Memoirs of the
American Philosophical Society
Held at Philadelphia
For Promoting Useful Knowledge
Volume 170

A SUPPLEMENT TO THE TUCKERMAN TABLES

A Supplement to the Tuckerman Tables

Michael A. Houlden
University of Liverpool

and

F. Richard Stephenson
University of Durham

American Philosophical Society
Independence Square Philadelphia
1986

Library of Congress Catalog Card No. 86-071019
International Standard Book No. 0-87169-170-1
US ISSN 0065-9738

INTRODUCTION

The present work is intended as a supplement to the well known planetary, lunar and solar tables produced by Bryant Tuckerman (1962, 1964). Since these tables appeared — as volumes 56 and 59 of the Memoirs of the Society — they have continued to prove an invaluable aid to historians of astronomy. An important usage is the dating of ancient and medieval astronomical observations but the tables also have wide application in determining the accuracy of early measurements and calculations.

Our supplementary volume owes its origin to the discovery by the authors — F. Richard Stephenson and Michael A. Houlden (1981) — of significant errors in Tuckerman's tabular positions of Mars. Following a query put to one of us (FRS) by the late Professor A.J. Sachs of Brown University, Providence, R.I. regarding the real accuracy of the tables, we made a systematic comparison between Tuckerman's positions for the Sun and planets and those computed from an integrated ephemeris. Only in the case of the longitude of Mars were errors found to be serious but these could amount to as much as 0.7 deg (considerably more than the Moon's apparent diameter). Before outlining the content of the present work, some remarks are necessary on various aspects of Tuckerman's original memoirs.

EXTENT AND PRECISION OF TUCKERMAN'S TABLES

In two remarkably compact volumes, positions of the five bright planets, together with the Moon and Sun, are tabulated over the entire period between 601 BC (or −600) and AD 1649 at intervals of 5 or 10 days. Co-ordinates are geocentric, relative to the ecliptic — i.e. longitude and latitude. These are more suitable than equatorial co-ordinates (right ascension and declination) for most planetary and lunar calculations and interpolation is much easier. In the case of the Moon and the more rapidly moving planets Mercury and Venus, positions are tabulated every five days, while for the Sun, Mars, Jupiter and Saturn the corresponding interval is 10 days. The hour selected is 16 h UT (4 p.m. Greenwich Civil Time or roughly 7 p.m. at both Babylon and Baghdad).

Tuckerman computed planetary and solar positions to the nearest 0.01 deg, which — provided this accuracy is realised — is more than sufficient for any practical purpose; not until the late seventeenth century was higher precision achieved in measurement. As he stated, it should be possible to interpolate satisfactorily all co-ordinates except the longitude of Mercury when this planet happened to be near inferior conjunction with the Sun. In the case of the Moon, which typically moves through more than 60 degrees every 5 days, there would have been little justification for tabulating positions to the same accuracy since interpolation would not be possible. Instead, Tuckerman rounded both the lunar longitude and latitude to the nearest 0.1 deg. At this level of precision, interpolation again becomes reasonably practicable. However, contrary to Tuckerman's suggestion, the lunar data are not really adequate for analysing such precise observations as eclipses and occultations. In our opinion, the lunar tables are mainly useful as a guide to the approximate location of the Moon; in any case many lunar calculations require a fairly substantial correction for parallax. Only the planetary and solar data are of value for detailed investigations, but here the applications are extensive. A knowledge of the real accuracy of the tabular co-ordinates is thus of prime importance.

In computing the data for the tables, Tuckerman rather surprisingly adopted what was in general outmoded orbital theory. He used the theory of Leverrier (1858–1861) for the Sun and inner planets, that of Gaillot (1904, 1913) for the outer planets and that of Hansen (1857) for the Moon; in each case he adopted modifications to certain of the orbital elements as derived by Schoch (1926). None of the theories cited had formed the basis of the American Ephemeris since well before 1930. However, despite these cautionary remarks, it is not our purpose to criticise Tuckerman's choice of orbital theory. Our main concern is the accuracy of the data in the published tables themselves.

The question of the true precision of the data tabulated by Tuckerman was first considered in detail by the present authors (Stephenson and Houlden, 1981). For this purpose, the then recently developed Long Export Ephemeris (code name DE 102) was used (Newhall et al., 1983). This ephemeris is based on a systematic numerical integration of the equations of motion of the planets.

The advantage of this type of ephemeris is that a "theory," such as was used to construct Tuckerman's tables, is not needed. Given precise masses and accurate starting positions and velocities for each planet, then in principle an ephemeris can be calculated at any time in the past or future. A typical step proceeds as follows. Starting with some moment at which the rectangular heliocentric co-ordinates and velocities of the planets are all accurately known, the total force on each planet due to the gravitational action of the Sun and the remaining planets is calculated. The position and velocity of each planet at some neighbouring moment (typically 12 hours away) is then obtained by integrating the equations of motion. The process is then repeated as often as required. DE 102 covers the entire period between 1411 BC (-1410) and AD 3002.

The time system adopted for DE 102 is ET (Ephemeris Time), whereas that used by Tuckerman was UT (Universal Time). In order to effect direct comparison between Tuckerman's tabular co-ordinates and the corresponding DE 102 values, we derived the following expression for ΔT (ET − UT) by comparing Tuckerman's adopted expression for the solar mean longitude with that deduced by Newcomb (1895) which defines ET:

$$\Delta T = +4.87 + 35.06\,T + 36.79\,T^2 \text{ seconds} \quad (1)$$

Here T is in Julian centuries of 36525 days, measured from the epoch 1900.0.

As might be expected, there is excellent accord between the planetary latitudes tabulated by Tuckerman and those calculated from DE 102. The latitudes of the planets are always fairly small; only Venus moves very far from the ecliptic (to a maximum of about 8 deg). However, the tabular longitudes require detailed discussion. For Mercury, Venus and the Sun, we found the agreement between the tabular and calculated longitudes to be very satisfactory. As far back as 601 BC (the epoch at which the tables commence) discrepancies of more than 0.02 deg are very rare and further are essentially random. In a sample of 200 positions for each object, the maximum error was found to be only 0.05 deg (for Venus). On the contrary, the deviations for Mars were disturbingly large, reaching 0.7 deg around 600 BC—see Fig 1. Huber (1983) pointed out that the cause of these discrepancies for Mars was Tuckerman's inadvertent choice of incorrect orbital elements for this planet; these corresponded to ET rather than UT. In particular, Huber demonstrated that the replacement of the quadratic term in Tuckerman's adopted expression for the mean longitude of Mars by its equivalent value in UT gave good accord with DE 102 (deviations as small as 0.05 deg).

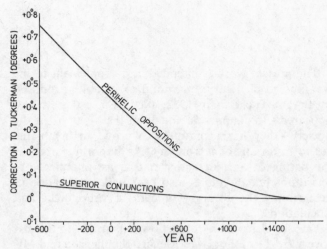

Fig 1. Deviations between Tuckerman's tabular longitudes of Mars and those computed from DE 102 at (i) perihelic oppositions (maximum discrepancy) and (ii) superior conjunctions (minimum discrepancy).

For the outer planets Jupiter and Saturn, the longitudes tabulated by Tuckerman are in very good agreement with the DE 102 data. In the case of Jupiter, over the entire period since about AD 200 the maximum discrepancy is as small as 0.03 deg. Although before that date deviations as large as 0.1 deg may occur, a smooth correction curve can be produced—see Fig 2. Use of this curve enables the tabular longitudes of Jupiter to be reduced to the equivalent DE 102 values with an accuracy of about 0.02 deg as far back as the beginning of the tables. It is evident from Fig 2 that the discrepancies are approximately periodic with increasing amplitude going backwards in time. For Saturn, we have deduced a similar curve (Fig 3) which allows the tabular longi-

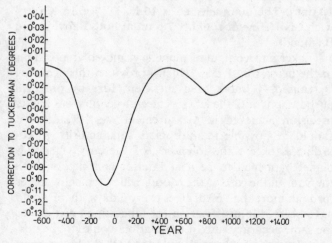

Fig 2. Mean corrections to Tuckerman's tabular longitudes of Jupiter in order to obtain best agreement with longitudes computed from DE 102.

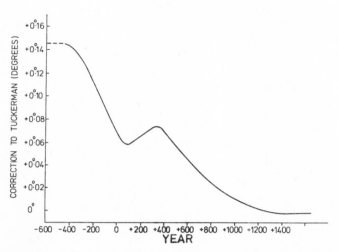

Fig 3. Mean corrections to Tuckerman's tabular longitudes of Saturn in order to obtain best agreement with longitudes computed from DE 102.

tudes to be corrected with errors as small as 0.03 deg at any time since about 300 BC. In earlier centuries the scatter of individual values becomes rather larger — approaching 0.1 deg — so that at this period only comparatively rough, but still useful estimates of the longitude of Saturn can be made from the tables.

OUTLINE OF THE PRESENT TABLES

In producing the present tables we have two main objectives — to make available revised positions for Mars throughout the period 601 BC to AD 1649 and to enable the apparent magnitude of each planet to be estimated at any time during this period.

Using the integrated ephemeris DE 102, we have computed the longitude and latitude of Mars to the nearest 0.01 deg at 10 day intervals for the same time of day as selected by Tuckerman (16 h UT). As a check on accuracy, we have compared a series of our calculated longitudes for the planet with those based on an orbital theory for Mars which until very recently was used in generating positions for the Astronomical Almanac. This is the theory of Newcomb (1898), with corrections derived by Ross (1917). Co-ordinates of Mars deduced from the Newcomb-Ross theory were kindly supplied by B. Emerson of the Royal Greenwich Observatory. The agreement between these and the equivalent DE 102 data is close to 0.01 deg as far back as 601 BC, sound evidence in favour of the reliability of our tabular data.

Brief remarks are needed on the question of the clock error ΔT. This arises from a gradual increase in the

length of the day due to a combination of lunar and solar tides and other causes. From an extensive study of historical observations, mainly of eclipses and occultations, Stephenson and Morrison (1984) deduced revised expressions for ΔT. The difference between values calculated from these formulas and figures based on equation (1) above is less than one hour at all periods back to 601 BC, which is negligible for the present purpose. For consistency, in producing the present tables we have used equation (1) to calculate ΔT values.

We have taken the opportunity to increase the versatility of the present volume by including — at 10-day intervals — the apparent magnitudes of each of the five bright planets Mercury, Venus, Mars, Jupiter and Saturn. We ourselves have often felt the need for readily accessible data of this kind. The values tabulated here should be especially useful in the case of Mercury and Mars, both of which fluctuate in brightness to a considerable degree. In not much more than a month, the magnitude of Mercury can vary from about -1.5 at superior conjunction to fainter than $+3$ at inferior conjunction. Changes in the brightness of Mars are much slower, but between opposition and conjunction the magnitude varies from about -2 to $+2$. Both planets revolve in rather elliptical orbits. As a result, the actual range in magnitude for Mercury at superior conjunction is from -0.8 to -1.6 whereas for Mars at opposition the range is between -1.2 and -2.6. At a close opposition, Mars can thus briefly outshine Jupiter. The brightness of Venus, Jupiter and Saturn is relatively steady, seldom varying over a range of much more than about one magnitude. The main factor in the case of Saturn is the visibility of the ring system; due to the changes in the aspect of the rings, the opposition magnitude of this planet varies between about $+0.7$ and -0.3. In the tables we have computed the apparent magnitudes of the planets from the formulae derived by Müller (1893). These formulae, which are based on numerous observations, formed the basis of the data in the Astronomical Almanac until 1983. The differences between these and the newer formulae employed — due to Harris (1961) — are trivial for all practical purposes.

USE OF TABLES

The accompanying pages of tables normally carry four years of data, the only exception being for the first page, covering -600 and -599 (i.e. 601 and 600 BC). This format has been chosen for convenience so that a typical single page of the present volume will correspond to an open double page of Tuckerman. Our tables actually extend to A. D. 1651, rather than 1649. Column by

column for each year we have: (i) the ecliptical longitude of Mars; (ii) the latitude of the planet; (iii) the apparent magnitude of Mars; (iv) the longitude of the Sun—given for reference; (v) the Julian Calendar date; (vi) to (ix) the apparent magnitudes of Mercury, Venus, Jupiter and Saturn.

Interpolation of the magnitude data should be accurate except in the rare instances for Mercury and Venus when these planets pass very close to the Sun at inferior conjunction—a transit across the solar disc being an extreme example.

The planetary symbols appearing at the top of the tables are:

♂ Mars ♀ Venus
☉ Sun ♃ Jupiter
☿ Mercury ♄ Saturn

REFERENCES

Gaillot, M.A. (1904, 1913). "Tables rectifiées du mouvement de Saturne", *Annales de l'Observatoire de Paris*, 24 (1904); "Tables rectifiées du mouvement de Jupiter", ibid., 31 (1913).

Hansen, P.A. (1857). *Tables de la Lune*, London, Eyre and Spottiswoode.

Harris, D.L. (1961). *In Planets and Satellites*, ed G.P. Kuiper and B.M. Middlehurst, Chicago, p. 272–342.

Huber, P.J. (1983). Review of "Ephemeriden von Sonne, Mond und hellen Planeten von -1000 bis -601". *Journal for the History of Astronomy*, 14, 228–229.

Leverrier, U.-J. (1858–61). "Théorie et tables du mouvement apparent du Soleil", *Annales de l'Observatoire imperial de Paris*, 4 (1858); "Théorie et tables du mouvement de Mercure", ibid., 5, 1–195 (1859); "Thé-orie et tables du mouvement de Venus", ibid., 6, 1–184 (1861); "Théorie et tables du mouvement de Mars", ibid., 6, 185–434 (1861).

Müller, G. (1893). "Helligkeitsbestimmungen der grosse Planeten und einiger Asteroiden", *Publicationem des Astrophysikalischen Observatoriums zu Potsdam*, 8, 197–389.

Newcomb, S. (1895). "Tables of the Sun", *Astronomical Papers of the American Ephemeris and Nautical Almanac*, 6, part 1.

Newcomb, S. (1898). "Tables of Mars", *Astronomical Papers of the American Ephemeris and Nautical Almanac*, 6, part 4.

Newhall, XX., Williams, J.G. and Standish, E.M. (1983), "DE 102: A Numerically Integrated Ephemeris of the Moon and Planets Spanning Forty-Four Centuries", *Astronomy and Astrophysics*, 125, 150–167.

Ross, F.E. (1917). "New Elements of Mars and Tables for Correcting the Heliocentric Positions". *Astronomical Papers of the American Ephemeris and Nautical Almanac*, 9, part 2.

Schoch, C. (1926). *Die säkulare Acceleration des Mondes und der Sonne*, Berlin-Steglitz, Selbstverlag.

Stephenson, F.R. and Houlden, M.A. (1981). "The Accuracy of Tuckerman's Solar and Planetary Tables", *Journal for the History of Astronomy*, 12, 133–138.

Stephenson, F.R. and Morrison, L.V. (1984). "Long-term Changes in the Rotation of the Earth: 700 BC to AD 1980". *Philosophical Transactions of the Royal Society*, A., 313, 47–70.

Tuckerman, B. (1962, 1964). *Planetary, Lunar and Solar Positions: 601 BC to AD 1*, Memoirs of the American Philosophical Society, 56 (1962); *Planetary, Lunar and Solar Positions: AD 2 to AD 1649*, ibid., 59 (1964).

♂ LONG	LAT	MAG	☉ LONG	16.00UT -600	☿	♀	♃	♄
							MAGNITUDES	
221.34	0.54	1.4	277.11	3JA	-0.0	-4.3	-1.5	0.8
228.25	0.42	1.4	287.23	13JA	-0.2	-4.3	-1.5	0.9
235.19	0.28	1.3	297.30	23JA	-0.5	-4.3	-1.5	0.9
242.15	0.13	1.2	307.32	2FE	-1.0	-4.1	-1.5	0.9
249.12	-0.03	1.1	317.27	12FE	-1.5	-3.6	-1.6	1.0
256.10	-0.22	0.9	327.17	22FE	-1.0	-3.3	-1.6	1.0
263.07	-0.42	0.8	337.01	3MR	0.1	-3.8	-1.6	1.0
270.03	-0.65	0.7	346.79	13MR	1.7	-4.1	-1.7	1.0
276.96	-0.89	0.6	356.52	23MR	3.6	-4.2	-1.7	1.0
283.84	-1.15	0.4	6.19	2AP	2.3	-4.2	-1.7	1.0
290.65	-1.44	0.3	15.82	12AP	1.3	-4.1	-1.8	0.9
297.35	-1.74	0.2	25.41	22AP	0.7	-4.0	-1.9	0.9
303.90	-2.06	0.0	34.96	2MY	-0.0	-3.9	-1.9	0.8
310.26	-2.40	-0.2	44.50	12MY	-0.9	-3.8	-2.0	0.8
316.37	-2.76	-0.3	54.02	22MY	-1.8	-3.7	-2.1	0.7
322.14	-3.13	-0.5	63.53	1JN	-1.1	-3.7	-2.1	0.7
327.48	-3.51	-0.7	73.05	11JN	-0.3	-3.6	-2.2	0.6
332.27	-3.91	-0.9	82.59	21JN	0.2	-3.5	-2.3	0.5
336.35	-4.31	-1.1	92.15	1JL	0.6	-3.5	-2.3	0.5
339.55	-4.70	-1.3	101.73	11JL	1.0	-3.4	-2.4	0.4
341.65	-5.07	-1.5	111.36	21JL	1.9	-3.4	-2.4	0.4
342.42	-5.37	-1.8	121.04	31JL	3.1	-3.4	-2.4	0.5
341.77	-5.55	-2.0	130.77	10AU	0.7	-3.4	-2.4	0.5
339.77	-5.55	-2.2	140.55	20AU	-0.7	-3.4	-2.4	0.6
336.81	-5.30	-2.3	150.39	30AU	-1.1	-3.4	-2.4	0.6
333.64	-4.80	-2.3	160.29	9SE	-1.1	-3.4	-2.3	0.7
331.03	-4.14	-2.0	170.26	19SE	-0.7	-3.4	-2.3	0.7
329.55	-3.41	-1.7	180.28	29SE	-0.4	-3.4	-2.2	0.8
329.40	-2.70	-1.3	190.34	9OC	-0.3	-3.4	-2.2	0.8
330.52	-2.06	-1.0	200.46	19OC	-0.3	-3.4	-2.1	0.9
332.74	-1.51	-0.7	210.62	29OC	-0.1	-3.4	-2.0	0.9
335.85	-1.05	-0.4	220.80	8NO	1.3	-3.4	-2.0	0.9
339.65	-0.66	-0.1	231.00	18NO	2.0	-3.5	-1.9	0.9
343.99	-0.33	0.1	241.22	28NO	0.2	-3.5	-1.8	0.9
348.75	-0.06	0.3	251.42	8DE	-0.1	-3.5	-1.8	0.9
353.83	0.17	0.6	261.62	18DE	-0.1	-3.5	-1.7	0.9
359.16	0.36	0.7	271.79	28DE	-0.3	-3.4	-1.7	0.9
				-599				
4.68	0.52	0.9	281.92	7JA	-0.6	-3.4	-1.6	0.9
10.35	0.65	1.1	292.02	17JA	-1.0	-3.4	-1.6	0.9
16.13	0.77	1.2	302.07	27JA	-1.3	-3.4	-1.6	1.0
22.00	0.86	1.3	312.05	6FE	-1.0	-3.3	-1.6	1.0
27.93	0.94	1.4	321.99	16FE	0.2	-3.3	-1.6	1.1
33.92	1.01	1.6	331.86	26FE	2.1	-3.3	-1.6	1.1
39.94	1.07	1.6	341.67	8MR	3.1	-3.3	-1.6	1.1
46.00	1.11	1.7	351.42	18MR	1.7	-3.3	-1.6	1.1
52.08	1.15	1.8	1.13	28MR	1.0	-3.4	-1.6	1.1
58.18	1.18	1.9	10.77	7AP	0.5	-3.4	-1.6	1.1
64.31	1.21	1.9	20.38	17AP	-0.1	-3.4	-1.7	1.1
70.46	1.22	1.9	29.96	27AP	-0.9	-3.4	-1.7	1.0
76.62	1.23	2.0	39.50	7MY	-1.8	-3.5	-1.8	1.0
82.80	1.24	2.0	49.03	17MY	-1.1	-3.5	-1.8	0.9
89.01	1.24	2.0	58.54	27MY	-0.2	-3.6	-1.9	0.9
95.24	1.23	2.0	68.06	6JN	0.4	-3.6	-1.9	0.8
101.51	1.22	2.0	77.59	16JN	0.8	-3.7	-2.0	0.7
107.80	1.21	2.0	87.13	26JN	1.4	-3.8	-2.1	0.7
114.13	1.19	2.0	96.70	6JL	2.5	-3.9	-2.1	0.6
120.50	1.17	2.0	106.31	16JL	2.7	-4.0	-2.2	0.5
126.91	1.14	2.0	115.96	26JL	0.6	-4.1	-2.3	0.5
133.36	1.11	1.9	125.66	5AU	-0.8	-4.2	-2.3	0.6
139.87	1.07	1.9	135.41	15AU	-1.2	-4.3	-2.4	0.6
146.42	1.03	1.9	145.22	25AU	-1.1	-4.3	-2.4	0.6
153.04	0.98	1.9	155.09	4SE	-0.6	-4.1	-2.4	0.7
159.71	0.93	1.9	165.03	14SE	-0.3	-3.7	-2.4	0.7
166.44	0.87	1.9	175.01	24SE	-0.2	-3.3	-2.4	0.8
173.24	0.81	1.8	185.06	4OC	-0.1	-3.8	-2.4	0.8
180.10	0.74	1.8	195.15	14OC	0.1	-4.2	-2.4	0.9
187.02	0.66	1.8	205.29	24OC	1.5	-4.4	-2.3	0.9
194.01	0.58	1.8	215.46	3NO	1.7	-4.4	-2.2	1.0
201.06	0.49	1.8	225.65	13NO	0.0	-4.3	-2.2	1.0
208.18	0.39	1.7	235.86	23NO	-0.2	-4.2	-2.1	1.0
215.36	0.29	1.7	246.07	3DE	-0.3	-4.1	-2.0	1.0
222.60	0.18	1.6	256.28	13DE	-0.4	-4.0	-2.0	1.0
229.90	0.06	1.6	266.46	23DE	-0.6	-3.9	-1.9	1.1

-598

♂ LONG	LAT	MAG	☉ LONG	16.00UT	☿	♀	♃	♄
237.26	-0.06	1.5	276.62	2JA	-1.0	-3.8	-1.8	1.1
244.66	-0.19	1.5	286.74	12JA	-1.2	-3.7	-1.8	1.0
252.11	-0.33	1.4	296.81	22JA	-0.9	-3.6	-1.7	1.0
259.60	-0.48	1.4	306.83	1FE	0.3	-3.6	-1.7	1.1
267.12	-0.62	1.3	316.79	11FE	2.5	-3.5	-1.6	1.1
274.66	-0.78	1.2	326.69	21FE	2.4	-3.4	-1.6	1.2
282.22	-0.93	1.2	336.54	3MR	1.2	-3.4	-1.6	1.2
289.78	-1.08	1.1	346.32	13MR	0.7	-3.4	-1.6	1.2
297.32	-1.23	1.1	356.05	23MR	0.3	-3.3	-1.5	1.2
304.84	-1.38	1.0	5.72	2AP	-0.2	-3.3	-1.5	1.2
312.31	-1.52	0.9	15.36	12AP	-0.9	-3.3	-1.5	1.2
319.72	-1.65	0.9	24.94	22AP	-1.8	-3.3	-1.6	1.2
327.05	-1.78	0.8	34.50	2MY	-1.1	-3.3	-1.6	1.2
334.28	-1.89	0.7	44.04	12MY	-0.2	-3.3	-1.6	1.2
341.37	-1.99	0.6	53.56	22MY	0.5	-3.3	-1.6	1.1
348.32	-2.08	0.6	63.07	1JN	1.1	-3.4	-1.7	1.1
355.08	-2.15	0.5	72.60	11JN	1.8	-3.4	-1.7	1.0
1.62	-2.20	0.4	82.12	21JN	3.1	-3.4	-1.8	1.0
7.90	-2.24	0.3	91.68	1JL	2.2	-3.5	-1.8	0.9
13.87	-2.25	0.2	101.27	11JL	0.5	-3.5	-1.9	0.8
19.46	-2.25	0.1	110.89	21JL	-0.8	-3.5	-1.9	0.7
24.60	-2.22	-0.1	120.57	31JL	-1.4	-3.4	-2.0	0.7
29.18	-2.17	-0.2	130.29	10AU	-1.1	-3.4	-2.1	0.7
33.08	-2.09	-0.4	140.07	20AU	-0.6	-3.4	-2.1	0.7
36.13	-1.97	-0.6	149.91	30AU	-0.2	-3.4	-2.2	0.7
38.14	-1.80	-0.8	159.81	9SE	-0.0	-3.3	-2.3	0.8
38.91	-1.57	-1.0	169.76	19SE	0.1	-3.3	-2.3	0.8
38.25	-1.27	-1.2	179.78	29SE	0.3	-3.3	-2.3	0.9
36.15	-0.89	-1.4	189.85	9OC	1.8	-3.3	-2.4	0.9
32.89	-0.44	-1.6	199.96	19OC	1.5	-3.3	-2.4	1.0
29.06	0.02	-1.6	210.12	29OC	-0.1	-3.4	-2.4	1.0
25.48	0.47	-1.4	220.30	8NO	-0.4	-3.4	-2.3	1.1
22.84	0.84	-1.1	230.50	18NO	-0.4	-3.4	-2.3	1.1
21.49	1.13	-0.8	240.72	28NO	-0.5	-3.4	-2.2	1.1
21.49	1.34	-0.5	250.93	8DE	-0.7	-3.5	-2.2	1.2
22.69	1.49	-0.2	261.12	18DE	-1.0	-3.5	-2.1	1.2
24.89	1.59	0.1	271.30	28DE	-1.0	-3.6	-2.0	1.2

-597

♂ LONG	LAT	MAG	☉ LONG	16.00UT	☿	♀	♃	♄
27.88	1.66	0.3	281.44	7JA	-0.8	-3.7	-2.0	1.2
31.50	1.69	0.6	291.53	17JA	0.4	-3.7	-1.9	1.2
35.61	1.71	0.8	301.58	27JA	2.9	-3.8	-1.8	1.2
40.11	1.72	0.9	311.57	6FE	1.8	-3.9	-1.8	1.2
44.91	1.71	1.1	321.51	16FE	0.9	-4.0	-1.7	1.3
49.94	1.70	1.2	331.38	26FE	0.5	-4.0	-1.6	1.3
55.18	1.68	1.4	341.20	8MR	0.2	-4.1	-1.6	1.4
60.57	1.65	1.5	350.95	18MR	-0.2	-4.2	-1.6	1.4
66.10	1.62	1.6	0.66	28MR	-0.9	-4.2	-1.5	1.4
71.73	1.59	1.7	10.31	7AP	-1.8	-4.1	-1.5	1.4
77.47	1.55	1.7	19.92	17AP	-1.1	-3.8	-1.5	1.4
83.28	1.51	1.8	29.49	27AP	-0.1	-3.1	-1.5	1.3
89.18	1.47	1.8	39.04	7MY	0.7	-3.2	-1.5	1.3
95.15	1.42	1.9	48.56	17MY	1.4	-3.8	-1.5	1.2
101.19	1.37	1.9	58.08	27MY	2.4	-4.1	-1.5	1.2
107.29	1.32	1.9	67.60	6JN	3.3	-4.2	-1.5	1.2
113.46	1.26	2.0	77.12	16JN	1.8	-4.2	-1.5	1.1
119.70	1.20	2.0	86.67	26JN	0.4	-4.1	-1.5	1.1
126.01	1.14	2.0	96.24	6JL	-0.9	-4.0	-1.6	1.0
132.39	1.08	2.0	105.84	16JL	-1.5	-3.9	-1.6	1.0
138.84	1.01	1.9	115.49	26JL	-1.1	-3.8	-1.7	0.9
145.36	0.94	1.9	125.19	5AU	-0.5	-3.7	-1.7	0.9
151.96	0.86	1.9	134.94	15AU	-0.1	-3.7	-1.8	0.8
158.65	0.78	1.9	144.75	25AU	0.1	-3.6	-1.8	0.8
165.41	0.70	1.8	154.61	4SE	0.2	-3.6	-1.9	0.9
172.25	0.62	1.8	164.54	14SE	0.6	-3.5	-2.0	0.9
179.18	0.53	1.7	174.53	24SE	2.2	-3.5	-2.0	1.0
186.19	0.43	1.7	184.57	4OC	1.3	-3.5	-2.1	1.0
193.28	0.33	1.6	194.66	14OC	-0.3	-3.4	-2.2	1.1
200.46	0.23	1.6	204.80	24OC	-0.6	-3.4	-2.2	1.1
207.72	0.13	1.6	214.96	3NO	-0.6	-3.4	-2.2	1.2
215.06	0.02	1.6	225.16	13NO	-0.6	-3.4	-2.3	1.2
222.48	-0.09	1.6	235.37	23NO	-0.8	-3.4	-2.3	1.3
229.96	-0.20	1.6	245.58	3DE	-0.8	-3.4	-2.2	1.3
237.51	-0.31	1.5	255.78	13DE	-0.9	-3.4	-2.2	1.3
245.13	-0.43	1.5	265.97	23DE	-0.7	-3.4	-2.2	1.4

-596

♂ LONG	LAT	MAG	☉ LONG	16.00UT	☿	♀	♃	♄
252.79	-0.54	1.5	276.13	2JA	0.6	-3.4	-2.1	1.3
260.49	-0.65	1.5	286.25	12JA	3.0	-3.4	-2.1	1.3
268.23	-0.76	1.5	296.32	22JA	1.3	-3.4	-2.0	1.3
276.00	-0.87	1.4	306.34	1FE	0.6	-3.4	-1.9	1.2
283.77	-0.96	1.4	316.31	11FE	0.4	-3.5	-1.9	1.2
291.55	-1.06	1.4	326.21	21FE	0.1	-3.5	-1.8	1.1
299.31	-1.14	1.4	336.05	2MR	-0.3	-3.5	-1.7	1.1
307.05	-1.21	1.3	345.84	12MR	-1.0	-3.4	-1.7	1.1
314.75	-1.27	1.3	355.57	22MR	-1.8	-3.4	-1.6	1.1
322.41	-1.33	1.3	5.25	1AP	-1.1	-3.4	-1.5	1.1
330.00	-1.36	1.3	14.88	11AP	-0.1	-3.4	-1.5	1.1
337.51	-1.39	1.3	24.47	21AP	0.9	-3.3	-1.5	1.1
344.94	-1.40	1.2	34.03	1MY	1.9	-3.3	-1.4	1.1
352.28	-1.39	1.2	43.57	11MY	3.2	-3.3	-1.4	1.0
359.50	-1.37	1.2	53.09	21MY	2.7	-3.3	-1.4	1.0
6.59	-1.34	1.2	62.60	31MY	1.4	-3.3	-1.4	0.9
13.56	-1.29	1.1	72.13	10JN	0.3	-3.4	-1.4	0.9
20.37	-1.23	1.1	81.66	20JN	-0.9	-3.4	-1.4	0.8
27.02	-1.15	1.0	91.21	30JN	-1.6	-3.4	-1.4	0.8
33.49	-1.06	1.0	100.80	10JL	-1.1	-3.4	-1.4	0.7
39.75	-0.96	0.9	110.43	20JL	-0.4	-3.5	-1.4	0.7
45.79	-0.83	0.9	120.09	30JL	-0.0	-3.5	-1.4	0.6
51.56	-0.70	0.8	129.82	9AU	0.2	-3.6	-1.5	0.6
57.02	-0.54	0.7	139.60	19AU	0.4	-3.7	-1.5	0.6
62.11	-0.36	0.6	149.43	29AU	0.9	-3.8	-1.6	0.5
66.77	-0.15	0.5	159.33	8SE	2.6	-3.9	-1.6	0.6
70.89	0.09	0.3	169.28	18SE	1.2	-4.0	-1.7	0.6
74.37	0.36	0.2	179.29	28SE	-0.4	-4.1	-1.7	0.7
77.03	0.67	-0.0	189.36	8OC	-0.7	-4.2	-1.8	0.8
78.70	1.03	-0.2	199.48	18OC	-0.7	-4.3	-1.9	0.8
79.20	1.43	-0.5	209.62	28OC	-0.7	-4.4	-1.9	0.9
78.34	1.88	-0.7	219.81	7NO	-0.7	-4.4	-2.0	1.0
76.11	2.33	-0.9	230.01	17NO	-0.7	-4.2	-2.1	1.0
72.74	2.74	-1.1	240.22	27NO	-0.7	-3.7	-2.1	1.0
68.78	3.06	-1.2	250.43	7DE	-0.5	-3.0	-2.1	1.1
65.01	3.24	-1.0	260.63	17DE	0.7	-3.8	-2.1	1.1
62.09	3.30	-0.7	270.80	27DE	2.9	-4.2	-2.1	1.1

-595

♂ LONG	LAT	MAG	☉ LONG	16.00UT	☿	♀	♃	♄
60.41	3.25	-0.4	280.95	6JA	1.0	-4.3	-2.1	1.1
60.06	3.13	-0.2	291.04	16JA	0.4	-4.3	-2.1	1.0
60.90	2.99	0.1	301.09	26JA	0.2	-4.2	-2.1	1.0
62.77	2.83	0.3	311.09	5FE	-0.0	-4.1	-2.0	1.0
65.46	2.68	0.5	321.03	15FE	-0.4	-4.0	-1.9	0.9
68.81	2.52	0.7	330.90	25FE	-1.0	-3.9	-1.9	0.9
72.68	2.38	0.9	340.72	7MR	-1.7	-3.8	-1.8	0.8
76.98	2.24	1.1	350.48	17MR	-1.1	-3.7	-1.7	0.8
81.62	2.11	1.2	0.18	27MR	-0.0	-3.6	-1.7	0.8
86.54	1.98	1.3	9.84	6AP	1.1	-3.6	-1.6	0.8
91.69	1.86	1.4	19.45	16AP	2.4	-3.5	-1.5	0.8
97.03	1.74	1.5	29.03	26AP	3.6	-3.4	-1.5	0.8
102.55	1.63	1.6	38.58	6MY	2.1	-3.4	-1.4	0.8
108.23	1.51	1.6	48.10	16MY	1.1	-3.4	-1.4	0.8
114.03	1.40	1.7	57.62	26MY	0.2	-3.3	-1.4	0.8
119.97	1.30	1.7	67.14	5JN	-0.9	-3.3	-1.3	0.7
126.02	1.19	1.8	76.66	15JN	-1.7	-3.3	-1.3	0.7
132.19	1.08	1.8	86.20	25JN	-1.1	-3.3	-1.3	0.6
138.47	0.97	1.8	95.78	5JL	-0.4	-3.3	-1.3	0.6
144.86	0.87	1.8	105.38	15JL	0.1	-3.3	-1.3	0.5
151.35	0.76	1.8	115.02	25JL	0.4	-3.4	-1.3	0.5
157.95	0.65	1.8	124.72	4AU	0.6	-3.4	-1.3	0.4
164.65	0.54	1.8	134.46	14AU	1.2	-3.4	-1.3	0.4
171.46	0.43	1.7	144.27	24AU	3.0	-3.4	-1.3	0.3
178.37	0.32	1.7	154.14	3SE	1.0	-3.5	-1.4	0.3
185.39	0.21	1.7	164.06	13SE	-0.5	-3.5	-1.4	0.2
192.50	0.10	1.7	174.04	23SE	-0.9	-3.5	-1.4	0.3
199.72	-0.01	1.6	184.08	3OC	-0.8	-3.5	-1.5	0.4
207.03	-0.12	1.6	194.17	13OC	-0.8	-3.4	-1.6	0.4
214.44	-0.23	1.5	204.30	23OC	-0.6	-3.4	-1.6	0.5
221.93	-0.34	1.5	214.47	2NO	-0.5	-3.4	-1.7	0.6
229.51	-0.44	1.5	224.66	12NO	-0.5	-3.4	-1.7	0.6
237.15	-0.54	1.4	234.87	22NO	-0.4	-3.3	-1.8	0.7
244.87	-0.64	1.4	245.08	2DE	0.9	-3.3	-1.9	0.7
252.63	-0.73	1.4	255.28	12DE	2.5	-3.3	-1.9	0.8
260.44	-0.81	1.4	265.47	22DE	0.7	-3.3	-2.0	0.8

♂ LONG	♂ LAT	♂ MAG	⊙ LONG	16.00UT	☿	♀	♃	♄
				-594	MAGNITUDES			
268.29	-0.89	1.4	275.63	1JA	0.2	-3.4	-2.0	0.8
276.15	-0.96	1.4	285.75	11JA	0.1	-3.4	-2.0	0.8
284.02	-1.01	1.4	295.83	21JA	-0.1	-3.4	-2.1	0.8
291.89	-1.06	1.4	305.86	31JA	-0.4	-3.4	-2.1	0.8
299.74	-1.10	1.4	315.82	10FE	-1.0	-3.4	-2.0	0.7
307.56	-1.12	1.4	325.73	20FE	-1.5	-3.5	-2.0	0.7
315.34	-1.14	1.4	335.58	2MR	-1.1	-3.5	-2.0	0.7
323.07	-1.14	1.4	345.37	12MR	0.1	-3.6	-1.9	0.6
330.74	-1.13	1.4	355.10	22MR	1.4	-3.6	-1.8	0.6
338.33	-1.10	1.5	4.79	1AP	3.2	-3.7	-1.8	0.6
345.85	-1.07	1.5	14.42	11AP	2.8	-3.8	-1.7	0.6
353.28	-1.02	1.5	24.01	21AP	1.6	-3.9	-1.6	0.6
0.61	-0.97	1.5	33.58	1MY	0.8	-3.9	-1.6	0.6
7.85	-0.90	1.5	43.11	11MY	0.1	-4.0	-1.5	0.6
14.98	-0.82	1.5	52.63	21MY	-0.9	-4.1	-1.5	0.6
22.01	-0.74	1.5	62.15	31MY	-1.7	-4.2	-1.4	0.6
28.93	-0.64	1.5	71.66	10JN	-1.1	-4.2	-1.4	0.6
35.73	-0.54	1.5	81.20	20JN	-0.3	-4.0	-1.3	0.5
42.41	-0.43	1.5	90.75	30JN	0.2	-3.5	-1.3	0.5
48.96	-0.31	1.5	100.33	10JL	0.5	-3.0	-1.3	0.4
55.38	-0.18	1.5	109.96	20JL	0.9	-3.6	-1.2	0.4
61.65	-0.05	1.4	119.63	30JL	1.6	-4.1	-1.2	0.3
67.77	0.10	1.4	129.34	9AU	3.2	-4.2	-1.2	0.3
73.71	0.25	1.3	139.12	19AU	0.9	-4.3	-1.2	0.2
79.46	0.42	1.3	148.95	29AU	-0.6	-4.2	-1.2	0.2
84.99	0.60	1.2	158.85	8SE	-1.0	-4.1	-1.2	0.1
90.25	0.80	1.1	168.80	18SE	-1.0	-4.0	-1.3	0.0
95.20	1.02	1.0	178.81	28SE	-0.8	-4.0	-1.3	0.0
99.79	1.26	0.9	188.87	8OC	-0.5	-3.9	-1.3	0.1
103.92	1.52	0.7	198.99	18OC	-0.4	-3.8	-1.4	0.1
107.50	1.82	0.5	209.13	28OC	-0.4	-3.7	-1.4	0.2
110.39	2.15	0.3	219.31	7NO	-0.2	-3.7	-1.5	0.3
112.43	2.53	0.1	229.52	17NO	1.1	-3.6	-1.5	0.3
113.43	2.93	-0.1	239.73	27NO	2.2	-3.5	-1.6	0.4
113.23	3.36	-0.3	249.94	7DE	0.4	-3.5	-1.7	0.5
111.70	3.79	-0.6	260.14	17DE	0.0	-3.4	-1.7	0.5
108.92	4.15	-0.8	270.31	27DE	-0.1	-3.4	-1.8	0.5
				-593				
105.23	4.40	-1.0	280.46	6JA	-0.2	-3.4	-1.9	0.6
101.27	4.47	-0.9	290.56	16JA	-0.5	-3.4	-1.9	0.6
97.79	4.38	-0.8	300.61	26JA	-1.0	-3.3	-2.0	0.6
95.32	4.15	-0.5	310.61	5FE	-1.4	-3.3	-2.0	0.6
94.13	3.85	-0.3	320.55	15FE	-1.0	-3.3	-2.0	0.6
94.23	3.53	-0.0	330.43	25FE	0.1	-3.3	-2.0	0.5
95.45	3.20	0.2	340.25	7MR	1.8	-3.3	-2.0	0.5
97.63	2.90	0.4	350.01	17MR	3.5	-3.3	-2.0	0.5
100.59	2.61	0.6	359.72	27MR	2.1	-3.4	-2.0	0.4
104.17	2.35	0.7	9.37	6AP	1.2	-3.4	-1.9	0.4
108.27	2.11	0.9	18.99	16AP	0.6	-3.4	-1.9	0.4
112.79	1.88	1.0	28.56	26AP	-0.0	-3.5	-1.8	0.4
117.66	1.68	1.1	38.11	6MY	-0.9	-3.5	-1.7	0.4
122.83	1.48	1.2	47.64	16MY	-1.8	-3.5	-1.7	0.5
128.25	1.30	1.3	57.16	26MY	-1.1	-3.4	-1.6	0.5
133.89	1.12	1.3	66.67	5JN	-0.3	-3.4	-1.5	0.4
139.74	0.96	1.4	76.20	15JN	0.3	-3.4	-1.5	0.4
145.77	0.80	1.4	85.74	25JN	0.7	-3.3	-1.4	0.4
151.97	0.64	1.5	95.30	5JL	1.1	-3.3	-1.4	0.4
158.33	0.49	1.5	104.91	15JL	2.1	-3.3	-1.3	0.3
164.84	0.35	1.5	114.55	25JL	3.0	-3.3	-1.3	0.3
171.49	0.21	1.5	124.24	4AU	0.7	-3.3	-1.3	0.2
178.28	0.08	1.5	133.99	14AU	-0.7	-3.3	-1.3	0.2
185.20	-0.05	1.5	143.79	24AU	-1.2	-3.4	-1.2	0.1
192.25	-0.18	1.5	153.65	3SE	-1.1	-3.4	-1.2	0.1
199.43	-0.30	1.5	163.57	13SE	-0.7	-3.4	-1.2	-0.0
206.72	-0.41	1.5	173.55	23SE	-0.4	-3.5	-1.2	-0.1
214.11	-0.52	1.5	183.59	3OC	-0.3	-3.5	-1.2	-0.1
221.61	-0.62	1.5	193.68	13OC	-0.2	-3.6	-1.3	-0.2
229.20	-0.71	1.5	203.81	23OC	-0.0	-3.7	-1.3	-0.1
236.87	-0.80	1.4	213.97	2NO	1.3	-3.7	-1.3	-0.0
244.61	-0.87	1.4	224.17	12NO	2.0	-3.8	-1.4	0.0
252.41	-0.94	1.4	234.37	22NO	0.2	-3.9	-1.4	0.1
260.26	-0.99	1.4	244.59	2DE	-0.1	-4.0	-1.5	0.2
268.14	-1.04	1.4	254.80	12DE	-0.2	-4.1	-1.5	0.2
276.03	-1.07	1.4	264.98	22DE	-0.3	-4.2	-1.6	0.3

♂ LONG	♂ LAT	♂ MAG	⊙ LONG	16.00UT	☿	♀	♃	♄
				-592	MAGNITUDES			
283.93	-1.09	1.4	275.14	1JA	-0.6	-4.3	-1.6	0.3
291.82	-1.10	1.3	285.27	11JA	-1.0	-4.3	-1.7	0.4
299.69	-1.09	1.3	295.34	21JA	-1.3	-4.3	-1.8	0.4
307.52	-1.08	1.3	305.37	31JA	-1.0	-4.1	-1.8	0.4
315.31	-1.05	1.3	315.34	10FE	0.2	-3.6	-1.9	0.4
323.04	-1.02	1.4	325.25	20FE	2.2	-3.3	-2.0	0.4
330.71	-0.97	1.4	335.11	1MR	2.8	-3.8	-2.0	0.4
338.30	-0.91	1.4	344.90	11MR	1.5	-4.1	-2.0	0.4
345.81	-0.85	1.5	354.63	21MR	0.9	-4.2	-2.0	0.4
353.24	-0.78	1.5	4.32	31MR	0.4	-4.2	-2.1	0.3
0.58	-0.70	1.6	13.95	10AP	-0.1	-4.1	-2.0	0.3
7.83	-0.61	1.6	23.55	20AP	-0.9	-4.0	-2.0	0.3
15.00	-0.52	1.6	33.11	30AP	-1.8	-3.9	-2.0	0.3
22.07	-0.42	1.7	42.65	10MY	-1.1	-3.8	-1.9	0.3
29.04	-0.32	1.7	52.17	20MY	-0.2	-3.7	-1.9	0.3
35.93	-0.22	1.7	61.69	30MY	0.4	-3.7	-1.8	0.3
42.73	-0.11	1.7	71.20	9JN	0.9	-3.6	-1.7	0.4
49.43	-0.00	1.8	80.73	19JN	1.5	-3.5	-1.7	0.3
56.05	0.11	1.8	90.29	29JN	2.7	-3.5	-1.6	0.3
62.57	0.22	1.8	99.87	9JL	2.5	-3.4	-1.6	0.3
69.01	0.34	1.8	109.49	19JL	0.6	-3.4	-1.5	0.3
75.35	0.46	1.8	119.16	29JL	-0.8	-3.4	-1.5	0.3
81.59	0.59	1.7	128.87	8AU	-1.3	-3.4	-1.4	0.2
87.73	0.72	1.7	138.64	18AU	-1.2	-3.4	-1.4	0.2
93.76	0.86	1.7	148.48	28AU	-0.6	-3.4	-1.3	0.1
99.67	1.00	1.6	158.37	7SE	-0.3	-3.4	-1.3	0.0
105.44	1.14	1.6	168.31	17SE	-0.1	-3.4	-1.3	-0.0
111.05	1.30	1.5	178.32	27SE	-0.0	-3.4	-1.3	-0.1
116.48	1.47	1.4	188.38	7OC	0.2	-3.4	-1.3	-0.2
121.70	1.65	1.3	198.49	17OC	1.6	-3.4	-1.3	-0.2
126.66	1.84	1.2	208.64	27OC	1.7	-3.4	-1.3	-0.3
131.31	2.05	1.0	218.82	6NO	-0.0	-3.4	-1.3	-0.2
135.59	2.28	0.9	229.02	16NO	-0.3	-3.5	-1.3	-0.1
139.39	2.53	0.7	239.24	26NO	-0.3	-3.5	-1.3	-0.1
142.61	2.81	0.5	249.44	6DE	-0.4	-3.5	-1.4	0.0
145.12	3.11	0.3	259.64	16DE	-0.7	-3.5	-1.4	0.1
146.72	3.44	0.0	269.82	26DE	-1.0	-3.4	-1.5	0.1
				-591				
147.26	3.77	-0.2	279.96	5JA	-1.1	-3.4	-1.5	0.2
146.57	4.10	-0.5	290.07	15JA	-0.9	-3.4	-1.6	0.3
144.58	4.36	-0.8	300.12	25JA	0.3	-3.4	-1.7	0.3
141.47	4.51	-1.0	310.12	4FE	2.6	-3.3	-1.7	0.3
137.68	4.50	-1.1	320.06	14FE	2.2	-3.3	-1.8	0.3
133.91	4.31	-1.0	329.95	24FE	1.1	-3.3	-1.9	0.3
130.85	3.97	-0.8	339.77	6MR	0.6	-3.3	-1.9	0.3
128.94	3.54	-0.6	349.54	16MR	0.3	-3.3	-2.0	0.3
128.34	3.09	-0.4	359.25	26MR	-0.2	-3.4	-2.0	0.3
129.00	2.65	-0.2	8.91	5AP	-0.9	-3.4	-2.1	0.3
130.75	2.24	0.0	18.52	15AP	-1.8	-3.4	-2.1	0.2
133.42	1.87	0.2	28.10	25AP	-1.2	-3.4	-2.1	0.2
136.85	1.53	0.3	37.65	5MY	-0.2	-3.5	-2.1	0.2
140.90	1.22	0.5	47.18	15MY	0.6	-3.5	-2.1	0.2
145.46	0.94	0.6	56.70	25MY	1.2	-3.6	-2.1	0.2
150.46	0.69	0.7	66.21	4JN	2.0	-3.6	-2.0	0.3
155.81	0.46	0.8	75.74	14JN	3.3	-3.7	-2.0	0.3
161.49	0.25	0.9	85.28	24JN	2.1	-3.8	-1.9	0.3
167.44	0.05	0.9	94.84	4JL	0.5	-3.9	-1.9	0.3
173.63	-0.13	1.0	104.44	14JL	-0.8	-4.0	-1.8	0.3
180.05	-0.29	1.1	114.09	24JL	-1.4	-4.1	-1.7	0.3
186.67	-0.45	1.1	123.77	3AU	-1.1	-4.2	-1.7	0.3
193.48	-0.59	1.1	133.51	13AU	-0.5	-4.3	-1.6	0.2
200.45	-0.71	1.2	143.31	23AU	-0.2	-4.3	-1.6	0.2
207.58	-0.83	1.2	153.17	2SE	0.0	-4.1	-1.5	0.1
214.84	-0.93	1.2	163.09	12SE	0.1	-3.6	-1.5	0.1
222.24	-1.01	1.2	173.07	22SE	0.4	-3.3	-1.4	-0.0
229.74	-1.09	1.3	183.10	2OC	1.9	-3.9	-1.4	-0.1
237.34	-1.14	1.3	193.18	12OC	1.5	-4.2	-1.4	-0.1
245.03	-1.19	1.3	203.31	22OC	-0.2	-4.4	-1.4	-0.2
252.78	-1.22	1.3	213.48	1NO	-0.5	-4.4	-1.4	-0.3
260.58	-1.24	1.3	223.67	11NO	-0.5	-4.3	-1.3	-0.3
268.41	-1.24	1.3	233.88	21NO	-0.5	-4.2	-1.4	-0.3
276.27	-1.23	1.4	244.09	1DE	-0.7	-4.1	-1.4	-0.2
284.12	-1.20	1.4	254.30	11DE	-0.9	-4.0	-1.4	-0.1
291.97	-1.17	1.4	264.49	21DE	-0.9	-3.9	-1.4	-0.0
299.79	-1.12	1.4	274.65	31DE	-0.8	-3.8	-1.4	0.0

-590

♂ LONG LAT MAG	☉ LONG	16.00UT	☿ ♀ ♃ ♄ MAGNITUDES
307.57 -1.06 1.4	284.78	10JA	0.4 -3.7 -1.5 0.1
315.30 -1.00 1.5	294.86	20JA	2.9 -3.6 -1.5 0.2
322.97 -0.92 1.5	304.89	30JA	1.6 -3.6 -1.6 0.2
330.58 -0.84 1.5	314.87	9FE	0.8 -3.5 -1.6 0.2
338.11 -0.76 1.5	324.78	19FE	0.5 -3.4 -1.7 0.3
345.56 -0.67 1.5	334.63	1MR	0.2 -3.4 -1.8 0.3
352.93 -0.57 1.6	344.43	11MR	-0.2 -3.4 -1.8 0.3
0.21 -0.48 1.6	354.17	21MR	-0.9 -3.3 -1.9 0.3
7.41 -0.38 1.6	3.85	31MR	-1.8 -3.3 -2.0 0.3
14.53 -0.28 1.6	13.49	10AP	-1.2 -3.3 -2.1 0.3
21.56 -0.18 1.6	23.09	20AP	-0.1 -3.3 -2.1 0.2
28.52 -0.07 1.7	32.65	30AP	0.7 -3.3 -2.2 0.2
35.39 0.03 1.7	42.19	10MY	1.6 -3.3 -2.2 0.2
42.19 0.13 1.8	51.71	20MY	2.7 -3.3 -2.2 0.2
48.91 0.23 1.8	61.23	30MY	3.2 -3.4 -2.2 0.2
55.57 0.33 1.9	70.75	9JN	1.7 -3.4 -2.2 0.2
62.16 0.43 1.9	80.28	19JN	0.4 -3.4 -2.2 0.3
68.69 0.52 1.9	89.83	29JN	-0.8 -3.5 -2.2 0.3
75.16 0.62 1.9	99.41	9JL	-1.5 -3.5 -2.2 0.3
81.57 0.71 1.9	109.02	19JL	-1.1 -3.5 -2.1 0.3
87.93 0.81 1.9	118.68	29JL	-0.5 -3.4 -2.0 0.3
94.23 0.91 1.9	128.40	8AU	-0.1 -3.4 -2.0 0.3
100.47 1.00 1.9	138.17	18AU	0.1 -3.4 -1.9 0.3
106.66 1.10 1.9	147.99	28AU	0.3 -3.4 -1.8 0.2
112.78 1.19 1.9	157.88	7SE	0.7 -3.3 -1.8 0.2
118.84 1.29 1.8	167.82	17SE	2.2 -3.3 -1.7 0.1
124.83 1.39 1.8	177.83	27SE	1.4 -3.3 -1.7 0.1
130.73 1.49 1.7	187.89	7OC	-0.3 -3.3 -1.6 0.0
136.54 1.60 1.7	197.99	17OC	-0.6 -3.3 -1.6 -0.1
142.24 1.70 1.6	208.14	27OC	-0.6 -3.4 -1.5 -0.1
147.81 1.81 1.5	218.32	6NO	-0.7 -3.4 -1.5 -0.2
153.22 1.93 1.4	228.52	16NO	-0.8 -3.4 -1.5 -0.3
158.45 2.05 1.2	238.73	26NO	-0.8 -3.4 -1.5 -0.3
163.44 2.18 1.1	248.95	6DE	-0.8 -3.5 -1.4 -0.2
168.16 2.31 0.9	259.14	16DE	-0.6 -3.5 -1.4 -0.2
172.54 2.45 0.7	269.32	26DE	0.5 -3.6 -1.5 -0.1

-589

♂ LONG LAT MAG	☉ LONG	16.00UT	☿ ♀ ♃ ♄ MAGNITUDES
176.48 2.59 0.5	279.47	5JA	3.0 -3.7 -1.5 -0.0
179.90 2.74 0.3	289.58	15JA	1.2 -3.7 -1.5 0.0
182.64 2.89 0.0	299.64	25JA	0.5 -3.8 -1.5 0.1
184.55 3.02 -0.2	309.64	4FE	0.3 -3.9 -1.6 0.2
185.45 3.13 -0.5	319.59	14FE	0.1 -4.0 -1.6 0.2
185.17 3.19 -0.8	329.48	24FE	-0.3 -4.1 -1.6 0.2
183.62 3.17 -1.1	339.30	6MR	-0.9 -4.1 -1.7 0.3
180.91 3.03 -1.4	349.07	16MR	-1.7 -4.2 -1.8 0.3
177.42 2.75 -1.6	358.78	26MR	-1.1 -4.2 -1.8 0.3
173.83 2.34 -1.5	8.45	5AP	-0.1 -4.1 -1.9 0.3
170.85 1.84 -1.4	18.06	15AP	1.0 -3.8 -2.0 0.3
168.98 1.32 -1.2	27.64	25AP	2.0 -3.1 -2.0 0.3
168.47 0.83 -0.9	37.19	5MY	3.5 -3.2 -2.1 0.3
169.28 0.38 -0.7	46.72	15MY	2.5 -3.9 -2.2 0.2
171.26 -0.01 -0.5	56.24	25MY	1.3 -4.1 -2.2 0.2
174.25 -0.34 -0.4	65.76	4JN	0.3 -4.2 -2.3 0.2
178.06 -0.63 -0.1	75.27	14JN	-0.8 -4.2 -2.3 0.2
182.54 -0.88 -0.1	84.82	24JN	-1.6 -4.1 -2.3 0.3
187.60 -1.08 0.1	94.38	4JL	-1.1 -4.0 -2.4 0.3
193.11 -1.26 0.2	103.97	14JL	-0.4 -3.9 -2.4 0.3
199.02 -1.40 0.3	113.62	24JL	0.0 -3.8 -2.3 0.4
205.27 -1.52 0.4	123.30	3AU	0.3 -3.7 -2.3 0.4
211.80 -1.61 0.5	133.04	13AU	0.5 -3.7 -2.3 0.4
218.57 -1.67 0.6	142.84	23AU	1.0 -3.6 -2.2 0.4
225.55 -1.71 0.6	152.69	2SE	2.6 -3.6 -2.1 0.3
232.70 -1.74 0.7	162.60	12SE	1.2 -3.5 -2.1 0.3
240.00 -1.74 0.8	172.58	22SE	-0.5 -3.5 -2.0 0.3
247.42 -1.72 0.9	182.61	2OC	-0.8 -3.5 -1.9 0.2
254.93 -1.68 0.9	192.69	12OC	-0.8 -3.4 -1.9 0.1
262.52 -1.63 1.0	202.82	22OC	-0.8 -3.4 -1.8 0.1
270.15 -1.56 1.0	212.98	1NO	-0.7 -3.4 -1.8 -0.0
277.82 -1.49 1.1	223.17	11NO	-0.6 -3.4 -1.7 -0.1
285.49 -1.40 1.2	233.38	21NO	-0.6 -3.4 -1.7 -0.1
293.16 -1.30 1.2	243.59	1DE	-0.5 -3.4 -1.6 -0.2
300.81 -1.19 1.3	253.80	11DE	0.7 -3.4 -1.6 -0.2
308.42 -1.08 1.4	263.99	21DE	2.8 -3.4 -1.6 -0.2
315.99 -0.97 1.4	274.16	31DE	0.9 -3.4 -1.5 -0.1

-588

♂ LONG LAT MAG	☉ LONG	16.00UT	☿ ♀ ♃ ♄ MAGNITUDES
323.50 -0.85 1.5	284.28	10JA	0.3 -3.4 -1.5 -0.0
330.95 -0.73 1.5	294.37	20JA	0.1 -3.4 -1.5 0.0
338.33 -0.61 1.6	304.40	30JA	-0.0 -3.4 -1.5 0.1
345.64 -0.49 1.6	314.38	9FE	-0.4 -3.5 -1.5 0.2
352.87 -0.38 1.7	324.30	19FE	-1.0 -3.5 -1.6 0.2
0.03 -0.26 1.7	334.15	29FE	-1.6 -3.5 -1.6 0.3
7.11 -0.15 1.7	343.95	10MR	-1.1 -3.4 -1.6 0.3
14.11 -0.04 1.8	353.70	20MR	-0.0 -3.4 -1.7 0.3
21.04 0.06 1.8	3.38	30MR	1.2 -3.4 -1.7 0.4
27.90 0.16 1.8	13.02	9AP	2.7 -3.4 -1.8 0.4
34.70 0.26 1.8	22.62	19AP	3.3 -3.3 -1.8 0.4
41.43 0.35 1.8	32.19	29AP	1.9 -3.3 -1.9 0.4
48.10 0.44 1.8	41.73	9MY	1.0 -3.3 -2.0 0.4
54.71 0.53 1.8	51.25	19MY	0.1 -3.3 -2.0 0.4
61.28 0.61 1.8	60.76	29MY	-0.9 -3.3 -2.1 0.3
67.80 0.69 1.9	70.28	8JN	-1.7 -3.4 -2.2 0.3
74.29 0.76 1.9	79.81	18JN	-1.1 -3.4 -2.2 0.3
80.73 0.83 1.9	89.36	28JN	-0.4 -3.4 -2.3 0.3
87.14 0.90 2.0	98.94	8JL	0.1 -3.4 -2.3 0.4
93.53 0.97 2.0	108.56	18JL	0.4 -3.5 -2.4 0.4
99.89 1.03 2.0	118.21	28JL	0.7 -3.5 -2.4 0.5
106.23 1.09 2.0	127.93	7AU	1.3 -3.6 -2.4 0.5
112.55 1.15 2.0	137.69	17AU	3.0 -3.7 -2.4 0.5
118.85 1.20 2.0	147.52	27AU	1.0 -3.8 -2.4 0.5
125.13 1.25 2.0	157.40	6SE	-0.6 -3.9 -2.4 0.5
131.40 1.30 2.0	167.35	16SE	-0.9 -4.0 -2.3 0.4
137.64 1.35 1.9	177.34	26SE	-0.9 -4.1 -2.3 0.4
143.87 1.39 1.9	187.40	6OC	-0.8 -4.2 -2.2 0.4
150.07 1.43 1.8	197.50	16OC	-0.6 -4.3 -2.2 0.3
156.24 1.47 1.8	207.65	26OC	-0.5 -4.4 -2.1 0.3
162.38 1.51 1.7	217.82	5NO	-0.5 -4.4 -2.0 0.2
168.47 1.54 1.6	228.02	15NO	-0.3 -4.2 -1.9 0.1
174.52 1.56 1.5	238.23	25NO	0.9 -3.7 -1.9 0.0
180.50 1.58 1.4	248.45	5DE	2.5 -3.0 -1.8 -0.0
186.40 1.60 1.3	258.65	15DE	0.6 -3.8 -1.8 -0.1
192.21 1.60 1.2	268.82	25DE	0.1 -4.2 -1.7 -0.1

-587

♂ LONG LAT MAG	☉ LONG	16.00UT	☿ ♀ ♃ ♄ MAGNITUDES
197.91 1.59 1.0	278.97	4JA	0.0 -4.4 -1.7 -0.0
203.46 1.58 0.9	289.08	14JA	-0.1 -4.3 -1.6 0.0
208.85 1.54 0.7	299.14	24JA	-0.4 -4.2 -1.6 0.1
214.02 1.48 0.5	309.15	3FE	-1.0 -4.1 -1.6 0.1
218.91 1.40 0.3	319.10	13FE	-1.5 -4.0 -1.6 0.2
223.48 1.29 0.1	328.99	23FE	-1.1 -3.9 -1.6 0.3
227.61 1.13 -0.2	338.82	5MR	0.0 -3.8 -1.6 0.3
231.21 0.92 -0.4	348.60	15MR	1.5 -3.7 -1.6 0.4
234.12 0.65 -0.7	358.31	25MR	3.4 -3.6 -1.6 0.4
236.18 0.29 -1.0	7.98	4AP	2.5 -3.6 -1.6 0.4
237.20 -0.16 -1.4	17.60	14AP	1.4 -3.5 -1.6 0.5
237.04 -0.71 -1.7	27.18	24AP	0.8 -3.4 -1.7 0.5
235.64 -1.34 -2.0	36.73	4MY	0.1 -3.4 -1.7 0.5
233.20 -2.02 -2.3	46.27	14MY	-0.9 -3.4 -1.8 0.5
230.23 -2.67 -2.3	55.78	24MY	-1.7 -3.3 -1.8 0.5
227.43 -3.21 -2.2	65.30	3JN	-1.2 -3.3 -1.9 0.5
225.49 -3.60 -2.0	74.82	13JN	-0.3 -3.3 -1.9 0.5
224.82 -3.82 -1.8	84.36	23JN	0.2 -3.3 -2.0 0.5
225.51 -3.91 -1.6	93.92	3JL	0.6 -3.3 -2.1 0.5
227.51 -3.90 -1.3	103.52	13JL	1.0 -3.3 -2.2 0.5
230.61 -3.82 -1.1	113.15	23JL	1.8 -3.4 -2.2 0.5
234.62 -3.69 -0.9	122.83	2AU	3.2 -3.4 -2.3 0.6
239.36 -3.52 -0.7	132.57	12AU	0.9 -3.4 -2.3 0.6
244.68 -3.34 -0.5	142.36	22AU	-0.6 -3.4 -2.4 0.6
250.47 -3.13 -0.3	152.21	1SE	-1.1 -3.5 -2.4 0.6
256.62 -2.91 -0.2	162.12	11SE	-1.1 -3.5 -2.4 0.6
263.04 -2.68 0.0	172.09	21SE	-0.7 -3.5 -2.4 0.6
269.69 -2.45 0.2	182.12	1OC	-0.4 -3.5 -2.4 0.6
276.50 -2.23 0.3	192.20	11OC	-0.3 -3.4 -2.4 0.6
283.42 -2.00 0.4	202.32	21OC	-0.3 -3.4 -2.3 0.5
290.43 -1.77 0.6	212.49	31OC	-0.1 -3.4 -2.3 0.5
297.50 -1.56 0.7	222.67	10NO	1.1 -3.4 -2.2 0.4
304.60 -1.35 0.8	232.88	20NO	2.2 -3.3 -2.1 0.4
311.71 -1.15 0.9	243.09	30NO	0.3 -3.3 -2.1 0.3
318.81 -0.95 1.1	253.30	10DE	-0.0 -3.3 -2.0 0.2
325.89 -0.77 1.2	263.49	20DE	-0.1 -3.3 -1.9 0.1
332.95 -0.60 1.3	273.66	30DE	-0.2 -3.4 -1.9 0.1

♂ LONG	♂ LAT	♂ MAG	☉ LONG	16.00UT	☿	♀	♃	♄
				-586	MAGNITUDES			
339.97	-0.44	1.4	283.79	9JA	-0.5	-3.4	-1.8	0.1
346.94	-0.29	1.4	293.87	19JA	-1.0	-3.4	-1.7	0.1
353.87	-0.15	1.5	303.91	29JA	-1.3	-3.4	-1.7	0.2
0.74	-0.02	1.6	313.89	8FE	-1.0	-3.4	-1.6	0.2
7.56	0.11	1.7	323.81	18FE	0.1	-3.5	-1.6	0.3
14.34	0.22	1.7	333.68	28FE	1.9	-3.5	-1.6	0.4
21.06	0.33	1.8	343.48	10MR	3.3	-3.6	-1.6	0.4
27.73	0.42	1.8	353.22	20MR	1.8	-3.6	-1.5	0.5
34.35	0.51	1.9	2.92	30MR	1.1	-3.7	-1.5	0.5
40.93	0.60	1.9	12.56	9AP	0.6	-3.8	-1.5	0.6
47.46	0.67	1.9	22.16	19AP	-0.0	-3.9	-1.5	0.6
53.96	0.75	2.0	31.73	29AP	-0.9	-3.9	-1.5	0.6
60.43	0.81	2.0	41.27	9MY	-1.8	-4.0	-1.6	0.7
66.87	0.87	2.0	50.79	19MY	-1.2	-4.1	-1.6	0.7
73.29	0.92	2.0	60.31	29MY	-0.3	-4.2	-1.6	0.7
79.68	0.97	2.0	69.83	8JN	0.3	-4.2	-1.7	0.7
86.07	1.01	2.0	79.35	18JN	0.8	-4.0	-1.7	0.7
92.44	1.05	2.0	88.90	28JN	1.3	-3.5	-1.8	0.7
98.81	1.09	1.9	98.48	8JL	2.3	-3.0	-1.8	0.7
105.18	1.12	2.0	108.09	18JL	2.9	-3.6	-1.9	0.7
111.55	1.14	2.0	117.75	28JL	0.7	-4.1	-1.9	0.7
117.93	1.17	2.0	127.46	7AU	-0.7	-4.2	-2.0	0.8
124.32	1.18	2.0	137.22	17AU	-1.2	-4.3	-2.1	0.8
130.73	1.20	2.0	147.04	27AU	-1.2	-4.2	-2.1	0.8
137.15	1.20	2.0	156.92	6SE	-0.7	-4.1	-2.2	0.9
143.59	1.21	2.0	166.86	16SE	-0.3	-4.0	-2.3	0.9
150.05	1.21	2.0	176.86	26SE	-0.2	-4.0	-2.3	0.9
156.53	1.20	2.0	186.91	6OC	-0.1	-3.9	-2.3	0.9
163.03	1.19	1.9	197.01	16OC	0.1	-3.8	-2.4	0.8
169.56	1.17	1.9	207.16	26OC	1.3	-3.7	-2.4	0.8
176.11	1.15	1.8	217.33	5NO	2.0	-3.7	-2.3	0.8
182.67	1.11	1.8	227.53	15NO	0.1	-3.6	-2.3	0.7
189.26	1.07	1.7	237.74	25NO	-0.2	-3.5	-2.3	0.7
195.86	1.02	1.6	247.95	5DE	-0.3	-3.5	-2.2	0.6
202.48	0.96	1.5	258.16	15DE	-0.4	-3.4	-2.1	0.5
209.11	0.88	1.5	268.34	25DE	-0.6	-3.4	-2.1	0.4
				-585				
215.74	0.79	1.4	278.48	4JA	-1.0	-3.4	-2.0	0.4
222.38	0.69	1.3	288.60	14JA	-1.2	-3.4	-1.9	0.3
229.02	0.57	1.1	298.66	24JA	-0.9	-3.3	-1.9	0.3
235.65	0.43	1.0	308.67	3FE	0.2	-3.3	-1.8	0.4
242.26	0.26	0.9	318.62	13FE	2.3	-3.3	-1.7	0.4
248.84	0.08	0.7	328.51	23FE	2.6	-3.3	-1.7	0.5
255.38	-0.14	0.6	338.34	5MR	1.3	-3.3	-1.6	0.5
261.86	-0.38	0.4	348.12	15MR	0.8	-3.3	-1.6	0.6
268.25	-0.66	0.3	357.84	25MR	0.4	-3.4	-1.5	0.6
274.53	-0.97	0.1	7.51	4AP	-0.1	-3.4	-1.5	0.7
280.65	-1.32	-0.1	17.13	14AP	-0.9	-3.4	-1.5	0.7
286.55	-1.71	-0.3	26.71	24AP	-1.8	-3.5	-1.5	0.8
292.18	-2.15	-0.5	36.27	4MY	-1.2	-3.5	-1.5	0.8
297.45	-2.63	-0.7	45.80	14MY	-0.2	-3.5	-1.4	0.8
302.23	-3.16	-0.9	55.32	24MY	0.4	-3.4	-1.5	0.9
306.39	-3.74	-1.2	64.83	3JN	1.0	-3.4	-1.5	0.9
309.76	-4.37	-1.4	74.36	13JN	1.7	-3.4	-1.5	0.9
312.12	-5.01	-1.7	83.89	23JN	2.9	-3.3	-1.5	0.9
313.28	-5.65	-1.9	93.45	3JL	2.4	-3.3	-1.5	0.9
313.08	-6.22	-2.2	103.05	13JL	0.6	-3.3	-1.6	0.9
311.58	-6.61	-2.4	112.68	23JL	-0.8	-3.3	-1.6	0.9
309.11	-6.74	-2.6	122.36	2AU	-1.3	-3.3	-1.7	0.9
306.30	-6.55	-2.5	132.10	12AU	-1.2	-3.3	-1.7	1.0
303.96	-6.08	-2.3	141.88	22AU	-0.6	-3.4	-1.8	1.0
302.64	-5.41	-2.0	151.73	1SE	-0.2	-3.4	-1.8	1.1
302.61	-4.68	-1.7	161.64	11SE	-0.1	-3.4	-1.9	1.1
303.87	-3.95	-1.4	171.61	21SE	0.0	-3.5	-2.0	1.1
306.24	-3.27	-1.1	181.63	1OC	0.3	-3.5	-2.0	1.1
309.52	-2.67	-0.8	191.71	11OC	1.6	-3.6	-2.1	1.1
313.52	-2.14	-0.5	201.83	21OC	1.7	-3.7	-2.2	1.1
318.08	-1.68	-0.2	211.99	31OC	-0.1	-3.7	-2.2	1.1
323.08	-1.29	0.0	222.18	10NO	-0.4	-3.8	-2.2	1.1
328.41	-0.94	0.2	232.38	20NO	-0.4	-3.9	-2.2	1.0
333.98	-0.64	0.4	242.60	30NO	-0.5	-4.0	-2.2	1.0
339.75	-0.39	0.6	252.81	10DE	-0.7	-4.1	-2.2	0.9
345.67	-0.16	0.8	263.00	20DE	-1.0	-4.2	-2.2	0.9
351.68	0.03	0.9	273.17	30DE	-1.0	-4.3	-2.1	0.8

♂ LONG	♂ LAT	♂ MAG	☉ LONG	16.00UT	☿	♀	♃	♄
				-584	MAGNITUDES			
357.79	0.20	1.1	283.30	9JA	-0.8	-4.3	-2.1	0.7
3.95	0.35	1.2	293.39	19JA	0.3	-4.3	-2.0	0.6
10.14	0.48	1.3	303.42	29JA	2.7	-4.1	-2.0	0.6
16.37	0.60	1.4	313.41	8FE	2.0	-3.6	-1.9	0.6
22.61	0.69	1.5	323.33	18FE	1.0	-3.3	-1.8	0.6
28.86	0.78	1.6	333.20	28FE	0.6	-3.8	-1.8	0.7
35.11	0.86	1.7	343.00	9MR	0.3	-4.1	-1.7	0.7
41.37	0.92	1.8	352.75	19MR	-0.2	-4.2	-1.6	0.8
47.62	0.98	1.8	2.44	29MR	-0.9	-4.2	-1.6	0.8
53.88	1.02	1.9	12.09	8AP	-1.8	-4.1	-1.5	0.9
60.13	1.06	1.9	21.69	18AP	-1.2	-4.0	-1.5	0.9
66.38	1.10	2.0	31.26	28AP	-0.2	-3.9	-1.4	1.0
72.64	1.13	2.0	40.80	8MY	0.6	-3.8	-1.4	1.0
78.90	1.15	2.0	50.33	18MY	1.3	-3.7	-1.4	1.1
85.17	1.16	2.0	59.85	28MY	2.3	-3.7	-1.4	1.1
91.45	1.17	2.0	69.36	7JN	3.4	-3.6	-1.4	1.1
97.74	1.18	2.0	78.89	17JN	1.9	-3.5	-1.3	1.2
104.06	1.18	2.0	88.44	27JN	0.5	-3.5	-1.4	1.2
110.40	1.18	2.0	98.02	7JL	-0.8	-3.4	-1.4	1.2
116.77	1.17	2.0	107.63	17JL	-1.4	-3.4	-1.4	1.2
123.16	1.16	2.0	117.28	27JL	-1.2	-3.4	-1.4	1.2
129.60	1.14	1.9	126.99	6AU	-0.5	-3.4	-1.4	1.2
136.07	1.12	1.9	136.75	16AU	-0.2	-3.4	-1.5	1.2
142.59	1.09	1.9	146.57	26AU	0.1	-3.4	-1.5	1.3
149.15	1.06	1.9	156.45	5SE	0.2	-3.4	-1.6	1.3
155.76	1.02	1.9	166.38	15SE	0.5	-3.4	-1.6	1.4
162.42	0.98	1.9	176.38	25SE	1.9	-3.4	-1.7	1.4
169.13	0.93	1.9	186.43	5OC	1.5	-3.4	-1.7	1.4
175.89	0.87	1.9	196.52	15OC	-0.2	-3.4	-1.8	1.4
182.72	0.81	1.8	206.66	25OC	-0.5	-3.4	-1.9	1.4
189.59	0.74	1.8	216.84	4NO	-0.5	-3.4	-1.9	1.3
196.52	0.66	1.8	227.03	14NO	-0.6	-3.5	-2.0	1.3
203.51	0.58	1.7	237.24	24NO	-0.8	-3.5	-2.1	1.2
210.55	0.48	1.7	247.46	4DE	-0.8	-3.5	-2.1	1.2
217.65	0.38	1.6	257.66	14DE	-0.9	-3.5	-2.1	1.1
224.80	0.27	1.6	267.84	24DE	-0.7	-3.4	-2.1	1.1
				-583				
232.00	0.14	1.5	277.99	3JA	0.4	-3.4	-2.1	1.0
239.24	0.01	1.4	288.10	13JA	3.0	-3.4	-2.1	1.0
246.53	-0.13	1.4	298.17	23JA	1.5	-3.4	-2.1	1.0
253.85	-0.28	1.3	308.18	2FE	0.7	-3.3	-2.0	0.9
261.21	-0.44	1.2	318.13	12FE	0.4	-3.3	-2.0	0.9
268.59	-0.61	1.1	328.03	22FE	0.1	-3.3	-1.9	0.9
275.98	-0.79	1.1	337.87	4MR	-0.3	-3.3	-1.8	0.9
283.38	-0.98	1.0	347.64	14MR	-0.9	-3.3	-1.8	1.0
290.77	-1.16	0.9	357.37	24MR	-1.8	-3.4	-1.7	1.0
298.13	-1.35	0.8	7.04	3AP	-1.2	-3.4	-1.6	1.1
305.44	-1.55	0.7	16.66	13AP	-0.1	-3.4	-1.6	1.1
312.69	-1.74	0.6	26.25	23AP	0.8	-3.4	-1.5	1.2
319.85	-1.93	0.5	35.80	3MY	1.7	-3.5	-1.5	1.3
326.89	-2.11	0.4	45.34	13MY	3.0	-3.5	-1.4	1.3
333.78	-2.28	0.3	54.86	23MY	2.9	-3.6	-1.4	1.3
340.47	-2.45	0.2	64.37	2JN	1.5	-3.6	-1.3	1.4
346.94	-2.60	0.1	73.89	12JN	0.4	-3.7	-1.3	1.4
353.11	-2.74	-0.0	83.43	22JN	-0.8	-3.8	-1.3	1.5
358.92	-2.87	-0.1	92.99	2JL	-1.5	-3.9	-1.3	1.5
4.29	-2.98	-0.3	102.58	12JL	-1.2	-4.0	-1.3	1.4
9.12	-3.07	-0.4	112.22	22JL	-0.5	-4.1	-1.3	1.4
13.26	-3.14	-0.6	121.89	1AU	-0.1	-4.2	-1.3	1.4
16.56	-3.17	-0.8	131.62	11AU	0.2	-4.3	-1.3	1.3
18.82	-3.16	-1.0	141.41	21AU	0.4	-4.3	-1.3	1.3
19.82	-3.09	-1.2	151.26	31AU	0.8	-4.1	-1.3	1.2
19.40	-2.93	-1.5	161.16	10SE	2.3	-3.6	-1.4	1.3
17.51	-2.66	-1.7	171.13	20SE	1.4	-3.3	-1.4	1.3
14.44	-2.26	-1.8	181.15	30SE	-0.4	-3.9	-1.5	1.2
10.78	-1.76	-1.9	191.22	10OC	-0.7	-4.2	-1.5	1.2
7.34	-1.21	-1.7	201.34	20OC	-0.7	-4.4	-1.6	1.2
4.86	-0.68	-1.3	211.50	30OC	-0.7	-4.4	-1.6	1.2
3.67	-0.21	-1.0	221.69	9NO	-0.7	-4.3	-1.7	1.2
3.83	0.18	-0.7	231.89	19NO	-0.7	-4.2	-1.8	1.1
5.19	0.49	-0.4	242.10	29NO	-0.7	-4.1	-1.8	1.1
7.55	0.73	-0.1	252.31	9DE	-0.6	-4.0	-1.9	1.1
10.70	0.92	0.2	262.51	19DE	0.5	-3.9	-1.9	1.0
14.47	1.06	0.4	272.67	29DE	3.0	-3.8	-2.0	1.0

♂ LONG	LAT	MAG	⊙ LONG	16.00UT	☿	♀	♃	♄
				-582	MAGNITUDES			
18.71	1.17	0.6	282.81	8JA	1.1	-3.7	-2.0	0.9
23.34	1.25	0.8	292.90	18JA	0.4	-3.6	-2.0	0.9
28.26	1.32	1.0	302.94	28JA	0.2	-3.6	-2.1	0.8
33.41	1.36	1.1	312.93	7FE	0.0	-3.5	-2.0	0.8
38.75	1.40	1.3	322.86	17FE	-0.3	-3.4	-2.0	0.8
44.23	1.42	1.4	332.72	27FE	-0.9	-3.4	-2.0	0.7
49.82	1.43	1.5	342.53	9MR	-1.7	-3.4	-1.9	0.8
55.52	1.44	1.6	352.28	19MR	-1.2	-3.3	-1.9	0.9
61.29	1.44	1.7	1.98	29MR	-0.1	-3.3	-1.8	0.9
67.14	1.43	1.8	11.63	8AP	1.0	-3.3	-1.8	1.0
73.04	1.42	1.8	21.23	18AP	2.3	-3.3	-1.7	1.1
79.00	1.40	1.9	30.80	28AP	3.8	-3.3	-1.6	1.1
85.02	1.38	1.9	40.35	8MY	2.3	-3.3	-1.6	1.2
91.08	1.36	2.0	49.87	18MY	1.2	-3.3	-1.5	1.2
97.19	1.33	2.0	59.38	28MY	0.2	-3.4	-1.4	1.3
103.36	1.29	2.0	68.90	7JN	-0.8	-3.4	-1.4	1.3
109.57	1.26	2.0	78.43	17JN	-1.6	-3.4	-1.4	1.3
115.83	1.21	2.0	87.97	27JN	-1.2	-3.5	-1.3	1.3
122.15	1.17	2.0	97.55	7JL	-0.4	-3.5	-1.3	1.3
128.53	1.12	2.0	107.16	17JL	0.0	-3.5	-1.3	1.3
134.97	1.07	2.0	116.81	27JL	0.3	-3.4	-1.2	1.3
141.47	1.01	1.9	126.51	6AU	0.6	-3.4	-1.2	1.2
148.04	0.95	1.9	136.27	16AU	1.1	-3.4	-1.2	1.2
154.68	0.88	1.9	146.08	26AU	2.7	-3.4	-1.2	1.1
161.39	0.81	1.8	155.96	5SE	1.2	-3.3	-1.2	1.1
168.17	0.74	1.8	165.89	15SE	-0.5	-3.3	-1.3	1.1
175.02	0.66	1.7	175.88	25SE	-0.8	-3.3	-1.3	1.1
181.96	0.58	1.7	185.94	5OC	-0.8	-3.3	-1.3	1.1
188.97	0.49	1.7	196.03	15OC	-0.8	-3.3	-1.3	1.1
196.05	0.39	1.7	206.17	25OC	-0.6	-3.4	-1.4	1.1
203.22	0.30	1.7	216.34	4NO	-0.5	-3.4	-1.4	1.1
210.45	0.19	1.7	226.54	14NO	-0.6	-3.4	-1.5	1.1
217.76	0.09	1.6	236.75	24NO	-0.4	-3.4	-1.6	1.0
225.14	-0.03	1.6	246.96	4DE	0.7	-3.5	-1.6	1.0
232.59	-0.14	1.6	257.16	14DE	2.8	-3.5	-1.7	1.0
240.09	-0.26	1.6	267.35	24DE	0.8	-3.6	-1.8	0.9
				-581				
247.66	-0.38	1.5	277.50	3JA	0.2	-3.7	-1.8	0.9
255.26	-0.50	1.5	287.62	13JA	0.1	-3.7	-1.9	0.8
262.91	-0.63	1.5	297.68	23JA	-0.1	-3.8	-1.9	0.8
270.59	-0.75	1.4	307.70	2FE	-0.4	-3.9	-2.0	0.7
278.29	-0.87	1.4	317.66	12FE	-0.9	-4.0	-2.0	0.7
286.01	-0.99	1.3	327.56	22FE	-1.6	-4.1	-2.0	0.6
293.72	-1.10	1.3	337.40	4MR	-1.1	-4.1	-2.0	0.6
301.41	-1.20	1.3	347.18	14MR	-0.0	-4.2	-2.0	0.6
309.08	-1.29	1.2	356.90	24MR	1.3	-4.2	-2.0	0.7
316.71	-1.38	1.2	6.58	3AP	2.9	-4.1	-1.9	0.7
324.28	-1.45	1.2	16.20	13AP	3.0	-3.8	-1.9	0.8
331.79	-1.51	1.1	25.79	23AP	1.7	-3.0	-1.8	0.9
339.21	-1.56	1.1	35.34	3MY	0.9	-3.2	-1.8	0.9
346.52	-1.59	1.0	44.88	13MY	0.1	-3.9	-1.7	1.0
353.73	-1.61	1.0	54.39	23MY	-0.8	-4.1	-1.7	1.0
0.80	-1.61	1.0	63.91	2JN	-1.7	-4.2	-1.6	1.1
7.72	-1.59	0.9	73.43	12JN	-1.2	-4.2	-1.5	1.1
14.47	-1.56	0.9	82.96	22JN	-0.4	-4.1	-1.5	1.1
21.03	-1.51	0.8	92.53	2JL	0.1	-4.0	-1.4	1.1
27.37	-1.44	0.7	102.11	12JL	0.5	-3.9	-1.4	1.2
33.46	-1.36	0.7	111.74	22JL	0.8	-3.8	-1.3	1.1
39.25	-1.26	0.6	121.42	1AU	1.5	-3.7	-1.3	1.1
44.69	-1.13	0.5	131.15	11AU	3.1	-3.7	-1.3	1.1
49.73	-0.98	0.3	140.93	21AU	1.0	-3.6	-1.2	1.1
54.26	-0.81	0.2	150.78	31AU	-0.6	-3.6	-1.2	1.0
58.19	-0.60	0.1	160.68	10SE	-1.0	-3.5	-1.2	1.0
61.36	-0.36	-0.1	170.64	20SE	-1.0	-3.5	-1.2	0.9
63.61	-0.07	-0.3	180.66	30SE	-0.8	-3.5	-1.2	1.0
64.73	0.27	-0.5	190.73	10OC	-0.5	-3.4	-1.2	1.0
64.55	0.67	-0.8	200.85	20OC	-0.4	-3.4	-1.3	1.0
62.94	1.10	-1.0	211.01	30OC	-0.4	-3.4	-1.3	1.0
60.04	1.55	-1.2	221.19	9NO	-0.2	-3.4	-1.3	1.0
56.26	1.96	-1.3	231.40	19NO	0.9	-3.4	-1.4	1.0
52.34	2.28	-1.2	241.61	29NO	2.5	-3.4	-1.4	1.0
49.05	2.49	-1.0	251.82	9DE	0.5	-3.4	-1.5	0.9
46.89	2.60	-0.7	262.02	19DE	0.1	-3.4	-1.5	0.9
46.07	2.62	-0.4	272.19	29DE	-0.0	-3.4	-1.6	0.9

♂ LONG	LAT	MAG	⊙ LONG	16.00UT	☿	♀	♃	♄
				-580	MAGNITUDES			
46.52	2.59	-0.1	282.32	8JA	-0.2	-3.4	-1.7	0.8
48.07	2.52	0.2	292.41	18JA	-0.5	-3.4	-1.7	0.8
50.52	2.45	0.4	302.46	28JA	-1.0	-3.4	-1.8	0.7
53.68	2.36	0.6	312.44	7FE	-1.4	-3.5	-1.9	0.7
57.41	2.27	0.8	322.38	17FE	-1.1	-3.5	-1.9	0.6
61.59	2.17	1.0	332.25	27FE	0.0	-3.5	-2.0	0.6
66.13	2.08	1.1	342.05	8MR	1.6	-3.4	-2.0	0.5
70.95	2.00	1.3	351.81	18MR	3.6	-3.4	-2.0	0.5
76.02	1.91	1.4	1.51	28MR	2.2	-3.4	-2.1	0.5
81.28	1.82	1.5	11.15	7AP	1.3	-3.4	-2.1	0.6
86.70	1.74	1.6	20.76	17AP	0.7	-3.3	-2.0	0.6
92.27	1.65	1.6	30.34	27AP	0.0	-3.3	-2.0	0.7
97.97	1.57	1.7	39.88	7MY	-0.8	-3.3	-2.0	0.8
103.78	1.48	1.8	49.41	17MY	-1.7	-3.3	-1.9	0.8
109.70	1.40	1.8	58.92	27MY	-1.2	-3.3	-1.9	0.9
115.72	1.31	1.8	68.44	6JN	-0.3	-3.4	-1.8	0.9
121.84	1.22	1.8	77.96	16JN	0.2	-3.4	-1.7	1.0
128.04	1.14	1.9	87.51	26JN	0.6	-3.4	-1.7	1.0
134.34	1.05	1.9	97.08	6JL	1.1	-3.4	-1.6	1.0
140.74	0.96	1.9	106.69	16JL	1.9	-3.5	-1.6	1.0
147.22	0.86	1.9	116.34	26JL	3.1	-3.6	-1.5	1.0
153.80	0.77	1.8	126.04	5AU	0.9	-3.6	-1.5	1.0
160.48	0.67	1.8	135.79	15AU	-0.6	-3.7	-1.4	1.0
167.24	0.57	1.8	145.61	25AU	-1.1	-3.8	-1.4	1.0
174.10	0.47	1.8	155.48	4SE	-1.1	-3.9	-1.3	1.0
181.06	0.37	1.7	165.41	14SE	-0.7	-4.0	-1.3	0.9
188.12	0.26	1.7	175.40	24SE	-0.4	-4.1	-1.3	0.9
195.26	0.16	1.6	185.44	4OC	-0.3	-4.2	-1.3	0.9
202.50	0.05	1.6	195.54	14OC	-0.2	-4.3	-1.3	0.9
209.83	-0.06	1.5	205.68	24OC	-0.1	-4.4	-1.3	0.9
217.24	-0.17	1.5	215.85	3NO	1.1	-4.4	-1.3	0.9
224.74	-0.27	1.5	226.04	13NO	2.2	-4.2	-1.3	0.9
232.31	-0.38	1.5	236.25	23NO	0.3	-3.6	-1.3	0.9
239.95	-0.49	1.5	246.46	3DE	-0.1	-3.0	-1.4	0.9
247.64	-0.59	1.4	256.67	13DE	-0.2	-3.9	-1.4	0.9
255.39	-0.69	1.4	266.85	23DE	-0.3	-4.2	-1.5	0.8
				-579				
263.18	-0.78	1.4	277.01	2JA	-0.5	-4.4	-1.5	0.8
271.00	-0.87	1.4	287.13	12JA	-1.0	-4.3	-1.6	0.8
278.83	-0.94	1.4	297.20	22JA	-1.3	-4.2	-1.6	0.7
286.68	-1.01	1.4	307.21	1FE	-1.0	-4.1	-1.7	0.7
294.51	-1.08	1.4	317.18	11FE	0.1	-4.0	-1.8	0.6
302.33	-1.13	1.4	327.08	21FE	2.0	-3.9	-1.8	0.6
310.12	-1.16	1.4	336.92	3MR	3.0	-3.8	-1.9	0.5
317.86	-1.19	1.4	346.71	13MR	1.6	-3.7	-2.0	0.5
325.55	-1.20	1.4	356.43	23MR	1.0	-3.6	-2.0	0.4
333.18	-1.21	1.4	6.11	2AP	0.5	-3.6	-2.1	0.4
340.73	-1.19	1.4	15.74	12AP	-0.0	-3.5	-2.1	0.4
348.20	-1.17	1.4	25.33	22AP	-0.8	-3.4	-2.1	0.5
355.58	-1.13	1.4	34.88	2MY	-1.8	-3.4	-2.1	0.6
2.85	-1.09	1.4	44.42	12MY	-1.2	-3.4	-2.1	0.6
10.02	-1.03	1.4	53.94	22MY	-0.3	-3.3	-2.1	0.7
17.08	-0.95	1.4	63.46	1JN	0.4	-3.3	-2.1	0.8
24.02	-0.87	1.4	72.98	11JN	0.8	-3.3	-2.0	0.8
30.83	-0.78	1.4	82.51	21JN	1.4	-3.3	-2.0	0.9
37.51	-0.67	1.3	92.07	1JL	2.5	-3.3	-1.9	0.9
44.03	-0.56	1.3	101.65	11JL	2.7	-3.3	-1.9	0.9
50.40	-0.43	1.3	111.28	21JL	0.7	-3.4	-1.8	0.9
56.60	-0.30	1.2	120.95	31JL	-0.7	-3.4	-1.7	0.9
62.61	-0.15	1.2	130.68	10AU	-1.3	-3.4	-1.7	0.9
68.39	0.02	1.1	140.45	20AU	-1.2	-3.4	-1.6	0.9
73.94	0.19	1.0	150.30	30AU	-0.6	-3.5	-1.6	0.9
79.18	0.39	0.9	160.20	9SE	-0.3	-3.5	-1.5	0.9
84.09	0.61	0.8	170.15	19SE	-0.2	-3.5	-1.5	0.9
88.58	0.85	0.7	180.17	29SE	-0.1	-3.5	-1.4	0.8
92.56	1.12	0.5	190.24	9OC	0.1	-3.4	-1.4	0.8
95.92	1.43	0.4	200.35	19OC	1.4	-3.4	-1.4	0.8
98.51	1.78	0.2	210.51	29OC	2.0	-3.4	-1.4	0.8
100.15	2.16	-0.0	220.69	8NO	0.1	-3.4	-1.4	0.8
100.66	2.59	-0.3	230.89	18NO	-0.3	-3.3	-1.4	0.9
99.87	3.04	-0.5	241.11	28NO	-0.3	-3.3	-1.4	0.9
97.73	3.47	-0.7	251.32	8DE	-0.4	-3.3	-1.4	0.9
94.46	3.83	-0.9	261.51	18DE	-0.6	-3.3	-1.4	0.8
90.53	4.07	-1.0	271.69	28DE	-1.0	-3.4	-1.4	0.8

Left Table

♂ LONG	LAT	MAG	☉ LONG	16.00UT -578	☿	♀	♃	♄
86.70	4.14	-0.9	281.83	7JA	-1.1	-3.4	-1.5	0.8
83.64	4.06	-0.7	291.92	17JA	-0.9	-3.4	-1.5	0.8
81.76	3.87	-0.4	301.97	27JA	0.2	-3.4	-1.6	0.7
81.20	3.62	-0.2	311.96	6FE	2.4	-3.4	-1.6	0.7
81.85	3.36	0.1	321.89	16FE	2.4	-3.5	-1.7	0.6
83.55	3.09	0.3	331.77	26FE	1.2	-3.5	-1.7	0.6
86.11	2.84	0.5	341.58	8MR	0.7	-3.6	-1.8	0.5
89.36	2.61	0.7	351.34	18MR	0.3	-3.6	-1.9	0.5
93.17	2.39	0.9	1.04	28MR	-0.1	-3.7	-1.9	0.4
97.43	2.19	1.0	10.70	7AP	-0.9	-3.8	-2.0	0.4
102.06	2.00	1.1	20.30	17AP	-1.8	-3.9	-2.1	0.3
106.99	1.82	1.2	29.88	27AP	-1.2	-4.0	-2.1	0.4
112.19	1.65	1.3	39.42	7MY	-0.3	-4.0	-2.2	0.5
117.61	1.50	1.4	48.95	17MY	0.5	-4.1	-2.2	0.5
123.23	1.34	1.5	58.47	27MY	1.1	-4.2	-2.3	0.6
129.03	1.20	1.5	67.98	6JN	1.9	-4.2	-2.3	0.7
134.99	1.05	1.6	77.51	16JN	3.2	-4.0	-2.3	0.7
141.10	0.92	1.6	87.05	26JN	2.3	-3.5	-2.2	0.8
147.36	0.78	1.6	96.62	6JL	0.6	-3.0	-2.2	0.8
153.74	0.65	1.6	106.23	16JL	-0.7	-3.7	-2.2	0.8
160.27	0.52	1.6	115.88	26JL	-1.4	-4.1	-2.1	0.9
166.91	0.39	1.6	125.57	5AU	-1.2	-4.2	-2.0	0.9
173.68	0.26	1.6	135.32	15AU	-0.6	-4.3	-2.0	0.9
180.58	0.14	1.6	145.13	25AU	-0.2	-4.2	-1.9	0.9
187.59	0.02	1.6	155.00	4SE	-0.0	-4.1	-1.8	0.9
194.71	-0.10	1.6	164.92	14SE	0.1	-4.0	-1.8	0.9
201.95	-0.21	1.6	174.91	24SE	0.4	-3.9	-1.7	0.8
209.29	-0.33	1.6	184.95	4OC	1.7	-3.9	-1.7	0.8
216.73	-0.43	1.5	195.04	14OC	1.8	-3.8	-1.6	0.8
224.27	-0.53	1.5	205.18	24OC	-0.1	-3.7	-1.6	0.7
231.89	-0.63	1.5	215.35	3NO	-0.4	-3.7	-1.5	0.8
239.58	-0.72	1.4	225.54	13NO	-0.5	-3.6	-1.5	0.8
247.34	-0.80	1.4	235.76	23NO	-0.5	-3.5	-1.5	0.8
255.15	-0.87	1.4	245.97	3DE	-0.7	-3.5	-1.5	0.8
263.00	-0.94	1.4	256.17	13DE	-0.9	-3.4	-1.5	0.8
270.88	-0.99	1.3	266.36	23DE	-1.0	-3.4	-1.5	0.8

-577

♂ LONG	LAT	MAG	☉ LONG	16.00UT	☿	♀	♃	♄
278.77	-1.03	1.3	276.52	2JA	-0.8	-3.4	-1.5	0.8
286.67	-1.06	1.3	286.64	12JA	0.3	-3.4	-1.5	0.8
294.56	-1.08	1.3	296.71	22JA	2.7	-3.3	-1.5	0.8
302.42	-1.09	1.3	306.73	1FE	1.8	-3.3	-1.5	0.7
310.24	-1.09	1.4	316.70	11FE	0.9	-3.3	-1.6	0.7
318.01	-1.07	1.4	326.60	21FE	0.5	-3.3	-1.6	0.6
325.73	-1.05	1.4	336.45	3MR	0.2	-3.3	-1.7	0.6
333.38	-1.01	1.5	346.23	13MR	-0.2	-3.3	-1.7	0.5
340.96	-0.96	1.5	355.97	23MR	-0.9	-3.4	-1.8	0.5
348.45	-0.91	1.5	5.64	2AP	-1.8	-3.4	-1.8	0.4
355.86	-0.84	1.5	15.27	12AP	-1.2	-3.4	-1.9	0.3
3.18	-0.77	1.6	24.86	22AP	-0.2	-3.5	-2.0	0.3
10.40	-0.69	1.6	34.42	2MY	0.7	-3.5	-2.1	0.3
17.53	-0.60	1.6	43.96	12MY	1.4	-3.5	-2.1	0.4
24.57	-0.50	1.6	53.48	22MY	2.5	-3.4	-2.2	0.4
31.50	-0.40	1.7	62.99	1JN	3.3	-3.4	-2.2	0.5
38.34	-0.30	1.7	72.51	11JN	1.8	-3.4	-2.3	0.6
45.08	-0.19	1.7	82.04	21JN	0.5	-3.3	-2.3	0.6
51.71	-0.07	1.7	91.60	1JL	-0.8	-3.3	-2.4	0.7
58.25	0.05	1.7	101.18	11JL	-1.5	-3.3	-2.4	0.7
64.67	0.17	1.7	110.81	21JL	-1.2	-3.3	-2.4	0.8
70.99	0.30	1.6	120.48	31JL	-0.5	-3.3	-2.3	0.8
77.18	0.44	1.6	130.20	10AU	-0.1	-3.3	-2.3	0.8
83.25	0.58	1.6	139.96	20AU	0.1	-3.4	-2.3	0.8
89.18	0.73	1.5	149.81	30AU	0.3	-3.4	-2.2	0.8
94.95	0.89	1.5	159.71	9SE	0.6	-3.4	-2.1	0.8
100.54	1.06	1.4	169.67	19SE	2.0	-3.5	-2.1	0.8
105.93	1.24	1.3	179.68	29SE	1.6	-3.5	-2.0	0.8
111.07	1.43	1.2	189.75	9OC	-0.3	-3.6	-1.9	0.8
115.92	1.65	1.1	199.86	19OC	-0.6	-3.7	-1.9	0.7
120.42	1.88	0.9	210.01	29OC	-0.6	-3.7	-1.8	0.7
124.49	2.14	0.8	220.20	8NO	-0.6	-3.8	-1.8	0.7
128.02	2.43	0.6	230.40	18NO	-0.8	-3.9	-1.7	0.7
130.89	2.74	0.4	240.61	28NO	-0.8	-4.0	-1.7	0.8
132.94	3.09	0.2	250.82	8DE	-0.8	-4.1	-1.6	0.8
133.99	3.46	-0.1	261.02	18DE	-0.7	-4.2	-1.6	0.8
133.87	3.85	-0.3	271.19	28DE	0.4	-4.3	-1.6	0.8

Right Table

♂ LONG	LAT	MAG	☉ LONG	16.00UT -576	☿	♀	♃	♄
132.44	4.21	-0.6	281.33	7JA	3.0	-4.4	-1.5	0.8
129.77	4.49	-0.8	291.43	17JA	1.3	-4.3	-1.5	0.8
126.18	4.64	-1.0	301.48	27JA	0.6	-4.1	-1.5	0.8
122.26	4.61	-1.0	311.48	6FE	0.3	-3.6	-1.5	0.7
118.76	4.41	-0.8	321.41	16FE	0.1	-3.3	-1.6	0.7
116.24	4.08	-0.6	331.29	26FE	-0.3	-3.8	-1.6	0.7
114.98	3.69	-0.4	341.11	7MR	-0.9	-4.2	-1.6	0.6
115.02	3.28	-0.2	350.87	17MR	-1.7	-4.2	-1.6	0.6
116.21	2.88	0.0	0.57	27MR	-1.2	-4.2	-1.7	0.5
118.39	2.51	0.2	10.23	6AP	-0.2	-4.1	-1.7	0.4
121.38	2.18	0.4	19.84	16AP	0.9	-4.0	-1.8	0.4
125.04	1.87	0.6	29.41	26AP	1.9	-3.9	-1.8	0.3
129.24	1.59	0.7	38.96	6MY	3.3	-3.8	-1.9	0.3
133.89	1.33	0.8	48.49	16MY	2.7	-3.7	-2.0	0.3
138.91	1.10	0.9	58.00	26MY	1.4	-3.7	-2.0	0.4
144.26	0.88	1.0	67.52	5JN	0.3	-3.6	-2.1	0.4
149.89	0.68	1.1	77.05	15JN	-0.8	-3.5	-2.2	0.5
155.77	0.48	1.1	86.59	25JN	-1.6	-3.5	-2.2	0.6
161.87	0.30	1.2	96.16	5JL	-1.2	-3.4	-2.3	0.6
168.18	0.14	1.2	105.76	15JL	-0.5	-3.4	-2.4	0.7
174.67	-0.02	1.3	115.41	25JL	-0.0	-3.4	-2.4	0.7
181.34	-0.17	1.3	125.10	4AU	0.2	-3.4	-2.4	0.7
188.17	-0.31	1.3	134.85	14AU	0.4	-3.4	-2.4	0.8
195.15	-0.44	1.3	144.66	24AU	0.9	-3.4	-2.4	0.8
202.27	-0.57	1.3	154.52	3SE	2.4	-3.4	-2.4	0.8
209.52	-0.68	1.4	164.44	13SE	1.4	-3.4	-2.4	0.8
216.89	-0.78	1.4	174.43	23SE	-0.4	-3.4	-2.3	0.8
224.38	-0.87	1.4	184.47	3OC	-0.8	-3.4	-2.3	0.8
231.96	-0.95	1.4	194.55	13OC	-0.7	-3.4	-2.2	0.8
239.63	-1.01	1.4	204.69	23OC	-0.8	-3.4	-2.1	0.7
247.37	-1.07	1.4	214.86	2NO	-0.7	-3.4	-2.1	0.7
255.17	-1.11	1.4	225.05	12NO	-0.6	-3.5	-2.0	0.7
263.01	-1.14	1.4	235.26	22NO	-0.6	-3.5	-1.9	0.7
270.88	-1.15	1.4	245.47	2DE	-0.5	-3.5	-1.9	0.7
278.77	-1.16	1.4	255.67	12DE	0.5	-3.5	-1.8	0.8
286.65	-1.15	1.4	265.86	22DE	3.0	-3.4	-1.7	0.8

-575

♂ LONG	LAT	MAG	☉ LONG	16.00UT	☿	♀	♃	♄
294.52	-1.13	1.4	276.02	1JA	1.0	-3.4	-1.7	0.8
302.36	-1.09	1.4	286.14	11JA	0.4	-3.4	-1.7	0.8
310.16	-1.05	1.4	296.21	21JA	0.2	-3.4	-1.6	0.8
317.90	-1.00	1.4	306.24	31JA	-0.0	-3.3	-1.6	0.8
325.59	-0.94	1.4	316.21	10FE	-0.3	-3.3	-1.6	0.8
333.20	-0.87	1.5	326.12	20FE	-0.9	-3.3	-1.6	0.7
340.74	-0.79	1.5	335.97	2MR	-1.6	-3.3	-1.6	0.7
348.20	-0.71	1.5	345.75	12MR	-1.2	-3.3	-1.6	0.6
355.58	-0.62	1.5	355.49	22MR	-0.1	-3.4	-1.6	0.6
2.87	-0.53	1.5	5.17	1AP	1.1	-3.4	-1.6	0.5
10.07	-0.43	1.6	14.80	11AP	2.5	-3.4	-1.6	0.5
17.18	-0.33	1.6	24.40	21AP	3.6	-3.4	-1.6	0.4
24.21	-0.24	1.7	33.96	1MY	2.1	-3.5	-1.7	0.3
31.15	-0.13	1.7	43.49	11MY	1.1	-3.5	-1.7	0.3
38.02	-0.03	1.8	53.02	21MY	0.2	-3.6	-1.8	0.3
44.80	0.07	1.8	62.53	31MY	-0.8	-3.6	-1.8	0.3
51.50	0.18	1.8	72.05	10JN	-1.6	-3.7	-1.9	0.4
58.13	0.28	1.8	81.59	20JN	-1.2	-3.8	-2.0	0.4
64.69	0.39	1.9	91.14	30JN	-0.4	-3.9	-2.0	0.5
71.18	0.49	1.9	100.72	10JL	0.1	-4.0	-2.1	0.6
77.59	0.60	1.9	110.35	20JL	0.4	-4.1	-2.2	0.6
83.94	0.70	1.9	120.01	30JL	0.6	-4.2	-2.2	0.7
90.21	0.81	1.9	129.73	9AU	1.2	-4.3	-2.3	0.7
96.42	0.92	1.9	139.51	19AU	2.9	-4.2	-2.3	0.7
102.54	1.03	1.8	149.34	29AU	1.2	-4.1	-2.4	0.8
108.58	1.15	1.8	159.23	8SE	-0.5	-3.6	-2.4	0.8
114.53	1.26	1.7	169.19	18SE	-0.9	-3.3	-2.4	0.8
120.38	1.39	1.7	179.20	28SE	-0.9	-3.9	-2.4	0.8
126.11	1.51	1.6	189.26	8OC	-0.8	-4.3	-2.4	0.8
131.71	1.64	1.5	199.37	18OC	-0.6	-4.4	-2.4	0.8
137.15	1.78	1.4	209.52	28OC	-0.5	-4.4	-2.3	0.8
142.40	1.93	1.3	219.70	7NO	-0.5	-4.3	-2.3	0.7
147.42	2.09	1.2	229.91	17NO	-0.4	-4.2	-2.2	0.7
152.16	2.26	1.0	240.12	27NO	0.7	-4.1	-2.1	0.7
156.57	2.44	0.9	250.32	7DE	2.7	-4.0	-2.1	0.7
160.55	2.64	0.7	260.53	17DE	0.7	-3.9	-2.0	0.8
164.00	2.84	0.5	270.70	27DE	0.2	-3.8	-1.9	0.8

♂ LONG	♂ LAT	♂ MAG	⊙ LONG	16.00UT -574	☿	♀	♃	♄
					MAGNITUDES			
166.80	3.07	0.2	280.84	6JA	0.0	-3.7	-1.8	0.8
168.78	3.29	-0.0	290.94	16JA	-0.1	-3.6	-1.8	0.8
169.77	3.52	-0.3	300.99	26JA	-0.4	-3.6	-1.7	0.8
169.60	3.72	-0.6	310.99	5FE	-0.9	-3.5	-1.7	0.8
168.14	3.85	-0.9	320.93	15FE	-1.5	-3.4	-1.6	0.8
165.48	3.88	-1.2	330.81	25FE	-1.1	-3.4	-1.6	0.8
161.97	3.76	-1.4	340.63	7MR	-0.0	-3.4	-1.6	0.7
158.22	3.48	-1.3	350.40	17MR	1.4	-3.3	-1.5	0.7
154.97	3.07	-1.2	0.10	27MR	3.2	-3.3	-1.5	0.6
152.76	2.58	-1.0	9.76	6AP	2.7	-3.3	-1.5	0.6
151.86	2.08	-0.8	19.38	16AP	1.5	-3.3	-1.5	0.5
152.29	1.61	-0.5	28.95	26AP	0.8	-3.3	-1.5	0.4
153.89	1.18	-0.3	38.50	6MY	0.1	-3.3	-1.5	0.4
156.50	0.79	-0.2	48.03	16MY	-0.8	-3.3	-1.6	0.3
159.95	0.45	0.0	57.55	26MY	-1.7	-3.4	-1.6	0.3
164.09	0.15	0.1	67.06	5JN	-1.2	-3.4	-1.6	0.3
168.80	-0.12	0.3	76.59	15JN	-0.4	-3.4	-1.7	0.3
173.98	-0.35	0.4	86.13	25JN	0.2	-3.5	-1.7	0.4
179.57	-0.56	0.5	95.69	5JL	0.5	-3.5	-1.8	0.4
185.50	-0.74	0.6	105.30	15JL	0.9	-3.5	-1.8	0.5
191.73	-0.90	0.7	114.94	25JL	1.6	-3.4	-1.9	0.6
198.22	-1.04	0.7	124.63	4AU	3.2	-3.4	-1.9	0.6
204.95	-1.16	0.8	134.37	14AU	1.1	-3.4	-2.0	0.7
211.87	-1.26	0.8	144.17	24AU	-0.6	-3.4	-2.1	0.7
218.97	-1.34	0.9	154.04	3SE	-1.0	-3.3	-2.1	0.7
226.23	-1.40	0.9	163.96	13SE	-1.0	-3.3	-2.2	0.8
233.61	-1.44	1.0	173.94	23SE	-0.8	-3.3	-2.3	0.8
241.11	-1.47	1.0	183.97	3OC	-0.5	-3.3	-2.3	0.8
248.70	-1.47	1.1	194.06	13OC	-0.4	-3.3	-2.3	0.8
256.37	-1.46	1.1	204.19	23OC	-0.3	-3.4	-2.3	0.8
264.08	-1.44	1.2	214.36	2NO	-0.2	-3.4	-2.3	0.8
271.84	-1.40	1.2	224.55	12NO	0.9	-3.4	-2.3	0.7
279.61	-1.35	1.3	234.76	22NO	2.5	-3.5	-2.3	0.7
287.38	-1.29	1.3	244.97	2DE	0.4	-3.5	-2.2	0.7
295.13	-1.21	1.3	255.18	12DE	-0.0	-3.5	-2.2	0.7
302.86	-1.13	1.4	265.37	22DE	-0.1	-3.6	-2.1	0.8
				-573				
310.54	-1.04	1.4	275.53	1JA	-0.2	-3.7	-2.0	0.8
318.18	-0.94	1.5	285.65	11JA	-0.5	-3.7	-2.0	0.8
325.75	-0.84	1.5	295.73	21JA	-1.0	-3.8	-1.9	0.8
333.25	-0.74	1.6	305.76	31JA	-1.4	-3.9	-1.8	0.8
340.69	-0.64	1.6	315.73	10FE	-1.1	-4.0	-1.8	0.8
348.04	-0.53	1.6	325.64	20FE	0.0	-4.1	-1.7	0.8
355.31	-0.42	1.7	335.49	2MR	1.7	-4.2	-1.6	0.8
2.51	-0.31	1.7	345.28	12MR	3.5	-4.2	-1.6	0.8
9.62	-0.21	1.7	355.02	22MR	2.0	-4.2	-1.6	0.7
16.66	-0.10	1.7	4.70	1AP	1.2	-4.1	-1.5	0.7
23.61	0.00	1.7	14.34	11AP	0.6	-3.8	-1.5	0.6
30.49	0.10	1.8	23.93	21AP	0.0	-3.0	-1.5	0.6
37.31	0.20	1.8	33.49	1MY	-0.8	-3.3	-1.4	0.5
44.05	0.29	1.8	43.03	11MY	-1.7	-3.9	-1.4	0.5
50.73	0.39	1.8	52.55	21MY	-1.2	-4.1	-1.4	0.4
57.36	0.48	1.8	62.07	31MY	-0.3	-4.2	-1.4	0.3
63.93	0.57	1.9	71.59	10JN	0.3	-4.1	-1.5	0.3
70.45	0.65	1.9	81.12	20JN	0.7	-4.1	-1.5	0.3
76.92	0.73	1.9	90.67	30JN	1.2	-4.0	-1.5	0.4
83.35	0.81	2.0	100.26	10JL	2.1	-3.9	-1.5	0.4
89.74	0.89	2.0	109.87	20JL	3.0	-3.8	-1.6	0.5
96.10	0.97	2.0	119.54	30JL	0.9	-3.7	-1.6	0.5
102.42	1.04	2.0	129.26	9AU	-0.6	-3.7	-1.7	0.6
108.71	1.11	2.0	139.03	19AU	-1.2	-3.6	-1.7	0.6
114.96	1.18	2.0	148.86	29AU	-1.2	-3.6	-1.8	0.7
121.18	1.25	2.0	158.75	8SE	-0.7	-3.5	-1.8	0.7
127.37	1.32	1.9	168.70	18SE	-0.4	-3.5	-1.9	0.7
133.52	1.38	1.9	178.71	28SE	-0.2	-3.4	-2.0	0.8
139.62	1.45	1.8	188.77	8OC	-0.2	-3.4	-2.0	0.8
145.68	1.51	1.8	198.88	18OC	0.0	-3.4	-2.1	0.8
151.68	1.58	1.7	209.03	28OC	1.1	-3.4	-2.2	0.8
157.61	1.64	1.6	219.21	7NO	2.2	-3.4	-2.2	0.8
163.46	1.70	1.5	229.41	17NO	0.2	-3.4	-2.2	0.8
169.22	1.76	1.4	239.62	27NO	-0.2	-3.4	-2.2	0.8
174.86	1.82	1.3	249.83	7DE	-0.2	-3.4	-2.2	0.7
180.37	1.87	1.2	260.03	17DE	-0.3	-3.4	-2.2	0.7
185.70	1.92	1.0	270.21	27DE	-0.6	-3.4	-2.2	0.8

♂ LONG	♂ LAT	♂ MAG	⊙ LONG	16.00UT -572	☿	♀	♃	♄
					MAGNITUDES			
190.82	1.96	0.9	280.35	6JA	-1.0	-3.4	-2.1	0.8
195.69	1.99	0.7	290.45	16JA	-1.2	-3.4	-2.1	0.8
200.24	2.02	0.5	300.51	26JA	-1.0	-3.4	-2.0	0.9
204.38	2.02	0.2	310.51	5FE	0.1	-3.5	-1.9	0.9
208.03	2.00	-0.0	320.45	15FE	2.1	-3.5	-1.9	0.9
211.03	1.95	-0.3	330.33	25FE	2.8	-3.5	-1.8	0.9
213.25	1.85	-0.6	340.15	6MR	1.5	-3.4	-1.7	0.9
214.50	1.70	-0.9	349.92	16MR	0.9	-3.4	-1.7	0.8
214.60	1.45	-1.2	359.63	26MR	0.4	-3.4	-1.6	0.8
213.46	1.11	-1.5	9.29	5AP	-0.1	-3.4	-1.5	0.8
211.14	0.67	-1.8	18.90	15AP	-0.8	-3.3	-1.5	0.7
208.01	0.14	-2.1	28.48	25AP	-1.7	-3.3	-1.5	0.7
204.72	-0.43	-2.0	38.03	5MY	-1.2	-3.3	-1.4	0.6
202.00	-0.97	-1.8	47.56	15MY	-0.3	-3.3	-1.4	0.5
200.39	-1.44	-1.6	57.08	25MY	0.4	-3.3	-1.4	0.5
200.16	-1.82	-1.4	66.59	4JN	0.9	-3.4	-1.3	0.4
201.27	-2.11	-1.2	76.12	14JN	1.6	-3.4	-1.3	0.3
203.60	-2.32	-1.0	85.66	24JN	2.7	-3.4	-1.3	0.3
206.96	-2.47	-0.8	95.22	4JL	2.6	-3.4	-1.3	0.3
211.15	-2.56	-0.6	104.83	14JL	0.7	-3.5	-1.4	0.4
216.03	-2.61	-0.4	114.47	24JL	-0.7	-3.6	-1.4	0.5
221.46	-2.62	-0.3	124.16	3AU	-1.3	-3.6	-1.4	0.5
227.34	-2.59	-0.1	133.90	13AU	-1.2	-3.7	-1.4	0.6
233.60	-2.54	0.0	143.70	23AU	-0.6	-3.8	-1.5	0.6
240.14	-2.47	0.1	153.56	2SE	-0.3	-3.9	-1.5	0.7
246.92	-2.38	0.3	163.48	12SE	-0.1	-4.0	-1.6	0.7
253.90	-2.26	0.4	173.46	22SE	-0.0	-4.1	-1.6	0.7
261.02	-2.14	0.5	183.49	2OC	0.2	-4.2	-1.7	0.8
268.25	-2.00	0.6	193.57	12OC	1.4	-4.3	-1.8	0.8
275.56	-1.85	0.7	203.70	22OC	2.0	-4.4	-1.8	0.8
282.92	-1.70	0.8	213.86	1NO	0.0	-4.4	-1.9	0.8
290.31	-1.54	0.9	224.06	11NO	-0.3	-4.2	-1.9	0.8
297.71	-1.38	1.0	234.26	21NO	-0.4	-3.6	-2.0	0.8
305.11	-1.22	1.1	244.47	1DE	-0.4	-3.1	-2.1	0.8
312.48	-1.06	1.2	254.68	11DE	-0.6	-3.9	-2.1	0.8
319.81	-0.91	1.3	264.87	21DE	-1.0	-4.3	-2.1	0.8
327.11	-0.75	1.3	275.03	31DE	-1.1	-4.4	-2.1	0.8
				-571				
334.35	-0.61	1.4	285.16	10JA	-0.9	-4.3	-2.1	0.8
341.53	-0.47	1.5	295.24	20JA	0.2	-4.2	-2.1	0.9
348.65	-0.33	1.6	305.27	30JA	2.5	-4.1	-2.1	0.9
355.71	-0.20	1.6	315.25	9FE	2.1	-4.0	-2.0	0.9
2.71	-0.08	1.7	325.16	19FE	1.1	-3.9	-1.9	0.9
9.63	0.04	1.7	335.01	1MR	0.6	-3.8	-1.9	0.9
16.50	0.15	1.8	344.81	11MR	0.3	-3.7	-1.8	0.9
23.30	0.25	1.8	354.55	21MR	-0.1	-3.6	-1.7	0.9
30.04	0.35	1.9	4.23	31MR	-0.8	-3.6	-1.7	0.9
36.73	0.44	1.9	13.87	10AP	-1.7	-3.5	-1.6	0.9
43.36	0.52	1.9	23.47	20AP	-1.2	-3.4	-1.6	0.8
49.95	0.61	1.9	33.03	30AP	-0.3	-3.4	-1.5	0.8
56.49	0.68	1.9	42.57	10MY	0.5	-3.4	-1.4	0.7
63.00	0.75	1.9	52.10	20MY	1.2	-3.3	-1.4	0.6
69.47	0.82	1.9	61.61	30MY	2.1	-3.3	-1.4	0.6
75.91	0.88	1.9	71.13	9JN	3.4	-3.3	-1.3	0.5
82.33	0.93	1.9	80.66	19JN	2.1	-3.3	-1.3	0.5
88.73	0.98	1.9	90.21	29JN	0.6	-3.3	-1.3	0.4
95.11	1.03	2.0	99.79	9JL	-0.7	-3.3	-1.3	0.4
101.49	1.08	2.0	109.41	19JL	-1.4	-3.4	-1.3	0.4
107.85	1.12	2.0	119.07	29JL	-1.2	-3.4	-1.3	0.5
114.22	1.15	2.0	128.79	8AU	-0.6	-3.4	-1.3	0.5
120.58	1.19	2.0	138.56	18AU	-0.2	-3.4	-1.3	0.6
126.94	1.21	2.0	148.38	28AU	0.0	-3.5	-1.3	0.6
133.31	1.24	2.0	158.27	7SE	0.2	-3.5	-1.3	0.7
139.68	1.26	2.0	168.22	17SE	0.4	-3.5	-1.4	0.7
146.06	1.27	2.0	178.22	27SE	1.7	-3.5	-1.4	0.8
152.45	1.29	1.9	188.28	7OC	1.8	-3.4	-1.5	0.8
158.84	1.29	1.9	198.38	17OC	-0.1	-3.4	-1.5	0.8
165.24	1.29	1.9	208.53	27OC	-0.5	-3.4	-1.6	0.9
171.63	1.29	1.8	218.71	6NO	-0.5	-3.4	-1.6	0.9
178.03	1.28	1.7	228.91	16NO	-0.6	-3.3	-1.7	0.9
184.42	1.26	1.7	239.12	26NO	-0.7	-3.3	-1.8	0.9
190.81	1.23	1.6	249.33	6DE	-0.9	-3.3	-1.8	0.9
197.18	1.19	1.5	259.53	16DE	-0.9	-3.3	-1.9	0.9
203.53	1.14	1.4	269.71	26DE	-0.8	-3.4	-1.9	0.9

-570

♂ LONG	LAT	MAG	☉ LONG	16.00UT	☿	♀	♃	♄
209.85	1.08	1.2	279.86	5JA	0.3	-3.4	-2.0	0.8
216.14	1.00	1.1	289.96	15JA	2.8	-3.4	-2.0	0.9
222.38	0.90	1.0	300.02	25JA	1.6	-3.4	-2.0	0.9
228.56	0.77	0.8	310.02	4FE	0.8	-3.4	-2.0	1.0
234.66	0.63	0.7	319.97	14FE	0.4	-3.5	-2.0	1.0
240.66	0.45	0.5	329.86	24FE	0.2	-3.5	-2.0	1.0
246.52	0.24	0.3	339.68	6MR	-0.2	-3.6	-2.0	1.0
252.22	-0.01	0.1	349.45	16MR	-0.9	-3.6	-1.9	1.0
257.71	-0.31	-0.1	359.16	26MR	-1.7	-3.7	-1.9	1.0
262.91	-0.67	-0.3	8.83	5AP	-1.2	-3.8	-1.8	1.0
267.75	-1.09	-0.5	18.44	15AP	-0.2	-3.9	-1.7	1.0
272.13	-1.58	-0.8	28.03	25AP	0.7	-4.0	-1.7	1.0
275.89	-2.15	-1.1	37.58	5MY	1.6	-4.0	-1.6	0.9
278.89	-2.80	-1.4	47.10	15MY	2.8	-4.1	-1.5	0.9
280.91	-3.54	-1.7	56.62	25MY	3.1	-4.2	-1.5	0.8
281.78	-4.33	-1.9	66.14	4JN	1.7	-4.1	-1.4	0.7
281.39	-5.12	-2.2	75.66	14JN	0.4	-4.1	-1.4	0.7
279.79	-5.82	-2.5	85.20	24JN	-0.7	-3.5	-1.3	0.6
277.37	-6.32	-2.6	94.77	4JL	-1.5	-3.0	-1.3	0.5
274.79	-6.53	-2.6	104.36	14JL	-1.2	-3.7	-1.3	0.5
272.75	-6.43	-2.4	114.00	24JL	-0.5	-4.1	-1.2	0.5
271.82	-6.08	-2.1	123.69	3AU	-0.1	-4.2	-1.2	0.5
272.19	-5.59	-1.8	133.42	13AU	0.1	-4.2	-1.2	0.6
273.84	-5.04	-1.5	143.22	23AU	0.3	-4.2	-1.2	0.6
276.60	-4.47	-1.3	153.08	2SE	0.7	-4.1	-1.2	0.7
280.27	-3.92	-1.0	162.99	12SE	2.1	-4.0	-1.2	0.7
284.65	-3.40	-0.7	172.97	22SE	1.6	-3.9	-1.3	0.8
289.60	-2.93	-0.5	183.00	2OC	-0.3	-3.9	-1.3	0.8
294.98	-2.49	-0.3	193.08	12OC	-0.7	-3.8	-1.3	0.8
300.68	-2.10	-0.1	203.21	22OC	-0.7	-3.7	-1.4	0.9
306.64	-1.74	0.1	213.37	1NO	-0.7	-3.6	-1.4	0.9
312.78	-1.42	0.3	223.56	11NO	-0.8	-3.6	-1.5	0.9
319.05	-1.13	0.5	233.77	21NO	-0.7	-3.5	-1.5	1.0
325.44	-0.87	0.7	243.98	1DE	-0.7	-3.5	-1.6	1.0
331.88	-0.63	0.8	254.19	11DE	-0.6	-3.4	-1.6	1.0
338.38	-0.42	1.0	264.38	21DE	0.4	-3.4	-1.7	1.0
344.90	-0.23	1.1	274.54	31DE	3.0	-3.4	-1.8	1.0

-569

♂ LONG	LAT	MAG	☉ LONG	16.00UT	☿	♀	♃	♄
351.43	-0.06	1.2	284.67	10JA	1.2	-3.4	-1.8	1.0
357.97	0.09	1.3	294.76	20JA	0.5	-3.3	-1.9	1.0
4.50	0.23	1.4	304.79	30JA	0.3	-3.3	-1.9	1.0
11.01	0.35	1.5	314.76	9FE	0.1	-3.3	-2.0	1.1
17.51	0.47	1.6	324.68	19FE	-0.3	-3.3	-2.0	1.1
23.98	0.57	1.7	334.54	1MR	-0.9	-3.3	-2.0	1.1
30.43	0.65	1.8	344.34	11MR	-1.7	-3.3	-2.0	1.2
36.86	0.73	1.8	354.08	21MR	-1.2	-3.4	-2.0	1.2
43.26	0.80	1.9	3.77	31MR	-0.2	-3.4	-2.0	1.2
49.65	0.87	1.9	13.41	10AP	0.9	-3.4	-1.9	1.2
56.02	0.92	2.0	23.01	20AP	2.1	-3.5	-1.9	1.1
62.37	0.97	2.0	32.57	30AP	3.7	-3.5	-1.8	1.1
68.71	1.01	2.0	42.11	10MY	2.5	-3.5	-1.8	1.1
75.04	1.05	2.0	51.64	20MY	1.3	-3.4	-1.7	1.0
81.36	1.08	2.0	61.15	30MY	0.3	-3.4	-1.6	1.0
87.69	1.11	2.0	70.67	9JN	-0.8	-3.4	-1.6	0.9
94.02	1.13	2.0	80.20	19JN	-1.6	-3.3	-1.5	0.9
100.36	1.14	2.0	89.74	29JN	-1.2	-3.3	-1.5	0.8
106.71	1.16	2.0	99.32	9JL	-0.5	-3.3	-1.4	0.7
113.07	1.16	2.0	108.94	19JL	-0.0	-3.3	-1.4	0.6
119.46	1.16	1.9	118.60	29JL	0.3	-3.3	-1.3	0.6
125.88	1.16	1.9	128.31	8AU	0.5	-3.3	-1.3	0.6
132.32	1.15	2.0	138.08	18AU	1.0	-3.4	-1.3	0.6
138.79	1.14	2.0	147.90	28AU	2.5	-3.4	-1.2	0.7
145.30	1.12	2.0	157.79	7SE	1.4	-3.4	-1.2	0.7
151.85	1.10	2.0	167.73	17SE	-0.4	-3.5	-1.2	0.8
158.43	1.07	1.9	177.73	27SE	-0.8	-3.5	-1.2	0.8
165.06	1.04	1.9	187.79	7OC	-0.8	-3.6	-1.2	0.9
171.73	1.00	1.9	197.90	17OC	-0.8	-3.7	-1.3	0.9
178.45	0.95	1.9	208.04	27OC	-0.7	-3.7	-1.3	1.0
185.21	0.89	1.8	218.22	6NO	-0.6	-3.8	-1.3	1.0
192.01	0.83	1.8	228.42	16NO	-0.6	-3.9	-1.3	1.0
198.85	0.76	1.7	238.63	26NO	-0.5	-4.0	-1.4	1.1
205.74	0.68	1.7	248.84	6DE	-0.4	-4.1	-1.4	1.1
212.67	0.59	1.6	259.04	16DE	3.0	-4.2	-1.5	1.1
219.65	0.48	1.5	269.22	26DE	0.9	-4.3	-1.6	1.1

-568

♂ LONG	LAT	MAG	☉ LONG	16.00UT	☿	♀	♃	♄
226.66	0.37	1.5	279.37	5JA	0.3	-4.4	-1.6	1.1
233.70	0.24	1.4	289.48	15JA	0.1	-4.3	-1.7	1.1
240.78	0.10	1.3	299.53	25JA	-0.0	-4.1	-1.8	1.1
247.89	-0.06	1.2	309.54	4FE	-0.4	-3.5	-1.8	1.1
255.04	-0.23	1.1	319.49	14FE	-0.9	-3.3	-1.9	1.2
262.17	-0.41	1.0	329.38	24FE	-1.6	-3.9	-1.9	1.2
269.32	-0.61	0.9	339.21	5MR	-1.2	-4.2	-2.0	1.3
276.46	-0.82	0.8	348.98	15MR	-0.1	-4.2	-2.0	1.3
283.59	-1.05	0.7	358.69	25MR	1.2	-4.2	-2.1	1.3
290.68	-1.29	0.6	8.36	4AP	2.7	-4.1	-2.1	1.3
297.71	-1.54	0.4	17.98	14AP	3.2	-4.0	-2.1	1.3
304.65	-1.80	0.3	27.56	24AP	1.9	-3.9	-2.0	1.3
311.47	-2.07	0.2	37.11	4MY	1.0	-3.8	-2.0	1.3
318.12	-2.35	0.1	46.65	14MY	0.2	-3.7	-2.0	1.3
324.58	-2.63	-0.1	56.16	24MY	-0.8	-3.7	-1.9	1.2
330.76	-2.91	-0.2	65.68	3JN	-1.6	-3.6	-1.9	1.2
336.59	-3.20	-0.4	75.20	13JN	-1.2	-3.5	-1.8	1.1
342.00	-3.48	-0.6	84.74	23JN	-0.4	-3.5	-1.7	1.1
346.85	-3.76	-0.7	94.30	3JL	0.1	-3.4	-1.7	1.0
350.99	-4.02	-0.9	103.90	13JL	0.4	-3.4	-1.6	0.9
354.27	-4.27	-1.1	113.53	23JL	0.7	-3.4	-1.5	0.9
356.45	-4.48	-1.4	123.22	2AU	1.3	-3.4	-1.5	0.8
357.33	-4.63	-1.6	132.96	12AU	3.0	-3.4	-1.4	0.7
356.78	-4.66	-1.8	142.75	22AU	1.2	-3.4	-1.4	0.8
354.80	-4.53	-2.0	152.60	1SE	-0.5	-3.4	-1.4	0.8
351.77	-4.21	-2.1	162.51	11SE	-1.0	-3.4	-1.3	0.8
348.37	-3.69	-2.1	172.48	21SE	-1.0	-3.4	-1.3	0.9
345.41	-3.05	-1.9	182.52	1OC	-0.8	-3.4	-1.3	0.9
343.52	-2.38	-1.6	192.59	11OC	-0.5	-3.4	-1.3	1.0
342.96	-1.75	-1.2	202.72	21OC	-0.4	-3.4	-1.3	1.0
343.70	-1.19	-0.9	212.88	31OC	-0.4	-3.4	-1.3	1.1
345.60	-0.72	-0.6	223.07	10NO	-0.3	-3.5	-1.3	1.1
348.41	-0.34	-0.3	233.27	20NO	0.7	-3.5	-1.3	1.2
351.96	-0.02	-0.0	243.49	30NO	2.7	-3.5	-1.4	1.2
356.08	0.24	0.2	253.69	10DE	0.6	-3.5	-1.4	1.3
0.64	0.45	0.5	263.88	20DE	0.1	-3.4	-1.4	1.3
5.54	0.62	0.7	274.05	30DE	-0.0	-3.4	-1.5	1.3

-567

♂ LONG	LAT	MAG	☉ LONG	16.00UT	☿	♀	♃	♄
10.71	0.76	0.8	284.18	9JA	-0.2	-3.4	-1.5	1.3
16.08	0.88	1.0	294.26	19JA	-0.4	-3.4	-1.6	1.3
21.61	0.97	1.2	304.30	29JA	-0.9	-3.3	-1.7	1.3
27.27	1.05	1.3	314.28	8FE	-1.4	-3.3	-1.7	1.3
33.02	1.12	1.4	324.20	18FE	-1.1	-3.3	-1.8	1.3
38.85	1.17	1.5	334.06	28FE	-0.1	-3.3	-1.9	1.3
44.75	1.21	1.6	343.86	10MR	1.5	-3.3	-1.9	1.3
50.69	1.24	1.7	353.61	20MR	3.4	-3.4	-2.0	1.2
56.68	1.26	1.8	3.30	30MR	2.4	-3.4	-2.0	1.2
62.70	1.28	1.8	12.94	9AP	1.4	-3.4	-2.1	1.2
68.75	1.29	1.9	22.54	19AP	0.8	-3.4	-2.1	1.2
74.84	1.29	1.9	32.11	29AP	0.1	-3.5	-2.2	1.2
80.96	1.29	2.0	41.65	9MY	-0.8	-3.5	-2.2	1.1
87.10	1.28	2.0	51.17	19MY	-1.7	-3.6	-2.2	1.1
93.28	1.27	2.0	60.69	29MY	-1.2	-3.6	-2.1	1.1
99.50	1.26	2.0	70.21	8JN	-0.4	-3.7	-2.1	1.0
105.75	1.24	2.0	79.74	18JN	0.2	-3.8	-2.0	1.0
112.03	1.21	2.0	89.29	28JN	0.6	-3.9	-2.0	0.9
118.36	1.18	2.0	98.86	8JL	1.0	-4.0	-1.9	0.9
124.74	1.15	2.0	108.48	18JL	1.8	-4.1	-1.9	0.8
131.17	1.11	2.0	118.14	28JL	3.2	-4.2	-1.8	0.8
137.65	1.07	1.9	127.84	7AU	1.0	-4.3	-1.7	0.7
144.18	1.02	1.9	137.61	17AU	-0.6	-4.2	-1.7	0.7
150.77	0.97	1.9	147.43	27AU	-1.1	-4.0	-1.6	0.7
157.43	0.91	1.8	157.31	6SE	-1.1	-3.6	-1.6	0.7
164.15	0.85	1.8	167.25	16SE	-0.7	-3.3	-1.5	0.8
170.93	0.79	1.8	177.25	26SE	-0.4	-3.9	-1.5	0.9
177.79	0.71	1.8	187.30	6OC	-0.3	-4.3	-1.4	0.9
184.71	0.64	1.8	197.40	16OC	-0.3	-4.4	-1.4	1.0
191.70	0.55	1.8	207.55	26OC	-0.1	-4.3	-1.4	1.1
198.76	0.46	1.8	217.72	5NO	0.9	-4.3	-1.4	1.1
205.90	0.37	1.7	227.92	15NO	2.5	-4.2	-1.4	1.1
213.09	0.26	1.7	238.13	25NO	0.3	-4.1	-1.4	1.2
220.36	0.15	1.7	248.34	5DE	-0.1	-4.0	-1.4	1.2
227.69	0.04	1.6	258.55	15DE	-0.2	-3.9	-1.4	1.2
235.07	-0.08	1.6	268.73	25DE	-0.3	-3.8	-1.4	1.2

Left table

♂ LONG	♂ LAT	♂ MAG	☉ LONG	16.00UT	☿	♀	♃	♄
				-566				
242.52	-0.21	1.5	278.88	4JA	-0.5	-3.7	-1.5	1.2
250.01	-0.34	1.5	288.99	14JA	-1.0	-3.6	-1.5	1.2
257.55	-0.47	1.4	299.05	24JA	-1.3	-3.6	-1.5	1.1
265.12	-0.61	1.4	309.06	3FE	-1.1	-3.5	-1.6	1.1
272.72	-0.75	1.3	319.01	13FE	-0.0	-3.4	-1.6	1.0
280.34	-0.89	1.3	328.91	23FE	1.8	-3.4	-1.7	1.0
287.96	-1.02	1.2	338.74	5MR	3.3	-3.4	-1.8	1.0
295.58	-1.16	1.2	348.51	15MR	1.8	-3.3	-1.8	1.0
303.18	-1.29	1.1	358.23	25MR	1.0	-3.3	-1.9	1.0
310.74	-1.41	1.1	7.90	4AP	0.6	-3.3	-2.0	1.0
318.26	-1.52	1.0	17.52	14AP	0.0	-3.3	-2.0	0.9
325.71	-1.63	0.9	27.10	24AP	-0.8	-3.3	-2.1	0.9
333.07	-1.72	0.9	36.65	4MY	-1.7	-3.3	-2.2	0.9
340.33	-1.80	0.8	46.19	14MY	-1.3	-3.3	-2.2	0.9
347.47	-1.86	0.8	55.71	24MY	-0.4	-3.4	-2.2	0.9
354.46	-1.91	0.7	65.22	3JN	0.3	-3.4	-2.3	0.8
1.27	-1.94	0.6	74.74	13JN	0.8	-3.4	-2.3	0.8
7.88	-1.95	0.6	84.28	23JN	1.3	-3.5	-2.3	0.7
14.26	-1.94	0.5	93.84	3JL	2.3	-3.5	-2.2	0.7
20.36	-1.92	0.4	103.43	13JL	2.9	-3.5	-2.2	0.6
26.13	-1.88	0.3	113.07	23JL	0.9	-3.4	-2.2	0.6
31.51	-1.81	0.2	122.75	2AU	-0.6	-3.4	-2.1	0.5
36.41	-1.72	0.0	132.48	12AU	-1.2	-3.4	-2.0	0.5
40.73	-1.60	-0.1	142.27	22AU	-1.2	-3.4	-2.0	0.4
44.34	-1.44	-0.3	152.12	1SE	-0.7	-3.3	-1.9	0.4
47.06	-1.25	-0.5	162.03	11SE	-0.3	-3.3	-1.8	0.4
48.71	-1.00	-0.7	171.99	21SE	-0.2	-3.3	-1.8	0.4
49.08	-0.70	-0.9	182.02	1OC	-0.1	-3.3	-1.7	0.5
48.02	-0.33	-1.1	192.10	11OC	0.1	-3.3	-1.7	0.6
45.56	0.10	-1.3	202.22	21OC	1.2	-3.4	-1.6	0.6
42.05	0.55	-1.5	212.38	31OC	2.2	-3.4	-1.6	0.7
38.14	0.98	-1.5	222.57	10NO	0.1	-3.4	-1.5	0.8
34.66	1.34	-1.2	232.77	20NO	-0.2	-3.5	-1.5	0.8
32.21	1.61	-0.9	242.98	30NO	-0.3	-3.5	-1.5	0.9
31.09	1.79	-0.6	253.20	10DE	-0.4	-3.5	-1.5	0.9
31.29	1.90	-0.3	263.39	20DE	-0.6	-3.6	-1.5	0.9
32.64	1.96	-0.0	273.56	30DE	-1.0	-3.7	-1.5	0.9
				-565				
34.95	1.98	0.2	283.69	9JA	-1.1	-3.7	-1.5	0.9
38.03	1.98	0.5	293.78	19JA	-1.0	-3.8	-1.5	0.9
41.69	1.96	0.7	303.82	29JA	0.1	-3.9	-1.5	0.9
45.83	1.94	0.9	313.80	8FE	2.2	-4.0	-1.6	0.8
50.34	1.90	1.0	323.72	18FE	2.6	-4.1	-1.6	0.8
55.15	1.86	1.2	333.59	28FE	1.3	-4.2	-1.6	0.7
60.19	1.82	1.3	343.40	10MR	0.8	-4.2	-1.7	0.7
65.43	1.77	1.4	353.14	20MR	0.4	-4.2	-1.7	0.7
70.83	1.72	1.5	2.83	30MR	-0.1	-4.1	-1.8	0.7
76.36	1.67	1.6	12.48	9AP	-0.8	-3.7	-1.9	0.7
82.01	1.61	1.7	22.08	19AP	-1.7	-3.0	-1.9	0.7
87.76	1.56	1.8	31.65	29AP	-1.3	-3.3	-2.0	0.7
93.60	1.50	1.8	41.19	9MY	-0.3	-3.9	-2.1	0.7
99.53	1.44	1.9	50.71	19MY	-0.4	-4.1	-2.1	0.7
105.54	1.38	1.9	60.23	29MY	1.0	-4.2	-2.2	0.7
111.63	1.31	1.9	69.75	8JN	1.7	-4.1	-2.3	0.6
117.79	1.24	1.9	79.27	18JN	3.0	-4.1	-2.3	0.6
124.03	1.17	1.9	88.82	28JN	2.4	-4.0	-2.4	0.6
130.35	1.10	1.9	98.40	8JL	0.7	-3.9	-2.4	0.5
136.74	1.03	1.9	108.01	18JL	-0.7	-3.8	-2.4	0.5
143.21	0.95	1.9	117.67	28JL	-1.3	-3.7	-2.4	0.4
149.77	0.87	1.9	127.37	7AU	-1.2	-3.7	-2.4	0.3
156.41	0.78	1.9	137.13	17AU	-0.6	-3.6	-2.3	0.3
163.13	0.70	1.8	146.95	27AU	-0.2	-3.6	-2.3	0.2
169.94	0.61	1.8	156.83	6SE	-0.1	-3.5	-2.2	0.2
176.84	0.51	1.8	166.76	16SE	0.1	-3.5	-2.1	0.1
183.82	0.42	1.7	176.76	26SE	0.3	-3.4	-2.1	0.1
190.89	0.32	1.7	186.81	6OC	1.5	-3.4	-2.0	0.2
198.05	0.22	1.6	196.91	16OC	2.0	-3.4	-1.9	0.3
205.30	0.11	1.6	207.05	26OC	-0.0	-3.4	-1.9	0.3
212.62	0.00	1.6	217.22	5NO	-0.4	-3.4	-1.8	0.4
220.0	-0.11	1.6	227.42	15NO	-0.4	-3.4	-1.8	0.5
227.51	-0.22	1.5	237.63	25NO	-0.5	-3.4	-1.7	0.5
235.06	-0.33	1.5	247.84	5DE	-0.7	-3.4	-1.7	0.6
242.68	-0.44	1.5	258.04	15DE	-0.9	-3.4	-1.6	0.6
250.35	-0.55	1.5	268.23	25DE	-1.0	-3.4	-1.6	0.6

Right table

♂ LONG	♂ LAT	♂ MAG	☉ LONG	16.00UT	☿	♀	♃	♄
				-564				
258.07	-0.65	1.5	278.38	4JA	-0.9	-3.4	-1.6	0.7
265.82	-0.75	1.5	288.49	14JA	0.2	-3.4	-1.6	0.7
273.61	-0.85	1.4	298.56	24JA	2.5	-3.4	-1.5	0.7
281.41	-0.94	1.4	308.57	3FE	2.0	-3.5	-1.5	0.7
289.21	-1.03	1.4	318.52	13FE	0.9	-3.5	-1.6	0.6
297.01	-1.10	1.4	328.42	23FE	0.5	-3.5	-1.6	0.6
304.78	-1.17	1.4	338.26	4MR	0.3	-3.4	-1.6	0.6
312.53	-1.22	1.4	348.03	14MR	-0.2	-3.4	-1.6	0.5
320.23	-1.26	1.3	357.76	24MR	-0.8	-3.4	-1.7	0.5
327.87	-1.29	1.3	7.42	3AP	-1.7	-3.4	-1.7	0.5
335.44	-1.31	1.3	17.05	13AP	-1.3	-3.3	-1.7	0.5
342.94	-1.31	1.3	26.64	23AP	-0.3	-3.3	-1.8	0.5
350.34	-1.30	1.3	36.19	3MY	0.6	-3.3	-1.9	0.5
357.64	-1.28	1.3	45.72	13MY	1.3	-3.3	-1.9	0.5
4.84	-1.24	1.3	55.24	23MY	2.3	-3.3	-2.0	0.5
11.91	-1.19	1.2	64.76	2JN	3.4	-3.4	-2.1	0.5
18.85	-1.12	1.2	74.28	12JN	2.0	-3.4	-2.1	0.5
25.65	-1.04	1.2	83.82	22JN	0.6	-3.4	-2.2	0.5
32.29	-0.95	1.2	93.37	2JL	-0.7	-3.4	-2.3	0.4
38.77	-0.85	1.1	102.96	12JL	-1.4	-3.5	-2.3	0.4
45.05	-0.73	1.1	112.60	22JL	-1.2	-3.6	-2.4	0.3
51.13	-0.60	1.0	122.28	1AU	-0.6	-3.6	-2.4	0.3
56.96	-0.45	0.9	132.01	11AU	-0.2	-3.7	-2.4	0.3
62.52	-0.28	0.8	141.80	21AU	0.1	-3.8	-2.4	0.2
67.76	-0.10	0.7	151.64	31AU	0.2	-3.9	-2.4	0.1
72.61	0.11	0.6	161.55	10SE	0.5	-4.0	-2.4	0.0
76.99	0.35	0.5	171.52	20SE	1.8	-4.1	-2.4	-0.0
80.80	0.62	0.3	181.53	30SE	1.8	-4.2	-2.3	-0.1
83.91	0.92	0.2	191.61	10OC	-0.2	-4.3	-2.3	-0.1
86.16	1.27	-0.0	201.73	20OC	-0.6	-4.4	-2.2	-0.0
87.35	1.67	-0.3	211.88	30OC	-0.6	-4.4	-2.1	0.1
87.30	2.10	-0.5	222.07	9NO	-0.6	-4.4	-2.1	0.1
85.88	2.55	-0.7	232.28	19NO	-0.8	-3.5	-2.0	0.3
83.16	2.99	-0.9	242.49	29NO	-0.8	-3.1	-1.9	0.3
79.47	3.34	-1.1	252.70	9DE	-0.8	-3.9	-1.9	0.3
75.49	3.57	-1.0	262.89	19DE	-0.7	-4.3	-1.8	0.4
71.97	3.65	-0.8	273.06	29DE	0.3	-4.4	-1.7	0.4
				-563				
69.47	3.60	-0.6	283.19	8JA	2.8	-4.3	-1.7	0.5
68.27	3.46	-0.3	293.28	18JA	1.5	-4.2	-1.7	0.5
68.37	3.28	-0.0	303.32	28JA	0.7	-4.1	-1.6	0.5
69.59	3.09	0.2	313.31	7FE	0.4	-4.0	-1.6	0.5
71.76	2.89	0.4	323.24	17FE	0.1	-3.9	-1.6	0.5
74.70	2.70	0.6	333.10	27FE	-0.2	-3.8	-1.6	0.5
78.24	2.51	0.8	342.92	9MR	-0.8	-3.7	-1.6	0.4
82.28	2.34	1.0	352.67	19MR	-1.7	-3.6	-1.6	0.4
86.71	2.18	1.1	2.36	29MR	-1.3	-3.6	-1.6	0.4
91.46	2.03	1.2	12.01	8AP	-0.2	-3.5	-1.6	0.3
96.49	1.89	1.3	21.62	18AP	0.8	-3.4	-1.6	0.3
101.73	1.75	1.4	31.19	28AP	1.7	-3.4	-1.7	0.3
107.18	1.62	1.5	40.73	8MY	3.1	-3.4	-1.7	0.4
112.79	1.49	1.6	50.26	18MY	2.9	-3.3	-1.7	0.4
118.56	1.37	1.6	59.77	28MY	1.5	-3.3	-1.8	0.4
124.48	1.25	1.7	69.29	7JN	0.4	-3.3	-1.8	0.4
130.52	1.13	1.7	78.82	17JN	-0.7	-3.3	-1.9	0.4
136.69	1.01	1.7	88.36	27JN	-1.5	-3.3	-2.0	0.4
142.98	0.89	1.7	97.94	7JL	-1.2	-3.3	-2.0	0.3
149.39	0.77	1.7	107.55	17JL	-0.5	-3.4	-2.1	0.3
155.91	0.66	1.7	117.20	27JL	-0.1	-3.4	-2.2	0.3
162.54	0.54	1.7	126.90	6AU	0.2	-3.4	-2.2	0.2
169.29	0.43	1.7	136.66	16AU	0.4	-3.4	-2.3	0.2
176.14	0.31	1.7	146.47	26AU	0.8	-3.5	-2.3	0.1
183.11	0.20	1.7	156.35	5SE	2.2	-3.5	-2.4	0.0
190.18	0.08	1.7	166.28	15SE	1.6	-3.5	-2.4	-0.0
197.35	-0.03	1.6	176.27	25SE	-0.3	-3.5	-2.4	-0.1
204.63	-0.14	1.6	186.32	5OC	-0.7	-3.4	-2.4	-0.1
212.01	-0.25	1.6	196.42	15OC	-0.7	-3.4	-2.4	-0.2
219.47	-0.36	1.5	206.55	25OC	-0.7	-3.4	-2.4	-0.2
227.03	-0.46	1.5	216.73	4NO	-0.7	-3.4	-2.3	-0.1
234.65	-0.56	1.4	226.92	14NO	-0.7	-3.3	-2.2	-0.1
242.35	-0.66	1.4	237.13	24NO	-0.7	-3.3	-2.2	0.0
250.11	-0.74	1.4	247.35	4DE	-0.6	-3.3	-2.1	0.1
257.92	-0.82	1.3	257.55	14DE	0.4	-3.3	-2.0	0.2
265.76	-0.90	1.3	267.73	24DE	3.0	-3.4	-2.0	0.2

♂			☉	16.00UT	☿	♀	♃	♄
LONG	LAT	MAG	LONG	-562	MAGNITUDES			
273.63	-0.96	1.3	277.89	3JA	1.1	-3.4	-1.9	0.3
281.51	-1.01	1.3	288.00	13JA	0.4	-3.4	-1.8	0.3
289.39	-1.06	1.4	298.06	23JA	0.2	-3.4	-1.8	0.3
297.26	-1.09	1.4	308.08	2FE	0.0	-3.5	-1.7	0.4
305.10	-1.11	1.4	318.04	12FE	-0.3	-3.5	-1.7	0.4
312.90	-1.12	1.4	327.94	22FE	-0.9	-3.5	-1.6	0.4
320.66	-1.11	1.4	337.78	4MR	-1.6	-3.6	-1.6	0.4
328.35	-1.10	1.4	347.56	14MR	-1.2	-3.6	-1.6	0.4
335.98	-1.07	1.5	357.28	24MR	-0.2	-3.7	-1.5	0.3
343.53	-1.03	1.5	6.96	3AP	1.0	-3.8	-1.5	0.3
351.00	-0.98	1.5	16.58	13AP	2.3	-3.9	-1.5	0.3
358.38	-0.92	1.5	26.17	23AP	3.8	-4.0	-1.5	0.2
5.66	-0.85	1.5	35.73	3MY	2.2	-4.0	-1.5	0.2
12.85	-0.78	1.5	45.26	13MY	1.2	-4.1	-1.5	0.3
19.93	-0.69	1.6	54.78	23MY	0.3	-4.2	-1.6	0.3
26.92	-0.60	1.6	64.30	2JN	-0.7	-4.1	-1.6	0.3
33.79	-0.50	1.6	73.82	12JN	-1.6	-3.9	-1.6	0.3
40.55	-0.39	1.6	83.35	22JN	-1.3	-3.4	-1.7	0.3
47.20	-0.27	1.6	92.91	2JL	-0.5	-3.0	-1.7	0.3
53.73	-0.15	1.5	102.50	12JL	0.0	-3.7	-1.8	0.3
60.14	-0.02	1.5	112.13	22JL	0.3	-4.1	-1.8	0.3
66.41	0.11	1.5	121.81	1AU	0.6	-4.2	-1.9	0.3
72.54	0.26	1.5	131.53	11AU	1.1	-4.2	-2.0	0.2
78.51	0.41	1.4	141.32	21AU	2.6	-4.2	-2.0	0.2
84.31	0.58	1.4	151.16	31AU	1.4	-4.1	-2.1	0.1
89.90	0.75	1.3	161.06	10SE	-0.4	-4.0	-2.1	0.1
95.26	0.94	1.2	171.03	20SE	-0.9	-3.9	-2.2	-0.0
100.35	1.15	1.1	181.05	30SE	-0.9	-3.9	-2.3	-0.1
105.10	1.38	1.0	191.12	10OC	-0.9	-3.8	-2.3	-0.2
109.47	1.63	0.8	201.24	20OC	-0.6	-3.7	-2.3	-0.2
113.34	1.91	0.7	211.39	30OC	-0.5	-3.6	-2.3	-0.3
116.61	2.22	0.5	221.58	9NO	-0.5	-3.6	-2.3	-0.3
119.14	2.56	0.3	231.78	19NO	-0.4	-3.5	-2.3	-0.2
120.75	2.94	0.1	242.00	29NO	0.6	-3.5	-2.3	-0.1
121.26	3.35	-0.2	252.20	9DE	3.0	-3.4	-2.2	-0.1
120.50	3.77	-0.4	262.40	19DE	0.8	-3.4	-2.2	0.0
118.43	4.15	-0.7	272.57	29DE	0.2	-3.4	-2.1	0.1
				-561				
115.22	4.44	-0.9	282.70	8JA	0.1	-3.4	-2.0	0.1
111.35	4.58	-1.0	292.80	18JA	-0.1	-3.3	-1.9	0.2
107.51	4.54	-0.9	302.84	28JA	-0.4	-3.3	-1.9	0.2
104.41	4.34	-0.7	312.83	7FE	-0.9	-3.3	-1.8	0.3
102.46	4.04	-0.4	322.76	17FE	-1.5	-3.3	-1.7	0.3
101.82	3.69	-0.2	332.63	27FE	-1.2	-3.3	-1.7	0.3
102.41	3.33	0.0	342.44	9MR	-0.1	-3.3	-1.6	0.3
104.06	2.98	0.2	352.19	19MR	1.3	-3.4	-1.6	0.3
106.60	2.66	0.4	1.89	29MR	2.9	-3.4	-1.5	0.3
109.86	2.37	0.6	11.54	8AP	2.9	-3.4	-1.5	0.3
113.70	2.10	0.7	21.15	18AP	1.7	-3.5	-1.5	0.2
118.03	1.85	0.9	30.72	28AP	0.9	-3.5	-1.4	0.2
122.75	1.62	1.0	40.26	8MY	0.2	-3.5	-1.4	0.2
127.80	1.40	1.1	49.79	18MY	-0.7	-3.4	-1.4	0.2
133.15	1.20	1.2	59.31	28MY	-1.6	-3.4	-1.4	0.2
138.74	1.02	1.2	68.82	7JN	-1.3	-3.4	-1.4	0.3
144.55	0.84	1.3	78.35	17JN	-0.4	-3.3	-1.4	0.3
150.57	0.67	1.3	87.90	27JN	0.1	-3.3	-1.5	0.3
156.77	0.51	1.4	97.47	7JL	0.5	-3.3	-1.5	0.3
163.14	0.35	1.4	107.08	17JL	0.8	-3.3	-1.5	0.3
169.67	0.20	1.4	116.73	27JL	1.5	-3.3	-1.6	0.3
176.35	0.06	1.5	126.43	6AU	3.1	-3.3	-1.6	0.3
183.19	-0.08	1.5	136.18	16AU	1.2	-3.4	-1.7	0.3
190.15	-0.21	1.5	145.99	26AU	-0.5	-3.4	-1.7	0.2
197.25	-0.33	1.5	155.86	5SE	-1.0	-3.4	-1.8	0.2
204.48	-0.45	1.5	165.80	15SE	-1.0	-3.5	-1.8	0.1
211.82	-0.56	1.5	175.79	25SE	-0.8	-3.5	-1.9	0.0
219.27	-0.66	1.5	185.83	5OC	-0.5	-3.6	-2.0	-0.0
226.81	-0.75	1.4	195.93	15OC	-0.4	-3.7	-2.0	-0.1
234.45	-0.84	1.4	206.06	25OC	-0.4	-3.7	-2.1	-0.2
242.16	-0.91	1.4	216.23	4NO	-0.2	-3.8	-2.1	-0.2
249.94	-0.97	1.4	226.43	14NO	0.7	-3.9	-2.2	-0.3
257.77	-1.02	1.4	236.64	24NO	2.7	-4.0	-2.2	-0.3
265.64	-1.06	1.4	246.85	4DE	0.5	-4.1	-2.2	-0.2
273.53	-1.09	1.4	257.06	14DE	0.0	-4.2	-2.2	-0.1
281.43	-1.11	1.4	267.24	24DE	-0.1	-4.3	-2.2	-0.1

♂			☉	16.00UT	☿	♀	♃	♄
LONG	LAT	MAG	LONG	-560	MAGNITUDES			
289.32	-1.11	1.4	277.39	3JA	-0.2	-4.4	-2.1	-0.0
297.20	-1.10	1.4	287.51	13JA	-0.5	-4.3	-2.1	0.1
305.04	-1.08	1.4	297.58	23JA	-0.9	-4.1	-2.0	0.1
312.84	-1.05	1.4	307.60	2FE	-1.4	-3.5	-2.0	0.2
320.59	-1.01	1.4	317.56	12FE	-1.1	-3.3	-1.9	0.2
328.27	-0.96	1.4	327.46	22FE	-0.1	-3.9	-1.8	0.3
335.89	-0.90	1.4	337.30	3MR	1.6	-4.2	-1.8	0.3
343.42	-0.83	1.4	347.09	13MR	3.5	-4.3	-1.7	0.3
350.88	-0.76	1.5	356.81	23MR	2.2	-4.2	-1.6	0.3
358.24	-0.67	1.5	6.48	2AP	1.2	-4.1	-1.6	0.3
5.52	-0.59	1.6	16.12	12AP	0.7	-4.0	-1.5	0.3
12.71	-0.49	1.6	25.70	22AP	0.1	-3.9	-1.5	0.3
19.82	-0.40	1.7	35.26	2MY	-0.8	-3.8	-1.4	0.2
26.83	-0.30	1.7	44.80	12MY	-1.7	-3.7	-1.4	0.2
33.75	-0.20	1.7	54.32	22MY	-1.3	-3.6	-1.4	0.2
40.59	-0.09	1.8	63.83	1JN	-0.4	-3.6	-1.3	0.2
47.34	0.01	1.8	73.36	11JN	0.2	-3.5	-1.3	0.2
54.00	0.12	1.8	82.89	21JN	0.6	-3.5	-1.3	0.3
60.59	0.23	1.8	92.45	1JL	1.1	-3.4	-1.3	0.3
67.08	0.35	1.8	102.04	11JL	2.0	-3.4	-1.3	0.3
73.50	0.46	1.8	111.66	21JL	3.2	-3.4	-1.3	0.3
79.83	0.58	1.8	121.34	31JL	1.0	-3.4	-1.4	0.3
86.07	0.70	1.8	131.07	10AU	-0.6	-3.4	-1.4	0.3
92.22	0.82	1.8	140.85	20AU	-1.2	-3.4	-1.4	0.3
98.27	0.95	1.7	150.69	30AU	-1.2	-3.4	-1.5	0.3
104.22	1.08	1.7	160.59	9SE	-0.7	-3.4	-1.5	0.3
110.04	1.22	1.6	170.54	19SE	-0.4	-3.4	-1.6	0.2
115.72	1.36	1.5	180.56	29SE	-0.3	-3.4	-1.6	0.1
121.24	1.52	1.5	190.63	9OC	-0.2	-3.4	-1.7	0.1
126.58	1.68	1.4	200.74	19OC	-0.0	-3.4	-1.8	0.0
131.69	1.86	1.3	210.90	29OC	1.0	-3.4	-1.8	-0.1
136.54	2.04	1.1	221.08	8NO	2.5	-3.5	-1.9	-0.1
141.05	2.25	1.0	231.28	18NO	0.3	-3.5	-1.9	-0.2
145.17	2.47	0.8	241.50	28NO	-0.1	-3.5	-2.0	-0.3
148.78	2.72	0.6	251.71	8DE	-0.2	-3.5	-2.0	-0.2
151.78	2.98	0.4	261.90	18DE	-0.3	-3.4	-2.1	-0.2
154.00	3.27	0.2	272.07	28DE	-0.5	-3.4	-2.1	-0.1
				-559				
155.27	3.57	-0.1	282.21	7JA	-1.0	-3.4	-2.1	-0.0
155.42	3.86	-0.4	292.30	17JA	-1.2	-3.4	-2.1	0.0
154.31	4.13	-0.7	302.35	27JA	-1.0	-3.3	-2.1	0.1
151.93	4.31	-0.9	312.34	6FE	-0.0	-3.3	-2.0	0.2
148.55	4.35	-1.1	322.27	16FE	1.9	-3.3	-2.0	0.2
144.71	4.22	-1.2	332.15	26FE	3.0	-3.3	-1.9	0.3
141.15	3.93	-1.0	341.96	8MR	1.6	-3.3	-1.9	0.3
138.49	3.52	-0.9	351.72	18MR	0.9	-3.4	-1.8	0.3
137.06	3.05	-0.6	1.42	28MR	0.5	-3.4	-1.7	0.3
136.96	2.58	-0.4	11.07	7AP	-0.0	-3.4	-1.7	0.3
138.07	2.14	-0.2	20.68	17AP	-0.8	-3.4	-1.6	0.3
140.22	1.74	-0.0	30.26	27AP	-1.7	-3.5	-1.5	0.3
143.25	1.37	0.1	39.80	7MY	-1.3	-3.5	-1.5	0.3
146.98	1.04	0.3	49.33	17MY	-0.4	-3.6	-1.4	0.3
151.31	0.75	0.4	58.85	27MY	0.3	-3.7	-1.4	0.3
156.14	0.48	0.5	68.36	6JN	0.9	-3.7	-1.3	0.2
161.37	0.24	0.6	77.89	16JN	1.5	-3.8	-1.3	0.3
166.95	0.02	0.7	87.44	26JN	2.5	-3.9	-1.3	0.3
172.85	-0.17	0.8	97.00	6JL	2.8	-4.0	-1.3	0.4
179.01	-0.35	0.9	106.61	16JL	0.9	-4.1	-1.3	0.4
185.42	-0.52	0.9	116.26	26JL	-0.6	-4.2	-1.2	0.4
192.04	-0.67	1.0	125.96	5AU	-1.3	-4.3	-1.3	0.4
198.85	-0.80	1.0	135.71	15AU	-1.2	-4.2	-1.3	0.4
205.85	-0.92	1.1	145.52	25AU	-0.7	-4.0	-1.3	0.4
213.00	-1.02	1.1	155.39	4SE	-0.3	-3.5	-1.3	0.4
220.29	-1.10	1.1	165.32	14SE	-0.1	-3.4	-1.3	0.4
227.71	-1.17	1.2	175.30	24SE	-0.0	-3.9	-1.4	0.3
235.24	-1.23	1.2	185.34	4OC	0.2	-4.3	-1.4	0.3
242.86	-1.27	1.2	195.44	14OC	1.2	-4.4	-1.5	0.2
250.56	-1.29	1.2	205.57	24OC	2.2	-4.3	-1.5	0.2
258.31	-1.30	1.3	215.74	3NO	0.1	-4.3	-1.6	0.1
266.11	-1.30	1.3	225.93	13NO	-0.3	-4.2	-1.6	0.0
273.94	-1.28	1.3	236.15	23NO	-0.4	-4.1	-1.7	-0.0
281.78	-1.25	1.4	246.35	3DE	-0.4	-4.0	-1.8	-0.1
289.61	-1.21	1.4	256.56	13DE	-0.6	-3.9	-1.8	-0.2
297.42	-1.15	1.4	266.75	23DE	-1.0	-3.8	-1.9	-0.1

Left table

♂ LONG LAT MAG	☉ LONG	16.00UT −558	☿ ♀ ♃ ♄ MAGNITUDES
305.19-1.09 1.4	276.90	2JA	-1.1 -3.7 -1.9 -0.1
312.92-1.02 1.5	287.02	12JA	-0.9 -3.6 -2.0 -0.0
320.60-0.94 1.5	297.10	22JA	0.1 -3.6 -2.0 0.0
328.21-0.85 1.5	307.12	1FE	2.2 -3.5 -2.0 0.1
335.75-0.76 1.5	317.08	11FE	2.3 -3.4 -2.0 0.2
343.21-0.66 1.6	326.99	21FE	1.2 -3.4 -2.0 0.2
350.59-0.56 1.6	336.83	3MR	0.7 -3.4 -2.0 0.3
357.89-0.46 1.6	346.62	13MR	0.3 -3.3 -1.9 0.3
5.11-0.36 1.6	356.35	23MR	-0.1 -3.3 -1.9 0.4
12.24-0.26 1.6	6.02	2AP	-0.8 -3.3 -1.8 0.4
19.29-0.16 1.7	15.65	12AP	-1.7 -3.3 -1.8 0.4
26.26-0.06 1.7	25.25	22AP	-1.3 -3.3 -1.7 0.4
33.15 0.04 1.7	34.80	2MY	-0.3 -3.3 -1.6 0.4
39.97 0.14 1.7	44.34	12MY	0.5 -3.3 -1.6 0.4
46.71 0.24 1.8	53.86	22MY	1.1 -3.4 -1.5 0.4
53.39 0.34 1.8	63.37	1JN	1.9 -3.4 -1.5 0.4
60.01 0.43 1.9	72.89	11JN	3.2 -3.4 -1.4 0.4
66.56 0.53 1.9	82.43	21JN	2.3 -3.5 -1.4 0.4
73.06 0.62 1.9	91.98	1JL	0.7 -3.5 -1.3 0.4
79.51 0.71 1.9	101.57	11JL	-0.6 -3.5 -1.3 0.4
85.91 0.80 2.0	111.19	21JL	-1.4 -3.4 -1.3 0.5
92.25 0.89 2.0	120.86	31JL	-1.3 -3.4 -1.2 0.5
98.56 0.98 2.0	130.59	10AU	-0.6 -3.4 -1.2 0.5
104.81 1.06 2.0	140.37	20AU	-0.2 -3.4 -1.2 0.6
111.01 1.15 1.9	150.20	30AU	-0.0 -3.3 -1.2 0.6
117.16 1.24 1.9	160.10	9SE	0.1 -3.3 -1.2 0.6
123.26 1.32 1.9	170.06	19SE	0.4 -3.3 -1.2 0.5
129.30 1.41 1.8	180.07	29SE	1.5 -3.3 -1.3 0.5
135.26 1.50 1.8	190.14	9OC	2.0 -3.3 -1.3 0.5
141.15 1.59 1.7	200.25	19OC	-0.1 -3.4 -1.3 0.4
146.95 1.68 1.6	210.40	29OC	-0.5 -3.4 -1.4 0.4
152.64 1.77 1.5	220.59	8NO	-0.5 -3.4 -1.4 0.3
158.20 1.86 1.4	230.79	18NO	-0.5 -3.5 -1.5 0.3
163.61 1.96 1.3	241.00	28NO	-0.7 -3.5 -1.5 0.2
168.84 2.06 1.1	251.21	8DE	-0.9 -3.5 -1.6 0.1
173.83 2.16 1.0	261.41	18DE	-0.9 -3.6 -1.7 0.0
178.54 2.27 0.8	271.58	28DE	-0.8 -3.7 -1.7 -0.0
		−557	
182.91 2.37 0.6	281.72	7JA	0.1 -3.7 -1.8 0.0
186.83 2.47 0.4	291.82	17JA	2.6 -3.8 -1.8 0.1
190.23 2.57 0.2	301.87	27JA	1.8 -3.9 -1.9 0.1
192.93 2.65 -0.1	311.87	6FE	0.8 -4.0 -2.0 0.2
194.80 2.71 -0.4	321.80	16FE	0.5 -4.1 -2.0 0.2
195.64 2.73 -0.7	331.67	26FE	-0.2 -4.2 -2.0 0.3
195.28 2.68 -1.0	341.49	8MR	-0.2 -4.2 -2.0 0.4
193.67 2.54 -1.3	351.25	18MR	-0.8 -4.2 -2.0 0.4
190.93 2.29 -1.6	0.95	28MR	-1.7 -4.1 -2.0 0.5
187.48 1.90 -1.7	10.61	7AP	-1.3 -3.7 -2.0 0.5
184.02 1.42 -1.6	20.22	17AP	-0.3 -3.0 -1.9 0.5
181.25 0.90 -1.5	29.79	27AP	0.6 -3.3 -1.9 0.6
179.65 0.38 -1.3	39.34	7MY	1.5 -3.9 -1.8 0.6
179.43 -0.08 -1.1	48.87	17MY	2.6 -4.1 -1.8 0.6
180.51 -0.49 -0.9	58.38	27MY	3.3 -4.2 -1.7 0.6
182.77 -0.82 -0.7	67.90	6JN	1.8 -4.1 -1.6 0.6
186.01 -1.11 -0.5	77.42	16JN	0.5 -4.1 -1.6 0.6
190.06 -1.34 -0.3	86.97	26JN	-0.7 -4.0 -1.5 0.6
194.78 -1.52 -0.2	96.54	6JL	-1.5 -3.9 -1.5 0.6
200.04 -1.67 -0.0	106.14	16JL	-1.3 -3.8 -1.4 0.6
205.76 -1.79 0.1	115.79	26JL	-0.6 -3.7 -1.4 0.6
211.86 -1.87 0.2	125.48	5AU	-0.1 -3.7 -1.3 0.7
218.28 -1.93 0.3	135.23	15AU	0.1 -3.6 -1.3 0.7
224.96 -1.96 0.4	145.04	25AU	0.3 -3.6 -1.3 0.7
231.87 -1.96 0.5	154.91	4SE	0.6 -3.5 -1.2 0.8
238.96 -1.95 0.6	164.83	14SE	1.9 -3.5 -1.2 0.8
246.20 -1.91 0.7	174.82	24SE	1.8 -3.4 -1.2 0.8
253.57 -1.86 0.7	184.86	4OC	-0.2 -3.4 -1.2 0.7
261.02 -1.79 0.8	194.94	14OC	-0.6 -3.4 -1.2 0.7
268.54 -1.70 0.9	205.08	24OC	-0.6 -3.4 -1.3 0.7
276.11 -1.60 1.0	215.25	3NO	-0.7 -3.4 -1.3 0.6
283.71 -1.50 1.1	225.44	13NO	-0.8 -3.4 -1.3 0.6
291.30 -1.38 1.1	235.65	23NO	-0.7 -3.4 -1.4 0.5
298.89 -1.26 1.2	245.86	3DE	-0.8 -3.4 -1.4 0.5
306.45 -1.14 1.3	256.07	13DE	-0.7 -3.4 -1.5 0.4
313.97 -1.01 1.3	266.26	23DE	0.3 -3.4 -1.5 0.3

Right table

♂ LONG LAT MAG	☉ LONG	16.00UT −556	☿ ♀ ♃ ♄ MAGNITUDES
321.45-0.88 1.4	276.41	2JA	2.9 -3.4 -1.6 0.2
328.87-0.75 1.5	286.53	12JA	1.3 -3.4 -1.7 0.2
336.22-0.62 1.5	296.61	22JA	0.6 -3.4 -1.7 0.2
343.51-0.50 1.6	306.63	1FE	0.3 -3.5 -1.8 0.3
350.73-0.37 1.6	316.60	11FE	0.1 -3.5 -1.9 0.3
357.87-0.25 1.7	326.51	21FE	-0.2 -3.5 -1.9 0.4
4.94-0.14 1.7	336.35	2MR	-0.8 -3.4 -2.0 0.4
11.93-0.03 1.8	346.14	12MR	-1.7 -3.4 -2.0 0.5
18.86 0.08 1.8	355.87	22MR	-1.3 -3.4 -2.0 0.6
25.72 0.18 1.8	5.55	1AP	-0.2 -3.4 -2.1 0.6
32.51 0.28 1.8	15.18	11AP	0.8 -3.3 -2.1 0.7
39.24 0.37 1.9	24.78	21AP	1.9 -3.3 -2.1 0.7
45.91 0.46 1.9	34.34	1MY	3.4 -3.3 -2.0 0.7
52.53 0.55 1.9	43.87	11MY	2.7 -3.3 -2.0 0.8
59.10 0.63 1.9	53.40	21MY	1.4 -3.3 -2.0 0.8
65.63 0.70 1.9	62.91	31MY	0.4 -3.4 -1.9 0.8
72.11 0.77 1.9	72.43	10JN	-0.7 -3.4 -1.8 0.8
78.57 0.84 1.9	81.96	20JN	-1.5 -3.4 -1.8 0.8
84.99 0.90 2.0	91.52	30JN	-1.3 -3.5 -1.7 0.8
91.39 0.97 2.0	101.10	10JL	-0.5 -3.5 -1.7 0.8
97.77 1.02 2.0	110.73	20JL	-0.0 -3.6 -1.6 0.8
104.13 1.08 2.0	120.39	30JL	0.2 -3.6 -1.5 0.8
110.47 1.13 2.0	130.11	9AU	0.5 -3.7 -1.5 0.9
116.81 1.17 2.0	139.89	19AU	0.9 -3.8 -1.4 0.9
123.13 1.22 2.0	149.72	29AU	2.3 -3.9 -1.4 1.0
129.44 1.26 2.0	159.62	8SE	1.6 -4.0 -1.4 1.0
135.74 1.30 2.0	169.57	18SE	-0.3 -4.1 -1.3 1.0
142.03 1.33 2.0	179.58	28SE	-0.8 -4.2 -1.3 1.0
148.32 1.36 1.9	189.65	8OC	-0.8 -4.3 -1.3 1.0
154.58 1.39 1.9	199.76	18OC	-0.8 -4.4 -1.3 1.0
160.84 1.41 1.8	209.91	28OC	-0.7 -4.3 -1.3 1.0
167.07 1.43 1.7	220.09	7NO	-0.6 -4.1 -1.3 0.9
173.27 1.45 1.7	230.30	17NO	-0.6 -3.5 -1.3 0.9
179.44 1.45 1.6	240.51	27NO	-0.5 -3.1 -1.3 0.8
185.57 1.45 1.5	250.72	7DE	0.4 -3.9 -1.4 0.8
191.65 1.44 1.4	260.92	17DE	3.0 -4.3 -1.4 0.7
197.66 1.42 1.2	271.09	27DE	1.0 -4.4 -1.5 0.6
		−555	
203.59 1.39 1.1	281.23	6JA	0.3 -4.3 -1.5 0.6
209.42 1.34 1.0	291.33	16JA	0.2 -4.2 -1.6 0.5
215.13 1.28 0.8	301.38	26JA	-0.0 -4.1 -1.6 0.4
220.69 1.19 0.6	311.38	5FE	-0.3 -4.0 -1.7 0.5
226.06 1.07 0.4	321.32	15FE	-0.9 -3.9 -1.8 0.5
231.19 0.92 0.2	331.20	25FE	-1.6 -3.8 -1.8 0.6
236.02 0.73 -0.0	341.02	7MR	-1.2 -3.7 -1.9 0.6
240.47 0.49 -0.3	350.78	17MR	-0.2 -3.6 -2.0 0.7
244.43 0.18 -0.5	0.49	27MR	1.1 -3.6 -2.0 0.7
247.78-0.20 -0.8	10.15	6AP	2.5 -3.5 -2.1 0.8
250.33-0.66 -1.1	19.76	16AP	3.5 -3.4 -2.1 0.8
251.93-1.23 -1.4	29.34	26AP	2.0 -3.4 -2.2 0.9
252.38-1.89 -1.7	38.88	6MY	1.1 -3.4 -2.2 0.9
251.59-2.62 -2.0	48.41	16MY	0.3 -3.3 -2.2 1.0
249.67-3.38 -2.3	57.93	26MY	-0.7 -3.3 -2.2 1.0
246.99-4.06 -2.5	67.45	5JN	-1.6 -3.3 -2.1 1.0
244.23-4.57 -2.4	76.97	15JN	-1.3 -3.3 -2.1 1.0
242.12-4.87 -2.2	86.51	25JN	-0.5 -3.3 -2.1 1.1
241.14-4.95 -2.0	96.08	5JL	0.0 -3.3 -2.0 1.1
241.53-4.88 -1.8	105.68	15JL	0.4 -3.4 -1.9 1.1
243.23-4.69 -1.5	115.32	25JL	0.7 -3.4 -1.9 1.1
246.07-4.43 -1.3	125.01	4AU	1.2 -3.4 -1.8 1.1
249.87-4.14 -1.0	134.76	14AU	2.8 -3.4 -1.7 1.1
254.43-3.80 -0.8	144.56	24AU	1.4 -3.5 -1.7 1.2
259.59-3.51 -0.6	154.42	3SE	-0.4 -3.5 -1.6 1.2
265.21-3.19 -0.4	164.35	13SE	-0.9 -3.5 -1.6 1.2
271.20-2.88 -0.2	174.32	23SE	-0.9 -3.5 -1.5 1.3
277.46-2.58 -0.0	184.36	3OC	-0.9 -3.4 -1.5 1.3
283.94-2.28 0.1	194.45	13OC	-0.6 -3.4 -1.5 1.3
290.56-2.00 0.3	204.58	23OC	-0.5 -3.4 -1.4 1.3
297.30-1.74 0.5	214.75	2NO	-0.4 -3.4 -1.4 1.3
304.13-1.48 0.6	224.94	12NO	-0.3 -3.3 -1.4 1.3
311.00-1.25 0.7	235.14	22NO	0.6 -3.3 -1.4 1.2
317.89-1.03 0.9	245.36	2DE	3.0 -3.3 -1.4 1.2
324.80-0.82 1.0	255.57	12DE	0.7 -3.3 -1.4 1.1
331.71-0.63 1.1	265.75	22DE	0.1 -3.4 -1.4 1.1

♂ LONG	♂ LAT	♂ MAG	☉ LONG	16.00UT	☿ MAG	♀ MAG	♃ MAG	♄ MAG
				-554				
338.59	-0.45	1.2	275.92	1JA	0.0	-3.4	-1.4	1.0
345.45	-0.28	1.3	286.04	11JA	-0.1	-3.4	-1.5	0.9
352.28	-0.13	1.4	296.12	21JA	-0.4	-3.4	-1.5	0.8
359.07	0.01	1.5	306.15	31JA	-0.9	-3.5	-1.6	0.8
5.82	0.14	1.6	316.12	10FE	-1.5	-3.5	-1.6	0.7
12.53	0.25	1.7	326.03	20FE	-1.2	-3.5	-1.7	0.8
19.19	0.36	1.7	335.88	2MR	-0.2	-3.6	-1.7	0.8
25.82	0.46	1.8	345.67	12MR	1.3	-3.7	-1.8	0.9
32.40	0.55	1.8	355.41	22MR	3.2	-3.7	-1.9	0.9
38.94	0.63	1.9	5.09	1AP	2.6	-3.8	-1.9	1.0
45.44	0.71	1.9	14.73	11AP	1.5	-3.9	-2.0	1.0
51.92	0.78	2.0	24.32	21AP	0.8	-4.0	-2.1	1.1
58.36	0.84	2.0	33.88	1MY	0.2	-4.1	-2.1	1.1
64.78	0.90	2.0	43.42	11MY	-0.7	-4.1	-2.2	1.2
71.18	0.95	2.0	52.94	21MY	-1.6	-4.2	-2.2	1.2
77.56	0.99	2.0	62.46	31MY	-1.3	-4.1	-2.3	1.3
83.93	1.03	2.0	71.98	10JN	-0.4	-3.9	-2.3	1.3
90.29	1.07	2.0	81.51	20JN	0.1	-3.4	-2.3	1.3
96.65	1.10	2.0	91.06	30JN	0.5	-3.0	-2.3	1.3
103.02	1.12	2.0	100.64	10JL	0.9	-3.7	-2.3	1.4
109.38	1.14	2.0	110.26	20JL	1.6	-4.1	-2.2	1.4
115.76	1.16	2.0	119.93	30JL	3.2	-4.2	-2.2	1.4
122.16	1.17	2.0	129.64	9AU	1.2	-4.2	-2.1	1.4
128.57	1.18	2.0	139.41	19AU	-0.5	-4.2	-2.0	1.3
135.00	1.18	2.0	149.25	29AU	-1.1	-4.1	-2.0	1.3
141.46	1.18	2.0	159.14	8SE	-1.1	-4.0	-1.9	1.3
147.94	1.17	2.0	169.08	18SE	-0.8	-3.9	-1.8	1.3
154.45	1.16	2.0	179.09	28SE	-0.5	-3.8	-1.8	1.3
160.99	1.14	1.9	189.15	8OC	-0.3	-3.8	-1.7	1.3
167.55	1.12	1.9	199.26	18OC	-0.3	-3.7	-1.7	1.3
174.15	1.09	1.9	209.41	28OC	-0.2	-3.6	-1.6	1.3
180.78	1.05	1.8	219.59	7NO	0.8	-3.6	-1.6	1.2
187.44	1.00	1.8	229.79	17NO	2.7	-3.5	-1.6	1.2
194.12	0.95	1.7	240.01	27NO	0.4	-3.5	-1.5	1.2
200.83	0.88	1.6	250.22	7DE	-0.0	-3.4	-1.5	1.1
207.57	0.81	1.6	260.42	17DE	-0.1	-3.4	-1.5	1.1
214.34	0.72	1.5	270.60	27DE	-0.2	-3.4	-1.5	1.0
				-553				
221.12	0.61	1.4	280.74	6JA	-0.5	-3.4	-1.5	1.0
227.92	0.50	1.3	290.84	16JA	-0.9	-3.3	-1.5	0.9
234.73	0.36	1.2	300.90	26JA	-1.3	-3.3	-1.5	0.9
241.55	0.21	1.1	310.90	5FE	-1.1	-3.3	-1.5	0.8
248.38	0.04	1.0	320.84	15FE	-0.1	-3.3	-1.6	0.8
255.20	-0.15	0.8	330.73	25FE	1.6	-3.3	-1.6	0.8
261.99	-0.37	0.7	340.55	7MR	3.4	-3.3	-1.7	0.9
268.75	-0.61	0.6	350.31	17MR	1.9	-3.4	-1.7	1.0
275.46	-0.87	0.4	0.02	27MR	1.1	-3.4	-1.8	1.0
282.09	-1.17	0.3	9.68	6AP	0.6	-3.4	-1.8	1.1
288.61	-1.48	0.1	19.30	16AP	0.1	-3.5	-1.9	1.2
294.98	-1.83	-0.1	28.88	26AP	-0.7	-3.5	-2.0	1.2
301.15	-2.21	-0.2	38.42	6MY	-1.7	-3.4	-2.0	1.3
307.05	-2.61	-0.4	47.95	16MY	-1.3	-3.4	-2.1	1.3
312.59	-3.05	-0.6	57.47	26MY	-0.4	-3.4	-2.2	1.3
317.69	-3.51	-0.8	66.98	5JN	0.2	-3.4	-2.2	1.4
322.19	-3.99	-1.0	76.51	15JN	0.7	-3.3	-2.3	1.4
325.92	-4.50	-1.3	86.05	25JN	1.2	-3.3	-2.3	1.4
328.71	-5.01	-1.5	95.61	5JL	2.1	-3.3	-2.4	1.4
330.32	-5.49	-1.7	105.21	15JL	3.1	-3.3	-2.4	1.3
330.57	-5.90	-2.0	114.86	25JL	1.0	-3.3	-2.4	1.3
329.43	-6.15	-2.2	124.54	4AU	-0.5	-3.3	-2.4	1.3
327.08	-6.18	-2.4	134.29	14AU	-1.2	-3.4	-2.4	1.2
324.11	-5.92	-2.5	144.09	24AU	-1.2	-3.4	-2.3	1.2
321.29	-5.39	-2.3	153.94	3SE	-0.7	-3.4	-2.3	1.1
319.31	-4.70	-2.0	163.86	13SE	-0.4	-3.5	-2.2	1.2
318.58	-3.94	-1.7	173.84	23SE	-0.2	-3.5	-2.1	1.2
319.17	-3.21	-1.3	183.87	3OC	-0.1	-3.6	-2.1	1.2
320.94	-2.55	-1.0	193.96	13OC	0.0	-3.7	-2.0	1.2
323.72	-1.97	-0.7	204.09	23OC	1.0	-3.7	-1.9	1.1
327.29	-1.48	-0.4	214.25	2NO	2.5	-3.8	-1.9	1.1
331.48	-1.06	-0.2	224.44	12NO	0.2	-3.9	-1.8	1.1
336.15	-0.70	0.1	234.65	22NO	-0.2	-4.0	-1.8	1.1
341.19	-0.40	0.3	244.86	2DE	-0.3	-4.1	-1.7	1.0
346.51	-0.14	0.5	255.07	12DE	-0.3	-4.2	-1.7	1.0
352.05	0.07	0.7	265.26	22DE	-0.6	-4.3	-1.6	1.0

♂ LONG	♂ LAT	♂ MAG	☉ LONG	16.00UT	☿ MAG	♀ MAG	♃ MAG	♄ MAG
				-552				
357.75	0.26	0.9	275.42	1JA	-1.0	-4.4	-1.6	0.9
3.57	0.42	1.0	285.55	11JA	-1.2	-4.3	-1.6	0.9
9.50	0.56	1.2	295.63	21JA	-1.0	-4.0	-1.6	0.8
15.49	0.67	1.3	305.66	31JA	-0.0	-3.5	-1.6	0.8
21.54	0.77	1.4	315.63	10FE	2.0	-3.4	-1.6	0.7
27.63	0.86	1.5	325.55	20FE	2.8	-3.9	-1.6	0.7
33.74	0.93	1.6	335.40	1MR	1.4	-4.2	-1.6	0.6
39.88	0.99	1.7	345.20	11MR	0.8	-4.3	-1.6	0.7
46.03	1.05	1.8	354.93	21MR	0.4	-4.2	-1.6	0.8
52.19	1.09	1.8	4.62	31MR	-0.0	-4.1	-1.7	0.8
58.37	1.12	1.9	14.26	10AP	-0.7	-4.0	-1.7	0.9
64.56	1.15	1.9	23.86	20AP	-1.7	-3.9	-1.8	1.0
70.75	1.18	2.0	33.42	30AP	-1.3	-3.8	-1.8	1.0
76.96	1.19	2.0	42.96	10MY	-0.4	-3.7	-1.9	1.1
83.19	1.20	2.0	52.48	20MY	0.4	-3.6	-1.9	1.1
89.42	1.21	2.0	62.00	30MY	0.9	-3.6	-2.0	1.2
95.68	1.21	2.0	71.52	9JN	1.6	-3.5	-2.1	1.2
101.97	1.20	2.0	81.05	19JN	2.8	-3.5	-2.1	1.2
108.28	1.20	2.0	90.60	29JN	2.6	-3.4	-2.2	1.2
114.62	1.18	2.0	100.18	9JL	0.8	-3.4	-2.3	1.2
120.99	1.16	2.0	109.80	19JL	-0.6	-3.4	-2.3	1.2
127.41	1.14	2.0	119.46	29JL	-1.3	-3.4	-2.4	1.2
133.86	1.11	1.9	129.18	8AU	-1.3	-3.4	-2.4	1.2
140.37	1.08	1.9	138.94	18AU	-0.7	-3.4	-2.4	1.1
146.92	1.04	1.9	148.77	28AU	-0.3	-3.4	-2.5	1.1
153.52	1.00	1.9	158.66	7SE	-0.1	-3.4	-2.4	1.0
160.17	0.95	1.9	168.61	17SE	0.0	-3.4	-2.4	1.0
166.89	0.90	1.9	178.61	27SE	0.2	-3.4	-2.4	1.0
173.66	0.84	1.9	188.67	7OC	1.3	-3.4	-2.3	1.0
180.49	0.78	1.8	198.77	17OC	2.3	-3.4	-2.3	1.0
187.38	0.70	1.8	208.92	27OC	0.1	-3.4	-2.2	1.0
194.34	0.62	1.8	219.10	6NO	-0.4	-3.5	-2.1	1.0
201.35	0.54	1.8	229.30	16NO	-0.4	-3.5	-2.0	1.0
208.42	0.44	1.7	239.51	26NO	-0.5	-3.5	-2.0	1.0
215.56	0.34	1.7	249.72	6DE	-0.6	-3.5	-1.9	1.0
222.75	0.23	1.6	259.92	16DE	-0.9	-3.4	-1.8	0.9
229.99	0.11	1.6	270.09	26DE	-1.0	-3.4	-1.8	0.9
				-551				
237.29	-0.01	1.5	280.24	5JA	-0.9	-3.4	-1.7	0.8
244.64	-0.15	1.5	290.34	15JA	0.0	-3.4	-1.7	0.8
252.03	-0.29	1.4	300.40	25JA	2.3	-3.3	-1.6	0.7
259.46	-0.44	1.3	310.41	4FE	2.1	-3.3	-1.6	0.7
266.92	-0.59	1.3	320.35	14FE	1.0	-3.3	-1.6	0.6
274.40	-0.75	1.2	330.24	24FE	0.6	-3.3	-1.6	0.6
281.89	-0.92	1.1	340.07	6MR	0.3	-3.4	-1.6	0.5
289.38	-1.09	1.0	349.83	16MR	-0.1	-3.4	-1.6	0.5
296.86	-1.25	1.0	359.55	26MR	-0.8	-3.4	-1.6	0.6
304.31	-1.42	0.9	9.21	5AP	-1.7	-3.4	-1.6	0.6
311.71	-1.58	0.8	18.83	15AP	-1.3	-3.4	-1.6	0.7
319.04	-1.73	0.7	28.41	25AP	-0.3	-3.5	-1.6	0.8
326.28	-1.88	0.7	37.96	5MY	0.5	-3.5	-1.7	0.8
333.41	-2.02	0.6	47.49	15MY	1.2	-3.6	-1.7	0.9
340.40	-2.15	0.5	57.01	25MY	2.1	-3.7	-1.7	0.9
347.21	-2.26	0.4	66.53	4JN	3.4	-3.7	-1.8	1.0
353.81	-2.36	0.3	76.05	14JN	2.1	-3.8	-1.8	1.0
0.16	-2.44	0.2	85.59	24JN	0.7	-3.9	-1.9	1.1
6.19	-2.50	0.1	95.15	4JL	-0.6	-4.0	-2.0	1.1
11.85	-2.55	-0.0	104.75	14JL	-1.4	-4.1	-2.0	1.1
17.05	-2.57	-0.2	114.39	24JL	-1.3	-4.2	-2.1	1.1
21.68	-2.57	-0.3	124.08	3AU	-0.6	-4.3	-2.2	1.1
25.62	-2.54	-0.5	133.81	13AU	-0.2	-4.2	-2.2	1.0
28.70	-2.48	-0.7	143.61	23AU	0.0	-4.0	-2.3	1.0
30.70	-2.36	-0.9	153.47	2SE	0.2	-3.5	-2.3	1.0
31.45	-2.19	-1.1	163.38	12SE	0.5	-3.4	-2.4	0.9
30.75	-1.93	-1.3	173.36	22SE	1.6	-4.0	-2.4	0.9
28.62	-1.58	-1.5	183.39	2OC	2.0	-4.3	-2.4	0.9
25.34	-1.14	-1.7	193.47	12OC	-0.1	-4.4	-2.4	0.9
21.56	-0.65	-1.7	203.60	22OC	-0.5	-4.3	-2.4	0.9
18.10	-0.16	-1.5	213.76	1NO	-0.6	-4.3	-2.3	0.9
15.63	0.28	-1.2	223.95	11NO	-0.6	-4.2	-2.3	0.9
14.47	0.64	-0.9	234.16	21NO	-0.7	-4.1	-2.2	0.9
14.66	0.91	-0.5	244.37	1DE	-0.8	-4.0	-2.2	0.9
16.02	1.12	-0.2	254.57	11DE	-0.9	-3.9	-2.1	0.9
18.36	1.27	0.0	264.77	21DE	-0.8	-3.8	-2.0	0.9
21.49	1.38	0.3	274.93	31DE	0.1	-3.7	-1.9	0.8

♂			⊙	16.00UT	☿	♀	♃	♄
LONG	LAT	MAG	LONG	-550	MAGNITUDES			
25.22	1.46	0.5	285.06	10JA	2.6	-3.6	-1.9	0.8
29.43	1.51	0.7	295.14	20JA	1.6	-3.6	-1.8	0.7
34.01	1.55	0.9	305.17	30JA	0.7	-3.5	-1.7	0.7
38.88	1.57	1.1	315.15	9FE	0.4	-3.4	-1.7	0.6
43.98	1.58	1.2	325.07	19FE	0.2	-3.4	-1.6	0.6
49.27	1.58	1.3	334.92	1MR	-0.2	-3.4	-1.6	0.5
54.70	1.57	1.5	344.72	11MR	-0.8	-3.3	-1.6	0.5
60.27	1.56	1.6	354.47	21MR	-1.6	-3.3	-1.5	0.4
65.93	1.54	1.7	4.15	31MR	-1.3	-3.3	-1.5	0.4
71.68	1.52	1.7	13.79	10AP	-0.3	-3.3	-1.5	0.5
77.52	1.49	1.8	23.39	20AP	0.7	-3.3	-1.5	0.6
83.42	1.46	1.8	32.96	30AP	1.6	-3.3	-1.5	0.6
89.38	1.42	1.9	42.50	10MY	2.9	-3.3	-1.5	0.7
95.41	1.38	1.9	52.02	20MY	3.1	-3.4	-1.5	0.8
101.49	1.34	2.0	61.54	30MY	1.7	-3.4	-1.6	0.8
107.63	1.30	2.0	71.05	9JN	0.5	-3.4	-1.6	0.9
113.84	1.25	2.0	80.59	19JN	-0.6	-3.5	-1.6	0.9
120.10	1.20	2.0	90.13	29JN	-1.5	-3.5	-1.7	0.9
126.43	1.14	2.0	99.71	9JL	-1.3	-3.5	-1.7	1.0
132.83	1.08	2.0	109.33	19JL	-0.6	-3.4	-1.8	1.0
139.29	1.02	1.9	118.99	29JL	-0.1	-3.4	-1.8	1.0
145.82	0.95	1.9	128.70	8AU	0.2	-3.4	-1.9	1.0
152.42	0.88	1.9	138.47	18AU	0.4	-3.4	-1.9	1.0
159.10	0.81	1.9	148.29	28AU	0.7	-3.3	-2.0	0.9
165.85	0.73	1.8	158.18	7SE	2.0	-3.3	-2.1	0.9
172.69	0.65	1.8	168.12	17SE	1.8	-3.3	-2.1	0.9
179.60	0.56	1.7	178.12	27SE	-0.2	-3.3	-2.2	0.8
186.60	0.47	1.7	188.18	7OC	-0.7	-3.3	-2.3	0.8
193.67	0.37	1.7	198.28	17OC	-0.7	-3.4	-2.3	0.8
200.83	0.28	1.7	208.42	27OC	-0.7	-3.4	-2.3	0.8
208.07	0.17	1.6	218.60	6NO	-0.7	-3.4	-2.3	0.9
215.38	0.07	1.6	228.80	16NO	-0.7	-3.5	-2.3	0.9
222.76	-0.04	1.6	239.01	26NO	-0.7	-3.5	-2.3	0.9
230.22	-0.16	1.6	249.22	6DE	-0.6	-3.6	-2.2	0.9
237.74	-0.27	1.6	259.43	16DE	0.3	-3.6	-2.2	0.9
245.32	-0.39	1.5	269.60	26DE	2.9	-3.7	-2.1	0.8
				-549				
252.95	-0.51	1.5	279.75	5JA	1.2	-3.7	-2.1	0.8
260.63	-0.62	1.5	289.86	15JA	0.5	-3.8	-2.0	0.8
268.33	-0.74	1.4	299.92	25JA	0.2	-3.9	-1.9	0.7
276.07	-0.85	1.4	309.93	4FE	0.1	-4.0	-1.8	0.7
283.82	-0.96	1.4	319.87	14FE	-0.3	-4.1	-1.8	0.6
291.57	-1.06	1.4	329.76	24FE	-0.8	-4.2	-1.7	0.6
299.31	-1.15	1.3	339.59	6MR	-1.6	-4.2	-1.7	0.5
307.03	-1.23	1.3	349.36	16MR	-1.3	-4.2	-1.6	0.5
314.71	-1.30	1.3	359.07	26MR	-0.3	-4.1	-1.6	0.4
322.34	-1.37	1.2	8.74	5AP	0.9	-3.7	-1.5	0.4
329.91	-1.41	1.2	18.36	15AP	2.1	-3.0	-1.5	0.4
337.41	-1.45	1.2	27.94	25AP	3.8	-3.4	-1.4	0.4
344.82	-1.47	1.2	37.50	5MY	2.4	-3.9	-1.4	0.5
352.12	-1.48	1.1	47.03	15MY	1.3	-4.1	-1.4	0.6
359.31	-1.47	1.1	56.54	25MY	0.4	-4.2	-1.4	0.6
6.37	-1.44	1.1	66.06	4JN	-0.7	-4.1	-1.4	0.7
13.29	-1.40	1.0	75.58	14JN	-1.6	-4.1	-1.4	0.7
20.04	-1.35	1.0	85.12	24JN	-1.3	-4.0	-1.4	0.8
26.62	-1.28	0.9	94.69	4JL	-0.5	-3.9	-1.5	0.8
32.99	-1.19	0.9	104.28	14JL	-0.0	-3.8	-1.5	0.9
39.13	-1.09	0.8	113.92	24JL	0.3	-3.7	-1.5	0.9
45.01	-0.97	0.7	123.61	3AU	0.5	-3.7	-1.6	0.9
50.57	-0.83	0.6	133.34	13AU	1.0	-3.6	-1.6	0.9
55.78	-0.67	0.5	143.13	23AU	2.4	-3.5	-1.7	0.9
60.55	-0.48	0.4	152.99	2SE	1.6	-3.5	-1.7	0.9
64.79	-0.26	0.3	162.90	12SE	-0.3	-3.5	-1.8	0.9
68.39	-0.01	0.1	172.87	22SE	-0.8	-3.4	-1.8	0.8
71.18	0.28	-0.1	182.90	2OC	-0.9	-3.4	-1.9	0.8
72.99	0.62	-0.3	192.98	12OC	-0.8	-3.4	-2.0	0.8
73.63	1.01	-0.5	203.10	22OC	-0.6	-3.4	-2.0	0.8
72.91	1.44	-0.7	213.27	1NO	-0.5	-3.4	-2.1	0.8
70.80	1.89	-1.0	223.45	11NO	-0.5	-3.4	-2.1	0.8
67.52	2.32	-1.1	233.66	21NO	-0.5	-3.4	-2.2	0.8
63.58	2.67	-1.2	243.88	1DE	0.4	-3.4	-2.2	0.8
59.78	2.90	-1.0	254.08	11DE	3.1	-3.4	-2.2	0.8
56.80	3.01	-0.8	264.27	21DE	0.9	-3.4	-2.2	0.8
55.04	3.01	-0.5	274.44	31DE	0.3	-3.4	-2.2	0.8

♂			⊙	16.00UT	☿	♀	♃	♄
LONG	LAT	MAG	LONG	-548	MAGNITUDES			
54.63	2.95	-0.2	284.57	10JA	0.1	-3.4	-2.1	0.8
55.42	2.84	0.0	294.65	20JA	-0.1	-3.4	-2.1	0.8
57.25	2.72	0.3	304.69	30JA	-0.3	-3.5	-2.0	0.7
59.91	2.59	0.5	314.66	9FE	-0.8	-3.5	-1.9	0.7
63.24	2.46	0.7	324.58	19FE	-1.5	-3.5	-1.9	0.6
67.10	2.34	0.9	334.45	29FE	-1.2	-3.4	-1.8	0.6
71.38	2.22	1.0	344.24	10MR	-0.2	-3.4	-1.7	0.5
76.00	2.10	1.2	353.99	20MR	1.1	-3.4	-1.7	0.5
80.90	1.99	1.3	3.68	30MR	2.7	-3.4	-1.6	0.4
86.03	1.88	1.4	13.32	9AP	3.2	-3.3	-1.5	0.4
91.35	1.78	1.5	22.92	19AP	1.8	-3.3	-1.5	0.3
96.84	1.68	1.6	32.49	29AP	1.0	-3.3	-1.4	0.4
102.48	1.57	1.6	42.03	9MY	0.3	-3.3	-1.4	0.4
108.24	1.47	1.7	51.55	19MY	-0.7	-3.3	-1.4	0.5
114.13	1.38	1.7	61.07	29MY	-1.6	-3.4	-1.3	0.5
120.13	1.28	1.8	70.59	8JN	-1.3	-3.4	-1.3	0.6
126.24	1.18	1.8	80.12	18JN	-0.5	-3.4	-1.3	0.7
132.45	1.08	1.8	89.67	28JN	0.1	-3.5	-1.3	0.7
138.76	0.98	1.8	99.24	8JL	0.4	-3.5	-1.3	0.8
145.17	0.88	1.8	108.86	18JL	0.8	-3.6	-1.3	0.8
151.69	0.78	1.8	118.52	28JL	1.4	-3.6	-1.3	0.8
158.30	0.67	1.8	128.23	7AU	2.9	-3.7	-1.4	0.8
165.01	0.57	1.8	137.99	17AU	1.4	-3.8	-1.4	0.9
171.82	0.46	1.8	147.82	27AU	-0.4	-3.9	-1.4	0.9
178.73	0.36	1.7	157.70	6SE	-1.0	-4.0	-1.5	0.9
185.75	0.25	1.7	167.64	16SE	-1.0	-4.1	-1.5	0.8
192.86	0.14	1.7	177.64	26SE	-0.8	-4.2	-1.6	0.8
200.06	0.03	1.6	187.69	6OC	-0.5	-4.3	-1.6	0.8
207.37	-0.08	1.6	197.79	16OC	-0.4	-4.4	-1.7	0.8
214.76	-0.19	1.5	207.93	26OC	-0.4	-4.3	-1.8	0.7
222.24	-0.29	1.5	218.11	5NO	-0.3	-4.1	-1.8	0.7
229.79	-0.40	1.4	228.31	15NO	0.6	-3.5	-1.9	0.8
237.42	-0.50	1.4	238.52	25NO	3.0	-3.2	-1.9	0.8
245.12	-0.60	1.4	248.73	5DE	0.6	-4.0	-2.0	0.8
252.87	-0.69	1.4	258.93	15DE	0.1	-4.3	-2.0	0.8
260.66	-0.78	1.4	269.11	25DE	-0.0	-4.4	-2.1	0.8
				-547				
268.49	-0.86	1.4	279.26	4JA	-0.2	-4.3	-2.1	0.8
276.34	-0.94	1.4	289.37	14JA	-0.4	-4.2	-2.1	0.8
284.20	-1.00	1.4	299.43	24JA	-0.9	-4.1	-2.1	0.8
292.06	-1.06	1.4	309.44	3FE	-1.4	-4.0	-2.0	0.7
299.91	-1.10	1.4	319.39	13FE	-1.2	-3.9	-2.0	0.7
307.72	-1.13	1.4	329.28	23FE	-0.2	-3.8	-1.9	0.7
315.50	-1.16	1.4	339.11	5MR	1.4	-3.7	-1.9	0.6
323.23	-1.16	1.4	348.89	15MR	3.4	-3.6	-1.8	0.5
330.89	-1.16	1.4	358.61	25MR	2.3	-3.6	-1.8	0.5
338.48	-1.14	1.4	8.27	4AP	1.3	-3.5	-1.7	0.4
346.00	-1.12	1.4	17.90	14AP	0.7	-3.4	-1.6	0.4
353.43	-1.08	1.4	27.48	24AP	0.1	-3.4	-1.6	0.3
0.76	-1.02	1.5	37.03	4MY	-0.7	-3.4	-1.5	0.3
8.00	-0.96	1.5	46.57	14MY	-1.6	-3.3	-1.4	0.3
15.13	-0.89	1.5	56.09	24MY	-1.3	-3.3	-1.4	0.4
22.14	-0.81	1.5	65.60	3JN	-0.5	-3.3	-1.4	0.5
29.05	-0.71	1.5	75.13	13JN	0.2	-3.3	-1.3	0.5
35.82	-0.61	1.4	84.66	23JN	0.6	-3.3	-1.3	0.6
42.47	-0.50	1.4	94.22	3JL	1.0	-3.3	-1.3	0.6
48.99	-0.38	1.4	103.82	13JL	1.8	-3.4	-1.3	0.7
55.35	-0.25	1.4	113.45	23JL	3.2	-3.4	-1.2	0.7
61.56	-0.11	1.3	123.13	2AU	1.2	-3.4	-1.2	0.8
67.59	0.04	1.3	132.87	12AU	-0.5	-3.4	-1.2	0.8
73.42	0.20	1.2	142.66	22AU	-1.1	-3.5	-1.3	0.8
79.04	0.38	1.1	152.51	1SE	-1.1	-3.5	-1.3	0.8
84.39	0.57	1.1	162.42	11SE	-0.8	-3.5	-1.3	0.8
89.43	0.79	1.0	172.39	21SE	-0.4	-3.5	-1.3	0.8
94.11	1.02	0.8	182.41	1OC	-0.3	-3.4	-1.4	0.8
98.34	1.28	0.7	192.49	11OC	-0.2	-3.4	-1.4	0.8
102.03	1.57	0.5	202.61	21OC	-0.1	-3.4	-1.5	0.7
105.04	1.90	0.3	212.77	31OC	0.8	-3.4	-1.5	0.7
107.22	2.27	0.1	222.96	10NO	2.8	-3.3	-1.6	0.7
108.38	2.67	-0.1	233.16	20NO	0.4	-3.3	-1.7	0.7
108.35	3.11	-0.3	243.37	30NO	-0.1	-3.3	-1.7	0.8
106.98	3.54	-0.6	253.58	10DE	-0.2	-3.3	-1.8	0.8
104.33	3.94	-0.8	263.77	20DE	-0.3	-3.4	-1.8	0.8
100.72	4.22	-1.0	273.94	30DE	-0.5	-3.4	-1.9	0.8

♂			☉	16.00UT	☿	♀	♃	♄
LONG	LAT	MAG	LONG	-546	MAGNITUDES			
96.74	4.35	-1.0	284.08	9JA	-0.9	-3.4	-2.0	0.8
93.16	4.30	-0.8	294.16	19JA	-1.3	-3.4	-2.0	0.8
90.55	4.12	-0.5	304.20	29JA	-1.1	-3.5	-2.0	0.8
89.21	3.86	-0.3	314.18	8FE	-0.1	-3.5	-2.0	0.8
89.16	3.56	-0.1	324.10	18FE	1.7	-3.5	-2.0	0.7
90.26	3.26	0.2	333.97	28FE	3.2	-3.6	-2.0	0.7
92.33	2.97	0.4	343.78	10MR	1.7	-3.7	-2.0	0.6
95.19	2.70	0.6	353.52	20MR	1.0	-3.7	-1.9	0.6
98.69	2.45	0.7	3.21	30MR	0.5	-3.8	-1.9	0.5
102.72	2.22	0.9	12.86	9AP	0.0	-3.9	-1.8	0.5
107.16	2.00	1.0	22.46	19AP	-0.7	-4.0	-1.8	0.4
111.95	1.81	1.1	32.03	29AP	-1.6	-4.1	-1.7	0.3
117.05	1.62	1.2	41.57	9MY	-1.4	-4.1	-1.6	0.3
122.39	1.44	1.3	51.10	19MY	-0.4	-4.2	-1.6	0.3
127.95	1.27	1.4	60.61	29MY	0.3	-4.1	-1.5	0.3
133.72	1.11	1.4	70.13	8JN	0.8	-3.9	-1.4	0.4
139.66	0.96	1.5	79.66	18JN	1.3	-3.4	-1.4	0.5
145.76	0.81	1.5	89.21	28JN	2.4	-3.0	-1.4	0.5
152.02	0.67	1.5	98.78	8JL	3.0	-3.7	-1.3	0.6
158.43	0.53	1.6	108.39	18JL	1.0	-4.1	-1.3	0.6
164.97	0.39	1.6	118.05	28JL	-0.5	-4.2	-1.3	0.7
171.66	0.25	1.6	127.76	7AU	-1.2	-4.2	-1.2	0.7
178.47	0.12	1.6	137.52	17AU	-1.3	-4.2	-1.2	0.8
185.41	-0.00	1.6	147.34	27AU	-0.7	-4.1	-1.2	0.8
192.47	-0.12	1.6	157.22	6SE	-0.3	-4.0	-1.2	0.8
199.65	-0.24	1.6	167.15	16SE	-0.2	-3.9	-1.2	0.8
206.94	-0.36	1.5	177.15	26SE	-0.1	-3.8	-1.2	0.8
214.33	-0.46	1.5	187.20	6OC	0.1	-3.8	-1.3	0.8
221.83	-0.57	1.5	197.30	16OC	1.1	-3.7	-1.3	0.8
229.41	-0.66	1.5	207.44	26OC	2.5	-3.6	-1.3	0.8
237.08	-0.75	1.5	217.62	5NO	0.2	-3.6	-1.4	0.7
244.81	-0.83	1.4	227.81	15NO	-0.3	-3.5	-1.4	0.7
252.61	-0.90	1.4	238.03	25NO	-0.3	-3.5	-1.5	0.7
260.45	-0.96	1.4	248.24	5DE	-0.4	-3.4	-1.5	0.7
268.32	-1.01	1.4	258.44	15DE	-0.6	-3.4	-1.6	0.8
276.22	-1.05	1.4	268.62	25DE	-0.9	-3.4	-1.7	0.8
				-545				
284.12	-1.07	1.3	278.77	4JA	-1.1	-3.4	-1.7	0.8
292.01	-1.09	1.3	288.88	14JA	-1.0	-3.3	-1.8	0.8
299.89	-1.09	1.3	298.95	24JA	-0.1	-3.3	-1.9	0.8
307.73	-1.08	1.3	308.96	3FE	2.0	-3.3	-1.9	0.8
315.52	-1.06	1.4	318.91	13FE	2.5	-3.3	-2.0	0.8
323.26	-1.03	1.4	328.81	23FE	1.3	-3.3	-2.0	0.8
330.94	-0.99	1.4	338.64	5MR	0.7	-3.3	-2.0	0.7
338.54	-0.94	1.5	348.42	15MR	0.4	-3.4	-2.0	0.7
346.06	-0.88	1.5	358.14	25MR	-0.0	-3.4	-2.0	0.6
353.51	-0.81	1.5	7.81	4AP	-0.7	-3.4	-2.0	0.6
0.86	-0.74	1.6	17.43	14AP	-1.6	-3.5	-2.0	0.5
8.12	-0.65	1.6	27.02	24AP	-1.4	-3.5	-1.9	0.4
15.29	-0.57	1.6	36.57	4MY	-0.4	-3.4	-1.9	0.4
22.37	-0.47	1.6	46.10	14MY	0.4	-3.4	-1.8	0.3
29.35	-0.37	1.7	55.62	24MY	1.0	-3.4	-1.7	0.3
36.24	-0.27	1.7	65.14	3JN	1.8	-3.4	-1.7	0.3
43.04	-0.16	1.7	74.66	13JN	3.0	-3.3	-1.6	0.4
49.74	-0.05	1.7	84.20	23JN	2.5	-3.3	-1.6	0.4
56.34	0.07	1.7	93.75	3JL	0.8	-3.3	-1.5	0.5
62.85	0.18	1.7	103.35	13JL	-0.6	-3.3	-1.4	0.5
69.26	0.31	1.7	112.98	23JL	-1.3	-3.3	-1.4	0.6
75.57	0.43	1.7	122.66	2AU	-1.3	-3.3	-1.4	0.6
81.77	0.57	1.7	132.39	12AU	-0.7	-3.4	-1.3	0.7
87.85	0.70	1.6	142.18	22AU	-0.2	-3.4	-1.3	0.7
93.81	0.85	1.6	152.02	1SE	-0.0	-3.4	-1.3	0.8
99.63	1.00	1.5	161.93	11SE	0.1	-3.5	-1.2	0.8
105.28	1.15	1.5	171.90	21SE	0.3	-3.5	-1.2	0.8
110.76	1.32	1.4	181.92	1OC	1.3	-3.6	-1.2	0.8
116.02	1.51	1.3	192.00	11OC	2.3	-3.7	-1.2	0.8
121.02	1.70	1.2	202.12	21OC	0.0	-3.7	-1.3	0.8
125.72	1.92	1.0	212.28	31OC	-0.4	-3.8	-1.3	0.8
130.04	2.15	0.9	222.46	10NO	-0.5	-3.9	-1.3	0.7
133.89	2.41	0.7	232.67	20NO	-0.5	-4.0	-1.3	0.7
137.17	2.70	0.5	242.88	30NO	-0.7	-4.1	-1.4	0.7
139.74	3.01	0.3	253.09	10DE	-0.9	-4.2	-1.4	0.7
141.42	3.35	0.1	263.29	20DE	-0.9	-4.3	-1.5	0.8
142.04	3.71	-0.2	273.45	30DE	-0.9	-4.4	-1.5	0.8

♂			☉	16.00UT	☿	♀	♃	♄
LONG	LAT	MAG	LONG	-544	MAGNITUDES			
141.43	4.06	-0.5	283.59	9JA	0.0	-4.3	-1.6	0.8
139.53	4.36	-0.7	293.68	19JA	2.4	-4.0	-1.7	0.8
136.47	4.55	-1.0	303.71	29JA	2.0	-3.5	-1.7	0.8
132.69	4.59	-1.1	313.70	8FE	0.9	-3.4	-1.8	0.8
128.87	4.45	-1.0	323.63	18FE	0.5	-3.9	-1.9	0.8
125.71	4.15	-0.8	333.49	28FE	0.2	-4.2	-1.9	0.8
123.68	3.75	-0.6	343.30	9MR	-0.1	-4.3	-2.0	0.8
122.95	3.32	-0.4	353.05	19MR	-0.8	-4.2	-2.0	0.7
123.48	2.89	-0.2	2.74	29MR	-1.6	-4.1	-2.1	0.7
125.11	2.49	0.1	12.39	8AP	-1.4	-4.0	-2.1	0.6
127.67	2.12	0.2	22.00	18AP	-0.4	-3.9	-2.1	0.6
130.99	1.78	0.4	31.57	28AP	0.6	-3.8	-2.1	0.5
134.93	1.48	0.5	41.11	8MY	1.4	-3.7	-2.0	0.4
139.39	1.20	0.7	50.64	18MY	2.4	-3.6	-2.0	0.4
144.27	0.95	0.8	60.15	28MY	3.5	-3.6	-2.0	0.3
149.51	0.72	0.9	69.67	7JN	2.0	-3.5	-1.9	0.3
155.08	0.50	0.9	79.20	17JN	0.6	-3.5	-1.8	0.3
160.91	0.30	1.0	88.74	27JN	-0.6	-3.4	-1.8	0.4
166.99	0.12	1.1	98.32	7JL	-1.4	-3.4	-1.7	0.4
173.29	-0.05	1.1	107.93	17JL	-1.3	-3.4	-1.7	0.5
179.79	-0.21	1.2	117.58	27JL	-0.6	-3.4	-1.6	0.6
186.48	-0.36	1.2	127.29	6AU	-0.2	-3.4	-1.5	0.6
193.34	-0.50	1.2	137.04	16AU	0.1	-3.4	-1.5	0.7
200.35	-0.63	1.2	146.86	26AU	0.3	-3.4	-1.4	0.7
207.51	-0.74	1.3	156.74	5SE	0.5	-3.4	-1.4	0.7
214.80	-0.84	1.3	166.67	15SE	1.7	-3.4	-1.4	0.8
222.22	-0.93	1.3	176.66	25SE	2.1	-3.4	-1.3	0.8
229.74	-1.01	1.3	186.71	5OC	-0.1	-3.4	-1.3	0.8
237.36	-1.07	1.3	196.81	15OC	-0.6	-3.4	-1.3	0.8
245.05	-1.12	1.3	206.94	25OC	-0.6	-3.4	-1.3	0.8
252.82	-1.16	1.3	217.12	4NO	-0.6	-3.5	-1.3	0.8
260.63	-1.19	1.4	227.32	14NO	-0.8	-3.5	-1.3	0.8
268.48	-1.20	1.4	237.52	24NO	-0.8	-3.5	-1.3	0.8
276.34	-1.20	1.4	247.74	4DE	-0.8	-3.5	-1.4	0.7
284.22	-1.18	1.4	257.94	14DE	-0.7	-3.4	-1.4	0.7
292.08	-1.15	1.4	268.12	24DE	0.1	-3.4	-1.4	0.8
				-543				
299.91	-1.11	1.4	278.28	3JA	2.7	-3.4	-1.5	0.8
307.71	-1.07	1.4	288.39	13JA	1.5	-3.4	-1.5	0.8
315.46	-1.01	1.5	298.46	23JA	0.6	-3.3	-1.6	0.9
323.16	-0.94	1.5	308.47	2FE	0.3	-3.3	-1.7	0.9
330.78	-0.87	1.5	318.43	12FE	0.1	-3.3	-1.7	0.9
338.34	-0.78	1.5	328.33	22FE	-0.2	-3.3	-1.8	0.9
345.81	-0.70	1.5	338.17	4MR	-0.8	-3.4	-1.9	0.8
353.20	-0.61	1.5	347.94	14MR	-1.6	-3.4	-1.9	0.8
0.51	-0.51	1.5	357.67	24MR	-1.3	-3.4	-2.0	0.8
7.73	-0.42	1.5	7.34	3AP	-0.3	-3.4	-2.1	0.7
14.87	-0.32	1.6	16.97	13AP	0.7	-3.4	-2.1	0.7
21.92	-0.22	1.6	26.55	23AP	1.8	-3.5	-2.1	0.6
28.89	-0.11	1.7	36.11	3MY	3.2	-3.5	-2.2	0.6
35.78	-0.01	1.7	45.64	13MY	2.9	-3.6	-2.2	0.5
42.59	0.09	1.8	55.16	23MY	1.5	-3.7	-2.2	0.4
49.32	0.19	1.8	64.68	2JN	0.5	-3.7	-2.2	0.4
55.99	0.29	1.8	74.20	12JN	-0.6	-3.8	-2.1	0.3
62.58	0.39	1.9	83.74	22JN	-1.5	-3.9	-2.1	0.3
69.11	0.49	1.9	93.30	2JL	-1.3	-4.0	-2.1	0.4
75.57	0.59	1.9	102.89	12JL	-0.6	-4.1	-2.0	0.4
81.97	0.69	1.9	112.52	22JL	-0.1	-4.2	-1.9	0.5
88.31	0.79	1.9	122.19	1AU	0.2	-4.3	-1.9	0.5
94.59	0.90	1.9	131.92	11AU	0.4	-4.2	-1.8	0.6
100.80	1.00	1.9	141.70	21AU	0.8	-4.0	-1.7	0.6
106.94	1.10	1.9	151.55	31AU	2.1	-3.5	-1.7	0.7
113.01	1.20	1.8	161.45	10SE	1.8	-3.4	-1.6	0.7
119.01	1.31	1.8	171.41	20SE	-0.2	-4.0	-1.6	0.8
124.92	1.42	1.7	181.43	30SE	-0.8	-4.3	-1.5	0.8
130.72	1.53	1.7	191.50	10OC	-0.8	-4.4	-1.5	0.8
136.42	1.65	1.6	201.62	20OC	-0.8	-4.3	-1.5	0.8
141.97	1.77	1.5	211.78	30OC	-0.7	-4.3	-1.4	0.8
147.37	1.90	1.4	221.96	9NO	-0.6	-4.2	-1.4	0.8
152.57	2.03	1.3	232.17	19NO	-0.6	-4.1	-1.4	0.8
157.54	2.18	1.1	242.38	29NO	-0.6	-4.0	-1.4	0.8
162.22	2.33	0.9	252.59	9DE	0.3	-3.9	-1.4	0.8
166.55	2.49	0.8	262.79	19DE	2.9	-3.8	-1.4	0.8
170.43	2.66	0.6	272.96	29DE	1.1	-3.7	-1.4	0.8

Left table (−542 / −541)

♂ LONG	♂ LAT	♂ MAG	☉ LONG	16.00UT	☿	♀	♃	♄
173.77	2.83	0.3	283.09	8JA	0.4	-3.6	-1.5	0.8
176.43	3.01	0.1	293.19	18JA	0.2	-3.6	-1.5	0.9
178.23	3.19	-0.2	303.23	28JA	0.0	-3.5	-1.5	0.9
179.01	3.35	-0.5	313.22	7FE	-0.3	-3.4	-1.6	0.9
178.59	3.47	-0.8	323.15	17FE	-0.8	-3.4	-1.6	0.9
176.89	3.50	-1.1	333.02	27FE	-1.6	-3.4	-1.7	0.9
174.05	3.42	-1.3	342.83	9MR	-1.3	-3.3	-1.8	0.9
170.47	3.19	-1.5	352.59	19MR	-0.3	-3.3	-1.8	0.9
166.84	2.82	-1.4	2.28	29MR	0.9	-3.3	-1.9	0.9
163.86	2.35	-1.2	11.93	8AP	2.3	-3.3	-2.0	0.8
162.01	1.84	-1.0	21.54	18AP	3.8	-3.3	-2.0	0.8
161.52	1.34	-0.8	31.11	28AP	2.2	-3.3	-2.1	0.7
162.32	0.88	-0.6	40.65	8MY	1.2	-3.3	-2.2	0.7
164.28	0.48	-0.4	50.18	18MY	0.3	-3.4	-2.2	0.6
167.22	0.12	-0.2	59.70	28MY	-0.6	-3.4	-2.3	0.5
170.97	-0.19	-0.1	69.21	7JN	-1.6	-3.4	-2.3	0.5
175.38	-0.45	0.1	78.74	17JN	-1.3	-3.5	-2.3	0.4
180.35	-0.69	0.2	88.28	27JN	-0.5	-3.5	-2.3	0.4
185.77	-0.89	0.3	97.85	7JL	0.0	-3.5	-2.3	0.4
191.59	-1.06	0.4	107.46	17JL	0.3	-3.4	-2.3	0.4
197.74	-1.20	0.5	117.11	27JL	0.6	-3.4	-2.2	0.5
204.17	-1.33	0.6	126.81	6AU	1.1	-3.4	-2.2	0.5
210.86	-1.42	0.6	136.57	16AU	2.6	-3.4	-2.1	0.6
217.76	-1.50	0.7	146.38	26AU	1.6	-3.3	-2.1	0.6
224.84	-1.56	0.8	156.25	5SE	-0.3	-3.3	-2.0	0.7
232.08	-1.59	0.8	166.18	15SE	-0.9	-3.3	-1.9	0.7
239.46	-1.61	0.9	176.17	25SE	-0.9	-3.3	-1.9	0.7
246.95	-1.60	1.0	186.21	5OC	-0.9	-3.3	-1.8	0.8
254.52	-1.58	1.0	196.31	15OC	-0.6	-3.4	-1.7	0.8
262.16	-1.55	1.1	206.45	25OC	-0.5	-3.4	-1.7	0.8
269.85	-1.50	1.1	216.62	4NO	-0.5	-3.4	-1.6	0.9
277.57	-1.43	1.2	226.82	14NO	-0.4	-3.5	-1.6	0.9
285.29	-1.36	1.2	237.02	24NO	0.4	-3.5	-1.6	0.9
293.01	-1.27	1.3	247.24	4DE	3.1	-3.6	-1.5	0.9
300.70	-1.18	1.3	257.44	14DE	0.8	-3.6	-1.5	0.9
308.36	-1.08	1.4	267.63	24DE	0.2	-3.7	-1.5	0.8

−541

♂ LONG	♂ LAT	♂ MAG	☉ LONG	16.00UT	☿	♀	♃	♄
315.98	-0.97	1.4	277.78	3JA	0.0	-3.7	-1.5	0.8
323.54	-0.86	1.5	287.90	13JA	-0.1	-3.8	-1.5	0.9
331.03	-0.75	1.5	297.97	23JA	-0.4	-3.9	-1.5	0.9
338.45	-0.64	1.6	307.99	2FE	-0.8	-4.0	-1.5	0.9
345.80	-0.53	1.6	317.95	12FE	-1.5	-4.1	-1.6	1.0
353.07	-0.41	1.6	327.85	22FE	-1.2	-4.2	-1.6	1.0
0.27	-0.30	1.7	337.69	4MR	-0.2	-4.2	-1.6	1.0
7.38	-0.19	1.7	347.48	14MR	1.2	-4.2	-1.7	1.0
14.42	-0.09	1.7	357.20	24MR	2.9	-4.1	-1.7	1.0
21.38	0.02	1.8	6.88	3AP	2.8	-3.7	-1.8	1.0
28.27	0.12	1.8	16.51	13AP	1.6	-3.0	-1.9	0.9
35.08	0.22	1.8	26.09	23AP	0.9	-3.4	-1.9	0.9
41.84	0.31	1.8	35.65	3MY	-0.3	-3.9	-2.0	0.9
48.53	0.40	1.8	45.19	13MY	-0.6	-4.2	-2.1	0.8
55.16	0.49	1.8	54.70	23MY	-1.6	-4.2	-2.1	0.7
61.74	0.58	1.8	64.22	2JN	-1.4	-4.1	-2.2	0.7
68.28	0.66	1.9	73.74	12JN	-0.5	-4.1	-2.3	0.6
74.76	0.74	1.9	83.27	22JN	0.1	-4.0	-2.3	0.5
81.21	0.81	2.0	92.83	2JL	0.5	-3.9	-2.4	0.5
87.63	0.89	2.0	102.42	12JL	0.8	-3.8	-2.4	0.4
94.01	0.96	2.0	112.05	22JL	1.5	-3.7	-2.4	0.5
100.36	1.02	2.0	121.72	1AU	3.0	-3.6	-2.4	0.5
106.69	1.09	2.0	131.45	11AU	1.4	-3.6	-2.4	0.5
112.99	1.15	2.0	141.23	21AU	-0.4	-3.5	-2.4	0.6
119.26	1.21	2.0	151.07	31AU	-1.0	-3.5	-2.3	0.6
125.51	1.27	2.0	160.97	10SE	-1.1	-3.5	-2.3	0.7
131.74	1.33	2.0	170.93	20SE	-0.8	-3.4	-2.2	0.7
137.94	1.38	1.9	180.95	30SE	-0.5	-3.4	-2.1	0.8
144.10	1.43	1.9	191.02	10OC	-0.4	-3.4	-2.1	0.8
150.23	1.48	1.8	201.13	20OC	-0.3	-3.4	-2.0	0.9
156.32	1.53	1.7	211.29	30OC	-0.2	-3.4	-1.9	0.9
162.36	1.57	1.7	221.47	9NO	0.6	-3.4	-1.9	0.9
168.33	1.62	1.6	231.67	19NO	3.0	-3.4	-1.8	0.9
174.24	1.65	1.5	241.89	29NO	0.5	-3.4	-1.8	0.9
180.06	1.69	1.4	252.09	9DE	-0.0	-3.4	-1.7	1.0
185.77	1.71	1.2	262.29	19DE	-0.1	-3.4	-1.7	0.9
191.35	1.73	1.1	272.46	29DE	-0.2	-3.4	-1.6	0.9

Right table (−540 / −539)

♂ LONG	♂ LAT	♂ MAG	☉ LONG	16.00UT	☿	♀	♃	♄
196.77	1.74	0.9	282.60	8JA	-0.4	-3.4	-1.6	0.9
201.99	1.74	0.8	292.69	18JA	-0.9	-3.4	-1.6	0.9
206.98	1.72	0.6	302.74	28JA	-1.3	-3.5	-1.6	1.0
211.66	1.68	0.3	312.73	7FE	-1.2	-3.5	-1.6	1.0
215.96	1.62	0.1	322.66	17FE	-0.2	-3.5	-1.6	1.1
219.78	1.52	-0.1	332.54	27FE	1.5	-3.4	-1.6	1.1
223.00	1.37	-0.4	342.35	8MR	3.4	-3.4	-1.6	1.1
225.46	1.17	-0.7	352.11	18MR	2.1	-3.4	-1.6	1.1
226.99	0.89	-1.0	1.81	28MR	1.2	-3.4	-1.6	1.1
227.40	0.51	-1.3	11.46	7AP	0.7	-3.3	-1.7	1.1
226.58	0.04	-1.7	21.07	17AP	0.1	-3.3	-1.7	1.1
224.56	-0.52	-2.0	30.64	27AP	-0.7	-3.3	-1.8	1.0
221.68	-1.14	-2.2	40.19	7MY	-1.6	-3.3	-1.8	1.0
218.53	-1.75	-2.2	49.71	17MY	-1.4	-3.3	-1.9	1.0
215.86	-2.28	-2.0	59.23	27MY	-0.5	-3.4	-2.0	0.9
214.26	-2.68	-1.8	68.75	6JN	0.2	-3.4	-2.0	0.8
214.01	-2.97	-1.6	78.27	16JN	0.7	-3.4	-2.1	0.8
215.12	-3.14	-1.4	87.82	26JN	1.1	-3.5	-2.2	0.7
217.47	-3.23	-1.1	97.39	6JL	2.0	-3.5	-2.2	0.6
220.85	-3.25	-0.9	106.99	16JL	3.2	-3.6	-2.3	0.6
225.08	-3.22	-0.7	116.65	26JL	1.2	-3.6	-2.3	0.5
230.01	-3.15	-0.6	126.34	5AU	-0.5	-3.7	-2.4	0.6
235.48	-3.05	-0.4	136.09	15AU	-1.2	-3.8	-2.4	0.6
241.40	-2.92	-0.2	145.90	25AU	-1.2	-3.9	-2.4	0.6
247.67	-2.78	-0.1	155.77	4SE	-0.8	-4.0	-2.4	0.7
254.21	-2.62	0.1	165.70	14SE	-0.4	-4.1	-2.4	0.8
260.98	-2.44	0.2	175.69	24SE	-0.2	-4.2	-2.4	0.8
267.91	-2.26	0.3	185.73	4OC	-0.2	-4.3	-2.4	0.9
274.96	-2.07	0.5	195.82	14OC	-0.0	-4.4	-2.3	0.9
282.10	-1.88	0.6	205.96	24OC	0.8	-4.3	-2.2	0.9
289.31	-1.69	0.7	216.12	3NO	2.8	-4.1	-2.2	1.0
296.54	-1.50	0.8	226.32	13NO	0.3	-3.4	-2.1	1.0
303.80	-1.32	0.9	236.53	23NO	-0.2	-3.2	-2.0	1.0
311.05	-1.13	1.0	246.74	3DE	-0.2	-4.0	-2.0	1.1
318.27	-0.96	1.1	256.94	13DE	-0.3	-4.3	-1.9	1.1
325.48	-0.79	1.2	267.13	23DE	-0.5	-4.4	-1.8	1.1

−539

♂ LONG	♂ LAT	♂ MAG	☉ LONG	16.00UT	☿	♀	♃	♄
332.64	-0.63	1.3	277.28	2JA	-0.9	-4.3	-1.8	1.1
339.75	-0.47	1.4	287.40	12JA	-1.2	-4.2	-1.7	1.1
346.81	-0.33	1.5	297.48	22JA	-1.1	-4.1	-1.7	1.1
353.82	-0.19	1.6	307.50	1FE	-0.1	-4.0	-1.6	1.1
0.77	-0.06	1.6	317.46	11FE	1.8	-3.9	-1.6	1.1
7.66	0.06	1.7	327.37	21FE	3.0	-3.8	-1.6	1.2
14.49	0.17	1.7	337.21	3MR	1.5	-3.7	-1.6	1.2
21.26	0.28	1.8	347.00	13MR	0.9	-3.6	-1.6	1.2
27.98	0.38	1.8	356.73	23MR	0.5	-3.6	-1.6	1.3
34.65	0.47	1.9	6.41	2AP	0.0	-3.5	-1.6	1.3
41.26	0.55	1.9	16.04	12AP	-0.7	-3.4	-1.6	1.3
47.84	0.63	1.9	25.63	22AP	-1.6	-3.4	-1.6	1.3
54.37	0.71	1.9	35.19	2MY	-1.4	-3.3	-1.6	1.2
60.86	0.77	2.0	44.73	12MY	-0.4	-3.3	-1.7	1.2
67.32	0.84	2.0	54.25	22MY	0.3	-3.3	-1.7	1.1
73.76	0.89	2.0	63.76	1JN	0.9	-3.3	-1.7	1.1
80.17	0.95	2.0	73.29	11JN	1.5	-3.3	-1.8	1.0
86.57	0.99	2.0	82.82	21JN	2.6	-3.3	-1.8	1.0
92.95	1.04	1.9	92.37	1JL	2.8	-3.3	-1.9	0.9
99.32	1.08	2.0	101.96	11JL	1.0	-3.4	-2.0	0.8
105.69	1.11	2.0	111.59	21JL	-0.5	-3.4	-2.0	0.8
112.06	1.14	2.0	121.25	31JL	-1.3	-3.4	-2.1	0.7
118.44	1.17	2.0	130.98	10AU	-1.3	-3.4	-2.2	0.7
124.82	1.19	2.0	140.76	20AU	-0.7	-3.5	-2.2	0.7
131.21	1.21	2.0	150.59	30AU	-0.3	-3.5	-2.3	0.7
137.61	1.22	2.0	160.49	9SE	-0.1	-3.5	-2.3	0.8
144.02	1.23	2.0	170.45	19SE	-0.0	-3.5	-2.4	0.8
150.45	1.24	2.0	180.46	29SE	0.2	-3.4	-2.4	0.9
156.89	1.24	1.9	190.53	9OC	1.1	-3.4	-2.4	0.9
163.35	1.23	1.9	200.64	19OC	2.5	-3.4	-2.4	1.0
169.82	1.22	1.9	210.79	29OC	0.1	-3.4	-2.4	1.0
176.31	1.20	1.8	220.97	8NO	-0.4	-3.3	-2.3	1.1
182.80	1.17	1.7	231.17	18NO	-0.4	-3.3	-2.3	1.1
189.31	1.14	1.7	241.38	28NO	-0.4	-3.3	-2.2	1.2
195.82	1.09	1.6	251.60	8DE	-0.6	-3.4	-2.1	1.2
202.33	1.04	1.5	261.79	18DE	-0.9	-3.4	-2.1	1.2
208.84	0.97	1.4	271.96	28DE	-1.0	-3.4	-2.0	1.2

Left table:

♂ LONG	LAT	MAG	☉ LONG	16.00UT −538	☿	♀	♃	♄
215.35	0.89	1.3	282.11	7JA	−0.9	−3.4	−1.9	1.2
221.84	0.79	1.2	292.20	17JA	−0.1	−3.4	−1.8	1.2
228.31	0.67	1.0	302.25	27JA	2.1	−3.5	−1.8	1.2
234.74	0.53	0.9	312.25	6FE	2.3	−3.5	−1.7	1.2
241.14	0.36	0.8	322.18	16FE	1.1	−3.5	−1.7	1.3
247.47	0.17	0.6	332.06	26FE	0.6	−3.6	−1.6	1.3
253.73	−0.05	0.4	341.88	8MR	0.3	−3.7	−1.6	1.4
259.88	−0.31	0.3	351.64	18MR	−0.1	−3.7	−1.5	1.4
265.89	−0.61	0.1	1.34	28MR	−0.7	−3.8	−1.5	1.4
271.71	−0.95	−0.1	11.00	7AP	−1.6	−3.9	−1.5	1.4
277.30	−1.35	−0.3	20.61	17AP	−1.4	−4.0	−1.5	1.3
282.56	−1.80	−0.5	30.18	27AP	−0.4	−4.1	−1.5	1.3
287.41	−2.31	−0.8	39.73	7MY	0.4	−4.1	−1.5	1.3
291.72	−2.88	−1.0	49.26	17MY	1.1	−4.2	−1.5	1.2
295.33	−3.52	−1.3	58.77	27MY	2.0	−4.1	−1.5	1.2
298.04	−4.22	−1.6	68.29	6JN	3.3	−3.9	−1.5	1.1
299.66	−4.95	−1.8	77.82	16JN	2.3	−3.3	−1.6	1.1
300.03	−5.68	−2.1	87.36	26JN	0.8	−3.0	−1.6	1.0
299.06	−6.31	−2.4	96.93	6JL	−0.5	−3.7	−1.7	1.0
297.01	−6.73	−2.6	106.53	16JL	−1.4	−4.1	−1.7	0.9
294.36	−6.86	−2.6	116.18	26JL	−1.3	−4.2	−1.8	0.9
291.88	−6.65	−2.5	125.87	5AU	−0.7	−4.2	−1.8	0.9
290.25	−6.18	−2.2	135.62	15AU	−0.2	−4.2	−1.9	0.8
289.83	−5.56	−1.9	145.43	25AU	0.0	−4.1	−1.9	0.8
290.72	−4.87	−1.6	155.29	4SE	0.2	−4.0	−2.0	0.9
292.81	−4.19	−1.3	165.22	14SE	0.4	−3.9	−2.1	0.9
295.88	−3.56	−1.0	175.20	24SE	1.4	−3.8	−2.1	1.0
299.76	−2.98	−0.7	185.24	4OC	2.3	−3.8	−2.2	1.0
304.27	−2.47	−0.5	195.33	14OC	−0.0	−3.7	−2.2	1.1
309.25	−2.02	−0.2	205.46	24OC	−0.5	−3.6	−2.3	1.1
314.61	−1.62	−0.0	215.63	3NO	−0.5	−3.6	−2.3	1.2
320.24	−1.26	0.2	225.82	13NO	−0.6	−3.5	−2.3	1.2
326.09	−0.95	0.4	236.03	23NO	−0.7	−3.5	−2.3	1.3
332.10	−0.68	0.6	246.25	3DE	−0.8	−3.4	−2.3	1.3
338.23	−0.44	0.7	256.45	13DE	−0.9	−3.4	−2.2	1.4
344.45	−0.22	0.9	266.64	23DE	−0.8	−3.4	−2.2	1.3

−537

♂ LONG	LAT	MAG	☉ LONG	16.00UT	☿	♀	♃	♄
350.72	−0.04	1.0	276.80	2JA	0.0	−3.4	−2.1	1.3
357.04	0.13	1.2	286.92	12JA	2.4	−3.3	−2.0	1.3
3.38	0.28	1.3	296.99	22JA	1.8	−3.3	−2.0	1.3
9.74	0.41	1.4	307.02	1FE	0.8	−3.3	−1.9	1.2
16.10	0.52	1.5	316.98	11FE	0.4	−3.3	−1.8	1.2
22.46	0.62	1.6	326.89	21FE	0.2	−3.3	−1.7	1.1
28.82	0.71	1.7	336.74	3MR	−0.1	−3.3	−1.7	1.1
35.16	0.79	1.8	346.52	13MR	−0.7	−3.4	−1.6	1.1
41.49	0.86	1.8	356.26	23MR	−1.6	−3.4	−1.6	1.1
47.82	0.92	1.9	5.94	2AP	−1.4	−3.4	−1.5	1.1
54.13	0.97	1.9	15.57	12AP	−0.4	−3.5	−1.5	1.1
60.43	1.02	2.0	25.16	22AP	0.6	−3.5	−1.5	1.1
66.73	1.06	2.0	34.73	2MY	1.5	−3.4	−1.4	1.0
73.02	1.09	2.0	44.26	12MY	2.7	−3.4	−1.4	1.0
79.31	1.11	2.0	53.78	22MY	3.3	−3.4	−1.4	1.0
85.61	1.14	2.0	63.30	1JN	1.8	−3.4	−1.4	0.9
91.91	1.15	2.0	72.82	11JN	0.6	−3.3	−1.4	0.9
98.22	1.16	2.0	82.35	21JN	−0.6	−3.3	−1.4	0.8
104.55	1.17	2.0	91.91	1JL	−1.4	−3.3	−1.4	0.8
110.90	1.17	2.0	101.49	11JL	−1.3	−3.3	−1.4	0.7
117.28	1.17	2.0	111.12	21JL	−0.6	−3.3	−1.5	0.7
123.68	1.16	1.9	120.78	31JL	−0.1	−3.3	−1.5	0.6
130.11	1.15	1.9	130.50	10AU	0.1	−3.4	−1.6	0.6
136.58	1.13	1.9	140.28	20AU	0.3	−3.4	−1.6	0.5
143.08	1.11	1.9	150.11	30AU	0.6	−3.4	−1.7	0.5
149.63	1.08	1.9	160.00	9SE	1.8	−3.5	−1.7	0.5
156.22	1.05	1.9	169.96	19SE	2.1	−3.5	−1.8	0.6
162.86	1.01	1.9	179.97	29SE	−0.1	−3.6	−1.8	0.7
169.54	0.96	1.9	190.03	9OC	−0.7	−3.7	−1.9	0.7
176.28	0.91	1.9	200.15	19OC	−0.7	−3.8	−2.0	0.8
183.06	0.85	1.9	210.30	29OC	−0.7	−3.8	−2.0	0.9
189.90	0.79	1.8	220.48	8NO	−0.8	−3.9	−2.1	0.9
196.79	0.71	1.8	230.68	18NO	−0.7	−4.0	−2.1	0.9
203.72	0.63	1.7	240.89	28NO	−0.7	−4.1	−2.2	1.0
210.71	0.54	1.7	251.10	8DE	−0.7	−4.2	−2.2	1.0
217.75	0.44	1.6	261.30	18DE	0.1	−4.3	−2.2	1.0
224.83	0.32	1.5	271.47	28DE	2.7	−4.4	−2.2	1.0

Right table:

♂ LONG	LAT	MAG	☉ LONG	16.00UT −536	☿	♀	♃	♄
231.96	0.20	1.5	281.61	7JA	1.4	−4.3	−2.1	1.0
239.13	0.07	1.4	291.72	17JA	0.5	−4.0	−2.1	1.0
246.34	−0.08	1.3	301.77	27JA	0.3	−3.5	−2.0	1.0
253.58	−0.23	1.2	311.76	6FE	0.1	−3.4	−2.0	0.9
260.85	−0.40	1.2	321.70	16FE	−0.2	−4.0	−1.9	0.9
268.14	−0.58	1.1	331.58	26FE	−0.8	−4.2	−1.8	0.9
275.44	−0.77	1.0	341.40	7MR	−1.6	−4.3	−1.8	0.8
282.74	−0.97	0.9	351.16	17MR	−1.3	−4.2	−1.7	0.8
290.02	−1.18	0.8	0.87	27MR	−0.3	−4.1	−1.6	0.8
297.27	−1.39	0.7	10.52	6AP	0.8	−4.0	−1.6	0.8
304.46	−1.61	0.6	20.14	16AP	1.9	−3.9	−1.5	0.8
311.58	−1.83	0.5	29.71	26AP	3.5	−3.8	−1.5	0.8
318.59	−2.05	0.4	39.26	6MY	2.6	−3.7	−1.4	0.8
325.46	−2.27	0.2	48.80	16MY	1.4	−3.6	−1.4	0.8
332.14	−2.48	0.1	58.31	26MY	0.5	−3.6	−1.4	0.7
338.60	−2.69	0.0	67.83	5JN	−0.6	−3.5	−1.3	0.7
344.77	−2.89	−0.1	77.36	15JN	−1.5	−3.5	−1.3	0.7
350.58	−3.08	−0.3	86.90	25JN	−1.4	−3.4	−1.3	0.6
355.94	−3.26	−0.4	96.47	5JL	−0.6	−3.4	−1.3	0.6
0.74	−3.43	−0.6	106.07	15JL	−0.1	−3.4	−1.3	0.5
4.83	−3.57	−0.8	115.71	25JL	0.3	−3.4	−1.3	0.5
8.05	−3.68	−1.0	125.41	4AU	0.5	−3.4	−1.3	0.4
10.18	−3.75	−1.2	135.15	14AU	0.9	−3.4	−1.4	0.4
11.03	−3.75	−1.4	144.95	24AU	2.2	−3.4	−1.4	0.3
10.42	−3.66	−1.6	154.82	3SE	1.8	−3.4	−1.4	0.3
8.38	−3.42	−1.8	164.74	13SE	−0.3	−3.4	−1.5	0.2
5.24	−3.04	−2.0	174.72	23SE	−0.8	−3.4	−1.5	0.3
1.66	−2.51	−2.0	184.76	3OC	−0.8	−3.4	−1.6	0.3
358.48	−1.92	−1.7	194.84	13OC	−0.8	−3.4	−1.6	0.4
356.34	−1.32	−1.4	204.97	23OC	−0.7	−3.4	−1.7	0.5
355.50	−0.79	−1.1	215.14	2NO	−0.6	−3.5	−1.8	0.5
356.04	−0.34	−0.8	225.33	12NO	−0.6	−3.5	−1.8	0.6
357.71	0.03	−0.4	235.54	22NO	−0.5	−3.5	−1.9	0.7
0.33	0.33	−0.2	245.75	2DE	0.3	−3.5	−2.0	0.7
3.71	0.56	0.1	255.96	12DE	3.0	−3.4	−2.0	0.7
7.67	0.75	0.3	266.14	22DE	1.0	−3.4	−2.0	0.8

−535

♂ LONG	LAT	MAG	☉ LONG	16.00UT	☿	♀	♃	♄
12.08	0.90	0.6	276.30	1JA	0.3	−3.4	−2.1	0.8
16.85	1.02	0.8	286.43	11JA	0.1	−3.4	−2.1	0.8
21.89	1.11	0.9	296.50	21JA	−0.0	−3.3	−2.1	0.8
27.15	1.19	1.1	306.53	31JA	−0.3	−3.3	−2.1	0.8
32.58	1.24	1.2	316.50	10FE	−0.8	−3.3	−2.0	0.7
38.13	1.29	1.4	326.41	20FE	−1.5	−3.3	−2.0	0.7
43.80	1.32	1.5	336.26	2MR	−1.3	−3.4	−1.9	0.6
49.56	1.34	1.6	346.05	12MR	−0.3	−3.4	−1.9	0.6
55.38	1.36	1.7	355.78	22MR	1.0	−3.4	−1.8	0.6
61.27	1.37	1.7	5.47	1AP	2.5	−3.4	−1.7	0.6
67.21	1.37	1.8	15.10	11AP	3.4	−3.4	−1.7	0.6
73.19	1.36	1.9	24.70	21AP	2.0	−3.5	−1.6	0.6
79.22	1.35	1.9	34.26	1MY	1.1	−3.5	−1.5	0.6
85.29	1.34	2.0	43.80	11MY	0.3	−3.6	−1.5	0.6
91.40	1.32	2.0	53.32	21MY	−0.6	−3.7	−1.4	0.6
97.55	1.30	2.0	62.84	31MY	−1.5	−3.7	−1.4	0.6
103.75	1.27	2.0	72.36	10JN	−1.4	−3.8	−1.3	0.5
109.99	1.24	2.0	81.89	20JN	−0.5	−3.9	−1.3	0.5
116.28	1.20	2.0	91.45	30JN	0.0	−4.0	−1.3	0.5
122.61	1.16	2.0	101.03	10JL	0.4	−4.1	−1.3	0.4
129.00	1.12	2.0	110.65	20JL	0.7	−4.2	−1.2	0.4
135.45	1.07	2.0	120.32	30JL	1.2	−4.2	−1.2	0.3
141.95	1.02	1.9	130.03	9AU	2.7	−4.2	−1.2	0.3
148.52	0.97	1.9	139.81	19AU	1.6	−3.9	−1.2	0.2
155.15	0.90	1.9	149.64	29AU	−0.3	−3.4	−1.3	0.2
161.85	0.84	1.8	159.53	8SE	−1.0	−3.4	−1.3	0.1
168.62	0.77	1.8	169.48	18SE	−1.0	−4.0	−1.3	0.0
175.46	0.69	1.8	179.49	28SE	−0.9	−4.3	−1.3	−0.0
182.38	0.61	1.8	189.55	8OC	−0.6	−4.4	−1.4	0.0
189.36	0.53	1.8	199.66	18OC	−0.4	−4.3	−1.4	0.1
196.42	0.44	1.7	209.81	28OC	−0.4	−4.2	−1.5	0.2
203.56	0.34	1.7	219.98	7NO	−0.3	−4.2	−1.5	0.3
210.76	0.24	1.7	230.19	17NO	0.5	−4.1	−1.6	0.3
218.04	0.13	1.7	240.40	27NO	3.1	−4.0	−1.7	0.4
225.38	0.02	1.6	250.61	7DE	0.7	−3.9	−1.7	0.4
232.79	−0.10	1.6	260.81	17DE	0.1	−3.8	−1.8	0.5
240.26	−0.22	1.6	270.98	27DE	−0.0	−3.7	−1.9	0.5

♂			☉	16.00UT	☿	♀	♃	♄
LONG	LAT	MAG	LONG	-534	MAGNITUDES			
247.78	-0.34	1.5	281.13	6JA	-0.1	-3.6	-1.9	0.6
255.35	-0.47	1.5	291.23	16JA	-0.4	-3.6	-2.0	0.6
262.96	-0.60	1.4	301.28	26JA	-0.8	-3.5	-2.0	0.6
270.60	-0.73	1.4	311.28	5FE	-1.4	-3.4	-2.0	0.6
278.27	-0.86	1.3	321.22	15FE	-1.2	-3.4	-2.0	0.5
285.94	-0.98	1.3	331.11	25FE	-0.3	-3.4	-2.0	0.5
293.62	-1.10	1.3	340.93	7MR	1.2	-3.3	-2.0	0.5
301.28	-1.22	1.2	350.69	17MR	3.2	-3.3	-2.0	0.4
308.92	-1.32	1.2	0.41	27MR	2.5	-3.3	-1.9	0.4
316.51	-1.42	1.1	10.06	6AP	1.5	-3.3	-1.9	0.4
324.06	-1.51	1.1	19.68	16AP	0.8	-3.3	-1.8	0.4
331.52	-1.58	1.0	29.26	26AP	0.2	-3.3	-1.7	0.4
338.90	-1.64	1.0	38.80	6MY	-0.6	-3.3	-1.7	0.4
346.18	-1.69	0.9	48.33	16MY	-1.6	-3.4	-1.6	0.4
353.33	-1.72	0.9	57.85	26MY	-1.4	-3.4	-1.5	0.4
0.34	-1.73	0.8	67.37	5JN	-0.5	-3.4	-1.5	0.4
7.19	-1.73	0.8	76.89	15JN	0.1	-3.5	-1.4	0.4
13.85	-1.71	0.7	86.43	25JN	0.5	-3.5	-1.4	0.4
20.29	-1.67	0.6	96.00	5JL	0.9	-3.5	-1.3	0.4
26.48	-1.62	0.6	105.60	15JL	1.7	-3.4	-1.3	0.3
32.38	-1.54	0.5	115.24	25JL	3.1	-3.4	-1.3	0.3
37.93	-1.45	0.4	124.93	4AU	1.4	-3.4	-1.2	0.2
43.07	-1.33	0.3	134.67	14AU	-0.4	-3.4	-1.2	0.2
47.71	-1.18	0.1	144.47	24AU	-1.1	-3.3	-1.2	0.1
51.74	-1.00	-0.0	154.33	3SE	-1.1	-3.3	-1.2	0.1
55.03	-0.79	-0.2	164.25	13SE	-0.8	-3.3	-1.2	-0.0
57.38	-0.53	-0.4	174.23	23SE	-0.5	-3.3	-1.2	-0.1
58.61	-0.21	-0.6	184.26	3OC	-0.3	-3.3	-1.2	-0.1
58.53	0.15	-0.8	194.35	13OC	-0.2	-3.4	-1.3	-0.2
57.01	0.57	-1.1	204.48	23OC	-0.1	-3.4	-1.3	-0.1
54.18	1.02	-1.2	214.64	2NO	0.7	-3.4	-1.3	-0.0
50.44	1.45	-1.4	224.84	12NO	3.0	-3.5	-1.4	0.0
46.54	1.81	-1.3	235.04	22NO	0.5	-3.5	-1.4	0.1
43.24	2.08	-1.0	245.25	2DE	-0.1	-3.6	-1.5	0.2
41.07	2.24	-0.7	255.46	12DE	-0.2	-3.6	-1.6	0.2
40.25	2.32	-0.4	265.65	22DE	-0.3	-3.7	-1.6	0.3
				-533				
40.71	2.34	-0.2	275.81	1JA	-0.5	-3.8	-1.7	0.3
42.27	2.32	0.1	285.94	11JA	-0.9	-3.8	-1.8	0.4
44.74	2.28	0.4	296.02	21JA	-1.3	-3.9	-1.8	0.4
47.92	2.22	0.6	306.04	31JA	-1.1	-4.0	-1.9	0.4
51.67	2.16	0.8	316.02	10FE	-0.2	-4.1	-1.9	0.4
55.87	2.09	1.0	325.93	20FE	1.5	-4.2	-2.0	0.4
60.42	2.02	1.1	335.78	2MR	3.4	-4.2	-2.0	0.4
65.26	1.95	1.2	345.58	12MR	1.9	-4.2	-2.0	0.4
70.33	1.88	1.4	355.32	22MR	1.1	-4.1	-2.0	0.3
75.59	1.81	1.5	5.00	1AP	0.6	-3.7	-2.0	0.3
81.01	1.74	1.6	14.64	11AP	0.1	-3.0	-2.0	0.3
86.57	1.66	1.6	24.24	21AP	-0.6	-3.4	-2.0	0.2
92.25	1.59	1.7	33.80	1MY	-1.6	-3.9	-1.9	0.3
98.04	1.52	1.8	43.34	11MY	-1.4	-4.2	-1.9	0.3
103.93	1.44	1.8	52.86	21MY	-0.5	-4.2	-1.8	0.3
109.91	1.37	1.8	62.38	31MY	0.2	-4.1	-1.7	0.3
115.98	1.29	1.9	71.90	10JN	0.7	-4.1	-1.7	0.3
122.14	1.21	1.9	81.43	20JN	1.3	-4.0	-1.6	0.3
128.38	1.13	1.9	90.98	30JN	2.2	-3.9	-1.5	0.3
134.71	1.05	1.9	100.56	10JL	3.1	-3.8	-1.5	0.3
141.12	0.96	1.9	110.18	20JL	1.1	-3.7	-1.4	0.3
147.62	0.88	1.9	119.84	30JL	-0.4	-3.6	-1.4	0.3
154.21	0.79	1.9	129.56	9AU	-1.2	-3.6	-1.4	0.2
160.88	0.70	1.8	139.33	19AU	-1.3	-3.5	-1.3	0.2
167.65	0.60	1.8	149.16	29AU	-0.7	-3.5	-1.3	0.1
174.51	0.50	1.8	159.05	8SE	-0.4	-3.5	-1.3	0.0
181.46	0.40	1.7	168.99	18SE	-0.2	-3.4	-1.3	-0.0
188.51	0.30	1.7	179.00	28SE	-0.1	-3.4	-1.2	-0.1
195.64	0.20	1.6	189.06	8OC	0.1	-3.4	-1.2	-0.2
202.86	0.09	1.6	199.16	18OC	0.9	-3.4	-1.3	-0.2
210.18	-0.01	1.5	209.31	28OC	2.8	-3.4	-1.3	-0.3
217.57	-0.12	1.5	219.49	7NO	0.3	-3.4	-1.3	-0.2
225.04	-0.23	1.5	229.69	17NO	-0.2	-3.4	-1.3	-0.1
232.59	-0.34	1.5	239.90	27NO	-0.3	-3.4	-1.4	-0.1
240.21	-0.45	1.5	250.12	7DE	-0.4	-3.4	-1.4	-0.0
247.88	-0.55	1.5	260.31	17DE	-0.6	-3.4	-1.5	0.1
255.61	-0.65	1.5	270.49	27DE	-0.9	-3.4	-1.5	0.1

♂			☉	16.00UT	☿	♀	♃	♄
LONG	LAT	MAG	LONG	-532	MAGNITUDES			
263.38	-0.75	1.5	280.64	6JA	-1.1	-3.4	-1.6	0.2
271.17	-0.84	1.4	290.74	16JA	-1.0	-3.4	-1.6	0.2
278.99	-0.93	1.4	300.80	26JA	-0.2	-3.5	-1.7	0.3
286.82	-1.01	1.4	310.80	5FE	1.8	-3.5	-1.8	0.3
294.64	-1.07	1.4	320.74	15FE	2.8	-3.5	-1.8	0.3
302.45	-1.13	1.4	330.63	25FE	1.4	-3.4	-1.9	0.3
310.22	-1.18	1.4	340.45	6MR	0.8	-3.4	-2.0	0.3
317.96	-1.21	1.4	350.22	16MR	0.4	-3.4	-2.0	0.3
325.64	-1.24	1.4	359.93	26MR	-0.0	-3.4	-2.1	0.3
333.26	-1.25	1.4	9.59	5AP	-0.7	-3.3	-2.1	0.3
340.81	-1.24	1.4	19.21	15AP	-1.6	-3.3	-2.1	0.2
348.27	-1.23	1.4	28.79	25AP	-1.4	-3.3	-2.1	0.2
355.64	-1.20	1.4	38.34	5MY	-0.5	-3.3	-2.1	0.2
2.91	-1.15	1.3	47.87	15MY	0.3	-3.3	-2.1	0.2
10.07	-1.10	1.3	57.39	25MY	1.0	-3.4	-2.0	0.2
17.11	-1.03	1.3	66.91	4JN	1.7	-3.4	-2.0	0.3
24.02	-0.95	1.3	76.43	14JN	2.8	-3.4	-1.9	0.3
30.80	-0.86	1.3	85.97	24JN	2.7	-3.5	-1.8	0.3
37.43	-0.76	1.3	95.53	4JL	0.9	-3.5	-1.8	0.3
43.90	-0.65	1.2	105.13	14JL	-0.5	-3.6	-1.7	0.3
50.20	-0.52	1.2	114.77	24JL	-1.3	-3.6	-1.7	0.3
56.30	-0.38	1.1	124.46	3AU	-1.3	-3.7	-1.6	0.3
62.19	-0.22	1.1	134.20	13AU	-0.7	-3.8	-1.5	0.2
67.83	-0.05	1.0	144.00	23AU	-0.3	-3.9	-1.5	0.2
73.18	0.13	0.9	153.85	2SE	-0.1	-4.0	-1.4	0.1
78.19	0.34	0.8	163.77	12SE	0.1	-4.1	-1.4	0.1
82.79	0.57	0.7	173.75	22SE	0.3	-4.2	-1.4	0.0
86.88	0.84	0.5	183.78	2OC	1.2	-4.3	-1.4	-0.1
90.37	1.13	0.3	193.86	12OC	2.6	-4.4	-1.3	-0.1
93.10	1.47	0.2	203.99	22OC	0.1	-4.3	-1.3	-0.2
94.89	1.85	-0.1	214.15	1NO	-0.4	-4.1	-1.3	-0.3
95.57	2.27	-0.3	224.34	11NO	-0.5	-3.4	-1.3	-0.3
94.95	2.71	-0.5	234.55	21NO	-0.5	-3.3	-1.3	-0.3
92.98	3.16	-0.7	244.76	1DE	-0.6	-4.0	-1.4	-0.2
89.81	3.55	-0.9	254.96	11DE	-0.9	-4.3	-1.4	-0.1
85.92	3.83	-1.1	265.16	21DE	-1.0	-4.4	-1.4	-0.0
82.03	3.95	-0.9	275.32	31DE	-0.9	-4.3	-1.5	0.0
				-531				
78.86	3.92	-0.7	285.45	10JA	-0.1	-4.2	-1.5	0.1
76.85	3.77	-0.4	295.53	20JA	2.1	-4.1	-1.6	0.1
76.15	3.57	-0.2	305.56	30JA	2.2	-4.0	-1.6	0.2
76.69	3.33	0.1	315.54	9FE	1.0	-3.9	-1.7	0.2
78.29	3.09	0.3	325.46	19FE	0.6	-3.8	-1.7	0.3
80.77	2.86	0.5	335.31	1MR	0.3	-3.7	-1.8	0.3
83.95	2.64	0.7	345.11	11MR	-0.1	-3.6	-1.9	0.3
87.70	2.44	0.9	354.85	21MR	-0.7	-3.6	-2.0	0.3
91.90	2.25	1.0	4.54	31MR	-1.6	-3.5	-2.0	0.3
96.47	2.07	1.1	14.18	10AP	-1.4	-3.4	-2.1	0.3
101.36	1.91	1.2	23.78	20AP	-0.4	-3.4	-2.1	0.2
106.50	1.75	1.3	33.34	30AP	0.5	-3.4	-2.2	0.2
111.86	1.60	1.4	42.88	10MY	1.2	-3.3	-2.2	0.2
117.42	1.46	1.5	52.41	20MY	2.2	-3.3	-2.2	0.2
123.15	1.32	1.5	61.92	30MY	3.5	-3.3	-2.2	0.2
129.03	1.18	1.6	71.44	9JN	2.1	-3.3	-2.2	0.2
135.06	1.05	1.6	80.97	19JN	0.8	-3.3	-2.2	0.3
141.23	0.92	1.7	90.52	29JN	-0.5	-3.3	-2.1	0.3
147.53	0.80	1.7	100.10	9JL	-1.4	-3.4	-2.1	0.3
153.96	0.67	1.7	109.72	19JL	-1.4	-3.4	-2.0	0.3
160.51	0.55	1.7	119.38	29JL	-0.7	-3.4	-1.9	0.3
167.17	0.42	1.7	129.09	8AU	-0.2	-3.4	-1.9	0.3
173.96	0.30	1.7	138.86	18AU	0.1	-3.5	-1.8	0.3
180.86	0.18	1.7	148.68	28AU	0.2	-3.5	-1.7	0.2
187.88	0.06	1.6	158.56	7SE	0.5	-3.5	-1.7	0.2
195.01	-0.05	1.6	168.51	17SE	1.5	-3.5	-1.6	0.1
202.24	-0.17	1.6	178.51	27SE	2.3	-3.4	-1.6	0.1
209.58	-0.28	1.6	188.56	7OC	-0.0	-3.4	-1.5	0.0
217.01	-0.38	1.5	198.67	17OC	-0.6	-3.4	-1.5	-0.1
224.54	-0.49	1.5	208.81	27OC	-0.6	-3.4	-1.5	-0.1
232.14	-0.58	1.5	218.99	6NO	-0.6	-3.3	-1.4	-0.2
239.83	-0.68	1.4	229.19	16NO	-0.7	-3.3	-1.4	-0.3
247.57	-0.76	1.4	239.40	26NO	-0.8	-3.3	-1.4	-0.3
255.37	-0.84	1.4	249.61	6DE	-0.8	-3.3	-1.4	-0.2
263.22	-0.91	1.3	259.81	16DE	-0.7	-3.4	-1.4	-0.2
271.09	-0.97	1.3	269.99	26DE	0.0	-3.4	-1.4	-0.1

Left table (year markers -530, -529):

♂ LONG	LAT	MAG	☉ LONG	16.00UT	☿	♀	♃	♄
				-530		MAGNITUDES		
278.98	-1.02	1.3	280.14	5JA	2.4	-3.4	-1.5	-0.0
286.87	-1.05	1.3	290.25	15JA	1.7	-3.4	-1.5	0.0
294.75	-1.08	1.3	300.31	25JA	0.7	-3.5	-1.5	0.1
302.61	-1.10	1.4	310.31	4FE	0.4	-3.5	-1.6	0.2
310.44	-1.10	1.4	320.26	14FE	0.2	-3.5	-1.6	0.2
318.22	-1.09	1.4	330.15	24FE	-0.2	-3.6	-1.7	0.2
325.94	-1.07	1.4	339.98	6MR	-0.7	-3.7	-1.7	0.3
333.60	-1.04	1.5	349.75	16MR	-1.6	-3.7	-1.8	0.3
341.18	-1.00	1.5	359.46	26MR	-1.4	-3.8	-1.9	0.3
348.68	-0.95	1.5	9.13	5AP	-0.4	-3.9	-1.9	0.3
356.10	-0.88	1.5	18.75	15AP	0.6	-4.0	-2.0	0.3
3.42	-0.81	1.6	28.33	25AP	1.6	-4.1	-2.1	0.3
10.66	-0.74	1.6	37.88	5MY	2.9	-4.1	-2.1	0.3
17.79	-0.65	1.6	47.41	15MY	3.1	-4.2	-2.2	0.2
24.82	-0.56	1.6	56.93	25MY	1.7	-4.1	-2.3	0.2
31.76	-0.46	1.6	66.45	4JN	-0.6	-3.9	-2.3	0.2
38.59	-0.35	1.6	75.97	14JN	-0.5	-3.3	-2.3	0.2
45.32	-0.24	1.6	85.51	24JN	-1.5	-3.0	-2.3	0.3
51.94	-0.12	1.6	95.07	4JL	-1.4	-3.7	-2.3	0.3
58.44	0.00	1.6	104.67	14JL	-0.6	-4.1	-2.3	0.3
64.84	0.13	1.6	114.30	24JL	-0.1	-4.2	-2.3	0.4
71.11	0.26	1.6	123.99	3AU	0.2	-4.2	-2.2	0.4
77.24	0.41	1.5	133.73	13AU	0.4	-4.2	-2.2	0.4
83.24	0.55	1.5	143.52	23AU	0.7	-4.1	-2.1	0.4
89.08	0.71	1.4	153.37	2SE	1.9	-4.0	-2.0	0.3
94.73	0.88	1.4	163.29	12SE	2.1	-3.9	-2.0	0.3
100.18	1.06	1.3	173.26	22SE	-0.2	-3.8	-1.9	0.3
105.38	1.26	1.2	183.29	2OC	-0.7	-3.8	-1.8	0.2
110.30	1.47	1.1	193.36	12OC	-0.7	-3.7	-1.8	0.2
114.86	1.71	0.9	203.49	22OC	-0.7	-3.6	-1.7	0.1
119.00	1.97	0.8	213.65	1NO	-0.7	-3.6	-1.7	0.0
122.62	2.25	0.6	223.84	11NO	-0.6	-3.5	-1.6	-0.1
125.58	2.57	0.4	234.05	21NO	-0.6	-3.5	-1.6	-0.1
127.74	2.93	0.2	244.26	1DE	-0.6	-3.4	-1.6	-0.2
128.91	3.31	-0.0	254.47	11DE	0.1	-3.4	-1.5	-0.2
128.92	3.71	-0.3	264.66	21DE	2.7	-3.4	-1.5	-0.2
127.62	4.09	-0.5	274.83	31DE	1.3	-3.4	-1.5	-0.1
				-529				
125.06	4.41	-0.8	284.95	10JA	0.4	-3.3	-1.5	-0.0
121.53	4.61	-1.0	295.04	20JA	0.2	-3.3	-1.5	0.0
117.59	4.63	-1.0	305.08	30JA	0.0	-3.3	-1.5	0.1
114.00	4.48	-0.8	315.05	9FE	-0.2	-3.3	-1.5	0.2
111.34	4.19	-0.6	324.97	19FE	-0.8	-3.3	-1.6	0.2
109.93	3.82	-0.4	334.84	1MR	-1.5	-3.3	-1.6	0.3
109.82	3.43	-0.2	344.63	11MR	-1.4	-3.4	-1.7	0.3
110.88	3.05	0.1	354.38	21MR	-0.4	-3.4	-1.7	0.3
112.94	2.69	0.3	4.07	31MR	0.8	-3.4	-1.8	0.4
115.82	2.37	0.4	13.71	10AP	2.1	-3.5	-1.8	0.4
119.37	2.07	0.6	23.31	20AP	3.9	-3.5	-1.9	0.4
123.47	1.79	0.7	32.88	30AP	2.3	-3.4	-1.9	0.4
128.02	1.54	0.9	42.42	10MY	1.3	-3.4	-2.0	0.4
132.94	1.31	1.0	51.94	20MY	0.4	-3.4	-2.1	0.4
138.19	1.09	1.1	61.46	30MY	-0.6	-3.4	-2.2	0.3
143.72	0.89	1.1	70.97	9JN	-1.5	-3.3	-2.2	0.3
149.49	0.70	1.2	80.50	19JN	-1.4	-3.3	-2.3	0.3
155.49	0.52	1.2	90.05	29JN	-0.6	-3.3	-2.3	0.3
161.68	0.35	1.3	99.63	9JL	-0.0	-3.3	-2.4	0.4
168.06	0.19	1.3	109.24	19JL	0.3	-3.3	-2.4	0.4
174.61	0.04	1.4	118.91	29JL	0.6	-3.3	-2.4	0.5
181.32	-0.11	1.4	128.61	8AU	1.0	-3.4	-2.4	0.5
188.19	-0.24	1.4	138.38	18AU	2.4	-3.4	-2.4	0.5
195.20	-0.37	1.4	148.20	28AU	1.8	-3.4	-2.4	0.5
202.34	-0.49	1.4	158.08	7SE	-0.3	-3.5	-2.3	0.5
209.61	-0.61	1.4	168.02	17SE	-0.9	-3.5	-2.3	0.5
217.00	-0.71	1.4	178.02	27SE	-0.9	-3.6	-2.2	0.4
224.49	-0.80	1.4	188.07	7OC	-0.9	-3.7	-2.1	0.4
232.08	-0.88	1.4	198.18	17OC	-0.6	-3.8	-2.1	0.3
239.75	-0.95	1.4	208.32	27OC	-0.5	-3.8	-2.0	0.3
247.50	-1.01	1.4	218.49	6NO	-0.5	-3.9	-1.9	0.1
255.30	-1.06	1.4	228.69	16NO	-0.4	-4.0	-1.9	0.1
263.15	-1.10	1.4	238.90	26NO	0.3	-4.1	-1.8	0.1
271.02	-1.12	1.4	249.11	6DE	3.0	-4.3	-1.7	-0.0
278.92	-1.13	1.4	259.32	16DE	0.9	-4.3	-1.7	-0.1
286.81	-1.13	1.4	269.50	26DE	0.2	-4.4	-1.7	-0.1

Right table (year markers -528, -527):

♂ LONG	LAT	MAG	☉ LONG	16.00UT	☿	♀	♃	♄
				-528		MAGNITUDES		
294.69	-1.12	1.4	279.64	5JA	0.1	-4.3	-1.6	-0.0
302.54	-1.09	1.4	289.76	15JA	-0.1	-4.0	-1.6	0.0
310.35	-1.06	1.4	299.82	25JA	-0.3	-3.4	-1.6	0.1
318.11	-1.01	1.4	309.82	4FE	-0.8	-3.4	-1.6	0.1
325.81	-0.95	1.4	319.78	14FE	-1.5	-4.0	-1.6	0.2
333.44	-0.89	1.4	329.67	24FE	-1.3	-4.2	-1.6	0.3
341.00	-0.82	1.4	339.50	5MR	-0.3	-4.3	-1.6	0.3
348.48	-0.74	1.4	349.28	15MR	1.1	-4.2	-1.6	0.4
355.87	-0.66	1.5	358.99	25MR	2.7	-4.1	-1.6	0.4
3.18	-0.57	1.5	8.66	4AP	3.1	-4.0	-1.6	0.4
10.40	-0.47	1.6	18.28	14AP	1.8	-3.9	-1.7	0.5
17.53	-0.38	1.6	27.87	24AP	1.0	-3.8	-1.7	0.5
24.57	-0.28	1.7	37.42	4MY	0.3	-3.7	-1.8	0.5
31.52	-0.18	1.7	46.95	14MY	-0.6	-3.6	-1.8	0.5
38.39	-0.07	1.8	56.47	24MY	-1.5	-3.6	-1.9	0.5
45.18	0.03	1.8	65.99	3JN	-1.4	-3.5	-2.0	0.5
51.89	0.14	1.8	75.51	13JN	-0.5	-3.5	-2.0	0.5
58.52	0.25	1.8	85.05	23JN	0.1	-3.4	-2.1	0.5
65.07	0.35	1.8	94.61	3JL	0.4	-3.4	-2.2	0.5
71.55	0.46	1.9	104.21	13JL	0.8	-3.4	-2.2	0.5
77.94	0.57	1.9	113.84	23JL	1.4	-3.4	-2.3	0.6
84.27	0.68	1.8	123.52	2AU	2.8	-3.4	-2.4	0.6
90.51	0.80	1.8	133.26	12AU	1.6	-3.3	-2.4	0.6
96.67	0.91	1.8	143.05	22AU	-0.3	-3.4	-2.4	0.6
102.75	1.03	1.8	152.90	1SE	-1.0	-3.4	-2.4	0.7
108.72	1.16	1.7	162.81	11SE	-1.0	-3.4	-2.4	0.7
114.59	1.28	1.7	172.77	21SE	-0.9	-3.4	-2.4	0.6
120.34	1.42	1.6	182.80	1OC	-0.5	-3.4	-2.4	0.6
125.96	1.55	1.5	192.88	11OC	-0.4	-3.4	-2.3	0.6
131.40	1.70	1.4	203.00	21OC	-0.3	-3.5	-2.3	0.6
136.66	1.86	1.3	213.16	31OC	-0.3	-3.5	-2.2	0.5
141.69	2.02	1.2	223.35	10NO	0.5	-3.5	-2.2	0.4
146.43	2.21	1.1	233.55	20NO	3.2	-3.5	-2.1	0.4
150.83	2.40	0.9	243.76	30NO	0.7	-3.5	-2.0	0.3
154.80	2.61	0.7	253.97	10DE	0.0	-3.4	-1.9	0.2
158.25	2.84	0.5	264.16	20DE	-0.1	-3.4	-1.9	0.2
161.04	3.08	0.3	274.33	30DE	-0.2	-3.4	-1.8	0.1
				-527				
163.00	3.34	0.0	284.46	9JA	-0.4	-3.4	-1.8	0.1
163.98	3.60	-0.2	294.54	19JA	-0.8	-3.3	-1.7	0.1
163.77	3.84	-0.5	304.58	29JA	-1.4	-3.3	-1.7	0.2
162.29	4.02	-0.8	314.56	8FE	-1.2	-3.3	-1.6	0.2
159.61	4.11	-1.1	324.49	18FE	-0.3	-3.3	-1.6	0.3
156.05	4.04	-1.3	334.35	28FE	1.3	-3.4	-1.6	0.4
152.87	3.80	-1.2	344.16	10MR	3.3	-3.4	-1.6	0.4
148.97	3.42	-1.1	353.90	20MR	2.3	-3.4	-1.5	0.5
146.70	2.96	-0.9	3.60	30MR	1.3	-3.4	-1.5	0.5
145.75	2.47	-0.7	13.24	9AP	0.7	-3.4	-1.6	0.6
146.10	2.00	-0.5	22.84	19AP	0.2	-3.5	-1.6	0.6
147.62	1.56	-0.3	32.41	29AP	-0.6	-3.5	-1.6	0.6
150.15	1.17	-0.1	41.96	9MY	-1.6	-3.6	-1.6	0.7
153.50	0.82	0.1	51.48	19MY	-1.4	-3.7	-1.7	0.7
157.54	0.51	0.2	61.00	29MY	-0.5	-3.7	-1.7	0.7
162.14	0.23	0.4	70.52	8JN	0.1	-3.8	-1.7	0.7
167.22	-0.01	0.5	80.04	18JN	0.6	-3.9	-1.8	0.7
172.69	-0.23	0.6	89.59	28JN	1.0	-4.0	-1.8	0.7
178.51	-0.43	0.7	99.17	8JL	1.8	-4.1	-1.9	0.7
184.63	-0.61	0.7	108.78	18JL	3.2	-4.2	-2.0	0.7
191.00	-0.77	0.8	118.44	28JL	1.3	-4.2	-2.1	0.8
197.62	-0.91	0.9	128.15	7AU	-0.4	-4.2	-2.1	0.8
204.43	-1.03	0.9	137.90	17AU	-1.1	-3.9	-2.2	0.8
211.43	-1.13	1.0	147.72	27AU	-1.2	-3.4	-2.2	0.8
218.60	-1.21	1.0	157.60	6SE	-0.8	-3.5	-2.3	0.9
225.90	-1.28	1.0	167.54	16SE	-0.4	-4.0	-2.3	0.9
233.34	-1.33	1.1	177.54	26SE	-0.3	-4.3	-2.4	0.9
240.88	-1.37	1.1	187.59	6OC	-0.2	-4.4	-2.4	0.9
248.50	-1.39	1.2	197.68	16OC	-0.1	-4.3	-2.4	0.9
256.20	-1.39	1.2	207.83	26OC	0.7	-4.2	-2.4	0.8
263.95	-1.38	1.2	218.00	5NO	3.1	-4.1	-2.3	0.8
271.73	-1.35	1.3	228.20	15NO	0.4	-4.0	-2.3	0.7
279.54	-1.31	1.3	238.41	25NO	-0.1	-3.9	-2.2	0.7
287.34	-1.26	1.3	248.62	5DE	-0.2	-3.9	-2.2	0.6
295.12	-1.19	1.4	258.82	15DE	-0.3	-3.8	-2.1	0.5
302.88	-1.12	1.4	269.01	25DE	-0.5	-3.7	-2.0	0.5

♂ LONG LAT MAG	☉ LONG	16.00UT −526	☿	♀	♃	♄ MAGNITUDES
310.60 -1.04 1.4	279.16	4JA	-0.9	-3.6	-2.0	0.4
318.27 -0.95 1.5	289.27	14JA	-1.2	-3.6	-1.9	0.3
325.88 -0.86 1.5	299.33	24JA	-1.1	-3.5	-1.8	0.3
333.41 -0.76 1.5	309.34	3FE	-0.2	-3.4	-1.8	0.4
340.88 -0.66 1.6	319.30	13FE	1.6	-3.4	-1.7	0.4
348.27 -0.56 1.6	329.19	23FE	3.2	-3.4	-1.6	0.5
355.57 -0.46 1.6	339.03	5MR	1.7	-3.3	-1.6	0.5
2.79 -0.35 1.7	348.80	15MR	1.0	-3.3	-1.6	0.6
9.94 -0.25 1.7	358.53	25MR	0.5	-3.3	-1.5	0.6
17.00 -0.14 1.7	8.19	4AP	0.1	-3.3	-1.5	0.7
23.98 -0.04 1.7	17.82	14AP	-0.6	-3.3	-1.5	0.7
30.88 0.06 1.7	27.41	24AP	-1.6	-3.3	-1.5	0.8
37.72 0.16 1.7	36.96	4MY	-1.4	-3.3	-1.5	0.8
44.48 0.26 1.8	46.49	14MY	-0.5	-3.4	-1.5	0.9
51.18 0.35 1.8	56.01	24MY	0.3	-3.4	-1.5	0.9
57.81 0.45 1.8	65.53	3JN	0.8	-3.4	-1.5	0.9
64.39 0.54 1.9	75.05	13JN	1.4	-3.5	-1.5	0.9
70.91 0.62 1.9	84.59	23JN	2.4	-3.5	-1.6	0.9
77.39 0.71 1.9	94.14	3JL	3.0	-3.5	-1.6	0.9
83.82 0.79 2.0	103.74	13JL	1.1	-3.4	-1.6	0.9
90.20 0.88 2.0	113.37	23JL	-0.4	-3.4	-1.7	0.9
96.54 0.96 2.0	123.05	2AU	-1.2	-3.4	-1.8	0.9
102.85 1.04 2.0	132.78	12AU	-1.3	-3.3	-1.8	1.0
109.11 1.11 2.0	142.57	22AU	-0.7	-3.3	-1.9	1.0
115.33 1.19 2.0	152.41	1SE	-0.3	-3.3	-1.9	1.1
121.51 1.27 1.9	162.32	11SE	-0.1	-3.3	-2.0	1.1
127.65 1.34 1.9	172.29	21SE	-0.0	-3.3	-2.1	1.1
133.74 1.42 1.9	182.31	10C	0.1	-3.3	-2.1	1.2
139.77 1.49 1.8	192.38	110C	1.0	-3.4	-2.2	1.2
145.74 1.56 1.7	202.50	210C	2.8	-3.4	-2.2	1.2
151.63 1.64 1.7	212.66	310C	0.2	-3.4	-2.3	1.1
157.45 1.71 1.6	222.85	10NO	-0.3	-3.5	-2.3	1.1
163.15 1.79 1.5	233.06	20NO	-0.4	-3.5	-2.3	1.1
168.73 1.86 1.3	243.26	30NO	-0.4	-3.6	-2.3	1.0
174.16 1.94 1.2	253.47	10DE	-0.6	-3.6	-2.2	0.9
179.40 2.01 1.1	263.67	20DE	-0.9	-3.7	-2.2	0.9
184.42 2.08 0.9	273.84	30DE	-1.1	-3.8	-2.1	0.8
		−525				
189.17 2.14 0.7	283.97	9JA	-1.0	-3.8	-2.1	0.7
193.57 2.20 0.5	294.06	19JA	-0.2	-3.9	-2.0	0.7
197.54 2.25 0.3	304.10	29JA	1.9	-4.0	-1.9	0.6
200.97 2.28 0.0	314.08	8FE	2.6	-4.1	-1.9	0.6
203.71 2.29 -0.2	324.01	18FE	1.2	-4.2	-1.8	0.7
205.62 2.25 -0.5	333.87	28FE	0.7	-4.2	-1.7	0.7
206.51 2.17 -0.8	343.68	10MR	0.4	-4.2	-1.7	0.7
206.20 2.00 -1.2	353.43	20MR	-0.0	-4.1	-1.6	0.8
204.64 1.74 -1.5	3.13	30MR	-0.7	-3.7	-1.5	0.9
201.96 1.36 -1.8	12.77	9AP	-1.6	-3.0	-1.5	0.9
198.61 0.88 -1.9	22.38	19AP	-1.4	-3.4	-1.5	1.0
195.29 0.35 -1.8	31.95	29AP	-0.5	-4.0	-1.4	1.0
192.68 -0.19 -1.6	41.49	9MY	0.4	-4.2	-1.4	1.1
191.28 -0.67 -1.4	51.02	19MY	1.1	-4.2	-1.4	1.1
191.26 -1.08 -1.2	60.53	29MY	1.8	-4.1	-1.4	1.1
192.56 -1.42 -1.0	70.05	8JN	3.1	-4.0	-1.4	1.2
195.02 -1.69 -0.8	79.58	18JN	2.5	-4.0	-1.4	1.2
198.47 -1.89 -0.6	89.13	28JN	0.9	-3.9	-1.4	1.2
202.71 -2.05 -0.4	98.70	8JL	-0.5	-3.8	-1.4	1.2
207.62 -2.16 -0.3	108.31	18JL	-1.3	-3.7	-1.4	1.2
213.06 -2.23 -0.1	117.97	28JL	-1.4	-3.6	-1.5	1.2
218.94 -2.27 -0.0	127.67	7AU	-0.7	-3.6	-1.5	1.2
225.20 -2.28 0.1	137.43	17AU	-0.3	-3.5	-1.5	1.2
231.74 -2.26 0.2	147.24	27AU	-0.0	-3.5	-1.6	1.3
238.53 -2.22 0.3	157.12	6SE	0.1	-3.5	-1.7	1.4
245.52 -2.16 0.4	167.06	16SE	0.3	-3.4	-1.7	1.4
252.67 -2.08 0.5	177.05	26SE	1.3	-3.4	-1.8	1.4
259.95 -1.98 0.6	187.10	6OC	2.6	-3.4	-1.8	1.4
267.32 -1.87 0.7	197.20	16OC	0.1	-3.4	-1.9	1.4
274.75 -1.75 0.8	207.33	26OC	-0.5	-3.4	-2.0	1.3
282.23 -1.62 0.9	217.51	5NO	-0.5	-3.4	-2.0	1.3
289.72 -1.49 1.0	227.70	15NO	-0.5	-3.4	-2.1	1.3
297.22 -1.35 1.1	237.91	25NO	-0.7	-3.4	-2.1	1.2
304.71 -1.20 1.2	248.13	5DE	-0.8	-3.4	-2.2	1.2
312.16 -1.06 1.2	258.33	15DE	-0.9	-3.4	-2.2	1.1
319.58 -0.91 1.3	268.51	25DE	-0.8	-3.4	-2.2	1.1

♂ LONG LAT MAG	☉ LONG	16.00UT −524	☿	♀	♃	♄ MAGNITUDES
326.95 -0.77 1.4	278.66	4JA	-0.1	-3.4	-2.1	1.0
334.26 -0.63 1.5	288.78	14JA	2.2	-3.4	-2.1	1.0
341.51 -0.50 1.5	298.84	24JA	2.0	-3.5	-2.1	1.0
348.69 -0.37 1.6	308.86	3FE	0.9	-3.5	-2.0	0.9
355.81 -0.24 1.6	318.81	13FE	0.5	-3.4	-1.9	0.9
2.85 -0.12 1.7	328.71	23FE	0.2	-3.4	-1.9	0.9
9.83 -0.01 1.7	338.55	4MR	-0.1	-3.4	-1.8	1.0
16.74 0.10 1.8	348.33	14MR	-0.7	-3.4	-1.7	1.0
23.58 0.20 1.8	358.05	24MR	-1.5	-3.4	-1.7	1.1
30.36 0.30 1.8	7.72	3AP	-1.4	-3.3	-1.6	1.1
37.08 0.40 1.9	17.35	13AP	-0.4	-3.3	-1.5	1.2
43.75 0.48 1.9	26.93	23AP	0.5	-3.3	-1.5	1.2
50.36 0.57 1.9	36.49	3MY	1.4	-3.3	-1.4	1.3
56.93 0.64 1.9	46.02	13MY	2.5	-3.3	-1.4	1.3
63.45 0.72 1.9	55.54	23MY	3.5	-3.4	-1.4	1.4
69.94 0.79 1.9	65.06	2JN	2.0	-3.4	-1.3	1.4
76.40 0.85 1.9	74.58	12JN	0.7	-3.4	-1.3	1.4
82.83 0.91 1.9	84.12	22JN	-0.5	-3.5	-1.3	1.5
89.23 0.97 2.0	93.68	2JL	-1.4	-3.5	-1.3	1.4
95.62 1.02 2.0	103.27	12JL	-1.4	-3.6	-1.3	1.4
101.99 1.07 2.0	112.90	22JL	-0.7	-3.6	-1.3	1.4
108.35 1.11 2.0	122.58	1AU	-0.2	-3.7	-1.3	1.3
114.70 1.15 2.0	132.31	11AU	0.1	-3.8	-1.3	1.3
121.05 1.19 2.0	142.09	21AU	0.3	-3.9	-1.4	1.2
127.40 1.23 2.0	151.94	31AU	0.6	-4.0	-1.4	1.2
133.74 1.26 2.0	161.84	10SE	1.6	-4.1	-1.4	1.2
140.08 1.28 2.0	171.80	20SE	2.3	-4.2	-1.5	1.2
146.42 1.31 2.0	181.83	30SE	-0.1	-4.3	-1.5	1.2
152.76 1.32 1.9	191.89	100C	-0.6	-4.4	-1.6	1.2
159.10 1.34 1.9	202.01	200C	-0.7	-4.3	-1.6	1.2
165.43 1.35 1.8	212.17	300C	-0.7	-4.0	-1.7	1.2
171.75 1.35 1.8	222.35	9NO	-0.8	-3.3	-1.8	1.2
178.06 1.35 1.7	232.56	19NO	-0.7	-3.3	-1.8	1.1
184.36 1.34 1.6	242.77	29NO	-0.7	-4.0	-1.9	1.1
190.62 1.32 1.5	252.98	9DE	-0.7	-4.3	-2.0	1.0
196.86 1.29 1.4	263.17	19DE	0.0	-4.4	-2.0	1.0
203.06 1.24 1.3	273.34	29DE	2.5	-4.3	-2.0	1.0
		−523				
209.21 1.19 1.2	283.48	8JA	1.5	-4.2	-2.1	0.9
215.29 1.12 1.0	293.57	18JA	0.6	-4.1	-2.1	0.9
221.29 1.02 0.9	303.61	28JA	0.3	-4.0	-2.1	0.8
227.19 0.91 0.7	313.60	7FE	0.1	-3.9	-2.0	0.8
232.97 0.76 0.5	323.53	17FE	-0.2	-3.8	-2.0	0.7
238.58 0.59 0.3	333.40	27FE	-0.7	-3.7	-2.0	0.7
243.99 0.37 0.1	343.21	9MR	-1.5	-3.6	-1.9	0.8
249.13 0.10 -0.1	352.97	19MR	-1.4	-3.6	-1.8	0.8
253.95 -0.22 -0.3	2.66	29MR	-0.4	-3.5	-1.8	0.9
258.33 -0.60 -0.6	12.31	8AP	0.7	-3.4	-1.7	1.0
262.17 -1.07 -0.9	21.92	18AP	1.8	-3.4	-1.6	1.1
265.30 -1.62 -1.1	31.49	28AP	3.3	-3.4	-1.6	1.1
267.55 -2.26 -1.4	41.03	8MY	2.8	-3.3	-1.5	1.2
268.72 -3.00 -1.8	50.56	18MY	1.5	-3.3	-1.5	1.2
268.67 -3.79 -2.1	60.08	28MY	0.5	-3.3	-1.4	1.2
267.40 -4.58 -2.3	69.59	7JN	-0.5	-3.3	-1.3	1.3
265.14 -5.27 -2.6	79.12	17JN	-1.5	-3.3	-1.3	1.3
262.48 -5.74 -2.6	88.67	27JN	-1.4	-3.3	-1.3	1.3
260.14 -5.95 -2.4	98.24	7JL	-0.6	-3.4	-1.3	1.3
258.73 -5.88 -2.2	107.85	17JL	-0.1	-3.4	-1.2	1.3
258.61 -5.63 -1.9	117.50	27JL	0.2	-3.4	-1.2	1.2
259.81 -5.25 -1.7	127.20	6AU	0.5	-3.4	-1.2	1.2
262.22 -4.81 -1.4	136.96	16AU	0.8	-3.5	-1.2	1.2
265.64 -4.36 -1.1	146.77	26AU	2.0	-3.5	-1.2	1.1
269.86 -3.91 -0.9	156.64	5SE	2.1	-3.5	-1.3	1.1
274.71 -3.47 -0.7	166.57	15SE	-0.2	-3.5	-1.3	1.1
280.07 -3.06 -0.4	176.56	25SE	-0.8	-3.4	-1.3	1.1
285.79 -2.67 -0.2	186.61	5OC	-0.8	-3.4	-1.3	1.1
291.81 -2.31 -0.0	196.70	15OC	-0.8	-3.4	-1.4	1.1
298.04 -1.97 0.1	206.84	25OC	-0.7	-3.4	-1.4	1.1
304.44 -1.66 0.3	217.01	4NO	-0.6	-3.3	-1.5	1.1
310.95 -1.38 0.5	227.21	14NO	-0.6	-3.3	-1.6	1.0
317.55 -1.12 0.7	237.41	24NO	-0.5	-3.3	-1.6	1.0
324.19 -0.88 0.8	247.63	4DE	0.2	-3.3	-1.7	1.0
330.88 -0.66 0.9	257.83	14DE	2.8	-3.4	-1.8	1.0
337.57 -0.46 1.1	268.01	24DE	1.2	-3.4	-1.8	0.9

♂			⊙	16.00UT	☿	♀	♃	♄
LONG	LAT	MAG	LONG	-522	MAGNITUDES			
344.27	-0.28	1.2	278.17	3JA	0.4	-3.4	-1.9	0.9
350.96	-0.12	1.3	288.29	13JA	0.2	-3.4	-1.9	0.8
357.63	0.03	1.4	298.35	23JA	0.0	-3.5	-2.0	0.8
4.28	0.17	1.5	308.37	2FE	-0.3	-3.5	-2.0	0.7
10.90	0.29	1.6	318.33	12FE	-0.8	-3.6	-2.0	0.7
17.48	0.40	1.7	328.23	22FE	-1.5	-3.6	-2.0	0.6
24.04	0.50	1.7	338.08	4MR	-1.3	-3.7	-2.0	0.6
30.56	0.60	1.8	347.86	14MR	-0.4	-3.7	-2.0	0.6
37.05	0.68	1.8	357.58	24MR	0.9	-3.8	-2.0	0.7
43.51	0.75	1.9	7.26	3AP	2.3	-3.9	-1.9	0.7
49.95	0.82	1.9	16.89	13AP	3.7	-4.0	-1.8	0.8
56.36	0.88	2.0	26.48	23AP	2.1	-4.1	-1.8	0.9
62.75	0.93	2.0	36.03	3MY	1.2	-4.1	-1.7	0.9
69.12	0.98	2.0	45.57	13MY	0.4	-4.2	-1.7	1.0
75.47	1.02	2.0	55.09	23MY	-0.5	-4.1	-1.6	1.0
81.82	1.05	2.0	64.60	2JN	-1.5	-3.9	-1.5	1.1
88.17	1.08	2.0	74.13	12JN	-1.4	-3.3	-1.5	1.1
94.51	1.11	2.0	83.66	22JN	-0.6	-3.0	-1.4	1.1
100.86	1.13	2.0	93.22	2JL	-0.0	-3.8	-1.4	1.1
107.22	1.15	2.0	102.81	12JL	0.4	-4.1	-1.3	1.1
113.59	1.16	2.0	112.43	22JL	0.6	-4.2	-1.3	1.1
119.98	1.16	2.0	122.11	1AU	1.1	-4.2	-1.3	1.1
126.39	1.17	2.0	131.83	11AU	2.5	-4.2	-1.2	1.1
132.82	1.16	2.0	141.61	21AU	1.8	-4.1	-1.2	1.1
139.29	1.16	2.0	151.46	31AU	-0.2	-4.0	-1.2	1.0
145.78	1.14	2.0	161.36	10SE	-0.9	-3.9	-1.2	1.0
152.30	1.13	2.0	171.32	20SE	-1.0	-3.8	-1.2	0.9
158.86	1.10	2.0	181.34	30SE	-0.9	-3.8	-1.2	1.0
165.46	1.07	1.9	191.41	10OC	-0.6	-3.7	-1.3	1.0
172.09	1.04	1.9	201.52	20OC	-0.5	-3.6	-1.3	1.0
178.76	1.00	1.9	211.68	30OC	-0.4	-3.6	-1.3	1.0
185.47	0.94	1.8	221.86	9NO	-0.4	-3.5	-1.4	1.0
192.22	0.89	1.8	232.06	19NO	0.3	-3.5	-1.4	1.0
199.00	0.82	1.7	242.28	29NO	3.0	-3.4	-1.5	0.9
205.82	0.74	1.6	252.49	9DE	0.9	-3.4	-1.5	0.9
212.68	0.65	1.6	262.68	19DE	0.2	-3.4	-1.6	0.9
219.57	0.55	1.5	272.86	29DE	0.0	-3.4	-1.7	0.9
				-521				
226.49	0.44	1.4	282.99	8JA	-0.1	-3.3	-1.7	0.8
233.44	0.31	1.3	293.08	18JA	-0.4	-3.3	-1.8	0.8
240.42	0.17	1.2	303.13	28JA	-0.8	-3.3	-1.9	0.7
247.41	0.01	1.1	313.12	7FE	-1.4	-3.3	-1.9	0.7
254.42	-0.17	1.0	323.05	17FE	-1.3	-3.3	-2.0	0.6
261.43	-0.36	0.9	332.93	27FE	-0.4	-3.3	-2.0	0.6
268.43	-0.57	0.8	342.74	9MR	1.1	-3.4	-2.0	0.5
275.43	-0.80	0.7	352.49	19MR	2.9	-3.4	-2.0	0.5
282.38	-1.05	0.5	2.20	29MR	2.8	-3.4	-2.0	0.5
289.28	-1.32	0.4	11.85	8AP	1.6	-3.5	-2.0	0.6
296.09	-1.60	0.3	21.45	18AP	0.9	-3.5	-2.0	0.6
302.79	-1.90	0.1	31.03	28AP	0.3	-3.4	-2.0	0.7
309.33	-2.22	-0.0	40.57	8MY	-0.6	-3.4	-1.9	0.8
315.67	-2.55	-0.2	50.10	18MY	-1.5	-3.4	-1.8	0.8
321.73	-2.89	-0.3	59.62	28MY	-1.5	-3.4	-1.8	0.9
327.43	-3.25	-0.5	69.13	7JN	-0.6	-3.3	-1.7	0.9
332.69	-3.61	-0.7	78.66	17JN	0.1	-3.3	-1.7	1.0
337.36	-3.98	-0.9	88.20	27JN	0.5	-3.3	-1.6	1.0
341.29	-4.35	-1.1	97.77	7JL	0.9	-3.3	-1.5	1.0
344.30	-4.70	-1.3	107.38	17JL	1.5	-3.3	-1.5	1.0
346.15	-5.02	-1.6	117.03	27JL	3.0	-3.3	-1.4	1.0
346.66	-5.26	-1.8	126.72	6AU	1.5	-3.4	-1.4	1.0
345.72	-5.38	-2.0	136.48	16AU	-0.3	-3.4	-1.3	1.0
343.46	-5.29	-2.2	146.29	26AU	-1.1	-3.4	-1.3	1.0
340.38	-4.96	-2.3	156.16	5SE	-1.1	-3.5	-1.3	1.0
337.21	-4.41	-2.2	166.09	15SE	-0.8	-3.5	-1.3	0.9
334.72	-3.73	-1.9	176.08	25SE	-0.5	-3.6	-1.3	0.9
333.44	-3.01	-1.6	186.12	5OC	-0.3	-3.7	-1.3	0.8
333.49	-2.33	-1.2	196.21	15OC	-0.3	-3.8	-1.3	0.9
334.79	-1.72	-0.9	206.35	25OC	-0.2	-3.8	-1.3	0.9
337.15	-1.21	-0.6	216.52	4NO	0.5	-3.9	-1.3	0.9
340.36	-0.78	-0.3	226.71	14NO	3.2	-4.0	-1.3	0.9
344.24	-0.42	-0.1	236.92	24NO	0.6	-4.2	-1.3	0.9
348.64	-0.12	0.2	247.13	4DE	-0.0	-4.3	-1.4	0.9
353.44	0.12	0.4	257.34	14DE	-0.1	-4.3	-1.4	0.9
358.55	0.33	0.6	267.53	24DE	-0.2	-4.4	-1.5	0.8

♂			⊙	16.00UT	☿	♀	♃	♄
LONG	LAT	MAG	LONG	-520	MAGNITUDES			
3.89	0.50	0.8	277.68	3JA	-0.4	-4.3	-1.5	0.8
9.41	0.65	1.0	287.80	13JA	-0.8	-4.0	-1.6	0.8
15.08	0.77	1.1	297.87	23JA	-1.3	-3.4	-1.7	0.7
20.86	0.87	1.3	307.89	2FE	-1.2	-3.5	-1.7	0.7
26.72	0.95	1.4	317.85	12FE	-0.3	-4.0	-1.8	0.6
32.64	1.02	1.5	327.76	22FE	1.4	-4.2	-1.9	0.6
38.62	1.08	1.6	337.60	3MR	3.4	-4.3	-1.9	0.5
44.64	1.13	1.7	347.38	13MR	2.0	-4.2	-2.0	0.5
50.68	1.17	1.7	357.11	23MR	1.2	-4.1	-2.0	0.4
56.76	1.20	1.8	6.79	2AP	0.7	-4.0	-2.1	0.4
62.86	1.22	1.9	16.42	12AP	0.1	-3.9	-2.1	0.4
68.98	1.24	1.9	26.01	22AP	-0.6	-3.8	-2.1	0.5
75.12	1.25	2.0	35.57	2MY	-1.5	-3.7	-2.1	0.6
81.28	1.25	2.0	45.11	12MY	-1.5	-3.6	-2.1	0.6
87.47	1.25	2.0	54.63	22MY	-0.5	-3.6	-2.1	0.7
93.68	1.24	2.0	64.14	1JN	0.2	-3.5	-2.0	0.7
99.92	1.23	2.0	73.67	11JN	0.7	-3.5	-2.0	0.8
106.20	1.22	2.0	83.20	21JN	1.2	-3.4	-1.9	0.8
112.50	1.20	2.0	92.75	1JL	2.0	-3.4	-1.8	0.9
118.85	1.18	2.0	102.34	11JL	3.2	-3.4	-1.8	0.9
125.23	1.15	2.0	111.97	21JL	1.3	-3.4	-1.7	0.9
131.66	1.12	2.0	121.64	31JL	-0.4	-3.3	-1.7	0.9
138.14	1.08	1.9	131.36	10AU	-1.2	-3.3	-1.6	0.9
144.67	1.04	1.9	141.14	20AU	-1.3	-3.4	-1.5	0.9
151.26	0.99	1.9	150.98	30AU	-0.8	-3.4	-1.5	0.9
157.90	0.94	1.8	160.88	9SE	-0.4	-3.4	-1.4	0.9
164.61	0.88	1.8	170.83	19SE	-0.2	-3.4	-1.4	0.9
171.38	0.82	1.8	180.85	29SE	-0.1	-3.4	-1.4	0.8
178.21	0.75	1.8	190.92	9OC	0.0	-3.4	-1.4	0.8
185.10	0.67	1.8	201.03	19OC	0.8	-3.5	-1.3	0.8
192.07	0.59	1.8	211.18	29OC	3.1	-3.5	-1.3	0.8
199.10	0.51	1.8	221.37	8NO	0.4	-3.5	-1.3	0.8
206.19	0.41	1.7	231.57	18NO	-0.2	-3.5	-1.3	0.8
213.35	0.31	1.7	241.78	28NO	-0.3	-3.5	-1.4	0.9
220.57	0.20	1.7	251.99	8DE	-0.3	-3.4	-1.4	0.9
227.85	0.09	1.6	262.19	18DE	-0.5	-3.4	-1.4	0.8
235.19	-0.03	1.6	272.36	28DE	-0.9	-3.4	-1.4	0.8
				-519				
242.59	-0.16	1.5	282.50	7JA	-1.1	-3.4	-1.5	0.8
250.03	-0.29	1.5	292.59	17JA	-1.1	-3.3	-1.5	0.8
257.51	-0.43	1.4	302.64	27JA	-0.3	-3.3	-1.6	0.7
265.03	-0.58	1.3	312.64	6FE	1.6	-3.3	-1.6	0.7
272.58	-0.73	1.3	322.57	16FE	3.0	-3.3	-1.7	0.6
280.15	-0.87	1.2	332.45	26FE	1.5	-3.4	-1.8	0.6
287.72	-1.02	1.2	342.26	8MR	0.9	-3.4	-1.8	0.5
295.28	-1.17	1.1	352.02	18MR	0.5	-3.4	-1.9	0.5
302.83	-1.31	1.0	1.72	28MR	0.0	-3.4	-2.0	0.4
310.34	-1.45	1.0	11.38	7AP	-0.6	-3.5	-2.0	0.4
317.80	-1.58	0.9	20.99	17AP	-1.5	-3.5	-2.1	0.3
325.19	-1.71	0.8	30.56	27AP	-1.5	-3.5	-2.2	0.4
332.49	-1.82	0.8	40.11	7MY	-0.5	-3.6	-2.2	0.5
339.68	-1.91	0.7	49.64	17MY	0.3	-3.7	-2.2	0.5
346.73	-2.00	0.6	59.16	27MY	0.9	-3.7	-2.2	0.6
353.61	-2.07	0.6	68.68	6JN	1.5	-3.8	-2.2	0.6
0.30	-2.12	0.5	78.20	16JN	2.6	-3.9	-2.2	0.7
6.76	-2.15	0.4	87.74	26JN	2.8	-4.0	-2.2	0.8
12.95	-2.17	0.3	97.32	6JL	1.1	-4.1	-2.1	0.8
18.80	-2.16	0.2	106.92	16JL	-0.4	-4.2	-2.1	0.8
24.27	-2.14	0.1	116.56	26JL	-1.3	-4.2	-2.0	0.9
29.25	-2.09	-0.1	126.26	5AU	-1.4	-4.2	-1.9	0.9
33.65	-2.01	-0.2	136.01	15AU	-0.7	-3.9	-1.9	0.9
37.33	-1.90	-0.4	145.81	25AU	-0.3	-3.4	-1.8	0.9
40.11	-1.75	-0.6	155.68	4SE	-0.1	-3.5	-1.7	0.9
41.80	-1.55	-0.8	165.60	14SE	0.0	-4.0	-1.7	0.9
42.20	-1.28	-1.0	175.59	24SE	0.2	-4.3	-1.6	0.8
41.15	-0.94	-1.2	185.63	4OC	1.0	-4.3	-1.6	0.8
38.71	-0.54	-1.4	195.72	14OC	2.9	-4.3	-1.5	0.8
35.21	-0.08	-1.6	205.85	24OC	0.2	-4.2	-1.5	0.7
31.35	0.38	-1.6	216.02	3NO	-0.4	-4.1	-1.5	0.8
27.95	0.79	-1.3	226.21	13NO	-0.4	-4.0	-1.5	0.8
25.59	1.12	-1.0	236.42	23NO	-0.5	-3.9	-1.4	0.8
24.58	1.37	-0.7	246.64	3DE	-0.6	-3.9	-1.4	0.8
24.89	1.54	-0.4	256.84	13DE	-0.9	-3.8	-1.4	0.8
26.35	1.65	-0.1	267.03	23DE	-1.0	-3.7	-1.4	0.8

♂ (Mars) LONG	LAT	MAG	⊙ (Sun) LONG	16.00UT	☿ (Mercury) MAGNITUDES	♀ (Venus)	♃ (Jupiter)	♄ (Saturn)
				-510				
180.53	2.59	0.4	285.34	10JA	1.7	-3.5	-1.9	1.0
183.81	2.73	0.2	295.43	20JA	0.7	-3.5	-1.9	1.0
186.37	2.85	-0.0	305.46	30JA	0.4	-3.4	-2.0	1.0
188.06	2.96	-0.3	315.44	9FE	0.1	-3.4	-2.0	1.1
188.70	3.04	-0.6	325.36	19FE	-0.2	-3.4	-2.0	1.1
188.12	3.06	-0.9	335.22	1MR	-0.7	-3.3	-2.0	1.1
186.28	2.99	-1.2	345.02	11MR	-1.5	-3.3	-2.0	1.2
183.34	2.79	-1.5	354.76	21MR	-1.4	-3.3	-2.0	1.2
179.77	2.45	-1.6	4.46	31MR	-0.5	-3.3	-1.9	1.2
176.29	2.00	-1.5	14.09	10AP	0.6	-3.3	-1.9	1.2
173.56	1.50	-1.3	23.70	20AP	1.7	-3.3	-1.8	1.2
172.04	0.98	-1.1	33.26	30AP	3.0	-3.3	-1.8	1.1
171.88	0.51	-0.9	42.80	10MY	3.0	-3.4	-1.7	1.1
173.02	0.08	-0.7	52.32	20MY	1.7	-3.4	-1.6	1.1
175.29	-0.28	-0.5	61.84	30MY	0.6	-3.4	-1.6	1.0
178.52	-0.59	-0.3	71.36	9JN	-0.4	-3.5	-1.5	0.9
182.53	-0.86	-0.2	80.89	19JN	-1.4	-3.5	-1.5	0.9
187.20	-1.08	-0.0	90.44	29JN	-1.5	-3.5	-1.4	0.8
192.40	-1.27	0.1	100.01	9JL	-0.7	-3.4	-1.4	0.7
198.05	-1.42	0.2	109.63	19JL	-0.1	-3.4	-1.3	0.7
204.08	-1.54	0.3	119.29	29JL	0.2	-3.4	-1.3	0.6
210.42	-1.64	0.4	129.00	8AU	0.4	-3.3	-1.3	0.6
217.04	-1.71	0.5	138.76	18AU	0.8	-3.3	-1.2	0.6
223.89	-1.76	0.6	148.59	28AU	1.8	-3.3	-1.2	0.7
230.93	-1.78	0.6	158.47	7SE	2.3	-3.3	-1.2	0.7
238.13	-1.78	0.7	168.41	17SE	-0.1	-3.3	-1.2	0.8
245.48	-1.77	0.8	178.41	27SE	-0.8	-3.3	-1.2	0.8
252.93	-1.73	0.9	188.46	7OC	-0.8	-3.4	-1.2	0.9
260.47	-1.68	0.9	198.57	17OC	-0.8	-3.4	-1.3	0.9
268.06	-1.62	1.0	208.71	27OC	-0.7	-3.4	-1.3	1.0
275.70	-1.54	1.1	218.88	6NO	-0.6	-3.5	-1.3	1.0
283.36	-1.45	1.1	229.09	16NO	-0.6	-3.5	-1.4	1.1
291.02	-1.35	1.2	239.30	26NO	-0.6	-3.6	-1.4	1.1
298.67	-1.24	1.3	249.51	6DE	0.1	-3.6	-1.5	1.1
306.29	-1.13	1.3	259.71	16DE	2.5	-3.7	-1.5	1.1
313.87	-1.01	1.4	269.89	26DE	1.3	-3.8	-1.6	1.1
				-509				
321.40	-0.89	1.4	280.04	5JA	0.4	-3.8	-1.7	1.1
328.87	-0.77	1.5	290.15	15JA	0.2	-3.9	-1.7	1.1
336.28	-0.65	1.5	300.21	25JA	0.0	-4.0	-1.8	1.1
343.61	-0.53	1.6	310.21	4FE	-0.2	-4.1	-1.9	1.1
350.88	-0.41	1.6	320.17	14FE	-0.7	-4.2	-1.9	1.2
358.06	-0.29	1.7	330.06	24FE	-1.5	-4.3	-2.0	1.2
5.17	-0.18	1.7	339.89	6MR	-1.4	-4.2	-2.0	1.3
12.21	-0.07	1.7	349.66	16MR	-0.5	-4.1	-2.0	1.3
19.17	0.04	1.8	359.38	26MR	0.8	-3.7	-2.0	1.3
26.06	0.14	1.8	9.04	5AP	2.1	-3.0	-2.0	1.4
32.88	0.24	1.8	18.67	15AP	3.9	-3.5	-2.0	1.4
39.63	0.33	1.8	28.25	25AP	2.3	-4.0	-2.0	1.4
46.33	0.42	1.8	37.80	5MY	1.3	-4.2	-1.9	1.3
52.97	0.51	1.9	47.33	15MY	0.5	-4.2	-1.9	1.3
59.56	0.59	1.9	56.85	25MY	-0.5	-4.1	-1.8	1.3
66.10	0.67	1.9	66.37	4JN	-1.4	-4.0	-1.8	1.2
72.60	0.75	1.9	75.89	14JN	-1.5	-3.9	-1.7	1.2
79.06	0.82	1.9	85.43	24JN	-0.6	-3.9	-1.7	1.1
85.49	0.89	2.0	94.99	4JL	-0.0	-3.8	-1.6	1.0
91.89	0.95	2.0	104.59	14JL	0.3	-3.7	-1.5	1.0
98.26	1.01	2.0	114.22	24JL	0.6	-3.6	-1.5	0.9
104.61	1.07	2.0	123.90	3AU	1.0	-3.6	-1.4	0.8
110.94	1.13	2.0	133.64	13AU	2.3	-3.5	-1.4	0.8
117.26	1.18	2.0	143.43	23AU	2.0	-3.5	-1.3	0.8
123.55	1.23	2.0	153.28	2SE	-0.2	-3.5	-1.3	0.8
129.84	1.28	2.0	163.19	12SE	-0.9	-3.4	-1.3	0.9
136.10	1.32	2.0	173.16	22SE	-0.9	-3.4	-1.3	0.9
142.35	1.36	1.9	183.19	2OC	-0.9	-3.4	-1.3	1.0
148.57	1.40	1.9	193.27	12OC	-0.6	-3.4	-1.3	1.0
154.77	1.44	1.8	203.39	22OC	-0.5	-3.4	-1.3	1.1
160.95	1.47	1.8	213.55	1NO	-0.4	-3.4	-1.3	1.1
167.09	1.50	1.7	223.74	11NO	-0.4	-3.4	-1.3	1.2
173.18	1.52	1.6	233.94	21NO	0.2	-3.4	-1.3	1.2
179.23	1.54	1.5	244.16	1DE	2.8	-3.4	-1.4	1.2
185.21	1.55	1.4	254.37	11DE	1.0	-3.4	-1.4	1.3
191.11	1.55	1.3	264.56	21DE	0.2	-3.4	-1.4	1.3
196.92	1.55	1.1	274.72	31DE	0.0	-3.4	-1.5	1.3

♂ (Mars) LONG	LAT	MAG	⊙ (Sun) LONG	16.00UT	☿ (Mercury) MAGNITUDES	♀ (Venus)	♃ (Jupiter)	♄ (Saturn)
				-508				
202.61	1.53	1.0	284.86	10JA	-0.1	-3.5	-1.6	1.3
208.16	1.49	0.8	294.94	20JA	-0.3	-3.5	-1.6	1.3
213.53	1.44	0.7	304.98	30JA	-0.8	-3.5	-1.7	1.3
218.68	1.36	0.5	314.96	9FE	-1.4	-3.4	-1.8	1.3
223.55	1.26	0.2	324.88	19FE	-1.3	-3.4	-1.8	1.2
228.08	1.12	0.0	334.74	29FE	-0.4	-3.4	-1.9	1.2
232.16	0.93	-0.2	344.55	10MR	1.0	-3.4	-2.0	1.2
235.69	0.68	-0.5	354.29	20MR	2.7	-3.4	-2.0	1.2
238.52	0.36	-0.8	3.98	30MR	3.0	-3.3	-2.1	1.2
240.45	-0.04	-1.1	13.63	9AP	1.7	-3.3	-2.1	1.2
241.34	-0.54	-1.4	23.23	19AP	1.0	-3.3	-2.1	1.2
241.01	-1.14	-1.8	32.80	29AP	0.3	-3.3	-2.1	1.1
239.45	-1.81	-2.1	42.34	9MY	-0.5	-3.3	-2.1	1.1
236.93	-2.51	-2.3	51.86	19MY	-1.5	-3.4	-2.1	1.1
233.97	-3.14	-2.4	61.38	29MY	-1.5	-3.4	-2.1	1.0
231.30	-3.64	-2.2	70.90	8JN	-0.6	-3.4	-2.0	1.0
229.59	-3.96	-2.0	80.42	18JN	0.0	-3.5	-2.0	0.9
229.16	-4.12	-1.8	89.97	28JN	0.5	-3.5	-1.9	0.9
230.11	-4.14	-1.5	99.55	8JL	0.8	-3.6	-1.8	0.8
232.32	-4.07	-1.3	109.16	18JL	1.4	-3.6	-1.8	0.8
235.60	-3.94	-1.1	118.82	28JL	2.8	-3.7	-1.7	0.7
239.76	-3.77	-0.9	128.53	7AU	1.7	-3.8	-1.7	0.7
244.62	-3.56	-0.7	138.29	17AU	-0.2	-3.9	-1.6	0.7
250.03	-3.34	-0.5	148.11	27AU	-1.0	-4.0	-1.5	0.6
255.88	-3.10	-0.3	157.99	6SE	-1.1	-4.1	-1.5	0.7
262.07	-2.86	-0.1	167.92	16SE	-0.9	-4.2	-1.5	0.8
268.53	-2.61	0.0	177.92	26SE	-0.5	-4.3	-1.4	0.8
275.19	-2.37	0.2	187.97	6OC	-0.3	-4.4	-1.4	0.9
281.99	-2.13	0.3	198.07	16OC	-0.3	-4.3	-1.4	1.0
288.91	-1.89	0.5	208.22	26OC	-0.2	-4.0	-1.4	1.0
295.90	-1.66	0.6	218.39	5NO	0.4	-3.2	-1.3	1.1
302.94	-1.44	0.7	228.59	15NO	3.1	-3.4	-1.3	1.1
310.01	-1.23	0.9	238.80	25NO	0.7	-4.1	-1.4	1.1
317.08	-1.02	1.0	249.01	5DE	0.0	-4.3	-1.4	1.2
324.14	-0.83	1.1	259.21	15DE	-0.1	-4.4	-1.4	1.2
331.18	-0.65	1.2	269.40	25DE	-0.2	-4.3	-1.4	1.2
				-507				
338.20	-0.49	1.3	279.55	4JA	-0.4	-4.2	-1.5	1.2
345.17	-0.33	1.4	289.66	14JA	-0.8	-4.1	-1.5	1.1
352.10	-0.18	1.5	299.72	24JA	-1.3	-4.0	-1.6	1.1
358.98	-0.04	1.5	309.73	3FE	-1.2	-3.9	-1.6	1.1
5.81	0.08	1.6	319.69	13FE	-0.4	-3.8	-1.7	1.0
12.59	0.20	1.7	329.58	23FE	1.2	-3.7	-1.7	1.0
19.32	0.31	1.7	339.42	5MR	3.3	-3.6	-1.8	0.9
26.00	0.41	1.8	349.19	15MR	2.2	-3.5	-1.9	0.9
32.64	0.50	1.8	358.92	25MR	1.3	-3.5	-1.9	0.9
39.23	0.59	1.9	8.58	4AP	0.7	-3.4	-2.0	0.9
45.77	0.66	1.9	18.21	14AP	0.2	-3.4	-2.1	0.9
52.29	0.74	1.9	27.79	24AP	-0.5	-3.4	-2.1	0.9
58.76	0.80	2.0	37.35	4MY	-1.5	-3.3	-2.2	0.9
65.21	0.86	2.0	46.88	14MY	-1.5	-3.3	-2.2	0.9
71.63	0.92	2.0	56.40	24MY	-0.6	-3.3	-2.2	0.8
78.03	0.96	2.0	65.91	3JN	0.1	-3.3	-2.2	0.8
84.42	1.01	2.0	75.43	13JN	0.6	-3.3	-2.2	0.8
90.79	1.05	2.0	84.97	23JN	1.1	-3.3	-2.2	0.7
97.16	1.08	2.0	94.53	3JL	1.9	-3.4	-2.2	0.7
103.53	1.11	1.9	104.12	13JL	3.2	-3.4	-2.1	0.6
109.90	1.14	2.0	113.76	23JL	1.5	-3.4	-2.1	0.6
116.28	1.16	2.0	123.43	2AU	-0.3	-3.4	-2.0	0.5
122.67	1.18	2.0	133.17	12AU	-1.1	-3.5	-1.9	0.5
129.07	1.19	2.0	142.96	22AU	-1.2	-3.5	-1.9	0.4
135.48	1.20	2.0	152.80	1SE	-0.8	-3.5	-1.8	0.4
141.92	1.20	2.0	162.71	11SE	-0.4	-3.5	-1.7	0.3
148.38	1.20	2.0	172.67	21SE	-0.2	-3.4	-1.7	0.4
154.85	1.19	2.0	182.69	1OC	-0.1	-3.4	-1.6	0.5
161.35	1.18	1.9	192.77	11OC	-0.0	-3.4	-1.6	0.5
167.88	1.16	1.9	202.89	21OC	0.6	-3.3	-1.5	0.6
174.43	1.13	1.9	213.05	31OC	3.4	-3.3	-1.5	0.7
181.00	1.10	1.8	223.23	10NO	0.5	-3.3	-1.5	0.7
187.59	1.06	1.7	233.44	20NO	-0.2	-3.3	-1.5	0.8
194.20	1.01	1.7	243.65	30NO	-0.3	-3.3	-1.4	0.8
200.83	0.95	1.6	253.86	10DE	-0.3	-3.4	-1.4	0.9
207.47	0.88	1.5	264.06	20DE	-0.5	-3.4	-1.4	0.9
214.13	0.80	1.4	274.22	30DE	-0.9	-3.4	-1.5	0.9

♂ LONG	♂ LAT	♂ MAG	⊙ LONG	16.00UT -506	☿	♀	♃	♄
					MAGNITUDES			
220.80	0.70	1.3	284.36	9JA	-1.2	-3.4	-1.5	0.9
227.47	0.58	1.2	294.45	19JA	-1.1	-3.5	-1.5	0.9
234.14	0.45	1.1	304.49	29JA	-0.3	-3.5	-1.5	0.9
240.80	0.30	1.0	314.48	8FE	1.4	-3.6	-1.6	0.8
247.45	0.12	0.8	324.40	18FE	3.2	-3.6	-1.6	0.8
254.06	-0.08	0.7	334.27	28FE	1.6	-3.7	-1.7	0.7
260.63	-0.31	0.6	344.08	10MR	0.9	-3.7	-1.7	0.7
267.14	-0.57	0.4	353.83	20MR	0.5	-3.8	-1.8	0.7
273.56	-0.85	0.2	3.52	30MR	0.1	-3.9	-1.8	0.7
279.85	-1.18	0.1	13.17	9AP	-0.6	-4.0	-1.9	0.7
285.98	-1.54	-0.1	22.77	19AP	-1.5	-4.1	-2.0	0.7
291.89	-1.94	-0.3	32.34	29AP	-1.5	-4.1	-2.1	0.7
297.51	-2.38	-0.5	41.89	9MY	-0.6	-4.2	-2.1	0.7
302.74	-2.87	-0.7	51.41	19MY	0.2	-4.1	-2.2	0.7
307.48	-3.40	-0.9	60.92	29MY	0.8	-3.8	-2.2	0.6
311.56	-3.97	-1.2	70.44	8JN	1.4	-3.2	-2.3	0.6
314.81	-4.57	-1.4	79.97	18JN	2.5	-3.1	-2.3	0.6
317.03	-5.19	-1.7	89.51	28JN	3.0	-3.8	-2.4	0.5
318.00	-5.79	-2.0	99.09	8JL	1.2	-4.1	-2.4	0.5
317.60	-6.29	-2.2	108.70	18JL	-0.3	-4.2	-2.3	0.4
315.92	-6.60	-2.4	118.35	28JL	-1.2	-4.2	-2.3	0.4
313.31	-6.64	-2.5	128.06	7AU	-1.4	-4.1	-2.3	0.3
310.50	-6.36	-2.5	137.82	17AU	-0.8	-4.1	-2.2	0.3
308.23	-5.81	-2.2	147.63	27AU	-0.3	-4.0	-2.2	0.2
307.04	-5.11	-1.9	157.51	6SE	-0.1	-3.9	-2.1	0.2
307.17	-4.35	-1.6	167.44	16SE	0.0	-3.8	-2.0	0.1
308.56	-3.63	-1.3	177.43	26SE	0.2	-3.7	-2.0	0.1
311.03	-2.97	-1.0	187.49	6OC	0.9	-3.7	-1.9	0.2
314.39	-2.39	-0.7	197.58	16OC	3.2	-3.6	-1.8	0.2
318.44	-1.88	-0.4	207.72	26OC	0.3	-3.6	-1.8	0.3
323.04	-1.44	-0.2	217.89	5NO	-0.3	-3.5	-1.7	0.4
328.05	-1.06	0.1	228.09	15NO	-0.4	-3.5	-1.7	0.4
333.38	-0.74	0.3	238.30	25NO	-0.4	-3.4	-1.6	0.5
338.96	-0.46	0.5	248.51	5DE	-0.6	-3.4	-1.6	0.6
344.72	-0.22	0.7	258.71	15DE	-0.9	-3.4	-1.6	0.6
350.61	-0.01	0.8	268.90	25DE	-1.0	-3.4	-1.5	0.6
				-505				
356.62	0.17	1.0	279.05	4JA	-1.0	-3.3	-1.5	0.6
2.70	0.33	1.1	289.16	14JA	-0.3	-3.3	-1.5	0.7
8.83	0.47	1.3	299.23	24JA	1.7	-3.3	-1.5	0.7
15.01	0.59	1.4	309.25	3FE	2.6	-3.3	-1.5	0.6
21.21	0.69	1.5	319.20	13FE	1.2	-3.3	-1.6	0.6
27.43	0.78	1.6	329.10	23FE	0.7	-3.4	-1.6	0.6
33.66	0.86	1.7	338.94	5MR	0.4	-3.4	-1.6	0.6
39.89	0.92	1.7	348.72	15MR	-0.0	-3.4	-1.6	0.5
46.12	0.98	1.8	358.44	25MR	-0.6	-3.4	-1.7	0.5
52.36	1.03	1.9	8.12	4AP	-1.5	-3.5	-1.7	0.5
58.60	1.07	1.9	17.74	14AP	-1.5	-3.5	-1.8	0.5
64.84	1.11	2.0	27.33	24AP	-0.6	-3.4	-1.8	0.5
71.09	1.13	2.0	36.88	4MY	0.3	-3.4	-1.9	0.5
77.33	1.15	2.0	46.41	14MY	1.1	-3.4	-2.0	0.5
83.59	1.17	2.0	55.93	24MY	1.9	-3.4	-2.1	0.5
89.86	1.18	2.0	65.45	3JN	3.2	-3.3	-2.1	0.5
96.14	1.19	2.0	74.97	13JN	2.5	-3.3	-2.2	0.5
102.44	1.19	2.0	84.51	23JN	1.0	-3.3	-2.3	0.5
108.77	1.18	2.0	94.07	3JL	-0.4	-3.3	-2.3	0.4
115.12	1.18	2.0	103.65	13JL	-1.3	-3.3	-2.4	0.4
121.50	1.16	2.0	113.29	23JL	-1.4	-3.3	-2.4	0.3
127.92	1.14	2.0	122.97	2AU	-0.7	-3.4	-2.4	0.3
134.37	1.12	1.9	132.69	12AU	-0.3	-3.4	-2.4	0.2
140.87	1.09	1.9	142.48	22AU	0.0	-3.4	-2.4	0.2
147.41	1.06	1.9	152.32	1SE	0.2	-3.5	-2.4	0.1
154.00	1.02	1.9	162.22	11SE	0.4	-3.5	-2.4	0.0
160.64	0.98	1.9	172.19	21SE	1.2	-3.6	-2.3	-0.0
167.33	0.93	1.9	182.21	1OC	2.9	-3.7	-2.3	-0.1
174.07	0.88	1.9	192.28	11OC	0.2	-3.8	-2.2	-0.1
180.87	0.82	1.9	202.40	21OC	-0.5	-3.9	-2.1	-0.0
187.73	0.75	1.8	212.56	31OC	-0.6	-4.0	-2.0	0.0
194.64	0.67	1.8	222.74	10NO	-0.6	-4.1	-2.0	0.1
201.62	0.59	1.8	232.94	20NO	-0.7	-4.2	-1.9	0.2
208.64	0.50	1.7	243.16	30NO	-0.8	-4.3	-1.8	0.3
215.73	0.40	1.7	253.36	10DE	-0.8	-4.4	-1.8	0.3
222.86	0.29	1.6	263.56	20DE	-0.8	-4.4	-1.7	0.4
230.05	0.17	1.5	273.73	30DE	-0.2	-4.3	-1.7	0.4

♂ LONG	♂ LAT	♂ MAG	⊙ LONG	16.00UT -504	☿	♀	♃	♄
					MAGNITUDES			
237.29	0.04	1.5	283.86	9JA	2.0	-4.0	-1.6	0.4
244.57	-0.10	1.4	293.96	19JA	2.0	-3.4	-1.6	0.5
251.89	-0.24	1.3	304.00	29JA	0.8	-3.5	-1.6	0.5
259.25	-0.40	1.3	313.98	8FE	0.5	-4.0	-1.6	0.5
266.64	-0.56	1.2	323.92	18FE	0.2	-4.2	-1.6	0.5
274.04	-0.73	1.1	333.79	28FE	-0.1	-4.3	-1.6	0.5
281.46	-0.91	1.0	343.59	9MR	-0.6	-4.2	-1.6	0.4
288.87	-1.09	1.0	353.35	19MR	-1.4	-4.1	-1.6	0.4
296.27	-1.27	0.9	3.05	29MR	-1.5	-4.0	-1.6	0.4
303.63	-1.46	0.8	12.70	8AP	-0.5	-3.9	-1.6	0.3
310.94	-1.64	0.7	22.31	18AP	0.5	-3.8	-1.7	0.3
318.17	-1.82	0.6	31.88	28AP	1.4	-3.7	-1.7	0.3
325.30	-2.00	0.5	41.42	8MY	2.5	-3.6	-1.7	0.4
332.31	-2.17	0.4	50.95	18MY	3.5	-3.6	-1.8	0.4
339.14	-2.33	0.3	60.47	28MY	2.0	-3.5	-1.9	0.4
345.78	-2.47	0.2	69.98	7JN	0.8	-3.5	-1.9	0.4
352.16	-2.61	0.1	79.51	17JN	-0.4	-3.4	-2.0	0.4
358.24	-2.73	-0.0	89.06	27JN	-1.4	-3.4	-2.1	0.4
3.94	-2.83	-0.2	98.63	7JL	-1.5	-3.4	-2.1	0.3
9.17	-2.91	-0.3	108.24	17JL	-0.7	-3.4	-2.2	0.3
13.82	-2.97	-0.5	117.89	27JL	-0.2	-3.3	-2.3	0.3
17.77	-3.01	-0.6	127.59	6AU	0.1	-3.3	-2.3	0.2
20.82	-3.01	-0.8	137.35	16AU	0.3	-3.3	-2.4	0.2
22.78	-2.95	-1.0	147.16	26AU	0.6	-3.4	-2.4	0.1
23.44	-2.84	-1.3	157.03	5SE	1.5	-3.4	-2.4	0.0
22.65	-2.63	-1.5	166.97	15SE	2.6	-3.4	-2.4	-0.0
20.43	-2.31	-1.7	176.95	25SE	0.0	-3.4	-2.4	-0.1
17.13	-1.87	-1.8	187.00	5OC	-0.7	-3.4	-2.4	-0.2
13.42	-1.36	-1.8	197.09	15OC	-0.7	-3.5	-2.4	-0.2
10.15	-0.82	-1.6	207.23	25OC	-0.7	-3.5	-2.3	-0.2
7.91	-0.32	-1.2	217.40	4NO	-0.8	-3.5	-2.3	-0.1
7.02	0.10	-0.9	227.60	14NO	-0.7	-3.5	-2.2	-0.1
7.46	0.45	-0.6	237.80	24NO	-0.7	-3.4	-2.1	-0.0
9.04	0.72	-0.3	248.01	4DE	-0.7	-3.4	-2.0	0.1
11.58	0.92	-0.0	258.22	14DE	-0.1	-3.4	-2.0	0.1
14.87	1.08	0.2	268.40	24DE	2.3	-3.4	-1.9	0.2
				-503				
18.74	1.20	0.5	278.55	3JA	1.6	-3.4	-1.8	0.3
23.07	1.29	0.7	288.67	13JA	0.6	-3.3	-1.8	0.3
27.76	1.35	0.9	298.74	23JA	0.3	-3.3	-1.7	0.3
32.72	1.40	1.0	308.75	2FE	0.1	-3.3	-1.7	0.4
37.90	1.44	1.2	318.72	12FE	-0.2	-3.3	-1.6	0.4
43.26	1.46	1.3	328.61	22FE	-0.7	-3.4	-1.6	0.4
48.76	1.47	1.4	338.46	4MR	-1.4	-3.4	-1.6	0.4
54.37	1.47	1.5	348.24	14MR	-1.4	-3.4	-1.6	0.3
60.08	1.47	1.6	357.96	24MR	-0.5	-3.4	-1.5	0.3
65.86	1.46	1.7	7.64	3AP	0.6	-3.5	-1.5	0.3
71.72	1.45	1.8	17.27	13AP	1.8	-3.5	-1.5	0.2
77.63	1.43	1.8	26.86	23AP	3.3	-3.5	-1.5	0.2
83.60	1.41	1.9	36.42	3MY	2.7	-3.6	-1.6	0.2
89.63	1.38	1.9	45.96	13MY	1.5	-3.7	-1.6	0.2
95.71	1.35	2.0	55.47	23MY	0.6	-3.7	-1.6	0.3
101.83	1.32	2.0	64.99	2JN	-0.4	-3.8	-1.7	0.3
108.01	1.28	2.0	74.51	12JN	-1.4	-3.9	-1.7	0.3
114.25	1.23	2.0	84.05	22JN	-1.5	-4.0	-1.7	0.3
120.53	1.19	2.0	93.61	2JL	-0.7	-4.1	-1.8	0.3
126.88	1.14	2.0	103.20	12JL	-0.1	-4.2	-1.9	0.3
133.29	1.08	2.0	112.82	22JL	0.2	-4.2	-1.9	0.3
139.75	1.03	2.0	122.50	1AU	0.5	-4.1	-2.0	0.3
146.29	0.97	1.9	132.23	11AU	0.9	-3.8	-2.1	0.2
152.89	0.90	1.9	142.00	21AU	2.0	-3.3	-2.1	0.2
159.57	0.83	1.8	151.85	31AU	2.3	-3.5	-2.2	0.1
166.31	0.76	1.8	161.75	10SE	-0.1	-4.0	-2.2	0.1
173.14	0.68	1.7	171.71	20SE	-0.8	-4.3	-2.3	-0.0
180.04	0.59	1.7	181.73	30SE	-0.9	-4.3	-2.3	-0.1
187.01	0.51	1.7	191.79	10OC	-0.8	-4.3	-2.4	-0.2
194.07	0.41	1.7	201.91	20OC	-0.7	-4.2	-2.4	-0.2
201.20	0.32	1.7	212.06	30OC	-0.5	-4.1	-2.4	-0.3
208.41	0.22	1.7	222.25	9NO	-0.5	-4.0	-2.3	-0.3
215.69	0.11	1.7	232.45	19NO	-0.5	-3.9	-2.3	-0.2
223.05	0.00	1.6	242.66	29NO	0.1	-3.8	-2.2	-0.1
230.47	-0.11	1.6	252.87	9DE	2.5	-3.8	-2.2	-0.1
237.96	-0.23	1.6	263.07	19DE	1.2	-3.7	-2.1	0.0
245.50	-0.35	1.5	273.24	29DE	0.3	-3.6	-2.0	0.1

♂ LONG LAT MAG	☉ LONG	16.00UT -502	☿ ♀ ♃ ♄ MAGNITUDES
253.10-0.47 1.5	283.38	8JA	0.1 -3.5 -2.0 0.1
260.74-0.59 1.5	293.47	18JA	-0.0 -3.5 -1.9 0.2
268.41-0.71 1.4	303.52	28JA	-0.3 -3.4 -1.8 0.2
276.11-0.83 1.4	313.51	7FE	-0.7 -3.4 -1.8 0.3
283.83-0.95 1.4	323.44	17FE	-1.4 -3.4 -1.7 0.3
291.55-1.06 1.3	333.31	27FE	-1.4 -3.3 -1.7 0.3
299.26-1.16 1.3	343.12	9MR	-0.5 -3.3 -1.6 0.3
306.96-1.25 1.3	352.88	19MR	0.8 -3.3 -1.6 0.3
314.61-1.34 1.2	2.58	29MR	2.3 -3.3 -1.5 0.3
322.22-1.41 1.2	12.23	8AP	3.6 -3.3 -1.5 0.3
329.77-1.47 1.2	21.84	18AP	2.0 -3.3 -1.5 0.2
337.24-1.52 1.1	31.41	28AP	1.2 -3.3 -1.5 0.2
344.61-1.55 1.1	40.96	8MY	0.4 -3.4 -1.5 0.2
351.88-1.57 1.0	50.48	18MY	-0.5 -3.4 -1.5 0.2
359.03-1.57 1.0	60.00	28MY	-1.4 -3.4 -1.5 0.2
6.04-1.56 1.0	69.52	7JN	-1.5 -3.5 -1.5 0.3
12.90-1.53 0.9	79.04	17JN	-0.7 -3.5 -1.5 0.3
19.58-1.48 0.9	88.59	27JN	-0.0 -3.5 -1.5 0.3
26.05-1.42 0.8	98.16	7JL	0.4 -3.4 -1.6 0.3
32.30-1.34 0.7	107.77	17JL	0.7 -3.4 -1.6 0.3
38.29-1.24 0.6	117.42	27JL	1.2 -3.4 -1.6 0.3
43.96-1.12 0.6	127.12	6AU	2.5 -3.3 -1.7 0.3
49.28-0.98 0.5	136.87	16AU	2.0 -3.3 -1.8 0.3
54.16-0.82 0.3	146.68	26AU	-0.1 -3.3 -1.8 0.2
58.52-0.62 0.2	156.55	5SE	-0.9 -3.3 -1.9 0.2
62.23-0.40 0.0	166.48	15SE	-1.0 -3.3 -1.9 0.1
65.15-0.13 -0.2	176.47	25SE	-0.9 -3.3 -2.0 0.1
67.09 0.18 -0.4	186.51	5OC	-0.6 -3.4 -2.1 -0.0
67.85 0.55 -0.6	196.60	15OC	-0.4 -3.4 -2.1 -0.1
67.26 0.96 -0.8	206.74	25OC	-0.4 -3.4 -2.2 -0.2
65.27 1.41 -1.0	216.90	4NO	-0.3 -3.5 -2.2 -0.2
62.06 1.85 -1.2	227.10	14NO	0.3 -3.5 -2.2 -0.3
58.15 2.23 -1.3	237.31	24NO	2.8 -3.6 -2.3 -0.3
54.34 2.51 -1.1	247.52	4DE	0.9 -3.6 -2.2 -0.2
51.31 2.67 -0.9	257.72	14DE	0.1 -3.7 -2.2 -0.2
49.51 2.73 -0.6	267.91	24DE	-0.0 -3.8 -2.2 -0.1
		-501	
49.05 2.72 -0.3	278.06	3JA	-0.1 -3.9 -2.1 -0.0
49.81 2.66 -0.0	288.18	13JA	-0.3 -3.9 -2.1 0.1
51.62 2.57 0.2	298.26	23JA	-0.8 -4.0 -2.0 0.1
54.28 2.48 0.5	308.27	2FE	-1.4 -4.1 -1.9 0.2
57.60 2.38 0.7	318.24	12FE	-1.3 -4.2 -1.9 0.2
61.46 2.27 0.9	328.14	22FE	-0.5 -4.3 -1.8 0.3
65.74 2.17 1.0	337.98	4MR	1.0 -4.2 -1.7 0.3
70.36 2.08 1.2	347.77	14MR	2.9 -4.1 -1.7 0.3
75.26 1.98 1.3	357.50	24MR	2.7 -3.7 -1.6 0.3
80.37 1.89 1.4	7.17	3AP	1.5 -3.0 -1.5 0.3
85.68 1.80 1.5	16.80	13AP	0.9 -3.5 -1.5 0.3
91.15 1.71 1.6	26.39	23AP	0.3 -4.0 -1.5 0.3
96.76 1.62 1.7	35.95	3MY	-0.5 -4.2 -1.4 0.2
102.49 1.53 1.7	45.49	13MY	-1.4 -4.2 -1.4 0.2
108.34 1.44 1.8	55.01	23MY	-1.5 -4.1 -1.4 0.2
114.30 1.35 1.8	64.52	2JN	-0.6 -4.0 -1.4 0.2
120.35 1.26 1.8	74.05	12JN	0.1 -3.9 -1.4 0.2
126.51 1.17 1.8	83.58	22JN	0.5 -3.9 -1.4 0.3
132.75 1.08 1.9	93.14	2JL	0.9 -3.8 -1.4 0.3
139.09 0.99 1.9	102.73	12JL	1.6 -3.7 -1.4 0.3
145.53 0.89 1.9	112.35	22JL	3.0 -3.6 -1.4 0.3
152.05 0.80 1.8	122.02	1AU	1.7 -3.6 -1.4 0.3
158.68 0.70 1.8	131.75	11AU	-0.2 -3.5 -1.5 0.3
165.40 0.60 1.8	141.53	21AU	-1.1 -3.5 -1.5 0.3
172.21 0.50 1.8	151.37	31AU	-1.2 -3.5 -1.5 0.3
179.12 0.39 1.7	161.27	10SE	-0.9 -3.4 -1.6 0.3
186.13 0.29 1.7	171.22	20SE	-0.5 -3.4 -1.7 0.2
193.23 0.18 1.7	181.24	30SE	-0.3 -3.4 -1.7 0.2
200.43 0.08 1.6	191.31	10OC	-0.2 -3.4 -1.8 0.1
207.72-0.03 1.6	201.42	20OC	-0.2 -3.4 -1.8 0.0
215.09-0.14 1.5	211.57	30OC	0.5 -3.4 -1.9 -0.0
222.55-0.25 1.5	221.76	9NO	3.2 -3.4 -2.0 -0.1
230.09-0.36 1.4	231.96	19NO	0.7 -3.4 -2.0 -0.2
237.70-0.46 1.4	242.17	29NO	-0.1 -3.4 -2.1 -0.3
245.38-0.56 1.4	252.38	9DE	-0.2 -3.4 -2.1 -0.2
253.11-0.66 1.4	262.57	19DE	-0.2 -3.4 -2.1 -0.2
260.88-0.75 1.4	272.74	29DE	-0.4 -3.4 -2.1 -0.1

♂ LONG LAT MAG	☉ LONG	16.00UT -500	☿ ♀ ♃ ♄ MAGNITUDES
268.70-0.84 1.4	282.88	8JA	-0.8 -3.5 -2.1 -0.0
276.53-0.92 1.4	292.98	18JA	-1.2 -3.5 -2.1 0.0
284.38-0.99 1.4	303.03	28JA	-1.2 -3.5 -2.1 0.1
292.22-1.05 1.4	313.02	7FE	-0.4 -3.4 -2.0 0.2
300.06-1.11 1.4	322.95	17FE	1.2 -3.4 -1.9 0.2
307.86-1.15 1.4	332.83	27FE	3.3 -3.4 -1.9 0.3
315.64-1.17 1.4	342.65	8MR	2.0 -3.4 -1.8 0.3
323.36-1.19 1.4	352.40	18MR	1.1 -3.4 -1.7 0.3
331.02-1.19 1.4	2.10	28MR	0.6 -3.3 -1.7 0.3
338.61-1.19 1.4	11.76	7AP	0.2 -3.3 -1.6 0.3
346.12-1.16 1.4	21.37	17AP	-0.5 -3.3 -1.6 0.3
353.55-1.13 1.4	30.94	27AP	-1.4 -3.3 -1.5 0.3
0.88-1.08 1.4	40.49	7MY	-1.6 -3.3 -1.4 0.3
8.11-1.03 1.4	50.02	17MY	-0.6 -3.4 -1.4 0.3
15.23-0.96 1.4	59.54	27MY	0.1 -3.4 -1.4 0.3
22.23-0.88 1.4	69.05	6JN	0.7 -3.4 -1.3 0.3
29.11-0.79 1.4	78.58	16JN	1.2 -3.5 -1.3 0.3
35.86-0.69 1.4	88.12	26JN	2.1 -3.5 -1.3 0.3
42.47-0.57 1.3	97.70	6JL	3.2 -3.6 -1.3 0.4
48.94-0.45 1.3	107.30	16JL	1.4 -3.6 -1.3 0.4
55.24-0.32 1.3	116.95	26JL	-0.3 -3.7 -1.3 0.4
61.37-0.18 1.2	126.65	5AU	-1.2 -3.8 -1.3 0.4
67.29-0.02 1.2	136.39	15AU	-1.3 -3.9 -1.3 0.4
73.00 0.15 1.1	146.20	25AU	-0.8 -4.0 -1.3 0.4
78.44 0.34 1.0	156.07	4SE	-0.4 -4.1 -1.3 0.4
83.58 0.54 0.9	165.99	14SE	-0.2 -4.2 -1.4 0.4
88.36 0.77 0.8	175.98	24SE	-0.1 -4.3 -1.4 0.4
92.69 1.02 0.7	186.02	4OC	0.0 -4.4 -1.5 0.3
96.49 1.31 0.5	196.11	14OC	0.7 -4.3 -1.5 0.3
99.63 1.63 0.3	206.24	24OC	3.4 -4.0 -1.6 0.2
101.95 1.99 0.1	216.41	3NO	0.5 -3.2 -1.7 0.1
103.28 2.39 -0.1	226.60	13NO	-0.2 -3.5 -1.7 0.0
103.41 2.82 -0.3	236.81	23NO	-0.3 -4.1 -1.8 -0.0
102.22 3.27 -0.6	247.02	3DE	-0.4 -4.3 -1.9 -0.1
99.71 3.68 -0.8	257.23	13DE	-0.5 -4.4 -1.9 -0.2
96.18 4.00 -1.0	267.41	23DE	-0.9 -4.3 -2.0 -0.1
		-499	
92.20 4.18 -1.0	277.57	2JA	-1.1 -4.2 -2.0 -0.1
88.54 4.19 -0.8	287.69	12JA	-1.1 -4.1 -2.0 -0.0
85.79 4.06 -0.6	297.77	22JA	-0.4 -4.0 -2.0 0.0
84.30 3.83 -0.3	307.79	1FE	1.5 -3.9 -2.0 0.1
84.11 3.56 -0.1	317.75	11FE	3.0 -3.8 -2.0 0.2
85.08 3.29 0.2	327.66	21FE	1.5 -3.7 -2.0 0.2
87.05 3.02 0.4	337.51	3MR	0.8 -3.6 -2.0 0.3
89.83 2.76 0.6	347.30	13MR	0.5 -3.5 -1.9 0.3
93.25 2.53 0.7	357.03	23MR	0.1 -3.5 -1.9 0.4
97.20 2.31 0.9	6.71	2AP	-0.6 -3.4 -1.8 0.4
101.58 2.10 1.0	16.34	12AP	-1.4 -3.4 -1.7 0.4
106.31 1.91 1.2	25.94	22AP	-1.6 -3.4 -1.7 0.4
111.34 1.74 1.3	35.50	2MY	-0.6 -3.3 -1.6 0.4
116.61 1.57 1.3	45.03	12MY	0.2 -3.3 -1.5 0.4
122.10 1.41 1.4	54.55	22MY	0.9 -3.3 -1.5 0.4
127.79 1.25 1.5	64.07	1JN	1.6 -3.3 -1.4 0.4
133.64 1.11 1.5	73.59	11JN	2.7 -3.3 -1.4 0.4
139.67 0.96 1.6	83.12	21JN	2.9 -3.3 -1.3 0.4
145.84 0.82 1.6	92.68	1JL	1.2 -3.4 -1.3 0.4
152.14 0.69 1.6	102.26	11JL	-0.3 -3.4 -1.3 0.5
158.59 0.55 1.6	111.89	21JL	-1.2 -3.4 -1.2 0.5
165.17 0.42 1.6	121.56	31JL	-1.4 -3.4 -1.2 0.5
171.87 0.30 1.6	131.28	10AU	-0.8 -3.5 -1.2 0.6
178.70 0.17 1.6	141.06	20AU	-0.3 -3.5 -1.2 0.6
185.65 0.05 1.6	150.89	30AU	-0.1 -3.5 -1.2 0.6
192.72-0.07 1.6	160.78	9SE	0.1 -3.4 -1.2 0.6
199.90-0.19 1.6	170.74	19SE	0.2 -3.4 -1.3 0.6
207.19-0.30 1.6	180.75	29SE	1.0 -3.4 -1.3 0.5
214.58-0.41 1.5	190.81	9OC	3.2 -3.4 -1.3 0.5
222.07-0.51 1.5	200.93	19OC	0.3 -3.3 -1.4 0.5
229.65-0.61 1.5	211.07	29OC	-0.4 -3.3 -1.4 0.4
237.31-0.70 1.5	221.26	8NO	-0.5 -3.3 -1.5 0.3
245.04-0.78 1.4	231.46	18NO	-0.5 -3.3 -1.5 0.3
252.83-0.86 1.4	241.67	28NO	-0.6 -3.3 -1.6 0.2
260.66-0.92 1.4	251.88	8DE	-0.9 -3.4 -1.6 0.1
268.53-0.98 1.4	262.08	18DE	-0.9 -3.4 -1.7 0.1
276.42-1.02 1.3	272.25	28DE	-0.9 -3.4 -1.8 -0.0

♂			☉	16.00UT	☿ ♀ ♃	♄
LONG	LAT	MAG	LONG	-498	MAGNITUDES	

♂ LONG	LAT	MAG	☉ LONG	16.00UT -498	☿	♀	♃	♄
284.32	-1.06	1.3	282.39	7JA	-0.3	-3.4	-1.8	0.0
292.21	-1.08	1.3	292.49	17JA	1.7	-3.5	-1.9	0.1
300.09	-1.09	1.3	302.54	27JA	2.4	-3.5	-1.9	0.1
307.93	-1.09	1.3	312.54	6FE	1.1	-3.6	-2.0	0.2
315.73	-1.08	1.4	322.48	16FE	0.6	-3.6	-2.0	0.3
323.48	-1.05	1.4	332.35	26FE	0.3	-3.7	-2.0	0.3
331.16	-1.02	1.4	342.17	8MR	-0.0	-3.7	-2.0	0.4
338.78	-0.97	1.5	351.94	18MR	-0.6	-3.8	-2.0	0.4
346.31	-0.92	1.5	1.64	28MR	-1.4	-3.9	-2.0	0.5
353.76	-0.85	1.5	11.30	7AP	-1.5	-4.0	-1.9	0.5
1.13	-0.78	1.6	20.91	17AP	-0.6	-4.1	-1.9	0.5
8.40	-0.70	1.6	30.48	27AP	0.4	-4.2	-1.8	0.6
15.58	-0.61	1.6	40.03	7MY	1.2	-4.2	-1.7	0.6
22.66	-0.52	1.6	49.56	17MY	2.1	-4.1	-1.7	0.6
29.65	-0.42	1.6	59.08	27MY	3.4	-3.8	-1.6	0.6
36.53	-0.32	1.7	68.60	6JN	2.3	-3.2	-1.6	0.6
43.33	-0.21	1.7	78.12	16JN	0.9	-3.1	-1.5	0.6
50.01	-0.10	1.7	87.66	26JN	-0.3	-3.8	-1.4	0.6
56.61	0.02	1.7	97.23	6JL	-1.3	-4.1	-1.4	0.6
63.09	0.14	1.7	106.83	16JL	-1.5	-4.2	-1.4	0.6
69.47	0.27	1.7	116.48	26JL	-0.7	-4.2	-1.3	0.6
75.74	0.40	1.6	126.17	5AU	-0.2	-4.1	-1.3	0.7
81.88	0.54	1.6	135.92	15AU	0.0	-4.1	-1.3	0.7
87.90	0.68	1.6	145.72	25AU	0.2	-4.0	-1.2	0.8
93.78	0.84	1.5	155.59	4SE	0.4	-3.9	-1.2	0.8
99.49	1.00	1.5	165.51	14SE	1.3	-3.8	-1.2	0.8
105.01	1.17	1.4	175.49	24SE	2.9	-3.7	-1.2	0.8
110.33	1.35	1.3	185.53	4OC	0.2	-3.7	-1.2	0.8
115.38	1.55	1.2	195.62	14OC	-0.6	-3.6	-1.3	0.7
120.13	1.77	1.0	205.75	24OC	-0.6	-3.6	-1.3	0.7
124.51	2.01	0.9	215.92	3NO	-0.6	-3.5	-1.3	0.7
128.42	2.27	0.7	226.11	13NO	-0.7	-3.5	-1.3	0.6
131.78	2.56	0.5	236.32	23NO	-0.8	-3.4	-1.4	0.5
134.42	2.88	0.3	246.53	3DE	-0.8	-3.4	-1.4	0.5
136.20	3.23	0.1	256.74	13DE	-0.8	-3.4	-1.5	0.4
136.93	3.60	-0.1	266.92	23DE	-0.2	-3.4	-1.6	0.3

-497

♂ LONG	LAT	MAG	☉ LONG	16.00UT -497	☿	♀	♃	♄
136.43	3.98	-0.4	277.08	2JA	2.0	-3.3	-1.6	0.3
134.63	4.31	-0.7	287.21	12JA	1.9	-3.3	-1.7	0.2
131.65	4.55	-0.9	297.28	22JA	0.7	-3.3	-1.8	0.2
127.89	4.64	-1.1	307.31	1FE	0.4	-3.3	-1.8	0.3
124.03	4.54	-1.0	317.27	11FE	0.2	-3.3	-1.9	0.3
120.76	4.29	-0.8	327.18	21FE	-0.1	-3.4	-1.9	0.4
118.59	3.92	-0.6	337.03	3MR	-0.6	-3.4	-2.0	0.4
117.72	3.51	-0.3	346.82	13MR	-1.4	-3.4	-2.0	0.5
118.11	3.10	-0.1	356.56	23MR	-1.5	-3.4	-2.0	0.6
119.62	2.71	0.1	6.24	2AP	-0.6	-3.5	-2.1	0.6
122.06	2.34	0.3	15.87	12AP	0.5	-3.5	-2.0	0.7
125.26	2.01	0.4	25.47	22AP	1.5	-3.4	-2.0	0.7
129.10	1.71	0.6	35.03	2MY	2.8	-3.4	-2.0	0.7
133.45	1.44	0.7	44.57	12MY	3.2	-3.4	-1.9	0.8
138.23	1.19	0.8	54.09	22MY	1.8	-3.4	-1.9	0.8
143.37	0.95	0.9	63.61	1JN	0.7	-3.3	-1.8	0.8
148.82	0.74	1.0	73.12	11JN	-0.4	-3.3	-1.8	0.8
154.54	0.54	1.1	82.65	21JN	-1.4	-3.3	-1.7	0.8
160.51	0.35	1.1	92.21	1JL	-1.5	-3.3	-1.6	0.8
166.70	0.18	1.2	101.79	11JL	-0.7	-3.3	-1.6	0.8
173.08	0.01	1.2	111.41	21JL	-0.2	-3.4	-1.5	0.8
179.65	-0.14	1.3	121.08	31JL	0.2	-3.4	-1.5	0.8
186.39	-0.29	1.3	130.80	10AU	0.4	-3.4	-1.4	0.9
193.29	-0.42	1.3	140.57	20AU	0.7	-3.4	-1.4	0.9
200.33	-0.55	1.3	150.41	30AU	1.7	-3.5	-1.3	1.0
207.52	-0.66	1.3	160.30	9SE	2.6	-3.5	-1.3	1.0
214.83	-0.76	1.3	170.25	19SE	0.0	-3.6	-1.3	1.0
222.26	-0.86	1.4	180.26	29SE	-0.7	-3.7	-1.3	1.0
229.79	-0.94	1.4	190.32	9OC	-0.8	-3.8	-1.3	1.0
237.42	-1.01	1.4	200.43	19OC	-0.8	-3.9	-1.3	1.0
245.13	-1.07	1.4	210.58	29OC	-0.8	-4.0	-1.3	1.0
252.90	-1.11	1.4	220.76	8NO	-0.6	-4.1	-1.3	1.0
260.72	-1.14	1.4	230.96	18NO	-0.6	-4.2	-1.3	0.9
268.58	-1.16	1.4	241.18	28NO	-0.6	-4.3	-1.3	0.9
276.45	-1.17	1.4	251.38	8DE	-0.0	-4.4	-1.4	0.8
284.34	-1.16	1.4	261.58	18DE	2.3	-4.4	-1.4	0.7
292.21	-1.14	1.4	271.76	28DE	1.5	-4.3	-1.5	0.7

♂			☉	16.00UT	☿ ♀ ♃	♄
LONG	LAT	MAG	LONG	-496	MAGNITUDES	

♂ LONG	LAT	MAG	☉ LONG	16.00UT -496	☿	♀	♃	♄
300.06	-1.11	1.4	281.90	7JA	0.5	-3.9	-1.5	0.6
307.88	-1.07	1.4	292.00	17JA	0.2	-3.3	-1.6	0.5
315.65	-1.02	1.4	302.06	27JA	0.1	-3.5	-1.7	0.5
323.36	-0.95	1.4	312.05	6FE	-0.2	-4.0	-1.7	0.5
331.01	-0.89	1.5	321.99	16FE	-0.7	-4.3	-1.8	0.5
338.58	-0.81	1.5	331.88	26FE	-1.4	-4.3	-1.9	0.6
346.07	-0.73	1.5	341.70	7MR	-1.4	-4.2	-1.9	0.6
353.49	-0.64	1.5	351.46	17MR	-0.5	-4.1	-2.0	0.7
0.82	-0.55	1.5	1.17	27MR	0.7	-4.0	-2.0	0.8
8.06	-0.45	1.6	10.83	6AP	2.0	-3.9	-2.1	0.8
15.21	-0.36	1.6	20.44	16AP	3.7	-3.8	-2.1	0.9
22.28	-0.26	1.7	30.02	26AP	2.5	-3.7	-2.1	0.9
29.26	-0.16	1.7	39.57	6MY	1.4	-3.6	-2.1	0.9
36.16	-0.05	1.7	49.10	16MY	0.5	-3.6	-2.1	1.0
42.98	0.05	1.8	58.62	26MY	-0.4	-3.5	-2.1	1.0
49.73	0.15	1.8	68.13	5JN	-1.4	-3.5	-2.1	1.0
56.39	0.26	1.8	77.66	15JN	-1.5	-3.4	-2.0	1.1
62.99	0.36	1.9	87.20	25JN	-0.7	-3.4	-2.0	1.1
69.51	0.46	1.9	96.77	5JL	-0.1	-3.4	-1.9	1.1
75.97	0.57	1.9	106.37	15JL	0.3	-3.4	-1.8	1.1
82.35	0.67	1.9	116.01	25JL	0.5	-3.3	-1.8	1.1
88.67	0.78	1.9	125.70	4AU	1.0	-3.3	-1.7	1.1
94.92	0.89	1.9	135.45	14AU	2.1	-3.3	-1.7	1.1
101.09	0.99	1.8	145.25	24AU	2.2	-3.4	-1.6	1.2
107.18	1.10	1.8	155.11	3SE	-0.0	-3.4	-1.5	1.2
113.20	1.22	1.8	165.03	13SE	-0.9	-3.4	-1.5	1.3
119.12	1.33	1.7	175.01	23SE	-0.9	-3.4	-1.5	1.3
124.93	1.45	1.7	185.04	3OC	-0.9	-3.4	-1.4	1.3
130.63	1.58	1.6	195.13	13OC	-0.7	-3.5	-1.4	1.3
136.19	1.71	1.5	205.26	23OC	-0.5	-3.5	-1.4	1.3
141.58	1.84	1.4	215.42	2NO	-0.5	-3.5	-1.4	1.3
146.77	1.99	1.3	225.61	12NO	-0.4	-3.5	-1.4	1.3
151.72	2.15	1.1	235.82	22NO	0.1	-3.4	-1.4	1.2
156.38	2.31	1.0	246.03	2DE	2.6	-3.4	-1.4	1.2
160.68	2.49	0.8	256.24	12DE	1.1	-3.4	-1.4	1.1
164.54	2.68	0.6	266.43	22DE	0.3	-3.4	-1.4	1.1

-495

♂ LONG	LAT	MAG	☉ LONG	16.00UT -495	☿	♀	♃	♄
167.84	2.88	0.4	276.59	1JA	0.1	-3.4	-1.4	1.0
170.44	3.10	0.1	286.71	11JA	-0.1	-3.3	-1.5	0.9
172.18	3.31	-0.1	296.79	21JA	-0.3	-3.3	-1.5	0.9
172.83	3.51	-0.4	306.82	31JA	-0.7	-3.3	-1.6	0.8
172.38	3.68	-0.7	316.79	10FE	-1.4	-3.3	-1.6	0.8
170.60	3.77	-1.0	326.70	20FE	-1.4	-3.4	-1.7	0.8
167.68	3.75	-1.2	336.55	2MR	-0.5	-3.4	-1.8	0.8
164.04	3.56	-1.4	346.35	12MR	0.8	-3.4	-1.8	0.9
160.37	3.23	-1.3	356.09	22MR	2.5	-3.4	-1.9	0.9
157.36	2.78	-1.1	5.77	1AP	3.2	-3.5	-2.0	1.0
155.52	2.29	-0.9	15.41	11AP	1.8	-3.5	-2.0	1.0
154.97	1.79	-0.7	25.01	21AP	1.0	-3.5	-2.1	1.1
155.74	1.33	-0.5	34.57	1MY	0.4	-3.6	-2.2	1.1
157.64	0.92	-0.3	44.11	11MY	-0.4	-3.7	-2.2	1.2
160.52	0.55	-0.1	53.63	21MY	-1.4	-3.7	-2.2	1.2
164.18	0.23	0.0	63.15	31MY	-1.6	-3.8	-2.3	1.3
168.50	-0.06	0.2	72.67	10JN	-0.7	-3.9	-2.3	1.3
173.37	-0.31	0.3	82.20	20JN	-0.0	-4.0	-2.2	1.3
178.69	-0.53	0.4	91.75	30JN	0.4	-4.1	-2.2	1.4
184.40	-0.72	0.5	101.33	10JL	0.8	-4.2	-2.2	1.4
190.44	-0.89	0.6	110.95	20JL	1.3	-4.2	-2.1	1.4
196.77	-1.04	0.7	120.62	30JL	2.6	-4.1	-2.1	1.4
203.36	-1.16	0.7	130.33	9AU	1.9	-3.8	-2.0	1.4
210.16	-1.27	0.8	140.10	19AU	-0.1	-3.3	-1.9	1.3
217.15	-1.35	0.8	149.93	29AU	-1.0	-3.6	-1.9	1.3
224.32	-1.42	0.9	159.82	8SE	-1.1	-4.1	-1.8	1.3
231.63	-1.46	0.9	169.76	18SE	-0.9	-4.3	-1.8	1.3
239.07	-1.49	1.0	179.77	28SE	-0.6	-4.3	-1.7	1.3
246.61	-1.50	1.0	189.83	8OC	-0.4	-4.3	-1.6	1.3
254.23	-1.49	1.1	199.94	18OC	-0.3	-4.2	-1.6	1.3
261.91	-1.47	1.1	210.08	28OC	-0.3	-4.1	-1.6	1.2
269.65	-1.43	1.2	220.27	7NO	0.3	-4.0	-1.5	1.2
277.40	-1.38	1.2	230.46	17NO	2.9	-3.9	-1.5	1.2
285.17	-1.32	1.3	240.68	27NO	0.9	-3.8	-1.5	1.1
292.93	-1.25	1.3	250.89	7DE	0.1	-3.8	-1.5	1.1
300.66	-1.16	1.4	261.09	17DE	-0.1	-3.7	-1.5	1.1
308.36	-1.07	1.4	271.27	27DE	-0.2	-3.6	-1.5	1.0

♂			⊙	16.00UT	☿	♀	♃	♄
LONG	LAT	MAG	LONG	-494	MAGNITUDES			
316.01	-0.98	1.4	281.41	6JA	-0.4	-3.5	-1.5	1.0
323.61	-0.88	1.5	291.51	16JA	-0.8	-3.5	-1.5	0.9
331.14	-0.77	1.5	301.57	26JA	-1.3	-3.4	-1.5	0.9
338.60	-0.67	1.6	311.58	5FE	-1.3	-3.4	-1.6	0.8
345.99	-0.56	1.6	321.52	15FE	-0.5	-3.4	-1.6	0.8
353.30	-0.45	1.6	331.40	25FE	1.1	-3.3	-1.6	0.8
0.52	-0.34	1.7	341.23	7MR	3.1	-3.3	-1.7	0.9
7.67	-0.23	1.7	350.99	17MR	2.4	-3.3	-1.7	0.9
14.74	-0.13	1.7	0.71	27MR	1.4	-3.3	-1.8	1.0
21.73	-0.02	1.7	10.37	6AP	0.8	-3.3	-1.9	1.1
28.64	0.08	1.8	19.98	16AP	0.3	-3.3	-1.9	1.1
35.48	0.18	1.8	29.56	26AP	-0.5	-3.3	-2.0	1.2
42.25	0.28	1.8	39.11	6MY	-1.4	-3.4	-2.1	1.3
48.96	0.37	1.8	48.64	16MY	-1.6	-3.4	-2.2	1.3
55.61	0.46	1.8	58.16	26MY	-0.7	-3.4	-2.2	1.3
62.20	0.55	1.9	67.68	5JN	0.1	-3.5	-2.3	1.3
68.75	0.63	1.9	77.20	15JN	0.6	-3.5	-2.3	1.4
75.24	0.71	1.9	86.74	25JN	1.0	-3.5	-2.4	1.4
81.69	0.79	2.0	96.30	5JL	1.7	-3.4	-2.4	1.3
88.10	0.87	2.0	105.90	15JL	3.1	-3.4	-2.4	1.3
94.48	0.94	2.0	115.54	25JL	1.6	-3.4	-2.4	1.3
100.82	1.02	2.0	125.23	4AU	-0.2	-3.3	-2.3	1.3
107.13	1.09	2.0	134.97	14AU	-1.1	-3.3	-2.3	1.2
113.41	1.16	2.0	144.77	24AU	-1.2	-3.3	-2.2	1.2
119.65	1.22	2.0	154.62	3SE	-0.9	-3.3	-2.2	1.1
125.86	1.29	2.0	164.54	13SE	-0.5	-3.3	-2.1	1.1
132.04	1.35	1.9	174.52	23SE	-0.3	-3.3	-2.0	1.1
138.18	1.41	1.9	184.55	3OC	-0.2	-3.4	-2.0	1.1
144.28	1.47	1.8	194.63	13OC	-0.1	-3.4	-1.9	1.1
150.33	1.53	1.8	204.76	23OC	0.5	-3.4	-1.8	1.1
156.32	1.59	1.7	214.92	2NO	3.2	-3.5	-1.8	1.1
162.25	1.65	1.6	225.11	12NO	0.6	-3.5	-1.7	1.1
168.09	1.70	1.5	235.32	22NO	-0.1	-3.6	-1.7	1.1
173.83	1.75	1.4	245.53	2DE	-0.2	-3.6	-1.6	1.0
179.46	1.80	1.3	255.74	12DE	-0.3	-3.7	-1.6	1.0
184.94	1.84	1.1	265.93	22DE	-0.5	-3.8	-1.6	0.9
				-493				
190.25	1.88	1.0	276.09	1JA	-0.8	-3.9	-1.6	0.9
195.34	1.91	0.8	286.22	11JA	-1.2	-3.9	-1.5	0.9
200.16	1.93	0.6	296.30	21JA	-1.1	-4.0	-1.5	0.8
204.64	1.93	0.4	306.33	31JA	-0.4	-4.1	-1.5	0.8
208.72	1.91	0.2	316.31	10FE	1.3	-4.2	-1.6	0.7
212.26	1.87	-0.1	326.23	20FE	3.3	-4.3	-1.6	0.7
215.15	1.78	-0.4	336.08	2MR	1.8	-4.3	-1.6	0.6
217.21	1.65	-0.7	345.88	12MR	1.0	-4.1	-1.6	0.7
218.26	1.44	-1.0	355.62	22MR	0.6	-3.6	-1.7	0.7
218.15	1.15	-1.3	5.31	1AP	0.1	-3.0	-1.7	0.8
216.78	0.76	-1.6	14.94	11AP	-0.5	-3.6	-1.8	0.9
214.28	0.26	-1.9	24.55	21AP	-1.4	-4.0	-1.8	0.9
211.10	-0.30	-2.1	34.11	1MY	-1.6	-4.2	-1.9	1.0
207.90	-0.87	-2.0	43.65	11MY	-0.6	-4.2	-1.9	1.0
205.39	-1.39	-1.8	53.17	21MY	0.2	-4.1	-2.0	1.1
204.09	-1.83	-1.6	62.68	31MY	0.7	-4.0	-2.1	1.1
204.15	-2.16	-1.4	72.20	10JN	1.3	-3.9	-2.1	1.2
205.56	-2.40	-1.2	81.74	20JN	2.3	-3.8	-2.2	1.2
208.14	-2.57	-1.0	91.28	30JN	3.2	-3.8	-2.3	1.2
211.70	-2.67	-0.8	100.87	10JL	1.4	-3.7	-2.3	1.2
216.07	-2.73	-0.6	110.49	20JL	-0.2	-3.6	-2.4	1.2
221.09	-2.74	-0.4	120.14	30JL	-1.2	-3.6	-2.4	1.2
226.64	-2.72	-0.3	129.86	9AU	-1.4	-3.5	-2.4	1.1
232.63	-2.67	-0.1	139.63	19AU	-0.8	-3.5	-2.4	1.1
238.96	-2.59	0.0	149.45	29AU	-0.4	-3.5	-2.4	1.1
245.56	-2.49	0.2	159.34	8SE	-0.1	-3.4	-2.4	1.0
252.39	-2.37	0.3	169.29	18SE	-0.0	-3.4	-2.4	1.0
259.40	-2.24	0.4	179.29	28SE	0.1	-3.4	-2.3	1.0
266.53	-2.10	0.5	189.34	8OC	0.8	-3.4	-2.2	1.0
273.77	-1.94	0.6	199.45	18OC	3.5	-3.4	-2.2	1.0
281.08	-1.78	0.7	209.59	28OC	0.4	-3.4	-2.1	1.0
288.43	-1.62	0.8	219.77	7NO	-0.3	-3.4	-2.0	1.0
295.81	-1.45	0.9	229.97	17NO	-0.4	-3.4	-2.0	1.0
303.18	-1.29	1.0	240.18	27NO	-0.4	-3.4	-1.9	1.0
310.54	-1.12	1.1	250.39	7DE	-0.6	-3.4	-1.8	0.9
317.88	-0.96	1.2	260.59	17DE	-0.9	-3.4	-1.8	0.9
325.17	-0.80	1.3	270.76	27DE	-1.0	-3.4	-1.7	0.9

♂			⊙	16.00UT	☿	♀	♃	♄
LONG	LAT	MAG	LONG	-492	MAGNITUDES			
332.42	-0.65	1.4	280.91	6JA	-1.0	-3.5	-1.7	0.8
339.62	-0.51	1.4	291.02	16JA	-0.4	-3.5	-1.6	0.8
346.75	-0.37	1.5	301.08	26JA	1.5	-3.5	-1.6	0.7
353.83	-0.23	1.6	311.08	5FE	2.8	-3.4	-1.6	0.7
0.84	-0.11	1.6	321.03	15FE	1.3	-3.4	-1.6	0.6
7.79	0.01	1.7	330.92	25FE	0.7	-3.4	-1.6	0.6
14.68	0.12	1.8	340.75	6MR	0.4	-3.4	-1.6	0.5
21.50	0.23	1.8	350.52	16MR	0.0	-3.4	-1.6	0.5
28.26	0.33	1.8	0.23	26MR	-0.5	-3.3	-1.6	0.6
34.97	0.42	1.9	9.90	5AP	-1.4	-3.3	-1.6	0.6
41.62	0.51	1.9	19.51	15AP	-1.6	-3.3	-1.6	0.7
48.22	0.59	1.9	29.09	25AP	-0.6	-3.3	-1.7	0.8
54.78	0.67	1.9	38.65	5MY	0.3	-3.3	-1.7	0.8
61.30	0.74	1.9	48.18	15MY	1.0	-3.4	-1.8	0.9
67.78	0.80	1.9	57.70	25MY	1.8	-3.4	-1.8	0.9
74.24	0.87	1.9	67.22	4JN	3.0	-3.4	-1.9	1.0
80.66	0.92	1.9	76.74	14JN	2.7	-3.5	-1.9	1.0
87.07	0.97	1.9	86.27	24JN	1.1	-3.5	-2.0	1.0
93.46	1.02	1.9	95.84	4JL	-0.3	-3.6	-2.1	1.1
99.83	1.07	2.0	105.44	14JL	-1.3	-3.6	-2.1	1.1
106.20	1.10	2.0	115.07	24JL	-1.5	-3.7	-2.2	1.1
112.57	1.14	2.0	124.76	3AU	-0.8	-3.8	-2.3	1.1
118.93	1.17	2.0	134.50	13AU	-0.3	-3.9	-2.3	1.0
125.30	1.20	2.0	144.29	23AU	-0.0	-4.0	-2.4	1.0
131.67	1.22	2.0	154.15	2SE	0.1	-4.1	-2.4	1.0
138.05	1.24	2.0	164.06	12SE	0.3	-4.2	-2.4	0.9
144.43	1.26	2.0	174.03	22SE	1.0	-4.3	-2.4	0.9
150.83	1.27	2.0	184.07	2OC	3.2	-4.4	-2.4	0.9
157.22	1.28	1.9	194.14	12OC	0.3	-4.3	-2.4	0.9
163.63	1.28	1.9	204.27	22OC	-0.5	-3.9	-2.4	0.9
170.05	1.27	1.8	214.43	1NO	-0.5	-3.2	-2.3	0.9
176.46	1.26	1.8	224.62	11NO	-0.5	-3.5	-2.2	0.9
182.88	1.24	1.7	234.82	21NO	-0.7	-4.1	-2.2	0.9
189.29	1.21	1.6	245.04	1DE	-0.8	-4.3	-2.1	0.9
195.70	1.17	1.5	255.24	11DE	-0.9	-4.4	-2.0	0.9
202.09	1.13	1.4	265.43	21DE	-0.9	-4.3	-2.0	0.9
208.47	1.06	1.3	275.60	31DE	-0.3	-4.2	-1.9	0.8
				-491				
214.81	0.99	1.2	285.73	10JA	1.8	-4.1	-1.8	0.8
221.13	0.89	1.1	295.81	20JA	2.2	-4.0	-1.8	0.7
227.39	0.78	0.9	305.85	30JA	0.9	-3.9	-1.7	0.7
233.60	0.64	0.8	315.82	9FE	0.5	-3.8	-1.7	0.6
239.72	0.47	0.6	325.74	19FE	0.3	-3.7	-1.6	0.6
245.75	0.28	0.5	335.60	1MR	-0.1	-3.6	-1.6	0.5
251.64	0.05	0.3	345.40	11MR	-0.6	-3.5	-1.6	0.5
257.36	-0.23	0.1	355.15	21MR	-1.4	-3.5	-1.5	0.4
262.86	-0.55	-0.1	4.84	31MR	-1.5	-3.4	-1.5	0.4
268.07	-0.93	-0.4	14.48	10AP	-0.6	-3.4	-1.5	0.5
272.91	-1.38	-0.6	24.08	20AP	0.4	-3.4	-1.5	0.5
277.26	-1.89	-0.9	33.65	30AP	1.3	-3.3	-1.5	0.6
280.99	-2.49	-1.1	43.19	10MY	2.3	-3.3	-1.6	0.7
283.93	-3.16	-1.4	52.71	20MY	3.6	-3.3	-1.6	0.7
285.86	-3.91	-1.7	62.23	30MY	2.1	-3.3	-1.6	0.8
286.62	-4.70	-2.0	71.75	9JN	0.9	-3.3	-1.6	0.8
286.10	-5.47	-2.3	81.28	19JN	-0.3	-3.3	-1.7	0.9
284.39	-6.12	-2.5	90.83	29JN	-1.3	-3.3	-1.7	0.9
281.93	-6.54	-2.6	100.40	9JL	-1.5	-3.4	-1.8	0.9
279.36	-6.65	-2.6	110.02	19JL	-0.8	-3.4	-1.9	1.0
277.42	-6.46	-2.3	119.68	29JL	-0.2	-3.4	-1.9	1.0
276.62	-6.04	-2.1	129.39	8AU	0.1	-3.5	-2.0	1.0
277.13	-5.49	-1.8	139.15	18AU	0.3	-3.5	-2.1	1.0
278.90	-4.89	-1.5	148.98	28AU	0.5	-3.5	-2.1	0.9
281.75	-4.29	-1.2	158.85	7SE	1.4	-3.4	-2.2	0.9
285.48	-3.73	-0.9	168.80	17SE	2.9	-3.4	-2.2	0.9
289.91	-3.20	-0.7	178.80	27SE	0.1	-3.4	-2.3	0.8
294.87	-2.73	-0.4	188.85	7OC	-0.6	-3.4	-2.3	0.8
300.25	-2.29	-0.2	198.95	17OC	-0.7	-3.3	-2.3	0.8
305.96	-1.90	0.0	209.10	27OC	-0.7	-3.3	-2.4	0.8
311.89	-1.55	0.2	219.27	6NO	-0.8	-3.3	-2.3	0.9
318.01	-1.24	0.4	229.47	16NO	-0.7	-3.3	-2.3	0.9
324.26	-0.96	0.6	239.68	26NO	-0.7	-3.3	-2.3	0.9
330.61	-0.71	0.7	249.89	6DE	-0.7	-3.4	-2.2	0.9
337.02	-0.48	0.9	260.09	16DE	-0.2	-3.4	-2.2	0.9
343.48	-0.28	1.0	270.27	26DE	2.0	-3.4	-2.1	0.8

Left table

♂ LONG	♂ LAT	♂ MAG	☉ LONG	16.00UT −490	☿	♀	♃	♄
349.97	-0.10	1.1	280.42	5JA	1.8	-3.4	-2.0	0.8
356.46	0.06	1.3	290.53	15JA	0.6	-3.5	-1.9	0.8
2.96	0.21	1.4	300.59	25JA	0.3	-3.5	-1.9	0.7
9.46	0.34	1.5	310.60	4FE	0.1	-3.6	-1.8	0.7
15.94	0.45	1.6	320.55	14FE	-0.1	-3.6	-1.7	0.6
22.40	0.55	1.6	330.44	24FE	-0.6	-3.7	-1.7	0.6
28.84	0.65	1.7	340.27	6MR	-1.4	-3.7	-1.6	0.5
35.26	0.73	1.8	350.05	16MR	-1.5	-3.8	-1.6	0.5
41.67	0.80	1.8	359.76	26MR	-0.6	-3.9	-1.5	0.4
48.05	0.87	1.9	9.43	5AP	0.5	-4.0	-1.5	0.4
54.41	0.92	1.9	19.05	15AP	1.7	-4.1	-1.5	0.4
60.76	0.97	2.0	28.63	25AP	3.1	-4.2	-1.5	0.4
67.10	1.01	2.0	38.19	5MY	2.9	-4.2	-1.4	0.5
73.42	1.05	2.0	47.72	15MY	1.6	-4.1	-1.4	0.6
79.74	1.08	2.0	57.24	25MY	0.7	-3.8	-1.4	0.6
86.06	1.11	2.0	66.75	4JN	-0.3	-3.1	-1.4	0.7
92.39	1.13	2.0	76.28	14JN	-1.3	-3.1	-1.5	0.7
98.72	1.15	2.0	85.82	24JN	-1.6	-3.8	-1.5	0.8
105.06	1.16	2.0	95.38	4JL	-0.7	-4.1	-1.5	0.8
111.42	1.16	2.0	104.98	14JL	-0.1	-4.2	-1.6	0.9
117.79	1.16	2.0	114.61	24JL	0.2	-4.2	-1.6	0.9
124.19	1.16	1.9	124.29	3AU	0.4	-4.1	-1.6	0.9
130.62	1.15	1.9	134.03	13AU	0.8	-4.0	-1.7	0.9
137.08	1.14	1.9	143.82	23AU	1.8	-4.0	-1.8	0.9
143.57	1.12	2.0	153.67	2SE	2.5	-3.9	-1.8	0.9
150.10	1.10	2.0	163.58	12SE	0.0	-3.8	-1.9	0.9
156.67	1.07	2.0	173.55	22SE	-0.8	-3.7	-1.9	0.8
163.29	1.04	1.9	183.57	2OC	-0.8	-3.7	-2.0	0.8
169.94	1.00	1.9	193.65	12OC	-0.8	-3.6	-2.1	0.8
176.64	0.95	1.9	203.77	22OC	-0.7	-3.6	-2.1	0.7
183.39	0.90	1.9	213.93	1NO	-0.6	-3.5	-2.2	0.8
190.18	0.84	1.8	224.12	11NO	-0.6	-3.5	-2.2	0.8
197.01	0.77	1.8	234.33	21NO	-0.5	-3.4	-2.2	0.8
203.90	0.69	1.7	244.54	1DE	-0.0	-3.4	-2.2	0.8
210.82	0.60	1.6	254.75	11DE	2.3	-3.4	-2.2	0.8
217.79	0.50	1.6	264.94	21DE	1.4	-3.4	-2.2	0.8
224.80	0.39	1.5	275.11	31DE	0.4	-3.3	-2.2	0.8
				−489				
231.85	0.27	1.4	285.24	10JA	0.2	-3.3	-2.1	0.8
238.94	0.13	1.3	295.32	20JA	0.0	-3.3	-2.0	0.8
246.06	-0.02	1.3	305.36	30JA	-0.2	-3.3	-2.0	0.7
253.21	-0.18	1.2	315.34	9FE	-0.7	-3.3	-1.9	0.7
260.37	-0.36	1.1	325.26	19FE	-1.4	-3.4	-1.8	0.6
267.56	-0.55	1.0	335.13	1MR	-1.4	-3.4	-1.8	0.6
274.74	-0.75	0.9	344.93	11MR	-0.6	-3.4	-1.7	0.5
281.91	-0.96	0.8	354.67	21MR	0.7	-3.4	-1.6	0.5
289.07	-1.19	0.7	4.37	31MR	2.1	-3.5	-1.6	0.4
296.17	-1.43	0.5	14.01	10AP	3.8	-3.5	-1.5	0.4
303.20	-1.68	0.4	23.61	20AP	2.2	-3.4	-1.5	0.3
310.14	-1.93	0.3	33.18	30AP	1.3	-3.4	-1.4	0.3
316.95	-2.19	0.2	42.72	10MY	0.5	-3.4	-1.4	0.4
323.58	-2.45	0.0	52.25	20MY	-0.4	-3.4	-1.4	0.5
329.99	-2.72	-0.1	61.76	30MY	-1.4	-3.3	-1.4	0.5
336.11	-2.98	-0.2	71.28	9JN	-1.6	-3.3	-1.3	0.6
341.87	-3.25	-0.4	80.81	19JN	-0.7	-3.3	-1.3	0.7
347.17	-3.50	-0.6	90.36	29JN	-0.1	-3.3	-1.3	0.7
351.87	-3.75	-0.7	99.94	9JL	0.3	-3.3	-1.3	0.8
355.86	-3.98	-0.9	109.55	19JL	0.6	-3.4	-1.4	0.8
358.92	-4.19	-1.1	119.21	29JL	1.1	-3.4	-1.4	0.8
0.84	-4.35	-1.4	128.92	8AU	2.3	-3.4	-1.4	0.8
1.44	-4.44	-1.6	138.68	18AU	2.2	-3.4	-1.5	0.9
0.57	-4.41	-1.8	148.50	28AU	-0.0	-3.5	-1.5	0.9
358.32	-4.21	-2.0	158.38	7SE	-0.9	-3.5	-1.5	0.9
355.12	-3.83	-2.1	168.31	17SE	-1.0	-3.6	-1.6	0.8
351.70	-3.27	-2.1	178.31	27SE	-0.9	-3.7	-1.7	0.8
348.88	-2.63	-1.8	188.36	7OC	-0.6	-3.8	-1.7	0.8
347.19	-1.98	-1.5	198.46	17OC	-0.4	-3.9	-1.8	0.8
346.85	-1.38	-1.1	208.61	27OC	-0.4	-4.0	-1.8	0.7
347.80	-0.87	-0.8	218.78	6NO	-0.4	-4.1	-1.9	0.7
349.85	-0.44	-0.5	228.97	16NO	0.2	-4.2	-2.0	0.7
352.79	-0.09	-0.2	239.19	26NO	2.6	-4.3	-2.0	0.8
356.43	0.19	0.1	249.40	6DE	1.1	-4.4	-2.1	0.8
0.62	0.42	0.3	259.60	16DE	0.2	-4.3	-2.1	0.8
5.23	0.61	0.5	269.78	26DE	-0.0	-4.3	-2.1	0.8

Right table

♂ LONG	♂ LAT	♂ MAG	☉ LONG	16.00UT −488	☿	♀	♃	♄
10.16	0.76	0.7	279.93	5JA	-0.1	-3.9	-2.1	0.8
15.35	0.89	0.9	290.04	15JA	-0.3	-3.3	-2.1	0.8
20.73	0.99	1.1	300.10	25JA	-0.7	-3.6	-2.1	0.8
26.27	1.07	1.2	310.11	4FE	-1.3	-4.1	-2.0	0.7
31.92	1.13	1.3	320.06	14FE	-1.3	-4.3	-2.0	0.7
37.67	1.19	1.4	329.96	24FE	-0.5	-4.3	-1.9	0.7
43.50	1.23	1.6	339.79	5MR	0.9	-4.2	-1.9	0.6
49.39	1.26	1.6	349.57	15MR	2.7	-4.1	-1.8	0.5
55.33	1.28	1.7	359.29	25MR	2.9	-4.0	-1.7	0.5
61.32	1.30	1.8	8.96	4AP	1.6	-3.9	-1.6	0.4
67.33	1.31	1.9	18.58	14AP	0.9	-3.8	-1.6	0.4
73.39	1.31	1.9	28.17	24AP	0.3	-3.7	-1.5	0.3
79.48	1.31	1.9	37.72	4MY	-0.4	-3.6	-1.5	0.3
85.60	1.30	2.0	47.26	14MY	-1.4	-3.6	-1.4	0.3
91.75	1.29	2.0	56.78	24MY	-1.6	-3.5	-1.4	0.4
97.94	1.27	2.0	66.29	3JN	-0.7	-3.5	-1.3	0.5
104.16	1.25	2.0	75.82	13JN	0.0	-3.4	-1.3	0.5
110.43	1.23	2.0	85.36	23JN	0.5	-3.4	-1.3	0.6
116.73	1.20	2.0	94.91	3JL	0.8	-3.4	-1.3	0.6
123.08	1.16	2.0	104.51	13JL	1.4	-3.4	-1.3	0.7
129.48	1.12	2.0	114.15	23JL	2.8	-3.3	-1.3	0.7
135.93	1.08	2.0	123.82	2AU	1.9	-3.3	-1.3	0.8
142.44	1.03	1.9	133.56	12AU	-0.1	-3.3	-1.3	0.8
149.00	0.98	1.9	143.35	22AU	-1.0	-3.4	-1.3	0.8
155.63	0.93	1.8	153.19	1SE	-1.1	-3.4	-1.3	0.8
162.32	0.86	1.8	163.10	11SE	-0.9	-3.4	-1.3	0.8
169.07	0.80	1.8	173.07	21SE	-0.5	-3.4	-1.4	0.8
175.89	0.73	1.8	183.09	1OC	-0.3	-3.4	-1.4	0.8
182.79	0.65	1.8	193.17	11OC	-0.3	-3.5	-1.5	0.8
189.75	0.57	1.8	203.29	21OC	-0.2	-3.5	-1.5	0.8
196.78	0.48	1.8	213.44	31OC	0.3	-3.5	-1.6	0.7
203.89	0.39	1.7	223.63	10NO	2.9	-3.5	-1.7	0.7
211.06	0.28	1.7	233.83	20NO	0.8	-3.4	-1.7	0.7
218.30	0.18	1.7	244.04	30NO	-0.0	-3.4	-1.8	0.8
225.60	0.07	1.6	254.25	10DE	-0.2	-3.4	-1.9	0.8
232.97	-0.05	1.6	264.44	20DE	-0.2	-3.4	-1.9	0.8
240.39	-0.17	1.6	274.61	30DE	-0.4	-3.4	-2.0	0.8
				−487				
247.87	-0.30	1.5	284.74	9JA	-0.8	-3.3	-2.0	0.8
255.40	-0.43	1.5	294.83	19JA	-1.2	-3.3	-2.0	0.8
262.96	-0.57	1.4	304.87	29JA	-1.2	-3.3	-2.0	0.8
270.56	-0.70	1.4	314.86	8FE	-0.5	-3.3	-2.0	0.8
278.18	-0.84	1.3	324.78	18FE	1.1	-3.4	-2.0	0.7
285.82	-0.97	1.3	334.64	28FE	3.2	-3.4	-2.0	0.7
293.45	-1.11	1.2	344.45	10MR	2.2	-3.4	-1.9	0.6
301.08	-1.23	1.2	354.20	20MR	1.2	-3.4	-1.9	0.6
308.67	-1.35	1.1	3.90	30MR	0.7	-3.5	-1.8	0.5
316.23	-1.47	1.0	13.55	9AP	0.2	-3.5	-1.8	0.5
323.72	-1.57	1.0	23.15	19AP	-0.5	-3.6	-1.7	0.4
331.15	-1.66	0.9	32.72	29AP	-1.4	-3.6	-1.6	0.3
338.48	-1.74	0.9	42.26	9MY	-1.6	-3.7	-1.6	0.3
345.70	-1.80	0.8	51.79	19MY	-0.7	-3.8	-1.5	0.3
352.79	-1.85	0.8	61.30	29MY	0.1	-3.8	-1.5	0.3
359.72	-1.88	0.7	70.82	8JN	0.6	-3.9	-1.4	0.4
6.46	-1.89	0.6	80.35	18JN	1.1	-4.0	-1.4	0.4
13.00	-1.89	0.6	89.90	28JN	1.9	-4.1	-1.3	0.5
19.29	-1.86	0.5	99.48	8JL	3.2	-4.2	-1.3	0.6
25.28	-1.82	0.4	109.09	18JL	1.6	-4.2	-1.3	0.6
30.94	-1.76	0.3	118.74	28JL	-0.2	-4.1	-1.2	0.7
36.18	-1.67	0.2	128.45	7AU	-1.1	-3.8	-1.2	0.7
40.92	-1.56	0.0	138.21	17AU	-1.3	-3.3	-1.2	0.8
45.05	-1.41	-0.1	148.02	27AU	-0.9	-3.6	-1.2	0.8
48.42	-1.23	-0.3	157.90	6SE	-0.4	-4.1	-1.2	0.8
50.86	-1.01	-0.5	167.83	16SE	-0.2	-4.3	-1.2	0.8
52.18	-0.73	-0.7	177.83	26SE	-0.1	-4.3	-1.3	0.8
52.16	-0.39	-0.9	187.88	6OC	-0.0	-4.3	-1.3	0.8
50.70	0.00	-1.1	197.97	16OC	0.6	-4.2	-1.3	0.8
47.92	0.44	-1.3	208.11	26OC	3.2	-4.1	-1.4	0.8
44.21	0.89	-1.5	218.29	5NO	0.6	-4.0	-1.4	0.7
40.34	1.29	-1.4	228.48	15NO	-0.2	-3.9	-1.5	0.7
37.06	1.61	-1.1	238.69	25NO	-0.3	-3.8	-1.5	0.7
34.94	1.83	-0.8	248.91	5DE	-0.3	-3.8	-1.6	0.7
34.16	1.97	-0.5	259.11	15DE	-0.5	-3.7	-1.7	0.8
34.67	2.05	-0.2	269.29	25DE	-0.8	-3.6	-1.7	0.8

♂ LONG	♂ LAT	♂ MAG	⊙ LONG	16.00UT	☿	♀	♃	♄
				-486	MAGNITUDES			
36.28	2.07	0.1	279.44	4JA	-1.1	-3.5	-1.8	0.8
38.79	2.07	0.3	289.55	14JA	-1.1	-3.5	-1.9	0.8
42.02	2.05	0.5	299.62	24JA	-0.4	-3.4	-1.9	0.8
45.81	2.02	0.7	309.64	3FE	1.3	-3.4	-2.0	0.8
50.05	1.98	0.9	319.59	13FE	3.1	-3.4	-2.0	0.8
54.63	1.93	1.1	329.49	23FE	1.6	-3.3	-2.0	0.8
59.49	1.88	1.2	339.33	5MR	0.9	-3.3	-2.0	0.7
64.58	1.83	1.4	349.10	15MR	0.5	-3.3	-2.0	0.7
69.86	1.78	1.5	358.82	25MR	0.1	-3.3	-2.0	0.6
75.29	1.72	1.6	8.50	4AP	-0.5	-3.3	-2.0	0.6
80.85	1.66	1.6	18.12	14AP	-1.4	-3.3	-1.9	0.5
86.52	1.60	1.7	27.71	24AP	-1.6	-3.3	-1.9	0.4
92.30	1.54	1.8	37.27	4MY	-0.7	-3.4	-1.8	0.4
98.17	1.47	1.8	46.80	14MY	0.2	-3.4	-1.7	0.3
104.12	1.41	1.9	56.32	24MY	0.8	-3.4	-1.7	0.3
110.16	1.34	1.9	65.83	3JN	1.5	-3.5	-1.6	0.3
116.27	1.27	1.9	75.35	13JN	2.5	-3.5	-1.5	0.4
122.47	1.20	1.9	84.89	23JN	3.0	-3.5	-1.5	0.4
128.74	1.13	1.9	94.45	3JL	1.3	-3.4	-1.4	0.5
135.09	1.05	1.9	104.04	13JL	-0.2	-3.4	-1.4	0.5
141.52	0.97	1.9	113.67	23JL	-1.2	-3.4	-1.3	0.6
148.03	0.89	1.9	123.35	2AU	-1.4	-3.3	-1.3	0.6
154.63	0.81	1.9	133.08	12AU	-0.8	-3.3	-1.3	0.7
161.31	0.72	1.8	142.86	22AU	-0.4	-3.3	-1.3	0.7
168.07	0.63	1.8	152.71	1SE	-0.1	-3.3	-1.2	0.8
174.93	0.54	1.8	162.61	11SE	0.0	-3.3	-1.2	0.8
181.87	0.44	1.7	172.58	21SE	0.2	-3.3	-1.2	0.8
188.90	0.34	1.7	182.60	1OC	0.8	-3.4	-1.2	0.8
196.02	0.24	1.6	192.67	11OC	3.5	-3.4	-1.2	0.8
203.23	0.13	1.6	202.79	21OC	0.4	-3.4	-1.3	0.8
210.52	0.03	1.6	212.95	31OC	-0.4	-3.5	-1.3	0.8
217.89	-0.08	1.6	223.13	10NO	-0.4	-3.5	-1.3	0.8
225.34	-0.19	1.5	233.34	20NO	-0.5	-3.6	-1.4	0.7
232.87	-0.30	1.5	243.55	30NO	-0.6	-3.6	-1.4	0.7
240.45	-0.41	1.5	253.76	10DE	-0.9	-3.7	-1.5	0.7
248.11	-0.52	1.5	263.95	20DE	-1.0	-3.8	-1.5	0.8
255.81	-0.62	1.5	274.12	30DE	-0.9	-3.9	-1.6	0.8
				-485				
263.55	-0.72	1.5	284.26	9JA	-0.4	-4.0	-1.7	0.8
271.33	-0.82	1.5	294.35	19JA	1.5	-4.0	-1.7	0.8
279.13	-0.91	1.4	304.39	29JA	2.6	-4.1	-1.8	0.8
286.93	-1.00	1.4	314.37	8FE	1.2	-4.2	-1.9	0.8
294.74	-1.07	1.4	324.31	18FE	0.6	-4.3	-1.9	0.8
302.53	-1.14	1.4	334.17	28FE	0.4	-4.3	-2.0	0.8
310.29	-1.20	1.4	343.98	10MR	0.0	-4.1	-2.0	0.8
318.02	-1.24	1.4	353.73	20MR	-0.5	-3.6	-2.0	0.7
325.69	-1.27	1.3	3.43	30MR	-1.4	-3.0	-2.1	0.7
333.30	-1.29	1.3	13.08	9AP	-1.6	-3.6	-2.1	0.6
340.84	-1.30	1.3	22.69	19AP	-0.7	-4.0	-2.1	0.6
348.30	-1.29	1.3	32.26	29AP	0.3	-4.2	-2.0	0.5
355.65	-1.26	1.3	41.80	9MY	1.1	-4.2	-2.0	0.4
2.91	-1.23	1.3	51.33	19MY	2.0	-4.1	-1.9	0.4
10.05	-1.18	1.3	60.84	29MY	3.3	-4.0	-1.9	0.3
17.06	-1.12	1.3	70.36	8JN	2.5	-3.9	-1.8	0.3
23.95	-1.04	1.2	79.89	18JN	1.0	-3.8	-1.8	0.3
30.68	-0.95	1.2	89.43	28JN	-0.2	-3.8	-1.7	0.4
37.26	-0.85	1.2	99.01	8JL	-1.3	-3.7	-1.6	0.4
43.66	-0.74	1.1	108.62	18JL	-1.5	-3.6	-1.6	0.5
49.87	-0.61	1.1	118.27	28JL	-0.8	-3.6	-1.5	0.5
55.86	-0.47	1.0	127.97	7AU	-0.3	-3.5	-1.5	0.6
61.60	-0.31	0.9	137.73	17AU	0.0	-3.5	-1.4	0.7
67.05	-0.13	0.8	147.54	27AU	0.2	-3.4	-1.4	0.7
72.17	0.06	0.7	157.41	6SE	0.4	-3.4	-1.3	0.7
76.88	0.28	0.6	167.35	16SE	1.1	-3.4	-1.3	0.8
81.09	0.53	0.5	177.34	26SE	3.2	-3.4	-1.3	0.8
84.70	0.82	0.3	187.39	6OC	0.3	-3.4	-1.3	0.8
87.57	1.14	0.1	197.48	16OC	-0.5	-3.4	-1.3	0.8
89.51	1.51	-0.1	207.62	26OC	-0.6	-3.4	-1.3	0.8
90.36	1.92	-0.3	217.79	5NO	-0.6	-3.4	-1.3	0.8
89.90	2.36	-0.5	227.99	15NO	-0.7	-3.4	-1.3	0.8
88.08	2.81	-0.8	238.20	25NO	-0.8	-3.4	-1.3	0.8
85.03	3.23	-1.0	248.41	5DE	-0.8	-3.4	-1.4	0.7
81.18	3.54	-1.1	258.61	15DE	-0.8	-3.4	-1.4	0.7
77.26	3.71	-1.0	268.79	25DE	-0.3	-3.4	-1.4	0.8

♂ LONG	♂ LAT	♂ MAG	⊙ LONG	16.00UT	☿	♀	♃	♄
				-484	MAGNITUDES			
73.99	3.73	-0.8	278.95	4JA	1.8	-3.5	-1.5	0.8
71.85	3.64	-0.5	289.07	14JA	2.1	-3.5	-1.6	0.8
71.04	3.47	-0.2	299.13	24JA	0.8	-3.5	-1.6	0.9
71.47	3.27	0.0	309.15	3FE	0.4	-3.4	-1.7	0.9
72.98	3.06	0.3	319.11	13FE	0.2	-3.4	-1.8	0.9
75.39	2.85	0.5	329.00	23FE	-0.1	-3.4	-1.8	0.9
78.51	2.65	0.7	338.85	4MR	-0.6	-3.4	-1.9	0.9
82.21	2.47	0.9	348.63	14MR	-1.4	-3.4	-2.0	0.8
86.37	2.29	1.0	358.35	24MR	-1.5	-3.3	-2.0	0.8
90.90	2.13	1.1	8.03	3AP	-0.6	-3.3	-2.1	0.8
95.74	1.97	1.3	17.65	13AP	0.4	-3.3	-2.1	0.7
100.83	1.83	1.4	27.24	23AP	1.4	-3.3	-2.1	0.6
106.14	1.69	1.4	36.80	3MY	2.6	-3.3	-2.2	0.6
111.65	1.55	1.5	46.34	13MY	3.4	-3.4	-2.2	0.5
117.32	1.42	1.6	55.85	23MY	1.9	-3.4	-2.1	0.5
123.14	1.30	1.6	65.37	2JN	0.8	-3.4	-2.1	0.4
129.10	1.17	1.7	74.89	12JN	-0.3	-3.5	-2.1	0.3
135.20	1.05	1.7	84.43	22JN	-1.3	-3.5	-2.0	0.3
141.41	0.93	1.7	93.99	2JL	-1.6	-3.6	-2.0	0.4
147.75	0.81	1.7	103.58	12JL	-0.8	-3.6	-1.9	0.4
154.21	0.69	1.7	113.20	22JL	-0.2	-3.7	-1.8	0.5
160.78	0.57	1.7	122.88	1AU	0.1	-3.8	-1.8	0.5
167.47	0.46	1.7	132.61	11AU	0.3	-3.9	-1.7	0.6
174.27	0.34	1.7	142.39	21AU	0.6	-4.0	-1.7	0.6
181.18	0.22	1.7	152.23	31AU	1.5	-4.1	-1.6	0.7
188.19	0.11	1.7	162.13	10SE	2.8	-4.2	-1.5	0.7
195.32	-0.00	1.6	172.09	20SE	0.2	-4.3	-1.5	0.8
202.55	-0.12	1.6	182.11	30SE	-0.7	-4.4	-1.5	0.8
209.88	-0.23	1.6	192.18	10OC	-0.8	-4.3	-1.4	0.8
217.30	-0.33	1.5	202.30	20OC	-0.7	-3.9	-1.4	0.8
224.82	-0.44	1.5	212.45	30OC	-0.8	-3.1	-1.4	0.8
232.41	-0.54	1.5	222.64	9NO	-0.6	-3.6	-1.4	0.8
240.08	-0.63	1.4	232.84	19NO	-0.6	-4.1	-1.4	0.8
247.82	-0.72	1.4	243.05	29NO	-0.6	-4.4	-1.4	0.8
255.61	-0.80	1.4	253.26	9DE	-0.1	-4.4	-1.4	0.8
263.44	-0.88	1.3	263.46	19DE	2.0	-4.3	-1.4	0.8
271.30	-0.94	1.3	273.63	29DE	1.7	-4.2	-1.4	0.8
				-483				
279.18	-1.00	1.3	283.77	8JA	0.5	-4.1	-1.5	0.8
287.06	-1.04	1.3	293.86	18JA	0.3	-4.0	-1.5	0.9
294.94	-1.08	1.4	303.91	28JA	0.1	-3.9	-1.6	0.9
302.80	-1.10	1.4	313.90	7FE	-0.2	-3.8	-1.6	0.9
310.62	-1.11	1.4	323.83	17FE	-0.6	-3.7	-1.7	0.9
318.41	-1.11	1.4	333.70	27FE	-1.4	-3.6	-1.7	0.9
326.13	-1.09	1.4	343.51	9MR	-1.5	-3.5	-1.8	0.9
333.79	-1.07	1.5	353.26	19MR	-0.6	-3.5	-1.9	0.9
341.38	-1.03	1.5	2.97	29MR	0.6	-3.4	-1.9	0.9
348.89	-0.99	1.5	12.62	8AP	1.8	-3.4	-2.0	0.8
356.31	-0.93	1.5	22.23	18AP	3.4	-3.4	-2.1	0.8
3.64	-0.86	1.5	31.80	28AP	2.7	-3.3	-2.1	0.7
10.88	-0.79	1.5	41.34	8MY	1.5	-3.3	-2.2	0.7
18.01	-0.70	1.6	50.87	18MY	0.6	-3.3	-2.2	0.6
25.05	-0.61	1.6	60.39	28MY	-0.3	-3.3	-2.3	0.6
31.98	-0.51	1.6	69.90	7JN	-1.3	-3.3	-2.3	0.5
38.80	-0.41	1.6	79.43	17JN	-1.6	-3.3	-2.3	0.4
45.51	-0.30	1.6	88.98	27JN	-0.7	-3.3	-2.3	0.4
52.11	-0.18	1.6	98.55	7JL	-0.1	-3.4	-2.2	0.4
58.58	-0.05	1.6	108.15	17JL	0.2	-3.4	-2.2	0.4
64.94	0.08	1.5	117.81	27JL	0.5	-3.4	-2.1	0.5
71.15	0.22	1.5	127.50	6AU	0.9	-3.5	-2.1	0.5
77.22	0.37	1.5	137.25	16AU	1.9	-3.5	-2.0	0.6
83.13	0.53	1.4	147.07	26AU	2.5	-3.5	-1.9	0.6
88.85	0.69	1.3	156.93	5SE	0.1	-3.4	-1.9	0.7
94.37	0.87	1.3	166.86	15SE	-0.8	-3.4	-1.8	0.7
99.65	1.07	1.2	176.85	25SE	-0.9	-3.4	-1.8	0.8
104.63	1.28	1.1	186.89	5OC	-0.9	-3.4	-1.7	0.8
109.28	1.52	0.9	196.98	15OC	-0.7	-3.3	-1.6	0.8
113.50	1.77	0.8	207.12	25OC	-0.5	-3.3	-1.6	0.9
117.20	2.06	0.6	217.29	4NO	-0.5	-3.3	-1.6	0.9
120.27	2.38	0.4	227.48	14NO	-0.5	-3.3	-1.5	0.9
122.54	2.74	0.2	237.69	24NO	0.0	-3.3	-1.5	0.9
123.84	3.12	-0.0	247.90	4DE	2.3	-3.4	-1.5	0.9
123.99	3.53	-0.3	258.11	14DE	1.3	-3.4	-1.5	0.9
122.84	3.94	-0.5	268.30	24DE	0.3	-3.4	-1.5	0.8

♂ LONG	♂ LAT	♂ MAG	⊙ LONG	16.00UT	☿	♀	♃	♄
				-482	MAGNITUDES			
120.41	4.29	-0.8	278.45	3JA	0.1	-3.4	-1.5	0.8
116.95	4.53	-1.0	288.57	13JA	-0.0	-3.5	-1.5	0.9
113.01	4.60	-1.0	298.64	23JA	-0.3	-3.5	-1.5	0.9
109.33	4.50	-0.8	308.66	2FE	-0.7	-3.6	-1.5	1.0
106.54	4.25	-0.6	318.63	12FE	-1.3	-3.6	-1.6	1.0
104.97	3.92	-0.4	328.53	22FE	-1.4	-3.7	-1.6	1.0
104.71	3.55	-0.1	338.37	4MR	-0.6	-3.8	-1.7	1.0
105.63	3.19	0.1	348.16	14MR	0.7	-3.8	-1.7	1.0
107.57	2.84	0.3	357.89	24MR	2.3	-3.9	-1.8	1.0
110.34	2.53	0.5	7.56	3AP	3.5	-4.0	-1.8	1.0
113.78	2.23	0.6	17.19	13AP	2.0	-4.1	-1.9	1.0
117.79	1.97	0.8	26.79	23AP	1.1	-4.2	-2.0	0.9
122.25	1.72	0.9	36.34	3MY	0.5	-4.2	-2.0	0.9
127.08	1.49	1.0	45.88	13MY	-0.4	-4.1	-2.1	0.8
132.24	1.28	1.1	55.40	23MY	-1.4	-3.8	-2.2	0.8
137.67	1.08	1.2	64.91	2JN	-1.6	-3.1	-2.2	0.7
143.34	0.90	1.2	74.43	12JN	-0.7	-3.1	-2.3	0.6
149.24	0.72	1.3	83.97	22JN	-0.1	-3.8	-2.3	0.6
155.32	0.55	1.3	93.52	2JL	0.4	-4.1	-2.4	0.5
161.59	0.39	1.4	103.11	12JL	0.7	-4.2	-2.4	0.4
168.03	0.24	1.4	112.74	22JL	1.2	-4.2	-2.4	0.5
174.63	0.09	1.4	122.41	1AU	2.4	-4.1	-2.4	0.5
181.38	-0.05	1.4	132.14	11AU	2.2	-4.0	-2.3	0.5
188.27	-0.18	1.5	141.91	21AU	-0.0	-4.0	-2.3	0.6
195.30	-0.31	1.5	151.75	31AU	-0.9	-3.9	-2.2	0.7
202.46	-0.43	1.5	161.65	10SE	-1.1	-3.8	-2.2	0.7
209.74	-0.54	1.5	171.61	20SE	-1.0	-3.7	-2.1	0.8
217.14	-0.64	1.4	181.62	30SE	-0.6	-3.7	-2.0	0.8
224.63	-0.74	1.4	191.69	10OC	-0.4	-3.6	-2.0	0.8
232.23	-0.82	1.4	201.80	20OC	-0.3	-3.6	-1.9	0.9
239.90	-0.90	1.4	211.95	30OC	-0.3	-3.5	-1.8	0.9
247.65	-0.96	1.4	222.14	9NO	0.2	-3.5	-1.8	0.9
255.45	-1.02	1.4	232.34	19NO	2.6	-3.4	-1.7	0.9
263.30	-1.06	1.4	242.55	29NO	1.0	-3.4	-1.7	1.0
271.18	-1.09	1.4	252.76	9DE	0.1	-3.4	-1.6	1.0
279.08	-1.11	1.4	262.96	19DE	-0.1	-3.4	-1.6	1.0
286.98	-1.11	1.4	273.13	29DE	-0.1	-3.3	-1.6	1.0
				-481				
294.87	-1.11	1.4	283.27	8JA	-0.3	-3.3	-1.6	0.9
302.73	-1.09	1.4	293.37	18JA	-0.7	-3.3	-1.6	0.9
310.55	-1.06	1.4	303.41	28JA	-1.3	-3.3	-1.5	1.0
318.32	-1.02	1.4	313.41	7FE	-1.3	-3.3	-1.6	1.0
326.04	-0.97	1.4	323.34	17FE	-0.6	-3.4	-1.6	1.1
333.68	-0.91	1.4	333.22	27FE	0.9	-3.4	-1.6	1.1
341.26	-0.85	1.4	343.04	9MR	2.9	-3.4	-1.6	1.1
348.75	-0.77	1.5	352.79	19MR	2.6	-3.4	-1.6	1.1
356.16	-0.69	1.5	2.49	29MR	1.5	-3.5	-1.7	1.1
3.48	-0.61	1.6	12.15	8AP	0.9	-3.5	-1.7	1.1
10.71	-0.51	1.6	21.76	18AP	0.3	-3.4	-1.8	1.1
17.85	-0.42	1.6	31.33	28AP	-0.4	-3.4	-1.8	1.1
24.91	-0.32	1.7	40.88	8MY	-1.4	-3.4	-1.9	1.0
31.87	-0.22	1.7	50.41	18MY	-1.6	-3.4	-2.0	1.0
38.75	-0.12	1.7	59.92	28MY	-0.7	-3.3	-2.0	0.9
45.54	-0.01	1.8	69.44	7JN	0.0	-3.3	-2.1	0.9
52.25	0.10	1.8	78.96	17JN	0.5	-3.3	-2.2	0.8
58.88	0.21	1.8	88.51	27JN	0.9	-3.3	-2.2	0.7
65.42	0.32	1.8	98.08	7JL	1.6	-3.3	-2.3	0.7
71.88	0.43	1.8	107.68	17JL	3.0	-3.4	-2.4	0.6
78.26	0.55	1.8	117.33	27JL	1.8	-3.4	-2.4	0.5
84.56	0.66	1.8	127.03	6AU	-0.1	-3.4	-2.4	0.6
90.76	0.78	1.8	136.78	16AU	-1.0	-3.4	-2.4	0.6
96.88	0.91	1.8	146.59	26AU	-1.2	-3.5	-2.4	0.7
102.89	1.03	1.7	156.45	5SE	-0.9	-3.5	-2.4	0.7
108.79	1.17	1.7	166.38	15SE	-0.5	-3.6	-2.4	0.8
114.57	1.30	1.6	176.36	25SE	-0.3	-3.7	-2.4	0.8
120.20	1.45	1.5	186.41	5OC	-0.2	-3.8	-2.3	0.9
125.67	1.60	1.4	196.49	15OC	-0.1	-3.9	-2.2	0.9
130.95	1.76	1.3	206.63	25OC	0.4	-4.0	-2.2	0.9
135.99	1.94	1.2	216.80	4NO	2.9	-4.1	-2.1	1.0
140.75	2.13	1.1	226.99	14NO	0.8	-4.2	-2.0	1.0
145.16	2.33	0.9	237.20	24NO	-0.1	-4.3	-2.0	1.0
149.15	2.56	0.7	247.41	4DE	-0.2	-4.4	-1.9	1.1
152.61	2.80	0.5	257.61	14DE	-0.3	-4.4	-1.8	1.1
155.41	3.07	0.3	267.80	24DE	-0.4	-4.3	-1.8	1.1

♂ LONG	♂ LAT	♂ MAG	⊙ LONG	16.00UT	☿	♀	♃	♄
				-480	MAGNITUDES			
157.38	3.35	0.1	277.96	3JA	-0.8	-3.9	-1.7	1.1
158.37	3.64	-0.2	288.07	13JA	-1.2	-3.3	-1.7	1.1
158.18	3.91	-0.5	298.15	23JA	-1.2	-3.6	-1.6	1.1
156.71	4.14	-0.7	308.17	2FE	-0.5	-4.1	-1.6	1.1
154.03	4.27	-1.0	318.14	12FE	1.1	-4.3	-1.6	1.1
150.46	4.25	-1.2	328.04	22FE	3.3	-4.3	-1.6	1.2
146.64	4.06	-1.2	337.89	3MR	1.9	-4.2	-1.6	1.2
143.28	3.72	-1.0	347.68	13MR	1.1	-4.1	-1.6	1.3
140.94	3.28	-0.8	357.41	23MR	0.6	-4.0	-1.6	1.3
139.90	2.80	-0.6	7.09	2AP	0.2	-3.9	-1.6	1.3
140.15	2.34	-0.4	16.72	12AP	-0.5	-3.8	-1.6	1.3
141.58	1.91	-0.2	26.32	22AP	-1.4	-3.7	-1.6	1.3
144.01	1.51	0.0	35.88	2MY	-1.6	-3.6	-1.7	1.3
147.27	1.16	0.2	45.41	12MY	-0.7	-3.6	-1.7	1.2
151.20	0.84	0.3	54.94	22MY	0.1	-3.5	-1.8	1.2
155.70	0.56	0.4	64.46	1JN	0.7	-3.5	-1.8	1.1
160.66	0.31	0.6	73.97	11JN	1.2	-3.4	-1.9	1.1
166.02	0.07	0.7	83.51	21JN	2.1	-3.4	-1.9	1.0
171.72	-0.13	0.7	93.06	1JL	3.3	-3.4	-2.0	0.9
177.72	-0.32	0.8	102.65	11JL	1.5	-3.3	-2.1	0.9
183.98	-0.49	0.9	112.28	21JL	-0.1	-3.3	-2.1	0.8
190.47	-0.65	0.9	121.94	31JL	-1.1	-3.3	-2.2	0.7
197.17	-0.79	1.0	131.66	10AU	-1.3	-3.3	-2.3	0.7
204.06	-0.91	1.0	141.44	20AU	-0.9	-3.4	-2.3	0.7
211.13	-1.01	1.1	151.28	30AU	-0.4	-3.4	-2.4	0.8
218.34	-1.10	1.1	161.17	9SE	-0.2	-3.4	-2.4	0.8
225.69	-1.18	1.1	171.13	19SE	-0.1	-3.4	-2.4	0.9
233.16	-1.24	1.2	181.14	29SE	0.1	-3.4	-2.4	0.9
240.73	-1.28	1.2	191.20	9OC	0.6	-3.5	-2.4	1.0
248.38	-1.31	1.2	201.31	19OC	3.2	-3.5	-2.4	1.0
256.10	-1.32	1.2	211.46	29OC	0.6	-3.5	-2.3	1.1
263.88	-1.32	1.3	221.64	8NO	-0.3	-3.5	-2.3	1.1
271.69	-1.30	1.3	231.85	18NO	-0.4	-3.4	-2.2	1.1
279.51	-1.27	1.3	242.05	28NO	-0.4	-3.4	-2.1	1.2
287.34	-1.23	1.4	252.26	8DE	-0.5	-3.4	-2.1	1.2
295.16	-1.18	1.4	262.46	18DE	-0.8	-3.4	-2.0	1.2
302.94	-1.12	1.4	272.63	28DE	-1.0	-3.4	-1.9	1.2
				-479				
310.69	-1.04	1.5	282.77	7JA	-1.0	-3.3	-1.9	1.2
318.39	-0.96	1.5	292.87	17JA	-0.5	-3.3	-1.8	1.3
326.02	-0.88	1.5	302.92	27JA	1.3	-3.3	-1.7	1.3
333.59	-0.79	1.5	312.92	6FE	3.0	-3.3	-1.7	1.2
341.08	-0.69	1.6	322.86	16FE	1.4	-3.4	-1.6	1.3
348.50	-0.59	1.6	332.73	26FE	0.8	-3.4	-1.6	1.3
355.83	-0.49	1.6	342.55	8MR	0.5	-3.4	-1.6	1.4
3.09	-0.39	1.6	352.32	18MR	0.1	-3.4	-1.5	1.4
10.25	-0.28	1.6	2.02	28MR	-0.5	-3.5	-1.5	1.4
17.34	-0.18	1.7	11.68	7AP	-1.3	-3.5	-1.5	1.3
24.34	-0.08	1.7	21.29	17AP	-1.6	-3.6	-1.5	1.3
31.27	0.02	1.7	30.87	27AP	-0.7	-3.6	-1.5	1.3
38.12	0.12	1.7	40.42	7MY	0.2	-3.7	-1.5	1.3
44.90	0.22	1.8	49.95	17MY	0.9	-3.8	-1.5	1.2
51.61	0.32	1.8	59.46	27MY	1.6	-3.8	-1.6	1.2
58.25	0.41	1.9	68.98	6JN	2.8	-3.9	-1.6	1.1
64.84	0.51	1.9	78.51	16JN	2.9	-4.0	-1.6	1.1
71.36	0.60	1.9	88.05	26JN	1.2	-4.1	-1.7	1.0
77.84	0.69	1.9	97.62	6JL	-0.2	-4.2	-1.7	1.0
84.26	0.78	2.0	107.22	16JL	-1.2	-4.2	-1.8	0.9
90.64	0.86	2.0	116.87	26JL	-1.5	-4.1	-1.9	0.9
96.97	0.95	2.0	126.56	5AU	-0.8	-3.8	-1.9	0.8
103.25	1.03	2.0	136.31	15AU	-0.3	-3.3	-2.0	0.8
109.48	1.12	2.0	146.11	25AU	-0.1	-3.6	-2.0	0.8
115.67	1.20	1.9	155.98	4SE	0.1	-4.1	-2.1	0.9
121.80	1.28	1.9	165.90	14SE	0.3	-4.3	-2.2	0.9
127.88	1.37	1.9	175.88	24SE	0.9	-4.3	-2.2	1.0
133.90	1.45	1.8	185.92	4OC	3.4	-4.3	-2.3	1.0
139.85	1.53	1.7	196.01	14OC	0.4	-4.2	-2.3	1.1
145.72	1.62	1.7	206.13	24OC	-0.4	-4.1	-2.3	1.1
151.50	1.71	1.6	216.30	3NO	-0.5	-4.0	-2.3	1.2
157.17	1.80	1.5	226.50	13NO	-0.5	-3.9	-2.3	1.2
162.70	1.88	1.4	236.70	23NO	-0.6	-3.8	-2.3	1.3
168.07	1.98	1.2	246.92	3DE	-0.8	-3.8	-2.2	1.3
173.25	2.07	1.1	257.12	13DE	-0.9	-3.7	-2.2	1.3
178.19	2.16	0.9	267.31	23DE	-0.9	-3.6	-2.1	1.3

♂ LONG	♂ LAT	♂ MAG	⊙ LONG	16.00UT -478	☿ MAGNITUDES	♀ MAGNITUDES	♃ MAGNITUDES	♄ MAGNITUDES
182.83	2.26	0.8	277.47	2JA	-0.4	-3.5	-2.1	1.3
187.11	2.35	0.6	287.59	12JA	1.5	-3.5	-2.0	1.3
190.94	2.43	0.3	297.66	22JA	2.4	-3.4	-1.9	1.2
194.19	2.51	0.1	307.69	1FE	1.0	-3.4	-1.8	1.2
196.73	2.57	-0.2	317.66	11FE	0.6	-3.4	-1.8	1.1
198.39	2.60	-0.5	327.57	21FE	0.3	-3.3	-1.7	1.1
198.98	2.58	-0.8	337.42	3MR	-0.0	-3.3	-1.7	1.1
198.35	2.49	-1.1	347.21	13MR	-0.5	-3.3	-1.6	1.1
196.47	2.29	-1.4	356.94	23MR	-1.3	-3.3	-1.6	1.1
193.53	1.98	-1.7	6.63	2AP	-1.6	-3.3	-1.5	1.1
190.04	1.55	-1.8	16.26	12AP	-0.7	-3.3	-1.5	1.1
186.70	1.04	-1.6	25.85	22AP	0.3	-3.3	-1.5	1.0
184.19	0.52	-1.5	35.42	2MY	1.2	-3.4	-1.4	1.0
182.93	0.02	-1.3	44.95	12MY	2.2	-3.4	-1.4	1.0
183.05	-0.42	-1.0	54.47	22MY	3.5	-3.4	-1.4	1.0
184.45	-0.79	-0.8	63.99	1JN	2.3	-3.5	-1.4	0.9
186.99	-1.10	-0.6	73.51	11JN	1.0	-3.5	-1.4	0.9
190.46	-1.35	-0.5	83.04	21JN	-0.2	-3.5	-1.5	0.8
194.71	-1.55	-0.3	92.60	1JL	-1.3	-3.4	-1.5	0.8
199.60	-1.71	-0.2	102.18	11JL	-1.6	-3.4	-1.5	0.7
205.01	-1.83	-0.0	111.80	21JL	-0.8	-3.4	-1.5	0.7
210.85	-1.92	0.1	121.47	31JL	-0.3	-3.3	-1.6	0.6
217.06	-1.99	0.2	131.19	10AU	0.1	-3.3	-1.6	0.6
223.57	-2.02	0.3	140.96	20AU	0.2	-3.3	-1.7	0.5
230.33	-2.03	0.4	150.80	30AU	0.5	-3.3	-1.7	0.5
237.31	-2.01	0.5	160.69	9SE	1.2	-3.3	-1.8	0.5
244.45	-1.98	0.6	170.64	19SE	3.1	-3.3	-1.9	0.6
251.73	-1.92	0.7	180.65	29SE	0.3	-3.4	-1.9	0.6
259.12	-1.85	0.8	190.71	9OC	-0.6	-3.4	-2.0	0.7
266.59	-1.77	0.8	200.82	19OC	-0.7	-3.4	-2.1	0.8
274.13	-1.67	0.9	210.97	29OC	-0.7	-3.5	-2.1	0.8
281.69	-1.56	1.0	221.15	8NO	-0.7	-3.5	-2.2	0.9
289.27	-1.44	1.1	231.35	18NO	-0.7	-3.6	-2.2	0.9
296.85	-1.31	1.1	241.56	28NO	-0.7	-3.6	-2.2	1.0
304.41	-1.19	1.2	251.77	8DE	-0.7	-3.7	-2.2	1.0
311.94	-1.05	1.3	261.97	18DE	-0.2	-3.8	-2.2	1.0
319.42	-0.92	1.4	272.15	28DE	1.8	-3.9	-2.2	1.0
				-477				
326.85	-0.79	1.4	282.29	7JA	2.0	-4.0	-2.1	1.0
334.23	-0.66	1.5	292.39	17JA	0.7	-4.0	-2.1	1.0
341.53	-0.53	1.5	302.44	27JA	0.4	-4.1	-2.0	1.0
348.77	-0.41	1.6	312.44	6FE	0.2	-4.2	-1.9	0.9
355.94	-0.28	1.6	322.38	16FE	-0.1	-4.3	-1.9	0.9
3.03	-0.17	1.7	332.26	26FE	-0.6	-4.3	-1.8	0.8
10.05	-0.05	1.7	342.08	8MR	-1.3	-4.1	-1.7	0.8
17.00	0.06	1.8	351.84	18MR	-1.5	-3.6	-1.7	0.8
23.88	0.16	1.8	1.55	28MR	-0.7	-3.1	-1.6	0.8
30.70	0.26	1.8	11.21	7AP	0.4	-3.6	-1.5	0.8
37.45	0.35	1.9	20.83	17AP	1.5	-4.0	-1.5	0.8
44.14	0.44	1.9	30.40	27AP	2.9	-4.2	-1.5	0.8
50.78	0.53	1.9	39.95	7MY	3.2	-4.2	-1.4	0.8
57.37	0.61	1.9	49.48	17MY	1.8	-4.1	-1.4	0.8
63.91	0.69	1.9	59.00	27MY	0.8	-4.0	-1.4	0.7
70.42	0.76	1.9	68.52	6JN	-0.3	-3.9	-1.3	0.7
76.88	0.83	1.9	78.04	16JN	-1.3	-3.8	-1.3	0.7
83.32	0.89	1.9	87.59	26JN	-1.6	-3.8	-1.3	0.6
89.73	0.95	2.0	97.15	6JL	-0.8	-3.7	-1.3	0.6
96.12	1.01	2.0	106.76	16JL	-0.2	-3.6	-1.3	0.5
102.49	1.06	2.0	116.40	26JL	0.2	-3.6	-1.4	0.5
108.84	1.11	2.0	126.09	5AU	0.4	-3.5	-1.4	0.4
115.18	1.16	2.0	135.84	15AU	0.7	-3.5	-1.4	0.4
121.51	1.20	2.0	145.64	25AU	1.6	-3.4	-1.4	0.3
127.83	1.24	2.0	155.50	4SE	2.8	-3.4	-1.5	0.3
134.15	1.28	2.0	165.42	14SE	0.2	-3.4	-1.5	0.2
140.45	1.31	2.0	175.40	24SE	-0.7	-3.4	-1.6	0.2
146.75	1.34	1.9	185.43	4OC	-0.8	-3.4	-1.7	0.3
153.04	1.36	1.9	195.52	14OC	-0.8	-3.4	-1.7	0.4
159.31	1.39	1.8	205.65	24OC	-0.7	-3.4	-1.8	0.5
165.57	1.40	1.8	215.81	3NO	-0.6	-3.4	-1.9	0.5
171.81	1.41	1.7	226.00	13NO	-0.6	-3.4	-1.9	0.6
178.02	1.42	1.6	236.21	23NO	-0.6	-3.4	-2.0	0.6
184.20	1.42	1.5	246.42	3DE	-0.1	-3.4	-2.0	0.7
190.34	1.41	1.4	256.63	13DE	2.0	-3.4	-2.1	0.7
196.42	1.39	1.3	266.81	23DE	1.6	-3.4	-2.1	0.7

♂ LONG	♂ LAT	♂ MAG	⊙ LONG	16.00UT -476	☿ MAGNITUDES	♀ MAGNITUDES	♃ MAGNITUDES	♄ MAGNITUDES
202.44	1.36	1.2	276.97	2JA	0.5	-3.5	-2.1	0.8
208.38	1.31	1.1	287.10	12JA	0.2	-3.5	-2.1	0.8
214.21	1.25	0.9	297.18	22JA	0.0	-3.5	-2.1	0.8
219.93	1.17	0.7	307.20	1FE	-0.2	-3.4	-2.0	0.7
225.48	1.06	0.6	317.18	11FE	-0.6	-3.4	-2.0	0.7
230.85	0.92	0.4	327.09	21FE	-1.3	-3.4	-1.9	0.7
235.97	0.74	0.1	336.94	2MR	-1.5	-3.4	-1.9	0.6
240.78	0.52	-0.1	346.73	12MR	-0.7	-3.3	-1.8	0.6
245.20	0.25	-0.3	356.47	22MR	0.6	-3.3	-1.8	0.5
249.12	-0.09	-0.6	6.15	1AP	2.0	-3.3	-1.7	0.6
252.39	-0.51	-0.9	15.79	11AP	3.8	-3.3	-1.6	0.6
254.86	-1.02	-1.2	25.38	21AP	2.4	-3.3	-1.6	0.6
256.34	-1.63	-1.5	34.95	1MY	1.4	-3.4	-1.5	0.6
256.65	-2.33	-1.8	44.49	11MY	0.6	-3.4	-1.4	0.6
255.73	-3.09	-2.1	54.01	21MY	-0.3	-3.4	-1.4	0.6
253.70	-3.85	-2.4	63.53	31MY	-1.3	-3.4	-1.4	0.6
251.00	-4.50	-2.5	73.05	10JN	-1.6	-3.5	-1.3	0.5
248.33	-4.95	-2.4	82.58	20JN	-0.8	-3.5	-1.3	0.5
246.38	-5.17	-2.2	92.13	30JN	-0.1	-3.6	-1.3	0.5
245.62	-5.17	-2.0	101.72	10JL	0.3	-3.6	-1.3	0.4
246.23	-5.02	-1.7	111.33	20JL	0.6	-3.7	-1.3	0.4
248.12	-4.77	-1.5	121.00	30JL	1.0	-3.8	-1.2	0.3
251.13	-4.47	-1.2	130.72	9AU	2.1	-3.9	-1.3	0.3
255.06	-4.13	-1.0	140.49	19AU	2.4	-4.0	-1.3	0.2
259.71	-3.79	-0.8	150.32	29AU	0.1	-4.1	-1.3	0.1
264.94	-3.44	-0.6	160.21	8SE	-0.9	-4.2	-1.3	0.1
270.61	-3.10	-0.4	170.16	18SE	-1.0	-4.3	-1.3	0.0
276.62	-2.77	-0.2	180.16	28SE	-0.9	-4.3	-1.4	-0.0
282.90	-2.45	0.0	190.22	8OC	-0.6	-4.2	-1.4	0.0
289.37	-2.15	0.2	200.33	18OC	-0.5	-3.9	-1.5	0.1
295.99	-1.86	0.3	210.48	28OC	-0.4	-3.1	-1.5	0.2
302.72	-1.60	0.5	220.66	7NO	-0.4	-3.6	-1.6	0.2
309.51	-1.34	0.6	230.85	17NO	0.1	-4.1	-1.7	0.3
316.35	-1.11	0.8	241.06	27NO	2.3	-4.4	-1.7	0.4
323.21	-0.89	0.9	251.28	7DE	1.2	-4.4	-1.8	0.4
330.08	-0.69	1.0	261.47	17DE	0.2	-4.3	-1.9	0.5
336.94	-0.50	1.2	271.65	27DE	0.0	-4.2	-1.9	0.5
				-475				
343.79	-0.33	1.3	281.80	6JA	-0.1	-4.1	-2.0	0.5
350.61	-0.17	1.4	291.90	16JA	-0.3	-4.0	-2.0	0.6
357.39	-0.02	1.5	301.95	26JA	-0.7	-3.9	-2.0	0.6
4.14	0.11	1.5	311.96	5FE	-1.3	-3.8	-2.0	0.6
10.85	0.23	1.6	321.90	15FE	-1.4	-3.7	-2.0	0.5
17.52	0.35	1.7	331.79	25FE	-0.6	-3.6	-2.0	0.5
24.15	0.45	1.8	341.61	7MR	0.8	-3.5	-2.0	0.5
30.73	0.54	1.8	351.37	17MR	2.5	-3.5	-1.9	0.4
37.28	0.63	1.9	1.09	27MR	3.1	-3.4	-1.9	0.4
43.79	0.70	1.9	10.75	6AP	1.8	-3.4	-1.8	0.4
50.27	0.77	1.9	20.36	16AP	1.0	-3.4	-1.7	0.4
56.72	0.84	2.0	29.95	26AP	0.4	-3.3	-1.7	0.4
63.14	0.89	2.0	39.50	6MY	-0.4	-3.3	-1.6	0.4
69.54	0.94	2.0	49.02	16MY	-1.3	-3.3	-1.6	0.4
75.92	0.99	2.0	58.55	26MY	-1.7	-3.3	-1.5	0.4
82.29	1.03	2.0	68.06	5JN	-0.7	-3.3	-1.4	0.4
88.65	1.06	2.0	77.58	15JN	-0.0	-3.3	-1.4	0.4
95.01	1.09	2.0	87.13	25JN	0.4	-3.3	-1.3	0.4
101.37	1.12	2.0	96.69	5JL	0.8	-3.4	-1.3	0.4
107.73	1.14	2.0	106.29	15JL	1.3	-3.4	-1.3	0.3
114.11	1.15	2.0	115.93	25JL	2.6	-3.4	-1.2	0.3
120.49	1.17	2.0	125.62	4AU	2.1	-3.5	-1.2	0.2
126.90	1.17	2.0	135.36	14AU	0.0	-3.5	-1.2	0.2
133.32	1.18	2.0	145.16	24AU	-1.0	-3.5	-1.2	0.1
139.77	1.17	2.0	155.02	3SE	-1.1	-3.4	-1.2	0.1
146.24	1.17	2.0	164.93	13SE	-1.0	-3.4	-1.2	-0.0
152.74	1.15	2.0	174.91	23SE	-0.6	-3.4	-1.2	-0.1
159.27	1.13	2.0	184.94	3OC	-0.4	-3.4	-1.3	-0.1
165.83	1.11	1.9	195.02	13OC	-0.3	-3.3	-1.3	-0.2
172.43	1.08	1.9	205.15	23OC	-0.2	-3.3	-1.3	-0.1
179.05	1.04	1.9	215.31	2NO	0.2	-3.3	-1.4	-0.1
185.71	1.00	1.8	225.50	12NO	2.6	-3.3	-1.4	0.0
192.39	0.94	1.7	235.71	22NO	0.9	-3.3	-1.5	0.1
199.11	0.88	1.7	245.92	2DE	0.0	-3.4	-1.6	0.2
205.85	0.81	1.6	256.13	12DE	-0.1	-3.4	-1.6	0.2
212.62	0.72	1.5	266.32	22DE	-0.2	-3.4	-1.7	0.3

Left page (-474, -473):

♂ LONG	LAT	MAG	☉ LONG	16.00UT	☿	♀	♃	♄
				-474		MAGNITUDES		
219.42	0.62	1.4	276.48	1JA	-0.4	-3.4	-1.8	0.3
226.24	0.51	1.4	286.61	11JA	-0.7	-3.5	-1.8	0.4
233.08	0.39	1.3	296.69	21JA	-1.3	-3.5	-1.9	0.4
239.92	0.24	1.1	306.72	31JA	-1.3	-3.6	-1.9	0.4
246.78	0.08	1.0	316.69	10FE	-0.6	-3.6	-2.0	0.4
253.64	-0.10	0.9	326.61	20FE	0.9	-3.7	-2.0	0.4
260.49	-0.31	0.8	336.47	2MR	3.0	-3.8	-2.0	0.4
267.32	-0.53	0.7	346.26	12MR	2.4	-3.8	-2.0	0.4
274.11	-0.78	0.5	356.01	22MR	1.3	-3.9	-2.0	0.3
280.84	-1.05	0.4	5.69	1AP	0.8	-4.0	-2.0	0.3
287.49	-1.35	0.2	15.33	11AP	0.3	-4.1	-1.9	0.3
294.02	-1.67	0.1	24.93	21AP	-0.4	-4.2	-1.9	0.2
300.39	-2.02	-0.1	34.49	1MY	-1.3	-4.2	-1.8	0.3
306.54	-2.39	-0.3	44.03	11MY	-1.7	-4.1	-1.8	0.3
312.41	-2.79	-0.4	53.56	21MY	-0.7	-3.8	-1.7	0.3
317.91	-3.22	-0.6	63.07	31MY	0.0	-3.1	-1.7	0.3
322.93	-3.66	-0.8	72.59	10JN	0.6	-3.1	-1.6	0.3
327.31	-4.13	-1.1	82.12	20JN	1.0	-3.8	-1.5	0.3
330.92	-4.60	-1.3	91.67	30JN	1.8	-4.1	-1.5	0.3
333.52	-5.07	-1.5	101.25	10JL	3.1	-4.2	-1.4	0.3
334.90	-5.51	-1.8	110.87	20JL	1.8	-4.2	-1.4	0.3
334.91	-5.85	-2.0	120.53	30JL	-0.1	-4.1	-1.3	0.3
333.52	-6.03	-2.2	130.24	9AU	-1.1	-4.0	-1.3	0.2
330.99	-5.96	-2.4	140.01	19AU	-1.3	-4.0	-1.3	0.2
327.95	-5.62	-2.4	149.84	29AU	-0.9	-3.9	-1.3	0.1
325.17	-5.04	-2.2	159.73	8SE	-0.5	-3.8	-1.2	0.0
323.34	-4.32	-1.9	169.67	18SE	-0.2	-3.7	-1.2	-0.0
322.78	-3.56	-1.6	179.67	28SE	-0.1	-3.7	-1.2	-0.1
323.53	-2.85	-1.2	189.73	8OC	-0.0	-3.6	-1.2	-0.2
325.46	-2.22	-0.9	199.84	18OC	0.5	-3.6	-1.3	-0.2
328.34	-1.67	-0.6	209.98	28OC	2.9	-3.5	-1.3	-0.3
331.99	-1.21	-0.3	220.16	7NO	0.7	-3.5	-1.3	-0.2
336.24	-0.82	-0.1	230.36	17NO	-0.2	-3.4	-1.3	-0.2
340.94	-0.49	0.2	240.57	27NO	-0.3	-3.4	-1.4	-0.1
346.00	-0.21	0.4	250.78	7DE	-0.3	-3.4	-1.4	-0.0
351.33	0.03	0.6	260.98	17DE	-0.5	-3.4	-1.5	0.1
356.87	0.23	0.8	271.16	27DE	-0.8	-3.3	-1.6	0.1
				-473				
2.57	0.40	0.9	281.31	6JA	-1.1	-3.3	-1.6	0.2
8.39	0.54	1.1	291.41	16JA	-1.1	-3.3	-1.7	0.2
14.30	0.67	1.2	301.47	26JA	-0.5	-3.3	-1.8	0.3
20.28	0.77	1.3	311.47	5FE	1.1	-3.3	-1.8	0.3
26.32	0.86	1.5	321.42	15FE	3.2	-3.4	-1.9	0.3
32.39	0.94	1.6	331.31	25FE	1.7	-3.4	-1.9	0.3
38.49	1.00	1.6	341.14	7MR	1.0	-3.4	-2.0	0.3
44.61	1.05	1.7	350.90	17MR	0.6	-3.4	-2.0	0.3
50.75	1.10	1.8	0.62	27MR	0.2	-3.5	-2.1	0.3
56.90	1.14	1.9	10.28	6AP	-0.4	-3.5	-2.1	0.3
63.07	1.16	1.9	19.90	16AP	-1.3	-3.4	-2.1	0.2
69.24	1.19	1.9	29.48	26AP	-1.7	-3.4	-2.1	0.2
75.44	1.20	2.0	39.03	6MY	-0.7	-3.4	-2.0	0.2
81.64	1.21	2.0	48.56	16MY	0.1	-3.4	-2.0	0.2
87.86	1.22	2.0	58.08	26MY	0.8	-3.3	-1.9	0.2
94.11	1.22	2.0	67.60	5JN	1.4	-3.3	-1.9	0.3
100.37	1.21	2.0	77.12	15JN	2.3	-3.3	-1.8	0.3
106.66	1.20	2.0	86.66	25JN	3.2	-3.3	-1.8	0.3
112.98	1.19	2.0	96.22	5JL	1.4	-3.3	-1.7	0.3
119.34	1.17	2.0	105.82	15JL	-0.1	-3.4	-1.6	0.3
125.73	1.15	2.0	115.46	25JL	-1.1	-3.4	-1.6	0.3
132.16	1.12	2.0	125.15	4AU	-1.4	-3.4	-1.5	0.3
138.64	1.09	1.9	134.88	14AU	-0.9	-3.4	-1.5	0.2
145.17	1.05	1.9	144.68	24AU	-0.4	-3.5	-1.4	0.2
151.75	1.01	1.9	154.53	3SE	-0.1	-3.5	-1.4	0.1
158.38	0.96	1.9	164.45	13SE	0.0	-3.6	-1.4	0.1
165.07	0.91	1.9	174.42	23SE	0.1	-3.7	-1.3	0.0
171.81	0.85	1.9	184.45	3OC	0.7	-3.8	-1.3	-0.1
178.62	0.79	1.9	194.53	13OC	3.3	-3.9	-1.3	-0.1
185.49	0.71	1.8	204.66	23OC	0.5	-4.0	-1.3	-0.2
192.42	0.64	1.8	214.82	2NO	-0.3	-4.1	-1.3	-0.3
199.41	0.55	1.8	225.01	12NO	-0.4	-4.2	-1.3	-0.3
206.47	0.46	1.7	235.21	22NO	-0.4	-4.3	-1.3	-0.3
213.58	0.36	1.7	245.43	2DE	-0.6	-4.4	-1.4	-0.2
220.76	0.26	1.7	255.63	12DE	-0.8	-4.4	-1.4	-0.2
227.99	0.14	1.6	265.82	22DE	-1.0	-4.3	-1.4	-0.1

Right page (-472, -471):

♂ LONG	LAT	MAG	☉ LONG	16.00UT	☿	♀	♃	♄
				-472		MAGNITUDES		
235.28	0.02	1.5	275.99	1JA	-1.0	-3.9	-1.5	0.0
242.61	-0.11	1.5	286.12	11JA	-0.4	-3.3	-1.5	0.1
250.00	-0.25	1.4	296.20	21JA	1.3	-3.6	-1.6	0.1
257.43	-0.40	1.4	306.24	31JA	2.8	-4.1	-1.6	0.2
264.89	-0.55	1.3	316.21	10FE	1.3	-4.3	-1.7	0.2
272.37	-0.70	1.2	326.13	20FE	0.7	-4.3	-1.8	0.3
279.88	-0.86	1.2	335.99	1MR	0.4	-4.2	-1.9	0.3
287.39	-1.02	1.1	345.79	11MR	0.1	-4.1	-1.9	0.3
294.89	-1.18	1.0	355.53	21MR	-0.5	-4.0	-2.0	0.3
302.37	-1.34	1.0	5.22	31MR	-1.3	-3.9	-2.0	0.3
309.82	-1.50	0.9	14.86	10AP	-1.6	-3.8	-2.1	0.2
317.21	-1.65	0.8	24.46	20AP	-0.7	-3.7	-2.1	0.2
324.52	-1.80	0.7	34.03	30AP	0.2	-3.6	-2.2	0.2
331.74	-1.93	0.6	43.57	10MY	1.0	-3.6	-2.2	0.2
338.83	-2.05	0.6	53.10	20MY	1.8	-3.5	-2.2	0.1
345.76	-2.16	0.5	62.61	30MY	3.1	-3.4	-2.2	0.2
352.51	-2.26	0.4	72.13	9JN	2.7	-3.4	-2.1	0.3
359.04	-2.33	0.3	81.66	19JN	1.2	-3.4	-2.1	0.3
5.29	-2.40	0.2	91.21	29JN	-0.2	-3.4	-2.0	0.3
11.22	-2.44	0.1	100.79	9JL	-1.2	-3.3	-2.0	0.3
16.75	-2.46	-0.0	110.41	19JL	-1.5	-3.3	-1.9	0.3
21.80	-2.46	-0.2	120.07	29JL	-0.8	-3.3	-1.8	0.3
26.25	-2.43	-0.3	129.77	8AU	-0.3	-3.3	-1.8	0.3
29.97	-2.37	-0.5	139.54	18AU	-0.0	-3.4	-1.7	0.3
32.78	-2.27	-0.7	149.36	28AU	0.1	-3.4	-1.7	0.2
34.48	-2.12	-0.9	159.24	7SE	0.3	-3.4	-1.6	0.2
34.86	-1.90	-1.1	169.19	17SE	1.0	-3.4	-1.6	0.2
33.78	-1.60	-1.4	179.19	27SE	3.4	-3.4	-1.5	0.1
31.32	-1.21	-1.5	189.24	7OC	0.4	-3.5	-1.5	0.0
27.82	-0.76	-1.7	199.34	17OC	-0.5	-3.5	-1.4	-0.0
24.03	-0.27	-1.6	209.49	27OC	-0.6	-3.5	-1.4	-0.1
20.75	0.20	-1.4	219.66	6NO	-0.6	-3.5	-1.4	-0.2
18.55	0.59	-1.1	229.86	16NO	-0.7	-3.4	-1.4	-0.3
17.72	0.90	-0.8	240.07	26NO	-0.8	-3.4	-1.4	-0.3
18.19	1.14	-0.2	250.28	6DE	-0.8	-3.4	-1.4	-0.2
19.79	1.30	-0.2	260.48	16DE	-0.8	-3.4	-1.4	-0.2
22.33	1.43	0.1	270.66	26DE	-0.3	-3.4	-1.4	-0.1
				-471				
25.59	1.51	0.4	280.81	5JA	1.6	-3.3	-1.5	-0.0
29.43	1.57	0.6	290.92	15JA	2.3	-3.3	-1.5	0.0
33.73	1.60	0.8	300.98	25JA	0.9	-3.3	-1.5	0.1
38.37	1.62	1.0	310.99	4FE	0.5	-3.3	-1.6	0.1
43.29	1.63	1.1	320.94	14FE	0.2	-3.4	-1.6	0.2
48.43	1.63	1.3	330.83	24FE	-0.0	-3.4	-1.7	0.2
53.75	1.62	1.4	340.66	6MR	-0.5	-3.4	-1.8	0.3
59.21	1.60	1.5	350.43	16MR	-1.3	-3.4	-1.8	0.3
64.79	1.58	1.6	0.15	26MR	-1.6	-3.5	-1.9	0.3
70.47	1.55	1.7	9.81	5AP	-0.7	-3.5	-2.0	0.3
76.24	1.52	1.7	19.44	15AP	0.3	-3.6	-2.0	0.3
82.09	1.49	1.8	29.02	25AP	1.3	-3.6	-2.1	0.3
88.00	1.45	1.9	38.57	5MY	2.4	-3.7	-2.2	0.3
93.99	1.41	1.9	48.11	15MY	3.6	-3.8	-2.2	0.3
100.03	1.37	1.9	57.62	25MY	2.1	-3.8	-2.3	0.2
106.14	1.32	2.0	67.14	4JN	0.9	-3.9	-2.3	0.2
112.30	1.27	2.0	76.67	14JN	-0.2	-4.0	-2.3	0.3
118.53	1.22	2.0	86.20	24JN	-1.3	-4.1	-2.3	0.3
124.82	1.16	2.0	95.76	4JL	-1.6	-4.2	-2.3	0.3
131.18	1.10	2.0	105.36	14JL	-0.8	-4.2	-2.2	0.3
137.60	1.04	2.0	115.00	24JL	-0.2	-4.1	-2.2	0.4
144.09	0.97	1.9	124.68	3AU	0.1	-3.7	-2.1	0.4
150.66	0.90	1.9	134.42	13AU	0.3	-3.2	-2.1	0.4
157.30	0.83	1.9	144.20	23AU	0.6	-3.6	-2.0	0.4
164.02	0.75	1.8	154.06	2SE	1.3	-4.1	-1.9	0.3
170.81	0.67	1.8	163.97	12SE	3.1	-4.3	-1.9	0.3
177.69	0.58	1.7	173.93	22SE	0.3	-4.3	-1.8	0.3
184.65	0.49	1.7	183.96	2OC	-0.6	-4.3	-1.8	0.2
191.69	0.40	1.7	194.04	12OC	-0.7	-4.2	-1.7	0.2
198.80	0.30	1.7	204.16	22OC	-0.7	-4.1	-1.6	0.1
206.01	0.20	1.6	214.32	1NO	-0.8	-4.0	-1.6	-0.0
213.29	0.09	1.6	224.51	11NO	-0.7	-3.9	-1.6	-0.0
220.64	-0.02	1.6	234.71	21NO	-0.7	-3.8	-1.5	-0.1
228.07	-0.13	1.6	244.93	1DE	-0.7	-3.7	-1.5	-0.2
235.56	-0.24	1.6	255.14	11DE	-0.2	-3.7	-1.5	-0.2
243.12	-0.36	1.5	265.33	21DE	1.8	-3.6	-1.5	-0.2
250.74	-0.47	1.5	275.50	31DE	1.8	-3.5	-1.5	-0.1

♂ LONG	LAT	MAG	⊙ LONG	16.00UT	☿	♀	♃	♄
				-470	MAGNITUDES			
258.40	-0.59	1.5	285.63	10JA	0.6	-3.5	-1.5	-0.0
266.10	-0.70	1.5	295.71	20JA	0.3	-3.4	-1.5	0.0
273.83	-0.81	1.4	305.75	30JA	0.1	-3.4	-1.5	0.1
281.58	-0.92	1.4	315.73	9FE	-0.1	-3.4	-1.6	0.1
289.34	-1.02	1.4	325.65	19FE	-0.6	-3.3	-1.6	0.2
297.09	-1.11	1.3	335.52	1MR	-1.3	-3.3	-1.6	0.3
304.83	-1.20	1.3	345.32	11MR	-1.5	-3.3	-1.7	0.3
312.54	-1.27	1.3	355.06	21MR	-0.7	-3.3	-1.7	0.3
320.20	-1.33	1.3	4.76	31MR	0.5	-3.3	-1.8	0.4
327.81	-1.38	1.2	14.40	10AP	1.7	-3.3	-1.9	0.4
335.35	-1.42	1.2	24.00	20AP	3.2	-3.3	-1.9	0.4
342.82	-1.44	1.2	33.57	30AP	2.9	-3.4	-2.0	0.4
350.18	-1.45	1.2	43.11	10MY	1.6	-3.4	-2.1	0.4
357.44	-1.44	1.1	52.63	20MY	0.7	-3.4	-2.1	0.4
4.57	-1.42	1.1	62.15	30MY	-0.3	-3.5	-2.2	0.3
11.57	-1.38	1.1	71.67	9JN	-1.3	-3.5	-2.3	0.3
18.43	-1.33	1.0	81.20	19JN	-1.6	-3.5	-2.3	0.3
25.11	-1.26	1.0	90.75	29JN	-0.8	-3.4	-2.4	0.4
31.61	-1.18	0.9	100.32	9JL	-0.2	-3.4	-2.4	0.4
37.90	-1.08	0.9	109.94	19JL	0.2	-3.4	-2.4	0.4
43.95	-0.97	0.8	119.60	29JL	0.5	-3.3	-2.4	0.5
49.72	-0.83	0.7	129.30	8AU	0.8	-3.3	-2.4	0.5
55.18	-0.68	0.6	139.06	18AU	1.8	-3.3	-2.3	0.5
60.25	-0.50	0.5	148.88	28AU	2.7	-3.3	-2.3	0.5
64.86	-0.30	0.4	158.76	7SE	0.2	-3.3	-2.2	0.5
68.92	-0.07	0.2	168.70	17SE	-0.8	-3.3	-2.2	0.5
72.29	0.20	0.1	178.70	27SE	-0.9	-3.4	-2.1	0.4
74.81	0.52	-0.1	188.75	7OC	-0.9	-3.4	-2.0	0.4
76.30	0.88	-0.3	198.85	17OC	-0.7	-3.4	-2.0	0.3
76.55	1.28	-0.6	208.99	27OC	-0.5	-3.5	-1.9	0.3
75.43	1.73	-0.8	219.16	6NO	-0.5	-3.5	-1.8	0.2
72.95	2.18	-1.0	229.36	16NO	-0.5	-3.6	-1.8	0.2
69.41	2.59	-1.2	239.57	26NO	-0.1	-3.6	-1.7	0.1
65.43	2.90	-1.2	249.78	6DE	2.0	-3.7	-1.7	0.0
61.79	3.08	-1.0	259.98	16DE	1.5	-3.8	-1.6	-0.1
59.12	3.13	-0.7	270.17	26DE	0.4	-3.9	-1.6	-0.1
				-469				
57.74	3.10	-0.4	280.32	5JA	0.1	-4.0	-1.6	-0.0
57.67	3.00	-0.1	290.43	15JA	-0.0	-4.1	-1.6	0.0
58.77	2.87	0.1	300.49	25JA	-0.2	-4.1	-1.6	0.1
60.84	2.74	0.4	310.50	4FE	-0.6	-4.2	-1.6	0.1
63.70	2.59	0.6	320.45	14FE	-1.3	-4.3	-1.6	0.2
67.19	2.46	0.8	330.35	24FE	-1.4	-4.3	-1.6	0.3
71.17	2.32	0.9	340.18	6MR	-0.7	-4.1	-1.6	0.3
75.55	2.20	1.1	349.96	16MR	0.6	-3.6	-1.6	0.4
80.26	2.08	1.2	359.68	26MR	2.1	-3.1	-1.6	0.4
85.23	1.96	1.3	9.35	5AP	3.7	-3.6	-1.7	0.4
90.41	1.85	1.4	18.97	15AP	2.2	-4.0	-1.7	0.5
95.79	1.74	1.5	28.56	25AP	1.2	-4.2	-1.8	0.5
101.32	1.63	1.6	38.11	5MY	0.5	-4.2	-1.8	0.5
107.00	1.53	1.7	47.64	15MY	-0.3	-4.1	-1.9	0.5
112.81	1.42	1.7	57.16	25MY	-1.3	-4.0	-2.0	0.5
118.74	1.32	1.7	66.68	4JN	-1.7	-3.9	-2.0	0.5
124.78	1.22	1.8	76.20	14JN	-0.8	-3.8	-2.1	0.5
130.93	1.12	1.8	85.74	24JN	-0.1	-3.8	-2.2	0.5
137.19	1.01	1.8	95.30	4JL	0.3	-3.7	-2.2	0.5
143.54	0.91	1.8	104.89	14JL	0.6	-3.6	-2.3	0.5
150.00	0.81	1.8	114.53	24JL	1.1	-3.6	-2.4	0.6
156.56	0.70	1.8	124.21	3AU	2.2	-3.5	-2.4	0.6
163.22	0.60	1.8	133.94	13AU	2.4	-3.5	-2.4	0.6
169.98	0.49	1.8	143.73	23AU	0.1	-3.4	-2.4	0.7
176.84	0.38	1.7	153.58	2SE	-0.9	-3.4	-2.4	0.7
183.81	0.28	1.7	163.49	12SE	-1.0	-3.4	-2.4	0.7
190.87	0.17	1.7	173.46	22SE	-1.0	-3.4	-2.4	0.7
198.03	0.06	1.6	183.48	2OC	-0.6	-3.4	-2.3	0.6
205.29	-0.05	1.6	193.55	12OC	-0.4	-3.4	-2.3	0.6
212.64	-0.16	1.6	203.68	22OC	-0.4	-3.4	-2.2	0.6
220.08	-0.27	1.5	213.83	1NO	-0.3	-3.4	-2.2	0.5
227.60	-0.37	1.5	224.02	11NO	0.1	-3.4	-2.1	0.5
235.20	-0.48	1.4	234.22	21NO	2.3	-3.4	-2.0	0.4
242.86	-0.58	1.4	244.43	1DE	1.2	-3.4	-1.9	0.3
250.59	-0.67	1.4	254.64	11DE	0.2	-3.4	-1.9	0.3
258.37	-0.76	1.4	264.83	21DE	-0.0	-3.4	-1.8	0.2
266.18	-0.84	1.4	275.00	31DE	-0.1	-3.5	-1.8	0.1

♂ LONG	LAT	MAG	⊙ LONG	16.00UT	☿	♀	♃	♄
				-468	MAGNITUDES			
274.03	-0.92	1.4	285.13	10JA	-0.3	-3.5	-1.7	0.1
281.89	-0.98	1.4	295.22	20JA	-0.7	-3.5	-1.7	0.1
289.75	-1.04	1.4	305.25	30JA	-1.3	-3.4	-1.6	0.2
297.60	-1.09	1.4	315.24	9FE	-1.3	-3.4	-1.6	0.3
305.43	-1.12	1.4	325.17	19FE	-0.6	-3.4	-1.6	0.3
313.23	-1.14	1.4	335.03	29FE	0.8	-3.4	-1.6	0.4
320.98	-1.15	1.4	344.84	10MR	2.7	-3.3	-1.6	0.4
328.68	-1.15	1.4	354.59	20MR	2.8	-3.3	-1.6	0.5
336.31	-1.14	1.4	4.28	30MR	1.6	-3.3	-1.6	0.5
343.87	-1.11	1.4	13.93	9AP	0.9	-3.3	-1.6	0.6
351.34	-1.08	1.5	23.53	19AP	0.4	-3.3	-1.6	0.6
358.72	-1.03	1.5	33.10	29AP	-0.4	-3.4	-1.6	0.6
6.01	-0.97	1.5	42.65	9MY	-1.3	-3.4	-1.7	0.7
13.19	-0.90	1.5	52.17	19MY	-1.7	-3.4	-1.7	0.7
20.27	-0.81	1.5	61.69	29MY	-0.8	-3.4	-1.8	0.7
27.23	-0.72	1.5	71.21	8JN	-0.0	-3.5	-1.8	0.7
34.08	-0.63	1.5	80.74	18JN	0.5	-3.5	-1.9	0.7
40.80	-0.52	1.4	90.28	28JN	0.9	-3.6	-1.9	0.7
47.39	-0.40	1.4	99.86	8JL	1.5	-3.6	-2.0	0.7
53.85	-0.27	1.4	109.47	18JL	2.8	-3.7	-2.1	0.7
60.15	-0.14	1.4	119.12	28JL	2.0	-3.8	-2.1	0.7
66.29	0.01	1.3	128.83	7AU	0.0	-3.9	-2.2	0.8
72.25	0.16	1.3	138.59	17AU	-1.0	-4.0	-2.3	0.8
78.01	0.33	1.2	148.41	27AU	-1.2	-4.1	-2.3	0.9
83.53	0.51	1.1	158.29	6SE	-0.9	-4.2	-2.4	0.9
88.78	0.72	1.0	168.22	16SE	-0.5	-4.3	-2.4	0.9
93.72	0.93	0.9	178.21	26SE	-0.3	-4.3	-2.4	0.9
98.26	1.18	0.8	188.27	6OC	-0.2	-4.2	-2.4	0.9
102.33	1.45	0.6	198.36	16OC	-0.2	-3.8	-2.4	0.9
105.83	1.75	0.5	208.50	26OC	0.3	-3.1	-2.4	0.8
108.60	2.09	0.3	218.67	5NO	2.6	-3.6	-2.3	0.8
110.49	2.47	0.1	228.87	15NO	0.9	-4.2	-2.3	0.8
111.32	2.89	-0.2	239.08	25NO	-0.0	-4.4	-2.2	0.7
110.89	3.32	-0.4	249.29	5DE	-0.2	-4.4	-2.1	0.6
109.13	3.75	-0.6	259.49	15DE	-0.2	-4.3	-2.0	0.6
106.15	4.11	-0.9	269.67	25DE	-0.4	-4.2	-2.0	0.5
				-467				
102.35	4.34	-1.0	279.83	4JA	-0.7	-4.1	-1.9	0.4
98.42	4.40	-0.9	289.94	14JA	-1.2	-4.0	-1.8	0.4
95.08	4.30	-0.7	300.00	24JA	-1.2	-3.9	-1.8	0.3
92.83	4.07	-0.5	310.02	3FE	-0.6	-3.8	-1.7	0.4
91.88	3.78	-0.2	319.97	13FE	1.0	-3.7	-1.7	0.4
92.18	3.47	0.0	329.87	23FE	3.1	-3.6	-1.6	0.5
93.58	3.16	0.2	339.71	5MR	2.1	-3.5	-1.6	0.5
95.91	2.87	0.4	349.48	15MR	1.2	-3.5	-1.6	0.6
98.97	2.60	0.6	359.21	25MR	0.7	-3.4	-1.5	0.7
102.64	2.35	0.8	8.88	4AP	0.2	-3.4	-1.5	0.7
106.81	2.12	0.9	18.50	14AP	-0.4	-3.4	-1.5	0.8
111.37	1.90	1.0	28.09	24AP	-1.3	-3.3	-1.5	0.8
116.26	1.71	1.2	37.65	4MY	-1.7	-3.3	-1.5	0.8
121.44	1.52	1.2	47.18	14MY	-0.8	-3.3	-1.5	0.9
126.86	1.34	1.3	56.70	24MY	0.1	-3.3	-1.5	0.9
132.50	1.17	1.4	66.22	3JN	0.6	-3.3	-1.6	0.9
138.33	1.01	1.4	75.74	13JN	1.1	-3.3	-1.6	0.9
144.34	0.86	1.5	85.28	23JN	2.0	-3.3	-1.6	1.0
150.51	0.71	1.5	94.84	3JL	3.2	-3.4	-1.7	1.0
156.83	0.57	1.5	104.43	13JL	1.7	-3.4	-1.7	1.0
163.30	0.43	1.6	114.06	23JL	-0.0	-3.4	-1.8	1.0
169.91	0.29	1.6	123.74	2AU	-1.1	-3.5	-1.9	1.0
176.65	0.16	1.6	133.47	12AU	-1.3	-3.5	-1.9	1.0
183.53	0.03	1.6	143.25	22AU	-0.9	-3.5	-2.0	1.1
190.53	-0.10	1.6	153.10	1SE	-0.4	-3.4	-2.0	1.1
197.65	-0.22	1.6	163.00	11SE	-0.2	-3.4	-2.1	1.1
204.88	-0.33	1.5	172.97	21SE	-0.1	-3.4	-2.2	1.2
212.23	-0.44	1.5	182.99	1OC	0.0	-3.4	-2.2	1.2
219.68	-0.55	1.5	193.06	11OC	0.5	-3.3	-2.3	1.2
227.22	-0.64	1.5	203.18	21OC	2.9	-3.3	-2.3	1.2
234.85	-0.73	1.5	213.33	31OC	0.7	-3.3	-2.3	1.2
242.56	-0.81	1.4	223.52	10NO	-0.2	-3.3	-2.3	1.1
250.33	-0.89	1.4	233.72	20NO	-0.3	-3.3	-2.3	1.1
258.15	-0.95	1.4	243.93	30NO	-0.4	-3.4	-2.3	1.0
266.01	-1.00	1.4	254.14	10DE	-0.5	-3.4	-2.2	1.0
273.90	-1.04	1.4	264.34	20DE	-0.8	-3.4	-2.2	0.9
281.80	-1.07	1.3	274.51	30DE	-1.1	-3.4	-2.1	0.8

Left table

♂ LONG	LAT	MAG	☉ LONG	16.00UT -466	☿	♀	♃	♄
289.70	-1.09	1.3	284.64	9JA	-1.1	-3.5	-2.0	0.8
297.58	-1.09	1.3	294.73	19JA	-0.5	-3.5	-2.0	0.7
305.44	-1.09	1.3	304.77	29JA	1.1	-3.6	-1.9	0.6
313.25	-1.07	1.3	314.76	8FE	3.1	-3.6	-1.8	0.6
321.02	-1.04	1.4	324.69	18FE	1.6	-3.7	-1.7	0.7
328.73	-1.00	1.4	334.56	28FE	0.9	-3.8	-1.7	0.7
336.36	-0.95	1.4	344.36	10MR	0.5	-3.8	-1.6	0.8
343.93	-0.89	1.5	354.12	20MR	-0.3	-3.9	-1.6	0.8
351.40	-0.83	1.5	3.81	30MR	-0.4	-4.0	-1.5	0.9
358.80	-0.75	1.5	13.46	9AP	-1.3	-4.1	-1.5	0.9
6.10	-0.67	1.6	23.07	19AP	-1.7	-4.2	-1.5	1.0
13.32	-0.58	1.6	32.64	29AP	-0.8	-4.2	-1.4	1.0
20.44	-0.49	1.6	42.18	9MY	0.1	-4.1	-1.4	1.1
27.46	-0.39	1.7	51.71	19MY	0.8	-3.8	-1.4	1.1
34.40	-0.29	1.7	61.23	29MY	1.5	-3.1	-1.4	1.1
41.24	-0.18	1.7	70.74	8JN	2.6	-3.2	-1.4	1.2
47.99	-0.07	1.7	80.28	18JN	3.1	-3.8	-1.4	1.2
54.64	0.04	1.7	89.82	28JN	1.4	-4.1	-1.4	1.2
61.20	0.16	1.7	99.39	8JL	-0.1	-4.2	-1.5	1.2
67.67	0.28	1.7	109.01	18JL	-1.2	-4.2	-1.5	1.2
74.03	0.40	1.7	118.66	28JL	-1.5	-4.1	-1.5	1.3
80.29	0.53	1.7	128.36	7AU	-0.9	-4.0	-1.6	1.2
86.45	0.66	1.7	138.12	17AU	-0.4	-3.9	-1.6	1.3
92.48	0.80	1.6	147.93	27AU	-0.1	-3.9	-1.7	1.3
98.39	0.94	1.6	157.80	6SE	0.1	-3.8	-1.7	1.4
104.16	1.09	1.5	167.74	16SE	0.2	-3.7	-1.8	1.4
109.76	1.25	1.5	177.73	26SE	0.8	-3.7	-1.9	1.4
115.17	1.43	1.4	187.77	6OC	3.3	-3.6	-1.9	1.4
120.35	1.61	1.3	197.87	16OC	0.5	-3.6	-2.0	1.3
125.26	1.81	1.1	208.00	26OC	-0.4	-3.5	-2.1	1.3
129.86	2.03	1.0	218.18	5NO	-0.5	-3.5	-2.1	1.3
134.05	2.27	0.8	228.37	15NO	-0.5	-3.4	-2.2	1.3
137.74	2.53	0.7	238.58	25NO	-0.6	-3.4	-2.2	1.2
140.83	2.82	0.5	248.79	5DE	-0.8	-3.4	-2.2	1.2
143.16	3.13	0.2	259.00	15DE	-0.9	-3.4	-2.2	1.1
144.56	3.47	-0.0	269.18	25DE	-0.9	-3.3	-2.2	1.1

-465

♂ LONG	LAT	MAG	☉ LONG	16.00UT	☿	♀	♃	♄
144.84	3.82	-0.3	279.33	4JA	-0.4	-3.3	-2.1	1.0
143.85	4.15	-0.5	289.45	14JA	1.4	-3.3	-2.1	1.0
141.60	4.41	-0.8	299.52	24JA	2.7	-3.3	-2.0	0.9
138.27	4.55	-1.0	309.53	3FE	1.1	-3.3	-2.0	0.9
134.41	4.52	-1.1	319.49	13FE	0.6	-3.4	-1.9	0.9
130.73	4.32	-1.0	329.39	23FE	0.3	-3.4	-1.8	0.9
127.86	3.97	-0.8	339.23	5MR	0.0	-3.4	-1.8	1.0
126.20	3.55	-0.5	349.01	15MR	-0.5	-3.4	-1.7	1.0
125.86	3.12	-0.3	358.73	25MR	-1.3	-3.5	-1.6	1.1
126.72	2.69	-0.1	8.41	4AP	-1.6	-3.5	-1.6	1.1
128.65	2.29	0.1	18.04	14AP	-0.7	-3.4	-1.5	1.2
131.46	1.93	0.3	27.62	24AP	0.2	-3.4	-1.5	1.2
134.98	1.61	0.4	37.18	4MY	1.1	-3.4	-1.4	1.3
139.09	1.31	0.6	46.72	14MY	2.0	-3.4	-1.4	1.3
143.69	1.04	0.7	56.23	24MY	3.3	-3.3	-1.4	1.4
148.70	0.79	0.8	65.75	3JN	2.5	-3.3	-1.3	1.4
154.07	0.57	0.9	75.27	13JN	1.1	-3.3	-1.3	1.4
159.73	0.36	1.0	84.81	23JN	-0.1	-3.3	-1.3	1.4
165.66	0.17	1.0	94.37	3JL	-1.2	-3.3	-1.3	1.4
171.83	-0.01	1.1	103.96	13JL	-1.6	-3.4	-1.3	1.4
178.21	-0.18	1.1	113.59	23JL	-0.8	-3.4	-1.3	1.4
184.79	-0.33	1.2	123.26	2AU	-0.3	-3.4	-1.3	1.3
191.55	-0.48	1.2	132.99	12AU	0.0	-3.4	-1.4	1.3
198.48	-0.61	1.2	142.77	22AU	0.2	-3.5	-1.4	1.2
205.56	-0.72	1.2	152.62	1SE	0.4	-3.6	-1.4	1.2
212.79	-0.83	1.3	162.52	11SE	1.1	-3.6	-1.5	1.2
220.14	-0.92	1.3	172.48	21SE	3.3	-3.7	-1.5	1.2
227.60	-1.01	1.3	182.50	1OC	0.4	-3.8	-1.6	1.2
235.17	-1.07	1.3	192.57	11OC	-0.5	-3.9	-1.7	1.2
242.83	-1.13	1.3	202.68	21OC	-0.6	-4.0	-1.7	1.2
250.56	-1.17	1.3	212.84	31OC	-0.6	-4.1	-1.8	1.2
258.35	-1.19	1.3	223.02	10NO	-0.7	-4.2	-1.9	1.1
266.18	-1.21	1.4	233.23	20NO	-0.7	-4.3	-1.9	1.1
274.04	-1.21	1.4	243.44	30NO	-0.8	-4.4	-2.0	1.1
281.92	-1.19	1.4	253.65	10DE	-0.8	-4.4	-2.0	1.0
289.78	-1.17	1.4	263.84	20DE	-0.3	-4.3	-2.1	1.0
297.63	-1.13	1.4	274.01	30DE	1.6	-3.9	-2.1	0.9

Right table

♂ LONG	LAT	MAG	☉ LONG	16.00UT -464	☿	♀	♃	♄
305.45	-1.08	1.4	284.15	9JA	2.2	-3.2	-2.1	0.9
313.22	-1.03	1.5	294.24	19JA	0.8	-3.7	-2.1	0.9
320.95	-0.96	1.5	304.29	29JA	0.4	-4.1	-2.1	0.8
328.60	-0.89	1.5	314.27	8FE	0.2	-4.3	-2.0	0.8
336.19	-0.81	1.5	324.20	18FE	-0.1	-4.3	-2.0	0.7
343.70	-0.72	1.5	334.08	28FE	-0.5	-4.2	-1.9	0.7
351.13	-0.63	1.5	343.89	9MR	-1.3	-4.1	-1.9	0.8
358.47	-0.53	1.5	353.64	19MR	-1.6	-4.0	-1.8	0.8
5.73	-0.44	1.6	3.34	29MR	-0.7	-3.9	-1.7	0.8
12.91	-0.34	1.6	12.99	8AP	0.4	-3.8	-1.7	1.0
20.00	-0.24	1.6	22.60	18AP	1.4	-3.7	-1.6	1.0
27.00	-0.14	1.7	32.18	28AP	2.7	-3.6	-1.5	1.1
33.93	-0.03	1.7	41.72	8MY	3.4	-3.6	-1.5	1.1
40.77	0.07	1.8	51.25	18MY	1.9	-3.5	-1.4	1.2
47.54	0.17	1.8	60.77	28MY	0.9	-3.4	-1.4	1.2
54.24	0.27	1.8	70.28	7JN	-0.2	-3.4	-1.3	1.2
60.86	0.37	1.9	79.81	17JN	-1.2	-3.4	-1.3	1.3
67.42	0.47	1.9	89.36	27JN	-1.6	-3.4	-1.3	1.3
73.92	0.57	1.9	98.93	7JL	-0.8	-3.3	-1.3	1.3
80.35	0.67	1.9	108.54	17JL	-0.2	-3.3	-1.2	1.3
86.72	0.77	1.9	118.19	27JL	0.1	-3.3	-1.2	1.2
93.04	0.86	1.9	127.89	6AU	0.4	-3.3	-1.2	1.2
99.29	0.96	1.9	137.64	16AU	0.6	-3.4	-1.3	1.2
105.47	1.06	1.9	147.46	26AU	1.5	-3.4	-1.3	1.1
111.59	1.16	1.9	157.32	5SE	3.0	-3.4	-1.3	1.1
117.64	1.26	1.8	167.25	15SE	0.3	-3.4	-1.3	1.1
123.61	1.37	1.8	177.24	25SE	-0.7	-3.4	-1.4	1.1
129.49	1.48	1.7	187.28	5OC	-0.8	-3.5	-1.4	1.1
135.27	1.59	1.6	197.38	15OC	-0.8	-3.5	-1.4	1.1
140.92	1.70	1.6	207.52	25OC	-0.8	-3.5	-1.5	1.1
146.44	1.82	1.5	217.68	4NO	-0.6	-3.5	-1.6	1.1
151.79	1.94	1.3	227.88	14NO	-0.6	-3.4	-1.6	1.0
156.94	2.07	1.2	238.09	24NO	-0.6	-3.4	-1.7	1.0
161.84	2.21	1.1	248.29	4DE	-0.2	-3.4	-1.8	1.0
166.44	2.36	0.9	258.50	14DE	1.8	-3.4	-1.8	0.9
170.67	2.51	0.7	268.68	24DE	1.7	-3.3	-1.9	0.9

-463

♂ LONG	LAT	MAG	☉ LONG	16.00UT	☿	♀	♃	♄
174.43	2.67	0.5	278.84	3JA	0.5	-3.3	-1.9	0.9
177.62	2.84	0.3	288.95	13JA	0.2	-3.3	-2.0	0.8
180.08	3.01	0.0	299.03	23JA	0.1	-3.3	-2.0	0.8
181.65	3.16	-0.3	309.04	2FE	-0.2	-3.3	-2.0	0.7
182.14	3.30	-0.6	319.01	12FE	-0.6	-3.4	-2.0	0.7
181.41	3.37	-0.9	328.91	22FE	-1.3	-3.4	-2.0	0.6
179.42	3.36	-1.1	338.75	4MR	-1.5	-3.4	-2.0	0.6
176.34	3.22	-1.4	348.54	14MR	-0.7	-3.4	-2.0	0.6
172.70	2.93	-1.5	358.27	24MR	0.5	-3.5	-1.9	0.6
169.18	2.52	-1.4	7.94	3AP	1.8	-3.5	-1.9	0.7
166.46	2.02	-1.2	17.57	13AP	3.5	-3.6	-1.8	0.8
164.98	1.51	-1.0	27.16	23AP	2.6	-3.6	-1.7	0.8
164.84	1.03	-0.8	36.72	3MY	1.5	-3.7	-1.7	0.9
165.97	0.59	-0.6	46.26	13MY	0.6	-3.8	-1.6	1.0
168.22	0.21	-0.4	55.78	23MY	-0.2	-3.8	-1.5	1.0
171.40	-0.12	-0.2	65.29	2JN	-1.3	-3.9	-1.5	1.0
175.35	-0.41	-0.1	74.82	12JN	-1.7	-4.0	-1.4	1.1
179.95	-0.66	0.1	84.35	22JN	-0.8	-4.1	-1.4	1.1
185.06	-0.87	0.2	93.91	2JL	-0.2	-4.2	-1.3	1.1
190.62	-1.06	0.3	103.50	12JL	0.2	-4.2	-1.3	1.1
196.56	-1.21	0.4	113.12	22JL	0.5	-4.1	-1.3	1.1
202.81	-1.34	0.5	122.79	1AU	0.9	-3.7	-1.2	1.1
209.34	-1.44	0.6	132.52	11AU	1.9	-3.6	-1.2	1.1
216.10	-1.52	0.6	142.30	21AU	2.7	-3.6	-1.2	1.1
223.07	-1.58	0.7	152.14	31AU	0.2	-4.1	-1.2	1.0
230.22	-1.62	0.8	162.04	10SE	-0.8	-4.3	-1.2	1.0
237.52	-1.64	0.8	172.00	20SE	-0.9	-4.3	-1.2	0.9
244.93	-1.64	0.9	182.01	30SE	-0.9	-4.3	-1.3	0.9
252.24	-1.62	1.0	192.08	10OC	-0.7	-4.2	-1.3	1.0
260.06	-1.59	1.0	202.19	20OC	-0.5	-4.1	-1.3	1.0
267.71	-1.54	1.1	212.35	30OC	-0.5	-4.0	-1.4	1.0
275.41	-1.47	1.1	222.53	9NO	-0.4	-3.9	-1.4	1.0
283.12	-1.40	1.2	232.73	19NO	-0.0	-3.8	-1.5	1.0
290.83	-1.31	1.2	242.94	29NO	2.0	-3.7	-1.5	0.9
298.53	-1.22	1.3	253.16	9DE	1.4	-3.7	-1.6	0.9
306.20	-1.12	1.3	263.35	19DE	0.3	-3.6	-1.6	0.9
313.83	-1.01	1.4	273.52	29DE	0.1	-3.5	-1.7	0.8

♂ LONG	♂ LAT	♂ MAG	⊙ LONG	16.00UT	☿	♀	♃	♄
				-462	MAGNITUDES			
321.41	-0.90	1.5	283.66	8JA	-0.1	-3.5	-1.8	0.8
328.93	-0.79	1.5	293.76	18JA	-0.3	-3.4	-1.8	0.8
336.37	-0.67	1.5	303.80	28JA	-0.6	-3.4	-1.9	0.7
343.75	-0.56	1.6	313.80	7FE	-1.3	-3.4	-1.9	0.7
351.06	-0.44	1.6	323.73	17FE	-1.4	-3.3	-2.0	0.6
358.28	-0.33	1.7	333.60	27FE	-0.7	-3.3	-2.0	0.6
5.43	-0.22	1.7	343.42	9MR	0.6	-3.3	-2.0	0.5
12.50	-0.11	1.7	353.18	19MR	2.3	-3.3	-2.0	0.5
19.49	-0.00	1.8	2.88	29MR	3.4	-3.3	-2.0	0.5
26.41	0.10	1.8	12.53	8AP	1.9	-3.3	-2.0	0.5
33.26	0.20	1.8	22.14	18AP	1.1	-3.3	-1.9	0.6
40.04	0.29	1.8	31.72	28AP	0.5	-3.4	-1.9	0.7
46.76	0.39	1.8	41.26	8MY	-0.3	-3.4	-1.8	0.7
53.41	0.47	1.8	50.79	18MY	-1.3	-3.4	-1.8	0.8
60.02	0.56	1.8	60.31	28MY	-1.7	-3.5	-1.7	0.9
66.57	0.64	1.9	69.82	7JN	-0.8	-3.5	-1.6	0.9
73.08	0.72	1.9	79.35	17JN	-0.1	-3.5	-1.6	0.9
79.55	0.80	1.9	88.89	27JN	0.4	-3.4	-1.5	1.0
85.98	0.87	2.0	98.46	7JL	0.7	-3.4	-1.5	1.0
92.37	0.94	2.0	108.07	17JL	1.2	-3.4	-1.4	1.0
98.74	1.00	2.0	117.72	27JL	2.4	-3.3	-1.4	1.0
105.08	1.07	2.0	127.41	6AU	2.3	-3.3	-1.3	1.0
111.40	1.13	2.0	137.16	16AU	0.1	-3.3	-1.3	1.0
117.69	1.19	2.0	146.97	26AU	-0.9	-3.3	-1.3	1.0
123.96	1.24	2.0	156.84	5SE	-1.1	-3.3	-1.3	0.9
130.20	1.30	2.0	166.76	15SE	-1.0	-3.3	-1.2	0.9
136.43	1.35	1.9	176.75	25SE	-0.6	-3.4	-1.2	0.9
142.62	1.40	1.9	186.79	5OC	-0.4	-3.4	-1.2	0.8
148.78	1.45	1.8	196.88	15OC	-0.3	-3.4	-1.3	0.9
154.91	1.49	1.8	207.02	25OC	-0.3	-3.5	-1.3	0.9
161.00	1.53	1.7	217.19	4NO	0.2	-3.5	-1.3	0.9
167.03	1.57	1.6	227.38	14NO	2.3	-3.6	-1.3	0.9
173.01	1.60	1.5	237.59	24NO	1.1	-3.6	-1.4	0.9
178.90	1.63	1.4	247.80	4DE	0.1	-3.7	-1.4	0.9
184.71	1.66	1.3	258.01	14DE	-0.1	-3.8	-1.5	0.9
190.42	1.68	1.2	268.19	24DE	-0.2	-3.9	-1.5	0.8
				-461				
195.98	1.68	1.0	278.35	3JA	-0.3	-4.0	-1.6	0.8
201.38	1.68	0.9	288.47	13JA	-0.7	-4.1	-1.6	0.8
206.59	1.66	0.7	298.54	23JA	-1.2	-4.2	-1.7	0.7
211.53	1.62	0.5	308.57	2FE	-1.3	-4.2	-1.8	0.7
216.17	1.56	0.3	318.53	12FE	-0.7	-4.3	-1.8	0.6
220.42	1.47	0.0	328.44	22FE	0.8	-4.3	-1.9	0.6
224.16	1.34	-0.2	338.28	4MR	2.8	-4.1	-2.0	0.5
227.29	1.16	-0.5	348.06	14MR	2.6	-3.6	-2.0	0.5
229.63	0.91	-0.8	357.80	24MR	1.4	-3.1	-2.1	0.4
230.99	0.58	-1.1	7.48	3AP	0.8	-3.7	-2.1	0.4
231.22	0.15	-1.4	17.11	13AP	0.3	-4.1	-2.1	0.4
230.21	-0.37	-1.7	26.70	23AP	-0.3	-4.2	-2.1	0.5
228.04	-0.98	-2.0	36.26	3MY	-1.3	-4.2	-2.1	0.5
225.10	-1.61	-2.3	45.79	13MY	-1.7	-4.1	-2.0	0.6
222.01	-2.21	-2.2	55.32	23MY	-0.8	-4.0	-2.0	0.7
219.54	-2.70	-2.0	64.83	2JN	-0.0	-3.9	-1.9	0.7
218.20	-3.06	-1.8	74.35	12JN	0.5	-3.8	-1.9	0.8
218.22	-3.28	-1.6	83.89	22JN	1.0	-3.7	-1.8	0.8
219.61	-3.40	-1.3	93.44	2JL	1.6	-3.7	-1.8	0.9
222.18	-3.43	-1.1	103.03	12JL	3.0	-3.6	-1.7	0.9
225.76	-3.41	-0.9	112.66	22JL	1.9	-3.6	-1.6	0.9
230.16	-3.34	-0.7	122.32	1AU	0.0	-3.5	-1.6	0.9
235.21	-3.23	-0.5	132.04	11AU	-1.0	-3.5	-1.5	0.9
240.79	-3.10	-0.4	141.83	21AU	-1.2	-3.4	-1.5	0.9
246.79	-2.94	-0.2	151.66	31AU	-0.9	-3.4	-1.4	0.9
253.11	-2.77	-0.0	161.55	10SE	-0.5	-3.4	-1.4	0.9
259.70	-2.58	0.1	171.51	20SE	-0.3	-3.4	-1.4	0.9
266.49	-2.39	0.2	181.52	30SE	-0.2	-3.4	-1.3	0.8
273.43	-2.19	0.4	191.59	10OC	-0.1	-3.4	-1.3	0.8
280.48	-1.99	0.5	201.70	20OC	0.4	-3.4	-1.3	0.8
287.62	-1.79	0.6	211.85	30OC	2.6	-3.4	-1.3	0.8
294.80	-1.59	0.7	222.04	9NO	0.9	-3.4	-1.3	0.8
302.02	-1.40	0.9	232.24	19NO	-0.1	-3.4	-1.3	0.8
309.24	-1.21	1.0	242.45	29NO	-0.3	-3.4	-1.3	0.8
316.45	-1.02	1.1	252.66	9DE	-0.3	-3.4	-1.4	0.8
323.64	-0.85	1.2	262.86	19DE	-0.4	-3.4	-1.4	0.8
330.79	-0.68	1.3	273.03	29DE	-0.8	-3.5	-1.4	0.8

♂ LONG	♂ LAT	♂ MAG	⊙ LONG	16.00UT	☿	♀	♃	♄
				-460	MAGNITUDES			
337.91	-0.52	1.3	283.17	8JA	-1.1	-3.5	-1.5	0.8
344.98	-0.37	1.4	293.27	18JA	-1.2	-3.5	-1.6	0.8
351.99	-0.23	1.5	303.32	28JA	-0.6	-3.4	-1.6	0.7
358.95	-0.09	1.6	313.31	7FE	1.0	-3.4	-1.7	0.7
5.86	0.03	1.6	323.25	17FE	3.2	-3.4	-1.7	0.6
12.70	0.15	1.7	333.12	27FE	1.9	-3.4	-1.8	0.6
19.49	0.26	1.8	342.94	8MR	1.0	-3.3	-1.9	0.5
26.23	0.36	1.8	352.70	18MR	0.6	-3.3	-2.0	0.5
32.91	0.45	1.9	2.41	28MR	0.2	-3.3	-2.0	0.4
39.54	0.54	1.9	12.06	7AP	-0.4	-3.3	-2.1	0.4
46.13	0.62	1.9	21.68	17AP	-1.3	-3.3	-2.1	0.3
52.67	0.70	1.9	31.25	27AP	-1.7	-3.4	-2.2	0.4
59.17	0.76	2.0	40.80	7MY	-0.8	-3.4	-2.2	0.4
65.65	0.83	2.0	50.33	17MY	0.1	-3.4	-2.2	0.5
72.09	0.89	2.0	59.84	27MY	0.7	-3.4	-2.2	0.6
78.51	0.94	2.0	69.36	6JN	1.3	-3.5	-2.2	0.6
84.91	0.99	2.0	78.89	16JN	2.2	-3.5	-2.1	0.7
91.30	1.03	2.0	88.43	26JN	3.3	-3.6	-2.1	0.7
97.68	1.07	1.9	98.00	6JL	1.6	-3.6	-2.0	0.8
104.05	1.10	2.0	107.61	16JL	-0.0	-3.7	-2.0	0.8
110.42	1.13	2.0	117.25	26JL	-1.1	-3.8	-1.9	0.9
116.79	1.16	2.0	126.94	5AU	-1.4	-3.9	-1.8	0.9
123.17	1.18	2.0	136.69	15AU	-0.9	-4.0	-1.8	0.9
129.56	1.20	2.0	146.49	25AU	-0.4	-4.1	-1.7	0.9
135.96	1.21	2.0	156.36	4SE	-0.2	-4.2	-1.7	0.9
142.37	1.22	2.0	166.28	14SE	-0.0	-4.3	-1.6	0.9
148.79	1.22	2.0	176.26	24SE	0.1	-4.3	-1.6	0.8
155.24	1.22	2.0	186.30	4OC	0.6	-4.2	-1.5	0.8
161.70	1.22	1.9	196.39	14OC	2.9	-3.8	-1.5	0.8
168.17	1.20	1.9	206.52	24OC	0.7	-3.1	-1.4	0.7
174.67	1.19	1.8	216.69	3NO	-0.3	-3.7	-1.4	0.8
181.17	1.16	1.8	226.88	13NO	-0.4	-4.2	-1.4	0.8
187.69	1.13	1.7	237.09	23NO	-0.4	-4.4	-1.4	0.8
194.22	1.08	1.6	247.31	3DE	-0.5	-4.4	-1.4	0.8
200.75	1.03	1.6	257.51	13DE	-0.8	-4.3	-1.4	0.8
207.29	0.96	1.5	267.70	23DE	-1.0	-4.2	-1.4	0.8
				-459				
213.83	0.88	1.4	277.86	2JA	-1.0	-4.1	-1.4	0.8
220.36	0.79	1.3	287.98	12JA	-0.5	-4.0	-1.5	0.8
226.88	0.68	1.1	298.06	22JA	1.2	-3.9	-1.5	0.8
233.38	0.55	1.0	308.08	1FE	3.0	-3.8	-1.6	0.7
239.84	0.39	0.9	318.05	11FE	1.4	-3.7	-1.6	0.7
246.27	0.21	0.7	327.96	21FE	0.8	-3.6	-1.7	0.6
252.63	0.00	0.6	337.81	3MR	0.4	-3.5	-1.7	0.6
258.91	-0.24	0.4	347.60	13MR	0.1	-3.5	-1.8	0.5
265.09	-0.51	0.2	357.33	23MR	-0.4	-3.4	-1.9	0.5
271.12	-0.83	0.0	7.02	2AP	-1.3	-3.4	-1.9	0.4
276.96	-1.19	-0.2	16.65	12AP	-1.7	-3.4	-2.0	0.4
282.55	-1.60	-0.4	26.24	22AP	-0.8	-3.3	-2.1	0.3
287.81	-2.06	-0.6	35.81	2MY	0.2	-3.3	-2.1	0.3
292.63	-2.59	-0.8	45.34	12MY	0.9	-3.3	-2.2	0.4
296.90	-3.17	-1.1	54.86	22MY	1.7	-3.3	-2.2	0.4
300.43	-3.81	-1.3	64.38	1JN	2.9	-3.3	-2.3	0.5
303.05	-4.50	-1.6	73.90	11JN	2.9	-3.3	-2.3	0.6
304.54	-5.22	-1.9	83.43	21JN	1.3	-3.3	-2.3	0.6
304.73	-5.90	-2.1	92.99	1JL	-0.1	-3.4	-2.3	0.7
303.62	-6.47	-2.4	102.57	11JL	-1.2	-3.4	-2.3	0.7
301.44	-6.81	-2.6	112.19	21JL	-1.5	-3.4	-2.2	0.8
298.75	-6.83	-2.6	121.86	31JL	-0.9	-3.5	-2.2	0.8
296.33	-6.54	-2.4	131.58	10AU	-0.3	-3.5	-2.1	0.8
294.80	-6.00	-2.2	141.35	20AU	-0.0	-3.5	-2.1	0.8
294.52	-5.32	-1.9	151.18	30AU	0.1	-3.4	-2.0	0.8
295.54	-4.61	-1.5	161.07	9SE	0.3	-3.4	-1.9	0.8
297.73	-3.93	-1.2	171.02	19SE	0.9	-3.4	-1.9	0.8
300.89	-3.30	-0.9	181.03	29SE	3.3	-3.4	-1.8	0.8
304.83	-2.73	-0.7	191.09	9OC	0.5	-3.3	-1.8	0.8
309.36	-2.23	-0.4	201.20	19OC	-0.4	-3.3	-1.7	0.8
314.36	-1.79	-0.2	211.35	29OC	-0.6	-3.3	-1.6	0.7
319.72	-1.40	0.1	221.53	8NO	-0.6	-3.3	-1.6	0.7
325.35	-1.06	0.3	231.73	18NO	-0.6	-3.3	-1.6	0.7
331.19	-0.77	0.5	241.95	28NO	-0.8	-3.4	-1.5	0.8
337.18	-0.51	0.6	252.16	8DE	-0.8	-3.4	-1.5	0.8
343.28	-0.28	0.8	262.35	18DE	-0.9	-3.4	-1.5	0.8
349.47	-0.08	1.0	272.53	28DE	-0.4	-3.4	-1.5	0.8

♂ / ☉ / 16.00UT / ☿ ♀ ♃ ♄ (-458, -457)

♂ LONG	LAT	MAG	☉ LONG	16.00UT	☿	♀	♃	♄
355.72	0.10	1.1	282.67	7JA	1.4	-3.5	-1.5	0.8
2.01	0.25	1.2	292.78	17JA	2.5	-3.5	-1.5	0.8
8.32	0.39	1.3	302.83	27JA	1.0	-3.6	-1.5	0.8
14.65	0.51	1.4	312.83	6FE	0.5	-3.6	-1.5	0.8
20.98	0.61	1.5	322.77	16FE	0.3	-3.7	-1.6	0.7
27.32	0.71	1.6	332.65	26FE	-0.0	-3.8	-1.6	0.7
33.64	0.79	1.7	342.47	8MR	-0.5	-3.8	-1.6	0.6
39.96	0.86	1.8	352.24	18MR	-1.3	-3.9	-1.7	0.6
46.28	0.92	1.8	1.95	28MR	-1.6	-4.0	-1.8	0.5
52.58	0.97	1.9	11.60	7AP	-0.8	-4.1	-1.8	0.4
58.87	1.02	1.9	21.22	17AP	0.3	-4.2	-1.9	0.4
65.16	1.06	2.0	30.80	27AP	1.2	-4.2	-1.9	0.3
71.44	1.09	2.0	40.34	7MY	2.2	-4.1	-2.0	0.3
77.73	1.12	2.0	49.87	17MY	3.6	-3.8	-2.1	0.3
84.02	1.14	2.0	59.39	27MY	2.3	-3.0	-2.2	0.4
90.31	1.16	2.0	68.90	6JN	1.0	-3.2	-2.2	0.4
96.61	1.17	2.0	78.43	16JN	-0.1	-3.8	-2.3	0.5
102.93	1.17	2.0	87.97	26JN	-1.2	-4.1	-2.3	0.5
109.27	1.17	2.0	97.54	6JL	-1.6	-4.2	-2.4	0.6
115.63	1.17	2.0	107.14	16JL	-0.9	-4.2	-2.4	0.7
122.01	1.16	2.0	116.78	26JL	-0.3	-4.1	-2.4	0.7
128.43	1.15	1.9	126.47	5AU	0.1	-4.0	-2.4	0.7
134.88	1.13	1.9	136.22	15AU	0.3	-3.9	-2.4	0.8
141.37	1.11	1.9	146.02	25AU	0.5	-3.9	-2.3	0.8
147.90	1.08	1.9	155.88	4SE	1.2	-3.8	-2.3	0.8
154.47	1.05	1.9	165.80	14SE	3.3	-3.7	-2.2	0.8
161.09	1.01	1.9	175.78	24SE	0.4	-3.7	-2.2	0.8
167.76	0.96	1.9	185.81	4OC	-0.6	-3.6	-2.1	0.8
174.47	0.91	1.9	195.90	14OC	-0.7	-3.5	-2.0	0.8
181.24	0.86	1.9	206.03	24OC	-0.7	-3.5	-2.0	0.8
188.06	0.79	1.8	216.19	3NO	-0.8	-3.5	-1.9	0.7
194.93	0.72	1.8	226.39	13NO	-0.7	-3.4	-1.8	0.7
201.86	0.64	1.8	236.59	23NO	-0.7	-3.4	-1.8	0.7
208.83	0.55	1.7	246.80	3DE	-0.7	-3.4	-1.7	0.7
215.86	0.45	1.6	257.01	13DE	-0.3	-3.4	-1.7	0.8
222.93	0.35	1.6	267.20	23DE	1.6	-3.3	-1.6	0.8
				-457				
230.05	0.23	1.5	277.36	2JA	2.0	-3.3	-1.6	0.8
237.22	0.10	1.4	287.49	12JA	0.7	-3.3	-1.6	0.8
244.43	-0.04	1.4	297.56	22JA	0.3	-3.3	-1.6	0.8
251.67	-0.19	1.3	307.59	1FE	0.2	-3.3	-1.6	0.8
258.95	-0.35	1.2	317.56	11FE	-0.1	-3.4	-1.6	0.8
266.25	-0.53	1.1	327.48	21FE	-0.5	-3.4	-1.6	0.7
273.57	-0.71	1.0	337.33	3MR	-1.2	-3.4	-1.6	0.7
280.90	-0.90	0.9	347.12	13MR	-1.6	-3.4	-1.6	0.7
288.21	-1.10	0.9	356.86	23MR	-0.8	-3.5	-1.6	0.6
295.50	-1.30	0.8	6.54	2AP	0.4	-3.5	-1.7	0.5
302.76	-1.51	0.7	16.18	12AP	1.5	-3.4	-1.7	0.5
309.95	-1.72	0.6	25.78	22AP	2.9	-3.4	-1.7	0.4
317.05	-1.93	0.5	35.34	2MY	3.1	-3.4	-1.8	0.4
324.03	-2.14	0.3	44.88	12MY	1.8	-3.4	-1.9	0.3
330.86	-2.34	0.2	54.40	22MY	0.8	-3.3	-1.9	0.3
337.50	-2.54	0.1	63.91	1JN	-0.2	-3.3	-2.0	0.3
343.89	-2.73	-0.0	73.44	11JN	-1.2	-3.3	-2.0	0.4
349.97	-2.91	-0.1	82.96	21JN	-1.7	-3.3	-2.1	0.4
355.68	-3.07	-0.3	92.52	1JL	-0.8	-3.3	-2.2	0.5
0.90	-3.23	-0.4	102.10	11JL	-0.2	-3.4	-2.3	0.6
5.54	-3.36	-0.6	111.72	21JL	0.2	-3.4	-2.3	0.6
9.44	-3.47	-0.8	121.39	31JL	0.4	-3.4	-2.4	0.7
12.41	-3.54	-1.0	131.10	10AU	0.7	-3.4	-2.4	0.7
14.26	-3.57	-1.2	140.87	20AU	1.6	-3.5	-2.4	0.7
14.78	-3.52	-1.4	150.70	30AU	3.0	-3.6	-2.4	0.8
13.82	-3.36	-1.6	160.59	9SE	0.3	-3.6	-2.4	0.8
11.47	-3.08	-1.8	170.54	19SE	-0.7	-3.7	-2.4	0.8
8.14	-2.64	-2.0	180.55	29SE	-0.9	-3.8	-2.4	0.8
4.56	-2.10	-1.9	190.61	9OC	-0.8	-3.9	-2.3	0.8
1.54	-1.51	-1.6	200.71	19OC	-0.7	-4.0	-2.3	0.8
359.64	-0.95	-1.3	210.86	29OC	-0.6	-4.1	-2.2	0.8
359.10	-0.45	-1.0	221.04	8NO	-0.5	-4.2	-2.1	0.7
359.86	-0.05	-0.7	231.24	18NO	-0.5	-4.3	-2.1	0.7
1.73	0.28	-0.4	241.45	28NO	-0.2	-4.4	-2.0	0.7
4.51	0.54	-0.1	251.66	8DE	1.8	-4.4	-1.9	0.7
8.00	0.75	0.2	261.86	18DE	1.7	-4.2	-1.9	0.8
12.05	0.91	0.4	272.04	28DE	0.4	-3.8	-1.8	0.8

♂ / ☉ / 16.00UT / ☿ ♀ ♃ ♄ (-456, -455)

♂ LONG	LAT	MAG	☉ LONG	16.00UT	☿	♀	♃	♄
16.53	1.04	0.6	282.18	7JA	0.2	-3.2	-1.7	0.8
21.34	1.14	0.8	292.28	17JA	0.0	-3.7	-1.7	0.8
26.42	1.21	1.0	302.34	27JA	-0.2	-4.1	-1.7	0.8
31.70	1.27	1.1	312.34	6FE	-0.6	-4.3	-1.6	0.8
37.14	1.32	1.3	322.28	16FE	-1.2	-4.3	-1.6	0.8
42.71	1.35	1.4	332.17	26FE	-1.5	-4.2	-1.6	0.8
48.39	1.37	1.5	341.99	7MR	-0.7	-4.1	-1.6	0.7
54.15	1.39	1.6	351.76	17MR	0.5	-4.0	-1.6	0.7
59.98	1.39	1.7	1.47	27MR	1.9	-3.9	-1.6	0.7
65.87	1.39	1.8	11.13	6AP	3.8	-3.8	-1.6	0.6
71.81	1.39	1.8	20.75	16AP	2.3	-3.7	-1.6	0.5
77.80	1.38	1.9	30.33	26AP	1.3	-3.6	-1.6	0.5
83.84	1.36	1.9	39.88	6MY	0.6	-3.6	-1.6	0.4
89.92	1.34	2.0	49.41	16MY	-0.2	-3.5	-1.7	0.3
96.04	1.32	2.0	58.93	26MY	-1.2	-3.4	-1.7	0.3
102.21	1.29	2.0	68.45	5JN	-1.7	-3.4	-1.8	0.3
108.42	1.26	2.0	77.97	15JN	-0.8	-3.4	-1.8	0.3
114.67	1.22	2.0	87.51	25JN	-0.1	-3.4	-1.9	0.4
120.98	1.18	2.0	97.08	5JL	0.3	-3.3	-1.9	0.4
127.34	1.14	2.0	106.68	15JL	0.6	-3.3	-2.0	0.5
133.76	1.09	2.0	116.32	25JL	1.0	-3.3	-2.1	0.6
140.23	1.04	2.0	126.01	4AU	2.0	-3.3	-2.1	0.6
146.70	0.98	1.9	135.75	14AU	2.6	-3.4	-2.2	0.7
153.37	0.92	1.9	145.55	24AU	0.2	-3.4	-2.3	0.7
160.04	0.85	1.8	155.40	3SE	-0.8	-3.4	-2.3	0.7
166.77	0.78	1.8	165.32	13SE	-1.0	-3.4	-2.4	0.8
173.58	0.71	1.8	175.30	23SE	-1.0	-3.4	-2.4	0.8
180.47	0.63	1.8	185.33	3OC	-0.6	-3.5	-2.4	0.8
187.42	0.54	1.7	195.41	13OC	-0.4	-3.5	-2.4	0.8
194.45	0.46	1.7	205.54	23OC	-0.4	-3.5	-2.4	0.8
201.56	0.36	1.7	215.70	2NO	-0.4	-3.5	-2.3	0.8
208.74	0.26	1.7	225.89	12NO	0.0	-3.4	-2.3	0.8
215.99	0.16	1.7	236.10	22NO	2.0	-3.4	-2.2	0.7
223.31	0.05	1.6	246.31	2DE	1.3	-3.4	-2.2	0.7
230.70	-0.07	1.6	256.51	12DE	0.2	-3.4	-2.1	0.7
238.15	-0.19	1.6	266.70	22DE	-0.0	-3.3	-2.0	0.8
				-455				
245.65	-0.31	1.5	276.86	1JA	-0.1	-3.3	-2.0	0.8
253.21	-0.43	1.5	286.99	11JA	-0.3	-3.3	-1.9	0.8
260.81	-0.56	1.5	297.07	21JA	-0.6	-3.3	-1.8	0.8
268.44	-0.69	1.4	307.10	31JA	-1.2	-3.3	-1.8	0.9
276.11	-0.81	1.4	317.07	10FE	-1.4	-3.4	-1.7	0.9
283.79	-0.94	1.3	326.99	20FE	-0.7	-3.4	-1.6	0.8
291.48	-1.06	1.3	336.84	2MR	0.7	-3.4	-1.6	0.8
299.16	-1.17	1.2	346.64	12MR	2.4	-3.4	-1.6	0.8
306.82	-1.28	1.2	356.38	22MR	3.1	-3.5	-1.5	0.8
314.44	-1.37	1.2	6.07	1AP	1.7	-3.5	-1.5	0.7
322.02	-1.46	1.1	15.71	11AP	1.0	-3.6	-1.5	0.7
329.53	-1.53	1.1	25.31	21AP	0.4	-3.6	-1.5	0.6
336.97	-1.59	1.0	34.87	1MY	-0.3	-3.7	-1.5	0.5
344.31	-1.64	1.0	44.42	11MY	-1.2	-3.8	-1.5	0.5
351.53	-1.67	0.9	53.94	21MY	-1.7	-3.8	-1.5	0.4
358.63	-1.69	0.9	63.46	31MY	-0.8	-3.9	-1.5	0.3
5.58	-1.69	0.8	72.98	10JN	-0.1	-4.0	-1.6	0.3
12.35	-1.67	0.8	82.51	20JN	0.4	-4.1	-1.6	0.3
18.93	-1.64	0.7	92.06	30JN	0.8	-4.2	-1.6	0.4
25.28	-1.58	0.6	101.64	10JL	1.4	-4.2	-1.7	0.4
31.37	-1.51	0.6	111.26	20JL	2.6	-4.1	-1.7	0.5
37.16	-1.42	0.5	120.92	30JL	2.2	-3.7	-1.8	0.5
42.59	-1.31	0.4	130.63	9AU	0.1	-3.2	-1.9	0.6
47.57	-1.17	0.2	140.40	19AU	-1.0	-3.7	-1.9	0.6
52.04	-1.00	0.1	150.23	29AU	-1.2	-4.1	-2.0	0.7
55.86	-0.80	-0.1	160.11	8SE	-1.0	-4.3	-2.0	0.7
58.89	-0.56	-0.2	170.06	18SE	-0.6	-4.3	-2.1	0.8
60.94	-0.28	-0.4	180.06	28SE	-0.3	-4.3	-2.2	0.8
61.81	0.06	-0.6	190.12	8OC	-0.2	-4.2	-2.2	0.8
61.32	0.46	-0.9	200.22	18OC	-0.2	-4.1	-2.3	0.8
59.42	0.89	-1.1	210.37	28OC	0.2	-4.0	-2.3	0.8
56.27	1.34	-1.3	220.55	7NO	2.3	-3.9	-2.3	0.8
52.40	1.75	-1.4	230.75	17NO	1.1	-3.8	-2.3	0.8
48.59	2.07	-1.2	240.96	27NO	0.0	-3.7	-2.3	0.8
45.55	2.29	-0.9	251.17	7DE	-0.2	-3.7	-2.2	0.8
43.74	2.41	-0.6	261.37	17DE	-0.2	-3.6	-2.2	0.7
43.27	2.45	-0.4	271.54	27DE	-0.4	-3.5	-2.1	0.8

♂ / ☉ / Planetary Magnitudes — Year −454 / −453

♂ LONG	LAT	MAG	☉ LONG	16.00UT	☿	♀	♃	♄
44.04	2.43	-0.1	281.69	6JA	-0.7	-3.5	-2.1	0.8
45.86	2.39	0.2	291.80	16JA	-1.2	-3.4	-2.0	0.8
48.53	2.33	0.4	301.85	26JA	-1.2	-3.4	-1.9	0.9
51.87	2.26	0.6	311.86	5FE	-0.7	-3.4	-1.9	0.9
55.74	2.18	0.8	321.80	15FE	0.8	-3.3	-1.8	0.9
60.04	2.11	1.0	331.69	25FE	3.0	-3.3	-1.7	0.9
64.67	2.03	1.2	341.52	7MR	2.3	-3.3	-1.7	0.9
69.57	1.95	1.3	351.29	17MR	1.3	-3.3	-1.6	0.9
74.69	1.87	1.4	1.00	27MR	0.7	-3.3	-1.6	0.8
79.99	1.80	1.5	10.67	6AP	0.3	-3.3	-1.5	0.8
85.45	1.72	1.6	20.28	16AP	-0.3	-3.3	-1.5	0.7
91.05	1.64	1.7	29.86	26AP	-1.2	-3.4	-1.4	0.7
96.76	1.56	1.7	39.42	6MY	-1.7	-3.4	-1.4	0.6
102.58	1.49	1.8	48.95	16MY	-0.8	-3.4	-1.4	0.6
108.50	1.41	1.8	58.47	26MY	0.0	-3.5	-1.4	0.5
114.52	1.33	1.8	67.98	5JN	0.6	-3.5	-1.4	0.4
120.62	1.24	1.9	77.51	15JN	1.1	-3.5	-1.4	0.4
126.81	1.16	1.9	87.04	25JN	1.8	-3.4	-1.4	0.3
133.09	1.08	1.9	96.61	5JL	3.1	-3.4	-1.4	0.4
139.45	0.99	1.9	106.21	15JL	1.9	-3.3	-1.5	0.4
145.90	0.90	1.9	115.85	25JL	0.1	-3.3	-1.5	0.5
152.45	0.81	1.9	125.54	4AU	-1.0	-3.3	-1.5	0.5
159.08	0.72	1.8	135.27	14AU	-1.3	-3.3	-1.6	0.6
165.80	0.63	1.8	145.07	24AU	-0.9	-3.3	-1.6	0.6
172.61	0.53	1.8	154.92	3SE	-0.5	-3.3	-1.7	0.7
179.52	0.43	1.8	164.84	13SE	-0.2	-3.3	-1.7	0.7
186.52	0.33	1.7	174.81	23SE	-0.1	-3.4	-1.8	0.8
193.61	0.22	1.7	184.84	3OC	-0.0	-3.4	-1.9	0.8
200.79	0.12	1.6	194.92	13OC	0.4	-3.4	-1.9	0.8
208.07	0.01	1.6	205.04	23OC	2.6	-3.5	-2.0	0.8
215.42	-0.10	1.5	215.21	2NO	0.9	-3.5	-2.1	0.8
222.86	-0.21	1.5	225.39	12NO	-0.2	-3.6	-2.1	0.8
230.38	-0.31	1.5	235.60	22NO	-0.3	-3.6	-2.2	0.8
237.97	-0.42	1.5	245.82	2DE	-0.3	-3.7	-2.2	0.8
245.63	-0.52	1.5	256.02	12DE	-0.5	-3.8	-2.2	0.8
253.34	-0.63	1.5	266.21	22DE	-0.8	-3.9	-2.2	0.8

−453

♂ LONG	LAT	MAG	☉ LONG	16.00UT	☿	♀	♃	♄
261.09	-0.72	1.5	276.38	1JA	-1.1	-4.0	-2.2	0.8
268.88	-0.82	1.4	286.50	11JA	-1.1	-4.1	-2.1	0.8
276.70	-0.90	1.4	296.59	21JA	-0.6	-4.2	-2.1	0.9
284.53	-0.98	1.4	306.62	31JA	1.0	-4.2	-2.0	0.9
292.36	-1.05	1.4	316.59	10FE	3.2	-4.3	-1.9	0.9
300.18	-1.11	1.4	326.51	20FE	1.7	-4.3	-1.9	1.0
307.98	-1.16	1.4	336.37	2MR	0.9	-4.1	-1.8	1.0
315.74	-1.20	1.4	346.17	12MR	0.5	-3.6	-1.7	1.0
323.46	-1.22	1.4	355.91	22MR	0.2	-3.1	-1.7	0.9
331.12	-1.23	1.4	5.60	1AP	-0.4	-3.7	-1.6	0.9
338.70	-1.23	1.4	15.24	11AP	-1.2	-4.1	-1.6	0.9
346.21	-1.22	1.4	24.84	21AP	-1.7	-4.2	-1.5	0.9
353.63	-1.19	1.4	34.41	1MY	-0.8	-4.2	-1.5	0.8
0.95	-1.15	1.4	43.95	11MY	0.1	-4.1	-1.4	0.7
8.17	-1.10	1.4	53.47	21MY	0.8	-4.0	-1.4	0.7
15.27	-1.03	1.3	62.99	31MY	1.4	-3.9	-1.3	0.6
22.26	-0.96	1.3	72.51	10JN	2.4	-3.8	-1.3	0.6
29.11	-0.87	1.3	82.04	20JN	3.2	-3.7	-1.3	0.5
35.82	-0.77	1.3	91.59	30JN	1.5	-3.7	-1.3	0.4
42.39	-0.66	1.3	101.17	10JL	0.0	-3.6	-1.3	0.4
48.79	-0.53	1.2	110.79	20JL	-1.1	-3.6	-1.3	0.4
55.02	-0.40	1.2	120.45	30JL	-1.4	-3.5	-1.3	0.5
61.04	-0.25	1.1	130.16	9AU	-0.9	-3.5	-1.3	0.5
66.84	-0.09	1.1	139.93	19AU	-0.4	-3.4	-1.4	0.6
72.38	0.09	1.0	149.75	29AU	-0.1	-3.4	-1.4	0.6
77.62	0.28	0.9	159.63	8SE	0.0	-3.4	-1.4	0.7
82.50	0.50	0.8	169.58	18SE	0.2	-3.4	-1.5	0.7
86.95	0.75	0.6	179.58	28SE	0.7	-3.4	-1.6	0.8
90.87	1.02	0.5	189.63	8OC	3.0	-3.4	-1.6	0.8
94.13	1.33	0.3	199.74	18OC	0.7	-3.4	-1.7	0.8
96.59	1.68	0.1	209.88	28OC	-0.3	-3.4	-1.7	0.9
98.08	2.08	-0.1	220.05	7NO	-0.5	-3.4	-1.8	0.9
98.38	2.51	-0.3	230.25	17NO	-0.5	-3.4	-1.9	0.9
97.35	2.96	-0.6	240.46	27NO	-0.6	-3.4	-1.9	0.9
95.00	3.39	-0.8	250.67	7DE	-0.8	-3.4	-2.0	0.9
91.55	3.75	-1.0	260.87	17DE	-0.9	-3.4	-2.0	0.9
87.59	3.97	-1.0	271.05	27DE	-0.9	-3.5	-2.1	0.9

♂ / ☉ / Planetary Magnitudes — Year −452 / −451

♂ LONG	LAT	MAG	☉ LONG	16.00UT	☿	♀	♃	♄
83.85	4.03	-0.9	281.20	6JA	-0.5	-3.5	-2.1	0.9
80.98	3.95	-0.6	291.31	16JA	1.2	-3.5	-2.1	0.9
79.35	3.76	-0.4	301.36	26JA	2.9	-3.4	-2.1	0.9
79.02	3.53	-0.1	311.37	5FE	1.3	-3.4	-2.0	1.0
79.88	3.28	0.1	321.32	15FE	0.7	-3.4	-2.0	1.0
81.76	3.03	0.4	331.21	25FE	0.4	-3.4	-2.0	1.0
84.46	2.80	0.6	341.04	6MR	0.1	-3.3	-1.9	1.1
87.81	2.58	0.7	350.81	16MR	-0.4	-3.3	-1.8	1.1
91.71	2.37	0.9	0.53	26MR	-1.2	-3.3	-1.8	1.1
96.02	2.18	1.0	10.19	5AP	-1.7	-3.3	-1.7	1.0
100.70	2.00	1.2	19.82	15AP	-0.8	-3.3	-1.6	1.0
105.67	1.83	1.3	29.40	25AP	0.2	-3.4	-1.6	1.0
110.88	1.67	1.4	38.95	5MY	1.0	-3.4	-1.5	0.9
116.31	1.52	1.4	48.48	15MY	1.9	-3.4	-1.5	0.9
121.93	1.38	1.5	58.00	25MY	3.1	-3.4	-1.4	0.8
127.72	1.24	1.6	67.52	4JN	2.7	-3.5	-1.4	0.8
133.66	1.10	1.6	77.04	14JN	1.2	-3.5	-1.3	0.7
139.75	0.97	1.6	86.58	24JN	-0.1	-3.6	-1.3	0.6
145.97	0.84	1.7	96.14	4JL	-1.1	-3.6	-1.3	0.6
152.32	0.71	1.7	105.74	14JL	-1.6	-3.7	-1.3	0.5
158.80	0.58	1.7	115.38	24JL	-0.9	-3.8	-1.2	0.5
165.41	0.46	1.7	125.06	3AU	-0.3	-3.9	-1.2	0.5
172.13	0.33	1.7	134.80	13AU	-0.0	-4.0	-1.2	0.6
178.97	0.21	1.7	144.59	23AU	0.2	-4.1	-1.3	0.6
185.93	0.09	1.6	154.45	2SE	0.4	-4.2	-1.3	0.7
193.00	-0.02	1.6	164.36	12SE	1.0	-4.3	-1.3	0.7
200.18	-0.14	1.6	174.33	22SE	3.3	-4.3	-1.3	0.8
207.47	-0.25	1.6	184.35	2OC	0.5	-4.2	-1.4	0.8
214.86	-0.36	1.6	194.43	12OC	-0.5	-3.8	-1.4	0.9
222.34	-0.46	1.5	204.55	22OC	-0.6	-3.1	-1.5	0.9
229.91	-0.56	1.5	214.71	1NO	-0.6	-3.7	-1.5	0.9
237.56	-0.66	1.5	224.90	11NO	-0.7	-4.2	-1.6	1.0
245.27	-0.74	1.4	235.11	21NO	-0.8	-4.4	-1.6	1.0
253.05	-0.82	1.4	245.32	1DE	-0.8	-4.4	-1.7	1.0
260.88	-0.89	1.4	255.53	11DE	-0.8	-4.3	-1.8	1.0
268.74	-0.95	1.3	265.72	21DE	-0.4	-4.2	-1.8	1.0
276.62	-1.00	1.3	275.88	31DE	1.4	-4.1	-1.9	1.0

−451

♂ LONG	LAT	MAG	☉ LONG	16.00UT	☿	♀	♃	♄
284.52	-1.04	1.3	286.01	10JA	2.4	-4.0	-1.9	1.0
292.41	-1.07	1.3	296.10	20JA	0.9	-3.9	-2.0	1.0
300.28	-1.09	1.3	306.13	30JA	0.4	-3.8	-2.0	1.0
308.11	-1.10	1.4	316.12	9FE	0.2	-3.7	-2.0	1.1
315.93	-1.09	1.4	326.03	19FE	-0.0	-3.6	-2.0	1.1
323.68	-1.07	1.4	335.90	1MR	-0.5	-3.5	-2.0	1.2
331.38	-1.04	1.4	345.70	11MR	-1.2	-3.5	-2.0	1.2
339.00	-1.00	1.5	355.44	21MR	-1.6	-3.4	-1.9	1.2
346.54	-0.95	1.5	5.14	31MR	-0.8	-3.4	-1.9	1.2
354.00	-0.89	1.5	14.78	10AP	0.3	-3.4	-1.8	1.2
1.37	-0.83	1.5	24.38	20AP	1.3	-3.3	-1.8	1.2
8.65	-0.75	1.6	33.95	30AP	2.5	-3.3	-1.7	1.2
15.84	-0.66	1.6	43.49	10MY	3.6	-3.3	-1.6	1.1
22.92	-0.57	1.6	53.02	20MY	1.7	-3.3	-1.6	1.1
29.91	-0.47	1.6	62.54	30MY	1.0	-3.3	-1.5	1.0
36.79	-0.37	1.6	72.05	9JN	-0.1	-3.3	-1.5	1.0
43.58	-0.26	1.6	81.58	19JN	-1.2	-3.3	-1.4	0.9
50.26	-0.15	1.6	91.13	29JN	-1.7	-3.4	-1.4	0.8
56.82	-0.03	1.6	100.71	9JL	-0.9	-3.4	-1.3	0.8
63.28	0.10	1.6	110.32	19JL	-0.3	-3.4	-1.3	0.7
69.63	0.23	1.6	119.98	29JL	0.1	-3.5	-1.3	0.6
75.84	0.37	1.6	129.69	8AU	0.3	-3.5	-1.2	0.6
81.93	0.51	1.5	139.45	18AU	0.6	-3.5	-1.2	0.7
87.87	0.66	1.5	149.27	28AU	1.3	-3.4	-1.2	0.7
93.64	0.82	1.4	159.15	7SE	3.2	-3.4	-1.2	0.7
99.23	1.00	1.4	169.09	17SE	0.4	-3.4	-1.2	0.8
104.60	1.18	1.3	179.09	27SE	-0.6	-3.4	-1.2	0.8
109.71	1.38	1.2	189.14	7OC	-0.8	-3.3	-1.3	0.9
114.53	1.60	1.0	199.24	17OC	-0.8	-3.3	-1.3	0.9
118.97	1.84	0.9	209.38	27OC	-0.8	-3.3	-1.3	1.0
122.96	2.11	0.7	219.56	6NO	-0.6	-3.3	-1.4	1.0
126.39	2.40	0.6	229.75	16NO	-0.6	-3.3	-1.4	1.1
129.13	2.73	0.4	239.97	26NO	-0.6	-3.4	-1.5	1.1
131.02	3.08	0.1	250.18	6DE	-0.3	-3.4	-1.5	1.1
131.86	3.47	-0.1	260.37	16DE	1.6	-3.4	-1.6	1.1
131.49	3.86	-0.4	270.56	26DE	1.9	-3.4	-1.7	1.2

♂			⊙	16.00UT	☿	♀	♃	♄
LONG	LAT	MAG	LONG	-450	MAGNITUDES			
129.82	4.22	-0.6	280.71	5JA	0.6	-3.5	-1.7	1.2
126.94	4.50	-0.9	290.81	15JA	0.3	-3.5	-1.8	1.2
123.21	4.64	-1.0	300.88	25JA	0.1	-3.6	-1.9	1.2
119.31	4.59	-1.0	310.89	4FE	-0.1	-3.6	-1.9	1.2
115.94	4.38	-0.8	320.84	14FE	-0.5	-3.7	-2.0	1.2
113.63	4.05	-0.6	330.74	24FE	-1.2	-3.8	-2.0	1.3
112.61	3.67	-0.3	340.57	6MR	-1.5	-3.8	-2.0	1.3
112.86	3.27	-0.1	350.34	16MR	-0.8	-3.9	-2.0	1.3
114.23	2.89	0.1	0.06	26MR	0.4	-4.0	-2.0	1.4
116.55	2.54	0.3	9.73	5AP	1.7	-4.1	-2.0	1.4
119.65	2.21	0.5	19.35	15AP	3.2	-4.2	-2.0	1.4
123.38	1.92	0.6	28.94	25AP	2.8	-4.2	-1.9	1.4
127.63	1.65	0.8	38.49	5MY	1.6	-4.1	-1.9	1.4
132.31	1.40	0.9	48.02	15MY	0.7	-3.7	-1.8	1.3
137.35	1.17	1.0	57.55	25MY	-0.2	-3.0	-1.8	1.3
142.70	0.96	1.1	67.06	4JN	-1.2	-3.2	-1.7	1.2
148.32	0.76	1.1	76.58	14JN	-1.7	-3.9	-1.6	1.2
154.18	0.57	1.2	86.12	24JN	-0.9	-4.1	-1.6	1.1
160.25	0.40	1.2	95.68	4JL	-0.2	-4.2	-1.5	1.1
166.52	0.23	1.3	105.27	14JL	0.2	-4.2	-1.5	1.0
172.97	0.07	1.3	114.91	24JL	0.5	-4.1	-1.4	0.9
179.60	-0.08	1.3	124.59	3AU	0.8	-4.0	-1.4	0.8
186.38	-0.22	1.4	134.32	13AU	1.7	-3.9	-1.3	0.8
193.31	-0.35	1.4	144.11	23AU	2.9	-3.9	-1.3	0.8
200.38	-0.47	1.4	153.96	2SE	0.3	-3.8	-1.3	0.8
207.59	-0.59	1.4	163.87	12SE	-0.8	-3.7	-1.3	0.9
214.91	-0.69	1.4	173.84	22SE	-0.9	-3.6	-1.2	0.9
222.35	-0.79	1.4	183.86	2OC	-0.9	-3.6	-1.2	1.0
229.89	-0.87	1.4	193.94	12OC	-0.7	-3.5	-1.2	1.0
237.53	-0.95	1.4	204.06	22OC	-0.5	-3.5	-1.3	1.1
245.24	-1.01	1.4	214.22	1NO	-0.5	-3.5	-1.3	1.1
253.01	-1.06	1.4	224.41	11NO	-0.5	-3.4	-1.3	1.2
260.84	-1.10	1.4	234.62	21NO	-0.1	-3.4	-1.3	1.2
268.70	-1.12	1.4	244.82	1DE	1.8	-3.4	-1.4	1.2
276.59	-1.14	1.4	255.03	11DE	1.6	-3.4	-1.4	1.3
284.48	-1.14	1.4	265.23	21DE	0.4	-3.3	-1.5	1.3
292.36	-1.13	1.4	275.39	31DE	0.1	-3.3	-1.5	1.3
				-449				
300.23	-1.10	1.4	285.53	10JA	-0.0	-3.3	-1.6	1.3
308.06	-1.07	1.4	295.61	20JA	-0.2	-3.3	-1.7	1.3
315.84	-1.02	1.4	305.65	30JA	-0.6	-3.3	-1.7	1.3
323.57	-0.97	1.4	315.63	9FE	-1.2	-3.4	-1.8	1.3
331.23	-0.91	1.4	325.56	19FE	-1.4	-3.4	-1.9	1.2
338.82	-0.84	1.4	335.42	1MR	-0.8	-3.4	-1.9	1.2
346.34	-0.76	1.4	345.23	11MR	0.5	-3.5	-2.0	1.2
353.77	-0.68	1.5	354.97	21MR	2.1	-3.5	-2.0	1.2
1.11	-0.59	1.5	4.67	31MR	3.6	-3.5	-2.1	1.2
8.37	-0.49	1.6	14.31	10AP	2.1	-3.4	-2.1	1.2
15.54	-0.40	1.6	23.92	20AP	1.2	-3.4	-2.1	1.2
22.63	-0.30	1.7	33.48	30AP	0.5	-3.4	-2.1	1.1
29.62	-0.20	1.7	43.03	10MY	-0.2	-3.4	-2.1	1.1
36.53	-0.10	1.7	52.55	20MY	-1.2	-3.3	-2.0	1.1
43.36	0.01	1.8	62.07	30MY	-1.8	-3.3	-2.0	1.0
50.11	0.11	1.8	71.59	9JN	-0.8	-3.3	-1.9	1.0
56.78	0.22	1.8	81.12	19JN	-0.1	-3.3	-1.9	0.9
63.37	0.33	1.8	90.66	29JN	0.3	-3.3	-1.8	0.9
69.89	0.43	1.9	100.24	9JL	0.7	-3.4	-1.7	0.8
76.33	0.54	1.9	109.85	19JL	1.1	-3.4	-1.7	0.8
82.69	0.65	1.9	119.50	29JL	2.2	-3.4	-1.6	0.7
88.99	0.76	1.8	129.21	8AU	2.5	-3.4	-1.6	0.7
95.20	0.88	1.8	138.97	18AU	0.2	-3.5	-1.5	0.6
101.33	0.99	1.8	148.79	28AU	-0.9	-3.6	-1.5	0.6
107.37	1.11	1.8	158.67	7SE	-1.1	-3.6	-1.4	0.7
113.31	1.23	1.7	168.60	17SE	-1.0	-3.7	-1.4	0.7
119.15	1.36	1.7	178.60	27SE	-0.6	-3.8	-1.4	0.8
124.86	1.49	1.6	188.65	7OC	-0.4	-3.9	-1.3	0.9
130.42	1.63	1.5	198.75	17OC	-0.3	-4.0	-1.3	0.9
135.82	1.77	1.4	208.89	27OC	-0.3	-4.1	-1.3	1.0
141.02	1.93	1.3	219.06	6NO	0.1	-4.2	-1.3	1.1
145.97	2.09	1.2	229.26	16NO	2.0	-4.3	-1.3	1.1
150.63	2.27	1.0	239.47	26NO	1.3	-4.4	-1.3	1.1
154.92	2.47	0.8	249.68	6DE	0.1	-4.4	-1.4	1.1
158.76	2.67	0.6	259.88	16DE	-0.1	-4.2	-1.4	1.2
162.04	2.90	0.4	270.06	26DE	-0.1	-3.8	-1.4	1.2

♂			⊙	16.00UT	☿	♀	♃	♄
LONG	LAT	MAG	LONG	-448	MAGNITUDES			
164.62	3.14	0.2	280.22	5JA	-0.3	-3.2	-1.5	1.1
166.34	3.38	-0.1	290.33	15JA	-0.7	-3.7	-1.5	1.1
167.02	3.63	-0.3	300.39	25JA	-1.2	-4.2	-1.6	1.1
166.47	3.84	-0.6	310.41	4FE	-1.3	-4.3	-1.6	1.0
164.66	3.98	-0.9	320.36	14FE	-0.7	-4.3	-1.7	1.0
161.69	4.01	-1.2	330.26	24FE	0.7	-4.2	-1.8	0.9
158.01	3.87	-1.3	340.09	5MR	2.6	-4.1	-1.8	0.9
154.30	3.58	-1.2	349.87	15MR	2.8	-4.0	-1.9	0.9
151.25	3.16	-1.0	359.59	25MR	1.5	-3.9	-2.0	0.9
149.34	2.68	-0.8	9.26	4AP	0.9	-3.8	-2.0	0.9
148.76	2.19	-0.6	18.89	14AP	0.4	-3.7	-2.1	0.9
149.45	1.73	-0.4	28.48	24AP	-0.3	-3.6	-2.1	0.9
151.28	1.31	-0.2	38.03	4MY	-1.2	-3.6	-2.2	0.9
154.07	0.94	-0.0	47.56	14MY	-1.8	-3.5	-2.2	0.9
157.64	0.60	0.1	57.09	24MY	-0.8	-3.4	-2.2	0.8
161.87	0.31	0.3	66.60	3JN	-0.1	-3.4	-2.2	0.8
166.63	0.04	0.4	76.12	13JN	0.5	-3.4	-2.2	0.8
171.84	-0.19	0.5	85.66	23JN	0.9	-3.4	-2.1	0.7
177.44	-0.40	0.6	95.22	3JL	1.5	-3.3	-2.1	0.7
183.37	-0.59	0.7	104.81	13JL	2.8	-3.3	-2.0	0.6
189.59	-0.75	0.7	114.44	23JL	2.1	-3.3	-2.0	0.6
196.06	-0.90	0.8	124.12	2AU	0.2	-3.3	-1.9	0.5
202.75	-1.02	0.9	133.85	12AU	-1.0	-3.4	-1.8	0.4
209.65	-1.13	0.9	143.64	22AU	-1.2	-3.4	-1.8	0.4
216.72	-1.22	1.0	153.49	1SE	-1.0	-3.4	-1.7	0.4
223.95	-1.29	1.0	163.39	11SE	-0.5	-3.4	-1.7	0.3
231.31	-1.35	1.0	173.35	21SE	-0.3	-3.4	-1.6	0.4
238.79	-1.39	1.1	183.37	1OC	-0.2	-3.5	-1.6	0.5
246.37	-1.41	1.1	193.44	11OC	-0.1	-3.5	-1.5	0.5
254.03	-1.41	1.2	203.57	21OC	0.3	-3.5	-1.5	0.6
261.75	-1.40	1.2	213.72	31OC	2.3	-3.5	-1.5	0.7
269.51	-1.38	1.2	223.90	10NO	1.0	-3.4	-1.4	0.7
277.30	-1.34	1.3	234.11	20NO	-0.1	-3.4	-1.4	0.8
285.10	-1.29	1.3	244.32	30NO	-0.2	-3.4	-1.4	0.8
292.89	-1.23	1.3	254.53	10DE	-0.3	-3.4	-1.4	0.8
300.66	-1.15	1.4	264.73	20DE	-0.4	-3.3	-1.4	0.9
308.39	-1.07	1.4	274.90	30DE	-0.7	-3.3	-1.4	0.9
				-447				
316.08	-0.98	1.5	285.03	9JA	-1.1	-3.3	-1.5	0.9
323.71	-0.89	1.5	295.12	19JA	-1.2	-3.3	-1.5	0.9
331.28	-0.79	1.5	305.16	29JA	-0.7	-3.3	-1.5	0.8
338.77	-0.69	1.6	315.15	8FE	0.8	-3.4	-1.6	0.8
346.19	-0.59	1.6	325.08	18FE	3.1	-3.4	-1.6	0.8
353.53	-0.48	1.6	334.94	28FE	2.1	-3.4	-1.7	0.7
0.79	-0.38	1.6	344.75	10MR	1.1	-3.4	-1.8	0.7
7.96	-0.27	1.7	354.50	20MR	0.7	-3.5	-1.8	0.7
15.06	-0.17	1.7	4.20	30MR	0.3	-3.5	-1.9	0.7
22.07	-0.06	1.7	13.85	9AP	-0.3	-3.6	-2.0	0.7
29.01	0.04	1.7	23.46	19AP	-1.2	-3.6	-2.0	0.7
35.87	0.14	1.7	33.03	29AP	-1.8	-3.7	-2.1	0.7
42.66	0.24	1.7	42.57	9MY	-0.8	-3.8	-2.2	0.7
49.39	0.33	1.8	52.10	19MY	0.0	-3.9	-2.2	0.7
56.05	0.43	1.8	61.61	29MY	0.6	-3.9	-2.3	0.6
62.66	0.52	1.9	71.13	8JN	1.2	-4.0	-2.3	0.6
69.21	0.60	1.9	80.66	18JN	2.0	-4.1	-2.3	0.6
75.71	0.69	1.9	90.21	28JN	3.3	-4.2	-2.3	0.5
82.16	0.77	2.0	99.78	8JL	1.8	-4.2	-2.3	0.5
88.57	0.85	2.0	109.39	18JL	0.1	-4.1	-2.3	0.4
94.93	0.93	2.0	119.04	28JL	-1.0	-3.7	-2.2	0.4
101.26	1.01	2.0	128.74	7AU	-1.4	-3.2	-2.2	0.3
107.55	1.09	2.0	138.50	17AU	-1.0	-3.7	-2.1	0.3
113.80	1.16	2.0	148.31	27AU	-0.5	-4.1	-2.1	0.2
120.01	1.23	2.0	158.19	6SE	-0.2	-4.3	-2.0	0.2
126.18	1.31	1.9	168.12	16SE	-0.0	-4.3	-1.9	0.1
132.30	1.38	1.9	178.11	26SE	0.1	-4.3	-1.9	0.1
138.38	1.45	1.8	188.16	6OC	0.5	-4.2	-1.8	0.1
144.41	1.52	1.8	198.26	16OC	2.7	-4.1	-1.8	0.2
150.36	1.59	1.7	208.39	26OC	0.8	-4.0	-1.7	0.3
156.25	1.66	1.6	218.56	5NO	-0.2	-3.9	-1.7	0.4
162.04	1.73	1.5	228.76	15NO	-0.4	-3.8	-1.6	0.4
167.72	1.79	1.4	238.97	25NO	-0.4	-3.7	-1.6	0.5
173.28	1.86	1.3	249.18	5DE	-0.5	-3.7	-1.5	0.5
178.68	1.93	1.2	259.39	15DE	-0.8	-3.6	-1.5	0.6
183.89	1.99	1.0	269.57	25DE	-1.0	-3.5	-1.5	0.6

♂ LONG	♂ LAT	♂ MAG	⊙ LONG	16.00UT -446	☿	♀	♃	♄
					MAGNITUDES			
188.86	2.05	0.8	279.72	4JA	-1.0	-3.5	-1.5	0.6
193.55	2.10	0.6	289.84	14JA	-0.6	-3.4	-1.5	0.6
197.87	2.14	0.4	299.90	24JA	1.0	-3.4	-1.5	0.6
201.75	2.17	0.2	309.92	3FE	3.1	-3.4	-1.5	0.6
205.06	2.18	-0.1	319.88	13FE	1.5	-3.3	-1.6	0.6
207.66	2.15	-0.3	329.78	23FE	0.8	-3.3	-1.6	0.6
209.38	2.08	-0.6	339.62	5MR	0.5	-3.3	-1.6	0.6
210.04	1.95	-0.9	349.40	15MR	0.1	-3.3	-1.7	0.5
209.49	1.74	-1.3	359.13	25MR	-0.4	-3.3	-1.7	0.5
207.68	1.42	-1.6	8.80	4AP	-1.2	-3.3	-1.8	0.4
204.83	0.99	-1.8	18.43	14AP	-1.7	-3.3	-1.8	0.5
201.46	0.48	-1.9	28.01	24AP	-0.8	-3.4	-1.9	0.5
198.25	-0.07	-1.8	37.57	4MY	0.1	-3.4	-2.0	0.5
195.90	-0.59	-1.6	47.11	14MY	0.8	-3.4	-2.0	0.5
194.82	-1.04	-1.4	56.62	24MY	1.6	-3.5	-2.1	0.5
195.12	-1.42	-1.2	66.14	3JN	2.7	-3.5	-2.2	0.5
196.72	-1.72	-1.0	75.66	13JN	3.1	-3.5	-2.2	0.5
199.44	-1.94	-0.8	85.20	23JN	1.4	-3.4	-2.3	0.4
203.10	-2.11	-0.6	94.76	3JL	0.0	-3.4	-2.4	0.4
207.53	-2.23	-0.4	104.35	13JL	-1.1	-3.4	-2.4	0.4
212.59	-2.31	-0.3	113.97	23JL	-1.5	-3.3	-2.4	0.3
218.16	-2.36	-0.1	123.65	2AU	-0.9	-3.3	-2.4	0.3
224.15	-2.37	-0.0	133.38	12AU	-0.4	-3.3	-2.4	0.2
230.49	-2.35	0.1	143.16	22AU	-0.1	-3.3	-2.4	0.2
237.11	-2.31	0.2	153.00	1SE	0.1	-3.3	-2.3	0.1
243.96	-2.25	0.3	162.91	11SE	0.2	-3.3	-2.3	0.0
250.99	-2.17	0.4	172.86	21SE	0.8	-3.4	-2.2	-0.0
258.18	-2.07	0.6	182.88	1OC	3.0	-3.4	-2.2	-0.1
265.47	-1.96	0.6	192.95	11OC	0.7	-3.4	-2.1	-0.1
272.85	-1.83	0.7	203.07	21OC	-0.4	-3.5	-2.0	-0.0
280.28	-1.70	0.8	213.23	31OC	-0.5	-3.5	-2.0	0.0
287.75	-1.56	0.9	223.41	10NO	-0.5	-3.6	-1.9	0.1
295.23	-1.41	1.0	233.61	20NO	-0.6	-3.6	-1.8	0.2
302.70	-1.26	1.1	243.83	30NO	-0.8	-3.7	-1.8	0.2
310.16	-1.11	1.2	254.03	10DE	-0.9	-3.8	-1.7	0.3
317.58	-0.97	1.3	264.23	20DE	-0.9	-3.9	-1.7	0.4
324.96	-0.82	1.3	274.40	30DE	-0.5	-4.0	-1.6	0.4
				-445				
332.28	-0.68	1.4	284.54	9JA	1.2	-4.1	-1.6	0.4
339.55	-0.54	1.5	294.63	19JA	2.7	-4.2	-1.6	0.5
346.75	-0.40	1.5	304.68	29JA	1.1	-4.2	-1.6	0.5
353.89	-0.28	1.6	314.66	8FE	0.6	-4.3	-1.6	0.5
0.96	-0.15	1.7	324.59	18FE	0.3	-4.3	-1.6	0.5
7.96	-0.03	1.7	334.47	28FE	0.0	-4.1	-1.6	0.4
14.89	0.08	1.8	344.28	10MR	-0.4	-3.6	-1.6	0.4
21.76	0.18	1.8	354.03	20MR	-1.2	-3.1	-1.6	0.4
28.56	0.28	1.8	3.73	30MR	-1.7	-3.7	-1.6	0.3
35.30	0.38	1.9	13.38	9AP	-0.8	-4.1	-1.7	0.3
41.99	0.47	1.9	22.99	19AP	0.2	-4.2	-1.7	0.3
48.62	0.55	1.9	32.57	29AP	1.1	-4.2	-1.8	0.3
55.21	0.63	1.9	42.11	9MY	2.1	-4.1	-1.8	0.3
61.75	0.71	1.9	51.63	19MY	3.4	-4.0	-1.9	0.4
68.25	0.77	1.9	61.15	29MY	2.5	-3.9	-1.9	0.4
74.72	0.84	1.9	70.67	8JN	1.1	-3.8	-2.0	0.4
81.15	0.90	1.9	80.20	18JN	-0.0	-3.7	-2.1	0.4
87.57	0.95	1.9	89.74	28JN	-1.1	-3.7	-2.1	0.4
93.96	1.01	2.0	99.31	8JL	-1.6	-3.6	-2.2	0.3
100.34	1.05	2.0	108.92	18JL	-0.9	-3.5	-2.3	0.3
106.71	1.10	2.0	118.58	28JL	-0.3	-3.5	-2.3	0.3
113.07	1.14	2.0	128.27	7AU	0.0	-3.5	-2.4	0.2
119.42	1.18	2.0	138.03	17AU	0.2	-3.4	-2.4	0.2
125.77	1.21	2.0	147.84	27AU	0.4	-3.4	-2.4	0.1
132.12	1.24	2.0	157.71	6SE	1.1	-3.4	-2.4	0.1
138.47	1.26	2.0	167.64	16SE	3.3	-3.4	-2.4	-0.0
144.82	1.29	2.0	177.63	26SE	0.5	-3.4	-2.4	-0.1
151.17	1.30	1.9	187.67	6OC	-0.5	-3.4	-2.4	-0.2
157.52	1.32	1.9	197.77	16OC	-0.7	-3.4	-2.3	-0.2
163.87	1.32	1.9	207.90	26OC	-0.7	-3.4	-2.3	-0.2
170.22	1.33	1.8	218.07	5NO	-0.7	-3.4	-2.2	-0.2
176.56	1.32	1.7	228.27	15NO	-0.7	-3.4	-2.1	-0.1
182.88	1.31	1.7	238.47	25NO	-0.7	-3.4	-2.0	-0.0
189.19	1.29	1.6	248.68	5DE	-0.7	-3.4	-2.0	0.1
195.48	1.26	1.5	258.89	15DE	-0.4	-3.4	-1.9	0.1
201.73	1.22	1.4	269.07	25DE	1.4	-3.5	-1.8	0.2

♂ LONG	♂ LAT	♂ MAG	⊙ LONG	16.00UT -444	☿	♀	♃	♄
					MAGNITUDES			
207.95	1.17	1.2	279.22	4JA	2.3	-3.5	-1.8	0.2
214.12	1.10	1.1	289.34	14JA	0.8	-3.5	-1.7	0.3
220.22	1.01	1.0	299.41	24JA	0.4	-3.4	-1.7	0.3
226.24	0.90	0.8	309.43	3FE	0.2	-3.4	-1.6	0.3
232.16	0.77	0.7	319.39	13FE	-0.1	-3.4	-1.6	0.4
237.95	0.60	0.5	329.29	23FE	-0.5	-3.4	-1.6	0.4
243.57	0.40	0.3	339.13	4MR	-1.2	-3.3	-1.6	0.4
248.99	0.16	0.1	348.92	14MR	-1.6	-3.3	-1.6	0.3
254.14	-0.13	-0.1	358.65	24MR	-0.8	-3.3	-1.6	0.3
258.96	-0.48	-0.4	8.32	3AP	0.3	-3.3	-1.6	0.3
263.33	-0.90	-0.6	17.96	13AP	1.4	-3.3	-1.6	0.2
267.13	-1.40	-0.9	27.55	23AP	2.7	-3.4	-1.6	0.2
270.21	-1.98	-1.2	37.10	3MY	3.3	-3.4	-1.6	0.2
272.38	-2.66	-1.5	46.64	13MY	1.9	-3.4	-1.6	0.2
273.45	-3.42	-1.8	56.17	23MY	0.9	-3.4	-1.7	0.3
273.29	-4.22	-2.1	65.68	2JN	-0.1	-3.5	-1.7	0.3
271.90	-5.00	-2.4	75.20	12JN	-1.2	-3.5	-1.8	0.3
269.59	-5.64	-2.6	84.74	22JN	-1.7	-3.6	-1.8	0.3
266.95	-6.04	-2.6	94.29	2JL	-0.9	-3.6	-1.9	0.3
264.69	-6.15	-2.4	103.88	12JL	-0.3	-3.7	-1.9	0.3
263.44	-6.00	-2.2	113.51	22JL	0.1	-3.8	-2.0	0.3
263.48	-5.67	-1.9	123.18	1AU	0.4	-3.9	-2.1	0.3
264.83	-5.23	-1.6	132.91	11AU	0.7	-4.0	-2.1	0.2
267.37	-4.74	-1.4	142.69	21AU	1.4	-4.1	-2.2	0.2
270.87	-4.25	-1.1	152.53	31AU	3.2	-4.2	-2.3	0.1
275.16	-3.78	-0.8	162.43	10SE	0.4	-4.3	-2.3	0.1
280.06	-3.32	-0.6	172.38	20SE	-0.7	-4.3	-2.4	-0.0
285.43	-2.90	-0.4	182.40	30SE	-0.8	-4.2	-2.4	-0.1
291.16	-2.50	-0.2	192.47	10OC	-0.8	-3.8	-2.4	-0.1
297.18	-2.14	0.0	202.58	20OC	-0.8	-3.1	-2.4	-0.2
303.39	-1.81	0.2	212.73	30OC	-0.6	-3.7	-2.4	-0.3
309.76	-1.50	0.4	222.92	9NO	-0.6	-4.2	-2.3	-0.3
316.25	-1.22	0.5	233.12	19NO	-0.6	-4.4	-2.3	-0.2
322.81	-0.96	0.7	243.33	29NO	-0.2	-4.4	-2.2	-0.2
329.42	-0.73	0.8	253.54	9DE	1.6	-4.3	-2.1	-0.1
336.07	-0.52	1.0	263.73	19DE	1.9	-4.2	-2.1	-0.0
342.72	-0.33	1.1	273.90	29DE	0.5	-4.1	-2.0	0.1
				-443				
349.38	-0.16	1.2	284.05	8JA	0.2	-4.0	-1.9	0.1
356.03	0.00	1.3	294.14	18JA	0.1	-3.9	-1.9	0.2
2.66	0.14	1.4	304.18	28JA	-0.2	-3.8	-1.8	0.2
9.27	0.27	1.5	314.18	7FE	-0.5	-3.7	-1.7	0.3
15.85	0.39	1.6	324.11	17FE	-1.2	-3.6	-1.7	0.3
22.40	0.49	1.7	333.99	27FE	-1.5	-3.5	-1.6	0.3
28.92	0.59	1.8	343.80	9MR	-0.8	-3.5	-1.6	0.3
35.42	0.67	1.8	353.56	19MR	0.4	-3.4	-1.6	0.3
41.88	0.75	1.9	3.26	29MR	1.8	-3.4	-1.5	0.3
48.31	0.81	1.9	12.92	8AP	3.5	-3.4	-1.5	0.3
54.73	0.87	1.9	22.53	18AP	2.5	-3.3	-1.5	0.2
61.11	0.93	2.0	32.10	28AP	1.4	-3.3	-1.5	0.2
67.49	0.98	2.0	41.65	8MY	0.7	-3.3	-1.5	0.2
73.84	1.02	2.0	51.18	18MY	-0.2	-3.3	-1.5	0.2
80.19	1.05	2.0	60.70	28MY	-1.2	-3.3	-1.5	0.2
86.53	1.08	2.0	70.22	7JN	-1.8	-3.3	-1.5	0.2
92.87	1.11	2.0	79.74	17JN	-0.9	-3.3	-1.6	0.3
99.22	1.13	2.0	89.28	27JN	-0.2	-3.4	-1.6	0.3
105.57	1.15	2.0	98.86	7JL	0.2	-3.4	-1.6	0.3
111.93	1.16	2.0	108.46	17JL	0.5	-3.4	-1.7	0.3
118.31	1.16	1.9	118.11	27JL	0.9	-3.5	-1.7	0.3
124.71	1.16	2.0	127.81	6AU	1.9	-3.5	-1.8	0.3
131.13	1.16	2.0	137.55	16AU	2.8	-3.5	-1.9	0.3
137.58	1.15	2.0	147.36	26AU	0.3	-3.4	-1.9	0.2
144.06	1.14	2.0	157.23	5SE	-0.8	-3.4	-2.0	0.2
150.57	1.12	2.0	167.15	15SE	-1.0	-3.4	-2.0	0.1
157.12	1.10	2.0	177.14	25SE	-1.0	-3.4	-2.1	0.1
163.70	1.07	1.9	187.18	5OC	-0.7	-3.3	-2.2	-0.0
170.33	1.04	1.9	197.27	15OC	-0.5	-3.3	-2.2	-0.1
176.99	0.99	1.9	207.40	25OC	-0.4	-3.3	-2.3	-0.2
183.69	0.95	1.8	217.57	4NO	-0.4	-3.3	-2.3	-0.2
190.43	0.89	1.8	227.76	14NO	-0.1	-3.3	-2.3	-0.3
197.21	0.82	1.7	237.97	24NO	1.8	-3.4	-2.3	-0.3
204.03	0.75	1.7	248.19	4DE	1.5	-3.4	-2.3	-0.2
210.89	0.66	1.6	258.39	14DE	0.3	-3.4	-2.2	-0.2
217.78	0.56	1.5	268.58	24DE	0.0	-3.4	-2.2	-0.1

♂ LONG	♂ LAT	♂ MAG	☉ LONG	16.00UT	☿	♀	♃	♄
				-442	MAGNITUDES			
224.71	0.46	1.5	278.73	3JA	-0.1	-3.5	-2.1	-0.0
231.67	0.33	1.4	288.85	13JA	-0.3	-3.5	-2.0	0.0
238.66	0.20	1.3	298.93	23JA	-0.6	-3.6	-2.0	0.1
245.67	0.04	1.2	308.95	2FE	-1.2	-3.6	-1.9	0.2
252.70	-0.12	1.1	318.91	12FE	-1.4	-3.7	-1.8	0.2
259.75	-0.31	1.0	328.82	22FE	-0.8	-3.8	-1.8	0.3
266.80	-0.51	0.9	338.66	4MR	0.5	-3.9	-1.7	0.3
273.84	-0.72	0.8	348.45	14MR	2.2	-3.9	-1.6	0.3
280.86	-0.96	0.6	358.18	24MR	3.3	-4.0	-1.6	0.3
287.84	-1.21	0.5	7.86	3AP	1.9	-4.1	-1.5	0.3
294.76	-1.48	0.4	17.49	13AP	1.1	-4.2	-1.5	0.3
301.59	-1.76	0.2	27.09	23AP	0.5	-4.2	-1.5	0.3
308.28	-2.05	0.1	36.65	3MY	-0.2	-4.1	-1.4	0.3
314.81	-2.36	-0.0	46.18	13MY	-1.2	-3.7	-1.4	0.2
321.12	-2.68	-0.2	55.71	23MY	-1.8	-3.0	-1.4	0.2
327.13	-3.01	-0.4	65.22	2JN	-0.9	-3.2	-1.4	0.2
332.77	-3.34	-0.5	74.74	12JN	-0.1	-3.9	-1.4	0.2
337.93	-3.68	-0.7	84.28	22JN	0.4	-4.1	-1.4	0.3
342.48	-4.02	-0.9	93.83	2JL	0.7	-4.2	-1.4	0.3
346.26	-4.36	-1.1	103.42	12JL	1.3	-4.2	-1.4	0.3
349.06	-4.67	-1.3	113.04	22JL	2.4	-4.1	-1.5	0.3
350.67	-4.93	-1.6	122.71	1AU	2.4	-4.0	-1.5	0.3
350.91	-5.11	-1.8	132.43	11AU	0.3	-3.9	-1.5	0.3
349.68	-5.15	-2.0	142.21	21AU	-0.9	-3.8	-1.6	0.3
347.20	-4.99	-2.2	152.05	31AU	-1.1	-3.8	-1.6	0.3
344.00	-4.59	-2.3	161.94	10SE	-1.0	-3.7	-1.7	0.3
340.85	-4.00	-2.1	171.90	20SE	-0.6	-3.6	-1.7	0.2
338.52	-3.31	-1.8	181.91	30SE	-0.4	-3.6	-1.8	0.2
337.42	-2.60	-1.5	191.98	10OC	-0.3	-3.5	-1.9	0.1
337.66	-1.96	-1.2	202.09	20OC	-0.2	-3.5	-1.9	0.0
339.12	-1.39	-0.8	212.24	30OC	0.1	-3.5	-2.0	-0.0
341.61	-0.92	-0.5	222.42	9NO	2.1	-3.4	-2.1	-0.1
344.92	-0.52	-0.2	232.63	19NO	1.2	-3.4	-2.1	-0.2
348.87	-0.19	0.0	242.83	29NO	0.1	-3.4	-2.1	-0.3
353.31	0.08	0.3	253.04	9DE	-0.1	-3.4	-2.2	-0.2
358.14	0.30	0.5	263.24	19DE	-0.2	-3.3	-2.2	-0.2
3.27	0.49	0.7	273.41	29DE	-0.3	-3.3	-2.2	-0.1
				-441				
8.61	0.64	0.9	283.55	8JA	-0.7	-3.3	-2.1	-0.0
14.14	0.77	1.0	293.65	18JA	-1.2	-3.3	-2.1	0.0
19.81	0.87	1.2	303.70	28JA	-1.3	-3.3	-2.0	0.1
25.58	0.96	1.3	313.69	7FE	-0.7	-3.4	-2.0	0.1
31.43	1.03	1.4	323.63	17FE	0.7	-3.4	-1.9	0.2
37.34	1.09	1.5	333.51	27FE	2.7	-3.4	-1.8	0.2
43.31	1.14	1.6	343.33	9MR	2.5	-3.5	-1.8	0.3
49.32	1.18	1.7	353.09	19MR	1.4	-3.5	-1.7	0.3
55.36	1.21	1.8	2.79	29MR	0.8	-3.5	-1.6	0.3
61.42	1.23	1.8	12.45	8AP	0.3	-3.4	-1.6	0.3
67.52	1.25	1.9	22.06	18AP	-0.3	-3.4	-1.5	0.4
73.63	1.26	1.9	31.63	28AP	-1.2	-3.4	-1.5	0.3
79.77	1.27	2.0	41.18	8MY	-1.8	-3.4	-1.4	0.3
85.94	1.26	2.0	50.71	18MY	-0.9	-3.3	-1.4	0.3
92.13	1.26	2.0	60.23	28MY	-0.1	-3.3	-1.4	0.3
98.35	1.25	2.0	69.75	7JN	0.5	-3.3	-1.3	0.3
104.60	1.23	2.0	79.27	17JN	1.0	-3.3	-1.3	0.3
110.88	1.21	2.0	88.81	27JN	1.7	-3.3	-1.3	0.3
117.20	1.19	2.0	98.38	7JL	3.0	-3.4	-1.3	0.4
123.57	1.16	2.0	107.99	17JL	2.0	-3.4	-1.3	0.4
129.97	1.13	2.0	117.63	27JL	0.2	-3.4	-1.3	0.4
136.43	1.09	2.0	127.33	6AU	-1.0	-3.4	-1.3	0.4
142.93	1.05	1.9	137.08	16AU	-1.3	-3.5	-1.3	0.4
149.49	1.00	1.9	146.88	26AU	-1.0	-3.6	-1.4	0.4
156.11	0.95	1.8	156.75	5SE	-0.5	-3.6	-1.4	0.4
162.78	0.89	1.8	166.67	15SE	-0.2	-3.7	-1.4	0.4
169.52	0.83	1.8	176.65	25SE	-0.1	-3.8	-1.5	0.4
176.33	0.76	1.8	186.70	5OC	-0.1	-3.9	-1.6	0.3
183.19	0.69	1.8	196.78	15OC	0.3	-4.0	-1.6	0.3
190.13	0.61	1.8	206.91	25OC	2.4	-4.1	-1.7	0.2
197.13	0.52	1.8	217.08	4NO	1.0	-4.2	-1.7	0.1
204.20	0.43	1.7	227.27	14NO	-0.1	-4.3	-1.8	0.1
211.34	0.33	1.7	237.48	24NO	-0.3	-4.4	-1.9	-0.0
218.54	0.23	1.7	247.69	4DE	-0.3	-4.4	-1.9	-0.1
225.80	0.12	1.6	257.90	14DE	-0.4	-4.2	-2.0	-0.1
233.12	-0.00	1.6	268.08	24DE	-0.7	-3.8	-2.0	-0.1

♂ LONG	♂ LAT	♂ MAG	☉ LONG	16.00UT	☿	♀	♃	♄
				-440	MAGNITUDES			
240.50	-0.13	1.5	278.24	3JA	-1.1	-3.2	-2.1	-0.1
247.93	-0.26	1.5	288.36	13JA	-1.1	-3.7	-2.1	-0.0
255.41	-0.39	1.4	298.43	23JA	-0.7	-4.2	-2.1	0.0
262.92	-0.53	1.4	308.46	2FE	0.8	-4.3	-2.1	0.1
270.47	-0.68	1.3	318.43	12FE	3.1	-4.3	-2.0	0.2
278.04	-0.82	1.3	328.33	22FE	1.9	-4.2	-2.0	0.2
285.63	-0.97	1.2	338.19	3MR	1.0	-4.1	-1.9	0.3
293.21	-1.11	1.1	347.98	13MR	0.6	-4.0	-1.9	0.3
300.79	-1.25	1.1	357.71	23MR	0.2	-3.9	-1.8	0.4
308.33	-1.39	1.0	7.39	2AP	-0.3	-3.8	-1.7	0.4
315.84	-1.52	1.0	17.03	12AP	-1.2	-3.7	-1.7	0.4
323.28	-1.64	0.9	26.62	22AP	-1.8	-3.6	-1.6	0.4
330.65	-1.75	0.8	36.18	2MY	-0.9	-3.6	-1.5	0.5
337.91	-1.84	0.8	45.72	12MY	0.0	-3.5	-1.5	0.5
345.06	-1.93	0.7	55.24	22MY	0.7	-3.4	-1.4	0.4
352.06	-1.99	0.6	64.76	1JN	1.3	-3.4	-1.4	0.4
358.88	-2.04	0.6	74.28	11JN	2.2	-3.4	-1.3	0.4
5.50	-2.08	0.5	83.81	21JN	3.3	-3.4	-1.3	0.4
11.88	-2.09	0.4	93.37	1JL	1.7	-3.3	-1.3	0.4
17.96	-2.09	0.3	102.95	11JL	0.1	-3.3	-1.3	0.5
23.71	-2.06	0.2	112.58	21JL	-1.0	-3.3	-1.2	0.5
29.04	-2.02	0.1	122.25	31JL	-1.4	-3.3	-1.2	0.5
33.86	-1.94	-0.1	131.96	10AU	-1.0	-3.4	-1.2	0.6
38.07	-1.84	-0.2	141.74	20AU	-0.4	-3.4	-1.2	0.6
41.52	-1.70	-0.4	151.58	30AU	-0.1	-3.4	-1.2	0.6
44.02	-1.52	-0.6	161.47	9SE	0.0	-3.4	-1.3	0.6
45.39	-1.28	-0.8	171.42	19SE	0.1	-3.4	-1.3	0.6
45.40	-0.98	-1.0	181.43	29SE	0.6	-3.5	-1.3	0.5
43.98	-0.61	-1.2	191.49	9OC	2.7	-3.5	-1.4	0.5
41.21	-0.18	-1.4	201.60	19OC	0.8	-3.5	-1.4	0.5
37.53	0.28	-1.6	211.75	29OC	-0.3	-3.5	-1.5	0.4
33.71	0.72	-1.5	221.93	8NO	-0.4	-3.4	-1.5	0.4
30.50	1.10	-1.2	232.13	18NO	-0.5	-3.4	-1.6	0.3
28.46	1.38	-0.9	242.34	28NO	-0.6	-3.4	-1.6	0.2
27.78	1.58	-0.6	252.55	8DE	-0.8	-3.4	-1.7	0.1
28.38	1.71	-0.3	262.74	18DE	-0.9	-3.3	-1.8	0.1
30.08	1.79	-0.0	272.92	28DE	-1.0	-3.3	-1.8	0.0
				-439				
32.67	1.84	0.3	283.06	7JA	-0.6	-3.3	-1.9	0.0
35.97	1.85	0.5	293.16	17JA	1.0	-3.3	-1.9	0.1
39.82	1.86	0.7	303.21	27JA	3.0	-3.3	-2.0	0.1
44.11	1.84	0.9	313.21	6FE	1.4	-3.4	-2.0	0.2
48.74	1.82	1.0	323.15	16FE	0.7	-3.4	-2.0	0.3
53.65	1.79	1.2	333.03	26FE	0.4	-3.4	-2.0	0.3
58.78	1.76	1.3	342.85	8MR	0.1	-3.4	-2.0	0.4
64.08	1.72	1.4	352.61	18MR	-0.4	-3.5	-2.0	0.4
69.53	1.68	1.5	2.32	28MR	-1.2	-3.5	-1.9	0.5
75.10	1.64	1.6	11.98	7AP	-1.7	-3.6	-1.9	0.5
80.78	1.59	1.7	21.60	17AP	-0.9	-3.6	-1.8	0.5
86.56	1.54	1.8	31.18	27AP	0.1	-3.7	-1.8	0.6
92.42	1.49	1.8	40.72	7MY	0.9	-3.8	-1.7	0.6
98.36	1.43	1.9	50.25	17MY	1.7	-3.9	-1.6	0.6
104.37	1.38	1.9	59.77	27MY	2.9	-3.9	-1.6	0.6
110.45	1.32	1.9	69.29	6JN	2.9	-4.0	-1.5	0.6
116.61	1.26	1.9	78.81	16JN	1.3	-4.1	-1.4	0.6
122.83	1.19	2.0	88.36	26JN	0.0	-4.2	-1.4	0.6
129.13	1.12	2.0	97.92	6JL	-1.1	-4.2	-1.3	0.6
135.50	1.05	1.9	107.53	16JL	-1.5	-4.0	-1.3	0.6
141.94	0.98	1.9	117.17	26JL	-0.9	-3.6	-1.3	0.7
148.46	0.91	1.9	126.86	5AU	-0.4	-3.2	-1.3	0.7
155.06	0.83	1.9	136.61	15AU	-0.0	-3.7	-1.2	0.7
161.74	0.74	1.9	146.41	25AU	0.1	-4.1	-1.2	0.8
168.51	0.66	1.8	156.27	4SE	0.3	-4.3	-1.2	0.8
175.35	0.57	1.8	166.19	14SE	0.9	-4.3	-1.2	0.8
182.29	0.47	1.7	176.17	24SE	3.0	-4.3	-1.2	0.8
189.30	0.38	1.7	186.21	4OC	0.7	-4.2	-1.2	0.8
196.41	0.28	1.6	196.29	14OC	-0.5	-4.1	-1.3	0.8
203.59	0.18	1.6	206.42	24OC	-0.6	-4.0	-1.3	0.7
210.86	0.07	1.6	216.59	3NO	-0.6	-3.9	-1.3	0.7
218.21	-0.04	1.6	226.78	13NO	-0.7	-3.8	-1.4	0.6
225.64	-0.15	1.6	236.99	23NO	-0.8	-3.7	-1.4	0.6
233.13	-0.26	1.6	247.20	3DE	-0.8	-3.7	-1.5	0.5
240.70	-0.37	1.5	257.41	13DE	-0.8	-3.6	-1.5	0.4
248.32	-0.48	1.5	267.59	23DE	-0.5	-3.5	-1.6	0.4

Left table (−438 / −437)

♂ LONG	LAT	MAG	☉ LONG	16.00UT −438	☿	♀	♃	♄
256.00	-0.59	1.5	277.75	2JA	1.2	-3.5	-1.7	0.3
263.71	-0.70	1.5	287.88	12JA	2.6	-3.4	-1.7	0.2
271.46	-0.80	1.5	297.95	22JA	1.0	-3.4	-1.8	0.2
279.24	-0.90	1.4	307.98	1FE	0.5	-3.3	-1.9	0.3
287.03	-0.99	1.4	317.95	11FE	0.3	-3.3	-1.9	0.3
294.81	-1.08	1.4	327.86	21FE	0.0	-3.3	-2.0	0.4
302.58	-1.15	1.4	337.71	3MR	-0.4	-3.3	-2.0	0.5
310.33	-1.22	1.3	347.51	13MR	-1.2	-3.3	-2.0	0.5
318.04	-1.27	1.3	357.24	23MR	-1.7	-3.3	-2.0	0.6
325.71	-1.31	1.3	6.93	2AP	-0.9	-3.3	-2.0	0.6
333.31	-1.34	1.3	16.56	12AP	0.2	-3.3	-2.0	0.7
340.83	-1.35	1.3	26.16	22AP	1.2	-3.4	-2.0	0.7
348.27	-1.35	1.3	35.72	2MY	2.3	-3.4	-1.9	0.7
355.62	-1.34	1.2	45.26	12MY	3.6	-3.4	-1.9	0.8
2.85	-1.31	1.2	54.78	22MY	2.2	-3.5	-1.8	0.8
9.96	-1.27	1.2	64.30	1JN	1.1	-3.5	-1.8	0.8
16.95	-1.21	1.2	73.82	11JN	-0.0	-3.5	-1.7	0.8
23.79	-1.14	1.1	83.35	21JN	-1.1	-3.4	-1.6	0.8
30.47	-1.06	1.1	92.90	1JL	-1.6	-3.4	-1.6	0.8
36.98	-0.96	1.1	102.49	11JL	-0.9	-3.4	-1.5	0.8
43.29	-0.85	1.0	112.10	21JL	-0.3	-3.3	-1.5	0.8
49.39	-0.72	0.9	121.77	31JL	0.1	-3.3	-1.4	0.8
55.24	-0.57	0.9	131.49	10AU	0.3	-3.3	-1.4	0.9
60.80	-0.41	0.8	141.26	20AU	0.5	-3.3	-1.3	0.9
66.02	-0.23	0.7	151.09	30AU	1.2	-3.3	-1.3	1.0
70.84	-0.02	0.6	160.98	9SE	3.3	-3.3	-1.3	1.0
75.18	0.22	0.4	170.93	19SE	0.6	-3.4	-1.3	1.0
78.92	0.49	0.3	180.94	29SE	-0.6	-3.4	-1.3	1.0
81.91	0.79	0.1	191.00	9OC	-0.8	-3.4	-1.2	1.0
84.01	1.14	-0.1	201.10	19OC	-0.7	-3.5	-1.3	1.0
85.01	1.54	-0.3	211.25	29OC	-0.8	-3.5	-1.3	1.0
84.72	1.98	-0.6	221.43	8NO	-0.7	-3.6	-1.3	1.0
83.05	2.43	-0.8	231.63	18NO	-0.6	-3.7	-1.3	0.9
80.11	2.86	-1.0	241.84	28NO	-0.7	-3.7	-1.4	0.9
76.31	3.21	-1.1	252.06	8DE	-0.3	-3.8	-1.4	0.8
72.37	3.43	-1.0	262.25	18DE	1.4	-3.9	-1.4	0.8
69.01	3.51	-0.8	272.43	28DE	2.2	-4.0	-1.5	0.7
				−437				
66.77	3.46	-0.6	282.57	7JA	0.7	-4.1	-1.6	0.6
65.85	3.34	-0.3	292.67	17JA	0.3	-4.2	-1.6	0.5
66.19	3.17	-0.0	302.73	27JA	0.1	-4.3	-1.7	0.5
67.63	2.99	0.2	312.73	6FE	-0.1	-4.3	-1.8	0.5
69.98	2.81	0.5	322.67	16FE	-0.5	-4.3	-1.8	0.6
73.05	2.63	0.7	332.56	26FE	-1.2	-4.1	-1.9	0.6
76.71	2.47	0.8	342.38	8MR	-1.6	-3.5	-2.0	0.7
80.84	2.31	1.0	352.14	18MR	-0.9	-3.2	-2.0	0.7
85.33	2.16	1.1	1.85	28MR	0.3	-3.7	-2.1	0.8
90.14	2.02	1.3	11.51	7AP	1.5	-4.1	-2.1	0.8
95.20	1.88	1.4	21.13	17AP	3.0	-4.2	-2.1	0.9
100.47	1.76	1.5	30.71	27AP	3.0	-4.2	-2.1	0.9
105.91	1.63	1.5	40.26	7MY	1.7	-4.1	-2.1	1.0
111.55	1.51	1.6	49.79	17MY	0.8	-4.0	-2.1	1.0
117.32	1.39	1.6	59.31	27MY	-0.1	-3.9	-2.0	1.0
123.22	1.28	1.7	68.83	6JN	-1.1	-3.8	-2.0	1.1
129.24	1.16	1.7	78.35	16JN	-1.7	-3.7	-1.9	1.1
135.39	1.05	1.7	87.89	26JN	-0.9	-3.7	-1.9	1.1
141.65	0.94	1.8	97.46	6JL	-0.2	-3.6	-1.8	1.1
148.02	0.83	1.8	107.06	16JL	0.2	-3.5	-1.7	1.1
154.51	0.71	1.8	116.70	26JL	0.4	-3.5	-1.7	1.1
161.10	0.60	1.8	126.39	5AU	0.8	-3.5	-1.6	1.1
167.80	0.49	1.8	136.13	15AU	1.6	-3.4	-1.6	1.1
174.60	0.38	1.7	145.93	25AU	3.1	-3.4	-1.5	1.2
181.52	0.26	1.7	155.79	4SE	0.5	-3.4	-1.5	1.2
188.53	0.15	1.7	165.71	14SE	-0.7	-3.4	-1.4	1.3
195.66	0.04	1.7	175.69	24SE	-0.9	-3.4	-1.4	1.3
202.88	-0.07	1.6	185.72	4OC	-0.9	-3.4	-1.4	1.3
210.20	-0.18	1.6	195.80	14OC	-0.7	-3.4	-1.3	1.3
217.61	-0.29	1.5	205.93	24OC	-0.5	-3.4	-1.3	1.3
225.11	-0.39	1.5	216.09	3NO	-0.5	-3.4	-1.3	1.3
232.69	-0.49	1.5	226.28	13NO	-0.5	-3.4	-1.3	1.3
240.35	-0.59	1.4	236.49	23NO	-0.2	-3.4	-1.3	1.3
248.07	-0.68	1.4	246.70	3DE	1.6	-3.4	-1.4	1.2
255.84	-0.77	1.3	256.91	13DE	1.8	-3.4	-1.4	1.2
263.66	-0.85	1.4	267.10	23DE	0.4	-3.5	-1.4	1.1

Right table (−436 / −435)

♂ LONG	LAT	MAG	☉ LONG	16.00UT −436	☿	♀	♃	♄
271.51	-0.92	1.4	277.26	2JA	0.1	-3.5	-1.5	1.0
279.38	-0.98	1.4	287.38	12JA	0.0	-3.5	-1.5	1.0
287.26	-1.03	1.4	297.47	22JA	-0.2	-3.4	-1.6	0.9
295.12	-1.07	1.4	307.49	1FE	-0.6	-3.4	-1.6	0.8
302.98	-1.10	1.4	317.46	11FE	-1.2	-3.4	-1.7	0.8
310.80	-1.12	1.4	327.38	21FE	-1.5	-3.4	-1.7	0.8
318.58	-1.13	1.4	337.23	2MR	-0.8	-3.3	-1.8	0.9
326.31	-1.12	1.4	347.03	12MR	0.4	-3.3	-1.9	0.9
333.97	-1.10	1.4	356.77	22MR	1.9	-3.3	-1.9	0.9
341.56	-1.07	1.5	6.45	1AP	3.8	-3.3	-2.0	1.0
349.07	-1.03	1.5	16.09	11AP	2.3	-3.3	-2.1	1.1
356.50	-0.98	1.5	25.69	21AP	1.3	-3.4	-2.1	1.1
3.83	-0.92	1.5	35.26	1MY	0.6	-3.4	-2.2	1.2
11.07	-0.84	1.5	44.80	11MY	-0.2	-3.4	-2.2	1.2
18.21	-0.76	1.5	54.32	21MY	-1.1	-3.4	-2.2	1.3
25.24	-0.67	1.5	63.84	31MY	-1.8	-3.5	-2.2	1.3
32.16	-0.57	1.5	73.36	10JN	-0.9	-3.5	-2.2	1.3
38.97	-0.47	1.5	82.89	20JN	-0.2	-3.6	-2.2	1.4
45.66	-0.36	1.5	92.44	30JN	0.3	-3.6	-2.1	1.4
52.23	-0.23	1.5	102.02	10JL	0.6	-3.7	-2.1	1.4
58.67	-0.11	1.5	111.64	20JL	1.0	-3.8	-2.0	1.4
64.97	0.03	1.5	121.30	30JL	2.0	-3.9	-2.0	1.4
71.12	0.18	1.4	131.02	9AU	2.7	-4.0	-1.9	1.4
77.11	0.33	1.4	140.79	19AU	-0.4	-4.1	-1.8	1.3
82.91	0.50	1.3	150.61	29AU	-0.8	-4.2	-1.8	1.3
88.51	0.67	1.2	160.50	8SE	-1.1	-4.3	-1.7	1.3
93.86	0.87	1.1	170.45	18SE	-1.0	-4.3	-1.7	1.3
98.93	1.08	1.0	180.45	28SE	-0.7	-4.2	-1.6	1.3
103.65	1.31	0.9	190.51	8OC	-0.4	-3.7	-1.6	1.3
107.96	1.56	0.8	200.61	18OC	-0.4	-3.1	-1.5	1.3
111.76	1.85	0.6	210.75	28OC	-0.3	-3.8	-1.5	1.2
114.94	2.17	0.4	220.93	7NO	-0.0	-4.2	-1.5	1.2
117.34	2.52	0.2	231.13	17NO	1.8	-4.4	-1.4	1.2
118.78	2.91	0.0	241.34	27NO	1.5	-4.4	-1.4	1.1
119.09	3.33	-0.2	251.56	7DE	0.2	-4.3	-1.4	1.1
118.10	3.75	-0.5	261.76	17DE	-0.0	-4.2	-1.4	1.0
115.81	4.13	-0.7	271.93	27DE	-0.1	-4.1	-1.4	1.0
				−435				
112.43	4.41	-0.9	282.08	6JA	-0.3	-4.0	-1.5	0.9
108.50	4.54	-1.0	292.19	16JA	-0.6	-3.9	-1.5	0.9
104.74	4.48	-0.8	302.24	26JA	-1.2	-3.8	-1.5	0.9
101.81	4.28	-0.6	312.25	5FE	-1.3	-3.7	-1.6	0.8
100.09	3.98	-0.4	322.19	15FE	-0.8	-3.6	-1.6	0.8
99.67	3.64	-0.1	332.08	25FE	0.5	-3.5	-1.7	0.8
100.45	3.29	0.1	341.91	7MR	2.4	-3.5	-1.7	0.8
102.27	2.97	0.3	351.68	17MR	3.0	-3.4	-1.8	0.9
104.93	2.66	0.5	1.39	27MR	1.7	-3.4	-1.9	1.0
108.28	2.38	0.7	11.05	6AP	1.0	-3.4	-1.9	1.1
112.20	2.12	0.8	20.67	16AP	0.4	-3.3	-2.0	1.1
116.57	1.88	0.9	30.25	26AP	-0.2	-3.3	-2.1	1.2
121.32	1.66	1.0	39.81	6MY	-1.1	-3.3	-2.1	1.2
126.39	1.45	1.1	49.34	16MY	-1.8	-3.3	-2.2	1.3
131.73	1.26	1.2	58.85	26MY	-0.9	-3.3	-2.2	1.3
137.31	1.08	1.3	68.37	5JN	-0.1	-3.3	-2.3	1.3
143.11	0.90	1.3	77.89	15JN	0.4	-3.3	-2.3	1.3
149.10	0.74	1.4	87.43	25JN	0.8	-3.4	-2.3	1.3
155.26	0.58	1.4	97.00	5JL	1.4	-3.4	-2.3	1.3
161.60	0.43	1.5	106.59	15JL	2.6	-3.4	-2.3	1.3
168.09	0.28	1.5	116.23	25JL	2.3	-3.5	-2.3	1.3
174.72	0.14	1.5	125.92	4AU	0.3	-3.5	-2.3	1.3
181.50	0.01	1.5	135.66	14AU	-0.9	-3.5	-2.2	1.2
188.42	-0.12	1.5	145.45	24AU	-1.2	-3.4	-2.1	1.2
195.46	-0.25	1.5	155.31	3SE	-1.0	-3.4	-2.1	1.1
202.63	-0.37	1.5	165.22	13SE	-0.6	-3.4	-2.0	1.1
209.92	-0.48	1.5	175.19	23SE	-0.3	-3.4	-1.9	1.1
217.32	-0.58	1.5	185.23	3OC	-0.2	-3.3	-1.9	1.1
224.82	-0.68	1.5	195.30	13OC	-0.2	-3.3	-1.8	1.1
232.41	-0.77	1.5	205.43	23OC	0.2	-3.3	-1.8	1.1
240.08	-0.85	1.4	215.59	2NO	2.1	-3.3	-1.7	1.1
247.83	-0.92	1.4	225.78	12NO	1.2	-3.3	-1.7	1.1
255.63	-0.98	1.4	235.99	22NO	-0.0	-3.4	-1.6	1.1
263.48	-1.02	1.4	246.20	2DE	-0.2	-3.4	-1.6	1.0
271.36	-1.06	1.4	256.40	12DE	-0.2	-3.4	-1.6	1.0
279.26	-1.09	1.4	266.60	22DE	-0.4	-3.4	-1.5	0.9

Left table (-434 / -433):

♂ LONG	LAT	MAG	☉ LONG	16.00UT -434	☿	♀	♃	♄
287.16	-1.10	1.4	276.76	1JA	-0.7	-3.5	-1.5	0.9
295.05	-1.10	1.4	286.89	11JA	-1.1	-3.5	-1.5	0.8
302.92	-1.09	1.4	296.97	21JA	-1.2	-3.6	-1.5	0.8
310.75	-1.07	1.3	307.01	31JA	-0.7	-3.6	-1.5	0.7
318.54	-1.03	1.3	316.98	10FE	0.7	-3.7	-1.6	0.7
326.26	-0.99	1.4	326.90	20FE	2.9	-3.8	-1.6	0.7
333.92	-0.94	1.4	336.76	2MR	2.3	-3.9	-1.6	0.6
341.51	-0.88	1.4	346.56	12MR	1.2	-3.9	-1.6	0.6
349.02	-0.81	1.5	356.30	22MR	0.7	-4.0	-1.7	0.7
356.44	-0.73	1.5	5.99	1AP	0.3	-4.1	-1.7	0.8
3.77	-0.65	1.6	15.63	11AP	-0.3	-4.2	-1.8	0.8
11.02	-0.56	1.6	25.23	21AP	-1.1	-4.2	-1.9	0.9
18.18	-0.46	1.6	34.80	1MY	-1.8	-4.1	-1.9	1.0
25.24	-0.37	1.7	44.34	11MY	-0.9	-3.7	-2.0	1.0
32.21	-0.26	1.7	53.86	21MY	-0.0	-2.9	-2.1	1.1
39.10	-0.16	1.7	63.38	31MY	0.6	-3.2	-2.1	1.1
45.89	-0.05	1.7	72.90	10JN	1.1	-3.9	-2.2	1.1
52.60	0.06	1.8	82.43	20JN	1.9	-4.1	-2.3	1.2
59.22	0.17	1.8	91.98	30JN	3.2	-4.2	-2.3	1.2
65.75	0.28	1.8	101.55	10JL	1.9	-4.2	-2.4	1.2
72.19	0.40	1.8	111.17	20JL	0.2	-4.1	-2.4	1.2
78.55	0.52	1.8	120.83	30JL	-1.0	-4.0	-2.4	1.2
84.81	0.64	1.8	130.54	9AU	-1.3	-3.9	-2.4	1.1
90.97	0.77	1.7	140.31	19AU	-1.0	-3.8	-2.4	1.1
97.03	0.90	1.7	150.14	29AU	-0.5	-3.8	-2.4	1.1
102.97	1.04	1.6	160.02	8SE	-0.2	-3.7	-2.3	1.0
108.78	1.18	1.6	169.96	18SE	-0.1	-3.6	-2.3	1.0
114.45	1.33	1.5	179.96	28SE	0.0	-3.6	-2.2	1.0
119.95	1.48	1.4	190.02	8OC	0.4	-3.5	-2.2	1.0
125.26	1.65	1.3	200.12	18OC	2.4	-3.5	-2.1	1.0
130.33	1.83	1.2	210.26	28OC	1.0	-3.5	-2.0	1.0
135.11	2.03	1.1	220.44	7NO	-0.2	-3.4	-2.0	1.0
139.55	2.24	0.9	230.64	17NO	-0.4	-3.4	-1.9	1.0
143.57	2.48	0.8	240.85	27NO	-0.4	-3.4	-1.8	1.0
147.06	2.74	0.6	251.06	7DE	-0.5	-3.4	-1.8	0.9
149.89	3.02	0.4	261.26	17DE	-0.8	-3.3	-1.7	0.9
151.91	3.32	0.1	271.44	27DE	-1.0	-3.3	-1.7	0.9
				-433				
152.94	3.63	-0.1	281.58	6JA	-1.1	-3.3	-1.6	0.8
152.80	3.94	-0.4	291.69	16JA	-0.7	-3.3	-1.6	0.8
151.38	4.21	-0.7	301.75	26JA	0.8	-3.3	-1.6	0.7
148.73	4.38	-0.9	311.76	5FE	3.1	-3.4	-1.6	0.7
145.17	4.41	-1.1	321.71	15FE	1.7	-3.4	-1.6	0.6
141.31	4.27	-1.3	331.60	25FE	0.9	-3.4	-1.6	0.6
137.88	3.96	-1.0	341.43	7MR	0.5	-3.5	-1.6	0.5
135.45	3.55	-0.8	351.20	17MR	0.2	-3.5	-1.6	0.5
134.30	3.10	-0.5	0.92	27MR	-0.3	-3.5	-1.6	0.5
134.44	2.64	-0.3	10.58	6AP	-1.1	-3.4	-1.6	0.6
135.76	2.21	-0.1	20.20	16AP	-1.8	-3.4	-1.7	0.7
138.08	1.82	0.1	29.78	26AP	-0.9	-3.4	-1.7	0.7
141.23	1.47	0.2	39.34	6MY	0.0	-3.4	-1.8	0.8
145.05	1.15	0.4	48.87	16MY	0.8	-3.3	-1.8	0.9
149.44	0.86	0.5	58.39	26MY	1.4	-3.3	-1.9	0.9
154.29	0.60	0.6	67.90	5JN	2.5	-3.3	-1.9	1.0
159.54	0.37	0.7	77.43	15JN	3.2	-3.3	-2.0	1.0
165.12	0.15	0.8	86.96	25JN	1.6	-3.3	-2.1	1.0
171.00	-0.05	0.9	96.53	5JL	0.1	-3.4	-2.2	1.0
177.14	-0.23	0.9	106.13	15JL	-1.0	-3.4	-2.2	1.1
183.51	-0.39	1.0	115.76	25JL	-1.5	-3.4	-2.3	1.1
190.10	-0.55	1.0	125.44	4AU	-1.0	-3.5	-2.3	1.0
196.87	-0.68	1.1	135.18	14AU	-0.4	-3.5	-2.4	1.0
203.82	-0.80	1.1	144.97	24AU	-0.1	-3.6	-2.4	1.0
210.93	-0.91	1.2	154.83	3SE	0.1	-3.6	-2.4	1.0
218.19	-1.01	1.2	164.74	13SE	0.2	-3.7	-2.4	0.9
225.57	-1.09	1.2	174.71	23SE	0.7	-3.8	-2.4	0.9
233.06	-1.15	1.2	184.74	3OC	2.7	-3.9	-2.4	0.9
240.65	-1.20	1.3	194.82	13OC	0.8	-4.0	-2.4	0.9
248.33	-1.24	1.3	204.94	23OC	-0.4	-4.1	-2.3	0.9
256.07	-1.26	1.3	215.10	2NO	-0.5	-4.2	-2.3	0.9
263.87	-1.27	1.3	225.29	12NO	-0.5	-4.3	-2.2	0.9
271.70	-1.26	1.3	235.49	22NO	-0.6	-4.4	-2.1	0.9
279.54	-1.24	1.4	245.70	2DE	-0.8	-4.4	-2.0	0.9
287.39	-1.21	1.4	255.91	12DE	-0.9	-4.2	-1.9	0.9
295.23	-1.16	1.4	266.10	22DE	-0.9	-3.7	-1.9	0.9

Right table (-432 / -431):

♂ LONG	LAT	MAG	☉ LONG	16.00UT -432	☿	♀	♃	♄
303.04	-1.11	1.4	276.27	1JA	-0.6	-3.2	-1.8	0.8
310.81	-1.05	1.5	286.40	11JA	1.0	-3.8	-1.8	0.8
318.53	-0.97	1.5	296.48	21JA	2.9	-4.2	-1.7	0.7
326.19	-0.89	1.5	306.51	31JA	1.2	-4.3	-1.7	0.7
333.79	-0.81	1.5	316.50	10FE	0.6	-4.3	-1.6	0.6
341.31	-0.72	1.5	326.41	20FE	0.4	-4.2	-1.6	0.6
348.75	-0.62	1.6	336.28	1MR	0.1	-4.1	-1.6	0.5
356.11	-0.52	1.6	346.08	11MR	-0.4	-4.0	-1.6	0.5
3.39	-0.42	1.6	355.82	21MR	-1.1	-3.9	-1.6	0.4
10.58	-0.32	1.6	5.52	31MR	-1.7	-3.8	-1.5	0.4
17.69	-0.22	1.6	15.16	10AP	-0.9	-3.7	-1.6	0.5
24.71	-0.12	1.6	24.76	20AP	0.1	-3.6	-1.6	0.5
31.65	-0.02	1.7	34.34	30AP	1.0	-3.5	-1.6	0.6
38.52	0.08	1.7	43.88	10MY	1.9	-3.5	-1.6	0.7
45.31	0.18	1.8	53.40	20MY	3.2	-3.4	-1.6	0.7
52.04	0.28	1.8	62.92	30MY	2.6	-3.4	-1.7	0.8
58.69	0.38	1.9	72.44	9JN	1.3	-3.4	-1.7	0.8
65.28	0.48	1.9	81.97	19JN	0.0	-3.4	-1.8	0.9
71.81	0.57	1.9	91.52	29JN	-1.1	-3.3	-1.8	0.9
78.28	0.66	1.9	101.10	9JL	-1.6	-3.3	-1.9	0.9
84.69	0.76	1.9	110.71	19JL	-1.0	-3.3	-2.0	1.0
91.06	0.85	1.9	120.37	29JL	-0.4	-3.3	-2.0	1.0
97.37	0.94	1.9	130.08	8AU	-0.0	-3.4	-2.1	1.0
103.62	1.03	1.9	139.84	18AU	0.2	-3.4	-2.2	1.0
109.82	1.12	1.9	149.66	28AU	0.4	-3.4	-2.2	0.9
115.97	1.21	1.9	159.54	7SE	1.0	-3.4	-2.3	0.9
122.05	1.30	1.9	169.48	17SE	3.1	-3.4	-2.3	0.9
128.07	1.39	1.8	179.48	27SE	0.7	-3.5	-2.4	0.8
134.01	1.49	1.8	189.53	7OC	-0.5	-3.5	-2.4	0.8
139.87	1.58	1.7	199.63	17OC	-0.7	-3.5	-2.4	0.8
145.63	1.68	1.6	209.77	27OC	-0.6	-3.5	-2.4	0.8
151.26	1.78	1.5	219.94	6NO	-0.7	-3.4	-2.3	0.8
156.76	1.88	1.4	230.14	16NO	-0.7	-3.4	-2.3	0.9
162.09	1.99	1.3	240.35	26NO	-0.7	-3.4	-2.2	0.9
167.22	2.10	1.1	250.56	6DE	-0.7	-3.4	-2.2	0.9
172.10	2.22	1.0	260.76	16DE	-0.4	-3.3	-2.1	0.8
176.67	2.34	0.8	270.94	26DE	1.2	-3.3	-2.0	0.8
				-431				
180.86	2.46	0.6	281.09	5JA	2.5	-3.3	-2.0	0.8
184.57	2.58	0.4	291.20	15JA	0.9	-3.3	-1.9	0.8
187.69	2.70	0.1	301.26	25JA	0.4	-3.3	-1.8	0.7
190.07	2.80	-0.1	311.27	4FE	0.2	-3.4	-1.8	0.7
191.54	2.89	-0.4	321.22	14FE	-0.0	-3.4	-1.7	0.6
191.91	2.93	-0.7	331.12	24FE	-0.5	-3.4	-1.7	0.6
191.04	2.90	-1.0	340.95	6MR	-1.1	-3.4	-1.6	0.5
188.93	2.78	-1.3	350.72	16MR	-1.6	-3.5	-1.6	0.5
185.80	2.52	-1.6	0.44	26MR	-0.9	-3.5	-1.5	0.4
182.20	2.14	-1.6	10.11	5AP	0.2	-3.6	-1.5	0.4
178.86	1.66	-1.5	19.73	15AP	1.3	-3.6	-1.5	0.4
176.41	1.14	-1.3	29.32	25AP	2.5	-3.7	-1.5	0.4
175.24	0.64	-1.1	38.87	5MY	3.5	-3.8	-1.5	0.5
175.43	0.18	-0.9	48.41	15MY	2.0	-3.9	-1.5	0.5
176.88	-0.21	-0.7	57.93	25MY	1.0	-3.9	-1.5	0.6
179.43	-0.55	-0.5	67.45	4JN	-0.0	-4.0	-1.5	0.7
182.89	-0.84	-0.3	76.97	14JN	-1.1	-4.1	-1.5	0.7
187.11	-1.08	-0.2	86.51	24JN	-1.7	-4.2	-1.5	0.8
191.94	-1.28	-0.0	96.07	4JL	-0.9	-4.2	-1.6	0.8
197.29	-1.44	0.1	105.66	14JL	-0.3	-4.0	-1.6	0.9
203.07	-1.57	0.2	115.30	24JL	0.1	-3.6	-1.7	0.9
209.20	-1.67	0.3	124.98	3AU	0.3	-3.2	-1.7	0.9
215.65	-1.75	0.4	134.71	13AU	0.6	-3.7	-1.8	0.9
222.35	-1.80	0.5	144.50	23AU	1.3	-4.1	-1.9	0.9
229.27	-1.83	0.6	154.35	2SE	3.2	-4.3	-1.9	0.9
236.37	-1.83	0.7	164.26	12SE	0.6	-4.3	-2.0	0.9
243.63	-1.82	0.7	174.23	22SE	-0.6	-4.2	-2.0	0.8
251.01	-1.79	0.8	184.25	2OC	-0.8	-4.2	-2.1	0.8
258.49	-1.74	0.9	194.33	12OC	-0.8	-4.1	-2.2	0.8
266.05	-1.67	0.9	204.45	22OC	-0.8	-4.0	-2.2	0.7
273.65	-1.59	1.0	214.60	1NO	-0.6	-3.9	-2.2	0.8
281.29	-1.50	1.1	224.79	11NO	-0.6	-3.8	-2.3	0.8
288.94	-1.40	1.1	235.00	21NO	-0.6	-3.7	-2.3	0.8
296.58	-1.29	1.2	245.21	1DE	-0.3	-3.7	-2.3	0.8
304.20	-1.17	1.3	255.42	11DE	1.3	-3.6	-2.3	0.8
311.79	-1.05	1.3	265.61	21DE	2.1	-3.5	-2.2	0.8
319.33	-0.93	1.4	275.78	31DE	0.6	-3.5	-2.1	0.8

♂			⊙	16.00UT	☿	♀	♃	♄
LONG	LAT	MAG	LONG	-430	MAGNITUDES			
326.82	-0.81	1.5	285.91	10JA	0.2	-3.4	-2.1	0.8
334.25	-0.68	1.5	296.00	20JA	0.1	-3.4	-2.0	0.8
341.60	-0.56	1.6	306.03	30JA	-0.1	-3.4	-1.9	0.7
348.89	-0.44	1.6	316.02	9FE	-0.5	-3.3	-1.9	0.7
356.10	-0.32	1.7	325.94	19FE	-1.1	-3.3	-1.8	0.6
3.24	-0.21	1.7	335.80	1MR	-1.5	-3.3	-1.7	0.6
10.30	-0.09	1.7	345.61	11MR	-0.9	-3.3	-1.7	0.5
17.29	0.01	1.8	355.36	21MR	0.3	-3.3	-1.6	0.5
24.20	0.12	1.8	5.05	31MR	1.6	-3.3	-1.6	0.4
31.05	0.22	1.8	14.70	10AP	3.3	-3.3	-1.5	0.4
37.83	0.31	1.8	24.30	20AP	2.7	-3.4	-1.5	0.3
44.55	0.41	1.8	33.87	30AP	1.6	-3.4	-1.4	0.3
51.21	0.49	1.9	43.41	10MY	0.7	-3.4	-1.4	0.4
57.82	0.58	1.9	52.94	20MY	-0.1	-3.5	-1.4	0.5
64.38	0.66	1.9	62.45	30MY	-1.1	-3.5	-1.4	0.5
70.90	0.73	1.9	71.97	9JN	-1.8	-3.5	-1.4	0.6
77.37	0.80	1.9	81.50	19JN	-0.9	-3.4	-1.4	0.6
83.81	0.87	1.9	91.05	29JN	-0.2	-3.4	-1.4	0.7
90.23	0.93	2.0	100.63	9JL	0.2	-3.4	-1.4	0.7
96.61	1.00	2.0	110.24	19JL	0.5	-3.3	-1.4	0.8
102.98	1.05	2.0	119.89	29JL	0.9	-3.3	-1.4	0.8
109.32	1.11	2.0	129.60	8AU	1.7	-3.3	-1.5	0.8
115.65	1.16	2.0	139.36	18AU	3.0	-3.3	-1.5	0.9
121.96	1.21	2.0	149.18	28AU	0.5	-3.3	-1.6	0.9
128.25	1.25	2.0	159.06	7SE	-0.8	-3.3	-1.6	0.9
134.53	1.30	2.0	168.99	17SE	-1.0	-3.4	-1.7	0.8
140.80	1.34	2.0	178.99	27SE	-0.9	-3.4	-1.7	0.8
147.05	1.37	1.9	189.04	7OC	-0.7	-3.4	-1.8	0.8
153.27	1.41	1.9	199.13	17OC	-0.5	-3.5	-1.9	0.8
159.48	1.44	1.8	209.27	27OC	-0.4	-3.5	-1.9	0.7
165.66	1.46	1.7	219.45	6NO	-0.4	-3.6	-2.0	0.7
171.80	1.48	1.7	229.64	16NO	-0.1	-3.7	-2.1	0.8
177.90	1.50	1.6	239.85	26NO	1.6	-3.7	-2.1	0.8
183.94	1.51	1.5	250.07	6DE	1.7	-3.8	-2.1	0.8
189.92	1.51	1.4	260.27	16DE	0.3	-3.9	-2.1	0.8
195.83	1.50	1.2	270.45	26DE	0.1	-4.0	-2.1	0.8
				-429				
201.63	1.48	1.1	280.60	5JA	-0.0	-4.1	-2.1	0.8
207.32	1.45	0.9	290.71	15JA	-0.2	-4.2	-2.1	0.8
212.86	1.40	0.8	300.78	25JA	-0.6	-4.3	-2.1	0.8
218.22	1.33	0.6	310.79	4FE	-1.2	-4.3	-2.0	0.7
223.35	1.23	0.4	320.74	14FE	-1.4	-4.3	-1.9	0.7
228.20	1.10	0.2	330.64	24FE	-0.9	-4.0	-1.9	0.7
232.69	0.93	-0.1	340.48	6MR	0.4	-3.5	-1.8	0.6
236.72	0.71	-0.3	350.25	16MR	2.1	-3.2	-1.7	0.6
240.18	0.42	-0.6	359.97	26MR	3.5	-3.8	-1.7	0.5
242.90	0.06	-0.9	9.65	5AP	2.0	-4.1	-1.6	0.4
244.72	-0.39	-1.2	19.27	15AP	1.2	-4.2	-1.6	0.4
245.45	-0.94	-1.5	28.85	25AP	0.6	-4.2	-1.5	0.3
244.95	-1.59	-1.8	38.41	5MY	-0.2	-4.1	-1.4	0.3
243.26	-2.29	-2.1	47.94	15MY	-1.1	-4.0	-1.4	0.3
240.66	-2.99	-2.4	57.47	25MY	-1.8	-3.9	-1.4	0.4
237.73	-3.60	-2.4	66.98	4JN	-0.9	-3.8	-1.3	0.4
235.22	-4.04	-2.2	76.50	14JN	-0.2	-3.7	-1.3	0.5
233.72	-4.29	-2.0	86.04	24JN	0.3	-3.7	-1.3	0.6
233.55	-4.38	-1.8	95.60	4JL	0.7	-3.6	-1.3	0.6
234.75	-4.34	-1.5	105.19	14JL	1.2	-3.5	-1.3	0.7
237.16	-4.21	-1.3	114.83	24JL	2.2	-3.5	-1.3	0.7
240.63	-4.03	-1.1	124.51	3AU	2.6	-3.5	-1.3	0.8
244.92	-3.81	-0.9	134.24	13AU	0.4	-3.4	-1.3	0.8
249.89	-3.57	-0.6	144.03	23AU	-0.8	-3.4	-1.3	0.8
255.39	-3.32	-0.5	153.87	2SE	-1.1	-3.4	-1.4	0.8
261.30	-3.06	-0.3	163.78	12SE	-1.0	-3.4	-1.4	0.8
267.53	-2.79	-0.1	173.75	22SE	-0.6	-3.4	-1.4	0.8
274.01	-2.53	0.1	183.77	2OC	-0.4	-3.4	-1.5	0.8
280.68	-2.27	0.2	193.84	12OC	-0.3	-3.4	-1.6	0.8
287.48	-2.02	0.4	203.96	22OC	-0.3	-3.4	-1.6	0.8
294.39	-1.77	0.5	214.11	1NO	0.1	-3.4	-1.7	0.7
301.36	-1.54	0.6	224.30	11NO	1.8	-3.4	-1.7	0.7
308.38	-1.31	0.8	234.50	21NO	1.4	-3.4	-1.8	0.7
315.41	-1.10	0.9	244.71	1DE	0.1	-3.4	-1.9	0.8
322.44	-0.90	1.0	254.92	11DE	-0.1	-3.4	-1.9	0.8
329.47	-0.71	1.1	265.11	21DE	-0.2	-3.5	-2.0	0.8
336.47	-0.53	1.2	275.28	31DE	-0.3	-3.5	-2.0	0.8

♂			⊙	16.00UT	☿	♀	♃	♄
LONG	LAT	MAG	LONG	-428	MAGNITUDES			
343.43	-0.37	1.3	285.41	10JA	-0.6	-3.5	-2.0	0.8
350.36	-0.22	1.4	295.51	20JA	-1.2	-3.4	-2.1	0.8
357.24	-0.07	1.5	305.55	30JA	-1.3	-3.4	-2.1	0.8
4.08	0.06	1.6	315.53	9FE	-0.8	-3.4	-2.0	0.8
10.87	0.18	1.6	325.46	19FE	0.5	-3.4	-2.0	0.7
17.61	0.29	1.7	335.32	29FE	2.5	-3.3	-2.0	0.7
24.30	0.39	1.8	345.13	10MR	2.7	-3.3	-1.9	0.6
30.94	0.49	1.8	354.88	20MR	1.5	-3.3	-1.8	0.6
37.54	0.58	1.9	4.58	30MR	0.9	-3.3	-1.8	0.5
44.10	0.66	1.9	14.23	9AP	0.4	-3.3	-1.7	0.5
50.61	0.73	1.9	23.84	19AP	-0.2	-3.4	-1.7	0.4
57.10	0.79	2.0	33.41	29AP	-1.1	-3.4	-1.6	0.3
63.55	0.85	2.0	42.95	9MY	-1.8	-3.4	-1.5	0.3
69.98	0.91	2.0	52.48	19MY	-0.9	-3.4	-1.5	0.3
76.38	0.96	2.0	61.99	29MY	-0.1	-3.5	-1.4	0.3
82.77	1.00	2.0	71.51	8JN	0.5	-3.5	-1.4	0.4
89.15	1.04	2.0	81.04	18JN	0.9	-3.6	-1.3	0.5
95.52	1.08	2.0	90.59	28JN	1.5	-3.7	-1.3	0.5
101.88	1.11	2.0	100.16	8JL	2.8	-3.7	-1.3	0.6
108.25	1.13	2.0	109.78	18JL	2.2	-3.8	-1.2	0.6
114.62	1.15	2.0	119.43	28JL	0.3	-3.9	-1.2	0.7
121.01	1.17	2.0	129.13	7AU	-0.9	-4.0	-1.2	0.7
127.40	1.18	2.0	138.89	17AU	-1.3	-4.1	-1.2	0.8
133.81	1.19	2.0	148.70	27AU	-1.0	-4.2	-1.2	0.8
140.24	1.19	2.0	158.58	6SE	-0.5	-4.3	-1.2	0.8
146.70	1.19	2.0	168.51	16SE	-0.3	-4.3	-1.3	0.8
153.17	1.18	2.0	178.50	26SE	-0.2	-4.2	-1.3	0.8
159.67	1.17	2.0	188.55	6OC	-0.1	-3.7	-1.3	0.8
166.19	1.15	1.9	198.65	16OC	0.3	-3.2	-1.4	0.8
172.73	1.13	1.9	208.78	26OC	2.1	-3.8	-1.4	0.8
179.31	1.09	1.8	218.96	5NO	1.2	-4.2	-1.5	0.7
185.90	1.05	1.8	229.15	15NO	-0.1	-4.4	-1.5	0.7
192.52	1.01	1.7	239.36	25NO	-0.3	-4.4	-1.6	0.7
199.16	0.95	1.6	249.57	5DE	-0.3	-4.3	-1.7	0.7
205.82	0.88	1.6	259.77	15DE	-0.4	-4.2	-1.7	0.8
212.49	0.80	1.5	269.95	25DE	-0.7	-4.1	-1.8	0.8
				-427				
219.18	0.70	1.4	280.11	4JA	-1.1	-4.0	-1.9	0.8
225.88	0.60	1.3	290.23	14JA	-1.2	-3.9	-1.9	0.8
232.58	0.47	1.2	300.29	24JA	-0.7	-3.8	-2.0	0.8
239.29	0.32	1.1	310.31	3FE	0.7	-3.7	-2.0	0.8
245.98	0.16	0.9	320.27	13FE	3.0	-3.6	-2.0	0.8
252.66	-0.03	0.8	330.16	23FE	2.0	-3.5	-2.0	0.8
259.31	-0.24	0.7	340.00	5MR	1.1	-3.5	-2.0	0.7
265.91	-0.48	0.5	349.79	15MR	0.6	-3.4	-2.0	0.7
272.44	-0.75	0.4	359.51	25MR	0.3	-3.4	-2.0	0.6
278.88	-1.05	0.2	9.18	4AP	-0.3	-3.4	-1.9	0.6
285.19	-1.39	0.0	18.81	14AP	-1.1	-3.3	-1.9	0.5
291.33	-1.75	-0.2	28.40	24AP	-1.8	-3.3	-1.8	0.5
297.24	-2.16	-0.3	37.96	4MY	-0.9	-3.3	-1.7	0.4
302.83	-2.60	-0.5	47.49	14MY	-0.0	-3.3	-1.7	0.3
308.03	-3.09	-0.8	57.01	24MY	0.6	-3.3	-1.6	0.3
312.70	-3.61	-1.0	66.53	3JN	1.2	-3.3	-1.5	0.3
316.69	-4.17	-1.2	76.05	13JN	2.1	-3.3	-1.5	0.3
319.82	-4.75	-1.5	85.58	23JN	3.3	-3.4	-1.4	0.4
321.87	-5.34	-1.7	95.14	3JL	1.8	-3.4	-1.4	0.5
322.64	-5.88	-2.0	104.73	13JL	0.2	-3.4	-1.3	0.5
322.04	-6.31	-2.2	114.36	23JL	-1.0	-3.5	-1.3	0.6
320.16	-6.54	-2.4	124.04	2AU	-1.4	-3.5	-1.3	0.6
317.44	-6.48	-2.5	133.77	12AU	-1.0	-3.5	-1.2	0.7
314.63	-6.11	-2.4	143.55	22AU	-0.5	-3.4	-1.2	0.7
312.44	-5.51	-2.2	153.39	1SE	-0.2	-3.4	-1.2	0.8
311.41	-4.77	-1.9	163.30	11SE	-0.0	-3.4	-1.2	0.8
311.68	-4.02	-1.5	173.25	21SE	0.1	-3.4	-1.2	0.8
313.20	-3.30	-1.2	183.28	1OC	0.5	-3.3	-1.2	0.8
315.78	-2.66	-0.9	193.35	11OC	2.4	-3.3	-1.2	0.8
319.22	-2.10	-0.6	203.46	21OC	1.0	-3.3	-1.3	0.8
323.32	-1.62	-0.4	213.62	31OC	-0.2	-3.3	-1.3	0.8
327.96	-1.20	-0.1	223.80	10NO	-0.4	-3.3	-1.3	0.8
332.99	-0.84	0.1	234.00	20NO	-0.4	-3.4	-1.4	0.7
338.33	-0.54	0.3	244.22	30NO	-0.5	-3.4	-1.4	0.7
343.90	-0.28	0.5	254.43	10DE	-0.8	-3.4	-1.5	0.7
349.65	-0.05	0.7	264.62	20DE	-1.0	-3.4	-1.6	0.8
355.53	0.14	0.9	274.79	30DE	-1.0	-3.5	-1.6	0.8

♂ LONG	♂ LAT	♂ MAG	⊙ LONG	16.00UT	☿	♀	♃	♄
				-426	MAGNITUDES			
1.52	0.31	1.0	284.93	9JA	-0.7	-3.5	-1.7	0.8
7.58	0.45	1.2	295.02	19JA	0.8	-3.6	-1.8	0.8
13.70	0.58	1.3	305.06	29JA	3.1	-3.6	-1.8	0.8
19.85	0.69	1.4	315.05	8FE	1.5	-3.7	-1.9	0.8
26.03	0.78	1.5	324.98	18FE	0.8	-3.8	-1.9	0.8
32.23	0.86	1.6	334.85	28FE	0.5	-3.9	-2.0	0.8
38.44	0.93	1.7	344.66	10MR	0.1	-3.9	-2.0	0.8
44.65	0.99	1.8	354.42	20MR	-0.3	-4.0	-2.0	0.7
50.87	1.04	1.8	4.12	30MR	-1.1	-4.1	-2.0	0.7
57.10	1.08	1.9	13.77	9AP	-1.8	-4.2	-2.0	0.6
63.32	1.11	1.9	23.37	19AP	-0.9	-4.2	-2.0	0.6
69.55	1.14	2.0	32.95	29AP	0.0	-4.1	-2.0	0.5
75.79	1.16	2.0	42.49	9MY	0.8	-3.7	-1.9	0.5
82.03	1.18	2.0	52.02	19MY	1.6	-2.9	-1.9	0.4
88.28	1.19	2.0	61.54	29MY	2.7	-3.3	-1.8	0.3
94.55	1.19	2.0	71.05	8JN	3.0	-3.9	-1.7	0.3
100.84	1.20	2.0	80.58	18JN	1.5	-4.1	-1.7	0.3
107.15	1.19	2.0	90.13	28JN	0.1	-4.2	-1.6	0.4
113.48	1.18	2.0	99.70	8JL	-1.0	-4.2	-1.6	0.4
119.84	1.17	2.0	109.31	18JL	-1.5	-4.1	-1.5	0.5
126.24	1.15	2.0	118.96	28JL	-1.0	-4.0	-1.4	0.5
132.67	1.13	1.9	128.66	7AU	-0.4	-3.9	-1.4	0.6
139.15	1.10	1.9	138.41	17AU	-0.1	-3.8	-1.4	0.7
145.67	1.07	1.9	148.22	27AU	0.1	-3.8	-1.3	0.7
152.23	1.03	1.9	158.09	6SE	0.3	-3.7	-1.3	0.7
158.85	0.99	1.9	168.03	16SE	0.8	-3.6	-1.3	0.8
165.52	0.94	1.9	178.02	26SE	2.8	-3.6	-1.3	0.8
172.24	0.88	1.9	188.06	6OC	0.8	-3.5	-1.3	0.8
179.02	0.82	1.9	198.15	16OC	-0.4	-3.5	-1.3	0.8
185.86	0.76	1.8	208.29	26OC	-0.6	-3.5	-1.3	0.8
192.76	0.68	1.8	218.46	5NO	-0.6	-3.4	-1.3	0.8
199.71	0.60	1.8	228.66	15NO	-0.6	-3.4	-1.3	0.8
206.72	0.51	1.7	238.87	25NO	-0.8	-3.4	-1.3	0.8
213.79	0.41	1.7	249.08	5DE	-0.8	-3.4	-1.4	0.7
220.91	0.31	1.6	259.28	15DE	-0.8	-3.3	-1.4	0.7
228.09	0.19	1.6	269.47	25DE	-0.5	-3.3	-1.5	0.8
				-425				
235.31	0.07	1.5	279.62	4JA	1.0	-3.3	-1.5	0.8
242.59	-0.06	1.5	289.74	14JA	2.8	-3.3	-1.6	0.8
249.91	-0.20	1.4	299.81	24JA	1.1	-3.4	-1.7	0.9
257.27	-0.35	1.3	309.82	3FE	0.5	-3.4	-1.7	0.9
264.66	-0.51	1.2	319.78	13FE	0.3	-3.4	-1.8	0.9
272.08	-0.68	1.2	329.69	23FE	0.0	-3.4	-1.9	0.9
279.51	-0.85	1.1	339.53	5MR	-0.4	-3.5	-1.9	0.9
286.95	-1.02	1.0	349.31	15MR	-1.1	-3.5	-2.0	0.8
294.37	-1.20	0.9	359.04	25MR	-1.7	-3.5	-2.0	0.8
301.77	-1.38	0.9	8.71	4AP	-0.9	-3.4	-2.1	0.8
309.14	-1.56	0.8	18.34	14AP	0.1	-3.4	-2.1	0.7
316.43	-1.73	0.7	27.93	24AP	1.1	-3.4	-2.1	0.7
323.65	-1.90	0.6	37.49	4MY	2.1	-3.4	-2.1	0.6
330.75	-2.06	0.5	47.03	14MY	3.5	-3.3	-2.1	0.5
337.71	-2.21	0.4	56.55	24MY	2.4	-3.3	-2.1	0.5
344.50	-2.35	0.3	66.06	3JN	1.2	-3.3	-2.0	0.4
351.06	-2.48	0.2	75.58	13JN	0.0	-3.3	-2.0	0.3
357.36	-2.59	0.1	85.12	23JN	-1.1	-3.3	-1.9	0.3
3.34	-2.69	-0.0	94.67	3JL	-1.6	-3.4	-1.9	0.4
8.92	-2.77	-0.2	104.26	13JL	-1.0	-3.4	-1.8	0.4
14.00	-2.82	-0.3	113.89	23JL	-0.3	-3.4	-1.7	0.5
18.49	-2.85	-0.5	123.56	2AU	0.0	-3.5	-1.7	0.5
22.21	-2.86	-0.6	133.29	12AU	0.3	-3.5	-1.6	0.6
25.00	-2.82	-0.8	143.07	22AU	0.5	-3.6	-1.6	0.6
26.66	-2.73	-1.1	152.91	1SE	1.1	-3.6	-1.5	0.7
26.97	-2.56	-1.3	162.81	11SE	3.1	-3.7	-1.5	0.7
25.81	-2.31	-1.5	172.77	21SE	0.7	-3.8	-1.4	0.8
23.28	-1.94	-1.7	182.78	1OC	-0.5	-3.9	-1.4	0.8
19.78	-1.48	-1.8	192.86	11OC	-0.7	-4.0	-1.4	0.8
16.09	-0.96	-1.7	202.97	21OC	-0.7	-4.1	-1.4	0.8
12.99	-0.44	-1.5	213.12	31OC	-0.8	-4.2	-1.3	0.8
11.03	0.02	-1.1	223.31	10NO	-0.7	-4.3	-1.3	0.8
10.44	0.40	-0.8	233.51	20NO	-0.7	-4.4	-1.3	0.8
11.13	0.70	-0.5	243.72	30NO	-0.7	-4.4	-1.4	0.8
12.94	0.93	-0.2	253.93	10DE	-0.4	-4.2	-1.4	0.8
15.65	1.10	0.1	264.12	20DE	1.2	-3.7	-1.4	0.8
19.06	1.23	0.3	274.30	30DE	2.4	-3.2	-1.4	0.8

♂ LONG	♂ LAT	♂ MAG	⊙ LONG	16.00UT	☿	♀	♃	♄
				-424	MAGNITUDES			
23.04	1.32	0.5	284.44	9JA	0.8	-3.8	-1.5	0.8
27.44	1.39	0.7	294.53	19JA	0.3	-4.2	-1.5	0.9
32.18	1.44	0.9	304.57	29JA	0.2	-4.3	-1.6	0.9
37.19	1.48	1.1	314.57	8FE	-0.1	-4.3	-1.6	0.9
42.40	1.50	1.2	324.50	18FE	-0.5	-4.2	-1.7	0.9
47.78	1.51	1.4	334.37	28FE	-1.1	-4.1	-1.8	0.9
53.30	1.51	1.5	344.19	9MR	-1.6	-4.0	-1.8	0.9
58.92	1.51	1.6	353.94	19MR	-0.9	-3.9	-1.9	0.9
64.64	1.50	1.7	3.65	29MR	0.2	-3.8	-2.0	0.9
70.44	1.48	1.7	13.30	8AP	1.4	-3.7	-2.0	0.9
76.30	1.46	1.8	22.91	18AP	2.8	-3.6	-2.1	0.8
82.23	1.44	1.9	32.49	28AP	3.3	-3.5	-2.2	0.8
88.22	1.41	1.9	42.03	8MY	1.9	-3.5	-2.2	0.7
94.25	1.37	1.9	51.56	18MY	0.9	-3.4	-2.2	0.6
100.34	1.34	2.0	61.08	28MY	-0.0	-3.4	-2.2	0.6
106.49	1.30	2.0	70.60	7JN	-1.1	-3.4	-2.2	0.5
112.69	1.25	2.0	80.12	17JN	-1.7	-3.3	-2.2	0.4
118.94	1.21	2.0	89.67	27JN	-1.0	-3.3	-2.2	0.4
125.25	1.16	2.0	99.24	7JL	-0.3	-3.3	-2.1	0.4
131.62	1.10	2.0	108.84	17JL	0.1	-3.3	-2.1	0.4
138.06	1.04	2.0	118.49	27JL	0.4	-3.3	-2.0	0.5
144.56	0.98	1.9	128.19	6AU	0.7	-3.4	-2.0	0.5
151.12	0.92	1.9	137.94	16AU	1.4	-3.4	-1.9	0.6
157.76	0.85	1.9	147.75	26AU	3.2	-3.4	-1.8	0.6
164.48	0.77	1.8	157.62	5SE	0.6	-3.4	-1.8	0.7
171.26	0.70	1.8	167.54	15SE	-0.7	-3.4	-1.7	0.7
178.12	0.61	1.7	177.53	25SE	-0.9	-3.5	-1.7	0.8
185.07	0.53	1.7	187.57	5OC	-0.9	-3.5	-1.6	0.8
192.08	0.43	1.7	197.66	15OC	-0.8	-3.5	-1.6	0.8
199.18	0.34	1.7	207.79	25OC	-0.6	-3.5	-1.5	0.9
206.36	0.24	1.7	217.96	4NO	-0.5	-3.4	-1.5	0.9
213.61	0.13	1.7	228.15	14NO	-0.5	-3.4	-1.5	0.9
220.94	0.03	1.6	238.36	24NO	-0.3	-3.4	-1.5	0.9
228.33	-0.09	1.6	248.58	4DE	1.3	-3.4	-1.4	0.9
235.79	-0.20	1.6	258.78	14DE	2.0	-3.3	-1.4	0.9
243.32	-0.32	1.6	268.97	24DE	0.5	-3.3	-1.4	0.9
				-423				
250.90	-0.44	1.5	279.12	3JA	0.2	-3.3	-1.5	0.8
258.53	-0.56	1.5	289.24	13JA	0.0	-3.3	-1.5	0.9
266.20	-0.68	1.5	299.31	23JA	-0.2	-3.4	-1.5	0.9
273.89	-0.79	1.4	309.34	2FE	-0.5	-3.4	-1.5	1.0
281.61	-0.91	1.4	319.30	12FE	-1.1	-3.4	-1.6	1.0
289.34	-1.02	1.4	329.21	22FE	-1.5	-3.4	-1.6	1.0
297.07	-1.12	1.3	339.05	4MR	-0.9	-3.4	-1.7	1.0
304.78	-1.21	1.3	348.84	14MR	0.3	-3.5	-1.7	1.0
312.46	-1.30	1.2	358.57	24MR	1.8	-3.5	-1.8	1.0
320.11	-1.37	1.2	8.25	3AP	3.6	-3.6	-1.9	1.0
327.69	-1.43	1.2	17.88	13AP	2.4	-3.6	-2.0	1.0
335.21	-1.48	1.2	27.47	23AP	1.4	-3.7	-2.0	0.9
342.64	-1.51	1.1	37.03	3MY	0.7	-3.8	-2.1	0.9
349.98	-1.53	1.1	46.57	13MY	-0.1	-3.9	-2.2	0.8
357.20	-1.54	1.0	56.09	23MY	-1.1	-4.0	-2.2	0.8
4.30	-1.53	1.0	65.61	2JN	-1.8	-4.0	-2.3	0.7
11.25	-1.50	1.0	75.13	12JN	-1.0	-4.1	-2.3	0.6
18.04	-1.46	0.9	84.66	22JN	-0.2	-4.2	-2.3	0.6
24.64	-1.40	0.9	94.22	2JL	0.2	-4.2	-2.4	0.5
31.03	-1.32	0.8	103.80	12JL	0.6	-4.0	-2.4	0.5
37.19	-1.23	0.7	113.43	22JL	1.0	-3.6	-2.3	0.5
43.07	-1.11	0.6	123.10	1AU	1.9	-3.2	-2.3	0.5
48.64	-0.98	0.5	132.82	11AU	2.9	-3.7	-2.3	0.6
53.82	-0.83	0.4	142.60	21AU	0.5	-4.1	-2.2	0.6
58.55	-0.64	0.3	152.43	31AU	-0.8	-4.3	-2.2	0.7
62.72	-0.43	0.2	162.33	10SE	-1.0	-4.3	-2.1	0.7
66.21	-0.18	-0.0	172.29	20SE	-1.0	-4.2	-2.0	0.8
68.85	0.11	-0.2	182.30	30SE	-0.7	-4.2	-1.9	0.8
70.47	0.44	-0.4	192.36	10OC	-0.5	-4.1	-1.9	0.8
70.86	0.83	-0.6	202.48	20OC	-0.4	-4.0	-1.8	0.9
69.87	1.26	-0.8	212.63	30OC	-0.4	-3.9	-1.8	0.9
67.50	1.72	-1.1	222.81	9NO	-0.1	-3.8	-1.7	0.9
64.02	2.14	-1.2	233.01	19NO	1.6	-3.7	-1.7	1.0
60.06	2.49	-1.3	243.22	29NO	1.7	-3.7	-1.6	1.0
56.39	2.72	-1.0	253.43	9DE	0.3	-3.6	-1.6	1.0
53.67	2.83	-0.8	263.63	19DE	-0.0	-3.5	-1.6	1.0
52.24	2.85	-0.5	273.80	29DE	-0.1	-3.5	-1.5	1.0

-422

♂ LONG	♂ LAT	♂ MAG	☉ LONG	16.00UT	☿	♀	♃	♄
52.12	2.80	-0.2	283.94	8JA	-0.3	-3.4	-1.5	1.0
53.19	2.71	0.1	294.04	18JA	-0.6	-3.4	-1.5	1.0
55.24	2.61	0.3	304.09	28JA	-1.1	-3.4	-1.5	1.0
58.09	2.50	0.5	314.08	7FE	-1.4	-3.3	-1.5	1.0
61.57	2.39	0.7	324.02	17FE	-0.9	-3.3	-1.6	1.1
65.55	2.28	0.9	333.90	27FE	-0.3	-3.3	-1.6	1.1
69.92	2.17	1.1	343.72	9MR	2.2	-3.3	-1.6	1.1
74.62	2.06	1.2	353.48	19MR	3.2	-3.3	-1.7	1.1
79.58	1.96	1.3	3.18	29MR	1.8	-3.3	-1.7	1.1
84.75	1.86	1.4	12.84	8AP	1.1	-3.3	-1.8	1.1
90.11	1.77	1.5	22.45	18AP	0.5	-3.4	-1.8	1.1
95.62	1.67	1.6	32.02	28AP	-0.2	-3.4	-1.9	1.1
101.21	1.58	1.7	41.57	8MY	-1.1	-3.4	-2.0	1.0
107.05	1.48	1.7	51.10	18MY	-1.8	-3.5	-2.0	1.0
112.93	1.39	1.8	60.62	28MY	-1.0	-3.5	-2.1	0.9
118.92	1.30	1.8	70.13	7JN	-0.2	-3.5	-2.2	0.9
125.02	1.21	1.8	79.66	17JN	0.4	-3.4	-2.2	0.8
131.21	1.11	1.8	89.20	27JN	0.8	-3.4	-2.3	0.7
137.49	1.02	1.9	98.77	7JL	1.3	-3.4	-2.3	0.7
143.88	0.92	1.9	108.38	17JL	2.4	-3.3	-2.4	0.6
150.35	0.82	1.8	118.02	27JL	2.5	-3.3	-2.4	0.6
156.93	0.72	1.8	127.72	6AU	0.4	-3.3	-2.4	0.6
163.59	0.62	1.8	137.47	16AU	-0.9	-3.3	-2.4	0.6
170.36	0.52	1.8	147.27	26AU	-1.2	-3.3	-2.4	0.7
177.22	0.42	1.8	157.14	5SE	-1.1	-3.3	-2.4	0.7
184.18	0.31	1.7	167.06	15SE	-0.6	-3.4	-2.3	0.8
191.24	0.21	1.7	177.04	25SE	-0.3	-3.4	-2.3	0.8
198.39	0.10	1.6	187.08	5OC	-0.2	-3.4	-2.2	0.9
205.64	-0.01	1.6	197.17	15OC	-0.2	-3.5	-2.2	0.9
212.97	-0.12	1.6	207.30	25OC	0.1	-3.5	-2.1	1.0
220.39	-0.22	1.5	217.47	4NO	1.8	-3.6	-2.0	1.0
227.90	-0.33	1.4	227.66	14NO	1.4	-3.7	-1.9	1.0
235.48	-0.43	1.4	237.86	24NO	0.1	-3.7	-1.9	1.1
243.12	-0.54	1.4	248.08	4DE	-0.2	-3.8	-1.8	1.1
250.83	-0.64	1.4	258.28	14DE	-0.2	-3.9	-1.8	1.1
258.59	-0.73	1.4	268.47	24DE	-0.4	-4.0	-1.7	1.1

-421

♂ LONG	♂ LAT	♂ MAG	☉ LONG	16.00UT	☿	♀	♃	♄
266.39	-0.82	1.4	278.63	3JA	-0.6	-4.1	-1.7	1.1
274.22	-0.90	1.4	288.75	13JA	-1.1	-4.2	-1.6	1.1
282.06	-0.97	1.4	298.82	23JA	-1.2	-4.3	-1.6	1.1
289.91	-1.03	1.4	308.85	2FE	-0.8	-4.3	-1.6	1.1
297.75	-1.09	1.4	318.81	12FE	0.5	-4.3	-1.6	1.2
305.58	-1.13	1.4	328.72	22FE	2.7	-4.0	-1.6	1.2
313.37	-1.16	1.4	338.57	4MR	2.5	-3.5	-1.6	1.2
321.12	-1.18	1.4	348.36	14MR	1.3	-3.2	-1.6	1.3
328.81	-1.18	1.4	358.09	24MR	0.8	-3.8	-1.6	1.3
336.44	-1.18	1.4	7.78	3AP	0.4	-4.1	-1.6	1.3
344.00	-1.16	1.4	17.41	13AP	-0.2	-4.2	-1.6	1.3
351.47	-1.13	1.4	27.00	23AP	-1.1	-4.2	-1.7	1.3
358.85	-1.08	1.4	36.57	3MY	-1.8	-4.1	-1.7	1.3
6.13	-1.03	1.4	46.10	13MY	-1.0	-4.0	-1.8	1.2
13.31	-0.96	1.4	55.63	23MY	-0.1	-3.9	-1.8	1.2
20.38	-0.88	1.4	65.15	2JN	0.5	-3.8	-1.9	1.1
27.32	-0.80	1.4	74.66	12JN	1.0	-3.7	-2.0	1.1
34.15	-0.70	1.4	84.20	22JN	1.7	-3.7	-2.0	1.1
40.84	-0.59	1.4	93.75	2JL	3.0	-3.6	-2.1	1.0
47.39	-0.47	1.4	103.34	12JL	2.1	-3.5	-2.2	0.9
53.79	-0.34	1.3	112.96	22JL	0.3	-3.5	-2.2	0.8
60.02	-0.21	1.3	122.63	1AU	-0.9	-3.5	-2.3	0.7
66.07	-0.06	1.2	132.35	11AU	-1.3	-3.4	-2.3	0.7
71.92	0.11	1.2	142.12	21AU	-1.0	-3.4	-2.4	0.7
77.54	0.29	1.1	151.96	31AU	-0.5	-3.4	-2.4	0.8
82.88	0.48	1.0	161.85	10SE	-0.2	-3.4	-2.4	0.8
87.91	0.69	0.9	171.80	20SE	-0.1	-3.4	-2.4	0.9
92.55	0.93	0.8	181.82	30SE	-0.0	-3.4	-2.4	0.9
96.73	1.19	0.6	191.88	10OC	0.3	-3.4	-2.4	1.0
100.34	1.49	0.5	201.99	20OC	2.1	-3.4	-2.4	1.0
103.25	1.82	0.3	212.14	30OC	1.2	-3.4	-2.3	1.1
105.28	2.20	0.1	222.31	9NO	-0.1	-3.4	-2.2	1.1
106.27	2.61	-0.2	232.52	19NO	-0.3	-3.4	-2.2	1.1
106.01	3.05	-0.4	242.73	29NO	-0.4	-3.4	-2.1	1.2
104.42	3.49	-0.6	252.93	9DE	-0.5	-3.5	-2.0	1.2
101.57	3.88	-0.9	263.13	19DE	-0.7	-3.5	-1.9	1.2
97.83	4.15	-1.0	273.31	29DE	-1.0	-3.5	-1.9	1.3

-420

♂ LONG	♂ LAT	♂ MAG	☉ LONG	16.00UT	☿	♀	♃	♄
93.87	4.26	-0.9	283.44	8JA	-1.1	-3.5	-1.8	1.3
90.42	4.21	-0.7	293.54	18JA	-0.7	-3.4	-1.8	1.3
88.03	4.03	-0.5	303.60	28JA	0.7	-3.4	-1.7	1.3
86.93	3.78	-0.3	313.59	7FE	3.0	-3.4	-1.7	1.3
87.09	3.49	-0.0	323.53	17FE	1.8	-3.4	-1.6	1.3
88.38	3.21	0.2	333.41	27FE	1.0	-3.3	-1.6	1.3
90.60	2.93	0.4	343.23	8MR	0.6	-3.3	-1.6	1.4
93.58	2.68	0.6	353.00	18MR	0.2	-3.3	-1.5	1.4
97.18	2.44	0.8	2.71	28MR	-0.3	-3.3	-1.5	1.4
101.27	2.22	0.9	12.36	7AP	-1.1	-3.3	-1.5	1.3
105.76	2.02	1.1	21.98	17AP	-1.8	-3.4	-1.5	1.3
110.59	1.83	1.2	31.56	27AP	-1.0	-3.4	-1.6	1.3
115.70	1.65	1.3	41.10	7MY	-0.0	-3.4	-1.6	1.2
121.05	1.48	1.4	50.64	17MY	0.7	-3.4	-1.6	1.2
126.61	1.32	1.4	60.16	27MY	1.3	-3.5	-1.6	1.2
132.36	1.16	1.5	69.67	6JN	2.3	-3.5	-1.7	1.1
138.28	1.01	1.5	79.20	16JN	3.3	-3.6	-1.7	1.1
144.37	0.87	1.6	88.74	26JN	1.7	-3.7	-1.8	1.0
150.59	0.73	1.6	98.31	6JL	0.2	-3.7	-1.8	1.0
156.97	0.59	1.6	107.91	16JL	-1.0	-3.8	-1.9	0.9
163.47	0.46	1.6	117.56	26JL	-1.4	-3.9	-2.0	0.9
170.11	0.33	1.6	127.25	5AU	-1.0	-4.0	-2.0	0.8
176.88	0.20	1.6	136.99	15AU	-0.5	-4.1	-2.1	0.8
183.76	0.08	1.6	146.80	25AU	-0.1	-4.2	-2.2	0.8
190.77	-0.05	1.6	156.66	4SE	0.0	-4.3	-2.2	0.9
197.90	-0.16	1.6	166.58	14SE	0.2	-4.3	-2.3	0.9
205.14	-0.28	1.6	176.56	24SE	0.6	-4.1	-2.3	1.0
212.48	-0.39	1.5	186.59	4OC	2.4	-3.7	-2.3	1.1
219.92	-0.49	1.5	196.68	14OC	1.0	-3.2	-2.4	1.1
227.46	-0.59	1.5	206.81	24OC	-0.3	-3.8	-2.4	1.2
235.08	-0.68	1.5	216.97	3NO	-0.5	-4.2	-2.4	1.2
242.78	-0.77	1.4	227.17	13NO	-0.5	-4.4	-2.3	1.3
250.54	-0.84	1.4	237.37	23NO	-0.6	-4.4	-2.3	1.3
258.36	-0.91	1.4	247.58	3DE	-0.8	-4.3	-2.2	1.3
266.22	-0.97	1.4	257.79	13DE	-0.9	-4.2	-2.2	1.3
274.10	-1.02	1.3	267.98	23DE	-0.9	-4.1	-2.1	1.3

-419

♂ LONG	♂ LAT	♂ MAG	☉ LONG	16.00UT	☿	♀	♃	♄
281.99	-1.05	1.3	278.13	2JA	-0.6	-4.0	-2.0	1.3
289.89	-1.08	1.3	288.26	12JA	0.8	-3.9	-1.9	1.3
297.78	-1.09	1.3	298.34	22JA	3.0	-3.8	-1.9	1.2
305.63	-1.09	1.3	308.36	1FE	1.4	-3.7	-1.8	1.2
313.46	-1.08	1.4	318.33	11FE	0.7	-3.6	-1.7	1.1
321.23	-1.06	1.4	328.24	21FE	0.4	-3.5	-1.7	1.1
328.94	-1.02	1.4	338.09	3MR	0.1	-3.5	-1.6	1.1
336.59	-0.98	1.4	347.89	13MR	-0.4	-3.4	-1.6	1.1
344.16	-0.93	1.5	357.62	23MR	-1.1	-3.4	-1.5	1.1
351.65	-0.86	1.5	7.31	2AP	-1.8	-3.4	-1.5	1.1
359.06	-0.79	1.5	16.95	12AP	-1.0	-3.3	-1.5	1.1
6.37	-0.72	1.6	26.54	22AP	0.0	-3.3	-1.5	1.0
13.59	-0.63	1.6	36.11	2MY	0.9	-3.3	-1.5	1.0
20.72	-0.54	1.6	45.65	12MY	1.8	-3.3	-1.5	1.0
27.75	-0.44	1.6	55.17	22MY	3.0	-3.3	-1.5	0.9
34.69	-0.34	1.7	64.68	1JN	2.8	-3.3	-1.5	0.9
41.53	-0.23	1.7	74.21	11JN	1.4	-3.3	-1.5	0.9
48.27	-0.12	1.7	83.74	21JN	0.1	-3.4	-1.5	0.8
54.91	-0.01	1.7	93.29	1JL	-1.0	-3.4	-1.5	0.8
61.46	0.11	1.7	102.88	11JL	-1.6	-3.4	-1.6	0.7
67.90	0.24	1.7	112.49	21JL	-1.0	-3.5	-1.6	0.7
74.23	0.37	1.7	122.16	31JL	-0.4	-3.5	-1.7	0.6
80.45	0.50	1.6	131.88	10AU	-0.0	-3.5	-1.7	0.6
86.54	0.64	1.6	141.65	20AU	0.2	-3.4	-1.8	0.5
92.51	0.78	1.6	151.48	30AU	0.3	-3.4	-1.8	0.5
98.33	0.94	1.5	161.37	9SE	0.9	-3.4	-1.9	0.5
103.98	1.10	1.4	171.32	19SE	2.8	-3.4	-2.0	0.6
109.44	1.28	1.4	181.32	29SE	0.9	-3.3	-2.0	0.6
114.67	1.46	1.3	191.38	9OC	-0.4	-3.3	-2.1	0.7
119.64	1.66	1.1	201.49	19OC	-0.6	-3.3	-2.2	0.8
124.28	1.89	1.0	211.64	29OC	-0.6	-3.3	-2.2	0.8
128.52	2.13	0.9	221.82	8NO	-0.7	-3.3	-2.2	0.9
132.29	2.40	0.7	232.01	18NO	-0.8	-3.4	-2.3	0.9
135.44	2.69	0.5	242.23	28NO	-0.7	-3.4	-2.3	1.0
137.85	3.02	0.3	252.44	8DE	-0.8	-3.4	-2.2	1.0
139.34	3.37	0.0	262.63	18DE	-0.5	-3.4	-2.2	1.0
139.72	3.74	-0.2	272.81	28DE	1.0	-3.5	-2.2	1.0

♂ LONG	♂ LAT	♂ MAG	☉ LONG	16.00UT -418	☿	♀	♃	♄
					MAGNITUDES			
138.85	4.09	-0.5	282.96	7JA	2.7	-3.5	-2.1	1.0
136.68	4.40	-0.7	293.06	17JA	1.0	-3.6	-2.0	1.0
133.42	4.58	-1.0	303.11	27JA	0.5	-3.6	-2.0	1.0
129.56	4.60	-1.1	313.11	6FE	0.2	-3.7	-1.9	0.9
125.81	4.44	-0.9	323.05	16FE	0.0	-3.8	-1.8	0.9
122.83	4.14	-0.7	332.94	26FE	-0.4	-3.9	-1.8	0.8
121.04	3.75	-0.5	342.76	8MR	-1.1	-4.0	-1.7	0.8
120.55	3.33	-0.3	352.52	18MR	-1.7	-4.0	-1.6	0.8
121.29	2.91	-0.1	2.24	28MR	-1.0	-4.1	-1.6	0.8
123.09	2.52	0.1	11.90	7AP	0.1	-4.2	-1.5	0.8
125.78	2.17	0.3	21.51	17AP	1.2	-4.2	-1.5	0.8
129.19	1.84	0.5	31.09	27AP	2.3	-4.1	-1.4	0.8
133.20	1.55	0.6	40.64	7MY	3.7	-3.7	-1.4	0.8
137.69	1.28	0.7	50.17	17MY	2.2	-2.9	-1.4	0.8
142.60	1.04	0.8	59.69	27MY	1.1	-3.3	-1.4	0.7
147.86	0.81	0.9	69.21	6JN	0.0	-3.9	-1.4	0.7
153.41	0.60	1.0	78.73	16JN	-1.0	-4.1	-1.4	0.7
159.23	0.41	1.1	88.28	26JN	-1.7	-4.2	-1.4	0.6
165.28	0.22	1.1	97.84	6JL	-1.0	-4.2	-1.4	0.6
171.55	0.05	1.2	107.44	16JL	-0.3	-4.1	-1.4	0.5
178.01	-0.11	1.2	117.09	26JL	0.1	-4.0	-1.4	0.5
184.65	-0.26	1.3	126.78	5AU	0.3	-3.9	-1.4	0.4
191.46	-0.40	1.3	136.52	15AU	0.6	-3.8	-1.5	0.4
198.43	-0.52	1.3	146.32	25AU	1.2	-3.8	-1.5	0.3
205.55	-0.64	1.3	156.18	4SE	3.1	-3.7	-1.6	0.3
212.79	-0.75	1.3	166.09	14SE	0.7	-3.6	-1.6	0.2
220.16	-0.85	1.3	176.07	24SE	-0.6	-3.6	-1.7	0.2
227.65	-0.93	1.3	186.10	4OC	-0.8	-3.5	-1.7	0.3
235.23	-1.00	1.4	196.19	14OC	-0.8	-3.5	-1.8	0.4
242.90	-1.06	1.4	206.32	24OC	-0.8	-3.5	-1.9	0.4
250.64	-1.11	1.4	216.48	3NO	-0.6	-3.4	-1.9	0.5
258.43	-1.14	1.4	226.67	13NO	-0.6	-3.4	-2.0	0.6
266.28	-1.17	1.4	236.88	23NO	-0.6	-3.4	-2.1	0.6
274.15	-1.17	1.4	247.09	3DE	-0.4	-3.4	-2.1	0.7
282.03	-1.17	1.4	257.29	13DE	1.1	-3.4	-2.1	0.7
289.91	-1.15	1.4	267.48	23DE	2.3	-3.3	-2.1	0.7
				-417				
297.77	-1.12	1.4	277.64	2JA	0.7	-3.3	-2.1	0.8
305.61	-1.08	1.4	287.77	12JA	0.3	-3.3	-2.1	0.8
313.40	-1.03	1.4	297.85	22JA	0.1	-3.4	-2.1	0.8
321.14	-0.97	1.4	307.88	1FE	-0.1	-3.4	-2.0	0.7
328.81	-0.91	1.5	317.85	11FE	-0.5	-3.4	-2.0	0.7
336.42	-0.83	1.5	327.77	21FE	-1.1	-3.4	-1.9	0.7
343.95	-0.75	1.5	337.62	3MR	-1.6	-3.5	-1.9	0.6
351.40	-0.66	1.5	347.41	13MR	-1.0	-3.5	-1.8	0.6
358.77	-0.57	1.5	357.15	23MR	0.2	-3.5	-1.7	0.5
6.04	-0.48	1.5	6.84	2AP	1.5	-3.4	-1.6	0.6
13.24	-0.38	1.6	16.47	12AP	3.1	-3.4	-1.6	0.6
20.34	-0.28	1.6	26.07	22AP	2.9	-3.4	-1.5	0.6
27.36	-0.18	1.7	35.64	2MY	1.7	-3.4	-1.5	0.6
34.30	-0.08	1.7	45.18	12MY	0.8	-3.3	-1.4	0.6
41.16	0.03	1.8	54.70	22MY	-0.0	-3.3	-1.4	0.6
47.93	0.13	1.8	64.22	1JN	-1.0	-3.3	-1.3	0.6
54.64	0.23	1.8	73.74	11JN	-1.8	-3.3	-1.3	0.5
61.27	0.34	1.9	83.27	21JN	-1.0	-3.3	-1.3	0.5
67.82	0.44	1.9	92.82	1JL	-0.3	-3.4	-1.3	0.5
74.31	0.54	1.9	102.40	11JL	0.2	-3.4	-1.3	0.4
80.74	0.64	1.9	112.02	21JL	0.5	-3.4	-1.3	0.4
87.09	0.75	1.9	121.69	31JL	0.8	-3.5	-1.3	0.3
93.38	0.85	1.9	131.40	10AU	1.6	-3.5	-1.3	0.3
99.60	0.96	1.9	141.17	20AU	3.2	-3.6	-1.3	0.2
105.74	1.06	1.8	151.00	30AU	0.6	-3.6	-1.3	0.1
111.81	1.17	1.8	160.89	9SE	-0.7	-3.7	-1.4	0.1
117.79	1.28	1.8	170.83	19SE	-0.9	-3.8	-1.4	0.0
123.68	1.40	1.7	180.84	29SE	-0.9	-3.9	-1.5	-0.0
129.47	1.51	1.6	190.90	9OC	-0.7	-4.0	-1.5	0.0
135.12	1.64	1.6	201.00	19OC	-0.5	-4.1	-1.6	0.1
140.64	1.77	1.5	211.14	29OC	-0.5	-4.2	-1.6	0.2
145.98	1.90	1.4	221.32	8NO	-0.4	-4.3	-1.7	0.2
151.11	2.04	1.2	231.52	18NO	-0.2	-4.4	-1.8	0.3
155.99	2.20	1.1	241.73	28NO	1.3	-4.4	-1.8	0.4
160.57	2.36	0.9	251.94	8DE	1.9	-4.2	-1.9	0.4
164.76	2.53	0.7	262.14	18DE	0.4	-3.7	-1.9	0.5
168.48	2.72	0.5	272.32	28DE	0.1	-3.2	-2.0	0.5

♂ LONG	♂ LAT	♂ MAG	☉ LONG	16.00UT -416	☿	♀	♃	♄
					MAGNITUDES			
171.61	2.91	0.3	282.46	7JA	-0.0	-3.8	-2.0	0.5
174.01	3.12	0.1	292.57	17JA	-0.2	-4.2	-2.0	0.5
175.51	3.31	-0.2	302.62	27JA	-0.5	-4.3	-2.1	0.6
175.91	3.49	-0.5	312.63	6FE	-1.1	-4.3	-2.0	0.5
175.08	3.62	-0.8	322.57	16FE	-1.5	-4.2	-2.0	0.5
172.99	3.67	-1.1	332.46	26FE	-0.9	-4.1	-2.0	0.5
169.83	3.58	-1.3	342.29	7MR	0.3	-4.0	-1.9	0.5
166.12	3.34	-1.4	352.05	17MR	1.9	-3.9	-1.9	0.4
162.57	2.96	-1.3	1.77	27MR	3.7	-3.8	-1.8	0.4
159.84	2.49	-1.1	11.43	6AP	2.2	-3.7	-1.8	0.3
158.34	1.99	-0.9	21.05	16AP	1.3	-3.6	-1.7	0.4
158.18	1.50	-0.7	30.63	26AP	0.6	-3.5	-1.6	0.4
159.28	1.06	-0.5	40.18	6MY	-0.1	-3.5	-1.6	0.4
161.48	0.66	-0.3	49.71	16MY	-1.1	-3.4	-1.5	0.4
164.60	0.31	-0.1	59.23	26MY	-1.8	-3.4	-1.4	0.4
168.47	0.00	0.0	68.75	5JN	-1.0	-3.4	-1.4	0.4
172.97	-0.26	0.2	78.27	15JN	-0.2	-3.3	-1.4	0.4
177.99	-0.50	0.3	87.81	25JN	0.3	-3.3	-1.3	0.4
183.45	-0.70	0.4	97.38	5JL	0.6	-3.3	-1.3	0.4
189.28	-0.88	0.5	106.98	15JL	1.1	-3.3	-1.3	0.3
195.43	-1.04	0.6	116.62	25JL	2.0	-3.3	-1.2	0.3
201.85	-1.17	0.7	126.31	4AU	2.9	-3.3	-1.2	0.2
208.52	-1.28	0.7	136.05	14AU	0.5	-3.4	-1.2	0.2
215.40	-1.37	0.8	145.84	24AU	-0.8	-3.4	-1.2	0.1
222.47	-1.43	0.8	155.70	3SE	-1.1	-3.4	-1.2	0.1
229.70	-1.48	0.9	165.61	13SE	-1.0	-3.4	-1.2	0.0
237.06	-1.52	1.0	175.59	23SE	-0.7	-3.5	-1.3	-0.1
244.54	-1.53	1.0	185.62	3OC	-0.4	-3.5	-1.3	-0.1
252.11	-1.52	1.0	195.70	13OC	-0.3	-3.5	-1.3	-0.2
259.76	-1.50	1.1	205.82	23OC	-0.3	-3.5	-1.4	-0.1
267.46	-1.47	1.1	215.99	2NO	-0.0	-3.4	-1.4	-0.1
275.20	-1.42	1.2	226.17	12NO	1.6	-3.4	-1.5	-0.0
282.95	-1.36	1.2	236.38	22NO	1.6	-3.4	-1.5	0.1
290.71	-1.28	1.3	246.59	2DE	0.2	-3.4	-1.6	0.1
298.45	-1.20	1.3	256.79	12DE	-0.1	-3.3	-1.7	0.2
306.16	-1.11	1.4	266.98	22DE	-0.1	-3.3	-1.7	0.3
				-415				
313.83	-1.01	1.4	277.15	1JA	-0.3	-3.3	-1.8	0.3
321.45	-0.91	1.5	287.27	11JA	-0.6	-3.3	-1.9	0.4
329.01	-0.81	1.5	297.36	21JA	-1.1	-3.4	-1.9	0.4
336.50	-0.70	1.5	307.39	31JA	-1.3	-3.4	-2.0	0.4
343.92	-0.59	1.6	317.36	10FE	-0.9	-3.4	-2.0	0.4
351.26	-0.48	1.6	327.28	20FE	0.4	-3.4	-2.0	0.4
358.52	-0.37	1.7	337.14	2MR	2.3	-3.4	-2.0	0.4
5.70	-0.26	1.7	346.94	12MR	2.9	-3.5	-2.0	0.4
12.80	-0.15	1.7	356.68	22MR	1.6	-3.5	-2.0	0.3
19.82	-0.04	1.7	6.37	1AP	0.9	-3.6	-1.9	0.3
26.77	0.06	1.8	16.01	11AP	0.5	-3.6	-1.9	0.3
33.64	0.16	1.8	25.61	21AP	-0.2	-3.7	-1.8	0.2
40.44	0.26	1.8	35.18	1MY	-1.1	-3.8	-1.8	0.3
47.18	0.35	1.8	44.72	11MY	-1.9	-3.9	-1.7	0.3
53.85	0.44	1.8	54.24	21MY	-1.0	-4.0	-1.7	0.3
60.47	0.53	1.8	63.76	31MY	-0.2	-4.0	-1.6	0.3
67.03	0.61	1.9	73.28	10JN	0.4	-4.1	-1.5	0.3
73.55	0.69	1.9	82.81	20JN	0.9	-4.2	-1.5	0.3
80.02	0.77	1.9	92.36	30JN	1.4	-4.2	-1.4	0.3
86.45	0.85	2.0	101.94	10JL	2.6	-4.0	-1.4	0.3
92.85	0.92	2.0	111.56	20JL	2.4	-3.5	-1.3	0.3
99.21	0.99	2.0	121.22	30JL	0.4	-3.2	-1.3	0.3
105.53	1.06	2.0	130.93	9AU	-0.9	-3.7	-1.3	0.2
111.83	1.13	2.0	140.70	19AU	-1.2	-4.1	-1.2	0.2
118.10	1.19	2.0	150.52	29AU	-1.1	-4.3	-1.2	0.1
124.33	1.26	2.0	160.40	8SE	-0.6	-4.3	-1.2	0.0
130.54	1.32	1.9	170.35	18SE	-0.3	-4.2	-1.2	-0.0
136.71	1.38	1.9	180.35	28SE	-0.2	-4.2	-1.2	-0.1
142.85	1.44	1.9	190.40	8OC	-0.1	-4.1	-1.2	-0.2
148.94	1.49	1.8	200.51	18OC	0.2	-4.0	-1.3	-0.2
154.98	1.55	1.7	210.65	28OC	1.8	-3.9	-1.3	-0.3
160.97	1.60	1.7	220.83	7NO	1.4	-3.8	-1.3	-0.2
166.88	1.65	1.6	231.03	17NO	-0.0	-3.7	-1.4	-0.2
172.71	1.70	1.5	241.24	27NO	-0.2	-3.7	-1.4	-0.1
178.44	1.74	1.4	251.45	7DE	-0.3	-3.6	-1.5	-0.0
184.05	1.78	1.2	261.65	17DE	-0.4	-3.5	-1.5	0.0
189.51	1.81	1.1	271.83	27DE	-0.7	-3.5	-1.6	0.1

♂ LONG	♂ LAT	♂ MAG	☉ LONG	16.00UT	☿	♀	♃	♄
				-414	MAGNITUDES			
194.79	1.84	0.9	281.97	6JA	-1.1	-3.4	-1.7	0.2
199.84	1.85	0.7	292.08	16JA	-1.2	-3.4	-1.7	0.2
204.61	1.85	0.5	302.14	26JA	-0.8	-3.4	-1.8	0.3
209.05	1.84	0.3	312.15	5FE	0.5	-3.3	-1.9	0.3
213.04	1.79	0.1	322.10	15FE	2.8	-3.3	-1.9	0.3
216.49	1.72	-0.2	331.99	25FE	2.2	-3.3	-2.0	0.3
219.25	1.60	-0.5	341.81	7MR	1.2	-3.3	-2.0	0.3
221.15	1.42	-0.8	351.59	17MR	0.7	-3.3	-2.0	0.3
222.01	1.17	-1.1	1.30	27MR	0.3	-3.3	-2.1	0.3
221.68	0.82	-1.4	10.97	6AP	-0.2	-3.3	-2.1	0.3
220.10	0.37	-1.7	20.59	16AP	-1.1	-3.4	-2.0	0.2
217.46	-0.16	-2.0	30.17	26AP	-1.9	-3.4	-2.0	0.2
214.24	-0.75	-2.1	39.72	6MY	-1.0	-3.4	-2.0	0.2
211.15	-1.31	-2.0	49.25	16MY	-0.1	-3.5	-1.9	0.2
208.89	-1.81	-1.8	58.77	26MY	0.6	-3.5	-1.9	0.2
207.87	-2.20	-1.6	68.29	5JN	1.1	-3.5	-1.8	0.3
208.24	-2.48	-1.4	77.81	15JN	1.9	-3.4	-1.7	0.3
209.93	-2.68	-1.2	87.35	25JN	3.2	-3.4	-1.7	0.3
212.75	-2.80	-0.9	96.91	5JL	2.0	-3.4	-1.6	0.3
216.51	-2.86	-0.8	106.51	15JL	0.3	-3.3	-1.6	0.3
221.05	-2.88	-0.6	116.14	25JL	-0.9	-3.3	-1.5	0.3
226.21	-2.85	-0.4	125.83	4AU	-1.4	-3.3	-1.4	0.3
231.88	-2.80	-0.2	135.57	14AU	-1.1	-3.3	-1.4	0.2
237.96	-2.72	-0.1	145.36	24AU	-0.5	-3.3	-1.4	0.2
244.35	-2.62	0.0	155.21	3SE	-0.2	-3.3	-1.3	0.1
251.02	-2.49	0.2	165.13	13SE	-0.0	-3.4	-1.3	0.1
257.89	-2.35	0.3	175.10	23SE	0.1	-3.4	-1.3	0.0
264.91	-2.20	0.4	185.12	3OC	0.4	-3.4	-1.3	-0.1
272.07	-2.04	0.5	195.21	13OC	2.1	-3.5	-1.3	-0.1
279.31	-1.88	0.6	205.33	23OC	1.2	-3.5	-1.3	-0.2
286.61	-1.71	0.7	215.49	2NO	-0.2	-3.6	-1.3	-0.3
293.94	-1.53	0.9	225.68	12NO	-0.4	-3.7	-1.3	-0.3
301.29	-1.36	1.0	235.88	22NO	-0.4	-3.7	-1.3	-0.3
308.63	-1.19	1.1	246.10	2DE	-0.5	-3.8	-1.4	-0.2
315.96	-1.02	1.1	256.31	12DE	-0.7	-3.9	-1.4	-0.1
323.25	-0.86	1.2	266.49	22DE	-1.0	-4.0	-1.4	-0.1
				-413				
330.51	-0.70	1.3	276.66	1JA	-1.0	-4.1	-1.5	0.0
337.71	-0.55	1.4	286.79	11JA	-0.7	-4.2	-1.6	0.1
344.86	-0.41	1.5	296.87	21JA	0.7	-4.3	-1.6	0.1
351.95	-0.27	1.5	306.91	31JA	3.1	-4.3	-1.7	0.2
358.98	-0.14	1.6	316.89	10FE	1.7	-4.3	-1.8	0.2
5.95	-0.02	1.7	326.81	20FE	0.9	-4.0	-1.8	0.3
12.85	0.10	1.7	336.67	2MR	0.5	-3.5	-1.9	0.3
19.70	0.21	1.8	346.47	12MR	0.2	-3.2	-2.0	0.3
26.48	0.31	1.8	356.21	22MR	-0.3	-3.8	-2.0	0.3
33.20	0.41	1.9	5.90	1AP	-1.1	-4.1	-2.1	0.3
39.87	0.50	1.9	15.55	11AP	-1.8	-4.2	-2.1	0.3
46.49	0.58	1.9	25.15	21AP	-1.0	-4.2	-2.1	0.3
53.07	0.66	1.9	34.72	1MY	-0.0	-4.1	-2.1	0.2
59.60	0.73	1.9	44.26	11MY	0.7	-4.0	-2.1	0.2
66.09	0.79	2.0	53.78	21MY	1.5	-3.9	-2.1	0.1
72.56	0.86	2.0	63.30	31MY	2.5	-3.8	-2.1	0.2
79.00	0.91	2.0	72.82	10JN	3.2	-3.7	-2.0	0.2
85.41	0.96	1.9	82.35	20JN	1.6	-3.7	-2.0	0.3
91.80	1.01	1.9	91.90	30JN	0.2	-3.6	-1.9	0.3
98.19	1.05	2.0	101.48	10JL	-1.0	-3.5	-1.9	0.3
104.56	1.09	2.0	111.09	20JL	-1.5	-3.5	-1.8	0.3
110.93	1.13	2.0	120.75	30JL	-1.0	-3.5	-1.7	0.3
117.29	1.16	2.0	130.46	9AU	-0.4	-3.4	-1.7	0.3
123.66	1.19	2.0	140.22	19AU	-0.1	-3.4	-1.6	0.3
130.03	1.21	2.0	150.04	29AU	0.1	-3.4	-1.6	0.3
136.41	1.23	2.0	159.92	8SE	0.2	-3.4	-1.5	0.2
142.80	1.24	2.0	169.86	18SE	0.7	-3.4	-1.5	0.2
149.19	1.25	2.0	179.87	28SE	2.5	-3.4	-1.4	0.1
155.60	1.26	2.0	189.92	8OC	1.0	-3.4	-1.4	0.0
162.01	1.26	1.9	200.02	18OC	-0.3	-3.4	-1.4	-0.0
168.43	1.25	1.9	210.16	28OC	-0.6	-3.4	-1.4	-0.1
174.86	1.24	1.8	220.33	7NO	-0.5	-3.4	-1.4	-0.2
181.30	1.22	1.7	230.53	17NO	-0.6	-3.4	-1.4	-0.3
187.73	1.19	1.7	240.74	27NO	-0.8	-3.4	-1.4	-0.3
194.16	1.16	1.6	250.95	7DE	-0.8	-3.5	-1.4	-0.3
200.59	1.11	1.5	261.15	17DE	-0.9	-3.5	-1.4	-0.2
207.01	1.05	1.4	271.34	27DE	-0.6	-3.5	-1.4	-0.1

♂ LONG	♂ LAT	♂ MAG	☉ LONG	16.00UT	☿	♀	♃	♄
				-412	MAGNITUDES			
213.40	0.98	1.3	281.48	6JA	0.8	-3.5	-1.5	-0.0
219.78	0.89	1.2	291.59	16JA	3.0	-3.4	-1.5	0.0
226.11	0.78	1.0	301.66	26JA	1.2	-3.4	-1.6	0.1
232.40	0.65	0.9	311.66	5FE	0.6	-3.4	-1.6	0.1
238.63	0.50	0.7	321.61	15FE	0.3	-3.4	-1.7	0.2
244.78	0.32	0.6	331.51	25FE	0.1	-3.3	-1.7	0.2
250.83	0.10	0.4	341.34	6MR	-0.4	-3.3	-1.8	0.3
256.75	-0.15	0.2	351.11	16MR	-1.1	-3.3	-1.9	0.3
262.48	-0.45	0.0	0.83	26MR	-1.7	-3.3	-1.9	0.3
267.99	-0.80	-0.2	10.50	5AP	-1.0	-3.3	-2.0	0.3
273.21	-1.20	-0.4	20.12	15AP	0.0	-3.4	-2.1	0.3
278.04	-1.67	-0.7	29.71	25AP	1.0	-3.4	-2.1	0.3
282.37	-2.20	-0.9	39.26	5MY	1.9	-3.4	-2.2	0.3
286.06	-2.82	-1.2	48.79	15MY	3.3	-3.4	-2.2	0.3
288.92	-3.51	-1.5	58.31	25MY	2.6	-3.5	-2.2	0.2
290.76	-4.26	-1.7	67.83	4JN	1.3	-3.5	-2.3	0.2
291.39	-5.04	-2.0	77.35	14JN	0.1	-3.6	-2.3	0.2
290.73	-5.78	-2.3	86.89	24JN	-1.0	-3.7	-2.2	0.3
288.92	-6.37	-2.5	96.45	4JL	-1.6	-3.7	-2.2	0.3
286.40	-6.70	-2.7	106.05	14JL	-1.0	-3.8	-2.2	0.3
283.86	-6.72	-2.5	115.68	24JL	-0.4	-3.9	-2.1	0.4
282.04	-6.43	-2.3	125.36	3AU	0.0	-4.0	-2.0	0.4
281.37	-5.94	-2.0	135.10	13AU	0.2	-4.1	-2.0	0.4
282.02	-5.34	-1.7	144.89	23AU	0.4	-4.2	-1.9	0.4
283.91	-4.71	-1.4	154.74	2SE	1.0	-4.3	-1.8	0.4
286.85	-4.09	-1.2	164.65	12SE	2.9	-4.3	-1.8	0.3
290.65	-3.52	-0.9	174.61	22SE	0.9	-4.1	-1.7	0.3
295.12	-2.99	-0.6	184.64	2OC	-0.5	-3.6	-1.7	0.2
300.11	-2.51	-0.4	194.71	12OC	-0.7	-3.2	-1.6	0.2
305.50	-2.09	-0.2	204.83	22OC	-0.7	-3.9	-1.6	0.1
311.19	-1.70	0.1	214.99	1NO	-0.7	-4.3	-1.5	0.0
317.12	-1.36	0.3	225.18	11NO	-0.7	-4.4	-1.5	-0.0
323.22	-1.06	0.4	235.39	21NO	-0.7	-4.4	-1.5	-0.1
329.44	-0.79	0.6	245.59	1DE	-0.7	-4.3	-1.5	-0.2
335.76	-0.55	0.8	255.81	11DE	-0.5	-4.2	-1.5	-0.2
342.14	-0.33	0.9	266.00	21DE	1.0	-4.1	-1.5	-0.2
348.57	-0.14	1.1	276.16	31DE	2.6	-4.0	-1.5	-0.1
				-411				
355.02	0.03	1.2	286.30	10JA	0.9	-3.9	-1.5	-0.1
1.48	0.18	1.3	296.39	20JA	0.4	-3.8	-1.5	0.0
7.94	0.32	1.4	306.42	30JA	0.2	-3.7	-1.5	0.1
14.40	0.44	1.5	316.41	9FE	-0.0	-3.6	-1.6	0.1
20.85	0.54	1.6	326.33	19FE	-0.4	-3.5	-1.6	0.2
27.28	0.64	1.7	336.19	1MR	-1.1	-3.5	-1.7	0.3
33.69	0.72	1.7	346.00	11MR	-1.6	-3.4	-1.7	0.3
40.09	0.80	1.8	355.75	21MR	-1.0	-3.4	-1.8	0.3
46.47	0.86	1.9	5.44	31MR	0.1	-3.4	-1.8	0.4
52.83	0.92	1.9	15.09	10AP	1.3	-3.3	-1.9	0.4
59.17	0.97	1.9	24.69	20AP	2.6	-3.3	-2.0	0.4
65.50	1.02	2.0	34.26	30AP	3.5	-3.3	-2.0	0.4
71.83	1.05	2.0	43.81	10MY	2.0	-3.3	-2.1	0.4
78.14	1.09	2.0	53.33	20MY	1.0	-3.3	-2.2	0.4
84.46	1.11	2.0	62.84	30MY	0.0	-3.3	-2.2	0.4
90.77	1.13	2.0	72.37	9JN	-1.0	-3.3	-2.3	0.3
97.09	1.15	2.0	81.89	19JN	-1.7	-3.4	-2.3	0.3
103.43	1.16	2.0	91.44	29JN	-1.0	-3.4	-2.4	0.4
109.77	1.16	2.0	101.02	9JL	-0.3	-3.4	-2.4	0.4
116.14	1.17	2.0	110.63	19JL	0.1	-3.5	-2.4	0.4
122.53	1.16	2.0	120.28	29JL	0.4	-3.5	-2.3	0.5
128.94	1.15	1.9	129.99	8AU	0.6	-3.5	-2.3	0.5
135.39	1.14	1.9	139.75	18AU	1.3	-3.4	-2.3	0.5
141.87	1.12	2.0	149.56	28AU	3.2	-3.4	-2.2	0.5
148.38	1.10	2.0	159.44	7SE	0.7	-3.4	-2.1	0.5
154.94	1.07	2.0	169.38	17SE	-0.6	-3.4	-2.1	0.5
161.53	1.04	1.9	179.37	27SE	-0.9	-3.3	-2.0	0.4
168.17	1.00	1.9	189.42	7OC	-0.8	-3.3	-1.9	0.4
174.86	0.95	1.9	199.52	17OC	-0.8	-3.3	-1.9	0.4
181.59	0.90	1.9	209.66	27OC	-0.6	-3.3	-1.8	0.3
188.37	0.84	1.8	219.83	6NO	-0.5	-3.3	-1.8	0.2
195.19	0.77	1.8	230.03	16NO	-0.5	-3.4	-1.7	0.2
202.06	0.69	1.7	240.24	26NO	-0.3	-3.4	-1.7	0.1
208.98	0.61	1.7	250.45	6DE	1.1	-3.4	-1.6	0.0
215.94	0.51	1.6	260.65	16DE	2.2	-3.5	-1.6	-0.0
222.95	0.41	1.6	270.83	26DE	0.6	-3.5	-1.6	-0.1

Left table (−410 / −409)

♂ LONG	LAT	MAG	☉ LONG	16.00UT	☿	♀	♃	♄
230.00	0.29	1.5	280.99	5JA	0.2	-3.5	-1.5	-0.0
237.08	0.16	1.4	291.10	15JA	0.1	-3.6	-1.5	0.0
244.21	0.02	1.3	301.16	25JA	-0.1	-3.7	-1.5	0.1
251.36	-0.14	1.2	311.18	4FE	-0.5	-3.7	-1.5	0.1
258.54	-0.31	1.1	321.13	14FE	-1.1	-3.8	-1.6	0.2
265.75	-0.49	1.0	331.03	24FE	-1.5	-3.9	-1.6	0.3
272.96	-0.68	0.9	340.87	6MR	-1.0	-4.0	-1.6	0.3
280.17	-0.89	0.8	350.64	16MR	0.2	-4.0	-1.6	0.4
287.37	-1.10	0.7	0.37	26MR	1.6	-4.1	-1.7	0.4
294.53	-1.33	0.6	10.04	5AP	3.3	-4.2	-1.7	0.4
301.64	-1.56	0.5	19.66	15AP	2.7	-4.2	-1.8	0.5
308.67	-1.80	0.4	29.25	25AP	1.5	-4.1	-1.8	0.5
315.60	-2.05	0.3	38.80	5MY	0.8	-3.7	-1.9	0.5
322.38	-2.30	0.2	48.33	15MY	-0.0	-2.8	-2.0	0.5
328.98	-2.55	0.0	57.86	25MY	-1.0	-3.3	-2.0	0.5
335.33	-2.80	-0.1	67.37	4JN	-1.8	-3.9	-2.1	0.5
341.38	-3.04	-0.3	76.89	14JN	-1.0	-4.2	-2.2	0.5
347.04	-3.28	-0.4	86.43	24JN	-0.3	-4.2	-2.2	0.5
352.21	-3.51	-0.6	95.99	4JL	0.2	-4.2	-2.3	0.5
356.78	-3.72	-0.8	105.58	14JL	0.5	-4.1	-2.4	0.5
0.57	-3.92	-1.0	115.22	24JL	0.9	-4.0	-2.4	0.6
3.39	-4.08	-1.2	124.90	3AU	1.7	-3.9	-2.4	0.6
5.04	-4.19	-1.4	134.62	13AU	3.1	-3.8	-2.4	0.6
5.32	-4.23	-1.6	144.41	23AU	0.6	-3.8	-2.4	0.7
4.12	-4.13	-1.8	154.26	2SE	-0.7	-3.7	-2.4	0.7
1.61	-3.87	-2.0	164.16	12SE	-1.0	-3.6	-2.4	0.7
358.26	-3.43	-2.1	174.13	22SE	-1.0	-3.6	-2.3	0.7
354.87	-2.85	-2.0	184.15	2OC	-0.7	-3.5	-2.3	0.7
352.22	-2.21	-1.7	194.22	12OC	-0.5	-3.5	-2.2	0.6
350.76	-1.58	-1.4	204.34	22OC	-0.4	-3.5	-2.1	0.6
350.67	-1.03	-1.0	214.50	1NO	-0.4	-3.4	-2.1	0.5
351.83	-0.56	-0.7	224.68	11NO	-0.1	-3.4	-2.0	0.5
354.05	-0.17	-0.4	234.89	21NO	1.4	-3.4	-1.9	0.4
357.12	0.14	-0.1	245.10	1DE	1.9	-3.4	-1.9	0.4
0.86	0.39	0.1	255.31	11DE	0.3	-3.4	-1.8	0.3
5.11	0.60	0.4	265.50	21DE	0.0	-3.3	-1.8	0.2
9.78	0.76	0.6	275.67	31DE	-0.1	-3.3	-1.7	0.1

−409

♂ LONG	LAT	MAG	☉ LONG	16.00UT	☿	♀	♃	♄
14.74	0.89	0.8	285.80	10JA	-0.2	-3.3	-1.7	0.1
19.95	1.00	0.9	295.89	20JA	-0.5	-3.4	-1.6	0.1
25.35	1.09	1.1	305.93	30JA	-1.1	-3.4	-1.6	0.2
30.90	1.15	1.2	315.92	9FE	-1.4	-3.4	-1.6	0.3
36.56	1.21	1.4	325.85	19FE	-0.9	-3.4	-1.6	0.3
42.31	1.25	1.5	335.71	1MR	0.3	-3.5	-1.6	0.4
48.14	1.28	1.6	345.52	11MR	2.0	-3.5	-1.6	0.4
54.03	1.30	1.7	355.27	21MR	3.5	-3.5	-1.6	0.5
59.97	1.32	1.7	4.97	31MR	2.0	-3.4	-1.6	0.5
65.95	1.33	1.8	14.61	10AP	1.1	-3.4	-1.6	0.6
71.97	1.33	1.9	24.22	20AP	0.6	-3.4	-1.6	0.6
78.03	1.33	1.9	33.79	30AP	-0.1	-3.4	-1.7	0.7
84.12	1.32	2.0	43.34	10MY	-1.0	-3.3	-1.7	0.7
90.24	1.31	2.0	52.86	20MY	-1.8	-3.3	-1.8	0.7
96.41	1.29	2.0	62.38	30MY	-1.0	-3.3	-1.8	0.7
102.60	1.27	2.0	71.90	9JN	-0.2	-3.3	-1.9	0.7
108.84	1.24	2.0	81.43	19JN	0.3	-3.3	-2.0	0.7
115.12	1.21	2.0	90.97	29JN	0.7	-3.4	-2.0	0.7
121.44	1.18	2.0	100.55	9JL	1.2	-3.4	-2.1	0.7
127.82	1.14	2.0	110.16	19JL	2.2	-3.4	-2.2	0.7
134.24	1.09	2.0	119.81	29JL	2.8	-3.5	-2.2	0.7
140.72	1.05	1.9	129.51	8AU	0.5	-3.5	-2.3	0.8
147.25	0.99	1.9	139.27	18AU	-0.8	-3.6	-2.3	0.8
153.85	0.94	1.9	149.09	28AU	-1.2	-3.6	-2.4	0.9
160.51	0.88	1.8	158.96	7SE	-1.1	-3.7	-2.4	0.9
167.24	0.81	1.8	168.90	17SE	-0.6	-3.8	-2.4	0.9
174.03	0.74	1.8	178.89	27SE	-0.4	-3.9	-2.4	0.9
180.89	0.66	1.8	188.94	7OC	-0.3	-4.0	-2.4	0.9
187.83	0.58	1.8	199.03	17OC	-0.2	-4.1	-2.4	0.9
194.83	0.50	1.8	209.17	27OC	0.0	-4.2	-2.3	0.9
201.91	0.40	1.7	219.34	6NO	1.6	-4.3	-2.3	0.8
209.06	0.31	1.7	229.54	16NO	1.6	-4.4	-2.2	0.8
216.27	0.20	1.7	239.74	26NO	0.1	-4.2	-2.1	0.7
223.55	0.09	1.7	249.95	6DE	-0.1	-4.2	-2.1	0.7
230.90	-0.02	1.6	260.16	16DE	-0.2	-3.6	-2.0	0.6
238.31	-0.14	1.6	270.34	26DE	-0.3	-3.2	-1.9	0.5

Right table (−408 / −407)

♂ LONG	LAT	MAG	☉ LONG	16.00UT	☿	♀	♃	♄
245.77	-0.27	1.5	280.49	5JA	-0.6	-3.8	-1.9	0.4
253.29	-0.40	1.5	290.61	15JA	-1.1	-4.2	-1.8	0.4
260.84	-0.53	1.4	300.67	25JA	-1.3	-4.3	-1.7	0.4
268.44	-0.66	1.4	310.69	4FE	-0.9	-4.3	-1.7	0.4
276.06	-0.79	1.3	320.65	14FE	0.4	-4.2	-1.6	0.4
283.70	-0.93	1.3	330.54	24FE	2.4	-4.1	-1.6	0.5
291.34	-1.06	1.2	340.38	5MR	2.7	-4.0	-1.6	0.5
298.98	-1.18	1.2	350.16	15MR	1.4	-3.9	-1.6	0.6
306.60	-1.30	1.1	359.89	25MR	0.8	-3.8	-1.5	0.7
314.19	-1.41	1.1	9.56	4AP	0.4	-3.7	-1.5	0.7
321.73	-1.51	1.0	19.19	14AP	-0.2	-3.6	-1.5	0.8
329.20	-1.60	1.0	28.78	24AP	-1.0	-3.5	-1.5	0.8
336.59	-1.68	0.9	38.34	4MY	-1.9	-3.5	-1.5	0.8
343.88	-1.74	0.9	47.88	14MY	-1.0	-3.4	-1.6	0.9
351.05	-1.79	0.8	57.39	24MY	-0.2	-3.4	-1.6	0.9
358.08	-1.82	0.8	66.91	3JN	0.5	-3.4	-1.6	0.9
4.94	-1.84	0.7	76.44	13JN	0.9	-3.3	-1.7	1.0
11.62	-1.83	0.6	85.97	23JN	1.6	-3.3	-1.7	1.0
18.07	-1.81	0.6	95.53	3JL	2.8	-3.3	-1.8	1.0
24.26	-1.77	0.5	105.12	13JL	2.3	-3.3	-1.8	1.0
30.15	-1.71	0.4	114.75	23JL	0.4	-3.3	-1.9	1.0
35.67	-1.63	0.3	124.43	2AU	-0.9	-3.3	-1.9	1.0
40.76	-1.52	0.2	134.16	12AU	-1.3	-3.4	-2.0	1.0
45.33	-1.39	0.0	143.94	22AU	-1.1	-3.4	-2.1	1.1
49.25	-1.22	-0.1	153.78	1SE	-0.6	-3.4	-2.1	1.1
52.37	-1.01	-0.3	163.68	11SE	-0.3	-3.4	-2.2	1.1
54.52	-0.76	-0.5	173.64	21SE	-0.1	-3.5	-2.3	1.2
55.47	-0.45	-0.7	183.66	1OC	-0.0	-3.5	-2.3	1.2
55.06	-0.09	-1.0	193.73	11OC	0.3	-3.5	-2.3	1.2
53.22	0.33	-1.2	203.85	21OC	1.9	-3.5	-2.4	1.2
50.12	0.78	-1.4	214.01	31OC	1.4	-3.4	-2.4	1.2
46.29	1.22	-1.5	224.19	10NO	-0.1	-3.4	-2.3	1.1
42.50	1.58	-1.3	234.39	20NO	-0.3	-3.4	-2.3	1.1
39.48	1.86	-1.0	244.60	30NO	-0.3	-3.4	-2.3	1.1
37.70	2.03	-0.7	254.81	10DE	-0.4	-3.3	-2.2	1.0
37.26	2.13	-0.4	265.00	20DE	-0.7	-3.3	-2.1	0.9
38.07	2.17	-0.1	275.17	30DE	-1.0	-3.3	-2.1	0.9

−407

♂ LONG	LAT	MAG	☉ LONG	16.00UT	☿	♀	♃	♄
39.93	2.17	0.1	285.31	9JA	-1.1	-3.3	-2.0	0.8
42.63	2.15	0.4	295.40	19JA	-0.8	-3.4	-1.9	0.7
46.01	2.11	0.6	305.44	29JA	0.5	-3.4	-1.8	0.7
49.92	2.07	0.8	315.43	8FE	2.9	-3.4	-1.8	0.6
54.25	2.01	1.0	325.36	18FE	2.0	-3.4	-1.7	0.7
58.90	1.96	1.1	335.23	28FE	1.1	-3.4	-1.7	0.7
63.83	1.90	1.3	345.04	10MR	0.6	-3.5	-1.6	0.8
68.96	1.84	1.4	354.79	20MR	0.3	-3.5	-1.6	0.8
74.28	1.77	1.5	4.50	30MR	-0.2	-3.6	-1.5	0.9
79.74	1.71	1.6	14.15	9AP	-1.0	-3.6	-1.5	0.9
85.33	1.65	1.7	23.75	19AP	-1.9	-3.7	-1.5	1.0
91.04	1.58	1.7	33.33	29AP	-1.0	-3.8	-1.5	1.0
96.84	1.51	1.8	42.87	9MY	-0.1	-3.9	-1.4	1.1
102.74	1.45	1.8	52.40	19MY	0.6	-4.0	-1.4	1.1
108.72	1.38	1.9	61.92	29MY	1.2	-4.1	-1.4	1.2
114.78	1.30	1.9	71.44	8JN	2.1	-4.1	-1.5	1.2
120.93	1.23	1.9	80.97	18JN	3.3	-4.2	-1.5	1.2
127.15	1.16	1.9	90.52	28JN	1.9	-4.2	-1.5	1.2
133.46	1.08	1.9	100.09	8JL	0.3	-4.0	-1.5	1.3
139.84	1.00	1.9	109.70	18JL	-0.9	-3.5	-1.6	1.3
146.31	0.92	1.9	119.35	28JL	-1.4	-3.2	-1.6	1.3
152.86	0.83	1.9	129.05	7AU	-1.1	-3.7	-1.7	1.3
159.50	0.74	1.8	138.80	17AU	-0.5	-4.1	-1.7	1.3
166.22	0.65	1.8	148.61	27AU	-0.2	-4.3	-1.8	1.3
173.03	0.56	1.8	158.48	6SE	0.0	-4.3	-1.8	1.4
179.93	0.46	1.8	168.41	16SE	0.1	-4.2	-1.9	1.4
186.92	0.36	1.7	178.41	26SE	0.5	-4.1	-2.0	1.4
194.00	0.26	1.7	188.45	6OC	2.2	-4.1	-2.0	1.4
201.17	0.16	1.6	198.54	16OC	1.2	-4.0	-2.1	1.3
208.43	0.05	1.5	208.68	26OC	-0.2	-3.9	-2.2	1.3
215.76	-0.05	1.5	218.85	5NO	-0.5	-3.8	-2.2	1.3
223.18	-0.16	1.5	229.04	15NO	-0.5	-3.7	-2.2	1.2
230.68	-0.27	1.5	239.25	25NO	-0.5	-3.7	-2.2	1.2
238.24	-0.38	1.5	249.46	5DE	-0.8	-3.6	-2.2	1.2
245.87	-0.49	1.5	259.66	15DE	-0.9	-3.5	-2.2	1.1
253.56	-0.59	1.5	269.85	25DE	-1.0	-3.5	-2.2	1.1

♂ LONG	♂ LAT	♂ MAG	☉ LONG	16.00UT -406	☿	♀	♃	♄
261.29	-0.70	1.5	280.00	4JA	-0.7	-3.4	-2.1	1.0
269.06	-0.79	1.5	290.12	14JA	0.7	-3.4	-2.1	1.0
276.86	-0.89	1.4	300.19	24JA	3.1	-3.4	-2.0	0.9
284.67	-0.97	1.4	310.20	3FE	1.5	-3.3	-1.9	0.9
292.48	-1.05	1.4	320.17	13FE	0.8	-3.3	-1.9	0.9
300.29	-1.12	1.4	330.07	23FE	0.4	-3.3	-1.8	0.9
308.07	-1.17	1.4	339.91	5MR	0.1	-3.3	-1.7	1.0
315.82	-1.22	1.4	349.69	15MR	-0.3	-3.3	-1.7	1.0
323.53	-1.25	1.4	359.42	25MR	-1.0	-3.3	-1.6	1.1
331.18	-1.27	1.4	9.09	4AP	-1.8	-3.3	-1.5	1.1
338.76	-1.28	1.3	18.72	14AP	-1.0	-3.4	-1.5	1.2
346.25	-1.27	1.3	28.31	24AP	-0.0	-3.4	-1.5	1.2
353.66	-1.25	1.3	37.87	4MY	0.8	-3.4	-1.4	1.3
0.98	-1.22	1.3	47.41	14MY	1.6	-3.5	-1.4	1.3
8.18	-1.17	1.3	56.93	24MY	2.8	-3.5	-1.4	1.4
15.26	-1.11	1.3	66.44	3JN	3.0	-3.4	-1.4	1.4
22.22	-1.04	1.3	75.97	13JN	1.5	-3.4	-1.3	1.4
29.04	-0.96	1.2	85.50	23JN	0.2	-3.4	-1.3	1.4
35.70	-0.86	1.2	95.06	3JL	-1.0	-3.4	-1.3	1.4
42.21	-0.75	1.2	104.65	13JL	-1.5	-3.3	-1.4	1.4
48.54	-0.63	1.1	114.28	23JL	-1.1	-3.3	-1.4	1.4
54.67	-0.49	1.1	123.95	2AU	-0.4	-3.3	-1.4	1.3
60.57	-0.34	1.0	133.68	12AU	-0.1	-3.3	-1.4	1.3
66.21	-0.17	0.9	143.46	22AU	0.1	-3.3	-1.5	1.2
71.55	0.02	0.8	153.30	1SE	0.3	-3.3	-1.5	1.2
76.54	0.23	0.7	163.20	11SE	0.8	-3.4	-1.6	1.2
81.10	0.46	0.6	173.16	21SE	2.6	-3.4	-1.6	1.2
85.14	0.72	0.4	183.17	1OC	1.0	-3.4	-1.7	1.2
88.53	1.02	0.3	193.24	11OC	-0.4	-3.5	-1.8	1.2
91.14	1.36	0.1	203.36	21OC	-0.6	-3.5	-1.8	1.2
92.77	1.74	-0.1	213.51	31OC	-0.6	-3.6	-1.9	1.2
93.24	2.17	-0.4	223.70	10NO	-0.7	-3.7	-1.9	1.1
92.39	2.62	-0.6	233.90	20NO	-0.8	-3.7	-2.0	1.1
90.19	3.06	-0.8	244.11	30NO	-0.8	-3.8	-2.0	1.1
86.84	3.45	-1.0	254.32	10DE	-0.8	-3.9	-2.1	1.0
82.89	3.71	-1.1	264.51	20DE	-0.6	-4.0	-2.1	1.0
79.09	3.83	-0.9	274.68	30DE	0.8	-4.1	-2.1	0.9
				-405				
76.11	3.79	-0.7	284.82	9JA	2.9	-4.2	-2.1	0.9
74.35	3.66	-0.4	294.91	19JA	1.1	-4.3	-2.1	0.8
73.90	3.46	-0.1	304.96	29JA	0.5	-4.3	-2.1	0.8
74.66	3.24	0.1	314.95	8FE	0.3	-4.3	-2.0	0.8
76.45	3.02	0.3	324.88	18FE	0.0	-4.0	-2.0	0.7
79.07	2.80	0.5	334.75	28FE	-0.4	-3.5	-1.9	0.7
82.38	2.60	0.7	344.57	10MR	-1.0	-3.3	-1.8	0.7
86.22	2.41	0.9	354.32	20MR	-1.7	-3.8	-1.8	0.8
90.49	2.23	1.0	4.03	30MR	-1.0	-4.1	-1.7	0.9
95.12	2.07	1.2	13.68	9AP	0.0	-4.2	-1.6	0.9
100.04	1.91	1.3	23.29	19AP	1.1	-4.2	-1.6	1.0
105.20	1.76	1.4	32.86	29AP	2.2	-4.1	-1.5	1.1
110.58	1.62	1.5	42.41	9MY	3.6	-4.0	-1.5	1.1
116.14	1.48	1.5	51.94	19MY	2.4	-3.9	-1.4	1.2
121.86	1.35	1.6	61.46	29MY	1.2	-3.8	-1.4	1.2
127.74	1.22	1.6	70.98	8JN	0.1	-3.7	-1.3	1.2
133.75	1.10	1.7	80.50	18JN	-1.0	-3.7	-1.3	1.2
139.89	0.97	1.7	90.05	28JN	-1.6	-3.6	-1.3	1.3
146.16	0.85	1.7	99.62	8JL	-1.0	-3.5	-1.3	1.3
152.55	0.73	1.7	109.23	18JL	-0.4	-3.5	-1.3	1.2
159.06	0.61	1.7	118.88	28JL	0.0	-3.5	-1.3	1.2
165.69	0.49	1.7	128.58	7AU	0.3	-3.4	-1.3	1.2
172.42	0.37	1.7	138.33	17AU	0.5	-3.4	-1.3	1.2
179.28	0.25	1.7	148.14	27AU	1.1	-3.4	-1.3	1.1
186.24	0.14	1.7	158.01	6SE	2.9	-3.4	-1.3	1.1
193.31	0.02	1.7	167.93	16SE	0.9	-3.4	-1.4	1.0
200.49	-0.09	1.6	177.92	26SE	-0.5	-3.4	-1.4	1.1
207.77	-0.20	1.6	187.96	6OC	-0.8	-3.4	-1.5	1.1
215.15	-0.31	1.6	198.05	16OC	-0.7	-3.4	-1.5	1.1
222.62	-0.42	1.5	208.19	26OC	-0.8	-3.4	-1.6	1.1
230.18	-0.52	1.5	218.36	5NO	-0.7	-3.4	-1.6	1.0
237.82	-0.61	1.4	228.55	15NO	-0.6	-3.4	-1.7	1.0
245.52	-0.70	1.4	238.76	25NO	-0.6	-3.4	-1.8	1.0
253.29	-0.78	1.4	248.97	5DE	-0.4	-3.5	-1.8	1.0
261.10	-0.86	1.3	259.17	15DE	1.0	-3.5	-1.9	0.9
268.96	-0.93	1.3	269.35	25DE	2.5	-3.5	-1.9	0.9

♂ LONG	♂ LAT	♂ MAG	☉ LONG	16.00UT -404	☿	♀	♃	♄
276.83	-0.98	1.3	279.51	4JA	0.8	-3.5	-2.0	0.9
284.72	-1.03	1.3	289.62	14JA	0.3	-3.4	-2.0	0.8
292.61	-1.07	1.3	299.70	24JA	0.1	-3.4	-2.0	0.8
300.48	-1.09	1.4	309.72	3FE	-0.1	-3.4	-2.0	0.7
308.32	-1.10	1.4	319.68	13FE	-0.4	-3.4	-2.0	0.7
316.13	-1.10	1.4	329.59	23FE	-1.0	-3.3	-2.0	0.6
323.88	-1.09	1.4	339.43	4MR	-1.6	-3.3	-2.0	0.6
331.58	-1.07	1.4	349.21	14MR	-1.0	-3.3	-1.9	0.6
339.20	-1.04	1.5	358.95	24MR	0.1	-3.3	-1.9	0.6
346.75	-0.99	1.5	8.62	3AP	1.4	-3.3	-1.8	0.7
354.22	-0.94	1.5	18.25	13AP	2.8	-3.4	-1.7	0.7
1.60	-0.87	1.5	27.85	23AP	3.2	-3.4	-1.7	0.8
8.88	-0.80	1.5	37.41	3MY	1.8	-3.4	-1.6	0.9
16.07	-0.72	1.6	46.94	13MY	0.9	-3.4	-1.5	0.9
23.16	-0.63	1.6	56.47	23MY	0.0	-3.5	-1.5	1.0
30.14	-0.53	1.6	65.98	2JN	-1.0	-3.5	-1.4	1.0
37.02	-0.43	1.6	75.50	12JN	-1.7	-3.6	-1.4	1.1
43.79	-0.32	1.6	85.04	22JN	-1.0	-3.7	-1.3	1.1
50.45	-0.20	1.6	94.60	2JL	-0.3	-3.7	-1.3	1.1
57.00	-0.08	1.6	104.18	12JL	0.1	-3.8	-1.3	1.1
63.42	0.05	1.6	113.81	22JL	0.4	-3.9	-1.2	1.1
69.72	0.19	1.5	123.48	1AU	0.7	-4.0	-1.2	1.1
75.88	0.33	1.5	133.21	11AU	1.4	-4.1	-1.2	1.1
81.89	0.48	1.5	142.99	21AU	3.2	-4.2	-1.2	1.0
87.73	0.64	1.4	152.82	31AU	0.8	-4.3	-1.2	1.0
93.39	0.81	1.3	162.72	10SE	-0.6	-4.3	-1.2	1.0
98.83	1.00	1.2	172.68	20SE	-0.9	-4.1	-1.3	0.9
104.01	1.20	1.1	182.69	30SE	-0.9	-3.6	-1.3	0.9
108.90	1.42	1.0	192.76	10OC	-0.8	-3.2	-1.3	0.9
113.42	1.66	0.9	202.87	20OC	-0.5	-3.9	-1.3	1.0
117.49	1.92	0.7	213.02	30OC	-0.5	-4.3	-1.4	1.0
121.02	2.22	0.6	223.20	9NO	-0.5	-4.4	-1.4	1.0
123.86	2.54	0.4	233.40	19NO	-0.3	-4.4	-1.5	0.9
125.86	2.91	0.2	243.61	29NO	1.1	-4.3	-1.6	0.9
126.84	3.30	-0.1	253.82	9DE	2.2	-4.2	-1.6	0.9
126.62	3.70	-0.3	264.02	19DE	0.5	-4.1	-1.7	0.9
125.09	4.09	-0.6	274.19	29DE	0.1	-4.0	-1.8	0.8
				-403				
122.31	4.41	-0.8	284.33	8JA	0.0	-3.9	-1.8	0.8
118.64	4.59	-1.0	294.43	18JA	-0.2	-3.8	-1.9	0.8
114.71	4.60	-1.0	304.47	28JA	-0.5	-3.7	-1.9	0.7
111.23	4.43	-0.8	314.47	7FE	-1.1	-3.6	-2.0	0.7
108.78	4.14	-0.6	324.41	17FE	-1.5	-3.5	-2.0	0.6
107.60	3.79	-0.3	334.28	27FE	-1.0	-3.5	-2.0	0.6
107.70	3.41	-0.1	344.10	9MR	0.2	-3.4	-2.0	0.5
108.94	3.04	0.1	353.86	19MR	1.7	-3.4	-2.0	0.5
111.14	2.70	0.3	3.56	29MR	3.6	-3.4	-2.0	0.5
114.13	2.38	0.5	13.22	8AP	2.4	-3.3	-1.9	0.5
117.76	2.10	0.7	22.83	18AP	1.4	-3.3	-1.9	0.6
121.91	1.83	0.8	32.40	28AP	0.7	-3.3	-1.8	0.7
126.50	1.59	0.9	41.95	8MY	-0.0	-3.3	-1.8	0.7
131.44	1.37	1.0	51.48	18MY	-1.0	-3.3	-1.7	0.8
136.70	1.16	1.1	61.00	28MY	-1.8	-3.3	-1.6	0.8
142.22	0.96	1.2	70.52	7JN	-1.0	-3.3	-1.6	0.9
147.98	0.78	1.2	80.04	17JN	-0.3	-3.4	-1.5	0.9
153.95	0.60	1.3	89.58	27JN	0.2	-3.4	-1.5	1.0
160.11	0.44	1.3	99.16	7JL	0.6	-3.4	-1.4	1.0
166.45	0.28	1.4	108.76	17JL	1.0	-3.5	-1.4	1.0
172.96	0.13	1.4	118.40	27JL	1.9	-3.5	-1.3	1.0
179.62	-0.02	1.4	128.10	6AU	3.0	-3.5	-1.3	1.0
186.44	-0.15	1.4	137.85	16AU	0.6	-3.4	-1.3	1.0
193.40	-0.28	1.4	147.65	26AU	-0.7	-3.4	-1.2	1.0
200.49	-0.40	1.4	157.52	5SE	-1.1	-3.4	-1.2	0.9
207.71	-0.52	1.4	167.45	15SE	-1.0	-3.4	-1.2	0.9
215.04	-0.63	1.4	177.43	25SE	-0.7	-3.3	-1.2	0.9
222.49	-0.72	1.4	187.47	5OC	-0.4	-3.3	-1.2	0.8
230.04	-0.81	1.4	197.56	15OC	-0.3	-3.3	-1.3	0.9
237.67	-0.89	1.4	207.69	25OC	-0.3	-3.3	-1.3	0.9
245.38	-0.96	1.4	217.86	4NO	-0.1	-3.3	-1.3	0.9
253.16	-1.01	1.4	228.05	14NO	1.4	-3.4	-1.3	0.9
260.99	-1.06	1.4	238.26	24NO	1.9	-3.4	-1.4	0.9
268.85	-1.09	1.4	248.47	4DE	0.3	-3.4	-1.4	0.9
276.74	-1.11	1.4	258.68	14DE	-0.0	-3.5	-1.5	0.9
284.64	-1.12	1.4	268.86	24DE	-0.1	-3.5	-1.6	0.8

♂ LONG	LAT	MAG	☉ LONG	16.00UT	☿	♀	♃	♄
				-402	MAGNITUDES			
292.53	-1.11	1.4	279.02	3JA	-0.3	-3.5	-1.6	0.8
300.41	-1.10	1.4	289.14	13JA	-0.6	-3.6	-1.7	0.8
308.25	-1.07	1.4	299.21	23JA	-1.1	-3.7	-1.8	0.7
316.04	-1.03	1.4	309.24	2FE	-1.3	-3.7	-1.8	0.7
323.79	-0.99	1.4	319.20	12FE	-0.9	-3.8	-1.9	0.6
331.46	-0.93	1.4	329.11	22FE	0.3	-3.9	-1.9	0.6
339.07	-0.86	1.4	338.96	4MR	2.1	-4.0	-2.0	0.5
346.60	-0.79	1.4	348.75	14MR	3.2	-4.1	-2.0	0.5
354.05	-0.71	1.5	358.48	24MR	1.8	-4.1	-2.1	0.4
1.41	-0.62	1.5	8.16	3AP	1.0	-4.2	-2.1	0.4
8.69	-0.53	1.6	17.79	13AP	0.5	-4.2	-2.1	0.4
15.87	-0.44	1.6	27.39	23AP	-0.1	-4.1	-2.0	0.5
22.97	-0.34	1.7	36.95	3MY	-1.0	-3.7	-2.0	0.5
29.98	-0.24	1.7	46.49	13MY	-1.8	-2.8	-2.0	0.6
36.90	-0.14	1.7	56.01	23MY	-1.1	-3.3	-1.9	0.7
43.73	-0.03	1.8	65.53	2JN	-0.2	-3.9	-1.9	0.7
50.48	0.07	1.8	75.04	12JN	0.4	-4.2	-1.8	0.8
57.15	0.18	1.8	84.58	22JN	0.8	-4.2	-1.7	0.8
63.74	0.29	1.8	94.13	2JL	1.3	-4.2	-1.7	0.9
70.24	0.40	1.8	103.72	12JL	2.4	-4.1	-1.6	0.9
76.67	0.52	1.8	113.34	22JL	2.6	-4.0	-1.5	0.9
83.01	0.63	1.8	123.01	1AU	0.5	-3.9	-1.5	0.9
89.27	0.75	1.8	132.73	11AU	-0.8	-3.8	-1.4	0.9
95.45	0.87	1.8	142.51	21AU	-1.2	-3.7	-1.4	0.9
101.52	0.99	1.7	152.34	31AU	-1.1	-3.7	-1.4	0.9
107.50	1.12	1.7	162.23	10SE	-0.6	-3.6	-1.3	0.9
113.36	1.25	1.7	172.19	20SE	-0.3	-3.6	-1.3	0.9
119.10	1.38	1.6	182.20	30SE	-0.2	-3.5	-1.3	0.8
124.68	1.53	1.5	192.26	10OC	-0.1	-3.5	-1.3	0.8
130.10	1.68	1.4	202.37	20OC	0.1	-3.5	-1.3	0.8
135.31	1.84	1.3	212.53	30OC	1.6	-3.4	-1.3	0.8
140.27	2.02	1.2	222.70	9NO	1.6	-3.4	-1.3	0.8
144.94	2.21	1.0	232.91	19NO	0.1	-3.4	-1.3	0.8
149.24	2.41	0.9	243.12	29NO	-0.2	-3.4	-1.3	0.8
153.10	2.64	0.7	253.33	9DE	-0.3	-3.4	-1.4	0.8
156.39	2.88	0.5	263.53	19DE	-0.4	-3.3	-1.4	0.8
158.97	3.14	0.2	273.70	29DE	-0.6	-3.3	-1.5	0.8
				-401				
160.70	3.41	-0.0	283.84	8JA	-1.1	-3.3	-1.5	0.8
161.38	3.69	-0.3	293.94	18JA	-1.2	-3.4	-1.6	0.8
160.84	3.94	-0.6	303.99	28JA	-0.9	-3.4	-1.6	0.7
159.03	4.13	-0.8	313.98	7FE	0.4	-3.4	-1.7	0.7
156.06	4.21	-1.1	323.93	17FE	2.6	-3.4	-1.8	0.6
152.36	4.12	-1.2	333.80	27FE	2.4	-3.5	-1.9	0.6
148.61	3.87	-1.1	343.62	9MR	1.3	-3.5	-1.9	0.5
145.49	3.49	-1.0	353.38	19MR	0.8	-3.5	-2.0	0.5
143.50	3.03	-0.8	3.09	29MR	0.4	-3.4	-2.0	0.4
142.84	2.55	-0.6	12.75	8AP	-0.2	-3.4	-2.1	0.4
143.44	2.09	-0.3	22.36	18AP	-1.0	-3.4	-2.1	0.3
145.18	1.67	-0.1	31.94	28AP	-1.9	-3.4	-2.1	0.4
147.87	1.29	0.0	41.49	8MY	-1.1	-3.3	-2.2	0.4
151.34	0.95	0.2	51.02	18MY	-0.2	-3.3	-2.1	0.5
155.46	0.65	0.3	60.54	28MY	0.5	-3.3	-2.1	0.6
160.11	0.38	0.5	70.05	7JN	1.0	-3.3	-2.1	0.6
165.21	0.13	0.6	79.58	17JN	1.8	-3.3	-2.0	0.7
170.70	-0.09	0.7	89.12	27JN	3.0	-3.4	-2.0	0.7
176.51	-0.29	0.7	98.69	7JL	2.2	-3.4	-1.9	0.8
182.61	-0.47	0.8	108.29	17JL	0.4	-3.4	-1.9	0.8
188.96	-0.63	0.9	117.93	27JL	-0.9	-3.5	-1.8	0.8
195.54	-0.77	0.9	127.62	6AU	-1.3	-3.5	-1.7	0.9
202.32	-0.90	1.0	137.37	16AU	-1.1	-3.6	-1.7	0.9
209.29	-1.01	1.0	147.18	26AU	-0.5	-3.6	-1.6	0.9
216.42	-1.11	1.1	157.04	5SE	-0.2	-3.7	-1.6	0.9
223.69	-1.18	1.1	166.96	15SE	-0.1	-3.8	-1.5	0.9
231.10	-1.25	1.1	176.94	25SE	0.0	-3.9	-1.5	0.8
238.61	-1.29	1.2	186.98	5OC	0.3	-4.0	-1.4	0.8
246.22	-1.32	1.2	197.07	15OC	1.9	-4.1	-1.4	0.8
253.91	-1.34	1.2	207.19	25OC	1.4	-4.2	-1.4	0.7
261.66	-1.34	1.3	217.36	4NO	-0.1	-4.3	-1.4	0.7
269.45	-1.33	1.3	227.55	14NO	-0.4	-4.4	-1.4	0.8
277.26	-1.30	1.3	237.76	24NO	-0.4	-4.4	-1.4	0.8
285.09	-1.26	1.3	247.97	4DE	-0.5	-4.2	-1.4	0.8
292.91	-1.21	1.4	258.18	14DE	-0.7	-3.6	-1.4	0.8
300.70	-1.14	1.4	268.36	24DE	-1.0	-3.2	-1.4	0.8

♂ LONG	LAT	MAG	☉ LONG	16.00UT	☿	♀	♃	♄
				-400	MAGNITUDES			
308.47	-1.07	1.4	278.52	3JA	-1.0	-3.9	-1.4	0.8
316.18	-0.99	1.5	288.65	13JA	-0.8	-4.2	-1.5	0.8
323.84	-0.90	1.5	298.72	23JA	0.5	-4.3	-1.5	0.8
331.44	-0.81	1.5	308.75	2FE	2.9	-4.3	-1.6	0.7
338.96	-0.72	1.6	318.72	12FE	1.8	-4.2	-1.6	0.7
346.41	-0.62	1.6	328.63	22FE	0.9	-4.1	-1.7	0.7
353.78	-0.52	1.6	338.48	3MR	0.5	-4.0	-1.8	0.6
1.07	-0.41	1.6	348.28	13MR	0.2	-3.9	-1.8	0.5
8.27	-0.31	1.6	358.01	23MR	-0.3	-3.8	-1.9	0.5
15.39	-0.21	1.7	7.69	2AP	-1.0	-3.7	-2.0	0.4
22.43	-0.10	1.7	17.33	12AP	-1.8	-3.6	-2.0	0.4
29.39	-0.00	1.7	26.93	22AP	-1.1	-3.5	-2.1	0.3
36.27	0.10	1.7	36.49	2MY	-0.1	-3.5	-2.2	0.3
43.08	0.20	1.7	46.03	12MY	0.7	-3.4	-2.2	0.4
49.82	0.30	1.8	55.55	22MY	1.4	-3.4	-2.2	0.4
56.49	0.39	1.8	65.07	1JN	2.3	-3.4	-2.3	0.5
63.11	0.48	1.9	74.59	11JN	3.3	-3.3	-2.3	0.5
69.66	0.58	1.9	84.12	21JN	1.8	-3.3	-2.3	0.6
76.16	0.66	1.9	93.67	1JL	0.3	-3.3	-2.2	0.7
82.61	0.75	2.0	103.26	11JL	-0.9	-3.3	-2.2	0.7
89.01	0.84	2.0	112.88	21JL	-1.5	-3.3	-2.2	0.8
95.37	0.92	2.0	122.55	31JL	-1.1	-3.3	-2.1	0.8
101.68	1.00	2.0	132.26	10AU	-0.5	-3.4	-2.0	0.8
107.94	1.09	2.0	142.03	20AU	-0.1	-3.4	-2.0	0.8
114.16	1.17	1.9	151.86	30AU	0.1	-3.4	-1.9	0.8
120.34	1.25	1.9	161.76	9SE	0.2	-3.4	-1.8	0.8
126.46	1.33	1.9	171.70	19SE	0.6	-3.5	-1.8	0.8
132.52	1.41	1.8	181.71	29SE	2.2	-3.5	-1.7	0.8
138.53	1.49	1.8	191.77	9OC	1.2	-3.5	-1.7	0.8
144.46	1.57	1.7	201.88	19OC	-0.3	-3.5	-1.6	0.8
150.32	1.65	1.6	212.02	29OC	-0.5	-3.4	-1.6	0.7
156.07	1.73	1.6	222.20	8NO	-0.5	-3.4	-1.5	0.7
161.71	1.81	1.5	232.40	18NO	-0.6	-3.4	-1.5	0.7
167.22	1.90	1.3	242.61	28NO	-0.8	-3.4	-1.5	0.8
172.55	1.98	1.2	252.83	8DE	-0.9	-3.3	-1.5	0.8
177.68	2.07	1.0	263.02	18DE	-0.9	-3.3	-1.5	0.8
182.57	2.15	0.9	273.20	28DE	-0.7	-3.3	-1.5	0.8
				-399				
187.14	2.23	0.7	283.34	7JA	0.6	-3.3	-1.5	0.8
191.34	2.31	0.5	293.44	17JA	3.1	-3.4	-1.5	0.8
195.06	2.38	0.3	303.50	27JA	1.4	-3.4	-1.5	0.8
198.17	2.44	0.0	313.50	6FE	0.7	-3.4	-1.5	0.8
200.54	2.47	-0.3	323.44	16FE	0.4	-3.4	-1.6	0.7
201.99	2.46	-0.6	333.33	26FE	0.1	-3.4	-1.6	0.7
202.32	2.40	-0.9	343.15	8MR	-0.3	-3.5	-1.7	0.6
201.43	2.26	-1.2	352.91	18MR	-1.0	-3.5	-1.7	0.6
199.29	2.01	-1.5	2.63	28MR	-1.8	-3.6	-1.8	0.5
196.19	1.65	-1.7	12.29	7AP	-1.1	-3.6	-1.9	0.5
192.69	1.18	-1.8	21.90	17AP	-0.0	-3.7	-1.9	0.4
189.50	0.66	-1.6	31.48	27AP	0.9	-3.8	-2.0	0.3
187.28	0.14	-1.4	41.03	7MY	1.8	-3.9	-2.1	0.3
186.36	-0.34	-1.2	50.56	17MY	3.1	-4.0	-2.1	0.3
186.80	-0.74	-1.0	60.08	27MY	2.8	-4.1	-2.2	0.4
188.51	-1.08	-0.8	69.60	6JN	1.4	-4.1	-2.3	0.4
191.30	-1.36	-0.6	79.12	16JN	0.2	-4.2	-2.3	0.5
195.00	-1.58	-0.5	88.66	26JN	-0.9	-4.2	-2.4	0.5
199.43	-1.75	-0.3	98.23	6JL	-1.6	-4.0	-2.4	0.6
204.48	-1.89	-0.2	107.83	16JL	-1.1	-3.5	-2.4	0.6
210.02	-1.98	-0.0	117.47	26JL	-0.4	-3.2	-2.4	0.7
215.98	-2.05	0.1	127.16	5AU	-0.0	-3.8	-2.4	0.7
222.29	-2.09	0.2	136.90	15AU	0.2	-4.1	-2.3	0.8
228.88	-2.10	0.3	146.70	25AU	0.4	-4.3	-2.3	0.8
235.71	-2.08	0.4	156.56	4SE	0.9	-4.3	-2.2	0.8
242.74	-2.05	0.5	166.48	14SE	2.6	-4.2	-2.1	0.8
249.93	-1.99	0.6	176.46	24SE	1.1	-4.1	-2.1	0.8
257.25	-1.92	0.7	186.49	4OC	-0.4	-4.0	-2.0	0.8
264.66	-1.83	0.8	196.57	14OC	-0.7	-4.0	-1.9	0.8
272.14	-1.73	0.9	206.70	24OC	-0.7	-3.9	-1.9	0.8
279.67	-1.62	0.9	216.86	3NO	-0.7	-3.8	-1.8	0.7
287.23	-1.50	1.0	227.05	13NO	-0.7	-3.7	-1.8	0.7
294.80	-1.37	1.1	237.26	23NO	-0.7	-3.6	-1.7	0.7
302.35	-1.24	1.2	247.47	3DE	-0.7	-3.6	-1.7	0.7
309.88	-1.11	1.2	257.68	13DE	-0.5	-3.5	-1.6	0.8
317.38	-0.97	1.3	267.87	23DE	0.8	-3.5	-1.6	0.8

♂			☉	16.00UT	☿	♀	♃	♄
LONG	LAT	MAG	LONG	-398	MAGNITUDES			
324.82	-0.83	1.4	278.03	2JA	2.8	-3.4	-1.6	0.8
332.21	-0.70	1.4	288.16	12JA	1.0	-3.4	-1.6	0.8
339.54	-0.57	1.5	298.24	22JA	0.4	-3.4	-1.5	0.8
346.80	-0.44	1.6	308.26	1FE	0.2	-3.3	-1.5	0.8
353.99	-0.31	1.6	318.24	11FE	0.0	-3.3	-1.5	0.8
1.11	-0.19	1.7	328.16	21FE	-0.4	-3.3	-1.6	0.8
8.16	-0.08	1.7	338.01	3MR	-1.0	-3.3	-1.6	0.7
15.14	0.03	1.8	347.80	13MR	-1.7	-3.3	-1.6	0.7
22.04	0.14	1.8	357.54	23MR	-1.1	-3.3	-1.6	0.6
28.88	0.24	1.8	7.23	2AP	0.0	-3.3	-1.7	0.6
35.66	0.34	1.8	16.87	12AP	1.1	-3.4	-1.7	0.5
42.37	0.43	1.9	26.47	22AP	2.4	-3.4	-1.8	0.4
49.03	0.51	1.9	36.03	2MY	3.7	-3.4	-1.9	0.4
55.64	0.60	1.9	45.57	12MY	2.2	-3.5	-1.9	0.3
62.20	0.67	1.9	55.09	22MY	1.1	-3.5	-2.0	0.3
68.72	0.74	1.9	64.61	1JN	0.1	-3.4	-2.1	0.3
75.20	0.81	1.9	74.13	11JN	-0.9	-3.4	-2.1	0.4
81.65	0.88	1.9	83.66	21JN	-1.7	-3.4	-2.2	0.4
88.07	0.94	1.9	93.21	1JL	-1.1	-3.4	-2.3	0.5
94.47	0.99	2.0	102.79	11JL	-0.4	-3.3	-2.3	0.5
100.85	1.05	2.0	112.41	21JL	0.1	-3.3	-2.4	0.6
107.21	1.09	2.0	122.07	31JL	0.3	-3.3	-2.4	0.6
113.56	1.14	2.0	131.79	10AU	0.6	-3.3	-2.4	0.7
119.90	1.18	2.0	141.56	20AU	1.2	-3.3	-2.4	0.7
126.23	1.22	2.0	151.38	30AU	3.0	-3.3	-2.4	0.8
132.55	1.26	2.0	161.27	9SE	0.9	-3.4	-2.4	0.8
138.87	1.29	2.0	171.22	19SE	-0.6	-3.4	-2.4	0.8
145.18	1.32	2.0	181.22	29SE	-0.8	-3.4	-2.3	0.8
151.49	1.34	1.9	191.28	9OC	-0.8	-3.5	-2.3	0.8
157.78	1.36	1.9	201.39	19OC	-0.8	-3.5	-2.2	0.8
164.07	1.38	1.8	211.53	29OC	-0.6	-3.6	-2.1	0.8
170.34	1.39	1.8	221.71	8NO	-0.6	-3.7	-2.1	0.7
176.59	1.39	1.7	231.91	18NO	-0.6	-3.7	-2.0	0.7
182.81	1.39	1.6	242.12	28NO	-0.4	-3.8	-1.9	0.7
189.00	1.38	1.5	252.33	8DE	1.0	-3.9	-1.9	0.7
195.15	1.36	1.4	262.53	18DE	2.4	-4.0	-1.8	0.8
201.24	1.33	1.3	272.70	28DE	0.7	-4.1	-1.7	0.8
				-397				
207.27	1.28	1.2	282.85	7JA	0.2	-4.2	-1.7	0.8
213.22	1.22	1.0	292.95	17JA	0.1	-4.3	-1.7	0.8
219.07	1.14	0.9	303.01	27JA	-0.1	-4.3	-1.6	0.8
224.79	1.04	0.7	313.01	6FE	-0.4	-4.3	-1.6	0.8
230.35	0.91	0.5	322.96	16FE	-1.0	-4.0	-1.6	0.8
235.72	0.75	0.3	332.84	26FE	-1.6	-3.5	-1.6	0.8
240.84	0.55	0.1	342.67	8MR	-1.0	-3.3	-1.6	0.8
245.63	0.30	-0.1	352.44	18MR	0.1	-3.8	-1.6	0.7
250.03	-0.01	-0.4	2.15	28MR	1.5	-4.1	-1.6	0.7
253.91	-0.39	-0.7	11.82	7AP	3.1	-4.2	-1.6	0.6
257.13	-0.85	-1.0	21.43	17AP	2.9	-4.2	-1.6	0.5
259.52	-1.40	-1.3	31.01	27AP	1.6	-4.1	-1.7	0.5
260.89	-2.05	-1.6	40.57	7MY	0.8	-4.0	-1.7	0.4
261.08	-2.78	-1.9	50.10	17MY	0.0	-3.9	-1.7	0.4
260.03	-3.56	-2.2	59.62	27MY	-1.0	-3.8	-1.8	0.3
257.91	-4.31	-2.4	69.14	6JN	-1.7	-3.7	-1.8	0.3
255.21	-4.91	-2.5	78.66	16JN	-1.1	-3.6	-1.9	0.3
252.62	-5.29	-2.4	88.20	26JN	-0.3	-3.6	-2.0	0.4
250.83	-5.43	-2.2	97.77	6JL	0.2	-3.5	-2.0	0.4
250.28	-5.36	-2.0	107.36	16JL	0.5	-3.5	-2.1	0.5
251.07	-5.13	-1.7	117.00	26JL	0.8	-3.4	-2.2	0.6
253.13	-4.82	-1.5	126.69	5AU	1.6	-3.4	-2.2	0.6
256.28	-4.47	-1.2	136.43	15AU	3.2	-3.4	-2.3	0.7
260.31	-4.09	-1.0	146.23	25AU	0.8	-3.4	-2.3	0.7
265.04	-3.71	-0.7	156.09	4SE	-0.7	-3.4	-2.4	0.7
270.32	-3.34	-0.5	166.00	14SE	-1.0	-3.4	-2.4	0.8
276.02	-2.98	-0.3	175.97	24SE	-1.0	-3.4	-2.4	0.8
282.06	-2.64	-0.1	186.01	4OC	-0.8	-3.4	-2.4	0.8
288.33	-2.31	0.0	196.08	14OC	-0.5	-3.4	-2.4	0.8
294.80	-2.01	0.2	206.21	24OC	-0.4	-3.4	-2.4	0.8
301.40	-1.72	0.4	216.37	3NO	-0.4	-3.4	-2.3	0.8
308.10	-1.45	0.5	226.56	13NO	-0.2	-3.4	-2.3	0.8
314.86	-1.20	0.7	236.77	23NO	1.2	-3.4	-2.2	0.7
321.66	-0.97	0.8	246.98	3DE	2.1	-3.5	-2.1	0.7
328.49	-0.75	1.0	257.18	13DE	0.4	-3.5	-2.0	0.7
335.31	-0.56	1.1	267.37	23DE	0.1	-3.5	-2.0	0.8

♂			☉	16.00UT	☿	♀	♃	♄
LONG	LAT	MAG	LONG	-396	MAGNITUDES			
342.14	-0.37	1.2	277.54	2JA	-0.0	-3.5	-1.9	0.8
348.94	-0.21	1.3	287.66	12JA	-0.2	-3.4	-1.8	0.8
355.71	-0.06	1.4	297.74	22JA	-0.5	-3.4	-1.8	0.8
2.46	0.08	1.5	307.77	1FE	-1.1	-3.4	-1.7	0.9
9.16	0.21	1.6	317.75	11FE	-1.4	-3.4	-1.7	0.9
15.83	0.33	1.6	327.67	21FE	-1.0	-3.3	-1.6	0.8
22.46	0.43	1.7	337.53	2MR	0.2	-3.3	-1.6	0.8
29.05	0.53	1.8	347.32	12MR	1.8	-3.3	-1.6	0.8
35.61	0.62	1.8	357.06	22MR	3.6	-3.3	-1.5	0.8
42.13	0.69	1.9	6.76	1AP	2.1	-3.3	-1.5	0.7
48.61	0.77	1.9	16.39	11AP	1.2	-3.4	-1.5	0.7
55.06	0.83	2.0	26.00	21AP	0.6	-3.4	-1.5	0.6
61.49	0.89	2.0	35.56	1MY	-0.0	-3.4	-1.5	0.6
67.90	0.94	2.0	45.10	11MY	-1.0	-3.4	-1.5	0.5
74.28	0.98	2.0	54.63	21MY	-1.8	-3.5	-1.6	0.4
80.65	1.02	2.0	64.15	31MY	-1.1	-3.5	-1.6	0.4
87.01	1.06	2.0	73.66	10JN	-0.3	-3.6	-1.6	0.3
93.37	1.09	2.0	83.20	20JN	0.3	-3.7	-1.7	0.3
99.72	1.11	2.0	92.75	30JN	0.7	-3.7	-1.7	0.4
106.08	1.14	2.0	102.33	10JL	1.1	-3.8	-1.8	0.4
112.45	1.15	2.0	111.95	20JL	2.0	-3.9	-1.8	0.5
118.83	1.16	2.0	121.61	30JL	2.9	-4.0	-1.9	0.5
125.22	1.17	2.0	131.32	9AU	0.7	-4.1	-2.0	0.6
131.64	1.17	2.0	141.09	19AU	-0.7	-4.2	-2.0	0.6
138.08	1.17	2.0	150.91	29AU	-1.1	-4.3	-2.1	0.7
144.54	1.16	2.0	160.79	8SE	-1.1	-4.3	-2.1	0.7
151.03	1.15	2.0	170.74	18SE	-0.7	-4.1	-2.2	0.8
157.55	1.13	2.0	180.74	28SE	-0.4	-3.6	-2.3	0.8
164.11	1.11	1.9	190.79	8OC	-0.3	-3.3	-2.3	0.8
170.69	1.08	1.9	200.90	18OC	-0.2	-3.9	-2.3	0.8
177.31	1.04	1.9	211.04	28OC	-0.0	-4.3	-2.3	0.8
183.96	1.00	1.8	221.21	7NO	1.4	-4.4	-2.3	0.8
190.65	0.94	1.8	231.42	17NO	1.8	-4.4	-2.3	0.8
197.37	0.88	1.7	241.62	27NO	0.2	-4.3	-2.3	0.8
204.12	0.81	1.7	251.83	7DE	-0.1	-4.2	-2.2	0.8
210.90	0.73	1.6	262.04	17DE	-0.2	-4.1	-2.2	0.7
217.70	0.64	1.5	272.21	27DE	-0.3	-4.0	-2.1	0.8
				-395				
224.54	0.53	1.4	282.36	6JA	-0.6	-3.9	-2.0	0.8
231.39	0.41	1.3	292.47	16JA	-1.1	-3.8	-2.0	0.8
238.27	0.27	1.2	302.53	26JA	-1.3	-3.7	-1.9	0.9
245.16	0.11	1.1	312.53	5FE	-0.9	-3.6	-1.8	0.9
252.06	-0.06	1.0	322.48	15FE	0.3	-3.5	-1.7	0.9
258.95	-0.25	0.9	332.37	25FE	2.2	-3.5	-1.7	0.9
265.84	-0.46	0.8	342.20	7MR	2.9	-3.4	-1.6	0.9
272.70	-0.70	0.6	351.97	17MR	1.6	-3.4	-1.6	0.9
279.52	-0.95	0.5	1.69	27MR	0.9	-3.4	-1.5	0.8
286.29	-1.23	0.3	11.35	6AP	0.5	-3.3	-1.5	0.8
292.95	-1.53	0.2	20.97	16AP	-0.1	-3.3	-1.5	0.8
299.49	-1.85	0.0	30.56	26AP	-1.0	-3.3	-1.5	0.7
305.86	-2.20	-0.1	40.11	6MY	-1.8	-3.3	-1.4	0.6
312.00	-2.57	-0.3	49.64	16MY	-1.1	-3.3	-1.4	0.6
317.83	-2.96	-0.5	59.16	26MY	-0.2	-3.3	-1.4	0.5
323.28	-3.36	-0.7	68.68	5JN	0.4	-3.3	-1.4	0.4
328.22	-3.79	-0.9	78.20	15JN	0.9	-3.4	-1.4	0.4
332.50	-4.23	-1.1	87.74	25JN	1.5	-3.4	-1.5	0.3
335.96	-4.67	-1.3	97.30	5JL	2.6	-3.4	-1.5	0.4
338.37	-5.10	-1.5	106.90	15JL	2.5	-3.5	-1.5	0.4
339.54	-5.47	-1.8	116.54	25JL	0.5	-3.5	-1.6	0.5
339.31	-5.75	-2.0	126.22	4AU	-0.8	-3.5	-1.6	0.5
337.67	-5.85	-2.2	135.96	14AU	-1.3	-3.4	-1.7	0.6
334.98	-5.70	-2.4	145.75	24AU	-1.1	-3.4	-1.7	0.6
331.88	-5.28	-2.4	155.60	3SE	-0.6	-3.4	-1.8	0.7
329.17	-4.66	-2.1	165.52	13SE	-0.3	-3.4	-1.8	0.7
327.48	-3.92	-1.8	175.49	23SE	-0.1	-3.3	-1.9	0.8
327.08	-3.18	-1.5	185.51	3OC	-0.1	-3.3	-2.0	0.8
327.98	-2.50	-1.2	195.59	13OC	0.2	-3.3	-2.0	0.8
330.03	-1.89	-0.8	205.71	23OC	1.6	-3.3	-2.1	0.8
333.01	-1.38	-0.5	215.87	2NO	1.6	-3.3	-2.1	0.8
336.73	-0.95	-0.3	226.06	12NO	-0.0	-3.4	-2.2	0.9
341.02	-0.58	-0.0	236.27	22NO	-0.3	-3.4	-2.2	0.9
345.75	-0.28	0.2	246.48	2DE	-0.3	-3.4	-2.2	0.8
350.82	-0.02	0.4	256.69	12DE	-0.4	-3.5	-2.2	0.8
356.16	0.20	0.6	266.88	22DE	-0.7	-3.5	-2.2	0.8

♂			☉	16.00UT	☿	♀	♃	♄
LONG	LAT	MAG	LONG	-394	MAGNITUDES			
1.69	0.38	0.8	277.04	1JA	-1.0	-3.5	-2.2	0.8
7.38	0.53	1.0	287.17	11JA	-1.1	-3.6	-2.1	0.8
13.19	0.66	1.1	297.26	21JA	-0.9	-3.7	-2.0	0.9
19.09	0.77	1.3	307.29	31JA	0.4	-3.7	-2.0	0.9
25.06	0.87	1.4	317.27	10FE	2.7	-3.8	-1.9	0.9
31.07	0.94	1.5	327.19	20FE	2.2	-3.9	-1.8	1.0
37.13	1.01	1.6	337.05	2MR	1.1	-4.0	-1.8	1.0
43.21	1.06	1.7	346.85	12MR	0.7	-4.1	-1.7	1.0
49.32	1.11	1.7	356.60	22MR	0.3	-4.1	-1.6	1.0
55.45	1.15	1.8	6.29	1AP	-0.2	-4.2	-1.6	0.9
61.59	1.18	1.9	15.93	11AP	-1.0	-4.2	-1.5	0.9
67.75	1.20	1.9	25.53	21AP	-1.9	-4.1	-1.5	0.9
73.92	1.21	2.0	35.10	1MY	-1.1	-3.6	-1.4	0.8
80.11	1.22	2.0	44.64	11MY	-0.2	-2.8	-1.4	0.8
86.31	1.23	2.0	54.17	21MY	0.6	-3.4	-1.4	0.7
92.54	1.23	2.0	63.69	31MY	1.1	-3.9	-1.4	0.6
98.78	1.22	2.0	73.20	10JN	2.0	-4.2	-1.3	0.6
105.05	1.21	2.0	82.73	20JN	3.2	-4.2	-1.3	0.5
111.36	1.20	2.0	92.28	30JN	2.1	-4.1	-1.3	0.4
117.69	1.18	2.0	101.86	10JL	0.4	-4.1	-1.3	0.4
124.06	1.16	2.0	111.48	20JL	-0.9	-4.0	-1.4	0.4
130.47	1.13	2.0	121.14	30JL	-1.4	-3.9	-1.4	0.5
136.93	1.10	1.9	130.85	9AU	-1.1	-3.8	-1.4	0.5
143.43	1.06	1.9	140.61	19AU	-0.5	-3.7	-1.4	0.6
149.98	1.02	1.9	150.43	29AU	-0.2	-3.7	-1.5	0.6
156.59	0.97	1.9	160.31	8SE	-0.0	-3.6	-1.5	0.7
163.25	0.92	1.9	170.25	18SE	0.1	-3.6	-1.6	0.7
169.97	0.86	1.9	180.25	28SE	0.4	-3.5	-1.6	0.8
176.75	0.80	1.9	190.30	8OC	2.0	-3.5	-1.7	0.8
183.60	0.73	1.8	200.40	18OC	1.4	-3.5	-1.8	0.8
190.50	0.65	1.8	210.55	28OC	-0.2	-3.4	-1.8	0.9
197.47	0.57	1.8	220.72	7NO	-0.4	-3.4	-1.9	0.9
204.50	0.48	1.8	230.92	17NO	-0.4	-3.4	-1.9	0.9
211.60	0.38	1.7	241.13	27NO	-0.5	-3.4	-2.0	0.9
218.75	0.28	1.7	251.34	7DE	-0.7	-3.4	-2.0	0.9
225.97	0.17	1.6	261.54	17DE	-0.9	-3.3	-2.1	0.9
233.24	0.05	1.6	271.72	27DE	-1.0	-3.3	-2.1	0.9
				-393				
240.57	-0.08	1.5	281.87	6JA	-0.8	-3.3	-2.1	0.9
247.94	-0.21	1.5	291.98	16JA	0.5	-3.4	-2.1	0.9
255.36	-0.35	1.4	302.04	26JA	3.0	-3.4	-2.1	1.0
262.82	-0.50	1.3	312.05	5FE	1.7	-3.4	-2.0	1.0
270.31	-0.65	1.3	322.00	15FE	0.8	-3.4	-2.0	1.0
277.83	-0.81	1.2	331.89	25FE	0.5	-3.5	-1.9	1.1
285.35	-0.96	1.1	341.72	7MR	0.2	-3.5	-1.9	1.1
292.88	-1.12	1.1	351.49	17MR	-0.3	-3.5	-1.8	1.1
300.40	-1.28	1.0	1.21	27MR	-1.0	-3.4	-1.7	1.1
307.88	-1.43	0.9	10.88	6AP	-1.8	-3.4	-1.7	1.1
315.32	-1.58	0.9	20.50	16AP	-1.1	-3.4	-1.6	1.0
322.70	-1.72	0.8	30.09	26AP	-0.1	-3.4	-1.5	1.0
329.99	-1.85	0.7	39.64	6MY	0.7	-3.3	-1.5	1.0
337.17	-1.96	0.6	49.17	16MY	1.5	-3.3	-1.4	0.9
344.22	-2.07	0.6	58.69	26MY	2.6	-3.3	-1.4	0.9
351.10	-2.16	0.5	68.21	5JN	3.2	-3.3	-1.3	0.8
357.79	-2.24	0.4	77.73	15JN	1.7	-3.3	-1.3	0.7
4.23	-2.30	0.3	87.27	25JN	0.3	-3.4	-1.3	0.7
10.39	-2.34	0.2	96.83	5JL	-0.9	-3.4	-1.3	0.6
16.21	-2.36	0.1	106.43	15JL	-1.5	-3.4	-1.3	0.5
21.61	-2.36	-0.0	116.07	25JL	-1.1	-3.5	-1.3	0.5
26.51	-2.33	-0.2	125.75	4AU	-0.5	-3.5	-1.3	0.5
30.78	-2.27	-0.4	135.48	14AU	-0.1	-3.6	-1.3	0.6
34.27	-2.18	-0.5	145.28	24AU	0.1	-3.6	-1.3	0.6
36.80	-2.05	-0.7	155.12	3SE	0.3	-3.7	-1.3	0.7
38.18	-1.86	-0.9	165.04	13SE	0.7	-3.8	-1.3	0.7
38.19	-1.60	-1.2	175.01	23SE	2.3	-3.9	-1.4	0.8
36.75	-1.26	-1.4	185.03	3OC	1.2	-4.0	-1.4	0.8
33.96	-0.85	-1.6	195.11	13OC	-0.3	-4.1	-1.5	0.9
30.31	-0.37	-1.7	205.23	23OC	-0.6	-4.2	-1.5	0.9
26.56	0.10	-1.6	215.38	2NO	-0.6	-4.3	-1.6	0.9
23.47	0.53	-1.3	225.57	12NO	-0.6	-4.4	-1.6	1.0
21.58	0.88	-1.0	235.78	22NO	-0.8	-4.4	-1.7	1.0
21.05	1.15	-0.7	245.98	2DE	-0.8	-4.1	-1.8	1.0
21.79	1.34	-0.4	256.19	12DE	-0.8	-3.6	-1.8	1.0
23.62	1.47	-0.1	266.38	22DE	-0.6	-3.2	-1.9	1.0

♂			☉	16.00UT	☿	♀	♃	♄
LONG	LAT	MAG	LONG	-392	MAGNITUDES			
26.33	1.57	0.2	276.55	1JA	0.6	-3.9	-2.0	1.0
29.73	1.63	0.4	286.68	11JA	3.0	-4.2	-2.0	1.0
33.68	1.66	0.6	296.77	21JA	1.2	-4.3	-2.0	1.0
38.05	1.68	0.8	306.80	31JA	0.6	-4.3	-2.0	1.0
42.75	1.68	1.0	316.79	10FE	0.3	-4.2	-2.0	1.1
47.71	1.68	1.2	326.71	20FE	0.1	-4.1	-2.0	1.1
52.89	1.67	1.3	336.57	1MR	-0.3	-4.0	-2.0	1.2
58.23	1.65	1.4	346.38	11MR	-1.0	-3.9	-2.0	1.2
63.72	1.62	1.5	356.13	21MR	-1.7	-3.8	-1.9	1.2
69.32	1.60	1.6	5.82	31MR	-1.1	-3.7	-1.8	1.2
75.02	1.56	1.7	15.47	10AP	-0.0	-3.6	-1.8	1.2
80.80	1.53	1.8	25.07	20AP	1.0	-3.5	-1.7	1.2
86.66	1.49	1.8	34.64	30AP	2.0	-3.5	-1.7	1.2
92.60	1.44	1.9	44.18	10MY	3.4	-3.4	-1.6	1.1
98.60	1.40	1.9	53.71	20MY	2.6	-3.4	-1.5	1.1
104.66	1.35	1.9	63.22	30MY	1.3	-3.4	-1.5	1.0
110.78	1.30	2.0	72.75	9JN	0.2	-3.3	-1.4	1.0
116.97	1.24	2.0	82.27	19JN	-0.9	-3.3	-1.4	0.9
123.23	1.18	2.0	91.82	29JN	-1.6	-3.3	-1.3	0.8
129.54	1.12	2.0	101.40	9JL	-1.1	-3.3	-1.3	0.8
135.93	1.06	2.0	111.01	19JL	-0.4	-3.3	-1.3	0.7
142.39	0.99	1.9	120.67	29JL	-0.0	-3.3	-1.2	0.6
148.91	0.92	1.9	130.38	8AU	0.2	-3.4	-1.2	0.6
155.52	0.85	1.9	140.14	18AU	0.5	-3.4	-1.2	0.7
162.19	0.77	1.9	149.96	28AU	1.0	-3.4	-1.2	0.7
168.95	0.68	1.8	159.84	7SE	2.7	-3.4	-1.2	0.8
175.79	0.60	1.8	169.77	17SE	1.1	-3.5	-1.2	0.8
182.71	0.51	1.7	179.77	27SE	-0.5	-3.5	-1.3	0.9
189.71	0.42	1.7	189.82	7OC	-0.7	-3.5	-1.3	0.9
196.79	0.32	1.7	199.91	17OC	-0.7	-3.5	-1.3	1.0
203.96	0.22	1.6	210.05	27OC	-0.8	-3.4	-1.4	1.0
211.21	0.11	1.6	220.23	6NO	-0.7	-3.4	-1.4	1.0
218.53	0.01	1.6	230.42	16NO	-0.6	-3.4	-1.5	1.1
225.93	-0.10	1.6	240.63	26NO	-0.7	-3.4	-1.5	1.1
233.39	-0.21	1.6	250.84	6DE	-0.5	-3.3	-1.6	1.1
240.93	-0.33	1.6	261.04	16DE	0.8	-3.3	-1.7	1.2
248.52	-0.44	1.5	271.22	26DE	2.7	-3.3	-1.7	1.2
				-391				
256.17	-0.56	1.5	281.37	5JA	0.9	-3.3	-1.8	1.2
263.85	-0.67	1.5	291.48	15JA	0.4	-3.4	-1.8	1.2
271.58	-0.78	1.5	301.55	25JA	0.2	-3.4	-1.9	1.2
279.33	-0.88	1.4	311.56	4FE	-0.0	-3.4	-2.0	1.2
287.08	-0.98	1.4	321.51	14FE	-0.4	-3.4	-2.0	1.2
294.85	-1.08	1.4	331.41	24FE	-1.0	-3.4	-2.0	1.3
302.60	-1.16	1.3	341.25	6MR	-1.6	-3.5	-2.0	1.3
310.33	-1.24	1.3	351.02	16MR	-1.1	-3.5	-2.0	1.3
318.03	-1.30	1.3	0.74	26MR	0.0	-3.6	-2.0	1.4
325.67	-1.35	1.3	10.42	5AP	1.2	-3.6	-2.0	1.4
333.25	-1.39	1.2	20.04	15AP	2.6	-3.7	-1.9	1.4
340.76	-1.41	1.2	29.63	25AP	3.4	-3.8	-1.9	1.4
348.18	-1.42	1.2	39.18	5MY	2.0	-3.9	-1.8	1.4
355.50	-1.42	1.2	48.72	15MY	1.0	-4.0	-1.8	1.3
2.71	-1.40	1.1	58.24	25MY	0.1	-4.1	-1.7	1.3
9.79	-1.37	1.1	67.76	4JN	-0.9	-4.1	-1.6	1.2
16.74	-1.32	1.1	77.27	14JN	-1.7	-4.2	-1.6	1.2
23.52	-1.25	1.0	86.81	24JN	-1.1	-4.2	-1.5	1.1
30.14	-1.17	1.0	96.37	4JL	-0.4	-3.9	-1.5	1.1
36.56	-1.08	0.9	105.96	14JL	0.1	-3.4	-1.4	1.0
42.77	-0.97	0.9	115.60	24JL	0.4	-3.2	-1.4	0.9
48.72	-0.84	0.8	125.28	3AU	0.7	-3.8	-1.3	0.9
54.39	-0.69	0.7	135.01	13AU	1.3	-4.1	-1.3	0.8
59.73	-0.53	0.6	144.80	23AU	3.1	-4.3	-1.3	0.8
64.66	-0.33	0.5	154.65	2SE	0.9	-4.3	-1.2	0.8
69.11	-0.12	0.4	164.55	12SE	-0.6	-4.2	-1.2	0.9
72.97	0.14	0.2	174.52	22SE	-0.9	-4.1	-1.2	0.9
76.10	0.42	0.0	184.54	2OC	-0.9	-4.0	-1.2	1.0
78.34	0.76	-0.2	194.61	12OC	-0.8	-4.0	-1.2	1.0
79.48	1.14	-0.4	204.73	22OC	-0.6	-3.9	-1.3	1.1
79.34	1.56	-0.6	214.89	1NO	-0.5	-3.8	-1.3	1.1
77.82	2.02	-0.8	225.08	11NO	-0.5	-3.7	-1.3	1.2
74.99	2.46	-1.0	235.28	21NO	-0.3	-3.6	-1.4	1.2
71.25	2.84	-1.2	245.49	1DE	1.0	-3.6	-1.4	1.3
67.29	3.10	-1.1	255.70	11DE	2.4	-3.5	-1.5	1.3
63.87	3.23	-0.9	265.90	21DE	0.6	-3.5	-1.5	1.3
61.54	3.24	-0.6	276.06	31DE	0.2	-3.4	-1.6	1.3

♂ LONG	♂ LAT	♂ MAG	⊙ LONG	16.00UT -390	☿ MAGNITUDES	♀	♃	♄
60.53	3.16	-0.3	286.19	10JA	0.0	-3.4	-1.6	1.4
60.80	3.04	-0.1	296.28	20JA	-0.1	-3.4	-1.7	1.4
62.19	2.89	0.2	306.32	30JA	-0.5	-3.3	-1.8	1.3
64.50	2.74	0.4	316.31	9FE	-1.0	-3.3	-1.8	1.2
67.54	2.59	0.6	326.23	19FE	-1.5	-3.3	-1.9	1.2
71.17	2.44	0.8	336.10	1MR	-1.0	-3.3	-2.0	1.2
75.27	2.30	1.0	345.91	11MR	0.1	-3.3	-2.0	1.2
79.75	2.17	1.1	355.66	21MR	1.6	-3.3	-2.0	1.2
84.53	2.04	1.3	5.35	31MR	3.4	-3.3	-2.1	1.2
89.56	1.92	1.4	15.00	10AP	2.6	-3.4	-2.1	1.2
94.81	1.80	1.5	24.61	20AP	1.5	-3.4	-2.1	1.2
100.23	1.69	1.5	34.18	30AP	0.8	-3.4	-2.0	1.1
105.81	1.58	1.6	43.72	10MY	0.0	-3.5	-2.0	1.1
111.54	1.47	1.7	53.25	20MY	-0.9	-3.5	-2.0	1.1
117.39	1.37	1.7	62.76	30MY	-1.8	-3.4	-1.9	1.0
123.36	1.26	1.8	72.28	9JN	-1.1	-3.4	-1.9	1.0
129.44	1.15	1.8	81.81	19JN	-0.3	-3.4	-1.8	0.9
135.63	1.05	1.8	91.35	29JN	0.2	-3.4	-1.7	0.9
141.93	0.94	1.8	100.93	9JL	0.5	-3.3	-1.7	0.8
148.33	0.84	1.8	110.54	19JL	0.9	-3.3	-1.6	0.8
154.83	0.73	1.8	120.19	29JL	1.7	-3.3	-1.5	0.7
161.44	0.63	1.8	129.90	8AU	3.2	-3.3	-1.5	0.7
168.15	0.52	1.8	139.66	18AU	0.8	-3.3	-1.4	0.6
174.96	0.41	1.8	149.47	28AU	-0.7	-3.3	-1.4	0.6
181.87	0.30	1.7	159.35	7SE	-1.0	-3.4	-1.4	0.6
188.89	0.19	1.7	169.28	17SE	-1.0	-3.4	-1.3	0.7
196.00	0.08	1.7	179.27	27SE	-0.7	-3.4	-1.3	0.8
203.22	-0.03	1.6	189.32	7OC	-0.5	-3.5	-1.3	0.9
210.53	-0.14	1.6	199.42	17OC	-0.4	-3.5	-1.3	0.9
217.92	-0.24	1.5	209.56	27OC	-0.3	-3.6	-1.3	1.0
225.41	-0.35	1.5	219.73	6NO	-0.1	-3.7	-1.3	1.0
232.98	-0.45	1.4	229.93	16NO	1.2	-3.7	-1.3	1.1
240.61	-0.55	1.4	240.14	26NO	2.1	-3.8	-1.3	1.1
248.32	-0.65	1.4	250.35	6DE	0.3	-3.9	-1.4	1.1
256.08	-0.74	1.4	260.55	16DE	-0.0	-4.0	-1.4	1.1
263.88	-0.82	1.4	270.73	26DE	-0.1	-4.1	-1.4	1.1
				-389				
271.71	-0.90	1.4	280.89	5JA	-0.2	-4.2	-1.5	1.1
279.57	-0.96	1.4	291.00	15JA	-0.5	-4.3	-1.5	1.1
287.43	-1.02	1.4	301.06	25JA	-1.0	-4.3	-1.6	1.1
295.30	-1.07	1.4	311.08	4FE	-1.4	-4.3	-1.7	1.0
303.14	-1.11	1.4	321.04	14FE	-1.0	-4.0	-1.7	1.0
310.96	-1.13	1.4	330.93	24FE	0.2	-3.5	-1.8	0.9
318.74	-1.14	1.4	340.77	6MR	1.9	-3.3	-1.9	0.9
326.46	-1.15	1.4	350.55	16MR	3.4	-3.9	-1.9	0.9
334.13	-1.13	1.4	0.27	26MR	1.9	-4.2	-2.0	0.9
341.72	-1.11	1.4	9.95	5AP	1.1	-4.2	-2.1	0.9
349.24	-1.08	1.5	19.57	15AP	0.6	-4.2	-2.1	0.9
356.66	-1.03	1.5	29.16	25AP	-0.1	-4.1	-2.1	0.9
4.00	-0.97	1.5	38.72	5MY	-0.9	-4.0	-2.2	0.9
11.24	-0.90	1.5	48.25	15MY	-1.8	-3.9	-2.2	0.9
18.37	-0.82	1.5	57.77	25MY	-1.1	-3.8	-2.2	0.8
25.39	-0.74	1.5	67.29	4JN	-0.3	-3.7	-2.1	0.8
32.30	-0.64	1.5	76.81	14JN	0.3	-3.6	-2.1	0.7
39.09	-0.53	1.5	86.35	24JN	0.7	-3.6	-2.0	0.7
45.76	-0.42	1.5	95.91	4JL	1.2	-3.5	-2.0	0.7
52.29	-0.30	1.4	105.50	14JL	2.2	-3.5	-1.9	0.6
58.68	-0.17	1.4	115.13	24JL	2.8	-3.4	-1.9	0.5
64.92	-0.03	1.4	124.81	3AU	0.7	-3.4	-1.8	0.5
70.99	0.12	1.3	134.54	13AU	-0.7	-3.4	-1.7	0.4
76.88	0.29	1.3	144.32	23AU	-1.2	-3.4	-1.7	0.4
82.56	0.46	1.2	154.17	2SE	-1.1	-3.4	-1.6	0.3
88.00	0.65	1.1	164.07	12SE	-0.7	-3.4	-1.6	0.3
93.15	0.86	1.0	174.03	22SE	-0.4	-3.4	-1.5	0.4
97.96	1.08	0.9	184.05	2OC	-0.2	-3.4	-1.5	0.4
102.37	1.34	0.8	194.12	12OC	-0.2	-3.4	-1.4	0.5
106.28	1.62	0.6	204.24	22OC	0.0	-3.4	-1.4	0.6
109.56	1.93	0.4	214.40	1NO	1.4	-3.4	-1.4	0.6
112.09	2.28	0.2	224.58	11NO	1.8	-3.4	-1.4	0.7
113.68	2.67	0.0	234.78	21NO	0.1	-3.4	-1.4	0.8
114.15	3.09	-0.2	245.00	1DE	-0.2	-3.5	-1.4	0.8
113.33	3.53	-0.5	255.20	11DE	-0.2	-3.5	-1.4	0.8
111.19	3.93	-0.7	265.40	21DE	-0.3	-3.5	-1.4	0.9
107.91	4.25	-0.9	275.57	31DE	-0.6	-3.5	-1.4	0.9

♂ LONG	♂ LAT	♂ MAG	⊙ LONG	16.00UT -388	☿ MAGNITUDES	♀	♃	♄
104.00	4.42	-1.0	285.70	10JA	-1.1	-3.4	-1.5	0.9
100.18	4.42	-0.9	295.79	20JA	-1.2	-3.4	-1.5	0.9
97.11	4.27	-0.6	305.84	30JA	-0.9	-3.4	-1.6	0.8
95.24	4.00	-0.4	315.82	9FE	0.3	-3.4	-1.6	0.8
94.67	3.69	-0.2	325.75	19FE	2.3	-3.3	-1.7	0.8
95.31	3.37	0.1	335.62	29FE	2.6	-3.3	-1.7	0.7
97.01	3.06	0.3	345.43	10MR	1.4	-3.3	-1.8	0.7
99.57	2.76	0.5	355.18	20MR	0.8	-3.3	-1.9	0.6
102.83	2.49	0.7	4.88	30MR	0.4	-3.3	-1.9	0.7
106.66	2.24	0.8	14.53	9AP	-0.1	-3.4	-2.0	0.7
110.95	2.01	1.0	24.14	19AP	-0.9	-3.4	-2.1	0.7
115.62	1.80	1.1	33.71	29AP	-1.8	-3.4	-2.1	0.7
120.61	1.60	1.2	43.26	9MY	-1.1	-3.4	-2.2	0.7
125.87	1.42	1.3	52.78	19MY	-0.2	-3.5	-2.2	0.7
131.38	1.24	1.3	62.31	29MY	0.4	-3.5	-2.3	0.6
137.08	1.07	1.4	71.82	8JN	1.0	-3.6	-2.3	0.6
142.98	0.91	1.4	81.35	18JN	1.6	-3.7	-2.3	0.6
149.05	0.76	1.5	90.90	28JN	2.9	-3.7	-2.3	0.5
155.29	0.61	1.5	100.47	8JL	2.4	-3.8	-2.3	0.5
161.67	0.46	1.5	110.08	18JL	0.5	-3.9	-2.2	0.4
168.20	0.32	1.5	119.73	28JL	-0.8	-4.0	-2.2	0.4
174.87	0.19	1.6	129.43	7AU	-1.3	-4.1	-2.1	0.3
181.68	0.06	1.6	139.19	17AU	-1.1	-4.2	-2.0	0.3
188.61	-0.07	1.6	149.00	27AU	-0.6	-4.3	-2.0	0.2
195.67	-0.19	1.5	158.87	6SE	-0.3	-4.3	-1.9	0.2
202.84	-0.31	1.5	168.80	16SE	-0.1	-4.1	-1.8	0.1
210.14	-0.42	1.5	178.79	26SE	-0.0	-3.5	-1.8	0.1
217.53	-0.52	1.5	188.83	6OC	0.3	-3.3	-1.7	0.1
225.03	-0.62	1.5	198.93	16OC	1.7	-3.9	-1.7	0.2
232.62	-0.71	1.5	209.07	26OC	1.6	-4.3	-1.6	0.3
240.29	-0.80	1.5	219.23	5NO	-0.1	-4.4	-1.6	0.3
248.03	-0.87	1.4	229.43	15NO	-0.3	-4.4	-1.5	0.4
255.83	-0.94	1.4	239.64	25NO	-0.4	-4.3	-1.5	0.5
263.68	-0.99	1.4	249.85	5DE	-0.4	-4.2	-1.5	0.5
271.55	-1.03	1.4	260.05	15DE	-0.7	-4.1	-1.5	0.6
279.45	-1.07	1.4	270.24	25DE	-1.0	-4.0	-1.5	0.6
				-387				
287.35	-1.09	1.3	280.39	4JA	-1.1	-3.9	-1.5	0.6
295.25	-1.09	1.3	290.51	14JA	-0.8	-3.8	-1.5	0.6
303.12	-1.09	1.3	300.58	24JA	0.4	-3.7	-1.5	0.6
310.96	-1.07	1.3	310.59	3FE	2.7	-3.6	-1.5	0.6
318.75	-1.05	1.3	320.56	13FE	2.0	-3.5	-1.6	0.6
326.49	-1.01	1.4	330.46	23FE	1.0	-3.5	-1.6	0.6
334.16	-0.96	1.4	340.30	5MR	0.6	-3.4	-1.7	0.5
341.76	-0.91	1.5	350.08	15MR	0.3	-3.4	-1.7	0.5
349.27	-0.84	1.5	359.81	25MR	-0.2	-3.4	-1.8	0.5
356.71	-0.77	1.5	9.48	4AP	-0.9	-3.3	-1.8	0.4
4.06	-0.69	1.6	19.12	14AP	-1.9	-3.3	-1.9	0.5
11.31	-0.60	1.6	28.71	24AP	-1.1	-3.3	-2.0	0.5
18.48	-0.51	1.6	38.26	4MY	-0.2	-3.3	-2.0	0.5
25.55	-0.41	1.7	47.80	14MY	0.6	-3.3	-2.1	0.5
32.53	-0.31	1.7	57.32	24MY	1.3	-3.3	-2.2	0.5
39.42	-0.21	1.7	66.83	3JN	2.2	-3.3	-2.2	0.5
46.21	-0.10	1.7	76.36	13JN	3.4	-3.4	-2.3	0.5
52.91	0.01	1.7	85.89	23JN	1.9	-3.4	-2.3	0.4
59.52	0.13	1.7	95.45	3JL	0.4	-3.4	-2.4	0.4
66.04	0.25	1.7	105.04	13JL	-0.8	-3.5	-2.4	0.4
72.46	0.37	1.7	114.67	23JL	-1.4	-3.5	-2.4	0.3
78.78	0.49	1.7	124.34	2AU	-1.1	-3.5	-2.4	0.3
85.01	0.62	1.7	134.07	12AU	-0.5	-3.4	-2.4	0.2
91.11	0.75	1.7	143.85	22AU	-0.2	-3.4	-2.3	0.2
97.11	0.89	1.6	153.68	1SE	0.0	-3.4	-2.3	0.1
102.97	1.04	1.6	163.59	11SE	0.2	-3.4	-2.2	0.0
108.68	1.19	1.5	173.54	21SE	0.5	-3.3	-2.1	-0.0
114.22	1.35	1.4	183.56	1OC	2.0	-3.3	-2.1	-0.1
119.56	1.53	1.3	193.63	11OC	1.4	-3.3	-2.0	-0.1
124.67	1.71	1.2	203.74	21OC	-0.2	-3.3	-1.9	-0.1
129.49	1.91	1.1	213.89	31OC	-0.5	-3.3	-1.9	0.0
133.97	2.13	1.0	224.08	10NO	-0.5	-3.4	-1.8	0.1
138.03	2.37	0.8	234.28	20NO	-0.6	-3.4	-1.8	0.2
141.57	2.64	0.6	244.49	30NO	-0.8	-3.4	-1.7	0.2
144.45	2.93	0.4	254.70	10DE	-0.9	-3.5	-1.7	0.3
146.53	3.25	0.2	264.90	20DE	-0.9	-3.5	-1.6	0.3
147.63	3.58	-0.1	275.07	30DE	-0.7	-3.5	-1.6	0.4

Left table (−386 / −385)

♂ LONG	♂ LAT	♂ MAG	☉ LONG	16.00UT	☿	♀	♃	♄
147.57	3.92	-0.4	285.21	9JA	0.5	-3.6	-1.6	0.4
146.22	4.22	-0.6	295.30	19JA	3.0	-3.7	-1.6	0.5
143.63	4.44	-0.9	305.35	29JA	1.5	-3.7	-1.5	0.5
140.09	4.52	-1.1	315.34	8FE	0.7	-3.8	-1.6	0.5
136.21	4.43	-1.1	325.27	18FE	0.4	-3.9	-1.6	0.5
132.70	4.16	-0.9	335.14	28FE	0.1	-4.0	-1.6	0.4
130.16	3.78	-0.7	344.96	10MR	-0.3	-4.1	-1.6	0.4
128.89	3.35	-0.5	354.71	20MR	-1.0	-4.1	-1.6	0.4
128.90	2.90	-0.3	4.42	30MR	-1.8	-4.2	-1.7	0.3
130.11	2.48	-0.1	14.07	9AP	-1.1	-4.2	-1.7	0.3
132.31	2.10	0.1	23.68	19AP	-0.1	-4.1	-1.8	0.3
135.34	1.74	0.3	33.25	29AP	-0.3	-4.0	-1.8	0.3
139.06	1.43	0.5	42.80	9MY	1.7	-2.8	-1.9	0.3
143.34	1.14	0.6	52.33	19MY	2.9	-3.4	-1.9	0.4
148.08	0.87	0.7	61.84	29MY	3.0	-3.9	-2.0	0.4
153.22	0.64	0.8	71.36	8JN	1.5	-4.2	-2.1	0.4
158.69	0.42	0.9	80.89	18JN	0.3	-4.2	-2.1	0.4
164.45	0.21	1.0	90.43	28JN	-0.9	-4.1	-2.2	0.4
170.48	0.03	1.0	100.01	8JL	-1.5	-4.1	-2.3	0.3
176.73	-0.15	1.1	109.61	18JL	-1.1	-4.0	-2.3	0.3
183.19	-0.31	1.1	119.26	28JL	-0.5	-3.9	-2.4	0.3
189.85	-0.45	1.2	128.96	7AU	-0.1	-3.8	-2.4	0.2
196.69	-0.59	1.2	138.71	17AU	0.2	-3.7	-2.4	0.2
203.68	-0.71	1.2	148.52	27AU	0.3	-3.7	-2.4	0.1
210.83	-0.82	1.2	158.39	6SE	0.8	-3.6	-2.4	0.1
218.11	-0.92	1.3	168.32	16SE	2.4	-3.6	-2.4	-0.0
225.52	-1.00	1.3	178.31	26SE	1.2	-3.5	-2.4	-0.1
233.03	-1.07	1.3	188.35	6OC	-0.4	-3.5	-2.3	-0.1
240.64	-1.13	1.3	198.44	16OC	-0.7	-3.5	-2.3	-0.2
248.34	-1.17	1.3	208.57	26OC	-0.6	-3.4	-2.2	-0.2
256.09	-1.20	1.3	218.74	5NO	-0.7	-3.4	-2.1	-0.2
263.90	-1.22	1.3	228.93	15NO	-0.8	-3.4	-2.0	-0.1
271.75	-1.22	1.4	239.14	25NO	-0.7	-3.4	-2.0	-0.0
279.61	-1.21	1.4	249.35	5DE	-0.8	-3.4	-1.9	0.0
287.48	-1.19	1.4	259.55	15DE	-0.6	-3.3	-1.8	0.1
295.33	-1.15	1.4	269.74	25DE	0.6	-3.3	-1.8	0.2

−385

♂ LONG	♂ LAT	♂ MAG	☉ LONG	16.00UT	☿	♀	♃	♄
303.16	-1.10	1.4	279.90	4JA	3.0	-3.4	-1.7	0.2
310.95	-1.05	1.4	290.01	14JA	1.1	-3.4	-1.7	0.3
318.70	-0.98	1.5	300.09	24JA	0.5	-3.4	-1.6	0.3
326.38	-0.91	1.5	310.10	3FE	0.3	-3.4	-1.6	0.3
334.00	-0.83	1.5	320.07	13FE	0.0	-3.4	-1.6	0.4
341.54	-0.74	1.5	329.97	23FE	-0.3	-3.5	-1.6	0.4
349.01	-0.65	1.5	339.82	5MR	-1.0	-3.5	-1.6	0.4
356.39	-0.56	1.5	349.60	15MR	-1.7	-3.5	-1.6	0.3
3.69	-0.46	1.6	359.33	25MR	-1.1	-3.4	-1.6	0.3
10.90	-0.36	1.6	9.01	4AP	-0.0	-3.4	-1.6	0.3
18.03	-0.26	1.6	18.64	14AP	1.0	-3.4	-1.6	0.2
25.07	-0.16	1.6	28.24	24AP	2.2	-3.4	-1.6	0.2
32.03	-0.06	1.7	37.79	4MY	3.7	-3.3	-1.7	0.2
38.91	0.04	1.7	47.33	14MY	2.3	-3.3	-1.7	0.2
45.72	0.15	1.8	56.85	24MY	1.2	-3.3	-1.7	0.3
52.45	0.25	1.8	66.37	3JN	0.2	-3.3	-1.8	0.3
59.11	0.35	1.9	75.89	13JN	-0.9	-3.3	-1.8	0.3
65.70	0.44	1.9	85.43	23JN	-1.6	-3.4	-1.9	0.3
72.23	0.54	1.9	94.98	3JL	-1.1	-3.4	-2.0	0.3
78.70	0.64	1.9	104.57	13JL	-0.4	-3.4	-2.0	0.3
85.10	0.74	1.9	114.20	23JL	0.0	-3.5	-2.1	0.3
91.45	0.83	1.9	123.87	2AU	0.3	-3.5	-2.2	0.3
97.73	0.93	1.9	133.59	12AU	0.5	-3.6	-2.2	0.2
103.96	1.03	1.9	143.37	22AU	1.1	-3.6	-2.3	0.2
110.12	1.12	1.9	153.21	1SE	2.8	-3.7	-2.3	0.1
116.22	1.22	1.9	163.10	11SE	1.1	-3.8	-2.4	0.1
122.25	1.32	1.8	173.06	21SE	-0.5	-3.9	-2.4	0.0
128.19	1.42	1.8	183.07	1OC	-0.8	-4.0	-2.4	-0.1
134.04	1.53	1.7	193.14	11OC	-0.8	-4.1	-2.4	-0.1
139.79	1.64	1.6	203.25	21OC	-0.8	-4.2	-2.4	-0.2
145.41	1.75	1.5	213.40	31OC	-0.6	-4.3	-2.3	-0.3
150.89	1.86	1.4	223.59	10NO	-0.6	-4.4	-2.3	-0.3
156.19	1.98	1.3	233.79	20NO	-0.6	-4.4	-2.2	-0.2
161.29	2.11	1.2	244.00	30NO	-0.4	-4.1	-2.2	-0.2
166.12	2.24	1.0	254.21	10DE	0.8	-3.5	-2.1	-0.1
170.64	2.38	0.8	264.40	20DE	2.7	-3.2	-2.0	-0.0
174.76	2.53	0.6	274.57	30DE	0.8	-3.9	-1.9	0.1

Right table (−384 / −383)

♂ LONG	♂ LAT	♂ MAG	☉ LONG	16.00UT	☿	♀	♃	♄
178.40	2.68	0.4	284.71	9JA	0.3	-4.3	-1.9	0.1
181.42	2.84	0.2	294.81	19JA	0.1	-4.3	-1.8	0.1
183.69	2.99	-0.1	304.85	29JA	-0.1	-4.3	-1.8	0.2
185.02	3.12	-0.4	314.85	8FE	-0.4	-4.2	-1.7	0.3
185.22	3.22	-0.7	324.79	18FE	-1.0	-4.1	-1.6	0.3
184.18	3.26	-1.0	334.66	28FE	-1.6	-4.0	-1.6	0.3
181.90	3.19	-1.2	344.48	9MR	-1.1	-3.9	-1.6	0.3
178.64	2.99	-1.5	354.24	19MR	0.0	-3.8	-1.5	0.3
174.97	2.65	-1.5	3.94	29MR	1.3	-3.7	-1.5	0.3
171.61	2.20	-1.4	13.60	8AP	2.9	-3.6	-1.5	0.3
169.20	1.69	-1.2	23.21	18AP	3.1	-3.5	-1.5	0.2
168.06	1.19	-1.0	32.79	28AP	1.8	-3.5	-1.5	0.2
168.27	0.72	-0.8	42.34	8MY	0.9	-3.4	-1.5	0.2
169.73	0.31	-0.6	51.87	18MY	0.1	-3.4	-1.5	0.2
172.26	-0.05	-0.4	61.38	28MY	-0.9	-3.4	-1.6	0.2
175.68	-0.36	-0.2	70.90	7JN	-1.7	-3.3	-1.6	0.2
179.83	-0.63	-0.1	80.43	17JN	-1.1	-3.3	-1.6	0.3
184.59	-0.86	0.1	89.97	27JN	-0.4	-3.3	-1.7	0.3
189.86	-1.05	0.2	99.54	7JL	0.1	-3.3	-1.7	0.3
195.55	-1.22	0.3	109.15	17JL	0.4	-3.3	-1.8	0.3
201.60	-1.36	0.4	118.79	27JL	0.8	-3.3	-1.8	0.3
207.95	-1.47	0.5	128.49	6AU	1.4	-3.4	-1.9	0.3
214.58	-1.55	0.6	138.24	16AU	3.1	-3.4	-2.0	0.3
221.42	-1.62	0.6	148.04	26AU	0.9	-3.4	-2.0	0.2
228.46	-1.66	0.7	157.91	5SE	-0.6	-3.4	-2.1	0.2
235.67	-1.68	0.8	167.84	15SE	-1.0	-3.5	-2.1	0.1
243.01	-1.68	0.8	177.82	25SE	-0.9	-3.5	-2.2	0.1
250.47	-1.67	0.9	187.86	5OC	-0.8	-3.5	-2.3	-0.0
258.02	-1.63	1.0	197.94	15OC	-0.5	-3.5	-2.3	-0.1
265.63	-1.58	1.0	208.07	25OC	-0.4	-3.4	-2.3	-0.1
273.30	-1.52	1.1	218.24	4NO	-0.4	-3.4	-2.3	-0.2
280.99	-1.44	1.1	228.44	14NO	-0.3	-3.4	-2.3	-0.3
288.69	-1.36	1.2	238.64	24NO	1.0	-3.4	-2.3	-0.3
296.39	-1.26	1.3	248.85	4DE	2.4	-3.3	-2.3	-0.2
304.06	-1.16	1.3	259.06	14DE	0.5	-3.3	-2.2	-0.2
311.70	-1.05	1.4	269.24	24DE	0.1	-3.3	-2.1	-0.1

−383

♂ LONG	♂ LAT	♂ MAG	☉ LONG	16.00UT	☿	♀	♃	♄
319.29	-0.94	1.4	279.40	3JA	-0.0	-3.3	-2.1	-0.0
326.83	-0.82	1.5	289.52	13JA	-0.2	-3.4	-2.0	0.0
334.30	-0.71	1.5	299.59	23JA	-0.5	-3.4	-1.9	0.1
341.71	-0.59	1.6	309.62	2FE	-1.0	-3.4	-1.9	0.2
349.03	-0.47	1.6	319.58	12FE	-1.5	-3.4	-1.8	0.2
356.29	-0.36	1.7	329.49	22FE	-1.1	-3.4	-1.7	0.2
3.46	-0.25	1.7	339.34	4MR	0.1	-3.5	-1.7	0.3
10.56	-0.13	1.7	349.13	14MR	1.7	-3.5	-1.6	0.3
17.58	-0.03	1.7	358.86	24MR	3.6	-3.6	-1.6	0.3
24.53	0.08	1.8	8.54	3AP	2.3	-3.6	-1.5	0.3
31.41	0.18	1.8	18.18	13AP	1.3	-3.7	-1.5	0.3
38.21	0.27	1.8	27.77	23AP	0.7	-3.8	-1.5	0.3
44.96	0.37	1.8	37.34	3MY	0.0	-3.9	-1.4	0.3
51.64	0.46	1.8	46.87	13MY	-0.9	-4.0	-1.4	0.2
58.26	0.54	1.8	56.39	23MY	-1.8	-4.1	-1.4	0.2
64.84	0.62	1.8	65.91	2JN	-1.1	-4.1	-1.4	0.2
71.37	0.70	1.9	75.43	12JN	-0.3	-4.2	-1.4	0.2
77.85	0.78	1.9	84.97	22JN	0.2	-4.1	-1.4	0.3
84.30	0.85	2.0	94.52	2JL	0.6	-3.9	-1.5	0.3
90.71	0.92	2.0	104.11	12JL	1.0	-3.4	-1.5	0.3
97.10	0.98	2.0	113.73	22JL	1.9	-3.2	-1.5	0.3
103.45	1.05	2.0	123.40	1AU	3.1	-3.8	-1.6	0.4
109.78	1.11	2.0	133.12	11AU	0.8	-4.2	-1.6	0.3
116.09	1.16	2.0	142.90	21AU	-0.7	-4.3	-1.7	0.3
122.38	1.22	2.0	152.73	31AU	-1.1	-4.3	-1.7	0.3
128.64	1.27	2.0	162.62	10SE	-1.1	-4.2	-1.8	0.3
134.89	1.32	2.0	172.58	20SE	-0.7	-4.1	-1.8	0.2
141.11	1.37	1.9	182.59	30SE	-0.4	-4.0	-1.9	0.2
147.30	1.41	1.9	192.65	10OC	-0.3	-3.9	-2.0	0.1
153.46	1.45	1.8	202.76	20OC	-0.3	-3.9	-2.0	0.0
159.59	1.49	1.8	212.91	30OC	-0.1	-3.8	-2.1	-0.0
165.67	1.53	1.7	223.09	9NO	1.2	-3.7	-2.1	-0.1
171.70	1.56	1.6	233.29	19NO	2.1	-3.6	-2.2	-0.2
177.67	1.59	1.5	243.50	29NO	0.3	-3.6	-2.2	-0.2
183.56	1.61	1.4	253.71	9DE	-0.1	-3.5	-2.2	-0.2
189.37	1.62	1.3	263.91	19DE	-0.2	-3.5	-2.2	-0.2
195.05	1.63	1.1	274.08	29DE	-0.3	-3.4	-2.2	-0.1

♂			⊙	16.00UT	☿	♀	♃	♄
LONG	LAT	MAG	LONG	-382	MAGNITUDES			
200.61	1.62	1.0	284.22	8JA	-0.5	-3.4	-2.1	-0.1
205.99	1.61	0.8	294.32	18JA	-1.0	-3.4	-2.1	0.0
211.17	1.57	0.6	304.38	28JA	-1.3	-3.3	-2.0	0.1
216.09	1.51	0.4	314.37	7FE	-1.0	-3.3	-1.9	0.1
220.68	1.43	0.2	324.31	17FE	0.2	-3.3	-1.9	0.2
224.87	1.31	-0.0	334.19	27FE	2.0	-3.3	-1.8	0.2
228.54	1.15	-0.3	344.01	9MR	3.1	-3.3	-1.7	0.3
231.56	0.92	-0.6	353.77	19MR	1.7	-3.3	-1.7	0.3
233.76	0.63	-0.9	3.48	29MR	1.0	-3.3	-1.6	0.3
234.98	0.25	-1.2	13.13	8AP	0.5	-3.4	-1.6	0.4
235.02	-0.23	-1.5	22.75	18AP	-0.1	-3.4	-1.5	0.4
233.82	-0.80	-1.8	32.32	28AP	-0.9	-3.4	-1.5	0.4
231.52	-1.44	-2.1	41.87	8MY	-1.8	-3.5	-1.4	0.3
228.52	-2.09	-2.3	51.40	18MY	-1.1	-3.5	-1.4	0.3
225.54	-2.66	-2.2	60.92	28MY	-0.3	-3.4	-1.4	0.3
223.27	-3.11	-2.0	70.44	7JN	0.4	-3.4	-1.3	0.3
222.20	-3.40	-1.8	79.96	17JN	0.8	-3.4	-1.3	0.3
222.51	-3.57	-1.6	89.51	27JN	1.4	-3.4	-1.3	0.3
224.15	-3.63	-1.3	99.07	7JL	2.4	-3.3	-1.3	0.4
226.95	-3.62	-1.1	108.68	17JL	2.7	-3.3	-1.3	0.4
230.72	-3.54	-0.9	118.32	27JL	0.7	-3.3	-1.3	0.4
235.27	-3.43	-0.7	128.01	6AU	-0.7	-3.3	-1.4	0.4
240.44	-3.29	-0.5	137.76	16AU	-1.2	-3.3	-1.4	0.4
246.12	-3.12	-0.3	147.57	26AU	-1.2	-3.4	-1.4	0.4
252.19	-2.94	-0.2	157.43	5SE	-0.6	-3.4	-1.5	0.4
258.57	-2.74	-0.0	167.35	15SE	-0.3	-3.4	-1.5	0.4
265.20	-2.54	0.1	177.33	25SE	-0.2	-3.4	-1.6	0.4
272.01	-2.32	0.3	187.37	5OC	-0.1	-3.5	-1.6	0.3
278.96	-2.11	0.4	197.45	15OC	0.1	-3.5	-1.7	0.3
286.01	-1.90	0.5	207.58	25OC	1.4	-3.6	-1.8	0.2
293.13	-1.69	0.6	217.75	4NO	1.8	-3.7	-1.8	0.1
300.30	-1.48	0.8	227.94	14NO	0.1	-3.7	-1.9	0.1
307.48	-1.28	0.9	238.15	24NO	-0.2	-3.8	-1.9	0.0
314.67	-1.09	1.0	248.36	4DE	-0.3	-3.9	-2.0	-0.1
321.84	-0.91	1.1	258.56	14DE	-0.4	-4.0	-2.0	-0.1
328.99	-0.73	1.2	268.75	24DE	-0.6	-4.1	-2.1	-0.1
				-381				
336.10	-0.57	1.3	278.91	3JA	-1.0	-4.2	-2.1	-0.1
343.17	-0.41	1.4	289.03	13JA	-1.2	-4.3	-2.1	-0.0
350.19	-0.26	1.5	299.11	23JA	-0.9	-4.3	-2.1	0.0
357.16	-0.12	1.5	309.13	2FE	0.3	-4.3	-2.1	0.1
4.08	0.01	1.6	319.11	12FE	2.4	-4.0	-2.0	0.2
10.93	0.13	1.7	329.01	22FE	2.4	-3.4	-2.0	0.2
17.74	0.24	1.7	338.86	4MR	1.2	-3.4	-1.9	0.3
24.48	0.34	1.8	348.66	14MR	0.7	-3.9	-1.8	0.3
31.18	0.44	1.8	358.39	24MR	0.4	-4.2	-1.8	0.4
37.82	0.53	1.9	8.07	3AP	-0.1	-4.2	-1.7	0.4
44.42	0.61	1.9	17.71	13AP	-0.9	-4.2	-1.6	0.4
50.98	0.69	1.9	27.30	23AP	-1.8	-4.1	-1.6	0.4
57.49	0.75	2.0	36.87	3MY	-1.2	-4.0	-1.5	0.5
63.97	0.82	2.0	46.41	13MY	-0.2	-3.9	-1.5	0.5
70.42	0.88	2.0	55.93	23MY	0.5	-3.8	-1.4	0.5
76.85	0.93	2.0	65.45	2JN	1.1	-3.7	-1.4	0.4
83.26	0.98	2.0	74.97	12JN	1.8	-3.6	-1.3	0.4
89.64	1.02	2.0	84.50	22JN	3.1	-3.6	-1.3	0.4
96.02	1.06	2.0	94.06	2JL	2.3	-3.5	-1.3	0.4
102.40	1.09	1.9	103.64	12JL	0.5	-3.5	-1.3	0.5
108.76	1.12	2.0	113.26	22JL	-0.8	-3.4	-1.2	0.5
115.14	1.15	2.0	122.93	1AU	-1.4	-3.4	-1.2	0.5
121.51	1.17	2.0	132.65	11AU	-1.2	-3.4	-1.2	0.6
127.90	1.19	2.0	142.42	21AU	-0.6	-3.4	-1.3	0.6
134.29	1.20	2.0	152.26	31AU	-0.2	-3.4	-1.3	0.6
140.71	1.21	2.0	162.15	10SE	-0.0	-3.4	-1.3	0.6
147.13	1.21	2.0	172.09	20SE	0.1	-3.4	-1.3	0.6
153.57	1.21	2.0	182.10	30SE	0.3	-3.4	-1.4	0.6
160.03	1.20	1.9	192.17	10OC	1.7	-3.4	-1.4	0.5
166.51	1.19	1.9	202.27	20OC	1.6	-3.4	-1.5	0.5
173.01	1.17	1.9	212.42	30OC	-0.1	-3.4	-1.5	0.4
179.52	1.15	1.8	222.60	9NO	-0.4	-3.4	-1.6	0.4
186.05	1.11	1.8	232.80	19NO	-0.4	-3.4	-1.7	0.3
192.59	1.07	1.7	243.01	29NO	-0.5	-3.5	-1.7	0.2
199.14	1.02	1.6	253.22	9DE	-0.7	-3.5	-1.8	0.2
205.70	0.96	1.5	263.41	19DE	-0.9	-3.5	-1.8	0.1
212.27	0.88	1.4	273.59	29DE	-1.0	-3.5	-1.9	0.0

♂			⊙	16.00UT	☿	♀	♃	♄
LONG	LAT	MAG	LONG	-380	MAGNITUDES			
218.83	0.79	1.3	283.73	8JA	-0.8	-3.4	-2.0	0.0
225.39	0.69	1.2	293.83	18JA	0.4	-3.4	-2.0	0.1
231.94	0.56	1.1	303.89	28JA	2.8	-3.4	-2.0	0.1
238.47	0.42	1.0	313.88	7FE	1.8	-3.4	-2.0	0.2
244.96	0.25	0.8	323.82	17FE	0.9	-3.3	-2.0	0.3
251.42	0.05	0.7	333.71	27FE	0.5	-3.3	-2.0	0.3
257.81	-0.17	0.5	343.53	8MR	0.2	-3.3	-2.0	0.4
264.12	-0.43	0.4	353.29	18MR	-0.2	-3.3	-1.9	0.4
270.32	-0.72	0.2	3.01	28MR	-0.9	-3.3	-1.9	0.5
276.37	-1.05	-0.0	12.67	7AP	-1.8	-3.4	-1.8	0.5
282.23	-1.43	-0.2	22.28	17AP	-1.2	-3.4	-1.8	0.6
287.82	-1.85	-0.4	31.86	27AP	-0.2	-3.4	-1.7	0.6
293.06	-2.32	-0.6	41.41	7MY	0.7	-3.4	-1.6	0.6
297.86	-2.85	-0.9	50.94	17MY	1.4	-3.5	-1.6	0.6
302.06	-3.44	-1.1	60.46	27MY	2.4	-3.5	-1.5	0.6
305.51	-4.08	-1.4	69.98	6JN	3.4	-3.6	-1.5	0.6
308.02	-4.76	-1.6	79.50	16JN	1.8	-3.7	-1.4	0.6
309.35	-5.44	-1.9	89.05	26JN	0.4	-3.7	-1.4	0.6
309.37	-6.08	-2.2	98.61	6JL	-0.8	-3.8	-1.3	0.6
308.09	-6.58	-2.4	108.21	16JL	-1.5	-3.9	-1.3	0.6
305.78	-6.83	-2.6	117.86	26JL	-1.1	-4.0	-1.3	0.7
303.06	-6.75	-2.6	127.55	5AU	-0.5	-4.1	-1.2	0.7
300.69	-6.37	-2.4	137.29	15AU	-0.1	-4.2	-1.2	0.7
299.28	-5.77	-2.1	147.09	25AU	0.1	-4.3	-1.2	0.8
299.15	-5.06	-1.8	156.95	4SE	0.2	-4.3	-1.2	0.8
300.30	-4.33	-1.5	166.87	14SE	0.6	-4.0	-1.2	0.8
302.60	-3.65	-1.2	176.85	24SE	2.1	-3.5	-1.2	0.8
305.84	-3.02	-0.9	186.88	4OC	1.4	-3.3	-1.3	0.8
309.83	-2.47	-0.6	196.97	14OC	-0.3	-4.0	-1.3	0.8
314.41	-1.98	-0.3	207.10	24OC	-0.6	-4.3	-1.3	0.7
319.43	-1.56	-0.1	217.26	3NO	-0.6	-4.4	-1.4	0.7
324.79	-1.19	0.1	227.45	13NO	-0.6	-4.4	-1.4	0.6
330.42	-0.86	0.3	237.66	23NO	-0.8	-4.3	-1.5	0.6
336.25	-0.58	0.5	247.86	3DE	-0.8	-4.2	-1.5	0.5
342.22	-0.34	0.7	258.07	13DE	-0.8	-4.1	-1.6	0.4
348.30	-0.12	0.8	268.26	23DE	-0.7	-4.0	-1.7	0.4
				-379				
354.47	0.06	1.0	278.42	2JA	0.5	-3.9	-1.7	0.3
0.69	0.23	1.1	288.54	12JA	3.0	-3.8	-1.8	0.2
6.95	0.37	1.3	298.63	22JA	1.4	-3.7	-1.9	0.3
13.24	0.50	1.4	308.65	1FE	0.6	-3.6	-1.9	0.3
19.54	0.61	1.5	318.62	11FE	0.4	-3.5	-2.0	0.3
25.85	0.70	1.6	328.54	21FE	0.1	-3.5	-2.0	0.4
32.15	0.79	1.7	338.39	3MR	-0.3	-3.4	-2.0	0.5
38.46	0.86	1.7	348.19	13MR	-0.9	-3.4	-2.0	0.5
44.75	0.92	1.8	357.93	23MR	-1.8	-3.4	-2.0	0.6
51.04	0.98	1.9	7.61	2AP	-1.2	-3.3	-2.0	0.6
57.33	1.02	1.9	17.25	12AP	-0.1	-3.3	-2.0	0.7
63.61	1.06	1.9	26.85	22AP	0.9	-3.3	-1.9	0.7
69.88	1.10	2.0	36.41	2MY	1.8	-3.3	-1.9	0.8
76.16	1.12	2.0	45.95	12MY	3.2	-3.3	-1.8	0.8
82.43	1.15	2.0	55.48	22MY	2.8	-3.3	-1.7	0.8
88.72	1.16	2.0	64.99	1JN	1.4	-3.3	-1.7	0.8
95.01	1.17	2.0	74.51	11JN	0.3	-3.4	-1.6	0.8
101.31	1.18	2.0	84.04	21JN	-0.8	-3.4	-1.6	0.8
107.64	1.18	2.0	93.59	1JL	-1.6	-3.4	-1.5	0.9
113.98	1.17	2.0	103.18	11JL	-1.1	-3.5	-1.4	0.8
120.35	1.17	2.0	112.80	21JL	-0.5	-3.5	-1.4	0.8
126.75	1.15	2.0	122.46	31JL	-0.0	-3.5	-1.4	0.9
133.18	1.13	1.9	132.17	10AU	0.2	-3.4	-1.3	0.9
139.65	1.11	1.9	141.94	20AU	0.4	-3.4	-1.3	1.0
146.16	1.08	1.9	151.77	30AU	0.9	-3.4	-1.3	1.0
152.72	1.05	1.9	161.66	9SE	2.5	-3.4	-1.2	1.0
159.32	1.01	1.9	171.61	19SE	1.3	-3.3	-1.2	1.0
165.96	0.97	1.9	181.61	29SE	-0.4	-3.3	-1.2	1.1
172.66	0.92	1.9	191.67	9OC	-0.7	-3.3	-1.2	1.1
179.41	0.86	1.9	201.78	19OC	-0.7	-3.3	-1.3	1.0
186.21	0.80	1.9	211.92	29OC	-0.7	-3.3	-1.3	1.0
193.07	0.73	1.8	222.10	8NO	-0.7	-3.4	-1.3	1.0
199.98	0.65	1.8	232.30	18NO	-0.7	-3.4	-1.3	1.0
206.94	0.56	1.7	242.51	28NO	-0.7	-3.4	-1.4	0.9
213.95	0.47	1.7	252.72	8DE	-0.5	-3.5	-1.4	0.8
221.02	0.37	1.6	262.92	18DE	0.6	-3.5	-1.5	0.8
228.13	0.25	1.6	273.10	28DE	2.9	-3.5	-1.5	0.7

Left table

♂ LONG	LAT	MAG	☉ LONG	16.00UT -378	☿	♀	♃	♄ MAGNITUDES
235.30	0.13	1.5	283.24	7JA	1.0	-3.6	-1.6	0.6
242.50	-0.01	1.4	293.35	17JA	0.4	-3.7	-1.7	0.6
249.75	-0.15	1.3	303.40	27JA	0.2	-3.7	-1.7	0.5
257.03	-0.31	1.3	313.41	6FE	0.0	-3.8	-1.8	0.5
264.34	-0.47	1.2	323.35	16FE	-0.4	-3.9	-1.9	0.6
271.67	-0.65	1.1	333.23	26FE	-0.9	-4.0	-1.9	0.6
279.02	-0.83	1.0	343.06	8MR	-1.7	-4.1	-2.0	0.7
286.36	-1.02	0.9	352.83	18MR	-1.1	-4.2	-2.0	0.7
293.70	-1.22	0.8	2.54	28MR	-0.0	-4.2	-2.1	0.8
301.00	-1.42	0.7	12.20	7AP	1.1	-4.2	-2.1	0.8
308.25	-1.62	0.6	21.82	17AP	2.4	-4.0	-2.1	0.9
315.43	-1.82	0.5	31.40	27AP	3.7	-3.6	-2.1	0.9
322.52	-2.02	0.4	40.95	7MY	2.1	-2.8	-2.0	1.0
329.47	-2.21	0.3	50.48	17MY	1.1	-3.4	-2.0	1.0
336.26	-2.40	0.2	60.00	27MY	0.2	-4.0	-2.0	1.0
342.84	-2.58	0.1	69.52	6JN	-0.9	-4.2	-1.9	1.1
349.15	-2.75	-0.0	79.04	16JN	-1.7	-4.2	-1.8	1.1
355.15	-2.90	-0.1	88.58	26JN	-1.1	-4.1	-1.8	1.1
0.73	-3.04	-0.3	98.15	6JL	-0.4	-4.1	-1.7	1.1
5.82	-3.17	-0.4	107.75	16JL	0.1	-4.0	-1.7	1.1
10.29	-3.27	-0.6	117.39	26JL	0.3	-3.9	-1.6	1.1
13.98	-3.34	-0.8	127.08	5AU	0.6	-3.8	-1.5	1.1
16.70	-3.38	-1.0	136.81	15AU	1.2	-3.7	-1.5	1.1
18.25	-3.36	-1.2	146.61	25AU	2.9	-3.7	-1.4	1.2
18.41	-3.26	-1.4	156.47	4SE	1.1	-3.6	-1.4	1.2
17.11	-3.05	-1.7	166.38	14SE	-0.5	-3.6	-1.4	1.3
14.47	-2.71	-1.8	176.36	24SE	-0.9	-3.5	-1.3	1.3
10.98	-2.24	-2.0	186.39	4OC	-0.9	-3.5	-1.3	1.3
7.44	-1.68	-1.8	196.47	14OC	-0.8	-3.4	-1.3	1.4
4.59	-1.11	-1.5	206.60	24OC	-0.6	-3.4	-1.3	1.4
2.96	-0.58	-1.2	216.77	3NO	-0.5	-3.4	-1.3	1.3
2.70	-0.14	-0.9	226.95	13NO	-0.5	-3.4	-1.3	1.3
3.69	0.23	-0.6	237.16	23NO	-0.4	-3.4	-1.3	1.3
5.76	0.52	-0.3	247.37	3DE	0.8	-3.4	-1.4	1.2
8.69	0.74	0.0	257.58	13DE	2.6	-3.3	-1.4	1.2
12.30	0.92	0.3	267.77	23DE	0.7	-3.3	-1.4	1.1

-377

♂ LONG	LAT	MAG	☉ LONG	16.00UT	☿	♀	♃	♄
16.43	1.05	0.5	277.93	2JA	0.2	-3.4	-1.5	1.1
20.97	1.16	0.7	288.05	12JA	0.1	-3.4	-1.5	1.0
25.83	1.24	0.9	298.14	22JA	-0.1	-3.4	-1.6	0.9
30.94	1.30	1.0	308.17	1FE	-0.4	-3.4	-1.6	0.9
36.25	1.35	1.2	318.14	11FE	-1.0	-3.4	-1.7	0.8
41.71	1.38	1.3	328.06	21FE	-1.5	-3.5	-1.8	0.8
47.29	1.40	1.4	337.92	3MR	-1.1	-3.5	-1.8	0.9
52.98	1.42	1.5	347.71	13MR	0.0	-3.5	-1.9	0.9
58.74	1.42	1.6	357.45	23MR	1.4	-3.4	-2.0	1.0
64.58	1.42	1.7	7.14	2AP	3.1	-3.4	-2.0	1.0
70.48	1.42	1.8	16.78	12AP	2.8	-3.4	-2.1	1.1
76.42	1.40	1.8	26.38	22AP	1.6	-3.4	-2.1	1.1
82.42	1.39	1.9	35.95	2MY	0.8	-3.3	-2.2	1.2
88.47	1.37	1.9	45.48	12MY	0.1	-3.3	-2.2	1.2
94.56	1.34	2.0	55.01	22MY	-0.9	-3.3	-2.2	1.3
100.69	1.31	2.0	64.53	1JN	-1.7	-3.3	-2.2	1.3
106.87	1.28	2.0	74.04	11JN	-1.2	-3.3	-2.1	1.3
113.10	1.24	2.0	83.58	21JN	-0.4	-3.4	-2.1	1.4
119.37	1.20	2.0	93.13	1JL	0.2	-3.4	-2.1	1.4
125.70	1.15	2.0	102.71	11JL	0.5	-3.4	-2.0	1.4
132.09	1.10	2.0	112.33	21JL	0.8	-3.5	-1.9	1.4
138.53	1.05	2.0	121.99	31JL	1.6	-3.5	-1.9	1.4
145.03	1.00	1.9	131.70	10AU	3.2	-3.6	-1.8	1.3
151.60	0.93	1.9	141.47	20AU	1.0	-3.6	-1.7	1.3
158.23	0.87	1.9	151.29	30AU	-0.6	-3.7	-1.7	1.3
164.94	0.80	1.8	161.18	9SE	-1.0	-3.8	-1.6	1.3
171.71	0.72	1.8	171.12	19SE	-1.0	-3.9	-1.6	1.3
178.56	0.65	1.8	181.12	29SE	-0.8	-4.0	-1.5	1.3
185.48	0.56	1.7	191.18	9OC	-0.5	-4.1	-1.5	1.3
192.48	0.47	1.7	201.28	19OC	-0.4	-4.2	-1.5	1.2
199.56	0.38	1.7	211.43	29OC	-0.4	-4.3	-1.4	1.2
206.70	0.28	1.7	221.60	8NO	-0.2	-4.4	-1.4	1.2
213.93	0.18	1.7	231.80	18NO	1.0	-4.4	-1.4	1.2
221.22	0.07	1.7	242.01	28NO	2.3	-4.1	-1.4	1.1
228.58	-0.04	1.6	252.22	8DE	0.4	-3.5	-1.4	1.1
236.01	-0.16	1.6	262.42	18DE	0.0	-3.2	-1.4	1.0
243.50	-0.28	1.6	272.60	28DE	-0.1	-3.9	-1.4	1.0

Right table

♂ LONG	LAT	MAG	☉ LONG	16.00UT -376	☿	♀	♃	♄ MAGNITUDES
251.04	-0.40	1.5	282.75	7JA	-0.2	-4.3	-1.5	0.9
258.63	-0.52	1.5	292.86	17JA	-0.5	-4.3	-1.5	0.9
266.26	-0.65	1.4	302.91	27JA	-1.0	-4.3	-1.5	0.8
273.92	-0.77	1.4	312.92	6FE	-1.4	-4.2	-1.6	0.8
281.61	-0.89	1.4	322.87	16FE	-1.1	-4.1	-1.6	0.8
289.30	-1.01	1.3	332.76	26FE	0.1	-4.0	-1.7	0.8
297.00	-1.12	1.3	342.58	7MR	1.8	-3.9	-1.8	0.8
304.68	-1.23	1.2	352.36	17MR	3.6	-3.8	-1.8	0.9
312.33	-1.33	1.2	2.07	27MR	2.1	-3.7	-1.9	1.0
319.94	-1.41	1.2	11.74	6AP	1.2	-3.6	-2.0	1.0
327.50	-1.49	1.1	21.36	16AP	0.6	-3.5	-2.0	1.1
334.98	-1.55	1.1	30.94	26AP	-0.0	-3.5	-2.1	1.2
342.38	-1.60	1.0	40.49	6MY	-0.9	-3.4	-2.2	1.2
349.68	-1.63	1.0	50.03	16MY	-1.8	-3.4	-2.2	1.3
356.86	-1.65	0.9	59.54	26MY	-1.2	-3.4	-2.3	1.3
3.90	-1.65	0.9	69.06	5JN	-0.3	-3.3	-2.3	1.3
10.78	-1.63	0.8	78.59	15JN	0.3	-3.3	-2.3	1.3
17.48	-1.60	0.8	88.12	25JN	0.7	-3.3	-2.3	1.3
23.99	-1.55	0.7	97.69	5JL	1.1	-3.3	-2.3	1.3
30.25	-1.48	0.6	107.29	15JL	2.1	-3.3	-2.3	1.3
36.24	-1.39	0.6	116.92	25JL	3.0	-3.3	-2.2	1.3
41.91	-1.29	0.5	126.61	4AU	0.8	-3.4	-2.2	1.2
47.20	-1.15	0.4	136.35	14AU	-0.7	-3.4	-2.1	1.2
52.04	-1.00	0.2	146.14	24AU	-1.2	-3.4	-2.0	1.2
56.32	-0.81	0.1	155.99	3SE	-1.1	-3.4	-1.9	1.1
59.92	-0.59	-0.1	165.91	13SE	-0.7	-3.5	-1.9	1.1
62.68	-0.33	-0.3	175.87	23SE	-0.4	-3.5	-1.8	1.1
64.42	-0.02	-0.5	185.90	3OC	-0.3	-3.5	-1.8	1.1
64.92	0.35	-0.7	195.98	13OC	-0.2	-3.5	-1.7	1.1
64.04	0.76	-0.9	206.10	23OC	-0.0	-3.4	-1.7	1.1
61.75	1.21	-1.1	216.26	2NO	1.2	-3.4	-1.6	1.1
58.33	1.65	-1.3	226.45	12NO	2.1	-3.4	-1.6	1.1
54.40	2.03	-1.3	236.65	22NO	0.2	-3.4	-1.6	1.0
50.72	2.31	-1.1	246.87	2DE	-0.1	-3.3	-1.5	1.0
47.98	2.48	-0.8	257.07	12DE	-0.2	-3.3	-1.5	1.0
46.53	2.55	-0.6	267.26	22DE	-0.3	-3.3	-1.5	0.9

-375

♂ LONG	LAT	MAG	☉ LONG	16.00UT	☿	♀	♃	♄
46.39	2.56	-0.3	277.43	1JA	-0.6	-3.3	-1.5	0.9
47.46	2.52	0.0	287.56	11JA	-1.0	-3.4	-1.5	0.8
49.52	2.45	0.3	297.64	21JA	-1.2	-3.4	-1.5	0.8
52.37	2.37	0.5	307.68	31JA	-1.0	-3.4	-1.5	0.7
55.87	2.29	0.7	317.66	10FE	0.2	-3.4	-1.6	0.7
59.85	2.20	0.9	327.58	20FE	2.1	-3.5	-1.6	0.7
64.24	2.11	1.1	337.44	2MR	2.9	-3.5	-1.6	0.6
68.95	2.03	1.2	347.24	12MR	1.5	-3.5	-1.7	0.6
73.91	1.94	1.3	356.98	22MR	0.9	-3.6	-1.7	0.7
79.08	1.86	1.4	6.68	1AP	0.5	-3.6	-1.8	0.8
84.43	1.78	1.5	16.32	11AP	-0.1	-3.7	-1.9	0.8
89.93	1.69	1.6	25.92	21AP	-0.9	-3.8	-1.9	0.9
95.56	1.61	1.7	35.49	1MY	-1.8	-3.9	-2.0	1.0
101.31	1.53	1.7	45.03	11MY	-1.2	-4.0	-2.1	1.0
107.16	1.45	1.8	54.55	21MY	-0.3	-4.1	-2.1	1.1
113.11	1.36	1.8	64.07	31MY	0.4	-4.1	-2.2	1.1
119.16	1.28	1.9	73.59	10JN	0.9	-4.2	-2.3	1.1
125.29	1.19	1.9	83.12	20JN	1.5	-4.1	-2.3	1.2
131.52	1.11	1.9	92.67	30JN	2.7	-3.9	-2.4	1.2
137.84	1.02	1.9	102.25	10JL	2.6	-3.4	-2.4	1.2
144.24	0.93	1.9	111.86	20JL	0.7	-3.2	-2.4	1.2
150.74	0.84	1.9	121.53	30JL	-0.7	-3.8	-2.4	1.1
157.32	0.75	1.9	131.23	9AU	-1.3	-4.2	-2.4	1.1
163.99	0.65	1.8	140.99	19AU	-1.2	-4.3	-2.4	1.1
170.76	0.55	1.8	150.82	29AU	-0.6	-4.3	-2.3	1.0
177.62	0.45	1.8	160.70	8SE	-0.3	-4.2	-2.3	1.0
184.57	0.35	1.7	170.64	18SE	-0.1	-4.1	-2.2	1.0
191.62	0.25	1.7	180.64	28SE	-0.0	-4.0	-2.1	1.0
198.76	0.14	1.6	190.69	8OC	0.2	-3.9	-2.1	1.0
205.99	0.04	1.6	200.79	18OC	1.5	-3.9	-2.0	1.0
213.31	-0.07	1.5	210.93	28OC	1.8	-3.8	-1.9	1.0
220.71	-0.18	1.5	221.11	7NO	0.0	-3.7	-1.9	1.0
228.20	-0.29	1.5	231.30	17NO	-0.3	-3.6	-1.8	1.0
235.76	-0.39	1.5	241.52	27NO	-0.3	-3.6	-1.8	1.0
243.38	-0.50	1.5	251.72	7DE	-0.4	-3.5	-1.7	0.9
251.07	-0.60	1.5	261.92	17DE	-0.6	-3.5	-1.7	0.9
258.81	-0.70	1.5	272.11	27DE	-1.0	-3.4	-1.6	0.9

-374

♂ LONG	LAT	MAG	☉ LONG	16.00UT	☿	♀	♃	♄
266.59	-0.79	1.4	282.25	6JA	-1.1	-3.4	-1.6	0.8
274.40	-0.88	1.4	292.36	16JA	-0.9	-3.4	-1.6	0.8
282.23	-0.96	1.4	302.43	26JA	0.3	-3.3	-1.6	0.7
290.06	-1.03	1.4	312.43	5FE	2.5	-3.3	-1.6	0.7
297.89	-1.09	1.4	322.38	15FE	2.2	-3.3	-1.6	0.6
305.71	-1.14	1.4	332.28	25FE	1.1	-3.3	-1.6	0.6
313.49	-1.18	1.4	342.11	7MR	0.6	-3.3	-1.6	0.5
321.23	-1.20	1.4	351.88	17MR	0.3	-3.3	-1.6	0.5
328.92	-1.22	1.4	1.60	27MR	-0.2	-3.3	-1.6	0.5
336.55	-1.22	1.4	11.27	6AP	-0.9	-3.4	-1.7	0.6
344.10	-1.21	1.4	20.89	16AP	-1.8	-3.4	-1.7	0.6
351.57	-1.18	1.4	30.48	26AP	-1.2	-3.4	-1.8	0.7
358.94	-1.14	1.4	40.03	6MY	-0.2	-3.5	-1.8	0.8
6.21	-1.09	1.4	49.56	16MY	0.5	-3.5	-1.9	0.8
13.38	-1.03	1.4	59.08	26MY	1.2	-3.4	-2.0	0.9
20.43	-0.96	1.3	68.60	5JN	2.0	-3.4	-2.0	0.9
27.36	-0.87	1.3	78.12	15JN	3.3	-3.4	-2.1	1.0
34.15	-0.78	1.3	87.66	25JN	2.1	-3.4	-2.2	1.0
40.80	-0.67	1.3	97.22	5JL	0.5	-3.3	-2.2	1.0
47.30	-0.55	1.3	106.82	15JL	-0.8	-3.3	-2.3	1.0
53.64	-0.42	1.2	116.45	25JL	-1.4	-3.3	-2.3	1.0
59.78	-0.28	1.2	126.13	4AU	-1.2	-3.3	-2.4	1.0
65.73	-0.12	1.1	135.87	14AU	-0.6	-3.3	-2.4	1.0
71.44	0.05	1.0	145.66	24AU	-0.2	-3.4	-2.5	1.0
76.88	0.23	1.0	155.51	3SE	0.0	-3.4	-2.5	1.0
82.01	0.44	0.9	165.42	13SE	0.1	-3.4	-2.4	0.9
86.76	0.67	0.7	175.39	23SE	0.4	-3.4	-2.4	0.9
91.05	0.92	0.6	185.41	3OC	1.8	-3.5	-2.4	0.9
94.78	1.21	0.4	195.49	13OC	1.6	-3.5	-2.3	0.9
97.81	1.54	0.3	205.61	23OC	-0.2	-3.6	-2.3	0.9
100.00	1.90	0.1	215.77	2NO	-0.5	-3.7	-2.2	0.9
101.15	2.31	-0.2	225.96	12NO	-0.5	-3.8	-2.1	0.9
101.07	2.75	-0.4	236.16	22NO	-0.5	-3.8	-2.0	0.9
99.65	3.20	-0.6	246.37	2DE	-0.7	-3.9	-2.0	0.9
96.93	3.61	-0.9	256.58	12DE	-0.9	-4.0	-1.9	0.9
93.26	3.92	-1.0	266.77	22DE	-0.9	-4.1	-1.8	0.9

-373

♂ LONG	LAT	MAG	☉ LONG	16.00UT	☿	♀	♃	♄
89.28	4.08	-1.0	276.93	1JA	-0.8	-4.2	-1.8	0.8
85.74	4.09	-0.8	287.07	11JA	0.4	-4.3	-1.7	0.8
83.21	3.95	-0.5	297.15	21JA	2.9	-4.3	-1.7	0.7
81.96	3.74	-0.3	307.19	31JA	1.6	-4.3	-1.6	0.7
81.99	3.48	-0.0	317.17	10FE	0.8	-4.0	-1.6	0.6
83.17	3.22	0.2	327.09	20FE	0.5	-3.4	-1.6	0.6
85.30	2.97	0.4	336.95	2MR	0.2	-3.4	-1.6	0.5
88.20	2.73	0.6	346.76	12MR	-0.2	-3.9	-1.6	0.5
91.73	2.50	0.8	356.51	22MR	-0.9	-4.2	-1.6	0.4
95.76	2.30	0.9	6.20	1AP	-1.8	-4.2	-1.6	0.4
100.19	2.10	1.1	15.85	11AP	-1.2	-4.2	-1.6	0.4
104.96	1.92	1.2	25.45	21AP	-0.2	-4.1	-1.6	0.5
110.01	1.75	1.3	35.02	1MY	0.7	-4.0	-1.6	0.6
115.30	1.59	1.4	44.57	11MY	1.5	-3.9	-1.7	0.6
120.80	1.44	1.4	54.09	21MY	2.7	-3.8	-1.7	0.7
126.48	1.29	1.5	63.61	31MY	3.2	-3.7	-1.7	0.8
132.32	1.15	1.6	73.13	10JN	1.7	-3.6	-1.8	0.8
138.32	1.01	1.6	82.66	20JN	0.4	-3.6	-1.9	0.9
144.47	0.88	1.6	92.21	30JN	-0.8	-3.5	-1.9	0.9
150.75	0.75	1.6	101.79	10JL	-1.5	-3.5	-2.0	0.9
157.16	0.62	1.7	111.40	20JL	-1.2	-3.4	-2.0	0.9
163.69	0.49	1.7	121.05	30JL	-0.5	-3.4	-2.1	1.0
170.36	0.37	1.7	130.76	9AU	-0.1	-3.4	-2.2	1.0
177.14	0.24	1.7	140.52	19AU	0.1	-3.4	-2.2	0.9
184.04	0.12	1.6	150.34	29AU	0.3	-3.4	-2.3	0.9
191.05	0.00	1.6	160.22	8SE	0.7	-3.4	-2.4	0.9
198.18	-0.11	1.6	170.16	18SE	2.2	-3.4	-2.4	0.9
205.41	-0.23	1.6	180.15	28SE	1.4	-3.4	-2.4	0.8
212.76	-0.34	1.6	190.21	8OC	-0.3	-3.4	-2.4	0.8
220.19	-0.45	1.5	200.30	18OC	-0.6	-3.4	-2.4	0.8
227.72	-0.54	1.5	210.44	28OC	-0.6	-3.4	-2.4	0.8
235.34	-0.64	1.5	220.62	7NO	-0.7	-3.4	-2.3	0.8
243.02	-0.72	1.4	230.81	17NO	-0.8	-3.4	-2.3	0.9
250.77	-0.80	1.4	241.02	27NO	-0.7	-3.5	-2.2	0.9
258.58	-0.88	1.4	251.23	7DE	-0.8	-3.5	-2.1	0.9
266.43	-0.94	1.3	261.43	17DE	-0.6	-3.5	-2.1	0.8
274.30	-0.99	1.3	271.61	27DE	0.5	-3.5	-2.0	0.8

-372

♂ LONG	LAT	MAG	☉ LONG	16.00UT	☿	♀	♃	♄
282.19	-1.03	1.3	281.76	6JA	3.0	-3.4	-1.9	0.8
290.09	-1.07	1.3	291.87	16JA	1.2	-3.4	-1.9	0.8
297.97	-1.09	1.3	301.93	26JA	0.5	-3.4	-1.8	0.7
305.83	-1.09	1.3	311.94	5FE	0.3	-3.4	-1.7	0.7
313.65	-1.09	1.4	321.89	15FE	0.1	-3.3	-1.7	0.6
321.43	-1.07	1.4	331.79	25FE	-0.3	-3.3	-1.6	0.6
329.15	-1.05	1.4	341.63	6MR	-0.9	-3.3	-1.6	0.5
336.80	-1.01	1.5	351.40	16MR	-1.7	-3.3	-1.6	0.5
344.38	-0.96	1.5	1.12	26MR	-1.2	-3.3	-1.5	0.4
351.88	-0.90	1.5	10.80	5AP	-0.1	-3.4	-1.5	0.4
359.30	-0.84	1.5	20.42	15AP	0.9	-3.4	-1.5	0.3
6.62	-0.76	1.6	30.01	25AP	2.0	-3.4	-1.5	0.4
13.85	-0.68	1.6	39.56	5MY	3.5	-3.4	-1.5	0.5
20.98	-0.59	1.6	49.10	15MY	2.5	-3.5	-1.5	0.5
28.02	-0.49	1.6	58.62	25MY	1.3	-3.5	-1.5	0.6
34.96	-0.39	1.6	68.14	4JN	0.3	-3.6	-1.6	0.7
41.79	-0.28	1.6	77.66	14JN	-0.8	-3.7	-1.6	0.7
48.53	-0.17	1.6	87.20	24JN	-1.6	-3.7	-1.6	0.8
55.15	-0.05	1.6	96.76	4JL	-1.2	-3.8	-1.7	0.8
61.67	0.07	1.6	106.35	14JL	-0.5	-3.9	-1.7	0.8
68.08	0.20	1.6	115.99	24JL	-0.0	-4.0	-1.8	0.9
74.38	0.33	1.6	125.67	3AU	0.3	-4.1	-1.8	0.9
80.54	0.47	1.6	135.40	13AU	0.5	-4.2	-1.9	0.9
86.57	0.62	1.5	145.19	23AU	1.0	-4.3	-2.0	0.9
92.45	0.77	1.5	155.03	2SE	2.6	-4.2	-2.0	0.9
98.16	0.93	1.4	164.94	12SE	1.3	-4.0	-2.1	0.9
103.68	1.11	1.3	174.91	22SE	-0.4	-3.5	-2.1	0.8
108.97	1.30	1.2	184.93	2OC	-0.8	-3.4	-2.2	0.8
114.00	1.50	1.1	195.00	12OC	-0.8	-4.0	-2.3	0.8
118.70	1.73	1.0	205.12	22OC	-0.8	-4.3	-2.3	0.7
123.01	1.97	0.9	215.28	1NO	-0.7	-4.4	-2.3	0.8
126.85	2.24	0.7	225.46	11NO	-0.6	-4.3	-2.3	0.8
130.09	2.54	0.5	235.67	21NO	-0.6	-4.3	-2.3	0.8
132.59	2.87	0.3	245.88	1DE	-0.5	-4.2	-2.3	0.8
134.19	3.24	0.1	256.08	11DE	0.6	-4.1	-2.2	0.8
134.69	3.62	-0.2	266.28	21DE	2.9	-4.0	-2.2	0.8
133.94	4.00	-0.4	276.45	31DE	0.9	-3.9	-2.1	0.8

-371

♂ LONG	LAT	MAG	☉ LONG	16.00UT	☿	♀	♃	♄
131.90	4.33	-0.7	286.58	10JA	0.3	-3.8	-2.0	0.8
128.71	4.56	-0.9	296.67	20JA	0.1	-3.7	-2.0	0.8
124.87	4.64	-1.0	306.71	30JA	-0.0	-3.6	-1.9	0.7
121.06	4.53	-0.9	316.69	9FE	-0.4	-3.5	-1.8	0.7
117.95	4.26	-0.7	326.62	19FE	-0.9	-3.5	-1.8	0.6
116.02	3.90	-0.5	336.48	1MR	-1.6	-3.4	-1.7	0.6
115.38	3.50	-0.3	346.29	11MR	-1.2	-3.4	-1.6	0.5
115.98	3.10	-0.0	356.04	21MR	-0.1	-3.4	-1.6	0.5
117.66	2.72	0.2	5.74	31MR	1.2	-3.3	-1.5	0.4
120.23	2.37	0.4	15.38	10AP	2.7	-3.3	-1.5	0.4
123.53	2.06	0.5	24.99	20AP	3.3	-3.3	-1.5	0.3
127.44	1.77	0.7	34.56	30AP	1.9	-3.3	-1.4	0.3
131.83	1.50	0.8	44.10	10MY	1.0	-3.3	-1.4	0.4
136.63	1.26	0.9	53.63	20MY	0.2	-3.3	-1.4	0.4
141.79	1.03	1.0	63.15	30MY	-0.8	-3.3	-1.4	0.5
147.23	0.83	1.1	72.67	9JN	-1.7	-3.4	-1.4	0.6
152.95	0.63	1.1	82.20	19JN	-1.2	-3.4	-1.4	0.6
158.89	0.45	1.2	91.74	29JN	-0.4	-3.4	-1.4	0.7
165.04	0.27	1.2	101.32	9JL	0.1	-3.5	-1.4	0.7
171.39	0.11	1.3	110.93	19JL	0.4	-3.5	-1.5	0.8
177.92	-0.04	1.3	120.58	29JL	0.7	-3.5	-1.5	0.8
184.61	-0.19	1.3	130.29	8AU	1.3	-3.4	-1.6	0.8
191.46	-0.32	1.4	140.05	18AU	3.0	-3.4	-1.6	0.8
198.46	-0.45	1.4	149.86	28AU	1.1	-3.4	-1.7	0.9
205.60	-0.57	1.4	159.74	7SE	-0.5	-3.4	-1.7	0.8
212.86	-0.68	1.4	169.67	17SE	-0.9	-3.3	-1.8	0.8
220.25	-0.77	1.4	179.66	27SE	-0.9	-3.3	-1.8	0.8
227.74	-0.86	1.4	189.71	7OC	-0.8	-3.3	-1.9	0.8
235.33	-0.94	1.4	199.81	17OC	-0.6	-3.3	-2.0	0.8
243.00	-1.00	1.4	209.94	27OC	-0.5	-3.4	-2.1	0.7
250.75	-1.06	1.4	220.12	6NO	-0.5	-3.4	-2.1	0.7
258.56	-1.10	1.4	230.31	16NO	-0.3	-3.4	-2.1	0.7
266.40	-1.13	1.4	240.52	26NO	0.8	-3.4	-2.2	0.8
274.28	-1.14	1.4	250.73	6DE	2.6	-3.5	-2.2	0.8
282.17	-1.15	1.4	260.94	16DE	0.6	-3.5	-2.2	0.8
290.06	-1.14	1.4	271.12	26DE	0.1	-3.6	-2.2	0.8

♂ LONG	♂ LAT	♂ MAG	⊙ LONG	16.00UT	☿	♀	♃	♄
				-370	MAGNITUDES			
297.93	-1.11	1.4	281.27	5JA	0.0	-3.6	-2.1	0.8
305.78	-1.08	1.4	291.38	15JA	-0.1	-3.7	-2.1	0.8
313.59	-1.04	1.4	301.45	25JA	-0.4	-3.7	-2.0	0.8
321.34	-0.99	1.4	311.46	4FE	-1.0	-3.8	-2.0	0.7
329.04	-0.92	1.4	321.42	14FE	-1.5	-3.9	-1.9	0.7
336.66	-0.85	1.4	331.32	24FE	-1.1	-4.0	-1.8	0.7
344.21	-0.78	1.4	341.16	6MR	0.0	-4.1	-1.8	0.6
351.68	-0.69	1.5	350.94	16MR	1.5	-4.2	-1.7	0.6
359.06	-0.61	1.5	0.66	26MR	3.4	-4.2	-1.6	0.5
6.36	-0.51	1.5	10.33	5AP	2.5	-4.2	-1.6	0.4
13.57	-0.42	1.6	19.96	15AP	1.4	-4.0	-1.5	0.4
20.69	-0.32	1.6	29.54	25AP	0.8	-3.6	-1.5	0.3
27.72	-0.22	1.7	39.10	5MY	0.1	-2.8	-1.4	0.3
34.67	-0.12	1.7	48.64	15MY	-0.8	-3.4	-1.4	0.3
41.54	-0.01	1.8	58.15	25MY	-1.7	-4.0	-1.4	0.4
48.32	0.09	1.8	67.67	4JN	-1.2	-4.2	-1.3	0.4
55.03	0.20	1.8	77.20	14JN	-0.4	-4.2	-1.3	0.5
61.66	0.30	1.8	86.73	24JN	0.2	-4.1	-1.3	0.6
68.21	0.41	1.9	96.29	4JL	0.6	-4.1	-1.3	0.6
74.69	0.51	1.9	105.88	14JL	0.9	-4.0	-1.3	0.7
81.10	0.62	1.9	115.51	24JL	1.7	-3.9	-1.3	0.7
87.44	0.73	1.9	125.20	3AU	3.2	-3.8	-1.3	0.8
93.69	0.84	1.8	134.92	13AU	1.0	-3.7	-1.4	0.8
99.87	0.95	1.8	144.71	23AU	-0.6	-3.7	-1.4	0.8
105.97	1.07	1.8	154.56	2SE	-1.1	-3.6	-1.4	0.8
111.98	1.18	1.8	164.46	12SE	-1.1	-3.6	-1.5	0.8
117.89	1.30	1.7	174.42	22SE	-0.7	-3.5	-1.5	0.8
123.69	1.43	1.6	184.44	2OC	-0.5	-3.5	-1.6	0.8
129.35	1.56	1.6	194.51	12OC	-0.3	-3.4	-1.6	0.8
134.88	1.69	1.5	204.63	22OC	-0.3	-3.4	-1.7	0.8
140.22	1.84	1.4	214.79	1NO	-0.1	-3.4	-1.8	0.7
145.35	1.99	1.3	224.97	11NO	1.0	-3.4	-1.8	0.7
150.23	2.16	1.1	235.17	21NO	2.3	-3.4	-1.9	0.7
154.79	2.33	1.0	245.38	1DE	0.3	-3.4	-1.9	0.8
158.97	2.52	0.8	255.59	11DE	-0.0	-3.3	-2.0	0.8
162.68	2.73	0.6	265.78	21DE	-0.1	-3.3	-2.0	0.8
165.79	2.95	0.4	275.95	31DE	-0.2	-3.4	-2.1	0.8
				-369				
168.16	3.18	0.1	286.09	10JA	-0.5	-3.4	-2.1	0.8
169.62	3.41	-0.1	296.18	20JA	-1.0	-3.4	-2.1	0.8
169.98	3.64	-0.4	306.22	30JA	-1.3	-3.4	-2.1	0.8
169.10	3.81	-0.7	316.21	9FE	-1.0	-3.4	-2.0	0.8
166.96	3.91	-1.0	326.13	19FE	0.1	-3.5	-2.0	0.7
163.75	3.88	-1.2	336.01	1MR	1.9	-3.5	-1.9	0.7
160.00	3.68	-1.3	345.81	11MR	3.3	-3.5	-1.9	0.7
156.41	3.34	-1.2	355.56	21MR	1.8	-3.4	-1.8	0.6
153.64	2.89	-1.0	5.27	31MR	1.1	-3.4	-1.7	0.5
152.10	2.40	-0.8	14.91	10AP	0.6	-3.4	-1.7	0.5
151.88	1.92	-0.6	24.52	20AP	-0.0	-3.4	-1.6	0.4
152.92	1.47	-0.4	34.09	30AP	-0.8	-3.3	-1.5	0.4
155.05	1.07	-0.2	43.64	10MY	-1.8	-3.3	-1.5	0.3
158.07	0.71	-0.0	53.16	20MY	-1.2	-3.3	-1.4	0.3
161.86	0.39	0.1	62.68	30MY	-0.3	-3.3	-1.4	0.3
166.26	0.11	0.3	72.20	9JN	0.3	-3.3	-1.3	0.4
171.17	-0.14	0.4	81.73	19JN	0.7	-3.4	-1.3	0.4
176.52	-0.36	0.5	91.28	29JN	1.3	-3.4	-1.3	0.5
182.24	-0.56	0.6	100.85	9JL	2.3	-3.4	-1.3	0.6
188.27	-0.73	0.7	110.46	19JL	2.9	-3.5	-1.2	0.6
194.59	-0.89	0.7	120.11	29JL	0.8	-3.5	-1.2	0.7
201.15	-1.02	0.8	129.81	8AU	-0.7	-3.6	-1.2	0.7
207.93	-1.13	0.9	139.57	18AU	-1.2	-3.6	-1.2	0.8
214.90	-1.23	0.9	149.39	28AU	-1.2	-3.7	-1.3	0.8
222.03	-1.30	1.0	159.26	7SE	-0.7	-3.8	-1.3	0.8
229.32	-1.36	1.0	169.19	17SE	-0.3	-3.9	-1.3	0.8
236.74	-1.40	1.0	179.18	27SE	-0.2	-4.0	-1.3	0.8
244.26	-1.43	1.1	189.22	7OC	-0.1	-4.1	-1.4	0.8
251.88	-1.44	1.1	199.32	17OC	0.1	-4.2	-1.4	0.8
259.56	-1.43	1.2	209.46	27OC	1.2	-4.3	-1.5	0.8
267.30	-1.41	1.2	219.62	6NO	2.1	-4.4	-1.5	0.7
275.07	-1.37	1.2	229.82	16NO	0.1	-4.3	-1.6	0.7
282.86	-1.32	1.3	240.03	26NO	-0.2	-4.1	-1.7	0.7
290.65	-1.26	1.3	250.24	6DE	-0.3	-3.4	-1.7	0.7
298.43	-1.19	1.4	260.44	16DE	-0.4	-3.3	-1.8	0.8
306.18	-1.10	1.4	270.62	26DE	-0.6	-4.0	-1.9	0.8

♂ LONG	♂ LAT	♂ MAG	⊙ LONG	16.00UT	☿	♀	♃	♄
				-368	MAGNITUDES			
313.88	-1.02	1.4	280.77	5JA	-1.0	-4.3	-1.9	0.8
321.54	-0.92	1.5	290.89	15JA	-1.2	-4.3	-2.0	0.8
329.13	-0.82	1.5	300.96	25JA	-1.0	-4.3	-2.0	0.8
336.66	-0.72	1.5	310.97	4FE	0.2	-4.2	-2.0	0.8
344.11	-0.62	1.6	320.94	14FE	2.2	-4.1	-2.0	0.8
351.48	-0.51	1.6	330.84	24FE	2.6	-4.0	-2.0	0.8
358.77	-0.40	1.6	340.68	5MR	1.4	-3.9	-2.0	0.7
5.98	-0.30	1.7	350.46	15MR	0.8	-3.8	-2.0	0.7
13.11	-0.19	1.7	0.19	25MR	0.4	-3.7	-1.9	0.6
20.16	-0.08	1.7	9.86	4AP	-0.1	-3.6	-1.9	0.6
27.13	0.02	1.7	19.49	14AP	-0.9	-3.5	-1.8	0.5
34.03	0.12	1.7	29.08	24AP	-1.8	-3.5	-1.7	0.5
40.85	0.22	1.7	38.64	4MY	-1.2	-3.4	-1.7	0.4
47.61	0.31	1.8	48.18	14MY	-0.3	-3.4	-1.6	0.3
54.30	0.41	1.8	57.70	24MY	0.4	-3.4	-1.5	0.3
60.93	0.50	1.8	67.21	3JN	1.0	-3.3	-1.5	0.3
67.50	0.58	1.9	76.74	13JN	1.7	-3.3	-1.4	0.3
74.02	0.67	1.9	86.27	23JN	2.9	-3.3	-1.4	0.4
80.50	0.75	1.9	95.83	3JL	2.4	-3.3	-1.3	0.5
86.93	0.83	2.0	105.42	13JL	0.7	-3.3	-1.3	0.5
93.31	0.91	2.0	115.05	23JL	-0.7	-3.3	-1.3	0.6
99.66	0.99	2.0	124.72	2AU	-1.3	-3.4	-1.2	0.6
105.97	1.06	2.0	134.45	12AU	-1.2	-3.4	-1.2	0.7
112.25	1.13	2.0	144.23	22AU	-0.6	-3.4	-1.2	0.7
118.48	1.20	2.0	154.07	1SE	-0.3	-3.4	-1.2	0.8
124.69	1.27	1.9	163.98	11SE	-0.1	-3.5	-1.2	0.8
130.84	1.34	1.9	173.93	21SE	0.0	-3.5	-1.2	0.8
136.96	1.41	1.9	183.95	1OC	0.3	-3.5	-1.2	0.8
143.03	1.48	1.8	194.02	11OC	1.5	-3.5	-1.3	0.8
149.04	1.54	1.8	204.13	21OC	1.8	-3.4	-1.3	0.8
154.98	1.61	1.7	214.29	31OC	-0.0	-3.4	-1.3	0.8
160.85	1.67	1.6	224.47	10NO	-0.4	-3.4	-1.4	0.8
166.63	1.73	1.5	234.67	20NO	-0.4	-3.4	-1.4	0.7
172.29	1.79	1.4	244.88	30NO	-0.5	-3.3	-1.5	0.7
177.83	1.85	1.3	255.09	10DE	-0.7	-3.3	-1.6	0.7
183.19	1.91	1.1	265.29	20DE	-1.0	-3.3	-1.6	0.8
188.37	1.96	1.0	275.46	30DE	-1.0	-3.3	-1.7	0.8
				-367				
193.30	2.01	0.8	285.59	9JA	-0.9	-3.4	-1.8	0.8
197.92	2.04	0.6	295.69	19JA	0.2	-3.4	-1.8	0.8
202.18	2.07	0.4	305.73	29JA	2.6	-3.4	-1.9	0.8
205.96	2.07	0.1	315.72	8FE	2.0	-3.4	-1.9	0.8
209.15	2.05	-0.1	325.65	18FE	1.0	-3.5	-2.0	0.8
211.60	2.00	-0.4	335.53	28FE	0.6	-3.5	-2.0	0.8
213.13	1.89	-0.7	345.34	10MR	0.3	-3.5	-2.0	0.8
213.56	1.71	-1.0	355.10	20MR	-0.2	-3.6	-2.0	0.7
212.77	1.44	-1.4	4.80	30MR	-0.9	-3.7	-2.0	0.7
210.74	1.07	-1.7	14.45	9AP	-1.8	-3.7	-2.0	0.7
207.76	0.60	-1.9	24.06	19AP	-1.2	-3.8	-2.0	0.6
204.38	0.06	-1.9	33.64	29AP	-0.2	-3.9	-1.9	0.5
201.33	-0.48	-1.8	43.18	9MY	0.6	-4.0	-1.9	0.5
199.26	-0.98	-1.6	52.71	19MY	1.3	-4.1	-1.8	0.4
198.49	-1.41	-1.4	62.23	29MY	2.2	-4.1	-1.7	0.3
199.11	-1.74	-1.2	71.75	8JN	3.4	-4.2	-1.7	0.3
200.99	-2.00	-1.0	81.27	18JN	2.0	-4.1	-1.6	0.3
203.95	-2.19	-0.8	90.82	28JN	0.5	-3.9	-1.5	0.4
207.82	-2.32	-0.6	100.39	8JL	-0.8	-3.4	-1.5	0.4
212.42	-2.41	-0.4	109.99	18JL	-1.4	-3.2	-1.4	0.5
217.62	-2.46	-0.3	119.64	28JL	-1.2	-3.8	-1.4	0.5
223.32	-2.47	-0.1	129.34	7AU	-0.6	-4.2	-1.3	0.6
229.41	-2.45	0.0	139.09	17AU	-0.2	-4.3	-1.3	0.6
235.83	-2.41	0.1	148.91	27AU	0.1	-4.3	-1.3	0.7
242.52	-2.35	0.2	158.78	6SE	0.2	-4.2	-1.3	0.7
249.42	-2.26	0.4	168.70	16SE	0.5	-4.1	-1.2	0.8
256.50	-2.16	0.5	178.69	26SE	1.9	-4.0	-1.2	0.8
263.71	-2.04	0.6	188.73	6OC	1.6	-3.9	-1.2	0.8
271.02	-1.92	0.7	198.82	16OC	-0.2	-3.9	-1.2	0.8
278.40	-1.78	0.8	208.96	26OC	-0.5	-3.8	-1.3	0.8
285.83	-1.63	0.9	219.13	5NO	-0.5	-3.7	-1.3	0.8
293.28	-1.48	1.0	229.32	15NO	-0.6	-3.6	-1.3	0.8
300.73	-1.33	1.0	239.53	25NO	-0.8	-3.6	-1.4	0.8
308.18	-1.17	1.1	249.75	5DE	-0.8	-3.5	-1.4	0.7
315.60	-1.02	1.2	259.95	15DE	-0.9	-3.5	-1.4	0.7
322.98	-0.87	1.3	270.13	25DE	-0.7	-3.4	-1.5	0.8

♂ LONG	♂ LAT	♂ MAG	⊙ LONG	16.00UT	☿	♀	♃	♄
				-366	MAGNITUDES			
330.32	-0.72	1.4	280.29	4JA	0.3	-3.4	-1.6	0.8
337.60	-0.58	1.4	290.40	14JA	2.9	-3.4	-1.6	0.8
344.82	-0.44	1.5	300.48	24JA	1.5	-3.3	-1.7	0.9
351.97	-0.31	1.6	310.50	3FE	0.7	-3.3	-1.8	0.9
359.06	-0.18	1.6	320.46	13FE	0.4	-3.3	-1.8	0.9
6.09	-0.06	1.7	330.36	23FE	0.1	-3.3	-1.9	0.9
13.04	0.05	1.7	340.21	5MR	-0.2	-3.3	-2.0	0.9
19.93	0.16	1.8	349.99	15MR	-0.9	-3.3	-2.0	0.8
26.76	0.27	1.8	359.72	25MR	-1.8	-3.3	-2.0	0.8
33.52	0.36	1.8	9.40	4AP	-1.2	-3.4	-2.1	0.8
40.23	0.45	1.9	19.03	14AP	-0.2	-3.4	-2.1	0.7
46.88	0.54	1.9	28.62	24AP	0.8	-3.4	-2.1	0.7
53.48	0.62	1.9	38.18	4MY	1.7	-3.5	-2.1	0.6
60.04	0.69	1.9	47.71	14MY	3.0	-3.5	-2.0	0.5
66.56	0.76	1.9	57.24	24MY	3.0	-3.4	-2.0	0.5
73.04	0.83	1.9	66.75	3JN	1.6	-3.4	-2.0	0.4
79.49	0.89	1.9	76.27	13JN	0.4	-3.4	-1.9	0.4
85.91	0.94	1.9	85.81	23JN	-0.8	-3.4	-1.8	0.3
92.31	1.00	1.9	95.36	3JL	-1.5	-3.3	-1.8	0.4
98.70	1.04	2.0	104.95	13JL	-1.2	-3.3	-1.7	0.4
105.07	1.09	2.0	114.58	23JL	-0.5	-3.3	-1.7	0.5
111.43	1.13	2.0	124.25	2AU	-0.1	-3.3	-1.6	0.5
117.79	1.16	2.0	133.97	12AU	0.2	-3.3	-1.5	0.6
124.15	1.19	2.0	143.75	22AU	0.4	-3.4	-1.5	0.6
130.50	1.22	2.0	153.59	1SE	0.8	-3.4	-1.4	0.7
136.85	1.25	2.0	163.49	11SE	2.3	-3.4	-1.4	0.7
143.21	1.27	2.0	173.45	21SE	1.5	-3.4	-1.4	0.8
149.57	1.28	2.0	183.46	1OC	-0.3	-3.5	-1.3	0.8
155.93	1.30	1.9	193.53	11OC	-0.7	-3.5	-1.3	0.8
162.29	1.30	1.9	203.64	21OC	-0.7	-3.6	-1.3	0.8
168.65	1.30	1.8	213.79	31OC	-0.7	-3.7	-1.3	0.8
175.01	1.30	1.8	223.97	10NO	-0.7	-3.8	-1.3	0.8
181.37	1.29	1.7	234.18	20NO	-0.7	-3.8	-1.3	0.8
187.71	1.27	1.6	244.39	30NO	-0.7	-3.9	-1.3	0.8
194.03	1.24	1.5	254.60	10DE	-0.6	-4.0	-1.4	0.8
200.33	1.20	1.4	264.80	20DE	0.5	-4.1	-1.4	0.8
206.61	1.15	1.3	274.97	30DE	3.0	-4.2	-1.4	0.8
				-365				
212.84	1.08	1.2	285.11	9JA	1.1	-4.3	-1.5	0.8
219.02	1.00	1.1	295.20	19JA	0.4	-4.3	-1.5	0.9
225.14	0.90	0.9	305.25	29JA	0.2	-4.3	-1.6	0.9
231.18	0.77	0.8	315.24	8FE	0.0	-4.0	-1.7	0.9
237.12	0.62	0.6	325.18	18FE	-0.3	-3.4	-1.7	0.9
242.92	0.44	0.4	335.05	28FE	-0.9	-3.4	-1.8	0.9
248.57	0.21	0.2	344.87	10MR	-1.7	-3.9	-1.9	0.9
253.99	-0.05	0.0	354.63	20MR	-1.2	-4.2	-1.9	0.9
259.15	-0.37	-0.2	4.33	30MR	-0.1	-4.2	-2.0	0.9
263.96	-0.75	-0.4	13.99	9AP	1.0	-4.2	-2.1	0.9
268.30	-1.20	-0.7	23.60	19AP	2.2	-4.1	-2.1	0.8
272.07	-1.73	-1.0	33.17	29AP	3.8	-4.0	-2.2	0.8
275.09	-2.35	-1.3	42.72	9MY	2.3	-3.9	-2.2	0.7
277.18	-3.05	-1.6	52.25	19MY	1.2	-3.8	-2.2	0.7
278.15	-3.83	-1.9	61.77	29MY	0.3	-3.7	-2.2	0.6
277.86	-4.63	-2.2	71.29	8JN	-0.8	-3.6	-2.2	0.5
276.37	-5.38	-2.4	80.81	18JN	-1.6	-3.6	-2.2	0.5
274.01	-5.96	-2.6	90.35	28JN	-1.2	-3.5	-2.1	0.4
271.38	-6.28	-2.6	99.93	8JL	-0.5	-3.5	-2.1	0.4
269.23	-6.30	-2.4	109.53	18JL	0.0	-3.4	-2.0	0.4
268.13	-6.06	-2.1	119.18	28JL	0.3	-3.4	-1.9	0.5
268.33	-5.66	-1.9	128.88	7AU	0.6	-3.4	-1.9	0.5
269.84	-5.16	-1.6	138.62	17AU	1.1	-3.4	-1.8	0.6
272.49	-4.64	-1.3	148.43	27AU	2.7	-3.4	-1.7	0.6
276.09	-4.12	-1.1	158.30	6SE	1.3	-3.4	-1.7	0.7
280.45	-3.62	-0.8	168.22	16SE	-0.4	-3.4	-1.6	0.7
285.38	-3.16	-0.6	178.20	26SE	-0.8	-3.4	-1.6	0.8
290.78	-2.72	-0.3	188.25	6OC	-0.8	-3.4	-1.5	0.8
296.52	-2.33	-0.1	198.33	16OC	-0.8	-3.4	-1.5	0.8
302.52	-1.96	0.1	208.46	26OC	-0.6	-3.4	-1.5	0.9
308.73	-1.63	0.3	218.63	5NO	-0.5	-3.4	-1.4	0.9
315.08	-1.33	0.4	228.83	15NO	-0.5	-3.4	-1.4	0.9
321.53	-1.06	0.6	239.03	25NO	-0.4	-3.5	-1.4	0.9
328.06	-0.81	0.7	249.25	5DE	0.6	-3.5	-1.4	0.9
334.64	-0.58	0.9	259.45	15DE	2.9	-3.5	-1.4	0.9
341.24	-0.38	1.0	269.63	25DE	0.8	-3.5	-1.4	0.9

♂ LONG	♂ LAT	♂ MAG	⊙ LONG	16.00UT	☿	♀	♃	♄
				-364	MAGNITUDES			
347.86	-0.20	1.1	279.79	4JA	0.2	-3.4	-1.5	0.9
354.48	-0.03	1.3	289.91	14JA	0.1	-3.4	-1.5	0.9
1.09	0.12	1.4	299.98	24JA	-0.1	-3.4	-1.5	0.9
7.68	0.25	1.5	310.01	3FE	-0.4	-3.4	-1.6	1.0
14.25	0.37	1.6	319.97	13FE	-0.9	-3.3	-1.6	1.0
20.79	0.48	1.6	329.88	23FE	-1.6	-3.3	-1.7	1.0
27.31	0.58	1.7	339.73	4MR	-1.2	-3.3	-1.7	1.0
33.80	0.66	1.8	349.52	14MR	-0.1	-3.3	-1.8	1.0
40.26	0.74	1.8	359.25	24MR	1.3	-3.3	-1.9	1.0
46.70	0.81	1.9	8.93	3AP	2.9	-3.4	-1.9	1.0
53.11	0.87	1.9	18.56	13AP	3.0	-3.4	-2.0	1.0
59.50	0.93	2.0	28.16	23AP	1.7	-3.4	-2.1	0.9
65.87	0.98	2.0	37.72	3MY	0.9	-3.4	-2.1	0.9
72.23	1.02	2.0	47.26	13MY	0.2	-3.5	-2.2	0.9
78.57	1.05	2.0	56.78	23MY	-0.8	-3.5	-2.2	0.8
84.92	1.08	2.0	66.30	2JN	-1.7	-3.6	-2.3	0.7
91.25	1.11	2.0	75.82	12JN	-1.2	-3.7	-2.3	0.7
97.59	1.13	2.0	85.35	22JN	-0.4	-3.7	-2.3	0.6
103.93	1.14	2.0	94.91	2JL	0.1	-3.8	-2.3	0.5
110.29	1.16	2.0	104.49	12JL	0.5	-3.9	-2.3	0.5
116.66	1.16	2.0	114.12	22JL	0.8	-4.0	-2.3	0.5
123.05	1.16	1.9	123.79	1AU	1.4	-4.1	-2.2	0.5
129.46	1.16	2.0	133.51	11AU	3.1	-4.2	-2.2	0.6
135.90	1.15	2.0	143.28	21AU	1.1	-4.3	-2.1	0.6
142.37	1.14	2.0	153.12	31AU	-0.5	-4.2	-2.0	0.7
148.86	1.12	2.0	163.01	10SE	-1.0	-4.0	-2.0	0.7
155.40	1.10	2.0	172.96	20SE	-1.0	-3.5	-1.9	0.8
161.97	1.07	2.0	182.98	30SE	-0.8	-3.4	-1.8	0.8
168.58	1.03	1.9	193.04	10OC	-0.5	-4.0	-1.8	0.9
175.23	0.99	1.9	203.15	20OC	-0.4	-4.3	-1.7	0.9
181.92	0.94	1.9	213.30	30OC	-0.4	-4.4	-1.7	0.9
188.65	0.89	1.8	223.48	9NO	-0.3	-4.3	-1.6	0.9
195.43	0.83	1.8	233.68	19NO	0.8	-4.3	-1.6	1.0
202.24	0.75	1.7	243.89	29NO	2.6	-4.2	-1.6	1.0
209.09	0.67	1.7	254.10	9DE	0.5	-4.1	-1.5	1.0
215.98	0.58	1.6	264.30	19DE	0.1	-3.9	-1.5	1.0
222.91	0.47	1.5	274.47	29DE	-0.0	-3.9	-1.5	1.0
				-363				
229.87	0.35	1.4	284.61	8JA	-0.2	-3.8	-1.5	1.0
236.87	0.22	1.4	294.71	18JA	-0.5	-3.7	-1.5	1.0
243.90	0.08	1.3	304.76	28JA	-1.0	-3.6	-1.5	1.0
250.95	-0.08	1.2	314.76	7FE	-1.4	-3.5	-1.5	1.1
258.02	-0.25	1.1	324.70	17FE	-1.1	-3.5	-1.6	1.1
265.10	-0.45	1.0	334.58	27FE	-0.0	-3.4	-1.6	1.1
272.18	-0.65	0.8	344.40	9MR	1.6	-3.4	-1.7	1.1
279.25	-0.87	0.7	354.16	19MR	3.5	-3.4	-1.7	1.2
286.29	-1.11	0.6	3.87	29MR	2.2	-3.3	-1.8	1.2
293.29	-1.36	0.5	13.52	8AP	1.3	-3.3	-1.8	1.1
300.21	-1.63	0.4	23.14	18AP	0.7	-3.3	-1.9	1.1
307.04	-1.91	0.2	32.72	28AP	0.1	-3.3	-1.9	1.1
313.73	-2.19	0.1	42.26	8MY	-0.8	-3.3	-2.0	1.1
320.23	-2.49	-0.1	51.79	18MY	-1.7	-3.3	-2.1	1.0
326.49	-2.80	-0.2	61.31	28MY	-1.2	-3.3	-2.2	1.0
332.45	-3.11	-0.4	70.83	7JN	-0.4	-3.4	-2.2	0.9
338.01	-3.42	-0.5	80.35	17JN	0.2	-3.4	-2.3	0.8
343.06	-3.74	-0.7	89.90	27JN	0.6	-3.4	-2.3	0.8
347.48	-4.04	-0.9	99.46	7JL	1.1	-3.5	-2.4	0.7
351.07	-4.34	-1.1	109.07	17JL	1.9	-3.5	-2.4	0.6
353.65	-4.61	-1.4	118.71	27JL	3.2	-3.5	-2.4	0.6
355.00	-4.82	-1.6	128.40	6AU	1.0	-3.4	-2.4	0.6
354.94	-4.93	-1.8	138.15	16AU	-0.6	-3.4	-2.4	0.6
353.43	-4.90	-2.0	147.95	26AU	-1.1	-3.4	-2.4	0.7
350.74	-4.66	-2.2	157.81	5SE	-1.1	-3.3	-2.3	0.7
347.44	-4.20	-2.2	167.74	15SE	-0.7	-3.3	-2.3	0.8
344.36	-3.58	-2.0	177.72	25SE	-0.4	-3.3	-2.2	0.8
342.19	-2.89	-1.7	187.75	5OC	-0.3	-3.3	-2.1	0.9
341.30	-2.21	-1.4	197.84	15OC	-0.2	-3.3	-2.1	0.9
341.75	-1.60	-1.1	207.97	25OC	-0.1	-3.3	-2.0	1.0
343.39	-1.07	-0.7	218.13	4NO	1.0	-3.4	-1.9	1.0
346.01	-0.63	-0.4	228.33	14NO	2.3	-3.4	-1.9	1.0
349.42	-0.27	-0.2	238.53	24NO	0.3	-3.4	-1.8	1.1
353.44	0.02	0.1	248.74	4DE	-0.1	-3.5	-1.8	1.1
357.94	0.27	0.3	258.95	14DE	-0.2	-3.5	-1.7	1.1
2.81	0.47	0.5	269.14	24DE	-0.3	-3.6	-1.7	1.1

♂			☉	16.00UT	☿	♀	♃	♄
LONG	LAT	MAG	LONG	-362	MAGNITUDES			
7.95	0.63	0.7	279.29	3JA	-0.5	-3.6	-1.6	1.1
13.31	0.77	0.9	289.42	13JA	-1.0	-3.7	-1.6	1.1
18.85	0.88	1.1	299.50	23JA	-1.3	-3.7	-1.6	1.1
24.51	0.97	1.2	309.52	2FE	-1.0	-3.8	-1.6	1.1
30.28	1.05	1.3	319.49	12FE	0.1	-3.9	-1.6	1.2
36.13	1.11	1.5	329.40	22FE	1.9	-4.0	-1.6	1.2
42.03	1.16	1.6	339.25	4MR	3.1	-4.1	-1.6	1.3
47.99	1.20	1.6	349.05	14MR	1.6	-4.2	-1.6	1.3
53.99	1.23	1.7	358.78	24MR	1.0	-4.2	-1.6	1.3
60.02	1.25	1.8	8.46	3AP	0.5	-4.2	-1.7	1.3
66.09	1.27	1.9	18.10	13AP	-0.0	-4.0	-1.7	1.3
72.18	1.28	1.9	27.70	23AP	-0.8	-3.6	-1.7	1.3
78.29	1.28	1.9	37.26	3MY	-1.7	-2.8	-1.8	1.3
84.44	1.28	2.0	46.80	13MY	-1.2	-3.5	-1.8	1.3
90.60	1.27	2.0	56.32	23MY	-0.3	-4.0	-1.9	1.2
96.80	1.26	2.0	65.84	2JN	0.3	-4.2	-2.0	1.2
103.03	1.25	2.0	75.36	12JN	0.8	-4.2	-2.0	1.1
109.29	1.22	2.0	84.89	22JN	1.4	-4.1	-2.1	1.1
115.59	1.20	2.0	94.44	2JL	2.5	-4.1	-2.2	1.0
121.93	1.17	2.0	104.03	12JL	2.8	-4.0	-2.2	0.9
128.31	1.14	2.0	113.65	22JL	0.8	-3.9	-2.3	0.8
134.74	1.10	2.0	123.32	1AU	-0.7	-3.8	-2.4	0.8
141.22	1.06	1.9	133.04	11AU	-1.3	-3.7	-2.4	0.7
147.75	1.01	1.9	142.81	21AU	-1.2	-3.7	-2.4	0.7
154.34	0.96	1.9	152.64	31AU	-0.7	-3.6	-2.5	0.8
160.99	0.90	1.8	162.53	10SE	-0.3	-3.6	-2.5	0.8
167.70	0.84	1.8	172.48	20SE	-0.1	-3.5	-2.4	0.9
174.48	0.77	1.8	182.49	30SE	-0.1	-3.5	-2.4	0.9
181.32	0.70	1.8	192.55	10OC	0.1	-3.4	-2.4	1.0
188.23	0.62	1.8	202.66	20OC	1.3	-3.4	-2.3	1.0
195.21	0.54	1.8	212.81	30OC	2.1	-3.4	-2.2	1.1
202.25	0.45	1.8	222.99	9NO	0.1	-3.4	-2.2	1.1
209.36	0.35	1.7	233.18	19NO	-0.3	-3.4	-2.1	1.2
216.54	0.25	1.7	243.40	29NO	-0.3	-3.4	-2.0	1.2
223.78	0.14	1.7	253.60	9DE	-0.4	-3.4	-2.0	1.2
231.08	0.02	1.6	263.80	19DE	-0.6	-3.4	-1.9	1.3
238.45	-0.10	1.6	273.98	29DE	-1.0	-3.4	-1.8	1.3
				-361				
245.86	-0.22	1.5	284.12	8JA	-1.1	-3.4	-1.8	1.3
253.33	-0.36	1.5	294.22	18JA	-0.9	-3.4	-1.7	1.3
260.84	-0.49	1.4	304.27	28JA	0.1	-3.4	-1.7	1.3
268.38	-0.63	1.4	314.27	7FE	2.3	-3.4	-1.6	1.3
275.95	-0.78	1.3	324.21	17FE	2.4	-3.5	-1.6	1.3
283.55	-0.92	1.2	334.09	27FE	1.2	-3.5	-1.6	1.4
291.14	-1.06	1.2	343.92	9MR	0.7	-3.5	-1.6	1.4
298.74	-1.20	1.1	353.68	19MR	0.4	-3.4	-1.6	1.4
306.31	-1.33	1.1	3.39	29MR	-0.1	-3.4	-1.6	1.3
313.85	-1.46	1.0	13.05	8AP	-0.8	-3.4	-1.6	1.3
321.34	-1.57	1.0	22.66	18AP	-1.7	-3.4	-1.6	1.3
328.76	-1.68	0.9	32.25	28AP	-1.2	-3.3	-1.6	1.3
336.09	-1.78	0.8	41.80	8MY	-0.3	-3.3	-1.6	1.2
343.31	-1.86	0.8	51.32	18MY	0.5	-3.3	-1.7	1.2
350.41	-1.92	0.7	60.85	28MY	1.1	-3.3	-1.7	1.2
357.35	-1.97	0.6	70.36	7JN	1.9	-3.3	-1.7	1.1
4.10	-2.01	0.6	79.89	17JN	3.1	-3.4	-1.8	1.1
10.64	-2.02	0.5	89.43	27JN	2.3	-3.4	-1.9	1.0
16.92	-2.02	0.4	99.00	7JL	0.6	-3.4	-1.9	1.0
22.90	-1.99	0.3	108.60	17JL	-0.7	-3.5	-2.0	0.9
28.52	-1.95	0.2	118.24	27JL	-1.4	-3.5	-2.0	0.9
33.71	-1.88	0.1	127.93	6AU	-1.2	-3.6	-2.1	0.8
38.36	-1.78	-0.1	137.67	16AU	-0.6	-3.6	-2.2	0.8
42.36	-1.66	-0.3	147.48	26AU	-0.2	-3.7	-2.2	0.8
45.56	-1.49	-0.4	157.34	5SE	-0.0	-3.8	-2.3	0.8
47.77	-1.28	-0.6	167.26	15SE	0.1	-3.9	-2.3	0.9
48.78	-1.01	-0.8	177.24	25SE	0.4	-4.0	-2.4	1.0
48.42	-0.67	-1.1	187.27	5OC	1.6	-4.1	-2.4	1.0
46.61	-0.27	-1.3	197.35	15OC	1.9	-4.2	-2.4	1.1
43.53	0.18	-1.5	207.48	25OC	-0.1	-4.3	-2.4	1.2
39.73	0.63	-1.6	217.64	4NO	-0.4	-4.4	-2.4	1.2
35.98	1.04	-1.4	227.83	14NO	-0.5	-4.3	-2.3	1.3
33.03	1.37	-1.1	238.04	24NO	-0.5	-4.1	-2.3	1.3
31.33	1.61	-0.8	248.25	4DE	-0.7	-3.4	-2.2	1.3
30.97	1.77	-0.5	258.45	14DE	-0.9	-3.3	-2.1	1.3
31.86	1.87	-0.2	268.64	24DE	-1.0	-4.0	-2.0	1.3

♂			☉	16.00UT	☿	♀	♃	♄
LONG	LAT	MAG	LONG	-360	MAGNITUDES			
33.80	1.92	0.1	278.80	3JA	-0.8	-4.3	-2.0	1.3
36.57	1.94	0.3	288.93	13JA	0.2	-4.4	-1.9	1.3
40.02	1.93	0.6	299.01	23JA	2.7	-4.3	-1.8	1.2
43.98	1.92	0.8	309.03	2FE	1.8	-4.2	-1.8	1.2
48.36	1.89	0.9	319.00	12FE	0.9	-4.1	-1.7	1.1
53.06	1.86	1.1	328.92	22FE	0.5	-4.0	-1.7	1.1
58.02	1.82	1.2	338.77	3MR	0.2	-3.9	-1.6	1.1
63.18	1.78	1.4	348.57	13MR	-0.2	-3.8	-1.6	1.1
68.52	1.73	1.5	358.31	23MR	-0.8	-3.7	-1.5	1.1
74.00	1.69	1.6	7.99	2AP	-1.7	-3.6	-1.5	1.1
79.60	1.63	1.7	17.63	12AP	-1.2	-3.5	-1.5	1.0
85.31	1.58	1.7	27.23	22AP	-0.2	-3.5	-1.5	1.0
91.10	1.53	1.8	36.79	2MY	0.6	-3.4	-1.5	1.0
96.98	1.47	1.8	46.34	12MY	1.4	-3.4	-1.5	1.0
102.95	1.41	1.9	55.86	22MY	2.5	-3.4	-1.5	0.9
108.98	1.35	1.9	65.38	1JN	3.4	-3.3	-1.5	0.9
115.09	1.29	1.9	74.90	11JN	1.8	-3.3	-1.5	0.9
121.27	1.22	1.9	84.43	21JN	0.5	-3.3	-1.6	0.8
127.53	1.15	1.9	93.98	1JL	-0.7	-3.3	-1.6	0.8
133.86	1.08	1.9	103.57	11JL	-1.5	-3.3	-1.7	0.7
140.26	1.00	1.9	113.19	21JL	-1.2	-3.3	-1.7	0.7
146.74	0.93	1.9	122.85	31JL	-0.5	-3.4	-1.8	0.6
153.30	0.85	1.9	132.57	10AU	-0.1	-3.4	-1.8	0.6
159.94	0.76	1.9	142.34	20AU	0.1	-3.4	-1.9	0.5
166.66	0.68	1.8	152.16	30AU	0.3	-3.4	-1.9	0.5
173.47	0.59	1.8	162.05	9SE	0.6	-3.5	-2.0	0.5
180.36	0.50	1.7	172.00	19SE	1.9	-3.5	-2.1	0.5
187.34	0.40	1.7	182.00	29SE	1.7	-3.5	-2.1	0.6
194.40	0.30	1.6	192.06	9OC	-0.2	-3.5	-2.2	0.7
201.55	0.20	1.6	202.16	19OC	-0.6	-3.4	-2.2	0.7
208.79	0.10	1.6	212.31	29OC	-0.6	-3.4	-2.3	0.8
216.10	-0.01	1.6	222.49	8NO	-0.6	-3.4	-2.3	0.9
223.50	-0.12	1.6	232.69	18NO	-0.8	-3.4	-2.3	0.9
230.97	-0.23	1.6	242.89	28NO	-0.8	-3.3	-2.3	1.0
238.51	-0.34	1.5	253.11	8DE	-0.8	-3.3	-2.2	1.0
246.11	-0.45	1.5	263.30	18DE	-0.7	-3.3	-2.2	1.0
253.77	-0.56	1.5	273.48	28DE	0.3	-3.3	-2.1	1.0
				-359				
261.48	-0.67	1.5	283.62	7JA	2.9	-3.4	-2.1	1.0
269.22	-0.77	1.5	293.73	17JA	1.4	-3.4	-2.0	1.0
276.99	-0.87	1.4	303.78	27JA	0.6	-3.4	-1.9	1.0
284.78	-0.96	1.4	313.79	6FE	0.3	-3.4	-1.9	0.9
292.57	-1.05	1.4	323.73	16FE	0.1	-3.5	-1.8	0.9
300.36	-1.12	1.4	333.61	26FE	-0.3	-3.5	-1.7	0.8
308.13	-1.19	1.4	343.44	8MR	-0.9	-3.5	-1.7	0.8
315.87	-1.24	1.3	353.21	18MR	-1.7	-3.6	-1.6	0.8
323.56	-1.28	1.3	2.92	28MR	-1.2	-3.7	-1.6	0.8
331.20	-1.31	1.3	12.58	7AP	-0.2	-3.7	-1.5	0.8
338.76	-1.33	1.3	22.20	17AP	0.8	-3.8	-1.5	0.8
346.25	-1.33	1.3	31.78	27AP	1.9	-3.9	-1.4	0.8
353.65	-1.32	1.3	41.34	7MY	3.3	-4.0	-1.4	0.8
0.94	-1.30	1.2	50.87	17MY	2.7	-4.1	-1.4	0.8
8.12	-1.26	1.2	60.39	27MY	1.4	-4.1	-1.4	0.7
15.18	-1.20	1.2	69.91	6JN	0.4	-4.2	-1.4	0.7
22.10	-1.14	1.2	79.43	16JN	-0.7	-4.1	-1.4	0.7
28.87	-1.05	1.1	88.97	26JN	-1.6	-3.9	-1.4	0.6
35.48	-0.96	1.1	98.54	6JL	-1.2	-3.3	-1.4	0.6
41.92	-0.85	1.1	108.13	16JL	-0.5	-3.2	-1.4	0.5
48.15	-0.73	1.0	117.78	26JL	-0.0	-3.8	-1.5	0.5
54.16	-0.59	0.9	127.47	5AU	0.2	-4.2	-1.5	0.4
59.90	-0.43	0.9	137.20	15AU	0.4	-4.3	-1.6	0.4
65.36	-0.26	0.8	147.00	25AU	0.9	-4.3	-1.6	0.3
70.45	-0.07	0.7	156.86	4SE	2.4	-4.2	-1.7	0.2
75.12	0.16	0.5	166.77	14SE	1.5	-4.1	-1.7	0.2
79.28	0.41	0.4	176.75	24SE	-0.4	-4.0	-1.8	0.2
82.80	0.69	0.2	186.78	4OC	-0.8	-3.9	-1.8	0.3
85.55	1.02	0.0	196.86	14OC	-0.8	-3.8	-1.9	0.3
87.33	1.39	-0.2	206.99	24OC	-0.8	-3.8	-2.0	0.4
87.96	1.80	-0.4	217.15	3NO	-0.7	-3.7	-2.0	0.5
87.28	2.25	-0.6	227.34	13NO	-0.6	-3.6	-2.1	0.6
85.22	2.70	-0.8	237.54	23NO	-0.6	-3.6	-2.1	0.6
81.96	3.11	-1.0	247.76	3DE	-0.5	-3.5	-2.2	0.7
78.05	3.41	-1.1	257.96	13DE	0.5	-3.5	-2.2	0.7
74.20	3.58	-1.0	268.15	23DE	3.0	-3.4	-2.2	0.7

♂ LONG	♂ LAT	♂ MAG	⊙ LONG	16.00UT	☿	♀	♃	♄
				-358	MAGNITUDES			
71.12	3.60	-0.7	278.31	2JA	1.0	-3.4	-2.1	0.7
69.25	3.51	-0.5	288.44	12JA	0.4	-3.4	-2.1	0.7
68.70	3.35	-0.2	298.52	22JA	0.2	-3.3	-2.1	0.7
69.37	3.17	0.1	308.55	1FE	-0.0	-3.3	-2.0	0.7
71.09	2.97	0.3	318.52	11FE	-0.3	-3.3	-2.0	0.7
73.65	2.78	0.5	328.44	21FE	-0.9	-3.3	-1.9	0.7
76.91	2.60	0.7	338.30	3MR	-1.6	-3.3	-1.8	0.6
80.71	2.43	0.9	348.09	13MR	-1.2	-3.3	-1.7	0.6
84.94	2.26	1.0	357.84	23MR	-0.1	-3.3	-1.7	0.5
89.54	2.11	1.2	7.52	2AP	1.1	-3.4	-1.6	0.5
94.42	1.97	1.3	17.16	12AP	2.5	-3.4	-1.6	0.6
99.54	1.83	1.4	26.76	22AP	3.6	-3.4	-1.5	0.6
104.88	1.70	1.5	36.33	2MY	2.1	-3.5	-1.5	0.6
110.39	1.57	1.5	45.87	12MY	1.1	-3.5	-1.4	0.6
116.06	1.45	1.6	55.39	22MY	0.2	-3.4	-1.4	0.6
121.88	1.33	1.7	64.91	1JN	-0.8	-3.4	-1.3	0.6
127.83	1.21	1.7	74.43	11JN	-1.6	-3.4	-1.3	0.5
133.90	1.09	1.7	83.96	21JN	-1.2	-3.4	-1.3	0.5
140.10	0.98	1.7	93.51	1JL	-0.5	-3.3	-1.3	0.5
146.40	0.86	1.8	103.09	11JL	0.1	-3.3	-1.3	0.4
152.82	0.75	1.8	112.71	21JL	0.4	-3.3	-1.3	0.4
159.36	0.63	1.8	122.38	31JL	0.6	-3.3	-1.3	0.3
166.00	0.52	1.8	132.09	10AU	1.2	-3.3	-1.3	0.3
172.75	0.41	1.7	141.86	20AU	2.8	-3.4	-1.4	0.2
179.61	0.29	1.7	151.68	30AU	1.3	-3.4	-1.4	0.1
186.57	0.18	1.7	161.57	9SE	-0.5	-3.4	-1.4	0.1
193.64	0.07	1.7	171.51	19SE	-0.9	-3.4	-1.5	0.0
200.81	-0.05	1.6	181.51	29SE	-0.9	-3.5	-1.5	-0.0
208.09	-0.16	1.6	191.57	9OC	-0.9	-3.5	-1.6	-0.0
215.46	-0.26	1.6	201.67	19OC	-0.6	-3.6	-1.6	0.1
222.92	-0.37	1.5	211.82	29OC	-0.5	-3.7	-1.7	0.1
230.46	-0.47	1.5	221.99	8NO	-0.5	-3.8	-1.8	0.2
238.09	-0.57	1.4	232.19	18NO	-0.4	-3.8	-1.8	0.3
245.78	-0.66	1.4	242.40	28NO	0.6	-3.9	-1.9	0.3
253.53	-0.75	1.4	252.61	8DE	2.8	-4.0	-2.0	0.4
261.33	-0.83	1.3	262.81	18DE	0.7	-4.1	-2.0	0.5
269.17	-0.90	1.3	272.99	28DE	0.2	-4.2	-2.0	0.5
				-357				
277.04	-0.96	1.4	283.13	7JA	0.0	-4.3	-2.1	0.5
284.92	-1.02	1.4	293.24	17JA	-0.1	-4.3	-2.1	0.5
292.80	-1.06	1.4	303.30	27JA	-0.4	-4.3	-2.1	0.5
300.66	-1.09	1.4	313.30	6FE	-0.9	-4.0	-2.0	0.5
308.51	-1.11	1.4	323.25	16FE	-1.5	-3.4	-2.0	0.5
316.31	-1.12	1.4	333.14	26FE	-1.2	-3.4	-2.0	0.5
324.07	-1.12	1.4	342.96	8MR	-0.1	-3.9	-1.9	0.5
331.77	-1.10	1.4	352.74	18MR	1.4	-4.2	-1.8	0.4
339.40	-1.07	1.5	2.45	28MR	3.1	-4.2	-1.8	0.4
346.95	-1.03	1.5	12.11	7AP	2.7	-4.2	-1.7	0.3
354.42	-0.98	1.5	21.74	17AP	1.6	-4.1	-1.6	0.4
1.80	-0.92	1.5	31.32	27AP	0.8	-4.0	-1.6	0.4
9.09	-0.85	1.5	40.87	7MY	0.1	-3.9	-1.5	0.4
16.28	-0.77	1.5	50.40	17MY	-0.8	-3.8	-1.5	0.4
23.36	-0.69	1.5	59.92	27MY	-1.7	-3.7	-1.4	0.4
30.34	-0.59	1.5	69.44	6JN	-1.2	-3.6	-1.4	0.4
37.21	-0.49	1.5	78.97	16JN	-0.4	-3.6	-1.3	0.4
43.97	-0.38	1.5	88.50	26JN	0.1	-3.5	-1.3	0.4
50.61	-0.26	1.5	98.07	6JL	0.5	-3.5	-1.3	0.4
57.12	-0.13	1.5	107.67	16JL	0.9	-3.4	-1.2	0.3
63.50	-0.00	1.5	117.31	26JL	1.6	-3.4	-1.2	0.3
69.74	0.14	1.5	126.99	5AU	3.2	-3.4	-1.2	0.3
75.83	0.29	1.4	136.73	15AU	1.1	-3.4	-1.2	0.2
81.75	0.45	1.4	146.53	25AU	-0.5	-3.4	-1.2	0.1
87.49	0.62	1.3	156.38	4SE	-1.0	-3.4	-1.3	0.1
93.00	0.80	1.2	166.30	14SE	-1.0	-3.4	-1.3	0.0
98.26	1.00	1.1	176.27	24SE	-0.8	-3.4	-1.3	-0.1
103.22	1.22	1.0	186.30	4OC	-0.5	-3.4	-1.3	-0.1
107.82	1.45	0.9	196.38	14OC	-0.4	-3.4	-1.4	-0.2
111.98	1.72	0.7	206.50	24OC	-0.3	-3.4	-1.4	-0.2
115.61	2.01	0.6	216.66	3NO	-0.2	-3.4	-1.5	-0.1
118.56	2.34	0.4	226.85	13NO	0.8	-3.4	-1.5	-0.0
120.69	2.70	0.2	237.05	23NO	2.6	-3.5	-1.6	0.1
121.82	3.10	-0.1	247.26	3DE	0.4	-3.5	-1.7	0.1
121.75	3.52	-0.3	257.47	13DE	-0.0	-3.5	-1.7	0.2
120.37	3.93	-0.6	267.65	23DE	-0.1	-3.5	-1.8	0.3

♂ LONG	♂ LAT	♂ MAG	⊙ LONG	16.00UT	☿	♀	♃	♄
				-356	MAGNITUDES			
117.72	4.28	-0.8	277.82	2JA	-0.2	-3.4	-1.9	0.3
114.12	4.50	-1.0	287.95	12JA	-0.5	-3.4	-1.9	0.3
110.17	4.56	-1.0	298.03	22JA	-0.9	-3.4	-2.0	0.4
106.59	4.45	-0.8	308.06	1FE	-1.4	-3.4	-2.0	0.4
104.00	4.20	-0.6	318.04	11FE	-1.1	-3.3	-2.0	0.4
102.66	3.87	-0.3	327.96	21FE	-0.0	-3.3	-2.0	0.4
102.61	3.52	-0.1	337.82	2MR	1.7	-3.3	-2.0	0.4
103.71	3.17	0.1	347.62	12MR	3.5	-3.3	-2.0	0.4
105.80	2.84	0.3	357.36	22MR	2.0	-3.4	-1.9	0.3
108.68	2.53	0.5	7.05	1AP	1.2	-3.4	-1.9	0.3
112.21	2.25	0.7	16.70	11AP	0.6	-3.4	-1.8	0.3
116.28	1.99	0.8	26.30	21AP	0.0	-3.4	-1.8	0.2
120.77	1.76	1.0	35.87	1MY	-0.8	-3.4	-1.7	0.3
125.63	1.54	1.1	45.41	11MY	-1.7	-3.5	-1.7	0.3
130.80	1.34	1.2	54.93	21MY	-1.3	-3.5	-1.6	0.3
136.23	1.14	1.2	64.45	31MY	-0.4	-3.6	-1.5	0.3
141.89	0.96	1.3	73.97	10JN	0.3	-3.7	-1.5	0.3
147.76	0.79	1.4	83.50	20JN	0.7	-3.7	-1.4	0.3
153.82	0.63	1.4	93.05	30JN	1.2	-3.8	-1.4	0.3
160.06	0.47	1.4	102.63	10JL	2.1	-3.9	-1.3	0.3
166.46	0.32	1.5	112.25	20JL	3.1	-4.0	-1.3	0.3
173.01	0.18	1.5	121.91	30JL	1.0	-4.1	-1.3	0.3
179.72	0.04	1.5	131.62	9AU	-0.6	-4.2	-1.2	0.2
186.56	-0.09	1.5	141.38	19AU	-1.2	-4.3	-1.2	0.2
193.54	-0.22	1.5	151.21	29AU	-1.2	-4.2	-1.2	0.1
200.64	-0.34	1.5	161.09	8SE	-0.7	-4.0	-1.2	0.1
207.87	-0.46	1.5	171.03	18SE	-0.4	-3.4	-1.2	-0.0
215.21	-0.56	1.5	181.03	28SE	-0.2	-3.4	-1.2	-0.1
222.66	-0.66	1.5	191.08	8OC	-0.2	-4.0	-1.3	-0.2
230.21	-0.75	1.5	201.18	18OC	0.0	-4.3	-1.3	-0.2
237.84	-0.83	1.4	211.33	28OC	1.1	-4.4	-1.3	-0.3
245.55	-0.91	1.4	221.50	7NO	2.3	-4.3	-1.4	-0.2
253.33	-0.97	1.4	231.70	17NO	0.2	-4.3	-1.4	-0.2
261.16	-1.02	1.4	241.91	27NO	-0.2	-4.2	-1.5	-0.1
269.02	-1.06	1.4	252.12	7DE	-0.2	-4.0	-1.5	-0.0
276.91	-1.09	1.4	262.32	17DE	-0.3	-3.9	-1.6	0.0
284.82	-1.10	1.4	272.50	27DE	-0.6	-3.8	-1.6	0.1
				-355				
292.71	-1.10	1.4	282.64	6JA	-1.0	-3.8	-1.7	0.2
300.59	-1.10	1.4	292.75	16JA	-1.2	-3.7	-1.8	0.2
308.44	-1.08	1.4	302.82	26JA	-1.0	-3.6	-1.8	0.3
316.25	-1.04	1.4	312.82	5FE	0.0	-3.5	-1.9	0.3
324.00	-1.00	1.4	322.77	15FE	2.0	-3.5	-2.0	0.3
331.70	-0.95	1.4	332.66	25FE	2.8	-3.4	-2.0	0.3
339.32	-0.89	1.4	342.49	7MR	1.5	-3.4	-2.0	0.3
346.86	-0.82	1.5	352.27	17MR	0.9	-3.4	-2.0	0.3
354.33	-0.75	1.5	1.99	27MR	0.5	-3.3	-2.0	0.3
1.70	-0.66	1.5	11.65	6AP	-0.0	-3.3	-2.0	0.3
8.99	-0.58	1.6	21.27	16AP	-0.8	-3.3	-2.0	0.2
16.19	-0.48	1.6	30.86	26AP	-1.7	-3.3	-2.0	0.2
23.30	-0.39	1.7	40.41	6MY	-1.3	-3.3	-1.9	0.2
30.32	-0.29	1.7	49.95	16MY	-0.3	-3.3	-1.8	0.2
37.24	-0.18	1.7	59.47	26MY	0.4	-3.3	-1.8	0.2
44.08	-0.08	1.7	68.98	5JN	0.9	-3.4	-1.7	0.3
50.84	0.03	1.8	78.50	15JN	1.6	-3.4	-1.7	0.3
57.50	0.14	1.8	88.04	25JN	2.7	-3.5	-1.6	0.3
64.08	0.26	1.8	97.60	5JL	2.6	-3.5	-1.5	0.3
70.57	0.37	1.8	107.20	15JL	0.8	-3.5	-1.5	0.3
76.98	0.49	1.8	116.84	25JL	-0.6	-3.5	-1.4	0.3
83.29	0.61	1.8	126.52	4AU	-1.3	-3.4	-1.4	0.3
89.51	0.73	1.8	136.25	14AU	-1.2	-3.4	-1.3	0.2
95.64	0.86	1.7	146.05	24AU	-0.6	-3.4	-1.3	0.2
101.65	0.99	1.7	155.89	3SE	-0.3	-3.3	-1.3	0.1
107.55	1.12	1.6	165.81	13SE	-0.1	-3.3	-1.3	0.1
113.32	1.26	1.6	175.78	23SE	-0.0	-3.3	-1.3	0.0
118.94	1.41	1.5	185.80	3OC	0.2	-3.3	-1.2	-0.0
124.38	1.57	1.4	195.88	13OC	1.3	-3.3	-1.3	-0.1
129.62	1.74	1.3	206.00	23OC	2.1	-3.3	-1.3	-0.2
134.61	1.92	1.2	216.16	2NO	0.0	-3.4	-1.3	-0.3
139.30	2.12	1.1	226.35	12NO	-0.4	-3.4	-1.3	-0.3
143.63	2.34	0.9	236.55	22NO	-0.4	-3.4	-1.3	-0.3
147.50	2.57	0.7	246.76	2DE	-0.4	-3.5	-1.4	-0.2
150.82	2.83	0.5	256.97	12DE	-0.6	-3.5	-1.4	-0.1
153.44	3.11	0.3	267.16	22DE	-1.0	-3.6	-1.5	-0.1

♂ LONG	♂ LAT	♂ MAG	⊙ LONG	16.00UT -354	☿	♀	♃	♄
					MAGNITUDES			
155.20	3.40	0.0	277.33	1JA	-1.1	-3.6	-1.5	0.0
155.92	3.71	-0.2	287.46	11JA	-0.9	-3.7	-1.6	0.1
155.42	4.00	-0.5	297.55	21JA	0.1	-3.7	-1.7	0.1
153.65	4.23	-0.8	307.58	31JA	2.4	-3.8	-1.7	0.2
150.69	4.35	-1.0	317.56	10FE	2.2	-3.9	-1.8	0.2
146.99	4.32	-1.2	327.48	20FE	1.1	-4.0	-1.9	0.3
143.19	4.11	-1.1	337.35	2MR	0.6	-4.1	-1.9	0.3
140.00	3.76	-0.9	347.15	12MR	0.3	-4.2	-2.0	0.3
137.92	3.32	-0.7	356.90	22MR	-0.1	-4.2	-2.0	0.3
137.14	2.86	-0.5	6.59	1AP	-0.8	-4.2	-2.1	0.3
137.64	2.41	-0.3	16.24	11AP	-1.7	-4.0	-2.1	0.3
139.28	1.99	-0.1	25.84	21AP	-1.3	-3.6	-2.1	0.3
141.86	1.61	0.1	35.40	1MY	-0.3	-2.8	-2.1	0.2
145.22	1.27	0.3	44.95	11MY	0.5	-3.5	-2.1	0.2
149.24	0.96	0.4	54.47	21MY	1.2	-4.0	-2.1	0.1
153.78	0.68	0.5	63.99	31MY	2.1	-4.2	-2.0	0.2
158.77	0.43	0.6	73.51	10JN	3.3	-4.2	-2.0	0.2
164.14	0.20	0.7	83.04	20JN	2.1	-4.1	-1.9	0.3
169.83	-0.00	0.8	92.59	30JN	0.6	-4.0	-1.8	0.3
175.81	-0.19	0.9	102.17	10JL	-0.7	-4.0	-1.8	0.3
182.05	-0.37	0.9	111.78	20JL	-1.4	-3.9	-1.7	0.3
188.51	-0.52	1.0	121.43	30JL	-1.2	-3.8	-1.7	0.3
195.18	-0.66	1.0	131.14	9AU	-0.6	-3.7	-1.6	0.3
202.03	-0.79	1.1	140.90	19AU	-0.2	-3.7	-1.5	0.3
209.05	-0.90	1.1	150.72	29AU	0.0	-3.6	-1.5	0.3
216.22	-1.00	1.1	160.61	8SE	0.2	-3.6	-1.4	0.2
223.54	-1.09	1.2	170.54	18SE	0.4	-3.5	-1.4	0.2
230.97	-1.15	1.2	180.54	28SE	1.7	-3.5	-1.4	0.1
238.52	-1.21	1.2	190.59	8OC	1.9	-3.4	-1.4	0.0
246.15	-1.25	1.2	200.69	18OC	-0.1	-3.4	-1.3	-0.0
253.86	-1.27	1.3	210.83	28OC	-0.5	-3.4	-1.3	-0.1
261.63	-1.28	1.3	221.01	7NO	-0.5	-3.4	-1.3	-0.2
269.44	-1.28	1.3	231.20	17NO	-0.6	-3.4	-1.3	-0.2
277.27	-1.26	1.3	241.41	27NO	-0.7	-3.4	-1.3	-0.3
285.12	-1.23	1.4	251.63	7DE	-0.8	-3.4	-1.4	-0.3
292.96	-1.19	1.4	261.82	17DE	-0.9	-3.4	-1.4	-0.2
300.78	-1.13	1.4	272.00	27DE	-0.8	-3.4	-1.4	-0.1
				-353				
308.57	-1.07	1.4	282.16	6JA	0.2	-3.4	-1.5	-0.1
316.31	-1.00	1.5	292.26	16JA	2.7	-3.4	-1.5	0.0
324.00	-0.92	1.5	302.33	26JA	1.7	-3.4	-1.6	0.1
331.62	-0.83	1.5	312.34	5FE	0.8	-3.4	-1.6	0.1
339.17	-0.74	1.5	322.29	15FE	0.4	-3.5	-1.7	0.2
346.64	-0.65	1.6	332.18	25FE	0.2	-3.5	-1.8	0.2
354.04	-0.55	1.6	342.02	7MR	-0.2	-3.4	-1.8	0.3
1.35	-0.45	1.6	351.79	17MR	-0.8	-3.4	-1.9	0.3
8.58	-0.35	1.6	1.51	27MR	-1.7	-3.4	-2.0	0.3
15.72	-0.24	1.6	11.18	6AP	-1.3	-3.4	-2.0	0.3
22.78	-0.14	1.6	20.80	16AP	-0.3	-3.4	-2.1	0.3
29.76	-0.04	1.7	30.39	26AP	0.7	-3.3	-2.1	0.3
36.66	0.06	1.7	39.95	6MY	1.6	-3.3	-2.2	0.3
43.48	0.16	1.8	49.48	16MY	2.8	-3.3	-2.2	0.3
50.24	0.26	1.8	59.00	26MY	3.2	-3.3	-2.2	0.2
56.92	0.36	1.8	68.52	5JN	1.7	-3.3	-2.2	0.2
63.54	0.45	1.9	78.04	15JN	0.5	-3.4	-2.2	0.2
70.10	0.55	1.9	87.58	25JN	-0.7	-3.4	-2.2	0.3
76.60	0.64	1.9	97.14	5JL	-1.5	-3.4	-2.1	0.3
83.05	0.73	1.9	106.73	15JL	-1.2	-3.5	-2.1	0.4
89.44	0.82	2.0	116.37	25JL	-0.5	-3.5	-2.0	0.4
95.78	0.91	2.0	126.05	4AU	-0.1	-3.6	-1.9	0.4
102.07	1.00	1.9	135.78	14AU	0.1	-3.6	-1.9	0.4
108.31	1.09	1.9	145.57	24AU	0.3	-3.7	-1.8	0.4
114.49	1.17	1.9	155.42	3SE	0.7	-3.8	-1.7	0.4
120.62	1.26	1.9	165.32	13SE	2.0	-3.9	-1.7	0.3
126.68	1.35	1.8	175.29	23SE	1.7	-4.0	-1.6	0.3
132.68	1.44	1.8	185.31	3OC	-0.3	-4.1	-1.6	0.2
138.60	1.53	1.7	195.38	13OC	-0.7	-4.2	-1.5	0.2
144.44	1.62	1.7	205.51	23OC	-0.7	-4.3	-1.5	0.1
150.17	1.72	1.6	215.66	2NO	-0.7	-4.4	-1.5	0.1
155.78	1.81	1.5	225.85	12NO	-0.8	-4.3	-1.5	-0.0
161.24	1.91	1.4	236.06	22NO	-0.7	-4.0	-1.4	-0.1
166.53	2.01	1.2	246.27	2DE	-0.7	-3.4	-1.4	-0.2
171.60	2.12	1.1	256.47	12DE	-0.6	-3.3	-1.4	-0.2
176.42	2.22	0.9	266.67	22DE	0.3	-4.0	-1.4	-0.2

♂ LONG	♂ LAT	♂ MAG	⊙ LONG	16.00UT -352	☿	♀	♃	♄
					MAGNITUDES			
180.91	2.33	0.7	276.83	1JA	3.0	-4.3	-1.5	-0.1
185.00	2.44	0.5	286.96	11JA	1.2	-4.4	-1.5	-0.1
188.59	2.55	0.3	297.06	21JA	0.5	-4.3	-1.5	0.0
191.55	2.65	0.1	307.09	31JA	0.3	-4.2	-1.5	0.1
193.74	2.73	-0.2	317.08	10FE	0.1	-4.1	-1.6	0.1
194.97	2.79	-0.5	327.01	20FE	-0.3	-4.0	-1.6	0.2
195.06	2.79	-0.8	336.87	1MR	-0.9	-3.9	-1.7	0.3
193.90	2.72	-1.1	346.68	11MR	-1.7	-3.8	-1.7	0.3
191.52	2.54	-1.4	356.43	21MR	-1.3	-3.7	-1.8	0.3
188.24	2.23	-1.7	6.12	31MR	-0.2	-3.6	-1.9	0.4
184.65	1.80	-1.6	15.77	10AP	0.9	-3.5	-2.0	0.4
181.48	1.30	-1.5	25.38	20AP	2.1	-3.5	-2.0	0.4
179.35	0.78	-1.3	34.95	30AP	3.6	-3.4	-2.1	0.4
178.52	0.30	-1.1	44.49	10MY	2.5	-3.4	-2.2	0.4
179.05	-0.13	-0.9	54.02	20MY	1.3	-3.4	-2.2	0.4
180.82	-0.50	-0.7	63.53	30MY	0.3	-3.3	-2.3	0.4
183.63	-0.81	-0.5	73.05	9JN	-0.7	-3.3	-2.3	0.3
187.31	-1.07	-0.3	82.58	19JN	-1.6	-3.3	-2.3	0.3
191.72	-1.29	-0.2	92.13	29JN	-1.3	-3.3	-2.3	0.4
196.71	-1.46	-0.0	101.70	9JL	-0.5	-3.3	-2.3	0.4
202.20	-1.60	0.1	111.32	19JL	-0.0	-3.3	-2.3	0.4
208.10	-1.71	0.2	120.97	29JL	0.3	-3.4	-2.3	0.5
214.34	-1.79	0.3	130.67	8AU	0.5	-3.4	-2.2	0.5
220.88	-1.85	0.4	140.44	18AU	1.0	-3.4	-2.2	0.5
227.66	-1.88	0.5	150.25	28AU	2.5	-3.4	-2.1	0.5
234.65	-1.89	0.6	160.12	7SE	1.5	-3.5	-2.1	0.5
241.81	-1.87	0.7	170.06	17SE	-0.4	-3.5	-2.0	0.5
249.11	-1.84	0.7	180.05	27SE	-0.8	-3.5	-1.9	0.5
256.52	-1.79	0.8	190.10	7OC	-0.8	-3.5	-1.9	0.4
264.03	-1.73	0.9	200.19	17OC	-0.8	-3.4	-1.8	0.4
271.60	-1.64	1.0	210.33	27OC	-0.7	-3.4	-1.7	0.3
279.21	-1.55	1.0	220.50	6NO	-0.6	-3.4	-1.7	0.2
286.84	-1.45	1.1	230.70	16NO	-0.6	-3.4	-1.6	0.2
294.47	-1.34	1.2	240.91	26NO	-0.5	-3.3	-1.6	0.1
302.09	-1.22	1.2	251.12	6DE	0.5	-3.3	-1.6	0.0
309.69	-1.10	1.3	261.32	16DE	3.0	-3.3	-1.5	-0.0
317.24	-0.97	1.4	271.50	26DE	0.9	-3.3	-1.5	-0.1
				-351				
324.75	-0.85	1.4	281.65	5JA	0.3	-3.4	-1.5	-0.0
332.19	-0.72	1.5	291.77	15JA	0.1	-3.4	-1.5	0.0
339.58	-0.60	1.5	301.83	25JA	-0.0	-3.4	-1.5	0.1
346.89	-0.47	1.6	311.85	4FE	-0.3	-3.4	-1.5	0.1
354.13	-0.35	1.6	321.81	14FE	-0.9	-3.5	-1.6	0.2
1.29	-0.23	1.7	331.70	24FE	-1.6	-3.5	-1.6	0.3
8.38	-0.12	1.7	341.54	6MR	-1.2	-3.5	-1.6	0.3
15.40	-0.01	1.7	351.32	16MR	-0.2	-3.6	-1.7	0.4
22.34	0.10	1.8	1.05	26MR	1.1	-3.7	-1.7	0.4
29.21	0.20	1.8	10.72	5AP	2.7	-3.7	-1.8	0.4
36.02	0.30	1.8	20.35	15AP	3.3	-3.8	-1.8	0.5
42.76	0.39	1.8	29.93	25AP	1.9	-3.9	-1.9	0.5
49.45	0.48	1.9	39.49	5MY	1.0	-4.0	-2.0	0.5
56.08	0.56	1.9	49.03	15MY	0.2	-4.1	-2.0	0.5
62.65	0.64	1.9	58.54	25MY	-0.7	-4.1	-2.1	0.5
69.19	0.71	1.9	68.06	4JN	-1.6	-4.2	-2.2	0.5
75.68	0.79	1.9	77.59	14JN	-1.3	-4.1	-2.2	0.5
82.14	0.85	1.9	87.12	24JN	-0.5	-3.9	-2.3	0.5
88.57	0.92	2.0	96.68	4JL	0.1	-3.3	-2.4	0.5
94.97	0.98	2.0	106.27	14JL	0.4	-3.2	-2.4	0.5
101.34	1.04	2.0	115.90	24JL	0.7	-3.8	-2.4	0.6
107.70	1.09	2.0	125.58	3AU	1.3	-4.2	-2.4	0.6
114.03	1.14	2.0	135.31	13AU	2.9	-4.3	-2.4	0.6
120.36	1.19	2.0	145.09	23AU	1.3	-4.2	-2.4	0.7
126.67	1.23	2.0	154.94	2SE	-0.5	-4.2	-2.4	0.7
132.96	1.27	2.0	164.84	12SE	-1.0	-4.1	-2.3	0.7
139.24	1.31	2.0	174.80	22SE	-1.0	-4.0	-2.3	0.7
145.51	1.35	1.9	184.83	2OC	-0.8	-3.9	-2.2	0.7
151.76	1.38	1.9	194.90	12OC	-0.5	-3.8	-2.1	0.6
158.00	1.41	1.8	205.01	22OC	-0.4	-3.8	-2.1	0.6
164.21	1.43	1.8	215.17	1NO	-0.4	-3.7	-2.0	0.6
170.39	1.45	1.7	225.35	11NO	-0.3	-3.6	-1.9	0.5
176.54	1.46	1.6	235.55	21NO	0.7	-3.6	-1.9	0.4
182.64	1.47	1.5	245.77	1DE	2.8	-3.5	-1.8	0.4
188.69	1.47	1.4	255.97	11DE	0.6	-3.5	-1.7	0.3
194.68	1.46	1.3	266.17	21DE	0.1	-3.4	-1.7	0.2
200.59	1.44	1.2	276.34	31DE	-0.0	-3.4	-1.7	0.2

Left table (−350 / −349)

♂ LONG	LAT	MAG	☉ LONG	16.00UT	☿	♀	♃	♄
				−350				
206.39	1.41	1.0	286.47	10JA	−0.2	−3.4	−1.6	0.1
212.08	1.36	0.9	296.56	20JA	−0.4	−3.3	−1.6	0.2
217.62	1.29	0.7	306.61	30JA	−0.9	−3.3	−1.6	0.2
222.96	1.20	0.5	316.59	9FE	−1.4	−3.3	−1.6	0.3
228.09	1.08	0.3	326.52	19FE	−1.2	−3.3	−1.6	0.3
232.91	0.93	0.1	336.39	1MR	−0.1	−3.3	−1.6	0.4
237.36	0.72	−0.1	346.20	11MR	1.4	−3.3	−1.6	0.4
241.35	0.47	−0.4	355.95	21MR	3.4	−3.3	−1.6	0.5
244.74	0.14	−0.7	5.66	31MR	2.4	−3.4	−1.6	0.5
247.37	−0.26	−1.0	15.30	10AP	1.4	−3.4	−1.7	0.6
249.08	−0.76	−1.3	24.91	20AP	0.8	−3.4	−1.7	0.6
249.66	−1.36	−1.6	34.49	30AP	0.1	−3.5	−1.7	0.7
249.02	−2.04	−1.9	44.03	10MY	−0.7	−3.4	−1.8	0.7
247.20	−2.77	−2.2	53.55	20MY	−1.7	−3.4	−1.8	0.7
244.53	−3.47	−2.4	63.07	30MY	−1.3	−3.4	−1.9	0.7
241.66	−4.03	−2.4	72.59	9JN	−0.4	−3.4	−2.0	0.7
239.30	−4.41	−2.2	82.12	19JN	0.2	−3.4	−2.0	0.7
238.02	−4.59	−2.0	91.66	29JN	0.6	−3.3	−2.1	0.7
238.10	−4.61	−1.8	101.24	9JL	1.0	−3.3	−2.2	0.7
239.50	−4.51	−1.5	110.85	19JL	1.8	−3.3	−2.2	0.7
242.12	−4.32	−1.3	120.50	29JL	3.2	−3.3	−2.3	0.7
245.73	−4.09	−1.0	130.20	8AU	1.1	−3.3	−2.4	0.8
250.14	−3.83	−0.8	139.96	18AU	−0.5	−3.4	−2.4	0.8
255.21	−3.56	−0.6	149.77	28AU	−1.1	−3.4	−2.4	0.9
260.77	−3.27	−0.4	159.64	7SE	−1.1	−3.4	−2.4	0.9
266.72	−2.99	−0.2	169.57	17SE	−0.8	−3.4	−2.4	0.9
272.99	−2.70	−0.1	179.57	27SE	−0.4	−3.5	−2.4	0.9
279.48	−2.43	0.1	189.61	7OC	−0.3	−3.5	−2.4	0.9
286.15	−2.15	0.2	199.70	17OC	−0.3	−3.6	−2.3	0.9
292.94	−1.89	0.4	209.84	27OC	−0.1	−3.7	−2.3	0.9
299.83	−1.64	0.5	220.01	6NO	0.9	−3.8	−2.2	0.8
306.78	−1.41	0.7	230.21	16NO	2.6	−3.8	−2.1	0.8
313.76	−1.18	0.8	240.41	26NO	0.4	−3.9	−2.1	0.7
320.76	−0.97	0.9	250.62	6DE	−0.1	−4.0	−2.0	0.7
327.75	−0.77	1.0	260.83	16DE	−0.2	−4.1	−1.9	0.6
334.73	−0.59	1.2	271.01	26DE	−0.3	−4.2	−1.9	0.5
				−349				
341.69	−0.42	1.3	281.16	5JA	−0.5	−4.3	−1.8	0.5
348.61	−0.26	1.4	291.28	15JA	−0.9	−4.3	−1.7	0.4
355.49	−0.11	1.4	301.35	25JA	−1.3	−4.3	−1.7	0.4
2.33	0.03	1.5	311.36	4FE	−1.1	−3.9	−1.7	0.4
9.13	0.15	1.6	321.32	14FE	−0.0	−3.4	−1.6	0.4
15.87	0.27	1.7	331.22	24FE	1.8	−3.5	−1.6	0.5
22.57	0.38	1.7	341.06	6MR	3.3	−4.0	−1.6	0.5
29.23	0.47	1.8	350.85	16MR	1.8	−4.2	−1.6	0.6
35.84	0.56	1.8	0.57	26MR	1.0	−4.2	−1.5	0.7
42.40	0.65	1.9	10.25	5AP	0.6	−4.2	−1.5	0.7
48.93	0.72	1.9	19.88	15AP	0.0	−4.1	−1.6	0.8
55.42	0.79	2.0	29.47	25AP	−0.8	−4.0	−1.6	0.8
61.88	0.85	2.0	39.02	5MY	−1.7	−3.9	−1.6	0.9
68.32	0.90	2.0	48.56	15MY	−1.3	−3.8	−1.6	0.9
74.73	0.95	2.0	58.08	25MY	−0.4	−3.7	−1.7	0.9
81.12	1.00	2.0	67.60	4JN	0.3	−3.6	−1.7	0.9
87.50	1.04	2.0	77.13	14JN	0.8	−3.6	−1.7	1.0
93.87	1.07	2.0	86.66	24JN	1.3	−3.5	−1.8	1.0
100.23	1.10	2.0	96.22	4JL	2.3	−3.5	−1.9	1.0
106.60	1.13	2.0	105.81	14JL	3.0	−3.4	−1.9	1.0
112.97	1.15	2.0	115.44	24JL	0.9	−3.4	−2.0	1.0
119.34	1.16	2.0	125.11	3AU	−0.6	−3.4	−2.0	1.0
125.73	1.17	2.0	134.84	13AU	−1.2	−3.4	−2.1	1.0
132.14	1.18	2.0	144.62	23AU	−1.2	−3.4	−2.2	1.1
138.56	1.18	2.0	154.46	2SE	−0.7	−3.4	−2.2	1.1
145.01	1.18	2.0	164.37	12SE	−0.3	−3.4	−2.3	1.2
151.48	1.17	2.0	174.32	22SE	−0.2	−3.4	−2.3	1.2
157.97	1.16	2.0	184.34	2OC	−0.1	−3.4	−2.4	1.2
164.49	1.14	1.9	194.41	12OC	0.1	−3.4	−2.4	1.2
171.03	1.12	1.9	204.52	22OC	1.1	−3.4	−2.4	1.2
177.60	1.09	1.9	214.68	1NO	2.3	−3.4	−2.4	1.2
184.20	1.05	1.8	224.86	11NO	0.2	−3.4	−2.3	1.2
190.82	1.00	1.8	235.06	21NO	−0.3	−3.5	−2.3	1.1
197.47	0.95	1.7	245.27	1DE	−0.3	−3.5	−2.2	1.1
204.14	0.88	1.6	255.48	11DE	−0.4	−3.5	−2.2	1.0
210.83	0.80	1.5	265.67	21DE	−0.6	−3.5	−2.1	0.9
217.54	0.71	1.5	275.84	31DE	−1.0	−3.4	−2.0	0.9

Right table (−348 / −347)

♂ LONG	LAT	MAG	☉ LONG	16.00UT	☿	♀	♃	♄
				−348				
224.27	0.61	1.4	285.98	10JA	−1.1	−3.4	−1.9	0.8
231.00	0.49	1.3	296.07	20JA	−1.0	−3.4	−1.9	0.7
237.74	0.35	1.1	306.11	30JA	0.0	−3.4	−1.8	0.7
244.48	0.19	1.0	316.11	9FE	2.1	−3.3	−1.7	0.6
251.22	0.01	0.9	326.04	19FE	2.6	−3.3	−1.7	0.7
257.93	−0.19	0.8	335.91	29FE	1.3	−3.3	−1.6	0.7
264.62	−0.41	0.6	345.72	10MR	0.8	−3.3	−1.6	0.8
271.25	−0.66	0.5	355.48	20MR	0.4	−3.4	−1.6	0.8
277.82	−0.94	0.3	5.18	30MR	−0.1	−3.4	−1.5	0.9
284.28	−1.25	0.2	14.83	9AP	−0.8	−3.4	−1.5	0.9
290.61	−1.59	−0.0	24.44	19AP	−1.7	−3.4	−1.5	1.0
296.76	−1.96	−0.2	34.02	29AP	−1.3	−3.4	−1.5	1.0
302.66	−2.37	−0.4	43.57	9MY	−0.3	−3.5	−1.5	1.1
308.24	−2.81	−0.6	53.09	19MY	0.4	−3.5	−1.5	1.1
313.40	−3.29	−0.8	62.61	29MY	1.0	−3.6	−1.5	1.2
318.00	−3.80	−1.0	72.13	8JN	1.7	−3.7	−1.5	1.2
321.90	−4.34	−1.2	81.66	18JN	3.0	−3.8	−1.5	1.2
324.89	−4.89	−1.5	91.21	28JN	2.5	−3.8	−1.6	1.2
326.77	−5.43	−1.7	100.78	8JL	0.8	−3.9	−1.6	1.3
327.35	−5.92	−2.0	110.38	18JL	−0.6	−4.0	−1.6	1.3
326.53	−6.28	−2.2	120.04	28JL	−1.3	−4.1	−1.7	1.3
324.46	−6.42	−2.4	129.74	7AU	−1.3	−4.2	−1.8	1.3
321.66	−6.27	−2.5	139.49	17AU	−0.6	−4.3	−1.8	1.3
318.83	−5.83	−2.4	149.30	27AU	−0.3	−4.2	−1.9	1.3
316.74	−5.17	−2.1	159.17	6SE	−0.0	−3.9	−1.9	1.4
315.84	−4.42	−1.8	169.09	16SE	0.1	−3.4	−2.0	1.4
316.25	−3.67	−1.5	179.08	26SE	0.3	−3.5	−2.1	1.4
317.89	−2.97	−1.1	189.12	6OC	1.4	−4.0	−2.1	1.3
320.56	−2.35	−0.8	199.21	16OC	2.1	−4.3	−2.2	1.3
324.06	−1.82	−0.6	209.35	26OC	−0.0	−4.4	−2.2	1.3
328.22	−1.36	−0.3	219.52	5NO	−0.4	−4.3	−2.3	1.3
332.87	−0.96	−0.0	229.71	15NO	−0.4	−4.2	−2.3	1.2
337.92	−0.63	0.2	239.92	25NO	−0.5	−4.1	−2.3	1.2
343.26	−0.35	0.4	250.13	5DE	−0.7	−4.0	−2.3	1.1
348.82	−0.10	0.6	260.33	15DE	−0.9	−3.9	−2.2	1.1
354.56	0.10	0.8	270.52	25DE	−1.0	−3.8	−2.2	1.0
				−347				
0.43	0.28	0.9	280.67	4JA	−0.9	−3.8	−2.1	1.0
6.40	0.44	1.1	290.79	14JA	0.1	−3.7	−2.0	1.0
12.44	0.57	1.2	300.86	24JA	2.5	−3.6	−2.0	0.9
18.53	0.68	1.3	310.88	3FE	2.0	−3.5	−1.9	0.9
24.66	0.78	1.5	320.84	13FE	0.9	−3.5	−1.8	0.8
30.82	0.86	1.6	330.75	23FE	0.5	−3.4	−1.8	0.9
37.00	0.93	1.6	340.59	5MR	0.3	−3.4	−1.7	0.9
43.19	0.99	1.7	350.37	15MR	−0.1	−3.4	−1.6	1.0
49.39	1.04	1.8	0.11	25MR	−0.8	−3.3	−1.6	1.1
55.59	1.09	1.8	9.78	4AP	−1.7	−3.3	−1.5	1.1
61.80	1.12	1.9	19.41	14AP	−1.3	−3.3	−1.5	1.2
68.02	1.15	1.9	29.01	24AP	−0.3	−3.3	−1.5	1.3
74.24	1.17	2.0	38.57	4MY	0.6	−3.3	−1.4	1.3
80.47	1.19	2.0	48.10	14MY	1.3	−3.3	−1.4	1.4
86.71	1.20	2.0	57.63	24MY	2.3	−3.3	−1.4	1.4
92.96	1.20	2.0	67.14	3JN	3.4	−3.4	−1.4	1.4
99.23	1.20	2.0	76.66	13JN	2.0	−3.4	−1.4	1.4
105.52	1.20	2.0	86.20	23JN	0.6	−3.5	−1.4	1.4
111.84	1.19	2.0	95.75	3JL	−0.7	−3.5	−1.4	1.4
118.19	1.17	2.0	105.34	13JL	−1.4	−3.5	−1.4	1.4
124.56	1.16	2.0	114.97	23JL	−1.3	−3.5	−1.4	1.4
130.98	1.13	2.0	124.64	2AU	−0.6	−3.4	−1.5	1.3
137.43	1.11	1.9	134.37	12AU	−0.2	−3.4	−1.5	1.3
143.93	1.07	1.9	144.15	22AU	0.1	−3.4	−1.5	1.2
150.47	1.04	1.9	153.98	1SE	0.2	−3.3	−1.6	1.2
157.07	0.99	1.9	163.88	11SE	0.5	−3.3	−1.7	1.2
163.71	0.95	1.9	173.84	21SE	1.7	−3.3	−1.7	1.2
170.41	0.89	1.9	183.85	1OC	1.9	−3.3	−1.8	1.2
177.17	0.83	1.9	193.92	11OC	−0.2	−3.3	−1.8	1.2
183.99	0.77	1.9	204.03	21OC	−0.6	−3.3	−1.9	1.2
190.86	0.69	1.8	214.18	31OC	−0.6	−3.4	−2.0	1.2
197.79	0.61	1.8	224.36	10NO	−0.6	−3.4	−2.0	1.1
204.78	0.53	1.8	234.57	20NO	−0.8	−3.4	−2.1	1.1
211.83	0.43	1.7	244.77	30NO	−0.8	−3.5	−2.1	1.1
218.94	0.33	1.7	254.98	10DE	−0.8	−3.5	−2.1	1.0
226.10	0.22	1.6	265.18	20DE	−0.7	−3.6	−2.2	1.0
233.32	0.10	1.6	275.35	30DE	0.2	−3.6	−2.1	0.9

♂ LONG	♂ LAT	♂ MAG	☉ LONG	16.00UT	☿	♀	♃	♄
LONG	LAT	MAG	LONG	-346	MAGNITUDES			
240.59	-0.03	1.5	285.49	9JA	2.8	-3.7	-2.1	0.9
247.90	-0.17	1.4	295.59	19JA	1.5	-3.8	-2.1	0.8
255.26	-0.31	1.4	305.63	29JA	0.7	-3.8	-2.0	0.8
262.65	-0.46	1.3	315.63	8FE	0.4	-3.9	-2.0	0.7
270.08	-0.62	1.2	325.56	18FE	0.1	-4.0	-1.9	0.7
277.52	-0.79	1.1	335.43	28FE	-0.2	-4.1	-1.9	0.7
284.98	-0.96	1.1	345.25	10MR	-0.8	-4.2	-1.8	0.7
292.43	-1.13	1.0	355.01	20MR	-1.7	-4.2	-1.7	0.8
299.87	-1.30	0.9	4.71	30MR	-1.3	-4.2	-1.6	0.9
307.28	-1.47	0.8	14.37	9AP	-0.3	-4.0	-1.6	0.9
314.64	-1.64	0.8	23.98	19AP	0.7	-3.6	-1.5	1.0
321.93	-1.80	0.7	33.55	29AP	1.7	-2.8	-1.5	1.0
329.12	-1.96	0.6	43.10	9MY	3.1	-3.5	-1.4	1.1
336.19	-2.11	0.5	52.63	19MY	2.9	-4.0	-1.4	1.1
343.11	-2.24	0.4	62.15	29MY	1.6	-4.2	-1.4	1.2
349.84	-2.36	0.3	71.67	8JN	0.5	-4.2	-1.3	1.2
356.33	-2.47	0.2	81.19	18JN	-0.7	-4.1	-1.3	1.2
2.55	-2.56	0.1	90.74	28JN	-1.5	-4.0	-1.3	1.2
8.42	-2.63	-0.0	100.31	8JL	-1.3	-4.0	-1.3	1.2
13.87	-2.68	-0.2	109.92	18JL	-0.5	-3.9	-1.3	1.2
18.81	-2.71	-0.3	119.56	28JL	-0.1	-3.8	-1.3	1.2
23.11	-2.72	-0.5	129.26	7AU	0.2	-3.7	-1.3	1.2
26.61	-2.68	-0.7	139.01	17AU	0.4	-3.6	-1.3	1.1
29.13	-2.61	-0.9	148.82	27AU	0.8	-3.6	-1.3	1.1
30.47	-2.48	-1.1	158.69	6SE	2.1	-3.5	-1.4	1.0
30.42	-2.27	-1.3	168.61	16SE	1.7	-3.5	-1.4	1.0
28.91	-1.96	-1.5	178.60	26SE	-0.3	-3.5	-1.5	1.0
26.07	-1.56	-1.7	188.64	6OC	-0.7	-3.4	-1.5	1.0
22.44	-1.08	-1.8	198.73	16OC	-0.7	-3.4	-1.6	1.0
18.80	-0.56	-1.6	208.86	26OC	-0.7	-3.4	-1.6	1.0
15.90	-0.08	-1.4	219.03	5NO	-0.7	-3.4	-1.7	1.0
14.23	0.34	-1.0	229.22	15NO	-0.6	-3.4	-1.8	1.0
13.93	0.67	-0.7	239.42	25NO	-0.7	-3.4	-1.8	1.0
14.87	0.93	-0.4	249.64	5DE	-0.6	-3.4	-1.9	1.0
16.89	1.12	-0.1	259.84	15DE	0.3	-3.4	-2.0	0.9
19.75	1.26	0.1	270.02	25DE	3.0	-3.4	-2.0	0.9
				-345				
23.29	1.36	0.4	280.18	4JA	1.1	-3.4	-2.0	0.8
27.35	1.44	0.6	290.30	14JA	0.4	-3.4	-2.1	0.8
31.82	1.49	0.8	300.37	24JA	0.2	-3.4	-2.1	0.8
36.61	1.52	1.0	310.39	3FE	0.0	-3.4	-2.1	0.7
41.66	1.54	1.1	320.36	13FE	-0.3	-3.5	-2.0	0.7
46.90	1.55	1.3	330.26	23FE	-0.8	-3.5	-2.0	0.6
52.30	1.55	1.4	340.11	5MR	-1.6	-3.4	-1.9	0.6
57.83	1.55	1.5	349.90	15MR	-1.3	-3.4	-1.9	0.5
63.47	1.53	1.6	359.63	25MR	-0.2	-3.4	-1.8	0.6
69.20	1.52	1.7	9.31	4AP	1.0	-3.4	-1.8	0.7
75.01	1.49	1.8	18.94	14AP	2.3	-3.3	-1.7	0.7
80.88	1.47	1.8	28.54	24AP	3.8	-3.3	-1.6	0.8
86.82	1.44	1.9	38.10	4MY	2.2	-3.3	-1.6	0.9
92.82	1.40	1.9	47.63	14MY	1.2	-3.3	-1.5	0.9
98.87	1.36	1.9	57.16	24MY	0.3	-3.3	-1.4	1.0
104.98	1.32	2.0	66.68	3JN	-0.7	-3.3	-1.4	1.0
111.14	1.28	2.0	76.19	13JN	-1.6	-3.4	-1.4	1.1
117.36	1.23	2.0	85.73	23JN	-1.3	-3.4	-1.3	1.1
123.64	1.18	2.0	95.29	3JL	-0.5	-3.4	-1.3	1.1
129.98	1.12	2.0	104.87	13JL	0.0	-3.5	-1.3	1.1
136.38	1.06	2.0	114.50	23JL	0.3	-3.5	-1.2	1.1
142.84	1.00	1.9	124.17	2AU	0.6	-3.6	-1.2	1.1
149.37	0.93	1.9	133.89	12AU	1.1	-3.6	-1.2	1.1
155.98	0.86	1.9	143.67	22AU	2.6	-3.7	-1.2	1.0
162.65	0.79	1.8	153.51	1SE	1.5	-3.8	-1.2	1.0
169.40	0.71	1.8	163.40	11SE	-0.4	-3.9	-1.3	1.0
176.23	0.63	1.8	173.36	21SE	-0.9	-4.0	-1.3	0.9
183.13	0.54	1.7	183.37	1OC	-0.9	-4.1	-1.3	0.9
190.12	0.45	1.7	193.43	11OC	-0.9	-4.2	-1.3	0.9
197.18	0.36	1.7	203.54	21OC	-0.6	-4.3	-1.4	0.9
204.32	0.26	1.7	213.69	31OC	-0.5	-4.4	-1.4	0.9
211.54	0.16	1.7	223.87	10NO	-0.5	-4.3	-1.5	0.9
218.84	0.05	1.6	234.07	20NO	-0.4	-4.0	-1.6	0.9
226.21	-0.06	1.6	244.28	30NO	0.5	-3.3	-1.6	0.9
233.64	-0.17	1.6	254.49	10DE	3.0	-3.4	-1.7	0.9
241.15	-0.29	1.6	264.68	20DE	0.8	-4.0	-1.8	0.9
248.70	-0.41	1.5	274.86	30DE	0.2	-4.3	-1.8	0.8

♂ LONG	♂ LAT	♂ MAG	☉ LONG	16.00UT	☿	♀	♃	♄
LONG	LAT	MAG	LONG	-344	MAGNITUDES			
256.32	-0.52	1.5	285.00	9JA	0.1	-4.4	-1.9	0.8
263.97	-0.64	1.5	295.10	19JA	-0.1	-4.3	-1.9	0.8
271.66	-0.76	1.4	305.15	29JA	-0.4	-4.2	-2.0	0.7
279.38	-0.87	1.4	315.14	8FE	-0.9	-4.1	-2.0	0.7
287.11	-0.98	1.4	325.08	18FE	-1.5	-4.0	-2.0	0.6
294.85	-1.08	1.3	334.96	28FE	-1.2	-3.9	-2.0	0.6
302.58	-1.17	1.3	344.77	9MR	-0.2	-3.8	-2.0	0.5
310.28	-1.26	1.3	354.54	19MR	1.2	-3.7	-2.0	0.5
317.95	-1.33	1.2	4.25	29MR	2.9	-3.6	-1.9	0.5
325.58	-1.40	1.2	13.90	8AP	2.9	-3.5	-1.9	0.5
333.14	-1.45	1.2	23.52	18AP	1.7	-3.5	-1.8	0.6
340.62	-1.48	1.2	33.09	28AP	0.9	-3.4	-1.8	0.6
348.02	-1.50	1.1	42.64	8MY	0.2	-3.4	-1.7	0.7
355.31	-1.51	1.1	52.17	18MY	-0.7	-3.4	-1.6	0.8
2.48	-1.50	1.0	61.69	28MY	-1.6	-3.3	-1.6	0.8
9.52	-1.47	1.0	71.21	7JN	-1.3	-3.3	-1.5	0.9
16.41	-1.43	1.0	80.74	17JN	-0.5	-3.3	-1.5	0.9
23.13	-1.38	0.9	90.28	27JN	0.1	-3.3	-1.4	1.0
29.66	-1.30	0.9	99.85	7JL	0.5	-3.3	-1.4	1.0
35.97	-1.21	0.8	109.45	17JL	0.8	-3.3	-1.3	1.0
42.04	-1.11	0.7	119.10	27JL	1.5	-3.4	-1.3	1.0
47.82	-0.98	0.6	128.79	6AU	3.0	-3.4	-1.3	1.0
53.26	-0.83	0.5	138.54	16AU	1.3	-3.4	-1.2	1.0
58.31	-0.66	0.4	148.34	26AU	-0.5	-3.4	-1.2	1.0
62.87	-0.46	0.3	158.20	5SE	-1.0	-3.5	-1.2	0.9
66.85	-0.23	0.1	168.13	15SE	-1.0	-3.5	-1.2	0.9
70.11	0.04	-0.0	178.11	25SE	-0.8	-3.5	-1.2	0.9
72.47	0.35	-0.2	188.15	5OC	-0.5	-3.5	-1.2	0.8
73.75	0.71	-0.4	198.24	15OC	-0.4	-3.4	-1.3	0.8
73.76	1.12	-0.7	208.36	25OC	-0.3	-3.4	-1.3	0.9
72.36	1.57	-0.9	218.53	4NO	-0.2	-3.4	-1.3	0.9
69.63	2.02	-1.1	228.72	14NO	0.7	-3.4	-1.4	0.9
65.95	2.42	-1.2	238.93	24NO	2.8	-3.3	-1.4	0.9
61.99	2.73	-1.2	249.14	4DE	0.5	-3.3	-1.5	0.9
58.52	2.91	-1.0	259.34	14DE	0.0	-3.3	-1.5	0.9
56.14	2.97	-0.7	269.53	24DE	-0.1	-3.3	-1.6	0.8
				-343				
55.07	2.94	-0.4	279.68	3JA	-0.2	-3.4	-1.7	0.8
55.29	2.86	-0.1	289.81	13JA	-0.4	-3.4	-1.7	0.8
56.64	2.75	0.1	299.88	23JA	-0.9	-3.4	-1.8	0.7
58.92	2.63	0.4	309.91	2FE	-1.4	-3.4	-1.9	0.7
61.95	2.51	0.6	319.88	12FE	-1.1	-3.5	-1.9	0.6
65.57	2.39	0.8	329.79	22FE	-0.1	-3.5	-2.0	0.6
69.66	2.27	1.0	339.64	4MR	1.5	-3.5	-2.0	0.5
74.13	2.15	1.1	349.43	14MR	3.5	-3.6	-2.0	0.5
78.90	2.04	1.2	359.16	24MR	2.2	-3.7	-2.0	0.4
83.92	1.94	1.4	8.85	3AP	1.2	-3.7	-2.0	0.4
89.15	1.83	1.5	18.48	13AP	0.7	-3.8	-2.0	0.4
94.55	1.73	1.6	28.08	23AP	0.1	-3.9	-2.0	0.5
100.10	1.63	1.6	37.64	3MY	-0.7	-4.0	-1.9	0.5
105.79	1.53	1.7	47.18	13MY	-1.7	-4.1	-1.9	0.6
111.60	1.44	1.7	56.70	23MY	-1.3	-4.1	-1.8	0.6
117.53	1.34	1.8	66.22	2JN	-0.4	-4.2	-1.8	0.7
123.56	1.24	1.8	75.74	12JN	0.2	-4.1	-1.7	0.8
129.69	1.15	1.8	85.27	22JN	0.6	-3.8	-1.7	0.8
135.92	1.05	1.8	94.83	2JL	1.1	-3.3	-1.6	0.9
142.25	0.95	1.8	104.41	12JL	1.9	-3.2	-1.5	0.9
148.67	0.85	1.8	114.03	22JL	3.2	-3.9	-1.5	0.9
155.19	0.75	1.8	123.70	1AU	1.1	-4.2	-1.4	0.9
161.81	0.65	1.8	133.42	11AU	-0.5	-4.3	-1.4	0.9
168.52	0.55	1.8	143.19	21AU	-1.1	-4.2	-1.3	0.9
175.34	0.44	1.8	153.02	31AU	-1.2	-4.2	-1.3	0.9
182.25	0.34	1.7	162.92	10SE	-0.7	-4.1	-1.3	0.9
189.26	0.23	1.7	172.87	20SE	-0.4	-4.0	-1.3	0.9
196.37	0.13	1.7	182.88	30SE	-0.2	-3.9	-1.3	0.8
203.57	0.02	1.6	192.94	10OC	-0.2	-3.8	-1.3	0.8
210.87	-0.09	1.6	203.05	20OC	-0.0	-3.8	-1.3	0.8
218.25	-0.20	1.5	213.20	30OC	0.9	-3.7	-1.3	0.8
225.72	-0.30	1.5	223.38	9NO	2.6	-3.6	-1.3	0.8
233.27	-0.41	1.4	233.57	19NO	0.3	-3.6	-1.3	0.8
240.89	-0.51	1.4	243.79	29NO	-0.2	-3.5	-1.4	0.8
248.57	-0.61	1.4	254.00	9DE	-0.2	-3.5	-1.4	0.8
256.31	-0.70	1.4	264.19	19DE	-0.3	-3.4	-1.4	0.8
264.09	-0.79	1.4	274.37	29DE	-0.5	-3.4	-1.5	0.8

Left table (−342 / −341):

♂ LONG	LAT	MAG	☉ LONG	16.00UT	☿	♀	♃	♄
				−342		MAGNITUDES		
271.91	-0.87	1.4	284.51	8JA	-0.9	-3.4	-1.6	0.8
279.75	-0.95	1.4	294.61	18JA	-1.2	-3.3	-1.6	0.8
287.61	-1.01	1.4	304.66	28JA	-1.1	-3.3	-1.7	0.7
295.46	-1.07	1.4	314.66	7FE	-0.1	-3.3	-1.8	0.7
303.29	-1.11	1.4	324.60	17FE	1.8	-3.3	-1.8	0.6
311.11	-1.14	1.4	334.48	27FE	3.0	-3.3	-1.9	0.6
318.88	-1.17	1.4	344.31	9MR	1.6	-3.3	-2.0	0.5
326.60	-1.17	1.4	354.07	19MR	0.9	-3.3	-2.0	0.5
334.27	-1.17	1.4	3.78	29MR	0.5	-3.4	-2.1	0.4
341.86	-1.15	1.4	13.44	8AP	0.0	-3.4	-2.1	0.4
349.37	-1.12	1.4	23.05	18AP	-0.7	-3.4	-2.1	0.3
356.80	-1.08	1.4	32.63	28AP	-1.7	-3.5	-2.1	0.4
4.13	-1.03	1.4	42.18	8MY	-1.3	-3.5	-2.1	0.4
11.36	-0.97	1.4	51.71	18MY	-0.4	-3.4	-2.1	0.5
18.49	-0.89	1.4	61.23	28MY	0.3	-3.4	-2.1	0.5
25.50	-0.81	1.4	70.75	7JN	0.8	-3.4	-2.0	0.6
32.39	-0.71	1.4	80.27	17JN	1.4	-3.4	-2.0	0.7
39.16	-0.61	1.4	89.81	27JN	2.5	-3.3	-1.9	0.7
45.79	-0.49	1.4	99.38	7JL	2.8	-3.3	-1.8	0.8
52.28	-0.37	1.4	108.98	17JL	0.9	-3.3	-1.8	0.8
58.61	-0.23	1.3	118.63	27JL	-0.6	-3.3	-1.7	0.8
64.78	-0.09	1.3	128.31	6AU	-1.3	-3.3	-1.7	0.9
70.76	0.07	1.2	138.06	16AU	-1.3	-3.4	-1.6	0.9
76.53	0.24	1.2	147.86	26AU	-0.7	-3.4	-1.5	0.9
82.05	0.42	1.1	157.72	5SE	-0.3	-3.4	-1.5	0.9
87.30	0.62	1.0	167.64	15SE	-0.1	-3.4	-1.5	0.9
92.20	0.85	0.9	177.62	25SE	-0.0	-3.5	-1.4	0.8
96.71	1.09	0.7	187.65	5OC	0.2	-3.5	-1.4	0.8
100.73	1.37	0.6	197.74	15OC	1.2	-3.6	-1.4	0.8
104.13	1.68	0.4	207.87	25OC	2.4	-3.7	-1.3	0.7
106.79	2.02	0.2	218.03	4NO	0.1	-3.8	-1.3	0.7
108.54	2.41	0.0	228.22	14NO	-0.3	-3.9	-1.3	0.8
109.17	2.83	-0.2	238.43	24NO	-0.4	-3.9	-1.3	0.8
108.52	3.27	-0.5	248.64	4DE	-0.4	-4.0	-1.4	0.8
106.54	3.70	-0.7	258.85	14DE	-0.6	-4.2	-1.4	0.8
103.37	4.05	-0.9	269.04	24DE	-1.0	-4.2	-1.4	0.8
				−341				
99.49	4.27	-1.0	279.20	3JA	-1.1	-4.3	-1.5	0.8
95.61	4.32	-0.9	289.32	13JA	-0.9	-4.3	-1.5	0.8
92.43	4.21	-0.7	299.40	23JA	-0.3	-4.3	-1.5	0.8
90.41	3.99	-0.4	309.42	2FE	2.2	-3.9	-1.6	0.7
89.68	3.71	-0.2	319.40	12FE	2.4	-3.4	-1.7	0.7
90.19	3.41	0.1	329.31	22FE	1.2	-3.5	-1.7	0.7
91.77	3.12	0.3	339.16	4MR	0.7	-4.0	-1.8	0.6
94.23	2.84	0.5	348.96	14MR	0.3	-4.2	-1.9	0.5
97.41	2.58	0.7	358.69	24MR	-0.1	-4.2	-1.9	0.5
101.16	2.35	0.8	8.38	3AP	-0.8	-4.2	-2.0	0.4
105.38	2.13	1.0	18.02	13AP	-1.7	-4.1	-2.1	0.4
109.99	1.92	1.1	27.61	23AP	-1.3	-4.0	-2.1	0.3
114.91	1.73	1.2	37.18	3MY	-0.4	-3.9	-2.2	0.3
120.10	1.55	1.3	46.72	13MY	0.5	-3.8	-2.2	0.3
125.52	1.38	1.4	56.24	23MY	1.1	-3.7	-2.2	0.4
131.15	1.22	1.4	65.76	2JN	1.9	-3.6	-2.2	0.5
136.97	1.07	1.5	75.28	12JN	3.2	-3.6	-2.2	0.5
142.96	0.92	1.5	84.81	22JN	2.3	-3.5	-2.2	0.6
149.09	0.77	1.6	94.36	2JL	0.7	-3.5	-2.2	0.7
155.38	0.63	1.6	103.95	12JL	-0.6	-3.4	-2.1	0.7
161.81	0.50	1.6	113.57	22JL	-1.4	-3.4	-2.1	0.8
168.38	0.36	1.6	123.23	1AU	-1.3	-3.4	-2.0	0.8
175.07	0.23	1.6	132.95	11AU	-0.6	-3.4	-1.9	0.8
181.90	0.11	1.6	142.72	21AU	-0.2	-3.4	-1.9	0.8
188.84	-0.02	1.6	152.55	31AU	-0.0	-3.4	-1.8	0.8
195.91	-0.14	1.6	162.44	10SE	0.1	-3.4	-1.7	0.8
203.09	-0.25	1.6	172.39	20SE	0.4	-3.4	-1.7	0.8
210.38	-0.36	1.6	182.39	30SE	1.5	-3.4	-1.6	0.8
217.78	-0.47	1.5	192.45	10OC	2.1	-3.4	-1.6	0.8
225.27	-0.57	1.5	202.55	20OC	-0.0	-3.4	-1.5	0.8
232.85	-0.66	1.5	212.70	30OC	-0.5	-3.4	-1.5	0.7
240.52	-0.75	1.5	222.88	9NO	-0.5	-3.4	-1.5	0.7
248.25	-0.83	1.4	233.08	19NO	-0.5	-3.5	-1.5	0.7
256.04	-0.90	1.4	243.28	29NO	-0.7	-3.5	-1.4	0.8
263.88	-0.96	1.4	253.50	9DE	-0.9	-3.5	-1.4	0.8
271.76	-1.01	1.4	263.69	19DE	-0.9	-3.5	-1.4	0.8
279.65	-1.04	1.3	273.87	29DE	-0.8	-3.4	-1.4	0.8

Right table (−340 / −339):

♂ LONG	LAT	MAG	☉ LONG	16.00UT	☿	♀	♃	♄
				−340		MAGNITUDES		
287.55	-1.07	1.3	284.02	8JA	0.1	-3.4	-1.5	0.8
295.45	-1.09	1.3	294.12	18JA	2.5	-3.4	-1.5	0.8
303.32	-1.09	1.3	304.17	28JA	1.8	-3.4	-1.5	0.8
311.16	-1.08	1.3	314.18	7FE	0.8	-3.3	-1.6	0.8
318.96	-1.06	1.4	324.12	17FE	0.5	-3.3	-1.6	0.7
326.70	-1.03	1.4	334.00	27FE	0.2	-3.3	-1.7	0.7
334.38	-0.99	1.4	343.83	8MR	-0.2	-3.3	-1.7	0.6
341.99	-0.94	1.5	353.59	18MR	-0.8	-3.4	-1.8	0.6
349.52	-0.88	1.5	3.31	28MR	-1.7	-3.4	-1.8	0.5
356.97	-0.81	1.5	12.97	7AP	-1.3	-3.4	-1.9	0.5
4.32	-0.73	1.6	22.59	17AP	-0.3	-3.4	-2.0	0.4
11.59	-0.65	1.6	32.17	27AP	0.6	-3.5	-2.0	0.3
18.77	-0.56	1.6	41.72	7MY	1.5	-3.5	-2.1	0.3
25.84	-0.46	1.6	51.25	17MY	2.6	-3.5	-2.2	0.3
32.83	-0.36	1.7	60.77	27MY	3.3	-3.6	-2.2	0.3
39.72	-0.25	1.7	70.29	6JN	1.8	-3.7	-2.3	0.4
46.51	-0.14	1.7	79.81	16JN	0.6	-3.8	-2.3	0.5
53.20	-0.03	1.7	89.35	26JN	-0.6	-3.8	-2.3	0.5
59.80	0.09	1.7	98.92	6JL	-1.5	-3.9	-2.4	0.6
66.30	0.21	1.7	108.52	16JL	-1.3	-4.0	-2.3	0.6
72.69	0.33	1.7	118.16	26JL	-0.6	-4.1	-2.3	0.7
78.98	0.46	1.7	127.85	5AU	-0.1	-4.2	-2.3	0.7
85.15	0.60	1.6	137.59	15AU	0.1	-4.3	-2.2	0.8
91.20	0.74	1.6	147.39	25AU	0.3	-4.2	-2.2	0.8
97.11	0.89	1.6	157.25	4SE	0.6	-3.9	-2.1	0.8
102.88	1.04	1.5	167.16	14SE	1.8	-3.4	-2.0	0.8
108.47	1.21	1.4	177.14	24SE	1.9	-3.5	-2.0	0.8
113.86	1.38	1.3	187.17	4OC	-0.2	-4.1	-1.9	0.8
119.01	1.57	1.2	197.25	14OC	-0.6	-4.3	-1.8	0.8
123.89	1.78	1.1	207.37	24OC	-0.6	-4.4	-1.8	0.8
128.41	2.00	1.0	217.54	3NO	-0.7	-4.3	-1.7	0.7
132.53	2.25	0.8	227.72	13NO	-0.8	-4.2	-1.7	0.7
136.13	2.52	0.6	237.93	23NO	-0.7	-4.1	-1.6	0.7
139.08	2.82	0.4	248.14	3DE	-0.6	-4.0	-1.6	0.7
141.24	3.15	0.2	258.35	13DE	-0.7	-3.9	-1.6	0.8
142.43	3.50	-0.0	268.54	23DE	0.2	-3.8	-1.5	0.8
				−339				
142.47	3.86	-0.3	278.70	2JA	2.8	-3.8	-1.5	0.8
141.22	4.19	-0.6	288.83	12JA	1.4	-3.7	-1.5	0.8
138.71	4.45	-0.8	298.91	22JA	0.6	-3.6	-1.5	0.8
135.22	4.58	-1.0	308.94	1FE	0.3	-3.5	-1.5	0.8
131.32	4.54	-1.0	318.91	11FE	0.1	-3.5	-1.5	0.8
127.73	4.32	-0.9	328.83	21FE	-0.2	-3.4	-1.6	0.8
125.07	3.97	-0.7	338.69	3MR	-0.8	-3.4	-1.6	0.7
123.66	3.56	-0.5	348.49	13MR	-1.6	-3.4	-1.6	0.7
123.54	3.13	-0.2	358.23	23MR	-1.3	-3.3	-1.7	0.6
124.61	2.72	-0.0	7.92	2AP	-0.3	-3.3	-1.7	0.6
126.69	2.34	0.2	17.55	12AP	0.8	-3.3	-1.8	0.5
129.61	1.99	0.4	27.16	22AP	1.9	-3.3	-1.8	0.4
133.22	1.67	0.5	36.72	2MY	3.4	-3.3	-1.9	0.4
137.39	1.39	0.6	46.26	12MY	2.7	-3.3	-2.0	0.3
142.03	1.13	0.8	55.79	22MY	1.4	-3.3	-2.0	0.3
147.06	0.89	0.9	65.30	1JN	0.4	-3.4	-2.1	0.3
152.42	0.66	0.9	74.82	11JN	-0.6	-3.4	-2.2	0.4
158.07	0.46	1.0	84.35	21JN	-1.5	-3.5	-2.3	0.4
163.98	0.27	1.1	93.90	1JL	-1.3	-3.5	-2.3	0.5
170.12	0.09	1.1	103.48	11JL	-0.5	-3.5	-2.4	0.5
176.46	-0.07	1.2	113.10	21JL	-0.0	-3.5	-2.4	0.6
183.00	-0.23	1.2	122.76	31JL	0.2	-3.4	-2.4	0.6
189.72	-0.37	1.2	132.47	10AU	0.5	-3.4	-2.4	0.7
196.60	-0.50	1.3	142.24	20AU	0.9	-3.4	-2.4	0.7
203.63	-0.63	1.3	152.07	30AU	2.3	-3.3	-2.4	0.8
210.81	-0.74	1.3	161.95	9SE	1.7	-3.3	-2.4	0.8
218.11	-0.84	1.3	171.90	19SE	-0.3	-3.3	-2.3	0.8
225.54	-0.92	1.3	181.90	29SE	-0.8	-3.3	-2.3	0.8
233.06	-1.00	1.3	191.95	9OC	-0.8	-3.3	-2.2	0.8
240.69	-1.06	1.3	202.06	19OC	-0.8	-3.3	-2.1	0.8
248.39	-1.11	1.4	212.20	29OC	-0.7	-3.4	-2.0	0.8
256.16	-1.15	1.4	222.38	8NO	-0.6	-3.4	-2.0	0.7
263.98	-1.17	1.4	232.58	18NO	-0.6	-3.4	-1.9	0.7
271.84	-1.18	1.4	242.79	28NO	-0.5	-3.5	-1.8	0.7
279.71	-1.18	1.4	253.00	8DE	0.3	-3.5	-1.8	0.7
287.59	-1.17	1.4	263.20	18DE	3.0	-3.6	-1.7	0.8
295.46	-1.14	1.4	273.37	28DE	1.0	-3.6	-1.7	0.8

♂ LONG	LAT	MAG	☉ LONG	16.00UT	☿	♀	♃	♄
				-338	MAGNITUDES			
303.31	-1.10	1.4	283.52	7JA	0.3	-3.7	-1.7	0.8
311.12	-1.05	1.4	293.63	17JA	0.1	-3.8	-1.6	0.8
318.88	-0.99	1.4	303.68	27JA	-0.0	-3.8	-1.6	0.8
326.59	-0.93	1.5	313.69	6FE	-0.3	-3.9	-1.6	0.8
334.23	-0.85	1.5	323.64	16FE	-0.8	-4.0	-1.6	0.8
341.79	-0.77	1.5	333.52	26FE	-1.6	-4.1	-1.6	0.8
349.28	-0.68	1.5	343.35	8MR	-1.3	-4.2	-1.6	0.8
356.68	-0.59	1.5	353.13	18MR	-0.2	-4.2	-1.6	0.7
4.00	-0.50	1.5	2.84	28MR	1.0	-4.2	-1.6	0.7
11.23	-0.40	1.5	12.50	7AP	2.5	-4.0	-1.6	0.6
18.38	-0.30	1.6	22.13	17AP	3.5	-3.5	-1.7	0.6
25.44	-0.20	1.7	31.71	27AP	2.0	-2.8	-1.7	0.5
32.41	-0.10	1.7	41.26	7MY	1.1	-3.5	-1.7	0.4
39.31	0.00	1.7	50.79	17MY	0.3	-4.0	-1.8	0.4
46.12	0.11	1.8	60.31	27MY	-0.7	-4.2	-1.9	0.3
52.86	0.21	1.8	69.83	6JN	-1.6	-4.2	-1.9	0.3
59.53	0.31	1.8	79.35	16JN	-1.3	-4.1	-2.0	0.3
66.12	0.41	1.9	88.89	26JN	-0.5	-4.0	-2.0	0.4
72.64	0.52	1.9	98.45	6JL	0.0	-3.9	-2.1	0.4
79.10	0.62	1.9	108.05	16JL	0.4	-3.9	-2.2	0.5
85.49	0.72	1.9	117.69	26JL	0.7	-3.8	-2.3	0.6
91.82	0.82	1.9	127.38	5AU	1.2	-3.7	-2.3	0.6
98.08	0.92	1.9	137.12	15AU	2.7	-3.6	-2.4	0.7
104.27	1.03	1.9	146.91	25AU	1.5	-3.6	-2.4	0.7
110.39	1.13	1.8	156.77	4SE	-0.4	-3.5	-2.4	0.7
116.43	1.24	1.8	166.68	14SE	-0.9	-3.5	-2.4	0.8
122.39	1.35	1.8	176.65	24SE	-0.9	-3.5	-2.4	0.8
128.25	1.46	1.7	186.68	4OC	-0.9	-3.4	-2.4	0.8
133.99	1.57	1.6	196.76	14OC	-0.6	-3.4	-2.4	0.8
139.61	1.69	1.5	206.88	24OC	-0.5	-3.4	-2.3	0.8
145.09	1.82	1.4	217.05	3NO	-0.4	-3.4	-2.3	0.8
150.37	1.95	1.3	227.23	13NO	-0.3	-3.4	-2.2	0.8
155.45	2.09	1.2	237.44	23NO	0.5	-3.4	-2.1	0.7
160.26	2.24	1.0	247.65	3DE	3.0	-3.4	-2.1	0.7
164.74	2.40	0.9	257.85	13DE	0.7	-3.4	-2.0	0.7
168.83	2.57	0.7	268.04	23DE	0.1	-3.4	-1.9	0.8
				-337				
172.42	2.75	0.5	278.21	2JA	0.0	-3.4	-1.8	0.8
175.38	2.93	0.2	288.33	12JA	-0.1	-3.4	-1.8	0.8
177.57	3.12	-0.0	298.41	22JA	-0.4	-3.4	-1.7	0.8
178.81	3.30	-0.3	308.45	1FE	-0.9	-3.4	-1.7	0.9
178.91	3.45	-0.6	318.43	11FE	-1.5	-3.5	-1.6	0.9
177.76	3.54	-0.9	328.34	21FE	-1.2	-3.5	-1.6	0.9
175.37	3.53	-1.2	338.20	3MR	-0.2	-3.4	-1.6	0.8
172.02	3.39	-1.4	348.00	13MR	1.3	-3.4	-1.6	0.8
168.30	3.09	-1.4	357.75	23MR	3.1	-3.4	-1.5	0.8
164.92	2.67	-1.3	7.44	2AP	2.6	-3.4	-1.5	0.7
162.50	2.18	-1.1	17.08	12AP	1.5	-3.3	-1.5	0.7
161.36	1.68	-0.9	26.68	22AP	0.8	-3.3	-1.5	0.6
161.55	1.21	-0.7	36.25	2MY	0.2	-3.3	-1.6	0.6
162.98	0.78	-0.5	45.79	12MY	-0.7	-3.3	-1.6	0.5
165.45	0.40	-0.3	55.32	22MY	-1.6	-3.3	-1.6	0.4
168.80	0.07	-0.1	64.84	1JN	-1.3	-3.3	-1.7	0.4
172.88	-0.21	0.0	74.35	11JN	-0.5	-3.4	-1.7	0.3
177.54	-0.46	0.2	83.88	21JN	0.1	-3.4	-1.7	0.3
182.71	-0.68	0.3	93.44	1JL	0.5	-3.4	-1.8	0.4
188.30	-0.87	0.4	103.02	11JL	0.9	-3.5	-1.9	0.4
194.25	-1.03	0.5	112.63	21JL	1.6	-3.5	-1.9	0.5
200.50	-1.17	0.6	122.29	31JL	3.1	-3.6	-2.0	0.5
207.02	-1.29	0.7	132.00	10AU	1.3	-3.6	-2.1	0.6
213.77	-1.38	0.7	141.77	20AU	-0.5	-3.7	-2.1	0.6
220.73	-1.46	0.8	151.59	30AU	-1.1	-3.8	-2.2	0.7
227.86	-1.51	0.8	161.47	9SE	-1.1	-3.9	-2.2	0.7
235.14	-1.54	0.9	171.41	19SE	-0.8	-4.0	-2.3	0.8
242.55	-1.56	1.0	181.42	29SE	-0.5	-4.1	-2.3	0.8
250.07	-1.56	1.0	191.47	9OC	-0.3	-4.2	-2.4	0.8
257.67	-1.54	1.1	201.57	19OC	-0.3	-4.3	-2.4	0.8
265.34	-1.51	1.1	211.71	29OC	-0.2	-4.4	-2.4	0.8
273.05	-1.46	1.2	221.89	8NO	0.7	-4.3	-2.3	0.8
280.79	-1.40	1.2	232.08	18NO	2.9	-4.0	-2.3	0.8
288.53	-1.32	1.3	242.29	28NO	0.5	-3.3	-2.3	0.8
296.27	-1.24	1.3	252.50	8DE	-0.0	-3.4	-2.2	0.8
303.99	-1.15	1.3	262.70	18DE	-0.1	-4.1	-2.1	0.7
311.67	-1.05	1.4	272.88	28DE	-0.2	-4.3	-2.1	0.8

♂ LONG	LAT	MAG	☉ LONG	16.00UT	☿	♀	♃	♄
				-336	MAGNITUDES			
319.31	-0.95	1.4	283.03	7JA	-0.5	-4.4	-2.0	0.8
326.88	-0.84	1.5	293.13	17JA	-0.9	-4.3	-1.9	0.9
334.40	-0.73	1.5	303.20	27JA	-1.3	-4.2	-1.8	0.9
341.84	-0.62	1.6	313.20	6FE	-1.1	-4.1	-1.8	0.9
349.21	-0.51	1.6	323.15	16FE	-0.1	-4.0	-1.7	0.9
356.50	-0.39	1.6	333.04	26FE	1.6	-3.9	-1.7	0.9
3.71	-0.28	1.7	342.87	7MR	3.4	-3.8	-1.6	0.9
10.85	-0.17	1.7	352.65	17MR	1.9	-3.7	-1.6	0.9
17.90	-0.07	1.7	2.37	27MR	1.1	-3.6	-1.5	0.9
24.88	0.04	1.8	12.03	6AP	0.6	-3.5	-1.5	0.8
31.78	0.14	1.8	21.66	16AP	0.1	-3.5	-1.5	0.8
38.61	0.23	1.8	31.24	26AP	-0.7	-3.4	-1.5	0.7
45.37	0.33	1.8	40.80	6MY	-1.6	-3.4	-1.5	0.7
52.07	0.42	1.8	50.33	16MY	-1.4	-3.4	-1.5	0.6
58.72	0.51	1.8	59.85	26MY	-0.4	-3.3	-1.5	0.5
65.30	0.59	1.9	69.37	5JN	0.2	-3.3	-1.5	0.5
71.84	0.68	1.9	78.89	15JN	0.7	-3.3	-1.5	0.4
78.34	0.75	1.9	88.43	25JN	1.2	-3.3	-1.5	0.3
84.79	0.83	2.0	97.99	5JL	2.1	-3.3	-1.6	0.4
91.20	0.90	2.0	107.59	15JL	3.1	-3.3	-1.6	0.4
97.58	0.97	2.0	117.23	25JL	1.1	-3.4	-1.6	0.5
103.92	1.04	2.0	126.91	4AU	-0.5	-3.4	-1.7	0.5
110.24	1.11	2.0	136.65	14AU	-1.2	-3.4	-1.8	0.6
116.52	1.17	2.0	146.44	24AU	-1.2	-3.4	-1.8	0.6
122.78	1.23	2.0	156.28	3SE	-0.7	-3.5	-1.9	0.7
129.01	1.29	2.0	166.20	13SE	-0.4	-3.5	-1.9	0.7
135.21	1.35	1.9	176.17	23SE	-0.2	-3.5	-2.0	0.8
141.38	1.40	1.9	186.19	3OC	-0.1	-3.5	-2.1	0.8
147.51	1.45	1.8	196.27	13OC	0.0	-3.4	-2.1	0.8
153.59	1.50	1.8	206.39	23OC	0.9	-3.4	-2.2	0.8
159.63	1.55	1.7	216.54	2NO	2.6	-3.4	-2.2	0.9
165.61	1.60	1.6	226.73	12NO	0.3	-3.4	-2.2	0.9
171.51	1.64	1.5	236.94	22NO	-0.2	-3.3	-2.3	0.9
177.33	1.68	1.4	247.15	2DE	-0.3	-3.3	-2.3	0.9
183.05	1.72	1.3	257.36	12DE	-0.3	-3.3	-2.2	0.8
188.64	1.75	1.2	267.55	22DE	-0.5	-3.3	-2.2	0.8
				-335				
194.08	1.77	1.0	277.71	1JA	-0.9	-3.4	-2.1	0.8
199.33	1.78	0.9	287.84	11JA	-1.2	-3.4	-2.1	0.9
204.34	1.78	0.7	297.93	21JA	-1.0	-3.4	-2.0	0.9
209.07	1.76	0.5	307.96	31JA	-0.1	-3.4	-1.9	0.9
213.44	1.72	0.2	317.94	10FE	1.9	-3.5	-1.9	1.0
217.36	1.66	0.0	327.87	20FE	2.8	-3.5	-1.8	1.0
220.71	1.55	-0.3	337.73	2MR	1.4	-3.6	-1.7	1.0
223.34	1.40	-0.5	347.53	12MR	0.8	-3.6	-1.7	1.0
225.08	1.17	-0.9	357.28	22MR	0.4	-3.7	-1.6	1.0
225.75	0.87	-1.2	6.97	1AP	-0.0	-3.7	-1.6	0.9
225.20	0.47	-1.5	16.62	11AP	-0.7	-3.8	-1.5	0.9
223.43	-0.03	-1.8	26.22	21AP	-1.6	-3.9	-1.5	0.9
220.66	-0.60	-2.1	35.79	1MY	-1.4	-4.0	-1.4	0.8
217.43	-1.20	-2.1	45.33	11MY	-0.4	-4.1	-1.4	0.8
214.50	-1.75	-2.0	54.86	21MY	0.4	-4.2	-1.4	0.7
212.48	-2.21	-1.8	64.38	31MY	0.9	-4.2	-1.4	0.7
211.76	-2.55	-1.6	73.90	10JN	1.6	-4.1	-1.4	0.6
212.44	-2.79	-1.4	83.43	20JN	2.8	-3.8	-1.4	0.5
214.39	-2.93	-1.1	92.97	30JN	2.7	-3.2	-1.4	0.5
217.44	-3.01	-0.9	102.55	10JL	0.9	-3.2	-1.4	0.4
221.40	-3.03	-0.7	112.17	20JL	-0.5	-3.9	-1.4	0.4
226.10	-3.01	-0.6	121.82	30JL	-1.3	-4.2	-1.4	0.5
231.39	-2.95	-0.4	131.53	9AU	-1.3	-4.3	-1.5	0.5
237.17	-2.87	-0.2	141.29	19AU	-0.7	-4.2	-1.5	0.6
243.33	-2.76	-0.1	151.11	29AU	-0.3	-4.2	-1.5	0.6
249.80	-2.63	0.1	160.99	8SE	-0.1	-4.1	-1.6	0.7
256.51	-2.48	0.2	170.93	18SE	0.0	-4.0	-1.7	0.7
263.41	-2.32	0.3	180.93	28SE	0.2	-3.9	-1.7	0.8
270.47	-2.15	0.4	190.98	8OC	1.2	-3.8	-1.8	0.8
277.63	-1.98	0.5	201.08	18OC	2.4	-3.8	-1.8	0.9
284.86	-1.80	0.7	211.22	28OC	0.1	-3.7	-1.9	0.9
292.15	-1.62	0.8	221.39	7NO	-0.4	-3.6	-2.0	0.9
299.47	-1.44	0.9	231.59	17NO	-0.4	-3.6	-2.0	0.9
306.79	-1.26	1.0	241.80	27NO	-0.5	-3.5	-2.1	0.9
314.10	-1.08	1.1	252.01	7DE	-0.6	-3.5	-2.1	0.9
321.39	-0.91	1.2	262.21	17DE	-0.9	-3.4	-2.1	0.9
328.64	-0.75	1.3	272.39	27DE	-1.0	-3.4	-2.1	0.9

♂			☉	16.00UT	☿	♀	♃	♄
LONG	LAT	MAG	LONG	-334	MAGNITUDES			
335.85	-0.59	1.3	282.54	6JA	-0.9	-3.4	-2.1	0.9
343.00	-0.44	1.4	292.65	16JA	-0.0	-3.3	-2.1	0.9
350.11	-0.30	1.5	302.71	26JA	2.2	-3.3	-2.1	1.0
357.15	-0.17	1.6	312.72	5FE	2.2	-3.3	-2.0	1.0
4.13	-0.04	1.6	322.67	15FE	1.0	-3.3	-2.0	1.0
11.05	0.08	1.7	332.57	25FE	0.6	-3.3	-1.9	1.1
17.91	0.19	1.7	342.40	7MR	0.3	-3.3	-1.8	1.1
24.71	0.29	1.8	352.18	17MR	-0.1	-3.3	-1.8	1.1
31.45	0.39	1.8	1.90	27MR	-0.7	-3.4	-1.7	1.1
38.14	0.48	1.9	11.57	6AP	-1.6	-3.4	-1.6	1.1
44.77	0.57	1.9	21.19	16AP	-1.4	-3.4	-1.6	1.1
51.36	0.64	1.9	30.78	26AP	-0.4	-3.5	-1.5	1.0
57.90	0.72	1.9	40.33	6MY	0.5	-3.5	-1.5	1.0
64.41	0.78	2.0	49.86	16MY	1.2	-3.4	-1.4	0.9
70.88	0.85	2.0	59.39	26MY	2.1	-3.4	-1.4	0.9
77.33	0.90	2.0	68.90	5JN	3.4	-3.4	-1.3	0.8
83.75	0.96	2.0	78.42	15JN	2.2	-3.4	-1.3	0.7
90.15	1.00	1.9	87.96	25JN	0.7	-3.3	-1.3	0.7
96.53	1.05	1.9	97.53	5JL	-0.6	-3.3	-1.3	0.6
102.91	1.08	2.0	107.12	15JL	-1.4	-3.3	-1.3	0.5
109.28	1.12	2.0	116.75	25JL	-1.3	-3.3	-1.3	0.5
115.65	1.15	2.0	126.44	4AU	-0.6	-3.3	-1.3	0.5
122.01	1.18	2.0	136.17	14AU	-0.2	-3.4	-1.3	0.6
128.39	1.20	2.0	145.96	24AU	0.0	-3.4	-1.3	0.6
134.77	1.22	2.0	155.81	3SE	0.2	-3.4	-1.3	0.7
141.15	1.23	2.0	165.71	13SE	0.5	-3.4	-1.4	0.7
147.55	1.24	2.0	175.68	23SE	1.6	-3.5	-1.4	0.8
153.96	1.24	2.0	185.70	3OC	2.2	-3.6	-1.5	0.8
160.38	1.24	1.9	195.78	13OC	-0.1	-3.6	-1.5	0.9
166.81	1.24	1.9	205.90	23OC	-0.5	-3.7	-1.6	0.9
173.25	1.22	1.8	216.05	2NO	-0.6	-3.8	-1.7	1.0
179.69	1.21	1.8	226.24	12NO	-0.6	-3.9	-1.7	1.0
186.15	1.18	1.7	236.44	22NO	-0.7	-4.0	-1.8	1.0
192.60	1.14	1.6	246.65	2DE	-0.8	-4.1	-1.8	1.0
199.06	1.10	1.6	256.86	12DE	-0.8	-4.2	-1.9	1.0
205.51	1.04	1.5	267.05	22DE	-0.8	-4.3	-2.0	1.0
				-333				
211.94	0.97	1.4	277.22	1JA	0.1	-4.3	-2.0	1.0
218.36	0.89	1.2	287.35	11JA	2.6	-4.4	-2.0	1.0
224.76	0.78	1.1	297.44	21JA	1.7	-4.3	-2.1	1.0
231.12	0.66	1.0	307.48	31JA	0.7	-3.9	-2.1	1.1
237.44	0.52	0.8	317.46	10FE	0.4	-3.3	-2.0	1.1
243.70	0.35	0.7	327.39	20FE	0.2	-3.5	-2.0	1.1
249.87	0.15	0.5	337.25	2MR	-0.2	-4.0	-2.0	1.2
255.94	-0.09	0.4	347.06	12MR	-0.8	-4.2	-1.9	1.2
261.87	-0.36	0.2	356.81	22MR	-1.6	-4.2	-1.9	1.2
267.63	-0.68	-0.0	6.50	1AP	-1.4	-4.2	-1.8	1.2
273.15	-1.05	-0.2	16.15	11AP	-0.3	-4.1	-1.7	1.2
278.36	-1.47	-0.5	25.76	21AP	0.7	-4.0	-1.7	1.2
283.18	-1.96	-0.7	35.33	1MY	1.6	-3.9	-1.6	1.2
287.48	-2.51	-1.0	44.87	11MY	2.8	-3.8	-1.5	1.2
291.11	-3.14	-1.2	54.40	21MY	3.1	-3.7	-1.5	1.1
293.89	-3.84	-1.5	63.91	31MY	1.7	-3.6	-1.4	1.1
295.62	-4.59	-1.8	73.43	10JN	0.5	-3.6	-1.4	1.0
296.11	-5.35	-2.1	82.97	20JN	-0.6	-3.5	-1.3	0.9
295.31	-6.05	-2.3	92.51	30JN	-1.5	-3.5	-1.3	0.9
293.39	-6.57	-2.6	102.09	10JL	-1.3	-3.4	-1.3	0.8
290.81	-6.81	-2.6	111.70	20JL	-0.6	-3.4	-1.2	0.7
288.33	-6.73	-2.5	121.36	30JL	-0.1	-3.4	-1.2	0.7
286.60	-6.36	-2.3	131.06	9AU	0.2	-3.4	-1.2	0.6
286.08	-5.80	-2.0	140.82	19AU	0.4	-3.4	-1.2	0.7
286.88	-5.15	-1.7	150.64	29AU	0.7	-3.4	-1.2	0.7
288.88	-4.50	-1.4	160.52	8SE	1.9	-3.4	-1.2	0.8
291.92	-3.87	-1.1	170.45	18SE	1.9	-3.4	-1.3	0.8
295.78	-3.29	-0.8	180.44	28SE	-0.2	-3.4	-1.3	0.9
300.29	-2.76	-0.6	190.49	8OC	-0.7	-3.4	-1.3	0.9
305.31	-2.29	-0.3	200.59	18OC	-0.7	-3.4	-1.4	1.0
310.71	-1.87	-0.1	210.73	28OC	-0.7	-3.4	-1.4	1.0
316.40	-1.50	0.1	220.90	7NO	-0.7	-3.5	-1.5	1.1
322.31	-1.17	0.3	231.10	17NO	-0.7	-3.5	-1.5	1.1
328.39	-0.88	0.5	241.30	27NO	-0.7	-3.5	-1.6	1.1
334.59	-0.62	0.7	251.51	7DE	-0.6	-3.5	-1.6	1.1
340.89	-0.39	0.8	261.71	17DE	0.2	-3.5	-1.7	1.2
347.24	-0.19	1.0	271.89	27DE	2.8	-3.4	-1.8	1.2

♂			☉	16.00UT	☿	♀	♃	♄
LONG	LAT	MAG	LONG	-332	MAGNITUDES			
353.63	-0.01	1.1	282.04	6JA	1.3	-3.4	-1.8	1.2
0.05	0.15	1.2	292.16	16JA	0.5	-3.4	-1.9	1.2
6.48	0.30	1.3	302.22	26JA	0.2	-3.4	-1.9	1.2
12.91	0.42	1.4	312.23	5FE	0.1	-3.3	-2.0	1.2
19.33	0.53	1.5	322.19	15FE	-0.3	-3.3	-2.0	1.2
25.75	0.63	1.6	332.08	25FE	-0.8	-3.3	-2.0	1.3
32.15	0.72	1.7	341.92	6MR	-1.6	-3.3	-2.0	1.3
38.54	0.80	1.8	351.70	16MR	-1.3	-3.4	-2.0	1.4
44.91	0.86	1.8	1.42	26MR	-0.3	-3.4	-2.0	1.4
51.26	0.92	1.9	11.10	5AP	0.9	-3.4	-1.9	1.4
57.60	0.97	1.9	20.73	15AP	2.1	-3.4	-1.9	1.4
63.93	1.02	2.0	30.31	25AP	3.7	-3.5	-1.8	1.4
70.24	1.06	2.0	39.87	5MY	2.4	-3.5	-1.8	1.4
76.55	1.09	2.0	49.41	15MY	1.3	-3.5	-1.7	1.3
82.86	1.11	2.0	58.93	25MY	0.4	-3.6	-1.6	1.3
89.17	1.13	2.0	68.45	4JN	-0.6	-3.7	-1.6	1.2
95.48	1.15	2.0	77.97	14JN	-1.5	-3.8	-1.5	1.2
101.80	1.16	2.0	87.50	24JN	-1.3	-3.8	-1.5	1.1
108.14	1.17	2.0	97.07	4JL	-0.5	-3.9	-1.4	1.1
114.49	1.17	2.0	106.66	14JL	-0.0	-4.0	-1.4	1.0
120.87	1.16	2.0	116.29	24JL	0.3	-4.1	-1.3	1.0
127.27	1.16	1.9	125.97	3AU	0.5	-4.2	-1.3	0.9
133.70	1.14	1.9	135.70	13AU	1.0	-4.3	-1.3	0.8
140.16	1.12	1.9	145.48	23AU	2.4	-4.2	-1.2	0.8
146.66	1.10	2.0	155.33	2SE	1.7	-3.9	-1.2	0.8
153.20	1.07	2.0	165.23	12SE	-0.3	-3.4	-1.2	0.9
159.78	1.04	1.9	175.20	22SE	-0.8	-3.5	-1.2	0.9
166.41	1.00	1.9	185.22	2OC	-0.9	-4.1	-1.2	1.0
173.07	0.96	1.9	195.29	12OC	-0.8	-4.3	-1.2	1.0
179.79	0.90	1.9	205.41	22OC	-0.6	-4.4	-1.3	1.1
186.56	0.84	1.9	215.56	1NO	-0.5	-4.3	-1.3	1.1
193.37	0.78	1.8	225.75	11NO	-0.5	-4.2	-1.3	1.2
200.23	0.70	1.8	235.95	21NO	-0.5	-4.1	-1.4	1.2
207.13	0.62	1.7	246.16	1DE	0.4	-4.0	-1.4	1.3
214.09	0.53	1.7	256.37	11DE	3.0	-3.9	-1.5	1.3
221.09	0.42	1.6	266.56	21DE	0.9	-3.8	-1.6	1.3
228.13	0.31	1.5	276.73	31DE	0.3	-3.8	-1.6	1.4
				-331				
235.22	0.19	1.5	286.86	10JA	0.1	-3.7	-1.7	1.4
242.35	0.05	1.4	296.95	20JA	-0.1	-3.6	-1.8	1.3
249.51	-0.10	1.3	307.00	30JA	-0.3	-3.5	-1.8	1.3
256.70	-0.26	1.2	316.98	9FE	-0.8	-3.5	-1.9	1.2
263.92	-0.43	1.1	326.91	19FE	-1.5	-3.4	-1.9	1.2
271.15	-0.62	1.0	336.78	1MR	-1.3	-3.4	-2.0	1.2
278.39	-0.81	0.9	346.59	11MR	-0.3	-3.4	-2.0	1.2
285.62	-1.02	0.8	356.34	21MR	1.1	-3.3	-2.0	1.2
292.84	-1.24	0.7	6.04	31MR	2.7	-3.3	-2.1	1.2
300.01	-1.46	0.6	15.69	10AP	3.2	-3.3	-2.0	1.2
307.13	-1.69	0.5	25.30	20AP	1.8	-3.3	-2.0	1.1
314.15	-1.92	0.4	34.87	30AP	1.0	-3.3	-2.0	1.1
321.06	-2.16	0.3	44.41	10MY	0.3	-3.3	-1.9	1.1
327.81	-2.39	0.1	53.94	20MY	-0.6	-3.3	-1.9	1.0
334.36	-2.62	0.0	63.46	30MY	-1.6	-3.4	-1.8	1.0
340.65	-2.85	-0.1	72.98	9JN	-1.4	-3.4	-1.8	1.0
346.62	-3.07	-0.3	82.50	19JN	-0.5	-3.5	-1.7	0.9
352.18	-3.29	-0.4	92.05	29JN	0.1	-3.5	-1.6	0.9
357.22	-3.49	-0.6	101.62	9JL	0.4	-3.5	-1.6	0.8
1.62	-3.67	-0.8	111.23	19JL	0.8	-3.5	-1.5	0.8
5.20	-3.83	-1.0	120.89	29JL	1.3	-3.4	-1.5	0.7
7.79	-3.95	-1.2	130.59	8AU	2.9	-3.4	-1.4	0.7
9.15	-4.01	-1.4	140.34	18AU	1.5	-3.4	-1.4	0.6
9.09	-3.98	-1.6	150.16	28AU	-0.4	-3.3	-1.3	0.6
7.58	-3.82	-1.8	160.03	7SE	-1.0	-3.3	-1.3	0.6
4.81	-3.50	-2.0	169.96	17SE	-1.0	-3.3	-1.3	0.7
1.35	-3.02	-2.1	179.96	27SE	-0.9	-3.3	-1.3	0.8
358.03	-2.42	-1.9	190.00	7OC	-0.5	-3.3	-1.3	0.8
355.55	-1.79	-1.6	200.09	17OC	-0.4	-3.3	-1.3	0.9
354.35	-1.20	-1.3	210.23	27OC	-0.4	-3.4	-1.3	1.0
354.50	-0.69	-0.9	220.40	6NO	-0.3	-3.4	-1.3	1.0
355.86	-0.26	-0.6	230.60	16NO	0.5	-3.4	-1.3	1.1
358.25	0.08	-0.3	240.81	26NO	3.1	-3.5	-1.3	1.1
1.45	0.36	-0.0	251.02	6DE	0.6	-3.5	-1.4	1.1
5.28	0.58	0.2	261.22	16DE	0.1	-3.6	-1.4	1.1
9.61	0.76	0.4	271.40	26DE	-0.1	-3.6	-1.5	1.1

Left table

♂ LONG	LAT	MAG	☉ LONG	16.00UT	☿	♀	♃	♄
				-330				
14.32	0.90	0.6	281.55	5JA	-0.2	-3.7	-1.5	1.1
19.32	1.02	0.8	291.67	15JA	-0.4	-3.8	-1.6	1.1
24.56	1.11	1.0	301.74	25JA	-0.9	-3.8	-1.6	1.1
29.97	1.18	1.2	311.75	4FE	-1.4	-3.9	-1.7	1.0
35.52	1.23	1.3	321.71	14FE	-1.2	-4.0	-1.8	1.0
41.19	1.27	1.4	331.61	24FE	-0.2	-4.1	-1.9	0.9
46.94	1.31	1.5	341.45	6MR	1.4	-4.2	-1.9	0.9
52.77	1.33	1.6	351.24	16MR	3.3	-4.2	-2.0	0.9
58.66	1.34	1.7	0.96	26MR	2.4	-4.2	-2.0	0.9
64.60	1.35	1.8	10.63	5AP	1.3	-4.0	-2.1	0.9
70.58	1.35	1.8	20.27	15AP	0.8	-3.5	-2.1	0.9
76.61	1.35	1.9	29.85	25AP	0.2	-2.8	-2.1	0.9
82.67	1.34	1.9	39.41	5MY	-0.7	-3.6	-2.1	0.9
88.76	1.33	2.0	48.95	15MY	-1.6	-4.0	-2.1	0.8
94.90	1.31	2.0	58.47	25MY	-1.4	-4.2	-2.1	0.8
101.07	1.28	2.0	67.98	4JN	-0.5	-4.2	-2.1	0.8
107.28	1.26	2.0	77.51	14JN	0.2	-4.1	-2.0	0.7
113.53	1.23	2.0	87.04	24JN	0.6	-4.0	-2.0	0.7
119.83	1.19	2.0	96.60	4JL	1.0	-3.9	-1.9	0.7
126.17	1.15	2.0	106.19	14JL	1.8	-3.9	-1.8	0.6
132.57	1.11	2.0	115.82	24JL	3.2	-3.8	-1.8	0.5
139.01	1.06	2.0	125.49	3AU	1.3	-3.7	-1.7	0.5
145.52	1.01	1.9	135.22	13AU	-0.4	-3.6	-1.7	0.4
152.09	0.95	1.9	145.01	23AU	-1.1	-3.6	-1.6	0.4
158.71	0.89	1.8	154.85	2SE	-1.2	-3.5	-1.5	0.3
165.41	0.83	1.8	164.75	12SE	-0.8	-3.5	-1.5	0.3
172.17	0.76	1.8	174.71	22SE	-0.4	-3.5	-1.5	0.4
179.00	0.68	1.8	184.73	2OC	-0.3	-3.4	-1.4	0.4
185.91	0.60	1.8	194.80	12OC	-0.2	-3.4	-1.4	0.5
192.88	0.51	1.8	204.91	22OC	-0.1	-3.4	-1.4	0.6
199.93	0.42	1.7	215.07	1NO	0.8	-3.4	-1.4	0.6
207.05	0.33	1.7	225.25	11NO	2.9	-3.4	-1.4	0.7
214.24	0.22	1.7	235.45	21NO	0.4	-3.4	-1.4	0.7
221.50	0.12	1.7	245.66	1DE	-0.1	-3.4	-1.4	0.8
228.82	0.00	1.6	255.88	11DE	-0.2	-3.4	-1.4	0.8
236.21	-0.11	1.6	266.07	21DE	-0.3	-3.4	-1.4	0.8
243.65	-0.24	1.6	276.24	31DE	-0.5	-3.4	-1.4	0.9
				-329				
251.16	-0.36	1.5	286.37	10JA	-0.9	-3.4	-1.5	0.9
258.70	-0.49	1.5	296.47	20JA	-1.2	-3.4	-1.5	0.8
266.29	-0.62	1.4	306.51	30JA	-1.1	-3.4	-1.6	0.8
273.91	-0.75	1.4	316.50	9FE	-0.2	-3.5	-1.6	0.8
281.55	-0.88	1.3	326.43	19FE	1.6	-3.5	-1.7	0.8
289.21	-1.01	1.3	336.30	1MR	3.2	-3.4	-1.8	0.7
296.87	-1.13	1.2	346.11	11MR	1.7	-3.4	-1.8	0.7
304.51	-1.25	1.2	355.87	21MR	1.0	-3.4	-1.9	0.6
312.13	-1.36	1.1	5.57	31MR	0.6	-3.4	-2.0	0.6
319.70	-1.46	1.1	15.22	10AP	0.1	-3.3	-2.0	0.7
327.22	-1.55	1.0	24.83	20AP	-0.7	-3.3	-2.1	0.7
334.66	-1.63	1.0	34.40	30AP	-1.6	-3.3	-2.2	0.7
342.02	-1.69	0.9	43.95	10MY	-1.4	-3.3	-2.2	0.7
349.26	-1.74	0.9	53.47	20MY	-0.5	-3.3	-2.2	0.6
356.38	-1.77	0.8	62.99	30MY	0.3	-3.3	-2.2	0.6
3.35	-1.78	0.8	72.51	9JN	0.8	-3.4	-2.3	0.6
10.15	-1.78	0.7	82.04	19JN	1.3	-3.4	-2.2	0.6
16.75	-1.76	0.6	91.59	29JN	2.3	-3.4	-2.2	0.5
23.11	-1.73	0.6	101.16	9JL	3.0	-3.5	-2.2	0.5
29.20	-1.67	0.5	110.77	19JL	1.1	-3.5	-2.1	0.4
34.98	-1.59	0.4	120.42	29JL	-0.5	-3.6	-2.1	0.4
40.37	-1.49	0.3	130.12	8AU	-1.2	-3.6	-2.0	0.3
45.31	-1.36	0.1	139.87	18AU	-1.3	-3.7	-1.9	0.3
49.70	-1.21	-0.0	149.68	28AU	-0.7	-3.8	-1.9	0.2
53.39	-1.02	-0.2	159.55	7SE	-0.3	-3.9	-1.8	0.2
56.26	-0.78	-0.4	169.48	17SE	-0.1	-4.0	-1.7	0.1
58.09	-0.50	-0.6	179.47	27SE	-0.1	-4.1	-1.7	0.1
58.68	-0.17	-0.8	189.51	7OC	0.1	-4.3	-1.6	0.1
57.88	0.23	-1.0	199.60	17OC	1.0	-4.3	-1.6	0.2
55.65	0.67	-1.2	209.74	27OC	2.7	-4.4	-1.5	0.3
52.28	1.12	-1.4	219.91	6NO	0.2	-4.3	-1.5	0.3
48.38	1.53	-1.4	230.10	16NO	-0.3	-4.0	-1.5	0.4
44.71	1.86	-1.2	240.31	26NO	-0.3	-3.2	-1.5	0.5
41.99	2.08	-0.9	250.52	6DE	-0.4	-3.5	-1.5	0.5
40.56	2.21	-0.6	260.72	16DE	-0.6	-4.1	-1.5	0.6
40.45	2.27	-0.3	270.91	26DE	-0.9	-4.3	-1.5	0.6

Right table

♂ LONG	LAT	MAG	☉ LONG	16.00UT	☿	♀	♃	♄
				-328				
41.55	2.28	-0.0	281.06	5JA	-1.1	-4.4	-1.5	0.6
43.64	2.26	0.2	291.18	15JA	-1.0	-4.3	-1.5	0.6
46.53	2.21	0.5	301.25	25JA	-0.1	-4.2	-1.5	0.6
50.05	2.16	0.7	311.27	4FE	2.0	-4.1	-1.5	0.6
54.07	2.10	0.9	321.23	14FE	2.6	-4.0	-1.6	0.6
58.48	2.04	1.0	331.14	24FE	1.3	-3.9	-1.6	0.6
63.21	1.97	1.2	340.98	5MR	0.7	-3.8	-1.7	0.5
68.18	1.90	1.3	350.76	15MR	0.4	-3.7	-1.7	0.5
73.37	1.84	1.4	0.49	25MR	-0.0	-3.6	-1.8	0.5
78.72	1.77	1.5	10.17	4AP	-0.7	-3.5	-1.9	0.4
84.22	1.70	1.6	19.80	14AP	-1.6	-3.5	-1.9	0.5
89.84	1.63	1.7	29.39	24AP	-1.4	-3.4	-2.0	0.5
95.58	1.56	1.7	38.95	4MY	-0.4	-3.4	-2.1	0.5
101.41	1.49	1.8	48.49	14MY	0.4	-3.4	-2.1	0.5
107.33	1.41	1.8	58.01	24MY	1.0	-3.3	-2.2	0.5
113.34	1.34	1.9	67.53	3JN	1.8	-3.3	-2.3	0.5
119.44	1.26	1.9	77.05	13JN	3.0	-3.3	-2.3	0.5
125.62	1.19	1.9	86.59	23JN	2.5	-3.3	-2.3	0.4
131.87	1.11	1.9	96.14	3JL	0.9	-3.3	-2.4	0.4
138.21	1.02	1.9	105.73	13JL	-0.5	-3.3	-2.4	0.4
144.64	0.94	1.9	115.36	23JL	-1.3	-3.4	-2.4	0.3
151.14	0.85	1.9	125.03	2AU	-1.3	-3.4	-2.3	0.3
157.73	0.77	1.9	134.75	12AU	-0.7	-3.5	-2.3	0.2
164.41	0.68	1.8	144.54	22AU	-0.3	-3.5	-2.2	0.2
171.18	0.58	1.8	154.37	1SE	-0.0	-3.5	-2.2	0.1
178.03	0.49	1.8	164.27	11SE	0.1	-3.5	-2.1	0.0
184.98	0.39	1.7	174.23	21SE	0.3	-3.5	-2.0	-0.0
192.02	0.29	1.7	184.24	1OC	1.3	-3.4	-2.0	-0.1
199.14	0.18	1.6	194.30	11OC	2.4	-3.4	-1.9	-0.1
206.36	0.08	1.6	204.42	21OC	0.0	-3.4	-1.8	-0.1
213.66	-0.03	1.5	214.57	31OC	-0.5	-3.4	-1.8	0.0
221.04	-0.14	1.5	224.75	10NO	-0.5	-3.3	-1.7	0.1
228.50	-0.25	1.5	234.95	20NO	-0.5	-3.3	-1.7	0.2
236.04	-0.35	1.5	245.16	30NO	-0.7	-3.3	-1.6	0.2
243.64	-0.46	1.5	255.37	10DE	-0.9	-3.3	-1.6	0.3
251.30	-0.57	1.5	265.57	20DE	-0.9	-3.3	-1.6	0.3
259.02	-0.67	1.5	275.73	30DE	-0.9	-3.4	-1.6	0.4
				-327				
266.78	-0.77	1.5	285.87	9JA	-0.0	-3.4	-1.5	0.4
274.57	-0.86	1.4	295.97	19JA	2.3	-3.4	-1.5	0.4
282.37	-0.94	1.4	306.02	29JA	2.0	-3.4	-1.5	0.5
290.19	-1.02	1.4	316.01	8FE	0.9	-3.5	-1.5	0.5
298.01	-1.09	1.4	325.95	18FE	0.5	-3.5	-1.6	0.5
305.81	-1.15	1.4	335.82	28FE	0.3	-3.6	-1.6	0.4
313.58	-1.20	1.4	345.64	10MR	-0.1	-3.6	-1.6	0.4
321.32	-1.23	1.4	355.40	20MR	-0.7	-3.7	-1.7	0.4
329.00	-1.25	1.4	5.10	30MR	-1.6	-3.7	-1.7	0.3
336.62	-1.26	1.4	14.76	9AP	-1.4	-3.8	-1.7	0.3
344.16	-1.26	1.3	24.37	19AP	-0.4	-3.9	-1.8	0.3
351.62	-1.24	1.3	33.94	29AP	0.5	-4.0	-1.9	0.3
358.99	-1.21	1.3	43.49	9MY	1.3	-4.1	-1.9	0.3
6.25	-1.17	1.3	53.02	19MY	2.4	-4.2	-2.0	0.4
13.40	-1.11	1.3	62.54	29MY	3.5	-4.2	-2.1	0.4
20.43	-1.04	1.3	72.06	8JN	2.0	-4.1	-2.1	0.4
27.32	-0.96	1.3	81.58	18JN	0.7	-3.8	-2.2	0.4
34.08	-0.86	1.2	91.13	28JN	-0.6	-3.2	-2.3	0.4
40.68	-0.76	1.2	100.70	8JL	-1.4	-3.2	-2.3	0.3
47.12	-0.64	1.2	110.30	18JL	-1.3	-3.9	-2.4	0.3
53.37	-0.51	1.1	119.95	28JL	-0.6	-4.2	-2.4	0.3
59.42	-0.36	1.1	129.65	7AU	-0.2	-4.3	-2.4	0.2
65.23	-0.20	1.0	139.40	17AU	0.1	-4.2	-2.4	0.2
70.77	-0.03	0.9	149.20	27AU	0.3	-4.2	-2.4	0.1
76.00	0.17	0.8	159.07	6SE	0.5	-4.1	-2.4	0.1
80.85	0.39	0.7	169.00	16SE	1.6	-4.0	-2.4	-0.0
85.26	0.64	0.6	178.98	26SE	2.2	-3.9	-2.3	-0.1
89.11	0.91	0.4	189.02	6OC	-0.1	-3.8	-2.2	-0.1
92.18	1.23	0.2	199.11	16OC	-0.6	-3.8	-2.2	-0.2
94.61	1.58	0.0	209.24	26OC	-0.6	-3.7	-2.1	-0.2
95.92	1.98	-0.2	219.41	5NO	-0.6	-3.6	-2.0	-0.2
96.01	2.42	-0.4	229.60	15NO	-0.8	-3.6	-2.0	-0.1
94.76	2.87	-0.7	239.81	25NO	-0.7	-3.5	-1.9	-0.0
92.17	3.30	-0.9	250.02	5DE	-0.8	-3.5	-1.8	0.0
88.58	3.65	-1.0	260.22	15DE	-0.7	-3.4	-1.8	0.1
84.60	3.86	-1.0	270.41	25DE	0.1	-3.4	-1.7	0.2

Left table

♂ LONG	LAT	MAG	⊙ LONG	16.00UT	☿	♀	♃	♄
						MAGNITUDES		
				-326				
80.97	3.91	-0.8	280.57	4JA	2.6	-3.4	-1.7	0.2
78.32	3.83	-0.6	290.68	14JA	1.5	-3.3	-1.6	0.3
76.94	3.66	-0.3	300.76	24JA	0.6	-3.3	-1.6	0.3
76.86	3.44	-0.1	310.78	3FE	0.3	-3.3	-1.6	0.3
77.93	3.20	0.2	320.74	13FE	0.1	-3.3	-1.6	0.4
79.98	2.97	0.4	330.65	23FE	-0.2	-3.3	-1.6	0.4
82.81	2.75	0.6	340.50	5MR	-0.8	-3.3	-1.6	0.4
86.28	2.54	0.8	350.29	15MR	-1.6	-3.3	-1.6	0.3
90.25	2.35	0.9	0.02	25MR	-1.4	-3.4	-1.6	0.3
94.64	2.17	1.1	9.70	4AP	-0.4	-3.4	-1.6	0.3
99.36	2.00	1.2	19.33	14AP	0.7	-3.5	-1.6	0.2
104.36	1.84	1.3	28.93	24AP	1.8	-3.5	-1.7	0.2
109.60	1.69	1.4	38.49	4MY	3.2	-3.5	-1.7	0.2
115.04	1.55	1.5	48.02	14MY	2.9	-3.4	-1.8	0.2
120.65	1.41	1.5	57.55	24MY	1.6	-3.4	-1.8	0.3
126.44	1.27	1.6	67.07	3JN	0.5	-3.4	-1.9	0.3
132.36	1.14	1.6	76.58	13JN	-0.6	-3.4	-1.9	0.3
138.43	1.01	1.7	86.12	23JN	-1.5	-3.3	-2.0	0.3
144.63	0.89	1.7	95.68	3JL	-1.4	-3.3	-2.1	0.3
150.95	0.76	1.7	105.26	13JL	-0.6	-3.3	-2.1	0.3
157.39	0.64	1.7	114.88	23JL	-0.1	-3.3	-2.2	0.3
163.96	0.52	1.7	124.56	2AU	0.2	-3.3	-2.3	0.3
170.63	0.40	1.7	134.28	12AU	0.4	-3.4	-2.3	0.2
177.43	0.28	1.7	144.05	22AU	0.8	-3.4	-2.4	0.2
184.34	0.16	1.7	153.89	1SE	2.0	-3.4	-2.4	0.1
191.35	0.05	1.7	163.78	11SE	1.9	-3.4	-2.4	0.1
198.48	-0.07	1.6	173.74	21SE	-0.2	-3.5	-2.4	0.0
205.71	-0.18	1.6	183.75	10C	-0.8	-3.6	-2.4	-0.1
213.04	-0.29	1.6	193.81	11OC	-0.8	-3.6	-2.4	-0.1
220.47	-0.39	1.5	203.93	21OC	-0.8	-3.7	-2.4	-0.2
227.99	-0.49	1.5	214.08	31OC	-0.7	-3.8	-2.3	-0.3
235.59	-0.59	1.5	224.25	10NO	-0.6	-3.9	-2.2	-0.3
243.27	-0.68	1.4	234.46	20NO	-0.6	-4.0	-2.2	-0.2
251.01	-0.77	1.4	244.67	30NO	-0.6	-4.1	-2.1	-0.2
258.80	-0.84	1.4	254.87	10DE	0.2	-4.2	-2.0	-0.1
266.64	-0.91	1.3	265.07	20DE	2.9	-4.3	-2.0	-0.0
274.51	-0.97	1.3	275.24	30DE	1.1	-4.3	-1.9	0.0
				-325				
282.39	-1.02	1.3	285.38	9JA	0.4	-4.4	-1.8	0.1
290.28	-1.06	1.3	295.48	19JA	0.2	-4.2	-1.8	0.2
298.16	-1.08	1.4	305.53	29JA	0.0	-3.9	-1.7	0.2
306.02	-1.10	1.4	315.52	8FE	-0.3	-3.3	-1.7	0.3
313.85	-1.10	1.4	325.46	18FE	-0.8	-3.5	-1.6	0.3
321.62	-1.09	1.4	335.34	28FE	-1.5	-4.0	-1.6	0.3
329.35	-1.07	1.4	345.16	10MR	-1.3	-4.2	-1.6	0.3
337.01	-1.04	1.5	354.92	20MR	-0.9	-4.2	-1.5	0.3
344.60	-1.00	1.5	4.63	30MR	0.9	-4.2	-1.5	0.3
352.10	-0.94	1.5	14.28	9AP	2.3	-4.1	-1.5	0.3
359.52	-0.88	1.5	23.90	19AP	3.8	-4.0	-1.5	0.2
6.85	-0.81	1.5	33.48	29AP	2.2	-3.9	-1.5	0.2
14.09	-0.73	1.6	43.02	9MY	1.2	-3.8	-1.6	0.2
21.23	-0.64	1.6	52.56	19MY	0.4	-3.7	-1.6	0.2
28.26	-0.55	1.6	62.08	29MY	-0.6	-3.6	-1.6	0.2
35.19	-0.45	1.6	71.59	8JN	-1.5	-3.6	-1.7	0.2
42.02	-0.34	1.6	81.12	18JN	-1.4	-3.5	-1.7	0.3
48.74	-0.22	1.6	90.66	28JN	-0.6	-3.5	-1.7	0.3
55.35	-0.10	1.6	100.23	8JL	-0.0	-3.4	-1.8	0.3
61.85	0.02	1.6	109.84	18JL	0.3	-3.4	-1.9	0.3
68.22	0.15	1.6	119.51	28JL	0.6	-3.4	-1.9	0.3
74.46	0.29	1.5	129.18	7AU	1.1	-3.4	-2.0	0.3
80.56	0.44	1.5	138.93	17AU	2.5	-3.4	-2.1	0.3
86.51	0.59	1.4	148.73	27AU	1.7	-3.4	-2.1	0.2
92.29	0.76	1.4	158.59	6SE	-0.3	-3.4	-2.2	0.2
97.88	0.93	1.3	168.52	16SE	-0.9	-3.4	-2.2	0.1
103.23	1.12	1.2	178.50	26SE	-0.9	-3.4	-2.3	0.1
108.33	1.32	1.1	188.53	6OC	-0.9	-3.4	-2.3	0.0
113.10	1.55	1.0	198.62	16OC	-0.6	-3.4	-2.4	-0.1
117.49	1.79	0.9	208.75	26OC	-0.5	-3.4	-2.4	-0.1
121.41	2.07	0.7	218.92	5NO	-0.5	-3.5	-2.4	-0.2
124.74	2.37	0.5	229.11	15NO	-0.4	-3.5	-2.3	-0.3
127.34	2.70	0.3	239.31	25NO	0.4	-3.5	-2.3	-0.2
129.06	3.07	0.1	249.52	5DE	3.1	-3.5	-2.2	-0.2
129.70	3.47	-0.2	259.73	15DE	0.8	-3.4	-2.2	-0.2
129.10	3.86	-0.4	269.91	25DE	0.2	-3.4	-2.1	-0.1

Right table

♂ LONG	LAT	MAG	⊙ LONG	16.00UT	☿	♀	♃	♄
						MAGNITUDES		
				-324				
127.18	4.23	-0.7	280.07	4JA	0.0	-3.4	-2.0	-0.0
124.09	4.50	-0.9	290.19	14JA	-0.1	-3.4	-2.0	0.0
120.28	4.62	-1.0	300.27	24JA	-0.4	-3.4	-1.9	0.1
116.41	4.56	-0.9	310.29	3FE	-0.8	-3.3	-1.8	0.2
113.19	4.34	-0.7	320.26	13FE	-1.5	-3.3	-1.8	0.2
111.11	4.02	-0.5	330.17	23FE	-1.3	-3.3	-1.7	0.2
110.31	3.64	-0.3	340.02	4MR	-0.3	-3.3	-1.6	0.3
110.77	3.26	-0.0	349.81	14MR	1.1	-3.4	-1.6	0.3
112.31	2.89	0.2	359.54	24MR	2.9	-3.4	-1.5	0.3
114.76	2.55	0.4	9.22	3AP	2.9	-3.4	-1.5	0.3
117.96	2.24	0.5	18.86	13AP	1.6	-3.4	-1.5	0.3
121.77	1.96	0.7	28.46	23AP	0.9	-3.5	-1.5	0.3
126.06	1.70	0.8	38.02	3MY	0.2	-3.5	-1.5	0.3
130.77	1.46	0.9	47.56	13MY	-0.6	-3.5	-1.4	0.2
135.83	1.24	1.0	57.08	23MY	-1.6	-3.6	-1.4	0.2
141.18	1.03	1.1	66.60	2JN	-1.4	-3.7	-1.5	0.2
146.79	0.84	1.2	76.13	12JN	-0.5	-3.8	-1.5	0.2
152.63	0.66	1.3	85.66	22JN	0.1	-3.8	-1.5	0.3
158.68	0.48	1.3	95.21	2JL	0.5	-3.9	-1.5	0.3
164.92	0.32	1.3	104.80	12JL	0.8	-4.0	-1.6	0.3
171.33	0.16	1.4	114.42	22JL	1.5	-4.1	-1.6	0.3
177.91	0.02	1.4	124.09	1AU	3.0	-4.2	-1.6	0.4
184.65	-0.12	1.4	133.81	11AU	1.5	-4.3	-1.7	0.4
191.53	-0.26	1.4	143.58	21AU	-0.4	-4.2	-1.8	0.3
198.55	-0.38	1.4	153.42	31AU	-1.0	-3.9	-1.8	0.3
205.71	-0.50	1.4	163.31	10SE	-1.1	-3.3	-1.9	0.3
212.98	-0.61	1.4	173.26	20SE	-0.8	-3.6	-1.9	0.2
220.38	-0.71	1.4	183.27	30SE	-0.5	-4.1	-2.0	0.2
227.88	-0.80	1.4	193.33	100C	-0.3	-4.3	-2.1	0.1
235.47	-0.88	1.4	203.43	200C	-0.3	-4.4	-2.1	0.1
243.15	-0.95	1.4	213.58	300C	-0.2	-4.3	-2.2	-0.0
250.90	-1.01	1.4	223.76	9NO	0.6	-4.2	-2.2	-0.1
258.70	-1.05	1.4	233.96	19NO	3.1	-4.1	-2.2	-0.2
266.55	-1.09	1.4	244.17	29NO	0.6	-4.0	-2.2	-0.2
274.43	-1.11	1.4	254.38	9DE	-0.0	-3.9	-2.2	-0.3
282.33	-1.12	1.4	264.58	19DE	-0.1	-3.8	-2.2	-0.2
290.23	-1.12	1.4	274.75	29DE	-0.2	-3.7	-2.2	-0.1
				-323				
298.11	-1.11	1.4	284.90	8JA	-0.4	-3.7	-2.1	-0.1
305.96	-1.08	1.4	294.99	18JA	-0.9	-3.6	-2.1	0.0
313.78	-1.05	1.4	305.05	28JA	-1.3	-3.5	-2.0	0.1
321.55	-1.00	1.4	315.05	7FE	-1.2	-3.5	-1.9	0.1
329.26	-0.94	1.4	324.98	17FE	-0.2	-3.4	-1.8	0.2
336.90	-0.88	1.4	334.87	27FE	1.4	-3.4	-1.8	0.2
344.46	-0.81	1.4	344.69	9MR	3.4	-3.4	-1.7	0.3
351.95	-0.73	1.5	354.45	19MR	2.1	-3.3	-1.6	0.3
359.35	-0.64	1.5	4.16	29MR	1.2	-3.3	-1.6	0.3
6.66	-0.55	1.6	13.82	8AP	0.7	-3.3	-1.5	0.4
13.89	-0.46	1.6	23.43	18AP	0.1	-3.3	-1.5	0.4
21.02	-0.36	1.6	33.02	28AP	-0.6	-3.3	-1.4	0.4
28.07	-0.26	1.7	42.56	8MY	-1.6	-3.3	-1.4	0.3
35.03	-0.16	1.7	52.09	18MY	-1.4	-3.4	-1.4	0.3
41.91	-0.06	1.7	61.62	28MY	-0.5	-3.4	-1.4	0.3
48.70	0.05	1.8	71.13	7JN	0.2	-3.4	-1.4	0.3
55.40	0.16	1.8	80.66	17JN	0.7	-3.5	-1.3	0.3
62.03	0.27	1.8	90.20	27JN	1.1	-3.5	-1.3	0.3
68.58	0.38	1.8	99.77	7JL	2.0	-3.5	-1.4	0.4
75.05	0.49	1.8	109.37	17JL	3.2	-3.5	-1.4	0.4
81.44	0.60	1.8	119.02	27JL	1.2	-3.4	-1.4	0.4
87.74	0.71	1.8	128.70	6AU	-0.4	-3.4	-1.4	0.4
93.97	0.83	1.8	138.45	16AU	-1.2	-3.4	-1.5	0.5
100.10	0.95	1.8	148.25	26AU	-1.2	-3.3	-1.5	0.4
106.14	1.07	1.7	158.11	5SE	-0.8	-3.3	-1.5	0.4
112.08	1.19	1.7	168.03	15SE	-0.4	-3.3	-1.6	0.4
117.90	1.32	1.6	178.01	25SE	-0.2	-3.3	-1.7	0.4
123.59	1.46	1.6	188.04	5OC	-0.2	-3.3	-1.7	0.3
129.13	1.60	1.5	198.13	15OC	-0.0	-3.3	-1.8	0.3
134.49	1.76	1.4	208.26	25OC	0.8	-3.4	-1.8	0.2
139.63	1.92	1.3	218.42	4NO	2.9	-3.4	-1.9	0.2
144.52	2.09	1.1	228.61	14NO	0.4	-3.4	-2.0	0.1
149.09	2.28	1.0	238.82	24NO	-0.2	-3.5	-2.0	0.1
153.28	2.49	0.8	249.03	4DE	-0.3	-3.5	-2.1	-0.1
156.99	2.71	0.6	259.23	14DE	-0.3	-3.6	-2.1	-0.1
160.09	2.95	0.4	269.42	24DE	-0.5	-3.6	-2.1	-0.1

♂ LONG	♂ LAT	♂ MAG	☉ LONG	16.00UT -322	☿	♀	♃	♄
					MAGNITUDES			
162.46	3.20	0.2	279.58	3JA	-0.9	-3.7	-2.1	-0.1
163.92	3.47	-0.1	289.70	13JA	-1.2	-3.8	-2.1	-0.0
164.28	3.73	-0.4	299.78	23JA	-1.1	-3.8	-2.1	0.0
163.40	3.95	-0.6	309.81	2FE	-0.2	-3.9	-2.0	0.1
161.24	4.10	-0.9	319.78	12FE	1.7	-4.0	-2.0	0.2
158.02	4.11	-1.2	329.69	22FE	3.0	-4.1	-1.9	0.2
154.24	3.97	-1.2	339.54	4MR	1.6	-4.2	-1.9	0.3
150.60	3.66	-1.1	349.34	14MR	0.9	-4.2	-1.8	0.3
147.77	3.24	-0.9	359.08	24MR	0.5	-4.2	-1.7	0.4
146.16	2.76	-0.7	8.76	3AP	0.0	-4.0	-1.7	0.4
145.86	2.29	-0.5	18.40	13AP	-0.7	-3.5	-1.6	0.4
146.81	1.84	-0.3	28.00	23AP	-1.6	-2.9	-1.5	0.5
148.85	1.43	-0.1	37.56	3MY	-1.4	-3.6	-1.5	0.5
151.78	1.07	0.1	47.10	13MY	-0.5	-4.0	-1.4	0.5
155.47	0.74	0.2	56.62	23MY	0.3	-4.2	-1.4	0.5
159.76	0.45	0.4	66.14	2JN	0.9	-4.2	-1.3	0.5
164.57	0.19	0.5	75.66	12JN	1.5	-4.1	-1.3	0.4
169.81	-0.04	0.6	85.19	22JN	2.6	-4.0	-1.3	0.4
175.41	-0.25	0.7	94.74	2JL	2.9	-3.9	-1.3	0.4
181.33	-0.44	0.7	104.33	12JL	1.0	-3.9	-1.3	0.5
187.53	-0.61	0.8	113.95	22JL	-0.5	-3.8	-1.3	0.5
193.97	-0.76	0.9	123.62	1AU	-1.3	-3.7	-1.3	0.5
200.63	-0.89	0.9	133.34	11AU	-1.3	-3.6	-1.3	0.6
207.50	-1.01	1.0	143.11	21AU	-0.7	-3.6	-1.3	0.6
214.53	-1.11	1.0	152.93	31AU	-0.3	-3.5	-1.3	0.6
221.73	-1.19	1.1	162.83	10SE	-0.1	-3.5	-1.3	0.6
229.06	-1.26	1.1	172.77	20SE	-0.0	-3.5	-1.4	0.6
236.52	-1.31	1.1	182.78	30SE	0.2	-3.4	-1.4	0.6
244.08	-1.34	1.2	192.84	10OC	1.1	-3.4	-1.5	0.5
251.73	-1.36	1.2	202.95	20OC	2.7	-3.4	-1.5	0.5
259.44	-1.36	1.2	213.09	30OC	0.2	-3.4	-1.6	0.4
267.21	-1.35	1.3	223.27	9NO	-0.4	-3.4	-1.7	0.4
275.01	-1.32	1.3	233.47	19NO	-0.4	-3.4	-1.7	0.3
282.83	-1.28	1.3	243.68	29NO	-0.4	-3.4	-1.8	0.2
290.65	-1.23	1.3	253.89	9DE	-0.6	-3.4	-1.9	0.2
298.46	-1.17	1.4	264.08	19DE	-0.9	-3.4	-1.9	0.1
306.23	-1.10	1.4	274.26	29DE	-1.0	-3.4	-2.0	0.0
				-321				
313.97	-1.02	1.4	284.40	8JA	-0.9	-3.4	-2.0	0.0
321.65	-0.93	1.5	294.50	18JA	-0.1	-3.4	-2.0	0.1
329.28	-0.84	1.5	304.56	28JA	2.0	-3.4	-2.0	0.1
336.83	-0.74	1.5	314.56	7FE	2.4	-3.5	-2.0	0.2
344.31	-0.64	1.6	324.50	17FE	1.1	-3.5	-2.0	0.3
351.72	-0.54	1.6	334.38	27FE	0.6	-3.4	-2.0	0.3
359.04	-0.44	1.6	344.21	9MR	0.3	-3.4	-1.9	0.4
6.28	-0.33	1.6	353.98	19MR	-0.1	-3.4	-1.9	0.4
13.43	-0.23	1.7	3.69	29MR	-0.7	-3.4	-1.8	0.5
20.51	-0.12	1.7	13.35	8AP	-1.6	-3.3	-1.8	0.5
27.50	-0.02	1.7	22.97	18AP	-1.4	-3.3	-1.7	0.6
34.41	0.08	1.7	32.55	28AP	-0.4	-3.3	-1.6	0.6
41.26	0.18	1.7	42.10	8MY	0.4	-3.3	-1.6	0.6
48.03	0.28	1.8	51.63	18MY	1.1	-3.3	-1.5	0.6
54.73	0.37	1.8	61.15	28MY	2.0	-3.3	-1.5	0.6
61.37	0.46	1.9	70.67	7JN	3.3	-3.4	-1.4	0.6
67.95	0.55	1.9	80.19	17JN	2.3	-3.4	-1.4	0.6
74.48	0.64	1.9	89.73	27JN	0.8	-3.4	-1.3	0.6
80.95	0.73	1.9	99.30	7JL	-0.5	-3.5	-1.3	0.6
87.38	0.81	2.0	108.90	17JL	-1.4	-3.5	-1.3	0.6
93.76	0.90	2.0	118.54	27JL	-1.4	-3.6	-1.2	0.7
100.09	0.98	2.0	128.23	6AU	-0.7	-3.6	-1.2	0.7
106.38	1.06	2.0	137.97	16AU	-0.2	-3.7	-1.2	0.7
112.63	1.14	2.0	147.77	26AU	0.0	-3.8	-1.2	0.8
118.84	1.21	1.9	157.63	5SE	0.2	-3.9	-1.2	0.8
124.99	1.29	1.9	167.55	15SE	0.4	-4.0	-1.2	0.8
131.10	1.37	1.9	177.53	25SE	1.4	-4.1	-1.3	0.8
137.15	1.44	1.8	187.56	5OC	2.4	-4.3	-1.3	0.8
143.14	1.52	1.8	197.64	15OC	0.0	-4.3	-1.3	0.8
149.06	1.60	1.7	207.77	25OC	-0.5	-4.4	-1.4	0.8
154.89	1.67	1.6	217.93	4NO	-0.5	-4.3	-1.4	0.7
160.63	1.75	1.5	228.12	14NO	-0.6	-3.9	-1.5	0.7
166.24	1.83	1.4	238.32	24NO	-0.7	-3.2	-1.5	0.6
171.71	1.90	1.3	248.54	4DE	-0.8	-3.5	-1.6	0.5
177.01	1.98	1.2	258.74	14DE	-0.9	-4.1	-1.7	0.5
182.10	2.06	1.0	268.93	24DE	-0.8	-4.3	-1.7	0.4

♂ LONG	♂ LAT	♂ MAG	☉ LONG	16.00UT -320	☿	♀	♃	♄
					MAGNITUDES			
186.93	2.13	0.8	279.09	3JA	-0.0	-4.4	-1.8	0.3
191.43	2.20	0.6	289.21	13JA	2.3	-4.3	-1.9	0.3
195.54	2.26	0.4	299.29	23JA	1.8	-4.2	-1.9	0.3
199.14	2.31	0.2	309.32	2FE	0.8	-4.1	-2.0	0.3
202.12	2.35	-0.1	319.30	12FE	0.4	-4.0	-2.0	0.4
204.31	2.35	-0.4	329.21	22FE	0.2	-3.9	-2.0	0.4
205.53	2.31	-0.7	339.07	3MR	-0.1	-3.8	-2.0	0.5
205.62	2.20	-1.0	348.86	13MR	-0.7	-3.7	-2.0	0.5
204.45	2.01	-1.3	358.61	23MR	-1.6	-3.6	-2.0	0.6
202.08	1.70	-1.6	8.30	2AP	-1.4	-3.5	-2.0	0.6
198.86	1.29	-1.8	17.93	12AP	-0.4	-3.5	-1.9	0.7
195.38	0.79	-1.8	27.54	22AP	0.6	-3.4	-1.9	0.7
192.39	0.26	-1.6	37.10	2MY	1.5	-3.4	-1.8	0.8
190.47	-0.24	-1.4	46.64	12MY	2.6	-3.4	-1.7	0.8
189.88	-0.69	-1.2	56.17	22MY	3.3	-3.3	-1.7	0.8
190.66	-1.06	-1.0	65.68	1JN	1.8	-3.3	-1.6	0.8
192.66	-1.36	-0.8	75.20	11JN	0.6	-3.3	-1.5	0.9
195.69	-1.61	-0.6	84.73	21JN	-0.5	-3.3	-1.5	0.9
199.60	-1.80	-0.4	94.29	1JL	-1.4	-3.3	-1.4	0.9
204.22	-1.94	-0.3	103.87	11JL	-1.4	-3.3	-1.4	0.9
209.41	-2.05	-0.1	113.49	21JL	-0.6	-3.4	-1.3	0.9
215.09	-2.12	-0.0	123.15	31JL	-0.1	-3.4	-1.3	0.9
221.16	-2.16	0.1	132.86	10AU	0.1	-3.4	-1.3	0.9
227.55	-2.17	0.2	142.63	20AU	0.3	-3.5	-1.3	1.0
234.23	-2.16	0.3	152.46	30AU	0.6	-3.5	-1.2	1.0
241.12	-2.12	0.4	162.34	9SE	1.7	-3.5	-1.2	1.0
248.20	-2.07	0.5	172.29	19SE	2.2	-3.5	-1.2	1.1
255.43	-1.99	0.6	182.29	29SE	-0.1	-3.4	-1.2	1.1
262.77	-1.91	0.7	192.34	9OC	-0.7	-3.4	-1.2	1.1
270.20	-1.80	0.8	202.45	19OC	-0.7	-3.4	-1.3	1.1
277.69	-1.69	0.9	212.59	29OC	-0.7	-3.4	-1.3	1.0
285.21	-1.56	1.0	222.77	8NO	-0.8	-3.3	-1.3	1.0
292.76	-1.43	1.0	232.97	18NO	-0.7	-3.3	-1.4	1.0
300.30	-1.30	1.1	243.18	28NO	-0.7	-3.3	-1.4	0.9
307.83	-1.16	1.2	253.39	8DE	-0.7	-3.3	-1.5	0.9
315.33	-1.02	1.3	263.59	18DE	0.1	-3.3	-1.5	0.8
322.79	-0.88	1.3	273.76	28DE	2.6	-3.4	-1.6	0.7
				-319				
330.19	-0.74	1.4	283.91	7JA	1.4	-3.4	-1.6	0.6
337.54	-0.61	1.5	294.02	17JA	0.5	-3.4	-1.7	0.6
344.82	-0.47	1.5	304.07	27JA	0.3	-3.4	-1.8	0.5
352.04	-0.35	1.6	314.07	6FE	0.1	-3.5	-1.8	0.5
359.18	-0.22	1.6	324.02	16FE	-0.2	-3.5	-1.9	0.6
6.26	-0.10	1.7	333.91	26FE	-0.7	-3.6	-2.0	0.6
13.26	0.01	1.7	343.74	8MR	-1.5	-3.6	-2.0	0.7
20.19	0.12	1.8	353.51	18MR	-1.4	-3.7	-2.0	0.7
27.06	0.22	1.8	3.22	28MR	-0.4	-3.7	-2.1	0.8
33.86	0.32	1.8	12.89	7AP	0.8	-3.8	-2.1	0.8
40.60	0.41	1.9	22.51	17AP	1.9	-3.9	-2.0	0.9
47.28	0.50	1.9	32.09	27AP	3.5	-4.0	-2.0	0.9
53.90	0.58	1.9	41.64	7MY	2.6	-4.1	-2.0	1.0
60.48	0.66	1.9	51.18	17MY	1.4	-4.2	-1.9	1.0
67.02	0.73	1.9	60.69	27MY	0.5	-4.2	-1.9	1.0
73.51	0.80	1.9	70.21	6JN	-0.6	-4.1	-1.8	1.1
79.98	0.86	1.9	79.74	16JN	-1.5	-3.8	-1.8	1.1
86.41	0.92	1.9	89.27	26JN	-1.4	-3.1	-1.7	1.1
92.82	0.98	2.0	98.84	6JL	-0.6	-3.3	-1.6	1.1
99.20	1.03	2.0	108.44	16JL	-0.1	-3.9	-1.6	1.1
105.57	1.08	2.0	118.07	26JL	0.3	-4.2	-1.5	1.1
111.93	1.12	2.0	127.76	5AU	0.5	-4.2	-1.5	1.1
118.28	1.17	2.0	137.50	15AU	0.9	-4.2	-1.4	1.2
124.61	1.20	2.0	147.29	25AU	2.2	-4.2	-1.4	1.2
130.95	1.24	2.0	157.15	4SE	1.9	-4.1	-1.3	1.3
137.28	1.27	2.0	167.07	14SE	-0.2	-4.0	-1.3	1.3
143.60	1.29	2.0	177.04	24SE	-0.8	-3.9	-1.3	1.3
149.92	1.32	2.0	187.07	4OC	-0.8	-3.8	-1.3	1.4
156.23	1.34	1.9	197.15	14OC	-0.8	-3.7	-1.3	1.4
162.53	1.35	1.9	207.27	24OC	-0.7	-3.7	-1.3	1.4
168.83	1.36	1.8	217.43	3NO	-0.6	-3.6	-1.3	1.3
175.10	1.36	1.7	227.62	13NO	-0.6	-3.6	-1.3	1.3
181.36	1.36	1.7	237.83	23NO	-0.5	-3.5	-1.3	1.3
187.60	1.35	1.6	248.04	3DE	0.2	-3.5	-1.4	1.2
193.80	1.33	1.5	258.25	13DE	2.9	-3.4	-1.4	1.2
199.95	1.30	1.4	268.43	23DE	1.1	-3.4	-1.4	1.1

♂ LONG	♂ LAT	♂ MAG	☉ LONG	16.00UT	☿	♀	♃	♄
				-318	MAGNITUDES			
206.06	1.26	1.2	278.60	2JA	0.3	-3.4	-1.5	1.1
212.10	1.20	1.1	288.73	12JA	0.1	-3.3	-1.5	1.0
218.06	1.13	1.0	298.81	22JA	-0.0	-3.3	-1.6	0.9
223.91	1.03	0.8	308.84	1FE	-0.3	-3.3	-1.7	0.9
229.64	0.91	0.6	318.82	11FE	-0.8	-3.3	-1.7	0.8
235.21	0.76	0.4	328.74	21FE	-1.5	-3.3	-1.8	0.8
240.58	0.58	0.2	338.60	3MR	-1.3	-3.3	-1.9	0.9
245.69	0.35	0.0	348.40	13MR	-0.4	-3.4	-1.9	0.9
250.48	0.07	-0.2	358.14	23MR	1.0	-3.4	-2.0	1.0
254.85	-0.28	-0.5	7.83	2AP	2.5	-3.4	-2.1	1.0
258.69	-0.69	-0.7	17.47	12AP	3.4	-3.5	-2.1	1.1
261.85	-1.19	-1.0	27.07	22AP	2.0	-3.5	-2.1	1.1
264.16	-1.78	-1.3	36.64	2MY	1.1	-3.5	-2.1	1.2
265.42	-2.47	-1.6	46.18	12MY	0.3	-3.4	-2.1	1.2
265.49	-3.23	-2.0	55.70	22MY	-0.6	-3.4	-2.1	1.3
264.31	-4.02	-2.2	65.22	1JN	-1.5	-3.4	-2.1	1.3
262.12	-4.74	-2.5	74.74	11JN	-1.4	-3.4	-2.1	1.3
259.42	-5.29	-2.6	84.27	21JN	-0.6	-3.3	-2.0	1.4
256.92	-5.60	-2.4	93.82	1JL	0.0	-3.3	-2.0	1.4
255.31	-5.65	-2.2	103.40	11JL	0.4	-3.3	-1.9	1.4
254.95	-5.50	-2.0	113.01	21JL	0.7	-3.3	-1.8	1.4
255.93	-5.20	-1.7	122.68	31JL	1.2	-3.3	-1.8	1.4
258.17	-4.84	-1.4	132.39	10AU	2.7	-3.4	-1.7	1.3
261.44	-4.44	-1.2	142.15	20AU	1.7	-3.4	-1.6	1.3
265.57	-4.03	-0.9	151.98	30AU	-0.3	-3.4	-1.6	1.3
270.38	-3.62	-0.7	161.86	9SE	-1.0	-3.4	-1.5	1.3
275.71	-3.23	-0.5	171.80	19SE	-1.0	-3.5	-1.5	1.3
281.44	-2.85	-0.3	181.80	29SE	-0.9	-3.6	-1.5	1.3
287.48	-2.50	-0.1	191.85	9OC	-0.6	-3.6	-1.4	1.2
293.75	-2.17	0.1	201.95	19OC	-0.4	-3.7	-1.4	1.2
300.21	-1.86	0.3	212.10	29OC	-0.4	-3.8	-1.4	1.2
306.79	-1.57	0.4	222.27	8NO	-0.3	-3.9	-1.4	1.2
313.46	-1.30	0.6	232.47	18NO	0.4	-4.0	-1.4	1.2
320.19	-1.05	0.7	242.68	28NO	3.1	-4.1	-1.4	1.1
326.96	-0.82	0.9	252.89	8DE	0.8	-4.2	-1.4	1.1
333.74	-0.62	1.0	263.09	18DE	0.1	-4.3	-1.4	1.0
340.53	-0.42	1.1	273.27	28DE	-0.0	-4.3	-1.4	1.0
				-317				
347.31	-0.25	1.2	283.42	7JA	-0.1	-4.4	-1.5	0.9
354.07	-0.09	1.3	293.53	17JA	-0.4	-4.2	-1.5	0.9
0.80	0.06	1.4	303.59	27JA	-0.8	-3.9	-1.5	0.8
7.50	0.19	1.5	313.59	6FE	-1.4	-3.3	-1.6	0.8
14.17	0.31	1.6	323.54	16FE	-1.2	-3.6	-1.7	0.8
20.80	0.42	1.7	333.44	26FE	-0.3	-4.0	-1.7	0.7
27.39	0.52	1.7	343.26	8MR	1.2	-4.2	-1.8	0.8
33.95	0.61	1.8	353.04	18MR	3.1	-4.2	-1.9	0.9
40.48	0.69	1.9	2.76	28MR	2.6	-4.2	-1.9	0.9
46.96	0.76	1.9	12.42	7AP	1.5	-4.1	-2.0	1.0
53.42	0.83	1.9	22.04	17AP	0.8	-4.0	-2.1	1.1
59.85	0.88	2.0	31.63	27AP	0.2	-3.9	-2.1	1.1
66.26	0.94	2.0	41.18	7MY	-0.6	-3.8	-2.2	1.2
72.65	0.98	2.0	50.71	17MY	-1.5	-3.7	-2.2	1.2
79.02	1.02	2.0	60.24	27MY	-1.4	-3.6	-2.3	1.3
85.38	1.06	2.0	69.75	6JN	-0.5	-3.6	-2.3	1.3
91.73	1.09	2.0	79.28	16JN	0.1	-3.5	-2.3	1.3
98.09	1.11	2.0	88.82	26JN	0.5	-3.5	-2.3	1.3
104.44	1.13	2.0	98.38	6JL	0.9	-3.4	-2.2	1.3
110.80	1.15	2.0	107.98	16JL	1.6	-3.4	-2.2	1.3
117.18	1.16	1.9	117.61	26JL	3.1	-3.4	-2.1	1.3
123.56	1.17	2.0	127.29	5AU	1.4	-3.4	-2.1	1.2
129.97	1.17	2.0	137.03	15AU	-0.4	-3.4	-2.0	1.2
136.40	1.16	2.0	146.82	25AU	-1.1	-3.4	-1.9	1.1
142.85	1.16	2.0	156.67	4SE	-1.1	-3.4	-1.9	1.1
149.33	1.14	2.0	166.59	14SE	-0.8	-3.4	-1.8	1.1
155.85	1.13	2.0	176.55	24SE	-0.5	-3.4	-1.7	1.1
162.39	1.10	2.0	186.58	4OC	-0.3	-3.4	-1.7	1.1
168.97	1.07	1.9	196.66	14OC	-0.2	-3.4	-1.6	1.1
175.58	1.04	1.9	206.78	24OC	-0.1	-3.4	-1.6	1.1
182.22	0.99	1.9	216.93	3NO	0.6	-3.5	-1.6	1.1
188.90	0.94	1.8	227.12	13NO	3.1	-3.5	-1.5	1.1
195.62	0.88	1.8	237.33	23NO	0.5	-3.5	-1.5	1.0
202.37	0.81	1.7	247.54	3DE	-0.1	-3.5	-1.5	1.0
209.15	0.73	1.6	257.75	13DE	-0.2	-3.4	-1.5	1.0
215.96	0.64	1.6	267.94	23DE	-0.3	-3.4	-1.5	0.9

♂ LONG	♂ LAT	♂ MAG	☉ LONG	16.00UT	☿	♀	♃	♄
				-316	MAGNITUDES			
222.80	0.54	1.5	278.10	2JA	-0.5	-3.4	-1.5	0.9
229.67	0.43	1.4	288.23	12JA	-0.9	-3.4	-1.5	0.8
236.56	0.29	1.3	298.32	22JA	-1.3	-3.4	-1.5	0.8
243.47	0.15	1.2	308.35	1FE	-1.1	-3.3	-1.5	0.7
250.40	-0.02	1.1	318.33	11FE	-0.3	-3.3	-1.6	0.7
257.33	-0.20	1.0	328.26	21FE	1.5	-3.3	-1.6	0.6
264.26	-0.40	0.8	338.12	2MR	3.4	-3.3	-1.7	0.6
271.18	-0.62	0.7	347.92	12MR	1.9	-3.4	-1.7	0.6
278.07	-0.86	0.6	357.67	22MR	1.1	-3.4	-1.8	0.7
284.92	-1.12	0.5	7.36	1AP	0.6	-3.4	-1.8	0.7
291.70	-1.40	0.3	17.01	11AP	0.1	-3.4	-1.9	0.8
298.37	-1.70	0.2	26.61	21AP	-0.6	-3.5	-2.0	0.9
304.92	-2.02	0.0	36.18	1MY	-1.6	-3.5	-2.0	0.9
311.27	-2.37	-0.1	45.72	11MY	-1.4	-3.6	-2.1	1.0
317.38	-2.72	-0.3	55.25	21MY	-0.5	-3.6	-2.1	1.0
323.18	-3.10	-0.5	64.76	31MY	0.2	-3.7	-2.2	1.1
328.55	-3.49	-0.7	74.29	10JN	0.7	-3.8	-2.3	1.1
333.39	-3.90	-0.9	83.81	20JN	1.2	-3.8	-2.3	1.1
337.56	-4.31	-1.1	93.36	30JN	2.2	-3.9	-2.4	1.2
340.84	-4.71	-1.3	102.94	10JL	3.1	-4.0	-2.4	1.2
343.05	-5.09	-1.6	112.56	20JL	1.2	-4.1	-2.4	1.1
343.98	-5.41	-1.8	122.21	30JL	-0.4	-4.2	-2.4	1.1
343.46	-5.62	-2.0	131.92	9AU	-1.2	-4.3	-2.3	1.1
341.60	-5.63	-2.2	141.68	19AU	-1.3	-4.2	-2.3	1.1
338.74	-5.40	-2.4	151.50	29AU	-0.8	-3.8	-2.2	1.0
335.61	-4.92	-2.3	161.38	8SE	-0.4	-3.3	-2.2	1.0
332.99	-4.25	-2.0	171.32	18SE	-0.2	-3.6	-2.1	0.9
331.46	-3.52	-1.7	181.32	28SE	-0.1	-4.1	-2.0	1.0
331.25	-2.79	-1.4	191.37	8OC	0.0	-4.3	-2.0	1.0
332.33	-2.14	-1.1	201.46	18OC	0.9	-4.4	-1.9	1.0
334.51	-1.57	-0.8	211.61	28OC	2.9	-4.3	-1.8	1.0
337.60	-1.09	-0.5	221.78	7NO	0.3	-4.2	-1.8	1.0
341.39	-0.69	-0.2	231.97	17NO	-0.3	-4.1	-1.7	1.0
345.73	-0.36	0.1	242.18	27NO	-0.3	-4.0	-1.7	0.9
350.50	-0.08	0.3	252.40	7DE	-0.4	-3.9	-1.6	0.9
355.60	0.16	0.5	262.59	17DE	-0.5	-3.8	-1.6	0.9
0.94	0.35	0.7	272.78	27DE	-0.9	-3.7	-1.6	0.9
				-315				
6.48	0.52	0.9	282.93	6JA	-1.1	-3.7	-1.6	0.8
12.17	0.66	1.0	293.03	16JA	-1.0	-3.6	-1.6	0.8
17.97	0.77	1.2	303.10	26JA	-0.2	-3.5	-1.5	0.7
23.85	0.87	1.3	313.11	5FE	1.8	-3.5	-1.5	0.7
29.81	0.95	1.4	323.06	15FE	2.8	-3.4	-1.6	0.6
35.81	1.02	1.5	332.96	25FE	1.4	-3.4	-1.6	0.6
41.86	1.08	1.6	342.79	7MR	0.8	-3.4	-1.6	0.5
47.93	1.12	1.7	352.57	17MR	0.4	-3.3	-1.6	0.5
54.02	1.16	1.8	2.29	27MR	0.0	-3.3	-1.7	0.5
60.14	1.19	1.8	11.96	6AP	-0.6	-3.3	-1.7	0.6
66.27	1.21	1.9	21.58	16AP	-1.6	-3.3	-1.8	0.6
72.43	1.23	1.9	31.17	26AP	-1.5	-3.3	-1.8	0.7
78.59	1.24	2.0	40.73	6MY	-0.5	-3.3	-1.9	0.8
84.78	1.24	2.0	50.26	16MY	0.3	-3.3	-2.0	0.8
90.99	1.24	2.0	59.78	26MY	0.9	-3.4	-2.0	0.9
97.22	1.24	2.0	69.29	5JN	1.6	-3.4	-2.1	0.9
103.47	1.23	2.0	78.81	15JN	2.8	-3.5	-2.2	1.0
109.75	1.21	2.0	88.35	25JN	2.7	-3.5	-2.2	1.0
116.06	1.19	2.0	97.92	5JL	1.0	-3.5	-2.3	1.0
122.41	1.17	2.0	107.51	15JL	-0.4	-3.5	-2.4	1.0
128.80	1.14	2.0	117.14	25JL	-1.3	-3.4	-2.4	1.0
135.24	1.11	2.0	126.82	4AU	-1.4	-3.4	-2.4	1.0
141.71	1.07	1.9	136.55	14AU	-0.7	-3.4	-2.4	1.0
148.24	1.03	1.9	146.35	24AU	-0.3	-3.3	-2.5	1.0
154.83	0.98	1.9	156.19	3SE	-0.1	-3.3	-2.4	1.0
161.46	0.93	1.9	166.10	13SE	0.1	-3.3	-2.4	0.9
168.16	0.87	1.9	176.07	23SE	0.3	-3.3	-2.4	0.9
174.92	0.81	1.9	186.09	3OC	1.1	-3.3	-2.3	0.9
181.73	0.74	1.8	196.16	13OC	2.7	-3.3	-2.2	0.9
188.62	0.66	1.8	206.29	23OC	0.1	-3.4	-2.2	0.9
195.56	0.58	1.8	216.44	2NO	-0.4	-3.4	-2.1	0.9
202.57	0.49	1.8	226.62	12NO	-0.5	-3.4	-2.0	0.9
209.64	0.40	1.7	236.83	22NO	-0.5	-3.5	-2.0	0.9
216.78	0.30	1.7	247.04	2DE	-0.6	-3.5	-1.9	0.9
223.97	0.19	1.7	257.25	12DE	-0.9	-3.6	-1.8	0.9
231.23	0.07	1.6	267.44	22DE	-1.0	-3.6	-1.8	0.8

Left Table

♂ LONG	LAT	MAG	☉ LONG	16.00UT	☿	♀	♃	♄
				-314				
238.54	-0.05	1.6	277.61	1JA	-0.9	-3.7	-1.7	0.8
245.91	-0.18	1.5	287.74	11JA	-0.1	-3.8	-1.7	0.8
253.32	-0.32	1.4	297.83	21JA	2.1	-3.9	-1.6	0.7
260.77	-0.46	1.4	307.86	31JA	2.2	-3.9	-1.6	0.7
268.26	-0.60	1.3	317.85	10FE	1.0	-4.0	-1.6	0.6
275.78	-0.75	1.3	327.78	20FE	0.6	-4.1	-1.6	0.6
283.31	-0.91	1.2	337.64	2MR	0.3	-4.2	-1.6	0.5
290.85	-1.06	1.1	347.45	12MR	-0.1	-4.2	-1.6	0.5
298.39	-1.21	1.1	357.20	22MR	-0.7	-4.2	-1.6	0.4
305.90	-1.36	1.0	6.89	1AP	-1.5	-4.0	-1.6	0.4
313.39	-1.51	0.9	16.54	11AP	-1.4	-3.5	-1.6	0.4
320.81	-1.64	0.9	26.15	21AP	-0.5	-2.9	-1.6	0.5
328.16	-1.77	0.8	35.71	1MY	0.5	-3.6	-1.7	0.6
335.42	-1.88	0.7	45.26	11MY	1.2	-4.0	-1.7	0.6
342.56	-1.99	0.6	54.79	21MY	2.2	-4.2	-1.8	0.7
349.56	-2.08	0.6	64.30	31MY	3.5	-4.2	-1.8	0.7
356.38	-2.15	0.5	73.82	10JN	2.2	-4.1	-1.9	0.8
2.99	-2.21	0.4	83.35	20JN	0.8	-4.0	-1.9	0.8
9.35	-2.25	0.3	92.90	30JN	-0.5	-3.9	-2.0	0.9
15.41	-2.27	0.2	102.47	10JL	-1.4	-3.8	-2.1	0.9
21.11	-2.26	0.1	112.09	20JL	-1.4	-3.8	-2.1	0.9
26.37	-2.24	-0.1	121.74	30JL	-0.7	-3.7	-2.2	0.9
31.09	-2.19	-0.2	131.45	9AU	-0.2	-3.6	-2.3	0.9
35.15	-2.10	-0.4	141.21	19AU	0.1	-3.6	-2.3	0.9
38.40	-1.98	-0.6	151.02	29AU	0.2	-3.5	-2.4	0.9
40.65	-1.82	-0.8	160.90	8SE	0.5	-3.5	-2.4	0.9
41.67	-1.59	-1.0	170.84	18SE	1.5	-3.5	-2.4	0.9
41.31	-1.29	-1.2	180.83	28SE	2.4	-3.4	-2.4	0.8
39.48	-0.92	-1.4	190.88	8OC	-0.0	-3.4	-2.4	0.8
36.41	-0.48	-1.6	200.98	18OC	-0.6	-3.4	-2.4	0.8
32.64	-0.00	-1.7	211.11	28OC	-0.6	-3.4	-2.4	0.8
28.97	0.45	-1.5	221.29	7NO	-0.6	-3.4	-2.3	0.8
26.14	0.84	-1.2	231.48	17NO	-0.7	-3.4	-2.2	0.8
24.57	1.15	-0.9	241.69	27NO	-0.8	-3.4	-2.2	0.9
24.35	1.37	-0.6	251.90	7DE	-0.8	-3.4	-2.1	0.8
25.38	1.52	-0.3	262.10	17DE	-0.8	-3.4	-2.0	0.8
27.43	1.62	0.0	272.28	27DE	-0.0	-3.4	-1.9	0.8
				-313				
30.31	1.69	0.3	282.43	6JA	2.4	-3.4	-1.9	0.8
33.85	1.73	0.5	292.54	16JA	1.7	-3.4	-1.8	0.8
37.89	1.74	0.7	302.61	26JA	0.7	-3.4	-1.8	0.7
42.34	1.75	0.9	312.62	5FE	0.4	-3.5	-1.7	0.7
47.10	1.74	1.1	322.58	15FE	0.2	-3.5	-1.7	0.6
52.11	1.72	1.2	332.47	25FE	-0.2	-3.4	-1.6	0.6
57.32	1.70	1.3	342.31	7MR	-0.7	-3.4	-1.6	0.5
62.70	1.67	1.5	352.09	17MR	-1.5	-3.4	-1.6	0.5
68.21	1.64	1.6	1.81	27MR	-1.4	-3.4	-1.5	0.4
73.83	1.60	1.6	11.48	6AP	-0.4	-3.3	-1.5	0.4
79.55	1.56	1.7	21.11	16AP	0.6	-3.3	-1.5	0.3
85.35	1.52	1.8	30.69	26AP	1.6	-3.3	-1.5	0.4
91.23	1.48	1.8	40.26	6MY	2.9	-3.3	-1.5	0.5
97.18	1.43	1.9	49.79	16MY	3.1	-3.3	-1.6	0.5
103.20	1.38	1.9	59.31	26MY	1.7	-3.4	-1.6	0.6
109.28	1.32	1.9	68.83	5JN	0.6	-3.4	-1.6	0.6
115.43	1.27	2.0	78.35	15JN	-0.5	-3.4	-1.7	0.7
121.65	1.21	2.0	87.89	25JN	-1.4	-3.4	-1.7	0.8
127.93	1.15	2.0	97.45	5JL	-1.4	-3.5	-1.7	0.8
134.28	1.08	2.0	107.04	15JL	-0.6	-3.5	-1.8	0.8
140.69	1.01	1.9	116.67	25JL	-0.1	-3.6	-1.9	0.9
147.18	0.94	1.9	126.36	4AU	0.2	-3.6	-1.9	0.9
153.75	0.87	1.9	136.08	14AU	0.4	-3.7	-2.0	0.9
160.39	0.79	1.9	145.87	24AU	0.7	-3.8	-2.1	0.9
167.11	0.71	1.8	155.72	3SE	1.9	-3.9	-2.1	0.9
173.91	0.62	1.8	165.62	13SE	2.2	-4.0	-2.2	0.9
180.79	0.53	1.7	175.58	23SE	-0.1	-4.1	-2.2	0.8
187.76	0.44	1.7	185.61	3OC	-0.7	-4.3	-2.3	0.8
194.80	0.34	1.6	195.68	13OC	-0.8	-4.3	-2.3	0.8
201.93	0.24	1.6	205.79	23OC	-0.7	-4.4	-2.3	0.7
209.15	0.14	1.6	215.95	2NO	-0.7	-4.3	-2.3	0.8
216.44	0.03	1.6	226.13	12NO	-0.6	-3.9	-2.3	0.8
223.81	-0.08	1.6	236.33	22NO	-0.6	-3.1	-2.3	0.8
231.25	-0.19	1.6	246.55	2DE	-0.6	-3.5	-2.3	0.8
238.76	-0.30	1.6	256.75	12DE	0.1	-4.1	-2.2	0.8
246.34	-0.41	1.5	266.94	22DE	2.7	-4.3	-2.1	0.8

Right Table

♂ LONG	LAT	MAG	☉ LONG	16.00UT	☿	♀	♃	♄
				-312				
253.97	-0.53	1.5	277.11	1JA	1.3	-4.4	-2.1	0.8
261.64	-0.64	1.5	287.25	11JA	0.4	-4.3	-2.0	0.8
269.36	-0.75	1.5	297.34	21JA	0.2	-4.2	-1.9	0.8
277.10	-0.85	1.4	307.38	31JA	0.0	-4.1	-1.9	0.7
284.86	-0.95	1.4	317.36	10FE	-0.2	-4.0	-1.8	0.7
292.63	-1.04	1.4	327.29	20FE	-0.7	-3.9	-1.7	0.6
300.40	-1.13	1.4	337.16	1MR	-1.5	-3.8	-1.7	0.6
308.15	-1.20	1.3	346.97	11MR	-1.4	-3.7	-1.6	0.5
315.87	-1.27	1.3	356.72	21MR	-0.4	-3.6	-1.6	0.5
323.54	-1.32	1.3	6.42	31MR	0.8	-3.5	-1.5	0.4
331.16	-1.36	1.3	16.07	10AP	2.1	-3.5	-1.5	0.4
338.71	-1.39	1.2	25.68	20AP	3.8	-3.4	-1.5	0.3
346.18	-1.40	1.2	35.25	30AP	2.4	-3.4	-1.5	0.3
353.56	-1.40	1.2	44.80	10MY	1.3	-3.4	-1.4	0.4
0.83	-1.38	1.2	54.32	20MY	0.4	-3.3	-1.4	0.4
7.98	-1.35	1.1	63.85	30MY	-0.5	-3.3	-1.4	0.5
15.00	-1.30	1.1	73.36	9JN	-1.5	-3.3	-1.4	0.6
21.88	-1.24	1.1	82.89	19JN	-1.4	-3.3	-1.5	0.6
28.60	-1.16	1.0	92.44	29JN	-0.6	-3.3	-1.5	0.7
35.14	-1.07	1.0	102.01	9JL	-0.0	-3.3	-1.5	0.7
41.48	-0.97	0.9	111.62	19JL	0.3	-3.4	-1.5	0.8
47.59	-0.85	0.8	121.28	29JL	0.6	-3.4	-1.6	0.8
53.45	-0.71	0.8	130.98	8AU	1.0	-3.4	-1.6	0.8
59.01	-0.55	0.7	140.74	18AU	2.3	-3.5	-1.7	0.8
64.21	-0.37	0.6	150.55	28AU	1.9	-3.5	-1.8	0.8
69.00	-0.16	0.5	160.42	7SE	-0.2	-3.5	-1.8	0.8
73.28	0.07	0.3	170.35	17SE	-0.9	-3.5	-1.9	0.8
76.92	0.34	0.2	180.34	27SE	-0.9	-3.4	-1.9	0.8
79.80	0.65	-0.0	190.39	7OC	-0.9	-3.4	-2.0	0.8
81.74	1.00	-0.2	200.48	17OC	-0.6	-3.4	-2.1	0.8
82.52	1.40	-0.4	210.62	27OC	-0.5	-3.4	-2.1	0.7
81.99	1.84	-0.7	220.79	6NO	-0.5	-3.3	-2.2	0.7
80.06	2.30	-0.9	230.98	16NO	-0.4	-3.3	-2.2	0.7
76.91	2.73	-1.1	241.19	26NO	0.3	-3.3	-2.2	0.8
73.03	3.07	-1.2	251.40	6DE	2.9	-3.3	-2.2	0.8
69.14	3.28	-1.0	261.60	16DE	1.0	-3.3	-2.2	0.8
65.99	3.35	-0.8	271.79	26DE	0.2	-3.4	-2.2	0.8
				-311				
64.03	3.31	-0.5	281.94	5JA	0.1	-3.4	-2.1	0.8
63.39	3.20	-0.2	292.05	15JA	-0.1	-3.4	-2.1	0.8
63.99	3.06	0.0	302.12	25JA	-0.3	-3.4	-2.0	0.8
65.65	2.89	0.3	312.13	4FE	-0.8	-3.5	-2.0	0.7
68.18	2.73	0.5	322.10	14FE	-1.5	-3.5	-1.9	0.7
71.40	2.57	0.7	331.99	24FE	-1.3	-3.6	-1.8	0.7
75.17	2.42	0.9	341.83	6MR	-0.4	-3.6	-1.7	0.6
79.38	2.27	1.0	351.62	16MR	1.0	-3.7	-1.7	0.6
83.95	2.13	1.2	1.34	26MR	2.7	-3.7	-1.6	0.5
88.81	2.00	1.3	11.01	5AP	3.1	-3.8	-1.6	0.4
93.90	1.88	1.4	20.65	15AP	1.8	-3.9	-1.5	0.4
99.21	1.76	1.5	30.24	25AP	1.0	-4.0	-1.5	0.3
104.68	1.64	1.6	39.79	5MY	0.3	-4.1	-1.4	0.3
110.31	1.53	1.6	49.33	15MY	-0.6	-4.2	-1.4	0.3
116.08	1.41	1.7	58.85	25MY	-1.5	-4.2	-1.4	0.4
121.97	1.30	1.7	68.37	4JN	-1.5	-4.1	-1.3	0.4
127.99	1.20	1.8	77.89	14JN	-0.6	-3.8	-1.3	0.5
134.11	1.09	1.8	87.43	24JN	0.0	-3.1	-1.3	0.6
140.34	0.98	1.8	96.99	4JL	0.4	-3.3	-1.3	0.6
146.69	0.87	1.8	106.58	14JL	0.8	-3.9	-1.3	0.7
153.13	0.76	1.8	116.21	24JL	1.4	-4.2	-1.4	0.7
159.68	0.66	1.8	125.88	3AU	2.8	-4.2	-1.4	0.8
166.34	0.55	1.8	135.61	13AU	1.7	-4.2	-1.4	0.8
173.10	0.44	1.8	145.39	23AU	-0.3	-4.1	-1.5	0.8
179.96	0.33	1.7	155.24	2SE	-1.1	-4.0	-1.5	0.8
186.92	0.22	1.7	165.14	12SE	-1.1	-4.0	-1.5	0.8
193.99	0.11	1.7	175.10	22SE	-0.9	-3.9	-1.6	0.8
201.16	-0.00	1.6	185.12	2OC	-0.5	-3.8	-1.7	0.8
208.42	-0.11	1.6	195.19	12OC	-0.4	-3.7	-1.7	0.8
215.78	-0.22	1.6	205.30	22OC	-0.3	-3.7	-1.8	0.8
223.23	-0.32	1.5	215.45	1NO	-0.3	-3.6	-1.9	0.7
230.76	-0.43	1.5	225.64	11NO	0.4	-3.6	-1.9	0.7
238.37	-0.53	1.4	235.84	21NO	3.2	-3.5	-2.0	0.7
246.04	-0.62	1.4	246.05	1DE	0.7	-3.5	-2.0	0.7
253.78	-0.71	1.4	256.26	11DE	0.0	-3.4	-2.1	0.8
261.57	-0.80	1.4	266.45	21DE	-0.1	-3.4	-2.1	0.8
269.39	-0.88	1.4	276.62	31DE	-0.2	-3.4	-2.1	0.8

Left table — ☉ –310 / –309

♂ LONG	LAT	MAG	☉ LONG	16.00UT	☿	♀	♃	♄
				-310				
277.25	-0.95	1.4	286.76	10JA	-0.4	-3.3	-2.1	0.8
285.11	-1.00	1.4	296.85	20JA	-0.8	-3.3	-2.1	0.8
292.98	-1.05	1.4	306.89	30JA	-1.3	-3.3	-2.1	0.8
300.84	-1.09	1.4	316.88	9FE	-1.2	-3.3	-2.0	0.8
308.68	-1.12	1.4	326.81	19FE	-0.3	-3.3	-2.0	0.7
316.48	-1.14	1.4	336.68	1MR	1.3	-3.3	-1.9	0.7
324.24	-1.14	1.4	346.50	11MR	3.3	-3.4	-1.8	0.7
331.94	-1.13	1.4	356.25	21MR	2.3	-3.4	-1.8	0.6
339.57	-1.11	1.4	5.95	31MR	1.3	-3.4	-1.7	0.5
347.13	-1.07	1.5	15.60	10AP	0.7	-3.5	-1.6	0.5
354.60	-1.03	1.5	25.21	20AP	0.2	-3.5	-1.6	0.4
1.98	-0.97	1.5	34.78	30AP	-0.6	-3.5	-1.5	0.4
9.27	-0.91	1.5	44.33	10MY	-1.5	-3.4	-1.5	0.3
16.45	-0.83	1.5	53.86	20MY	-1.5	-3.4	-1.4	0.3
23.53	-0.75	1.5	63.38	30MY	-0.5	-3.4	-1.4	0.3
30.50	-0.65	1.5	72.90	9JN	0.1	-3.4	-1.3	0.4
37.36	-0.55	1.5	82.42	19JN	0.6	-3.3	-1.3	0.4
44.09	-0.44	1.5	91.97	29JN	1.0	-3.3	-1.3	0.5
50.70	-0.32	1.5	101.54	9JL	1.8	-3.3	-1.3	0.6
57.17	-0.19	1.4	111.15	19JL	3.2	-3.3	-1.3	0.6
63.50	-0.06	1.4	120.80	29JL	1.4	-3.3	-1.3	0.7
69.68	0.09	1.4	130.50	8AU	-0.3	-3.4	-1.3	0.7
75.68	0.24	1.3	140.26	18AU	-1.1	-3.4	-1.3	0.7
81.50	0.41	1.3	150.07	28AU	-1.2	-3.4	-1.3	0.8
87.09	0.59	1.2	159.94	7SE	-0.8	-3.5	-1.3	0.8
92.44	0.79	1.1	169.87	17SE	-0.4	-3.5	-1.3	0.8
97.49	1.00	1.0	179.86	27SE	-0.2	-3.6	-1.4	0.8
102.17	1.24	0.9	189.90	7OC	-0.2	-3.6	-1.4	0.8
106.43	1.50	0.7	199.99	17OC	-0.1	-3.7	-1.5	0.8
110.16	1.79	0.6	210.13	27OC	0.7	-3.8	-1.5	0.8
113.23	2.11	0.4	220.30	6NO	3.2	-3.9	-1.6	0.7
115.50	2.47	0.2	230.49	16NO	0.5	-4.0	-1.7	0.7
116.77	2.87	-0.1	240.70	26NO	-0.1	-4.1	-1.7	0.7
116.86	3.30	-0.3	250.91	6DE	-0.2	-4.2	-1.8	0.7
115.65	3.72	-0.5	261.11	16DE	-0.3	-4.3	-1.9	0.8
113.14	4.10	-0.8	271.29	26DE	-0.5	-4.3	-1.9	0.8
				-309				
109.61	4.37	-1.0	281.45	5JA	-0.9	-4.4	-2.0	0.8
105.66	4.48	-1.0	291.56	15JA	-1.2	-4.2	-2.0	0.8
101.99	4.42	-0.8	301.63	25JA	-1.1	-3.9	-2.0	0.8
99.26	4.21	-0.6	311.65	4FE	-0.3	-3.3	-2.0	0.8
97.76	3.92	-0.3	321.61	14FE	1.5	-3.6	-2.0	0.8
97.56	3.59	-0.1	331.52	24FE	3.2	-4.1	-2.0	0.8
98.53	3.26	0.1	341.36	6MR	1.7	-4.2	-2.0	0.7
100.50	2.94	0.3	351.14	16MR	1.0	-4.3	-1.9	0.7
103.28	2.65	0.5	0.87	26MR	0.5	-4.2	-1.9	0.6
106.73	2.38	0.7	10.55	5AP	0.1	-4.1	-1.8	0.6
110.70	2.13	0.9	20.18	15AP	-0.6	-4.0	-1.7	0.5
115.12	1.90	1.0	29.77	25AP	-1.5	-3.9	-1.7	0.5
119.90	1.69	1.1	39.33	5MY	-1.5	-3.8	-1.6	0.4
124.99	1.50	1.2	48.86	15MY	-0.5	-3.7	-1.6	0.3
130.33	1.31	1.3	58.39	25MY	0.2	-3.6	-1.5	0.3
135.91	1.13	1.3	67.91	4JN	0.8	-3.6	-1.4	0.3
141.69	0.97	1.4	77.43	14JN	1.4	-3.5	-1.4	0.3
147.66	0.81	1.4	86.97	24JN	2.4	-3.5	-1.3	0.4
153.80	0.65	1.5	96.52	4JL	3.0	-3.4	-1.3	0.5
160.09	0.50	1.5	106.11	14JL	1.2	-3.4	-1.3	0.5
166.54	0.36	1.5	115.74	24JL	-0.4	-3.4	-1.2	0.6
173.13	0.22	1.5	125.41	3AU	-1.2	-3.4	-1.2	0.6
179.86	0.09	1.5	135.14	13AU	-1.3	-3.4	-1.2	0.7
186.73	-0.04	1.5	144.92	23AU	-0.8	-3.4	-1.2	0.7
193.72	-0.16	1.5	154.76	2SE	-0.3	-3.4	-1.2	0.7
200.84	-0.28	1.5	164.66	12SE	-0.1	-3.4	-1.2	0.8
208.07	-0.40	1.5	174.62	22SE	-0.0	-3.4	-1.2	0.8
215.42	-0.50	1.5	184.63	2OC	0.1	-3.4	-1.3	0.8
222.87	-0.60	1.5	194.70	12OC	0.9	-3.4	-1.3	0.8
230.41	-0.70	1.5	204.81	22OC	3.0	-3.4	-1.3	0.8
238.04	-0.78	1.5	214.96	1NO	0.3	-3.5	-1.4	0.8
245.75	-0.86	1.4	225.15	11NO	-0.3	-3.5	-1.4	0.8
253.52	-0.93	1.4	235.35	21NO	-0.4	-3.5	-1.5	0.7
261.35	-0.98	1.4	245.55	1DE	-0.4	-3.5	-1.5	0.7
269.21	-1.03	1.4	255.76	11DE	-0.6	-3.4	-1.6	0.7
277.10	-1.06	1.4	265.96	21DE	-0.9	-3.4	-1.7	0.8
285.00	-1.08	1.4	276.13	31DE	-1.0	-3.4	-1.7	0.8

Right table — ☉ –308 / –307

♂ LONG	LAT	MAG	☉ LONG	16.00UT	☿	♀	♃	♄
				-308				
292.90	-1.09	1.3	286.26	10JA	-1.0	-3.4	-1.8	0.8
300.79	-1.09	1.3	296.36	20JA	-0.2	-3.4	-1.9	0.8
308.64	-1.08	1.3	306.40	30JA	1.8	-3.3	-1.9	0.8
316.46	-1.05	1.3	316.40	9FE	2.6	-3.3	-2.0	0.8
324.22	-1.02	1.4	326.33	19FE	1.2	-3.3	-2.0	0.8
331.92	-0.97	1.4	336.20	29FE	0.7	-3.3	-2.0	0.8
339.56	-0.92	1.4	346.02	10MR	0.4	-3.4	-2.0	0.8
347.12	-0.85	1.5	355.78	20MR	-0.0	-3.4	-2.0	0.8
354.59	-0.78	1.5	5.48	30MR	-0.6	-3.4	-2.0	0.7
1.98	-0.70	1.5	15.14	9AP	-1.5	-3.4	-1.9	0.7
9.28	-0.62	1.6	24.75	19AP	-1.5	-3.5	-1.9	0.6
16.49	-0.53	1.6	34.32	29AP	-0.5	-3.5	-1.8	0.5
23.61	-0.43	1.6	43.87	9MY	0.4	-3.6	-1.8	0.5
30.64	-0.33	1.7	53.40	19MY	1.0	-3.6	-1.7	0.4
37.57	-0.23	1.7	62.92	29MY	1.8	-3.7	-1.7	0.3
44.42	-0.12	1.7	72.44	8JN	3.1	-3.8	-1.6	0.3
51.17	-0.01	1.7	81.96	18JN	2.5	-3.9	-1.5	0.3
57.82	0.10	1.7	91.51	28JN	0.9	-3.9	-1.5	0.4
64.39	0.22	1.8	101.08	8JL	-0.4	-4.1	-1.4	0.4
70.87	0.34	1.8	110.69	18JL	-1.3	-4.1	-1.4	0.5
77.25	0.46	1.7	120.33	28JL	-1.4	-4.2	-1.3	0.5
83.53	0.58	1.7	130.04	7AU	-0.7	-4.2	-1.3	0.6
89.71	0.71	1.7	139.78	17AU	-0.3	-4.2	-1.3	0.6
95.77	0.85	1.7	149.59	27AU	-0.0	-3.8	-1.2	0.7
101.72	0.99	1.6	159.46	6SE	0.1	-3.3	-1.2	0.7
107.53	1.13	1.6	169.39	16SE	0.3	-3.6	-1.2	0.8
113.19	1.28	1.5	179.37	26SE	1.2	-4.1	-1.2	0.8
118.67	1.45	1.4	189.41	6OC	2.7	-4.3	-1.2	0.8
123.94	1.62	1.3	199.50	16OC	0.1	-4.4	-1.2	0.8
128.97	1.81	1.2	209.63	26OC	-0.5	-4.3	-1.3	0.8
133.70	2.01	1.1	219.81	5NO	-0.5	-4.2	-1.3	0.8
138.06	2.24	0.9	230.00	15NO	-0.5	-4.1	-1.3	0.8
141.98	2.48	0.7	240.20	25NO	-0.7	-4.0	-1.4	0.8
145.34	2.75	0.5	250.42	5DE	-0.8	-3.9	-1.4	0.8
148.01	3.04	0.3	260.62	15DE	-0.9	-3.8	-1.5	0.7
149.83	3.36	0.1	270.80	25DE	-0.8	-3.7	-1.5	0.8
				-307				
150.61	3.68	-0.2	280.96	4JA	-0.1	-3.7	-1.6	0.8
150.19	4.00	-0.4	291.08	14JA	2.1	-3.6	-1.7	0.8
148.48	4.27	-0.7	301.15	24JA	2.0	-3.5	-1.7	0.9
145.57	4.45	-0.9	311.17	3FE	0.9	-3.5	-1.8	0.9
141.88	4.46	-1.1	321.14	13FE	0.5	-3.4	-1.9	0.9
138.04	4.30	-1.0	331.04	23FE	0.2	-3.4	-1.9	0.9
134.76	3.99	-0.9	340.89	5MR	-0.1	-3.4	-2.0	0.9
132.57	3.58	-0.7	350.67	15MR	-0.7	-3.3	-2.0	0.9
131.67	3.13	-0.4	0.41	25MR	-1.5	-3.3	-2.0	0.8
132.05	2.69	-0.2	10.09	4AP	-1.5	-3.3	-2.1	0.8
133.57	2.27	-0.0	19.72	14AP	-0.5	-3.3	-2.1	0.7
136.03	1.90	0.2	29.31	24AP	0.5	-3.3	-2.0	0.7
139.29	1.55	0.3	38.87	4MY	1.4	-3.3	-2.0	0.6
143.19	1.24	0.5	48.41	14MY	2.4	-3.3	-2.0	0.6
147.62	0.96	0.6	57.93	24MY	3.5	-3.4	-1.9	0.5
152.50	0.71	0.7	67.45	3JN	2.0	-3.4	-1.9	0.4
157.76	0.48	0.8	76.97	13JN	0.7	-3.5	-1.8	0.4
163.34	0.27	0.9	86.50	23JN	-0.5	-3.5	-1.8	0.3
169.20	0.07	1.0	96.06	3JL	-1.4	-3.5	-1.7	0.4
175.32	-0.11	1.0	105.64	13JL	-1.4	-3.5	-1.6	0.4
181.66	-0.28	1.1	115.27	23JL	-0.7	-3.4	-1.6	0.5
188.21	-0.43	1.1	124.94	2AU	-0.2	-3.4	-1.5	0.5
194.94	-0.57	1.1	134.66	12AU	0.1	-3.4	-1.5	0.6
201.85	-0.69	1.2	144.44	22AU	0.3	-3.3	-1.4	0.6
208.91	-0.81	1.2	154.28	1SE	0.6	-3.3	-1.4	0.7
216.12	-0.91	1.2	164.17	11SE	1.6	-3.3	-1.3	0.7
223.46	-1.00	1.2	174.13	21SE	2.4	-3.3	-1.3	0.8
230.92	-1.07	1.3	184.14	1OC	-0.0	-3.3	-1.3	0.8
238.48	-1.13	1.3	194.20	11OC	-0.6	-3.3	-1.3	0.8
246.13	-1.18	1.3	204.32	21OC	-0.7	-3.4	-1.3	0.8
253.86	-1.21	1.3	214.47	31OC	-0.7	-3.4	-1.3	0.8
261.64	-1.23	1.3	224.65	10NO	-0.8	-3.4	-1.3	0.8
269.47	-1.23	1.3	234.85	20NO	-0.7	-3.5	-1.3	0.8
277.32	-1.22	1.4	245.06	30NO	-0.7	-3.5	-1.3	0.8
285.19	-1.20	1.4	255.27	10DE	-0.7	-3.6	-1.4	0.8
293.04	-1.17	1.4	265.47	20DE	-0.0	-3.6	-1.4	0.8
300.89	-1.12	1.4	275.64	30DE	2.4	-3.7	-1.5	0.8

♂ LONG	♂ LAT	♂ MAG	⊙ LONG	16.00UT -306	☿	♀	♃	♄
					MAGNITUDES			
308.69	-1.07	1.4	285.78	9JA	1.6	-3.8	-1.5	0.8
316.46	-1.01	1.5	295.88	19JA	0.6	-3.9	-1.6	0.9
324.17	-0.93	1.5	305.93	29JA	0.3	-3.9	-1.6	0.9
331.81	-0.85	1.5	315.92	8FE	0.1	-4.0	-1.7	0.9
339.39	-0.77	1.5	325.86	18FE	-0.2	-4.1	-1.8	0.9
346.89	-0.68	1.5	335.73	28FE	-0.7	-4.2	-1.8	1.0
354.31	-0.58	1.5	345.55	10MR	-1.5	-4.2	-1.9	0.9
1.64	-0.48	1.6	355.31	20MR	-1.4	-4.2	-2.0	0.9
8.89	-0.39	1.6	5.02	30MR	-0.5	-4.0	-2.0	0.9
16.06	-0.28	1.6	14.67	9AP	0.7	-3.5	-2.1	0.9
23.14	-0.18	1.6	24.29	19AP	1.8	-2.9	-2.1	0.8
30.13	-0.08	1.7	33.86	29AP	3.2	-3.6	-2.2	0.8
37.05	0.02	1.7	43.41	9MY	2.8	-4.0	-2.2	0.7
43.89	0.12	1.8	52.94	19MY	1.5	-4.2	-2.2	0.7
50.66	0.22	1.8	62.46	29MY	0.6	-4.2	-2.1	0.6
57.35	0.32	1.8	71.97	8JN	-0.5	-4.1	-2.1	0.5
63.98	0.42	1.9	81.50	18JN	-1.4	-4.0	-2.1	0.5
70.54	0.52	1.9	91.04	28JN	-1.4	-3.9	-2.0	0.4
77.04	0.62	1.9	100.61	8JL	-0.6	-3.8	-2.0	0.4
83.47	0.71	1.9	110.22	18JL	-0.1	-3.8	-1.9	0.4
89.85	0.81	1.9	119.87	28JL	0.2	-3.7	-1.8	0.5
96.17	0.90	1.9	129.56	7AU	0.5	-3.6	-1.8	0.5
102.44	0.99	1.9	139.31	17AU	0.8	-3.6	-1.7	0.6
108.64	1.09	1.9	149.11	27AU	2.0	-3.5	-1.6	0.6
114.78	1.18	1.9	158.98	6SE	2.2	-3.5	-1.6	0.7
120.86	1.28	1.8	168.90	16SE	-0.1	-3.5	-1.5	0.7
126.86	1.38	1.8	178.88	26SE	-0.8	-3.4	-1.5	0.8
132.78	1.47	1.7	188.92	6OC	-0.8	-3.4	-1.5	0.8
138.61	1.58	1.7	199.01	16OC	-0.8	-3.4	-1.4	0.8
144.32	1.68	1.6	209.14	26OC	-0.7	-3.4	-1.4	0.9
149.91	1.79	1.5	219.31	5NO	-0.6	-3.4	-1.4	0.9
155.35	1.90	1.4	229.50	15NO	-0.6	-3.4	-1.4	0.9
160.61	2.02	1.3	239.71	25NO	-0.5	-3.4	-1.4	0.9
165.64	2.14	1.1	249.92	5DE	0.1	-3.4	-1.4	0.9
170.41	2.27	1.0	260.12	15DE	2.7	-3.4	-1.4	0.9
174.84	2.40	0.8	270.31	25DE	1.2	-3.4	-1.4	0.9
				-305				
178.86	2.54	0.6	280.47	4JA	0.4	-3.4	-1.4	0.9
182.35	2.68	0.4	290.59	14JA	0.2	-3.4	-1.5	0.9
185.21	2.82	0.1	300.66	24JA	0.0	-3.5	-1.5	0.9
187.27	2.95	-0.2	310.68	3FE	-0.3	-3.5	-1.6	1.0
188.34	3.06	-0.4	320.65	13FE	-0.7	-3.5	-1.6	1.0
188.25	3.12	-0.8	330.56	23FE	-1.5	-3.4	-1.7	1.0
186.91	3.11	-1.1	340.41	5MR	-1.4	-3.4	-1.8	1.0
184.35	2.99	-1.3	350.20	15MR	-0.4	-3.4	-1.8	1.0
180.94	2.73	-1.6	359.93	25MR	0.8	-3.4	-1.9	1.0
177.29	2.34	-1.5	9.61	4AP	2.3	-3.3	-2.0	1.0
174.14	1.87	-1.4	19.25	14AP	3.7	-3.3	-2.0	1.0
172.04	1.36	-1.2	28.84	24AP	2.1	-3.3	-2.1	1.0
171.26	0.87	-1.0	38.41	4MY	1.2	-3.3	-2.2	0.9
171.82	0.42	-0.7	47.95	14MY	0.4	-3.3	-2.2	0.9
173.59	0.03	-0.5	57.47	24MY	-0.5	-3.3	-2.2	0.8
176.37	-0.31	-0.4	66.98	3JN	-1.5	-3.4	-2.3	0.7
180.02	-0.60	-0.2	76.51	13JN	-1.5	-3.4	-2.3	0.7
184.36	-0.84	-0.0	86.04	23JN	-0.6	-3.4	-2.3	0.6
189.28	-1.05	0.1	95.59	3JL	-0.0	-3.5	-2.3	0.5
194.69	-1.23	0.2	105.18	13JL	0.3	-3.5	-2.2	0.5
200.51	-1.37	0.3	114.80	23JL	0.6	-3.6	-2.2	0.5
206.66	-1.49	0.4	124.47	2AU	1.1	-3.7	-2.1	0.5
213.12	-1.58	0.5	134.19	12AU	2.5	-3.7	-2.1	0.6
219.82	-1.65	0.6	143.96	22AU	1.9	-3.8	-2.0	0.6
226.74	-1.70	0.6	153.80	1SE	-0.2	-3.9	-1.9	0.7
233.85	-1.72	0.7	163.69	11SE	-0.9	-4.0	-1.9	0.7
241.11	-1.72	0.8	173.64	21SE	-1.0	-4.1	-1.8	0.8
248.50	-1.71	0.9	183.65	1OC	-0.9	-4.3	-1.8	0.8
255.99	-1.68	0.9	193.71	11OC	-0.6	-4.3	-1.7	0.9
263.56	-1.63	1.0	203.82	21OC	-0.4	-4.4	-1.6	0.9
271.19	-1.57	1.0	213.97	31OC	-0.4	-4.3	-1.6	0.9
278.87	-1.49	1.1	224.15	10NO	-0.4	-3.9	-1.6	1.0
286.55	-1.40	1.2	234.35	20NO	0.3	-3.1	-1.5	1.0
294.24	-1.31	1.2	244.56	30NO	3.0	-3.6	-1.5	1.0
301.92	-1.20	1.3	254.77	10DE	0.9	-4.1	-1.5	1.0
309.57	-1.09	1.3	264.96	20DE	0.2	-4.3	-1.5	1.0
317.18	-0.98	1.4	275.14	30DE	0.0	-4.4	-1.5	1.0

♂ LONG	♂ LAT	♂ MAG	⊙ LONG	16.00UT -304	☿	♀	♃	♄
					MAGNITUDES			
324.73	-0.86	1.4	285.28	9JA	-0.1	-4.3	-1.5	1.0
332.23	-0.74	1.5	295.38	19JA	-0.3	-4.2	-1.5	1.0
339.66	-0.62	1.5	305.44	29JA	-0.8	-4.1	-1.5	1.0
347.01	-0.50	1.6	315.43	8FE	-1.4	-4.0	-1.6	1.1
354.29	-0.39	1.6	325.37	18FE	-1.3	-3.9	-1.6	1.1
1.50	-0.27	1.7	335.26	28FE	-0.4	-3.8	-1.6	1.1
8.63	-0.16	1.7	345.08	9MR	1.1	-3.7	-1.7	1.2
15.68	-0.05	1.7	354.84	19MR	2.9	-3.6	-1.7	1.2
22.66	0.05	1.8	4.55	29MR	2.8	-3.5	-1.8	1.2
29.56	0.16	1.8	14.21	8AP	1.6	-3.5	-1.9	1.2
36.40	0.25	1.8	23.82	18AP	0.9	-3.4	-1.9	1.1
43.16	0.35	1.8	33.41	28AP	0.3	-3.4	-2.0	1.1
49.87	0.44	1.8	42.95	8MY	-0.5	-3.4	-2.1	1.1
56.52	0.52	1.8	52.48	18MY	-1.5	-3.3	-2.1	1.0
63.12	0.61	1.8	62.00	28MY	-1.5	-3.3	-2.2	1.0
69.66	0.69	1.9	71.52	7JN	-0.6	-3.3	-2.3	0.9
76.17	0.76	1.9	81.04	17JN	0.1	-3.3	-2.3	0.9
82.63	0.83	1.9	90.59	27JN	0.5	-3.3	-2.4	0.8
89.06	0.90	2.0	100.16	7JL	0.9	-3.3	-2.4	0.7
95.46	0.97	2.0	109.76	17JL	1.5	-3.4	-2.4	0.7
101.83	1.03	2.0	119.40	27JL	3.0	-3.4	-2.4	0.6
108.18	1.09	2.0	129.09	6AU	1.6	-3.4	-2.4	0.6
114.50	1.14	2.0	138.84	16AU	-0.3	-3.5	-2.3	0.6
120.80	1.20	2.0	148.64	26AU	-1.0	-3.5	-2.3	0.7
127.08	1.25	2.0	158.50	5SE	-1.1	-3.5	-2.2	0.7
133.35	1.29	2.0	168.42	15SE	-0.9	-3.5	-2.2	0.8
139.59	1.34	2.0	178.40	25SE	-0.5	-3.4	-2.1	0.8
145.81	1.38	1.9	188.43	5OC	-0.3	-3.4	-2.0	0.9
152.00	1.42	1.9	198.51	15OC	-0.3	-3.4	-2.0	0.9
158.16	1.46	1.8	208.64	25OC	-0.2	-3.4	-1.9	1.0
164.29	1.49	1.7	218.80	4NO	0.5	-3.3	-1.8	1.0
170.38	1.52	1.7	229.00	14NO	3.2	-3.3	-1.8	1.0
176.41	1.54	1.6	239.20	24NO	0.6	-3.3	-1.7	1.1
182.37	1.56	1.5	249.41	4DE	-0.0	-3.3	-1.7	1.1
188.27	1.57	1.4	259.62	14DE	-0.1	-3.3	-1.6	1.1
194.06	1.58	1.2	269.81	24DE	-0.2	-3.4	-1.6	1.1
				-303				
199.75	1.57	1.1	279.96	3JA	-0.4	-3.4	-1.6	1.1
205.29	1.55	0.9	290.09	13JA	-0.8	-3.4	-1.6	1.1
210.66	1.52	0.8	300.17	23JA	-1.3	-3.4	-1.6	1.1
215.81	1.46	0.6	310.19	2FE	-1.2	-3.5	-1.6	1.1
220.71	1.39	0.4	320.17	12FE	-0.4	-3.5	-1.6	1.2
225.26	1.28	0.1	330.08	22FE	1.3	-3.6	-1.6	1.2
229.39	1.13	-0.1	339.93	4MR	3.4	-3.6	-1.6	1.3
232.99	0.93	-0.4	349.73	14MR	2.1	-3.7	-1.6	1.3
235.91	0.67	-0.7	359.47	24MR	1.2	-3.8	-1.6	1.3
237.99	0.33	-1.0	9.15	3AP	0.7	-3.8	-1.7	1.3
239.05	-0.10	-1.3	18.79	13AP	0.2	-3.9	-1.7	1.3
238.91	-0.63	-1.6	28.39	23AP	-0.6	-4.0	-1.8	1.3
237.55	-1.25	-1.9	37.95	3MY	-1.5	-4.1	-1.8	1.3
235.12	-1.92	-2.2	47.49	13MY	-1.5	-4.2	-1.9	1.3
232.11	-2.56	-2.3	57.01	23MY	-0.6	-4.2	-2.0	1.2
229.24	-3.10	-2.2	66.53	2JN	0.2	-4.1	-2.0	1.2
227.17	-3.49	-2.0	76.05	12JN	0.7	-3.7	-2.1	1.1
226.37	-3.73	-1.8	85.58	22JN	1.2	-3.1	-2.2	1.1
226.95	-3.83	-1.5	95.13	2JL	2.0	-3.3	-2.2	1.0
228.82	-3.84	-1.3	104.72	12JL	3.2	-3.9	-2.3	0.9
231.83	-3.77	-1.1	114.34	22JL	1.4	-4.2	-2.4	0.9
235.76	-3.65	-0.9	124.00	1AU	-0.3	-4.2	-2.4	0.8
240.44	-3.50	-0.7	133.72	11AU	-1.2	-4.2	-2.4	0.7
245.72	-3.32	-0.5	143.49	21AU	-1.3	-4.1	-2.5	0.8
251.48	-3.12	-0.3	153.32	31AU	-0.8	-4.1	-2.5	0.8
257.61	-2.91	-0.2	163.21	10SE	-0.4	-4.0	-2.4	0.8
264.03	-2.69	-0.0	173.16	20SE	-0.2	-3.9	-2.4	0.9
270.68	-2.47	0.1	183.16	30SE	-0.1	-3.8	-2.4	0.9
277.51	-2.24	0.3	193.22	10OC	0.0	-3.7	-2.3	1.0
284.46	-2.02	0.4	203.33	20OC	0.7	-3.7	-2.2	1.0
291.50	-1.80	0.6	213.47	30OC	3.2	-3.6	-2.2	1.1
298.61	-1.58	0.7	223.66	9NO	0.4	-3.6	-2.1	1.1
305.75	-1.37	0.8	233.85	19NO	-0.2	-3.5	-2.0	1.2
312.90	-1.17	0.9	244.06	29NO	-0.3	-3.5	-2.0	1.2
320.05	-0.97	1.0	254.27	9DE	-0.3	-3.4	-1.9	1.2
327.19	-0.79	1.1	264.47	19DE	-0.5	-3.4	-1.8	1.3
334.29	-0.62	1.2	274.64	29DE	-0.9	-3.4	-1.8	1.3

♂			☉	16.00UT	☿	♀	♃	♄
LONG	LAT	MAG	LONG	-302	MAGNITUDES			
341.36	-0.45	1.3	284.79	8JA	-1.1	-3.3	-1.7	1.3
348.39	-0.30	1.4	294.89	18JA	-1.1	-3.3	-1.7	1.3
355.36	-0.16	1.5	304.94	28JA	-0.3	-3.3	-1.6	1.3
2.29	-0.02	1.6	314.95	7FE	1.6	-3.3	-1.6	1.3
9.16	0.10	1.6	324.89	17FE	3.0	-3.3	-1.6	1.3
15.97	0.22	1.7	334.77	27FE	1.5	-3.3	-1.6	1.4
22.73	0.32	1.8	344.60	9MR	0.9	-3.4	-1.6	1.3
29.44	0.42	1.8	354.37	19MR	0.5	-3.4	-1.6	1.3
36.10	0.51	1.9	4.08	29MR	0.1	-3.4	-1.6	1.3
42.71	0.60	1.9	13.74	8AP	-0.6	-3.5	-1.6	1.3
49.28	0.68	1.9	23.36	18AP	-1.5	-3.5	-1.6	1.3
55.81	0.75	1.9	32.94	28AP	-1.5	-3.5	-1.6	1.3
62.30	0.81	2.0	42.49	8MY	-0.5	-3.4	-1.7	1.2
68.76	0.87	2.0	52.02	18MY	0.3	-3.4	-1.7	1.2
75.19	0.92	2.0	61.54	28MY	0.9	-3.4	-1.8	1.1
81.60	0.97	2.0	71.06	7JN	1.5	-3.4	-1.8	1.1
87.99	1.01	2.0	80.58	17JN	2.6	-3.3	-1.9	1.0
94.37	1.05	2.0	90.12	27JN	2.9	-3.3	-1.9	1.0
100.75	1.09	1.9	99.69	7JL	1.1	-3.3	-2.0	0.9
107.12	1.12	2.0	109.29	17JL	-0.4	-3.3	-2.1	0.9
113.49	1.14	2.0	118.93	27JL	-1.3	-3.3	-2.1	0.9
119.86	1.16	2.0	128.62	6AU	-1.4	-3.4	-2.2	0.8
126.24	1.18	2.0	138.36	16AU	-0.8	-3.4	-2.3	0.8
132.64	1.19	2.0	148.16	26AU	-0.3	-3.4	-2.3	0.8
139.04	1.20	2.0	158.02	5SE	-0.1	-3.5	-2.4	0.8
145.47	1.20	2.0	167.93	15SE	0.0	-3.5	-2.4	0.9
151.91	1.20	2.0	177.91	25SE	0.2	-3.6	-2.4	1.0
158.37	1.19	2.0	187.94	5OC	1.0	-3.6	-2.4	1.0
164.85	1.18	1.9	198.02	15OC	3.0	-3.7	-2.4	1.1
171.35	1.16	1.9	208.15	25OC	0.2	-3.8	-2.4	1.1
177.87	1.14	1.8	218.32	4NO	-0.4	-3.9	-2.3	1.2
184.40	1.10	1.8	228.50	14NO	-0.4	-4.0	-2.3	1.2
190.96	1.06	1.7	238.71	24NO	-0.5	-4.1	-2.2	1.3
197.53	1.01	1.7	248.92	4DE	-0.6	-4.2	-2.1	1.3
204.10	0.95	1.6	259.12	14DE	-0.9	-4.3	-2.1	1.3
210.70	0.88	1.5	269.31	24DE	-1.0	-4.4	-2.0	1.3
				-301				
217.29	0.79	1.4	279.48	3JA	-0.9	-4.4	-1.9	1.3
223.89	0.69	1.3	289.60	13JA	-0.2	-4.2	-1.9	1.2
230.48	0.57	1.2	299.68	23JA	1.8	-3.9	-1.8	1.2
237.07	0.44	1.1	309.71	2FE	2.4	-3.3	-1.7	1.2
243.63	0.28	0.9	319.68	12FE	1.1	-3.6	-1.7	1.1
250.16	0.10	0.8	329.60	22FE	0.6	-4.1	-1.6	1.1
256.65	-0.11	0.6	339.45	4MR	0.3	-4.2	-1.6	1.0
263.08	-0.35	0.5	349.25	14MR	-0.0	-4.3	-1.6	1.0
269.42	-0.62	0.3	358.99	24MR	-0.6	-4.2	-1.5	1.0
275.65	-0.93	0.1	8.68	3AP	-1.5	-4.1	-1.5	1.0
281.72	-1.27	-0.1	18.32	13AP	-1.5	-4.0	-1.5	1.0
287.59	-1.66	-0.2	27.92	23AP	-0.5	-3.9	-1.5	1.0
293.18	-2.09	-0.5	37.49	3MY	0.4	-3.8	-1.5	1.0
298.42	-2.58	-0.7	47.02	13MY	1.1	-3.7	-1.5	1.0
303.18	-3.11	-0.9	56.55	23MY	2.0	-3.6	-1.5	0.9
307.33	-3.69	-1.1	66.07	2JN	3.3	-3.6	-1.6	0.9
310.70	-4.32	-1.4	75.59	12JN	2.3	-3.5	-1.6	0.9
313.08	-4.98	-1.7	85.12	22JN	0.9	-3.5	-1.6	0.8
314.27	-5.63	-1.9	94.68	2JL	-0.4	-3.4	-1.7	0.8
314.13	-6.21	-2.2	104.26	12JL	-1.3	-3.4	-1.7	0.7
312.67	-6.63	-2.4	113.88	22JL	-1.4	-3.4	-1.8	0.7
310.25	-6.79	-2.6	123.54	1AU	-0.7	-3.4	-1.9	0.6
307.50	-6.62	-2.5	133.25	11AU	-0.2	-3.4	-1.9	0.6
305.17	-6.16	-2.3	143.02	21AU	0.0	-3.4	-2.0	0.5
303.88	-5.50	-2.0	152.85	31AU	0.2	-3.4	-2.1	0.5
303.87	-4.77	-1.7	162.73	10SE	0.4	-3.4	-2.1	0.5
305.13	-4.03	-1.4	172.68	20SE	1.3	-3.4	-2.2	0.5
307.53	-3.35	-1.1	182.68	30SE	2.7	-3.4	-2.2	0.6
310.83	-2.74	-0.8	192.74	10OC	0.1	-3.4	-2.3	0.7
314.86	-2.20	-0.5	202.84	20OC	-0.5	-3.4	-2.3	0.7
319.46	-1.73	-0.3	212.98	30OC	-0.6	-3.5	-2.3	0.8
324.49	-1.33	-0.0	223.16	9NO	-0.6	-3.5	-2.3	0.8
329.85	-0.98	0.2	233.36	19NO	-0.7	-3.5	-2.3	0.9
335.47	-0.67	0.4	243.57	29NO	-0.8	-3.5	-2.3	0.9
341.28	-0.41	0.6	253.77	9DE	-0.8	-3.4	-2.2	1.0
347.23	-0.18	0.7	263.98	19DE	-0.8	-3.4	-2.2	1.0
353.29	0.02	0.9	274.15	29DE	-0.1	-3.4	-2.1	1.0

♂			☉	16.00UT	☿	♀	♃	♄
LONG	LAT	MAG	LONG	-300	MAGNITUDES			
359.43	0.20	1.0	284.29	8JA	2.1	-3.4	-2.0	1.0
5.63	0.35	1.2	294.40	18JA	1.9	-3.4	-2.0	1.0
11.86	0.48	1.3	304.45	28JA	0.8	-3.3	-1.9	0.9
18.12	0.60	1.4	314.46	7FE	0.4	-3.3	-1.8	0.9
24.39	0.70	1.5	324.41	17FE	0.2	-3.3	-1.8	0.9
30.67	0.78	1.6	334.29	27FE	-0.1	-3.3	-1.7	0.8
36.96	0.86	1.7	344.12	8MR	-0.7	-3.4	-1.6	0.8
43.24	0.93	1.8	353.89	18MR	-1.5	-3.4	-1.6	0.8
49.51	0.98	1.8	3.60	28MR	-1.5	-3.4	-1.5	0.8
55.79	1.03	1.9	13.27	7AP	-0.5	-3.4	-1.5	0.8
62.06	1.07	1.9	22.89	17AP	0.5	-3.5	-1.5	0.8
68.32	1.10	2.0	32.47	27AP	1.5	-3.5	-1.4	0.8
74.59	1.13	2.0	42.02	7MY	2.7	-3.6	-1.4	0.8
80.85	1.15	2.0	51.56	17MY	3.3	-3.6	-1.4	0.7
87.13	1.17	2.0	61.08	27MY	1.8	-3.7	-1.4	0.7
93.41	1.18	2.0	70.60	6JN	0.7	-3.8	-1.4	0.7
99.70	1.18	2.0	80.12	16JN	-0.4	-3.9	-1.4	0.7
106.01	1.18	2.0	89.66	26JN	-1.4	-4.0	-1.4	0.6
112.34	1.18	2.0	99.23	6JL	-1.5	-4.1	-1.5	0.6
118.70	1.17	2.0	108.83	16JL	-0.7	-4.2	-1.5	0.5
125.08	1.16	2.0	118.47	26JL	-0.2	-4.2	-1.5	0.5
131.49	1.14	1.9	128.15	5AU	0.1	-4.2	-1.6	0.4
137.95	1.12	1.9	137.90	15AU	0.3	-4.1	-1.6	0.4
144.44	1.09	1.9	147.69	25AU	0.6	-3.8	-1.7	0.3
150.97	1.05	1.9	157.54	4SE	1.7	-3.3	-1.7	0.2
157.55	1.02	1.9	167.46	14SE	2.4	-3.6	-1.8	0.2
164.18	0.97	1.9	177.43	24SE	-0.0	-4.1	-1.9	0.2
170.85	0.92	1.9	187.46	4OC	-0.7	-4.3	-1.9	0.3
177.58	0.87	1.9	197.54	14OC	-0.7	-4.4	-2.0	0.3
184.37	0.81	1.9	207.66	24OC	-0.7	-4.3	-2.1	0.4
191.20	0.74	1.8	217.82	3NO	-0.8	-4.2	-2.1	0.5
198.10	0.66	1.8	228.01	13NO	-0.7	-4.1	-2.2	0.5
205.04	0.58	1.8	238.21	23NO	-0.7	-4.0	-2.2	0.6
212.04	0.49	1.7	248.43	3DE	-0.6	-3.9	-2.2	0.6
219.10	0.38	1.7	258.63	13DE	-0.0	-3.8	-2.2	0.7
226.20	0.27	1.6	268.82	23DE	2.4	-3.7	-2.2	0.7
				-299				
233.36	0.15	1.5	278.98	2JA	1.5	-3.7	-2.2	0.7
240.56	0.02	1.5	289.11	12JA	0.5	-3.6	-2.1	0.7
247.80	-0.12	1.4	299.19	22JA	0.2	-3.5	-2.0	0.7
255.09	-0.27	1.3	309.23	1FE	0.1	-3.5	-2.0	0.7
262.41	-0.42	1.2	319.20	11FE	-0.2	-3.4	-1.9	0.7
269.75	-0.59	1.2	329.12	21FE	-0.7	-3.4	-1.8	0.7
277.12	-0.77	1.1	338.98	3MR	-1.5	-3.4	-1.8	0.6
284.49	-0.95	1.0	348.78	13MR	-1.4	-3.3	-1.7	0.6
291.86	-1.14	0.9	358.52	23MR	-0.5	-3.3	-1.6	0.5
299.21	-1.33	0.8	8.21	2AP	0.7	-3.3	-1.6	0.5
306.52	-1.52	0.7	17.86	12AP	1.9	-3.3	-1.5	0.6
313.77	-1.72	0.6	27.45	22AP	3.6	-3.3	-1.5	0.6
320.94	-1.91	0.5	37.02	2MY	2.5	-3.3	-1.4	0.6
328.00	-2.09	0.4	46.57	12MY	1.4	-3.3	-1.4	0.6
334.93	-2.27	0.3	56.09	22MY	0.5	-3.4	-1.4	0.6
341.67	-2.44	0.2	65.61	1JN	-0.5	-3.4	-1.3	0.6
348.18	-2.60	0.1	75.13	11JN	-1.4	-3.5	-1.3	0.5
354.42	-2.75	-0.0	84.66	21JN	-1.5	-3.5	-1.3	0.5
0.32	-2.88	-0.2	94.21	1JL	-0.6	-3.5	-1.3	0.5
5.79	-2.99	-0.3	103.79	11JL	-0.1	-3.5	-1.3	0.4
10.74	-3.09	-0.5	113.41	21JL	0.3	-3.4	-1.3	0.4
15.02	-3.16	-0.6	123.07	31JL	0.5	-3.4	-1.4	0.3
18.49	-3.19	-0.8	132.78	10AU	0.9	-3.4	-1.4	0.3
20.96	-3.19	-1.0	142.54	20AU	2.1	-3.3	-1.4	0.2
22.19	-3.12	-1.2	152.37	30AU	2.1	-3.3	-1.4	0.2
22.02	-2.97	-1.5	162.25	9SE	-0.1	-3.3	-1.5	0.1
20.38	-2.71	-1.7	172.19	19SE	-0.8	-3.3	-1.5	0.0
17.47	-2.32	-1.8	182.19	29SE	-0.9	-3.3	-1.6	-0.0
13.88	-1.82	-1.9	192.25	9OC	-0.9	-3.4	-1.7	-0.0
10.40	-1.27	-1.7	202.35	19OC	-0.7	-3.4	-1.7	0.1
7.76	-0.72	-1.4	212.49	29OC	-0.5	-3.4	-1.8	0.1
6.40	-0.24	-1.1	222.66	8NO	-0.5	-3.4	-1.9	0.2
6.39	0.16	-0.8	232.86	18NO	-0.5	-3.5	-1.9	0.3
7.61	0.48	-0.5	243.07	28NO	0.1	-3.5	-2.0	0.3
9.86	0.73	-0.2	253.28	8DE	2.7	-3.6	-2.0	0.4
12.92	0.93	0.1	263.48	18DE	1.1	-3.6	-2.1	0.4
16.63	1.07	0.3	273.66	28DE	0.3	-3.7	-2.1	0.5

♂			☉	16.00UT	☿	♀	♃	♄
LONG	LAT	MAG	LONG	-298	MAGNITUDES			
20.84	1.19	0.6	283.80	7JA	0.1	-3.8	-2.1	0.5
25.44	1.27	0.8	293.91	17JA	-0.0	-3.9	-2.1	0.5
30.34	1.34	0.9	303.97	27JA	-0.3	-3.9	-2.1	0.5
35.48	1.38	1.1	313.98	6FE	-0.7	-4.0	-2.0	0.5
40.80	1.42	1.2	323.93	16FE	-1.4	-4.1	-2.0	0.5
46.28	1.44	1.4	333.82	26FE	-1.3	-4.2	-1.9	0.5
51.87	1.45	1.5	343.65	8MR	-0.5	-4.2	-1.9	0.5
57.56	1.46	1.6	353.42	18MR	0.9	-4.2	-1.8	0.4
63.33	1.45	1.7	3.14	28MR	2.5	-4.0	-1.7	0.4
69.18	1.44	1.7	12.80	7AP	3.3	-3.5	-1.7	0.3
75.08	1.43	1.8	22.42	17AP	1.9	-3.0	-1.6	0.4
81.03	1.41	1.9	32.01	27AP	1.1	-3.6	-1.5	0.4
87.04	1.39	1.9	41.56	7MY	0.4	-4.0	-1.5	0.4
93.09	1.36	1.9	51.09	17MY	-0.5	-4.2	-1.4	0.4
99.19	1.33	2.0	60.62	27MY	-1.5	-4.2	-1.4	0.4
105.34	1.30	2.0	70.13	6JN	-1.5	-4.1	-1.3	0.4
111.54	1.26	2.0	79.66	16JN	-0.6	-4.0	-1.3	0.4
117.78	1.22	2.0	89.20	26JN	0.0	-3.9	-1.3	0.4
124.08	1.17	2.0	98.76	6JL	0.4	-3.8	-1.3	0.4
130.43	1.12	2.0	108.36	16JL	0.7	-3.8	-1.2	0.3
136.84	1.07	2.0	118.00	26JL	1.3	-3.7	-1.2	0.3
143.31	1.01	2.0	127.68	5AU	2.6	-3.6	-1.2	0.3
149.85	0.95	1.9	137.42	15AU	1.9	-3.6	-1.2	0.2
156.45	0.88	1.9	147.21	25AU	-0.2	-3.5	-1.3	0.1
163.12	0.82	1.8	157.06	4SE	-1.0	-3.5	-1.3	0.1
169.86	0.74	1.8	166.98	14SE	-1.0	-3.5	-1.3	0.0
176.68	0.66	1.7	176.95	24SE	-0.9	-3.4	-1.3	-0.1
183.56	0.58	1.7	186.97	4OC	-0.6	-3.4	-1.4	-0.1
190.53	0.49	1.7	197.05	14OC	-0.4	-3.4	-1.4	-0.2
197.57	0.40	1.7	207.17	24OC	-0.4	-3.4	-1.5	-0.2
204.69	0.30	1.7	217.33	3NO	-0.3	-3.4	-1.6	-0.1
211.88	0.20	1.7	227.52	13NO	0.3	-3.4	-1.6	-0.0
219.14	0.10	1.7	237.72	23NO	3.0	-3.4	-1.7	0.1
226.48	-0.01	1.6	247.93	3DE	0.8	-3.4	-1.7	0.1
233.88	-0.13	1.6	258.14	13DE	0.1	-3.4	-1.8	0.2
241.35	-0.25	1.6	268.33	23DE	-0.1	-3.4	-1.9	0.2
				-297				
248.87	-0.37	1.5	278.49	2JA	-0.2	-3.4	-1.9	0.3
256.45	-0.49	1.5	288.62	12JA	-0.4	-3.4	-2.0	0.3
264.07	-0.61	1.5	298.70	22JA	-0.8	-3.5	-2.0	0.4
271.72	-0.73	1.4	308.74	1FE	-1.3	-3.5	-2.0	0.4
279.41	-0.85	1.4	318.72	11FE	-1.3	-3.5	-2.0	0.4
287.10	-0.97	1.3	328.64	21FE	-0.4	-3.4	-2.0	0.4
294.81	-1.08	1.3	338.50	3MR	1.1	-3.4	-2.0	0.4
302.51	-1.19	1.3	348.30	13MR	3.1	-3.4	-2.0	0.4
310.18	-1.28	1.2	358.05	23MR	2.5	-3.4	-1.9	0.3
317.83	-1.37	1.2	7.74	2AP	1.4	-3.3	-1.9	0.3
325.42	-1.44	1.2	17.39	12AP	0.8	-3.3	-1.8	0.3
332.96	-1.51	1.1	26.99	22AP	0.2	-3.3	-1.7	0.2
340.41	-1.56	1.1	36.55	2MY	-0.5	-3.3	-1.7	0.2
347.77	-1.59	1.0	46.10	12MY	-1.5	-3.3	-1.6	0.3
355.02	-1.61	1.0	55.62	22MY	-1.5	-3.3	-1.5	0.3
2.15	-1.61	0.9	65.14	1JN	-0.6	-3.4	-1.5	0.3
9.13	-1.60	0.9	74.66	11JN	0.1	-3.4	-1.4	0.3
15.95	-1.57	0.8	84.19	21JN	0.5	-3.4	-1.4	0.3
22.58	-1.52	0.8	93.74	1JL	1.0	-3.5	-1.3	0.3
29.00	-1.45	0.7	103.32	11JL	1.7	-3.5	-1.3	0.3
35.17	-1.37	0.6	112.94	21JL	3.1	-3.6	-1.3	0.3
41.06	-1.27	0.6	122.60	31JL	1.6	-3.7	-1.2	0.3
46.62	-1.14	0.5	132.31	10AU	-0.2	-3.7	-1.2	0.2
51.77	-1.00	0.3	142.07	20AU	-1.1	-3.8	-1.2	0.2
56.45	-0.82	0.2	151.89	30AU	-1.2	-3.9	-1.2	0.1
60.54	-0.61	0.1	161.77	9SE	-0.9	-4.0	-1.2	0.1
63.91	-0.37	-0.1	171.71	19SE	-0.5	-4.1	-1.2	-0.0
66.40	-0.08	-0.3	181.71	29SE	-0.3	-4.3	-1.2	-0.1
67.79	0.25	-0.5	191.76	9OC	-0.2	-4.3	-1.3	-0.1
67.92	0.64	-0.7	201.86	19OC	-0.1	-4.4	-1.3	-0.2
66.63	1.08	-1.0	212.00	29OC	0.5	-4.3	-1.3	-0.3
63.99	1.53	-1.2	222.17	8NO	3.3	-3.9	-1.4	-0.3
60.35	1.96	-1.3	232.37	18NO	0.6	-3.0	-1.4	-0.2
56.40	2.30	-1.3	242.58	28NO	-0.1	-3.6	-1.5	-0.1
52.92	2.53	-1.0	252.79	8DE	-0.2	-4.2	-1.6	-0.0
50.51	2.65	-0.8	262.98	18DE	-0.3	-4.4	-1.6	0.0
49.41	2.68	-0.5	273.17	28DE	-0.5	-4.4	-1.7	0.1

♂			☉	16.00UT	☿	♀	♃	♄
LONG	LAT	MAG	LONG	-296	MAGNITUDES			
49.61	2.65	-0.2	283.32	7JA	-0.8	-4.3	-1.8	0.2
50.95	2.58	0.1	293.42	17JA	-1.2	-4.2	-1.8	0.2
53.23	2.50	0.3	303.48	27JA	-1.1	-4.1	-1.9	0.3
56.27	2.40	0.6	313.50	6FE	-0.4	-4.0	-1.9	0.3
59.90	2.31	0.8	323.44	16FE	1.3	-3.9	-2.0	0.3
63.99	2.21	0.9	333.34	26FE	3.3	-3.8	-2.0	0.3
68.47	2.12	1.1	343.17	7MR	1.8	-3.7	-2.0	0.3
73.24	2.02	1.2	352.95	17MR	1.0	-3.6	-2.0	0.3
78.26	1.93	1.4	2.67	27MR	0.6	-3.5	-2.0	0.3
83.48	1.84	1.5	12.34	6AP	0.1	-3.5	-2.0	0.3
88.87	1.75	1.6	21.96	16AP	-0.5	-3.4	-1.9	0.2
94.40	1.66	1.6	31.55	26AP	-1.5	-3.4	-1.9	0.2
100.07	1.58	1.7	41.11	6MY	-1.5	-3.4	-1.8	0.2
105.85	1.49	1.7	50.64	16MY	-0.6	-3.3	-1.8	0.2
111.74	1.40	1.8	60.16	26MY	0.2	-3.3	-1.7	0.2
117.73	1.32	1.8	69.68	5JN	0.7	-3.3	-1.6	0.3
123.81	1.23	1.9	79.20	15JN	1.3	-3.3	-1.6	0.3
129.98	1.14	1.9	88.74	25JN	2.2	-3.3	-1.5	0.3
136.24	1.05	1.9	98.30	5JL	3.2	-3.3	-1.5	0.3
142.60	0.96	1.9	107.89	15JL	1.3	-3.4	-1.4	0.3
149.04	0.87	1.9	117.53	25JL	-0.3	-3.4	-1.4	0.3
155.58	0.77	1.9	127.21	4AU	-1.2	-3.4	-1.3	0.3
162.20	0.68	1.8	136.94	14AU	-1.3	-3.5	-1.3	0.2
168.92	0.58	1.8	146.74	24AU	-0.8	-3.5	-1.3	0.2
175.74	0.48	1.8	156.58	3SE	-0.4	-3.5	-1.3	0.2
182.65	0.38	1.8	166.49	13SE	-0.2	-3.5	-1.2	0.1
189.65	0.27	1.7	176.46	23SE	-0.1	-3.4	-1.2	0.0
196.75	0.17	1.7	186.48	3OC	0.1	-3.4	-1.2	-0.0
203.93	0.06	1.6	196.55	13OC	0.8	-3.4	-1.2	-0.1
211.22	-0.05	1.6	206.68	23OC	3.3	-3.4	-1.3	-0.2
218.58	-0.15	1.5	216.83	2NO	0.4	-3.3	-1.3	-0.3
226.03	-0.26	1.5	227.02	12NO	-0.3	-3.3	-1.3	-0.3
233.56	-0.37	1.5	237.22	22NO	-0.4	-3.3	-1.4	-0.3
241.16	-0.47	1.5	247.43	2DE	-0.4	-3.3	-1.4	-0.2
248.82	-0.57	1.5	257.64	12DE	-0.6	-3.3	-1.4	-0.1
256.54	-0.67	1.5	267.83	22DE	-0.9	-3.4	-1.5	-0.1
				-295				
264.30	-0.77	1.4	278.00	1JA	-1.1	-3.4	-1.6	-0.0
272.10	-0.85	1.4	288.13	11JA	-1.0	-3.4	-1.6	0.1
279.93	-0.93	1.4	298.22	21JA	-0.3	-3.4	-1.7	0.1
287.76	-1.00	1.4	308.25	31JA	1.6	-3.5	-1.8	0.2
295.60	-1.07	1.4	318.23	10FE	2.8	-3.5	-1.8	0.2
303.43	-1.12	1.4	328.16	20FE	1.4	-3.6	-1.9	0.3
311.23	-1.16	1.4	338.03	2MR	0.8	-3.6	-2.0	0.3
319.00	-1.19	1.4	347.83	12MR	0.4	-3.7	-2.0	0.3
326.72	-1.20	1.4	357.58	22MR	0.0	-3.8	-2.0	0.3
334.37	-1.21	1.4	7.28	1AP	-0.6	-3.8	-2.1	0.3
341.97	-1.20	1.4	16.92	11AP	-1.5	-3.9	-2.1	0.3
349.48	-1.18	1.4	26.53	21AP	-1.5	-4.0	-2.1	0.3
356.90	-1.14	1.4	36.10	1MY	-0.6	-4.1	-2.1	0.2
4.23	-1.09	1.4	45.64	11MY	0.3	-4.2	-2.0	0.2
11.45	-1.03	1.4	55.17	21MY	1.0	-4.2	-2.0	0.2
18.56	-0.96	1.4	64.69	31MY	1.7	-4.1	-1.9	0.2
25.56	-0.88	1.4	74.20	10JN	2.9	-3.7	-1.9	0.2
32.43	-0.79	1.3	83.74	20JN	2.7	-3.0	-1.8	0.3
39.16	-0.68	1.3	93.28	30JN	1.1	-3.3	-1.8	0.3
45.75	-0.57	1.3	102.86	10JL	-0.3	-3.9	-1.7	0.3
52.18	-0.44	1.3	112.47	20JL	-1.3	-4.2	-1.6	0.3
58.45	-0.31	1.2	122.13	30JL	-1.4	-4.2	-1.6	0.3
64.52	-0.16	1.2	131.83	9AU	-0.8	-4.2	-1.5	0.3
70.39	0.01	1.1	141.59	19AU	-0.3	-4.1	-1.5	0.3
76.01	0.18	1.0	151.41	29AU	-0.0	-4.1	-1.4	0.3
81.35	0.38	0.9	161.29	8SE	0.1	-4.0	-1.4	0.2
86.36	0.59	0.8	171.22	18SE	0.3	-3.9	-1.4	0.2
90.97	0.83	0.7	181.22	28SE	1.1	-3.8	-1.3	0.1
95.10	1.10	0.6	191.27	8OC	3.0	-3.7	-1.3	0.1
98.62	1.40	0.4	201.37	18OC	0.2	-3.7	-1.3	-0.0
101.42	1.74	0.2	211.50	28OC	-0.4	-3.6	-1.3	-0.1
103.32	2.12	0.0	221.68	7NO	-0.5	-3.6	-1.3	-0.2
104.11	2.54	-0.2	231.87	17NO	-0.5	-3.5	-1.3	-0.2
103.64	2.99	-0.5	242.08	27NO	-0.6	-3.5	-1.3	-0.3
101.82	3.43	-0.7	252.29	7DE	-0.9	-3.4	-1.4	-0.3
98.77	3.81	-0.9	262.50	17DE	-0.9	-3.4	-1.4	-0.2
94.94	4.07	-1.0	272.67	27DE	-0.9	-3.4	-1.4	-0.1

♂			⊙	16.00UT	☿	♀	♃	♄
LONG	LAT	MAG	LONG	-294	MAGNITUDES			
91.01	4.17	-0.9	282.83	6JA	-0.2	-3.3	-1.5	-0.1
87.72	4.11	-0.7	292.94	16JA	1.9	-3.3	-1.5	0.0
85.55	3.94	-0.5	303.00	26JA	2.2	-3.3	-1.6	0.1
84.68	3.69	-0.2	313.01	5FE	1.0	-3.3	-1.7	0.1
85.07	3.42	0.0	322.97	15FE	0.5	-3.3	-1.7	0.2
86.54	3.15	0.3	332.86	25FE	0.3	-3.3	-1.8	0.2
88.91	2.89	0.5	342.70	7MR	-0.1	-3.4	-1.9	0.3
92.01	2.65	0.7	352.48	17MR	-0.6	-3.4	-1.9	0.3
95.69	2.42	0.8	2.20	27MR	-1.5	-3.4	-2.0	0.3
99.85	2.22	1.0	11.87	6AP	-1.5	-3.5	-2.1	0.3
104.39	2.02	1.1	21.50	16AP	-0.6	-3.5	-2.1	0.3
109.25	1.84	1.2	31.08	26AP	0.4	-3.5	-2.1	0.3
114.38	1.67	1.3	40.64	6MY	1.3	-3.4	-2.2	0.3
119.74	1.51	1.4	50.17	16MY	2.3	-3.4	-2.2	0.3
125.30	1.36	1.5	59.69	26MY	3.5	-3.4	-2.2	0.2
131.04	1.21	1.5	69.21	5JN	2.2	-3.4	-2.2	0.2
136.95	1.06	1.6	78.73	15JN	0.9	-3.3	-2.1	0.2
143.00	0.92	1.6	88.27	25JN	-0.4	-3.3	-2.1	0.3
149.20	0.79	1.6	97.83	5JL	-1.3	-3.3	-2.0	0.3
155.54	0.66	1.6	107.43	15JL	-1.5	-3.3	-2.0	0.4
162.00	0.53	1.7	117.06	25JL	-0.7	-3.3	-1.9	0.4
168.60	0.40	1.7	126.74	4AU	-0.2	-3.4	-1.8	0.4
175.31	0.27	1.7	136.47	14AU	0.1	-3.4	-1.8	0.4
182.15	0.15	1.6	146.25	24AU	0.2	-3.4	-1.7	0.4
189.11	0.03	1.6	156.10	3SE	0.5	-3.5	-1.7	0.4
196.18	-0.09	1.6	166.00	13SE	1.4	-3.5	-1.6	0.3
203.36	-0.20	1.6	175.97	23SE	2.7	-3.6	-1.6	0.3
210.65	-0.31	1.6	185.99	3OC	0.1	-3.6	-1.5	0.3
218.04	-0.42	1.5	196.06	13OC	-0.6	-3.7	-1.5	0.2
225.53	-0.52	1.5	206.18	23OC	-0.7	-3.8	-1.4	0.1
233.11	-0.62	1.5	216.34	2NO	-0.6	-3.9	-1.4	0.1
240.76	-0.71	1.5	226.52	12NO	-0.7	-4.0	-1.4	-0.0
248.49	-0.79	1.4	236.72	22NO	-0.7	-4.1	-1.4	-0.1
256.27	-0.86	1.4	246.94	2DE	-0.7	-4.2	-1.4	-0.2
264.10	-0.93	1.4	257.14	12DE	-0.7	-4.3	-1.4	-0.2
271.97	-0.98	1.3	267.34	22DE	-0.1	-4.4	-1.4	-0.2
				-293				
279.86	-1.03	1.3	277.51	1JA	2.1	-4.4	-1.4	-0.1
287.75	-1.06	1.3	287.64	11JA	1.7	-4.2	-1.5	-0.1
295.64	-1.08	1.3	297.73	21JA	0.7	-3.8	-1.5	0.0
303.52	-1.09	1.3	307.77	31JA	0.3	-3.3	-1.6	0.1
311.36	-1.09	1.4	317.75	10FE	0.2	-3.7	-1.6	0.1
319.16	-1.08	1.4	327.68	20FE	-0.1	-4.1	-1.7	0.2
326.91	-1.05	1.4	337.55	2MR	-0.7	-4.3	-1.7	0.3
334.60	-1.02	1.4	347.36	12MR	-1.4	-4.3	-1.8	0.3
342.22	-0.97	1.5	357.11	22MR	-1.5	-4.2	-1.9	0.3
349.76	-0.91	1.5	6.81	1AP	-0.5	-4.1	-1.9	0.4
357.21	-0.85	1.5	16.45	11AP	0.6	-4.0	-2.0	0.4
4.58	-0.77	1.5	26.06	21AP	1.6	-3.9	-2.1	0.4
11.85	-0.69	1.6	35.64	1MY	3.0	-3.8	-2.1	0.4
19.03	-0.61	1.6	45.18	11MY	3.0	-3.7	-2.2	0.4
26.12	-0.51	1.6	54.71	21MY	1.7	-3.6	-2.2	0.4
33.10	-0.41	1.6	64.23	31MY	0.7	-3.5	-2.3	0.4
39.99	-0.30	1.6	73.74	10JN	-0.4	-3.5	-2.3	0.3
46.78	-0.19	1.7	83.27	20JN	-1.4	-3.4	-2.3	0.3
53.46	-0.08	1.7	92.82	30JN	-1.5	-3.4	-2.3	0.4
60.04	0.04	1.7	102.40	10JL	-0.7	-3.4	-2.3	0.4
66.51	0.17	1.6	112.01	20JL	-0.1	-3.4	-2.2	0.4
72.87	0.30	1.6	121.66	30JL	0.2	-3.4	-2.2	0.5
79.11	0.43	1.6	131.36	9AU	0.4	-3.4	-2.1	0.5
85.23	0.57	1.6	141.12	19AU	0.7	-3.3	-2.1	0.5
91.21	0.72	1.5	150.94	29AU	1.8	-3.4	-2.0	0.5
97.03	0.88	1.5	160.81	8SE	2.4	-3.4	-1.9	0.5
102.68	1.04	1.4	170.74	18SE	-0.0	-3.4	-1.9	0.5
108.12	1.22	1.3	180.73	28SE	-0.7	-3.4	-1.8	0.5
113.33	1.41	1.2	190.78	8OC	-0.8	-3.4	-1.8	0.4
118.26	1.62	1.1	200.87	18OC	-0.8	-3.4	-1.7	0.4
122.85	1.85	1.0	211.01	28OC	-0.7	-3.5	-1.7	0.3
127.04	2.10	0.8	221.18	7NO	-0.6	-3.5	-1.6	0.3
130.71	2.38	0.6	231.37	17NO	-0.6	-3.5	-1.6	0.2
133.74	2.68	0.4	241.58	27NO	-0.6	-3.5	-1.5	0.1
136.00	3.02	0.2	251.79	7DE	0.0	-3.4	-1.5	0.0
137.30	3.38	-0.0	261.99	17DE	2.4	-3.4	-1.5	-0.0
137.45	3.76	-0.3	272.17	27DE	1.4	-3.4	-1.5	-0.1

♂			⊙	16.00UT	☿	♀	♃	♄
LONG	LAT	MAG	LONG	-292	MAGNITUDES			
136.33	4.12	-0.5	282.32	6JA	0.4	-3.4	-1.5	-0.0
133.92	4.42	-0.8	292.44	16JA	0.2	-3.4	-1.5	0.0
130.49	4.60	-1.0	302.51	26JA	0.0	-3.3	-1.5	0.1
126.58	4.60	-1.0	312.52	5FE	-0.2	-3.3	-1.5	0.1
122.91	4.43	-0.9	322.48	15FE	-0.7	-3.3	-1.6	0.2
120.13	4.12	-0.7	332.38	25FE	-1.4	-3.3	-1.6	0.3
118.57	3.74	-0.4	342.22	6MR	-1.4	-3.4	-1.6	0.3
118.30	3.33	-0.2	352.00	16MR	-0.5	-3.4	-1.7	0.4
119.24	2.93	0.0	1.73	26MR	0.7	-3.4	-1.8	0.4
121.20	2.55	0.2	11.40	5AP	2.1	-3.4	-1.8	0.5
124.00	2.21	0.4	21.03	15AP	3.9	-3.5	-1.9	0.5
127.51	1.90	0.5	30.62	25AP	2.3	-3.5	-1.9	0.5
131.57	1.61	0.7	40.18	5MY	1.3	-3.6	-2.0	0.5
136.10	1.35	0.8	49.72	15MY	0.5	-3.6	-2.1	0.5
141.03	1.12	0.9	59.24	25MY	-0.4	-3.7	-2.2	0.5
146.28	0.89	1.0	68.75	4JN	-1.4	-3.8	-2.2	0.5
151.83	0.69	1.1	78.28	14JN	-1.5	-3.9	-2.3	0.5
157.63	0.50	1.1	87.81	24JN	-0.7	-4.0	-2.3	0.5
163.66	0.32	1.2	97.37	4JL	-0.1	-4.1	-2.4	0.5
169.89	0.15	1.2	106.97	14JL	0.3	-4.2	-2.4	0.5
176.31	-0.01	1.3	116.60	24JL	0.6	-4.2	-2.4	0.6
182.91	-0.16	1.3	126.27	3AU	1.0	-4.2	-2.4	0.6
189.67	-0.30	1.3	136.00	13AU	2.3	-4.1	-2.4	0.6
196.59	-0.43	1.3	145.78	23AU	2.1	-3.8	-2.3	0.7
203.65	-0.55	1.4	155.62	2SE	-0.1	-3.3	-2.3	0.7
210.85	-0.66	1.4	165.53	12SE	-0.9	-3.7	-2.2	0.7
218.17	-0.76	1.4	175.49	22SE	-1.0	-4.1	-2.2	0.7
225.61	-0.85	1.4	185.50	2OC	-0.9	-4.3	-2.1	0.7
233.15	-0.93	1.4	195.58	12OC	-0.6	-4.3	-2.0	0.7
240.78	-1.00	1.4	205.69	22OC	-0.5	-4.3	-2.0	0.6
248.49	-1.06	1.4	215.84	1NO	-0.4	-4.2	-1.9	0.6
256.27	-1.10	1.4	226.03	11NO	-0.4	-4.1	-1.8	0.5
264.09	-1.13	1.4	236.23	21NO	0.2	-4.0	-1.8	0.5
271.95	-1.15	1.4	246.44	1DE	2.7	-3.9	-1.7	0.4
279.84	-1.15	1.4	256.65	11DE	1.0	-3.8	-1.7	0.3
287.73	-1.15	1.4	266.84	21DE	0.2	-3.7	-1.6	0.2
295.61	-1.13	1.4	277.01	31DE	0.0	-3.7	-1.6	0.2
				-291				
303.47	-1.10	1.4	287.15	10JA	-0.1	-3.6	-1.6	0.1
311.30	-1.05	1.4	297.24	20JA	-0.3	-3.5	-1.6	0.2
319.08	-1.00	1.4	307.28	30JA	-0.7	-3.5	-1.6	0.2
326.80	-0.94	1.4	317.27	9FE	-1.4	-3.4	-1.6	0.3
334.45	-0.87	1.4	327.20	19FE	-1.3	-3.4	-1.6	0.3
342.04	-0.80	1.5	337.07	1MR	-0.5	-3.4	-1.6	0.4
349.55	-0.72	1.5	346.89	11MR	0.9	-3.3	-1.6	0.4
356.97	-0.63	1.5	356.64	21MR	2.7	-3.3	-1.6	0.5
4.30	-0.54	1.5	6.34	31MR	3.0	-3.3	-1.7	0.5
11.56	-0.44	1.6	15.99	10AP	1.7	-3.3	-1.7	0.6
18.72	-0.34	1.6	25.60	20AP	1.0	-3.3	-1.7	0.6
25.79	-0.24	1.7	35.18	30AP	0.3	-3.3	-1.8	0.7
32.78	-0.14	1.7	44.72	10MY	-0.5	-3.3	-1.9	0.7
39.69	-0.04	1.7	54.25	20MY	-1.4	-3.4	-1.9	0.7
46.51	0.07	1.8	63.77	30MY	-1.5	-3.4	-2.0	0.7
53.26	0.17	1.8	73.29	9JN	-0.6	-3.5	-2.1	0.7
59.92	0.28	1.8	82.81	19JN	0.0	-3.5	-2.1	0.7
66.52	0.38	1.9	92.36	29JN	0.4	-3.5	-2.2	0.7
73.03	0.49	1.9	101.93	9JL	0.8	-3.5	-2.3	0.7
79.48	0.59	1.9	111.54	19JL	1.4	-3.4	-2.3	0.7
85.86	0.70	1.9	121.19	29JL	2.8	-3.4	-2.4	0.8
92.16	0.81	1.9	130.89	8AU	1.8	-3.4	-2.4	0.8
98.39	0.91	1.9	140.64	18AU	-0.2	-3.3	-2.4	0.8
104.54	1.02	1.8	150.46	28AU	-1.0	-3.3	-2.5	0.9
110.60	1.14	1.8	160.33	7SE	-1.1	-3.3	-2.4	0.9
116.58	1.25	1.7	170.25	17SE	-0.9	-3.3	-2.4	0.9
122.46	1.37	1.7	180.25	27SE	-0.5	-3.3	-2.4	0.9
128.21	1.49	1.6	190.29	7OC	-0.3	-3.4	-2.3	0.9
133.84	1.62	1.5	200.38	17OC	-0.3	-3.4	-2.3	0.9
139.32	1.76	1.5	210.52	27OC	-0.2	-3.4	-2.2	0.9
144.60	1.90	1.3	220.68	6NO	0.4	-3.4	-2.1	0.9
149.67	2.05	1.2	230.87	16NO	3.0	-3.5	-2.1	0.8
154.47	2.22	1.1	241.08	26NO	0.8	-3.5	-2.0	0.8
158.94	2.39	0.9	251.29	6DE	0.0	-3.6	-1.9	0.7
163.01	2.58	0.7	261.49	16DE	-0.1	-3.6	-1.9	0.6
166.57	2.78	0.5	271.68	26DE	-0.2	-3.7	-1.8	0.5

♂ LONG	♂ LAT	♂ MAG	⊙ LONG	16.00UT	☿	♀	♃	♄
				-290	MAGNITUDES			
169.50	2.99	0.3	281.83	5JA	-0.4	-3.8	-1.8	0.5
171.65	3.21	0.0	291.95	15JA	-0.8	-3.9	-1.7	0.4
172.84	3.43	-0.2	302.02	25JA	-1.3	-4.0	-1.7	0.4
172.89	3.62	-0.5	312.04	4FE	-1.2	-4.0	-1.6	0.4
171.68	3.77	-0.8	322.00	14FE	-0.4	-4.1	-1.6	0.4
169.23	3.81	-1.1	331.90	24FE	1.1	-4.2	-1.6	0.5
165.83	3.72	-1.3	341.74	6MR	3.2	-4.3	-1.6	0.6
162.06	3.47	-1.3	351.53	16MR	2.2	-4.2	-1.6	0.6
158.64	3.08	-1.2	1.26	26MR	1.3	-4.0	-1.6	0.7
156.19	2.61	-1.0	10.93	5AP	0.7	-3.5	-1.6	0.7
155.01	2.11	-0.8	20.56	15AP	0.2	-3.0	-1.6	0.8
155.15	1.64	-0.6	30.16	25AP	-0.5	-3.7	-1.6	0.8
156.52	1.21	-0.4	39.72	5MY	-1.4	-4.1	-1.6	0.9
158.92	0.82	-0.2	49.25	15MY	-1.6	-4.2	-1.7	0.9
162.19	0.48	-0.0	58.78	25MY	-0.6	-4.2	-1.7	0.9
166.17	0.18	0.1	68.29	4JN	0.1	-4.1	-1.8	1.0
170.74	-0.09	0.3	77.81	14JN	0.6	-4.0	-1.8	1.0
175.81	-0.33	0.4	87.35	24JN	1.1	-3.9	-1.9	1.0
181.29	-0.53	0.5	96.91	4JL	1.8	-3.8	-1.9	1.0
187.12	-0.72	0.6	106.50	14JL	3.2	-3.8	-2.0	1.0
193.26	-0.88	0.7	116.13	24JL	1.5	-3.7	-2.1	1.0
199.67	-1.02	0.7	125.80	3AU	-0.2	-3.6	-2.1	1.0
206.32	-1.14	0.8	135.52	13AU	-1.1	-3.6	-2.2	1.0
213.18	-1.24	0.9	145.31	23AU	-1.2	-3.5	-2.3	1.1
220.21	-1.32	0.9	155.14	2SE	-0.8	-3.5	-2.3	1.1
227.41	-1.38	1.0	165.04	12SE	-0.4	-3.5	-2.4	1.2
234.76	-1.43	1.0	175.00	22SE	-0.2	-3.4	-2.4	1.2
242.22	-1.45	1.0	185.02	2OC	-0.1	-3.4	-2.4	1.2
249.78	-1.46	1.1	195.09	12OC	-0.0	-3.4	-2.4	1.2
257.43	-1.46	1.1	205.20	22OC	0.6	-3.4	-2.4	1.2
265.13	-1.44	1.2	215.35	1NO	3.4	-3.4	-2.4	1.2
272.88	-1.40	1.2	225.53	11NO	0.5	-3.4	-2.3	1.2
280.66	-1.35	1.3	235.74	21NO	-0.2	-3.4	-2.2	1.1
288.44	-1.29	1.3	245.94	1DE	-0.3	-3.4	-2.2	1.1
296.21	-1.22	1.3	256.15	11DE	-0.3	-3.4	-2.1	1.0
303.97	-1.14	1.4	266.35	21DE	-0.5	-3.4	-2.0	1.0
311.69	-1.05	1.4	276.51	31DE	-0.8	-3.4	-2.0	0.9
				-289				
319.36	-0.95	1.5	286.65	10JA	-1.1	-3.4	-1.9	0.8
326.97	-0.85	1.5	296.75	20JA	-1.1	-3.5	-1.8	0.8
334.52	-0.75	1.5	306.79	30JA	-0.4	-3.5	-1.8	0.7
342.00	-0.65	1.6	316.78	9FE	1.4	-3.5	-1.7	0.7
349.41	-0.54	1.6	326.72	19FE	3.2	-3.4	-1.7	0.7
356.73	-0.43	1.6	336.59	1MR	1.7	-3.4	-1.6	0.7
3.98	-0.32	1.7	346.40	11MR	0.9	-3.4	-1.6	0.8
11.14	-0.21	1.7	356.16	21MR	0.5	-3.4	-1.5	0.8
18.22	-0.11	1.7	5.86	31MR	0.1	-3.3	-1.5	0.9
25.22	-0.01	1.7	15.52	10AP	-0.5	-3.3	-1.5	0.9
32.15	0.10	1.7	25.13	20AP	-1.4	-3.3	-1.5	1.0
39.00	0.20	1.7	34.70	30AP	-1.6	-3.3	-1.5	1.0
45.79	0.29	1.7	44.25	10MY	-0.6	-3.3	-1.5	1.1
52.51	0.39	1.8	53.78	20MY	0.2	-3.4	-1.5	1.1
59.16	0.48	1.8	63.30	30MY	0.8	-3.4	-1.5	1.2
65.76	0.56	1.9	72.82	9JN	1.4	-3.4	-1.6	1.2
72.31	0.65	1.9	82.35	19JN	2.4	-3.4	-1.6	1.2
78.80	0.73	1.9	91.89	29JN	3.1	-3.5	-1.6	1.3
85.26	0.81	2.0	101.47	9JL	1.3	-3.5	-1.7	1.3
91.67	0.89	2.0	111.07	19JL	-0.3	-3.6	-1.7	1.3
98.03	0.96	2.0	120.72	29JL	-1.2	-3.7	-1.8	1.3
104.37	1.04	2.0	130.42	8AU	-1.4	-3.7	-1.9	1.3
110.67	1.11	2.0	140.17	18AU	-0.8	-3.8	-1.9	1.3
116.93	1.18	2.0	149.98	28AU	-0.4	-3.9	-2.0	1.3
123.16	1.24	2.0	159.85	7SE	-0.1	-4.0	-2.0	1.4
129.34	1.31	1.9	169.77	17SE	0.0	-4.2	-2.1	1.4
135.49	1.37	1.9	179.76	27SE	0.2	-4.3	-2.2	1.3
141.60	1.44	1.9	189.80	7OC	0.9	-4.3	-2.2	1.3
147.66	1.50	1.8	199.89	17OC	3.3	-4.4	-2.3	1.3
153.66	1.56	1.7	210.02	27OC	0.4	-4.2	-2.3	1.3
159.59	1.62	1.7	220.19	6NO	-0.3	-3.8	-2.3	1.3
165.45	1.68	1.6	230.38	16NO	-0.4	-3.0	-2.3	1.2
171.21	1.73	1.5	240.59	26NO	-0.4	-3.6	-2.3	1.2
176.85	1.79	1.3	250.80	6DE	-0.6	-4.2	-2.3	1.1
182.36	1.84	1.2	261.00	16DE	-0.9	-4.4	-2.2	1.1
187.70	1.88	1.1	271.18	26DE	-1.0	-4.4	-2.1	1.0

♂ LONG	♂ LAT	♂ MAG	⊙ LONG	16.00UT	☿	♀	♃	♄
				-288	MAGNITUDES			
192.83	1.92	0.9	281.34	5JA	-1.0	-4.3	-2.1	1.0
197.71	1.96	0.7	291.46	15JA	-0.3	-4.2	-2.0	0.9
202.29	1.98	0.5	301.53	25JA	1.6	-4.1	-1.9	0.9
206.46	1.98	0.3	311.55	4FE	2.6	-4.0	-1.9	0.9
210.14	1.97	0.0	321.52	14FE	1.2	-3.9	-1.8	0.8
213.21	1.92	-0.2	331.42	24FE	0.7	-3.8	-1.7	0.9
215.50	1.83	-0.5	341.27	5MR	0.4	-3.7	-1.7	0.9
216.84	1.68	-0.8	351.05	15MR	0.0	-3.6	-1.6	1.0
217.05	1.45	-1.1	0.79	25MR	-0.6	-3.5	-1.6	1.1
216.01	1.13	-1.5	10.47	4AP	-1.4	-3.5	-1.5	1.1
213.78	0.70	-1.7	20.10	14AP	-1.5	-3.4	-1.5	1.2
210.68	0.19	-2.0	29.69	24AP	-0.6	-3.4	-1.5	1.3
207.33	-0.36	-1.9	39.26	4MY	0.3	-3.3	-1.4	1.3
204.48	-0.90	-1.8	48.79	14MY	1.1	-3.3	-1.4	1.3
202.69	-1.37	-1.6	58.32	24MY	1.9	-3.3	-1.4	1.4
202.25	-1.76	-1.4	67.84	3JN	3.2	-3.3	-1.4	1.4
203.18	-2.05	-1.2	77.35	13JN	2.5	-3.3	-1.4	1.4
205.33	-2.26	-1.0	86.89	23JN	1.0	-3.3	-1.4	1.4
208.54	-2.41	-0.8	96.45	3JL	-0.3	-3.3	-1.4	1.4
212.61	-2.51	-0.6	106.03	13JL	-1.3	-3.4	-1.5	1.4
217.38	-2.57	-0.4	115.66	23JL	-1.5	-3.4	-1.5	1.3
222.73	-2.58	-0.3	125.33	2AU	-0.8	-3.4	-1.5	1.3
228.54	-2.57	-0.1	135.05	12AU	-0.3	-3.5	-1.6	1.3
234.72	-2.52	0.0	144.83	22AU	0.0	-3.5	-1.6	1.2
241.23	-2.46	0.1	154.67	1SE	0.2	-3.5	-1.7	1.2
247.98	-2.37	0.3	164.56	11SE	0.4	-3.5	-1.7	1.2
254.93	-2.26	0.4	174.52	21SE	1.2	-3.4	-1.8	1.2
262.04	-2.14	0.5	184.53	1OC	3.0	-3.4	-1.9	1.2
269.27	-2.01	0.6	194.59	11OC	0.2	-3.4	-1.9	1.2
276.58	-1.86	0.7	204.70	21OC	-0.5	-3.4	-2.0	1.2
283.97	-1.71	0.8	214.85	31OC	-0.6	-3.3	-2.1	1.1
291.38	-1.55	0.9	225.03	10NO	-0.6	-3.3	-2.1	1.1
298.81	-1.40	1.0	235.23	20NO	-0.7	-3.3	-2.2	1.1
306.25	-1.24	1.1	245.44	30NO	-0.8	-3.3	-2.2	1.1
313.66	-1.08	1.2	255.65	10DE	-0.8	-3.3	-2.2	1.0
321.04	-0.92	1.2	265.85	20DE	-0.8	-3.4	-2.2	1.0
328.38	-0.77	1.3	276.02	30DE	-0.2	-3.4	-2.2	0.9
				-287				
335.67	-0.62	1.4	286.16	9JA	1.9	-3.4	-2.1	0.9
342.90	-0.48	1.5	296.26	19JA	2.1	-3.4	-2.1	0.8
350.08	-0.34	1.5	306.31	29JA	0.9	-3.5	-2.0	0.8
357.18	-0.21	1.6	316.30	8FE	0.5	-3.5	-2.0	0.7
4.23	-0.09	1.7	326.24	18FE	0.2	-3.6	-1.9	0.7
11.20	0.03	1.7	336.12	28FE	-0.1	-3.6	-1.8	0.7
18.11	0.14	1.8	345.93	10MR	-0.6	-3.7	-1.7	0.7
24.96	0.25	1.8	355.69	20MR	-1.4	-3.8	-1.7	0.8
31.74	0.34	1.8	5.40	30MR	-1.5	-3.8	-1.6	0.8
38.46	0.44	1.9	15.05	9AP	-0.6	-3.9	-1.6	0.9
45.13	0.52	1.9	24.67	19AP	0.4	-4.0	-1.5	1.0
51.75	0.60	1.9	34.25	29AP	1.4	-4.1	-1.5	1.0
58.32	0.68	1.9	43.79	9MY	2.5	-4.2	-1.4	1.1
64.85	0.75	1.9	53.33	19MY	3.5	-4.2	-1.4	1.1
71.35	0.82	1.9	62.84	29MY	2.0	-4.1	-1.4	1.2
77.81	0.88	1.9	72.36	8JN	0.8	-3.7	-1.3	1.2
84.24	0.93	1.9	81.89	18JN	-0.3	-3.0	-1.3	1.2
90.65	0.98	1.9	91.43	28JN	-1.3	-3.3	-1.3	1.2
97.04	1.03	2.0	101.00	8JL	-1.5	-3.9	-1.3	1.2
103.42	1.07	2.0	110.61	18JL	-0.7	-4.2	-1.3	1.2
109.79	1.11	2.0	120.25	28JL	-0.2	-4.2	-1.3	1.2
116.15	1.15	2.0	129.95	7AU	0.1	-4.2	-1.3	1.2
122.51	1.18	2.0	139.70	17AU	0.3	-4.1	-1.4	1.1
128.86	1.21	2.0	149.50	27AU	0.6	-4.0	-1.4	1.1
135.22	1.23	2.0	159.37	6SE	1.5	-4.0	-1.4	1.0
141.59	1.25	2.0	169.29	16SE	2.7	-3.9	-1.5	1.0
147.95	1.27	2.0	179.27	26SE	0.1	-3.8	-1.5	1.0
154.32	1.28	2.0	189.31	6OC	-0.7	-3.7	-1.6	1.0
160.69	1.28	1.9	199.40	16OC	-0.7	-3.7	-1.7	1.0
167.07	1.28	1.9	209.53	26OC	-0.7	-3.6	-1.7	1.0
173.44	1.28	1.8	219.69	5NO	-0.8	-3.6	-1.8	1.0
179.81	1.27	1.8	229.89	15NO	-0.7	-3.5	-1.9	1.0
186.18	1.25	1.7	240.09	25NO	-0.7	-3.5	-1.9	1.0
192.54	1.22	1.6	250.30	5DE	-0.7	-3.4	-2.0	1.0
198.88	1.18	1.5	260.51	15DE	-0.1	-3.4	-2.0	0.9
205.20	1.13	1.4	270.69	25DE	2.2	-3.4	-2.1	0.9

♂ LONG	♂ LAT	♂ MAG	⊙ LONG	16.00UT	☿	♀	♃	♄
				-286	MAGNITUDES			
211.49	1.07	1.3	280.85	4JA	1.6	-3.3	-2.1	0.8
217.74	0.99	1.2	290.97	14JA	0.6	-3.3	-2.1	0.8
223.94	0.89	1.0	301.05	24JA	0.3	-3.3	-2.1	0.7
230.08	0.78	0.9	311.07	3FE	0.1	-3.3	-2.1	0.7
236.14	0.63	0.7	321.04	13FE	-0.2	-3.3	-2.0	0.6
242.09	0.46	0.5	330.94	23FE	-0.6	-3.3	-2.0	0.6
247.91	0.26	0.4	340.79	5MR	-1.4	-3.4	-1.9	0.6
253.56	0.01	0.2	350.59	15MR	-1.5	-3.4	-1.8	0.5
259.00	-0.28	-0.0	0.32	25MR	-0.6	-3.4	-1.8	0.6
264.16	-0.62	-0.3	10.00	4AP	0.6	-3.5	-1.7	0.6
268.95	-1.03	-0.5	19.64	14AP	1.8	-3.5	-1.6	0.7
273.28	-1.51	-0.8	29.23	24AP	3.3	-3.5	-1.6	0.8
277.01	-2.07	-1.0	38.79	4MY	2.7	-3.4	-1.5	0.8
279.96	-2.71	-1.3	48.33	14MY	1.5	-3.4	-1.5	0.9
281.96	-3.44	-1.6	57.85	24MY	0.6	-3.4	-1.4	1.0
282.80	-4.23	-1.9	67.37	3JN	-0.4	-3.4	-1.4	1.0
282.38	-5.02	-2.2	76.89	13JN	-1.4	-3.3	-1.3	1.0
280.79	-5.73	-2.5	86.42	23JN	-1.5	-3.3	-1.3	1.1
278.36	-6.25	-2.6	95.98	3JL	-0.7	-3.3	-1.3	1.1
275.77	-6.47	-2.6	105.57	13JL	-0.1	-3.3	-1.3	1.1
273.74	-6.40	-2.4	115.19	23JL	0.2	-3.3	-1.2	1.1
272.79	-6.08	-2.1	124.86	2AU	0.5	-3.4	-1.2	1.1
273.16	-5.61	-1.8	134.58	12AU	0.8	-3.4	-1.2	1.1
274.82	-5.06	-1.6	144.35	22AU	1.9	-3.4	-1.2	1.0
277.59	-4.50	-1.3	154.19	1SE	2.4	-3.5	-1.3	1.0
281.28	-3.96	-1.0	164.08	11SE	-0.0	-3.5	-1.3	1.0
285.69	-3.45	-0.8	174.03	21SE	-0.8	-3.6	-1.3	0.9
290.67	-2.97	-0.5	184.04	1OC	-0.9	-3.6	-1.4	0.9
296.09	-2.54	-0.3	194.10	11OC	-0.8	-3.7	-1.4	0.9
301.84	-2.14	-0.1	204.21	21OC	-0.7	-3.8	-1.4	0.9
307.83	-1.78	0.1	214.36	31OC	-0.5	-3.9	-1.5	0.9
314.03	-1.45	0.3	224.54	10NO	-0.5	-4.0	-1.6	0.9
320.35	-1.16	0.5	234.74	20NO	-0.5	-4.1	-1.6	0.9
326.78	-0.89	0.6	244.95	30NO	0.0	-4.2	-1.7	0.9
333.28	-0.65	0.8	255.16	10DE	2.4	-4.3	-1.8	0.9
339.82	-0.44	0.9	265.35	20DE	1.3	-4.4	-1.8	0.9
346.39	-0.24	1.1	275.53	30DE	0.3	-4.4	-1.9	0.8
				-285				
352.97	-0.07	1.2	285.67	9JA	0.1	-4.2	-1.9	0.8
359.55	0.09	1.3	295.77	19JA	-0.0	-3.8	-2.0	0.8
6.12	0.23	1.4	305.82	29JA	-0.3	-3.3	-2.0	0.7
12.68	0.35	1.5	315.82	8FE	-0.7	-3.7	-2.0	0.6
19.21	0.46	1.6	325.75	18FE	-1.4	-4.1	-2.0	0.6
25.72	0.57	1.7	335.64	28FE	-1.4	-4.3	-2.0	0.5
32.21	0.66	1.7	345.46	10MR	-0.5	-4.3	-2.0	0.5
38.67	0.74	1.8	355.22	20MR	0.8	-4.2	-1.9	0.5
45.10	0.81	1.9	4.93	30MR	2.3	-4.1	-1.9	0.4
51.51	0.87	1.9	14.59	9AP	3.6	-4.0	-1.8	0.5
57.90	0.93	1.9	24.20	19AP	2.1	-3.9	-1.8	0.6
64.27	0.98	2.0	33.78	29AP	1.2	-3.8	-1.7	0.6
70.62	1.02	2.0	43.33	9MY	0.4	-3.7	-1.6	0.7
76.97	1.05	2.0	52.86	19MY	-0.4	-3.6	-1.6	0.8
83.30	1.08	2.0	62.38	29MY	-1.4	-3.5	-1.5	0.8
89.63	1.11	2.0	71.90	8JN	-1.6	-3.5	-1.5	0.9
95.96	1.13	2.0	81.43	18JN	-0.7	-3.4	-1.4	0.9
102.30	1.15	2.0	90.97	28JN	-0.0	-3.4	-1.4	0.9
108.65	1.16	2.0	100.54	8JL	0.4	-3.4	-1.3	1.0
115.01	1.16	2.0	110.14	18JL	0.7	-3.4	-1.3	1.0
121.39	1.16	2.0	119.79	28JL	1.2	-3.4	-1.3	1.0
127.78	1.16	1.9	129.48	7AU	2.4	-3.3	-1.2	1.0
134.21	1.15	2.0	139.23	17AU	2.1	-3.3	-1.2	1.0
140.67	1.14	2.0	149.03	27AU	-0.1	-3.4	-1.2	1.0
147.15	1.12	2.0	158.89	6SE	-0.9	-3.4	-1.2	0.9
153.67	1.10	2.0	168.81	16SE	-1.0	-3.4	-1.2	0.9
160.23	1.07	2.0	178.79	26SE	-0.9	-3.4	-1.2	0.9
166.83	1.03	1.9	188.82	6OC	-0.6	-3.4	-1.3	0.8
173.47	0.99	1.9	198.91	16OC	-0.4	-3.4	-1.3	0.8
180.14	0.95	1.9	209.04	26OC	-0.4	-3.5	-1.3	0.9
186.86	0.89	1.8	219.20	5NO	-0.3	-3.5	-1.4	0.9
193.63	0.83	1.8	229.39	15NO	0.2	-3.5	-1.4	0.9
200.44	0.76	1.8	239.60	25NO	2.7	-3.5	-1.5	0.9
207.28	0.68	1.7	249.81	5DE	1.0	-3.4	-1.5	0.9
214.17	0.59	1.6	260.01	15DE	0.1	-3.4	-1.6	0.9
221.10	0.49	1.6	270.20	25DE	-0.0	-3.4	-1.6	0.8

♂ LONG	♂ LAT	♂ MAG	⊙ LONG	16.00UT	☿	♀	♃	♄
				-284	MAGNITUDES			
228.06	0.37	1.5	280.36	4JA	-0.1	-3.4	-1.7	0.8
235.07	0.25	1.4	290.48	14JA	-0.3	-3.4	-1.8	0.8
242.10	0.11	1.3	300.56	24JA	-0.7	-3.3	-1.8	0.7
249.16	-0.04	1.2	310.58	3FE	-1.3	-3.3	-1.9	0.7
256.25	-0.21	1.1	320.55	13FE	-1.3	-3.3	-2.0	0.6
263.35	-0.39	1.0	330.47	23FE	-0.5	-3.3	-2.0	0.6
270.46	-0.58	0.9	340.31	4MR	1.0	-3.4	-2.0	0.5
277.58	-0.80	0.8	350.11	14MR	2.9	-3.4	-2.0	0.5
284.67	-1.02	0.7	359.85	24MR	2.7	-3.4	-2.0	0.4
291.73	-1.26	0.6	9.53	3AP	1.5	-3.4	-2.0	0.4
298.74	-1.51	0.5	19.17	13AP	0.9	-3.5	-2.0	0.4
305.67	-1.77	0.3	28.77	23AP	0.3	-3.5	-1.9	0.4
312.49	-2.04	0.2	38.33	3MY	-0.5	-3.6	-1.9	0.5
319.16	-2.32	0.1	47.87	13MY	-1.4	-3.6	-1.8	0.6
325.63	-2.61	-0.1	57.40	23MY	-1.6	-3.7	-1.8	0.6
331.85	-2.90	-0.2	66.91	2JN	-0.7	-3.8	-1.7	0.7
337.73	-3.19	-0.4	76.43	12JN	0.0	-3.9	-1.6	0.7
343.20	-3.48	-0.6	85.97	22JN	0.5	-4.0	-1.6	0.8
348.13	-3.76	-0.7	95.52	2JL	0.9	-4.1	-1.5	0.8
352.39	-4.04	-0.9	105.11	12JL	1.5	-4.2	-1.5	0.9
355.78	-4.30	-1.1	114.73	22JL	2.9	-4.2	-1.4	0.9
358.13	-4.52	-1.4	124.39	1AU	1.8	-4.2	-1.4	0.9
359.19	-4.67	-1.6	134.11	11AU	-0.2	-4.1	-1.3	0.9
358.82	-4.72	-1.8	143.88	21AU	-1.0	-3.7	-1.3	0.9
357.04	-4.61	-2.0	153.71	31AU	-1.2	-3.3	-1.3	0.9
354.12	-4.30	-2.2	163.60	10SE	-0.9	-3.7	-1.3	0.9
350.77	-3.80	-2.2	173.55	20SE	-0.5	-4.1	-1.2	0.9
347.78	-3.16	-1.9	183.55	30SE	-0.3	-4.3	-1.2	0.8
345.80	-2.47	-1.6	193.62	10OC	-0.2	-4.3	-1.2	0.8
345.14	-1.82	-1.3	203.72	20OC	-0.2	-4.3	-1.3	0.8
345.80	-1.25	-1.0	213.87	30OC	0.4	-4.2	-1.3	0.8
347.61	-0.77	-0.7	224.05	9NO	3.1	-4.1	-1.3	0.8
350.38	-0.37	-0.4	234.25	19NO	0.7	-4.0	-1.3	0.8
353.89	-0.04	-0.1	244.46	29NO	-0.1	-3.9	-1.4	0.8
357.99	0.23	0.2	254.67	9DE	-0.2	-3.8	-1.4	0.8
2.54	0.45	0.4	264.86	19DE	-0.2	-3.7	-1.5	0.8
7.44	0.62	0.6	275.04	29DE	-0.4	-3.7	-1.5	0.8
				-283				
12.61	0.77	0.8	285.18	8JA	-0.8	-3.6	-1.6	0.8
17.99	0.89	1.0	295.28	18JA	-1.2	-3.5	-1.7	0.8
23.53	0.98	1.1	305.34	28JA	-1.2	-3.5	-1.7	0.7
29.20	1.06	1.3	315.34	7FE	-0.5	-3.4	-1.8	0.7
34.96	1.13	1.4	325.28	17FE	1.2	-3.4	-1.9	0.6
40.80	1.18	1.5	335.16	27FE	3.3	-3.4	-1.9	0.6
46.71	1.22	1.6	344.99	9MR	2.0	-3.3	-2.0	0.5
52.66	1.25	1.7	354.75	19MR	1.1	-3.3	-2.0	0.5
58.66	1.27	1.8	4.46	29MR	0.7	-3.3	-2.1	0.4
64.69	1.29	1.8	14.13	8AP	0.2	-3.3	-2.1	0.4
70.75	1.30	1.9	23.74	18AP	-0.5	-3.3	-2.1	0.3
76.83	1.30	1.9	33.32	28AP	-1.4	-3.3	-2.1	0.4
82.95	1.30	2.0	42.88	8MY	-1.6	-3.3	-2.1	0.4
89.09	1.29	2.0	52.40	18MY	-0.6	-3.4	-2.0	0.5
95.27	1.28	2.0	61.92	28MY	0.1	-3.4	-2.0	0.5
101.47	1.26	2.0	71.44	7JN	0.7	-3.5	-1.9	0.6
107.71	1.24	2.0	80.97	17JN	1.2	-3.5	-1.9	0.7
113.98	1.21	2.0	90.51	27JN	2.0	-3.5	-1.8	0.7
120.30	1.18	2.0	100.08	7JL	3.2	-3.5	-1.7	0.8
126.65	1.15	2.0	109.67	17JL	1.5	-3.4	-1.7	0.8
133.05	1.11	2.0	119.32	27JL	-0.2	-3.4	-1.6	0.8
139.51	1.07	2.0	129.01	6AU	-1.1	-3.4	-1.6	0.9
146.01	1.02	1.9	138.75	16AU	-1.3	-3.3	-1.5	0.9
152.57	0.97	1.9	148.55	26AU	-0.8	-3.3	-1.5	0.9
159.19	0.91	1.8	158.41	5SE	-0.4	-3.3	-1.4	0.9
165.88	0.85	1.8	168.32	15SE	-0.2	-3.3	-1.4	0.8
172.62	0.79	1.8	178.30	25SE	-0.1	-3.3	-1.4	0.8
179.44	0.71	1.8	188.33	5OC	0.0	-3.4	-1.3	0.8
186.32	0.64	1.8	198.41	15OC	0.7	-3.4	-1.3	0.8
193.27	0.55	1.8	208.54	25OC	3.4	-3.4	-1.3	0.7
200.29	0.47	1.8	218.71	4NO	0.5	-3.4	-1.3	0.7
207.37	0.37	1.7	228.89	14NO	-0.2	-3.5	-1.3	0.8
214.53	0.27	1.7	239.10	24NO	-0.3	-3.5	-1.3	0.8
221.75	0.16	1.7	249.31	4DE	-0.4	-3.6	-1.4	0.8
229.03	0.05	1.6	259.52	14DE	-0.5	-3.6	-1.4	0.8
236.37	-0.07	1.6	269.71	24DE	-0.8	-3.7	-1.4	0.8

♂			☉	16.00UT	☿	♀	♃	♄
LONG	LAT	MAG	LONG	-282	MAGNITUDES			
243.78	-0.19	1.5	279.87	3JA	-1.1	-3.8	-1.5	0.8
251.23	-0.32	1.5	289.99	13JA	-1.1	-3.9	-1.5	0.8
258.73	-0.45	1.4	300.07	23JA	-0.4	-4.0	-1.6	0.8
266.27	-0.59	1.4	310.10	2FE	1.4	-4.1	-1.6	0.7
273.84	-0.73	1.3	320.07	12FE	3.0	-4.1	-1.7	0.7
281.44	-0.87	1.3	329.99	22FE	1.5	-4.2	-1.8	0.7
289.05	-1.01	1.2	339.85	4MR	0.8	-4.3	-1.8	0.6
296.66	-1.14	1.2	349.64	14MR	0.5	-4.2	-1.9	0.5
304.26	-1.27	1.1	359.38	24MR	0.1	-4.0	-2.0	0.5
311.83	-1.40	1.1	9.07	3AP	-0.5	-3.4	-2.0	0.4
319.36	-1.51	1.0	18.70	13AP	-1.4	-3.0	-2.1	0.4
326.83	-1.62	0.9	28.30	23AP	-1.6	-3.7	-2.1	0.3
334.22	-1.71	0.9	37.87	3MY	-0.6	-4.1	-2.2	0.3
341.52	-1.79	0.8	47.41	13MY	0.2	-4.2	-2.2	0.3
348.70	-1.86	0.8	56.93	23MY	0.9	-4.2	-2.2	0.4
355.74	-1.91	0.7	66.45	2JN	1.6	-4.1	-2.2	0.5
2.62	-1.94	0.6	75.97	12JN	2.7	-4.0	-2.2	0.5
9.30	-1.96	0.6	85.50	22JN	2.9	-3.9	-2.1	0.6
15.75	-1.95	0.5	95.06	2JL	1.2	-3.8	-2.1	0.6
21.94	-1.93	0.4	104.64	12JL	-0.2	-3.7	-2.0	0.7
27.81	-1.89	0.3	114.26	22JL	-1.2	-3.7	-2.0	0.7
33.30	-1.82	0.2	123.92	1AU	-1.4	-3.6	-1.9	0.8
38.33	-1.73	0.0	133.63	11AU	-0.8	-3.6	-1.8	0.8
42.80	-1.61	-0.1	143.41	21AU	-0.3	-3.5	-1.8	0.8
46.59	-1.46	-0.3	153.23	31AU	-0.1	-3.5	-1.7	0.8
49.53	-1.26	-0.5	163.12	10SE	0.1	-3.4	-1.7	0.8
51.42	-1.02	-0.7	173.07	20SE	0.2	-3.4	-1.6	0.8
52.07	-0.72	-0.9	183.07	30SE	0.9	-3.4	-1.6	0.8
51.32	-0.35	-1.1	193.12	10OC	3.3	-3.4	-1.5	0.8
49.13	0.07	-1.3	203.23	20OC	0.3	-3.4	-1.5	0.8
45.79	0.53	-1.5	213.37	30OC	-0.4	-3.4	-1.4	0.7
41.92	0.97	-1.5	223.55	9NO	-0.5	-3.4	-1.4	0.7
38.30	1.35	-1.3	233.75	19NO	-0.5	-3.4	-1.4	0.7
35.64	1.63	-1.0	243.96	29NO	-0.6	-3.4	-1.4	0.8
34.27	1.82	-0.7	254.17	9DE	-0.9	-3.4	-1.4	0.8
34.24	1.94	-0.4	264.37	19DE	-0.9	-3.4	-1.4	0.8
35.41	2.00	-0.1	274.54	29DE	-0.9	-3.4	-1.4	0.8
				-281				
37.56	2.02	0.2	284.69	8JA	-0.3	-3.4	-1.5	0.8
40.51	2.02	0.4	294.80	18JA	1.7	-3.5	-1.5	0.8
44.10	2.00	0.6	304.85	28JA	2.4	-3.5	-1.5	0.8
48.16	1.97	0.8	314.85	7FE	1.1	-3.5	-1.6	0.8
52.62	1.93	1.0	324.80	17FE	0.6	-3.4	-1.6	0.7
57.39	1.89	1.1	334.68	27FE	0.3	-3.4	-1.7	0.7
62.39	1.84	1.3	344.51	9MR	-0.0	-3.4	-1.7	0.6
67.60	1.79	1.4	354.28	19MR	-0.6	-3.4	-1.8	0.6
72.97	1.74	1.5	3.99	29MR	-1.4	-3.3	-1.9	0.5
78.48	1.68	1.6	13.66	8AP	-1.6	-3.3	-2.0	0.5
84.11	1.63	1.7	23.28	18AP	-0.6	-3.3	-2.0	0.4
89.84	1.57	1.7	32.86	28AP	0.3	-3.3	-2.1	0.3
95.66	1.51	1.8	42.41	8MY	1.2	-3.3	-2.2	0.3
101.57	1.45	1.8	51.94	18MY	2.1	-3.4	-2.2	0.3
107.55	1.38	1.9	61.46	28MY	3.4	-3.4	-2.3	0.3
113.62	1.32	1.9	70.98	7JN	2.3	-3.4	-2.3	0.4
119.75	1.25	1.9	80.51	17JN	1.0	-3.4	-2.3	0.5
125.96	1.18	1.9	90.04	27JN	-0.3	-3.5	-2.3	0.5
132.25	1.10	1.9	99.61	7JL	-1.3	-3.5	-2.3	0.6
138.61	1.03	1.9	109.21	17JL	-1.5	-3.6	-2.3	0.6
145.05	0.95	1.9	118.85	27JL	-0.8	-3.7	-2.2	0.7
151.56	0.87	1.9	128.54	6AU	-0.3	-3.7	-2.2	0.7
158.16	0.79	1.9	138.28	16AU	0.0	-3.8	-2.1	0.8
164.84	0.70	1.9	148.07	26AU	0.2	-3.9	-2.1	0.8
171.61	0.61	1.8	157.93	5SE	0.4	-4.0	-2.0	0.8
178.46	0.52	1.8	167.84	15SE	1.2	-4.2	-1.9	0.8
185.39	0.42	1.7	177.81	25SE	3.0	-4.3	-1.9	0.8
192.42	0.32	1.7	187.85	5OC	0.2	-4.3	-1.8	0.8
199.53	0.22	1.6	197.93	15OC	-0.6	-4.4	-1.8	0.8
206.72	0.12	1.6	208.05	25OC	-0.6	-4.2	-1.7	0.8
214.01	0.01	1.6	218.21	4NO	-0.6	-3.8	-1.7	0.7
221.37	-0.09	1.6	228.40	14NO	-0.7	-2.9	-1.6	0.7
228.80	-0.20	1.6	238.60	24NO	-0.8	-3.7	-1.6	0.7
236.32	-0.31	1.5	248.81	4DE	-0.8	-4.2	-1.5	0.7
243.89	-0.42	1.5	259.02	14DE	-0.8	-4.4	-1.5	0.8
251.53	-0.53	1.5	269.21	24DE	-0.2	-4.4	-1.5	0.8

♂			☉	16.00UT	☿	♀	♃	♄
LONG	LAT	MAG	LONG	-280	MAGNITUDES			
259.22	-0.64	1.5	279.37	3JA	1.9	-4.3	-1.5	0.8
266.95	-0.74	1.5	289.50	13JA	1.9	-4.2	-1.5	0.8
274.72	-0.84	1.5	299.58	23JA	0.7	-4.1	-1.5	0.8
282.51	-0.93	1.4	309.61	2FE	0.4	-4.0	-1.5	0.8
290.30	-1.02	1.4	319.59	12FE	0.2	-3.9	-1.6	0.8
298.10	-1.09	1.4	329.51	22FE	-0.1	-3.8	-1.6	0.8
305.89	-1.16	1.4	339.37	3MR	-0.6	-3.7	-1.6	0.7
313.64	-1.22	1.4	349.17	13MR	-1.4	-3.6	-1.7	0.7
321.36	-1.26	1.3	358.91	23MR	-1.5	-3.5	-1.7	0.6
329.04	-1.29	1.3	8.60	2AP	-0.6	-3.5	-1.8	0.6
336.64	-1.31	1.3	18.24	12AP	0.5	-3.4	-1.8	0.5
344.17	-1.31	1.3	27.84	22AP	1.5	-3.4	-1.9	0.4
351.62	-1.31	1.3	37.41	2MY	2.8	-3.3	-2.0	0.4
358.97	-1.28	1.3	46.95	12MY	3.2	-3.3	-2.0	0.3
6.22	-1.25	1.3	56.48	22MY	1.8	-3.3	-2.1	0.3
13.35	-1.19	1.2	66.00	1JN	0.7	-3.3	-2.2	0.3
20.34	-1.13	1.2	75.52	11JN	-0.3	-3.3	-2.2	0.3
27.21	-1.05	1.2	85.04	21JN	-1.3	-3.3	-2.3	0.4
33.91	-0.96	1.1	94.60	1JL	-1.5	-3.3	-2.4	0.5
40.45	-0.86	1.1	104.18	11JL	-0.7	-3.4	-2.4	0.5
46.81	-0.74	1.1	113.79	21JL	-0.2	-3.4	-2.4	0.6
52.96	-0.61	1.0	123.46	31JL	0.2	-3.4	-2.4	0.6
58.88	-0.46	0.9	133.17	10AU	0.4	-3.5	-2.4	0.7
64.53	-0.29	0.8	142.93	20AU	0.7	-3.5	-2.4	0.7
69.87	-0.11	0.8	152.76	30AU	1.6	-3.5	-2.3	0.8
74.83	0.10	0.6	162.64	9SE	2.7	-3.5	-2.3	0.8
79.35	0.33	0.5	172.58	19SE	0.1	-3.4	-2.2	0.8
83.32	0.60	0.4	182.58	29SE	-0.7	-3.4	-2.2	0.8
86.62	0.90	0.2	192.63	9OC	-0.8	-3.4	-2.1	0.8
89.10	1.24	-0.0	202.73	19OC	-0.8	-3.4	-2.0	0.8
90.56	1.63	-0.2	212.88	29OC	-0.8	-3.3	-2.0	0.8
90.81	2.06	-0.4	223.05	8NO	-0.6	-3.3	-1.9	0.7
89.72	2.52	-0.7	233.25	18NO	-0.6	-3.3	-1.8	0.7
87.28	2.96	-0.9	243.46	28NO	-0.6	-3.3	-1.8	0.7
83.77	3.34	-1.1	253.66	8DE	-0.1	-3.3	-1.7	0.7
79.79	3.59	-1.1	263.86	18DE	2.2	-3.4	-1.7	0.8
76.10	3.70	-0.9	274.04	28DE	1.5	-3.4	-1.6	0.8
				-279				
73.34	3.66	-0.6	284.19	7JA	0.5	-3.4	-1.6	0.8
71.84	3.53	-0.4	294.30	17JA	0.2	-3.4	-1.6	0.8
71.65	3.35	-0.1	304.36	27JA	0.1	-3.5	-1.6	0.8
72.64	3.15	0.1	314.36	6FE	-0.2	-3.5	-1.6	0.8
74.62	2.94	0.4	324.31	16FE	-0.6	-3.6	-1.6	0.8
77.39	2.74	0.6	334.21	26FE	-1.4	-3.6	-1.6	0.8
80.82	2.56	0.8	344.03	8MR	-1.5	-3.7	-1.6	0.8
84.75	2.38	0.9	353.81	18MR	-0.6	-3.8	-1.6	0.7
89.09	2.21	1.1	3.53	28MR	0.6	-3.8	-1.6	0.7
93.78	2.06	1.2	13.19	7AP	1.9	-3.9	-1.7	0.6
98.73	1.91	1.3	22.81	17AP	3.7	-4.0	-1.7	0.6
103.93	1.77	1.4	32.40	27AP	2.5	-4.1	-1.8	0.5
109.32	1.63	1.5	41.95	7MY	1.4	-4.2	-1.8	0.4
114.89	1.50	1.6	51.49	17MY	0.6	-4.2	-1.9	0.4
120.61	1.38	1.6	61.01	27MY	-0.4	-4.1	-1.9	0.3
126.48	1.26	1.7	70.52	6JN	-1.4	-3.7	-2.0	0.3
132.48	1.13	1.7	80.05	16JN	-1.6	-3.0	-2.1	0.3
138.60	1.02	1.7	89.59	26JN	-0.7	-3.3	-2.1	0.4
144.84	0.90	1.7	99.15	6JL	-0.1	-3.9	-2.2	0.4
151.20	0.78	1.8	108.75	16JL	0.3	-4.2	-2.3	0.5
157.67	0.67	1.8	118.38	26JL	0.5	-4.2	-2.3	0.5
164.25	0.55	1.8	128.06	5AU	0.9	-4.2	-2.4	0.6
170.95	0.43	1.7	137.80	15AU	2.1	-4.1	-2.4	0.7
177.75	0.32	1.7	147.60	25AU	2.3	-4.0	-2.4	0.7
184.66	0.21	1.7	157.45	4SE	-0.0	-4.0	-2.4	0.7
191.68	0.09	1.7	167.36	14SE	-0.8	-3.9	-2.4	0.8
198.80	-0.02	1.6	177.33	24SE	-0.9	-3.8	-2.4	0.8
206.03	-0.13	1.6	187.36	4OC	-0.9	-3.7	-2.4	0.8
213.35	-0.24	1.6	197.44	14OC	-0.7	-3.7	-2.3	0.8
220.77	-0.35	1.5	207.56	24OC	-0.5	-3.6	-2.3	0.8
228.28	-0.45	1.5	217.71	3NO	-0.5	-3.6	-2.2	0.8
235.87	-0.55	1.5	227.90	13NO	-0.4	-3.5	-2.1	0.8
243.53	-0.64	1.4	238.11	23NO	0.1	-3.5	-2.1	0.8
251.26	-0.73	1.4	248.32	3DE	2.5	-3.4	-2.0	0.7
259.04	-0.81	1.3	258.52	13DE	1.2	-3.4	-1.9	0.7
266.86	-0.88	1.3	268.71	23DE	0.3	-3.4	-1.9	0.8

Left column — year -278 / -277

♂ LONG	LAT	MAG	☉ LONG	16.00UT	☿	♀	♃	♄
				-278				
274.72	-0.95	1.3	278.88	2JA	0.1	-3.4	-1.8	0.8
282.59	-1.00	1.4	289.01	12JA	-0.1	-3.3	-1.7	0.8
290.47	-1.05	1.4	299.09	22JA	-0.3	-3.3	-1.7	0.8
298.35	-1.08	1.4	309.12	1FE	-0.7	-3.3	-1.6	0.9
306.20	-1.10	1.4	319.11	11FE	-1.4	-3.3	-1.6	0.9
314.03	-1.11	1.4	329.03	21FE	-1.4	-3.3	-1.6	0.9
321.81	-1.11	1.4	338.89	3MR	-0.6	-3.4	-1.6	0.8
329.54	-1.10	1.4	348.69	13MR	0.8	-3.4	-1.6	0.8
337.20	-1.07	1.5	358.43	23MR	2.5	-3.4	-1.6	0.8
344.79	-1.04	1.5	8.13	2AP	3.2	-3.5	-1.6	0.7
352.31	-0.99	1.5	17.77	12AP	1.8	-3.5	-1.6	0.7
359.73	-0.93	1.5	27.37	22AP	1.1	-3.5	-1.6	0.7
7.06	-0.86	1.5	36.94	2MY	0.4	-3.4	-1.6	0.6
14.30	-0.78	1.5	46.49	12MY	-0.4	-3.4	-1.6	0.5
21.44	-0.70	1.5	56.01	22MY	-1.4	-3.4	-1.7	0.4
28.47	-0.61	1.5	65.53	1JN	-1.6	-3.4	-1.7	0.4
35.40	-0.50	1.6	75.05	11JN	-0.7	-3.3	-1.8	0.3
42.22	-0.40	1.6	84.58	21JN	-0.0	-3.3	-1.8	0.3
48.92	-0.28	1.5	94.13	1JL	0.4	-3.3	-1.9	0.4
55.50	-0.16	1.5	103.71	11JL	0.7	-3.3	-1.9	0.4
61.96	-0.03	1.5	113.32	21JL	1.3	-3.3	-2.0	0.5
68.29	0.11	1.5	122.98	31JL	2.6	-3.4	-2.1	0.5
74.47	0.25	1.5	132.69	10AU	2.0	-3.4	-2.1	0.6
80.49	0.40	1.4	142.45	20AU	-0.1	-3.4	-2.2	0.6
86.35	0.57	1.4	152.27	30AU	-1.0	-3.5	-2.3	0.7
92.01	0.74	1.3	162.16	9SE	-1.1	-3.5	-2.3	0.7
97.44	0.93	1.2	172.09	19SE	-0.9	-3.6	-2.4	0.8
102.61	1.13	1.1	182.09	29SE	-0.6	-3.6	-2.4	0.8
107.46	1.35	1.0	192.15	9OC	-0.4	-3.7	-2.4	0.8
111.93	1.60	0.9	202.24	19OC	-0.3	-3.8	-2.4	0.8
115.94	1.87	0.7	212.38	29OC	-0.3	-3.9	-2.4	0.8
119.37	2.17	0.5	222.56	8NO	0.3	-4.0	-2.3	0.8
122.09	2.51	0.3	232.75	18NO	2.8	-4.1	-2.3	0.8
123.94	2.88	0.1	242.96	28NO	0.9	-4.2	-2.2	0.8
124.72	3.28	-0.1	253.17	8DE	0.1	-4.3	-2.2	0.8
124.28	3.70	-0.4	263.37	18DE	-0.1	-4.4	-2.1	0.8
122.51	4.09	-0.6	273.55	28DE	-0.2	-4.4	-2.0	0.8
				-277				
119.53	4.40	-0.9	283.70	7JA	-0.4	-4.2	-1.9	0.8
115.75	4.57	-1.0	293.81	17JA	-0.7	-3.8	-1.9	0.9
111.84	4.56	-0.9	303.87	27JA	-1.3	-3.3	-1.8	0.9
108.51	4.38	-0.7	313.88	6FE	-1.3	-3.7	-1.7	0.9
106.27	4.09	-0.5	323.83	16FE	-0.5	-4.1	-1.7	0.9
105.32	3.75	-0.3	333.72	26FE	1.0	-4.3	-1.6	0.9
105.63	3.38	-0.0	343.56	8MR	3.0	-4.3	-1.6	0.9
107.04	3.03	0.2	353.33	18MR	2.4	-4.2	-1.6	0.9
109.38	2.70	0.4	3.05	28MR	1.4	-4.1	-1.5	0.9
112.48	2.40	0.6	12.72	7AP	0.8	-4.0	-1.5	0.8
116.19	2.12	0.7	22.34	17AP	0.3	-3.9	-1.5	0.8
120.39	1.87	0.9	31.93	27AP	-0.4	-3.8	-1.5	0.7
125.01	1.64	1.0	41.49	7MY	-1.4	-3.7	-1.5	0.7
129.98	1.42	1.1	51.02	17MY	-1.6	-3.6	-1.5	0.6
135.24	1.22	1.2	60.54	27MY	-0.7	-3.5	-1.5	0.5
140.76	1.03	1.2	70.06	6JN	0.1	-3.5	-1.5	0.5
146.50	0.85	1.3	79.58	16JN	0.6	-3.4	-1.6	0.4
152.45	0.68	1.4	89.13	26JN	1.0	-3.4	-1.6	0.4
158.58	0.52	1.4	98.69	6JL	1.7	-3.4	-1.6	0.4
164.89	0.36	1.4	108.28	16JL	3.1	-3.4	-1.7	0.4
171.35	0.21	1.4	117.92	26JL	1.7	-3.4	-1.7	0.5
177.98	0.07	1.5	127.60	5AU	-0.1	-3.3	-1.8	0.5
184.74	-0.06	1.5	137.33	15AU	-1.1	-3.3	-1.9	0.6
191.65	-0.19	1.5	147.13	25AU	-1.2	-3.4	-1.9	0.6
198.69	-0.32	1.5	156.97	4SE	-0.9	-3.4	-2.0	0.7
205.86	-0.43	1.5	166.88	14SE	-0.5	-3.4	-2.0	0.7
213.15	-0.54	1.5	176.85	24SE	-0.3	-3.4	-2.1	0.8
220.55	-0.65	1.5	186.87	4OC	-0.2	-3.4	-2.2	0.8
228.05	-0.74	1.5	196.94	14OC	-0.1	-3.4	-2.2	0.8
235.64	-0.82	1.4	207.07	24OC	0.5	-3.5	-2.3	0.8
243.32	-0.90	1.4	217.22	3NO	3.1	-3.5	-2.3	0.9
251.06	-0.96	1.4	227.41	13NO	0.7	-3.5	-2.3	0.9
258.87	-1.01	1.4	237.61	23NO	-0.1	-3.5	-2.3	0.9
266.72	-1.05	1.4	247.82	3DE	-0.2	-3.4	-2.3	0.9
274.60	-1.08	1.4	258.02	13DE	-0.3	-3.4	-2.2	0.8
282.50	-1.10	1.4	268.22	23DE	-0.5	-3.4	-2.2	0.8

Right column — year -276 / -275

♂ LONG	LAT	MAG	☉ LONG	16.00UT	☿	♀	♃	♄
				-276				
290.40	-1.11	1.4	278.38	2JA	-0.8	-3.4	-2.1	0.8
298.29	-1.10	1.4	288.51	12JA	-1.2	-3.3	-2.0	0.9
306.15	-1.08	1.4	298.60	22JA	-1.1	-3.3	-2.0	0.9
313.98	-1.05	1.4	308.64	1FE	-0.5	-3.3	-1.9	0.9
321.76	-1.01	1.4	318.62	11FE	1.2	-3.3	-1.8	1.0
329.48	-0.96	1.4	328.54	21FE	3.3	-3.3	-1.8	1.0
337.13	-0.91	1.4	338.41	2MR	1.8	-3.4	-1.7	1.0
344.71	-0.84	1.4	348.21	12MR	1.0	-3.4	-1.6	1.0
352.21	-0.76	1.5	357.96	22MR	0.6	-3.4	-1.6	1.0
359.63	-0.68	1.5	7.65	1AP	0.2	-3.4	-1.5	1.0
6.95	-0.59	1.6	17.30	11AP	-0.5	-3.5	-1.5	0.9
14.20	-0.50	1.6	26.91	21AP	-1.4	-3.5	-1.5	0.9
21.34	-0.41	1.6	36.48	1MY	-1.6	-3.6	-1.4	0.8
28.40	-0.31	1.7	46.02	11MY	-0.7	-3.6	-1.4	0.8
35.37	-0.20	1.7	55.55	21MY	0.1	-3.7	-1.4	0.7
42.25	-0.10	1.7	65.07	31MY	0.7	-3.8	-1.4	0.7
49.05	0.01	1.8	74.59	10JN	1.3	-3.9	-1.4	0.6
55.76	0.12	1.8	84.12	20JN	2.3	-4.0	-1.4	0.5
62.38	0.23	1.8	93.67	30JN	3.2	-4.1	-1.4	0.5
68.92	0.34	1.8	103.25	10JL	1.4	-4.2	-1.4	0.4
75.37	0.46	1.8	112.86	20JL	-0.2	-4.2	-1.5	0.4
81.74	0.57	1.8	122.52	30JL	-1.2	-4.2	-1.5	0.5
88.01	0.69	1.8	132.22	9AU	-1.4	-4.1	-1.5	0.5
94.20	0.82	1.8	141.99	19AU	-0.8	-3.7	-1.6	0.6
100.28	0.94	1.7	151.80	29AU	-0.4	-3.3	-1.6	0.6
106.26	1.07	1.7	161.68	8SE	-0.1	-3.7	-1.7	0.7
112.11	1.21	1.6	171.62	18SE	-0.0	-4.1	-1.7	0.7
117.83	1.35	1.6	181.61	28SE	0.1	-4.3	-1.8	0.8
123.40	1.50	1.5	191.66	8OC	0.7	-4.3	-1.9	0.8
128.78	1.66	1.4	201.75	18OC	3.4	-4.3	-1.9	0.9
133.95	1.83	1.3	211.89	28OC	0.5	-4.2	-2.0	0.9
138.86	2.01	1.2	222.06	7NO	-0.3	-4.1	-2.1	0.9
143.45	2.21	1.0	232.26	17NO	-0.4	-4.0	-2.1	0.9
147.66	2.42	0.8	242.47	27NO	-0.4	-3.9	-2.1	0.9
151.39	2.66	0.7	252.68	7DE	-0.6	-3.8	-2.2	0.9
154.52	2.91	0.4	262.88	17DE	-0.9	-3.7	-2.2	0.9
156.92	3.19	0.2	273.06	27DE	-1.0	-3.7	-2.2	0.9
				-275				
158.40	3.48	-0.0	283.21	6JA	-1.0	-3.6	-2.1	0.9
158.79	3.77	-0.3	293.32	16JA	-0.4	-3.5	-2.1	0.9
157.95	4.03	-0.6	303.38	26JA	1.4	-3.5	-2.1	1.0
155.82	4.23	-0.9	313.40	5FE	2.8	-3.4	-2.0	1.0
152.60	4.29	-1.1	323.35	15FE	1.3	-3.4	-1.9	1.0
148.80	4.20	-1.2	333.25	25FE	0.7	-3.4	-1.9	1.1
145.11	3.93	-1.1	343.08	7MR	0.4	-3.3	-1.8	1.1
142.21	3.54	-0.9	352.86	17MR	0.0	-3.3	-1.7	1.1
140.49	3.08	-0.7	2.58	27MR	-0.5	-3.3	-1.7	1.1
140.10	2.62	-0.4	12.26	6AP	-1.4	-3.3	-1.6	1.1
140.95	2.17	-0.2	21.88	16AP	-1.6	-3.3	-1.5	1.1
142.88	1.77	-0.0	31.47	26AP	-0.7	-3.3	-1.5	1.0
145.71	1.40	0.1	41.03	6MY	0.2	-3.3	-1.4	1.0
149.29	1.07	0.3	50.56	16MY	1.0	-3.4	-1.4	0.9
153.47	0.77	0.4	60.08	26MY	1.7	-3.4	-1.4	0.9
158.17	0.50	0.5	69.60	5JN	3.0	-3.5	-1.3	0.8
163.30	0.26	0.7	79.12	15JN	2.7	-3.5	-1.3	0.8
168.78	0.04	0.7	88.66	25JN	1.1	-3.5	-1.3	0.7
174.58	-0.15	0.8	98.22	5JL	-0.2	-3.5	-1.3	0.6
180.66	-0.33	0.9	107.81	15JL	-1.2	-3.4	-1.3	0.6
186.99	-0.50	0.9	117.44	25JL	-1.5	-3.4	-1.3	0.5
193.53	-0.65	1.0	127.13	4AU	-0.8	-3.3	-1.3	0.6
200.28	-0.78	1.0	136.86	14AU	-0.3	-3.3	-1.3	0.6
207.21	-0.90	1.1	146.64	24AU	-0.0	-3.3	-1.4	0.6
214.30	-1.00	1.1	156.49	3SE	0.1	-3.3	-1.4	0.7
221.54	-1.09	1.1	166.40	13SE	0.3	-3.3	-1.4	0.7
228.91	-1.16	1.2	176.36	23SE	1.0	-3.3	-1.5	0.8
236.39	-1.22	1.2	186.38	3OC	3.3	-3.4	-1.5	0.8
243.98	-1.26	1.2	196.45	13OC	0.3	-3.4	-1.6	0.9
251.65	-1.29	1.2	206.57	23OC	-0.5	-3.4	-1.7	0.9
259.39	-1.30	1.3	216.73	2NO	-0.5	-3.4	-1.7	1.0
267.18	-1.30	1.3	226.91	12NO	-0.5	-3.5	-1.8	1.0
275.00	-1.28	1.3	237.11	22NO	-0.7	-3.5	-1.9	1.0
282.85	-1.25	1.3	247.33	2DE	-0.8	-3.6	-1.9	1.0
290.69	-1.21	1.4	257.53	12DE	-0.9	-3.6	-2.0	1.0
298.52	-1.16	1.4	267.72	22DE	-0.9	-3.7	-2.0	1.1

♂ LONG	♂ LAT	♂ MAG	⊙ LONG	16.00UT -274	☿	♀	♃	♄
					MAGNITUDES			
306.32	-1.09	1.4	277.89	1JA	-0.3	-3.8	-2.1	1.1
314.08	-1.02	1.5	288.02	11JA	1.7	-3.9	-2.1	1.0
321.79	-0.94	1.5	298.11	21JA	2.3	-4.0	-2.1	1.0
329.45	-0.86	1.5	308.16	31JA	0.9	-4.1	-2.1	1.1
337.03	-0.77	1.5	318.14	10FE	0.5	-4.1	-2.0	1.1
344.54	-0.67	1.6	328.07	20FE	0.3	-4.2	-2.0	1.2
351.97	-0.57	1.6	337.94	2MR	-0.0	-4.3	-1.9	1.2
359.31	-0.47	1.6	347.74	12MR	-0.6	-4.2	-1.9	1.2
6.58	-0.37	1.6	357.49	22MR	-1.4	-4.0	-1.8	1.2
13.76	-0.27	1.6	7.19	1AP	-1.6	-3.4	-1.8	1.2
20.85	-0.16	1.6	16.84	11AP	-0.6	-3.0	-1.7	1.2
27.87	-0.06	1.6	26.44	21AP	0.4	-3.7	-1.6	1.2
34.80	0.04	1.7	36.02	1MY	1.3	-4.1	-1.6	1.2
41.66	0.14	1.7	45.56	11MY	2.3	-4.2	-1.5	1.2
48.45	0.24	1.8	55.09	21MY	3.6	-4.2	-1.4	1.1
55.16	0.34	1.8	64.61	31MY	2.2	-4.1	-1.4	1.1
61.81	0.43	1.9	74.12	10JN	0.9	-4.0	-1.4	1.0
68.40	0.53	1.9	83.65	20JN	-0.3	-3.9	-1.3	1.0
74.93	0.62	1.9	93.20	30JN	-1.3	-3.8	-1.3	0.9
81.40	0.71	1.9	102.78	10JL	-1.5	-3.7	-1.3	0.8
87.82	0.80	2.0	112.39	20JL	-0.8	-3.7	-1.2	0.7
94.19	0.88	2.0	122.05	30JL	-0.2	-3.6	-1.2	0.7
100.50	0.97	2.0	131.75	9AU	0.1	-3.6	-1.2	0.7
106.77	1.06	2.0	141.51	19AU	0.3	-3.5	-1.2	0.7
112.99	1.14	1.9	151.32	29AU	0.5	-3.5	-1.2	0.7
119.15	1.23	1.9	161.20	8SE	1.4	-3.4	-1.3	0.8
125.26	1.31	1.9	171.13	18SE	3.0	-3.4	-1.3	0.8
131.31	1.40	1.8	181.13	28SE	0.2	-3.4	-1.3	0.9
137.29	1.48	1.8	191.17	8OC	-0.6	-3.4	-1.4	0.9
143.19	1.57	1.7	201.27	18OC	-0.7	-3.4	-1.4	1.0
149.00	1.66	1.6	211.40	28OC	-0.7	-3.4	-1.5	1.0
154.71	1.74	1.5	221.57	7NO	-0.8	-3.4	-1.5	1.1
160.29	1.84	1.4	231.77	17NO	-0.7	-3.4	-1.6	1.1
165.71	1.93	1.3	241.98	27NO	-0.7	-3.4	-1.6	1.1
170.96	2.03	1.2	252.18	7DE	-0.7	-3.4	-1.7	1.2
175.98	2.12	1.0	262.39	17DE	-0.2	-3.4	-1.8	1.2
180.73	2.22	0.9	272.57	27DE	1.9	-3.4	-1.8	1.2
				-273				
185.14	2.32	0.7	282.72	6JA	1.8	-3.4	-1.9	1.2
189.13	2.42	0.5	292.83	16JA	0.6	-3.5	-1.9	1.2
192.59	2.51	0.2	302.90	26JA	0.3	-3.5	-2.0	1.2
195.40	2.59	-0.0	312.91	5FE	0.1	-3.5	-2.0	1.2
197.39	2.65	-0.3	322.87	15FE	-0.1	-3.4	-2.0	1.2
198.37	2.67	-0.6	332.77	25FE	-0.6	-3.4	-2.0	1.3
198.20	2.63	-0.9	342.60	7MR	-1.4	-3.4	-2.0	1.3
196.75	2.51	-1.2	352.39	17MR	-1.5	-3.4	-2.0	1.4
194.13	2.27	-1.5	2.11	27MR	-0.6	-3.3	-1.9	1.4
190.73	1.91	-1.7	11.78	6AP	0.5	-3.3	-1.9	1.4
187.19	1.44	-1.6	21.41	16AP	1.6	-3.3	-1.8	1.4
184.25	0.93	-1.5	31.00	26AP	3.1	-3.3	-1.8	1.4
182.43	0.42	-1.3	40.56	6MY	3.0	-3.3	-1.7	1.3
181.95	-0.04	-1.1	50.10	16MY	1.7	-3.4	-1.6	1.3
182.82	-0.44	-0.9	59.62	26MY	0.7	-3.4	-1.6	1.3
184.87	-0.78	-0.7	69.13	5JN	-0.3	-3.4	-1.5	1.2
187.93	-1.06	-0.5	78.66	15JN	-1.3	-3.4	-1.5	1.2
191.83	-1.30	-0.3	88.19	25JN	-1.6	-3.5	-1.4	1.1
196.41	-1.48	-0.2	97.75	5JL	-0.7	-3.5	-1.4	1.1
201.56	-1.64	-0.0	107.35	15JL	-0.2	-3.6	-1.3	1.0
207.19	-1.75	0.1	116.98	25JL	0.2	-3.7	-1.3	1.0
213.20	-1.84	0.2	126.65	4AU	0.4	-3.7	-1.3	0.9
219.54	-1.90	0.3	136.38	14AU	0.8	-3.8	-1.2	0.8
226.16	-1.93	0.4	146.17	24AU	1.7	-3.9	-1.2	0.8
233.01	-1.94	0.5	156.01	3SE	2.6	-4.0	-1.2	0.9
240.06	-1.93	0.6	165.92	13SE	0.1	-4.2	-1.2	0.9
247.27	-1.90	0.7	175.88	23SE	-0.8	-4.3	-1.2	0.9
254.61	-1.85	0.7	185.89	3OC	-0.8	-4.3	-1.2	1.0
262.06	-1.78	0.8	195.97	13OC	-0.8	-4.4	-1.3	1.0
269.58	-1.70	0.9	206.08	23OC	-0.7	-4.2	-1.3	1.1
277.15	-1.61	1.0	216.23	2NO	-0.6	-3.8	-1.3	1.2
284.76	-1.51	1.0	226.42	12NO	-0.6	-2.9	-1.4	1.2
292.38	-1.39	1.1	236.62	22NO	-0.5	-3.7	-1.4	1.2
299.99	-1.27	1.2	246.83	2DE	-0.0	-4.2	-1.5	1.3
307.59	-1.15	1.2	257.04	12DE	2.2	-4.4	-1.5	1.3
315.15	-1.02	1.3	267.23	22DE	1.4	-4.4	-1.6	1.3

♂ LONG	♂ LAT	♂ MAG	⊙ LONG	16.00UT -272	☿	♀	♃	♄
					MAGNITUDES			
322.67	-0.89	1.4	277.40	1JA	0.4	-4.3	-1.7	1.4
330.14	-0.76	1.4	287.54	11JA	0.2	-4.2	-1.7	1.4
337.54	-0.63	1.5	297.62	21JA	0.0	-4.1	-1.8	1.3
344.88	-0.51	1.6	307.67	31JA	-0.2	-4.0	-1.9	1.3
352.15	-0.38	1.6	317.66	10FE	-0.7	-3.9	-1.9	1.2
359.34	-0.26	1.6	327.59	20FE	-1.3	-3.8	-2.0	1.2
6.46	-0.15	1.7	337.46	1MR	-1.4	-3.7	-2.0	1.2
13.50	-0.03	1.7	347.27	11MR	-0.6	-3.6	-2.0	1.2
20.47	0.07	1.8	357.02	21MR	0.7	-3.5	-2.0	1.2
27.37	0.18	1.8	6.72	31MR	2.1	-3.5	-2.0	1.2
34.21	0.28	1.8	16.38	10AP	3.9	-3.4	-2.0	1.2
40.97	0.37	1.8	25.98	20AP	2.2	-3.4	-2.0	1.1
47.68	0.46	1.9	35.56	30AP	1.3	-3.3	-1.9	1.1
54.33	0.54	1.9	45.11	10MY	0.5	-3.3	-1.9	1.1
60.93	0.62	1.9	54.63	20MY	-0.4	-3.3	-1.8	1.0
67.49	0.70	1.9	64.15	30MY	-1.3	-3.3	-1.8	1.0
74.00	0.77	1.9	73.67	9JN	-1.6	-3.3	-1.7	1.0
80.47	0.84	1.9	83.20	19JN	-0.7	-3.3	-1.6	0.9
86.91	0.90	1.9	92.74	29JN	-0.1	-3.3	-1.6	0.9
93.32	0.96	2.0	102.32	9JL	0.3	-3.4	-1.5	0.8
99.71	1.02	2.0	111.92	19JL	0.6	-3.4	-1.5	0.8
106.07	1.07	2.0	121.58	29JL	1.1	-3.4	-1.4	0.7
112.42	1.12	2.0	131.28	8AU	2.2	-3.5	-1.4	0.7
118.75	1.17	2.0	141.03	18AU	2.3	-3.5	-1.3	0.6
125.07	1.21	2.0	150.84	28AU	0.0	-3.5	-1.3	0.6
131.38	1.25	2.0	160.71	7SE	-0.9	-3.5	-1.3	0.6
137.68	1.29	2.0	170.64	17SE	-1.0	-3.4	-1.3	0.7
143.96	1.32	2.0	180.63	27SE	-0.9	-3.4	-1.2	0.7
150.23	1.35	1.9	190.68	7OC	-0.6	-3.4	-1.2	0.8
156.49	1.38	1.9	200.77	17OC	-0.4	-3.4	-1.3	0.9
162.73	1.40	1.8	210.91	27OC	-0.4	-3.3	-1.3	0.9
168.94	1.42	1.8	221.07	6NO	-0.4	-3.3	-1.3	1.0
175.13	1.43	1.7	231.27	16NO	0.1	-3.3	-1.3	1.0
181.28	1.44	1.6	241.48	26NO	2.5	-3.3	-1.3	1.1
187.39	1.44	1.5	251.69	6DE	1.1	-3.3	-1.4	1.1
193.45	1.43	1.4	261.89	16DE	0.2	-3.4	-1.4	1.1
199.44	1.41	1.3	272.07	26DE	-0.0	-3.4	-1.5	1.1
				-271				
205.35	1.37	1.1	282.23	5JA	-0.1	-3.4	-1.6	1.1
211.16	1.33	1.0	292.34	15JA	-0.3	-3.4	-1.6	1.1
216.85	1.26	0.8	302.41	25JA	-0.7	-3.5	-1.7	1.1
222.38	1.18	0.7	312.43	4FE	-1.3	-3.5	-1.8	1.0
227.72	1.07	0.5	322.39	14FE	-1.3	-3.6	-1.8	1.0
232.83	0.92	0.3	332.29	24FE	-0.6	-3.6	-1.9	0.9
237.63	0.74	0.0	342.13	6MR	0.8	-3.7	-1.9	0.9
242.06	0.51	-0.2	351.92	16MR	2.7	-3.8	-2.0	0.9
246.00	0.21	-0.5	1.65	26MR	2.9	-3.8	-2.0	0.9
249.32	-0.15	-0.7	11.32	5AP	1.7	-3.9	-2.1	0.9
251.87	-0.60	-1.1	20.95	15AP	1.0	-4.0	-2.1	0.9
253.45	-1.15	-1.4	30.55	25AP	0.4	-4.1	-2.1	0.9
253.90	-1.79	-1.7	40.10	5MY	-0.4	-4.2	-2.1	0.9
253.12	-2.51	-2.0	49.64	15MY	-1.4	-4.2	-2.1	0.8
251.18	-3.25	-2.3	59.16	25MY	-1.6	-4.1	-2.0	0.8
248.48	-3.93	-2.5	68.68	4JN	-0.7	-3.7	-2.0	0.8
245.67	-4.45	-2.4	78.20	14JN	-0.0	-2.9	-1.9	0.7
243.47	-4.76	-2.2	87.74	24JN	0.5	-3.4	-1.9	0.7
242.42	-4.86	-2.0	97.29	4JL	0.8	-3.9	-1.8	0.6
242.72	-4.81	-1.7	106.88	14JL	1.4	-4.2	-1.7	0.6
244.34	-4.64	-1.5	116.51	24JL	2.8	-4.2	-1.7	0.5
247.14	-4.40	-1.3	126.18	3AU	2.0	-4.2	-1.6	0.5
250.89	-4.12	-1.0	135.91	13AU	-0.1	-4.1	-1.6	0.4
255.42	-3.82	-0.8	145.69	23AU	-1.0	-4.0	-1.5	0.4
260.57	-3.52	-0.6	155.53	2SE	-1.1	-3.9	-1.5	0.3
266.18	-3.21	-0.4	165.43	12SE	-0.9	-3.9	-1.4	0.3
272.18	-2.90	-0.2	175.39	22SE	-0.5	-3.8	-1.4	0.3
278.46	-2.60	-0.0	185.40	2OC	-0.3	-3.7	-1.4	0.4
284.96	-2.31	0.1	195.47	12OC	-0.3	-3.7	-1.3	0.5
291.63	-2.03	0.3	205.59	22OC	-0.2	-3.6	-1.3	0.5
298.41	-1.76	0.4	215.74	1NO	0.3	-3.5	-1.3	0.6
305.27	-1.51	0.6	225.92	11NO	2.8	-3.5	-1.3	0.7
312.19	-1.27	0.7	236.13	21NO	0.8	-3.5	-1.3	0.7
319.14	-1.05	0.8	246.33	1DE	-0.0	-3.4	-1.3	0.8
326.09	-0.84	1.0	256.54	11DE	-0.2	-3.4	-1.4	0.8
333.05	-0.64	1.1	266.74	21DE	-0.2	-3.4	-1.4	0.8
339.98	-0.46	1.2	276.91	31DE	-0.4	-3.4	-1.4	0.8

♂ LONG	♂ LAT	♂ MAG	⊙ LONG	16.00UT	☿	♀	♃	♄
				-270	MAGNITUDES			
346.89	-0.30	1.3	287.05	10JA	-0.8	-3.3	-1.5	0.8
353.77	-0.14	1.4	297.14	20JA	-1.2	-3.3	-1.5	0.8
0.61	-0.00	1.5	307.18	30JA	-1.2	-3.3	-1.6	0.8
7.41	0.13	1.6	317.18	9FE	-0.5	-3.3	-1.7	0.8
14.16	0.25	1.6	327.11	19FE	1.0	-3.3	-1.7	0.8
20.87	0.36	1.7	336.98	1MR	3.2	-3.4	-1.8	0.7
27.53	0.46	1.8	346.80	11MR	2.2	-3.4	-1.9	0.7
34.15	0.55	1.8	356.55	21MR	1.2	-3.4	-1.9	0.6
40.72	0.64	1.9	6.25	31MR	0.7	-3.5	-2.0	0.6
47.26	0.71	1.9	15.91	10AP	0.2	-3.5	-2.1	0.7
53.76	0.78	1.9	25.52	20AP	-0.4	-3.5	-2.1	0.7
60.23	0.84	2.0	35.09	30AP	-1.4	-3.4	-2.2	0.7
66.67	0.90	2.0	44.64	10MY	-1.6	-3.4	-2.2	0.7
73.08	0.95	2.0	54.17	20MY	-0.7	-3.4	-2.2	0.6
79.48	0.99	2.0	63.69	30MY	0.1	-3.4	-2.2	0.6
85.86	1.03	2.0	73.21	9JN	0.6	-3.3	-2.2	0.6
92.23	1.07	2.0	82.73	19JN	1.1	-3.3	-2.2	0.6
98.59	1.10	2.0	92.28	29JN	1.9	-3.3	-2.1	0.5
104.96	1.12	2.0	101.85	9JL	3.2	-3.3	-2.1	0.5
111.32	1.14	2.0	111.46	19JL	1.6	-3.3	-2.0	0.4
117.70	1.16	2.0	121.10	29JL	-0.1	-3.4	-2.0	0.4
124.08	1.17	2.0	130.80	8AU	-1.1	-3.4	-1.9	0.3
130.48	1.17	2.0	140.55	18AU	-1.3	-3.4	-1.8	0.3
136.90	1.18	2.0	150.36	28AU	-0.9	-3.5	-1.8	0.2
143.33	1.17	2.0	160.23	7SE	-0.4	-3.5	-1.7	0.2
149.80	1.17	2.0	170.16	17SE	-0.2	-3.6	-1.7	0.1
156.28	1.15	2.0	180.14	27SE	-0.1	-3.6	-1.6	0.1
162.79	1.14	2.0	190.19	7OC	-0.0	-3.7	-1.6	0.1
169.33	1.11	1.9	200.27	17OC	0.5	-3.8	-1.5	0.2
175.90	1.08	1.9	210.41	27OC	3.1	-3.9	-1.5	0.2
182.50	1.04	1.8	220.58	6NO	0.6	-4.0	-1.5	0.3
189.12	1.00	1.8	230.77	16NO	-0.2	-4.1	-1.4	0.4
195.77	0.94	1.7	240.98	26NO	-0.3	-4.2	-1.4	0.4
202.45	0.88	1.7	251.19	6DE	-0.3	-4.3	-1.4	0.5
209.15	0.80	1.6	261.39	16DE	-0.5	-4.4	-1.4	0.5
215.87	0.72	1.5	271.58	26DE	-0.8	-4.4	-1.4	0.6
				-269				
222.61	0.62	1.4	281.73	5JA	-1.1	-4.2	-1.5	0.6
229.37	0.50	1.3	291.85	15JA	-1.1	-3.8	-1.5	0.6
236.14	0.37	1.2	301.92	25JA	-0.5	-3.3	-1.5	0.6
242.91	0.22	1.1	311.95	4FE	1.2	-3.7	-1.6	0.6
249.69	0.06	1.0	321.91	14FE	3.2	-4.1	-1.6	0.6
256.46	-0.13	0.9	331.81	24FE	1.6	-4.3	-1.7	0.6
263.21	-0.34	0.7	341.66	6MR	0.9	-4.3	-1.7	0.5
269.92	-0.58	0.6	351.44	16MR	0.5	-4.2	-1.8	0.5
276.59	-0.84	0.4	1.18	26MR	0.1	-4.1	-1.8	0.5
283.18	-1.13	0.3	10.86	5AP	-0.5	-4.0	-1.9	0.4
289.67	-1.44	0.1	20.49	15AP	-1.3	-3.9	-2.0	0.4
296.01	-1.79	-0.0	30.08	25AP	-1.6	-3.8	-2.1	0.5
302.15	-2.16	-0.2	39.64	5MY	-0.7	-3.7	-2.1	0.5
308.04	-2.57	-0.4	49.18	15MY	0.2	-3.6	-2.2	0.5
313.59	-3.01	-0.6	58.70	25MY	0.8	-3.5	-2.2	0.5
318.69	-3.47	-0.8	68.22	4JN	1.5	-3.5	-2.3	0.5
323.22	-3.97	-1.0	77.74	14JN	2.5	-3.4	-2.3	0.5
327.01	-4.48	-1.3	87.28	24JN	3.1	-3.4	-2.3	0.4
329.85	-5.00	-1.5	96.84	4JL	1.3	-3.4	-2.3	0.4
331.55	-5.50	-1.8	106.42	14JL	-0.2	-3.4	-2.3	0.4
331.91	-5.93	-2.0	116.05	24JL	-1.2	-3.4	-2.3	0.3
330.86	-6.21	-2.2	125.72	3AU	-1.4	-3.3	-2.3	0.3
328.62	-6.26	-2.4	135.44	13AU	-0.8	-3.3	-2.2	0.2
325.71	-6.02	-2.5	145.22	23AU	-0.4	-3.4	-2.1	0.2
322.91	-5.51	-2.3	155.06	2SE	-0.1	-3.4	-2.1	0.1
320.94	-4.81	-2.0	164.95	12SE	0.0	-3.4	-2.0	0.0
320.18	-4.05	-1.7	174.91	22SE	0.2	-3.4	-1.9	-0.0
320.75	-3.31	-1.4	184.92	2OC	0.8	-3.4	-1.9	-0.1
322.52	-2.63	-1.1	194.98	12OC	3.4	-3.4	-1.8	-0.1
325.29	-2.04	-0.8	205.09	22OC	0.5	-3.5	-1.8	-0.1
328.87	-1.53	-0.5	215.24	1NO	-0.4	-3.5	-1.7	-0.0
333.07	-1.10	-0.2	225.42	11NO	-0.5	-3.5	-1.7	0.1
337.75	-0.73	0.0	235.62	21NO	-0.5	-3.5	-1.6	0.1
342.82	-0.42	0.3	245.83	1DE	-0.6	-3.4	-1.6	0.2
348.16	-0.16	0.5	256.04	11DE	-0.9	-3.4	-1.6	0.3
353.72	0.06	0.7	266.24	21DE	-0.9	-3.4	-1.5	0.3
359.45	0.25	0.8	276.41	31DE	-0.9	-3.4	-1.5	0.4

♂ LONG	♂ LAT	♂ MAG	⊙ LONG	16.00UT	☿	♀	♃	♄
				-268	MAGNITUDES			
5.31	0.42	1.0	286.54	10JA	-0.4	-3.3	-1.5	0.4
11.26	0.56	1.1	296.65	20JA	1.5	-3.3	-1.5	0.4
17.28	0.68	1.3	306.69	30JA	2.6	-3.3	-1.5	0.5
23.36	0.78	1.4	316.69	9FE	1.2	-3.3	-1.5	0.5
29.47	0.86	1.5	326.63	19FE	0.6	-3.3	-1.6	0.5
35.61	0.94	1.6	336.50	29FE	0.4	-3.4	-1.6	0.4
41.77	1.00	1.7	346.32	10MR	0.0	-3.4	-1.6	0.4
47.94	1.05	1.7	356.08	20MR	-0.5	-3.4	-1.7	0.4
54.12	1.10	1.8	5.79	30MR	-1.3	-3.4	-1.7	0.3
60.31	1.13	1.9	15.44	9AP	-1.6	-3.5	-1.8	0.3
66.51	1.16	1.9	25.06	19AP	-0.7	-3.5	-1.9	0.3
72.72	1.18	2.0	34.63	29AP	0.3	-3.6	-1.9	0.3
78.93	1.20	2.0	44.18	9MY	1.1	-3.6	-2.0	0.3
85.16	1.21	2.0	53.71	19MY	1.9	-3.7	-2.1	0.4
91.40	1.21	2.0	63.23	29MY	3.2	-3.8	-2.1	0.4
97.65	1.21	2.0	72.75	8JN	2.5	-3.9	-2.2	0.4
103.93	1.21	2.0	82.28	18JN	1.1	-4.0	-2.3	0.4
110.23	1.20	2.0	91.82	28JN	-0.2	-4.1	-2.3	0.4
116.56	1.18	2.0	101.39	8JL	-1.2	-4.2	-2.4	0.3
122.92	1.16	2.0	111.00	18JL	-1.5	-4.2	-2.4	0.3
129.31	1.14	2.0	120.64	28JL	-0.8	-4.2	-2.4	0.3
135.75	1.11	1.9	130.34	7AU	-0.3	-4.1	-2.4	0.2
142.22	1.08	1.9	140.09	17AU	0.0	-3.7	-2.4	0.2
148.75	1.04	1.9	149.89	27AU	0.2	-3.3	-2.4	0.1
155.32	1.00	1.9	159.76	6SE	0.4	-3.7	-2.3	0.1
161.94	0.95	1.9	169.68	16SE	1.1	-4.2	-2.3	-0.0
168.62	0.90	1.9	179.66	26SE	3.3	-4.3	-2.2	-0.1
175.35	0.84	1.9	189.70	6OC	0.3	-4.3	-2.2	-0.1
182.15	0.78	1.9	199.79	16OC	-0.5	-4.3	-2.1	-0.2
189.00	0.70	1.8	209.92	26OC	-0.6	-4.2	-2.0	-0.3
195.91	0.63	1.8	220.08	5NO	-0.6	-4.1	-2.0	-0.2
202.88	0.54	1.8	230.28	15NO	-0.7	-4.0	-1.9	-0.1
209.91	0.45	1.7	240.48	25NO	-0.8	-3.9	-1.8	-0.0
217.00	0.35	1.7	250.69	5DE	-0.8	-3.8	-1.8	0.0
224.15	0.24	1.6	260.90	15DE	-0.8	-3.7	-1.7	0.1
231.35	0.12	1.6	271.08	25DE	-0.3	-3.7	-1.7	0.2
				-267				
238.60	0.00	1.5	281.24	4JA	1.7	-3.6	-1.6	0.2
245.91	-0.13	1.5	291.36	14JA	2.1	-3.5	-1.6	0.3
253.26	-0.27	1.4	301.43	24JA	0.8	-3.5	-1.6	0.3
260.65	-0.42	1.3	311.46	3FE	0.4	-3.4	-1.6	0.3
268.08	-0.57	1.3	321.43	13FE	0.2	-3.4	-1.6	0.3
275.53	-0.73	1.2	331.33	23FE	-0.1	-3.4	-1.6	0.4
283.00	-0.90	1.1	341.18	5MR	-0.6	-3.3	-1.6	0.4
290.47	-1.06	1.1	350.98	15MR	-1.3	-3.3	-1.6	0.3
297.94	-1.23	1.0	0.71	25MR	-1.6	-3.3	-1.6	0.3
305.38	-1.40	0.9	10.39	4AP	-0.7	-3.3	-1.6	0.3
312.79	-1.56	0.8	20.03	14AP	0.4	-3.3	-1.7	0.3
320.13	-1.72	0.7	29.62	24AP	1.4	-3.3	-1.7	0.2
327.40	-1.87	0.7	39.18	4MY	2.6	-3.3	-1.8	0.2
334.56	-2.01	0.6	48.72	14MY	3.4	-3.4	-1.8	0.2
341.59	-2.14	0.5	58.24	24MY	2.0	-3.4	-1.9	0.3
348.46	-2.26	0.4	67.76	3JN	0.8	-3.5	-1.9	0.3
355.11	-2.36	0.3	77.28	13JN	-0.3	-3.5	-2.0	0.3
1.53	-2.44	0.2	86.81	23JN	-1.3	-3.5	-2.1	0.3
7.65	-2.51	0.1	96.37	3JL	-1.6	-3.4	-2.1	0.3
13.41	-2.56	-0.0	105.96	13JL	-0.8	-3.4	-2.2	0.3
18.73	-2.59	-0.2	115.58	23JL	-0.2	-3.4	-2.3	0.3
23.50	-2.59	-0.3	125.25	2AU	0.1	-3.4	-2.3	0.3
27.59	-2.56	-0.5	134.97	12AU	0.3	-3.3	-2.4	0.2
30.86	-2.50	-0.7	144.74	22AU	0.6	-3.3	-2.4	0.2
33.10	-2.39	-0.9	154.58	1SE	1.5	-3.3	-2.4	0.1
34.09	-2.21	-1.1	164.47	11SE	2.9	-3.3	-2.4	0.1
33.68	-1.96	-1.3	174.42	21SE	0.2	-3.3	-2.4	0.0
31.80	-1.61	-1.5	184.43	1OC	-0.7	-3.4	-2.4	-0.1
28.70	-1.18	-1.7	194.49	11OC	-0.8	-3.4	-2.4	-0.1
24.98	-0.68	-1.8	204.60	21OC	-0.7	-3.4	-2.3	-0.2
21.42	-0.19	-1.6	214.75	31OC	-0.8	-3.4	-2.3	-0.3
18.77	0.26	-1.3	224.93	10NO	-0.6	-3.5	-2.2	-0.3
17.41	0.63	-0.9	235.13	20NO	-0.6	-3.5	-2.1	-0.2
17.39	0.92	-0.6	245.34	30NO	-0.6	-3.6	-2.0	-0.2
18.60	1.14	-0.3	255.55	10DE	-0.2	-3.7	-2.0	-0.1
20.82	1.29	-0.0	265.74	20DE	1.9	-3.7	-1.9	-0.0
23.84	1.41	0.2	275.92	30DE	1.7	-3.8	-1.8	0.0

Left table — year −266 / −265

δ LONG	LAT	MAG	☉ LONG	16.00UT	☿	♀	♃	♄
				−266		MAGNITUDES		
27.50	1.49	0.5	286.06	9JA	0.6	-3.9	-1.8	0.1
31.65	1.54	0.7	296.15	19JA	0.3	-4.0	-1.7	0.2
36.19	1.57	0.9	306.21	29JA	0.1	-4.1	-1.7	0.2
41.03	1.59	1.0	316.20	8FE	-0.2	-4.2	-1.6	0.2
46.11	1.60	1.2	326.14	18FE	-0.6	-4.2	-1.6	0.3
51.38	1.60	1.3	336.02	28FE	-1.3	-4.3	-1.6	0.3
56.81	1.59	1.4	345.84	10MR	-1.5	-4.2	-1.6	0.3
62.36	1.58	1.5	355.60	20MR	-0.7	-4.0	-1.6	0.3
68.01	1.55	1.6	5.32	30MR	0.5	-3.4	-1.5	0.3
73.76	1.53	1.7	14.97	9AP	1.8	-3.1	-1.6	0.3
79.58	1.50	1.8	24.59	19AP	3.4	-3.7	-1.6	0.3
85.47	1.47	1.8	34.17	29AP	2.7	-4.1	-1.6	0.2
91.42	1.43	1.9	43.72	9MY	1.5	-4.2	-1.6	0.2
97.43	1.39	1.9	53.25	19MY	0.6	-4.2	-1.6	0.2
103.50	1.35	1.9	62.77	29MY	-0.3	-4.1	-1.7	0.2
109.63	1.30	2.0	72.29	8JN	-1.3	-4.0	-1.7	0.2
115.81	1.25	2.0	81.81	18JN	-1.6	-3.9	-1.8	0.3
122.05	1.20	2.0	91.36	28JN	-0.8	-3.8	-1.8	0.3
128.35	1.14	2.0	100.93	8JL	-0.1	-3.7	-1.9	0.3
134.72	1.08	2.0	110.53	18JL	0.2	-3.7	-2.0	0.3
141.15	1.02	2.0	120.18	28JL	0.5	-3.6	-2.0	0.3
147.65	0.95	1.9	129.87	7AU	0.9	-3.6	-2.1	0.3
154.21	0.88	1.9	139.61	17AU	1.9	-3.5	-2.2	0.3
160.85	0.81	1.9	149.42	27AU	2.6	-3.5	-2.2	0.2
167.57	0.73	1.8	159.28	6SE	0.1	-3.4	-2.3	0.2
174.36	0.65	1.8	169.20	16SE	-0.8	-3.4	-2.3	0.1
181.23	0.56	1.7	179.18	26SE	-0.9	-3.4	-2.4	0.1
188.18	0.47	1.7	189.21	6OC	-0.9	-3.4	-2.4	0.0
195.21	0.38	1.7	199.30	16OC	-0.7	-3.4	-2.4	-0.1
202.32	0.28	1.7	209.43	26OC	-0.5	-3.4	-2.4	-0.1
209.51	0.18	1.7	219.59	5NO	-0.5	-3.4	-2.4	-0.2
216.77	0.08	1.6	229.78	15NO	-0.5	-3.4	-2.3	-0.3
224.12	-0.03	1.6	239.99	25NO	-0.0	-3.4	-2.3	-0.3
231.53	-0.14	1.6	250.20	5DE	2.2	-3.4	-2.2	-0.2
239.01	-0.26	1.6	260.40	15DE	1.3	-3.4	-2.1	-0.2
246.55	-0.37	1.6	270.59	25DE	0.3	-3.4	-2.1	-0.1
				−265				
254.14	-0.49	1.5	280.74	4JA	0.1	-3.4	-2.0	-0.0
261.79	-0.61	1.5	290.87	14JA	-0.0	-3.5	-1.9	0.0
269.47	-0.72	1.5	300.94	24JA	-0.2	-3.5	-1.8	0.1
277.18	-0.83	1.4	310.97	3FE	-0.7	-3.5	-1.8	0.2
284.92	-0.94	1.4	320.94	13FE	-1.3	-3.4	-1.7	0.2
292.66	-1.04	1.4	330.85	23FE	-1.4	-3.4	-1.7	0.2
300.40	-1.14	1.3	340.70	5MR	-0.6	-3.4	-1.6	0.3
308.13	-1.22	1.3	350.49	15MR	0.7	-3.4	-1.6	0.3
315.82	-1.30	1.3	0.23	25MR	2.3	-3.3	-1.5	0.3
323.47	-1.36	1.2	9.91	4AP	3.5	-3.3	-1.5	0.3
331.07	-1.41	1.2	19.55	14AP	2.0	-3.3	-1.5	0.3
338.60	-1.45	1.2	29.15	24AP	1.1	-3.3	-1.5	0.3
346.05	-1.47	1.2	38.71	4MY	0.5	-3.3	-1.5	0.3
353.40	-1.48	1.1	48.25	14MY	-0.3	-3.4	-1.5	0.2
0.64	-1.47	1.1	57.78	24MY	-1.3	-3.4	-1.5	0.2
7.76	-1.45	1.1	67.29	3JN	-1.6	-3.4	-1.5	0.2
14.73	-1.41	1.0	76.82	13JN	-0.7	-3.4	-1.5	0.2
21.55	-1.36	1.0	86.35	23JN	-0.1	-3.5	-1.6	0.3
28.20	-1.29	0.9	95.90	3JL	0.4	-3.5	-1.6	0.3
34.64	-1.20	0.9	105.49	13JL	0.7	-3.6	-1.6	0.3
40.87	-1.10	0.8	115.11	23JL	1.2	-3.7	-1.7	0.4
46.83	-0.98	0.7	124.78	2AU	2.4	-3.7	-1.7	0.4
52.50	-0.84	0.6	134.50	12AU	2.2	-3.8	-1.8	0.4
57.82	-0.68	0.5	144.27	22AU	0.0	-3.9	-1.9	0.3
62.72	-0.49	0.4	154.10	1SE	-0.9	-4.0	-1.9	0.3
67.11	-0.28	0.3	163.99	11SE	-1.1	-4.2	-2.0	0.3
70.88	-0.03	0.1	173.94	21SE	-1.0	-4.3	-2.0	0.2
73.89	0.26	-0.1	183.94	1OC	-0.6	-4.3	-2.1	0.2
75.95	0.60	-0.3	194.01	11OC	-0.4	-4.4	-2.2	0.1
76.88	0.98	-0.5	204.11	21OC	-0.3	-4.2	-2.2	0.1
76.49	1.41	-0.7	214.26	31OC	-0.3	-3.7	-2.3	-0.0
74.68	1.86	-0.9	224.44	10NO	0.2	-3.2	-2.3	-0.1
71.63	2.30	-1.1	234.63	20NO	2.5	-3.7	-2.3	-0.2
67.79	2.68	-1.2	244.84	30NO	1.0	-4.2	-2.3	-0.2
63.88	2.94	-1.1	255.05	10DE	0.1	-4.4	-2.2	-0.3
60.68	3.06	-0.9	265.25	20DE	-0.1	-4.4	-2.2	-0.2
58.65	3.08	-0.6	275.42	30DE	-0.2	-4.3	-2.1	-0.1

Right table — year −264 / −263

δ LONG	LAT	MAG	☉ LONG	16.00UT	☿	♀	♃	♄
				−264		MAGNITUDES		
57.94	3.01	-0.3	285.57	9JA	-0.3	-4.2	-2.1	-0.1
58.50	2.91	-0.0	295.67	19JA	-0.7	-4.1	-2.0	0.0
60.12	2.78	0.2	305.72	29JA	-1.3	-4.0	-1.9	0.1
62.62	2.64	0.5	315.72	8FE	-1.3	-3.9	-1.9	0.1
65.83	2.51	0.7	325.66	18FE	-0.6	-3.8	-1.8	0.2
69.58	2.38	0.8	335.54	28FE	0.9	-3.7	-1.7	0.2
73.79	2.25	1.0	345.37	9MR	2.8	-3.6	-1.7	0.3
78.34	2.13	1.2	355.13	19MR	2.6	-3.5	-1.6	0.3
83.18	2.02	1.3	4.84	29MR	1.5	-3.5	-1.6	0.3
88.26	1.90	1.4	14.51	8AP	0.9	-3.4	-1.5	0.4
93.54	1.79	1.5	24.12	18AP	0.3	-3.4	-1.5	0.4
98.99	1.69	1.6	33.71	28AP	-0.4	-3.3	-1.4	0.4
104.59	1.59	1.6	43.26	8MY	-1.3	-3.3	-1.4	0.4
110.32	1.48	1.7	52.79	18MY	-1.7	-3.3	-1.4	0.3
116.17	1.38	1.7	62.31	28MY	-0.7	-3.3	-1.4	0.3
122.13	1.28	1.8	71.83	7JN	0.0	-3.3	-1.4	0.3
128.20	1.18	1.8	81.35	17JN	0.5	-3.3	-1.4	0.3
134.37	1.09	1.8	90.90	27JN	0.9	-3.3	-1.4	0.3
140.64	0.98	1.8	100.47	7JL	1.6	-3.4	-1.4	0.4
147.01	0.88	1.8	110.06	17JL	2.9	-3.4	-1.4	0.4
153.48	0.78	1.8	119.71	27JL	1.9	-3.4	-1.5	0.4
160.04	0.68	1.8	129.40	6AU	-0.0	-3.5	-1.5	0.4
166.71	0.58	1.8	139.14	16AU	-1.0	-3.5	-1.5	0.5
173.47	0.47	1.8	148.94	26AU	-1.2	-3.5	-1.6	0.5
180.34	0.37	1.8	158.80	5SE	-0.9	-3.5	-1.6	0.4
187.30	0.26	1.7	168.71	15SE	-0.5	-3.4	-1.7	0.4
194.36	0.15	1.7	178.69	25SE	-0.3	-3.4	-1.7	0.4
201.52	0.04	1.6	188.72	5OC	-0.2	-3.4	-1.8	0.4
208.77	-0.07	1.6	198.80	15OC	-0.1	-3.4	-1.9	0.3
216.11	-0.17	1.6	208.93	25OC	0.4	-3.3	-1.9	0.2
223.55	-0.28	1.5	219.09	4NO	2.8	-3.3	-2.0	0.2
231.06	-0.38	1.5	229.28	14NO	0.8	-3.3	-2.1	0.1
238.65	-0.49	1.4	239.49	24NO	-0.1	-3.3	-2.1	0.0
246.31	-0.59	1.4	249.70	4DE	-0.2	-3.3	-2.1	-0.0
254.03	-0.68	1.4	259.90	14DE	-0.3	-3.4	-2.2	-0.1
261.80	-0.77	1.4	270.09	24DE	-0.4	-3.4	-2.2	-0.1
				−263				
269.61	-0.85	1.4	280.25	3JA	-0.8	-3.4	-2.1	-0.1
277.44	-0.93	1.4	290.37	13JA	-1.2	-3.4	-2.1	-0.0
285.30	-0.99	1.4	300.46	23JA	-1.2	-3.5	-2.1	0.0
293.16	-1.05	1.4	310.48	2FE	-0.6	-3.5	-2.0	0.1
301.01	-1.10	1.4	320.45	12FE	1.1	-3.6	-2.0	0.2
308.84	-1.13	1.4	330.37	22FE	3.2	-3.6	-1.9	0.2
316.64	-1.15	1.4	340.23	4MR	2.0	-3.7	-1.8	0.3
324.39	-1.16	1.4	350.02	14MR	1.1	-3.8	-1.8	0.3
332.09	-1.16	1.4	359.76	24MR	0.6	-3.9	-1.7	0.4
339.72	-1.15	1.4	9.45	3AP	0.2	-3.9	-1.6	0.4
347.27	-1.12	1.4	19.09	13AP	-0.4	-4.0	-1.6	0.4
354.75	-1.08	1.4	28.69	23AP	-1.3	-4.1	-1.5	0.5
2.13	-1.03	1.4	38.25	3MY	-1.7	-4.2	-1.5	0.5
9.41	-0.97	1.4	47.79	13MY	-0.7	-4.2	-1.4	0.5
16.59	-0.90	1.4	57.32	23MY	0.1	-4.0	-1.4	0.5
23.66	-0.81	1.4	66.84	2JN	0.7	-3.7	-1.3	0.5
30.62	-0.72	1.4	76.35	12JN	1.2	-2.9	-1.3	0.4
37.46	-0.62	1.4	85.89	22JN	2.1	-3.4	-1.3	0.4
44.16	-0.51	1.4	95.44	2JL	3.3	-3.9	-1.3	0.4
50.73	-0.39	1.4	105.02	12JL	1.6	-4.2	-1.3	0.5
57.16	-0.26	1.4	114.65	22JL	-0.1	-4.2	-1.3	0.5
63.42	-0.12	1.3	124.31	1AU	-1.1	-4.2	-1.3	0.5
69.52	0.03	1.3	134.02	11AU	-1.3	-4.1	-1.3	0.6
75.42	0.19	1.2	143.79	21AU	-0.9	-4.0	-1.3	0.6
81.11	0.37	1.1	153.62	31AU	-0.4	-3.9	-1.4	0.6
86.54	0.56	1.1	163.51	10SE	-0.2	-3.9	-1.4	0.6
91.68	0.77	1.0	173.46	20SE	-0.0	-3.8	-1.4	0.6
96.46	1.00	0.8	183.46	30SE	0.1	-3.7	-1.5	0.6
100.82	1.26	0.7	193.51	10OC	0.6	-3.7	-1.6	0.6
104.66	1.54	0.6	203.62	20OC	3.1	-3.6	-1.6	0.5
107.85	1.86	0.4	213.76	30OC	0.6	-3.5	-1.7	0.5
110.25	2.22	0.2	223.94	9NO	-0.3	-3.5	-1.7	0.4
111.67	2.62	-0.0	234.14	19NO	-0.4	-3.5	-1.8	0.3
111.94	3.05	-0.3	244.35	29NO	-0.4	-3.4	-1.9	0.3
110.90	3.49	-0.5	254.56	9DE	-0.5	-3.4	-1.9	0.2
108.53	3.89	-0.8	264.76	19DE	-0.8	-3.4	-2.0	0.1
105.10	4.20	-0.9	274.93	29DE	-1.0	-3.4	-2.0	0.1

♂			⊙	16.00UT	☿	♀	♃	♄
LONG	LAT	MAG	LONG	-262	MAGNITUDES			
101.14	4.36	-1.0	285.07	8JA	-1.0	-3.3	-2.0	0.0
97.40	4.34	-0.8	295.18	18JA	-0.5	-3.3	-2.1	0.1
94.53	4.18	-0.6	305.23	28JA	1.3	-3.3	-2.1	0.1
92.88	3.93	-0.4	315.24	7FE	3.0	-3.3	-2.0	0.2
92.53	3.63	-0.1	325.18	17FE	1.4	-3.3	-2.0	0.3
93.38	3.32	0.1	335.07	27FE	0.8	-3.4	-2.0	0.3
95.23	3.02	0.3	344.89	9MR	0.5	-3.4	-1.9	0.4
97.92	2.74	0.5	354.66	19MR	0.1	-3.4	-1.9	0.4
101.28	2.48	0.7	4.37	29MR	-0.5	-3.5	-1.8	0.5
105.18	2.25	0.9	14.04	8AP	-1.3	-3.5	-1.7	0.5
109.53	2.03	1.0	23.66	18AP	-1.6	-3.5	-1.7	0.6
114.23	1.82	1.1	33.24	28AP	-0.7	-3.4	-1.6	0.6
119.24	1.63	1.2	42.79	8MY	0.2	-3.4	-1.5	0.6
124.52	1.46	1.3	52.32	18MY	0.9	-3.4	-1.5	0.6
130.02	1.29	1.4	61.84	28MY	1.6	-3.4	-1.4	0.6
135.72	1.12	1.4	71.36	7JN	2.8	-3.3	-1.4	0.6
141.60	0.97	1.5	80.89	17JN	2.9	-3.3	-1.3	0.6
147.64	0.82	1.5	90.42	27JN	1.3	-3.3	-1.3	0.6
153.84	0.68	1.6	99.99	7JL	-0.1	-3.3	-1.3	0.6
160.19	0.54	1.6	109.59	17JL	-1.2	-3.3	-1.2	0.6
166.68	0.40	1.6	119.23	27JL	-1.5	-3.4	-1.2	0.7
173.31	0.27	1.6	128.92	6AU	-0.8	-3.4	-1.2	0.7
180.06	0.14	1.6	138.66	16AU	-0.3	-3.4	-1.2	0.8
186.94	0.01	1.6	148.46	26AU	-0.1	-3.5	-1.2	0.8
193.95	-0.11	1.6	158.32	5SE	0.1	-3.5	-1.2	0.8
201.07	-0.23	1.6	168.23	15SE	0.3	-3.6	-1.3	0.8
208.31	-0.34	1.6	178.20	25SE	0.9	-3.6	-1.3	0.8
215.65	-0.45	1.5	188.24	5OC	3.4	-3.7	-1.3	0.8
223.10	-0.55	1.5	198.32	15OC	0.4	-3.8	-1.4	0.8
230.64	-0.65	1.5	208.44	25OC	-0.4	-3.9	-1.4	0.8
238.26	-0.73	1.5	218.60	4NO	-0.5	-4.0	-1.5	0.7
245.97	-0.81	1.4	228.79	14NO	-0.5	-4.1	-1.5	0.7
253.73	-0.89	1.4	238.99	24NO	-0.6	-4.2	-1.6	0.6
261.55	-0.95	1.4	249.21	4DE	-0.8	-4.3	-1.7	0.5
269.41	-1.00	1.4	259.41	14DE	-0.9	-4.4	-1.7	0.5
277.30	-1.04	1.4	269.60	24DE	-0.9	-4.4	-1.8	0.4
				-261				
285.20	-1.07	1.3	279.76	3JA	-0.4	-4.2	-1.8	0.3
293.10	-1.08	1.3	289.89	13JA	1.5	-3.7	-1.9	0.3
300.98	-1.09	1.3	299.96	23JA	2.5	-3.3	-2.0	0.3
308.85	-1.08	1.3	310.00	2FE	1.0	-3.8	-2.0	0.3
316.67	-1.07	1.3	319.97	12FE	0.6	-4.2	-2.0	0.4
324.44	-1.04	1.4	329.89	22FE	0.3	-4.3	-2.0	0.4
332.15	-1.00	1.4	339.75	4MR	-0.0	-4.3	-2.0	0.5
339.80	-0.95	1.4	349.55	14MR	-0.5	-4.2	-2.0	0.5
347.36	-0.89	1.5	359.29	24MR	-1.3	-4.1	-2.0	0.6
354.85	-0.82	1.5	8.98	3AP	-1.6	-4.0	-1.9	0.6
2.25	-0.75	1.5	18.62	13AP	-0.7	-3.9	-1.9	0.7
9.57	-0.66	1.6	28.22	23AP	0.3	-3.8	-1.8	0.7
16.79	-0.57	1.6	37.79	3MY	1.2	-3.7	-1.7	0.8
23.91	-0.48	1.6	47.33	13MY	2.2	-3.6	-1.7	0.8
30.95	-0.38	1.7	56.85	23MY	3.5	-3.5	-1.6	0.8
37.88	-0.28	1.7	66.38	2JN	2.3	-3.5	-1.5	0.8
44.72	-0.17	1.7	75.89	12JN	1.0	-3.4	-1.5	0.9
51.47	-0.06	1.7	85.42	22JN	-0.2	-3.4	-1.4	0.9
58.12	0.06	1.7	94.98	2JL	-1.2	-3.4	-1.4	0.9
64.67	0.18	1.7	104.56	12JL	-1.6	-3.4	-1.3	0.9
71.12	0.30	1.7	114.18	22JL	-0.8	-3.3	-1.3	0.9
77.47	0.43	1.7	123.84	1AU	-0.3	-3.3	-1.3	0.9
83.71	0.56	1.7	133.55	11AU	0.1	-3.3	-1.2	0.9
89.84	0.69	1.6	143.32	21AU	0.2	-3.3	-1.2	1.0
95.84	0.84	1.6	153.15	31AU	0.5	-3.4	-1.2	1.0
101.70	0.98	1.5	163.03	10SE	1.2	-3.4	-1.2	1.0
107.41	1.14	1.5	172.97	20SE	3.2	-3.4	-1.2	1.1
112.94	1.31	1.4	182.97	30SE	0.3	-3.4	-1.2	1.1
118.25	1.49	1.3	193.03	10OC	-0.6	-3.4	-1.2	1.1
123.33	1.68	1.2	203.13	20OC	-0.7	-3.5	-1.3	1.1
128.10	1.89	1.1	213.27	30OC	-0.7	-3.5	-1.3	1.1
132.51	2.11	0.9	223.45	9NO	-0.7	-3.5	-1.3	1.0
136.49	2.36	0.8	233.64	19NO	-0.7	-3.5	-1.4	1.0
139.91	2.64	0.6	243.85	29NO	-0.7	-3.4	-1.4	0.9
142.65	2.94	0.4	254.06	9DE	-0.7	-3.4	-1.5	0.9
144.55	3.27	0.1	264.26	19DE	-0.3	-3.4	-1.6	0.8
145.42	3.62	-0.1	274.44	29DE	1.7	-3.4	-1.6	0.7

♂			⊙	16.00UT	☿	♀	♃	♄
LONG	LAT	MAG	LONG	-260	MAGNITUDES			
145.09	3.97	-0.4	284.58	8JA	2.0	-3.3	-1.7	0.7
143.48	4.27	-0.7	294.69	18JA	0.7	-3.3	-1.8	0.6
140.64	4.49	-0.9	304.75	28JA	0.4	-3.3	-1.8	0.5
136.97	4.56	-1.1	314.75	7FE	0.2	-3.3	-1.9	0.5
133.09	4.45	-1.0	324.70	17FE	-0.1	-3.3	-1.9	0.6
129.72	4.17	-0.8	334.59	27FE	-0.6	-3.4	-2.0	0.6
127.41	3.79	-0.6	344.42	8MR	-1.3	-3.4	-2.0	0.7
126.37	3.36	-0.4	354.19	18MR	-1.6	-3.4	-2.0	0.7
126.62	2.93	-0.2	3.91	28MR	-0.7	-3.4	-2.0	0.8
128.01	2.53	0.0	13.57	7AP	0.4	-3.5	-2.0	0.8
130.35	2.15	0.2	23.19	17AP	1.5	-3.5	-2.0	0.9
133.50	1.81	0.4	32.78	27AP	2.9	-3.6	-2.0	0.9
137.29	1.50	0.5	42.33	7MY	3.2	-3.6	-1.9	1.0
141.62	1.22	0.7	51.87	17MY	1.8	-3.7	-1.9	1.0
146.39	0.97	0.8	61.39	27MY	0.8	-3.8	-1.8	1.1
151.54	0.74	0.9	70.90	6JN	-0.2	-3.9	-1.7	1.1
157.00	0.52	1.0	80.43	16JN	-1.3	-4.0	-1.7	1.1
162.75	0.32	1.0	89.97	26JN	-1.6	-4.1	-1.6	1.1
168.75	0.14	1.1	99.53	6JL	-0.8	-4.2	-1.6	1.1
174.97	-0.04	1.1	109.13	16JL	-0.2	-4.2	-1.5	1.1
181.40	-0.20	1.2	118.77	26JL	0.2	-4.2	-1.4	1.2
188.02	-0.34	1.2	128.45	5AU	0.4	-4.1	-1.4	1.1
194.81	-0.48	1.2	138.19	15AU	0.7	-3.7	-1.4	1.2
201.76	-0.61	1.3	147.98	25AU	1.6	-3.3	-1.3	1.2
208.86	-0.72	1.3	157.83	4SE	2.9	-3.8	-1.3	1.3
216.09	-0.82	1.3	167.75	14SE	0.2	-4.2	-1.3	1.3
223.46	-0.92	1.3	177.72	24SE	-0.7	-4.3	-1.3	1.3
230.93	-0.99	1.3	187.75	4OC	-0.8	-4.3	-1.3	1.4
238.51	-1.06	1.3	197.83	14OC	-0.8	-4.3	-1.3	1.4
246.17	-1.11	1.3	207.95	24OC	-0.8	-4.2	-1.3	1.4
253.91	-1.15	1.3	218.11	3NO	-0.6	-4.1	-1.3	1.3
261.70	-1.18	1.4	228.30	13NO	-0.6	-4.0	-1.3	1.3
269.54	-1.19	1.4	238.50	23NO	-0.6	-3.9	-1.3	1.2
277.40	-1.19	1.4	248.71	3DE	-0.1	-3.8	-1.4	1.2
285.28	-1.18	1.4	258.92	13DE	1.9	-3.7	-1.4	1.2
293.16	-1.15	1.4	269.11	23DE	1.6	-3.7	-1.5	1.1
				-259				
301.01	-1.12	1.4	279.27	2JA	0.5	-3.6	-1.5	1.1
308.84	-1.07	1.4	289.40	12JA	0.2	-3.5	-1.6	1.0
316.62	-1.01	1.4	299.48	22JA	0.0	-3.5	-1.6	1.0
324.35	-0.95	1.5	309.52	1FE	-0.2	-3.4	-1.7	0.9
332.02	-0.87	1.5	319.50	11FE	-0.6	-3.4	-1.8	0.8
339.62	-0.79	1.5	329.42	21FE	-1.3	-3.4	-1.8	0.8
347.14	-0.71	1.5	339.28	3MR	-1.5	-3.3	-1.9	0.9
354.58	-0.62	1.5	349.08	13MR	-0.7	-3.3	-2.0	0.9
1.94	-0.52	1.5	358.82	23MR	0.6	-3.3	-2.0	1.0
9.21	-0.42	1.5	8.51	2AP	1.9	-3.3	-2.1	1.0
16.39	-0.32	1.6	18.16	12AP	3.7	-3.3	-2.1	1.1
23.49	-0.22	1.6	27.76	22AP	2.4	-3.3	-2.1	1.1
30.51	-0.12	1.7	37.33	2MY	1.4	-3.3	-2.1	1.2
37.44	-0.02	1.7	46.87	12MY	0.6	-3.4	-2.1	1.2
44.29	0.08	1.8	56.40	22MY	-0.3	-3.4	-2.1	1.3
51.07	0.19	1.8	65.91	1JN	-1.3	-3.5	-2.0	1.3
57.77	0.29	1.8	75.44	11JN	-1.7	-3.5	-2.0	1.4
64.39	0.39	1.9	84.96	21JN	-0.8	-3.5	-1.9	1.4
70.96	0.49	1.9	94.51	1JL	-0.1	-3.4	-1.9	1.4
77.45	0.59	1.9	104.09	11JL	0.3	-3.4	-1.8	1.4
83.88	0.69	1.9	113.71	21JL	0.6	-3.4	-1.7	1.4
90.24	0.79	1.9	123.37	31JL	1.0	-3.4	-1.7	1.4
96.54	0.89	1.9	133.08	10AU	2.0	-3.3	-1.6	1.3
102.77	0.99	1.9	142.84	20AU	2.5	-3.3	-1.6	1.3
108.94	1.09	1.9	152.66	30AU	0.1	-3.3	-1.5	1.2
115.03	1.19	1.8	162.54	9SE	-0.8	-3.3	-1.5	1.3
121.05	1.30	1.8	172.48	19SE	-1.0	-3.3	-1.4	1.3
126.97	1.40	1.7	182.48	29SE	-0.9	-3.4	-1.4	1.2
132.80	1.51	1.7	192.53	9OC	-0.7	-3.4	-1.4	1.2
138.52	1.63	1.6	202.63	19OC	-0.5	-3.4	-1.3	1.2
144.10	1.75	1.5	212.77	29OC	-0.4	-3.4	-1.3	1.2
149.52	1.87	1.4	222.95	8NO	-0.4	-3.5	-1.3	1.2
154.76	2.00	1.3	233.14	18NO	0.0	-3.5	-1.3	1.1
159.77	2.14	1.1	243.35	28NO	2.2	-3.6	-1.3	1.1
164.50	2.28	1.0	253.57	8DE	1.3	-3.7	-1.4	1.1
168.90	2.43	0.8	263.76	18DE	0.2	-3.7	-1.4	1.0
172.87	2.60	0.6	273.95	28DE	0.0	-3.8	-1.4	1.0

♂			☉	16.00UT	☿	♀	♃	♄
LONG	LAT	MAG	LONG	-258	MAGNITUDES			
176.30	2.77	0.4	284.09	7JA	-0.1	-3.9	-1.5	0.9
179.09	2.94	0.2	294.20	17JA	-0.3	-4.0	-1.5	0.9
181.05	3.11	-0.1	304.26	27JA	-0.7	-4.1	-1.6	0.8
182.01	3.27	-0.4	314.27	6FE	-1.3	-4.2	-1.6	0.8
181.81	3.39	-0.7	324.22	16FE	-1.4	-4.2	-1.7	0.8
180.34	3.43	-1.0	334.12	26FE	-0.7	-4.3	-1.8	0.7
177.66	3.37	-1.2	343.95	8MR	0.7	-4.2	-1.8	0.8
174.17	3.16	-1.5	353.72	18MR	2.4	-4.0	-1.9	0.9
170.47	2.82	-1.4	3.44	28MR	3.2	-3.4	-2.0	0.9
167.30	2.36	-1.2	13.11	7AP	1.8	-3.1	-2.0	1.0
165.21	1.86	-1.0	22.73	17AP	1.0	-3.8	-2.1	1.1
164.43	1.37	-0.8	32.32	27AP	0.4	-4.1	-2.1	1.1
164.99	0.92	-0.6	41.87	7MY	-0.3	-4.2	-2.2	1.2
166.72	0.51	-0.4	51.40	17MY	-1.3	-4.2	-2.2	1.2
169.46	0.15	-0.2	60.93	27MY	-1.7	-4.1	-2.2	1.3
173.04	-0.16	-0.1	70.44	6JN	-0.8	-4.0	-2.2	1.3
177.30	-0.42	0.1	79.96	16JN	-0.1	-3.9	-2.2	1.3
182.14	-0.66	0.2	89.51	26JN	0.4	-3.8	-2.2	1.3
187.45	-0.86	0.3	99.07	6JL	0.8	-3.7	-2.1	1.3
193.17	-1.03	0.4	108.66	16JL	1.3	-3.7	-2.1	1.3
199.22	-1.18	0.5	118.30	26JL	2.6	-3.6	-2.0	1.3
205.58	-1.30	0.6	127.98	5AU	2.2	-3.6	-2.0	1.2
212.19	-1.40	0.7	137.71	15AU	0.1	-3.5	-1.9	1.2
219.02	-1.48	0.7	147.51	25AU	-1.0	-3.5	-1.8	1.1
226.05	-1.54	0.8	157.36	4SE	-1.1	-3.4	-1.8	1.1
233.24	-1.57	0.8	167.26	14SE	-1.0	-3.4	-1.7	1.1
240.58	-1.59	0.9	177.23	24SE	-0.6	-3.4	-1.7	1.1
248.04	-1.59	1.0	187.26	4OC	-0.3	-3.4	-1.6	1.1
255.59	-1.58	1.0	197.33	14OC	-0.3	-3.4	-1.6	1.1
263.22	-1.54	1.1	207.46	24OC	-0.2	-3.4	-1.5	1.1
270.90	-1.50	1.1	217.61	3NO	0.2	-3.4	-1.5	1.1
278.62	-1.44	1.2	227.80	13NO	2.5	-3.4	-1.5	1.1
286.35	-1.36	1.2	238.00	23NO	1.0	-3.4	-1.4	1.0
294.09	-1.28	1.3	248.21	3DE	0.0	-3.4	-1.4	1.0
301.81	-1.19	1.3	258.42	13DE	-0.1	-3.4	-1.4	1.0
309.51	-1.09	1.4	268.61	23DE	-0.2	-3.4	-1.4	0.9
				-257				
317.16	-0.98	1.4	278.78	2JA	-0.4	-3.4	-1.5	0.9
324.75	-0.87	1.5	288.91	12JA	-0.7	-3.5	-1.5	0.8
332.29	-0.76	1.5	298.99	22JA	-1.2	-3.5	-1.5	0.8
339.76	-0.65	1.5	309.03	1FE	-1.3	-3.5	-1.5	0.7
347.16	-0.54	1.6	319.01	11FE	-0.6	-3.4	-1.6	0.7
354.48	-0.42	1.6	328.94	21FE	0.9	-3.4	-1.6	0.6
1.73	-0.31	1.7	338.80	3MR	3.0	-3.4	-1.7	0.6
8.89	-0.20	1.7	348.60	13MR	2.4	-3.4	-1.7	0.6
15.97	-0.09	1.7	358.35	23MR	1.3	-3.3	-1.8	0.7
22.98	0.01	1.7	8.04	2AP	0.8	-3.3	-1.9	0.7
29.91	0.12	1.8	17.69	12AP	0.3	-3.3	-1.9	0.8
36.77	0.21	1.8	27.30	22AP	-0.4	-3.3	-2.0	0.9
43.56	0.31	1.8	36.86	2MY	-1.3	-3.3	-2.1	0.9
50.29	0.40	1.8	46.41	12MY	-1.7	-3.4	-2.1	1.0
56.96	0.49	1.8	55.94	22MY	-0.8	-3.4	-2.2	1.0
63.57	0.58	1.8	65.45	1JN	0.0	-3.4	-2.3	1.1
70.13	0.66	1.9	74.97	11JN	0.6	-3.4	-2.3	1.1
76.64	0.74	1.9	84.50	21JN	1.0	-3.5	-2.3	1.1
83.11	0.81	1.9	94.05	1JL	1.7	-3.5	-2.4	1.1
89.54	0.88	2.0	103.63	11JL	3.1	-3.6	-2.4	1.1
95.94	0.95	2.0	113.24	21JL	1.8	-3.7	-2.3	1.1
102.30	1.02	2.0	122.90	31JL	-0.0	-3.7	-2.3	1.1
108.64	1.08	2.0	132.61	10AU	-1.1	-3.8	-2.3	1.1
114.94	1.15	2.0	142.37	20AU	-1.3	-3.9	-2.2	1.1
121.22	1.20	2.0	152.18	30AU	-0.9	-4.0	-2.1	1.0
127.48	1.26	2.0	162.06	9SE	-0.5	-4.2	-2.1	1.0
133.70	1.32	2.0	172.00	19SE	-0.2	-4.3	-2.0	0.9
139.90	1.37	1.9	181.99	29SE	-0.1	-4.3	-1.9	1.0
146.06	1.42	1.9	192.04	9OC	-0.0	-4.3	-1.9	1.0
152.18	1.47	1.8	202.14	19OC	0.4	-4.2	-1.8	1.0
158.27	1.51	1.8	212.28	29OC	2.8	-3.7	-1.8	1.0
164.30	1.55	1.7	222.45	8NO	0.8	-3.0	-1.7	1.0
170.27	1.59	1.6	232.65	18NO	-0.2	-3.8	-1.7	1.0
176.17	1.63	1.5	242.85	28NO	-0.3	-4.2	-1.6	0.9
181.98	1.66	1.4	253.07	8DE	-0.3	-4.4	-1.6	0.9
187.68	1.69	1.3	263.27	18DE	-0.5	-4.4	-1.6	0.9
193.26	1.71	1.1	273.44	28DE	-0.8	-4.3	-1.5	0.9

♂			☉	16.00UT	☿	♀	♃	♄
LONG	LAT	MAG	LONG	-256	MAGNITUDES			
198.68	1.72	1.0	283.60	7JA	-1.1	-4.2	-1.5	0.8
203.90	1.71	0.8	293.71	17JA	-1.1	-4.1	-1.5	0.8
208.88	1.70	0.6	303.77	27JA	-0.6	-4.0	-1.5	0.7
213.57	1.66	0.4	313.79	6FE	1.1	-3.9	-1.5	0.7
217.87	1.60	0.2	323.74	16FE	3.2	-3.8	-1.6	0.6
221.71	1.50	-0.1	333.63	26FE	1.8	-3.7	-1.6	0.6
224.96	1.36	-0.3	343.47	7MR	1.0	-3.6	-1.6	0.5
227.46	1.17	-0.6	353.25	17MR	0.6	-3.5	-1.7	0.5
229.04	0.90	-0.9	2.97	27MR	0.2	-3.5	-1.7	0.5
229.51	0.54	-1.3	12.65	6AP	-0.4	-3.4	-1.8	0.6
228.75	0.09	-1.6	22.27	16AP	-1.3	-3.4	-1.8	0.6
226.80	-0.45	-1.9	31.86	26AP	-1.7	-3.3	-1.9	0.7
223.92	-1.06	-2.1	41.42	6MY	-0.8	-3.3	-1.9	0.7
220.72	-1.65	-2.1	50.95	16MY	0.1	-3.3	-2.0	0.8
217.95	-2.18	-2.0	60.47	26MY	0.7	-3.3	-2.1	0.9
216.18	-2.59	-1.8	69.99	5JN	1.4	-3.3	-2.2	0.9
215.77	-2.88	-1.6	79.51	15JN	2.3	-3.3	-2.2	1.0
216.73	-3.07	-1.3	89.05	25JN	3.2	-3.3	-2.3	1.0
218.93	-3.16	-1.1	98.61	5JL	1.5	-3.4	-2.3	1.0
222.20	-3.19	-0.9	108.20	15JL	-0.1	-3.4	-2.4	1.0
226.34	-3.17	-0.7	117.83	25JL	-1.1	-3.4	-2.4	1.0
231.19	-3.11	-0.5	127.52	4AU	-1.4	-3.5	-2.4	1.0
236.60	-3.02	-0.4	137.24	14AU	-0.9	-3.5	-2.4	1.0
242.47	-2.91	-0.2	147.03	24AU	-0.4	-3.5	-2.4	1.0
248.70	-2.77	-0.1	156.88	3SE	-0.1	-3.5	-2.4	1.0
255.23	-2.62	0.1	166.78	13SE	0.0	-3.4	-2.3	0.9
261.98	-2.45	0.2	176.75	23SE	0.1	-3.4	-2.3	0.9
268.91	-2.27	0.3	186.77	3OC	0.7	-3.4	-2.2	0.8
275.98	-2.09	0.5	196.84	13OC	3.1	-3.4	-2.2	0.9
283.14	-1.90	0.6	206.96	23OC	0.6	-3.3	-2.1	0.9
290.37	-1.71	0.7	217.11	2NO	-0.3	-3.3	-2.0	0.9
297.64	-1.52	0.8	227.30	12NO	-0.4	-3.3	-2.0	0.9
304.93	-1.33	0.9	237.50	22NO	-0.4	-3.3	-1.9	0.9
312.23	-1.15	1.0	247.71	2DE	-0.6	-3.3	-1.8	0.9
319.50	-0.98	1.1	257.91	12DE	-0.8	-3.4	-1.8	0.9
326.75	-0.81	1.2	268.11	22DE	-1.0	-3.4	-1.7	0.8
				-255				
333.96	-0.64	1.3	278.28	1JA	-1.0	-3.4	-1.7	0.8
341.13	-0.49	1.4	288.41	11JA	-0.5	-3.4	-1.6	0.8
348.24	-0.34	1.5	298.50	21JA	1.3	-3.5	-1.6	0.7
355.29	-0.20	1.5	308.54	31JA	2.9	-3.5	-1.6	0.7
2.29	-0.07	1.6	318.52	10FE	1.3	-3.6	-1.6	0.6
9.23	0.05	1.7	328.45	20FE	0.7	-3.6	-1.6	0.6
16.11	0.17	1.7	338.32	2MR	0.4	-3.7	-1.6	0.5
22.92	0.27	1.8	348.13	12MR	0.1	-3.8	-1.6	0.5
29.68	0.37	1.8	357.88	22MR	-0.5	-3.9	-1.6	0.4
36.39	0.47	1.9	7.58	1AP	-1.3	-3.9	-1.6	0.4
43.04	0.55	1.9	17.23	11AP	-1.7	-4.0	-1.6	0.4
49.64	0.63	1.9	26.84	21AP	-0.8	-4.1	-1.7	0.5
56.20	0.71	1.9	36.41	1MY	0.2	-4.2	-1.7	0.5
62.72	0.77	2.0	45.95	11MY	1.0	-4.2	-1.8	0.6
69.20	0.84	2.0	55.48	21MY	1.8	-4.0	-1.8	0.7
75.66	0.89	2.0	65.00	31MY	3.0	-3.6	-1.9	0.7
82.08	0.95	2.0	74.51	10JN	2.7	-2.8	-2.0	0.8
88.49	0.99	2.0	84.04	20JN	1.2	-3.4	-2.0	0.8
94.88	1.04	1.9	93.59	30JN	-0.1	-4.0	-2.1	0.9
101.26	1.08	1.9	103.16	10JL	-1.2	-4.2	-2.2	0.9
107.63	1.11	2.0	112.78	20JL	-1.5	-4.2	-2.2	0.9
114.00	1.14	2.0	122.43	30JL	-0.9	-4.2	-2.3	0.9
120.37	1.17	2.0	132.13	9AU	-0.3	-4.1	-2.3	0.9
126.74	1.19	2.0	141.89	19AU	-0.0	-4.0	-2.4	0.9
133.12	1.20	2.0	151.71	29AU	0.2	-3.9	-2.4	0.9
139.51	1.22	2.0	161.58	8SE	0.3	-3.9	-2.4	0.9
145.91	1.23	2.0	171.52	18SE	1.0	-3.8	-2.5	0.9
152.32	1.23	2.0	181.51	28SE	3.4	-3.7	-2.4	0.8
158.74	1.23	2.0	191.55	8OC	0.4	-3.6	-2.4	0.8
165.18	1.22	1.9	201.65	18OC	-0.5	-3.6	-2.4	0.8
171.63	1.21	1.9	211.79	28OC	-0.6	-3.5	-2.3	0.8
178.08	1.19	1.8	221.96	7NO	-0.6	-3.5	-2.2	0.8
184.55	1.16	1.8	232.15	17NO	-0.7	-3.5	-2.2	0.8
191.03	1.13	1.7	242.36	27NO	-0.8	-3.4	-2.1	0.8
197.51	1.09	1.6	252.57	7DE	-0.8	-3.4	-2.0	0.8
203.99	1.03	1.5	262.77	17DE	-0.8	-3.4	-2.0	0.8
210.46	0.96	1.4	272.95	27DE	-0.4	-3.4	-1.9	0.8

Left Table

♂ LONG	LAT	MAG	☉ LONG	16.00UT -254	☿	♀	♃	♄ MAGNITUDES
216.93	0.88	1.3	283.10	6JA	1.5	-3.3	-1.8	0.8
223.38	0.79	1.2	293.22	16JA	2.3	-3.3	-1.8	0.8
229.80	0.67	1.1	303.28	26JA	0.9	-3.3	-1.7	0.7
236.19	0.54	0.9	313.30	5FE	0.5	-3.3	-1.7	0.7
242.54	0.38	0.8	323.26	15FE	0.3	-3.3	-1.6	0.6
248.83	0.19	0.6	333.15	25FE	-0.0	-3.4	-1.6	0.6
255.03	-0.03	0.5	342.99	7MR	-0.5	-3.4	-1.6	0.5
261.13	-0.28	0.3	352.77	17MR	-1.3	-3.4	-1.6	0.5
267.09	-0.57	0.1	2.50	27MR	-1.6	-3.5	-1.5	0.4
272.86	-0.91	-0.1	12.17	6AP	-0.7	-3.5	-1.5	0.4
278.40	-1.30	-0.3	21.80	16AP	0.3	-3.5	-1.6	0.3
283.61	-1.74	-0.5	31.39	26AP	1.3	-3.4	-1.6	0.4
288.42	-2.24	-0.8	40.95	6MY	2.4	-3.4	-1.6	0.4
292.69	-2.81	-1.0	50.49	16MY	3.6	-3.4	-1.6	0.5
296.27	-3.45	-1.3	60.00	26MY	2.1	-3.4	-1.6	0.6
298.97	-4.15	-1.6	69.52	5JN	0.9	-3.3	-1.7	0.6
300.58	-4.89	-1.8	79.05	15JN	-0.2	-3.3	-1.7	0.7
300.94	-5.62	-2.1	88.58	25JN	-1.2	-3.3	-1.8	0.7
300.00	-6.27	-2.4	98.14	5JL	-1.6	-3.3	-1.8	0.8
297.95	-6.71	-2.6	107.73	15JL	-0.8	-3.3	-1.9	0.8
295.35	-6.86	-2.6	117.36	25JL	-0.3	-3.4	-2.0	0.8
292.89	-6.68	-2.5	127.04	4AU	0.1	-3.4	-2.0	0.9
291.27	-6.23	-2.2	136.77	14AU	0.3	-3.4	-2.1	0.9
290.88	-5.62	-1.9	146.55	24AU	0.5	-3.5	-2.2	0.9
291.79	-4.94	-1.6	156.40	3SE	1.3	-3.5	-2.2	0.9
293.90	-4.26	-1.3	166.30	13SE	3.2	-3.6	-2.3	0.9
297.01	-3.63	-1.0	176.26	23SE	0.3	-3.6	-2.3	0.8
300.92	-3.05	-0.8	186.28	3OC	-0.6	-3.7	-2.4	0.8
305.46	-2.53	-0.5	196.35	13OC	-0.7	-3.8	-2.4	0.8
310.49	-2.07	-0.3	206.47	23OC	-0.7	-3.9	-2.4	0.7
315.89	-1.66	-0.0	216.62	2NO	-0.8	-4.0	-2.4	0.7
321.57	-1.30	0.2	226.81	12NO	-0.7	-4.1	-2.3	0.8
327.47	-0.98	0.4	237.00	22NO	-0.7	-4.2	-2.3	0.8
333.52	-0.70	0.5	247.22	2DE	-0.7	-4.3	-2.2	0.8
339.70	-0.46	0.7	257.42	12DE	-0.2	-4.4	-2.2	0.8
345.96	-0.24	0.9	267.61	22DE	1.7	-4.4	-2.1	0.8

-253

♂ LONG	LAT	MAG	☉ LONG	16.00UT	☿	♀	♃	♄
352.28	-0.05	1.0	277.78	1JA	1.9	-4.2	-2.0	0.8
358.64	0.12	1.1	287.92	11JA	0.6	-3.7	-2.0	0.8
5.03	0.27	1.3	298.01	21JA	0.3	-3.3	-1.9	0.8
11.42	0.40	1.4	308.05	31JA	0.1	-3.8	-1.8	0.7
17.82	0.52	1.5	318.04	10FE	-0.1	-4.2	-1.8	0.7
24.22	0.62	1.6	327.97	20FE	-0.6	-4.3	-1.7	0.6
30.60	0.71	1.7	337.84	2MR	-1.3	-4.3	-1.6	0.6
36.98	0.79	1.7	347.65	12MR	-1.5	-4.2	-1.6	0.5
43.34	0.86	1.8	357.40	22MR	-0.7	-4.1	-1.6	0.5
49.69	0.92	1.9	7.10	1AP	0.4	-4.0	-1.5	0.4
56.02	0.98	1.9	16.76	11AP	1.6	-3.9	-1.5	0.4
62.34	1.02	1.9	26.36	21AP	3.2	-3.8	-1.5	0.3
68.65	1.06	2.0	35.94	1MY	2.9	-3.7	-1.5	0.3
74.96	1.09	2.0	45.49	11MY	1.6	-3.6	-1.5	0.4
81.26	1.12	2.0	55.01	21MY	0.7	-3.5	-1.5	0.4
87.56	1.14	2.0	64.54	31MY	-0.2	-3.5	-1.5	0.5
93.87	1.15	2.0	74.06	10JN	-1.2	-3.4	-1.5	0.5
100.18	1.16	2.0	83.58	20JN	-1.7	-3.4	-1.5	0.6
106.50	1.17	2.0	93.13	30JN	-0.8	-3.4	-1.5	0.7
112.84	1.17	2.0	102.71	10JL	-0.2	-3.4	-1.6	0.7
119.20	1.17	2.0	112.31	20JL	0.2	-3.3	-1.6	0.8
125.59	1.16	2.0	121.97	30JL	0.5	-3.3	-1.7	0.8
132.01	1.14	1.9	131.67	9AU	0.8	-3.3	-1.7	0.8
138.45	1.13	1.9	141.42	19AU	1.7	-3.3	-1.8	0.8
144.94	1.10	1.9	151.24	29AU	2.8	-3.4	-1.9	0.8
151.46	1.08	1.9	161.11	8SE	0.2	-3.4	-1.9	0.8
158.02	1.04	1.9	171.03	18SE	-0.8	-3.4	-2.0	0.8
164.63	1.00	1.9	181.03	28SE	-0.9	-3.4	-2.0	0.8
171.28	0.96	1.9	191.07	8OC	-0.9	-3.4	-2.1	0.8
177.98	0.91	1.9	201.16	18OC	-0.7	-3.5	-2.2	0.7
184.73	0.85	1.9	211.30	28OC	-0.5	-3.5	-2.2	0.7
191.52	0.78	1.8	221.46	7NO	-0.5	-3.5	-2.2	0.7
198.37	0.71	1.8	231.65	17NO	-0.5	-3.4	-2.3	0.7
205.27	0.63	1.7	241.86	27NO	-0.1	-3.4	-2.3	0.8
212.21	0.54	1.7	252.07	7DE	1.9	-3.4	-2.3	0.8
219.21	0.44	1.6	262.27	17DE	1.5	-3.4	-2.2	0.8
226.25	0.33	1.6	272.46	27DE	0.4	-3.4	-2.2	0.8

Right Table

♂ LONG	LAT	MAG	☉ LONG	16.00UT -252	☿	♀	♃	♄ MAGNITUDES
233.33	0.21	1.5	282.61	6JA	0.1	-3.3	-2.1	0.8
240.46	0.08	1.4	292.72	16JA	-0.0	-3.3	-2.1	0.8
247.63	-0.06	1.3	302.79	26JA	-0.2	-3.3	-2.0	0.8
254.83	-0.22	1.3	312.81	5FE	-0.6	-3.3	-1.9	0.7
262.06	-0.38	1.2	322.77	15FE	-1.3	-3.3	-1.8	0.7
269.31	-0.56	1.1	332.67	25FE	-1.4	-3.4	-1.8	0.7
276.58	-0.75	1.0	342.51	6MR	-0.7	-3.4	-1.7	0.6
283.85	-0.94	0.9	352.30	16MR	0.6	-3.4	-1.7	0.6
291.11	-1.15	0.8	2.03	26MR	2.1	-3.4	-1.6	0.5
298.34	-1.36	0.7	11.70	5AP	3.7	-3.5	-1.5	0.4
305.52	-1.58	0.6	21.33	15AP	2.2	-3.5	-1.5	0.4
312.64	-1.80	0.5	30.93	25AP	1.2	-3.6	-1.5	0.3
319.65	-2.03	0.4	40.48	5MY	0.5	-3.6	-1.4	0.3
326.54	-2.25	0.2	50.02	15MY	-0.3	-3.7	-1.4	0.3
333.26	-2.47	0.1	59.55	25MY	-1.3	-3.8	-1.4	0.4
339.76	-2.68	0.0	69.06	4JN	-1.7	-3.9	-1.4	0.4
345.99	-2.89	-0.1	78.59	14JN	-0.8	-4.0	-1.4	0.5
351.87	-3.09	-0.3	88.13	24JN	-0.1	-4.1	-1.4	0.5
357.31	-3.27	-0.4	97.68	4JL	0.3	-4.2	-1.4	0.6
2.22	-3.44	-0.6	107.27	14JL	0.6	-4.2	-1.4	0.7
6.44	-3.59	-0.8	116.90	24JL	1.1	-4.2	-1.4	0.7
9.82	-3.71	-1.0	126.58	3AU	2.2	-4.0	-1.4	0.7
12.15	-3.79	-1.2	136.30	13AU	2.5	-3.6	-1.5	0.8
13.20	-3.80	-1.4	146.09	23AU	0.1	-3.3	-1.5	0.8
12.82	-3.71	-1.6	155.92	2SE	-0.9	-3.8	-1.6	0.8
10.99	-3.49	-1.8	165.82	12SE	-1.0	-4.2	-1.6	0.8
7.99	-3.11	-2.0	175.78	22SE	-1.0	-4.3	-1.7	0.8
4.47	-2.60	-2.0	185.79	2OC	-0.6	-4.3	-1.7	0.8
1.23	-1.99	-1.8	195.86	12OC	-0.4	-4.3	-1.8	0.8
358.96	-1.39	-1.5	205.98	22OC	-0.4	-4.2	-1.9	0.8
358.01	-0.84	-1.2	216.13	1NO	-0.3	-4.1	-1.9	0.7
358.38	-0.37	-0.8	226.31	11NO	0.1	-4.0	-2.0	0.7
359.95	0.01	-0.5	236.51	21NO	2.2	-3.9	-2.1	0.7
2.49	0.32	-0.2	246.72	1DE	1.2	-3.8	-2.1	0.7
5.80	0.56	0.0	256.93	11DE	0.2	-3.7	-2.1	0.8
9.72	0.76	0.3	267.12	21DE	-0.0	-3.7	-2.1	0.8
14.11	0.91	0.5	277.29	31DE	-0.1	-3.6	-2.1	0.8

-251

♂ LONG	LAT	MAG	☉ LONG	16.00UT	☿	♀	♃	♄
18.86	1.03	0.7	287.43	10JA	-0.3	-3.5	-2.1	0.8
23.89	1.13	0.9	297.53	20JA	-0.7	-3.5	-2.1	0.8
29.15	1.20	1.1	307.57	30JA	-1.3	-3.4	-2.0	0.8
34.57	1.26	1.2	317.56	9FE	-1.3	-3.4	-2.0	0.8
40.14	1.30	1.3	327.49	19FE	-0.7	-3.4	-1.9	0.7
45.80	1.33	1.4	337.37	1MR	0.7	-3.3	-1.9	0.7
51.56	1.36	1.6	347.18	11MR	2.6	-3.3	-1.8	0.7
57.39	1.37	1.6	356.94	21MR	2.9	-3.3	-1.7	0.6
63.28	1.38	1.7	6.64	31MR	1.6	-3.3	-1.7	0.5
69.22	1.38	1.8	16.29	10AP	0.9	-3.3	-1.6	0.5
75.21	1.37	1.9	25.90	20AP	0.4	-3.3	-1.5	0.4
81.23	1.36	1.9	35.48	30AP	-0.3	-3.3	-1.5	0.4
87.30	1.35	1.9	45.03	10MY	-1.3	-3.4	-1.4	0.3
93.40	1.33	2.0	54.56	20MY	-1.7	-3.4	-1.4	0.3
99.54	1.30	2.0	64.07	30MY	-0.8	-3.5	-1.4	0.3
105.73	1.27	2.0	73.59	9JN	-0.0	-3.5	-1.3	0.4
111.95	1.24	2.0	83.12	19JN	0.5	-3.5	-1.3	0.4
118.22	1.21	2.0	92.66	29JN	0.9	-3.4	-1.3	0.5
124.53	1.17	2.0	102.24	9JL	1.5	-3.4	-1.3	0.5
130.90	1.12	2.0	111.85	19JL	2.7	-3.4	-1.3	0.6
137.32	1.07	2.0	121.49	29JL	2.1	-3.4	-1.3	0.7
143.79	1.02	1.9	131.20	8AU	0.1	-3.3	-1.3	0.7
150.33	0.97	1.9	140.95	18AU	-1.0	-3.3	-1.3	0.7
156.93	0.90	1.9	150.75	28AU	-1.2	-3.3	-1.3	0.8
163.59	0.84	1.8	160.63	7SE	-1.0	-3.3	-1.4	0.8
170.32	0.77	1.8	170.55	17SE	-0.5	-3.3	-1.4	0.8
177.12	0.70	1.8	180.54	27SE	-0.3	-3.4	-1.4	0.8
183.99	0.62	1.8	190.58	7OC	-0.2	-3.4	-1.5	0.8
190.93	0.53	1.8	200.67	17OC	-0.2	-3.4	-1.6	0.8
197.95	0.44	1.8	210.80	27OC	0.3	-3.4	-1.6	0.8
205.04	0.35	1.7	220.97	6NO	2.5	-3.5	-1.7	0.7
212.20	0.25	1.7	231.16	16NO	1.0	-3.5	-1.7	0.7
219.43	0.14	1.7	241.37	26NO	-0.0	-3.6	-1.8	0.7
226.73	0.03	1.7	251.58	6DE	-0.2	-3.7	-1.9	0.7
234.09	-0.08	1.6	261.78	16DE	-0.2	-3.7	-1.9	0.8
241.52	-0.20	1.6	271.96	26DE	-0.4	-3.8	-2.0	0.8

Left table (−250 / −249)

♂ LONG	LAT	MAG	☉ LONG	16.00UT	☿	♀	♃	♄
				−250				
249.01	−0.33	1.5	282.12	5JA	−0.7	−3.9	−2.0	0.8
256.54	−0.45	1.5	292.24	15JA	−1.2	−4.0	−2.0	0.8
264.12	−0.58	1.5	302.31	25JA	−1.2	−4.1	−2.1	0.8
271.74	−0.71	1.4	312.33	4FE	−0.6	−4.2	−2.0	0.8
279.38	−0.84	1.4	322.29	14FE	0.9	−4.2	−2.0	0.8
287.04	−0.96	1.3	332.20	24FE	3.1	−4.3	−2.0	0.8
294.71	−1.09	1.3	342.04	6MR	2.1	−4.2	−1.9	0.7
302.37	−1.20	1.2	351.83	16MR	1.2	−4.0	−1.9	0.7
310.02	−1.31	1.2	1.56	26MR	0.7	−3.4	−1.8	0.7
317.62	−1.41	1.1	11.24	5AP	0.3	−3.1	−1.8	0.6
325.18	−1.50	1.1	20.87	15AP	−0.4	−3.8	−1.7	0.5
332.68	−1.58	1.0	30.46	25AP	−1.3	−4.1	−1.6	0.5
340.10	−1.64	1.0	40.02	5MY	−1.7	−4.2	−1.6	0.4
347.41	−1.69	0.9	49.56	15MY	−0.8	−4.2	−1.5	0.3
354.62	−1.72	0.9	59.08	25MY	0.0	−4.1	−1.5	0.3
1.68	−1.74	0.8	68.60	4JN	0.6	−4.0	−1.4	0.3
8.59	−1.74	0.8	78.12	14JN	1.1	−3.9	−1.4	0.3
15.32	−1.72	0.7	87.66	24JN	1.9	−3.8	−1.3	0.4
21.84	−1.68	0.6	97.22	4JL	3.2	−3.7	−1.3	0.4
28.11	−1.63	0.6	106.80	14JL	1.7	−3.7	−1.3	0.5
34.10	−1.55	0.5	116.43	24JL	0.0	−3.6	−1.2	0.6
39.76	−1.46	0.4	126.11	3AU	−1.1	−3.5	−1.2	0.6
45.02	−1.34	0.3	135.83	13AU	−1.3	−3.5	−1.2	0.7
49.80	−1.19	0.1	145.61	23AU	−0.9	−3.5	−1.2	0.7
54.00	−1.01	−0.0	155.45	2SE	−0.5	−3.4	−1.2	0.7
57.47	−0.80	−0.2	165.34	12SE	−0.2	−3.4	−1.2	0.8
60.06	−0.54	−0.4	175.30	22SE	−0.1	−3.4	−1.3	0.8
61.55	−0.23	−0.6	185.31	2OC	0.0	−3.4	−1.3	0.8
61.77	0.13	−0.8	195.37	12OC	0.5	−3.4	−1.3	0.8
60.57	0.55	−1.0	205.49	22OC	2.8	−3.4	−1.4	0.8
57.99	1.00	−1.2	215.64	1NO	0.8	−3.4	−1.4	0.8
54.40	1.44	−1.4	225.82	11NO	−0.2	−3.4	−1.5	0.8
50.47	1.82	−1.4	236.02	21NO	−0.3	−3.4	−1.5	0.7
47.00	2.11	−1.1	246.23	1DE	−0.4	−3.4	−1.6	0.7
44.59	2.28	−0.8	256.43	11DE	−0.5	−3.4	−1.7	0.7
43.50	2.37	−0.5	266.63	21DE	−0.8	−3.4	−1.7	0.8
43.72	2.39	−0.2	276.80	31DE	−1.0	−3.4	−1.8	0.8
				−249				
45.09	2.37	0.0	286.94	10JA	−1.1	−3.5	−1.9	0.8
47.39	2.32	0.3	297.04	20JA	−0.6	−3.5	−1.9	0.8
50.46	2.27	0.5	307.08	30JA	1.1	−3.5	−2.0	0.8
54.11	2.20	0.7	317.07	9FE	3.2	−3.4	−2.0	0.8
58.23	2.13	0.9	327.01	19FE	1.6	−3.4	−2.0	0.8
62.73	2.05	1.1	336.89	1MR	0.9	−3.4	−2.0	0.8
67.52	1.98	1.2	346.70	11MR	0.5	−3.4	−2.0	0.8
72.55	1.90	1.3	356.46	21MR	0.1	−3.3	−2.0	0.8
77.78	1.83	1.5	6.17	31MR	−0.4	−3.3	−1.9	0.7
83.17	1.75	1.5	15.82	10AP	−1.3	−3.3	−1.9	0.7
88.70	1.68	1.6	25.44	20AP	−1.7	−3.3	−1.8	0.6
94.35	1.60	1.7	35.01	30AP	−0.8	−3.3	−1.8	0.5
100.11	1.53	1.8	44.56	10MY	0.1	−3.4	−1.7	0.5
105.97	1.45	1.8	54.09	20MY	0.8	−3.4	−1.7	0.4
111.93	1.38	1.8	63.61	30MY	1.5	−3.4	−1.6	0.4
117.97	1.30	1.9	73.13	9JN	2.6	−3.4	−1.5	0.3
124.09	1.22	1.9	82.66	19JN	3.1	−3.5	−1.5	0.3
130.30	1.14	1.9	92.20	29JN	1.4	−3.5	−1.4	0.4
136.60	1.05	1.9	101.77	9JL	−0.0	−3.6	−1.4	0.4
142.97	0.97	1.9	111.38	19JL	−1.1	−3.7	−1.3	0.5
149.43	0.88	1.9	121.03	29JL	−1.5	−3.7	−1.3	0.5
155.98	0.79	1.9	130.72	8AU	−0.9	−3.8	−1.3	0.6
162.62	0.70	1.9	140.47	18AU	−0.4	−3.9	−1.2	0.6
169.34	0.61	1.8	150.28	28AU	−0.1	−4.0	−1.2	0.7
176.15	0.51	1.8	160.14	7SE	0.1	−4.2	−1.2	0.7
183.05	0.41	1.8	170.07	17SE	0.2	−4.3	−1.2	0.8
190.05	0.31	1.7	180.05	27SE	0.8	−4.3	−1.2	0.8
197.13	0.21	1.7	190.09	7OC	3.1	−4.3	−1.2	0.8
204.31	0.10	1.6	200.18	17OC	0.6	−4.2	−1.3	0.8
211.57	−0.00	1.6	210.31	27OC	−0.4	−3.7	−1.3	0.8
218.91	−0.11	1.5	220.46	6NO	−0.5	−3.0	−1.3	0.8
226.34	−0.22	1.5	230.67	16NO	−0.5	−3.8	−1.4	0.8
233.85	−0.33	1.5	240.87	26NO	−0.6	−4.0	−1.4	0.8
241.42	−0.43	1.5	251.08	6DE	−0.8	−4.4	−1.5	0.8
249.06	−0.54	1.5	261.29	16DE	−0.9	−4.4	−1.5	0.7
256.76	−0.64	1.5	271.47	26DE	−0.9	−4.3	−1.6	0.8

Right table (−248 / −247)

♂ LONG	LAT	MAG	☉ LONG	16.00UT	☿	♀	♃	♄
				−248				
264.50	−0.74	1.5	281.63	5JA	−0.5	−4.2	−1.6	0.8
272.28	−0.83	1.5	291.75	15JA	1.3	−4.1	−1.7	0.8
280.08	−0.92	1.4	301.82	25JA	2.7	−4.0	−1.8	0.9
287.90	−1.00	1.4	311.84	4FE	1.2	−3.9	−1.8	0.9
295.73	−1.06	1.4	321.81	14FE	0.6	−3.8	−1.9	0.9
303.54	−1.12	1.4	331.72	24FE	0.3	−3.7	−2.0	0.9
311.33	−1.17	1.4	341.57	5MR	0.0	−3.6	−2.0	0.9
319.09	−1.21	1.4	351.36	15MR	−0.5	−3.5	−2.0	0.9
326.80	−1.24	1.4	1.09	25MR	−1.2	−3.5	−2.0	0.8
334.45	−1.25	1.4	10.77	4AP	−1.7	−3.4	−2.0	0.8
342.04	−1.25	1.4	20.41	14AP	−0.8	−3.4	−2.0	0.7
349.54	−1.23	1.4	30.00	24AP	0.2	−3.3	−2.0	0.7
356.96	−1.20	1.3	39.56	4MY	1.1	−3.3	−2.0	0.6
4.27	−1.16	1.3	49.10	14MY	2.0	−3.3	−1.9	0.6
11.49	−1.11	1.3	58.63	24MY	3.3	−3.3	−1.8	0.5
18.58	−1.04	1.3	68.14	3JN	2.5	−3.3	−1.8	0.4
25.55	−0.96	1.3	77.67	13JN	1.1	−3.3	−1.7	0.4
32.39	−0.87	1.3	87.20	23JN	−0.1	−3.3	−1.7	0.3
39.09	−0.77	1.2	96.75	3JL	−1.2	−3.4	−1.6	0.4
45.62	−0.65	1.2	106.34	13JL	−1.6	−3.4	−1.5	0.4
51.99	−0.53	1.2	115.96	23JL	−0.9	−3.4	−1.5	0.5
58.17	−0.39	1.1	125.63	2AU	−0.3	−3.5	−1.4	0.5
64.13	−0.23	1.1	135.35	12AU	0.0	−3.5	−1.4	0.6
69.85	−0.06	1.0	145.13	22AU	0.2	−3.5	−1.4	0.6
75.29	0.12	0.9	154.96	1SE	0.4	−3.5	−1.3	0.7
80.40	0.33	0.8	164.86	11SE	1.1	−3.4	−1.3	0.7
85.12	0.56	0.7	174.81	21SE	3.3	−3.4	−1.3	0.8
89.36	0.82	0.5	184.82	1OC	0.5	−3.4	−1.3	0.8
93.02	1.11	0.4	194.88	11OC	−0.5	−3.4	−1.3	0.8
95.95	1.44	0.2	204.99	21OC	−0.7	−3.3	−1.3	0.8
97.99	1.81	−0.0	215.14	31OC	−0.6	−3.3	−1.3	0.8
98.95	2.22	−0.2	225.32	10NO	−0.7	−3.3	−1.3	0.9
98.66	2.67	−0.5	235.52	20NO	−0.7	−3.3	−1.3	0.8
97.00	3.12	−0.7	245.73	30NO	−0.7	−3.3	−1.3	0.8
94.08	3.53	−0.9	255.94	10DE	−0.8	−3.4	−1.4	0.8
90.30	3.83	−1.1	266.13	20DE	−0.4	−3.4	−1.4	0.8
86.34	3.98	−1.0	276.31	30DE	1.5	−3.4	−1.5	0.8
				−247				
82.95	3.97	−0.8	286.45	9JA	2.2	−3.4	−1.5	0.8
80.65	3.84	−0.5	296.55	19JA	0.8	−3.5	−1.6	0.9
79.65	3.64	−0.2	306.60	29JA	0.4	−3.5	−1.7	0.9
79.92	3.40	0.0	316.59	8FE	0.2	−3.6	−1.7	0.9
81.29	3.15	0.2	326.53	18FE	−0.1	−3.6	−1.8	1.0
83.58	2.91	0.5	336.41	28FE	−0.5	−3.7	−1.9	1.0
86.61	2.69	0.7	346.23	10MR	−1.2	−3.8	−1.9	1.0
90.23	2.48	0.8	355.99	20MR	−1.6	−3.9	−2.0	0.9
94.34	2.28	1.0	5.70	30MR	−0.8	−3.9	−2.0	0.9
98.83	2.10	1.1	15.36	9AP	0.3	−4.0	−2.1	0.9
103.63	1.93	1.2	24.97	19AP	1.4	−4.1	−2.1	0.8
108.71	1.77	1.3	34.56	29AP	2.6	−4.2	−2.1	0.8
114.02	1.62	1.4	44.10	9MY	3.4	−4.2	−2.1	0.7
119.51	1.47	1.5	53.63	19MY	1.9	−4.0	−2.1	0.7
125.19	1.33	1.5	63.16	29MY	0.9	−3.6	−2.1	0.6
131.03	1.19	1.6	72.67	8JN	−0.2	−2.8	−2.0	0.5
137.01	1.06	1.6	82.20	18JN	−1.2	−3.4	−2.0	0.5
143.13	0.93	1.7	91.74	28JN	−1.7	−4.0	−1.9	0.4
149.37	0.80	1.7	101.31	8JL	−0.8	−4.2	−1.9	0.4
155.75	0.68	1.7	110.91	18JL	−0.2	−4.2	−1.8	0.4
162.25	0.55	1.7	120.56	28JL	0.1	−4.2	−1.7	0.5
168.86	0.43	1.7	130.25	7AU	0.4	−4.1	−1.7	0.5
175.59	0.31	1.7	139.99	17AU	0.6	−4.0	−1.6	0.6
182.44	0.19	1.7	149.80	27AU	1.4	−3.9	−1.6	0.6
189.41	0.08	1.7	159.66	6SE	3.1	−3.8	−1.5	0.7
196.48	−0.04	1.6	169.58	16SE	0.3	−3.8	−1.5	0.7
203.66	−0.15	1.6	179.56	26SE	−0.7	−3.7	−1.4	0.8
210.94	−0.26	1.6	189.60	6OC	−0.8	−3.6	−1.4	0.8
218.33	−0.37	1.6	199.68	16OC	−0.8	−3.6	−1.4	0.8
225.81	−0.47	1.5	209.82	26OC	−0.8	−3.5	−1.4	0.9
233.37	−0.57	1.5	219.98	5NO	−0.6	−3.5	−1.3	0.9
241.02	−0.66	1.5	230.17	15NO	−0.6	−3.5	−1.3	0.9
248.73	−0.75	1.4	240.38	25NO	−0.6	−3.4	−1.4	0.9
256.50	−0.83	1.4	250.59	5DE	−0.2	−3.4	−1.4	0.9
264.32	−0.90	1.3	260.79	15DE	1.7	−3.4	−1.4	0.9
272.18	−0.96	1.3	270.98	25DE	1.8	−3.4	−1.4	0.9

♂ LONG	LAT	MAG	⊙ LONG	16.00UT	☿	♀	♃	♄
				-246	MAGNITUDES			
280.06	-1.01	1.3	281.14	4JA	0.5	-3.3	-1.4	0.9
287.95	-1.05	1.3	291.26	14JA	0.2	-3.3	-1.5	0.9
295.84	-1.08	1.3	301.34	24JA	0.1	-3.3	-1.5	0.9
303.71	-1.09	1.4	311.36	3FE	-0.2	-3.3	-1.6	1.0
311.56	-1.10	1.4	321.33	13FE	-0.6	-3.3	-1.7	1.0
319.36	-1.09	1.4	331.24	23FE	-1.2	-3.4	-1.7	1.0
327.11	-1.07	1.4	341.09	5MR	-1.5	-3.4	-1.8	1.0
334.81	-1.04	1.4	350.88	15MR	-0.8	-3.4	-1.9	1.1
342.43	-1.00	1.5	0.62	25MR	0.5	-3.5	-1.9	1.0
349.97	-0.95	1.5	10.30	4AP	1.8	-3.5	-2.0	1.0
357.44	-0.89	1.5	19.94	14AP	3.5	-3.5	-2.1	1.0
4.81	-0.82	1.5	29.54	24AP	2.6	-3.4	-2.1	1.0
12.09	-0.74	1.6	39.10	4MY	1.5	-3.4	-2.2	0.9
19.28	-0.66	1.6	48.64	14MY	0.7	-3.4	-2.2	0.9
26.36	-0.56	1.6	58.16	24MY	-0.2	-3.4	-2.2	0.8
33.35	-0.46	1.6	67.68	3JN	-1.2	-3.3	-2.2	0.8
40.23	-0.36	1.6	77.20	13JN	-1.7	-3.3	-2.2	0.7
47.01	-0.25	1.6	86.73	23JN	-0.8	-3.3	-2.2	0.6
53.68	-0.13	1.6	96.29	3JL	-0.2	-3.3	-2.2	0.6
60.23	-0.01	1.6	105.87	13JL	0.2	-3.3	-2.1	0.5
66.67	0.12	1.6	115.49	23JL	0.5	-3.4	-2.1	0.5
72.99	0.26	1.6	125.16	2AU	0.9	-3.4	-2.0	0.5
79.18	0.40	1.5	134.88	12AU	1.9	-3.4	-2.0	0.6
85.23	0.55	1.5	144.65	22AU	2.7	-3.5	-1.9	0.6
91.12	0.70	1.4	154.48	1SE	0.2	-3.5	-1.8	0.7
96.83	0.87	1.4	164.37	11SE	-0.8	-3.6	-1.8	0.7
102.34	1.05	1.3	174.32	21SE	-1.0	-3.6	-1.7	0.8
107.62	1.24	1.2	184.33	1OC	-0.9	-3.7	-1.7	0.8
112.62	1.45	1.1	194.39	11OC	-0.7	-3.8	-1.6	0.9
117.28	1.68	1.0	204.50	21OC	-0.5	-3.9	-1.6	0.9
121.53	1.93	0.8	214.64	31OC	-0.4	-4.0	-1.5	0.9
125.29	2.21	0.7	224.82	10NO	-0.4	-4.1	-1.5	1.0
128.41	2.52	0.5	235.02	20NO	-0.1	-4.2	-1.5	1.0
130.78	2.86	0.3	245.23	30NO	1.9	-4.3	-1.5	1.0
132.20	3.23	0.0	255.44	10DE	1.4	-4.4	-1.5	1.0
132.49	3.63	-0.2	265.64	20DE	0.3	-4.4	-1.5	1.0
131.50	4.01	-0.5	275.81	30DE	0.1	-4.2	-1.5	1.0
				-245				
129.22	4.35	-0.7	285.96	9JA	-0.1	-3.7	-1.5	1.0
125.86	4.57	-0.9	296.06	19JA	-0.2	-3.3	-1.5	1.0
121.96	4.62	-1.0	306.11	29JA	-0.6	-3.8	-1.5	1.0
118.21	4.50	-0.9	316.11	8FE	-1.2	-4.2	-1.6	1.1
115.30	4.23	-0.7	326.05	18FE	-1.4	-4.3	-1.6	1.1
113.58	3.87	-0.4	335.93	28FE	-0.7	-4.3	-1.7	1.1
113.17	3.48	-0.2	345.76	10MR	0.6	-4.2	-1.7	1.2
113.97	3.10	0.0	355.52	20MR	2.2	-4.1	-1.8	1.2
115.80	2.73	0.2	5.23	30MR	3.4	-4.0	-1.8	1.2
118.49	2.40	0.4	14.90	9AP	1.9	-3.9	-1.9	1.2
121.89	2.09	0.6	24.51	19AP	1.1	-3.8	-2.0	1.2
125.85	1.81	0.7	34.09	29AP	0.5	-3.7	-2.0	1.1
130.29	1.56	0.9	43.64	9MY	-0.3	-3.6	-2.1	1.1
135.12	1.32	1.0	53.17	19MY	-1.2	-3.5	-2.2	1.1
140.28	1.11	1.1	62.69	29MY	-1.7	-3.5	-2.2	1.0
145.72	0.90	1.1	72.21	8JN	-0.8	-3.4	-2.3	0.9
151.42	0.71	1.2	81.74	18JN	-0.1	-3.4	-2.3	0.9
157.34	0.53	1.3	91.28	28JN	0.4	-3.4	-2.4	0.8
163.46	0.36	1.3	100.85	8JL	0.7	-3.4	-2.4	0.7
169.77	0.20	1.3	110.45	18JL	1.2	-3.3	-2.4	0.7
176.26	0.05	1.4	120.09	28JL	2.4	-3.3	-2.3	0.6
182.91	-0.09	1.4	129.79	7AU	2.4	-3.3	-2.3	0.6
189.71	-0.23	1.4	139.53	17AU	0.2	-3.3	-2.3	0.6
196.65	-0.36	1.4	149.33	27AU	-0.9	-3.4	-2.2	0.7
203.74	-0.48	1.4	159.19	6SE	-1.1	-3.4	-2.1	0.7
210.95	-0.59	1.4	169.10	16SE	-1.0	-3.4	-2.1	0.8
218.28	-0.69	1.4	179.08	26SE	-0.6	-3.4	-2.0	0.8
225.73	-0.79	1.4	189.11	6OC	-0.4	-3.4	-1.9	0.9
233.27	-0.87	1.4	199.19	16OC	-0.3	-3.5	-1.9	0.9
240.91	-0.94	1.4	209.32	26OC	-0.3	-3.5	-1.8	1.0
248.62	-1.00	1.4	219.48	5NO	0.1	-3.5	-1.8	1.0
256.40	-1.05	1.4	229.67	15NO	2.2	-3.5	-1.7	1.1
264.23	-1.09	1.4	239.88	25NO	1.2	-3.4	-1.7	1.1
272.10	-1.11	1.4	250.09	5DE	0.1	-3.4	-1.6	1.1
279.99	-1.13	1.4	260.29	15DE	-0.1	-3.4	-1.6	1.1
287.89	-1.13	1.4	270.48	25DE	-0.2	-3.4	-1.6	1.1

♂ LONG	LAT	MAG	⊙ LONG	16.00UT	☿	♀	♃	♄
				-244	MAGNITUDES			
295.78	-1.12	1.4	280.64	4JA	-0.3	-3.3	-1.5	1.1
303.65	-1.09	1.4	290.76	14JA	-0.7	-3.3	-1.5	1.1
311.48	-1.06	1.4	300.84	24JA	-1.2	-3.3	-1.5	1.1
319.28	-1.01	1.4	310.87	3FE	-1.3	-3.3	-1.5	1.1
327.01	-0.96	1.4	320.84	13FE	-0.7	-3.3	-1.6	1.2
334.69	-0.90	1.4	330.76	23FE	0.7	-3.4	-1.6	1.2
342.29	-0.83	1.4	340.61	4MR	2.8	-3.4	-1.6	1.3
349.81	-0.75	1.4	350.41	14MR	2.6	-3.4	-1.6	1.3
357.25	-0.66	1.5	0.15	24MR	1.4	-3.4	-1.7	1.3
4.60	-0.57	1.5	9.84	3AP	0.8	-3.5	-1.7	1.3
11.87	-0.48	1.6	19.47	13AP	0.3	-3.5	-1.8	1.3
19.05	-0.38	1.6	29.08	23AP	-0.3	-3.6	-1.8	1.3
26.14	-0.28	1.7	38.64	3MY	-1.2	-3.6	-1.9	1.3
33.14	-0.18	1.7	48.18	13MY	-1.7	-3.7	-2.0	1.3
40.06	-0.08	1.7	57.71	23MY	-0.8	-3.8	-2.0	1.3
46.89	0.03	1.8	67.22	2JN	-0.0	-3.9	-2.1	1.2
53.64	0.13	1.8	76.74	12JN	0.5	-4.0	-2.2	1.2
60.30	0.24	1.8	86.28	22JN	1.0	-4.1	-2.2	1.1
66.89	0.35	1.8	95.83	2JL	1.6	-4.2	-2.3	1.0
73.40	0.46	1.8	105.41	12JL	2.9	-4.2	-2.4	1.0
79.84	0.57	1.8	115.03	22JL	2.0	-4.2	-2.4	0.9
86.19	0.68	1.8	124.70	1AU	0.1	-4.0	-2.4	0.8
92.46	0.79	1.8	134.41	11AU	-1.0	-3.6	-2.4	0.8
98.65	0.91	1.8	144.18	21AU	-1.3	-3.3	-2.4	0.8
104.75	1.02	1.8	154.01	31AU	-1.0	-3.8	-2.4	0.8
110.76	1.14	1.7	163.89	10SE	-0.5	-4.2	-2.4	0.8
116.66	1.27	1.7	173.84	20SE	-0.3	-4.3	-2.3	0.9
122.44	1.40	1.6	183.84	30SE	-0.2	-4.3	-2.3	0.9
128.09	1.53	1.5	193.90	10OC	-0.1	-4.3	-2.2	1.0
133.58	1.68	1.5	204.01	20OC	0.3	-4.2	-2.2	1.0
138.88	1.83	1.4	214.15	30OC	2.5	-4.1	-2.1	1.1
143.96	1.99	1.2	224.33	9NO	0.9	-4.0	-2.0	1.1
148.76	2.16	1.1	234.53	19NO	-0.1	-3.9	-1.9	1.2
153.23	2.35	0.9	244.73	29NO	-0.3	-3.8	-1.9	1.2
157.30	2.55	0.8	254.94	9DE	-0.3	-3.7	-1.8	1.3
160.86	2.77	0.6	265.14	19DE	-0.4	-3.6	-1.8	1.3
163.79	3.01	0.3	275.32	29DE	-0.7	-3.6	-1.7	1.3
				-243				
165.93	3.26	0.1	285.46	8JA	-1.1	-3.5	-1.7	1.3
167.11	3.51	-0.2	295.57	18JA	-1.1	-3.5	-1.6	1.3
167.15	3.75	-0.5	305.62	28JA	-0.6	-3.4	-1.6	1.3
165.93	3.94	-0.7	315.62	7FE	0.9	-3.4	-1.6	1.3
163.45	4.03	-1.0	325.57	17FE	3.2	-3.4	-1.6	1.3
160.03	3.99	-1.2	335.45	27FE	1.9	-3.3	-1.6	1.3
156.22	3.78	-1.2	345.28	9MR	1.0	-3.3	-1.6	1.3
152.75	3.43	-1.1	355.05	19MR	0.6	-3.3	-1.6	1.3
150.24	2.98	-0.9	4.76	29MR	0.2	-3.3	-1.6	1.3
148.99	2.50	-0.7	14.43	8AP	-0.4	-3.3	-1.6	1.3
149.06	2.03	-0.5	24.05	18AP	-1.2	-3.3	-1.7	1.3
150.34	1.59	-0.3	33.63	28AP	-1.7	-3.3	-1.7	1.2
152.65	1.20	-0.1	43.18	8MY	-0.8	-3.4	-1.7	1.2
155.83	0.85	0.1	52.72	18MY	0.0	-3.4	-1.8	1.2
159.71	0.54	0.2	62.23	28MY	0.7	-3.5	-1.8	1.1
164.17	0.26	0.4	71.75	7JN	1.3	-3.5	-1.9	1.1
169.13	0.01	0.5	81.28	17JN	2.1	-3.5	-2.0	1.0
174.50	-0.21	0.6	90.81	27JN	3.3	-3.4	-2.0	1.0
180.21	-0.41	0.7	100.38	7JL	1.7	-3.4	-2.1	0.9
186.24	-0.59	0.7	109.98	17JL	0.0	-3.4	-2.2	0.9
192.53	-0.74	0.8	119.62	27JL	-1.1	-3.4	-2.2	0.8
199.07	-0.88	0.9	129.31	6AU	-1.4	-3.3	-2.3	0.8
205.81	-1.00	0.9	139.05	16AU	-0.9	-3.3	-2.4	0.8
212.75	-1.11	1.0	148.84	26AU	-0.4	-3.3	-2.4	0.7
219.85	-1.20	1.0	158.70	5SE	-0.2	-3.3	-2.4	0.8
227.11	-1.27	1.1	168.62	15SE	-0.0	-3.3	-2.4	0.9
234.49	-1.32	1.1	178.59	25SE	0.1	-3.4	-2.4	0.9
241.99	-1.36	1.1	188.62	5OC	0.6	-3.4	-2.4	1.0
249.59	-1.38	1.2	198.70	15OC	2.8	-3.4	-2.4	1.1
257.27	-1.38	1.2	208.82	25OC	0.7	-3.4	-2.3	1.1
265.00	-1.37	1.2	218.99	4NO	-0.3	-3.5	-2.3	1.2
272.78	-1.35	1.3	229.18	14NO	-0.4	-3.5	-2.2	1.2
280.59	-1.31	1.3	239.38	24NO	-0.4	-3.6	-2.2	1.2
288.40	-1.26	1.3	249.59	4DE	-0.5	-3.7	-2.1	1.3
296.21	-1.20	1.4	259.80	14DE	-0.8	-3.7	-2.0	1.3
303.99	-1.13	1.4	269.98	24DE	-1.0	-3.8	-1.9	1.3

♂ LONG	LAT	MAG	☉ LONG	16.00UT	☿	♀	♃	♄
				-242	MAGNITUDES			
311.74	-1.05	1.4	280.15	3JA	-1.0	-3.9	-1.9	1.3
319.45	-0.96	1.5	290.27	13JA	-0.6	-4.0	-1.8	1.2
327.09	-0.87	1.5	300.35	23JA	1.1	-4.1	-1.7	1.2
334.68	-0.77	1.5	310.39	2FE	3.0	-4.2	-1.7	1.2
342.19	-0.67	1.6	320.36	12FE	1.4	-4.2	-1.7	1.1
349.62	-0.57	1.6	330.27	22FE	0.8	-4.3	-1.6	1.1
356.98	-0.46	1.6	340.14	4MR	0.4	-4.2	-1.6	1.0
4.25	-0.36	1.6	349.93	14MR	0.1	-4.0	-1.6	1.0
11.44	-0.25	1.7	359.67	24MR	-0.4	-3.4	-1.5	1.0
18.55	-0.15	1.7	9.37	3AP	-1.2	-3.2	-1.5	1.0
25.58	-0.05	1.7	19.01	13AP	-1.7	-3.8	-1.5	1.0
32.53	0.06	1.7	28.61	23AP	-0.8	-4.1	-1.5	1.0
39.40	0.16	1.7	38.18	3MY	0.1	-4.2	-1.6	1.0
46.21	0.25	1.7	47.72	13MY	0.9	-4.2	-1.6	1.0
52.94	0.35	1.8	57.24	23MY	1.7	-4.1	-1.6	0.9
59.61	0.44	1.8	66.76	2JN	2.8	-4.0	-1.6	0.9
66.22	0.53	1.9	76.28	12JN	2.9	-3.9	-1.7	0.8
72.77	0.62	1.9	85.81	22JN	1.3	-3.8	-1.7	0.8
79.27	0.71	1.9	95.37	2JL	-0.0	-3.7	-1.8	0.8
85.72	0.79	2.0	104.94	12JL	-1.1	-3.7	-1.8	0.7
92.12	0.87	2.0	114.56	22JL	-1.5	-3.6	-1.9	0.6
98.48	0.95	2.0	124.23	1AU	-0.9	-3.5	-2.0	0.6
104.80	1.03	2.0	133.94	11AU	-0.4	-3.5	-2.0	0.5
111.08	1.11	2.0	143.70	21AU	-0.0	-3.5	-2.1	0.5
117.31	1.18	2.0	153.53	31AU	0.1	-3.4	-2.2	0.5
123.50	1.26	1.9	163.41	10SE	0.3	-3.4	-2.2	0.4
129.64	1.33	1.9	173.36	20SE	0.9	-3.4	-2.3	0.5
135.74	1.40	1.9	183.36	30SE	3.2	-3.4	-2.3	0.6
141.78	1.48	1.8	193.41	10OC	0.6	-3.4	-2.3	0.6
147.75	1.55	1.7	203.52	20OC	-0.4	-3.4	-2.4	0.7
153.65	1.62	1.7	213.66	30OC	-0.6	-3.4	-2.4	0.8
159.47	1.69	1.6	223.83	9NO	-0.6	-3.4	-2.3	0.8
165.18	1.76	1.5	234.03	19NO	-0.6	-3.4	-2.3	0.9
170.77	1.83	1.4	244.24	29NO	-0.8	-3.4	-2.3	0.9
176.21	1.90	1.2	254.45	9DE	-0.8	-3.4	-2.2	1.0
181.48	1.97	1.1	264.65	19DE	-0.8	-3.4	-2.1	1.0
186.52	2.04	0.9	274.82	29DE	-0.5	-3.4	-2.1	1.0
				-241				
191.29	2.10	0.8	284.97	8JA	1.3	-3.5	-2.0	1.0
195.72	2.16	0.6	295.07	18JA	2.6	-3.5	-1.9	1.0
199.74	2.20	0.3	305.13	28JA	1.0	-3.5	-1.9	0.9
203.23	2.23	0.1	315.13	7FE	0.5	-3.4	-1.8	0.9
206.06	2.24	-0.2	325.08	17FE	0.3	-3.4	-1.7	0.9
208.07	2.21	-0.5	334.97	27FE	0.0	-3.4	-1.7	0.8
209.08	2.13	-0.8	344.80	9MR	-0.5	-3.4	-1.6	0.8
208.92	1.98	-1.1	354.57	19MR	-1.2	-3.3	-1.6	0.8
207.49	1.73	-1.4	4.29	29MR	-1.6	-3.3	-1.5	0.8
204.93	1.37	-1.7	13.95	8AP	-0.8	-3.3	-1.5	0.8
201.60	0.91	-1.9	23.58	18AP	0.2	-3.3	-1.5	0.8
198.19	0.39	-1.8	33.16	28AP	1.2	-3.3	-1.5	0.8
195.43	-0.13	-1.6	42.71	8MY	2.2	-3.4	-1.5	0.8
193.81	-0.62	-1.4	52.25	18MY	3.6	-3.4	-1.4	0.7
193.56	-1.03	-1.2	61.77	28MY	2.3	-3.4	-1.5	0.7
194.65	-1.37	-1.0	71.29	7JN	1.1	-3.4	-1.5	0.7
196.92	-1.64	-0.8	80.81	17JN	-0.1	-3.5	-1.5	0.7
200.21	-1.85	-0.6	90.35	27JN	-1.2	-3.5	-1.5	0.6
204.32	-2.00	-0.4	99.92	7JL	-1.6	-3.6	-1.5	0.6
209.10	-2.12	-0.3	109.52	17JL	-0.9	-3.7	-1.6	0.5
214.44	-2.19	-0.1	119.16	27JL	-0.3	-3.7	-1.6	0.5
220.24	-2.24	-0.0	128.84	6AU	0.1	-3.8	-1.7	0.4
226.41	-2.25	0.1	138.58	16AU	0.3	-3.9	-1.7	0.3
232.90	-2.24	0.2	148.37	26AU	0.5	-4.1	-1.8	0.3
239.64	-2.21	0.3	158.22	5SE	1.2	-4.2	-1.8	0.2
246.60	-2.15	0.4	168.14	15SE	3.3	-4.3	-1.9	0.2
253.72	-2.08	0.5	178.11	25SE	0.5	-4.3	-2.0	0.2
260.98	-1.98	0.6	188.13	5OC	-0.6	-4.3	-2.0	0.2
268.34	-1.88	0.7	198.21	15OC	-0.7	-4.2	-2.1	0.3
275.78	-1.76	0.8	208.33	25OC	-0.7	-3.6	-2.2	0.4
283.27	-1.63	0.9	218.49	4NO	-0.8	-3.0	-2.2	0.5
290.79	-1.50	1.0	228.68	14NO	-0.7	-3.8	-2.2	0.5
298.32	-1.36	1.1	238.88	24NO	-0.7	-4.3	-2.2	0.6
305.84	-1.22	1.1	249.09	4DE	-0.7	-4.4	-2.2	0.6
313.33	-1.07	1.2	259.30	14DE	-0.3	-4.4	-2.2	0.7
320.79	-0.93	1.3	269.49	24DE	1.5	-4.3	-2.2	0.7

♂ LONG	LAT	MAG	☉ LONG	16.00UT	☿	♀	♃	♄
				-240	MAGNITUDES			
328.21	-0.79	1.4	279.65	3JA	2.1	-4.2	-2.1	0.7
335.57	-0.65	1.4	289.78	13JA	0.7	-4.1	-2.1	0.7
342.87	-0.51	1.5	299.87	23JA	0.3	-4.0	-2.0	0.7
350.10	-0.38	1.6	309.90	2FE	0.2	-3.9	-2.0	0.7
357.27	-0.25	1.6	319.88	12FE	-0.1	-3.8	-1.9	0.7
4.36	-0.13	1.7	329.80	22FE	-0.5	-3.7	-1.8	0.7
11.39	-0.02	1.7	339.66	3MR	-1.2	-3.6	-1.7	0.6
18.34	0.10	1.8	349.46	13MR	-1.6	-3.5	-1.7	0.6
25.23	0.20	1.8	359.21	23MR	-0.8	-3.5	-1.6	0.5
32.05	0.30	1.8	8.90	2AP	0.3	-3.4	-1.6	0.5
38.81	0.39	1.9	18.54	12AP	1.5	-3.4	-1.5	0.5
45.51	0.48	1.9	28.14	22AP	2.9	-3.3	-1.5	0.6
52.16	0.57	1.9	37.71	2MY	3.1	-3.3	-1.4	0.6
58.76	0.64	1.9	47.26	12MY	1.8	-3.3	-1.4	0.6
65.31	0.72	1.9	56.78	22MY	0.8	-3.3	-1.4	0.6
71.82	0.79	1.9	66.30	1JN	-0.2	-3.3	-1.4	0.6
78.29	0.85	1.9	75.82	11JN	-1.2	-3.3	-1.3	0.5
84.74	0.91	1.9	85.35	21JN	-1.7	-3.3	-1.3	0.5
91.15	0.97	1.9	94.90	1JL	-0.9	-3.4	-1.3	0.5
97.55	1.02	2.0	104.48	11JL	-0.2	-3.4	-1.4	0.4
103.93	1.07	2.0	114.10	21JL	0.2	-3.4	-1.4	0.4
110.29	1.11	2.0	123.76	31JL	0.4	-3.5	-1.4	0.3
116.65	1.15	2.0	133.47	10AU	0.7	-3.5	-1.4	0.3
122.99	1.19	2.0	143.23	20AU	1.6	-3.5	-1.5	0.2
129.33	1.22	2.0	153.05	30AU	3.0	-3.5	-1.5	0.2
135.67	1.25	2.0	162.93	9SE	0.4	-3.4	-1.6	0.1
142.00	1.27	2.0	172.87	19SE	-0.7	-3.4	-1.6	0.0
148.33	1.30	2.0	182.87	29SE	-0.9	-3.4	-1.7	-0.0
154.65	1.31	1.9	192.92	9OC	-0.8	-3.4	-1.7	-0.0
160.97	1.33	1.9	203.02	19OC	-0.7	-3.3	-1.8	0.0
167.29	1.33	1.8	213.16	29OC	-0.6	-3.3	-1.9	0.1
173.59	1.34	1.8	223.34	8NO	-0.5	-3.3	-1.9	0.2
179.88	1.33	1.7	233.53	18NO	-0.5	-3.3	-2.0	0.3
186.15	1.32	1.6	243.74	28NO	-0.2	-3.3	-2.0	0.3
192.39	1.30	1.5	253.95	8DE	1.7	-3.4	-2.1	0.4
198.60	1.27	1.4	264.15	18DE	1.7	-3.4	-2.1	0.4
204.77	1.23	1.3	274.33	28DE	0.4	-3.4	-2.1	0.5
				-239				
210.89	1.18	1.2	284.48	7JA	0.2	-3.4	-2.1	0.5
216.94	1.11	1.1	294.58	17JA	0.0	-3.5	-2.1	0.5
222.91	1.02	0.9	304.65	27JA	-0.2	-3.5	-2.1	0.5
228.78	0.91	0.7	314.66	6FE	-0.6	-3.6	-2.0	0.5
234.51	0.77	0.6	324.60	16FE	-1.2	-3.6	-2.0	0.5
240.09	0.60	0.4	334.50	26FE	-1.5	-3.7	-1.9	0.5
245.46	0.39	0.2	344.33	8MR	-0.8	-3.8	-1.8	0.5
250.57	0.13	-0.0	354.10	18MR	0.5	-3.9	-1.8	0.4
255.34	-0.18	-0.3	3.82	28MR	1.9	-3.9	-1.7	0.4
259.69	-0.55	-0.5	13.49	7AP	3.8	-4.0	-1.6	0.3
263.49	-1.00	-0.8	23.11	17AP	2.3	-4.1	-1.6	0.4
266.59	-1.54	-1.1	32.70	27AP	1.3	-4.2	-1.5	0.4
268.80	-2.17	-1.4	42.26	7MY	0.6	-4.2	-1.5	0.4
269.96	-2.89	-1.7	51.79	17MY	-0.2	-4.0	-1.4	0.4
269.89	-3.68	-2.0	61.31	27MY	-1.2	-3.6	-1.4	0.4
268.60	-4.46	-2.3	70.83	6JN	-1.7	-2.8	-1.3	0.4
266.34	-5.15	-2.5	80.35	16JN	-0.9	-3.4	-1.3	0.4
263.64	-5.64	-2.6	89.89	26JN	-0.2	-4.0	-1.3	0.4
261.26	-5.86	-2.4	99.45	6JL	0.3	-4.2	-1.3	0.4
259.82	-5.83	-2.2	109.05	16JL	0.6	-4.2	-1.3	0.3
259.65	-5.59	-1.9	118.69	26JL	1.0	-4.2	-1.3	0.3
260.83	-5.23	-1.7	128.37	5AU	2.0	-4.1	-1.3	0.3
263.22	-4.82	-1.4	138.10	15AU	2.7	-4.0	-1.3	0.2
266.62	-4.37	-1.1	147.90	25AU	0.3	-3.9	-1.3	0.1
270.85	-3.93	-0.9	157.75	4SE	-0.8	-3.8	-1.3	0.1
275.72	-3.50	-0.7	167.65	14SE	-1.0	-3.8	-1.4	0.0
281.09	-3.09	-0.5	177.62	24SE	-1.0	-3.7	-1.4	-0.1
286.85	-2.70	-0.2	187.65	4OC	-0.7	-3.6	-1.5	-0.1
292.90	-2.34	-0.1	197.72	14OC	-0.4	-3.6	-1.5	-0.2
299.17	-2.01	0.1	207.85	24OC	-0.4	-3.5	-1.6	-0.2
305.61	-1.70	0.3	218.00	3NO	-0.4	-3.5	-1.6	-0.1
312.17	-1.41	0.5	228.19	13NO	-0.0	-3.5	-1.7	-0.0
318.81	-1.14	0.6	238.39	23NO	1.9	-3.4	-1.8	0.0
325.51	-0.90	0.8	248.60	3DE	1.4	-3.4	-1.8	0.1
332.24	-0.68	0.9	258.81	13DE	0.2	-3.4	-1.9	0.2
338.99	-0.48	1.0	269.00	23DE	-0.0	-3.4	-1.9	0.2

♂			⊙	16.00UT	☿	♀	♃	♄
LONG	LAT	MAG	LONG	-238	MAGNITUDES			
345.73	-0.29	1.2	279.16	2JA	-0.1	-3.3	-2.0	0.3
352.47	-0.13	1.3	289.29	12JA	-0.3	-3.3	-2.0	0.3
359.19	0.03	1.4	299.38	22JA	-0.6	-3.3	-2.0	0.4
5.88	0.16	1.5	309.41	1FE	-1.2	-3.3	-2.0	0.4
12.54	0.29	1.5	319.40	11FE	-1.4	-3.3	-2.0	0.4
19.17	0.40	1.6	329.32	21FE	-0.8	-3.4	-2.0	0.4
25.76	0.50	1.7	339.18	3MR	0.6	-3.4	-2.0	0.4
32.32	0.60	1.8	348.99	13MR	2.4	-3.4	-1.9	0.4
38.84	0.68	1.8	358.74	23MR	3.1	-3.5	-1.9	0.3
45.34	0.75	1.9	8.43	2AP	1.7	-3.5	-1.8	0.3
51.80	0.82	1.9	18.07	12AP	1.0	-3.5	-1.7	0.3
58.23	0.88	1.9	27.68	22AP	0.4	-3.4	-1.7	0.2
64.64	0.93	2.0	37.25	2MY	-0.3	-3.4	-1.6	0.2
71.02	0.98	2.0	46.79	12MY	-1.2	-3.4	-1.5	0.3
77.40	1.02	2.0	56.32	22MY	-1.8	-3.3	-1.5	0.3
83.76	1.06	2.0	65.83	1JN	-0.8	-3.3	-1.4	0.3
90.11	1.09	2.0	75.35	11JN	-0.1	-3.3	-1.4	0.3
96.46	1.11	2.0	84.89	21JN	0.4	-3.3	-1.3	0.3
102.80	1.13	2.0	94.43	1JL	0.8	-3.3	-1.3	0.3
109.16	1.15	2.0	104.01	11JL	1.3	-3.3	-1.3	0.3
115.53	1.16	2.0	113.63	21JL	2.6	-3.4	-1.2	0.3
121.90	1.16	2.0	123.28	31JL	2.3	-3.4	-1.2	0.3
128.30	1.16	2.0	132.99	10AU	0.2	-3.4	-1.2	0.2
134.72	1.16	2.0	142.75	20AU	-0.9	-3.5	-1.2	0.2
141.16	1.15	2.0	152.57	30AU	-1.2	-3.5	-1.2	0.1
147.63	1.14	2.0	162.45	9SE	-1.0	-3.6	-1.2	0.1
154.14	1.12	2.0	172.39	19SE	-0.6	-3.6	-1.2	0.0
160.67	1.10	2.0	182.38	29SE	-0.3	-3.7	-1.3	-0.1
167.24	1.07	1.9	192.43	9OC	-0.2	-3.8	-1.3	-0.1
173.84	1.03	1.9	202.53	19OC	-0.2	-3.9	-1.3	-0.2
180.48	0.99	1.9	212.67	29OC	0.2	-4.0	-1.4	-0.3
187.15	0.94	1.8	222.84	8NO	2.2	-4.1	-1.4	-0.3
193.86	0.88	1.8	233.04	18NO	1.1	-4.2	-1.5	-0.2
200.61	0.82	1.7	243.25	28NO	0.0	-4.3	-1.6	-0.1
207.39	0.74	1.7	253.46	8DE	-0.2	-4.4	-1.6	-0.0
214.20	0.65	1.6	263.66	18DE	-0.2	-4.4	-1.7	0.0
221.05	0.55	1.5	273.83	28DE	-0.4	-4.2	-1.8	0.1
				-237				
227.93	0.44	1.4	283.99	7JA	-0.7	-3.7	-1.8	0.2
234.83	0.32	1.4	294.10	17JA	-1.2	-3.3	-1.9	0.2
241.76	0.18	1.3	304.16	27JA	-1.2	-3.8	-1.9	0.3
248.71	0.02	1.2	314.17	6FE	-0.7	-4.2	-2.0	0.3
255.67	-0.15	1.0	324.12	16FE	0.8	-4.3	-2.0	0.3
262.64	-0.34	0.9	334.02	26FE	2.9	-4.3	-2.0	0.3
269.60	-0.55	0.8	343.85	8MR	2.3	-4.2	-2.0	0.3
276.55	-0.77	0.7	353.63	18MR	1.3	-4.1	-2.0	0.3
283.47	-1.02	0.6	3.35	28MR	0.7	-4.0	-2.0	0.3
290.34	-1.28	0.4	13.03	7AP	0.3	-3.9	-1.9	0.3
297.13	-1.57	0.3	22.65	17AP	-0.3	-3.8	-1.9	0.2
303.81	-1.87	0.1	32.24	27AP	-1.2	-3.7	-1.8	0.2
310.35	-2.18	-0.0	41.80	7MY	-1.8	-3.6	-1.8	0.2
316.68	-2.52	-0.2	51.33	17MY	-0.8	-3.5	-1.7	0.2
322.76	-2.87	-0.3	60.85	27MY	-0.0	-3.5	-1.6	0.2
328.50	-3.23	-0.5	70.37	6JN	0.6	-3.4	-1.6	0.3
333.80	-3.60	-0.7	79.89	16JN	1.1	-3.4	-1.5	0.3
338.54	-3.98	-0.9	89.43	26JN	1.8	-3.4	-1.5	0.3
342.55	-4.36	-1.1	98.99	6JL	3.1	-3.4	-1.4	0.3
345.66	-4.72	-1.3	108.59	16JL	1.9	-3.3	-1.4	0.3
347.65	-5.05	-1.6	118.22	26JL	0.1	-3.3	-1.3	0.3
348.31	-5.31	-1.8	127.90	5AU	-1.0	-3.3	-1.3	0.3
347.53	-5.44	-2.0	137.63	15AU	-1.3	-3.3	-1.3	0.3
345.42	-5.38	-2.2	147.42	25AU	-1.0	-3.4	-1.2	0.2
342.41	-5.07	-2.3	157.27	4SE	-0.5	-3.4	-1.2	0.2
339.28	-4.53	-2.2	167.17	14SE	-0.2	-3.4	-1.2	0.1
336.77	-3.84	-2.0	177.14	24SE	-0.1	-3.4	-1.2	0.0
335.42	-3.11	-1.6	187.16	4OC	-0.0	-3.4	-1.2	-0.0
335.41	-2.41	-1.3	197.23	14OC	0.4	-3.5	-1.2	-0.1
336.66	-1.79	-1.0	207.35	24OC	2.5	-3.5	-1.3	-0.2
338.98	-1.26	-0.7	217.51	3NO	0.9	-3.5	-1.3	-0.2
342.17	-0.82	-0.4	227.69	13NO	-0.2	-3.5	-1.3	-0.3
346.03	-0.45	-0.1	237.89	23NO	-0.3	-3.4	-1.4	-0.3
350.43	-0.14	0.1	248.11	3DE	-0.4	-3.4	-1.4	-0.2
355.24	0.11	0.4	258.31	13DE	-0.5	-3.4	-1.5	-0.1
0.35	0.33	0.6	268.50	23DE	-0.8	-3.4	-1.5	-0.1

♂			⊙	16.00UT	☿	♀	♃	♄
LONG	LAT	MAG	LONG	-236	MAGNITUDES			
5.71	0.50	0.7	278.67	2JA	-1.1	-3.3	-1.6	-0.0
11.25	0.65	0.9	288.80	12JA	-1.1	-3.3	-1.7	0.1
16.93	0.77	1.1	298.89	22JA	-0.6	-3.3	-1.7	0.1
22.73	0.88	1.2	308.93	1FE	0.9	-3.3	-1.8	0.2
28.61	0.96	1.3	318.91	11FE	3.2	-3.4	-1.9	0.2
34.55	1.03	1.5	328.84	21FE	1.7	-3.4	-1.9	0.3
40.54	1.09	1.6	338.71	2MR	0.9	-3.4	-2.0	0.3
46.57	1.14	1.6	348.51	12MR	0.5	-3.4	-2.0	0.3
52.63	1.18	1.7	358.26	22MR	0.2	-3.4	-2.0	0.3
58.72	1.21	1.8	7.96	1AP	-0.4	-3.5	-2.1	0.3
64.83	1.23	1.9	17.61	11AP	-1.2	-3.5	-2.1	0.3
70.96	1.24	1.9	27.22	21AP	-1.7	-3.6	-2.0	0.3
77.11	1.25	1.9	36.79	1MY	-0.8	-3.6	-2.0	0.2
83.27	1.26	2.0	46.33	11MY	0.1	-3.7	-2.0	0.2
89.46	1.25	2.0	55.86	21MY	0.7	-3.8	-1.9	0.2
95.67	1.25	2.0	65.38	31MY	1.4	-3.9	-1.9	0.2
101.90	1.24	2.0	74.90	10JN	2.4	-4.0	-1.8	0.2
108.16	1.22	2.0	84.43	20JN	3.2	-4.1	-1.7	0.3
114.45	1.20	2.0	93.98	30JN	1.6	-4.2	-1.7	0.3
120.78	1.18	2.0	103.55	10JL	0.0	-4.2	-1.6	0.3
127.15	1.15	2.0	113.17	20JL	-1.1	-4.2	-1.5	0.3
133.56	1.12	2.0	122.82	30JL	-1.4	-4.0	-1.5	0.3
140.01	1.08	1.9	132.52	9AU	-0.9	-3.6	-1.4	0.3
146.51	1.04	1.9	142.28	19AU	-0.4	-3.3	-1.4	0.3
153.07	0.99	1.9	152.10	29AU	-0.1	-3.8	-1.4	0.3
159.68	0.94	1.9	161.97	8SE	0.0	-4.2	-1.3	0.2
166.35	0.88	1.9	171.91	18SE	0.2	-4.3	-1.3	0.2
173.08	0.82	1.9	181.90	28SE	0.7	-4.3	-1.3	0.1
179.87	0.75	1.8	191.94	8OC	2.9	-4.3	-1.3	0.1
186.73	0.68	1.8	202.04	18OC	0.7	-4.2	-1.3	-0.0
193.65	0.60	1.8	212.18	28OC	-0.3	-4.1	-1.3	-0.1
200.63	0.51	1.8	222.35	7NO	-0.5	-4.0	-1.3	-0.1
207.68	0.42	1.8	232.55	17NO	-0.5	-3.9	-1.3	-0.2
214.80	0.32	1.7	242.75	27NO	-0.6	-3.8	-1.3	-0.3
221.98	0.21	1.7	252.96	7DE	-0.8	-3.7	-1.4	-0.3
229.22	0.10	1.6	263.17	17DE	-0.9	-3.6	-1.4	-0.2
236.51	-0.02	1.6	273.35	27DE	-0.9	-3.6	-1.5	-0.1
				-235				
243.86	-0.15	1.5	283.50	6JA	-0.5	-3.5	-1.5	-0.1
251.27	-0.28	1.5	293.61	16JA	1.1	-3.5	-1.6	0.0
258.71	-0.42	1.4	303.68	26JA	2.9	-3.4	-1.6	0.1
266.20	-0.56	1.4	313.69	5FE	1.3	-3.4	-1.7	0.1
273.72	-0.71	1.3	323.65	15FE	0.7	-3.4	-1.8	0.2
281.26	-0.85	1.2	333.55	25FE	0.4	-3.3	-1.8	0.2
288.82	-1.00	1.2	343.38	7MR	0.1	-3.3	-1.9	0.3
296.37	-1.15	1.1	353.17	17MR	-0.4	-3.3	-2.0	0.3
303.91	-1.30	1.0	2.89	27MR	-1.2	-3.3	-2.0	0.3
311.43	-1.44	1.0	12.56	6AP	-1.7	-3.3	-2.1	0.3
318.90	-1.57	0.9	22.19	16AP	-0.8	-3.3	-2.1	0.3
326.31	-1.70	0.8	31.78	26AP	0.1	-3.3	-2.1	0.3
333.64	-1.81	0.8	41.33	6MY	1.0	-3.4	-2.1	0.3
340.86	-1.91	0.7	50.87	16MY	1.8	-3.4	-2.1	0.3
347.95	-2.00	0.6	60.39	26MY	3.1	-3.5	-2.1	0.3
354.90	-2.07	0.6	69.91	5JN	2.7	-3.5	-2.1	0.2
1.65	-2.12	0.5	79.43	15JN	1.3	-3.5	-2.0	0.2
8.18	-2.16	0.4	88.97	25JN	-0.0	-3.4	-2.0	0.3
14.45	-2.18	0.3	98.53	5JL	-1.1	-3.4	-1.9	0.3
20.40	-2.18	0.2	108.12	15JL	-1.6	-3.4	-1.9	0.4
25.97	-2.15	0.1	117.75	25JL	-0.9	-3.4	-1.8	0.4
31.09	-2.10	-0.1	127.43	4AU	-0.3	-3.3	-1.7	0.4
35.63	-2.03	-0.2	137.16	14AU	-0.0	-3.3	-1.7	0.4
39.49	-1.92	-0.4	146.94	24AU	0.2	-3.3	-1.6	0.4
42.48	-1.77	-0.6	156.78	3SE	0.4	-3.3	-1.6	0.4
44.41	-1.57	-0.8	166.69	13SE	1.0	-3.3	-1.5	0.3
45.09	-1.30	-1.0	176.65	23SE	3.2	-3.4	-1.5	0.3
44.34	-0.97	-1.2	186.67	3OC	0.6	-3.4	-1.4	0.3
42.15	-0.56	-1.4	196.74	13OC	-0.5	-3.4	-1.4	0.2
38.83	-0.11	-1.6	206.85	23OC	-0.6	-3.4	-1.4	0.1
34.99	0.36	-1.6	217.01	2NO	-0.6	-3.5	-1.4	0.1
31.45	0.79	-1.4	227.20	12NO	-0.7	-3.5	-1.4	0.0
28.90	1.13	-1.1	237.40	22NO	-0.8	-3.6	-1.4	-0.1
27.65	1.39	-0.8	247.61	2DE	-0.8	-3.7	-1.4	-0.1
27.75	1.57	-0.5	257.82	12DE	-0.8	-3.7	-1.4	-0.2
29.03	1.68	-0.2	268.01	22DE	-0.4	-3.8	-1.4	-0.2

♂ LONG	♂ LAT	♂ MAG	☉ LONG	16.00UT -234	☿	♀	♃	♄
					MAGNITUDES			
31.29	1.76	0.1	278.18	1JA	1.3	-3.9	-1.4	-0.1
34.34	1.80	0.4	288.31	11JA	2.4	-4.0	-1.5	-0.1
38.00	1.81	0.6	298.40	21JA	0.9	-4.1	-1.5	0.0
42.14	1.81	0.8	308.45	31JA	0.4	-4.2	-1.6	0.1
46.66	1.80	1.0	318.43	10FE	0.2	-4.3	-1.6	0.1
51.48	1.78	1.1	328.36	20FE	-0.0	-4.3	-1.7	0.2
56.53	1.76	1.3	338.23	2MR	-0.5	-4.2	-1.8	0.3
61.78	1.73	1.4	348.04	12MR	-1.2	-3.9	-1.8	0.3
67.18	1.69	1.5	357.79	22MR	-1.6	-3.4	-1.9	0.3
72.71	1.65	1.6	7.49	1AP	-0.8	-3.2	-2.0	0.4
78.36	1.61	1.7	17.15	11AP	0.2	-3.8	-2.0	0.4
84.09	1.56	1.7	26.75	21AP	1.3	-4.1	-2.1	0.4
89.92	1.51	1.8	36.33	1MY	2.4	-4.2	-2.2	0.4
95.82	1.46	1.8	45.87	11MY	3.6	-4.2	-2.2	0.4
101.79	1.41	1.9	55.40	21MY	2.1	-4.1	-2.2	0.4
107.83	1.35	1.9	64.92	31MY	1.0	-4.0	-2.3	0.4
113.94	1.30	1.9	74.44	10JN	-0.1	-3.9	-2.3	0.4
120.11	1.23	2.0	83.96	20JN	-1.2	-3.8	-2.3	0.3
126.35	1.17	2.0	93.51	30JN	-1.7	-3.7	-2.2	0.4
132.66	1.10	2.0	103.09	10JL	-0.9	-3.7	-2.2	0.4
139.03	1.03	2.0	112.70	20JL	-0.3	-3.6	-2.2	0.4
145.49	0.96	1.9	122.35	30JL	0.1	-3.5	-2.1	0.5
152.01	0.89	1.9	132.05	9AU	0.3	-3.5	-2.0	0.5
158.61	0.81	1.9	141.81	19AU	0.6	-3.5	-2.0	0.5
165.29	0.73	1.9	151.62	29AU	1.3	-3.4	-1.9	0.5
172.05	0.64	1.8	161.49	8SE	3.3	-3.4	-1.8	0.5
178.89	0.55	1.8	171.42	18SE	0.5	-3.4	-1.8	0.5
185.82	0.46	1.7	181.41	28SE	-0.6	-3.4	-1.7	0.5
192.83	0.36	1.7	191.46	8OC	-0.8	-3.4	-1.7	0.4
199.92	0.26	1.6	201.55	18OC	-0.8	-3.4	-1.6	0.4
207.10	0.16	1.6	211.69	28OC	-0.8	-3.4	-1.6	0.3
214.35	0.06	1.6	221.85	7NO	-0.6	-3.4	-1.5	0.3
221.69	-0.05	1.6	232.04	17NO	-0.6	-3.4	-1.5	0.2
229.10	-0.16	1.6	242.26	27NO	-0.6	-3.4	-1.5	0.1
236.59	-0.27	1.6	252.46	7DE	-0.3	-3.4	-1.5	0.1
244.14	-0.38	1.5	262.66	17DE	1.5	-3.4	-1.5	-0.0
251.75	-0.49	1.5	272.85	27DE	2.0	-3.4	-1.5	-0.1
				-233				
259.41	-0.61	1.5	283.00	6JA	0.6	-3.5	-1.5	-0.0
267.11	-0.71	1.5	293.11	16JA	0.3	-3.5	-1.5	0.0
274.85	-0.82	1.5	303.19	26JA	0.1	-3.5	-1.5	0.1
282.61	-0.92	1.4	313.20	5FE	-0.1	-3.4	-1.5	0.1
290.39	-1.01	1.4	323.16	15FE	-0.5	-3.4	-1.6	0.2
298.16	-1.10	1.4	333.06	25FE	-1.2	-3.4	-1.6	0.3
305.93	-1.17	1.4	342.90	7MR	-1.5	-3.4	-1.7	0.3
313.67	-1.24	1.3	352.69	17MR	-0.8	-3.3	-1.7	0.4
321.37	-1.29	1.3	2.42	27MR	0.4	-3.3	-1.8	0.4
329.03	-1.33	1.3	12.09	6AP	1.6	-3.3	-1.9	0.5
336.62	-1.36	1.3	21.72	16AP	3.2	-3.3	-1.9	0.5
344.14	-1.38	1.2	31.31	26AP	2.8	-3.3	-2.0	0.5
351.57	-1.37	1.2	40.87	6MY	1.6	-3.4	-2.1	0.5
358.90	-1.36	1.2	50.41	16MY	0.7	-3.4	-2.1	0.5
6.12	-1.33	1.2	59.93	26MY	-0.1	-3.4	-2.2	0.5
13.22	-1.29	1.1	69.45	5JN	-1.2	-3.4	-2.3	0.5
20.18	-1.23	1.1	78.97	15JN	-1.7	-3.5	-2.3	0.5
26.99	-1.16	1.1	88.51	25JN	-0.9	-3.5	-2.4	0.5
33.64	-1.07	1.0	98.06	5JL	-0.2	-3.6	-2.4	0.5
40.10	-0.97	1.0	107.66	15JL	0.2	-3.7	-2.4	0.5
46.36	-0.85	0.9	117.29	25JL	0.5	-3.7	-2.4	0.6
52.39	-0.72	0.9	126.96	4AU	0.8	-3.8	-2.4	0.6
58.15	-0.57	0.8	136.69	14AU	1.7	-3.9	-2.3	0.7
63.59	-0.39	0.7	146.47	24AU	3.0	-4.1	-2.3	0.7
68.67	-0.20	0.6	156.31	3SE	0.4	-4.2	-2.2	0.7
73.30	0.02	0.5	166.21	13SE	-0.8	-4.3	-2.2	0.7
77.39	0.27	0.3	176.17	23SE	-0.9	-4.3	-2.1	0.7
80.82	0.55	0.1	186.18	3OC	-0.9	-4.3	-2.0	0.7
83.44	0.88	-0.0	196.25	13OC	-0.7	-4.1	-1.9	0.7
85.04	1.26	-0.3	206.37	23OC	-0.5	-3.6	-1.9	0.6
85.46	1.67	-0.5	216.51	2NO	-0.5	-3.1	-1.8	0.6
84.52	2.12	-0.7	226.70	12NO	-0.5	-3.9	-1.8	0.5
82.21	2.58	-0.9	236.90	22NO	-0.1	-4.3	-1.7	0.5
78.79	2.98	-1.1	247.11	2DE	1.7	-4.4	-1.7	0.4
74.82	3.28	-1.2	257.32	12DE	1.6	-4.4	-1.6	0.3
71.08	3.43	-1.0	267.51	22DE	0.4	-4.3	-1.6	0.3

♂ LONG	♂ LAT	♂ MAG	☉ LONG	16.00UT -232	☿	♀	♃	♄
					MAGNITUDES			
68.24	3.45	-0.7	277.68	1JA	0.1	-4.2	-1.6	0.2
66.64	3.37	-0.4	287.82	11JA	-0.0	-4.1	-1.6	0.1
66.36	3.23	-0.2	297.91	21JA	-0.2	-4.0	-1.5	0.2
67.28	3.06	0.1	307.95	31JA	-0.6	-3.8	-1.5	0.2
69.19	2.88	0.3	317.95	10FE	-1.2	-3.8	-1.6	0.3
71.93	2.71	0.6	327.88	20FE	-1.4	-3.7	-1.6	0.3
75.31	2.54	0.7	337.75	1MR	-0.8	-3.6	-1.6	0.4
79.22	2.38	0.9	347.57	11MR	0.5	-3.5	-1.6	0.4
83.54	2.23	1.1	357.33	21MR	2.1	-3.5	-1.7	0.5
88.19	2.09	1.2	7.03	31MR	3.7	-3.4	-1.7	0.5
93.12	1.96	1.3	16.68	10AP	2.1	-3.4	-1.7	0.6
98.28	1.83	1.4	26.29	20AP	1.2	-3.3	-1.8	0.6
103.64	1.70	1.5	35.87	30AP	0.6	-3.3	-1.9	0.7
109.16	1.58	1.6	45.42	10MY	-0.2	-3.3	-1.9	0.7
114.84	1.47	1.6	54.94	20MY	-1.2	-3.3	-2.0	0.7
120.65	1.35	1.7	64.46	30MY	-1.8	-3.3	-2.1	0.7
126.59	1.24	1.7	73.98	9JN	-0.9	-3.3	-2.1	0.7
132.65	1.13	1.8	83.51	19JN	-0.1	-3.3	-2.2	0.7
138.82	1.02	1.8	93.05	29JN	0.3	-3.4	-2.3	0.7
145.10	0.91	1.8	102.63	9JL	0.7	-3.4	-2.3	0.7
151.49	0.80	1.8	112.23	19JL	1.1	-3.4	-2.4	0.7
157.98	0.69	1.8	121.88	29JL	2.2	-3.5	-2.4	0.8
164.58	0.58	1.8	131.58	8AU	2.6	-3.5	-2.4	0.8
171.29	0.47	1.8	141.33	18AU	0.3	-3.5	-2.5	0.9
178.10	0.36	1.7	151.14	28AU	-0.9	-3.5	-2.4	0.9
185.01	0.25	1.7	161.01	7SE	-1.1	-3.4	-2.4	0.9
192.03	0.13	1.7	170.94	17SE	-1.0	-3.4	-2.4	0.9
199.14	0.02	1.7	180.92	27SE	-0.6	-3.4	-2.3	0.9
206.36	-0.08	1.6	190.97	7OC	-0.4	-3.4	-2.3	0.9
213.68	-0.19	1.6	201.05	17OC	-0.3	-3.3	-2.2	0.9
221.08	-0.30	1.5	211.19	27OC	-0.3	-3.3	-2.1	0.9
228.57	-0.40	1.5	221.36	6NO	0.1	-3.3	-2.1	0.9
236.15	-0.50	1.4	231.55	16NO	1.9	-3.3	-2.0	0.8
243.79	-0.60	1.4	241.75	26NO	1.3	-3.3	-1.9	0.8
251.51	-0.69	1.4	251.96	6DE	0.1	-3.4	-1.9	0.7
259.27	-0.78	1.4	262.16	16DE	-0.1	-3.4	-1.8	0.6
267.08	-0.86	1.4	272.35	26DE	-0.1	-3.4	-1.8	0.6
				-231				
274.92	-0.93	1.4	282.51	5JA	-0.3	-3.5	-1.7	0.5
282.79	-0.99	1.4	292.62	15JA	-0.6	-3.5	-1.7	0.4
290.66	-1.04	1.4	302.69	25JA	-1.2	-3.5	-1.6	0.4
298.53	-1.08	1.4	312.71	4FE	-1.3	-3.6	-1.6	0.4
306.38	-1.11	1.4	322.68	14FE	-0.8	-3.6	-1.6	0.5
314.20	-1.13	1.4	332.58	24FE	0.6	-3.7	-1.6	0.5
321.98	-1.13	1.4	342.43	6MR	2.6	-3.8	-1.6	0.6
329.71	-1.13	1.4	352.21	16MR	2.8	-3.9	-1.6	0.6
337.38	-1.11	1.4	1.94	26MR	1.5	-4.0	-1.6	0.7
344.97	-1.08	1.5	11.63	5AP	0.9	-4.0	-1.6	0.7
352.49	-1.03	1.5	21.26	15AP	0.4	-4.1	-1.6	0.8
359.92	-0.98	1.5	30.85	25AP	-0.3	-4.2	-1.7	0.8
7.25	-0.92	1.5	40.41	5MY	-1.2	-4.2	-1.7	0.9
14.49	-0.84	1.5	49.95	15MY	-1.8	-4.0	-1.7	0.9
21.63	-0.76	1.5	59.47	25MY	-0.9	-3.6	-1.8	0.9
28.65	-0.67	1.5	68.99	4JN	-0.1	-2.7	-1.8	1.0
35.57	-0.57	1.5	78.51	14JN	0.5	-3.4	-1.9	1.0
42.37	-0.46	1.5	88.05	24JN	0.9	-4.0	-2.0	1.0
49.05	-0.34	1.5	97.60	4JL	1.5	-4.2	-2.0	1.0
55.60	-0.22	1.5	107.19	14JL	2.8	-4.2	-2.1	1.0
62.01	-0.09	1.4	116.82	24JL	2.2	-4.2	-2.2	1.0
68.28	0.05	1.4	126.49	3AU	0.2	-4.1	-2.2	1.0
74.39	0.20	1.4	136.21	13AU	-1.0	-4.0	-2.3	1.0
80.33	0.36	1.3	145.99	23AU	-1.2	-3.9	-2.4	1.1
86.07	0.54	1.3	155.83	2SE	-1.0	-3.8	-2.4	1.1
91.58	0.72	1.2	165.72	12SE	-0.5	-3.8	-2.4	1.2
96.83	0.92	1.1	175.68	22SE	-0.3	-3.7	-2.4	1.2
101.77	1.14	1.0	185.70	2OC	-0.2	-3.6	-2.4	1.2
106.33	1.39	0.8	195.76	12OC	-0.1	-3.6	-2.4	1.2
110.43	1.66	0.7	205.87	22OC	0.3	-3.5	-2.4	1.2
113.97	1.96	0.5	216.02	1NO	2.2	-3.5	-2.3	1.2
116.81	2.29	0.3	226.20	11NO	1.1	-3.5	-2.3	1.2
118.80	2.66	0.1	236.41	21NO	-0.1	-3.4	-2.2	1.2
119.73	3.07	-0.1	246.62	1DE	-0.2	-3.4	-2.1	1.1
119.45	3.49	-0.4	256.82	11DE	-0.3	-3.4	-2.1	1.0
117.84	3.91	-0.6	267.02	21DE	-0.4	-3.4	-2.0	1.0
114.98	4.25	-0.8	277.19	31DE	-0.7	-3.3	-1.9	0.9

♂ LONG	LAT	MAG	☉ LONG	16.00UT	☿	♀	♃	♄
				-230	MAGNITUDES			
111.26	4.47	-1.0	287.32	10JA	-1.1	-3.3	-1.8	0.8
107.31	4.51	-0.9	297.42	20JA	-1.2	-3.3	-1.8	0.8
103.88	4.39	-0.7	307.47	30JA	-0.7	-3.3	-1.7	0.7
101.49	4.13	-0.5	317.46	9FE	0.8	-3.3	-1.7	0.7
100.38	3.82	-0.3	327.40	19FE	3.0	-3.4	-1.6	0.7
100.54	3.47	-0.0	337.27	1MR	2.1	-3.4	-1.6	0.7
101.83	3.14	0.2	347.09	11MR	1.1	-3.4	-1.6	0.8
104.05	2.82	0.4	356.85	21MR	0.7	-3.5	-1.5	0.8
107.05	2.53	0.6	6.55	31MR	0.3	-3.5	-1.5	0.9
110.67	2.26	0.7	16.21	10AP	-0.3	-3.5	-1.5	1.0
114.79	2.02	0.9	25.82	20AP	-1.2	-3.4	-1.5	1.0
119.33	1.79	1.0	35.40	30AP	-1.8	-3.4	-1.5	1.1
124.21	1.58	1.1	44.95	10MY	-0.9	-3.4	-1.5	1.1
129.39	1.39	1.2	54.48	20MY	-0.0	-3.3	-1.6	1.1
134.82	1.20	1.3	64.00	30MY	0.6	-3.3	-1.6	1.2
140.47	1.03	1.3	73.51	9JN	1.2	-3.3	-1.6	1.2
146.33	0.86	1.4	83.04	19JN	2.0	-3.3	-1.7	1.2
152.36	0.70	1.4	92.59	29JN	3.3	-3.3	-1.7	1.3
158.57	0.55	1.5	102.16	9JL	1.8	-3.3	-1.8	1.3
164.93	0.40	1.5	111.77	19JL	0.1	-3.4	-1.8	1.3
171.45	0.26	1.5	121.41	29JL	-1.0	-3.4	-1.9	1.3
178.10	0.12	1.5	131.11	8AU	-1.4	-3.4	-2.0	1.3
184.90	-0.01	1.5	140.86	18AU	-1.0	-3.5	-2.0	1.3
191.82	-0.14	1.5	150.66	28AU	-0.5	-3.5	-2.1	1.3
198.88	-0.26	1.5	160.53	7SE	-0.2	-3.6	-2.2	1.3
206.06	-0.37	1.5	170.46	17SE	-0.0	-3.6	-2.2	1.3
213.35	-0.48	1.5	180.44	27SE	0.1	-3.7	-2.3	1.3
220.75	-0.59	1.5	190.48	7OC	0.5	-3.8	-2.3	1.3
228.25	-0.68	1.5	200.57	17OC	2.5	-3.9	-2.3	1.3
235.84	-0.77	1.5	210.69	27OC	0.9	-4.0	-2.3	1.3
243.51	-0.85	1.4	220.86	6NO	-0.2	-4.1	-2.3	1.2
251.26	-0.92	1.4	231.06	16NO	-0.4	-4.2	-2.3	1.2
259.06	-0.97	1.4	241.26	26NO	-0.4	-4.3	-2.3	1.2
266.91	-1.02	1.4	251.47	6DE	-0.5	-4.4	-2.2	1.1
274.79	-1.06	1.4	261.67	16DE	-0.8	-4.4	-2.2	1.1
282.68	-1.08	1.4	271.86	26DE	-1.0	-4.1	-2.1	1.0
				-229				
290.59	-1.09	1.4	282.01	5JA	-1.0	-3.6	-2.0	1.0
298.48	-1.10	1.3	292.13	15JA	-0.6	-3.3	-2.0	0.9
306.35	-1.08	1.3	302.20	25JA	0.9	-3.9	-1.9	0.9
314.18	-1.06	1.3	312.23	4FE	3.1	-4.2	-1.8	0.9
321.97	-1.03	1.3	322.19	14FE	1.6	-4.3	-1.8	0.8
329.70	-0.98	1.4	332.10	24FE	0.8	-4.3	-1.7	0.8
337.37	-0.93	1.4	341.95	6MR	0.5	-4.2	-1.6	0.9
344.96	-0.87	1.4	351.74	16MR	0.1	-4.1	-1.6	1.0
352.47	-0.80	1.5	1.47	26MR	-0.4	-4.0	-1.6	1.0
359.90	-0.72	1.5	11.15	5AP	-1.2	-3.9	-1.5	1.1
7.24	-0.64	1.6	20.79	15AP	-1.7	-3.8	-1.5	1.2
14.49	-0.55	1.6	30.38	25AP	-0.9	-3.7	-1.5	1.2
21.66	-0.45	1.6	39.94	5MY	0.1	-3.6	-1.4	1.3
28.72	-0.35	1.7	49.48	15MY	0.8	-3.5	-1.4	1.3
35.70	-0.25	1.7	59.01	25MY	1.5	-3.5	-1.4	1.4
42.59	-0.14	1.7	68.53	4JN	2.6	-3.4	-1.4	1.4
49.38	-0.03	1.7	78.05	14JN	3.1	-3.4	-1.5	1.4
56.09	0.08	1.8	87.58	24JN	1.5	-3.4	-1.5	1.4
62.70	0.19	1.8	97.14	4JL	0.1	-3.4	-1.5	1.4
69.23	0.31	1.8	106.73	14JL	-1.1	-3.3	-1.5	1.4
75.66	0.43	1.8	116.35	24JL	-1.5	-3.3	-1.6	1.3
82.00	0.55	1.7	126.02	3AU	-0.9	-3.3	-1.6	1.3
88.24	0.67	1.7	135.75	13AU	-0.4	-3.3	-1.7	1.2
94.37	0.80	1.7	145.52	23AU	-0.1	-3.4	-1.7	1.2
100.40	0.94	1.7	155.36	2SE	0.1	-3.4	-1.8	1.1
106.30	1.08	1.6	165.25	12SE	0.2	-3.4	-1.8	1.2
112.06	1.22	1.6	175.20	22SE	0.8	-3.4	-1.9	1.2
117.66	1.38	1.5	185.21	2OC	2.9	-3.4	-2.0	1.2
123.08	1.54	1.4	195.27	12OC	0.7	-3.5	-2.0	1.2
128.28	1.72	1.3	205.38	22OC	-0.4	-3.5	-2.1	1.1
133.22	1.90	1.2	215.53	1NO	-0.5	-3.5	-2.1	1.1
137.85	2.11	1.0	225.71	11NO	-0.5	-3.5	-2.2	1.1
142.09	2.33	0.9	235.90	21NO	-0.6	-3.4	-2.2	1.1
145.86	2.58	0.7	246.12	1DE	-0.8	-3.4	-2.2	1.0
149.03	2.85	0.5	256.32	11DE	-0.8	-3.4	-2.2	1.0
151.48	3.14	0.3	266.52	21DE	-0.9	-3.4	-2.2	1.0
153.02	3.45	0.0	276.69	31DE	-0.5	-3.3	-2.2	0.9

♂ LONG	LAT	MAG	☉ LONG	16.00UT	☿	♀	♃	♄
				-228	MAGNITUDES			
153.48	3.77	-0.3	286.83	10JA	1.1	-3.3	-2.1	0.9
152.69	4.07	-0.5	296.93	20JA	2.8	-3.3	-2.1	0.8
150.62	4.30	-0.8	306.98	30JA	1.1	-3.3	-2.0	0.8
147.44	4.42	-1.0	316.97	9FE	0.6	-3.4	-1.9	0.7
143.63	4.37	-1.1	326.91	19FE	0.3	-3.4	-1.9	0.7
139.89	4.15	-1.0	336.79	29FE	0.0	-3.4	-1.8	0.7
136.89	3.79	-0.8	346.61	10MR	-0.4	-3.4	-1.7	0.7
135.06	3.36	-0.6	356.37	20MR	-1.2	-3.4	-1.6	0.7
134.55	2.91	-0.4	6.08	30MR	-1.7	-3.5	-1.6	0.8
135.28	2.47	-0.2	15.74	9AP	-0.9	-3.5	-1.5	0.9
137.09	2.06	0.0	25.35	19AP	0.2	-3.6	-1.5	0.9
139.81	1.69	0.2	34.94	29AP	1.1	-3.6	-1.4	1.0
143.28	1.36	0.4	44.48	9MY	2.0	-3.7	-1.4	1.1
147.36	1.06	0.5	54.01	19MY	3.4	-3.8	-1.4	1.1
151.95	0.79	0.6	63.54	29MY	2.5	-3.9	-1.4	1.2
156.96	0.55	0.7	73.05	8JN	1.2	-4.0	-1.3	1.2
162.33	0.32	0.8	82.58	18JN	-0.0	-4.1	-1.3	1.2
168.02	0.12	0.9	92.13	28JN	-1.1	-4.2	-1.3	1.2
173.98	-0.07	1.0	101.70	8JL	-1.6	-4.2	-1.3	1.2
180.18	-0.24	1.0	111.30	18JL	-0.9	-4.2	-1.4	1.2
186.61	-0.40	1.1	120.95	28JL	-0.3	-4.0	-1.4	1.2
193.24	-0.55	1.1	130.64	7AU	0.0	-3.5	-1.4	1.2
200.05	-0.68	1.1	140.39	17AU	0.2	-3.3	-1.4	1.1
207.03	-0.80	1.2	150.19	27AU	0.4	-3.8	-1.5	1.1
214.16	-0.90	1.2	160.05	6SE	1.1	-4.2	-1.5	1.0
221.43	-0.99	1.2	169.97	16SE	3.2	-4.3	-1.6	1.0
228.83	-1.07	1.2	179.96	26SE	0.6	-4.3	-1.6	1.0
236.34	-1.13	1.3	189.99	6OC	-0.5	-4.2	-1.7	1.0
243.95	-1.18	1.3	200.07	16OC	-0.7	-4.2	-1.8	1.0
251.64	-1.22	1.3	210.21	26OC	-0.7	-4.1	-1.8	1.0
259.39	-1.24	1.3	220.37	5NO	-0.7	-4.0	-1.9	1.0
267.20	-1.25	1.3	230.56	15NO	-0.7	-3.9	-1.9	1.0
275.04	-1.24	1.3	240.77	25NO	-0.7	-3.8	-2.0	1.0
282.90	-1.22	1.4	250.97	5DE	-0.7	-3.7	-2.0	1.0
290.76	-1.19	1.4	261.18	15DE	-0.4	-3.6	-2.1	0.9
298.61	-1.15	1.4	271.37	25DE	1.3	-3.6	-2.1	0.9
				-227				
306.43	-1.09	1.4	281.52	4JA	2.3	-3.5	-2.1	0.8
314.22	-1.03	1.4	291.65	14JA	0.8	-3.5	-2.1	0.8
321.96	-0.96	1.5	301.72	24JA	0.4	-3.4	-2.1	0.7
329.63	-0.88	1.5	311.74	3FE	0.2	-3.4	-2.0	0.7
337.24	-0.79	1.5	321.72	13FE	-0.1	-3.4	-2.0	0.6
344.77	-0.70	1.5	331.63	23FE	-0.5	-3.3	-1.9	0.6
352.22	-0.61	1.5	341.47	5MR	-1.2	-3.3	-1.9	0.6
359.59	-0.51	1.6	351.27	15MR	-1.6	-3.3	-1.8	0.5
6.88	-0.41	1.6	1.00	25MR	-0.9	-3.3	-1.7	0.6
14.08	-0.31	1.6	10.68	4AP	0.3	-3.3	-1.7	0.6
21.20	-0.20	1.6	20.32	14AP	1.4	-3.3	-1.6	0.7
28.23	-0.10	1.6	29.92	24AP	2.7	-3.3	-1.5	0.8
35.19	-0.00	1.7	39.48	4MY	3.3	-3.4	-1.5	0.8
42.06	0.10	1.7	49.02	14MY	1.9	-3.4	-1.4	0.9
48.86	0.20	1.8	58.55	24MY	0.9	-3.5	-1.4	0.9
55.58	0.30	1.8	68.06	3JN	-0.1	-3.5	-1.3	1.0
62.24	0.40	1.9	77.59	13JN	-1.1	-3.5	-1.3	1.0
68.83	0.50	1.9	87.12	23JN	-1.7	-3.4	-1.3	1.1
75.36	0.59	1.9	96.67	3JL	-0.9	-3.4	-1.3	1.1
81.83	0.69	1.9	106.26	13JL	-0.3	-3.4	-1.3	1.1
88.24	0.78	1.9	115.88	23JL	0.1	-3.4	-1.3	1.1
94.59	0.87	1.9	125.55	2AU	0.4	-3.3	-1.3	1.1
100.89	0.96	1.9	135.27	12AU	0.7	-3.3	-1.3	1.1
107.13	1.06	1.9	145.04	22AU	1.4	-3.3	-1.3	1.0
113.31	1.15	1.9	154.87	1SE	3.2	-3.3	-1.3	1.0
119.43	1.24	1.9	164.76	11SE	0.5	-3.3	-1.3	0.9
125.48	1.33	1.8	174.71	21SE	-0.7	-3.4	-1.4	0.9
131.46	1.43	1.8	184.72	1OC	-0.8	-3.4	-1.4	0.9
137.36	1.52	1.7	194.78	11OC	-0.8	-3.4	-1.5	0.9
143.15	1.62	1.6	204.89	21OC	-0.8	-3.4	-1.5	0.9
148.84	1.72	1.6	215.03	31OC	-0.6	-3.5	-1.6	0.9
154.40	1.82	1.5	225.21	10NO	-0.6	-3.5	-1.6	0.9
159.79	1.93	1.3	235.41	20NO	-0.6	-3.6	-1.7	0.9
165.00	2.04	1.2	245.62	30NO	-0.2	-3.7	-1.8	0.9
169.97	2.16	1.1	255.83	10DE	1.5	-3.7	-1.8	0.9
174.66	2.28	0.9	266.03	20DE	1.9	-3.8	-1.9	0.9
179.00	2.41	0.7	276.20	30DE	0.5	-3.9	-1.9	0.8

♂			☉	16.00UT	☿	♀	♃	♄
LONG	LAT	MAG	LONG	-226	MAGNITUDES			
182.91	2.53	0.5	286.34	9JA	0.2	-4.0	-2.0	0.8
186.27	2.66	0.3	296.44	19JA	0.1	-4.1	-2.0	0.7
188.95	2.78	0.0	306.49	29JA	-0.2	-4.2	-2.0	0.7
190.78	2.89	-0.3	316.50	8FE	-0.5	-4.3	-2.0	0.6
191.59	2.97	-0.5	326.43	18FE	-1.2	-4.3	-2.0	0.6
191.21	2.99	-0.8	336.32	28FE	-1.5	-4.2	-2.0	0.5
189.55	2.94	-1.1	346.14	10MR	-0.9	-3.9	-2.0	0.5
186.76	2.76	-1.4	355.90	20MR	0.4	-3.3	-1.9	0.4
183.24	2.45	-1.6	5.61	30MR	1.8	-3.2	-1.8	0.4
179.66	2.02	-1.5	15.28	9AP	3.5	-3.8	-1.8	0.5
176.75	1.52	-1.3	24.89	19AP	2.5	-4.1	-1.7	0.5
174.98	1.02	-1.1	34.47	29AP	1.4	-4.2	-1.7	0.6
174.56	0.54	-0.9	44.02	9MY	0.7	-4.2	-1.6	0.7
175.46	0.12	-0.7	53.55	19MY	-0.1	-4.1	-1.5	0.7
177.52	-0.25	-0.5	63.07	29MY	-1.1	-4.0	-1.5	0.8
180.57	-0.56	-0.3	72.59	8JN	-1.8	-3.9	-1.4	0.8
184.42	-0.82	-0.2	82.12	18JN	-0.9	-3.8	-1.4	0.9
188.95	-1.05	-0.0	91.66	28JN	-0.2	-3.7	-1.3	0.9
194.03	-1.23	0.1	101.23	8JL	0.2	-3.7	-1.3	1.0
199.58	-1.39	0.2	110.83	18JL	0.5	-3.6	-1.3	1.0
205.51	-1.51	0.3	120.47	28JL	0.9	-3.5	-1.2	1.0
211.77	-1.61	0.4	130.17	7AU	1.8	-3.5	-1.2	1.0
218.32	-1.68	0.5	139.91	17AU	2.9	-3.5	-1.2	1.0
225.10	-1.73	0.6	149.71	27AU	0.4	-3.4	-1.2	0.9
232.09	-1.76	0.7	159.57	6SE	-0.8	-3.4	-1.2	0.9
239.26	-1.77	0.7	169.49	16SE	-1.0	-3.4	-1.2	0.9
246.56	-1.76	0.8	179.47	26SE	-1.0	-3.4	-1.3	0.8
253.99	-1.73	0.9	189.51	6OC	-0.7	-3.4	-1.3	0.8
261.51	-1.68	0.9	199.59	16OC	-0.5	-3.4	-1.3	0.8
269.10	-1.62	1.0	209.71	26OC	-0.4	-3.4	-1.4	0.8
276.75	-1.54	1.1	219.88	5NO	-0.4	-3.4	-1.4	0.9
284.42	-1.45	1.1	230.06	15NO	-0.1	-3.4	-1.5	0.9
292.10	-1.35	1.2	240.27	25NO	1.7	-3.4	-1.5	0.9
299.77	-1.25	1.2	250.48	5DE	1.6	-3.4	-1.6	0.9
307.43	-1.14	1.3	260.68	15DE	0.3	-3.4	-1.6	0.9
315.04	-1.02	1.4	270.87	25DE	0.0	-3.4	-1.7	0.8
				-225				
322.62	-0.90	1.4	281.03	4JA	-0.1	-3.5	-1.8	0.8
330.13	-0.78	1.5	291.15	14JA	-0.2	-3.5	-1.8	0.8
337.59	-0.66	1.5	301.23	24JA	-0.6	-3.5	-1.9	0.7
344.97	-0.54	1.6	311.26	3FE	-1.2	-3.4	-1.9	0.7
352.28	-0.42	1.6	321.23	13FE	-1.4	-3.4	-2.0	0.6
359.52	-0.30	1.7	331.14	23FE	-0.8	-3.4	-2.0	0.6
6.68	-0.19	1.7	341.00	5MR	0.5	-3.4	-2.0	0.5
13.76	-0.08	1.7	350.79	15MR	2.2	-3.3	-2.0	0.5
20.77	0.03	1.8	0.53	25MR	3.3	-3.3	-2.0	0.4
27.70	0.14	1.8	10.22	4AP	1.9	-3.3	-2.0	0.4
34.56	0.23	1.8	19.85	14AP	1.1	-3.3	-1.9	0.4
41.36	0.33	1.8	29.45	24AP	0.5	-3.3	-1.9	0.4
48.09	0.42	1.8	39.02	4MY	-0.2	-3.4	-1.8	0.5
54.76	0.51	1.8	48.56	14MY	-1.1	-3.4	-1.8	0.6
61.38	0.59	1.8	58.08	24MY	-1.8	-3.4	-1.7	0.6
67.95	0.67	1.8	67.60	3JN	-0.9	-3.4	-1.6	0.7
74.47	0.75	1.9	77.12	13JN	-0.1	-3.5	-1.6	0.7
80.95	0.82	1.9	86.66	23JN	0.4	-3.5	-1.5	0.8
87.40	0.88	2.0	96.21	3JL	0.7	-3.6	-1.5	0.8
93.81	0.95	2.0	105.79	13JL	1.2	-3.7	-1.4	0.9
100.20	1.01	2.0	115.41	23JL	2.4	-3.8	-1.4	0.9
106.56	1.07	2.0	125.08	2AU	2.5	-3.8	-1.3	0.9
112.89	1.12	2.0	134.79	12AU	0.3	-3.9	-1.3	0.9
119.21	1.17	2.0	144.57	22AU	-0.9	-4.1	-1.3	0.9
125.51	1.22	2.0	154.39	1SE	-1.1	-4.2	-1.2	0.9
131.79	1.27	2.0	164.28	11SE	-1.0	-4.3	-1.2	0.9
138.05	1.31	2.0	174.23	21SE	-0.6	-4.3	-1.2	0.8
144.29	1.35	1.9	184.23	1OC	-0.4	-4.3	-1.2	0.8
150.51	1.39	1.9	194.29	11OC	-0.3	-4.1	-1.2	0.8
156.70	1.42	1.8	204.40	21OC	-0.2	-3.6	-1.3	0.7
162.86	1.46	1.8	214.54	31OC	0.1	-3.1	-1.3	0.8
168.99	1.48	1.7	224.72	10NO	2.0	-3.9	-1.3	0.8
175.08	1.50	1.6	234.92	20NO	1.3	-4.3	-1.3	0.8
181.11	1.52	1.5	245.13	30NO	0.1	-4.4	-1.4	0.8
187.08	1.53	1.4	255.34	10DE	-0.1	-4.4	-1.4	0.8
192.96	1.53	1.3	265.54	20DE	-0.2	-4.3	-1.5	0.8
198.76	1.53	1.2	275.71	30DE	-0.3	-4.2	-1.6	0.8

♂			☉	16.00UT	☿	♀	♃	♄
LONG	LAT	MAG	LONG	-224	MAGNITUDES			
204.43	1.51	1.0	285.85	9JA	-0.7	-4.1	-1.6	0.8
209.97	1.47	0.9	295.96	19JA	-1.2	-3.9	-1.7	0.8
215.32	1.42	0.7	306.01	29JA	-1.3	-3.8	-1.8	0.7
220.46	1.35	0.5	316.01	8FE	-0.8	-3.8	-1.8	0.7
225.31	1.25	0.3	325.96	18FE	0.6	-3.7	-1.9	0.6
229.83	1.11	0.1	335.84	28FE	2.7	-3.6	-2.0	0.6
233.91	0.93	-0.2	345.67	9MR	2.5	-3.5	-2.0	0.5
237.43	0.69	-0.5	355.44	19MR	1.4	-3.5	-2.0	0.5
240.26	0.39	-0.7	5.15	29MR	0.8	-3.4	-2.1	0.4
242.22	0.00	-1.1	14.81	8AP	0.4	-3.4	-2.1	0.4
243.11	-0.48	-1.4	24.43	18AP	-0.3	-3.3	-2.1	0.3
242.81	-1.06	-1.7	34.01	28AP	-1.1	-3.3	-2.0	0.3
241.28	-1.71	-2.0	43.57	8MY	-1.8	-3.3	-2.0	0.4
238.75	-2.39	-2.3	53.10	18MY	-0.9	-3.3	-2.0	0.5
235.76	-3.02	-2.3	62.62	28MY	-0.1	-3.3	-1.9	0.5
233.00	-3.52	-2.2	72.14	7JN	0.5	-3.3	-1.9	0.6
231.17	-3.86	-2.0	81.66	17JN	1.0	-3.3	-1.8	0.6
230.63	-4.03	-1.8	91.20	27JN	1.7	-3.4	-1.7	0.7
231.46	-4.07	-1.5	100.77	7JL	3.0	-3.4	-1.7	0.7
233.57	-4.01	-1.3	110.37	17JL	2.1	-3.4	-1.6	0.8
236.77	-3.90	-1.1	120.01	27JL	0.2	-3.5	-1.5	0.8
240.85	-3.73	-0.9	129.70	6AU	-1.0	-3.5	-1.5	0.8
245.66	-3.54	-0.7	139.44	16AU	-1.3	-3.5	-1.4	0.9
251.04	-3.33	-0.5	149.23	26AU	-1.0	-3.5	-1.4	0.9
256.87	-3.10	-0.3	159.09	5SE	-0.5	-3.4	-1.4	0.9
263.06	-2.87	-0.1	169.00	15SE	-0.2	-3.4	-1.3	0.8
269.51	-2.63	0.0	178.98	25SE	-0.1	-3.4	-1.3	0.8
276.18	-2.39	0.2	189.01	5OC	-0.0	-3.4	-1.3	0.8
283.01	-2.15	0.3	199.09	15OC	0.3	-3.3	-1.3	0.8
289.96	-1.91	0.5	209.21	25OC	2.2	-3.3	-1.3	0.7
296.99	-1.68	0.6	219.38	4NO	1.1	-3.3	-1.3	0.7
304.07	-1.46	0.7	229.57	14NO	-0.1	-3.3	-1.3	0.8
311.18	-1.25	0.8	239.77	24NO	-0.3	-3.3	-1.3	0.8
318.30	-1.04	0.9	249.98	4DE	-0.3	-3.4	-1.4	0.8
325.41	-0.85	1.1	260.19	14DE	-0.4	-3.4	-1.4	0.8
332.50	-0.67	1.2	270.37	24DE	-0.7	-3.4	-1.4	0.8
				-223				
339.57	-0.50	1.3	280.54	3JA	-1.1	-3.5	-1.5	0.8
346.59	-0.34	1.4	290.66	13JA	-1.1	-3.5	-1.5	0.8
353.57	-0.19	1.4	300.74	23JA	-0.7	-3.5	-1.6	0.8
0.50	-0.05	1.5	310.78	2FE	0.8	-3.6	-1.7	0.7
7.38	0.08	1.6	320.75	12FE	3.1	-3.7	-1.7	0.7
14.21	0.20	1.7	330.67	22FE	1.9	-3.7	-1.8	0.7
20.98	0.31	1.7	340.53	4MR	1.0	-3.8	-1.9	0.6
27.70	0.41	1.8	350.32	14MR	0.6	-3.9	-1.9	0.5
34.38	0.50	1.8	0.06	24MR	0.2	-4.0	-2.0	0.5
41.00	0.59	1.9	9.76	3AP	-0.3	-4.0	-2.1	0.4
47.58	0.67	1.9	19.40	13AP	-1.1	-4.1	-2.1	0.4
54.12	0.74	1.9	28.99	23AP	-1.8	-4.2	-2.1	0.3
60.62	0.80	2.0	38.56	3MY	-0.9	-4.2	-2.2	0.3
67.09	0.86	2.0	48.10	13MY	-0.0	-4.0	-2.2	0.3
73.53	0.92	2.0	57.63	23MY	0.7	-3.6	-2.2	0.4
79.95	0.97	2.0	67.15	2JN	1.3	-2.7	-2.1	0.5
86.34	1.01	2.0	76.67	12JN	2.2	-3.5	-2.1	0.5
92.73	1.05	2.0	86.20	22JN	3.3	-4.0	-2.0	0.6
99.10	1.08	2.0	95.75	2JL	1.7	-4.2	-2.0	0.6
105.47	1.11	1.9	105.33	12JL	0.1	-4.2	-1.9	0.7
111.84	1.14	2.0	114.95	22JL	-1.0	-4.2	-1.9	0.7
118.21	1.16	2.0	124.61	1AU	-1.4	-4.1	-1.8	0.8
124.59	1.17	2.0	134.32	11AU	-1.0	-4.0	-1.7	0.8
130.98	1.18	2.0	144.09	21AU	-0.4	-3.9	-1.7	0.8
137.38	1.19	2.0	153.92	31AU	-0.1	-3.8	-1.6	0.8
143.80	1.19	2.0	163.80	10SE	0.0	-3.8	-1.6	0.8
150.24	1.19	2.0	173.74	20SE	0.1	-3.7	-1.5	0.8
156.70	1.18	2.0	183.75	30SE	0.6	-3.6	-1.5	0.8
163.17	1.17	2.0	193.80	10OC	2.6	-3.6	-1.4	0.8
169.67	1.15	1.9	203.90	20OC	0.9	-3.5	-1.4	0.8
176.19	1.13	1.9	214.05	30OC	-0.3	-3.5	-1.4	0.7
182.73	1.10	1.8	224.22	9NO	-0.5	-3.5	-1.4	0.7
189.29	1.05	1.8	234.42	19NO	-0.5	-3.4	-1.4	0.7
195.87	1.01	1.7	244.63	29NO	-0.6	-3.4	-1.4	0.8
202.47	0.95	1.6	254.84	9DE	-0.8	-3.4	-1.4	0.8
209.07	0.88	1.5	265.04	19DE	-0.9	-3.4	-1.4	0.8
215.69	0.80	1.5	275.22	29DE	-1.0	-3.3	-1.4	0.8

♂			☉	16.00UT	☿	♀	♃	♄
LONG	LAT	MAG	LONG	-222	MAGNITUDES			
222.32	0.70	1.4	285.36	8JA	-0.6	-3.3	-1.5	0.8
228.95	0.59	1.2	295.47	18JA	0.9	-3.3	-1.5	0.8
235.57	0.46	1.1	305.53	28JA	3.0	-3.3	-1.5	0.8
242.19	0.31	1.0	315.53	7FE	1.4	-3.3	-1.6	0.8
248.79	0.14	0.9	325.48	17FE	0.7	-3.4	-1.7	0.7
255.35	-0.06	0.7	335.37	27FE	0.4	-3.4	-1.7	0.7
261.87	-0.28	0.6	345.19	9MR	0.1	-3.4	-1.8	0.6
268.33	-0.53	0.4	354.97	19MR	-0.4	-3.5	-1.8	0.6
274.70	-0.82	0.3	4.68	29MR	-1.1	-3.5	-1.9	0.5
280.95	-1.14	0.1	14.34	8AP	-1.7	-3.5	-2.0	0.5
287.05	-1.49	-0.1	23.97	18AP	-0.9	-3.4	-2.1	0.4
292.92	-1.89	-0.3	33.55	28AP	0.1	-3.4	-2.1	0.3
298.51	-2.33	-0.5	43.10	8MY	0.9	-3.4	-2.2	0.3
303.72	-2.81	-0.7	52.64	18MY	1.7	-3.3	-2.2	0.3
308.44	-3.35	-0.9	62.16	28MY	2.9	-3.3	-2.3	0.3
312.53	-3.92	-1.2	71.67	7JN	2.9	-3.3	-2.3	0.4
315.80	-4.54	-1.4	81.20	17JN	1.4	-3.3	-2.3	0.5
318.04	-5.17	-1.7	90.74	27JN	0.1	-3.3	-2.3	0.5
319.07	-5.77	-2.0	100.30	7JL	-1.1	-3.3	-2.2	0.6
318.74	-6.30	-2.2	109.90	17JL	-1.5	-3.4	-2.2	0.6
317.11	-6.63	-2.4	119.54	27JL	-1.0	-3.4	-2.2	0.7
314.59	-6.69	-2.6	129.22	6AU	-0.4	-3.4	-2.1	0.7
311.80	-6.44	-2.5	138.96	16AU	-0.0	-3.5	-2.0	0.8
309.56	-5.90	-2.3	148.76	26AU	0.1	-3.5	-2.0	0.8
308.39	-5.21	-2.0	158.61	5SE	0.3	-3.6	-1.9	0.8
308.51	-4.45	-1.7	168.52	15SE	0.8	-3.6	-1.8	0.8
309.90	-3.72	-1.3	178.49	25SE	2.9	-3.7	-1.8	0.8
312.39	-3.05	-1.0	188.52	5OC	0.8	-3.8	-1.7	0.8
315.76	-2.45	-0.7	198.60	15OC	-0.4	-3.9	-1.7	0.8
319.85	-1.94	-0.5	208.72	25OC	-0.6	-4.0	-1.6	0.8
324.47	-1.49	-0.2	218.88	4NO	-0.6	-4.1	-1.6	0.7
329.51	-1.10	0.0	229.07	14NO	-0.7	-4.2	-1.5	0.7
334.88	-0.77	0.2	239.27	24NO	-0.8	-4.3	-1.5	0.7
340.49	-0.48	0.4	249.48	4DE	-0.8	-4.4	-1.5	0.7
346.28	-0.23	0.6	259.69	14DE	-0.8	-4.4	-1.5	0.8
352.22	-0.02	0.8	269.88	24DE	-0.5	-4.1	-1.5	0.8
				-221				
358.26	0.17	0.9	280.04	3JA	1.1	-3.6	-1.5	0.8
4.38	0.33	1.1	290.17	13JA	2.6	-3.3	-1.5	0.8
10.55	0.47	1.2	300.25	23JA	1.0	-3.9	-1.5	0.8
16.75	0.59	1.3	310.29	2FE	0.5	-4.2	-1.5	0.8
22.99	0.69	1.5	320.27	12FE	0.3	-4.3	-1.6	0.8
29.24	0.78	1.5	330.19	22FE	0.0	-4.3	-1.6	0.8
35.49	0.86	1.6	340.05	4MR	-0.4	-4.2	-1.6	0.7
41.75	0.93	1.7	349.85	14MR	-1.1	-4.1	-1.7	0.7
48.01	0.99	1.8	359.59	24MR	-1.7	-4.0	-1.8	0.6
54.27	1.04	1.8	9.28	3AP	-0.9	-3.9	-1.8	0.6
60.52	1.08	1.9	18.93	13AP	0.2	-3.8	-1.9	0.5
66.78	1.11	1.9	28.53	23AP	1.2	-3.7	-2.0	0.4
73.03	1.14	2.0	38.10	3MY	2.3	-3.6	-2.0	0.4
79.29	1.16	2.0	47.65	13MY	3.6	-3.5	-2.1	0.3
85.55	1.17	2.0	57.17	23MY	2.3	-3.5	-2.2	0.3
91.82	1.18	2.0	66.69	2JN	1.1	-3.4	-2.2	0.3
98.10	1.19	2.0	76.21	12JN	-0.0	-3.4	-2.3	0.3
104.40	1.19	2.0	85.74	22JN	-1.1	-3.4	-2.3	0.4
110.72	1.18	2.0	95.29	2JL	-1.6	-3.4	-2.4	0.5
117.06	1.18	2.0	104.87	12JL	-0.9	-3.3	-2.4	0.5
123.42	1.16	2.0	114.49	22JL	-0.3	-3.3	-2.4	0.6
129.82	1.14	2.0	124.15	1AU	0.1	-3.3	-2.4	0.6
136.26	1.12	1.9	133.86	11AU	0.3	-3.3	-2.4	0.7
142.73	1.09	1.9	143.62	21AU	0.5	-3.4	-2.3	0.7
149.24	1.06	1.9	153.44	31AU	1.2	-3.4	-2.3	0.8
155.80	1.02	1.9	163.32	10SE	3.2	-3.4	-2.2	0.8
162.41	0.98	1.9	173.26	20SE	0.6	-3.4	-2.1	0.8
169.07	0.93	1.9	183.26	30SE	-0.6	-3.4	-2.1	0.8
175.78	0.88	1.9	193.31	10OC	-0.8	-3.5	-2.0	0.8
182.54	0.81	1.9	203.41	20OC	-0.7	-3.5	-1.9	0.8
189.36	0.75	1.8	213.55	30OC	-0.8	-3.5	-1.9	0.8
196.23	0.67	1.8	223.73	9NO	-0.7	-3.5	-1.8	0.7
203.16	0.59	1.8	233.92	19NO	-0.6	-3.4	-1.8	0.7
210.14	0.50	1.7	244.13	29NO	-0.6	-3.4	-1.7	0.7
217.18	0.40	1.7	254.34	9DE	-0.4	-3.4	-1.7	0.7
224.27	0.30	1.6	264.53	19DE	1.3	-3.4	-1.6	0.8
231.42	0.18	1.6	274.71	29DE	2.2	-3.3	-1.6	0.8

♂			☉	16.00UT	☿	♀	♃	♄
LONG	LAT	MAG	LONG	-220	MAGNITUDES			
238.61	0.05	1.5	284.86	8JA	0.7	-3.3	-1.6	0.8
245.85	-0.08	1.4	294.97	18JA	0.3	-3.3	-1.6	0.8
253.13	-0.23	1.4	305.03	28JA	0.1	-3.3	-1.6	0.8
260.45	-0.38	1.3	315.04	7FE	-0.1	-3.4	-1.6	0.8
267.80	-0.54	1.2	324.99	17FE	-0.5	-3.4	-1.6	0.8
275.18	-0.71	1.1	334.88	27FE	-1.1	-3.4	-1.6	0.8
282.57	-0.89	1.1	344.72	8MR	-1.6	-3.4	-1.6	0.8
289.96	-1.07	1.0	354.49	18MR	-0.9	-3.4	-1.6	0.7
297.34	-1.25	0.9	4.21	28MR	0.3	-3.5	-1.7	0.7
304.69	-1.44	0.8	13.88	7AP	1.5	-3.5	-1.7	0.6
312.00	-1.62	0.7	23.50	17AP	3.0	-3.6	-1.8	0.6
319.25	-1.81	0.6	33.09	27AP	3.0	-3.6	-1.8	0.5
326.40	-1.98	0.5	42.65	7MY	1.7	-3.7	-1.9	0.4
333.43	-2.16	0.4	52.18	17MY	0.8	-3.8	-1.9	0.4
340.31	-2.32	0.3	61.70	27MY	-0.1	-3.9	-2.0	0.3
347.00	-2.47	0.2	71.22	6JN	-1.1	-4.0	-2.1	0.3
353.44	-2.61	0.1	80.74	16JN	-1.7	-4.1	-2.1	0.3
359.60	-2.73	-0.0	90.28	26JN	-0.9	-4.2	-2.2	0.4
5.38	-2.84	-0.2	99.85	6JL	-0.3	-4.2	-2.3	0.4
10.73	-2.92	-0.3	109.44	16JL	0.2	-4.2	-2.3	0.5
15.51	-2.99	-0.5	119.08	26JL	0.4	-4.0	-2.4	0.5
19.61	-3.03	-0.6	128.76	5AU	0.8	-3.5	-2.4	0.6
22.84	-3.03	-0.8	138.49	15AU	1.5	-3.3	-2.4	0.6
25.02	-2.98	-1.0	148.29	25AU	3.2	-3.8	-2.4	0.7
25.92	-2.87	-1.3	158.13	4SE	0.5	-4.2	-2.4	0.7
25.39	-2.67	-1.5	168.04	14SE	-0.7	-4.3	-2.4	0.8
23.40	-2.36	-1.7	178.01	24SE	-0.9	-4.3	-2.4	0.8
20.26	-1.93	-1.8	188.04	4OC	-0.9	-4.2	-2.3	0.8
16.60	-1.41	-1.9	198.11	14OC	-0.7	-4.2	-2.3	0.8
13.22	-0.87	-1.6	208.23	24OC	-0.5	-4.1	-2.2	0.8
10.83	-0.35	-1.3	218.39	3NO	-0.5	-4.0	-2.1	0.8
9.76	0.09	-1.0	228.57	13NO	-0.5	-3.9	-2.1	0.8
10.02	0.44	-0.7	238.78	23NO	-0.2	-3.8	-2.0	0.8
11.48	0.72	-0.4	248.99	3DE	1.5	-3.7	-1.9	0.7
13.90	0.93	-0.1	259.19	13DE	1.8	-3.6	-1.9	0.7
17.11	1.09	0.2	269.39	23DE	0.4	-3.6	-1.8	0.8
				-219				
20.93	1.22	0.4	279.55	2JA	0.1	-3.5	-1.7	0.8
25.21	1.31	0.6	289.68	12JA	0.0	-3.5	-1.7	0.8
29.87	1.37	0.8	299.77	22JA	-0.2	-3.4	-1.7	0.8
34.82	1.42	1.0	309.80	1FE	-0.5	-3.4	-1.6	0.9
39.98	1.46	1.1	319.78	11FE	-1.2	-3.4	-1.6	0.9
45.33	1.48	1.3	329.71	21FE	-1.5	-3.3	-1.6	0.9
50.82	1.49	1.4	339.57	3MR	-0.9	-3.3	-1.6	0.8
56.43	1.49	1.5	349.37	13MR	0.4	-3.3	-1.6	0.8
62.13	1.49	1.6	359.12	23MR	1.9	-3.3	-1.6	0.8
67.91	1.48	1.7	8.82	2AP	3.7	-3.3	-1.6	0.8
73.76	1.46	1.8	18.46	12AP	2.3	-3.3	-1.6	0.7
79.67	1.44	1.8	28.07	22AP	1.3	-3.3	-1.6	0.6
85.64	1.42	1.9	37.64	2MY	0.6	-3.4	-1.7	0.6
91.65	1.39	1.9	47.18	12MY	-0.1	-3.4	-1.7	0.5
97.72	1.36	2.0	56.71	22MY	-1.1	-3.5	-1.7	0.5
103.83	1.32	2.0	66.23	1JN	-1.8	-3.5	-1.8	0.4
110.00	1.28	2.0	75.75	11JN	-0.9	-3.5	-1.8	0.3
116.21	1.24	2.0	85.28	21JN	-0.2	-3.4	-1.9	0.3
122.48	1.19	2.0	94.82	1JL	0.3	-3.4	-2.0	0.4
128.80	1.14	2.0	104.40	11JL	0.6	-3.4	-2.0	0.4
135.18	1.09	2.0	114.02	21JL	1.0	-3.4	-2.1	0.5
141.61	1.03	2.0	123.67	31JL	2.0	-3.3	-2.2	0.5
148.12	0.97	1.9	133.38	10AU	2.8	-3.3	-2.2	0.6
154.69	0.90	1.9	143.14	20AU	0.4	-3.3	-2.3	0.6
161.32	0.83	1.9	152.96	30AU	-0.8	-3.3	-2.4	0.7
168.03	0.76	1.8	162.84	9SE	-1.1	-3.3	-2.4	0.7
174.81	0.68	1.8	172.78	19SE	-1.0	-3.4	-2.4	0.8
181.67	0.60	1.7	182.77	29SE	-0.7	-3.4	-2.4	0.8
188.60	0.51	1.7	192.82	9OC	-0.4	-3.4	-2.4	0.8
195.61	0.42	1.7	202.92	19OC	-0.3	-3.4	-2.4	0.8
202.69	0.32	1.7	213.06	29OC	-0.3	-3.5	-2.4	0.8
209.86	0.22	1.7	223.23	8NO	-0.0	-3.5	-2.3	0.8
217.09	0.12	1.7	233.43	18NO	1.7	-3.6	-2.2	0.8
224.40	0.01	1.6	243.63	28NO	1.5	-3.7	-2.2	0.8
231.78	-0.10	1.6	253.84	8DE	0.2	-3.7	-2.1	0.8
239.23	-0.22	1.6	264.04	18DE	-0.0	-3.8	-2.0	0.8
246.73	-0.34	1.6	274.22	28DE	-0.1	-3.9	-2.0	0.8

♂ LONG	LAT	MAG	☉ LONG	16.00UT	☿	♀	♃	♄
				-218	MAGNITUDES			
254.29	-0.46	1.5	284.37	7JA	-0.3	-4.0	-1.9	0.8
261.90	-0.58	1.5	294.48	17JA	-0.6	-4.1	-1.8	0.9
269.55	-0.70	1.5	304.54	27JA	-1.2	-4.2	-1.8	0.9
277.23	-0.81	1.4	314.56	6FE	-1.3	-4.3	-1.7	0.9
284.93	-0.93	1.4	324.51	16FE	-0.8	-4.3	-1.7	0.9
292.64	-1.04	1.3	334.40	26FE	0.5	-4.2	-1.6	0.9
300.35	-1.14	1.3	344.24	8MR	2.3	-3.9	-1.6	0.9
308.04	-1.24	1.3	354.02	18MR	3.0	-3.3	-1.6	0.9
315.71	-1.33	1.2	3.73	28MR	1.7	-3.3	-1.5	0.9
323.34	-1.40	1.2	13.41	7AP	1.0	-3.9	-1.5	0.8
330.91	-1.47	1.1	23.03	17AP	0.5	-4.1	-1.5	0.8
338.41	-1.52	1.1	32.62	27AP	-0.2	-4.2	-1.5	0.7
345.83	-1.55	1.1	42.18	7MY	-1.1	-4.2	-1.5	0.7
353.15	-1.57	1.0	51.71	17MY	-1.8	-4.1	-1.5	0.6
0.35	-1.58	1.0	61.23	27MY	-0.9	-4.0	-1.6	0.6
7.42	-1.56	0.9	70.76	6JN	-0.1	-3.9	-1.6	0.5
14.33	-1.54	0.9	80.28	16JN	0.4	-3.8	-1.6	0.4
21.08	-1.49	0.8	89.82	26JN	0.8	-3.7	-1.7	0.4
27.63	-1.43	0.8	99.38	6JL	1.4	-3.6	-1.7	0.4
33.96	-1.35	0.7	108.97	16JL	2.6	-3.6	-1.8	0.4
40.04	-1.25	0.6	118.61	26JL	2.4	-3.5	-1.8	0.5
45.81	-1.13	0.6	128.29	5AU	0.3	-3.5	-1.9	0.5
51.24	-0.99	0.4	138.02	15AU	-0.9	-3.5	-2.0	0.6
56.25	-0.83	0.3	147.81	25AU	-1.2	-3.4	-2.0	0.6
60.76	-0.64	0.2	157.66	4SE	-1.0	-3.4	-2.1	0.7
64.64	-0.41	0.0	167.56	14SE	-0.6	-3.4	-2.1	0.7
67.77	-0.15	-0.1	177.53	24SE	-0.3	-3.4	-2.2	0.8
69.95	0.17	-0.3	187.55	4OC	-0.2	-3.4	-2.3	0.8
71.00	0.53	-0.5	197.62	14OC	-0.1	-3.4	-2.3	0.8
70.73	0.94	-0.8	207.74	24OC	0.2	-3.4	-2.3	0.8
69.03	1.39	-1.0	217.90	3NO	2.0	-3.4	-2.3	0.9
66.06	1.84	-1.2	228.08	13NO	1.3	-3.4	-2.3	0.9
62.25	2.24	-1.3	238.28	23NO	0.0	-3.4	-2.3	0.9
58.35	2.54	-1.2	248.50	3DE	-0.2	-3.4	-2.3	0.9
55.12	2.72	-0.9	258.70	13DE	-0.3	-3.4	-2.2	0.9
53.05	2.79	-0.7	268.89	23DE	-0.4	-3.5	-2.2	0.8
				-217				
52.31	2.78	-0.4	279.06	2JA	-0.7	-3.5	-2.1	0.8
52.85	2.72	-0.1	289.19	12JA	-1.1	-3.5	-2.0	0.9
54.46	2.63	0.2	299.27	22JA	-1.2	-3.5	-1.9	0.9
56.96	2.53	0.4	309.31	1FE	-0.8	-3.4	-1.9	0.9
60.16	2.42	0.6	319.29	11FE	0.6	-3.4	-1.8	1.0
63.92	2.31	0.8	329.22	21FE	2.8	-3.4	-1.7	1.0
68.13	2.21	1.0	339.09	3MR	2.3	-3.4	-1.7	1.0
72.68	2.11	1.1	348.89	13MR	1.2	-3.3	-1.6	1.0
77.52	2.01	1.3	358.64	23MR	0.7	-3.3	-1.6	1.0
82.60	1.91	1.4	8.34	2AP	0.3	-3.3	-1.5	1.0
87.87	1.81	1.5	17.99	12AP	-0.3	-3.3	-1.5	0.9
93.30	1.72	1.6	27.60	22AP	-1.1	-3.3	-1.5	0.9
98.88	1.63	1.6	37.17	2MY	-1.8	-3.4	-1.4	0.9
104.58	1.54	1.7	46.71	12MY	-0.9	-3.4	-1.4	0.8
110.40	1.45	1.8	56.24	22MY	-0.1	-3.4	-1.4	0.7
116.32	1.36	1.8	65.76	1JN	0.6	-3.4	-1.4	0.7
122.34	1.27	1.8	75.28	11JN	1.1	-3.5	-1.4	0.6
128.46	1.18	1.9	84.81	21JN	1.8	-3.5	-1.4	0.5
134.67	1.08	1.9	94.36	1JL	3.1	-3.6	-1.5	0.5
140.97	0.99	1.9	103.94	11JL	2.0	-3.7	-1.5	0.4
147.36	0.90	1.9	113.55	21JL	0.2	-3.8	-1.5	0.4
153.85	0.80	1.9	123.21	31JL	-1.0	-3.8	-1.6	0.5
160.43	0.70	1.8	132.91	10AU	-1.3	-3.9	-1.6	0.5
167.10	0.60	1.8	142.67	20AU	-1.0	-4.1	-1.7	0.6
173.87	0.50	1.8	152.49	30AU	-0.5	-4.2	-1.7	0.6
180.73	0.40	1.8	162.36	9SE	-0.2	-4.3	-1.8	0.7
187.69	0.30	1.7	172.30	19SE	-0.1	-4.3	-1.8	0.7
194.74	0.19	1.7	182.29	29SE	0.0	-4.3	-1.9	0.8
201.89	0.08	1.6	192.33	9OC	0.4	-4.1	-2.0	0.8
209.13	-0.02	1.6	202.43	19OC	2.3	-3.5	-2.0	0.9
216.46	-0.13	1.5	212.57	29OC	1.1	-3.2	-2.1	0.9
223.87	-0.24	1.5	222.73	8NO	-0.2	-3.9	-2.1	0.9
231.37	-0.34	1.5	232.93	18NO	-0.4	-4.3	-2.2	0.9
238.94	-0.45	1.5	243.14	28NO	-0.4	-4.4	-2.2	0.9
246.57	-0.55	1.5	253.35	8DE	-0.5	-4.4	-2.2	0.9
254.27	-0.65	1.4	263.55	18DE	-0.7	-4.3	-2.2	0.9
262.02	-0.74	1.4	273.73	28DE	-1.0	-4.2	-2.2	0.9

♂ LONG	LAT	MAG	☉ LONG	16.00UT	☿	♀	♃	♄
				-216	MAGNITUDES			
269.81	-0.83	1.4	283.88	7JA	-1.1	-4.1	-2.1	0.9
277.64	-0.91	1.4	293.99	17JA	-0.7	-3.9	-2.1	0.9
285.47	-0.98	1.4	304.06	27JA	0.8	-3.8	-2.0	1.0
293.32	-1.04	1.4	314.07	6FE	3.1	-3.8	-2.0	1.0
301.16	-1.10	1.4	324.03	16FE	1.7	-3.7	-1.9	1.1
308.98	-1.14	1.4	333.93	26FE	0.9	-3.6	-1.8	1.1
316.77	-1.17	1.4	343.76	7MR	0.5	-3.5	-1.8	1.1
324.51	-1.19	1.4	353.55	17MR	0.2	-3.5	-1.7	1.1
332.21	-1.20	1.4	3.27	27MR	-0.3	-3.4	-1.6	1.1
339.83	-1.19	1.4	12.94	6AP	-1.1	-3.4	-1.6	1.1
347.39	-1.17	1.4	22.57	16AP	-1.8	-3.3	-1.5	1.1
354.86	-1.14	1.4	32.16	26AP	-0.9	-3.3	-1.5	1.0
2.23	-1.09	1.4	41.72	6MY	0.0	-3.3	-1.4	1.0
9.51	-1.03	1.4	51.26	16MY	0.7	-3.3	-1.4	1.0
16.68	-0.97	1.4	60.78	26MY	1.4	-3.3	-1.4	0.9
23.74	-0.89	1.4	70.29	5JN	2.4	-3.3	-1.3	0.8
30.68	-0.80	1.4	79.82	15JN	3.2	-3.3	-1.3	0.8
37.49	-0.69	1.4	89.35	25JN	1.6	-3.4	-1.3	0.7
44.16	-0.58	1.3	98.92	5JL	0.2	-3.4	-1.3	0.6
50.68	-0.46	1.3	108.51	15JL	-1.0	-3.4	-1.3	0.6
57.05	-0.33	1.3	118.14	25JL	-1.5	-3.5	-1.3	0.5
63.24	-0.19	1.2	127.82	4AU	-1.0	-3.5	-1.4	0.6
69.24	-0.03	1.2	137.55	14AU	-0.4	-3.5	-1.4	0.6
75.01	0.14	1.1	147.33	24AU	-0.1	-3.5	-1.4	0.6
80.54	0.32	1.0	157.18	3SE	0.1	-3.4	-1.5	0.7
85.77	0.53	0.9	167.08	13SE	0.2	-3.4	-1.5	0.7
90.66	0.75	0.8	177.04	23SE	0.7	-3.4	-1.6	0.8
95.12	1.00	0.7	187.06	3OC	2.6	-3.4	-1.6	0.8
99.07	1.28	0.5	197.13	13OC	0.9	-3.3	-1.7	0.9
102.39	1.59	0.4	207.24	23OC	-0.3	-3.3	-1.8	0.9
104.93	1.95	0.2	217.40	2NO	-0.5	-3.3	-1.8	1.0
106.51	2.34	-0.1	227.58	12NO	-0.5	-3.3	-1.9	1.0
106.95	2.77	-0.3	237.78	22NO	-0.6	-3.3	-1.9	1.0
106.08	3.22	-0.5	247.99	2DE	-0.8	-3.4	-2.0	1.0
103.86	3.64	-0.8	258.20	12DE	-0.9	-3.4	-2.0	1.1
100.53	3.98	-0.9	268.39	22DE	-0.9	-3.4	-2.1	1.1
				-215				
96.59	4.19	-1.0	278.56	1JA	-0.6	-3.5	-2.1	1.1
92.78	4.23	-0.9	288.70	11JA	0.9	-3.5	-2.1	1.1
89.79	4.12	-0.6	298.78	21JA	2.9	-3.5	-2.1	1.0
87.99	3.90	-0.4	308.83	31JA	1.3	-3.6	-2.1	1.1
87.50	3.63	-0.1	318.82	10FE	0.6	-3.7	-2.0	1.1
88.22	3.35	0.1	328.74	20FE	0.4	-3.7	-2.0	1.2
89.97	3.07	0.3	338.61	2MR	0.1	-3.8	-1.9	1.2
92.57	2.81	0.5	348.43	12MR	-0.4	-3.9	-1.8	1.2
95.86	2.56	0.7	358.18	22MR	-1.1	-4.0	-1.8	1.2
99.69	2.34	0.9	7.88	1AP	-1.7	-4.1	-1.7	1.3
103.97	2.13	1.0	17.53	11AP	-0.9	-4.1	-1.6	1.3
108.61	1.94	1.1	27.13	21AP	0.1	-4.2	-1.6	1.2
113.56	1.75	1.2	36.71	1MY	1.0	-4.2	-1.5	1.2
118.77	1.58	1.3	46.26	11MY	1.9	-4.0	-1.5	1.2
124.20	1.42	1.4	55.78	21MY	3.2	-3.5	-1.4	1.2
129.83	1.27	1.5	65.30	31MY	2.7	-2.7	-1.4	1.1
135.63	1.12	1.5	74.82	10JN	1.3	-3.5	-1.3	1.0
141.59	0.97	1.6	84.35	20JN	0.1	-4.0	-1.3	1.0
147.71	0.83	1.6	93.90	30JN	-1.0	-4.2	-1.3	0.9
153.96	0.70	1.6	103.47	10JL	-1.6	-4.2	-1.3	0.8
160.35	0.56	1.6	113.08	20JL	-1.0	-4.2	-1.2	0.8
166.88	0.43	1.6	122.74	30JL	-0.4	-4.1	-1.2	0.7
173.53	0.31	1.6	132.44	9AU	-0.0	-4.0	-1.2	0.7
180.30	0.18	1.6	142.19	19AU	0.2	-3.9	-1.3	0.7
187.20	0.06	1.6	152.01	29AU	0.4	-3.8	-1.3	0.7
194.21	-0.06	1.6	161.88	8SE	0.9	-3.8	-1.3	0.8
201.33	-0.18	1.6	171.81	18SE	3.0	-3.7	-1.3	0.8
208.57	-0.29	1.6	181.80	28SE	0.8	-3.6	-1.4	0.9
215.91	-0.40	1.6	191.85	8OC	-0.5	-3.6	-1.4	0.9
223.35	-0.50	1.5	201.94	18OC	-0.7	-3.5	-1.5	1.0
230.89	-0.60	1.5	212.08	28OC	-0.7	-3.5	-1.5	1.0
238.51	-0.69	1.5	222.24	7NO	-0.7	-3.5	-1.6	1.1
246.20	-0.77	1.4	232.44	17NO	-0.7	-3.4	-1.6	1.1
253.96	-0.85	1.4	242.65	27NO	-0.7	-3.4	-1.7	1.1
261.77	-0.91	1.4	252.85	7DE	-0.7	-3.4	-1.8	1.2
269.62	-0.97	1.4	263.05	17DE	-0.5	-3.4	-1.8	1.2
277.51	-1.02	1.3	273.24	27DE	1.1	-3.3	-1.9	1.2

♂ LONG	LAT	MAG	⊙ LONG	16.00UT	☿	♀	♃	♄
				-214	MAGNITUDES			
285.40	-1.05	1.3	283.39	6JA	2.5	-3.3	-2.0	1.2
293.30	-1.08	1.3	293.50	16JA	0.9	-3.3	-2.0	1.2
301.19	-1.09	1.3	303.57	26JA	0.4	-3.3	-2.0	1.2
309.05	-1.09	1.3	313.59	5FE	0.2	-3.4	-2.0	1.2
316.87	-1.08	1.4	323.55	15FE	-0.0	-3.4	-2.0	1.3
324.65	-1.06	1.4	333.45	25FE	-0.4	-3.4	-2.0	1.3
332.37	-1.02	1.4	343.29	7MR	-1.1	-3.4	-2.0	1.3
340.02	-0.98	1.4	353.07	17MR	-1.6	-3.5	-1.9	1.4
347.60	-0.92	1.5	2.80	27MR	-0.9	-3.5	-1.9	1.4
355.10	-0.86	1.5	12.47	6AP	0.2	-3.5	-1.8	1.4
2.51	-0.79	1.5	22.10	16AP	1.3	-3.4	-1.8	1.4
9.83	-0.71	1.6	31.69	26AP	2.5	-3.4	-1.7	1.4
17.06	-0.62	1.6	41.25	6MY	3.6	-3.4	-1.6	1.3
24.19	-0.53	1.6	50.79	16MY	2.1	-3.3	-1.6	1.3
31.23	-0.43	1.6	60.31	26MY	1.0	-3.3	-1.5	1.2
38.16	-0.33	1.6	69.83	5JN	-0.0	-3.3	-1.5	1.2
45.00	-0.22	1.7	79.35	15JN	-1.1	-3.3	-1.4	1.2
51.74	-0.10	1.7	88.89	25JN	-1.7	-3.3	-1.4	1.1
58.38	0.01	1.7	98.44	5JL	-1.0	-3.3	-1.3	1.1
64.91	0.14	1.7	108.04	15JL	-0.3	-3.4	-1.3	1.0
71.33	0.26	1.7	117.67	25JL	0.1	-3.4	-1.3	1.0
77.64	0.40	1.6	127.34	4AU	0.3	-3.4	-1.2	0.9
83.84	0.53	1.6	137.07	14AU	0.6	-3.5	-1.2	0.9
89.90	0.68	1.6	146.85	24AU	1.3	-3.5	-1.2	0.8
95.82	0.83	1.5	156.69	3SE	3.2	-3.6	-1.2	0.9
101.59	0.98	1.5	166.60	13SE	0.6	-3.6	-1.2	0.9
107.17	1.15	1.4	176.56	23SE	-0.6	-3.7	-1.2	1.0
112.54	1.33	1.3	186.57	3OC	-0.8	-3.8	-1.3	1.0
117.67	1.53	1.2	196.64	13OC	-0.8	-3.9	-1.3	1.1
122.50	1.74	1.1	206.76	23OC	-0.8	-4.0	-1.3	1.1
126.97	1.97	0.9	216.91	2NO	-0.6	-4.1	-1.4	1.2
131.01	2.23	0.8	227.09	12NO	-0.6	-4.2	-1.4	1.2
134.50	2.51	0.6	237.29	22NO	-0.6	-4.3	-1.5	1.3
137.33	2.82	0.4	247.50	2DE	-0.3	-4.4	-1.5	1.3
139.32	3.16	0.2	257.71	12DE	1.3	-4.4	-1.6	1.3
140.30	3.52	-0.1	267.90	22DE	2.1	-4.1	-1.7	1.4
				-213				
140.09	3.89	-0.3	278.07	1JA	0.6	-3.6	-1.7	1.4
138.59	4.23	-0.6	288.20	11JA	0.2	-3.3	-1.8	1.4
135.84	4.49	-0.8	298.30	21JA	0.1	-3.9	-1.9	1.3
132.21	4.60	-1.0	308.34	31JA	-0.1	-4.2	-1.9	1.3
128.30	4.54	-1.0	318.33	10FE	-0.5	-4.3	-2.0	1.2
124.84	4.31	-0.8	328.26	20FE	-1.1	-4.3	-2.0	1.2
122.39	3.96	-0.6	338.13	2MR	-1.5	-4.2	-2.0	1.2
121.21	3.56	-0.4	347.95	12MR	-0.9	-4.1	-2.0	1.2
121.32	3.14	-0.2	357.71	22MR	0.3	-4.0	-2.0	1.2
122.57	2.74	0.1	7.41	1AP	1.6	-3.9	-2.0	1.2
124.80	2.38	0.3	17.06	11AP	3.3	-3.8	-2.0	1.1
127.84	2.04	0.4	26.67	21AP	2.7	-3.7	-1.9	1.1
131.52	1.74	0.6	36.24	1MY	1.6	-3.6	-1.9	1.1
135.74	1.46	0.7	45.80	11MY	0.8	-3.5	-1.8	1.1
140.41	1.20	0.8	55.32	21MY	-0.1	-3.5	-1.7	1.0
145.45	0.97	0.9	64.84	31MY	-1.1	-3.4	-1.7	1.0
150.81	0.76	1.0	74.36	10JN	-1.8	-3.4	-1.6	0.9
156.46	0.56	1.1	83.89	20JN	-1.0	-3.4	-1.6	0.9
162.34	0.37	1.1	93.43	30JN	-0.2	-3.3	-1.5	0.8
168.45	0.19	1.2	103.01	10JL	0.2	-3.3	-1.4	0.8
174.76	0.03	1.2	112.62	20JL	0.5	-3.3	-1.4	0.7
181.26	-0.13	1.3	122.27	30JL	0.9	-3.3	-1.4	0.7
187.93	-0.27	1.3	131.97	9AU	1.7	-3.3	-1.3	0.6
194.77	-0.40	1.3	141.72	19AU	3.1	-3.4	-1.3	0.6
201.75	-0.53	1.3	151.53	29AU	0.5	-3.4	-1.3	0.6
208.88	-0.64	1.3	161.40	8SE	-0.7	-3.4	-1.2	0.6
216.13	-0.75	1.4	171.33	18SE	-1.0	-3.4	-1.2	0.7
223.51	-0.84	1.4	181.31	28SE	-0.9	-3.4	-1.2	0.7
231.00	-0.92	1.4	191.36	8OC	-0.7	-3.5	-1.2	0.8
238.58	-0.99	1.4	201.45	18OC	-0.5	-3.5	-1.2	0.9
246.25	-1.05	1.4	211.58	28OC	-0.4	-3.5	-1.3	0.9
254.00	-1.10	1.4	221.75	7NO	-0.4	-3.5	-1.3	1.0
261.80	-1.13	1.4	231.94	17NO	-0.1	-3.4	-1.3	1.0
269.64	-1.15	1.4	242.15	27NO	1.5	-3.4	-1.4	1.1
277.52	-1.16	1.4	252.36	7DE	1.8	-3.4	-1.4	1.1
285.40	-1.16	1.4	262.56	17DE	0.3	-3.4	-1.5	1.1
293.29	-1.14	1.4	272.74	27DE	0.1	-3.3	-1.5	1.1

♂ LONG	LAT	MAG	⊙ LONG	16.00UT	☿	♀	♃	♄
				-212	MAGNITUDES			
301.16	-1.11	1.4	282.90	6JA	-0.0	-3.3	-1.6	1.1
309.00	-1.07	1.4	293.01	16JA	-0.2	-3.3	-1.7	1.1
316.80	-1.02	1.4	303.08	26JA	-0.6	-3.3	-1.7	1.0
324.55	-0.96	1.4	313.10	5FE	-1.1	-3.4	-1.8	1.0
332.23	-0.89	1.4	323.06	15FE	-1.4	-3.4	-1.9	1.0
339.85	-0.82	1.5	332.97	25FE	-0.9	-3.4	-1.9	0.9
347.39	-0.74	1.5	342.81	6MR	0.4	-3.4	-2.0	0.9
354.85	-0.65	1.5	352.60	16MR	2.0	-3.4	-2.0	0.9
2.23	-0.56	1.5	2.33	26MR	3.6	-3.5	-2.1	0.9
9.52	-0.46	1.5	12.01	5AP	2.0	-3.5	-2.1	0.9
16.72	-0.36	1.6	21.64	15AP	1.2	-3.6	-2.1	0.9
23.84	-0.26	1.6	31.23	25AP	0.6	-3.6	-2.1	0.9
30.87	-0.16	1.7	40.79	5MY	-0.1	-3.7	-2.0	0.9
37.81	-0.06	1.7	50.33	15MY	-1.1	-3.8	-2.0	0.8
44.68	0.04	1.8	59.85	25MY	-1.8	-3.9	-2.0	0.8
51.46	0.15	1.8	69.37	4JN	-1.0	-4.0	-1.9	0.8
58.17	0.25	1.8	78.89	14JN	-0.2	-4.1	-1.8	0.7
64.80	0.36	1.8	88.43	24JN	0.3	-4.2	-1.8	0.7
71.36	0.46	1.9	97.99	4JL	0.7	-4.2	-1.7	0.6
77.84	0.56	1.9	107.57	14JL	1.2	-4.2	-1.7	0.6
84.26	0.67	1.9	117.20	24JL	2.2	-4.0	-1.6	0.5
90.60	0.77	1.9	126.88	3AU	2.7	-3.5	-1.5	0.5
96.87	0.88	1.9	136.60	13AU	0.4	-3.3	-1.5	0.4
103.07	0.99	1.8	146.38	23AU	-0.8	-3.9	-1.4	0.4
109.19	1.09	1.8	156.22	2SE	-1.1	-4.2	-1.4	0.3
115.23	1.21	1.8	166.11	12SE	-1.1	-4.3	-1.4	0.3
121.18	1.32	1.7	176.07	22SE	-0.6	-4.3	-1.3	0.3
127.02	1.44	1.7	186.08	2OC	-0.4	-4.2	-1.3	0.4
132.74	1.56	1.6	196.15	12OC	-0.3	-4.1	-1.3	0.5
138.33	1.68	1.5	206.26	22OC	-0.2	-4.1	-1.3	0.5
143.75	1.82	1.4	216.41	1NO	0.0	-4.0	-1.3	0.6
148.98	1.96	1.3	226.59	11NO	1.7	-3.9	-1.3	0.7
153.99	2.11	1.2	236.80	21NO	1.5	-3.8	-1.3	0.7
158.70	2.27	1.0	247.01	1DE	0.1	-3.7	-1.3	0.8
163.07	2.44	0.9	257.21	11DE	-0.1	-3.6	-1.4	0.8
167.02	2.62	0.7	267.41	21DE	-0.2	-3.6	-1.4	0.8
170.42	2.82	0.4	277.58	31DE	-0.3	-3.5	-1.5	0.8
				-211				
173.17	3.02	0.2	287.72	10JA	-0.6	-3.5	-1.5	0.8
175.08	3.23	-0.0	297.82	20JA	-1.1	-3.4	-1.6	0.8
175.99	3.43	-0.3	307.86	30JA	-1.3	-3.4	-1.6	0.8
175.72	3.59	-0.6	317.85	9FE	-0.8	-3.4	-1.7	0.8
174.17	3.70	-0.9	327.79	19FE	0.5	-3.3	-1.8	0.7
171.43	3.69	-1.2	337.66	1MR	2.5	-3.3	-1.8	0.7
167.88	3.53	-1.4	347.48	11MR	2.7	-3.3	-1.9	0.7
164.13	3.22	-1.3	357.24	21MR	1.5	-3.3	-2.0	0.6
160.94	2.80	-1.1	6.94	31MR	0.9	-3.3	-2.0	0.6
158.82	2.31	-0.9	16.60	10AP	0.4	-3.3	-2.1	0.6
158.01	1.82	-0.7	26.21	20AP	-0.2	-3.3	-2.1	0.6
158.51	1.36	-0.5	35.79	30AP	-1.1	-3.4	-2.2	0.6
160.19	0.95	-0.3	45.33	10MY	-1.9	-3.4	-2.2	0.6
162.85	0.58	-0.1	54.86	20MY	-1.0	-3.5	-2.2	0.6
166.35	0.25	0.0	64.38	30MY	-0.1	-3.5	-2.2	0.6
170.52	-0.03	0.2	73.90	9JN	0.5	-3.5	-2.1	0.6
175.26	-0.28	0.3	83.43	19JN	0.9	-3.4	-2.1	0.6
180.47	-0.50	0.4	92.97	29JN	1.5	-3.4	-2.1	0.5
186.07	-0.70	0.5	102.54	9JL	2.8	-3.4	-2.0	0.5
192.02	-0.87	0.6	112.15	19JL	2.3	-3.4	-1.9	0.4
198.27	-1.01	0.7	121.80	29JL	0.3	-3.3	-1.9	0.4
204.77	-1.14	0.7	131.49	8AU	-0.9	-3.3	-1.8	0.3
211.50	-1.24	0.8	141.24	18AU	-1.3	-3.3	-1.7	0.3
218.44	-1.33	0.9	151.05	28AU	-1.1	-3.3	-1.7	0.2
225.54	-1.40	0.9	160.91	7SE	-0.6	-3.3	-1.6	0.1
232.81	-1.45	1.0	170.84	17SE	-0.3	-3.4	-1.6	0.1
240.20	-1.48	1.0	180.82	27SE	-0.2	-3.4	-1.5	0.0
247.71	-1.49	1.0	190.86	7OC	-0.1	-3.4	-1.5	0.1
255.31	-1.49	1.1	200.95	17OC	0.2	-3.4	-1.5	0.2
262.98	-1.47	1.1	211.08	27OC	2.0	-3.5	-1.4	0.2
270.70	-1.43	1.2	221.25	6NO	1.3	-3.5	-1.4	0.3
278.46	-1.39	1.2	231.44	16NO	-0.1	-3.6	-1.4	0.4
286.23	-1.32	1.3	241.65	26NO	-0.3	-3.7	-1.4	0.4
294.00	-1.25	1.3	251.86	6DE	-0.3	-3.7	-1.4	0.5
301.77	-1.17	1.3	262.07	16DE	-0.4	-3.8	-1.4	0.5
309.50	-1.08	1.4	272.25	26DE	-0.7	-3.9	-1.4	0.6

♂			☉	16.00UT	☿	♀	♃	♄		♂			☉	16.00UT	☿	♀	♃	♄
LONG	LAT	MAG	LONG	-210		MAGN	ITUDES			LONG	LAT	MAG	LONG	-208		MAGN	ITUDES	
317.18	-0.99	1.4	282.41	5JA	-1.1	-4.0	-1.4	0.6		333.73	-0.67	1.3	281.91	5JA	0.9	-4.1	-1.6	0.2
324.82	-0.89	1.5	292.53	15JA	-1.1	-4.1	-1.5	0.6		340.98	-0.52	1.4	292.03	15JA	2.8	-3.9	-1.6	0.2
332.40	-0.78	1.5	302.60	25JA	-0.8	-4.2	-1.5	0.6		348.17	-0.38	1.5	302.11	25JA	1.1	-3.8	-1.6	0.3
339.90	-0.68	1.5	312.62	4FE	0.6	-4.3	-1.6	0.6		355.29	-0.24	1.6	312.13	4FE	0.5	-3.7	-1.6	0.3
347.34	-0.57	1.6	322.59	14FE	2.9	-4.3	-1.6	0.6		2.35	-0.12	1.6	322.10	14FE	0.3	-3.7	-1.6	0.3
354.69	-0.46	1.6	332.49	24FE	2.1	-4.2	-1.7	0.6		9.35	0.00	1.7	332.01	24FE	0.0	-3.6	-1.6	0.4
1.97	-0.35	1.6	342.34	6MR	1.1	-3.9	-1.7	0.5		16.28	0.12	1.7	341.86	5MR	-0.4	-3.5	-1.6	0.4
9.17	-0.24	1.7	352.13	16MR	0.6	-3.3	-1.8	0.5		23.15	0.23	1.8	351.66	15MR	-1.1	-3.5	-1.6	0.3
16.28	-0.13	1.7	1.86	26MR	0.3	-3.3	-1.9	0.4		29.95	0.33	1.8	1.39	25MR	-1.7	-3.4	-1.6	0.3
23.32	-0.03	1.7	11.54	5AP	-0.3	-3.9	-2.0	0.4		36.70	0.42	1.9	11.08	4AP	-1.0	-3.4	-1.7	0.3
30.27	0.07	1.7	21.18	15AP	-1.1	-4.1	-2.0	0.4		43.38	0.51	1.9	20.72	14AP	0.1	-3.3	-1.7	0.3
37.16	0.17	1.7	30.77	25AP	-1.8	-4.2	-2.1	0.5		50.02	0.59	1.9	30.31	24AP	1.0	-3.3	-1.8	0.2
43.97	0.27	1.8	40.33	5MY	-1.0	-4.2	-2.2	0.5		56.61	0.67	1.9	39.87	4MY	2.1	-3.3	-1.8	0.2
50.72	0.37	1.8	49.87	15MY	-0.1	-4.1	-2.2	0.5		63.15	0.74	1.9	49.41	14MY	3.5	-3.3	-1.9	0.2
57.40	0.46	1.8	59.39	25MY	0.6	-4.0	-2.2	0.5		69.66	0.80	1.9	58.94	24MY	2.4	-3.3	-1.9	0.3
64.02	0.54	1.8	68.91	4JN	1.2	-3.9	-2.3	0.5		76.13	0.87	1.9	68.45	3JN	1.2	-3.3	-2.0	0.3
70.59	0.63	1.9	78.43	14JN	2.0	-3.8	-2.3	0.5		82.57	0.92	1.9	77.98	13JN	0.1	-3.3	-2.1	0.3
77.11	0.71	1.9	87.96	24JN	3.3	-3.7	-2.3	0.4		88.99	0.97	1.9	87.51	23JN	-1.0	-3.4	-2.2	0.3
83.59	0.79	1.9	97.52	4JL	1.9	-3.6	-2.3	0.4		95.39	1.02	1.9	97.06	3JL	-1.6	-3.4	-2.2	0.3
90.02	0.87	2.0	107.11	14JL	0.3	-3.6	-2.3	0.4		101.77	1.06	2.0	106.65	13JL	-1.0	-3.4	-2.3	0.3
96.41	0.94	2.0	116.73	24JL	-1.0	-3.5	-2.2	0.3		108.15	1.10	2.0	116.27	23JL	-0.4	-3.5	-2.3	0.3
102.76	1.01	2.0	126.41	3AU	-1.4	-3.5	-2.2	0.3		114.51	1.14	2.0	125.94	2AU	0.0	-3.5	-2.4	0.3
109.08	1.08	2.0	136.13	13AU	-1.0	-3.5	-2.1	0.2		120.87	1.17	2.0	135.66	12AU	0.3	-3.5	-2.4	0.2
115.37	1.15	2.0	145.9C	23AU	-0.5	-3.4	-2.0	0.2		127.23	1.19	2.0	145.43	22AU	0.5	-3.5	-2.4	0.2
121.62	1.21	2.0	155.74	2SE	-0.2	-3.4	-2.0	0.1		133.60	1.22	2.0	155.26	1SE	1.0	-3.4	-2.5	0.1
127.84	1.28	2.0	165.63	12SE	-0.0	-3.4	-1.9	0.0		139.96	1.24	2.0	165.15	11SE	3.0	-3.4	-2.4	0.1
134.02	1.34	1.9	175.58	22SE	0.1	-3.4	-1.8	-0.0		146.33	1.25	2.0	175.10	21SE	0.8	-3.4	-2.4	0.0
140.16	1.40	1.9	185.60	2OC	0.5	-3.4	-1.8	-0.1		152.71	1.26	2.0	185.11	1OC	-0.5	-3.4	-2.4	-0.1
146.26	1.46	1.8	195.66	12OC	2.3	-3.4	-1.7	-0.1		159.09	1.27	1.9	195.17	11OC	-0.7	-3.3	-2.3	-0.1
152.31	1.52	1.8	205.77	22OC	1.1	-3.4	-1.7	-0.1		165.48	1.27	1.9	205.27	21OC	-0.7	-3.3	-2.3	-0.2
158.30	1.57	1.7	215.92	1NO	-0.2	-3.4	-1.6	-0.0		171.87	1.26	1.9	215.42	31OC	-0.8	-3.3	-2.2	-0.3
164.23	1.63	1.6	226.10	11NO	-0.4	-3.4	-1.6	0.1		178.26	1.25	1.8	225.60	10NO	-0.7	-3.3	-2.1	-0.3
170.07	1.68	1.5	236.30	21NO	-0.4	-3.4	-1.6	0.1		184.65	1.23	1.7	235.80	20NO	-0.7	-3.3	-2.0	-0.3
175.81	1.73	1.4	246.51	1DE	-0.5	-3.4	-1.5	0.2		191.04	1.20	1.6	246.00	30NO	-0.7	-3.4	-2.0	-0.2
181.44	1.77	1.3	256.71	11DE	-0.8	-3.4	-1.5	0.3		197.41	1.16	1.6	256.22	10DE	-0.4	-3.4	-1.9	-0.1
186.93	1.81	1.2	266.91	21DE	-1.0	-3.5	-1.5	0.3		203.77	1.12	1.5	266.41	20DE	1.1	-3.4	-1.8	-0.0
192.24	1.85	1.0	277.08	31DE	-1.0	-3.5	-1.5	0.4		210.12	1.06	1.4	276.58	30DE	2.4	-3.5	-1.8	0.0
				-209										-207				
197.33	1.88	0.8	287.22	10JA	-0.7	-3.5	-1.5	0.4		216.43	0.98	1.2	286.73	9JA	0.8	-3.5	-1.7	0.1
202.17	1.89	0.7	297.32	20JA	0.8	-3.5	-1.5	0.4		222.70	0.89	1.1	296.83	19JA	0.3	-3.5	-1.7	0.2
206.68	1.90	0.4	307.37	30JA	3.1	-3.4	-1.5	0.4		228.93	0.78	1.0	306.88	29JA	0.2	-3.6	-1.6	0.2
210.77	1.88	0.2	317.37	9FE	1.5	-3.4	-1.5	0.5		235.09	0.65	0.8	316.88	8FE	-0.1	-3.7	-1.6	0.2
214.36	1.84	-0.0	327.30	19FE	0.8	-3.4	-1.6	0.5		241.17	0.49	0.7	326.82	18FE	-0.4	-3.7	-1.6	0.3
217.30	1.76	-0.3	337.18	1MR	0.5	-3.4	-1.6	0.4		247.15	0.29	0.5	336.70	28FE	-1.1	-3.8	-1.6	0.3
219.43	1.63	-0.6	347.00	11MR	0.2	-3.3	-1.7	0.4		252.99	0.07	0.3	346.53	10MR	-1.6	-3.9	-1.6	0.3
220.58	1.44	-0.9	356.76	21MR	-0.3	-3.3	-1.7	0.4		258.66	-0.20	0.1	356.29	20MR	-1.0	-4.0	-1.6	0.3
220.56	1.16	-1.2	6.47	31MR	-1.1	-3.3	-1.8	0.3		264.11	-0.51	-0.1	6.00	30MR	0.2	-4.1	-1.6	0.3
219.29	0.78	-1.5	16.13	10AP	-1.8	-3.3	-1.8	0.3		269.27	-0.88	-0.3	15.67	9AP	1.4	-4.1	-1.6	0.3
216.88	0.31	-1.8	25.74	20AP	-1.0	-3.3	-1.9	0.3		274.06	-1.32	-0.6	25.28	19AP	2.8	-4.2	-1.6	0.3
213.69	-0.24	-2.0	35.32	30AP	0.0	-3.4	-2.0	0.3		278.38	-1.82	-0.8	34.86	29AP	3.3	-4.2	-1.6	0.2
210.42	-0.80	-1.9	44.87	10MY	0.8	-3.4	-2.0	0.3		282.06	-2.41	-1.1	44.42	9MY	1.9	-4.0	-1.7	0.2
207.77	-1.32	-1.8	54.40	20MY	1.6	-3.4	-2.1	0.3		284.95	-3.07	-1.4	53.94	19MY	0.9	-3.5	-1.7	0.2
206.27	-1.75	-1.6	63.92	30MY	2.7	-3.4	-2.2	0.4		286.86	-3.81	-1.7	63.46	29MY	0.0	-2.6	-1.7	0.2
206.15	-2.09	-1.4	73.44	9JN	3.1	-3.5	-2.2	0.4		287.59	-4.60	-2.0	72.98	8JN	-1.0	-3.5	-1.8	0.2
207.37	-2.34	-1.2	82.97	19JN	1.5	-3.5	-2.3	0.4		287.06	-5.38	-2.3	82.51	18JN	-1.7	-4.0	-1.9	0.3
209.79	-2.51	-0.9	92.51	29JN	0.2	-3.6	-2.4	0.4		285.35	-6.04	-2.5	92.05	28JN	-1.0	-4.2	-1.9	0.3
213.22	-2.62	-0.8	102.08	9JL	-1.0	-3.7	-2.4	0.3		282.87	-6.48	-2.6	101.62	8JL	-0.3	-4.2	-2.0	0.3
217.46	-2.68	-0.6	111.68	19JL	-1.5	-3.8	-2.4	0.3		280.32	-6.61	-2.6	111.22	18JL	0.1	-4.2	-2.0	0.3
222.39	-2.70	-0.4	121.33	29JL	-1.0	-3.8	-2.4	0.3		278.38	-6.45	-2.3	120.86	28JL	0.4	-4.1	-2.1	0.3
227.87	-2.69	-0.3	131.02	8AU	-0.4	-4.0	-2.4	0.2		277.58	-6.05	-2.1	130.55	7AU	0.7	-4.0	-2.2	0.3
233.78	-2.64	-0.1	140.77	18AU	-0.1	-4.1	-2.4	0.2		278.10	-5.51	-1.8	140.30	17AU	1.4	-3.9	-2.2	0.3
240.06	-2.57	0.0	150.58	28AU	0.1	-4.2	-2.3	0.1		279.87	-4.93	-1.5	150.10	27AU	3.2	-3.8	-2.3	0.2
246.63	-2.48	0.2	160.44	7SE	0.3	-4.3	-2.3	0.1		282.74	-4.34	-1.2	159.96	6SE	0.7	-3.7	-2.4	0.2
253.43	-2.37	0.3	170.36	17SE	0.7	-4.3	-2.2	-0.0		286.50	-3.78	-1.0	169.88	16SE	-0.7	-3.7	-2.4	0.2
260.42	-2.24	0.4	180.34	27SE	2.7	-4.3	-2.1	-0.1		290.96	-3.25	-0.7	179.86	26SE	-0.9	-3.6	-2.4	0.1
267.56	-2.10	0.5	190.37	7OC	0.9	-4.1	-2.1	-0.1		295.97	-2.77	-0.5	189.89	6OC	-0.9	-3.6	-2.4	0.0
274.80	-1.95	0.6	200.46	17OC	-0.4	-3.5	-2.0	-0.2		301.39	-2.34	-0.2	199.97	16OC	-0.8	-3.5	-2.4	-0.0
282.12	-1.80	0.7	210.59	27OC	-0.6	-3.2	-1.9	-0.3		307.14	-1.95	-0.0	210.10	26OC	-0.6	-3.5	-2.4	-0.1
289.50	-1.63	0.8	220.75	6NO	-0.6	-3.9	-1.9	-0.2		313.12	-1.59	0.2	220.26	5NO	-0.5	-3.5	-2.3	-0.2
296.90	-1.47	0.9	230.94	16NO	-0.6	-4.3	-1.8	-0.1		319.29	-1.27	0.3	230.45	15NO	-0.5	-3.4	-2.3	-0.3
304.31	-1.30	1.0	241.15	26NO	-0.8	-4.4	-1.8	-0.1		325.59	-0.99	0.5	240.66	25NO	-0.3	-3.4	-2.2	-0.3
311.71	-1.14	1.1	251.36	6DE	-0.8	-4.4	-1.7	0.0		331.99	-0.73	0.7	250.87	5DE	1.3	-3.4	-2.2	-0.3
319.09	-0.98	1.2	261.56	16DE	-0.8	-4.3	-1.7	0.1		338.45	-0.50	0.8	261.07	15DE	2.1	-3.4	-2.1	-0.2
326.43	-0.82	1.3	271.75	26DE	-0.6	-4.2	-1.6	0.2		344.96	-0.29	1.0	271.26	25DE	0.5	-3.3	-2.0	-0.1

-206

♂ LONG	LAT	MAG	☉ LONG	16.00UT	☿	♀	♃	♄
351.49	-0.11	1.1	281.42	4JA	0.2	-3.3	-1.9	-0.0
358.04	0.05	1.2	291.54	14JA	0.0	-3.3	-1.9	0.0
4.58	0.20	1.3	301.62	24JA	-0.2	-3.3	-1.8	0.1
11.11	0.33	1.4	311.64	3FE	-0.5	-3.4	-1.7	0.2
17.63	0.45	1.5	321.61	13FE	-1.1	-3.4	-1.7	0.2
24.13	0.55	1.6	331.53	23FE	-1.5	-3.4	-1.6	0.2
30.61	0.65	1.7	341.38	5MR	-0.9	-3.4	-1.6	0.3
37.06	0.73	1.8	351.18	15MR	0.3	-3.5	-1.6	0.3
43.50	0.80	1.8	0.92	25MR	1.7	-3.5	-1.5	0.3
49.91	0.87	1.9	10.60	4AP	3.5	-3.5	-1.5	0.3
56.29	0.93	1.9	20.24	14AP	2.5	-3.4	-1.5	0.3
62.66	0.98	2.0	29.84	24AP	1.4	-3.4	-1.5	0.3
69.01	1.02	2.0	39.41	4MY	0.7	-3.4	-1.5	0.3
75.35	1.06	2.0	48.94	14MY	-0.1	-3.3	-1.5	0.2
81.69	1.09	2.0	58.47	24MY	-1.0	-3.3	-1.5	0.2
88.01	1.11	2.0	67.99	3JN	-1.8	-3.3	-1.6	0.2
94.34	1.13	2.0	77.51	13JN	-1.0	-3.3	-1.6	0.2
100.67	1.15	2.0	87.04	23JN	-0.2	-3.3	-1.6	0.3
107.01	1.16	2.0	96.60	3JL	0.2	-3.3	-1.7	0.3
113.36	1.16	2.0	106.18	13JL	0.6	-3.4	-1.7	0.3
119.72	1.16	2.0	115.80	23JL	1.0	-3.4	-1.8	0.4
126.11	1.16	1.9	125.47	2AU	1.8	-3.4	-1.8	0.4
132.52	1.15	1.9	135.18	12AU	3.0	-3.5	-1.9	0.4
138.97	1.14	1.9	144.95	22AU	0.6	-3.5	-2.0	0.4
145.44	1.12	2.0	154.78	1SE	-0.8	-3.6	-2.0	0.3
151.94	1.10	2.0	164.67	11SE	-1.0	-3.6	-2.1	0.3
158.49	1.07	2.0	174.62	21SE	-1.0	-3.7	-2.1	0.3
165.07	1.03	1.9	184.62	1OC	-0.7	-3.8	-2.2	0.2
171.70	0.99	1.9	194.68	11OC	-0.5	-3.9	-2.3	0.1
178.36	0.95	1.9	204.78	21OC	-0.4	-4.0	-2.3	0.1
185.07	0.89	1.9	214.93	31OC	-0.3	-4.1	-2.3	0.0
191.83	0.83	1.8	225.10	10NO	-0.1	-4.2	-2.3	-0.1
198.62	0.76	1.8	235.30	20NO	1.5	-4.3	-2.3	-0.1
205.46	0.69	1.7	245.51	30NO	1.7	-4.4	-2.3	-0.2
212.35	0.60	1.7	255.72	10DE	0.3	-4.3	-2.2	-0.3
219.28	0.50	1.6	265.92	20DE	-0.0	-4.1	-2.2	-0.2
226.24	0.39	1.5	276.09	30DE	-0.1	-3.5	-2.1	-0.1

-205

♂ LONG	LAT	MAG	☉ LONG	16.00UT	☿	♀	♃	♄
233.25	0.27	1.5	286.23	9JA	-0.3	-3.3	-2.1	-0.1
240.29	0.14	1.4	296.34	19JA	-0.6	-3.9	-2.0	0.0
247.37	-0.00	1.3	306.39	29JA	-1.1	-4.2	-1.9	0.1
254.47	-0.16	1.2	316.39	8FE	-1.4	-4.3	-1.8	0.1
261.60	-0.34	1.1	326.34	18FE	-0.9	-4.3	-1.8	0.2
268.74	-0.52	1.0	336.22	28FE	0.4	-4.2	-1.7	0.2
275.89	-0.72	0.9	346.05	10MR	2.1	-4.1	-1.6	0.3
283.03	-0.94	0.8	355.82	20MR	3.3	-4.0	-1.6	0.3
290.15	-1.16	0.7	5.53	30MR	1.8	-3.9	-1.5	0.3
297.23	-1.40	0.6	15.19	9AP	1.1	-3.8	-1.5	0.4
304.25	-1.65	0.4	24.81	19AP	0.5	-3.7	-1.5	0.4
311.19	-1.90	0.3	34.39	29AP	-0.1	-3.6	-1.4	0.4
318.00	-2.17	0.2	43.95	9MY	-1.1	-3.5	-1.4	0.3
324.65	-2.43	0.0	53.48	19MY	-1.8	-3.5	-1.4	0.3
331.08	-2.70	-0.1	63.00	29MY	-1.0	-3.4	-1.4	0.3
337.25	-2.97	-0.2	72.52	8JN	-0.2	-3.4	-1.4	0.3
343.06	-3.24	-0.4	82.05	18JN	0.4	-3.4	-1.4	0.3
348.43	-3.50	-0.6	91.59	28JN	0.8	-3.3	-1.4	0.3
353.23	-3.76	-0.8	101.16	8JL	1.3	-3.3	-1.5	0.4
357.32	-4.00	-0.9	110.76	18JL	2.4	-3.3	-1.5	0.4
0.53	-4.21	-1.2	120.40	28JL	2.6	-3.3	-1.5	0.4
2.62	-4.39	-1.4	130.09	7AU	0.5	-3.3	-1.6	0.5
3.40	-4.49	-1.6	139.83	17AU	-0.8	-3.4	-1.6	0.5
2.74	-4.47	-1.8	149.62	27AU	-1.2	-3.4	-1.7	0.5
0.67	-4.29	-2.0	159.48	6SE	-1.1	-3.4	-1.7	0.5
357.58	-3.92	-2.2	169.40	16SE	-0.6	-3.4	-1.8	0.4
354.20	-3.37	-2.1	179.37	26SE	-0.3	-3.4	-1.8	0.4
351.31	-2.72	-1.9	189.40	6OC	-0.2	-3.5	-1.9	0.4
349.53	-2.06	-1.5	199.48	16OC	-0.2	-3.5	-2.0	0.3
349.08	-1.44	-1.2	209.60	26OC	0.1	-3.5	-2.0	0.3
349.93	-0.91	-0.9	219.77	5NO	1.7	-3.5	-2.1	0.2
351.91	-0.47	-0.6	229.95	15NO	1.5	-3.4	-2.1	0.1
354.79	-0.11	-0.3	240.16	25NO	0.1	-3.4	-2.2	0.0
358.39	0.18	-0.0	250.37	5DE	-0.2	-3.4	-2.2	-0.0
2.56	0.42	0.2	260.57	15DE	-0.2	-3.4	-2.2	-0.1
7.15	0.61	0.5	270.76	25DE	-0.4	-3.3	-2.2	-0.1

-204

♂ LONG	LAT	MAG	☉ LONG	16.00UT	☿	♀	♃	♄
12.08	0.77	0.7	280.92	4JA	-0.6	-3.3	-2.2	-0.1
17.27	0.89	0.8	291.05	14JA	-1.1	-3.3	-2.1	-0.0
22.66	1.00	1.0	301.12	24JA	-1.2	-3.3	-2.1	0.0
28.20	1.08	1.2	311.16	3FE	-0.8	-3.4	-2.0	0.1
33.87	1.15	1.3	321.13	13FE	0.5	-3.4	-1.9	0.2
39.63	1.20	1.4	331.05	23FE	2.6	-3.4	-1.9	0.2
45.47	1.24	1.5	340.91	4MR	2.5	-3.4	-1.8	0.3
51.37	1.27	1.6	350.70	14MR	1.3	-3.4	-1.7	0.3
57.32	1.29	1.7	0.44	24MR	0.8	-3.5	-1.7	0.4
63.31	1.31	1.8	10.13	3AP	0.4	-3.5	-1.6	0.4
69.33	1.32	1.8	19.77	13AP	-0.2	-3.6	-1.5	0.4
75.39	1.32	1.9	29.37	23AP	-1.1	-3.6	-1.5	0.5
81.48	1.32	1.9	38.94	3MY	-1.9	-3.7	-1.4	0.5
87.60	1.31	2.0	48.49	13MY	-1.0	-3.8	-1.4	0.5
93.75	1.29	2.0	58.01	23MY	-0.1	-3.9	-1.4	0.5
99.93	1.28	2.0	67.53	2JN	0.5	-4.0	-1.3	0.5
106.14	1.25	2.0	77.05	12JN	1.0	-4.1	-1.3	0.5
112.39	1.23	2.0	86.58	22JN	1.7	-4.2	-1.3	0.4
118.68	1.20	2.0	96.14	2JL	3.0	-4.2	-1.3	0.4
125.01	1.16	2.0	105.72	12JL	2.2	-4.2	-1.3	0.5
131.38	1.12	2.0	115.34	22JL	0.4	-3.9	-1.3	0.5
137.81	1.08	2.0	125.00	1AU	-0.9	-3.4	-1.3	0.6
144.29	1.03	1.9	134.71	11AU	-1.3	-3.4	-1.4	0.6
150.82	0.98	1.9	144.48	21AU	-1.1	-3.9	-1.4	0.6
157.41	0.93	1.9	154.31	31AU	-0.5	-4.2	-1.4	0.6
164.07	0.86	1.8	164.19	10SE	-0.2	-4.3	-1.5	0.6
170.78	0.80	1.8	174.14	20SE	-0.1	-4.3	-1.5	0.6
177.57	0.73	1.8	184.14	30SE	-0.0	-4.2	-1.6	0.6
184.42	0.65	1.8	194.19	10OC	0.3	-4.1	-1.6	0.6
191.34	0.57	1.8	204.29	20OC	2.0	-4.0	-1.7	0.5
198.33	0.48	1.8	214.44	30OC	1.3	-4.0	-1.8	0.5
205.38	0.39	1.8	224.61	9NO	-0.1	-3.9	-1.8	0.4
212.51	0.29	1.7	234.81	19NO	-0.3	-3.8	-1.9	0.3
219.71	0.19	1.7	245.02	29NO	-0.4	-3.7	-1.9	0.3
226.97	0.08	1.7	255.22	9DE	-0.5	-3.6	-2.0	0.2
234.29	-0.04	1.6	265.42	19DE	-0.7	-3.6	-2.0	0.1
241.68	-0.16	1.6	275.60	29DE	-1.0	-3.5	-2.1	0.1

-203

♂ LONG	LAT	MAG	☉ LONG	16.00UT	☿	♀	♃	♄
249.12	-0.29	1.5	285.74	8JA	-1.1	-3.5	-2.1	0.0
256.61	-0.42	1.5	295.85	18JA	-0.8	-3.4	-2.1	0.1
264.14	-0.55	1.4	305.91	28JA	0.6	-3.4	-2.1	0.1
271.71	-0.69	1.4	315.91	7FE	3.0	-3.4	-2.0	0.2
279.31	-0.82	1.3	325.86	17FE	1.9	-3.3	-2.0	0.3
286.93	-0.96	1.3	335.75	27FE	1.0	-3.3	-1.9	0.3
294.55	-1.09	1.2	345.58	9MR	0.6	-3.3	-1.9	0.4
302.17	-1.22	1.2	355.35	19MR	0.2	-3.3	-1.8	0.4
309.77	-1.34	1.1	5.06	29MR	-0.3	-3.3	-1.7	0.5
317.34	-1.45	1.1	14.73	8AP	-1.1	-3.3	-1.7	0.5
324.86	-1.56	1.0	24.35	18AP	-1.8	-3.3	-1.6	0.6
332.30	-1.65	0.9	33.93	28AP	-1.0	-3.4	-1.6	0.6
339.67	-1.73	0.8	43.48	8MY	-0.1	-3.4	-1.5	0.6
346.93	-1.80	0.8	53.02	18MY	0.7	-3.5	-1.4	0.6
354.07	-1.85	0.8	62.54	28MY	1.3	-3.5	-1.4	0.6
1.05	-1.88	0.7	72.06	7JN	2.3	-3.5	-1.4	0.7
7.87	-1.90	0.7	81.58	17JN	3.3	-3.4	-1.3	0.7
14.47	-1.89	0.6	91.12	27JN	1.8	-3.4	-1.3	0.6
20.85	-1.87	0.5	100.69	7JL	0.3	-3.4	-1.3	0.6
26.94	-1.83	0.4	110.29	17JL	-1.0	-3.4	-1.2	0.6
32.69	-1.77	0.3	119.93	27JL	-1.4	-3.3	-1.2	0.7
38.05	-1.68	0.2	129.61	6AU	-1.0	-3.3	-1.2	0.7
42.93	-1.57	0.0	139.35	16AU	-0.5	-3.3	-1.2	0.8
47.21	-1.43	-0.1	149.15	26AU	-0.1	-3.3	-1.3	0.8
50.78	-1.25	-0.3	159.00	5SE	0.0	-3.3	-1.3	0.8
53.45	-1.03	-0.5	168.91	15SE	0.2	-3.4	-1.3	0.8
55.02	-0.75	-0.7	178.89	25SE	0.6	-3.4	-1.3	0.8
55.31	-0.42	-0.9	188.91	5OC	2.4	-3.4	-1.4	0.8
54.15	-0.02	-1.1	198.99	15OC	1.1	-3.4	-1.4	0.8
51.62	0.42	-1.3	209.11	25OC	-0.3	-3.5	-1.5	0.8
48.06	0.88	-1.5	219.27	4NO	-0.5	-3.5	-1.5	0.7
44.16	1.29	-1.4	229.46	14NO	-0.5	-3.6	-1.6	0.7
40.73	1.63	-1.2	239.66	24NO	-0.6	-3.7	-1.7	0.6
38.37	1.87	-0.9	249.87	4DE	-0.8	-3.8	-1.7	0.6
37.33	2.01	-0.6	260.08	14DE	-0.9	-3.8	-1.8	0.5
37.61	2.09	-0.3	270.27	24DE	-0.9	-3.9	-1.8	0.4

-202

♂ LONG	LAT	MAG	☉ LONG	16.00UT	☿	♀	♃	♄
39.04	2.12	-0.0	280.43	3JA	-0.7	-4.0	-1.9	0.4
41.40	2.12	0.2	290.56	13JA	0.7	-4.1	-2.0	0.3
44.52	2.09	0.5	300.64	23JA	3.1	-4.2	-2.0	0.3
48.22	2.06	0.7	310.67	2FE	1.4	-4.3	-2.0	0.3
52.38	2.01	0.9	320.65	12FE	0.7	-4.3	-2.0	0.4
56.91	1.96	1.0	330.57	22FE	0.4	-4.2	-2.0	0.4
61.73	1.91	1.2	340.43	4MR	0.1	-3.9	-2.0	0.5
66.79	1.85	1.3	350.23	14MR	-0.3	-3.3	-2.0	0.5
72.04	1.80	1.4	359.97	24MR	-1.1	-3.3	-1.9	0.6
77.44	1.74	1.5	9.66	3AP	-1.8	-3.9	-1.9	0.6
82.98	1.68	1.6	19.31	13AP	-1.0	-4.1	-1.8	0.7
88.63	1.61	1.7	28.91	23AP	0.0	-4.2	-1.7	0.7
94.38	1.55	1.8	38.48	3MY	0.9	-4.2	-1.7	0.8
100.23	1.48	1.8	48.02	13MY	1.7	-4.1	-1.6	0.8
106.16	1.42	1.9	57.55	23MY	2.9	-4.0	-1.6	0.8
112.17	1.35	1.9	67.07	2JN	2.9	-3.9	-1.5	0.9
118.26	1.28	1.9	76.59	12JN	1.4	-3.8	-1.4	0.9
124.43	1.21	1.9	86.12	22JN	0.2	-3.7	-1.4	0.9
130.67	1.13	1.9	95.67	2JL	-1.0	-3.6	-1.3	0.9
136.98	1.05	1.9	105.25	12JL	-1.6	-3.6	-1.3	0.9
143.38	0.98	1.9	114.87	22JL	-1.0	-3.5	-1.3	0.9
149.85	0.89	1.9	124.53	1AU	-0.4	-3.5	-1.2	0.9
156.41	0.81	1.9	134.24	11AU	-0.0	-3.5	-1.2	0.9
163.05	0.72	1.9	144.00	21AU	0.2	-3.4	-1.2	1.0
169.77	0.63	1.8	153.83	31AU	0.3	-3.4	-1.2	1.0
176.58	0.54	1.8	163.71	10SE	0.8	-3.4	-1.2	1.1
183.47	0.45	1.8	173.65	20SE	2.7	-3.4	-1.2	1.1
190.46	0.35	1.7	183.65	30SE	0.9	-3.4	-1.2	1.1
197.53	0.25	1.7	193.71	10OC	-0.4	-3.4	-1.3	1.1
204.68	0.14	1.6	203.80	20OC	-0.6	-3.4	-1.3	1.1
211.93	0.04	1.6	213.95	30OC	-0.6	-3.4	-1.3	1.1
219.25	-0.07	1.6	224.12	9NO	-0.7	-3.4	-1.4	1.0
226.66	-0.18	1.6	234.32	19NO	-0.8	-3.4	-1.4	1.0
234.14	-0.29	1.5	244.52	29NO	-0.7	-3.4	-1.5	1.0
241.69	-0.39	1.5	254.73	9DE	-0.8	-3.4	-1.5	0.9
249.30	-0.50	1.5	264.93	19DE	-0.5	-3.5	-1.6	0.8
256.97	-0.61	1.5	275.11	29DE	0.9	-3.5	-1.7	0.8

-201

♂ LONG	LAT	MAG	☉ LONG	16.00UT	☿	♀	♃	♄
264.69	-0.71	1.5	285.26	8JA	2.7	-3.5	-1.7	0.7
272.44	-0.81	1.5	295.36	18JA	1.0	-3.5	-1.8	0.6
280.23	-0.90	1.4	305.42	28JA	0.5	-3.4	-1.9	0.6
288.00	-0.99	1.4	315.43	7FE	0.2	-3.4	-1.9	0.5
295.83	-1.06	1.4	325.37	17FE	0.0	-3.4	-2.0	0.6
303.63	-1.13	1.4	335.27	27FE	-0.4	-3.4	-2.0	0.6
311.40	-1.19	1.4	345.10	9MR	-1.1	-3.3	-2.0	0.7
319.15	-1.24	1.4	354.87	19MR	-1.7	-3.3	-2.0	0.7
326.85	-1.27	1.3	4.59	29MR	-1.0	-3.3	-2.0	0.8
334.49	-1.29	1.3	14.26	8AP	0.1	-3.3	-2.0	0.8
342.06	-1.30	1.3	23.88	18AP	1.1	-3.3	-2.0	0.9
349.56	-1.29	1.3	33.47	28AP	2.3	-3.4	-1.9	0.9
356.96	-1.27	1.3	43.02	8MY	3.7	-3.4	-1.9	1.0
4.27	-1.24	1.3	52.55	18MY	2.2	-3.4	-1.8	1.0
11.46	-1.19	1.2	62.08	28MY	1.1	-3.4	-1.7	1.1
18.54	-1.13	1.2	71.60	7JN	0.1	-3.5	-1.7	1.1
25.48	-1.05	1.2	81.12	17JN	-1.0	-3.5	-1.6	1.1
32.28	-0.96	1.2	90.66	27JN	-1.7	-3.6	-1.5	1.1
38.92	-0.86	1.1	100.22	7JL	-1.0	-3.7	-1.5	1.2
45.39	-0.75	1.1	109.82	17JL	-0.3	-3.8	-1.4	1.2
51.67	-0.62	1.1	119.46	27JL	0.1	-3.9	-1.4	1.2
57.74	-0.48	1.0	129.14	6AU	0.3	-4.0	-1.3	1.2
63.57	-0.32	0.9	138.88	16AU	0.6	-4.1	-1.3	1.2
69.12	-0.14	0.8	148.67	26AU	1.2	-4.2	-1.3	1.2
74.33	0.05	0.7	158.52	5SE	3.1	-4.3	-1.3	1.3
79.16	0.27	0.6	168.43	15SE	0.8	-4.3	-1.2	1.3
83.52	0.52	0.5	178.40	25SE	-0.6	-4.3	-1.2	1.4
87.30	0.80	0.3	188.43	5OC	-0.8	-4.1	-1.2	1.4
90.37	1.12	0.2	198.50	15OC	-0.8	-3.5	-1.2	1.4
92.56	1.48	-0.0	208.63	25OC	-0.8	-3.3	-1.3	1.4
93.68	1.88	-0.3	218.78	4NO	-0.6	-4.0	-1.3	1.3
93.56	2.32	-0.5	228.97	14NO	-0.6	-4.3	-1.3	1.3
92.05	2.78	-0.7	239.17	24NO	-0.4	-4.4	-1.3	1.2
89.26	3.21	-0.9	249.38	4DE	-0.4	-4.4	-1.4	1.2
85.54	3.54	-1.1	259.58	14DE	1.1	-4.3	-1.4	1.1
81.56	3.74	-1.0	269.78	24DE	2.4	-4.2	-1.5	1.1

-200

♂ LONG	LAT	MAG	☉ LONG	16.00UT	☿	♀	♃	♄
78.09	3.79	-0.8	279.94	3JA	0.7	-4.0	-1.5	1.1
75.67	3.71	-0.6	290.07	13JA	0.3	-3.9	-1.6	1.0
74.55	3.54	-0.3	300.16	23JA	0.1	-3.8	-1.7	1.0
74.71	3.34	-0.0	310.19	2FE	-0.1	-3.7	-1.7	0.9
76.00	3.12	0.2	320.17	12FE	-0.5	-3.7	-1.8	0.8
78.22	2.90	0.4	330.10	22FE	-1.1	-3.6	-1.9	0.9
81.19	2.70	0.6	339.96	3MR	-1.6	-3.5	-1.9	0.9
84.77	2.51	0.8	349.76	13MR	-1.0	-3.5	-2.0	0.9
88.83	2.32	1.0	359.51	23MR	0.2	-3.4	-2.0	1.0
93.27	2.16	1.1	9.20	2AP	1.5	-3.4	-2.1	1.0
98.04	2.00	1.2	18.85	12AP	3.0	-3.3	-2.1	1.1
103.07	1.85	1.3	28.45	22AP	3.0	-3.3	-2.1	1.1
108.33	1.70	1.4	38.02	2MY	1.7	-3.3	-2.1	1.2
113.78	1.57	1.5	47.57	12MY	0.8	-3.3	-2.0	1.2
119.40	1.43	1.6	57.09	22MY	-0.0	-3.3	-2.0	1.3
125.18	1.31	1.6	66.61	1JN	-1.0	-3.3	-2.0	1.3
131.09	1.18	1.7	76.13	11JN	-1.7	-3.3	-1.9	1.4
137.14	1.06	1.7	85.66	21JN	-1.0	-3.4	-1.8	1.4
143.31	0.94	1.7	95.21	1JL	-0.3	-3.4	-1.8	1.4
149.60	0.82	1.7	104.79	11JL	0.2	-3.4	-1.7	1.4
156.01	0.70	1.7	114.40	21JL	0.5	-3.5	-1.6	1.4
162.53	0.58	1.7	124.06	31JL	0.8	-3.5	-1.6	1.4
169.16	0.46	1.7	133.77	10AU	1.5	-3.5	-1.5	1.3
175.91	0.35	1.7	143.53	20AU	3.2	-3.5	-1.5	1.3
182.76	0.23	1.7	153.34	30AU	0.7	-3.4	-1.4	1.2
189.73	0.12	1.7	163.23	9SE	-0.7	-3.4	-1.4	1.2
196.80	0.01	1.7	173.16	19SE	-1.0	-3.4	-1.4	1.2
203.98	-0.11	1.6	183.16	29SE	-0.9	-3.4	-1.3	1.2
211.26	-0.21	1.6	193.21	9OC	-0.8	-3.3	-1.3	1.2
218.63	-0.32	1.6	203.31	19OC	-0.5	-3.3	-1.3	1.2
226.10	-0.43	1.5	213.45	29OC	-0.4	-3.3	-1.3	1.2
233.65	-0.52	1.5	223.62	8NO	-0.4	-3.3	-1.3	1.2
241.28	-0.62	1.4	233.82	18NO	-0.2	-3.4	-1.3	1.1
248.98	-0.71	1.4	244.02	28NO	1.3	-3.4	-1.3	1.1
256.74	-0.79	1.4	254.24	8DE	2.0	-3.4	-1.4	1.1
264.55	-0.87	1.3	264.44	18DE	0.4	-3.4	-1.4	1.0
272.40	-0.93	1.3	274.61	28DE	0.1	-3.5	-1.4	1.0

-199

♂ LONG	LAT	MAG	☉ LONG	16.00UT	☿	♀	♃	♄
280.27	-0.99	1.3	284.77	7JA	-0.0	-3.5	-1.5	0.9
288.15	-1.04	1.4	294.87	17JA	-0.2	-3.5	-1.5	0.9
296.03	-1.07	1.4	304.93	27JA	-0.5	-3.6	-1.6	0.8
303.90	-1.09	1.4	314.95	6FE	-1.1	-3.7	-1.7	0.8
311.74	-1.11	1.4	324.90	16FE	-1.5	-3.7	-1.7	0.7
319.55	-1.11	1.4	334.79	26FE	-1.8	-3.8	-1.8	0.7
327.30	-1.10	1.4	344.63	8MR	0.3	-3.9	-1.9	0.8
335.00	-1.07	1.4	354.41	18MR	1.8	-4.0	-1.9	0.8
342.63	-1.04	1.5	4.12	28MR	3.7	-4.1	-2.0	0.9
350.18	-0.99	1.5	13.80	7AP	2.2	-4.1	-2.1	1.0
357.65	-0.94	1.5	23.42	17AP	1.3	-4.2	-2.1	1.0
5.03	-0.87	1.5	33.01	27AP	0.6	-4.2	-2.1	1.1
12.31	-0.80	1.5	42.57	7MY	-0.1	-4.0	-2.2	1.2
19.50	-0.71	1.5	52.10	17MY	-1.0	-3.5	-2.2	1.2
26.58	-0.62	1.6	61.62	27MY	-1.8	-2.6	-2.2	1.2
33.57	-0.52	1.6	71.14	6JN	-1.0	-3.5	-2.2	1.3
40.44	-0.42	1.6	80.66	16JN	-0.2	-4.0	-2.1	1.3
47.20	-0.30	1.6	90.20	26JN	0.3	-4.2	-2.1	1.3
53.85	-0.18	1.6	99.76	6JL	0.6	-4.2	-2.0	1.3
60.38	-0.06	1.5	109.35	16JL	1.1	-4.1	-2.0	1.3
66.79	0.07	1.5	118.99	26JL	2.0	-4.1	-1.9	1.2
73.06	0.21	1.5	128.67	5AU	2.9	-4.0	-1.9	1.2
79.18	0.36	1.5	138.40	15AU	0.6	-3.9	-1.8	1.2
85.14	0.52	1.4	148.19	25AU	-0.8	-3.8	-1.7	1.1
90.93	0.68	1.3	158.04	4SE	-1.1	-3.7	-1.7	1.1
96.51	0.86	1.3	167.94	14SE	-1.1	-3.7	-1.6	1.1
101.86	1.05	1.2	177.91	24SE	-0.7	-3.6	-1.6	1.1
106.93	1.26	1.1	187.94	4OC	-0.4	-3.6	-1.5	1.1
111.67	1.49	1.0	198.01	14OC	-0.3	-3.5	-1.5	1.0
116.01	1.74	0.8	208.13	24OC	-0.3	-3.5	-1.5	1.0
119.85	2.02	0.7	218.29	3NO	-0.0	-3.4	-1.4	1.1
123.08	2.33	0.5	228.47	13NO	1.5	-3.4	-1.4	1.0
125.57	2.68	0.3	238.67	23NO	1.7	-3.4	-1.4	1.0
127.11	3.06	0.0	248.89	3DE	0.2	-3.4	-1.4	1.0
127.55	3.46	-0.2	259.09	13DE	-0.1	-3.4	-1.4	1.0
126.72	3.86	-0.5	269.28	23DE	-0.2	-3.3	-1.4	0.9

♂ LONG	LAT	MAG	☉ LONG	16.00UT	☿	♀	♃	♄
				-198	MAGNITUDES			
124.57	4.23	-0.7	279.45	2JA	-0.3	-3.3	-1.4	0.9
121.31	4.49	-0.9	289.58	12JA	-0.6	-3.3	-1.5	0.8
117.41	4.60	-1.0	299.67	22JA	-1.1	-3.3	-1.5	0.8
113.60	4.53	-0.9	309.71	1FE	-1.3	-3.4	-1.5	0.7
110.56	4.30	-0.7	319.69	11FE	-0.9	-3.4	-1.6	0.7
108.69	3.98	-0.4	329.61	21FE	0.4	-3.4	-1.6	0.6
108.12	3.61	-0.2	339.48	3MR	2.3	-3.4	-1.7	0.6
108.78	3.24	0.0	349.28	13MR	3.0	-3.5	-1.8	0.6
110.48	2.89	0.2	359.04	23MR	1.6	-3.5	-1.8	0.6
113.07	2.56	0.4	8.73	2AP	1.0	-3.5	-1.9	0.7
116.36	2.26	0.6	18.38	12AP	0.5	-3.4	-2.0	0.8
120.23	1.99	0.8	27.98	22AP	-0.2	-3.4	-2.1	0.8
124.57	1.74	0.9	37.56	2MY	-1.0	-3.4	-2.1	0.9
129.31	1.51	1.0	47.10	12MY	-1.9	-3.3	-2.2	1.0
134.38	1.30	1.1	56.63	22MY	-1.0	-3.3	-2.2	1.0
139.73	1.10	1.2	66.15	1JN	-0.2	-3.3	-2.3	1.1
145.33	0.91	1.2	75.66	11JN	0.4	-3.3	-2.3	1.1
151.15	0.73	1.3	85.19	21JN	0.8	-3.3	-2.3	1.1
157.17	0.56	1.4	94.74	1JL	1.4	-3.3	-2.3	1.1
163.37	0.40	1.4	104.32	11JL	2.6	-3.4	-2.3	1.1
169.75	0.25	1.4	113.93	21JL	2.5	-3.4	-2.3	1.1
176.28	0.11	1.4	123.59	31JL	0.5	-3.4	-2.2	1.1
182.97	-0.03	1.5	133.29	10AU	-0.8	-3.5	-2.2	1.1
189.80	-0.17	1.5	143.05	20AU	-1.2	-3.5	-2.1	1.1
196.77	-0.29	1.5	152.87	30AU	-1.1	-3.6	-2.0	1.0
203.87	-0.41	1.5	162.74	9SE	-0.6	-3.6	-2.0	1.0
211.10	-0.52	1.5	172.68	19SE	-0.3	-3.7	-1.9	0.9
218.44	-0.63	1.5	182.67	29SE	-0.2	-3.8	-1.8	0.9
225.89	-0.72	1.5	192.72	9OC	-0.1	-3.9	-1.8	1.0
233.44	-0.81	1.4	202.82	19OC	0.2	-4.0	-1.7	1.0
241.07	-0.89	1.4	212.95	29OC	1.8	-4.1	-1.7	1.0
248.79	-0.95	1.4	223.12	8NO	1.5	-4.2	-1.6	1.0
256.57	-1.01	1.4	233.32	18NO	-0.0	-4.3	-1.6	1.0
264.40	-1.05	1.4	243.53	28NO	-0.2	-4.4	-1.6	0.9
272.27	-1.08	1.4	253.73	8DE	-0.3	-4.3	-1.5	0.9
280.16	-1.10	1.4	263.94	18DE	-0.4	-4.1	-1.5	0.9
288.06	-1.11	1.4	274.12	28DE	-0.7	-3.5	-1.5	0.8
				-197				
295.96	-1.11	1.4	284.27	7JA	-1.1	-3.3	-1.5	0.8
303.84	-1.09	1.4	294.38	17JA	-1.2	-4.0	-1.5	0.8
311.68	-1.06	1.4	304.45	27JA	-0.8	-4.3	-1.5	0.7
319.48	-1.03	1.4	314.46	6FE	0.5	-4.3	-1.5	0.7
327.23	-0.98	1.4	324.42	16FE	2.7	-4.3	-1.6	0.6
334.92	-0.92	1.4	334.31	26FE	2.3	-4.2	-1.6	0.6
342.53	-0.85	1.4	344.15	8MR	1.2	-4.1	-1.6	0.5
350.07	-0.78	1.4	353.93	18MR	0.7	-4.0	-1.7	0.5
357.53	-0.70	1.5	3.66	28MR	0.3	-3.9	-1.7	0.5
4.90	-0.61	1.5	13.33	7AP	-0.2	-3.8	-1.8	0.5
12.18	-0.52	1.6	22.96	17AP	-1.0	-3.7	-1.9	0.6
19.37	-0.43	1.6	32.54	27AP	-1.9	-3.6	-1.9	0.7
26.47	-0.33	1.7	42.10	7MY	-1.0	-3.5	-2.0	0.7
33.49	-0.23	1.7	51.64	17MY	-0.1	-3.5	-2.1	0.8
40.41	-0.12	1.7	61.16	27MY	0.5	-3.4	-2.1	0.8
47.25	-0.02	1.8	70.68	6JN	1.1	-3.4	-2.2	0.9
54.00	0.09	1.8	80.20	16JN	1.9	-3.4	-2.3	0.9
60.66	0.20	1.8	89.74	26JN	3.2	-3.3	-2.3	1.0
67.24	0.31	1.8	99.30	6JL	2.1	-3.3	-2.4	1.0
73.74	0.43	1.8	108.90	16JL	0.4	-3.3	-2.4	1.0
80.16	0.54	1.8	118.53	26JL	-0.9	-3.3	-2.4	1.0
86.49	0.66	1.8	128.20	5AU	-1.4	-3.3	-2.4	1.0
92.73	0.78	1.8	137.94	15AU	-1.1	-3.4	-2.4	1.0
98.87	0.90	1.8	147.72	25AU	-0.5	-3.4	-2.4	1.0
104.92	1.02	1.7	157.56	4SE	-0.2	-3.4	-2.3	0.9
110.85	1.15	1.7	167.47	14SE	-0.0	-3.4	-2.3	0.9
116.67	1.29	1.6	177.43	24SE	0.1	-3.4	-2.2	0.9
122.34	1.43	1.5	187.45	4OC	0.4	-3.5	-2.1	0.8
127.85	1.58	1.5	197.52	14OC	2.1	-3.5	-2.1	0.9
133.18	1.74	1.4	207.63	24OC	1.3	-3.5	-2.0	0.9
138.27	1.91	1.2	217.79	3NO	-0.2	-3.5	-1.9	0.9
143.10	2.09	1.1	227.97	13NO	-0.4	-3.4	-1.9	0.9
147.59	2.29	1.0	238.17	23NO	-0.4	-3.4	-1.8	0.9
151.68	2.50	0.8	248.38	3DE	-0.5	-3.4	-1.8	0.9
155.25	2.74	0.6	258.59	13DE	-0.7	-3.4	-1.7	0.9
158.20	2.99	0.4	268.78	23DE	-1.0	-3.3	-1.7	0.8

♂ LONG	LAT	MAG	☉ LONG	16.00UT	☿	♀	♃	♄
				-196	MAGNITUDES			
160.36	3.26	0.1	278.95	2JA	-1.0	-3.3	-1.6	0.8
161.57	3.54	-0.1	289.08	12JA	-0.7	-3.3	-1.6	0.8
161.63	3.82	-0.4	299.17	22JA	0.6	-3.3	-1.6	0.7
160.43	4.05	-0.7	309.21	1FE	3.0	-3.4	-1.6	0.7
157.97	4.20	-0.9	319.20	11FE	1.7	-3.4	-1.6	0.6
154.55	4.20	-1.2	329.13	21FE	0.9	-3.4	-1.6	0.6
150.71	4.04	-1.2	339.00	2MR	0.5	-3.4	-1.6	0.5
147.19	3.73	-1.0	348.81	12MR	0.2	-3.4	-1.6	0.5
144.59	3.30	-0.8	358.56	22MR	-0.3	-3.5	-1.6	0.4
143.25	2.84	-0.6	8.26	1AP	-1.0	-3.5	-1.6	0.4
143.22	2.37	-0.4	17.92	11AP	-1.8	-3.6	-1.7	0.4
144.40	1.94	-0.2	27.52	21AP	-1.0	-3.6	-1.7	0.5
146.61	1.54	-0.0	37.10	1MY	-0.1	-3.7	-1.8	0.5
149.68	1.19	0.2	46.64	11MY	0.7	-3.8	-1.8	0.6
153.45	0.87	0.3	56.17	21MY	1.5	-3.9	-1.9	0.7
157.81	0.58	0.4	65.69	31MY	2.5	-4.0	-2.0	0.7
162.66	0.33	0.6	75.21	10JN	3.2	-4.1	-2.0	0.8
167.91	0.10	0.7	84.74	20JN	1.7	-4.2	-2.1	0.8
173.51	-0.11	0.7	94.29	30JN	0.3	-4.2	-2.2	0.9
179.42	-0.30	0.8	103.86	10JL	-0.9	-4.2	-2.2	0.9
185.59	-0.47	0.9	113.47	20JL	-1.5	-3.9	-2.3	0.9
192.00	-0.63	0.9	123.12	30JL	-1.1	-3.4	-2.4	0.9
198.63	-0.77	1.0	132.83	9AU	-0.5	-3.3	-2.4	0.9
205.45	-0.89	1.0	142.58	19AU	-0.1	-3.9	-2.4	0.9
212.45	-0.99	1.1	152.40	29AU	0.1	-4.2	-2.5	0.9
219.61	-1.09	1.1	162.27	8SE	0.2	-4.3	-2.5	0.9
226.90	-1.16	1.1	172.20	18SE	0.6	-4.3	-2.4	0.9
234.33	-1.22	1.2	182.19	28SE	2.4	-4.2	-2.4	0.8
241.86	-1.27	1.2	192.23	8OC	1.1	-4.1	-2.4	0.8
249.48	-1.30	1.2	202.32	18OC	-0.3	-4.0	-2.3	0.8
257.18	-1.31	1.2	212.46	28OC	-0.6	-3.9	-2.2	0.8
264.94	-1.31	1.3	222.63	7NO	-0.6	-3.9	-2.2	0.8
272.75	-1.30	1.3	232.82	17NO	-0.6	-3.8	-2.1	0.8
280.58	-1.27	1.3	243.03	27NO	-0.8	-3.7	-2.0	0.8
288.41	-1.23	1.4	253.24	7DE	-0.8	-3.6	-2.0	0.8
296.24	-1.18	1.4	263.44	17DE	-0.9	-3.6	-1.9	0.8
304.06	-1.12	1.4	273.62	27DE	-0.6	-3.5	-1.8	0.8
				-195				
311.83	-1.05	1.4	283.77	6JA	0.7	-3.5	-1.8	0.8
319.56	-0.97	1.5	293.89	16JA	3.0	-3.4	-1.7	0.8
327.24	-0.89	1.5	303.96	26JA	1.2	-3.4	-1.7	0.7
334.85	-0.79	1.5	313.97	5FE	0.6	-3.4	-1.6	0.7
342.39	-0.70	1.5	323.93	15FE	0.3	-3.3	-1.6	0.6
349.86	-0.60	1.6	333.84	25FE	0.1	-3.3	-1.6	0.6
357.24	-0.50	1.6	343.67	7MR	-0.3	-3.3	-1.6	0.5
4.54	-0.40	1.6	353.46	17MR	-1.0	-3.3	-1.6	0.5
11.76	-0.29	1.6	3.19	27MR	-1.7	-3.3	-1.6	0.4
18.89	-0.19	1.6	12.86	6AP	-1.0	-3.3	-1.6	0.4
25.94	-0.09	1.7	22.49	16AP	0.0	-3.3	-1.6	0.3
32.91	0.02	1.7	32.08	26AP	1.0	-3.4	-1.6	0.4
39.80	0.12	1.7	41.64	6MY	1.9	-3.4	-1.6	0.4
46.62	0.22	1.7	51.18	16MY	3.3	-3.5	-1.7	0.5
53.37	0.31	1.8	60.70	26MY	2.6	-3.5	-1.7	0.6
60.05	0.41	1.8	70.22	5JN	1.3	-3.5	-1.8	0.6
66.66	0.50	1.9	79.74	15JN	0.2	-3.4	-1.8	0.7
73.22	0.59	1.9	89.28	25JN	-1.0	-3.4	-1.9	0.7
79.72	0.68	1.9	98.83	5JL	-1.6	-3.4	-1.9	0.8
86.17	0.77	1.9	108.42	15JL	-1.0	-3.4	-2.0	0.8
92.56	0.86	2.0	118.06	25JL	-0.4	-3.3	-2.1	0.8
98.91	0.94	2.0	127.73	4AU	-0.0	-3.3	-2.1	0.9
105.21	1.03	2.0	137.46	14AU	0.2	-3.3	-2.2	0.9
111.46	1.11	2.0	147.24	24AU	0.4	-3.3	-2.3	0.9
117.65	1.19	1.9	157.08	3SE	0.9	-3.3	-2.3	0.9
123.80	1.27	1.9	166.98	13SE	2.8	-3.4	-2.4	0.8
129.90	1.35	1.9	176.94	23SE	1.0	-3.4	-2.4	0.8
135.92	1.44	1.8	186.96	3OC	-0.5	-3.4	-2.4	0.8
141.89	1.52	1.8	197.03	13OC	-0.7	-3.4	-2.4	0.8
147.77	1.60	1.7	207.14	23OC	-0.7	-3.5	-2.4	0.7
153.55	1.68	1.6	217.29	2NO	-0.7	-3.5	-2.4	0.7
159.23	1.77	1.5	227.48	12NO	-0.7	-3.6	-2.3	0.8
164.78	1.86	1.4	237.68	22NO	-0.7	-3.7	-2.3	0.8
170.17	1.94	1.3	247.89	2DE	-0.7	-3.8	-2.2	0.8
175.37	2.03	1.1	258.09	12DE	-0.5	-3.8	-2.1	0.8
180.34	2.12	1.0	268.29	22DE	0.9	-3.9	-2.1	0.8

♂ LONG	LAT	MAG	⊙ LONG	16.00UT -194	☿	♀	♃	♄
185.03	2.21	0.8	278.45	1JA	2.7	-4.0	-2.0	0.8
189.36	2.30	0.6	288.59	11JA	0.9	-4.1	-1.9	0.8
193.24	2.38	0.4	298.68	21JA	0.4	-4.2	-1.8	0.8
196.58	2.45	0.1	308.73	31JA	0.2	-4.3	-1.8	0.7
199.22	2.51	-0.1	318.72	10FE	-0.0	-4.3	-1.7	0.7
201.00	2.54	-0.4	328.65	20FE	-0.4	-4.2	-1.7	0.6
201.75	2.53	-0.7	338.52	2MR	-1.0	-3.9	-1.6	0.6
201.29	2.44	-1.0	348.33	12MR	-1.6	-3.3	-1.6	0.5
199.57	2.27	-1.3	358.09	22MR	-1.0	-3.4	-1.6	0.5
196.75	1.97	-1.6	7.79	1AP	0.1	-3.9	-1.5	0.4
193.27	1.56	-1.7	17.44	11AP	1.2	-4.2	-1.5	0.4
189.83	1.07	-1.6	27.05	21AP	2.6	-4.2	-1.5	0.3
187.14	0.56	-1.5	36.63	1MY	3.5	-4.2	-1.5	0.3
185.64	0.07	-1.3	46.18	11MY	2.0	-4.1	-1.5	0.3
185.51	-0.37	-1.0	55.71	21MY	1.0	-4.0	-1.5	0.4
186.70	-0.74	-0.8	65.22	31MY	0.1	-3.9	-1.5	0.5
189.03	-1.05	-0.6	74.75	10JN	-1.0	-3.8	-1.6	0.5
192.34	-1.31	-0.5	84.27	20JN	-1.7	-3.7	-1.6	0.6
196.44	-1.51	-0.3	93.82	30JN	-1.0	-3.6	-1.6	0.7
201.20	-1.67	-0.2	103.40	10JL	-0.3	-3.6	-1.7	0.7
206.50	-1.80	-0.0	113.00	20JL	0.1	-3.5	-1.7	0.7
212.25	-1.89	0.1	122.65	30JL	0.4	-3.5	-1.8	0.8
218.37	-1.96	0.2	132.36	9AU	0.6	-3.4	-1.8	0.8
224.81	-2.00	0.3	142.11	19AU	1.3	-3.4	-1.9	0.8
231.51	-2.01	0.4	151.92	29AU	3.1	-3.4	-2.0	0.8
238.44	-2.00	0.5	161.79	8SE	0.8	-3.4	-2.0	0.8
245.54	-1.97	0.6	171.72	18SE	-0.6	-3.4	-2.1	0.8
252.79	-1.92	0.7	181.70	28SE	-0.9	-3.4	-2.1	0.8
260.17	-1.85	0.8	191.75	8OC	-0.8	-3.4	-2.2	0.8
267.63	-1.77	0.8	201.83	18OC	-0.8	-3.4	-2.3	0.7
275.16	-1.67	0.9	211.97	28OC	-0.6	-3.4	-2.3	0.7
282.74	-1.57	1.0	222.14	7NO	-0.5	-3.4	-2.3	0.7
290.34	-1.45	1.1	232.33	17NO	-0.5	-3.4	-2.3	0.7
297.94	-1.33	1.1	242.53	27NO	-0.3	-3.4	-2.3	0.8
305.53	-1.20	1.2	252.75	7DE	1.1	-3.4	-2.3	0.8
313.10	-1.07	1.3	262.95	17DE	2.3	-3.5	-2.2	0.8
320.62	-0.94	1.3	273.13	27DE	0.6	-3.5	-2.2	0.8
				-193				
328.10	-0.80	1.4	283.28	6JA	0.2	-3.5	-2.1	0.8
335.52	-0.67	1.5	293.40	16JA	0.1	-3.5	-2.0	0.8
342.88	-0.54	1.5	303.47	26JA	-0.1	-3.4	-2.0	0.8
350.16	-0.42	1.6	313.49	5FE	-0.5	-3.4	-1.9	0.7
357.38	-0.29	1.6	323.45	15FE	-1.1	-3.4	-1.8	0.7
4.52	-0.17	1.7	333.35	25FE	-1.5	-3.4	-1.7	0.7
11.59	-0.06	1.7	343.20	7MR	-1.0	-3.3	-1.7	0.6
18.59	0.05	1.8	352.98	17MR	0.2	-3.3	-1.6	0.6
25.52	0.16	1.8	2.71	27MR	1.6	-3.3	-1.6	0.5
32.37	0.26	1.8	12.39	6AP	3.3	-3.3	-1.5	0.4
39.17	0.35	1.8	22.02	16AP	2.7	-3.3	-1.5	0.4
45.90	0.44	1.9	31.61	26AP	1.5	-3.4	-1.5	0.3
52.57	0.53	1.9	41.17	6MY	0.8	-3.4	-1.4	0.3
59.19	0.61	1.9	50.71	16MY	-0.0	-3.4	-1.4	0.3
65.76	0.69	1.9	60.23	26MY	-1.0	-3.4	-1.4	0.3
72.29	0.76	1.9	69.75	5JN	-1.8	-3.5	-1.4	0.4
78.77	0.83	1.9	79.27	15JN	-1.0	-3.5	-1.4	0.5
85.23	0.89	1.9	88.81	25JN	-0.3	-3.6	-1.4	0.5
91.65	0.95	2.0	98.37	5JL	0.2	-3.7	-1.4	0.6
98.05	1.00	2.0	107.96	15JL	0.5	-3.8	-1.4	0.6
104.43	1.06	2.0	117.59	25JL	0.9	-3.9	-1.5	0.7
110.79	1.11	2.0	127.26	4AU	1.7	-4.0	-1.5	0.7
117.13	1.15	2.0	136.99	14AU	3.2	-4.1	-1.6	0.8
123.46	1.19	2.0	146.77	24AU	0.7	-4.2	-1.6	0.8
129.78	1.23	2.0	156.61	3SE	-0.7	-4.3	-1.7	0.8
136.08	1.27	2.0	166.50	13SE	-1.0	-4.3	-1.7	0.8
142.38	1.30	2.0	176.46	23SE	-1.0	-4.3	-1.8	0.8
148.67	1.33	2.0	186.47	3OC	-0.7	-4.0	-1.8	0.8
154.94	1.35	1.9	196.54	13OC	-0.5	-3.4	-1.9	0.8
161.21	1.37	1.9	206.65	23OC	-0.4	-3.3	-2.0	0.8
167.45	1.39	1.8	216.80	2NO	-0.4	-4.0	-2.0	0.7
173.67	1.40	1.7	226.98	12NO	-0.2	-4.3	-2.1	0.7
179.87	1.40	1.7	237.18	22NO	1.3	-4.4	-2.1	0.7
186.02	1.40	1.6	247.39	2DE	2.0	-4.3	-2.2	0.7
192.14	1.39	1.5	257.60	12DE	0.3	-4.3	-2.2	0.8
198.20	1.37	1.4	267.79	22DE	0.0	-4.2	-2.2	0.8

♂ LONG	LAT	MAG	⊙ LONG	16.00UT -192	☿	♀	♃	♄
204.20	1.34	1.2	277.96	1JA	-0.1	-4.0	-2.2	0.8
210.11	1.30	1.1	288.10	11JA	-0.2	-3.9	-2.1	0.8
215.92	1.24	0.9	298.20	21JA	-0.5	-3.8	-2.1	0.8
221.61	1.16	0.6	308.24	31JA	-1.1	-3.7	-2.0	0.8
227.14	1.05	0.6	318.23	10FE	-1.4	-3.7	-2.0	0.8
232.48	0.92	0.4	328.17	20FE	-1.0	-3.6	-1.9	0.7
237.57	0.75	0.2	338.05	1MR	0.3	-3.5	-1.8	0.7
242.36	0.54	-0.0	347.86	11MR	2.0	-3.5	-1.8	0.7
246.75	0.27	-0.3	357.62	21MR	3.5	-3.4	-1.7	0.6
250.64	-0.06	-0.5	7.33	31MR	2.0	-3.4	-1.6	0.5
253.90	-0.46	-0.8	16.98	10AP	1.1	-3.3	-1.6	0.5
256.35	-0.96	-1.1	26.59	20AP	0.6	-3.3	-1.5	0.4
257.81	-1.55	-1.4	36.17	30AP	-0.1	-3.3	-1.5	0.4
258.13	-2.23	-1.8	45.72	10MY	-1.0	-3.3	-1.4	0.3
257.19	-2.98	-2.1	55.25	20MY	-1.8	-3.3	-1.4	0.2
255.16	-3.72	-2.3	64.77	30MY	-1.1	-3.3	-1.4	0.3
252.43	-4.37	-2.5	74.29	9JN	-0.2	-3.3	-1.3	0.4
249.70	-4.83	-2.4	83.82	19JN	0.3	-3.4	-1.3	0.4
247.68	-5.07	-2.2	93.36	29JN	0.7	-3.4	-1.3	0.5
246.85	-5.10	-2.0	102.93	9JL	1.2	-3.4	-1.3	0.5
247.38	-4.97	-1.7	112.54	19JL	2.2	-3.5	-1.3	0.6
249.21	-4.74	-1.5	122.18	29JL	2.8	-3.5	-1.3	0.6
252.17	-4.45	-1.2	131.88	8AU	0.6	-3.5	-1.3	0.7
256.07	-4.12	-1.0	141.63	18AU	-0.8	-3.5	-1.4	0.7
260.71	-3.79	-0.8	151.44	28AU	-1.2	-3.4	-1.4	0.8
265.92	-3.45	-0.6	161.30	7SE	-1.1	-3.4	-1.4	0.8
271.60	-3.12	-0.4	171.23	17SE	-0.6	-3.4	-1.5	0.8
277.63	-2.79	-0.2	181.21	27SE	-0.4	-3.4	-1.5	0.8
283.93	-2.48	-0.0	191.25	7OC	-0.3	-3.3	-1.6	0.8
290.44	-2.18	0.2	201.34	17OC	-0.2	-3.3	-1.6	0.8
297.09	-1.89	0.3	211.47	27OC	0.0	-3.3	-1.7	0.8
303.86	-1.62	0.5	221.64	6NO	1.5	-3.3	-1.8	0.7
310.70	-1.37	0.6	231.83	16NO	1.7	-3.4	-1.8	0.7
317.59	-1.13	0.8	242.03	26NO	0.1	-3.4	-1.9	0.7
324.50	-0.91	0.9	252.25	6DE	-0.1	-3.4	-1.9	0.7
331.42	-0.71	1.0	262.45	16DE	-0.2	-3.4	-2.0	0.8
338.33	-0.52	1.1	272.63	26DE	-0.3	-3.5	-2.0	0.8
				-191				
345.22	-0.34	1.2	282.79	5JA	-0.6	-3.5	-2.1	0.8
352.09	-0.18	1.3	292.91	15JA	-1.1	-3.6	-2.1	0.8
358.92	-0.03	1.4	302.98	25JA	-1.3	-3.6	-2.1	0.8
5.72	0.10	1.5	313.00	4FE	-0.9	-3.7	-2.0	0.8
12.47	0.23	1.6	322.97	14FE	0.4	-3.7	-2.0	0.8
19.18	0.34	1.7	332.87	24FE	2.4	-3.8	-2.0	0.8
25.85	0.45	1.7	342.72	6MR	2.7	-3.9	-1.9	0.7
32.47	0.54	1.8	352.51	16MR	1.4	-4.0	-1.9	0.7
39.06	0.63	1.8	2.24	26MR	0.9	-4.1	-1.8	0.7
45.60	0.70	1.9	11.93	5AP	0.4	-4.1	-1.7	0.6
52.11	0.77	1.9	21.56	15AP	-0.2	-4.2	-1.7	0.5
58.58	0.84	2.0	31.15	25AP	-1.0	-4.2	-1.6	0.5
65.02	0.89	2.0	40.72	5MY	-1.9	-4.0	-1.5	0.4
71.44	0.94	2.0	50.25	15MY	-1.1	-3.5	-1.5	0.3
77.84	0.99	2.0	59.77	25MY	-0.2	-2.6	-1.4	0.3
84.22	1.03	2.0	69.29	4JN	0.4	-3.5	-1.4	0.3
90.59	1.06	2.0	78.82	14JN	0.9	-4.0	-1.3	0.3
96.95	1.09	2.0	88.35	24JN	1.6	-4.2	-1.3	0.4
103.31	1.12	2.0	97.91	4JL	2.8	-4.2	-1.3	0.4
109.67	1.14	2.0	107.49	14JL	2.4	-4.1	-1.3	0.5
116.04	1.15	2.0	117.12	24JL	0.5	-4.1	-1.2	0.6
122.42	1.16	2.0	126.79	3AU	-0.8	-4.0	-1.2	0.6
128.81	1.17	2.0	136.51	13AU	-1.3	-3.9	-1.2	0.7
135.22	1.17	2.0	146.29	23AU	-1.1	-3.8	-1.2	0.7
141.65	1.17	2.0	156.13	2SE	-0.6	-3.7	-1.3	0.7
148.11	1.16	2.0	166.02	12SE	-0.3	-3.7	-1.3	0.8
154.58	1.15	2.0	175.97	22SE	-0.1	-3.6	-1.3	0.8
161.09	1.13	2.0	185.99	2OC	-0.0	-3.6	-1.3	0.8
167.62	1.10	1.9	196.05	12OC	0.2	-3.5	-1.4	0.8
174.19	1.07	1.9	206.16	22OC	1.8	-3.5	-1.4	0.8
180.78	1.04	1.9	216.31	1NO	1.5	-3.4	-1.5	0.8
187.40	0.99	1.8	226.49	11NO	-0.1	-3.4	-1.5	0.8
194.05	0.94	1.8	236.69	21NO	-0.3	-3.4	-1.6	0.7
200.73	0.88	1.7	246.90	1DE	-0.3	-3.4	-1.7	0.7
207.44	0.81	1.6	257.11	11DE	-0.4	-3.4	-1.7	0.7
214.17	0.72	1.6	267.30	21DE	-0.7	-3.3	-1.8	0.8
220.92	0.63	1.5	277.47	31DE	-1.0	-3.3	-1.9	0.8

Left page (years −190 and −189):

♂ LONG	LAT	MAG	☉ LONG	16.00UT	☿	♀	♃	♄
				−190				
227.70	0.52	1.4	287.61	10JA	−1.1	−3.3	−1.9	0.8
234.49	0.39	1.3	297.71	20JA	−0.8	−3.3	−2.0	0.8
241.29	0.25	1.2	307.76	30JA	0.5	−3.4	−2.0	0.8
248.10	0.09	1.1	317.75	9FE	2.8	−3.4	−2.0	0.8
254.92	−0.08	0.9	327.69	19FE	2.0	−3.4	−2.0	0.8
261.72	−0.28	0.8	337.57	1MR	1.1	−3.4	−2.0	0.8
268.50	−0.50	0.7	347.39	11MR	0.6	−3.5	−2.0	0.8
275.25	−0.75	0.5	357.15	21MR	0.3	−3.5	−2.0	0.8
281.94	−1.02	0.4	6.86	31MR	−0.2	−3.5	−1.9	0.7
288.55	−1.31	0.2	16.51	10AP	−1.0	−3.4	−1.8	0.7
295.06	−1.63	0.1	26.13	20AP	−1.9	−3.4	−1.8	0.6
301.40	−1.98	−0.1	35.71	30AP	−1.1	−3.4	−1.7	0.5
307.54	−2.35	−0.3	45.25	10MY	−0.1	−3.3	−1.7	0.5
313.41	−2.75	−0.4	54.78	20MY	0.6	−3.3	−1.6	0.4
318.91	−3.18	−0.6	64.30	30MY	1.2	−3.3	−1.5	0.4
323.95	−3.64	−0.8	73.82	9JN	2.1	−3.3	−1.5	0.3
328.39	−4.11	−1.1	83.35	19JN	3.3	−3.3	−1.4	0.3
332.04	−4.59	−1.3	92.89	29JN	1.9	−3.3	−1.4	0.4
334.72	−5.08	−1.5	102.46	9JL	0.4	−3.4	−1.3	0.4
336.22	−5.53	−1.8	112.07	19JL	−0.9	−3.4	−1.3	0.5
336.33	−5.89	−2.0	121.72	29JL	−1.4	−3.4	−1.3	0.5
335.06	−6.09	−2.2	131.41	8AU	−1.1	−3.5	−1.2	0.6
332.64	−6.05	−2.4	141.16	18AU	−0.5	−3.5	−1.2	0.6
329.65	−5.73	−2.5	150.96	28AU	−0.2	−3.6	−1.2	0.7
326.90	−5.15	−2.3	160.82	7SE	0.0	−3.7	−1.2	0.7
325.05	−4.43	−2.0	170.75	17SE	0.1	−3.7	−1.2	0.8
324.46	−3.67	−1.6	180.73	27SE	0.5	−3.8	−1.2	0.8
325.19	−2.94	−1.3	190.77	7OC	2.1	−3.9	−1.2	0.8
327.09	−2.30	−1.0	200.85	17OC	1.3	−4.0	−1.3	0.8
329.97	−1.74	−0.7	210.98	27OC	−0.2	−4.1	−1.3	0.8
333.62	−1.26	−0.4	221.15	6NO	−0.5	−4.2	−1.3	0.8
337.87	−0.85	−0.1	231.34	16NO	−0.5	−4.3	−1.4	0.8
342.60	−0.51	0.1	241.54	26NO	−0.5	−4.4	−1.4	0.8
347.67	−0.22	0.3	251.75	6DE	−0.8	−4.5	−1.5	0.8
353.02	0.02	0.5	261.96	16DE	−0.9	−4.1	−1.6	0.7
358.59	0.22	0.7	272.14	26DE	−1.0	−3.5	−1.6	0.8
				−189				
4.31	0.40	0.9	282.30	5JA	−0.7	−3.4	−1.7	0.8
10.16	0.55	1.0	292.42	15JA	0.6	−4.0	−1.8	0.8
16.10	0.67	1.2	302.49	25JA	3.1	−4.3	−1.8	0.9
22.10	0.78	1.3	312.52	4FE	1.5	−4.3	−1.9	0.9
28.16	0.87	1.4	322.49	14FE	0.8	−4.3	−1.9	0.9
34.26	0.94	1.5	332.40	24FE	0.4	−4.2	−2.0	0.9
40.38	1.01	1.6	342.24	6MR	0.2	−4.1	−2.0	0.9
46.52	1.06	1.7	352.04	16MR	−0.3	−4.0	−2.0	0.9
52.68	1.11	1.8	1.77	26MR	−1.0	−3.9	−2.0	0.8
58.84	1.14	1.8	11.45	5AP	−1.8	−3.8	−2.0	0.8
65.02	1.17	1.9	21.09	15AP	−1.1	−3.7	−2.0	0.8
71.21	1.19	1.9	30.69	25AP	−0.1	−3.6	−1.9	0.7
77.41	1.21	2.0	40.25	5MY	0.8	−3.5	−1.9	0.6
83.62	1.22	2.0	49.80	15MY	1.6	−3.5	−1.8	0.6
89.84	1.22	2.0	59.32	25MY	2.8	−3.4	−1.8	0.5
96.08	1.22	2.0	68.83	4JN	3.1	−3.4	−1.7	0.4
102.34	1.22	2.0	78.36	14JN	1.6	−3.4	−1.7	0.4
108.62	1.21	2.0	87.89	24JN	0.3	−3.3	−1.6	0.3
114.93	1.19	2.0	97.45	4JL	−0.9	−3.3	−1.5	0.4
121.27	1.17	2.0	107.03	14JL	−1.5	−3.3	−1.5	0.4
127.65	1.15	2.0	116.66	24JL	−1.1	−3.3	−1.4	0.5
134.06	1.12	2.0	126.32	3AU	−0.4	−3.3	−1.4	0.5
140.52	1.09	1.9	136.04	13AU	−0.1	−3.4	−1.3	0.6
147.01	1.05	1.9	145.82	23AU	−0.1	−3.4	−1.3	0.6
153.56	1.01	1.9	155.65	2SE	0.3	−3.4	−1.3	0.7
160.16	0.96	1.9	165.54	12SE	0.7	−3.4	−1.3	0.7
166.82	0.91	1.9	175.49	22SE	2.5	−3.5	−1.2	0.8
173.53	0.85	1.9	185.50	2OC	1.1	−3.5	−1.2	0.8
180.30	0.79	1.9	195.56	12OC	−0.4	−3.5	−1.2	0.8
187.12	0.72	1.8	205.67	22OC	−0.6	−3.5	−1.3	0.8
194.01	0.64	1.8	215.81	1NO	−0.6	−3.5	−1.3	0.8
200.96	0.56	1.8	225.99	11NO	−0.7	−3.4	−1.3	0.9
207.97	0.47	1.8	236.19	21NO	−0.8	−3.4	−1.3	0.8
215.04	0.37	1.7	246.40	1DE	−0.8	−3.4	−1.4	0.8
222.17	0.26	1.7	256.61	11DE	−0.8	−3.4	−1.4	0.8
229.36	0.15	1.6	266.80	21DE	−0.6	−3.3	−1.5	0.8
236.60	0.03	1.6	276.98	31DE	0.7	−3.3	−1.5	0.8

Right page (years −188 and −187):

♂ LONG	LAT	MAG	☉ LONG	16.00UT	☿	♀	♃	♄
				−188				
243.90	−0.10	1.5	287.12	10JA	2.9	−3.3	−1.6	0.8
251.24	−0.23	1.4	297.22	20JA	1.1	−3.3	−1.6	0.9
258.63	−0.38	1.4	307.27	30JA	0.5	−3.4	−1.7	0.9
266.06	−0.53	1.3	317.27	9FE	0.3	−3.4	−1.8	0.9
273.51	−0.68	1.2	327.21	19FE	0.0	−3.4	−1.8	1.0
280.99	−0.84	1.2	337.09	29FE	−0.4	−3.4	−1.9	1.0
288.48	−1.00	1.1	346.91	10MR	−1.0	−3.5	−2.0	1.0
295.97	−1.16	1.0	356.68	20MR	−1.7	−3.5	−2.0	0.9
303.45	−1.32	1.0	6.39	30MR	−1.1	−3.5	−2.1	0.9
310.89	−1.48	0.9	16.05	9AP	0.0	−3.6	−2.1	0.9
318.29	−1.64	0.8	25.66	19AP	1.0	−3.7	−2.1	0.9
325.62	−1.78	0.7	35.24	29AP	2.1	−3.7	−2.1	0.8
332.86	−1.92	0.6	44.80	9MY	3.6	−3.8	−2.1	0.8
339.99	−2.05	0.6	54.33	19MY	2.4	−3.9	−2.0	0.7
346.97	−2.16	0.5	63.85	29MY	1.2	−4.0	−2.0	0.6
353.77	−2.26	0.4	73.37	8JN	0.2	−4.1	−2.0	0.6
0.36	−2.34	0.3	82.89	18JN	−0.9	−4.2	−1.9	0.5
6.70	−2.40	0.2	92.44	28JN	−1.6	−4.2	−1.8	0.4
12.71	−2.45	0.1	102.01	8JL	−1.1	−4.1	−1.8	0.4
18.35	−2.47	−0.1	111.60	18JL	−0.4	−3.9	−1.7	0.4
23.52	−2.48	−0.2	121.25	28JL	0.0	−3.4	−1.6	0.5
28.12	−2.45	−0.3	130.94	7AU	0.3	−3.3	−1.6	0.5
32.01	−2.39	−0.5	140.68	17AU	0.5	−3.9	−1.5	0.6
35.05	−2.30	−0.7	150.49	27AU	1.0	−4.2	−1.5	0.6
36.95	−2.15	−0.9	160.35	6SE	2.9	−4.3	−1.4	0.7
37.60	−1.93	−1.1	170.26	16SE	1.0	−4.3	−1.4	0.7
36.80	−1.64	−1.4	180.24	26SE	−0.5	−4.2	−1.4	0.8
34.57	−1.25	−1.5	190.28	6OC	−0.8	−4.1	−1.3	0.8
31.25	−0.79	−1.7	200.36	16OC	−0.8	−4.0	−1.3	0.8
27.46	−0.30	−1.7	210.49	26OC	−0.8	−3.9	−1.3	0.9
24.06	0.17	−1.5	220.65	5NO	−0.7	−3.9	−1.3	0.9
21.68	0.58	−1.2	230.84	15NO	−0.6	−3.8	−1.3	0.9
20.62	0.91	−0.8	241.05	25NO	−0.6	−3.7	−1.3	0.9
20.90	1.15	−0.5	251.26	5DE	−0.4	−3.6	−1.4	0.9
22.35	1.33	−0.2	261.46	15DE	0.9	−3.6	−1.4	0.9
24.76	1.45	0.0	271.65	25DE	2.6	−3.5	−1.4	0.9
				−187				
27.94	1.54	0.3	281.81	4JA	0.8	−3.5	−1.5	0.9
31.71	1.59	0.5	291.93	14JA	0.3	−3.4	−1.5	0.9
35.95	1.63	0.7	302.01	24JA	0.1	−3.4	−1.6	0.9
40.56	1.65	0.9	312.04	3FE	−0.1	−3.4	−1.6	1.0
45.45	1.65	1.1	322.01	13FE	−0.4	−3.3	−1.7	1.0
50.57	1.65	1.2	331.92	23FE	−1.0	−3.3	−1.8	1.0
55.87	1.64	1.4	341.78	5MR	−1.6	−3.3	−1.8	1.1
61.32	1.62	1.5	351.57	15MR	−1.1	−3.3	−1.9	1.1
66.89	1.60	1.6	1.31	25MR	0.1	−3.3	−2.0	1.1
72.56	1.57	1.7	10.99	4AP	1.3	−3.3	−2.0	1.0
78.32	1.54	1.7	20.63	14AP	2.8	−3.3	−2.1	1.0
84.15	1.50	1.8	30.23	24AP	3.2	−3.4	−2.1	1.0
90.06	1.46	1.8	39.79	4MY	1.8	−3.4	−2.2	0.9
96.02	1.42	1.9	49.33	14MY	0.9	−3.5	−2.2	0.9
102.05	1.38	1.9	58.86	24MY	0.1	−3.5	−2.2	0.8
108.14	1.33	2.0	68.38	3JN	−1.0	−3.5	−2.2	0.8
114.28	1.28	2.0	77.89	13JN	−1.7	−3.4	−2.2	0.7
120.49	1.22	2.0	87.43	23JN	−1.1	−3.4	−2.2	0.6
126.75	1.16	2.0	96.98	3JL	−0.3	−3.4	−2.1	0.6
133.08	1.10	2.0	106.56	13JL	0.1	−3.4	−2.1	0.6
139.48	1.04	2.0	116.19	23JL	0.4	−3.3	−2.0	0.5
145.94	0.97	1.9	125.85	2AU	0.7	−3.3	−1.9	0.5
152.46	0.90	1.9	135.57	12AU	1.4	−3.3	−1.9	0.6
159.07	0.83	1.9	145.34	22AU	3.2	−3.3	−1.8	0.6
165.74	0.75	1.8	155.17	1SE	0.8	−3.4	−1.7	0.7
172.50	0.67	1.8	165.05	11SE	−0.6	−3.4	−1.7	0.7
179.33	0.58	1.8	175.00	21SE	−0.9	−3.4	−1.6	0.8
186.24	0.49	1.7	185.01	1OC	−0.9	−3.4	−1.6	0.8
193.24	0.40	1.7	195.06	11OC	−0.8	−3.5	−1.5	0.9
200.31	0.30	1.7	205.17	21OC	−0.5	−3.5	−1.5	0.9
207.46	0.20	1.7	215.32	31OC	−0.5	−3.6	−1.5	0.9
214.70	0.10	1.7	225.49	10NO	−0.5	−3.6	−1.4	1.0
222.01	−0.01	1.6	235.69	20NO	−0.3	−3.7	−1.4	1.0
229.39	−0.12	1.6	245.90	30NO	1.1	−3.8	−1.4	1.0
236.84	−0.23	1.6	256.11	10DE	2.2	−3.8	−1.4	1.0
244.36	−0.34	1.5	266.31	20DE	0.5	−3.9	−1.4	1.0
251.94	−0.46	1.5	276.48	30DE	0.1	−4.0	−1.4	1.0

♂			☉	16.00UT	☿	♀	♃	♄
LONG	LAT	MAG	LONG	-186	MAGNITUDES			
259.57	-0.57	1.5	286.63	9JA	0.0	-4.1	-1.5	1.0
267.24	-0.69	1.5	296.73	19JA	-0.2	-4.2	-1.5	1.0
274.95	-0.80	1.4	306.78	29JA	-0.5	-4.3	-1.5	1.0
282.69	-0.90	1.4	316.79	8FE	-1.0	-4.3	-1.6	1.1
290.44	-1.01	1.4	326.73	18FE	-1.5	-4.2	-1.6	1.1
298.18	-1.10	1.4	336.61	28FE	-1.0	-3.9	-1.7	1.1
305.93	-1.19	1.3	346.44	10MR	0.2	-3.3	-1.7	1.2
313.65	-1.26	1.3	356.21	20MR	1.7	-3.4	-1.8	1.2
321.33	-1.33	1.3	5.92	30MR	3.6	-3.9	-1.9	1.2
328.96	-1.38	1.2	15.58	9AP	2.4	-4.2	-1.9	1.2
336.53	-1.42	1.2	25.20	19AP	1.4	-4.2	-2.0	1.2
344.03	-1.44	1.2	34.78	29AP	0.7	-4.2	-2.1	1.2
351.44	-1.45	1.2	44.33	9MY	-0.0	-4.1	-2.1	1.1
358.74	-1.45	1.1	53.87	19MY	-1.0	-4.0	-2.2	1.1
5.93	-1.43	1.1	63.38	29MY	-1.8	-3.9	-2.3	1.0
12.99	-1.39	1.1	72.90	8JN	-1.1	-3.8	-2.3	1.0
19.91	-1.34	1.0	82.43	18JN	-0.3	-3.7	-2.3	0.9
26.66	-1.27	1.0	91.97	28JN	0.2	-3.6	-2.3	0.8
33.23	-1.19	0.9	101.54	8JL	0.6	-3.6	-2.3	0.8
39.59	-1.09	0.9	111.14	18JL	1.0	-3.5	-2.3	0.7
45.73	-0.98	0.8	120.78	28JL	1.9	-3.5	-2.3	0.6
51.60	-0.84	0.7	130.47	7AU	3.1	-3.4	-2.2	0.6
57.15	-0.69	0.6	140.21	17AU	0.7	-3.4	-2.2	0.7
62.34	-0.51	0.5	150.01	27AU	-0.7	-3.4	-2.1	0.7
67.08	-0.31	0.4	159.87	6SE	-1.1	-3.4	-2.0	0.7
71.29	-0.08	0.2	169.79	16SE	-1.0	-3.4	-2.0	0.8
74.85	0.19	0.1	179.76	26SE	-0.7	-3.4	-1.9	0.8
77.59	0.50	-0.1	189.79	6OC	-0.4	-3.4	-1.8	0.9
79.34	0.85	-0.3	199.87	16OC	-0.3	-3.4	-1.8	0.9
79.90	1.26	-0.5	209.99	26OC	-0.3	-3.4	-1.7	1.0
79.10	1.70	-0.8	220.16	5NO	-0.1	-3.4	-1.7	1.0
76.91	2.16	-1.0	230.34	15NO	1.3	-3.4	-1.6	1.1
73.57	2.58	-1.1	240.55	25NO	1.9	-3.4	-1.6	1.1
69.62	2.92	-1.2	250.76	5DE	0.3	-3.4	-1.6	1.1
65.85	3.12	-1.0	260.96	15DE	-0.0	-3.5	-1.5	1.1
62.94	3.19	-0.8	271.15	25DE	-0.1	-3.5	-1.5	1.2
				-185				
61.27	3.16	-0.5	281.31	4JA	-0.3	-3.5	-1.5	1.2
60.93	3.07	-0.2	291.44	14JA	-0.5	-3.5	-1.5	1.2
61.79	2.94	0.1	301.52	24JA	-1.1	-3.4	-1.5	1.2
63.67	2.79	0.3	311.55	3FE	-1.3	-3.4	-1.5	1.1
66.38	2.64	0.5	321.52	13FE	-1.0	-3.4	-1.6	1.2
69.75	2.50	0.7	331.44	23FE	0.3	-3.4	-1.6	1.2
73.63	2.36	0.9	341.30	5MR	2.1	-3.3	-1.6	1.3
77.94	2.23	1.1	351.09	15MR	3.2	-3.3	-1.7	1.3
82.58	2.10	1.2	0.83	25MR	1.8	-3.3	-1.7	1.3
87.49	1.98	1.3	10.52	4AP	1.0	-3.3	-1.8	1.4
92.63	1.87	1.4	20.16	14AP	0.5	-3.3	-1.8	1.4
97.96	1.75	1.5	29.76	24AP	-0.1	-3.4	-1.9	1.4
103.46	1.64	1.6	39.33	4MY	-1.0	-3.4	-2.0	1.3
109.10	1.54	1.7	48.87	14MY	-1.8	-3.4	-2.0	1.3
114.87	1.43	1.7	58.39	24MY	-1.1	-3.4	-2.1	1.3
120.76	1.33	1.7	67.92	3JN	-0.2	-3.5	-2.2	1.2
126.77	1.22	1.8	77.43	13JN	0.4	-3.5	-2.2	1.2
132.87	1.12	1.8	86.97	23JN	0.8	-3.6	-2.3	1.1
139.09	1.02	1.8	96.52	3JL	1.3	-3.7	-2.3	1.1
145.40	0.92	1.8	106.10	13JL	2.4	-3.8	-2.4	1.0
151.81	0.81	1.8	115.72	23JL	2.7	-3.9	-2.4	0.9
158.33	0.71	1.8	125.39	2AU	0.6	-4.0	-2.4	0.8
164.94	0.60	1.8	135.10	12AU	-0.8	-4.1	-2.4	0.8
171.65	0.50	1.8	144.87	22AU	-1.2	-4.2	-2.4	0.8
178.47	0.39	1.8	154.70	1SE	-1.1	-4.3	-2.4	0.8
185.38	0.28	1.7	164.58	11SE	-0.6	-4.3	-2.3	0.9
192.39	0.18	1.7	174.52	21SE	-0.3	-4.3	-2.3	0.9
199.50	0.07	1.7	184.53	1OC	-0.2	-4.0	-2.2	1.0
206.71	-0.04	1.6	194.58	11OC	-0.1	-3.4	-2.1	1.0
214.01	-0.15	1.6	204.68	21OC	0.1	-3.4	-2.1	1.1
221.40	-0.25	1.5	214.83	31OC	1.5	-4.0	-2.0	1.1
228.88	-0.36	1.5	225.00	10NO	1.7	-4.3	-1.9	1.2
236.43	-0.46	1.4	235.20	20NO	0.1	-4.4	-1.9	1.2
244.06	-0.56	1.4	245.41	30NO	-0.2	-4.3	-1.8	1.2
251.76	-0.66	1.4	255.61	10DE	-0.3	-4.3	-1.7	1.3
259.51	-0.75	1.4	265.81	20DE	-0.4	-4.2	-1.7	1.3
267.30	-0.83	1.4	275.99	30DE	-0.6	-4.0	-1.7	1.3

♂			☉	16.00UT	☿	♀	♃	♄
LONG	LAT	MAG	LONG	-184	MAGNITUDES			
275.13	-0.91	1.4	286.13	9JA	-1.1	-3.9	-1.6	1.3
282.98	-0.97	1.4	296.24	19JA	-1.2	-3.8	-1.6	1.3
290.84	-1.03	1.4	306.30	29JA	-0.9	-3.7	-1.6	1.3
298.70	-1.08	1.4	316.30	8FE	0.4	-3.7	-1.6	1.3
306.54	-1.12	1.4	326.25	18FE	2.5	-3.6	-1.6	1.3
314.36	-1.14	1.4	336.14	28FE	2.5	-3.5	-1.6	1.3
322.13	-1.15	1.4	345.96	9MR	1.3	-3.5	-1.6	1.3
329.86	-1.15	1.4	355.74	19MR	0.8	-3.4	-1.6	1.3
337.53	-1.14	1.4	5.45	29MR	0.4	-3.4	-1.6	1.3
345.12	-1.12	1.4	15.12	8AP	-0.2	-3.3	-1.7	1.3
352.64	-1.08	1.4	24.74	18AP	-1.0	-3.3	-1.7	1.3
0.07	-1.03	1.4	34.32	28AP	-1.9	-3.3	-1.7	1.2
7.40	-0.97	1.5	43.87	8MY	-1.1	-3.3	-1.8	1.2
14.64	-0.90	1.5	53.41	18MY	-0.2	-3.3	-1.8	1.2
21.77	-0.82	1.5	62.93	28MY	0.5	-3.3	-1.9	1.1
28.79	-0.73	1.5	72.45	7JN	1.0	-3.3	-2.0	1.1
35.69	-0.63	1.4	81.97	17JN	1.8	-3.4	-2.0	1.0
42.46	-0.53	1.4	91.51	27JN	3.0	-3.4	-2.1	1.0
49.11	-0.41	1.4	101.08	7JL	2.2	-3.4	-2.2	0.9
55.62	-0.28	1.4	110.68	17JL	0.5	-3.5	-2.3	0.9
61.99	-0.15	1.4	120.31	27JL	-0.8	-3.5	-2.3	0.8
68.19	-0.00	1.3	130.00	6AU	-1.3	-3.5	-2.4	0.8
74.21	0.15	1.3	139.74	16AU	-1.1	-3.5	-2.4	0.7
80.04	0.32	1.2	149.53	26AU	-0.6	-3.4	-2.4	0.7
85.64	0.50	1.1	159.39	5SE	-0.2	-3.4	-2.5	0.8
90.98	0.70	1.0	169.30	15SE	-0.1	-3.4	-2.5	0.8
96.00	0.92	0.9	179.27	25SE	0.0	-3.3	-2.4	0.9
100.66	1.16	0.8	189.30	5OC	0.3	-3.3	-2.4	1.0
104.86	1.42	0.7	199.38	15OC	1.8	-3.3	-2.4	1.0
108.51	1.72	0.5	209.50	25OC	1.5	-3.3	-2.3	1.1
111.48	2.05	0.3	219.66	4NO	-0.1	-3.3	-2.2	1.2
113.61	2.42	0.1	229.85	14NO	-0.4	-3.4	-2.2	1.2
114.70	2.83	-0.1	240.05	24NO	-0.4	-3.4	-2.1	1.2
114.59	3.26	-0.4	250.26	4DE	-0.5	-3.4	-2.0	1.2
113.14	3.69	-0.6	260.47	14DE	-0.7	-3.4	-1.9	1.3
110.42	4.06	-0.8	270.65	24DE	-1.0	-3.5	-1.9	1.3
				-183				
106.77	4.32	-1.0	280.81	3JA	-1.0	-3.5	-1.8	1.2
102.80	4.42	-1.0	290.94	13JA	-0.8	-3.6	-1.8	1.2
99.26	4.34	-0.8	301.03	23JA	0.5	-3.6	-1.7	1.2
96.72	4.14	-0.5	311.06	2FE	2.9	-3.7	-1.7	1.1
95.46	3.85	-0.3	321.04	12FE	1.8	-3.7	-1.6	1.1
95.48	3.53	-0.0	330.96	22FE	0.9	-3.8	-1.6	1.0
96.64	3.22	0.2	340.81	4MR	0.6	-3.9	-1.6	1.0
98.76	2.92	0.4	350.62	14MR	0.2	-4.0	-1.6	1.0
101.66	2.64	0.6	0.36	24MR	-0.2	-4.1	-1.6	1.0
105.19	2.38	0.8	10.05	3AP	-1.0	-4.2	-1.6	1.0
109.24	2.14	0.9	19.70	13AP	-1.9	-4.2	-1.6	1.0
113.71	1.93	1.0	29.30	23AP	-1.1	-4.2	-1.6	1.0
118.51	1.72	1.1	38.87	3MY	-0.1	-4.0	-1.6	1.0
123.62	1.53	1.2	48.41	13MY	0.7	-3.5	-1.6	0.9
128.97	1.36	1.3	57.94	23MY	1.4	-2.6	-1.7	0.9
134.54	1.19	1.4	67.45	2JN	2.3	-3.6	-1.7	0.9
140.31	1.02	1.4	76.98	12JN	3.4	-4.0	-1.7	0.8
146.26	0.87	1.5	86.50	22JN	1.8	-4.2	-1.8	0.8
152.37	0.72	1.5	96.06	2JL	0.4	-4.2	-1.9	0.7
158.64	0.58	1.5	105.64	12JL	-0.9	-4.1	-1.9	0.7
165.05	0.44	1.6	115.25	22JL	-1.5	-4.1	-2.0	0.6
171.60	0.30	1.6	124.91	1AU	-1.1	-4.0	-2.1	0.6
178.28	0.17	1.6	134.63	11AU	-0.5	-3.9	-2.1	0.5
185.10	0.04	1.6	144.39	21AU	-0.1	-3.8	-2.2	0.5
192.04	-0.08	1.6	154.21	31AU	0.1	-3.7	-2.3	0.4
199.11	-0.20	1.6	164.10	10SE	0.2	-3.7	-2.3	0.4
206.29	-0.32	1.6	174.04	20SE	0.6	-3.6	-2.3	0.5
213.58	-0.43	1.5	184.04	30SE	2.2	-3.6	-2.4	0.5
220.98	-0.53	1.5	194.09	10OC	1.3	-3.5	-2.4	0.6
228.47	-0.63	1.5	204.19	20OC	-0.3	-3.5	-2.4	0.7
236.06	-0.72	1.5	214.33	30OC	-0.5	-3.4	-2.4	0.8
243.73	-0.80	1.5	224.51	9NO	-0.5	-3.4	-2.3	0.8
251.47	-0.87	1.4	234.70	19NO	-0.6	-3.4	-2.3	0.9
259.26	-0.94	1.4	244.91	29NO	-0.8	-3.4	-2.2	0.9
267.11	-0.99	1.4	255.12	9DE	-0.8	-3.4	-2.2	0.9
274.98	-1.03	1.4	265.32	19DE	-0.9	-3.3	-2.1	1.0
282.88	-1.06	1.3	275.50	29DE	-0.7	-3.3	-2.0	1.0

Left Table

♂ LONG	LAT	MAG	☉ LONG	16.00UT −182	☿	♀	♃	♄
290.78	-1.08	1.3	285.64	8JA	0.6	-3.3	-2.0	1.0
298.67	-1.09	1.3	295.75	18JA	3.1	-3.3	-1.9	1.0
306.54	-1.09	1.3	305.81	28JA	1.4	-3.4	-1.8	0.9
314.38	-1.07	1.3	315.81	7FE	0.7	-3.4	-1.8	0.9
322.18	-1.04	1.4	325.76	17FE	0.4	-3.4	-1.7	0.9
329.92	-1.01	1.4	335.65	27FE	0.1	-3.4	-1.6	0.8
337.59	-0.96	1.4	345.48	9MR	-0.3	-3.5	-1.6	0.8
345.20	-0.90	1.5	355.25	19MR	-1.0	-3.5	-1.6	0.7
352.72	-0.83	1.5	4.98	29MR	-1.8	-3.4	-1.5	0.8
0.16	-0.76	1.5	14.64	8AP	-1.1	-3.4	-1.5	0.8
7.51	-0.68	1.6	24.26	18AP	-0.1	-3.4	-1.5	0.8
14.78	-0.59	1.6	33.85	28AP	0.9	-3.4	-1.5	0.8
21.95	-0.50	1.6	43.41	8MY	1.8	-3.3	-1.5	0.7
29.03	-0.40	1.6	52.94	18MY	3.1	-3.3	-1.5	0.7
36.01	-0.30	1.7	62.46	28MY	2.8	-3.3	-1.5	0.7
42.90	-0.19	1.7	71.98	7JN	1.4	-3.3	-1.5	0.7
49.69	-0.08	1.7	81.50	17JN	0.3	-3.3	-1.5	0.6
56.39	0.03	1.7	91.05	27JN	-0.9	-3.3	-1.6	0.6
62.99	0.15	1.7	100.61	7JL	-1.6	-3.4	-1.6	0.6
69.50	0.27	1.7	110.20	17JL	-1.1	-3.4	-1.7	0.5
75.91	0.39	1.7	119.84	27JL	-0.4	-3.4	-1.7	0.5
82.21	0.52	1.7	129.53	6AU	-0.0	-3.5	-1.8	0.4
88.41	0.65	1.7	139.26	16AU	0.2	-3.5	-1.8	0.3
94.49	0.79	1.6	149.06	26AU	0.4	-3.6	-1.9	0.3
100.44	0.93	1.6	158.91	5SE	0.8	-3.7	-2.0	0.2
106.25	1.08	1.5	168.82	15SE	2.6	-3.7	-2.0	0.2
111.90	1.24	1.5	178.79	25SE	1.1	-3.8	-2.1	0.2
117.36	1.41	1.4	188.81	5OC	-0.4	-3.9	-2.1	0.2
122.60	1.59	1.3	198.89	15OC	-0.7	-4.0	-2.2	0.3
127.59	1.78	1.2	209.01	25OC	-0.7	-4.1	-2.2	0.4
132.26	1.99	1.0	219.17	4NO	-0.7	-4.3	-2.3	0.4
136.55	2.22	0.9	229.35	14NO	-0.7	-4.4	-2.3	0.5
140.37	2.48	0.7	239.56	24NO	-0.7	-4.4	-2.3	0.6
143.61	2.76	0.5	249.76	4DE	-0.7	-4.3	-2.3	0.6
146.12	3.06	0.3	259.97	14DE	-0.5	-4.1	-2.2	0.7
147.74	3.39	0.1	270.16	24DE	0.7	-3.4	-2.2	0.7

−181

♂ LONG	LAT	MAG	☉ LONG	16.00UT	☿	♀	♃	♄
148.28	3.73	-0.2	280.32	3JA	2.9	-3.4	-2.1	0.7
147.59	4.06	-0.5	290.45	13JA	1.0	-4.0	-2.1	0.7
145.60	4.33	-0.7	300.54	23JA	0.4	-4.3	-2.0	0.7
142.48	4.50	-1.0	310.57	2FE	0.2	-4.3	-1.9	0.7
138.67	4.50	-1.1	320.55	12FE	0.0	-4.3	-1.8	0.7
134.87	4.32	-1.0	330.48	22FE	-0.4	-4.2	-1.8	0.7
131.78	4.00	-0.8	340.33	4MR	-1.0	-4.1	-1.7	0.6
129.82	3.59	-0.6	350.14	14MR	-1.7	-4.0	-1.7	0.6
129.17	3.16	-0.4	359.89	24MR	-1.1	-3.9	-1.6	0.5
129.78	2.73	-0.1	9.58	3AP	0.0	-3.8	-1.5	0.5
131.47	2.33	0.1	19.23	13AP	1.1	-3.7	-1.5	0.5
134.07	1.96	0.3	28.83	23AP	2.4	-3.6	-1.5	0.6
137.43	1.63	0.4	38.40	3MY	3.7	-3.5	-1.4	0.6
141.40	1.33	0.6	47.95	13MY	2.2	-3.5	-1.4	0.6
145.88	1.06	0.7	57.48	23MY	1.1	-3.4	-1.4	0.6
150.78	0.81	0.8	66.99	2JN	0.2	-3.4	-1.4	0.5
156.04	0.58	0.9	76.51	12JN	-0.9	-3.4	-1.4	0.5
161.62	0.38	1.0	86.05	22JN	-1.7	-3.3	-1.4	0.5
167.47	0.18	1.0	95.59	2JL	-1.1	-3.3	-1.4	0.5
173.55	0.00	1.1	105.18	12JL	-0.4	-3.3	-1.4	0.4
179.86	-0.16	1.1	114.79	22JL	0.1	-3.3	-1.4	0.4
186.37	-0.32	1.2	124.45	1AU	0.3	-3.3	-1.5	0.3
193.06	-0.46	1.2	134.16	11AU	0.6	-3.4	-1.5	0.3
199.93	-0.59	1.2	143.92	21AU	1.2	-3.4	-1.5	0.2
206.95	-0.71	1.3	153.74	31AU	3.0	-3.4	-1.6	0.2
214.11	-0.81	1.3	163.62	10SE	1.0	-3.4	-1.7	0.1
221.41	-0.91	1.3	173.56	20SE	-0.5	-3.5	-1.7	0.0
228.83	-0.99	1.3	183.55	30SE	-0.8	-3.5	-1.8	-0.0
236.36	-1.06	1.3	193.60	10OC	-0.8	-3.5	-1.8	-0.0
243.98	-1.11	1.3	203.70	20OC	-0.8	-3.5	-1.9	0.0
251.68	-1.16	1.3	213.83	30OC	-0.6	-3.5	-2.0	0.1
259.45	-1.19	1.3	224.01	9NO	-0.6	-3.4	-2.0	0.2
267.26	-1.20	1.4	234.20	19NO	-0.6	-3.4	-2.1	0.2
275.12	-1.20	1.4	244.41	29NO	-0.4	-3.4	-2.1	0.3
282.99	-1.19	1.4	254.62	9DE	0.9	-3.4	-2.1	0.4
290.86	-1.17	1.4	264.82	19DE	2.5	-3.3	-2.2	0.4
298.73	-1.13	1.4	275.00	29DE	0.7	-3.3	-2.2	0.5

Right Table

♂ LONG	LAT	MAG	☉ LONG	16.00UT −180	☿	♀	♃	♄
306.57	-1.09	1.4	285.15	8JA	0.2	-3.3	-2.1	0.5
314.37	-1.03	1.4	295.26	18JA	0.1	-3.3	-2.1	0.5
322.13	-0.97	1.5	305.32	28JA	-0.1	-3.4	-2.1	0.5
329.82	-0.89	1.5	315.33	7FE	-0.4	-3.4	-2.0	0.5
337.45	-0.81	1.5	325.28	17FE	-1.0	-3.4	-1.9	0.5
345.01	-0.73	1.5	335.17	27FE	-1.6	-3.4	-1.9	0.5
352.48	-0.64	1.5	345.01	8MR	-1.1	-3.5	-1.8	0.5
359.88	-0.54	1.5	354.78	18MR	0.1	-3.5	-1.7	0.4
7.18	-0.45	1.5	4.50	28MR	1.4	-3.5	-1.7	0.4
14.41	-0.35	1.5	14.18	7AP	3.1	-3.6	-1.6	0.3
21.54	-0.25	1.6	23.80	17AP	2.9	-3.7	-1.5	0.4
28.59	-0.14	1.6	33.39	27AP	1.6	-3.7	-1.5	0.4
35.56	-0.04	1.7	42.95	7MY	0.9	-3.8	-1.4	0.4
42.45	0.06	1.7	52.48	17MY	0.1	-3.9	-1.4	0.4
49.26	0.16	1.8	62.00	27MY	-0.9	-4.0	-1.4	0.4
56.00	0.26	1.8	71.52	6JN	-1.7	-4.1	-1.3	0.4
62.66	0.37	1.8	81.05	16JN	-1.1	-4.2	-1.3	0.4
69.25	0.47	1.9	90.58	26JN	-0.3	-4.2	-1.3	0.4
75.78	0.56	1.9	100.15	6JL	0.2	-4.1	-1.3	0.4
82.24	0.66	1.9	109.74	16JL	0.5	-3.9	-1.3	0.3
88.64	0.76	1.9	119.38	26JL	0.8	-3.3	-1.3	0.3
94.97	0.86	1.9	129.06	5AU	1.5	-3.3	-1.3	0.3
101.24	0.96	1.9	138.79	15AU	3.2	-3.9	-1.3	0.2
107.45	1.05	1.9	148.58	25AU	0.9	-4.2	-1.3	0.1
113.59	1.15	1.9	158.43	4SE	-0.6	-4.3	-1.4	0.1
119.66	1.25	1.8	168.34	14SE	-1.0	-4.3	-1.4	0.0
125.65	1.36	1.8	178.30	24SE	-1.0	-4.2	-1.5	-0.0
131.55	1.46	1.7	188.33	4OC	-0.8	-4.1	-1.5	-0.1
137.34	1.57	1.7	198.40	14OC	-0.5	-4.0	-1.6	-0.2
143.03	1.68	1.6	208.52	24OC	-0.4	-3.9	-1.6	-0.2
148.57	1.79	1.5	218.68	3NO	-0.4	-3.9	-1.7	-0.1
153.95	1.91	1.4	228.86	13NO	-0.2	-3.8	-1.8	-0.0
159.13	2.04	1.2	239.06	23NO	1.1	-3.7	-1.8	0.0
164.07	2.17	1.1	249.27	3DE	2.2	-3.6	-1.9	0.1
168.73	2.31	0.9	259.48	13DE	0.4	-3.6	-2.0	0.2
173.02	2.46	0.8	269.67	23DE	0.1	-3.5	-2.0	0.2

−179

♂ LONG	LAT	MAG	☉ LONG	16.00UT	☿	♀	♃	♄
176.87	2.61	0.6	279.84	2JA	-0.0	-3.5	-2.0	0.3
180.15	2.77	0.3	289.96	12JA	-0.2	-3.4	-2.1	0.3
182.75	2.93	0.1	300.05	22JA	-0.5	-3.4	-2.1	0.4
184.48	3.09	-0.2	310.09	1FE	-1.0	-3.4	-2.1	0.4
185.17	3.22	-0.5	320.07	11FE	-1.4	-3.3	-2.0	0.4
184.65	3.30	-0.8	330.00	21FE	-1.0	-3.3	-2.0	0.4
182.85	3.30	-1.1	339.87	3MR	0.2	-3.3	-1.9	0.4
179.94	3.18	-1.3	349.67	13MR	1.8	-3.3	-1.9	0.4
176.33	2.91	-1.5	359.42	23MR	3.6	-3.3	-1.8	0.4
172.71	2.52	-1.4	9.12	2AP	2.1	-3.3	-1.8	0.3
169.80	2.05	-1.2	18.76	12AP	1.2	-3.3	-1.7	0.3
168.04	1.54	-1.0	28.37	22AP	0.6	-3.4	-1.6	0.2
167.64	1.06	-0.8	37.94	2MY	-0.0	-3.4	-1.6	0.2
168.53	0.63	-0.6	47.48	12MY	-0.9	-3.5	-1.5	0.3
170.56	0.24	-0.4	57.01	22MY	-1.8	-3.5	-1.5	0.3
173.56	-0.09	-0.2	66.53	1JN	-1.1	-3.5	-1.4	0.3
177.35	-0.38	-0.1	76.05	11JN	-0.3	-3.4	-1.4	0.3
181.80	-0.63	0.1	85.58	21JN	0.3	-3.4	-1.3	0.3
186.80	-0.84	0.2	95.13	1JL	0.7	-3.4	-1.3	0.3
192.25	-1.03	0.3	104.70	11JL	1.1	-3.4	-1.3	0.3
198.08	-1.18	0.4	114.32	21JL	2.0	-3.3	-1.2	0.3
204.25	-1.31	0.5	123.97	31JL	3.0	-3.3	-1.2	0.3
210.70	-1.42	0.6	133.68	10AU	0.7	-3.3	-1.2	0.2
217.40	-1.50	0.7	143.44	20AU	-0.7	-3.3	-1.2	0.2
224.31	-1.56	0.7	153.26	30AU	-1.1	-3.3	-1.2	0.1
231.40	-1.61	0.8	163.13	9SE	-1.1	-3.4	-1.3	0.1
238.65	-1.63	0.8	173.07	19SE	-0.7	-3.4	-1.3	0.0
246.04	-1.63	0.9	183.06	29SE	-0.4	-3.4	-1.3	-0.1
253.53	-1.62	1.0	193.11	9OC	-0.3	-3.5	-1.3	-0.1
261.11	-1.58	1.0	203.21	19OC	-0.2	-3.5	-1.4	-0.2
268.76	-1.54	1.1	213.34	29OC	-0.0	-3.6	-1.4	-0.3
276.46	-1.48	1.1	223.51	8NO	1.3	-3.6	-1.5	-0.3
284.18	-1.40	1.2	233.71	18NO	1.9	-3.7	-1.6	-0.2
291.91	-1.32	1.2	243.92	28NO	0.2	-3.8	-1.6	-0.1
299.63	-1.23	1.3	254.13	8DE	-0.1	-3.8	-1.7	-0.1
307.33	-1.13	1.3	264.33	18DE	-0.2	-3.9	-1.7	0.0
315.00	-1.02	1.4	274.51	28DE	-0.3	-4.0	-1.8	0.1

♂ LONG	♂ LAT	♂ MAG	☉ LONG	16.00UT -178	☿	♀	♃	♄
					MAGNITUDES			
322.61	-0.91	1.4	284.66	7JA	-0.6	-4.1	-1.9	0.2
330.18	-0.80	1.5	294.77	17JA	-1.0	-4.2	-1.9	0.2
337.67	-0.68	1.5	304.83	27JA	-1.3	-4.3	-2.0	0.2
345.10	-0.57	1.6	314.84	6FE	-1.0	-4.3	-2.0	0.3
352.45	-0.45	1.6	324.80	16FE	0.2	-4.2	-2.0	0.3
359.73	-0.34	1.6	334.70	26FE	2.2	-3.9	-2.0	0.3
6.92	-0.23	1.7	344.53	8MR	2.9	-3.3	-2.0	0.3
14.04	-0.12	1.7	354.31	18MR	1.6	-3.4	-2.0	0.3
21.08	-0.01	1.7	4.04	28MR	0.9	-3.9	-1.9	0.3
28.04	0.09	1.8	13.71	7AP	0.5	-4.2	-1.9	0.3
34.93	0.19	1.8	23.34	17AP	-0.1	-4.2	-1.8	0.3
41.75	0.29	1.8	32.92	27AP	-0.9	-4.2	-1.8	0.2
48.51	0.38	1.8	42.48	7MY	-1.8	-4.1	-1.7	0.2
55.20	0.47	1.8	52.02	17MY	-1.1	-4.0	-1.6	0.2
61.83	0.56	1.8	61.54	27MY	-0.2	-3.9	-1.6	0.2
68.41	0.64	1.8	71.06	6JN	0.4	-3.8	-1.5	0.3
74.95	0.72	1.9	80.58	16JN	0.9	-3.7	-1.5	0.3
81.44	0.79	1.9	90.12	26JN	1.5	-3.6	-1.4	0.3
87.89	0.87	2.0	99.68	6JL	2.6	-3.6	-1.4	0.3
94.30	0.93	2.0	109.28	16JL	2.6	-3.5	-1.3	0.3
100.68	1.00	2.0	118.91	26JL	0.6	-3.5	-1.3	0.3
107.03	1.06	2.0	128.59	5AU	-0.8	-3.4	-1.3	0.3
113.35	1.12	2.0	138.32	15AU	-1.3	-3.4	-1.2	0.3
119.65	1.18	2.0	148.10	25AU	-1.1	-3.4	-1.2	0.2
125.92	1.24	2.0	157.95	4SE	-0.6	-3.4	-1.2	0.2
132.17	1.29	2.0	167.86	14SE	-0.3	-3.4	-1.2	0.1
138.39	1.34	1.9	177.82	24SE	-0.1	-3.4	-1.2	0.1
144.58	1.39	1.9	187.84	4OC	-0.1	-3.4	-1.2	-0.0
150.74	1.43	1.9	197.91	14OC	0.2	-3.4	-1.3	-0.1
156.86	1.47	1.8	208.03	24OC	1.6	-3.4	-1.3	-0.2
162.94	1.51	1.7	218.18	3NO	1.7	-3.4	-1.3	-0.2
168.97	1.55	1.7	228.37	13NO	0.0	-3.4	-1.4	-0.3
174.93	1.58	1.6	238.57	23NO	-0.3	-3.4	-1.4	-0.3
180.83	1.61	1.5	248.78	3DE	-0.3	-3.4	-1.5	-0.2
186.63	1.63	1.4	258.98	13DE	-0.4	-3.5	-1.5	-0.2
192.32	1.65	1.2	269.17	23DE	-0.6	-3.5	-1.6	-0.1
				-177				
197.88	1.66	1.1	279.34	2JA	-1.0	-3.5	-1.7	-0.0
203.28	1.65	0.9	289.47	12JA	-1.1	-3.5	-1.7	0.1
208.47	1.64	0.7	299.56	22JA	-0.9	-3.4	-1.8	0.1
213.43	1.60	0.5	309.60	1FE	0.3	-3.4	-1.8	0.2
218.06	1.54	0.3	319.59	11FE	2.6	-3.4	-1.9	0.2
222.32	1.45	0.1	329.51	21FE	2.2	-3.4	-2.0	0.3
226.08	1.33	-0.2	339.38	3MR	1.1	-3.3	-2.0	0.3
229.22	1.15	-0.4	349.19	13MR	0.7	-3.3	-2.0	0.3
231.60	0.92	-0.7	358.94	23MR	0.3	-3.3	-2.0	0.3
233.02	0.60	-1.0	8.64	2AP	-0.2	-3.3	-2.0	0.3
233.30	0.19	-1.4	18.30	12AP	-0.9	-3.3	-2.0	0.3
232.36	-0.31	-1.7	27.90	22AP	-1.8	-3.4	-2.0	0.3
230.23	-0.90	-2.0	37.48	2MY	-1.1	-3.4	-1.9	0.2
227.28	-1.52	-2.2	47.02	12MY	-0.2	-3.4	-1.9	0.2
224.14	-2.11	-2.1	56.55	22MY	0.5	-3.4	-1.8	0.2
221.54	-2.60	-2.0	66.07	1JN	1.1	-3.5	-1.8	0.2
220.05	-2.96	-1.8	75.59	11JN	1.9	-3.5	-1.7	0.2
219.92	-3.20	-1.6	85.12	21JN	3.2	-3.6	-1.6	0.3
221.15	-3.33	-1.3	94.67	1JL	2.1	-3.7	-1.6	0.3
223.60	-3.37	-1.1	104.24	11JL	0.5	-3.8	-1.5	0.3
227.06	-3.36	-0.9	113.85	21JL	-0.8	-3.9	-1.5	0.3
231.37	-3.30	-0.7	123.51	31JL	-1.4	-4.0	-1.4	0.3
236.35	-3.20	-0.5	133.21	10AU	-1.1	-4.1	-1.4	0.3
241.87	-3.08	-0.4	142.96	20AU	-0.5	-4.2	-1.3	0.3
247.83	-2.93	-0.2	152.78	30AU	-0.2	-4.3	-1.3	0.3
254.13	-2.76	-0.1	162.65	9SE	-0.0	-4.3	-1.3	0.2
260.70	-2.59	0.1	172.58	19SE	0.1	-4.3	-1.3	0.2
267.49	-2.40	0.2	182.58	29SE	0.4	-4.0	-1.3	0.1
274.44	-2.21	0.4	192.62	9OC	1.9	-3.4	-1.3	0.1
281.51	-2.01	0.5	202.71	19OC	1.5	-3.4	-1.3	0.0
288.67	-1.81	0.6	212.85	29OC	-0.2	-4.0	-1.3	-0.1
295.88	-1.61	0.7	223.02	8NO	-0.4	-4.3	-1.3	-0.1
303.13	-1.42	0.8	233.21	18NO	-0.5	-4.4	-1.3	-0.2
310.40	-1.23	0.9	243.42	28NO	-0.5	-4.3	-1.3	-0.3
317.65	-1.04	1.0	253.63	8DE	-0.7	-4.3	-1.4	-0.3
324.89	-0.86	1.1	263.83	18DE	-0.9	-4.1	-1.4	-0.2
332.10	-0.69	1.2	274.02	28DE	-1.0	-4.0	-1.5	-0.1

♂ LONG	♂ LAT	♂ MAG	☉ LONG	16.00UT -176	☿	♀	♃	♄
					MAGNITUDES			
339.26	-0.53	1.3	284.17	7JA	-0.8	-3.9	-1.5	-0.1
346.38	-0.38	1.4	294.28	17JA	0.4	-3.8	-1.6	0.0
353.45	-0.24	1.5	304.35	27JA	2.9	-3.7	-1.7	0.1
0.46	-0.10	1.6	314.36	6FE	1.7	-3.7	-1.7	0.1
7.41	0.03	1.6	324.32	16FE	0.8	-3.6	-1.8	0.2
14.31	0.14	1.7	334.23	26FE	0.5	-3.5	-1.9	0.2
21.14	0.25	1.7	344.06	7MR	0.2	-3.5	-1.9	0.3
27.91	0.36	1.8	353.85	17MR	-0.2	-3.4	-2.0	0.3
34.64	0.45	1.8	3.57	27MR	-0.9	-3.4	-2.0	0.3
41.30	0.54	1.9	13.25	6AP	-1.8	-3.3	-2.1	0.3
47.92	0.62	1.9	22.88	16AP	-1.1	-3.3	-2.1	0.3
54.50	0.70	1.9	32.47	26AP	-0.1	-3.3	-2.1	0.3
61.03	0.76	1.9	42.03	6MY	0.7	-3.3	-2.1	0.3
67.52	0.83	2.0	51.56	16MY	1.5	-3.3	-2.1	0.3
73.99	0.89	2.0	61.09	26MY	2.6	-3.3	-2.1	0.3
80.42	0.94	2.0	70.60	5JN	3.2	-3.3	-2.0	0.2
86.84	0.99	2.0	80.13	15JN	1.7	-3.4	-2.0	0.2
93.23	1.03	2.0	89.66	25JN	0.4	-3.4	-1.9	0.3
99.62	1.07	1.9	99.22	5JL	-0.9	-3.4	-1.8	0.3
105.99	1.10	2.0	108.81	15JL	-1.5	-3.5	-1.8	0.4
112.36	1.13	2.0	118.44	25JL	-1.1	-3.5	-1.7	0.4
118.73	1.16	2.0	128.11	4AU	-0.5	-3.5	-1.6	0.4
125.10	1.18	2.0	137.84	14AU	-0.1	-3.5	-1.6	0.4
131.48	1.19	2.0	147.63	24AU	0.1	-3.4	-1.5	0.4
137.86	1.21	2.0	157.47	3SE	0.3	-3.4	-1.5	0.4
144.26	1.21	2.0	167.37	13SE	0.7	-3.4	-1.4	0.4
150.67	1.22	2.0	177.33	23SE	2.2	-3.3	-1.4	0.3
157.10	1.22	2.0	187.34	3OC	1.3	-3.3	-1.4	0.3
163.54	1.21	1.9	197.41	13OC	-0.3	-3.3	-1.4	0.2
169.99	1.20	1.9	207.53	23OC	-0.6	-3.3	-1.3	0.2
176.45	1.18	1.9	217.68	2NO	-0.6	-3.3	-1.3	0.1
182.93	1.15	1.8	227.87	12NO	-0.6	-3.4	-1.3	0.0
189.42	1.12	1.7	238.07	22NO	-0.8	-3.4	-1.3	-0.1
195.92	1.07	1.7	248.28	2DE	-0.8	-3.4	-1.4	-0.1
202.42	1.02	1.6	258.49	12DE	-0.8	-3.4	-1.4	-0.2
208.92	0.96	1.5	268.68	22DE	-0.6	-3.5	-1.4	-0.2
				-175				
215.42	0.88	1.4	278.85	1JA	0.6	-3.5	-1.4	-0.1
221.91	0.79	1.3	288.99	11JA	3.0	-3.6	-1.5	-0.1
228.39	0.68	1.2	299.08	21JA	1.2	-3.6	-1.5	0.0
234.84	0.55	1.0	309.12	31JA	0.6	-3.7	-1.6	0.1
241.27	0.40	0.9	319.11	10FE	0.3	-3.7	-1.7	0.1
247.64	0.23	0.8	329.04	20FE	0.1	-3.8	-1.7	0.2
253.96	0.02	0.6	338.91	2MR	-0.3	-3.9	-1.8	0.3
260.20	-0.21	0.4	348.73	12MR	-1.0	-4.0	-1.9	0.3
266.32	-0.48	0.3	358.48	22MR	-1.7	-4.1	-1.9	0.3
272.30	-0.79	0.1	8.18	1AP	-1.1	-4.2	-2.0	0.4
278.10	-1.14	-0.1	17.84	11AP	-0.1	-4.2	-2.1	0.4
283.64	-1.55	-0.3	27.44	21AP	0.9	-4.2	-2.1	0.4
288.86	-2.01	-0.6	37.02	1MY	2.0	-4.0	-2.2	0.4
293.65	-2.52	-0.8	46.57	11MY	3.4	-3.4	-2.2	0.4
297.88	-3.10	-1.1	56.09	21MY	2.6	-2.7	-2.2	0.4
301.40	-3.74	-1.3	65.61	31MY	1.3	-3.6	-2.2	0.4
304.00	-4.44	-1.6	75.13	10JN	0.2	-4.0	-2.2	0.4
305.50	-5.16	-1.9	84.66	20JN	-0.9	-4.2	-2.2	0.3
305.72	-5.86	-2.1	94.20	30JN	-1.6	-4.2	-2.2	0.4
304.62	-6.44	-2.4	103.78	10JL	-1.1	-4.1	-2.1	0.4
302.48	-6.81	-2.6	113.39	20JL	-0.4	-4.0	-2.1	0.5
299.83	-6.86	-2.6	123.04	30JL	-0.0	-4.0	-2.0	0.5
297.41	-6.59	-2.4	132.74	9AU	0.2	-3.9	-1.9	0.5
295.91	-6.06	-2.2	142.49	19AU	0.5	-3.8	-1.9	0.5
295.64	-5.40	-1.9	152.30	29AU	0.9	-3.7	-1.8	0.5
296.68	-4.69	-1.6	162.17	8SE	2.6	-3.7	-1.7	0.5
298.89	-4.00	-1.3	172.10	18SE	1.2	-3.6	-1.7	0.5
302.08	-3.37	-1.0	182.09	28SE	-0.4	-3.6	-1.6	0.5
306.04	-2.79	-0.7	192.13	8OC	-0.8	-3.5	-1.6	0.5
310.62	-2.28	-0.5	202.22	18OC	-0.7	-3.5	-1.5	0.4
315.66	-1.84	-0.2	212.35	28OC	-0.8	-3.4	-1.5	0.4
321.06	-1.44	0.0	222.53	7NO	-0.7	-3.4	-1.5	0.3
326.73	-1.10	0.2	232.72	17NO	-0.6	-3.4	-1.5	0.2
332.61	-0.79	0.4	242.92	27NO	-0.7	-3.4	-1.4	0.2
338.65	-0.53	0.6	253.14	7DE	-0.5	-3.4	-1.4	0.1
344.80	-0.29	0.8	263.34	17DE	0.7	-3.3	-1.4	0.0
351.03	-0.09	0.9	273.52	27DE	2.8	-3.3	-1.4	-0.1

Left table (years -174 and -173):

♂ LONG	LAT	MAG	☉ LONG	16.00UT -174	☿	♀	♃	♄
357.32	0.09	1.1	283.68	6JA	0.9	-3.3	-1.5	-0.0
3.65	0.25	1.2	293.79	16JA	0.4	-3.3	-1.5	0.0
10.00	0.39	1.3	303.86	26JA	0.2	-3.4	-1.5	0.1
16.37	0.51	1.4	313.88	5FE	-0.0	-3.4	-1.6	0.1
22.73	0.62	1.5	323.84	15FE	-0.4	-3.4	-1.6	0.2
29.10	0.71	1.6	333.74	25FE	-1.0	-3.4	-1.6	0.3
35.46	0.79	1.7	343.59	7MR	-1.6	-3.5	-1.7	0.3
41.80	0.86	1.8	353.37	17MR	-1.1	-3.5	-1.8	0.4
48.14	0.93	1.8	3.10	27MR	-0.0	-3.4	-1.8	0.4
54.47	0.98	1.9	12.78	6AP	1.2	-3.4	-1.9	0.5
60.78	1.03	1.9	22.41	16AP	2.6	-3.4	-2.0	0.5
67.08	1.06	2.0	32.00	26AP	3.4	-3.4	-2.0	0.5
73.38	1.10	2.0	41.56	6MY	2.0	-3.3	-2.1	0.5
79.68	1.12	2.0	51.10	16MY	1.0	-3.3	-2.2	0.6
85.97	1.14	2.0	60.62	26MY	0.1	-3.3	-2.2	0.6
92.27	1.16	2.0	70.14	5JN	-0.9	-3.3	-2.3	0.6
98.57	1.17	2.0	79.66	15JN	-1.7	-3.3	-2.3	0.5
104.89	1.17	2.0	89.20	25JN	-1.1	-3.3	-2.3	0.5
111.22	1.17	2.0	98.76	5JL	-0.4	-3.4	-2.4	0.5
117.56	1.17	2.0	108.34	15JL	0.1	-3.4	-2.3	0.5
123.94	1.16	2.0	117.97	25JL	0.4	-3.4	-2.3	0.6
130.34	1.15	1.9	127.65	4AU	0.7	-3.5	-2.3	0.6
136.77	1.13	1.9	137.37	14AU	1.3	-3.5	-2.2	0.7
143.24	1.11	1.9	147.15	24AU	3.0	-3.6	-2.2	0.7
149.74	1.08	1.9	156.99	3SE	1.0	-3.7	-2.1	0.7
156.29	1.04	1.9	166.89	13SE	-0.6	-3.7	-2.0	0.7
162.88	1.01	1.9	176.85	23SE	-0.9	-3.8	-2.0	0.7
169.51	0.96	1.9	186.86	3OC	-0.9	-3.9	-1.9	0.7
176.19	0.91	1.9	196.92	13OC	-0.8	-4.0	-1.8	0.7
182.92	0.85	1.9	207.04	23OC	-0.6	-4.1	-1.8	0.6
189.70	0.79	1.9	217.19	2NO	-0.5	-4.3	-1.7	0.6
196.54	0.72	1.8	227.37	12NO	-0.5	-4.3	-1.7	0.5
203.42	0.64	1.8	237.57	22NO	-0.3	-4.4	-1.6	0.5
210.35	0.55	1.7	247.78	2DE	0.9	-4.3	-1.6	0.4
217.33	0.46	1.7	257.98	12DE	2.5	-4.0	-1.6	0.3
224.37	0.35	1.6	268.18	22DE	0.6	-3.4	-1.5	0.3

-173

♂ LONG	LAT	MAG	☉ LONG	16.00UT -173	☿	♀	♃	♄
231.44	0.24	1.5	278.35	1JA	0.2	-3.4	-1.5	0.2
238.56	0.11	1.5	288.49	11JA	0.0	-4.0	-1.5	0.1
245.73	-0.03	1.4	298.58	21JA	-0.1	-4.3	-1.5	0.2
252.93	-0.18	1.3	308.63	31JA	-0.4	-4.3	-1.5	0.2
260.17	-0.34	1.2	318.62	10FE	-1.0	-4.3	-1.5	0.3
267.43	-0.50	1.1	328.56	20FE	-1.5	-4.2	-1.6	0.3
274.72	-0.68	1.1	338.43	2MR	-1.1	-4.1	-1.6	0.4
282.01	-0.87	1.0	348.25	12MR	0.1	-4.0	-1.6	0.4
289.30	-1.07	0.9	358.01	22MR	1.5	-3.8	-1.7	0.5
296.58	-1.27	0.8	7.71	1AP	3.3	-3.8	-1.7	0.6
303.82	-1.48	0.7	17.37	11AP	2.6	-3.7	-1.8	0.6
311.01	-1.69	0.6	26.98	21AP	1.5	-3.6	-1.8	0.6
318.11	-1.91	0.5	36.56	1MY	0.8	-3.5	-1.9	0.7
325.12	-2.12	0.3	46.11	11MY	0.0	-3.5	-2.0	0.7
331.97	-2.33	0.2	55.64	21MY	-0.9	-3.4	-2.0	0.7
338.65	-2.53	0.1	65.16	31MY	-1.7	-3.4	-2.1	0.7
345.09	-2.72	-0.0	74.67	10JN	-1.1	-3.4	-2.2	0.8
351.24	-2.91	-0.1	84.20	20JN	-0.3	-3.3	-2.3	0.8
357.02	-3.08	-0.3	93.75	30JN	0.2	-3.3	-2.3	0.8
2.35	-3.24	-0.4	103.32	10JL	0.5	-3.3	-2.4	0.8
7.11	-3.38	-0.6	112.93	20JL	0.9	-3.3	-2.4	0.7
11.15	-3.49	-0.8	122.57	30JL	1.7	-3.3	-2.4	0.8
14.30	-3.57	-1.0	132.27	9AU	3.2	-3.4	-2.4	0.8
16.34	-3.60	-1.2	142.02	19AU	0.9	-3.4	-2.4	0.9
17.08	-3.56	-1.4	151.83	29AU	-0.6	-3.4	-2.4	0.9
16.37	-3.42	-1.7	161.70	8SE	-1.0	-3.4	-2.4	0.9
14.22	-3.14	-1.9	171.62	18SE	-1.0	-3.5	-2.3	1.0
11.03	-2.72	-2.0	181.60	28SE	-0.7	-3.5	-2.3	1.0
7.46	-2.17	-2.0	191.64	8OC	-0.5	-3.5	-2.2	1.0
4.35	-1.58	-1.7	201.73	18OC	-0.4	-3.5	-2.1	0.9
2.32	-1.00	-1.4	211.86	28OC	-0.3	-3.5	-2.1	0.9
1.62	-0.49	-1.1	222.03	7NO	-0.2	-3.4	-2.0	0.9
2.24	-0.07	-0.7	232.22	17NO	1.1	-3.4	-1.9	0.8
4.01	0.27	-0.4	242.42	27NO	2.2	-3.4	-1.9	0.8
6.70	0.54	-0.1	252.63	7DE	0.3	-3.4	-1.8	0.7
10.13	0.75	0.1	262.84	17DE	-0.0	-3.3	-1.7	0.7
14.14	0.92	0.4	273.02	27DE	-0.1	-3.3	-1.7	0.6

Right table (years -172 and -171):

♂ LONG	LAT	MAG	☉ LONG	16.00UT -172	☿	♀	♃	♄
18.59	1.05	0.6	283.17	6JA	-0.2	-3.3	-1.7	0.5
23.38	1.15	0.8	293.29	16JA	-0.5	-3.3	-1.6	0.4
28.45	1.23	0.9	303.36	26JA	-1.0	-3.4	-1.6	0.4
33.72	1.29	1.1	313.39	5FE	-1.4	-3.4	-1.6	0.4
39.17	1.33	1.2	323.35	15FE	-1.0	-3.4	-1.6	0.5
44.74	1.36	1.4	333.26	25FE	0.1	-3.4	-1.6	0.5
50.41	1.39	1.5	343.11	6MR	1.9	-3.5	-1.6	0.6
56.18	1.40	1.6	352.90	16MR	3.4	-3.5	-1.6	0.6
62.01	1.41	1.7	2.63	26MR	1.9	-3.5	-1.6	0.7
67.90	1.41	1.7	12.31	5AP	1.1	-3.6	-1.6	0.7
73.84	1.40	1.8	21.95	15AP	0.6	-3.7	-1.7	0.8
79.83	1.39	1.9	31.54	25AP	-0.0	-3.7	-1.7	0.8
85.86	1.37	1.9	41.10	5MY	-0.9	-3.8	-1.7	0.9
91.94	1.35	2.0	50.64	15MY	-1.8	-3.9	-1.8	0.9
98.05	1.32	2.0	60.16	25MY	-1.1	-4.0	-1.9	0.9
104.20	1.29	2.0	69.69	4JN	-0.3	-4.1	-1.9	1.0
110.40	1.26	2.0	79.21	14JN	0.3	-4.2	-2.0	1.0
116.64	1.22	2.0	88.74	24JN	0.7	-4.2	-2.1	1.0
122.93	1.18	2.0	98.30	4JL	1.2	-4.1	-2.1	1.0
129.26	1.14	2.0	107.89	14JL	2.2	-3.9	-2.2	1.0
135.65	1.09	2.0	117.51	24JL	2.9	-3.3	-2.3	1.0
142.10	1.04	2.0	127.18	3AU	0.7	-3.4	-2.3	1.0
148.60	0.98	1.9	136.90	13AU	-0.7	-3.9	-2.4	1.0
155.17	0.92	1.9	146.68	23AU	-1.2	-4.2	-2.4	1.1
161.80	0.85	1.9	156.52	2SE	-1.1	-4.3	-2.4	1.1
168.50	0.78	1.8	166.41	12SE	-0.7	-4.3	-2.5	1.2
175.27	0.71	1.8	176.36	22SE	-0.4	-4.2	-2.4	1.2
182.11	0.63	1.8	186.38	2OC	-0.2	-4.1	-2.4	1.2
189.02	0.55	1.8	196.44	12OC	-0.2	-4.0	-2.4	1.2
196.01	0.46	1.8	206.55	22OC	0.0	-3.9	-2.3	1.2
203.07	0.37	1.7	216.70	1NO	1.3	-3.8	-2.3	1.2
210.20	0.27	1.7	226.87	11NO	1.9	-3.8	-2.2	1.2
217.41	0.16	1.7	237.07	21NO	0.1	-3.7	-2.1	1.2
224.68	0.06	1.7	247.28	1DE	-0.2	-3.6	-2.1	1.1
232.02	-0.06	1.6	257.49	11DE	-0.2	-3.6	-2.0	1.1
239.43	-0.17	1.6	267.68	21DE	-0.3	-3.5	-1.9	1.0
246.90	-0.30	1.6	277.86	31DE	-0.6	-3.5	-1.9	0.9

-171

♂ LONG	LAT	MAG	☉ LONG	16.00UT -171	☿	♀	♃	♄
254.42	-0.42	1.5	288.00	10JA	-1.0	-3.4	-1.8	0.9
261.98	-0.54	1.5	298.09	20JA	-1.2	-3.4	-1.7	0.8
269.59	-0.67	1.4	308.14	30JA	-0.9	-3.4	-1.7	0.7
277.23	-0.80	1.4	318.14	9FE	0.2	-3.3	-1.6	0.7
284.90	-0.92	1.3	328.07	19FE	2.3	-3.3	-1.6	0.7
292.57	-1.04	1.3	337.95	1MR	2.7	-3.3	-1.6	0.8
300.24	-1.15	1.3	347.77	11MR	1.4	-3.3	-1.6	0.8
307.91	-1.26	1.2	357.53	21MR	0.8	-3.3	-1.6	0.9
315.54	-1.36	1.2	7.24	31MR	0.4	-3.3	-1.5	0.9
323.14	-1.45	1.1	16.90	10AP	-0.1	-3.3	-1.5	1.0
330.67	-1.53	1.1	26.51	20AP	-0.9	-3.4	-1.6	1.0
338.14	-1.59	1.0	36.09	30AP	-1.8	-3.4	-1.6	1.1
345.52	-1.64	1.0	45.64	10MY	-1.2	-3.5	-1.6	1.1
352.79	-1.67	0.9	55.17	20MY	-0.2	-3.5	-1.6	1.2
359.94	-1.69	0.9	64.69	30MY	0.4	-3.5	-1.7	1.2
6.95	-1.69	0.8	74.21	9JN	1.0	-3.4	-1.7	1.2
13.79	-1.68	0.8	83.74	19JN	1.6	-3.4	-1.7	1.3
20.44	-1.65	0.7	93.28	29JN	2.8	-3.4	-1.8	1.3
26.87	-1.59	0.6	102.85	9JL	2.4	-3.4	-1.9	1.3
33.05	-1.52	0.6	112.46	19JL	0.6	-3.3	-1.9	1.3
38.94	-1.43	0.5	122.10	29JL	-0.8	-3.3	-2.0	1.3
44.47	-1.32	0.4	131.80	8AU	-1.3	-3.3	-2.1	1.3
49.59	-1.18	0.2	141.54	18AU	-1.2	-3.3	-2.1	1.3
54.20	-1.01	0.1	151.35	28AU	-0.6	-3.3	-2.2	1.3
58.19	-0.81	-0.0	161.21	7SE	-0.3	-3.4	-2.2	1.3
61.43	-0.58	-0.2	171.14	17SE	-0.1	-3.4	-2.3	1.3
63.72	-0.29	-0.4	181.12	27SE	-0.0	-3.4	-2.3	1.3
64.87	0.04	-0.6	191.15	7OC	0.3	-3.5	-2.4	1.3
64.70	0.43	-0.9	201.24	17OC	1.6	-3.5	-2.4	1.3
63.09	0.87	-1.1	211.37	27OC	1.7	-3.6	-2.4	1.3
60.19	1.32	-1.3	221.53	6NO	-0.0	-3.6	-2.4	1.2
56.42	1.75	-1.4	231.72	16NO	-0.3	-3.7	-2.3	1.2
52.53	2.09	-1.3	241.93	26NO	-0.4	-3.8	-2.3	1.2
49.30	2.33	-1.0	252.14	6DE	-0.4	-3.9	-2.2	1.1
47.22	2.46	-0.7	262.34	16DE	-0.7	-3.9	-2.1	1.1
46.49	2.50	-0.4	272.53	26DE	-1.0	-4.0	-2.1	1.0

♂ LONG	LAT	MAG	⊙ LONG	16.00UT	☿	♀	♃	♄
				-170	MAGNITUDES			
47.03	2.49	-0.1	282.68	5JA	-1.1	-4.1	-2.0	1.0
48.65	2.44	0.1	292.80	15JA	-0.8	-4.2	-1.9	0.9
51.17	2.38	0.4	302.88	25JA	0.3	-4.3	-1.9	0.9
54.40	2.30	0.6	312.90	4FE	2.7	-4.3	-1.8	0.8
58.18	2.22	0.8	322.87	14FE	2.0	-4.2	-1.7	0.8
62.40	2.14	1.0	332.78	24FE	1.0	-3.9	-1.7	0.8
66.98	2.06	1.1	342.63	6MR	0.6	-3.3	-1.6	0.9
71.83	1.98	1.3	352.42	16MR	0.3	-3.5	-1.6	0.9
76.91	1.90	1.4	2.15	26MR	-0.2	-4.0	-1.5	1.0
82.18	1.81	1.5	11.84	5AP	-0.9	-4.2	-1.5	1.1
87.61	1.73	1.6	21.47	15AP	-1.8	-4.2	-1.5	1.2
93.18	1.65	1.6	31.07	25AP	-1.2	-4.2	-1.5	1.2
98.86	1.57	1.7	40.63	5MY	-0.2	-4.1	-1.5	1.3
104.66	1.49	1.8	50.18	15MY	0.6	-4.0	-1.5	1.3
110.55	1.41	1.8	59.70	25MY	1.3	-3.9	-1.5	1.3
116.53	1.33	1.8	69.22	4JN	2.2	-3.8	-1.5	1.4
122.60	1.25	1.9	78.74	14JN	3.4	-3.7	-1.5	1.4
128.76	1.17	1.9	88.28	24JN	2.0	-3.6	-1.5	1.4
135.01	1.08	1.9	97.83	4JL	0.5	-3.6	-1.6	1.4
141.34	1.00	1.9	107.42	14JL	-0.8	-3.5	-1.6	1.3
147.75	0.91	1.9	117.04	24JL	-1.4	-3.5	-1.6	1.3
154.25	0.82	1.9	126.71	3AU	-1.2	-3.4	-1.7	1.3
160.84	0.72	1.9	136.43	13AU	-0.5	-3.4	-1.8	1.2
167.52	0.63	1.8	146.21	23AU	-0.2	-3.4	-1.8	1.2
174.28	0.53	1.8	156.04	2SE	0.0	-3.4	-1.9	1.1
181.14	0.43	1.8	165.93	12SE	0.2	-3.4	-1.9	1.1
188.10	0.33	1.7	175.88	22SE	0.5	-3.4	-2.0	1.2
195.14	0.23	1.7	185.89	2OC	1.9	-3.4	-2.1	1.2
202.27	0.13	1.6	195.95	12OC	1.5	-3.4	-2.1	1.1
209.50	0.02	1.6	206.06	22OC	-0.2	-3.4	-2.2	1.1
216.81	-0.09	1.5	216.20	1NO	-0.5	-3.4	-2.2	1.1
224.20	-0.19	1.5	226.38	11NO	-0.5	-3.4	-2.3	1.1
231.68	-0.30	1.5	236.58	21NO	-0.6	-3.4	-2.3	1.1
239.22	-0.41	1.5	246.79	1DE	-0.7	-3.4	-2.3	1.0
246.84	-0.51	1.5	257.00	11DE	-0.9	-3.5	-2.2	1.0
254.52	-0.61	1.5	267.19	21DE	-0.9	-3.5	-2.2	1.0
262.24	-0.71	1.5	277.36	31DE	-0.7	-3.5	-2.2	0.9
				-169				
270.01	-0.80	1.5	287.50	10JA	0.4	-3.5	-2.1	0.9
277.81	-0.89	1.4	297.60	20JA	3.0	-3.4	-2.0	0.8
285.63	-0.97	1.4	307.65	30JA	1.5	-3.4	-2.0	0.8
293.46	-1.04	1.4	317.65	9FE	0.7	-3.4	-1.9	0.7
301.28	-1.10	1.4	327.59	19FE	0.4	-3.4	-1.8	0.7
309.09	-1.15	1.4	337.47	1MR	0.2	-3.3	-1.7	0.6
316.87	-1.19	1.4	347.29	11MR	-0.3	-3.3	-1.7	0.7
324.61	-1.22	1.4	357.05	21MR	-0.9	-3.3	-1.6	0.7
332.29	-1.23	1.4	6.76	31MR	-1.8	-3.3	-1.6	0.8
339.92	-1.23	1.4	16.43	10AP	-1.2	-3.3	-1.5	0.9
347.46	-1.22	1.4	26.04	20AP	-0.1	-3.4	-1.5	0.9
354.93	-1.19	1.4	35.62	30AP	0.8	-3.4	-1.4	1.0
2.30	-1.16	1.4	45.18	10MY	1.7	-3.4	-1.4	1.0
9.56	-1.10	1.3	54.71	20MY	2.9	-3.5	-1.4	1.1
16.72	-1.04	1.3	64.23	30MY	3.0	-3.5	-1.4	1.1
23.76	-0.96	1.3	73.75	9JN	1.6	-3.5	-1.4	1.2
30.67	-0.88	1.3	83.27	19JN	0.3	-3.6	-1.4	1.2
37.45	-0.78	1.3	92.82	29JN	-0.8	-3.7	-1.4	1.2
44.08	-0.67	1.3	102.39	9JL	-1.5	-3.8	-1.4	1.2
50.54	-0.54	1.2	111.99	19JL	-1.1	-3.9	-1.4	1.2
56.84	-0.41	1.2	121.64	29JL	-0.5	-4.0	-1.4	1.2
62.93	-0.26	1.1	131.33	8AU	-0.1	-4.1	-1.5	1.1
68.81	-0.10	1.1	141.07	18AU	0.2	-4.2	-1.5	1.1
74.44	0.08	1.0	150.88	28AU	0.3	-4.3	-1.5	1.1
79.77	0.27	0.9	160.74	7SE	0.7	-4.3	-1.6	1.0
84.76	0.49	0.8	170.66	17SE	2.3	-4.2	-1.7	1.0
89.33	0.73	0.6	180.64	27SE	1.3	-4.0	-1.7	1.0
93.40	1.00	0.5	190.67	7OC	-0.4	-3.4	-1.8	1.0
96.85	1.31	0.3	200.75	17OC	-0.7	-3.5	-1.8	1.0
99.53	1.65	0.1	210.88	27OC	-0.7	-4.1	-1.9	1.0
101.26	2.04	-0.1	221.04	6NO	-0.7	-4.3	-2.0	1.0
101.87	2.46	-0.3	231.23	16NO	-0.8	-4.4	-2.0	1.0
101.17	2.91	-0.5	241.43	26NO	-0.7	-4.3	-2.1	1.0
99.11	3.35	-0.8	251.64	6DE	-0.7	-4.2	-2.1	0.9
95.89	3.73	-1.0	261.85	16DE	-0.6	-4.1	-2.1	0.9
91.97	3.98	-1.1	272.03	26DE	0.6	-4.0	-2.1	0.9

♂ LONG	LAT	MAG	⊙ LONG	16.00UT	☿	♀	♃	♄
				-168	MAGNITUDES			
88.11	4.07	-0.9	282.19	5JA	3.0	-3.9	-2.1	0.8
85.00	4.01	-0.7	292.31	15JA	1.1	-3.8	-2.1	0.8
83.07	3.83	-0.4	302.39	25JA	0.5	-3.7	-2.1	0.7
82.45	3.60	-0.2	312.42	4FE	0.3	-3.7	-2.0	0.7
83.05	3.35	0.1	322.39	14FE	0.0	-3.6	-2.0	0.6
84.71	3.09	0.3	332.30	24FE	-0.3	-3.5	-1.9	0.6
87.23	2.85	0.5	342.16	5MR	-0.9	-3.5	-1.8	0.5
90.44	2.62	0.7	351.95	15MR	-1.7	-3.4	-1.8	0.5
94.22	2.41	0.9	1.69	25MR	-1.2	-3.4	-1.7	0.5
98.44	2.21	1.0	11.37	4AP	-0.1	-3.3	-1.6	0.6
103.03	2.02	1.1	21.01	14AP	1.0	-3.3	-1.6	0.7
107.93	1.85	1.3	30.61	24AP	2.2	-3.3	-1.5	0.7
113.08	1.69	1.4	40.18	4MY	3.7	-3.3	-1.5	0.8
118.45	1.54	1.4	49.71	14MY	2.4	-3.3	-1.4	0.9
124.01	1.39	1.5	59.24	24MY	1.2	-3.3	-1.4	0.9
129.75	1.25	1.6	68.76	3JN	0.2	-3.3	-1.3	1.0
135.64	1.11	1.6	78.28	13JN	-0.8	-3.4	-1.3	1.0
141.67	0.98	1.6	87.82	23JN	-1.6	-3.4	-1.3	1.0
147.84	0.84	1.7	97.37	3JL	-1.1	-3.4	-1.3	1.1
154.14	0.72	1.7	106.95	13JL	-0.4	-3.5	-1.3	1.1
160.57	0.59	1.7	116.58	23JL	0.0	-3.5	-1.3	1.1
167.12	0.47	1.7	126.24	2AU	0.3	-3.5	-1.3	1.1
173.79	0.34	1.7	135.96	12AU	0.5	-3.5	-1.3	1.0
180.58	0.22	1.7	145.73	22AU	1.0	-3.4	-1.3	1.0
187.48	0.10	1.7	155.56	1SE	2.7	-3.4	-1.3	1.0
194.50	-0.01	1.6	165.45	11SE	1.2	-3.4	-1.4	0.9
201.63	-0.13	1.6	175.40	21SE	-0.5	-3.3	-1.4	0.9
208.86	-0.24	1.6	185.40	1OC	-0.8	-3.3	-1.5	0.9
216.20	-0.35	1.6	195.46	11OC	-0.8	-3.3	-1.5	0.9
223.63	-0.45	1.5	205.56	21OC	-0.8	-3.3	-1.6	0.9
231.16	-0.55	1.5	215.71	31OC	-0.6	-3.3	-1.6	0.9
238.77	-0.64	1.5	225.88	10NO	-0.6	-3.4	-1.7	0.9
246.45	-0.73	1.4	236.08	20NO	-0.6	-3.4	-1.8	0.9
254.20	-0.81	1.4	246.29	30NO	-0.4	-3.4	-1.8	0.9
262.00	-0.88	1.4	256.50	10DE	0.7	-3.4	-1.9	0.9
269.84	-0.94	1.3	266.70	20DE	2.8	-3.5	-2.0	0.9
277.72	-0.99	1.3	276.87	30DE	0.8	-3.5	-2.0	0.8
				-167				
285.61	-1.04	1.3	287.01	9JA	0.3	-3.6	-2.0	0.8
293.50	-1.07	1.3	297.12	19JA	0.1	-3.6	-2.0	0.7
301.39	-1.09	1.3	307.17	29JA	-0.1	-3.7	-2.1	0.7
309.25	-1.09	1.4	317.17	8FE	-0.4	-3.7	-2.0	0.6
317.08	-1.09	1.4	327.11	18FE	-1.0	-3.8	-2.0	0.6
324.86	-1.07	1.4	337.00	28FE	-1.6	-3.9	-2.0	0.5
332.58	-1.05	1.4	346.82	10MR	-1.1	-4.0	-1.9	0.5
340.24	-1.01	1.5	356.59	20MR	-0.0	-4.1	-1.9	0.4
347.83	-0.96	1.5	6.30	30MR	1.3	-4.2	-1.8	0.4
355.33	-0.90	1.5	15.96	9AP	2.8	-4.2	-1.7	0.5
2.75	-0.83	1.5	25.58	19AP	3.1	-4.2	-1.7	0.5
10.08	-0.76	1.5	35.16	29AP	1.8	-4.0	-1.6	0.6
17.32	-0.67	1.6	44.71	9MY	0.9	-3.4	-1.5	0.7
24.45	-0.58	1.6	54.25	19MY	0.1	-2.7	-1.5	0.7
31.49	-0.48	1.6	63.76	29MY	-0.9	-3.6	-1.4	0.8
38.42	-0.38	1.6	73.29	8JN	-1.7	-4.0	-1.4	0.8
45.25	-0.27	1.6	82.81	18JN	-1.2	-4.2	-1.3	0.9
51.98	-0.15	1.6	92.35	28JN	-0.4	-4.2	-1.3	0.9
58.60	-0.03	1.6	101.92	8JL	0.1	-4.1	-1.3	0.9
65.11	0.09	1.6	111.52	18JL	0.4	-4.0	-1.2	1.0
71.50	0.22	1.6	121.16	28JL	0.8	-4.0	-1.2	1.0
77.76	0.36	1.6	130.86	7AU	1.4	-3.9	-1.2	1.0
83.90	0.50	1.5	140.60	17AU	3.1	-3.8	-1.2	1.0
89.89	0.65	1.5	150.39	27AU	1.0	-3.7	-1.2	0.9
95.71	0.81	1.4	160.26	6SE	-0.6	-3.7	-1.2	0.9
101.36	0.98	1.4	170.17	16SE	-1.0	-3.6	-1.3	0.9
106.79	1.17	1.3	180.15	26SE	-0.9	-3.6	-1.3	0.8
111.99	1.36	1.2	190.18	6OC	-0.8	-3.5	-1.3	0.8
116.88	1.58	1.1	200.26	16OC	-0.5	-3.5	-1.4	0.8
121.42	1.81	0.9	210.39	26OC	-0.4	-3.4	-1.4	0.8
125.53	2.07	0.8	220.55	5NO	-0.4	-3.4	-1.4	0.8
129.11	2.35	0.6	230.74	15NO	-0.3	-3.4	-1.5	0.9
132.03	2.67	0.4	240.94	25NO	0.9	-3.4	-1.6	0.9
134.13	3.02	0.2	251.15	5DE	2.4	-3.4	-1.6	0.9
135.23	3.39	-0.1	261.36	15DE	0.5	-3.3	-1.7	0.8
135.16	3.78	-0.3	271.54	25DE	0.1	-3.3	-1.8	0.8

-166

♂ LONG	LAT	MAG	☉ LONG	16.00UT	☿	♀	♃	♄
133.79	4.14	-0.6	281.70	4JA	-0.0	-3.3	-1.8	0.8
131.16	4.44	-0.8	291.83	14JA	-0.2	-3.3	-1.9	0.8
127.58	4.60	-1.0	301.90	24JA	-0.5	-3.4	-1.9	0.7
123.64	4.59	-1.0	311.94	3FE	-1.0	-3.4	-2.0	0.7
120.09	4.41	-0.8	321.91	13FE	-1.4	-3.4	-2.0	0.6
117.50	4.10	-0.6	331.82	23FE	-1.1	-3.4	-2.0	0.6
116.17	3.72	-0.4	341.68	5MR	0.1	-3.5	-2.0	0.5
116.13	3.32	-0.1	351.47	15MR	1.6	-3.5	-2.0	0.5
117.25	2.93	0.1	1.21	25MR	3.5	-3.4	-2.0	0.4
119.36	2.57	0.3	10.90	4AP	2.3	-3.4	-1.9	0.4
122.28	2.24	0.5	20.54	14AP	1.3	-3.4	-1.9	0.4
125.86	1.94	0.6	30.14	24AP	0.7	-3.4	-1.8	0.4
129.98	1.67	0.8	39.71	4MY	0.0	-3.3	-1.8	0.5
134.55	1.42	0.9	49.25	14MY	-0.9	-3.3	-1.7	0.5
139.49	1.19	1.0	58.77	24MY	-1.7	-3.3	-1.6	0.6
144.75	0.97	1.1	68.29	3JN	-1.2	-3.3	-1.6	0.7
150.29	0.77	1.1	77.81	13JN	-0.3	-3.3	-1.5	0.7
156.07	0.59	1.2	87.34	23JN	0.2	-3.3	-1.5	0.8
162.07	0.41	1.3	96.90	3JL	0.6	-3.4	-1.4	0.8
168.27	0.24	1.3	106.48	13JL	1.0	-3.4	-1.4	0.9
174.66	0.09	1.3	116.10	23JL	1.9	-3.4	-1.3	0.9
181.21	-0.06	1.4	125.77	2AU	3.1	-3.5	-1.3	0.9
187.93	-0.20	1.4	135.48	12AU	0.9	-3.5	-1.3	0.9
194.80	-0.33	1.4	145.25	22AU	-0.7	-3.6	-1.2	0.9
201.81	-0.46	1.4	155.08	1SE	-1.1	-3.7	-1.2	0.9
208.95	-0.57	1.4	164.96	11SE	-1.1	-3.7	-1.2	0.9
216.23	-0.68	1.4	174.91	21SE	-0.7	-3.8	-1.2	0.8
223.62	-0.77	1.4	184.91	1OC	-0.4	-3.9	-1.2	0.8
231.11	-0.86	1.4	194.97	11OC	-0.3	-4.0	-1.2	0.8
238.70	-0.93	1.4	205.07	21OC	-0.3	-4.2	-1.3	0.7
246.38	-1.00	1.4	215.22	31OC	-0.1	-4.3	-1.3	0.8
254.12	-1.05	1.4	225.39	10NO	1.1	-4.4	-1.3	0.8
261.93	-1.09	1.4	235.59	20NO	2.2	-4.4	-1.4	0.8
269.78	-1.12	1.4	245.80	30NO	0.3	-4.3	-1.4	0.8
277.66	-1.13	1.4	256.00	10DE	-0.1	-4.0	-1.5	0.8
285.55	-1.14	1.4	266.20	20DE	-0.2	-3.3	-1.6	0.8
293.45	-1.13	1.4	276.38	30DE	-0.3	-3.4	-1.6	0.8

-165

♂ LONG	LAT	MAG	☉ LONG	16.00UT	☿	♀	♃	♄
301.33	-1.10	1.4	286.52	9JA	-0.5	-4.0	-1.7	0.8
309.18	-1.07	1.4	296.63	19JA	-1.0	-4.3	-1.7	0.8
316.99	-1.03	1.4	306.68	29JA	-1.3	-4.3	-1.8	0.7
324.75	-0.98	1.4	316.68	8FE	-1.0	-4.3	-1.9	0.7
332.46	-0.91	1.4	326.63	18FE	0.1	-4.2	-1.9	0.6
340.09	-0.84	1.4	336.52	28FE	2.0	-4.1	-2.0	0.6
347.65	-0.77	1.4	346.34	10MR	3.1	-4.0	-2.0	0.5
355.13	-0.68	1.4	356.12	20MR	1.7	-3.8	-2.0	0.5
2.52	-0.60	1.5	5.83	30MR	1.0	-3.8	-2.0	0.4
9.83	-0.50	1.5	15.49	9AP	0.5	-3.7	-2.0	0.4
17.05	-0.41	1.6	25.12	19AP	-0.0	-3.6	-2.0	0.3
24.18	-0.31	1.6	34.70	29AP	-0.9	-3.5	-2.0	0.3
31.23	-0.20	1.7	44.25	9MY	-1.8	-3.5	-1.9	0.4
38.18	-0.10	1.7	53.79	19MY	-1.2	-3.4	-1.9	0.4
45.06	0.00	1.8	63.31	29MY	-0.3	-3.4	-1.8	0.5
51.85	0.11	1.8	72.83	8JN	0.3	-3.4	-1.8	0.6
58.56	0.22	1.8	82.35	18JN	0.8	-3.3	-1.7	0.6
65.18	0.32	1.8	91.89	28JN	1.4	-3.3	-1.6	0.7
71.74	0.43	1.8	101.46	8JL	2.4	-3.3	-1.6	0.7
78.25	0.54	1.8	111.06	18JL	2.8	-3.3	-1.5	0.8
84.61	0.65	1.8	120.70	28JL	0.7	-3.3	-1.5	0.8
90.93	0.76	1.8	130.38	7AU	-0.7	-3.4	-1.4	0.8
97.17	0.87	1.8	140.13	17AU	-1.2	-3.4	-1.4	0.8
103.33	0.98	1.8	149.92	27AU	-1.2	-3.4	-1.3	0.9
109.40	1.10	1.8	159.77	6SE	-0.6	-3.4	-1.3	0.8
115.37	1.22	1.7	169.69	16SE	-0.3	-3.5	-1.3	0.8
121.23	1.34	1.7	179.66	26SE	-0.2	-3.5	-1.3	0.8
126.98	1.47	1.6	189.69	6OC	-0.1	-3.5	-1.3	0.8
132.58	1.61	1.5	199.77	16OC	0.1	-3.5	-1.3	0.8
138.01	1.75	1.4	209.89	26OC	1.4	-3.5	-1.3	0.7
143.25	1.90	1.3	220.05	5NO	1.9	-3.4	-1.3	0.7
148.26	2.06	1.2	230.24	15NO	0.1	-3.4	-1.3	0.7
152.98	2.23	1.1	240.44	25NO	-0.3	-3.4	-1.3	0.8
157.35	2.42	0.9	250.65	5DE	-0.3	-3.4	-1.4	0.8
161.29	2.62	0.7	260.86	15DE	-0.4	-3.3	-1.4	0.8
164.69	2.83	0.5	271.05	25DE	-0.6	-3.3	-1.5	0.8

-164

♂ LONG	LAT	MAG	☉ LONG	16.00UT	☿	♀	♃	♄
167.42	3.06	0.3	281.21	4JA	-1.0	-3.3	-1.5	0.8
169.32	3.30	0.0	291.34	14JA	-1.2	-3.3	-1.6	0.8
170.21	3.53	-0.3	301.42	24JA	-0.9	-3.4	-1.6	0.8
169.92	3.75	-0.5	311.45	3FE	0.2	-3.4	-1.7	0.7
168.34	3.90	-0.8	321.43	13FE	2.4	-3.4	-1.8	0.7
165.58	3.94	-1.1	331.34	23FE	2.4	-3.4	-1.8	0.7
161.99	3.84	-1.3	341.20	4MR	1.2	-3.5	-1.9	0.6
158.21	3.57	-1.2	351.00	14MR	0.7	-3.5	-2.0	0.5
154.97	3.18	-1.1	0.75	24MR	0.4	-3.5	-2.0	0.5
152.78	2.71	-0.9	10.44	3AP	-0.1	-3.6	-2.1	0.4
151.91	2.22	-0.6	20.08	13AP	-0.9	-3.7	-2.1	0.4
152.34	1.76	-0.4	29.68	23AP	-1.8	-3.7	-2.1	0.3
153.93	1.34	-0.2	39.25	3MY	-1.2	-3.8	-2.1	0.3
156.51	0.96	-0.1	48.80	13MY	-0.2	-3.9	-2.1	0.3
159.91	0.63	0.1	58.32	23MY	0.5	-4.0	-2.1	0.4
163.98	0.33	0.3	67.84	2JN	1.1	-4.1	-2.1	0.4
168.62	0.07	0.4	77.36	12JN	1.8	-4.2	-2.1	0.5
173.71	-0.17	0.5	86.89	22JN	3.1	-4.2	-2.0	0.6
179.20	-0.38	0.6	96.44	2JL	2.3	-4.1	-1.9	0.6
185.04	-0.56	0.7	106.02	12JL	0.6	-3.8	-1.8	0.7
191.17	-0.73	0.7	115.64	22JL	-0.8	-3.3	-1.8	0.7
197.55	-0.87	0.8	125.30	1AU	-1.4	-3.4	-1.7	0.8
204.18	-1.00	0.9	135.01	11AU	-1.2	-3.9	-1.6	0.8
211.00	-1.11	0.9	144.77	21AU	-0.6	-4.2	-1.6	0.8
218.01	-1.20	1.0	154.60	31AU	-0.2	-4.3	-1.5	0.8
225.18	-1.28	1.0	164.48	10SE	-0.0	-4.3	-1.5	0.8
232.50	-1.33	1.1	174.42	20SE	0.1	-4.2	-1.5	0.8
239.94	-1.37	1.1	184.42	30SE	0.3	-4.1	-1.4	0.8
247.48	-1.40	1.1	194.48	10OC	1.7	-4.0	-1.4	0.8
255.12	-1.41	1.2	204.58	20OC	1.7	-3.9	-1.4	0.8
262.82	-1.40	1.2	214.72	30OC	-0.1	-3.8	-1.4	0.7
270.57	-1.38	1.2	224.90	9NO	-0.4	-3.8	-1.3	0.7
278.36	-1.34	1.3	235.09	19NO	-0.4	-3.7	-1.3	0.7
286.17	-1.29	1.3	245.30	29NO	-0.5	-3.6	-1.4	0.7
293.97	-1.23	1.3	255.51	9DE	-0.7	-3.6	-1.4	0.8
301.76	-1.16	1.4	265.71	19DE	-0.9	-3.5	-1.4	0.8
309.53	-1.08	1.4	275.89	29DE	-1.0	-3.5	-1.4	0.8

-163

♂ LONG	LAT	MAG	☉ LONG	16.00UT	☿	♀	♃	♄
317.25	-0.99	1.4	286.04	8JA	-0.8	-3.4	-1.5	0.8
324.92	-0.90	1.5	296.14	18JA	0.3	-3.4	-1.5	0.8
332.53	-0.80	1.5	306.20	28JA	2.8	-3.4	-1.6	0.8
340.07	-0.70	1.5	316.21	7FE	1.8	-3.3	-1.6	0.8
347.53	-0.60	1.6	326.15	17FE	0.9	-3.3	-1.7	0.7
354.92	-0.49	1.6	336.05	27FE	0.5	-3.3	-1.7	0.7
2.23	-0.38	1.6	345.88	9MR	0.2	-3.3	-1.8	0.6
9.45	-0.28	1.7	355.65	19MR	-0.2	-3.3	-1.9	0.6
16.60	-0.17	1.7	5.37	29MR	-0.9	-3.3	-2.0	0.5
23.66	-0.07	1.7	15.03	8AP	-1.8	-3.3	-2.0	0.5
30.64	0.03	1.7	24.65	18AP	-1.2	-3.4	-2.1	0.4
37.55	0.13	1.7	34.24	28AP	-0.2	-3.4	-2.1	0.3
44.38	0.23	1.7	43.79	8MY	0.6	-3.5	-2.2	0.3
51.14	0.33	1.8	53.33	18MY	1.4	-3.5	-2.2	0.3
57.84	0.42	1.8	62.85	28MY	2.4	-3.5	-2.2	0.3
64.48	0.51	1.9	72.37	7JN	3.4	-3.4	-2.2	0.4
71.05	0.60	1.9	81.89	17JN	1.8	-3.4	-2.2	0.4
77.58	0.69	1.9	91.43	27JN	0.5	-3.4	-2.2	0.5
84.05	0.77	1.9	100.99	7JL	-0.8	-3.3	-2.1	0.6
90.48	0.85	2.0	110.59	17JL	-1.5	-3.3	-2.1	0.6
96.87	0.93	2.0	120.23	27JL	-1.2	-3.3	-2.1	0.7
103.21	1.01	2.0	129.91	6AU	-0.5	-3.3	-2.0	0.7
109.51	1.08	2.0	139.65	16AU	-0.1	-3.3	-1.9	0.7
115.77	1.15	2.0	149.44	26AU	0.1	-3.3	-1.9	0.8
121.99	1.23	2.0	159.29	5SE	0.2	-3.4	-1.8	0.8
128.17	1.30	1.9	169.20	15SE	0.6	-3.4	-1.7	0.8
134.30	1.37	1.9	179.17	25SE	2.0	-3.4	-1.7	0.8
140.38	1.43	1.8	189.20	5OC	1.5	-3.5	-1.6	0.8
146.41	1.50	1.8	199.27	15OC	-0.3	-3.5	-1.6	0.8
152.37	1.57	1.7	209.40	25OC	-0.6	-3.6	-1.5	0.8
158.26	1.64	1.6	219.55	4NO	-0.6	-3.6	-1.5	0.7
164.06	1.70	1.6	229.74	14NO	-0.6	-3.7	-1.5	0.7
169.75	1.77	1.5	239.95	24NO	-0.8	-3.8	-1.5	0.7
175.31	1.83	1.3	250.15	4DE	-0.8	-3.9	-1.5	0.7
180.72	1.89	1.2	260.36	14DE	-0.8	-3.9	-1.4	0.8
185.95	1.95	1.1	270.55	24DE	-0.7	-4.0	-1.5	0.8

♂ LONG	♂ LAT	♂ MAG	⊙ LONG	16.00UT -162	☿	♀	♃	♄
					MAGNITUDES			
190.94	2.01	0.9	280.71	3JA	0.4	-4.1	-1.5	0.8
195.65	2.06	0.7	290.84	13JA	3.0	-4.2	-1.5	0.8
200.01	2.10	0.5	300.93	23JA	1.4	-4.3	-1.5	0.8
203.93	2.12	0.3	310.96	2FE	0.6	-4.3	-1.5	0.8
207.31	2.13	0.0	320.94	12FE	0.4	-4.2	-1.6	0.8
209.98	2.11	-0.3	330.87	22FE	0.1	-3.9	-1.6	0.8
211.81	2.05	-0.5	340.73	4MR	-0.3	-3.3	-1.7	0.7
212.59	1.92	-0.9	350.53	14MR	-0.9	-3.5	-1.7	0.7
212.17	1.72	-1.2	0.28	24MR	-1.8	-4.0	-1.8	0.6
210.50	1.42	-1.5	9.97	3AP	-1.2	-4.2	-1.9	0.6
207.73	1.01	-1.8	19.61	13AP	-0.1	-4.2	-1.9	0.5
204.35	0.52	-1.9	29.22	23AP	0.8	-4.2	-2.0	0.5
201.05	-0.01	-1.8	38.79	3MY	1.8	-4.1	-2.1	0.4
198.52	-0.53	-1.6	48.33	13MY	3.2	-4.0	-2.1	0.3
197.23	-0.98	-1.4	57.86	23MY	2.8	-3.9	-2.2	0.3
197.31	-1.36	-1.2	67.38	2JN	1.5	-3.8	-2.3	0.3
198.70	-1.66	-1.0	76.90	12JN	0.3	-3.7	-2.3	0.3
201.25	-1.89	-0.8	86.43	22JN	-0.8	-3.6	-2.3	0.4
204.76	-2.07	-0.6	95.98	2JL	-1.6	-3.6	-2.4	0.5
209.06	-2.19	-0.4	105.56	12JL	-1.2	-3.5	-2.4	0.5
214.01	-2.28	-0.3	115.17	22JL	-0.5	-3.5	-2.4	0.6
219.49	-2.32	-0.1	124.83	1AU	-0.0	-3.4	-2.3	0.6
225.40	-2.34	0.0	134.54	11AU	0.2	-3.4	-2.3	0.7
231.67	-2.33	0.1	144.30	21AU	0.4	-3.4	-2.2	0.7
238.24	-2.30	0.2	154.12	31AU	0.8	-3.4	-2.2	0.7
245.05	-2.24	0.3	164.00	10SE	2.4	-3.4	-2.1	0.8
252.06	-2.16	0.4	173.94	20SE	1.3	-3.4	-2.0	0.8
259.22	-2.07	0.5	183.94	30SE	-0.4	-3.4	-2.0	0.8
266.51	-1.96	0.6	193.99	10OC	-0.7	-3.4	-1.9	0.8
273.89	-1.84	0.7	204.09	20OC	-0.7	-3.4	-1.8	0.8
281.33	-1.71	0.8	214.22	30OC	-0.7	-3.4	-1.8	0.8
288.82	-1.57	0.9	224.40	9NO	-0.7	-3.4	-1.7	0.8
296.32	-1.42	1.0	234.59	19NO	-0.7	-3.4	-1.7	0.7
303.83	-1.28	1.1	244.80	29NO	-0.7	-3.4	-1.6	0.7
311.32	-1.13	1.2	255.01	9DE	-0.5	-3.5	-1.6	0.7
318.78	-0.98	1.2	265.21	19DE	0.6	-3.5	-1.6	0.8
326.20	-0.83	1.3	275.38	29DE	3.0	-3.5	-1.6	0.8
				-161				
333.58	-0.69	1.4	285.54	8JA	1.0	-3.4	-1.5	0.8
340.89	-0.55	1.5	295.64	18JA	0.4	-3.4	-1.5	0.8
348.15	-0.42	1.5	305.70	28JA	0.2	-3.4	-1.5	0.8
355.33	-0.28	1.6	315.72	7FE	0.0	-3.4	-1.5	0.8
2.45	-0.16	1.6	325.67	17FE	-0.3	-3.4	-1.6	0.8
9.50	-0.04	1.7	335.56	27FE	-0.9	-3.3	-1.6	0.8
16.48	0.07	1.7	345.40	9MR	-1.7	-3.3	-1.6	0.8
23.39	0.18	1.8	355.17	19MR	-1.2	-3.3	-1.7	0.7
30.24	0.28	1.8	4.89	29MR	-0.1	-3.3	-1.7	0.7
37.02	0.38	1.8	14.57	8AP	1.1	-3.4	-1.7	0.6
43.74	0.47	1.9	24.19	18AP	2.4	-3.4	-1.8	0.6
50.41	0.55	1.9	33.77	28AP	3.7	-3.4	-1.9	0.5
57.02	0.63	1.9	43.33	8MY	2.1	-3.4	-1.9	0.4
63.59	0.70	1.9	52.87	18MY	1.1	-3.5	-2.0	0.4
70.12	0.77	1.9	62.39	28MY	0.2	-3.5	-2.1	0.3
76.61	0.84	1.9	71.91	7JN	-0.8	-3.5	-2.1	0.3
83.06	0.90	1.9	81.43	17JN	-1.6	-3.6	-2.2	0.3
89.49	0.95	1.9	90.97	27JN	-1.2	-3.7	-2.3	0.4
95.90	1.01	2.0	100.54	7JL	-0.4	-3.8	-2.3	0.4
102.28	1.05	2.0	110.13	17JL	0.1	-3.9	-2.4	0.5
108.65	1.10	2.0	119.76	27JL	0.3	-4.0	-2.4	0.5
115.02	1.14	2.0	129.45	6AU	0.6	-4.1	-2.4	0.6
121.37	1.17	2.0	139.18	16AU	1.2	-4.2	-2.4	0.6
127.71	1.20	2.0	148.97	26AU	2.8	-4.3	-2.4	0.7
134.06	1.23	2.0	158.82	5SE	1.2	-4.3	-2.4	0.7
140.39	1.26	2.0	168.72	15SE	-0.5	-4.2	-2.4	0.8
146.73	1.28	2.0	178.69	25SE	-0.9	-3.9	-2.3	0.8
153.07	1.29	2.0	188.71	5OC	-0.9	-3.3	-2.3	0.8
159.41	1.31	1.9	198.78	15OC	-0.9	-3.5	-2.2	0.8
165.74	1.31	1.9	208.90	25OC	-0.6	-4.1	-2.1	0.8
172.06	1.31	1.8	219.06	4NO	-0.5	-4.3	-2.0	0.8
178.38	1.31	1.8	229.24	14NO	-0.5	-4.4	-2.0	0.8
184.68	1.30	1.7	239.45	24NO	-0.4	-4.3	-1.9	0.8
190.97	1.28	1.6	249.66	4DE	0.7	-4.2	-1.8	0.7
197.23	1.25	1.5	259.86	14DE	2.7	-4.1	-1.8	0.7
203.45	1.21	1.4	270.05	24DE	0.7	-4.0	-1.7	0.8

♂ LONG	♂ LAT	♂ MAG	⊙ LONG	16.00UT -160	☿	♀	♃	♄
					MAGNITUDES			
209.64	1.16	1.3	280.22	3JA	0.2	-3.9	-1.7	0.8
215.77	1.09	1.2	290.35	13JA	0.1	-3.8	-1.6	0.8
221.84	1.01	1.0	300.44	23JA	-0.1	-3.7	-1.6	0.8
227.82	0.90	0.9	310.48	2FE	-0.4	-3.7	-1.6	0.9
233.70	0.77	0.7	320.46	12FE	-0.9	-3.6	-1.6	0.9
239.46	0.61	0.5	330.39	22FE	-1.5	-3.5	-1.6	0.9
245.04	0.42	0.3	340.25	3MR	-1.1	-3.5	-1.6	0.9
250.41	0.18	0.1	350.06	13MR	-0.0	-3.4	-1.6	0.8
255.52	-0.10	-0.1	359.81	23MR	1.4	-3.4	-1.6	0.8
260.28	-0.44	-0.3	9.51	2AP	3.1	-3.3	-1.6	0.8
264.61	-0.85	-0.6	19.15	12AP	2.8	-3.3	-1.6	0.7
268.37	-1.33	-0.9	28.76	22AP	1.6	-3.3	-1.7	0.7
271.41	-1.90	-1.2	38.33	2MY	0.9	-3.3	-1.7	0.6
273.54	-2.57	-1.5	47.87	12MY	0.1	-3.3	-1.8	0.5
274.59	-3.31	-1.8	57.40	22MY	-0.8	-3.3	-1.8	0.5
274.39	-4.11	-2.1	66.92	1JN	-1.7	-3.3	-1.9	0.4
273.00	-4.88	-2.4	76.44	11JN	-1.2	-3.4	-1.9	0.3
270.66	-5.53	-2.6	85.97	21JN	-0.4	-3.4	-2.0	0.3
267.99	-5.94	-2.6	95.52	1JL	0.1	-3.4	-2.1	0.3
265.72	-6.08	-2.4	105.09	11JL	0.5	-3.5	-2.1	0.4
264.43	-5.95	-2.2	114.71	21JL	0.8	-3.5	-2.2	0.5
264.45	-5.64	-1.9	124.36	31JL	1.6	-3.5	-2.3	0.5
265.79	-5.22	-1.6	134.07	10AU	3.2	-3.5	-2.3	0.6
268.31	-4.75	-1.4	143.83	20AU	1.0	-3.4	-2.4	0.6
271.83	-4.28	-1.1	153.64	30AU	-0.6	-3.4	-2.4	0.7
276.13	-3.81	-0.9	163.52	9SE	-1.0	-3.4	-2.4	0.7
281.05	-3.36	-0.6	173.46	19SE	-1.0	-3.3	-2.4	0.7
286.45	-2.94	-0.4	183.45	29SE	-0.8	-3.3	-2.4	0.8
292.22	-2.54	-0.2	193.49	9OC	-0.5	-3.3	-2.4	0.8
298.27	-2.18	-0.0	203.59	19OC	-0.4	-3.3	-2.4	0.8
304.54	-1.84	0.2	213.73	29OC	-0.4	-3.3	-2.3	0.8
310.96	-1.53	0.3	223.90	8NO	-0.2	-3.4	-2.3	0.8
317.49	-1.25	0.5	234.10	18NO	0.9	-3.4	-2.2	0.8
324.10	-0.99	0.7	244.30	28NO	2.4	-3.4	-2.1	0.8
330.77	-0.75	0.8	254.51	8DE	0.4	-3.4	-2.0	0.8
337.46	-0.54	0.9	264.71	18DE	0.0	-3.5	-2.0	0.8
344.17	-0.34	1.1	274.89	28DE	-0.1	-3.5	-1.9	0.8
				-159				
350.87	-0.17	1.2	285.04	7JA	-0.2	-3.6	-1.8	0.8
357.57	-0.01	1.3	295.15	17JA	-0.5	-3.6	-1.8	0.9
4.25	0.14	1.4	305.22	27JA	-1.0	-3.7	-1.7	0.9
10.90	0.27	1.5	315.23	6FE	-1.4	-3.8	-1.7	0.9
17.52	0.38	1.6	325.19	16FE	-1.1	-3.8	-1.6	0.9
24.12	0.49	1.7	335.08	26FE	0.0	-3.9	-1.6	0.9
30.67	0.59	1.7	344.92	8MR	1.7	-4.0	-1.6	0.9
37.20	0.67	1.8	354.70	18MR	3.6	-4.1	-1.6	0.9
43.69	0.75	1.8	4.42	28MR	2.1	-4.2	-1.5	0.9
50.16	0.82	1.9	14.10	7AP	1.2	-4.2	-1.5	0.8
56.60	0.88	1.9	23.73	17AP	0.6	-4.2	-1.5	0.8
63.01	0.93	2.0	33.31	27AP	0.0	-4.0	-1.5	0.7
69.39	0.98	2.0	42.87	7MY	-0.8	-3.4	-1.6	0.7
75.77	1.02	2.0	52.41	17MY	-1.7	-2.7	-1.6	0.6
82.13	1.06	2.0	61.93	27MY	-1.2	-3.6	-1.6	0.6
88.48	1.09	2.0	71.45	6JN	-0.3	-4.0	-1.7	0.5
94.82	1.11	2.0	80.97	16JN	0.3	-4.2	-1.7	0.4
101.17	1.13	2.0	90.51	26JN	0.7	-4.2	-1.7	0.4
107.51	1.15	2.0	100.07	6JL	1.1	-4.1	-1.8	0.4
113.87	1.16	2.0	109.66	16JL	2.0	-4.0	-1.9	0.4
120.24	1.16	2.0	119.29	26JL	3.1	-4.0	-1.9	0.5
126.63	1.16	2.0	128.97	5AU	0.9	-3.9	-2.0	0.5
133.04	1.16	2.0	138.71	15AU	-0.6	-3.8	-2.1	0.6
139.47	1.15	2.0	148.49	25AU	-1.2	-3.7	-2.1	0.6
145.93	1.14	2.0	158.34	4SE	-1.1	-3.7	-2.2	0.7
152.42	1.12	2.0	168.24	14SE	-0.7	-3.6	-2.2	0.7
158.94	1.10	2.0	178.20	24SE	-0.4	-3.6	-2.3	0.8
165.50	1.07	2.0	188.22	4OC	-0.2	-3.5	-2.3	0.8
172.09	1.03	1.9	198.30	14OC	-0.2	-3.5	-2.4	0.8
178.72	0.99	1.9	208.41	24OC	-0.0	-3.4	-2.4	0.8
185.39	0.94	1.9	218.57	3NO	1.1	-3.4	-2.4	0.9
192.10	0.88	1.8	228.75	13NO	2.2	-3.4	-2.3	0.9
198.84	0.82	1.8	238.95	23NO	0.2	-3.4	-2.3	0.9
205.62	0.75	1.7	249.16	3DE	-0.2	-3.4	-2.2	0.9
212.44	0.66	1.6	259.37	13DE	-0.2	-3.3	-2.2	0.9
219.29	0.57	1.6	269.56	23DE	-0.3	-3.3	-2.1	0.8

Left table

♂ LONG	LAT	MAG	☉ LONG	16.00UT	☿	♀	♃	♄
				-158				
226.17	0.46	1.5	279.73	2JA	-0.6	-3.3	-2.0	0.8
233.09	0.34	1.4	289.86	12JA	-1.0	-3.4	-2.0	0.9
240.03	0.21	1.3	299.95	22JA	-1.2	-3.4	-1.9	0.9
247.00	0.06	1.2	309.99	1FE	-1.0	-3.4	-1.8	0.9
253.99	-0.10	1.1	319.98	11FE	0.1	-3.4	-1.8	1.0
260.99	-0.28	1.0	329.90	21FE	2.1	-3.4	-1.7	1.0
268.00	-0.48	0.9	339.77	3MR	2.9	-3.5	-1.6	1.0
275.00	-0.70	0.8	349.58	13MR	1.5	-3.5	-1.6	1.0
281.98	-0.93	0.7	359.33	23MR	0.9	-3.4	-1.6	1.0
288.93	-1.18	0.5	9.03	2AP	0.5	-3.4	-1.5	1.0
295.82	-1.44	0.4	18.68	12AP	-0.1	-3.4	-1.5	1.0
302.62	-1.72	0.3	28.29	22AP	-0.8	-3.4	-1.5	0.9
309.31	-2.02	0.1	37.86	2MY	-1.8	-3.3	-1.5	0.9
315.83	-2.33	-0.0	47.41	12MY	-1.2	-3.3	-1.5	0.8
322.15	-2.65	-0.2	56.93	22MY	-0.3	-3.3	-1.5	0.8
328.19	-2.99	-0.4	66.46	1JN	0.4	-3.3	-1.5	0.7
333.86	-3.33	-0.5	75.98	11JN	0.9	-3.3	-1.5	0.6
339.08	-3.68	-0.7	85.50	21JN	1.5	-3.3	-1.5	0.6
343.70	-4.03	-0.9	95.05	1JL	2.6	-3.4	-1.5	0.5
347.57	-4.37	-1.1	104.63	11JL	2.6	-3.4	-1.6	0.4
350.49	-4.69	-1.3	114.24	21JL	0.7	-3.4	-1.6	0.4
352.24	-4.97	-1.6	123.90	31JL	-0.7	-3.5	-1.6	0.5
352.64	-5.17	-1.8	133.60	10AU	-1.3	-3.5	-1.7	0.5
351.59	-5.22	-2.0	143.35	20AU	-1.2	-3.6	-1.8	0.6
349.24	-5.08	-2.2	153.17	30AU	-0.6	-3.7	-1.8	0.6
346.12	-4.70	-2.3	163.04	9SE	-0.3	-3.7	-1.9	0.7
343.00	-4.12	-2.2	172.97	19SE	-0.1	-3.8	-1.9	0.7
340.61	-3.42	-1.9	182.97	29SE	-0.0	-3.9	-2.0	0.8
339.45	-2.70	-1.6	193.01	9OC	0.2	-4.0	-2.1	0.8
339.61	-2.03	-1.2	203.10	19OC	1.4	-4.2	-2.1	0.9
341.02	-1.45	-0.9	213.24	29OC	1.9	-4.3	-2.2	0.9
343.47	-0.96	-0.6	223.41	8NO	0.0	-4.4	-2.2	0.9
346.74	-0.55	-0.3	233.60	18NO	-0.3	-4.4	-2.2	0.9
350.68	-0.21	-0.0	243.81	28NO	-0.4	-4.3	-2.2	1.0
355.12	0.06	0.2	254.02	8DE	-0.4	-4.0	-2.2	1.0
359.95	0.30	0.4	264.22	18DE	-0.6	-3.3	-2.2	1.0
5.09	0.49	0.6	274.40	28DE	-1.0	-3.5	-2.2	0.9
				-157				
10.45	0.64	0.8	284.55	7JA	-1.1	-4.1	-2.1	0.9
15.99	0.77	1.0	294.66	17JA	-0.9	-4.3	-2.1	0.9
21.67	0.88	1.1	304.73	27JA	0.2	-4.3	-2.0	1.0
27.46	0.97	1.3	314.74	6FE	2.5	-4.3	-1.9	1.0
33.33	1.04	1.4	324.70	16FE	2.2	-4.2	-1.9	1.1
39.26	1.10	1.5	334.60	26FE	1.1	-4.1	-1.8	1.1
45.24	1.15	1.6	344.44	8MR	0.7	-4.0	-1.7	1.1
51.26	1.19	1.7	354.22	18MR	0.3	-3.8	-1.7	1.1
57.32	1.22	1.8	3.95	28MR	-0.1	-3.7	-1.6	1.1
63.40	1.24	1.8	13.63	7AP	-0.9	-3.7	-1.5	1.1
69.50	1.26	1.9	23.26	17AP	-1.8	-3.6	-1.5	1.1
75.62	1.27	1.9	32.85	27AP	-1.2	-3.5	-1.5	1.1
81.76	1.27	2.0	42.41	7MY	-0.2	-3.5	-1.4	1.0
87.93	1.27	2.0	51.94	17MY	0.5	-3.4	-1.4	1.0
94.12	1.26	2.0	61.47	27MY	1.2	-3.4	-1.4	0.9
100.33	1.25	2.0	70.99	6JN	2.0	-3.4	-1.4	0.9
106.57	1.23	2.0	80.51	16JN	3.3	-3.3	-1.4	0.8
112.84	1.21	2.0	90.05	26JN	2.2	-3.3	-1.4	0.7
119.15	1.19	2.0	99.61	6JL	0.6	-3.3	-1.4	0.7
125.49	1.16	2.0	109.20	16JL	-0.7	-3.3	-1.4	0.6
131.88	1.13	2.0	118.83	26JL	-1.4	-3.3	-1.4	0.5
138.31	1.09	2.0	128.51	5AU	-1.2	-3.4	-1.4	0.6
144.79	1.05	1.9	138.24	15AU	-0.6	-3.4	-1.5	0.6
151.32	1.00	1.9	148.02	25AU	-0.2	-3.4	-1.5	0.6
157.90	0.95	1.9	157.86	4SE	0.0	-3.4	-1.5	0.7
164.54	0.89	1.8	167.76	14SE	0.1	-3.5	-1.6	0.7
171.24	0.83	1.8	177.72	24SE	0.4	-3.5	-1.7	0.8
178.01	0.76	1.8	187.74	4OC	1.7	-3.5	-1.7	0.8
184.84	0.69	1.8	197.81	14OC	1.7	-3.5	-1.8	0.9
191.73	0.61	1.8	207.92	24OC	-0.1	-3.4	-1.8	0.9
198.69	0.53	1.8	218.07	3NO	-0.5	-3.4	-1.9	1.0
205.72	0.44	1.8	228.25	13NO	-0.5	-3.4	-2.0	1.0
212.81	0.34	1.7	238.45	23NO	-0.5	-3.4	-2.0	1.0
219.96	0.24	1.7	248.66	3DE	-0.7	-3.4	-2.1	1.1
227.18	0.12	1.7	258.87	13DE	-0.9	-3.3	-2.1	1.1
234.46	0.01	1.6	269.06	23DE	-0.9	-3.3	-2.1	1.1

Right table

♂ LONG	LAT	MAG	☉ LONG	16.00UT	☿	♀	♃	♄
				-156				
241.79	-0.11	1.6	279.23	2JA	-0.8	-3.3	-2.1	1.1
249.19	-0.24	1.5	289.37	12JA	0.3	-3.3	-2.1	1.1
256.62	-0.38	1.5	299.46	22JA	2.8	-3.4	-2.1	1.1
264.11	-0.52	1.4	309.50	1FE	1.7	-3.4	-2.1	1.1
271.63	-0.66	1.3	319.49	11FE	0.8	-3.4	-2.0	1.1
279.17	-0.80	1.3	329.42	21FE	0.5	-3.4	-1.9	1.2
286.74	-0.95	1.2	339.29	2MR	0.2	-3.5	-1.9	1.2
294.31	-1.09	1.2	349.11	12MR	-0.2	-3.5	-1.8	1.2
301.88	-1.24	1.1	358.86	22MR	-0.9	-3.5	-1.7	1.3
309.43	-1.37	1.0	8.56	1AP	-1.8	-3.6	-1.7	1.3
316.94	-1.50	1.0	18.22	11AP	-1.2	-3.7	-1.6	1.3
324.40	-1.63	0.9	27.83	21AP	-0.2	-3.7	-1.5	1.3
331.79	-1.74	0.8	37.40	1MY	0.7	-3.8	-1.5	1.2
339.09	-1.84	0.8	46.95	11MY	1.5	-3.9	-1.4	1.2
346.28	-1.92	0.7	56.48	21MY	2.7	-4.0	-1.4	1.2
353.33	-1.99	0.6	66.00	31MY	3.2	-4.1	-1.4	1.1
0.21	-2.05	0.6	75.52	10JN	1.7	-4.2	-1.3	1.1
6.89	-2.08	0.5	85.05	20JN	0.4	-4.2	-1.3	1.0
13.35	-2.10	0.4	94.59	30JN	-0.8	-4.1	-1.3	0.9
19.52	-2.10	0.3	104.17	10JL	-1.5	-3.8	-1.3	0.9
25.36	-2.08	0.2	113.78	20JL	-1.2	-3.2	-1.3	0.8
30.81	-2.03	0.1	123.43	30JL	-0.5	-3.4	-1.3	0.7
35.77	-1.96	-0.0	133.13	9AU	-0.1	-3.9	-1.3	0.7
40.13	-1.85	-0.2	142.88	19AU	0.1	-4.2	-1.3	0.7
43.76	-1.72	-0.4	152.69	29AU	0.3	-4.3	-1.3	0.7
46.48	-1.54	-0.6	162.57	8SE	0.7	-4.3	-1.3	0.8
48.10	-1.30	-0.8	172.49	18SE	2.1	-4.2	-1.4	0.8
48.41	-1.00	-1.0	182.48	28SE	1.5	-4.1	-1.4	0.9
47.27	-0.64	-1.2	192.52	8OC	-0.3	-4.0	-1.5	0.9
44.75	-0.20	-1.6	202.61	18OC	-0.6	-3.9	-1.5	1.0
41.21	0.26	-1.6	212.75	28OC	-0.6	-3.8	-1.6	1.0
37.36	0.71	-1.5	222.92	7NO	-0.7	-3.8	-1.7	1.1
34.01	1.10	-1.3	233.11	17NO	-0.8	-3.7	-1.7	1.1
31.74	1.40	-1.0	243.31	27NO	-0.7	-3.6	-1.8	1.2
30.82	1.61	-0.7	253.52	7DE	-0.8	-3.6	-1.9	1.2
31.21	1.75	-0.4	263.72	17DE	-0.6	-3.5	-1.9	1.2
32.74	1.83	-0.1	273.91	27DE	0.4	-3.5	-2.0	1.2
				-155				
35.20	1.87	0.2	284.06	6JA	3.0	-3.4	-2.0	1.2
38.39	1.89	0.4	294.17	16JA	1.2	-3.4	-2.0	1.2
42.17	1.89	0.6	304.25	26JA	0.5	-3.4	-2.0	1.2
46.40	1.87	0.8	314.27	5FE	0.3	-3.3	-2.0	1.2
50.99	1.85	1.0	324.22	15FE	0.1	-3.3	-2.0	1.3
55.85	1.82	1.2	334.13	25FE	-0.3	-3.3	-2.0	1.3
60.95	1.78	1.3	343.97	7MR	-0.9	-3.3	-2.0	1.4
66.23	1.74	1.4	353.75	17MR	-1.7	-3.3	-1.9	1.4
71.66	1.70	1.5	3.48	27MR	-1.2	-3.3	-1.8	1.4
77.22	1.65	1.6	13.16	6AP	-0.2	-3.4	-1.8	1.4
82.88	1.60	1.7	22.79	16AP	0.9	-3.4	-1.7	1.4
88.64	1.55	1.8	32.38	26AP	2.0	-3.4	-1.7	1.3
94.48	1.50	1.8	41.95	6MY	3.5	-3.5	-1.6	1.3
100.40	1.44	1.9	51.48	16MY	2.5	-3.5	-1.5	1.3
106.39	1.38	1.9	61.00	26MY	1.3	-3.5	-1.5	1.2
112.46	1.32	1.9	70.52	5JN	0.3	-3.4	-1.4	1.2
118.58	1.26	1.9	80.04	15JN	-0.8	-3.4	-1.4	1.1
124.78	1.20	2.0	89.58	25JN	-1.6	-3.4	-1.3	1.1
131.05	1.13	2.0	99.14	5JL	-1.2	-3.4	-1.3	1.0
137.39	1.06	2.0	108.73	15JL	-0.5	-3.3	-1.3	1.0
143.80	0.98	1.9	118.36	25JL	-0.0	-3.3	-1.2	0.9
150.29	0.91	1.9	128.03	4AU	0.3	-3.3	-1.2	0.9
156.85	0.83	1.9	137.76	14AU	0.5	-3.3	-1.2	0.9
163.49	0.75	1.9	147.54	24AU	0.9	-3.3	-1.2	0.9
170.21	0.66	1.8	157.38	3SE	2.5	-3.4	-1.2	0.9
177.01	0.57	1.8	167.28	13SE	1.4	-3.4	-1.2	0.9
183.90	0.48	1.7	177.24	23SE	-0.4	-3.4	-1.3	1.0
190.87	0.38	1.7	187.25	3OC	-0.8	-3.5	-1.3	1.0
197.92	0.29	1.6	197.31	13OC	-0.8	-3.5	-1.3	1.1
205.06	0.18	1.6	207.43	23OC	-0.8	-3.6	-1.4	1.1
212.28	0.08	1.6	217.58	2NO	-0.7	-3.6	-1.4	1.2
219.58	-0.03	1.6	227.76	12NO	-0.6	-3.7	-1.5	1.2
226.96	-0.13	1.6	237.96	22NO	-0.5	-3.8	-1.5	1.3
234.42	-0.24	1.6	248.17	2DE	-0.5	-3.9	-1.6	1.3
241.94	-0.35	1.6	258.38	12DE	0.6	-4.0	-1.7	1.3
249.53	-0.47	1.5	268.57	22DE	2.9	-4.1	-1.7	1.4

♂ LONG	LAT	MAG	☉ LONG	16.00UT -154	☿	♀	♃	♄
					MAGNITUDES			
257.17	-0.57	1.5	278.74	1JA	0.9	-4.2	-1.8	1.4
264.86	-0.68	1.5	288.87	11JA	0.3	-4.2	-1.8	1.3
272.59	-0.79	1.5	298.97	21JA	0.1	-4.3	-1.9	1.3
280.35	-0.89	1.4	309.02	31JA	-0.0	-4.3	-2.0	1.3
288.12	-0.98	1.4	319.01	10FE	-0.4	-4.2	-2.0	1.2
295.90	-1.06	1.4	328.94	20FE	-0.9	-3.8	-2.0	1.1
303.68	-1.14	1.4	338.82	2MR	-1.6	-3.3	-2.0	1.1
311.44	-1.21	1.3	348.63	12MR	-1.2	-3.5	-2.0	1.1
319.17	-1.26	1.3	358.39	22MR	-0.1	-4.0	-2.0	1.1
326.85	-1.31	1.3	8.09	1AP	1.2	-4.2	-2.0	1.1
334.48	-1.34	1.3	17.75	11AP	2.6	-4.2	-1.9	1.1
342.05	-1.35	1.3	27.36	21AP	3.4	-4.2	-1.9	1.1
349.53	-1.36	1.2	36.93	1MY	1.9	-4.1	-1.8	1.1
356.92	-1.34	1.2	46.48	11MY	1.0	-4.0	-1.7	1.1
4.20	-1.32	1.2	56.02	21MY	0.2	-3.9	-1.7	1.0
11.37	-1.28	1.2	65.53	31MY	-0.8	-3.8	-1.6	1.0
18.41	-1.22	1.2	75.05	10JN	-1.7	-3.7	-1.6	0.9
25.31	-1.15	1.1	84.58	20JN	-1.2	-3.6	-1.5	0.9
32.06	-1.07	1.1	94.13	30JN	-0.4	-3.6	-1.4	0.8
38.64	-0.97	1.0	103.70	10JL	0.1	-3.5	-1.4	0.8
45.03	-0.86	1.0	113.31	20JL	0.4	-3.5	-1.3	0.7
51.20	-0.73	0.9	122.96	30JL	0.7	-3.4	-1.3	0.7
57.14	-0.58	0.9	132.66	9AU	1.3	-3.4	-1.3	0.6
62.79	-0.42	0.8	142.41	19AU	2.9	-3.4	-1.3	0.6
68.12	-0.24	0.7	152.22	29AU	1.2	-3.4	-1.2	0.6
73.06	-0.03	0.6	162.08	8SE	-0.5	-3.4	-1.2	0.6
77.54	0.20	0.4	172.01	18SE	-0.9	-3.4	-1.2	0.6
81.44	0.47	0.3	182.00	28SE	-0.9	-3.4	-1.2	0.7
84.64	0.77	0.1	192.04	8OC	-0.8	-3.4	-1.2	0.8
86.97	1.12	-0.1	202.13	18OC	-0.6	-3.4	-1.3	0.8
88.25	1.51	-0.3	212.26	28OC	-0.5	-3.4	-1.3	0.9
88.28	1.94	-0.5	222.42	7NO	-0.5	-3.4	-1.3	1.0
86.93	2.40	-0.8	232.62	17NO	-0.3	-3.4	-1.3	1.0
84.26	2.84	-1.0	242.82	27NO	0.7	-3.4	-1.4	1.0
80.61	3.22	-1.1	253.03	7DE	2.7	-3.5	-1.4	1.1
76.62	3.46	-1.1	263.23	17DE	0.6	-3.5	-1.5	1.1
73.08	3.56	-0.9	273.41	27DE	0.1	-3.5	-1.6	1.1

-153

♂ LONG	LAT	MAG	☉ LONG	16.00UT	☿	♀	♃	♄
70.56	3.53	-0.6	283.57	6JA	0.0	-3.4	-1.6	1.1
69.34	3.41	-0.3	293.69	16JA	-0.1	-3.4	-1.7	1.1
69.42	3.24	-0.1	303.75	26JA	-0.4	-3.4	-1.8	1.0
70.63	3.05	0.2	313.78	5FE	-0.9	-3.4	-1.8	1.0
72.79	2.87	0.4	323.74	15FE	-1.5	-3.4	-1.9	1.0
75.72	2.68	0.6	333.64	25FE	-1.1	-3.3	-2.0	0.9
79.26	2.51	0.8	343.49	7MR	-0.0	-3.3	-2.0	0.8
83.29	2.34	1.0	353.28	17MR	1.5	-3.3	-2.0	0.9
87.71	2.19	1.1	3.01	27MR	3.3	-3.3	-2.0	0.9
92.44	2.04	1.2	12.69	6AP	2.5	-3.4	-2.1	0.9
97.45	1.90	1.3	22.33	16AP	1.4	-3.4	-2.0	0.9
102.67	1.77	1.4	31.92	26AP	0.8	-3.4	-2.0	0.9
108.08	1.65	1.5	41.48	6MY	0.1	-3.4	-2.0	0.8
113.65	1.52	1.6	51.02	16MY	-0.8	-3.5	-1.9	0.8
119.38	1.40	1.6	60.54	26MY	-1.7	-3.6	-1.9	0.8
125.24	1.29	1.7	70.06	5JN	-1.2	-3.6	-1.8	0.8
131.22	1.17	1.7	79.59	15JN	-0.4	-3.6	-1.8	0.7
137.32	1.06	1.8	89.12	25JN	0.2	-3.7	-1.7	0.7
143.54	0.94	1.8	98.68	5JL	0.6	-3.8	-1.6	0.6
149.87	0.83	1.8	108.27	15JL	0.9	-3.9	-1.6	0.6
156.30	0.72	1.8	117.89	25JL	1.7	-4.0	-1.5	0.5
162.84	0.61	1.8	127.57	4AU	3.2	-4.1	-1.5	0.5
169.49	0.50	1.8	137.29	14AU	1.0	-4.2	-1.4	0.4
176.25	0.38	1.8	147.06	24AU	-0.6	-4.3	-1.4	0.4
183.11	0.27	1.7	156.90	3SE	-1.1	-4.3	-1.3	0.3
190.08	0.16	1.7	166.80	13SE	-1.1	-4.2	-1.3	0.3
197.14	0.05	1.7	176.75	23SE	-0.8	-3.9	-1.3	0.3
204.32	-0.06	1.6	186.76	3OC	-0.5	-3.3	-1.3	0.4
211.58	-0.17	1.6	196.83	13OC	-0.3	-3.5	-1.3	0.4
218.95	-0.28	1.6	206.94	23OC	-0.3	-4.1	-1.3	0.5
226.40	-0.38	1.5	217.09	2NO	-0.1	-4.3	-1.3	0.6
233.94	-0.48	1.5	227.27	12NO	0.9	-4.4	-1.3	0.6
241.56	-0.58	1.4	237.47	22NO	2.4	-4.3	-1.3	0.7
249.24	-0.67	1.4	247.68	2DE	0.4	-4.2	-1.3	0.7
256.98	-0.76	1.4	257.88	12DE	-0.1	-4.1	-1.4	0.8
264.78	-0.84	1.4	268.08	22DE	-0.1	-4.0	-1.4	0.8

♂ LONG	LAT	MAG	☉ LONG	16.00UT -152	☿	♀	♃	♄
					MAGNITUDES			
272.61	-0.91	1.4	278.25	1JA	-0.2	-3.9	-1.5	0.8
280.47	-0.97	1.4	288.39	11JA	-0.5	-3.8	-1.5	0.8
288.34	-1.02	1.4	298.48	21JA	-1.0	-3.7	-1.6	0.8
296.22	-1.07	1.4	308.54	31JA	-1.3	-3.7	-1.7	0.8
304.08	-1.10	1.4	318.53	10FE	-1.1	-3.6	-1.7	0.8
311.92	-1.12	1.4	328.46	20FE	0.0	-3.5	-1.8	0.7
319.72	-1.13	1.4	338.34	1MR	1.8	-3.5	-1.9	0.7
327.47	-1.12	1.4	348.16	11MR	3.4	-3.4	-1.9	0.6
335.17	-1.10	1.4	357.92	21MR	1.9	-3.4	-2.0	0.6
342.80	-1.08	1.5	7.63	31MR	1.1	-3.3	-2.0	0.6
350.36	-1.04	1.5	17.29	10AP	0.6	-3.3	-2.1	0.6
357.83	-0.98	1.5	26.90	20AP	0.0	-3.3	-2.1	0.6
5.21	-0.92	1.5	36.48	30AP	-0.8	-3.3	-2.1	0.6
12.50	-0.85	1.5	46.03	10MY	-1.7	-3.3	-2.1	0.6
19.69	-0.77	1.5	55.56	20MY	-1.2	-3.3	-2.1	0.6
26.77	-0.68	1.5	65.08	30MY	-0.3	-3.3	-2.1	0.6
33.74	-0.58	1.5	74.59	9JN	0.3	-3.4	-2.1	0.6
40.61	-0.48	1.5	84.12	19JN	0.7	-3.4	-2.0	0.6
47.35	-0.36	1.5	93.67	29JN	1.3	-3.4	-2.0	0.5
53.97	-0.24	1.5	103.24	9JL	2.2	-3.5	-1.9	0.5
60.47	-0.11	1.5	112.84	19JL	3.0	-3.5	-1.8	0.4
66.83	0.02	1.4	122.49	29JL	0.9	-3.5	-1.8	0.4
73.03	0.17	1.4	132.18	8AU	-0.6	-3.5	-1.7	0.3
79.08	0.32	1.4	141.93	18AU	-1.2	-3.4	-1.6	0.3
84.95	0.49	1.3	151.74	28AU	-1.2	-3.4	-1.6	0.2
90.61	0.66	1.2	161.60	7SE	-0.7	-3.4	-1.5	0.1
96.04	0.85	1.2	171.52	17SE	-0.4	-3.3	-1.5	0.1
101.19	1.06	1.1	181.51	27SE	-0.2	-3.3	-1.5	0.0
106.01	1.29	0.9	191.54	7OC	-0.1	-3.3	-1.4	0.1
110.44	1.54	0.8	201.63	17OC	0.1	-3.3	-1.4	0.1
114.38	1.81	0.7	211.76	27OC	1.2	-3.3	-1.4	0.2
117.72	2.12	0.5	221.92	6NO	2.2	-3.4	-1.4	0.3
120.33	2.47	0.3	232.11	16NO	0.2	-3.4	-1.4	0.3
122.01	2.85	0.1	242.32	26NO	-0.2	-3.4	-1.4	0.4
122.61	3.26	-0.2	252.53	6DE	-0.3	-3.4	-1.4	0.5
121.93	3.68	-0.4	262.73	16DE	-0.4	-3.5	-1.4	0.5
119.93	4.07	-0.7	272.92	26DE	-0.6	-3.5	-1.4	0.6

-151

♂ LONG	LAT	MAG	☉ LONG	16.00UT	☿	♀	♃	♄
116.78	4.37	-0.9	283.07	5JA	-1.0	-3.6	-1.4	0.6
112.91	4.53	-1.0	293.20	15JA	-1.2	-3.6	-1.5	0.6
109.04	4.51	-0.9	303.27	25JA	-1.0	-3.7	-1.5	0.6
105.88	4.33	-0.7	313.29	4FE	0.1	-3.8	-1.6	0.6
103.85	4.04	-0.4	323.26	14FE	2.2	-3.9	-1.6	0.6
103.13	3.70	-0.2	333.17	24FE	2.6	-3.9	-1.7	0.6
103.64	3.35	0.0	343.02	6MR	1.4	-4.0	-1.8	0.5
105.22	3.01	0.3	352.81	16MR	0.8	-4.1	-1.8	0.5
107.69	2.70	0.5	2.55	26MR	0.4	-4.2	-1.9	0.4
110.89	2.41	0.6	12.23	5AP	-0.1	-4.2	-2.0	0.4
114.67	2.14	0.8	21.86	15AP	-0.8	-4.2	-2.1	0.4
118.93	1.90	0.9	31.46	25AP	-1.7	-3.9	-2.1	0.4
123.58	1.68	1.0	41.02	5MY	-1.2	-3.4	-2.2	0.5
128.57	1.47	1.1	50.56	15MY	-0.3	-2.8	-2.2	0.5
133.84	1.27	1.2	60.09	25MY	0.4	-3.6	-2.2	0.5
139.34	1.09	1.3	69.60	4JN	1.0	-4.0	-2.3	0.5
145.07	0.92	1.4	79.13	14JN	1.7	-4.2	-2.3	0.5
151.00	0.75	1.4	88.66	24JN	2.9	-4.2	-2.2	0.4
157.10	0.59	1.4	98.21	4JL	2.5	-4.1	-2.2	0.4
163.37	0.44	1.5	107.80	14JL	0.7	-4.0	-2.2	0.4
169.80	0.29	1.5	117.43	24JL	-0.7	-3.9	-2.1	0.3
176.37	0.15	1.5	127.09	3AU	-1.3	-3.9	-2.1	0.3
183.09	0.02	1.5	136.82	13AU	-1.2	-3.8	-2.0	0.2
189.95	-0.11	1.5	146.59	23AU	-0.6	-3.7	-1.9	0.2
196.94	-0.23	1.5	156.42	2SE	-0.3	-3.7	-1.9	0.1
204.05	-0.35	1.5	166.32	12SE	-0.1	-3.6	-1.8	0.0
211.29	-0.46	1.5	176.26	22SE	0.0	-3.5	-1.7	-0.0
218.63	-0.57	1.5	186.27	2OC	0.3	-3.5	-1.7	-0.1
226.08	-0.66	1.5	196.34	12OC	1.5	-3.5	-1.6	-0.1
233.63	-0.75	1.5	206.44	22OC	2.0	-3.4	-1.6	-0.1
241.27	-0.83	1.5	216.59	1NO	-0.0	-3.4	-1.6	-0.0
248.98	-0.90	1.4	226.77	11NO	-0.4	-3.4	-1.5	0.0
256.76	-0.97	1.4	236.97	21NO	-0.4	-3.4	-1.5	0.1
264.58	-1.02	1.4	247.18	1DE	-0.5	-3.4	-1.5	0.2
272.45	-1.05	1.4	257.39	11DE	-0.7	-3.4	-1.5	0.3
280.34	-1.08	1.4	267.58	21DE	-0.9	-3.3	-1.5	0.3
288.25	-1.10	1.4	277.75	31DE	-1.0	-3.3	-1.5	0.4

♂ (Mars) LONG	LAT	MAG	☉ (Sun) LONG	16.00UT -150	☿ (Mercury) MAG	♀ (Venus) MAG	♃ (Jupiter) MAG	♄ (Saturn) MAG
296.15	-1.10	1.4	287.90	10JA	-0.9	-3.4	-1.5	0.4
304.03	-1.09	1.3	297.99	20JA	0.2	-3.4	-1.5	0.4
311.88	-1.07	1.3	308.05	30JA	2.5	-3.4	-1.5	0.4
319.69	-1.04	1.3	318.05	9FE	2.0	-3.4	-1.6	0.4
327.45	-1.00	1.3	327.98	19FE	1.0	-3.4	-1.6	0.4
335.15	-0.94	1.4	337.86	1MR	0.6	-3.5	-1.6	0.4
342.78	-0.88	1.4	347.69	11MR	0.3	-3.5	-1.7	0.4
350.33	-0.81	1.5	357.45	21MR	-0.2	-3.4	-1.8	0.4
357.80	-0.74	1.5	7.16	31MR	-0.8	-3.4	-1.8	0.3
5.18	-0.65	1.5	16.82	10AP	-1.7	-3.4	-1.9	0.3
12.47	-0.56	1.6	26.43	20AP	-1.3	-3.4	-2.0	0.3
19.68	-0.47	1.6	36.01	30AP	-0.3	-3.3	-2.0	0.3
26.79	-0.37	1.7	45.56	10MY	0.6	-3.3	-2.1	0.3
33.81	-0.27	1.7	55.09	20MY	1.3	-3.3	-2.2	0.3
40.74	-0.17	1.7	64.61	30MY	2.2	-3.3	-2.2	0.4
47.58	-0.06	1.7	74.13	9JN	3.4	-3.3	-2.3	0.4
54.33	0.05	1.8	83.66	19JN	2.0	-3.3	-2.3	0.4
60.99	0.16	1.8	93.20	29JN	0.6	-3.4	-2.4	0.4
67.57	0.28	1.8	102.77	9JL	-0.7	-3.4	-2.4	0.3
74.05	0.40	1.8	112.37	19JL	-1.4	-3.4	-2.4	0.3
80.44	0.51	1.8	122.02	29JL	-1.2	-3.5	-2.4	0.3
86.74	0.64	1.8	131.71	8AU	-0.6	-3.5	-2.3	0.2
92.94	0.76	1.7	141.45	18AU	-0.2	-3.6	-2.3	0.2
99.03	0.89	1.7	151.26	28AU	0.1	-3.7	-2.2	0.1
105.01	1.02	1.7	161.12	7SE	0.2	-3.7	-2.2	0.1
110.86	1.16	1.6	171.04	17SE	0.5	-3.8	-2.1	0.0
116.58	1.31	1.5	181.02	27SE	1.8	-3.9	-2.0	-0.1
122.12	1.47	1.5	191.05	7OC	1.8	-4.0	-2.0	-0.1
127.48	1.63	1.4	201.13	17OC	-0.2	-4.2	-1.9	-0.2
132.61	1.81	1.3	211.26	27OC	-0.5	-4.3	-1.8	-0.3
137.46	2.00	1.1	221.43	6NO	-0.6	-4.4	-1.8	-0.2
141.98	2.20	1.0	231.61	16NO	-0.6	-4.4	-1.7	-0.1
146.10	2.43	0.8	241.82	26NO	-0.7	-4.3	-1.7	-0.1
149.71	2.68	0.6	252.03	6DE	-0.8	-4.0	-1.6	0.0
152.70	2.94	0.4	262.23	16DE	-0.9	-3.2	-1.6	0.1
154.91	3.24	0.2	272.42	26DE	-0.7	-3.5	-1.6	0.1
				-149				
156.16	3.54	-0.1	282.58	5JA	0.3	-4.1	-1.6	0.2
156.29	3.84	-0.3	292.70	15JA	2.9	-4.3	-1.5	0.3
155.14	4.11	-0.6	302.78	25JA	1.5	-4.3	-1.5	0.3
152.73	4.31	-0.9	312.81	4FE	0.7	-4.3	-1.5	0.3
149.32	4.36	-1.1	322.77	14FE	0.4	-4.2	-1.6	0.3
145.46	4.25	-1.1	332.69	24FE	0.1	-4.1	-1.6	0.4
141.87	3.97	-1.0	342.54	6MR	-0.2	-4.0	-1.6	0.4
139.18	3.58	-0.8	352.33	16MR	-0.9	-3.8	-1.6	0.3
137.73	3.13	-0.6	2.08	26MR	-1.7	-3.7	-1.7	0.3
137.59	2.68	-0.3	11.76	5AP	-1.2	-3.7	-1.7	0.3
138.65	2.24	-0.1	21.40	15AP	-0.2	-3.6	-1.8	0.3
140.75	1.85	0.1	31.00	25AP	0.8	-3.5	-1.8	0.2
143.71	1.49	0.2	40.56	5MY	1.7	-3.5	-1.9	0.2
147.38	1.17	0.4	50.10	15MY	3.0	-3.4	-1.9	0.2
151.63	0.88	0.5	59.63	25MY	3.0	-3.4	-2.0	0.3
156.36	0.62	0.6	69.15	4JN	1.6	-3.4	-2.1	0.3
161.50	0.38	0.7	78.67	14JN	0.4	-3.3	-2.2	0.3
166.98	0.17	0.8	88.20	24JN	-0.7	-3.3	-2.2	0.3
172.77	-0.03	0.9	97.76	4JL	-1.5	-3.3	-2.3	0.3
178.83	-0.21	1.0	107.34	14JL	-1.2	-3.3	-2.3	0.3
185.12	-0.38	1.0	116.96	24JL	-0.5	-3.3	-2.4	0.3
191.63	-0.53	1.1	126.63	3AU	-0.1	-3.4	-2.4	0.3
198.34	-0.66	1.1	136.34	13AU	0.2	-3.4	-2.4	0.2
205.22	-0.78	1.1	146.12	23AU	0.4	-3.4	-2.4	0.2
212.27	-0.89	1.2	155.95	2SE	0.8	-3.4	-2.4	0.2
219.46	-0.99	1.2	165.83	12SE	2.2	-3.5	-2.4	0.1
226.79	-1.07	1.2	175.78	22SE	1.6	-3.5	-2.4	0.0
234.24	-1.14	1.2	185.79	2OC	-0.3	-3.5	-2.3	-0.0
241.80	-1.19	1.3	195.84	12OC	-0.7	-3.5	-2.2	-0.1
249.44	-1.23	1.3	205.95	22OC	-0.7	-3.4	-2.2	-0.2
257.16	-1.25	1.3	216.09	1NO	-0.7	-3.4	-2.1	-0.3
264.94	-1.26	1.3	226.27	11NO	-0.7	-3.4	-2.0	-0.3
272.76	-1.26	1.3	236.47	21NO	-0.7	-3.4	-2.0	-0.3
280.61	-1.24	1.4	246.67	1DE	-0.7	-3.4	-1.9	-0.2
288.46	-1.21	1.4	256.88	11DE	-0.6	-3.3	-1.8	-0.1
296.32	-1.17	1.4	267.08	21DE	0.4	-3.3	-1.8	-0.0
304.15	-1.11	1.4	277.25	31DE	3.0	-3.3	-1.7	0.0

♂ (Mars) LONG	LAT	MAG	☉ (Sun) LONG	16.00UT -148	☿ (Mercury) MAG	♀ (Venus) MAG	♃ (Jupiter) MAG	♄ (Saturn) MAG
311.95	-1.05	1.4	287.40	10JA	1.1	-3.3	-1.7	0.1
319.71	-0.98	1.5	297.50	20JA	0.5	-3.4	-1.6	0.2
327.41	-0.90	1.5	307.55	30JA	0.2	-3.4	-1.6	0.2
335.04	-0.82	1.5	317.55	9FE	0.0	-3.4	-1.6	0.2
342.61	-0.73	1.5	327.50	19FE	-0.3	-3.4	-1.6	0.3
350.10	-0.63	1.5	337.38	29FE	-0.9	-3.5	-1.6	0.3
357.51	-0.53	1.6	347.21	10MR	-1.7	-3.5	-1.6	0.3
4.83	-0.43	1.6	356.98	20MR	-1.2	-3.6	-1.6	0.3
12.07	-0.33	1.6	6.69	30MR	-0.2	-3.6	-1.6	0.3
19.23	-0.23	1.6	16.35	9AP	1.0	-3.7	-1.6	0.3
26.30	-0.13	1.6	25.97	19AP	2.2	-3.7	-1.6	0.3
33.29	-0.02	1.7	35.55	29AP	3.8	-3.8	-1.7	0.2
40.20	0.08	1.7	45.10	9MY	2.3	-3.9	-1.7	0.2
47.03	0.18	1.8	54.64	19MY	1.2	-4.0	-1.8	0.2
53.79	0.28	1.8	64.16	29MY	0.3	-4.1	-1.8	0.2
60.48	0.38	1.8	73.68	8JN	-0.8	-4.2	-1.9	0.2
67.10	0.47	1.9	83.20	18JN	-1.6	-4.2	-1.9	0.3
73.66	0.57	1.9	92.74	28JN	-1.2	-4.1	-2.0	0.3
80.16	0.66	1.9	102.31	8JL	-0.5	-3.8	-2.1	0.3
86.60	0.75	1.9	111.91	18JL	0.0	-3.2	-2.1	0.3
92.98	0.84	1.9	121.55	28JL	0.3	-3.4	-2.2	0.3
99.31	0.93	1.9	131.24	7AU	0.6	-4.0	-2.3	0.3
105.59	1.02	1.9	140.99	17AU	1.1	-4.2	-2.3	0.3
111.81	1.11	1.9	150.78	27AU	2.6	-4.3	-2.4	0.3
117.97	1.20	1.9	160.64	6SE	1.4	-4.3	-2.4	0.2
124.07	1.29	1.9	170.56	16SE	-0.4	-4.2	-2.4	0.2
130.10	1.38	1.8	180.53	26SE	-0.9	-4.1	-2.4	0.1
136.06	1.47	1.8	190.57	6OC	-0.8	-4.0	-2.4	0.0
141.92	1.56	1.7	200.65	16OC	-0.8	-3.9	-2.4	-0.0
147.70	1.66	1.6	210.77	26OC	-0.6	-3.8	-2.4	-0.1
153.36	1.76	1.5	220.94	5NO	-0.5	-3.8	-2.3	-0.2
158.87	1.86	1.4	231.12	15NO	-0.5	-3.7	-2.2	-0.3
164.23	1.96	1.3	241.33	25NO	-0.4	-3.6	-2.2	-0.3
169.39	2.07	1.2	251.54	5DE	0.6	-3.6	-2.1	-0.3
174.30	2.17	1.0	261.74	15DE	2.9	-3.5	-2.0	-0.2
178.92	2.29	0.8	271.93	25DE	0.8	-3.5	-1.9	-0.1
				-147				
183.17	2.40	0.7	282.09	4JA	0.2	-3.4	-1.9	-0.0
186.96	2.52	0.4	292.21	14JA	0.1	-3.4	-1.8	0.0
190.18	2.63	0.2	302.29	24JA	-0.1	-3.4	-1.8	0.1
192.67	2.74	-0.1	312.32	3FE	-0.4	-3.3	-1.7	0.1
194.29	2.82	-0.3	322.29	13FE	-0.9	-3.3	-1.7	0.2
194.84	2.86	-0.6	332.21	23FE	-1.5	-3.3	-1.6	0.2
194.16	2.85	-0.9	342.07	5MR	-1.2	-3.3	-1.6	0.3
192.22	2.73	-1.2	351.86	15MR	-0.1	-3.3	-1.6	0.3
189.22	2.50	-1.5	1.60	25MR	1.2	-3.3	-1.5	0.3
185.62	2.14	-1.6	11.29	4AP	2.9	-3.4	-1.5	0.3
182.17	1.68	-1.5	20.93	14AP	3.0	-3.4	-1.5	0.3
179.51	1.17	-1.3	30.53	24AP	1.7	-3.4	-1.5	0.3
178.09	0.68	-1.1	40.10	4MY	0.9	-3.5	-1.5	0.3
178.03	0.22	-0.9	49.64	14MY	0.2	-3.5	-1.6	0.3
179.25	-0.18	-0.7	59.16	24MY	-0.8	-3.5	-1.6	0.2
181.59	-0.52	-0.5	68.68	3JN	-1.7	-3.4	-1.6	0.2
184.88	-0.80	-0.3	78.20	13JN	-1.2	-3.4	-1.7	0.2
188.94	-1.04	-0.2	87.73	23JN	-0.4	-3.4	-1.7	0.3
193.64	-1.24	-0.0	97.29	3JL	0.1	-3.4	-1.7	0.3
198.88	-1.41	0.1	106.87	13JL	0.5	-3.3	-1.8	0.3
204.55	-1.54	0.2	116.49	23JL	0.8	-3.3	-1.9	0.4
210.60	-1.65	0.3	126.16	2AU	1.4	-3.3	-1.9	0.4
216.97	-1.72	0.4	135.87	12AU	3.0	-3.3	-2.0	0.4
223.60	-1.78	0.5	145.64	22AU	1.2	-3.3	-2.1	0.4
230.46	-1.81	0.6	155.46	1SE	-0.5	-3.4	-2.1	0.3
237.52	-1.82	0.7	165.35	11SE	-1.0	-3.4	-2.2	0.3
244.73	-1.81	0.7	175.29	21SE	-1.0	-3.4	-2.2	0.3
252.09	-1.78	0.8	185.30	1OC	-0.8	-3.5	-2.3	0.2
259.55	-1.73	0.9	195.35	11OC	-0.5	-3.5	-2.3	0.2
267.09	-1.67	0.9	205.46	21OC	-0.4	-3.6	-2.3	0.1
274.69	-1.59	1.0	215.60	31OC	-0.4	-3.6	-2.4	0.0
282.34	-1.50	1.1	225.78	10NO	-0.3	-3.7	-2.3	-0.1
290.00	-1.40	1.1	235.97	20NO	0.8	-3.8	-2.3	-0.1
297.67	-1.30	1.2	246.18	30NO	2.7	-3.9	-2.3	-0.2
305.32	-1.18	1.3	256.39	10DE	0.5	-4.0	-2.2	-0.2
312.94	-1.07	1.3	266.59	20DE	0.1	-4.1	-2.2	-0.2
320.52	-0.94	1.4	276.77	30DE	-0.1	-4.2	-2.1	-0.1

♂			⊙	16.00UT	☿	♀	♃	♄
LONG	LAT	MAG	LONG	-146	MAGNITUDES			
328.05	-0.82	1.4	286.91	9JA	-0.2	-4.2	-2.0	-0.1
335.53	-0.70	1.5	297.01	19JA	-0.4	-4.3	-1.9	0.0
342.93	-0.57	1.5	307.07	29JA	-0.9	-4.3	-1.9	0.1
350.27	-0.45	1.6	317.07	8FE	-1.4	-4.2	-1.8	0.1
357.53	-0.33	1.6	327.02	18FE	-1.1	-3.8	-1.7	0.2
4.72	-0.21	1.7	336.90	28FE	-0.1	-3.3	-1.7	0.2
11.83	-0.10	1.7	346.73	10MR	1.5	-3.6	-1.6	0.3
18.86	0.01	1.7	356.50	20MR	3.5	-4.0	-1.6	0.3
25.83	0.11	1.8	6.21	30MR	2.3	-4.2	-1.5	0.3
32.71	0.21	1.8	15.88	9AP	1.3	-4.2	-1.5	0.4
39.54	0.31	1.8	25.50	19AP	0.7	-4.2	-1.5	0.4
46.29	0.40	1.8	35.08	29AP	0.1	-4.1	-1.5	0.4
52.99	0.49	1.8	44.63	9MY	-0.8	-4.0	-1.4	0.4
59.63	0.57	1.9	54.17	19MY	-1.7	-3.9	-1.4	0.4
66.22	0.65	1.9	63.69	29MY	-1.3	-3.8	-1.4	0.3
72.76	0.73	1.9	73.21	8JN	-0.4	-3.7	-1.5	0.3
79.26	0.80	1.9	82.74	18JN	0.2	-3.6	-1.5	0.3
85.72	0.87	1.9	92.28	28JN	0.6	-3.6	-1.5	0.3
92.15	0.93	2.0	101.84	8JL	1.0	-3.5	-1.5	0.4
98.55	0.99	2.0	111.45	18JL	1.9	-3.5	-1.6	0.4
104.92	1.05	2.0	121.09	28JL	3.2	-3.4	-1.6	0.4
111.27	1.10	2.0	130.77	7AU	1.0	-3.4	-1.6	0.5
117.60	1.15	2.0	140.52	17AU	-0.6	-3.4	-1.7	0.5
123.91	1.20	2.0	150.31	27AU	-1.1	-3.4	-1.8	0.5
130.20	1.25	2.0	160.16	6SE	-1.1	-3.4	-1.8	0.5
136.48	1.29	2.0	170.08	16SE	-0.7	-3.4	-1.9	0.4
142.74	1.33	2.0	180.05	26SE	-0.4	-3.4	-1.9	0.4
148.98	1.36	1.9	190.08	6OC	-0.3	-3.4	-2.0	0.4
155.20	1.39	1.9	200.16	16OC	-0.2	-3.4	-2.1	0.3
161.40	1.42	1.8	210.28	26OC	-0.1	-3.4	-2.1	0.3
167.56	1.45	1.8	220.44	5NO	1.0	-3.4	-2.2	0.2
173.69	1.47	1.7	230.63	15NO	2.4	-3.4	-2.2	0.1
179.78	1.48	1.6	240.83	25NO	0.3	-3.4	-2.2	0.1
185.81	1.49	1.5	251.04	5DE	-0.1	-3.5	-2.2	-0.0
191.78	1.49	1.4	261.25	15DE	-0.2	-3.5	-2.2	-0.1
197.66	1.48	1.3	271.43	25DE	-0.3	-3.5	-2.2	-0.1
				-145				
203.45	1.46	1.1	281.59	4JA	-0.5	-3.4	-2.1	-0.1
209.12	1.43	1.0	291.72	14JA	-1.0	-3.4	-2.1	-0.0
214.64	1.38	0.8	301.80	24JA	-1.3	-3.4	-2.0	0.0
219.99	1.31	0.6	311.83	3FE	-1.0	-3.4	-2.0	0.1
225.10	1.22	0.4	321.81	13FE	0.0	-3.4	-1.9	0.2
229.93	1.09	0.2	331.72	23FE	1.9	-3.3	-1.8	0.2
234.41	0.93	-0.0	341.58	5MR	3.1	-3.3	-1.8	0.3
238.44	0.72	-0.3	351.38	15MR	1.7	-3.3	-1.7	0.3
241.89	0.44	-0.5	1.13	25MR	1.0	-3.3	-1.6	0.4
244.63	0.10	-0.8	10.82	4AP	0.5	-3.4	-1.6	0.4
246.45	-0.34	-1.1	20.46	14AP	-0.0	-3.4	-1.5	0.4
247.21	-0.87	-1.5	30.06	24AP	-0.8	-3.4	-1.5	0.5
246.74	-1.49	-1.8	39.63	4MY	-1.7	-3.4	-1.4	0.5
245.06	-2.18	-2.1	49.18	14MY	-1.3	-3.5	-1.4	0.5
242.45	-2.87	-2.3	58.70	24MY	-0.4	-3.5	-1.4	0.5
239.48	-3.48	-2.3	68.22	3JN	0.3	-3.6	-1.4	0.5
236.87	-3.93	-2.2	77.74	13JN	0.8	-3.6	-1.3	0.5
235.27	-4.20	-2.0	87.27	23JN	1.4	-3.7	-1.3	0.4
234.98	-4.30	-1.8	96.83	3JL	2.5	-3.8	-1.3	0.4
236.06	-4.28	-1.5	106.41	13JL	2.8	-3.9	-1.3	0.5
238.39	-4.16	-1.3	116.03	23JL	0.9	-4.0	-1.4	0.5
241.77	-4.00	-1.1	125.69	2AU	-0.6	-4.1	-1.4	0.6
246.01	-3.79	-0.9	135.40	12AU	-1.3	-4.2	-1.4	0.6
250.94	-3.56	-0.7	145.17	22AU	-1.2	-4.3	-1.5	0.6
256.40	-3.32	-0.5	154.99	1SE	-0.7	-4.3	-1.5	0.6
262.30	-3.06	-0.3	164.87	11SE	-0.3	-4.2	-1.5	0.6
268.53	-2.80	-0.1	174.81	21SE	-0.1	-3.9	-1.6	0.6
275.02	-2.55	0.0	184.81	1OC	-0.1	-3.3	-1.7	0.6
281.71	-2.29	0.2	194.87	11OC	0.1	-3.6	-1.7	0.6
288.54	-2.04	0.3	204.97	21OC	1.2	-4.1	-1.8	0.5
295.48	-1.79	0.5	215.11	31OC	2.2	-4.3	-1.9	0.5
302.49	-1.56	0.6	225.28	10NO	0.1	-4.4	-1.9	0.4
309.54	-1.33	0.7	235.48	20NO	-0.3	-4.3	-2.0	0.4
316.62	-1.12	0.9	245.69	30NO	-0.3	-4.2	-2.0	0.3
323.71	-0.92	1.0	255.90	10DE	-0.4	-4.1	-2.1	0.2
330.78	-0.73	1.1	266.09	20DE	-0.6	-4.0	-2.1	0.2
337.82	-0.55	1.2	276.27	30DE	-1.0	-3.9	-2.1	0.1

♂			⊙	16.00UT	☿	♀	♃	♄
LONG	LAT	MAG	LONG	-144	MAGNITUDES			
344.84	-0.38	1.3	286.42	9JA	-1.1	-3.8	-2.1	0.1
351.82	-0.23	1.4	296.52	19JA	-0.9	-3.7	-2.1	0.1
358.75	-0.08	1.5	306.58	29JA	0.1	-3.7	-2.1	0.2
5.64	0.05	1.6	316.59	8FE	2.2	-3.6	-2.0	0.2
12.47	0.17	1.6	326.54	18FE	2.4	-3.5	-2.0	0.3
19.25	0.29	1.7	336.43	28FE	1.2	-3.5	-1.9	0.3
25.99	0.39	1.7	346.26	9MR	0.7	-3.4	-1.8	0.4
32.67	0.49	1.8	356.03	19MR	0.4	-3.4	-1.8	0.4
39.30	0.58	1.8	5.75	29MR	-0.1	-3.3	-1.7	0.5
45.89	0.66	1.9	15.42	8AP	-0.8	-3.3	-1.6	0.5
52.44	0.73	1.9	25.04	18AP	-1.7	-3.3	-1.6	0.6
58.95	0.80	2.0	34.63	28AP	-1.3	-3.3	-1.5	0.6
65.43	0.86	2.0	44.18	8MY	-0.3	-3.3	-1.5	0.6
71.87	0.91	2.0	53.71	18MY	0.5	-3.3	-1.4	0.6
78.29	0.96	2.0	63.24	28MY	1.1	-3.3	-1.4	0.7
84.70	1.00	2.0	72.75	7JN	1.9	-3.4	-1.3	0.7
91.08	1.04	2.0	82.28	17JN	3.1	-3.4	-1.3	0.7
97.46	1.08	2.0	91.82	27JN	2.3	-3.4	-1.3	0.7
103.83	1.11	2.0	101.38	7JL	0.7	-3.5	-1.3	0.6
110.19	1.13	2.0	110.98	17JL	-0.7	-3.5	-1.3	0.6
116.56	1.15	2.0	120.62	27JL	-1.4	-3.5	-1.3	0.7
122.94	1.17	2.0	130.30	6AU	-1.2	-3.5	-1.3	0.7
129.32	1.18	2.0	140.04	16AU	-0.6	-3.4	-1.3	0.8
135.72	1.18	2.0	149.83	26AU	-0.2	-3.4	-1.3	0.8
142.14	1.19	2.0	159.68	5SE	-0.0	-3.4	-1.3	0.8
148.57	1.18	2.0	169.59	15SE	0.1	-3.3	-1.3	0.8
155.02	1.17	2.0	179.57	25SE	0.3	-3.3	-1.4	0.8
161.50	1.16	2.0	189.59	5OC	1.5	-3.3	-1.4	0.8
167.99	1.14	1.9	199.66	15OC	2.0	-3.3	-1.5	0.8
174.51	1.12	1.9	209.79	25OC	-0.1	-3.3	-1.5	0.8
181.05	1.09	1.9	219.94	4NO	-0.5	-3.4	-1.6	0.8
187.62	1.05	1.8	230.13	14NO	-0.5	-3.4	-1.7	0.7
194.20	1.00	1.7	240.34	24NO	-0.5	-3.4	-1.7	0.7
200.80	0.94	1.7	250.54	4DE	-0.7	-3.4	-1.8	0.6
207.43	0.88	1.6	260.75	14DE	-0.9	-3.5	-1.9	0.5
214.06	0.80	1.5	270.94	24DE	-1.0	-3.5	-1.9	0.4
				-143				
220.71	0.70	1.4	281.10	3JA	-0.8	-3.6	-2.0	0.4
227.37	0.60	1.3	291.23	13JA	0.2	-3.6	-2.0	0.3
234.03	0.48	1.2	301.32	23JA	2.6	-3.7	-2.0	0.3
240.68	0.33	1.1	311.35	2FE	1.8	-3.8	-2.0	0.3
247.33	0.17	1.0	321.33	12FE	0.9	-3.8	-2.0	0.4
253.96	-0.01	0.8	331.25	22FE	0.5	-3.9	-2.0	0.4
260.56	-0.22	0.7	341.11	4MR	0.2	-4.0	-2.0	0.5
267.12	-0.45	0.5	350.92	14MR	-0.2	-4.1	-1.9	0.5
273.60	-0.72	0.4	0.66	24MR	-0.8	-4.2	-1.9	0.6
280.00	-1.01	0.2	10.35	3AP	-1.7	-4.2	-1.8	0.6
286.27	-1.34	0.0	20.00	13AP	-1.3	-4.2	-1.8	0.7
292.37	-1.71	-0.1	29.60	23AP	-0.3	-3.9	-1.7	0.7
298.25	-2.11	-0.3	39.17	3MY	0.6	-3.4	-1.6	0.8
303.83	-2.56	-0.5	48.72	13MY	1.4	-2.8	-1.6	0.8
309.01	-3.04	-0.7	58.24	23MY	2.5	-3.7	-1.5	0.8
313.68	-3.57	-1.0	67.76	2JN	3.4	-4.1	-1.4	0.9
317.69	-4.13	-1.2	77.28	12JN	1.9	-4.2	-1.4	0.9
320.84	-4.72	-1.5	86.81	22JN	0.5	-4.2	-1.4	0.9
322.94	-5.32	-1.7	96.36	2JL	-0.7	-4.1	-1.3	0.9
323.78	-5.88	-2.0	105.94	12JL	-1.5	-4.0	-1.3	0.9
323.25	-6.34	-2.2	115.56	22JL	-1.2	-3.9	-1.3	0.9
321.46	-6.59	-2.4	125.22	1AU	-0.6	-3.9	-1.2	0.9
318.80	-6.55	-2.6	134.93	11AU	-0.1	-3.8	-1.2	0.9
316.02	-6.21	-2.5	144.69	21AU	0.1	-3.7	-1.2	1.0
313.86	-5.61	-2.2	154.51	31AU	0.3	-3.6	-1.2	1.0
312.82	-4.88	-1.9	164.39	10SE	0.6	-3.6	-1.2	1.1
313.10	-4.12	-1.6	174.33	20SE	1.9	-3.5	-1.2	1.1
314.62	-3.39	-1.3	184.33	30SE	1.8	-3.5	-1.3	1.1
317.20	-2.74	-1.0	194.38	10OC	-0.2	-3.5	-1.3	1.1
320.66	-2.16	-0.7	204.48	20OC	-0.6	-3.4	-1.3	1.1
324.79	-1.67	-0.4	214.62	30OC	-0.6	-3.4	-1.4	1.1
329.44	-1.24	-0.2	224.79	9NO	-0.6	-3.4	-1.4	1.1
334.51	-0.88	0.1	234.99	19NO	-0.8	-3.4	-1.5	1.0
339.88	-0.56	0.3	245.19	29NO	-0.8	-3.4	-1.5	1.0
345.48	-0.30	0.5	255.40	9DE	-0.8	-3.4	-1.6	0.9
351.27	-0.07	0.7	265.60	19DE	-0.7	-3.3	-1.7	0.9
357.19	0.13	0.8	275.78	29DE	0.3	-3.3	-1.7	0.8

♂ LONG	♂ LAT	♂ MAG	⊙ LONG	16.00UT -142	☿ MAGNITUDES	♀	♃	♄
3.21	0.30	1.0	285.93	8JA	2.9	-3.4	-1.8	0.7
9.30	0.45	1.1	296.03	18JA	1.4	-3.4	-1.9	0.6
15.45	0.58	1.3	306.09	28JA	0.6	-3.4	-1.9	0.6
21.63	0.69	1.4	316.11	7FE	0.3	-3.4	-2.0	0.6
27.84	0.78	1.5	326.06	17FE	0.1	-3.4	-2.0	0.6
34.07	0.86	1.6	335.95	27FE	-0.2	-3.5	-2.0	0.6
40.30	0.93	1.7	345.78	9MR	-0.8	-3.5	-2.0	0.7
46.54	0.99	1.7	355.56	19MR	-1.7	-3.4	-2.0	0.7
52.78	1.04	1.8	5.28	29MR	-1.3	-3.4	-2.0	0.8
59.02	1.08	1.9	14.95	8AP	-0.2	-3.4	-1.9	0.9
65.26	1.12	1.9	24.57	18AP	0.8	-3.4	-1.9	0.9
71.50	1.15	2.0	34.16	28AP	1.9	-3.3	-1.8	1.0
77.74	1.17	2.0	43.71	8MY	3.3	-3.3	-1.8	1.0
83.99	1.18	2.0	53.25	18MY	2.7	-3.3	-1.7	1.0
90.25	1.19	2.0	62.77	28MY	1.5	-3.3	-1.7	1.1
96.52	1.20	2.0	72.29	7JN	0.4	-3.3	-1.6	1.1
102.80	1.20	2.0	81.81	17JN	-0.7	-3.3	-1.5	1.1
109.10	1.19	2.0	91.35	27JN	-1.5	-3.4	-1.5	1.2
115.43	1.18	2.0	100.92	7JL	-1.3	-3.4	-1.4	1.2
121.78	1.17	2.0	110.51	17JL	-0.5	-3.4	-1.4	1.2
128.16	1.15	2.0	120.14	27JL	-0.0	-3.5	-1.3	1.2
134.57	1.13	2.0	129.83	6AU	0.2	-3.5	-1.3	1.2
141.02	1.10	1.9	139.56	16AU	0.4	-3.6	-1.3	1.2
147.52	1.06	1.9	149.35	26AU	0.9	-3.7	-1.2	1.2
154.06	1.03	1.9	159.20	5SE	2.3	-3.7	-1.2	1.3
160.64	0.98	1.9	169.11	15SE	1.6	-3.8	-1.2	1.3
167.28	0.94	1.9	179.08	25SE	-0.3	-3.9	-1.2	1.4
173.97	0.88	1.9	189.10	5OC	-0.8	-4.0	-1.2	1.4
180.71	0.82	1.9	199.18	15OC	-0.8	-4.2	-1.2	1.4
187.51	0.76	1.9	209.30	25OC	-0.8	-4.3	-1.3	1.3
194.37	0.68	1.8	219.45	4NO	-0.7	-4.4	-1.3	1.3
201.27	0.60	1.8	229.64	14NO	-0.6	-4.4	-1.3	1.3
208.24	0.52	1.8	239.84	24NO	-0.6	-4.3	-1.4	1.2
215.27	0.42	1.7	250.05	4DE	-0.5	-3.9	-1.4	1.2
222.34	0.32	1.7	260.25	14DE	0.4	-3.2	-1.5	1.1
229.48	0.20	1.6	270.44	24DE	3.0	-3.5	-1.5	1.1
				-141				
236.66	0.08	1.5	280.61	3JA	1.0	-4.1	-1.6	1.0
243.89	-0.05	1.5	290.74	13JA	0.4	-4.3	-1.7	1.0
251.17	-0.19	1.4	300.83	23JA	0.2	-4.3	-1.7	1.0
258.49	-0.34	1.3	310.86	2FE	-0.0	-4.3	-1.8	0.9
265.84	-0.49	1.3	320.84	12FE	-0.3	-4.2	-1.9	0.9
273.23	-0.65	1.2	330.77	22FE	-0.9	-4.1	-1.9	0.9
280.63	-0.82	1.1	340.64	4MR	-1.6	-3.9	-2.0	0.9
288.04	-1.00	1.0	350.44	14MR	-1.2	-3.8	-2.0	0.9
295.45	-1.18	0.9	0.19	24MR	-0.2	-3.7	-2.0	1.0
302.84	-1.36	0.9	9.89	3AP	1.0	-3.7	-2.1	1.0
310.20	-1.54	0.8	19.53	13AP	2.4	-3.6	-2.1	1.1
317.51	-1.71	0.7	29.14	23AP	3.6	-3.5	-2.0	1.2
324.73	-1.88	0.6	38.71	3MY	2.1	-3.5	-2.0	1.2
331.86	-2.05	0.5	48.26	13MY	1.1	-3.4	-2.0	1.3
338.86	-2.20	0.4	57.79	23MY	0.3	-3.4	-1.9	1.3
345.69	-2.35	0.3	67.30	2JN	-0.7	-3.4	-1.9	1.3
352.31	-2.48	0.2	76.82	12JN	-1.6	-3.3	-1.8	1.4
358.68	-2.60	0.1	86.36	22JN	-1.3	-3.3	-1.8	1.4
4.74	-2.70	-0.0	95.90	2JL	-0.5	-3.3	-1.7	1.4
10.41	-2.78	-0.2	105.48	12JL	0.0	-3.3	-1.6	1.4
15.62	-2.84	-0.3	115.10	22JL	0.4	-3.3	-1.6	1.4
20.23	-2.87	-0.5	124.75	1AU	0.6	-3.4	-1.5	1.4
24.12	-2.88	-0.7	134.46	11AU	1.2	-3.4	-1.5	1.3
27.11	-2.84	-0.9	144.22	21AU	2.7	-3.4	-1.4	1.3
28.99	-2.76	-1.1	154.03	31AU	1.4	-3.4	-1.4	1.2
29.56	-2.60	-1.3	163.91	10SE	-0.4	-3.5	-1.3	1.2
28.66	-2.35	-1.5	173.85	20SE	-0.9	-3.5	-1.3	1.2
26.35	-1.99	-1.7	183.84	30SE	-0.9	-3.5	-1.3	1.2
23.00	-1.53	-1.8	193.89	10OC	-0.9	-3.5	-1.3	1.2
19.30	-1.01	-1.8	203.99	20OC	-0.6	-3.4	-1.3	1.2
16.08	-0.48	-1.5	214.12	30OC	-0.5	-3.4	-1.3	1.2
13.96	-0.01	-1.2	224.29	9NO	-0.5	-3.4	-1.3	1.2
13.17	0.39	-0.9	234.49	19NO	-0.4	-3.4	-1.3	1.1
13.70	0.70	-0.6	244.70	29NO	0.6	-3.4	-1.3	1.1
15.38	0.94	-0.3	254.90	9DE	2.9	-3.3	-1.4	1.0
17.98	1.12	-0.0	265.11	19DE	0.7	-3.3	-1.4	1.0
21.32	1.25	0.2	275.28	29DE	0.2	-3.3	-1.4	1.0

♂ LONG	♂ LAT	♂ MAG	⊙ LONG	16.00UT -140	☿ MAGNITUDES	♀	♃	♄
25.24	1.35	0.5	285.43	8JA	0.0	-3.3	-1.5	0.9
29.60	1.42	0.7	295.55	18JA	-0.1	-3.4	-1.6	0.9
34.31	1.47	0.9	305.61	28JA	-0.4	-3.4	-1.6	0.8
39.29	1.50	1.0	315.62	7FE	-0.9	-3.4	-1.7	0.8
44.49	1.52	1.2	325.58	17FE	-1.5	-3.4	-1.8	0.7
49.87	1.53	1.3	335.47	27FE	-1.2	-3.5	-1.8	0.7
55.37	1.53	1.4	345.31	8MR	-0.1	-3.5	-1.9	0.8
60.99	1.52	1.5	355.09	18MR	1.3	-3.6	-2.0	0.8
66.71	1.51	1.6	4.81	28MR	3.1	-3.6	-2.0	0.9
72.50	1.49	1.7	14.48	7AP	2.7	-3.7	-2.1	1.0
78.36	1.47	1.8	24.11	17AP	1.6	-3.7	-2.1	1.0
84.28	1.45	1.8	33.70	27AP	0.8	-3.8	-2.1	1.1
90.25	1.42	1.9	43.26	7MY	0.2	-3.9	-2.2	1.1
96.28	1.38	1.9	52.79	17MY	-0.7	-4.0	-2.1	1.2
102.36	1.34	2.0	62.31	27MY	-1.7	-4.1	-2.1	1.2
108.49	1.30	2.0	71.83	6JN	-1.3	-4.2	-2.1	1.2
114.67	1.26	2.0	81.36	16JN	-0.4	-4.2	-2.1	1.3
120.90	1.21	2.0	90.89	26JN	0.1	-4.1	-2.0	1.3
127.19	1.16	2.0	100.46	6JL	0.5	-3.8	-2.0	1.3
133.53	1.10	2.0	110.05	16JL	0.9	-3.2	-1.9	1.3
139.94	1.05	2.0	119.68	26JL	1.6	-3.4	-1.8	1.2
146.41	0.98	2.0	129.36	5AU	3.1	-4.0	-1.8	1.2
152.94	0.92	1.9	139.09	15AU	1.2	-4.2	-1.7	1.2
159.54	0.85	1.9	148.88	25AU	-0.5	-4.3	-1.6	1.1
166.21	0.78	1.8	158.72	4SE	-1.0	-4.2	-1.6	1.1
172.96	0.70	1.8	168.63	14SE	-1.1	-4.2	-1.5	1.1
179.78	0.62	1.7	178.59	24SE	-0.8	-4.1	-1.5	1.1
186.68	0.53	1.7	188.61	4OC	-0.5	-4.0	-1.5	1.1
193.65	0.44	1.7	198.69	14OC	-0.3	-3.9	-1.4	1.1
200.70	0.34	1.7	208.80	24OC	-0.3	-3.8	-1.4	1.1
207.83	0.25	1.7	218.96	3NO	-0.2	-3.8	-1.4	1.1
215.04	0.14	1.7	229.14	13NO	0.8	-3.7	-1.4	1.0
222.32	0.04	1.7	239.34	23NO	2.7	-3.6	-1.4	1.0
229.67	-0.07	1.6	249.55	3DE	0.5	-3.6	-1.4	1.0
237.09	-0.19	1.6	259.76	13DE	-0.0	-3.5	-1.4	0.9
244.58	-0.30	1.6	269.95	23DE	-0.1	-3.5	-1.4	0.9
				-139				
252.12	-0.42	1.5	280.12	2JA	-0.2	-3.4	-1.4	0.9
259.71	-0.54	1.5	290.25	12JA	-0.5	-3.4	-1.5	0.8
267.35	-0.66	1.5	300.34	22JA	-0.9	-3.4	-1.5	0.8
275.03	-0.78	1.4	310.38	1FE	-1.3	-3.3	-1.6	0.7
282.73	-0.89	1.4	320.37	11FE	-1.1	-3.3	-1.6	0.7
290.45	-1.00	1.4	330.29	21FE	-0.1	-3.3	-1.7	0.6
298.17	-1.10	1.3	340.16	3MR	1.6	-3.3	-1.7	0.6
305.88	-1.20	1.3	349.97	13MR	3.5	-3.3	-1.8	0.6
313.57	-1.29	1.3	359.72	23MR	2.0	-3.3	-1.9	0.6
321.23	-1.36	1.2	9.42	2AP	1.2	-3.4	-1.9	0.7
328.84	-1.43	1.2	19.07	12AP	0.6	-3.4	-2.0	0.8
336.38	-1.48	1.1	28.67	22AP	0.1	-3.4	-2.1	0.8
343.86	-1.51	1.1	38.25	2MY	-0.8	-3.5	-2.1	0.9
351.23	-1.54	1.1	47.80	12MY	-1.7	-3.5	-2.2	0.9
358.50	-1.54	1.0	57.32	22MY	-1.3	-3.5	-2.2	1.0
5.65	-1.53	1.0	66.84	1JN	-0.4	-3.4	-2.3	1.0
12.66	-1.51	1.0	76.36	11JN	0.2	-3.4	-2.3	1.1
19.51	-1.47	0.9	85.89	21JN	0.7	-3.4	-2.3	1.1
26.19	-1.41	0.8	95.44	1JL	1.2	-3.4	-2.3	1.1
32.66	-1.33	0.8	105.01	11JL	2.1	-3.3	-2.2	1.1
38.90	-1.24	0.7	114.62	21JL	3.1	-3.3	-2.2	1.1
44.87	-1.13	0.6	124.28	31JL	1.0	-3.3	-2.1	1.1
50.54	-0.99	0.5	133.98	10AU	-0.6	-3.3	-2.1	1.1
55.84	-0.84	0.4	143.74	20AU	-1.2	-3.3	-2.0	1.1
60.70	-0.66	0.3	153.55	30AU	-1.2	-3.4	-1.9	1.0
65.02	-0.45	0.2	163.43	9SE	-0.7	-3.4	-1.9	1.0
68.70	-0.20	0.0	173.36	19SE	-0.4	-3.4	-1.8	0.9
71.56	0.09	-0.2	183.35	29SE	-0.2	-3.5	-1.7	0.9
73.44	0.42	-0.4	193.39	9OC	-0.2	-3.5	-1.7	0.9
74.13	0.81	-0.6	203.49	19OC	0.0	-3.6	-1.6	1.0
73.45	1.24	-0.8	213.63	29OC	1.0	-3.6	-1.6	1.0
71.37	1.69	-1.0	223.79	8NO	2.4	-3.7	-1.6	1.0
68.11	2.14	-1.2	233.99	18NO	0.2	-3.8	-1.5	0.9
64.19	2.50	-1.3	244.20	28NO	-0.2	-3.9	-1.5	0.9
60.40	2.76	-1.1	254.41	8DE	-0.2	-4.0	-1.5	0.9
57.44	2.89	-0.8	264.61	18DE	-0.3	-4.1	-1.5	0.9
55.73	2.91	-0.6	274.79	28DE	-0.5	-4.2	-1.5	0.8

♂			⊙	16.00UT	☿	♀	♃	♄
LONG	LAT	MAG	LONG	-138	MAGNITUDES			
55.35	2.86	-0.3	284.94	7JA	-1.0	-4.3	-1.5	0.8
56.18	2.78	-0.0	295.05	17JA	-1.2	-4.3	-1.5	0.8
58.05	2.67	0.3	305.12	27JA	-1.0	-4.3	-1.5	0.7
60.75	2.55	0.5	315.14	6FE	-0.0	-4.2	-1.5	0.7
64.11	2.43	0.7	325.09	16FE	2.0	-3.8	-1.6	0.6
68.00	2.31	0.9	334.99	26FE	2.9	-3.3	-1.6	0.6
72.30	2.20	1.0	344.83	8MR	1.5	-3.6	-1.7	0.5
76.94	2.09	1.2	354.61	18MR	0.9	-4.0	-1.7	0.5
81.85	1.99	1.3	4.34	28MR	0.5	-4.2	-1.8	0.5
86.98	1.88	1.4	14.01	7AP	-0.0	-4.2	-1.8	0.5
92.29	1.78	1.5	23.64	17AP	-0.8	-4.2	-1.9	0.6
97.77	1.69	1.6	33.24	27AP	-1.7	-4.1	-2.0	0.6
103.39	1.59	1.7	42.79	7MY	-1.3	-4.0	-2.1	0.7
109.13	1.49	1.7	52.33	17MY	-0.4	-3.9	-2.1	0.8
114.98	1.40	1.8	61.85	27MY	0.4	-3.8	-2.2	0.8
120.94	1.31	1.8	71.37	6JN	0.9	-3.7	-2.3	0.9
126.99	1.21	1.8	80.89	16JN	1.5	-3.6	-2.3	0.9
133.15	1.12	1.9	90.43	26JN	2.7	-3.6	-2.3	1.0
139.40	1.02	1.9	99.99	6JL	2.7	-3.5	-2.4	1.0
145.74	0.93	1.9	109.59	16JL	0.8	-3.5	-2.4	1.0
152.17	0.83	1.9	119.22	26JL	-0.6	-3.4	-2.4	1.0
158.70	0.73	1.8	128.89	5AU	-1.3	-3.4	-2.4	1.0
165.33	0.63	1.8	138.62	15AU	-1.3	-3.4	-2.3	1.0
172.05	0.53	1.8	148.41	25AU	-0.7	-3.4	-2.3	1.0
178.86	0.43	1.8	158.25	4SE	-0.3	-3.4	-2.2	0.9
185.77	0.32	1.7	168.15	14SE	-0.1	-3.4	-2.2	0.9
192.78	0.22	1.7	178.11	24SE	-0.0	-3.4	-2.1	0.9
199.88	0.11	1.7	188.13	4OC	0.2	-3.4	-2.0	0.8
207.07	0.00	1.6	198.20	14OC	1.3	-3.4	-2.0	0.8
214.36	-0.10	1.6	208.31	24OC	2.2	-3.4	-1.9	0.9
221.73	-0.21	1.5	218.46	3NO	0.1	-3.4	-1.8	0.9
229.19	-0.32	1.5	228.64	13NO	-0.4	-3.4	-1.8	0.9
236.73	-0.42	1.5	238.84	23NO	-0.4	-3.4	-1.7	0.9
244.34	-0.52	1.4	249.05	3DE	-0.4	-3.5	-1.7	0.9
252.01	-0.62	1.4	259.26	13DE	-0.6	-3.5	-1.6	0.9
259.74	-0.72	1.4	269.45	23DE	-0.9	-3.5	-1.6	0.8
				-137				
267.52	-0.80	1.4	279.62	2JA	-1.0	-3.4	-1.6	0.8
275.33	-0.89	1.4	289.75	12JA	-0.9	-3.4	-1.6	0.8
283.16	-0.96	1.4	299.85	22JA	0.1	-3.4	-1.6	0.7
291.01	-1.02	1.4	309.89	1FE	2.3	-3.4	-1.6	0.7
298.85	-1.08	1.4	319.88	11FE	2.2	-3.4	-1.6	0.6
306.69	-1.12	1.4	329.81	21FE	1.1	-3.3	-1.6	0.6
314.50	-1.16	1.4	339.68	3MR	0.6	-3.3	-1.6	0.5
322.27	-1.18	1.4	349.49	13MR	0.3	-3.3	-1.6	0.5
329.99	-1.18	1.4	359.25	23MR	-0.1	-3.3	-1.6	0.4
337.65	-1.18	1.4	8.95	2AP	-0.8	-3.4	-1.7	0.4
345.25	-1.16	1.4	18.60	12AP	-1.7	-3.4	-1.7	0.4
352.76	-1.13	1.4	28.21	22AP	-1.3	-3.4	-1.8	0.5
0.19	-1.09	1.4	37.78	2MY	-0.3	-3.4	-1.8	0.5
7.52	-1.04	1.4	47.33	12MY	0.5	-3.5	-1.9	0.6
14.75	-0.97	1.4	56.86	22MY	1.2	-3.5	-2.0	0.6
21.87	-0.89	1.4	66.38	1JN	2.1	-3.6	-2.0	0.7
28.87	-0.80	1.4	75.90	11JN	3.3	-3.6	-2.1	0.8
35.75	-0.71	1.4	85.43	21JN	2.2	-3.7	-2.2	0.8
42.50	-0.60	1.4	94.98	1JL	0.7	-3.8	-2.2	0.8
49.11	-0.48	1.3	104.55	11JL	-0.6	-3.9	-2.3	0.9
55.57	-0.35	1.3	114.16	21JL	-1.4	-4.0	-2.4	0.9
61.87	-0.21	1.3	123.81	31JL	-1.3	-4.1	-2.4	0.9
67.99	-0.06	1.2	133.52	10AU	-0.6	-4.2	-2.4	0.9
73.91	0.10	1.2	143.27	20AU	-0.2	-4.3	-2.5	0.9
79.60	0.28	1.1	153.08	30AU	0.0	-4.3	-2.5	0.9
85.03	0.47	1.0	162.95	9SE	0.2	-4.2	-2.4	0.9
90.16	0.68	0.9	172.88	19SE	0.4	-3.9	-2.4	0.8
94.91	0.91	0.8	182.87	29SE	1.6	-3.3	-2.4	0.8
99.22	1.17	0.7	192.91	9OC	2.0	-3.6	-2.3	0.8
102.99	1.46	0.5	203.00	19OC	-0.1	-4.1	-2.2	0.8
106.09	1.79	0.3	213.13	29OC	-0.5	-4.3	-2.2	0.8
108.36	2.15	0.1	223.30	8NO	-0.5	-4.4	-2.1	0.8
109.61	2.56	-0.1	233.49	18NO	-0.6	-4.3	-2.0	0.8
109.67	3.00	-0.3	243.70	28NO	-0.7	-4.2	-2.0	0.8
108.40	3.44	-0.6	253.91	8DE	-0.8	-4.1	-1.9	0.8
105.82	3.84	-0.8	264.11	18DE	-0.9	-4.0	-1.8	0.8
102.24	4.14	-1.0	274.29	28DE	-0.8	-3.9	-1.8	0.8

♂			⊙	16.00UT	☿	♀	♃	♄
LONG	LAT	MAG	LONG	-136	MAGNITUDES			
98.26	4.28	-1.0	284.45	7JA	0.2	-3.8	-1.7	0.8
94.64	4.26	-0.8	294.56	17JA	2.7	-3.7	-1.7	0.8
91.96	4.10	-0.6	304.63	27JA	1.7	-3.7	-1.6	0.7
90.55	3.85	-0.3	314.65	6FE	0.8	-3.6	-1.6	0.7
90.43	3.56	-0.1	324.61	16FE	0.4	-3.5	-1.6	0.6
91.46	3.27	0.2	334.51	26FE	0.2	-3.5	-1.6	0.6
93.48	2.98	0.4	344.36	7MR	-0.2	-3.4	-1.6	0.5
96.30	2.72	0.6	354.14	17MR	-0.8	-3.4	-1.6	0.5
99.76	2.47	0.8	3.87	27MR	-1.7	-3.3	-1.6	0.4
103.74	2.25	0.9	13.55	6AP	-1.3	-3.3	-1.6	0.3
108.13	2.04	1.0	23.18	16AP	-0.3	-3.3	-1.6	0.3
112.88	1.84	1.2	32.77	26AP	0.7	-3.3	-1.6	0.4
117.92	1.66	1.3	42.34	6MY	1.6	-3.3	-1.7	0.4
123.20	1.49	1.3	51.87	16MY	2.7	-3.3	-1.7	0.5
128.70	1.33	1.4	61.40	26MY	3.2	-3.3	-1.8	0.5
134.39	1.17	1.5	70.91	5JN	1.7	-3.4	-1.8	0.6
140.26	1.02	1.5	80.43	15JN	0.5	-3.4	-1.9	0.7
146.28	0.88	1.6	89.97	25JN	-0.7	-3.4	-1.9	0.7
152.46	0.74	1.6	99.53	5JL	-1.5	-3.5	-2.0	0.8
158.77	0.60	1.6	109.12	15JL	-1.3	-3.5	-2.1	0.8
165.22	0.47	1.6	118.75	25JL	-0.6	-3.5	-2.2	0.8
171.80	0.34	1.6	128.42	4AU	-0.1	-3.5	-2.2	0.9
178.51	0.21	1.6	138.14	14AU	0.1	-3.4	-2.3	0.9
185.34	0.09	1.6	147.93	24AU	0.3	-3.4	-2.3	0.9
192.29	-0.03	1.6	157.77	3SE	0.7	-3.4	-2.4	0.9
199.36	-0.15	1.6	167.66	13SE	2.0	-3.3	-2.4	0.8
206.55	-0.27	1.6	177.62	23SE	1.8	-3.3	-2.4	0.8
213.84	-0.37	1.6	187.64	3OC	-0.2	-3.3	-2.4	0.8
221.23	-0.48	1.5	197.70	13OC	-0.7	-3.3	-2.4	0.7
228.72	-0.58	1.5	207.82	23OC	-0.7	-3.3	-2.4	0.7
236.30	-0.67	1.5	217.97	2NO	-0.7	-3.4	-2.3	0.7
243.96	-0.75	1.5	228.15	12NO	-0.8	-3.4	-2.3	0.8
251.69	-0.83	1.4	238.35	22NO	-0.7	-3.4	-2.2	0.8
259.48	-0.90	1.4	248.56	2DE	-0.7	-3.4	-2.2	0.8
267.31	-0.96	1.4	258.76	12DE	-0.6	-3.5	-2.1	0.8
275.18	-1.01	1.3	268.96	22DE	0.3	-3.5	-2.0	0.8
				-135				
283.08	-1.04	1.3	279.13	1JA	2.9	-3.6	-1.9	0.8
290.97	-1.07	1.3	289.26	11JA	1.3	-3.6	-1.9	0.8
298.87	-1.09	1.3	299.36	21JA	0.5	-3.7	-1.8	0.8
306.74	-1.09	1.3	309.40	31JA	0.3	-3.8	-1.7	0.7
314.58	-1.08	1.3	319.39	10FE	0.1	-3.8	-1.7	0.7
322.39	-1.06	1.4	329.33	20FE	-0.3	-3.9	-1.6	0.6
330.13	-1.03	1.4	339.20	2MR	-0.8	-4.0	-1.6	0.6
337.82	-0.99	1.4	349.01	12MR	-1.6	-4.1	-1.6	0.5
345.43	-0.93	1.5	358.78	22MR	-1.3	-4.2	-1.6	0.5
352.96	-0.87	1.5	8.48	1AP	-0.2	-4.2	-1.5	0.4
0.42	-0.80	1.5	18.13	11AP	0.9	-4.2	-1.5	0.4
7.78	-0.72	1.6	27.75	21AP	2.0	-3.9	-1.5	0.3
15.05	-0.64	1.6	37.32	1MY	3.6	-3.3	-1.5	0.3
22.23	-0.55	1.6	46.87	11MY	2.5	-2.9	-1.5	0.3
29.31	-0.45	1.6	56.40	21MY	1.3	-3.7	-1.6	0.4
36.30	-0.35	1.6	65.92	31MY	0.4	-4.1	-1.6	0.5
43.19	-0.24	1.7	75.44	10JN	-0.7	-4.2	-1.6	0.5
49.98	-0.13	1.7	84.97	20JN	-1.6	-4.2	-1.7	0.6
56.67	-0.01	1.7	94.51	30JN	-1.3	-4.1	-1.7	0.6
63.26	0.11	1.7	104.09	10JL	-0.5	-4.0	-1.7	0.7
69.74	0.23	1.7	113.70	20JL	-0.0	-3.9	-1.8	0.7
76.12	0.36	1.7	123.34	30JL	0.3	-3.9	-1.9	0.8
82.38	0.49	1.6	133.04	9AU	0.5	-3.8	-1.9	0.8
88.52	0.63	1.6	142.79	19AU	1.0	-3.7	-2.0	0.8
94.53	0.78	1.6	152.60	29AU	2.4	-3.6	-2.1	0.8
100.40	0.93	1.5	162.47	8SE	1.6	-3.6	-2.1	0.8
106.10	1.09	1.4	172.40	18SE	-0.3	-3.5	-2.2	0.8
111.62	1.26	1.4	182.38	28SE	-0.8	-3.5	-2.2	0.8
116.92	1.44	1.3	192.42	8OC	-0.8	-3.5	-2.3	0.8
121.95	1.64	1.2	202.51	18OC	-0.8	-3.4	-2.3	0.7
126.68	1.85	1.0	212.64	28OC	-0.7	-3.4	-2.3	0.7
131.03	2.09	0.9	222.81	7NO	-0.6	-3.4	-2.3	0.7
134.91	2.35	0.7	233.00	17NO	-0.6	-3.4	-2.3	0.7
138.22	2.64	0.5	243.21	27NO	-0.5	-3.4	-2.3	0.7
140.82	2.95	0.3	253.42	7DE	0.4	-3.4	-2.3	0.8
142.53	3.29	0.1	263.62	17DE	3.0	-3.3	-2.2	0.8
143.18	3.65	-0.2	273.80	27DE	0.9	-3.3	-2.1	0.8

♂ LONG	♂ LAT	♂ MAG	⊙ LONG	16.00UT	☿	♀	♃	♄
				-134	MAGNITUDES			
142.61	4.01	-0.4	283.96	6JA	0.3	-3.4	-2.1	0.8
140.72	4.32	-0.7	294.07	16JA	0.1	-3.4	-2.0	0.8
137.68	4.53	-0.9	304.14	26JA	-0.0	-3.4	-1.9	0.8
133.89	4.58	-1.1	314.16	5FE	-0.3	-3.4	-1.8	0.7
130.04	4.45	-1.0	324.13	15FE	-0.9	-3.4	-1.8	0.7
126.84	4.17	-0.8	334.03	25FE	-1.6	-3.5	-1.7	0.7
124.74	3.79	-0.6	343.88	7MR	-1.2	-3.5	-1.7	0.6
123.95	3.37	-0.3	353.66	17MR	-0.2	-3.4	-1.6	0.6
124.42	2.95	-0.1	3.39	27MR	1.1	-3.4	-1.6	0.5
125.98	2.56	0.1	13.07	6AP	2.6	-3.4	-1.5	0.4
128.47	2.20	0.3	22.71	16AP	3.3	-3.4	-1.5	0.4
131.71	1.87	0.5	32.30	26AP	1.9	-3.3	-1.5	0.3
135.58	1.57	0.6	41.86	6MY	1.0	-3.3	-1.4	0.3
139.95	1.30	0.7	51.40	16MY	0.3	-3.3	-1.4	0.3
144.75	1.05	0.8	60.92	26MY	-0.7	-3.3	-1.4	0.3
149.91	0.83	0.9	70.45	5JN	-1.6	-3.3	-1.4	0.4
155.37	0.62	1.0	79.97	15JN	-1.3	-3.3	-1.4	0.5
161.11	0.42	1.1	89.50	25JN	-0.5	-3.4	-1.5	0.5
167.08	0.24	1.1	99.06	5JL	0.1	-3.4	-1.5	0.6
173.28	0.07	1.2	108.65	15JL	0.4	-3.4	-1.5	0.6
179.67	-0.09	1.2	118.28	25JL	0.7	-3.5	-1.5	0.7
186.24	-0.24	1.3	127.95	4AU	1.3	-3.5	-1.6	0.7
192.99	-0.38	1.3	137.67	14AU	2.9	-3.6	-1.6	0.8
199.89	-0.51	1.3	147.45	24AU	1.4	-3.7	-1.7	0.8
206.94	-0.63	1.3	157.29	3SE	-0.4	-3.7	-1.8	0.8
214.13	-0.73	1.3	167.18	13SE	-1.0	-3.8	-1.8	0.8
221.45	-0.83	1.3	177.14	23SE	-1.0	-3.9	-1.9	0.8
228.88	-0.92	1.4	187.15	3OC	-0.8	-4.1	-1.9	0.8
236.42	-0.99	1.4	197.21	13OC	-0.5	-4.2	-2.0	0.8
244.05	-1.05	1.4	207.32	23OC	-0.4	-4.3	-2.1	0.7
251.76	-1.10	1.4	217.47	2NO	-0.4	-4.4	-2.1	0.7
259.54	-1.14	1.4	227.65	12NO	-0.3	-4.4	-2.2	0.7
267.36	-1.16	1.4	237.85	22NO	0.6	-4.3	-2.2	0.7
275.23	-1.17	1.4	248.06	2DE	2.9	-3.9	-2.2	0.7
283.11	-1.17	1.4	258.27	12DE	0.6	-3.1	-2.2	0.8
290.99	-1.15	1.4	268.46	22DE	0.1	-3.6	-2.2	0.8
				-133				
298.87	-1.12	1.4	278.63	1JA	-0.0	-4.1	-2.2	0.8
306.73	-1.09	1.4	288.77	11JA	-0.1	-4.3	-2.1	0.8
314.54	-1.04	1.4	298.87	21JA	-0.4	-4.3	-2.1	0.8
322.32	-0.98	1.4	308.92	31JA	-0.9	-4.3	-2.0	0.8
330.03	-0.91	1.4	318.91	10FE	-1.4	-4.2	-1.9	0.8
337.68	-0.84	1.5	328.84	20FE	-1.2	-4.1	-1.9	0.7
345.25	-0.76	1.5	338.72	2MR	-0.2	-3.9	-1.8	0.7
352.75	-0.67	1.5	348.54	12MR	1.4	-3.8	-1.7	0.7
0.16	-0.58	1.5	358.30	22MR	3.3	-3.7	-1.7	0.6
7.49	-0.48	1.5	8.01	1AP	2.4	-3.7	-1.6	0.5
14.73	-0.39	1.6	17.66	11AP	1.4	-3.6	-1.5	0.5
21.88	-0.29	1.6	27.28	21AP	0.8	-3.5	-1.5	0.4
28.95	-0.18	1.7	36.86	1MY	0.1	-3.5	-1.4	0.4
35.93	-0.08	1.7	46.41	11MY	-0.7	-3.4	-1.4	0.3
42.84	0.02	1.7	55.94	21MY	-1.6	-3.4	-1.4	0.2
49.66	0.12	1.8	65.46	31MY	-1.3	-3.4	-1.4	0.3
56.40	0.23	1.8	74.98	10JN	-0.4	-3.3	-1.3	0.3
63.06	0.33	1.8	84.51	20JN	0.2	-3.3	-1.3	0.4
69.66	0.43	1.9	94.05	30JN	0.6	-3.3	-1.3	0.5
76.18	0.54	1.9	103.62	10JL	1.0	-3.3	-1.3	0.5
82.63	0.64	1.9	113.23	20JL	1.7	-3.3	-1.3	0.6
89.01	0.74	1.9	122.88	30JL	3.2	-3.4	-1.4	0.6
95.33	0.85	1.9	132.57	9AU	1.2	-3.4	-1.4	0.7
101.57	0.95	1.9	142.32	19AU	-0.5	-3.4	-1.4	0.7
107.74	1.06	1.8	152.13	29AU	-1.1	-3.4	-1.5	0.8
113.83	1.16	1.8	161.99	8SE	-1.1	-3.5	-1.5	0.8
119.84	1.27	1.8	171.91	18SE	-0.8	-3.5	-1.5	0.8
125.75	1.38	1.7	181.90	28SE	-0.4	-3.5	-1.6	0.8
131.56	1.50	1.7	191.93	8OC	-0.3	-3.5	-1.7	0.8
137.24	1.62	1.6	202.02	18OC	-0.3	-3.4	-1.7	0.8
142.79	1.74	1.5	212.15	28OC	-0.1	-3.4	-1.8	0.8
148.16	1.87	1.4	222.31	7NO	0.8	-3.4	-1.9	0.7
153.34	2.01	1.3	232.50	17NO	2.7	-3.4	-1.9	0.7
158.27	2.16	1.1	242.71	27NO	0.4	-3.3	-2.0	0.7
162.90	2.31	1.0	252.91	7DE	-0.1	-3.3	-2.0	0.7
167.17	2.48	0.8	263.12	17DE	-0.2	-3.3	-2.1	0.7
170.98	2.66	0.6	273.30	27DE	-0.3	-3.3	-2.1	0.8

♂ LONG	♂ LAT	♂ MAG	⊙ LONG	16.00UT	☿	♀	♃	♄
				-132	MAGNITUDES			
174.23	2.84	0.4	283.46	6JA	-0.5	-3.3	-2.1	0.8
176.77	3.04	0.1	293.58	16JA	-0.9	-3.4	-2.1	0.8
178.44	3.23	-0.1	303.65	26JA	-1.3	-3.4	-2.1	0.8
179.06	3.40	-0.4	313.67	5FE	-1.1	-3.4	-2.0	0.8
178.46	3.54	-0.7	323.64	15FE	-0.1	-3.4	-2.0	0.8
176.58	3.60	-1.0	333.55	25FE	1.7	-3.5	-1.9	0.8
173.59	3.53	-1.2	343.40	6MR	3.3	-3.5	-1.9	0.7
169.92	3.32	-1.4	353.19	16MR	1.8	-3.6	-1.8	0.7
166.27	2.96	-1.3	2.93	26MR	1.0	-3.6	-1.7	0.7
163.33	2.50	-1.1	12.61	5AP	0.6	-3.7	-1.7	0.6
161.55	2.01	-0.9	22.25	15AP	0.0	-3.7	-1.6	0.5
161.12	1.53	-0.7	31.84	25AP	-0.7	-3.8	-1.6	0.5
161.97	1.09	-0.5	41.40	5MY	-1.7	-3.9	-1.5	0.4
163.94	0.69	-0.3	50.95	15MY	-1.3	-4.0	-1.4	0.4
166.87	0.34	-0.1	60.47	25MY	-0.4	-4.1	-1.4	0.3
170.58	0.03	0.0	69.99	4JN	0.3	-4.2	-1.4	0.3
174.93	-0.24	0.2	79.51	14JN	0.8	-4.2	-1.3	0.3
179.83	-0.47	0.3	89.04	24JN	1.3	-4.1	-1.3	0.4
185.17	-0.68	0.4	98.60	4JL	2.3	-3.7	-1.3	0.4
190.90	-0.86	0.5	108.19	14JL	3.0	-3.1	-1.3	0.5
196.96	-1.01	0.6	117.81	24JL	1.0	-3.4	-1.2	0.5
203.30	-1.14	0.7	127.48	3AU	-0.6	-4.0	-1.2	0.6
209.90	-1.25	0.7	137.20	13AU	-1.2	-4.2	-1.3	0.6
216.71	-1.34	0.8	146.98	23AU	-1.2	-4.3	-1.3	0.7
223.72	-1.42	0.9	156.81	2SE	-0.7	-4.2	-1.3	0.7
230.89	-1.47	0.9	166.71	12SE	-0.3	-4.2	-1.3	0.8
238.21	-1.50	1.0	176.65	22SE	-0.2	-4.1	-1.3	0.8
245.65	-1.52	1.0	186.66	2OC	-0.1	-4.0	-1.4	0.8
253.20	-1.52	1.0	196.73	12OC	0.1	-3.9	-1.4	0.8
260.83	-1.50	1.1	206.83	22OC	1.1	-3.8	-1.5	0.8
268.52	-1.47	1.1	216.98	1NO	2.5	-3.7	-1.5	0.8
276.26	-1.42	1.2	227.16	11NO	0.2	-3.7	-1.6	0.8
284.02	-1.36	1.2	237.36	21NO	-0.3	-3.6	-1.7	0.7
291.79	-1.29	1.3	247.57	1DE	-0.3	-3.6	-1.7	0.7
299.56	-1.21	1.3	257.78	11DE	-0.4	-3.5	-1.8	0.7
307.30	-1.12	1.4	267.97	21DE	-0.6	-3.5	-1.9	0.7
315.00	-1.02	1.4	278.14	31DE	-1.0	-3.4	-1.9	0.8
				-131				
322.66	-0.92	1.4	288.28	10JA	-1.1	-3.4	-2.0	0.8
330.26	-0.82	1.5	298.38	20JA	-1.0	-3.4	-2.0	0.8
337.79	-0.71	1.5	308.43	30JA	-0.0	-3.3	-2.0	0.8
345.26	-0.60	1.6	318.43	9FE	2.0	-3.3	-2.0	0.8
352.65	-0.49	1.6	328.37	19FE	2.6	-3.3	-2.0	0.8
359.96	-0.37	1.6	338.25	1MR	1.3	-3.3	-2.0	0.8
7.19	-0.26	1.7	348.07	11MR	0.8	-3.3	-2.0	0.8
14.33	-0.16	1.7	357.83	21MR	0.4	-3.3	-1.9	0.8
21.40	-0.05	1.7	7.54	31MR	-0.0	-3.4	-1.9	0.7
28.39	0.05	1.7	17.20	10AP	-0.7	-3.4	-1.8	0.7
35.31	0.15	1.8	26.81	20AP	-1.7	-3.4	-1.7	0.6
42.15	0.25	1.8	36.39	30AP	-1.3	-3.5	-1.7	0.6
48.93	0.35	1.8	45.95	10MY	-0.4	-3.5	-1.6	0.5
55.63	0.44	1.8	55.47	20MY	0.4	-3.5	-1.5	0.4
62.28	0.53	1.8	65.00	30MY	1.0	-3.4	-1.5	0.4
68.88	0.61	1.9	74.52	9JN	1.7	-3.4	-1.4	0.3
75.42	0.69	1.9	84.04	19JN	2.9	-3.4	-1.4	0.3
81.91	0.77	1.9	93.58	29JN	2.5	-3.3	-1.3	0.3
88.36	0.85	2.0	103.15	9JL	0.8	-3.3	-1.3	0.4
94.77	0.92	2.0	112.76	19JL	-0.6	-3.3	-1.3	0.5
101.15	0.99	2.0	122.40	29JL	-1.3	-3.3	-1.2	0.5
107.49	1.06	2.0	132.10	8AU	-1.3	-3.3	-1.2	0.6
113.79	1.12	2.0	141.84	18AU	-0.7	-3.3	-1.2	0.6
120.07	1.19	2.0	151.65	28AU	-0.3	-3.4	-1.2	0.7
126.31	1.25	2.0	161.51	7SE	-0.0	-3.4	-1.2	0.7
132.52	1.31	2.0	171.43	17SE	0.1	-3.4	-1.2	0.7
138.69	1.37	1.9	181.41	27SE	0.3	-3.5	-1.2	0.8
144.83	1.42	1.9	191.44	7OC	1.3	-3.5	-1.3	0.8
150.92	1.48	1.8	201.53	17OC	2.2	-3.6	-1.3	0.8
156.96	1.53	1.8	211.66	27OC	0.0	-3.6	-1.3	0.8
162.94	1.58	1.7	221.82	6NO	-0.4	-3.7	-1.4	0.8
168.85	1.63	1.6	232.01	16NO	-0.4	-3.8	-1.4	0.8
174.68	1.67	1.5	242.22	26NO	-0.5	-3.9	-1.5	0.8
180.41	1.71	1.4	252.42	6DE	-0.7	-4.0	-1.6	0.8
186.02	1.75	1.3	262.62	16DE	-0.9	-4.1	-1.6	0.7
191.48	1.78	1.1	272.81	26DE	-1.0	-4.2	-1.7	0.8

☿ = Mercury, ♀ = Venus, ♃ = Jupiter, ♄ = Saturn — Left table (−130, −129)

♂ LONG	♂ LAT	♂ MAG	☉ LONG	16.00UT	☿	♀	♃	♄
				−130				
196.76	1.80	1.0	282.97	5JA	-0.9	-4.3	-1.7	0.8
201.82	1.82	0.8	293.09	15JA	0.1	-4.3	-1.8	0.8
206.61	1.82	0.6	303.17	25JA	2.4	-4.3	-1.9	0.9
211.06	1.80	0.4	313.19	4FE	2.0	-4.2	-1.9	0.9
215.08	1.76	0.1	323.16	14FE	0.9	-3.8	-2.0	0.9
218.57	1.69	-0.1	333.08	24FE	0.5	-3.2	-2.0	0.9
221.37	1.58	-0.4	342.93	6MR	0.3	-3.6	-2.0	0.9
223.35	1.41	-0.7	352.72	16MR	-0.1	-4.0	-2.0	0.9
224.29	1.17	-1.0	2.46	26MR	-0.8	-4.2	-2.0	0.8
224.05	0.84	-1.3	12.14	5AP	-1.7	-4.2	-2.0	0.8
222.57	0.41	-1.6	21.78	15AP	-1.3	-4.2	-1.9	0.8
219.98	-0.11	-1.9	31.38	25AP	-0.3	-4.1	-1.9	0.7
216.76	-0.67	-2.1	40.94	5MY	0.5	-4.0	-1.8	0.6
213.58	-1.23	-2.0	50.48	15MY	1.3	-3.9	-1.8	0.6
211.15	-1.73	-1.8	60.01	25MY	2.3	-3.8	-1.7	0.5
209.96	-2.12	-1.6	69.52	4JN	3.4	-3.7	-1.6	0.5
210.14	-2.41	-1.4	79.05	14JN	2.0	-3.6	-1.6	0.4
211.65	-2.61	-1.1	88.58	24JN	0.6	-3.6	-1.5	0.3
214.33	-2.74	-0.9	98.14	4JL	-0.6	-3.5	-1.5	0.4
217.96	-2.81	-0.7	107.72	14JL	-1.4	-3.5	-1.4	0.4
222.39	-2.84	-0.6	117.35	24JL	-1.3	-3.4	-1.4	0.5
227.46	-2.82	-0.4	127.01	3AU	-0.6	-3.4	-1.3	0.5
233.05	-2.77	-0.2	136.73	13AU	-0.2	-3.4	-1.3	0.6
239.07	-2.70	-0.1	146.50	23AU	0.1	-3.4	-1.3	0.6
245.43	-2.60	0.0	156.33	2SE	0.2	-3.4	-1.2	0.7
252.06	-2.49	0.2	166.22	12SE	0.5	-3.4	-1.2	0.7
258.91	-2.35	0.3	176.17	22SE	1.7	-3.4	-1.2	0.8
265.93	-2.21	0.4	186.18	2OC	2.0	-3.4	-1.2	0.8
273.08	-2.05	0.5	196.24	12OC	-0.1	-3.4	-1.2	0.8
280.34	-1.89	0.6	206.34	22OC	-0.6	-3.4	-1.2	0.8
287.66	-1.72	0.7	216.49	1NO	-0.6	-3.4	-1.3	0.8
295.02	-1.55	0.8	226.67	11NO	-0.6	-3.4	-1.3	0.9
302.40	-1.38	0.9	236.87	21NO	-0.8	-3.5	-1.3	0.9
309.78	-1.21	1.0	247.07	1DE	-0.8	-3.5	-1.4	0.8
317.15	-1.04	1.1	257.28	11DE	-0.8	-3.5	-1.4	0.8
324.50	-0.87	1.2	267.48	21DE	-0.7	-3.5	-1.5	0.8
331.80	-0.72	1.3	277.65	31DE	0.2	-3.4	-1.5	0.8
				−129				
339.05	-0.56	1.4	287.79	10JA	2.7	-3.4	-1.6	0.8
346.25	-0.42	1.4	297.89	20JA	1.5	-3.4	-1.7	0.9
353.39	-0.28	1.5	307.94	30JA	0.7	-3.4	-1.7	0.9
0.47	-0.15	1.6	317.94	9FE	0.4	-3.4	-1.8	0.9
7.49	-0.02	1.6	327.89	19FE	0.1	-3.3	-1.9	1.0
14.44	0.10	1.7	337.77	1MR	-0.2	-3.3	-1.9	1.0
21.33	0.21	1.8	347.59	11MR	-0.8	-3.3	-2.0	1.0
28.16	0.31	1.8	357.36	21MR	-1.6	-3.3	-2.0	1.0
34.92	0.40	1.8	7.07	31MR	-1.3	-3.4	-2.1	0.9
41.63	0.49	1.9	16.73	10AP	-0.3	-3.4	-2.1	0.9
48.28	0.58	1.9	26.35	20AP	0.7	-3.4	-2.1	0.9
54.89	0.66	1.9	35.93	30AP	1.7	-3.4	-2.0	0.8
61.45	0.73	1.9	45.48	10MY	3.0	-3.5	-2.0	0.8
67.97	0.79	1.9	55.02	20MY	2.9	-3.5	-2.0	0.7
74.45	0.86	1.9	64.54	30MY	1.6	-3.6	-1.9	0.6
80.91	0.91	1.9	74.06	9JN	0.5	-3.6	-1.9	0.6
87.33	0.96	1.9	83.58	19JN	-0.6	-3.7	-1.8	0.5
93.74	1.01	1.9	93.12	29JN	-1.5	-3.8	-1.7	0.4
100.13	1.05	1.9	102.69	9JL	-1.3	-3.9	-1.7	0.4
106.50	1.09	2.0	112.30	19JL	-0.6	-4.0	-1.6	0.4
112.87	1.13	2.0	121.94	29JL	-0.1	-4.1	-1.6	0.5
119.24	1.16	2.0	131.63	8AU	0.2	-4.2	-1.5	0.5
125.60	1.18	2.0	141.37	18AU	0.4	-4.3	-1.5	0.6
131.96	1.20	2.0	151.17	28AU	0.8	-4.3	-1.4	0.6
138.33	1.22	2.0	161.03	7SE	2.1	-4.2	-1.4	0.7
144.71	1.24	2.0	170.95	17SE	1.8	-3.8	-1.3	0.7
151.09	1.25	2.0	180.92	27SE	-0.3	-3.3	-1.3	0.8
157.47	1.25	2.0	190.96	7OC	-0.7	-3.6	-1.3	0.8
163.87	1.25	1.9	201.04	17OC	-0.7	-4.1	-1.3	0.8
170.27	1.24	1.9	211.16	27OC	-0.7	-4.3	-1.3	0.9
176.67	1.23	1.8	221.33	6NO	-0.7	-4.4	-1.3	0.9
183.08	1.21	1.8	231.51	16NO	-0.6	-4.3	-1.3	0.9
189.49	1.18	1.7	241.72	26NO	-0.7	-4.2	-1.3	0.9
195.89	1.15	1.6	251.93	6DE	-0.6	-4.1	-1.3	0.9
202.29	1.10	1.5	262.13	16DE	0.3	-4.0	-1.4	0.9
208.67	1.04	1.4	272.32	26DE	3.0	-3.9	-1.4	0.9

Right table (−128, −127)

♂ LONG	♂ LAT	♂ MAG	☉ LONG	16.00UT	☿	♀	♃	♄
				−128				
215.03	0.97	1.3	282.48	5JA	1.1	-3.8	-1.5	0.9
221.37	0.89	1.2	292.61	15JA	0.4	-3.7	-1.5	0.9
227.67	0.78	1.1	302.68	25JA	0.2	-3.7	-1.6	0.9
233.92	0.66	0.9	312.72	4FE	0.0	-3.6	-1.6	1.0
240.10	0.51	0.8	322.69	14FE	-0.3	-3.5	-1.7	1.0
246.21	0.33	0.6	332.60	24FE	-0.8	-3.5	-1.8	1.0
252.21	0.12	0.4	342.46	5MR	-1.6	-3.4	-1.9	1.1
258.07	-0.13	0.3	352.25	15MR	-1.3	-3.4	-1.9	1.1
263.76	-0.42	0.1	1.99	25MR	-0.3	-3.3	-2.0	1.1
269.22	-0.75	-0.2	11.68	4AP	0.9	-3.3	-2.0	1.1
274.38	-1.15	-0.4	21.32	14AP	2.2	-3.3	-2.1	1.0
279.17	-1.61	-0.6	30.92	24AP	3.8	-3.3	-2.1	1.0
283.45	-2.13	-0.9	40.49	4MY	2.3	-3.3	-2.2	1.0
287.10	-2.74	-1.2	50.03	14MY	1.2	-3.3	-2.2	0.9
289.93	-3.42	-1.4	59.55	24MY	0.4	-3.3	-2.2	0.9
291.73	-4.17	-1.7	69.07	3JN	-0.7	-3.4	-2.1	0.8
292.35	-4.95	-2.0	78.59	13JN	-1.6	-3.4	-2.1	0.7
291.69	-5.70	-2.3	88.12	23JN	-1.3	-3.4	-2.1	0.7
289.87	-6.31	-2.5	97.68	3JL	-0.5	-3.5	-2.0	0.6
287.37	-6.66	-2.7	107.26	13JL	0.0	-3.5	-2.0	0.5
284.84	-6.70	-2.5	116.88	23JL	0.3	-3.5	-1.9	0.5
283.01	-6.44	-2.3	126.54	2AU	0.6	-3.4	-1.8	0.5
282.36	-5.97	-2.0	136.25	12AU	1.1	-3.4	-1.8	0.6
283.01	-5.38	-1.8	146.02	22AU	2.5	-3.4	-1.7	0.6
284.92	-4.76	-1.5	155.85	1SE	1.6	-3.4	-1.6	0.7
287.89	-4.15	-1.2	165.73	11SE	-0.4	-3.3	-1.6	0.7
291.71	-3.57	-0.9	175.68	21SE	-0.9	-3.3	-1.5	0.8
296.22	-3.04	-0.7	185.69	1OC	-0.9	-3.3	-1.5	0.8
301.25	-2.56	-0.4	195.74	11OC	-0.9	-3.3	-1.5	0.9
306.68	-2.13	-0.2	205.84	21OC	-0.6	-3.3	-1.4	0.9
312.42	-1.75	0.0	215.99	31OC	-0.5	-3.4	-1.4	0.9
318.39	-1.40	0.2	226.16	10NO	-0.5	-3.4	-1.4	1.0
324.54	-1.09	0.4	236.36	20NO	-0.4	-3.4	-1.4	1.0
330.81	-0.81	0.6	246.57	30NO	0.4	-3.4	-1.4	1.0
337.18	-0.57	0.7	256.78	10DE	3.1	-3.5	-1.4	1.0
343.61	-0.35	0.9	266.98	20DE	0.8	-3.5	-1.4	1.0
350.08	-0.15	1.0	277.16	30DE	0.2	-3.6	-1.4	1.0
				−127				
356.58	0.02	1.1	287.30	9JA	0.1	-3.6	-1.5	1.0
3.08	0.17	1.3	297.40	19JA	-0.1	-3.7	-1.5	1.0
9.59	0.31	1.4	307.46	29JA	-0.4	-3.8	-1.5	1.0
16.09	0.44	1.5	317.46	8FE	-0.9	-3.9	-1.6	1.1
22.57	0.54	1.6	327.41	18FE	-1.5	-3.9	-1.6	1.1
29.04	0.64	1.6	337.30	28FE	-1.2	-4.0	-1.7	1.2
35.49	0.73	1.7	347.12	10MR	-0.2	-4.1	-1.8	1.2
41.91	0.80	1.8	356.89	20MR	1.2	-4.2	-1.8	1.2
48.32	0.87	1.8	6.61	30MR	2.9	-4.2	-1.9	1.2
54.70	0.93	1.9	16.27	9AP	3.0	-4.2	-2.0	1.2
61.07	0.98	1.9	25.89	19AP	1.7	-3.9	-2.0	1.2
67.42	1.02	2.0	35.47	29AP	0.9	-3.3	-2.1	1.2
73.76	1.06	2.0	45.02	9MY	0.2	-2.9	-2.2	1.1
80.08	1.09	2.0	54.56	19MY	-0.7	-3.7	-2.2	1.1
86.41	1.11	2.0	64.08	29MY	-1.6	-4.1	-2.3	1.0
92.73	1.13	2.0	73.60	8JN	-1.3	-4.2	-2.3	1.0
99.05	1.15	2.0	83.12	18JN	-0.5	-4.2	-2.3	0.9
105.38	1.16	2.0	92.66	28JN	0.1	-4.1	-2.3	0.8
111.72	1.16	2.0	102.23	8JL	0.5	-4.0	-2.3	0.8
118.08	1.16	2.0	111.83	18JL	0.8	-3.9	-2.2	0.7
124.46	1.16	2.0	121.47	28JL	1.4	-3.8	-2.2	0.6
130.86	1.15	1.9	131.16	7AU	3.0	-3.8	-2.1	0.6
137.28	1.14	1.9	140.90	17AU	1.4	-3.7	-2.1	0.7
143.74	1.12	2.0	150.69	27AU	-0.4	-3.6	-2.0	0.7
150.23	1.10	2.0	160.55	6SE	-1.0	-3.6	-1.9	0.7
156.76	1.07	2.0	170.46	16SE	-1.0	-3.5	-1.9	0.8
163.33	1.03	2.0	180.44	26SE	-0.8	-3.5	-1.8	0.8
169.94	0.99	1.9	190.46	6OC	-0.5	-3.5	-1.8	0.9
176.59	0.95	1.9	200.55	16OC	-0.4	-3.4	-1.7	0.9
183.29	0.90	1.9	210.67	26OC	-0.3	-3.4	-1.6	1.0
190.03	0.84	1.9	220.83	5NO	-0.2	-3.4	-1.6	1.0
196.82	0.77	1.8	231.02	15NO	0.6	-3.4	-1.6	1.1
203.65	0.69	1.8	241.22	25NO	2.9	-3.4	-1.5	1.1
210.52	0.61	1.7	251.43	5DE	0.6	-3.4	-1.5	1.1
217.44	0.52	1.6	261.64	15DE	0.0	-3.3	-1.5	1.1
224.40	0.41	1.6	271.82	25DE	-0.1	-3.4	-1.5	1.2

♂ LONG	♂ LAT	♂ MAG	☉ LONG	16.00UT	☿	♀	♃	♄
				-126				
231.41	0.30	1.5	281.98	4JA	-0.2	-3.4	-1.5	1.2
238.45	0.17	1.4	292.11	14JA	-0.4	-3.4	-1.5	1.2
245.53	0.03	1.3	302.19	24JA	-0.9	-3.4	-1.5	1.2
252.64	-0.12	1.3	312.22	3FE	-1.4	-3.4	-1.5	1.2
259.78	-0.29	1.2	322.20	13FE	-1.2	-3.4	-1.6	1.2
266.94	-0.47	1.1	332.12	23FE	-0.2	-3.5	-1.6	1.3
274.12	-0.66	1.0	341.98	5MR	1.5	-3.5	-1.6	1.3
281.30	-0.86	0.9	351.78	15MR	3.5	-3.4	-1.7	1.3
288.46	-1.08	0.8	1.52	25MR	2.2	-3.4	-1.7	1.4
295.60	-1.30	0.6	11.21	4AP	1.3	-3.4	-1.8	1.4
302.70	-1.54	0.5	20.85	14AP	0.7	-3.4	-1.9	1.4
309.72	-1.78	0.4	30.45	24AP	0.1	-3.3	-1.9	1.4
316.66	-2.03	0.3	40.02	4MY	-0.7	-3.3	-2.0	1.4
323.45	-2.28	0.2	49.56	14MY	-1.6	-3.3	-2.1	1.3
330.07	-2.53	0.0	59.09	24MY	-1.4	-3.3	-2.2	1.3
336.46	-2.78	-0.1	68.61	3JN	-0.5	-3.3	-2.2	1.3
342.55	-3.03	-0.3	78.13	13JN	0.2	-3.3	-2.3	1.2
348.28	-3.28	-0.4	87.66	23JN	0.6	-3.4	-2.3	1.1
353.54	-3.51	-0.6	97.21	3JL	1.1	-3.4	-2.4	1.1
358.21	-3.74	-0.8	106.79	13JL	1.9	-3.4	-2.4	1.0
2.12	-3.94	-1.0	116.41	23JL	3.2	-3.5	-2.4	0.9
5.11	-4.11	-1.2	126.07	2AU	1.2	-3.5	-2.4	0.9
6.93	-4.23	-1.4	135.79	12AU	-0.5	-3.6	-2.4	0.8
7.42	-4.28	-1.6	145.55	22AU	-1.1	-3.7	-2.3	0.8
6.43	-4.19	-1.8	155.38	1SE	-1.2	-3.7	-2.3	0.8
4.09	-3.95	-2.0	165.26	11SE	-0.8	-3.8	-2.2	0.9
0.85	-3.52	-2.2	175.20	21SE	-0.4	-3.9	-2.2	0.9
357.46	-2.94	-2.0	185.20	1OC	-0.2	-4.1	-2.1	1.0
354.72	-2.29	-1.8	195.26	11OC	-0.2	-4.2	-2.0	1.0
353.16	-1.66	-1.4	205.35	21OC	-0.0	-4.3	-2.0	1.1
352.94	-1.08	-1.1	215.50	31OC	0.9	-4.4	-1.9	1.1
354.00	-0.60	-0.8	225.67	10NO	2.7	-4.4	-1.8	1.2
356.15	-0.20	-0.5	235.86	20NO	0.3	-4.3	-1.8	1.2
359.16	0.13	-0.2	246.07	30NO	-0.2	-3.9	-1.7	1.2
2.86	0.39	0.1	256.28	10DE	-0.2	-3.1	-1.7	1.3
7.09	0.60	0.3	266.48	20DE	-0.3	-3.6	-1.6	1.3
11.73	0.77	0.5	276.66	30DE	-0.5	-4.1	-1.6	1.3
				-125				
16.70	0.90	0.7	286.80	9JA	-0.9	-4.3	-1.6	1.3
21.91	1.01	0.9	296.91	19JA	-1.2	-4.3	-1.6	1.4
27.31	1.10	1.1	306.97	29JA	-1.1	-4.3	-1.6	1.4
32.86	1.17	1.2	316.97	8FE	-0.1	-4.2	-1.6	1.3
38.53	1.22	1.3	326.92	18FE	1.8	-4.1	-1.6	1.3
44.29	1.26	1.5	336.81	28FE	3.1	-3.9	-1.6	1.3
50.13	1.29	1.6	346.65	10MR	1.6	-3.8	-1.6	1.3
56.02	1.32	1.6	356.42	20MR	0.9	-3.7	-1.6	1.3
61.97	1.33	1.7	6.14	30MR	0.5	-3.7	-1.7	1.3
67.96	1.34	1.8	15.80	9AP	0.0	-3.6	-1.7	1.3
73.98	1.34	1.9	25.43	19AP	-0.7	-3.5	-1.7	1.2
80.04	1.34	1.9	35.01	29AP	-1.6	-3.5	-1.8	1.2
86.13	1.33	1.9	44.57	9MY	-1.4	-3.4	-1.9	1.2
92.25	1.31	2.0	54.10	19MY	-0.4	-3.4	-1.9	1.1
98.40	1.29	2.0	63.62	29MY	0.3	-3.4	-2.0	1.1
104.59	1.27	2.0	73.14	8JN	0.8	-3.3	-2.1	1.1
110.82	1.24	2.0	82.67	18JN	1.4	-3.3	-2.1	1.0
117.08	1.21	2.0	92.21	28JN	2.5	-3.3	-2.2	1.0
123.39	1.18	2.0	101.77	8JL	2.9	-3.3	-2.3	0.9
129.74	1.14	2.0	111.37	18JL	1.0	-3.3	-2.3	0.9
136.14	1.09	2.0	121.01	28JL	-0.5	-3.4	-2.4	0.8
142.59	1.05	2.0	130.69	7AU	-1.3	-3.4	-2.4	0.8
149.09	0.99	1.9	140.43	17AU	-1.3	-3.4	-2.4	0.7
155.66	0.94	1.9	150.22	27AU	-0.7	-3.4	-2.5	0.7
162.28	0.88	1.8	160.07	6SE	-0.3	-3.5	-2.5	0.7
168.97	0.81	1.8	169.98	16SE	-0.1	-3.5	-2.4	0.8
175.73	0.74	1.8	179.96	26SE	-0.0	-3.5	-2.4	0.9
182.55	0.67	1.8	189.98	6OC	0.1	-3.5	-2.4	0.9
189.44	0.59	1.8	200.05	16OC	1.1	-3.4	-2.3	1.0
196.40	0.50	1.8	210.18	26OC	2.5	-3.4	-2.2	1.1
203.43	0.41	1.8	220.33	5NO	0.1	-3.4	-2.2	1.1
210.53	0.31	1.7	230.52	15NO	-0.3	-3.4	-2.1	1.2
217.70	0.21	1.7	240.72	25NO	-0.4	-3.3	-2.0	1.2
224.94	0.10	1.7	250.93	5DE	-0.4	-3.3	-1.9	1.2
232.24	-0.01	1.6	261.13	15DE	-0.6	-3.3	-1.9	1.2
239.61	-0.13	1.6	271.32	25DE	-0.9	-3.3	-1.8	1.2

♂ LONG	♂ LAT	♂ MAG	☉ LONG	16.00UT	☿	♀	♃	♄
				-124				
247.03	-0.25	1.6	281.48	4JA	-1.1	-3.3	-1.8	1.2
254.50	-0.38	1.5	291.61	14JA	-1.0	-3.4	-1.7	1.2
262.03	-0.51	1.5	301.70	24JA	-0.0	-3.4	-1.7	1.2
269.59	-0.64	1.4	311.73	3FE	2.1	-3.4	-1.6	1.1
277.19	-0.78	1.4	321.71	13FE	2.4	-3.4	-1.6	1.1
284.81	-0.91	1.3	331.63	23FE	1.2	-3.5	-1.6	1.0
292.44	-1.04	1.3	341.49	4MR	0.7	-3.5	-1.6	1.0
300.07	-1.17	1.2	351.30	14MR	0.4	-3.6	-1.6	1.0
307.69	-1.29	1.1	1.05	24MR	-0.1	-3.6	-1.6	1.0
315.29	-1.40	1.1	10.74	3AP	-0.7	-3.7	-1.6	1.0
322.84	-1.50	1.0	20.39	13AP	-1.6	-3.8	-1.6	1.0
330.34	-1.60	1.0	29.99	23AP	-1.4	-3.8	-1.6	1.0
337.76	-1.68	0.9	39.56	3MY	-0.4	-3.9	-1.6	1.0
345.09	-1.74	0.9	49.11	13MY	0.4	-4.0	-1.7	0.9
352.30	-1.79	0.8	58.63	23MY	1.1	-4.1	-1.7	0.9
359.38	-1.83	0.8	68.15	2JN	1.9	-4.2	-1.8	0.9
6.31	-1.84	0.7	77.67	12JN	3.2	-4.2	-1.8	0.8
13.05	-1.84	0.6	87.20	22JN	2.4	-4.1	-1.9	0.8
19.58	-1.82	0.6	96.75	2JL	0.8	-3.7	-2.0	0.7
25.86	-1.78	0.5	106.33	12JL	-0.6	-3.1	-2.0	0.7
31.84	-1.72	0.4	115.95	22JL	-1.4	-3.4	-2.1	0.6
37.47	-1.64	0.3	125.61	1AU	-1.3	-4.0	-2.2	0.6
42.69	-1.54	0.1	135.32	11AU	-0.6	-4.2	-2.2	0.5
47.40	-1.40	0.0	145.08	21AU	-0.2	-4.3	-2.3	0.5
51.49	-1.24	-0.1	154.90	31AU	-0.0	-4.2	-2.3	0.4
54.81	-1.03	-0.3	164.78	10SE	0.1	-4.2	-2.4	0.4
57.18	-0.78	-0.5	174.72	20SE	0.4	-4.1	-2.4	0.5
58.42	-0.47	-0.7	184.71	30SE	1.4	-4.0	-2.4	0.5
58.31	-0.11	-0.9	194.77	10OC	2.3	-3.9	-2.4	0.6
56.77	0.31	-1.2	204.87	20OC	-0.0	-3.8	-2.4	0.7
53.90	0.76	-1.3	215.00	30OC	-0.5	-3.7	-2.4	0.7
50.16	1.21	-1.5	225.18	9NO	-0.5	-3.7	-2.3	0.8
46.31	1.60	-1.4	235.37	19NO	-0.5	-3.6	-2.3	0.8
43.11	1.89	-1.1	245.58	29NO	-0.7	-3.6	-2.2	0.9
41.07	2.07	-0.8	255.79	9DE	-0.9	-3.5	-2.1	0.9
40.39	2.18	-0.5	265.99	19DE	-0.9	-3.5	-2.1	0.9
40.97	2.22	-0.2	276.16	29DE	-0.8	-3.4	-2.0	0.9
				-123				
42.65	2.22	0.1	286.31	8JA	0.0	-3.4	-1.9	0.9
45.22	2.20	0.3	296.42	18JA	2.4	-3.4	-1.8	0.9
48.49	2.15	0.5	306.48	28JA	1.9	-3.3	-1.8	0.9
52.32	2.10	0.7	316.49	7FE	0.8	-3.3	-1.7	0.9
56.58	2.05	0.9	326.44	17FE	0.5	-3.3	-1.7	0.8
61.18	1.99	1.1	336.33	27FE	0.2	-3.3	-1.6	0.8
66.06	1.92	1.2	346.17	9MR	-0.1	-3.3	-1.6	0.7
71.17	1.86	1.4	355.94	19MR	-0.8	-3.3	-1.6	0.7
76.45	1.79	1.5	5.66	29MR	-1.6	-3.4	-1.5	0.7
81.89	1.73	1.6	15.33	8AP	-1.4	-3.4	-1.5	0.7
87.46	1.66	1.6	24.96	18AP	-0.4	-3.4	-1.5	0.7
93.14	1.59	1.7	34.54	28AP	0.6	-3.5	-1.5	0.7
98.92	1.52	1.8	44.10	8MY	1.4	-3.5	-1.5	0.7
104.79	1.45	1.8	53.63	18MY	2.5	-3.5	-1.5	0.7
110.75	1.38	1.9	63.16	28MY	3.4	-3.4	-1.5	0.7
116.79	1.31	1.9	72.68	7JN	1.9	-3.4	-1.6	0.7
122.91	1.24	1.9	82.20	17JN	0.6	-3.4	-1.6	0.6
129.10	1.16	1.9	91.74	27JN	-0.6	-3.3	-1.6	0.6
135.37	1.08	1.9	101.31	7JL	-1.4	-3.3	-1.7	0.6
141.72	1.00	1.9	110.90	17JL	-1.3	-3.3	-1.7	0.5
148.15	0.92	1.9	120.53	27JL	-0.6	-3.3	-1.8	0.5
154.67	0.83	1.9	130.22	6AU	-0.1	-3.3	-1.9	0.4
161.26	0.75	1.9	139.95	16AU	0.1	-3.3	-1.9	0.3
167.94	0.66	1.8	149.74	26AU	0.3	-3.4	-2.0	0.3
174.71	0.56	1.8	159.59	5SE	0.6	-3.4	-2.1	0.2
181.56	0.47	1.8	169.50	15SE	1.8	-3.4	-2.1	0.2
188.51	0.37	1.7	179.47	25SE	2.0	-3.5	-2.2	0.1
195.54	0.27	1.7	189.49	5OC	-0.2	-3.5	-2.2	0.2
202.66	0.17	1.6	199.56	15OC	-0.6	-3.6	-2.3	0.3
209.86	0.06	1.6	209.68	25OC	-0.7	-3.6	-2.3	0.3
217.16	-0.04	1.6	219.84	4NO	-0.7	-3.7	-2.3	0.4
224.53	-0.15	1.6	230.02	14NO	-0.8	-3.8	-2.3	0.5
231.98	-0.26	1.5	240.23	24NO	-0.7	-3.9	-2.3	0.5
239.51	-0.37	1.5	250.44	4DE	-0.7	-4.0	-2.3	0.6
247.10	-0.47	1.5	260.64	14DE	-0.7	-4.1	-2.3	0.6
254.75	-0.58	1.5	270.83	24DE	0.2	-4.2	-2.2	0.7

♂ / ☉ / Magnitudes — -122 / -121

♂ LONG	LAT	MAG	☉ LONG	16.00UT	☿	♀	♃	♄
262.45	-0.68	1.5	281.00	3JA	2.7	-4.3	-2.1	0.7
270.19	-0.78	1.5	291.12	13JA	1.4	-4.3	-2.0	0.7
277.97	-0.87	1.4	301.21	23JA	0.6	-4.3	-2.0	0.7
285.77	-0.96	1.4	311.25	2FE	0.3	-4.2	-1.9	0.7
293.58	-1.04	1.4	321.23	12FE	0.1	-3.8	-1.8	0.7
301.39	-1.11	1.4	331.15	22FE	-0.2	-3.2	-1.7	0.6
309.18	-1.16	1.4	341.02	4MR	-0.8	-3.6	-1.7	0.6
316.94	-1.21	1.4	350.82	14MR	-1.6	-4.1	-1.6	0.6
324.67	-1.25	1.4	0.57	24MR	-1.3	-4.2	-1.6	0.5
332.35	-1.27	1.3	10.27	3AP	-0.3	-4.2	-1.5	0.5
339.96	-1.28	1.3	19.91	13AP	0.8	-4.2	-1.5	0.5
347.50	-1.28	1.3	29.52	23AP	1.9	-4.1	-1.5	0.5
354.95	-1.26	1.3	39.09	3MY	3.4	-4.0	-1.4	0.5
2.31	-1.23	1.3	48.63	13MY	2.7	-3.9	-1.4	0.6
9.56	-1.18	1.3	58.17	23MY	1.5	-3.8	-1.4	0.5
16.70	-1.12	1.3	67.69	2JN	0.5	-3.7	-1.4	0.5
23.71	-1.05	1.2	77.20	12JN	-0.6	-3.6	-1.4	0.5
30.59	-0.97	1.2	86.74	22JN	-1.5	-3.5	-1.4	0.5
37.32	-0.87	1.2	96.29	2JL	-1.3	-3.5	-1.4	0.5
43.90	-0.76	1.2	105.87	12JL	-0.6	-3.5	-1.5	0.4
50.29	-0.64	1.1	115.48	22JL	-0.1	-3.4	-1.5	0.4
56.49	-0.50	1.1	125.14	1AU	0.2	-3.4	-1.5	0.3
62.48	-0.35	1.0	134.84	11AU	0.5	-3.4	-1.6	0.3
68.21	-0.18	0.9	144.61	21AU	0.9	-3.4	-1.6	0.2
73.64	0.01	0.8	154.43	31AU	2.2	-3.4	-1.7	0.2
78.74	0.21	0.7	164.30	10SE	1.8	-3.4	-1.8	0.1
83.42	0.44	0.6	174.24	20SE	-0.3	-3.4	-1.8	0.0
87.61	0.70	0.5	184.23	30SE	-0.8	-3.4	-1.9	-0.0
91.18	1.00	0.3	194.28	10OC	-0.8	-3.4	-1.9	-0.1
94.00	1.33	0.1	204.38	20OC	-0.8	-3.4	-2.0	0.0
95.89	1.71	-0.1	214.51	30OC	-0.7	-3.4	-2.1	0.1
96.66	2.13	-0.3	224.68	9NO	-0.6	-3.4	-2.1	0.2
96.12	2.58	-0.6	234.86	19NO	-0.6	-3.5	-2.2	0.2
94.22	3.03	-0.8	245.08	29NO	-0.5	-3.5	-2.2	0.3
91.11	3.43	-1.0	255.29	9DE	0.3	-3.5	-2.2	0.4
87.23	3.72	-1.1	265.49	19DE	3.0	-3.5	-2.2	0.4
83.33	3.86	-1.0	275.67	29DE	1.0	-3.4	-2.2	0.5
				-121				
80.12	3.85	-0.7	285.82	8JA	0.3	-3.4	-2.1	0.5
78.06	3.73	-0.5	295.93	18JA	0.1	-3.4	-2.1	0.5
77.33	3.53	-0.2	305.99	28JA	-0.0	-3.4	-2.0	0.5
77.82	3.31	0.0	316.00	7FE	-0.3	-3.4	-2.0	0.5
79.40	3.08	0.3	325.96	17FE	-0.8	-3.3	-1.9	0.5
81.85	2.85	0.5	335.85	27FE	-1.6	-3.3	-1.8	0.5
85.01	2.64	0.7	345.69	9MR	-1.3	-3.3	-1.8	0.5
88.73	2.45	0.9	355.47	19MR	-0.3	-3.3	-1.7	0.4
92.91	2.26	1.0	5.19	29MR	1.0	-3.4	-1.6	0.4
97.46	2.09	1.2	14.86	8AP	2.4	-3.4	-1.6	0.3
102.31	1.93	1.3	24.49	18AP	3.5	-3.4	-1.5	0.3
107.42	1.78	1.4	34.08	28AP	2.0	-3.4	-1.5	0.4
112.74	1.63	1.5	43.63	8MY	1.1	-3.5	-1.4	0.4
118.25	1.49	1.5	53.17	18MY	0.3	-3.5	-1.4	0.4
123.92	1.36	1.6	62.69	28MY	-0.6	-3.6	-1.4	0.4
129.75	1.23	1.6	72.21	7JN	-1.6	-3.6	-1.3	0.4
135.71	1.10	1.7	81.74	17JN	-1.4	-3.7	-1.3	0.4
141.81	0.98	1.7	91.28	27JN	-0.5	-3.8	-1.3	0.4
148.03	0.86	1.7	100.84	7JL	0.0	-3.9	-1.3	0.4
154.37	0.73	1.7	110.44	17JL	0.4	-4.0	-1.3	0.3
160.83	0.62	1.7	120.07	27JL	-0.7	-4.1	-1.3	0.3
167.40	0.50	1.7	129.75	6AU	1.2	-4.2	-1.3	0.3
174.09	0.38	1.7	139.48	16AU	2.7	-4.3	-1.4	0.2
180.89	0.26	1.7	149.27	26AU	1.6	-4.3	-1.4	0.2
187.80	0.15	1.7	159.12	5SE	-0.4	-4.2	-1.4	0.1
194.82	0.03	1.7	169.02	15SE	-0.9	-3.8	-1.5	0.0
201.94	-0.08	1.6	178.98	25SE	-1.0	-3.3	-1.5	-0.0
209.17	-0.19	1.6	189.00	5OC	-0.9	-3.7	-1.6	-0.1
216.50	-0.30	1.6	199.08	15OC	-0.6	-4.2	-1.7	-0.2
223.93	-0.40	1.5	209.19	25OC	-0.4	-4.3	-1.7	-0.2
231.44	-0.50	1.5	219.35	4NO	-0.4	-4.4	-1.8	-0.1
239.04	-0.60	1.5	229.53	14NO	-0.4	-4.3	-1.9	-0.1
246.71	-0.69	1.4	239.73	24NO	0.5	-4.2	-1.9	0.0
254.44	-0.77	1.4	249.94	4DE	3.1	-4.1	-2.0	0.1
262.23	-0.85	1.3	260.15	14DE	0.7	-4.0	-2.0	0.2
270.07	-0.92	1.3	270.34	24DE	0.1	-3.9	-2.1	0.2

♂ / ☉ / Magnitudes — -120 / -119

♂ LONG	LAT	MAG	☉ LONG	16.00UT	☿	♀	♃	♄
277.93	-0.97	1.3	280.50	3JA	0.0	-3.8	-2.1	0.3
285.81	-1.02	1.3	290.64	13JA	-0.1	-3.7	-2.1	0.3
293.70	-1.06	1.3	300.72	23JA	-0.4	-3.6	-2.1	0.4
301.58	-1.09	1.4	310.77	2FE	-0.8	-3.6	-2.1	0.4
309.44	-1.10	1.4	320.75	12FE	-1.4	-3.5	-2.0	0.4
317.27	-1.10	1.4	330.67	22FE	-1.2	-3.5	-2.0	0.4
325.05	-1.09	1.4	340.55	3MR	-0.2	-3.4	-1.9	0.4
332.78	-1.07	1.4	350.35	13MR	1.2	-3.4	-1.9	0.4
340.44	-1.04	1.5	0.10	23MR	3.1	-3.3	-1.8	0.4
348.04	-1.00	1.5	9.80	2AP	2.7	-3.3	-1.7	0.3
355.55	-0.94	1.5	19.45	12AP	1.5	-3.3	-1.7	0.3
2.97	-0.88	1.5	29.06	22AP	0.8	-3.3	-1.6	0.2
10.31	-0.81	1.5	38.63	2MY	0.2	-3.3	-1.5	0.2
17.54	-0.72	1.5	48.18	12MY	-0.6	-3.3	-1.5	0.3
24.68	-0.64	1.6	57.70	22MY	-1.6	-3.3	-1.4	0.3
31.72	-0.54	1.6	67.23	1JN	-1.4	-3.4	-1.4	0.3
38.64	-0.43	1.6	76.75	11JN	-0.5	-3.4	-1.3	0.3
45.47	-0.32	1.6	86.27	21JN	0.1	-3.4	-1.3	0.3
52.18	-0.21	1.6	95.83	1JL	0.5	-3.5	-1.3	0.3
58.77	-0.09	1.6	105.40	11JL	0.9	-3.5	-1.3	0.3
65.25	0.04	1.6	115.01	21JL	1.6	-3.5	-1.2	0.3
71.60	0.18	1.5	124.67	31JL	3.1	-3.4	-1.2	0.3
77.81	0.32	1.5	134.37	10AU	1.4	-3.4	-1.2	0.2
83.87	0.47	1.5	144.13	20AU	-0.4	-3.4	-1.2	0.2
89.77	0.63	1.4	153.95	30AU	-1.1	-3.4	-1.3	0.1
95.49	0.80	1.3	163.82	9SE	-1.1	-3.3	-1.3	0.1
100.99	0.98	1.3	173.75	19SE	-0.8	-3.3	-1.3	0.0
106.25	1.18	1.2	183.74	29SE	-0.5	-3.3	-1.4	-0.1
111.22	1.39	1.1	193.79	9OC	-0.3	-3.3	-1.4	-0.1
115.84	1.63	0.9	203.88	19OC	-0.3	-3.3	-1.4	-0.2
120.03	1.89	0.8	214.02	29OC	-0.2	-3.4	-1.5	-0.3
123.70	2.17	0.6	224.19	8NO	0.7	-3.4	-1.6	-0.3
126.73	2.49	0.4	234.38	18NO	3.0	-3.4	-1.6	-0.2
128.94	2.84	0.2	244.59	28NO	0.5	-3.4	-1.7	-0.1
130.18	3.23	-0.0	254.80	8DE	-0.0	-3.5	-1.8	-0.1
130.26	3.63	-0.3	265.00	18DE	-0.1	-3.5	-1.8	0.0
129.03	4.02	-0.5	275.18	28DE	-0.2	-3.6	-1.9	0.1
				-119				
126.52	4.35	-0.8	285.33	7JA	-0.5	-3.6	-1.9	0.1
123.01	4.56	-1.0	295.44	17JA	-0.9	-3.7	-2.0	0.2
119.07	4.60	-1.0	305.51	27JA	-1.3	-3.8	-2.0	0.2
115.43	4.47	-0.8	315.52	6FE	-1.1	-3.9	-2.0	0.3
112.70	4.19	-0.6	325.48	16FE	-0.2	-3.9	-2.0	0.3
111.21	3.84	-0.4	335.38	26FE	1.5	-4.0	-2.0	0.3
111.02	3.46	-0.1	345.22	8MR	3.4	-4.1	-2.0	0.3
112.00	3.09	0.1	355.00	18MR	2.0	-4.2	-1.9	0.3
113.99	2.74	0.3	4.73	28MR	1.1	-4.2	-1.9	0.3
116.80	2.42	0.5	14.40	7AP	0.6	-4.2	-1.8	0.3
120.28	2.12	0.6	24.03	17AP	0.1	-3.9	-1.8	0.3
124.31	1.86	0.8	33.62	27AP	-0.7	-3.3	-1.7	0.2
128.78	1.61	0.9	43.17	7MY	-1.6	-2.9	-1.6	0.2
133.63	1.38	1.0	52.71	17MY	-1.4	-3.7	-1.6	0.2
138.80	1.17	1.1	62.23	27MY	-0.5	-4.1	-1.5	0.2
144.24	0.97	1.2	71.75	6JN	0.2	-4.2	-1.5	0.3
149.93	0.79	1.3	81.27	16JN	0.7	-4.2	-1.4	0.3
155.83	0.61	1.3	90.81	26JN	1.2	-4.1	-1.4	0.3
161.92	0.45	1.4	100.37	6JL	2.1	-4.0	-1.3	0.3
168.19	0.29	1.4	109.97	16JL	3.1	-3.9	-1.3	0.3
174.64	0.14	1.4	119.60	26JL	1.1	-3.8	-1.3	0.3
181.24	-0.00	1.4	129.27	5AU	-0.5	-3.8	-1.2	0.3
187.99	-0.14	1.4	139.01	15AU	-1.2	-3.7	-1.2	0.3
194.89	-0.27	1.5	148.79	25AU	-1.2	-3.6	-1.2	0.2
201.92	-0.39	1.5	158.63	4SE	-0.7	-3.6	-1.2	0.2
209.08	-0.50	1.5	168.54	14SE	-0.4	-3.5	-1.2	0.1
216.37	-0.61	1.5	178.50	24SE	-0.2	-3.5	-1.2	0.1
223.76	-0.71	1.5	188.52	4OC	-0.1	-3.5	-1.2	-0.0
231.26	-0.80	1.4	198.59	14OC	0.0	-3.4	-1.3	-0.1
238.86	-0.88	1.4	208.70	24OC	0.9	-3.4	-1.3	-0.1
246.53	-0.94	1.4	218.85	3NO	2.8	-3.4	-1.3	-0.2
254.28	-1.00	1.4	229.04	13NO	0.3	-3.4	-1.4	-0.3
262.09	-1.05	1.4	239.24	23NO	-0.2	-3.4	-1.5	-0.3
269.94	-1.08	1.4	249.45	3DE	-0.3	-3.4	-1.5	-0.2
277.82	-1.10	1.4	259.66	13DE	-0.3	-3.4	-1.6	-0.2
285.72	-1.12	1.4	269.85	23DE	-0.5	-3.4	-1.6	-0.1

♂ LONG	♂ LAT	♂ MAG	☉ LONG	16.00UT	☿	♀	♃	♄
				-118	MAGNITUDES			
293.62	-1.11	1.4	280.01	2JA	-0.9	-3.4	-1.7	-0.0
301.51	-1.10	1.4	290.15	12JA	-1.2	-3.4	-1.8	0.1
309.37	-1.07	1.4	300.24	22JA	-1.0	-3.4	-1.8	0.1
317.19	-1.04	1.4	310.28	1FE	-0.1	-3.4	-1.9	0.2
324.97	-0.99	1.4	320.27	11FE	1.8	-3.4	-1.9	0.2
332.68	-0.94	1.4	330.19	21FE	2.8	-3.5	-2.0	0.3
340.33	-0.87	1.4	340.06	3MR	1.4	-3.5	-2.0	0.3
347.91	-0.80	1.4	349.88	13MR	0.8	-3.4	-2.0	0.3
355.40	-0.72	1.5	359.63	23MR	0.5	-3.4	-2.0	0.3
2.81	-0.63	1.5	9.33	2AP	0.0	-3.4	-2.0	0.3
10.13	-0.54	1.6	18.98	12AP	-0.7	-3.4	-2.0	0.3
17.37	-0.45	1.6	28.59	22AP	-1.6	-3.3	-1.9	0.3
24.51	-0.35	1.6	38.16	2MY	-1.4	-3.3	-1.9	0.3
31.57	-0.25	1.7	47.71	12MY	-0.4	-3.3	-1.8	0.2
38.54	-0.14	1.7	57.24	22MY	0.3	-3.3	-1.8	0.2
45.42	-0.04	1.7	66.76	1JN	0.9	-3.3	-1.7	0.2
52.21	0.07	1.8	76.28	11JN	1.6	-3.3	-1.6	0.2
58.92	0.18	1.8	85.81	21JN	2.7	-3.4	-1.6	0.3
65.54	0.29	1.8	95.35	1JL	2.7	-3.4	-1.5	0.3
72.09	0.40	1.8	104.93	11JL	0.9	-3.4	-1.5	0.3
78.55	0.51	1.8	114.54	21JL	-0.5	-3.5	-1.4	0.3
84.93	0.62	1.8	124.19	31JL	-1.3	-3.5	-1.4	0.3
91.22	0.74	1.8	133.90	10AU	-1.3	-3.6	-1.3	0.3
97.42	0.86	1.8	143.65	20AU	-0.7	-3.7	-1.3	0.3
103.53	0.98	1.8	153.46	30AU	-0.3	-3.8	-1.3	0.3
109.54	1.10	1.7	163.34	9SE	-0.1	-3.8	-1.2	0.3
115.43	1.23	1.7	173.27	19SE	0.0	-3.9	-1.2	0.2
121.21	1.37	1.6	183.25	29SE	0.2	-4.1	-1.2	0.2
126.83	1.51	1.5	193.30	9OC	1.2	-4.2	-1.2	0.1
132.29	1.66	1.4	203.39	19OC	2.5	-4.3	-1.2	0.0
137.55	1.82	1.3	213.53	29OC	0.1	-4.4	-1.3	-0.0
142.57	1.99	1.2	223.70	8NO	-0.4	-4.4	-1.3	-0.1
147.30	2.17	1.1	233.89	18NO	-0.4	-4.3	-1.3	-0.2
151.69	2.37	0.9	244.09	28NO	-0.5	-3.9	-1.4	-0.3
155.64	2.58	0.7	254.30	8DE	-0.6	-3.0	-1.4	-0.3
159.06	2.81	0.5	264.50	18DE	-0.9	-3.6	-1.5	-0.2
161.80	3.06	0.3	274.68	28DE	-1.0	-4.2	-1.5	-0.1
				-117				
163.72	3.33	0.1	284.84	7JA	-0.9	-4.3	-1.6	-0.1
164.62	3.60	-0.2	294.95	17JA	-0.1	-4.3	-1.6	0.0
164.35	3.85	-0.5	305.02	27JA	2.2	-4.3	-1.7	0.1
162.79	4.04	-0.8	315.04	6FE	2.2	-4.2	-1.8	0.1
160.03	4.14	-1.0	325.00	16FE	1.0	-4.1	-1.8	0.2
156.43	4.09	-1.2	334.90	26FE	0.6	-3.9	-1.9	0.2
152.61	3.86	-1.2	344.74	8MR	0.3	-3.8	-2.0	0.3
149.30	3.50	-1.0	354.52	18MR	-0.1	-3.7	-2.0	0.3
147.04	3.05	-0.8	4.25	28MR	-0.7	-3.7	-2.0	0.3
146.08	2.58	-0.6	13.93	7AP	-1.6	-3.6	-2.1	0.3
146.41	2.12	-0.4	23.56	17AP	-1.4	-3.5	-2.1	0.3
147.91	1.70	-0.2	33.16	27AP	-0.4	-3.5	-2.1	0.3
150.39	1.32	0.0	42.72	7MY	0.5	-3.4	-2.1	0.3
153.69	0.97	0.2	52.25	17MY	1.2	-3.4	-2.0	0.3
157.65	0.67	0.3	61.78	27MY	2.1	-3.3	-2.0	0.3
162.18	0.40	0.5	71.30	6JN	3.4	-3.3	-1.9	0.2
167.16	0.15	0.6	80.82	16JN	2.2	-3.3	-1.9	0.3
172.54	-0.07	0.7	90.36	26JN	0.8	-3.3	-1.8	0.3
178.26	-0.27	0.8	99.91	6JL	-0.5	-3.3	-1.7	0.3
184.27	-0.45	0.8	109.50	16JL	-1.4	-3.3	-1.7	0.4
190.54	-0.61	0.9	119.13	26JL	-1.3	-3.4	-1.6	0.4
197.04	-0.75	0.9	128.81	5AU	-0.6	-3.4	-1.6	0.4
203.75	-0.88	1.0	138.53	15AU	-0.2	-3.4	-1.5	0.4
210.65	-0.99	1.0	148.32	25AU	0.0	-3.4	-1.5	0.4
217.72	-1.09	1.1	158.15	4SE	0.2	-3.5	-1.4	0.4
224.94	-1.17	1.1	168.05	14SE	0.4	-3.5	-1.4	0.4
232.29	-1.23	1.1	178.01	24SE	1.5	-3.5	-1.4	0.3
239.77	-1.28	1.2	188.03	4OC	2.3	-3.5	-1.3	0.3
247.34	-1.31	1.2	198.09	14OC	-0.0	-3.4	-1.3	0.2
255.00	-1.33	1.2	208.21	24OC	-0.5	-3.4	-1.3	0.2
262.73	-1.33	1.2	218.36	3NO	-0.6	-3.4	-1.3	0.1
270.51	-1.32	1.3	228.54	13NO	-0.6	-3.4	-1.3	0.0
278.32	-1.30	1.3	238.74	23NO	-0.7	-3.3	-1.3	-0.0
286.15	-1.26	1.3	248.95	3DE	-0.8	-3.3	-1.3	-0.1
293.98	-1.21	1.4	259.15	13DE	-0.8	-3.3	-1.4	-0.2
301.80	-1.15	1.4	269.35	23DE	-0.8	-3.3	-1.4	-0.2

♂ LONG	♂ LAT	♂ MAG	☉ LONG	16.00UT	☿	♀	♃	♄
				-116	MAGNITUDES			
309.59	-1.08	1.4	279.52	2JA	0.0	-3.3	-1.5	-0.1
317.34	-1.00	1.4	289.65	12JA	2.5	-3.4	-1.5	-0.0
325.04	-0.91	1.5	299.75	22JA	1.7	-3.4	-1.6	0.0
332.68	-0.82	1.5	309.79	1FE	0.7	-3.4	-1.6	0.1
340.25	-0.73	1.5	319.78	11FE	0.4	-3.4	-1.7	0.1
347.74	-0.63	1.6	329.72	21FE	0.2	-3.5	-1.8	0.2
355.16	-0.52	1.6	339.59	2MR	-0.2	-3.5	-1.8	0.3
2.49	-0.42	1.6	349.40	12MR	-0.7	-3.6	-1.9	0.3
9.75	-0.32	1.6	359.16	22MR	-1.6	-3.6	-2.0	0.3
16.92	-0.21	1.6	8.86	1AP	-1.4	-3.7	-2.0	0.4
24.00	-0.11	1.7	18.52	11AP	-0.4	-3.8	-2.1	0.4
31.01	-0.01	1.7	28.13	21AP	0.6	-3.8	-2.1	0.4
37.93	0.10	1.7	37.71	1MY	1.6	-3.9	-2.2	0.4
44.78	0.19	1.7	47.26	11MY	2.8	-4.0	-2.2	0.4
51.56	0.29	1.8	56.79	21MY	3.2	-4.1	-2.2	0.4
58.27	0.39	1.8	66.30	31MY	1.7	-4.2	-2.2	0.4
64.92	0.48	1.9	75.82	10JN	0.6	-4.2	-2.2	0.4
71.51	0.57	1.9	85.35	20JN	-0.6	-4.1	-2.1	0.4
78.03	0.66	1.9	94.90	30JN	-1.5	-3.7	-2.1	0.4
84.51	0.75	1.9	104.47	10JL	-1.4	-3.1	-2.0	0.4
90.93	0.83	2.0	114.08	20JL	-0.6	-3.5	-2.0	0.5
97.30	0.92	2.0	123.73	30JL	-0.1	-4.0	-1.9	0.5
103.63	1.00	2.0	133.43	9AU	0.2	-4.2	-1.8	0.5
109.91	1.08	2.0	143.18	19AU	0.4	-4.3	-1.8	0.5
116.14	1.16	1.9	152.98	29AU	0.7	-4.2	-1.7	0.5
122.33	1.24	1.9	162.85	8SE	1.9	-4.2	-1.7	0.5
128.46	1.32	1.9	172.78	18SE	2.0	-4.1	-1.6	0.5
134.53	1.39	1.9	182.76	28SE	-0.2	-4.0	-1.5	0.5
140.55	1.47	1.8	192.81	8OC	-0.7	-3.9	-1.5	0.5
146.49	1.55	1.7	202.90	18OC	-0.7	-3.8	-1.5	0.4
152.35	1.63	1.7	213.03	28OC	-0.7	-3.7	-1.4	0.4
158.12	1.71	1.6	223.20	7NO	-0.7	-3.7	-1.4	0.3
163.77	1.79	1.5	233.39	17NO	-0.7	-3.6	-1.4	0.2
169.29	1.87	1.4	243.59	27NO	-0.7	-3.6	-1.4	0.2
174.64	1.95	1.2	253.81	7DE	-0.6	-3.5	-1.4	0.1
179.80	2.03	1.1	264.01	17DE	0.2	-3.5	-1.4	0.0
184.71	2.11	0.9	274.19	27DE	2.8	-3.4	-1.4	-0.0
				-115				
189.32	2.19	0.7	284.35	6JA	1.3	-3.4	-1.5	-0.0
193.56	2.26	0.5	294.47	16JA	0.5	-3.4	-1.5	0.0
197.34	2.33	0.3	304.53	26JA	0.2	-3.3	-1.5	0.1
200.54	2.38	0.1	314.56	5FE	0.1	-3.3	-1.6	0.1
203.01	2.41	-0.2	324.52	15FE	-0.2	-3.3	-1.6	0.2
204.58	2.41	-0.5	334.42	25FE	-0.8	-3.3	-1.7	0.3
205.07	2.36	-0.8	344.27	7MR	-1.6	-3.3	-1.7	0.3
204.33	2.23	-1.1	354.06	17MR	-1.3	-3.3	-1.8	0.4
202.35	2.00	-1.4	3.78	27MR	-0.3	-3.4	-1.9	0.4
199.33	1.65	-1.7	13.47	6AP	0.8	-3.4	-1.9	0.5
195.81	1.20	-1.7	23.10	16AP	2.1	-3.4	-2.0	0.5
192.52	0.69	-1.6	32.69	26AP	3.7	-3.5	-2.1	0.5
190.08	0.18	-1.4	42.25	6MY	2.4	-3.5	-2.1	0.6
188.93	-0.29	-1.2	51.79	16MY	1.3	-3.5	-2.2	0.6
189.15	-0.70	-1.0	61.31	26MY	0.4	-3.4	-2.3	0.6
190.64	-1.04	-0.8	70.83	5JN	-0.6	-3.4	-2.3	0.6
193.25	-1.31	-0.6	80.35	15JN	-1.5	-3.4	-2.3	0.6
196.79	-1.54	-0.4	89.89	25JN	-1.4	-3.3	-2.3	0.6
201.08	-1.71	-0.3	99.45	5JL	-0.6	-3.3	-2.3	0.5
206.01	-1.85	-0.1	109.04	15JL	-0.0	-3.3	-2.3	0.5
211.45	-1.95	-0.0	118.66	25JL	0.3	-3.3	-2.2	0.6
217.32	-2.02	0.1	128.34	4AU	0.5	-3.3	-2.2	0.6
223.56	-2.06	0.2	138.06	14AU	1.0	-3.3	-2.1	0.7
230.08	-2.07	0.3	147.83	24AU	2.3	-3.4	-2.1	0.7
236.86	-2.07	0.4	157.67	3SE	1.8	-3.4	-2.0	0.7
243.85	-2.04	0.5	167.57	13SE	-0.3	-3.4	-1.9	0.7
251.00	-1.99	0.6	177.52	23SE	-0.8	-3.5	-1.9	0.7
258.30	-1.92	0.7	187.54	3OC	-0.9	-3.5	-1.8	0.7
265.70	-1.83	0.8	197.60	13OC	-0.9	-3.6	-1.8	0.7
273.18	-1.74	0.8	207.71	23OC	-0.6	-3.6	-1.7	0.7
280.72	-1.63	0.9	217.86	2NO	-0.5	-3.7	-1.7	0.6
288.30	-1.51	1.0	228.04	12NO	-0.5	-3.8	-1.6	0.6
295.88	-1.38	1.1	238.24	22NO	-0.5	-3.9	-1.6	0.5
303.47	-1.25	1.2	248.45	2DE	0.3	-4.0	-1.5	0.4
311.03	-1.12	1.2	258.66	12DE	3.0	-4.1	-1.5	0.4
318.57	-0.98	1.3	268.85	22DE	0.9	-4.2	-1.5	0.3

♂ LONG	♂ LAT	♂ MAG	⊙ LONG	16.00UT -114	☿	♀	♃	♄ MAGNITUDES
326.06	-0.85	1.4	279.02	1JA	0.3	-4.3	-1.5	0.2
333.49	-0.71	1.4	289.16	11JA	0.1	-4.3	-1.5	0.2
340.87	-0.58	1.5	299.26	21JA	-0.1	-4.3	-1.5	0.2
348.18	-0.45	1.5	309.31	31JA	-0.3	-4.2	-1.5	0.2
355.42	-0.32	1.6	319.30	10FE	-0.8	-3.8	-1.6	0.3
2.59	-0.20	1.6	329.23	20FE	-1.5	-3.2	-1.6	0.3
9.69	-0.08	1.7	339.11	2MR	-1.3	-3.7	-1.6	0.4
16.71	0.03	1.7	348.93	12MR	-0.3	-4.1	-1.7	0.5
23.66	0.13	1.8	358.69	22MR	1.0	-4.2	-1.7	0.5
30.55	0.24	1.8	8.40	1AP	2.7	-4.2	-1.8	0.6
37.36	0.33	1.8	18.05	11AP	3.2	-4.2	-1.8	0.6
44.11	0.43	1.9	27.67	21AP	1.8	-4.1	-1.9	0.7
50.81	0.51	1.9	37.25	1MY	1.0	-4.0	-2.0	0.7
57.45	0.59	1.9	46.79	11MY	0.3	-3.9	-2.0	0.7
64.04	0.67	1.9	56.32	21MY	-0.6	-3.8	-2.1	0.7
70.58	0.74	1.9	65.85	31MY	-1.6	-3.7	-2.2	0.8
77.08	0.81	1.9	75.36	10JN	-1.4	-3.6	-2.2	0.8
83.55	0.88	1.9	84.89	20JN	-0.5	-3.5	-2.3	0.8
89.99	0.93	1.9	94.44	30JN	0.1	-3.5	-2.3	0.8
96.40	0.99	2.0	104.01	10JL	0.4	-3.5	-2.4	0.8
102.79	1.04	2.0	113.62	20JL	0.7	-3.4	-2.4	0.8
109.16	1.09	2.0	123.26	30JL	1.3	-3.4	-2.4	0.8
115.51	1.14	2.0	132.96	9AU	2.8	-3.4	-2.4	0.8
121.85	1.18	2.0	142.71	19AU	1.6	-3.4	-2.4	0.9
128.18	1.21	2.0	152.52	29AU	-0.3	-3.4	-2.3	0.9
134.49	1.25	2.0	162.38	8SE	-1.0	-3.4	-2.3	0.9
140.81	1.28	2.0	172.30	18SE	-1.0	-3.4	-2.2	1.0
147.11	1.31	2.0	182.29	28SE	-0.9	-3.4	-2.2	1.0
153.40	1.33	1.9	192.32	8OC	-0.5	-3.4	-2.1	1.0
159.69	1.35	1.9	202.41	18OC	-0.4	-3.4	-2.0	1.0
165.95	1.36	1.8	212.54	28OC	-0.4	-3.4	-2.0	0.9
172.21	1.37	1.8	222.70	7NO	-0.3	-3.4	-1.9	0.9
178.44	1.37	1.7	232.89	17NO	0.5	-3.5	-1.8	0.9
184.64	1.37	1.6	243.10	27NO	3.1	-3.5	-1.8	0.8
190.81	1.36	1.5	253.30	7DE	0.7	-3.5	-1.7	0.8
196.93	1.34	1.4	263.51	17DE	0.1	-3.5	-1.7	0.7
203.00	1.31	1.3	273.69	27DE	-0.1	-3.4	-1.6	0.6
				-113				
209.01	1.27	1.2	283.84	6JA	-0.2	-3.4	-1.6	0.5
214.93	1.21	1.1	293.97	16JA	-0.4	-3.4	-1.6	0.5
220.75	1.13	0.9	304.04	26JA	-0.8	-3.4	-1.6	0.4
226.44	1.04	0.7	314.06	5FE	-1.4	-3.4	-1.6	0.4
231.97	0.91	0.5	324.03	15FE	-1.2	-3.3	-1.6	0.5
237.30	0.76	0.3	333.94	25FE	-0.3	-3.3	-1.6	0.5
242.39	0.56	0.1	343.79	7MR	1.3	-3.3	-1.6	0.6
247.15	0.32	-0.1	353.58	17MR	3.3	-3.3	-1.6	0.6
251.51	0.02	-0.3	3.31	27MR	2.4	-3.4	-1.6	0.7
255.36	-0.34	-0.6	12.99	6AP	1.4	-3.4	-1.7	0.7
258.56	-0.79	-0.9	22.63	16AP	0.8	-3.4	-1.7	0.8
260.92	-1.32	-1.2	32.23	26AP	0.2	-3.4	-1.8	0.8
262.28	-1.96	-1.5	41.79	6MY	-0.6	-3.5	-1.8	0.9
262.45	-2.67	-1.8	51.33	16MY	-1.6	-3.5	-1.9	0.9
261.39	-3.44	-2.1	60.86	26MY	-1.4	-3.6	-1.9	1.0
259.26	-4.18	-2.4	70.37	5JN	-0.5	-3.6	-2.0	1.0
256.51	-4.79	-2.5	79.90	15JN	0.1	-3.7	-2.1	1.0
253.88	-5.18	-2.4	89.43	25JN	0.6	-3.8	-2.1	1.0
252.03	-5.34	-2.2	98.99	5JL	1.0	-3.9	-2.2	1.0
251.40	-5.29	-2.0	108.58	15JL	1.8	-4.0	-2.3	1.0
252.14	-5.09	-1.7	118.20	25JL	3.2	-4.1	-2.3	1.0
254.16	-4.79	-1.5	127.87	4AU	1.3	-4.2	-2.4	1.0
257.27	-4.45	-1.2	137.59	14AU	-0.4	-4.3	-2.4	1.1
261.29	-4.09	-1.0	147.37	24AU	-1.1	-4.3	-2.4	1.1
266.01	-3.73	-0.7	157.20	3SE	-1.2	-4.2	-2.5	1.2
271.29	-3.36	-0.5	167.09	13SE	-0.8	-3.8	-2.5	1.2
277.01	-3.01	-0.3	177.04	23SE	-0.4	-3.3	-2.4	1.2
283.06	-2.67	-0.2	187.05	3OC	-0.3	-3.7	-2.4	1.3
289.37	-2.34	0.0	197.11	13OC	-0.2	-4.2	-2.3	1.3
295.88	-2.04	0.2	207.22	23OC	-0.1	-4.3	-2.3	1.3
302.51	-1.75	0.4	217.37	2NO	0.7	-4.4	-2.2	1.3
309.26	-1.48	0.5	227.55	12NO	3.0	-4.3	-2.1	1.2
316.07	-1.22	0.7	237.74	22NO	0.4	-4.2	-2.1	1.2
322.92	-0.99	0.8	247.95	2DE	-0.1	-4.1	-2.0	1.2
329.80	-0.77	0.9	258.16	12DE	-0.2	-4.0	-1.9	1.1
336.68	-0.57	1.0	268.35	22DE	-0.3	-3.9	-1.9	1.0

♂ LONG	♂ LAT	♂ MAG	⊙ LONG	16.00UT -112	☿	♀	♃	♄ MAGNITUDES
343.55	-0.39	1.2	278.53	1JA	-0.5	-3.8	-1.8	1.0
350.41	-0.22	1.3	288.67	11JA	-0.9	-3.7	-1.7	0.9
357.23	-0.06	1.4	298.76	21JA	-1.2	-3.6	-1.7	0.8
4.02	0.08	1.5	308.82	31JA	-1.1	-3.6	-1.7	0.8
10.78	0.21	1.5	318.82	10FE	-0.2	-3.5	-1.6	0.7
17.49	0.32	1.6	328.75	20FE	1.6	-3.5	-1.6	0.7
24.16	0.43	1.7	338.63	1MR	3.3	-3.4	-1.6	0.8
30.79	0.53	1.8	348.46	11MR	1.8	-3.4	-1.6	0.8
37.38	0.62	1.8	358.22	21MR	1.0	-3.3	-1.6	0.9
43.93	0.70	1.9	7.93	31MR	0.6	-3.3	-1.6	0.9
50.44	0.77	1.9	17.59	10AP	0.1	-3.3	-1.6	1.0
56.92	0.83	1.9	27.20	20AP	-0.7	-3.3	-1.6	1.0
63.37	0.89	2.0	36.79	30AP	-1.6	-3.3	-1.6	1.1
69.80	0.94	2.0	46.34	10MY	-1.4	-3.3	-1.6	1.1
76.20	0.99	2.0	55.87	20MY	-0.5	-3.3	-1.7	1.2
82.58	1.03	2.0	65.39	30MY	0.2	-3.4	-1.7	1.2
88.95	1.06	2.0	74.91	9JN	0.8	-3.4	-1.8	1.2
95.31	1.09	2.0	84.43	19JN	1.3	-3.4	-1.8	1.3
101.67	1.11	2.0	93.98	29JN	2.3	-3.5	-1.9	1.3
108.03	1.13	2.0	103.55	9JL	3.0	-3.5	-2.0	1.3
114.39	1.15	2.0	113.15	19JL	1.1	-3.5	-2.0	1.3
120.77	1.16	2.0	122.79	29JL	-0.4	-3.4	-2.1	1.3
127.15	1.17	2.0	132.49	8AU	-1.2	-3.4	-2.2	1.3
133.55	1.17	2.0	142.23	18AU	-1.3	-3.4	-2.2	1.3
139.98	1.16	2.0	152.03	28AU	-0.7	-3.4	-2.3	1.3
146.42	1.16	2.0	161.90	7SE	-0.3	-3.3	-2.3	1.3
152.89	1.14	2.0	171.81	17SE	-0.1	-3.3	-2.4	1.3
159.39	1.12	2.0	181.80	27SE	-0.1	-3.3	-2.4	1.3
165.92	1.10	2.0	191.83	7OC	0.1	-3.3	-2.4	1.3
172.47	1.07	1.9	201.91	17OC	1.0	-3.3	-2.4	1.3
179.06	1.03	1.9	212.04	27OC	2.8	-3.4	-2.4	1.3
185.68	0.99	1.9	222.21	6NO	0.2	-3.4	-2.3	1.2
192.33	0.94	1.8	232.39	16NO	-0.3	-3.4	-2.3	1.2
199.02	0.88	1.7	242.60	26NO	-0.3	-3.4	-2.2	1.2
205.73	0.81	1.7	252.81	6DE	-0.4	-3.5	-2.2	1.1
212.47	0.73	1.6	263.01	16DE	-0.6	-3.5	-2.1	1.1
219.23	0.64	1.5	273.20	26DE	-0.9	-3.6	-2.0	1.0
				-111				
226.02	0.53	1.4	283.36	5JA	-1.1	-3.6	-2.0	1.0
232.83	0.41	1.4	293.48	15JA	-1.0	-3.7	-1.9	0.9
239.66	0.28	1.3	303.55	25JA	-0.1	-3.8	-1.8	0.9
246.51	0.13	1.1	313.58	4FE	1.9	-3.9	-1.8	0.8
253.36	-0.04	1.0	323.55	14FE	2.6	-3.9	-1.7	0.8
260.21	-0.23	0.9	333.46	24FE	1.3	-4.0	-1.6	0.8
267.05	-0.44	0.8	343.31	6MR	0.7	-4.1	-1.6	0.9
273.87	-0.67	0.7	353.10	16MR	0.4	-4.2	-1.6	0.9
280.65	-0.92	0.5	2.84	26MR	-0.0	-4.2	-1.5	1.0
287.37	-1.19	0.4	12.53	5AP	-0.7	-4.2	-1.5	1.1
294.00	-1.49	0.2	22.16	15AP	-1.6	-3.9	-1.5	1.1
300.52	-1.81	0.1	31.77	25AP	-1.4	-3.3	-1.5	1.2
306.86	-2.16	-0.1	41.33	5MY	-0.5	-3.0	-1.5	1.2
313.00	-2.53	-0.3	50.87	15MY	0.4	-3.7	-1.5	1.3
318.84	-2.92	-0.5	60.40	25MY	1.0	-4.1	-1.5	1.3
324.29	-3.34	-0.7	69.91	4JN	1.8	-4.2	-1.5	1.3
329.27	-3.77	-0.9	79.43	14JN	3.0	-4.2	-1.6	1.4
333.60	-4.22	-1.1	88.97	24JN	2.5	-4.1	-1.6	1.4
337.12	-4.67	-1.3	98.52	4JL	0.9	-4.0	-1.6	1.4
339.63	-5.11	-1.5	108.11	14JL	-0.5	-3.9	-1.7	1.3
340.90	-5.51	-1.8	117.73	24JL	-1.3	-3.8	-1.7	1.3
340.78	-5.80	-2.0	127.40	3AU	-1.4	-3.8	-1.8	1.3
339.28	-5.92	-2.2	137.12	13AU	-0.7	-3.7	-1.9	1.2
336.68	-5.79	-2.4	146.89	23AU	-0.3	-3.6	-1.9	1.2
333.64	-5.39	-2.4	156.72	2SE	-0.0	-3.6	-2.0	1.1
330.94	-4.77	-2.2	166.61	12SE	0.1	-3.5	-2.1	1.1
329.21	-4.03	-1.9	176.56	22SE	0.3	-3.5	-2.1	1.1
328.79	-3.28	-1.6	186.56	2OC	1.2	-3.5	-2.2	1.1
329.66	-2.58	-1.2	196.62	12OC	2.5	-3.4	-2.2	1.1
331.69	-1.96	-0.9	206.73	22OC	0.1	-3.4	-2.3	1.1
334.66	-1.43	-0.6	216.87	1NO	-0.5	-3.4	-2.3	1.1
338.37	-0.99	-0.3	227.05	11NO	-0.5	-3.4	-2.3	1.1
342.67	-0.61	-0.1	237.25	21NO	-0.5	-3.4	-2.3	1.1
347.42	-0.30	0.2	247.46	1DE	-0.7	-3.4	-2.3	1.0
352.51	-0.03	0.4	257.67	11DE	-0.9	-3.4	-2.2	1.0
357.87	0.19	0.6	267.86	21DE	-0.9	-3.4	-2.2	0.9
3.43	0.37	0.8	278.03	31DE	-0.9	-3.4	-2.1	0.9

Left table (-110 / -109):

♂ LONG	LAT	MAG	☉ LONG	16.00UT	☿	♀	♃	♄
				-110		MAGNITUDES		
9.15	0.53	0.9	288.18	10JA	-0.1	-3.4	-2.1	0.9
14.98	0.67	1.1	298.28	20JA	2.2	-3.4	-2.0	0.8
20.90	0.78	1.2	308.33	30JA	2.0	-3.4	-1.9	0.8
26.89	0.87	1.3	318.33	9FE	0.9	-3.5	-1.8	0.7
32.93	0.95	1.5	328.27	19FE	0.5	-3.5	-1.8	0.7
39.01	1.02	1.6	338.15	1MR	0.3	-3.5	-1.7	0.6
45.12	1.07	1.6	347.98	11MR	-0.1	-3.4	-1.7	0.6
51.24	1.12	1.7	357.74	21MR	-0.7	-3.4	-1.6	0.7
57.39	1.16	1.8	7.45	31MR	-1.6	-3.4	-1.6	0.8
63.54	1.18	1.9	17.11	10AP	-1.4	-3.4	-1.5	0.8
69.71	1.21	1.9	26.73	20AP	-0.4	-3.3	-1.5	0.9
75.89	1.22	1.9	36.31	30AP	0.5	-3.3	-1.4	1.0
82.08	1.23	2.0	45.87	10MY	1.3	-3.3	-1.4	1.0
88.29	1.23	2.0	55.40	20MY	2.4	-3.3	-1.4	1.1
94.51	1.23	2.0	64.92	30MY	3.5	-3.3	-1.4	1.1
100.76	1.23	2.0	74.44	9JN	2.0	-3.4	-1.4	1.2
107.02	1.22	2.0	83.97	19JN	0.7	-3.4	-1.4	1.2
113.31	1.20	2.0	93.51	29JN	-0.5	-3.4	-1.4	1.2
119.63	1.18	2.0	103.08	9JL	-1.4	-3.4	-1.4	1.2
125.99	1.16	2.0	112.68	19JL	-1.4	-3.5	-1.5	1.2
132.38	1.13	2.0	122.32	29JL	-0.6	-3.5	-1.5	1.2
138.81	1.10	2.0	132.02	8AU	-0.2	-3.6	-1.5	1.1
145.29	1.06	1.9	141.76	18AU	0.1	-3.7	-1.6	1.1
151.81	1.02	1.9	151.56	28AU	0.3	-3.8	-1.6	1.1
158.39	0.97	1.9	161.42	7SE	0.5	-3.8	-1.7	1.0
165.02	0.92	1.9	171.34	17SE	1.6	-3.9	-1.7	1.0
171.70	0.86	1.9	181.31	27SE	2.3	-4.1	-1.8	1.0
178.45	0.80	1.9	191.35	7OC	-0.1	-4.2	-1.9	1.0
185.25	0.73	1.9	201.43	17OC	-0.6	-4.3	-1.9	1.0
192.12	0.65	1.8	211.55	27OC	-0.6	-4.4	-2.0	1.0
199.05	0.57	1.8	221.71	6NO	-0.6	-4.4	-2.1	1.0
206.03	0.48	1.8	231.90	16NO	-0.8	-4.3	-2.1	1.0
213.08	0.39	1.7	242.10	26NO	-0.7	-3.8	-2.1	1.0
220.20	0.28	1.7	252.31	6DE	-0.8	-3.0	-2.2	0.9
227.37	0.17	1.6	262.52	16DE	-0.7	-3.7	-2.2	0.9
234.60	0.06	1.6	272.70	26DE	0.0	-4.2	-2.2	0.9
				-109				
241.88	-0.07	1.5	282.86	5JA	2.5	-4.3	-2.1	0.8
249.21	-0.20	1.5	292.99	15JA	1.6	-4.3	-2.1	0.8
256.60	-0.34	1.4	303.06	25JA	0.6	-4.3	-2.1	0.7
264.02	-0.48	1.4	313.09	4FE	0.3	-4.2	-2.0	0.7
271.48	-0.63	1.3	323.06	14FE	0.1	-4.1	-1.9	0.6
278.97	-0.79	1.2	332.98	24FE	-0.2	-3.9	-1.9	0.6
286.47	-0.94	1.2	342.84	6MR	-0.7	-3.8	-1.8	0.5
293.98	-1.10	1.1	352.63	16MR	-1.6	-3.7	-1.7	0.5
301.49	-1.26	1.0	2.37	26MR	-1.4	-3.7	-1.7	0.5
308.97	-1.41	0.9	12.06	5AP	-0.4	-3.6	-1.6	0.6
316.41	-1.56	0.9	21.70	15AP	0.7	-3.5	-1.5	0.7
323.80	-1.70	0.8	31.30	25AP	1.7	-3.5	-1.5	0.7
331.12	-1.84	0.7	40.87	5MY	3.1	-3.4	-1.4	0.8
338.33	-1.96	0.6	50.41	15MY	2.9	-3.4	-1.4	0.9
345.42	-2.07	0.6	59.93	25MY	1.6	-3.3	-1.4	0.9
352.35	-2.16	0.5	69.46	4JN	0.5	-3.3	-1.3	1.0
359.10	-2.24	0.4	78.98	14JN	-0.5	-3.3	-1.3	1.0
5.62	-2.30	0.3	88.51	24JN	-1.5	-3.3	-1.3	1.0
11.86	-2.35	0.2	98.07	4JL	-1.4	-3.3	-1.3	1.0
17.77	-2.37	0.1	107.65	14JL	-0.6	-3.3	-1.3	1.1
23.28	-2.37	-0.1	117.27	24JL	-0.1	-3.3	-1.3	1.1
28.30	-2.35	-0.2	126.94	3AU	0.2	-3.4	-1.3	1.0
32.72	-2.29	-0.4	136.65	13AU	0.4	-3.4	-1.3	1.0
36.39	-2.20	-0.5	146.42	23AU	0.8	-3.4	-1.4	1.0
39.14	-2.07	-0.7	156.25	2SE	2.0	-3.5	-1.4	1.0
40.76	-1.89	-0.9	166.13	12SE	2.0	-3.5	-1.4	0.9
41.06	-1.63	-1.2	176.08	22SE	-0.2	-3.5	-1.5	0.9
39.88	-1.33	-1.4	186.08	2OC	-0.8	-3.5	-1.5	0.9
37.34	-0.88	-1.6	196.13	12OC	-0.8	-3.4	-1.6	0.9
33.81	-0.41	-1.7	206.24	22OC	-0.8	-3.4	-1.7	0.9
30.02	0.08	-1.6	216.38	1NO	-0.7	-3.4	-1.7	0.9
26.80	0.52	-1.4	226.55	11NO	-0.6	-3.4	-1.8	0.9
24.70	0.88	-1.1	236.75	21NO	-0.6	-3.3	-1.9	0.9
23.95	1.16	-0.7	246.96	1DE	-0.6	-3.3	-1.9	0.9
24.51	1.36	-0.4	257.17	11DE	0.2	-3.3	-2.0	0.9
26.18	1.50	-0.1	267.36	21DE	2.8	-3.3	-2.0	0.9
28.77	1.60	0.1	277.54	31DE	1.2	-3.3	-2.1	0.8

Right table (-108 / -107):

♂ LONG	LAT	MAG	☉ LONG	16.00UT	☿	♀	♃	♄
				-108		MAGNITUDES		
32.08	1.66	0.4	287.68	10JA	0.4	-3.4	-2.1	0.8
35.96	1.69	0.6	297.79	20JA	0.2	-3.4	-2.1	0.7
40.28	1.71	0.8	307.84	30JA	0.0	-3.4	-2.1	0.7
44.94	1.71	1.0	317.84	9FE	-0.3	-3.4	-2.0	0.6
49.88	1.71	1.1	327.79	19FE	-0.8	-3.5	-2.0	0.6
55.03	1.69	1.3	337.68	29FE	-1.5	-3.5	-1.9	0.5
60.36	1.67	1.4	347.50	10MR	-1.3	-3.6	-1.9	0.5
65.83	1.64	1.5	357.27	20MR	-0.4	-3.6	-1.8	0.4
71.42	1.61	1.6	6.99	30MR	0.9	-3.7	-1.8	0.4
77.10	1.58	1.7	16.65	9AP	2.3	-3.8	-1.7	0.4
82.88	1.54	1.8	26.27	19AP	3.8	-3.8	-1.6	0.5
88.73	1.50	1.8	35.86	29AP	2.2	-3.9	-1.6	0.6
94.64	1.45	1.9	45.41	9MY	1.2	-4.0	-1.5	0.6
100.63	1.41	1.9	54.94	19MY	0.4	-4.1	-1.4	0.7
106.67	1.35	1.9	64.46	29MY	-0.6	-4.2	-1.4	0.8
112.78	1.30	2.0	73.98	8JN	-1.5	-4.2	-1.4	0.8
118.95	1.25	2.0	83.51	18JN	-1.4	-4.0	-1.3	0.9
125.18	1.19	2.0	93.05	28JN	-0.6	-3.7	-1.3	0.9
131.47	1.13	2.0	102.62	8JL	-0.0	-3.1	-1.3	0.9
137.82	1.06	2.0	112.22	18JL	0.3	-3.5	-1.2	0.9
144.25	0.99	2.0	121.86	28JL	0.6	-4.0	-1.2	1.0
150.74	0.92	1.9	131.54	7AU	1.1	-4.2	-1.2	1.0
157.30	0.85	1.9	141.29	17AU	2.5	-4.3	-1.2	1.0
163.94	0.77	1.9	151.08	27AU	1.8	-4.2	-1.2	0.9
170.66	0.69	1.8	160.94	6SE	-0.3	-4.1	-1.3	0.9
177.46	0.60	1.8	170.86	16SE	-0.9	-4.1	-1.3	0.9
184.33	0.51	1.7	180.83	26SE	-0.9	-4.0	-1.3	0.8
191.28	0.42	1.7	190.86	6OC	-0.9	-3.9	-1.4	0.8
198.32	0.33	1.7	200.94	16OC	-0.6	-3.8	-1.4	0.8
205.44	0.23	1.7	211.06	26OC	-0.5	-3.7	-1.5	0.8
212.64	0.12	1.7	221.22	5NO	-0.5	-3.7	-1.5	0.8
219.92	0.02	1.6	231.41	15NO	-0.4	-3.6	-1.6	0.8
227.27	-0.09	1.6	241.61	25NO	0.3	-3.6	-1.6	0.9
234.69	-0.20	1.6	251.82	5DE	3.0	-3.5	-1.7	0.8
242.19	-0.31	1.6	262.03	15DE	0.9	-3.5	-1.8	0.8
249.74	-0.43	1.5	272.21	25DE	0.2	-3.4	-1.8	0.8
				-107				
257.35	-0.54	1.5	282.37	4JA	0.0	-3.4	-1.9	0.8
265.01	-0.65	1.5	292.50	14JA	-0.1	-3.4	-1.9	0.8
272.71	-0.76	1.5	302.58	24JA	-0.3	-3.3	-2.0	0.7
280.44	-0.87	1.4	312.61	3FE	-0.8	-3.3	-2.0	0.7
288.19	-0.97	1.4	322.59	13FE	-1.4	-3.3	-2.0	0.6
295.95	-1.06	1.4	332.50	23FE	-1.3	-3.3	-2.0	0.6
303.70	-1.15	1.3	342.36	5MR	-0.3	-3.3	-2.0	0.5
311.44	-1.23	1.3	352.16	15MR	1.1	-3.3	-2.0	0.5
319.15	-1.29	1.3	1.90	25MR	2.9	-3.4	-1.9	0.4
326.82	-1.35	1.3	11.59	4AP	2.9	-3.4	-1.9	0.4
334.43	-1.39	1.2	21.24	14AP	1.6	-3.4	-1.8	0.3
341.97	-1.41	1.2	30.83	24AP	0.9	-3.5	-1.8	0.4
349.43	-1.43	1.2	40.40	4MY	0.3	-3.5	-1.7	0.5
356.80	-1.42	1.2	49.95	14MY	-0.6	-3.5	-1.6	0.5
4.06	-1.41	1.1	59.47	24MY	-1.5	-3.4	-1.6	0.6
11.20	-1.37	1.1	68.99	3JN	-1.4	-3.4	-1.5	0.7
18.20	-1.33	1.1	78.51	13JN	-0.5	-3.4	-1.5	0.7
25.05	-1.26	1.0	88.04	23JN	0.1	-3.3	-1.4	0.8
31.73	-1.18	1.0	97.59	3JL	0.5	-3.3	-1.4	0.8
38.23	-1.09	0.9	107.18	13JL	0.8	-3.3	-1.3	0.8
44.51	-0.98	0.9	116.79	23JL	1.5	-3.3	-1.3	0.9
50.56	-0.85	0.8	126.46	2AU	3.0	-3.3	-1.3	0.9
56.32	-0.70	0.7	136.17	12AU	1.5	-3.3	-1.2	0.9
61.76	-0.54	0.6	145.94	22AU	-0.3	-3.4	-1.2	0.9
66.81	-0.35	0.5	155.76	1SE	-1.0	-3.4	-1.2	0.9
71.40	-0.13	0.4	165.65	11SE	-1.1	-3.4	-1.2	0.9
75.43	0.12	0.2	175.59	21SE	-0.8	-3.5	-1.2	0.8
78.76	0.41	0.1	185.59	1OC	-0.5	-3.5	-1.2	0.8
81.22	0.74	-0.1	195.65	11OC	-0.3	-3.6	-1.3	0.8
82.65	1.11	-0.4	205.75	21OC	-0.3	-3.6	-1.3	0.7
82.83	1.54	-0.6	215.89	31OC	-0.2	-3.7	-1.3	0.8
81.62	1.99	-0.8	226.06	10NO	0.5	-3.8	-1.4	0.8
79.07	2.44	-1.0	236.26	20NO	3.2	-3.9	-1.4	0.8
75.48	2.84	-1.2	246.47	30NO	0.6	-4.0	-1.5	0.8
71.50	3.13	-1.2	256.68	10DE	-0.0	-4.1	-1.5	0.8
67.91	3.28	-1.0	266.87	20DE	-0.1	-4.2	-1.6	0.8
65.31	3.30	-0.7	277.05	30DE	-0.2	-4.3	-1.7	0.8

♂ LONG	♂ LAT	♂ MAG	⊙ LONG	16.00UT -106	☿ MAGNITUDES	♀	♃	♄
64.01	3.23	-0.4	287.19	9JA	-0.4	-4.3	-1.7	0.8
64.02	3.11	-0.1	297.30	19JA	-0.8	-4.3	-1.8	0.8
65.18	2.95	0.1	307.36	29JA	-1.3	-4.2	-1.9	0.7
67.30	2.79	0.4	317.36	8FE	-1.2	-3.7	-1.9	0.7
70.20	2.64	0.6	327.31	18FE	-0.3	-3.3	-2.0	0.6
73.72	2.48	0.8	337.20	28FE	1.4	-3.7	-2.0	0.6
77.73	2.34	0.9	347.03	10MR	3.4	-4.1	-2.0	0.5
82.13	2.20	1.1	356.80	20MR	2.1	-4.2	-2.0	0.5
86.84	2.07	1.2	6.52	30MR	1.2	-4.2	-2.0	0.4
91.82	1.94	1.3	16.18	9AP	0.7	-4.2	-2.0	0.4
97.02	1.82	1.4	25.80	19AP	0.1	-4.1	-2.0	0.3
102.40	1.71	1.5	35.39	29AP	-0.6	-4.0	-1.9	0.3
107.94	1.59	1.6	44.94	9MY	-1.5	-3.9	-1.9	0.4
113.62	1.48	1.7	54.48	19MY	-1.4	-3.8	-1.8	0.4
119.44	1.37	1.7	64.00	29MY	-0.5	-3.7	-1.8	0.5
125.37	1.27	1.8	73.52	8JN	0.2	-3.6	-1.7	0.6
131.41	1.16	1.8	83.04	18JN	0.6	-3.5	-1.6	0.6
137.56	1.06	1.8	92.59	28JN	1.1	-3.5	-1.6	0.7
143.82	0.95	1.8	102.15	8JL	1.9	-3.5	-1.5	0.7
150.17	0.84	1.8	111.75	18JL	3.2	-3.4	-1.5	0.8
156.63	0.74	1.8	121.39	28JL	1.3	-3.4	-1.4	0.8
163.19	0.63	1.8	131.07	7AU	-0.4	-3.4	-1.4	0.8
169.85	0.53	1.8	140.81	17AU	-1.2	-3.4	-1.3	0.8
176.62	0.42	1.8	150.61	27AU	-1.2	-3.4	-1.3	0.8
183.48	0.31	1.7	160.46	6SE	-0.8	-3.4	-1.3	0.8
190.44	0.20	1.7	170.37	16SE	-0.4	-3.4	-1.3	0.8
197.51	0.09	1.7	180.35	26SE	-0.2	-3.4	-1.2	0.8
204.67	-0.02	1.6	190.37	6OC	-0.1	-3.4	-1.2	0.8
211.93	-0.12	1.6	200.45	16OC	-0.0	-3.4	-1.2	0.7
219.28	-0.23	1.6	210.57	26OC	0.8	-3.4	-1.3	0.7
226.72	-0.34	1.5	220.73	5NO	3.0	-3.4	-1.3	0.7
234.24	-0.44	1.5	230.91	15NO	0.4	-3.5	-1.3	0.7
241.84	-0.54	1.4	241.12	25NO	-0.2	-3.5	-1.3	0.8
249.50	-0.63	1.4	251.32	5DE	-0.3	-3.5	-1.4	0.8
257.23	-0.72	1.4	261.53	15DE	-0.3	-3.5	-1.4	0.8
265.01	-0.81	1.4	271.72	25DE	-0.5	-3.4	-1.5	0.8
				-105				
272.82	-0.89	1.4	281.88	4JA	-0.9	-3.4	-1.5	0.8
280.67	-0.95	1.4	292.01	14JA	-1.2	-3.4	-1.6	0.8
288.53	-1.01	1.4	302.09	24JA	-1.1	-3.4	-1.7	0.8
296.39	-1.06	1.4	312.12	3FE	-0.2	-3.4	-1.7	0.7
304.24	-1.10	1.4	322.10	13FE	1.6	-3.3	-1.8	0.7
312.08	-1.13	1.4	332.02	23FE	3.0	-3.3	-1.9	0.7
319.87	-1.14	1.4	341.88	5MR	1.6	-3.3	-1.9	0.6
327.63	-1.15	1.4	351.68	15MR	0.9	-3.3	-2.0	0.6
335.33	-1.14	1.4	1.43	25MR	0.5	-3.4	-2.0	0.5
342.96	-1.11	1.4	11.12	4AP	0.0	-3.4	-2.1	0.4
350.51	-1.08	1.4	20.77	14AP	-0.6	-3.4	-2.1	0.4
357.99	-1.03	1.5	30.37	24AP	-1.6	-3.4	-2.1	0.3
5.37	-0.98	1.5	39.94	4MY	-1.5	-3.5	-2.1	0.3
12.66	-0.91	1.5	49.48	14MY	-0.5	-3.5	-2.1	0.3
19.84	-0.83	1.5	59.01	24MY	0.3	-3.6	-2.0	0.4
26.92	-0.75	1.5	68.53	3JN	0.9	-3.6	-2.0	0.4
33.88	-0.65	1.5	78.05	13JN	1.5	-3.7	-1.9	0.5
40.73	-0.54	1.5	87.58	23JN	2.5	-3.8	-1.9	0.6
47.45	-0.43	1.4	97.13	3JL	2.9	-3.9	-1.8	0.6
54.04	-0.31	1.4	106.72	13JL	1.1	-4.0	-1.7	0.7
60.49	-0.17	1.4	116.33	23JL	-0.4	-4.1	-1.7	0.7
66.79	-0.03	1.4	125.99	2AU	-1.3	-4.2	-1.6	0.7
72.93	0.12	1.3	135.70	12AU	-1.3	-4.3	-1.6	0.8
78.88	0.28	1.3	145.47	22AU	-0.7	-4.3	-1.5	0.8
84.63	0.45	1.2	155.28	1SE	-0.3	-4.1	-1.5	0.8
90.14	0.64	1.1	165.17	11SE	-0.1	-3.8	-1.4	0.8
95.38	0.84	1.0	175.11	21SE	0.0	-3.3	-1.4	0.8
100.30	1.07	0.9	185.10	1OC	0.2	-3.7	-1.4	0.8
104.82	1.31	0.8	195.16	11OC	1.0	-4.2	-1.3	0.8
108.86	1.59	0.6	205.26	21OC	2.8	-4.3	-1.3	0.7
112.32	1.90	0.5	215.39	31OC	0.2	-4.4	-1.3	0.7
115.05	2.24	0.3	225.57	10NO	-0.4	-4.3	-1.3	0.7
116.88	2.62	0.1	235.76	20NO	-0.4	-4.2	-1.3	0.7
117.64	3.03	-0.2	245.97	30NO	-0.4	-4.1	-1.3	0.7
117.13	3.46	-0.4	256.18	10DE	-0.6	-4.0	-1.4	0.8
115.30	3.88	-0.7	266.38	20DE	-0.9	-3.9	-1.4	0.8
112.26	4.22	-0.9	276.56	30DE	-1.0	-3.8	-1.4	0.8

♂ LONG	♂ LAT	♂ MAG	⊙ LONG	16.00UT -104	☿ MAGNITUDES	♀	♃	♄
108.43	4.42	-1.0	286.71	9JA	-0.9	-3.7	-1.5	0.8
104.52	4.45	-0.9	296.81	19JA	-0.2	-3.6	-1.5	0.8
101.23	4.32	-0.7	306.87	29JA	1.9	-3.6	-1.6	0.8
99.05	4.07	-0.5	316.88	8FE	2.4	-3.5	-1.6	0.8
98.17	3.76	-0.2	326.83	18FE	1.1	-3.5	-1.7	0.7
98.54	3.43	0.0	336.72	28FE	0.6	-3.4	-1.8	0.7
100.00	3.11	0.3	346.56	9MR	0.3	-3.4	-1.9	0.6
102.37	2.81	0.5	356.33	19MR	-0.0	-3.3	-1.9	0.6
105.47	2.53	0.6	6.05	29MR	-0.7	-3.3	-2.0	0.5
109.17	2.27	0.8	15.72	8AP	-1.5	-3.3	-2.1	0.5
113.35	2.04	0.9	25.35	18AP	-1.4	-3.3	-2.1	0.4
117.93	1.82	1.1	34.93	28AP	-0.5	-3.3	-2.2	0.3
122.84	1.62	1.2	44.49	8MY	0.4	-3.3	-2.2	0.3
128.03	1.43	1.3	54.02	18MY	1.1	-3.3	-2.2	0.3
133.46	1.25	1.3	63.54	28MY	2.0	-3.4	-2.2	0.3
139.10	1.08	1.4	73.06	7JN	3.2	-3.4	-2.2	0.4
144.94	0.92	1.4	82.59	17JN	2.4	-3.4	-2.2	0.4
150.95	0.77	1.5	92.12	27JN	0.9	-3.5	-2.1	0.5
157.12	0.62	1.5	101.69	7JL	-0.5	-3.5	-2.1	0.6
163.45	0.47	1.5	111.28	17JL	-1.3	-3.5	-2.0	0.6
169.92	0.33	1.6	120.92	27JL	-1.4	-3.4	-2.0	0.7
176.53	0.20	1.6	130.60	6AU	-0.7	-3.4	-1.9	0.7
183.28	0.07	1.6	140.34	16AU	-0.2	-3.4	-1.8	0.7
190.15	-0.06	1.6	150.13	26AU	0.0	-3.4	-1.8	0.8
197.15	-0.18	1.6	159.98	5SE	0.2	-3.3	-1.7	0.8
204.27	-0.29	1.6	169.88	15SE	0.4	-3.3	-1.7	0.8
211.51	-0.41	1.5	179.85	25SE	1.3	-3.3	-1.6	0.8
218.86	-0.51	1.5	189.88	5OC	2.5	-3.3	-1.6	0.8
226.31	-0.61	1.5	199.95	15OC	0.0	-3.3	-1.5	0.8
233.85	-0.70	1.5	210.07	25OC	-0.5	-3.4	-1.5	0.8
241.48	-0.78	1.5	220.23	4NO	-0.6	-3.4	-1.5	0.7
249.19	-0.86	1.4	230.41	14NO	-0.6	-3.4	-1.4	0.7
256.96	-0.93	1.4	240.62	24NO	-0.7	-3.4	-1.4	0.7
264.79	-0.98	1.4	250.83	4DE	-0.8	-3.5	-1.4	0.7
272.65	-1.03	1.4	261.03	14DE	-0.9	-3.5	-1.4	0.8
280.54	-1.06	1.4	271.22	24DE	-0.8	-3.6	-1.4	0.8
				-103				
288.44	-1.08	1.3	281.39	3JA	-0.1	-3.6	-1.4	0.8
296.34	-1.09	1.3	291.52	13JA	2.2	-3.7	-1.5	0.8
304.23	-1.09	1.3	301.61	23JA	1.9	-3.8	-1.5	0.8
312.09	-1.08	1.3	311.64	2FE	0.8	-3.9	-1.5	0.8
319.90	-1.05	1.3	321.62	12FE	0.4	-4.0	-1.6	0.8
327.67	-1.01	1.4	331.55	22FE	0.2	-4.0	-1.6	0.8
335.38	-0.97	1.4	341.41	4MR	-0.1	-4.1	-1.7	0.7
343.02	-0.91	1.4	351.21	14MR	-0.7	-4.2	-1.8	0.7
350.58	-0.85	1.5	0.96	24MR	-1.5	-4.2	-1.8	0.6
358.06	-0.78	1.5	10.66	3AP	-1.4	-4.2	-1.9	0.6
5.46	-0.70	1.5	20.30	13AP	-0.4	-3.9	-2.0	0.5
12.76	-0.61	1.6	29.91	23AP	0.5	-3.3	-2.0	0.5
19.98	-0.52	1.6	39.48	3MY	1.5	-3.0	-2.1	0.4
27.10	-0.42	1.6	49.02	13MY	2.6	-3.7	-2.2	0.3
34.13	-0.32	1.7	58.55	23MY	3.3	-4.1	-2.2	0.3
41.06	-0.21	1.7	68.07	2JN	1.9	-4.2	-2.3	0.3
47.90	-0.10	1.7	77.59	12JN	0.7	-4.2	-2.3	0.3
54.65	0.01	1.7	87.12	22JN	-0.5	-4.1	-2.3	0.4
61.30	0.12	1.7	96.67	2JL	-1.4	-4.0	-2.3	0.4
67.86	0.24	1.7	106.25	12JL	-1.4	-3.9	-2.3	0.5
74.33	0.36	1.7	115.86	22JL	-0.6	-3.8	-2.3	0.6
80.69	0.49	1.7	125.52	1AU	-0.2	-3.8	-2.3	0.6
86.95	0.61	1.7	135.23	11AU	0.1	-3.7	-2.2	0.7
93.10	0.75	1.7	144.99	21AU	0.3	-3.6	-2.1	0.7
99.13	0.88	1.6	154.81	31AU	0.6	-3.6	-2.1	0.7
105.03	1.03	1.6	164.68	10SE	1.7	-3.5	-2.0	0.8
110.79	1.18	1.5	174.62	20SE	2.3	-3.5	-1.9	0.8
116.38	1.34	1.4	184.62	30SE	-0.1	-3.5	-1.9	0.8
121.78	1.50	1.4	194.66	10OC	-0.7	-3.4	-1.8	0.8
126.95	1.69	1.3	204.76	20OC	-0.7	-3.4	-1.8	0.8
131.84	1.88	1.1	214.90	30OC	-0.7	-3.4	-1.7	0.8
136.41	2.10	1.0	225.07	9NO	-0.8	-3.4	-1.7	0.8
140.57	2.33	0.8	235.27	19NO	-0.7	-3.4	-1.6	0.7
144.24	2.58	0.7	245.47	29NO	-0.7	-3.4	-1.6	0.7
147.28	2.87	0.5	255.68	9DE	-0.7	-3.4	-1.5	0.7
149.56	3.17	0.2	265.88	19DE	0.1	-3.4	-1.5	0.7
150.89	3.50	-0.0	276.06	29DE	2.5	-3.4	-1.5	0.8

♂ LONG	♂ LAT	♂ MAG	☉ LONG	16.00UT	☿	♀	♃	♄
				-102	MAGNITUDES			
151.10	3.83	-0.3	286.21	8JA	1.4	-3.4	-1.5	0.8
150.03	4.13	-0.6	296.32	18JA	0.5	-3.4	-1.5	0.8
147.70	4.37	-0.8	306.38	28JA	0.3	-3.4	-1.5	0.8
144.33	4.47	-1.0	316.39	7FE	0.1	-3.5	-1.5	0.8
140.45	4.41	-1.1	326.35	17FE	-0.2	-3.5	-1.6	0.8
136.80	4.18	-0.9	336.24	27FE	-0.7	-3.5	-1.6	0.8
134.00	3.81	-0.7	346.08	9MR	-1.5	-3.4	-1.6	0.8
132.41	3.39	-0.5	355.86	19MR	-1.4	-3.4	-1.7	0.7
132.14	2.94	-0.3	5.58	29MR	-0.4	-3.4	-1.7	0.7
133.08	2.52	-0.1	15.25	8AP	0.7	-3.4	-1.8	0.6
135.06	2.13	0.1	24.88	18AP	1.9	-3.3	-1.8	0.6
137.91	1.77	0.3	34.46	28AP	3.5	-3.3	-1.9	0.5
141.46	1.45	0.4	44.02	8MY	2.6	-3.3	-2.0	0.5
145.61	1.16	0.6	53.56	18MY	1.4	-3.3	-2.1	0.4
150.23	0.89	0.7	63.08	28MY	0.5	-3.3	-2.1	0.3
155.25	0.65	0.8	72.60	7JN	-0.5	-3.4	-2.2	0.3
160.63	0.43	0.9	82.12	17JN	-1.5	-3.4	-2.3	0.3
166.31	0.23	1.0	91.66	27JN	-1.4	-3.4	-2.3	0.4
172.24	0.04	1.0	101.22	7JL	-0.6	-3.4	-2.4	0.4
178.42	-0.13	1.1	110.82	17JL	-0.1	-3.5	-2.4	0.5
184.81	-0.29	1.1	120.45	27JL	0.3	-3.5	-2.4	0.5
191.40	-0.44	1.2	130.13	6AU	0.5	-3.6	-2.4	0.6
198.16	-0.57	1.2	139.87	16AU	0.9	-3.7	-2.4	0.6
205.09	-0.69	1.2	149.65	26AU	2.1	-3.8	-2.4	0.7
212.18	-0.80	1.2	159.50	5SE	2.0	-3.8	-2.3	0.7
219.41	-0.90	1.3	169.41	15SE	-0.2	-4.0	-2.3	0.8
226.76	-0.99	1.3	179.37	25SE	-0.8	-4.1	-2.2	0.8
234.23	-1.06	1.3	189.39	5OC	-0.8	-4.2	-2.2	0.8
241.80	-1.12	1.3	199.46	15OC	-0.8	-4.3	-2.1	0.8
249.46	-1.16	1.3	209.58	25OC	-0.7	-4.4	-2.0	0.8
257.19	-1.19	1.3	219.73	4NO	-0.5	-4.4	-2.0	0.8
264.99	-1.21	1.3	229.92	14NO	-0.5	-4.2	-1.9	0.8
272.82	-1.22	1.4	240.12	24NO	-0.5	-3.8	-1.8	0.8
280.68	-1.21	1.4	250.33	4DE	0.2	-2.9	-1.8	0.7
288.56	-1.19	1.4	260.53	14DE	2.8	-3.7	-1.7	0.7
296.42	-1.15	1.4	270.72	24DE	1.1	-4.2	-1.7	0.8
				-101				
304.28	-1.11	1.4	280.89	3JA	0.3	-4.3	-1.6	0.8
312.10	-1.05	1.4	291.02	13JA	0.1	-4.3	-1.6	0.8
319.87	-0.99	1.4	301.11	23JA	-0.0	-4.3	-1.6	0.8
327.59	-0.92	1.5	311.15	2FE	-0.3	-4.2	-1.6	0.9
335.25	-0.84	1.5	321.14	12FE	-0.8	-4.1	-1.6	0.9
342.84	-0.75	1.5	331.06	22FE	-1.5	-3.9	-1.6	0.9
350.35	-0.66	1.5	340.93	4MR	-1.3	-3.8	-1.6	0.9
357.78	-0.57	1.5	350.74	14MR	-0.4	-3.7	-1.6	0.8
5.13	-0.47	1.5	0.49	24MR	0.9	-3.7	-1.6	0.8
12.39	-0.37	1.6	10.19	3AP	2.5	-3.6	-1.6	0.8
19.57	-0.27	1.6	19.84	13AP	3.4	-3.5	-1.7	0.7
26.66	-0.17	1.6	29.45	23AP	2.0	-3.5	-1.7	0.7
33.66	-0.06	1.7	39.02	3MY	1.1	-3.4	-1.8	0.6
40.59	0.04	1.7	48.57	13MY	0.4	-3.4	-1.8	0.5
47.43	0.14	1.8	58.09	23MY	-0.5	-3.3	-1.9	0.5
54.21	0.24	1.8	67.62	2JN	-1.5	-3.3	-1.9	0.4
60.90	0.34	1.8	77.13	12JN	-1.4	-3.3	-2.0	0.3
67.53	0.44	1.9	86.66	22JN	-0.6	-3.3	-2.1	0.3
74.09	0.54	1.9	96.21	2JL	0.0	-3.3	-2.1	0.3
80.58	0.64	1.9	105.79	12JL	0.4	-3.3	-2.2	0.4
87.01	0.73	1.9	115.40	22JL	0.7	-3.3	-2.3	0.4
93.39	0.83	1.9	125.06	1AU	1.2	-3.4	-2.3	0.5
99.69	0.92	1.9	134.76	11AU	2.6	-3.4	-2.4	0.6
105.94	1.02	1.9	144.52	21AU	1.8	-3.4	-2.4	0.6
112.13	1.12	1.9	154.33	31AU	-0.3	-3.5	-2.4	0.7
118.24	1.21	1.9	164.21	10SE	-1.0	-3.5	-2.5	0.7
124.28	1.31	1.8	174.14	20SE	-1.0	-3.5	-2.4	0.7
130.25	1.41	1.8	184.13	30SE	-0.9	-3.5	-2.4	0.8
136.12	1.51	1.7	194.17	10OC	-0.6	-3.4	-2.4	0.8
141.89	1.62	1.6	204.27	20OC	-0.4	-3.4	-2.3	0.8
147.53	1.72	1.5	214.41	30OC	-0.4	-3.4	-2.3	0.8
153.03	1.83	1.4	224.57	9NO	-0.3	-3.4	-2.2	0.8
158.37	1.95	1.3	234.77	19NO	0.4	-3.3	-2.1	0.8
163.50	2.07	1.2	244.97	29NO	3.1	-3.3	-2.0	0.8
168.38	2.20	1.1	255.18	9DE	0.8	-3.3	-2.0	0.8
172.95	2.33	0.9	265.38	19DE	0.1	-3.3	-1.9	0.8
177.14	2.47	0.7	275.56	29DE	-0.0	-3.3	-1.8	0.8

♂ LONG	♂ LAT	♂ MAG	☉ LONG	16.00UT	☿	♀	♃	♄
				-100	MAGNITUDES			
180.87	2.62	0.5	285.71	8JA	-0.1	-3.4	-1.8	0.8
184.00	2.77	0.3	295.82	18JA	-0.4	-3.4	-1.7	0.9
186.40	2.91	-0.0	305.89	28JA	-0.8	-3.4	-1.7	0.9
187.90	3.04	-0.3	315.90	7FE	-1.4	-3.4	-1.6	0.9
188.31	3.14	-0.6	325.86	17FE	-1.3	-3.5	-1.6	0.9
187.47	3.18	-0.9	335.76	27FE	-0.4	-3.5	-1.6	0.9
185.39	3.13	-1.2	345.60	8MR	1.2	-3.6	-1.6	0.9
182.26	2.96	-1.4	355.38	18MR	3.1	-3.6	-1.6	0.9
178.59	2.64	-1.5	5.11	28MR	2.6	-3.7	-1.6	0.9
175.11	2.21	-1.4	14.78	7AP	1.5	-3.8	-1.6	0.9
172.46	1.72	-1.2	24.41	17AP	0.8	-3.8	-1.6	0.8
171.06	1.22	-1.0	34.01	27AP	0.2	-3.9	-1.6	0.8
171.02	0.76	-0.8	43.56	7MY	-0.6	-4.0	-1.6	0.7
172.23	0.34	-0.6	53.10	17MY	-1.5	-4.1	-1.6	0.6
174.55	-0.02	-0.4	62.63	27MY	-1.5	-4.2	-1.7	0.6
177.79	-0.33	-0.2	72.14	6JN	-0.6	-4.2	-1.7	0.5
181.78	-0.60	-0.1	81.67	16JN	0.1	-4.0	-1.8	0.4
186.41	-0.83	0.1	91.20	26JN	0.5	-3.6	-1.8	0.4
191.55	-1.03	0.2	100.76	6JL	0.9	-3.0	-1.9	0.4
197.14	-1.19	0.3	110.36	16JL	1.6	-3.5	-2.0	0.4
203.09	-1.33	0.4	119.99	26JL	3.1	-4.0	-2.0	0.5
209.36	-1.44	0.5	129.66	5AU	1.5	-4.2	-2.1	0.5
215.90	-1.53	0.6	139.39	15AU	-0.3	-4.3	-2.2	0.6
222.69	-1.59	0.7	149.18	25AU	-1.1	-4.2	-2.2	0.6
229.67	-1.64	0.7	159.02	4SE	-1.1	-4.1	-2.2	0.7
236.82	-1.66	0.8	168.92	14SE	-0.8	-4.1	-2.3	0.7
244.13	-1.67	0.8	178.89	24SE	-0.5	-4.0	-2.4	0.8
251.55	-1.66	0.9	188.90	4OC	-0.3	-3.9	-2.4	0.8
259.08	-1.63	1.0	198.97	14OC	-0.2	-3.8	-2.4	0.8
266.69	-1.58	1.0	209.09	24OC	-0.1	-3.7	-2.4	0.8
274.35	-1.52	1.1	219.24	3NO	0.6	-3.7	-2.4	0.9
282.04	-1.45	1.1	229.42	13NO	3.2	-3.6	-2.3	0.9
289.76	-1.36	1.2	239.62	23NO	0.5	-3.5	-2.3	0.9
297.48	-1.27	1.2	249.83	3DE	-0.1	-3.5	-2.2	0.9
305.18	-1.17	1.3	260.04	13DE	-0.2	-3.5	-2.1	0.9
312.85	-1.06	1.4	270.23	23DE	-0.3	-3.4	-2.1	0.9
				-99				
320.48	-0.95	1.4	280.40	2JA	-0.5	-3.4	-2.0	0.8
328.06	-0.84	1.5	290.53	12JA	-0.8	-3.4	-1.9	0.9
335.58	-0.72	1.5	300.62	22JA	-1.3	-3.3	-1.9	0.9
343.03	-0.60	1.5	310.66	1FE	-1.2	-3.3	-1.8	0.9
350.41	-0.48	1.6	320.65	11FE	-0.3	-3.3	-1.7	1.0
357.71	-0.37	1.6	330.58	21FE	1.4	-3.3	-1.7	1.0
4.94	-0.25	1.7	340.45	3MR	3.4	-3.3	-1.6	1.0
12.09	-0.14	1.7	350.26	13MR	1.9	-3.3	-1.6	1.0
19.16	-0.03	1.7	0.02	23MR	1.1	-3.4	-1.5	1.0
26.15	0.07	1.8	9.71	2AP	0.6	-3.4	-1.5	1.0
33.07	0.17	1.8	19.37	12AP	0.1	-3.4	-1.5	1.0
39.92	0.27	1.8	28.98	22AP	-0.6	-3.5	-1.5	0.9
46.70	0.36	1.8	38.55	2MY	-1.5	-3.5	-1.5	0.9
53.42	0.45	1.8	48.10	12MY	-1.5	-3.5	-1.5	0.8
60.07	0.54	1.8	57.63	22MY	-0.5	-3.4	-1.5	0.8
66.68	0.62	1.8	67.15	1JN	0.2	-3.4	-1.5	0.7
73.23	0.70	1.9	76.67	11JN	0.7	-3.4	-1.5	0.6
79.74	0.78	1.9	86.20	21JN	1.2	-3.3	-1.6	0.6
86.21	0.85	1.9	95.74	1JL	2.1	-3.3	-1.6	0.5
92.64	0.92	2.0	105.32	11JL	3.2	-3.3	-1.6	0.5
99.03	0.98	2.0	114.93	21JL	1.3	-3.3	-1.7	0.4
105.40	1.04	2.0	124.58	31JL	-0.4	-3.3	-1.7	0.5
111.74	1.10	2.0	134.29	10AU	-1.2	-3.3	-1.8	0.5
118.05	1.16	2.0	144.04	20AU	-1.3	-3.4	-1.9	0.6
124.34	1.21	2.0	153.85	30AU	-0.8	-3.4	-1.9	0.6
130.61	1.26	2.0	163.72	9SE	-0.4	-3.4	-2.0	0.7
136.85	1.31	2.0	173.65	19SE	-0.2	-3.5	-2.1	0.7
143.07	1.36	1.9	183.64	29SE	-0.1	-3.5	-2.1	0.8
149.25	1.40	1.9	193.69	9OC	0.0	-3.6	-2.2	0.8
155.41	1.44	1.8	203.78	19OC	0.8	-3.6	-2.2	0.9
161.53	1.48	1.8	213.91	29OC	3.1	-3.7	-2.3	0.9
167.61	1.51	1.7	224.08	8NO	0.3	-3.8	-2.3	0.9
173.63	1.54	1.6	234.27	18NO	-0.3	-3.9	-2.3	0.9
179.59	1.57	1.5	244.48	28NO	-0.3	-4.0	-2.3	1.0
185.48	1.59	1.4	254.69	8DE	-0.4	-4.1	-2.3	1.0
191.27	1.60	1.3	264.89	18DE	-0.5	-4.2	-2.2	1.0
196.95	1.60	1.2	275.07	28DE	-0.9	-4.3	-2.2	1.0

♂ LONG	♂ LAT	♂ MAG	⊙ LONG	16.00UT -98	☿	♀	♃	♄
					MAGNITUDES			
202.50	1.60	1.0	285.22	7JA	-1.1	-4.3	-2.1	0.9
207.88	1.58	0.9	295.33	17JA	-1.0	-4.3	-2.0	0.9
213.05	1.55	0.7	305.40	27JA	-0.2	-4.2	-2.0	1.0
217.96	1.49	0.5	315.42	6FE	1.7	-3.7	-1.9	1.0
222.56	1.41	0.3	325.38	16FE	2.8	-3.3	-1.8	1.1
226.76	1.30	0.0	335.28	26FE	1.4	-3.7	-1.8	1.1
230.44	1.14	-0.2	345.12	8MR	0.8	-4.1	-1.7	1.1
233.49	0.93	-0.5	354.90	18MR	0.4	-4.2	-1.6	1.1
235.73	0.65	-0.8	4.63	28MR	0.0	-4.2	-1.6	1.1
236.99	0.28	-1.1	14.31	7AP	-0.6	-4.2	-1.5	1.1
237.09	-0.18	-1.5	23.94	17AP	-1.5	-4.1	-1.5	1.1
235.95	-0.73	-1.8	33.54	27AP	-1.5	-4.0	-1.5	1.1
233.67	-1.35	-2.1	43.10	7MY	-0.5	-3.8	-1.4	1.0
230.68	-1.99	-2.2	52.63	17MY	0.3	-3.8	-1.4	1.0
227.61	-2.56	-2.1	62.16	27MY	0.9	-3.7	-1.4	0.9
225.22	-3.01	-2.0	71.68	6JN	1.6	-3.6	-1.4	0.9
224.00	-3.32	-1.8	81.20	16JN	2.8	-3.5	-1.4	0.8
224.15	-3.49	-1.5	90.74	26JN	2.7	-3.5	-1.4	0.7
225.66	-3.57	-1.3	100.30	6JL	1.0	-3.4	-1.4	0.7
228.33	-3.56	-1.1	109.89	16JL	-0.4	-3.4	-1.4	0.6
231.99	-3.50	-0.9	119.52	26JL	-1.3	-3.4	-1.5	0.6
236.46	-3.40	-0.7	129.20	5AU	-1.4	-3.4	-1.5	0.6
241.57	-3.26	-0.5	138.92	15AU	-0.7	-3.4	-1.5	0.6
247.20	-3.11	-0.3	148.71	25AU	-0.3	-3.4	-1.6	0.6
253.24	-2.93	-0.2	158.55	4SE	-0.1	-3.4	-1.6	0.7
259.60	-2.74	-0.0	168.44	14SE	0.1	-3.4	-1.7	0.7
266.21	-2.54	0.1	178.40	24SE	0.2	-3.4	-1.7	0.8
273.03	-2.34	0.2	188.42	4OC	1.1	-3.4	-1.8	0.8
279.99	-2.13	0.4	198.48	14OC	2.8	-3.4	-1.9	0.9
287.06	-1.92	0.5	208.60	24OC	0.2	-3.4	-1.9	0.9
294.21	-1.71	0.6	218.75	3NO	-0.4	-3.4	-2.0	1.0
301.41	-1.50	0.7	228.93	13NO	-0.5	-3.5	-2.1	1.0
308.63	-1.30	0.9	239.13	23NO	-0.5	-3.5	-2.1	1.0
315.86	-1.11	1.0	249.34	3DE	-0.6	-3.5	-2.1	1.1
323.08	-0.93	1.1	259.54	13DE	-0.9	-3.5	-2.2	1.1
330.28	-0.75	1.2	269.74	23DE	-0.9	-3.4	-2.2	1.1
				-97				
337.44	-0.58	1.3	279.90	2JA	-0.9	-3.4	-2.2	1.1
344.56	-0.42	1.4	290.04	12JA	-0.2	-3.4	-2.1	1.1
351.63	-0.27	1.4	300.13	22JA	2.0	-3.4	-2.1	1.1
358.65	-0.13	1.5	310.18	1FE	2.2	-3.3	-2.0	1.1
5.61	-0.00	1.6	320.16	11FE	1.0	-3.3	-2.0	1.1
12.52	0.12	1.7	330.10	21FE	0.6	-3.3	-1.9	1.2
19.37	0.23	1.7	339.97	3MR	0.3	-3.3	-1.8	1.2
26.16	0.34	1.8	349.78	13MR	-0.1	-3.3	-1.8	1.3
32.89	0.44	1.8	359.54	23MR	-0.7	-3.4	-1.7	1.3
39.57	0.53	1.9	9.24	2AP	-1.5	-3.4	-1.6	1.3
46.21	0.61	1.9	18.90	12AP	-1.5	-3.4	-1.6	1.3
52.79	0.69	1.9	28.51	22AP	-0.5	-3.4	-1.5	1.3
59.34	0.76	1.9	38.09	2MY	0.4	-3.5	-1.5	1.3
65.84	0.82	2.0	47.64	12MY	1.2	-3.5	-1.4	1.2
72.31	0.88	2.0	57.17	22MY	2.2	-3.6	-1.4	1.2
78.76	0.93	2.0	66.69	1JN	3.5	-3.6	-1.3	1.1
85.18	0.98	2.0	76.21	11JN	2.2	-3.7	-1.3	1.1
91.58	1.02	2.0	85.74	21JN	0.8	-3.8	-1.3	1.0
97.96	1.06	2.0	95.28	1JL	-0.4	-3.9	-1.3	1.0
104.34	1.09	1.9	104.86	11JL	-1.4	-4.0	-1.3	0.9
110.71	1.12	2.0	114.47	21JL	-1.4	-4.1	-1.3	0.8
117.08	1.15	2.0	124.12	31JL	-0.7	-4.2	-1.3	0.7
123.45	1.17	2.0	133.82	10AU	-0.2	-4.3	-1.3	0.7
129.83	1.18	2.0	143.57	20AU	0.1	-4.3	-1.3	0.7
136.21	1.20	2.0	153.38	30AU	0.2	-4.1	-1.4	0.8
142.61	1.20	2.0	163.25	9SE	0.5	-3.7	-1.4	0.8
149.02	1.21	2.0	173.18	19SE	1.4	-3.3	-1.4	0.8
155.44	1.20	2.0	183.16	29SE	2.5	-3.8	-1.5	0.9
161.88	1.20	2.0	193.20	9OC	0.0	-4.2	-1.6	0.9
168.34	1.18	1.9	203.29	19OC	-0.6	-4.3	-1.6	1.0
174.81	1.17	1.9	213.42	29OC	-0.6	-4.4	-1.7	1.0
181.29	1.14	1.8	223.59	8NO	-0.6	-4.3	-1.7	1.1
187.79	1.11	1.8	233.78	18NO	-0.7	-4.2	-1.8	1.1
194.30	1.06	1.7	243.98	28NO	-0.8	-4.1	-1.9	1.2
200.82	1.01	1.6	254.19	8DE	-0.8	-4.0	-1.9	1.2
207.34	0.95	1.6	264.40	18DE	-0.8	-3.9	-2.0	1.2
213.87	0.88	1.5	274.58	28DE	-0.1	-3.8	-2.0	1.2

♂ LONG	♂ LAT	♂ MAG	⊙ LONG	16.00UT -96	☿	♀	♃	♄
					MAGNITUDES			
220.40	0.79	1.4	284.73	7JA	2.3	-3.7	-2.1	1.2
226.92	0.69	1.2	294.85	17JA	1.7	-3.6	-2.1	1.3
233.42	0.57	1.1	304.92	27JA	0.7	-3.6	-2.1	1.3
239.91	0.42	1.0	314.94	6FE	0.4	-3.5	-2.0	1.2
246.36	0.26	0.9	324.91	16FE	0.2	-3.5	-2.0	1.3
252.76	0.07	0.7	334.81	26FE	-0.1	-3.4	-2.0	1.3
259.11	-0.15	0.6	344.65	7MR	-0.7	-3.4	-1.9	1.4
265.37	-0.40	0.4	354.44	17MR	-1.5	-3.3	-1.9	1.4
271.52	-0.68	0.2	4.17	27MR	-1.4	-3.3	-1.8	1.4
277.52	-1.01	0.0	13.85	6AP	-0.5	-3.3	-1.7	1.4
283.33	-1.38	-0.2	23.48	16AP	0.6	-3.3	-1.7	1.4
288.88	-1.80	-0.4	33.08	26AP	1.6	-3.3	-1.6	1.3
294.09	-2.27	-0.6	42.64	6MY	2.9	-3.3	-1.5	1.3
298.85	-2.79	-0.8	52.18	16MY	3.1	-3.3	-1.5	1.3
303.04	-3.38	-1.1	61.70	26MY	1.7	-3.4	-1.4	1.2
306.48	-4.02	-1.4	71.22	5JN	0.6	-3.4	-1.4	1.2
308.98	-4.71	-1.6	80.74	15JN	-0.5	-3.4	-1.3	1.1
310.35	-5.40	-1.9	90.28	25JN	-1.4	-3.5	-1.3	1.1
310.40	-6.06	-2.2	99.84	5JL	-1.4	-3.5	-1.3	1.0
309.15	-6.57	-2.4	109.42	15JL	-0.7	-3.5	-1.3	1.0
306.89	-6.85	-2.6	119.05	25JL	-0.1	-3.4	-1.2	0.9
304.20	-6.80	-2.6	128.73	4AU	0.2	-3.4	-1.2	0.9
301.86	-6.44	-2.4	138.45	14AU	0.4	-3.4	-1.2	0.8
300.47	-5.85	-2.1	148.23	24AU	0.7	-3.4	-1.2	0.8
300.34	-5.14	-1.8	158.07	3SE	1.8	-3.3	-1.2	0.9
301.52	-4.42	-1.5	167.96	13SE	2.3	-3.3	-1.3	0.9
303.83	-3.72	-1.2	177.91	23SE	-0.1	-3.3	-1.3	1.0
307.10	-3.09	-0.9	187.93	3OC	-0.7	-3.3	-1.3	1.0
311.12	-2.53	-0.7	197.99	13OC	-0.8	-3.3	-1.4	1.1
315.72	-2.04	-0.4	208.10	23OC	-0.8	-3.4	-1.4	1.1
320.78	-1.60	-0.2	218.25	2NO	-0.7	-3.4	-1.5	1.2
326.19	-1.22	0.1	228.43	12NO	-0.6	-3.4	-1.5	1.2
331.85	-0.90	0.3	238.63	22NO	-0.6	-3.4	-1.6	1.3
337.72	-0.61	0.5	248.84	2DE	-0.6	-3.5	-1.6	1.3
343.74	-0.36	0.6	259.05	12DE	0.1	-3.5	-1.7	1.4
349.86	-0.14	0.8	269.24	22DE	2.6	-3.6	-1.8	1.4
				-95				
356.07	0.05	1.0	279.41	1JA	1.3	-3.6	-1.8	1.3
2.33	0.22	1.1	289.55	11JA	0.4	-3.7	-1.9	1.3
8.63	0.37	1.2	299.64	21JA	0.2	-3.8	-1.9	1.3
14.95	0.50	1.3	309.69	31JA	0.0	-3.9	-2.0	1.2
21.29	0.61	1.4	319.69	10FE	-0.2	-4.0	-2.0	1.2
27.63	0.71	1.5	329.62	20FE	-0.7	-4.0	-2.0	1.1
33.97	0.79	1.6	339.50	2MR	-1.5	-4.1	-2.0	1.1
40.30	0.86	1.7	349.31	12MR	-1.4	-4.2	-2.0	1.1
46.62	0.93	1.8	359.07	22MR	-0.4	-4.2	-2.0	1.1
52.93	0.98	1.8	8.78	1AP	0.8	-4.2	-1.9	1.1
59.24	1.03	1.9	18.44	11AP	2.1	-3.9	-1.9	1.1
65.53	1.07	1.9	28.05	21AP	3.8	-3.2	-1.8	1.1
71.82	1.10	2.0	37.63	1MY	2.4	-3.0	-1.7	1.1
78.11	1.13	2.0	47.18	11MY	1.3	-3.8	-1.7	1.0
84.39	1.15	2.0	56.71	21MY	0.5	-4.1	-1.6	1.0
90.68	1.16	2.0	66.23	31MY	-0.5	-4.2	-1.6	1.0
96.97	1.17	2.0	75.74	10JN	-1.5	-4.2	-1.5	0.9
103.27	1.18	2.0	85.27	20JN	-1.5	-4.1	-1.4	0.9
109.59	1.18	2.0	94.82	30JN	-0.6	-4.0	-1.4	0.8
115.93	1.17	2.0	104.39	10JL	-0.1	-3.9	-1.3	0.8
122.29	1.16	2.0	114.00	20JL	0.3	-3.8	-1.3	0.7
128.67	1.15	2.0	123.65	30JL	0.6	-3.7	-1.3	0.7
135.09	1.13	1.9	133.34	9AU	1.0	-3.7	-1.2	0.6
141.54	1.11	1.9	143.09	19AU	2.3	-3.6	-1.2	0.6
148.02	1.08	1.9	152.90	29AU	2.0	-3.6	-1.2	0.5
154.55	1.05	1.9	162.76	8SE	-0.2	-3.5	-1.2	0.5
161.12	1.01	1.9	172.69	18SE	-0.9	-3.5	-1.2	0.6
167.74	0.97	1.9	182.68	28SE	-0.9	-3.5	-1.2	0.7
174.40	0.92	1.9	192.71	8OC	-0.9	-3.4	-1.2	0.7
181.12	0.86	1.9	202.80	18OC	-0.6	-3.4	-1.3	0.8
187.88	0.80	1.9	212.93	28OC	-0.5	-3.4	-1.3	0.9
194.70	0.73	1.8	223.09	7NO	-0.5	-3.4	-1.3	0.9
201.56	0.65	1.8	233.29	17NO	-0.4	-3.4	-1.4	1.0
208.48	0.57	1.8	243.49	27NO	0.2	-3.4	-1.4	1.0
215.45	0.47	1.7	253.70	7DE	2.8	-3.4	-1.5	1.0
222.48	0.37	1.6	263.90	17DE	1.0	-3.4	-1.6	1.1
229.54	0.26	1.6	274.09	27DE	0.2	-3.4	-1.6	1.1

♂ LONG	♂ LAT	♂ MAG	☉ LONG	16.00UT	☿ MAGNITUDES	♀	♃	♄
				-94				
236.66	0.14	1.5	284.24	6JA	0.1	-3.4	-1.7	1.1
243.82	0.00	1.4	294.36	16JA	-0.1	-3.4	-1.8	1.0
251.03	-0.14	1.4	304.43	26JA	-0.3	-3.4	-1.8	1.0
258.27	-0.29	1.3	314.45	5FE	-0.8	-3.5	-1.9	1.0
265.54	-0.45	1.2	324.42	15FE	-1.4	-3.5	-1.9	0.9
272.83	-0.63	1.1	334.33	25FE	-1.3	-3.5	-2.0	0.9
280.15	-0.81	1.0	344.17	7MR	-0.4	-3.4	-2.0	0.8
287.47	-1.00	0.9	353.97	17MR	1.0	-3.4	-2.0	0.8
294.78	-1.19	0.8	3.70	27MR	2.7	-3.4	-2.0	0.8
302.06	-1.39	0.7	13.38	6AP	3.1	-3.4	-2.0	0.8
309.31	-1.59	0.6	23.02	16AP	1.8	-3.3	-2.0	0.8
316.49	-1.80	0.5	32.61	26AP	1.0	-3.3	-2.0	0.8
323.59	-2.00	0.4	42.17	6MY	0.3	-3.3	-1.9	0.8
330.57	-2.20	0.3	51.71	16MY	-0.5	-3.3	-1.9	0.8
337.38	-2.39	0.2	61.24	26MY	-1.5	-3.3	-1.8	0.8
344.01	-2.57	0.1	70.75	5JN	-1.5	-3.4	-1.7	0.8
350.38	-2.75	-0.0	80.28	15JN	-0.6	-3.4	-1.7	0.7
356.45	-2.91	-0.2	89.81	25JN	0.0	-3.4	-1.6	0.7
2.13	-3.05	-0.3	99.37	5JL	0.4	-3.4	-1.6	0.6
7.32	-3.18	-0.5	108.96	15JL	0.8	-3.5	-1.5	0.6
11.91	-3.29	-0.6	118.58	25JL	1.4	-3.5	-1.4	0.5
15.76	-3.37	-0.8	128.25	4AU	2.8	-3.6	-1.4	0.5
18.66	-3.41	-1.0	137.98	14AU	1.7	-3.7	-1.4	0.4
20.41	-3.39	-1.2	147.75	24AU	-0.2	-3.8	-1.3	0.4
20.82	-3.30	-1.5	157.58	3SE	-1.0	-3.9	-1.3	0.3
19.75	-3.10	-1.7	167.48	13SE	-1.1	-4.0	-1.3	0.3
17.32	-2.77	-1.9	177.43	23SE	-0.9	-4.1	-1.3	0.3
13.94	-2.31	-2.0	187.44	3OC	-0.5	-4.2	-1.3	0.3
10.38	-1.75	-1.9	197.51	13OC	-0.4	-4.3	-1.2	0.4
7.44	-1.17	-1.6	207.61	23OC	-0.3	-4.4	-1.3	0.5
5.66	-0.63	-1.3	217.76	2NO	-0.3	-4.4	-1.3	0.5
5.23	-0.16	-1.0	227.94	12NO	0.4	-4.2	-1.3	0.6
6.09	0.21	-0.6	238.14	22NO	3.1	-3.8	-1.3	0.7
8.05	0.51	-0.3	248.35	2DE	0.7	-2.9	-1.4	0.7
10.90	0.75	-0.1	258.56	12DE	0.0	-3.7	-1.4	0.8
14.45	0.93	0.2	268.75	22DE	-0.1	-4.2	-1.5	0.8
				-93				
18.54	1.07	0.4	278.92	1JA	-0.2	-4.4	-1.5	0.8
23.05	1.18	0.6	289.06	11JA	-0.4	-4.3	-1.6	0.8
27.89	1.26	0.8	299.16	21JA	-0.8	-4.3	-1.6	0.8
32.99	1.32	1.0	309.21	31JA	-1.3	-4.2	-1.7	0.8
38.29	1.37	1.2	319.20	10FE	-1.2	-4.0	-1.8	0.8
43.75	1.40	1.3	329.14	20FE	-0.4	-3.9	-1.8	0.7
49.33	1.42	1.4	339.02	2MR	1.2	-3.8	-1.9	0.7
55.02	1.43	1.5	348.84	12MR	3.2	-3.7	-2.0	0.6
60.78	1.44	1.6	358.60	22MR	2.3	-3.7	-2.0	0.6
66.62	1.43	1.7	8.31	1AP	1.3	-3.6	-2.1	0.6
72.51	1.43	1.8	17.97	11AP	0.7	-3.5	-2.1	0.6
78.46	1.41	1.8	27.59	21AP	0.2	-3.5	-2.1	0.6
84.45	1.40	1.9	37.17	1MY	-0.5	-3.4	-2.1	0.6
90.49	1.37	1.9	46.72	11MY	-1.5	-3.4	-2.1	0.6
96.57	1.35	2.0	56.25	21MY	-1.5	-3.3	-2.1	0.6
102.70	1.32	2.0	65.77	31MY	-0.6	-3.3	-2.0	0.6
108.86	1.28	2.0	75.29	10JN	0.1	-3.3	-2.0	0.6
115.07	1.24	2.0	84.82	20JN	0.6	-3.3	-1.9	0.6
121.33	1.20	2.0	94.36	30JN	1.0	-3.3	-1.9	0.5
127.64	1.16	2.0	103.93	10JL	1.8	-3.3	-1.8	0.5
134.00	1.11	2.0	113.53	20JL	3.2	-3.3	-1.7	0.4
140.41	1.05	2.0	123.18	30JL	1.5	-3.4	-1.7	0.4
146.89	1.00	2.0	132.88	9AU	-0.3	-3.4	-1.6	0.3
153.42	0.93	1.9	142.62	19AU	-1.1	-3.4	-1.6	0.3
160.02	0.87	1.9	152.42	29AU	-1.2	-3.5	-1.5	0.2
166.68	0.80	1.8	162.29	8SE	-0.8	-3.5	-1.5	0.1
173.42	0.73	1.8	172.20	18SE	-0.4	-3.5	-1.4	0.1
180.23	0.65	1.8	182.19	28SE	-0.2	-3.5	-1.4	0.0
187.11	0.57	1.8	192.22	8OC	-0.2	-3.4	-1.4	0.0
194.06	0.48	1.8	202.30	18OC	-0.1	-3.4	-1.3	0.1
201.09	0.39	1.7	212.43	28OC	0.6	-3.4	-1.3	0.2
208.19	0.29	1.7	222.60	7NO	3.3	-3.4	-1.3	0.3
215.37	0.19	1.7	232.78	17NO	0.5	-3.3	-1.3	0.3
222.62	0.08	1.7	242.99	27NO	-0.1	-3.3	-1.3	0.4
229.94	-0.03	1.6	253.20	7DE	-0.2	-3.3	-1.4	0.5
237.32	-0.15	1.6	263.40	17DE	-0.3	-3.3	-1.4	0.5
244.77	-0.26	1.6	273.59	27DE	-0.5	-3.3	-1.4	0.5

♂ LONG	♂ LAT	♂ MAG	☉ LONG	16.00UT	☿ MAGNITUDES	♀	♃	♄
				-92				
252.27	-0.38	1.5	283.75	6JA	-0.9	-3.4	-1.5	0.6
259.83	-0.51	1.5	293.87	16JA	-1.2	-3.4	-1.5	0.6
267.43	-0.63	1.5	303.95	26JA	-1.1	-3.4	-1.6	0.6
275.06	-0.76	1.4	313.97	5FE	-0.3	-3.4	-1.6	0.6
282.73	-0.88	1.4	323.94	15FE	1.5	-3.5	-1.7	0.6
290.41	-1.00	1.3	333.85	25FE	3.2	-3.5	-1.7	0.6
298.09	-1.11	1.3	343.70	6MR	1.7	-3.6	-1.8	0.5
305.77	-1.22	1.2	353.49	16MR	1.0	-3.6	-1.9	0.5
313.43	-1.32	1.2	3.23	26MR	0.5	-3.7	-2.0	0.4
321.06	-1.40	1.2	12.92	5AP	0.1	-3.8	-2.0	0.4
328.64	-1.48	1.1	22.55	15AP	-0.6	-3.8	-2.1	0.4
336.15	-1.55	1.1	32.15	25AP	-1.5	-3.9	-2.1	0.4
343.58	-1.60	1.0	41.72	5MY	-1.5	-4.0	-2.2	0.5
350.92	-1.63	1.0	51.26	15MY	-0.6	-4.1	-2.2	0.5
358.15	-1.65	0.9	60.78	25MY	0.2	-4.2	-2.2	0.5
5.24	-1.65	0.9	70.30	4JN	0.8	-4.2	-2.2	0.5
12.19	-1.64	0.8	79.82	14JN	1.4	-4.0	-2.2	0.5
18.96	-1.61	0.8	89.36	24JN	2.4	-3.6	-2.2	0.4
25.53	-1.56	0.7	98.91	4JL	3.1	-3.0	-2.1	0.4
31.88	-1.49	0.6	108.50	14JL	1.2	-3.5	-2.1	0.4
37.96	-1.41	0.6	118.12	24JL	-0.3	-4.0	-2.0	0.3
43.73	-1.30	0.5	127.78	3AU	-1.2	-4.2	-2.0	0.3
49.14	-1.17	0.4	137.50	13AU	-1.3	-4.3	-1.9	0.2
54.10	-1.01	0.2	147.28	23AU	-0.8	-4.2	-1.8	0.2
58.54	-0.83	0.1	157.11	2SE	-0.4	-4.1	-1.8	0.1
62.32	-0.61	-0.1	167.00	12SE	-0.1	-4.0	-1.7	0.0
65.29	-0.34	-0.3	176.95	22SE	-0.0	-4.0	-1.7	-0.0
67.28	-0.04	-0.5	186.95	2OC	0.1	-3.9	-1.6	-0.1
68.08	0.33	-0.7	197.01	12OC	0.9	-3.8	-1.6	-0.1
67.50	0.74	-0.9	207.12	22OC	3.1	-3.7	-1.5	-0.1
65.51	1.19	-1.1	217.26	1NO	0.3	-3.7	-1.5	-0.0
62.30	1.64	-1.3	227.44	11NO	-0.3	-3.6	-1.5	0.0
58.41	2.04	-1.4	237.64	21NO	-0.4	-3.5	-1.4	0.1
54.63	2.34	-1.2	247.85	1DE	-0.4	-3.5	-1.4	0.2
51.66	2.53	-0.9	258.06	11DE	-0.6	-3.5	-1.4	0.2
49.94	2.61	-0.6	268.25	21DE	-0.9	-3.4	-1.4	0.3
49.56	2.61	-0.3	278.43	31DE	-1.0	-3.4	-1.4	0.3
				-91				
50.40	2.57	-0.1	288.57	10JA	-1.0	-3.4	-1.5	0.4
52.28	2.50	0.2	298.67	20JA	-0.3	-3.3	-1.5	0.4
54.99	2.42	0.4	308.72	30JA	1.7	-3.3	-1.5	0.4
58.37	2.33	0.7	318.72	9FE	2.6	-3.3	-1.6	0.4
62.28	2.24	0.8	328.67	19FE	1.2	-3.3	-1.6	0.4
66.60	2.15	1.0	338.54	1MR	0.7	-3.3	-1.7	0.4
71.24	2.06	1.2	348.37	11MR	0.4	-3.3	-1.7	0.4
76.16	1.97	1.3	358.14	21MR	0.0	-3.4	-1.8	0.4
81.29	1.88	1.4	7.84	31MR	-0.6	-3.4	-1.9	0.3
86.61	1.79	1.5	17.51	10AP	-1.5	-3.4	-1.9	0.3
92.08	1.71	1.6	27.12	20AP	-1.5	-3.5	-2.0	0.3
97.68	1.62	1.7	36.70	30AP	-0.5	-3.5	-2.1	0.3
103.39	1.54	1.7	46.26	10MY	0.3	-3.5	-2.1	0.3
109.22	1.46	1.8	55.79	20MY	1.0	-3.4	-2.2	0.3
115.14	1.37	1.8	65.31	30MY	1.8	-3.4	-2.3	0.4
121.16	1.29	1.9	74.83	9JN	3.1	-3.4	-2.3	0.4
127.26	1.20	1.9	84.35	19JN	2.6	-3.3	-2.3	0.4
133.46	1.11	1.9	93.89	29JN	1.0	-3.3	-2.3	0.4
139.74	1.02	1.9	103.47	9JL	-0.4	-3.3	-2.3	0.3
146.10	0.93	1.9	113.07	19JL	-1.3	-3.3	-2.3	0.3
152.55	0.84	1.9	122.71	29JL	-1.4	-3.3	-2.3	0.3
159.10	0.75	1.9	132.40	8AU	-0.7	-3.3	-2.3	0.2
165.73	0.65	1.8	142.15	18AU	-0.3	-3.4	-2.2	0.2
172.45	0.56	1.8	151.94	28AU	-0.0	-3.4	-2.1	0.1
179.26	0.46	1.8	161.80	7SE	0.1	-3.4	-2.1	0.1
186.17	0.36	1.8	171.72	17SE	0.3	-3.5	-2.0	0.0
193.17	0.25	1.7	181.70	27SE	1.2	-3.5	-1.9	-0.1
200.26	0.15	1.7	191.73	7OC	2.8	-3.6	-1.9	-0.1
207.44	0.05	1.6	201.81	17OC	0.2	-3.6	-1.8	-0.2
214.71	-0.06	1.6	211.94	27OC	-0.5	-3.7	-1.8	-0.3
222.07	-0.17	1.5	222.10	6NO	-0.5	-3.8	-1.7	-0.2
229.51	-0.27	1.5	232.29	16NO	-0.5	-3.9	-1.7	-0.1
237.02	-0.38	1.5	242.49	26NO	-0.7	-4.0	-1.6	-0.1
244.61	-0.48	1.5	252.70	6DE	-0.8	-4.1	-1.6	-0.0
252.26	-0.59	1.5	262.90	16DE	-0.9	-4.2	-1.6	0.1
259.97	-0.68	1.5	273.09	26DE	-0.8	-4.3	-1.5	0.1

♂ / ⊙ — MAGNITUDES (☿ ♀ ♃ ♄), 16.00UT, −90 / −89

♂ LONG	LAT	MAG	⊙ LONG	16.00UT	☿	♀	♃	♄
267.73	-0.78	1.5	283.25	5JA	-0.2	-4.4	-1.5	0.2
275.52	-0.86	1.4	293.37	15JA	2.0	-4.3	-1.5	0.2
283.33	-0.94	1.4	303.45	25JA	2.1	-4.2	-1.5	0.3
291.16	-1.02	1.4	313.48	4FE	0.9	-3.7	-1.5	0.3
298.99	-1.08	1.4	323.45	14FE	0.5	-3.3	-1.6	0.3
306.82	-1.13	1.4	333.37	24FE	0.2	-3.7	-1.6	0.3
314.62	-1.17	1.4	343.22	6MR	-0.1	-4.1	-1.6	0.3
322.38	-1.20	1.4	353.02	16MR	-0.6	-4.2	-1.7	0.3
330.09	-1.22	1.4	2.76	26MR	-1.5	-4.2	-1.7	0.3
337.75	-1.22	1.4	12.45	5AP	-1.5	-4.2	-1.8	0.3
345.34	-1.21	1.4	22.08	15AP	-0.5	-4.1	-1.8	0.3
352.85	-1.19	1.4	31.68	25AP	0.5	-3.9	-1.9	0.2
0.27	-1.15	1.4	41.25	5MY	1.3	-3.8	-1.9	0.2
7.59	-1.10	1.4	50.79	15MY	2.4	-3.8	-2.0	0.2
14.81	-1.04	1.4	60.32	25MY	3.5	-3.7	-2.1	0.2
21.92	-0.97	1.3	69.84	4JN	2.0	-3.6	-2.1	0.3
28.90	-0.88	1.3	79.36	14JN	0.8	-3.5	-2.2	0.3
35.75	-0.79	1.3	88.89	24JN	-0.4	-3.5	-2.3	0.3
42.46	-0.68	1.3	98.45	4JL	-1.4	-3.4	-2.3	0.3
49.02	-0.56	1.3	108.03	14JL	-1.5	-3.4	-2.4	0.3
55.43	-0.43	1.2	117.65	24JL	-0.7	-3.4	-2.4	0.3
61.64	-0.29	1.2	127.32	3AU	-0.2	-3.4	-2.4	0.3
67.66	-0.13	1.1	137.03	13AU	0.1	-3.4	-2.4	0.3
73.45	0.04	1.0	146.81	23AU	0.3	-3.4	-2.4	0.2
78.98	0.22	1.0	156.63	2SE	0.6	-3.4	-2.4	0.2
84.20	0.43	0.9	166.52	12SE	1.5	-3.4	-2.3	0.1
89.07	0.65	0.7	176.47	22SE	2.5	-3.4	-2.3	0.0
93.49	0.90	0.6	186.47	2OC	0.0	-3.4	-2.2	-0.0
97.37	1.19	0.5	196.52	12OC	-0.6	-3.4	-2.2	-0.1
100.60	1.51	0.3	206.63	22OC	-0.7	-3.4	-2.1	-0.2
103.01	1.87	0.1	216.77	1NO	-0.7	-3.4	-2.0	-0.2
104.43	2.27	-0.1	226.94	11NO	-0.8	-3.5	-1.9	-0.3
104.66	2.70	-0.4	237.14	21NO	-0.7	-3.5	-1.9	-0.3
103.55	3.15	-0.6	247.35	1DE	-0.7	-3.5	-1.8	-0.2
101.13	3.57	-0.8	257.55	11DE	-0.7	-3.5	-1.8	-0.1
97.63	3.91	-1.0	267.75	21DE	-0.1	-3.4	-1.7	-0.1
93.66	4.10	-1.0	277.93	31DE	2.3	-3.4	-1.7	0.0

−89

♂ LONG	LAT	MAG	⊙ LONG	16.00UT	☿	♀	♃	♄
89.95	4.13	-0.8	288.07	10JA	1.6	-3.4	-1.6	0.1
87.15	4.02	-0.6	298.17	20JA	0.6	-3.4	-1.6	0.1
85.60	3.81	-0.4	308.23	30JA	0.3	-3.3	-1.6	0.2
85.34	3.55	-0.1	318.23	9FE	0.1	-3.3	-1.6	0.2
86.27	3.28	0.2	328.18	19FE	-0.2	-3.3	-1.6	0.3
88.20	3.02	0.4	338.06	1MR	-0.7	-3.3	-1.6	0.3
90.93	2.77	0.6	347.89	11MR	-1.5	-3.3	-1.6	0.3
94.32	2.54	0.8	357.66	21MR	-1.4	-3.4	-1.6	0.3
98.24	2.33	0.9	7.37	31MR	-0.5	-3.4	-1.6	0.3
102.58	2.13	1.1	17.03	10AP	0.6	-3.4	-1.6	0.3
107.27	1.94	1.2	26.66	20AP	1.7	-3.4	-1.7	0.3
112.25	1.77	1.3	36.24	30AP	3.2	-3.5	-1.7	0.2
117.48	1.61	1.4	45.79	10MY	2.8	-3.5	-1.8	0.2
122.91	1.45	1.4	55.33	20MY	1.6	-3.6	-1.8	0.2
128.54	1.30	1.5	64.85	30MY	0.6	-3.6	-1.9	0.2
134.33	1.16	1.6	74.37	9JN	-0.4	-3.7	-1.9	0.2
140.28	1.02	1.6	83.90	19JN	-1.4	-3.8	-2.0	0.3
146.37	0.89	1.6	93.44	29JN	-1.5	-3.9	-2.1	0.3
152.60	0.76	1.7	103.00	9JL	-0.7	-4.0	-2.2	0.3
158.96	0.63	1.7	112.61	19JL	-0.1	-4.1	-2.2	0.3
165.44	0.50	1.7	122.25	29JL	0.2	-4.2	-2.3	0.3
172.04	0.37	1.7	131.93	8AU	0.5	-4.3	-2.3	0.3
178.77	0.25	1.7	141.68	18AU	0.8	-4.3	-2.4	0.3
185.62	0.13	1.7	151.47	28AU	1.9	-4.1	-2.4	0.3
192.58	0.01	1.6	161.33	7SE	2.3	-3.7	-2.4	0.2
199.65	-0.10	1.6	171.25	17SE	-0.1	-3.3	-2.4	0.2
206.83	-0.21	1.6	181.22	27SE	-0.8	-3.8	-2.4	0.1
214.12	-0.32	1.6	191.24	7OC	-0.8	-4.2	-2.4	0.0
221.51	-0.43	1.6	201.33	17OC	-0.8	-4.3	-2.4	-0.0
228.99	-0.53	1.5	211.45	27OC	-0.7	-4.4	-2.3	-0.1
236.56	-0.62	1.5	221.61	6NO	-0.6	-4.3	-2.2	-0.2
244.21	-0.71	1.5	231.80	16NO	-0.6	-4.2	-2.2	-0.2
251.93	-0.79	1.4	242.00	26NO	-0.5	-4.1	-2.1	-0.3
259.70	-0.86	1.4	252.21	6DE	0.1	-4.0	-2.0	-0.3
267.53	-0.93	1.4	262.41	16DE	2.6	-3.9	-2.0	-0.2
275.39	-0.98	1.3	272.60	26DE	1.2	-3.8	-1.9	-0.1

♂ / ⊙ — MAGNITUDES (☿ ♀ ♃ ♄), 16.00UT, −88 / −87

♂ LONG	LAT	MAG	⊙ LONG	16.00UT	☿	♀	♃	♄
283.28	-1.03	1.3	282.76	5JA	0.4	-3.7	-1.8	-0.0
291.17	-1.06	1.3	292.89	15JA	0.1	-3.6	-1.8	0.0
299.06	-1.08	1.3	302.96	25JA	0.0	-3.6	-1.7	0.1
306.94	-1.09	1.3	312.99	4FE	-0.3	-3.5	-1.7	0.1
314.78	-1.09	1.4	322.97	14FE	-0.7	-3.5	-1.6	0.2
322.58	-1.08	1.4	332.89	24FE	-1.4	-3.4	-1.6	0.2
330.34	-1.05	1.4	342.75	5MR	-1.4	-3.4	-1.6	0.3
338.03	-1.01	1.4	352.55	15MR	-0.5	-3.3	-1.6	0.3
345.65	-0.97	1.5	2.29	25MR	0.8	-3.3	-1.6	0.3
353.19	-0.91	1.5	11.98	4AP	2.3	-3.3	-1.5	0.3
0.65	-0.84	1.5	21.62	14AP	3.7	-3.3	-1.6	0.3
8.03	-0.77	1.5	31.22	24AP	2.1	-3.3	-1.6	0.3
15.31	-0.69	1.6	40.79	4MY	1.2	-3.3	-1.6	0.3
22.49	-0.60	1.6	50.34	14MY	0.4	-3.3	-1.6	0.3
29.58	-0.50	1.6	59.86	24MY	-0.5	-3.4	-1.6	0.2
36.56	-0.40	1.6	69.38	3JN	-1.4	-3.4	-1.7	0.2
43.45	-0.29	1.6	78.90	13JN	-1.5	-3.4	-1.7	0.2
50.23	-0.18	1.6	88.43	23JN	-0.6	-3.5	-1.8	0.3
56.91	-0.06	1.6	97.99	3JL	-0.0	-3.5	-1.8	0.3
63.48	0.06	1.6	107.57	13JL	0.3	-3.5	-1.9	0.3
69.93	0.19	1.6	117.18	23JL	0.6	-3.4	-2.0	0.4
76.27	0.32	1.6	126.85	2AU	1.1	-3.4	-2.0	0.4
82.48	0.46	1.6	136.56	12AU	2.4	-3.4	-2.1	0.4
88.56	0.61	1.5	146.32	22AU	2.0	-3.4	-2.2	0.4
94.50	0.76	1.5	156.15	1SE	-0.2	-3.3	-2.2	0.3
100.26	0.92	1.4	166.04	11SE	-0.9	-3.3	-2.3	0.3
105.84	1.10	1.3	175.98	21SE	-1.0	-3.3	-2.3	0.3
111.20	1.28	1.3	185.98	1OC	-0.9	-3.3	-2.4	0.2
116.29	1.48	1.2	196.03	11OC	-0.6	-3.3	-2.4	0.2
121.09	1.70	1.0	206.13	21OC	-0.4	-3.4	-2.4	0.1
125.50	1.94	0.9	216.28	31OC	-0.4	-3.4	-2.4	0.0
129.46	2.20	0.7	226.45	10NO	-0.4	-3.4	-2.3	-0.0
132.85	2.49	0.6	236.64	20NO	0.3	-3.4	-2.3	-0.1
135.54	2.81	0.3	246.85	30NO	2.9	-3.5	-2.2	-0.2
137.37	3.16	0.1	257.06	10DE	0.9	-3.5	-2.2	-0.2
138.14	3.54	-0.1	267.26	20DE	0.2	-3.6	-2.1	-0.2
137.69	3.91	-0.4	277.44	30DE	-0.0	-3.7	-2.0	-0.1

−87

♂ LONG	LAT	MAG	⊙ LONG	16.00UT	☿	♀	♃	♄
135.93	4.26	-0.6	287.58	9JA	-0.1	-3.7	-2.0	-0.1
132.98	4.51	-0.9	297.68	19JA	-0.3	-3.8	-1.9	0.0
129.23	4.61	-1.0	307.74	29JA	-0.8	-3.9	-1.8	0.1
125.34	4.53	-0.9	317.75	8FE	-1.4	-4.0	-1.8	0.1
122.02	4.30	-0.8	327.69	18FE	-1.3	-4.1	-1.7	0.2
119.78	3.95	-0.5	337.58	28FE	-0.4	-4.1	-1.7	0.2
118.84	3.55	-0.3	347.41	10MR	1.0	-4.2	-1.6	0.3
119.16	3.15	-0.1	357.18	20MR	2.9	-4.2	-1.6	0.3
120.60	2.76	0.1	6.90	30MR	2.8	-4.2	-1.5	0.4
122.97	2.41	0.3	16.57	9AP	1.6	-3.9	-1.5	0.4
126.10	2.08	0.5	26.19	19AP	0.9	-3.2	-1.5	0.4
129.86	1.79	0.6	35.78	29AP	0.3	-3.1	-1.5	0.4
134.14	1.52	0.8	45.33	9MY	-0.5	-3.8	-1.5	0.4
138.84	1.28	0.9	54.86	19MY	-1.5	-4.1	-1.5	0.4
143.89	1.05	1.0	64.39	29MY	-1.5	-4.2	-1.5	0.3
149.26	0.84	1.1	73.90	8JN	-0.6	-4.2	-1.5	0.3
154.89	0.64	1.1	83.43	18JN	0.1	-4.1	-1.5	0.3
160.75	0.46	1.2	92.97	28JN	0.5	-4.0	-1.6	0.3
166.84	0.29	1.3	102.54	8JL	0.9	-3.9	-1.6	0.4
173.11	0.13	1.3	112.14	18JL	1.5	-3.8	-1.6	0.4
179.57	-0.03	1.3	121.78	28JL	2.9	-3.7	-1.7	0.4
186.20	-0.17	1.3	131.46	7AU	1.7	-3.7	-1.7	0.5
192.99	-0.31	1.4	141.20	17AU	-0.2	-3.6	-1.8	0.5
199.79	-0.43	1.4	151.00	27AU	-1.0	-3.6	-1.9	0.5
207.00	-0.55	1.4	160.85	6SE	-1.1	-3.5	-1.9	0.5
214.21	-0.66	1.4	170.76	16SE	-0.9	-3.5	-2.0	0.5
221.54	-0.76	1.4	180.73	26SE	-0.5	-3.5	-2.1	0.4
228.99	-0.85	1.4	190.76	6OC	-0.3	-3.4	-2.1	0.4
236.53	-0.93	1.4	200.83	16OC	-0.3	-3.4	-2.2	0.3
244.17	-0.99	1.4	210.96	26OC	-0.2	-3.4	-2.2	0.3
251.88	-1.05	1.4	221.11	5NO	0.5	-3.4	-2.2	0.2
259.66	-1.09	1.4	231.30	15NO	3.2	-3.4	-2.3	0.2
267.50	-1.12	1.4	241.51	25NO	0.7	-3.4	-2.3	0.1
275.36	-1.14	1.4	251.71	5DE	-0.0	-3.4	-2.3	0.0
283.25	-1.14	1.4	261.92	15DE	-0.1	-3.4	-2.2	-0.1
291.14	-1.14	1.4	272.11	25DE	-0.2	-3.4	-2.2	-0.1

-86

♂ LONG	LAT	MAG	⊙ LONG	16.00UT -86	☿	♀	♃	♄
299.03	-1.12	1.4	282.27	4JA	-0.4	-3.4	-2.1	-0.1
306.90	-1.09	1.4	292.39	14JA	-0.8	-3.4	-2.1	-0.0
314.73	-1.04	1.4	302.48	24JA	-1.3	-3.4	-2.0	0.0
322.51	-0.99	1.4	312.51	3FE	-1.2	-3.5	-1.9	0.1
330.24	-0.93	1.4	322.49	13FE	-0.4	-3.5	-1.9	0.2
337.91	-0.86	1.4	332.41	23FE	1.3	-3.5	-1.8	0.2
345.50	-0.79	1.4	342.26	5MR	3.3	-3.4	-1.7	0.3
353.01	-0.70	1.4	352.07	15MR	2.1	-3.4	-1.7	0.3
0.44	-0.62	1.5	1.81	25MR	1.2	-3.4	-1.6	0.4
7.79	-0.52	1.5	11.50	4AP	0.7	-3.4	-1.5	0.4
15.04	-0.43	1.6	21.15	14AP	0.2	-3.3	-1.5	0.5
22.22	-0.33	1.6	30.75	24AP	-0.5	-3.3	-1.5	0.5
29.30	-0.23	1.7	40.32	4MY	-1.5	-3.3	-1.4	0.5
36.29	-0.12	1.7	49.86	14MY	-1.5	-3.3	-1.4	0.5
43.21	-0.02	1.7	59.39	24MY	-0.6	-3.3	-1.4	0.5
50.03	0.09	1.8	68.91	3JN	0.1	-3.4	-1.4	0.5
56.78	0.19	1.8	78.43	13JN	0.7	-3.4	-1.4	0.5
63.45	0.30	1.8	87.97	23JN	1.1	-3.4	-1.4	0.5
70.04	0.40	1.8	97.52	3JL	2.0	-3.4	-1.4	0.4
76.55	0.51	1.9	107.10	13JL	3.2	-3.5	-1.4	0.5
82.99	0.62	1.9	116.72	23JL	1.4	-3.5	-1.4	0.5
89.36	0.72	1.9	126.38	2AU	-0.3	-3.6	-1.4	0.6
95.65	0.83	1.8	136.09	12AU	-1.1	-3.7	-1.5	0.6
101.85	0.94	1.8	145.85	22AU	-1.3	-3.8	-1.5	0.6
107.98	1.06	1.8	155.67	1SE	-0.8	-3.9	-1.6	0.6
114.01	1.17	1.8	165.56	11SE	-0.4	-4.0	-1.6	0.6
119.95	1.29	1.7	175.50	21SE	-0.2	-4.1	-1.7	0.6
125.78	1.41	1.7	185.49	1OC	-0.1	-4.2	-1.8	0.6
131.48	1.54	1.6	195.55	11OC	0.0	-4.3	-1.8	0.6
137.03	1.67	1.5	205.64	21OC	0.7	-4.4	-1.9	0.6
142.41	1.81	1.4	215.78	31OC	3.4	-4.4	-1.9	0.5
147.59	1.96	1.3	225.96	10NO	0.5	-4.2	-2.0	0.5
152.52	2.12	1.2	236.15	20NO	-0.2	-3.7	-2.1	0.4
157.15	2.29	1.0	246.36	30NO	-0.3	-2.9	-2.1	0.3
161.41	2.47	0.8	256.57	10DE	-0.3	-3.8	-2.1	0.3
165.21	2.67	0.6	266.76	20DE	-0.5	-4.2	-2.1	0.2
168.45	2.88	0.4	276.94	30DE	-0.9	-4.4	-2.1	0.1

-85

♂ LONG	LAT	MAG	⊙ LONG	16.00UT	☿	♀	♃	♄
170.97	3.10	0.2	287.09	9JA	-1.1	-4.3	-2.1	0.1
172.61	3.33	-0.1	297.19	19JA	-1.1	-4.3	-2.1	0.1
173.20	3.54	-0.4	307.25	29JA	-0.3	-4.2	-2.1	0.2
172.57	3.73	-0.6	317.26	8FE	1.5	-4.0	-2.0	0.2
170.66	3.83	-0.9	327.21	18FE	3.0	-3.9	-1.9	0.3
167.63	3.82	-1.2	337.10	28FE	1.5	-3.8	-1.9	0.3
163.91	3.66	-1.3	346.94	10MR	0.9	-3.7	-1.8	0.4
160.23	3.34	-1.2	356.71	20MR	0.5	-3.7	-1.7	0.4
157.24	2.91	-1.0	6.43	30MR	0.1	-3.6	-1.7	0.5
155.41	2.43	-0.8	16.10	9AP	-0.6	-3.5	-1.6	0.5
154.92	1.95	-0.6	25.73	19AP	-1.5	-3.5	-1.5	0.6
155.70	1.50	-0.4	35.31	29AP	-1.5	-3.4	-1.5	0.6
157.59	1.09	-0.2	44.87	9MY	-0.6	-3.4	-1.4	0.6
160.44	0.73	-0.0	54.40	19MY	0.2	-3.3	-1.4	0.7
164.05	0.41	0.1	63.93	29MY	0.9	-3.3	-1.4	0.7
168.31	0.13	0.3	73.45	8JN	1.5	-3.3	-1.3	0.7
173.10	-0.12	0.4	82.97	18JN	2.6	-3.3	-1.3	0.7
178.33	-0.34	0.5	92.51	28JN	2.9	-3.3	-1.3	0.7
183.95	-0.54	0.6	102.08	8JL	1.2	-3.3	-1.3	0.7
189.89	-0.71	0.7	111.67	18JL	-0.3	-3.3	-1.3	0.6
196.12	-0.86	0.7	121.31	28JL	-1.2	-3.4	-1.3	0.7
202.61	-1.00	0.8	130.99	7AU	-1.4	-3.4	-1.3	0.7
209.31	-1.11	0.9	140.73	17AU	-0.8	-3.4	-1.3	0.8
216.21	-1.21	0.9	150.52	27AU	-0.3	-3.5	-1.3	0.8
223.29	-1.29	1.0	160.37	6SE	-0.1	-3.5	-1.4	0.8
230.53	-1.35	1.0	170.28	16SE	0.0	-3.5	-1.4	0.9
237.90	-1.39	1.0	180.25	26SE	0.2	-3.5	-1.5	0.9
245.39	-1.42	1.1	190.27	6OC	1.0	-3.4	-1.5	0.9
252.97	-1.43	1.1	200.34	16OC	3.1	-3.4	-1.6	0.8
260.64	-1.42	1.2	210.46	26OC	0.3	-3.4	-1.6	0.8
268.37	-1.40	1.2	220.62	5NO	-0.4	-3.4	-1.7	0.8
276.14	-1.37	1.2	230.80	15NO	-0.4	-3.3	-1.7	0.7
283.93	-1.32	1.3	241.00	25NO	-0.5	-3.3	-1.8	0.7
291.74	-1.26	1.3	251.21	5DE	-0.6	-3.3	-1.9	0.6
299.53	-1.19	1.3	261.42	15DE	-0.9	-3.3	-1.9	0.5
307.30	-1.11	1.4	271.61	25DE	-1.0	-3.3	-2.0	0.5

-84

♂ LONG	LAT	MAG	⊙ LONG	16.00UT -84	☿	♀	♃	♄
315.04	-1.03	1.4	281.77	4JA	-0.9	-3.4	-2.0	0.4
322.73	-0.93	1.5	291.90	14JA	-0.3	-3.4	-2.0	0.3
330.37	-0.83	1.5	301.99	24JA	1.8	-3.4	-2.1	0.3
337.94	-0.73	1.5	312.02	3FE	2.4	-3.4	-2.1	0.3
345.44	-0.63	1.6	322.00	13FE	1.1	-3.5	-2.0	0.4
352.86	-0.52	1.6	331.93	23FE	0.6	-3.5	-2.0	0.4
0.20	-0.41	1.6	341.79	4MR	0.3	-3.6	-2.0	0.5
7.46	-0.30	1.6	351.59	14MR	-0.0	-3.6	-1.9	0.6
14.64	-0.20	1.7	1.35	24MR	-0.6	-3.7	-1.8	0.6
21.73	-0.09	1.7	11.04	3AP	-1.5	-3.8	-1.8	0.7
28.75	0.01	1.7	20.68	13AP	-1.5	-3.9	-1.7	0.7
35.69	0.11	1.7	30.29	23AP	-0.6	-3.9	-1.6	0.8
42.55	0.21	1.7	39.86	3MY	0.4	-4.0	-1.6	0.8
49.34	0.31	1.7	49.41	13MY	1.1	-4.1	-1.5	0.8
56.07	0.40	1.8	58.94	23MY	2.0	-4.2	-1.5	0.9
62.73	0.49	1.8	68.45	2JN	3.3	-4.2	-1.4	0.9
69.33	0.58	1.9	77.97	12JN	2.4	-4.0	-1.4	0.9
75.88	0.67	1.9	87.51	22JN	0.9	-3.6	-1.3	0.9
82.38	0.75	1.9	97.06	2JL	-0.4	-3.0	-1.3	0.9
88.83	0.83	2.0	106.63	12JL	-1.3	-3.5	-1.3	0.9
95.24	0.91	2.0	116.25	22JL	-1.5	-4.0	-1.2	0.9
101.60	0.98	2.0	125.91	1AU	-0.7	-4.2	-1.2	0.9
107.93	1.05	2.0	135.61	11AU	-0.2	-4.2	-1.2	1.0
114.21	1.13	2.0	145.38	21AU	0.0	-4.2	-1.2	1.0
120.46	1.20	2.0	155.20	31AU	0.2	-4.1	-1.2	1.1
126.67	1.26	2.0	165.07	10SE	0.4	-4.0	-1.2	1.1
132.83	1.33	1.9	175.01	20SE	1.3	-4.0	-1.3	1.1
138.95	1.40	1.9	185.01	30SE	2.8	-3.9	-1.3	1.1
145.02	1.46	1.8	195.05	10OC	0.1	-3.8	-1.3	1.1
151.04	1.52	1.8	205.15	20OC	-0.5	-3.7	-1.4	1.1
156.98	1.59	1.7	215.29	30OC	-0.6	-3.7	-1.4	1.1
162.86	1.65	1.6	225.46	9NO	-0.6	-3.6	-1.5	1.1
168.64	1.71	1.5	235.66	19NO	-0.7	-3.5	-1.5	1.0
174.31	1.77	1.4	245.86	29NO	-0.8	-3.5	-1.6	1.0
179.85	1.82	1.3	256.07	9DE	-0.8	-3.5	-1.7	0.9
185.23	1.88	1.2	266.27	19DE	-0.8	-3.4	-1.7	0.9
190.41	1.93	1.0	276.45	29DE	-0.2	-3.4	-1.8	0.8

-83

♂ LONG	LAT	MAG	⊙ LONG	16.00UT	☿	♀	♃	♄
195.36	1.97	0.8	286.60	8JA	2.0	-3.4	-1.9	0.7
200.01	2.00	0.6	296.71	18JA	1.9	-3.3	-1.9	0.7
204.29	2.03	0.4	306.77	28JA	0.8	-3.3	-2.0	0.6
208.12	2.03	0.2	316.78	7FE	0.4	-3.3	-2.0	0.6
211.37	2.02	-0.1	326.74	17FE	0.2	-3.3	-2.0	0.6
213.89	1.96	-0.3	336.63	27FE	-0.1	-3.3	-2.0	0.7
215.52	1.86	-0.6	346.46	9MR	-0.6	-3.3	-2.0	0.7
216.07	1.69	-1.0	356.24	19MR	-1.4	-3.4	-2.0	0.8
215.40	1.44	-1.3	5.96	29MR	-1.5	-3.4	-1.9	0.8
213.49	1.09	-1.6	15.63	8AP	-0.5	-3.4	-1.9	0.9
210.56	0.63	-1.8	25.26	18AP	0.5	-3.5	-1.8	0.9
207.16	0.11	-1.9	34.85	28AP	1.5	-3.5	-1.8	1.0
204.00	-0.42	-1.8	44.40	8MY	2.7	-3.5	-1.7	1.0
201.74	-0.92	-1.6	53.94	18MY	3.3	-3.4	-1.7	1.1
200.77	-1.34	-1.4	63.46	28MY	1.9	-3.4	-1.6	1.1
201.18	-1.68	-1.2	72.98	7JN	0.7	-3.4	-1.5	1.1
202.87	-1.94	-1.0	82.51	17JN	-0.4	-3.3	-1.5	1.2
205.67	-2.14	-0.8	92.04	27JN	-1.4	-3.3	-1.4	1.2
209.40	-2.27	-0.6	101.61	7JL	-1.5	-3.3	-1.4	1.2
213.88	-2.37	-0.4	111.20	17JL	-0.7	-3.3	-1.3	1.2
218.98	-2.42	-0.3	120.83	27JL	-0.2	-3.3	-1.3	1.2
224.59	-2.44	-0.1	130.52	6AU	0.1	-3.3	-1.3	1.2
230.61	-2.43	0.0	140.25	16AU	0.3	-3.4	-1.2	1.2
236.97	-2.39	0.1	150.04	26AU	0.6	-3.4	-1.2	1.3
243.61	-2.33	0.2	159.89	5SE	1.6	-3.4	-1.2	1.3
250.48	-2.26	0.4	169.79	15SE	2.5	-3.5	-1.2	1.4
257.53	-2.16	0.5	179.76	25SE	0.0	-3.5	-1.2	1.4
264.73	-2.05	0.6	189.78	5OC	-0.7	-3.6	-1.2	1.4
272.04	-1.92	0.7	199.85	15OC	-0.7	-3.6	-1.2	1.4
279.43	-1.79	0.8	209.97	25OC	-0.7	-3.7	-1.3	1.3
286.87	-1.64	0.8	220.13	4NO	-0.8	-3.8	-1.3	1.3
294.35	-1.49	0.9	230.31	14NO	-0.6	-3.9	-1.3	1.3
301.83	-1.34	1.0	240.51	24NO	-0.7	-4.0	-1.4	1.2
309.32	-1.19	1.1	250.72	4DE	-0.6	-4.1	-1.4	1.2
316.78	-1.04	1.2	260.93	14DE	-0.0	-4.2	-1.5	1.1
324.20	-0.88	1.3	271.11	24DE	2.3	-4.3	-1.6	1.1

♂			☉	16.00UT	☿	♀	♃	♄
LONG	LAT	MAG	LONG	-82	MAGNITUDES			
331.59	-0.74	1.3	281.28	3JA	1.5	-4.4	-1.6	1.0
338.92	-0.59	1.4	291.41	13JA	0.5	-4.3	-1.7	1.0
346.19	-0.45	1.5	301.50	23JA	0.2	-4.2	-1.8	0.9
353.40	-0.32	1.5	311.54	2FE	0.1	-3.7	-1.8	0.9
0.54	-0.19	1.6	321.52	12FE	-0.2	-3.3	-1.9	0.9
7.61	-0.07	1.7	331.45	22FE	-0.7	-3.8	-1.9	0.9
14.62	0.05	1.7	341.32	4MR	-1.4	-4.1	-2.0	0.9
21.55	0.16	1.8	351.12	14MR	-1.4	-4.3	-2.0	1.0
28.42	0.26	1.8	0.87	24MR	-0.5	-4.2	-2.0	1.0
35.23	0.36	1.8	10.57	3AP	0.7	-4.2	-2.0	1.1
41.97	0.45	1.9	20.22	13AP	1.9	-4.1	-2.0	1.1
48.66	0.54	1.9	29.83	23AP	3.6	-3.9	-2.0	1.2
55.29	0.62	1.9	39.40	3MY	2.6	-3.8	-2.0	1.2
61.88	0.69	1.9	48.95	13MY	1.4	-3.8	-1.9	1.3
68.42	0.76	1.9	58.47	23MY	0.6	-3.7	-1.9	1.3
74.92	0.83	1.9	68.00	2JN	-0.4	-3.6	-1.8	1.4
81.39	0.89	1.9	77.51	12JN	-1.4	-3.5	-1.7	1.4
87.83	0.94	1.9	87.04	22JN	-1.5	-3.5	-1.7	1.4
94.24	0.99	1.9	96.60	2JL	-0.7	-3.4	-1.6	1.4
100.64	1.04	2.0	106.17	12JL	-0.1	-3.4	-1.5	1.4
107.02	1.08	2.0	115.78	22JL	0.3	-3.4	-1.5	1.4
113.38	1.12	2.0	125.44	1AU	0.5	-3.4	-1.4	1.3
119.74	1.16	2.0	135.14	11AU	0.9	-3.4	-1.4	1.3
126.09	1.19	2.0	144.90	21AU	2.1	-3.4	-1.4	1.2
132.44	1.22	2.0	154.72	31AU	2.2	-3.4	-1.3	1.2
138.78	1.24	2.0	164.59	10SE	-0.1	-3.4	-1.3	1.2
145.13	1.26	2.0	174.53	20SE	-0.8	-3.4	-1.3	1.2
151.48	1.27	2.0	184.52	30SE	-0.9	-3.4	-1.3	1.2
157.82	1.29	1.9	194.57	10OC	-0.9	-3.4	-1.3	1.2
164.17	1.29	1.9	204.66	20OC	-0.7	-3.4	-1.3	1.2
170.51	1.29	1.9	214.80	30OC	-0.5	-3.4	-1.3	1.2
176.85	1.29	1.8	224.97	9NO	-0.5	-3.5	-1.3	1.1
183.18	1.28	1.7	235.16	19NO	-0.5	-3.5	-1.3	1.1
189.49	1.26	1.7	245.37	29NO	0.1	-3.5	-1.3	1.1
195.79	1.23	1.6	255.58	9DE	2.6	-3.5	-1.4	1.0
202.07	1.19	1.5	265.78	19DE	1.1	-3.4	-1.4	1.0
208.31	1.14	1.4	275.96	29DE	0.3	-3.4	-1.5	0.9
				-81				
214.51	1.07	1.2	286.11	8JA	0.1	-3.4	-1.5	0.9
220.66	0.99	1.1	296.22	18JA	-0.0	-3.4	-1.6	0.9
226.74	0.90	1.0	306.28	28JA	-0.3	-3.3	-1.7	0.8
232.74	0.77	0.8	316.29	7FE	-0.7	-3.3	-1.7	0.8
238.64	0.63	0.6	326.25	17FE	-1.4	-3.3	-1.8	0.7
244.40	0.45	0.5	336.15	27FE	-1.4	-3.3	-1.9	0.7
250.00	0.23	0.3	345.99	9MR	-0.5	-3.3	-1.9	0.7
255.38	-0.03	0.1	355.77	19MR	0.9	-3.4	-2.0	0.8
260.49	-0.34	-0.2	5.50	29MR	2.4	-3.4	-2.0	0.9
265.25	-0.71	-0.4	15.17	8AP	3.4	-3.4	-2.1	0.9
269.55	-1.14	-0.7	24.80	18AP	1.9	-3.4	-2.1	1.0
273.27	-1.66	-0.9	34.39	28AP	1.1	-3.5	-2.1	1.1
276.26	-2.27	-1.2	43.94	8MY	0.4	-3.5	-2.1	1.1
278.31	-2.96	-1.5	53.48	18MY	-0.5	-3.6	-2.1	1.2
279.24	-3.72	-1.8	63.01	28MY	-1.4	-3.6	-2.1	1.2
278.94	-4.52	-2.1	72.52	7JN	-1.5	-3.7	-2.0	1.2
277.42	-5.27	-2.4	82.05	17JN	-0.7	-3.8	-2.0	1.3
275.03	-5.87	-2.6	91.59	27JN	-0.0	-3.9	-1.9	1.3
272.40	-6.20	-2.6	101.15	7JL	0.4	-4.0	-1.9	1.3
270.22	-6.24	-2.4	110.74	17JL	0.7	-4.1	-1.8	1.2
269.10	-6.03	-2.1	120.37	27JL	1.2	-4.2	-1.7	1.2
269.29	-5.65	-1.9	130.05	6AU	2.6	-4.2	-1.7	1.2
270.79	-5.17	-1.6	139.78	16AU	1.9	-4.2	-1.6	1.2
273.44	-4.66	-1.3	149.57	26AU	-0.1	-4.1	-1.6	1.1
277.05	-4.15	-1.1	159.41	5SE	-1.0	-3.7	-1.5	1.1
281.42	-3.66	-0.8	169.31	15SE	-1.0	-3.3	-1.5	1.0
286.40	-3.20	-0.6	179.27	25SE	-0.9	-3.8	-1.4	1.1
291.82	-2.76	-0.4	189.29	5OC	-0.6	-4.2	-1.4	1.1
297.60	-2.37	-0.2	199.36	15OC	-0.4	-4.3	-1.4	1.1
303.65	-2.00	0.0	209.48	25OC	-0.3	-4.4	-1.3	1.1
309.90	-1.67	0.2	219.63	4NO	-0.3	-4.3	-1.3	1.0
316.30	-1.36	0.4	229.81	14NO	0.3	-4.2	-1.3	1.0
322.80	-1.08	0.6	240.02	24NO	2.9	-4.1	-1.3	1.0
329.38	-0.83	0.7	250.22	4DE	0.9	-4.0	-1.4	1.0
336.01	-0.60	0.9	260.43	14DE	0.1	-3.9	-1.4	0.9
342.67	-0.40	1.0	270.62	24DE	-0.1	-3.8	-1.4	0.9

♂			☉	16.00UT	☿	♀	♃	♄
LONG	LAT	MAG	LONG	-80	MAGNITUDES			
349.33	-0.21	1.1	280.79	3JA	-0.2	-3.7	-1.4	0.9
356.00	-0.04	1.2	290.93	13JA	-0.4	-3.6	-1.5	0.8
2.66	0.11	1.3	301.02	23JA	-0.8	-3.6	-1.5	0.8
9.29	0.24	1.4	311.06	2FE	-1.3	-3.5	-1.6	0.7
15.91	0.37	1.5	321.05	12FE	-1.3	-3.5	-1.6	0.7
22.49	0.48	1.6	330.97	22FE	-0.5	-3.4	-1.7	0.6
29.05	0.58	1.7	340.84	3MR	1.1	-3.4	-1.8	0.6
35.58	0.66	1.8	350.66	13MR	3.0	-3.3	-1.8	0.5
42.07	0.74	1.8	0.41	23MR	2.5	-3.3	-1.9	0.6
48.53	0.81	1.9	10.11	2AP	1.4	-3.3	-2.0	0.7
54.97	0.88	1.9	19.76	12AP	0.8	-3.3	-2.0	0.7
61.39	0.93	1.9	29.37	22AP	0.3	-3.3	-2.1	0.8
67.78	0.98	2.0	38.94	2MY	-0.5	-3.3	-2.2	0.9
74.15	1.02	2.0	48.49	12MY	-1.4	-3.3	-2.2	0.9
80.51	1.06	2.0	58.02	22MY	-1.6	-3.4	-2.2	1.0
86.86	1.09	2.0	67.54	1JN	-0.6	-3.4	-2.2	1.0
93.20	1.11	2.0	77.06	11JN	0.1	-3.4	-2.2	1.1
99.54	1.13	2.0	86.59	21JN	0.5	-3.5	-2.2	1.1
105.88	1.14	2.0	96.13	1JL	1.0	-3.5	-2.2	1.1
112.24	1.15	2.0	105.71	11JL	1.7	-3.5	-2.1	1.1
118.60	1.16	2.0	115.32	21JL	3.1	-3.4	-2.1	1.1
124.98	1.16	1.9	124.97	31JL	1.6	-3.4	-2.0	1.1
131.38	1.16	2.0	134.67	10AU	-0.2	-3.4	-2.0	1.1
137.80	1.15	2.0	144.43	20AU	-1.1	-3.4	-1.9	1.0
144.24	1.14	2.0	154.24	30AU	-1.2	-3.3	-1.8	1.0
150.72	1.12	2.0	164.11	9SE	-0.9	-3.3	-1.8	1.0
157.23	1.09	2.0	174.04	19SE	-0.5	-3.3	-1.7	0.9
163.78	1.06	2.0	184.03	29SE	-0.3	-3.3	-1.7	0.9
170.36	1.03	1.9	194.07	9OC	-0.2	-3.3	-1.6	0.9
176.98	0.99	1.9	204.16	19OC	-0.1	-3.4	-1.6	0.9
183.63	0.94	1.9	214.30	29OC	0.5	-3.4	-1.5	0.9
190.33	0.89	1.8	224.47	8NO	3.2	-3.4	-1.5	0.9
197.07	0.82	1.8	234.66	18NO	0.6	-3.4	-1.5	0.9
203.84	0.75	1.7	244.87	28NO	-0.1	-3.5	-1.5	0.9
210.65	0.67	1.7	255.08	8DE	-0.2	-3.5	-1.4	0.9
217.50	0.58	1.6	265.28	18DE	-0.3	-3.6	-1.4	0.9
224.39	0.48	1.5	275.46	28DE	-0.5	-3.7	-1.4	0.8
				-79				
231.31	0.36	1.5	285.62	7JA	-0.8	-3.7	-1.5	0.8
238.26	0.23	1.4	295.73	17JA	-1.2	-3.8	-1.5	0.8
245.24	0.09	1.3	305.80	27JA	-1.2	-3.9	-1.5	0.7
252.25	-0.06	1.2	315.82	6FE	-0.4	-4.0	-1.5	0.6
259.27	-0.24	1.1	325.77	16FE	1.3	-4.1	-1.6	0.6
266.31	-0.42	1.0	335.68	26FE	3.3	-4.1	-1.6	0.5
273.35	-0.63	0.9	345.52	8MR	1.9	-4.2	-1.7	0.5
280.38	-0.85	0.7	355.30	18MR	1.0	-4.2	-1.8	0.5
287.39	-1.08	0.6	5.03	28MR	0.6	-4.2	-1.8	0.4
294.36	-1.33	0.5	14.71	7AP	0.1	-3.9	-1.9	0.5
301.27	-1.60	0.4	24.33	17AP	-0.5	-3.2	-2.0	0.6
308.08	-1.87	0.2	33.93	27AP	-1.4	-3.1	-2.0	0.6
314.76	-2.17	0.1	43.49	7MY	-1.6	-3.8	-2.1	0.7
321.27	-2.47	-0.1	53.02	17MY	-0.6	-4.1	-2.2	0.8
327.56	-2.78	-0.2	62.55	27MY	0.2	-4.2	-2.2	0.8
333.54	-3.09	-0.4	72.07	6JN	0.7	-4.2	-2.3	0.9
339.15	-3.41	-0.5	81.59	16JN	1.3	-4.1	-2.3	0.9
344.27	-3.73	-0.7	91.13	26JN	2.2	-4.0	-2.3	0.9
348.76	-4.05	-0.9	100.69	6JL	3.2	-3.9	-2.4	1.0
352.46	-4.36	-1.1	110.27	16JL	1.4	-3.8	-2.4	1.0
355.18	-4.64	-1.4	119.91	26JL	-0.3	-3.7	-2.3	1.0
356.68	-4.86	-1.6	129.58	5AU	-1.2	-3.7	-2.3	1.0
356.80	-4.99	-1.8	139.31	15AU	-1.3	-3.6	-2.3	1.0
355.46	-4.97	-2.0	149.09	25AU	-0.8	-3.6	-2.2	1.0
352.89	-4.75	-2.2	158.93	4SE	-0.4	-3.5	-2.1	0.9
349.68	-4.31	-2.3	168.83	14SE	-0.2	-3.5	-2.1	0.9
346.58	-3.69	-2.1	178.79	24SE	-0.1	-3.4	-2.0	0.9
344.36	-2.99	-1.8	188.81	4OC	0.1	-3.4	-1.9	0.8
343.39	-2.30	-1.5	198.87	14OC	0.7	-3.4	-1.9	0.8
343.75	-1.67	-1.1	208.99	24OC	3.4	-3.4	-1.8	0.8
345.33	-1.12	-0.8	219.14	3NO	0.4	-3.4	-1.8	0.9
347.91	-0.67	-0.5	229.32	13NO	-0.3	-3.4	-1.7	0.9
351.29	-0.30	-0.2	239.52	23NO	-0.4	-3.4	-1.7	0.9
355.30	0.01	0.0	249.73	3DE	-0.4	-3.4	-1.6	0.9
359.79	0.26	0.3	259.93	13DE	-0.5	-3.4	-1.6	0.8
4.65	0.47	0.5	270.13	23DE	-0.9	-3.4	-1.6	0.8

♂ LONG	♂ LAT	♂ MAG	⊙ LONG	16.00UT	☿	♀	♃	♄
				-78	MAGNITUDES			
9.80	0.63	0.7	280.29	2JA	-1.1	-3.4	-1.5	0.8
15.18	0.77	0.9	290.43	12JA	-1.0	-3.4	-1.5	0.8
20.72	0.89	1.0	300.52	22JA	-0.3	-3.4	-1.5	0.7
26.40	0.98	1.2	310.57	1FE	1.5	-3.5	-1.5	0.7
32.18	1.06	1.3	320.56	11FE	2.8	-3.5	-1.5	0.6
38.05	1.12	1.4	330.49	21FE	1.4	-3.5	-1.6	0.6
43.97	1.17	1.5	340.36	3MR	0.8	-3.4	-1.6	0.5
49.94	1.21	1.6	350.17	13MR	0.4	-3.4	-1.6	0.5
55.96	1.24	1.7	359.93	23MR	0.0	-3.4	-1.7	0.4
62.00	1.26	1.8	9.63	2AP	-0.6	-3.4	-1.7	0.4
68.07	1.28	1.8	19.29	12AP	-1.4	-3.3	-1.8	0.4
74.17	1.29	1.9	28.90	22AP	-1.6	-3.3	-1.8	0.4
80.29	1.29	1.9	38.48	2MY	-0.6	-3.3	-1.9	0.5
86.43	1.29	2.0	48.02	12MY	0.3	-3.3	-2.0	0.6
92.60	1.28	2.0	57.56	22MY	0.9	-3.3	-2.0	0.6
98.79	1.27	2.0	67.07	1JN	1.7	-3.4	-2.1	0.7
105.01	1.25	2.0	76.59	11JN	2.9	-3.4	-2.2	0.7
111.26	1.23	2.0	86.12	21JN	2.7	-3.4	-2.2	0.8
117.54	1.20	2.0	95.67	1JL	1.1	-3.4	-2.3	0.8
123.86	1.17	2.0	105.24	11JL	-0.3	-3.5	-2.4	0.9
130.23	1.14	2.0	114.85	21JL	-1.3	-3.5	-2.4	0.9
136.63	1.10	2.0	124.50	31JL	-1.4	-3.6	-2.4	0.9
143.09	1.06	2.0	134.20	10AU	-0.8	-3.7	-2.4	0.9
149.59	1.01	1.9	143.96	20AU	-0.3	-3.8	-2.4	0.9
156.15	0.96	1.9	153.76	30AU	-0.0	-3.9	-2.4	0.9
162.77	0.90	1.8	163.63	9SE	0.1	-4.0	-2.4	0.9
169.44	0.84	1.8	173.56	19SE	0.3	-4.1	-2.3	0.8
176.18	0.77	1.8	183.55	29SE	1.0	-4.2	-2.3	0.8
182.98	0.70	1.8	193.59	9OC	3.1	-4.3	-2.2	0.8
189.85	0.62	1.8	203.68	19OC	0.3	-4.4	-2.2	0.7
196.78	0.54	1.8	213.81	29OC	-0.4	-4.4	-2.1	0.8
203.78	0.45	1.8	223.98	8NO	-0.5	-4.2	-2.0	0.8
210.85	0.36	1.7	234.17	18NO	-0.5	-3.7	-1.9	0.8
217.98	0.26	1.7	244.37	28NO	-0.6	-2.9	-1.9	0.8
225.18	0.15	1.7	254.58	8DE	-0.8	-3.8	-1.8	0.8
232.44	0.03	1.6	264.78	18DE	-0.9	-4.2	-1.8	0.8
239.76	-0.09	1.6	274.96	28DE	-0.9	-4.4	-1.7	0.8
				-77				
247.13	-0.21	1.5	285.12	7JA	-0.3	-4.3	-1.7	0.8
254.56	-0.34	1.5	295.23	17JA	1.8	-4.3	-1.6	0.8
262.03	-0.48	1.4	305.30	27JA	2.3	-4.2	-1.6	0.7
269.55	-0.62	1.4	315.32	6FE	1.0	-4.0	-1.6	0.7
277.09	-0.76	1.3	325.29	16FE	0.5	-3.9	-1.6	0.6
284.66	-0.90	1.3	335.19	26FE	0.3	-3.8	-1.6	0.6
292.24	-1.04	1.2	345.04	8MR	-0.0	-3.7	-1.6	0.5
299.82	-1.18	1.1	354.83	18MR	-0.6	-3.7	-1.6	0.5
307.39	-1.31	1.1	4.56	28MR	-1.4	-3.6	-1.6	0.4
314.94	-1.44	1.0	14.24	7AP	-1.5	-3.5	-1.6	0.3
322.44	-1.56	1.0	23.87	17AP	-0.6	-3.5	-1.7	0.3
329.88	-1.67	0.9	33.46	27AP	0.4	-3.4	-1.7	0.3
337.25	-1.77	0.8	43.03	7MY	1.2	-3.4	-1.7	0.4
344.51	-1.85	0.8	52.57	17MY	2.2	-3.3	-1.8	0.5
351.65	-1.92	0.7	62.09	27MY	3.5	-3.3	-1.8	0.5
358.65	-1.98	0.6	71.61	6JN	2.2	-3.3	-1.9	0.6
5.46	-2.01	0.5	81.13	16JN	0.9	-3.3	-2.0	0.6
12.07	-2.03	0.5	90.67	26JN	-0.3	-3.3	-2.0	0.7
18.44	-2.03	0.4	100.23	6JL	-1.3	-3.3	-2.1	0.7
24.51	-2.01	0.3	109.82	16JL	-1.5	-3.3	-2.2	0.8
30.24	-1.96	0.2	119.44	26JL	-0.7	-3.4	-2.2	0.8
35.54	-1.90	0.0	129.12	5AU	-0.2	-3.4	-2.3	0.8
40.33	-1.80	-0.1	138.84	15AU	0.1	-3.4	-2.4	0.9
44.50	-1.67	-0.3	148.62	25AU	0.2	-3.5	-2.4	0.9
47.89	-1.51	-0.4	158.46	4SE	0.5	-3.5	-2.4	0.9
50.33	-1.29	-0.6	168.35	14SE	1.4	-3.5	-2.4	0.8
51.62	-1.03	-0.8	178.30	24SE	2.8	-3.5	-2.4	0.8
51.54	-0.69	-1.1	188.32	4OC	0.1	-3.4	-2.4	0.8
50.02	-0.30	-1.3	198.38	14OC	-0.6	-3.4	-2.4	0.7
47.18	0.15	-1.5	208.49	24OC	-0.7	-3.4	-2.4	0.7
43.46	0.62	-1.6	218.64	3NO	-0.7	-3.4	-2.3	0.7
39.66	1.04	-1.5	228.82	13NO	-0.7	-3.3	-2.2	0.7
36.53	1.39	-1.2	239.02	23NO	-0.7	-3.3	-2.2	0.8
34.58	1.64	-0.9	249.23	3DE	-0.7	-3.3	-2.1	0.8
34.00	1.81	-0.6	259.43	13DE	-0.7	-3.3	-2.0	0.8
34.68	1.91	-0.3	269.62	23DE	-0.2	-3.3	-1.9	0.8

♂ LONG	♂ LAT	♂ MAG	⊙ LONG	16.00UT	☿	♀	♃	♄
				-76	MAGNITUDES			
36.45	1.96	0.0	279.80	2JA	2.1	-3.4	-1.9	0.8
39.10	1.98	0.3	289.93	12JA	1.8	-3.4	-1.8	0.8
42.44	1.97	0.5	300.03	22JA	0.7	-3.4	-1.8	0.8
46.33	1.95	0.7	310.08	1FE	0.3	-3.4	-1.7	0.7
50.65	1.92	0.9	320.07	11FE	0.2	-3.5	-1.7	0.7
55.30	1.89	1.1	330.00	21FE	-0.1	-3.5	-1.6	0.6
60.23	1.85	1.2	339.88	2MR	-0.6	-3.6	-1.6	0.6
65.37	1.80	1.3	349.70	12MR	-1.4	-3.6	-1.6	0.5
70.68	1.75	1.4	359.46	22MR	-1.5	-3.7	-1.6	0.5
76.14	1.70	1.6	9.17	1AP	-0.6	-3.8	-1.5	0.4
81.72	1.65	1.6	18.82	11AP	0.5	-3.9	-1.5	0.4
87.41	1.60	1.7	28.44	21AP	1.6	-3.9	-1.6	0.3
93.19	1.54	1.8	38.02	1MY	3.0	-4.0	-1.6	0.3
99.05	1.48	1.8	47.56	11MY	3.0	-4.1	-1.6	0.3
104.99	1.42	1.9	57.10	21MY	1.7	-4.2	-1.6	0.4
111.01	1.36	1.9	66.62	31MY	0.7	-4.2	-1.6	0.4
117.09	1.29	1.9	76.14	10JN	-0.4	-4.0	-1.7	0.5
123.25	1.22	1.9	85.66	20JN	-1.4	-3.6	-1.7	0.6
129.48	1.15	2.0	95.21	30JN	-1.5	-2.9	-1.8	0.6
135.77	1.08	2.0	104.78	10JL	-0.7	-3.5	-1.8	0.7
142.14	1.01	1.9	114.39	20JL	-0.1	-4.0	-1.9	0.7
148.59	0.93	1.9	124.04	30JL	0.2	-4.2	-2.0	0.8
155.11	0.85	1.9	133.73	9AU	0.4	-4.2	-2.0	0.8
161.71	0.77	1.9	143.48	19AU	0.7	-4.2	-2.1	0.8
168.39	0.68	1.9	153.29	29AU	1.8	-4.1	-2.2	0.8
175.15	0.59	1.8	163.15	8SE	2.5	-4.0	-2.2	0.8
182.00	0.50	1.8	173.08	18SE	0.0	-3.9	-2.3	0.8
188.93	0.41	1.7	183.06	28SE	-0.7	-3.9	-2.3	0.8
195.95	0.31	1.7	193.10	8OC	-0.8	-3.8	-2.4	0.8
203.05	0.21	1.6	203.19	18OC	-0.8	-3.7	-2.4	0.7
210.24	0.10	1.6	213.32	28OC	-0.7	-3.7	-2.4	0.7
217.51	-0.00	1.6	223.48	7NO	-0.6	-3.6	-2.4	0.7
224.86	-0.11	1.6	233.67	17NO	-0.6	-3.5	-2.3	0.7
232.29	-0.22	1.6	243.88	27NO	-0.6	-3.5	-2.3	0.7
239.78	-0.33	1.6	254.08	7DE	-0.0	-3.5	-2.2	0.8
247.35	-0.44	1.5	264.29	17DE	2.3	-3.4	-2.2	0.8
254.97	-0.55	1.5	274.47	27DE	1.4	-3.4	-2.1	0.8
				-75				
262.64	-0.65	1.5	284.63	6JA	0.4	-3.4	-2.0	0.8
270.36	-0.76	1.5	294.75	16JA	0.2	-3.3	-1.9	0.8
278.11	-0.85	1.5	304.82	26JA	0.0	-3.3	-1.9	0.8
285.89	-0.95	1.4	314.84	5FE	-0.2	-3.3	-1.8	0.7
293.68	-1.03	1.4	324.81	15FE	-0.7	-3.3	-1.7	0.7
301.46	-1.11	1.4	334.71	25FE	-1.4	-3.3	-1.7	0.7
309.24	-1.18	1.4	344.56	7MR	-1.4	-3.3	-1.6	0.6
316.99	-1.24	1.3	354.35	17MR	-0.5	-3.4	-1.6	0.6
324.70	-1.28	1.3	4.08	27MR	0.7	-3.4	-1.5	0.5
332.36	-1.31	1.3	13.76	6AP	2.1	-3.4	-1.5	0.4
339.96	-1.33	1.3	23.40	16AP	3.9	-3.5	-1.5	0.4
347.48	-1.34	1.3	32.99	26AP	2.3	-3.5	-1.5	0.3
354.92	-1.33	1.3	42.56	6MY	1.3	-3.5	-1.5	0.3
2.27	-1.30	1.2	52.10	16MY	0.5	-3.4	-1.5	0.3
9.50	-1.26	1.2	61.62	26MY	-0.4	-3.4	-1.5	0.3
16.61	-1.21	1.2	71.14	5JN	-1.4	-3.4	-1.5	0.4
23.59	-1.14	1.2	80.67	15JN	-1.6	-3.3	-1.5	0.4
30.43	-1.06	1.1	90.20	25JN	-0.7	-3.3	-1.5	0.5
37.11	-0.97	1.1	99.76	5JL	-0.1	-3.3	-1.5	0.6
43.61	-0.86	1.0	109.35	15JL	0.3	-3.3	-1.6	0.6
49.91	-0.74	1.0	118.97	25JL	0.6	-3.3	-1.6	0.7
56.00	-0.60	0.9	128.64	4AU	1.0	-3.3	-1.7	0.7
61.84	-0.45	0.9	138.36	14AU	2.2	-3.4	-1.7	0.7
67.39	-0.27	0.8	148.14	24AU	2.2	-3.4	-1.8	0.8
72.59	-0.08	0.7	157.97	3SE	-0.1	-3.4	-1.9	0.8
77.39	0.14	0.5	167.87	13SE	-0.9	-3.5	-1.9	0.8
81.69	0.39	0.4	177.82	23SE	-1.0	-3.5	-2.0	0.8
85.39	0.67	0.2	187.83	3OC	-0.9	-3.6	-2.1	0.8
88.34	0.99	0.1	197.89	13OC	-0.6	-3.6	-2.1	0.8
90.38	1.36	-0.1	208.00	23OC	-0.5	-3.7	-2.2	0.7
91.31	1.77	-0.4	218.15	2NO	-0.4	-3.8	-2.2	0.7
90.94	2.21	-0.6	228.33	12NO	-0.4	-3.9	-2.2	0.7
89.19	2.67	-0.8	238.52	22NO	0.1	-4.0	-2.3	0.7
86.19	3.09	-1.0	248.73	2DE	2.6	-4.1	-2.3	0.7
82.36	3.42	-1.1	258.94	12DE	1.1	-4.2	-2.2	0.7
78.43	3.61	-1.0	269.13	22DE	0.2	-4.3	-2.2	0.8

♂ LONG	♂ LAT	♂ MAG	⊙ LONG	16.00UT	☿	♀	♃	♄
				-74	MAGNITUDES			
75.13	3.65	-0.8	279.30	1JA	0.0	-4.4	-2.2	0.8
72.97	3.58	-0.5	289.45	11JA	-0.1	-4.3	-2.1	0.8
72.12	3.42	-0.3	299.54	21JA	-0.3	-4.1	-2.0	0.8
72.53	3.23	0.0	309.59	31JA	-0.7	-3.7	-2.0	0.8
74.04	3.03	0.2	319.59	10FE	-1.4	-3.3	-1.9	0.8
76.43	2.83	0.5	329.52	20FE	-1.3	-3.8	-1.8	0.7
79.55	2.64	0.7	339.40	2MR	-0.5	-4.1	-1.8	0.7
83.23	2.46	0.9	349.22	12MR	0.9	-4.3	-1.7	0.7
87.38	2.30	1.0	358.98	22MR	2.6	-4.2	-1.6	0.6
91.89	2.14	1.1	8.69	1AP	3.0	-4.2	-1.6	0.6
96.71	1.99	1.3	18.35	11AP	1.7	-4.1	-1.5	0.5
101.78	1.85	1.4	27.97	21AP	1.0	-3.9	-1.5	0.4
107.06	1.71	1.5	37.55	1MY	0.4	-3.8	-1.4	0.4
112.53	1.58	1.5	47.10	11MY	-0.4	-3.8	-1.4	0.3
118.16	1.46	1.6	56.63	21MY	-1.4	-3.7	-1.4	0.2
123.93	1.33	1.7	66.15	31MY	-1.6	-3.6	-1.4	0.3
129.83	1.22	1.7	75.67	10JN	-0.7	-3.5	-1.4	0.3
135.86	1.10	1.7	85.20	20JN	0.0	-3.5	-1.4	0.4
142.01	0.98	1.8	94.75	30JN	0.4	-3.4	-1.4	0.5
148.27	0.87	1.8	104.32	10JL	0.8	-3.4	-1.4	0.5
154.65	0.75	1.8	113.92	20JL	1.4	-3.4	-1.4	0.6
161.13	0.64	1.8	123.57	30JL	2.7	-3.4	-1.4	0.6
167.72	0.53	1.8	133.27	9AU	1.9	-3.4	-1.4	0.7
174.42	0.41	1.8	143.01	19AU	-0.1	-3.4	-1.5	0.7
181.23	0.30	1.7	152.82	29AU	-1.0	-3.4	-1.5	0.7
188.14	0.19	1.7	162.68	8SE	-1.1	-3.4	-1.6	0.8
195.16	0.08	1.7	172.60	18SE	-0.9	-3.4	-1.6	0.8
202.28	-0.03	1.7	182.58	28SE	-0.5	-3.4	-1.7	0.8
209.51	-0.14	1.6	192.61	8OC	-0.3	-3.4	-1.8	0.8
216.83	-0.25	1.6	202.70	18OC	-0.3	-3.4	-1.8	0.8
224.24	-0.36	1.5	212.83	28OC	-0.2	-3.4	-1.9	0.8
231.74	-0.46	1.5	222.99	7NO	0.3	-3.5	-1.9	0.7
239.32	-0.56	1.5	233.17	17NO	2.9	-3.5	-2.0	0.7
246.98	-0.65	1.4	243.38	27NO	0.8	-3.5	-2.1	0.7
254.70	-0.74	1.4	253.59	7DE	0.0	-3.5	-2.1	0.7
262.48	-0.82	1.3	263.79	17DE	-0.1	-3.4	-2.1	0.7
270.30	-0.89	1.4	273.98	27DE	-0.2	-3.4	-2.1	0.8
				-73				
278.15	-0.95	1.4	284.13	6JA	-0.4	-3.4	-2.1	0.8
286.02	-1.01	1.4	294.25	16JA	-0.8	-3.4	-2.1	0.8
293.90	-1.05	1.4	304.33	26JA	-1.3	-3.3	-2.1	0.8
301.77	-1.09	1.4	314.35	5FE	-1.2	-3.3	-2.0	0.8
309.63	-1.11	1.4	324.32	15FE	-0.5	-3.3	-2.0	0.8
317.45	-1.12	1.4	334.23	25FE	1.1	-3.3	-1.9	0.8
325.23	-1.12	1.4	344.08	7MR	3.2	-3.3	-1.8	0.7
332.96	-1.10	1.4	353.87	17MR	2.3	-3.4	-1.8	0.7
340.63	-1.08	1.4	3.61	27MR	1.3	-3.4	-1.7	0.7
348.23	-1.04	1.5	13.29	6AP	0.7	-3.4	-1.6	0.6
355.74	-0.99	1.5	22.93	16AP	0.2	-3.4	-1.6	0.5
3.17	-0.93	1.5	32.53	26AP	-0.5	-3.5	-1.5	0.5
10.50	-0.86	1.5	42.09	6MY	-1.4	-3.5	-1.5	0.4
17.74	-0.78	1.5	51.63	16MY	-1.6	-3.6	-1.4	0.4
24.88	-0.69	1.5	61.16	26MY	-0.7	-3.6	-1.4	0.3
31.91	-0.60	1.5	70.68	5JN	0.1	-3.7	-1.3	0.3
38.83	-0.49	1.5	80.20	15JN	0.6	-3.8	-1.3	0.3
45.64	-0.38	1.5	89.74	25JN	1.1	-3.9	-1.3	0.4
52.33	-0.27	1.5	99.29	5JL	1.8	-4.0	-1.3	0.4
58.90	-0.14	1.5	108.88	15JL	3.2	-4.1	-1.3	0.5
65.33	-0.01	1.5	118.51	25JL	1.6	-4.2	-1.3	0.5
71.63	0.13	1.5	128.17	4AU	-0.2	-4.2	-1.3	0.6
77.78	0.28	1.4	137.89	14AU	-1.1	-4.2	-1.3	0.6
83.76	0.44	1.4	147.67	24AU	-1.3	-4.1	-1.3	0.7
89.55	0.61	1.3	157.50	3SE	-0.9	-3.6	-1.3	0.7
95.14	0.79	1.2	167.39	13SE	-0.4	-3.3	-1.4	0.8
100.47	0.98	1.1	177.34	23SE	-0.2	-3.8	-1.4	0.8
105.52	1.20	1.0	187.34	3OC	-0.1	-4.2	-1.5	0.8
110.22	1.43	0.9	197.40	13OC	-0.0	-4.3	-1.5	0.8
114.51	1.69	0.8	207.51	23OC	0.6	-4.3	-1.6	0.8
118.27	1.97	0.6	217.65	2NO	3.3	-4.3	-1.6	0.8
121.41	2.29	0.4	227.83	12NO	0.6	-4.2	-1.7	0.8
123.75	2.65	0.2	238.03	22NO	-0.2	-4.1	-1.8	0.7
125.13	3.03	-0.0	248.24	2DE	-0.3	-4.0	-1.8	0.7
125.36	3.45	-0.3	258.45	12DE	-0.3	-3.9	-1.9	0.7
124.30	3.86	-0.5	268.64	22DE	-0.5	-3.8	-1.9	0.7

♂ LONG	♂ LAT	♂ MAG	⊙ LONG	16.00UT	☿	♀	♃	♄
				-72	MAGNITUDES			
121.93	4.22	-0.7	278.81	1JA	-0.8	-3.7	-2.0	0.8
118.50	4.47	-0.9	288.96	11JA	-1.1	-3.6	-2.0	0.8
114.56	4.57	-1.0	299.06	21JA	-1.1	-3.6	-2.0	0.8
110.84	4.48	-0.8	309.11	31JA	-0.4	-3.5	-2.0	0.8
107.97	4.25	-0.6	319.11	10FE	1.3	-3.5	-2.0	0.8
106.33	3.93	-0.4	329.05	20FE	3.2	-3.4	-2.0	0.8
105.98	3.57	-0.1	338.93	1MR	1.7	-3.4	-2.0	0.8
106.82	3.22	0.1	348.75	11MR	0.9	-3.3	-1.9	0.8
108.69	2.88	0.3	358.52	21MR	0.5	-3.3	-1.9	0.8
111.39	2.57	0.5	8.23	31MR	0.1	-3.3	-1.8	0.7
114.78	2.28	0.7	17.89	10AP	-0.5	-3.3	-1.8	0.7
118.72	2.02	0.8	27.51	20AP	-1.4	-3.3	-1.7	0.6
123.11	1.78	0.9	37.09	30AP	-1.6	-3.3	-1.6	0.6
127.87	1.56	1.1	46.64	10MY	-0.6	-3.3	-1.6	0.5
132.96	1.35	1.1	56.17	20MY	0.2	-3.4	-1.5	0.4
138.31	1.16	1.2	65.69	30MY	0.8	-3.4	-1.4	0.4
143.90	0.98	1.3	75.21	9JN	1.4	-3.4	-1.4	0.3
149.70	0.80	1.4	84.74	19JN	2.4	-3.5	-1.3	0.3
155.70	0.64	1.4	94.28	29JN	3.1	-3.5	-1.3	0.3
161.87	0.48	1.4	103.85	9JL	1.3	-3.5	-1.3	0.4
168.21	0.33	1.5	113.45	19JL	-0.2	-3.4	-1.3	0.4
174.70	0.19	1.5	123.10	29JL	-1.2	-3.4	-1.2	0.5
181.34	0.05	1.5	132.79	8AU	-1.4	-3.4	-1.2	0.6
188.12	-0.08	1.5	142.53	18AU	-0.8	-3.4	-1.2	0.6
195.04	-0.21	1.5	152.33	28AU	-0.4	-3.3	-1.2	0.7
202.09	-0.33	1.5	162.20	7SE	-0.1	-3.3	-1.2	0.7
209.26	-0.44	1.5	172.11	17SE	0.0	-3.3	-1.2	0.7
216.55	-0.55	1.5	182.09	27SE	0.2	-3.3	-1.3	0.8
223.95	-0.65	1.5	192.12	7OC	0.8	-3.3	-1.3	0.8
231.45	-0.74	1.5	202.21	17OC	3.5	-3.4	-1.3	0.8
239.04	-0.82	1.5	212.33	27OC	0.4	-3.4	-1.4	0.8
246.72	-0.89	1.4	222.50	6NO	-0.3	-3.4	-1.4	0.8
254.46	-0.96	1.4	232.68	16NO	-0.4	-3.4	-1.5	0.8
262.27	-1.01	1.4	242.88	26NO	-0.4	-3.5	-1.5	0.8
270.12	-1.05	1.4	253.10	6DE	-0.6	-3.5	-1.6	0.8
278.00	-1.08	1.4	263.30	16DE	-0.9	-3.6	-1.7	0.7
285.90	-1.10	1.4	273.48	26DE	-1.0	-3.7	-1.7	0.8
				-71				
293.80	-1.10	1.4	283.64	5JA	-1.0	-3.7	-1.8	0.8
301.70	-1.10	1.4	293.76	15JA	-0.4	-3.8	-1.9	0.8
309.56	-1.08	1.3	303.84	25JA	1.6	-3.9	-1.9	0.9
317.40	-1.05	1.3	313.87	4FE	2.7	-4.0	-2.0	0.9
325.18	-1.01	1.3	323.84	14FE	1.2	-4.1	-2.0	0.9
332.91	-0.96	1.4	333.75	24FE	0.7	-4.2	-2.0	0.9
340.58	-0.90	1.4	343.61	6MR	0.4	-4.2	-2.0	0.9
348.16	-0.83	1.4	353.40	16MR	0.0	-4.2	-2.0	0.9
355.67	-0.75	1.5	3.14	26MR	-0.6	-4.2	-2.0	0.8
3.10	-0.67	1.5	12.83	5AP	-1.4	-3.9	-1.9	0.8
10.43	-0.58	1.6	22.47	15AP	-1.6	-3.2	-1.9	0.8
17.68	-0.49	1.6	32.07	25AP	-0.6	-3.1	-1.8	0.7
24.84	-0.39	1.6	41.64	5MY	0.3	-3.8	-1.8	0.7
31.90	-0.29	1.7	51.17	15MY	1.0	-4.1	-1.7	0.6
38.88	-0.19	1.7	60.70	25MY	1.9	-4.2	-1.6	0.5
45.77	-0.08	1.7	70.22	4JN	3.1	-4.2	-1.6	0.5
52.56	0.03	1.7	79.74	14JN	2.6	-4.1	-1.5	0.4
59.27	0.14	1.8	89.27	24JN	1.1	-4.0	-1.5	0.3
65.89	0.25	1.8	98.83	4JL	-0.3	-3.9	-1.4	0.3
72.42	0.37	1.8	108.41	14JL	-1.3	-3.8	-1.4	0.4
78.86	0.48	1.8	118.03	24JL	-1.5	-3.7	-1.3	0.4
85.21	0.60	1.8	127.70	3AU	-0.8	-3.7	-1.3	0.5
91.47	0.72	1.8	137.42	13AU	-0.3	-3.6	-1.3	0.6
97.63	0.85	1.7	147.19	23AU	-0.0	-3.6	-1.2	0.6
103.68	0.98	1.7	157.02	2SE	0.2	-3.5	-1.2	0.7
109.62	1.11	1.6	166.90	12SE	0.4	-3.5	-1.2	0.7
115.43	1.25	1.6	176.85	22SE	1.1	-3.4	-1.2	0.7
121.09	1.40	1.5	186.86	2OC	3.1	-3.4	-1.2	0.8
126.57	1.55	1.4	196.91	12OC	0.3	-3.4	-1.2	0.8
131.87	1.72	1.3	207.02	22OC	-0.5	-3.4	-1.3	0.8
136.91	1.89	1.2	217.17	1NO	-0.6	-3.4	-1.3	0.8
141.68	2.08	1.1	227.34	11NO	-0.6	-3.4	-1.3	0.9
146.09	2.29	0.9	237.54	21NO	-0.7	-3.4	-1.4	0.9
150.07	2.52	0.8	247.75	1DE	-0.8	-3.4	-1.4	0.8
153.52	2.77	0.6	257.95	11DE	-0.8	-3.4	-1.5	0.8
156.31	3.03	0.3	268.15	21DE	-0.8	-3.4	-1.5	0.8
158.26	3.32	0.1	278.32	31DE	-0.3	-3.4	-1.6	0.8

♂ LONG	LAT	MAG	☉ LONG	16.00UT -70	☿	♀	♃	♄
159.22	3.62	-0.2	288.46	10JA	1.8	-3.4	-1.7	0.8
158.99	3.90	-0.4	298.57	20JA	2.1	-3.4	-1.7	0.9
157.48	4.14	-0.7	308.62	30JA	0.9	-3.5	-1.8	0.9
154.76	4.28	-0.9	318.62	9FE	0.5	-3.5	-1.9	0.9
151.16	4.28	-1.1	328.57	19FE	0.2	-3.5	-1.9	1.0
147.31	4.10	-1.1	338.45	1MR	-0.1	-3.4	-2.0	1.0
143.93	3.78	-0.9	348.27	11MR	-0.6	-3.4	-2.0	1.0
141.57	3.35	-0.7	358.04	21MR	-1.4	-3.4	-2.0	1.0
140.50	2.89	-0.5	7.76	31MR	-1.5	-3.4	-2.0	0.9
140.72	2.44	-0.3	17.42	10AP	-0.6	-3.3	-2.0	0.9
142.10	2.02	-0.1	27.04	20AP	0.4	-3.3	-2.0	0.9
144.48	1.64	0.1	36.62	30AP	1.4	-3.3	-2.0	0.8
147.66	1.29	0.3	46.17	10MY	2.5	-3.3	-2.0	0.8
151.52	0.98	0.4	55.71	20MY	3.5	-3.3	-1.9	0.7
155.93	0.70	0.5	65.23	30MY	2.0	-3.4	-1.8	0.7
160.80	0.45	0.6	74.75	9JN	0.8	-3.4	-1.8	0.6
166.07	0.22	0.7	84.27	19JN	-0.3	-3.4	-1.7	0.5
171.67	0.01	0.8	93.82	29JN	-1.3	-3.4	-1.7	0.5
177.56	-0.17	0.9	103.38	9JL	-1.5	-3.5	-1.6	0.4
183.71	-0.35	1.0	112.99	19JL	-0.7	-3.5	-1.5	0.4
190.09	-0.50	1.0	122.63	29JL	-0.2	-3.6	-1.5	0.5
196.68	-0.65	1.1	132.31	8AU	0.1	-3.7	-1.4	0.5
203.47	-0.77	1.1	142.06	18AU	0.3	-3.8	-1.4	0.6
210.42	-0.89	1.1	151.86	28AU	0.6	-3.9	-1.4	0.6
217.54	-0.98	1.2	161.71	7SE	1.5	-4.0	-1.3	0.7
224.79	-1.07	1.2	171.63	17SE	2.8	-4.1	-1.3	0.7
232.18	-1.14	1.2	181.60	27SE	0.1	-4.2	-1.3	0.8
239.68	-1.20	1.2	191.63	7OC	-0.7	-4.3	-1.3	0.8
247.28	-1.24	1.3	201.72	17OC	-0.7	-4.4	-1.3	0.8
254.96	-1.26	1.3	211.84	27OC	-0.7	-4.4	-1.3	0.9
262.71	-1.28	1.3	222.00	6NO	-0.8	-4.2	-1.3	0.9
270.51	-1.27	1.3	232.19	16NO	-0.7	-3.7	-1.3	0.9
278.34	-1.26	1.3	242.39	26NO	-0.7	-2.9	-1.3	0.9
286.19	-1.23	1.4	252.60	6DE	-0.7	-3.8	-1.4	0.9
294.04	-1.19	1.4	262.81	16DE	-0.1	-4.2	-1.4	0.9
301.88	-1.14	1.4	272.99	26DE	2.1	-4.4	-1.4	0.9

♂ LONG	LAT	MAG	☉ LONG	16.00UT -69	☿	♀	♃	♄
309.69	-1.08	1.4	283.15	5JA	1.7	-4.3	-1.5	0.9
317.47	-1.01	1.4	293.28	15JA	0.6	-4.3	-1.6	0.9
325.19	-0.93	1.5	303.36	25JA	0.3	-4.2	-1.6	0.9
332.85	-0.84	1.5	313.39	4FE	0.1	-4.0	-1.7	1.0
340.45	-0.75	1.5	323.36	14FE	-0.2	-3.9	-1.8	1.0
347.97	-0.66	1.5	333.28	24FE	-0.6	-3.8	-1.8	1.1
355.41	-0.56	1.6	343.13	6MR	-1.4	-3.7	-1.9	1.1
2.78	-0.46	1.6	352.93	16MR	-1.5	-3.6	-2.0	1.1
10.05	-0.35	1.6	2.67	26MR	-0.6	-3.6	-2.0	1.1
17.24	-0.25	1.6	12.36	5AP	0.6	-3.5	-2.1	1.1
24.35	-0.15	1.6	22.01	15AP	1.8	-3.5	-2.1	1.0
31.38	-0.05	1.6	31.61	25AP	3.3	-3.4	-2.1	1.0
38.32	0.06	1.7	41.18	5MY	2.8	-3.4	-2.1	1.0
45.19	0.16	1.7	50.72	15MY	1.5	-3.3	-2.1	0.9
51.98	0.26	1.8	60.24	25MY	0.6	-3.3	-2.1	0.9
58.71	0.35	1.8	69.76	4JN	-0.4	-3.3	-2.1	0.8
65.36	0.45	1.9	79.29	14JN	-1.4	-3.3	-2.0	0.7
71.95	0.54	1.9	88.82	24JN	-1.6	-3.3	-2.0	0.7
78.48	0.64	1.9	98.37	4JL	-0.7	-3.3	-1.9	0.6
84.95	0.73	1.9	107.95	14JL	-0.1	-3.3	-1.9	0.5
91.37	0.82	1.9	117.57	24JL	0.2	-3.4	-1.8	0.5
97.73	0.91	1.9	127.23	3AU	0.5	-3.4	-1.7	0.5
104.03	0.99	1.9	136.95	13AU	0.8	-3.4	-1.7	0.6
110.29	1.08	1.9	146.71	23AU	1.9	-3.5	-1.6	0.6
116.49	1.17	1.9	156.54	2SE	2.5	-3.5	-1.6	0.7
122.63	1.25	1.9	166.42	12SE	0.0	-3.5	-1.5	0.7
128.71	1.34	1.9	176.36	22SE	-0.8	-3.5	-1.5	0.8
134.72	1.43	1.8	186.37	2OC	-0.9	-3.4	-1.4	0.8
140.65	1.51	1.7	196.42	12OC	-0.8	-3.4	-1.4	0.9
146.50	1.60	1.7	206.52	22OC	-0.7	-3.4	-1.4	0.9
152.25	1.69	1.6	216.67	1NO	-0.5	-3.4	-1.4	0.9
157.87	1.79	1.5	226.84	11NO	-0.5	-3.3	-1.4	1.0
163.36	1.88	1.4	237.03	21NO	-0.5	-3.3	-1.4	1.0
168.67	1.98	1.3	247.24	1DE	0.0	-3.3	-1.4	1.0
173.77	2.08	1.1	257.45	11DE	2.3	-3.4	-1.4	1.0
178.63	2.18	1.0	267.65	21DE	1.3	-3.4	-1.4	1.0
183.16	2.28	0.8	277.83	31DE	0.3	-3.4	-1.4	1.0

♂ LONG	LAT	MAG	☉ LONG	16.00UT -68	☿	♀	♃	♄
187.31	2.39	0.6	287.97	10JA	0.1	-3.4	-1.5	1.0
190.98	2.49	0.4	298.07	20JA	-0.0	-3.4	-1.5	1.0
194.05	2.59	0.1	308.14	30JA	-0.2	-3.4	-1.6	1.0
196.35	2.67	-0.1	318.14	9FE	-0.7	-3.5	-1.6	1.1
197.74	2.72	-0.4	328.08	19FE	-1.4	-3.5	-1.7	1.1
198.01	2.73	-0.7	337.98	29FE	-1.4	-3.6	-1.7	1.2
197.03	2.67	-1.0	347.80	10MR	-0.6	-3.6	-1.8	1.2
194.82	2.50	-1.3	357.57	20MR	0.7	-3.7	-1.9	1.2
191.62	2.21	-1.6	7.29	30MR	2.3	-3.8	-1.9	1.2
188.02	1.81	-1.6	16.96	9AP	3.6	-3.9	-2.0	1.2
184.72	1.33	-1.5	26.58	19AP	2.1	-3.9	-2.1	1.2
182.35	0.82	-1.3	36.17	29AP	1.2	-4.0	-2.1	1.2
181.28	0.34	-1.1	45.72	9MY	0.5	-4.1	-2.2	1.2
181.56	-0.09	-0.9	55.25	19MY	-0.4	-4.2	-2.2	1.1
183.10	-0.46	-0.7	64.78	29MY	-1.4	-4.2	-2.2	1.1
185.72	-0.77	-0.5	74.29	8JN	-1.6	-4.0	-2.3	1.0
189.24	-1.04	-0.3	83.82	18JN	-0.7	-3.5	-2.2	0.9
193.50	-1.25	-0.2	93.36	28JN	-0.1	-2.9	-2.2	0.9
198.37	-1.43	-0.0	102.92	8JL	0.4	-3.6	-2.2	0.8
203.74	-1.57	0.1	112.52	18JL	0.7	-4.0	-2.1	0.7
209.55	-1.68	0.2	122.16	28JL	1.1	-4.2	-2.1	0.7
215.71	-1.77	0.3	131.85	7AU	2.4	-4.2	-2.0	0.6
222.17	-1.83	0.4	141.58	17AU	2.1	-4.2	-2.0	0.7
228.89	-1.86	0.6	151.38	27AU	-0.0	-4.1	-1.9	0.7
235.82	-1.87	0.6	161.23	6SE	-0.9	-4.0	-1.8	0.8
242.93	-1.86	0.7	171.14	16SE	-1.0	-3.9	-1.8	0.8
250.21	-1.83	0.7	181.12	26SE	-0.9	-3.9	-1.7	0.9
257.60	-1.79	0.8	191.14	6OC	-0.6	-3.8	-1.7	0.9
265.08	-1.72	0.9	201.22	16OC	-0.4	-3.7	-1.6	1.0
272.65	-1.65	0.9	211.34	26OC	-0.4	-3.7	-1.6	1.0
280.26	-1.56	1.0	221.50	5NO	-0.3	-3.6	-1.5	1.0
287.90	-1.46	1.1	231.69	15NO	0.2	-3.5	-1.5	1.1
295.56	-1.35	1.1	241.89	25NO	2.6	-3.5	-1.5	1.1
303.21	-1.23	1.2	252.10	5DE	1.0	-3.5	-1.5	1.1
310.83	-1.11	1.3	262.31	15DE	0.1	-3.4	-1.5	1.2
318.43	-0.99	1.3	272.50	25DE	-0.0	-3.4	-1.5	1.2

♂ LONG	LAT	MAG	☉ LONG	16.00UT -67	☿	♀	♃	♄
325.97	-0.86	1.4	282.66	4JA	-0.1	-3.4	-1.5	1.2
333.47	-0.73	1.5	292.79	14JA	-0.3	-3.3	-1.5	1.2
340.90	-0.61	1.5	302.87	24JA	-0.7	-3.3	-1.5	1.2
348.26	-0.48	1.6	312.90	3FE	-1.3	-3.3	-1.5	1.2
355.55	-0.36	1.6	322.88	13FE	-1.3	-3.3	-1.6	1.2
2.76	-0.24	1.7	332.80	23FE	-0.5	-3.3	-1.6	1.3
9.90	-0.13	1.7	342.66	5MR	0.9	-3.3	-1.7	1.3
16.96	-0.02	1.7	352.46	15MR	2.8	-3.4	-1.7	1.3
23.95	0.09	1.8	2.21	25MR	2.7	-3.4	-1.8	1.4
30.87	0.19	1.8	11.90	4AP	1.5	-3.4	-1.8	1.4
37.72	0.29	1.8	21.54	14AP	0.9	-3.5	-1.9	1.4
44.50	0.39	1.8	31.14	24AP	0.3	-3.5	-2.0	1.4
51.22	0.47	1.8	40.71	4MY	-0.4	-3.5	-2.1	1.4
57.88	0.56	1.9	50.26	14MY	-1.4	-3.4	-2.1	1.4
64.49	0.64	1.9	59.78	24MY	-1.6	-3.4	-2.2	1.3
71.05	0.71	1.9	69.30	3JN	-0.7	-3.4	-2.3	1.3
77.57	0.79	1.9	78.82	13JN	0.0	-3.3	-2.3	1.2
84.04	0.85	1.9	88.35	23JN	0.5	-3.3	-2.3	1.2
90.49	0.92	1.9	97.90	3JL	0.9	-3.3	-2.4	1.1
96.90	0.98	2.0	107.49	13JL	1.5	-3.3	-2.4	1.0
103.28	1.03	2.0	117.10	23JL	2.9	-3.3	-2.4	1.0
109.65	1.09	2.0	126.76	2AU	1.8	-3.3	-2.3	0.9
115.99	1.14	2.0	136.47	12AU	-0.1	-3.4	-2.3	0.8
122.31	1.18	2.0	146.24	22AU	-1.0	-3.4	-2.3	0.8
128.62	1.23	2.0	156.06	1SE	-1.2	-3.4	-2.2	0.8
134.92	1.27	2.0	165.94	11SE	-0.9	-3.5	-2.1	0.9
141.19	1.30	2.0	175.88	21SE	-0.5	-3.5	-2.1	0.9
147.45	1.34	2.0	185.88	1OC	-0.3	-3.6	-2.0	1.0
153.70	1.37	1.9	195.93	11OC	-0.2	-3.6	-1.9	1.0
159.92	1.39	1.9	206.03	21OC	-0.1	-3.7	-1.9	1.1
166.12	1.42	1.8	216.17	31OC	-0.4	-3.8	-1.8	1.1
172.29	1.43	1.7	226.34	10NO	2.9	-3.9	-1.8	1.2
178.42	1.45	1.7	236.54	20NO	0.8	-4.0	-1.7	1.2
184.51	1.45	1.6	246.74	30NO	-0.1	-4.1	-1.7	1.3
190.55	1.45	1.5	256.95	10DE	-0.2	-4.2	-1.6	1.3
196.51	1.44	1.4	267.15	20DE	-0.2	-4.3	-1.6	1.3
202.40	1.42	1.2	277.33	30DE	-0.4	-4.4	-1.6	1.3

♂ LONG	LAT	MAG	☉ LONG	16.00UT -66	☿	♀	♃	♄
						MAGNITUDES		
208.19	1.39	1.1	287.48	9JA	-0.8	-4.3	-1.6	1.4
213.86	1.35	0.9	297.58	19JA	-1.2	-4.1	-1.5	1.4
219.38	1.28	0.8	307.64	29JA	-1.2	-3.6	-1.5	1.4
224.71	1.19	0.6	317.65	8FE	-0.5	-3.3	-1.5	1.3
229.80	1.08	0.4	327.60	18FE	1.1	-3.8	-1.6	1.3
234.61	0.93	0.2	337.49	28FE	3.3	-4.2	-1.6	1.3
239.05	0.73	-0.1	347.33	10MR	2.0	-4.3	-1.6	1.3
243.02	0.48	-0.3	357.10	20MR	1.1	-4.2	-1.6	1.3
246.40	0.17	-0.6	6.82	30MR	0.7	-4.2	-1.7	1.2
249.03	-0.22	-0.9	16.49	9AP	0.2	-4.0	-1.7	1.2
250.74	-0.70	-1.2	26.11	19AP	-0.5	-3.9	-1.8	1.2
251.34	-1.28	-1.5	35.70	29AP	-1.4	-3.8	-1.8	1.2
250.71	-1.94	-1.9	45.26	9MY	-1.6	-3.7	-1.9	1.2
248.90	-2.66	-2.2	54.79	19MY	-0.7	-3.7	-2.0	1.1
246.22	-3.34	-2.4	64.31	29MY	0.1	-3.6	-2.1	1.1
243.28	-3.91	-2.3	73.83	8JN	0.7	-3.5	-2.1	1.0
240.84	-4.30	-2.2	83.36	18JN	1.2	-3.5	-2.2	1.0
239.46	-4.50	-2.0	92.90	28JN	2.0	-3.4	-2.3	0.9
239.42	-4.53	-1.7	102.46	8JL	3.2	-3.4	-2.3	0.9
240.75	-4.45	-1.5	112.06	18JL	1.5	-3.4	-2.4	0.8
243.27	-4.28	-1.3	121.70	28JL	-0.2	-3.4	-2.4	0.8
246.82	-4.06	-1.0	131.38	7AU	-1.1	-3.4	-2.4	0.7
251.20	-3.82	-0.8	141.11	17AU	-1.3	-3.4	-2.5	0.7
256.22	-3.55	-0.6	150.91	27AU	-0.9	-3.4	-2.4	0.7
261.77	-3.28	-0.4	160.76	6SE	-0.4	-3.4	-2.4	0.7
267.72	-3.00	-0.3	170.67	16SE	-0.2	-3.4	-2.4	0.8
273.98	-2.72	-0.1	180.63	26SE	-0.1	-3.4	-2.3	0.8
280.50	-2.45	0.1	190.66	6OC	0.0	-3.4	-2.3	0.9
287.19	-2.18	0.2	200.73	16OC	0.6	-3.4	-2.2	1.0
294.01	-1.92	0.4	210.85	26OC	3.3	-3.4	-2.1	1.0
300.94	-1.67	0.5	221.01	5NO	0.5	-3.5	-2.1	1.1
307.93	-1.43	0.7	231.19	15NO	-0.2	-3.5	-2.0	1.1
314.95	-1.20	0.8	241.39	25NO	-0.3	-3.5	-1.9	1.2
322.00	-0.99	0.9	251.60	5DE	-0.4	-3.5	-1.9	1.2
329.05	-0.79	1.0	261.80	15DE	-0.5	-3.4	-1.8	1.2
336.08	-0.60	1.1	271.99	25DE	-0.8	-3.4	-1.8	1.2
				-65				
343.08	-0.43	1.2	282.16	4JA	-1.1	-3.4	-1.7	1.2
350.06	-0.27	1.3	292.28	14JA	-1.1	-3.4	-1.7	1.2
356.99	-0.12	1.4	302.37	24JA	-0.4	-3.3	-1.6	1.1
3.88	0.02	1.5	312.41	3FE	1.3	-3.3	-1.6	1.1
10.72	0.15	1.6	322.39	13FE	3.0	-3.3	-1.6	1.1
17.51	0.27	1.7	332.31	23FE	1.5	-3.3	-1.6	1.0
24.25	0.37	1.7	342.18	5MR	0.8	-3.3	-1.6	1.0
30.95	0.47	1.8	351.98	15MR	0.5	-3.4	-1.6	1.0
37.59	0.56	1.8	1.73	25MR	0.1	-3.4	-1.6	1.0
44.20	0.65	1.9	11.43	4AP	-0.5	-3.4	-1.6	1.0
50.75	0.72	1.9	21.07	14AP	-1.4	-3.4	-1.6	1.0
57.27	0.79	1.9	30.68	24AP	-1.6	-3.5	-1.7	1.0
63.76	0.85	2.0	40.25	4MY	-0.7	-3.5	-1.7	0.9
70.21	0.91	2.0	49.79	14MY	0.2	-3.6	-1.7	0.9
76.64	0.96	2.0	59.33	24MY	0.9	-3.6	-1.8	0.9
83.05	1.00	2.0	68.84	3JN	1.6	-3.7	-1.8	0.9
89.44	1.04	2.0	78.36	13JN	2.7	-3.8	-1.9	0.8
95.81	1.07	2.0	87.90	23JN	2.9	-3.9	-2.0	0.8
102.18	1.10	2.0	97.45	3JL	1.2	-4.0	-2.0	0.7
108.55	1.12	2.0	107.02	13JL	-0.2	-4.1	-2.1	0.7
114.92	1.14	2.0	116.64	23JL	-1.2	-4.2	-2.2	0.6
121.29	1.16	2.0	126.30	2AU	-1.4	-4.2	-2.2	0.6
127.67	1.17	2.0	136.00	12AU	-0.8	-4.2	-2.3	0.5
134.06	1.18	2.0	145.77	22AU	-0.3	-4.0	-2.4	0.5
140.47	1.18	2.0	155.59	1SE	-0.1	-3.6	-2.4	0.4
146.90	1.18	2.0	165.46	11SE	0.1	-3.3	-2.4	0.4
153.35	1.17	2.0	175.40	21SE	0.2	-3.8	-2.4	0.4
159.82	1.15	2.0	185.40	1OC	0.9	-4.2	-2.4	0.5
166.31	1.14	2.0	195.44	11OC	3.4	-4.3	-2.4	0.6
172.83	1.11	1.9	205.54	21OC	0.4	-4.3	-2.4	0.6
179.37	1.08	1.9	215.68	31OC	-0.4	-4.3	-2.3	0.7
185.94	1.04	1.8	225.85	10NO	-0.5	-4.2	-2.3	0.8
192.53	0.99	1.8	236.05	20NO	-0.5	-4.1	-2.2	0.8
199.14	0.94	1.7	246.25	30NO	-0.6	-4.0	-2.1	0.9
205.78	0.87	1.6	256.46	10DE	-0.9	-3.9	-2.1	0.9
212.43	0.80	1.6	266.66	20DE	-0.9	-3.8	-2.0	0.9
219.10	0.71	1.5	276.84	30DE	-0.9	-3.7	-1.9	0.9

♂ LONG	LAT	MAG	☉ LONG	16.00UT -64	☿	♀	♃	♄
						MAGNITUDES		
225.78	0.61	1.4	286.98	9JA	-0.3	-3.6	-1.9	0.9
232.47	0.49	1.3	297.10	19JA	1.6	-3.6	-1.8	0.9
239.17	0.36	1.2	307.16	29JA	2.5	-3.5	-1.7	0.9
245.86	0.20	1.1	317.17	8FE	1.1	-3.5	-1.7	0.9
252.55	0.03	0.9	327.12	18FE	0.6	-3.4	-1.6	0.8
259.21	-0.16	0.8	337.02	28FE	0.3	-3.4	-1.6	0.8
265.85	-0.38	0.7	346.85	9MR	-0.0	-3.3	-1.6	0.7
272.44	-0.63	0.5	356.63	19MR	-0.5	-3.3	-1.6	0.7
278.96	-0.91	0.3	6.35	29MR	-1.4	-3.3	-1.5	0.7
285.38	-1.21	0.2	16.02	8AP	-1.6	-3.3	-1.5	0.7
291.67	-1.55	0.0	25.65	18AP	-0.7	-3.3	-1.5	0.7
297.78	-1.92	-0.2	35.24	28AP	0.3	-3.3	-1.5	0.7
303.66	-2.33	-0.4	44.80	8MY	1.1	-3.3	-1.6	0.7
309.22	-2.77	-0.6	54.33	18MY	2.1	-3.4	-1.6	0.7
314.37	-3.25	-0.8	63.85	28MY	3.4	-3.4	-1.6	0.7
318.98	-3.76	-1.0	73.37	7JN	2.4	-3.4	-1.6	0.7
322.90	-4.31	-1.2	82.90	17JN	1.0	-3.5	-1.7	0.6
325.93	-4.87	-1.5	92.44	27JN	-0.3	-3.5	-1.7	0.6
327.87	-5.43	-1.7	102.00	7JL	-1.3	-3.5	-1.8	0.5
328.52	-5.94	-2.0	111.60	17JL	-1.5	-3.4	-1.8	0.5
327.79	-6.32	-2.2	121.23	27JL	-0.8	-3.4	-1.9	0.4
325.81	-6.48	-2.4	130.91	6AU	-0.3	-3.4	-2.0	0.4
323.06	-6.36	-2.5	140.64	16AU	0.0	-3.4	-2.0	0.3
320.28	-5.93	-2.4	150.43	26AU	0.2	-3.3	-2.1	0.3
318.20	-5.28	-2.2	160.28	5SE	0.4	-3.3	-2.2	0.2
317.29	-4.53	-1.8	170.18	15SE	1.2	-3.3	-2.2	0.2
317.71	-3.77	-1.5	180.14	25SE	3.1	-3.3	-2.3	0.1
319.34	-3.06	-1.2	190.17	5OC	0.2	-3.3	-2.3	0.2
322.02	-2.42	-0.9	200.24	15OC	-0.6	-3.4	-2.3	0.2
325.54	-1.88	-0.6	210.36	25OC	-0.6	-3.4	-2.4	0.3
329.71	-1.40	-0.3	220.51	4NO	-0.6	-3.4	-2.4	0.4
334.40	-1.00	-0.1	230.70	14NO	-0.7	-3.4	-2.3	0.5
339.47	-0.66	0.1	240.90	24NO	-0.7	-3.5	-2.3	0.5
344.84	-0.37	0.4	251.11	4DE	-0.8	-3.5	-2.3	0.6
350.44	-0.12	0.5	261.31	14DE	-0.8	-3.6	-2.2	0.6
356.21	0.09	0.7	271.50	24DE	-0.2	-3.7	-2.1	0.7
				-63				
2.11	0.28	0.9	281.67	3JA	1.8	-3.7	-2.1	0.7
8.11	0.43	1.0	291.80	13JA	2.0	-3.8	-2.0	0.7
14.18	0.57	1.2	301.89	23JA	0.8	-3.9	-1.9	0.7
20.31	0.68	1.3	311.93	2FE	0.4	-4.0	-1.8	0.7
26.47	0.78	1.4	321.91	12FE	0.2	-4.1	-1.8	0.7
32.66	0.87	1.5	331.83	22FE	-0.1	-4.2	-1.7	0.6
38.86	0.94	1.6	341.70	4MR	-0.6	-4.2	-1.7	0.6
45.07	1.00	1.7	351.51	14MR	-1.4	-4.2	-1.6	0.6
51.29	1.05	1.8	1.26	24MR	-1.5	-4.2	-1.6	0.5
57.51	1.09	1.8	10.96	3AP	-0.6	-3.9	-1.5	0.5
63.74	1.13	1.9	20.61	13AP	0.4	-3.2	-1.5	0.5
69.97	1.16	1.9	30.21	23AP	1.5	-3.2	-1.5	0.5
76.20	1.18	2.0	39.79	3MY	2.8	-3.8	-1.5	0.5
82.43	1.19	2.0	49.33	13MY	3.3	-4.1	-1.4	0.5
88.68	1.20	2.0	58.86	23MY	1.8	-4.2	-1.4	0.5
94.93	1.21	2.0	68.38	2JN	0.8	-4.2	-1.5	0.5
101.20	1.21	2.0	77.90	12JN	-0.3	-4.1	-1.5	0.5
107.49	1.20	2.0	87.43	22JN	-1.3	-4.0	-1.5	0.5
113.80	1.19	2.0	96.98	2JL	-1.6	-3.9	-1.5	0.5
120.13	1.18	2.0	106.56	12JL	-0.8	-3.8	-1.5	0.4
126.49	1.16	2.0	116.17	22JL	-0.2	-3.7	-1.6	0.4
132.89	1.13	2.0	125.83	1AU	0.2	-3.7	-1.6	0.3
139.32	1.10	1.9	135.53	11AU	0.4	-3.6	-1.7	0.3
145.80	1.07	1.9	145.29	21AU	0.7	-3.6	-1.7	0.2
152.32	1.03	1.9	155.11	31AU	1.6	-3.5	-1.8	0.2
158.88	0.99	1.9	164.98	10SE	2.8	-3.5	-1.9	0.1
165.49	0.94	1.9	174.92	20SE	0.1	-3.4	-1.9	0.0
172.16	0.89	1.9	184.91	30SE	-0.7	-3.4	-2.0	-0.0
178.88	0.83	1.9	194.96	10OC	-0.8	-3.4	-2.0	-0.1
185.66	0.77	1.9	205.05	20OC	-0.8	-3.4	-2.1	-0.0
192.49	0.69	1.8	215.19	30OC	-0.8	-3.4	-2.2	0.1
199.38	0.62	1.8	225.36	9NO	-0.6	-3.4	-2.2	0.1
206.33	0.53	1.8	235.55	19NO	-0.6	-3.4	-2.2	0.2
213.34	0.44	1.7	245.76	29NO	-0.6	-3.4	-2.2	0.3
220.40	0.34	1.7	255.96	9DE	-0.1	-3.4	-2.2	0.3
227.52	0.23	1.6	266.16	19DE	2.1	-3.4	-2.2	0.4
234.70	0.11	1.6	276.35	29DE	1.6	-3.4	-2.2	0.4

Left table (year −62 / −61):

♂ LONG LAT MAG	☉ LONG	16.00UT	☿ ♀ ♃ ♄ MAGNITUDES
241.92-0.02 1.5	286.49	8JA	0.5 -3.4 -2.1 0.5
249.19-0.15 1.5	296.60	18JA	0.2 -3.4 -2.1 0.5
256.51-0.29 1.4	306.67	28JA	0.1 -3.5 -2.0 0.5
263.87-0.45 1.3	316.68	7FE	-0.2 -3.5 -1.9 0.5
271.26-0.60 1.2	326.64	17FE	-0.6 -3.5 -1.9 0.5
278.67-0.77 1.2	336.54	27FE	-1.3 -3.4 -1.8 0.5
286.10-0.94 1.1	346.37	9MR	-1.5 -3.4 -1.7 0.5
293.54-1.11 1.0	356.15	19MR	-0.6 -3.4 -1.7 0.4
300.97-1.28 0.9	5.88	29MR	0.6 -3.4 -1.6 0.4
308.37-1.45 0.8	15.55	8AP	1.9 -3.3 -1.5 0.3
315.73-1.62 0.8	25.18	18AP	3.6 -3.3 -1.5 0.3
323.03-1.79 0.7	34.77	28AP	2.5 -3.3 -1.5 0.4
330.24-1.95 0.6	44.33	8MY	1.4 -3.3 -1.4 0.4
337.34-2.10 0.5	53.86	18MY	0.6 -3.3 -1.4 0.4
344.29-2.24 0.4	63.39	28MY	-0.3 -3.4 -1.4 0.4
351.07-2.36 0.3	72.91	7JN	-1.3 -3.4 -1.4 0.4
357.63-2.47 0.2	82.43	17JN	-1.6 -3.4 -1.3 0.4
3.92-2.57 0.1	91.97	27JN	-0.7 -3.4 -1.3 0.4
9.88-2.64 -0.0	101.53	7JL	-0.1 -3.5 -1.3 0.4
15.43-2.70 -0.2	111.13	17JL	0.3 -3.5 -1.4 0.3
20.49-2.73 -0.3	120.76	27JL	0.5 -3.6 -1.4 0.3
24.93-2.73 -0.5	130.44	6AU	0.9 -3.7 -1.4 0.3
28.61-2.71 -0.7	140.17	16AU	2.0 -3.8 -1.4 0.2
31.33-2.64 -0.9	149.96	26AU	2.4 -3.9 -1.5 0.2
32.91-2.51 -1.1	159.80	5SE	0.0 -4.0 -1.5 0.1
33.12-2.30 -1.3	169.71	15SE	-0.8 -4.1 -1.6 0.0
31.86-2.01 -1.5	179.67	25SE	-0.9 -4.2 -1.6 -0.0
29.25-1.61 -1.7	189.68	5OC	-0.9 -4.3 -1.7 -0.1
25.72-1.13 -1.8	199.75	15OC	-0.7 -4.4 -1.8 -0.2
22.04-0.61 -1.7	209.87	25OC	-0.5 -4.4 -1.8 -0.2
19.01-0.11 -1.4	220.02	4NO	-0.5 -4.2 -1.9 -0.1
17.15 0.32 -1.1	230.20	14NO	-0.4 -3.6 -1.9 -0.1
16.66 0.67 -0.8	240.40	24NO	0.1 -3.0 -2.0 0.0
17.44 0.94 -0.5	250.61	4DE	2.3 -3.9 -2.1 0.1
19.32 1.14 -0.2	260.82	14DE	1.2 -4.4 -2.1 0.1
22.09 1.28 0.1	271.01	24DE	0.3 -4.4 -2.1 0.2
		−61	
25.55 1.39 0.3	281.17	3JA	0.1 -4.3 -2.1 0.3
29.56 1.46 0.5	291.31	13JA	-0.1 -4.3 -2.1 0.3
33.99 1.51 0.7	301.40	23JA	-0.3 -4.2 -2.1 0.3
38.75 1.55 0.9	311.44	2FE	-0.7 -4.0 -2.1 0.4
43.77 1.57 1.1	321.43	12FE	-1.3 -3.9 -2.0 0.4
49.00 1.57 1.2	331.35	22FE	-1.4 -3.8 -1.9 0.4
54.39 1.57 1.4	341.22	4MR	-0.6 -3.7 -1.9 0.4
59.91 1.56 1.5	351.04	14MR	0.8 -3.6 -1.8 0.4
65.55 1.55 1.6	0.79	24MR	2.4 -3.6 -1.7 0.4
71.27 1.53 1.7	10.49	3AP	3.3 -3.5 -1.7 0.3
77.07 1.51 1.7	20.14	13AP	1.9 -3.5 -1.6 0.3
82.94 1.48 1.8	29.75	23AP	1.1 -3.4 -1.6 0.2
88.87 1.45 1.9	39.32	3MY	0.4 -3.4 -1.5 0.2
94.86 1.41 1.9	48.87	13MY	-0.4 -3.3 -1.4 0.3
100.90 1.37 1.9	58.40	23MY	-1.3 -3.3 -1.4 0.3
106.99 1.33 2.0	67.92	2JN	-1.6 -3.3 -1.4 0.3
113.14 1.28 2.0	77.44	12JN	-0.7 -3.3 -1.3 0.3
119.33 1.23 2.0	86.97	22JN	-0.0 -3.3 -1.3 0.3
125.59 1.18 2.0	96.52	2JL	0.4 -3.3 -1.3 0.3
131.90 1.12 2.0	106.10	12JL	0.7 -3.3 -1.3 0.3
138.27 1.06 2.0	115.71	22JL	1.3 -3.4 -1.3 0.3
144.71 1.00 2.0	125.36	1AU	2.5 -3.4 -1.3 0.3
151.21 0.94 1.9	135.07	11AU	2.1 -3.4 -1.3 0.3
157.77 0.87 1.9	144.82	21AU	-0.0 -3.5 -1.3 0.2
164.41 0.79 1.9	154.63	31AU	-1.0 -3.5 -1.3 0.2
171.12 0.72 1.8	164.51	10SE	-1.1 -3.5 -1.3 0.1
177.90 0.63 1.8	174.43	20SE	-0.9 -3.5 -1.4 0.0
184.76 0.55 1.7	184.42	30SE	-0.6 -3.4 -1.4 -0.0
191.70 0.46 1.7	194.47	10OC	-0.4 -3.4 -1.5 -0.1
198.72 0.37 1.7	204.56	20OC	-0.3 -3.4 -1.5 -0.2
205.82 0.27 1.7	214.69	30OC	-0.3 -3.4 -1.6 -0.2
212.99 0.17 1.7	224.86	9NO	0.2 -3.3 -1.6 -0.3
220.24 0.06 1.7	235.05	19NO	2.6 -3.3 -1.7 -0.2
227.56-0.05 1.6	245.26	29NO	0.9 -3.3 -1.8 -0.1
234.96-0.16 1.6	255.45	9DE	0.1 -3.3 -1.8 -0.1
242.42-0.27 1.6	265.66	19DE	-0.1 -3.4 -1.9 0.0
249.94-0.39 1.6	275.85	29DE	-0.2 -3.4 -1.9 0.1

Right table (year −60 / −59):

♂ LONG LAT MAG	☉ LONG	16.00UT	☿ ♀ ♃ ♄ MAGNITUDES
257.52-0.51 1.5	286.00	8JA	-0.4 -3.4 -2.0 0.1
265.14-0.62 1.5	296.11	18JA	-0.7 -3.4 -2.0 0.2
272.81-0.74 1.5	306.18	28JA	-1.3 -3.4 -2.0 0.2
280.51-0.85 1.4	316.20	7FE	-1.3 -3.5 -2.0 0.3
288.22-0.96 1.4	326.15	17FE	-0.6 -3.5 -2.0 0.3
295.95-1.07 1.3	336.06	27FE	0.9 -3.6 -2.0 0.3
303.68-1.16 1.3	345.90	8MR	3.0 -3.6 -2.0 0.3
311.39-1.25 1.3	355.68	18MR	2.5 -3.7 -1.9 0.3
319.08-1.32 1.2	5.41	28MR	1.4 -3.8 -1.9 0.3
326.72-1.39 1.2	15.09	7AP	0.8 -3.9 -1.8 0.3
334.31-1.44 1.2	24.72	17AP	0.3 -3.9 -1.7 0.3
341.83-1.48 1.1	34.31	27AP	-0.4 -4.0 -1.7 0.2
349.27-1.50 1.1	43.87	7MY	-1.3 -4.1 -1.6 0.2
356.60-1.51 1.1	53.41	17MY	-1.6 -4.2 -1.5 0.2
3.83-1.51 1.0	62.93	27MY	-0.7 -4.2 -1.5 0.2
10.92-1.48 1.0	72.45	6JN	0.0 -4.0 -1.4 0.3
17.87-1.44 1.0	81.97	16JN	0.5 -3.5 -1.4 0.3
24.66-1.39 0.9	91.51	26JN	1.0 -2.9 -1.3 0.3
31.26-1.31 0.9	101.07	6JL	1.7 -3.6 -1.3 0.3
37.65-1.22 0.8	110.66	16JL	3.1 -4.0 -1.3 0.3
43.80-1.12 0.7	120.29	26JL	1.8 -4.2 -1.2 0.3
49.67-0.99 0.6	129.97	5AU	-0.1 -4.2 -1.2 0.3
55.22-0.84 0.5	139.69	15AU	-1.1 -4.2 -1.2 0.3
60.39-0.67 0.4	149.48	25AU	-1.2 -4.1 -1.2 0.2
65.09-0.47 0.3	159.32	4SE	-0.9 -4.0 -1.2 0.2
69.23-0.25 0.2	169.22	14SE	-0.5 -3.9 -1.2 0.1
72.68 0.02 -0.0	179.18	24SE	-0.3 -3.9 -1.2 0.1
75.27 0.33 -0.2	189.20	4OC	-0.2 -3.8 -1.3 0.0
76.83 0.69 -0.4	199.26	14OC	-0.1 -3.7 -1.3 -0.1
77.13 1.09 -0.6	209.38	24OC	0.5 -3.6 -1.3 -0.1
76.06 1.54 -0.8	219.53	3NO	3.0 -3.6 -1.4 -0.2
73.61 2.00 -1.1	229.71	13NO	0.7 -3.5 -1.4 -0.3
70.08 2.42 -1.2	239.91	23NO	-0.1 -3.5 -1.5 -0.3
66.12 2.75 -1.2	250.12	3DE	-0.2 -3.5 -1.6 -0.2
62.50 2.95 -1.0	260.32	13DE	-0.3 -3.4 -1.6 -0.2
59.85 3.03 -0.8	270.52	23DE	-0.5 -3.4 -1.7 -0.1
		−59	
58.50 3.01 -0.5	280.69	2JA	-0.8 -3.4 -1.8 -0.0
58.46 2.93 -0.2	290.82	12JA	-1.2 -3.3 -1.8 0.0
59.59 2.82 0.1	300.91	22JA	-1.1 -3.3 -1.9 0.1
61.70 2.69 0.3	310.95	1FE	-0.5 -3.3 -1.9 0.2
64.59 2.56 0.6	320.94	11FE	1.2 -3.3 -2.0 0.2
68.10 2.43 0.7	330.88	21FE	3.3 -3.3 -2.0 0.3
72.10 2.31 0.9	340.75	3MR	1.8 -3.3 -2.0 0.3
76.50 2.19 1.1	350.56	13MR	1.0 -3.4 -2.0 0.3
81.21 2.07 1.2	0.32	23MR	0.6 -3.4 -2.0 0.3
86.18 1.96 1.3	10.02	2AP	0.2 -3.4 -2.0 0.3
91.36 1.85 1.4	19.67	12AP	-0.5 -3.5 -1.9 0.3
96.72 1.75 1.5	29.29	22AP	-1.3 -3.5 -1.9 0.3
102.24 1.64 1.6	38.86	2MY	-1.6 -3.5 -1.8 0.3
107.89 1.54 1.7	48.41	12MY	-0.7 -3.4 -1.8 0.2
113.67 1.45 1.7	57.94	22MY	0.1 -3.4 -1.7 0.2
119.56 1.35 1.8	67.46	1JN	0.7 -3.4 -1.6 0.2
125.55 1.25 1.8	76.98	11JN	1.3 -3.3 -1.6 0.2
131.65 1.15 1.8	86.51	21JN	2.2 -3.3 -1.5 0.3
137.84 1.06 1.8	96.05	1JL	3.2 -3.3 -1.5 0.3
144.13 0.96 1.9	105.63	11JL	1.5 -3.3 -1.4 0.3
150.51 0.86 1.9	115.24	21JL	-0.1 -3.3 -1.4 0.3
156.99 0.76 1.8	124.88	31JL	-1.1 -3.3 -1.3 0.3
163.56 0.66 1.8	134.59	10AU	-1.4 -3.4 -1.3 0.3
170.23 0.55 1.8	144.34	20AU	-0.9 -3.4 -1.3 0.3
177.00 0.45 1.8	154.15	30AU	-0.4 -3.4 -1.2 0.3
183.86 0.35 1.8	164.02	9SE	-0.1 -3.5 -1.2 0.3
190.82 0.24 1.7	173.95	19SE	-0.0 -3.5 -1.2 0.2
197.88 0.13 1.7	183.93	29SE	0.1 -3.6 -1.2 0.2
205.03 0.03 1.6	193.98	9OC	0.7 -3.7 -1.2 0.1
212.28-0.08 1.6	204.07	19OC	3.3 -3.7 -1.2 0.0
219.61-0.19 1.5	214.20	29OC	0.5 -3.8 -1.3 -0.0
227.03-0.29 1.5	224.37	8NO	-0.3 -3.9 -1.3 -0.1
234.54-0.40 1.4	234.56	18NO	-0.4 -4.0 -1.3 -0.2
242.11-0.50 1.4	244.76	28NO	-0.4 -4.1 -1.4 -0.2
249.76-0.60 1.4	254.97	8DE	-0.6 -4.2 -1.4 -0.3
257.47-0.69 1.4	265.18	18DE	-0.8 -4.3 -1.5 -0.2
265.23-0.78 1.4	275.36	28DE	-1.0 -4.4 -1.6 -0.1

♂ LONG	LAT	MAG	☉ LONG	16.00UT -58	☿	♀	♃	♄
273.03	-0.86	1.4	285.51	7JA	-1.0	-4.3	-1.6	-0.1
280.85	-0.94	1.4	295.63	17JA	-0.4	-4.1	-1.7	0.0
288.70	-1.00	1.4	305.69	27JA	1.4	-3.6	-1.8	0.1
296.55	-1.06	1.4	315.71	6FE	2.9	-3.3	-1.8	0.1
304.39	-1.11	1.4	325.68	16FE	1.3	-3.8	-1.9	0.2
312.22	-1.14	1.4	335.58	26FE	0.7	-4.2	-1.9	0.2
320.01	-1.16	1.4	345.42	8MR	0.4	-4.3	-2.0	0.3
327.76	-1.17	1.4	355.21	18MR	0.1	-4.2	-2.0	0.3
335.46	-1.17	1.4	4.94	28MR	-0.5	-4.2	-2.0	0.3
343.09	-1.16	1.4	14.62	7AP	-1.3	-4.0	-2.1	0.3
350.64	-1.13	1.4	24.25	17AP	-1.6	-3.9	-2.1	0.3
358.11	-1.09	1.4	33.84	27AP	-0.7	-3.8	-2.0	0.3
5.50	-1.04	1.4	43.41	7MY	0.2	-3.7	-2.0	0.3
12.78	-0.97	1.4	52.95	17MY	1.0	-3.7	-1.9	0.3
19.96	-0.90	1.4	62.47	27MY	1.7	-3.6	-1.9	0.3
27.02	-0.81	1.4	71.99	6JN	2.9	-3.5	-1.8	0.3
33.97	-0.72	1.4	81.51	16JN	2.7	-3.5	-1.8	0.3
40.79	-0.61	1.4	91.05	26JN	1.2	-3.4	-1.7	0.3
47.48	-0.50	1.4	100.61	6JL	-0.2	-3.4	-1.7	0.3
54.03	-0.38	1.3	110.20	16JL	-1.2	-3.4	-1.6	0.4
60.43	-0.24	1.3	119.82	26JL	-1.5	-3.4	-1.5	0.4
66.66	-0.10	1.3	129.50	5AU	-0.8	-3.4	-1.5	0.4
72.71	0.06	1.2	139.22	15AU	-0.3	-3.3	-1.4	0.4
78.55	0.23	1.2	149.00	25AU	-0.0	-3.4	-1.4	0.4
84.15	0.41	1.1	158.84	4SE	0.1	-3.4	-1.4	0.4
89.48	0.61	1.0	168.74	14SE	0.3	-3.4	-1.3	0.4
94.49	0.83	0.9	178.69	24SE	1.0	-3.4	-1.3	0.4
99.12	1.07	0.8	188.71	4OC	3.4	-3.4	-1.3	0.3
103.27	1.34	0.6	198.77	14OC	0.4	-3.4	-1.3	0.3
106.84	1.65	0.5	208.88	24OC	-0.5	-3.4	-1.3	0.2
109.70	1.98	0.3	219.04	3NO	-0.5	-3.5	-1.3	0.1
111.69	2.36	0.1	229.21	13NO	-0.6	-3.5	-1.3	0.1
112.61	2.78	-0.2	239.41	23NO	-0.7	-3.5	-1.3	-0.0
112.27	3.22	-0.4	249.62	3DE	-0.8	-3.5	-1.3	-0.1
110.60	3.65	-0.7	259.83	13DE	-0.8	-3.4	-1.4	-0.2
107.69	4.02	-0.9	270.02	23DE	-0.9	-3.4	-1.4	-0.2

-57

♂ LONG	LAT	MAG	☉ LONG	16.00UT	☿	♀	♃	♄
103.91	4.27	-1.0	280.19	2JA	-0.3	-3.4	-1.5	-0.1
99.97	4.35	-0.9	290.33	12JA	1.6	-3.4	-1.5	-0.0
96.57	4.26	-0.7	300.42	22JA	2.3	-3.3	-1.6	0.0
94.24	4.06	-0.5	310.47	1FE	1.0	-3.3	-1.6	0.1
93.22	3.78	-0.2	320.46	11FE	0.5	-3.3	-1.7	0.1
93.45	3.48	0.0	330.39	21FE	0.3	-3.3	-1.8	0.2
94.79	3.17	0.2	340.27	3MR	-0.0	-3.3	-1.9	0.3
97.07	2.89	0.4	350.08	13MR	-0.5	-3.4	-1.9	0.3
100.08	2.62	0.6	359.84	23MR	-1.3	-3.4	-2.0	0.4
103.70	2.38	0.8	9.55	2AP	-1.6	-3.4	-2.0	0.4
107.82	2.15	0.9	19.21	12AP	-0.7	-3.4	-2.1	0.4
112.32	1.94	1.1	28.82	22AP	0.3	-3.5	-2.1	0.4
117.17	1.75	1.2	38.40	2MY	1.2	-3.5	-2.1	0.4
122.29	1.57	1.3	47.95	12MY	2.3	-3.6	-2.2	0.4
127.65	1.40	1.4	57.48	22MY	3.6	-3.6	-2.2	0.4
133.22	1.23	1.4	67.00	1JN	2.2	-3.7	-2.1	0.4
138.97	1.08	1.5	76.52	11JN	0.9	-3.8	-2.1	0.4
144.90	0.93	1.5	86.05	21JN	-0.2	-3.9	-2.0	0.4
150.98	0.78	1.6	95.60	1JL	-1.3	-4.0	-2.0	0.4
157.21	0.64	1.6	105.17	11JL	-1.6	-4.1	-1.9	0.4
163.59	0.51	1.6	114.77	21JL	-0.8	-4.2	-1.9	0.5
170.09	0.37	1.6	124.42	31JL	-0.2	-4.2	-1.8	0.5
176.73	0.24	1.6	134.12	10AU	0.1	-4.2	-1.7	0.5
183.50	0.12	1.6	143.87	20AU	0.3	-4.0	-1.7	0.6
190.39	-0.01	1.6	153.68	30AU	0.5	-3.6	-1.6	0.6
197.40	-0.13	1.6	163.54	9SE	1.3	-3.3	-1.6	0.6
204.53	-0.24	1.6	173.47	19SE	3.1	-3.9	-1.5	0.6
211.77	-0.35	1.6	183.45	29SE	0.2	-4.2	-1.5	0.5
219.11	-0.46	1.5	193.49	9OC	-0.6	-4.3	-1.4	0.5
226.56	-0.56	1.5	203.58	19OC	-0.7	-4.3	-1.4	0.5
234.10	-0.65	1.5	213.71	29OC	-0.7	-4.3	-1.4	0.4
241.72	-0.74	1.5	223.87	8NO	-0.8	-4.2	-1.4	0.3
249.42	-0.82	1.4	234.06	18NO	-0.7	-4.1	-1.4	0.3
257.18	-0.89	1.4	244.27	28NO	-0.7	-4.0	-1.4	0.2
265.00	-0.95	1.4	254.48	8DE	-0.7	-3.9	-1.4	0.1
272.86	-1.00	1.4	264.68	18DE	-0.2	-3.8	-1.4	0.1
280.74	-1.04	1.3	274.87	28DE	1.8	-3.7	-1.4	-0.0

♂ LONG	LAT	MAG	☉ LONG	16.00UT -56	☿	♀	♃	♄
288.64	-1.07	1.3	285.02	7JA	1.9	-3.6	-1.4	-0.0
296.54	-1.08	1.3	295.14	17JA	0.7	-3.6	-1.5	0.0
304.43	-1.09	1.3	305.21	27JA	0.3	-3.5	-1.5	0.1
312.29	-1.08	1.3	315.23	6FE	0.1	-3.5	-1.6	0.2
320.11	-1.06	1.4	325.20	16FE	-0.1	-3.4	-1.6	0.2
327.88	-1.03	1.4	335.11	26FE	-0.6	-3.4	-1.7	0.3
335.60	-0.99	1.4	344.95	7MR	-1.3	-3.3	-1.8	0.3
343.25	-0.94	1.4	354.74	17MR	-1.5	-3.3	-1.8	0.4
350.82	-0.88	1.5	4.47	27MR	-0.7	-3.3	-1.9	0.4
358.31	-0.81	1.5	14.15	6AP	0.5	-3.3	-2.0	0.5
5.72	-0.74	1.5	23.79	16AP	1.6	-3.3	-2.0	0.5
13.03	-0.65	1.6	33.39	26AP	3.1	-3.3	-2.1	0.5
20.26	-0.56	1.6	42.95	6MY	3.0	-3.3	-2.2	0.6
27.39	-0.47	1.6	52.49	16MY	1.7	-3.4	-2.2	0.6
34.42	-0.37	1.6	62.01	26MY	0.7	-3.4	-2.2	0.6
41.35	-0.26	1.7	71.53	5JN	-0.3	-3.4	-2.3	0.6
48.19	-0.15	1.7	81.05	15JN	-1.3	-3.5	-2.3	0.6
54.93	-0.04	1.7	90.59	25JN	-1.6	-3.5	-2.3	0.6
61.58	0.08	1.7	100.14	5JL	-0.8	-3.5	-2.2	0.6
68.12	0.20	1.7	109.73	15JL	-0.2	-3.4	-2.2	0.6
74.56	0.33	1.7	119.36	25JL	0.2	-3.4	-2.2	0.6
80.89	0.46	1.7	129.03	4AU	0.4	-3.4	-2.1	0.7
87.10	0.59	1.6	138.75	14AU	0.8	-3.4	-2.0	0.7
93.19	0.73	1.6	148.52	24AU	1.7	-3.3	-2.0	0.7
99.16	0.88	1.6	158.36	3SE	2.7	-3.3	-1.9	0.7
104.97	1.03	1.5	168.25	13SE	0.1	-3.3	-1.8	0.8
110.61	1.19	1.4	178.20	23SE	-0.8	-3.3	-1.8	0.8
116.06	1.36	1.3	188.21	3OC	-0.9	-3.3	-1.7	0.7
121.28	1.55	1.2	198.28	13OC	-0.8	-3.4	-1.7	0.7
126.22	1.75	1.1	208.39	23OC	-0.7	-3.4	-1.6	0.7
130.84	1.97	1.0	218.53	2NO	-0.6	-3.4	-1.6	0.6
135.06	2.21	0.8	228.72	12NO	-0.5	-3.4	-1.5	0.6
138.79	2.47	0.7	238.91	22NO	-0.5	-3.5	-1.5	0.5
141.91	2.76	0.5	249.12	2DE	-0.1	-3.5	-1.5	0.5
144.26	3.08	0.3	259.33	12DE	2.1	-3.6	-1.5	0.4
145.69	3.42	0.0	269.52	22DE	1.5	-3.7	-1.5	0.3

-55

♂ LONG	LAT	MAG	☉ LONG	16.00UT	☿	♀	♃	♄
146.00	3.77	-0.2	279.70	1JA	0.4	-3.7	-1.5	0.3
145.04	4.11	-0.5	289.84	11JA	0.1	-3.8	-1.5	0.2
142.80	4.38	-0.8	299.93	21JA	0.0	-3.9	-1.5	0.2
139.48	4.54	-1.0	309.98	31JA	-0.2	-4.0	-1.5	0.2
135.61	4.52	-1.1	319.98	10FE	-0.6	-4.1	-1.6	0.3
131.89	4.33	-0.9	329.92	20FE	-1.3	-4.2	-1.6	0.3
128.97	4.01	-0.7	339.80	2MR	-1.4	-4.2	-1.6	0.4
127.25	3.60	-0.5	349.62	12MR	-0.7	-4.3	-1.7	0.5
126.84	3.17	-0.3	359.38	22MR	0.6	-4.2	-1.7	0.5
127.64	2.76	-0.1	9.09	1AP	2.1	-3.8	-1.8	0.6
129.50	2.37	0.1	18.74	11AP	3.9	-3.2	-1.9	0.6
132.24	2.02	0.3	28.36	21AP	2.2	-3.2	-1.9	0.7
135.68	1.70	0.5	37.94	1MY	1.3	-3.8	-2.0	0.7
139.72	1.41	0.6	47.49	11MY	0.5	-4.1	-2.1	0.7
144.23	1.14	0.8	57.02	21MY	-0.3	-4.2	-2.2	0.8
149.16	0.90	0.9	66.54	31MY	-1.3	-4.2	-2.2	0.8
154.42	0.68	0.9	76.06	10JN	-1.6	-4.1	-2.3	0.8
159.99	0.48	1.0	85.58	20JN	-0.8	-4.0	-2.3	0.8
165.81	0.28	1.1	95.13	30JN	-0.1	-3.9	-2.4	0.8
171.88	0.11	1.1	104.70	10JL	0.3	-3.8	-2.4	0.8
178.15	-0.06	1.2	114.30	20JL	0.6	-3.7	-2.4	0.8
184.62	-0.21	1.2	123.95	30JL	1.1	-3.7	-2.4	0.8
191.27	-0.35	1.3	133.65	9AU	2.2	-3.6	-2.4	0.8
198.08	-0.49	1.3	143.39	19AU	2.4	-3.6	-2.3	0.9
205.05	-0.61	1.3	153.20	29AU	0.1	-3.5	-2.3	0.9
212.17	-0.72	1.3	163.06	8SE	-0.9	-3.5	-2.2	1.0
219.41	-0.82	1.3	172.98	18SE	-1.0	-3.4	-2.1	1.0
226.79	-0.91	1.3	182.97	28SE	-1.0	-3.4	-2.1	1.0
234.27	-0.98	1.3	193.00	8OC	-0.6	-3.4	-2.0	1.0
241.86	-1.05	1.4	203.08	18OC	-0.4	-3.4	-1.9	1.0
249.53	-1.10	1.4	213.21	28OC	-0.4	-3.4	-1.9	1.0
257.27	-1.14	1.4	223.38	7NO	-0.0	-3.4	-1.8	0.9
265.07	-1.17	1.4	233.56	17NO	0.1	-3.4	-1.8	0.9
272.92	-1.18	1.4	243.77	27NO	2.3	-3.4	-1.7	0.8
280.79	-1.18	1.4	253.98	7DE	1.2	-3.4	-1.7	0.8
288.67	-1.16	1.4	264.18	17DE	0.2	-3.4	-1.6	0.7
296.55	-1.14	1.4	274.37	27DE	-0.0	-3.4	-1.6	0.6

♂ LONG	♂ LAT	♂ MAG	⊙ LONG	16.00UT	☿	♀	♃	♄
				-54	MAGNITUDES			
304.42	-1.10	1.4	284.52	6JA	-0.1	-3.4	-1.6	0.6
312.25	-1.06	1.4	294.64	16JA	-0.3	-3.4	-1.6	0.5
320.05	-1.00	1.4	304.72	26JA	-0.7	-3.5	-1.6	0.4
327.79	-0.93	1.4	314.74	5FE	-1.3	-3.5	-1.6	0.5
335.47	-0.86	1.5	324.71	15FE	-1.3	-3.5	-1.6	0.5
343.08	-0.78	1.5	334.62	25FE	-0.6	-3.4	-1.6	0.5
350.61	-0.69	1.5	344.47	7MR	0.8	-3.4	-1.6	0.6
358.06	-0.60	1.5	354.26	17MR	2.6	-3.4	-1.6	0.7
5.43	-0.51	1.5	4.00	27MR	3.0	-3.4	-1.7	0.7
12.71	-0.41	1.5	13.68	6AP	1.7	-3.3	-1.7	0.8
19.90	-0.31	1.6	23.32	16AP	1.0	-3.3	-1.8	0.8
27.01	-0.21	1.6	32.92	26AP	0.4	-3.3	-1.8	0.9
34.03	-0.10	1.7	42.48	6MY	-0.4	-3.3	-1.9	0.9
40.97	-0.00	1.7	52.02	16MY	-1.3	-3.3	-1.9	0.9
47.83	0.10	1.8	61.55	26MY	-1.7	-3.4	-2.0	1.0
54.60	0.20	1.8	71.07	5JN	-0.7	-3.4	-2.1	1.0
61.31	0.31	1.8	80.59	15JN	-0.0	-3.4	-2.1	1.0
67.94	0.41	1.9	90.12	25JN	0.4	-3.4	-2.2	1.0
74.49	0.51	1.9	99.68	5JL	0.8	-3.5	-2.3	1.1
80.98	0.61	1.9	109.27	15JL	1.4	-3.5	-2.3	1.1
87.40	0.71	1.9	118.89	25JL	2.7	-3.6	-2.4	1.1
93.75	0.81	1.9	128.56	4AU	2.0	-3.7	-2.4	1.1
100.04	0.92	1.9	138.28	14AU	-0.0	-3.8	-2.4	1.1
106.26	1.02	1.9	148.05	24AU	-1.0	-3.9	-2.5	1.1
112.40	1.12	1.8	157.88	3SE	-1.2	-4.0	-2.4	1.2
118.46	1.23	1.8	167.78	13SE	-0.9	-4.1	-2.4	1.2
124.44	1.33	1.8	177.73	23SE	-0.5	-4.2	-2.4	1.3
130.32	1.44	1.7	187.73	3OC	-0.3	-4.3	-2.3	1.3
136.10	1.55	1.6	197.79	13OC	-0.2	-4.4	-2.3	1.3
141.74	1.67	1.6	207.90	23OC	-0.2	-4.3	-2.2	1.3
147.24	1.79	1.5	218.04	2NO	0.3	-4.1	-2.1	1.3
152.57	1.92	1.4	228.22	12NO	2.6	-3.6	-2.1	1.3
157.68	2.06	1.2	238.42	22NO	0.9	-3.0	-2.0	1.2
162.54	2.20	1.1	248.62	2DE	-0.0	-3.9	-1.9	1.2
167.09	2.35	0.9	258.83	12DE	-0.2	-4.3	-1.9	1.1
171.25	2.51	0.7	269.03	22DE	-0.2	-4.4	-1.8	1.1
				-53				
174.93	2.68	0.5	279.20	1JA	-0.4	-4.3	-1.7	1.0
178.02	2.86	0.3	289.34	11JA	-0.7	-4.3	-1.7	0.9
180.35	3.04	0.1	299.44	21JA	-1.2	-4.2	-1.7	0.9
181.77	3.22	-0.2	309.49	31JA	-1.2	-4.0	-1.6	0.8
182.09	3.37	-0.5	319.49	10FE	-0.6	-3.9	-1.6	0.7
181.17	3.46	-0.8	329.43	20FE	1.0	-3.8	-1.6	0.8
178.99	3.47	-1.1	339.31	2MR	3.1	-3.7	-1.6	0.8
175.77	3.35	-1.3	349.14	12MR	2.2	-3.6	-1.6	0.8
172.05	3.07	-1.4	358.90	22MR	1.2	-3.6	-1.6	0.9
168.54	2.68	-1.3	8.61	1AP	0.7	-3.5	-1.6	0.9
165.87	2.20	-1.1	18.28	11AP	0.2	-3.5	-1.6	1.0
164.46	1.71	-0.9	27.90	21AP	-0.4	-3.4	-1.6	1.0
164.39	1.24	-0.7	37.47	1MY	-1.3	-3.4	-1.7	1.1
165.57	0.81	-0.5	47.03	11MY	-1.7	-3.3	-1.7	1.1
167.83	0.43	-0.3	56.56	21MY	-0.7	-3.3	-1.7	1.2
171.00	0.10	-0.1	66.08	31MY	0.1	-3.3	-1.8	1.2
174.91	-0.19	0.1	75.60	10JN	0.6	-3.3	-1.9	1.3
179.45	-0.44	0.2	85.13	20JN	1.1	-3.3	-1.9	1.3
184.50	-0.66	0.3	94.67	30JN	1.9	-3.3	-2.0	1.3
189.97	-0.84	0.4	104.24	10JL	3.2	-3.3	-2.0	1.3
195.82	-1.01	0.5	113.84	20JL	1.7	-3.4	-2.1	1.4
201.99	-1.15	0.6	123.49	30JL	-0.1	-3.4	-2.2	1.4
208.43	-1.26	0.7	133.18	9AU	-1.1	-3.4	-2.2	1.4
215.11	-1.36	0.7	142.92	19AU	-1.3	-3.5	-2.3	1.4
222.00	-1.44	0.8	152.72	29AU	-0.9	-3.5	-2.4	1.3
229.07	-1.49	0.9	162.58	8SE	-0.4	-3.5	-2.4	1.3
236.31	-1.53	0.9	172.50	18SE	-0.2	-3.5	-2.4	1.3
243.67	-1.55	1.0	182.48	28SE	-0.1	-3.4	-2.4	1.3
251.16	-1.55	1.0	192.51	8OC	-0.0	-3.4	-2.4	1.3
258.74	-1.53	1.1	202.59	18OC	0.5	-3.4	-2.4	1.3
266.39	-1.50	1.1	212.72	28OC	3.0	-3.4	-2.4	1.2
274.10	-1.46	1.1	222.88	7NO	0.7	-3.3	-2.3	1.2
281.84	-1.40	1.2	233.07	17NO	-0.2	-3.3	-2.3	1.2
289.59	-1.33	1.2	243.27	27NO	-0.3	-3.3	-2.2	1.1
297.35	-1.25	1.3	253.48	7DE	-0.3	-3.3	-2.1	1.1
305.10	-1.16	1.3	263.68	17DE	-0.5	-3.4	-2.0	1.1
312.81	-1.06	1.4	273.87	27DE	-0.8	-3.4	-2.0	1.0

♂ LONG	♂ LAT	♂ MAG	⊙ LONG	16.00UT	☿	♀	♃	♄
				-52	MAGNITUDES			
320.48	-0.96	1.4	284.03	6JA	-1.1	-3.4	-1.9	1.0
328.10	-0.85	1.5	294.15	16JA	-1.1	-3.4	-1.8	0.9
335.66	-0.74	1.5	304.23	26JA	-0.5	-3.5	-1.8	0.9
343.16	-0.63	1.5	314.26	5FE	1.2	-3.5	-1.7	0.8
350.57	-0.52	1.6	324.22	15FE	3.2	-3.5	-1.7	0.8
357.91	-0.40	1.6	334.14	25FE	1.6	-3.6	-1.6	0.8
5.18	-0.29	1.7	343.99	6MR	0.9	-3.6	-1.6	0.8
12.36	-0.18	1.7	353.79	16MR	0.5	-3.7	-1.6	0.9
19.46	-0.07	1.7	3.53	26MR	0.1	-3.8	-1.5	1.0
26.48	0.03	1.7	13.22	5AP	-0.5	-3.9	-1.5	1.0
33.43	0.13	1.8	22.85	15AP	-1.3	-4.0	-1.5	1.1
40.30	0.23	1.8	32.45	25AP	-1.7	-4.0	-1.5	1.2
47.11	0.33	1.8	42.02	5MY	-0.7	-4.1	-1.5	1.2
53.84	0.42	1.8	51.56	15MY	0.1	-4.2	-1.5	1.3
60.52	0.51	1.8	61.09	25MY	0.8	-4.2	-1.6	1.3
67.14	0.59	1.8	70.61	4JN	1.5	-4.0	-1.6	1.3
73.70	0.67	1.9	80.13	14JN	2.5	-3.5	-1.6	1.3
80.22	0.75	1.9	89.66	24JN	3.1	-2.9	-1.7	1.3
86.69	0.83	1.9	99.22	4JL	1.4	-3.6	-1.7	1.3
93.12	0.90	2.0	108.80	14JL	-0.1	-4.0	-1.8	1.3
99.51	0.97	2.0	118.42	24JL	-1.2	-4.2	-1.8	1.3
105.87	1.04	2.0	128.09	3AU	-1.4	-4.2	-1.9	1.3
112.19	1.10	2.0	137.80	13AU	-0.9	-4.2	-2.0	1.2
118.49	1.16	2.0	147.58	23AU	-0.4	-4.1	-2.0	1.2
124.75	1.22	2.0	157.41	2SE	-0.1	-4.0	-2.1	1.1
130.98	1.28	2.0	167.29	12SE	0.0	-3.9	-2.2	1.1
137.19	1.33	1.9	177.24	22SE	0.2	-3.8	-2.2	1.1
143.35	1.39	1.9	187.24	2OC	0.8	-3.8	-2.3	1.1
149.48	1.44	1.9	197.30	12OC	3.3	-3.7	-2.3	1.1
155.57	1.49	1.8	207.41	22OC	0.5	-3.6	-2.3	1.1
161.60	1.54	1.7	217.55	1NO	-0.4	-3.6	-2.3	1.1
167.57	1.58	1.7	227.72	11NO	-0.5	-3.5	-2.3	1.1
173.48	1.62	1.6	237.92	21NO	-0.5	-3.5	-2.3	1.1
179.29	1.66	1.5	248.13	1DE	-0.6	-3.5	-2.3	1.0
185.01	1.69	1.3	258.34	11DE	-0.8	-3.4	-2.2	1.0
190.60	1.72	1.2	268.54	21DE	-0.9	-3.4	-2.2	0.9
196.04	1.74	1.1	278.71	31DE	-0.9	-3.4	-2.1	0.9
				-51				
201.29	1.75	0.9	288.85	10JA	-0.4	-3.3	-2.0	0.8
206.32	1.75	0.7	298.95	20JA	1.4	-3.3	-2.0	0.8
211.06	1.73	0.5	309.01	30JA	2.7	-3.3	-1.9	0.8
215.44	1.70	0.3	319.01	9FE	1.2	-3.3	-1.8	0.7
219.39	1.63	0.1	328.95	19FE	0.6	-3.3	-1.7	0.7
222.77	1.53	-0.2	338.84	1MR	0.4	-3.3	-1.7	0.6
225.46	1.38	-0.5	348.66	11MR	0.0	-3.4	-1.6	0.6
227.26	1.17	-0.8	358.43	21MR	-0.5	-3.4	-1.6	0.7
228.01	0.88	-1.1	8.14	31MR	-1.3	-3.4	-1.5	0.8
227.57	0.50	-1.4	17.80	10AP	-1.6	-3.5	-1.5	0.8
225.87	0.02	-1.7	27.43	20AP	-0.7	-3.5	-1.5	0.9
223.14	-0.54	-2.0	37.01	30AP	0.2	-3.5	-1.5	1.0
219.90	-1.12	-2.1	46.56	10MY	1.0	-3.4	-1.4	1.0
216.85	-1.67	-2.0	56.09	20MY	1.9	-3.4	-1.4	1.1
214.68	-2.13	-1.8	65.61	30MY	3.2	-3.4	-1.4	1.1
213.78	-2.48	-1.6	75.13	9JN	2.5	-3.3	-1.4	1.1
214.27	-2.72	-1.3	84.66	19JN	1.1	-3.3	-1.4	1.2
216.06	-2.87	-1.1	94.20	29JN	-0.2	-3.3	-1.5	1.2
218.97	-2.96	-0.9	103.77	9JL	-1.2	-3.3	-1.5	1.2
222.80	-2.99	-0.7	113.37	19JL	-1.5	-3.3	-1.5	1.2
227.40	-2.97	-0.6	123.01	29JL	-0.8	-3.3	-1.6	1.2
232.61	-2.92	-0.4	132.70	8AU	-0.3	-3.4	-1.6	1.1
238.32	-2.84	-0.2	142.45	18AU	0.0	-3.4	-1.7	1.1
244.43	-2.74	-0.1	152.24	28AU	0.2	-3.4	-1.7	1.1
250.86	-2.62	0.0	162.10	7SE	0.4	-3.5	-1.8	1.0
257.55	-2.48	0.2	172.02	17SE	1.1	-3.5	-1.8	1.0
264.44	-2.33	0.3	181.99	27SE	3.3	-3.6	-1.9	1.0
271.48	-2.16	0.4	192.02	7OC	0.4	-3.7	-2.0	1.0
278.65	-1.99	0.5	202.10	17OC	-0.5	-3.7	-2.0	1.0
285.91	-1.81	0.6	212.22	27OC	-0.6	-3.8	-2.1	1.0
293.22	-1.63	0.8	222.38	6NO	-0.6	-3.9	-2.1	1.0
300.56	-1.46	0.9	232.57	16NO	-0.7	-4.0	-2.2	1.0
307.92	-1.28	1.0	242.78	26NO	-0.8	-4.1	-2.2	1.0
315.27	-1.10	1.1	252.98	6DE	-0.8	-4.2	-2.2	0.9
322.60	-0.93	1.1	263.19	16DE	-0.8	-4.3	-2.2	0.9
329.90	-0.77	1.2	273.37	26DE	-0.3	-4.4	-2.2	0.9

♂ LONG	♂ LAT	♂ MAG	☉ LONG	16.00UT -50	☿	♀	♃	♄
337.16	-0.61	1.3	283.54	5JA	1.6	-4.3	-2.1	0.8
344.37	-0.46	1.4	293.66	15JA	2.2	-4.1	-2.1	0.8
351.52	-0.31	1.5	303.74	25JA	0.8	-3.6	-2.0	0.7
358.61	-0.18	1.5	313.77	4FE	0.4	-3.3	-2.0	0.7
5.65	-0.05	1.6	323.74	14FE	0.2	-3.9	-1.9	0.6
12.61	0.07	1.7	333.66	24FE	-0.1	-4.2	-1.8	0.6
19.52	0.18	1.7	343.51	6MR	-0.5	-4.3	-1.8	0.5
26.36	0.29	1.8	353.31	16MR	-1.3	-4.2	-1.7	0.5
33.14	0.39	1.8	3.05	26MR	-1.6	-4.2	-1.6	0.5
39.87	0.48	1.9	12.74	5AP	-0.7	-4.0	-1.6	0.6
46.54	0.57	1.9	22.39	15AP	0.4	-3.9	-1.5	0.6
53.16	0.64	1.9	31.99	25AP	1.4	-3.8	-1.5	0.7
59.73	0.72	1.9	41.56	5MY	2.6	-3.7	-1.4	0.8
66.27	0.78	1.9	51.10	15MY	3.5	-3.7	-1.4	0.8
72.76	0.85	2.0	60.62	25MY	2.0	-3.6	-1.4	0.9
79.23	0.90	2.0	70.15	4JN	0.9	-3.5	-1.3	0.9
85.66	0.95	1.9	79.67	14JN	-0.2	-3.5	-1.3	1.0
92.08	1.00	1.9	89.20	24JN	-1.3	-3.4	-1.3	1.0
98.47	1.04	1.9	98.76	4JL	-1.6	-3.4	-1.3	1.0
104.85	1.08	2.0	108.34	14JL	-0.8	-3.4	-1.3	1.0
111.22	1.12	2.0	117.96	24JL	-0.2	-3.4	-1.3	1.0
117.59	1.15	2.0	127.62	3AU	0.1	-3.4	-1.4	1.0
123.96	1.17	2.0	137.34	13AU	0.3	-3.3	-1.4	1.0
130.32	1.19	2.0	147.10	23AU	0.6	-3.3	-1.4	1.0
136.69	1.21	2.0	156.93	2SE	1.4	-3.4	-1.5	1.0
143.07	1.22	2.0	166.82	12SE	3.0	-3.4	-1.5	0.9
149.45	1.23	2.0	176.76	22SE	0.3	-3.4	-1.6	0.9
155.84	1.23	2.0	186.76	2OC	-0.7	-3.4	-1.6	0.9
162.24	1.23	2.0	196.81	12OC	-0.8	-3.4	-1.7	0.9
168.65	1.23	1.9	206.91	22OC	-0.7	-3.4	-1.8	0.9
175.06	1.21	1.9	217.06	1NO	-0.8	-3.5	-1.8	0.9
181.49	1.20	1.8	227.23	11NO	-0.6	-3.5	-1.9	0.9
187.91	1.17	1.7	237.42	21NO	-0.6	-3.5	-1.9	0.9
194.34	1.13	1.7	247.63	1DE	-0.6	-3.5	-2.0	0.9
200.76	1.09	1.6	257.84	11DE	-0.2	-3.4	-2.0	0.9
207.18	1.03	1.5	268.04	21DE	1.8	-3.4	-2.1	0.8
213.58	0.97	1.4	278.21	31DE	1.8	-3.4	-2.1	0.8
				-49				
219.96	0.88	1.3	288.36	10JA	0.6	-3.4	-2.1	0.8
226.32	0.78	1.2	298.46	20JA	0.3	-3.3	-2.1	0.7
232.64	0.67	1.0	308.52	30JA	0.1	-3.3	-2.1	0.7
238.91	0.52	0.9	318.52	9FE	-0.2	-3.3	-2.0	0.6
245.13	0.36	0.7	328.47	19FE	-0.6	-3.3	-2.0	0.6
251.26	0.16	0.6	338.36	1MR	-1.3	-3.4	-1.9	0.5
257.28	-0.06	0.4	348.18	11MR	-1.5	-3.4	-1.9	0.5
263.16	-0.33	0.2	357.95	21MR	-0.7	-3.4	-1.8	0.4
268.87	-0.64	0.0	7.67	31MR	0.5	-3.4	-1.7	0.4
274.34	-1.00	-0.2	17.34	10AP	1.8	-3.4	-1.7	0.4
279.51	-1.41	-0.4	26.96	20AP	3.4	-3.5	-1.6	0.5
284.28	-1.89	-0.7	36.55	30AP	2.7	-3.5	-1.5	0.6
288.53	-2.44	-0.9	46.10	10MY	1.5	-3.6	-1.5	0.6
292.13	-3.06	-1.2	55.63	20MY	0.7	-3.6	-1.4	0.7
294.88	-3.76	-1.5	65.16	30MY	-0.3	-3.7	-1.4	0.7
296.59	-4.51	-1.8	74.68	9JN	-1.3	-3.8	-1.3	0.8
297.08	-5.28	-2.1	84.20	19JN	-1.7	-3.9	-1.3	0.8
296.28	-5.98	-2.3	93.75	29JN	-0.8	-4.0	-1.3	0.9
294.36	-6.52	-2.6	103.31	9JL	-0.2	-4.1	-1.3	0.9
291.81	-6.79	-2.7	112.91	19JL	0.2	-4.2	-1.3	0.9
289.32	-6.73	-2.5	122.55	29JL	0.5	-4.2	-1.2	0.9
287.62	-6.38	-2.3	132.24	8AU	0.9	-4.2	-1.2	1.0
287.10	-5.84	-2.0	141.98	18AU	1.8	-4.0	-1.3	0.9
287.91	-5.21	-1.7	151.77	28AU	2.7	-3.5	-1.3	0.9
289.94	-4.55	-1.4	161.62	7SE	0.2	-3.3	-1.3	0.9
292.99	-3.93	-1.1	171.54	17SE	-0.8	-3.9	-1.3	0.9
296.89	-3.35	-0.9	181.51	27SE	-0.9	-4.2	-1.4	0.8
301.44	-2.82	-0.6	191.54	7OC	-0.9	-4.3	-1.4	0.8
306.49	-2.34	-0.4	201.61	17OC	-0.7	-4.3	-1.5	0.8
311.94	-1.92	-0.1	211.74	27OC	-0.5	-4.3	-1.5	0.8
317.68	-1.54	0.1	221.89	6NO	-0.5	-4.2	-1.6	0.8
323.63	-1.21	0.3	232.08	16NO	-0.5	-4.1	-1.6	0.8
329.76	-0.91	0.4	242.28	26NO	-0.0	-4.0	-1.7	0.8
336.01	-0.64	0.6	252.49	6DE	2.1	-3.9	-1.8	0.8
342.35	-0.41	0.8	262.70	16DE	1.4	-3.8	-1.8	0.8
348.75	-0.20	0.9	272.89	26DE	0.3	-3.7	-1.9	0.8

♂ LONG	♂ LAT	♂ MAG	☉ LONG	16.00UT -48	☿	♀	♃	♄
355.18	-0.02	1.1	283.04	5JA	0.1	-3.6	-1.9	0.8
1.64	0.15	1.2	293.17	15JA	-0.0	-3.6	-2.0	0.8
8.12	0.29	1.3	303.26	25JA	-0.2	-3.5	-2.0	0.7
14.59	0.42	1.4	313.29	4FE	-0.6	-3.5	-2.0	0.7
21.05	0.53	1.5	323.27	14FE	-1.3	-3.4	-2.0	0.6
27.50	0.63	1.6	333.19	24FE	-1.4	-3.4	-2.0	0.6
33.94	0.72	1.7	343.04	5MR	-0.7	-3.3	-2.0	0.5
40.35	0.80	1.7	352.85	15MR	0.6	-3.3	-1.9	0.5
46.75	0.87	1.8	2.59	25MR	2.2	-3.3	-1.9	0.4
53.13	0.93	1.9	12.28	4AP	3.5	-3.3	-1.8	0.4
59.49	0.98	1.9	21.93	14AP	2.0	-3.3	-1.8	0.3
65.84	1.02	2.0	31.53	24AP	1.1	-3.3	-1.7	0.4
72.17	1.06	2.0	41.10	4MY	0.5	-3.3	-1.6	0.4
78.49	1.09	2.0	50.64	14MY	-0.3	-3.4	-1.6	0.5
84.81	1.12	2.0	60.17	24MY	-1.3	-3.4	-1.5	0.6
91.12	1.14	2.0	69.68	3JN	-1.7	-3.4	-1.5	0.6
97.44	1.15	2.0	79.21	13JN	-0.8	-3.5	-1.4	0.7
103.76	1.16	2.0	88.74	23JN	-0.1	-3.5	-1.4	0.7
110.09	1.17	2.0	98.29	3JL	0.4	-3.5	-1.3	0.8
116.44	1.17	2.0	107.87	13JL	0.7	-3.4	-1.3	0.8
122.80	1.16	2.0	117.49	23JL	1.2	-3.4	-1.3	0.9
129.19	1.15	2.0	127.15	2AU	2.3	-3.4	-1.2	0.9
135.60	1.14	1.9	136.86	12AU	2.3	-3.4	-1.2	0.9
142.05	1.12	1.9	146.63	22AU	0.1	-3.3	-1.2	0.9
148.52	1.10	2.0	156.45	1SE	-0.9	-3.3	-1.2	0.9
155.04	1.07	2.0	166.33	11SE	-1.1	-3.3	-1.2	0.9
161.59	1.04	2.0	176.27	21SE	-1.0	-3.3	-1.2	0.8
168.18	1.00	1.9	186.27	1OC	-0.6	-3.3	-1.3	0.8
174.82	0.95	1.9	196.32	11OC	-0.4	-3.4	-1.3	0.8
181.50	0.90	1.9	206.42	21OC	-0.3	-3.4	-1.3	0.7
188.23	0.84	1.9	216.56	31OC	-0.3	-3.4	-1.4	0.7
195.01	0.78	1.8	226.74	10NO	0.1	-3.4	-1.4	0.8
201.82	0.70	1.8	236.93	20NO	2.4	-3.5	-1.5	0.8
208.69	0.62	1.7	247.14	30NO	1.1	-3.5	-1.5	0.8
215.60	0.53	1.7	257.35	10DE	0.1	-3.6	-1.6	0.8
222.56	0.43	1.6	267.55	20DE	-0.1	-3.7	-1.6	0.8
229.56	0.32	1.6	277.72	30DE	-0.2	-3.7	-1.7	0.8
				-47				
236.60	0.20	1.5	287.87	9JA	-0.3	-3.8	-1.8	0.8
243.68	0.06	1.4	297.97	19JA	-0.7	-3.9	-1.8	0.8
250.80	-0.08	1.3	308.03	29JA	-1.3	-4.0	-1.9	0.7
257.95	-0.24	1.2	318.04	8FE	-1.3	-4.1	-1.9	0.7
265.12	-0.41	1.1	327.99	18FE	-0.6	-4.2	-2.0	0.6
272.32	-0.60	1.0	337.88	28FE	0.8	-4.2	-2.0	0.6
279.52	-0.79	0.9	347.72	10MR	2.8	-4.3	-2.0	0.5
286.73	-0.99	0.8	357.49	20MR	2.7	-4.2	-2.0	0.5
293.92	-1.21	0.7	7.20	30MR	1.5	-3.8	-2.0	0.4
301.07	-1.43	0.6	16.88	9AP	0.9	-3.2	-2.0	0.4
308.17	-1.66	0.5	26.50	19AP	0.3	-3.2	-1.9	0.3
315.20	-1.90	0.4	36.08	29AP	-0.4	-3.9	-1.9	0.3
322.12	-2.14	0.3	45.64	9MY	-1.3	-4.1	-1.8	0.4
328.88	-2.37	0.1	55.17	19MY	-1.7	-4.2	-1.7	0.4
335.47	-2.61	0.0	64.69	29MY	-0.8	-4.2	-1.7	0.5
341.80	-2.84	-0.1	74.22	8JN	-0.0	-4.1	-1.6	0.5
347.82	-3.07	-0.3	83.74	18JN	0.5	-4.0	-1.6	0.6
353.46	-3.29	-0.4	93.28	28JN	0.9	-3.9	-1.5	0.7
358.59	-3.50	-0.6	102.85	8JL	1.6	-3.8	-1.4	0.7
3.10	-3.69	-0.8	112.44	18JL	2.9	-3.7	-1.4	0.8
6.83	-3.85	-1.0	122.08	28JL	2.0	-3.7	-1.4	0.8
9.57	-3.98	-1.2	131.77	7AU	0.0	-3.6	-1.3	0.8
11.12	-4.05	-1.4	141.50	17AU	-1.0	-3.5	-1.3	0.8
11.29	-4.04	-1.6	151.29	27AU	-1.2	-3.5	-1.3	0.8
9.97	-3.89	-1.9	161.15	6SE	-0.9	-3.5	-1.2	0.8
7.38	-3.58	-2.0	171.05	16SE	-0.5	-3.4	-1.2	0.8
4.00	-3.11	-2.1	181.03	26SE	-0.3	-3.4	-1.2	0.8
0.65	-2.51	-2.0	191.05	6OC	-0.2	-3.4	-1.2	0.8
358.10	-1.87	-1.7	201.12	16OC	-0.1	-3.4	-1.2	0.7
356.76	-1.26	-1.3	211.25	26OC	0.4	-3.4	-1.3	0.7
356.79	-0.73	-1.0	221.41	5NO	2.7	-3.4	-1.3	0.7
358.06	-0.29	-0.7	231.59	15NO	0.9	-3.4	-1.3	0.7
0.36	0.07	-0.4	241.79	25NO	-0.1	-3.4	-1.4	0.8
3.51	0.35	-0.1	252.00	5DE	-0.2	-3.4	-1.4	0.8
7.30	0.58	0.1	262.20	15DE	-0.3	-3.4	-1.5	0.8
11.60	0.77	0.4	272.39	25DE	-0.4	-3.4	-1.5	0.8

♂ LONG	LAT	MAG	☉ LONG	16.00UT	☿ ♀ 2+ ♄ MAGNITUDES			
				-46				
16.30	0.91	0.6	282.56	4JA	-0.8	-3.4	-1.6	0.8
21.30	1.03	0.8	292.68	14JA	-1.2	-3.4	-1.6	0.8
26.53	1.12	1.0	302.77	24JA	-1.2	-3.5	-1.7	0.8
31.94	1.19	1.1	312.80	3FE	-0.6	-3.5	-1.8	0.7
37.50	1.25	1.3	322.78	13FE	1.0	-3.5	-1.8	0.7
43.17	1.29	1.4	332.70	23FE	3.2	-3.4	-1.9	0.7
48.94	1.32	1.5	342.57	5MR	2.0	-3.4	-2.0	0.6
54.77	1.34	1.6	352.37	15MR	1.1	-3.4	-2.0	0.6
60.67	1.36	1.7	2.12	25MR	0.6	-3.4	-2.0	0.5
66.61	1.36	1.8	11.81	4AP	0.2	-3.3	-2.1	0.4
72.60	1.36	1.8	21.45	14AP	-0.4	-3.3	-2.1	0.4
78.62	1.36	1.9	31.06	24AP	-1.3	-3.3	-2.1	0.3
84.68	1.35	1.9	40.63	4MY	-1.7	-3.3	-2.0	0.3
90.77	1.33	2.0	50.17	14MY	-0.8	-3.3	-2.0	0.3
96.90	1.31	2.0	59.71	24MY	0.1	-3.4	-2.0	0.4
103.07	1.29	2.0	69.22	3JN	0.7	-3.4	-1.9	0.4
109.26	1.26	2.0	78.74	13JN	1.2	-3.4	-1.8	0.5
115.50	1.23	2.0	88.28	23JN	2.1	-3.4	-1.8	0.5
121.78	1.19	2.0	97.83	3JL	3.3	-3.5	-1.7	0.6
128.10	1.15	2.0	107.41	13JL	1.6	-3.5	-1.7	0.7
134.48	1.11	2.0	117.02	23JL	-0.0	-3.6	-1.6	0.7
140.90	1.06	2.0	126.68	2AU	-1.1	-3.7	-1.5	0.7
147.37	1.01	1.9	136.39	12AU	-1.4	-3.8	-1.5	0.8
153.91	0.95	1.9	146.15	22AU	-0.9	-3.9	-1.4	0.8
160.50	0.89	1.9	155.97	1SE	-0.4	-4.0	-1.4	0.8
167.16	0.83	1.8	165.85	11SE	-0.2	-4.1	-1.4	0.8
173.89	0.76	1.8	175.79	21SE	-0.0	-4.2	-1.3	0.8
180.68	0.68	1.8	185.78	1OC	0.1	-4.3	-1.3	0.8
187.54	0.60	1.8	195.83	11OC	0.6	-4.4	-1.3	0.8
194.47	0.52	1.8	205.93	21OC	3.0	-4.3	-1.3	0.7
201.47	0.43	1.8	216.07	31OC	0.7	-4.1	-1.3	0.7
208.55	0.33	1.7	226.24	10NO	-0.3	-3.5	-1.3	0.7
215.69	0.23	1.7	236.44	20NO	-0.4	-3.1	-1.3	0.7
222.90	0.13	1.7	246.64	30NO	-0.4	-3.9	-1.3	0.7
230.18	0.01	1.7	256.85	10DE	-0.5	-4.3	-1.4	0.8
237.53	-0.10	1.6	267.05	20DE	-0.8	-4.4	-1.4	0.8
244.93	-0.22	1.6	277.23	30DE	-1.0	-4.3	-1.4	0.8
				-45				
252.40	-0.35	1.5	287.38	9JA	-1.0	-4.3	-1.5	0.8
259.91	-0.47	1.5	297.49	19JA	-0.5	-4.1	-1.6	0.8
267.46	-0.60	1.4	307.55	29JA	1.2	-4.0	-1.6	0.8
275.06	-0.73	1.4	317.56	8FE	3.0	-3.9	-1.7	0.8
282.68	-0.86	1.3	327.51	18FE	1.5	-3.8	-1.7	0.7
290.31	-0.99	1.3	337.40	28FE	0.8	-3.7	-1.8	0.7
297.96	-1.12	1.2	347.24	10MR	0.5	-3.6	-1.9	0.7
305.60	-1.24	1.2	357.02	20MR	0.1	-3.6	-2.0	0.6
313.22	-1.35	1.1	6.74	30MR	-0.5	-3.5	-2.0	0.5
320.81	-1.45	1.1	16.41	9AP	-1.3	-3.4	-2.1	0.5
328.35	-1.54	1.0	26.04	19AP	-1.7	-3.4	-2.1	0.4
335.82	-1.62	1.0	35.62	29AP	-0.8	-3.4	-2.1	0.3
343.21	-1.69	0.9	45.18	9MY	0.2	-3.3	-2.2	0.3
350.50	-1.74	0.9	54.72	19MY	0.9	-3.3	-2.2	0.2
357.66	-1.77	0.8	64.24	29MY	1.6	-3.3	-2.2	0.3
4.69	-1.79	0.8	73.76	8JN	2.7	-3.3	-2.1	0.4
11.55	-1.79	0.7	83.29	18JN	2.9	-3.3	-2.1	0.4
18.21	-1.77	0.6	92.82	28JN	1.3	-3.3	-2.0	0.5
24.66	-1.74	0.6	102.39	8JL	-0.1	-3.3	-2.0	0.5
30.84	-1.68	0.5	111.98	18JL	-1.2	-3.4	-1.9	0.6
36.71	-1.60	0.4	121.61	28JL	-1.5	-3.4	-1.9	0.6
42.22	-1.50	0.3	131.30	7AU	-0.9	-3.4	-1.8	0.7
47.29	-1.38	0.1	141.03	17AU	-0.4	-3.5	-1.7	0.7
51.82	-1.22	-0.0	150.82	27AU	-0.1	-3.5	-1.7	0.8
55.70	-1.03	-0.2	160.66	6SE	0.1	-3.5	-1.6	0.8
58.77	-0.80	-0.3	170.57	16SE	0.3	-3.5	-1.6	0.8
60.85	-0.52	-0.5	180.53	26SE	0.9	-3.4	-1.5	0.8
61.73	-0.19	-0.8	190.56	6OC	3.3	-3.4	-1.5	0.8
61.23	0.20	-1.0	200.63	16OC	0.5	-3.4	-1.4	0.8
59.30	0.64	-1.2	210.74	26OC	-0.4	-3.4	-1.4	0.8
56.13	1.10	-1.4	220.90	5NO	-0.5	-3.3	-1.4	0.7
52.28	1.53	-1.5	231.09	15NO	-0.5	-3.3	-1.4	0.7
48.52	1.88	-1.3	241.29	25NO	-0.6	-3.3	-1.4	0.7
45.57	2.12	-1.0	251.50	5DE	-0.8	-3.3	-1.4	0.7
43.88	2.26	-0.7	261.70	15DE	-0.9	-3.4	-1.4	0.7
43.54	2.32	-0.4	271.89	25DE	-0.9	-3.4	-1.4	0.8

♂ LONG	LAT	MAG	☉ LONG	16.00UT	☿ ♀ 2+ ♄ MAGNITUDES			
				-44				
44.42	2.33	-0.1	282.06	4JA	-0.4	-3.4	-1.4	0.8
46.34	2.31	0.1	292.19	14JA	1.4	-3.4	-1.5	0.8
49.10	2.26	0.4	302.28	24JA	2.5	-3.5	-1.5	0.8
52.52	2.20	0.6	312.32	3FE	1.1	-3.5	-1.6	0.8
56.46	2.14	0.8	322.30	13FE	0.6	-3.5	-1.6	0.8
60.81	2.07	1.0	332.22	23FE	0.3	-3.6	-1.7	0.8
65.48	2.00	1.1	342.09	4MR	0.0	-3.6	-1.7	0.7
70.42	1.93	1.3	351.90	14MR	-0.5	-3.7	-1.8	0.7
75.57	1.86	1.4	1.65	24MR	-1.3	-3.8	-1.9	0.6
80.90	1.79	1.5	11.35	3AP	-1.6	-3.9	-1.9	0.6
86.37	1.71	1.6	21.00	13AP	-0.8	-4.0	-2.0	0.5
91.97	1.64	1.7	30.60	23AP	0.3	-4.0	-2.1	0.5
97.68	1.57	1.7	40.18	3MY	1.1	-4.1	-2.1	0.4
103.48	1.50	1.8	49.72	13MY	2.1	-4.2	-2.1	0.3
109.38	1.42	1.8	59.25	23MY	3.5	-4.2	-2.2	0.3
115.37	1.35	1.9	68.77	2JN	2.3	-4.0	-2.3	0.3
121.43	1.27	1.9	78.29	12JN	1.0	-3.5	-2.3	0.3
127.58	1.19	1.9	87.82	22JN	-0.2	-2.9	-2.3	0.4
133.81	1.11	1.9	97.37	2JL	-1.2	-3.6	-2.3	0.4
140.11	1.03	1.9	106.94	12JL	-1.6	-4.0	-2.3	0.5
146.50	0.94	1.9	116.56	22JL	-0.8	-4.2	-2.2	0.6
152.97	0.86	1.9	126.22	1AU	-0.3	-4.2	-2.2	0.6
159.52	0.77	1.9	135.92	11AU	0.0	-4.2	-2.1	0.7
166.15	0.68	1.9	145.68	21AU	0.2	-4.1	-2.0	0.7
172.88	0.59	1.8	155.50	31AU	0.5	-4.0	-2.0	0.7
179.69	0.49	1.8	165.37	10SE	1.2	-3.9	-1.9	0.8
186.58	0.39	1.8	175.30	20SE	3.3	-3.8	-1.8	0.8
193.57	0.29	1.7	185.30	30SE	0.4	-3.8	-1.8	0.8
200.65	0.19	1.7	195.34	10OC	-0.6	-3.7	-1.7	0.8
207.82	0.09	1.6	205.44	20OC	-0.7	-3.6	-1.7	0.8
215.07	-0.02	1.6	215.58	30OC	-0.7	-3.6	-1.6	0.8
222.41	-0.13	1.5	225.74	9NO	-0.7	-3.5	-1.6	0.8
229.82	-0.23	1.5	235.94	19NO	-0.7	-3.5	-1.5	0.7
237.32	-0.34	1.5	246.15	29NO	-0.7	-3.5	-1.5	0.7
244.88	-0.45	1.5	256.35	9DE	-0.7	-3.4	-1.5	0.7
252.51	-0.55	1.5	266.55	19DE	-0.3	-3.4	-1.5	0.7
260.20	-0.65	1.5	276.74	29DE	1.6	-3.4	-1.5	0.8
				-43				
267.92	-0.75	1.5	286.88	8JA	2.1	-3.3	-1.5	0.8
275.69	-0.84	1.5	297.00	18JA	0.7	-3.3	-1.5	0.8
283.49	-0.93	1.4	307.06	28JA	0.4	-3.3	-1.5	0.8
291.30	-1.01	1.4	317.07	7FE	0.2	-3.3	-1.5	0.8
299.12	-1.08	1.4	327.03	17FE	-0.1	-3.3	-1.6	0.8
306.92	-1.14	1.4	336.93	27FE	-0.5	-3.3	-1.6	0.8
314.71	-1.19	1.4	346.76	9MR	-1.3	-3.4	-1.7	0.8
322.46	-1.23	1.4	356.54	19MR	-1.6	-3.4	-1.7	0.7
330.17	-1.25	1.4	6.27	29MR	-0.7	-3.4	-1.8	0.7
337.81	-1.26	1.3	15.94	8AP	0.4	-3.5	-1.8	0.6
345.40	-1.26	1.3	25.57	18AP	1.5	-3.5	-1.9	0.6
352.90	-1.25	1.3	35.16	28AP	2.8	-3.5	-2.0	0.5
0.31	-1.22	1.3	44.72	8MY	3.2	-3.4	-2.0	0.5
7.62	-1.17	1.3	54.25	18MY	1.8	-3.4	-2.1	0.4
14.82	-1.12	1.3	63.78	28MY	0.8	-3.4	-2.2	0.3
21.90	-1.05	1.3	73.29	7JN	-0.2	-3.3	-2.2	0.3
28.86	-0.97	1.3	82.82	17JN	-1.2	-3.3	-2.3	0.3
35.68	-0.87	1.2	92.36	27JN	-1.7	-3.3	-2.3	0.3
42.34	-0.77	1.2	101.92	7JL	-0.8	-3.3	-2.4	0.4
48.85	-0.65	1.2	111.51	17JL	-0.2	-3.3	-2.4	0.5
55.17	-0.52	1.1	121.15	27JL	0.2	-3.3	-2.4	0.5
61.29	-0.37	1.1	130.82	6AU	0.4	-3.4	-2.4	0.6
67.18	-0.21	1.0	140.56	16AU	0.7	-3.4	-2.4	0.6
72.81	-0.04	0.9	150.34	26AU	1.6	-3.4	-2.3	0.7
78.14	0.16	0.8	160.18	5SE	3.0	-3.5	-2.3	0.7
83.11	0.38	0.7	170.09	15SE	0.3	-3.5	-2.2	0.7
87.64	0.62	0.6	180.05	25SE	-0.7	-3.6	-2.1	0.8
91.65	0.89	0.4	190.07	5OC	-0.8	-3.7	-2.1	0.8
95.01	1.20	0.3	200.14	15OC	-0.8	-3.7	-2.0	0.8
97.56	1.55	0.1	210.25	25OC	-0.8	-3.8	-1.9	0.8
99.14	1.95	-0.1	220.41	4NO	-0.6	-3.9	-1.9	0.8
99.53	2.38	-0.4	230.59	14NO	-0.6	-4.0	-1.8	0.8
98.60	2.83	-0.6	240.79	24NO	-0.6	-4.1	-1.8	0.7
96.32	3.27	-0.8	251.00	4DE	-0.2	-4.2	-1.7	0.7
92.92	3.64	-1.0	261.21	14DE	1.8	-4.3	-1.7	0.7
88.96	3.88	-1.1	271.40	24DE	1.7	-4.4	-1.6	0.8

♂			☉	16.00UT	☿	♀	♃	♄
LONG	LAT	MAG	LONG	-42	MAGNITUDES			
85.19	3.96	-0.9	281.56	3JA	0.5	-4.3	-1.6	0.8
82.27	3.89	-0.7	291.70	13JA	0.2	-4.1	-1.6	0.8
80.59	3.73	-0.4	301.79	23JA	0.0	-3.6	-1.6	0.8
80.22	3.51	-0.1	311.82	2FE	-0.2	-3.3	-1.6	0.9
81.04	3.27	0.1	321.81	12FE	-0.6	-3.9	-1.6	0.9
82.89	3.03	0.4	331.74	22FE	-1.3	-4.2	-1.6	0.9
85.56	2.80	0.6	341.61	4MR	-1.5	-4.3	-1.6	0.9
88.89	2.58	0.7	351.42	14MR	-0.7	-4.2	-1.6	0.8
92.76	2.38	0.9	1.17	24MR	0.5	-4.2	-1.6	0.8
97.05	2.20	1.1	10.87	3AP	1.9	-4.0	-1.7	0.8
101.69	2.02	1.2	20.53	13AP	3.7	-3.9	-1.7	0.7
106.63	1.86	1.3	30.14	23AP	2.4	-3.8	-1.8	0.7
111.80	1.71	1.4	39.71	3MY	1.4	-3.7	-1.8	0.6
117.19	1.56	1.5	49.26	13MY	0.6	-3.7	-1.9	0.5
122.76	1.42	1.5	58.79	23MY	-0.3	-3.6	-1.9	0.5
128.48	1.28	1.6	68.31	2JN	-1.3	-3.5	-2.0	0.4
134.37	1.15	1.6	77.83	12JN	-1.7	-3.5	-2.1	0.4
140.38	1.02	1.7	87.36	22JN	-0.8	-3.4	-2.2	0.3
146.53	0.90	1.7	96.91	2JL	-0.1	-3.4	-2.2	0.3
152.80	0.77	1.7	106.48	12JL	0.3	-3.4	-2.3	0.4
159.19	0.65	1.7	116.09	22JL	0.6	-3.4	-2.3	0.4
165.70	0.53	1.7	125.75	1AU	1.0	-3.4	-2.4	0.5
172.33	0.41	1.7	135.45	11AU	2.0	-3.3	-2.4	0.6
179.07	0.29	1.7	145.21	21AU	2.6	-3.3	-2.4	0.6
185.93	0.17	1.7	155.02	31AU	0.2	-3.4	-2.5	0.7
192.89	0.06	1.7	164.90	10SE	-0.8	-3.4	-2.4	0.7
199.96	-0.05	1.7	174.82	20SE	-1.0	-3.4	-2.4	0.7
207.14	-0.17	1.6	184.81	30SE	-0.9	-3.4	-2.4	0.8
214.42	-0.27	1.6	194.86	10OC	-0.7	-3.4	-2.3	0.8
221.80	-0.38	1.6	204.95	20OC	-0.5	-3.4	-2.3	0.8
229.27	-0.48	1.5	215.08	30OC	-0.4	-3.5	-2.2	0.8
236.83	-0.58	1.5	225.25	9NO	-0.4	-3.5	-2.1	0.8
244.47	-0.67	1.4	235.44	19NO	0.0	-3.5	-2.0	0.8
252.18	-0.75	1.4	245.65	29NO	2.1	-3.5	-2.0	0.8
259.94	-0.83	1.4	255.86	9DE	1.3	-3.4	-1.9	0.8
267.76	-0.90	1.3	266.05	19DE	0.2	-3.4	-1.8	0.8
275.61	-0.96	1.3	276.23	29DE	0.0	-3.4	-1.8	0.8
				-41				
283.48	-1.01	1.3	286.39	8JA	-0.1	-3.4	-1.7	0.8
291.37	-1.05	1.3	296.50	18JA	-0.3	-3.3	-1.7	0.9
299.26	-1.08	1.4	306.56	28JA	-0.6	-3.3	-1.6	0.9
307.13	-1.10	1.4	316.58	7FE	-1.3	-3.3	-1.6	0.9
314.97	-1.10	1.4	326.54	17FE	-1.4	-3.3	-1.6	0.9
322.78	-1.09	1.4	336.44	27FE	-0.7	-3.4	-1.6	0.9
330.53	-1.07	1.4	346.28	9MR	0.7	-3.4	-1.6	0.9
338.23	-1.04	1.4	356.06	19MR	2.4	-3.4	-1.6	0.9
345.86	-1.00	1.5	5.79	29MR	3.2	-3.4	-1.6	0.9
353.41	-0.95	1.5	15.47	8AP	1.8	-3.4	-1.6	0.9
0.88	-0.89	1.5	25.10	18AP	1.0	-3.5	-1.6	0.8
8.26	-0.82	1.5	34.70	28AP	0.4	-3.5	-1.6	0.8
15.54	-0.74	1.5	44.26	8MY	-0.3	-3.6	-1.7	0.7
22.73	-0.65	1.6	53.79	18MY	-1.3	-3.6	-1.7	0.7
29.82	-0.56	1.6	63.32	28MY	-1.7	-3.7	-1.7	0.6
36.80	-0.45	1.6	72.84	7JN	-0.8	-3.8	-1.8	0.5
43.68	-0.35	1.6	82.36	17JN	-0.1	-3.9	-1.9	0.5
50.45	-0.23	1.6	91.90	27JN	0.4	-4.0	-1.9	0.4
57.11	-0.11	1.6	101.46	7JL	0.8	-4.1	-2.0	0.4
63.65	0.01	1.6	111.05	17JL	1.3	-4.2	-2.0	0.4
70.07	0.15	1.6	120.68	27JL	2.5	-4.2	-2.1	0.5
76.37	0.28	1.5	130.36	6AU	2.2	-4.2	-2.2	0.5
82.52	0.43	1.5	140.08	16AU	0.1	-4.0	-2.2	0.6
88.53	0.58	1.4	149.87	26AU	-0.9	-3.5	-2.3	0.6
94.36	0.75	1.4	159.71	5SE	-1.1	-3.3	-2.4	0.7
100.01	0.92	1.3	169.61	15SE	-1.0	-3.9	-2.4	0.7
105.44	1.10	1.2	179.57	25SE	-0.6	-4.2	-2.4	0.8
110.60	1.30	1.1	189.58	5OC	-0.3	-4.3	-2.4	0.8
115.47	1.52	1.0	199.65	15OC	-0.3	-4.3	-2.4	0.8
119.96	1.76	0.9	209.77	25OC	-0.2	-4.3	-2.4	0.9
124.00	2.03	0.7	219.92	4NO	0.2	-4.2	-2.3	0.9
127.49	2.32	0.6	230.10	14NO	2.4	-4.1	-2.3	0.9
130.29	2.65	0.4	240.30	24NO	1.1	-4.0	-2.2	0.9
132.23	3.01	0.1	250.51	4DE	0.0	-3.9	-2.2	0.9
133.14	3.39	-0.1	260.71	14DE	-0.1	-3.8	-2.1	0.9
132.84	3.79	-0.4	270.91	24DE	-0.2	-3.7	-2.0	0.9

♂			☉	16.00UT	☿	♀	♃	♄
LONG	LAT	MAG	LONG	-40	MAGNITUDES			
131.22	4.16	-0.6	281.07	3JA	-0.4	-3.6	-1.9	0.8
128.38	4.45	-0.8	291.20	13JA	-0.7	-3.6	-1.9	0.9
124.67	4.60	-1.0	301.30	23JA	-1.2	-3.5	-1.8	0.9
120.75	4.57	-0.9	311.34	2FE	-1.3	-3.5	-1.7	0.9
117.33	4.38	-0.8	321.33	12FE	-0.7	-3.4	-1.7	1.0
114.94	4.07	-0.5	331.26	22FE	0.8	-3.4	-1.6	1.0
113.85	3.69	-0.3	341.13	3MR	2.9	-3.3	-1.6	1.0
114.01	3.31	-0.1	350.95	13MR	2.4	-3.3	-1.6	1.0
115.31	2.94	0.1	0.71	23MR	1.3	-3.3	-1.5	1.0
117.57	2.59	0.3	10.41	2AP	0.8	-3.3	-1.5	1.0
120.59	2.27	0.5	20.06	12AP	0.3	-3.3	-1.5	1.0
124.25	1.98	0.7	29.67	22AP	-0.4	-3.3	-1.5	0.9
128.43	1.72	0.8	39.25	2MY	-1.3	-3.3	-1.5	0.9
133.03	1.48	0.9	48.80	12MY	-1.7	-3.4	-1.5	0.8
138.00	1.25	1.0	58.33	22MY	-0.8	-3.4	-1.5	0.8
143.27	1.04	1.1	67.85	1JN	0.0	-3.4	-1.6	0.7
148.80	0.85	1.2	77.37	11JN	0.5	-3.5	-1.6	0.7
154.56	0.67	1.3	86.90	21JN	1.0	-3.5	-1.6	0.6
160.54	0.50	1.3	96.44	1JL	1.7	-3.5	-1.7	0.5
166.71	0.33	1.3	106.02	11JL	3.1	-3.4	-1.7	0.5
173.06	0.18	1.4	115.63	21JL	1.9	-3.4	-1.8	0.5
179.57	0.03	1.4	125.28	31JL	0.0	-3.4	-1.8	0.5
186.24	-0.11	1.4	134.98	10AU	-1.0	-3.4	-1.9	0.5
193.06	-0.24	1.4	144.73	20AU	-1.3	-3.3	-2.0	0.6
200.02	-0.37	1.4	154.54	30AU	-0.9	-3.3	-2.0	0.6
207.12	-0.48	1.4	164.41	9SE	-0.5	-3.3	-2.1	0.7
214.34	-0.59	1.4	174.34	19SE	-0.2	-3.3	-2.2	0.7
221.68	-0.69	1.4	184.32	29SE	-0.1	-3.3	-2.2	0.8
229.13	-0.78	1.4	194.36	9OC	-0.0	-3.4	-2.3	0.8
236.68	-0.87	1.4	204.46	19OC	0.4	-3.4	-2.3	0.9
244.32	-0.94	1.4	214.59	29OC	2.7	-3.4	-2.3	0.9
252.04	-1.00	1.4	224.76	8NO	0.8	-3.5	-2.3	0.9
259.82	-1.05	1.4	234.95	18NO	-0.2	-3.5	-2.3	1.0
267.65	-1.08	1.4	245.15	28NO	-0.3	-3.5	-2.3	1.0
275.52	-1.11	1.4	255.36	8DE	-0.3	-3.6	-2.2	1.0
283.41	-1.12	1.4	265.56	18DE	-0.5	-3.7	-2.2	1.0
291.31	-1.12	1.4	275.74	28DE	-0.8	-3.7	-2.1	1.0
				-39				
299.20	-1.11	1.4	285.90	7JA	-1.1	-3.8	-2.1	1.0
307.08	-1.08	1.4	296.01	17JA	-1.1	-3.9	-2.0	0.9
314.92	-1.05	1.4	306.08	27JA	-0.6	-4.0	-1.9	1.0
322.72	-1.00	1.4	316.10	6FE	1.0	-4.1	-1.9	1.0
330.46	-0.95	1.4	326.06	16FE	3.2	-4.2	-1.8	1.1
338.14	-0.89	1.4	335.96	26FE	1.8	-4.2	-1.7	1.1
345.75	-0.82	1.4	345.81	8MR	1.0	-4.3	-1.7	1.1
353.28	-0.74	1.4	355.59	18MR	0.6	-4.2	-1.6	1.1
0.72	-0.65	1.5	5.32	28MR	0.2	-3.8	-1.6	1.1
8.09	-0.56	1.5	15.00	7AP	-0.4	-3.2	-1.5	1.1
15.36	-0.47	1.6	24.63	17AP	-1.3	-3.3	-1.5	1.1
22.54	-0.37	1.6	34.23	27AP	-1.7	-3.9	-1.5	1.1
29.64	-0.27	1.7	43.79	7MY	-0.8	-4.1	-1.5	1.1
36.65	-0.17	1.7	53.33	17MY	0.1	-4.2	-1.4	1.0
43.57	-0.06	1.7	62.85	27MY	0.7	-4.1	-1.4	1.0
50.40	0.04	1.8	72.37	6JN	1.3	-4.1	-1.4	0.9
57.15	0.15	1.8	81.90	16JN	2.3	-4.0	-1.4	0.8
63.82	0.26	1.8	91.43	26JN	3.2	-3.9	-1.4	0.8
70.41	0.37	1.8	100.99	6JL	1.5	-3.8	-1.5	0.7
76.91	0.48	1.8	110.58	16JL	-0.0	-3.7	-1.5	0.6
83.33	0.59	1.8	120.21	26JL	-1.1	-3.7	-1.5	0.6
89.68	0.71	1.8	129.89	5AU	-1.4	-3.6	-1.6	0.6
95.93	0.82	1.8	139.61	15AU	-0.9	-3.5	-1.6	0.6
102.10	0.94	1.8	149.39	25AU	-0.4	-3.5	-1.7	0.7
108.17	1.06	1.7	159.23	4SE	-0.1	-3.5	-1.7	0.7
114.14	1.18	1.7	169.13	14SE	0.0	-3.4	-1.8	0.8
120.00	1.31	1.6	179.08	24SE	0.1	-3.4	-1.8	0.8
125.72	1.44	1.6	189.10	4OC	0.7	-3.4	-1.9	0.9
131.30	1.58	1.5	199.16	14OC	3.0	-3.4	-2.0	0.9
136.70	1.73	1.4	209.27	24OC	0.7	-3.4	-2.0	0.9
141.89	1.89	1.3	219.43	3NO	-0.3	-3.4	-2.1	1.0
146.83	2.06	1.2	229.60	13NO	-0.4	-3.4	-2.1	1.0
151.48	2.24	1.0	239.80	23NO	-0.5	-3.4	-2.2	1.0
155.75	2.44	0.9	250.01	3DE	-0.6	-3.4	-2.2	1.1
159.56	2.65	0.7	260.22	13DE	-0.8	-3.4	-2.2	1.1
162.81	2.88	0.5	270.41	23DE	-1.0	-3.4	-2.2	1.1

♂			☉	16.00UT	☿	♀	♃	♄
LONG	LAT	MAG	LONG	-38	MAGNITUDES			
165.34	3.13	0.2	280.58	2JA	-1.0	-3.4	-2.2	1.1
166.99	3.38	-0.0	290.71	12JA	-0.5	-3.4	-2.1	1.1
167.59	3.63	-0.3	300.81	22JA	1.2	-3.5	-2.1	1.1
166.96	3.86	-0.6	310.85	1FE	2.9	-3.5	-2.0	1.1
165.06	4.01	-0.8	320.84	11FE	1.3	-3.5	-1.9	1.1
162.02	4.05	-1.1	330.78	21FE	0.7	-3.4	-1.9	1.2
158.29	3.94	-1.2	340.65	3MR	0.4	-3.4	-1.8	1.2
154.56	3.66	-1.1	350.47	13MR	0.1	-3.4	-1.7	1.3
151.50	3.26	-1.0	0.23	23MR	-0.4	-3.4	-1.7	1.3
149.60	2.79	-0.7	9.93	2AP	-1.2	-3.3	-1.6	1.3
149.02	2.32	-0.5	19.59	12AP	-1.7	-3.3	-1.5	1.3
149.70	1.87	-0.3	29.20	22AP	-0.8	-3.3	-1.5	1.3
151.51	1.46	-0.1	38.78	2MY	0.2	-3.3	-1.4	1.3
154.25	1.09	0.1	48.33	12MY	1.0	-3.3	-1.4	1.3
157.76	0.77	0.2	57.86	22MY	1.8	-3.4	-1.4	1.2
161.91	0.47	0.4	67.38	1JN	3.0	-3.4	-1.4	1.2
166.59	0.21	0.5	76.90	11JN	2.7	-3.4	-1.3	1.1
171.71	-0.02	0.6	86.43	21JN	1.2	-3.4	-1.3	1.1
177.22	-0.23	0.7	95.98	1JL	-0.1	-3.5	-1.3	1.0
183.04	-0.42	0.8	105.55	11JL	-1.2	-3.5	-1.3	0.9
189.15	-0.59	0.8	115.16	21JL	-1.5	-3.6	-1.3	0.8
195.52	-0.74	0.9	124.81	31JL	-0.9	-3.7	-1.3	0.8
202.11	-0.87	0.9	134.50	10AU	-0.3	-3.8	-1.4	0.7
208.90	-0.99	1.0	144.26	20AU	-0.0	-3.9	-1.4	0.7
215.87	-1.09	1.0	154.07	30AU	0.2	-4.0	-1.4	0.8
223.01	-1.17	1.1	163.93	9SE	0.3	-4.1	-1.5	0.8
230.29	-1.24	1.1	173.86	19SE	1.0	-4.2	-1.5	0.9
237.70	-1.29	1.1	183.84	29SE	3.3	-4.3	-1.6	0.9
245.22	-1.33	1.2	193.88	9OC	0.5	-4.4	-1.6	1.0
252.84	-1.35	1.2	203.97	19OC	-0.5	-4.3	-1.7	1.0
260.54	-1.35	1.2	214.10	29OC	-0.6	-4.1	-1.8	1.1
268.29	-1.34	1.3	224.26	8NO	-0.6	-3.5	-1.8	1.1
276.08	-1.32	1.3	234.45	18NO	-0.7	-3.1	-1.9	1.1
283.91	-1.28	1.3	244.66	28NO	-0.8	-3.9	-2.0	1.2
291.73	-1.24	1.3	254.86	8DE	-0.8	-4.3	-2.0	1.2
299.56	-1.18	1.4	265.07	18DE	-0.8	-4.4	-2.0	1.2
307.36	-1.11	1.4	275.25	28DE	-0.4	-4.3	-2.1	1.3
				-37				
315.13	-1.03	1.4	285.40	7JA	1.4	-4.3	-2.1	1.3
322.85	-0.94	1.5	295.52	17JA	2.4	-4.1	-2.1	1.3
330.51	-0.85	1.5	305.59	27JA	0.9	-4.0	-2.1	1.3
338.11	-0.75	1.5	315.61	6FE	0.5	-3.9	-2.0	1.3
345.64	-0.65	1.6	325.58	16FE	0.3	-3.8	-2.0	1.3
353.09	-0.55	1.6	335.49	26FE	-0.0	-3.7	-2.0	1.3
0.46	-0.45	1.6	345.33	8MR	-0.5	-3.6	-1.9	1.4
7.75	-0.34	1.6	355.12	18MR	-1.2	-3.6	-1.8	1.4
14.95	-0.24	1.6	4.86	28MR	-1.6	-3.5	-1.8	1.4
22.07	-0.13	1.7	14.54	7AP	-0.8	-3.4	-1.7	1.4
29.11	-0.03	1.7	24.17	17AP	0.3	-3.4	-1.6	1.3
36.07	0.07	1.7	33.77	27AP	1.2	-3.4	-1.6	1.3
42.95	0.17	1.7	43.33	7MY	2.4	-3.3	-1.5	1.3
49.76	0.27	1.7	52.87	17MY	3.6	-3.3	-1.5	1.2
56.50	0.37	1.8	62.39	27MY	2.1	-3.3	-1.4	1.2
63.18	0.46	1.8	71.91	6JN	1.0	-3.3	-1.4	1.1
69.79	0.55	1.9	81.44	16JN	-0.1	-3.3	-1.3	1.1
76.34	0.64	1.9	90.97	26JN	-1.2	-3.3	-1.3	1.0
82.84	0.73	1.9	100.53	6JL	-1.6	-3.3	-1.3	1.0
89.29	0.81	2.0	110.12	16JL	-0.9	-3.4	-1.3	0.9
95.69	0.89	2.0	119.75	26JL	-0.3	-3.4	-1.2	0.9
102.04	0.97	2.0	129.42	5AU	0.1	-3.4	-1.2	0.9
108.35	1.05	2.0	139.14	15AU	0.3	-3.5	-1.2	0.8
114.61	1.13	2.0	148.92	25AU	0.5	-3.5	-1.3	0.8
120.82	1.21	1.9	158.75	4SE	1.3	-3.5	-1.3	0.9
126.99	1.28	1.9	168.65	14SE	3.2	-3.5	-1.3	0.9
133.11	1.36	1.9	178.60	24SE	0.4	-3.4	-1.3	1.0
139.17	1.43	1.8	188.61	4OC	-0.6	-3.4	-1.4	1.0
145.17	1.50	1.8	198.67	14OC	-0.7	-3.4	-1.4	1.1
151.09	1.58	1.7	208.78	24OC	-0.7	-3.4	-1.5	1.1
156.93	1.65	1.6	218.93	3NO	-0.8	-3.3	-1.5	1.2
162.68	1.72	1.5	229.11	13NO	-0.7	-3.3	-1.6	1.2
168.30	1.80	1.4	239.30	23NO	-0.6	-3.3	-1.7	1.3
173.79	1.87	1.3	249.51	3DE	-0.7	-3.3	-1.7	1.3
179.10	1.95	1.2	259.72	13DE	-0.3	-3.4	-1.8	1.3
184.21	2.02	1.0	269.91	23DE	1.6	-3.4	-1.8	1.3

♂			☉	16.00UT	☿	♀	♃	♄
LONG	LAT	MAG	LONG	-36	MAGNITUDES			
189.07	2.09	0.9	280.08	2JA	2.0	-3.4	-1.9	1.3
193.61	2.16	0.7	290.22	12JA	0.6	-3.4	-2.0	1.3
197.76	2.21	0.5	300.32	22JA	0.3	-3.5	-2.0	1.3
201.43	2.26	0.2	310.37	1FE	0.1	-3.5	-2.0	1.2
204.48	2.29	-0.0	320.36	11FE	-0.1	-3.5	-2.0	1.2
206.78	2.30	-0.3	330.30	21FE	-0.5	-3.6	-2.0	1.1
208.13	2.26	-0.6	340.18	2MR	-1.2	-3.7	-2.0	1.1
208.36	2.16	-0.9	350.00	12MR	-1.5	-3.7	-2.0	1.1
207.35	1.98	-1.2	359.76	22MR	-0.8	-3.8	-1.9	1.1
205.12	1.70	-1.5	9.47	1AP	0.4	-3.9	-1.9	1.1
201.95	1.30	-1.8	19.13	11AP	1.6	-4.0	-1.8	1.1
198.44	0.82	-1.7	28.74	21AP	3.1	-4.1	-1.8	1.1
195.30	0.31	-1.6	38.32	1MY	2.9	-4.1	-1.7	1.0
193.17	-0.19	-1.4	47.87	11MY	1.6	-4.2	-1.6	1.0
192.36	-0.64	-1.2	57.40	21MY	0.7	-4.2	-1.6	1.0
192.90	-1.01	-1.0	66.92	31MY	-0.2	-4.0	-1.5	0.9
194.70	-1.32	-0.8	76.44	10JN	-1.2	-3.4	-1.4	0.9
197.57	-1.56	-0.6	85.97	20JN	-1.7	-2.9	-1.4	0.9
201.32	-1.75	-0.4	95.52	30JN	-0.8	-3.6	-1.3	0.8
205.81	-1.90	-0.3	105.09	10JL	-0.2	-4.1	-1.3	0.8
210.89	-2.01	-0.1	114.69	20JL	0.2	-4.2	-1.3	0.7
216.47	-2.08	-0.0	124.34	30JL	0.5	-4.2	-1.2	0.7
222.46	-2.13	0.1	134.03	9AU	0.8	-4.2	-1.2	0.6
228.78	-2.15	0.2	143.78	19AU	1.7	-4.1	-1.2	0.6
235.39	-2.14	0.3	153.59	29AU	2.9	-4.0	-1.2	0.5
242.24	-2.11	0.4	163.45	8SE	0.3	-3.9	-1.2	0.5
249.29	-2.06	0.5	173.37	18SE	-0.8	-3.8	-1.2	0.6
256.49	-1.99	0.6	183.36	28SE	-0.9	-3.8	-1.2	0.6
263.81	-1.91	0.7	193.39	8OC	-0.9	-3.7	-1.3	0.7
271.24	-1.81	0.8	203.47	18OC	-0.7	-3.6	-1.3	0.8
278.73	-1.70	0.9	213.61	28OC	-0.5	-3.6	-1.3	0.8
286.27	-1.57	0.9	223.77	7NO	-0.5	-3.5	-1.4	0.9
293.84	-1.45	1.0	233.96	17NO	-0.5	-3.5	-1.4	0.9
301.41	-1.31	1.1	244.16	27NO	-0.1	-3.4	-1.5	1.0
308.97	-1.17	1.2	254.37	7DE	1.8	-3.4	-1.5	1.0
316.51	-1.03	1.2	264.57	17DE	1.6	-3.4	-1.6	1.0
324.01	-0.89	1.3	274.76	27DE	0.4	-3.4	-1.7	1.0
				-35				
331.46	-0.76	1.4	284.91	6JA	0.1	-3.3	-1.7	1.0
338.86	-0.62	1.4	295.03	16JA	-0.0	-3.3	-1.8	1.0
346.19	-0.49	1.5	305.11	26JA	-0.2	-3.3	-1.9	1.0
353.45	-0.36	1.6	315.13	5FE	-0.6	-3.3	-1.9	1.0
0.65	-0.23	1.6	325.10	15FE	-1.2	-3.3	-2.0	0.9
7.77	-0.11	1.7	335.01	25FE	-1.4	-3.3	-2.0	0.9
14.82	0.00	1.7	344.86	7MR	-0.7	-3.4	-2.0	0.8
21.80	0.11	1.8	354.65	17MR	0.5	-3.4	-2.0	0.8
28.71	0.22	1.8	4.39	27MR	2.1	-3.4	-2.0	0.8
35.55	0.32	1.8	14.07	6AP	3.8	-3.5	-2.0	0.8
42.33	0.41	1.8	23.70	16AP	2.2	-3.5	-2.0	0.8
49.05	0.50	1.9	33.30	26AP	1.2	-3.5	-1.9	0.8
55.71	0.58	1.9	42.87	6MY	0.6	-3.4	-1.9	0.8
62.31	0.66	1.9	52.40	16MY	-0.2	-3.4	-1.8	0.8
68.88	0.73	1.9	61.93	26MY	-1.2	-3.4	-1.7	0.8
75.40	0.80	1.9	71.45	5JN	-1.7	-3.3	-1.7	0.7
81.88	0.86	1.9	80.97	15JN	-0.8	-3.3	-1.6	0.7
88.33	0.92	1.9	90.51	25JN	-0.1	-3.3	-1.5	0.7
94.75	0.98	1.9	100.06	5JL	0.3	-3.3	-1.5	0.6
101.14	1.03	2.0	109.65	15JL	0.6	-3.3	-1.4	0.6
107.52	1.08	2.0	119.27	25JL	1.1	-3.3	-1.4	0.5
113.88	1.12	2.0	128.94	4AU	2.2	-3.4	-1.3	0.5
120.23	1.16	2.0	138.66	14AU	2.5	-3.4	-1.3	0.4
126.57	1.20	2.0	148.44	24AU	0.2	-3.4	-1.3	0.3
132.89	1.23	2.0	158.27	3SE	-0.9	-3.5	-1.3	0.3
139.22	1.26	2.0	168.16	13SE	-1.0	-3.5	-1.2	0.2
145.53	1.29	2.0	178.11	23SE	-1.0	-3.6	-1.2	0.2
151.84	1.31	2.0	188.12	3OC	-0.6	-3.7	-1.2	0.3
158.14	1.32	1.9	198.18	13OC	-0.4	-3.7	-1.2	0.4
164.43	1.34	1.9	208.29	23OC	-0.4	-3.8	-1.3	0.4
170.70	1.35	1.8	218.43	2NO	-0.3	-3.9	-1.3	0.5
176.96	1.35	1.8	228.61	12NO	0.1	-4.0	-1.3	0.6
183.20	1.34	1.7	238.81	22NO	2.1	-4.1	-1.3	0.6
189.42	1.33	1.6	249.02	2DE	1.3	-4.2	-1.4	0.7
195.59	1.31	1.5	259.23	12DE	0.2	-4.3	-1.4	0.7
201.73	1.28	1.4	269.42	22DE	-0.0	-4.4	-1.5	0.8

♂ LONG	♂ LAT	♂ MAG	☉ LONG	16.00UT	☿	♀	♃	♄
				-34	MAGNITUDES			
207.81	1.24	1.3	279.59	1JA	-0.1	-4.3	-1.5	0.8
213.82	1.19	1.1	289.73	11JA	-0.3	-4.1	-1.6	0.8
219.75	1.12	1.0	299.83	21JA	-0.7	-3.5	-1.7	0.8
225.58	1.02	0.8	309.88	31JA	-1.2	-3.4	-1.7	0.8
231.28	0.91	0.7	319.88	10FE	-1.3	-3.9	-1.8	0.7
236.81	0.76	0.5	329.82	20FE	-0.7	-4.2	-1.9	0.7
242.14	0.58	0.3	339.70	2MR	0.7	-4.3	-1.9	0.7
247.22	0.36	0.1	349.52	12MR	2.6	-4.2	-2.0	0.6
251.97	0.09	-0.2	359.29	22MR	2.9	-4.2	-2.0	0.6
256.31	-0.24	-0.4	9.00	1AP	1.6	-4.0	-2.1	0.6
260.11	-0.64	-0.7	18.66	11AP	0.9	-3.9	-2.1	0.6
263.24	-1.12	-1.0	28.28	21AP	0.4	-3.8	-2.1	0.6
265.52	-1.70	-1.3	37.86	1MY	-0.3	-3.7	-2.1	0.6
266.76	-2.37	-1.6	47.41	11MY	-1.2	-3.7	-2.0	0.6
266.81	-3.12	-1.9	56.94	21MY	-1.7	-3.6	-2.0	0.6
265.63	-3.90	-2.2	66.46	31MY	-0.8	-3.5	-2.0	0.6
263.41	-4.62	-2.5	75.98	10JN	-0.1	-3.5	-1.9	0.6
260.67	-5.18	-2.5	85.51	20JN	0.4	-3.4	-1.8	0.5
258.14	-5.50	-2.4	95.05	30JN	0.9	-3.4	-1.8	0.5
256.45	-5.57	-2.2	104.63	10JL	1.4	-3.4	-1.7	0.5
256.04	-5.44	-1.9	114.23	20JL	2.7	-3.4	-1.6	0.4
256.98	-5.17	-1.7	123.87	30JL	2.2	-3.3	-1.6	0.4
259.17	-4.82	-1.4	133.57	9AU	0.1	-3.3	-1.5	0.3
262.43	-4.43	-1.2	143.31	19AU	-1.0	-3.3	-1.5	0.3
266.55	-4.03	-0.9	153.11	29AU	-1.2	-3.4	-1.4	0.2
271.35	-3.64	-0.7	162.97	8SE	-1.0	-3.4	-1.4	0.1
276.69	-3.25	-0.5	172.89	18SE	-0.5	-3.4	-1.4	0.1
282.44	-2.88	-0.3	182.87	28SE	-0.3	-3.4	-1.3	0.0
288.51	-2.53	-0.1	192.90	8OC	-0.2	-3.4	-1.3	0.0
294.82	-2.20	0.1	202.98	18OC	-0.1	-3.4	-1.3	0.1
301.31	-1.89	0.2	213.11	28OC	0.3	-3.5	-1.3	0.2
307.93	-1.60	0.4	223.27	7NO	2.4	-3.5	-1.3	0.2
314.65	-1.33	0.5	233.46	17NO	1.0	-3.5	-1.3	0.3
321.43	-1.08	0.7	243.66	27NO	-0.0	-3.5	-1.3	0.4
328.25	-0.85	0.8	253.88	7DE	-0.2	-3.4	-1.4	0.4
335.09	-0.64	1.0	264.08	17DE	-0.3	-3.4	-1.4	0.5
341.93	-0.44	1.1	274.26	27DE	-0.4	-3.4	-1.4	0.5
				-33				
348.76	-0.26	1.2	284.42	6JA	-0.7	-3.4	-1.5	0.5
355.57	-0.10	1.3	294.54	16JA	-1.2	-3.3	-1.5	0.6
2.35	0.05	1.4	304.62	26JA	-1.2	-3.3	-1.6	0.6
9.10	0.18	1.5	314.65	5FE	-0.7	-3.3	-1.6	0.6
15.81	0.30	1.6	324.61	15FE	0.8	-3.3	-1.7	0.6
22.48	0.42	1.6	334.53	25FE	3.1	-3.4	-1.8	0.5
29.12	0.52	1.7	344.38	7MR	2.2	-3.4	-1.8	0.5
35.71	0.61	1.8	354.18	17MR	1.2	-3.4	-1.9	0.5
42.27	0.69	1.8	3.91	27MR	0.7	-3.4	-2.0	0.4
48.79	0.76	1.9	13.60	6AP	0.3	-3.4	-2.0	0.4
55.27	0.83	1.9	23.24	16AP	-0.4	-3.5	-2.1	0.4
61.73	0.89	2.0	32.84	26AP	-1.2	-3.5	-2.1	0.4
68.16	0.94	2.0	42.41	6MY	-1.7	-3.6	-2.2	0.4
74.56	0.98	2.0	51.95	16MY	-0.8	-3.6	-2.2	0.4
80.95	1.02	2.0	61.47	26MY	0.0	-3.7	-2.2	0.5
87.32	1.06	2.0	71.00	5JN	0.6	-3.8	-2.2	0.5
93.68	1.09	2.0	80.52	15JN	1.1	-3.9	-2.1	0.4
100.04	1.11	2.0	90.05	25JN	1.9	-4.0	-2.1	0.4
106.40	1.13	2.0	99.61	5JL	3.2	-4.1	-2.0	0.4
112.75	1.15	2.0	109.19	15JL	1.8	-4.2	-2.0	0.4
119.12	1.16	2.0	118.81	25JL	0.1	-4.2	-1.9	0.3
125.50	1.16	2.0	128.48	4AU	-1.0	-4.2	-1.9	0.3
131.89	1.16	2.0	138.19	14AU	-1.3	-4.0	-1.8	0.2
138.31	1.16	2.0	147.97	24AU	-0.9	-3.5	-1.7	0.2
144.74	1.15	2.0	157.79	3SE	-0.5	-3.4	-1.7	0.1
151.20	1.14	2.0	167.68	13SE	-0.2	-3.9	-1.6	0.0
157.69	1.12	2.0	177.63	23SE	-0.1	-4.2	-1.6	-0.0
164.21	1.10	2.0	187.63	3OC	0.0	-4.3	-1.5	-0.1
170.76	1.07	1.9	197.69	13OC	0.5	-4.3	-1.5	-0.1
177.34	1.03	1.9	207.79	23OC	2.7	-4.3	-1.5	-0.1
183.96	0.99	1.9	217.94	2NO	0.8	-4.2	-1.4	-0.1
190.60	0.94	1.8	228.11	12NO	-0.2	-4.1	-1.4	0.0
197.28	0.88	1.8	238.31	22NO	-0.4	-4.0	-1.4	0.1
203.99	0.81	1.7	248.52	2DE	-0.4	-3.9	-1.4	0.1
210.73	0.73	1.7	258.73	12DE	-0.5	-3.8	-1.4	0.2
217.51	0.65	1.6	268.93	22DE	-0.8	-3.7	-1.4	0.3

♂ LONG	♂ LAT	♂ MAG	☉ LONG	16.00UT	☿	♀	♃	♄
				-32	MAGNITUDES			
224.30	0.54	1.5	279.10	1JA	-1.0	-3.6	-1.4	0.3
231.13	0.43	1.4	289.24	11JA	-1.1	-3.6	-1.5	0.4
237.98	0.30	1.3	299.35	21JA	-0.6	-3.5	-1.5	0.4
244.84	0.16	1.2	309.40	31JA	1.0	-3.5	-1.5	0.4
251.72	0.00	1.1	319.40	10FE	3.2	-3.4	-1.6	0.4
258.61	-0.18	1.0	329.35	20FE	1.6	-3.4	-1.6	0.4
265.50	-0.37	0.9	339.23	1MR	0.9	-3.3	-1.7	0.4
272.37	-0.59	0.7	349.05	11MR	0.5	-3.3	-1.8	0.4
279.23	-0.83	0.6	358.82	21MR	0.1	-3.3	-1.8	0.4
286.03	-1.09	0.5	8.54	31MR	-0.4	-3.3	-1.9	0.3
292.78	-1.37	0.3	18.20	10AP	-1.2	-3.3	-2.0	0.3
299.43	-1.67	0.2	27.82	20AP	-1.7	-3.3	-2.0	0.3
305.95	-1.99	0.0	37.40	30AP	-0.8	-3.3	-2.1	0.3
312.30	-2.33	-0.1	46.95	10MY	0.1	-3.4	-2.2	0.3
318.41	-2.69	-0.3	56.49	20MY	0.8	-3.4	-2.2	0.3
324.21	-3.07	-0.5	66.01	30MY	1.5	-3.4	-2.3	0.3
329.62	-3.47	-0.7	75.52	9JN	2.5	-3.5	-2.3	0.4
334.50	-3.88	-0.9	85.05	19JN	3.1	-3.5	-2.3	0.4
338.72	-4.30	-1.1	94.59	29JN	1.5	-3.5	-2.3	0.4
342.10	-4.72	-1.3	104.16	9JL	-0.0	-3.4	-2.3	0.3
344.41	-5.11	-1.6	113.76	19JL	-1.1	-3.4	-2.3	0.3
345.45	-5.45	-1.8	123.40	29JL	-1.5	-3.4	-2.2	0.3
345.09	-5.67	-2.0	133.09	8AU	-0.9	-3.4	-2.2	0.3
343.35	-5.71	-2.3	142.84	18AU	-0.4	-3.3	-2.1	0.2
340.60	-5.50	-2.4	152.63	28AU	-0.1	-3.3	-2.0	0.1
337.51	-5.03	-2.4	162.49	7SE	0.1	-3.3	-2.0	0.1
334.88	-4.37	-2.1	172.41	17SE	0.2	-3.3	-1.9	0.0
333.32	-3.63	-1.8	182.38	27SE	0.7	-3.3	-1.8	-0.1
333.06	-2.89	-1.5	192.41	7OC	3.0	-3.4	-1.8	-0.1
334.09	-2.22	-1.1	202.49	17OC	0.6	-3.4	-1.7	-0.2
336.25	-1.64	-0.8	212.61	27OC	-0.4	-3.4	-1.7	-0.3
339.31	-1.14	-0.5	222.77	6NO	-0.5	-3.5	-1.6	-0.2
343.10	-0.73	-0.3	232.96	16NO	-0.5	-3.5	-1.6	-0.2
347.45	-0.38	0.0	243.16	26NO	-0.6	-3.6	-1.6	-0.1
352.22	-0.09	0.2	253.37	6DE	-0.8	-3.6	-1.5	-0.0
357.33	0.15	0.4	263.58	16DE	-0.9	-3.7	-1.5	0.1
2.70	0.35	0.6	273.77	26DE	-0.9	-3.7	-1.5	0.1
				-31				
8.26	0.52	0.8	283.93	5JA	-0.5	-3.8	-1.5	0.2
13.97	0.66	1.0	294.05	15JA	1.2	-3.9	-1.5	0.2
19.79	0.78	1.1	304.13	25JA	2.7	-4.0	-1.5	0.3
25.70	0.88	1.3	314.16	4FE	1.2	-4.1	-1.5	0.3
31.68	0.96	1.4	324.14	14FE	0.6	-4.2	-1.6	0.3
37.70	1.03	1.5	334.05	24FE	0.3	-4.2	-1.6	0.3
43.76	1.09	1.6	343.91	6MR	0.0	-4.3	-1.6	0.3
49.85	1.13	1.7	353.71	16MR	-0.4	-4.2	-1.7	0.3
55.97	1.17	1.8	3.45	26MR	-1.2	-3.8	-1.7	0.3
62.10	1.20	1.8	13.14	5AP	-1.7	-3.1	-1.8	0.3
68.24	1.22	1.9	22.78	15AP	-0.8	-3.3	-1.9	0.3
74.40	1.24	1.9	32.38	25AP	0.2	-3.9	-1.9	0.2
80.58	1.24	2.0	41.95	5MY	1.0	-4.1	-2.0	0.2
86.77	1.25	2.0	51.49	15MY	2.0	-4.2	-2.1	0.2
92.97	1.25	2.0	61.01	25MY	3.3	-4.1	-2.1	0.2
99.20	1.24	2.0	70.53	4JN	2.5	-4.1	-2.2	0.3
105.45	1.23	2.0	80.06	14JN	1.2	-4.0	-2.3	0.3
111.72	1.21	2.0	89.59	24JN	-0.1	-3.9	-2.3	0.3
118.02	1.19	2.0	99.14	4JL	-1.2	-3.8	-2.4	0.3
124.36	1.17	2.0	108.73	14JL	-1.6	-3.7	-2.4	0.3
130.73	1.14	2.0	118.34	24JL	-0.9	-3.6	-2.4	0.3
137.14	1.10	2.0	128.01	3AU	-0.3	-3.6	-2.4	0.3
143.59	1.07	1.9	137.72	13AU	0.0	-3.5	-2.4	0.3
150.09	1.02	1.9	147.49	23AU	0.2	-3.5	-2.4	0.2
156.65	0.98	1.9	157.32	2SE	0.4	-3.5	-2.3	0.2
163.25	0.93	1.9	167.20	12SE	1.1	-3.4	-2.3	0.1
169.91	0.87	1.9	177.15	22SE	3.3	-3.4	-2.2	0.0
176.63	0.81	1.9	187.15	2OC	0.5	-3.4	-2.1	-0.0
183.41	0.74	1.9	197.20	12OC	-0.5	-3.4	-2.1	-0.1
190.25	0.66	1.8	207.30	22OC	-0.7	-3.4	-2.0	-0.2
197.16	0.58	1.8	217.45	1NO	-0.6	-3.4	-1.9	-0.2
204.12	0.50	1.8	227.62	11NO	-0.7	-3.4	-1.9	-0.3
211.15	0.40	1.8	237.81	21NO	-0.7	-3.4	-1.8	-0.3
218.24	0.30	1.7	248.02	1DE	-0.7	-3.4	-1.7	-0.2
225.39	0.20	1.7	258.23	11DE	-0.8	-3.4	-1.7	-0.1
232.60	0.08	1.6	268.42	21DE	-0.4	-3.4	-1.7	-0.1
239.87	-0.04	1.6	278.60	31DE	1.4	-3.4	-1.6	0.0

♂			☉	16.00UT	☿	♀	♃	♄
LONG	LAT	MAG	LONG	-30	MAGNITUDES			
247.19	-0.17	1.5	288.75	10JA	2.3	-3.4	-1.6	0.1
254.57	-0.30	1.5	298.85	20JA	0.8	-3.5	-1.6	0.1
261.98	-0.44	1.4	308.91	30JA	0.4	-3.5	-1.6	0.2
269.44	-0.59	1.3	318.91	9FE	0.2	-3.5	-1.6	0.2
276.93	-0.74	1.3	328.86	19FE	-0.1	-3.4	-1.6	0.3
284.44	-0.89	1.2	338.75	1MR	-0.5	-3.4	-1.6	0.3
291.96	-1.04	1.1	348.57	11MR	-1.2	-3.4	-1.6	0.3
299.48	-1.19	1.1	358.34	21MR	-1.6	-3.4	-1.6	0.3
306.99	-1.34	1.0	8.06	31MR	-0.8	-3.3	-1.6	0.3
314.47	-1.49	0.9	17.73	10AP	0.3	-3.3	-1.7	0.3
321.91	-1.63	0.9	27.35	20AP	1.4	-3.3	-1.7	0.3
329.29	-1.76	0.8	36.93	30AP	2.6	-3.3	-1.8	0.2
336.57	-1.88	0.7	46.49	10MY	3.4	-3.3	-1.8	0.2
343.75	-1.98	0.6	56.02	20MY	2.0	-3.4	-1.9	0.2
350.79	-2.08	0.6	65.55	30MY	0.9	-3.4	-2.0	0.2
357.67	-2.15	0.5	75.06	9JN	-0.1	-3.4	-2.0	0.2
4.34	-2.21	0.4	84.59	19JN	-1.2	-3.4	-2.1	0.3
10.78	-2.25	0.3	94.13	29JN	-1.7	-3.5	-2.2	0.3
16.92	-2.28	0.2	103.70	9JL	-0.9	-3.5	-2.2	0.3
22.73	-2.28	0.1	113.30	19JL	-0.3	-3.6	-2.3	0.3
28.10	-2.25	-0.1	122.94	29JL	0.1	-3.7	-2.4	0.3
32.96	-2.20	-0.2	132.62	8AU	0.4	-3.8	-2.4	0.3
37.19	-2.12	-0.4	142.36	18AU	0.6	-3.9	-2.4	0.3
40.62	-2.00	-0.6	152.16	28AU	1.4	-4.0	-2.5	0.3
43.09	-1.84	-0.7	162.01	7SE	3.2	-4.1	-2.5	0.2
44.38	-1.62	-1.0	171.93	17SE	0.4	-4.2	-2.4	0.2
44.29	-1.32	-1.2	181.90	27SE	-0.7	-4.3	-2.4	0.1
42.75	-0.95	-1.4	191.92	7OC	-0.8	-4.4	-2.4	0.1
39.89	-0.51	-1.6	202.00	17OC	-0.8	-4.3	-2.3	-0.0
36.20	-0.03	-1.7	212.13	27OC	-0.8	-4.1	-2.2	-0.1
32.48	0.44	-1.5	222.28	6NO	-0.6	-3.5	-2.2	-0.2
29.47	0.84	-1.3	232.47	16NO	-0.6	-3.2	-2.1	-0.2
27.68	1.16	-1.0	242.67	26NO	-0.6	-4.0	-2.0	-0.3
27.26	1.39	-0.6	252.88	6DE	-0.2	-4.3	-2.0	-0.3
28.08	1.55	-0.3	263.08	16DE	1.6	-4.4	-1.9	-0.2
29.99	1.66	-0.0	273.27	26DE	1.9	-4.3	-1.8	-0.1
				-29				
32.76	1.72	0.2	283.43	5JA	0.5	-4.3	-1.8	-0.1
36.20	1.76	0.4	293.56	15JA	0.2	-4.1	-1.7	0.0
40.19	1.77	0.7	303.64	25JA	0.1	-4.0	-1.7	0.1
44.58	1.78	0.9	313.67	4FE	-0.1	-3.9	-1.6	0.1
49.31	1.77	1.0	323.65	14FE	-0.6	-3.8	-1.6	0.2
54.29	1.75	1.2	333.57	24FE	-1.2	-3.7	-1.6	0.2
59.48	1.72	1.3	343.43	6MR	-1.5	-3.6	-1.6	0.3
64.83	1.69	1.4	353.23	16MR	-0.8	-3.6	-1.6	0.3
70.33	1.66	1.5	2.96	26MR	0.4	-3.5	-1.6	0.3
75.94	1.62	1.6	12.67	5AP	1.7	-3.4	-1.6	0.3
81.64	1.58	1.7	22.31	15AP	3.4	-3.4	-1.6	0.3
87.43	1.53	1.8	31.92	25AP	2.6	-3.4	-1.6	0.3
93.30	1.49	1.8	41.48	5MY	1.5	-3.3	-1.6	0.3
99.23	1.44	1.9	51.03	15MY	0.7	-3.3	-1.7	0.3
105.24	1.38	1.9	60.56	25MY	-0.2	-3.3	-1.7	0.2
111.30	1.33	1.9	70.08	4JN	-1.2	-3.3	-1.7	0.2
117.43	1.27	2.0	79.60	14JN	-1.7	-3.3	-1.8	0.2
123.62	1.21	2.0	89.13	24JN	-0.9	-3.3	-1.9	0.3
129.87	1.15	2.0	98.68	4JL	-0.2	-3.3	-1.9	0.3
136.19	1.08	2.0	108.27	14JL	0.2	-3.4	-2.0	0.3
142.58	1.01	2.0	117.88	24JL	0.5	-3.4	-2.1	0.4
149.03	0.94	1.9	127.54	3AU	0.9	-3.4	-2.1	0.4
155.56	0.87	1.9	137.25	13AU	1.8	-3.5	-2.2	0.4
162.17	0.79	1.9	147.02	23AU	2.8	-3.5	-2.3	0.4
168.84	0.71	1.9	156.84	2SE	0.3	-3.5	-2.3	0.4
175.60	0.62	1.8	166.72	12SE	-0.8	-3.5	-2.4	0.3
182.44	0.53	1.8	176.66	22SE	-1.0	-3.4	-2.4	0.3
189.36	0.44	1.7	186.66	2OC	-0.9	-3.4	-2.4	0.2
196.36	0.35	1.7	196.71	12OC	-0.7	-3.4	-2.4	0.2
203.44	0.25	1.7	206.81	22OC	-0.5	-3.4	-2.4	0.1
210.61	0.15	1.6	216.95	1NO	-0.4	-3.3	-2.4	0.1
217.86	0.04	1.6	227.12	11NO	-0.4	-3.3	-2.3	-0.0
225.18	-0.07	1.6	237.32	21NO	-0.1	-3.3	-2.3	-0.1
232.58	-0.18	1.6	247.52	1DE	1.8	-3.3	-2.2	-0.2
240.05	-0.29	1.6	257.73	11DE	1.5	-3.4	-2.1	-0.2
247.58	-0.40	1.6	267.93	21DE	0.3	-3.4	-2.1	-0.2
255.17	-0.51	1.5	278.10	31DE	0.1	-3.4	-2.0	-0.1

♂			☉	16.00UT	☿	♀	♃	♄
LONG	LAT	MAG	LONG	-28	MAGNITUDES			
262.82	-0.62	1.5	288.25	10JA	-0.1	-3.4	-1.9	-0.1
270.50	-0.73	1.5	298.36	20JA	-0.2	-3.5	-1.9	0.0
278.23	-0.84	1.5	308.42	30JA	-0.6	-3.5	-1.8	0.1
285.98	-0.94	1.4	318.43	9FE	-1.2	-3.5	-1.7	0.1
293.74	-1.03	1.4	328.37	19FE	-1.4	-3.6	-1.7	0.2
301.50	-1.12	1.4	338.26	29FE	-0.8	-3.7	-1.6	0.2
309.25	-1.19	1.3	348.10	10MR	0.5	-3.7	-1.6	0.3
316.98	-1.26	1.3	357.87	20MR	2.2	-3.8	-1.6	0.3
324.68	-1.32	1.3	7.59	30MR	3.4	-3.9	-1.5	0.4
332.32	-1.36	1.3	17.26	9AP	1.9	-4.0	-1.5	0.4
339.90	-1.39	1.2	26.88	19AP	1.1	-4.1	-1.5	0.4
347.41	-1.40	1.2	36.47	29AP	0.5	-4.1	-1.5	0.4
354.83	-1.40	1.2	46.03	9MY	-0.2	-4.2	-1.5	0.4
2.15	-1.39	1.2	55.56	19MY	-1.2	-4.2	-1.5	0.4
9.35	-1.36	1.1	65.08	29MY	-1.8	-3.9	-1.5	0.4
16.43	-1.31	1.1	74.61	8JN	-0.9	-3.4	-1.6	0.3
23.37	-1.25	1.1	84.13	18JN	-0.1	-2.9	-1.6	0.3
30.16	-1.18	1.0	93.67	28JN	0.4	-3.6	-1.6	0.3
36.76	-1.08	1.0	103.24	8JL	0.7	-4.1	-1.7	0.4
43.18	-0.98	0.9	112.83	18JL	1.2	-4.2	-1.7	0.4
49.38	-0.86	0.9	122.47	28JL	2.3	-4.2	-1.8	0.5
55.32	-0.72	0.8	132.16	7AU	2.4	-4.2	-1.8	0.5
60.97	-0.56	0.7	141.89	17AU	0.2	-4.1	-1.9	0.5
66.29	-0.38	0.6	151.68	27AU	-0.9	-4.0	-2.0	0.5
71.20	-0.17	0.5	161.53	6SE	-1.1	-3.9	-2.0	0.5
75.62	0.06	0.3	171.44	16SE	-1.0	-3.8	-2.1	0.5
79.45	0.32	0.2	181.41	26SE	-0.6	-3.8	-2.2	0.4
82.53	0.63	0.0	191.44	6OC	-0.4	-3.7	-2.2	0.4
84.70	0.98	-0.2	201.51	16OC	-0.3	-3.6	-2.3	0.4
85.78	1.38	-0.4	211.63	26OC	-0.3	-3.6	-2.3	0.3
85.56	1.81	-0.6	221.79	5NO	0.1	-3.5	-2.3	0.2
83.95	2.27	-0.8	231.97	15NO	2.1	-3.5	-2.3	0.2
81.05	2.71	-1.0	242.18	25NO	1.2	-3.4	-2.3	0.1
77.27	3.08	-1.2	252.39	5DE	0.1	-3.4	-2.3	0.0
73.33	3.31	-1.1	262.59	15DE	-0.1	-3.4	-2.2	-0.0
69.96	3.41	-0.9	272.78	25DE	-0.2	-3.4	-2.2	-0.1
				-27				
67.72	3.38	-0.6	282.94	4JA	-0.3	-3.3	-2.1	-0.1
66.79	3.27	-0.3	293.07	14JA	-0.7	-3.3	-2.0	-0.0
67.13	3.12	-0.0	303.15	24JA	-1.2	-3.3	-2.0	0.1
68.59	2.95	0.2	313.19	3FE	-1.3	-3.3	-1.9	0.1
70.94	2.78	0.4	323.16	13FE	-0.7	-3.3	-1.8	0.2
74.02	2.62	0.6	333.09	23FE	0.7	-3.3	-1.8	0.2
77.69	2.46	0.8	342.95	5MR	2.7	-3.4	-1.7	0.3
81.81	2.30	1.0	352.75	15MR	2.6	-3.4	-1.6	0.3
86.31	2.16	1.1	2.50	25MR	1.4	-3.4	-1.6	0.4
91.10	2.03	1.3	12.19	4AP	0.8	-3.5	-1.5	0.4
96.15	1.90	1.4	21.84	14AP	0.4	-3.5	-1.5	0.5
101.40	1.77	1.5	31.45	24AP	-0.3	-3.5	-1.5	0.5
106.83	1.65	1.6	41.02	4MY	-1.2	-3.4	-1.4	0.5
112.42	1.54	1.6	50.56	14MY	-1.8	-3.4	-1.4	0.5
118.15	1.42	1.7	60.09	24MY	-0.9	-3.4	-1.4	0.5
124.00	1.31	1.7	69.61	3JN	-0.0	-3.3	-1.4	0.5
129.98	1.20	1.8	79.13	13JN	0.5	-3.3	-1.4	0.5
136.06	1.09	1.8	88.66	23JN	0.9	-3.3	-1.4	0.5
142.26	0.99	1.8	98.21	3JL	1.6	-3.3	-1.4	0.5
148.55	0.88	1.8	107.79	13JL	2.9	-3.3	-1.5	0.5
154.96	0.77	1.8	117.41	23JL	2.1	-3.3	-1.5	0.5
161.46	0.66	1.8	127.07	2AU	0.1	-3.4	-1.5	0.6
168.07	0.55	1.8	136.78	12AU	-1.0	-3.4	-1.6	0.6
174.78	0.45	1.8	146.54	22AU	-1.3	-3.4	-1.6	0.6
181.59	0.34	1.8	156.36	1SE	-1.0	-3.5	-1.7	0.7
188.50	0.23	1.7	166.24	11SE	-0.5	-3.5	-1.7	0.7
195.52	0.12	1.7	176.18	21SE	-0.3	-3.6	-1.8	0.7
202.64	0.01	1.7	186.17	1OC	-0.2	-3.7	-1.9	0.6
209.85	-0.10	1.6	196.22	11OC	-0.1	-3.7	-1.9	0.6
217.16	-0.21	1.6	206.32	21OC	0.3	-3.8	-2.0	0.6
224.56	-0.31	1.5	216.46	31OC	2.4	-3.9	-2.0	0.5
232.04	-0.41	1.5	226.63	10NO	1.0	-4.0	-2.1	0.5
239.61	-0.51	1.4	236.83	20NO	-0.1	-4.1	-2.1	0.4
247.25	-0.61	1.4	247.03	30NO	-0.3	-4.2	-2.2	0.4
254.96	-0.70	1.4	257.24	10DE	-0.3	-4.3	-2.2	0.3
262.72	-0.79	1.4	267.44	20DE	-0.4	-4.4	-2.2	0.2
270.52	-0.86	1.4	277.61	30DE	-0.7	-4.3	-2.2	0.1

Left section:

♂ LONG	LAT	MAG	☉ LONG	16.00UT −26	☿	♀	♃	♄ MAGNITUDES
278.36	-0.93	1.4	287.76	9JA	-1.1	-4.1	-2.1	0.1
286.22	-1.00	1.4	297.87	19JA	-1.1	-3.5	-2.1	0.1
294.08	-1.05	1.4	307.93	29JA	-0.7	-3.4	-2.0	0.2
301.95	-1.09	1.4	317.94	8FE	0.8	-3.9	-2.0	0.2
309.80	-1.12	1.4	327.89	18FE	3.1	-4.2	-1.9	0.3
317.62	-1.13	1.4	337.78	28FE	1.9	-4.3	-1.8	0.3
325.40	-1.14	1.4	347.62	10MR	1.1	-4.2	-1.8	0.4
333.13	-1.13	1.4	357.40	20MR	0.6	-4.1	-1.7	0.5
340.79	-1.11	1.4	7.12	30MR	0.2	-4.0	-1.6	0.5
348.39	-1.08	1.4	16.79	9AP	-0.3	-3.9	-1.6	0.6
355.91	-1.04	1.5	26.42	19AP	-1.2	-3.8	-1.5	0.6
3.33	-0.98	1.5	36.00	29AP	-1.8	-3.7	-1.5	0.6
10.67	-0.92	1.5	45.56	9MY	-0.9	-3.7	-1.4	0.7
17.91	-0.84	1.5	55.10	19MY	0.0	-3.6	-1.4	0.7
25.04	-0.76	1.5	64.62	29MY	0.7	-3.5	-1.4	0.7
32.06	-0.66	1.5	74.14	8JN	1.2	-3.5	-1.3	0.7
38.97	-0.56	1.5	83.67	18JN	2.1	-3.4	-1.3	0.7
45.76	-0.45	1.5	93.21	28JN	3.3	-3.4	-1.3	0.7
52.42	-0.33	1.5	102.77	8JL	1.7	-3.4	-1.3	0.7
58.95	-0.20	1.4	112.37	18JL	0.1	-3.4	-1.3	0.7
65.34	-0.06	1.4	122.00	28JL	-1.1	-3.3	-1.3	0.7
71.58	0.08	1.4	131.69	7AU	-1.4	-3.3	-1.3	0.8
77.64	0.24	1.3	141.42	17AU	-1.0	-3.3	-1.4	0.8
83.52	0.40	1.3	151.21	27AU	-0.4	-3.4	-1.4	0.8
89.19	0.58	1.2	161.06	6SE	-0.2	-3.4	-1.4	0.9
94.61	0.77	1.1	170.97	16SE	-0.0	-3.4	-1.5	0.9
99.75	0.98	1.0	180.93	26SE	0.1	-3.4	-1.5	0.9
104.54	1.21	0.9	190.95	6OC	0.6	-3.4	-1.6	0.9
108.92	1.47	0.8	201.02	16OC	2.7	-3.4	-1.6	0.9
112.79	1.75	0.6	211.14	26OC	0.8	-3.5	-1.7	0.8
116.04	2.07	0.4	221.30	5NO	-0.3	-3.5	-1.8	0.8
118.51	2.42	0.2	231.48	15NO	-0.4	-3.5	-1.8	0.8
120.05	2.81	-0.0	241.68	25NO	-0.4	-3.5	-1.9	0.7
120.44	3.23	-0.2	251.89	5DE	-0.5	-3.4	-2.0	0.6
119.54	3.66	-0.5	262.09	15DE	-0.8	-3.4	-2.0	0.6
117.32	4.05	-0.7	272.28	25DE	-1.0	-3.4	-2.0	0.5

−25

♂ LONG	LAT	MAG	☉ LONG	16.00UT	☿	♀	♃	♄
113.99	4.34	-0.9	282.45	4JA	-1.0	-3.4	-2.1	0.4
110.06	4.49	-1.0	292.57	14JA	-0.6	-3.3	-2.1	0.4
106.27	4.45	-0.8	302.66	24JA	1.0	-3.3	-2.1	0.3
103.27	4.26	-0.6	312.70	3FE	3.1	-3.3	-2.1	0.4
101.47	3.98	-0.4	322.68	13FE	1.4	-3.3	-2.0	0.4
100.97	3.65	-0.1	332.60	23FE	0.8	-3.4	-2.0	0.5
101.68	3.31	0.1	342.47	5MR	0.4	-3.4	-1.9	0.5
103.43	2.99	0.3	352.28	15MR	0.1	-3.4	-1.9	0.6
106.03	2.69	0.5	2.03	25MR	-0.4	-3.4	-1.8	0.6
109.32	2.41	0.7	11.73	4AP	-1.2	-3.4	-1.7	0.7
113.18	2.16	0.8	21.37	14AP	-1.7	-3.5	-1.7	0.7
117.49	1.93	1.0	30.98	24AP	-0.9	-3.5	-1.6	0.8
122.18	1.71	1.1	40.56	4MY	0.1	-3.6	-1.5	0.8
127.18	1.51	1.2	50.10	14MY	0.9	-3.7	-1.5	0.9
132.46	1.32	1.3	59.63	24MY	1.7	-3.7	-1.4	0.9
137.96	1.15	1.3	69.15	3JN	2.8	-3.8	-1.4	0.9
143.68	0.98	1.4	78.67	13JN	2.9	-3.9	-1.3	0.9
149.58	0.82	1.4	88.20	23JN	1.4	-4.0	-1.3	0.9
155.66	0.66	1.5	97.75	3JL	0.0	-4.1	-1.3	0.9
161.90	0.51	1.5	107.33	13JL	-1.1	-4.2	-1.3	0.9
168.28	0.37	1.5	116.94	23JL	-1.5	-4.2	-1.2	0.9
174.82	0.23	1.6	126.60	2AU	-0.9	-4.2	-1.2	0.9
181.49	0.10	1.6	136.31	12AU	-0.4	-4.0	-1.2	1.0
188.29	-0.03	1.6	146.07	22AU	-0.0	-3.5	-1.2	1.0
195.23	-0.15	1.6	155.89	1SE	0.1	-3.4	-1.3	1.1
202.29	-0.27	1.6	165.76	11SE	0.3	-3.9	-1.3	1.1
209.47	-0.38	1.5	175.70	21SE	0.8	-4.2	-1.3	1.1
216.76	-0.49	1.5	185.69	1OC	3.1	-4.3	-1.3	1.2
224.16	-0.59	1.5	195.74	11OC	0.7	-4.3	-1.4	1.2
231.66	-0.68	1.5	205.83	21OC	-0.4	-4.2	-1.4	1.2
239.25	-0.77	1.5	215.97	31OC	-0.6	-4.2	-1.5	1.1
246.92	-0.85	1.5	226.14	10NO	-0.6	-4.1	-1.5	1.1
254.66	-0.91	1.4	236.33	20NO	-0.6	-4.0	-1.6	1.1
262.46	-0.97	1.4	246.54	30NO	-0.8	-3.9	-1.7	1.0
270.31	-1.02	1.4	256.74	10DE	-0.8	-3.8	-1.7	1.0
278.19	-1.05	1.4	266.94	20DE	-0.8	-3.7	-1.8	0.9
286.09	-1.08	1.4	277.13	30DE	-0.5	-3.6	-1.9	0.8

Right section:

♂ LONG	LAT	MAG	☉ LONG	16.00UT −24	☿	♀	♃	♄ MAGNITUDES
293.99	-1.09	1.3	287.27	9JA	1.2	-3.6	-1.9	0.8
301.89	-1.09	1.3	297.38	19JA	2.6	-3.5	-2.0	0.7
309.76	-1.08	1.3	307.45	29JA	1.0	-3.5	-2.0	0.6
317.60	-1.06	1.3	317.46	8FE	0.5	-3.4	-2.0	0.6
325.40	-1.02	1.3	327.42	18FE	0.3	-3.4	-2.0	0.6
333.14	-0.98	1.4	337.31	28FE	0.0	-3.3	-2.0	0.7
340.81	-0.92	1.4	347.15	9MR	-0.5	-3.3	-2.0	0.7
348.41	-0.86	1.5	356.93	19MR	-1.2	-3.3	-2.0	0.8
355.93	-0.79	1.5	6.65	29MR	-1.7	-3.3	-1.9	0.8
3.37	-0.71	1.5	16.32	8AP	-0.8	-3.3	-1.9	0.9
10.72	-0.63	1.6	25.95	18AP	0.2	-3.3	-1.8	0.9
17.98	-0.54	1.6	35.54	28AP	1.1	-3.3	-1.7	1.0
25.15	-0.44	1.6	45.10	8MY	2.2	-3.4	-1.7	1.0
32.22	-0.34	1.7	54.64	18MY	3.6	-3.4	-1.6	1.1
39.20	-0.23	1.7	64.16	28MY	2.3	-3.4	-1.5	1.1
46.09	-0.13	1.7	73.68	7JN	1.1	-3.5	-1.5	1.2
52.89	-0.02	1.7	83.20	17JN	-0.1	-3.5	-1.4	1.2
59.59	0.10	1.7	92.74	27JN	-1.1	-3.5	-1.4	1.2
66.20	0.21	1.7	102.30	7JL	-1.6	-3.4	-1.3	1.2
72.71	0.33	1.7	111.90	17JL	-0.9	-3.4	-1.3	1.2
79.13	0.45	1.7	121.53	27JL	-0.3	-3.4	-1.3	1.2
85.45	0.58	1.7	131.21	6AU	0.1	-3.4	-1.2	1.2
91.67	0.71	1.7	140.94	16AU	0.3	-3.3	-1.2	1.2
97.78	0.84	1.7	150.73	26AU	0.5	-3.3	-1.2	1.3
103.76	0.98	1.6	160.57	5SE	1.2	-3.3	-1.2	1.3
109.62	1.12	1.6	170.48	15SE	3.3	-3.3	-1.2	1.4
115.32	1.27	1.5	180.44	25SE	0.5	-3.3	-1.2	1.4
120.85	1.43	1.4	190.46	5OC	-0.6	-3.4	-1.2	1.4
126.18	1.60	1.3	200.53	15OC	-0.7	-3.4	-1.3	1.3
131.26	1.78	1.2	210.65	25OC	-0.7	-3.4	-1.3	1.3
136.07	1.98	1.1	220.80	4NO	-0.8	-3.5	-1.3	1.3
140.52	2.19	1.0	230.99	14NO	-0.7	-3.5	-1.4	1.3
144.55	2.43	0.8	241.19	24NO	-0.7	-3.6	-1.4	1.2
148.04	2.69	0.6	251.39	4DE	-0.7	-3.6	-1.5	1.2
150.88	2.97	0.4	261.60	14DE	-0.3	-3.7	-1.6	1.1
152.90	3.27	0.2	271.79	24DE	1.4	-3.8	-1.6	1.1

−23

♂ LONG	LAT	MAG	☉ LONG	16.00UT	☿	♀	♃	♄
153.93	3.59	-0.1	281.95	3JA	2.2	-3.8	-1.7	1.0
153.78	3.91	-0.4	292.09	13JA	0.7	-3.9	-1.8	1.0
152.35	4.19	-0.6	302.18	23JA	0.3	-4.0	-1.8	0.9
149.68	4.38	-0.9	312.22	2FE	0.2	-4.1	-1.9	0.9
146.10	4.42	-1.1	322.20	12FE	-0.1	-4.2	-1.9	0.9
142.23	4.29	-1.1	332.13	22FE	-0.5	-4.3	-2.0	0.9
138.76	4.01	-0.9	342.00	4MR	-1.2	-4.3	-2.0	0.9
136.29	3.61	-0.7	351.81	14MR	-1.6	-4.2	-2.0	1.0
135.10	3.17	-0.5	1.56	24MR	-0.8	-3.8	-2.0	1.0
135.19	2.72	-0.3	11.26	3AP	0.3	-3.1	-2.0	1.1
136.45	2.31	-0.0	20.91	13AP	1.5	-3.3	-2.0	1.1
138.71	1.92	0.1	30.52	23AP	2.9	-3.9	-1.9	1.2
141.79	1.58	0.3	40.09	3MY	3.1	-4.1	-1.9	1.2
145.54	1.27	0.5	49.64	13MY	1.8	-4.2	-1.8	1.3
149.84	0.98	0.6	59.17	23MY	0.8	-4.1	-1.8	1.3
154.60	0.73	0.7	68.69	2JN	-0.1	-4.1	-1.7	1.4
159.76	0.50	0.8	78.21	12JN	-1.2	-4.0	-1.7	1.4
165.24	0.28	0.9	87.74	22JN	-1.7	-3.9	-1.6	1.4
171.02	0.09	1.0	97.29	2JL	-0.9	-3.8	-1.5	1.4
177.05	-0.09	1.0	106.86	12JL	-0.2	-3.7	-1.5	1.4
183.31	-0.26	1.1	116.47	22JL	0.2	-3.6	-1.4	1.4
189.78	-0.41	1.1	126.13	1AU	0.4	-3.6	-1.4	1.3
196.45	-0.55	1.2	135.84	11AU	0.7	-3.5	-1.3	1.3
203.29	-0.68	1.2	145.59	21AU	1.5	-3.5	-1.3	1.2
210.29	-0.79	1.2	155.41	31AU	3.1	-3.5	-1.3	1.2
217.44	-0.89	1.2	165.28	10SE	0.4	-3.4	-1.3	1.2
224.73	-0.98	1.3	175.21	20SE	-0.7	-3.4	-1.2	1.2
232.14	-1.06	1.3	185.20	30SE	-0.9	-3.4	-1.2	1.2
239.66	-1.12	1.3	195.25	10OC	-0.8	-3.4	-1.2	1.2
247.27	-1.17	1.3	205.34	20OC	-0.8	-3.4	-1.2	1.2
254.97	-1.20	1.3	215.48	30OC	-0.6	-3.4	-1.3	1.2
262.73	-1.22	1.3	225.65	9NO	-0.5	-3.4	-1.3	1.1
270.54	-1.23	1.3	235.84	19NO	-0.5	-3.4	-1.3	1.1
278.39	-1.22	1.4	246.04	29NO	-0.2	-3.4	-1.4	1.0
286.26	-1.20	1.4	256.25	9DE	1.6	-3.4	-1.4	1.0
294.12	-1.17	1.4	266.45	19DE	1.8	-3.4	-1.4	1.0
301.98	-1.13	1.4	276.63	29DE	0.5	-3.4	-1.5	0.9

♂			⊙	16.00UT	☿	♀	♃	♄
LONG	LAT	MAG	LONG	-22	MAGNITUDES			
309.81	-1.07	1.4	286.78	8JA	0.2	-3.4	-1.6	0.9
317.61	-1.01	1.4	296.89	18JA	0.0	-3.5	-1.6	0.8
325.35	-0.94	1.5	306.96	28JA	-0.2	-3.5	-1.7	0.8
333.04	-0.86	1.5	316.98	7FE	-0.6	-3.5	-1.8	0.8
340.66	-0.78	1.5	326.93	17FE	-1.2	-3.4	-1.8	0.7
348.20	-0.69	1.5	336.83	27FE	-1.5	-3.4	-1.9	0.7
355.67	-0.59	1.5	346.67	9MR	-0.8	-3.4	-2.0	0.7
3.05	-0.49	1.5	356.45	19MR	0.4	-3.4	-2.0	0.8
10.35	-0.39	1.6	6.18	29MR	1.9	-3.3	-2.0	0.8
17.57	-0.29	1.6	15.86	8AP	3.7	-3.3	-2.1	0.9
24.69	-0.19	1.6	25.48	18AP	2.4	-3.3	-2.1	1.0
31.74	-0.09	1.6	35.08	28AP	1.3	-3.3	-2.1	1.0
38.70	0.02	1.7	44.64	8MY	0.6	-3.3	-2.1	1.1
45.58	0.12	1.7	54.17	18MY	-0.2	-3.4	-2.0	1.1
52.39	0.22	1.8	63.70	28MY	-1.2	-3.4	-2.0	1.2
59.12	0.32	1.8	73.22	7JN	-1.8	-3.4	-1.9	1.2
65.78	0.42	1.9	82.74	17JN	-0.9	-3.4	-1.9	1.2
72.38	0.52	1.9	92.28	27JN	-0.2	-3.5	-1.8	1.2
78.91	0.61	1.9	101.84	7JL	0.3	-3.5	-1.8	1.2
85.37	0.71	1.9	111.43	17JL	0.6	-3.6	-1.7	1.2
91.78	0.80	1.9	121.06	27JL	1.0	-3.7	-1.6	1.2
98.12	0.89	1.9	130.74	6AU	2.0	-3.8	-1.6	1.2
104.41	0.99	1.9	140.47	16AU	2.7	-3.9	-1.5	1.2
110.63	1.08	1.9	150.25	26AU	0.3	-4.0	-1.5	1.1
116.79	1.17	1.9	160.09	5SE	-0.8	-4.1	-1.4	1.1
122.88	1.27	1.8	169.99	15SE	-1.0	-4.2	-1.4	1.0
128.90	1.36	1.8	179.96	25SE	-1.0	-4.3	-1.4	1.0
134.84	1.46	1.8	189.97	5OC	-0.7	-4.4	-1.3	1.0
140.68	1.56	1.7	200.04	15OC	-0.4	-4.3	-1.3	1.0
146.42	1.66	1.6	210.16	25OC	-0.4	-4.1	-1.3	1.0
152.03	1.76	1.5	220.31	4NO	-0.4	-3.4	-1.3	1.0
157.49	1.87	1.4	230.49	14NO	-0.0	-3.2	-1.3	1.0
162.78	1.98	1.3	240.69	24NO	1.8	-4.0	-1.3	1.0
167.85	2.10	1.2	250.90	4DE	1.5	-4.3	-1.3	1.0
172.66	2.22	1.0	261.10	14DE	0.2	-4.4	-1.4	0.9
177.15	2.35	0.8	271.30	24DE	-0.0	-4.3	-1.4	0.9
				-21				
181.23	2.48	0.6	281.46	3JA	-0.1	-4.3	-1.4	0.9
184.82	2.61	0.4	291.60	13JA	-0.3	-4.1	-1.5	0.8
187.79	2.75	0.2	301.69	23JA	-0.6	-4.0	-1.5	0.8
189.98	2.87	-0.1	311.73	2FE	-1.2	-3.9	-1.6	0.7
191.23	2.98	-0.4	321.72	12FE	-1.4	-3.8	-1.7	0.7
191.33	3.05	-0.7	331.65	22FE	-0.8	-3.7	-1.7	0.6
190.19	3.05	-1.0	341.52	4MR	0.6	-3.6	-1.8	0.6
187.82	2.94	-1.3	351.34	14MR	2.4	-3.6	-1.9	0.5
184.50	2.71	-1.5	1.09	24MR	3.1	-3.5	-1.9	0.6
180.82	2.34	-1.5	10.79	3AP	1.7	-3.4	-2.0	0.6
177.51	1.88	-1.3	20.45	13AP	1.0	-3.4	-2.1	0.7
175.17	1.39	-1.2	30.06	23AP	0.5	-3.4	-2.1	0.8
174.14	0.90	-1.0	39.63	3MY	-0.2	-3.3	-2.2	0.8
174.44	0.45	-0.7	49.18	13MY	-1.2	-3.3	-2.2	0.9
175.98	0.06	-0.5	58.71	23MY	-1.8	-3.3	-2.2	1.0
178.58	-0.28	-0.4	68.23	2JN	-0.9	-3.3	-2.2	1.0
182.04	-0.57	-0.2	77.75	12JN	-0.1	-3.3	-2.2	1.0
186.24	-0.81	-0.0	87.28	22JN	0.4	-3.3	-2.1	1.1
191.04	-1.02	0.1	96.83	2JL	0.8	-3.3	-2.1	1.1
196.33	-1.20	0.2	106.40	12JL	1.3	-3.4	-2.0	1.1
202.04	-1.34	0.3	116.01	22JL	2.5	-3.4	-2.0	1.1
208.11	-1.46	0.4	125.66	1AU	2.4	-3.4	-1.9	1.1
214.48	-1.56	0.5	135.36	11AU	0.2	-3.5	-1.9	1.1
221.12	-1.63	0.6	145.12	21AU	-0.9	-3.5	-1.8	1.0
227.98	-1.68	0.7	154.92	31AU	-1.2	-3.5	-1.7	1.0
235.03	-1.70	0.7	164.80	10SE	-1.0	-3.5	-1.7	1.0
242.24	-1.71	0.8	174.73	20SE	-0.6	-3.4	-1.6	0.9
249.60	-1.70	0.9	184.71	30SE	-0.3	-3.4	-1.6	0.9
257.06	-1.67	0.9	194.75	10OC	-0.2	-3.4	-1.5	0.9
264.62	-1.63	1.0	204.84	20OC	-0.2	-3.4	-1.5	0.9
272.24	-1.57	1.0	214.97	30OC	0.2	-3.3	-1.5	0.9
279.92	-1.50	1.1	225.14	9NO	2.1	-3.3	-1.4	0.9
287.62	-1.41	1.1	235.34	19NO	1.2	-3.3	-1.4	0.9
295.32	-1.32	1.2	245.54	29NO	0.0	-3.3	-1.4	0.9
303.02	-1.21	1.3	255.75	9DE	-0.2	-3.4	-1.4	0.9
310.70	-1.10	1.3	265.95	19DE	-0.2	-3.4	-1.4	0.9
318.34	-0.99	1.4	276.13	29DE	-0.4	-3.4	-1.4	0.8

♂			⊙	16.00UT	☿	♀	♃	♄
LONG	LAT	MAG	LONG	-20	MAGNITUDES			
325.94	-0.87	1.4	286.29	8JA	-0.7	-3.4	-1.4	0.8
333.48	-0.75	1.5	296.40	18JA	-1.2	-3.5	-1.5	0.7
340.96	-0.63	1.5	306.47	28JA	-1.2	-3.5	-1.5	0.7
348.36	-0.52	1.6	316.49	7FE	-0.7	-3.5	-1.6	0.6
355.70	-0.40	1.6	326.45	17FE	0.7	-3.6	-1.6	0.6
2.95	-0.28	1.6	336.35	27FE	2.9	-3.7	-1.7	0.5
10.13	-0.17	1.7	346.20	8MR	2.3	-3.7	-1.7	0.5
17.23	-0.06	1.7	355.99	18MR	1.3	-3.8	-1.8	0.5
24.25	0.05	1.7	5.71	28MR	0.8	-3.9	-1.9	0.4
31.20	0.15	1.8	15.40	7AP	0.3	-4.0	-1.9	0.5
38.08	0.25	1.8	25.03	17AP	-0.3	-4.1	-2.0	0.5
44.89	0.35	1.8	34.62	27AP	-1.2	-4.1	-2.1	0.6
51.63	0.44	1.8	44.18	7MY	-1.8	-4.2	-2.1	0.7
58.31	0.52	1.8	53.72	17MY	-0.9	-4.1	-2.2	0.7
64.94	0.61	1.8	63.24	27MY	-0.0	-3.9	-2.2	0.8
71.51	0.68	1.8	72.76	6JN	0.5	-3.4	-2.3	0.9
78.04	0.76	1.9	82.28	16JN	1.0	-2.9	-2.3	0.9
84.53	0.83	1.9	91.82	26JN	1.8	-3.6	-2.3	0.9
90.98	0.90	2.0	101.38	6JL	3.1	-4.1	-2.3	1.0
97.39	0.96	2.0	110.97	16JL	2.0	-4.2	-2.3	1.0
103.77	1.02	2.0	120.60	26JL	0.2	-4.2	-2.3	1.0
110.13	1.08	2.0	130.27	5AU	-1.0	-4.2	-2.2	1.0
116.46	1.14	2.0	140.00	15AU	-1.3	-4.1	-2.2	1.0
122.76	1.19	2.0	149.78	25AU	-1.0	-4.0	-2.1	1.0
129.05	1.24	2.0	159.62	4SE	-0.5	-3.9	-2.0	0.9
135.31	1.28	2.0	169.51	14SE	-0.2	-3.8	-2.0	0.9
141.55	1.33	2.0	179.47	24SE	-0.1	-3.8	-1.9	0.8
147.76	1.37	1.9	189.48	4OC	-0.0	-3.7	-1.8	0.8
153.95	1.41	1.9	199.55	14OC	0.4	-3.6	-1.8	0.8
160.11	1.44	1.8	209.66	24OC	2.4	-3.6	-1.7	0.8
166.22	1.47	1.8	219.81	3NO	1.0	-3.5	-1.7	0.9
172.30	1.50	1.7	229.99	13NO	-0.2	-3.5	-1.6	0.9
178.32	1.52	1.6	240.19	23NO	-0.3	-3.4	-1.6	0.9
184.28	1.54	1.5	250.40	3DE	-0.4	-3.4	-1.6	0.9
190.16	1.55	1.4	260.60	13DE	-0.5	-3.4	-1.5	0.8
195.95	1.56	1.3	270.80	23DE	-0.7	-3.4	-1.5	0.8
				-19				
201.62	1.55	1.1	280.97	2JA	-1.1	-3.3	-1.5	0.8
207.15	1.53	1.0	291.10	12JA	-1.1	-3.3	-1.5	0.8
212.51	1.50	0.8	301.20	22JA	-0.7	-3.3	-1.5	0.7
217.66	1.45	0.6	311.25	1FE	0.9	-3.3	-1.5	0.7
222.54	1.37	0.4	321.24	11FE	3.2	-3.3	-1.5	0.6
227.09	1.26	0.2	331.17	21FE	1.7	-3.3	-1.6	0.6
231.22	1.12	-0.0	341.05	3MR	0.9	-3.4	-1.6	0.5
234.83	0.93	-0.3	350.86	13MR	0.5	-3.4	-1.7	0.5
237.78	0.68	-0.6	0.62	23MR	0.2	-3.4	-1.7	0.4
239.88	0.35	-0.9	10.33	2AP	-0.3	-3.5	-1.8	0.4
240.98	-0.06	-1.2	19.98	12AP	-1.2	-3.5	-1.8	0.4
240.89	-0.57	-1.5	29.59	22AP	-1.8	-3.5	-1.9	0.4
239.56	-1.17	-1.9	39.17	2MY	-0.9	-3.4	-1.9	0.5
237.16	-1.82	-2.1	48.72	12MY	0.0	-3.4	-2.0	0.6
234.12	-2.45	-2.3	58.25	22MY	0.7	-3.4	-2.1	0.6
231.16	-2.99	-2.2	67.77	1JN	1.4	-3.3	-2.2	0.7
228.99	-3.39	-2.0	77.29	11JN	2.4	-3.3	-2.2	0.7
228.03	-3.64	-1.8	86.82	21JN	3.2	-3.3	-2.3	0.8
228.47	-3.76	-1.5	96.36	1JL	1.6	-3.3	-2.3	0.8
230.23	-3.77	-1.3	105.93	11JL	0.1	-3.3	-2.4	0.9
233.12	-3.72	-1.1	115.54	21JL	-1.1	-3.3	-2.4	0.9
236.97	-3.61	-0.9	125.19	31JL	-1.4	-3.4	-2.4	0.9
241.58	-3.47	-0.7	134.89	10AU	-1.0	-3.4	-2.4	0.9
246.80	-3.30	-0.5	144.64	20AU	-0.4	-3.4	-2.4	0.9
252.52	-3.11	-0.3	154.45	30AU	-0.1	-3.5	-2.4	0.9
258.63	-2.91	-0.2	164.31	9SE	0.0	-3.5	-2.3	0.9
265.04	-2.70	-0.0	174.24	19SE	0.2	-3.6	-2.3	0.8
271.69	-2.48	0.1	184.23	29SE	0.6	-3.7	-2.2	0.8
278.52	-2.26	0.3	194.26	9OC	2.7	-3.7	-2.1	0.8
285.49	-2.04	0.4	204.35	19OC	0.8	-3.8	-2.1	0.7
292.57	-1.82	0.5	214.48	29OC	-0.3	-3.9	-2.0	0.8
299.70	-1.60	0.7	224.65	8NO	-0.5	-4.0	-1.9	0.8
306.88	-1.39	0.8	234.84	18NO	-0.5	-4.1	-1.9	0.8
314.08	-1.19	0.9	245.04	28NO	-0.6	-4.2	-1.8	0.8
321.27	-0.99	1.0	255.25	8DE	-0.8	-4.3	-1.7	0.8
328.45	-0.81	1.1	265.45	18DE	-0.9	-4.4	-1.7	0.8
335.61	-0.63	1.2	275.64	28DE	-0.9	-4.3	-1.7	0.8

♂ LONG	♂ LAT	♂ MAG	⊙ LONG	16.00UT -18	☿	♀	♃	♄
					MAGNITUDES			
342.73	-0.47	1.3	285.79	7JA	-0.6	-4.1	-1.6	0.8
349.80	-0.31	1.4	295.91	17JA	1.0	-3.5	-1.6	0.8
356.83	-0.17	1.5	305.98	27JA	2.9	-3.4	-1.6	0.7
3.80	-0.03	1.5	316.00	6FE	1.3	-4.0	-1.6	0.7
10.72	0.10	1.6	325.97	16FE	0.7	-4.2	-1.6	0.6
17.59	0.21	1.7	335.87	26FE	0.4	-4.3	-1.6	0.6
24.39	0.32	1.7	345.72	8MR	0.1	-4.2	-1.6	0.5
31.14	0.42	1.8	355.51	18MR	-0.4	-4.1	-1.6	0.5
37.84	0.51	1.8	5.24	28MR	-1.2	-4.0	-1.6	0.4
44.48	0.60	1.9	14.92	7AP	-1.7	-3.9	-1.7	0.3
51.09	0.68	1.9	24.56	17AP	-0.9	-3.8	-1.7	0.3
57.64	0.75	1.9	34.15	27AP	0.1	-3.7	-1.7	0.3
64.16	0.81	2.0	43.72	7MY	1.0	-3.7	-1.8	0.4
70.64	0.87	2.0	53.26	17MY	1.8	-3.6	-1.8	0.4
77.09	0.92	2.0	62.78	27MY	3.1	-3.5	-1.9	0.5
83.52	0.97	2.0	72.30	6JN	2.7	-3.5	-2.0	0.6
89.93	1.02	2.0	81.83	16JN	1.3	-3.4	-2.0	0.6
96.31	1.05	2.0	91.36	26JN	0.0	-3.4	-2.1	0.7
102.69	1.09	1.9	100.92	6JL	-1.1	-3.4	-2.2	0.7
109.07	1.12	2.0	110.51	16JL	-1.6	-3.4	-2.3	0.8
115.43	1.14	2.0	120.13	26JL	-0.9	-3.3	-2.3	0.8
121.80	1.16	2.0	129.80	5AU	-0.4	-3.3	-2.4	0.8
128.18	1.18	2.0	139.53	15AU	-0.0	-3.3	-2.4	0.8
134.56	1.19	2.0	149.30	25AU	0.2	-3.4	-2.4	0.9
140.96	1.19	2.0	159.14	4SE	0.4	-3.4	-2.5	0.8
147.37	1.20	2.0	169.04	14SE	0.9	-3.4	-2.5	0.8
153.79	1.19	2.0	178.99	24SE	3.1	-3.4	-2.4	0.8
160.23	1.19	2.0	189.00	4OC	0.7	-3.4	-2.4	0.8
166.69	1.17	1.9	199.06	14OC	-0.5	-3.4	-2.4	0.7
173.16	1.15	1.9	209.17	24OC	-0.6	-3.5	-2.3	0.7
179.65	1.13	1.9	219.32	3NO	-0.6	-3.5	-2.2	0.7
186.16	1.10	1.8	229.50	13NO	-0.7	-3.5	-2.2	0.7
192.68	1.06	1.8	239.69	23NO	-0.8	-3.5	-2.1	0.8
199.22	1.01	1.7	249.90	3DE	-0.8	-3.4	-2.0	0.8
205.76	0.95	1.6	260.11	13DE	-0.8	-3.4	-1.9	0.8
212.32	0.87	1.5	270.30	23DE	-0.4	-3.4	-1.9	0.8
				-17				
218.87	0.79	1.4	280.47	2JA	1.2	-3.4	-1.8	0.8
225.43	0.69	1.3	290.61	12JA	2.5	-3.3	-1.8	0.8
231.98	0.58	1.2	300.70	22JA	0.9	-3.3	-1.7	0.8
238.52	0.44	1.1	310.75	1FE	0.4	-3.3	-1.7	0.7
245.03	0.29	1.0	320.75	11FE	0.2	-3.3	-1.6	0.7
251.52	0.11	0.8	330.68	21FE	-0.0	-3.4	-1.6	0.6
257.96	-0.09	0.7	340.56	3MR	-0.5	-3.4	-1.6	0.6
264.33	-0.32	0.5	350.38	13MR	-1.2	-3.4	-1.6	0.5
270.63	-0.59	0.3	0.14	23MR	-1.6	-3.4	-1.6	0.5
276.81	-0.89	0.2	9.85	2AP	-0.9	-3.4	-1.6	0.4
282.83	-1.23	-0.0	19.51	12AP	0.2	-3.5	-1.6	0.4
288.65	-1.61	-0.2	29.13	22AP	1.2	-3.5	-1.6	0.3
294.20	-2.04	-0.4	38.71	2MY	2.4	-3.6	-1.6	0.3
299.40	-2.52	-0.7	48.26	12MY	3.6	-3.7	-1.6	0.3
304.14	-3.05	-0.9	57.79	22MY	2.1	-3.7	-1.7	0.4
308.27	-3.64	-1.1	67.31	1JN	1.0	-3.8	-1.7	0.4
311.63	-4.27	-1.4	76.83	11JN	-0.0	-3.9	-1.8	0.5
314.02	-4.94	-1.7	86.36	21JN	-1.1	-4.0	-1.8	0.6
315.23	-5.60	-1.9	95.91	1JL	-1.7	-4.1	-1.9	0.6
315.11	-6.20	-2.2	105.48	11JL	-0.9	-4.2	-1.9	0.7
313.72	-6.64	-2.4	115.08	21JL	-0.3	-4.2	-2.0	0.7
311.33	-6.82	-2.6	124.73	31JL	0.1	-4.2	-2.1	0.8
308.62	-6.68	-2.6	134.43	10AU	0.3	-3.9	-2.1	0.8
306.33	-6.24	-2.4	144.17	20AU	0.6	-3.4	-2.2	0.8
305.05	-5.59	-2.1	153.98	30AU	1.3	-3.4	-2.3	0.8
305.06	-4.85	-1.8	163.84	9SE	3.3	-3.9	-2.3	0.8
306.35	-4.12	-1.5	173.76	19SE	0.5	-4.2	-2.4	0.8
308.76	-3.43	-1.2	183.74	29SE	-0.6	-4.3	-2.4	0.8
312.10	-2.81	-0.9	193.78	9OC	-0.8	-4.3	-2.4	0.8
316.16	-2.26	-0.6	203.86	19OC	-0.8	-4.2	-2.4	0.7
320.80	-1.79	-0.3	213.99	29OC	-0.8	-4.1	-2.4	0.7
325.87	-1.37	-0.1	224.16	8NO	-0.6	-4.0	-2.4	0.7
331.27	-1.01	0.1	234.34	18NO	-0.6	-4.0	-2.3	0.7
336.93	-0.70	0.3	244.55	28NO	-0.6	-3.9	-2.3	0.7
342.78	-0.43	0.5	254.76	8DE	-0.3	-3.8	-2.2	0.8
348.77	-0.19	0.7	264.96	18DE	1.4	-3.7	-2.1	0.8
354.88	0.01	0.9	275.15	28DE	2.1	-3.6	-2.0	0.8

♂ LONG	♂ LAT	♂ MAG	⊙ LONG	16.00UT -16	☿	♀	♃	♄
					MAGNITUDES			
1.06	0.19	1.0	285.30	7JA	0.6	-3.6	-2.0	0.8
7.29	0.35	1.1	295.42	17JA	0.3	-3.5	-1.9	0.8
13.56	0.48	1.3	305.50	27JA	0.1	-3.5	-1.8	0.8
19.86	0.60	1.4	315.52	6FE	-0.1	-3.4	-1.8	0.7
26.16	0.70	1.5	325.48	16FE	-0.5	-3.4	-1.7	0.7
32.48	0.79	1.6	335.40	26FE	-1.2	-3.3	-1.7	0.7
38.79	0.87	1.7	345.24	7MR	-1.6	-3.3	-1.6	0.6
45.10	0.93	1.7	355.04	17MR	-0.9	-3.3	-1.6	0.6
51.40	0.99	1.8	4.77	27MR	0.3	-3.3	-1.5	0.5
57.69	1.04	1.9	14.46	6AP	1.6	-3.3	-1.5	0.4
63.98	1.08	1.9	24.09	16AP	3.2	-3.3	-1.5	0.4
70.26	1.11	2.0	33.69	26AP	2.8	-3.3	-1.5	0.3
76.53	1.14	2.0	43.25	6MY	1.6	-3.4	-1.5	0.3
82.81	1.16	2.0	52.80	16MY	0.8	-3.4	-1.5	0.3
89.09	1.17	2.0	62.32	26MY	-0.1	-3.4	-1.5	0.3
95.37	1.18	2.0	71.84	5JN	-1.1	-3.5	-1.5	0.4
101.66	1.18	2.0	81.36	15JN	-1.7	-3.5	-1.6	0.4
107.97	1.18	2.0	90.90	25JN	-0.9	-3.5	-1.6	0.5
114.29	1.18	2.0	100.45	5JL	-0.2	-3.4	-1.6	0.6
120.64	1.17	2.0	110.04	15JL	0.2	-3.4	-1.7	0.6
127.01	1.16	2.0	119.67	25JL	0.5	-3.4	-1.7	0.7
133.41	1.14	2.0	129.33	4AU	0.8	-3.4	-1.8	0.7
139.84	1.11	1.9	139.05	14AU	1.7	-3.3	-1.8	0.7
146.31	1.09	1.9	148.83	24AU	3.0	-3.3	-1.9	0.8
152.81	1.05	1.9	158.66	3SE	0.4	-3.3	-2.0	0.8
159.37	1.01	1.9	168.55	13SE	-0.7	-3.3	-2.0	0.8
165.96	0.97	1.9	178.50	23SE	-0.9	-3.3	-2.1	0.8
172.61	0.92	1.9	188.51	3OC	-0.9	-3.4	-2.2	0.8
179.31	0.87	1.9	198.57	13OC	-0.7	-3.4	-2.2	0.8
186.05	0.81	1.9	208.68	23OC	-0.5	-3.4	-2.3	0.7
192.85	0.74	1.9	218.82	2NO	-0.5	-3.5	-2.3	0.7
199.70	0.66	1.8	229.00	12NO	-0.5	-3.5	-2.3	0.7
206.61	0.58	1.8	239.20	22NO	-0.1	-3.6	-2.3	0.7
213.56	0.49	1.7	249.40	2DE	1.6	-3.6	-2.3	0.7
220.58	0.39	1.7	259.61	12DE	1.7	-3.7	-2.2	0.7
227.64	0.28	1.6	269.81	22DE	0.4	-3.8	-2.2	0.8
				-15				
234.75	0.16	1.6	279.98	1JA	0.1	-3.8	-2.1	0.8
241.91	0.04	1.5	290.12	11JA	-0.0	-3.9	-2.1	0.8
249.11	-0.10	1.4	300.22	21JA	-0.2	-4.0	-2.0	0.8
256.35	-0.25	1.3	310.27	31JA	-0.6	-4.1	-1.9	0.8
263.63	-0.41	1.3	320.27	10FE	-1.2	-4.2	-1.9	0.8
270.94	-0.57	1.2	330.21	20FE	-1.4	-4.3	-1.8	0.7
278.27	-0.75	1.1	340.08	2MR	-0.8	-4.3	-1.7	0.7
285.62	-0.93	1.0	349.91	12MR	0.4	-4.2	-1.7	0.7
292.96	-1.12	0.9	359.67	22MR	2.0	-3.8	-1.6	0.6
300.29	-1.31	0.8	9.38	1AP	3.7	-3.1	-1.6	0.6
307.59	-1.50	0.7	19.04	11AP	2.1	-3.4	-1.5	0.5
314.84	-1.70	0.6	28.66	21AP	1.2	-3.9	-1.5	0.4
322.02	-1.89	0.5	38.24	1MY	0.6	-4.2	-1.4	0.4
329.10	-2.08	0.4	47.79	11MY	-0.2	-4.2	-1.4	0.3
336.04	-2.26	0.3	57.33	21MY	-1.1	-4.1	-1.4	0.2
342.83	-2.44	0.2	66.85	31MY	-1.8	-4.1	-1.4	0.3
349.39	-2.60	0.1	76.37	10JN	-0.9	-4.0	-1.4	0.3
355.69	-2.75	-0.0	85.90	20JN	-0.2	-3.9	-1.4	0.4
1.67	-2.89	-0.2	95.44	30JN	0.3	-3.8	-1.4	0.4
7.23	-3.01	-0.3	105.01	10JL	0.7	-3.7	-1.4	0.5
12.29	-3.10	-0.5	114.62	20JL	1.1	-3.6	-1.4	0.6
16.71	-3.18	-0.6	124.26	30JL	2.2	-3.6	-1.5	0.6
20.34	-3.22	-0.8	133.96	9AU	2.7	-3.5	-1.5	0.7
22.99	-3.22	-1.0	143.70	19AU	0.3	-3.5	-1.6	0.7
24.45	-3.16	-1.2	153.50	29AU	-0.8	-3.5	-1.6	0.7
24.51	-3.02	-1.5	163.36	8SE	-1.1	-3.4	-1.7	0.8
23.11	-2.76	-1.7	173.28	18SE	-1.0	-3.4	-1.7	0.8
20.39	-2.38	-1.9	183.26	28SE	-0.6	-3.4	-1.8	0.8
16.87	-1.89	-2.0	193.29	8OC	-0.4	-3.4	-1.9	0.8
13.36	-1.33	-1.8	203.38	18OC	-0.3	-3.4	-1.9	0.8
10.60	-0.77	-1.5	213.50	28OC	-0.3	-3.4	-2.0	0.8
9.07	-0.27	-1.2	223.67	7NO	0.0	-3.4	-2.0	0.7
8.91	0.15	-0.9	233.85	17NO	1.8	-3.4	-2.1	0.7
10.00	0.48	-0.5	244.05	27NO	1.4	-3.4	-2.1	0.7
12.14	0.74	-0.2	254.27	7DE	0.1	-3.4	-2.2	0.7
15.13	0.94	0.0	264.47	17DE	-0.1	-3.4	-2.2	0.7
18.78	1.09	0.3	274.65	27DE	-0.2	-3.4	-2.2	0.8

♂ LONG	♂ LAT	♂ MAG	☉ LONG	16.00UT -14	☿	♀	♃	♄
22.95	1.20	0.5	284.81	6JA	-0.3	-3.4	-2.2	0.8
27.52	1.29	0.7	294.93	16JA	-0.6	-3.5	-2.1	0.8
32.40	1.36	0.9	305.00	26JA	-1.2	-3.5	-2.1	0.8
37.53	1.40	1.1	315.03	5FE	-1.3	-3.5	-2.0	0.8
42.85	1.44	1.2	325.00	15FE	-0.8	-3.4	-1.9	0.8
48.32	1.46	1.3	334.91	25FE	0.6	-3.4	-1.9	0.8
53.91	1.47	1.4	344.77	7MR	2.5	-3.4	-1.8	0.8
59.60	1.47	1.6	354.56	17MR	2.8	-3.4	-1.7	0.7
65.37	1.47	1.6	4.30	27MR	1.6	-3.3	-1.7	0.7
71.22	1.46	1.7	13.98	6AP	0.9	-3.3	-1.6	0.6
77.11	1.44	1.8	23.62	16AP	0.4	-3.3	-1.6	0.6
83.07	1.42	1.9	33.22	26AP	-0.2	-3.3	-1.5	0.5
89.07	1.40	1.9	42.79	6MY	-1.1	-3.3	-1.4	0.4
95.11	1.37	1.9	52.33	16MY	-1.8	-3.4	-1.4	0.4
101.20	1.34	2.0	61.85	26MY	-0.9	-3.4	-1.4	0.3
107.34	1.30	2.0	71.38	5JN	-0.1	-3.4	-1.3	0.2
113.52	1.26	2.0	80.90	15JN	0.4	-3.4	-1.3	0.3
119.75	1.22	2.0	90.43	25JN	0.9	-3.5	-1.3	0.3
126.03	1.17	2.0	99.99	5JL	1.5	-3.6	-1.3	0.4
132.35	1.12	2.0	109.57	15JL	2.7	-3.6	-1.3	0.5
138.74	1.07	2.0	119.20	25JL	2.3	-3.7	-1.3	0.5
145.18	1.01	2.0	128.87	4AU	0.3	-3.8	-1.3	0.6
151.68	0.95	1.9	138.58	14AU	-0.9	-3.9	-1.3	0.6
158.25	0.89	1.9	148.36	24AU	-1.2	-4.0	-1.4	0.7
164.88	0.82	1.9	158.19	3SE	-1.0	-4.1	-1.4	0.7
171.58	0.74	1.8	168.07	13SE	-0.6	-4.2	-1.4	0.7
178.36	0.67	1.8	178.02	23SE	-0.3	-4.3	-1.5	0.8
185.20	0.58	1.8	188.03	3OC	-0.2	-4.4	-1.5	0.8
192.12	0.50	1.7	198.08	13OC	-0.1	-4.3	-1.6	0.8
199.12	0.41	1.7	208.19	23OC	0.2	-4.0	-1.6	0.8
206.19	0.31	1.7	218.33	2NO	2.1	-3.4	-1.7	0.8
213.34	0.21	1.7	228.50	12NO	1.2	-3.3	-1.8	0.8
220.56	0.10	1.7	238.70	22NO	-0.0	-4.0	-1.8	0.7
227.85	-0.00	1.7	248.91	2DE	-0.2	-4.3	-1.9	0.7
235.21	-0.12	1.6	259.11	12DE	-0.3	-4.4	-2.0	0.7
242.63	-0.23	1.6	269.31	22DE	-0.4	-4.3	-2.0	0.7

-13

♂ LONG	♂ LAT	♂ MAG	☉ LONG	16.00UT	☿	♀	♃	♄
250.12	-0.35	1.6	279.49	1JA	-0.7	-4.2	-2.0	0.8
257.66	-0.47	1.5	289.63	11JA	-1.1	-4.1	-2.1	0.8
265.25	-0.60	1.5	299.73	21JA	-1.2	-4.0	-2.1	0.8
272.87	-0.72	1.4	309.78	31JA	-0.7	-3.9	-2.1	0.8
280.54	-0.84	1.4	319.78	10FE	0.7	-3.8	-2.0	0.8
288.22	-0.95	1.4	329.73	20FE	3.0	-3.7	-2.0	0.8
295.91	-1.07	1.3	339.61	2MR	2.1	-3.6	-2.0	0.8
303.61	-1.17	1.3	349.43	12MR	1.1	-3.6	-1.9	0.8
311.29	-1.27	1.2	359.20	22MR	0.7	-3.5	-1.8	0.8
318.95	-1.36	1.2	8.92	1AP	0.3	-3.4	-1.8	0.7
326.56	-1.44	1.2	18.58	11AP	-0.3	-3.4	-1.7	0.7
334.12	-1.50	1.1	28.20	21AP	-1.1	-3.4	-1.6	0.6
341.61	-1.55	1.1	37.78	1MY	-1.8	-3.3	-1.6	0.6
349.01	-1.59	1.0	47.33	11MY	-0.9	-3.3	-1.5	0.5
356.31	-1.61	1.0	56.87	21MY	-0.0	-3.3	-1.5	0.4
3.48	-1.62	0.9	66.39	31MY	0.6	-3.3	-1.4	0.4
10.53	-1.60	0.9	75.91	10JN	1.2	-3.3	-1.4	0.3
17.41	-1.58	0.8	85.44	20JN	2.0	-3.3	-1.3	0.3
24.11	-1.53	0.8	94.98	30JN	3.2	-3.3	-1.3	0.3
30.60	-1.47	0.7	104.55	10JL	1.9	-3.4	-1.3	0.4
36.86	-1.38	0.6	114.15	20JL	0.2	-3.4	-1.2	0.4
42.84	-1.28	0.6	123.79	30JL	-1.0	-3.4	-1.2	0.5
48.50	-1.16	0.5	133.48	9AU	-1.4	-3.5	-1.2	0.6
53.77	-1.01	0.3	143.23	19AU	-1.0	-3.5	-1.2	0.6
58.58	-0.83	0.2	153.02	29AU	-0.5	-3.5	-1.2	0.7
62.84	-0.63	0.1	162.88	8SE	-0.2	-3.5	-1.3	0.7
66.39	-0.39	-0.1	172.80	18SE	-0.0	-3.4	-1.3	0.7
69.10	-0.10	-0.3	182.77	28SE	0.1	-3.4	-1.3	0.8
70.77	0.23	-0.5	192.80	8OC	0.5	-3.4	-1.3	0.8
71.19	0.62	-0.7	202.89	18OC	2.4	-3.4	-1.4	0.8
70.22	1.05	-0.9	213.01	28OC	1.0	-3.3	-1.4	0.8
67.86	1.51	-1.1	223.17	7NO	-0.2	-3.3	-1.5	0.8
64.39	1.95	-1.3	233.36	17NO	-0.4	-3.3	-1.6	0.8
60.45	2.32	-1.3	243.56	27NO	-0.4	-3.3	-1.6	0.8
56.82	2.57	-1.1	253.76	7DE	-0.5	-3.4	-1.7	0.8
54.15	2.70	-0.8	263.97	17DE	-0.8	-3.4	-1.7	0.7
52.78	2.74	-0.5	274.15	27DE	-1.0	-3.4	-1.8	0.8

♂ LONG	♂ LAT	♂ MAG	☉ LONG	16.00UT -12	☿	♀	♃	♄
52.73	2.71	-0.2	284.31	6JA	-1.0	-3.4	-1.9	0.8
53.86	2.64	0.0	294.44	16JA	-0.7	-3.5	-1.9	0.8
55.97	2.55	0.3	304.52	26JA	0.9	-3.5	-2.0	0.9
58.87	2.45	0.5	314.55	5FE	3.1	-3.5	-2.0	0.9
62.39	2.35	0.7	324.52	15FE	1.6	-3.6	-2.0	0.9
66.41	2.25	0.9	334.43	25FE	0.8	-3.7	-2.0	0.9
70.81	2.15	1.1	344.29	6MR	0.5	-3.7	-2.0	0.9
75.54	2.05	1.2	354.09	16MR	0.2	-3.8	-2.0	0.9
80.51	1.95	1.3	3.83	26MR	-0.4	-3.9	-1.9	0.9
85.69	1.86	1.4	13.52	5AP	-1.1	-4.0	-1.9	0.8
91.05	1.77	1.5	23.16	15AP	-1.8	-4.1	-1.8	0.8
96.55	1.68	1.6	32.76	25AP	-0.9	-4.1	-1.8	0.7
102.18	1.59	1.7	42.33	5MY	0.0	-4.1	-1.7	0.7
107.94	1.50	1.7	51.87	15MY	0.8	-4.1	-1.6	0.6
113.79	1.41	1.8	61.40	25MY	1.5	-3.9	-1.6	0.5
119.75	1.32	1.8	70.92	4JN	2.6	-3.4	-1.5	0.5
125.80	1.24	1.9	80.44	14JN	3.1	-2.9	-1.5	0.4
131.93	1.15	1.9	89.97	24JN	1.5	-3.7	-1.4	0.3
138.16	1.06	1.9	99.53	4JL	0.1	-4.1	-1.4	0.3
144.48	0.96	1.9	109.11	14JL	-1.0	-4.2	-1.3	0.4
150.88	0.87	1.9	118.73	24JL	-1.5	-4.2	-1.3	0.4
157.38	0.78	1.9	128.39	3AU	-1.0	-4.2	-1.3	0.5
163.96	0.68	1.9	138.11	13AU	-0.4	-4.1	-1.2	0.5
170.64	0.58	1.8	147.88	23AU	-0.1	-4.0	-1.2	0.6
177.41	0.48	1.8	157.71	2SE	0.1	-3.9	-1.2	0.7
184.27	0.38	1.8	167.59	12SE	0.2	-3.8	-1.2	0.7
191.22	0.28	1.7	177.53	22SE	0.7	-3.7	-1.2	0.7
198.27	0.17	1.7	187.54	2OC	2.8	-3.7	-1.2	0.8
205.41	0.07	1.6	197.59	12OC	0.8	-3.6	-1.2	0.8
212.64	-0.04	1.6	207.69	22OC	-0.4	-3.6	-1.3	0.8
219.96	-0.14	1.5	217.84	1NO	-0.5	-3.5	-1.3	0.8
227.36	-0.25	1.5	228.01	11NO	-0.5	-3.5	-1.4	0.9
234.84	-0.36	1.5	238.21	21NO	-0.6	-3.4	-1.4	0.9
242.40	-0.46	1.5	248.42	1DE	-0.8	-3.4	-1.5	0.9
250.02	-0.56	1.5	258.62	11DE	-0.8	-3.4	-1.5	0.8
257.71	-0.66	1.5	268.82	21DE	-0.9	-3.4	-1.6	0.8
265.45	-0.75	1.5	279.00	31DE	-0.5	-3.3	-1.6	0.8

-11

♂ LONG	♂ LAT	♂ MAG	☉ LONG	16.00UT	☿	♀	♃	♄
273.23	-0.84	1.4	289.14	10JA	1.0	-3.3	-1.7	0.8
281.03	-0.92	1.4	299.24	20JA	2.8	-3.3	-1.8	0.9
288.86	-0.99	1.4	309.30	30JA	1.1	-3.3	-1.8	0.9
296.70	-1.06	1.4	319.30	9FE	0.6	-3.3	-1.9	0.9
304.53	-1.11	1.4	329.24	19FE	0.3	-3.4	-1.9	1.0
312.35	-1.15	1.4	339.13	1MR	0.0	-3.4	-2.0	1.0
320.13	-1.18	1.4	348.96	11MR	-0.4	-3.4	-2.0	1.0
327.87	-1.20	1.4	358.73	21MR	-1.1	-3.4	-2.0	1.0
335.56	-1.21	1.4	8.45	31MR	-1.7	-3.5	-2.0	1.0
343.19	-1.20	1.4	18.11	10AP	-0.9	-3.5	-2.0	0.9
350.74	-1.18	1.4	27.73	20AP	0.1	-3.4	-2.0	0.9
358.21	-1.15	1.4	37.32	30AP	1.0	-3.4	-1.9	0.8
5.58	-1.10	1.4	46.87	10MY	2.0	-3.4	-1.9	0.8
12.86	-1.04	1.4	56.40	20MY	3.4	-3.4	-1.8	0.7
20.03	-0.97	1.4	65.93	30MY	2.5	-3.3	-1.8	0.7
27.07	-0.89	1.4	75.44	9JN	1.2	-3.3	-1.7	0.6
34.00	-0.80	1.3	84.97	19JN	0.0	-3.3	-1.6	0.5
40.80	-0.69	1.3	94.51	29JN	-1.1	-3.3	-1.6	0.5
47.45	-0.58	1.3	104.08	9JL	-1.6	-3.3	-1.5	0.4
53.95	-0.45	1.3	113.68	19JL	-1.0	-3.3	-1.5	0.4
60.28	-0.31	1.2	123.32	29JL	-0.3	-3.4	-1.4	0.5
66.42	-0.17	1.2	133.01	8AU	0.0	-3.4	-1.4	0.5
72.36	-0.00	1.1	142.75	18AU	0.2	-3.4	-1.3	0.6
78.06	0.17	1.0	152.54	28AU	0.4	-3.5	-1.3	0.6
83.49	0.37	0.9	162.40	7SE	1.0	-3.5	-1.3	0.7
88.60	0.58	0.8	172.31	17SE	3.1	-3.6	-1.3	0.7
93.33	0.82	0.7	182.29	27SE	0.7	-3.7	-1.2	0.8
97.59	1.08	0.6	192.31	7OC	-0.5	-3.7	-1.2	0.8
101.29	1.38	0.4	202.39	17OC	-0.7	-3.8	-1.2	0.8
104.28	1.71	0.2	212.52	27OC	-0.7	-3.9	-1.3	0.9
106.42	2.08	0.0	222.67	6NO	-0.7	-4.0	-1.3	0.9
107.50	2.49	-0.2	232.86	16NO	-0.7	-4.1	-1.3	0.9
107.34	2.93	-0.4	243.07	26NO	-0.7	-4.3	-1.3	0.9
105.84	3.38	-0.7	253.27	6DE	-0.7	-4.3	-1.4	0.9
103.05	3.78	-0.9	263.48	16DE	-0.4	-4.4	-1.4	0.9
99.34	4.07	-1.0	273.66	26DE	1.2	-4.3	-1.5	0.9

♂			☉	16.00UT	☿	♀	♃	♄
LONG	LAT	MAG	LONG	-10	MAGNITUDES			
95.38	4.20	-1.0	283.82	5JA	2.4	-4.0	-1.5	0.9
91.88	4.17	-0.8	293.95	15JA	0.8	-3.5	-1.6	0.9
89.42	4.00	-0.5	304.03	25JA	0.4	-3.4	-1.7	0.9
88.25	3.76	-0.3	314.06	4FE	0.2	-4.0	-1.7	1.0
88.35	3.49	-0.0	324.04	14FE	-0.1	-4.2	-1.8	1.0
89.59	3.21	0.2	333.96	24FE	-0.5	-4.3	-1.9	1.1
91.77	2.94	0.4	343.81	6MR	-1.1	-4.2	-1.9	1.1
94.70	2.69	0.6	353.61	16MR	-1.6	-4.1	-2.0	1.1
98.26	2.46	0.8	3.36	26MR	-0.9	-4.0	-2.0	1.1
102.31	2.25	1.0	13.05	5AP	0.2	-3.9	-2.1	1.1
106.76	2.05	1.1	22.69	15AP	1.3	-3.8	-2.1	1.1
111.55	1.86	1.2	32.30	25AP	2.7	-3.7	-2.1	1.0
116.60	1.69	1.3	41.86	5MY	3.4	-3.7	-2.1	1.0
121.90	1.52	1.4	51.41	15MY	1.9	-3.6	-2.1	0.9
127.41	1.37	1.5	60.94	25MY	0.9	-3.5	-2.0	0.9
133.09	1.22	1.5	70.46	4JN	-0.0	-3.5	-2.0	0.8
138.94	1.07	1.6	79.98	14JN	-1.1	-3.4	-1.9	0.8
144.95	0.93	1.6	89.51	24JN	-1.7	-3.4	-1.9	0.7
151.09	0.80	1.6	99.06	4JL	-0.9	-3.4	-1.8	0.6
157.37	0.66	1.7	108.65	14JL	-0.3	-3.4	-1.8	0.6
163.78	0.53	1.7	118.27	24JL	0.1	-3.3	-1.7	0.5
170.32	0.41	1.7	127.93	3AU	0.4	-3.3	-1.6	0.5
176.98	0.28	1.7	137.64	13AU	0.7	-3.3	-1.6	0.6
183.77	0.16	1.7	147.40	23AU	1.4	-3.4	-1.5	0.6
190.67	0.04	1.7	157.23	2SE	3.2	-3.4	-1.5	0.7
197.68	-0.08	1.6	167.11	12SE	0.6	-3.4	-1.4	0.7
204.81	-0.19	1.6	177.05	22SE	-0.7	-3.4	-1.4	0.8
212.05	-0.30	1.6	187.05	2OC	-0.9	-3.4	-1.4	0.8
219.39	-0.41	1.6	197.10	12OC	-0.8	-3.5	-1.3	0.9
226.83	-0.51	1.5	207.20	22OC	-0.8	-3.5	-1.3	0.9
234.36	-0.60	1.5	217.34	1NO	-0.6	-3.5	-1.3	0.9
241.97	-0.69	1.5	227.52	11NO	-0.5	-3.5	-1.3	1.0
249.66	-0.78	1.4	237.71	21NO	-0.5	-3.5	-1.3	1.0
257.42	-0.85	1.4	247.92	1DE	-0.3	-3.4	-1.3	1.0
265.22	-0.92	1.4	258.13	11DE	1.4	-3.4	-1.4	1.0
273.07	-0.97	1.3	268.32	21DE	2.0	-3.4	-1.4	1.0
280.95	-1.02	1.3	278.50	31DE	0.5	-3.4	-1.4	1.0
				-9				
288.84	-1.05	1.3	288.65	10JA	0.2	-3.3	-1.5	1.0
296.74	-1.08	1.3	298.75	20JA	0.1	-3.3	-1.5	1.0
304.62	-1.09	1.3	308.81	30JA	-0.1	-3.3	-1.6	1.0
312.48	-1.09	1.3	318.82	9FE	-0.5	-3.3	-1.6	1.1
320.31	-1.08	1.4	328.76	19FE	-1.1	-3.4	-1.7	1.1
328.09	-1.05	1.4	338.65	1MR	-1.5	-3.4	-1.8	1.2
335.81	-1.02	1.4	348.49	11MR	-0.9	-3.4	-1.8	1.2
343.46	-0.98	1.5	358.26	21MR	0.3	-3.4	-1.9	1.2
351.05	-0.92	1.5	7.98	31MR	1.7	-3.4	-2.0	1.2
358.55	-0.86	1.5	17.65	10AP	3.5	-3.5	-2.0	1.2
5.96	-0.78	1.5	27.27	20AP	2.5	-3.5	-2.1	1.2
13.29	-0.70	1.6	36.85	30AP	1.5	-3.6	-2.1	1.2
20.52	-0.61	1.6	46.41	10MY	0.7	-3.7	-2.2	1.2
27.65	-0.52	1.6	55.95	20MY	-0.1	-3.7	-2.2	1.1
34.69	-0.42	1.6	65.47	30MY	-1.1	-3.8	-2.2	1.1
41.62	-0.31	1.6	74.99	9JN	-1.8	-3.9	-2.2	1.0
48.46	-0.20	1.6	84.51	19JN	-0.9	-4.0	-2.2	1.0
55.19	-0.09	1.6	94.06	29JN	-0.2	-4.1	-2.1	0.9
61.82	0.03	1.6	103.62	9JL	0.2	-4.2	-2.1	0.8
68.34	0.16	1.6	113.22	19JL	0.5	-4.2	-2.0	0.8
74.75	0.29	1.6	122.86	29JL	0.9	-4.2	-2.0	0.7
81.03	0.42	1.6	132.54	8AU	1.8	-3.9	-1.9	0.6
87.20	0.56	1.6	142.28	18AU	3.0	-3.4	-1.9	0.7
93.22	0.71	1.5	152.07	28AU	0.5	-3.4	-1.8	0.7
99.10	0.87	1.5	161.92	7SE	-0.8	-4.0	-1.7	0.8
104.80	1.03	1.4	171.83	17SE	-1.0	-4.2	-1.7	0.8
110.31	1.21	1.3	181.80	27SE	-1.0	-4.3	-1.6	0.9
115.58	1.39	1.2	191.83	7OC	-0.7	-4.3	-1.6	0.9
120.59	1.60	1.1	201.90	17OC	-0.5	-4.2	-1.5	1.0
125.28	1.82	1.0	212.02	27OC	-0.4	-4.1	-1.5	1.0
129.56	2.06	0.9	222.18	6NO	-0.4	-4.0	-1.5	1.0
133.37	2.33	0.7	232.36	16NO	-0.1	-3.9	-1.4	1.1
136.57	2.63	0.5	242.57	26NO	1.6	-3.9	-1.4	1.1
139.02	2.95	0.3	252.78	6DE	1.7	-3.8	-1.4	1.1
140.55	3.31	0.0	262.98	16DE	0.3	-3.7	-1.4	1.2
140.98	3.68	-0.2	273.17	26DE	0.0	-3.6	-1.4	1.2

♂			☉	16.00UT	☿	♀	♃	♄
LONG	LAT	MAG	LONG	-8	MAGNITUDES			
140.16	4.04	-0.5	283.33	5JA	-0.1	-3.6	-1.4	1.2
138.03	4.35	-0.7	293.46	15JA	-0.2	-3.5	-1.5	1.2
134.79	4.55	-0.9	303.55	25JA	-0.6	-3.5	-1.5	1.2
130.93	4.59	-1.0	313.58	4FE	-1.2	-3.4	-1.5	1.2
127.14	4.45	-0.9	323.56	14FE	-1.4	-3.4	-1.6	1.2
124.10	4.16	-0.7	333.48	24FE	-0.9	-3.3	-1.6	1.3
122.24	3.78	-0.5	343.34	5MR	0.4	-3.3	-1.7	1.3
121.68	3.37	-0.2	353.15	15MR	2.2	-3.3	-1.8	1.4
122.34	2.97	-0.0	2.89	25MR	3.4	-3.3	-1.8	1.4
124.07	2.59	0.2	12.59	4AP	1.9	-3.3	-1.9	1.4
126.69	2.24	0.4	22.23	14AP	1.1	-3.3	-2.0	1.4
130.03	1.92	0.5	31.84	24AP	0.5	-3.3	-2.0	1.4
133.96	1.63	0.7	41.41	4MY	-0.2	-3.4	-2.1	1.4
138.37	1.37	0.8	50.95	14MY	-1.1	-3.4	-2.2	1.4
143.19	1.13	0.9	60.48	24MY	-1.8	-3.4	-2.2	1.3
148.36	0.91	1.0	70.00	3JN	-0.9	-3.5	-2.3	1.3
153.82	0.70	1.1	79.52	13JN	-0.2	-3.5	-2.3	1.2
159.54	0.51	1.1	89.05	23JN	0.4	-3.5	-2.3	1.2
165.49	0.33	1.2	98.60	3JL	0.7	-3.4	-2.3	1.1
171.65	0.16	1.3	108.18	13JL	1.2	-3.4	-2.3	1.1
178.00	0.01	1.3	117.80	23JL	2.3	-3.4	-2.3	1.0
184.53	-0.14	1.3	127.46	2AU	2.6	-3.4	-2.3	0.9
191.23	-0.28	1.3	137.16	12AU	0.4	-3.3	-2.2	0.8
198.08	-0.41	1.4	146.93	22AU	-0.9	-3.3	-2.2	0.8
205.08	-0.53	1.4	156.75	1SE	-1.2	-3.3	-2.1	0.9
212.22	-0.64	1.4	166.62	11SE	-1.1	-3.3	-2.0	0.9
219.49	-0.75	1.4	176.57	21SE	-0.6	-3.3	-2.0	0.9
226.88	-0.84	1.4	186.56	1OC	-0.4	-3.4	-1.9	1.0
234.37	-0.92	1.4	196.61	11OC	-0.3	-3.4	-1.8	1.0
241.96	-0.99	1.4	206.71	21OC	-0.2	-3.4	-1.8	1.1
249.64	-1.04	1.4	216.85	31OC	0.1	-3.5	-1.7	1.1
257.39	-1.09	1.4	227.02	10NO	1.9	-3.5	-1.7	1.2
265.19	-1.12	1.4	237.21	20NO	1.4	-3.6	-1.6	1.2
273.04	-1.14	1.4	247.42	30NO	0.1	-3.6	-1.6	1.3
280.92	-1.15	1.4	257.63	10DE	-0.1	-3.7	-1.6	1.3
288.82	-1.14	1.4	267.83	20DE	-0.2	-3.8	-1.5	1.3
296.71	-1.13	1.4	278.00	30DE	-0.3	-3.8	-1.5	1.4
				-7				
304.58	-1.10	1.4	288.15	9JA	-0.6	-3.9	-1.5	1.4
312.43	-1.06	1.4	298.26	19JA	-1.1	-4.0	-1.5	1.4
320.24	-1.01	1.4	308.32	29JA	-1.3	-4.1	-1.5	1.3
328.00	-0.95	1.4	318.33	8FE	-0.8	-4.2	-1.5	1.3
335.69	-0.88	1.4	328.29	18FE	0.6	-4.3	-1.6	1.2
343.32	-0.81	1.4	338.18	28FE	2.6	-4.3	-1.6	1.2
350.87	-0.72	1.4	348.01	10MR	2.6	-4.2	-1.6	1.2
358.34	-0.64	1.5	357.79	20MR	1.4	-3.8	-1.7	1.2
5.72	-0.54	1.5	7.51	30MR	0.8	-3.1	-1.7	1.2
13.02	-0.45	1.5	17.18	9AP	0.4	-3.4	-1.8	1.2
20.23	-0.35	1.6	26.81	19AP	-0.2	-3.9	-1.8	1.2
27.35	-0.25	1.6	36.39	29AP	-1.1	-4.2	-1.9	1.2
34.39	-0.15	1.7	45.95	9MY	-1.8	-4.2	-2.0	1.1
41.34	-0.04	1.7	55.49	19MY	-0.9	-4.1	-2.0	1.1
48.21	0.06	1.8	65.01	29MY	-0.1	-4.1	-2.1	1.1
55.00	0.17	1.8	74.53	8JN	0.5	-4.0	-2.2	1.0
61.70	0.27	1.8	84.05	18JN	1.0	-3.9	-2.2	1.0
68.33	0.38	1.8	93.59	28JN	1.6	-3.8	-2.3	0.9
74.89	0.48	1.9	103.16	8JL	2.9	-3.7	-2.4	0.9
81.36	0.59	1.9	112.75	18JL	2.2	-3.6	-2.4	0.8
87.77	0.69	1.9	122.39	28JL	0.3	-3.6	-2.4	0.8
94.10	0.80	1.9	132.07	7AU	-0.9	-3.5	-2.4	0.7
100.36	0.91	1.8	141.81	17AU	-1.3	-3.5	-2.4	0.7
106.53	1.02	1.8	151.59	27AU	-1.0	-3.5	-2.4	0.6
112.63	1.13	1.8	161.44	6SE	-0.5	-3.4	-2.4	0.7
118.63	1.24	1.8	171.35	16SE	-0.2	-3.4	-2.3	0.7
124.53	1.36	1.7	181.32	26SE	-0.1	-3.4	-2.3	0.8
130.32	1.48	1.6	191.34	6OC	-0.0	-3.4	-2.2	0.9
135.98	1.60	1.6	201.41	16OC	0.3	-3.4	-2.1	0.9
141.49	1.73	1.5	211.53	26OC	2.1	-3.4	-2.1	1.0
146.82	1.87	1.4	221.69	5NO	1.2	-3.4	-2.0	1.0
151.94	2.02	1.2	231.87	15NO	-0.1	-3.4	-1.9	1.1
156.79	2.17	1.1	242.07	25NO	-0.3	-3.4	-1.9	1.1
161.33	2.34	1.0	252.28	5DE	-0.3	-3.4	-1.8	1.2
165.48	2.52	0.8	262.48	15DE	-0.4	-3.4	-1.7	1.2
169.15	2.71	0.6	272.67	25DE	-0.7	-3.4	-1.7	1.2

Left half (–6):

♂ LONG	♂ LAT	♂ MAG	☉ LONG	16.00UT	☿	♀	♃	♄
							MAGNITUDES	
172.21	2.92	0.4	282.84	4JA	-1.1	-3.5	-1./	1.2
174.52	3.13	0.1	292.96	14JA	-1.1	-3.5	-1.6	1.1
175.91	3.34	-0.2	303.05	24JA	-0.7	-3.5	-1.6	1.1
176.19	3.53	-0.4	313.09	3FE	0.7	-3.5	-1.6	1.1
175.22	3.68	-0.7	323.07	13FE	3.0	-3.4	-1.6	1.0
173.00	3.74	-1.0	332.99	23FE	1.9	-3.4	-1.6	1.0
169.73	3.67	-1.2	342.86	5MR	1.0	-3.4	-1.6	0.9
165.97	3.45	-1.3	352.67	15MR	0.6	-3.4	-1.6	0.9
162.41	3.08	-1.2	2.41	25MR	0.2	-3.3	-1.6	0.9
159.71	2.63	-1.0	12.11	4AP	-0.3	-3.3	-1.6	0.9
158.25	2.14	-0.8	21.76	14AP	-1.1	-3.3	-1.7	0.9
158.12	1.67	-0.6	31.37	24AP	-1.8	-3.3	-1.7	0.9
159.23	1.24	-0.4	40.94	4MY	-0.9	-3.3	-1.8	0.9
161.42	0.85	-0.2	50.49	14MY	-0.0	-3.4	-1.8	0.9
164.50	0.50	-0.0	60.01	24MY	0.7	-3.4	-1.9	0.9
168.32	0.20	0.1	69.54	3JN	1.3	-3.4	-1.9	0.8
172.75	-0.07	0.3	79.06	13JN	2.2	-3.5	-2.0	0.8
177.69	-0.30	0.4	88.59	23JN	3.3	-3.5	-2.1	0.8
183.06	-0.51	0.5	98.14	3JL	1.8	-3.6	-2.1	0.7
188.80	-0.69	0.6	107.72	13JL	0.2	-3.6	-2.2	0.7
194.84	-0.85	0.7	117.33	23JL	-1.0	-3.7	-2.3	0.6
201.17	-0.99	0.7	126.99	2AU	-1.4	-3.8	-2.3	0.5
207.74	-1.11	0.8	136.69	12AU	-1.0	-3.9	-2.4	0.5
214.52	-1.22	0.9	146.45	22AU	-0.5	-4.0	-2.4	0.4
221.50	-1.30	0.9	156.27	1SE	-0.1	-4.1	-2.4	0.4
228.65	-1.36	1.0	166.15	11SE	0.0	-4.2	-2.5	0.4
235.94	-1.41	1.0	176.08	21SE	0.1	-4.3	-2.5	0.4
243.36	-1.44	1.0	186.08	1OC	0.6	-4.4	-2.4	0.4
250.89	-1.45	1.1	196.12	11OC	2.5	-4.3	-2.4	0.5
258.51	-1.45	1.1	206.22	21OC	1.0	-4.0	-2.3	0.6
266.20	-1.43	1.2	216.36	31OC	-0.3	-3.3	-2.3	0.7
273.94	-1.40	1.2	226.52	10NO	-0.5	-3.3	-2.2	0.7
281.71	-1.35	1.2	236.72	20NO	-0.5	-4.0	-2.1	0.8
289.51	-1.29	1.3	246.92	30NO	-0.6	-4.3	-2.1	0.8
297.30	-1.23	1.3	257.13	10DE	-0.8	-4.4	-2.0	0.9
305.08	-1.15	1.4	267.33	20DE	-0.9	-4.3	-1.9	0.9
312.83	-1.06	1.4	277.51	30DE	-1.0	-4.2	-1.9	0.9

–5

♂ LONG	♂ LAT	♂ MAG	☉ LONG	16.00UT	☿	♀	♃	♄
320.54	-0.96	1.4	287.66	9JA	-0.6	-4.1	-1.8	0.9
328.19	-0.87	1.5	297.77	19JA	0.8	-4.0	-1.7	0.9
335.79	-0.76	1.5	307.83	29JA	3.1	-3.9	-1.7	0.9
343.31	-0.66	1.5	317.84	8FE	1.4	-3.8	-1.7	0.8
350.77	-0.55	1.6	327.80	18FE	0.7	-3.7	-1.6	0.8
358.14	-0.44	1.6	337.70	28FE	0.4	-3.6	-1.6	0.8
5.43	-0.33	1.6	347.53	10MR	0.1	-3.6	-1.6	0.7
12.65	-0.22	1.7	357.31	20MR	-0.4	-3.5	-1.6	0.7
19.78	-0.12	1.7	7.04	30MR	-1.1	-3.4	-1.6	0.7
26.83	-0.01	1.7	16.71	9AP	-1.8	-3.4	-1.6	0.7
33.80	0.09	1.7	26.34	19AP	-0.9	-3.4	-1.6	0.7
40.69	0.19	1.7	35.93	29AP	0.1	-3.3	-1.6	0.7
47.52	0.29	1.7	45.49	9MY	0.9	-3.3	-1.6	0.7
54.27	0.38	1.8	55.03	19MY	1.7	-3.3	-1.6	0.7
60.96	0.47	1.8	64.55	29MY	2.9	-3.3	-1.7	0.7
67.59	0.56	1.8	74.07	8JN	2.9	-3.3	-1.7	0.6
74.17	0.65	1.9	83.59	18JN	1.4	-3.3	-1.8	0.6
80.69	0.73	1.9	93.13	28JN	0.1	-3.3	-1.8	0.6
87.16	0.81	1.9	102.69	8JL	-1.0	-3.4	-1.9	0.5
93.59	0.88	2.0	112.29	18JL	-1.5	-3.4	-1.9	0.5
99.98	0.96	2.0	121.92	28JL	-1.0	-3.4	-2.0	0.4
106.32	1.03	2.0	131.60	7AU	-0.4	-3.5	-2.1	0.4
112.63	1.10	2.0	141.33	17AU	-0.0	-3.5	-2.1	0.3
118.91	1.17	2.0	151.12	27AU	0.1	-3.5	-2.2	0.3
125.14	1.23	2.0	160.96	6SE	0.3	-3.5	-2.3	0.2
131.34	1.30	1.9	170.87	16SE	0.8	-3.4	-2.3	0.2
137.49	1.36	1.9	180.83	26SE	2.8	-3.4	-2.4	0.1
143.60	1.42	1.9	190.85	6OC	0.8	-3.4	-2.4	0.1
149.66	1.48	1.8	200.92	16OC	-0.4	-3.3	-2.4	0.2
155.66	1.54	1.8	211.03	26OC	-0.6	-3.3	-2.4	0.3
161.60	1.60	1.7	221.19	5NO	-0.6	-3.3	-2.4	0.4
167.46	1.65	1.6	231.37	15NO	-0.7	-3.3	-2.3	0.4
173.22	1.71	1.5	241.57	25NO	-0.8	-3.3	-2.3	0.5
178.87	1.76	1.4	251.78	5DE	-0.8	-3.4	-2.2	0.5
184.39	1.81	1.3	261.99	15DE	-0.8	-3.4	-2.2	0.6
189.73	1.85	1.1	272.17	25DE	-0.5	-3.4	-2.1	0.6

Right half (–4):

♂ LONG	♂ LAT	♂ MAG	☉ LONG	16.00UT	☿	♀	♃	♄
							MAGNITUDES	
194.88	1.89	0.9	282.34	4JA	1.0	-3.4	-2.0	0.7
199.78	1.92	0.8	292.47	14JA	2.7	-3.5	-1.9	0.7
204.38	1.94	0.6	302.56	24JA	1.0	-3.5	-1.9	0.7
208.59	1.94	0.4	312.60	3FE	0.5	-3.6	-1.8	0.7
212.32	1.93	0.1	322.59	13FE	0.3	-3.6	-1.7	0.6
215.44	1.88	-0.1	332.51	23FE	0.0	-3.7	-1.7	0.6
217.82	1.80	-0.4	342.38	4MR	-0.4	-3.7	-1.6	0.6
219.25	1.65	-0.7	352.19	14MR	-1.1	-3.8	-1.6	0.6
219.58	1.44	-1.1	1.94	24MR	-1.7	-3.9	-1.6	0.5
218.67	1.13	-1.4	11.64	3AP	-0.9	-4.0	-1.5	0.5
216.54	0.73	-1.7	21.30	13AP	0.1	-4.1	-1.5	0.5
213.49	0.23	-1.9	30.90	23AP	1.1	-4.1	-1.5	0.5
210.10	-0.31	-1.9	40.48	3MY	2.2	-4.2	-1.5	0.5
207.11	-0.84	-1.8	50.03	13MY	3.6	-4.1	-1.5	0.5
205.13	-1.31	-1.6	59.56	23MY	2.3	-3.9	-1.5	0.5
204.48	-1.69	-1.4	69.08	2JN	1.1	-3.3	-1.5	0.5
205.20	-1.99	-1.2	78.60	12JN	0.0	-2.9	-1.5	0.5
207.18	-2.21	-0.9	88.13	22JN	-1.1	-3.7	-1.5	0.5
210.22	-2.37	-0.8	97.68	2JL	-1.6	-4.1	-1.6	0.5
214.16	-2.47	-0.6	107.25	12JL	-1.0	-4.2	-1.6	0.4
218.81	-2.53	-0.4	116.86	22JL	-0.3	-4.2	-1.7	0.4
224.06	-2.55	-0.3	126.52	1AU	0.1	-4.1	-1.7	0.3
229.79	-2.54	0.0	136.22	11AU	0.3	-4.1	-1.8	0.3
235.91	-2.50	0.0	145.98	21AU	0.5	-4.0	-1.8	0.2
242.36	-2.44	0.1	155.79	31AU	1.1	-3.9	-1.9	0.1
249.07	-2.36	0.3	165.67	10SE	3.2	-3.8	-2.0	0.1
255.99	-2.26	0.4	175.60	20SE	0.7	-3.7	-2.0	0.0
263.08	-2.14	0.5	185.59	30SE	-0.6	-3.7	-2.1	-0.0
270.30	-2.01	0.6	195.63	10OC	-0.8	-3.6	-2.1	-0.1
277.62	-1.87	0.7	205.72	20OC	-0.7	-3.6	-2.2	-0.0
285.01	-1.72	0.8	215.86	30OC	-0.8	-3.5	-2.2	0.0
292.45	-1.57	0.9	226.03	9NO	-0.7	-3.5	-2.3	0.1
299.91	-1.41	1.0	236.22	19NO	-0.6	-3.4	-2.3	0.2
307.37	-1.25	1.0	246.43	29NO	-0.6	-3.4	-2.3	0.2
314.83	-1.09	1.1	256.64	9DE	-0.4	-3.4	-2.3	0.3
322.25	-0.94	1.2	266.84	19DE	1.2	-3.4	-2.2	0.4
329.63	-0.78	1.3	277.02	29DE	2.3	-3.3	-2.2	0.4

–3

♂ LONG	♂ LAT	♂ MAG	☉ LONG	16.00UT	☿	♀	♃	♄
336.97	-0.64	1.4	287.17	8JA	0.7	-3.3	-2.1	0.5
344.26	-0.49	1.4	297.28	18JA	0.3	-3.3	-2.0	0.5
351.48	-0.35	1.5	307.35	28JA	0.1	-3.3	-2.0	0.5
358.64	-0.22	1.6	317.36	7FE	-0.1	-3.3	-1.9	0.5
5.73	-0.10	1.6	327.32	17FE	-0.5	-3.4	-1.8	0.5
12.75	0.02	1.7	337.22	27FE	-1.1	-3.4	-1.8	0.5
19.71	0.14	1.7	347.06	9MR	-1.6	-3.4	-1.7	0.5
26.60	0.24	1.8	356.84	19MR	-0.9	-3.4	-1.6	0.4
33.42	0.34	1.8	6.57	29MR	0.2	-3.5	-1.6	0.4
40.19	0.43	1.8	16.24	8AP	1.5	-3.5	-1.5	0.3
46.89	0.52	1.9	25.87	18AP	3.0	-3.4	-1.5	0.3
53.55	0.60	1.9	35.46	28AP	3.1	-3.4	-1.5	0.3
60.15	0.68	1.9	45.02	8MY	1.7	-3.4	-1.4	0.4
66.71	0.75	1.9	54.56	18MY	0.9	-3.4	-1.4	0.4
73.22	0.82	1.9	64.08	28MY	-0.0	-3.3	-1.4	0.4
79.70	0.88	1.9	73.60	7JN	-1.1	-3.3	-1.4	0.4
86.15	0.93	1.9	83.12	17JN	-1.7	-3.3	-1.4	0.4
92.58	0.98	1.9	92.67	27JN	-1.0	-3.3	-1.4	0.4
98.98	1.03	1.9	102.23	7JL	-0.3	-3.3	-1.4	0.4
105.36	1.07	2.0	111.82	17JL	0.2	-3.3	-1.4	0.3
111.74	1.11	2.0	121.45	27JL	0.4	-3.4	-1.4	0.3
118.10	1.15	2.0	131.13	6AU	0.8	-3.4	-1.5	0.3
124.45	1.18	2.0	140.85	16AU	1.5	-3.4	-1.5	0.2
130.81	1.20	2.0	150.64	26AU	3.2	-3.5	-1.6	0.2
137.16	1.23	2.0	160.48	5SE	0.6	-3.5	-1.6	0.1
143.51	1.24	2.0	170.38	15SE	-0.7	-3.6	-1.7	0.0
149.86	1.26	2.0	180.35	25SE	-0.9	-3.7	-1.7	-0.0
156.21	1.27	2.0	190.36	5OC	-0.9	-3.7	-1.8	-0.1
162.57	1.27	1.9	200.43	15OC	-0.8	-3.8	-1.9	-0.2
168.93	1.27	1.9	210.54	25OC	-0.5	-3.9	-1.9	-0.2
175.28	1.27	1.8	220.69	4NO	-0.5	-4.0	-2.0	-0.2
181.63	1.25	1.8	230.87	14NO	-0.5	-4.2	-2.0	-0.1
187.98	1.24	1.7	241.08	24NO	-0.2	-4.3	-2.1	-0.0
194.30	1.21	1.6	251.28	4DE	1.4	-4.4	-2.1	0.1
200.62	1.17	1.5	261.49	14DE	1.9	-4.4	-2.1	0.1
206.91	1.12	1.4	271.68	24DE	0.4	-4.3	-2.2	0.2

♂ LONG	♂ LAT	♂ MAG	☉ LONG	16.00UT	☿	♀	♃	♄
				-2	MAGNITUDES			
213.17	1.06	1.3	281.85	3JA	0.1	-4.0	-2.1	0.2
219.39	0.98	1.2	291.98	13JA	0.0	-3.4	-2.1	0.3
225.56	0.89	1.1	302.07	23JA	-0.2	-3.5	-2.1	0.3
231.66	0.78	0.9	312.11	2FE	-0.5	-4.0	-2.0	0.4
237.68	0.64	0.8	322.10	12FE	-1.1	-4.2	-2.0	0.4
243.59	0.47	0.6	332.03	22FE	-1.5	-4.3	-1.9	0.4
249.36	0.27	0.4	341.90	4MR	-0.9	-4.2	-1.8	0.4
254.98	0.04	0.2	351.72	14MR	0.3	-4.1	-1.8	0.4
260.36	-0.25	-0.0	1.47	24MR	1.8	-4.0	-1.7	0.3
265.47	-0.58	-0.2	11.17	3AP	3.7	-3.9	-1.6	0.3
270.23	-0.98	-0.5	20.83	13AP	2.3	-3.8	-1.6	0.3
274.51	-1.45	-0.7	30.44	23AP	1.3	-3.7	-1.5	0.2
278.19	-1.99	-1.0	40.01	3MY	0.6	-3.6	-1.5	0.2
281.11	-2.63	-1.3	49.56	13MY	-0.1	-3.6	-1.4	0.3
283.07	-3.34	-1.6	59.10	23MY	-1.1	-3.5	-1.4	0.3
283.90	-4.12	-1.9	68.61	2JN	-1.8	-3.5	-1.3	0.3
283.47	-4.92	-2.2	78.14	12JN	-1.0	-3.4	-1.3	0.3
281.85	-5.64	-2.4	87.67	22JN	-0.2	-3.4	-1.3	0.3
279.43	-6.16	-2.6	97.21	2JL	0.3	-3.4	-1.3	0.3
276.81	-6.41	-2.6	106.79	12JL	0.6	-3.4	-1.3	0.3
274.76	-6.36	-2.4	116.40	22JL	1.0	-3.3	-1.3	0.3
273.80	-6.07	-2.1	126.05	1AU	2.0	-3.3	-1.3	0.3
274.16	-5.61	-1.8	135.76	11AU	2.9	-3.3	-1.3	0.3
275.81	-5.09	-1.6	145.51	21AU	0.5	-3.4	-1.3	0.2
278.59	-4.54	-1.3	155.32	31AU	-0.8	-3.4	-1.4	0.2
282.29	-4.00	-1.0	165.19	10SE	-1.1	-3.4	-1.4	0.1
286.73	-3.49	-0.8	175.12	20SE	-1.0	-3.4	-1.4	0.0
291.74	-3.02	-0.5	185.10	30SE	-0.7	-3.4	-1.5	-0.0
297.19	-2.58	-0.3	195.15	10OC	-0.4	-3.5	-1.5	-0.1
302.98	-2.18	-0.1	205.24	20OC	-0.3	-3.5	-1.6	-0.2
309.02	-1.82	0.1	215.37	30OC	-0.3	-3.5	-1.6	-0.2
315.25	-1.49	0.3	225.54	9NO	-0.0	-3.5	-1.7	-0.3
321.63	-1.19	0.4	235.73	19NO	1.6	-3.5	-1.8	-0.2
328.11	-0.92	0.6	245.93	29NO	1.6	-3.4	-1.8	-0.2
334.65	-0.67	0.8	256.14	9DE	0.2	-3.4	-1.9	-0.1
341.25	-0.46	0.9	266.34	19DE	-0.0	-3.4	-2.0	-0.0
347.86	-0.26	1.0	276.52	29DE	-0.1	-3.4	-2.0	0.1
				-1				
354.49	-0.08	1.2	286.67	8JA	-0.3	-3.3	-2.0	0.1
1.12	0.08	1.3	296.79	18JA	-0.6	-3.3	-2.1	0.2
7.73	0.22	1.4	306.85	28JA	-1.1	-3.3	-2.1	0.2
14.33	0.35	1.5	316.87	7FE	-1.3	-3.3	-2.0	0.3
20.91	0.46	1.6	326.83	17FE	-0.9	-3.4	-2.0	0.3
27.45	0.57	1.6	336.73	27FE	0.4	-3.4	-2.0	0.3
33.97	0.66	1.7	346.58	9MR	2.3	-3.4	-1.9	0.3
40.47	0.74	1.8	356.37	19MR	3.1	-3.4	-1.9	0.3
46.93	0.81	1.8	6.09	29MR	1.7	-3.5	-1.8	0.3
53.37	0.87	1.9	15.78	8AP	1.0	-3.5	-1.7	0.3
59.78	0.93	1.9	25.41	18AP	0.5	-3.5	-1.7	0.3
66.17	0.98	2.0	35.00	28AP	-0.2	-3.6	-1.6	0.2
72.54	1.02	2.0	44.56	8MY	-1.1	-3.7	-1.6	0.2
78.90	1.06	2.0	54.10	18MY	-1.8	-3.7	-1.5	0.2
85.24	1.09	2.0	63.62	28MY	-1.0	-3.8	-1.4	0.2
91.58	1.11	2.0	73.15	7JN	-0.2	-3.9	-1.4	0.3
97.92	1.13	2.0	82.67	17JN	0.4	-4.0	-1.3	0.3
104.26	1.15	2.0	92.21	27JN	0.8	-4.1	-1.3	0.3
110.60	1.16	2.0	101.77	7JL	1.4	-4.2	-1.3	0.3
116.96	1.16	2.0	111.36	17JL	2.5	-4.2	-1.3	0.3
123.32	1.16	2.0	120.99	27JL	2.5	-4.2	-1.2	0.3
129.71	1.16	1.9	130.66	6AU	0.4	-3.9	-1.2	0.3
136.12	1.15	2.0	140.39	16AU	-0.9	-3.4	-1.2	0.3
142.56	1.13	2.0	150.17	26AU	-1.2	-3.4	-1.2	0.3
149.02	1.12	2.0	160.01	5SE	-1.1	-4.0	-1.2	0.2
155.52	1.09	2.0	169.91	15SE	-0.6	-4.3	-1.3	0.2
162.05	1.06	2.0	179.86	25SE	-0.3	-4.3	-1.3	0.1
168.62	1.03	2.0	189.88	5OC	-0.2	-4.3	-1.3	0.0
175.23	0.99	1.9	199.94	15OC	-0.1	-4.2	-1.3	-0.0
181.87	0.94	1.9	210.05	25OC	0.2	-4.1	-1.4	-0.1
188.56	0.89	1.9	220.21	4NO	1.9	-4.0	-1.4	-0.2
195.29	0.83	1.8	230.38	14NO	1.4	-3.9	-1.5	-0.3
202.05	0.76	1.8	240.58	24NO	0.0	-3.9	-1.6	-0.3
208.86	0.68	1.7	250.79	4DE	-0.2	-3.8	-1.6	-0.2
215.71	0.59	1.7	261.00	14DE	-0.3	-3.7	-1.7	-0.2
222.59	0.49	1.6	271.19	24DE	-0.4	-3.6	-1.8	-0.1

♂ LONG	♂ LAT	♂ MAG	☉ LONG	16.00UT	☿	♀	♃	♄
				0	MAGNITUDES			
229.52	0.38	1.5	281.36	3JA	-0.7	-3.6	-1.8	-0.0
236.47	0.26	1.4	291.49	13JA	-1.1	-3.5	-1.9	0.0
243.46	0.12	1.3	301.59	23JA	-1.2	-3.5	-1.9	0.1
250.48	-0.03	1.3	311.63	2FE	-0.8	-3.4	-2.0	0.2
257.52	-0.19	1.2	321.62	12FE	0.6	-3.4	-2.0	0.2
264.58	-0.37	1.1	331.56	22FE	2.8	-3.3	-2.0	0.3
271.65	-0.56	0.9	341.43	3MR	2.3	-3.3	-2.0	0.3
278.73	-0.77	0.8	351.25	13MR	1.2	-3.3	-2.0	0.3
285.79	-0.99	0.7	1.01	23MR	0.7	-3.3	-2.0	0.3
292.82	-1.23	0.6	10.71	2AP	0.3	-3.3	-1.9	0.3
299.81	-1.48	0.5	20.36	12AP	-0.2	-3.3	-1.9	0.3
306.72	-1.74	0.3	29.98	22AP	-1.1	-3.3	-1.8	0.3
313.53	-2.01	0.2	39.56	2MY	-1.8	-3.4	-1.8	0.3
320.21	-2.30	0.1	49.10	12MY	-1.0	-3.4	-1.7	0.2
326.69	-2.59	-0.1	58.64	22MY	-0.1	-3.4	-1.6	0.2
332.93	-2.88	-0.2	68.16	1JN	0.5	-3.5	-1.6	0.2
338.86	-3.18	-0.4	77.67	11JN	1.1	-3.5	-1.5	0.2
344.39	-3.47	-0.6	87.20	21JN	1.8	-3.5	-1.5	0.3
349.39	-3.77	-0.7	96.75	1JL	3.1	-3.4	-1.4	0.3
353.75	-4.05	-0.9	106.32	11JL	2.0	-3.4	-1.4	0.3
357.27	-4.32	-1.2	115.93	21JL	0.3	-3.4	-1.3	0.3
359.75	-4.55	-1.4	125.58	31JL	-0.9	-3.3	-1.3	0.4
1.00	-4.72	-1.6	135.28	10AU	-1.3	-3.3	-1.3	0.4
0.82	-4.78	-1.8	145.03	20AU	-1.0	-3.3	-1.2	0.3
359.20	-4.69	-2.1	154.84	30AU	-0.5	-3.3	-1.2	0.3
356.43	-4.40	-2.2	164.70	9SE	-0.2	-3.3	-1.2	0.3
353.13	-3.90	-2.3	174.63	19SE	-0.1	-3.3	-1.2	0.2
350.11	-3.26	-2.0	184.62	29SE	0.0	-3.4	-1.2	0.2
348.06	-2.56	-1.7	194.66	9OC	0.4	-3.4	-1.2	0.1
347.30	-1.90	-1.4	204.75	19OC	2.2	-3.4	-1.3	0.1
347.87	-1.31	-1.0	214.88	29OC	1.2	-3.5	-1.3	-0.0
349.62	-0.81	-0.7	225.04	8NO	-0.2	-3.5	-1.3	-0.1
352.33	-0.40	-0.4	235.23	18NO	-0.4	-3.6	-1.4	-0.2
355.81	-0.06	-0.1	245.44	28NO	-0.4	-3.6	-1.4	-0.2
359.89	0.22	0.1	255.65	8DE	-0.5	-3.7	-1.5	-0.3
4.43	0.44	0.3	265.85	18DE	-0.7	-3.8	-1.5	-0.2
9.33	0.63	0.6	276.03	28DE	-1.0	-3.8	-1.6	-0.1
				1				
14.50	0.77	0.7	286.18	7JA	-1.1	-3.9	-1.7	-0.1
19.89	0.90	0.9	296.30	17JA	-0.7	-4.0	-1.7	0.0
25.44	0.99	1.1	306.37	27JA	0.7	-4.1	-1.8	0.1
31.12	1.07	1.2	316.39	6FE	3.1	-4.2	-1.9	0.1
36.90	1.14	1.3	326.36	16FE	1.7	-4.3	-1.9	0.2
42.75	1.19	1.5	336.26	26FE	0.9	-4.3	-2.0	0.2
48.67	1.23	1.6	346.11	8MR	0.5	-4.1	-2.0	0.3
54.63	1.26	1.7	355.90	18MR	0.2	-3.8	-2.0	0.3
60.64	1.28	1.7	5.63	28MR	-0.3	-3.1	-2.0	0.3
66.68	1.30	1.8	15.31	7AP	-1.1	-3.4	-2.0	0.3
72.74	1.31	1.9	24.94	17AP	-1.8	-4.0	-2.0	0.4
78.84	1.31	1.9	34.54	27AP	-1.0	-4.2	-2.0	0.4
84.95	1.30	2.0	44.10	7MY	-0.0	-4.2	-1.9	0.3
91.10	1.29	2.0	53.64	17MY	0.7	-4.1	-1.9	0.3
97.26	1.28	2.0	63.16	27MY	1.4	-4.1	-1.8	0.3
103.46	1.26	2.0	72.68	6JN	2.4	-4.0	-1.8	0.3
109.69	1.24	2.0	82.21	16JN	3.3	-3.9	-1.7	0.3
115.95	1.22	2.0	91.74	26JN	1.7	-3.8	-1.6	0.3
122.25	1.18	2.0	101.30	6JL	0.2	-3.7	-1.6	0.3
128.59	1.15	2.0	110.89	16JL	-1.0	-3.6	-1.5	0.4
134.97	1.11	2.0	120.52	26JL	-1.5	-3.6	-1.5	0.4
141.39	1.07	2.0	130.19	5AU	-1.0	-3.5	-1.4	0.4
147.87	1.02	1.9	139.92	15AU	-0.4	-3.5	-1.4	0.4
154.40	0.97	1.9	149.69	25AU	-0.1	-3.5	-1.3	0.4
160.99	0.91	1.9	159.53	4SE	0.1	-3.4	-1.3	0.4
167.64	0.85	1.8	169.43	14SE	0.2	-3.4	-1.3	0.4
174.35	0.79	1.8	179.38	24SE	0.6	-3.4	-1.3	0.4
181.13	0.71	1.8	189.39	4OC	2.5	-3.4	-1.3	0.3
187.97	0.64	1.8	199.46	14OC	1.0	-3.4	-1.3	0.3
194.87	0.56	1.8	209.56	24OC	-0.3	-3.4	-1.3	0.2
201.85	0.47	1.8	219.71	3NO	-0.5	-3.4	-1.3	0.2
208.89	0.38	1.8	229.89	13NO	-0.5	-3.4	-1.3	0.1
216.00	0.28	1.7	240.09	23NO	-0.6	-3.4	-1.3	0.0
223.17	0.17	1.7	250.30	3DE	-0.8	-3.4	-1.4	-0.1
230.41	0.06	1.7	260.50	13DE	-0.9	-3.4	-1.4	-0.1
237.71	-0.06	1.6	270.69	23DE	-0.9	-3.4	-1.4	-0.2

♂			☉	16.00UT	☿	♀	♃	♄
LONG	LAT	MAG	LONG	²	MAGNITUDES			
245.07	-0.18	1.6	280.87	2JA	-0.6	-3.5	-1.5	-0.1
252.48	-0.31	1.5	291.00	12JA	0.8	-3.5	-1.6	-0.0
259.95	-0.44	1.5	301.10	22JA	3.0	-3.5	-1.6	0.0
267.45	-0.57	1.4	311.15	1FE	1.3	-3.5	-1.7	0.1
275.00	-0.71	1.4	321.14	11FE	0.6	-3.4	-1.8	0.2
282.57	-0.85	1.3	331.07	21FE	0.4	-3.4	-1.8	0.2
290.16	-0.99	1.2	340.95	3MR	0.1	-3.4	-1.9	0.3
297.75	-1.12	1.2	350.77	13MR	-0.4	-3.4	-2.0	0.3
305.34	-1.26	1.1	0.53	23MR	-1.1	-3.3	-2.0	0.4
312.92	-1.38	1.1	10.24	2AP	-1.7	-3.3	-2.1	0.4
320.46	-1.50	1.0	19.90	12AP	-1.0	-3.3	-2.1	0.4
327.94	-1.61	0.9	29.51	22AP	0.1	-3.3	-2.1	0.4
335.36	-1.71	0.9	39.09	2MY	0.9	-3.3	-2.1	0.4
342.69	-1.79	0.8	48.64	12MY	1.9	-3.4	-2.1	0.5
349.92	-1.86	0.8	58.17	22MY	3.2	-3.4	-2.1	0.4
357.01	-1.91	0.7	67.69	1JN	2.7	-3.4	-2.0	0.4
3.94	-1.95	0.6	77.21	11JN	1.3	-3.4	-2.0	0.4
10.69	-1.97	0.5	86.74	21JN	0.1	-3.5	-1.9	0.4
17.22	-1.97	0.5	96.29	1JL	-1.0	-3.6	-1.9	0.4
23.49	-1.94	0.4	105.86	11JL	-1.6	-3.6	-1.8	0.4
29.46	-1.90	0.3	115.46	21JL	-1.0	-3.7	-1.8	0.5
35.06	-1.84	0.2	125.11	31JL	-0.4	-3.8	-1.7	0.5
40.22	-1.75	0.0	134.81	10AU	-0.0	-3.9	-1.6	0.5
44.84	-1.63	-0.1	144.56	20AU	0.2	-4.0	-1.6	0.6
48.80	-1.48	-0.3	154.36	30AU	0.4	-4.1	-1.5	0.6
51.94	-1.28	-0.5	164.23	9SE	0.9	-4.2	-1.5	0.6
54.09	-1.04	-0.7	174.15	19SE	2.9	-4.3	-1.4	0.6
55.01	-0.74	-0.9	184.13	29SE	0.8	-4.3	-1.4	0.6
54.55	-0.38	-1.1	194.17	9OC	-0.5	-4.3	-1.4	0.5
52.65	0.05	-1.3	204.25	19OC	-0.7	-4.0	-1.4	0.5
49.51	0.51	-1.5	214.39	29OC	-0.7	-3.3	-1.3	0.4
45.69	0.96	-1.6	224.55	8NO	-0.7	-3.4	-1.3	0.4
41.98	1.35	-1.4	234.74	18NO	-0.7	-4.0	-1.3	0.3
39.10	1.66	-1.1	244.94	28NO	-0.7	-4.3	-1.3	0.2
37.50	1.86	-0.8	255.15	8DE	-0.7	-4.4	-1.4	0.2
37.23	1.98	-0.5	265.35	18DE	-0.5	-4.3	-1.4	0.1
38.20	2.05	-0.2	275.54	28DE	1.0	-4.2	-1.4	0.0
				3				
40.20	2.07	0.1	285.69	7JA	2.6	-4.1	-1.5	0.0
43.03	2.06	0.3	295.81	17JA	0.9	-4.0	-1.5	0.1
46.52	2.04	0.6	305.89	27JA	0.4	-3.9	-1.5	0.1
50.52	2.01	0.8	315.91	6FE	0.2	-3.8	-1.6	0.2
54.92	1.97	0.9	325.88	16FE	-0.0	-3.7	-1.7	0.2
59.64	1.92	1.1	335.79	26FE	-0.4	-3.6	-1.7	0.3
64.61	1.87	1.2	345.63	8MR	-1.1	-3.6	-1.8	0.3
69.79	1.81	1.4	355.43	18MR	-1.7	-3.5	-1.9	0.4
75.14	1.76	1.5	5.16	28MR	-1.0	-3.4	-1.9	0.4
80.63	1.70	1.6	14.84	7AP	0.1	-3.4	-2.0	0.5
86.23	1.64	1.7	24.48	17AP	1.2	-3.4	-2.1	0.5
91.94	1.58	1.7	34.08	27AP	2.5	-3.3	-2.1	0.6
97.75	1.52	1.8	43.64	7MY	3.6	-3.3	-2.2	0.6
103.63	1.45	1.8	53.18	17MY	2.1	-3.3	-2.2	0.6
109.60	1.39	1.9	62.71	27MY	1.0	-3.3	-2.2	0.6
115.63	1.32	1.9	72.23	6JN	0.0	-3.3	-2.2	0.6
121.74	1.25	1.9	81.75	16JN	-1.0	-3.3	-2.2	0.6
127.93	1.18	1.9	91.29	26JN	-1.7	-3.3	-2.2	0.6
134.18	1.11	1.9	100.84	6JL	-1.0	-3.4	-2.2	0.6
140.51	1.03	1.9	110.43	16JL	-0.3	-3.4	-2.1	0.6
146.91	0.95	1.9	120.05	26JL	0.1	-3.4	-2.1	0.6
153.39	0.87	1.9	129.72	5AU	0.3	-3.5	-2.0	0.7
159.95	0.79	1.9	139.44	15AU	0.6	-3.5	-1.9	0.7
166.59	0.70	1.9	149.22	25AU	1.3	-3.5	-1.9	0.7
173.31	0.61	1.8	159.05	4SE	3.2	-3.5	-1.8	0.8
180.12	0.52	1.8	168.94	14SE	0.7	-3.4	-1.7	0.8
187.01	0.43	1.7	178.89	24SE	-0.6	-3.4	-1.7	0.8
193.98	0.33	1.7	188.90	4OC	-0.8	-3.4	-1.6	0.8
201.05	0.23	1.6	198.96	14OC	-0.8	-3.3	-1.6	0.8
208.20	0.13	1.6	209.07	24OC	-0.8	-3.3	-1.5	0.7
215.43	0.02	1.6	219.21	3NO	-0.6	-3.3	-1.5	0.7
222.74	-0.08	1.6	229.39	13NO	-0.6	-3.3	-1.5	0.6
230.14	-0.19	1.6	239.59	23NO	-0.6	-3.3	-1.5	0.6
237.60	-0.30	1.6	249.79	3DE	-0.3	-3.4	-1.4	0.5
245.14	-0.41	1.5	260.00	13DE	1.2	-3.4	-1.4	0.4
252.74	-0.52	1.5	270.20	23DE	2.2	-3.4	-1.4	0.4

♂ LONG	♂ LAT	♂ MAG	☉ LONG	16.00UT	☿	♀	♃	♄
				4	MAGNITUDES			
260.40	-0.62	1.5	280.37	2JA	0.6	-3.4	-1.5	0.3
268.11	-0.73	1.5	290.51	12JA	0.2	-3.5	-1.5	0.2
275.85	-0.82	1.5	300.61	22JA	0.1	-3.5	-1.5	0.2
283.62	-0.92	1.4	310.66	1FE	-0.1	-3.6	-1.5	0.3
291.41	-1.00	1.4	320.66	11FE	-0.5	-3.6	-1.6	0.3
299.21	-1.08	1.4	330.60	21FE	-1.1	-3.7	-1.6	0.4
306.99	-1.15	1.4	340.48	2MR	-1.5	-3.7	-1.7	0.4
314.77	-1.21	1.4	350.30	12MR	-1.0	-3.8	-1.7	0.5
322.50	-1.26	1.3	0.07	22MR	0.2	-3.9	-1.8	0.5
330.20	-1.29	1.3	9.77	1AP	1.6	-4.0	-1.8	0.6
337.83	-1.31	1.3	19.44	11AP	3.2	-4.1	-1.9	0.6
345.40	-1.32	1.3	29.05	21AP	2.8	-4.1	-2.0	0.7
352.89	-1.31	1.3	38.63	1MY	1.6	-4.2	-2.1	0.7
0.29	-1.29	1.3	48.18	11MY	0.8	-4.1	-2.1	0.8
7.58	-1.25	1.2	57.72	21MY	-0.0	-3.9	-2.2	0.8
14.76	-1.20	1.2	67.23	31MY	-1.0	-3.3	-2.3	0.8
21.82	-1.14	1.2	76.76	10JN	-1.8	-2.9	-2.3	0.8
28.74	-1.06	1.2	86.28	20JN	-1.0	-3.7	-2.3	0.8
35.51	-0.97	1.1	95.83	30JN	-0.3	-4.1	-2.4	0.8
42.12	-0.87	1.1	105.40	10JL	0.2	-4.2	-2.4	0.8
48.54	-0.75	1.0	115.00	20JL	0.5	-4.2	-2.3	0.8
54.77	-0.62	1.0	124.64	30JL	0.8	-4.1	-2.3	0.8
60.77	-0.47	0.9	134.34	9AU	1.7	-4.1	-2.3	0.9
66.51	-0.30	0.8	144.08	19AU	3.1	-4.0	-2.2	0.9
71.95	-0.12	0.8	153.88	29AU	0.6	-3.9	-2.2	1.0
77.02	0.09	0.6	163.75	8SE	-0.7	-3.8	-2.1	1.0
81.67	0.32	0.5	173.67	18SE	-1.0	-3.7	-2.0	1.0
85.80	0.58	0.4	183.64	28SE	-0.9	-3.7	-2.0	1.0
89.28	0.88	0.2	193.68	8OC	-0.7	-3.6	-1.9	1.0
91.97	1.22	0.0	203.76	18OC	-0.5	-3.6	-1.8	1.0
93.70	1.60	-0.2	213.89	28OC	-0.4	-3.5	-1.8	1.0
94.26	2.02	-0.4	224.05	7NO	-0.4	-3.5	-1.7	1.0
93.49	2.48	-0.6	234.24	17NO	-0.2	-3.4	-1.7	0.9
91.35	2.93	-0.9	244.44	27NO	1.4	-3.4	-1.6	0.9
88.04	3.33	-1.0	254.65	7DE	1.9	-3.4	-1.6	0.8
84.11	3.61	-1.1	264.85	17DE	0.4	-3.4	-1.6	0.8
80.29	3.74	-1.0	275.04	27DE	0.0	-3.3	-1.6	0.7
				5				
77.27	3.73	-0.7	285.20	6JA	-0.1	-3.3	-1.5	0.6
75.49	3.61	-0.4	295.32	16JA	-0.2	-3.3	-1.5	0.5
75.01	3.42	-0.2	305.40	26JA	-0.5	-3.3	-1.5	0.5
75.75	3.22	0.1	315.42	5FE	-1.1	-3.3	-1.5	0.5
77.52	3.00	0.3	325.39	15FE	-1.4	-3.4	-1.6	0.5
80.13	2.79	0.5	335.30	25FE	-0.9	-3.4	-1.6	0.6
83.42	2.60	0.7	345.16	7MR	0.3	-3.4	-1.6	0.6
87.25	2.41	0.9	354.95	17MR	2.0	-3.4	-1.6	0.7
91.51	2.24	1.1	4.69	27MR	3.6	-3.5	-1.7	0.7
96.12	2.08	1.2	14.38	6AP	2.0	-3.5	-1.7	0.8
101.01	1.93	1.3	24.01	16AP	1.2	-3.4	-1.8	0.8
106.15	1.79	1.4	33.61	26AP	0.6	-3.4	-1.9	0.9
111.49	1.65	1.5	43.18	6MY	-0.1	-3.4	-1.9	0.9
117.01	1.52	1.6	52.72	16MY	-1.0	-3.4	-2.0	1.0
122.69	1.39	1.6	62.24	26MY	-1.8	-3.3	-2.1	1.0
128.51	1.26	1.7	71.76	5JN	-1.0	-3.3	-2.1	1.0
134.46	1.14	1.7	81.28	15JN	-0.2	-3.3	-2.2	1.1
140.54	1.02	1.7	90.82	25JN	0.3	-3.3	-2.3	1.1
146.74	0.90	1.7	100.38	5JL	0.7	-3.3	-2.3	1.1
153.05	0.79	1.8	109.96	15JL	1.1	-3.3	-2.4	1.1
159.47	0.67	1.8	119.58	25JL	2.2	-3.4	-2.4	1.1
166.00	0.56	1.8	129.25	4AU	2.8	-3.4	-2.4	1.1
172.64	0.44	1.8	138.97	14AU	0.5	-3.4	-2.4	1.1
179.40	0.33	1.7	148.74	24AU	-0.8	-3.5	-2.4	1.2
186.26	0.21	1.7	158.57	3SE	-1.1	-3.5	-2.4	1.2
193.22	0.10	1.7	168.46	13SE	-1.1	-3.6	-2.4	1.2
200.29	-0.01	1.7	178.41	23SE	-0.6	-3.7	-2.3	1.3
207.47	-0.12	1.6	188.41	3OC	-0.4	-3.8	-2.3	1.3
214.74	-0.23	1.6	198.47	13OC	-0.3	-3.8	-2.2	1.3
222.11	-0.33	1.6	208.57	23OC	-0.2	-3.9	-2.1	1.3
229.57	-0.44	1.5	218.72	2NO	0.0	-4.0	-2.0	1.3
237.11	-0.53	1.5	228.89	12NO	1.6	-4.2	-2.0	1.3
244.74	-0.63	1.4	239.09	22NO	1.6	-4.3	-1.9	1.3
252.43	-0.72	1.4	249.30	2DE	0.1	-4.4	-1.8	1.2
260.18	-0.80	1.4	259.50	12DE	-0.1	-4.4	-1.8	1.2
267.98	-0.87	1.3	269.70	22DE	-0.2	-4.3	-1.7	1.1

♂ LONG	♂ LAT	♂ MAG	☉ LONG	16.00UT	☿	♀	♃	♄
				6	MAGNITUDES			
275.82	-0.94	1.3	279.87	1JA	-0.3	-4.0	-1.7	1.0
283.69	-0.99	1.4	290.01	11JA	-0.6	-3.4	-1.7	1.0
291.57	-1.04	1.4	300.11	21JA	-1.1	-3.5	-1.6	0.9
299.44	-1.08	1.4	310.17	31JA	-1.3	-4.0	-1.6	0.8
307.31	-1.10	1.4	320.16	10FE	-0.9	-4.3	-1.6	0.8
315.15	-1.11	1.4	330.11	20FE	0.4	-4.3	-1.6	0.8
322.96	-1.11	1.4	339.99	2MR	2.4	-4.2	-1.6	0.8
330.72	-1.10	1.4	349.82	12MR	2.8	-4.1	-1.6	0.9
338.42	-1.08	1.4	359.59	22MR	1.5	-4.0	-1.6	0.9
346.05	-1.04	1.5	9.30	1AP	0.9	-3.9	-1.6	1.0
353.60	-0.99	1.5	18.96	11AP	0.4	-3.8	-1.6	1.0
1.08	-0.94	1.5	28.59	21AP	-0.2	-3.7	-1.7	1.1
8.46	-0.87	1.5	38.17	1MY	-1.0	-3.6	-1.7	1.1
15.75	-0.79	1.5	47.72	11MY	-1.9	-3.6	-1.8	1.2
22.94	-0.71	1.5	57.26	21MY	-1.0	-3.5	-1.8	1.2
30.02	-0.61	1.5	66.78	31MY	-0.1	-3.5	-1.9	1.3
37.00	-0.51	1.5	76.29	10JN	0.4	-3.4	-1.9	1.3
43.87	-0.40	1.5	85.83	20JN	0.9	-3.4	-2.0	1.3
50.62	-0.29	1.5	95.37	30JN	1.5	-3.4	-2.1	1.4
57.26	-0.17	1.5	104.93	10JL	2.7	-3.3	-2.1	1.4
63.77	-0.04	1.5	114.54	20JL	2.3	-3.3	-2.2	1.4
70.15	0.10	1.5	124.18	30JL	0.4	-3.3	-2.3	1.4
76.39	0.24	1.5	133.87	9AU	-0.9	-3.3	-2.3	1.4
82.47	0.39	1.4	143.62	19AU	-1.3	-3.3	-2.4	1.3
88.38	0.56	1.4	153.41	29AU	-1.1	-3.4	-2.4	1.3
94.11	0.73	1.3	163.27	8SE	-0.6	-3.4	-2.4	1.3
99.61	0.91	1.2	173.19	18SE	-0.3	-3.4	-2.5	1.3
104.85	1.11	1.1	183.16	28SE	-0.1	-3.4	-2.4	1.3
109.80	1.33	1.0	193.19	8OC	-0.1	-3.5	-2.4	1.3
114.38	1.57	0.9	203.27	18OC	0.2	-3.5	-2.4	1.2
118.51	1.84	0.7	213.40	28OC	1.9	-3.5	-2.3	1.2
122.10	2.13	0.6	223.56	7NO	1.3	-3.5	-2.3	1.2
125.01	2.46	0.4	233.74	17NO	-0.1	-3.5	-2.2	1.2
127.08	2.82	0.2	243.94	27NO	-0.3	-3.4	-2.1	1.1
128.14	3.21	-0.1	254.15	7DE	-0.3	-3.4	-2.1	1.1
127.99	3.62	-0.3	264.36	17DE	-0.4	-3.4	-2.0	1.0
126.53	4.02	-0.6	274.54	27DE	-0.7	-3.4	-1.9	1.0
				7				
123.82	4.35	-0.8	284.70	6JA	-1.1	-3.3	-1.8	0.9
120.17	4.54	-1.0	294.82	16JA	-1.1	-3.3	-1.8	0.9
116.22	4.57	-0.9	304.90	26JA	-0.8	-3.3	-1.7	0.9
112.69	4.42	-0.8	314.93	5FE	0.6	-3.3	-1.7	0.8
110.16	4.15	-0.5	324.90	15FE	2.9	-3.4	-1.6	0.8
108.91	3.80	-0.3	334.82	25FE	2.1	-3.4	-1.6	0.7
108.92	3.44	-0.1	344.67	7MR	1.1	-3.4	-1.6	0.8
110.09	3.08	0.2	354.47	17MR	0.6	-3.4	-1.6	0.9
112.23	2.74	0.4	4.21	27MR	0.3	-3.5	-1.5	0.9
115.15	2.43	0.5	13.90	6AP	-0.2	-3.5	-1.5	1.0
118.71	2.15	0.7	23.55	16AP	-1.0	-3.5	-1.5	1.1
122.80	1.89	0.8	33.14	26AP	-1.9	-3.6	-1.6	1.1
127.32	1.65	1.0	42.71	6MY	-1.0	-3.7	-1.6	1.2
132.19	1.44	1.1	52.26	16MY	-0.1	-3.7	-1.6	1.2
137.37	1.23	1.2	61.78	26MY	0.6	-3.8	-1.6	1.3
142.81	1.04	1.2	71.31	5JN	1.2	-3.9	-1.7	1.3
148.49	0.86	1.3	80.83	15JN	2.0	-4.0	-1.7	1.3
154.36	0.69	1.4	90.36	25JN	3.3	-4.1	-1.8	1.3
160.43	0.53	1.4	99.92	5JL	1.9	-4.2	-1.8	1.3
166.67	0.37	1.4	109.50	15JL	0.3	-4.2	-1.9	1.3
173.08	0.22	1.5	119.12	25JL	-0.9	-4.1	-1.9	1.3
179.64	0.08	1.5	128.79	4AU	-1.4	-3.9	-2.0	1.3
186.34	-0.05	1.5	138.50	14AU	-1.0	-3.3	-2.1	1.2
193.19	-0.18	1.5	148.27	24AU	-0.5	-3.5	-2.1	1.2
200.17	-0.30	1.5	158.09	3SE	-0.2	-4.0	-2.2	1.1
207.28	-0.42	1.5	167.98	13SE	-0.0	-4.3	-2.3	1.1
214.51	-0.53	1.5	177.92	23SE	0.1	-4.3	-2.3	1.1
221.86	-0.63	1.5	187.93	3OC	0.5	-4.3	-2.3	1.1
229.31	-0.72	1.5	197.98	13OC	2.2	-4.2	-2.4	1.1
236.86	-0.81	1.5	208.08	23OC	1.2	-4.1	-2.4	1.1
244.50	-0.88	1.4	218.23	2NO	-0.2	-4.0	-2.4	1.1
252.21	-0.95	1.4	228.40	12NO	-0.4	-3.9	-2.3	1.1
259.99	-1.00	1.4	238.60	22NO	-0.4	-3.8	-2.3	1.0
267.82	-1.05	1.4	248.81	2DE	-0.5	-3.8	-2.3	1.0
275.69	-1.08	1.4	259.01	12DE	-0.8	-3.7	-2.2	1.0
283.58	-1.10	1.4	269.21	22DE	-0.9	-3.6	-2.1	0.9

Left section (years 8–9):

♂ LONG	LAT	MAG	☉ LONG	16.00UT	☿	♀	♃	♄
				8				
291.48	-1.10	1.4	279.38	1JA	-1.0	-3.6	-2.1	0.9
299.38	-1.10	1.4	289.52	11JA	-0.7	-3.5	-2.0	0.8
307.26	-1.08	1.4	299.63	21JA	0.7	-3.5	-1.9	0.8
315.11	-1.06	1.4	309.68	31JA	3.1	-3.4	-1.8	0.7
322.92	-1.02	1.3	319.69	10FE	1.6	-3.4	-1.8	0.7
330.67	-0.97	1.3	329.63	20FE	0.8	-3.3	-1.7	0.7
338.37	-0.91	1.4	339.52	1MR	0.5	-3.3	-1.7	0.6
345.99	-0.85	1.4	349.34	11MR	0.2	-3.3	-1.6	0.6
353.53	-0.77	1.5	359.11	21MR	-0.3	-3.3	-1.6	0.7
0.99	-0.69	1.5	8.83	31MR	-1.1	-3.3	-1.5	0.7
8.37	-0.60	1.5	18.50	10AP	-1.8	-3.3	-1.5	0.8
15.66	-0.51	1.6	28.12	20AP	-1.0	-3.3	-1.5	0.9
22.86	-0.41	1.6	37.70	30AP	-0.0	-3.4	-1.5	0.9
29.97	-0.31	1.7	47.26	10MY	0.8	-3.4	-1.5	1.0
36.98	-0.21	1.7	56.79	20MY	1.6	-3.4	-1.5	1.0
43.91	-0.11	1.7	66.31	30MY	2.7	-3.5	-1.5	1.1
50.75	0.00	1.7	75.83	9JN	3.1	-3.5	-1.5	1.1
57.50	0.11	1.8	85.36	19JN	1.6	-3.5	-1.5	1.1
64.17	0.22	1.8	94.90	29JN	0.2	-3.4	-1.5	1.2
70.74	0.34	1.8	104.47	9JL	-1.0	-3.4	-1.6	1.2
77.23	0.45	1.8	114.07	19JL	-1.5	-3.4	-1.6	1.2
83.64	0.57	1.8	123.71	29JL	-1.0	-3.3	-1.7	1.1
89.95	0.69	1.8	133.40	8AU	-0.4	-3.3	-1.7	1.1
96.17	0.81	1.8	143.14	18AU	-0.1	-3.3	-1.8	1.1
102.29	0.93	1.7	152.93	28AU	0.1	-3.3	-1.8	1.0
108.30	1.06	1.7	162.79	7SE	0.3	-3.3	-1.9	1.0
114.19	1.19	1.6	172.70	17SE	0.7	-3.3	-2.0	1.0
119.95	1.33	1.6	182.68	27SE	2.6	-3.4	-2.0	1.0
125.56	1.48	1.5	192.70	7OC	1.0	-3.4	-2.1	1.0
130.99	1.63	1.4	202.78	17OC	-0.4	-3.4	-2.1	1.0
136.21	1.80	1.3	212.90	27OC	-0.6	-3.5	-2.2	1.0
141.18	1.98	1.2	223.06	6NO	-0.6	-3.5	-2.2	1.0
145.85	2.17	1.0	233.25	16NO	-0.6	-3.6	-2.3	1.0
150.14	2.38	0.9	243.45	26NO	-0.8	-3.6	-2.3	1.0
153.98	2.60	0.7	253.66	6DE	-0.8	-3.7	-2.3	0.9
157.25	2.85	0.5	263.86	16DE	-0.8	-3.8	-2.2	0.9
159.81	3.11	0.3	274.05	26DE	-0.6	-3.9	-2.2	0.9
				9				
161.51	3.39	0.0	284.21	5JA	0.8	-3.9	-2.1	0.8
162.15	3.68	-0.2	294.34	15JA	2.9	-4.0	-2.1	0.8
161.56	3.94	-0.5	304.42	25JA	1.1	-4.1	-2.0	0.7
159.70	4.14	-0.8	314.45	4FE	0.5	-4.2	-1.9	0.7
156.68	4.23	-1.0	324.42	14FE	0.3	-4.3	-1.9	0.6
152.94	4.16	-1.2	334.34	24FE	-0.0	-4.3	-1.8	0.6
149.16	3.93	-1.1	344.20	6MR	-0.4	-4.1	-1.7	0.5
146.03	3.56	-0.9	354.00	16MR	-1.1	-3.8	-1.7	0.5
144.03	3.11	-0.7	3.74	26MR	-1.7	-3.1	-1.6	0.5
143.34	2.65	-0.5	13.43	5AP	-1.0	-3.5	-1.6	0.6
143.92	2.20	-0.3	23.08	15AP	0.1	-4.0	-1.5	0.6
145.61	1.79	-0.1	32.68	25AP	1.0	-4.2	-1.5	0.7
148.25	1.42	0.1	42.25	5MY	2.1	-4.2	-1.4	0.8
151.65	1.09	0.3	51.79	15MY	3.5	-4.1	-1.4	0.8
155.70	0.79	0.4	61.32	25MY	2.5	-4.1	-1.4	0.9
160.27	0.52	0.5	70.84	4JN	1.2	-4.0	-1.4	0.9
165.28	0.28	0.7	80.36	14JN	0.1	-3.9	-1.4	1.0
170.67	0.06	0.7	89.90	24JN	-1.0	-3.8	-1.4	1.0
176.37	-0.14	0.8	99.45	4JL	-1.6	-3.7	-1.4	1.0
182.37	-0.32	0.9	109.03	14JL	-1.0	-3.6	-1.4	1.0
188.61	-0.48	1.0	118.65	24JL	-0.4	-3.6	-1.4	1.0
195.08	-0.63	1.0	128.31	3AU	0.0	-3.5	-1.4	1.0
201.76	-0.76	1.1	138.03	13AU	0.3	-3.5	-1.5	1.0
208.62	-0.88	1.1	147.79	23AU	0.5	-3.4	-1.5	1.0
215.64	-0.98	1.1	157.62	2SE	1.0	-3.4	-1.5	1.0
222.83	-1.07	1.2	167.50	12SE	3.0	-3.4	-1.6	0.9
230.15	-1.14	1.2	177.44	22SE	0.9	-3.4	-1.7	0.9
237.59	-1.20	1.2	187.44	2OC	-0.5	-3.4	-1.7	0.8
245.14	-1.25	1.2	197.50	12OC	-0.7	-3.4	-1.8	0.9
252.78	-1.28	1.3	207.59	22OC	-0.7	-3.4	-1.9	0.9
260.49	-1.29	1.3	217.73	1NO	-0.8	-3.4	-1.9	0.9
268.26	-1.29	1.3	227.91	11NO	-0.7	-3.4	-2.0	0.9
276.08	-1.28	1.3	238.10	21NO	-0.6	-3.4	-2.0	0.9
283.92	-1.25	1.3	248.31	1DE	-0.7	-3.4	-2.1	0.9
291.77	-1.21	1.4	258.52	11DE	-0.4	-3.4	-2.1	0.9
299.62	-1.16	1.4	268.71	21DE	1.0	-3.4	-2.1	0.8
307.44	-1.10	1.4	278.89	31DE	2.5	-3.5	-2.1	0.8

Right section (years 10–11):

♂ LONG	LAT	MAG	☉ LONG	16.00UT	☿	♀	♃	♄
				10				
315.23	-1.03	1.4	289.03	10JA	0.8	-3.5	-2.1	0.8
322.98	-0.95	1.5	299.14	20JA	0.3	-3.5	-2.1	0.7
330.67	-0.87	1.5	309.19	30JA	0.2	-3.4	-2.1	0.7
338.29	-0.78	1.5	319.20	9FE	-0.1	-3.4	-2.0	0.6
345.84	-0.68	1.5	329.15	19FE	-0.4	-3.4	-1.9	0.6
353.32	-0.58	1.6	339.04	1MR	-1.1	-3.4	-1.9	0.5
0.72	-0.48	1.6	348.87	11MR	-1.6	-3.4	-1.8	0.5
8.03	-0.38	1.6	358.64	21MR	-1.0	-3.3	-1.7	0.4
15.26	-0.28	1.6	8.36	31MR	0.1	-3.3	-1.7	0.4
22.40	-0.17	1.6	18.03	10AP	1.3	-3.3	-1.6	0.4
29.46	-0.07	1.6	27.65	20AP	2.7	-3.3	-1.6	0.5
36.44	0.03	1.7	37.23	30AP	3.3	-3.3	-1.5	0.5
43.34	0.13	1.7	46.79	10MY	1.9	-3.4	-1.4	0.6
50.17	0.23	1.8	56.32	20MY	0.9	-3.4	-1.4	0.7
56.92	0.33	1.8	65.85	30MY	0.0	-3.4	-1.4	0.7
63.61	0.43	1.8	75.37	9JN	-1.0	-3.5	-1.3	0.8
70.22	0.52	1.9	84.89	19JN	-1.7	-3.5	-1.3	0.8
76.78	0.61	1.9	94.44	29JN	-1.0	-3.6	-1.3	0.9
83.28	0.70	1.9	104.00	9JL	-0.3	-3.6	-1.3	0.9
89.72	0.79	1.9	113.60	19JL	0.1	-3.7	-1.3	0.9
96.12	0.88	2.0	123.24	29JL	0.4	-3.8	-1.3	0.9
102.45	0.97	2.0	132.93	8AU	0.7	-3.9	-1.3	0.9
108.74	1.05	1.9	142.66	18AU	1.4	-4.0	-1.3	0.9
114.97	1.13	1.9	152.46	28AU	3.2	-4.1	-1.3	0.9
121.15	1.22	1.9	162.31	7SE	0.7	-4.2	-1.3	0.9
127.27	1.30	1.9	172.22	17SE	-0.6	-4.3	-1.4	0.9
133.33	1.38	1.8	182.19	27SE	-0.9	-4.3	-1.4	0.8
139.32	1.47	1.8	192.22	7OC	-0.9	-4.3	-1.5	0.8
145.24	1.55	1.7	202.29	17OC	-0.8	-4.0	-1.5	0.8
151.06	1.63	1.7	212.41	27OC	-0.6	-3.3	-1.6	0.8
156.78	1.72	1.6	222.57	6NO	-0.5	-3.4	-1.7	0.8
162.37	1.81	1.5	232.75	16NO	-0.5	-4.1	-1.7	0.8
167.82	1.90	1.4	242.96	26NO	-0.3	-4.3	-1.8	0.8
173.09	1.99	1.2	253.16	6DE	1.2	-4.4	-1.8	0.8
178.14	2.08	1.1	263.37	16DE	2.1	-4.3	-1.9	0.8
182.93	2.18	0.9	273.56	26DE	0.5	-4.2	-2.0	0.8
				11				
187.38	2.27	0.7	283.72	5JA	0.2	-4.1	-2.0	0.8
191.43	2.36	0.5	293.84	15JA	0.0	-4.0	-2.0	0.8
194.97	2.45	0.3	303.93	25JA	-0.1	-3.9	-2.0	0.7
197.87	2.53	0.0	313.96	4FE	-0.5	-3.8	-2.0	0.7
199.99	2.58	-0.2	323.94	14FE	-1.1	-3.7	-2.0	0.6
201.12	2.61	-0.5	333.86	24FE	-1.5	-3.4	-2.0	0.6
201.11	2.58	-0.8	343.73	6MR	-1.0	-3.6	-2.0	0.5
199.84	2.46	-1.1	353.53	16MR	0.2	-3.5	-1.9	0.5
197.37	2.24	-1.4	3.28	26MR	1.7	-3.4	-1.8	0.4
194.03	1.90	-1.7	12.97	5AP	3.5	-3.4	-1.8	0.4
190.44	1.46	-1.6	22.61	15AP	2.5	-3.4	-1.7	0.3
187.33	0.96	-1.5	32.22	25AP	1.4	-3.3	-1.7	0.4
185.29	0.46	-1.3	41.79	5MY	0.7	-3.3	-1.6	0.4
184.57	0.00	-1.1	51.33	15MY	-0.0	-3.3	-1.5	0.5
185.20	-0.40	-0.9	60.86	25MY	-1.0	-3.3	-1.5	0.6
187.04	-0.74	-0.7	70.38	4JN	-1.8	-3.3	-1.4	0.6
189.92	-1.02	-0.5	79.90	14JN	-1.0	-3.3	-1.4	0.7
193.66	-1.26	-0.3	89.44	24JN	-0.3	-3.3	-1.3	0.7
198.11	-1.45	-0.1	98.99	4JL	0.2	-3.4	-1.3	0.8
203.14	-1.60	-0.0	108.57	14JL	0.6	-3.4	-1.3	0.8
208.66	-1.72	0.1	118.19	24JL	0.9	-3.4	-1.2	0.8
214.59	-1.81	0.2	127.84	3AU	1.8	-3.5	-1.2	0.9
220.85	-1.87	0.3	137.55	13AU	3.1	-3.5	-1.2	0.9
227.40	-1.91	0.4	147.32	23AU	0.6	-3.4	-1.2	0.9
234.20	-1.93	0.5	157.14	2SE	-0.7	-3.5	-1.2	0.9
241.19	-1.92	0.6	167.02	12SE	-1.0	-3.4	-1.2	0.9
248.37	-1.89	0.7	176.96	22SE	-1.0	-3.4	-1.3	0.8
255.68	-1.85	0.7	186.95	2OC	-0.7	-3.4	-1.3	0.8
263.11	-1.78	0.8	197.00	12OC	-0.5	-3.3	-1.3	0.8
270.62	-1.71	0.9	207.10	22OC	-0.4	-3.3	-1.4	0.7
278.20	-1.62	1.0	217.24	1NO	-0.3	-3.3	-1.4	0.7
285.81	-1.51	1.0	227.41	11NO	-0.1	-3.3	-1.5	0.8
293.45	-1.40	1.1	237.61	21NO	1.4	-3.3	-1.5	0.8
301.09	-1.28	1.2	247.81	1DE	1.8	-3.4	-1.6	0.8
308.72	-1.16	1.2	258.02	11DE	0.3	-3.4	-1.6	0.8
316.32	-1.03	1.3	268.22	21DE	-0.0	-3.4	-1.7	0.8
323.88	-0.90	1.4	278.39	31DE	-0.1	-3.4	-1.8	0.8

♂ LONG	♂ LAT	♂ MAG	⊙ LONG	16.00UT 12	☿	♀	♃	♄
					MAGNITUDES			
331.39	-0.77	1.4	288.54	10JA	-0.2	-3.5	-1.8	0.8
338.84	-0.65	1.5	298.65	20JA	-0.6	-3.5	-1.9	0.8
346.23	-0.52	1.5	308.71	30JA	-1.1	-3.6	-1.9	0.7
353.55	-0.39	1.6	318.72	9FE	-1.4	-3.6	-2.0	0.7
0.79	-0.27	1.6	328.67	19FE	-0.9	-3.7	-2.0	0.6
7.96	-0.15	1.7	338.56	29FE	0.3	-3.7	-2.0	0.6
15.05	-0.04	1.7	348.40	10MR	2.1	-3.8	-2.0	0.5
22.07	0.07	1.7	358.17	20MR	3.3	-3.9	-2.0	0.5
29.01	0.17	1.8	7.89	30MR	1.8	-4.0	-2.0	0.4
35.89	0.27	1.8	17.56	9AP	1.1	-4.1	-1.9	0.4
42.70	0.37	1.8	27.19	19AP	0.5	-4.2	-1.9	0.3
49.44	0.46	1.8	36.77	29AP	-0.1	-4.2	-1.8	0.3
56.12	0.54	1.9	46.33	9MY	-1.0	-4.1	-1.7	0.3
62.76	0.62	1.9	55.87	19MY	-1.8	-3.9	-1.7	0.4
69.33	0.70	1.9	65.39	29MY	-1.0	-3.3	-1.6	0.5
75.87	0.77	1.9	74.91	8JN	-0.2	-2.9	-1.6	0.5
82.36	0.84	1.9	84.44	18JN	0.3	-3.7	-1.5	0.6
88.82	0.90	1.9	93.97	28JN	0.8	-4.1	-1.4	0.7
95.25	0.96	2.0	103.54	8JL	1.3	-4.2	-1.4	0.7
101.64	1.02	2.0	113.14	18JL	2.3	-4.2	-1.3	0.7
108.02	1.07	2.0	122.77	28JL	2.7	-4.1	-1.3	0.8
114.37	1.12	2.0	132.45	7AU	0.5	-4.1	-1.3	0.8
120.71	1.16	2.0	142.19	17AU	-0.8	-4.0	-1.2	0.8
127.03	1.21	2.0	151.98	27AU	-1.2	-3.9	-1.2	0.8
133.33	1.25	2.0	161.83	6SE	-1.1	-3.8	-1.2	0.8
139.62	1.28	2.0	171.74	16SE	-0.6	-3.7	-1.2	0.8
145.90	1.31	2.0	181.70	26SE	-0.3	-3.7	-1.2	0.8
152.17	1.34	1.9	191.73	6OC	-0.2	-3.6	-1.2	0.8
158.41	1.37	1.9	201.80	16OC	-0.2	-3.6	-1.2	0.7
164.64	1.39	1.8	211.92	26OC	0.1	-3.5	-1.3	0.7
170.84	1.40	1.8	222.08	5NO	1.6	-3.5	-1.3	0.7
177.01	1.41	1.7	232.26	15NO	1.6	-3.4	-1.3	0.7
183.15	1.42	1.6	242.46	25NO	0.1	-3.4	-1.4	0.8
189.25	1.42	1.5	252.67	5DE	-0.2	-3.4	-1.4	0.8
195.28	1.41	1.4	262.88	15DE	-0.2	-3.4	-1.5	0.8
201.25	1.39	1.3	273.06	25DE	-0.4	-3.3	-1.6	0.8
				13				
207.14	1.36	1.2	283.23	4JA	-0.6	-3.3	-1.6	0.8
212.93	1.31	1.0	293.36	14JA	-1.1	-3.3	-1.7	0.8
218.60	1.25	0.9	303.44	24JA	-1.2	-3.3	-1.8	0.8
224.11	1.17	0.7	313.48	3FE	-0.9	-3.3	-1.8	0.7
229.43	1.06	0.5	323.46	13FE	0.4	-3.4	-1.9	0.7
234.51	0.92	0.3	333.38	23FE	2.5	-3.4	-1.9	0.7
239.29	0.74	0.1	343.25	5MR	2.5	-3.4	-2.0	0.6
243.69	0.52	-0.1	353.05	15MR	1.3	-3.4	-2.0	0.6
247.62	0.24	-0.4	2.80	25MR	0.8	-3.5	-2.0	0.5
250.93	-0.11	-0.7	12.50	4AP	0.4	-3.5	-2.0	0.4
253.46	-0.55	-1.0	22.15	14AP	-0.2	-3.4	-2.0	0.4
255.05	-1.08	-1.3	31.75	24AP	-1.0	-3.4	-2.0	0.3
255.50	-1.70	-1.6	41.33	4MY	-1.9	-3.4	-2.0	0.3
254.72	-2.40	-1.9	50.87	14MY	-1.0	-3.4	-1.9	0.3
252.79	-3.13	-2.2	60.40	24MY	-0.1	-3.3	-1.9	0.3
250.06	-3.80	-2.4	69.92	3JN	0.5	-3.3	-1.8	0.4
247.20	-4.33	-2.4	79.44	13JN	1.0	-3.3	-1.8	0.5
244.92	-4.65	-2.2	88.97	23JN	1.7	-3.3	-1.7	0.5
243.76	-4.78	-2.0	98.52	3JL	3.0	-3.3	-1.6	0.6
243.97	-4.74	-1.7	108.10	13JL	2.2	-3.3	-1.6	0.6
245.51	-4.59	-1.5	117.71	23JL	0.4	-3.4	-1.5	0.7
248.23	-4.36	-1.2	127.37	2AU	-0.9	-3.4	-1.5	0.7
251.94	-4.10	-1.0	137.07	12AU	-1.3	-3.4	-1.4	0.8
256.43	-3.82	-0.8	146.84	22AU	-1.1	-3.5	-1.4	0.8
261.55	-3.52	-0.6	156.66	1SE	-0.5	-3.5	-1.3	0.8
267.17	-3.22	-0.4	166.53	11SE	-0.2	-3.6	-1.3	0.8
273.16	-2.92	-0.2	176.47	21SE	-0.1	-3.7	-1.3	0.8
279.46	-2.62	-0.1	186.46	1OC	-0.0	-3.8	-1.3	0.8
285.98	-2.33	0.1	196.51	11OC	0.3	-3.8	-1.3	0.8
292.67	-2.06	0.3	206.61	21OC	1.9	-3.9	-1.3	0.7
299.49	-1.79	0.4	216.75	31OC	1.3	-4.1	-1.3	0.7
306.39	-1.54	0.6	226.91	10NO	-0.1	-4.2	-1.3	0.7
313.35	-1.30	0.7	237.11	20NO	-0.3	-4.3	-1.3	0.7
320.35	-1.07	0.8	247.32	30NO	-0.4	-4.4	-1.3	0.7
327.36	-0.86	0.9	257.52	10DE	-0.5	-4.4	-1.4	0.8
334.36	-0.66	1.1	267.72	20DE	-0.7	-4.3	-1.4	0.8
341.35	-0.48	1.2	277.90	30DE	-1.0	-4.0	-1.5	0.8

♂ LONG	♂ LAT	♂ MAG	⊙ LONG	16.00UT 14	☿	♀	♃	♄
					MAGNITUDES			
348.31	-0.31	1.3	288.05	9JA	-1.1	-3.4	-1.5	0.8
355.24	-0.15	1.4	298.16	19JA	-0.8	-3.5	-1.6	0.8
2.13	-0.01	1.5	308.22	29JA	0.6	-4.0	-1.6	0.8
8.98	0.12	1.5	318.23	8FE	2.9	-4.3	-1.7	0.8
15.78	0.25	1.6	328.19	18FE	1.9	-4.3	-1.8	0.7
22.53	0.36	1.7	338.09	28FE	1.0	-4.2	-1.9	0.7
29.23	0.46	1.7	347.92	10MR	0.6	-4.1	-1.9	0.7
35.89	0.55	1.8	357.70	20MR	0.2	-4.0	-2.0	0.6
42.50	0.64	1.8	7.42	30MR	-0.3	-3.9	-2.0	0.5
49.07	0.71	1.9	17.09	9AP	-1.0	-3.8	-2.1	0.5
55.60	0.78	1.9	26.72	19AP	-1.8	-3.7	-2.1	0.4
62.09	0.84	2.0	36.32	29AP	-1.0	-3.6	-2.1	0.4
68.55	0.90	2.0	45.87	9MY	-0.1	-3.6	-2.1	0.3
74.99	0.95	2.0	55.41	19MY	0.7	-3.5	-2.1	0.2
81.40	0.99	2.0	64.93	29MY	1.3	-3.5	-2.1	0.3
87.79	1.03	2.0	74.45	8JN	2.3	-3.4	-2.1	0.3
94.17	1.07	2.0	83.98	18JN	3.3	-3.4	-2.0	0.4
100.54	1.10	2.0	93.52	28JN	1.8	-3.4	-2.0	0.5
106.91	1.12	2.0	103.08	8JL	0.3	-3.3	-1.9	0.5
113.27	1.14	2.0	112.68	18JL	-0.9	-3.3	-1.8	0.6
119.64	1.15	2.0	122.31	28JL	-1.4	-3.3	-1.8	0.6
126.02	1.17	2.0	131.99	7AU	-1.1	-3.3	-1.7	0.7
132.41	1.17	2.0	141.72	17AU	-0.5	-3.3	-1.6	0.7
138.81	1.17	2.0	151.51	27AU	-0.1	-3.4	-1.6	0.8
145.23	1.17	2.0	161.35	6SE	0.0	-3.4	-1.5	0.8
151.67	1.16	2.0	171.26	16SE	0.2	-3.4	-1.5	0.8
158.14	1.15	2.0	181.22	26SE	0.5	-3.4	-1.4	0.8
164.63	1.13	2.0	191.24	6OC	2.3	-3.5	-1.4	0.8
171.14	1.10	1.9	201.31	16OC	1.2	-3.5	-1.4	0.8
177.68	1.07	1.9	211.42	26OC	-0.3	-3.5	-1.4	0.8
184.25	1.04	1.9	221.58	5NO	-0.5	-3.5	-1.4	0.7
190.84	0.99	1.8	231.76	15NO	-0.5	-3.5	-1.4	0.7
197.45	0.94	1.8	241.96	25NO	-0.6	-3.4	-1.4	0.7
204.09	0.87	1.7	252.17	5DE	-0.8	-3.4	-1.4	0.7
210.76	0.80	1.6	262.38	15DE	-0.9	-3.4	-1.4	0.7
217.44	0.72	1.5	272.57	25DE	-0.9	-3.4	-1.4	0.8
				15				
224.14	0.62	1.4	282.73	4JA	-0.7	-3.3	-1.4	0.8
230.85	0.51	1.4	292.87	14JA	0.7	-3.3	-1.5	0.8
237.58	0.38	1.2	302.95	24JA	3.1	-3.3	-1.5	0.8
244.31	0.23	1.1	312.99	3FE	1.4	-3.3	-1.6	0.8
251.04	0.07	1.0	322.98	13FE	0.7	-3.4	-1.6	0.8
257.76	-0.11	0.9	332.90	23FE	0.4	-3.4	-1.7	0.8
264.46	-0.32	0.8	342.77	5MR	0.1	-3.4	-1.8	0.7
271.13	-0.55	0.6	352.58	15MR	-0.3	-3.4	-1.8	0.7
277.75	-0.81	0.5	2.33	25MR	-1.0	-3.5	-1.9	0.6
284.30	-1.09	0.3	12.03	4AP	-1.8	-3.5	-2.0	0.6
290.75	-1.40	0.1	21.69	14AP	-1.0	-3.5	-2.0	0.5
297.05	-1.75	-0.0	31.29	24AP	-0.0	-3.6	-2.1	0.5
303.17	-2.12	-0.2	40.87	4MY	0.9	-3.7	-2.2	0.4
309.04	-2.53	-0.4	50.42	14MY	1.7	-3.7	-2.2	0.3
314.58	-2.97	-0.6	59.94	24MY	3.0	-3.8	-2.2	0.3
319.69	-3.44	-0.8	69.46	3JN	2.9	-3.9	-2.3	0.3
324.23	-3.94	-1.0	78.99	13JN	1.4	-4.0	-2.3	0.3
328.04	-4.46	-1.3	88.51	23JN	0.2	-4.1	-2.2	0.4
330.95	-4.99	-1.5	98.06	3JL	-1.0	-4.2	-2.2	0.4
332.70	-5.51	-1.8	107.64	13JL	-1.6	-4.2	-2.2	0.5
333.15	-5.95	-2.0	117.25	23JL	-1.1	-4.1	-2.1	0.5
332.21	-6.26	-2.3	126.91	2AU	-0.4	-3.8	-2.1	0.6
330.04	-6.33	-2.4	136.61	12AU	-0.0	-3.3	-2.0	0.6
327.21	-6.11	-2.5	146.37	22AU	0.2	-3.5	-1.9	0.7
324.44	-5.62	-2.4	156.18	1SE	0.4	-4.0	-1.9	0.7
322.47	-4.93	-2.1	166.06	11SE	0.8	-4.3	-1.8	0.8
321.71	-4.16	-1.8	175.99	21SE	2.6	-4.3	-1.7	0.8
322.26	-3.40	-1.4	185.98	1OC	1.0	-4.3	-1.7	0.8
324.03	-2.71	-1.1	196.02	11OC	-0.4	-4.2	-1.6	0.8
326.80	-2.11	-0.8	206.11	21OC	-0.7	-4.1	-1.6	0.8
330.39	-1.59	-0.5	216.25	31OC	-0.6	-4.0	-1.5	0.8
334.61	-1.14	-0.3	226.42	10NO	-0.7	-3.9	-1.5	0.8
339.32	-0.77	-0.0	236.61	20NO	-0.8	-3.8	-1.5	0.7
344.40	-0.45	0.2	246.82	30NO	-0.7	-3.8	-1.5	0.7
349.78	-0.18	0.4	257.03	10DE	-0.8	-3.7	-1.5	0.7
355.37	0.05	0.6	267.23	20DE	-0.5	-3.6	-1.5	0.7
1.13	0.25	0.8	277.41	30DE	0.8	-3.6	-1.5	0.8

Left table (16 / 17)

♂ LONG	♂ LAT	♂ MAG	☉ LONG	16.00UT	☿	♀	♃	♄
7.02	0.42	0.9	287.56	9JA	2.8	-3.5	-1.5	0.8
13.00	0.56	1.1	297.67	19JA	1.0	-3.5	-1.5	0.8
19.05	0.68	1.2	307.74	29JA	0.5	-3.4	-1.5	0.8
25.16	0.78	1.3	317.75	8FE	0.2	-3.4	-1.5	0.8
31.30	0.87	1.5	327.71	18FE	0.0	-3.3	-1.6	0.8
37.46	0.94	1.6	337.61	28FE	-0.4	-3.3	-1.6	0.8
43.64	1.01	1.6	347.45	9MR	-1.0	-3.3	-1.7	0.8
49.84	1.06	1.7	357.23	19MR	-1.7	-3.3	-1.7	0.8
56.04	1.10	1.8	6.96	29MR	-1.0	-3.3	-1.8	0.7
62.25	1.14	1.9	16.63	8AP	0.1	-3.3	-1.9	0.7
68.46	1.17	1.9	26.26	18AP	1.1	-3.3	-1.9	0.6
74.68	1.19	1.9	35.86	28AP	2.3	-3.4	-2.0	0.5
80.90	1.20	2.0	45.41	8MY	3.7	-3.4	-2.1	0.5
87.13	1.21	2.0	54.95	18MY	2.2	-3.4	-2.2	0.4
93.37	1.22	2.0	64.48	28MY	1.1	-3.5	-2.2	0.3
99.63	1.21	2.0	73.99	7JN	0.1	-3.5	-2.3	0.3
105.90	1.21	2.0	83.52	17JN	-1.0	-3.5	-2.3	0.3
112.19	1.20	2.0	93.06	27JN	-1.7	-3.4	-2.4	0.3
118.51	1.18	2.0	102.62	7JL	-1.0	-3.4	-2.4	0.4
124.85	1.16	2.0	112.21	17JL	-0.4	-3.4	-2.4	0.5
131.23	1.14	2.0	121.84	27JL	0.1	-3.3	-2.4	0.5
137.65	1.11	2.0	131.52	6AU	0.3	-3.3	-2.3	0.6
144.10	1.08	1.9	141.24	16AU	0.6	-3.3	-2.3	0.6
150.60	1.04	1.9	151.03	26AU	1.1	-3.3	-2.2	0.7
157.14	1.00	1.9	160.87	5SE	3.0	-3.3	-2.2	0.7
163.73	0.95	1.9	170.77	15SE	0.9	-3.3	-2.1	0.7
170.38	0.90	1.9	180.73	25SE	-0.6	-3.4	-2.0	0.8
177.08	0.84	1.9	190.75	5OC	-0.8	-3.4	-2.0	0.8
183.83	0.77	1.9	200.82	15OC	-0.8	-3.4	-1.9	0.8
190.64	0.70	1.9	210.93	25OC	-0.8	-3.5	-1.8	0.8
197.51	0.63	1.8	221.08	4NO	-0.6	-3.5	-1.8	0.8
204.44	0.54	1.8	231.26	14NO	-0.6	-3.6	-1.7	0.8
211.43	0.45	1.8	241.46	24NO	-0.6	-3.6	-1.7	0.8
218.47	0.35	1.7	251.67	4DE	-0.4	-3.7	-1.6	0.7
225.58	0.25	1.7	261.88	14DE	1.0	-3.8	-1.6	0.7
232.73	0.13	1.6	272.07	24DE	2.4	-3.9	-1.6	0.7

17

♂ LONG	♂ LAT	♂ MAG	☉ LONG	16.00UT	☿	♀	♃	♄
239.95	0.01	1.6	282.24	3JA	0.7	-3.9	-1.6	0.8
247.21	-0.12	1.5	292.37	13JA	0.3	-4.0	-1.5	0.8
254.52	-0.26	1.4	302.47	23JA	0.1	-4.1	-1.5	0.8
261.87	-0.40	1.4	312.50	2FE	-0.1	-4.2	-1.5	0.9
269.26	-0.55	1.3	322.49	12FE	-0.4	-4.3	-1.6	0.9
276.68	-0.71	1.2	332.43	22FE	-1.0	-4.3	-1.6	0.9
284.12	-0.88	1.1	342.30	4MR	-1.6	-4.1	-1.6	0.9
291.58	-1.04	1.1	352.11	14MR	-1.0	-3.8	-1.6	0.8
299.03	-1.21	1.0	1.87	24MR	0.1	-3.1	-1.7	0.8
306.46	-1.38	0.9	11.56	3AP	1.4	-3.5	-1.7	0.7
313.87	-1.54	0.8	21.22	13AP	3.0	-4.0	-1.8	0.7
321.22	-1.70	0.7	30.83	23AP	3.0	-4.2	-1.8	0.7
328.50	-1.86	0.7	40.40	3MY	1.7	-4.2	-1.9	0.6
335.69	-2.00	0.6	49.95	13MY	0.9	-4.1	-1.9	0.6
342.76	-2.13	0.5	59.48	23MY	0.0	-4.0	-2.0	0.5
349.67	-2.26	0.4	69.00	2JN	-1.0	-4.0	-2.1	0.4
356.39	-2.36	0.3	78.52	12JN	-1.7	-3.9	-2.2	0.4
2.87	-2.45	0.2	88.05	22JN	-1.0	-3.8	-2.2	0.3
9.07	-2.52	0.1	97.60	2JL	-0.3	-3.7	-2.3	0.3
14.92	-2.57	-0.1	107.18	12JL	0.2	-3.6	-2.3	0.4
20.34	-2.60	-0.2	116.79	22JL	0.5	-3.6	-2.4	0.4
25.25	-2.61	-0.3	126.44	1AU	0.8	-3.5	-2.4	0.5
29.50	-2.58	-0.5	136.14	11AU	1.5	-3.5	-2.4	0.5
32.94	-2.52	-0.7	145.90	21AU	3.2	-3.4	-2.4	0.6
35.39	-2.41	-0.9	155.71	31AU	0.8	-3.4	-2.4	0.7
36.64	-2.24	-1.1	165.58	10SE	-0.7	-3.4	-2.4	0.7
36.48	-2.00	-1.3	175.51	20SE	-1.0	-3.4	-2.4	0.7
34.87	-1.66	-1.5	185.49	30SE	-0.9	-3.4	-2.3	0.8
31.96	-1.22	-1.7	195.54	10OC	-0.8	-3.4	-2.2	0.8
28.31	-0.73	-1.8	205.63	20OC	-0.5	-3.4	-2.2	0.8
24.70	-0.22	-1.6	215.76	30OC	-0.4	-3.4	-2.1	0.8
21.88	0.24	-1.3	225.93	9NO	-0.4	-3.4	-2.0	0.8
20.33	0.63	-1.0	236.12	19NO	-0.2	-3.4	-2.0	0.8
20.12	0.93	-0.7	246.32	29NO	1.2	-3.4	-1.9	0.8
21.16	1.15	-0.4	256.53	9DE	2.1	-3.4	-1.8	0.8
23.25	1.32	-0.1	266.73	19DE	0.4	-3.4	-1.8	0.8
26.18	1.43	0.2	276.91	29DE	0.1	-3.5	-1.7	0.8

Right table (18 / 19)

♂ LONG	♂ LAT	♂ MAG	☉ LONG	16.00UT	☿	♀	♃	♄
29.76	1.51	0.4	287.06	8JA	-0.0	-3.5	-1.7	0.8
33.86	1.57	0.6	297.18	18JA	-0.2	-3.5	-1.6	0.9
38.36	1.60	0.8	307.24	28JA	-0.5	-3.4	-1.6	0.9
43.18	1.62	1.0	317.26	7FE	-1.1	-3.4	-1.6	0.9
48.24	1.62	1.1	327.22	17FE	-1.5	-3.4	-1.6	0.9
53.49	1.62	1.3	337.12	27FE	-1.0	-3.4	-1.6	0.9
58.91	1.61	1.4	346.97	9MR	0.2	-3.4	-1.6	0.9
64.45	1.59	1.5	356.75	19MR	1.8	-3.3	-1.6	0.9
70.10	1.57	1.6	6.48	29MR	3.7	-3.3	-1.6	0.9
75.83	1.54	1.7	16.16	8AP	2.2	-3.3	-1.6	0.9
81.65	1.51	1.8	25.79	18AP	1.3	-3.3	-1.6	0.8
87.52	1.48	1.8	35.38	28AP	0.6	-3.3	-1.7	0.8
93.47	1.44	1.9	44.95	8MY	-0.1	-3.4	-1.7	0.7
99.47	1.40	1.9	54.49	18MY	-1.0	-3.4	-1.8	0.7
105.52	1.35	1.9	64.01	28MY	-1.8	-3.4	-1.8	0.6
111.63	1.31	2.0	73.53	7JN	-1.1	-3.5	-1.9	0.5
117.80	1.26	2.0	83.06	17JN	-0.3	-3.5	-1.9	0.5
124.02	1.20	2.0	92.59	27JN	0.3	-3.6	-2.0	0.4
130.30	1.14	2.0	102.15	7JL	0.6	-3.6	-2.1	0.4
136.63	1.08	2.0	111.74	17JL	1.1	-3.7	-2.1	0.4
143.03	1.02	2.0	121.37	27JL	2.0	-3.8	-2.2	0.4
149.50	0.95	1.9	131.05	6AU	3.0	-3.9	-2.3	0.5
156.03	0.88	1.9	140.77	16AU	0.6	-4.0	-2.3	0.6
162.63	0.81	1.9	150.56	26AU	-0.8	-4.1	-2.4	0.6
169.31	0.73	1.8	160.40	5SE	-1.1	-4.2	-2.4	0.7
176.06	0.65	1.8	170.29	15SE	-1.1	-4.3	-2.4	0.7
182.89	0.57	1.8	180.25	25SE	-0.7	-4.3	-2.4	0.8
189.79	0.48	1.7	190.27	5OC	-0.4	-4.3	-2.4	0.8
196.78	0.39	1.7	200.33	15OC	-0.3	-3.9	-2.4	0.8
203.84	0.29	1.7	210.44	25OC	-0.3	-3.2	-2.4	0.9
210.98	0.19	1.7	220.59	4NO	-0.0	-3.5	-2.3	0.9
218.20	0.08	1.7	230.77	14NO	1.4	-4.1	-2.2	0.9
225.49	-0.02	1.6	240.97	24NO	1.8	-4.3	-2.2	0.9
232.86	-0.13	1.6	251.18	4DE	0.2	-4.4	-2.1	0.9
240.30	-0.25	1.6	261.38	14DE	-0.1	-4.3	-2.0	0.9
247.80	-0.36	1.6	271.57	24DE	-0.2	-4.2	-2.0	0.9

19

♂ LONG	♂ LAT	♂ MAG	☉ LONG	16.00UT	☿	♀	♃	♄
255.36	-0.48	1.5	281.74	3JA	-0.3	-4.1	-1.9	0.9
262.97	-0.59	1.5	291.88	13JA	-0.6	-4.0	-1.8	0.9
270.62	-0.71	1.5	301.97	23JA	-1.1	-3.9	-1.8	0.9
278.31	-0.82	1.4	312.02	2FE	-1.3	-3.8	-1.7	1.0
286.03	-0.93	1.4	322.01	12FE	-0.9	-3.7	-1.7	1.0
293.76	-1.03	1.4	331.94	22FE	0.3	-3.6	-1.6	1.0
301.50	-1.12	1.3	341.82	4MR	2.2	-3.6	-1.6	1.0
309.22	-1.21	1.3	351.63	14MR	3.0	-3.5	-1.6	1.0
316.93	-1.29	1.3	1.39	24MR	1.6	-3.4	-1.5	1.0
324.60	-1.35	1.2	11.10	3AP	1.0	-3.4	-1.5	1.0
332.22	-1.41	1.2	20.75	13AP	0.5	-3.4	-1.5	1.0
339.78	-1.45	1.2	30.37	23AP	-0.1	-3.3	-1.5	1.0
347.27	-1.47	1.1	39.94	3MY	-1.0	-3.3	-1.5	0.9
354.46	-1.48	1.1	49.49	13MY	-1.9	-3.3	-1.6	0.9
1.95	-1.48	1.1	59.02	23MY	-1.1	-3.3	-1.6	0.8
9.12	-1.46	1.0	68.55	2JN	-0.2	-3.3	-1.6	0.7
16.15	-1.42	1.0	78.06	12JN	0.4	-3.3	-1.7	0.7
23.04	-1.37	1.0	87.60	22JN	0.8	-3.3	-1.7	0.6
29.75	-1.30	0.9	97.14	2JL	1.4	-3.4	-1.8	0.5
36.27	-1.21	0.9	106.71	12JL	2.6	-3.4	-1.8	0.5
42.58	-1.11	0.8	116.32	22JL	2.5	-3.4	-1.9	0.5
48.63	-0.99	0.7	125.97	1AU	0.5	-3.5	-1.9	0.5
54.40	-0.85	0.6	135.67	11AU	-0.8	-3.5	-2.0	0.5
59.83	-0.69	0.5	145.42	21AU	-1.2	-3.5	-2.1	0.6
64.85	-0.50	0.4	155.23	31AU	-1.1	-3.5	-2.1	0.6
69.39	-0.29	0.3	165.09	10SE	-0.6	-3.4	-2.2	0.7
73.33	-0.04	0.1	175.02	20SE	-0.3	-3.4	-2.2	0.7
76.53	0.24	-0.0	185.01	30SE	-0.2	-3.4	-2.3	0.8
78.84	0.58	-0.2	195.04	10OC	-0.1	-3.3	-2.3	0.8
80.05	0.96	-0.5	205.13	20OC	0.2	-3.3	-2.4	0.9
79.97	1.38	-0.7	215.26	30OC	1.7	-3.3	-2.4	0.9
78.49	1.84	-0.9	225.43	9NO	1.6	-3.3	-2.4	0.9
75.68	2.29	-1.1	235.62	19NO	0.0	-3.3	-2.3	1.0
71.96	2.68	-1.2	245.82	29NO	-0.2	-3.4	-2.3	1.0
68.01	2.97	-1.2	256.03	9DE	-0.3	-3.4	-2.2	1.0
64.60	3.11	-0.9	266.23	19DE	-0.4	-3.4	-2.2	1.0
62.30	3.14	-0.7	276.42	29DE	-0.6	-3.4	-2.1	1.0

♂			☉	16.00UT	☿	♀	♃	♄
LONG	LAT	MAG	LONG	20	MAGNITUDES			
61.31	3.08	-0.4	286.57	8JA	-1.1	-3.5	-2.0	1.0
61.61	2.97	-0.1	296.69	18JA	-1.2	-3.5	-2.0	1.0
63.03	2.84	0.2	306.76	28JA	-0.9	-3.6	-1.9	1.0
65.36	2.70	0.4	316.77	7FE	0.4	-3.6	-1.8	1.0
68.43	2.56	0.6	326.74	17FE	2.7	-3.7	-1.7	1.1
72.09	2.42	0.8	336.65	27FE	2.3	-3.7	-1.7	1.1
76.21	2.29	1.0	346.49	8MR	1.2	-3.8	-1.6	1.1
80.69	2.16	1.1	356.28	18MR	0.7	-3.9	-1.6	1.2
85.48	2.04	1.3	6.01	28MR	0.3	-4.0	-1.5	1.2
90.51	1.92	1.4	15.69	7AP	-0.2	-4.1	-1.5	1.2
95.75	1.81	1.5	25.33	17AP	-1.0	-4.2	-1.5	1.1
101.16	1.70	1.6	34.92	27AP	-1.9	-4.2	-1.5	1.1
106.72	1.60	1.6	44.48	7MY	-1.1	-4.1	-1.5	1.1
112.41	1.49	1.7	54.03	17MY	-0.1	-3.9	-1.4	1.0
118.23	1.39	1.7	63.55	27MY	0.5	-3.3	-1.4	1.0
124.15	1.29	1.8	73.07	6JN	1.1	-2.9	-1.5	0.9
130.18	1.19	1.8	82.60	16JN	1.9	-3.7	-1.5	0.9
136.32	1.09	1.8	92.13	26JN	3.2	-4.1	-1.5	0.8
142.55	0.99	1.8	101.69	6JL	2.1	-4.2	-1.5	0.7
148.88	0.89	1.8	111.28	16JL	0.4	-4.2	-1.6	0.7
155.30	0.79	1.8	120.90	26JL	-0.9	-4.1	-1.6	0.6
161.83	0.68	1.8	130.58	5AU	-1.4	-4.0	-1.6	0.6
168.45	0.58	1.8	140.30	15AU	-1.1	-4.0	-1.7	0.6
175.16	0.48	1.8	150.08	25AU	-0.5	-3.9	-1.8	0.7
181.98	0.37	1.8	159.92	4SE	-0.2	-3.8	-1.8	0.7
188.89	0.27	1.7	169.81	14SE	-0.0	-3.7	-1.9	0.8
195.90	0.16	1.7	179.76	24SE	0.1	-3.7	-2.0	0.8
203.01	0.05	1.7	189.78	4OC	0.4	-3.6	-2.0	0.9
210.21	-0.06	1.6	199.84	14OC	2.0	-3.6	-2.1	0.9
217.51	-0.16	1.6	209.95	24OC	1.4	-3.5	-2.1	1.0
224.89	-0.27	1.5	220.10	3NO	-0.2	-3.5	-2.2	1.0
232.36	-0.37	1.5	230.28	13NO	-0.4	-3.4	-2.2	1.0
239.91	-0.47	1.4	240.47	23NO	-0.4	-3.4	-2.2	1.1
247.53	-0.57	1.4	250.68	3DE	-0.5	-3.4	-2.2	1.1
255.22	-0.67	1.4	260.89	13DE	-0.7	-3.4	-2.2	1.1
262.96	-0.76	1.4	271.08	23DE	-1.0	-3.4	-2.2	1.1
				21				
270.74	-0.84	1.4	281.25	2JA	-1.0	-3.3	-2.2	1.1
278.57	-0.92	1.4	291.39	12JA	-0.8	-3.3	-2.1	1.1
286.41	-0.98	1.4	301.48	22JA	0.5	-3.3	-2.0	1.1
294.26	-1.04	1.4	311.53	1FE	3.0	-3.3	-2.0	1.1
302.12	-1.09	1.4	321.53	11FE	1.7	-3.4	-1.9	1.2
309.96	-1.13	1.4	331.46	21FE	0.9	-3.4	-1.8	1.2
317.77	-1.15	1.4	341.34	3MR	0.5	-3.4	-1.8	1.2
325.54	-1.16	1.4	351.16	13MR	0.2	-3.4	-1.7	1.3
333.27	-1.16	1.4	0.91	23MR	-0.3	-3.5	-1.6	1.3
340.93	-1.15	1.4	10.62	2AP	-1.0	-3.5	-1.6	1.3
348.53	-1.12	1.4	20.28	12AP	-1.8	-3.4	-1.5	1.3
356.04	-1.09	1.4	29.89	22AP	-1.1	-3.4	-1.5	1.3
3.47	-1.04	1.4	39.47	2MY	-0.1	-3.4	-1.4	1.3
10.80	-0.98	1.4	49.02	12MY	0.7	-3.4	-1.4	1.3
18.04	-0.90	1.4	58.55	22MY	1.5	-3.3	-1.4	1.2
25.16	-0.82	1.4	68.08	1JN	2.5	-3.3	-1.4	1.2
32.17	-0.73	1.4	77.60	11JN	3.3	-3.3	-1.3	1.1
39.06	-0.63	1.4	87.12	21JN	1.7	-3.3	-1.3	1.1
45.82	-0.52	1.4	96.67	1JL	0.3	-3.3	-1.3	1.0
52.45	-0.40	1.4	106.24	11JL	-0.9	-3.3	-1.4	0.9
58.94	-0.27	1.4	115.85	21JL	-1.5	-3.4	-1.4	0.9
65.27	-0.13	1.3	125.50	31JL	-1.1	-3.4	-1.4	0.8
71.43	0.02	1.3	135.20	10AU	-0.5	-3.4	-1.4	0.7
77.40	0.19	1.2	144.94	20AU	-0.1	-3.5	-1.5	0.7
83.16	0.36	1.2	154.75	30AU	0.1	-3.5	-1.5	0.8
88.67	0.55	1.1	164.62	9SE	0.2	-3.6	-1.5	0.8
93.90	0.76	1.0	174.54	19SE	0.6	-3.7	-1.6	0.9
98.78	0.98	0.9	184.52	29SE	2.3	-3.8	-1.7	0.9
103.26	1.23	0.7	194.56	9OC	1.2	-3.8	-1.7	1.0
107.25	1.51	0.6	204.64	19OC	-0.3	-3.9	-1.8	1.0
110.62	1.83	0.4	214.77	29OC	-0.6	-4.1	-1.9	1.1
113.23	2.18	0.2	224.94	8NO	-0.6	-4.2	-1.9	1.1
114.91	2.57	-0.0	235.12	18NO	-0.6	-4.3	-2.0	1.2
115.47	2.99	-0.2	245.33	28NO	-0.8	-4.4	-2.0	1.2
114.74	3.43	-0.5	255.54	8DE	-0.8	-4.4	-2.1	1.2
112.68	3.84	-0.7	265.74	18DE	-0.9	-4.3	-2.1	1.2
109.45	4.17	-0.9	275.92	28DE	-0.6	-4.0	-2.1	1.3

♂			☉	16.00UT	☿	♀	♃	♄
LONG	LAT	MAG	LONG	22	MAGNITUDES			
105.56	4.36	-1.0	286.08	7JA	0.7	-3.3	-2.1	1.3
101.70	4.38	-0.9	296.19	17JA	3.0	-3.5	-2.1	1.3
98.58	4.24	-0.7	306.27	27JA	1.3	-4.1	-2.1	1.3
96.63	3.99	-0.4	316.29	6FE	0.6	-4.3	-2.0	1.3
95.98	3.69	-0.2	326.26	16FE	0.3	-4.3	-2.0	1.3
96.56	3.38	0.1	336.17	26FE	0.1	-4.2	-1.9	1.4
98.20	3.07	0.3	346.01	8MR	-0.3	-4.1	-1.9	1.4
100.70	2.79	0.5	355.80	18MR	-1.0	-4.0	-1.8	1.4
103.91	2.52	0.7	5.54	28MR	-1.8	-3.9	-1.7	1.3
107.69	2.28	0.8	15.22	7AP	-1.1	-3.8	-1.7	1.3
111.93	2.05	1.0	24.86	17AP	-0.0	-3.7	-1.6	1.3
116.55	1.84	1.1	34.46	27AP	0.9	-3.6	-1.5	1.3
121.48	1.65	1.2	44.02	7MY	1.9	-3.6	-1.5	1.2
126.68	1.47	1.3	53.56	17MY	3.3	-3.5	-1.4	1.2
132.12	1.30	1.4	63.09	27MY	2.7	-3.5	-1.4	1.2
137.76	1.14	1.4	72.61	6JN	1.3	-3.4	-1.3	1.1
143.57	0.98	1.5	82.13	16JN	0.2	-3.4	-1.3	1.1
149.56	0.83	1.5	91.67	26JN	-0.9	-3.4	-1.3	1.0
155.71	0.68	1.6	101.23	6JL	-1.6	-3.3	-1.3	1.0
162.00	0.54	1.6	110.81	16JL	-1.1	-3.3	-1.3	0.9
168.43	0.41	1.6	120.44	26JL	-0.4	-3.3	-1.3	0.9
175.00	0.28	1.6	130.11	5AU	-0.0	-3.3	-1.3	0.8
181.69	0.15	1.6	139.83	15AU	0.2	-3.3	-1.3	0.8
188.52	0.02	1.6	149.61	25AU	0.4	-3.4	-1.3	0.8
195.47	-0.10	1.6	159.44	4SE	0.9	-3.4	-1.3	0.8
202.53	-0.22	1.6	169.33	14SE	2.7	-3.4	-1.4	0.9
209.72	-0.33	1.6	179.29	24SE	1.0	-3.4	-1.4	0.9
217.01	-0.44	1.6	189.29	4OC	-0.5	-3.5	-1.4	1.0
224.41	-0.54	1.5	199.35	14OC	-0.7	-3.5	-1.5	1.1
231.90	-0.63	1.5	209.46	24OC	-0.7	-3.5	-1.5	1.1
239.48	-0.72	1.5	219.60	3NO	-0.7	-3.5	-1.6	1.2
247.15	-0.80	1.5	229.78	13NO	-0.7	-3.4	-1.7	1.2
254.89	-0.87	1.4	239.98	23NO	-0.7	-3.4	-1.7	1.3
262.68	-0.94	1.4	250.18	3DE	-0.7	-3.4	-1.8	1.3
270.52	-0.99	1.4	260.39	13DE	-0.5	-3.4	-1.9	1.3
278.40	-1.03	1.4	270.58	23DE	0.8	-3.4	-1.9	1.3
				23				
286.29	-1.06	1.3	280.75	2JA	2.7	-3.3	-2.0	1.3
294.19	-1.08	1.3	290.89	12JA	0.9	-3.3	-2.0	1.3
302.09	-1.09	1.3	300.99	22JA	0.4	-3.3	-2.0	1.2
309.97	-1.08	1.3	311.04	1FE	0.2	-3.3	-2.0	1.2
317.81	-1.07	1.3	321.04	11FE	-0.0	-3.4	-2.0	1.1
325.61	-1.04	1.4	330.98	21FE	-0.4	-3.4	-2.0	1.1
333.36	-1.00	1.4	340.86	3MR	-1.0	-3.4	-2.0	1.1
341.04	-0.95	1.4	350.68	13MR	-1.6	-3.4	-1.9	1.1
348.65	-0.90	1.5	0.44	23MR	-1.1	-3.5	-1.9	1.1
356.18	-0.83	1.5	10.15	2AP	0.1	-3.5	-1.8	1.1
3.63	-0.75	1.5	19.82	12AP	1.2	-3.5	-1.8	1.0
10.99	-0.67	1.6	29.43	22AP	2.5	-3.6	-1.7	1.0
18.26	-0.58	1.6	39.01	2MY	3.5	-3.7	-1.6	1.0
25.44	-0.49	1.6	48.57	12MY	2.0	-3.7	-1.6	1.0
32.52	-0.39	1.6	58.10	22MY	1.0	-3.8	-1.5	1.0
39.51	-0.28	1.7	67.62	1JN	0.1	-3.9	-1.5	0.9
46.40	-0.17	1.7	77.14	11JN	-0.9	-4.0	-1.4	0.9
53.19	-0.06	1.7	86.67	21JN	-1.7	-4.1	-1.4	0.8
59.88	0.05	1.7	96.21	1JL	-1.1	-4.2	-1.3	0.8
66.48	0.17	1.7	105.78	11JL	-0.4	-4.2	-1.3	0.7
72.97	0.30	1.7	115.39	21JL	0.1	-4.1	-1.3	0.7
79.37	0.42	1.7	125.03	31JL	0.4	-3.8	-1.2	0.6
85.65	0.55	1.7	134.73	10AU	0.6	-3.3	-1.2	0.6
91.82	0.69	1.6	144.47	20AU	1.3	-3.5	-1.2	0.5
97.86	0.83	1.6	154.27	30AU	3.1	-4.0	-1.2	0.5
103.77	0.97	1.6	164.14	9SE	0.9	-4.3	-1.2	0.5
109.52	1.13	1.5	174.06	19SE	-0.6	-4.3	-1.2	0.5
115.10	1.29	1.4	184.04	29SE	-0.9	-4.3	-1.3	0.6
120.48	1.47	1.3	194.07	9OC	-0.8	-4.2	-1.3	0.7
125.61	1.65	1.2	204.15	19OC	-0.8	-4.1	-1.3	0.7
130.47	1.86	1.1	214.28	29OC	-0.6	-4.0	-1.4	0.8
134.97	2.08	1.0	224.45	8NO	-0.5	-3.9	-1.4	0.9
139.06	2.32	0.8	234.63	18NO	-0.5	-3.8	-1.5	0.9
142.62	2.59	0.6	244.83	28NO	-0.3	-3.8	-1.5	0.9
145.52	2.88	0.4	255.05	8DE	1.0	-3.7	-1.6	1.0
147.63	3.20	0.2	265.25	18DE	2.4	-3.6	-1.7	1.0
148.75	3.53	-0.1	275.43	28DE	0.6	-3.6	-1.7	1.0

Year 24

♂ LONG	LAT	MAG	☉ LONG	16.00UT 24	☿	♀	♃	♄
148.70	3.88	-0.3	285.59	7JA	0.2	-3.5	-1.8	1.0
147.37	4.19	-0.6	295.71	17JA	0.1	-3.5	-1.9	1.0
144.79	4.42	-0.8	305.78	27JA	-0.1	-3.4	-1.9	1.0
141.25	4.52	-1.0	315.81	6FE	-0.5	-3.4	-2.0	0.9
137.35	4.44	-1.0	325.78	16FE	-1.0	-3.3	-2.0	0.9
133.81	4.19	-0.9	335.69	26FE	-1.5	-3.3	-2.0	0.9
131.21	3.83	-0.7	345.54	7MR	-1.0	-3.3	-2.0	0.8
129.88	3.40	-0.4	355.33	17MR	0.1	-3.3	-2.0	0.8
129.83	2.97	-0.2	5.07	27MR	1.5	-3.3	-2.0	0.8
130.97	2.56	-0.0	14.76	6AP	3.3	-3.3	-2.0	0.8
133.11	2.18	0.2	24.40	16AP	2.7	-3.3	-1.9	0.8
136.07	1.84	0.4	34.00	26AP	1.5	-3.4	-1.9	0.8
139.71	1.52	0.5	43.56	6MY	0.8	-3.4	-1.8	0.8
143.90	1.24	0.7	53.10	16MY	0.0	-3.4	-1.7	0.8
148.56	0.99	0.8	62.63	26MY	-1.0	-3.5	-1.7	0.7
153.60	0.75	0.9	72.15	5JN	-1.8	-3.5	-1.6	0.7
158.98	0.53	1.0	81.67	15JN	-1.1	-3.5	-1.5	0.7
164.64	0.34	1.0	91.20	25JN	-0.3	-3.4	-1.5	0.6
170.56	0.15	1.1	100.76	5JL	0.2	-3.4	-1.4	0.6
176.70	-0.02	1.1	110.34	15JL	0.5	-3.4	-1.4	0.5
183.06	-0.18	1.2	119.97	25JL	0.9	-3.3	-1.3	0.5
189.60	-0.33	1.2	129.64	4AU	1.7	-3.3	-1.3	0.4
196.32	-0.47	1.3	139.35	14AU	3.2	-3.3	-1.3	0.4
203.21	-0.59	1.3	149.13	24AU	0.8	-3.3	-1.2	0.3
210.25	-0.71	1.3	158.96	3SE	-0.7	-3.3	-1.2	0.3
217.43	-0.81	1.3	168.84	13SE	-1.0	-3.3	-1.2	0.2
224.73	-0.90	1.3	178.80	23SE	-1.0	-3.4	-1.2	0.2
232.16	-0.98	1.3	188.80	3OC	-0.7	-3.4	-1.2	0.3
239.69	-1.05	1.3	198.86	13OC	-0.5	-3.4	-1.2	0.3
247.32	-1.10	1.3	208.97	23OC	-0.4	-3.5	-1.3	0.4
255.03	-1.14	1.4	219.11	2NO	-0.4	-3.5	-1.3	0.5
262.80	-1.17	1.4	229.29	12NO	-0.2	-3.6	-1.3	0.5
270.62	-1.19	1.4	239.49	22NO	1.2	-3.6	-1.4	0.6
278.48	-1.19	1.4	249.69	2DE	2.1	-3.7	-1.4	0.7
286.36	-1.18	1.4	259.90	12DE	0.4	-3.8	-1.5	0.7
294.24	-1.15	1.4	270.10	22DE	0.0	-3.9	-1.5	0.7

Year 25

♂ LONG	LAT	MAG	☉ LONG	16.00UT 25	☿	♀	♃	♄
302.11	-1.12	1.4	280.27	1JA	-0.1	-4.0	-1.6	0.7
309.96	-1.07	1.4	290.41	11JA	-0.2	-4.0	-1.6	0.8
317.77	-1.02	1.4	300.51	21JA	-0.5	-4.1	-1.7	0.8
325.54	-0.95	1.4	310.56	31JA	-1.1	-4.2	-1.8	0.7
333.24	-0.88	1.5	320.56	10FE	-1.4	-4.3	-1.8	0.7
340.88	-0.80	1.5	330.50	20FE	-1.0	-4.3	-1.9	0.7
348.45	-0.72	1.5	340.38	2MR	0.2	-4.1	-2.0	0.7
355.94	-0.62	1.5	350.21	12MR	1.9	-3.7	-2.0	0.6
3.34	-0.53	1.5	359.97	22MR	3.5	-3.1	-2.0	0.6
10.66	-0.43	1.5	9.69	1AP	2.0	-3.5	-2.1	0.6
17.90	-0.33	1.5	19.35	11AP	1.1	-4.0	-2.1	0.6
25.04	-0.23	1.6	28.97	21AP	0.6	-4.2	-2.0	0.6
32.11	-0.13	1.7	38.55	1MY	-0.1	-4.2	-2.0	0.6
39.08	-0.02	1.7	48.10	11MY	-1.0	-4.1	-2.0	0.6
45.98	0.08	1.7	57.64	21MY	-1.8	-4.0	-1.9	0.6
52.80	0.18	1.8	67.16	31MY	-1.1	-3.9	-1.9	0.6
59.53	0.28	1.8	76.67	10JN	-0.3	-3.9	-1.8	0.6
66.20	0.38	1.8	86.20	20JN	0.3	-3.8	-1.8	0.5
72.79	0.49	1.9	95.75	30JN	0.7	-3.7	-1.7	0.5
79.32	0.59	1.9	105.32	10JL	1.2	-3.6	-1.6	0.5
85.78	0.69	1.9	114.92	20JL	2.2	-3.6	-1.6	0.4
92.17	0.79	1.9	124.56	30JL	2.9	-3.5	-1.5	0.4
98.49	0.88	1.9	134.25	9AU	0.6	-3.5	-1.5	0.3
104.75	0.98	1.9	144.00	19AU	-0.8	-3.4	-1.4	0.2
110.94	1.08	1.9	153.80	29AU	-1.2	-3.4	-1.4	0.2
117.06	1.18	1.8	163.66	8SE	-1.1	-3.4	-1.3	0.1
123.09	1.29	1.8	173.58	18SE	-0.7	-3.4	-1.3	0.1
129.04	1.39	1.8	183.55	28SE	-0.4	-3.4	-1.3	0.0
134.89	1.50	1.7	193.58	8OC	-0.3	-3.4	-1.3	-0.0
140.63	1.61	1.6	203.67	18OC	-0.2	-3.4	-1.3	0.1
146.24	1.72	1.5	213.79	28OC	0.0	-3.4	-1.3	0.1
151.70	1.84	1.4	223.95	7NO	1.4	-3.4	-1.3	0.2
156.97	1.97	1.3	234.14	17NO	1.8	-3.4	-1.3	0.3
162.02	2.10	1.2	244.34	27NO	0.1	-3.4	-1.3	0.3
166.81	2.24	1.0	254.55	7DE	-0.2	-3.4	-1.4	0.4
171.26	2.38	0.9	264.75	17DE	-0.2	-3.4	-1.4	0.5
175.31	2.54	0.7	274.94	27DE	-0.3	-3.5	-1.4	0.5

Year 26

♂ LONG	LAT	MAG	☉ LONG	16.00UT 26	☿	♀	♃	♄
178.85	2.70	0.5	285.10	6JA	-0.6	-3.5	-1.5	0.5
181.75	2.87	0.2	295.22	16JA	-1.1	-3.5	-1.5	0.5
183.88	3.03	-0.0	305.29	26JA	-1.3	-3.4	-1.6	0.6
185.03	3.18	-0.3	315.32	5FE	-0.9	-3.4	-1.7	0.6
185.04	3.30	-0.6	325.30	15FE	0.3	-3.4	-1.7	0.6
183.80	3.36	-0.9	335.21	25FE	2.3	-3.4	-1.8	0.5
181.32	3.31	-1.2	345.06	7MR	2.7	-3.4	-1.9	0.5
177.92	3.13	-1.4	354.86	17MR	1.5	-3.3	-1.9	0.5
174.19	2.81	-1.4	4.60	27MR	0.9	-3.3	-2.0	0.4
170.86	2.38	-1.2	14.29	6AP	0.4	-3.3	-2.1	0.4
168.51	1.89	-1.1	23.93	16AP	-0.1	-3.3	-2.1	0.4
167.47	1.40	-0.8	33.53	26AP	-1.0	-3.3	-2.1	0.4
167.75	0.95	-0.6	43.10	6MY	-1.9	-3.4	-2.1	0.4
169.25	0.54	-0.4	52.64	16MY	-1.1	-3.4	-2.1	0.4
171.79	0.18	-0.2	62.16	26MY	-0.2	-3.4	-2.1	0.4
175.18	-0.13	-0.1	71.69	5JN	0.4	-3.5	-2.1	0.4
179.30	-0.40	0.1	81.21	15JN	0.9	-3.5	-2.1	0.4
184.00	-0.63	0.2	90.74	25JN	1.6	-3.6	-2.0	0.4
189.20	-0.83	0.3	100.30	5JL	2.8	-3.6	-2.0	0.4
194.81	-1.00	0.4	109.88	15JL	2.4	-3.7	-1.9	0.4
200.77	-1.15	0.5	119.50	25JL	0.5	-3.8	-1.8	0.3
207.04	-1.28	0.6	129.17	4AU	-0.8	-3.9	-1.8	0.3
213.57	-1.38	0.7	138.88	14AU	-1.3	-4.0	-1.7	0.2
220.34	-1.46	0.7	148.65	24AU	-1.1	-4.1	-1.6	0.2
227.30	-1.52	0.8	158.48	3SE	-0.6	-4.2	-1.6	0.1
234.44	-1.56	0.9	168.37	13SE	-0.3	-4.3	-1.5	0.0
241.73	-1.58	0.9	178.31	23SE	-0.1	-4.3	-1.5	-0.0
249.15	-1.58	1.0	188.32	3OC	-0.0	-4.3	-1.5	-0.1
256.68	-1.57	1.0	198.37	13OC	0.2	-3.9	-1.4	-0.1
264.29	-1.54	1.1	208.47	23OC	1.7	-3.2	-1.4	-0.2
271.96	-1.50	1.1	218.62	2NO	1.6	-3.5	-1.4	-0.1
279.68	-1.44	1.2	228.79	12NO	-0.0	-4.1	-1.4	-0.0
287.42	-1.37	1.2	238.99	22NO	-0.3	-4.3	-1.4	0.1
295.17	-1.29	1.3	249.20	2DE	-0.3	-4.4	-1.4	0.1
302.92	-1.20	1.3	259.40	12DE	-0.4	-4.3	-1.4	0.2
310.64	-1.10	1.3	269.60	22DE	-0.7	-4.2	-1.4	0.3

Year 27

♂ LONG	LAT	MAG	☉ LONG	16.00UT 27	☿	♀	♃	♄
318.32	-0.99	1.4	279.78	1JA	-1.0	-4.1	-1.4	0.3
325.96	-0.89	1.4	289.92	11JA	-1.1	-4.0	-1.5	0.4
333.54	-0.77	1.5	300.02	21JA	-0.8	-3.9	-1.5	0.4
341.06	-0.66	1.5	310.08	31JA	0.4	-3.8	-1.6	0.4
348.51	-0.55	1.6	320.08	10FE	2.7	-3.7	-1.6	0.4
355.88	-0.43	1.6	330.02	20FE	2.1	-3.6	-1.7	0.4
3.17	-0.32	1.6	339.91	2MR	1.1	-3.6	-1.7	0.4
10.38	-0.21	1.7	349.74	12MR	0.6	-3.5	-1.8	0.4
17.52	-0.10	1.7	359.51	22MR	0.3	-3.4	-1.9	0.4
24.57	0.01	1.7	9.22	1AP	-0.2	-3.4	-1.9	0.3
31.55	0.11	1.7	18.89	11AP	-1.0	-3.4	-2.0	0.3
38.45	0.21	1.8	28.51	21AP	-1.9	-3.3	-2.1	0.3
45.28	0.31	1.8	38.09	1MY	-1.1	-3.3	-2.1	0.3
52.05	0.40	1.8	47.65	11MY	-0.1	-3.3	-2.2	0.3
58.75	0.49	1.8	57.18	21MY	0.6	-3.3	-2.2	0.3
65.39	0.57	1.8	66.70	31MY	1.2	-3.3	-2.3	0.3
71.98	0.66	1.9	76.22	10JN	2.1	-3.3	-2.3	0.3
78.52	0.73	1.9	85.75	20JN	3.3	-3.3	-2.3	0.4
85.01	0.81	1.9	95.29	30JN	2.0	-3.4	-2.2	0.4
91.46	0.88	2.0	104.86	10JL	0.4	-3.4	-2.2	0.3
97.87	0.95	2.0	114.46	20JL	-0.9	-3.4	-2.2	0.3
104.25	1.02	2.0	124.10	30JL	-1.4	-3.5	-2.1	0.3
110.59	1.08	2.0	133.78	9AU	-1.1	-3.5	-2.1	0.3
116.91	1.14	2.0	143.52	19AU	-0.5	-3.5	-2.0	0.2
123.19	1.20	2.0	153.32	29AU	-0.2	-3.4	-1.9	0.2
129.45	1.25	2.0	163.17	8SE	0.0	-3.4	-1.9	0.1
135.68	1.31	2.0	173.09	18SE	0.1	-3.4	-1.8	0.0
141.87	1.36	1.9	183.06	28SE	0.5	-3.4	-1.7	-0.0
148.03	1.41	1.9	193.09	8OC	2.0	-3.3	-1.7	-0.1
154.16	1.45	1.8	203.17	18OC	1.4	-3.3	-1.6	-0.2
160.24	1.50	1.8	213.29	28OC	-0.2	-3.3	-1.6	-0.2
166.26	1.54	1.7	223.45	7NO	-0.5	-3.3	-1.6	-0.2
172.23	1.57	1.6	233.63	17NO	-0.5	-3.3	-1.5	-0.2
178.13	1.61	1.5	243.84	27NO	-0.5	-3.4	-1.5	-0.1
183.93	1.64	1.4	254.05	7DE	-0.7	-3.4	-1.5	-0.0
189.63	1.66	1.3	264.25	17DE	-0.9	-3.4	-1.5	0.0
195.20	1.68	1.2	274.44	27DE	-0.9	-3.4	-1.5	0.1

♂			⊙	16.00UT	☿	♀	♃	♄
LONG	LAT	MAG	LONG	28	MAGNITUDES			
200.62	1.69	1.0	284.60	6JA	-0.7	-3.5	-1.5	0.2
205.85	1.69	0.8	294.73	16JA	0.5	-3.5	-1.5	0.2
210.83	1.67	0.7	304.81	26JA	3.0	-3.6	-1.5	0.3
215.52	1.63	0.5	314.84	5FE	1.5	-3.6	-1.5	0.3
219.85	1.57	0.2	324.82	15FE	0.8	-3.7	-1.6	0.3
223.71	1.48	-0.0	334.73	25FE	0.4	-3.8	-1.6	0.3
226.99	1.35	-0.3	344.59	6MR	0.2	-3.8	-1.7	0.3
229.54	1.16	-0.6	354.39	16MR	-0.3	-3.9	-1.7	0.3
231.18	0.91	-0.9	4.14	26MR	-1.0	-4.0	-1.8	0.3
231.74	0.57	-1.2	13.82	5AP	-1.8	-4.1	-1.8	0.3
231.06	0.13	-1.5	23.47	15AP	-1.1	-4.2	-1.9	0.3
229.17	-0.39	-1.8	33.07	25AP	-0.1	-4.2	-2.0	0.2
226.32	-0.98	-2.1	42.64	5MY	0.8	-4.1	-2.0	0.2
223.08	-1.57	-2.1	52.18	15MY	1.6	-3.9	-2.1	0.2
220.19	-2.09	-2.0	61.71	25MY	2.8	-3.2	-2.2	0.2
218.28	-2.51	-1.8	71.23	4JN	3.1	-2.9	-2.2	0.3
217.68	-2.81	-1.6	80.75	14JN	1.6	-3.7	-2.3	0.3
218.47	-3.00	-1.3	90.28	24JN	0.3	-4.1	-2.3	0.3
220.52	-3.11	-1.1	99.83	4JL	-0.9	-4.2	-2.4	0.3
223.66	-3.14	-0.9	109.42	14JL	-1.5	-4.2	-2.4	0.3
227.69	-3.13	-0.7	119.04	24JL	-1.1	-4.1	-2.4	0.3
232.44	-3.08	-0.5	128.70	3AU	-0.5	-4.0	-2.4	0.3
237.78	-3.00	-0.4	138.41	13AU	-0.1	-4.0	-2.3	0.3
243.60	-2.89	-0.2	148.18	23AU	0.1	-3.9	-2.3	0.2
249.79	-2.76	-0.1	158.00	2SE	0.3	-3.8	-2.2	0.2
256.28	-2.61	0.1	167.89	12SE	0.7	-3.7	-2.2	0.1
263.02	-2.45	0.2	177.83	22SE	2.4	-3.7	-2.1	0.1
269.94	-2.28	0.3	187.82	2OC	1.2	-3.6	-2.0	-0.0
277.01	-2.10	0.4	197.88	12OC	-0.4	-3.6	-2.0	-0.1
284.19	-1.91	0.6	207.98	22OC	-0.6	-3.5	-1.9	-0.2
291.44	-1.73	0.7	218.12	1NO	-0.6	-3.5	-1.8	-0.2
298.74	-1.54	0.8	228.29	11NO	-0.7	-3.4	-1.8	-0.3
306.07	-1.35	0.9	238.49	21NO	-0.8	-3.4	-1.7	-0.3
313.39	-1.17	1.0	248.69	1DE	-0.8	-3.4	-1.7	-0.2
320.72	-0.99	1.1	258.90	11DE	-0.8	-3.4	-1.6	-0.1
328.01	-0.82	1.2	269.10	21DE	-0.6	-3.4	-1.6	-0.1
335.27	-0.66	1.3	279.28	31DE	0.7	-3.3	-1.6	-0.0
				29				
342.48	-0.50	1.3	289.42	10JA	3.0	-3.3	-1.6	0.1
349.65	-0.35	1.4	299.53	20JA	1.1	-3.3	-1.5	0.1
356.75	-0.21	1.5	309.58	30JA	0.5	-3.3	-1.5	0.2
3.80	-0.08	1.6	319.59	9FE	0.3	-3.4	-1.5	0.2
10.78	0.05	1.6	329.54	19FE	0.0	-3.4	-1.6	0.3
17.71	0.16	1.7	339.43	1MR	-0.3	-3.4	-1.6	0.3
24.57	0.27	1.7	349.26	11MR	-1.0	-3.4	-1.6	0.3
31.37	0.37	1.8	359.03	21MR	-1.7	-3.5	-1.6	0.3
38.11	0.47	1.8	8.75	31MR	-1.1	-3.5	-1.7	0.3
44.80	0.55	1.9	18.42	10AP	-0.0	-3.4	-1.7	0.3
51.44	0.63	1.9	28.04	20AP	1.0	-3.4	-1.8	0.3
58.03	0.71	1.9	37.62	30AP	2.1	-3.4	-1.8	0.2
64.57	0.78	1.9	47.18	10MY	3.6	-3.4	-1.9	0.2
71.08	0.84	2.0	56.71	20MY	2.4	-3.3	-2.0	0.2
77.55	0.89	2.0	66.24	30MY	1.2	-3.3	-2.0	0.2
84.00	0.95	2.0	75.76	9JN	0.2	-3.3	-2.1	0.2
90.42	0.99	2.0	85.28	19JN	-0.9	-3.3	-2.2	0.3
96.82	1.04	1.9	94.82	29JN	-1.6	-3.3	-2.2	0.3
103.21	1.07	1.9	104.39	9JL	-1.1	-3.3	-2.3	0.3
109.58	1.11	2.0	113.99	19JL	-0.4	-3.4	-2.4	0.3
115.95	1.14	2.0	123.63	29JL	0.0	-3.4	-2.4	0.3
122.32	1.16	2.0	133.31	8AU	0.3	-3.4	-2.4	0.3
128.68	1.18	2.0	143.05	18AU	0.5	-3.5	-2.4	0.3
135.06	1.20	2.0	152.84	28AU	1.0	-3.5	-2.4	0.3
141.43	1.21	2.0	162.70	7SE	2.8	-3.6	-2.4	0.2
147.82	1.22	2.0	172.61	17SE	1.0	-3.7	-2.4	0.2
154.21	1.22	2.0	182.58	27SE	-0.5	-3.8	-2.4	0.1
160.62	1.22	2.0	192.60	7OC	-0.8	-3.9	-2.3	0.1
167.03	1.21	1.9	202.68	17OC	-0.8	-4.0	-2.2	0.0
173.46	1.20	1.9	212.80	27OC	-0.8	-4.1	-2.2	-0.1
179.89	1.18	1.8	222.96	6NO	-0.7	-4.2	-2.1	-0.1
186.33	1.15	1.8	233.14	16NO	-0.6	-4.3	-2.0	-0.2
192.78	1.12	1.7	243.34	26NO	-0.6	-4.4	-2.0	-0.3
199.23	1.08	1.6	253.55	6DE	-0.4	-4.4	-1.9	-0.3
205.67	1.02	1.6	263.75	16DE	0.8	-4.3	-1.8	-0.2
212.11	0.96	1.5	273.94	26DE	2.7	-3.9	-1.8	-0.1

♂			⊙	16.00UT	☿	♀	♃	♄
LONG	LAT	MAG	LONG	30	MAGNITUDES			
218.54	0.88	1.4	284.10	5JA	0.8	-3.3	-1.7	-0.1
224.95	0.78	1.2	294.23	15JA	0.3	-3.6	-1.7	0.0
231.34	0.67	1.1	304.31	25JA	0.1	-4.1	-1.6	0.1
237.68	0.54	1.0	314.35	4FE	-0.1	-4.3	-1.6	0.1
243.99	0.39	0.8	324.32	14FE	-0.4	-4.3	-1.6	0.2
250.23	0.20	0.7	334.25	24FE	-1.0	-4.2	-1.6	0.2
256.38	-0.01	0.5	344.11	6MR	-1.6	-4.1	-1.6	0.3
262.43	-0.25	0.3	353.91	16MR	-1.1	-4.0	-1.6	0.3
268.34	-0.54	0.2	3.66	26MR	0.0	-3.9	-1.6	0.3
274.06	-0.87	-0.0	13.35	5AP	1.3	-3.8	-1.6	0.3
279.54	-1.25	-0.3	23.00	15AP	2.8	-3.7	-1.6	0.3
284.71	-1.68	-0.5	32.61	25AP	3.2	-3.6	-1.6	0.3
289.47	-2.18	-0.7	42.18	5MY	1.8	-3.6	-1.7	0.3
293.70	-2.74	-1.0	51.72	15MY	0.9	-3.5	-1.7	0.3
297.24	-3.38	-1.3	61.25	25MY	0.1	-3.5	-1.8	0.3
299.91	-4.08	-1.5	70.77	4JN	-0.9	-3.4	-1.8	0.2
301.51	-4.82	-1.8	80.29	14JN	-1.7	-3.4	-1.9	0.2
301.86	-5.56	-2.1	89.83	24JN	-1.1	-3.4	-1.9	0.3
300.93	-6.22	-2.4	99.38	4JL	-0.4	-3.3	-2.0	0.3
298.91	-6.69	-2.6	108.96	14JL	0.1	-3.3	-2.1	0.4
296.31	-6.86	-2.6	118.58	24JL	0.4	-3.3	-2.1	0.4
293.88	-6.70	-2.5	128.23	3AU	0.7	-3.3	-2.2	0.4
292.27	-6.27	-2.2	137.94	13AU	1.4	-3.3	-2.3	0.4
291.89	-5.67	-2.0	147.71	23AU	3.1	-3.4	-2.3	0.4
292.83	-5.00	-1.7	157.53	2SE	0.9	-3.4	-2.4	0.4
294.96	-4.32	-1.4	167.41	12SE	-0.6	-3.4	-2.4	0.3
298.10	-3.69	-1.1	177.35	22SE	-0.9	-3.4	-2.4	0.3
302.06	-3.11	-0.8	187.34	2OC	-0.9	-3.5	-2.4	0.3
306.64	-2.58	-0.6	197.39	12OC	-0.8	-3.5	-2.4	0.2
311.71	-2.12	-0.3	207.49	22OC	-0.5	-3.5	-2.4	0.1
317.15	-1.70	-0.1	217.63	1NO	-0.5	-3.5	-2.3	0.1
322.88	-1.34	0.1	227.80	11NO	-0.5	-3.4	-2.3	0.0
328.83	-1.01	0.3	237.99	21NO	-0.3	-3.4	-2.2	-0.1
334.93	-0.73	0.5	248.20	1DE	1.0	-3.4	-2.2	-0.1
341.15	-0.48	0.7	258.40	11DE	2.3	-3.4	-2.1	-0.2
347.46	-0.25	0.8	268.60	21DE	0.5	-3.4	-2.0	-0.2
353.83	-0.06	1.0	278.78	31DE	0.1	-3.3	-1.9	-0.1
				31				
0.23	0.11	1.1	288.92	10JA	0.0	-3.3	-1.9	-0.1
6.66	0.27	1.2	299.03	20JA	-0.2	-3.3	-1.8	0.0
13.10	0.40	1.3	309.09	30JA	-0.5	-3.3	-1.7	0.1
19.53	0.52	1.4	319.10	9FE	-1.0	-3.4	-1.7	0.1
25.97	0.63	1.5	329.05	19FE	-1.5	-3.4	-1.6	0.2
32.39	0.72	1.6	338.94	1MR	-1.0	-3.4	-1.6	0.2
38.79	0.80	1.7	348.78	11MR	0.1	-3.4	-1.6	0.3
45.18	0.87	1.8	358.56	21MR	1.6	-3.5	-1.6	0.3
51.56	0.93	1.8	8.27	31MR	3.5	-3.5	-1.5	0.4
57.91	0.98	1.9	17.95	10AP	2.4	-3.5	-1.5	0.4
64.25	1.03	1.9	27.58	20AP	1.4	-3.6	-1.5	0.4
70.58	1.06	2.0	37.16	30AP	0.7	-3.7	-1.5	0.4
76.90	1.10	2.0	46.72	10MY	0.0	-3.7	-1.5	0.4
83.21	1.12	2.0	56.26	20MY	-0.9	-3.8	-1.6	0.4
89.52	1.14	2.0	65.78	30MY	-1.8	-3.9	-1.6	0.4
95.83	1.16	2.0	75.30	9JN	-1.1	-4.0	-1.6	0.4
102.14	1.17	2.0	84.83	19JN	-0.3	-4.1	-1.7	0.3
108.46	1.17	2.0	94.37	29JN	0.2	-4.2	-1.7	0.3
114.80	1.17	2.0	103.93	9JL	0.6	-4.2	-1.8	0.4
121.15	1.17	2.0	113.53	19JL	1.0	-4.1	-1.8	0.4
127.53	1.16	2.0	123.16	29JL	1.8	-3.8	-1.9	0.5
133.93	1.14	1.9	132.85	8AU	3.1	-3.3	-1.9	0.5
140.36	1.12	1.9	142.58	18AU	0.8	-3.5	-2.0	0.5
146.82	1.10	1.9	152.37	28AU	-0.7	-4.0	-2.1	0.5
153.32	1.07	2.0	162.22	7SE	-1.1	-4.3	-2.1	0.5
159.85	1.04	2.0	172.13	17SE	-1.1	-4.3	-2.2	0.5
166.43	1.00	1.9	182.09	27SE	-0.7	-4.3	-2.2	0.5
173.05	0.95	1.9	192.12	7OC	-0.4	-4.2	-2.3	0.4
179.72	0.90	1.9	202.19	17OC	-0.3	-4.1	-2.3	0.4
186.43	0.85	1.9	212.31	27OC	-0.3	-4.0	-2.3	0.3
193.19	0.78	1.9	222.46	6NO	-0.1	-3.9	-2.3	0.3
200.00	0.71	1.8	232.65	16NO	1.2	-3.8	-2.3	0.2
206.85	0.63	1.8	242.85	26NO	2.0	-3.8	-2.3	0.1
213.76	0.54	1.7	253.06	6DE	0.3	-3.7	-2.3	0.1
220.71	0.45	1.7	263.26	16DE	-0.1	-3.6	-2.2	-0.0
227.71	0.34	1.6	273.45	26DE	-0.1	-3.6	-2.1	-0.1

32

♂ LONG	LAT	MAG	☉ LONG	16.00UT	☿	♀	♃	♄
234.75	0.22	1.5	283.62	5JA	-0.3	-3.5	-2.1	-0.1
241.83	0.09	1.4	293.74	15JA	-0.5	-3.5	-2.0	0.0
248.95	-0.05	1.4	303.83	25JA	-1.0	-3.4	-1.9	0.1
256.11	-0.20	1.3	313.87	4FE	-1.3	-3.4	-1.9	0.1
263.30	-0.36	1.2	323.85	14FE	-1.0	-3.3	-1.8	0.2
270.51	-0.54	1.1	333.77	24FE	0.2	-3.3	-1.7	0.2
277.74	-0.72	1.0	343.64	5MR	2.0	-3.3	-1.7	0.3
284.98	-0.92	0.9	353.44	15MR	3.2	-3.3	-1.6	0.4
292.21	-1.12	0.8	3.19	25MR	1.8	-3.3	-1.6	0.4
299.42	-1.34	0.7	12.89	4AP	1.0	-3.3	-1.5	0.4
306.59	-1.56	0.6	22.53	14AP	0.5	-3.3	-1.5	0.5
313.70	-1.78	0.5	32.14	24AP	-0.1	-3.4	-1.5	0.5
320.72	-2.01	0.4	41.71	4MY	-0.9	-3.4	-1.5	0.5
327.62	-2.23	0.3	51.26	14MY	-1.8	-3.4	-1.4	0.5
334.37	-2.45	0.1	60.79	24MY	-1.1	-3.5	-1.4	0.5
340.90	-2.67	-0.0	70.31	3JN	-0.3	-3.5	-1.4	0.5
347.17	-2.89	-0.1	79.83	13JN	0.3	-3.5	-1.4	0.5
353.12	-3.09	-0.3	89.36	23JN	0.8	-3.4	-1.5	0.5
358.65	-3.28	-0.4	98.91	3JL	1.3	-3.4	-1.5	0.5
3.65	-3.45	-0.6	108.49	13JL	2.4	-3.4	-1.5	0.5
8.01	-3.61	-0.8	118.10	23JL	2.7	-3.3	-1.6	0.6
11.53	-3.74	-1.0	127.76	2AU	0.6	-3.3	-1.6	0.6
14.02	-3.82	-1.2	137.47	12AU	-0.8	-3.3	-1.6	0.6
15.29	-3.84	-1.4	147.23	22AU	-1.2	-3.3	-1.7	0.7
15.12	-3.77	-1.7	157.05	1SE	-1.1	-3.3	-1.8	0.7
13.51	-3.56	-1.9	166.92	11SE	-0.6	-3.3	-1.8	0.7
10.67	-3.19	-2.0	176.86	21SE	-0.3	-3.4	-1.9	0.7
7.20	-2.68	-2.1	186.86	10C	-0.2	-3.4	-2.0	0.7
3.92	-2.08	-1.9	196.90	110C	-0.1	-3.4	-2.0	0.6
1.54	-1.46	-1.6	207.00	210C	0.1	-3.5	-2.1	0.6
0.45	-0.89	-1.2	217.14	310C	1.5	-3.5	-2.1	0.6
0.71	-0.40	-0.9	227.30	10NO	1.8	-3.6	-2.2	0.5
2.16	-0.01	-0.6	237.50	20NO	0.1	-3.6	-2.2	0.5
4.63	0.31	-0.3	247.70	30NO	-0.2	-3.7	-2.2	0.4
7.88	0.56	-0.0	257.91	10DE	-0.3	-3.8	-2.2	0.3
11.76	0.76	0.2	268.11	20DE	-0.4	-3.9	-2.2	0.2
16.12	0.92	0.5	278.29	30DE	-0.6	-4.0	-2.2	0.2

33

♂ LONG	LAT	MAG	☉ LONG	16.00UT	☿	♀	♃	♄
20.86	1.04	0.7	288.43	9JA	-1.0	-4.1	-2.1	0.1
25.88	1.14	0.8	298.55	19JA	-1.2	-4.1	-2.1	0.1
31.14	1.22	1.0	308.61	29JA	-0.9	-4.2	-2.0	0.2
36.56	1.27	1.2	318.62	8FE	0.3	-4.3	-1.9	0.2
42.13	1.32	1.3	328.57	18FE	2.4	-4.3	-1.9	0.3
47.81	1.35	1.4	338.47	28FE	2.5	-4.1	-1.8	0.4
53.57	1.37	1.5	348.30	10MR	1.3	-3.7	-1.7	0.4
59.40	1.38	1.6	358.08	20MR	0.8	-3.1	-1.7	0.5
65.30	1.39	1.7	7.81	30MR	0.4	-3.6	-1.6	0.5
71.24	1.39	1.8	17.48	9AP	-0.1	-4.0	-1.6	0.6
77.23	1.38	1.8	27.11	19AP	-0.9	-4.2	-1.5	0.6
83.25	1.37	1.9	36.70	29AP	-1.8	-4.2	-1.5	0.6
89.31	1.35	1.9	46.25	9MY	-1.1	-4.1	-1.4	0.7
95.41	1.33	2.0	55.79	19MY	-0.2	-4.0	-1.4	0.7
101.55	1.31	2.0	65.32	29MY	0.5	-3.9	-1.4	0.7
107.72	1.28	2.0	74.83	8JN	1.0	-3.9	-1.3	0.7
113.93	1.25	2.0	84.36	18JN	1.7	-3.8	-1.3	0.7
120.18	1.21	2.0	93.90	28JN	3.0	-3.7	-1.3	0.7
126.48	1.17	2.0	103.46	8JL	2.3	-3.6	-1.3	0.7
132.82	1.12	2.0	113.06	18JL	0.5	-3.6	-1.3	0.7
139.22	1.07	2.0	122.70	28JL	-0.8	-3.5	-1.4	0.7
145.67	1.02	2.0	132.38	7AU	-1.3	-3.5	-1.4	0.8
152.17	0.97	1.9	142.11	17AU	-1.1	-3.4	-1.4	0.8
158.73	0.90	1.9	151.90	27AU	-0.6	-3.4	-1.5	0.9
165.36	0.84	1.8	161.74	6SE	-0.2	-3.4	-1.5	0.9
172.05	0.77	1.8	171.65	16SE	-0.1	-3.4	-1.6	0.9
178.81	0.70	1.8	181.61	26SE	0.0	-3.4	-1.6	0.9
185.64	0.62	1.8	191.63	6OC	0.3	-3.4	-1.7	0.9
192.54	0.53	1.8	201.70	16OC	1.7	-3.4	-1.7	0.9
199.52	0.45	1.8	211.82	26OC	1.6	-3.4	-1.8	0.9
206.56	0.35	1.8	221.97	5NO	-0.1	-3.4	-1.9	0.8
213.68	0.25	1.7	232.16	15NO	-0.4	-3.4	-1.9	0.8
220.86	0.15	1.7	242.35	25NO	-0.4	-3.4	-2.0	0.7
228.12	0.04	1.7	252.56	5DE	-0.5	-3.4	-2.0	0.7
235.44	-0.07	1.6	262.77	15DE	-0.7	-3.4	-2.1	0.6
242.82	-0.19	1.6	272.95	25DE	-1.0	-3.5	-2.1	0.5

34

♂ LONG	LAT	MAG	☉ LONG	16.00UT	☿	♀	♃	♄
250.27	-0.31	1.6	283.12	4JA	-1.0	-3.5	-2.1	0.5
257.77	-0.44	1.5	293.25	14JA	-0.8	-3.5	-2.1	0.4
265.31	-0.56	1.5	303.34	24JA	0.4	-3.4	-2.1	0.3
272.90	-0.69	1.4	313.37	3FE	2.8	-3.4	-2.1	0.4
280.52	-0.82	1.4	323.36	13FE	1.9	-3.4	-2.0	0.4
288.16	-0.95	1.3	333.29	23FE	0.9	-3.4	-2.0	0.5
295.82	-1.07	1.3	343.15	5MR	0.6	-3.4	-1.9	0.5
303.48	-1.19	1.2	352.96	15MR	0.2	-3.3	-1.8	0.6
311.12	-1.30	1.2	2.71	25MR	-0.2	-3.3	-1.8	0.6
318.74	-1.40	1.1	12.41	4AP	-0.9	-3.3	-1.7	0.7
326.32	-1.49	1.1	22.06	14AP	-1.9	-3.3	-1.6	0.7
333.84	-1.57	1.0	31.67	24AP	-1.1	-3.4	-1.6	0.8
341.28	-1.64	1.0	41.24	4MY	-0.1	-3.4	-1.5	0.8
348.64	-1.69	0.9	50.79	14MY	0.6	-3.4	-1.5	0.9
355.89	-1.72	0.9	60.32	24MY	1.3	-3.4	-1.4	0.9
3.01	-1.74	0.8	69.84	3JN	2.3	-3.5	-1.4	0.9
9.98	-1.74	0.8	79.37	13JN	3.4	-3.5	-1.3	0.9
16.77	-1.73	0.7	88.90	23JN	1.9	-3.6	-1.3	1.0
23.36	-1.69	0.6	98.45	3JL	0.4	-3.6	-1.3	1.0
29.72	-1.64	0.6	108.03	13JL	-0.8	-3.7	-1.3	1.0
35.80	-1.57	0.5	117.64	23JL	-1.5	-3.8	-1.3	1.0
41.56	-1.47	0.4	127.29	2AU	-1.1	-3.9	-1.3	1.0
46.94	-1.35	0.3	137.00	12AU	-0.5	-4.0	-1.3	1.0
51.85	-1.20	0.1	146.76	22AU	-0.1	-4.1	-1.3	1.0
56.21	-1.03	-0.0	156.57	1SE	0.1	-4.2	-1.3	1.1
59.86	-0.82	-0.2	166.45	11SE	0.2	-4.3	-1.3	1.1
62.67	-0.56	-0.4	176.38	21SE	0.6	-4.3	-1.4	1.2
64.44	-0.25	-0.6	186.37	10C	2.1	-4.2	-1.4	1.2
64.94	0.11	-0.8	196.42	110C	1.4	-3.9	-1.4	1.2
64.05	0.53	-1.0	206.51	210C	-0.3	-3.2	-1.5	1.2
61.75	0.98	-1.2	216.64	310C	-0.5	-3.6	-1.6	1.2
58.33	1.43	-1.4	226.81	10NO	-0.5	-4.1	-1.6	1.2
54.41	1.83	-1.4	237.00	20NO	-0.6	-4.4	-1.7	1.1
50.80	2.13	-1.2	247.21	30NO	-0.8	-4.4	-1.7	1.1
48.15	2.33	-0.9	257.42	10DE	-0.8	-4.3	-1.8	1.0
46.80	2.42	-0.6	267.61	20DE	-0.9	-4.2	-1.9	1.0
46.77	2.45	-0.3	277.79	30DE	-0.7	-4.1	-1.9	0.9

35

♂ LONG	LAT	MAG	☉ LONG	16.00UT	☿	♀	♃	♄
47.94	2.42	-0.0	287.95	9JA	0.5	-4.0	-2.0	0.8
50.09	2.38	0.2	298.05	19JA	3.0	-3.9	-2.0	0.7
53.02	2.31	0.5	308.12	29JA	1.4	-3.8	-2.0	0.7
56.58	2.24	0.7	318.14	8FE	0.7	-3.7	-2.0	0.6
60.62	2.16	0.9	328.09	18FE	0.4	-3.6	-2.0	0.7
65.05	2.08	1.0	337.99	28FE	0.1	-3.5	-2.0	0.7
69.80	2.01	1.2	347.83	10MR	-0.3	-3.5	-2.0	0.7
74.79	1.93	1.3	357.61	20MR	-1.0	-3.4	-1.9	0.8
79.98	1.85	1.4	7.34	30MR	-1.8	-3.4	-1.9	0.9
85.34	1.77	1.5	17.02	9AP	-1.1	-3.4	-1.8	0.9
90.84	1.69	1.6	26.64	19AP	-0.1	-3.3	-1.7	1.0
96.47	1.62	1.7	36.24	29AP	0.8	-3.3	-1.7	1.0
102.21	1.54	1.7	45.80	9MY	1.8	-3.3	-1.6	1.1
108.04	1.46	1.8	55.33	19MY	3.1	-3.3	-1.6	1.1
113.97	1.38	1.8	64.86	29MY	2.9	-3.3	-1.5	1.1
119.99	1.30	1.9	74.38	8JN	1.5	-3.3	-1.4	1.2
126.08	1.22	1.9	83.90	18JN	0.3	-3.3	-1.4	1.2
132.26	1.14	1.9	93.44	28JN	-0.9	-3.4	-1.3	1.2
138.52	1.06	1.9	103.00	8JL	-1.6	-3.4	-1.3	1.2
144.86	0.97	1.9	112.59	18JL	-1.1	-3.4	-1.3	1.3
151.28	0.88	1.9	122.23	28JL	-0.5	-3.5	-1.2	1.3
157.79	0.79	1.9	131.90	7AU	-0.0	-3.5	-1.2	1.3
164.38	0.70	1.9	141.63	17AU	0.2	-3.5	-1.2	1.3
171.06	0.61	1.8	151.42	27AU	0.4	-3.4	-1.2	1.3
177.82	0.51	1.8	161.26	6SE	0.8	-3.4	-1.2	1.4
184.68	0.42	1.8	171.16	16SE	2.5	-3.4	-1.2	1.4
191.63	0.32	1.7	181.13	26SE	1.2	-3.4	-1.2	1.4
198.66	0.21	1.7	191.14	6OC	-0.4	-3.3	-1.3	1.3
205.78	0.11	1.6	201.21	16OC	-0.7	-3.3	-1.3	1.3
213.00	0.01	1.6	211.33	26OC	-0.7	-3.3	-1.3	1.3
220.30	-0.10	1.5	221.48	5NO	-0.7	-3.3	-1.4	1.3
227.68	-0.21	1.5	231.66	15NO	-0.7	-3.3	-1.4	1.2
235.14	-0.31	1.5	241.86	25NO	-0.7	-3.4	-1.5	1.2
242.68	-0.42	1.5	252.06	5DE	-0.7	-3.4	-1.6	1.1
250.28	-0.52	1.5	262.27	15DE	-0.5	-3.4	-1.6	1.1
257.94	-0.63	1.5	272.46	25DE	0.7	-3.4	-1.7	1.1

♂ LONG	LAT	MAG	☉ LONG	16.00UT	☿	♀	♃	♄
				36				
265.65	-0.72	1.5	282.63	4JA	2.9	-3.5	-1.7	1.0
273.41	-0.82	1.5	292.76	14JA	1.0	-3.5	-1.8	1.0
281.20	-0.90	1.4	302.85	24JA	0.4	-3.6	-1.9	0.9
289.01	-0.98	1.4	312.89	3FE	0.2	-3.6	-1.9	0.9
296.83	-1.06	1.4	322.88	13FE	0.0	-3.7	-2.0	0.9
304.64	-1.12	1.4	332.81	23FE	-0.4	-3.8	-2.0	0.8
312.44	-1.17	1.4	342.68	4MR	-1.0	-3.8	-2.0	0.9
320.22	-1.21	1.4	352.49	14MR	-1.7	-3.9	-2.0	1.0
327.95	-1.23	1.4	2.25	24MR	-1.1	-4.0	-2.0	1.1
335.63	-1.25	1.4	11.95	3AP	-0.0	-4.1	-2.0	1.1
343.25	-1.25	1.4	21.60	13AP	1.1	-4.2	-1.9	1.2
350.80	-1.23	1.3	31.21	23AP	2.3	-4.2	-1.9	1.2
358.26	-1.21	1.3	40.79	3MY	3.7	-4.1	-1.8	1.3
5.63	-1.17	1.3	50.33	13MY	2.2	-3.9	-1.8	1.3
12.89	-1.12	1.3	59.87	23MY	1.1	-3.2	-1.7	1.4
20.04	-1.05	1.3	69.38	2JN	0.2	-3.0	-1.6	1.4
27.07	-0.97	1.3	78.90	12JN	-0.9	-3.7	-1.6	1.4
33.97	-0.88	1.3	88.44	22JN	-1.6	-4.1	-1.5	1.4
40.72	-0.78	1.2	97.98	2JL	-1.1	-4.2	-1.5	1.4
47.32	-0.66	1.2	107.56	12JL	-0.4	-4.2	-1.4	1.4
53.76	-0.54	1.2	117.17	22JL	0.1	-4.1	-1.4	1.4
60.00	-0.40	1.1	126.82	1AU	0.3	-4.0	-1.3	1.3
66.05	-0.24	1.1	136.52	11AU	0.6	-3.9	-1.3	1.3
71.85	-0.07	1.0	146.28	21AU	1.1	-3.9	-1.3	1.2
77.38	0.11	0.9	156.09	31AU	2.9	-3.8	-1.2	1.2
82.60	0.32	0.8	165.96	10SE	1.1	-3.7	-1.2	1.2
87.43	0.54	0.7	175.90	20SE	-0.5	-3.7	-1.2	1.2
91.81	0.80	0.5	185.88	30SE	-0.8	-3.6	-1.2	1.2
95.63	1.09	0.4	195.93	10OC	-0.8	-3.6	-1.2	1.2
98.75	1.41	0.2	206.02	20OC	-0.8	-3.5	-1.2	1.2
101.03	1.78	0.0	216.15	30OC	-0.6	-3.5	-1.3	1.1
102.27	2.18	-0.2	226.32	9NO	-0.5	-3.4	-1.3	1.1
102.29	2.62	-0.4	236.51	19NO	-0.6	-3.4	-1.3	1.1
100.95	3.08	-0.7	246.71	29NO	-0.4	-3.4	-1.4	1.1
98.30	3.50	-0.9	256.93	9DE	0.8	-3.4	-1.4	1.0
94.67	3.83	-1.0	267.12	19DE	2.6	-3.4	-1.5	1.0
90.69	4.01	-1.0	277.30	29DE	0.7	-3.3	-1.5	0.9
				37				
87.10	4.03	-0.8	287.46	8JA	0.2	-3.3	-1.6	0.9
84.52	3.91	-0.6	297.57	18JA	0.1	-3.3	-1.7	0.8
83.22	3.71	-0.3	307.64	28JA	-0.1	-3.4	-1.7	0.8
83.21	3.47	-0.1	317.65	7FE	-0.4	-3.4	-1.8	0.7
84.34	3.21	0.2	327.61	17FE	-1.0	-3.4	-1.9	0.7
86.44	2.97	0.4	337.51	27FE	-1.6	-3.4	-1.9	0.7
89.31	2.73	0.6	347.36	9MR	-1.1	-3.4	-2.0	0.7
92.81	2.52	0.8	357.14	19MR	0.0	-3.5	-2.0	0.8
96.81	2.31	1.0	6.87	29MR	1.4	-3.5	-2.0	0.8
101.21	2.13	1.1	16.55	8AP	3.0	-3.4	-2.1	0.9
105.95	1.95	1.2	26.18	18AP	2.9	-3.4	-2.1	1.0
110.95	1.79	1.3	35.77	28AP	1.7	-3.4	-2.0	1.0
116.20	1.63	1.4	45.33	8MY	0.9	-3.3	-2.0	1.1
121.65	1.48	1.5	54.87	18MY	0.1	-3.3	-2.0	1.1
127.27	1.34	1.5	64.39	28MY	-0.9	-3.3	-1.9	1.2
133.05	1.20	1.6	73.91	7JN	-1.7	-3.3	-1.9	1.2
138.99	1.07	1.6	83.43	17JN	-1.1	-3.3	-1.8	1.2
145.05	0.94	1.7	92.97	27JN	-0.4	-3.3	-1.7	1.2
151.25	0.81	1.7	102.53	7JL	0.1	-3.3	-1.7	1.2
157.58	0.69	1.7	112.12	17JL	0.5	-3.4	-1.6	1.2
164.02	0.56	1.7	121.75	27JL	0.8	-3.4	-1.6	1.2
170.59	0.44	1.7	131.43	6AU	1.5	-3.4	-1.5	1.2
177.27	0.32	1.7	141.16	16AU	3.2	-3.5	-1.5	1.1
184.06	0.20	1.7	150.94	26AU	0.9	-3.5	-1.4	1.1
190.97	0.09	1.7	160.78	5SE	-0.6	-3.6	-1.4	1.1
197.99	-0.03	1.7	170.68	15SE	-1.0	-3.7	-1.3	1.0
205.12	-0.14	1.6	180.64	25SE	-1.0	-3.8	-1.3	1.0
212.35	-0.25	1.6	190.65	5OC	-0.8	-3.9	-1.3	1.0
219.68	-0.36	1.6	200.72	15OC	-0.5	-4.0	-1.3	1.0
227.11	-0.46	1.5	210.83	25OC	-0.4	-4.1	-1.3	1.0
234.63	-0.56	1.5	220.98	4NO	-0.4	-4.2	-1.3	1.0
242.24	-0.65	1.5	231.16	14NO	-0.2	-4.3	-1.3	1.0
249.91	-0.73	1.4	241.36	24NO	1.0	-4.4	-1.3	1.0
257.66	-0.81	1.4	251.57	4DE	2.3	-4.4	-1.3	1.0
265.45	-0.88	1.4	261.77	14DE	0.0	-4.3	-1.4	0.9
273.29	-0.95	1.3	271.97	24DE	0.0	-3.9	-1.4	0.9

♂ LONG	LAT	MAG	☉ LONG	16.00UT	☿	♀	♃	♄
				38				
281.16	-1.00	1.3	282.14	3JA	-0.1	-3.3	-1.5	0.8
289.04	-1.04	1.3	292.27	13JA	-0.2	-3.6	-1.5	0.8
296.93	-1.07	1.3	302.36	23JA	-0.5	-4.1	-1.6	0.8
304.81	-1.09	1.4	312.41	2FE	-1.0	-4.3	-1.6	0.7
312.67	-1.10	1.4	322.40	12FE	-1.4	-4.3	-1.7	0.7
320.50	-1.09	1.4	332.33	22FE	-1.0	-4.2	-1.8	0.6
328.28	-1.08	1.4	342.20	4MR	0.1	-4.1	-1.8	0.6
336.01	-1.05	1.4	352.02	14MR	1.7	-4.0	-1.9	0.5
343.67	-1.01	1.5	1.77	24MR	3.6	-3.9	-2.0	0.6
351.26	-0.96	1.5	11.48	3AP	2.2	-3.8	-2.0	0.6
358.76	-0.90	1.5	21.13	13AP	1.2	-3.7	-2.1	0.7
6.19	-0.83	1.5	30.75	23AP	0.6	-3.6	-2.1	0.8
13.52	-0.75	1.5	40.33	3MY	0.0	-3.6	-2.1	0.8
20.75	-0.67	1.6	49.87	13MY	-0.9	-3.5	-2.2	0.9
27.89	-0.57	1.6	59.41	23MY	-1.8	-3.5	-2.2	0.9
34.93	-0.47	1.6	68.93	2JN	-1.1	-3.4	-2.1	1.0
41.86	-0.37	1.6	78.45	12JN	-0.3	-3.4	-2.1	1.0
48.69	-0.25	1.6	87.98	22JN	0.3	-3.4	-2.1	1.1
55.41	-0.14	1.6	97.53	2JL	0.6	-3.3	-2.0	1.1
62.01	-0.01	1.6	107.10	12JL	1.1	-3.3	-1.9	1.1
68.51	0.11	1.6	116.71	22JL	2.0	-3.3	-1.9	1.1
74.88	0.25	1.6	126.36	1AU	3.0	-3.3	-1.8	1.1
81.11	0.39	1.5	136.05	11AU	0.8	-3.3	-1.8	1.1
87.21	0.54	1.5	145.81	21AU	-0.7	-3.4	-1.7	1.0
93.16	0.69	1.4	155.62	31AU	-1.1	-3.4	-1.6	1.0
98.93	0.86	1.4	165.48	10SE	-1.1	-3.4	-1.6	1.0
104.50	1.03	1.3	175.41	20SE	-0.7	-3.4	-1.5	0.9
109.85	1.22	1.2	185.40	30SE	-0.4	-3.5	-1.5	0.9
114.93	1.43	1.1	195.43	10OC	-0.3	-3.5	-1.5	0.9
119.68	1.65	1.0	205.52	20OC	-0.2	-3.5	-1.4	0.9
124.04	1.90	0.9	215.65	30OC	-0.0	-3.5	-1.4	0.9
127.93	2.17	0.7	225.82	9NO	1.2	-3.4	-1.4	0.9
131.22	2.47	0.5	236.01	19NO	2.0	-3.4	-1.4	0.9
133.78	2.80	0.3	246.21	29NO	0.2	-3.4	-1.4	0.9
135.44	3.16	0.1	256.42	9DE	-0.1	-3.4	-1.4	0.9
136.01	3.55	-0.2	266.63	19DE	-0.2	-3.4	-1.4	0.9
135.32	3.93	-0.4	276.81	29DE	-0.3	-3.3	-1.4	0.8
				39				
133.34	4.27	-0.7	286.96	8JA	-0.6	-3.3	-1.4	0.8
130.18	4.52	-0.9	297.08	18JA	-1.0	-3.3	-1.5	0.7
126.34	4.61	-1.0	307.15	28JA	-1.3	-3.3	-1.5	0.7
122.50	4.52	-0.9	317.16	7FE	-1.0	-3.4	-1.6	0.6
119.34	4.27	-0.7	327.13	17FE	0.2	-3.4	-1.6	0.6
117.33	3.92	-0.5	337.03	27FE	2.1	-3.4	-1.7	0.5
116.61	3.54	-0.2	346.88	9MR	2.9	-3.4	-1.8	0.5
117.13	3.15	-0.0	356.67	19MR	1.6	-3.5	-1.8	0.4
118.73	2.77	0.2	6.40	29MR	0.9	-3.5	-1.9	0.4
121.23	2.43	0.4	16.08	8AP	0.5	-3.6	-2.0	0.5
124.46	2.12	0.6	25.72	18AP	-0.1	-3.6	-2.0	0.5
128.29	1.84	0.7	35.31	28AP	-0.9	-3.7	-2.1	0.6
132.61	1.58	0.8	44.87	8MY	-1.8	-3.7	-2.2	0.7
137.34	1.34	1.0	54.41	18MY	-1.2	-3.8	-2.2	0.7
142.41	1.12	1.0	63.94	28MY	-0.3	-3.9	-2.2	0.8
147.77	0.92	1.1	73.46	7JN	0.4	-4.0	-2.3	0.8
153.39	0.72	1.2	82.98	17JN	0.9	-4.1	-2.3	0.9
159.23	0.55	1.3	92.52	27JN	1.5	-4.2	-2.3	0.9
165.28	0.38	1.3	102.07	7JL	2.6	-4.2	-2.3	0.9
171.53	0.22	1.3	111.67	17JL	2.6	-4.1	-2.2	1.0
177.94	0.06	1.4	121.29	27JL	0.6	-3.8	-2.2	1.0
184.53	-0.08	1.4	130.96	6AU	-0.8	-3.2	-2.1	1.0
191.27	-0.21	1.4	140.69	16AU	-1.3	-3.5	-2.1	1.0
198.15	-0.34	1.4	150.46	26AU	-1.2	-4.0	-2.0	0.9
205.17	-0.46	1.4	160.30	5SE	-0.6	-4.3	-1.9	0.9
212.33	-0.57	1.4	170.20	15SE	-0.3	-4.3	-1.9	0.9
219.61	-0.68	1.4	180.15	25SE	-0.1	-4.3	-1.8	0.8
227.00	-0.77	1.4	190.16	5OC	-0.1	-4.2	-1.7	0.8
234.51	-0.85	1.4	200.23	15OC	0.2	-4.1	-1.7	0.8
242.10	-0.93	1.4	210.33	25OC	1.5	-4.0	-1.6	0.8
249.78	-0.99	1.4	220.49	4NO	1.8	-3.9	-1.6	0.9
257.53	-1.04	1.4	230.67	14NO	0.0	-3.8	-1.6	0.9
265.34	-1.08	1.4	240.86	24NO	-0.3	-3.8	-1.5	0.9
273.19	-1.11	1.4	251.07	4DE	-0.3	-3.7	-1.5	0.9
281.08	-1.12	1.4	261.28	14DE	-0.4	-3.6	-1.5	0.8
288.97	-1.13	1.4	271.47	24DE	-0.6	-3.6	-1.5	0.8

♂ LONG LAT MAG	☉ LONG	16.00UT 40	☿ ♀ ♃ ♄ MAGNITUDES
296.87 -1.12 1.4	281.64	3JA	-1.0 -3.5 -1.5 0.8
304.76 -1.09 1.4	291.78	13JA	-1.1 -3.5 -1.5 0.8
312.62 -1.06 1.4	301.87	23JA	-0.9 -3.4 -1.5 0.7
320.44 -1.02 1.4	311.92	2FE	0.3 -3.4 -1.5 0.7
328.21 -0.97 1.4	321.92	12FE	2.5 -3.3 -1.5 0.6
335.92 -0.90 1.4	331.85	22FE	2.3 -3.3 -1.6 0.6
343.56 -0.83 1.4	341.73	3MR	1.2 -3.3 -1.6 0.5
351.12 -0.76 1.4	351.55	13MR	0.7 -3.3 -1.7 0.5
358.61 -0.67 1.5	1.31	23MR	0.3 -3.3 -1.7 0.4
6.01 -0.58 1.5	11.02	2AP	-0.2 -3.3 -1.8 0.4
13.33 -0.49 1.6	20.67	12AP	-0.9 -3.3 -1.9 0.3
20.55 -0.39 1.6	30.28	22AP	-1.8 -3.4 -1.9 0.4
27.69 -0.29 1.6	39.86	2MY	-1.2 -3.4 -2.0 0.5
34.74 -0.19 1.7	49.41	12MY	-0.2 -3.4 -2.1 0.5
41.70 -0.08 1.7	58.94	22MY	0.5 -3.5 -2.1 0.6
48.58 0.02 1.7	68.47	1JN	1.1 -3.5 -2.2 0.7
55.37 0.13 1.8	77.99	11JN	1.9 -3.5 -2.3 0.7
62.07 0.24 1.8	87.51	21JN	3.2 -3.5 -2.3 0.8
68.70 0.34 1.8	97.06	1JL	2.2 -3.4 -2.4 0.8
75.25 0.45 1.8	106.63	11JL	0.5 -3.4 -2.4 0.8
81.72 0.56 1.8	116.24	21JL	-0.8 -3.3 -2.4 0.9
88.10 0.67 1.8	125.89	31JL	-1.4 -3.3 -2.4 0.9
94.41 0.79 1.8	135.58	10AU	-1.2 -3.3 -2.4 0.9
100.63 0.90 1.8	145.33	20AU	-0.6 -3.3 -2.3 0.9
106.76 1.02 1.8	155.14	30AU	-0.2 -3.3 -2.3 0.9
112.80 1.13 1.7	165.00	9SE	-0.0 -3.3 -2.2 0.9
118.73 1.26 1.7	174.92	19SE	0.1 -3.4 -2.2 0.8
124.55 1.38 1.6	184.91	29SE	0.4 -3.4 -2.1 0.8
130.23 1.52 1.6	194.94	9OC	1.8 -3.4 -2.0 0.8
135.75 1.65 1.5	205.03	19OC	1.6 -3.5 -2.0 0.7
141.10 1.80 1.4	215.16	29OC	-0.2 -3.5 -1.9 0.8
146.22 1.96 1.3	225.32	8NO	-0.4 -3.6 -1.8 0.8
151.09 2.13 1.1	235.51	18NO	-0.5 -3.6 -1.8 0.8
155.64 2.31 1.0	245.72	28NO	-0.5 -3.7 -1.7 0.8
159.79 2.50 0.8	255.92	8DE	-0.7 -3.8 -1.7 0.8
163.46 2.71 0.6	266.12	18DE	-0.9 -3.9 -1.6 0.8
166.53 2.94 0.4	276.31	28DE	-1.0 -4.0 -1.6 0.8
		41	
168.84 3.18 0.2	286.47	7JA	-0.8 -4.1 -1.6 0.8
170.23 3.42 -0.1	296.58	17JA	-0.4 -4.2 -1.6 0.8
170.50 3.65 -0.4	306.66	27JA	2.9 -4.2 -1.6 0.7
169.54 3.85 -0.7	316.68	6FE	1.7 -4.3 -1.6 0.7
167.30 3.96 -0.9	326.65	16FE	0.8 -4.3 -1.6 0.6
164.02 3.94 -1.2	336.56	26FE	0.5 -4.1 -1.6 0.6
160.23 3.76 -1.2	346.40	8MR	0.2 -3.7 -1.6 0.5
156.62 3.43 -1.1	356.19	18MR	-0.2 -3.1 -1.6 0.5
153.86 3.00 -0.9	5.93	28MR	-0.9 -3.6 -1.7 0.4
152.33 2.53 -0.7	15.61	7AP	-1.8 -4.0 -1.7 0.3
152.11 2.06 -0.5	25.25	17AP	-1.2 -4.2 -1.7 0.3
153.14 1.62 -0.3	34.85	27AP	-0.2 -4.2 -1.8 0.3
155.23 1.22 -0.1	44.41	7MY	0.7 -4.1 -1.8 0.4
158.21 0.87 0.1	53.95	17MY	1.5 -4.0 -1.9 0.4
161.94 0.56 0.2	63.48	27MY	2.6 -3.9 -2.0 0.5
166.26 0.28 0.4	72.99	6JN	3.3 -3.8 -2.0 0.6
171.09 0.03 0.5	82.52	16JN	1.7 -3.8 -2.1 0.6
176.35 -0.19 0.6	92.05	26JN	0.4 -3.7 -2.2 0.7
181.97 -0.39 0.7	101.61	6JL	-0.8 -3.6 -2.3 0.7
187.90 -0.56 0.8	111.20	16JL	-1.5 -3.6 -2.3 0.8
194.11 -0.72 0.8	120.83	26JL	-1.2 -3.5 -2.4 0.8
200.56 -0.86 0.9	130.49	5AU	-0.5 -3.5 -2.4 0.8
207.24 -0.98 0.9	140.22	15AU	-0.1 -3.4 -2.4 0.8
214.10 -1.09 1.0	149.99	25AU	0.1 -3.4 -2.5 0.8
221.15 -1.18 1.0	159.83	4SE	0.3 -3.4 -2.5 0.8
228.35 -1.25 1.1	169.72	14SE	0.6 -3.4 -2.4 0.8
235.69 -1.30 1.1	179.67	24SE	2.2 -3.4 -2.4 0.8
243.15 -1.34 1.1	189.68	4OC	1.4 -3.4 -2.3 0.8
250.71 -1.37 1.2	199.74	14OC	-0.3 -3.4 -2.3 0.7
258.36 -1.38 1.2	209.85	24OC	-0.6 -3.4 -2.2 0.7
266.08 -1.37 1.2	219.99	3NO	-0.6 -3.4 -2.2 0.7
273.85 -1.35 1.3	230.17	13NO	-0.6 -3.4 -2.1 0.7
281.65 -1.31 1.3	240.37	23NO	-0.8 -3.4 -2.0 0.8
289.47 -1.26 1.3	250.57	3DE	-0.8 -3.4 -1.9 0.8
297.29 -1.21 1.4	260.78	13DE	-0.8 -3.4 -1.9 0.8
305.10 -1.14 1.4	270.97	23DE	-0.6 -3.5 -1.8 0.8

♂ LONG LAT MAG	☉ LONG	16.00UT 42	☿ ♀ ♃ ♄ MAGNITUDES
312.88 -1.06 1.4	281.14	2JA	0.5 -3.5 -1.8 0.8
320.62 -0.97 1.4	291.28	12JA	3.0 -3.5 -1.7 0.8
328.31 -0.88 1.5	301.38	22JA	1.3 -3.4 -1.7 0.8
335.93 -0.78 1.5	311.43	1FE	0.6 -3.4 -1.6 0.7
343.49 -0.68 1.5	321.43	11FE	0.3 -3.4 -1.6 0.7
350.97 -0.58 1.6	331.36	21FE	0.1 -3.4 -1.6 0.6
358.37 -0.47 1.6	341.24	3MR	-0.3 -3.4 -1.6 0.6
5.70 -0.37 1.6	351.07	13MR	-0.9 -3.3 -1.6 0.5
12.94 -0.26 1.6	0.83	23MR	-1.7 -3.3 -1.6 0.5
20.09 -0.16 1.7	10.54	2AP	-1.2 -3.3 -1.6 0.4
27.17 -0.05 1.7	20.20	12AP	-0.1 -3.3 -1.6 0.4
34.16 0.05 1.7	29.82	22AP	0.9 -3.4 -1.6 0.3
41.08 0.15 1.7	39.40	2MY	2.0 -3.4 -1.6 0.2
47.92 0.25 1.7	48.95	12MY	3.4 -3.4 -1.7 0.3
54.69 0.35 1.8	58.48	22MY	2.6 -3.4 -1.7 0.4
61.40 0.44 1.8	68.00	1JN	1.4 -3.5 -1.8 0.4
68.04 0.53 1.9	77.53	11JN	0.3 -3.5 -1.8 0.5
74.62 0.62 1.9	87.05	21JN	-0.8 -3.6 -1.9 0.5
81.15 0.70 1.9	96.60	1JL	-1.6 -3.6 -2.0 0.6
87.62 0.79 1.9	106.17	11JL	-1.1 -3.7 -2.0 0.7
94.05 0.87 2.0	115.77	21JL	-0.4 -3.8 -2.1 0.7
100.42 0.95 2.0	125.42	31JL	-0.0 -3.9 -2.2 0.7
106.76 1.03 2.0	135.12	10AU	0.2 -4.0 -2.2 0.8
113.05 1.10 2.0	144.86	20AU	0.5 -4.1 -2.3 0.8
119.29 1.18 2.0	154.66	30AU	0.9 -4.2 -2.3 0.8
125.49 1.25 1.9	164.53	9SE	2.6 -4.3 -2.4 0.8
131.64 1.32 1.9	174.45	19SE	1.2 -4.3 -2.4 0.8
137.75 1.39 1.9	184.42	29SE	-0.4 -4.2 -2.4 0.8
143.79 1.46 1.8	194.46	9OC	-0.8 -3.9 -2.4 0.7
149.78 1.53 1.8	204.54	19OC	-0.7 -3.2 -2.4 0.7
155.69 1.60 1.7	214.67	29OC	-0.8 -3.6 -2.4 0.7
161.51 1.67 1.6	224.83	8NO	-0.7 -4.1 -2.3 0.7
167.23 1.74 1.5	235.02	18NO	-0.6 -4.4 -2.3 0.7
172.83 1.80 1.4	245.22	28NO	-0.6 -4.4 -2.2 0.7
178.28 1.87 1.3	255.43	8DE	-0.5 -4.3 -2.1 0.8
183.56 1.94 1.1	265.63	18DE	0.7 -4.2 -2.1 0.8
188.63 2.00 1.0	275.81	28DE	2.9 -4.1 -2.0 0.8
		43	
193.43 2.06 0.8	285.97	7JA	0.9 -4.0 -1.9 0.8
197.90 2.11 0.6	296.09	17JA	0.4 -3.9 -1.8 0.8
201.96 2.16 0.4	306.17	27JA	0.2 -3.8 -1.8 0.8
205.51 2.18 0.2	316.20	6FE	-0.0 -3.7 -1.7 0.7
208.43 2.19 -0.1	326.16	16FE	-0.4 -3.6 -1.7 0.7
210.54 2.16 -0.4	336.07	26FE	-1.0 -3.6 -1.6 0.7
211.67 2.09 -0.7	345.93	8MR	-1.6 -3.5 -1.6 0.6
211.66 1.95 -1.0	355.72	18MR	-1.1 -3.4 -1.6 0.6
210.38 1.71 -1.3	5.46	28MR	-0.0 -3.4 -1.5 0.5
207.92 1.38 -1.6	15.15	7AP	1.2 -3.4 -1.5 0.5
204.64 0.94 -1.8	24.78	17AP	2.6 -3.3 -1.5 0.4
201.17 0.43 -1.8	34.38	27AP	3.5 -3.3 -1.5 0.3
198.24 -0.08 -1.6	43.95	7MY	2.0 -3.3 -1.5 0.3
196.42 -0.56 -1.4	53.49	17MY	1.0 -3.3 -1.5 0.2
195.94 -0.98 -1.2	63.02	27MY	0.2 -3.3 -1.6 0.3
196.81 -1.32 -1.0	72.54	6JN	-0.9 -3.3 -1.6 0.4
198.50 -1.59 -0.8	82.06	16JN	-1.7 -3.3 -1.6 0.4
202.01 -1.80 -0.6	91.59	26JN	-1.2 -3.4 -1.7 0.5
205.98 -1.96 -0.4	101.15	6JL	-0.4 -3.4 -1.7 0.5
210.64 -2.08 -0.3	110.73	16JL	0.1 -3.4 -1.8 0.6
215.87 -2.16 -0.1	120.36	26JL	0.4 -3.5 -1.8 0.6
221.58 -2.21 0.0	130.03	5AU	0.7 -3.5 -1.9 0.7
227.67 -2.23 0.1	139.74	15AU	1.3 -3.5 -1.9 0.7
234.09 -2.22 0.2	149.52	25AU	3.0 -3.4 -2.0 0.8
240.79 -2.19 0.3	159.35	4SE	1.1 -3.4 -2.1 0.8
247.70 -2.14 0.4	169.23	14SE	-0.5 -3.4 -2.1 0.8
254.79 -2.07 0.5	179.18	24SE	-0.9 -3.4 -2.2 0.8
262.03 -1.98 0.6	189.19	4OC	-0.9 -3.3 -2.2 0.8
269.38 -1.88 0.7	199.24	14OC	-0.8 -3.3 -2.3 0.8
276.82 -1.77 0.8	209.35	24OC	-0.6 -3.3 -2.3 0.7
284.32 -1.64 0.9	219.50	3NO	-0.5 -3.3 -2.3 0.7
291.85 -1.51 1.0	229.67	13NO	-0.5 -3.3 -2.3 0.7
299.40 -1.37 1.0	239.87	23NO	-0.3 -3.3 -2.3 0.7
306.95 -1.23 1.1	250.08	3DE	0.8 -3.4 -2.3 0.7
314.48 -1.09 1.2	260.28	13DE	2.6 -3.4 -2.2 0.7
321.98 -0.94 1.3	270.48	23DE	0.6 -3.4 -2.2 0.8

♂ LONG	♂ LAT	♂ MAG	☉ LONG	16.00UT	☿	♀	♃	♄
				44	MAGNITUDES			
329.44	-0.80	1.3	280.65	2JA	0.2	-3.5	-2.1	0.8
336.85	-0.66	1.4	290.79	12JA	0.0	-3.5	-2.0	0.8
344.20	-0.52	1.5	300.89	22JA	-0.1	-3.6	-2.0	0.8
351.48	-0.39	1.5	310.94	1FE	-0.4	-3.6	-1.9	0.8
358.70	-0.26	1.6	320.94	11FE	-1.0	-3.7	-1.8	0.8
5.85	-0.14	1.6	330.89	21FE	-1.5	-3.8	-1.8	0.7
12.92	-0.02	1.7	340.77	2MR	-1.1	-3.8	-1.7	0.7
19.92	0.09	1.7	350.59	12MR	0.0	-3.9	-1.6	0.7
26.86	0.20	1.8	0.36	22MR	1.5	-4.0	-1.6	0.6
33.72	0.30	1.8	10.07	1AP	3.3	-4.1	-1.5	0.6
40.52	0.39	1.8	19.73	11AP	2.6	-4.2	-1.5	0.5
47.26	0.48	1.9	29.35	21AP	1.5	-4.2	-1.5	0.4
53.94	0.56	1.9	38.93	1MY	0.8	-4.1	-1.5	0.4
60.57	0.64	1.9	48.49	11MY	0.1	-3.8	-1.4	0.3
67.15	0.72	1.9	58.02	21MY	-0.9	-3.2	-1.4	0.3
73.69	0.79	1.9	67.54	31MY	-1.7	-3.0	-1.4	0.3
80.18	0.85	1.9	77.06	10JN	-1.2	-3.8	-1.4	0.3
86.64	0.91	1.9	86.59	20JN	-0.4	-4.1	-1.4	0.4
93.08	0.96	1.9	96.13	30JN	0.2	-4.2	-1.5	0.4
99.48	1.02	2.0	105.70	10JL	0.5	-4.2	-1.5	0.5
105.87	1.06	2.0	115.31	20JL	0.9	-4.1	-1.5	0.5
112.24	1.11	2.0	124.95	30JL	1.7	-4.0	-1.6	0.6
118.59	1.15	2.0	134.64	9AU	3.2	-3.9	-1.6	0.7
124.94	1.18	2.0	144.39	19AU	0.9	-3.9	-1.6	0.7
131.27	1.21	2.0	154.18	29AU	-0.6	-3.8	-1.7	0.7
137.60	1.24	2.0	164.04	8SE	-1.1	-3.7	-1.8	0.8
143.93	1.27	2.0	173.96	18SE	-1.0	-3.7	-1.8	0.8
150.25	1.29	2.0	183.94	28SE	-0.7	-3.6	-1.9	0.8
156.56	1.30	2.0	193.97	8OC	-0.5	-3.5	-2.0	0.8
162.87	1.32	1.9	204.05	18OC	-0.4	-3.5	-2.0	0.8
169.16	1.32	1.9	214.17	28OC	-0.3	-3.5	-2.1	0.8
175.45	1.32	1.8	224.34	7NO	-0.2	-3.4	-2.1	0.7
181.72	1.32	1.7	234.53	17NO	1.0	-3.4	-2.2	0.7
187.97	1.31	1.7	244.73	27NO	2.3	-3.4	-2.2	0.7
194.19	1.29	1.6	254.94	7DE	0.4	-3.4	-2.2	0.7
200.38	1.26	1.5	265.14	17DE	-0.0	-3.4	-2.2	0.7
206.52	1.22	1.4	275.32	27DE	-0.1	-3.3	-2.2	0.8
				45				
212.61	1.17	1.2	285.48	6JA	-0.2	-3.3	-2.2	0.8
218.64	1.10	1.1	295.61	16JA	-0.5	-3.3	-2.1	0.8
224.57	1.01	1.0	305.68	26JA	-1.0	-3.3	-2.1	0.8
230.41	0.90	0.8	315.71	5FE	-1.4	-3.4	-2.0	0.8
236.11	0.77	0.6	325.68	15FE	-1.0	-3.4	-1.9	0.8
241.65	0.60	0.4	335.59	25FE	0.1	-3.4	-1.8	0.8
246.99	0.40	0.2	345.45	7MR	1.8	-3.4	-1.8	0.8
252.06	0.15	0.0	355.25	17MR	3.4	-3.5	-1.7	0.7
256.79	-0.15	-0.2	4.98	27MR	1.9	-3.5	-1.6	0.7
261.11	-0.51	-0.5	14.67	6AP	1.1	-3.4	-1.6	0.6
264.87	-0.95	-0.8	24.31	16AP	0.6	-3.4	-1.5	0.6
267.94	-1.47	-1.1	33.91	26AP	-0.0	-3.4	-1.5	0.5
270.14	-2.09	-1.4	43.48	6MY	-0.9	-3.4	-1.4	0.4
271.26	-2.79	-1.7	53.02	16MY	-1.8	-3.3	-1.4	0.4
271.19	-3.56	-2.0	62.55	26MY	-1.2	-3.3	-1.4	0.3
269.88	-4.34	-2.3	72.07	5JN	-0.3	-3.3	-1.3	0.2
267.60	-5.03	-2.5	81.59	15JN	0.3	-3.3	-1.3	0.3
264.88	-5.53	-2.5	91.12	25JN	0.7	-3.3	-1.3	0.3
262.44	-5.78	-2.4	100.68	5JL	1.2	-3.3	-1.3	0.4
260.95	-5.76	-2.2	110.27	15JL	2.2	-3.4	-1.3	0.5
260.74	-5.55	-1.9	119.89	25JL	2.9	-3.4	-1.3	0.5
261.87	-5.21	-1.7	129.55	4AU	0.8	-3.4	-1.4	0.6
264.23	-4.81	-1.4	139.27	14AU	-0.7	-3.5	-1.4	0.6
267.61	-4.38	-1.2	149.04	24AU	-1.2	-3.5	-1.4	0.7
271.83	-3.95	-0.9	158.87	3SE	-1.2	-3.6	-1.5	0.7
276.71	-3.53	-0.7	168.75	13SE	-0.7	-3.7	-1.5	0.7
282.10	-3.12	-0.5	178.70	23SE	-0.4	-3.8	-1.6	0.8
287.88	-2.74	-0.3	188.70	3OC	-0.2	-3.9	-1.6	0.8
293.96	-2.38	-0.1	198.76	13OC	-0.2	-4.0	-1.7	0.8
300.27	-2.04	0.1	208.86	23OC	0.0	-4.1	-1.7	0.8
306.75	-1.73	0.3	219.01	2NO	1.3	-4.2	-1.8	0.8
313.35	-1.44	0.4	229.18	12NO	2.0	-4.3	-1.9	0.8
320.04	-1.17	0.6	239.37	22NO	0.2	-4.4	-1.9	0.7
326.79	-0.93	0.7	249.58	2DE	-0.2	-4.4	-2.0	0.7
333.57	-0.70	0.9	259.79	12DE	-0.2	-4.3	-2.0	0.7
340.37	-0.50	1.0	269.98	22DE	-0.3	-3.9	-2.1	0.7

♂ LONG	♂ LAT	♂ MAG	☉ LONG	16.00UT	☿	♀	♃	♄
				46	MAGNITUDES			
347.17	-0.31	1.1	280.16	1JA	-0.6	-3.2	-2.1	0.8
353.95	-0.14	1.2	290.30	11JA	-1.0	-3.6	-2.1	0.8
0.72	0.02	1.3	300.40	21JA	-1.2	-4.1	-2.1	0.8
7.46	0.16	1.4	310.46	31JA	-1.0	-4.3	-2.1	0.8
14.16	0.28	1.5	320.46	10FE	0.2	-4.3	-2.0	0.8
20.83	0.40	1.6	330.40	20FE	2.2	-4.2	-2.0	0.8
27.47	0.50	1.7	340.29	2MR	2.7	-4.1	-1.9	0.8
34.06	0.60	1.7	350.12	12MR	1.4	-4.0	-1.9	0.8
40.62	0.68	1.8	359.88	22MR	0.8	-3.9	-1.8	0.8
47.15	0.76	1.9	9.60	1AP	0.4	-3.8	-1.7	0.7
53.63	0.82	1.9	19.27	11AP	-0.1	-3.7	-1.7	0.7
60.09	0.88	1.9	28.89	21AP	-0.9	-3.6	-1.6	0.6
66.53	0.94	2.0	38.47	1MY	-1.8	-3.6	-1.5	0.6
72.93	0.98	2.0	48.03	11MY	-1.2	-3.5	-1.5	0.5
79.32	1.02	2.0	57.56	21MY	-0.3	-3.5	-1.4	0.4
85.69	1.06	2.0	67.09	31MY	0.4	-3.4	-1.4	0.4
92.05	1.09	2.0	76.60	10JN	0.9	-3.4	-1.3	0.3
98.41	1.11	2.0	86.13	20JN	1.6	-3.4	-1.3	0.3
104.76	1.13	2.0	95.68	30JN	2.8	-3.3	-1.3	0.3
111.11	1.15	2.0	105.24	10JL	2.5	-3.3	-1.3	0.4
117.47	1.16	2.0	114.84	20JL	0.6	-3.3	-1.3	0.4
123.84	1.16	2.0	124.49	30JL	-0.7	-3.3	-1.2	0.5
130.23	1.16	2.0	134.17	9AU	-1.3	-3.3	-1.2	0.5
136.63	1.16	2.0	143.92	19AU	-1.2	-3.4	-1.3	0.6
143.06	1.15	2.0	153.71	29AU	-0.6	-3.4	-1.3	0.6
149.51	1.13	2.0	163.57	8SE	-0.3	-3.4	-1.3	0.7
155.99	1.12	2.0	173.48	18SE	-0.1	-3.4	-1.3	0.7
162.50	1.09	2.0	183.46	28SE	-0.0	-3.5	-1.4	0.8
169.04	1.06	2.0	193.48	8OC	0.2	-3.5	-1.4	0.8
175.61	1.03	1.9	203.56	18OC	1.5	-3.5	-1.4	0.8
182.22	0.99	1.9	213.69	28OC	1.8	-3.5	-1.5	0.8
188.86	0.94	1.9	223.84	7NO	-0.0	-3.4	-1.6	0.8
195.54	0.88	1.8	234.03	17NO	-0.4	-3.4	-1.6	0.8
202.24	0.81	1.8	244.23	27NO	-0.4	-3.4	-1.7	0.8
208.98	0.74	1.7	254.44	7DE	-0.4	-3.4	-1.8	0.8
215.76	0.65	1.6	264.64	17DE	-0.7	-3.4	-1.8	0.7
222.56	0.56	1.6	274.83	27DE	-1.0	-3.3	-1.9	0.8
				47				
229.39	0.45	1.5	284.98	6JA	-1.1	-3.3	-1.9	0.8
236.26	0.33	1.4	295.11	16JA	-0.9	-3.3	-2.0	0.8
243.14	0.19	1.3	305.19	26JA	0.3	-3.3	-2.0	0.9
250.04	0.04	1.2	315.22	5FE	2.6	-3.4	-2.0	0.9
256.96	-0.13	1.1	325.20	15FE	2.1	-3.4	-2.0	0.9
263.88	-0.32	1.0	335.12	25FE	1.0	-3.4	-2.0	0.9
270.80	-0.52	0.8	344.97	7MR	0.6	-3.4	-2.0	0.9
277.71	-0.75	0.7	354.77	17MR	0.3	-3.5	-2.0	0.9
284.59	-0.99	0.6	4.52	27MR	-0.2	-3.5	-1.9	0.9
291.42	-1.25	0.4	14.21	6AP	-0.9	-3.6	-1.8	0.8
298.19	-1.53	0.3	23.85	16AP	-1.8	-3.6	-1.8	0.8
304.85	-1.83	0.2	33.45	26AP	-1.2	-3.7	-1.7	0.7
311.37	-2.15	-0.0	43.02	6MY	-0.2	-3.8	-1.7	0.7
317.70	-2.49	-0.2	52.57	16MY	0.6	-3.8	-1.6	0.6
323.79	-2.84	-0.3	62.09	26MY	1.2	-3.9	-1.5	0.6
329.55	-3.21	-0.5	71.61	5JN	2.1	-4.0	-1.5	0.5
334.89	-3.58	-0.7	81.14	15JN	3.4	-4.1	-1.4	0.4
339.67	-3.97	-0.9	90.67	25JN	2.0	-4.2	-1.4	0.4
343.77	-4.36	-1.1	100.22	5JL	0.5	-4.2	-1.3	0.3
346.97	-4.74	-1.3	109.81	15JL	-0.8	-4.1	-1.3	0.4
349.07	-5.08	-1.6	119.42	25JL	-1.4	-3.7	-1.3	0.4
349.87	-5.36	-1.8	129.09	4AU	-1.2	-3.2	-1.2	0.5
349.24	-5.51	-2.1	138.80	14AU	-0.5	-3.5	-1.2	0.5
347.26	-5.46	-2.3	148.57	24AU	-0.2	-4.0	-1.2	0.6
344.37	-5.17	-2.4	158.39	3SE	0.0	-4.3	-1.2	0.6
341.25	-4.64	-2.3	168.28	13SE	0.2	-4.3	-1.2	0.7
338.73	-3.96	-2.0	178.22	23SE	0.5	-4.3	-1.2	0.7
337.34	-3.21	-1.7	188.22	3OC	1.9	-4.2	-1.2	0.8
337.27	-2.50	-1.4	198.27	13OC	1.6	-4.1	-1.3	0.8
338.47	-1.86	-1.1	208.37	23OC	-0.2	-4.0	-1.3	0.8
340.76	-1.32	-0.7	218.51	2NO	-0.5	-3.9	-1.3	0.8
343.93	-0.86	-0.4	228.69	12NO	-0.5	-3.8	-1.4	0.9
347.79	-0.48	-0.2	238.88	22NO	-0.6	-3.7	-1.4	0.9
352.18	-0.16	0.1	249.09	2DE	-0.7	-3.7	-1.5	0.9
357.00	0.10	0.3	259.30	12DE	-0.9	-3.6	-1.6	0.8
2.13	0.32	0.5	269.49	22DE	-0.9	-3.6	-1.6	0.8

♂ LONG	♂ LAT	♂ MAG	☉ LONG	16.00UT	☿	♀	♃	♄
				48	MAGNITUDES			
7.50	0.50	0.7	279.67	1JA	-0.7	-3.5	-1.7	0.8
13.06	0.65	0.9	289.81	11JA	0.4	-3.4	-1.8	0.8
18.77	0.78	1.0	299.92	21JA	2.9	-3.4	-1.8	0.9
24.58	0.88	1.2	309.97	31JA	1.5	-3.4	-1.9	0.9
30.48	0.97	1.3	319.98	10FE	0.7	-3.3	-1.9	0.9
36.44	1.04	1.4	329.93	20FE	0.4	-3.3	-2.0	1.0
42.45	1.10	1.5	339.82	1MR	0.2	-3.3	-2.0	1.0
48.50	1.15	1.6	349.65	11MR	-0.2	-3.3	-2.0	1.0
54.58	1.19	1.7	359.42	21MR	-0.9	-3.3	-2.0	1.0
60.68	1.21	1.8	9.14	31MR	-1.8	-3.3	-2.0	1.0
66.80	1.24	1.8	18.80	10AP	-1.2	-3.3	-2.0	0.9
72.94	1.25	1.9	28.42	20AP	-0.2	-3.4	-1.9	0.9
79.09	1.26	1.9	38.01	30AP	0.8	-3.4	-1.9	0.9
85.26	1.26	2.0	47.57	10MY	1.6	-3.4	-1.8	0.8
91.44	1.26	2.0	57.10	20MY	2.9	-3.5	-1.8	0.7
97.65	1.25	2.0	66.62	30MY	3.0	-3.5	-1.7	0.7
103.88	1.24	2.0	76.14	9JN	1.6	-3.5	-1.6	0.6
110.13	1.22	2.0	85.67	19JN	0.4	-3.4	-1.6	0.5
116.41	1.20	2.0	95.21	29JN	-0.8	-3.4	-1.5	0.5
122.73	1.18	2.0	104.78	9JL	-1.5	-3.4	-1.5	0.4
129.08	1.15	2.0	114.37	19JL	-1.2	-3.3	-1.4	0.4
135.47	1.11	2.0	124.01	29JL	-0.5	-3.3	-1.4	0.5
141.90	1.08	2.0	133.70	8AU	-0.1	-3.3	-1.3	0.5
148.37	1.03	1.9	143.44	18AU	0.2	-3.3	-1.3	0.6
154.90	0.99	1.9	153.23	28AU	0.3	-3.3	-1.3	0.6
161.48	0.93	1.9	163.09	7SE	0.7	-3.3	-1.2	0.7
168.12	0.88	1.9	172.99	17SE	2.2	-3.4	-1.2	0.7
174.81	0.82	1.9	182.97	27SE	1.4	-3.4	-1.2	0.8
181.57	0.75	1.9	192.99	7OC	-0.3	-3.4	-1.2	0.8
188.38	0.68	1.8	203.07	17OC	-0.7	-3.5	-1.2	0.8
195.26	0.60	1.8	213.19	27OC	-0.7	-3.5	-1.3	0.9
202.20	0.51	1.8	223.35	6NO	-0.7	-3.6	-1.3	0.9
209.21	0.42	1.8	233.53	16NO	-0.8	-3.6	-1.3	0.9
216.28	0.32	1.7	243.74	26NO	-0.7	-3.7	-1.3	0.9
223.41	0.22	1.7	253.95	6DE	-0.7	-3.8	-1.4	0.9
230.60	0.11	1.7	264.15	16DE	-0.6	-3.9	-1.4	0.9
237.86	-0.01	1.6	274.34	26DE	0.5	-4.0	-1.5	0.9
				49				
245.17	-0.13	1.6	284.50	5JA	3.0	-4.1	-1.6	0.9
252.53	-0.26	1.5	294.62	15JA	1.1	-4.2	-1.6	0.9
259.94	-0.40	1.4	304.71	25JA	0.5	-4.2	-1.7	1.0
267.39	-0.54	1.4	314.74	4FE	0.3	-4.3	-1.8	1.0
274.87	-0.69	1.3	324.72	14FE	0.0	-4.3	-1.8	1.0
282.39	-0.84	1.3	334.64	24FE	-0.3	-4.1	-1.9	1.1
289.92	-0.98	1.2	344.50	6MR	-0.9	-3.7	-2.0	1.1
297.46	-1.13	1.1	354.30	16MR	-1.7	-3.2	-2.0	1.1
305.00	-1.28	1.1	4.05	26MR	-1.2	-3.6	-2.0	1.1
312.51	-1.42	1.0	13.74	5AP	-0.1	-4.0	-2.1	1.1
319.99	-1.56	0.9	23.38	15AP	1.0	-4.2	-2.1	1.1
327.41	-1.68	0.8	32.99	25AP	2.2	-4.2	-2.1	1.0
334.76	-1.80	0.8	42.56	5MY	3.7	-4.1	-2.0	1.0
342.02	-1.91	0.7	52.10	15MY	2.4	-4.0	-2.0	1.0
349.15	-2.00	0.6	61.63	25MY	1.2	-3.9	-2.0	0.9
356.14	-2.07	0.5	71.15	4JN	0.3	-3.8	-1.9	0.8
2.96	-2.13	0.5	80.67	14JN	-0.8	-3.8	-1.9	0.8
9.56	-2.17	0.4	90.21	24JN	-1.6	-3.7	-1.8	0.7
15.91	-2.19	0.3	99.76	4JL	-1.2	-3.6	-1.7	0.6
21.95	-2.19	0.2	109.34	14JL	-0.4	-3.6	-1.7	0.6
27.63	-2.17	0.1	118.96	24JL	0.0	-3.5	-1.6	0.5
32.86	-2.12	-0.1	128.62	3AU	0.3	-3.5	-1.6	0.5
37.56	-2.05	-0.2	138.33	13AU	0.5	-3.4	-1.5	0.6
41.57	-1.94	-0.4	148.09	23AU	1.0	-3.4	-1.5	0.6
44.76	-1.79	-0.6	157.91	2SE	2.7	-3.4	-1.4	0.7
46.94	-1.59	-0.8	167.79	12SE	1.3	-3.4	-1.4	0.7
47.87	-1.33	-1.0	177.74	22SE	-0.5	-3.4	-1.3	0.8
47.41	-1.00	-1.2	187.73	2OC	-0.8	-3.4	-1.3	0.8
45.50	-0.60	-1.4	197.78	12OC	-0.8	-3.4	-1.3	0.9
42.35	-0.14	-1.6	207.88	22OC	-0.8	-3.4	-1.3	0.9
38.57	0.34	-1.7	218.02	1NO	-0.6	-3.4	-1.3	1.0
34.94	0.78	-1.5	228.19	11NO	-0.6	-3.4	-1.3	1.0
32.19	1.14	-1.2	238.39	21NO	-0.6	-3.4	-1.3	1.0
30.73	1.41	-0.9	248.59	1DE	-0.4	-3.4	-1.3	1.0
30.60	1.60	-0.5	258.80	11DE	0.7	-3.4	-1.4	1.0
31.71	1.72	-0.2	269.00	21DE	2.8	-3.5	-1.4	1.1
33.83	1.79	0.0	279.17	31DE	0.8	-3.5	-1.4	1.1

♂ LONG	♂ LAT	♂ MAG	☉ LONG	16.00UT	☿	♀	♃	♄
				50	MAGNITUDES			
36.77	1.83	0.3	289.32	10JA	0.3	-3.5	-1.5	1.1
40.35	1.85	0.5	299.43	20JA	0.1	-3.4	-1.5	1.0
44.43	1.85	0.7	309.49	30JA	-0.1	-3.4	-1.6	1.0
48.90	1.83	0.9	319.49	9FE	-0.4	-3.4	-1.7	1.1
53.68	1.81	1.1	329.45	19FE	-0.9	-3.4	-1.7	1.1
58.71	1.78	1.2	339.33	1MR	-1.6	-3.4	-1.8	1.2
63.93	1.75	1.3	349.17	11MR	-1.2	-3.3	-1.9	1.2
69.32	1.71	1.5	358.94	21MR	-0.1	-3.3	-1.9	1.2
74.84	1.67	1.6	8.66	31MR	1.3	-3.3	-2.0	1.2
80.46	1.62	1.6	18.33	10AP	2.8	-3.3	-2.1	1.2
86.19	1.57	1.7	27.96	20AP	3.1	-3.4	-2.1	1.2
92.00	1.52	1.8	37.54	30AP	1.8	-3.4	-2.1	1.2
97.88	1.47	1.8	47.10	10MY	1.0	-3.4	-2.2	1.2
103.84	1.42	1.9	56.64	20MY	0.2	-3.4	-2.2	1.2
109.86	1.36	1.9	66.16	30MY	-0.8	-3.5	-2.2	1.1
115.94	1.30	1.9	75.68	9JN	-1.7	-3.5	-2.1	1.0
122.09	1.24	2.0	85.21	19JN	-1.2	-3.6	-2.1	1.0
128.31	1.17	2.0	94.75	29JN	-0.4	-3.6	-2.1	0.9
134.59	1.11	2.0	104.31	9JL	0.1	-3.7	-2.0	0.8
140.94	1.04	2.0	113.91	19JL	0.4	-3.8	-1.9	0.8
147.35	0.96	2.0	123.55	29JL	0.7	-3.9	-1.9	0.7
153.84	0.89	1.9	133.23	8AU	1.4	-4.0	-1.8	0.6
160.41	0.81	1.9	142.97	18AU	3.1	-4.1	-1.8	0.7
167.04	0.73	1.9	152.76	28AU	1.1	-4.2	-1.7	0.7
173.76	0.64	1.8	162.61	7SE	-0.5	-4.3	-1.6	0.8
180.56	0.55	1.8	172.52	17SE	-1.0	-4.3	-1.6	0.8
187.44	0.46	1.7	182.48	27SE	-1.0	-4.2	-1.5	0.9
194.41	0.37	1.7	192.51	7OC	-0.8	-3.8	-1.5	0.9
201.45	0.27	1.6	202.58	17OC	-0.5	-3.2	-1.5	1.0
208.58	0.17	1.6	212.70	27OC	-0.4	-3.6	-1.4	1.0
215.79	0.07	1.6	222.85	6NO	-0.4	-4.2	-1.4	1.1
223.08	-0.04	1.6	233.04	16NO	-0.3	-4.4	-1.4	1.1
230.45	-0.15	1.6	243.24	26NO	0.8	-4.4	-1.4	1.1
237.89	-0.26	1.6	253.45	6DE	2.5	-4.3	-1.4	1.2
245.40	-0.37	1.6	263.65	16DE	0.5	-4.2	-1.4	1.2
252.97	-0.48	1.5	273.84	26DE	0.1	-4.1	-1.4	1.2
				51				
260.60	-0.59	1.5	284.01	5JA	-0.0	-4.0	-1.4	1.2
268.27	-0.70	1.5	294.13	15JA	-0.2	-3.9	-1.5	1.2
275.99	-0.80	1.5	304.22	25JA	-0.5	-3.8	-1.5	1.2
283.73	-0.90	1.4	314.26	4FE	-1.0	-3.7	-1.6	1.2
291.50	-1.00	1.4	324.24	14FE	-1.4	-3.6	-1.6	1.2
299.27	-1.08	1.4	334.16	24FE	-1.1	-3.6	-1.7	1.3
307.04	-1.16	1.4	344.03	6MR	0.0	-3.5	-1.7	1.3
314.78	-1.23	1.3	353.83	16MR	1.6	-3.4	-1.8	1.4
322.50	-1.29	1.3	3.58	26MR	3.5	-3.4	-1.9	1.4
330.18	-1.33	1.3	13.28	5AP	2.3	-3.4	-1.9	1.4
337.80	-1.36	1.3	22.92	15AP	1.3	-3.3	-2.0	1.4
345.35	-1.38	1.2	32.53	25AP	0.7	-3.3	-2.1	1.4
352.82	-1.38	1.2	42.10	5MY	0.1	-3.3	-2.1	1.4
0.20	-1.37	1.2	51.65	15MY	-0.8	-3.3	-2.2	1.3
7.47	-1.34	1.2	61.17	25MY	-1.7	-3.3	-2.2	1.3
14.62	-1.30	1.1	70.70	4JN	-1.2	-3.3	-2.3	1.2
21.64	-1.24	1.1	80.22	14JN	-0.4	-3.3	-2.3	1.2
28.52	-1.17	1.1	89.75	24JN	0.2	-3.4	-2.3	1.1
35.23	-1.08	1.0	99.30	4JL	0.6	-3.4	-2.3	1.1
41.77	-0.98	1.0	108.88	14JL	1.0	-3.4	-2.3	1.0
48.10	-0.86	0.9	118.49	24JL	1.8	-3.5	-2.2	1.0
54.21	-0.73	0.9	128.15	3AU	3.2	-3.5	-2.2	0.9
60.05	-0.58	0.8	137.86	13AU	0.9	-3.5	-2.1	0.9
65.60	-0.41	0.7	147.62	23AU	-0.6	-3.4	-2.1	0.8
70.79	-0.21	0.6	157.44	2SE	-1.1	-3.4	-2.0	0.9
75.55	0.00	0.5	167.31	12SE	-1.1	-3.4	-1.9	0.9
79.80	0.25	0.3	177.25	22SE	-0.7	-3.4	-1.9	0.9
83.41	0.53	0.2	187.24	2OC	-0.4	-3.3	-1.8	1.0
86.24	0.86	-0.0	197.29	12OC	-0.3	-3.3	-1.7	1.0
88.11	1.23	-0.2	207.38	22OC	-0.3	-3.3	-1.7	1.1
88.82	1.64	-0.4	217.52	1NO	-0.1	-3.3	-1.6	1.1
88.20	2.09	-0.7	227.69	11NO	1.0	-3.3	-1.6	1.2
86.20	2.55	-0.9	237.88	21NO	2.3	-3.4	-1.6	1.2
82.98	2.97	-1.1	248.09	1DE	0.3	-3.4	-1.5	1.3
79.08	3.29	-1.2	258.30	11DE	-0.1	-3.4	-1.5	1.3
75.23	3.47	-1.0	268.50	21DE	-0.2	-3.4	-1.5	1.4
72.14	3.51	-0.8	278.68	31DE	-0.3	-3.5	-1.5	1.4

♂			☉	16.00UT	☿	♀	♃	♄
LONG	LAT	MAG	LONG	52	MAGNITUDES			
70.26	3.44	-0.5	288.82	10JA	-0.5	-3.5	-1.5	1.4
69.69	3.30	-0.2	298.94	20JA	-1.0	-3.6	-1.5	1.4
70.36	3.13	0.0	309.00	30JA	-1.3	-3.6	-1.5	1.3
72.08	2.94	0.3	319.01	9FE	-1.0	-3.7	-1.5	1.3
74.66	2.76	0.5	328.97	19FE	0.1	-3.8	-1.6	1.2
77.91	2.59	0.7	338.86	29FE	1.9	-3.8	-1.6	1.2
81.71	2.42	0.9	348.70	10MR	3.2	-3.9	-1.7	1.2
85.95	2.26	1.0	358.48	20MR	1.7	-4.0	-1.7	1.2
90.53	2.12	1.2	8.20	30MR	1.0	-4.1	-1.8	1.2
95.40	1.98	1.3	17.87	9AP	0.5	-4.2	-1.8	1.2
100.51	1.85	1.4	27.50	19AP	-0.0	-4.2	-1.9	1.2
105.82	1.72	1.5	37.09	29AP	-0.8	-4.1	-2.0	1.1
111.30	1.60	1.6	46.64	9MY	-1.8	-3.8	-2.0	1.1
116.94	1.48	1.6	56.18	19MY	-1.2	-3.2	-2.1	1.1
122.71	1.36	1.7	65.70	29MY	-0.3	-3.0	-2.2	1.0
128.61	1.25	1.7	75.22	8JN	0.3	-3.8	-2.2	1.0
134.63	1.13	1.8	84.75	18JN	0.8	-4.1	-2.3	0.9
140.76	1.02	1.8	94.29	28JN	1.3	-4.2	-2.3	0.9
146.99	0.91	1.8	103.85	8JL	2.4	-4.2	-2.4	0.8
153.34	0.80	1.8	113.45	18JL	2.8	-4.1	-2.4	0.8
159.79	0.69	1.8	123.08	28JL	0.8	-4.0	-2.4	0.7
166.34	0.58	1.8	132.76	7AU	-0.7	-3.9	-2.4	0.7
172.99	0.47	1.8	142.49	17AU	-1.2	-3.9	-2.4	0.7
179.75	0.36	1.8	152.28	27AU	-1.2	-3.8	-2.3	0.6
186.61	0.25	1.7	162.13	6SE	-0.7	-3.7	-2.3	0.6
193.58	0.14	1.7	172.04	16SE	-0.3	-3.6	-2.2	0.7
200.65	0.03	1.7	182.00	26SE	-0.2	-3.6	-2.2	0.7
207.81	-0.07	1.6	192.02	6OC	-0.1	-3.5	-2.1	0.8
215.08	-0.18	1.6	202.09	16OC	0.1	-3.5	-2.0	0.9
222.43	-0.29	1.6	212.20	26OC	1.3	-3.5	-2.0	0.9
229.88	-0.39	1.5	222.36	5NO	2.0	-3.4	-1.9	1.0
237.41	-0.49	1.5	232.54	15NO	0.1	-3.4	-1.8	1.1
245.02	-0.59	1.4	242.74	25NO	-0.3	-3.4	-1.8	1.1
252.69	-0.68	1.4	252.95	5DE	-0.3	-3.4	-1.7	1.1
260.43	-0.77	1.4	263.16	15DE	-0.4	-3.4	-1.7	1.1
268.21	-0.84	1.4	273.34	25DE	-0.6	-3.3	-1.6	1.1
				53				
276.04	-0.92	1.4	283.51	4JA	-1.0	-3.3	-1.6	1.1
283.89	-0.98	1.4	293.64	14JA	-1.1	-3.3	-1.6	1.1
291.76	-1.03	1.4	303.73	24JA	-0.9	-3.3	-1.6	1.1
299.63	-1.07	1.4	313.76	3FE	0.2	-3.4	-1.6	1.1
307.49	-1.10	1.4	323.75	13FE	2.3	-3.4	-1.6	1.0
315.33	-1.12	1.4	333.68	23FE	2.5	-3.4	-1.6	1.0
323.13	-1.13	1.4	343.54	5MR	1.3	-3.4	-1.6	0.9
330.89	-1.13	1.4	353.35	15MR	0.7	-3.5	-1.6	0.9
338.59	-1.11	1.4	3.10	25MR	0.4	-3.5	-1.6	0.9
346.22	-1.08	1.4	12.80	4AP	-0.1	-3.4	-1.7	0.9
353.78	-1.04	1.5	22.45	14AP	-0.9	-3.4	-1.7	0.9
1.25	-0.99	1.5	32.06	24AP	-1.8	-3.4	-1.8	0.9
8.64	-0.92	1.5	41.63	4MY	-1.2	-3.4	-1.8	0.9
15.92	-0.85	1.5	51.18	14MY	-0.3	-3.3	-1.9	0.9
23.11	-0.77	1.5	60.71	24MY	0.5	-3.3	-1.9	0.8
30.19	-0.68	1.5	70.23	3JN	1.0	-3.3	-2.0	0.8
37.16	-0.58	1.5	79.75	13JN	1.8	-3.3	-2.1	0.8
44.01	-0.47	1.5	89.28	23JN	3.0	-3.3	-2.1	0.7
50.75	-0.35	1.5	98.83	3JL	2.3	-3.3	-2.2	0.7
57.35	-0.23	1.5	108.41	13JL	0.6	-3.4	-2.3	0.6
63.82	-0.09	1.4	118.02	23JL	-0.7	-3.4	-2.3	0.6
70.15	0.05	1.4	127.68	2AU	-1.4	-3.4	-2.4	0.5
76.32	0.20	1.4	137.38	12AU	-1.2	-3.5	-2.4	0.5
82.32	0.36	1.3	147.14	22AU	-0.6	-3.5	-2.4	0.4
88.13	0.53	1.3	156.96	1SE	-0.2	-3.6	-2.5	0.4
93.71	0.71	1.2	166.83	11SE	-0.0	-3.7	-2.5	0.3
99.04	0.91	1.1	176.76	21SE	0.1	-3.8	-2.4	0.3
104.07	1.12	1.0	186.76	1OC	0.3	-3.9	-2.4	0.4
108.74	1.36	0.9	196.80	11OC	1.6	-4.0	-2.3	0.5
112.97	1.63	0.7	206.89	21OC	1.8	-4.1	-2.3	0.5
116.66	1.92	0.6	217.03	31OC	-0.1	-4.2	-2.2	0.6
119.69	2.25	0.4	227.20	10NO	-0.4	-4.3	-2.1	0.7
121.90	2.61	0.2	237.39	20NO	-0.4	-4.4	-2.1	0.7
123.11	3.01	-0.1	247.60	30NO	-0.5	-4.4	-2.0	0.8
123.13	3.43	-0.3	257.80	10DE	-0.7	-4.3	-1.9	0.8
121.84	3.84	-0.6	268.00	20DE	-0.9	-3.9	-1.9	0.8
119.25	4.20	-0.8	278.18	30DE	-1.0	-3.2	-1.8	0.9

♂			☉	16.00UT	☿	♀	♃	♄
LONG	LAT	MAG	LONG	54	MAGNITUDES			
115.68	4.45	-1.0	288.33	9JA	-0.8	-3.6	-1.7	0.9
111.72	4.52	-1.0	298.44	19JA	0.3	-4.1	-1.7	0.9
108.10	4.43	-0.8	308.51	29JA	2.7	-4.3	-1.7	0.8
105.43	4.19	-0.6	318.52	8FE	1.9	-4.3	-1.6	0.8
104.01	3.88	-0.3	328.47	18FE	0.9	-4.2	-1.6	0.8
103.88	3.53	-0.1	338.38	28FE	0.5	-4.1	-1.6	0.7
104.92	3.19	0.2	348.21	10MR	0.2	-4.0	-1.6	0.7
106.94	2.87	0.4	357.99	20MR	-0.2	-3.9	-1.6	0.6
109.76	2.57	0.6	7.72	30MR	-0.9	-3.8	-1.6	0.7
113.24	2.30	0.7	17.40	9AP	-1.8	-3.7	-1.6	0.7
117.24	2.04	0.9	27.03	19AP	-1.2	-3.6	-1.6	0.7
121.67	1.81	1.0	36.62	29AP	-0.2	-3.6	-1.6	0.7
126.47	1.60	1.1	46.18	9MY	0.6	-3.5	-1.7	0.7
131.57	1.40	1.2	55.72	19MY	1.4	-3.5	-1.7	0.7
136.93	1.21	1.3	65.25	29MY	2.4	-3.4	-1.7	0.7
142.51	1.04	1.3	74.76	8JN	3.4	-3.4	-1.8	0.6
148.30	0.87	1.4	84.29	18JN	1.9	-3.4	-1.8	0.6
154.27	0.71	1.4	93.83	28JN	0.5	-3.3	-1.9	0.6
160.42	0.56	1.5	103.39	8JL	-0.8	-3.3	-2.0	0.5
166.72	0.41	1.5	112.98	18JL	-1.5	-3.3	-2.0	0.5
173.17	0.27	1.5	122.62	28JL	-1.2	-3.3	-2.1	0.4
179.77	0.13	1.5	132.29	7AU	-0.5	-3.3	-2.2	0.4
186.50	0.00	1.5	142.02	17AU	-0.1	-3.4	-2.2	0.3
193.37	-0.13	1.5	151.81	27AU	0.1	-3.4	-2.3	0.3
200.37	-0.25	1.5	161.65	6SE	0.2	-3.4	-2.3	0.2
207.49	-0.36	1.5	171.55	16SE	0.6	-3.4	-2.4	0.1
214.72	-0.47	1.5	181.52	26SE	1.9	-3.5	-2.4	0.1
222.07	-0.57	1.5	191.53	6OC	1.6	-3.5	-2.4	0.1
229.52	-0.67	1.5	201.60	16OC	-0.2	-3.5	-2.4	0.2
237.07	-0.76	1.5	211.71	26OC	-0.6	-3.5	-2.4	0.3
244.70	-0.83	1.5	221.86	5NO	-0.6	-3.4	-2.4	0.3
252.41	-0.90	1.4	232.04	15NO	-0.6	-3.4	-2.3	0.4
260.19	-0.96	1.4	242.25	25NO	-0.8	-3.4	-2.2	0.5
268.01	-1.01	1.4	252.45	5DE	-0.8	-3.4	-2.2	0.5
275.88	-1.05	1.4	262.65	15DE	-0.8	-3.3	-2.1	0.6
283.77	-1.08	1.4	272.85	25DE	-0.7	-3.3	-2.0	0.6
				55				
291.67	-1.09	1.4	283.01	4JA	0.4	-3.3	-2.0	0.6
299.57	-1.09	1.3	293.14	14JA	3.0	-3.3	-1.9	0.6
307.45	-1.08	1.3	303.24	24JA	1.4	-3.3	-1.8	0.6
315.31	-1.06	1.3	313.27	3FE	0.6	-3.4	-1.8	0.6
323.12	-1.03	1.3	323.26	13FE	0.4	-3.4	-1.7	0.6
330.89	-0.99	1.4	333.19	23FE	0.1	-3.4	-1.7	0.6
338.59	-0.94	1.4	343.06	5MR	-0.3	-3.4	-1.6	0.6
346.23	-0.88	1.4	352.87	15MR	-0.9	-3.5	-1.6	0.5
353.78	-0.81	1.5	2.63	25MR	-1.7	-3.5	-1.6	0.5
1.26	-0.73	1.5	12.33	4AP	-1.2	-3.6	-1.5	0.4
8.65	-0.64	1.5	21.98	14AP	-0.2	-3.6	-1.5	0.5
15.95	-0.55	1.6	31.60	24AP	0.8	-3.7	-1.5	0.5
23.16	-0.46	1.6	41.17	4MY	1.8	-3.8	-1.5	0.5
30.28	-0.36	1.6	50.72	14MY	3.2	-3.8	-1.5	0.5
37.31	-0.26	1.7	60.25	24MY	2.8	-3.9	-1.5	0.5
44.24	-0.15	1.7	69.77	3JN	1.5	-4.0	-1.5	0.5
51.08	-0.04	1.7	79.29	13JN	0.4	-4.1	-1.6	0.5
57.83	0.07	1.7	88.83	23JN	-0.8	-4.2	-1.6	0.5
64.49	0.18	1.7	98.37	3JL	-1.6	-4.2	-1.7	0.4
71.06	0.30	1.8	107.95	13JL	-1.2	-4.1	-1.7	0.4
77.53	0.42	1.8	117.56	23JL	-0.5	-3.7	-1.8	0.4
83.91	0.54	1.7	127.21	2AU	-0.0	-3.2	-1.8	0.3
90.19	0.67	1.7	136.92	12AU	0.2	-3.6	-1.9	0.3
96.36	0.79	1.7	146.67	22AU	0.4	-4.0	-1.9	0.2
102.42	0.93	1.7	156.48	1SE	0.8	-4.3	-2.0	0.1
108.36	1.06	1.6	166.35	11SE	2.3	-4.3	-2.1	0.1
114.16	1.21	1.6	176.28	21SE	1.4	-4.3	-2.1	0.0
119.81	1.36	1.5	186.27	1OC	-0.4	-4.2	-2.2	-0.0
125.28	1.52	1.4	196.31	11OC	-0.7	-4.1	-2.2	-0.1
130.53	1.69	1.3	206.41	21OC	-0.7	-4.0	-2.3	-0.1
135.54	1.87	1.2	216.54	31OC	-0.7	-3.9	-2.3	0.0
140.25	2.07	1.1	226.71	10NO	-0.7	-3.8	-2.3	0.1
144.58	2.29	0.9	236.90	20NO	-0.6	-3.7	-2.3	0.1
148.46	2.53	0.7	247.10	30NO	-0.7	-3.7	-2.3	0.2
151.78	2.79	0.5	257.31	10DE	-0.5	-3.6	-2.3	0.3
154.40	3.07	0.3	267.51	20DE	0.5	-3.5	-2.2	0.3
156.16	3.37	0.1	277.69	30DE	3.0	-3.5	-2.1	0.4

♂ LONG	♂ LAT	♂ MAG	⊙ LONG	16.00UT	☿	♀	♃	♄
				56	MAGNITUDES			
156.86	3.68	-0.2	287.84	9JA	1.0	-3.4	-2.1	0.4
156.35	3.98	-0.5	297.96	19JA	0.4	-3.4	-2.0	0.5
154.56	4.22	-0.7	308.02	29JA	0.2	-3.4	-1.9	0.5
151.58	4.36	-1.0	318.04	8FE	0.0	-3.3	-1.9	0.5
147.85	4.34	-1.1	328.00	18FE	-0.3	-3.3	-1.8	0.5
144.03	4.15	-1.0	337.90	28FE	-0.9	-3.3	-1.7	0.5
140.80	3.82	-0.9	347.74	9MR	-1.6	-3.3	-1.7	0.4
138.69	3.39	-0.6	357.52	19MR	-1.2	-3.3	-1.6	0.4
137.88	2.94	-0.4	7.25	29MR	-0.1	-3.3	-1.6	0.4
138.33	2.50	-0.2	16.93	8AP	1.1	-3.3	-1.5	0.3
139.91	2.09	-0.0	26.56	18AP	2.4	-3.4	-1.5	0.3
142.43	1.72	0.2	36.15	28AP	3.7	-3.4	-1.5	0.3
145.73	1.39	0.3	45.72	8MY	2.2	-3.4	-1.4	0.4
149.66	1.08	0.5	55.25	18MY	1.1	-3.5	-1.4	0.4
154.12	0.81	0.6	64.78	28MY	0.2	-3.5	-1.4	0.4
159.02	0.56	0.7	74.30	7JN	-0.8	-3.5	-1.4	0.4
164.30	0.34	0.8	83.82	17JN	-1.6	-3.4	-1.4	0.4
169.89	0.13	0.9	93.36	27JN	-1.2	-3.4	-1.4	0.4
175.76	-0.05	1.0	102.92	7JL	-0.4	-3.4	-1.4	0.4
181.89	-0.23	1.0	112.51	17JL	0.0	-3.3	-1.5	0.3
188.24	-0.38	1.1	122.14	27JL	0.3	-3.3	-1.5	0.3
194.79	-0.53	1.1	131.82	6AU	0.6	-3.3	-1.5	0.3
201.54	-0.66	1.2	141.55	16AU	1.2	-3.3	-1.6	0.2
208.45	-0.78	1.2	151.33	26AU	2.8	-3.3	-1.6	0.2
215.52	-0.88	1.2	161.17	5SE	1.3	-3.3	-1.7	0.1
222.74	-0.98	1.2	171.07	15SE	-0.5	-3.4	-1.8	0.0
230.08	-1.06	1.3	181.03	25SE	-0.9	-3.4	-1.8	-0.0
237.55	-1.12	1.3	191.04	5OC	-0.9	-3.4	-1.9	-0.1
245.12	-1.17	1.3	201.11	15OC	-0.9	-3.5	-2.0	-0.2
252.77	-1.21	1.3	211.22	25OC	-0.6	-3.5	-2.0	-0.2
260.51	-1.23	1.3	221.37	4NO	-0.5	-3.6	-2.1	-0.2
268.29	-1.24	1.3	231.55	14NO	-0.5	-3.6	-2.1	-0.1
276.12	-1.24	1.3	241.75	24NO	-0.4	-3.7	-2.2	-0.0
283.98	-1.22	1.4	251.96	4DE	0.7	-3.8	-2.2	0.0
291.85	-1.19	1.4	262.16	14DE	2.8	-3.9	-2.2	0.1
299.71	-1.15	1.4	272.35	24DE	0.7	-4.0	-2.2	0.2
				57				
307.56	-1.10	1.4	282.52	3JA	0.2	-4.1	-2.2	0.2
315.37	-1.03	1.4	292.66	13JA	0.1	-4.2	-2.1	0.3
323.13	-0.96	1.5	302.75	23JA	-0.1	-4.3	-2.1	0.3
330.85	-0.89	1.5	312.79	2FE	-0.4	-4.3	-2.0	0.3
338.49	-0.80	1.5	322.78	12FE	-0.9	-4.3	-2.0	0.4
346.07	-0.71	1.5	332.71	22FE	-1.5	-4.1	-1.9	0.4
353.57	-0.62	1.5	342.59	4MR	-1.2	-3.7	-1.8	0.4
0.99	-0.52	1.5	352.40	14MR	-0.1	-3.2	-1.7	0.4
8.32	-0.42	1.6	2.16	24MR	1.3	-3.6	-1.7	0.3
15.57	-0.32	1.6	11.86	3AP	3.1	-4.1	-1.6	0.3
22.74	-0.21	1.6	21.51	13AP	2.8	-4.2	-1.6	0.3
29.82	-0.11	1.6	31.13	23AP	1.6	-4.2	-1.5	0.2
36.82	-0.01	1.7	40.71	3MY	0.9	-4.1	-1.5	0.2
43.73	0.10	1.7	50.25	13MY	0.1	-4.0	-1.4	0.2
50.57	0.20	1.8	59.79	23MY	-0.8	-3.9	-1.4	0.3
57.34	0.30	1.8	69.31	2JN	-1.7	-3.8	-1.3	0.3
64.03	0.39	1.8	78.83	12JN	-1.2	-3.8	-1.3	0.3
70.66	0.49	1.9	88.36	22JN	-0.4	-3.7	-1.3	0.3
77.22	0.59	1.9	97.91	2JL	0.1	-3.6	-1.3	0.3
83.71	0.68	1.9	107.48	12JL	0.5	-3.6	-1.3	0.3
90.15	0.77	1.9	117.09	22JL	0.8	-3.5	-1.3	0.3
96.53	0.87	1.9	126.74	1AU	1.5	-3.5	-1.3	0.3
102.84	0.96	1.9	136.44	11AU	3.1	-3.4	-1.3	0.3
109.11	1.05	1.9	146.20	21AU	1.1	-3.4	-1.4	0.2
115.31	1.14	1.9	156.01	31AU	-0.6	-3.4	-1.4	0.2
121.44	1.23	1.9	165.87	10SE	-1.0	-3.4	-1.5	0.1
127.51	1.32	1.8	175.80	20SE	-1.0	-3.4	-1.5	0.0
133.51	1.41	1.8	185.79	30SE	-0.8	-3.4	-1.6	-0.0
139.42	1.51	1.7	195.83	10OC	-0.5	-3.4	-1.6	-0.1
145.24	1.60	1.7	205.92	20OC	-0.4	-3.4	-1.7	-0.2
150.94	1.70	1.6	216.05	30OC	-0.4	-3.4	-1.7	-0.2
156.52	1.80	1.5	226.21	9NO	-0.2	-3.4	-1.8	-0.3
161.94	1.90	1.4	236.40	19NO	0.9	-3.4	-1.9	-0.2
167.17	2.01	1.3	246.61	29NO	2.5	-3.4	-1.9	-0.2
172.19	2.12	1.1	256.81	9DE	0.5	-3.4	-2.0	-0.1
176.92	2.24	0.9	267.01	19DE	0.0	-3.5	-2.0	-0.0
181.32	2.35	0.8	277.19	29DE	-0.1	-3.5	-2.1	0.0

♂ LONG	♂ LAT	♂ MAG	⊙ LONG	16.00UT	☿	♀	♃	♄
				58	MAGNITUDES			
185.30	2.48	0.6	287.35	8JA	-0.2	-3.5	-2.1	0.1
188.75	2.60	0.3	297.46	18JA	-0.5	-3.4	-2.1	0.2
191.54	2.72	0.1	307.53	28JA	-1.0	-3.4	-2.1	0.2
193.52	2.82	-0.2	317.55	7FE	-1.4	-3.4	-2.0	0.3
194.50	2.90	-0.5	327.51	17FE	-1.1	-3.4	-2.0	0.3
194.32	2.93	-0.8	337.42	27FE	-0.0	-3.4	-2.0	0.3
192.87	2.88	-1.1	347.26	9MR	1.7	-3.3	-1.9	0.3
190.22	2.72	-1.3	357.05	19MR	3.6	-3.3	-1.8	0.3
186.77	2.43	-1.6	6.78	29MR	2.1	-3.3	-1.8	0.3
183.13	2.03	-1.5	16.46	8AP	1.2	-3.3	-1.7	0.3
180.03	1.55	-1.3	26.09	18AP	0.6	-3.4	-1.6	0.3
178.02	1.05	-1.1	35.69	28AP	0.0	-3.4	-1.6	0.2
177.34	0.58	-0.9	45.25	8MY	-0.8	-3.4	-1.5	0.2
177.99	0.15	-0.7	54.79	18MY	-1.7	-3.4	-1.5	0.2
179.84	-0.21	-0.5	64.31	28MY	-1.2	-3.5	-1.4	0.2
182.69	-0.53	-0.3	73.84	7JN	-0.4	-3.5	-1.4	0.2
186.38	-0.79	-0.2	83.36	17JN	0.2	-3.6	-1.3	0.3
190.77	-1.01	-0.0	92.90	27JN	0.7	-3.6	-1.3	0.3
195.73	-1.20	0.1	102.46	7JL	1.1	-3.7	-1.3	0.3
201.16	-1.36	0.2	112.05	17JL	2.0	-3.8	-1.2	0.3
207.00	-1.48	0.3	121.68	27JL	3.1	-3.9	-1.2	0.3
213.17	-1.58	0.4	131.35	6AU	0.9	-4.0	-1.2	0.3
219.65	-1.66	0.5	141.08	16AU	-0.6	-4.1	-1.2	0.3
226.36	-1.71	0.6	150.85	26AU	-1.2	-4.2	-1.2	0.3
233.30	-1.75	0.7	160.69	5SE	-1.2	-4.3	-1.3	0.2
240.41	-1.76	0.7	170.59	15SE	-0.7	-4.3	-1.3	0.2
247.68	-1.75	0.8	180.54	25SE	-0.4	-4.2	-1.3	0.1
255.08	-1.72	0.9	190.56	5OC	-0.2	-3.8	-1.4	0.0
262.58	-1.68	0.9	200.62	15OC	-0.2	-3.2	-1.4	-0.0
270.17	-1.62	1.0	210.73	25OC	-0.0	-3.7	-1.5	-0.1
277.80	-1.54	1.0	220.88	4NO	1.1	-4.2	-1.5	-0.2
285.48	-1.46	1.1	231.06	14NO	2.3	-4.4	-1.6	-0.2
293.18	-1.36	1.2	241.25	24NO	0.2	-4.4	-1.6	-0.3
300.88	-1.26	1.2	251.46	4DE	-0.2	-4.3	-1.7	-0.2
308.56	-1.15	1.3	261.67	14DE	-0.2	-4.2	-1.8	-0.2
316.21	-1.03	1.3	271.86	24DE	-0.3	-4.1	-1.8	-0.1
				59				
323.82	-0.91	1.4	282.03	3JA	-0.5	-4.0	-1.9	-0.0
331.38	-0.79	1.4	292.17	13JA	-1.0	-3.9	-1.9	0.0
338.88	-0.67	1.5	302.26	23JA	-1.2	-3.8	-2.0	0.1
346.32	-0.55	1.5	312.31	2FE	-1.0	-3.7	-2.0	0.2
353.68	-0.43	1.6	322.30	12FE	0.1	-3.6	-2.0	0.2
0.96	-0.31	1.6	332.24	22FE	2.0	-3.6	-2.0	0.2
8.17	-0.19	1.7	342.11	4MR	2.9	-3.5	-2.0	0.3
15.30	-0.08	1.7	351.93	14MR	1.5	-3.4	-2.0	0.3
22.36	0.03	1.7	1.69	24MR	0.9	-3.4	-1.9	0.3
29.34	0.13	1.8	11.40	3AP	0.5	-3.4	-1.9	0.3
36.24	0.23	1.8	21.06	13AP	-0.0	-3.3	-1.8	0.3
43.08	0.33	1.8	30.67	23AP	-0.8	-3.3	-1.8	0.3
49.85	0.42	1.8	40.25	3MY	-1.7	-3.3	-1.7	0.3
56.55	0.51	1.8	49.80	13MY	-1.2	-3.3	-1.6	0.3
63.20	0.59	1.8	59.33	23MY	-0.3	-3.3	-1.6	0.2
69.80	0.67	1.8	68.85	2JN	0.4	-3.3	-1.5	0.2
76.35	0.74	1.9	78.37	12JN	0.9	-3.3	-1.5	0.2
82.85	0.81	1.9	87.90	22JN	1.5	-3.4	-1.4	0.3
89.31	0.88	1.9	97.45	2JL	2.6	-3.4	-1.4	0.3
95.74	0.95	2.0	107.02	12JL	2.7	-3.4	-1.3	0.3
102.14	1.01	2.0	116.62	22JL	0.8	-3.5	-1.3	0.4
108.51	1.06	2.0	126.27	1AU	-0.7	-3.5	-1.3	0.4
114.85	1.12	2.0	135.97	11AU	-1.3	-3.5	-1.2	0.4
121.17	1.17	2.0	145.72	21AU	-1.2	-3.4	-1.2	0.4
127.47	1.22	2.0	155.53	31AU	-0.6	-3.4	-1.2	0.3
133.75	1.26	2.0	165.39	10SE	-0.3	-3.4	-1.2	0.3
140.01	1.30	2.0	175.31	20SE	-0.1	-3.4	-1.2	0.3
146.25	1.34	2.0	185.30	30SE	-0.0	-3.3	-1.2	0.2
152.46	1.38	1.9	195.34	10OC	0.2	-3.3	-1.3	0.2
158.64	1.41	1.9	205.42	20OC	1.3	-3.3	-1.3	0.1
164.80	1.44	1.8	215.55	30OC	2.0	-3.3	-1.3	0.0
170.92	1.46	1.7	225.72	9NO	0.0	-3.3	-1.4	-0.1
176.99	1.49	1.7	235.90	19NO	-0.3	-3.4	-1.4	-0.1
183.01	1.50	1.6	246.11	29NO	-0.4	-3.4	-1.5	-0.2
188.97	1.51	1.5	256.32	9DE	-0.4	-3.4	-1.5	-0.3
194.85	1.51	1.4	266.52	19DE	-0.6	-3.4	-1.6	-0.2
200.63	1.50	1.2	276.70	29DE	-1.0	-3.5	-1.7	-0.1

Left table (☿ ♀ ♃ ♄ Magnitudes — 60 / 61)

♂ LONG	LAT	MAG	☉ LONG	16.00UT	☿	♀	♃	♄
206.29	1.48	1.1	286.86	8JA	-1.1	-3.5	-1.7	-0.1
211.81	1.45	0.9	296.97	18JA	-0.9	-3.6	-1.8	-0.0
217.15	1.40	0.7	307.05	28JA	0.1	-3.6	-1.8	0.1
222.27	1.33	0.6	317.07	7FE	2.4	-3.7	-1.9	0.1
227.13	1.23	0.4	327.03	17FE	2.2	-3.8	-2.0	0.2
231.63	1.10	0.1	336.94	27FE	1.1	-3.8	-2.0	0.2
235.71	0.93	-0.1	346.79	8MR	0.7	-3.9	-2.0	0.3
239.24	0.70	-0.4	356.58	18MR	0.3	-4.0	-2.0	0.3
242.08	0.41	-0.7	6.32	28MR	-0.1	-4.1	-2.0	0.3
244.06	0.04	-1.0	16.00	7AP	-0.8	-4.2	-2.0	0.4
245.00	-0.42	-1.3	25.63	17AP	-1.7	-4.2	-2.0	0.4
244.73	-0.98	-1.6	35.23	27AP	-1.3	-4.1	-1.9	0.4
243.24	-1.62	-1.9	44.79	7MY	-0.3	-3.8	-1.9	0.4
240.71	-2.29	-2.2	54.33	17MY	0.5	-3.1	-1.8	0.3
237.67	-2.91	-2.3	63.86	27MY	1.2	-3.0	-1.8	0.3
234.85	-3.42	-2.2	73.38	6JN	2.0	-3.8	-1.7	0.3
232.88	-3.76	-2.0	82.90	16JN	3.3	-4.1	-1.6	0.3
232.20	-3.94	-1.7	92.44	26JN	2.2	-4.2	-1.6	0.3
232.91	-4.00	-1.5	101.99	6JL	0.6	-4.2	-1.5	0.4
234.90	-3.96	-1.3	111.58	16JL	-0.7	-4.1	-1.5	0.4
238.00	-3.85	-1.1	121.21	26JL	-1.4	-4.0	-1.4	0.4
242.01	-3.70	-0.9	130.88	5AU	-1.2	-3.9	-1.4	0.4
246.76	-3.52	-0.7	140.60	15AU	-0.6	-3.8	-1.3	0.5
252.10	-3.32	-0.5	150.38	25AU	-0.2	-3.8	-1.3	0.5
257.89	-3.10	-0.3	160.21	4SE	0.0	-3.7	-1.3	0.4
264.06	-2.87	-0.1	170.11	14SE	0.1	-3.6	-1.2	0.4
270.52	-2.64	0.0	180.06	24SE	0.4	-3.6	-1.2	0.4
277.19	-2.40	0.2	190.07	4OC	1.7	-3.5	-1.2	0.4
284.04	-2.17	0.3	200.13	14OC	1.8	-3.5	-1.2	0.3
291.01	-1.93	0.4	210.24	24OC	-0.1	-3.5	-1.2	0.2
298.07	-1.70	0.6	220.38	3NO	-0.5	-3.4	-1.3	0.2
305.19	-1.48	0.7	230.56	13NO	-0.5	-3.4	-1.3	0.1
312.34	-1.27	0.8	240.76	23NO	-0.5	-3.4	-1.3	0.0
319.50	-1.07	0.9	250.97	3DE	-0.7	-3.4	-1.4	-0.0
326.66	-0.87	1.0	261.18	13DE	-0.9	-3.4	-1.4	-0.1
333.81	-0.69	1.1	271.37	23DE	-0.9	-3.3	-1.5	-0.1

61

♂ LONG	LAT	MAG	☉ LONG	16.00UT	☿	♀	♃	♄
340.92	-0.51	1.2	281.54	2JA	-0.8	-3.3	-1.5	-0.1
347.99	-0.35	1.3	291.68	12JA	0.2	-3.3	-1.6	-0.0
355.03	-0.20	1.4	301.78	22JA	2.7	-3.4	-1.7	0.0
2.01	-0.06	1.5	311.82	1FE	1.7	-3.4	-1.7	0.1
8.93	0.07	1.6	321.82	11FE	0.8	-3.4	-1.8	0.2
15.81	0.19	1.6	331.76	21FE	0.5	-3.4	-1.9	0.2
22.63	0.30	1.7	341.63	3MR	0.2	-3.4	-1.9	0.3
29.39	0.41	1.8	351.46	13MR	-0.2	-3.5	-2.0	0.3
36.11	0.50	1.8	1.22	23MR	-0.8	-3.5	-2.0	0.4
42.77	0.59	1.9	10.92	2AP	-1.7	-3.4	-2.1	0.4
49.38	0.67	1.9	20.59	12AP	-1.3	-3.4	-2.1	0.4
55.95	0.74	1.9	30.20	22AP	-0.2	-3.4	-2.1	0.4
62.48	0.80	1.9	39.78	2MY	0.7	-3.4	-2.1	0.5
68.97	0.86	2.0	49.33	12MY	1.5	-3.3	-2.1	0.5
75.43	0.92	2.0	58.86	22MY	2.7	-3.3	-2.0	0.5
81.86	0.97	2.0	68.38	1JN	3.2	-3.3	-2.0	0.5
88.28	1.01	2.0	77.91	11JN	1.7	-3.3	-1.9	0.4
94.67	1.05	2.0	87.43	21JN	0.5	-3.3	-1.9	0.4
101.05	1.08	2.0	96.98	1JL	-0.7	-3.4	-1.8	0.4
107.42	1.11	1.9	106.55	11JL	-1.5	-3.4	-1.7	0.4
113.79	1.13	1.9	116.15	21JL	-1.2	-3.4	-1.7	0.5
120.16	1.15	2.0	125.80	31JL	-0.5	-3.4	-1.6	0.5
126.53	1.17	2.0	135.50	10AU	-0.1	-3.5	-1.6	0.6
132.91	1.18	2.0	145.24	20AU	0.1	-3.5	-1.5	0.6
139.31	1.19	2.0	155.05	30AU	0.3	-3.6	-1.5	0.6
145.71	1.19	2.0	164.91	9SE	0.7	-3.7	-1.4	0.6
152.13	1.18	2.0	174.83	19SE	2.0	-3.8	-1.4	0.6
158.57	1.18	2.0	184.81	29SE	1.6	-3.9	-1.3	0.6
165.02	1.16	2.0	194.85	9OC	-0.3	-4.0	-1.3	0.5
171.50	1.14	1.9	204.93	19OC	-0.6	-4.1	-1.3	0.5
177.99	1.12	1.9	215.06	29OC	-0.6	-4.2	-1.3	0.5
184.51	1.09	1.8	225.22	8NO	-0.7	-4.3	-1.3	0.4
191.03	1.05	1.8	235.41	18NO	-0.8	-4.4	-1.3	0.3
197.58	1.00	1.7	245.61	28NO	-0.7	-4.4	-1.3	0.3
204.14	0.94	1.7	255.82	8DE	-0.8	-4.3	-1.4	0.2
210.71	0.87	1.6	266.02	18DE	-0.6	-3.8	-1.4	0.1
217.29	0.79	1.5	276.21	28DE	0.4	-3.2	-1.4	0.1

Right table (☿ ♀ ♃ ♄ Magnitudes — 62 / 63)

♂ LONG	LAT	MAG	☉ LONG	16.00UT	☿	♀	♃	♄
223.88	0.70	1.4	286.37	7JA	3.0	-3.7	-1.5	0.0
230.46	0.59	1.3	296.48	17JA	1.3	-4.1	-1.5	0.1
237.05	0.46	1.2	306.56	27JA	0.5	-4.3	-1.6	0.1
243.62	0.32	1.0	316.59	6FE	0.3	-4.3	-1.6	0.2
250.16	0.15	0.9	326.55	16FE	0.1	-4.2	-1.7	0.2
256.69	-0.04	0.8	336.46	26FE	-0.3	-4.1	-1.8	0.3
263.16	-0.26	0.7	346.32	8MR	-0.9	-4.0	-1.8	0.4
269.57	-0.50	0.5	356.11	18MR	-1.7	-3.9	-1.9	0.4
275.89	-0.78	0.3	5.85	28MR	-1.2	-3.8	-2.0	0.5
282.09	-1.10	0.1	15.53	7AP	-0.2	-3.7	-2.0	0.5
288.14	-1.45	-0.1	25.17	17AP	0.9	-3.6	-2.1	0.5
293.97	-1.84	-0.3	34.77	27AP	2.0	-3.6	-2.1	0.6
299.52	-2.28	-0.5	44.34	7MY	3.5	-3.5	-2.2	0.6
304.71	-2.76	-0.7	53.87	17MY	2.6	-3.5	-2.2	0.6
309.41	-3.30	-0.9	63.40	27MY	1.4	-3.4	-2.2	0.6
313.49	-3.87	-1.2	72.92	6JN	0.3	-3.4	-2.2	0.6
316.76	-4.49	-1.4	82.44	16JN	-0.7	-3.4	-2.2	0.6
319.03	-5.14	-1.7	91.98	26JN	-1.6	-3.3	-2.1	0.6
320.08	-5.76	-2.0	101.54	6JL	-1.2	-3.3	-2.1	0.6
319.81	-6.30	-2.2	111.12	16JL	-0.5	-3.3	-2.0	0.6
318.24	-6.66	-2.4	120.75	26JL	-0.0	-3.3	-2.0	0.6
315.75	-6.75	-2.6	130.42	5AU	0.3	-3.3	-1.9	0.7
313.02	-6.51	-2.5	140.13	15AU	0.5	-3.4	-1.8	0.7
310.79	-5.99	-2.3	149.91	25AU	0.9	-3.4	-1.8	0.8
309.64	-5.30	-2.0	159.74	4SE	2.4	-3.4	-1.7	0.8
309.78	-4.54	-1.7	169.62	14SE	1.5	-3.4	-1.6	0.8
311.18	-3.81	-1.4	179.57	24SE	-0.4	-3.5	-1.6	0.8
313.69	-3.13	-1.1	189.58	4OC	-0.8	-3.5	-1.5	0.8
317.09	-2.52	-0.8	199.63	14OC	-0.8	-3.5	-1.5	0.8
321.20	-1.99	-0.5	209.74	24OC	-0.8	-3.5	-1.5	0.8
325.86	-1.54	-0.3	219.89	3NO	-0.7	-3.4	-1.4	0.7
330.94	-1.14	-0.0	230.06	13NO	-0.6	-3.4	-1.4	0.7
336.34	-0.80	0.2	240.26	23NO	-0.6	-3.4	-1.4	0.6
342.00	-0.50	0.4	250.47	3DE	-0.5	-3.4	-1.4	0.5
347.83	-0.25	0.6	260.67	13DE	0.5	-3.3	-1.4	0.5
353.80	-0.03	0.7	270.87	23DE	3.0	-3.3	-1.4	0.4

63

♂ LONG	LAT	MAG	☉ LONG	16.00UT	☿	♀	♃	♄
359.88	0.16	0.9	281.04	2JA	0.9	-3.3	-1.4	0.3
6.04	0.32	1.1	291.18	12JA	0.3	-3.3	-1.5	0.3
12.24	0.47	1.2	301.28	22JA	0.1	-3.3	-1.5	0.2
18.49	0.59	1.3	311.33	1FE	-0.0	-3.4	-1.5	0.3
24.75	0.70	1.4	321.33	11FE	-0.3	-3.4	-1.6	0.3
31.03	0.79	1.5	331.28	21FE	-0.6	-3.4	-1.6	0.4
37.32	0.87	1.6	341.16	3MR	-1.6	-3.4	-1.7	0.4
43.61	0.94	1.7	350.98	13MR	-1.2	-3.5	-1.8	0.5
49.89	0.99	1.8	0.75	23MR	-0.1	-3.5	-1.8	0.6
56.17	1.04	1.8	10.46	2AP	1.1	-3.6	-1.9	0.6
62.45	1.08	1.9	20.12	12AP	2.6	-3.6	-2.0	0.7
68.72	1.12	1.9	29.74	22AP	3.4	-3.7	-2.0	0.7
74.99	1.14	2.0	39.32	2MY	1.9	-3.8	-2.1	0.7
81.25	1.16	2.0	48.88	12MY	1.0	-3.9	-2.2	0.8
87.52	1.18	2.0	58.41	22MY	0.2	-3.9	-2.2	0.8
93.79	1.19	2.0	67.93	1JN	-0.8	-4.0	-2.3	0.8
100.08	1.19	2.0	77.45	11JN	-1.6	-4.1	-2.3	0.8
106.37	1.19	2.0	86.98	21JN	-1.2	-4.2	-2.3	0.8
112.68	1.19	2.0	96.52	1JL	-0.4	-4.2	-2.3	0.9
119.01	1.18	2.0	106.09	11JL	0.1	-4.1	-2.3	0.8
125.37	1.16	2.0	115.70	21JL	0.4	-3.7	-2.3	0.8
131.75	1.14	2.0	125.34	31JL	0.7	-3.2	-2.2	0.8
138.16	1.12	1.9	135.03	10AU	1.3	-3.6	-2.2	0.9
144.61	1.09	1.9	144.78	20AU	2.9	-4.1	-2.1	0.9
151.10	1.06	1.9	154.57	30AU	1.3	-4.3	-2.1	1.0
157.64	1.02	1.9	164.43	9SE	-0.5	-4.3	-2.0	1.0
164.21	0.98	1.9	174.35	19SE	-0.9	-4.2	-1.9	1.0
170.84	0.93	1.9	184.32	29SE	-0.9	-4.2	-1.9	1.1
177.51	0.87	1.9	194.36	9OC	-0.8	-4.1	-1.8	1.1
184.24	0.81	1.9	204.44	19OC	-0.6	-4.0	-1.7	1.1
191.02	0.75	1.9	214.56	29OC	-0.5	-3.9	-1.7	1.0
197.85	0.67	1.8	224.72	8NO	-0.4	-3.8	-1.6	1.0
204.74	0.59	1.8	234.91	18NO	-0.3	-3.7	-1.6	1.0
211.68	0.50	1.8	245.11	28NO	0.7	-3.7	-1.6	0.9
218.68	0.41	1.7	255.32	8DE	2.8	-3.6	-1.5	0.9
225.73	0.30	1.7	265.53	18DE	0.6	-3.5	-1.5	0.8
232.82	0.19	1.6	275.71	28DE	0.1	-3.5	-1.5	0.7

♂ LONG	LAT	MAG	☉ LONG	16.00UT 64	☿	♀	♃	♄
239.98	0.06	1.5	285.87	7JA	0.0	-3.4	-1.5	0.7
247.17	-0.07	1.5	296.00	17JA	-0.1	-3.4	-1.5	0.6
254.41	-0.21	1.4	306.07	27JA	-0.4	-3.4	-1.5	0.5
261.69	-0.36	1.3	316.10	6FE	-0.9	-3.3	-1.5	0.5
269.01	-0.52	1.2	326.07	16FE	-1.5	-3.3	-1.6	0.5
276.35	-0.69	1.2	335.99	26FE	-1.2	-3.3	-1.6	0.6
283.71	-0.86	1.1	345.84	7MR	-0.1	-3.3	-1.6	0.6
291.07	-1.04	1.0	355.64	17MR	1.4	-3.3	-1.7	0.7
298.43	-1.23	0.9	5.38	27MR	3.3	-3.3	-1.7	0.7
305.78	-1.42	0.8	15.07	6AP	2.5	-3.3	-1.8	0.8
313.08	-1.60	0.7	24.71	16AP	1.4	-3.4	-1.8	0.9
320.33	-1.79	0.6	34.30	26AP	0.8	-3.4	-1.9	0.9
327.50	-1.97	0.5	43.87	6MY	0.1	-3.4	-2.0	0.9
334.55	-2.14	0.4	53.42	16MY	-0.8	-3.5	-2.1	1.0
341.47	-2.31	0.3	62.94	26MY	-1.7	-3.5	-2.1	1.0
348.20	-2.47	0.2	72.46	5JN	-1.3	-3.5	-2.2	1.1
354.70	-2.61	0.1	81.98	15JN	-0.4	-3.4	-2.3	1.1
0.93	-2.74	-0.0	91.51	25JN	0.2	-3.4	-2.3	1.1
6.80	-2.85	-0.2	101.07	5JL	0.6	-3.4	-2.4	1.1
12.24	-2.94	-0.3	110.66	15JL	0.9	-3.3	-2.4	1.1
17.15	-3.01	-0.5	120.28	25JL	1.7	-3.3	-2.4	1.1
21.39	-3.05	-0.6	129.94	4AU	3.2	-3.3	-2.4	1.1
24.79	-3.06	-0.8	139.66	14AU	1.1	-3.3	-2.4	1.1
27.18	-3.01	-1.0	149.43	24AU	-0.6	-3.3	-2.4	1.2
28.32	-2.91	-1.3	159.26	3SE	-1.1	-3.3	-2.3	1.2
28.03	-2.71	-1.5	169.14	13SE	-1.1	-3.4	-2.3	1.3
26.29	-2.41	-1.7	179.09	23SE	-0.8	-3.4	-2.2	1.3
23.31	-1.99	-1.9	189.09	30C	-0.5	-3.4	-2.2	1.3
19.71	-1.47	-1.9	199.15	130C	-0.3	-3.5	-2.1	1.3
16.27	-0.92	-1.7	209.25	230C	-0.3	-3.5	-2.0	1.4
13.73	-0.39	-1.4	219.39	2NO	-0.1	-3.6	-2.0	1.3
12.48	0.06	-1.1	229.57	12NO	0.9	-3.6	-1.9	1.3
12.58	0.43	-0.8	239.76	22NO	2.5	-3.7	-1.8	1.3
13.90	0.72	-0.5	249.97	2DE	0.4	-3.8	-1.8	1.2
16.22	0.94	-0.2	260.18	12DE	-0.1	-3.9	-1.7	1.2
19.35	1.11	0.1	270.37	22DE	-0.1	-4.0	-1.7	1.1

65

♂ LONG	LAT	MAG	☉ LONG	16.00UT	☿	♀	♃	♄
23.10	1.24	0.3	280.55	1JA	-0.2	-4.1	-1.6	1.1
27.35	1.33	0.6	290.69	11JA	-0.5	-4.2	-1.6	1.0
31.98	1.40	0.8	300.79	21JA	-1.0	-4.3	-1.6	0.9
36.90	1.44	0.9	310.85	31JA	-1.3	-4.3	-1.6	0.9
42.06	1.48	1.1	320.85	10FE	-1.1	-4.3	-1.6	0.8
47.40	1.50	1.2	330.79	20FE	-0.0	-4.1	-1.6	0.8
52.89	1.51	1.4	340.68	2MR	1.8	-3.7	-1.6	0.8
58.49	1.51	1.5	350.50	12MR	3.4	-3.2	-1.6	0.9
64.19	1.50	1.6	0.27	22MR	1.9	-3.7	-1.6	0.9
69.97	1.49	1.7	9.99	1AP	1.1	-4.1	-1.6	1.0
75.82	1.47	1.7	19.65	11AP	0.6	-4.2	-1.7	1.0
81.72	1.45	1.8	29.27	21AP	0.0	-4.2	-1.7	1.1
87.68	1.43	1.9	38.86	1MY	-0.8	-4.1	-1.8	1.1
93.69	1.40	1.9	48.41	11MY	-1.7	-4.0	-1.8	1.2
99.75	1.36	1.9	57.95	21MY	-1.3	-3.9	-1.9	1.2
105.85	1.33	2.0	67.47	31MY	-0.4	-3.8	-1.9	1.3
112.00	1.28	2.0	76.99	10JN	0.3	-3.8	-2.0	1.3
118.19	1.24	2.0	86.52	20JN	0.7	-3.7	-2.1	1.4
124.44	1.19	2.0	96.06	30JN	1.2	-3.6	-2.1	1.4
130.74	1.14	2.0	105.63	10JL	2.2	-3.5	-2.2	1.4
137.09	1.09	2.0	115.23	20JL	3.0	-3.5	-2.3	1.4
143.50	1.03	2.0	124.87	30JL	0.9	-3.5	-2.3	1.4
149.97	0.97	2.0	134.56	9AU	-0.6	-3.4	-2.4	1.3
156.51	0.90	1.9	144.30	19AU	-1.2	-3.4	-2.4	1.3
163.11	0.83	1.9	154.10	29AU	-1.2	-3.4	-2.5	1.3
169.78	0.76	1.8	163.96	8SE	-0.7	-3.4	-2.5	1.3
176.52	0.68	1.8	173.87	18SE	-0.4	-3.4	-2.4	1.3
183.34	0.60	1.7	183.85	28SE	-0.2	-3.4	-2.4	1.3
190.22	0.51	1.7	193.87	80C	-0.1	-3.4	-2.4	1.2
197.19	0.42	1.7	203.95	180C	0.1	-3.4	-2.3	1.2
204.23	0.33	1.7	214.08	280C	1.1	-3.4	-2.3	1.2
211.34	0.23	1.7	224.23	7NO	2.3	-3.4	-2.2	1.2
218.53	0.13	1.7	234.42	17NO	0.2	-3.4	-2.1	1.2
225.80	0.02	1.7	244.62	27NO	-0.2	-3.4	-2.1	1.1
233.13	-0.09	1.6	254.83	7DE	-0.3	-3.4	-2.0	1.1
240.53	-0.20	1.6	265.03	17DE	-0.4	-3.5	-1.9	1.0
247.99	-0.32	1.6	275.22	27DE	-0.6	-3.5	-1.8	1.0

♂ LONG	LAT	MAG	☉ LONG	16.00UT 66	☿	♀	♃	♄
255.52	-0.44	1.5	285.37	6JA	-1.0	-3.5	-1.8	0.9
263.09	-0.56	1.5	295.50	16JA	-1.2	-3.4	-1.7	0.9
270.71	-0.68	1.5	305.58	26JA	-1.0	-3.4	-1.7	0.8
278.36	-0.80	1.4	315.60	5FE	0.1	-3.4	-1.7	0.8
286.04	-0.91	1.4	325.58	15FE	2.1	-3.4	-1.6	0.8
293.74	-1.03	1.3	335.50	25FE	2.7	-3.4	-1.6	0.7
301.41	-1.13	1.3	345.35	7MR	1.4	-3.3	-1.6	0.8
309.14	-1.23	1.3	355.16	17MR	0.8	-3.3	-1.6	0.8
316.82	-1.32	1.2	4.90	27MR	0.4	-3.3	-1.6	0.9
324.46	-1.40	1.2	14.59	6AP	-0.1	-3.3	-1.6	1.0
332.05	-1.46	1.1	24.23	16AP	-0.8	-3.4	-1.6	1.1
339.58	-1.51	1.1	33.84	26AP	-1.7	-3.4	-1.6	1.1
347.04	-1.55	1.1	43.40	6MY	-1.3	-3.4	-1.6	1.2
354.40	-1.57	1.0	52.95	16MY	-0.3	-3.4	-1.6	1.2
1.65	-1.58	1.0	62.48	26MY	0.4	-3.5	-1.7	1.3
8.77	-1.57	0.9	72.00	5JN	1.0	-3.5	-1.7	1.3
15.75	-1.55	0.9	81.52	15JN	1.7	-3.6	-1.8	1.3
22.56	-1.50	0.8	91.06	25JN	2.9	-3.6	-1.8	1.3
29.19	-1.44	0.8	100.61	5JL	2.5	-3.7	-1.9	1.3
35.60	-1.36	0.7	110.19	15JL	0.8	-3.8	-2.0	1.3
41.76	-1.27	0.6	119.81	25JL	-1.3	-4.0	-2.1	1.2
47.64	-1.15	0.5	129.47	4AU	-1.3	-4.0	-2.1	1.2
53.18	-1.01	0.4	139.19	14AU	-1.2	-4.1	-2.2	1.2
58.31	-0.84	0.3	148.96	24AU	-0.6	-4.2	-2.2	1.2
62.96	-0.65	0.2	158.78	3SE	-0.3	-4.3	-2.3	1.1
67.01	-0.43	0.0	168.67	13SE	-0.1	-4.3	-2.3	1.1
70.33	-0.16	-0.1	178.61	23SE	0.0	-4.2	-2.4	1.1
72.76	0.15	-0.3	188.61	30C	0.3	-3.8	-2.4	1.1
74.08	0.50	-0.5	198.66	130C	1.4	-3.2	-2.4	1.1
74.12	0.91	-0.8	208.76	230C	2.1	-3.7	-2.4	1.1
72.75	1.36	-1.0	218.90	2NO	-0.0	-4.2	-2.4	1.1
70.02	1.82	-1.2	229.08	12NO	-0.4	-4.4	-2.3	1.1
66.35	2.24	-1.3	239.27	22NO	-0.4	-4.3	-2.3	1.0
62.41	2.57	-1.3	249.47	2DE	-0.5	-4.3	-2.2	1.0
58.98	2.77	-1.0	259.68	12DE	-0.7	-4.2	-2.2	1.0
56.65	2.85	-0.7	269.88	22DE	-0.9	-4.1	-2.1	0.9

67

♂ LONG	LAT	MAG	☉ LONG	16.00UT	☿	♀	♃	♄
55.64	2.84	-0.4	280.05	1JA	-1.0	-4.0	-2.0	0.9
55.92	2.78	-0.2	290.20	11JA	-0.9	-3.9	-1.9	0.8
57.34	2.69	0.1	300.30	21JA	0.1	-3.8	-1.9	0.8
59.67	2.58	0.4	310.36	31JA	2.5	-3.7	-1.8	0.7
62.75	2.47	0.6	320.36	10FE	2.0	-3.6	-1.7	0.7
66.41	2.36	0.8	330.31	20FE	1.0	-3.6	-1.7	0.6
70.53	2.25	0.9	340.20	2MR	0.6	-3.5	-1.6	0.6
75.03	2.14	1.1	350.03	12MR	0.3	-3.4	-1.6	0.6
79.82	2.03	1.2	359.80	22MR	-0.1	-3.4	-1.6	0.6
84.85	1.93	1.4	9.52	1AP	-0.8	-3.4	-1.5	0.7
90.08	1.83	1.5	19.19	11AP	-1.7	-3.3	-1.5	0.8
95.48	1.74	1.6	28.81	21AP	-1.3	-3.3	-1.5	0.9
101.02	1.64	1.6	38.40	1MY	-0.3	-3.3	-1.5	0.9
106.69	1.55	1.7	47.95	11MY	0.6	-3.3	-1.5	1.0
112.48	1.46	1.8	57.49	21MY	1.3	-3.3	-1.5	1.0
118.37	1.36	1.8	67.01	31MY	2.2	-3.3	-1.5	1.1
124.36	1.27	1.8	76.53	10JN	3.4	-3.3	-1.5	1.1
130.44	1.18	1.9	86.06	20JN	2.0	-3.4	-1.6	1.1
136.61	1.09	1.9	95.60	30JN	0.6	-3.4	-1.6	1.1
142.88	0.99	1.9	105.17	10JL	-0.7	-3.4	-1.6	1.1
149.23	0.90	1.9	114.76	20JL	-1.4	-3.5	-1.7	1.1
155.68	0.80	1.9	124.40	30JL	-1.2	-3.5	-1.7	1.1
162.22	0.71	1.8	134.09	9AU	-0.6	-3.5	-1.8	1.1
168.84	0.61	1.8	143.83	19AU	-0.2	-3.4	-1.9	1.1
175.56	0.51	1.8	153.62	29AU	0.1	-3.4	-1.9	1.0
182.38	0.41	1.8	163.48	8SE	0.2	-3.4	-2.0	1.0
189.29	0.30	1.8	173.38	18SE	0.5	-3.4	-2.1	0.9
196.29	0.20	1.7	183.36	28SE	1.7	-3.3	-2.1	0.9
203.39	0.09	1.7	193.38	80C	1.9	-3.3	-2.2	1.0
210.58	-0.01	1.6	203.46	180C	-0.2	-3.3	-2.2	1.0
217.86	-0.12	1.6	213.58	280C	-0.5	-3.3	-2.3	1.0
225.23	-0.22	1.5	223.74	7NO	-0.6	-3.3	-2.3	1.0
232.68	-0.33	1.5	233.92	17NO	-0.6	-3.4	-2.3	1.0
240.21	-0.43	1.5	244.12	27NO	-0.7	-3.4	-2.3	0.9
247.81	-0.53	1.5	254.33	7DE	-0.8	-3.4	-2.3	0.9
255.47	-0.63	1.5	264.53	17DE	-0.9	-3.4	-2.2	0.9
263.19	-0.73	1.5	274.72	27DE	-0.7	-3.5	-2.2	0.9

Left (1968–1969)

♂ LONG	LAT	MAG	☉ LONG	16.00UT	☿	♀	♃	♄
270.96	-0.81	1.4	284.88	6JA	0.2	-3.5	-2.1	0.8
278.76	-0.90	1.4	295.01	16JA	2.8	-3.6	-2.0	0.8
286.58	-0.97	1.4	305.09	26JA	1.5	-3.6	-2.0	0.7
294.42	-1.04	1.4	315.12	5FE	0.7	-3.7	-1.9	0.7
302.26	-1.09	1.4	325.10	15FE	0.4	-3.8	-1.8	0.6
310.09	-1.13	1.4	335.02	25FE	0.2	-3.9	-1.8	0.6
317.89	-1.17	1.4	344.88	6MR	-0.2	-3.9	-1.7	0.5
325.66	-1.19	1.4	354.68	16MR	-0.8	-4.0	-1.6	0.5
333.38	-1.20	1.4	4.43	26MR	-1.7	-4.1	-1.6	0.5
341.04	-1.19	1.4	14.12	5AP	-1.3	-4.2	-1.5	0.5
348.63	-1.17	1.4	23.77	15AP	-0.2	-4.2	-1.5	0.6
356.15	-1.14	1.4	33.37	25AP	0.7	-4.1	-1.5	0.7
3.57	-1.10	1.4	42.94	5MY	1.7	-3.8	-1.4	0.7
10.90	-1.04	1.4	52.49	15MY	2.9	-3.1	-1.4	0.8
18.12	-0.97	1.4	62.02	25MY	3.0	-3.1	-1.4	0.9
25.23	-0.90	1.4	71.54	4JN	1.6	-3.8	-1.4	0.9
32.22	-0.81	1.4	81.06	14JN	0.5	-4.1	-1.4	0.9
39.09	-0.70	1.3	90.59	24JN	-0.7	-4.2	-1.4	1.0
45.82	-0.59	1.3	100.14	4JL	-1.5	-4.2	-1.4	1.0
52.41	-0.47	1.3	109.72	14JL	-1.3	-4.1	-1.4	1.0
58.84	-0.34	1.3	119.35	24JL	-0.5	-4.0	-1.5	1.0
65.09	-0.20	1.2	129.00	3AU	-0.1	-3.9	-1.5	1.0
71.16	-0.04	1.2	138.71	13AU	0.2	-3.8	-1.5	1.0
77.01	0.13	1.1	148.48	23AU	0.4	-3.8	-1.6	1.0
82.62	0.31	1.0	158.30	2SE	0.7	-3.7	-1.6	1.0
87.95	0.51	0.9	168.18	12SE	2.1	-3.6	-1.7	0.9
92.94	0.74	0.8	178.13	22SE	1.7	-3.6	-1.8	0.9
97.52	0.98	0.7	188.12	2OC	-0.3	-3.5	-1.8	0.8
101.62	1.26	0.6	198.17	12OC	-0.7	-3.5	-1.9	0.8
105.11	1.56	0.4	208.27	22OC	-0.7	-3.5	-2.0	0.9
107.86	1.91	0.2	218.41	1NO	-0.7	-3.4	-2.0	0.9
109.69	2.30	-0.0	228.58	11NO	-0.7	-3.4	-2.1	0.9
110.41	2.72	-0.2	238.78	21NO	-0.7	-3.4	-2.1	0.9
109.86	3.16	-0.5	248.98	1DE	-0.7	-3.4	-2.1	0.9
107.96	3.59	-0.7	259.19	11DE	-0.6	-3.4	-2.2	0.9
104.85	3.96	-0.9	269.39	21DE	0.4	-3.3	-2.2	0.8
100.99	4.20	-1.0	279.56	31DE	3.0	-3.3	-2.2	0.8
				69				
97.09	4.27	-0.9	289.71	10JA	1.1	-3.3	-2.1	0.8
93.85	4.18	-0.7	299.81	20JA	0.5	-3.4	-2.1	0.7
91.76	3.97	-0.5	309.87	30JA	0.2	-3.4	-2.0	0.7
90.97	3.70	-0.2	319.88	9FE	0.0	-3.4	-2.0	0.6
91.42	3.41	0.1	329.83	19FE	-0.3	-3.4	-1.9	0.6
92.95	3.13	0.3	339.72	1MR	-0.9	-3.5	-1.8	0.5
95.37	2.86	0.5	349.55	11MR	-1.7	-3.5	-1.8	0.5
98.51	2.60	0.7	359.33	21MR	-1.3	-3.5	-1.7	0.4
102.22	2.37	0.8	9.04	31MR	-0.2	-3.4	-1.6	0.4
106.39	2.16	1.0	18.71	10AP	0.9	-3.4	-1.6	0.4
110.95	1.96	1.1	28.34	20AP	2.2	-3.4	-1.5	0.5
115.83	1.77	1.2	37.92	30AP	3.8	-3.4	-1.5	0.5
120.97	1.60	1.3	47.48	10MY	2.3	-3.3	-1.4	0.6
126.34	1.43	1.4	57.02	20MY	1.2	-3.3	-1.4	0.7
131.90	1.28	1.5	66.54	30MY	0.3	-3.3	-1.4	0.7
137.65	1.13	1.5	76.06	9JN	-0.7	-3.3	-1.3	0.8
143.56	0.98	1.6	85.59	19JN	-1.6	-3.3	-1.3	0.8
149.62	0.84	1.6	95.13	29JN	-1.3	-3.4	-1.3	0.9
155.82	0.70	1.6	104.70	9JL	-0.5	-3.4	-1.3	0.9
162.16	0.57	1.6	114.30	19JL	0.0	-3.4	-1.3	0.9
168.63	0.44	1.7	123.93	29JL	0.3	-3.4	-1.3	0.9
175.22	0.31	1.7	133.62	8AU	0.6	-3.5	-1.3	0.9
181.94	0.19	1.7	143.35	18AU	1.0	-3.5	-1.3	0.9
188.78	0.07	1.6	153.14	28AU	2.6	-3.6	-1.4	0.9
195.73	-0.05	1.6	163.00	7SE	1.5	-3.7	-1.4	0.9
202.81	-0.16	1.6	172.91	17SE	-0.4	-3.8	-1.5	0.9
209.99	-0.28	1.6	182.87	27SE	-0.9	-3.9	-1.5	0.8
217.28	-0.38	1.6	192.90	7OC	-0.9	-4.0	-1.6	0.8
224.67	-0.49	1.5	202.97	17OC	-0.8	-4.1	-1.6	0.8
232.16	-0.58	1.5	213.09	27OC	-0.6	-4.2	-1.7	0.8
239.74	-0.67	1.5	223.24	6NO	-0.5	-4.3	-1.7	0.8
247.39	-0.76	1.5	233.43	16NO	-0.5	-4.4	-1.8	0.8
255.12	-0.83	1.4	243.63	26NO	-0.4	-4.4	-1.9	0.8
262.91	-0.90	1.4	253.84	6DE	0.5	-4.2	-1.9	0.8
270.74	-0.96	1.4	264.04	16DE	3.0	-3.8	-2.0	0.8
278.61	-1.01	1.3	274.22	26DE	0.8	-3.1	-2.0	0.8

Right (1970–1971)

♂ LONG	LAT	MAG	☉ LONG	16.00UT	☿	♀	♃	♄
286.50	-1.05	1.3	284.39	5JA	0.2	-3.7	-2.1	0.8
294.40	-1.07	1.3	294.52	15JA	0.1	-4.2	-2.1	0.8
302.29	-1.09	1.3	304.60	25JA	-0.1	-4.3	-2.1	0.7
310.17	-1.09	1.3	314.64	4FE	-0.4	-4.3	-2.1	0.7
318.01	-1.08	1.4	324.62	14FE	-0.9	-4.2	-2.0	0.6
325.82	-1.06	1.4	334.54	24FE	-1.5	-4.1	-2.0	0.6
333.57	-1.03	1.4	344.41	6MR	-1.2	-4.0	-1.9	0.5
341.26	-0.98	1.4	354.21	16MR	-0.2	-3.9	-1.9	0.5
348.88	-0.93	1.5	3.96	26MR	1.2	-3.8	-1.8	0.4
356.42	-0.87	1.5	13.66	5AP	2.8	-3.7	-1.7	0.4
3.88	-0.80	1.5	23.30	15AP	3.0	-3.6	-1.7	0.3
11.25	-0.72	1.5	32.91	25AP	1.7	-3.6	-1.6	0.4
18.53	-0.63	1.6	42.48	5MY	0.9	-3.5	-1.5	0.4
25.71	-0.54	1.6	52.03	15MY	0.2	-3.4	-1.5	0.5
32.79	-0.44	1.6	61.55	25MY	-0.7	-3.4	-1.4	0.5
39.78	-0.33	1.6	71.08	4JN	-1.6	-3.4	-1.4	0.6
46.67	-0.22	1.6	80.60	14JN	-1.3	-3.4	-1.3	0.7
53.46	-0.11	1.7	90.13	24JN	-0.5	-3.3	-1.3	0.7
60.14	0.01	1.7	99.68	4JL	0.1	-3.3	-1.3	0.8
66.72	0.13	1.7	109.26	14JL	0.5	-3.3	-1.3	0.8
73.19	0.26	1.6	118.88	24JL	0.8	-3.3	-1.2	0.8
79.54	0.39	1.6	128.54	3AU	1.4	-3.3	-1.2	0.9
85.78	0.52	1.6	138.24	13AU	3.0	-3.4	-1.2	0.9
91.89	0.67	1.6	148.01	23AU	1.3	-3.4	-1.2	0.9
97.86	0.82	1.5	157.83	2SE	-0.5	-3.4	-1.2	0.9
103.67	0.97	1.5	167.70	12SE	-1.0	-3.4	-1.3	0.8
109.31	1.14	1.4	177.64	22SE	-1.0	-3.5	-1.3	0.8
114.75	1.32	1.3	187.64	2OC	-0.8	-3.5	-1.3	0.8
119.94	1.51	1.2	197.68	12OC	-0.5	-3.5	-1.4	0.8
124.85	1.71	1.1	207.78	22OC	-0.4	-3.5	-1.4	0.7
129.42	1.94	1.0	217.92	1NO	-0.4	-3.4	-1.5	0.7
133.57	2.18	0.8	228.08	11NO	-0.3	-3.4	-1.5	0.8
137.20	2.46	0.6	238.28	21NO	0.7	-3.4	-1.6	0.8
140.19	2.76	0.4	248.48	1DE	2.8	-3.4	-1.6	0.8
142.40	3.09	0.2	258.69	11DE	0.6	-3.3	-1.7	0.8
143.63	3.44	-0.0	268.89	21DE	0.1	-3.3	-1.8	0.8
143.70	3.80	-0.3	279.07	31DE	-0.1	-3.3	-1.8	0.8
				71				
142.50	4.14	-0.5	289.21	10JA	-0.2	-3.3	-1.9	0.8
140.02	4.42	-0.8	299.32	20JA	-0.4	-3.3	-1.9	0.8
136.53	4.56	-1.0	309.38	30JA	-0.9	-3.4	-2.0	0.7
132.62	4.54	-1.0	319.39	9FE	-1.4	-3.4	-2.0	0.7
128.99	4.33	-0.9	329.35	19FE	-1.1	-3.4	-2.0	0.6
126.27	4.00	-0.6	339.24	1MR	-0.1	-3.4	-2.0	0.6
124.79	3.60	-0.4	349.07	11MR	1.5	-3.5	-2.0	0.5
124.60	3.19	-0.2	358.86	21MR	3.5	-3.5	-2.0	0.5
125.60	2.78	0.0	8.58	31MR	2.3	-3.6	-1.9	0.4
127.61	2.41	0.2	18.25	10AP	1.3	-3.6	-1.9	0.4
130.46	2.07	0.4	27.88	20AP	0.7	-3.7	-1.8	0.3
133.99	1.76	0.6	37.47	30AP	0.1	-3.8	-1.7	0.3
138.08	1.48	0.7	47.02	10MY	-0.7	-3.8	-1.7	0.3
142.64	1.22	0.8	56.56	20MY	-1.7	-3.9	-1.6	0.4
147.58	0.99	0.9	66.09	30MY	-1.3	-4.0	-1.6	0.5
152.85	0.77	1.0	75.60	9JN	-0.4	-4.1	-1.5	0.5
158.40	0.57	1.1	85.13	19JN	0.2	-4.2	-1.4	0.6
164.21	0.38	1.2	94.67	29JN	0.6	-4.2	-1.4	0.6
170.25	0.21	1.2	104.23	9JL	1.0	-4.0	-1.3	0.7
176.49	0.04	1.2	113.83	19JL	1.9	-3.7	-1.3	0.7
182.91	-0.11	1.3	123.47	29JL	3.2	-3.2	-1.3	0.8
189.52	-0.25	1.3	133.15	8AU	1.1	-3.6	-1.2	0.8
196.28	-0.39	1.3	142.88	18AU	-0.5	-4.1	-1.2	0.8
203.21	-0.51	1.4	152.67	28AU	-1.1	-4.3	-1.2	0.8
210.27	-0.63	1.4	162.51	7SE	-1.1	-4.3	-1.2	0.8
217.47	-0.73	1.4	172.42	17SE	-0.7	-4.2	-1.2	0.8
224.80	-0.83	1.4	182.39	27SE	-0.4	-4.2	-1.2	0.8
232.23	-0.91	1.4	192.41	7OC	-0.3	-4.1	-1.2	0.8
239.78	-0.98	1.4	202.48	17OC	-0.2	-4.0	-1.3	0.7
247.41	-1.04	1.4	212.60	27OC	-0.1	-3.9	-1.3	0.7
255.13	-1.09	1.4	222.75	6NO	0.9	-3.8	-1.3	0.7
262.90	-1.12	1.4	232.94	16NO	2.5	-3.7	-1.4	0.7
270.73	-1.15	1.4	243.13	26NO	0.3	-3.6	-1.4	0.7
278.60	-1.16	1.4	253.34	6DE	-0.1	-3.6	-1.5	0.8
286.48	-1.15	1.4	263.55	16DE	-0.2	-3.5	-1.5	0.8
294.38	-1.14	1.4	273.74	26DE	-0.3	-3.5	-1.6	0.8

72 / 73

♂ LONG	LAT	MAG	☉ LONG	16.00UT 72	☿	♀	♃	♄
302.26	-1.11	1.4	283.90	5JA	-0.5	-3.4	-1.7	0.8
310.12	-1.07	1.4	294.03	15JA	-0.9	-3.4	-1.7	0.8
317.95	-1.03	1.4	304.12	25JA	-1.3	-3.4	-1.8	0.8
325.73	-0.97	1.4	314.16	4FE	-1.1	-3.3	-1.9	0.7
333.45	-0.90	1.4	324.14	14FE	-0.0	-3.3	-1.9	0.7
341.11	-0.83	1.4	334.06	24FE	1.8	-3.3	-2.0	0.7
348.70	-0.75	1.4	343.93	5MR	3.1	-3.3	-2.0	0.6
356.20	-0.66	1.5	353.74	15MR	1.7	-3.3	-2.0	0.6
3.63	-0.57	1.5	3.49	25MR	1.0	-3.3	-2.0	0.5
10.97	-0.47	1.5	13.19	4AP	0.5	-3.3	-2.0	0.4
18.22	-0.37	1.6	22.84	14AP	0.0	-3.4	-2.0	0.4
25.38	-0.27	1.6	32.44	24AP	-0.8	-3.4	-2.0	0.3
32.46	-0.17	1.7	42.02	4MY	-1.7	-3.4	-1.9	0.3
39.45	-0.07	1.7	51.57	14MY	-1.3	-3.5	-1.9	0.3
46.36	0.04	1.7	61.09	24MY	-0.4	-3.5	-1.8	0.3
53.19	0.14	1.8	70.61	3JN	0.3	-3.5	-1.7	0.4
59.93	0.25	1.8	80.14	13JN	0.8	-3.4	-1.7	0.5
66.60	0.35	1.8	89.66	23JN	1.4	-3.4	-1.6	0.5
73.19	0.46	1.9	99.21	3JL	2.4	-3.4	-1.6	0.6
79.71	0.56	1.9	108.79	13JL	2.9	-3.3	-1.5	0.6
86.16	0.66	1.9	118.40	23JL	0.9	-3.3	-1.4	0.7
92.53	0.77	1.9	128.06	2AU	-0.6	-3.3	-1.4	0.7
98.83	0.87	1.9	137.77	12AU	-1.3	-3.3	-1.4	0.8
105.06	0.98	1.8	147.52	22AU	-1.2	-3.3	-1.3	0.8
111.20	1.09	1.8	157.34	1SE	-0.7	-3.4	-1.3	0.8
117.26	1.19	1.8	167.22	11SE	-0.3	-3.4	-1.3	0.8
123.23	1.31	1.7	177.15	21SE	-0.1	-3.4	-1.3	0.8
129.10	1.42	1.7	187.15	1OC	-0.1	-3.4	-1.2	0.8
134.85	1.54	1.6	197.19	11OC	0.1	-3.5	-1.2	0.8
140.47	1.66	1.5	207.28	21OC	1.2	-3.5	-1.2	0.7
145.91	1.79	1.4	217.42	31OC	2.3	-3.6	-1.3	0.7
151.20	1.93	1.3	227.59	10NO	0.1	-3.6	-1.3	0.7
156.25	2.07	1.2	237.78	20NO	-0.3	-3.7	-1.3	0.7
161.03	2.22	1.1	247.99	30NO	-0.3	-3.8	-1.3	0.7
165.47	2.39	0.9	258.20	10DE	-0.4	-3.9	-1.4	0.8
169.50	2.56	0.7	268.39	20DE	-0.6	-4.0	-1.4	0.8
173.02	2.75	0.5	278.58	30DE	-1.0	-4.1	-1.5	0.8
				73				
175.89	2.95	0.3	288.73	9JA	-1.1	-4.2	-1.6	0.8
177.98	3.15	0.0	298.83	19JA	-1.0	-4.3	-1.6	0.8
179.09	3.34	-0.2	308.90	29JA	0.0	-4.3	-1.7	0.8
179.04	3.51	-0.5	318.91	8FE	2.2	-4.3	-1.8	0.8
177.74	3.61	-0.8	328.87	18FE	2.4	-4.1	-1.8	0.7
175.20	3.62	-1.1	338.77	28FE	1.2	-3.7	-1.9	0.7
171.75	3.49	-1.3	348.60	10MR	0.7	-3.2	-2.0	0.7
167.98	3.21	-1.3	358.38	20MR	0.4	-3.7	-2.0	0.6
164.61	2.81	-1.1	8.11	30MR	-0.1	-4.1	-2.0	0.5
162.23	2.34	-1.0	17.78	9AP	-0.8	-4.2	-2.1	0.5
161.14	1.85	-0.7	27.41	19AP	-1.7	-4.2	-2.1	0.4
161.37	1.39	-0.5	37.00	29AP	-1.3	-4.1	-2.1	0.4
162.81	0.97	-0.3	46.56	9MY	-0.3	-4.0	-2.1	0.3
165.27	0.60	-0.1	56.10	19MY	0.4	-3.9	-2.1	0.2
168.59	0.28	0.0	65.63	29MY	1.1	-3.8	-2.0	0.3
172.61	-0.01	0.2	75.14	8JN	1.8	-3.7	-2.0	0.3
177.21	-0.26	0.3	84.67	18JN	3.1	-3.7	-1.9	0.4
182.30	-0.48	0.4	94.21	28JN	2.4	-3.6	-1.9	0.5
187.80	-0.67	0.5	103.77	8JL	0.7	-3.5	-1.8	0.5
193.65	-0.84	0.6	113.36	18JL	-0.6	-3.5	-1.7	0.6
199.81	-0.99	0.7	123.00	28JL	-1.4	-3.5	-1.7	0.6
206.23	-1.12	0.7	132.68	7AU	-1.3	-3.4	-1.6	0.7
212.89	-1.22	0.8	142.41	17AU	-0.6	-3.4	-1.6	0.7
219.76	-1.31	0.9	152.19	27AU	-0.2	-3.4	-1.5	0.7
226.80	-1.38	0.9	162.04	6SE	-0.0	-3.4	-1.5	0.8
234.01	-1.43	1.0	171.94	16SE	0.1	-3.4	-1.4	0.8
241.36	-1.46	1.0	181.90	26SE	0.3	-3.4	-1.4	0.8
248.83	-1.48	1.0	191.92	6OC	1.5	-3.4	-1.4	0.8
256.40	-1.48	1.1	201.99	16OC	2.1	-3.4	-1.3	0.8
264.05	-1.46	1.1	212.11	26OC	-0.0	-3.4	-1.3	0.8
271.76	-1.43	1.2	222.26	5NO	-0.5	-3.4	-1.3	0.7
279.51	-1.39	1.2	232.44	15NO	-0.5	-3.4	-1.3	0.7
287.30	-1.33	1.3	242.64	25NO	-0.5	-3.4	-1.3	0.7
295.08	-1.26	1.3	252.85	5DE	-0.7	-3.5	-1.3	0.7
302.86	-1.18	1.3	263.05	15DE	-0.9	-3.5	-1.4	0.7
310.62	-1.09	1.4	273.24	25DE	-0.9	-3.5	-1.4	0.8

74 / 75

♂ LONG	LAT	MAG	☉ LONG	16.00UT 74	☿	♀	♃	♄
318.34	-1.00	1.4	283.41	4JA	-0.8	-3.5	-1.4	0.8
326.02	-0.90	1.5	293.54	14JA	0.1	-3.4	-1.5	0.8
333.64	-0.79	1.5	303.63	24JA	2.5	-3.4	-1.5	0.8
341.19	-0.69	1.5	313.67	3FE	1.9	-3.4	-1.6	0.8
348.67	-0.58	1.6	323.66	13FE	0.9	-3.4	-1.7	0.8
356.07	-0.47	1.6	333.59	23FE	0.5	-3.3	-1.7	0.8
3.40	-0.36	1.6	343.45	5MR	0.2	-3.3	-1.8	0.7
10.64	-0.25	1.7	353.26	15MR	-0.2	-3.3	-1.9	0.7
17.81	-0.14	1.7	3.02	25MR	-0.8	-3.3	-1.9	0.7
24.89	-0.03	1.7	12.72	4AP	-1.7	-3.3	-2.0	0.6
31.90	0.07	1.7	22.37	14AP	-1.3	-3.4	-2.1	0.5
38.82	0.17	1.7	31.98	24AP	-0.3	-3.4	-2.1	0.5
45.68	0.27	1.7	41.55	4MY	0.6	-3.4	-2.2	0.4
52.46	0.36	1.7	51.10	14MY	1.4	-3.4	-2.2	0.3
59.18	0.45	1.8	60.63	24MY	2.5	-3.5	-2.2	0.3
65.84	0.54	1.8	70.15	3JN	3.4	-3.5	-2.2	0.2
72.44	0.63	1.9	79.68	13JN	1.9	-3.6	-2.2	0.3
78.99	0.71	1.9	89.21	23JN	0.6	-3.6	-2.2	0.4
85.48	0.79	1.9	98.75	3JL	-0.7	-3.7	-2.1	0.4
91.93	0.86	2.0	108.33	13JL	-1.5	-3.8	-2.1	0.5
98.34	0.94	2.0	117.94	23JL	-1.3	-3.9	-2.0	0.5
104.71	1.01	2.0	127.60	2AU	-0.6	-4.0	-2.0	0.6
111.04	1.08	2.0	137.30	12AU	-0.1	-4.1	-1.9	0.6
117.34	1.14	2.0	147.06	22AU	0.1	-4.2	-1.8	0.7
123.60	1.21	2.0	156.87	1SE	0.3	-4.3	-1.8	0.7
129.82	1.27	2.0	166.74	11SE	0.6	-4.3	-1.7	0.8
136.01	1.33	1.9	176.67	21SE	1.8	-4.2	-1.6	0.8
142.15	1.39	1.9	186.66	1OC	1.9	-3.8	-1.6	0.8
148.26	1.44	1.9	196.70	11OC	-0.2	-3.2	-1.6	0.8
154.31	1.50	1.8	206.79	21OC	-0.6	-3.7	-1.5	0.8
160.30	1.55	1.7	216.92	31OC	-0.6	-4.2	-1.5	0.8
166.22	1.60	1.6	227.09	10NO	-0.6	-4.4	-1.5	0.8
172.06	1.65	1.6	237.29	20NO	-0.8	-4.4	-1.4	0.7
177.81	1.70	1.5	247.49	30NO	-0.8	-4.3	-1.4	0.7
183.44	1.74	1.3	257.70	10DE	-0.8	-4.2	-1.4	0.7
188.93	1.78	1.2	267.90	20DE	-0.7	-4.1	-1.4	0.7
194.25	1.82	1.1	278.08	30DE	0.2	-4.0	-1.4	0.8
				75				
199.36	1.84	0.9	288.23	9JA	2.8	-3.9	-1.5	0.8
204.21	1.86	0.7	298.35	19JA	1.4	-3.8	-1.5	0.8
208.74	1.86	0.5	308.41	29JA	0.6	-3.7	-1.5	0.8
212.87	1.85	0.3	318.43	8FE	0.3	-3.6	-1.6	0.8
216.49	1.80	0.0	328.39	18FE	0.1	-3.6	-1.6	0.8
219.49	1.73	-0.2	338.29	28FE	-0.2	-3.5	-1.7	0.8
221.69	1.61	-0.5	348.14	10MR	-0.8	-3.4	-1.7	0.8
222.93	1.42	-0.8	357.92	20MR	-1.7	-3.4	-1.8	0.8
223.03	1.16	-1.2	7.64	30MR	-1.3	-3.4	-1.8	0.7
221.87	0.80	-1.5	17.32	9AP	-0.3	-3.3	-1.9	0.7
219.53	0.35	-1.8	26.95	19AP	0.8	-3.3	-2.0	0.6
216.38	-0.18	-2.0	36.54	29AP	1.8	-3.3	-2.1	0.5
213.03	-0.73	-1.9	46.11	9MY	3.3	-3.3	-2.1	0.5
210.25	-1.25	-1.8	55.64	19MY	2.8	-3.3	-2.2	0.4
208.56	-1.69	-1.6	65.17	29MY	1.5	-3.3	-2.2	0.4
208.23	-2.03	-1.4	74.69	8JN	0.4	-3.3	-2.3	0.3
209.27	-2.28	-1.1	84.21	18JN	-0.7	-3.4	-2.3	0.3
211.52	-2.46	-0.9	93.75	28JN	-1.5	-3.4	-2.3	0.3
214.80	-2.57	-0.7	103.31	8JL	-1.3	-3.4	-2.3	0.4
218.93	-2.64	-0.6	112.90	18JL	-0.5	-3.5	-2.3	0.4
223.75	-2.67	-0.4	122.53	28JL	-0.1	-3.5	-2.3	0.5
229.14	-2.66	-0.2	132.21	7AU	0.2	-3.5	-2.2	0.6
234.98	-2.62	-0.1	141.93	17AU	0.4	-3.4	-2.2	0.6
241.20	-2.55	0.0	151.71	27AU	0.8	-3.4	-2.1	0.7
247.73	-2.47	0.2	161.56	6SE	2.2	-3.4	-2.1	0.7
254.50	-2.36	0.3	171.45	16SE	1.7	-3.4	-2.0	0.7
261.46	-2.24	0.4	181.41	26SE	-0.3	-3.3	-1.9	0.8
268.59	-2.11	0.5	191.43	6OC	-0.8	-3.3	-1.9	0.8
275.83	-1.96	0.6	201.49	16OC	-0.8	-3.3	-1.8	0.8
283.16	-1.81	0.7	211.60	26OC	-0.8	-3.3	-1.7	0.8
290.56	-1.65	0.8	221.76	5NO	-0.7	-3.3	-1.7	0.8
297.98	-1.49	0.9	231.93	15NO	-0.6	-3.4	-1.6	0.8
305.42	-1.32	1.0	242.13	25NO	-0.6	-3.4	-1.6	0.8
312.86	-1.16	1.1	252.34	5DE	-0.5	-3.4	-1.6	0.8
320.28	-0.99	1.2	262.55	15DE	0.4	-3.4	-1.5	0.7
327.67	-0.84	1.2	272.74	25DE	3.0	-3.5	-1.5	0.7

♂ LONG	♂ LAT	♂ MAG	☉ LONG	16.00UT 76	☿	♀	♃	♄
					MAGNITUDES			
335.02	-0.68	1.3	282.91	4JA	1.0	-3.5	-1.5	0.8
342.31	-0.53	1.4	293.04	14JA	0.4	-3.6	-1.5	0.8
349.55	-0.39	1.5	303.14	24JA	0.2	-3.6	-1.5	0.8
356.73	-0.25	1.5	313.19	3FE	-0.0	-3.7	-1.5	0.9
3.84	-0.13	1.6	323.17	13FE	-0.3	-3.8	-1.5	0.9
10.88	-0.00	1.7	333.11	23FE	-0.8	-3.9	-1.6	0.9
17.86	0.11	1.7	342.98	4MR	-1.6	-3.9	-1.6	0.9
24.77	0.22	1.8	352.79	14MR	-1.3	-4.0	-1.6	0.9
31.62	0.32	1.8	2.55	24MR	-0.2	-4.1	-1.7	0.8
38.41	0.42	1.8	12.26	3AP	1.0	-4.2	-1.7	0.8
45.13	0.51	1.9	21.91	13AP	2.4	-4.2	-1.8	0.7
51.80	0.59	1.9	31.52	23AP	3.6	-4.1	-1.9	0.7
58.42	0.67	1.9	41.10	3MY	2.1	-3.8	-1.9	0.6
65.00	0.74	1.9	50.64	13MY	1.1	-3.1	-2.0	0.6
71.53	0.80	1.9	60.18	23MY	0.3	-3.1	-2.1	0.5
78.02	0.87	1.9	69.70	2JN	-0.7	-3.8	-2.1	0.4
84.48	0.92	1.9	79.21	12JN	-1.6	-4.1	-2.2	0.4
90.91	0.97	1.9	88.75	22JN	-1.3	-4.2	-2.3	0.3
97.32	1.02	1.9	98.29	2JL	-0.5	-4.2	-2.3	0.3
103.72	1.06	2.0	107.87	12JL	0.0	-4.1	-2.4	0.4
110.09	1.10	2.0	117.48	22JL	0.4	-4.0	-2.4	0.4
116.46	1.13	2.0	127.13	1AU	0.6	-3.9	-2.4	0.5
122.82	1.16	2.0	136.83	11AU	1.2	-3.8	-2.4	0.5
129.18	1.19	2.0	146.58	21AU	2.7	-3.8	-2.4	0.6
135.53	1.21	2.0	156.39	31AU	1.5	-3.7	-2.4	0.6
141.89	1.23	2.0	166.26	10SE	-0.4	-3.6	-2.3	0.7
148.25	1.24	2.0	176.19	20SE	-0.9	-3.6	-2.3	0.7
154.61	1.25	2.0	186.17	30SE	-0.9	-3.5	-2.2	0.8
160.98	1.26	2.0	196.21	10OC	-0.9	-3.5	-2.2	0.8
167.35	1.26	1.9	206.30	20OC	-0.6	-3.5	-2.1	0.8
173.71	1.25	1.9	216.43	30OC	-0.5	-3.4	-2.0	0.8
180.08	1.24	1.8	226.60	9NO	-0.5	-3.4	-1.9	0.8
186.45	1.22	1.7	236.79	19NO	-0.4	-3.4	-1.9	0.8
192.81	1.19	1.7	246.99	29NO	0.5	-3.4	-1.8	0.8
199.16	1.15	1.6	257.20	9DE	3.0	-3.4	-1.8	0.8
205.49	1.11	1.5	267.40	19DE	0.7	-3.3	-1.7	0.8
211.80	1.05	1.4	277.58	29DE	0.2	-3.3	-1.7	0.8
				77				
218.08	0.97	1.3	287.73	8JA	0.0	-3.3	-1.6	0.8
224.32	0.89	1.2	297.85	18JA	-0.1	-3.4	-1.6	0.9
230.50	0.78	1.0	307.92	28JA	-0.4	-3.4	-1.6	0.9
236.63	0.65	0.9	317.94	7FE	-0.9	-3.4	-1.6	0.9
242.67	0.49	0.7	327.91	17FE	-1.5	-3.4	-1.6	0.9
248.59	0.31	0.5	337.81	27FE	-1.2	-3.5	-1.6	0.9
254.39	0.09	0.4	347.65	9MR	-0.2	-3.5	-1.6	0.9
260.01	-0.17	0.2	357.44	19MR	1.3	-3.5	-1.6	0.9
265.41	-0.48	-0.1	7.17	29MR	3.1	-3.4	-1.6	0.9
270.53	-0.84	-0.3	16.85	8AP	2.7	-3.4	-1.6	0.9
275.26	-1.26	-0.5	26.49	18AP	1.6	-3.4	-1.7	0.8
279.52	-1.76	-0.8	36.08	28AP	0.9	-3.4	-1.7	0.8
283.17	-2.33	-1.1	45.64	8MY	0.2	-3.3	-1.8	0.7
286.01	-2.99	-1.4	55.18	18MY	-0.7	-3.3	-1.8	0.7
287.89	-3.72	-1.7	64.70	28MY	-1.6	-3.3	-1.9	0.6
288.60	-4.50	-2.0	74.22	7JN	-1.3	-3.3	-2.0	0.6
288.04	-5.28	-2.2	83.75	17JN	-0.5	-3.3	-2.0	0.5
286.32	-5.95	-2.5	93.28	27JN	0.1	-3.4	-2.1	0.4
283.84	-6.41	-2.6	102.84	7JL	0.5	-3.4	-2.2	0.4
281.27	-6.57	-2.5	112.44	17JL	0.9	-3.4	-2.2	0.4
279.33	-6.42	-2.3	122.06	27JL	1.6	-3.4	-2.3	0.4
278.52	-6.05	-2.1	131.74	6AU	3.1	-3.5	-2.3	0.5
279.04	-5.53	-1.8	141.46	16AU	1.3	-3.5	-2.4	0.5
280.83	-4.96	-1.5	151.24	26AU	-0.5	-3.6	-2.4	0.6
283.72	-4.38	-1.3	161.08	5SE	-1.0	-3.7	-2.4	0.7
287.50	-3.82	-1.0	170.98	15SE	-1.1	-3.8	-2.5	0.7
292.00	-3.30	-0.7	180.93	25SE	-0.8	-3.9	-2.4	0.7
297.03	-2.82	-0.5	190.94	5OC	-0.5	-4.0	-2.4	0.8
302.50	-2.38	-0.3	201.01	15OC	-0.3	-4.1	-2.4	0.8
308.29	-1.99	-0.1	211.11	25OC	-0.3	-4.2	-2.3	0.9
314.32	-1.63	0.1	221.26	4NO	-0.2	-4.3	-2.2	0.9
320.54	-1.31	0.3	231.44	14NO	0.7	-4.4	-2.2	0.9
326.89	-1.02	0.5	241.64	24NO	2.8	-4.4	-2.1	0.9
333.34	-0.75	0.6	251.85	4DE	0.5	-4.2	-2.0	0.9
339.86	-0.52	0.8	262.05	14DE	-0.0	-3.8	-2.0	0.9
346.41	-0.31	0.9	272.24	24DE	-0.1	-3.1	-1.9	0.9

♂ LONG	♂ LAT	♂ MAG	☉ LONG	16.00UT 78	☿	♀	♃	♄
					MAGNITUDES			
352.99	-0.12	1.1	282.41	3JA	-0.2	-3.7	-1.8	0.9
359.58	0.05	1.2	292.55	13JA	-0.5	-4.2	-1.8	0.9
6.17	0.20	1.3	302.64	23JA	-0.9	-4.3	-1.7	0.9
12.75	0.33	1.4	312.69	2FE	-1.3	-4.3	-1.7	1.0
19.31	0.45	1.5	322.68	12FE	-1.1	-4.2	-1.6	1.0
25.85	0.56	1.6	332.62	22FE	-0.1	-4.1	-1.6	1.0
32.37	0.65	1.7	342.50	4MR	1.6	-4.0	-1.6	1.0
38.86	0.73	1.7	352.31	14MR	3.5	-3.9	-1.6	1.0
45.32	0.81	1.8	2.07	24MR	2.0	-3.8	-1.6	1.0
51.75	0.87	1.9	11.78	3AP	1.2	-3.7	-1.6	1.0
58.17	0.93	1.9	21.44	13AP	0.6	-3.6	-1.6	1.0
64.56	0.98	1.9	31.05	23AP	0.1	-3.6	-1.6	1.0
70.93	1.02	2.0	40.64	3MY	-0.7	-3.5	-1.6	0.9
77.29	1.06	2.0	50.19	13MY	-1.7	-3.4	-1.6	0.9
83.63	1.09	2.0	59.72	23MY	-1.3	-3.4	-1.6	0.8
89.96	1.11	2.0	69.24	2JN	-0.4	-3.4	-1.7	0.8
96.30	1.13	2.0	78.76	12JN	0.2	-3.3	-1.7	0.7
102.63	1.15	2.0	88.29	22JN	0.7	-3.3	-1.8	0.6
108.97	1.16	2.0	97.84	2JL	1.2	-3.3	-1.8	0.6
115.31	1.16	2.0	107.41	12JL	2.1	-3.3	-1.9	0.5
121.67	1.16	2.0	117.01	22JL	3.1	-3.3	-2.0	0.5
128.05	1.16	2.0	126.67	1AU	1.1	-3.3	-2.0	0.5
134.45	1.15	1.9	136.36	11AU	-0.5	-3.4	-2.1	0.5
140.87	1.13	2.0	146.11	21AU	-1.2	-3.4	-2.2	0.6
147.32	1.12	2.0	155.92	31AU	-1.2	-3.4	-2.2	0.6
153.81	1.09	2.0	165.78	10SE	-0.7	-3.4	-2.3	0.7
160.33	1.06	2.0	175.71	20SE	-0.4	-3.5	-2.3	0.7
166.88	1.03	2.0	185.69	30SE	-0.2	-3.5	-2.4	0.8
173.48	0.99	1.9	195.72	10OC	-0.2	-3.5	-2.4	0.8
180.11	0.94	1.9	205.81	20OC	-0.0	-3.5	-2.4	0.9
186.79	0.89	1.9	215.94	30OC	1.0	-3.4	-2.4	0.9
193.51	0.83	1.8	226.10	9NO	2.6	-3.4	-2.4	0.9
200.27	0.76	1.8	236.29	19NO	0.3	-3.4	-2.3	1.0
207.07	0.69	1.8	246.50	29NO	-0.2	-3.4	-2.3	1.0
213.91	0.60	1.7	256.70	9DE	-0.3	-3.3	-2.2	1.0
220.79	0.51	1.6	266.90	19DE	-0.3	-3.3	-2.1	1.0
227.72	0.40	1.6	277.09	29DE	-0.5	-3.3	-2.1	1.0
				79				
234.68	0.28	1.5	287.24	8JA	-0.9	-3.3	-2.0	1.0
241.68	0.15	1.4	297.36	18JA	-1.2	-3.3	-1.9	1.0
248.71	0.01	1.3	307.43	28JA	-1.0	-3.4	-1.8	1.0
255.76	-0.15	1.2	317.45	7FE	-0.1	-3.4	-1.8	1.0
262.85	-0.32	1.1	327.42	17FE	1.9	-3.4	-1.7	1.1
269.95	-0.50	1.0	337.33	27FE	2.9	-3.4	-1.7	1.1
277.05	-0.70	0.9	347.17	9MR	1.5	-3.5	-1.6	1.1
284.16	-0.91	0.8	356.96	19MR	0.9	-3.5	-1.6	1.2
291.25	-1.14	0.7	6.70	29MR	0.5	-3.6	-1.5	1.2
298.30	-1.37	0.6	16.38	8AP	-0.0	-3.6	-1.5	1.2
305.31	-1.62	0.4	26.02	18AP	-0.7	-3.7	-1.5	1.2
312.23	-1.88	0.3	35.62	28AP	-1.7	-3.8	-1.5	1.1
319.03	-2.14	0.2	45.18	8MY	-1.3	-3.8	-1.5	1.1
325.69	-2.41	0.0	54.72	18MY	-0.4	-3.9	-1.5	1.1
332.15	-2.69	-0.1	64.25	28MY	0.3	-4.0	-1.5	1.0
338.34	-2.96	-0.2	73.77	7JN	0.9	-4.1	-1.5	0.9
344.20	-3.24	-0.4	83.29	17JN	1.5	-4.2	-1.5	0.9
349.63	-3.51	-0.6	92.83	27JN	2.7	-4.2	-1.6	0.8
354.52	-3.77	-0.8	102.39	7JL	2.7	-4.0	-1.6	0.7
358.72	-4.02	-1.0	111.98	17JL	0.9	-3.6	-1.6	0.7
2.04	-4.24	-1.2	121.60	27JL	-0.6	-3.1	-1.7	0.6
4.30	-4.43	-1.4	131.27	6AU	-1.3	-3.6	-1.7	0.6
5.26	-4.54	-1.6	140.99	16AU	-1.3	-4.1	-1.8	0.6
4.77	-4.54	-1.9	150.77	26AU	-0.7	-4.3	-1.9	0.7
2.88	-4.38	-2.1	160.60	5SE	-0.3	-4.3	-1.9	0.7
359.91	-4.02	-2.2	170.50	15SE	-0.1	-4.2	-2.0	0.8
356.56	-3.48	-2.2	180.45	25SE	0.0	-4.2	-2.1	0.8
353.65	-2.82	-1.9	190.45	5OC	0.2	-4.1	-2.1	0.9
351.78	-2.14	-1.6	200.52	15OC	1.2	-4.0	-2.2	0.9
351.24	-1.51	-1.3	210.63	25OC	2.3	-3.9	-2.2	1.0
352.00	-0.97	-0.9	220.77	4NO	0.1	-3.8	-2.3	1.0
353.91	-0.51	-0.6	230.95	14NO	-0.4	-3.7	-2.3	1.0
356.75	-0.13	-0.3	241.15	24NO	-0.4	-3.7	-2.3	1.1
0.32	0.17	-0.1	251.35	4DE	-0.4	-3.6	-2.3	1.1
4.46	0.42	0.2	261.56	14DE	-0.6	-3.5	-2.2	1.1
9.05	0.61	0.4	271.76	24DE	-0.9	-3.5	-2.2	1.1

♂			☉	16.00UT	☿	♀	♃	♄
LONG	LAT	MAG	LONG	80	MAGNITUDES			
13.98	0.77	0.6	281.92	3JA	-1.0	-3.4	-2.1	1.1
19.17	0.90	0.8	292.06	13JA	-0.9	-3.4	-2.1	1.1
24.57	1.01	1.0	302.16	23JA	0.0	-3.4	-2.0	1.1
30.12	1.09	1.1	312.21	2FE	2.2	-3.3	-1.9	1.1
35.80	1.16	1.3	322.20	12FE	2.2	-3.3	-1.9	1.2
41.58	1.21	1.4	332.14	22FE	1.1	-3.3	-1.8	1.2
47.43	1.25	1.5	342.02	3MR	0.6	-3.3	-1.7	1.3
53.34	1.28	1.6	351.84	13MR	0.3	-3.3	-1.7	1.3
59.30	1.30	1.7	1.60	23MR	-0.1	-3.3	-1.6	1.3
65.30	1.32	1.8	11.31	2AP	-0.8	-3.3	-1.6	1.3
71.33	1.33	1.8	20.97	12AP	-1.7	-3.4	-1.5	1.3
77.39	1.33	1.9	30.59	22AP	-1.3	-3.4	-1.5	1.3
83.48	1.32	1.9	40.17	2MY	-0.4	-3.4	-1.4	1.3
89.60	1.31	2.0	49.72	12MY	0.5	-3.5	-1.4	1.3
95.75	1.30	2.0	59.25	22MY	1.2	-3.5	-1.4	1.3
101.92	1.28	2.0	68.77	1JN	2.0	-3.5	-1.4	1.2
108.12	1.26	2.0	78.30	11JN	3.3	-3.4	-1.4	1.2
114.36	1.23	2.0	87.82	21JN	2.2	-3.4	-1.4	1.1
120.64	1.20	2.0	97.37	1JL	0.7	-3.4	-1.4	1.0
126.95	1.16	2.0	106.94	11JL	-0.6	-3.3	-1.4	1.0
133.31	1.12	2.0	116.55	21JL	-1.4	-3.3	-1.4	0.9
139.71	1.08	2.0	126.19	31JL	-1.3	-3.3	-1.5	0.8
146.16	1.03	2.0	135.89	10AU	-0.6	-3.3	-1.5	0.8
152.66	0.98	1.9	145.63	20AU	-0.2	-3.3	-1.5	0.8
159.22	0.92	1.9	155.44	30AU	0.0	-3.3	-1.6	0.8
165.84	0.86	1.8	165.30	9SE	0.2	-3.4	-1.6	0.8
172.52	0.80	1.8	175.22	19SE	0.4	-3.4	-1.7	0.9
179.27	0.73	1.8	185.20	29SE	1.5	-3.4	-1.8	0.9
186.08	0.65	1.8	195.24	9OC	2.1	-3.5	-1.8	1.0
192.96	0.57	1.8	205.32	19OC	-0.1	-3.5	-1.9	1.0
199.90	0.49	1.8	215.45	29OC	-0.5	-3.6	-2.0	1.1
206.92	0.40	1.8	225.61	8NO	-0.5	-3.6	-2.0	1.1
214.00	0.30	1.7	235.80	18NO	-0.6	-3.7	-2.1	1.2
221.15	0.20	1.7	246.00	28NO	-0.7	-3.8	-2.1	1.2
228.37	0.09	1.7	256.21	8DE	-0.8	-3.9	-2.1	1.2
235.65	-0.03	1.6	266.41	18DE	-0.9	-4.0	-2.2	1.3
242.99	-0.15	1.6	276.59	28DE	-0.8	-4.1	-2.2	1.3
				81				
250.39	-0.27	1.5	286.75	7JA	0.1	-4.2	-2.1	1.3
257.84	-0.40	1.5	296.87	17JA	2.6	-4.3	-2.1	1.3
265.34	-0.53	1.4	306.94	27JA	1.7	-4.3	-2.1	1.3
272.88	-0.67	1.4	316.97	6FE	0.8	-4.3	-2.0	1.3
280.46	-0.80	1.3	326.94	16FE	0.4	-4.1	-2.0	1.3
288.05	-0.94	1.3	336.85	26FE	0.2	-3.6	-1.9	1.3
295.66	-1.07	1.2	346.70	8MR	-0.2	-3.2	-1.8	1.3
303.27	-1.20	1.2	356.49	18MR	-0.8	-3.7	-1.7	1.3
310.87	-1.33	1.1	6.23	28MR	-1.6	-4.1	-1.7	1.3
318.44	-1.44	1.1	15.91	7AP	-1.3	-4.2	-1.6	1.3
325.97	-1.55	1.0	25.55	17AP	-0.3	-4.2	-1.6	1.3
333.45	-1.65	0.9	35.15	27AP	0.6	-4.1	-1.5	1.2
340.84	-1.73	0.9	44.71	7MY	1.5	-4.0	-1.5	1.2
348.14	-1.80	0.8	54.25	17MY	2.7	-3.9	-1.4	1.2
355.32	-1.85	0.8	63.78	27MY	3.2	-3.8	-1.4	1.1
2.36	-1.89	0.7	73.30	6JN	1.7	-3.7	-1.3	1.1
9.23	-1.90	0.6	82.82	16JN	0.5	-3.7	-1.3	1.0
15.91	-1.90	0.5	92.36	26JN	-0.6	-3.6	-1.3	1.0
22.36	-1.89	0.5	101.92	6JL	-1.5	-3.5	-1.3	0.9
28.53	-1.85	0.4	111.51	16JL	-1.3	-3.5	-1.3	0.9
34.39	-1.78	0.3	121.13	26JL	-0.6	-3.5	-1.3	0.8
39.86	-1.70	0.2	130.80	5AU	-0.1	-3.4	-1.3	0.8
44.87	-1.59	0.0	140.52	15AU	0.1	-3.4	-1.3	0.8
49.31	-1.44	-0.1	150.30	25AU	0.3	-3.4	-1.3	0.7
53.05	-1.27	-0.3	160.13	4SE	0.7	-3.4	-1.4	0.8
55.94	-1.04	-0.5	170.02	14SE	1.9	-3.4	-1.4	0.8
57.77	-0.77	-0.7	179.97	24SE	1.9	-3.4	-1.5	0.9
58.33	-0.44	-0.9	189.98	4OC	-0.2	-3.4	-1.5	1.0
57.49	-0.05	-1.1	200.03	14OC	-0.7	-3.4	-1.6	1.0
55.21	0.40	-1.3	210.14	24OC	-0.7	-3.4	-1.6	1.1
51.82	0.86	-1.5	220.28	3NO	-0.7	-3.4	-1.7	1.2
47.94	1.29	-1.5	230.46	13NO	-0.8	-3.4	-1.7	1.2
44.37	1.65	-1.3	240.65	23NO	-0.7	-3.4	-1.8	1.2
41.78	1.90	-1.0	250.86	3DE	-0.7	-3.5	-1.9	1.3
40.51	2.05	-0.7	261.06	13DE	-0.6	-3.5	-1.9	1.3
40.56	2.14	-0.4	271.26	23DE	0.2	-3.5	-2.0	1.3

♂			☉	16.00UT	☿	♀	♃	♄
LONG	LAT	MAG	LONG	82	MAGNITUDES			
41.80	2.17	-0.1	281.43	2JA	2.9	-3.5	-2.0	1.3
44.01	2.16	0.2	291.57	12JA	1.3	-3.4	-2.1	1.2
47.01	2.14	0.4	301.67	22JA	0.5	-3.4	-2.1	1.2
50.62	2.10	0.6	311.72	1FE	0.3	-3.4	-2.1	1.2
54.72	2.05	0.8	321.71	11FE	0.1	-3.4	-2.0	1.1
59.20	2.00	1.0	331.66	21FE	-0.3	-3.3	-2.0	1.1
63.98	1.94	1.2	341.54	3MR	-0.8	-3.3	-2.0	1.0
69.00	1.88	1.3	351.36	13MR	-1.6	-3.3	-1.9	1.0
74.22	1.82	1.4	1.13	23MR	-1.3	-3.3	-1.8	1.0
79.60	1.76	1.5	10.84	2AP	-0.3	-3.3	-1.8	1.0
85.11	1.69	1.6	20.50	12AP	0.8	-3.4	-1.7	1.0
90.74	1.63	1.7	30.12	22AP	2.0	-3.4	-1.7	1.0
96.48	1.56	1.7	39.70	2MY	3.6	-3.4	-1.6	1.0
102.30	1.49	1.8	49.25	12MY	2.5	-3.4	-1.5	1.0
108.21	1.42	1.8	58.79	22MY	1.4	-3.5	-1.5	0.9
114.20	1.36	1.9	68.31	1JN	0.4	-3.5	-1.4	0.9
120.26	1.28	1.9	77.83	11JN	-0.6	-3.6	-1.4	0.9
126.40	1.21	1.9	87.36	21JN	-1.5	-3.6	-1.3	0.8
132.61	1.13	1.9	96.91	1JL	-1.3	-3.7	-1.3	0.8
138.90	1.06	1.9	106.48	11JL	-0.5	-3.8	-1.3	0.7
145.26	0.98	1.9	116.08	21JL	-0.0	-3.9	-1.2	0.7
151.70	0.90	1.9	125.73	31JL	0.3	-4.0	-1.2	0.6
158.21	0.81	1.9	135.42	10AU	0.5	-4.1	-1.2	0.6
164.81	0.73	1.9	145.17	20AU	1.0	-4.2	-1.2	0.5
171.49	0.64	1.9	154.96	30AU	2.3	-4.3	-1.2	0.5
178.26	0.54	1.8	164.82	9SE	1.7	-4.3	-1.2	0.4
185.11	0.45	1.8	174.75	19SE	-0.3	-4.1	-1.3	0.5
192.04	0.35	1.7	184.72	29SE	-0.8	-3.7	-1.3	0.5
199.06	0.25	1.7	194.75	9OC	-0.8	-3.2	-1.3	0.6
206.17	0.15	1.6	204.84	19OC	-0.8	-3.8	-1.4	0.7
213.37	0.05	1.6	214.96	29OC	-0.7	-4.2	-1.4	0.8
220.65	-0.06	1.6	225.12	8NO	-0.6	-4.4	-1.5	0.8
228.01	-0.17	1.6	235.31	18NO	-0.6	-4.4	-1.5	0.9
235.44	-0.27	1.6	245.51	28NO	-0.5	-4.3	-1.6	0.9
242.95	-0.38	1.5	255.72	8DE	0.4	-4.2	-1.7	0.9
250.53	-0.49	1.5	265.92	18DE	3.0	-4.1	-1.7	1.0
258.16	-0.59	1.5	276.10	28DE	0.9	-4.0	-1.8	1.0
				83				
265.85	-0.70	1.5	286.26	7JA	0.3	-3.9	-1.9	1.0
273.58	-0.79	1.5	296.38	17JA	0.1	-3.8	-1.9	1.0
281.35	-0.89	1.5	306.46	27JA	-0.0	-3.7	-2.0	0.9
289.14	-0.97	1.4	316.49	6FE	-0.3	-3.6	-2.0	0.9
296.93	-1.05	1.4	326.46	16FE	-0.8	-3.6	-2.0	0.9
304.73	-1.12	1.4	336.37	26FE	-1.5	-3.5	-2.0	0.8
312.52	-1.18	1.4	346.23	8MR	-1.3	-3.4	-2.0	0.8
320.28	-1.23	1.4	356.02	18MR	-0.2	-3.4	-2.0	0.7
328.00	-1.27	1.3	5.76	28MR	1.1	-3.4	-2.0	0.8
335.67	-1.29	1.3	15.45	7AP	2.6	-3.3	-1.9	0.8
343.28	-1.30	1.3	25.09	17AP	3.3	-3.3	-1.9	0.8
350.81	-1.29	1.3	34.69	27AP	1.9	-3.3	-1.8	0.8
358.26	-1.28	1.3	44.26	7MY	1.0	-3.3	-1.7	0.8
5.61	-1.24	1.3	53.80	17MY	0.3	-3.3	-1.7	0.7
12.86	-1.20	1.2	63.32	27MY	-0.7	-3.3	-1.6	0.7
19.99	-1.14	1.2	72.85	6JN	-1.6	-3.3	-1.5	0.7
26.99	-1.06	1.2	82.37	16JN	-1.3	-3.4	-1.5	0.7
33.85	-0.97	1.2	91.90	26JN	-0.5	-3.4	-1.4	0.6
40.56	-0.87	1.1	101.46	6JL	0.1	-3.4	-1.4	0.6
47.09	-0.76	1.1	111.04	16JL	0.4	-3.5	-1.3	0.5
53.45	-0.63	1.0	120.66	26JL	0.7	-3.5	-1.3	0.5
59.59	-0.49	1.0	130.33	5AU	1.3	-3.5	-1.3	0.4
65.50	-0.33	0.9	140.05	15AU	2.8	-3.4	-1.2	0.4
71.14	-0.16	0.8	149.81	25AU	1.5	-3.4	-1.2	0.3
76.46	0.04	0.7	159.65	4SE	-0.4	-3.4	-1.2	0.3
81.41	0.26	0.6	169.53	14SE	-1.0	-3.4	-1.2	0.2
85.90	0.50	0.5	179.48	24SE	-1.0	-3.3	-1.2	0.2
89.84	0.78	0.3	189.48	4OC	-0.9	-3.3	-1.2	0.2
93.10	1.09	0.2	199.54	14OC	-0.5	-3.3	-1.2	0.3
95.53	1.45	-0.0	209.64	24OC	-0.4	-3.3	-1.3	0.4
96.92	1.84	-0.2	219.79	3NO	-0.4	-3.3	-1.3	0.4
97.10	2.28	-0.5	229.96	13NO	-0.3	-3.4	-1.3	0.5
95.93	2.74	-0.7	240.16	23NO	0.6	-3.4	-1.4	0.6
93.41	3.18	-0.9	250.37	3DE	3.0	-3.4	-1.4	0.6
89.86	3.54	-1.1	260.57	13DE	0.7	-3.4	-1.5	0.7
85.88	3.77	-1.1	270.76	23DE	0.1	-3.5	-1.6	0.7

♂ LONG	♂ LAT	♂ MAG	☉ LONG	16.00UT	☿	♀	♃	♄
				84	MAGNITUDES			
82.22	3.84	-0.9	280.94	2JA	-0.0	-3.5	-1.6	0.7
79.54	3.78	-0.6	291.08	12JA	-0.1	-3.6	-1.7	0.7
78.12	3.62	-0.4	301.18	22JA	-0.4	-3.6	-1.8	0.7
78.00	3.41	-0.1	311.24	1FE	-0.9	-3.7	-1.8	0.7
79.05	3.18	0.2	321.24	11FE	-1.4	-3.8	-1.9	0.7
81.07	2.96	0.4	331.18	21FE	-1.2	-3.9	-1.9	0.7
83.89	2.75	0.6	341.07	2MR	-0.2	-3.9	-2.0	0.6
87.35	2.55	0.8	350.89	12MR	1.3	-4.0	-2.0	0.6
91.30	2.36	0.9	0.66	22MR	3.3	-4.1	-2.0	0.5
95.67	2.18	1.1	10.38	1AP	2.5	-4.2	-2.0	0.5
100.36	2.02	1.2	20.04	11AP	1.4	-4.2	-2.0	0.5
105.33	1.87	1.3	29.66	21AP	0.8	-4.1	-2.0	0.6
110.53	1.72	1.4	39.24	1MY	0.2	-3.8	-2.0	0.6
115.93	1.58	1.5	48.80	11MY	-0.7	-3.1	-1.9	0.6
121.51	1.45	1.6	58.33	21MY	-1.6	-3.1	-1.9	0.6
127.24	1.32	1.6	67.85	31MY	-1.3	-3.8	-1.8	0.6
133.10	1.19	1.7	77.37	10JN	-0.5	-4.1	-1.7	0.5
139.10	1.06	1.7	86.90	20JN	0.2	-4.2	-1.7	0.5
145.23	0.94	1.7	96.44	30JN	0.6	-4.2	-1.6	0.5
151.47	0.82	1.7	106.01	10JL	1.0	-4.1	-1.5	0.4
157.83	0.71	1.8	115.61	20JL	1.7	-4.0	-1.5	0.4
164.31	0.59	1.8	125.26	30JL	3.2	-3.9	-1.4	0.4
170.89	0.47	1.8	134.94	9AU	1.2	-3.8	-1.4	0.3
177.58	0.36	1.7	144.69	19AU	-0.5	-3.8	-1.4	0.2
184.39	0.24	1.7	154.49	29AU	-1.1	-3.7	-1.3	0.2
191.30	0.13	1.7	164.34	8SE	-1.1	-3.6	-1.3	0.1
198.32	0.02	1.7	174.26	18SE	-0.8	-3.6	-1.3	0.1
205.45	-0.09	1.6	184.23	28SE	-0.4	-3.5	-1.3	0.0
212.67	-0.20	1.6	194.26	8OC	-0.3	-3.5	-1.3	-0.0
220.00	-0.31	1.6	204.34	18OC	-0.2	-3.5	-1.3	0.0
227.42	-0.41	1.5	214.47	28OC	-0.1	-3.4	-1.3	0.1
234.92	-0.51	1.5	224.62	7NO	0.8	-3.4	-1.3	0.2
242.51	-0.61	1.5	234.81	17NO	2.8	-3.4	-1.3	0.2
250.18	-0.70	1.4	245.02	27NO	0.4	-3.4	-1.3	0.3
257.90	-0.78	1.4	255.22	7DE	-0.1	-3.4	-1.4	0.4
265.69	-0.85	1.3	265.43	17DE	-0.2	-3.3	-1.4	0.4
273.51	-0.92	1.3	275.61	27DE	-0.3	-3.3	-1.5	0.5
				85				
281.37	-0.98	1.3	285.77	6JA	-0.5	-3.3	-1.5	0.5
289.24	-1.03	1.4	295.89	16JA	-0.9	-3.4	-1.6	0.5
297.13	-1.06	1.4	305.97	26JA	-1.3	-3.4	-1.6	0.5
305.00	-1.09	1.4	316.00	5FE	-1.1	-3.4	-1.7	0.5
312.86	-1.11	1.4	325.98	15FE	-0.1	-3.4	-1.8	0.5
320.68	-1.11	1.4	335.89	25FE	1.6	-3.5	-1.9	0.5
328.46	-1.10	1.4	345.75	7MR	3.3	-3.5	-1.9	0.5
336.19	-1.08	1.4	355.55	17MR	1.8	-3.5	-2.0	0.5
343.86	-1.04	1.4	5.29	27MR	1.0	-3.4	-2.0	0.4
351.45	-1.00	1.5	14.97	6AP	0.6	-3.4	-2.1	0.4
358.96	-0.94	1.5	24.62	16AP	0.1	-3.4	-2.1	0.4
6.39	-0.88	1.5	34.22	26AP	-0.7	-3.4	-2.1	0.4
13.72	-0.80	1.5	43.79	6MY	-1.6	-3.3	-2.1	0.4
20.96	-0.72	1.5	53.33	16MY	-1.4	-3.3	-2.1	0.4
28.10	-0.63	1.5	62.86	26MY	-0.4	-3.3	-2.1	0.4
35.14	-0.53	1.5	72.38	5JN	0.3	-3.3	-2.0	0.4
42.06	-0.42	1.6	81.90	15JN	0.8	-3.3	-2.0	0.4
48.88	-0.31	1.6	91.44	25JN	1.3	-3.4	-1.9	0.4
55.58	-0.19	1.5	100.99	5JL	2.3	-3.4	-1.9	0.4
62.16	-0.07	1.5	110.57	15JL	3.0	-3.4	-1.8	0.4
68.62	0.07	1.5	120.19	25JL	1.0	-3.4	-1.7	0.3
74.94	0.21	1.5	129.86	4AU	-0.5	-3.5	-1.7	0.3
81.12	0.35	1.5	139.57	14AU	-1.2	-3.5	-1.6	0.2
87.14	0.51	1.4	149.34	24AU	-1.3	-3.6	-1.6	0.2
92.99	0.67	1.3	159.16	3SE	-0.7	-3.7	-1.5	0.1
98.64	0.85	1.3	169.05	13SE	-0.4	-3.8	-1.5	0.0
104.06	1.04	1.2	178.99	23SE	-0.2	-3.9	-1.4	-0.0
109.21	1.24	1.1	188.99	3OC	-0.1	-4.0	-1.4	-0.1
114.05	1.47	1.0	199.05	13OC	0.1	-4.1	-1.4	-0.1
118.49	1.71	0.8	209.15	23OC	1.0	-4.2	-1.3	-0.2
122.47	1.99	0.7	219.29	2NO	2.6	-4.3	-1.3	-0.1
125.87	2.29	0.5	229.47	12NO	0.2	-4.4	-1.3	-0.0
128.54	2.62	0.3	239.66	22NO	-0.3	-4.4	-1.3	0.0
130.33	2.99	0.1	249.87	2DE	-0.3	-4.2	-1.3	0.1
131.04	3.39	-0.1	260.08	12DE	-0.4	-3.8	-1.4	0.2
130.52	3.79	-0.4	270.27	22DE	-0.6	-3.1	-1.4	0.2

♂ LONG	♂ LAT	♂ MAG	☉ LONG	16.00UT	☿	♀	♃	♄
				86	MAGNITUDES			
128.68	4.16	-0.6	280.44	1JA	-0.9	-3.8	-1.4	0.3
125.63	4.45	-0.9	290.59	11JA	-1.1	-4.2	-1.5	0.3
121.83	4.59	-1.0	300.69	21JA	-1.0	-4.3	-1.5	0.4
117.93	4.55	-0.9	310.75	31JA	-0.1	-4.3	-1.6	0.4
114.65	4.34	-0.7	320.75	10FE	2.0	-4.2	-1.6	0.4
112.49	4.03	-0.5	330.70	20FE	2.7	-4.1	-1.7	0.4
111.61	3.67	-0.2	340.59	2MR	1.3	-4.0	-1.8	0.4
111.99	3.29	-0.0	350.42	12MR	0.8	-3.9	-1.8	0.4
113.46	2.93	0.2	0.19	22MR	0.4	-3.8	-1.9	0.4
115.85	2.60	0.4	9.91	1AP	-0.0	-3.7	-2.0	0.3
118.98	2.29	0.6	19.57	11AP	-0.7	-3.6	-2.0	0.3
122.71	2.02	0.7	29.19	21AP	-1.6	-3.6	-2.1	0.3
126.94	1.76	0.9	38.78	1MY	-1.4	-3.5	-2.1	0.2
131.58	1.53	1.0	48.34	11MY	-0.4	-3.4	-2.2	0.3
136.56	1.31	1.1	57.87	21MY	0.4	-3.4	-2.2	0.3
141.83	1.11	1.2	67.40	31MY	1.0	-3.4	-2.2	0.3
147.35	0.92	1.2	76.92	10JN	1.7	-3.3	-2.2	0.3
153.10	0.74	1.3	86.44	20JN	2.9	-3.3	-2.2	0.3
159.05	0.58	1.4	95.99	30JN	2.6	-3.3	-2.2	0.3
165.19	0.42	1.4	105.55	10JL	0.9	-3.3	-2.1	0.3
171.50	0.26	1.4	115.15	20JL	-0.6	-3.3	-2.1	0.3
177.97	0.12	1.5	124.79	30JL	-1.3	-3.3	-2.0	0.3
184.60	-0.02	1.5	134.48	9AU	-1.3	-3.4	-2.0	0.3
191.37	-0.15	1.5	144.21	19AU	-0.7	-3.4	-1.9	0.2
198.28	-0.28	1.5	154.01	29AU	-0.3	-3.4	-1.8	0.2
205.32	-0.40	1.5	163.86	8SE	-0.0	-3.4	-1.8	0.1
212.49	-0.51	1.5	173.77	18SE	0.1	-3.5	-1.7	0.0
219.78	-0.61	1.5	183.74	28SE	0.3	-3.5	-1.6	-0.0
227.18	-0.71	1.5	193.77	8OC	1.3	-3.5	-1.6	-0.1
234.68	-0.80	1.5	203.84	18OC	2.4	-3.5	-1.6	-0.2
242.28	-0.87	1.4	213.97	28OC	0.0	-3.4	-1.5	-0.2
249.96	-0.94	1.4	224.13	7NO	-0.4	-3.4	-1.5	-0.3
257.70	-1.00	1.4	234.31	17NO	-0.5	-3.4	-1.5	-0.2
265.51	-1.04	1.4	244.51	27NO	-0.5	-3.4	-1.4	-0.1
273.37	-1.08	1.4	254.72	7DE	-0.7	-3.3	-1.4	-0.1
281.25	-1.10	1.4	264.92	17DE	-0.9	-3.3	-1.4	0.0
289.15	-1.11	1.4	275.11	27DE	-1.0	-3.3	-1.4	0.1
				87				
297.05	-1.11	1.4	285.27	6JA	-0.9	-3.3	-1.5	0.1
304.94	-1.09	1.4	295.40	16JA	0.0	-3.4	-1.5	0.2
312.81	-1.07	1.4	305.48	26JA	2.3	-3.4	-1.5	0.3
320.64	-1.03	1.4	315.51	5FE	2.1	-3.4	-1.5	0.3
328.42	-0.98	1.4	325.49	15FE	1.0	-3.4	-1.6	0.3
336.14	-0.93	1.4	335.41	25FE	0.5	-3.4	-1.6	0.3
343.80	-0.86	1.4	345.27	7MR	0.3	-3.5	-1.7	0.3
351.38	-0.79	1.4	355.07	17MR	-0.1	-3.5	-1.7	0.3
358.88	-0.71	1.5	4.82	27MR	-0.7	-3.6	-1.8	0.3
6.30	-0.62	1.5	14.51	6AP	-1.6	-3.6	-1.9	0.3
13.62	-0.53	1.6	24.16	16AP	-1.4	-3.7	-1.9	0.3
20.87	-0.44	1.6	33.76	26AP	-0.4	-3.8	-2.0	0.2
28.02	-0.34	1.6	43.33	6MY	0.5	-3.9	-2.1	0.2
35.08	-0.23	1.7	52.87	16MY	1.3	-3.9	-2.2	0.2
42.05	-0.13	1.7	62.41	26MY	2.3	-4.0	-2.2	0.2
48.93	-0.02	1.7	71.92	5JN	3.4	-4.1	-2.3	0.3
55.72	0.09	1.8	81.45	15JN	2.1	-4.2	-2.3	0.3
62.43	0.20	1.8	90.98	25JN	0.7	-4.2	-2.3	0.3
69.05	0.31	1.8	100.53	5JL	-0.6	-4.0	-2.4	0.3
75.59	0.42	1.8	110.11	15JL	-1.4	-3.6	-2.3	0.3
82.04	0.54	1.8	119.73	25JL	-1.3	-3.1	-2.3	0.3
88.40	0.65	1.8	129.39	4AU	-0.6	-3.6	-2.3	0.3
94.68	0.77	1.8	139.10	14AU	-0.2	-4.1	-2.3	0.3
100.86	0.89	1.8	148.87	24AU	0.1	-4.3	-2.2	0.2
106.94	1.01	1.7	158.69	3SE	0.2	-4.3	-2.1	0.2
112.91	1.14	1.7	168.57	13SE	0.5	-4.2	-2.1	0.1
118.76	1.27	1.6	178.51	23SE	1.6	-4.2	-2.0	0.1
124.47	1.41	1.6	188.51	3OC	2.1	-4.1	-1.9	0.0
130.03	1.56	1.5	198.56	13OC	-0.1	-4.0	-1.9	-0.1
135.40	1.71	1.4	208.66	23OC	-0.6	-3.9	-1.8	-0.1
140.55	1.88	1.3	218.79	2NO	-0.6	-3.8	-1.7	-0.2
145.44	2.06	1.1	228.97	12NO	-0.6	-3.7	-1.7	-0.3
150.01	2.25	1.0	239.16	22NO	-0.7	-3.7	-1.7	-0.3
154.18	2.45	0.8	249.37	2DE	-0.8	-3.6	-1.6	-0.2
157.88	2.68	0.6	259.57	12DE	-0.8	-3.5	-1.6	-0.2
160.97	2.92	0.4	269.77	22DE	-0.7	-3.5	-1.6	-0.1

Left (1988–1989)

♂ LONG	LAT	MAG	☉ LONG	16.00UT 88	☿	♀	♃	♄
163.31	3.18	0.2	279.95	1JA	0.1	-3.4	-1.5	-0.0
164.73	3.46	-0.1	290.10	11JA	2.6	-3.4	-1.5	0.1
165.04	3.72	-0.3	300.20	21JA	1.6	-3.4	-1.5	0.1
164.11	3.96	-0.6	310.26	31JA	0.7	-3.3	-1.5	0.2
161.90	4.12	-0.9	320.27	10FE	0.4	-3.3	-1.5	0.2
158.62	4.15	-1.1	330.22	20FE	0.1	-3.3	-1.6	0.3
154.81	4.02	-1.2	340.11	1MR	-0.2	-3.3	-1.6	0.3
151.15	3.73	-1.0	349.94	11MR	-0.8	-3.3	-1.6	0.3
148.31	3.32	-0.9	359.72	21MR	-1.6	-3.3	-1.7	0.3
146.68	2.87	-0.6	9.44	31MR	-1.4	-3.3	-1.7	0.3
146.36	2.40	-0.4	19.11	10AP	-0.3	-3.4	-1.8	0.3
147.28	1.97	-0.2	28.73	20AP	0.7	-3.4	-1.8	0.3
149.27	1.57	-0.0	38.32	30AP	1.7	-3.4	-1.9	0.3
152.15	1.21	0.2	47.88	10MY	3.0	-3.5	-2.0	0.2
155.76	0.89	0.3	57.41	20MY	3.0	-3.5	-2.0	0.2
159.98	0.60	0.4	66.93	30MY	1.6	-3.5	-2.1	0.2
164.70	0.35	0.6	76.45	9JN	0.5	-3.4	-2.2	0.2
169.84	0.11	0.7	85.98	19JN	-0.6	-3.4	-2.2	0.3
175.34	-0.09	0.8	95.52	29JN	-1.5	-3.4	-2.3	0.3
181.16	-0.28	0.8	105.08	9JL	-1.3	-3.3	-2.3	0.3
187.25	-0.45	0.9	114.68	19JL	-0.6	-3.3	-2.4	0.3
193.58	-0.61	1.0	124.32	29JL	-0.1	-3.3	-2.4	0.3
200.14	-0.75	1.0	134.00	8AU	0.2	-3.3	-2.4	0.3
206.89	-0.87	1.0	143.74	18AU	0.4	-3.3	-2.4	0.3
213.82	-0.98	1.1	153.53	28AU	0.8	-3.3	-2.4	0.3
220.92	-1.07	1.1	163.38	7SE	2.0	-3.4	-2.4	0.3
228.16	-1.15	1.1	173.29	17SE	1.9	-3.4	-2.3	0.2
235.53	-1.21	1.2	183.26	27SE	-0.2	-3.4	-2.3	0.2
243.02	-1.26	1.2	193.28	7OC	-0.7	-3.5	-2.2	0.1
250.61	-1.29	1.2	203.36	17OC	-0.7	-3.5	-2.1	-0.0
258.29	-1.31	1.3	213.47	27OC	-0.7	-3.6	-2.1	-0.0
266.03	-1.31	1.3	223.63	6NO	-0.7	-3.7	-2.0	-0.1
273.82	-1.30	1.3	233.81	16NO	-0.6	-3.7	-1.9	-0.2
281.65	-1.27	1.3	244.01	26NO	-0.6	-3.8	-1.9	-0.3
289.49	-1.24	1.3	254.22	6DE	-0.6	-3.9	-1.8	-0.3
297.34	-1.19	1.4	264.43	16DE	0.2	-4.0	-1.8	-0.2
305.17	-1.13	1.4	274.61	26DE	2.9	-4.1	-1.7	-0.1

89

♂ LONG	LAT	MAG	☉ LONG	16.00UT 89	☿	♀	♃	♄
312.97	-1.06	1.4	284.78	5JA	1.2	-4.2	-1.7	-0.1
320.74	-0.98	1.4	294.91	15JA	0.4	-4.3	-1.6	0.0
328.45	-0.90	1.5	304.99	25JA	0.2	-4.3	-1.6	0.1
336.10	-0.80	1.5	315.03	4FE	0.0	-4.3	-1.6	0.1
343.69	-0.71	1.5	325.01	14FE	-0.3	-4.1	-1.6	0.2
351.20	-0.61	1.5	334.93	24FE	-0.8	-3.6	-1.6	0.2
358.63	-0.51	1.6	344.79	6MR	-1.6	-3.2	-1.6	0.3
5.98	-0.40	1.6	354.60	16MR	-1.3	-3.7	-1.6	0.3
13.24	-0.30	1.6	4.34	26MR	-0.3	-4.1	-1.6	0.3
20.42	-0.20	1.6	14.04	5AP	0.9	-4.2	-1.6	0.3
27.52	-0.09	1.6	23.69	15AP	2.2	-4.2	-1.7	0.3
34.54	0.01	1.6	33.29	25AP	3.9	-4.1	-1.7	0.3
41.47	0.11	1.7	42.87	5MY	2.3	-4.0	-1.7	0.3
48.33	0.21	1.7	52.41	15MY	1.2	-3.9	-1.8	0.3
55.12	0.31	1.8	61.94	25MY	0.4	-3.8	-1.8	0.3
61.83	0.41	1.8	71.46	4JN	-0.6	-3.7	-1.9	0.2
68.48	0.50	1.9	80.98	14JN	-1.5	-3.7	-2.0	0.2
75.07	0.59	1.9	90.52	24JN	-1.4	-3.6	-2.0	0.3
81.60	0.68	1.9	100.07	4JL	-0.5	-3.5	-2.1	0.3
88.07	0.77	1.9	109.65	14JL	-0.0	-3.5	-2.2	0.4
94.49	0.85	2.0	119.26	24JL	0.3	-3.5	-2.2	0.4
100.86	0.94	2.0	128.93	3AU	0.6	-3.4	-2.3	0.4
107.17	1.02	2.0	138.63	13AU	1.1	-3.4	-2.4	0.4
113.44	1.10	2.0	148.39	23AU	2.5	-3.4	-2.4	0.4
119.65	1.18	1.9	158.22	2SE	1.7	-3.4	-2.4	0.4
125.81	1.26	1.9	168.09	12SE	-0.3	-3.4	-2.4	0.4
131.92	1.34	1.9	178.03	22SE	-0.9	-3.4	-2.4	0.3
137.96	1.42	1.8	188.03	2OC	-0.9	-3.4	-2.4	0.3
143.93	1.50	1.8	198.07	12OC	-0.9	-3.4	-2.4	0.2
149.82	1.58	1.7	208.17	22OC	-0.6	-3.4	-2.4	0.2
155.62	1.66	1.6	218.31	1NO	-0.5	-3.4	-2.3	0.1
161.32	1.74	1.5	228.47	11NO	-0.5	-3.4	-2.2	0.0
166.88	1.83	1.4	238.67	21NO	-0.4	-3.4	-2.2	-0.0
172.29	1.91	1.3	248.87	1DE	0.4	-3.5	-2.1	-0.1
177.52	2.00	1.2	259.08	11DE	3.1	-3.5	-2.0	-0.2
182.52	2.08	1.0	269.28	21DE	0.9	-3.5	-1.9	-0.2
187.24	2.17	0.9	279.45	31DE	0.2	-3.5	-1.9	-0.1

Right (1990–1991)

♂ LONG	LAT	MAG	☉ LONG	16.00UT 90	☿	♀	♃	♄
191.61	2.25	0.7	289.60	10JA	0.1	-3.4	-1.8	-0.1
195.57	2.33	0.4	299.71	20JA	-0.1	-3.4	-1.8	0.0
198.98	2.40	0.2	309.77	30JA	-0.3	-3.4	-1.7	0.1
201.72	2.45	-0.0	319.78	9FE	-0.8	-3.4	-1.7	0.1
203.64	2.48	-0.3	329.73	19FE	-1.5	-3.3	-1.6	0.2
204.53	2.47	-0.6	339.63	1MR	-1.3	-3.3	-1.6	0.2
204.26	2.40	-0.9	349.46	11MR	-0.3	-3.3	-1.6	0.3
202.71	2.23	-1.2	359.24	21MR	1.1	-3.3	-1.6	0.3
200.00	1.96	-1.5	8.96	31MR	2.8	-3.3	-1.5	0.4
196.57	1.57	-1.7	18.63	10AP	3.0	-3.4	-1.5	0.4
193.04	1.10	-1.6	28.26	20AP	1.7	-3.4	-1.6	0.4
190.17	0.59	-1.4	37.85	30AP	0.9	-3.4	-1.6	0.4
188.44	0.11	-1.3	47.41	10MY	0.2	-3.4	-1.6	0.4
188.06	-0.33	-1.0	56.95	20MY	-0.6	-3.5	-1.6	0.4
189.02	-0.70	-0.8	66.47	30MY	-1.6	-3.5	-1.6	0.4
191.15	-1.01	-0.5	75.99	9JN	-1.4	-3.6	-1.7	0.4
194.28	-1.27	-0.5	85.52	19JN	-0.5	-3.6	-1.7	0.3
198.23	-1.47	-0.3	95.06	29JN	0.1	-3.7	-1.8	0.4
202.85	-1.64	-0.1	104.62	9JL	0.5	-3.8	-1.8	0.4
208.04	-1.77	-0.0	114.22	19JL	0.8	-3.9	-1.9	0.4
213.69	-1.86	0.1	123.85	29JL	1.4	-4.0	-2.0	0.5
219.71	-1.93	0.2	133.53	8AU	3.0	-4.1	-2.0	0.5
226.09	-1.97	0.3	143.27	18AU	1.4	-4.2	-2.1	0.5
232.73	-1.99	0.4	153.06	28AU	-0.4	-4.3	-2.2	0.5
239.59	-1.98	0.5	162.90	7SE	-1.0	-4.3	-2.2	0.5
246.65	-1.96	0.6	172.81	17SE	-1.0	-4.1	-2.3	0.5
253.88	-1.91	0.7	182.78	27SE	-0.8	-3.7	-2.3	0.5
261.22	-1.85	0.8	192.79	7OC	-0.5	-3.2	-2.4	0.5
268.68	-1.77	0.8	202.87	17OC	-0.4	-3.8	-2.4	0.4
276.21	-1.68	0.9	212.98	27OC	-0.3	-4.2	-2.4	0.4
283.78	-1.57	1.0	223.14	6NO	-0.2	-4.4	-2.4	0.3
291.40	-1.46	1.0	233.32	16NO	0.6	-4.4	-2.3	0.2
299.02	-1.34	1.1	243.52	26NO	3.0	-4.3	-2.3	0.2
306.64	-1.21	1.2	253.73	6DE	0.6	-4.2	-2.2	0.1
314.24	-1.08	1.3	263.93	16DE	0.0	-4.1	-2.2	0.0
321.81	-0.95	1.3	274.12	26DE	-0.1	-4.0	-2.1	-0.0

91

♂ LONG	LAT	MAG	☉ LONG	16.00UT 91	☿	♀	♃	♄
329.33	-0.82	1.4	284.29	5JA	-0.2	-3.9	-2.0	-0.0
336.80	-0.68	1.4	294.42	15JA	-0.4	-3.8	-1.9	0.0
344.20	-0.55	1.5	304.50	25JA	-0.9	-3.7	-1.9	0.1
351.54	-0.43	1.6	314.54	4FE	-1.4	-3.6	-1.8	0.1
358.81	-0.30	1.6	324.53	14FE	-1.2	-3.6	-1.7	0.2
6.00	-0.18	1.7	334.45	24FE	-0.2	-3.5	-1.7	0.3
13.12	-0.07	1.7	344.32	6MR	1.4	-3.4	-1.6	0.3
20.17	0.05	1.7	354.13	16MR	3.4	-3.4	-1.6	0.4
27.14	0.15	1.8	3.87	26MR	2.2	-3.4	-1.6	0.4
34.04	0.25	1.8	13.57	5AP	1.3	-3.3	-1.5	0.5
40.87	0.35	1.8	23.23	15AP	0.7	-3.3	-1.5	0.5
47.64	0.44	1.8	32.83	25AP	0.1	-3.3	-1.5	0.5
54.35	0.53	1.9	42.41	5MY	-0.7	-3.3	-1.5	0.5
61.00	0.61	1.9	51.95	15MY	-1.6	-3.3	-1.5	0.5
67.60	0.68	1.9	61.48	25MY	-1.4	-3.3	-1.5	0.5
74.15	0.76	1.9	71.00	4JN	-0.5	-3.4	-1.5	0.5
80.67	0.82	1.9	80.53	14JN	0.2	-3.4	-1.5	0.5
87.14	0.89	1.9	90.05	24JN	0.6	-3.4	-1.5	0.5
93.58	0.95	1.9	99.61	4JL	1.1	-3.4	-1.6	0.5
99.99	1.00	2.0	109.19	14JL	1.9	-3.5	-1.6	0.5
106.37	1.05	2.0	118.80	24JL	3.2	-3.5	-1.6	0.6
112.74	1.10	2.0	128.45	3AU	1.2	-3.5	-1.7	0.6
119.09	1.15	2.0	138.16	13AU	-0.4	-3.4	-1.7	0.7
125.41	1.19	2.0	147.92	23AU	-1.1	-3.4	-1.8	0.7
131.73	1.23	2.0	157.73	2SE	-1.2	-3.4	-1.9	0.7
138.04	1.26	2.0	167.61	12SE	-0.8	-3.4	-1.9	0.7
144.33	1.29	2.0	177.54	22SE	-0.4	-3.3	-2.0	0.7
150.61	1.32	2.0	187.53	2OC	-0.2	-3.3	-2.1	0.7
156.87	1.34	1.9	197.58	12OC	-0.2	-3.3	-2.1	0.7
163.12	1.36	1.9	207.67	22OC	-0.1	-3.3	-2.2	0.6
169.35	1.37	1.8	217.81	1NO	0.8	-3.3	-2.2	0.6
175.56	1.38	1.8	227.98	11NO	2.8	-3.4	-2.2	0.5
181.74	1.39	1.7	238.17	21NO	0.4	-3.4	-2.3	0.5
187.88	1.39	1.6	248.38	1DE	-0.2	-3.4	-2.3	0.4
193.98	1.38	1.5	258.58	11DE	-0.2	-3.4	-2.2	0.3
200.02	1.36	1.4	268.78	21DE	-0.3	-3.5	-2.2	0.3
206.00	1.33	1.3	278.96	31DE	-0.5	-3.5	-2.2	0.2

♂			⊙	16.00UT	☿	♀	♃	♄
LONG	LAT	MAG	LONG	92	MAGNITUDES			
211.89	1.28	1.1	289.11	10JA	-0.9	-3.6	-2.1	0.1
217.68	1.22	1.0	299.22	20JA	-1.2	-3.6	-2.1	0.2
223.35	1.15	0.8	309.29	30JA	-1.1	-3.7	-2.0	0.2
228.85	1.04	0.7	319.30	9FE	-0.2	-3.8	-1.9	0.3
234.16	0.92	0.5	329.25	19FE	1.7	-3.9	-1.8	0.3
239.23	0.75	0.3	339.15	29FE	3.1	-4.0	-1.8	0.4
243.99	0.55	0.0	348.99	10MR	1.6	-4.0	-1.7	0.4
248.36	0.29	-0.2	358.77	20MR	0.9	-4.1	-1.6	0.5
252.24	-0.02	-0.5	8.50	30MR	0.5	-4.2	-1.6	0.5
255.48	-0.41	-0.8	18.17	9AP	0.0	-4.2	-1.5	0.6
257.93	-0.89	-1.1	27.80	19AP	-0.7	-4.1	-1.5	0.6
259.39	-1.46	-1.4	37.39	29AP	-1.6	-3.8	-1.5	0.7
259.71	-2.13	-1.7	46.95	9MY	-1.4	-3.0	-1.4	0.7
258.79	-2.86	-2.0	56.48	19MY	-0.4	-3.2	-1.4	0.7
256.74	-3.60	-2.3	66.01	29MY	0.3	-3.8	-1.4	0.7
253.99	-4.25	-2.5	75.53	8JN	0.8	-4.1	-1.4	0.7
251.21	-4.72	-2.4	85.05	18JN	1.4	-4.2	-1.4	0.7
249.09	-4.97	-2.2	94.60	28JN	2.5	-4.2	-1.4	0.7
248.18	-5.02	-2.0	104.16	8JL	2.9	-4.1	-1.4	0.7
248.62	-4.91	-1.7	113.75	18JL	1.0	-4.0	-1.4	0.7
250.37	-4.70	-1.5	123.39	28JL	-0.5	-3.9	-1.4	0.7
253.28	-4.42	-1.2	133.06	7AU	-1.3	-3.8	-1.5	0.8
257.12	-4.11	-1.0	142.79	17AU	-1.3	-3.8	-1.5	0.8
261.73	-3.79	-0.8	152.58	27AU	-0.7	-3.7	-1.5	0.9
266.93	-3.46	-0.6	162.42	6SE	-0.3	-3.6	-1.6	0.9
272.60	-3.14	-0.4	172.33	16SE	-0.1	-3.6	-1.6	0.9
278.64	-2.82	-0.2	182.29	26SE	-0.0	-3.5	-1.7	0.9
284.96	-2.50	-0.0	192.31	6OC	0.1	-3.5	-1.8	0.9
291.49	-2.21	0.1	202.38	16OC	1.1	-3.5	-1.8	0.9
298.18	-1.92	0.3	212.49	26OC	2.6	-3.4	-1.9	0.9
304.98	-1.65	0.4	222.64	5NO	0.2	-3.4	-2.0	0.9
311.86	-1.40	0.6	232.82	15NO	-0.3	-3.4	-2.0	0.8
318.80	-1.16	0.7	243.03	25NO	-0.4	-3.4	-2.1	0.8
325.76	-0.93	0.9	253.23	5DE	-0.4	-3.4	-2.1	0.7
332.72	-0.73	1.0	263.44	15DE	-0.6	-3.3	-2.1	0.7
339.69	-0.53	1.1	273.63	25DE	-0.9	-3.3	-2.1	0.6
				93				
346.63	-0.35	1.2	283.79	4JA	-1.1	-3.3	-2.1	0.5
353.55	-0.19	1.3	293.92	14JA	-1.0	-3.4	-2.1	0.4
0.43	-0.04	1.4	304.02	24JA	-0.1	-3.4	-2.1	0.4
7.27	0.10	1.5	314.05	3FE	2.0	-3.4	-2.0	0.4
14.07	0.22	1.6	324.04	13FE	2.4	-3.4	-2.0	0.4
20.83	0.34	1.6	333.97	23FE	1.2	-3.5	-1.9	0.5
27.54	0.44	1.7	343.84	5MR	0.7	-3.5	-1.9	0.5
34.20	0.54	1.8	353.65	15MR	0.4	-3.5	-1.8	0.6
40.82	0.63	1.8	3.40	25MR	-0.1	-3.4	-1.7	0.7
47.39	0.71	1.9	13.10	4AP	-0.7	-3.4	-1.7	0.7
53.93	0.78	1.9	22.75	14AP	-1.6	-3.4	-1.6	0.8
60.43	0.84	1.9	32.36	24AP	-1.4	-3.4	-1.5	0.8
66.90	0.90	2.0	41.94	4MY	-0.4	-3.3	-1.5	0.9
73.34	0.95	2.0	51.49	14MY	0.4	-3.3	-1.4	0.9
79.76	0.99	2.0	61.02	24MY	1.1	-3.3	-1.4	0.9
86.15	1.03	2.0	70.53	3JN	1.9	-3.3	-1.4	1.0
92.53	1.06	2.0	80.06	13JN	3.2	-3.3	-1.3	1.0
98.90	1.09	2.0	89.59	23JN	2.4	-3.4	-1.3	1.0
105.26	1.12	2.0	99.14	3JL	0.8	-3.4	-1.3	1.0
111.63	1.14	2.0	108.72	13JL	-0.5	-3.4	-1.3	1.0
117.99	1.15	2.0	118.33	23JL	-1.4	-3.4	-1.3	1.0
124.36	1.16	2.0	127.98	2AU	-1.3	-3.5	-1.3	1.0
130.75	1.17	2.0	137.69	12AU	-0.7	-3.5	-1.3	1.0
137.14	1.17	2.0	147.44	22AU	-0.2	-3.6	-1.3	1.1
143.56	1.16	2.0	157.26	1SE	-0.0	-3.7	-1.3	1.1
149.99	1.15	2.0	167.13	11SE	0.1	-3.8	-1.4	1.2
156.45	1.14	2.0	177.06	21SE	0.4	-3.9	-1.4	1.2
162.93	1.12	2.0	187.05	1OC	1.4	-4.0	-1.5	1.2
169.44	1.10	2.0	197.09	11OC	2.4	-4.1	-1.5	1.2
175.98	1.07	1.9	207.18	21OC	0.0	-4.2	-1.6	1.2
182.54	1.03	1.9	217.32	31OC	-0.5	-4.3	-1.6	1.2
189.13	0.99	1.8	227.49	10NO	-0.5	-4.4	-1.7	1.2
195.75	0.93	1.8	237.68	20NO	-0.5	-4.4	-1.8	1.2
202.39	0.87	1.7	247.88	30NO	-0.7	-4.2	-1.8	1.1
209.06	0.80	1.7	258.09	10DE	-0.8	-3.7	-1.9	1.1
215.75	0.72	1.6	268.29	20DE	-0.9	-3.1	-1.9	1.0
222.46	0.63	1.5	278.46	30DE	-0.8	-3.8	-2.0	0.9

♂			⊙	16.00UT	☿	♀	♃	♄
LONG	LAT	MAG	LONG	94	MAGNITUDES			
229.20	0.52	1.4	288.62	9JA	0.0	-4.2	-2.0	0.9
235.94	0.40	1.3	298.73	19JA	2.3	-4.3	-2.0	0.8
242.70	0.26	1.2	308.79	29JA	1.9	-4.3	-2.1	0.7
249.47	0.11	1.1	318.81	8FE	0.8	-4.2	-2.0	0.7
256.23	-0.07	1.0	328.77	18FE	0.5	-4.1	-2.0	0.7
262.99	-0.26	0.8	338.67	28FE	0.2	-4.0	-2.0	0.7
269.72	-0.48	0.7	348.51	10MR	-0.1	-3.9	-1.9	0.8
276.43	-0.72	0.6	358.30	20MR	-0.7	-3.8	-1.9	0.8
283.08	-0.98	0.4	8.02	30MR	-1.6	-3.7	-1.8	0.9
289.65	-1.27	0.3	17.70	9AP	-1.4	-3.6	-1.8	0.9
296.11	-1.59	0.1	27.33	19AP	-0.4	-3.6	-1.7	1.0
302.43	-1.94	-0.1	36.93	29AP	0.6	-3.5	-1.6	1.0
308.55	-2.31	-0.2	46.49	9MY	1.4	-3.4	-1.6	1.1
314.40	-2.72	-0.4	56.03	19MY	2.5	-3.4	-1.5	1.1
319.92	-3.15	-0.6	65.55	29MY	3.4	-3.4	-1.4	1.2
324.97	-3.61	-0.8	75.07	8JN	1.9	-3.3	-1.4	1.2
329.43	-4.09	-1.1	84.60	18JN	0.6	-3.3	-1.4	1.2
333.14	-4.58	-1.3	94.13	28JN	-0.6	-3.3	-1.3	1.3
335.88	-5.08	-1.5	103.70	8JL	-1.4	-3.3	-1.3	1.3
337.45	-5.54	-1.8	113.29	18JL	-1.4	-3.3	-1.3	1.3
337.68	-5.93	-2.0	122.92	28JL	-0.6	-3.3	-1.2	1.3
336.51	-6.15	-2.3	132.60	7AU	-0.2	-3.4	-1.2	1.3
334.19	-6.13	-2.4	142.32	17AU	0.1	-3.4	-1.2	1.3
331.27	-5.83	-2.5	152.11	27AU	0.3	-3.4	-1.2	1.3
328.53	-5.27	-2.3	161.95	6SE	0.6	-3.4	-1.2	1.4
326.69	-4.55	-2.0	171.85	16SE	1.7	-3.5	-1.2	1.4
326.08	-3.78	-1.7	181.81	26SE	2.1	-3.5	-1.3	1.3
326.78	-3.04	-1.4	191.82	6OC	-0.1	-3.5	-1.3	1.3
328.68	-2.37	-1.0	201.89	16OC	-0.6	-3.5	-1.3	1.3
331.55	-1.80	-0.7	212.00	26OC	-0.7	-3.4	-1.4	1.3
335.21	-1.31	-0.5	222.15	5NO	-0.7	-3.4	-1.4	1.3
339.48	-0.89	-0.2	232.33	15NO	-0.8	-3.4	-1.5	1.2
344.21	-0.54	0.0	242.53	25NO	-0.7	-3.4	-1.5	1.2
349.32	-0.24	0.3	252.74	5DE	-0.7	-3.3	-1.6	1.1
354.70	0.01	0.5	262.94	15DE	-0.7	-3.3	-1.7	1.1
0.28	0.22	0.7	273.13	25DE	0.1	-3.3	-1.7	1.1
				95				
6.04	0.39	0.8	283.30	4JA	2.7	-3.3	-1.8	1.0
11.91	0.55	1.0	293.43	14JA	1.4	-3.4	-1.9	1.0
17.88	0.67	1.1	303.52	24JA	0.6	-3.4	-1.9	0.9
23.91	0.78	1.3	313.57	3FE	0.3	-3.4	-2.0	0.9
29.99	0.87	1.4	323.55	13FE	0.1	-3.4	-2.0	0.8
36.11	0.95	1.5	333.49	23FE	-0.2	-3.4	-2.0	0.8
42.26	1.02	1.6	343.36	5MR	-0.8	-3.5	-2.0	0.9
48.42	1.07	1.7	353.17	15MR	-1.6	-3.5	-2.0	1.0
54.60	1.12	1.7	2.93	25MR	-1.4	-3.6	-2.0	1.0
60.78	1.15	1.8	12.64	4AP	-0.4	-3.6	-1.9	1.1
66.98	1.18	1.9	22.29	14AP	0.7	-3.7	-1.9	1.2
73.18	1.20	1.9	31.90	24AP	1.9	-3.8	-1.8	1.2
79.38	1.21	2.0	41.48	4MY	3.4	-3.9	-1.8	1.3
85.60	1.22	2.0	51.03	14MY	2.7	-3.9	-1.7	1.3
91.82	1.23	2.0	60.56	24MY	1.5	-4.0	-1.6	1.4
98.06	1.23	2.0	70.08	3JN	0.5	-4.1	-1.6	1.4
104.32	1.22	2.0	79.60	13JN	-0.6	-4.2	-1.5	1.4
110.60	1.21	2.0	89.13	23JN	-1.5	-4.2	-1.5	1.4
116.90	1.19	2.0	98.68	3JL	-1.4	-4.0	-1.4	1.4
123.22	1.17	2.0	108.25	13JL	-0.6	-3.6	-1.4	1.4
129.58	1.15	2.0	117.87	23JL	-0.1	-3.1	-1.3	1.3
135.98	1.12	2.0	127.52	2AU	0.2	-3.6	-1.3	1.3
142.41	1.09	1.9	137.21	12AU	0.5	-4.1	-1.3	1.3
148.88	1.05	1.9	146.97	22AU	0.9	-4.3	-1.2	1.2
155.41	1.01	1.9	156.78	1SE	2.1	-4.3	-1.2	1.2
161.97	0.96	1.9	166.65	11SE	1.9	-4.2	-1.2	1.2
168.60	0.91	1.9	176.58	21SE	-0.2	-4.1	-1.2	1.2
175.27	0.85	1.9	186.56	1OC	-0.8	-4.1	-1.2	1.2
182.00	0.78	1.9	196.60	11OC	-0.8	-4.0	-1.2	1.2
188.79	0.72	1.9	206.70	21OC	-0.8	-3.9	-1.3	1.2
195.64	0.64	1.8	216.83	31OC	-0.7	-3.8	-1.3	1.1
202.55	0.56	1.8	226.99	10NO	-0.6	-3.7	-1.3	1.1
209.52	0.47	1.8	237.18	20NO	-0.6	-3.7	-1.4	1.1
216.54	0.37	1.7	247.39	30NO	-0.5	-3.6	-1.4	1.1
223.63	0.27	1.7	257.59	10DE	0.3	-3.5	-1.5	1.0
230.77	0.16	1.6	267.80	20DE	2.9	-3.5	-1.5	1.0
237.97	0.04	1.6	277.98	30DE	1.1	-3.4	-1.6	0.9

Left table (1996–1997)

♂ LONG	LAT	MAG	☉ LONG	16.00UT	☿	♀	♃	♄
				96				
245.22	-0.09	1.5	288.13	9JA	0.3	-3.4	-1.7	0.9
252.53	-0.22	1.5	298.25	19JA	0.1	-3.4	-1.7	0.8
259.87	-0.36	1.4	308.31	29JA	-0.0	-3.3	-1.8	0.8
267.26	-0.51	1.3	318.33	8FE	-0.3	-3.3	-1.9	0.7
274.68	-0.66	1.3	328.29	18FE	-0.8	-3.3	-1.9	0.7
282.13	-0.82	1.2	338.20	28FE	-1.5	-3.3	-2.0	0.7
289.59	-0.98	1.1	348.04	9MR	-1.3	-3.3	-2.0	0.7
297.07	-1.14	1.0	357.83	19MR	-0.3	-3.3	-2.0	0.7
304.53	-1.31	1.0	7.56	29MR	1.0	-3.3	-2.0	0.8
311.97	-1.47	0.9	17.24	8AP	2.4	-3.4	-2.0	0.9
319.37	-1.62	0.8	26.87	18AP	3.6	-3.4	-2.0	0.9
326.71	-1.77	0.7	36.46	28AP	2.0	-3.4	-2.0	1.0
333.98	-1.91	0.6	46.03	8MY	1.1	-3.5	-2.0	1.1
341.14	-2.04	0.6	55.57	18MY	0.3	-3.5	-1.9	1.1
348.16	-2.16	0.5	65.09	28MY	-0.6	-3.5	-1.8	1.1
355.02	-2.26	0.4	74.61	7JN	-1.5	-3.4	-1.8	1.2
1.67	-2.35	0.3	84.13	17JN	-1.4	-3.4	-1.7	1.2
8.07	-2.41	0.2	93.67	27JN	-0.5	-3.4	-1.7	1.2
14.18	-2.46	0.1	103.23	7JL	0.0	-3.3	-1.6	1.2
19.91	-2.49	-0.1	112.82	17JL	0.4	-3.3	-1.5	1.2
25.20	-2.49	-0.2	122.45	27JL	0.7	-3.3	-1.5	1.2
29.94	-2.47	-0.3	132.12	6AU	1.2	-3.3	-1.4	1.2
33.99	-2.41	-0.5	141.85	16AU	2.6	-3.3	-1.4	1.1
37.19	-2.32	-0.7	151.63	26AU	1.7	-3.3	-1.3	1.1
39.35	-2.17	-0.9	161.47	5SE	-0.3	-3.4	-1.3	1.0
40.25	-1.96	-1.1	171.36	15SE	-0.9	-3.4	-1.3	1.0
39.74	-1.67	-1.4	181.32	25SE	-1.0	-3.4	-1.3	1.0
37.77	-1.29	-1.6	191.33	5OC	-0.9	-3.5	-1.3	1.0
34.60	-0.84	-1.7	201.40	15OC	-0.6	-3.5	-1.3	1.0
30.86	-0.34	-1.8	211.51	25OC	-0.4	-3.6	-1.3	1.0
27.36	0.15	-1.5	221.66	4NO	-0.4	-3.7	-1.3	1.0
24.80	0.58	-1.2	231.84	14NO	-0.3	-3.7	-1.3	1.0
23.55	0.91	-0.9	242.03	24NO	0.4	-3.8	-1.3	1.0
23.63	1.17	-0.6	252.24	4DE	3.1	-3.9	-1.3	1.0
24.92	1.35	-0.3	262.45	14DE	0.8	-4.0	-1.4	0.9
27.21	1.48	-0.0	272.64	24DE	0.1	-4.1	-1.4	0.9
				97				
30.29	1.57	0.2	282.81	3JA	-0.0	-4.2	-1.5	0.8
33.99	1.62	0.5	292.95	13JA	-0.1	-4.3	-1.5	0.8
38.18	1.66	0.7	303.04	23JA	-0.4	-4.3	-1.6	0.7
42.74	1.67	0.9	313.09	2FE	-0.8	-4.3	-1.7	0.7
47.61	1.68	1.0	323.08	12FE	-1.4	-4.1	-1.7	0.6
52.71	1.67	1.2	333.01	22FE	-1.2	-3.6	-1.8	0.6
57.99	1.66	1.3	342.89	4MR	-0.3	-3.2	-1.9	0.6
63.43	1.64	1.4	352.70	14MR	1.2	-3.8	-1.9	0.5
68.99	1.61	1.5	2.46	24MR	3.1	-4.1	-2.0	0.5
74.65	1.58	1.6	12.17	3AP	2.7	-4.2	-2.0	0.6
80.40	1.55	1.7	21.82	13AP	1.5	-4.2	-2.1	0.7
86.22	1.51	1.8	31.44	23AP	0.8	-4.1	-2.1	0.7
92.12	1.47	1.8	41.02	3MY	0.2	-4.0	-2.1	0.8
98.07	1.43	1.9	50.57	13MY	-0.6	-3.9	-2.1	0.9
104.09	1.38	1.9	60.10	23MY	-1.6	-3.8	-2.1	0.9
110.16	1.33	2.0	69.62	2JN	-1.4	-3.7	-2.1	1.0
116.29	1.28	2.0	79.14	12JN	-0.5	-3.7	-2.0	1.0
122.47	1.23	2.0	88.67	22JN	0.1	-3.6	-2.0	1.0
128.71	1.17	2.0	98.22	2JL	0.5	-3.5	-1.9	1.1
135.01	1.11	2.0	107.79	12JL	0.9	-3.5	-1.9	1.1
141.38	1.04	2.0	117.40	22JL	1.6	-3.4	-1.8	1.1
147.81	0.97	2.0	127.05	1AU	3.1	-3.4	-1.7	1.1
154.30	0.90	1.9	136.75	11AU	1.4	-3.4	-1.7	1.0
160.87	0.83	1.9	146.50	21AU	-0.4	-3.4	-1.6	1.0
167.51	0.75	1.9	156.31	31AU	-1.1	-3.4	-1.5	1.0
174.22	0.67	1.8	166.17	10SE	-1.1	-3.4	-1.5	1.0
181.01	0.59	1.8	176.09	20SE	-0.8	-3.4	-1.5	0.9
187.88	0.50	1.7	186.08	30SE	-0.5	-3.4	-1.4	0.9
194.83	0.41	1.7	196.12	10OC	-0.3	-3.4	-1.4	0.9
201.86	0.31	1.7	206.20	20OC	-0.3	-3.4	-1.4	0.9
208.96	0.21	1.7	216.34	30OC	-0.2	-3.4	-1.4	0.9
216.15	0.11	1.7	226.50	9NO	0.6	-3.4	-1.3	0.9
223.41	0.00	1.7	236.69	19NO	3.1	-3.4	-1.3	0.9
230.75	-0.11	1.6	246.89	29NO	0.5	-3.5	-1.3	0.9
238.16	-0.22	1.6	257.10	9DE	-0.0	-3.5	-1.4	0.9
245.64	-0.33	1.6	267.30	19DE	-0.1	-3.5	-1.4	0.9
253.18	-0.44	1.6	277.48	29DE	-0.2	-3.5	-1.4	0.8

Right table (1998–1999)

♂ LONG	LAT	MAG	☉ LONG	16.00UT	☿	♀	♃	♄
				98				
260.77	-0.56	1.5	287.63	8JA	-0.5	-3.4	-1.4	0.8
268.42	-0.67	1.5	297.75	18JA	-0.9	-3.4	-1.5	0.7
276.10	-0.78	1.5	307.83	28JA	-1.3	-3.4	-1.5	0.7
283.81	-0.89	1.4	317.84	7FE	-1.2	-3.4	-1.6	0.6
291.55	-0.99	1.4	327.81	17FE	-0.2	-3.3	-1.7	0.6
299.29	-1.09	1.4	337.72	27FE	1.5	-3.3	-1.7	0.5
307.03	-1.17	1.3	347.56	9MR	3.4	-3.3	-1.8	0.5
314.75	-1.25	1.3	357.35	19MR	2.0	-3.3	-1.9	0.4
322.45	-1.32	1.3	7.09	29MR	1.1	-3.3	-1.9	0.4
330.10	-1.37	1.2	16.77	8AP	0.6	-3.4	-2.0	0.5
337.71	-1.42	1.2	26.40	18AP	0.1	-3.4	-2.1	0.5
345.24	-1.44	1.2	36.00	28AP	-0.6	-3.4	-2.1	0.6
352.68	-1.45	1.1	45.56	8MY	-1.6	-3.4	-2.2	0.6
0.03	-1.45	1.1	55.10	18MY	-1.4	-3.5	-2.2	0.7
7.27	-1.43	1.1	64.63	28MY	-0.5	-3.5	-2.2	0.8
14.39	-1.40	1.0	74.15	7JN	0.2	-3.6	-2.2	0.8
21.37	-1.35	1.0	83.67	17JN	0.7	-3.6	-2.2	0.9
28.18	-1.29	1.0	93.21	27JN	1.2	-3.7	-2.2	0.9
34.82	-1.20	0.9	102.77	7JL	2.1	-3.8	-2.2	0.9
41.27	-1.11	0.9	112.36	17JL	3.2	-3.9	-2.1	1.0
47.48	-0.99	0.8	121.99	27JL	1.2	-4.0	-2.1	1.0
53.44	-0.86	0.7	131.66	6AU	-0.4	-4.1	-2.0	1.0
59.09	-0.70	0.6	141.38	16AU	-1.2	-4.2	-2.0	1.0
64.39	-0.53	0.5	151.16	26AU	-1.2	-4.3	-1.9	0.9
69.27	-0.33	0.4	160.99	5SE	-0.8	-4.3	-1.8	0.9
73.63	-0.10	0.3	170.88	15SE	-0.4	-4.1	-1.8	0.9
77.36	0.17	0.1	180.84	25SE	-0.2	-3.7	-1.7	0.8
80.32	0.48	-0.1	190.84	5OC	-0.1	-3.2	-1.6	0.8
82.33	0.83	-0.3	200.91	15OC	0.0	-3.8	-1.6	0.8
83.17	1.23	-0.5	211.02	25OC	0.9	-4.2	-1.6	0.8
82.70	1.67	-0.7	221.16	4NO	2.9	-4.4	-1.5	0.8
80.81	2.13	-0.9	231.34	14NO	0.3	-4.4	-1.5	0.8
77.69	2.57	-1.1	241.54	24NO	-0.2	-4.3	-1.5	0.9
73.82	2.93	-1.3	251.74	4DE	-0.3	-4.2	-1.5	0.8
69.95	3.16	-1.1	261.95	14DE	-0.3	-4.1	-1.4	0.8
66.81	3.25	-0.8	272.15	24DE	-0.5	-4.0	-1.4	0.8
				99				
64.86	3.23	-0.6	282.31	3JA	-0.9	-3.9	-1.5	0.8
64.23	3.14	-0.3	292.46	13JA	-1.1	-3.8	-1.5	0.8
64.86	3.00	-0.0	302.55	23JA	-1.0	-3.7	-1.5	0.7
66.55	2.85	0.2	312.60	2FE	-0.2	-3.6	-1.5	0.7
69.09	2.70	0.5	322.60	12FE	1.8	-3.6	-1.6	0.6
72.34	2.55	0.7	332.53	22FE	2.9	-3.5	-1.6	0.6
76.13	2.40	0.9	342.41	4MR	1.4	-3.4	-1.7	0.5
80.35	2.26	1.0	352.23	14MR	0.8	-3.4	-1.7	0.5
84.92	2.13	1.2	2.00	24MR	0.5	-3.4	-1.8	0.4
89.78	2.01	1.3	11.70	3AP	0.0	-3.3	-1.8	0.4
94.87	1.88	1.4	21.36	13AP	-0.7	-3.3	-1.9	0.3
100.16	1.77	1.5	30.98	23AP	-1.6	-3.3	-2.0	0.4
105.62	1.66	1.6	40.56	3MY	-1.4	-3.3	-2.0	0.5
111.22	1.55	1.6	50.11	13MY	-0.5	-3.3	-2.1	0.5
116.96	1.44	1.7	59.64	23MY	0.3	-3.3	-2.2	0.6
122.81	1.34	1.7	69.16	2JN	0.9	-3.3	-2.2	0.6
128.78	1.23	1.8	78.69	12JN	1.6	-3.4	-2.3	0.7
134.85	1.13	1.8	88.21	22JN	2.7	-3.4	-2.3	0.8
141.02	1.02	1.8	97.76	2JL	2.7	-3.4	-2.4	0.8
147.29	0.92	1.8	107.33	12JL	1.0	-3.5	-2.4	0.8
153.66	0.82	1.8	116.93	22JL	-0.5	-3.5	-2.4	0.9
160.13	0.71	1.8	126.58	1AU	-1.3	-3.5	-2.3	0.9
166.70	0.61	1.8	136.28	11AU	-1.3	-3.4	-2.3	0.9
173.36	0.50	1.8	146.02	21AU	-0.7	-3.4	-2.3	0.9
180.13	0.40	1.8	155.82	31AU	-0.3	-3.4	-2.2	0.9
186.99	0.29	1.8	165.69	10SE	-0.1	-3.4	-2.1	0.9
193.95	0.18	1.7	175.61	20SE	0.0	-3.3	-2.1	0.8
201.01	0.08	1.7	185.59	30SE	0.2	-3.3	-2.0	0.8
208.17	-0.03	1.6	195.62	10OC	1.1	-3.3	-1.9	0.8
215.42	-0.14	1.6	205.71	20OC	2.6	-3.3	-1.9	0.7
222.76	-0.24	1.6	215.83	30OC	0.1	-3.3	-1.8	0.7
230.20	-0.35	1.5	226.00	9NO	-0.4	-3.4	-1.7	0.8
237.71	-0.45	1.5	236.18	19NO	-0.4	-3.4	-1.7	0.8
245.30	-0.55	1.4	246.39	29NO	-0.5	-3.4	-1.6	0.8
252.96	-0.64	1.4	256.60	9DE	-0.6	-3.4	-1.6	0.8
260.67	-0.73	1.4	266.80	19DE	-0.9	-3.5	-1.6	0.8
268.44	-0.82	1.4	276.98	29DE	-1.0	-3.5	-1.6	0.8

Left table:

♂ LONG	LAT	MAG	☉ LONG	16.00UT	☿	♀	♃	♄
				100		MAGNITUDES		
276.25	-0.89	1.4	287.14	8JA	-0.9	-3.6	-1.5	0.8
284.09	-0.96	1.4	297.26	18JA	-0.1	-3.7	-1.5	0.8
291.94	-1.02	1.4	307.33	28JA	2.1	-3.7	-1.5	0.7
299.80	-1.07	1.4	317.36	7FE	2.3	-3.8	-1.5	0.7
307.65	-1.11	1.4	327.33	17FE	1.0	-3.9	-1.6	0.6
315.48	-1.14	1.4	337.24	27FE	0.6	-4.0	-1.6	0.6
323.28	-1.15	1.4	347.09	8MR	0.3	-4.0	-1.6	0.5
331.04	-1.15	1.4	356.88	18MR	-0.1	-4.1	-1.6	0.5
338.73	-1.14	1.4	6.62	28MR	-0.7	-4.2	-1.7	0.4
346.37	-1.12	1.4	16.30	7AP	-1.6	-4.2	-1.7	0.3
353.92	-1.09	1.4	25.94	17AP	-1.4	-4.1	-1.8	0.3
1.40	-1.04	1.4	35.54	27AP	-0.4	-3.8	-1.8	0.3
8.78	-0.98	1.4	45.10	7MY	0.5	-3.0	-1.9	0.4
16.07	-0.91	1.4	54.64	17MY	1.2	-3.2	-2.0	0.4
23.25	-0.83	1.4	64.17	27MY	2.1	-3.8	-2.0	0.5
30.32	-0.74	1.4	73.69	6JN	3.4	-4.1	-2.1	0.5
37.27	-0.64	1.4	83.21	16JN	2.2	-4.2	-2.2	0.6
44.11	-0.53	1.4	92.75	26JN	0.8	-4.2	-2.3	0.7
50.81	-0.42	1.4	102.31	6JL	-0.5	-4.1	-2.3	0.7
57.38	-0.29	1.4	111.89	16JL	-1.4	-4.0	-2.4	0.8
63.80	-0.16	1.4	121.52	26JL	-1.4	-3.9	-2.4	0.8
70.07	-0.01	1.3	131.19	5AU	-0.7	-3.8	-2.4	0.8
76.16	0.14	1.3	140.90	15AU	-0.2	-3.7	-2.4	0.8
82.05	0.31	1.2	150.68	25AU	0.0	-3.7	-2.4	0.8
87.72	0.49	1.1	160.51	4SE	0.2	-3.6	-2.4	0.8
93.14	0.69	1.1	170.40	14SE	0.4	-3.6	-2.4	0.8
98.27	0.90	1.0	180.35	24SE	1.5	-3.5	-2.3	0.8
103.03	1.14	0.8	190.36	4OC	2.4	-3.5	-2.3	0.8
107.36	1.40	0.7	200.42	14OC	-0.0	-3.5	-2.2	0.7
111.17	1.69	0.5	210.53	24OC	-0.6	-3.4	-2.1	0.7
114.32	2.01	0.4	220.67	3NO	-0.6	-3.4	-2.1	0.7
116.67	2.37	0.2	230.84	13NO	-0.6	-3.4	-2.0	0.7
118.03	2.77	-0.1	241.04	23NO	-0.7	-3.4	-1.9	0.8
118.22	3.20	-0.3	251.25	3DE	-0.8	-3.4	-1.9	0.8
117.10	3.63	-0.5	261.45	13DE	-0.8	-3.3	-1.8	0.8
114.65	4.02	-0.8	271.65	23DE	-0.8	-3.3	-1.7	0.8
				101				
111.17	4.30	-1.0	281.82	2JA	-0.0	-3.4	-1.7	0.8
107.21	4.43	-1.0	291.96	12JA	2.4	-3.4	-1.7	0.8
103.50	4.39	-0.8	302.06	22JA	1.7	-3.4	-1.6	0.8
100.70	4.20	-0.6	312.11	1FE	0.7	-3.4	-1.6	0.7
99.13	3.92	-0.3	322.10	11FE	0.4	-3.4	-1.6	0.7
98.85	3.60	-0.1	332.05	21FE	0.2	-3.5	-1.6	0.7
99.76	3.28	0.1	341.93	3MR	-0.2	-3.5	-1.6	0.6
101.67	2.97	0.4	351.75	13MR	-0.7	-3.5	-1.6	0.5
104.40	2.68	0.6	1.52	23MR	-1.6	-3.4	-1.6	0.5
107.79	2.42	0.7	11.23	2AP	-1.4	-3.4	-1.6	0.4
111.72	2.17	0.9	20.89	12AP	-0.4	-3.4	-1.6	0.4
116.08	1.95	1.0	30.51	22AP	0.6	-3.4	-1.7	0.3
120.81	1.74	1.1	40.09	2MY	1.6	-3.3	-1.7	0.3
125.83	1.55	1.2	49.64	12MY	2.8	-3.3	-1.7	0.3
131.12	1.37	1.3	59.18	22MY	3.2	-3.3	-1.8	0.3
136.62	1.20	1.4	68.70	1JN	1.7	-3.3	-1.8	0.4
142.33	1.03	1.4	78.22	11JN	0.6	-3.3	-1.9	0.5
148.22	0.88	1.5	87.75	21JN	-0.5	-3.4	-2.0	0.5
154.27	0.73	1.5	97.29	1JL	-1.4	-3.4	-2.0	0.6
160.47	0.58	1.6	106.86	11JL	-1.4	-3.4	-2.1	0.6
166.83	0.44	1.6	116.47	21JL	-0.6	-3.4	-2.2	0.7
173.32	0.31	1.6	126.11	31JL	-0.1	-3.5	-2.2	0.7
179.95	0.18	1.6	135.80	10AU	0.2	-3.5	-2.3	0.8
186.71	0.05	1.6	145.55	20AU	0.4	-3.6	-2.4	0.8
193.59	-0.07	1.6	155.35	30AU	0.7	-3.7	-2.4	0.8
200.60	-0.19	1.6	165.21	9SE	1.8	-3.8	-2.4	0.8
207.72	-0.31	1.6	175.13	19SE	2.1	-3.9	-2.4	0.8
214.96	-0.41	1.6	185.10	29SE	-0.1	-4.0	-2.4	0.8
222.31	-0.52	1.5	195.14	9OC	-0.7	-4.1	-2.4	0.8
229.76	-0.61	1.5	205.22	19OC	-0.7	-4.2	-2.4	0.7
237.30	-0.70	1.5	215.34	29OC	-0.7	-4.3	-2.3	0.7
244.92	-0.79	1.5	225.51	8NO	-0.7	-4.4	-2.3	0.7
252.63	-0.86	1.4	235.69	18NO	-0.7	-4.4	-2.2	0.7
260.39	-0.93	1.4	245.89	28NO	-0.7	-4.2	-2.1	0.7
268.21	-0.98	1.4	256.10	8DE	-0.6	-3.7	-2.1	0.7
276.08	-1.02	1.4	266.30	18DE	0.1	-3.1	-2.0	0.8
283.96	-1.06	1.4	276.49	28DE	2.7	-3.8	-1.9	0.8

Right table:

♂ LONG	LAT	MAG	☉ LONG	16.00UT	☿	♀	♃	♄
				102		MAGNITUDES		
291.86	-1.08	1.3	286.64	7JA	1.3	-4.2	-1.9	0.8
299.76	-1.09	1.3	296.77	17JA	0.5	-4.3	-1.8	0.8
307.65	-1.09	1.3	306.84	27JA	0.2	-4.3	-1.7	0.8
315.51	-1.07	1.3	316.87	6FE	0.1	-4.2	-1.7	0.8
323.33	-1.05	1.3	326.84	16FE	-0.2	-4.1	-1.6	0.7
331.10	-1.01	1.4	336.75	26FE	-0.8	-4.0	-1.6	0.7
338.81	-0.96	1.4	346.61	8MR	-1.5	-3.9	-1.6	0.6
346.46	-0.91	1.4	356.40	18MR	-1.4	-3.8	-1.6	0.6
354.03	-0.84	1.5	6.14	28MR	-0.4	-3.7	-1.5	0.5
1.52	-0.77	1.5	15.83	7AP	0.8	-3.6	-1.5	0.5
8.92	-0.69	1.5	25.47	17AP	2.0	-3.6	-1.5	0.4
16.23	-0.60	1.6	35.07	27AP	3.7	-3.5	-1.5	0.3
23.45	-0.51	1.6	44.64	7MY	2.4	-3.4	-1.6	0.3
30.58	-0.41	1.6	54.18	17MY	1.3	-3.4	-1.6	0.2
37.61	-0.30	1.7	63.71	27MY	0.5	-3.4	-1.6	0.3
44.55	-0.20	1.7	73.23	6JN	-0.6	-3.3	-1.6	0.3
51.39	-0.09	1.7	82.76	16JN	-1.5	-3.3	-1.7	0.4
58.13	0.03	1.7	92.29	26JN	-1.4	-3.3	-1.7	0.5
64.78	0.14	1.7	101.85	6JL	-0.6	-3.3	-1.8	0.5
71.33	0.26	1.7	111.43	16JL	-0.0	-3.3	-1.8	0.6
77.78	0.39	1.7	121.05	26JL	0.3	-3.3	-1.9	0.6
84.13	0.51	1.7	130.72	5AU	0.5	-3.4	-2.0	0.7
90.37	0.65	1.7	140.44	15AU	1.0	-3.4	-2.0	0.7
96.49	0.78	1.6	150.20	25AU	2.3	-3.4	-2.1	0.8
102.48	0.92	1.6	160.04	4SE	1.9	-3.4	-2.2	0.8
108.34	1.07	1.5	169.92	14SE	-0.2	-3.5	-2.2	0.8
114.03	1.22	1.5	179.87	24SE	-0.8	-3.5	-2.3	0.8
119.55	1.39	1.4	189.87	4OC	-0.9	-3.5	-2.3	0.8
124.85	1.56	1.3	199.93	14OC	-0.9	-3.5	-2.4	0.8
129.91	1.75	1.2	210.03	24OC	-0.6	-3.4	-2.4	0.7
134.66	1.96	1.1	220.17	3NO	-0.5	-3.4	-2.4	0.7
139.04	2.18	0.9	230.35	13NO	-0.5	-3.4	-2.4	0.7
142.98	2.43	0.8	240.54	23NO	-0.5	-3.4	-2.3	0.7
146.36	2.70	0.6	250.75	3DE	0.3	-3.3	-2.3	0.7
149.05	2.99	0.3	260.95	13DE	3.0	-3.3	-2.2	0.7
150.89	3.31	0.1	271.15	23DE	1.0	-3.3	-2.1	0.8
				103				
151.68	3.64	-0.1	281.32	2JA	0.3	-3.3	-2.1	0.8
151.28	3.97	-0.4	291.46	12JA	0.1	-3.4	-2.0	0.8
149.57	4.25	-0.7	301.56	22JA	-0.1	-3.4	-1.9	0.8
146.67	4.43	-0.9	311.62	1FE	-0.3	-3.4	-1.9	0.8
142.96	4.47	-1.1	321.62	11FE	-0.8	-3.4	-1.8	0.8
139.09	4.33	-1.0	331.56	21FE	-1.5	-3.4	-1.7	0.8
135.78	4.03	-0.8	341.45	3MR	-1.3	-3.5	-1.7	0.7
133.54	3.63	-0.6	351.27	13MR	-0.3	-3.5	-1.6	0.7
132.58	3.20	-0.4	1.04	23MR	1.0	-3.6	-1.6	0.6
132.91	2.77	-0.2	10.76	2AP	2.6	-3.6	-1.5	0.6
134.36	2.36	0.0	20.42	12AP	3.2	-3.7	-1.5	0.5
136.76	1.99	0.2	30.04	22AP	1.8	-3.8	-1.5	0.4
139.94	1.65	0.4	39.63	2MY	1.0	-3.9	-1.5	0.4
143.77	1.35	0.5	49.18	12MY	0.3	-3.9	-1.5	0.3
148.12	1.08	0.7	58.72	22MY	-0.6	-4.0	-1.5	0.3
152.91	0.83	0.8	68.24	1JN	-1.5	-4.1	-1.5	0.2
158.08	0.60	0.9	77.76	11JN	-1.4	-4.2	-1.5	0.3
163.56	0.39	1.0	87.29	21JN	-0.6	-4.2	-1.5	0.4
169.32	0.20	1.0	96.83	1JL	0.0	-4.0	-1.5	0.4
175.33	0.02	1.1	106.40	11JL	0.4	-3.6	-1.6	0.5
181.56	-0.15	1.1	116.00	21JL	0.7	-3.1	-1.6	0.5
188.00	-0.30	1.2	125.65	31JL	1.3	-3.7	-1.6	0.6
194.62	-0.44	1.2	135.33	10AU	2.8	-4.1	-1.7	0.6
201.42	-0.57	1.2	145.08	20AU	1.6	-4.3	-1.7	0.7
208.37	-0.69	1.3	154.88	30AU	-0.3	-4.3	-1.8	0.7
215.48	-0.80	1.3	164.73	9SE	-1.0	-4.2	-1.9	0.7
222.72	-0.89	1.3	174.65	19SE	-1.0	-4.1	-1.9	0.8
230.09	-0.98	1.3	184.62	29SE	-0.9	-4.0	-2.0	0.8
237.57	-1.05	1.3	194.65	9OC	-0.5	-4.0	-2.1	0.8
245.15	-1.10	1.3	204.73	19OC	-0.4	-3.9	-2.1	0.8
252.82	-1.15	1.3	214.85	29OC	-0.4	-3.8	-2.2	0.8
260.56	-1.18	1.4	225.01	8NO	-0.3	-3.7	-2.2	0.7
268.36	-1.20	1.4	235.20	18NO	0.4	-3.7	-2.2	0.7
276.20	-1.20	1.4	245.40	28NO	3.1	-3.6	-2.3	0.7
284.07	-1.19	1.4	255.61	8DE	0.7	-3.5	-2.2	0.7
291.95	-1.17	1.4	265.81	18DE	0.1	-3.5	-2.2	0.7
299.83	-1.14	1.4	276.00	28DE	-0.1	-3.4	-2.2	0.8

♂ LONG	♂ LAT	♂ MAG	☉ LONG	16.00UT 104	☿	♀	♃	♄
					MAGNITUDES			
307.69	-1.09	1.4	286.15	7JA	-0.2	-3.4	-2.1	0.8
315.52	-1.04	1.4	296.28	17JA	-0.4	-3.4	-2.1	0.8
323.30	-0.97	1.4	306.36	27JA	-0.8	-3.3	-2.0	0.8
331.03	-0.90	1.5	316.39	6FE	-1.4	-3.3	-2.0	0.8
338.70	-0.82	1.5	326.36	16FE	-1.2	-3.3	-1.9	0.8
346.30	-0.74	1.5	336.28	26FE	-0.3	-3.3	-1.8	0.8
353.82	-0.65	1.5	346.13	7MR	1.3	-3.3	-1.7	0.8
1.26	-0.55	1.5	355.93	17MR	3.3	-3.3	-1.7	0.7
8.62	-0.45	1.5	5.67	27MR	2.4	-3.3	-1.6	0.7
15.89	-0.35	1.5	15.36	6AP	1.4	-3.4	-1.6	0.6
23.07	-0.25	1.6	25.01	16AP	0.8	-3.4	-1.5	0.6
30.17	-0.15	1.6	34.61	26AP	0.2	-3.4	-1.5	0.5
37.18	-0.05	1.7	44.17	6MY	-0.6	-3.5	-1.4	0.4
44.12	0.06	1.7	53.72	16MY	-1.5	-3.5	-1.4	0.4
50.97	0.16	1.8	63.25	26MY	-1.4	-3.5	-1.4	0.3
57.74	0.26	1.8	72.76	5JN	-0.5	-3.4	-1.4	0.3
64.44	0.36	1.8	82.29	15JN	0.1	-3.4	-1.4	0.3
71.07	0.46	1.9	91.82	25JN	0.6	-3.4	-1.4	0.3
77.63	0.56	1.9	101.38	5JL	1.0	-3.3	-1.4	0.4
84.12	0.66	1.9	110.96	15JL	1.7	-3.3	-1.4	0.4
90.55	0.76	1.9	120.58	25JL	3.2	-3.3	-1.4	0.5
96.91	0.85	1.9	130.24	4AU	1.4	-3.3	-1.4	0.6
103.21	0.95	1.9	139.96	14AU	-0.4	-3.3	-1.4	0.6
109.44	1.05	1.9	149.73	24AU	-1.1	-3.3	-1.5	0.7
115.60	1.14	1.9	159.55	3SE	-1.2	-3.4	-1.5	0.7
121.69	1.24	1.8	169.44	13SE	-0.8	-3.4	-1.6	0.7
127.69	1.34	1.8	179.38	23SE	-0.4	-3.4	-1.6	0.8
133.61	1.44	1.7	189.38	3OC	-0.3	-3.5	-1.7	0.8
139.43	1.55	1.7	199.44	13OC	-0.2	-3.5	-1.8	0.8
145.14	1.66	1.6	209.54	23OC	-0.1	-3.6	-1.8	0.8
150.71	1.77	1.5	219.68	2NO	0.7	-3.7	-1.9	0.8
156.12	1.88	1.4	229.85	12NO	3.1	-3.7	-2.0	0.8
161.33	2.00	1.3	240.05	22NO	0.5	-3.8	-2.0	0.7
166.32	2.13	1.1	250.25	2DE	-0.1	-3.9	-2.1	0.7
171.03	2.27	1.0	260.46	12DE	-0.2	-4.0	-2.1	0.7
175.38	2.41	0.8	270.66	22DE	-0.3	-4.1	-2.1	0.7
				105				
179.31	2.55	0.6	280.83	1JA	-0.5	-4.2	-2.1	0.8
182.70	2.71	0.4	290.98	11JA	-0.9	-4.3	-2.1	0.8
185.42	2.86	0.1	301.08	21JA	-1.2	-4.3	-2.1	0.8
187.32	3.01	-0.1	311.13	31JA	-1.1	-4.3	-2.1	0.8
188.19	3.14	-0.4	321.14	10FE	-0.3	-4.1	-2.0	0.9
187.89	3.22	-0.7	331.08	20FE	1.5	-3.6	-2.0	0.8
186.32	3.23	-1.0	340.97	2MR	3.3	-3.3	-1.9	0.8
183.56	3.13	-1.3	350.80	12MR	1.8	-3.8	-1.8	0.8
180.02	2.89	-1.5	0.57	22MR	1.0	-4.1	-1.8	0.8
176.34	2.52	-1.4	10.29	1AP	0.6	-4.2	-1.7	0.7
173.23	2.06	-1.2	19.95	11AP	0.1	-4.2	-1.6	0.7
171.23	1.57	-1.0	29.57	21AP	-0.6	-4.1	-1.6	0.6
170.54	1.09	-0.8	39.16	1MY	-1.5	-4.0	-1.5	0.6
171.18	0.66	-0.6	48.72	11MY	-1.5	-3.9	-1.5	0.5
172.99	0.27	-0.4	58.25	21MY	-0.5	-3.8	-1.4	0.5
175.79	-0.06	-0.2	67.77	31MY	0.2	-3.7	-1.4	0.4
179.42	-0.35	-0.1	77.30	10JN	0.8	-3.7	-1.3	0.3
183.73	-0.60	0.1	86.82	20JN	1.3	-3.6	-1.3	0.3
188.59	-0.82	0.2	96.37	30JN	2.3	-3.5	-1.3	0.3
193.93	-1.00	0.3	105.94	10JL	3.1	-3.5	-1.3	0.4
199.67	-1.16	0.4	115.53	20JL	1.2	-3.4	-1.3	0.4
205.74	-1.29	0.5	125.18	30JL	-0.4	-3.4	-1.3	0.5
212.11	-1.39	0.6	134.87	9AU	-1.2	-3.4	-1.3	0.5
218.74	-1.48	0.7	144.60	19AU	-1.3	-3.4	-1.3	0.6
225.58	-1.54	0.7	154.40	29AU	-0.8	-3.4	-1.3	0.6
232.62	-1.59	0.8	164.26	8SE	-0.4	-3.4	-1.3	0.7
239.82	-1.61	0.9	174.17	18SE	-0.1	-3.4	-1.4	0.7
247.17	-1.62	0.9	184.14	28SE	-0.1	-3.4	-1.4	0.8
254.63	-1.61	1.0	194.17	8OC	0.1	-3.4	-1.5	0.8
262.19	-1.58	1.0	204.24	18OC	0.9	-3.4	-1.5	0.8
269.83	-1.54	1.1	214.36	28OC	2.9	-3.4	-1.6	0.8
277.52	-1.48	1.1	224.52	7NO	0.3	-3.4	-1.6	0.8
285.24	-1.41	1.2	234.70	17NO	-0.3	-3.4	-1.7	0.8
292.99	-1.33	1.2	244.91	27NO	-0.3	-3.5	-1.8	0.8
300.73	-1.24	1.3	255.11	7DE	-0.4	-3.5	-1.8	0.8
308.46	-1.14	1.3	265.31	17DE	-0.6	-3.5	-1.9	0.8
316.16	-1.03	1.4	275.50	27DE	-0.9	-3.5	-1.9	0.7

♂ LONG	♂ LAT	♂ MAG	☉ LONG	16.00UT 106	☿	♀	♃	♄
					MAGNITUDES			
323.81	-0.92	1.4	285.66	6JA	-1.1	-3.4	-2.0	0.8
331.41	-0.81	1.5	295.78	16JA	-1.0	-3.4	-2.0	0.8
338.96	-0.69	1.5	305.87	26JA	-0.2	-3.4	-2.0	0.9
346.43	-0.58	1.5	315.90	5FE	1.8	-3.4	-2.0	0.9
353.83	-0.46	1.6	325.87	15FE	2.7	-3.3	-2.0	0.9
1.16	-0.35	1.6	335.80	25FE	1.3	-3.3	-2.0	0.9
8.40	-0.23	1.7	345.65	7MR	0.7	-3.3	-2.0	0.9
15.57	-0.12	1.7	355.45	17MR	0.4	-3.3	-1.9	0.9
22.65	-0.02	1.7	5.20	27MR	-0.0	-3.3	-1.9	0.9
29.66	0.09	1.7	14.89	6AP	-0.6	-3.4	-1.8	0.8
36.59	0.19	1.8	24.53	16AP	-1.5	-3.4	-1.7	0.8
43.46	0.29	1.8	34.14	26AP	-1.5	-3.4	-1.7	0.7
50.25	0.38	1.8	43.71	6MY	-0.5	-3.4	-1.6	0.7
56.97	0.47	1.8	53.25	16MY	0.3	-3.5	-1.5	0.6
63.64	0.56	1.8	62.79	26MY	1.0	-3.5	-1.5	0.6
70.25	0.64	1.8	72.30	5JN	1.8	-3.6	-1.4	0.5
76.81	0.72	1.9	81.83	15JN	3.0	-3.6	-1.4	0.4
83.33	0.79	1.9	91.36	25JN	2.6	-3.7	-1.3	0.4
89.79	0.86	1.9	100.91	5JL	0.9	-3.8	-1.3	0.3
96.22	0.93	2.0	110.50	15JL	-0.4	-3.9	-1.3	0.4
102.62	1.00	2.0	120.12	25JL	-1.3	-4.0	-1.2	0.4
108.98	1.06	2.0	129.78	4AU	-1.4	-4.1	-1.2	0.5
115.31	1.12	2.0	139.49	14AU	-0.7	-4.2	-1.2	0.5
121.62	1.17	2.0	149.26	24AU	-0.3	-4.3	-1.2	0.6
127.89	1.23	2.0	159.08	3SE	-0.0	-4.3	-1.2	0.6
134.14	1.28	2.0	168.96	13SE	0.1	-4.1	-1.2	0.7
140.36	1.33	2.0	178.90	23SE	0.3	-3.6	-1.2	0.7
146.55	1.37	1.9	188.90	3OC	1.2	-3.2	-1.3	0.8
152.71	1.42	1.9	198.95	13OC	2.7	-3.8	-1.3	0.8
158.83	1.46	1.8	209.05	23OC	0.1	-4.2	-1.3	0.8
164.90	1.50	1.8	219.19	2NO	-0.5	-4.4	-1.4	0.8
170.92	1.53	1.7	229.36	12NO	-0.5	-4.4	-1.4	0.9
176.88	1.56	1.6	239.56	22NO	-0.5	-4.3	-1.5	0.9
182.77	1.59	1.5	249.76	2DE	-0.7	-4.2	-1.6	0.9
188.56	1.61	1.4	259.97	12DE	-0.9	-4.1	-1.6	0.9
194.25	1.63	1.3	270.17	22DE	-0.9	-4.0	-1.7	0.8
				107				
199.80	1.63	1.1	280.34	1JA	-0.9	-3.9	-1.8	0.8
205.20	1.63	1.0	290.49	11JA	-0.1	-3.8	-1.8	0.8
210.39	1.61	0.8	300.59	21JA	2.1	-3.7	-1.9	0.9
215.34	1.58	0.6	310.65	31JA	2.1	-3.6	-1.9	0.9
219.99	1.52	0.4	320.66	10FE	0.9	-3.6	-2.0	0.9
224.25	1.44	0.2	330.61	20FE	0.5	-3.5	-2.0	1.0
228.03	1.31	-0.1	340.50	2MR	0.3	-3.4	-2.0	1.0
231.21	1.15	-0.4	350.33	12MR	-0.1	-3.4	-2.0	1.0
233.61	0.92	-0.7	0.11	22MR	-0.7	-3.4	-2.0	1.0
235.09	0.62	-1.0	9.82	1AP	-1.5	-3.3	-2.0	1.0
235.44	0.23	-1.3	19.49	11AP	-1.4	-3.3	-1.9	0.9
234.55	-0.26	-1.6	29.12	21AP	-0.5	-3.3	-1.9	0.9
232.49	-0.82	-1.9	38.70	1MY	0.5	-3.3	-1.8	0.9
229.54	-1.43	-2.2	48.26	11MY	1.3	-3.3	-1.8	0.8
226.34	-2.01	-2.1	57.80	21MY	2.3	-3.3	-1.7	0.8
223.63	-2.51	-2.0	67.32	31MY	3.5	-3.3	-1.6	0.7
221.98	-2.88	-1.8	76.84	10JN	2.1	-3.4	-1.6	0.6
221.69	-3.12	-1.5	86.36	20JN	0.7	-3.4	-1.5	0.6
222.77	-3.26	-1.3	95.90	30JN	-0.5	-3.4	-1.5	0.5
225.07	-3.32	-1.1	105.47	10JL	-1.4	-3.5	-1.4	0.4
228.43	-3.31	-0.9	115.07	20JL	-1.4	-3.5	-1.4	0.4
232.63	-3.26	-0.7	124.70	30JL	-0.7	-3.5	-1.3	0.5
237.53	-3.17	-0.5	134.39	9AU	-0.2	-3.4	-1.3	0.5
243.00	-3.05	-0.4	144.13	19AU	0.1	-3.4	-1.3	0.6
248.90	-2.92	-0.2	153.92	29AU	0.3	-3.4	-1.2	0.6
255.17	-2.76	-0.1	163.77	8SE	0.5	-3.4	-1.2	0.7
261.72	-2.59	0.1	173.68	18SE	1.5	-3.3	-1.2	0.7
268.50	-2.41	0.2	183.65	28SE	2.4	-3.3	-1.2	0.8
275.45	-2.22	0.3	193.67	8OC	-0.0	-3.3	-1.2	0.8
282.53	-2.02	0.5	203.75	18OC	-0.6	-3.3	-1.2	0.8
289.71	-1.83	0.6	213.86	28OC	-0.6	-3.3	-1.3	0.9
296.95	-1.63	0.7	224.02	7NO	-0.6	-3.4	-1.3	0.9
304.24	-1.44	0.8	234.21	17NO	-0.8	-3.4	-1.3	0.9
311.54	-1.25	0.9	244.41	27NO	-0.7	-3.4	-1.4	0.9
318.84	-1.06	1.0	254.62	7DE	-0.8	-3.4	-1.4	0.9
326.12	-0.88	1.1	264.82	17DE	-0.7	-3.5	-1.5	0.9
333.38	-0.71	1.2	275.01	27DE	0.0	-3.5	-1.5	0.9

♂			⊙	16.00UT	☿	♀	♃	♄
LONG	LAT	MAG	LONG	108	MAGNITUDES			
340.60	-0.55	1.3	285.17	6JA	2.4	-3.6	-1.6	0.9
347.77	-0.39	1.4	295.30	16JA	1.6	-3.7	-1.7	0.9
354.88	-0.25	1.5	305.38	26JA	0.6	-3.7	-1.7	1.0
1.95	-0.11	1.5	315.42	5FE	0.3	-3.8	-1.8	1.0
8.95	0.02	1.6	325.40	15FE	0.1	-3.9	-1.9	1.0
15.89	0.14	1.7	335.32	25FE	-0.2	-4.0	-1.9	1.1
22.77	0.25	1.7	345.18	6MR	-0.7	-4.1	-2.0	1.1
29.59	0.35	1.8	354.99	16MR	-1.5	-4.1	-2.0	1.1
36.35	0.45	1.8	4.73	26MR	-1.4	-4.2	-2.0	1.1
43.06	0.54	1.9	14.43	5AP	-0.4	-4.2	-2.0	1.1
49.71	0.62	1.9	24.07	15AP	0.7	-4.1	-2.0	1.1
56.31	0.70	1.9	33.68	25AP	1.7	-3.8	-2.0	1.1
62.87	0.77	1.9	43.25	5MY	3.1	-3.0	-2.0	1.0
69.39	0.83	2.0	52.80	15MY	2.9	-3.2	-1.9	1.0
75.88	0.89	2.0	62.32	25MY	1.6	-3.9	-1.9	0.9
82.33	0.94	2.0	71.85	4JN	0.6	-4.1	-1.8	0.9
88.76	0.99	2.0	81.37	14JN	-0.5	-4.2	-1.8	0.8
95.17	1.03	2.0	90.90	24JN	-1.4	-4.2	-1.7	0.7
101.56	1.07	1.9	100.45	4JL	-1.4	-4.1	-1.7	0.7
107.94	1.10	2.0	110.03	14JL	-0.6	-4.0	-1.6	0.6
114.31	1.13	2.0	119.65	24JL	-0.1	-3.9	-1.5	0.5
120.68	1.15	2.0	129.31	3AU	0.2	-3.8	-1.5	0.5
127.04	1.17	2.0	139.01	13AU	0.4	-3.7	-1.4	0.6
133.41	1.19	2.0	148.78	23AU	0.8	-3.7	-1.4	0.6
139.79	1.20	2.0	158.60	2SE	1.9	-3.6	-1.3	0.7
146.18	1.21	2.0	168.48	12SE	2.1	-3.6	-1.3	0.7
152.57	1.21	2.0	178.42	22SE	-0.1	-3.5	-1.3	0.8
158.98	1.21	2.0	188.41	2OC	-0.8	-3.5	-1.3	0.8
165.40	1.20	2.0	198.46	12OC	-0.8	-3.5	-1.3	0.9
171.83	1.19	1.9	208.56	22OC	-0.8	-3.4	-1.3	0.9
178.27	1.17	1.9	218.70	1NO	-0.7	-3.4	-1.3	1.0
184.72	1.14	1.8	228.87	11NO	-0.6	-3.4	-1.3	1.0
191.18	1.11	1.8	239.06	21NO	-0.6	-3.9	-1.3	1.0
197.65	1.07	1.7	249.27	1DE	-0.6	-3.4	-1.3	1.0
204.12	1.01	1.6	259.47	11DE	0.1	-3.4	-1.4	1.1
210.59	0.95	1.5	269.67	21DE	2.7	-3.3	-1.4	1.1
217.05	0.88	1.4	279.85	31DE	1.2	-3.4	-1.5	1.1
				109				
223.50	0.79	1.3	289.99	10JA	0.4	-3.4	-1.5	1.1
229.94	0.68	1.2	300.10	20JA	0.2	-3.4	-1.6	1.1
236.35	0.56	1.1	310.17	30JA	0.0	-3.4	-1.6	1.1
242.73	0.41	0.9	320.17	9FE	-0.3	-3.4	-1.7	1.1
249.06	0.24	0.8	330.13	19FE	-0.7	-3.5	-1.8	1.2
255.33	0.04	0.6	340.02	1MR	-1.5	-3.5	-1.8	1.2
261.51	-0.19	0.5	349.85	11MR	-1.4	-3.5	-1.9	1.2
267.59	-0.45	0.3	359.63	21MR	-0.4	-3.4	-2.0	1.2
273.52	-0.75	0.1	9.35	31MR	0.8	-3.4	-2.0	1.3
279.26	-1.10	-0.1	19.02	10AP	2.2	-3.4	-2.1	1.3
284.75	-1.50	-0.3	28.65	20AP	3.8	-3.4	-2.1	1.3
289.92	-1.95	-0.5	38.24	30AP	2.2	-3.3	-2.1	1.2
294.67	-2.46	-0.8	47.79	10MY	1.2	-3.3	-2.1	1.2
298.87	-3.04	-1.0	57.33	20MY	0.4	-3.3	-2.1	1.2
302.35	-3.68	-1.3	66.86	30MY	-0.5	-3.3	-2.1	1.1
304.94	-4.37	-1.6	76.37	9JN	-1.5	-3.3	-2.1	1.1
306.42	-5.10	-1.9	85.90	19JN	-1.4	-3.4	-2.0	1.0
306.64	-5.81	-2.1	95.44	29JN	-0.6	-3.4	-2.0	0.9
305.57	-6.41	-2.4	105.00	9JL	-0.0	-3.4	-1.9	0.9
303.43	-6.80	-2.6	114.60	19JL	0.3	-3.4	-1.8	0.8
300.81	-6.87	-2.6	124.24	29JL	0.6	-3.5	-1.8	0.7
298.42	-6.62	-2.5	133.92	8AU	1.1	-3.6	-1.7	0.7
296.93	-6.11	-2.2	143.65	18AU	2.4	-3.6	-1.7	0.7
296.69	-5.46	-1.9	153.44	28AU	1.9	-3.7	-1.6	0.7
297.75	-4.76	-1.6	163.29	7SE	-0.2	-3.8	-1.5	0.8
299.98	-4.07	-1.3	173.20	17SE	-0.9	-3.9	-1.5	0.8
303.21	-3.43	-1.0	183.16	27SE	-0.9	-4.0	-1.5	0.9
307.21	-2.86	-0.8	193.18	7OC	-0.9	-4.1	-1.4	0.9
311.82	-2.34	-0.5	203.26	17OC	-0.6	-4.2	-1.4	1.0
316.91	-1.89	-0.3	213.37	27OC	-0.5	-4.3	-1.4	1.0
322.35	-1.48	-0.0	223.53	6NO	-0.4	-4.4	-1.4	1.1
328.07	-1.13	0.2	233.71	16NO	-0.4	-4.4	-1.4	1.1
334.00	-0.82	0.4	243.91	26NO	0.3	-4.2	-1.4	1.1
340.08	-0.55	0.5	254.12	6DE	3.0	-3.6	-1.4	1.2
346.27	-0.31	0.7	264.32	16DE	0.9	-3.1	-1.4	1.2
352.55	-0.10	0.9	274.51	26DE	0.2	-3.8	-1.4	1.2

♂			⊙	16.00UT	☿	♀	♃	♄
LONG	LAT	MAG	LONG	110	MAGNITUDES			
358.89	0.08	1.0	284.68	5JA	0.0	-4.2	-1.4	1.2
5.26	0.24	1.1	294.81	15JA	-0.1	-4.3	-1.5	1.2
11.65	0.38	1.3	304.89	25JA	-0.3	-4.3	-1.5	1.2
18.06	0.51	1.4	314.93	4FE	-0.8	-4.2	-1.6	1.2
24.47	0.62	1.5	324.92	14FE	-1.4	-4.1	-1.6	1.2
30.87	0.71	1.6	334.84	24FE	-1.3	-4.0	-1.7	1.3
37.26	0.80	1.7	344.71	6MR	-0.4	-3.9	-1.8	1.3
43.64	0.87	1.7	354.52	16MR	1.1	-3.8	-1.8	1.4
50.00	0.93	1.8	4.26	26MR	2.8	-3.7	-1.9	1.4
56.35	0.98	1.9	13.96	5AP	2.9	-3.6	-2.0	1.4
62.68	1.03	1.9	23.61	15AP	1.6	-3.6	-2.0	1.4
69.01	1.07	1.9	33.22	25AP	0.9	-3.5	-2.1	1.4
75.32	1.10	2.0	42.79	5MY	0.3	-3.4	-2.2	1.3
81.63	1.13	2.0	52.34	15MY	-0.6	-3.4	-2.2	1.3
87.93	1.15	2.0	61.87	25MY	-1.5	-3.4	-2.2	1.3
94.23	1.16	2.0	71.39	4JN	-1.5	-3.3	-2.3	1.2
100.54	1.17	2.0	80.91	14JN	-0.6	-3.3	-2.3	1.2
106.85	1.17	2.0	90.44	24JN	0.1	-3.3	-2.2	1.1
113.18	1.17	2.0	100.00	4JL	0.5	-3.3	-2.2	1.1
119.52	1.17	2.0	109.57	14JL	0.8	-3.3	-2.2	1.0
125.88	1.16	2.0	119.19	24JL	1.5	-3.3	-2.1	1.0
132.27	1.15	2.0	128.84	3AU	2.9	-3.4	-2.1	0.9
138.68	1.13	1.9	138.55	13AU	1.6	-3.4	-2.0	0.9
145.13	1.10	1.9	148.31	23AU	-0.3	-3.4	-2.0	0.8
151.61	1.07	1.9	158.12	2SE	-1.0	-3.4	-1.9	0.9
158.13	1.04	1.9	168.00	12SE	-1.1	-3.5	-1.8	0.9
164.69	1.00	1.9	177.93	22SE	-0.9	-3.5	-1.8	1.0
171.29	0.96	1.9	187.92	2OC	-0.5	-3.5	-1.7	1.0
177.94	0.91	1.9	197.97	12OC	-0.3	-3.5	-1.7	1.1
184.64	0.85	1.9	208.06	22OC	-0.3	-3.4	-1.6	1.1
191.38	0.79	1.9	218.20	1NO	-0.2	-3.4	-1.6	1.2
198.17	0.72	1.8	228.37	11NO	0.5	-3.4	-1.5	1.2
205.01	0.64	1.8	238.56	21NO	3.2	-3.4	-1.5	1.3
211.91	0.56	1.7	248.76	1DE	0.6	-3.3	-1.5	1.3
218.85	0.46	1.7	258.97	11DE	-0.0	-3.3	-1.5	1.3
225.83	0.36	1.6	269.17	21DE	-0.1	-3.3	-1.5	1.4
232.87	0.24	1.6	279.35	31DE	-0.2	-3.3	-1.5	1.4
				111				
239.95	0.12	1.5	289.50	10JA	-0.4	-3.4	-1.5	1.4
247.07	-0.01	1.4	299.61	20JA	-0.8	-3.4	-1.5	1.3
254.23	-0.16	1.3	309.68	30JA	-1.3	-3.4	-1.5	1.3
261.42	-0.32	1.3	319.69	9FE	-1.2	-3.4	-1.5	1.2
268.65	-0.48	1.2	329.64	19FE	-0.3	-3.4	-1.6	1.2
275.89	-0.66	1.1	339.54	1MR	1.3	-3.5	-1.6	1.2
283.15	-0.85	1.0	349.38	11MR	3.4	-3.5	-1.7	1.2
290.42	-1.05	0.9	359.16	21MR	2.1	-3.6	-1.7	1.2
297.67	-1.25	0.8	8.89	31MR	1.2	-3.6	-1.8	1.1
304.89	-1.46	0.7	18.56	10AP	0.7	-3.7	-1.9	1.1
312.07	-1.67	0.6	28.19	20AP	0.2	-3.8	-1.9	1.1
319.18	-1.89	0.5	37.78	30AP	-0.6	-3.9	-2.0	1.1
326.19	-2.10	0.4	47.34	10MY	-1.5	-4.0	-2.1	1.1
333.07	-2.31	0.2	56.88	20MY	-1.5	-4.0	-2.1	1.0
339.78	-2.52	0.1	66.40	30MY	-0.5	-4.1	-2.2	1.0
346.26	-2.72	-0.0	75.92	9JN	0.2	-4.2	-2.3	1.0
352.47	-2.91	-0.2	85.44	19JN	0.6	-4.2	-2.3	0.9
358.32	-3.09	-0.3	94.99	29JN	1.1	-4.0	-2.4	0.9
3.74	-3.25	-0.5	104.55	9JL	1.9	-3.5	-2.4	0.8
8.61	-3.39	-0.6	114.14	19JL	3.2	-3.1	-2.4	0.8
12.77	-3.51	-0.8	123.78	29JL	1.4	-3.7	-2.4	0.7
16.08	-3.60	-1.0	133.45	8AU	-0.3	-4.1	-2.3	0.7
18.32	-3.64	-1.2	143.18	18AU	-1.1	-4.3	-2.3	0.6
19.27	-3.61	-1.4	152.97	28AU	-1.2	-4.3	-2.3	0.6
18.78	-3.47	-1.7	162.81	7SE	-0.8	-4.2	-2.2	0.6
16.85	-3.21	-1.9	172.72	17SE	-0.4	-4.1	-2.1	0.6
13.78	-2.79	-2.0	182.68	27SE	-0.2	-4.0	-2.1	0.7
10.26	-2.25	-2.0	192.70	7OC	-0.1	-4.0	-2.0	0.8
7.08	-1.65	-1.8	202.76	17OC	-0.0	-3.9	-1.9	0.8
4.92	-1.06	-1.5	212.88	27OC	0.7	-3.8	-1.9	0.9
4.09	-0.53	-1.1	223.03	6NO	3.2	-3.7	-1.8	1.0
4.58	-0.09	-0.8	233.21	16NO	0.4	-3.7	-1.7	1.0
6.23	0.26	-0.5	243.41	26NO	-0.2	-3.6	-1.7	1.0
8.85	0.54	-0.2	253.62	6DE	-0.3	-3.5	-1.7	1.1
12.23	0.76	0.1	263.82	16DE	-0.3	-3.5	-1.6	1.1
16.19	0.93	0.3	274.02	26DE	-0.5	-3.4	-1.6	1.1

112

♂ LONG	LAT	MAG	☉ LONG	16.00UT	☿	♀	♃	♄
20.62	1.06	0.5	284.18	5JA	-0.9	-3.4	-1.6	1.1
25.40	1.17	0.7	294.31	15JA	-1.2	-3.4	-1.5	1.1
30.46	1.24	0.9	304.41	25JA	-1.1	-3.4	-1.5	1.1
35.73	1.30	1.1	314.44	4FE	-0.3	-3.3	-1.5	1.0
41.17	1.35	1.2	324.43	14FE	1.6	-3.3	-1.6	1.0
46.75	1.38	1.3	334.36	24FE	3.1	-3.3	-1.6	0.9
52.43	1.40	1.5	344.23	5MR	1.6	-3.3	-1.6	0.9
58.19	1.41	1.6	354.04	15MR	0.9	-3.3	-1.6	0.9
64.03	1.42	1.6	3.80	25MR	0.5	-3.3	-1.7	0.9
69.92	1.42	1.7	13.49	4AP	0.1	-3.4	-1.7	0.9
75.87	1.41	1.8	23.15	14AP	-0.6	-3.4	-1.7	0.9
81.86	1.40	1.9	32.76	24AP	-1.5	-3.4	-1.8	0.9
87.88	1.38	1.9	42.33	4MY	-1.5	-3.5	-1.9	0.9
93.95	1.36	1.9	51.88	14MY	-0.5	-3.5	-1.9	0.8
100.06	1.33	2.0	61.41	24MY	0.3	-3.5	-2.0	0.8
106.20	1.30	2.0	70.93	3JN	0.8	-3.4	-2.1	0.8
112.39	1.26	2.0	80.45	13JN	1.5	-3.4	-2.1	0.8
118.61	1.23	2.0	89.98	23JN	2.5	-3.4	-2.2	0.7
124.88	1.18	2.0	99.53	3JL	2.9	-3.3	-2.3	0.7
131.20	1.14	2.0	109.11	13JL	1.1	-3.3	-2.3	0.6
137.56	1.09	2.0	118.72	23JL	-0.4	-3.3	-2.4	0.6
143.98	1.04	2.0	128.37	2AU	-1.2	-3.3	-2.4	0.5
150.46	0.98	1.9	138.08	12AU	-1.4	-3.3	-2.4	0.5
156.99	0.92	1.9	147.83	22AU	-0.7	-3.3	-2.5	0.4
163.59	0.85	1.9	157.64	1SE	-0.3	-3.4	-2.4	0.4
170.25	0.79	1.8	167.52	11SE	-0.1	-3.4	-2.4	0.3
176.98	0.71	1.8	177.45	21SE	0.0	-3.4	-2.4	0.3
183.78	0.63	1.8	187.44	1OC	0.2	-3.5	-2.3	0.4
190.65	0.55	1.8	197.48	11OC	1.0	-3.5	-2.3	0.4
197.59	0.46	1.8	207.57	21OC	2.9	-3.6	-2.2	0.5
204.61	0.37	1.8	217.70	31OC	0.2	-3.7	-2.1	0.6
211.70	0.27	1.7	227.87	10NO	-0.4	-3.7	-2.1	0.6
218.85	0.17	1.7	238.06	20NO	-0.4	-3.8	-2.0	0.7
226.08	0.06	1.7	248.27	30NO	-0.4	-3.9	-1.9	0.7
233.38	-0.05	1.7	258.48	10DE	-0.6	-4.0	-1.9	0.8
240.74	-0.16	1.6	268.67	20DE	-0.9	-4.1	-1.8	0.8
248.17	-0.28	1.6	278.85	30DE	-1.0	-4.2	-1.7	0.8

113

♂ LONG	LAT	MAG	☉ LONG	16.00UT	☿	♀	♃	♄
255.65	-0.40	1.5	289.01	9JA	-1.0	-4.3	-1.7	0.8
263.18	-0.53	1.5	299.12	19JA	-0.2	-4.3	-1.7	0.8
270.76	-0.65	1.4	309.18	29JA	1.9	-4.3	-1.6	0.8
278.37	-0.78	1.4	319.20	8FE	2.5	-4.1	-1.6	0.8
286.01	-0.90	1.4	329.16	18FE	1.1	-3.6	-1.6	0.8
293.67	-1.02	1.3	339.06	28FE	0.6	-3.3	-1.6	0.7
301.34	-1.14	1.3	348.90	10MR	0.3	-3.8	-1.6	0.7
309.00	-1.25	1.2	358.68	20MR	-0.0	-4.1	-1.6	0.6
316.64	-1.35	1.2	8.41	30MR	-0.6	-4.2	-1.6	0.6
324.25	-1.44	1.1	18.09	9AP	-1.5	-4.2	-1.6	0.6
331.81	-1.52	1.1	27.72	19AP	-1.5	-4.1	-1.6	0.7
339.30	-1.59	1.0	37.31	29AP	-0.5	-4.0	-1.7	0.7
346.72	-1.64	1.0	46.88	9MY	0.4	-3.9	-1.7	0.7
354.03	-1.68	0.9	56.41	19MY	1.1	-3.8	-1.7	0.6
1.23	-1.70	0.9	65.94	29MY	2.0	-3.7	-1.8	0.6
8.29	-1.70	0.8	75.46	8JN	3.2	-3.7	-1.9	0.6
15.20	-1.69	0.8	84.98	18JN	2.4	-3.6	-1.9	0.6
21.92	-1.66	0.7	94.52	28JN	0.9	-3.5	-2.0	0.6
28.43	-1.61	0.6	104.09	8JL	-0.4	-3.5	-2.1	0.5
34.69	-1.54	0.5	113.68	18JL	-1.3	-3.4	-2.1	0.5
40.67	-1.44	0.5	123.31	28JL	-1.4	-3.4	-2.2	0.4
46.31	-1.33	0.4	132.99	7AU	-0.7	-3.4	-2.3	0.4
51.55	-1.19	0.2	142.71	17AU	-0.2	-3.4	-2.3	0.3
56.31	-1.03	0.0	152.50	27AU	0.0	-3.4	-2.4	0.2
60.46	-0.83	-0.0	162.34	6SE	0.2	-3.4	-2.4	0.2
63.89	-0.59	-0.2	172.24	16SE	0.4	-3.4	-2.4	0.1
66.42	-0.31	-0.4	182.20	26SE	1.3	-3.4	-2.4	0.1
67.84	0.02	-0.6	192.22	6OC	2.7	-3.4	-2.4	0.1
67.97	0.41	-0.8	202.28	16OC	0.1	-3.4	-2.4	0.1
66.67	0.85	-1.1	212.39	26OC	-0.5	-3.4	-2.4	0.2
64.01	1.31	-1.2	222.54	5NO	-0.6	-3.4	-2.3	0.3
60.38	1.75	-1.4	232.72	15NO	-0.6	-3.4	-2.3	0.4
56.45	2.11	-1.3	242.92	25NO	-0.7	-3.5	-2.2	0.4
53.04	2.37	-1.1	253.13	5DE	-0.8	-3.5	-2.1	0.5
50.72	2.51	-0.8	263.33	15DE	-0.8	-3.5	-2.0	0.5
49.72	2.56	-0.5	273.52	25DE	-0.8	-3.5	-2.0	0.6

114

♂ LONG	LAT	MAG	☉ LONG	16.00UT	☿	♀	♃	♄
50.03	2.55	-0.2	283.69	4JA	-0.1	-3.4	-1.9	0.6
51.47	2.50	0.1	293.82	14JA	2.2	-3.4	-1.8	0.6
53.83	2.43	0.3	303.91	24JA	1.9	-3.4	-1.8	0.6
56.94	2.35	0.5	313.95	3FE	0.8	-3.4	-1.7	0.6
60.63	2.26	0.7	323.94	13FE	0.4	-3.3	-1.7	0.6
64.77	2.18	0.9	333.87	23FE	0.2	-3.3	-1.6	0.6
69.29	2.09	1.1	343.74	5MR	-0.1	-3.3	-1.6	0.6
74.10	2.00	1.2	353.55	15MR	-0.7	-3.3	-1.6	0.5
79.14	1.92	1.3	3.31	25MR	-1.5	-3.3	-1.6	0.5
84.38	1.83	1.5	13.02	4AP	-1.5	-3.4	-1.5	0.4
89.78	1.75	1.6	22.67	14AP	-0.5	-3.4	-1.5	0.4
95.32	1.67	1.6	32.29	24AP	0.5	-3.4	-1.5	0.5
100.98	1.59	1.7	41.86	4MY	1.4	-3.4	-1.5	0.5
106.74	1.50	1.8	51.41	14MY	2.6	-3.5	-1.6	0.5
112.61	1.42	1.8	60.95	24MY	3.4	-3.5	-1.6	0.5
118.57	1.34	1.8	70.47	3JN	1.9	-3.6	-1.6	0.5
124.61	1.26	1.9	79.99	13JN	0.7	-3.6	-1.6	0.5
130.74	1.17	1.9	89.52	23JN	-0.5	-3.7	-1.7	0.5
136.95	1.09	1.9	99.07	3JL	-1.4	-3.8	-1.7	0.4
143.24	1.00	1.9	108.64	13JL	-1.4	-3.9	-1.8	0.4
149.62	0.91	1.9	118.26	23JL	-0.7	-4.0	-1.8	0.4
156.08	0.82	1.9	127.91	2AU	-0.2	-4.1	-1.9	0.3
162.63	0.73	1.9	137.61	12AU	0.1	-4.2	-2.0	0.3
169.26	0.63	1.9	147.36	22AU	0.3	-4.3	-2.0	0.2
175.99	0.54	1.8	157.17	1SE	0.6	-4.3	-2.1	0.1
182.80	0.44	1.8	167.04	11SE	1.7	-4.1	-2.2	0.1
189.70	0.34	1.8	176.97	21SE	2.4	-3.6	-2.2	0.0
196.70	0.24	1.7	186.95	1OC	-0.0	-3.3	-2.3	-0.0
203.78	0.13	1.7	196.99	11OC	-0.7	-3.9	-2.3	-0.1
210.96	0.03	1.6	207.08	21OC	-0.7	-4.2	-2.3	-0.1
218.22	-0.08	1.6	217.21	31OC	-0.7	-4.4	-2.4	-0.0
225.57	-0.18	1.5	227.38	10NO	-0.8	-4.4	-2.4	0.0
233.00	-0.29	1.5	237.57	20NO	-0.7	-4.3	-2.3	0.1
240.50	-0.39	1.5	247.77	30NO	-0.7	-4.2	-2.3	0.2
248.08	-0.50	1.5	257.98	10DE	-0.7	-4.1	-2.2	0.3
255.72	-0.60	1.5	268.18	20DE	0.0	-4.0	-2.2	0.3
263.42	-0.70	1.5	278.36	30DE	2.4	-3.9	-2.1	0.4

115

♂ LONG	LAT	MAG	☉ LONG	16.00UT	☿	♀	♃	♄
271.16	-0.79	1.5	288.51	9JA	1.5	-3.8	-2.0	0.4
278.94	-0.88	1.5	298.63	19JA	0.5	-3.7	-2.0	0.4
286.75	-0.96	1.4	308.70	29JA	0.3	-3.6	-1.9	0.5
294.57	-1.03	1.4	318.71	8FE	0.1	-3.6	-1.8	0.5
302.39	-1.09	1.4	328.68	18FE	-0.2	-3.5	-1.8	0.5
310.20	-1.15	1.4	338.58	28FE	-0.7	-3.4	-1.7	0.5
318.00	-1.19	1.4	348.42	10MR	-1.5	-3.4	-1.6	0.4
325.75	-1.22	1.4	358.21	20MR	-1.4	-3.4	-1.6	0.4
333.46	-1.23	1.4	7.94	30MR	-0.5	-3.3	-1.6	0.4
341.12	-1.23	1.4	17.62	9AP	0.7	-3.3	-1.5	0.3
348.70	-1.22	1.4	27.26	19AP	1.9	-3.3	-1.5	0.3
356.20	-1.20	1.3	36.85	29AP	3.4	-3.3	-1.5	0.3
3.62	-1.16	1.3	46.41	9MY	2.6	-3.4	-1.5	0.4
10.94	-1.11	1.3	55.95	19MY	1.5	-3.3	-1.4	0.4
18.15	-1.05	1.3	65.48	29MY	0.5	-3.3	-1.4	0.4
25.24	-0.97	1.3	75.00	8JN	-0.5	-3.4	-1.4	0.4
32.21	-0.89	1.3	84.52	18JN	-1.4	-3.4	-1.5	0.4
39.05	-0.79	1.3	94.06	28JN	-1.5	-3.4	-1.5	0.4
45.74	-0.68	1.2	103.62	8JL	-0.6	-3.5	-1.5	0.4
52.27	-0.55	1.2	113.21	18JL	-0.1	-3.5	-1.5	0.3
58.63	-0.42	1.2	122.84	28JL	0.2	-3.5	-1.6	0.3
64.80	-0.27	1.1	132.51	7AU	0.5	-3.4	-1.6	0.3
70.75	-0.11	1.0	142.24	17AU	0.9	-3.4	-1.7	0.2
76.46	0.07	1.0	152.02	27AU	2.1	-3.4	-1.7	0.2
81.89	0.26	0.9	161.86	6SE	2.1	-3.4	-1.8	0.1
86.98	0.48	0.8	171.76	16SE	-0.1	-3.3	-1.9	0.0
91.68	0.71	0.7	181.71	26SE	-0.8	-3.3	-1.9	-0.0
95.89	0.98	0.5	191.72	6OC	-0.9	-3.3	-2.0	-0.1
99.50	1.28	0.4	201.79	16OC	-0.8	-3.3	-2.1	-0.2
102.39	1.62	0.2	211.89	26OC	-0.7	-3.3	-2.1	-0.2
104.38	2.00	-0.0	222.04	5NO	-0.5	-3.4	-2.2	-0.2
105.26	2.42	-0.3	232.22	15NO	-0.5	-3.4	-2.2	-0.1
104.89	2.86	-0.5	242.42	25NO	-0.5	-3.4	-2.2	-0.1
103.14	3.31	-0.7	252.63	5DE	0.2	-3.5	-2.2	0.0
100.15	3.71	-0.9	262.84	15DE	2.7	-3.5	-2.2	0.1
96.35	3.98	-1.1	273.02	25DE	1.1	-3.5	-2.2	0.1

♂			⊙	16.00UT	☿	♀	♃	♄
LONG	LAT	MAG	LONG	116	MAGNITUDES			
92.40	4.11	-1.0	283.20	4JA	0.3	-3.6	-2.2	0.2
89.07	4.07	-0.7	293.33	14JA	0.1	-3.7	-2.1	0.3
86.84	3.90	-0.5	303.42	24JA	-0.0	-3.7	-2.1	0.3
85.92	3.67	-0.2	313.47	3FE	-0.3	-3.8	-2.0	0.3
86.26	3.41	0.0	323.46	13FE	-0.7	-3.9	-1.9	0.4
87.69	3.15	0.3	333.39	23FE	-1.5	-4.0	-1.8	0.4
90.03	2.90	0.5	343.27	4MR	-1.4	-4.1	-1.8	0.4
93.10	2.66	0.7	353.09	14MR	-0.4	-4.1	-1.7	0.4
96.75	2.44	0.8	2.84	24MR	0.9	-4.2	-1.6	0.3
100.88	2.24	1.0	12.55	3AP	2.4	-4.2	-1.6	0.3
105.39	2.05	1.1	22.21	13AP	3.5	-4.1	-1.5	0.3
110.21	1.87	1.2	31.82	23AP	2.0	-3.7	-1.5	0.2
115.30	1.71	1.3	41.40	3MY	1.1	-3.0	-1.4	0.2
120.61	1.55	1.4	50.95	13MY	0.4	-3.3	-1.4	0.2
126.12	1.40	1.5	60.48	23MY	-0.5	-3.9	-1.4	0.3
131.80	1.26	1.6	70.00	2JN	-1.5	-4.1	-1.4	0.3
137.64	1.12	1.6	79.52	12JN	-1.5	-4.2	-1.3	0.3
143.62	0.98	1.6	89.05	22JN	-0.6	-4.2	-1.3	0.3
149.75	0.85	1.7	98.60	2JL	-0.0	-4.1	-1.3	0.3
156.00	0.72	1.7	108.18	12JL	0.4	-4.0	-1.3	0.3
162.37	0.60	1.7	117.78	22JL	0.7	-3.9	-1.4	0.3
168.87	0.47	1.7	127.44	1AU	1.2	-3.8	-1.4	0.3
175.49	0.35	1.7	137.13	11AU	2.6	-3.7	-1.4	0.3
182.22	0.23	1.7	146.88	21AU	1.8	-3.7	-1.4	0.2
189.07	0.11	1.7	156.69	31AU	-0.2	-3.6	-1.5	0.2
196.04	-0.00	1.7	166.56	10SE	-1.0	-3.6	-1.5	0.1
203.11	-0.12	1.6	176.48	20SE	-1.0	-3.5	-1.6	0.1
210.29	-0.23	1.6	186.47	30SE	-0.9	-3.5	-1.6	-0.0
217.58	-0.33	1.6	196.51	10OC	-0.6	-3.5	-1.7	-0.1
224.96	-0.44	1.6	206.59	20OC	-0.4	-3.4	-1.8	-0.2
232.44	-0.54	1.5	216.73	30OC	-0.4	-3.4	-1.8	-0.2
240.01	-0.63	1.5	226.89	9NO	-0.3	-3.4	-1.9	-0.3
247.66	-0.72	1.5	237.08	19NO	0.3	-3.4	-2.0	-0.3
255.37	-0.80	1.4	247.28	29NO	3.0	-3.4	-2.0	-0.2
263.15	-0.87	1.4	257.49	9DE	0.8	-3.4	-2.1	-0.1
270.97	-0.93	1.3	267.69	19DE	0.1	-3.4	-2.1	-0.0
278.83	-0.99	1.3	277.87	29DE	-0.0	-3.4	-2.1	0.0
				117				
286.71	-1.03	1.3	288.02	8JA	-0.1	-3.4	-2.1	0.1
294.60	-1.06	1.3	298.14	18JA	-0.4	-3.4	-2.1	0.2
302.50	-1.08	1.3	308.21	28JA	-0.8	-3.4	-2.1	0.2
310.37	-1.09	1.4	318.23	7FE	-1.4	-3.4	-2.0	0.2
318.22	-1.09	1.4	328.19	17FE	-1.3	-3.5	-2.0	0.3
326.02	-1.08	1.4	338.10	27FE	-0.4	-3.5	-1.9	0.3
333.78	-1.05	1.4	347.94	9MR	1.1	-3.5	-1.9	0.3
341.47	-1.01	1.4	357.73	19MR	3.0	-3.4	-1.8	0.3
349.10	-0.97	1.5	7.47	29MR	2.6	-3.4	-1.7	0.3
356.65	-0.91	1.5	17.15	8AP	1.5	-3.4	-1.7	0.3
4.11	-0.84	1.5	26.78	18AP	0.8	-3.4	-1.6	0.3
11.49	-0.76	1.5	36.38	28AP	0.3	-3.3	-1.5	0.2
18.77	-0.68	1.6	45.94	8MY	-0.5	-3.3	-1.5	0.2
25.96	-0.59	1.6	55.48	18MY	-1.5	-3.3	-1.4	0.2
33.04	-0.49	1.6	65.01	28MY	-1.5	-3.3	-1.4	0.2
40.03	-0.39	1.6	74.53	7JN	-0.6	-3.3	-1.3	0.2
46.91	-0.28	1.6	84.05	17JN	0.1	-3.4	-1.3	0.3
53.69	-0.16	1.6	93.59	27JN	0.5	-3.4	-1.3	0.3
60.35	-0.04	1.6	103.15	7JL	0.9	-3.4	-1.3	0.3
66.91	0.08	1.6	112.74	17JL	1.6	-3.4	-1.3	0.3
73.35	0.22	1.6	122.37	27JL	3.1	-3.5	-1.3	0.3
79.66	0.35	1.6	132.04	6AU	1.6	-3.6	-1.3	0.3
85.85	0.50	1.5	141.77	16AU	-0.3	-3.6	-1.3	0.3
91.89	0.65	1.5	151.54	26AU	-1.1	-3.7	-1.3	0.3
97.77	0.80	1.4	161.38	5SE	-1.2	-3.8	-1.3	0.2
103.47	0.97	1.4	171.27	15SE	-0.8	-3.9	-1.3	0.2
108.97	1.15	1.3	181.23	25SE	-0.5	-4.0	-1.4	0.1
114.23	1.34	1.2	191.24	5OC	-0.3	-4.1	-1.4	0.1
119.21	1.55	1.1	201.30	15OC	-0.2	-4.2	-1.5	-0.0
123.85	1.78	1.0	211.41	25OC	-0.1	-4.3	-1.5	-0.1
128.08	2.03	0.8	221.55	4NO	0.5	-4.4	-1.6	-0.2
131.79	2.31	0.6	231.73	14NO	3.3	-4.4	-1.6	-0.2
134.88	2.61	0.5	241.93	24NO	0.6	-4.2	-1.7	-0.3
137.19	2.95	0.2	252.13	4DE	-0.1	-3.6	-1.8	-0.3
138.54	3.31	0.0	262.34	14DE	-0.2	-3.1	-1.8	-0.2
138.75	3.69	-0.2	272.53	24DE	-0.3	-3.9	-1.9	-0.1

♂			⊙	16.00UT	☿	♀	♃	♄
LONG	LAT	MAG	LONG	118	MAGNITUDES			
137.69	4.06	-0.5	282.70	3JA	-0.4	-4.2	-1.9	-0.0
135.33	4.37	-0.7	292.84	13JA	-0.8	-4.4	-2.0	0.0
131.92	4.56	-0.9	302.93	23JA	-1.2	-4.3	-2.0	0.1
128.01	4.59	-1.0	312.98	2FE	-1.2	-4.2	-2.0	0.1
124.30	4.43	-0.8	322.98	12FE	-0.3	-4.1	-2.0	0.2
121.45	4.14	-0.6	332.91	22FE	1.4	-4.0	-2.0	0.2
119.81	3.77	-0.4	342.79	4MR	3.4	-3.9	-2.0	0.3
119.47	3.37	-0.2	352.61	14MR	1.9	-3.8	-1.9	0.3
120.33	2.98	0.0	2.37	24MR	1.1	-3.7	-1.9	0.3
122.22	2.61	0.2	12.08	3AP	0.6	-3.6	-1.8	0.3
124.95	2.27	0.4	21.74	13AP	0.1	-3.6	-1.8	0.3
128.38	1.97	0.6	31.36	23AP	-0.6	-3.5	-1.7	0.3
132.37	1.69	0.7	40.94	3MY	-1.5	-3.4	-1.6	0.3
136.83	1.44	0.9	50.49	13MY	-1.5	-3.4	-1.6	0.3
141.67	1.20	1.0	60.02	23MY	-0.6	-3.4	-1.5	0.2
146.84	0.99	1.1	69.54	2JN	0.2	-3.3	-1.5	0.2
152.30	0.79	1.1	79.07	12JN	0.7	-3.3	-1.4	0.2
158.01	0.60	1.2	88.59	22JN	1.2	-3.3	-1.4	0.3
163.93	0.42	1.3	98.14	2JL	2.1	-3.3	-1.3	0.3
170.06	0.26	1.3	107.72	12JL	3.2	-3.3	-1.3	0.3
176.38	0.10	1.3	117.32	22JL	1.3	-3.3	-1.3	0.4
182.87	-0.05	1.4	126.97	1AU	-0.3	-3.4	-1.2	0.4
189.52	-0.19	1.4	136.67	11AU	-1.2	-3.4	-1.2	0.4
196.32	-0.32	1.4	146.41	21AU	-1.3	-3.4	-1.2	0.4
203.27	-0.44	1.4	156.22	31AU	-0.8	-3.4	-1.2	0.3
210.36	-0.56	1.4	166.08	10SE	-0.4	-3.5	-1.2	0.3
217.57	-0.66	1.4	176.00	20SE	-0.2	-3.5	-1.2	0.3
224.91	-0.76	1.4	185.98	30SE	-0.1	-3.5	-1.3	0.2
232.36	-0.84	1.4	196.02	10OC	0.0	-3.5	-1.3	0.2
239.91	-0.92	1.4	206.10	20OC	0.8	-3.4	-1.3	0.1
247.54	-0.98	1.4	216.23	30OC	3.2	-3.4	-1.4	0.0
255.26	-1.04	1.4	226.39	9NO	0.4	-3.4	-1.4	-0.0
263.04	-1.08	1.4	236.58	19NO	-0.3	-3.4	-1.5	-0.1
270.87	-1.11	1.4	246.78	29NO	-0.3	-3.3	-1.5	-0.2
278.74	-1.13	1.4	256.99	9DE	-0.4	-3.3	-1.6	-0.2
286.63	-1.13	1.4	267.19	19DE	-0.5	-3.3	-1.6	-0.2
294.53	-1.12	1.4	277.37	29DE	-0.9	-3.3	-1.7	-0.1
				119				
302.43	-1.11	1.4	287.53	8JA	-1.1	-3.4	-1.8	-0.1
310.30	-1.07	1.4	297.65	18JA	-1.0	-3.4	-1.8	-0.0
318.14	-1.03	1.4	307.72	28JA	-0.3	-3.4	-1.9	0.1
325.93	-0.98	1.4	317.75	7FE	1.6	-3.4	-1.9	0.1
333.67	-0.92	1.4	327.71	17FE	2.9	-3.4	-2.0	0.2
341.34	-0.85	1.4	337.62	27FE	1.4	-3.5	-2.0	0.2
348.95	-0.78	1.4	347.47	9MR	0.8	-3.5	-2.0	0.3
356.47	-0.69	1.4	357.26	19MR	0.4	-3.6	-2.0	0.3
3.91	-0.60	1.5	7.00	29MR	0.0	-3.6	-2.0	0.3
11.27	-0.51	1.5	16.69	8AP	-0.6	-3.7	-2.0	0.4
18.54	-0.41	1.6	26.32	18AP	-1.5	-3.8	-1.9	0.4
25.72	-0.31	1.6	35.92	28AP	-1.5	-3.9	-1.9	0.4
32.81	-0.21	1.7	45.49	8MY	-0.5	-4.0	-1.8	0.4
39.81	-0.11	1.7	55.03	18MY	0.3	-4.0	-1.7	0.4
46.73	-0.00	1.7	64.55	28MY	0.9	-4.1	-1.7	0.3
53.57	0.10	1.8	74.07	7JN	1.6	-4.2	-1.6	0.3
60.31	0.21	1.8	83.60	17JN	2.8	-4.2	-1.6	0.3
66.98	0.32	1.8	93.13	27JN	2.8	-4.0	-1.5	0.3
73.57	0.43	1.8	102.69	7JL	1.1	-3.5	-1.4	0.4
80.08	0.53	1.8	112.28	17JL	-0.4	-3.1	-1.4	0.4
86.51	0.64	1.8	121.90	27JL	-1.3	-3.7	-1.3	0.4
92.87	0.75	1.8	131.57	6AU	-1.4	-4.1	-1.3	0.5
99.14	0.86	1.8	141.29	16AU	-0.7	-4.2	-1.3	0.5
105.32	0.98	1.8	151.07	26AU	-0.3	-4.3	-1.3	0.5
111.42	1.09	1.8	160.90	5SE	-0.1	-4.2	-1.2	0.5
117.42	1.21	1.7	170.79	15SE	0.1	-4.1	-1.2	0.4
123.32	1.33	1.7	180.74	25SE	0.2	-4.0	-1.2	0.4
129.09	1.45	1.6	190.75	5OC	1.1	-3.9	-1.2	0.4
134.72	1.58	1.5	200.81	15OC	3.0	-3.9	-1.2	0.3
140.20	1.72	1.5	210.92	25OC	0.2	-3.8	-1.3	0.3
145.48	1.87	1.3	221.06	4NO	-0.4	-3.7	-1.3	0.2
150.54	2.02	1.2	231.24	14NO	-0.5	-3.6	-1.3	0.1
155.32	2.19	1.1	241.44	24NO	-0.5	-3.6	-1.4	0.1
159.77	2.37	0.9	251.64	4DE	-0.6	-3.5	-1.4	-0.0
163.80	2.56	0.8	261.85	14DE	-0.9	-3.5	-1.4	-0.1
167.32	2.77	0.5	272.04	24DE	-0.9	-3.4	-1.5	-0.1

♂ / ☉ — 16.00UT — MAGNITUDES (☿ ♀ ♃ ♄)

120

♂ LONG	LAT	MAG	☉ LONG	16.00UT	☿	♀	♃	♄
170.19	2.99	0.3	282.21	3JA	-0.9	-3.4	-1.6	-0.1
172.27	3.22	0.1	292.35	13JA	-0.2	-3.4	-1.6	-0.0
173.37	3.44	-0.2	302.45	23JA	1.9	-3.4	-1.7	0.0
173.32	3.65	-0.5	312.50	2FE	2.3	-3.3	-1.8	0.1
172.01	3.81	-0.8	322.50	12FE	1.0	-3.3	-1.8	0.2
169.45	3.87	-1.0	332.44	22FE	0.6	-3.3	-1.9	0.2
165.98	3.80	-1.2	342.32	3MR	0.3	-3.3	-2.0	0.3
162.17	3.56	-1.2	352.14	13MR	-0.1	-3.3	-2.0	0.3
158.75	3.18	-1.1	1.91	23MR	-0.6	-3.3	-2.0	0.4
156.31	2.73	-0.9	11.61	2AP	-1.5	-3.4	-2.1	0.4
155.15	2.25	-0.7	21.27	12AP	-1.5	-3.4	-2.1	0.4
155.30	1.79	-0.5	30.90	22AP	-0.5	-3.4	-2.0	0.5
156.65	1.37	-0.2	40.47	2MY	0.4	-3.5	-2.0	0.5
159.02	0.99	-0.1	50.03	12MY	1.2	-3.5	-2.0	0.5
162.24	0.65	0.1	59.56	22MY	2.2	-3.5	-1.9	0.5
166.16	0.35	0.3	69.08	1JN	3.4	-3.4	-1.9	0.5
170.65	0.09	0.4	78.60	11JN	2.2	-3.4	-1.8	0.5
175.63	-0.15	0.5	88.13	21JN	0.9	-3.4	-1.8	0.4
181.02	-0.35	0.6	97.67	1JL	-0.4	-3.3	-1.7	0.4
186.75	-0.54	0.7	107.24	11JL	-1.3	-3.3	-1.6	0.5
192.79	-0.71	0.8	116.85	21JL	-1.4	-3.3	-1.6	0.5
199.10	-0.85	0.8	126.49	31JL	-0.7	-3.3	-1.5	0.5
205.64	-0.98	0.9	136.19	10AU	-0.2	-3.3	-1.5	0.6
212.40	-1.09	0.9	145.93	20AU	0.1	-3.3	-1.4	0.6
219.34	-1.18	1.0	155.73	30AU	0.2	-3.4	-1.4	0.6
226.45	-1.26	1.0	165.59	9SE	0.5	-3.4	-1.4	0.6
233.71	-1.32	1.1	175.52	19SE	1.4	-3.4	-1.3	0.6
241.11	-1.36	1.1	185.49	29SE	2.7	-3.5	-1.3	0.6
248.61	-1.39	1.1	195.52	9OC	0.1	-3.5	-1.3	0.6
256.22	-1.40	1.2	205.61	19OC	-0.6	-3.6	-1.3	0.5
263.90	-1.39	1.2	215.73	29OC	-0.6	-3.7	-1.3	0.5
271.64	-1.37	1.2	225.90	8NO	-0.6	-3.7	-1.3	0.4
279.43	-1.34	1.3	236.09	18NO	-0.7	-3.8	-1.3	0.4
287.23	-1.29	1.3	246.29	28NO	-0.8	-3.9	-1.3	0.3
295.05	-1.24	1.3	256.50	8DE	-0.8	-4.0	-1.4	0.2
302.86	-1.17	1.4	266.70	18DE	-0.8	-4.1	-1.4	0.2
310.65	-1.09	1.4	276.88	28DE	-0.1	-4.2	-1.4	0.1

121

♂ LONG	LAT	MAG	☉ LONG	16.00UT	☿	♀	♃	♄
318.40	-1.00	1.4	287.04	7JA	2.2	-4.3	-1.5	0.0
326.11	-0.91	1.5	297.16	17JA	1.8	-4.3	-1.5	0.1
333.76	-0.81	1.5	307.23	27JA	0.7	-4.3	-1.6	0.1
341.34	-0.71	1.5	317.26	6FE	0.4	-4.1	-1.7	0.2
348.86	-0.61	1.6	327.23	16FE	0.2	-3.6	-1.7	0.2
356.29	-0.50	1.6	337.14	26FE	-0.1	-3.3	-1.8	0.3
3.65	-0.39	1.6	347.00	8MR	-0.7	-3.8	-1.9	0.4
10.92	-0.29	1.6	356.79	18MR	-1.5	-4.2	-1.9	0.4
18.11	-0.18	1.7	6.53	28MR	-1.5	-4.2	-2.0	0.5
25.23	-0.08	1.7	16.22	7AP	-0.5	-4.2	-2.1	0.5
32.25	0.03	1.7	25.86	17AP	0.6	-4.1	-2.1	0.6
39.20	0.13	1.7	35.45	27AP	1.6	-4.0	-2.1	0.6
46.08	0.23	1.7	45.02	7MY	2.9	-3.9	-2.2	0.6
52.88	0.33	1.7	54.57	17MY	3.1	-3.8	-2.2	0.6
59.62	0.42	1.8	64.09	27MY	1.7	-3.7	-2.1	0.6
66.29	0.51	1.8	73.61	6JN	0.7	-3.7	-2.1	0.7
72.90	0.60	1.9	83.14	16JN	-0.4	-3.6	-2.1	0.7
79.45	0.68	1.9	92.67	26JN	-1.4	-3.5	-2.0	0.6
85.95	0.77	1.9	102.23	6JL	-1.5	-3.5	-2.0	0.6
92.40	0.85	2.0	111.81	16JL	-0.7	-3.4	-1.9	0.6
98.80	0.92	2.0	121.43	26JL	-0.1	-3.4	-1.9	0.6
105.16	1.00	2.0	131.10	5AU	0.2	-3.4	-1.8	0.7
111.48	1.07	2.0	140.82	15AU	0.4	-3.4	-1.7	0.8
117.75	1.15	2.0	150.59	25AU	0.7	-3.4	-1.7	0.8
123.98	1.22	2.0	160.42	4SE	1.8	-3.4	-1.6	0.8
130.17	1.29	1.9	170.31	14SE	2.4	-3.4	-1.6	0.8
136.31	1.35	1.9	180.26	24SE	-0.0	-3.4	-1.5	0.8
142.40	1.42	1.9	190.26	4OC	-0.7	-3.4	-1.5	0.8
148.43	1.49	1.8	200.32	14OC	-0.8	-3.4	-1.4	0.8
154.40	1.55	1.7	210.42	24OC	-0.8	-3.4	-1.4	0.8
160.29	1.62	1.7	220.57	3NO	-0.7	-3.4	-1.4	0.8
166.09	1.68	1.6	230.74	13NO	-0.6	-3.4	-1.4	0.7
171.79	1.74	1.5	240.93	23NO	-0.6	-3.5	-1.4	0.6
177.37	1.80	1.4	251.14	3DE	-0.6	-3.5	-1.4	0.6
182.79	1.86	1.2	261.35	13DE	0.0	-3.5	-1.4	0.5
188.03	1.92	1.1	271.54	23DE	2.5	-3.5	-1.4	0.4

122

♂ LONG	LAT	MAG	☉ LONG	16.00UT	☿	♀	♃	♄
193.04	1.97	0.9	281.72	2JA	1.4	-3.4	-1.4	0.4
197.78	2.02	0.8	291.86	12JA	0.5	-3.4	-1.5	0.3
202.18	2.06	0.5	301.96	22JA	0.2	-3.4	-1.5	0.3
206.15	2.08	0.3	312.01	1FE	0.0	-3.4	-1.5	0.3
209.58	2.09	0.1	322.01	11FE	-0.2	-3.3	-1.6	0.3
212.34	2.07	-0.2	331.95	21FE	-0.7	-3.3	-1.7	0.4
214.26	2.01	-0.5	341.84	3MR	-1.5	-3.3	-1.7	0.5
215.17	1.89	-0.8	351.66	13MR	-1.4	-3.3	-1.8	0.5
214.90	1.70	-1.1	1.43	23MR	-0.5	-3.3	-1.9	0.6
213.36	1.42	-1.4	11.15	2AP	0.7	-3.4	-1.9	0.6
210.70	1.03	-1.7	20.81	12AP	2.1	-3.4	-2.0	0.7
207.33	0.55	-1.9	30.43	22AP	3.8	-3.4	-2.1	0.7
203.94	0.03	-1.8	40.01	2MY	2.4	-3.4	-2.1	0.8
201.25	-0.48	-1.6	49.57	12MY	1.3	-3.5	-2.2	0.8
199.73	-0.93	-1.4	59.10	22MY	0.5	-3.5	-2.2	0.8
199.59	-1.31	-1.2	68.63	1JN	-0.5	-3.6	-2.3	0.8
200.79	-1.61	-1.0	78.14	11JN	-1.4	-3.6	-2.3	0.9
203.14	-1.85	-0.8	87.67	21JN	-1.5	-3.7	-2.3	0.9
206.50	-2.02	-0.6	97.22	1JL	-0.6	-3.8	-2.3	0.9
210.66	-2.15	-0.4	106.78	11JL	-0.1	-3.9	-2.2	0.9
215.49	-2.24	-0.3	116.39	21JL	0.3	-4.0	-2.2	0.9
220.87	-2.29	-0.1	126.03	31JL	0.6	-4.1	-2.1	0.9
226.70	-2.31	0.0	135.72	10AU	1.0	-4.2	-2.1	0.9
232.90	-2.31	0.1	145.46	20AU	2.2	-4.3	-2.0	1.0
239.41	-2.28	0.2	155.26	30AU	2.1	-4.3	-2.0	1.0
246.17	-2.23	0.3	165.12	9SE	-0.1	-4.1	-1.9	1.0
253.14	-2.16	0.4	175.03	19SE	-0.9	-3.6	-1.8	1.1
260.28	-2.07	0.5	185.01	29SE	-0.9	-3.3	-1.8	1.1
267.55	-1.96	0.6	195.03	9OC	-0.9	-3.9	-1.7	1.1
274.93	-1.84	0.7	205.12	19OC	-0.6	-4.3	-1.7	1.1
282.38	-1.72	0.8	215.24	29OC	-0.5	-4.4	-1.6	1.1
289.88	-1.58	0.9	225.40	8NO	-0.5	-4.4	-1.6	1.0
297.40	-1.44	1.0	235.59	18NO	-0.4	-4.3	-1.5	1.0
304.94	-1.29	1.1	245.79	28NO	0.2	-4.2	-1.5	1.0
312.46	-1.14	1.1	255.99	8DE	2.8	-4.1	-1.5	0.9
319.97	-1.00	1.2	266.20	18DE	1.0	-4.0	-1.5	0.8
327.43	-0.85	1.3	276.39	28DE	0.2	-3.9	-1.5	0.8

123

♂ LONG	LAT	MAG	☉ LONG	16.00UT	☿	♀	♃	♄
334.85	-0.70	1.4	286.54	7JA	0.1	-3.8	-1.5	0.7
342.22	-0.56	1.4	296.67	17JA	-0.1	-3.7	-1.5	0.6
349.52	-0.43	1.5	306.75	27JA	-0.3	-3.6	-1.5	0.6
356.76	-0.30	1.5	316.78	6FE	-0.7	-3.4	-1.5	0.5
3.93	-0.17	1.6	326.75	16FE	-1.4	-3.5	-1.6	0.6
11.03	-0.05	1.7	336.67	26FE	-1.3	-3.4	-1.6	0.6
18.05	0.07	1.7	346.52	8MR	-0.5	-3.4	-1.7	0.7
25.01	0.17	1.8	356.33	18MR	0.9	-3.4	-1.7	0.7
31.90	0.28	1.8	6.07	28MR	2.6	-3.3	-1.8	0.8
38.72	0.37	1.8	15.75	7AP	3.1	-3.3	-1.8	0.8
45.48	0.46	1.9	25.40	17AP	1.8	-3.3	-1.9	0.9
52.19	0.55	1.9	35.00	27AP	1.0	-3.3	-2.0	0.9
58.83	0.63	1.9	44.56	7MY	0.3	-3.3	-2.0	1.0
65.43	0.70	1.9	54.11	17MY	-0.5	-3.3	-2.1	1.0
71.98	0.77	1.9	63.64	27MY	-1.5	-3.4	-2.2	1.1
78.49	0.84	1.9	73.15	6JN	-1.5	-3.4	-2.2	1.1
84.97	0.90	1.9	82.68	16JN	-0.6	-3.4	-2.3	1.1
91.41	0.95	1.9	92.21	26JN	0.0	-3.4	-2.3	1.1
97.83	1.00	1.9	101.77	6JL	0.4	-3.5	-2.4	1.1
104.23	1.05	2.0	111.35	16JL	0.8	-3.5	-2.4	1.2
110.61	1.09	2.0	120.97	26JL	1.3	-3.5	-2.4	1.2
116.97	1.13	2.0	130.63	5AU	2.7	-3.4	-2.4	1.2
123.32	1.17	2.0	140.35	15AU	1.8	-3.4	-2.4	1.2
129.66	1.20	2.0	150.12	25AU	-0.2	-3.4	-2.3	1.2
136.00	1.23	2.0	159.94	4SE	-1.0	-3.4	-2.3	1.3
142.33	1.25	2.0	169.83	14SE	-1.1	-3.3	-2.2	1.3
148.66	1.27	2.0	179.77	24SE	-0.9	-3.3	-2.1	1.3
154.99	1.28	2.0	189.77	4OC	-0.5	-3.3	-2.1	1.4
161.31	1.29	1.9	199.83	14OC	-0.4	-3.3	-2.0	1.4
167.62	1.30	1.9	209.92	24OC	-0.3	-3.3	-1.9	1.4
173.93	1.30	1.8	220.07	3NO	-0.2	-3.4	-1.9	1.3
180.23	1.30	1.8	230.24	13NO	0.4	-3.4	-1.8	1.3
186.51	1.28	1.7	240.43	23NO	3.1	-3.4	-1.7	1.3
192.77	1.26	1.6	250.64	3DE	0.8	-3.5	-1.7	1.2
199.01	1.24	1.5	260.85	13DE	0.0	-3.5	-1.7	1.2
205.21	1.20	1.4	271.04	23DE	-0.1	-3.5	-1.6	1.1

♂			☉	16.00UT	☿	♀	♃	♄
LONG	LAT	MAG	LONG	124	MAGNITUDES			
211.37	1.15	1.3	281.22	2JA	-0.2	-3.6	-1.6	1.1
217.47	1.08	1.2	291.36	12JA	-0.4	-3.7	-1.6	1.0
223.50	1.00	1.1	301.47	22JA	-0.8	-3.7	-1.6	1.0
229.46	0.90	0.9	311.52	1FE	-1.3	-3.8	-1.6	0.9
235.30	0.77	0.7	321.53	11FE	-1.2	-3.9	-1.6	0.9
241.01	0.62	0.6	331.47	21FE	-0.4	-4.0	-1.6	0.8
246.56	0.43	0.4	341.36	2MR	1.2	-4.1	-1.6	0.9
251.89	0.20	0.2	351.19	12MR	3.2	-4.1	-1.6	0.9
256.96	-0.07	-0.1	0.96	22MR	2.3	-4.2	-1.6	1.0
261.68	-0.40	-0.3	10.68	1AP	1.3	-4.2	-1.7	1.0
265.96	-0.79	-0.6	20.35	11AP	0.8	-4.1	-1.7	1.1
269.68	-1.27	-0.8	29.97	21AP	0.2	-3.7	-1.8	1.1
272.69	-1.82	-1.1	39.55	1MY	-0.5	-3.0	-1.8	1.2
274.79	-2.47	-1.4	49.11	11MY	-1.5	-3.3	-1.9	1.2
275.81	-3.21	-1.7	58.64	21MY	-1.5	-3.9	-1.9	1.3
275.60	-3.99	-2.0	68.16	31MY	-0.6	-4.1	-2.0	1.3
274.17	-4.76	-2.3	77.69	10JN	0.1	-4.2	-2.1	1.4
271.83	-5.42	-2.5	87.21	20JN	0.6	-4.1	-2.2	1.4
269.12	-5.85	-2.5	96.75	30JN	1.0	-4.1	-2.2	1.4
266.80	-6.00	-2.4	106.32	10JL	1.8	-4.0	-2.3	1.4
265.49	-5.90	-2.2	115.92	20JL	3.2	-3.9	-2.3	1.4
265.46	-5.61	-1.9	125.56	30JL	1.5	-3.8	-2.4	1.4
266.78	-5.21	-1.6	135.25	9AU	-0.3	-3.7	-2.4	1.3
269.29	-4.76	-1.4	144.99	19AU	-1.1	-3.7	-2.5	1.3
272.79	-4.29	-1.1	154.79	29AU	-1.2	-3.6	-2.5	1.2
277.10	-3.83	-0.9	164.64	8SE	-0.8	-3.6	-2.4	1.2
282.04	-3.39	-0.7	174.55	18SE	-0.4	-3.5	-2.4	1.2
287.47	-2.97	-0.4	184.53	28SE	-0.2	-3.5	-2.4	1.2
293.27	-2.58	-0.2	194.55	8OC	-0.2	-3.4	-2.3	1.2
299.36	-2.21	-0.0	204.63	18OC	-0.1	-3.4	-2.3	1.2
305.66	-1.88	0.1	214.75	28OC	0.6	-3.4	-2.2	1.2
312.13	-1.56	0.3	224.91	7NO	3.3	-3.4	-2.1	1.2
318.71	-1.28	0.5	235.09	17NO	0.5	-3.4	-2.0	1.1
325.37	-1.01	0.6	245.29	27NO	-0.1	-3.4	-2.0	1.1
332.09	-0.78	0.8	255.50	7DE	-0.2	-3.4	-1.9	1.1
338.83	-0.56	0.9	265.70	17DE	-0.3	-3.4	-1.8	1.0
345.59	-0.36	1.0	275.89	27DE	-0.5	-3.4	-1.8	1.0
				125				
352.35	-0.18	1.2	286.05	6JA	-0.8	-3.4	-1.7	0.9
359.09	-0.02	1.3	296.17	16JA	-1.2	-3.4	-1.7	0.9
5.82	0.13	1.4	306.26	26JA	-1.1	-3.4	-1.7	0.8
12.52	0.26	1.5	316.29	5FE	-0.4	-3.4	-1.6	0.8
19.18	0.38	1.6	326.26	15FE	1.4	-3.5	-1.6	0.8
25.81	0.49	1.6	336.18	25FE	3.2	-3.5	-1.6	0.7
32.41	0.59	1.7	346.04	7MR	1.7	-3.5	-1.6	0.8
38.97	0.67	1.8	355.84	17MR	1.0	-3.4	-1.6	0.8
45.50	0.75	1.8	5.59	27MR	0.6	-3.4	-1.6	0.9
51.99	0.82	1.9	15.28	6AP	0.1	-3.4	-1.6	1.0
58.46	0.88	1.9	24.92	16AP	-0.6	-3.4	-1.6	1.0
64.89	0.93	2.0	34.53	26AP	-1.5	-3.3	-1.6	1.1
71.30	0.98	2.0	44.10	6MY	-1.5	-3.3	-1.7	1.1
77.69	1.02	2.0	53.64	16MY	-0.6	-3.3	-1.7	1.2
84.06	1.06	2.0	63.17	26MY	0.2	-3.3	-1.8	1.2
90.42	1.09	2.0	72.69	5JN	0.8	-3.3	-1.8	1.3
96.77	1.11	2.0	82.21	15JN	1.4	-3.4	-1.9	1.3
103.12	1.13	2.0	91.75	25JN	2.3	-3.4	-1.9	1.3
109.47	1.15	2.0	101.30	5JL	3.1	-3.4	-2.0	1.3
115.83	1.15	2.0	110.88	15JL	1.3	-3.4	-2.1	1.3
122.19	1.16	2.0	120.50	25JL	-0.3	-3.5	-2.1	1.3
128.57	1.16	2.0	130.16	4AU	-1.2	-3.6	-2.2	1.2
134.97	1.16	2.0	139.88	14AU	-1.3	-3.6	-2.3	1.2
141.39	1.15	2.0	149.64	24AU	-0.8	-3.7	-2.3	1.1
147.83	1.13	2.0	159.46	3SE	-0.4	-3.8	-2.4	1.1
154.30	1.11	2.0	169.35	13SE	-0.1	-3.9	-2.4	1.1
160.80	1.09	2.0	179.29	23SE	-0.0	-4.0	-2.4	1.1
167.33	1.06	2.0	189.29	3OC	0.1	-4.1	-2.4	1.1
173.89	1.03	1.9	199.34	13OC	0.8	-4.2	-2.4	1.1
180.49	0.98	1.9	209.44	23OC	3.2	-4.3	-2.4	1.1
187.12	0.94	1.9	219.58	2NO	0.3	-4.4	-2.4	1.1
193.80	0.88	1.8	229.75	12NO	-0.3	-4.4	-2.3	1.0
200.50	0.82	1.8	239.94	22NO	-0.4	-4.1	-2.2	1.0
207.24	0.74	1.7	250.15	2DE	-0.4	-3.6	-2.2	1.0
214.02	0.66	1.7	260.35	12DE	-0.6	-3.1	-2.1	1.0
220.83	0.57	1.6	270.55	22DE	-0.9	-3.9	-2.0	0.9

♂			☉	16.00UT	☿	♀	♃	♄
LONG	LAT	MAG	LONG	126	MAGNITUDES			
227.67	0.46	1.5	280.72	1JA	-1.0	-4.3	-2.0	0.9
234.54	0.35	1.4	290.87	11JA	-1.0	-4.4	-1.9	0.8
241.44	0.22	1.3	300.98	21JA	-0.3	-4.3	-1.8	0.8
248.36	0.07	1.2	311.03	31JA	1.7	-4.2	-1.8	0.7
255.30	-0.09	1.1	321.04	10FE	2.7	-4.1	-1.7	0.7
262.26	-0.27	1.0	330.99	20FE	1.3	-4.0	-1.7	0.6
269.22	-0.46	0.9	340.88	2MR	0.7	-3.9	-1.6	0.6
276.18	-0.67	0.8	350.71	12MR	0.4	-3.8	-1.6	0.6
283.12	-0.90	0.7	0.49	22MR	0.0	-3.7	-1.6	0.6
290.03	-1.15	0.5	10.20	1AP	-0.6	-3.6	-1.5	0.7
296.89	-1.41	0.4	19.88	11AP	-1.5	-3.6	-1.5	0.7
303.66	-1.69	0.3	29.50	21AP	-1.5	-3.5	-1.5	0.8
310.33	-1.99	0.1	39.09	1MY	-0.6	-3.4	-1.5	0.9
316.85	-2.30	-0.0	48.65	11MY	0.3	-3.4	-1.5	0.9
323.17	-2.63	-0.2	58.18	21MY	1.0	-3.4	-1.5	1.0
329.22	-2.97	-0.4	67.71	31MY	1.8	-3.3	-1.6	1.0
334.93	-3.31	-0.5	77.23	10JN	3.0	-3.3	-1.6	1.1
340.18	-3.67	-0.7	86.76	20JN	2.6	-3.3	-1.6	1.1
344.87	-4.03	-0.9	96.30	30JN	1.0	-3.3	-1.7	1.1
348.81	-4.38	-1.1	105.86	10JL	-0.3	-3.3	-1.7	1.1
351.83	-4.72	-1.4	115.46	20JL	-1.3	-3.3	-1.8	1.1
353.72	-5.01	-1.6	125.10	30JL	-1.4	-3.4	-1.8	1.1
354.26	-5.22	-1.8	134.78	9AU	-0.7	-3.4	-1.9	1.1
353.36	-5.30	-2.1	144.52	19AU	-0.3	-3.4	-2.0	1.1
351.16	-5.17	-2.3	154.31	29AU	-0.0	-3.4	-2.0	1.0
348.11	-4.81	-2.4	164.16	8SE	0.1	-3.5	-2.1	1.0
345.01	-4.23	-2.2	174.07	18SE	0.3	-3.5	-2.2	0.9
342.61	-3.53	-1.9	184.04	28SE	1.1	-3.5	-2.2	0.9
341.38	-2.80	-1.6	194.06	8OC	3.0	-3.5	-2.3	0.9
341.50	-2.12	-1.3	204.14	18OC	0.2	-3.4	-2.3	1.0
342.85	-1.52	-1.0	214.25	28OC	-0.5	-3.4	-2.3	1.0
345.26	-1.01	-0.7	224.41	7NO	-0.5	-3.4	-2.3	1.0
348.53	-0.59	-0.4	234.59	17NO	-0.5	-3.4	-2.3	1.0
352.46	-0.24	-0.1	244.79	27NO	-0.7	-3.3	-2.3	0.9
356.90	0.05	0.1	255.00	7DE	-0.8	-3.3	-2.3	0.9
1.74	0.29	0.4	265.20	17DE	-0.9	-3.3	-2.2	0.9
6.88	0.48	0.6	275.39	27DE	-0.8	-3.3	-2.1	0.9
				127				
12.27	0.65	0.8	285.56	6JA	-0.2	-3.4	-2.1	0.8
17.83	0.78	0.9	295.68	16JA	1.9	-3.4	-2.0	0.8
23.53	0.89	1.1	305.76	26JA	2.1	-3.4	-1.9	0.7
29.33	0.98	1.2	315.80	5FE	0.9	-3.4	-1.9	0.7
35.22	1.05	1.3	325.78	15FE	0.5	-3.4	-1.8	0.6
41.17	1.11	1.5	335.70	25FE	0.2	-3.5	-1.7	0.6
47.17	1.16	1.6	345.57	7MR	-0.1	-3.5	-1.7	0.5
53.21	1.20	1.7	355.37	17MR	-0.6	-3.6	-1.6	0.5
59.28	1.23	1.7	5.11	27MR	-1.4	-3.6	-1.6	0.5
65.37	1.25	1.8	14.81	6AP	-1.5	-3.7	-1.5	0.5
71.48	1.27	1.9	24.46	16AP	-0.6	-3.8	-1.5	0.6
77.61	1.28	1.9	34.06	26AP	0.4	-3.9	-1.5	0.6
83.76	1.28	2.0	43.64	6MY	1.3	-4.0	-1.4	0.7
89.92	1.28	2.0	53.18	16MY	2.4	-4.0	-1.4	0.8
96.11	1.27	2.0	62.71	26MY	3.5	-4.1	-1.4	0.8
102.32	1.25	2.0	72.24	5JN	2.0	-4.2	-1.4	0.9
108.55	1.24	2.0	81.76	15JN	0.8	-4.2	-1.4	0.9
114.82	1.22	2.0	91.29	25JN	-0.4	-4.0	-1.4	1.0
121.11	1.19	2.0	100.84	5JL	-1.3	-3.5	-1.5	1.0
127.44	1.16	2.0	110.42	15JL	-1.5	-3.1	-1.5	1.0
133.80	1.13	2.0	120.04	25JL	-0.7	-3.7	-1.5	1.0
140.21	1.09	2.0	129.70	4AU	-0.2	-4.1	-1.6	1.0
146.66	1.04	1.9	139.41	14AU	0.1	-4.2	-1.6	1.0
153.17	1.00	1.9	149.17	24AU	0.3	-4.3	-1.7	1.0
159.72	0.95	1.9	158.99	3SE	0.6	-4.2	-1.7	0.9
166.33	0.89	1.9	168.87	13SE	1.5	-4.1	-1.8	0.9
173.00	0.83	1.9	178.81	23SE	2.7	-4.0	-1.9	0.9
179.73	0.76	1.9	188.80	3OC	0.1	-3.9	-1.9	0.8
186.51	0.69	1.8	198.85	13OC	-0.6	-3.9	-2.0	0.8
193.37	0.61	1.8	208.95	23OC	-0.7	-3.8	-2.1	0.9
200.29	0.53	1.8	219.08	2NO	-0.7	-3.7	-2.1	0.9
207.27	0.44	1.8	229.25	12NO	-0.8	-3.6	-2.2	0.9
214.32	0.34	1.8	239.45	22NO	-0.7	-3.6	-2.2	0.9
221.43	0.24	1.7	249.65	2DE	-0.7	-3.5	-2.2	0.9
228.60	0.13	1.7	259.86	12DE	-0.7	-3.5	-2.2	0.9
235.84	0.02	1.6	270.06	22DE	-0.1	-3.4	-2.2	0.8

128

♂ LONG	LAT	MAG	☉ LONG	16.00UT	☿	♀	♃	♄
243.13	-0.10	1.6	280.24	1JA	2.2	-3.4	-2.2	0.8
250.48	-0.23	1.5	290.38	11JA	1.7	-3.4	-2.1	0.8
257.88	-0.36	1.5	300.49	21JA	0.6	-3.4	-2.1	0.7
265.33	-0.50	1.4	310.55	31JA	0.3	-3.3	-2.0	0.7
272.81	-0.64	1.4	320.56	10FE	0.1	-3.3	-2.0	0.6
280.30	-0.79	1.3	330.51	20FE	-0.2	-3.3	-1.9	0.6
287.87	-0.93	1.2	340.40	1MR	-0.7	-3.3	-1.8	0.5
295.43	-1.08	1.2	350.24	11MR	-1.4	-3.3	-1.7	0.5
302.98	-1.22	1.1	0.02	21MR	-1.5	-3.3	-1.7	0.4
310.53	-1.36	1.0	9.74	31MR	-0.5	-3.4	-1.6	0.4
318.04	-1.49	1.0	19.41	10AP	0.6	-3.4	-1.6	0.4
325.52	-1.61	0.9	29.03	20AP	1.7	-3.4	-1.5	0.4
332.92	-1.73	0.8	38.62	30AP	3.2	-3.5	-1.5	0.5
340.25	-1.83	0.8	48.18	10MY	2.8	-3.5	-1.4	0.6
347.48	-1.92	0.7	57.72	20MY	1.6	-3.5	-1.4	0.6
354.57	-2.00	0.6	67.24	30MY	0.6	-3.4	-1.4	0.7
1.50	-2.05	0.5	76.76	9JN	-0.4	-3.4	-1.3	0.8
8.25	-2.09	0.5	86.29	19JN	-1.4	-3.4	-1.3	0.8
14.78	-2.11	0.4	95.83	29JN	-1.5	-3.3	-1.3	0.8
21.04	-2.11	0.3	105.39	9JL	-0.7	-3.3	-1.3	0.9
26.97	-2.09	0.2	114.99	19JL	-0.1	-3.3	-1.3	0.9
32.52	-2.05	0.0	124.63	29JL	0.2	-3.3	-1.3	0.9
37.61	-1.97	-0.1	134.31	8AU	0.5	-3.3	-1.4	0.9
42.13	-1.87	-0.2	144.04	18AU	0.8	-3.3	-1.4	0.9
45.93	-1.74	-0.4	153.83	28AU	1.9	-3.4	-1.4	0.9
48.87	-1.56	-0.6	163.68	7SE	2.4	-3.4	-1.5	0.9
50.73	-1.32	-0.8	173.59	17SE	-0.0	-3.4	-1.5	0.9
51.32	-1.03	-1.0	183.55	27SE	-0.8	-3.5	-1.6	0.8
50.47	-0.66	-1.2	193.57	7OC	-0.8	-3.5	-1.6	0.8
48.20	-0.23	-1.4	203.65	17OC	-0.8	-3.6	-1.7	0.8
44.82	0.24	-1.6	213.76	27OC	-0.7	-3.7	-1.8	0.8
40.98	0.70	-1.6	223.92	6NO	-0.6	-3.7	-1.8	0.8
37.49	1.10	-1.4	234.10	16NO	-0.6	-3.8	-1.9	0.8
35.03	1.42	-1.1	244.30	26NO	-0.3	-3.9	-2.0	0.8
33.87	1.64	-0.6	254.51	6DE	0.1	-4.0	-2.0	0.8
34.05	1.78	-0.4	264.71	16DE	2.5	-4.1	-2.1	0.8
35.41	1.87	-0.2	274.90	26DE	1.3	-4.2	-2.1	0.8

129

♂ LONG	LAT	MAG	☉ LONG	16.00UT	☿	♀	♃	♄
37.73	1.91	0.1	285.06	5JA	0.4	-4.3	-2.1	0.8
40.82	1.93	0.4	295.19	15JA	0.1	-4.3	-2.1	0.8
44.52	1.93	0.6	305.28	25JA	0.0	-4.3	-2.1	0.7
48.69	1.91	0.8	315.31	4FE	-0.2	-4.1	-2.1	0.7
53.23	1.88	1.0	325.30	14FE	-0.7	-3.5	-2.0	0.6
58.07	1.85	1.1	335.22	24FE	-1.4	-3.3	-2.0	0.6
63.13	1.81	1.3	345.09	6MR	-1.4	-3.9	-1.9	0.5
68.39	1.76	1.4	354.90	16MR	-0.5	-4.2	-1.8	0.5
73.80	1.72	1.5	4.64	26MR	0.8	-4.2	-1.8	0.4
79.34	1.67	1.6	14.34	5AP	2.2	-4.2	-1.7	0.4
84.99	1.62	1.7	23.99	15AP	3.7	-4.1	-1.6	0.3
90.73	1.56	1.7	33.60	25AP	2.1	-4.0	-1.6	0.3
96.56	1.51	1.8	43.17	5MY	1.2	-3.9	-1.5	0.4
102.46	1.45	1.9	52.72	15MY	0.5	-3.8	-1.5	0.5
108.43	1.39	1.9	62.25	25MY	-0.4	-3.7	-1.4	0.5
114.48	1.33	1.9	71.77	4JN	-1.4	-3.7	-1.4	0.6
120.59	1.27	1.9	81.29	14JN	-1.5	-3.6	-1.3	0.7
126.76	1.20	2.0	90.82	24JN	-0.7	-3.5	-1.3	0.7
133.00	1.13	2.0	100.38	4JL	-0.0	-3.5	-1.3	0.8
139.31	1.06	2.0	109.96	14JL	0.3	-3.4	-1.3	0.8
145.69	0.99	2.0	119.57	24JL	0.6	-3.4	-1.2	0.8
152.14	0.91	1.9	129.23	3AU	1.1	-3.4	-1.2	0.8
158.60	0.83	1.9	138.94	13AU	2.4	-3.4	-1.2	0.9
165.26	0.75	1.9	148.70	23AU	2.1	-3.4	-1.3	0.9
171.94	0.66	1.9	158.52	2SE	-0.1	-3.4	-1.3	0.9
178.70	0.58	1.8	168.39	12SE	-0.9	-3.4	-1.3	0.9
185.54	0.48	1.8	178.33	22SE	-1.0	-3.4	-1.3	0.8
192.47	0.39	1.7	188.32	2OC	-0.9	-3.4	-1.4	0.8
199.47	0.29	1.7	198.37	12OC	-0.6	-3.4	-1.4	0.8
206.56	0.19	1.6	208.46	22OC	-0.5	-3.4	-1.5	0.7
213.74	0.09	1.6	218.60	1NO	-0.4	-3.4	-1.5	0.7
220.99	-0.02	1.6	228.76	11NO	-0.4	-3.4	-1.6	0.7
228.33	-0.12	1.6	238.95	21NO	0.2	-3.5	-1.7	0.8
235.74	-0.23	1.6	249.16	1DE	2.8	-3.5	-1.7	0.8
243.22	-0.34	1.6	259.36	11DE	1.0	-3.5	-1.8	0.8
250.77	-0.45	1.5	269.56	21DE	0.2	-3.5	-1.8	0.8
258.37	-0.56	1.5	279.74	31DE	-0.0	-3.4	-1.9	0.8

130

♂ LONG	LAT	MAG	☉ LONG	16.00UT	☿	♀	♃	♄
266.03	-0.67	1.5	289.89	10JA	-0.1	-3.4	-2.0	0.8
273.74	-0.77	1.5	299.99	20JA	-0.3	-3.4	-2.0	0.8
281.48	-0.87	1.5	310.06	30JA	-0.7	-3.4	-2.0	0.7
289.24	-0.97	1.4	320.07	9FE	-1.4	-3.3	-2.0	0.7
297.01	-1.05	1.4	330.02	19FE	-1.3	-3.3	-2.0	0.6
304.79	-1.13	1.4	339.92	1MR	-0.5	-3.3	-2.0	0.6
312.55	-1.20	1.4	349.76	11MR	1.0	-3.3	-2.0	0.5
320.29	-1.26	1.3	359.54	21MR	2.8	-3.3	-1.9	0.5
328.00	-1.30	1.3	9.27	31MR	2.8	-3.4	-1.9	0.4
335.66	-1.33	1.3	18.94	10AP	1.6	-3.4	-1.8	0.4
343.25	-1.35	1.3	28.56	20AP	0.9	-3.4	-1.8	0.3
350.77	-1.36	1.2	38.16	30AP	0.3	-3.4	-1.7	0.3
358.20	-1.35	1.2	47.72	10MY	-0.5	-3.5	-1.6	0.3
5.54	-1.32	1.2	57.25	20MY	-1.4	-3.5	-1.6	0.4
12.76	-1.28	1.2	66.78	30MY	-1.6	-3.6	-1.5	0.4
19.86	-1.23	1.1	76.30	9JN	-0.6	-3.7	-1.4	0.5
26.82	-1.16	1.1	85.83	19JN	0.0	-3.7	-1.4	0.6
33.63	-1.08	1.1	95.37	29JN	0.5	-3.8	-1.3	0.6
40.28	-0.98	1.0	104.93	9JL	0.9	-3.9	-1.3	0.7
46.74	-0.87	1.0	114.53	19JL	1.5	-4.0	-1.3	0.7
52.99	-0.74	0.9	124.16	29JL	2.9	-4.1	-1.2	0.8
59.01	-0.59	0.9	133.84	8AU	1.8	-4.2	-1.2	0.8
64.76	-0.43	0.8	143.57	18AU	-0.2	-4.3	-1.2	0.8
70.19	-0.25	0.7	153.36	28AU	-1.0	-4.2	-1.2	0.8
75.25	-0.04	0.6	163.20	7SE	-1.1	-4.0	-1.2	0.8
79.86	0.19	0.4	173.11	17SE	-0.9	-3.5	-1.2	0.8
83.92	0.45	0.3	183.07	27SE	-0.5	-3.3	-1.2	0.8
87.31	0.75	0.1	193.09	7OC	-0.3	-3.9	-1.3	0.8
89.87	1.09	-0.1	203.16	17OC	-0.3	-4.3	-1.3	0.7
91.41	1.48	-0.3	213.28	27OC	-0.2	-4.4	-1.3	0.7
91.75	1.91	-0.5	223.43	6NO	0.4	-4.3	-1.4	0.7
90.73	2.37	-0.7	233.61	16NO	3.1	-4.3	-1.4	0.7
88.34	2.82	-0.9	243.81	26NO	0.7	-4.2	-1.5	0.7
84.87	3.21	-1.1	254.01	6DE	-0.0	-4.1	-1.5	0.8
80.90	3.48	-1.2	264.22	16DE	-0.2	-4.0	-1.6	0.8
77.19	3.61	-0.9	274.41	26DE	-0.2	-3.9	-1.7	0.8

131

♂ LONG	LAT	MAG	☉ LONG	16.00UT	☿	♀	♃	♄
74.42	3.59	-0.7	284.57	5JA	-0.4	-3.8	-1.7	0.8
72.90	3.48	-0.4	294.70	15JA	-0.8	-3.7	-1.8	0.8
72.70	3.31	-0.1	304.80	25JA	-1.3	-3.6	-1.9	0.8
73.68	3.12	0.1	314.83	4FE	-1.2	-3.6	-1.9	0.7
75.65	2.92	0.4	324.82	14FE	-0.4	-3.5	-2.0	0.7
78.43	2.73	0.6	334.75	24FE	1.2	-3.4	-2.0	0.7
81.84	2.55	0.8	344.61	6MR	3.3	-3.4	-2.0	0.6
85.77	2.38	0.9	354.42	16MR	2.1	-3.4	-2.0	0.6
90.11	2.22	1.1	4.18	26MR	1.2	-3.3	-2.0	0.5
94.78	2.07	1.2	13.88	5AP	0.7	-3.3	-2.0	0.5
99.72	1.93	1.3	23.53	15AP	0.2	-3.3	-2.0	0.4
104.89	1.79	1.4	33.14	25AP	-0.5	-3.3	-1.9	0.3
110.25	1.66	1.5	42.71	5MY	-1.4	-3.3	-1.9	0.3
115.78	1.53	1.6	52.26	15MY	-1.6	-3.3	-1.8	0.3
121.47	1.41	1.6	61.79	25MY	-0.6	-3.3	-1.7	0.3
127.28	1.29	1.7	71.31	4JN	0.1	-3.4	-1.7	0.4
133.22	1.18	1.7	80.83	14JN	0.6	-3.4	-1.6	0.4
139.29	1.06	1.8	90.37	24JN	1.1	-3.4	-1.5	0.5
145.46	0.95	1.8	99.91	4JL	2.0	-3.5	-1.5	0.6
151.74	0.84	1.8	109.49	14JL	3.2	-3.5	-1.4	0.6
158.13	0.72	1.8	119.10	24JL	1.5	-3.5	-1.4	0.7
164.62	0.61	1.8	128.76	3AU	-0.2	-3.4	-1.3	0.7
171.22	0.50	1.8	138.46	13AU	-1.1	-3.4	-1.3	0.7
177.93	0.39	1.8	148.22	23AU	-1.3	-3.4	-1.3	0.8
184.74	0.28	1.8	158.03	2SE	-0.8	-3.4	-1.3	0.8
191.65	0.17	1.7	167.91	12SE	-0.4	-3.3	-1.2	0.8
198.67	0.06	1.7	177.84	22SE	-0.2	-3.3	-1.2	0.8
205.79	-0.05	1.7	187.83	2OC	-0.1	-3.3	-1.2	0.8
213.01	-0.16	1.6	197.87	12OC	0.0	-3.3	-1.2	0.8
220.32	-0.26	1.6	207.96	22OC	0.6	-3.3	-1.2	0.8
227.73	-0.37	1.5	218.10	1NO	3.4	-3.4	-1.3	0.7
235.22	-0.47	1.5	228.27	11NO	0.5	-3.4	-1.3	0.7
242.80	-0.57	1.4	238.46	21NO	-0.2	-3.4	-1.3	0.7
250.44	-0.66	1.4	248.66	1DE	-0.3	-3.5	-1.4	0.7
258.15	-0.74	1.4	258.87	11DE	-0.3	-3.5	-1.4	0.8
265.92	-0.82	1.4	269.07	21DE	-0.5	-3.5	-1.5	0.8
273.73	-0.90	1.4	279.25	31DE	-0.8	-3.6	-1.5	0.8

132

♂ LONG	LAT	MAG	☉ LONG	16.00UT	☿	♀	♃	♄
281.57	-0.96	1.4	289.40	10JA	-1.1	-3.7	-1.6	0.8
289.44	-1.02	1.4	299.51	20JA	-1.1	-3.7	-1.7	0.8
297.31	-1.06	1.4	309.57	30JA	-0.4	-3.8	-1.7	0.8
305.18	-1.09	1.4	319.59	9FE	1.4	-3.9	-1.8	0.8
313.03	-1.11	1.4	329.55	19FE	3.1	-4.0	-1.9	0.7
320.85	-1.12	1.4	339.45	29FE	1.5	-4.1	-1.9	0.7
328.63	-1.12	1.4	349.29	10MR	0.9	-4.2	-2.0	0.7
336.36	-1.11	1.4	359.07	20MR	0.5	-4.2	-2.0	0.6
344.03	-1.08	1.4	8.80	30MR	0.1	-4.2	-2.1	0.6
351.63	-1.04	1.5	18.48	9AP	-0.5	-4.1	-2.1	0.5
359.14	-0.99	1.5	28.10	19AP	-1.4	-3.7	-2.1	0.4
6.57	-0.93	1.5	37.70	29AP	-1.6	-2.9	-2.1	0.4
13.91	-0.86	1.5	47.26	9MY	-0.6	-3.3	-2.0	0.3
21.15	-0.78	1.5	56.79	19MY	0.2	-3.9	-2.0	0.2
28.28	-0.69	1.5	66.32	29MY	0.9	-4.1	-1.9	0.3
35.31	-0.59	1.5	75.84	8JN	1.5	-4.2	-1.9	0.3
42.23	-0.49	1.5	85.36	18JN	2.6	-4.1	-1.8	0.4
49.03	-0.37	1.5	94.90	28JN	2.9	-4.1	-1.8	0.4
55.70	-0.25	1.5	104.47	8JL	1.2	-4.0	-1.7	0.5
62.25	-0.12	1.5	114.06	18JL	-0.3	-3.9	-1.6	0.6
68.67	0.01	1.4	123.69	28JL	-1.2	-3.8	-1.6	0.6
74.93	0.16	1.4	133.37	7AU	-1.4	-3.7	-1.5	0.7
81.04	0.31	1.4	143.09	17AU	-0.8	-3.7	-1.5	0.7
86.97	0.48	1.3	152.88	27AU	-0.3	-3.6	-1.4	0.7
92.70	0.65	1.2	162.72	6SE	-0.1	-3.6	-1.4	0.8
98.20	0.84	1.2	172.62	16SE	0.0	-3.5	-1.4	0.8
103.44	1.04	1.1	182.58	26SE	0.2	-3.5	-1.3	0.8
108.36	1.27	1.0	192.60	6OC	0.9	-3.4	-1.3	0.8
112.89	1.51	0.8	202.67	16OC	3.3	-3.4	-1.3	0.8
116.97	1.78	0.7	212.78	26OC	0.3	-3.4	-1.3	0.8
120.47	2.08	0.5	222.94	5NO	-0.4	-3.4	-1.3	0.7
123.27	2.42	0.3	233.11	15NO	-0.5	-3.4	-1.3	0.7
125.20	2.79	0.1	243.31	25NO	-0.5	-3.4	-1.3	0.7
126.06	3.19	-0.1	253.52	5DE	-0.6	-3.4	-1.3	0.7
125.71	3.61	-0.4	263.72	15DE	-0.9	-3.4	-1.4	0.7
124.02	4.01	-0.6	273.92	25DE	-1.0	-3.4	-1.4	0.8

133

♂ LONG	LAT	MAG	☉ LONG	16.00UT	☿	♀	♃	♄
121.09	4.33	-0.8	284.08	4JA	-0.9	-3.4	-1.5	0.8
117.34	4.52	-1.0	294.21	14JA	-0.3	-3.4	-1.5	0.8
113.40	4.53	-0.9	304.31	24JA	1.7	-3.4	-1.6	0.8
110.02	4.38	-0.7	314.35	3FE	2.5	-3.4	-1.6	0.8
107.70	4.10	-0.5	324.33	13FE	1.1	-3.5	-1.7	0.8
106.66	3.76	-0.2	334.27	23FE	0.6	-3.5	-1.8	0.8
106.90	3.41	-0.0	344.14	5MR	0.3	-3.5	-1.8	0.7
108.24	3.06	0.2	353.95	15MR	-0.0	-3.4	-1.9	0.7
110.51	2.74	0.4	3.70	25MR	-0.6	-3.4	-2.0	0.7
113.55	2.44	0.6	13.41	4AP	-1.4	-3.4	-2.0	0.6
117.19	2.17	0.8	23.06	14AP	-1.5	-3.4	-2.1	0.6
121.34	1.92	0.9	32.67	24AP	-0.6	-3.3	-2.1	0.5
125.90	1.70	1.0	42.25	4MY	0.3	-3.3	-2.2	0.4
130.79	1.48	1.1	51.79	14MY	1.1	-3.3	-2.2	0.4
135.98	1.29	1.2	61.33	24MY	2.0	-3.3	-2.2	0.3
141.42	1.10	1.3	70.85	3JN	3.3	-3.3	-2.2	0.2
147.08	0.93	1.4	80.37	13JN	2.4	-3.4	-2.1	0.3
152.94	0.76	1.4	89.90	23JN	1.0	-3.4	-2.1	0.3
158.98	0.60	1.4	99.45	3JL	-0.3	-3.4	-2.0	0.4
165.19	0.45	1.5	109.02	13JL	-1.3	-3.5	-2.0	0.5
171.55	0.31	1.5	118.64	23JL	-1.5	-3.5	-1.9	0.5
178.07	0.17	1.5	128.29	2AU	-0.8	-3.6	-1.9	0.6
184.73	0.03	1.5	137.99	12AU	-0.3	-3.6	-1.8	0.6
191.52	-0.10	1.5	147.74	22AU	0.0	-3.7	-1.7	0.7
198.45	-0.22	1.5	157.55	1SE	0.2	-3.8	-1.7	0.7
205.51	-0.34	1.5	167.42	11SE	0.4	-3.9	-1.6	0.7
212.69	-0.45	1.5	177.35	21SE	1.2	-4.0	-1.6	0.8
219.98	-0.55	1.5	187.34	1OC	3.0	-4.1	-1.5	0.8
227.38	-0.65	1.5	197.38	11OC	0.2	-4.2	-1.5	0.8
234.89	-0.74	1.5	207.47	21OC	-0.5	-4.3	-1.4	0.8
242.48	-0.82	1.5	217.60	31OC	-0.6	-4.4	-1.4	0.8
250.15	-0.89	1.4	227.77	10NO	-0.6	-4.4	-1.4	0.8
257.90	-0.95	1.4	237.96	20NO	-0.7	-4.1	-1.4	0.7
265.70	-1.01	1.4	248.16	30NO	-0.8	-3.5	-1.4	0.7
273.55	-1.05	1.4	258.37	10DE	-0.8	-3.2	-1.4	0.7
281.44	-1.07	1.4	268.57	20DE	-0.8	-3.9	-1.4	0.7
289.34	-1.09	1.4	278.75	30DE	-0.2	-4.3	-1.4	0.8

134

♂ LONG	LAT	MAG	☉ LONG	16.00UT	☿	♀	♃	♄
297.24	-1.10	1.4	288.90	9JA	1.9	-4.4	-1.5	0.8
305.13	-1.09	1.3	299.02	19JA	2.0	-4.3	-1.5	0.8
313.01	-1.07	1.3	309.09	29JA	0.8	-4.2	-1.5	0.8
320.84	-1.04	1.3	319.10	8FE	0.4	-4.1	-1.6	0.8
328.63	-1.00	1.3	329.07	18FE	0.2	-4.0	-1.6	0.8
336.36	-0.95	1.4	338.97	28FE	-0.1	-3.9	-1.7	0.8
344.03	-0.89	1.4	348.81	10MR	-0.6	-3.8	-1.8	0.8
351.63	-0.82	1.4	358.60	20MR	-1.4	-3.7	-1.8	0.8
359.14	-0.75	1.5	8.33	30MR	-1.5	-3.6	-1.9	0.7
6.57	-0.66	1.5	18.01	9AP	-0.6	-3.6	-2.0	0.7
13.91	-0.57	1.6	27.64	19AP	0.5	-3.6	-2.0	0.6
21.17	-0.48	1.6	37.24	29AP	1.5	-3.4	-2.1	0.6
28.33	-0.38	1.6	46.80	9MY	2.7	-3.4	-2.2	0.5
35.40	-0.28	1.7	56.34	19MY	3.3	-3.4	-2.2	0.4
42.38	-0.17	1.7	65.86	29MY	1.9	-3.3	-2.3	0.4
49.26	-0.07	1.7	75.38	8JN	0.8	-3.3	-2.3	0.3
56.06	0.05	1.7	84.91	18JN	-0.4	-3.3	-2.3	0.3
62.76	0.16	1.7	94.45	28JN	-1.3	-3.3	-2.3	0.3
69.38	0.27	1.8	104.01	8JL	-1.5	-3.3	-2.3	0.4
75.90	0.39	1.8	113.60	18JL	-0.7	-3.3	-2.2	0.4
82.33	0.51	1.8	123.23	28JL	-0.2	-3.4	-2.2	0.5
88.67	0.63	1.7	132.90	7AU	0.1	-3.4	-2.2	0.6
94.90	0.75	1.7	142.63	17AU	0.3	-3.4	-2.1	0.6
101.03	0.88	1.7	152.41	27AU	0.6	-3.4	-2.0	0.7
107.05	1.01	1.7	162.24	6SE	1.6	-3.5	-2.0	0.7
112.95	1.15	1.6	172.14	16SE	2.6	-3.5	-1.9	0.7
118.70	1.30	1.5	182.10	26SE	0.1	-3.5	-1.8	0.8
124.29	1.45	1.5	192.11	6OC	-0.7	-3.5	-1.8	0.8
129.70	1.61	1.4	202.17	16OC	-0.8	-3.5	-1.7	0.8
134.88	1.78	1.3	212.28	26OC	-0.7	-3.4	-1.7	0.8
139.81	1.96	1.2	222.43	5NO	-0.8	-3.4	-1.6	0.8
144.41	2.16	1.0	232.61	15NO	-0.6	-3.4	-1.6	0.8
148.62	2.38	0.9	242.81	25NO	-0.6	-3.3	-1.5	0.8
152.36	2.62	0.7	253.02	5DE	-0.6	-3.3	-1.5	0.8
155.49	2.88	0.5	263.22	15DE	-0.1	-3.3	-1.5	0.7
157.88	3.16	0.2	273.41	25DE	2.2	-3.3	-1.5	0.7

135

♂ LONG	LAT	MAG	☉ LONG	16.00UT	☿	♀	♃	♄
159.36	3.45	-0.0	283.58	4JA	1.5	-3.4	-1.5	0.8
159.73	3.75	-0.3	293.72	14JA	0.5	-3.4	-1.5	0.8
158.87	4.02	-0.5	303.81	24JA	0.2	-3.4	-1.5	0.9
156.72	4.23	-0.8	313.86	3FE	0.1	-3.4	-1.5	0.9
153.48	4.31	-1.0	323.85	13FE	-0.2	-3.5	-1.6	0.9
149.65	4.23	-1.1	333.78	23FE	-0.7	-3.5	-1.6	0.9
145.94	3.98	-1.0	343.66	5MR	-1.4	-3.5	-1.6	0.9
143.00	3.60	-0.8	353.48	15MR	-1.5	-3.6	-1.7	0.9
141.25	3.16	-0.6	3.23	25MR	-0.6	-3.6	-1.7	0.8
140.82	2.71	-0.4	12.94	4AP	0.6	-3.7	-1.8	0.8
141.62	2.28	-0.2	22.60	14AP	1.9	-3.8	-1.9	0.8
143.49	1.88	0.0	32.21	24AP	3.5	-3.9	-1.9	0.7
146.26	1.52	0.2	41.79	4MY	2.6	-4.0	-2.0	0.7
149.76	1.19	0.4	51.34	14MY	1.4	-4.1	-2.1	0.6
153.87	0.90	0.5	60.87	24MY	0.6	-4.1	-2.1	0.5
158.48	0.64	0.6	70.39	3JN	-0.4	-4.2	-2.2	0.5
163.51	0.40	0.7	79.91	13JN	-1.4	-4.1	-2.3	0.4
168.90	0.19	0.8	89.44	23JN	-1.5	-3.9	-2.3	0.3
174.60	-0.01	0.9	98.99	3JL	-0.7	-3.4	-2.4	0.3
180.57	-0.19	1.0	108.56	13JL	-0.1	-3.1	-2.4	0.4
186.78	-0.36	1.0	118.17	23JL	0.3	-3.7	-2.4	0.4
193.21	-0.51	1.1	127.82	2AU	0.5	-4.1	-2.4	0.5
199.85	-0.64	1.1	137.52	12AU	0.9	-4.2	-2.4	0.5
206.66	-0.77	1.1	147.27	22AU	2.0	-4.2	-2.4	0.6
213.65	-0.88	1.2	157.08	1SE	2.3	-4.2	-2.3	0.6
220.78	-0.97	1.2	166.95	11SE	-0.0	-4.1	-2.3	0.7
228.06	-1.05	1.2	176.87	21SE	-0.8	-4.0	-2.2	0.7
235.46	-1.12	1.2	186.86	1OC	-0.9	-3.9	-2.1	0.8
242.97	-1.18	1.3	196.89	11OC	-0.9	-3.9	-2.1	0.8
250.58	-1.22	1.3	206.98	21OC	-0.7	-3.8	-2.0	0.8
258.28	-1.24	1.3	217.11	31OC	-0.5	-3.7	-1.9	0.8
266.03	-1.26	1.3	227.27	10NO	-0.5	-3.6	-1.9	0.8
273.84	-1.25	1.3	237.46	20NO	-0.5	-3.6	-1.8	0.8
281.69	-1.24	1.3	247.67	30NO	0.1	-3.5	-1.7	0.8
289.55	-1.21	1.4	257.87	10DE	2.5	-3.4	-1.7	0.8
297.41	-1.17	1.4	268.07	20DE	1.2	-3.4	-1.7	0.8
305.26	-1.12	1.4	278.26	30DE	0.3	-3.4	-1.6	0.8

♂			☉	16.00UT		☿	♀	♃	♄
LONG LAT	MAG		LONG		136		MAGNITUDES		

♂ LONG LAT MAG	☉ LONG	16.00UT	136	☿	♀	♃	♄
313.09 -1.06 1.4	288.41	9JA		0.1	-3.4	-1.6	0.8
320.87 -0.99 1.4	298.53	19JA		-0.0	-3.4	-1.6	0.9
328.61 -0.91 1.5	308.60	29JA		-0.3	-3.3	-1.6	0.9
336.29 -0.83 1.5	318.62	8FE		-0.7	-3.3	-1.6	0.9
343.90 -0.73 1.5	328.58	18FE		-1.4	-3.3	-1.6	0.9
351.43 -0.64 1.5	338.49	28FE		-1.4	-3.3	-1.6	1.0
358.89 -0.54 1.5	348.34	9MR		-0.5	-3.3	-1.6	1.0
6.26 -0.44 1.6	358.13	19MR		0.8	-3.3	-1.6	0.9
13.55 -0.34 1.6	7.86	29MR		2.4	-3.4	-1.6	0.9
20.75 -0.24 1.6	17.54	8AP		3.4	-3.4	-1.7	0.9
27.87 -0.13 1.6	27.18	18AP		1.9	-3.4	-1.7	0.9
34.91 -0.03 1.6	36.77	28AP		1.1	-3.5	-1.8	0.8
41.86 0.07 1.7	46.33	8MY		0.4	-3.5	-1.8	0.8
48.73 0.17 1.7	55.88	18MY		-0.4	-3.4	-1.9	0.7
55.53 0.27 1.8	65.40	28MY		-1.4	-3.4	-2.0	0.6
62.26 0.37 1.8	74.92	7JN		-1.6	-3.4	-2.0	0.6
68.91 0.47 1.9	84.44	17JN		-0.7	-3.4	-2.1	0.5
75.51 0.56 1.9	93.98	27JN		-0.0	-3.3	-2.2	0.4
82.03 0.66 1.9	103.54	7JL		0.4	-3.3	-2.2	0.4
88.50 0.75 1.9	113.13	17JL		0.7	-3.3	-2.3	0.4
94.91 0.84 1.9	122.76	27JL		1.2	-3.3	-2.4	0.4
101.26 0.93 1.9	132.43	6AU		2.5	-3.3	-2.4	0.5
107.56 1.02 1.9	142.15	16AU		2.0	-3.3	-2.4	0.5
113.79 1.11 1.9	151.93	26AU		-0.1	-3.4	-2.5	0.6
119.97 1.19 1.9	161.76	5SE		-1.0	-3.4	-2.5	0.7
126.09 1.28 1.9	171.66	15SE		-1.1	-3.4	-2.4	0.7
132.13 1.37 1.8	181.61	25SE		-0.9	-3.5	-2.4	0.7
138.11 1.46 1.8	191.62	5OC		-0.6	-3.5	-2.4	0.8
144.00 1.55 1.7	201.69	15OC		-0.4	-3.6	-2.3	0.8
149.78 1.64 1.6	211.79	25OC		-0.3	-3.7	-2.2	0.9
155.46 1.73 1.6	221.94	4NO		-0.3	-3.7	-2.2	0.9
161.00 1.83 1.5	232.12	14NO		0.3	-3.8	-2.1	0.9
166.38 1.93 1.3	242.32	24NO		2.8	-3.9	-2.0	0.9
171.57 2.03 1.2	252.52	4DE		0.9	-4.0	-2.0	0.9
176.52 2.13 1.1	262.73	14DE		0.1	-4.1	-1.9	0.9
181.18 2.24 0.9	272.92	24DE		-0.1	-4.2	-1.8	0.9
				137			
185.49 2.35 0.7	283.09	3JA		-0.2	-4.3	-1.8	0.9
189.36 2.46 0.5	293.23	13JA		-0.4	-4.4	-1.7	0.9
192.67 2.57 0.3	303.32	23JA		-0.8	-4.3	-1.7	0.9
195.29 2.67 0.0	313.37	2FE		-1.3	-4.0	-1.6	1.0
197.05 2.75 -0.3	323.36	12FE		-1.3	-3.5	-1.6	1.0
197.77 2.79 -0.6	333.30	22FE		-0.5	-3.4	-1.6	1.0
197.29 2.78 -0.9	343.18	4MR		1.0	-3.9	-1.6	1.0
195.54 2.69 -1.2	353.00	14MR		3.0	-4.2	-1.6	1.1
192.67 2.47 -1.4	2.76	24MR		2.5	-4.2	-1.6	1.1
189.12 2.13 -1.6	12.47	3AP		1.4	-4.2	-1.6	1.0
185.56 1.69 -1.5	22.13	13AP		0.8	-4.1	-1.6	1.0
182.71 1.20 -1.3	31.74	23AP		0.3	-4.0	-1.6	1.0
181.03 0.71 -1.1	41.32	3MY		-0.5	-3.9	-1.6	1.0
180.71 0.26 -0.9	50.88	13MY		-1.4	-3.8	-1.7	0.9
181.70 -0.14 -0.7	60.41	23MY		-1.6	-3.7	-1.7	0.9
183.83 -0.48 -0.5	69.93	2JN		-0.7	-3.6	-1.8	0.8
186.94 -0.77 -0.3	79.45	12JN		0.1	-3.6	-1.8	0.7
190.84 -1.01 -0.2	88.98	22JN		0.5	-3.5	-1.9	0.7
195.41 -1.21 -0.0	98.53	2JL		1.0	-3.5	-1.9	0.6
200.52 -1.38 0.1	108.10	12JL		1.6	-3.4	-2.0	0.5
206.09 -1.51 0.2	117.71	22JL		3.0	-3.4	-2.1	0.5
212.04 -1.62 0.3	127.35	1AU		1.7	-3.4	-2.1	0.5
218.33 -1.70 0.4	137.05	11AU		-0.2	-3.4	-2.2	0.5
224.89 -1.76 0.5	146.80	21AU		-1.1	-3.4	-2.3	0.6
231.68 -1.79 0.6	156.60	31AU		-1.2	-3.4	-2.3	0.6
238.69 -1.80 0.7	166.47	10SE		-0.9	-3.4	-2.4	0.7
245.86 -1.80 0.7	176.39	20SE		-0.5	-3.4	-2.4	0.7
253.18 -1.77 0.8	186.37	30SE		-0.3	-3.4	-2.4	0.8
260.61 -1.73 0.9	196.41	10OC		-0.2	-3.4	-2.4	0.8
268.14 -1.67 0.9	206.49	20OC		-0.1	-3.4	-2.4	0.9
275.74 -1.60 1.0	216.62	30OC		0.5	-3.4	-2.4	0.9
283.38 -1.51 1.1	226.78	9NO		3.1	-3.4	-2.3	0.9
291.06 -1.41 1.1	236.97	19NO		0.7	-3.5	-2.3	1.0
298.74 -1.31 1.2	247.17	29NO		-0.1	-3.5	-2.2	1.0
306.42 -1.20 1.2	257.38	9DE		-0.2	-3.5	-2.1	1.0
314.08 -1.08 1.3	267.58	19DE		-0.3	-3.5	-2.1	1.0
321.70 -0.96 1.4	277.76	29DE		-0.4	-3.4	-2.0	1.0

♂ LONG LAT MAG	☉ LONG	16.00UT	138	☿	♀	♃	♄
329.27 -0.83 1.4	287.92	8JA		-0.8	-3.4	-1.9	1.0
336.79 -0.71 1.5	298.03	18JA		-1.2	-3.4	-1.9	1.0
344.25 -0.58 1.5	308.11	28JA		-1.2	-3.4	-1.8	1.0
351.63 -0.46 1.6	318.13	7FE		-0.5	-3.3	-1.7	1.1
358.94 -0.34 1.6	328.09	17FE		1.2	-3.3	-1.7	1.1
6.18 -0.22 1.7	338.00	27FE		3.3	-3.3	-1.6	1.1
13.34 -0.11 1.7	347.86	9MR		1.9	-3.3	-1.6	1.2
20.43 0.00 1.7	357.64	19MR		1.0	-3.3	-1.6	1.2
27.43 0.11 1.8	7.38	29MR		0.6	-3.4	-1.5	1.2
34.37 0.21 1.8	17.07	8AP		0.2	-3.4	-1.5	1.2
41.23 0.31 1.8	26.70	18AP		-0.5	-3.4	-1.5	1.2
48.03 0.40 1.8	36.30	28AP		-1.4	-3.4	-1.5	1.2
54.76 0.49 1.8	45.87	8MY		-1.6	-3.5	-1.5	1.1
61.43 0.57 1.8	55.41	18MY		-0.6	-3.5	-1.5	1.1
68.05 0.65 1.8	64.94	28MY		0.1	-3.6	-1.5	1.0
74.62 0.73 1.8	74.46	7JN		0.7	-3.7	-1.6	1.0
81.14 0.80 1.9	83.98	17JN		1.3	-3.7	-1.6	0.9
87.63 0.87 1.9	93.52	27JN		2.2	-3.8	-1.6	0.8
94.07 0.93 2.0	103.08	7JL		3.2	-3.9	-1.7	0.8
100.48 0.99 2.0	112.67	17JL		1.4	-4.0	-1.7	0.7
106.87 1.05 2.0	122.29	27JL		-0.2	-4.1	-1.8	0.6
113.22 1.10 2.0	131.96	6AU		-1.2	-4.2	-1.8	0.6
119.56 1.15 2.0	141.68	16AU		-1.3	-4.3	-1.9	0.6
125.87 1.20 2.0	151.46	26AU		-0.8	-4.2	-2.0	0.7
132.17 1.24 2.0	161.29	5SE		-0.4	-4.0	-2.0	0.7
138.44 1.28 2.0	171.18	15SE		-0.2	-3.5	-2.1	0.8
144.70 1.32 2.0	181.13	25SE		-0.0	-3.4	-2.2	0.8
150.93 1.35 1.9	191.14	5OC		0.1	-3.9	-2.2	0.9
157.15 1.38 1.9	201.19	15OC		0.7	-4.3	-2.3	0.9
163.33 1.41 1.8	211.30	25OC		3.5	-4.4	-2.3	1.0
169.49 1.43 1.8	221.45	4NO		0.5	-4.3	-2.3	1.0
175.61 1.45 1.7	231.62	14NO		-0.3	-4.3	-2.3	1.0
181.68 1.46 1.6	241.82	24NO		-0.4	-4.2	-2.3	1.1
187.70 1.47 1.5	252.03	4DE		-0.4	-4.1	-2.3	1.1
193.66 1.47 1.4	262.23	14DE		-0.5	-4.0	-2.2	1.1
199.53 1.46 1.3	272.43	24DE		-0.9	-3.9	-2.2	1.1
				139			
205.30 1.44 1.2	282.60	3JA		-1.0	-3.8	-2.1	1.2
210.96 1.41 1.0	292.74	13JA		-1.0	-3.7	-2.0	1.2
216.47 1.36 0.9	302.84	23JA		-0.4	-3.6	-2.0	1.1
221.80 1.30 0.7	312.89	2FE		1.5	-3.6	-1.9	1.1
226.90 1.21 0.5	322.88	12FE		2.9	-3.5	-1.8	1.2
231.72 1.09 0.3	332.82	22FE		1.4	-3.4	-1.8	1.2
236.19 0.93 0.1	342.70	4MR		0.8	-3.4	-1.7	1.3
240.22 0.72 -0.2	352.52	14MR		0.4	-3.4	-1.6	1.3
243.67 0.46 -0.5	2.29	24MR		0.1	-3.3	-1.6	1.3
246.41 0.13 -0.8	12.00	3AP		-0.5	-3.3	-1.5	1.4
248.26 -0.29 -1.1	21.66	13AP		-1.4	-3.3	-1.5	1.4
249.04 -0.80 -1.4	31.28	23AP		-1.6	-3.3	-1.5	1.4
248.61 -1.41 -1.7	40.86	3MY		-0.6	-3.3	-1.4	1.3
246.96 -2.08 -2.0	50.41	13MY		0.2	-3.3	-1.4	1.3
244.34 -2.76 -2.3	59.95	23MY		0.9	-3.3	-1.4	1.3
241.33 -3.36 -2.3	69.47	2JN		1.7	-3.4	-1.4	1.3
238.63 -3.82 -2.2	78.99	12JN		2.8	-3.4	-1.4	1.2
236.90 -4.10 -2.0	88.52	22JN		2.8	-3.4	-1.4	1.1
236.49 -4.22 -1.7	98.06	2JL		1.2	-3.5	-1.4	1.1
237.45 -4.21 -1.5	107.64	12JL		-0.3	-3.5	-1.5	1.0
239.68 -4.11 -1.3	117.24	22JL		-1.2	-3.5	-1.5	0.9
242.97 -3.96 -1.1	126.88	1AU		-1.5	-3.4	-1.5	0.9
247.14 -3.77 -0.8	136.58	11AU		-0.8	-3.4	-1.6	0.8
252.01 -3.55 -0.6	146.33	21AU		-0.3	-3.4	-1.6	0.8
257.45 -3.31 -0.5	156.12	31AU		-0.0	-3.4	-1.7	0.8
263.32 -3.07 -0.3	165.99	10SE		0.1	-3.3	-1.7	0.8
269.54 -2.81 -0.1	175.91	20SE		0.3	-3.3	-1.8	0.9
276.03 -2.56 0.0	185.88	30SE		1.0	-3.3	-1.9	0.9
282.72 -2.31 0.2	195.91	10OC		3.3	-3.3	-1.9	1.0
289.58 -2.06 0.3	206.00	20OC		-0.3	-3.3	-2.0	1.0
296.54 -1.82 0.5	216.12	30OC		-0.4	-3.4	-2.0	1.1
303.59 -1.58 0.6	226.28	9NO		-0.5	-3.4	-2.1	1.1
310.68 -1.36 0.7	236.47	19NO		-0.5	-3.4	-2.1	1.2
317.81 -1.14 0.8	246.67	29NO		-0.6	-3.5	-2.2	1.2
324.93 -0.94 1.0	256.88	9DE		-0.8	-3.5	-2.2	1.3
332.06 -0.75 1.1	267.08	19DE		-0.9	-3.6	-2.2	1.3
339.15 -0.57 1.2	277.27	29DE		-0.9	-3.6	-2.2	1.3

♂ LONG	LAT	MAG	☉ LONG	16.00UT 140	☿	♀	♃	♄
346.22	-0.40	1.3	287.43	8JA	-0.3	-3.7	-2.1	1.3
353.25	-0.24	1.4	297.55	18JA	1.7	-3.7	-2.1	1.3
0.23	-0.09	1.4	307.62	28JA	2.3	-3.8	-2.0	1.3
7.16	0.04	1.5	317.65	7FE	1.0	-3.9	-2.0	1.3
14.05	0.17	1.6	327.62	17FE	0.5	-4.0	-1.9	1.3
20.87	0.28	1.7	337.53	27FE	0.3	-4.1	-1.8	1.3
27.65	0.39	1.7	347.38	8MR	-0.0	-4.2	-1.8	1.3
34.37	0.49	1.8	357.18	18MR	-0.6	-4.2	-1.7	1.3
41.05	0.57	1.8	6.91	28MR	-1.4	-4.2	-1.6	1.3
47.67	0.66	1.9	16.60	7AP	-1.6	-4.1	-1.6	1.3
54.25	0.73	1.9	26.24	17AP	-0.6	-3.7	-1.5	1.2
60.79	0.80	1.9	35.84	27AP	0.4	-2.9	-1.5	1.2
67.29	0.86	2.0	45.41	7MY	1.2	-3.3	-1.4	1.2
73.76	0.91	2.0	54.95	17MY	2.2	-3.9	-1.4	1.2
80.20	0.96	2.0	64.48	27MY	3.5	-4.1	-1.4	1.1
86.62	1.00	2.0	74.00	6JN	2.2	-4.2	-1.3	1.1
93.02	1.04	2.0	83.52	16JN	0.9	-4.1	-1.3	1.0
99.40	1.08	2.0	93.06	26JN	-0.3	-4.1	-1.3	1.0
105.77	1.10	2.0	102.61	6JL	-1.3	-4.0	-1.3	0.9
112.14	1.13	1.9	112.20	16JL	-1.5	-3.9	-1.3	0.9
118.51	1.15	2.0	121.82	26JL	-0.8	-3.8	-1.3	0.8
124.88	1.16	2.0	131.49	5AU	-0.2	-3.7	-1.3	0.8
131.26	1.17	2.0	141.21	15AU	0.1	-3.7	-1.4	0.7
137.65	1.18	2.0	150.98	25AU	0.3	-3.6	-1.4	0.7
144.05	1.18	2.0	160.81	4SE	0.5	-3.6	-1.4	0.7
150.46	1.18	2.0	170.70	14SE	1.3	-3.5	-1.5	0.8
156.90	1.17	2.0	180.65	24SE	2.9	-3.5	-1.5	0.9
163.35	1.15	2.0	190.65	4OC	0.2	-3.4	-1.6	0.9
169.82	1.14	2.0	200.71	14OC	-0.6	-3.4	-1.6	1.0
176.32	1.11	1.9	210.81	24OC	-0.7	-3.4	-1.7	1.1
182.83	1.08	1.9	220.96	3NO	-0.7	-3.4	-1.8	1.1
189.36	1.04	1.8	231.13	13NO	-0.7	-3.4	-1.8	1.2
195.92	0.99	1.8	241.33	23NO	-0.7	-3.4	-1.9	1.2
202.49	0.94	1.7	251.53	3DE	-0.7	-3.4	-2.0	1.2
209.07	0.87	1.6	261.74	13DE	-0.7	-3.4	-2.0	1.2
215.67	0.79	1.5	271.93	23DE	-0.2	-3.4	-2.0	1.2
				141				
222.28	0.70	1.5	282.11	2JA	2.0	-3.4	-2.1	1.2
228.89	0.60	1.4	292.25	12JA	1.8	-3.4	-2.1	1.2
235.51	0.48	1.2	302.34	22JA	0.7	-3.4	-2.1	1.2
242.13	0.34	1.1	312.40	1FE	0.3	-3.4	-2.1	1.1
248.73	0.19	1.0	322.40	11FE	0.2	-3.5	-2.0	1.1
255.31	0.01	0.9	332.34	21FE	-0.1	-3.5	-2.0	1.0
261.86	-0.20	0.7	342.23	3MR	-0.6	-3.5	-1.9	1.0
268.37	-0.43	0.6	352.05	13MR	-1.4	-3.4	-1.9	1.0
274.81	-0.69	0.4	1.82	23MR	-1.5	-3.4	-1.8	1.0
281.16	-0.98	0.3	11.53	2AP	-0.6	-3.4	-1.7	1.0
287.38	-1.30	0.1	21.19	12AP	0.5	-3.4	-1.7	1.0
293.45	-1.66	-0.1	30.81	22AP	1.6	-3.3	-1.6	1.0
299.28	-2.06	-0.3	40.40	2MY	3.0	-3.3	-1.5	1.0
304.83	-2.51	-0.5	49.95	12MY	3.1	-3.3	-1.5	0.9
310.00	-2.99	-0.7	59.48	22MY	1.7	-3.3	-1.4	0.9
314.66	-3.52	-1.1	69.01	1JN	0.7	-3.3	-1.4	0.9
318.67	-4.09	-1.2	78.53	11JN	-0.3	-3.4	-1.3	0.8
321.84	-4.69	-1.5	88.05	21JN	-1.3	-3.4	-1.3	0.8
323.97	-5.30	-1.7	97.60	1JL	-1.6	-3.4	-1.3	0.7
324.86	-5.88	-2.0	107.17	11JL	-0.7	-3.5	-1.3	0.7
324.39	-6.35	-2.2	116.77	21JL	-0.2	-3.5	-1.2	0.6
322.65	-6.63	-2.5	126.42	31JL	0.2	-3.6	-1.2	0.6
320.07	-6.62	-2.6	136.11	10AU	0.4	-3.6	-1.2	0.5
317.32	-6.29	-2.5	145.85	20AU	0.7	-3.7	-1.2	0.5
315.18	-5.71	-2.3	155.65	30AU	1.7	-3.8	-1.3	0.4
314.16	-4.98	-2.0	165.51	9SE	2.6	-3.9	-1.3	0.4
314.43	-4.21	-1.6	175.43	19SE	0.1	-4.0	-1.3	0.4
315.97	-3.48	-1.3	185.40	29SE	-0.7	-4.1	-1.3	0.5
318.57	-2.81	-1.0	195.43	9OC	-0.8	-4.2	-1.4	0.6
322.04	-2.23	-0.7	205.51	19OC	-0.8	-4.3	-1.4	0.6
326.20	-1.72	-0.5	215.64	29OC	-0.7	-4.4	-1.5	0.7
330.89	-1.29	-0.2	225.79	8NO	-0.6	-4.3	-1.5	0.8
335.98	-0.91	0.0	235.98	18NO	-0.6	-4.1	-1.6	0.8
341.39	-0.59	0.2	246.18	28NO	-0.6	-3.5	-1.7	0.9
347.03	-0.32	0.4	256.39	8DE	-0.0	-3.2	-1.7	0.9
352.85	-0.08	0.6	266.59	18DE	2.2	-3.9	-1.8	0.9
358.81	0.12	0.8	276.77	28DE	1.4	-4.3	-1.9	0.9

♂ LONG	LAT	MAG	☉ LONG	16.00UT 142	☿	♀	♃	♄
4.87	0.30	1.0	286.93	7JA	0.4	-4.4	-1.9	0.9
10.99	0.45	1.1	297.05	17JA	0.2	-4.3	-2.0	0.9
17.18	0.58	1.2	307.13	27JA	0.0	-4.2	-2.0	0.9
23.39	0.69	1.3	317.16	6FE	-0.2	-4.1	-2.0	0.9
29.63	0.79	1.5	327.14	16FE	-0.7	-4.0	-2.0	0.9
35.89	0.87	1.6	337.05	26FE	-1.4	-3.9	-2.0	0.8
42.15	0.94	1.6	346.90	8MR	-1.4	-3.8	-2.0	0.8
48.41	1.00	1.7	356.71	18MR	-0.6	-3.7	-2.0	0.7
54.68	1.05	1.8	6.45	28MR	0.7	-3.6	-1.9	0.7
60.93	1.09	1.8	16.13	7AP	2.1	-3.6	-1.9	0.7
67.19	1.13	1.9	25.78	17AP	3.9	-3.5	-1.8	0.7
73.45	1.15	1.9	35.38	27AP	2.3	-3.4	-1.7	0.7
79.70	1.17	2.0	44.95	7MY	1.3	-3.4	-1.7	0.7
85.96	1.19	2.0	54.49	17MY	0.5	-3.4	-1.6	0.7
92.21	1.20	2.0	64.02	27MY	-0.4	-3.3	-1.5	0.7
98.49	1.20	2.0	73.54	6JN	-1.4	-3.3	-1.5	0.7
104.77	1.20	2.0	83.07	16JN	-1.6	-3.3	-1.4	0.6
111.07	1.19	2.0	92.60	26JN	-0.7	-3.3	-1.4	0.6
117.39	1.18	2.0	102.15	6JL	-0.1	-3.3	-1.3	0.6
123.73	1.17	2.0	111.74	16JL	0.3	-3.3	-1.3	0.5
130.09	1.15	2.0	121.36	26JL	0.6	-3.4	-1.3	0.5
136.49	1.12	2.0	131.02	5AU	1.0	-3.4	-1.2	0.4
142.92	1.10	1.9	140.74	15AU	2.2	-3.4	-1.2	0.4
149.39	1.06	1.9	150.51	25AU	2.3	-3.4	-1.2	0.3
155.90	1.02	1.9	160.33	4SE	-0.0	-3.5	-1.2	0.2
162.46	0.98	1.9	170.22	14SE	-0.9	-3.5	-1.2	0.2
169.07	0.93	1.9	180.16	24SE	-1.0	-3.5	-1.2	0.1
175.72	0.88	1.9	190.16	4OC	-0.9	-3.5	-1.2	0.2
182.43	0.82	1.9	200.22	14OC	-0.6	-3.4	-1.3	0.3
189.19	0.76	1.9	210.32	24OC	-0.5	-3.4	-1.3	0.3
196.00	0.68	1.8	220.46	3NO	-0.4	-3.4	-1.3	0.4
202.87	0.60	1.8	230.64	13NO	-0.4	-3.4	-1.4	0.5
209.80	0.52	1.8	240.83	23NO	0.1	-3.3	-1.4	0.5
216.78	0.42	1.7	251.03	3DE	2.5	-3.3	-1.5	0.6
223.81	0.32	1.7	261.24	13DE	1.1	-3.3	-1.6	0.6
230.90	0.21	1.6	271.44	23DE	0.2	-3.3	-1.6	0.7
				143				
238.04	0.09	1.6	281.61	2JA	0.0	-3.4	-1.7	0.7
245.23	-0.04	1.5	291.75	12JA	-0.1	-3.4	-1.8	0.7
252.46	-0.17	1.4	301.85	22JA	-0.3	-3.4	-1.8	0.7
259.74	-0.32	1.4	311.91	1FE	-0.7	-3.4	-1.9	0.7
267.05	-0.47	1.3	321.92	11FE	-1.3	-3.5	-1.9	0.7
274.40	-0.63	1.2	331.86	21FE	-1.3	-3.5	-2.0	0.7
281.77	-0.80	1.1	341.75	3MR	-0.6	-3.5	-2.0	0.6
289.15	-0.98	1.0	351.58	13MR	0.8	-3.6	-2.0	0.6
296.54	-1.15	1.0	1.35	23MR	2.6	-3.7	-2.0	0.5
303.92	-1.33	0.9	11.06	2AP	3.0	-3.7	-2.0	0.5
311.27	-1.52	0.8	20.73	12AP	1.7	-3.8	-2.0	0.5
318.57	-1.69	0.7	30.35	22AP	1.0	-3.9	-1.9	0.5
325.82	-1.87	0.6	39.94	2MY	0.4	-4.0	-1.9	0.5
332.96	-2.04	0.5	49.49	12MY	-0.4	-4.1	-1.8	0.6
339.98	-2.20	0.4	59.03	22MY	-1.4	-4.1	-1.8	0.6
346.86	-2.35	0.3	68.55	1JN	-1.6	-4.2	-1.7	0.5
353.53	-2.48	0.2	78.07	11JN	-0.7	-4.1	-1.7	0.5
359.96	-2.60	0.1	87.60	21JN	-0.0	-3.9	-1.6	0.5
6.10	-2.71	-0.0	97.14	1JL	0.4	-3.4	-1.5	0.5
11.86	-2.79	-0.2	106.71	11JL	0.8	-3.1	-1.5	0.4
17.18	-2.86	-0.3	116.31	21JL	1.4	-3.7	-1.4	0.4
21.92	-2.90	-0.5	125.95	31JL	2.7	-4.1	-1.4	0.3
25.96	-2.90	-0.7	135.64	10AU	2.0	-4.2	-1.3	0.3
29.14	-2.87	-0.9	145.38	20AU	-0.1	-4.2	-1.3	0.2
31.23	-2.79	-1.1	155.17	30AU	-1.0	-4.2	-1.3	0.2
32.03	-2.64	-1.3	165.03	9SE	-1.1	-4.1	-1.3	0.1
31.39	-2.39	-1.5	174.94	19SE	-0.9	-4.0	-1.2	0.0
29.31	-2.04	-1.7	184.92	29SE	-0.5	-3.9	-1.2	-0.0
26.10	-1.59	-1.9	194.94	9OC	-0.3	-3.8	-1.2	-0.1
22.44	-1.06	-1.9	205.02	19OC	-0.3	-3.8	-1.2	-0.0
19.12	-0.52	-1.6	215.14	29OC	-0.2	-3.7	-1.3	0.1
16.84	-0.03	-1.3	225.30	8NO	0.3	-3.6	-1.3	0.1
15.88	0.38	-1.0	235.48	18NO	2.8	-3.6	-1.3	0.2
16.25	0.70	-0.7	245.69	28NO	0.8	-3.5	-1.3	0.3
17.79	0.95	-0.4	255.90	8DE	0.0	-3.5	-1.4	0.3
20.29	1.13	-0.1	266.10	18DE	-0.1	-3.4	-1.4	0.4
23.56	1.27	0.2	276.29	28DE	-0.2	-3.4	-1.5	0.4

144

♂ LONG	LAT	MAG	☉ LONG	16.00UT	☿	♀	♃	♄
27.42	1.37	0.4	286.44	7JA	-0.4	-3.4	-1.6	0.5
31.74	1.44	0.6	296.57	17JA	-0.8	-3.4	-1.6	0.5
36.43	1.49	0.8	306.65	27JA	-1.3	-3.3	-1.7	0.5
41.39	1.52	1.0	316.68	6FE	-1.2	-3.3	-1.8	0.5
46.58	1.54	1.2	326.65	16FE	-0.5	-3.3	-1.8	0.5
51.94	1.55	1.3	336.58	26FE	1.0	-3.3	-1.9	0.5
57.45	1.55	1.4	346.43	7MR	3.2	-3.3	-1.9	0.5
63.06	1.54	1.5	356.23	17MR	-2.3	-3.3	-2.0	0.4
68.77	1.53	1.6	5.98	27MR	1.3	-3.4	-2.0	0.4
74.56	1.51	1.7	15.67	6AP	0.7	-3.4	-2.1	0.4
80.41	1.48	1.8	25.31	16AP	0.2	-3.4	-2.1	0.3
86.33	1.46	1.8	34.92	26AP	-0.5	-3.5	-2.1	0.4
92.29	1.42	1.9	44.48	6MY	-1.4	-3.5	-2.1	0.4
98.31	1.39	1.9	54.03	16MY	-1.6	-3.4	-2.0	0.4
104.38	1.35	2.0	63.56	26MY	-0.7	-3.4	-2.0	0.4
110.49	1.31	2.0	73.08	5JN	0.1	-3.4	-1.9	0.4
116.66	1.26	2.0	82.60	15JN	0.6	-3.4	-1.9	0.4
122.87	1.21	2.0	92.14	25JN	1.1	-3.3	-1.8	0.4
129.14	1.16	2.0	101.69	5JL	1.8	-3.3	-1.8	0.4
135.46	1.11	2.0	111.27	15JL	3.2	-3.3	-1.7	0.4
141.84	1.05	2.0	120.89	25JL	1.6	-3.3	-1.6	0.3
148.27	0.98	2.0	130.55	4AU	-0.1	-3.3	-1.6	0.3
154.77	0.92	1.9	140.26	14AU	-1.1	-3.3	-1.5	0.2
161.34	0.85	1.9	150.03	24AU	-1.3	-3.4	-1.5	0.2
167.98	0.78	1.9	159.85	3SE	-0.9	-3.4	-1.4	0.1
174.68	0.70	1.8	169.73	13SE	-0.4	-3.4	-1.4	0.0
181.46	0.62	1.8	179.68	23SE	-0.2	-3.5	-1.4	-0.0
188.32	0.53	1.7	189.67	3OC	-0.1	-3.5	-1.3	-0.1
195.25	0.44	1.7	199.73	13OC	-0.0	-3.6	-1.3	-0.2
202.26	0.35	1.7	209.83	23OC	0.5	-3.7	-1.3	-0.2
209.34	0.25	1.7	219.97	2NO	3.1	-3.7	-1.3	-0.1
216.50	0.15	1.7	230.14	12NO	0.6	-3.8	-1.3	-0.1
223.74	0.05	1.7	240.34	22NO	-0.2	-3.9	-1.3	0.0
231.04	-0.06	1.6	250.54	2DE	-0.3	-4.0	-1.3	0.1
238.42	-0.18	1.6	260.75	12DE	-0.3	-4.1	-1.4	0.1
245.86	-0.29	1.6	270.95	22DE	-0.5	-4.2	-1.4	0.2

145

♂ LONG	LAT	MAG	☉ LONG	16.00UT	☿	♀	♃	♄
253.36	-0.41	1.6	281.12	1JA	-0.8	-4.3	-1.4	0.3
260.92	-0.53	1.5	291.26	11JA	-1.1	-4.4	-1.5	0.3
268.53	-0.64	1.5	301.37	21JA	-1.1	-4.3	-1.5	0.3
276.18	-0.76	1.4	311.43	31JA	-0.5	-4.0	-1.6	0.4
283.86	-0.88	1.4	321.43	10FE	1.3	-3.5	-1.7	0.4
291.56	-0.99	1.4	331.38	20FE	3.2	-3.4	-1.7	0.4
299.27	-1.09	1.3	341.27	2MR	1.7	-3.9	-1.8	0.4
306.98	-1.19	1.3	351.10	12MR	0.9	-4.2	-1.9	0.4
314.67	-1.28	1.3	0.87	22MR	0.5	-4.3	-1.9	0.4
322.34	-1.36	1.2	10.59	1AP	0.1	-4.2	-2.0	0.3
329.97	-1.42	1.2	20.26	11AP	-0.5	-4.1	-2.1	0.3
337.54	-1.48	1.1	29.89	21AP	-1.4	-4.0	-2.1	0.2
345.04	-1.52	1.1	39.47	1MY	-1.6	-3.9	-2.2	0.2
352.46	-1.54	1.1	49.03	11MY	-0.7	-3.8	-2.2	0.3
359.78	-1.55	1.0	58.57	21MY	0.2	-3.7	-2.2	0.3
6.98	-1.54	1.0	68.09	31MY	0.8	-3.6	-2.2	0.3
14.05	-1.52	0.9	77.61	10JN	1.4	-3.6	-2.2	0.3
20.96	-1.48	0.9	87.14	20JN	2.4	-3.5	-2.1	0.3
27.70	-1.42	0.8	96.68	30JN	3.1	-3.5	-2.1	0.3
34.25	-1.34	0.8	106.25	10JL	1.4	-3.4	-2.0	0.3
40.57	-1.25	0.7	115.85	20JL	-0.2	-3.4	-2.0	0.3
46.63	-1.14	0.6	125.48	30JL	-1.2	-3.4	-1.9	0.3
52.40	-1.01	0.5	135.17	9AU	-1.4	-3.4	-1.9	0.3
57.81	-0.85	0.4	144.91	19AU	-0.8	-3.4	-1.8	0.2
62.80	-0.67	0.3	154.70	29AU	-0.4	-3.4	-1.7	0.2
67.27	-0.46	0.2	164.55	8SE	-0.1	-3.4	-1.7	0.1
71.12	-0.22	0.0	174.46	18SE	0.0	-3.4	-1.6	0.0
74.19	0.07	-0.2	184.43	28SE	0.2	-3.4	-1.6	-0.0
76.32	0.40	-0.4	194.46	8OC	-0.8	-3.4	-1.5	-0.1
77.29	0.78	-0.6	204.53	18OC	3.5	-3.4	-1.5	-0.2
76.93	1.21	-0.8	214.65	28OC	0.4	-3.4	-1.5	-0.2
75.15	1.67	-1.0	224.81	7NO	-0.3	-3.4	-1.4	-0.3
72.11	2.12	-1.2	234.99	17NO	-0.4	-3.5	-1.4	-0.2
68.28	2.51	-1.3	245.19	27NO	-0.4	-3.5	-1.4	-0.1
64.40	2.79	-1.2	255.40	7DE	-0.6	-3.5	-1.4	-0.1
61.23	2.94	-0.9	265.60	17DE	-0.9	-3.5	-1.4	-0.0
59.24	2.97	-0.6	275.78	27DE	-1.0	-3.4	-1.4	0.1

146

♂ LONG	LAT	MAG	☉ LONG	16.00UT	☿	♀	♃	♄
58.59	2.93	-0.3	285.95	6JA	-1.0	-3.4	-1.4	0.1
59.20	2.84	-0.1	296.07	16JA	-0.4	-3.4	-1.5	0.2
60.88	2.72	0.2	306.16	26JA	1.5	-3.4	-1.5	0.2
63.43	2.60	0.4	316.19	5FE	2.7	-3.3	-1.6	0.3
66.67	2.48	0.6	326.17	15FE	1.2	-3.3	-1.6	0.3
70.46	2.35	0.8	336.09	25FE	0.7	-3.3	-1.7	0.3
74.69	2.24	1.0	345.96	7MR	0.4	-3.3	-1.7	0.3
79.27	2.12	1.1	355.76	17MR	0.0	-3.4	-1.8	0.3
84.12	2.01	1.3	5.50	27MR	-0.5	-3.4	-1.8	0.3
89.21	1.90	1.4	15.20	6AP	-1.4	-3.4	-1.9	0.3
94.49	1.80	1.5	24.84	16AP	-1.6	-3.4	-2.0	0.3
99.93	1.70	1.6	34.45	26AP	-0.7	-3.4	-2.1	0.2
105.51	1.60	1.7	44.03	6MY	0.3	-3.5	-2.1	0.2
111.22	1.50	1.7	53.57	16MY	1.0	-3.5	-2.2	0.2
117.04	1.41	1.8	63.10	26MY	1.9	-3.6	-2.2	0.2
122.97	1.31	1.8	72.62	5JN	3.1	-3.7	-2.3	0.3
128.99	1.22	1.8	82.14	15JN	2.6	-3.7	-2.3	0.3
135.11	1.12	1.9	91.68	25JN	1.1	-3.8	-2.3	0.3
141.32	1.03	1.9	101.23	5JL	-0.2	-3.9	-2.3	0.3
147.62	0.93	1.9	110.81	15JL	-1.2	-4.0	-2.3	0.3
154.02	0.83	1.9	120.43	25JL	-1.5	-4.1	-2.3	0.3
160.51	0.73	1.9	130.09	4AU	-0.8	-4.2	-2.2	0.3
167.08	0.63	1.8	139.79	14AU	-0.3	-4.3	-2.2	0.3
173.76	0.53	1.8	149.56	24AU	-0.0	-4.2	-2.1	0.2
180.53	0.43	1.8	159.38	3SE	0.2	-4.0	-2.0	0.2
187.39	0.33	1.8	169.26	13SE	0.4	-3.5	-2.0	0.1
194.34	0.22	1.7	179.20	23SE	1.1	-3.4	-1.9	0.1
201.40	0.12	1.7	189.19	3OC	3.2	-4.0	-1.8	0.0
208.54	0.01	1.6	199.24	13OC	0.3	-4.3	-1.8	-0.1
215.78	-0.09	1.6	209.34	23OC	-0.5	-4.4	-1.7	-0.1
223.11	-0.20	1.5	219.47	2NO	-0.6	-4.3	-1.7	-0.2
230.52	-0.30	1.5	229.64	12NO	-0.6	-4.3	-1.6	-0.3
238.01	-0.41	1.5	239.84	22NO	-0.7	-4.2	-1.6	-0.3
245.58	-0.51	1.5	250.04	2DE	-0.8	-4.1	-1.5	-0.2
253.22	-0.61	1.5	260.25	12DE	-0.8	-4.0	-1.5	-0.2
260.92	-0.70	1.4	270.45	22DE	-0.8	-3.9	-1.5	-0.1

147

♂ LONG	LAT	MAG	☉ LONG	16.00UT	☿	♀	♃	♄
268.67	-0.79	1.4	280.62	1JA	-0.3	-3.8	-1.5	-0.0
276.46	-0.87	1.4	290.77	11JA	1.7	-3.7	-1.5	0.0
284.28	-0.95	1.4	300.88	21JA	2.2	-3.6	-1.5	0.1
292.12	-1.01	1.4	310.94	31JA	0.9	-3.5	-1.5	0.2
299.96	-1.07	1.4	320.95	10FE	0.5	-3.5	-1.5	0.2
307.80	-1.12	1.4	330.90	20FE	0.2	-3.4	-1.6	0.2
315.62	-1.15	1.4	340.80	2MR	-0.1	-3.4	-1.6	0.3
323.41	-1.17	1.4	350.63	12MR	-0.6	-3.4	-1.6	0.3
331.16	-1.18	1.4	0.41	22MR	-1.4	-3.3	-1.7	0.3
338.85	-1.18	1.4	10.13	1AP	-1.6	-3.3	-1.7	0.3
346.48	-1.17	1.4	19.80	11AP	-0.6	-3.3	-1.8	0.3
354.04	-1.14	1.4	29.43	21AP	0.4	-3.3	-1.9	0.3
1.51	-1.10	1.4	39.01	1MY	1.3	-3.3	-1.9	0.3
8.89	-1.04	1.4	48.57	11MY	2.5	-3.3	-2.0	0.2
16.17	-0.98	1.4	58.11	21MY	3.5	-3.3	-2.1	0.2
23.34	-0.90	1.4	67.63	31MY	2.0	-3.4	-2.2	0.2
30.40	-0.81	1.4	77.15	10JN	0.9	-3.4	-2.2	0.2
37.33	-0.72	1.4	86.68	20JN	-0.3	-3.4	-2.3	0.3
44.14	-0.61	1.4	96.22	30JN	-1.3	-3.5	-2.3	0.3
50.81	-0.49	1.3	105.78	10JL	-1.6	-3.5	-2.4	0.3
57.33	-0.36	1.3	115.38	20JL	-0.8	-3.5	-2.4	0.3
63.69	-0.22	1.3	125.02	30JL	-0.2	-3.4	-2.4	0.3
69.88	-0.07	1.2	134.70	9AU	0.1	-3.4	-2.4	0.3
75.87	0.09	1.2	144.43	19AU	0.3	-3.4	-2.4	0.3
81.64	0.26	1.1	154.22	29AU	0.6	-3.4	-2.4	0.3
87.15	0.46	1.0	164.07	8SE	1.4	-3.3	-2.3	0.3
92.37	0.66	0.9	173.98	18SE	2.9	-3.3	-2.3	0.2
97.23	0.89	0.8	183.94	28SE	0.2	-3.3	-2.2	0.2
101.67	1.15	0.7	193.96	8OC	-0.7	-3.3	-2.1	0.1
105.59	1.44	0.5	204.04	18OC	-0.7	-3.3	-2.1	0.0
108.87	1.76	0.3	214.15	28OC	-0.7	-3.4	-2.0	-0.0
111.36	2.11	0.1	224.30	7NO	-0.8	-3.4	-1.9	-0.1
112.88	2.51	-0.1	234.49	17NO	-0.7	-3.4	-1.9	-0.2
113.24	2.94	-0.3	244.69	27NO	-0.7	-3.5	-1.8	-0.2
112.29	3.38	-0.5	254.89	7DE	-0.7	-3.4	-1.7	-0.3
110.00	3.80	-0.8	265.10	17DE	-0.2	-3.6	-1.7	-0.2
106.61	4.12	-1.0	275.29	27DE	2.0	-3.6	-1.7	-0.1

♂			⊙	16.00UT	☿	♀	♃	♄
LONG	LAT	MAG	LONG	148	MAGNITUDES			
102.66	4.30	-1.0	285.45	6JA	1.7	-3.7	-1.6	-0.1
98.88	4.30	-0.9	295.58	16JA	0.6	-3.7	-1.6	-0.0
95.95	4.16	-0.6	305.67	26JA	0.3	-3.8	-1.6	0.1
94.23	3.92	-0.4	315.70	5FE	0.1	-3.9	-1.6	0.1
93.81	3.63	-0.1	325.69	15FE	-0.2	-4.0	-1.6	0.2
94.60	3.33	0.1	335.61	25FE	-0.6	-4.1	-1.6	0.2
96.41	3.04	0.3	345.48	6MR	-1.3	-4.2	-1.6	0.3
99.05	2.76	0.5	355.29	16MR	-1.5	-4.2	-1.6	0.3
102.37	2.51	0.7	5.03	26MR	-0.6	-4.2	-1.6	0.3
106.23	2.28	0.9	14.73	5AP	0.5	-4.1	-1.7	0.3
110.53	2.06	1.0	24.38	15AP	1.7	-3.7	-1.7	0.3
115.19	1.86	1.1	33.99	25AP	3.3	-2.9	-1.7	0.3
120.15	1.68	1.2	43.56	5MY	2.8	-3.4	-1.8	0.3
125.37	1.50	1.3	53.11	15MY	1.6	-3.9	-1.8	0.3
130.81	1.34	1.4	62.64	25MY	0.7	-4.2	-1.9	0.3
136.44	1.18	1.5	72.16	4JN	-0.3	-4.2	-2.0	0.3
142.25	1.03	1.5	81.68	14JN	-1.3	-4.1	-2.0	0.2
148.22	0.89	1.6	91.21	24JN	-1.6	-4.1	-2.1	0.3
154.34	0.75	1.6	100.76	4JL	-0.7	-4.0	-2.2	0.3
160.60	0.61	1.6	110.34	14JL	-0.1	-3.9	-2.2	0.4
167.00	0.48	1.6	119.96	24JL	0.2	-3.8	-2.3	0.4
173.52	0.35	1.6	129.62	3AU	0.5	-3.7	-2.4	0.4
180.18	0.22	1.6	139.32	13AU	0.8	-3.7	-2.4	0.4
186.95	0.10	1.6	149.08	23AU	1.8	-3.6	-2.4	0.4
193.85	-0.02	1.6	158.90	2SE	2.6	-3.5	-2.5	0.4
200.86	-0.14	1.6	168.78	12SE	0.1	-3.5	-2.5	0.4
207.99	-0.25	1.6	178.71	22SE	-0.8	-3.5	-2.4	0.3
215.23	-0.36	1.6	188.71	2OC	-0.9	-3.4	-2.4	0.3
222.57	-0.47	1.6	198.75	12OC	-0.9	-3.4	-2.4	0.3
230.01	-0.56	1.5	208.84	22OC	-0.7	-3.4	-2.3	0.2
237.55	-0.66	1.5	218.98	1NO	-0.5	-3.4	-2.2	0.1
245.17	-0.74	1.5	229.15	11NO	-0.5	-3.4	-2.2	0.1
252.86	-0.82	1.4	239.34	21NO	-0.5	-3.4	-2.1	-0.0
260.62	-0.89	1.4	249.55	1DE	-0.0	-3.4	-2.0	-0.1
268.43	-0.95	1.4	259.75	11DE	2.2	-3.4	-1.9	-0.2
276.28	-1.00	1.4	269.95	21DE	1.3	-3.4	-1.9	-0.2
284.17	-1.04	1.3	280.13	31DE	0.3	-3.4	-1.8	-0.1
				149				
292.06	-1.07	1.3	290.28	10JA	0.1	-3.4	-1.8	-0.1
299.96	-1.08	1.3	300.38	20JA	-0.0	-3.4	-1.7	0.0
307.85	-1.09	1.3	310.45	30JA	-0.2	-3.4	-1.7	0.1
315.71	-1.08	1.3	320.46	9FE	-0.7	-3.5	-1.6	0.1
323.53	-1.06	1.4	330.41	19FE	-1.3	-3.5	-1.6	0.2
331.31	-1.03	1.4	340.31	1MR	-1.4	-3.5	-1.6	0.2
339.03	-0.99	1.4	350.15	11MR	-0.6	-3.4	-1.6	0.3
346.68	-0.94	1.4	359.92	21MR	0.7	-3.4	-1.6	0.3
354.26	-0.88	1.5	9.65	31MR	2.2	-3.4	-1.6	0.4
1.76	-0.81	1.5	19.32	10AP	3.6	-3.4	-1.6	0.4
9.17	-0.73	1.5	28.95	20AP	2.1	-3.3	-1.6	0.4
16.50	-0.65	1.6	38.54	30AP	1.2	-3.3	-1.6	0.4
23.72	-0.56	1.6	48.10	10MY	0.5	-3.3	-1.6	0.4
30.86	-0.46	1.6	57.64	20MY	-0.4	-3.3	-1.7	0.4
37.89	-0.35	1.6	67.17	30MY	-1.3	-3.3	-1.7	0.4
44.83	-0.25	1.6	76.69	9JN	-1.6	-3.4	-1.8	0.4
51.67	-0.14	1.7	86.21	19JN	-0.7	-3.4	-1.8	0.4
58.41	-0.02	1.7	95.75	29JN	-0.1	-3.4	-1.9	0.4
65.04	0.10	1.7	105.32	9JL	0.3	-3.5	-1.9	0.4
71.57	0.22	1.7	114.91	19JL	0.7	-3.5	-2.0	0.5
77.99	0.35	1.7	124.55	29JL	1.1	-3.6	-2.1	0.5
84.30	0.49	1.6	134.23	8AU	2.3	-3.6	-2.1	0.5
90.49	0.62	1.6	143.96	18AU	2.2	-3.7	-2.2	0.5
96.55	0.77	1.6	153.75	28AU	0.0	-3.8	-2.3	0.5
102.46	0.92	1.5	163.59	7SE	-0.9	-3.9	-2.3	0.5
108.22	1.07	1.5	173.49	17SE	-1.0	-4.0	-2.4	0.5
113.79	1.24	1.4	183.46	27SE	-1.0	-4.1	-2.4	0.5
119.15	1.42	1.3	193.48	7OC	-0.6	-4.2	-2.4	0.5
124.26	1.61	1.2	203.55	17OC	-0.4	-4.3	-2.4	0.4
129.07	1.82	1.1	213.66	27OC	-0.4	-4.4	-2.4	0.4
133.51	2.05	0.9	223.81	6NO	-0.3	-4.3	-2.4	0.3
137.52	2.30	0.8	234.00	16NO	0.2	-4.1	-2.3	0.3
140.97	2.58	0.6	244.20	26NO	2.5	-3.4	-2.3	0.2
143.75	2.89	0.4	254.40	6DE	1.0	-3.2	-2.2	0.1
145.68	3.22	0.1	264.60	16DE	0.1	-4.0	-2.1	0.0
146.59	3.57	-0.1	274.80	26DE	-0.0	-4.3	-2.0	-0.0

♂			⊙	16.00UT	☿	♀	♃	♄
LONG	LAT	MAG	LONG	150	MAGNITUDES			
146.30	3.92	-0.4	284.96	5JA	-0.1	-4.4	-2.0	-0.0
144.71	4.23	-0.6	295.09	15JA	-0.3	-4.3	-1.9	0.0
141.89	4.46	-0.9	305.18	25JA	-0.7	-4.2	-1.8	0.1
138.22	4.55	-1.0	315.21	4FE	-1.3	-4.1	-1.8	0.1
134.32	4.45	-1.0	325.20	14FE	-1.3	-4.0	-1.7	0.2
130.91	4.20	-0.8	335.13	24FE	-0.6	-3.9	-1.7	0.3
128.53	3.83	-0.6	344.99	6MR	0.9	-3.8	-1.6	0.3
127.43	3.42	-0.4	354.81	16MR	2.8	-3.7	-1.6	0.4
127.61	3.00	-0.1	4.56	26MR	2.7	-3.6	-1.5	0.4
128.94	2.60	0.1	14.26	5AP	1.5	-3.6	-1.5	0.5
131.21	2.23	0.3	23.91	15AP	0.9	-3.5	-1.5	0.5
134.29	1.90	0.4	33.52	25AP	0.3	-3.4	-1.5	0.5
138.01	1.59	0.6	43.10	5MY	-0.4	-3.4	-1.5	0.5
142.25	1.32	0.7	52.65	15MY	-1.3	-3.4	-1.5	0.6
146.94	1.07	0.8	62.18	25MY	-1.6	-3.3	-1.5	0.6
152.00	0.84	0.9	71.70	4JN	-0.7	-3.3	-1.5	0.6
157.38	0.63	1.0	81.22	14JN	0.0	-3.3	-1.6	0.6
163.03	0.43	1.1	90.76	24JN	0.5	-3.3	-1.6	0.6
168.93	0.25	1.2	100.30	4JL	0.9	-3.3	-1.6	0.5
175.05	0.08	1.2	109.88	14JL	1.5	-3.3	-1.7	0.5
181.37	-0.08	1.2	119.50	24JL	2.9	-3.4	-1.7	0.6
187.87	-0.23	1.3	129.15	3AU	1.9	-3.4	-1.8	0.6
194.55	-0.36	1.3	138.85	13AU	-0.1	-3.4	-1.8	0.7
201.39	-0.49	1.3	148.61	23AU	-1.0	-3.4	-1.9	0.7
208.38	-0.61	1.3	158.42	2SE	-1.2	-3.5	-2.0	0.7
215.51	-0.72	1.4	168.30	12SE	-0.9	-3.5	-2.0	0.7
222.77	-0.82	1.4	178.23	22SE	-0.5	-3.5	-2.1	0.7
230.15	-0.90	1.4	188.22	2OC	-0.3	-3.5	-2.2	0.7
237.64	-0.98	1.4	198.26	12OC	-0.2	-3.4	-2.2	0.7
245.24	-1.04	1.4	208.35	22OC	-0.1	-3.4	-2.3	0.7
252.91	-1.09	1.4	218.48	1NO	0.4	-3.4	-2.3	0.6
260.66	-1.13	1.4	228.65	11NO	2.8	-3.4	-2.3	0.6
268.46	-1.15	1.4	238.84	21NO	0.8	-3.3	-2.3	0.5
276.31	-1.16	1.4	249.05	1DE	-0.1	-3.3	-2.3	0.5
284.19	-1.16	1.4	259.26	11DE	-0.2	-3.3	-2.3	0.4
292.08	-1.15	1.4	269.45	21DE	-0.3	-3.3	-2.2	0.3
299.96	-1.13	1.4	279.63	31DE	-0.4	-3.4	-2.1	0.2
				151				
307.84	-1.09	1.4	289.78	10JA	-0.8	-3.4	-2.1	0.2
315.68	-1.04	1.4	299.89	20JA	-1.2	-3.4	-2.0	0.2
323.48	-0.99	1.4	309.96	30JA	-1.2	-3.4	-1.9	0.2
331.23	-0.92	1.4	319.97	9FE	-0.5	-3.5	-1.9	0.3
338.92	-0.85	1.4	329.93	19FE	1.1	-3.5	-1.8	0.3
346.53	-0.77	1.5	339.83	1MR	3.3	-3.5	-1.7	0.4
354.07	-0.68	1.5	349.67	11MR	2.0	-3.6	-1.7	0.4
1.53	-0.59	1.5	359.45	21MR	1.1	-3.7	-1.6	0.5
8.91	-0.49	1.5	9.18	31MR	0.7	-3.7	-1.6	0.5
16.20	-0.39	1.5	18.86	10AP	0.2	-3.8	-1.5	0.6
23.40	-0.29	1.6	28.49	20AP	-0.4	-3.9	-1.5	0.6
30.52	-0.19	1.6	38.08	30AP	-1.3	-4.0	-1.5	0.7
37.55	-0.09	1.7	47.64	10MY	-1.6	-4.1	-1.4	0.7
44.49	0.01	1.7	57.18	20MY	-0.7	-4.1	-1.4	0.7
51.36	0.12	1.8	66.71	30MY	0.1	-4.2	-1.4	0.8
58.14	0.22	1.8	76.23	9JN	0.7	-4.1	-1.4	0.8
64.84	0.33	1.8	85.75	19JN	1.2	-3.9	-1.4	0.8
71.47	0.43	1.8	95.29	29JN	2.0	-3.4	-1.4	0.8
78.03	0.53	1.9	104.86	9JL	3.2	-3.1	-1.4	0.8
84.51	0.64	1.9	114.45	19JL	1.6	-3.7	-1.5	0.8
90.93	0.74	1.9	124.08	29JL	-0.1	-4.1	-1.5	0.8
97.27	0.84	1.9	133.76	8AU	-1.1	-4.2	-1.5	0.8
103.54	0.94	1.9	143.49	18AU	-1.3	-4.2	-1.6	0.9
109.73	1.05	1.8	153.27	28AU	-0.9	-4.2	-1.6	0.9
115.85	1.15	1.8	163.11	7SE	-0.4	-4.1	-1.7	0.9
121.88	1.26	1.8	173.01	17SE	-0.2	-4.0	-1.7	1.0
127.82	1.37	1.7	182.97	27SE	-0.1	-3.9	-1.8	1.0
133.65	1.48	1.7	192.99	7OC	0.0	-3.8	-1.9	1.0
139.36	1.60	1.6	203.05	17OC	0.6	-3.8	-1.9	1.0
144.94	1.72	1.5	213.17	27OC	3.2	-3.7	-2.0	0.9
150.35	1.84	1.4	223.32	6NO	0.6	-3.6	-2.0	0.9
155.56	1.98	1.3	233.50	16NO	-0.2	-3.6	-2.1	0.9
160.54	2.12	1.2	243.70	26NO	-0.3	-3.5	-2.1	0.8
165.23	2.27	1.0	253.91	6DE	-0.4	-3.5	-2.2	0.8
169.57	2.43	0.8	264.11	16DE	-0.5	-3.4	-2.2	0.7
173.48	2.60	0.7	274.30	26DE	-0.8	-3.4	-2.2	0.6

♂ LONG	♂ LAT	♂ MAG	⊙ LONG	16.00UT	☿	♀	♃	♄
				152	MAGNITUDES			
176.83	2.78	0.4	284.47	5JA	-1.1	-3.4	-2.2	0.6
179.52	2.96	0.2	294.60	15JA	-1.1	-3.4	-2.1	0.5
181.36	3.15	-0.1	304.69	25JA	-0.5	-3.3	-2.1	0.4
182.18	3.32	-0.3	314.73	4FE	1.3	-3.3	-2.0	0.4
181.81	3.46	-0.6	324.72	14FE	3.1	-3.3	-2.0	0.5
180.17	3.52	-0.9	334.65	24FE	1.5	-3.3	-1.9	0.5
177.35	3.48	-1.2	344.52	5MR	0.8	-3.3	-1.8	0.6
173.76	3.29	-1.4	354.33	15MR	0.5	-3.3	-1.8	0.6
170.03	2.96	-1.3	4.09	25MR	0.1	-3.4	-1.7	0.7
166.89	2.52	-1.1	13.79	4AP	-0.5	-3.4	-1.6	0.7
164.86	2.04	-0.9	23.44	14AP	-1.3	-3.4	-1.6	0.8
164.14	1.56	-0.7	33.06	24AP	-1.6	-3.5	-1.5	0.8
164.73	1.11	-0.5	42.63	4MY	-0.7	-3.5	-1.5	0.9
166.47	0.72	-0.3	52.18	14MY	0.2	-3.4	-1.4	0.9
169.20	0.36	-0.1	61.71	24MY	0.9	-3.4	-1.4	0.9
172.75	0.06	0.0	71.23	3JN	1.6	-3.4	-1.3	1.0
176.96	-0.21	0.2	80.75	13JN	2.7	-3.4	-1.3	1.0
181.72	-0.45	0.3	90.29	23JN	3.0	-3.3	-1.3	1.0
186.96	-0.65	0.4	99.83	3JL	1.3	-3.3	-1.3	1.0
192.58	-0.83	0.5	109.41	13JL	-0.2	-3.3	-1.3	1.0
198.55	-0.99	0.6	119.02	23JL	-1.2	-3.3	-1.3	1.0
204.81	-1.12	0.7	128.68	2AU	-1.4	-3.3	-1.3	1.0
211.32	-1.23	0.7	138.38	12AU	-0.8	-3.3	-1.3	1.0
218.07	-1.32	0.8	148.13	22AU	-0.3	-3.4	-1.4	1.1
225.01	-1.40	0.9	157.94	1SE	-0.1	-3.4	-1.4	1.1
232.13	-1.45	0.9	167.81	11SE	0.1	-3.4	-1.4	1.2
239.40	-1.49	1.0	177.75	21SE	0.2	-3.5	-1.5	1.2
246.80	-1.51	1.0	187.73	1OC	0.9	-3.5	-1.5	1.2
254.31	-1.51	1.1	197.77	11OC	3.5	-3.6	-1.6	1.3
261.92	-1.49	1.1	207.86	21OC	0.4	-3.7	-1.7	1.3
269.60	-1.47	1.1	217.99	31OC	-0.4	-3.8	-1.7	1.2
277.33	-1.42	1.2	228.16	10NO	-0.5	-3.8	-1.8	1.2
285.09	-1.36	1.2	238.35	20NO	-0.5	-3.9	-1.8	1.2
292.87	-1.30	1.3	248.55	30NO	-0.6	-4.0	-1.9	1.2
300.65	-1.22	1.3	258.76	10DE	-0.8	-4.1	-2.0	1.1
308.42	-1.13	1.3	268.96	20DE	-0.9	-4.2	-2.0	1.0
316.15	-1.03	1.4	279.14	30DE	-0.9	-4.3	-2.0	1.0
				153				
323.85	-0.93	1.4	289.29	9JA	-0.4	-4.4	-2.1	0.9
331.49	-0.83	1.5	299.41	19JA	1.5	-4.3	-2.1	0.8
339.07	-0.72	1.5	309.47	29JA	2.5	-4.0	-2.1	0.8
346.58	-0.61	1.5	319.49	8FE	1.1	-3.5	-2.0	0.7
354.02	-0.50	1.6	329.45	18FE	0.6	-3.4	-2.0	0.7
1.37	-0.38	1.6	339.35	28FE	0.3	-3.9	-2.0	0.8
8.65	-0.27	1.6	349.19	10MR	0.0	-4.2	-1.9	0.8
15.85	-0.16	1.7	358.98	20MR	-0.5	-4.3	-1.8	0.8
22.96	-0.06	1.7	8.71	30MR	-1.3	-4.2	-1.8	0.9
30.00	0.05	1.7	18.39	9AP	-1.6	-4.1	-1.7	1.0
36.96	0.15	1.7	28.02	19AP	-0.7	-4.0	-1.6	1.0
43.84	0.25	1.7	37.61	29AP	0.3	-3.9	-1.6	1.1
50.66	0.34	1.8	47.18	9MY	1.1	-3.8	-1.5	1.1
57.40	0.43	1.8	56.72	19MY	2.1	-3.7	-1.5	1.2
64.09	0.52	1.8	66.24	29MY	3.4	-3.6	-1.4	1.2
70.71	0.61	1.8	75.76	8JN	2.4	-3.6	-1.4	1.2
77.28	0.69	1.9	85.29	18JN	1.0	-3.5	-1.3	1.3
83.80	0.77	1.9	94.83	28JN	-0.2	-3.5	-1.3	1.3
90.27	0.84	1.9	104.39	8JL	-1.2	-3.4	-1.3	1.3
96.70	0.92	2.0	113.98	18JL	-1.5	-3.4	-1.3	1.3
103.09	0.99	2.0	123.61	28JL	-0.8	-3.4	-1.2	1.3
109.44	1.05	2.0	133.29	7AU	-0.3	-3.4	-1.2	1.3
115.76	1.12	2.0	143.02	17AU	0.0	-3.4	-1.2	1.3
122.04	1.18	2.0	152.80	27AU	0.2	-3.4	-1.2	1.3
128.29	1.24	2.0	162.64	6SE	0.4	-3.4	-1.3	1.3
134.50	1.30	2.0	172.54	16SE	1.2	-3.4	-1.3	1.3
140.68	1.35	1.9	182.49	26SE	3.2	-3.4	-1.3	1.3
146.82	1.41	1.9	192.51	6OC	0.3	-3.4	-1.3	1.3
152.91	1.46	1.8	202.57	16OC	-0.6	-3.4	-1.4	1.3
158.95	1.51	1.8	212.68	26OC	-0.6	-3.4	-1.4	1.3
164.93	1.56	1.7	222.83	5NO	-0.6	-3.4	-1.5	1.2
170.84	1.60	1.6	233.01	15NO	-0.7	-3.5	-1.6	1.2
176.67	1.65	1.5	243.20	25NO	-0.7	-3.5	-1.6	1.2
182.40	1.69	1.4	253.41	5DE	-0.8	-3.5	-1.7	1.1
188.01	1.72	1.3	263.62	15DE	-0.8	-3.5	-1.7	1.1
193.48	1.75	1.2	273.80	25DE	-0.3	-3.4	-1.8	1.0

♂ LONG	♂ LAT	♂ MAG	⊙ LONG	16.00UT	☿	♀	♃	♄
				154	MAGNITUDES			
198.77	1.77	1.0	283.97	4JA	1.7	-3.4	-1.9	1.0
203.83	1.79	0.8	294.11	14JA	2.0	-3.4	-1.9	0.9
208.64	1.79	0.6	304.20	24JA	0.8	-3.4	-2.0	0.9
213.10	1.77	0.4	314.24	3FE	0.4	-3.3	-2.0	0.9
217.15	1.73	0.2	324.24	13FE	0.2	-3.3	-2.0	0.8
220.68	1.66	-0.0	334.17	23FE	-0.1	-3.3	-2.0	0.8
223.54	1.56	-0.3	344.04	5MR	-0.6	-3.3	-2.0	0.9
225.58	1.40	-0.6	353.86	15MR	-1.3	-3.4	-2.0	0.9
226.62	1.17	-0.9	3.61	25MR	-1.6	-3.4	-1.9	1.0
226.48	0.85	-1.2	13.32	4AP	-0.7	-3.4	-1.9	1.1
225.09	0.44	-1.6	22.98	14AP	0.4	-3.4	-1.8	1.1
222.58	-0.05	-1.8	32.59	24AP	1.5	-3.4	-1.8	1.2
219.34	-0.61	-2.0	42.17	4MY	2.7	-3.5	-1.7	1.2
216.09	-1.16	-1.9	51.72	14MY	3.3	-3.5	-1.7	1.3
213.52	-1.65	-1.8	61.25	24MY	1.9	-3.6	-1.6	1.3
212.13	-2.05	-1.6	70.78	3JN	0.8	-3.7	-1.5	1.4
212.13	-2.35	-1.3	80.30	13JN	-0.3	-3.7	-1.5	1.4
213.46	-2.55	-1.1	89.82	23JN	-1.3	-3.8	-1.4	1.4
215.97	-2.69	-0.9	99.38	3JL	-1.6	-3.9	-1.4	1.4
219.47	-2.76	-0.7	108.95	13JL	-0.8	-4.0	-1.3	1.4
223.78	-2.79	-0.5	118.56	23JL	-0.2	-4.1	-1.3	1.3
228.76	-2.79	-0.4	128.21	2AU	0.1	-4.2	-1.3	1.3
234.28	-2.75	-0.2	137.91	12AU	0.4	-4.3	-1.2	1.3
240.22	-2.68	-0.1	147.66	22AU	0.7	-4.2	-1.2	1.2
246.53	-2.59	0.0	157.47	1SE	1.6	-4.0	-1.2	1.2
253.13	-2.48	0.2	167.34	11SE	2.9	-3.5	-1.2	1.1
259.95	-2.35	0.3	177.26	21SE	0.2	-3.4	-1.2	1.1
266.96	-2.21	0.4	187.25	1OC	-0.7	-4.0	-1.2	1.2
274.11	-2.06	0.5	197.28	11OC	-0.8	-4.3	-1.2	1.1
281.37	-1.90	0.6	207.37	21OC	-0.8	-4.4	-1.3	1.1
288.70	-1.73	0.7	217.51	31OC	-0.8	-4.3	-1.3	1.1
296.09	-1.57	0.8	227.67	10NO	-0.6	-4.3	-1.4	1.1
303.50	-1.39	0.9	237.86	20NO	-0.6	-4.2	-1.4	1.1
310.92	-1.22	1.0	248.06	30NO	-0.6	-4.1	-1.5	1.0
318.33	-1.06	1.1	258.27	10DE	-0.1	-4.0	-1.5	1.0
325.72	-0.89	1.2	268.47	20DE	2.0	-3.9	-1.6	1.0
333.07	-0.73	1.3	278.65	30DE	1.6	-3.8	-1.6	0.9
				155				
340.37	-0.58	1.3	288.80	9JA	0.5	-3.7	-1.7	0.9
347.62	-0.43	1.4	298.92	19JA	0.2	-3.6	-1.8	0.8
354.82	-0.29	1.5	308.99	29JA	0.1	-3.5	-1.8	0.8
1.95	-0.16	1.6	319.01	8FE	-0.2	-3.5	-1.9	0.7
9.01	-0.03	1.6	328.97	18FE	-0.6	-3.4	-1.9	0.7
16.01	0.09	1.7	338.88	28FE	-1.3	-3.4	-2.0	0.7
22.95	0.20	1.7	348.72	10MR	-1.5	-3.4	-2.0	0.6
29.82	0.31	1.8	358.51	20MR	-0.7	-3.3	-2.0	0.7
36.62	0.40	1.8	8.25	30MR	0.6	-3.3	-2.0	0.8
43.37	0.49	1.9	17.93	9AP	1.9	-3.3	-2.0	0.8
50.06	0.58	1.9	27.56	19AP	3.6	-3.3	-2.0	0.9
56.70	0.66	1.9	37.16	29AP	2.5	-3.3	-1.9	1.0
63.29	0.73	1.9	46.72	9MY	1.4	-3.3	-1.9	1.0
69.83	0.79	1.9	56.26	19MY	0.6	-3.3	-1.8	1.1
76.34	0.86	1.9	65.79	29MY	-0.3	-3.4	-1.8	1.1
82.81	0.91	1.9	75.30	8JN	-1.3	-3.4	-1.7	1.2
89.26	0.96	1.9	84.83	18JN	-1.6	-3.4	-1.6	1.2
95.67	1.01	1.9	94.37	28JN	-0.8	-3.5	-1.6	1.2
102.07	1.05	1.9	103.92	8JL	-0.1	-3.5	-1.5	1.2
108.46	1.09	2.0	113.52	18JL	0.3	-3.5	-1.5	1.2
114.83	1.12	2.0	123.14	28JL	0.5	-3.4	-1.4	1.2
121.19	1.15	2.0	132.81	7AU	0.9	-3.4	-1.4	1.2
127.55	1.18	2.0	142.54	17AU	2.0	-3.4	-1.3	1.1
133.91	1.20	2.0	152.32	27AU	2.5	-3.4	-1.3	1.1
140.27	1.22	2.0	162.15	6SE	0.1	-3.3	-1.3	1.0
146.63	1.23	2.0	172.05	16SE	-0.8	-3.3	-1.3	1.0
153.00	1.24	2.0	182.00	26SE	-0.9	-3.3	-1.2	1.0
159.37	1.24	2.0	192.01	6OC	-0.9	-3.3	-1.2	1.0
165.75	1.24	1.9	202.08	16OC	-0.7	-3.3	-1.2	1.0
172.13	1.23	1.9	212.19	26OC	-0.5	-3.4	-1.3	1.0
178.51	1.22	1.9	222.33	5NO	-0.4	-3.4	-1.3	1.0
184.89	1.20	1.8	232.51	15NO	-0.4	-3.4	-1.3	1.0
191.28	1.17	1.7	242.71	25NO	0.0	-3.5	-1.3	1.0
197.65	1.14	1.6	252.91	5DE	2.2	-3.5	-1.4	1.0
204.02	1.09	1.6	263.12	15DE	1.3	-3.6	-1.4	0.9
210.37	1.04	1.5	273.32	25DE	0.3	-3.6	-1.5	0.9

♂ LONG	♂ LAT	♂ MAG	☉ LONG	16.00UT	☿	♀	♃	♄
				156	MAGNITUDES			
216.70	0.97	1.4	283.48	4JA	0.1	-3.7	-1.5	0.8
223.00	0.88	1.2	293.62	14JA	-0.1	-3.7	-1.6	0.8
229.26	0.78	1.1	303.72	24JA	-0.3	-3.8	-1.6	0.7
235.47	0.66	1.0	313.76	3FE	-0.7	-3.9	-1.7	0.7
241.61	0.51	0.8	323.76	13FE	-1.3	-4.0	-1.8	0.6
247.67	0.34	0.7	333.69	23FE	-1.4	-4.1	-1.9	0.6
253.63	0.14	0.5	343.57	4MR	-0.6	-4.2	-1.9	0.6
259.44	-0.10	0.3	353.39	14MR	0.7	-4.2	-2.0	0.5
265.08	-0.38	0.1	3.15	24MR	2.4	-4.2	-2.0	0.5
270.49	-0.71	-0.1	12.86	3AP	3.3	-4.1	-2.1	0.6
275.60	-1.10	-0.3	22.52	13AP	1.9	-3.7	-2.1	0.6
280.34	-1.55	-0.6	32.13	23AP	1.1	-2.9	-2.1	0.7
284.58	-2.07	-0.8	41.71	3MY	0.4	-3.4	-2.1	0.8
288.17	-2.66	-1.1	51.26	13MY	-0.3	-3.9	-2.1	0.8
290.96	-3.34	-1.4	60.79	23MY	-1.3	-4.2	-2.0	0.9
292.74	-4.08	-1.7	70.31	2JN	-1.7	-4.2	-2.0	0.9
293.33	-4.86	-2.0	79.84	12JN	-0.7	-4.1	-1.9	1.0
292.65	-5.61	-2.3	89.36	22JN	-0.1	-4.1	-1.9	1.0
290.84	-6.24	-2.5	98.91	2JL	0.4	-4.0	-1.8	1.0
288.31	-6.61	-2.6	108.49	12JL	0.7	-3.9	-1.8	1.1
285.79	-6.67	-2.5	118.09	22JL	1.3	-3.8	-1.7	1.1
283.96	-6.43	-2.3	127.74	1AU	2.5	-3.7	-1.6	1.1
283.30	-5.98	-2.1	137.44	11AU	2.2	-3.7	-1.6	1.0
283.98	-5.41	-1.8	147.18	21AU	0.0	-3.6	-1.5	1.0
285.89	-4.80	-1.5	156.99	31AU	-1.0	-3.5	-1.5	1.0
288.88	-4.19	-1.2	166.86	10SE	-1.1	-3.5	-1.4	0.9
292.73	-3.62	-0.9	176.78	20SE	-1.0	-3.5	-1.4	0.9
297.27	-3.09	-0.7	186.76	30SE	-0.6	-3.4	-1.4	0.9
302.34	-2.61	-0.5	196.80	10OC	-0.4	-3.4	-1.3	0.9
307.82	-2.18	-0.2	206.88	20OC	-0.3	-3.4	-1.3	0.9
313.60	-1.79	-0.0	217.01	30OC	-0.3	-3.4	-1.3	0.9
319.62	-1.44	0.2	227.17	9NO	0.2	-3.4	-1.3	0.9
325.82	-1.12	0.4	237.36	19NO	2.5	-3.4	-1.3	0.9
332.14	-0.84	0.5	247.57	29NO	1.0	-3.4	-1.3	0.9
338.56	-0.59	0.7	257.77	9DE	0.1	-3.4	-1.4	0.9
345.04	-0.37	0.8	267.97	19DE	-0.1	-3.4	-1.4	0.9
351.56	-0.17	1.0	278.16	29DE	-0.2	-3.4	-1.4	0.8
				157				
358.11	0.01	1.1	288.31	8JA	-0.4	-3.4	-1.5	0.8
4.66	0.17	1.2	298.43	18JA	-0.7	-3.4	-1.5	0.7
11.21	0.31	1.3	308.50	28JA	-1.3	-3.4	-1.6	0.7
17.75	0.43	1.4	318.53	7FE	-1.3	-3.5	-1.6	0.7
24.28	0.54	1.5	328.49	17FE	-0.6	-3.5	-1.7	0.6
30.78	0.64	1.6	338.40	27FE	0.9	-3.5	-1.8	0.5
37.27	0.73	1.7	348.25	9MR	2.9	-3.4	-1.8	0.5
43.72	0.80	1.8	358.04	19MR	2.5	-3.4	-1.9	0.4
50.16	0.87	1.8	7.77	29MR	1.4	-3.4	-2.0	0.4
56.57	0.93	1.9	17.46	8AP	0.8	-3.4	-2.0	0.4
62.96	0.98	1.9	27.09	18AP	0.3	-3.3	-2.1	0.5
69.33	1.02	2.0	36.69	28AP	-0.4	-3.3	-2.1	0.6
75.69	1.06	2.0	46.26	8MY	-1.3	-3.3	-2.2	0.6
82.03	1.09	2.0	55.80	18MY	-1.7	-3.3	-2.2	0.7
88.36	1.12	2.0	65.32	28MY	-0.7	-3.3	-2.2	0.8
94.69	1.14	2.0	74.85	7JN	0.0	-3.4	-2.2	0.8
101.01	1.15	2.0	84.37	17JN	0.5	-3.4	-2.2	0.9
107.34	1.16	2.0	93.91	27JN	1.0	-3.4	-2.1	0.9
113.68	1.16	2.0	103.46	7JL	1.7	-3.5	-2.1	0.9
120.03	1.16	2.0	113.05	17JL	3.0	-3.5	-2.0	0.9
126.40	1.16	2.0	122.68	27JL	1.8	-3.6	-2.0	1.0
132.79	1.15	1.9	132.35	6AU	-0.0	-3.6	-1.9	1.0
139.20	1.14	2.0	142.07	16AU	-1.1	-3.7	-1.8	1.0
145.64	1.12	2.0	151.84	26AU	-1.2	-3.8	-1.8	0.9
152.11	1.09	2.0	161.68	5SE	-0.9	-3.9	-1.7	0.9
158.61	1.06	2.0	171.57	15SE	-0.5	-4.0	-1.7	0.9
165.16	1.03	2.0	181.52	25SE	-0.3	-4.1	-1.6	0.8
171.74	0.99	1.9	191.53	5OC	-0.2	-4.2	-1.6	0.8
178.36	0.95	1.9	201.58	15OC	-0.1	-4.3	-1.5	0.8
185.02	0.89	1.9	211.69	25OC	0.4	-4.4	-1.5	0.8
191.72	0.83	1.9	221.84	4NO	2.8	-4.3	-1.5	0.8
198.47	0.77	1.8	232.01	14NO	0.8	-4.1	-1.4	0.8
205.27	0.69	1.8	242.21	24NO	-0.1	-3.4	-1.4	0.8
212.10	0.61	1.7	252.42	4DE	-0.3	-3.3	-1.4	0.8
218.98	0.52	1.7	262.62	14DE	-0.3	-4.0	-1.4	0.8
225.90	0.42	1.6	272.82	24DE	-0.5	-4.3	-1.4	0.8

♂ LONG	♂ LAT	♂ MAG	☉ LONG	16.00UT	☿	♀	♃	♄
				158	MAGNITUDES			
232.86	0.30	1.5	282.99	3JA	-0.8	-4.4	-1.4	0.8
239.86	0.18	1.5	293.13	13JA	-1.1	-4.3	-1.5	0.8
246.89	0.04	1.4	303.23	23JA	-1.1	-4.2	-1.5	0.7
253.96	-0.11	1.3	313.28	2FE	-0.5	-4.1	-1.5	0.7
261.06	-0.27	1.2	323.27	12FE	1.1	-4.0	-1.6	0.6
268.18	-0.44	1.1	333.21	22FE	3.3	-3.9	-1.6	0.6
275.31	-0.63	1.0	343.09	4MR	1.8	-3.8	-1.7	0.5
282.45	-0.83	0.9	352.91	14MR	1.0	-3.7	-1.7	0.5
289.58	-1.05	0.8	2.68	24MR	0.6	-3.6	-1.8	0.4
296.70	-1.27	0.7	12.39	3AP	0.2	-3.6	-1.9	0.4
303.77	-1.51	0.5	22.05	13AP	-0.4	-3.5	-1.9	0.3
310.78	-1.75	0.4	31.67	23AP	-1.3	-3.4	-2.0	0.4
317.70	-2.00	0.3	41.25	3MY	-1.7	-3.4	-2.1	0.4
324.51	-2.26	0.2	50.80	13MY	-0.7	-3.4	-2.2	0.5
331.14	-2.51	0.0	60.34	23MY	0.1	-3.3	-2.2	0.6
337.56	-2.77	-0.1	69.86	2JN	0.7	-3.3	-2.3	0.6
343.70	-3.03	-0.3	79.38	12JN	1.3	-3.3	-2.3	0.7
349.48	-3.28	-0.4	88.91	22JN	2.2	-3.3	-2.3	0.7
354.81	-3.52	-0.6	98.45	2JL	3.2	-3.3	-2.3	0.8
359.57	-3.75	-0.8	108.03	12JL	1.5	-3.3	-2.3	0.8
3.60	-3.96	-1.0	117.63	22JL	-0.1	-3.4	-2.3	0.9
6.72	-4.14	-1.2	127.28	1AU	-1.1	-3.4	-2.3	0.9
8.72	-4.28	-1.4	136.97	11AU	-1.4	-3.4	-2.2	0.9
9.39	-4.33	-1.6	146.72	21AU	-0.9	-3.4	-2.2	0.9
8.60	-4.26	-1.9	156.51	31AU	-0.4	-3.5	-2.1	0.9
6.44	-4.03	-2.1	166.38	10SE	-0.1	-3.5	-2.0	0.9
3.30	-3.62	-2.2	176.30	20SE	-0.0	-3.5	-2.0	0.8
359.94	-3.04	-2.1	186.27	30SE	0.1	-3.5	-1.9	0.8
357.14	-2.38	-1.8	196.30	10OC	0.7	-3.4	-1.8	0.8
355.48	-1.73	-1.5	206.39	20OC	3.2	-3.4	-1.8	0.7
355.16	-1.14	-1.2	216.51	30OC	0.6	-3.4	-1.7	0.7
356.12	-0.64	-0.9	226.67	9NO	-0.3	-3.4	-1.7	0.8
358.19	-0.22	-0.5	236.86	19NO	-0.4	-3.3	-1.6	0.8
1.16	0.11	-0.3	247.06	29NO	-0.4	-3.3	-1.6	0.8
4.82	0.38	0.0	257.27	9DE	-0.6	-3.3	-1.6	0.8
9.03	0.60	0.3	267.47	19DE	-0.8	-3.3	-1.5	0.8
13.67	0.77	0.5	277.65	29DE	-1.0	-3.4	-1.5	0.8
				159				
18.62	0.91	0.7	287.81	8JA	-1.0	-3.4	-1.5	0.8
23.84	1.02	0.9	297.93	18JA	-0.5	-3.4	-1.5	0.8
29.25	1.11	1.0	308.01	28JA	1.3	-3.4	-1.5	0.7
34.81	1.18	1.2	318.04	7FE	2.9	-3.5	-1.5	0.7
40.49	1.24	1.3	328.01	17FE	1.3	-3.5	-1.6	0.6
46.26	1.28	1.4	337.92	27FE	0.7	-3.5	-1.6	0.6
52.10	1.31	1.5	347.77	9MR	0.4	-3.6	-1.6	0.5
58.01	1.33	1.6	357.57	19MR	0.1	-3.7	-1.7	0.5
63.96	1.34	1.7	7.30	29MR	-0.5	-3.7	-1.7	0.4
69.96	1.35	1.8	16.99	8AP	-1.3	-3.8	-1.8	0.4
75.99	1.35	1.8	26.63	18AP	-1.6	-3.9	-1.8	0.3
82.05	1.34	1.9	36.23	28AP	-0.7	-4.0	-1.9	0.3
88.14	1.33	1.9	45.80	8MY	0.2	-4.1	-2.0	0.3
94.26	1.32	2.0	55.34	18MY	0.9	-4.1	-2.0	0.4
100.41	1.30	2.0	64.87	28MY	1.7	-4.2	-2.1	0.5
106.59	1.27	2.0	74.39	7JN	2.9	-4.1	-2.2	0.5
112.81	1.25	2.0	83.91	17JN	2.8	-3.9	-2.2	0.6
119.06	1.21	2.0	93.45	27JN	1.2	-3.3	-2.3	0.7
125.35	1.18	2.0	103.01	7JL	-0.1	-3.1	-2.3	0.7
131.68	1.14	2.0	112.59	17JL	-1.2	-3.8	-2.4	0.7
138.05	1.09	2.0	122.21	27JL	-1.5	-4.1	-2.4	0.8
144.48	1.05	2.0	131.88	6AU	-0.8	-4.2	-2.4	0.8
150.96	0.99	1.9	141.60	16AU	-0.3	-4.2	-2.4	0.8
157.49	0.94	1.9	151.37	26AU	-0.0	-4.2	-2.4	0.8
164.08	0.88	1.9	161.20	5SE	0.1	-4.1	-2.4	0.8
170.73	0.81	1.8	171.09	15SE	0.3	-4.0	-2.3	0.8
177.45	0.74	1.8	181.03	25SE	1.0	-3.9	-2.2	0.8
184.23	0.67	1.8	191.04	5OC	3.4	-3.8	-2.2	0.8
191.08	0.59	1.8	201.10	15OC	0.4	-3.8	-2.1	0.7
198.00	0.50	1.8	211.20	25OC	-0.5	-3.7	-2.0	0.7
204.99	0.41	1.8	221.35	4NO	-0.6	-3.6	-2.0	0.7
212.04	0.32	1.8	231.52	14NO	-0.6	-3.6	-1.9	0.7
219.17	0.22	1.7	241.71	24NO	-0.7	-3.5	-1.8	0.7
226.36	0.11	1.7	251.92	4DE	-0.8	-3.5	-1.8	0.8
233.62	-0.00	1.7	262.12	14DE	-0.8	-3.4	-1.7	0.8
240.94	-0.12	1.6	272.32	24DE	-0.8	-3.4	-1.7	0.8

Left section (160 / 161):

♂ LONG	LAT	MAG	☉ LONG	16.00UT	☿	♀	♃	♄
248.32	-0.24	1.6	282.49	3JA	-0.4	-3.4	-1.6	0.8
255.75	-0.37	1.5	292.63	13JA	1.5	-3.4	-1.6	0.8
263.24	-0.50	1.5	302.73	23JA	2.4	-3.3	-1.6	0.8
270.77	-0.63	1.4	312.79	2FE	1.0	-3.3	-1.6	0.7
278.34	-0.76	1.4	322.79	12FE	0.5	-3.3	-1.6	0.7
285.94	-0.89	1.3	332.73	22FE	0.3	-3.3	-1.6	0.7
293.55	-1.02	1.3	342.62	3MR	-0.0	-3.3	-1.6	0.6
301.17	-1.15	1.2	352.44	13MR	-0.5	-3.3	-1.6	0.6
308.79	-1.27	1.2	2.21	23MR	-1.3	-3.4	-1.6	0.5
316.39	-1.39	1.1	11.92	2AP	-1.6	-3.4	-1.6	0.4
323.95	-1.49	1.0	21.58	12AP	-0.7	-3.4	-1.7	0.4
331.47	-1.59	1.0	31.20	22AP	0.3	-3.5	-1.7	0.3
338.92	-1.67	0.9	40.79	2MY	1.2	-3.5	-1.7	0.3
346.28	-1.74	0.9	50.34	12MY	2.3	-3.4	-1.8	0.3
353.54	-1.79	0.8	59.87	22MY	3.6	-3.4	-1.9	0.3
0.67	-1.83	0.8	69.40	1JN	2.2	-3.4	-1.9	0.4
7.65	-1.85	0.7	78.92	11JN	1.0	-3.4	-2.0	0.5
14.46	-1.85	0.6	88.44	21JN	-0.2	-3.3	-2.1	0.5
21.06	-1.83	0.5	97.99	1JL	-1.2	-3.3	-2.1	0.6
27.42	-1.80	0.5	107.56	11JL	-1.6	-3.3	-2.2	0.6
33.50	-1.74	0.4	117.16	21JL	-0.8	-3.3	-2.3	0.7
39.23	-1.66	0.3	126.81	31JL	-0.3	-3.3	-2.3	0.7
44.57	-1.55	0.1	136.49	10AU	0.1	-3.3	-2.4	0.8
49.42	-1.42	0.0	146.24	20AU	0.3	-3.4	-2.4	0.8
53.66	-1.25	-0.1	156.04	30AU	0.5	-3.4	-2.4	0.8
57.18	-1.05	-0.3	165.89	9SE	1.3	-3.4	-2.5	0.8
59.79	-0.80	-0.5	175.81	19SE	3.2	-3.5	-2.4	0.8
61.28	-0.50	-0.7	185.79	29SE	0.3	-3.5	-2.4	0.8
61.48	-0.13	-0.9	195.82	9OC	-0.6	-3.6	-2.4	0.8
60.23	0.29	-1.2	205.90	19OC	-0.7	-3.7	-2.3	0.7
57.61	0.74	-1.3	216.02	29OC	-0.7	-3.8	-2.3	0.7
54.01	1.20	-1.5	226.18	8NO	-0.8	-3.8	-2.2	0.7
50.12	1.60	-1.4	236.37	18NO	-0.7	-3.9	-2.1	0.7
46.75	1.91	-1.2	246.57	28NO	-0.7	-4.0	-2.1	0.7
44.48	2.11	-0.9	256.77	8DE	-0.7	-4.1	-2.0	0.7
43.54	2.22	-0.6	266.98	18DE	-0.2	-4.2	-1.9	0.8
43.91	2.27	-0.3	277.16	28DE	1.7	-4.3	-1.9	0.8

161

♂ LONG	LAT	MAG	☉ LONG	16.00UT	☿	♀	♃	♄
45.40	2.27	-0.0	287.32	7JA	1.9	-4.4	-1.8	0.8
47.83	2.24	0.3	297.44	17JA	0.7	-4.3	-1.7	0.8
50.99	2.20	0.5	307.52	27JA	0.3	-4.0	-1.7	0.8
54.73	2.14	0.7	317.54	6FE	0.1	-3.5	-1.7	0.8
58.92	2.08	0.9	327.52	16FE	-0.1	-3.4	-1.6	0.7
63.48	2.02	1.1	337.43	26FE	-0.6	-4.0	-1.6	0.7
68.31	1.95	1.2	347.29	8MR	-1.3	-4.2	-1.6	0.6
73.38	1.88	1.3	357.09	18MR	-1.5	-4.3	-1.6	0.6
78.64	1.82	1.4	6.83	28MR	-0.7	-4.2	-1.6	0.5
84.05	1.75	1.5	16.52	7AP	0.4	-4.1	-1.6	0.5
89.60	1.68	1.6	26.16	17AP	1.6	-4.0	-1.6	0.4
95.26	1.61	1.7	35.76	27AP	3.0	-3.9	-1.6	0.3
101.02	1.54	1.8	45.33	7MY	3.0	-3.8	-1.6	0.3
106.87	1.46	1.8	54.88	17MY	1.7	-3.7	-1.6	0.2
112.80	1.39	1.9	64.41	27MY	0.7	-3.6	-1.7	0.3
118.82	1.32	1.9	73.93	6JN	-0.2	-3.6	-1.7	0.3
124.91	1.24	1.9	83.45	16JN	-1.3	-3.5	-1.8	0.4
131.07	1.16	1.9	92.99	26JN	-1.6	-3.5	-1.8	0.5
137.32	1.08	1.9	102.54	6JL	-0.8	-3.4	-1.9	0.5
143.64	1.00	1.9	112.13	16JL	-0.2	-3.4	-1.9	0.6
150.03	0.92	1.9	121.75	26JL	0.2	-3.4	-2.0	0.6
156.51	0.84	1.9	131.41	5AU	0.4	-3.4	-2.1	0.7
163.06	0.75	1.9	141.13	15AU	0.8	-3.4	-2.1	0.7
169.70	0.66	1.9	150.90	25AU	1.7	-3.4	-2.2	0.7
176.42	0.57	1.8	160.72	4SE	2.8	-3.4	-2.3	0.8
183.23	0.47	1.8	170.61	14SE	0.2	-3.4	-2.3	0.8
190.13	0.38	1.8	180.55	24SE	-0.8	-3.4	-2.4	0.8
197.11	0.28	1.7	190.56	4OC	-0.9	-3.4	-2.4	0.8
204.18	0.17	1.6	200.61	14OC	-0.8	-3.4	-2.4	0.8
211.34	0.07	1.6	210.71	24OC	-0.7	-3.4	-2.4	0.7
218.58	-0.03	1.6	220.85	3NO	-0.6	-3.4	-2.3	0.7
225.91	-0.14	1.6	231.03	13NO	-0.5	-3.5	-2.3	0.7
233.31	-0.25	1.6	241.22	23NO	-0.5	-3.5	-2.3	0.7
240.80	-0.35	1.5	251.42	3DE	-0.1	-3.5	-2.2	0.7
248.35	-0.46	1.5	261.63	13DE	2.0	-3.4	-2.2	0.7
255.96	-0.57	1.5	271.82	23DE	1.5	-3.4	-2.1	0.8

Right section (162 / 163):

♂ LONG	LAT	MAG	☉ LONG	16.00UT	☿	♀	♃	♄
263.63	-0.67	1.5	282.00	2JA	0.4	-3.4	-2.0	0.8
271.35	-0.77	1.5	292.14	12JA	0.1	-3.4	-1.9	0.8
279.10	-0.86	1.5	302.24	22JA	0.0	-3.4	-1.9	0.8
286.89	-0.95	1.4	312.29	1FE	-0.2	-3.3	-1.8	0.8
294.69	-1.03	1.4	322.30	11FE	-0.6	-3.3	-1.7	0.8
302.49	-1.10	1.4	332.24	21FE	-1.3	-3.3	-1.7	0.8
310.29	-1.16	1.4	342.13	3MR	-1.5	-3.3	-1.6	0.7
318.06	-1.21	1.4	351.96	13MR	-0.7	-3.4	-1.6	0.7
325.81	-1.24	1.4	1.73	23MR	0.6	-3.4	-1.6	0.6
333.51	-1.27	1.3	11.44	2AP	2.0	-3.4	-1.5	0.6
341.15	-1.28	1.3	21.11	12AP	3.9	-3.4	-1.5	0.5
348.72	-1.28	1.3	30.73	22AP	2.3	-3.4	-1.5	0.5
356.22	-1.26	1.3	40.32	2MY	1.3	-3.5	-1.5	0.4
3.62	-1.23	1.3	49.88	12MY	0.6	-3.5	-1.5	0.3
10.93	-1.19	1.3	59.41	22MY	-0.3	-3.6	-1.5	0.3
18.12	-1.13	1.3	68.94	1JN	-1.3	-3.7	-1.5	0.2
25.19	-1.06	1.2	78.46	11JN	-1.7	-3.7	-1.5	0.3
32.13	-0.98	1.2	87.98	21JN	-0.8	-3.8	-1.6	0.4
38.93	-0.88	1.2	97.53	1JL	-0.1	-3.9	-1.6	0.4
45.56	-0.77	1.1	107.10	11JL	0.3	-4.0	-1.6	0.5
52.03	-0.65	1.1	116.70	21JL	0.6	-4.1	-1.7	0.5
58.30	-0.51	1.0	126.34	31JL	1.0	-4.2	-1.7	0.6
64.36	-0.36	1.0	136.03	10AU	2.1	-4.3	-1.8	0.6
70.18	-0.19	0.9	145.77	20AU	2.4	-4.2	-1.8	0.7
75.71	-0.01	0.8	155.57	30AU	0.1	-3.9	-1.9	0.7
80.91	0.20	0.7	165.42	9SE	-0.9	-3.4	-2.0	0.7
85.72	0.43	0.6	175.33	19SE	-1.0	-3.4	-2.0	0.8
90.05	0.69	0.5	185.31	29SE	-1.0	-4.0	-2.1	0.8
93.79	0.98	0.3	195.33	9OC	-0.6	-4.3	-2.2	0.8
96.81	1.30	0.1	205.41	19OC	-0.4	-4.4	-2.2	0.8
98.94	1.68	-0.1	215.53	29OC	-0.4	-4.3	-2.3	0.8
99.99	2.09	-0.3	225.69	8NO	-0.4	-4.2	-2.3	0.7
99.78	2.54	-0.5	235.87	18NO	0.1	-4.2	-2.3	0.7
98.20	2.99	-0.7	246.08	28NO	2.2	-4.0	-2.3	0.7
95.33	3.41	-0.9	256.28	8DE	1.2	-4.0	-2.3	0.7
91.58	3.73	-1.1	266.48	18DE	0.2	-3.9	-2.2	0.7
87.61	3.90	-1.0	276.67	28DE	-0.0	-3.8	-2.2	0.8

163

♂ LONG	LAT	MAG	☉ LONG	16.00UT	☿	♀	♃	♄
84.19	3.91	-0.8	286.83	7JA	-0.1	-3.7	-2.1	0.8
81.85	3.80	-0.5	296.95	17JA	-0.3	-3.6	-2.1	0.8
80.81	3.61	-0.3	307.04	27JA	-0.7	-3.5	-2.0	0.8
81.05	3.38	-0.0	317.07	6FE	-1.3	-3.5	-1.9	0.8
82.39	3.14	0.2	327.04	16FE	-1.3	-3.4	-1.8	0.8
84.66	2.91	0.5	336.96	26FE	-0.7	-3.4	-1.8	0.8
87.68	2.69	0.7	346.82	8MR	0.7	-3.4	-1.7	0.8
91.28	2.49	0.8	356.62	18MR	2.6	-3.3	-1.6	0.7
95.37	2.30	1.0	6.36	28MR	3.0	-3.3	-1.6	0.7
99.83	2.12	1.1	16.05	7AP	1.7	-3.3	-1.5	0.6
104.61	1.95	1.2	25.70	17AP	1.0	-3.3	-1.5	0.6
109.66	1.80	1.4	35.30	27AP	0.4	-3.3	-1.5	0.5
114.93	1.65	1.4	44.87	7MY	-0.3	-3.3	-1.4	0.5
120.38	1.51	1.5	54.41	17MY	-1.3	-3.3	-1.4	0.4
126.01	1.37	1.6	63.95	27MY	-1.7	-3.4	-1.4	0.3
131.79	1.24	1.6	73.46	6JN	-0.8	-3.4	-1.4	0.3
137.71	1.11	1.7	82.99	16JN	-0.0	-3.4	-1.4	0.3
143.76	0.99	1.7	92.52	26JN	0.4	-3.5	-1.4	0.3
149.93	0.86	1.7	102.08	6JL	0.8	-3.5	-1.4	0.4
156.22	0.74	1.7	111.66	16JL	1.4	-3.5	-1.4	0.4
162.64	0.62	1.7	121.28	26JL	2.7	-3.4	-1.4	0.5
169.16	0.50	1.7	130.94	5AU	2.1	-3.4	-1.5	0.5
175.79	0.39	1.7	140.65	15AU	0.0	-3.4	-1.5	0.6
182.54	0.27	1.7	150.42	25AU	-1.0	-3.3	-1.6	0.7
189.40	0.16	1.7	160.24	4SE	-1.2	-3.3	-1.6	0.7
196.36	0.04	1.7	170.13	14SE	-1.0	-3.3	-1.7	0.7
203.44	-0.07	1.7	180.07	24SE	-0.5	-3.3	-1.7	0.8
210.61	-0.18	1.6	190.06	4OC	-0.3	-3.3	-1.8	0.8
217.89	-0.29	1.6	200.12	14OC	-0.2	-3.3	-1.9	0.8
225.27	-0.39	1.6	210.22	24OC	-0.2	-3.4	-1.9	0.8
232.74	-0.49	1.5	220.36	3NO	0.3	-3.4	-2.0	0.8
240.29	-0.59	1.5	230.53	13NO	2.5	-3.4	-2.0	0.8
247.93	-0.68	1.4	240.72	23NO	0.9	-3.5	-2.1	0.8
255.63	-0.76	1.4	250.93	3DE	-0.0	-3.5	-2.1	0.7
263.39	-0.84	1.4	261.13	13DE	-0.2	-3.6	-2.2	0.7
271.20	-0.91	1.3	271.33	23DE	-0.2	-3.6	-2.2	0.7

♂ LONG	LAT	MAG	☉ LONG	16.00UT	☿	♀	♃	♄
				164				
279.05	-0.96	1.3	281.50	2JA	-0.4	-3.7	-2.2	0.8
286.92	-1.01	1.3	291.65	12JA	-0.7	-3.8	-2.1	0.8
294.81	-1.05	1.4	301.76	22JA	-1.2	-3.8	-2.1	0.8
302.69	-1.08	1.4	311.81	1FE	-1.2	-3.9	-2.1	0.8
310.56	-1.10	1.4	321.82	11FE	-0.6	-4.0	-2.0	0.9
318.41	-1.10	1.4	331.77	21FE	0.9	-4.1	-1.9	0.9
326.21	-1.10	1.4	341.65	2MR	3.1	-4.2	-1.9	0.8
333.97	-1.08	1.4	351.49	12MR	2.2	-4.2	-1.8	0.8
341.67	-1.05	1.4	1.26	22MR	1.2	-4.2	-1.7	0.8
349.30	-1.00	1.5	10.97	1AP	0.7	-4.1	-1.7	0.8
356.85	-0.95	1.5	20.65	11AP	0.3	-3.7	-1.6	0.7
4.32	-0.89	1.5	30.27	21AP	-0.4	-2.9	-1.5	0.7
11.70	-0.82	1.5	39.85	1MY	-1.3	-3.4	-1.5	0.6
18.99	-0.73	1.5	49.41	11MY	-1.7	-4.0	-1.4	0.5
26.18	-0.64	1.5	58.95	21MY	-0.8	-4.2	-1.4	0.5
33.26	-0.55	1.6	68.47	31MY	0.0	-4.2	-1.4	0.4
40.25	-0.44	1.6	77.99	10JN	0.6	-4.1	-1.3	0.3
47.12	-0.33	1.6	87.52	20JN	1.1	-4.1	-1.3	0.3
53.88	-0.22	1.6	97.06	30JN	1.9	-4.0	-1.3	0.3
60.53	-0.09	1.6	106.63	10JL	3.2	-3.9	-1.3	0.4
67.05	0.04	1.5	116.23	20JL	1.7	-3.8	-1.3	0.4
73.45	0.17	1.5	125.87	30JL	-0.0	-3.7	-1.3	0.5
79.72	0.31	1.5	135.56	9AU	-1.1	-3.6	-1.3	0.5
85.84	0.46	1.5	145.29	19AU	-1.3	-3.6	-1.3	0.6
91.80	0.62	1.4	155.09	29AU	-0.9	-3.5	-1.4	0.6
97.57	0.79	1.3	164.94	8SE	-0.5	-3.5	-1.4	0.7
103.14	0.97	1.3	174.85	18SE	-0.2	-3.5	-1.4	0.7
108.48	1.16	1.2	184.82	28SE	-0.1	-3.4	-1.5	0.8
113.53	1.37	1.1	194.85	8OC	-0.0	-3.4	-1.5	0.8
118.25	1.60	1.0	204.92	18OC	0.5	-3.4	-1.6	0.8
122.56	1.85	0.8	215.04	28OC	2.9	-3.4	-1.7	0.8
126.37	2.13	0.6	225.20	7NO	0.7	-3.4	-1.7	0.8
129.56	2.44	0.5	235.38	17NO	-0.2	-3.4	-1.8	0.8
131.98	2.78	0.3	245.58	27NO	-0.3	-3.4	-1.9	0.8
133.47	3.16	0.0	255.79	7DE	-0.4	-3.4	-1.9	0.8
133.83	3.55	-0.2	265.99	17DE	-0.5	-3.4	-2.0	0.8
132.92	3.94	-0.5	276.17	27DE	-0.8	-3.4	-2.0	0.7
				165				
130.69	4.29	-0.7	286.34	6JA	-1.1	-3.4	-2.0	0.8
127.37	4.52	-0.9	296.46	16JA	-1.1	-3.4	-2.1	0.8
123.47	4.60	-1.0	306.54	26JA	-0.5	-3.4	-2.1	0.9
119.68	4.49	-0.9	316.58	5FE	1.1	-3.5	-2.1	0.9
116.70	4.24	-0.6	326.56	15FE	3.2	-3.5	-2.0	0.9
114.91	3.90	-0.4	336.48	25FE	1.6	-3.5	-2.0	0.9
114.42	3.52	-0.2	346.34	7MR	0.9	-3.4	-1.9	0.9
115.14	3.14	0.1	356.14	17MR	0.5	-3.4	-1.9	0.9
116.90	2.78	0.3	5.89	27MR	0.1	-3.4	-1.8	0.9
119.52	2.45	0.5	15.58	6AP	-0.4	-3.4	-1.8	0.8
122.85	2.15	0.6	25.23	16AP	-1.3	-3.3	-1.7	0.8
126.75	1.88	0.8	34.83	26AP	-1.7	-3.3	-1.6	0.8
131.11	1.63	0.9	44.40	6MY	-0.8	-3.3	-1.6	0.7
135.87	1.40	1.0	53.95	16MY	0.1	-3.3	-1.5	0.6
140.95	1.19	1.1	63.48	26MY	0.8	-3.3	-1.4	0.6
146.31	0.99	1.2	73.00	5JN	1.4	-3.4	-1.4	0.5
151.92	0.80	1.3	82.52	15JN	2.5	-3.4	-1.4	0.4
157.74	0.63	1.3	92.06	25JN	3.1	-3.4	-1.3	0.4
163.77	0.46	1.4	101.61	5JL	1.4	-3.5	-1.3	0.3
169.98	0.30	1.4	111.19	15JL	-0.1	-3.5	-1.3	0.4
176.36	0.15	1.4	120.81	25JL	-1.1	-3.6	-1.2	0.4
182.90	0.01	1.4	130.47	4AU	-1.4	-3.6	-1.2	0.5
189.59	-0.13	1.5	140.18	14AU	-0.9	-3.7	-1.2	0.5
196.42	-0.25	1.5	149.95	24AU	-0.4	-3.8	-1.2	0.6
203.39	-0.38	1.5	159.77	3SE	-0.1	-3.9	-1.2	0.6
210.50	-0.49	1.5	169.65	13SE	0.0	-4.0	-1.3	0.7
217.72	-0.60	1.5	179.59	23SE	0.2	-4.1	-1.3	0.7
225.07	-0.69	1.5	189.58	3OC	0.8	-4.2	-1.3	0.8
232.52	-0.78	1.5	199.63	13OC	3.2	-4.3	-1.4	0.8
240.07	-0.86	1.4	209.73	23OC	0.6	-4.4	-1.4	0.8
247.71	-0.93	1.4	219.87	2NO	-0.4	-4.3	-1.4	0.9
255.43	-0.99	1.4	230.04	12NO	-0.5	-4.0	-1.5	0.9
263.21	-1.04	1.4	240.23	22NO	-0.5	-3.3	-1.6	0.9
271.04	-1.07	1.4	250.44	2DE	-0.6	-3.3	-1.6	0.9
278.91	-1.10	1.4	260.64	12DE	-0.8	-4.0	-1.7	0.9
286.80	-1.11	1.4	270.84	22DE	-0.9	-4.3	-1.8	0.8

♂ LONG	LAT	MAG	☉ LONG	16.00UT	☿	♀	♃	♄
				166				
294.71	-1.11	1.4	281.01	1JA	-0.9	-4.4	-1.8	0.8
302.61	-1.10	1.4	291.16	11JA	-0.5	-4.3	-1.9	0.8
310.49	-1.08	1.4	301.27	21JA	1.3	-4.2	-1.9	0.9
318.34	-1.04	1.4	311.33	31JA	2.7	-4.1	-2.0	0.9
326.14	-1.00	1.4	321.33	10FE	1.2	-4.0	-2.0	1.0
333.89	-0.94	1.4	331.29	20FE	0.6	-3.9	-2.0	1.0
341.58	-0.88	1.4	341.18	2MR	0.4	-3.8	-2.0	1.0
349.20	-0.81	1.4	351.01	12MR	0.0	-3.7	-2.0	1.0
356.74	-0.73	1.4	0.79	22MR	-0.5	-3.6	-2.0	1.0
4.20	-0.64	1.5	10.51	1AP	-1.3	-3.5	-1.9	1.0
11.57	-0.55	1.5	20.18	11AP	-1.6	-3.5	-1.9	1.0
18.85	-0.46	1.6	29.81	21AP	-0.8	-3.4	-1.8	0.9
26.05	-0.36	1.6	39.40	1MY	0.2	-3.4	-1.8	0.9
33.15	-0.26	1.7	48.95	11MY	1.0	-3.4	-1.7	0.8
40.16	-0.15	1.7	58.49	21MY	1.9	-3.3	-1.6	0.8
47.09	-0.05	1.7	68.01	31MY	3.2	-3.3	-1.6	0.7
53.93	0.06	1.8	77.53	10JN	2.6	-3.3	-1.5	0.7
60.68	0.17	1.8	87.06	20JN	1.1	-3.3	-1.5	0.6
67.34	0.28	1.8	96.60	30JN	-0.1	-3.3	-1.4	0.5
73.93	0.39	1.8	106.17	10JL	-1.2	-3.3	-1.4	0.5
80.43	0.51	1.8	115.77	20JL	-1.5	-3.4	-1.3	0.4
86.84	0.62	1.8	125.40	30JL	-0.9	-3.4	-1.3	0.5
93.16	0.73	1.8	135.08	9AU	-0.3	-3.4	-1.3	0.5
99.40	0.85	1.8	144.82	19AU	0.0	-3.4	-1.2	0.6
105.54	0.97	1.8	154.61	29AU	0.2	-3.5	-1.2	0.6
111.58	1.09	1.7	164.46	8SE	0.4	-3.5	-1.2	0.7
117.51	1.22	1.7	174.37	18SE	1.0	-3.5	-1.2	0.7
123.32	1.35	1.6	184.33	28SE	3.4	-3.5	-1.2	0.8
128.98	1.49	1.5	194.35	8OC	0.4	-3.4	-1.2	0.8
134.48	1.64	1.5	204.43	18OC	-0.5	-3.4	-1.3	0.8
139.79	1.79	1.4	214.54	28OC	-0.6	-3.4	-1.3	0.9
144.87	1.95	1.2	224.70	7NO	-0.6	-3.4	-1.3	0.9
149.67	2.13	1.1	234.88	17NO	-0.7	-3.3	-1.4	0.9
154.13	2.32	1.0	245.08	27NO	-0.8	-3.3	-1.4	0.9
158.18	2.53	0.8	255.29	7DE	-0.8	-3.3	-1.5	1.0
161.72	2.75	0.6	265.49	17DE	-0.8	-3.3	-1.5	1.0
164.61	2.99	0.4	275.68	27DE	-0.4	-3.4	-1.6	1.0
				167				
166.71	3.25	0.1	285.84	6JA	1.5	-3.4	-1.7	0.9
167.84	3.51	-0.1	295.97	16JA	2.2	-3.4	-1.7	0.9
167.82	3.75	-0.4	306.06	26JA	0.8	-3.4	-1.8	1.0
166.53	3.96	-0.7	316.09	5FE	0.4	-3.5	-1.9	1.0
163.99	4.07	-0.9	326.08	15FE	0.2	-3.5	-1.9	1.0
160.52	4.04	-1.2	336.00	25FE	-0.1	-3.5	-2.0	1.1
156.68	3.85	-1.2	345.86	7MR	-0.5	-3.6	-2.0	1.1
153.19	3.51	-1.0	355.67	17MR	-1.3	-3.7	-2.0	1.1
150.68	3.08	-0.8	5.42	27MR	-1.6	-3.7	-2.0	1.1
149.42	2.61	-0.6	15.12	6AP	-0.7	-3.8	-2.0	1.1
149.47	2.15	-0.4	24.77	16AP	0.3	-3.9	-2.0	1.1
150.72	1.73	-0.2	34.37	26AP	1.3	-4.0	-2.0	1.1
152.98	1.34	0.0	43.95	6MY	2.5	-4.1	-1.9	1.0
156.10	1.00	0.2	53.49	16MY	3.5	-4.1	-1.9	1.0
159.91	0.69	0.3	63.02	26MY	2.0	-4.2	-1.8	0.9
164.30	0.42	0.5	72.54	5JN	0.9	-4.1	-1.8	0.9
169.17	0.17	0.6	82.07	15JN	-0.2	-3.9	-1.7	0.8
174.44	-0.05	0.7	91.60	25JN	-1.2	-3.3	-1.6	0.8
180.05	-0.25	0.8	101.15	5JL	-1.6	-3.1	-1.6	0.7
185.97	-0.43	0.8	110.73	15JL	-0.8	-3.8	-1.5	0.6
192.16	-0.59	0.9	120.34	25JL	-0.2	-4.1	-1.5	0.6
198.58	-0.73	1.0	130.00	4AU	0.1	-4.2	-1.4	0.5
205.22	-0.86	1.0	139.71	14AU	0.3	-4.2	-1.4	0.6
212.05	-0.97	1.0	149.47	24AU	0.6	-4.2	-1.4	0.6
219.05	-1.07	1.1	159.29	3SE	1.4	-4.1	-1.3	0.7
226.22	-1.15	1.1	169.16	13SE	3.1	-4.0	-1.3	0.7
233.52	-1.22	1.1	179.10	23SE	0.3	-3.9	-1.3	0.8
240.95	-1.27	1.2	189.09	3OC	-0.7	-3.8	-1.2	0.8
248.49	-1.30	1.2	199.14	13OC	-0.8	-3.8	-1.2	0.9
256.11	-1.32	1.2	209.23	23OC	-0.8	-3.7	-1.3	0.9
263.82	-1.33	1.3	219.37	2NO	-0.8	-3.6	-1.3	1.0
271.59	-1.32	1.3	229.54	12NO	-0.6	-3.6	-1.3	1.0
279.39	-1.30	1.3	239.73	22NO	-0.6	-3.5	-1.3	1.0
287.22	-1.26	1.3	249.94	2DE	-0.6	-3.5	-1.3	1.0
295.07	-1.21	1.4	260.15	12DE	-0.2	-3.4	-1.4	1.1
302.90	-1.15	1.4	270.34	22DE	1.7	-3.4	-1.4	1.1

Left table — 168 / 169

♂ LONG	LAT	MAG	☉ LONG	16.00UT	☿	♀	♃	♄
310.71	-1.09	1.4	280.52	1JA	1.8	-3.4	-1.5	1.1
318.49	-1.01	1.4	290.67	11JA	0.6	-3.4	-1.5	1.1
326.23	-0.92	1.5	300.78	21JA	0.2	-3.3	-1.6	1.1
333.91	-0.83	1.5	310.84	31JA	0.1	-3.3	-1.7	1.1
341.52	-0.74	1.5	320.85	10FE	-0.1	-3.3	-1.7	1.1
349.06	-0.64	1.5	330.80	20FE	-0.6	-3.3	-1.8	1.2
356.52	-0.53	1.6	340.70	1MR	-1.3	-3.3	-1.9	1.2
3.91	-0.43	1.6	350.54	11MR	-1.5	-3.3	-1.9	1.2
11.21	-0.33	1.6	0.32	21MR	-0.7	-3.4	-2.0	1.3
18.43	-0.22	1.6	10.04	31MR	0.5	-3.4	-2.0	1.3
25.56	-0.12	1.6	19.71	10AP	1.7	-3.4	-2.1	1.3
32.61	-0.01	1.6	29.34	20AP	3.4	-3.5	-2.1	1.3
39.58	0.09	1.7	38.93	30AP	2.7	-3.5	-2.1	1.3
46.48	0.19	1.7	48.49	10MY	1.5	-3.4	-2.1	1.2
53.30	0.29	1.7	58.03	20MY	0.7	-3.4	-2.1	1.2
60.05	0.38	1.8	67.55	30MY	-0.2	-3.4	-2.0	1.2
66.73	0.48	1.8	77.07	9JN	-1.2	-3.4	-2.0	1.1
73.34	0.57	1.9	86.60	19JN	-1.7	-3.3	-1.9	1.0
79.90	0.66	1.9	96.14	29JN	-0.8	-3.3	-1.9	1.0
86.40	0.75	1.9	105.70	9JL	-0.2	-3.3	-1.8	0.9
92.85	0.83	1.9	115.29	19JL	0.2	-3.3	-1.8	0.8
99.24	0.91	2.0	124.93	29JL	0.5	-3.3	-1.7	0.8
105.59	0.99	2.0	134.61	8AU	0.9	-3.3	-1.6	0.7
111.88	1.07	2.0	144.34	18AU	1.8	-3.4	-1.6	0.7
118.13	1.15	2.0	154.13	28AU	2.7	-3.4	-1.5	0.7
124.33	1.23	1.9	163.97	7SE	0.2	-3.4	-1.5	0.8
130.47	1.31	1.9	173.88	17SE	-0.8	-3.5	-1.4	0.8
136.56	1.38	1.9	183.85	27SE	-0.9	-3.5	-1.4	0.9
142.58	1.46	1.8	193.86	7OC	-0.9	-3.6	-1.4	0.9
148.54	1.53	1.8	203.93	17OC	-0.7	-3.7	-1.3	1.0
154.41	1.61	1.7	214.05	27OC	-0.5	-3.8	-1.3	1.0
160.18	1.68	1.6	224.20	6NO	-0.5	-3.8	-1.3	1.1
165.85	1.76	1.5	234.39	16NO	-0.5	-3.9	-1.3	1.1
171.38	1.84	1.4	244.59	26NO	-0.1	-4.0	-1.3	1.2
176.75	1.92	1.3	254.79	6DE	2.0	-4.1	-1.3	1.2
181.93	1.99	1.1	265.00	16DE	1.4	-4.3	-1.4	1.2
186.87	2.07	1.0	275.19	26DE	0.3	-4.3	-1.4	1.2
				169				
191.53	2.14	0.8	285.35	5JA	0.1	-4.4	-1.4	1.2
195.82	2.21	0.6	295.48	15JA	-0.0	-4.3	-1.5	1.3
199.66	2.28	0.4	305.57	25JA	-0.2	-4.0	-1.5	1.3
202.93	2.33	0.1	315.61	4FE	-0.6	-3.4	-1.6	1.2
205.51	2.36	-0.1	325.59	14FE	-1.3	-3.5	-1.7	1.3
207.20	2.36	-0.4	335.52	24FE	-1.4	-4.0	-1.7	1.3
207.85	2.31	-0.7	345.39	6MR	-0.7	-4.2	-1.8	1.4
207.28	2.19	-1.0	355.20	16MR	0.6	-4.3	-1.9	1.4
205.45	1.97	-1.3	4.95	26MR	2.2	-4.2	-1.9	1.4
202.55	1.65	-1.6	14.65	5AP	3.5	-4.1	-2.0	1.4
199.04	1.22	-1.7	24.30	15AP	2.0	-4.0	-2.1	1.4
195.63	0.73	-1.6	33.91	25AP	1.2	-3.9	-2.1	1.3
193.01	0.22	-1.4	43.48	5MY	0.5	-3.8	-2.2	1.3
191.61	-0.25	-1.2	53.03	15MY	-0.3	-3.7	-2.2	1.3
191.60	-0.65	-1.0	62.56	25MY	-1.3	-3.6	-2.2	1.2
192.87	-0.99	-0.8	72.08	4JN	-1.7	-3.6	-2.2	1.2
195.28	-1.27	-0.6	81.61	14JN	-0.8	-3.5	-2.2	1.1
198.66	-1.50	-0.4	91.14	24JN	-0.1	-3.5	-2.2	1.1
202.81	-1.68	-0.3	100.69	4JL	0.3	-3.4	-2.1	1.0
207.61	-1.81	-0.1	110.27	14JL	-0.4	-3.4	-2.1	1.0
212.95	-1.92	-0.0	119.88	24JL	1.2	-3.4	-2.0	0.9
218.72	-1.99	0.1	129.53	3AU	2.3	-3.4	-2.0	0.9
224.87	-2.03	0.2	139.24	13AU	2.4	-3.4	-1.9	0.9
231.33	-2.05	0.3	149.00	23AU	0.1	-3.4	-1.8	0.8
238.05	-2.05	0.4	158.81	2SE	-0.9	-3.4	-1.8	0.9
244.98	-2.02	0.5	168.69	12SE	-1.1	-3.4	-1.7	0.9
252.10	-1.98	0.6	178.62	22SE	-1.0	-3.4	-1.7	1.0
259.37	-1.91	0.7	188.61	2OC	-0.6	-3.4	-1.6	1.0
266.75	-1.83	0.8	198.65	12OC	-0.4	-3.4	-1.6	1.1
274.23	-1.74	0.8	208.74	22OC	-0.3	-3.4	-1.5	1.1
281.77	-1.64	0.9	218.87	1NO	-0.3	-3.4	-1.5	1.2
289.35	-1.52	1.0	229.05	11NO	0.1	-3.5	-1.5	1.2
296.96	-1.40	1.1	239.24	21NO	2.2	-3.5	-1.4	1.3
304.57	-1.27	1.1	249.44	1DE	1.2	-3.5	-1.4	1.3
312.17	-1.13	1.2	259.65	11DE	0.1	-3.5	-1.4	1.4
319.74	-1.00	1.3	269.85	21DE	-0.1	-3.4	-1.4	1.4
327.27	-0.86	1.3	280.02	31DE	-0.2	-3.4	-1.4	1.4

Right table — 170 / 171

♂ LONG	LAT	MAG	☉ LONG	16.00UT	☿	♀	♃	♄
334.75	-0.73	1.4	290.18	10JA	-0.3	-3.4	-1.5	1.3
342.18	-0.59	1.5	300.29	20JA	-0.7	-3.4	-1.5	1.3
349.53	-0.46	1.5	310.35	30JA	-1.2	-3.3	-1.5	1.3
356.83	-0.33	1.6	320.37	9FE	-1.3	-3.3	-1.6	1.2
4.05	-0.21	1.6	330.32	19FE	-0.7	-3.3	-1.6	1.1
11.19	-0.09	1.7	340.22	1MR	0.8	-3.3	-1.6	1.1
18.27	0.02	1.7	350.06	11MR	2.7	-3.4	-1.7	1.1
25.26	0.13	1.8	359.84	21MR	2.7	-3.4	-1.8	1.1
32.19	0.23	1.8	9.57	31MR	1.5	-3.4	-1.8	1.1
39.05	0.33	1.8	19.25	10AP	0.9	-3.4	-1.9	1.1
45.84	0.42	1.8	28.88	20AP	0.4	-3.4	-2.0	1.1
52.57	0.51	1.9	38.47	30AP	-0.3	-3.5	-2.0	1.1
59.25	0.59	1.9	48.03	10MY	-1.3	-3.5	-2.1	1.0
65.87	0.67	1.9	57.57	20MY	-1.7	-3.6	-2.2	1.0
72.44	0.74	1.9	67.09	30MY	-0.8	-3.7	-2.2	1.0
78.97	0.81	1.9	76.62	9JN	-0.0	-3.7	-2.3	0.9
85.45	0.87	1.9	86.14	19JN	0.5	-3.8	-2.3	0.9
91.91	0.93	1.9	95.68	29JN	0.9	-3.9	-2.3	0.8
98.33	0.99	2.0	105.24	9JL	1.6	-4.0	-2.3	0.8
104.73	1.04	2.0	114.83	19JL	2.9	-4.1	-2.3	0.7
111.11	1.09	2.0	124.47	29JL	2.0	-4.2	-2.3	0.7
117.46	1.13	2.0	134.15	8AU	0.1	-4.3	-2.3	0.6
123.80	1.17	2.0	143.87	18AU	-1.0	-4.2	-2.2	0.6
130.13	1.21	2.0	153.66	28AU	-1.2	-3.9	-2.2	0.6
136.45	1.24	2.0	163.50	7SE	-1.0	-3.4	-2.1	0.5
142.75	1.27	2.0	173.40	17SE	-0.5	-3.5	-2.0	0.6
149.05	1.30	2.0	183.36	27SE	-0.3	-4.0	-2.0	0.6
155.33	1.32	2.0	193.38	7OC	-0.2	-4.3	-1.9	0.7
161.60	1.34	1.9	203.44	17OC	-0.1	-4.4	-1.8	0.8
167.86	1.35	1.9	213.56	27OC	0.3	-4.3	-1.8	0.9
174.10	1.36	1.8	223.71	6NO	2.5	-4.2	-1.7	0.9
180.31	1.36	1.7	233.89	16NO	0.9	-4.1	-1.7	1.0
186.50	1.36	1.7	244.09	26NO	-0.1	-4.0	-1.6	1.0
192.65	1.35	1.6	254.30	6DE	-0.2	-3.9	-1.6	1.0
198.75	1.33	1.5	264.50	16DE	-0.3	-3.9	-1.6	1.1
204.80	1.30	1.4	274.69	26DE	-0.4	-3.8	-1.5	1.1
				171				
210.78	1.25	1.2	284.86	5JA	-0.7	-3.7	-1.5	1.1
216.68	1.20	1.1	294.99	15JA	-1.2	-3.6	-1.5	1.1
222.47	1.12	0.9	305.08	25JA	-1.2	-3.5	-1.5	1.0
228.13	1.03	0.8	315.13	4FE	-0.6	-3.4	-1.5	1.0
233.63	0.91	0.6	325.11	14FE	0.9	-3.4	-1.5	1.0
238.94	0.76	0.4	335.04	24FE	3.2	-3.4	-1.6	0.9
243.99	0.57	0.2	344.92	6MR	2.0	-3.4	-1.6	0.9
248.73	0.34	-0.0	354.73	16MR	1.1	-3.3	-1.6	0.8
253.07	0.05	-0.3	4.48	26MR	0.6	-3.3	-1.7	0.8
256.88	-0.30	-0.6	14.19	5AP	0.2	-3.3	-1.7	0.8
260.06	-0.73	-0.8	23.84	15AP	-0.4	-3.3	-1.8	0.8
262.40	-1.25	-1.2	33.45	25AP	-1.3	-3.3	-1.9	0.8
263.74	-1.87	-1.5	43.03	5MY	-1.7	-3.3	-1.9	0.8
263.92	-2.57	-1.8	52.57	15MY	-0.8	-3.3	-2.0	0.8
262.84	-3.32	-2.1	62.11	25MY	0.1	-3.4	-2.1	0.8
260.69	-4.05	-2.4	71.63	4JN	0.7	-3.4	-2.1	0.8
257.93	-4.67	-2.5	81.15	14JN	1.2	-3.4	-2.2	0.7
255.22	-5.07	-2.4	90.68	24JN	2.1	-3.5	-2.3	0.7
253.31	-5.25	-2.2	100.23	4JL	3.3	-3.5	-2.3	0.6
252.61	-5.21	-1.9	109.80	14JL	1.7	-3.5	-2.4	0.6
253.27	-5.04	-1.7	119.41	24JL	0.0	-3.4	-2.4	0.5
255.24	-4.76	-1.4	129.06	3AU	-1.1	-3.4	-2.4	0.5
258.31	-4.44	-1.2	138.76	13AU	-1.4	-3.4	-2.4	0.4
262.29	-4.09	-1.0	148.52	23AU	-0.9	-3.3	-2.4	0.4
267.00	-3.73	-0.8	158.33	2SE	-0.4	-3.3	-2.4	0.3
272.28	-3.38	-0.6	168.20	12SE	-0.2	-3.3	-2.4	0.3
278.00	-3.03	-0.4	178.13	22SE	-0.0	-3.3	-2.3	0.3
284.08	-2.70	-0.2	188.12	2OC	0.1	-3.3	-2.2	0.3
290.41	-2.37	0.0	198.16	12OC	0.6	-3.3	-2.2	0.4
296.94	-2.07	0.2	208.25	22OC	2.9	-3.4	-2.1	0.5
303.62	-1.78	0.3	218.38	1NO	0.7	-3.4	-2.0	0.5
310.40	-1.51	0.5	228.55	11NO	-0.3	-3.4	-2.0	0.6
317.26	-1.25	0.6	238.74	21NO	-0.4	-3.5	-1.9	0.7
324.17	-1.01	0.8	248.94	1DE	-0.4	-3.5	-1.8	0.7
331.09	-0.79	0.9	259.15	11DE	-0.5	-3.6	-1.8	0.7
338.02	-0.59	1.0	269.35	21DE	-0.8	-3.6	-1.7	0.8
344.95	-0.40	1.1	279.53	31DE	-1.0	-3.7	-1.7	0.8

♂ LONG	LAT	MAG	☉ LONG	16.00UT 172	☿	♀	♃	♄
					MAGNITUDES			
351.85	-0.23	1.2	289.68	10JA	-1.0	-3.8	-1.6	0.8
358.72	-0.07	1.3	299.80	20JA	-0.5	-3.8	-1.6	0.8
5.56	0.07	1.4	309.86	30JA	1.1	-3.9	-1.6	0.8
12.36	0.20	1.5	319.88	9FE	3.1	-4.0	-1.6	0.8
19.12	0.32	1.6	329.84	19FE	1.5	-4.1	-1.6	0.7
25.83	0.43	1.7	339.74	29FE	0.8	-4.2	-1.6	0.7
32.50	0.53	1.7	349.59	10MR	0.5	-4.2	-1.6	0.7
39.13	0.62	1.8	359.37	20MR	0.1	-4.2	-1.6	0.6
45.72	0.70	1.8	9.10	30MR	-0.4	-4.1	-1.6	0.6
52.26	0.77	1.9	18.78	9AP	-1.3	-3.7	-1.6	0.6
58.77	0.83	1.9	28.42	19AP	-1.7	-2.9	-1.7	0.6
65.25	0.89	2.0	38.01	29AP	-0.8	-3.4	-1.7	0.6
71.69	0.94	2.0	47.57	9MY	0.1	-4.0	-1.8	0.6
78.11	0.99	2.0	57.11	19MY	0.9	-4.2	-1.8	0.6
84.51	1.03	2.0	66.63	29MY	1.6	-4.2	-1.8	0.6
90.89	1.06	2.0	76.15	8JN	2.7	-4.1	-1.9	0.6
97.26	1.09	2.0	85.68	18JN	3.0	-4.0	-2.0	0.6
103.62	1.11	2.0	95.22	28JN	1.4	-4.0	-2.1	0.5
109.98	1.13	2.0	104.78	8JL	-0.1	-3.9	-2.1	0.5
116.35	1.15	2.0	114.37	18JL	-1.1	-3.8	-2.2	0.5
122.71	1.16	2.0	124.00	28JL	-1.5	-3.7	-2.3	0.4
129.09	1.16	2.0	133.68	7AU	-0.9	-3.6	-2.3	0.3
135.48	1.16	2.0	143.40	17AU	-0.4	-3.6	-2.4	0.3
141.89	1.16	2.0	153.18	27AU	-0.1	-3.5	-2.4	0.2
148.32	1.15	2.0	163.02	6SE	0.1	-3.5	-2.4	0.2
154.77	1.14	2.0	172.92	16SE	0.3	-3.5	-2.5	0.1
161.25	1.12	2.0	182.88	26SE	0.8	-3.4	-2.4	0.1
167.75	1.09	2.0	192.89	6OC	3.2	-3.4	-2.4	0.0
174.28	1.06	1.9	202.96	16OC	0.6	-3.4	-2.4	0.1
180.84	1.03	1.9	213.07	26OC	-0.4	-3.4	-2.3	0.2
187.43	0.98	1.9	223.22	5NO	-0.5	-3.4	-2.3	0.2
194.05	0.93	1.8	233.40	15NO	-0.5	-3.4	-2.2	0.3
200.69	0.87	1.8	243.59	25NO	-0.6	-3.4	-2.1	0.4
207.37	0.81	1.7	253.80	5DE	-0.8	-3.4	-2.1	0.4
214.07	0.73	1.6	264.01	15DE	-0.9	-3.4	-2.0	0.5
220.79	0.64	1.6	274.19	25DE	-0.9	-3.4	-1.9	0.5
				173				
227.54	0.53	1.5	284.36	4JA	-0.4	-3.4	-1.8	0.6
234.31	0.42	1.4	294.50	14JA	1.3	-3.4	-1.8	0.6
241.09	0.29	1.3	304.59	24JA	2.6	-3.4	-1.7	0.6
247.89	0.14	1.2	314.63	3FE	1.1	-3.5	-1.7	0.6
254.69	-0.02	1.1	324.62	13FE	0.6	-3.5	-1.6	0.6
261.49	-0.21	0.9	334.55	23FE	0.3	-3.5	-1.6	0.6
268.29	-0.41	0.8	344.43	5MR	0.0	-3.4	-1.6	0.5
275.06	-0.64	0.7	354.25	15MR	-0.5	-3.4	-1.6	0.5
281.80	-0.89	0.5	4.00	25MR	-1.2	-3.4	-1.6	0.5
288.48	-1.16	0.4	13.71	4AP	-1.6	-3.4	-1.6	0.4
295.07	-1.46	0.2	23.37	14AP	-0.8	-3.3	-1.6	0.4
301.55	-1.78	0.1	32.98	24AP	0.2	-3.3	-1.6	0.4
307.88	-2.12	-0.1	42.56	4MY	1.1	-3.3	-1.6	0.5
313.98	-2.49	-0.3	52.11	14MY	2.1	-3.3	-1.6	0.5
319.82	-2.89	-0.5	61.64	24MY	3.5	-3.3	-1.6	0.5
325.29	-3.31	-0.7	71.16	3JN	2.4	-3.4	-1.7	0.5
330.27	-3.75	-0.9	80.68	13JN	1.1	-3.4	-1.7	0.5
334.64	-4.21	-1.1	90.21	23JN	-0.1	-3.4	-1.8	0.5
338.21	-4.67	-1.3	99.76	3JL	-1.2	-3.5	-1.8	0.4
340.79	-5.13	-1.6	109.34	13JL	-1.6	-3.5	-1.9	0.4
342.17	-5.54	-1.8	118.95	23JL	-0.9	-3.6	-1.9	0.4
342.15	-5.85	-2.1	128.60	2AU	-0.3	-3.6	-2.0	0.3
340.76	-5.99	-2.3	138.30	12AU	0.0	-3.7	-2.1	0.3
338.26	-5.89	-2.4	148.05	22AU	0.2	-3.8	-2.1	0.2
335.27	-5.51	-2.5	157.86	1SE	0.5	-3.9	-2.2	0.1
332.59	-4.89	-2.2	167.72	11SE	1.2	-4.0	-2.3	0.1
330.86	-4.15	-1.9	177.65	21SE	3.3	-4.1	-2.3	0.0
330.41	-3.38	-1.6	187.64	1OC	0.4	-4.2	-2.4	-0.1
331.28	-2.67	-1.3	197.67	11OC	-0.6	-4.3	-2.4	-0.1
333.29	-2.04	-1.0	207.76	21OC	-0.7	-4.4	-2.4	-0.1
336.25	-1.49	-0.7	217.89	31OC	-0.7	-4.3	-2.4	-0.1
339.98	-1.03	-0.4	228.05	10NO	-0.7	-4.0	-2.4	0.0
344.29	-0.65	-0.1	238.24	20NO	-0.7	-3.3	-2.3	0.1
349.06	-0.32	0.1	248.45	30NO	-0.7	-3.4	-2.3	0.2
354.18	-0.05	0.3	258.65	10DE	-0.7	-4.0	-2.2	0.2
359.56	0.18	0.5	268.85	20DE	-0.3	-4.3	-2.1	0.3
5.15	0.37	0.7	279.04	30DE	1.5	-4.4	-2.1	0.3

♂ LONG	LAT	MAG	☉ LONG	16.00UT 174	☿	♀	♃	♄
					MAGNITUDES			
10.89	0.53	0.9	289.19	9JA	2.1	-4.3	-2.0	0.4
16.75	0.67	1.0	299.30	19JA	0.7	-4.2	-1.9	0.4
22.70	0.78	1.2	309.37	29JA	0.4	-4.1	-1.9	0.4
28.72	0.88	1.3	319.39	8FE	0.2	-4.0	-1.8	0.5
34.80	0.96	1.4	329.35	18FE	-0.1	-3.9	-1.7	0.5
40.89	1.03	1.5	339.26	28FE	-0.5	-3.8	-1.7	0.4
47.01	1.08	1.6	349.11	10MR	-1.2	-3.7	-1.6	0.4
53.16	1.13	1.7	358.90	20MR	-1.6	-3.6	-1.6	0.4
59.32	1.16	1.8	8.63	30MR	-0.8	-3.5	-1.5	0.4
65.49	1.19	1.8	18.31	9AP	0.4	-3.5	-1.5	0.3
71.67	1.21	1.9	27.95	19AP	1.5	-3.4	-1.5	0.3
77.86	1.23	1.9	37.54	29AP	2.8	-3.4	-1.5	0.3
84.06	1.24	2.0	47.11	9MY	3.2	-3.4	-1.5	0.3
90.27	1.24	2.0	56.65	19MY	1.8	-3.3	-1.5	0.3
96.50	1.24	2.0	66.18	29MY	0.8	-3.3	-1.5	0.4
102.74	1.23	2.0	75.70	8JN	-0.2	-3.3	-1.5	0.4
109.00	1.22	2.0	85.22	18JN	-1.2	-3.3	-1.5	0.4
115.28	1.20	2.0	94.76	28JN	-1.7	-3.3	-1.5	0.4
121.59	1.18	2.0	104.32	8JL	-0.8	-3.3	-1.6	0.4
127.93	1.16	2.0	113.91	18JL	-0.2	-3.4	-1.6	0.3
134.31	1.13	2.0	123.54	28JL	0.2	-3.4	-1.7	0.3
140.72	1.09	2.0	133.21	7AU	0.4	-3.4	-1.7	0.3
147.17	1.06	1.9	142.93	17AU	0.7	-3.4	-1.8	0.2
153.67	1.01	1.9	152.71	27AU	1.5	-3.5	-1.8	0.2
160.22	0.97	1.9	162.54	6SE	3.0	-3.5	-1.9	0.1
166.82	0.91	1.9	172.44	16SE	0.3	-3.5	-2.0	0.0
173.47	0.86	1.9	182.40	26SE	-0.7	-3.5	-2.0	-0.0
180.17	0.79	1.9	192.40	6OC	-0.8	-3.4	-2.1	-0.1
186.94	0.73	1.9	202.47	16OC	-0.8	-3.4	-2.2	-0.2
193.77	0.65	1.8	212.57	26OC	-0.8	-3.4	-2.2	-0.2
200.65	0.57	1.8	222.72	5NO	-0.6	-3.4	-2.2	-0.2
207.60	0.49	1.8	232.90	15NO	-0.6	-3.3	-2.3	-0.2
214.61	0.39	1.8	243.10	25NO	-0.6	-3.3	-2.3	-0.1
221.67	0.29	1.7	253.30	5DE	-0.2	-3.3	-2.3	-0.0
228.80	0.18	1.7	263.51	15DE	1.7	-3.3	-2.2	0.1
235.99	0.07	1.6	273.70	25DE	1.7	-3.4	-2.2	0.1
				175				
243.23	-0.06	1.6	283.87	4JA	0.5	-3.4	-2.2	0.2
250.52	-0.19	1.5	294.01	14JA	0.2	-3.4	-2.1	0.2
257.86	-0.32	1.4	304.10	24JA	0.0	-3.4	-2.0	0.3
265.25	-0.46	1.4	314.14	3FE	-0.2	-3.5	-2.0	0.3
272.67	-0.61	1.3	324.14	13FE	-0.6	-3.5	-1.9	0.3
280.13	-0.77	1.2	334.08	23FE	-1.2	-3.5	-1.8	0.4
287.61	-0.92	1.2	343.95	5MR	-1.5	-3.6	-1.7	0.4
295.10	-1.08	1.1	353.77	15MR	-0.8	-3.7	-1.7	0.4
302.58	-1.24	1.0	3.53	25MR	0.5	-3.7	-1.6	0.3
310.06	-1.39	1.0	13.24	4AP	1.9	-3.8	-1.6	0.3
317.51	-1.54	0.9	22.90	14AP	3.7	-3.9	-1.5	0.3
324.90	-1.69	0.8	32.52	24AP	2.4	-4.0	-1.5	0.2
332.23	-1.83	0.7	42.09	4MY	1.4	-4.1	-1.4	0.2
339.47	-1.95	0.6	51.65	14MY	0.6	-4.1	-1.4	0.2
346.59	-2.06	0.6	61.18	24MY	-0.2	-4.2	-1.4	0.3
353.57	-2.16	0.5	70.70	3JN	-1.2	-4.1	-1.4	0.3
0.37	-2.25	0.4	80.23	13JN	-1.7	-3.9	-1.4	0.3
6.95	-2.31	0.3	89.75	23JN	-0.8	-3.3	-1.4	0.3
13.28	-2.36	0.2	99.30	3JL	-0.2	-3.1	-1.4	0.3
19.28	-2.38	0.1	108.88	13JL	0.3	-3.8	-1.4	0.3
24.89	-2.39	-0.1	118.48	23JL	0.6	-4.1	-1.4	0.3
30.03	-2.36	-0.2	128.13	2AU	1.0	-4.2	-1.4	0.3
34.59	-2.31	-0.4	137.83	12AU	2.0	-4.2	-1.5	0.3
38.43	-2.23	-0.5	147.58	22AU	2.7	-4.2	-1.5	0.2
41.38	-2.10	-0.7	157.38	1SE	0.2	-4.1	-1.6	0.2
43.23	-1.91	-0.9	167.25	11SE	-0.8	-4.0	-1.6	0.1
43.79	-1.66	-1.2	177.17	21SE	-1.0	-3.9	-1.7	0.1
42.90	-1.33	-1.4	187.15	1OC	-0.9	-3.8	-1.7	0.0
40.58	-0.92	-1.6	197.19	11OC	-0.7	-3.8	-1.8	-0.1
37.20	-0.45	-1.7	207.27	21OC	-0.5	-3.7	-1.9	-0.1
33.42	0.05	-1.7	217.40	31OC	-0.4	-3.6	-2.0	-0.2
30.07	0.50	-1.4	227.56	10NO	-0.4	-3.6	-2.0	-0.3
27.78	0.88	-1.1	237.75	20NO	-0.0	-3.5	-2.0	-0.3
26.82	1.17	-0.8	247.95	30NO	2.0	-3.5	-2.1	-0.2
27.20	1.38	-0.5	258.16	10DE	1.4	-3.4	-2.1	-0.1
28.72	1.53	-0.2	268.36	20DE	0.2	-3.4	-2.1	-0.1
31.19	1.63	0.1	278.54	30DE	0.0	-3.4	-2.1	0.0

♂			⊙	16.00UT	☿	♀	♃	♄
LONG	LAT	MAG	LONG	176	MAGNITUDES			
34.42	1.69	0.3	288.70	9JA	-0.1	-3.4	-2.1	0.1
38.23	1.72	0.5	298.81	19JA	-0.3	-3.3	-2.1	0.1
42.49	1.74	0.7	308.89	29JA	-0.6	-3.3	-2.1	0.2
47.12	1.74	0.9	318.91	8FE	-1.2	-3.3	-2.0	0.2
52.03	1.73	1.1	328.87	18FE	-1.4	-3.3	-2.0	0.3
57.17	1.71	1.2	338.78	28FE	-0.7	-3.3	-1.9	0.3
62.48	1.69	1.4	348.63	9MR	0.6	-3.3	-1.8	0.3
67.94	1.66	1.5	358.42	19MR	2.4	-3.4	-1.8	0.3
73.51	1.63	1.6	8.16	29MR	3.2	-3.4	-1.7	0.3
79.19	1.59	1.7	17.84	8AP	1.8	-3.4	-1.6	0.3
84.95	1.55	1.7	27.48	18AP	1.0	-3.5	-1.6	0.3
90.79	1.51	1.8	37.08	28AP	0.5	-3.5	-1.5	0.3
96.70	1.46	1.9	46.64	8MY	-0.3	-3.4	-1.5	0.2
102.67	1.41	1.9	56.18	18MY	-1.2	-3.4	-1.4	0.2
108.70	1.36	1.9	65.71	28MY	-1.7	-3.4	-1.4	0.2
114.79	1.31	2.0	75.23	7JN	-0.8	-3.4	-1.3	0.2
120.93	1.25	2.0	84.75	17JN	-0.1	-3.3	-1.3	0.3
127.14	1.19	2.0	94.29	27JN	0.4	-3.3	-1.3	0.3
133.41	1.13	2.0	103.85	7JL	0.8	-3.3	-1.3	0.3
139.73	1.06	2.0	113.44	17JL	1.3	-3.3	-1.3	0.3
146.13	0.99	2.0	123.07	27JL	2.5	-3.3	-1.3	0.3
152.59	0.92	1.9	132.74	6AU	2.3	-3.3	-1.3	0.3
159.12	0.85	1.9	142.45	16AU	0.2	-3.4	-1.3	0.3
165.72	0.77	1.9	152.23	26AU	-0.9	-3.4	-1.3	0.3
172.40	0.69	1.9	162.07	5SE	-1.1	-3.4	-1.4	0.2
179.15	0.61	1.8	171.96	15SE	-1.0	-3.5	-1.4	0.2
185.98	0.52	1.8	181.91	25SE	-0.6	-3.5	-1.4	0.1
192.89	0.43	1.7	191.92	5OC	-0.3	-3.6	-1.5	0.1
199.88	0.33	1.7	201.98	15OC	-0.3	-3.7	-1.5	0.0
206.95	0.23	1.7	212.09	25OC	-0.2	-3.8	-1.6	-0.1
214.10	0.13	1.7	222.23	4NO	0.2	-3.8	-1.7	-0.1
221.33	0.03	1.7	232.40	14NO	2.3	-3.9	-1.7	-0.2
228.64	-0.08	1.6	242.60	24NO	1.1	-4.0	-1.8	-0.3
236.02	-0.19	1.6	252.81	4DE	0.0	-4.2	-1.9	-0.3
243.48	-0.30	1.6	263.01	14DE	-0.1	-4.3	-1.9	-0.2
250.99	-0.41	1.6	273.21	24DE	-0.2	-4.3	-2.0	-0.1
				177				
258.57	-0.53	1.5	283.38	3JA	-0.4	-4.4	-2.0	-0.1
266.20	-0.64	1.5	293.51	13JA	-0.7	-4.3	-2.0	0.0
273.87	-0.75	1.5	303.61	23JA	-1.2	-4.0	-2.1	0.1
281.58	-0.85	1.4	313.66	2FE	-1.3	-3.4	-2.1	0.1
289.31	-0.96	1.4	323.65	12FE	-0.7	-3.5	-2.0	0.2
297.06	-1.05	1.4	333.60	22FE	0.8	-4.0	-2.0	0.2
304.81	-1.14	1.4	343.47	4MR	2.9	-4.2	-2.0	0.3
312.55	-1.22	1.3	353.29	14MR	2.4	-4.3	-1.9	0.3
320.27	-1.29	1.3	3.06	24MR	1.3	-4.2	-1.9	0.3
327.96	-1.34	1.3	12.77	3AP	0.8	-4.1	-1.8	0.3
335.59	-1.38	1.2	22.43	13AP	0.3	-4.0	-1.7	0.3
343.17	-1.41	1.2	32.05	23AP	-0.3	-3.9	-1.7	0.3
350.67	-1.43	1.2	41.63	3MY	-1.2	-3.8	-1.6	0.3
358.08	-1.43	1.1	51.18	13MY	-1.7	-3.7	-1.5	0.3
5.38	-1.41	1.1	60.72	23MY	-0.8	-3.6	-1.5	0.2
12.57	-1.38	1.1	70.24	2JN	-0.0	-3.6	-1.4	0.2
19.64	-1.33	1.0	79.76	12JN	0.5	-3.5	-1.4	0.2
26.55	-1.27	1.0	89.29	22JN	1.0	-3.5	-1.3	0.3
33.30	-1.19	1.0	98.84	2JL	1.7	-3.4	-1.3	0.3
39.87	-1.10	0.9	108.41	12JL	3.1	-3.4	-1.3	0.3
46.23	-0.99	0.9	118.02	22JL	1.9	-3.4	-1.2	0.4
52.35	-0.86	0.8	127.66	1AU	0.1	-3.4	-1.2	0.4
58.21	-0.72	0.7	137.36	11AU	-1.0	-3.4	-1.2	0.4
63.75	-0.55	0.6	147.11	21AU	-1.3	-3.3	-1.2	0.4
68.92	-0.36	0.5	156.91	31AU	-1.0	-3.4	-1.2	0.4
73.65	-0.14	0.4	166.77	10SE	-0.5	-3.4	-1.2	0.3
77.82	0.10	0.2	176.69	20SE	-0.2	-3.4	-1.3	0.3
81.34	0.39	0.1	186.67	30SE	-0.1	-3.4	-1.3	0.3
84.03	0.71	-0.1	196.70	10OC	-0.0	-3.4	-1.3	0.2
85.71	1.09	-0.3	206.79	20OC	0.4	-3.4	-1.4	0.1
86.20	1.51	-0.5	216.91	30OC	2.6	-3.5	-1.4	0.1
85.31	1.96	-0.8	227.07	9NO	0.9	-3.5	-1.5	-0.0
83.04	2.42	-1.0	237.26	19NO	-0.1	-3.5	-1.5	-0.1
79.65	2.84	-1.2	247.46	29NO	-0.3	-3.5	-1.6	-0.2
75.69	3.15	-1.2	257.67	9DE	-0.3	-3.5	-1.6	-0.2
71.95	3.32	-1.0	267.87	19DE	-0.5	-3.4	-1.7	-0.2
69.11	3.36	-0.8	278.05	29DE	-0.8	-3.4	-1.8	-0.1

♂			⊙	16.00UT	☿	♀	♃	♄
LONG	LAT	MAG	LONG	178	MAGNITUDES			
67.52	3.30	-0.5	288.20	8JA	-1.1	-3.4	-1.8	-0.1
67.26	3.17	-0.2	298.32	18JA	-1.1	-3.4	-1.9	-0.0
68.19	3.02	0.1	308.40	28JA	-0.6	-3.3	-1.9	0.1
70.12	2.85	0.3	318.42	7FE	0.9	-3.3	-2.0	0.1
72.88	2.69	0.5	328.39	17FE	3.2	-3.3	-2.0	0.2
76.28	2.53	0.7	338.30	27FE	1.8	-3.3	-2.0	0.2
80.19	2.37	0.9	348.15	9MR	1.0	-3.4	-2.0	0.3
84.51	2.23	1.1	357.95	19MR	0.6	-3.4	-2.0	0.3
89.17	2.09	1.2	7.68	29MR	0.2	-3.4	-2.0	0.3
94.09	1.96	1.3	17.37	8AP	-0.4	-3.4	-1.9	0.4
99.24	1.84	1.4	27.01	18AP	-1.2	-3.5	-1.9	0.4
104.58	1.72	1.5	36.61	28AP	-1.7	-3.5	-1.8	0.4
110.08	1.60	1.6	46.18	8MY	-0.8	-3.5	-1.7	0.4
115.73	1.49	1.7	55.72	18MY	0.1	-3.6	-1.7	0.4
121.50	1.38	1.7	65.25	28MY	0.7	-3.7	-1.6	0.4
127.40	1.27	1.8	74.77	7JN	1.3	-3.7	-1.6	0.3
133.40	1.17	1.8	84.29	17JN	2.3	-3.8	-1.5	0.3
139.51	1.06	1.8	93.83	27JN	3.2	-3.9	-1.4	0.3
145.73	0.96	1.8	103.39	7JL	1.6	-4.0	-1.4	0.4
152.04	0.85	1.8	112.97	17JL	0.0	-4.1	-1.3	0.4
158.46	0.74	1.8	122.60	27JL	-1.1	-4.2	-1.3	0.4
164.97	0.64	1.8	132.27	6AU	-1.4	-4.2	-1.3	0.5
171.59	0.53	1.8	141.99	16AU	-0.9	-4.2	-1.2	0.5
178.30	0.42	1.8	151.76	26AU	-0.4	-3.9	-1.2	0.5
185.11	0.32	1.8	161.59	5SE	-0.1	-3.4	-1.2	0.5
192.03	0.21	1.7	171.48	15SE	0.0	-3.5	-1.2	0.5
199.04	0.10	1.7	181.43	25SE	0.1	-4.0	-1.2	0.4
206.15	-0.01	1.7	191.43	5OC	0.6	-4.3	-1.2	0.4
213.36	-0.11	1.6	201.49	15OC	2.9	-4.4	-1.2	0.4
220.66	-0.22	1.6	211.59	25OC	0.7	-4.3	-1.3	0.3
228.05	-0.32	1.5	221.74	4NO	-0.3	-4.2	-1.3	0.2
235.53	-0.43	1.5	231.91	14NO	-0.4	-4.1	-1.3	0.2
243.08	-0.52	1.4	242.11	24NO	-0.5	-4.0	-1.4	0.1
250.71	-0.62	1.4	252.32	4DE	-0.6	-3.9	-1.4	0.0
258.40	-0.71	1.4	262.52	14DE	-0.8	-3.8	-1.5	-0.0
266.15	-0.80	1.4	272.71	24DE	-0.9	-3.8	-1.6	-0.1
				179				
273.95	-0.87	1.4	282.89	3JA	-1.0	-3.7	-1.6	-0.1
281.77	-0.94	1.4	293.03	13JA	-0.5	-3.6	-1.7	-0.0
289.63	-1.00	1.4	303.13	23JA	1.1	-3.5	-1.8	0.0
297.49	-1.06	1.4	313.18	2FE	2.9	-3.5	-1.8	0.1
305.34	-1.10	1.4	323.18	12FE	1.3	-3.4	-1.9	0.2
313.19	-1.12	1.4	333.12	22FE	0.7	-3.4	-1.9	0.2
321.01	-1.14	1.4	343.00	4MR	0.4	-3.4	-2.0	0.3
328.78	-1.15	1.4	352.82	14MR	0.1	-3.3	-2.0	0.3
336.51	-1.14	1.4	2.59	24MR	-0.4	-3.3	-2.0	0.4
344.18	-1.12	1.4	12.31	3AP	-1.2	-3.3	-2.0	0.4
351.77	-1.09	1.4	21.97	13AP	-1.7	-3.3	-2.0	0.4
359.29	-1.04	1.4	31.59	23AP	-0.8	-3.3	-2.0	0.5
6.72	-0.99	1.4	41.17	3MY	0.2	-3.3	-2.0	0.5
14.06	-0.92	1.4	50.72	13MY	0.9	-3.3	-1.9	0.5
21.30	-0.84	1.5	60.26	23MY	1.8	-3.4	-1.9	0.5
28.43	-0.75	1.5	69.78	2JN	3.0	-3.4	-1.8	0.5
35.44	-0.66	1.4	79.30	12JN	2.8	-3.4	-1.8	0.5
42.35	-0.55	1.4	88.83	22JN	1.3	-3.5	-1.7	0.5
49.12	-0.44	1.4	98.38	2JL	-0.0	-3.5	-1.6	0.4
55.77	-0.32	1.4	107.94	12JL	-1.1	-3.5	-1.6	0.5
62.28	-0.18	1.4	117.55	22JL	-1.5	-3.4	-1.5	0.5
68.64	-0.04	1.4	127.19	1AU	-0.9	-3.4	-1.5	0.6
74.84	0.11	1.3	136.88	11AU	-0.3	-3.4	-1.4	0.6
80.86	0.27	1.3	146.63	21AU	-0.0	-3.3	-1.4	0.6
86.67	0.44	1.2	156.43	31AU	0.2	-3.3	-1.3	0.6
92.26	0.63	1.1	166.28	10SE	0.3	-3.3	-1.3	0.6
97.59	0.83	1.0	176.20	20SE	0.9	-3.3	-1.3	0.6
102.60	1.05	0.9	186.18	30SE	3.2	-3.3	-1.3	0.6
107.23	1.29	0.8	196.21	10OC	0.6	-3.3	-1.3	0.6
111.41	1.56	0.7	206.29	20OC	-0.5	-3.4	-1.3	0.6
115.02	1.86	0.5	216.41	30OC	-0.6	-3.4	-1.3	0.5
117.95	2.19	0.3	226.57	9NO	-0.6	-3.4	-1.3	0.5
120.02	2.57	0.1	236.76	19NO	-0.7	-3.5	-1.3	0.4
121.04	2.97	-0.1	246.96	29NO	-0.8	-3.5	-1.3	0.3
120.86	3.40	-0.4	257.17	9DE	-0.8	-3.6	-1.4	0.3
119.33	3.82	-0.6	267.37	19DE	-0.8	-3.6	-1.4	0.2
116.54	4.18	-0.8	277.56	29DE	-0.4	-3.7	-1.5	0.1

♂ LONG	♂ LAT	♂ MAG	⊙ LONG	16.00UT	☿	♀	♃	♄
LONG	LAT	MAG	LONG	180	MAGNITUDES			
112.85	4.41	-1.0	287.71	8JA	1.3	-3.8	-1.5	0.1
108.88	4.47	-0.9	297.84	18JA	2.5	-3.8	-1.6	0.1
105.39	4.37	-0.8	307.92	28JA	0.9	-3.9	-1.6	0.1
102.93	4.13	-0.5	317.94	7FE	0.5	-4.0	-1.7	0.2
101.74	3.82	-0.3	327.92	17FE	0.3	-4.1	-1.8	0.3
101.83	3.49	-0.0	337.83	27FE	-0.0	-4.2	-1.8	0.3
103.05	3.16	0.2	347.68	8MR	-0.5	-4.2	-1.9	0.4
105.22	2.85	0.4	357.48	18MR	-1.2	-4.2	-2.0	0.4
108.16	2.57	0.6	7.22	28MR	-1.6	-4.1	-2.0	0.5
111.72	2.30	0.8	16.91	7AP	-0.8	-3.6	-2.1	0.5
115.79	2.06	0.9	26.55	17AP	0.3	-2.9	-2.1	0.6
120.27	1.84	1.0	36.15	27AP	1.2	-3.5	-2.1	0.6
125.10	1.64	1.2	45.72	7MY	2.4	-4.0	-2.1	0.6
130.21	1.44	1.2	55.26	17MY	3.6	-4.2	-2.1	0.7
135.57	1.26	1.3	64.79	27MY	2.2	-4.2	-2.1	0.7
141.15	1.09	1.4	74.31	6JN	1.0	-4.1	-2.0	0.7
146.93	0.93	1.4	83.83	16JN	-0.1	-4.0	-2.0	0.7
152.88	0.78	1.5	93.37	26JN	-1.2	-4.0	-1.9	0.7
158.99	0.63	1.5	102.92	6JL	-1.6	-3.9	-1.9	0.7
165.26	0.48	1.6	112.51	16JL	-0.9	-3.8	-1.8	0.7
171.67	0.34	1.6	122.13	26JL	-0.3	-3.7	-1.8	0.7
178.22	0.21	1.6	131.79	5AU	0.1	-3.6	-1.7	0.7
184.91	0.08	1.6	141.51	15AU	0.3	-3.6	-1.6	0.8
191.73	-0.04	1.6	151.28	25AU	0.5	-3.5	-1.6	0.8
198.67	-0.17	1.6	161.11	4SE	1.3	-3.5	-1.5	0.8
205.74	-0.28	1.6	171.00	14SE	3.3	-3.5	-1.5	0.9
212.92	-0.39	1.6	180.94	24SE	0.4	-3.4	-1.4	0.9
220.21	-0.50	1.5	190.94	4OC	-0.6	-3.4	-1.4	0.9
227.62	-0.60	1.5	201.00	14OC	-0.8	-3.4	-1.4	0.8
235.11	-0.69	1.5	211.10	24OC	-0.7	-3.4	-1.4	0.8
242.70	-0.77	1.5	221.24	3NO	-0.8	-3.4	-1.3	0.8
250.37	-0.85	1.5	231.42	13NO	-0.7	-3.4	-1.3	0.7
258.11	-0.91	1.4	241.61	23NO	-0.6	-3.4	-1.3	0.7
265.91	-0.97	1.4	251.82	3DE	-0.7	-3.4	-1.4	0.6
273.76	-1.02	1.4	262.03	13DE	-0.3	-3.4	-1.4	0.6
281.63	-1.05	1.4	272.22	23DE	1.5	-3.4	-1.4	0.5
				181				
289.53	-1.08	1.3	282.39	2JA	2.0	-3.4	-1.4	0.4
297.43	-1.09	1.3	292.54	12JA	0.6	-3.4	-1.5	0.3
305.33	-1.09	1.3	302.64	22JA	0.3	-3.4	-1.5	0.3
313.20	-1.08	1.3	312.69	1FE	0.1	-3.5	-1.6	0.3
321.05	-1.05	1.3	322.69	11FE	-0.1	-3.5	-1.6	0.4
328.84	-1.02	1.4	332.64	21FE	-0.5	-3.4	-1.7	0.4
336.58	-0.97	1.4	342.52	3MR	-1.2	-3.4	-1.8	0.5
344.26	-0.92	1.4	352.35	13MR	-1.6	-3.4	-1.8	0.5
351.86	-0.86	1.5	2.12	23MR	-0.8	-3.4	-1.9	0.6
359.39	-0.78	1.5	11.83	2AP	0.4	-3.4	-2.0	0.6
6.83	-0.70	1.5	21.50	12AP	1.6	-3.3	-2.0	0.7
14.19	-0.62	1.6	31.12	22AP	3.1	-3.3	-2.1	0.7
21.45	-0.53	1.6	40.70	2MY	2.9	-3.3	-2.1	0.8
28.62	-0.43	1.6	50.26	12MY	1.7	-3.3	-2.2	0.8
35.70	-0.33	1.6	59.79	22MY	0.8	-3.3	-2.2	0.8
42.68	-0.22	1.7	69.32	1JN	-0.2	-3.4	-2.2	0.9
49.57	-0.11	1.7	78.84	11JN	-1.2	-3.4	-2.2	0.9
56.37	0.00	1.7	88.36	21JN	-1.7	-3.4	-2.2	0.9
63.07	0.12	1.7	97.91	1JL	-0.9	-3.5	-2.2	0.9
69.67	0.23	1.7	107.48	11JL	-0.2	-3.5	-2.2	0.9
76.17	0.36	1.7	117.08	21JL	0.2	-3.6	-2.1	0.9
82.58	0.48	1.7	126.72	31JL	0.5	-3.6	-2.0	0.9
88.88	0.61	1.7	136.41	10AU	0.8	-3.7	-2.0	0.9
95.07	0.74	1.7	146.15	20AU	1.7	-3.8	-1.9	1.0
101.15	0.87	1.6	155.95	30AU	3.0	-3.9	-1.9	1.0
107.09	1.02	1.6	165.81	9SE	0.3	-4.0	-1.8	1.1
112.89	1.16	1.5	175.72	19SE	-0.7	-4.1	-1.7	1.1
118.54	1.32	1.5	185.69	29SE	-0.9	-4.2	-1.7	1.1
123.98	1.48	1.4	195.72	9OC	-0.9	-4.3	-1.6	1.1
129.22	1.66	1.3	205.80	19OC	-0.7	-4.4	-1.6	1.1
134.18	1.85	1.2	215.92	29OC	-0.5	-4.3	-1.5	1.1
138.83	2.06	1.0	226.08	8NO	-0.5	-4.0	-1.5	1.1
143.10	2.28	0.9	236.26	18NO	-0.5	-3.3	-1.5	1.1
146.88	2.53	0.7	246.46	28NO	-0.1	-3.4	-1.5	1.0
150.07	2.80	0.5	256.67	8DE	1.7	-4.1	-1.4	1.0
152.54	3.10	0.3	266.87	18DE	1.6	-4.3	-1.4	0.9
154.09	3.41	0.0	277.06	28DE	0.4	-4.4	-1.4	0.8

♂ LONG	♂ LAT	♂ MAG	⊙ LONG	16.00UT	☿	♀	♃	♄
LONG	LAT	MAG	LONG	182	MAGNITUDES			
154.56	3.74	-0.2	287.22	7JA	0.1	-4.3	-1.5	0.8
153.78	4.04	-0.5	297.34	17JA	-0.0	-4.2	-1.5	0.7
151.71	4.29	-0.7	307.43	27JA	-0.2	-4.1	-1.5	0.6
148.53	4.42	-1.0	317.46	6FE	-0.6	-4.0	-1.5	0.6
144.71	4.39	-1.1	327.43	16FE	-1.2	-3.9	-1.6	0.6
140.93	4.18	-1.0	337.35	26FE	-1.5	-3.8	-1.6	0.6
137.89	3.84	-0.8	347.21	8MR	-0.8	-3.7	-1.7	0.7
136.02	3.42	-0.6	357.01	18MR	0.5	-3.6	-1.7	0.7
135.45	2.98	-0.3	6.75	28MR	2.0	-3.5	-1.8	0.8
136.12	2.55	-0.1	16.44	7AP	3.8	-3.5	-1.9	0.8
137.87	2.16	0.1	26.09	17AP	2.2	-3.4	-1.9	0.9
140.52	1.80	0.3	35.69	27AP	1.3	-3.4	-2.0	1.0
143.92	1.47	0.4	45.26	7MY	0.6	-3.4	-2.1	1.0
147.91	1.18	0.6	54.80	17MY	-0.2	-3.3	-2.1	1.0
152.42	0.91	0.7	64.33	27MY	-1.2	-3.3	-2.2	1.1
157.34	0.67	0.8	73.85	6JN	-1.8	-3.3	-2.3	1.1
162.61	0.45	0.9	83.38	16JN	-0.9	-3.3	-2.3	1.1
168.20	0.25	1.0	92.91	26JN	-0.1	-3.3	-2.3	1.2
174.06	0.06	1.0	102.46	6JL	0.3	-3.3	-2.4	1.2
180.15	-0.11	1.1	112.05	16JL	0.6	-3.4	-2.4	1.2
186.47	-0.27	1.1	121.67	26JL	1.1	-3.4	-2.3	1.2
192.98	-0.42	1.2	131.33	5AU	2.1	-3.4	-2.3	1.2
199.68	-0.55	1.2	141.04	15AU	2.6	-3.4	-2.3	1.2
206.54	-0.68	1.2	150.81	25AU	0.3	-3.5	-2.2	1.2
213.57	-0.79	1.3	160.63	4SE	-0.9	-3.5	-2.2	1.3
220.73	-0.88	1.3	170.51	14SE	-1.1	-3.5	-2.1	1.3
228.04	-0.97	1.3	180.46	24SE	-1.0	-3.5	-2.0	1.4
235.46	-1.04	1.3	190.45	4OC	-0.6	-3.4	-2.0	1.4
242.99	-1.10	1.3	200.51	14OC	-0.4	-3.4	-1.9	1.4
250.61	-1.15	1.3	210.61	24OC	-0.3	-3.4	-1.8	1.3
258.32	-1.18	1.3	220.74	3NO	-0.3	-3.4	-1.8	1.3
266.09	-1.20	1.3	230.92	13NO	0.0	-3.3	-1.7	1.3
273.91	-1.21	1.4	241.11	23NO	2.0	-3.3	-1.7	1.2
281.76	-1.21	1.4	251.31	3DE	1.3	-3.3	-1.6	1.2
289.64	-1.19	1.4	261.52	13DE	0.2	-3.3	-1.6	1.2
297.52	-1.15	1.4	271.72	23DE	-0.0	-3.4	-1.6	1.1
				183				
305.38	-1.11	1.4	281.89	2JA	-0.1	-3.4	-1.5	1.1
313.23	-1.06	1.4	292.04	12JA	-0.3	-3.4	-1.5	1.0
321.03	-1.00	1.4	302.14	22JA	-0.6	-3.4	-1.5	1.0
328.79	-0.93	1.5	312.20	1FE	-1.2	-3.5	-1.5	0.9
336.49	-0.85	1.5	322.21	11FE	-1.3	-3.5	-1.5	0.9
344.12	-0.76	1.5	332.16	21FE	-0.7	-3.6	-1.6	0.9
351.67	-0.67	1.5	342.04	3MR	0.6	-3.6	-1.6	0.9
359.15	-0.58	1.5	351.88	13MR	2.5	-3.7	-1.6	0.9
6.55	-0.48	1.5	1.65	23MR	2.9	-3.7	-1.7	1.0
13.85	-0.38	1.5	11.37	2AP	1.6	-3.8	-1.7	1.0
21.08	-0.28	1.5	21.04	12AP	0.9	-3.9	-1.8	1.1
28.22	-0.17	1.6	30.66	22AP	0.4	-4.0	-1.8	1.1
35.27	-0.07	1.6	40.25	2MY	-0.3	-4.1	-1.9	1.2
42.24	0.03	1.7	49.80	12MY	-1.2	-4.1	-1.9	1.3
49.13	0.14	1.7	59.34	22MY	-1.8	-4.2	-2.0	1.3
55.94	0.24	1.8	68.86	1JN	-0.9	-4.1	-2.1	1.3
62.67	0.34	1.8	78.38	11JN	-0.1	-3.9	-2.2	1.4
69.34	0.44	1.8	87.91	21JN	0.4	-3.3	-2.2	1.4
75.93	0.54	1.9	97.45	1JL	0.9	-3.1	-2.3	1.4
82.46	0.63	1.9	107.02	11JL	1.4	-3.8	-2.3	1.4
88.91	0.73	1.9	116.62	21JL	2.7	-4.1	-2.4	1.4
95.31	0.82	1.9	126.25	31JL	2.2	-4.2	-2.4	1.4
101.65	0.92	1.9	135.94	10AU	0.2	-4.2	-2.4	1.3
107.92	1.01	1.9	145.68	20AU	-1.0	-4.2	-2.4	1.3
114.12	1.11	1.9	155.47	30AU	-1.2	-4.1	-2.4	1.2
120.26	1.20	1.9	165.33	9SE	-1.0	-4.0	-2.4	1.2
126.32	1.30	1.8	175.24	19SE	-0.5	-3.9	-2.4	1.2
132.30	1.40	1.8	185.21	29SE	-0.3	-3.8	-2.3	1.2
138.19	1.49	1.7	195.23	9OC	-0.2	-3.7	-2.2	1.2
143.98	1.60	1.7	205.31	19OC	-0.1	-3.7	-2.2	1.2
149.65	1.70	1.6	215.43	29OC	0.2	-3.6	-2.1	1.2
155.18	1.81	1.5	225.58	8NO	2.3	-3.6	-2.0	1.2
160.55	1.92	1.4	235.77	18NO	1.1	-3.5	-2.0	1.1
165.71	2.04	1.2	245.97	28NO	-0.0	-3.5	-1.9	1.1
170.63	2.16	1.1	256.18	8DE	-0.2	-3.4	-1.8	1.1
175.26	2.29	0.9	266.38	18DE	-0.3	-3.4	-1.8	1.0
179.52	2.42	0.7	276.56	28DE	-0.4	-3.4	-1.7	1.0

Left section (184 / 185):

♂ LONG	LAT	MAG	☉ LONG	16.00UT	☿	♀	♃	♄
				184				
183.33	2.56	0.5	286.73	7JA	-0.7	-3.4	-1.7	0.9
186.57	2.70	0.3	296.85	17JA	-1.2	-3.3	-1.6	0.9
189.10	2.84	0.1	306.93	27JA	-1.2	-3.3	-1.6	0.8
190.76	2.97	-0.2	316.97	6FE	-0.7	-3.3	-1.6	0.8
191.36	3.06	-0.5	326.94	16FE	0.8	-3.3	-1.6	0.7
190.75	3.11	-0.8	336.87	26FE	3.0	-3.3	-1.6	0.7
188.86	3.08	-1.1	346.73	7MR	2.2	-3.4	-1.6	0.7
185.87	2.92	-1.3	356.53	17MR	1.2	-3.4	-1.6	0.8
182.24	2.63	-1.5	6.28	27MR	0.7	-3.4	-1.6	0.9
178.65	2.22	-1.4	15.97	6AP	0.3	-3.4	-1.6	0.9
175.80	1.74	-1.2	25.62	16AP	-0.3	-3.5	-1.6	1.0
174.14	1.25	-1.0	35.22	26AP	-1.2	-3.5	-1.7	1.1
173.82	0.79	-0.8	44.80	6MY	-1.8	-3.4	-1.7	1.1
174.80	0.37	-0.6	54.34	16MY	-0.9	-3.4	-1.8	1.2
176.90	0.01	-0.4	63.87	26MY	-0.0	-3.4	-1.8	1.2
179.95	-0.30	-0.2	73.39	5JN	0.6	-3.4	-1.9	1.2
183.79	-0.57	-0.1	82.91	15JN	1.1	-3.3	-1.9	1.3
188.27	-0.80	0.1	92.44	25JN	1.9	-3.3	-2.0	1.3
193.30	-1.00	0.2	102.00	5JL	3.2	-3.3	-2.1	1.3
198.77	-1.16	0.3	111.58	15JL	1.8	-3.3	-2.1	1.3
204.62	-1.30	0.4	121.20	25JL	0.1	-3.3	-2.2	1.2
210.81	-1.41	0.5	130.86	4AU	-1.0	-3.3	-2.3	1.2
217.27	-1.51	0.6	140.56	14AU	-1.3	-3.4	-2.3	1.2
223.98	-1.57	0.7	150.33	24AU	-1.0	-3.4	-2.4	1.1
230.90	-1.62	0.7	160.15	3SE	-0.5	-3.4	-2.4	1.1
238.00	-1.65	0.8	170.03	13SE	-0.2	-3.5	-2.4	1.0
245.26	-1.66	0.9	179.97	23SE	-0.1	-3.5	-2.4	1.1
252.66	-1.65	0.9	189.97	3OC	0.0	-3.6	-2.4	1.1
260.16	-1.62	1.0	200.02	13OC	0.5	-3.7	-2.4	1.1
267.74	-1.58	1.0	210.12	23OC	2.6	-3.8	-2.4	1.1
275.40	-1.52	1.1	220.25	2NO	0.9	-3.9	-2.3	1.1
283.09	-1.45	1.1	230.42	12NO	-0.2	-3.9	-2.2	1.0
290.82	-1.37	1.2	240.62	22NO	-0.4	-4.1	-2.2	1.0
298.55	-1.28	1.2	250.82	2DE	-0.4	-4.2	-2.1	1.0
306.28	-1.18	1.3	261.03	12DE	-0.5	-4.3	-2.0	1.0
313.98	-1.07	1.3	271.22	22DE	-0.8	-4.3	-2.0	0.9
				185				
321.65	-0.96	1.4	281.40	1JA	-1.0	-4.4	-1.9	0.9
329.27	-0.85	1.4	291.54	11JA	-1.1	-4.3	-1.8	0.8
336.83	-0.73	1.5	301.65	21JA	-0.6	-4.0	-1.8	0.8
344.33	-0.61	1.5	311.71	31JA	0.9	-3.4	-1.7	0.7
351.76	-0.49	1.6	321.72	10FE	3.2	-3.5	-1.7	0.7
359.11	-0.38	1.6	331.67	20FE	1.6	-4.0	-1.6	0.6
6.39	-0.26	1.6	341.56	2MR	0.9	-4.2	-1.6	0.6
13.59	-0.15	1.7	351.39	12MR	0.5	-4.3	-1.6	0.6
20.71	-0.04	1.7	1.17	22MR	0.2	-4.2	-1.6	0.6
27.75	0.07	1.7	10.89	1AP	-0.4	-4.1	-1.5	0.7
34.71	0.17	1.8	20.56	11AP	-1.2	-4.0	-1.5	0.7
41.60	0.27	1.8	30.19	21AP	-1.7	-3.9	-1.5	0.8
48.42	0.36	1.8	39.78	1MY	-0.9	-3.8	-1.6	0.9
55.18	0.45	1.8	49.34	11MY	0.1	-3.7	-1.6	0.9
61.87	0.54	1.8	58.88	21MY	0.8	-3.6	-1.6	1.0
68.51	0.62	1.8	68.40	31MY	1.5	-3.6	-1.6	1.0
75.09	0.70	1.9	77.92	10JN	2.5	-3.5	-1.7	1.1
81.62	0.77	1.9	87.45	20JN	3.1	-3.5	-1.7	1.1
88.11	0.85	1.9	96.99	30JN	1.5	-3.4	-1.8	1.1
94.56	0.91	2.0	106.56	10JL	0.0	-3.4	-1.8	1.1
100.97	0.98	2.0	116.16	20JL	-1.1	-3.4	-1.9	1.1
107.35	1.04	2.0	125.79	30JL	-1.5	-3.4	-1.9	1.1
113.70	1.10	2.0	135.47	9AU	-0.9	-3.4	-2.0	1.1
120.02	1.15	2.0	145.21	19AU	-0.4	-3.3	-2.1	1.1
126.31	1.20	2.0	155.00	29AU	-0.1	-3.4	-2.1	1.0
132.58	1.25	2.0	164.85	8SE	0.1	-3.4	-2.2	1.0
138.83	1.30	2.0	174.76	18SE	0.2	-3.4	-2.3	0.9
145.04	1.35	1.9	184.72	28SE	0.7	-3.4	-2.3	0.9
151.23	1.39	1.9	194.74	8OC	2.9	-3.4	-2.3	0.9
157.38	1.43	1.9	204.82	18OC	0.7	-3.4	-2.4	0.9
163.49	1.46	1.8	214.93	28OC	-0.4	-3.5	-2.4	1.0
169.57	1.49	1.7	225.09	7NO	-0.5	-3.5	-2.4	1.0
175.58	1.52	1.7	235.27	17NO	-0.5	-3.5	-2.3	0.9
181.54	1.55	1.6	245.47	27NO	-0.6	-3.5	-2.3	0.9
187.42	1.56	1.5	255.68	7DE	-0.8	-3.5	-2.2	0.9
193.20	1.58	1.3	265.88	17DE	-0.9	-3.4	-2.2	0.9
198.88	1.58	1.2	276.07	27DE	-0.9	-3.4	-2.1	0.9

Right section (186 / 187):

♂ LONG	LAT	MAG	☉ LONG	16.00UT	☿	♀	♃	♄
				186				
204.42	1.57	1.1	286.23	6JA	-0.5	-3.4	-2.0	0.8
209.79	1.56	0.9	296.36	16JA	1.1	-3.4	-2.0	0.8
214.96	1.52	0.7	306.44	26JA	2.8	-3.3	-1.9	0.7
219.88	1.47	0.5	316.48	5FE	1.2	-3.3	-1.8	0.7
224.48	1.39	0.3	326.46	15FE	0.6	-3.3	-1.8	0.6
228.68	1.28	0.1	336.38	25FE	0.3	-3.3	-1.7	0.6
232.39	1.13	-0.2	346.25	7MR	0.0	-3.4	-1.6	0.5
235.46	0.93	-0.4	356.05	17MR	-0.4	-3.4	-1.6	0.5
237.75	0.66	-0.7	5.80	27MR	-1.2	-3.4	-1.6	0.4
239.06	0.31	-1.1	15.50	6AP	-1.7	-3.4	-1.5	0.5
239.22	-0.13	-1.4	25.15	16AP	-0.9	-3.5	-1.5	0.6
238.15	-0.66	-1.7	34.75	26AP	0.2	-3.5	-1.5	0.6
235.92	-1.27	-2.0	44.33	6MY	1.0	-3.5	-1.5	0.7
232.90	-1.89	-2.2	53.88	16MY	2.0	-3.6	-1.5	0.8
229.78	-2.46	-2.1	63.41	26MY	3.3	-3.7	-1.5	0.8
227.26	-2.91	-2.0	72.93	5JN	2.6	-3.7	-1.5	0.9
225.88	-3.23	-1.8	82.45	15JN	1.2	-3.8	-1.5	0.9
225.88	-3.42	-1.5	91.98	25JN	-0.0	-3.9	-1.5	0.9
227.22	-3.50	-1.3	101.54	5JL	-1.1	-4.0	-1.5	1.0
229.77	-3.51	-1.1	111.12	15JL	-1.6	-4.1	-1.6	1.0
233.33	-3.46	-0.9	120.73	25JL	-0.9	-4.2	-1.6	1.0
237.70	-3.36	-0.7	130.39	4AU	-0.3	-4.2	-1.7	1.0
242.74	-3.24	-0.5	140.10	14AU	0.0	-4.2	-1.7	1.0
248.31	-3.09	-0.3	149.86	24AU	0.2	-3.9	-1.8	1.0
254.30	-2.92	-0.2	159.68	3SE	0.4	-3.4	-1.8	0.9
260.63	-2.74	-0.0	169.56	13SE	1.0	-3.5	-1.9	0.9
267.23	-2.55	0.1	179.49	23SE	3.2	-4.0	-2.0	0.9
274.04	-2.35	0.2	189.48	3OC	0.6	-4.3	-2.0	0.8
281.01	-2.14	0.4	199.53	13OC	-0.5	-4.3	-2.1	0.8
288.10	-1.94	0.5	209.62	23OC	-0.7	-4.3	-2.1	0.8
295.27	-1.73	0.6	219.76	2NO	-0.6	-4.2	-2.2	0.9
302.50	-1.52	0.7	229.93	12NO	-0.7	-4.1	-2.2	0.9
309.76	-1.33	0.8	240.12	22NO	-0.7	-4.0	-2.2	0.9
317.03	-1.13	0.9	250.33	2DE	-0.7	-3.9	-2.3	0.9
324.30	-0.94	1.0	260.53	12DE	-0.7	-3.8	-2.2	0.9
331.54	-0.77	1.1	270.73	22DE	-0.4	-3.8	-2.2	0.8
				187				
338.75	-0.60	1.2	280.91	1JA	1.3	-3.7	-2.2	0.8
345.92	-0.44	1.3	291.06	11JA	2.3	-3.6	-2.1	0.8
353.04	-0.28	1.4	301.16	21JA	0.8	-3.5	-2.1	0.7
0.11	-0.14	1.5	311.23	31JA	0.4	-3.5	-2.0	0.7
7.13	-0.01	1.6	321.24	10FE	0.2	-3.4	-1.9	0.6
14.08	0.11	1.6	331.19	20FE	-0.0	-3.4	-1.8	0.6
20.97	0.23	1.7	341.09	2MR	-0.5	-3.4	-1.8	0.5
27.81	0.34	1.7	350.92	12MR	-1.2	-3.3	-1.7	0.5
34.58	0.43	1.8	0.70	22MR	-1.6	-3.3	-1.6	0.4
41.31	0.53	1.8	10.43	1AP	-0.8	-3.3	-1.6	0.4
47.98	0.61	1.9	20.10	11AP	0.3	-3.3	-1.5	0.4
54.59	0.69	1.9	29.73	21AP	1.3	-3.3	-1.5	0.4
61.17	0.76	1.9	39.32	1MY	2.6	-3.3	-1.5	0.5
67.70	0.82	2.0	48.88	11MY	3.4	-3.3	-1.4	0.6
74.20	0.88	2.0	58.41	21MY	2.0	-3.4	-1.4	0.6
80.66	0.93	2.0	67.94	31MY	0.9	-3.4	-1.4	0.7
87.10	0.98	2.0	77.46	10JN	-0.1	-3.4	-1.4	0.7
93.51	1.02	2.0	86.98	20JN	-1.2	-3.5	-1.4	0.8
99.91	1.06	2.0	96.53	30JN	-1.7	-3.5	-1.4	0.8
106.29	1.09	1.9	106.09	10JL	-0.9	-3.5	-1.4	0.9
112.66	1.12	2.0	115.69	20JL	-0.3	-3.4	-1.4	0.9
119.03	1.14	2.0	125.32	30JL	0.1	-3.4	-1.4	0.9
125.40	1.16	2.0	135.00	9AU	0.4	-3.4	-1.4	0.9
131.77	1.18	2.0	144.73	19AU	0.6	-3.3	-1.5	0.9
138.14	1.19	2.0	154.52	29AU	1.4	-3.3	-1.5	0.9
144.53	1.20	2.0	164.37	8SE	3.2	-3.3	-1.6	0.9
150.92	1.20	2.0	174.27	18SE	0.5	-3.3	-1.6	0.9
157.33	1.20	2.0	184.24	28SE	-0.7	-3.3	-1.7	0.8
163.75	1.19	2.0	194.25	8OC	-0.8	-3.3	-1.7	0.7
170.18	1.18	1.9	204.33	18OC	-0.8	-3.4	-1.8	0.7
176.63	1.16	1.9	214.44	28OC	-0.8	-3.4	-1.9	0.8
183.09	1.13	1.9	224.59	7NO	-0.6	-3.4	-1.9	0.8
189.56	1.10	1.8	234.78	17NO	-0.6	-3.5	-2.0	0.8
196.04	1.06	1.7	244.98	27NO	-0.6	-3.5	-2.0	0.8
202.53	1.01	1.7	255.18	7DE	-0.3	-3.6	-2.1	0.8
209.02	0.95	1.6	265.39	17DE	1.5	-3.6	-2.1	0.8
215.51	0.87	1.5	275.58	27DE	1.9	-3.7	-2.1	0.8

♂			⊙	16.00UT	☿	♀	♃	♄
LONG	LAT	MAG	LONG	188	MAGNITUDES			
222.00	0.79	1.4	285.74	6JA	0.5	-3.8	-2.1	0.8
228.48	0.69	1.3	295.87	16JA	0.2	-3.8	-2.1	0.8
234.94	0.57	1.2	305.96	26JA	0.1	-3.9	-2.1	0.7
241.38	0.43	1.0	315.99	5FE	-0.1	-4.0	-2.0	0.7
247.79	0.27	0.9	325.98	15FE	-0.5	-4.1	-2.0	0.6
254.15	0.09	0.7	335.91	25FE	-1.2	-4.2	-1.9	0.6
260.45	-0.13	0.6	345.77	6MR	-1.5	-4.3	-1.9	0.5
266.66	-0.37	0.4	355.58	16MR	-0.8	-4.2	-1.8	0.5
272.76	-0.65	0.2	5.34	26MR	0.4	-4.1	-1.7	0.4
278.71	-0.97	0.1	15.03	5AP	1.7	-3.6	-1.7	0.4
284.48	-1.33	-0.1	24.68	15AP	3.4	-2.9	-1.6	0.3
289.98	-1.74	-0.4	34.29	25AP	2.6	-3.5	-1.5	0.3
295.15	-2.21	-0.6	43.86	5MY	1.5	-4.0	-1.5	0.4
299.88	-2.73	-0.8	53.42	15MY	0.7	-4.2	-1.4	0.5
304.03	-3.32	-1.1	62.95	25MY	-0.2	-4.2	-1.4	0.5
307.46	-3.96	-1.3	72.46	4JN	-1.2	-4.1	-1.3	0.6
309.95	-4.65	-1.6	81.99	14JN	-1.7	-4.0	-1.3	0.6
311.31	-5.35	-1.9	91.52	24JN	-0.9	-3.9	-1.3	0.7
311.39	-6.02	-2.2	101.07	4JL	-0.2	-3.9	-1.3	0.7
310.16	-6.56	-2.4	110.65	14JL	0.2	-3.8	-1.3	0.8
307.93	-6.85	-2.6	120.27	24JL	0.5	-3.7	-1.3	0.8
305.28	-6.83	-2.6	129.92	3AU	0.9	-3.6	-1.3	0.8
302.94	-6.49	-2.4	139.63	13AU	1.8	-3.6	-1.3	0.9
301.58	-5.91	-2.2	149.39	23AU	2.9	-3.5	-1.3	0.9
301.47	-5.22	-1.9	159.20	2SE	0.4	-3.5	-1.3	0.9
302.65	-4.49	-1.6	169.08	12SE	-0.8	-3.5	-1.4	0.8
305.00	-3.80	-1.3	179.01	22SE	-1.0	-3.4	-1.4	0.8
308.30	-3.16	-1.0	189.00	2OC	-0.9	-3.4	-1.4	0.8
312.35	-2.59	-0.7	199.05	12OC	-0.7	-3.4	-1.5	0.8
316.99	-2.09	-0.4	209.14	22OC	-0.5	-3.4	-1.5	0.7
322.09	-1.65	-0.2	219.27	1NO	-0.4	-3.4	-1.6	0.7
327.54	-1.26	0.0	229.44	11NO	-0.4	-3.4	-1.7	0.7
333.25	-0.93	0.2	239.63	21NO	-0.1	-3.4	-1.7	0.8
339.16	-0.63	0.4	249.83	1DE	1.7	-3.4	-1.8	0.8
345.22	-0.38	0.6	260.04	11DE	1.6	-3.4	-1.9	0.8
351.39	-0.15	0.8	270.24	21DE	0.3	-3.4	-1.9	0.8
357.64	0.04	0.9	280.41	31DE	0.0	-3.4	-2.0	0.8
				189				
3.94	0.21	1.1	290.57	10JA	-0.1	-3.4	-2.0	0.8
10.28	0.36	1.2	300.67	20JA	-0.2	-3.4	-2.0	0.8
16.64	0.50	1.3	310.74	30JA	-0.6	-3.5	-2.0	0.7
23.02	0.61	1.4	320.75	9FE	-1.2	-3.5	-2.0	0.7
29.39	0.71	1.5	330.71	19FE	-1.4	-3.4	-2.0	0.7
35.76	0.79	1.6	340.61	1MR	-0.8	-3.4	-2.0	0.6
42.12	0.87	1.7	350.45	11MR	0.5	-3.4	-1.9	0.5
48.47	0.93	1.8	0.23	21MR	2.2	-3.4	-1.9	0.5
54.81	0.99	1.8	9.95	31MR	3.5	-3.4	-1.8	0.4
61.14	1.04	1.9	19.63	10AP	2.0	-3.3	-1.8	0.4
67.45	1.07	1.9	29.26	20AP	1.1	-3.3	-1.7	0.3
73.76	1.11	2.0	38.85	30AP	0.5	-3.3	-1.6	0.3
80.06	1.13	2.0	48.41	10MY	-0.2	-3.3	-1.6	0.3
86.35	1.15	2.0	57.95	20MY	-1.2	-3.3	-1.5	0.4
92.64	1.17	2.0	67.47	30MY	-1.8	-3.4	-1.5	0.4
98.94	1.18	2.0	77.00	9JN	-0.9	-3.4	-1.4	0.5
105.24	1.18	2.0	86.52	19JN	-0.1	-3.4	-1.4	0.6
111.56	1.18	2.0	96.06	29JN	0.3	-3.5	-1.3	0.6
117.89	1.17	2.0	105.63	9JL	0.7	-3.5	-1.3	0.7
124.24	1.16	2.0	115.22	19JL	1.2	-3.6	-1.3	0.7
130.61	1.15	2.0	124.85	29JL	2.3	-3.6	-1.2	0.8
137.01	1.13	1.9	134.53	8AU	2.5	-3.7	-1.2	0.8
143.44	1.11	1.9	144.26	18AU	0.3	-3.8	-1.2	0.8
149.90	1.08	1.9	154.05	28AU	-0.9	-3.9	-1.2	0.8
156.41	1.04	1.9	163.89	7SE	-1.1	-4.0	-1.2	0.8
162.95	1.01	1.9	173.79	17SE	-1.0	-4.1	-1.2	0.8
169.54	0.96	1.9	183.76	27SE	-0.6	-4.2	-1.3	0.8
176.17	0.91	1.9	193.77	7OC	-0.4	-4.3	-1.3	0.8
182.85	0.86	1.9	203.84	17OC	-0.3	-4.4	-1.3	0.8
189.58	0.80	1.9	213.95	27OC	-0.2	-4.3	-1.4	0.7
196.35	0.73	1.9	224.11	6NO	0.1	-4.0	-1.4	0.7
203.18	0.65	1.8	234.28	16NO	2.0	-3.2	-1.5	0.7
210.06	0.57	1.8	244.48	26NO	1.3	-3.4	-1.5	0.7
216.98	0.48	1.7	254.69	6DE	0.1	-4.1	-1.6	0.8
223.96	0.38	1.7	264.89	16DE	-0.1	-4.3	-1.7	0.8
230.99	0.27	1.6	275.08	26DE	-0.2	-4.4	-1.7	0.8

♂			⊙	16.00UT	☿	♀	♃	♄
LONG	LAT	MAG	LONG	190	MAGNITUDES			
238.06	0.15	1.5	285.25	5JA	-0.3	-4.3	-1.8	0.8
245.18	0.02	1.5	295.38	15JA	-0.7	-4.2	-1.9	0.8
252.34	-0.12	1.4	305.47	25JA	-1.2	-4.1	-1.9	0.8
259.53	-0.27	1.3	315.51	4FE	-1.3	-4.0	-2.0	0.8
266.76	-0.43	1.2	325.50	14FE	-0.8	-3.9	-2.0	0.7
274.02	-0.61	1.1	335.43	24FE	0.6	-3.8	-2.0	0.7
281.30	-0.79	1.0	345.30	6MR	2.7	-3.7	-2.0	0.6
288.59	-0.97	1.0	355.11	16MR	2.6	-3.6	-2.0	0.6
295.87	-1.17	0.9	4.87	26MR	1.4	-3.5	-2.0	0.5
303.14	-1.37	0.8	14.57	5AP	0.8	-3.5	-1.9	0.5
310.38	-1.57	0.7	24.22	15AP	0.4	-3.4	-1.9	0.4
317.56	-1.78	0.6	33.83	25AP	-0.3	-3.4	-1.8	0.3
324.66	-1.98	0.4	43.41	5MY	-1.2	-3.4	-1.8	0.3
331.65	-2.18	0.3	52.96	15MY	-1.8	-3.3	-1.7	0.3
338.50	-2.38	0.2	62.49	25MY	-0.9	-3.3	-1.7	0.3
345.16	-2.57	0.1	72.01	4JN	-0.1	-3.3	-1.6	0.4
351.59	-2.75	-0.0	81.53	14JN	0.5	-3.3	-1.5	0.4
357.72	-2.91	-0.2	91.07	24JN	0.9	-3.3	-1.5	0.5
3.47	-3.07	-0.3	100.61	4JL	1.6	-3.3	-1.4	0.6
8.77	-3.20	-0.5	110.19	14JL	2.9	-3.4	-1.4	0.6
13.48	-3.31	-0.6	119.80	24JL	2.1	-3.4	-1.3	0.7
17.46	-3.40	-0.8	129.45	3AU	0.2	-3.4	-1.3	0.7
20.54	-3.44	-1.0	139.16	13AU	-1.0	-3.4	-1.3	0.7
22.49	-3.43	-1.2	148.91	23AU	-1.3	-3.5	-1.2	0.8
23.12	-3.35	-1.5	158.72	2SE	-1.0	-3.5	-1.2	0.8
22.30	-3.16	-1.7	168.59	12SE	-0.5	-3.5	-1.2	0.8
20.05	-2.84	-1.9	178.53	22SE	-0.3	-3.4	-1.2	0.8
16.80	-2.38	-2.0	188.51	2OC	-0.1	-3.4	-1.2	0.8
13.26	-1.82	-2.0	198.55	12OC	-0.1	-3.4	-1.2	0.8
10.22	-1.23	-1.7	208.65	22OC	0.3	-3.4	-1.3	0.8
8.31	-0.67	-1.4	218.78	1NO	2.3	-3.3	-1.3	0.7
7.73	-0.19	-1.0	228.94	11NO	1.1	-3.3	-1.3	0.7
8.46	0.20	-0.7	239.13	21NO	-0.1	-3.3	-1.4	0.7
10.31	0.51	-0.4	249.34	1DE	-0.3	-3.3	-1.4	0.7
13.08	0.75	-0.1	259.54	11DE	-0.3	-3.3	-1.5	0.7
16.57	0.94	0.1	269.74	21DE	-0.4	-3.4	-1.5	0.8
20.63	1.08	0.4	279.92	31DE	-0.7	-3.4	-1.6	0.8
				191				
25.11	1.19	0.6	290.07	10JA	-1.1	-3.4	-1.6	0.8
29.94	1.28	0.8	300.19	20JA	-1.1	-3.4	-1.7	0.8
35.03	1.34	1.0	310.25	30JA	-0.7	-3.5	-1.8	0.8
40.32	1.38	1.1	320.27	9FE	0.8	-3.5	-1.8	0.8
45.78	1.42	1.3	330.23	19FE	3.1	-3.6	-1.9	0.8
51.36	1.44	1.4	340.13	1MR	2.0	-3.6	-2.0	0.7
57.05	1.45	1.5	349.97	11MR	1.1	-3.7	-2.0	0.7
62.82	1.45	1.6	359.76	21MR	0.6	-3.7	-2.0	0.6
68.66	1.45	1.7	9.49	31MR	0.2	-3.8	-2.0	0.6
74.55	1.44	1.8	19.17	10AP	-0.3	-3.9	-2.0	0.5
80.50	1.42	1.8	28.80	20AP	-1.2	-4.0	-2.0	0.4
86.49	1.40	1.9	38.39	30AP	-1.8	-4.1	-2.0	0.4
92.52	1.38	1.9	47.96	10MY	-0.9	-4.2	-2.0	0.3
98.60	1.35	2.0	57.50	20MY	0.0	-4.2	-1.9	0.3
104.71	1.32	2.0	67.02	30MY	0.7	-4.1	-1.9	0.3
110.86	1.29	2.0	76.54	9JN	1.2	-3.8	-1.8	0.3
117.06	1.25	2.0	86.07	19JN	2.1	-3.2	-1.7	0.4
123.30	1.20	2.0	95.60	29JN	3.3	-3.1	-1.7	0.4
129.59	1.16	2.0	105.16	9JL	1.7	-3.8	-1.6	0.5
135.93	1.11	2.0	114.76	19JL	0.1	-4.1	-1.6	0.6
142.31	1.05	2.0	124.38	29JL	-1.0	-4.2	-1.5	0.6
148.76	1.00	2.0	134.06	8AU	-1.4	-4.2	-1.5	0.7
155.26	0.93	1.9	143.79	18AU	-1.0	-4.1	-1.4	0.7
161.83	0.87	1.9	153.57	28AU	-0.5	-4.1	-1.4	0.7
168.46	0.80	1.9	163.41	7SE	-0.2	-4.0	-1.3	0.8
175.16	0.73	1.8	173.31	17SE	-0.0	-3.9	-1.3	0.8
181.92	0.65	1.8	183.27	27SE	0.1	-3.8	-1.3	0.8
188.76	0.57	1.8	193.28	7OC	0.5	-3.7	-1.3	0.8
195.67	0.48	1.8	203.35	17OC	2.6	-3.7	-1.3	0.8
202.66	0.39	1.8	213.46	27OC	0.9	-3.6	-1.3	0.8
209.72	0.29	1.7	223.61	6NO	-0.3	-3.6	-1.3	0.7
216.85	0.19	1.7	233.79	16NO	-0.4	-3.5	-1.3	0.7
224.05	0.09	1.7	243.99	26NO	-0.4	-3.5	-1.3	0.7
231.32	-0.02	1.7	254.20	6DE	-0.5	-3.4	-1.3	0.7
238.66	-0.13	1.6	264.40	16DE	-0.8	-3.4	-1.4	0.7
246.07	-0.25	1.6	274.59	26DE	-1.0	-3.4	-1.4	0.8

♂ LONG	♂ LAT	♂ MAG	☉ LONG	16.00UT	☿	♀	♃	♄
				192	MAGNITUDES			
253.53	-0.37	1.6	284.76	5JA	-1.0	-3.4	-1.5	0.8
261.05	-0.49	1.5	294.89	15JA	-0.6	-3.3	-1.5	0.8
268.61	-0.62	1.5	304.98	25JA	0.9	-3.3	-1.6	0.8
276.22	-0.74	1.4	315.03	4FE	3.1	-3.3	-1.7	0.8
283.86	-0.86	1.4	325.02	14FE	1.4	-3.3	-1.7	0.8
291.52	-0.98	1.3	334.95	24FE	0.8	-3.3	-1.8	0.8
299.20	-1.10	1.3	344.82	5MR	0.4	-3.4	-1.9	0.8
306.87	-1.20	1.3	354.64	15MR	0.1	-3.4	-1.9	0.7
314.53	-1.30	1.2	4.39	25MR	-0.4	-3.4	-2.0	0.7
322.17	-1.40	1.2	14.10	4AP	-1.2	-3.4	-2.1	0.6
329.76	-1.48	1.1	23.75	14AP	-1.7	-3.5	-2.1	0.6
337.30	-1.54	1.1	33.36	24AP	-0.9	-3.5	-2.1	0.5
344.77	-1.60	1.0	42.95	4MY	0.1	-3.4	-2.1	0.4
352.14	-1.63	1.0	52.49	14MY	0.9	-3.4	-2.1	0.4
359.42	-1.66	0.9	62.02	24MY	1.6	-3.4	-2.1	0.3
6.57	-1.66	0.9	71.55	3JN	2.8	-3.4	-2.1	0.3
13.57	-1.65	0.8	81.07	13JN	3.0	-3.3	-2.0	0.3
20.40	-1.62	0.8	90.60	23JN	1.4	-3.3	-2.0	0.3
27.05	-1.57	0.7	100.15	3JL	0.1	-3.3	-1.9	0.4
33.47	-1.51	0.6	109.72	13JL	-1.1	-3.3	-1.9	0.5
39.64	-1.42	0.5	119.33	23JL	-1.5	-3.3	-1.8	0.5
45.51	-1.31	0.5	128.98	2AU	-0.9	-3.3	-1.8	0.6
51.03	-1.18	0.4	138.68	12AU	-0.4	-3.4	-1.7	0.6
56.13	-1.03	0.2	148.43	22AU	-0.1	-3.4	-1.6	0.7
60.71	-0.84	0.1	158.24	1SE	0.1	-3.4	-1.6	0.7
64.66	-0.62	-0.1	168.11	11SE	0.3	-3.5	-1.5	0.7
67.84	-0.36	-0.2	178.04	21SE	0.8	-3.5	-1.5	0.8
70.07	-0.06	-0.4	188.02	1OC	3.0	-3.6	-1.4	0.8
71.15	0.30	-0.6	198.06	11OC	0.7	-3.7	-1.4	0.8
70.89	0.72	-0.9	208.15	21OC	-0.4	-3.8	-1.4	0.8
69.19	1.17	-1.1	218.28	31OC	-0.6	-3.9	-1.4	0.8
66.21	1.63	-1.3	228.44	10NO	-0.6	-4.0	-1.4	0.8
62.41	2.05	-1.4	238.64	20NO	-0.6	-4.1	-1.4	0.7
58.54	2.37	-1.3	248.84	30NO	-0.8	-4.2	-1.4	0.7
55.38	2.57	-1.0	259.05	10DE	-0.8	-4.3	-1.4	0.7
53.39	2.66	-0.7	269.25	20DE	-0.8	-4.3	-1.4	0.7
52.75	2.67	-0.4	279.43	30DE	-0.5	-4.4	-1.4	0.8
				193				
53.37	2.63	-0.1	289.58	9JA	1.1	-4.3	-1.5	0.8
55.06	2.56	0.1	299.70	19JA	2.7	-3.9	-1.5	0.8
57.63	2.47	0.4	309.77	29JA	1.0	-3.4	-1.5	0.8
60.90	2.38	0.6	319.78	8FE	0.5	-3.5	-1.6	0.8
64.72	2.28	0.8	329.75	18FE	0.3	-4.0	-1.7	0.8
68.97	2.18	1.0	339.65	28FE	0.0	-4.2	-1.7	0.8
73.56	2.09	1.1	349.50	10MR	-0.4	-4.3	-1.8	0.8
78.43	1.99	1.3	359.28	20MR	-1.2	-4.2	-1.9	0.8
83.52	1.90	1.4	9.02	30MR	-1.7	-4.1	-1.9	0.7
88.80	1.81	1.5	18.70	9AP	-0.9	-4.0	-2.0	0.7
94.24	1.73	1.6	28.33	19AP	0.2	-3.9	-2.1	0.6
99.82	1.64	1.7	37.93	29AP	1.1	-3.8	-2.1	0.6
105.51	1.55	1.7	47.49	9MY	2.2	-3.7	-2.2	0.5
111.30	1.46	1.8	57.03	19MY	3.6	-3.6	-2.2	0.4
117.20	1.38	1.8	66.56	29MY	2.3	-3.6	-2.2	0.4
123.18	1.29	1.9	76.08	8JN	1.1	-3.5	-2.3	0.3
129.26	1.21	1.9	85.61	18JN	-0.0	-3.5	-2.3	0.3
135.42	1.12	1.9	95.14	28JN	-1.1	-3.4	-2.2	0.3
141.66	1.03	1.9	104.70	8JL	-1.6	-3.4	-2.2	0.4
147.99	0.94	1.9	114.29	18JL	-0.9	-3.4	-2.2	0.4
154.41	0.85	1.9	123.92	28JL	-0.3	-3.4	-2.1	0.5
160.91	0.75	1.9	133.59	7AU	0.0	-3.3	-2.1	0.5
167.50	0.66	1.9	143.32	17AU	0.3	-3.3	-2.0	0.6
174.17	0.56	1.8	153.10	27AU	0.5	-3.3	-1.9	0.6
180.94	0.46	1.8	162.93	6SE	1.1	-3.4	-1.9	0.7
187.80	0.36	1.8	172.83	16SE	3.3	-3.4	-1.8	0.7
194.75	0.26	1.7	182.79	26SE	0.6	-3.4	-1.7	0.8
201.79	0.16	1.7	192.79	6OC	-0.6	-3.4	-1.7	0.8
208.92	0.05	1.6	202.86	16OC	-0.7	-3.4	-1.6	0.8
216.15	-0.05	1.6	212.97	26OC	-0.7	-3.5	-1.6	0.8
223.45	-0.16	1.5	223.11	5NO	-0.8	-3.5	-1.5	0.8
230.85	-0.26	1.5	233.29	15NO	-0.7	-3.5	-1.5	0.8
238.32	-0.37	1.5	243.49	25NO	-0.7	-3.5	-1.5	0.8
245.87	-0.47	1.5	253.69	5DE	-0.7	-3.4	-1.5	0.8
253.48	-0.57	1.5	263.90	15DE	-0.4	-3.4	-1.5	0.7
261.16	-0.67	1.5	274.09	25DE	1.3	-3.4	-1.5	0.7

♂ LONG	♂ LAT	♂ MAG	☉ LONG	16.00UT	☿	♀	♃	♄
				194	MAGNITUDES			
268.88	-0.76	1.5	284.26	4JA	2.2	-3.4	-1.5	0.8
276.65	-0.85	1.5	294.40	14JA	0.7	-3.4	-1.5	0.8
284.45	-0.93	1.4	304.49	24JA	0.3	-3.3	-1.5	0.9
292.27	-1.01	1.4	314.54	3FE	0.2	-3.3	-1.5	0.9
300.10	-1.07	1.4	324.53	13FE	-0.1	-3.3	-1.6	0.9
307.93	-1.12	1.4	334.47	23FE	-0.5	-3.3	-1.6	0.9
315.74	-1.17	1.4	344.34	5MR	-1.2	-3.4	-1.7	0.9
323.52	-1.20	1.4	354.16	15MR	-1.6	-3.4	-1.7	0.9
331.26	-1.22	1.4	3.92	25MR	-0.9	-3.4	-1.8	0.8
338.94	-1.22	1.4	13.63	4AP	0.3	-3.4	-1.8	0.8
346.57	-1.21	1.4	23.29	14AP	1.4	-3.5	-1.9	0.8
354.11	-1.19	1.4	32.90	24AP	2.9	-3.5	-2.0	0.7
1.58	-1.16	1.4	42.48	4MY	3.1	-3.5	-2.0	0.7
8.95	-1.11	1.3	52.04	14MY	1.8	-3.6	-2.1	0.6
16.22	-1.05	1.3	61.57	24MY	0.9	-3.7	-2.2	0.5
23.38	-0.98	1.3	71.09	3JN	-0.1	-3.7	-2.2	0.5
30.42	-0.89	1.3	80.61	13JN	-1.1	-3.8	-2.3	0.4
37.33	-0.80	1.3	90.14	23JN	-1.7	-3.9	-2.3	0.3
44.10	-0.69	1.3	99.69	3JL	-0.9	-4.0	-2.4	0.3
50.73	-0.57	1.2	109.26	13JL	-0.3	-4.1	-2.4	0.4
57.19	-0.44	1.2	118.87	23JL	0.2	-4.2	-2.4	0.4
63.48	-0.30	1.2	128.52	2AU	0.4	-4.2	-2.4	0.5
69.57	-0.14	1.1	138.22	12AU	0.7	-4.2	-2.3	0.5
75.44	0.02	1.0	147.96	22AU	1.5	-3.9	-2.3	0.6
81.05	0.21	1.0	157.77	1SE	3.2	-3.3	-2.2	0.6
86.37	0.41	0.9	167.64	11SE	0.5	-3.5	-2.2	0.7
91.34	0.64	0.8	177.56	21SE	-0.7	-4.0	-2.1	0.7
95.89	0.89	0.6	187.54	1OC	-0.9	-4.3	-2.0	0.8
99.92	1.16	0.5	197.58	11OC	-0.9	-4.3	-2.0	0.8
103.33	1.48	0.3	207.66	21OC	-0.8	-4.3	-1.9	0.8
105.96	1.83	0.1	217.79	31OC	-0.6	-4.2	-1.8	0.8
107.63	2.22	-0.1	227.95	10NO	-0.5	-4.1	-1.8	0.8
108.16	2.65	-0.3	238.14	20NO	-0.5	-4.0	-1.7	0.8
107.39	3.10	-0.5	248.34	30NO	-0.2	-3.9	-1.7	0.8
105.25	3.53	-0.8	258.55	10DE	1.5	-3.8	-1.6	0.8
101.97	3.89	-1.0	268.75	20DE	1.8	-3.8	-1.6	0.8
98.04	4.12	-1.1	278.93	30DE	0.5	-3.7	-1.6	0.8
				195				
94.20	4.18	-0.9	289.09	9JA	0.1	-3.6	-1.6	0.8
91.15	4.08	-0.7	299.20	19JA	0.0	-3.5	-1.5	0.9
89.30	3.88	-0.4	309.28	29JA	-0.2	-3.5	-1.5	0.9
88.75	3.62	-0.2	319.30	8FE	-0.5	-3.4	-1.5	0.9
89.43	3.35	0.1	329.27	18FE	-1.2	-3.4	-1.6	1.0
91.14	3.08	0.3	339.18	28FE	-1.5	-3.4	-1.6	1.0
93.70	2.82	0.5	349.03	10MR	-0.9	-3.3	-1.6	1.0
96.96	2.58	0.7	358.81	20MR	0.4	-3.3	-1.6	1.0
100.76	2.36	0.9	8.55	30MR	1.8	-3.3	-1.7	0.9
105.00	2.16	1.0	18.24	9AP	3.7	-3.3	-1.7	0.9
109.61	1.97	1.2	27.87	19AP	2.4	-3.3	-1.8	0.9
114.52	1.79	1.3	37.47	29AP	1.4	-3.3	-1.8	0.8
119.68	1.62	1.4	47.03	9MY	0.6	-3.3	-1.9	0.8
125.06	1.47	1.4	56.57	19MY	-0.2	-3.4	-2.0	0.7
130.63	1.31	1.5	66.10	29MY	-1.1	-3.4	-2.0	0.7
136.37	1.17	1.6	75.62	8JN	-1.8	-3.4	-2.1	0.6
142.27	1.03	1.6	85.14	18JN	-0.9	-3.5	-2.2	0.5
148.31	0.89	1.6	94.68	28JN	-0.2	-3.5	-2.2	0.5
154.48	0.76	1.7	104.24	8JL	0.3	-3.5	-2.3	0.4
160.79	0.63	1.7	113.83	18JL	0.6	-3.4	-2.4	0.4
167.22	0.51	1.7	123.45	28JL	1.0	-3.4	-2.4	0.4
173.77	0.38	1.7	133.12	7AU	2.0	-3.4	-2.4	0.5
180.44	0.26	1.7	142.84	17AU	2.8	-3.3	-2.4	0.5
187.23	0.14	1.7	152.62	27AU	0.4	-3.3	-2.4	0.6
194.14	0.02	1.7	162.45	6SE	-0.8	-3.3	-2.4	0.7
201.15	-0.09	1.6	172.35	16SE	-1.0	-3.3	-2.4	0.7
208.28	-0.20	1.6	182.30	26SE	-1.0	-3.3	-2.4	0.8
215.52	-0.31	1.6	192.31	6OC	-0.7	-3.3	-2.3	0.8
222.85	-0.42	1.6	202.36	16OC	-0.4	-3.4	-2.2	0.8
230.29	-0.52	1.5	212.47	26OC	-0.4	-3.4	-2.2	0.9
237.82	-0.61	1.5	222.62	5NO	-0.3	-3.4	-2.1	0.9
245.42	-0.70	1.5	232.79	15NO	-0.0	-3.5	-2.0	0.9
253.11	-0.78	1.4	242.99	25NO	1.7	-3.5	-2.0	0.9
260.86	-0.85	1.4	253.20	5DE	1.5	-3.6	-1.9	0.9
268.66	-0.92	1.4	263.40	15DE	0.2	-3.6	-1.8	0.9
276.50	-0.97	1.3	273.60	25DE	-0.0	-3.7	-1.8	0.9

Left page (196 / 197)

♂ LONG	LAT	MAG	☉ LONG	16.00UT 196	☿	♀	♃	♄
284.37	-1.02	1.3	283.76	4JA	-0.1	-3.8	-1.7	0.9
292.26	-1.05	1.3	293.90	14JA	-0.3	-3.9	-1.7	0.9
300.16	-1.08	1.3	304.00	24JA	-0.6	-3.9	-1.6	0.9
308.04	-1.09	1.3	314.05	3FE	-1.2	-4.0	-1.6	1.0
315.91	-1.09	1.4	324.04	13FE	-1.4	-4.1	-1.6	1.0
323.74	-1.08	1.4	333.99	23FE	-0.8	-4.2	-1.6	1.0
331.52	-1.05	1.4	343.86	4MR	0.5	-4.3	-1.6	1.1
339.24	-1.02	1.4	353.68	14MR	2.3	-4.2	-1.6	1.1
346.90	-0.97	1.5	3.45	24MR	3.1	-4.1	-1.6	1.1
354.49	-0.92	1.5	13.16	3AP	1.7	-3.6	-1.6	1.1
1.99	-0.85	1.5	22.82	13AP	1.0	-2.9	-1.6	1.0
9.42	-0.78	1.5	32.44	23AP	0.5	-3.5	-1.6	1.0
16.74	-0.70	1.5	42.02	3MY	-0.2	-4.0	-1.7	1.0
23.98	-0.61	1.6	51.57	13MY	-1.1	-4.2	-1.7	0.9
31.12	-0.51	1.6	61.11	23MY	-1.8	-4.2	-1.8	0.9
38.15	-0.41	1.6	70.63	2JN	-0.9	-4.1	-1.8	0.8
45.09	-0.30	1.6	80.15	12JN	-0.1	-4.0	-1.9	0.7
51.92	-0.19	1.6	89.68	22JN	0.4	-3.9	-1.9	0.7
58.64	-0.07	1.6	99.22	2JL	0.8	-3.9	-2.0	0.6
65.26	0.06	1.6	108.80	12JL	1.3	-3.8	-2.1	0.5
71.77	0.18	1.6	118.40	22JL	2.5	-3.7	-2.2	0.5
78.15	0.32	1.6	128.05	1AU	2.4	-3.6	-2.2	0.5
84.42	0.45	1.6	137.74	11AU	0.3	-3.6	-2.3	0.5
90.55	0.60	1.5	147.49	21AU	-0.9	-3.5	-2.3	0.6
96.53	0.75	1.5	157.29	31AU	-1.2	-3.5	-2.4	0.6
102.35	0.91	1.4	167.16	10SE	-1.0	-3.5	-2.4	0.7
107.98	1.08	1.4	177.08	20SE	-0.6	-3.4	-2.4	0.7
113.41	1.26	1.3	187.05	30SE	-0.3	-3.4	-2.4	0.8
118.58	1.46	1.2	197.09	10OC	-0.2	-3.4	-2.4	0.8
123.46	1.67	1.1	207.17	20OC	-0.2	-3.4	-2.4	0.9
127.97	1.90	0.9	217.30	30OC	0.2	-3.4	-2.3	0.9
132.06	2.16	0.8	227.46	9NO	2.0	-3.4	-2.3	1.0
135.59	2.44	0.6	237.65	19NO	1.3	-3.4	-2.2	1.0
138.46	2.75	0.4	247.85	29NO	0.0	-3.4	-2.2	1.0
140.51	3.09	0.2	258.06	9DE	-0.2	-3.4	-2.1	1.0
141.54	3.46	-0.1	268.26	19DE	-0.2	-3.4	-2.0	1.0
141.39	3.83	-0.3	278.44	29DE	-0.4	-3.4	-1.9	1.0

197

♂ LONG	LAT	MAG	☉ LONG	16.00UT	☿	♀	♃	♄
139.93	4.18	-0.6	288.59	8JA	-0.7	-3.4	-1.9	1.0
137.23	4.45	-0.8	298.71	18JA	-1.2	-3.5	-1.8	1.0
133.61	4.58	-1.0	308.78	28JA	-1.2	-3.5	-1.7	1.0
129.68	4.54	-1.0	318.81	7FE	-0.8	-3.5	-1.7	1.1
126.17	4.33	-0.8	328.78	17FE	0.6	-3.4	-1.7	1.1
123.65	3.99	-0.6	338.69	27FE	2.8	-3.4	-1.6	1.1
122.40	3.60	-0.3	348.54	9MR	2.4	-3.4	-1.6	1.2
122.43	3.19	-0.1	358.33	19MR	1.3	-3.4	-1.6	1.2
123.62	2.80	0.1	8.07	29MR	0.8	-3.4	-1.5	1.2
125.77	2.44	0.3	17.76	8AP	0.3	-3.3	-1.5	1.2
128.74	2.11	0.5	27.40	18AP	-0.3	-3.3	-1.5	1.2
132.35	1.81	0.6	36.99	28AP	-1.1	-3.3	-1.5	1.2
136.50	1.54	0.8	46.57	8MY	-1.8	-3.3	-1.5	1.1
141.09	1.29	0.9	56.11	18MY	-0.9	-3.3	-1.6	1.1
146.05	1.06	1.0	65.63	28MY	-0.1	-3.4	-1.6	1.1
151.32	0.85	1.1	75.16	7JN	0.5	-3.4	-1.6	1.0
156.88	0.66	1.1	84.68	17JN	1.0	-3.4	-1.7	0.9
162.67	0.47	1.2	94.21	27JN	1.8	-3.5	-1.7	0.9
168.68	0.30	1.3	103.78	7JL	3.1	-3.5	-1.8	0.8
174.88	0.14	1.3	113.36	17JL	2.0	-3.6	-1.8	0.7
181.27	-0.01	1.3	122.98	27JL	0.2	-3.6	-1.9	0.7
187.83	-0.16	1.4	132.66	6AU	-1.0	-3.7	-1.9	0.6
194.56	-0.29	1.4	142.37	16AU	-1.3	-3.8	-2.0	0.6
201.43	-0.42	1.4	152.15	26AU	-1.0	-3.9	-2.1	0.7
208.45	-0.54	1.4	161.98	5SE	-0.5	-4.0	-2.1	0.7
215.60	-0.65	1.4	171.87	15SE	-0.2	-4.1	-2.2	0.8
222.87	-0.74	1.4	181.81	25SE	-0.1	-4.2	-2.3	0.8
230.27	-0.83	1.4	191.82	5OC	-0.0	-4.3	-2.3	0.9
237.76	-0.91	1.4	201.88	15OC	0.4	-4.4	-2.3	0.9
245.36	-0.98	1.4	211.98	25OC	2.3	-4.3	-2.4	1.0
253.04	-1.04	1.4	222.13	4NO	1.0	-3.9	-2.4	1.0
260.79	-1.08	1.4	232.30	14NO	-0.2	-3.2	-2.3	1.1
268.60	-1.11	1.4	242.50	24NO	-0.3	-3.5	-2.3	1.1
276.45	-1.13	1.4	252.70	4DE	-0.4	-4.1	-2.3	1.1
284.33	-1.14	1.4	262.91	14DE	-0.5	-4.3	-2.2	1.1
292.23	-1.13	1.4	273.10	24DE	-0.7	-4.4	-2.1	1.2

Right page (198 / 199)

♂ LONG	LAT	MAG	☉ LONG	16.00UT 198	☿	♀	♃	♄
300.12	-1.12	1.4	283.27	3JA	-1.0	-4.3	-2.1	1.2
308.00	-1.09	1.4	293.41	13JA	-1.1	-4.2	-2.0	1.2
315.86	-1.05	1.4	303.51	23JA	-0.7	-4.1	-1.9	1.2
323.67	-1.00	1.4	313.56	2FE	0.8	-4.0	-1.9	1.2
331.43	-0.94	1.4	323.56	12FE	3.1	-3.9	-1.8	1.2
339.14	-0.87	1.4	333.50	22FE	1.8	-3.8	-1.7	1.2
346.77	-0.79	1.4	343.39	4MR	0.9	-3.7	-1.7	1.3
354.33	-0.71	1.4	353.21	14MR	0.6	-3.6	-1.6	1.3
1.81	-0.62	1.4	2.98	24MR	0.2	-3.5	-1.6	1.4
9.20	-0.53	1.5	12.69	3AP	-0.3	-3.5	-1.5	1.4
16.51	-0.44	1.5	22.35	13AP	-1.1	-3.4	-1.5	1.4
23.73	-0.34	1.6	31.97	23AP	-1.8	-3.4	-1.5	1.4
30.86	-0.23	1.6	41.56	3MY	-0.9	-3.4	-1.5	1.4
37.90	-0.13	1.7	51.11	13MY	0.0	-3.3	-1.4	1.4
44.86	-0.03	1.7	60.65	23MY	0.7	-3.3	-1.4	1.3
51.73	0.08	1.8	70.17	2JN	1.4	-3.3	-1.4	1.3
58.52	0.18	1.8	79.69	12JN	2.3	-3.3	-1.5	1.2
65.23	0.29	1.8	89.22	22JN	3.3	-3.3	-1.5	1.2
71.86	0.40	1.8	98.77	2JL	1.6	-3.3	-1.5	1.1
78.41	0.50	1.8	108.33	12JL	0.1	-3.4	-1.5	1.0
84.88	0.61	1.8	117.94	22JL	-1.0	-3.4	-1.6	1.0
91.28	0.72	1.8	127.58	1AU	-1.4	-3.4	-1.6	0.9
97.60	0.83	1.8	137.27	11AU	-1.0	-3.4	-1.7	0.8
103.84	0.94	1.8	147.02	21AU	-0.4	-3.5	-1.7	0.8
109.99	1.05	1.8	156.82	31AU	-0.1	-3.5	-1.8	0.8
116.05	1.16	1.8	166.67	10SE	0.0	-3.5	-1.8	0.9
122.02	1.28	1.7	176.59	20SE	0.2	-3.4	-1.9	0.9
127.87	1.40	1.7	186.57	30SE	0.6	-3.4	-2.0	0.9
133.60	1.52	1.6	196.60	10OC	2.6	-3.4	-2.0	1.0
139.19	1.65	1.5	206.68	20OC	0.9	-3.4	-2.1	1.0
144.61	1.79	1.4	216.80	30OC	-0.3	-3.3	-2.1	1.1
149.84	1.93	1.3	226.96	9NO	-0.5	-3.3	-2.2	1.1
154.82	2.08	1.2	237.15	19NO	-0.5	-3.3	-2.2	1.2
159.51	2.25	1.0	247.35	29NO	-0.6	-3.3	-2.2	1.2
163.85	2.42	0.9	257.55	9DE	-0.8	-3.3	-2.2	1.3
167.75	2.61	0.7	267.76	19DE	-0.9	-3.4	-2.2	1.3
171.10	2.81	0.5	277.94	29DE	-0.9	-3.4	-2.2	1.3

199

♂ LONG	LAT	MAG	☉ LONG	16.00UT	☿	♀	♃	♄
173.78	3.02	0.3	288.10	8JA	-0.6	-3.4	-2.1	1.3
175.61	3.24	-0.0	298.22	18JA	0.9	-3.4	-2.1	1.4
176.41	3.45	-0.3	308.30	28JA	3.0	-3.5	-2.0	1.4
176.03	3.64	-0.6	318.32	7FE	1.3	-3.5	-2.0	1.3
174.36	3.75	-0.8	328.30	17FE	0.7	-3.6	-1.9	1.3
171.51	3.76	-1.1	338.21	27FE	0.4	-3.6	-1.8	1.3
167.89	3.62	-1.3	348.06	9MR	0.1	-3.7	-1.7	1.3
164.12	3.33	-1.2	357.86	19MR	-0.4	-3.7	-1.7	1.3
160.93	2.92	-1.0	7.60	29MR	-1.1	-3.8	-1.6	1.2
158.84	2.45	-0.8	17.29	8AP	-1.7	-3.9	-1.6	1.2
158.05	1.98	-0.6	26.94	18AP	-0.9	-4.0	-1.5	1.2
158.57	1.53	-0.4	36.53	28AP	0.1	-4.1	-1.5	1.2
160.23	1.12	-0.2	46.10	8MY	0.9	-4.2	-1.4	1.2
162.87	0.76	-0.0	55.65	18MY	1.8	-4.2	-1.4	1.1
166.32	0.44	0.1	65.17	28MY	3.1	-4.1	-1.4	1.1
170.42	0.15	0.3	74.69	7JN	2.7	-3.8	-1.4	1.0
175.09	-0.10	0.4	84.22	17JN	1.3	-3.2	-1.3	1.0
180.21	-0.32	0.5	93.75	27JN	0.1	-3.1	-1.3	0.9
185.72	-0.51	0.6	103.31	7JL	-1.1	-3.8	-1.3	0.9
191.57	-0.69	0.7	112.90	17JL	-1.6	-4.1	-1.4	0.8
197.71	-0.84	0.8	122.52	27JL	-1.0	-4.2	-1.4	0.8
204.11	-0.98	0.8	132.18	6AU	-0.4	-4.2	-1.4	0.7
210.75	-1.09	0.9	141.90	16AU	-0.0	-4.1	-1.4	0.7
217.58	-1.19	0.9	151.67	26AU	0.2	-4.1	-1.5	0.7
224.60	-1.27	1.0	161.50	5SE	0.4	-4.0	-1.5	0.7
231.78	-1.33	1.0	171.39	15SE	0.9	-3.9	-1.6	0.7
239.10	-1.38	1.1	181.33	25SE	3.0	-3.8	-1.6	0.8
246.55	-1.41	1.1	191.33	5OC	0.7	-3.7	-1.7	0.9
254.10	-1.42	1.1	201.39	15OC	-0.5	-3.7	-1.7	0.9
261.74	-1.42	1.2	211.49	25OC	-0.6	-3.6	-1.8	1.0
269.45	-1.40	1.2	221.63	4NO	-0.6	-3.6	-1.9	1.1
277.22	-1.37	1.2	231.81	14NO	-0.7	-3.5	-1.9	1.1
285.01	-1.32	1.3	242.00	24NO	-0.8	-3.5	-2.0	1.2
292.82	-1.27	1.3	252.21	4DE	-0.7	-3.4	-2.0	1.2
300.63	-1.20	1.3	262.41	14DE	-0.8	-3.4	-2.1	1.2
308.43	-1.12	1.4	272.61	24DE	-0.5	-3.4	-2.1	1.2

♂ LONG	♂ LAT	♂ MAG	☉ LONG	16.00UT	☿	♀	♃	♄
				200	MAGNITUDES			
316.19	-1.03	1.4	282.78	3JA	1.1	-3.4	-2.1	1.2
323.92	-0.94	1.4	292.92	13JA	2.5	-3.3	-2.1	1.2
331.59	-0.84	1.5	303.02	23JA	0.9	-3.3	-2.1	1.1
339.20	-0.74	1.5	313.08	2FE	0.4	-3.3	-2.1	1.1
346.75	-0.64	1.5	323.08	12FE	0.2	-3.3	-2.0	1.1
354.22	-0.53	1.6	333.02	22FE	-0.0	-3.3	-2.0	1.0
1.61	-0.42	1.6	342.91	3MR	-0.4	-3.4	-1.9	1.0
8.91	-0.31	1.6	352.74	13MR	-1.1	-3.4	-1.8	1.0
16.14	-0.20	1.7	2.50	23MR	-1.7	-3.4	-1.8	1.0
23.28	-0.10	1.7	12.22	2AP	-0.9	-3.4	-1.7	1.0
30.34	0.01	1.7	21.88	12AP	0.2	-3.5	-1.6	1.0
37.33	0.11	1.7	31.50	22AP	1.2	-3.5	-1.6	1.0
44.23	0.21	1.7	41.09	2MY	2.4	-3.4	-1.5	0.9
51.07	0.30	1.7	50.65	12MY	3.6	-3.4	-1.5	0.9
57.83	0.40	1.8	60.18	22MY	2.1	-3.4	-1.4	0.9
64.53	0.49	1.8	69.70	1JN	1.0	-3.4	-1.4	0.9
71.16	0.58	1.9	79.22	11JN	-0.0	-3.3	-1.3	0.8
77.74	0.66	1.9	88.75	21JN	-1.1	-3.3	-1.3	0.8
84.26	0.75	1.9	98.30	1JL	-1.7	-3.3	-1.3	0.7
90.74	0.83	1.9	107.86	11JL	-0.9	-3.3	-1.3	0.7
97.16	0.90	2.0	117.46	21JL	-0.3	-3.3	-1.3	0.6
103.55	0.98	2.0	127.11	31JL	0.1	-3.4	-1.3	0.6
109.88	1.05	2.0	136.80	10AU	0.3	-3.4	-1.3	0.5
116.18	1.12	2.0	146.54	20AU	0.6	-3.4	-1.3	0.5
122.44	1.19	2.0	156.34	30AU	1.2	-3.4	-1.3	0.4
128.66	1.26	2.0	166.19	9SE	3.3	-3.5	-1.3	0.4
134.83	1.32	1.9	176.11	19SE	0.6	-3.5	-1.4	0.4
140.96	1.38	1.9	186.08	29SE	-0.6	-3.6	-1.4	0.5
147.03	1.45	1.8	196.11	9OC	-0.8	-3.7	-1.4	0.5
153.05	1.51	1.8	206.19	19OC	-0.8	-3.8	-1.5	0.6
159.01	1.57	1.7	216.31	29OC	-0.8	-3.9	-1.6	0.7
164.88	1.63	1.6	226.47	8NO	-0.6	-4.0	-1.6	0.7
170.67	1.68	1.6	236.65	18NO	-0.6	-4.1	-1.7	0.8
176.34	1.74	1.4	246.86	28NO	-0.6	-4.2	-1.7	0.8
181.89	1.79	1.3	257.06	8DE	-0.3	-4.3	-1.8	0.9
187.28	1.84	1.2	267.26	18DE	1.3	-4.4	-1.9	0.9
192.48	1.89	1.0	277.45	28DE	2.1	-4.4	-1.9	0.9
				201				
197.45	1.93	0.9	287.61	7JA	0.6	-4.3	-2.0	0.9
202.13	1.97	0.7	297.73	17JA	0.3	-3.9	-2.0	0.9
206.44	1.99	0.5	307.81	27JA	0.1	-3.3	-2.0	0.9
210.31	1.99	0.3	317.84	6FE	-0.1	-3.6	-2.0	0.9
213.62	1.97	0.0	327.81	16FE	-0.5	-4.0	-2.0	0.8
216.23	1.93	-0.3	337.73	26FE	-1.1	-4.3	-2.0	0.8
217.96	1.83	-0.6	347.59	8MR	-1.6	-4.3	-2.0	0.7
218.63	1.67	-0.9	357.39	18MR	-0.9	-4.2	-1.9	0.7
218.09	1.43	-1.2	7.13	28MR	0.3	-4.1	-1.9	0.7
216.30	1.10	-1.5	16.82	7AP	1.6	-4.0	-1.8	0.7
213.46	0.66	-1.8	26.47	17AP	3.2	-3.9	-1.7	0.7
210.04	0.16	-1.9	36.07	27AP	2.8	-3.8	-1.7	0.7
206.78	-0.37	-1.8	45.64	7MY	1.6	-3.7	-1.6	0.7
204.33	-0.86	-1.6	55.18	17MY	0.8	-3.6	-1.6	0.7
203.15	-1.29	-1.4	64.71	27MY	-0.1	-3.6	-1.5	0.7
203.35	-1.63	-1.2	74.23	6JN	-1.1	-3.5	-1.4	0.7
204.84	-1.89	-1.0	83.76	16JN	-1.8	-3.5	-1.4	0.6
207.47	-2.09	-0.8	93.29	26JN	-0.9	-3.4	-1.3	0.6
211.05	-2.23	-0.6	102.85	6JL	-0.2	-3.4	-1.3	0.6
215.40	-2.33	-0.4	112.43	16JL	0.2	-3.4	-1.3	0.5
220.40	-2.38	-0.3	122.05	26JL	0.5	-3.4	-1.2	0.5
225.91	-2.41	-0.1	131.72	5AU	0.8	-3.3	-1.2	0.4
231.86	-2.40	0.0	141.43	15AU	1.6	-3.3	-1.2	0.3
238.16	-2.37	0.1	151.20	25AU	3.1	-3.3	-1.2	0.3
244.75	-2.32	0.2	161.02	4SE	0.5	-3.4	-1.2	0.2
251.57	-2.25	0.4	170.91	14SE	-0.7	-3.4	-1.2	0.2
258.60	-2.15	0.5	180.85	24SE	-0.9	-3.4	-1.2	0.1
265.78	-2.05	0.6	190.85	4OC	-0.9	-3.4	-1.3	0.2
273.07	-1.93	0.6	200.90	14OC	-0.7	-3.4	-1.3	0.2
280.47	-1.79	0.7	211.00	24OC	-0.5	-3.5	-1.3	0.3
287.92	-1.65	0.8	221.14	3NO	-0.5	-3.5	-1.4	0.4
295.42	-1.51	0.9	231.31	13NO	-0.4	-3.5	-1.4	0.4
302.93	-1.36	1.0	241.51	23NO	-0.2	-3.5	-1.5	0.5
310.45	-1.20	1.1	251.71	3DE	1.5	-3.4	-1.5	0.5
317.95	-1.05	1.2	261.92	13DE	1.8	-3.4	-1.6	0.6
325.42	-0.90	1.2	272.11	23DE	0.4	-3.4	-1.7	0.6

♂ LONG	♂ LAT	♂ MAG	☉ LONG	16.00UT	☿	♀	♃	♄
				202	MAGNITUDES			
332.84	-0.75	1.3	282.28	2JA	0.1	-3.4	-1.7	0.7
340.22	-0.61	1.4	292.43	12JA	-0.0	-3.4	-1.8	0.7
347.55	-0.46	1.5	302.53	22JA	-0.2	-3.3	-1.9	0.7
354.80	-0.33	1.5	312.59	1FE	-0.6	-3.3	-1.9	0.7
2.00	-0.20	1.6	322.59	11FE	-1.2	-3.3	-2.0	0.7
9.12	-0.08	1.6	332.54	21FE	-1.4	-3.3	-2.0	0.6
16.17	0.04	1.7	342.43	3MR	-0.9	-3.4	-2.0	0.6
23.15	0.15	1.7	352.26	13MR	0.4	-3.4	-2.0	0.6
30.07	0.26	1.8	2.03	23MR	2.0	-3.4	-2.0	0.5
36.91	0.36	1.8	11.75	2AP	3.7	-3.4	-2.0	0.5
43.70	0.45	1.8	21.42	12AP	2.1	-3.5	-1.9	0.5
50.42	0.54	1.9	31.04	22AP	1.2	-3.5	-1.9	0.5
57.09	0.62	1.9	40.63	2MY	0.6	-3.5	-1.8	0.5
63.71	0.69	1.9	50.19	12MY	-0.2	-3.6	-1.8	0.5
70.28	0.76	1.9	59.72	22MY	-1.1	-3.7	-1.7	0.5
76.80	0.83	1.9	69.24	1JN	-1.8	-3.7	-1.6	0.5
83.29	0.89	1.9	78.77	11JN	-0.9	-3.8	-1.6	0.5
89.75	0.94	1.9	88.29	21JN	-0.2	-3.9	-1.5	0.5
96.18	0.99	1.9	97.84	1JL	0.3	-4.0	-1.5	0.5
102.58	1.04	2.0	107.41	11JL	0.7	-4.1	-1.4	0.4
108.97	1.08	2.0	117.00	21JL	1.1	-4.2	-1.4	0.4
115.33	1.12	2.0	126.64	31JL	2.1	-4.2	-1.3	0.3
121.69	1.15	2.0	136.33	10AU	2.7	-4.1	-1.3	0.3
128.04	1.18	2.0	146.07	20AU	0.4	-3.8	-1.3	0.2
134.39	1.21	2.0	155.86	30AU	-0.8	-3.3	-1.2	0.2
140.73	1.23	2.0	165.72	9SE	-1.1	-3.6	-1.2	0.1
147.06	1.25	2.0	175.63	19SE	-1.0	-4.1	-1.2	0.0
153.40	1.27	2.0	185.60	29SE	-0.6	-4.3	-1.2	-0.0
159.73	1.28	2.0	195.63	9OC	-0.4	-4.3	-1.2	-0.1
166.06	1.28	1.9	205.70	19OC	-0.3	-4.3	-1.2	-0.0
172.38	1.28	1.9	215.82	29OC	-0.3	-4.2	-1.3	0.0
178.70	1.28	1.8	225.98	8NO	0.0	-4.1	-1.3	0.1
185.01	1.26	1.8	236.16	18NO	1.8	-4.0	-1.3	0.2
191.30	1.24	1.7	246.36	28NO	1.5	-3.9	-1.4	0.2
197.58	1.21	1.6	256.57	8DE	0.2	-3.8	-1.4	0.3
203.83	1.18	1.5	266.77	18DE	-0.1	-3.8	-1.5	0.4
210.04	1.13	1.4	276.96	28DE	-0.2	-3.7	-1.5	0.4
				203				
216.21	1.07	1.3	287.12	7JA	-0.3	-3.6	-1.6	0.5
222.33	0.99	1.1	297.24	17JA	-0.6	-3.5	-1.7	0.5
228.38	0.89	1.0	307.33	27JA	-1.2	-3.5	-1.7	0.5
234.35	0.77	0.9	317.36	6FE	-1.3	-3.4	-1.8	0.5
240.20	0.63	0.7	327.34	16FE	-0.8	-3.4	-1.9	0.5
245.93	0.46	0.5	337.26	26FE	0.5	-3.4	-1.9	0.5
251.48	0.25	0.3	347.12	8MR	2.5	-3.3	-2.0	0.5
256.82	-0.00	0.1	356.92	18MR	2.8	-3.3	-2.0	0.4
261.88	-0.30	-0.1	6.67	28MR	1.6	-3.3	-2.0	0.4
266.59	-0.66	-0.4	16.36	7AP	0.9	-3.3	-2.1	0.4
270.85	-1.09	-0.6	26.01	17AP	0.4	-3.3	-2.1	0.3
274.53	-1.59	-0.9	35.61	27AP	-0.2	-3.3	-2.0	0.3
277.47	-2.18	-1.2	45.18	7MY	-1.1	-3.3	-2.0	0.4
279.48	-2.86	-1.5	54.73	17MY	-1.8	-3.4	-2.0	0.4
280.39	-3.62	-1.8	64.26	27MY	-0.9	-3.4	-1.9	0.4
280.05	-4.41	-2.1	73.78	6JN	-0.1	-3.4	-1.9	0.4
278.53	-5.16	-2.4	83.30	16JN	0.4	-3.5	-1.8	0.4
276.11	-5.76	-2.6	92.83	26JN	0.9	-3.5	-1.7	0.4
273.44	-6.12	-2.5	102.39	6JL	1.5	-3.5	-1.7	0.4
271.25	-6.18	-2.4	111.96	16JL	2.7	-3.4	-1.6	0.4
270.09	-5.99	-2.1	121.58	26JL	2.3	-3.4	-1.6	0.3
270.26	-5.63	-1.9	131.25	5AU	0.3	-3.4	-1.5	0.3
271.75	-5.17	-1.6	140.95	15AU	-0.9	-3.3	-1.5	0.2
274.39	-4.68	-1.3	150.72	25AU	-1.2	-3.3	-1.4	0.2
278.01	-4.18	-1.1	160.54	4SE	-1.0	-3.3	-1.4	0.1
282.40	-3.69	-0.8	170.42	14SE	-0.6	-3.3	-1.3	0.0
287.38	-3.23	-0.6	180.36	24SE	-0.3	-3.3	-1.3	-0.0
292.85	-2.80	-0.4	190.36	4OC	-0.2	-3.3	-1.3	-0.1
298.66	-2.41	-0.2	200.41	14OC	-0.1	-3.4	-1.3	-0.2
304.75	-2.04	0.0	210.51	24OC	0.2	-3.4	-1.3	-0.2
311.05	-1.70	0.2	220.65	3NO	2.0	-3.4	-1.3	-0.2
317.49	-1.39	0.4	230.81	13NO	1.2	-3.5	-1.3	-0.1
324.04	-1.11	0.5	241.01	23NO	-0.0	-3.5	-1.3	-0.0
330.68	-0.86	0.7	251.22	3DE	-0.2	-3.6	-1.3	0.0
337.36	-0.62	0.8	261.42	13DE	-0.3	-3.6	-1.4	0.1
344.06	-0.41	1.0	271.62	23DE	-0.4	-3.7	-1.4	0.2

♂ LONG	♂ LAT	♂ MAG	⊙ LONG	16.00UT 204	☿	♀	♃	♄
					MAGNITUDES			
350.78	-0.22	1.1	281.80	2JA	-0.7	-3.8	-1.5	0.2
357.50	-0.05	1.2	291.94	12JA	-1.1	-3.9	-1.5	0.3
4.20	0.10	1.3	302.05	22JA	-1.2	-3.9	-1.6	0.3
10.89	0.24	1.4	312.11	1FE	-0.8	-4.0	-1.6	0.4
17.55	0.36	1.5	322.11	11FE	0.6	-4.1	-1.7	0.4
24.17	0.48	1.6	332.06	21FE	2.9	-4.2	-1.8	0.4
30.77	0.58	1.7	341.96	2MR	2.1	-4.3	-1.8	0.4
37.33	0.67	1.7	351.79	12MR	1.1	-4.2	-1.9	0.4
43.86	0.75	1.8	1.56	22MR	0.7	-4.1	-2.0	0.4
50.36	0.82	1.8	11.28	1AP	0.3	-3.6	-2.0	0.3
56.82	0.88	1.9	20.95	11AP	-0.3	-3.0	-2.1	0.3
63.26	0.93	1.9	30.58	21AP	-1.1	-3.6	-2.1	0.2
69.68	0.98	2.0	40.17	1MY	-1.8	-4.0	-2.1	0.2
76.07	1.02	2.0	49.72	11MY	-0.9	-4.2	-2.2	0.2
82.44	1.06	2.0	59.26	21MY	-0.1	-4.2	-2.1	0.3
88.80	1.09	2.0	68.79	31MY	0.6	-4.1	-2.1	0.3
95.15	1.11	2.0	78.30	10JN	1.2	-4.0	-2.1	0.3
101.50	1.13	2.0	87.83	20JN	2.0	-3.9	-2.1	0.3
107.84	1.15	2.0	97.37	30JN	3.2	-3.8	-2.0	0.3
114.19	1.15	2.0	106.94	10JL	1.9	-3.8	-1.9	0.3
120.55	1.16	2.0	116.54	20JL	0.2	-3.7	-1.9	0.3
126.92	1.16	2.0	126.18	30JL	-1.0	-3.6	-1.8	0.3
133.31	1.15	2.0	135.86	9AU	-1.4	-3.6	-1.8	0.3
139.72	1.15	2.0	145.60	19AU	-1.0	-3.5	-1.7	0.2
146.15	1.13	2.0	155.39	29AU	-0.5	-3.5	-1.6	0.2
152.60	1.11	2.0	165.24	8SE	-0.2	-3.5	-1.6	0.1
159.09	1.09	2.0	175.15	18SE	-0.0	-3.4	-1.5	0.0
165.61	1.06	2.0	185.11	28SE	0.1	-3.4	-1.5	-0.0
172.16	1.02	2.0	195.13	8OC	0.5	-3.4	-1.4	-0.1
178.75	0.98	1.9	205.21	18OC	2.3	-3.4	-1.4	-0.2
185.38	0.94	1.9	215.33	28OC	1.0	-3.4	-1.4	-0.2
192.04	0.88	1.9	225.48	7NO	-0.2	-3.4	-1.4	-0.3
198.74	0.82	1.8	235.67	17NO	-0.4	-3.4	-1.4	-0.2
205.48	0.75	1.8	245.86	27NO	-0.4	-3.4	-1.4	-0.2
212.25	0.67	1.7	256.07	7DE	-0.5	-3.4	-1.4	-0.1
219.06	0.58	1.6	266.28	17DE	-0.8	-3.4	-1.4	-0.0
225.90	0.48	1.6	276.46	27DE	-1.0	-3.4	-1.4	0.0
				205				
232.78	0.37	1.5	286.63	6JA	-1.0	-3.4	-1.4	0.1
239.69	0.24	1.4	296.76	16JA	-0.7	-3.5	-1.5	0.2
246.62	0.10	1.3	306.84	26JA	0.8	-3.5	-1.5	0.2
253.58	-0.05	1.2	316.87	5FE	3.1	-3.5	-1.6	0.3
260.56	-0.22	1.1	326.86	15FE	1.6	-3.4	-1.6	0.3
267.55	-0.40	1.0	336.77	25FE	0.8	-3.4	-1.7	0.3
274.55	-0.60	0.9	346.64	7MR	0.5	-3.4	-1.7	0.3
281.54	-0.82	0.8	356.45	17MR	0.2	-3.4	-1.8	0.3
288.52	-1.05	0.6	6.19	27MR	-0.3	-3.4	-1.9	0.3
295.45	-1.30	0.5	15.89	6AP	-1.1	-3.3	-2.0	0.3
302.33	-1.56	0.4	25.54	16AP	-1.8	-3.3	-2.0	0.3
309.12	-1.84	0.2	35.14	26AP	-0.9	-3.3	-2.1	0.2
315.80	-2.14	0.1	44.72	6MY	0.0	-3.3	-2.2	0.2
322.30	-2.44	-0.1	54.27	16MY	0.8	-3.3	-2.2	0.2
328.60	-2.75	-0.2	63.79	26MY	1.5	-3.4	-2.2	0.2
334.61	-3.07	-0.4	73.32	5JN	2.6	-3.4	-2.3	0.2
340.25	-3.40	-0.6	82.84	15JN	3.1	-3.4	-2.3	0.3
345.42	-3.73	-0.7	92.37	25JN	1.5	-3.5	-2.3	0.3
349.99	-4.06	-0.9	101.92	5JL	0.1	-3.5	-2.2	0.3
353.78	-4.38	-1.2	111.50	15JL	-1.0	-3.6	-2.2	0.3
356.61	-4.67	-1.4	121.12	25JL	-1.5	-3.6	-2.2	0.3
358.25	-4.91	-1.6	130.78	4AU	-1.0	-3.7	-2.1	0.3
358.52	-5.05	-1.9	140.49	14AU	-0.4	-3.8	-2.1	0.3
357.35	-5.05	-2.1	150.25	24AU	-0.1	-3.9	-2.0	0.2
354.91	-4.85	-2.2	160.07	3SE	0.1	-4.0	-1.9	0.2
351.77	-4.42	-2.3	169.94	13SE	0.2	-4.1	-1.9	0.2
348.70	-3.81	-2.2	179.88	23SE	0.7	-4.2	-1.8	0.1
346.42	-3.10	-1.9	189.87	3OC	2.7	-4.3	-1.7	0.0
345.39	-2.39	-1.5	199.92	13OC	0.9	-4.4	-1.7	-0.0
345.69	-1.74	-1.2	210.01	23OC	-0.4	-4.3	-1.6	-0.1
347.21	-1.18	-0.9	220.15	2NO	-0.6	-3.9	-1.6	-0.2
349.75	-0.71	-0.6	230.32	12NO	-0.5	-3.1	-1.5	-0.3
353.11	-0.33	-0.3	240.51	22NO	-0.6	-3.5	-1.5	-0.3
357.10	-0.01	-0.0	250.72	2DE	-0.8	-4.1	-1.5	-0.3
1.60	0.25	0.2	260.92	12DE	-0.8	-4.3	-1.5	-0.2
6.47	0.46	0.4	271.12	22DE	-0.9	-4.4	-1.5	-0.1

♂ LONG	♂ LAT	♂ MAG	⊙ LONG	16.00UT 206	☿	♀	♃	♄
					MAGNITUDES			
11.63	0.64	0.6	281.30	1JA	-0.6	-4.3	-1.5	-0.0
17.02	0.78	0.8	291.45	11JA	0.9	-4.2	-1.5	0.0
22.58	0.90	1.0	301.55	21JA	2.9	-4.1	-1.5	0.1
28.28	0.99	1.1	311.62	31JA	1.2	-4.0	-1.5	0.1
34.08	1.07	1.3	321.63	10FE	0.6	-3.9	-1.5	0.2
39.96	1.13	1.4	331.58	20FE	0.3	-3.8	-1.6	0.2
45.90	1.18	1.5	341.48	2MR	0.1	-3.7	-1.6	0.3
51.89	1.22	1.6	351.31	12MR	-0.4	-3.6	-1.7	0.3
57.91	1.25	1.7	1.09	22MR	-1.1	-3.5	-1.7	0.3
63.97	1.27	1.8	10.82	1AP	-1.7	-3.5	-1.8	0.3
70.05	1.29	1.8	20.49	11AP	-0.9	-3.4	-1.9	0.3
76.16	1.29	1.9	30.12	21AP	0.1	-3.4	-1.9	0.3
82.28	1.30	1.9	39.71	1MY	1.0	-3.4	-2.0	0.3
88.43	1.29	2.0	49.27	11MY	2.0	-3.3	-2.1	0.2
94.59	1.28	2.0	58.81	21MY	3.4	-3.3	-2.1	0.2
100.78	1.27	2.0	68.33	31MY	2.5	-3.3	-2.2	0.2
107.00	1.25	2.0	77.85	10JN	1.2	-3.3	-2.3	0.2
113.24	1.23	2.0	87.38	20JN	0.1	-3.3	-2.3	0.3
119.51	1.20	2.0	96.92	30JN	-1.0	-3.3	-2.4	0.3
125.82	1.17	2.0	106.48	10JL	-1.6	-3.4	-2.4	0.3
132.16	1.14	2.0	116.08	20JL	-1.0	-3.4	-2.4	0.3
138.55	1.10	2.0	125.71	30JL	-0.4	-3.4	-2.4	0.3
144.98	1.06	2.0	135.39	9AU	0.0	-3.4	-2.4	0.3
151.46	1.01	1.9	145.13	19AU	0.2	-3.5	-2.3	0.3
157.99	0.96	1.9	154.92	29AU	0.4	-3.5	-2.3	0.3
164.57	0.90	1.9	164.76	8SE	1.0	-3.5	-2.2	0.3
171.21	0.84	1.9	174.66	18SE	3.1	-3.4	-2.2	0.2
177.91	0.77	1.9	184.63	28SE	0.7	-3.4	-2.1	0.2
184.67	0.70	1.8	194.64	8OC	-0.5	-3.4	-2.0	0.1
191.50	0.63	1.8	204.71	18OC	-0.7	-3.4	-2.0	0.1
198.39	0.54	1.8	214.83	28OC	-0.7	-3.3	-1.9	-0.0
205.35	0.46	1.8	224.98	7NO	-0.7	-3.3	-1.8	-0.1
212.37	0.36	1.8	235.16	17NO	-0.7	-3.3	-1.8	-0.2
219.46	0.26	1.7	245.36	27NO	-0.7	-3.3	-1.7	-0.2
226.61	0.16	1.7	255.57	7DE	-0.7	-3.3	-1.7	-0.3
233.82	0.04	1.7	265.77	17DE	-0.4	-3.4	-1.6	-0.2
241.10	-0.07	1.6	275.96	27DE	1.1	-3.4	-1.6	-0.2
				207				
248.43	-0.20	1.6	286.12	6JA	2.4	-3.4	-1.6	-0.1
255.82	-0.33	1.5	296.26	16JA	0.8	-3.4	-1.6	-0.0
263.25	-0.46	1.5	306.35	26JA	0.4	-3.5	-1.5	0.1
270.73	-0.60	1.4	316.38	5FE	0.2	-3.5	-1.5	0.1
278.25	-0.74	1.3	326.37	15FE	-0.0	-3.6	-1.6	0.2
285.79	-0.88	1.3	336.30	25FE	-0.4	-3.6	-1.6	0.2
293.35	-1.02	1.2	346.16	7MR	-1.1	-3.7	-1.6	0.3
300.93	-1.16	1.2	355.97	17MR	-1.6	-3.7	-1.6	0.3
308.49	-1.30	1.1	5.73	27MR	-0.9	-3.8	-1.6	0.3
316.03	-1.43	1.0	15.42	6AP	0.2	-3.9	-1.7	0.3
323.55	-1.55	1.0	25.08	16AP	1.3	-4.0	-1.7	0.3
331.00	-1.66	0.9	34.69	26AP	2.7	-4.1	-1.8	0.3
338.39	-1.76	0.8	44.26	6MY	3.4	-4.2	-1.8	0.3
345.69	-1.85	0.8	53.81	16MY	1.9	-4.2	-1.9	0.3
352.87	-1.92	0.7	63.34	26MY	0.9	-4.1	-2.0	0.3
359.92	-1.98	0.6	72.86	5JN	-0.0	-3.8	-2.0	0.3
6.79	-2.02	0.5	82.38	15JN	-1.1	-3.2	-2.1	0.2
13.47	-2.04	0.5	91.91	25JN	-1.7	-3.1	-2.2	0.3
19.91	-2.04	0.4	101.46	5JL	-1.0	-3.8	-2.3	0.3
26.08	-2.02	0.3	111.04	15JL	-0.3	-4.1	-2.3	0.4
31.90	-1.98	0.2	120.65	25JL	0.1	-4.2	-2.4	0.4
37.32	-1.91	0.0	130.31	4AU	0.4	-4.2	-2.4	0.4
42.25	-1.82	-0.1	140.02	14AU	0.7	-4.1	-2.4	0.4
46.57	-1.69	-0.3	149.77	24AU	1.4	-4.1	-2.5	0.4
50.15	-1.53	-0.4	159.59	3SE	3.2	-4.0	-2.4	0.4
52.81	-1.32	-0.6	169.47	13SE	0.6	-3.9	-2.4	0.4
54.34	-1.05	-0.8	179.40	23SE	-0.6	-3.8	-2.4	0.4
54.56	-0.72	-1.0	189.39	3OC	-0.9	-3.7	-2.3	0.3
53.32	-0.32	-1.3	199.43	13OC	-0.8	-3.7	-2.3	0.3
50.71	0.13	-1.5	209.52	23OC	-0.8	-3.6	-2.2	0.2
47.13	0.60	-1.6	219.66	2NO	-0.6	-3.6	-2.2	0.2
43.29	1.04	-1.5	229.83	12NO	-0.5	-3.5	-2.1	0.1
40.00	1.40	-1.3	240.02	22NO	-0.5	-3.5	-2.0	0.0
37.83	1.67	-1.0	250.22	2DE	-0.3	-3.4	-1.9	-0.1
37.01	1.84	-0.7	260.43	12DE	1.3	-3.4	-1.9	-0.1
37.50	1.95	-0.3	270.62	22DE	2.1	-3.4	-1.8	-0.2

♂ LONG	♂ LAT	♂ MAG	☉ LONG	16.00UT 208	☿	♀	♃	♄
39.10	2.00	-0.1	280.80	1JA	0.5	-3.4	-1.8	-0.1
41.62	2.02	0.2	290.95	11JA	0.2	-3.3	-1.7	-0.1
44.86	2.01	0.4	301.06	21JA	0.0	-3.3	-1.7	0.0
48.68	1.99	0.7	311.13	31JA	-0.1	-3.3	-1.6	0.1
52.94	1.96	0.8	321.14	10FE	-0.5	-3.3	-1.6	0.1
57.55	1.92	1.0	331.10	20FE	-1.1	-3.3	-1.6	0.2
62.44	1.87	1.2	340.99	1MR	-1.5	-3.4	-1.6	0.2
67.55	1.83	1.3	350.84	11MR	-0.9	-3.4	-1.6	0.3
72.84	1.77	1.4	0.62	21MR	0.3	-3.4	-1.6	0.3
78.28	1.72	1.5	10.34	31MR	1.7	-3.4	-1.6	0.4
83.84	1.67	1.6	20.02	10AP	3.5	-3.5	-1.6	0.4
89.51	1.61	1.7	29.65	20AP	2.6	-3.5	-1.6	0.4
95.28	1.55	1.8	39.24	30AP	1.5	-3.4	-1.6	0.4
101.12	1.49	1.8	48.80	10MY	0.7	-3.4	-1.7	0.4
107.05	1.43	1.9	58.34	20MY	-0.1	-3.4	-1.7	0.4
113.04	1.36	1.9	67.86	30MY	-1.1	-3.4	-1.8	0.4
119.11	1.30	1.9	77.39	9JN	-1.8	-3.3	-1.8	0.4
125.24	1.23	1.9	86.91	19JN	-1.0	-3.3	-1.9	0.4
131.44	1.16	2.0	96.45	29JN	-0.2	-3.3	-2.0	0.4
137.71	1.08	2.0	106.02	9JL	0.2	-3.3	-2.0	0.4
144.05	1.01	2.0	115.61	19JL	0.5	-3.3	-2.1	0.5
150.46	0.93	1.9	125.24	29JL	0.9	-3.4	-2.2	0.5
156.94	0.85	1.9	134.92	8AU	1.8	-3.4	-2.2	0.5
163.50	0.77	1.9	144.65	18AU	3.0	-3.4	-2.3	0.5
170.14	0.68	1.9	154.43	28AU	0.5	-3.4	-2.3	0.6
176.87	0.60	1.8	164.28	7SE	-0.8	-3.5	-2.4	0.6
183.67	0.51	1.8	174.18	17SE	-1.0	-3.5	-2.4	0.6
190.55	0.41	1.7	184.14	27SE	-1.0	-3.6	-2.4	0.5
197.53	0.31	1.7	194.16	7OC	-0.7	-3.7	-2.4	0.5
204.58	0.21	1.6	204.23	17OC	-0.5	-3.8	-2.4	0.5
211.72	0.11	1.6	214.34	27OC	-0.4	-3.9	-2.4	0.4
218.94	0.01	1.6	224.49	6NO	-0.4	-4.0	-2.3	0.4
226.24	-0.10	1.6	234.67	16NO	-0.1	-4.1	-2.3	0.3
233.62	-0.21	1.6	244.87	26NO	1.5	-4.2	-2.2	0.2
241.08	-0.31	1.6	255.08	6DE	1.7	-4.3	-2.1	0.2
248.60	-0.42	1.6	265.28	16DE	0.3	-4.4	-2.1	0.1
256.19	-0.53	1.5	275.47	26DE	0.0	-4.4	-2.0	0.0
				209				
263.83	-0.64	1.5	285.63	5JA	-0.1	-4.3	-1.9	-0.0
271.52	-0.74	1.5	295.76	15JA	-0.2	-3.9	-1.8	0.0
279.25	-0.84	1.5	305.85	25JA	-0.6	-3.3	-1.8	0.1
287.00	-0.93	1.4	315.89	4FE	-1.1	-3.6	-1.7	0.1
294.78	-1.02	1.4	325.88	14FE	-1.4	-4.1	-1.7	0.2
302.56	-1.10	1.4	335.81	24FE	-0.9	-4.3	-1.6	0.3
310.34	-1.17	1.4	345.68	6MR	0.4	-4.3	-1.6	0.3
318.09	-1.23	1.3	355.49	16MR	2.1	-4.2	-1.6	0.4
325.82	-1.28	1.3	5.25	26MR	3.4	-4.1	-1.5	0.4
333.51	-1.31	1.3	14.95	5AP	1.9	-4.0	-1.5	0.5
341.14	-1.33	1.3	24.60	15AP	1.1	-3.9	-1.5	0.5
348.70	-1.34	1.3	34.22	25AP	0.5	-3.8	-1.5	0.5
356.18	-1.33	1.2	43.79	5MY	-0.2	-3.7	-1.5	0.6
3.57	-1.31	1.2	53.34	15MY	-1.1	-3.6	-1.5	0.6
10.85	-1.27	1.2	62.88	25MY	-1.8	-3.6	-1.6	0.6
18.02	-1.22	1.2	72.40	4JN	-1.0	-3.5	-1.6	0.6
25.06	-1.16	1.1	81.92	14JN	-0.2	-3.5	-1.6	0.6
31.96	-1.08	1.1	91.45	24JN	0.3	-3.4	-1.7	0.6
38.70	-0.98	1.1	101.00	4JL	0.7	-3.4	-1.7	0.6
45.27	-0.87	1.0	110.58	14JL	1.2	-3.4	-1.8	0.5
51.66	-0.75	1.0	120.19	24JL	2.3	-3.4	-1.8	0.6
57.82	-0.61	0.9	129.85	3AU	2.6	-3.4	-1.9	0.6
63.75	-0.46	0.9	139.55	13AU	0.4	-3.3	-1.9	0.7
69.39	-0.28	0.8	149.30	23AU	-0.8	-3.3	-2.0	0.7
74.70	-0.09	0.7	159.12	2SE	-1.2	-3.4	-2.1	0.7
79.62	0.13	0.6	168.99	12SE	-1.1	-3.4	-2.1	0.8
84.06	0.37	0.4	178.92	22SE	-0.6	-3.4	-2.2	0.8
87.92	0.65	0.3	188.90	2OC	-0.4	-3.4	-2.3	0.7
91.08	0.97	0.1	198.94	12OC	-0.3	-3.4	-2.3	0.7
93.35	1.33	-0.1	209.04	22OC	-0.2	-3.5	-2.3	0.7
94.56	1.73	-0.3	219.17	1NO	0.1	-3.5	-2.3	0.7
94.51	2.18	-0.5	229.33	11NO	1.8	-3.5	-2.3	0.6
93.07	2.64	-0.8	239.52	21NO	1.5	-3.5	-2.3	0.6
90.33	3.07	-1.0	249.72	1DE	0.1	-3.4	-2.3	0.5
86.63	3.43	-1.1	259.93	11DE	-0.1	-3.4	-2.2	0.4
82.66	3.65	-1.1	270.13	21DE	-0.2	-3.4	-2.2	0.4
79.17	3.71	-0.9	280.31	31DE	-0.3	-3.4	-2.1	0.3

♂ LONG	♂ LAT	♂ MAG	☉ LONG	16.00UT 210	☿	♀	♃	♄
76.73	3.65	-0.6	290.46	10JA	-0.6	-3.4	-2.0	0.2
75.59	3.50	-0.3	300.57	20JA	-1.1	-3.3	-2.0	0.2
75.74	3.30	-0.1	310.63	30JA	-1.3	-3.3	-1.9	0.2
77.02	3.10	0.2	320.65	9FE	-0.8	-3.3	-1.8	0.3
79.24	2.89	0.4	330.61	19FE	0.5	-3.3	-1.8	0.3
82.21	2.69	0.6	340.51	1MR	2.6	-3.4	-1.7	0.4
85.78	2.50	0.8	350.35	11MR	2.6	-3.4	-1.6	0.5
89.83	2.33	1.0	0.14	21MR	1.4	-3.4	-1.6	0.5
94.27	2.17	1.1	9.87	31MR	0.8	-3.4	-1.5	0.6
99.02	2.01	1.2	19.55	10AP	0.4	-3.5	-1.5	0.6
104.04	1.87	1.4	29.18	20AP	-0.2	-3.5	-1.5	0.7
109.27	1.73	1.4	38.77	30AP	-1.1	-3.6	-1.5	0.7
114.68	1.60	1.5	48.34	10MY	-1.9	-3.6	-1.4	0.7
120.27	1.47	1.6	57.88	20MY	-1.0	-3.7	-1.4	0.8
126.00	1.34	1.7	67.40	30MY	-0.1	-3.8	-1.4	0.8
131.86	1.22	1.7	76.93	9JN	0.5	-3.8	-1.4	0.8
137.85	1.10	1.7	86.45	19JN	1.0	-3.9	-1.4	0.8
143.95	0.99	1.8	95.99	29JN	1.6	-4.0	-1.5	0.8
150.17	0.87	1.8	105.55	9JL	2.9	-4.1	-1.5	0.8
156.50	0.76	1.8	115.15	19JL	2.2	-4.2	-1.5	0.8
162.93	0.64	1.8	124.78	29JL	0.3	-4.2	-1.6	0.8
169.48	0.53	1.8	134.45	8AU	-0.9	-4.1	-1.6	0.8
176.13	0.42	1.8	144.18	18AU	-1.3	-3.8	-1.7	0.9
182.89	0.31	1.8	153.96	28AU	-1.0	-3.3	-1.7	0.9
189.75	0.20	1.7	163.80	7SE	-0.5	-3.6	-1.8	1.0
196.72	0.09	1.7	173.70	17SE	-0.3	-4.1	-1.8	1.0
203.78	-0.02	1.7	183.66	27SE	-0.1	-4.3	-1.9	1.0
210.96	-0.13	1.6	193.67	7OC	-0.0	-4.3	-2.0	1.0
218.23	-0.24	1.6	203.74	17OC	-0.3	-4.3	-2.0	1.0
225.59	-0.34	1.6	213.85	27OC	2.0	-4.2	-2.1	1.0
233.05	-0.45	1.5	224.00	6NO	1.2	-4.1	-2.1	1.0
240.59	-0.54	1.5	234.18	16NO	-0.1	-4.0	-2.2	0.9
248.20	-0.64	1.4	244.37	26NO	-0.3	-3.9	-2.2	0.9
255.89	-0.72	1.4	254.58	6DE	-0.3	-3.8	-2.2	0.8
263.64	-0.81	1.4	264.79	16DE	-0.4	-3.8	-2.2	0.7
271.43	-0.88	1.4	274.97	26DE	-0.7	-3.7	-2.2	0.7
				211				
279.27	-0.94	1.4	285.15	5JA	-1.1	-3.6	-2.2	0.6
287.13	-1.00	1.4	295.28	15JA	-1.1	-3.5	-2.1	0.5
295.00	-1.05	1.4	305.37	25JA	-0.8	-3.5	-2.1	0.5
302.88	-1.08	1.4	315.41	4FE	0.6	-3.4	-2.0	0.4
310.74	-1.11	1.4	325.40	14FE	3.0	-3.4	-1.9	0.5
318.58	-1.12	1.4	335.33	24FE	1.9	-3.4	-1.9	0.5
326.39	-1.12	1.4	345.21	6MR	1.0	-3.3	-1.8	0.6
334.15	-1.10	1.4	355.02	16MR	0.6	-3.3	-1.7	0.6
341.85	-1.08	1.4	4.78	26MR	0.2	-3.3	-1.7	0.7
349.48	-1.04	1.4	14.48	5AP	-0.3	-3.3	-1.6	0.7
357.04	-1.00	1.5	24.14	15AP	-1.1	-3.3	-1.5	0.8
4.51	-0.94	1.5	33.75	25AP	-1.8	-3.3	-1.5	0.9
11.89	-0.87	1.5	43.33	5MY	-1.0	-3.3	-1.4	0.9
19.18	-0.79	1.5	52.88	15MY	-0.0	-3.4	-1.4	0.9
26.37	-0.70	1.5	62.41	25MY	0.6	-3.4	-1.4	1.0
33.45	-0.61	1.5	71.94	4JN	1.3	-3.4	-1.4	1.0
40.42	-0.50	1.5	81.45	14JN	2.2	-3.5	-1.3	1.0
47.28	-0.39	1.5	90.98	24JN	3.3	-3.5	-1.3	1.0
54.03	-0.27	1.5	100.54	4JL	1.8	-3.5	-1.3	1.1
60.65	-0.15	1.5	110.11	14JL	0.2	-3.4	-1.3	1.1
67.14	-0.02	1.5	119.72	24JL	-1.0	-3.4	-1.3	1.1
73.49	0.12	1.4	129.37	3AU	-1.4	-3.4	-1.4	1.1
79.69	0.27	1.4	139.07	13AU	-1.0	-3.3	-1.4	1.1
85.73	0.43	1.4	148.82	23AU	-0.5	-3.3	-1.4	1.1
91.59	0.60	1.3	158.64	2SE	-0.1	-3.3	-1.5	1.2
97.24	0.78	1.2	168.50	12SE	0.0	-3.3	-1.5	1.2
102.66	0.97	1.2	178.43	22SE	0.1	-3.3	-1.6	1.2
107.79	1.18	1.1	188.42	2OC	0.5	-3.3	-1.6	1.3
112.59	1.41	0.9	198.45	12OC	2.4	-3.4	-1.7	1.3
116.99	1.66	0.8	208.54	22OC	1.0	-3.4	-1.7	1.3
120.90	1.94	0.6	218.67	1NO	-0.3	-3.4	-1.8	1.3
124.20	2.25	0.5	228.84	11NO	-0.5	-3.5	-1.9	1.3
126.75	2.59	0.3	239.03	21NO	-0.5	-3.5	-1.9	1.2
128.38	2.97	0.0	249.23	1DE	-0.6	-3.6	-2.0	1.2
128.89	3.38	-0.2	259.43	11DE	-0.8	-3.6	-2.0	1.1
128.14	3.78	-0.4	269.63	21DE	-0.9	-3.7	-2.1	1.1
126.06	4.16	-0.7	279.82	31DE	-1.0	-3.8	-2.1	1.0

Left table:

♂ LONG	LAT	MAG	☉ LONG	16.00UT	☿	♀	♃	♄
				212		MAGNITUDES		
122.85	4.44	-0.9	289.97	10JA	-0.7	-3.9	-2.1	1.0
118.96	4.56	-1.0	300.08	20JA	0.8	-3.9	-2.1	0.9
115.12	4.51	-0.9	310.15	30JA	3.1	-4.0	-2.1	0.8
112.01	4.30	-0.7	320.17	9FE	1.4	-4.1	-2.0	0.7
110.06	3.99	-0.4	330.13	19FE	0.7	-4.2	-2.0	0.7
109.41	3.63	-0.2	340.04	29FE	0.4	-4.3	-1.9	0.8
109.99	3.27	0.1	349.88	10MR	0.1	-4.2	-1.9	0.8
111.63	2.93	0.3	359.67	20MR	-0.3	-4.1	-1.8	0.9
114.15	2.61	0.5	9.40	30MR	-1.1	-3.6	-1.7	0.9
117.38	2.31	0.6	19.08	9AP	-1.8	-3.0	-1.7	1.0
121.19	2.04	0.8	28.71	19AP	-1.0	-3.6	-1.6	1.0
125.47	1.80	0.9	38.31	29AP	0.0	-4.0	-1.5	1.1
130.14	1.57	1.0	47.87	9MY	0.9	-4.2	-1.5	1.1
135.14	1.37	1.1	57.41	19MY	1.7	-4.2	-1.4	1.2
140.42	1.17	1.2	66.94	29MY	2.9	-4.1	-1.4	1.2
145.94	0.99	1.3	76.46	8JN	2.9	-4.0	-1.4	1.3
151.67	0.81	1.4	85.99	18JN	1.4	-3.9	-1.3	1.3
157.60	0.65	1.4	95.53	28JN	0.1	-3.8	-1.3	1.3
163.71	0.49	1.4	105.08	8JL	-1.0	-3.8	-1.3	1.3
169.98	0.34	1.5	114.68	18JL	-1.5	-3.7	-1.3	1.4
176.42	0.20	1.5	124.31	28JL	-1.0	-3.6	-1.2	1.4
182.99	0.06	1.5	133.98	7AU	-0.4	-3.6	-1.2	1.4
189.72	-0.07	1.5	143.71	17AU	-0.0	-3.5	-1.3	1.3
196.58	-0.19	1.5	153.49	27AU	0.1	-3.5	-1.3	1.3
203.56	-0.31	1.5	163.32	6SE	0.3	-3.5	-1.3	1.3
210.68	-0.43	1.5	173.22	16SE	0.8	-3.4	-1.3	1.3
217.92	-0.53	1.5	183.18	26SE	2.7	-3.4	-1.4	1.3
225.26	-0.63	1.5	193.19	6OC	0.9	-3.4	-1.4	1.3
232.72	-0.73	1.5	203.25	16OC	-0.4	-3.4	-1.5	1.3
240.27	-0.81	1.5	213.36	26OC	-0.6	-3.4	-1.5	1.3
247.90	-0.88	1.5	223.51	5NO	-0.6	-3.4	-1.6	1.2
255.62	-0.95	1.4	233.69	15NO	-0.7	-3.4	-1.6	1.2
263.39	-1.00	1.4	243.88	25NO	-0.8	-3.4	-1.7	1.2
271.22	-1.04	1.4	254.09	5DE	-0.8	-3.4	-1.8	1.1
279.09	-1.07	1.4	264.29	15DE	-0.8	-3.4	-1.8	1.1
286.99	-1.09	1.4	274.48	25DE	-0.5	-3.4	-1.9	1.0
				213				
294.89	-1.10	1.4	284.65	4JA	0.9	-3.4	-1.9	1.0
302.80	-1.09	1.4	294.79	14JA	2.8	-3.5	-2.0	0.9
310.68	-1.08	1.3	304.88	24JA	1.0	-3.5	-2.0	0.9
318.54	-1.05	1.3	314.92	3FE	0.5	-3.5	-2.0	0.8
326.35	-1.01	1.3	324.92	13FE	0.3	-3.4	-2.0	0.8
334.11	-0.96	1.3	334.85	23FE	0.0	-3.4	-2.0	0.8
341.81	-0.90	1.4	344.72	5MR	-0.4	-3.4	-2.0	0.8
349.44	-0.84	1.4	354.54	15MR	-1.1	-3.4	-2.0	0.9
357.00	-0.76	1.5	4.30	25MR	-1.7	-3.4	-1.9	1.0
4.47	-0.68	1.5	14.01	4AP	-1.0	-3.3	-1.9	1.0
11.85	-0.59	1.5	23.67	14AP	0.1	-3.3	-1.8	1.1
19.15	-0.50	1.6	33.28	24AP	1.1	-3.3	-1.7	1.2
26.36	-0.40	1.6	42.86	4MY	2.2	-3.3	-1.7	1.2
33.47	-0.30	1.6	52.42	14MY	3.6	-3.3	-1.6	1.3
40.49	-0.20	1.7	61.95	24MY	2.3	-3.4	-1.5	1.3
47.43	-0.09	1.7	71.47	3JN	1.1	-3.4	-1.5	1.3
54.27	0.02	1.7	80.99	13JN	0.1	-3.4	-1.4	1.3
61.02	0.13	1.7	90.52	23JN	-1.0	-3.5	-1.4	1.4
67.68	0.24	1.8	100.07	3JL	-1.6	-3.5	-1.3	1.3
74.25	0.36	1.8	109.64	13JL	-1.0	-3.6	-1.3	1.3
80.73	0.48	1.8	119.25	23JL	-0.3	-3.6	-1.3	1.3
87.12	0.60	1.8	128.90	2AU	0.1	-3.7	-1.2	1.3
93.42	0.72	1.7	138.60	12AU	0.3	-3.8	-1.2	1.2
99.61	0.84	1.7	148.35	22AU	0.5	-3.9	-1.2	1.2
105.70	0.97	1.7	158.16	1SE	1.1	-4.0	-1.2	1.1
111.68	1.10	1.7	168.03	11SE	3.1	-4.1	-1.2	1.1
117.52	1.24	1.6	177.95	21SE	0.8	-4.2	-1.2	1.1
123.22	1.38	1.5	187.93	1OC	-0.6	-4.3	-1.2	1.1
128.76	1.53	1.5	197.97	11OC	-0.8	-4.4	-1.3	1.1
134.10	1.69	1.4	208.05	21OC	-0.7	-4.3	-1.3	1.1
139.21	1.86	1.2	218.18	31OC	-0.8	-3.9	-1.3	1.1
144.04	2.05	1.1	228.35	10NO	-0.7	-3.1	-1.4	1.1
148.54	2.25	1.0	238.53	20NO	-0.6	-3.6	-1.4	1.1
152.62	2.47	0.8	248.74	30NO	-0.6	-4.1	-1.5	1.0
156.19	2.70	0.6	258.94	10DE	-0.4	-4.4	-1.6	1.0
159.13	2.96	0.4	269.14	20DE	1.1	-4.4	-1.6	1.0
161.28	3.24	0.2	279.32	30DE	2.4	-4.3	-1.7	0.9

Right table:

♂ LONG	LAT	MAG	☉ LONG	16.00UT	☿	♀	♃	♄
				214		MAGNITUDES		
162.46	3.53	-0.1	289.48	9JA	0.7	-4.2	-1.8	0.9
162.50	3.81	-0.4	299.59	19JA	0.3	-4.1	-1.8	0.8
161.26	4.05	-0.6	309.67	29JA	0.1	-4.0	-1.9	0.8
158.77	4.21	-0.9	319.69	8FE	-0.1	-3.9	-1.9	0.7
155.32	4.23	-1.1	329.65	18FE	-0.5	-3.8	-2.0	0.7
151.45	4.09	-1.1	339.56	28FE	-1.1	-3.7	-2.0	0.6
147.90	3.79	-1.0	349.41	10MR	-1.6	-3.6	-2.0	0.6
145.28	3.38	-0.8	359.20	20MR	-1.0	-3.5	-2.0	0.7
143.91	2.93	-0.5	8.93	30MR	0.2	-3.5	-2.0	0.7
143.85	2.47	-0.3	18.62	9AP	1.4	-3.4	-2.0	0.8
144.98	2.05	-0.1	28.25	19AP	2.9	-3.4	-1.9	0.9
147.14	1.66	0.1	37.85	29AP	3.1	-3.4	-1.9	0.9
150.15	1.31	0.2	47.42	9MY	1.8	-3.3	-1.8	0.9
153.85	1.00	0.4	56.96	19MY	0.9	-3.3	-1.8	1.1
158.12	0.72	0.5	66.48	29MY	-0.0	-3.3	-1.7	1.1
162.88	0.47	0.6	76.00	8JN	-1.0	-3.3	-1.6	1.1
168.04	0.24	0.7	85.53	18JN	-1.7	-3.3	-1.6	1.2
173.54	0.03	0.8	95.07	28JN	-1.0	-3.3	-1.5	1.2
179.34	-0.16	0.9	104.62	8JL	-0.3	-3.4	-1.5	1.2
185.41	-0.33	1.0	114.21	18JL	0.2	-3.4	-1.4	1.2
191.71	-0.48	1.0	123.84	28JL	0.4	-3.4	-1.4	1.2
198.23	-0.63	1.1	133.51	7AU	0.7	-3.4	-1.3	1.1
204.94	-0.75	1.1	143.23	17AU	1.5	-3.5	-1.3	1.1
211.83	-0.87	1.1	153.01	27AU	3.2	-3.5	-1.3	1.1
218.88	-0.97	1.2	162.84	6SE	0.6	-3.5	-1.2	1.0
226.08	-1.05	1.2	172.74	16SE	-0.7	-3.4	-1.2	1.0
233.41	-1.12	1.2	182.69	26SE	-0.9	-3.4	-1.2	1.0
240.87	-1.18	1.2	192.70	6OC	-0.9	-3.4	-1.2	1.0
248.43	-1.23	1.3	202.76	16OC	-0.8	-3.4	-1.2	1.0
256.80	-1.25	1.3	212.87	26OC	-0.5	-3.3	-1.3	1.0
263.80	-1.27	1.3	223.01	5NO	-0.5	-3.3	-1.3	1.0
271.58	-1.27	1.3	233.19	15NO	-0.5	-3.3	-1.3	1.0
279.41	-1.26	1.3	243.39	25NO	-0.2	-3.3	-1.3	1.0
287.26	-1.23	1.4	253.59	5DE	1.3	-3.3	-1.4	0.9
295.12	-1.19	1.4	263.79	15DE	2.0	-3.4	-1.4	0.9
302.97	-1.14	1.4	273.99	25DE	0.5	-3.4	-1.5	0.9
				215				
310.81	-1.08	1.4	284.16	4JA	0.1	-3.4	-1.6	0.8
318.61	-1.01	1.4	294.29	14JA	-0.0	-3.4	-1.6	0.8
326.37	-0.94	1.5	304.39	24JA	-0.2	-3.5	-1.7	0.7
334.07	-0.85	1.5	314.44	3FE	-0.5	-3.5	-1.8	0.7
341.71	-0.76	1.5	324.43	13FE	-1.1	-3.6	-1.8	0.6
349.28	-0.67	1.5	334.38	23FE	-1.5	-3.6	-1.9	0.6
356.77	-0.57	1.5	344.25	5MR	-0.9	-3.7	-1.9	0.5
4.18	-0.47	1.6	354.07	15MR	0.3	-3.8	-2.0	0.5
11.50	-0.36	1.6	3.84	25MR	1.8	-3.8	-2.0	0.5
18.74	-0.26	1.6	13.55	4AP	3.7	-3.9	-2.1	0.6
25.90	-0.16	1.6	23.21	14AP	2.3	-4.0	-2.1	0.6
32.97	-0.05	1.6	32.83	24AP	1.3	-4.1	-2.1	0.7
39.96	0.05	1.7	42.40	4MY	0.7	-4.2	-2.0	0.8
46.87	0.15	1.7	51.96	14MY	-0.1	-4.2	-2.0	0.8
53.71	0.25	1.8	61.49	24MY	-1.0	-4.1	-2.0	0.9
60.47	0.35	1.8	71.01	3JN	-1.8	-3.8	-1.9	0.9
67.16	0.45	1.8	80.53	13JN	-1.0	-3.1	-1.9	1.0
73.78	0.54	1.9	90.06	23JN	-0.2	-3.2	-1.8	1.0
80.34	0.63	1.9	99.61	3JL	0.3	-3.8	-1.7	1.0
86.84	0.72	1.9	109.18	13JL	0.6	-4.1	-1.7	1.0
93.28	0.81	1.9	118.79	23JL	2.0	-4.2	-1.6	1.0
99.66	0.90	1.9	128.43	2AU	2.0	-4.2	-1.5	1.0
105.99	0.99	1.9	138.13	12AU	2.9	-4.1	-1.5	1.0
112.27	1.07	1.9	147.88	22AU	0.5	-4.0	-1.4	1.0
118.48	1.16	1.9	157.68	1SE	-0.8	-4.0	-1.4	1.0
124.64	1.24	1.9	167.54	11SE	-1.1	-3.9	-1.4	0.9
130.73	1.33	1.9	177.46	21SE	-1.0	-3.8	-1.3	0.9
136.76	1.41	1.8	187.44	1OC	-0.7	-3.7	-1.3	0.9
142.71	1.50	1.8	197.48	11OC	-0.4	-3.7	-1.3	0.9
148.57	1.58	1.7	207.56	21OC	-0.3	-3.6	-1.3	0.9
154.33	1.67	1.6	217.69	31OC	-0.3	-3.6	-1.3	0.9
159.98	1.76	1.5	227.85	10NO	-0.0	-3.5	-1.3	0.9
165.48	1.85	1.4	238.04	20NO	1.5	-3.4	-1.3	0.9
170.82	1.94	1.3	248.24	30NO	1.7	-3.4	-1.3	0.9
175.95	2.04	1.2	258.45	10DE	0.2	-3.4	-1.4	0.9
180.84	2.14	1.0	268.65	20DE	-0.1	-3.4	-1.4	0.9
185.43	2.24	0.8	278.83	30DE	-0.1	-3.4	-1.4	0.8

♂			☉	16.00UT	☿	♀	♃	♄
LONG	LAT	MAG	LONG	216	MAGNITUDES			
189.64	2.34	0.6	288.99	9JA	-0.3	-3.3	-1.5	0.8
193.38	2.43	0.4	299.11	19JA	-0.6	-3.3	-1.5	0.8
196.54	2.53	0.2	309.18	29JA	-1.1	-3.3	-1.6	0.7
198.97	2.60	-0.1	319.21	8FE	-1.3	-3.3	-1.7	0.7
200.50	2.66	-0.4	329.17	18FE	-0.9	-3.3	-1.7	0.6
200.95	2.67	-0.7	339.08	28FE	0.4	-3.4	-1.8	0.5
200.17	2.62	-1.0	348.93	9MR	2.2	-3.4	-1.9	0.5
198.13	2.47	-1.3	358.73	19MR	3.1	-3.4	-1.9	0.4
195.05	2.20	-1.5	8.46	29MR	1.7	-3.4	-2.0	0.4
191.43	1.81	-1.6	18.15	8AP	1.0	-3.5	-2.1	0.4
188.01	1.35	-1.5	27.79	18AP	0.5	-3.5	-2.1	0.5
185.44	0.85	-1.3	37.38	28AP	-0.2	-3.4	-2.1	0.5
184.11	0.37	-1.1	46.95	8MY	-1.0	-3.4	-2.2	0.6
184.15	-0.06	-0.9	56.49	18MY	-1.9	-3.4	-2.2	0.7
185.46	-0.43	-0.7	66.02	28MY	-1.0	-3.4	-2.2	0.7
187.88	-0.74	-0.5	75.54	7JN	-0.2	-3.3	-2.1	0.8
191.23	-1.00	-0.3	85.06	17JN	0.4	-3.3	-2.1	0.8
195.33	-1.22	-0.2	94.60	27JN	0.8	-3.3	-2.1	0.9
200.07	-1.40	-0.0	104.16	7JL	1.4	-3.3	-2.0	0.9
205.34	-1.54	0.1	113.74	17JL	2.5	-3.3	-1.9	0.9
211.04	-1.65	0.2	123.37	27JL	2.5	-3.4	-1.9	0.9
217.11	-1.74	0.3	133.04	6AU	0.4	-3.4	-1.8	1.0
223.49	-1.80	0.4	142.76	16AU	-0.9	-3.4	-1.7	0.9
230.14	-1.84	0.5	152.53	26AU	-1.2	-3.4	-1.7	0.9
237.02	-1.85	0.6	162.36	5SE	-1.1	-3.5	-1.6	0.9
244.09	-1.85	0.7	172.25	15SE	-0.6	-3.5	-1.6	0.9
251.31	-1.82	0.7	182.20	25SE	-0.3	-3.6	-1.5	0.8
258.68	-1.78	0.8	192.21	5OC	-0.2	-3.7	-1.5	0.8
266.15	-1.72	0.9	202.26	15OC	-0.1	-3.8	-1.5	0.8
273.70	-1.65	0.9	212.37	25OC	0.2	-3.9	-1.4	0.8
281.31	-1.56	1.0	222.52	4NO	1.8	-4.0	-1.4	0.8
288.96	-1.47	1.1	232.69	14NO	1.4	-4.1	-1.4	0.8
296.63	-1.36	1.1	242.89	24NO	0.0	-4.2	-1.4	0.8
304.30	-1.24	1.2	253.10	4DE	-0.2	-4.3	-1.4	0.8
311.96	-1.12	1.3	263.30	14DE	-0.3	-4.4	-1.4	0.8
319.59	-1.00	1.3	273.49	24DE	-0.4	-4.4	-1.4	0.8
				217				
327.18	-0.87	1.4	283.67	3JA	-0.7	-4.3	-1.4	0.8
334.71	-0.75	1.4	293.80	13JA	-1.1	-3.9	-1.5	0.8
342.19	-0.62	1.5	303.90	23JA	-1.2	-3.3	-1.5	0.7
349.60	-0.49	1.5	313.96	2FE	-0.8	-3.6	-1.5	0.7
356.94	-0.37	1.6	323.95	12FE	0.5	-4.1	-1.6	0.6
4.20	-0.25	1.6	333.89	22FE	2.7	-4.3	-1.6	0.6
11.39	-0.14	1.7	343.78	4MR	2.3	-4.3	-1.7	0.5
18.50	-0.02	1.7	353.60	14MR	1.2	-4.2	-1.8	0.5
25.54	0.09	1.7	3.37	24MR	0.7	-4.1	-1.8	0.4
32.50	0.19	1.8	13.08	3AP	0.3	-4.0	-1.9	0.4
39.39	0.29	1.8	22.74	13AP	-0.2	-3.9	-2.0	0.3
46.21	0.38	1.8	32.36	23AP	-1.0	-3.8	-2.1	0.4
52.97	0.47	1.8	41.94	3MY	-1.9	-3.7	-2.1	0.4
59.67	0.56	1.8	51.49	13MY	-1.0	-3.6	-2.2	0.5
66.31	0.64	1.8	61.03	23MY	-0.1	-3.6	-2.2	0.6
72.90	0.71	1.8	70.56	2JN	0.5	-3.5	-2.3	0.6
79.44	0.78	1.9	80.07	12JN	1.1	-3.5	-2.3	0.7
85.94	0.85	1.9	89.60	22JN	1.8	-3.4	-2.3	0.7
92.40	0.91	1.9	99.15	2JL	3.1	-3.4	-2.3	0.8
98.83	0.97	2.0	108.72	12JL	2.1	-3.4	-2.3	0.8
105.23	1.03	2.0	118.32	22JL	0.3	-3.4	-2.2	0.8
111.60	1.08	2.0	127.97	1AU	-0.9	-3.3	-2.2	0.9
117.95	1.13	2.0	137.66	11AU	-1.3	-3.3	-2.1	0.9
124.28	1.18	2.0	147.41	21AU	-1.1	-3.3	-2.1	0.9
130.59	1.22	2.0	157.21	31AU	-0.5	-3.4	-2.0	0.9
136.88	1.26	2.0	167.06	10SE	-0.2	-3.4	-1.9	0.9
143.15	1.29	2.0	176.98	20SE	-0.1	-3.4	-1.9	0.8
149.41	1.33	2.0	186.96	30SE	0.0	-3.4	-1.8	0.8
155.64	1.35	1.9	196.99	10OC	0.4	-3.4	-1.7	0.8
161.86	1.38	1.9	207.07	20OC	2.1	-3.5	-1.7	0.7
168.05	1.40	1.8	217.19	30OC	1.2	-3.5	-1.6	0.7
174.21	1.42	1.8	227.35	9NO	-0.2	-3.5	-1.6	0.8
180.33	1.43	1.7	237.54	19NO	-0.4	-3.5	-1.5	0.8
186.40	1.44	1.6	247.74	29NO	-0.4	-3.4	-1.5	0.8
192.42	1.43	1.5	257.94	9DE	-0.5	-3.4	-1.5	0.8
198.38	1.42	1.4	268.15	19DE	-0.7	-3.4	-1.5	0.8
204.25	1.41	1.3	278.33	29DE	-1.0	-3.4	-1.5	0.8

♂			☉	16.00UT	☿	♀	♃	♄
LONG	LAT	MAG	LONG	218	MAGNITUDES			
210.02	1.37	1.1	288.49	8JA	-1.0	-3.4	-1.5	0.8
215.67	1.33	1.0	298.61	18JA	-0.7	-3.3	-1.5	0.8
221.17	1.26	0.8	308.69	28JA	0.6	-3.3	-1.5	0.7
226.48	1.18	0.6	318.71	7FE	3.1	-3.3	-1.5	0.7
231.56	1.07	0.4	328.69	17FE	1.7	-3.3	-1.6	0.6
236.35	0.92	0.2	338.60	27FE	0.9	-3.4	-1.6	0.6
240.78	0.74	-0.0	348.45	9MR	0.5	-3.4	-1.6	0.5
244.74	0.50	-0.3	358.25	19MR	0.2	-3.4	-1.7	0.5
248.11	0.20	-0.6	7.99	29MR	-0.3	-3.4	-1.7	0.4
250.76	-0.18	-0.9	17.68	8AP	-1.0	-3.5	-1.8	0.4
252.47	-0.64	-1.2	27.32	18AP	-1.8	-3.5	-1.9	0.3
253.09	-1.20	-1.5	36.92	28AP	-1.0	-3.6	-1.9	0.3
252.49	-1.85	-1.8	46.49	8MY	-0.0	-3.6	-2.0	0.3
250.68	-2.54	-2.1	56.04	18MY	0.7	-3.7	-2.1	0.4
247.99	-3.22	-2.3	65.56	28MY	1.4	-3.8	-2.2	0.5
245.01	-3.80	-2.3	75.08	7JN	2.4	-3.8	-2.2	0.5
242.47	-4.20	-2.2	84.61	17JN	3.3	-3.9	-2.3	0.6
240.98	-4.41	-2.0	94.14	27JN	1.7	-4.0	-2.3	0.6
240.83	-4.46	-1.7	103.70	7JL	0.2	-4.1	-2.4	0.7
242.04	-4.39	-1.5	113.29	17JL	-1.0	-4.2	-2.4	0.7
244.49	-4.23	-1.3	122.91	27JL	-1.5	-4.2	-2.4	0.8
247.96	-4.03	-1.0	132.57	6AU	-1.0	-4.1	-2.4	0.8
252.27	-3.80	-0.8	142.29	16AU	-0.5	-3.8	-2.4	0.8
257.27	-3.54	-0.6	152.06	26AU	-0.1	-3.3	-2.3	0.8
262.78	-3.28	-0.4	161.89	5SE	0.1	-3.6	-2.3	0.8
268.72	-3.01	-0.3	171.78	15SE	0.2	-4.1	-2.2	0.8
274.99	-2.74	-0.1	181.72	25SE	0.6	-4.3	-2.2	0.8
281.51	-2.47	0.1	191.72	5OC	2.4	-4.3	-2.1	0.8
288.22	-2.20	0.2	201.78	15OC	1.1	-4.3	-2.0	0.8
295.07	-1.94	0.4	211.88	25OC	-0.3	-4.2	-1.9	0.7
302.03	-1.69	0.5	222.02	4NO	-0.5	-4.1	-1.9	0.7
309.06	-1.46	0.6	232.20	14NO	-0.5	-4.0	-1.8	0.7
316.13	-1.23	0.8	242.39	24NO	-0.6	-3.9	-1.8	0.7
323.22	-1.01	0.9	252.59	4DE	-0.8	-3.8	-1.7	0.8
330.31	-0.81	1.0	262.80	14DE	-0.9	-3.7	-1.7	0.8
337.39	-0.62	1.1	272.99	24DE	-0.9	-3.7	-1.6	0.8
				219				
344.45	-0.44	1.2	283.17	3JA	-0.6	-3.6	-1.6	0.8
351.48	-0.28	1.3	293.31	13JA	0.8	-3.5	-1.6	0.8
358.46	-0.13	1.4	303.41	23JA	3.0	-3.5	-1.6	0.8
5.40	0.01	1.5	313.47	2FE	1.3	-3.4	-1.6	0.7
12.29	0.14	1.6	323.47	12FE	0.6	-3.4	-1.6	0.7
19.12	0.26	1.6	333.41	22FE	0.4	-3.4	-1.6	0.7
25.91	0.37	1.7	343.30	4MR	0.1	-3.3	-1.6	0.6
32.65	0.47	1.8	353.13	14MR	-0.3	-3.3	-1.6	0.6
39.33	0.56	1.8	2.90	24MR	-1.1	-3.3	-1.6	0.5
45.97	0.65	1.9	12.61	3AP	-1.8	-3.3	-1.7	0.4
52.56	0.72	1.9	22.28	13AP	-1.0	-3.3	-1.7	0.4
59.11	0.79	1.9	31.90	23AP	0.0	-3.3	-1.7	0.3
65.62	0.85	2.0	41.48	3MY	0.9	-3.3	-1.8	0.3
72.10	0.91	2.0	51.04	13MY	1.9	-3.4	-1.9	0.3
78.55	0.96	2.0	60.57	23MY	3.2	-3.4	-1.9	0.3
84.97	1.00	2.0	70.10	2JN	2.7	-3.4	-2.0	0.4
91.37	1.04	2.0	79.62	12JN	1.3	-3.5	-2.1	0.4
97.76	1.07	2.0	89.14	22JN	0.2	-3.5	-2.1	0.5
104.13	1.10	2.0	98.69	2JL	-1.0	-3.5	-2.2	0.6
110.50	1.12	2.0	108.26	12JL	-1.6	-3.4	-2.3	0.6
116.87	1.14	2.0	117.86	22JL	-1.0	-3.4	-2.3	0.7
123.24	1.16	2.0	127.50	1AU	-0.4	-3.4	-2.4	0.7
129.61	1.17	2.0	137.19	11AU	-0.0	-3.3	-2.4	0.8
136.00	1.17	2.0	146.93	21AU	0.2	-3.3	-2.4	0.8
142.39	1.17	2.0	156.73	31AU	0.4	-3.3	-2.5	0.8
148.81	1.17	2.0	166.58	10SE	0.9	-3.3	-2.4	0.8
155.24	1.16	2.0	176.50	20SE	2.8	-3.3	-2.4	0.8
161.69	1.15	2.0	186.47	30SE	0.9	-3.3	-2.4	0.8
168.16	1.13	2.0	196.50	10OC	-0.5	-3.4	-2.3	0.8
174.65	1.10	1.9	206.58	20OC	-0.7	-3.4	-2.3	0.7
181.17	1.07	1.9	216.70	30OC	-0.7	-3.4	-2.2	0.7
187.71	1.03	1.9	226.86	9NO	-0.7	-3.5	-2.1	0.7
194.26	0.99	1.8	237.04	19NO	-0.7	-3.5	-2.1	0.7
200.84	0.93	1.7	247.24	29NO	-0.7	-3.6	-2.0	0.7
207.44	0.87	1.7	257.45	9DE	-0.7	-3.6	-1.9	0.7
214.06	0.80	1.6	267.65	19DE	-0.5	-3.7	-1.9	0.8
220.69	0.71	1.5	277.84	29DE	0.9	-3.8	-1.8	0.8

♂			☉	16.00UT	☿	♀	♃	♄
LONG	LAT	MAG	LONG	220	MAGNITUDES			
227.32	0.61	1.4	288.00	8JA	2.7	-3.9	-1.7	0.8
233.97	0.50	1.3	298.12	18JA	0.9	-4.0	-1.7	0.8
240.62	0.36	1.2	308.20	28JA	0.4	-4.0	-1.7	0.8
247.27	0.22	1.1	318.23	7FE	0.2	-4.1	-1.6	0.8
253.91	0.05	1.0	328.20	17FE	-0.0	-4.2	-1.6	0.7
260.53	-0.14	0.8	338.12	27FE	-0.4	-4.3	-1.6	0.7
267.11	-0.36	0.7	347.98	8MR	-1.1	-4.2	-1.6	0.7
273.65	-0.60	0.5	357.78	18MR	-1.7	-4.1	-1.6	0.6
280.12	-0.87	0.4	7.52	28MR	-1.0	-3.6	-1.6	0.5
286.50	-1.17	0.2	17.21	7AP	0.1	-3.0	-1.6	0.5
292.75	-1.51	0.0	26.86	17AP	1.2	-3.6	-1.6	0.4
298.82	-1.87	-0.2	36.46	27AP	2.5	-4.0	-1.6	0.4
304.66	-2.28	-0.3	46.03	7MY	3.6	-4.2	-1.7	0.3
310.20	-2.72	-0.6	55.57	17MY	2.1	-4.2	-1.7	0.2
315.33	-3.21	-0.8	65.10	27MY	1.0	-4.1	-1.7	0.3
319.93	-3.73	-1.0	74.62	6JN	0.1	-4.0	-1.8	0.3
323.86	-4.28	-1.2	84.15	16JN	-1.0	-3.9	-1.8	0.4
326.91	-4.85	-1.5	93.68	26JN	-1.7	-3.8	-1.9	0.5
328.89	-5.43	-1.8	103.24	6JL	-1.0	-3.8	-2.0	0.5
329.60	-5.95	-2.0	112.82	16JL	-0.3	-3.7	-2.0	0.6
328.93	-6.35	-2.3	122.44	26JL	0.1	-3.6	-2.1	0.6
327.03	-6.54	-2.5	132.10	5AU	0.3	-3.6	-2.2	0.7
324.34	-6.44	-2.6	141.82	15AU	0.6	-3.5	-2.2	0.7
321.59	-6.03	-2.5	151.59	25AU	1.2	-3.5	-2.3	0.7
319.54	-5.39	-2.2	161.41	4SE	3.2	-3.4	-2.3	0.8
318.64	-4.63	-1.9	171.30	14SE	0.8	-3.4	-2.4	0.8
319.06	-3.86	-1.6	181.24	24SE	-0.6	-3.4	-2.4	0.8
320.71	-3.14	-1.3	191.24	4OC	-0.8	-3.4	-2.4	0.8
323.41	-2.50	-0.9	201.29	14OC	-0.8	-3.4	-2.4	0.8
326.95	-1.94	-0.7	211.39	24OC	-0.8	-3.4	-2.4	0.8
331.15	-1.45	-0.4	221.53	3NO	-0.6	-3.4	-2.4	0.7
335.86	-1.04	-0.1	231.70	13NO	-0.6	-3.4	-2.3	0.7
340.97	-0.69	0.1	241.90	23NO	-0.6	-3.4	-2.2	0.7
346.38	-0.39	0.3	252.10	3DE	-0.3	-3.4	-2.2	0.7
352.01	-0.14	0.5	262.31	13DE	1.1	-3.4	-2.1	0.7
357.82	0.08	0.7	272.50	23DE	2.3	-3.4	-2.0	0.8
				221				
3.76	0.27	0.8	282.67	2JA	0.6	-3.4	-2.0	0.8
9.80	0.43	1.0	292.82	12JA	0.2	-3.5	-1.9	0.8
15.90	0.57	1.1	302.92	22JA	0.1	-3.5	-1.8	0.8
22.06	0.69	1.3	312.97	1FE	-0.1	-3.5	-1.8	0.8
28.25	0.79	1.4	322.98	11FE	-0.5	-3.4	-1.7	0.8
34.47	0.87	1.5	332.93	21FE	-1.1	-3.4	-1.7	0.8
40.70	0.95	1.6	342.81	3MR	-1.5	-3.4	-1.6	0.7
46.94	1.01	1.7	352.64	13MR	-1.0	-3.4	-1.6	0.7
53.18	1.06	1.7	2.42	23MR	0.2	-3.3	-1.6	0.6
59.43	1.10	1.8	12.13	2AP	1.5	-3.3	-1.5	0.6
65.67	1.13	1.9	21.80	12AP	3.2	-3.3	-1.5	0.5
71.91	1.16	1.9	31.43	22AP	2.8	-3.3	-1.5	0.5
78.16	1.18	2.0	41.01	2MY	1.6	-3.3	-1.5	0.4
84.40	1.20	2.0	50.57	12MY	0.8	-3.3	-1.5	0.3
90.65	1.21	2.0	60.11	22MY	-0.0	-3.4	-1.5	0.3
96.91	1.21	2.0	69.63	1JN	-1.0	-3.4	-1.6	0.2
103.18	1.21	2.0	79.15	11JN	-1.8	-3.4	-1.6	0.3
109.47	1.20	2.0	88.68	21JN	-1.0	-3.5	-1.6	0.3
115.77	1.19	2.0	98.22	1JL	-0.3	-3.5	-1.7	0.4
122.09	1.18	2.0	107.79	11JL	0.2	-3.6	-1.7	0.5
128.44	1.16	2.0	117.39	21JL	0.5	-3.6	-1.8	0.5
134.82	1.13	2.0	127.03	31JL	0.8	-3.7	-1.8	0.6
141.24	1.10	2.0	136.72	10AU	1.6	-3.8	-1.9	0.6
147.69	1.07	1.9	146.46	20AU	3.2	-3.9	-1.9	0.7
154.18	1.03	1.9	156.25	30AU	0.7	-4.0	-2.0	0.7
160.72	0.99	1.9	166.11	9SE	-0.7	-4.1	-2.1	0.7
167.30	0.94	1.9	176.02	19SE	-1.0	-4.2	-2.1	0.8
173.94	0.89	1.9	185.99	29SE	-1.0	-4.3	-2.2	0.8
180.62	0.83	1.9	196.02	9OC	-0.7	-4.4	-2.3	0.8
187.36	0.76	1.9	206.09	19OC	-0.5	-4.2	-2.3	0.8
194.16	0.69	1.9	216.21	29OC	-0.4	-3.9	-2.3	0.8
201.01	0.62	1.8	226.37	8NO	-0.4	-3.0	-2.3	0.7
207.92	0.53	1.8	236.55	18NO	-0.2	-3.6	-2.3	0.7
214.88	0.44	1.8	246.75	28NO	1.3	-4.2	-2.3	0.7
221.90	0.34	1.7	256.96	8DE	1.9	-4.4	-2.3	0.7
228.98	0.23	1.7	267.16	18DE	0.4	-4.4	-2.2	0.7
236.11	0.12	1.6	277.34	28DE	0.0	-4.3	-2.2	0.8

♂			☉	16.00UT	☿	♀	♃	♄
LONG	LAT	MAG	LONG	222	MAGNITUDES			
243.29	-0.01	1.5	287.51	7JA	-0.1	-4.2	-2.1	0.8
250.52	-0.14	1.5	297.63	17JA	-0.2	-4.1	-2.0	0.8
257.79	-0.28	1.4	307.71	27JA	-0.5	-4.0	-1.9	0.8
265.11	-0.43	1.3	317.75	6FE	-1.1	-3.9	-1.9	0.8
272.46	-0.58	1.3	327.72	16FE	-1.4	-3.8	-1.8	0.8
279.84	-0.75	1.2	337.64	26FE	-0.9	-3.7	-1.7	0.8
287.24	-0.91	1.1	347.50	8MR	0.3	-3.6	-1.7	0.8
294.65	-1.09	1.0	357.30	18MR	1.9	-3.5	-1.6	0.7
302.06	-1.26	0.9	7.05	28MR	3.6	-3.5	-1.6	0.7
309.45	-1.43	0.9	16.75	7AP	2.1	-3.4	-1.5	0.7
316.81	-1.61	0.8	26.39	17AP	1.2	-3.4	-1.5	0.6
324.11	-1.77	0.7	36.00	27AP	0.6	-3.4	-1.5	0.5
331.34	-1.94	0.6	45.57	7MY	-0.1	-3.3	-1.4	0.5
338.46	-2.09	0.5	55.11	17MY	-1.0	-3.3	-1.4	0.4
345.45	-2.23	0.4	64.64	27MY	-1.8	-3.3	-1.4	0.3
352.27	-2.36	0.3	74.17	6JN	-1.0	-3.3	-1.4	0.3
358.89	-2.48	0.2	83.69	16JN	-0.2	-3.3	-1.4	0.3
5.25	-2.58	0.1	93.22	26JN	0.3	-3.3	-1.4	0.3
11.28	-2.66	-0.1	102.78	6JL	0.7	-3.4	-1.5	0.4
16.93	-2.71	-0.2	112.36	16JL	1.1	-3.4	-1.5	0.4
22.11	-2.75	-0.3	121.98	26JL	2.1	-3.4	-1.5	0.5
26.68	-2.76	-0.5	131.64	5AU	2.8	-3.4	-1.6	0.5
30.52	-2.73	-0.7	141.35	15AU	0.6	-3.5	-1.6	0.6
33.44	-2.67	-0.9	151.11	25AU	-0.8	-3.5	-1.7	0.6
35.23	-2.54	-1.1	160.94	4SE	-1.1	-3.5	-1.7	0.7
35.70	-2.34	-1.3	170.81	14SE	-1.1	-3.4	-1.8	0.7
34.70	-2.05	-1.5	180.75	24SE	-0.6	-3.4	-1.8	0.8
32.29	-1.66	-1.7	190.75	4OC	-0.4	-3.4	-1.9	0.8
28.90	-1.18	-1.9	200.80	14OC	-0.3	-3.4	-2.0	0.8
25.21	-0.65	-1.8	210.90	24OC	-0.2	-3.3	-2.0	0.8
22.06	-0.14	-1.5	221.04	3NO	0.0	-3.3	-2.1	0.8
20.04	0.31	-1.2	231.20	13NO	1.5	-3.3	-2.1	0.8
19.35	0.67	-0.9	241.40	23NO	1.7	-3.3	-2.2	0.8
19.98	0.95	-0.6	251.60	3DE	0.2	-3.3	-2.2	0.7
21.73	1.15	-0.3	261.81	13DE	-0.1	-3.4	-2.2	0.7
24.40	1.30	0.0	272.01	23DE	-0.2	-3.4	-2.2	0.7
				223				
27.79	1.41	0.3	282.18	2JA	-0.3	-3.4	-2.2	0.8
31.75	1.49	0.5	292.32	12JA	-0.6	-3.4	-2.1	0.8
36.14	1.54	0.7	302.43	22JA	-1.1	-3.5	-2.1	0.8
40.88	1.57	0.9	312.49	1FE	-1.3	-3.5	-2.0	0.8
45.88	1.59	1.0	322.50	11FE	-0.9	-3.6	-2.0	0.9
51.09	1.60	1.2	332.45	21FE	0.4	-3.6	-1.9	0.9
56.48	1.59	1.3	342.34	3MR	2.4	-3.7	-1.8	0.9
61.99	1.58	1.4	352.17	13MR	2.8	-3.8	-1.8	0.8
67.62	1.57	1.6	1.95	23MR	1.5	-3.8	-1.7	0.8
73.34	1.55	1.6	11.67	2AP	0.9	-3.9	-1.6	0.8
79.14	1.52	1.7	21.34	12AP	0.4	-4.0	-1.6	0.7
85.00	1.49	1.8	30.97	22AP	-0.2	-4.1	-1.5	0.7
90.92	1.45	1.8	40.55	2MY	-1.0	-4.2	-1.5	0.6
96.90	1.42	1.9	50.11	12MY	-1.9	-4.2	-1.4	0.6
102.93	1.38	1.9	59.65	22MY	-1.0	-4.1	-1.4	0.5
109.01	1.33	2.0	69.17	1JN	-0.2	-3.8	-1.4	0.4
115.14	1.29	2.0	78.69	11JN	0.4	-3.1	-1.3	0.4
121.32	1.24	2.0	88.22	21JN	0.9	-3.2	-1.3	0.3
127.55	1.18	2.0	97.76	1JL	1.5	-3.8	-1.3	0.3
133.84	1.13	2.0	107.33	11JL	2.7	-4.1	-1.3	0.4
140.19	1.07	2.0	116.93	21JL	2.4	-4.2	-1.3	0.4
146.59	1.00	2.0	126.56	31JL	0.4	-4.2	-1.3	0.5
153.06	0.94	2.0	136.25	10AU	-0.9	-4.1	-1.4	0.5
159.59	0.87	1.9	145.99	20AU	-1.3	-4.0	-1.4	0.6
166.19	0.79	1.9	155.78	30AU	-1.1	-4.0	-1.4	0.6
172.86	0.72	1.8	165.63	9SE	-0.6	-3.9	-1.5	0.7
179.61	0.64	1.8	175.54	19SE	-0.3	-3.8	-1.5	0.7
186.43	0.55	1.7	185.50	29SE	-0.1	-3.7	-1.6	0.8
193.32	0.46	1.7	195.53	9OC	-0.1	-3.7	-1.6	0.8
200.30	0.37	1.7	205.60	19OC	0.2	-3.6	-1.7	0.8
207.35	0.27	1.7	215.72	29OC	1.8	-3.6	-1.7	0.8
214.47	0.17	1.7	225.87	8NO	1.4	-3.5	-1.8	0.8
221.68	0.07	1.7	236.06	18NO	-0.0	-3.5	-1.9	0.8
228.95	-0.04	1.7	246.25	28NO	-0.3	-3.4	-1.9	0.8
236.31	-0.15	1.6	256.46	8DE	-0.3	-3.4	-2.0	0.8
243.72	-0.26	1.6	266.67	18DE	-0.4	-3.4	-2.0	0.8
251.20	-0.38	1.6	276.85	28DE	-0.7	-3.4	-2.1	0.7

224

♂ LONG	LAT	MAG	☉ LONG	16.00UT	☿	♀	♃	♄
258.75	-0.49	1.5	287.01	7JA	-1.1	-3.3	-2.1	0.8
266.34	-0.61	1.5	297.14	17JA	-1.1	-3.3	-2.1	0.8
273.98	-0.73	1.5	307.22	27JA	-0.8	-3.3	-2.1	0.9
281.65	-0.84	1.4	317.26	6FE	0.5	-3.3	-2.1	0.9
289.35	-0.95	1.4	327.24	16FE	2.8	-3.3	-2.0	0.9
297.07	-1.05	1.4	337.16	26FE	2.1	-3.4	-2.0	0.9
304.79	-1.15	1.3	347.02	7MR	1.1	-3.4	-1.9	0.9
312.51	-1.24	1.3	356.83	17MR	0.7	-3.4	-1.8	0.9
320.20	-1.32	1.2	6.58	27MR	0.3	-3.4	-1.8	0.9
327.86	-1.38	1.2	16.27	6AP	-0.2	-3.5	-1.7	0.9
335.47	-1.44	1.2	25.92	16AP	-1.0	-3.5	-1.6	0.8
343.02	-1.48	1.1	35.53	26AP	-1.9	-3.4	-1.6	0.8
350.49	-1.51	1.1	45.10	6MY	-1.0	-3.4	-1.5	0.7
357.87	-1.52	1.1	54.65	16MY	-0.1	-3.4	-1.5	0.7
5.14	-1.51	1.0	64.17	26MY	0.6	-3.4	-1.4	0.6
12.29	-1.49	1.0	73.70	5JN	1.2	-3.3	-1.4	0.5
19.30	-1.45	0.9	83.22	15JN	2.0	-3.3	-1.3	0.5
26.15	-1.40	0.9	92.75	25JN	3.3	-3.3	-1.3	0.4
32.82	-1.33	0.8	102.31	5JL	2.0	-3.3	-1.3	0.3
39.29	-1.24	0.8	111.89	15JL	0.3	-3.3	-1.3	0.4
45.52	-1.13	0.7	121.50	25JL	-0.9	-3.4	-1.2	0.4
51.49	-1.00	0.6	131.16	4AU	-1.4	-3.4	-1.2	0.5
57.14	-0.86	0.5	140.87	14AU	-1.1	-3.4	-1.2	0.5
62.42	-0.69	0.4	150.63	24AU	-0.5	-3.4	-1.3	0.6
67.26	-0.49	0.3	160.46	3SE	-0.2	-3.5	-1.3	0.6
71.55	-0.26	0.2	170.33	13SE	-0.0	-3.6	-1.3	0.7
75.18	0.00	-0.0	180.27	23SE	0.1	-3.6	-1.3	0.7
78.00	0.31	-0.2	190.27	3OC	0.5	-3.7	-1.4	0.8
79.81	0.66	-0.4	200.31	13OC	2.1	-3.8	-1.4	0.8
80.42	1.07	-0.6	210.41	23OC	1.2	-3.9	-1.5	0.8
79.66	1.51	-0.8	220.55	2NO	-0.2	-4.0	-1.5	0.9
77.50	1.98	-1.0	230.71	12NO	-0.4	-4.1	-1.6	0.9
74.17	2.41	-1.2	240.90	22NO	-0.4	-4.2	-1.6	0.9
70.24	2.77	-1.3	251.11	2DE	-0.5	-4.3	-1.7	0.9
66.48	2.99	-1.1	261.32	12DE	-0.7	-4.4	-1.8	0.9
63.60	3.08	-0.8	271.51	22DE	-0.9	-4.4	-1.8	0.9

225

♂ LONG	LAT	MAG	☉ LONG	16.00UT	☿	♀	♃	♄
61.96	3.07	-0.5	281.69	1JA	-1.0	-4.3	-1.9	0.8
61.67	2.99	-0.3	291.83	11JA	-0.7	-3.9	-1.9	0.8
62.57	2.88	0.0	301.94	21JA	0.6	-3.3	-2.0	0.9
64.50	2.75	0.3	312.00	31JA	3.1	-3.6	-2.0	0.9
67.24	2.61	0.5	322.01	10FE	1.6	-4.1	-2.0	1.0
70.64	2.48	0.7	331.96	20FE	0.8	-4.3	-2.0	1.0
74.55	2.34	0.9	341.86	2MR	0.5	-4.3	-2.0	1.0
78.88	2.22	1.0	351.69	12MR	0.2	-4.2	-2.0	1.0
83.53	2.10	1.2	1.47	22MR	-0.3	-4.1	-1.9	1.0
88.45	1.98	1.3	11.20	1AP	-1.0	-4.0	-1.9	1.0
93.59	1.87	1.4	20.87	11AP	-1.8	-3.9	-1.8	1.0
98.91	1.76	1.5	30.50	21AP	-1.0	-3.8	-1.8	0.9
104.40	1.66	1.6	40.09	1MY	-0.0	-3.7	-1.7	0.9
110.02	1.56	1.7	49.65	11MY	0.8	-3.6	-1.6	0.9
115.76	1.45	1.7	59.18	21MY	1.6	-3.6	-1.6	0.8
121.62	1.36	1.8	68.71	31MY	2.7	-3.5	-1.5	0.7
127.58	1.26	1.8	78.23	10JN	3.1	-3.5	-1.5	0.7
133.64	1.16	1.8	87.76	20JN	1.6	-3.4	-1.4	0.6
139.79	1.06	1.9	97.30	30JN	0.2	-3.4	-1.4	0.5
146.04	0.96	1.9	106.86	10JL	-0.9	-3.4	-1.3	0.5
152.39	0.86	1.9	116.46	20JL	-1.5	-3.3	-1.3	0.4
158.82	0.76	1.9	126.10	30JL	-1.1	-3.3	-1.3	0.5
165.35	0.66	1.9	135.78	9AU	-0.4	-3.3	-1.2	0.5
171.97	0.56	1.8	145.51	19AU	-0.1	-3.3	-1.2	0.6
178.69	0.46	1.8	155.31	29AU	0.1	-3.4	-1.2	0.6
185.51	0.35	1.8	165.15	8SE	0.3	-3.4	-1.2	0.7
192.42	0.25	1.7	175.06	18SE	0.7	-3.4	-1.2	0.7
199.43	0.14	1.7	185.02	28SE	2.5	-3.4	-1.2	0.8
206.53	0.04	1.7	195.04	8OC	1.1	-3.4	-1.3	0.8
213.72	-0.07	1.6	205.11	18OC	-0.4	-3.5	-1.3	0.9
221.01	-0.18	1.6	215.23	28OC	-0.6	-3.5	-1.3	0.9
228.38	-0.28	1.5	225.38	7NO	-0.6	-3.5	-1.4	0.9
235.84	-0.38	1.5	235.56	17NO	-0.6	-3.4	-1.4	0.9
243.38	-0.49	1.5	245.76	27NO	-0.8	-3.4	-1.5	1.0
250.98	-0.58	1.5	255.96	7DE	-0.8	-3.4	-1.5	1.0
258.66	-0.68	1.4	266.17	17DE	-0.8	-3.4	-1.6	1.0
266.39	-0.77	1.4	276.36	27DE	-0.6	-3.4	-1.7	1.0

226

♂ LONG	LAT	MAG	☉ LONG	16.00UT	☿	♀	♃	♄
274.16	-0.85	1.4	286.52	6JA	0.8	-3.4	-1.7	1.0
281.97	-0.93	1.4	296.65	16JA	2.9	-3.3	-1.8	0.9
289.81	-0.99	1.4	306.73	26JA	1.2	-3.3	-1.8	1.0
297.65	-1.05	1.4	316.77	5FE	0.5	-3.3	-1.9	1.0
305.50	-1.10	1.4	326.75	15FE	0.3	-3.3	-2.0	1.1
313.33	-1.14	1.4	336.68	25FE	0.1	-3.4	-2.0	1.1
321.14	-1.16	1.4	346.54	7MR	-0.4	-3.4	-2.0	1.1
328.92	-1.17	1.4	356.35	17MR	-1.0	-3.4	-2.0	1.1
336.64	-1.17	1.4	6.11	27MR	-1.7	-3.4	-2.0	1.1
344.30	-1.16	1.4	15.80	6AP	-1.0	-3.5	-2.0	1.1
351.90	-1.13	1.4	25.46	16AP	0.0	-3.5	-2.0	1.1
359.41	-1.10	1.4	35.07	26AP	1.0	-3.6	-1.9	1.1
6.84	-1.04	1.4	44.64	6MY	2.1	-3.6	-1.9	1.1
14.18	-0.98	1.4	54.19	16MY	3.5	-3.7	-1.8	1.0
21.40	-0.91	1.4	63.72	26MY	2.5	-3.8	-1.8	1.0
28.52	-0.82	1.4	73.24	5JN	1.2	-3.8	-1.7	0.9
35.53	-0.73	1.4	82.76	15JN	0.1	-3.9	-1.6	0.9
42.41	-0.62	1.4	92.30	25JN	-1.0	-4.0	-1.6	0.8
49.16	-0.51	1.4	101.85	5JL	-1.6	-4.1	-1.5	0.7
55.77	-0.38	1.3	111.43	15JL	-1.0	-4.2	-1.4	0.6
62.23	-0.25	1.3	121.04	25JL	-0.4	-4.2	-1.4	0.6
68.52	-0.11	1.3	130.70	4AU	0.0	-4.1	-1.4	0.6
74.64	0.05	1.2	140.40	14AU	0.3	-3.8	-1.3	0.6
80.55	0.22	1.2	150.16	24AU	0.5	-3.5	-1.3	0.6
86.23	0.40	1.1	159.98	3SE	1.0	-3.6	-1.3	0.7
91.65	0.60	1.0	169.86	13SE	2.9	-4.1	-1.2	0.7
96.76	0.82	0.9	179.79	23SE	0.9	-4.3	-1.2	0.8
101.49	1.05	0.8	189.78	3OC	-0.5	-4.3	-1.2	0.8
105.78	1.32	0.6	199.83	13OC	-0.7	-4.3	-1.2	0.9
109.51	1.62	0.5	209.92	23OC	-0.7	-4.2	-1.2	0.9
112.56	1.95	0.3	220.05	2NO	-0.8	-4.1	-1.3	1.0
114.78	2.32	0.1	230.22	12NO	-0.7	-4.0	-1.3	1.0
115.97	2.72	-0.1	240.41	22NO	-0.6	-3.9	-1.3	1.0
115.95	3.16	-0.4	250.61	2DE	-0.7	-3.8	-1.4	1.1
114.59	3.59	-0.6	260.82	12DE	-0.5	-3.7	-1.4	1.1
111.94	3.98	-0.8	271.02	22DE	0.9	-3.7	-1.5	1.1

227

♂ LONG	LAT	MAG	☉ LONG	16.00UT	☿	♀	♃	♄
108.32	4.25	-1.0	281.20	1JA	2.6	-3.6	-1.5	1.1
104.34	4.37	-1.0	291.35	11JA	0.8	-3.5	-1.6	1.1
100.76	4.31	-0.8	301.46	21JA	0.3	-3.5	-1.6	1.1
98.15	4.12	-0.6	311.52	31JA	0.2	-3.4	-1.7	1.1
96.82	3.85	-0.3	321.53	10FE	-0.0	-3.4	-1.8	1.1
96.77	3.54	-0.0	331.49	20FE	-0.4	-3.4	-1.9	1.2
97.87	3.23	0.2	341.39	2MR	-1.0	-3.3	-1.9	1.2
99.94	2.94	0.4	351.23	12MR	-1.6	-3.3	-2.0	1.2
102.80	2.66	0.6	1.01	22MR	-1.0	-3.3	-2.0	1.3
106.28	2.41	0.8	10.73	1AP	0.1	-3.3	-2.0	1.3
110.28	2.18	0.9	20.41	11AP	1.3	-3.3	-2.1	1.3
114.70	1.97	1.1	30.04	21AP	2.7	-3.3	-2.1	1.3
119.46	1.77	1.2	39.63	1MY	3.3	-3.3	-2.1	1.3
124.51	1.58	1.3	49.19	11MY	1.9	-3.4	-2.0	1.3
129.80	1.41	1.4	58.73	21MY	1.0	-3.4	-2.0	1.2
135.31	1.24	1.4	68.25	31MY	0.1	-3.4	-2.0	1.2
141.01	1.09	1.5	77.77	10JN	-1.0	-3.5	-1.9	1.1
146.88	0.94	1.5	87.30	20JN	-1.7	-3.5	-1.9	1.1
152.90	0.79	1.6	96.84	30JN	-1.0	-3.5	-1.8	1.0
159.08	0.65	1.6	106.40	10JL	-0.3	-3.4	-1.7	0.9
165.40	0.51	1.6	115.99	20JL	0.1	-3.4	-1.7	0.9
171.85	0.38	1.6	125.63	30JL	0.4	-3.4	-1.6	0.8
178.43	0.25	1.6	135.31	9AU	0.7	-3.3	-1.5	0.7
185.14	0.13	1.6	145.03	19AU	1.4	-3.3	-1.5	0.7
191.98	0.00	1.6	154.82	29AU	3.2	-3.3	-1.4	0.7
198.93	-0.11	1.6	164.67	8SE	0.8	-3.3	-1.4	0.8
206.00	-0.23	1.6	174.57	18SE	-0.6	-3.3	-1.4	0.8
213.19	-0.34	1.6	184.53	28SE	-0.9	-3.3	-1.3	0.9
220.48	-0.44	1.6	194.55	8OC	-0.9	-3.4	-1.3	0.9
227.87	-0.54	1.5	204.61	18OC	-0.8	-3.4	-1.3	1.0
235.37	-0.64	1.5	214.73	28OC	-0.6	-3.4	-1.3	1.0
242.95	-0.72	1.5	224.88	7NO	-0.5	-3.5	-1.3	1.1
250.61	-0.80	1.5	235.06	17NO	-0.5	-3.5	-1.3	1.1
258.34	-0.88	1.4	245.26	27NO	-0.3	-3.6	-1.3	1.2
266.13	-0.94	1.4	255.47	7DE	1.1	-3.6	-1.3	1.2
273.97	-0.99	1.4	265.67	17DE	2.2	-3.7	-1.4	1.2
281.84	-1.03	1.3	275.87	27DE	0.5	-3.8	-1.4	1.2

♂ LONG	♂ LAT	♂ MAG	⊙ LONG	16.00UT	☿	♀	♃	♄
				228	MAGNITUDES			
289.73	-1.06	1.3	286.03	6JA	0.2	-3.9	-1.5	1.3
297.63	-1.08	1.3	296.16	16JA	0.0	-4.0	-1.5	1.3
305.53	-1.09	1.3	306.25	26JA	-0.1	-4.1	-1.6	1.3
313.40	-1.08	1.3	316.29	5FE	-0.5	-4.1	-1.6	1.3
321.25	-1.07	1.3	326.27	15FE	-1.1	-4.2	-1.7	1.3
329.05	-1.04	1.4	336.21	25FE	-1.5	-4.3	-1.8	1.3
336.80	-1.00	1.4	346.07	6MR	-1.0	-4.2	-1.8	1.4
344.48	-0.95	1.4	355.88	16MR	0.2	-4.1	-1.9	1.4
352.10	-0.89	1.5	5.64	26MR	1.7	-3.6	-2.0	1.4
359.63	-0.82	1.5	15.34	5AP	3.5	-3.0	-2.0	1.4
7.09	-0.75	1.5	24.99	15AP	2.5	-3.6	-2.1	1.3
14.45	-0.66	1.6	34.60	25AP	1.4	-4.0	-2.1	1.3
21.72	-0.57	1.6	44.18	5MY	0.7	-4.2	-2.2	1.3
28.90	-0.48	1.6	53.72	15MY	-0.0	-4.2	-2.2	1.2
35.99	-0.38	1.6	63.26	25MY	-1.0	-4.1	-2.2	1.2
42.97	-0.27	1.6	72.78	4JN	-1.8	-4.0	-2.2	1.2
49.86	-0.16	1.7	82.30	14JN	-1.1	-3.9	-2.1	1.1
56.65	-0.05	1.7	91.83	24JN	-0.3	-3.8	-2.1	1.1
63.34	0.07	1.7	101.38	4JL	0.2	-3.8	-2.1	1.0
69.93	0.19	1.7	110.96	14JL	0.6	-3.7	-2.0	1.0
76.41	0.32	1.7	120.57	24JL	0.9	-3.6	-1.9	0.9
82.79	0.45	1.7	130.23	3AU	1.8	-3.6	-1.9	0.9
89.05	0.58	1.6	139.93	13AU	3.1	-3.5	-1.8	0.8
95.19	0.72	1.6	149.69	23AU	0.7	-3.5	-1.7	0.8
101.19	0.87	1.6	159.50	2SE	-0.7	-3.4	-1.7	0.8
107.05	1.02	1.5	169.37	12SE	-1.0	-3.4	-1.6	0.9
112.75	1.18	1.4	179.31	22SE	-1.0	-3.4	-1.6	1.0
118.25	1.35	1.4	189.29	2OC	-0.7	-3.4	-1.5	1.0
123.53	1.53	1.3	199.33	12OC	-0.5	-3.4	-1.5	1.1
128.56	1.72	1.2	209.43	22OC	-0.4	-3.4	-1.5	1.1
133.26	1.93	1.0	219.56	1NO	-0.3	-3.4	-1.4	1.2
137.59	2.17	0.9	229.72	11NO	-0.1	-3.4	-1.4	1.2
141.44	2.42	0.7	239.92	21NO	1.3	-3.4	-1.4	1.3
144.70	2.70	0.5	250.12	1DE	1.9	-3.4	-1.4	1.3
147.25	3.01	0.3	260.32	11DE	0.3	-3.4	-1.4	1.3
148.90	3.34	0.1	270.53	21DE	-0.0	-3.4	-1.4	1.3
149.48	3.68	-0.2	280.70	31DE	-0.1	-3.4	-1.4	1.3
				229				
148.82	4.02	-0.4	290.85	10JA	-0.2	-3.5	-1.4	1.3
146.85	4.30	-0.7	300.97	20JA	-0.5	-3.5	-1.5	1.3
143.74	4.48	-0.9	311.03	30JA	-1.1	-3.5	-1.5	1.2
139.93	4.50	-1.1	321.05	9FE	-1.4	-3.4	-1.6	1.2
136.10	4.34	-0.9	331.01	19FE	-1.0	-3.4	-1.6	1.1
132.96	4.04	-0.8	340.90	1MR	0.3	-3.4	-1.7	1.1
130.94	3.64	-0.5	350.75	11MR	2.1	-3.4	-1.7	1.1
130.23	3.22	-0.3	0.53	21MR	3.3	-3.3	-1.8	1.1
130.76	2.80	-0.1	10.26	31MR	1.8	-3.3	-1.9	1.1
132.38	2.41	0.1	19.94	10AP	1.1	-3.3	-1.9	1.1
134.92	2.05	0.3	29.57	20AP	0.5	-3.3	-2.0	1.1
138.20	1.72	0.5	39.16	30AP	-0.1	-3.3	-2.1	1.0
142.09	1.43	0.6	48.72	10MY	-1.0	-3.4	-2.1	1.0
146.49	1.16	0.7	58.26	20MY	-1.8	-3.4	-2.2	1.0
151.31	0.92	0.9	67.79	30MY	-1.1	-3.4	-2.2	1.0
156.48	0.69	0.9	77.31	9JN	-0.2	-3.4	-2.3	0.9
161.96	0.49	1.0	86.84	19JN	0.3	-3.5	-2.3	0.9
167.70	0.30	1.1	96.37	29JN	0.8	-3.5	-2.3	0.8
173.68	0.12	1.2	105.94	9JL	1.3	-3.6	-2.3	0.8
179.88	-0.04	1.2	115.53	19JL	2.3	-3.6	-2.3	0.7
186.28	-0.20	1.2	125.16	29JL	2.7	-3.7	-2.2	0.7
192.86	-0.34	1.3	134.84	8AU	0.6	-3.8	-2.2	0.6
199.61	-0.47	1.3	144.57	18AU	-0.8	-3.9	-2.1	0.6
206.51	-0.59	1.3	154.35	28AU	-1.2	-4.0	-2.1	0.5
213.57	-0.70	1.3	164.19	7SE	-1.1	-4.1	-2.0	0.5
220.76	-0.80	1.3	174.09	17SE	-0.6	-4.2	-1.9	0.5
228.08	-0.89	1.4	184.05	27SE	-0.3	-4.3	-1.9	0.6
235.52	-0.97	1.4	194.06	7OC	-0.2	-4.3	-1.8	0.7
243.06	-1.04	1.4	204.13	17OC	-0.2	-4.2	-1.7	0.7
250.69	-1.09	1.4	214.24	27OC	0.1	-3.8	-1.7	0.8
258.40	-1.13	1.4	224.39	6NO	1.6	-3.0	-1.6	0.9
266.18	-1.16	1.4	234.57	16NO	1.6	-3.6	-1.6	0.9
274.01	-1.17	1.4	244.76	26NO	0.1	-4.2	-1.6	1.0
281.87	-1.17	1.4	254.97	6DE	-0.2	-4.4	-1.5	1.0
289.76	-1.16	1.4	265.18	16DE	-0.2	-4.4	-1.5	1.0
297.65	-1.14	1.4	275.36	26DE	-0.3	-4.3	-1.5	1.0

♂ LONG	♂ LAT	♂ MAG	⊙ LONG	16.00UT	☿	♀	♃	♄
				230	MAGNITUDES			
305.53	-1.11	1.4	285.53	5JA	-0.6	-4.2	-1.5	1.0
313.39	-1.06	1.4	295.67	15JA	-1.1	-4.1	-1.5	1.0
321.20	-1.00	1.4	305.76	25JA	-1.2	-4.0	-1.5	1.0
328.98	-0.94	1.4	315.80	4FE	-0.9	-3.9	-1.5	1.0
336.70	-0.87	1.4	325.79	14FE	0.4	-3.8	-1.5	0.9
344.34	-0.79	1.5	335.72	24FE	2.5	-3.7	-1.6	0.9
351.92	-0.70	1.5	345.60	6MR	2.5	-3.6	-1.6	0.8
359.42	-0.61	1.5	355.41	16MR	1.3	-3.5	-1.7	0.8
6.83	-0.52	1.5	5.17	26MR	0.8	-3.5	-1.7	0.8
14.16	-0.42	1.5	14.88	5AP	0.4	-3.4	-1.8	0.8
21.40	-0.32	1.6	24.53	15AP	-0.2	-3.4	-1.8	0.8
28.56	-0.21	1.6	34.14	25AP	-1.0	-3.4	-1.9	0.8
35.63	-0.11	1.7	43.72	5MY	-1.9	-3.3	-2.0	0.8
42.61	-0.01	1.7	53.27	15MY	-1.1	-3.3	-2.0	0.8
49.51	0.10	1.7	62.80	25MY	-0.2	-3.3	-2.1	0.8
56.34	0.20	1.8	72.33	4JN	0.5	-3.3	-2.2	0.7
63.08	0.30	1.8	81.85	14JN	1.0	-3.3	-2.2	0.7
69.74	0.40	1.8	91.38	24JN	1.7	-3.3	-2.3	0.7
76.34	0.51	1.9	100.93	4JL	2.9	-3.4	-2.4	0.6
82.86	0.61	1.9	110.50	14JL	2.3	-3.4	-2.4	0.6
89.31	0.71	1.9	120.11	24JL	0.5	-3.4	-2.4	0.5
95.69	0.81	1.9	129.76	3AU	-0.9	-3.4	-2.4	0.5
102.00	0.91	1.9	139.46	13AU	-1.3	-3.5	-2.4	0.4
108.24	1.01	1.9	149.21	23AU	-1.1	-3.5	-2.4	0.4
114.41	1.11	1.8	159.03	2SE	-0.6	-3.5	-2.3	0.3
120.49	1.22	1.8	168.89	12SE	-0.2	-3.4	-2.3	0.3
126.50	1.32	1.8	178.82	22SE	-0.1	-3.4	-2.2	0.2
132.40	1.43	1.7	188.81	2OC	-0.0	-3.4	-2.2	0.3
138.20	1.54	1.7	198.84	12OC	0.3	-3.4	-2.1	0.3
143.88	1.65	1.6	208.93	22OC	1.8	-3.3	-2.0	0.4
149.41	1.77	1.5	219.06	1NO	1.4	-3.3	-1.9	0.5
154.77	1.89	1.4	229.22	11NO	-0.1	-3.3	-1.9	0.6
159.92	2.02	1.3	239.41	21NO	-0.3	-3.3	-1.8	0.6
164.83	2.16	1.1	249.62	1DE	-0.4	-3.3	-1.8	0.7
169.44	2.31	1.0	259.82	11DE	-0.5	-3.4	-1.7	0.7
173.68	2.46	0.8	270.02	21DE	-0.7	-3.4	-1.7	0.7
177.45	2.62	0.6	280.20	31DE	-1.0	-3.4	-1.6	0.8
				231				
180.65	2.79	0.4	290.35	10JA	-1.1	-3.4	-1.6	0.8
183.13	2.97	0.1	300.47	20JA	-0.8	-3.5	-1.6	0.8
184.73	3.13	-0.1	310.54	30JA	0.5	-3.5	-1.6	0.8
185.26	3.28	-0.4	320.56	9FE	2.9	-3.6	-1.6	0.8
184.57	3.38	-0.7	330.52	19FE	1.9	-3.6	-1.6	0.7
182.60	3.40	-1.0	340.43	1MR	1.0	-3.7	-1.6	0.7
179.54	3.30	-1.3	350.27	11MR	0.6	-3.8	-1.6	0.6
175.85	3.05	-1.4	0.06	21MR	0.2	-3.8	-1.6	0.6
172.22	2.68	-1.3	9.79	31MR	-0.2	-3.9	-1.6	0.6
169.35	2.22	-1.1	19.47	10AP	-1.0	-4.0	-1.7	0.6
167.66	1.73	-0.9	29.11	20AP	-1.9	-4.1	-1.7	0.6
167.32	1.27	-0.7	38.70	30AP	-1.1	-4.2	-1.8	0.6
168.25	0.84	-0.5	48.26	10MY	-0.1	-4.2	-1.8	0.6
170.29	0.46	-0.3	57.81	20MY	0.6	-4.1	-1.9	0.6
173.27	0.13	-0.1	67.33	30MY	1.3	-3.8	-1.9	0.6
177.02	-0.16	0.1	76.85	9JN	2.2	-3.0	-2.0	0.6
181.41	-0.41	0.2	86.38	19JN	3.3	-3.2	-2.1	0.6
186.34	-0.63	0.3	95.92	29JN	1.9	-3.9	-2.1	0.5
191.71	-0.82	0.4	105.47	9JL	0.3	-4.1	-2.2	0.5
197.45	-0.98	0.5	115.07	19JL	-0.9	-4.2	-2.3	0.4
203.53	-1.12	0.6	124.69	29JL	-1.4	-4.2	-2.3	0.4
209.88	-1.24	0.7	134.37	8AU	-1.1	-4.1	-2.4	0.3
216.49	-1.34	0.7	144.09	18AU	-0.5	-4.0	-2.4	0.3
223.31	-1.42	0.8	153.87	28AU	-0.1	-3.9	-2.4	0.2
230.32	-1.48	0.9	163.71	7SE	0.0	-3.9	-2.5	0.2
237.50	-1.51	0.9	173.61	17SE	0.2	-3.8	-2.4	0.1
244.83	-1.54	1.0	183.56	27SE	0.5	-3.7	-2.4	0.0
252.28	-1.54	1.0	193.57	7OC	2.2	-3.7	-2.4	0.0
259.83	-1.53	1.1	203.64	17OC	1.2	-3.6	-2.3	0.1
267.46	-1.50	1.1	213.75	27OC	-0.3	-3.6	-2.3	0.1
275.16	-1.46	1.1	223.89	6NO	-0.5	-3.5	-2.2	0.2
282.90	-1.40	1.2	234.07	16NO	-0.5	-3.5	-2.1	0.3
290.67	-1.33	1.2	244.27	26NO	-0.6	-3.4	-2.1	0.3
298.44	-1.25	1.3	254.47	6DE	-0.8	-3.4	-2.0	0.4
306.20	-1.17	1.3	264.68	16DE	-0.9	-3.4	-1.9	0.5
313.95	-1.07	1.4	274.87	26DE	-0.9	-3.4	-1.8	0.5

♂ / ☉ — 232, 233

♂ LONG	LAT	MAG	☉ LONG	16.00UT	☿	♀	♃	♄
321.65	-0.97	1.4	285.04	5JA	-0.7	-3.3	-1.8	0.5
329.31	-0.86	1.4	295.18	15JA	0.6	-3.3	-1.7	0.6
336.92	-0.75	1.5	305.27	25JA	3.1	-3.3	-1.7	0.6
344.45	-0.64	1.5	315.31	4FE	1.4	-3.3	-1.7	0.6
351.92	-0.53	1.6	325.31	14FE	0.7	-3.3	-1.6	0.6
359.31	-0.41	1.6	335.24	24FE	0.4	-3.4	-1.6	0.6
6.62	-0.30	1.6	345.11	5MR	0.1	-3.4	-1.6	0.5
13.85	-0.19	1.7	354.93	15MR	-0.3	-3.4	-1.6	0.5
21.00	-0.08	1.7	4.69	25MR	-1.0	-3.5	-1.6	0.5
28.07	0.02	1.7	14.40	4AP	-1.8	-3.5	-1.6	0.4
35.06	0.13	1.7	24.06	14AP	-1.1	-3.5	-1.6	0.4
41.98	0.23	1.7	33.67	24AP	-0.0	-3.4	-1.6	0.4
48.82	0.32	1.8	43.25	4MY	0.8	-3.4	-1.6	0.4
55.60	0.41	1.8	52.81	14MY	1.7	-3.4	-1.7	0.5
62.31	0.50	1.8	62.34	24MY	3.0	-3.4	-1.7	0.5
68.96	0.59	1.8	71.86	3JN	2.9	-3.3	-1.7	0.5
75.55	0.67	1.9	81.38	13JN	1.5	-3.3	-1.8	0.5
82.09	0.75	1.9	90.91	23JN	0.2	-3.3	-1.8	0.4
88.59	0.83	1.9	100.46	3JL	-0.9	-3.3	-1.9	0.4
95.04	0.90	2.0	110.03	13JL	-1.6	-3.3	-2.0	0.4
101.45	0.97	2.0	119.64	23JL	-1.1	-3.4	-2.0	0.4
107.82	1.03	2.0	129.29	2AU	-0.4	-3.4	-2.1	0.3
114.16	1.10	2.0	138.99	12AU	-0.0	-3.4	-2.2	0.3
120.46	1.16	2.0	148.74	22AU	0.2	-3.4	-2.2	0.2
126.73	1.21	2.0	158.54	1SE	0.4	-3.5	-2.3	0.1
132.97	1.27	2.0	168.41	11SE	0.8	-3.6	-2.3	0.1
139.17	1.32	2.0	178.33	21SE	2.6	-3.6	-2.4	0.0
145.34	1.38	1.9	188.32	1OC	1.1	-3.7	-2.4	-0.1
151.47	1.43	1.9	198.36	11OC	-0.4	-3.8	-2.4	-0.1
157.56	1.47	1.8	208.44	21OC	-0.7	-3.9	-2.4	-0.1
163.59	1.52	1.8	218.57	31OC	-0.6	-4.0	-2.4	-0.1
169.56	1.56	1.7	228.73	10NO	-0.7	-4.1	-2.3	-0.0
175.47	1.60	1.6	238.92	20NO	-0.8	-4.2	-2.3	0.1
181.28	1.63	1.5	249.12	30NO	-0.7	-4.3	-2.2	0.1
187.00	1.67	1.4	259.33	10DE	-0.8	-4.4	-2.2	0.2
192.59	1.69	1.3	269.53	20DE	-0.6	-4.4	-2.1	0.3
198.03	1.71	1.1	279.71	30DE	0.8	-4.2	-2.0	0.3

233

♂ LONG	LAT	MAG	☉ LONG	16.00UT	☿	♀	♃	♄
203.29	1.72	1.0	289.86	9JA	2.9	-3.8	-1.9	0.4
208.32	1.72	0.8	299.98	19JA	1.0	-3.3	-1.9	0.4
213.08	1.70	0.6	310.05	29JA	0.5	-3.7	-1.8	0.4
217.48	1.67	0.4	320.07	8FE	0.2	-4.1	-1.7	0.4
221.46	1.60	0.1	330.03	18FE	0.0	-4.3	-1.7	0.4
224.88	1.51	-0.1	339.94	28FE	-0.4	-4.3	-1.6	0.4
227.61	1.37	-0.4	349.79	10MR	-1.0	-4.2	-1.6	0.4
229.50	1.16	-0.7	359.58	20MR	-1.7	-4.1	-1.6	0.4
230.34	0.89	-1.0	9.32	30MR	-1.1	-4.0	-1.5	0.4
229.98	0.52	-1.3	19.00	9AP	0.0	-3.9	-1.5	0.3
228.38	0.06	-1.7	28.63	19AP	1.1	-3.8	-1.5	0.3
225.70	-0.48	-1.9	38.23	29AP	2.3	-3.7	-1.5	0.3
222.44	-1.05	-2.0	47.80	9MY	3.7	-3.6	-1.5	0.3
219.30	-1.59	-1.9	57.34	19MY	2.3	-3.6	-1.5	0.3
216.96	-2.05	-1.8	66.87	29MY	1.1	-3.5	-1.5	0.3
215.88	-2.40	-1.5	76.39	8JN	0.1	-3.4	-1.5	0.4
216.18	-2.65	-1.3	85.91	18JN	-0.9	-3.4	-1.6	0.4
217.80	-2.81	-1.1	95.46	28JN	-1.7	-3.4	-1.6	0.4
220.56	-2.90	-0.9	105.01	8JL	-1.1	-3.4	-1.7	0.4
224.27	-2.94	-0.7	114.60	18JL	-0.4	-3.3	-1.7	0.3
228.76	-2.93	-0.5	124.23	28JL	0.0	-3.3	-1.8	0.3
233.88	-2.89	-0.4	133.90	7AU	0.3	-3.3	-1.8	0.3
239.51	-2.82	-0.2	143.62	17AU	0.5	-3.3	-1.9	0.2
245.56	-2.72	-0.1	153.40	27AU	1.1	-3.4	-1.9	0.2
251.95	-2.61	0.1	163.23	6SE	3.0	-3.4	-2.0	0.1
258.60	-2.48	0.2	173.13	16SE	1.0	-3.4	-2.1	0.0
265.47	-2.33	0.3	183.08	26SE	-0.5	-3.4	-2.1	-0.0
272.51	-2.17	0.4	193.09	6OC	-0.8	-3.4	-2.2	-0.1
279.68	-2.00	0.5	203.15	16OC	-0.8	-3.5	-2.2	-0.2
286.94	-1.83	0.6	213.26	26OC	-0.8	-3.5	-2.3	-0.2
294.28	-1.65	0.7	223.40	5NO	-0.6	-3.5	-2.3	-0.2
301.64	-1.47	0.8	233.57	15NO	-0.6	-3.4	-2.3	-0.2
309.04	-1.30	0.9	243.77	25NO	-0.6	-3.4	-2.3	-0.1
316.43	-1.12	1.0	253.98	5DE	-0.4	-3.4	-2.3	-0.0
323.80	-0.95	1.1	264.18	15DE	0.9	-3.4	-2.2	0.0
331.15	-0.78	1.2	274.38	25DE	2.5	-3.4	-2.2	0.1

♂ / ☉ — 234, 235

♂ LONG	LAT	MAG	☉ LONG	16.00UT	☿	♀	♃	♄
338.45	-0.62	1.3	284.54	4JA	0.7	-3.4	-2.1	0.2
345.71	-0.47	1.4	294.68	14JA	0.3	-3.3	-2.1	0.2
352.91	-0.33	1.5	304.78	24JA	0.1	-3.3	-2.0	0.3
0.06	-0.19	1.5	314.82	3FE	-0.1	-3.3	-1.9	0.3
7.14	-0.06	1.6	324.82	13FE	-0.3	-3.3	-1.8	0.3
14.16	0.06	1.6	334.76	23FE	-1.0	-3.4	-1.8	0.3
21.11	0.18	1.7	344.63	5MR	-1.6	-3.4	-1.7	0.4
28.00	0.29	1.8	354.45	15MR	-1.0	-3.4	-1.6	0.3
34.82	0.39	1.8	4.22	25MR	0.1	-3.4	-1.6	0.3
41.59	0.48	1.8	13.93	4AP	1.4	-3.5	-1.5	0.3
48.30	0.56	1.9	23.59	14AP	3.0	-3.5	-1.5	0.3
54.95	0.64	1.9	33.21	24AP	3.0	-3.6	-1.5	0.2
61.56	0.72	1.9	42.79	4MY	1.7	-3.6	-1.4	0.2
68.12	0.78	1.9	52.34	14MY	0.9	-3.7	-1.4	0.2
74.64	0.85	1.9	61.88	24MY	0.1	-3.8	-1.4	0.2
81.12	0.90	1.9	71.40	3JN	-0.9	-3.8	-1.4	0.3
87.58	0.95	1.9	80.92	13JN	-1.7	-3.9	-1.4	0.3
94.01	1.00	1.9	90.45	23JN	-1.1	-4.0	-1.4	0.3
100.41	1.04	1.9	100.00	3JL	-0.3	-4.1	-1.4	0.3
106.80	1.08	2.0	109.57	13JL	0.1	-4.2	-1.4	0.3
113.18	1.11	2.0	119.18	23JL	0.5	-4.2	-1.5	0.3
119.54	1.14	2.0	128.82	2AU	0.8	-4.1	-1.5	0.3
125.91	1.17	2.0	138.52	12AU	1.5	-3.7	-1.5	0.3
132.27	1.19	2.0	148.27	22AU	3.2	-3.2	-1.6	0.2
138.63	1.20	2.0	158.07	1SE	0.8	-3.6	-1.6	0.2
144.99	1.22	2.0	167.93	11SE	-0.6	-4.1	-1.7	0.1
151.36	1.22	2.0	177.86	21SE	-1.0	-4.3	-1.8	0.1
157.74	1.23	2.0	187.83	1OC	-0.9	-4.3	-1.8	0.0
164.12	1.22	2.0	197.87	11OC	-0.8	-4.3	-1.9	-0.1
170.51	1.22	1.9	207.95	21OC	-0.5	-4.2	-2.0	-0.1
176.90	1.20	1.9	218.07	31OC	-0.4	-4.1	-2.0	-0.2
183.30	1.18	1.8	228.24	10NO	-0.4	-4.0	-2.1	-0.3
189.70	1.16	1.8	238.43	20NO	-0.2	-3.9	-2.1	-0.3
196.10	1.12	1.7	248.63	30NO	1.1	-3.8	-2.2	-0.2
202.50	1.08	1.6	258.84	10DE	2.2	-3.7	-2.2	-0.2
208.88	1.03	1.5	269.04	20DE	0.4	-3.7	-2.2	-0.1
215.25	0.96	1.4	279.22	30DE	0.1	-3.6	-2.2	-0.0

235

♂ LONG	LAT	MAG	☉ LONG	16.00UT	☿	♀	♃	♄
221.60	0.88	1.3	289.38	9JA	-0.0	-3.5	-2.1	0.1
227.92	0.78	1.2	299.49	19JA	-0.2	-3.5	-2.1	0.1
234.21	0.67	1.1	309.57	29JA	-0.5	-3.4	-2.1	0.2
240.44	0.53	0.9	319.59	8FE	-1.0	-3.4	-2.0	0.2
246.61	0.37	0.8	329.56	18FE	-1.4	-3.4	-1.9	0.3
252.69	0.18	0.6	339.47	28FE	-1.0	-3.3	-1.9	0.3
258.67	-0.04	0.4	349.32	10MR	0.2	-3.3	-1.8	0.3
264.50	-0.30	0.2	359.11	20MR	1.8	-3.3	-1.7	0.3
270.16	-0.60	0.0	8.85	30MR	3.7	-3.3	-1.7	0.3
275.58	-0.95	-0.2	18.54	9AP	2.2	-3.3	-1.6	0.3
280.70	-1.36	-0.4	28.17	19AP	1.3	-3.3	-1.5	0.3
285.43	-1.83	-0.6	37.77	29AP	0.7	-3.3	-1.5	0.3
289.64	-2.37	-0.9	47.34	9MY	-0.0	-3.4	-1.4	0.2
293.19	-2.99	-1.2	56.88	19MY	-1.0	-3.4	-1.4	0.2
295.92	-3.68	-1.5	66.41	29MY	-1.8	-3.4	-1.4	0.2
297.59	-4.43	-1.8	75.93	8JN	-1.1	-3.5	-1.3	0.3
298.07	-5.20	-2.1	85.45	18JN	-0.3	-3.5	-1.3	0.3
297.27	-5.91	-2.3	94.99	28JN	0.3	-3.5	-1.3	0.3
295.35	-6.47	-2.6	104.55	8JL	0.6	-3.4	-1.3	0.3
292.80	-6.76	-2.7	114.13	18JL	1.1	-3.4	-1.3	0.3
290.32	-6.72	-2.5	123.76	28JL	2.0	-3.4	-1.3	0.3
288.61	-6.40	-2.3	133.43	7AU	3.0	-3.3	-1.3	0.3
288.11	-5.87	-2.0	143.15	17AU	0.7	-3.3	-1.4	0.3
288.92	-5.25	-1.7	152.92	27AU	-0.7	-3.3	-1.4	0.3
290.96	-4.61	-1.4	162.76	6SE	-1.1	-3.3	-1.4	0.3
294.05	-3.98	-1.2	172.64	16SE	-1.1	-3.3	-1.5	0.2
297.97	-3.40	-0.9	182.59	26SE	-0.7	-3.3	-1.5	0.2
302.55	-2.87	-0.6	192.60	6OC	-0.4	-3.4	-1.6	0.1
307.65	-2.39	-0.4	202.66	16OC	-0.3	-3.4	-1.6	0.0
313.13	-1.97	-0.2	212.76	26OC	-0.3	-3.4	-1.7	-0.1
318.91	-1.58	0.0	222.91	5NO	-0.0	-3.5	-1.8	-0.1
324.92	-1.24	0.2	233.08	15NO	1.3	-3.5	-1.8	-0.2
331.09	-0.94	0.4	243.28	25NO	1.9	-3.6	-1.9	-0.3
337.39	-0.67	0.6	253.48	5DE	0.2	-3.6	-1.9	-0.3
343.78	-0.43	0.7	263.69	15DE	-0.1	-3.7	-2.0	-0.2
350.22	-0.22	0.9	273.88	25DE	-0.2	-3.8	-2.0	-0.1

♂ LONG	♂ LAT	♂ MAG	☉ LONG	16.00UT	☿	♀	♃	♄
				236	MAGNITUDES			
356.71	-0.03	1.0	284.05	4JA	-0.3	-3.9	-2.1	-0.1
3.21	0.14	1.2	294.19	14JA	-0.6	-4.0	-2.1	-0.0
9.73	0.29	1.3	304.29	24JA	-1.1	-4.1	-2.1	0.1
16.24	0.42	1.4	314.34	3FE	-1.3	-4.2	-2.1	0.1
22.75	0.53	1.5	324.34	13FE	-1.0	-4.2	-2.0	0.2
29.23	0.63	1.6	334.28	23FE	0.3	-4.3	-2.0	0.2
35.71	0.72	1.7	344.16	4MR	2.2	-4.3	-1.9	0.3
42.15	0.80	1.7	353.98	14MR	3.0	-4.1	-1.9	0.3
48.58	0.87	1.8	3.75	24MR	1.6	-3.6	-1.8	0.3
54.99	0.93	1.8	13.46	3AP	1.0	-3.1	-1.7	0.3
61.37	0.98	1.9	23.12	13AP	0.5	-3.7	-1.7	0.3
67.74	1.03	1.9	32.74	23AP	-0.1	-4.1	-1.6	0.3
74.09	1.06	2.0	42.33	3MY	-1.0	-4.2	-1.6	0.3
80.43	1.10	2.0	51.88	13MY	-1.8	-4.2	-1.5	0.3
86.76	1.12	2.0	61.41	23MY	-1.1	-4.1	-1.4	0.3
93.08	1.14	2.0	70.94	2JN	-0.2	-4.0	-1.4	0.2
99.40	1.15	2.0	80.46	12JN	0.4	-3.9	-1.3	0.2
105.72	1.16	2.0	89.99	22JN	0.8	-3.8	-1.3	0.3
112.05	1.17	2.0	99.53	2JL	1.4	-3.7	-1.3	0.3
118.39	1.17	2.0	109.10	12JL	2.5	-3.7	-1.3	0.3
124.75	1.16	2.0	118.71	22JL	2.6	-3.6	-1.2	0.4
131.13	1.15	2.0	128.36	1AU	0.6	-3.6	-1.2	0.4
137.52	1.14	1.9	138.05	11AU	-0.8	-3.5	-1.2	0.4
143.95	1.12	2.0	147.80	21AU	-1.2	-3.5	-1.2	0.4
150.41	1.09	2.0	157.60	31AU	-1.1	-3.4	-1.2	0.4
156.90	1.07	2.0	167.45	10SE	-0.6	-3.4	-1.3	0.3
163.43	1.03	2.0	177.38	20SE	-0.3	-3.4	-1.3	0.3
169.99	0.99	2.0	187.35	30SE	-0.2	-3.4	-1.3	0.3
176.60	0.95	1.9	197.38	10OC	-0.1	-3.4	-1.4	0.2
183.25	0.90	1.9	207.47	20OC	0.2	-3.4	-1.4	0.2
189.94	0.84	1.9	217.59	30OC	1.6	-3.4	-1.5	0.1
196.67	0.77	1.9	227.75	9NO	1.6	-3.4	-1.5	0.0
203.46	0.70	1.8	237.93	19NO	0.0	-3.4	-1.6	-0.1
210.28	0.62	1.8	248.14	29NO	-0.3	-3.4	-1.6	-0.1
217.15	0.53	1.7	258.34	9DE	-0.3	-3.4	-1.7	-0.2
224.06	0.43	1.6	268.54	19DE	-0.4	-3.4	-1.8	-0.2
231.02	0.32	1.6	278.72	29DE	-0.6	-3.4	-1.8	-0.1
				237				
238.02	0.20	1.5	288.88	8JA	-1.1	-3.5	-1.9	-0.1
245.05	0.07	1.4	299.00	18JA	-1.2	-3.5	-1.9	-0.0
252.12	-0.07	1.3	309.08	28JA	-0.9	-3.5	-2.0	0.1
259.23	-0.23	1.3	319.10	7FE	0.4	-3.4	-2.0	0.1
266.36	-0.39	1.2	329.07	17FE	2.6	-3.4	-2.0	0.2
273.52	-0.57	1.1	338.99	27FE	2.3	-3.4	-2.0	0.2
280.68	-0.77	1.0	348.84	9MR	1.2	-3.4	-2.0	0.3
287.85	-0.97	0.9	358.64	19MR	0.7	-3.3	-2.0	0.3
295.01	-1.18	0.7	8.38	29MR	0.3	-3.3	-1.9	0.4
302.15	-1.41	0.6	18.06	8AP	-0.2	-3.3	-1.9	0.4
309.23	-1.64	0.5	27.71	18AP	-1.0	-3.3	-1.8	0.4
316.25	-1.87	0.4	37.30	28AP	-1.9	-3.3	-1.7	0.4
323.17	-2.11	0.3	46.87	8MY	-1.1	-3.4	-1.7	0.4
329.95	-2.36	0.1	56.42	18MY	-0.2	-3.4	-1.6	0.4
336.55	-2.60	0.0	65.94	28MY	0.5	-3.4	-1.6	0.4
342.92	-2.84	-0.1	75.47	7JN	1.1	-3.4	-1.5	0.3
349.00	-3.07	-0.3	84.99	17JN	1.9	-3.5	-1.4	0.3
354.69	-3.30	-0.4	94.53	27JN	3.1	-3.5	-1.4	0.3
359.92	-3.51	-0.6	104.08	7JL	2.1	-3.6	-1.3	0.4
4.53	-3.71	-0.8	113.67	17JL	0.5	-3.6	-1.3	0.4
8.38	-3.88	-1.0	123.29	27JL	-0.8	-3.7	-1.3	0.4
11.28	-4.02	-1.2	132.96	6AU	-1.4	-3.8	-1.2	0.5
13.01	-4.10	-1.4	142.68	16AU	-1.1	-3.9	-1.2	0.5
13.37	-4.09	-1.7	152.45	26AU	-0.5	-4.0	-1.2	0.5
12.27	-3.96	-1.9	162.28	5SE	-0.2	-4.1	-1.2	0.5
9.83	-3.67	-2.1	172.17	15SE	-0.0	-4.2	-1.2	0.5
6.56	-3.20	-2.2	182.11	25SE	0.1	-4.3	-1.2	0.5
3.21	-2.60	-2.0	192.12	5OC	0.4	-4.3	-1.2	0.4
0.57	-1.95	-1.7	202.17	15OC	1.9	-4.2	-1.3	0.4
359.13	-1.33	-1.4	212.27	25OC	1.4	-3.8	-1.3	0.3
359.04	-0.78	-1.1	222.42	4NO	-0.2	-3.0	-1.3	0.3
0.22	-0.32	-0.8	232.59	14NO	-0.4	-3.7	-1.4	0.2
2.45	0.05	-0.5	242.78	24NO	-0.4	-4.2	-1.4	0.1
5.54	0.35	-0.2	252.99	4DE	-0.5	-4.4	-1.5	0.1
9.30	0.58	0.1	263.20	14DE	-0.7	-4.4	-1.5	-0.0
13.58	0.77	0.3	273.39	24DE	-0.9	-4.3	-1.6	-0.1

♂ LONG	♂ LAT	♂ MAG	☉ LONG	16.00UT	☿	♀	♃	♄
				238	MAGNITUDES			
18.26	0.92	0.5	283.56	3JA	-1.0	-4.2	-1.7	-0.1
23.25	1.04	0.7	293.70	13JA	-0.8	-4.1	-1.7	-0.0
28.49	1.13	0.9	303.80	23JA	0.5	-4.0	-1.8	0.0
33.91	1.21	1.1	313.86	2FE	2.9	-3.9	-1.9	0.1
39.47	1.26	1.2	323.86	12FE	1.7	-3.8	-1.9	0.2
45.15	1.30	1.3	333.80	22FE	0.9	-3.7	-2.0	0.2
50.92	1.33	1.5	343.69	4MR	0.5	-3.6	-2.0	0.3
56.77	1.36	1.6	353.51	14MR	0.2	-3.5	-2.0	0.3
62.67	1.37	1.7	3.28	24MR	-0.2	-3.5	-2.0	0.4
68.62	1.37	1.7	13.00	3AP	-1.0	-3.4	-2.0	0.4
74.61	1.37	1.8	22.66	13AP	-1.8	-3.4	-2.0	0.5
80.64	1.37	1.9	32.28	23AP	-1.1	-3.4	-2.0	0.5
86.70	1.36	1.9	41.87	3MY	-0.1	-3.3	-1.9	0.5
92.79	1.34	2.0	51.42	13MY	0.7	-3.3	-1.9	0.5
98.91	1.32	2.0	60.96	23MY	1.4	-3.3	-1.8	0.5
105.07	1.29	2.0	70.48	2JN	2.5	-3.3	-1.7	0.5
111.26	1.26	2.0	80.00	12JN	3.3	-3.3	-1.7	0.5
117.48	1.23	2.0	89.53	22JN	1.7	-3.3	-1.6	0.5
123.75	1.19	2.0	99.08	2JL	0.3	-3.4	-1.5	0.5
130.05	1.15	2.0	108.64	12JL	-0.9	-3.4	-1.5	0.5
136.40	1.11	2.0	118.24	22JL	-1.5	-3.4	-1.4	0.5
142.80	1.06	2.0	127.89	1AU	-1.1	-3.4	-1.4	0.6
149.25	1.01	2.0	137.58	11AU	-0.5	-3.5	-1.3	0.6
155.75	0.95	1.9	147.32	21AU	-0.1	-3.5	-1.3	0.6
162.32	0.89	1.9	157.12	31AU	0.1	-3.5	-1.3	0.7
168.94	0.83	1.8	166.97	10SE	0.2	-3.4	-1.3	0.7
175.63	0.76	1.8	176.89	20SE	0.6	-3.4	-1.2	0.7
182.38	0.68	1.8	186.86	30SE	2.2	-3.4	-1.2	0.7
189.20	0.60	1.8	196.89	10OC	1.3	-3.4	-1.2	0.6
196.09	0.52	1.8	206.97	20OC	-0.3	-3.3	-1.2	0.6
203.05	0.43	1.8	217.09	30OC	-0.6	-3.3	-1.3	0.6
210.08	0.34	1.8	227.25	9NO	-0.6	-3.3	-1.3	0.5
217.17	0.24	1.7	237.43	19NO	-0.6	-3.3	-1.3	0.4
224.34	0.13	1.7	247.64	29NO	-0.8	-3.3	-1.3	0.4
231.58	0.02	1.7	257.84	9DE	-0.8	-3.4	-1.4	0.3
238.88	-0.09	1.6	268.04	19DE	-0.8	-3.4	-1.4	0.2
246.24	-0.21	1.6	278.23	29DE	-0.6	-3.4	-1.5	0.2
				239				
253.66	-0.33	1.6	288.39	8JA	0.6	-3.4	-1.5	0.1
261.13	-0.46	1.5	298.51	18JA	3.0	-3.5	-1.6	0.1
268.66	-0.59	1.5	308.59	28JA	1.3	-3.5	-1.7	0.2
276.22	-0.72	1.4	318.62	7FE	0.6	-3.6	-1.7	0.2
283.81	-0.85	1.4	328.59	17FE	0.3	-3.6	-1.8	0.3
291.43	-0.98	1.3	338.51	27FE	0.1	-3.7	-1.9	0.3
299.06	-1.10	1.3	348.37	9MR	-0.3	-3.8	-1.9	0.4
306.69	-1.22	1.2	358.17	19MR	-1.0	-3.8	-2.0	0.4
314.31	-1.33	1.1	7.91	29MR	-1.8	-3.9	-2.0	0.5
321.91	-1.44	1.1	17.60	8AP	-1.1	-4.0	-2.1	0.5
329.46	-1.53	1.0	27.24	18AP	-0.0	-4.1	-2.1	0.6
336.96	-1.62	1.0	36.85	28AP	0.9	-4.2	-2.1	0.6
344.37	-1.69	0.9	46.41	8MY	1.9	-4.2	-2.1	0.7
351.70	-1.74	0.9	55.96	18MY	3.3	-4.1	-2.0	0.7
358.92	-1.78	0.8	65.49	28MY	2.7	-3.7	-2.0	0.7
5.99	-1.80	0.8	75.01	7JN	1.4	-3.0	-2.0	0.7
12.92	-1.80	0.7	84.53	17JN	0.2	-3.2	-1.9	0.7
19.65	-1.79	0.6	94.07	27JN	-0.9	-3.9	-1.8	0.7
26.17	-1.75	0.5	103.62	7JL	-1.6	-4.1	-1.8	0.7
32.44	-1.69	0.5	113.20	17JL	-1.1	-4.2	-1.7	0.7
38.41	-1.62	0.4	122.83	27JL	-0.4	-4.2	-1.7	0.7
44.02	-1.52	0.3	132.49	6AU	-0.0	-4.1	-1.6	0.7
49.22	-1.39	0.1	142.20	16AU	0.2	-4.0	-1.5	0.8
53.89	-1.24	-0.0	151.97	26AU	0.4	-3.9	-1.5	0.8
57.94	-1.05	-0.2	161.80	5SE	0.9	-3.9	-1.4	0.9
61.21	-0.82	-0.3	171.68	15SE	2.6	-3.8	-1.4	0.9
63.52	-0.54	-0.5	181.63	25SE	1.1	-3.7	-1.4	0.9
64.68	-0.21	-0.7	191.62	5OC	-0.4	-3.7	-1.3	0.9
64.48	0.18	-1.0	201.68	15OC	-0.7	-3.6	-1.3	0.9
62.84	0.62	-1.2	211.78	25OC	-0.7	-3.5	-1.3	0.9
59.90	1.08	-1.4	221.92	4NO	-0.7	-3.5	-1.3	0.8
56.13	1.53	-1.5	232.10	14NO	-0.7	-3.5	-1.3	0.8
52.29	1.89	-1.4	242.29	24NO	-0.7	-3.4	-1.3	0.7
49.16	2.15	-1.1	252.49	4DE	-0.7	-3.4	-1.3	0.7
47.22	2.30	-0.8	262.70	14DE	-0.5	-3.4	-1.4	0.6
46.63	2.37	-0.5	272.90	24DE	0.7	-3.4	-1.4	0.5

♂			⊙	16.00UT	☿	♀	♃	♄
LONG	LAT	MAG	LONG	240	MAGNITUDES			
47.30	2.38	-0.2	283.07	3JA	2.8	-3.3	-1.4	0.5
49.05	2.36	0.1	293.21	13JA	0.9	-3.3	-1.5	0.4
51.68	2.31	0.3	303.32	23JA	0.4	-3.3	-1.5	0.3
54.99	2.25	0.6	313.37	2FE	0.2	-3.3	-1.6	0.3
58.85	2.18	0.8	323.38	12FE	-0.0	-3.3	-1.7	0.4
63.14	2.10	0.9	333.32	22FE	-0.4	-3.4	-1.7	0.4
67.77	2.03	1.1	343.21	3MR	-1.0	-3.4	-1.8	0.5
72.66	1.96	1.2	353.04	13MR	-1.7	-3.4	-1.9	0.5
77.78	1.88	1.4	2.81	23MR	-1.1	-3.5	-1.9	0.6
83.08	1.81	1.5	12.52	2AP	0.0	-3.5	-2.0	0.7
88.52	1.73	1.6	22.19	12AP	1.2	-3.5	-2.1	0.7
94.10	1.66	1.7	31.82	22AP	2.5	-3.4	-2.1	0.8
99.78	1.58	1.7	41.40	2MY	3.5	-3.4	-2.2	0.8
105.57	1.51	1.8	50.96	12MY	2.1	-3.4	-2.2	0.8
111.44	1.43	1.8	60.49	22MY	1.1	-3.4	-2.2	0.9
117.40	1.35	1.9	70.01	1JN	0.1	-3.3	-2.2	0.9
123.44	1.27	1.9	79.54	11JN	-0.9	-3.3	-2.2	0.9
129.56	1.19	1.9	89.06	21JN	-1.7	-3.3	-2.2	0.9
135.76	1.11	1.9	98.61	1JL	-1.1	-3.3	-2.1	0.9
142.03	1.03	1.9	108.18	11JL	-0.4	-3.3	-2.1	0.9
148.38	0.95	1.9	117.78	21JL	0.1	-3.4	-2.0	0.9
154.81	0.86	1.9	127.41	31JL	0.4	-3.4	-1.9	0.9
161.33	0.77	1.9	137.11	10AU	0.6	-3.4	-1.9	1.0
167.92	0.68	1.9	146.84	20AU	1.2	-3.4	-1.8	1.0
174.60	0.59	1.8	156.64	30AU	3.0	-3.5	-1.8	1.1
181.37	0.50	1.8	166.49	9SE	1.0	-3.6	-1.7	1.1
188.22	0.40	1.8	176.40	19SE	-0.6	-3.6	-1.6	1.1
195.16	0.30	1.7	186.37	29SE	-0.9	-3.7	-1.6	1.1
202.19	0.20	1.7	196.40	9OC	-0.9	-3.8	-1.5	1.2
209.31	0.10	1.6	206.48	19OC	-0.8	-3.9	-1.5	1.2
216.51	-0.01	1.6	216.60	29OC	-0.6	-4.0	-1.5	1.1
223.80	-0.12	1.6	226.76	8NO	-0.5	-4.1	-1.4	1.1
231.17	-0.22	1.6	236.94	18NO	-0.5	-4.2	-1.4	1.1
238.62	-0.33	1.5	247.14	28NO	-0.3	-4.3	-1.4	1.0
246.15	-0.43	1.5	257.35	8DE	0.9	-4.4	-1.4	1.0
253.74	-0.54	1.5	267.55	18DE	2.5	-4.4	-1.4	0.9
261.39	-0.64	1.5	277.73	28DE	0.6	-4.2	-1.4	0.9
				241				
269.09	-0.74	1.5	287.90	7JA	0.2	-3.8	-1.4	0.8
276.83	-0.83	1.5	298.02	17JA	0.1	-3.3	-1.5	0.7
284.61	-0.92	1.4	308.10	27JA	-0.1	-3.7	-1.5	0.7
292.41	-1.00	1.4	318.14	6FE	-0.4	-4.1	-1.5	0.6
300.22	-1.07	1.4	328.11	16FE	-1.0	-4.3	-1.6	0.6
308.03	-1.13	1.4	338.03	26FE	-1.5	-4.3	-1.7	0.7
315.82	-1.18	1.4	347.89	8MR	-1.1	-4.2	-1.7	0.7
323.59	-1.22	1.4	357.69	18MR	0.1	-4.1	-1.8	0.8
331.32	-1.25	1.4	7.44	28MR	1.5	-4.0	-1.8	0.8
338.99	-1.26	1.3	17.13	7AP	3.2	-3.9	-1.9	0.9
346.61	-1.26	1.3	26.78	17AP	2.7	-3.8	-2.0	0.9
354.15	-1.25	1.3	36.38	27AP	1.5	-3.7	-2.1	1.0
1.60	-1.22	1.3	45.95	7MY	0.8	-3.6	-2.1	1.0
8.96	-1.18	1.3	55.50	17MY	0.0	-3.6	-2.2	1.1
16.21	-1.13	1.3	65.03	27MY	-0.9	-3.5	-2.2	1.1
23.35	-1.06	1.3	74.55	6JN	-1.7	-3.4	-2.3	1.1
30.37	-0.98	1.2	84.07	16JN	-1.1	-3.4	-2.3	1.2
37.24	-0.89	1.2	93.61	26JN	-0.3	-3.4	-2.3	1.2
43.97	-0.78	1.2	103.16	6JL	0.2	-3.4	-2.3	1.2
50.54	-0.66	1.1	112.74	16JL	0.5	-3.3	-2.3	1.2
56.94	-0.53	1.1	122.36	26JL	0.9	-3.3	-2.3	1.2
63.13	-0.38	1.0	132.03	5AU	1.6	-3.3	-2.2	1.2
69.10	-0.22	1.0	141.74	15AU	3.2	-3.3	-2.2	1.2
74.82	-0.05	0.9	151.50	25AU	0.8	-3.4	-2.1	1.3
80.24	0.15	0.8	161.32	4SE	-0.7	-3.4	-2.1	1.3
85.32	0.36	0.7	171.20	14SE	-1.0	-3.4	-2.0	1.4
89.98	0.60	0.6	181.14	24SE	-1.0	-3.4	-1.9	1.4
94.14	0.87	0.4	191.14	4OC	-0.7	-3.4	-1.9	1.4
97.67	1.18	0.3	201.19	14OC	-0.5	-3.5	-1.8	1.4
100.44	1.52	0.1	211.29	24OC	-0.4	-3.5	-1.7	1.3
102.27	1.91	-0.1	221.43	3NO	-0.4	-3.5	-1.7	1.3
102.96	2.33	-0.3	231.59	13NO	-0.2	-3.5	-1.6	1.3
102.35	2.79	-0.6	241.79	23NO	1.1	-3.4	-1.6	1.2
100.37	3.24	-0.8	251.99	3DE	2.1	-3.4	-1.6	1.2
97.19	3.62	-1.0	262.20	13DE	0.4	-3.4	-1.5	1.1
93.29	3.89	-1.1	272.39	23DE	0.0	-3.4	-1.5	1.1

♂			⊙	16.00UT	☿	♀	♃	♄
LONG	LAT	MAG	LONG	242	MAGNITUDES			
89.41	4.00	-1.0	282.57	2JA	-0.1	-3.4	-1.5	1.0
86.26	3.96	-0.7	292.71	12JA	-0.2	-3.3	-1.5	1.0
84.28	3.80	-0.5	302.82	22JA	-0.5	-3.3	-1.5	0.9
83.63	3.58	-0.2	312.88	1FE	-1.0	-3.3	-1.5	0.9
84.20	3.33	0.1	322.88	11FE	-1.4	-3.3	-1.5	0.9
85.83	3.09	0.3	332.84	21FE	-1.0	-3.4	-1.6	0.9
88.33	2.85	0.5	342.73	3MR	0.2	-3.4	-1.6	0.9
91.52	2.63	0.7	352.56	13MR	1.9	-3.4	-1.6	1.0
95.27	2.42	0.9	2.34	23MR	3.5	-3.4	-1.7	1.0
99.48	2.23	1.0	12.06	2AP	2.0	-3.5	-1.7	1.1
104.04	2.05	1.2	21.72	12AP	1.2	-3.5	-1.8	1.1
108.91	1.88	1.3	31.35	22AP	0.6	-3.6	-1.9	1.2
114.03	1.72	1.4	40.94	2MY	-0.0	-3.6	-1.9	1.2
119.35	1.57	1.5	50.50	12MY	-0.9	-3.7	-2.0	1.3
124.87	1.43	1.5	60.04	22MY	-1.8	-3.8	-2.1	1.3
130.55	1.29	1.6	69.56	1JN	-1.1	-3.9	-2.1	1.4
136.38	1.16	1.6	79.08	11JN	-0.3	-3.9	-2.2	1.4
142.36	1.03	1.7	88.61	21JN	0.3	-4.0	-2.3	1.4
148.45	0.90	1.7	98.15	1JL	0.7	-4.1	-2.3	1.4
154.68	0.78	1.7	107.72	11JL	1.2	-4.2	-2.4	1.4
161.02	0.66	1.7	117.32	21JL	2.1	-4.2	-2.4	1.4
167.48	0.54	1.7	126.95	31JL	2.9	-4.1	-2.4	1.3
174.06	0.42	1.7	136.64	10AU	0.7	-3.7	-2.4	1.3
180.74	0.30	1.7	146.38	20AU	-0.7	-3.2	-2.4	1.3
187.54	0.18	1.7	156.17	30AU	-1.2	-3.7	-2.4	1.2
194.45	0.07	1.7	166.02	9SE	-1.1	-4.1	-2.3	1.2
201.47	-0.04	1.7	175.93	19SE	-0.7	-4.3	-2.3	1.2
208.60	-0.15	1.6	185.89	29SE	-0.4	-4.3	-2.2	1.2
215.83	-0.26	1.6	195.91	9OC	-0.2	-4.3	-2.2	1.2
223.16	-0.37	1.6	205.99	19OC	-0.2	-4.2	-2.1	1.2
230.58	-0.47	1.5	216.10	29OC	0.0	-4.1	-2.0	1.2
238.10	-0.56	1.5	226.26	8NO	1.4	-4.0	-1.9	1.1
245.69	-0.66	1.5	236.44	18NO	1.9	-3.9	-1.9	1.1
253.36	-0.74	1.4	246.64	28NO	0.2	-3.8	-1.8	1.1
261.10	-0.82	1.4	256.85	8DE	-0.2	-3.7	-1.8	1.0
268.89	-0.89	1.3	267.05	18DE	-0.2	-3.7	-1.7	1.0
276.72	-0.95	1.3	277.24	28DE	-0.3	-3.6	-1.7	1.0
				243				
284.58	-1.00	1.3	287.40	7JA	-0.6	-3.5	-1.6	0.9
292.47	-1.04	1.3	297.53	17JA	-1.0	-3.5	-1.6	0.9
300.35	-1.07	1.4	307.61	27JA	-1.2	-3.4	-1.6	0.8
308.24	-1.09	1.4	317.65	6FE	-0.9	-3.4	-1.6	0.8
316.10	-1.10	1.4	327.63	16FE	0.3	-3.4	-1.6	0.7
323.92	-1.09	1.4	337.55	26FE	2.3	-3.3	-1.6	0.7
331.71	-1.08	1.4	347.42	8MR	2.8	-3.3	-1.6	0.7
339.44	-1.05	1.4	357.22	18MR	1.5	-3.3	-1.6	0.8
347.10	-1.01	1.5	6.97	28MR	0.9	-3.3	-1.6	0.8
354.70	-0.96	1.5	16.67	7AP	0.4	-3.3	-1.6	0.9
2.21	-0.90	1.5	26.31	17AP	-0.1	-3.3	-1.7	1.0
9.63	-0.83	1.5	35.92	27AP	-0.9	-3.3	-1.7	1.0
16.97	-0.75	1.5	45.49	7MY	-1.8	-3.4	-1.8	1.1
24.21	-0.66	1.5	55.04	17MY	-1.1	-3.4	-1.8	1.1
31.34	-0.56	1.6	64.57	27MY	-0.2	-3.4	-1.9	1.2
38.38	-0.46	1.6	74.09	6JN	0.4	-3.5	-2.0	1.2
45.31	-0.35	1.6	83.61	16JN	0.9	-3.5	-2.0	1.2
52.13	-0.24	1.6	93.14	26JN	1.6	-3.5	-2.1	1.2
58.84	-0.12	1.6	102.70	6JL	2.7	-3.4	-2.2	1.2
65.43	0.01	1.6	112.28	16JL	2.5	-3.4	-2.2	1.2
71.91	0.14	1.5	121.89	26JL	0.6	-3.4	-2.3	1.2
78.25	0.28	1.5	131.55	5AU	-0.8	-3.3	-2.4	1.2
84.46	0.42	1.5	141.26	15AU	-1.3	-3.3	-2.4	1.2
90.52	0.57	1.4	151.02	25AU	-1.1	-3.3	-2.4	1.1
96.41	0.73	1.4	160.84	4SE	-0.6	-3.3	-2.5	1.1
102.12	0.91	1.3	170.72	14SE	-0.3	-3.3	-2.5	1.0
107.61	1.09	1.3	180.66	24SE	-0.1	-3.3	-2.4	1.0
112.86	1.29	1.2	190.65	4OC	-0.0	-3.4	-2.4	1.0
117.81	1.50	1.0	200.70	14OC	0.2	-3.4	-2.4	1.1
122.40	1.73	0.9	210.79	24OC	1.6	-3.4	-2.3	1.0
126.57	1.99	0.8	220.93	3NO	1.7	-3.5	-2.3	1.0
130.20	2.28	0.6	231.10	13NO	-0.0	-3.5	-2.2	1.0
133.18	2.59	0.4	241.29	23NO	-0.3	-3.6	-2.1	1.0
135.34	2.94	0.2	251.50	3DE	-0.3	-3.6	-2.0	1.0
136.51	3.32	-0.0	261.70	13DE	-0.4	-3.7	-2.0	0.9
136.50	3.71	-0.3	271.90	23DE	-0.7	-3.8	-1.9	0.9

♂ LONG	♂ LAT	♂ MAG	⊙ LONG	16.00UT 244	☿ MAGNITUDES	♀	♃	♄
135.19	4.08	-0.5	282.08	2JA	-1.0	-3.9	-1.8	0.9
132.61	4.39	-0.8	292.22	12JA	-1.1	-4.0	-1.8	0.8
129.06	4.57	-1.0	302.33	22JA	-0.9	-4.1	-1.7	0.8
125.11	4.58	-1.0	312.39	1FE	0.4	-4.2	-1.7	0.7
121.51	4.41	-0.8	322.40	11FE	2.7	-4.2	-1.6	0.7
118.85	4.12	-0.6	332.35	21FE	2.1	-4.3	-1.6	0.6
117.44	3.75	-0.3	342.25	2MR	1.1	-4.3	-1.6	0.6
117.33	3.36	-0.1	352.08	12MR	0.6	-4.0	-1.6	0.5
118.37	2.98	0.1	1.86	22MR	0.3	-3.5	-1.6	0.6
120.41	2.63	0.3	11.58	1AP	-0.2	-3.1	-1.6	0.6
123.27	2.30	0.5	21.26	11AP	-0.9	-3.7	-1.6	0.7
126.78	2.01	0.7	30.88	21AP	-1.8	-4.1	-1.6	0.8
130.83	1.74	0.8	40.48	1MY	-1.1	-4.2	-1.6	0.8
135.33	1.49	0.9	50.03	11MY	-0.2	-4.2	-1.6	0.9
140.20	1.27	1.0	59.57	21MY	0.6	-4.1	-1.7	0.9
145.38	1.06	1.1	69.10	31MY	1.2	-4.0	-1.7	1.0
150.84	0.86	1.2	78.62	10JN	2.1	-3.9	-1.7	1.0
156.53	0.68	1.3	88.14	20JN	3.3	-3.8	-1.8	1.1
162.43	0.51	1.3	97.69	30JN	2.0	-3.7	-1.8	1.1
168.54	0.34	1.4	107.25	10JL	0.5	-3.7	-1.9	1.1
174.82	0.19	1.4	116.85	20JL	-0.8	-3.6	-2.0	1.1
181.27	0.04	1.4	126.49	30JL	-1.4	-3.6	-2.0	1.1
187.87	-0.10	1.4	136.17	9AU	-1.1	-3.5	-2.1	1.1
194.63	-0.23	1.5	145.90	19AU	-0.5	-3.5	-2.2	1.1
201.53	-0.35	1.5	155.69	29AU	-0.2	-3.4	-2.2	1.0
208.57	-0.47	1.5	165.54	8SE	0.0	-3.4	-2.3	1.0
215.73	-0.58	1.5	175.44	18SE	0.1	-3.4	-2.3	0.9
223.02	-0.68	1.5	185.41	28SE	0.5	-3.4	-2.4	0.9
230.42	-0.77	1.5	195.43	8OC	1.9	-3.4	-2.4	0.9
237.92	-0.85	1.4	205.50	18OC	1.5	-3.4	-2.4	0.9
245.52	-0.92	1.4	215.62	28OC	-0.2	-3.4	-2.4	0.9
253.20	-0.99	1.4	225.77	7NO	-0.5	-3.4	-2.4	0.9
260.95	-1.04	1.4	235.95	17NO	-0.5	-3.4	-2.3	0.9
268.76	-1.07	1.4	246.15	27NO	-0.5	-3.4	-2.3	0.9
276.61	-1.10	1.4	256.35	7DE	-0.7	-3.4	-2.2	0.9
284.49	-1.12	1.4	266.56	17DE	-0.9	-3.4	-2.1	0.9
292.39	-1.12	1.4	276.75	27DE	-0.9	-3.4	-2.1	0.8
				245				
300.29	-1.11	1.4	286.91	6JA	-0.7	-3.5	-2.0	0.8
308.18	-1.09	1.4	297.04	16JA	0.5	-3.5	-1.9	0.8
316.04	-1.05	1.4	307.12	26JA	3.0	-3.5	-1.8	0.7
323.86	-1.01	1.4	317.16	5FE	1.6	-3.4	-1.8	0.7
331.64	-0.96	1.4	327.14	15FE	0.8	-3.4	-1.7	0.6
339.36	-0.89	1.4	337.07	25FE	0.4	-3.4	-1.7	0.6
347.00	-0.82	1.4	346.93	7MR	0.2	-3.4	-1.6	0.5
354.58	-0.75	1.4	356.74	17MR	-0.3	-3.3	-1.6	0.5
2.07	-0.66	1.5	6.49	27MR	-0.9	-3.3	-1.5	0.4
9.48	-0.57	1.5	16.19	6AP	-1.8	-3.3	-1.5	0.5
16.81	-0.48	1.6	25.84	16AP	-1.1	-3.3	-1.5	0.5
24.04	-0.38	1.6	35.45	26AP	-0.1	-3.3	-1.5	0.6
31.19	-0.28	1.6	45.02	6MY	0.8	-3.4	-1.5	0.7
38.24	-0.17	1.7	54.57	16MY	1.6	-3.4	-1.5	0.7
45.21	-0.07	1.7	64.10	26MY	2.8	-3.4	-1.5	0.8
52.09	0.04	1.7	73.62	5JN	3.1	-3.4	-1.5	0.8
58.88	0.15	1.8	83.15	15JN	1.6	-3.5	-1.5	0.9
65.59	0.25	1.8	92.68	25JN	0.3	-3.5	-1.6	0.9
72.22	0.36	1.8	102.23	5JL	-0.9	-3.6	-1.6	1.0
78.76	0.48	1.8	111.81	15JL	-1.5	-3.6	-1.6	1.0
85.22	0.59	1.8	121.43	25JL	-1.1	-3.7	-1.7	1.0
91.59	0.70	1.8	131.08	4AU	-0.5	-3.8	-1.8	1.0
97.89	0.81	1.8	140.79	14AU	-0.1	-3.9	-1.8	1.0
104.08	0.93	1.8	150.55	24AU	0.1	-4.0	-1.9	1.0
110.19	1.05	1.7	160.37	3SE	0.3	-4.1	-1.9	0.9
116.19	1.17	1.7	170.24	13SE	0.7	-4.2	-2.0	0.9
122.08	1.30	1.7	180.18	23SE	2.3	-4.3	-2.1	0.9
127.84	1.43	1.6	190.17	3OC	1.3	-4.3	-2.1	0.8
133.45	1.56	1.5	200.21	13OC	-0.3	-4.2	-2.2	0.8
138.90	1.71	1.4	210.31	23OC	-0.6	-3.8	-2.2	0.8
144.14	1.86	1.3	220.44	2NO	-0.6	-3.0	-2.3	0.9
149.14	2.02	1.2	230.61	12NO	-0.7	-3.7	-2.3	0.9
153.85	2.20	1.1	240.80	22NO	-0.8	-4.2	-2.3	0.9
158.21	2.39	0.9	251.00	2DE	-0.7	-4.4	-2.3	0.9
162.12	2.59	0.7	261.21	12DE	-0.8	-4.4	-2.3	0.9
165.49	2.81	0.5	271.41	22DE	-0.6	-4.3	-2.2	0.8

♂ LONG	♂ LAT	♂ MAG	⊙ LONG	16.00UT 246	☿ MAGNITUDES	♀	♃	♄
168.18	3.05	0.3	281.58	1JA	0.6	-4.2	-2.2	0.8
170.02	3.30	0.0	291.73	11JA	3.0	-4.1	-2.1	0.8
170.85	3.54	-0.2	301.84	21JA	1.1	-4.0	-2.0	0.7
170.48	3.77	-0.5	311.90	31JA	0.5	-3.9	-2.0	0.7
168.83	3.93	-0.8	321.92	10FE	0.3	-3.8	-1.9	0.6
165.99	3.99	-1.0	331.87	20FE	0.1	-3.7	-1.8	0.6
162.35	3.90	-1.2	341.77	2MR	-0.3	-3.6	-1.7	0.5
158.54	3.65	-1.1	351.61	12MR	-1.0	-3.5	-1.7	0.5
155.29	3.27	-1.0	1.39	22MR	-1.7	-3.5	-1.6	0.4
153.11	2.82	-0.8	11.11	1AP	-1.1	-3.4	-1.6	0.4
152.24	2.35	-0.6	20.79	11AP	-0.1	-3.4	-1.5	0.4
152.66	1.90	-0.3	30.42	21AP	1.0	-3.4	-1.5	0.4
154.22	1.49	-0.1	40.01	1MY	2.1	-3.3	-1.4	0.5
156.76	1.12	0.0	49.57	11MY	3.6	-3.3	-1.4	0.5
160.11	0.79	0.2	59.11	21MY	2.4	-3.3	-1.4	0.6
164.11	0.49	0.3	68.64	31MY	1.3	-3.3	-1.4	0.7
168.66	0.23	0.5	78.16	10JN	0.2	-3.3	-1.4	0.7
173.67	-0.00	0.6	87.68	20JN	-0.9	-3.3	-1.4	0.8
179.06	-0.21	0.7	97.23	30JN	-1.6	-3.4	-1.4	0.8
184.80	-0.40	0.8	106.79	10JL	-1.1	-3.4	-1.4	0.9
190.82	-0.57	0.8	116.38	20JL	-0.4	-3.4	-1.4	0.9
197.10	-0.72	0.9	126.02	30JL	0.0	-3.4	-1.5	0.9
203.62	-0.85	0.9	135.70	9AU	0.3	-3.5	-1.5	0.9
210.34	-0.97	1.0	145.43	19AU	0.5	-3.5	-1.5	0.9
217.24	-1.07	1.0	155.21	29AU	1.0	-3.5	-1.6	0.9
224.32	-1.15	1.1	165.06	8SE	2.7	-3.4	-1.6	0.9
231.55	-1.22	1.1	174.96	18SE	1.1	-3.4	-1.7	0.9
238.91	-1.28	1.1	184.92	28SE	-0.5	-3.4	-1.8	0.8
246.39	-1.32	1.2	194.94	8OC	-0.8	-3.4	-1.8	0.8
253.97	-1.34	1.2	205.00	18OC	-0.8	-3.3	-1.9	0.7
261.64	-1.35	1.2	215.12	28OC	-0.8	-3.3	-2.0	0.8
269.38	-1.34	1.3	225.27	7NO	-0.7	-3.3	-2.0	0.8
277.16	-1.32	1.3	235.45	17NO	-0.6	-3.3	-2.1	0.8
284.98	-1.29	1.3	245.65	27NO	-0.6	-3.3	-2.1	0.8
292.82	-1.24	1.3	255.86	7DE	-0.5	-3.4	-2.1	0.8
300.65	-1.18	1.4	266.06	17DE	0.7	-3.4	-2.2	0.8
308.47	-1.11	1.4	276.25	27DE	2.7	-3.4	-2.2	0.8
				247				
316.27	-1.04	1.4	286.42	6JA	0.8	-3.4	-2.2	0.8
324.02	-0.95	1.4	296.54	16JA	0.3	-3.5	-2.1	0.8
331.72	-0.86	1.5	306.64	26JA	0.1	-3.5	-2.1	0.7
339.36	-0.76	1.5	316.67	5FE	-0.1	-3.6	-2.0	0.7
346.93	-0.66	1.5	326.66	15FE	-0.4	-3.6	-2.0	0.6
354.43	-0.56	1.6	336.59	25FE	-1.0	-3.7	-1.9	0.6
1.85	-0.46	1.6	346.46	7MR	-1.6	-3.8	-1.8	0.5
9.18	-0.35	1.6	356.27	17MR	-1.1	-3.8	-1.8	0.5
16.44	-0.24	1.6	6.02	27MR	0.0	-3.9	-1.7	0.4
23.60	-0.14	1.6	15.72	6AP	1.3	-4.0	-1.6	0.4
30.69	-0.03	1.7	25.37	16AP	2.8	-4.1	-1.6	0.3
37.70	0.07	1.7	34.99	26AP	3.2	-4.2	-1.5	0.3
44.62	0.17	1.7	44.56	6MY	1.9	-4.2	-1.5	0.4
51.47	0.27	1.7	54.11	16MY	1.0	-4.1	-1.4	0.4
58.25	0.36	1.8	63.64	26MY	0.1	-3.7	-1.4	0.5
64.96	0.46	1.8	73.16	5JN	-0.9	-3.0	-1.3	0.6
71.61	0.55	1.9	82.69	15JN	-1.7	-3.2	-1.3	0.6
78.19	0.64	1.9	92.22	25JN	-1.1	-3.9	-1.3	0.7
84.72	0.72	1.9	101.77	5JL	-0.4	-4.1	-1.3	0.7
91.19	0.81	1.9	111.35	15JL	0.1	-4.2	-1.3	0.8
97.61	0.89	2.0	120.96	25JL	0.4	-4.2	-1.3	0.8
103.98	0.97	2.0	130.61	4AU	0.7	-4.1	-1.3	0.8
110.31	1.05	2.0	140.32	14AU	1.4	-4.0	-1.3	0.8
116.58	1.12	2.0	150.08	24AU	3.1	-3.9	-1.3	0.9
122.81	1.20	1.9	159.89	3SE	1.0	-3.9	-1.4	0.9
128.99	1.27	1.9	169.76	13SE	-0.6	-3.8	-1.4	0.8
135.12	1.34	1.9	179.70	23SE	-0.9	-3.7	-1.5	0.8
141.19	1.42	1.9	189.68	3OC	-0.9	-3.6	-1.5	0.8
147.20	1.49	1.8	199.72	13OC	-0.8	-3.6	-1.6	0.8
153.13	1.56	1.7	209.82	23OC	-0.5	-3.5	-1.6	0.7
158.98	1.63	1.7	219.95	2NO	-0.5	-3.5	-1.7	0.7
164.74	1.70	1.6	230.11	12NO	-0.5	-3.5	-1.8	0.7
170.37	1.77	1.5	240.31	22NO	-0.3	-3.4	-1.8	0.8
175.87	1.84	1.4	250.51	2DE	0.9	-3.4	-1.9	0.8
181.21	1.91	1.2	260.71	12DE	2.4	-3.4	-1.9	0.8
186.34	1.98	1.1	270.91	22DE	0.5	-3.4	-2.0	0.8

♂ LONG	♂ LAT	♂ MAG	☉ LONG	16.00UT 248	☿	♀	♃	♄
191.22	2.05	0.9	281.09	1JA	0.1	-3.3	-2.0	0.8
195.80	2.11	0.7	291.24	11JA	-0.0	-3.3	-2.1	0.8
200.00	2.17	0.5	301.35	21JA	-0.2	-3.3	-2.1	0.8
203.73	2.21	0.3	311.42	31JA	-0.5	-3.3	-2.1	0.7
206.86	2.24	0.1	321.43	10FE	-1.0	-3.3	-2.0	0.7
209.25	2.25	-0.2	331.39	20FE	-1.5	-3.4	-2.0	0.7
210.74	2.22	-0.5	341.29	1MR	-1.1	-3.4	-2.0	0.6
211.12	2.13	-0.8	351.13	11MR	0.1	-3.4	-1.9	0.6
210.27	1.96	-1.1	0.92	21MR	1.6	-3.5	-1.9	0.5
208.18	1.69	-1.4	10.64	31MR	3.5	-3.5	-1.8	0.4
205.09	1.31	-1.7	20.32	10AP	2.4	-3.5	-1.7	0.4
201.55	0.85	-1.7	29.95	20AP	1.4	-3.4	-1.7	0.3
198.30	0.35	-1.6	39.54	30AP	0.7	-3.4	-1.6	0.3
195.96	-0.15	-1.4	49.11	10MY	0.0	-3.4	-1.5	0.3
194.91	-0.59	-1.2	58.65	20MY	-0.9	-3.4	-1.5	0.4
195.24	-0.97	-1.0	68.17	30MY	-1.8	-3.3	-1.4	0.4
196.82	-1.27	-0.8	77.69	9JN	-1.1	-3.3	-1.4	0.5
199.51	-1.52	-0.6	87.22	19JN	-0.3	-3.3	-1.3	0.5
203.11	-1.71	-0.4	96.75	29JN	0.2	-3.3	-1.3	0.6
207.45	-1.86	-0.3	106.32	9JL	0.6	-3.3	-1.3	0.7
212.42	-1.97	-0.1	115.91	19JL	1.0	-3.4	-1.2	0.7
217.90	-2.05	0.0	125.54	29JL	1.8	-3.4	-1.2	0.7
223.79	-2.10	0.1	135.22	8AU	3.2	-3.4	-1.2	0.8
230.05	-2.12	0.2	144.95	18AU	0.8	-3.4	-1.2	0.8
236.60	-2.12	0.3	154.73	28AU	-0.7	-3.5	-1.2	0.8
243.39	-2.10	0.4	164.58	7SE	-1.1	-3.6	-1.2	0.8
250.40	-2.05	0.5	174.48	17SE	-1.1	-3.6	-1.3	0.8
257.56	-1.99	0.6	184.44	27SE	-0.7	-3.7	-1.3	0.8
264.87	-1.91	0.7	194.45	7OC	-0.4	-3.8	-1.3	0.8
272.28	-1.81	0.8	204.52	17OC	-0.3	-3.9	-1.4	0.8
279.78	-1.70	0.9	214.63	27OC	-0.3	-4.0	-1.4	0.7
287.32	-1.58	0.9	224.78	6NO	-0.1	-4.1	-1.5	0.7
294.91	-1.46	1.0	234.96	16NO	1.1	-4.2	-1.5	0.7
302.50	-1.32	1.1	245.15	26NO	2.1	-4.3	-1.6	0.7
310.10	-1.19	1.2	255.36	6DE	0.3	-4.4	-1.7	0.8
317.67	-1.05	1.2	265.57	16DE	-0.1	-4.4	-1.7	0.8
325.21	-0.91	1.3	275.75	26DE	-0.1	-4.2	-1.8	0.8

♂ LONG	♂ LAT	♂ MAG	☉ LONG	16.00UT 249	☿	♀	♃	♄
332.70	-0.77	1.4	285.92	5JA	-0.3	-3.8	-1.9	0.8
340.15	-0.63	1.4	296.05	15JA	-0.5	-3.2	-1.9	0.8
347.53	-0.50	1.5	306.14	25JA	-1.0	-3.7	-2.0	0.8
354.84	-0.37	1.5	316.19	4FE	-1.3	-4.2	-2.0	0.8
2.09	-0.24	1.6	326.18	14FE	-1.0	-4.3	-2.0	0.7
9.26	-0.12	1.6	336.10	24FE	0.2	-4.3	-2.0	0.7
16.36	-0.00	1.7	345.98	6MR	2.0	-4.2	-2.0	0.6
23.39	0.11	1.7	355.79	16MR	3.2	-4.1	-2.0	0.6
30.34	0.21	1.8	5.55	26MR	1.8	-4.0	-2.0	0.5
37.23	0.31	1.8	15.26	5AP	1.0	-3.9	-1.9	0.5
44.04	0.41	1.8	24.91	15AP	0.5	-3.8	-1.9	0.4
50.80	0.49	1.9	34.52	25AP	-0.1	-3.7	-1.8	0.3
57.49	0.58	1.9	44.10	5MY	-0.9	-3.6	-1.7	0.3
64.13	0.66	1.9	53.65	15MY	-1.8	-3.6	-1.7	0.2
70.72	0.73	1.9	63.18	25MY	-1.1	-3.5	-1.6	0.3
77.27	0.80	1.9	72.71	4JN	-0.3	-3.4	-1.5	0.4
83.77	0.86	1.9	82.23	14JN	0.3	-3.4	-1.5	0.4
90.24	0.92	1.9	91.76	24JN	0.8	-3.4	-1.4	0.5
96.68	0.98	1.9	101.31	4JL	1.3	-3.4	-1.4	0.5
103.09	1.03	2.0	110.88	14JL	2.3	-3.3	-1.3	0.6
109.47	1.07	2.0	120.49	24JL	2.8	-3.3	-1.3	0.7
115.84	1.12	2.0	130.15	3AU	0.7	-3.3	-1.3	0.7
122.19	1.16	2.0	139.85	13AU	-0.7	-3.3	-1.2	0.7
128.53	1.19	2.0	149.60	23AU	-1.2	-3.4	-1.2	0.8
134.85	1.22	2.0	159.42	2SE	-1.2	-3.4	-1.2	0.8
141.17	1.25	2.0	169.28	12SE	-0.6	-3.4	-1.2	0.8
147.47	1.28	2.0	179.21	22SE	-0.3	-3.4	-1.2	0.8
153.77	1.30	2.0	189.20	2OC	-0.2	-3.4	-1.2	0.8
160.06	1.31	1.9	199.24	12OC	-0.1	-3.5	-1.2	0.8
166.34	1.33	1.9	209.32	22OC	0.1	-3.5	-1.3	0.8
172.60	1.33	1.8	219.46	1NO	1.4	-3.5	-1.3	0.7
178.84	1.33	1.8	229.62	11NO	1.9	-3.5	-1.3	0.7
185.07	1.33	1.7	239.81	21NO	0.1	-3.4	-1.4	0.7
191.26	1.32	1.6	250.01	1DE	-0.2	-3.4	-1.4	0.7
197.41	1.30	1.5	260.22	11DE	-0.3	-3.4	-1.5	0.7
203.53	1.27	1.4	270.42	21DE	-0.4	-3.4	-1.6	0.8
209.58	1.23	1.3	280.60	31DE	-0.6	-3.3	-1.6	0.8

♂ LONG	♂ LAT	♂ MAG	☉ LONG	16.00UT 250	☿	♀	♃	♄
215.57	1.17	1.2	290.75	10JA	-1.0	-3.3	-1.7	0.8
221.47	1.10	1.0	300.86	20JA	-1.2	-3.3	-1.8	0.8
227.27	1.01	0.9	310.93	30JA	-0.9	-3.3	-1.8	0.8
232.94	0.90	0.7	320.94	9FE	0.2	-3.3	-1.9	0.8
238.44	0.76	0.5	330.91	19FE	2.4	-3.4	-1.9	0.8
243.74	0.59	0.3	340.81	1MR	2.5	-3.4	-2.0	0.7
248.78	0.38	0.1	350.65	11MR	1.3	-3.4	-2.0	0.7
253.50	0.12	-0.1	0.44	21MR	0.8	-3.4	-2.0	0.6
257.80	-0.20	-0.4	10.18	31MR	0.4	-3.5	-2.0	0.6
261.58	-0.59	-0.6	19.85	10AP	-0.1	-3.5	-2.0	0.5
264.68	-1.06	-0.9	29.49	20AP	-0.9	-3.6	-2.0	0.5
266.93	-1.62	-1.2	39.09	30AP	-1.8	-3.6	-2.0	0.4
268.16	-2.28	-1.5	48.65	10MY	-1.2	-3.7	-1.9	0.3
268.19	-3.01	-1.9	58.19	20MY	-0.2	-3.8	-1.9	0.3
266.98	-3.77	-2.2	67.72	30MY	0.5	-3.9	-1.8	0.3
264.75	-4.49	-2.4	77.23	9JN	1.0	-3.9	-1.7	0.3
261.97	-5.06	-2.5	86.76	19JN	1.7	-4.0	-1.7	0.4
259.38	-5.39	-2.4	96.30	29JN	3.0	-4.1	-1.6	0.4
257.64	-5.49	-2.2	105.86	9JL	2.3	-4.2	-1.5	0.5
257.16	-5.37	-1.9	115.45	19JL	0.6	-4.2	-1.5	0.6
258.04	-5.12	-1.7	125.08	29JL	-0.8	-4.1	-1.4	0.6
260.19	-4.80	-1.4	134.75	8AU	-1.3	-3.7	-1.4	0.7
263.41	-4.42	-1.2	144.48	18AU	-1.2	-3.2	-1.3	0.7
267.52	-4.04	-0.9	154.26	28AU	-0.6	-3.7	-1.3	0.7
272.31	-3.65	-0.7	164.10	7SE	-0.2	-4.1	-1.3	0.8
277.65	-3.27	-0.5	174.00	17SE	-0.1	-4.3	-1.3	0.8
283.42	-2.91	-0.3	183.95	27SE	0.0	-4.3	-1.3	0.8
289.52	-2.56	-0.1	193.96	7OC	0.3	-4.3	-1.2	0.8
295.86	-2.23	0.0	204.03	17OC	1.7	-4.2	-1.2	0.8
302.39	-1.92	0.2	214.14	27OC	1.7	-4.1	-1.3	0.8
309.05	-1.63	0.4	224.29	6NO	-0.1	-4.0	-1.3	0.7
315.82	-1.36	0.5	234.47	16NO	-0.4	-3.9	-1.3	0.7
322.65	-1.10	0.7	244.66	26NO	-0.4	-3.8	-1.3	0.7
329.52	-0.87	0.8	254.87	6DE	-0.5	-3.7	-1.4	0.7
336.41	-0.65	0.9	265.08	16DE	-0.7	-3.7	-1.4	0.7
343.30	-0.46	1.0	275.27	26DE	-1.0	-3.6	-1.5	0.8

♂ LONG	♂ LAT	♂ MAG	☉ LONG	16.00UT 251	☿	♀	♃	♄
350.18	-0.28	1.2	285.43	5JA	-1.0	-3.5	-1.5	0.8
357.04	-0.11	1.3	295.57	15JA	-0.8	-3.5	-1.6	0.8
3.87	0.04	1.4	305.66	25JA	0.3	-3.4	-1.6	0.8
10.67	0.18	1.5	315.71	4FE	2.8	-3.4	-1.7	0.8
17.43	0.30	1.5	325.70	14FE	1.9	-3.4	-1.8	0.8
24.14	0.41	1.6	335.63	24FE	0.9	-3.3	-1.8	0.8
30.82	0.52	1.7	345.51	6MR	0.6	-3.3	-1.9	0.8
37.45	0.61	1.8	355.33	16MR	0.2	-3.3	-2.0	0.7
44.05	0.69	1.8	5.08	26MR	-0.2	-3.3	-2.0	0.7
50.60	0.76	1.9	14.79	5AP	-0.9	-3.3	-2.1	0.6
57.12	0.83	1.9	24.45	15AP	-1.8	-3.3	-2.1	0.6
63.60	0.89	1.9	34.06	25AP	-1.2	-3.3	-2.1	0.5
70.05	0.94	2.0	43.64	5MY	-0.2	-3.4	-2.1	0.4
76.47	0.99	2.0	53.19	15MY	0.6	-3.4	-2.1	0.4
82.88	1.03	2.0	62.72	25MY	1.3	-3.4	-2.1	0.3
89.26	1.06	2.0	72.25	4JN	2.3	-3.5	-2.0	0.3
95.63	1.09	2.0	81.77	14JN	3.4	-3.5	-2.0	0.3
101.99	1.11	2.0	91.30	24JN	1.9	-3.5	-1.9	0.3
108.35	1.13	2.0	100.84	4JL	0.4	-3.4	-1.8	0.4
114.71	1.15	2.0	110.42	14JL	-0.8	-3.4	-1.8	0.5
121.07	1.16	2.0	120.03	24JL	-1.5	-3.4	-1.7	0.5
127.44	1.16	2.0	129.68	3AU	-1.2	-3.3	-1.7	0.6
133.83	1.16	2.0	139.38	13AU	-0.5	-3.3	-1.6	0.6
140.23	1.16	2.0	149.12	23AU	-0.1	-3.3	-1.5	0.7
146.65	1.15	2.0	158.93	2SE	0.1	-3.3	-1.5	0.7
153.09	1.13	2.0	168.80	12SE	0.2	-3.3	-1.5	0.7
159.56	1.11	2.0	178.72	22SE	0.5	-3.3	-1.4	0.8
166.06	1.09	2.0	188.71	2OC	2.0	-3.4	-1.4	0.8
172.58	1.06	2.0	198.74	12OC	1.5	-3.4	-1.4	0.8
179.13	1.02	1.9	208.83	22OC	-0.2	-3.4	-1.3	0.8
185.72	0.98	1.9	218.96	1NO	-0.5	-3.5	-1.3	0.8
192.33	0.93	1.9	229.12	11NO	-0.5	-3.5	-1.3	0.8
198.97	0.87	1.8	239.31	21NO	-0.6	-3.6	-1.3	0.8
205.65	0.81	1.7	249.52	1DE	-0.8	-3.6	-1.3	0.7
212.35	0.73	1.7	259.72	11DE	-0.8	-3.7	-1.4	0.7
219.08	0.65	1.6	269.92	21DE	-0.9	-3.8	-1.4	0.7
225.84	0.55	1.5	280.11	31DE	-0.7	-3.9	-1.4	0.8

♂ / ☉ / Planet Magnitudes — 252 / 253

♂ LONG	LAT	MAG	☉ LONG	16.00UT	☿	♀	♃	♄
232.62	0.44	1.4	290.26	10JA	0.5	-4.0	-1.5	0.8
239.43	0.31	1.3	300.37	20JA	3.0	-4.1	-1.5	0.8
246.25	0.17	1.2	310.45	30JA	1.4	-4.2	-1.6	0.8
253.08	0.01	1.1	320.47	9FE	0.7	-4.2	-1.6	0.9
259.92	-0.16	1.0	330.43	19FE	0.4	-4.3	-1.7	0.9
266.76	-0.35	0.9	340.34	29FE	0.1	-4.3	-1.8	0.8
273.59	-0.57	0.8	350.19	10MR	-0.3	-4.0	-1.8	0.8
280.40	-0.80	0.6	359.97	20MR	-0.9	-3.5	-1.9	0.8
287.17	-1.05	0.5	9.71	30MR	-1.8	-3.1	-2.0	0.8
293.87	-1.33	0.4	19.39	9AP	-1.2	-3.7	-2.0	0.7
300.49	-1.63	0.2	29.02	19AP	-0.1	-4.1	-2.1	0.6
306.98	-1.95	0.0	38.62	29AP	0.8	-4.2	-2.1	0.6
313.31	-2.29	-0.1	48.19	9MY	1.8	-4.2	-2.2	0.5
319.41	-2.66	-0.3	57.73	19MY	3.1	-4.1	-2.2	0.5
325.22	-3.04	-0.5	67.26	29MY	2.9	-4.0	-2.2	0.4
330.63	-3.45	-0.7	76.78	8JN	1.5	-3.9	-2.2	0.3
335.55	-3.87	-0.9	86.30	18JN	0.3	-3.8	-2.2	0.3
339.81	-4.30	-1.1	95.84	28JN	-0.8	-3.7	-2.2	0.3
343.25	-4.73	-1.3	105.40	8JL	-1.6	-3.7	-2.1	0.4
345.65	-5.14	-1.6	114.99	18JL	-1.2	-3.6	-2.1	0.4
346.80	-5.49	-1.8	124.62	28JL	-0.5	-3.6	-2.0	0.5
346.55	-5.73	-2.1	134.29	7AU	-0.0	-3.5	-1.9	0.5
344.93	-5.79	-2.3	144.01	17AU	0.2	-3.5	-1.9	0.6
342.27	-5.60	-2.4	153.79	27AU	0.4	-3.4	-1.8	0.6
339.24	-5.15	-2.4	163.62	6SE	0.8	-3.4	-1.8	0.7
336.62	-4.49	-2.2	173.52	16SE	2.4	-3.4	-1.7	0.7
335.03	-3.74	-1.9	183.47	26SE	1.3	-3.4	-1.6	0.8
334.75	-2.99	-1.5	193.48	6OC	-0.4	-3.4	-1.6	0.8
335.76	-2.30	-1.2	203.54	16OC	-0.7	-3.4	-1.5	0.8
337.89	-1.70	-0.9	213.65	26OC	-0.7	-3.4	-1.5	0.8
340.96	-1.19	-0.6	223.79	5NO	-0.7	-3.4	-1.5	0.8
344.75	-0.77	-0.3	233.97	15NO	-0.7	-3.4	-1.5	0.8
349.10	-0.41	-0.1	244.17	25NO	-0.7	-3.4	-1.4	0.8
353.90	-0.11	0.2	254.37	5DE	-0.7	-3.4	-1.4	0.8
359.03	0.14	0.4	264.57	15DE	-0.6	-3.4	-1.4	0.8
4.42	0.34	0.6	274.77	25DE	0.6	-3.4	-1.4	0.7
				253				
10.00	0.52	0.8	284.94	4JA	3.0	-3.5	-1.4	0.8
15.73	0.66	0.9	295.07	14JA	1.0	-3.5	-1.5	0.8
21.58	0.78	1.1	305.17	24JA	0.4	-3.5	-1.5	0.9
27.52	0.88	1.2	315.22	3FE	0.2	-3.4	-1.5	0.9
33.52	0.97	1.3	325.21	13FE	-0.0	-3.4	-1.6	0.9
39.57	1.04	1.5	335.15	23FE	-0.3	-3.4	-1.6	0.9
45.65	1.09	1.6	345.03	5MR	-0.9	-3.4	-1.7	0.9
51.76	1.14	1.7	354.85	15MR	-1.7	-3.3	-1.7	0.9
57.90	1.18	1.7	4.61	25MR	-1.2	-3.3	-1.8	0.9
64.04	1.21	1.8	14.32	4AP	-0.1	-3.3	-1.9	0.8
70.20	1.23	1.9	23.98	14AP	1.1	-3.3	-1.9	0.8
76.38	1.24	1.9	33.60	24AP	2.3	-3.3	-2.0	0.7
82.56	1.25	2.0	43.17	4MY	3.7	-3.4	-2.1	0.7
88.75	1.25	2.0	52.73	14MY	2.2	-3.4	-2.2	0.6
94.96	1.25	2.0	62.26	24MY	1.2	-3.4	-2.2	0.6
101.19	1.24	2.0	71.78	3JN	0.2	-3.4	-2.3	0.5
107.43	1.23	2.0	81.31	13JN	-0.8	-3.5	-2.3	0.4
113.70	1.21	2.0	90.84	23JN	-1.6	-3.5	-2.3	0.4
119.99	1.19	2.0	100.38	3JL	-1.2	-3.6	-2.3	0.3
126.31	1.17	2.0	109.96	13JL	-0.4	-3.6	-2.3	0.4
132.67	1.14	2.0	119.57	23JL	0.0	-3.7	-2.3	0.4
139.06	1.10	2.0	129.21	2AU	0.3	-3.8	-2.3	0.5
145.49	1.07	2.0	138.91	12AU	0.6	-3.9	-2.2	0.5
151.97	1.02	1.9	148.66	22AU	1.1	-4.0	-2.2	0.6
158.49	0.98	1.9	158.46	1SE	2.8	-4.1	-2.1	0.6
165.06	0.92	1.9	168.32	11SE	1.1	-4.2	-2.1	0.7
171.69	0.87	1.9	178.25	21SE	-0.5	-4.3	-2.0	0.7
178.37	0.80	1.9	188.22	1OC	-0.9	-4.3	-1.9	0.8
185.12	0.74	1.9	198.26	11OC	-0.8	-4.2	-1.9	0.8
191.92	0.66	1.9	208.34	21OC	-0.8	-3.7	-1.8	0.8
198.78	0.59	1.8	218.46	31OC	-0.6	-3.0	-1.7	0.8
205.70	0.50	1.8	228.63	10NO	-0.5	-3.7	-1.7	0.9
212.69	0.41	1.8	238.81	20NO	-0.5	-4.2	-1.6	0.9
219.73	0.31	1.7	249.01	30NO	-0.4	-4.4	-1.6	0.9
226.84	0.20	1.7	259.22	10DE	0.7	-4.4	-1.6	0.8
234.01	0.09	1.6	269.42	20DE	2.7	-4.3	-1.5	0.8
241.23	-0.03	1.6	279.60	30DE	0.7	-4.2	-1.5	0.8

♂ / ☉ / Planet Magnitudes — 254 / 255

♂ LONG	LAT	MAG	☉ LONG	16.00UT	☿	♀	♃	♄
248.51	-0.15	1.5	289.76	9JA	0.2	-4.1	-1.5	0.8
255.84	-0.29	1.5	299.88	19JA	0.1	-4.0	-1.5	0.9
263.22	-0.42	1.4	309.95	29JA	-0.1	-3.9	-1.5	0.9
270.64	-0.57	1.4	319.98	8FE	-0.4	-3.8	-1.5	0.9
278.10	-0.72	1.3	329.95	18FE	-1.0	-3.7	-1.6	1.0
285.58	-0.87	1.2	339.86	28FE	-1.6	-3.6	-1.6	1.0
293.08	-1.02	1.2	349.71	10MR	-1.1	-3.5	-1.6	1.0
300.59	-1.17	1.1	359.50	20MR	-0.0	-3.5	-1.7	1.0
308.09	-1.33	1.0	9.24	30MR	1.4	-3.4	-1.7	1.0
315.57	-1.47	0.9	18.93	9AP	3.0	-3.4	-1.8	0.9
323.01	-1.61	0.9	28.57	19AP	2.9	-3.4	-1.8	0.9
330.40	-1.75	0.8	38.16	29AP	1.7	-3.3	-1.9	0.9
337.71	-1.87	0.7	47.73	9MY	0.9	-3.3	-2.0	0.8
344.92	-1.98	0.6	57.27	19MY	0.1	-3.3	-2.0	0.7
352.00	-2.08	0.5	66.80	29MY	-0.9	-3.3	-2.1	0.7
358.93	-2.16	0.5	76.32	8JN	-1.7	-3.3	-2.2	0.6
5.66	-2.22	0.4	85.85	18JN	-1.2	-3.3	-2.2	0.5
12.17	-2.27	0.3	95.38	28JN	-0.4	-3.3	-2.3	0.5
18.40	-2.29	0.2	104.94	8JL	0.1	-3.4	-2.3	0.4
24.29	-2.29	0.1	114.53	18JL	0.5	-3.4	-2.4	0.4
29.78	-2.27	-0.1	124.15	28JL	0.8	-3.4	-2.4	0.5
34.77	-2.22	-0.2	133.82	7AU	1.5	-3.5	-2.4	0.5
39.14	-2.14	-0.4	143.54	17AU	3.2	-3.5	-2.4	0.5
42.76	-2.03	-0.6	153.31	27AU	1.0	-3.5	-2.4	0.6
45.44	-1.86	-0.7	163.15	6SE	-0.6	-3.4	-2.4	0.7
46.98	-1.64	-1.0	173.05	16SE	-1.0	-3.4	-2.3	0.7
47.17	-1.35	-1.2	182.98	26SE	-1.0	-3.4	-2.3	0.8
45.89	-0.99	-1.4	192.99	6OC	-0.8	-3.4	-2.2	0.8
43.26	-0.54	-1.6	203.05	16OC	-0.5	-3.3	-2.1	0.8
39.69	-0.06	-1.7	213.15	26OC	-0.4	-3.3	-2.1	0.9
35.92	0.42	-1.6	223.30	5NO	-0.4	-3.3	-2.0	0.9
32.78	0.84	-1.4	233.47	15NO	-0.2	-3.3	-1.9	0.9
30.78	1.17	-1.0	243.66	25NO	0.9	-3.3	-1.9	0.9
30.13	1.41	-0.7	253.87	5DE	2.4	-3.4	-1.8	0.9
30.79	1.58	-0.4	264.07	15DE	0.5	-3.4	-1.8	0.9
32.54	1.69	-0.1	274.27	25DE	0.0	-3.4	-1.7	0.9
				255				
35.19	1.76	0.1	284.44	4JA	-0.1	-3.4	-1.7	0.9
38.55	1.79	0.4	294.58	14JA	-0.2	-3.5	-1.6	0.9
42.47	1.81	0.6	304.68	24JA	-0.5	-3.5	-1.6	0.9
46.82	1.81	0.8	314.73	3FE	-1.0	-3.6	-1.6	1.0
51.51	1.79	1.0	324.72	13FE	-1.4	-3.6	-1.6	1.0
56.46	1.77	1.1	334.67	23FE	-1.1	-3.7	-1.6	1.0
61.63	1.75	1.3	344.55	5MR	0.1	-3.8	-1.6	1.1
66.97	1.71	1.4	354.37	15MR	1.7	-3.9	-1.6	1.1
72.45	1.68	1.5	4.14	25MR	3.6	-3.9	-1.6	1.1
78.05	1.64	1.6	13.85	4AP	2.2	-4.0	-1.6	1.1
83.74	1.59	1.7	23.51	14AP	1.2	-4.1	-1.7	1.1
89.52	1.55	1.8	33.13	24AP	0.7	-4.2	-1.7	1.0
95.37	1.50	1.8	42.72	4MY	0.0	-4.2	-1.7	1.0
101.29	1.45	1.9	52.27	14MY	-0.9	-4.1	-1.8	1.0
107.28	1.39	1.9	61.81	24MY	-1.8	-3.7	-1.8	0.9
113.33	1.34	1.9	71.33	3JN	-1.2	-2.9	-1.9	0.8
119.44	1.28	2.0	80.85	13JN	-0.3	-3.3	-2.0	0.8
125.61	1.22	2.0	90.38	23JN	0.2	-3.9	-2.0	0.7
131.84	1.15	2.0	99.92	3JL	0.6	-4.1	-2.1	0.6
138.13	1.09	2.0	109.49	13JL	1.1	-4.2	-2.2	0.6
144.49	1.02	2.0	119.10	23JL	2.0	-4.2	-2.2	0.5
150.91	0.94	2.0	128.74	2AU	3.1	-4.1	-2.3	0.5
157.40	0.87	1.9	138.43	12AU	0.8	-4.0	-2.4	0.6
163.97	0.79	1.9	148.18	22AU	-0.7	-3.9	-2.4	0.6
170.61	0.71	1.9	157.98	1SE	-1.1	-3.8	-2.4	0.7
177.32	0.63	1.8	167.84	11SE	-1.1	-3.8	-2.5	0.7
184.12	0.54	1.8	177.76	21SE	-0.7	-3.7	-2.5	0.8
190.99	0.45	1.7	187.74	1OC	-0.4	-3.6	-2.4	0.8
197.95	0.35	1.7	197.77	11OC	-0.3	-3.6	-2.4	0.9
204.99	0.25	1.7	207.85	21OC	-0.2	-3.5	-2.4	0.9
212.10	0.15	1.7	217.98	31OC	-0.0	-3.5	-2.3	0.9
219.30	0.05	1.7	228.13	10NO	1.2	-3.5	-2.2	1.0
226.58	-0.06	1.6	238.32	20NO	2.1	-3.4	-2.2	1.0
233.93	-0.16	1.6	248.52	30NO	0.2	-3.4	-2.1	1.0
241.35	-0.27	1.6	258.73	10DE	-0.1	-3.4	-1.9	1.0
248.85	-0.39	1.6	268.93	20DE	-0.2	-3.4	-1.9	1.0
256.40	-0.50	1.5	279.11	30DE	-0.3	-3.3	-1.9	1.0

256 / 257

♂ LONG	LAT	MAG	☉ LONG	16.00UT 256	☿	♀	♃	♄
264.01	-0.61	1.5	289.27	9JA	-0.6	-3.3	-1.8	1.0
271.67	-0.72	1.5	299.39	19JA	-1.0	-3.3	-1.8	1.0
279.37	-0.82	1.5	309.46	29JA	-1.3	-3.3	-1.7	1.0
287.10	-0.92	1.4	319.49	8FE	-1.0	-3.3	-1.7	1.1
294.84	-1.02	1.4	329.46	18FE	0.1	-3.4	-1.6	1.1
302.60	-1.11	1.4	339.37	28FE	2.1	-3.4	-1.6	1.2
310.35	-1.18	1.3	349.23	9MR	3.0	-3.4	-1.6	1.2
318.09	-1.25	1.3	359.02	19MR	1.6	-3.5	-1.6	1.2
325.80	-1.31	1.3	8.76	29MR	0.9	-3.5	-1.6	1.2
333.46	-1.36	1.3	18.45	8AP	0.5	-3.5	-1.6	1.2
341.08	-1.39	1.2	28.09	18AP	-0.1	-3.4	-1.6	1.2
348.62	-1.40	1.2	37.69	28AP	-0.9	-3.4	-1.6	1.2
356.08	-1.41	1.2	47.26	8MY	-1.8	-3.4	-1.6	1.2
3.44	-1.39	1.2	56.81	18MY	-1.2	-3.4	-1.6	1.1
10.70	-1.36	1.1	66.33	28MY	-0.3	-3.3	-1.7	1.1
17.84	-1.32	1.1	75.85	7JN	0.4	-3.3	-1.7	1.0
24.84	-1.26	1.1	85.38	17JN	0.9	-3.3	-1.7	1.0
31.68	-1.19	1.0	94.91	27JN	1.4	-3.3	-1.8	0.9
38.36	-1.10	1.0	104.47	7JL	2.6	-3.3	-1.8	0.8
44.85	-0.99	0.9	114.06	17JL	2.7	-3.4	-1.9	0.8
51.12	-0.87	0.9	123.68	27JL	0.7	-3.4	-2.0	0.7
57.16	-0.73	0.8	133.35	6AU	-0.7	-3.4	-2.0	0.7
62.91	-0.57	0.7	143.07	16AU	-1.3	-3.5	-2.1	0.7
68.33	-0.39	0.6	152.83	26AU	-1.2	-3.5	-2.2	0.7
73.37	-0.19	0.5	162.66	5SE	-0.6	-3.6	-2.2	0.7
77.92	0.04	0.4	172.55	15SE	-0.3	-3.6	-2.3	0.8
81.91	0.31	0.2	182.50	25SE	-0.1	-3.7	-2.3	0.8
85.20	0.61	0.0	192.50	5OC	-0.1	-3.8	-2.4	0.9
87.60	0.95	-0.2	202.56	15OC	0.2	-3.9	-2.4	0.9
88.95	1.35	-0.4	212.66	25OC	1.4	-4.0	-2.4	1.0
89.05	1.78	-0.6	222.80	4NO	1.9	-4.1	-2.4	1.0
87.75	2.24	-0.8	232.98	14NO	0.0	-4.2	-2.3	1.1
85.12	2.70	-1.0	243.17	24NO	-0.3	-4.3	-2.3	1.1
81.49	3.08	-1.2	253.38	4DE	-0.3	-4.4	-2.2	1.1
77.51	3.34	-1.2	263.58	14DE	-0.4	-4.4	-2.2	1.2
73.97	3.46	-0.9	273.77	24DE	-0.6	-4.2	-2.1	1.2

257

♂ LONG	LAT	MAG	☉ LONG	16.00UT 257	☿	♀	♃	♄
71.46	3.45	-0.7	283.95	3JA	-1.0	-3.8	-2.0	1.2
70.24	3.35	-0.4	294.09	13JA	-1.1	-3.2	-2.0	1.2
70.33	3.19	-0.1	304.18	23JA	-0.9	-3.8	-1.9	1.2
71.56	3.02	0.1	314.24	2FE	0.2	-4.2	-1.8	1.2
73.74	2.84	0.4	324.24	12FE	2.5	-4.3	-1.8	1.2
76.68	2.66	0.6	334.18	22FE	2.3	-4.3	-1.7	1.3
80.23	2.50	0.8	344.07	4MR	1.2	-4.2	-1.6	1.3
84.27	2.34	1.0	353.89	14MR	0.7	-4.1	-1.6	1.3
88.69	2.19	1.1	3.66	24MR	0.3	-4.0	-1.6	1.4
93.42	2.05	1.2	13.38	3AP	-0.1	-3.9	-1.5	1.4
98.41	1.92	1.4	23.04	13AP	-0.9	-3.8	-1.5	1.4
103.62	1.79	1.5	32.66	23AP	-1.8	-3.7	-1.5	1.4
109.01	1.67	1.5	42.25	3MY	-1.2	-3.6	-1.5	1.4
114.56	1.55	1.6	51.81	13MY	-0.2	-3.5	-1.5	1.4
120.25	1.43	1.7	61.34	23MY	0.5	-3.5	-1.5	1.3
126.06	1.32	1.7	70.87	2JN	1.1	-3.4	-1.5	1.3
132.00	1.21	1.8	80.39	12JN	1.9	-3.4	-1.5	1.2
138.04	1.10	1.8	89.92	22JN	3.2	-3.4	-1.5	1.2
144.20	0.99	1.8	99.46	2JL	2.2	-3.4	-1.6	1.1
150.46	0.88	1.8	109.03	12JL	0.6	-3.3	-1.6	1.1
156.81	0.78	1.8	118.63	22JL	-0.8	-3.3	-1.6	1.0
163.27	0.67	1.8	128.28	1AU	-1.4	-3.3	-1.7	0.9
169.84	0.56	1.8	137.97	11AU	-1.2	-3.3	-1.7	0.9
176.50	0.45	1.8	147.71	21AU	-0.6	-3.4	-1.8	0.8
183.26	0.34	1.8	157.51	31AU	-0.2	-3.4	-1.9	0.8
190.12	0.23	1.8	167.37	10SE	-0.0	-3.4	-1.9	0.9
197.09	0.13	1.7	177.28	20SE	0.1	-3.4	-2.0	0.9
204.15	0.02	1.7	187.26	30SE	0.4	-3.4	-2.1	1.0
211.32	-0.09	1.6	197.28	10OC	1.7	-3.5	-2.1	1.0
218.58	-0.20	1.6	207.36	20OC	1.7	-3.5	-2.2	1.1
225.93	-0.30	1.6	217.48	30OC	-0.1	-3.5	-2.2	1.1
233.37	-0.40	1.5	227.64	9NO	-0.5	-3.5	-2.3	1.2
240.89	-0.50	1.5	237.82	19NO	-0.5	-3.4	-2.3	1.2
248.49	-0.60	1.4	248.03	29NO	-0.5	-3.4	-2.3	1.3
256.16	-0.69	1.4	258.23	9DE	-0.7	-3.4	-2.3	1.3
263.89	-0.77	1.4	268.43	19DE	-0.9	-3.4	-2.2	1.3
271.67	-0.85	1.4	278.62	29DE	-1.0	-3.3	-2.2	1.3

258 / 259

♂ LONG	LAT	MAG	☉ LONG	16.00UT 258	☿	♀	♃	♄
279.48	-0.92	1.4	288.77	8JA	-0.8	-3.3	-2.1	1.4
287.33	-0.99	1.4	298.89	18JA	0.3	-3.3	-2.1	1.4
295.19	-1.04	1.4	308.98	28JA	2.8	-3.3	-2.0	1.4
303.06	-1.08	1.4	319.00	7FE	1.7	-3.3	-1.9	1.3
310.91	-1.11	1.4	328.97	17FE	0.8	-3.4	-1.8	1.2
318.75	-1.13	1.4	338.89	27FE	0.5	-3.4	-1.8	1.2
326.55	-1.14	1.4	348.75	9MR	0.2	-3.4	-1.7	1.2
334.30	-1.13	1.4	358.55	19MR	-0.2	-3.4	-1.6	1.2
342.00	-1.12	1.4	8.29	29MR	-0.9	-3.5	-1.6	1.2
349.63	-1.09	1.4	17.98	8AP	-1.8	-3.5	-1.5	1.2
357.19	-1.04	1.4	27.62	18AP	-1.2	-3.6	-1.5	1.2
4.67	-0.99	1.4	37.23	28AP	-0.2	-3.6	-1.5	1.2
12.05	-0.93	1.5	46.80	8MY	0.7	-3.7	-1.4	1.1
19.34	-0.85	1.5	56.34	18MY	1.5	-3.8	-1.4	1.1
26.52	-0.77	1.5	65.87	28MY	2.6	-3.9	-1.4	1.1
33.60	-0.67	1.5	75.39	7JN	3.3	-3.9	-1.4	1.0
40.56	-0.57	1.5	84.92	17JN	1.8	-4.0	-1.4	1.0
47.40	-0.46	1.5	94.46	27JN	0.4	-4.1	-1.4	0.9
54.12	-0.34	1.4	104.01	7JL	-0.8	-4.2	-1.4	0.9
60.71	-0.21	1.4	113.60	17JL	-1.5	-4.2	-1.4	0.8
67.15	-0.07	1.4	123.22	27JL	-1.2	-4.0	-1.4	0.8
73.45	0.07	1.4	132.88	6AU	-0.5	-3.6	-1.5	0.7
79.58	0.23	1.3	142.60	16AU	-0.1	-3.2	-1.5	0.7
85.52	0.39	1.3	152.37	26AU	0.1	-3.7	-1.5	0.6
91.26	0.57	1.2	162.19	5SE	0.3	-4.1	-1.6	0.6
96.76	0.76	1.1	172.08	15SE	0.6	-4.3	-1.6	0.7
101.98	0.97	1.0	182.02	25SE	2.1	-4.3	-1.7	0.8
106.88	1.19	0.9	192.02	5OC	1.5	-4.3	-1.8	0.8
111.37	1.44	0.8	202.07	15OC	-0.3	-4.2	-1.8	0.9
115.38	1.72	0.6	212.17	25OC	-0.6	-4.1	-1.9	1.0
118.80	2.03	0.5	222.31	4NO	-0.6	-4.0	-2.0	1.0
121.48	2.38	0.3	232.48	14NO	-0.6	-3.9	-2.0	1.1
123.25	2.76	0.0	242.68	24NO	-0.8	-3.8	-2.1	1.1
123.93	3.17	-0.2	252.88	4DE	-0.8	-3.7	-2.1	1.1
123.35	3.59	-0.4	263.09	14DE	-0.8	-3.7	-2.1	1.2
121.43	3.99	-0.7	273.28	24DE	-0.7	-3.6	-2.2	1.2

259

♂ LONG	LAT	MAG	☉ LONG	16.00UT 259	☿	♀	♃	♄
118.33	4.31	-0.9	283.45	3JA	0.4	-3.5	-2.1	1.2
114.47	4.48	-1.0	293.60	13JA	3.0	-3.5	-2.1	1.1
110.58	4.48	-0.9	303.70	23JA	1.3	-3.4	-2.1	1.1
107.35	4.32	-0.7	313.75	2FE	0.6	-3.4	-2.0	1.1
105.25	4.04	-0.4	323.76	12FE	0.3	-3.4	-2.0	1.0
104.45	3.71	-0.2	333.71	22FE	0.1	-3.3	-1.9	1.0
104.89	3.37	0.0	343.59	4MR	-0.3	-3.3	-1.9	0.9
106.41	3.04	0.3	353.42	14MR	-0.9	-3.3	-1.8	0.9
108.82	2.73	0.5	3.19	24MR	-1.7	-3.3	-1.7	0.9
111.96	2.45	0.6	12.91	3AP	-1.2	-3.3	-1.7	0.9
115.69	2.19	0.8	22.58	13AP	-0.1	-3.3	-1.6	0.9
119.90	1.95	0.9	32.20	23AP	0.9	-3.3	-1.5	0.9
124.49	1.73	1.1	41.78	3MY	1.9	-3.4	-1.5	0.9
129.41	1.53	1.2	51.34	13MY	3.4	-3.4	-1.4	0.9
134.61	1.34	1.3	60.88	23MY	2.6	-3.4	-1.4	0.9
140.05	1.16	1.3	70.40	2JN	1.4	-3.5	-1.4	0.8
145.70	0.99	1.4	79.92	12JN	0.3	-3.5	-1.3	0.8
151.54	0.83	1.5	89.45	22JN	-0.8	-3.5	-1.3	0.8
157.56	0.67	1.5	98.99	2JL	-1.6	-3.4	-1.3	0.7
163.74	0.52	1.5	108.57	12JL	-1.2	-3.4	-1.3	0.7
170.07	0.38	1.5	118.16	22JL	-0.5	-3.4	-1.3	0.6
176.54	0.24	1.6	127.80	1AU	-0.0	-3.3	-1.3	0.6
183.15	0.11	1.6	137.50	11AU	0.2	-3.3	-1.3	0.5
189.90	-0.02	1.6	147.23	21AU	0.5	-3.3	-1.3	0.4
196.78	-0.14	1.6	157.03	31AU	0.9	-3.3	-1.3	0.4
203.78	-0.26	1.6	166.89	10SE	2.5	-3.3	-1.4	0.4
210.90	-0.37	1.6	176.80	20SE	1.3	-3.3	-1.4	0.4
218.14	-0.48	1.5	186.77	30SE	-0.4	-3.4	-1.5	0.4
225.49	-0.58	1.5	196.80	10OC	-0.8	-3.4	-1.5	0.5
232.94	-0.67	1.5	206.87	20OC	-0.8	-3.4	-1.6	0.6
240.49	-0.76	1.5	216.99	30OC	-0.8	-3.5	-1.6	0.6
248.12	-0.83	1.5	227.15	9NO	-0.7	-3.5	-1.7	0.7
255.83	-0.90	1.4	237.33	19NO	-0.6	-3.6	-1.8	0.7
263.60	-0.96	1.4	247.53	29NO	-0.6	-3.7	-1.8	0.8
271.43	-1.01	1.4	257.74	9DE	-0.5	-3.7	-1.9	0.8
279.30	-1.05	1.4	267.94	19DE	0.6	-3.8	-1.9	0.9
287.19	-1.07	1.4	278.12	29DE	2.9	-3.9	-2.0	0.9

♂ LONG	LAT	MAG	☉ LONG	16.00UT 260	☿	♀	♃	♄
295.09	-1.09	1.3	288.29	8JA	0.9	-4.0	-2.0	0.9
302.99	-1.09	1.3	298.41	18JA	0.4	-4.1	-2.1	0.9
310.88	-1.08	1.3	308.49	28JA	0.2	-4.2	-2.1	0.9
318.75	-1.06	1.3	318.52	7FE	-0.0	-4.3	-2.1	0.8
326.56	-1.03	1.3	328.50	17FE	-0.4	-4.3	-2.1	0.8
334.34	-0.98	1.4	338.42	27FE	-0.9	-4.3	-2.0	0.8
342.05	-0.93	1.4	348.28	8MR	-1.6	-4.0	-1.9	0.7
349.69	-0.87	1.4	358.07	18MR	-1.2	-3.5	-1.9	0.7
357.25	-0.80	1.5	7.82	28MR	-0.1	-3.1	-1.8	0.7
4.74	-0.72	1.5	17.51	7AP	1.1	-3.7	-1.8	0.7
12.13	-0.64	1.5	27.16	17AP	2.5	-4.1	-1.7	0.7
19.44	-0.54	1.6	36.76	27AP	3.5	-4.2	-1.6	0.7
26.66	-0.45	1.6	46.33	7MY	2.0	-4.2	-1.6	0.7
33.78	-0.35	1.6	55.88	17MY	1.1	-4.1	-1.5	0.7
40.81	-0.24	1.7	65.41	27MY	0.2	-4.0	-1.5	0.7
47.75	-0.13	1.7	74.93	6JN	-0.8	-3.9	-1.4	0.6
54.59	-0.02	1.7	84.45	16JN	-1.7	-3.8	-1.4	0.6
61.34	0.09	1.7	93.99	26JN	-1.2	-3.7	-1.3	0.6
67.99	0.21	1.7	103.54	6JL	-0.4	-3.7	-1.3	0.5
74.55	0.33	1.7	113.13	16JL	0.1	-3.6	-1.3	0.5
81.01	0.45	1.7	122.75	26JL	0.4	-3.5	-1.2	0.4
87.37	0.57	1.7	132.41	5AU	0.7	-3.5	-1.2	0.4
93.62	0.70	1.7	142.12	15AU	1.2	-3.5	-1.2	0.3
99.77	0.83	1.7	151.89	25AU	2.9	-3.4	-1.2	0.3
105.80	0.97	1.6	161.71	4SE	1.2	-3.4	-1.2	0.2
111.69	1.11	1.6	171.59	14SE	-0.5	-3.4	-1.2	0.2
117.44	1.26	1.5	181.54	24SE	-0.9	-3.4	-1.3	0.1
123.02	1.41	1.4	191.54	4OC	-0.9	-3.4	-1.3	0.1
128.40	1.58	1.4	201.58	14OC	-0.8	-3.4	-1.3	0.2
133.56	1.76	1.2	211.68	24OC	-0.6	-3.4	-1.4	0.3
138.43	1.95	1.1	221.82	3NO	-0.5	-3.4	-1.4	0.3
142.97	2.16	1.0	231.99	13NO	-0.5	-3.4	-1.5	0.4
147.10	2.38	0.8	242.19	23NO	-0.3	-3.4	-1.5	0.5
150.73	2.63	0.6	252.39	3DE	0.8	-3.4	-1.6	0.5
153.72	2.90	0.4	262.59	13DE	2.7	-3.4	-1.7	0.6
155.94	3.20	0.2	272.79	23DE	0.6	-3.4	-1.7	0.6

261

♂ LONG	LAT	MAG	☉ LONG	16.00UT	☿	♀	♃	♄
157.20	3.51	-0.0	282.96	2JA	0.2	-3.5	-1.8	0.6
157.32	3.82	-0.3	293.10	12JA	0.0	-3.5	-1.9	0.7
156.17	4.10	-0.6	303.21	22JA	-0.1	-3.5	-1.9	0.7
153.76	4.30	-0.8	313.27	1FE	-0.4	-3.4	-2.0	0.7
150.34	4.38	-1.0	323.27	11FE	-1.0	-3.4	-2.0	0.6
146.45	4.28	-1.1	333.22	21FE	-1.5	-3.4	-2.0	0.6
142.84	4.02	-0.9	343.11	3MR	-1.1	-3.4	-2.0	0.6
140.11	3.64	-0.7	352.94	13MR	-0.0	-3.3	-2.0	0.6
138.61	3.20	-0.5	2.72	23MR	1.4	-3.3	-2.0	0.5
138.41	2.76	-0.3	12.44	2AP	3.3	-3.3	-1.9	0.5
139.42	2.34	-0.1	22.11	12AP	2.6	-3.3	-1.9	0.5
141.46	1.95	0.1	31.73	22AP	1.5	-3.3	-1.8	0.5
144.35	1.60	0.3	41.32	2MY	0.8	-3.4	-1.8	0.5
147.94	1.29	0.5	50.88	12MY	0.1	-3.4	-1.7	0.5
152.11	1.00	0.6	60.42	22MY	-0.8	-3.4	-1.6	0.5
156.76	0.75	0.7	69.94	1JN	-1.7	-3.4	-1.6	0.5
161.80	0.51	0.8	79.46	11JN	-1.2	-3.5	-1.5	0.5
167.19	0.30	0.9	88.99	21JN	-0.4	-3.5	-1.5	0.5
172.88	0.10	1.0	98.53	1JL	0.2	-3.6	-1.4	0.5
178.82	-0.08	1.0	108.10	11JL	0.5	-3.7	-1.4	0.4
185.01	-0.24	1.1	117.70	21JL	0.9	-3.7	-1.3	0.4
191.41	-0.39	1.1	127.34	31JL	1.7	-3.8	-1.3	0.3
198.00	-0.53	1.2	137.02	10AU	3.2	-3.9	-1.3	0.3
204.77	-0.66	1.2	146.76	20AU	1.0	-4.0	-1.2	0.2
211.71	-0.77	1.2	156.55	30AU	-0.6	-4.1	-1.2	0.2
218.80	-0.88	1.3	166.40	9SE	-1.1	-4.2	-1.2	0.1
226.03	-0.97	1.3	176.32	19SE	-1.0	-4.3	-1.2	0.0
233.39	-1.04	1.3	186.28	29SE	-0.8	-4.3	-1.2	-0.0
240.86	-1.11	1.3	196.31	9OC	-0.5	-4.2	-1.2	-0.1
248.44	-1.15	1.3	206.38	19OC	-0.3	-3.7	-1.3	-0.1
256.10	-1.19	1.3	216.50	29OC	-0.3	-3.0	-1.3	0.0
263.83	-1.21	1.3	226.65	8NO	-0.2	-3.8	-1.3	0.1
271.63	-1.22	1.3	236.84	18NO	1.0	-4.2	-1.4	0.1
279.47	-1.22	1.4	247.04	28NO	2.4	-4.4	-1.4	0.2
287.33	-1.20	1.4	257.24	8DE	0.4	-4.4	-1.5	0.3
295.21	-1.17	1.4	267.45	18DE	-0.0	-4.3	-1.5	0.3
303.08	-1.13	1.4	277.63	28DE	-0.1	-4.2	-1.6	0.4

♂ LONG	LAT	MAG	☉ LONG	16.00UT 262	☿	♀	♃	♄
310.93	-1.08	1.4	287.79	7JA	-0.2	-4.1	-1.6	0.4
318.75	-1.02	1.4	297.92	17JA	-0.5	-4.0	-1.7	0.5
326.53	-0.95	1.4	308.00	27JA	-1.0	-3.9	-1.8	0.5
334.25	-0.87	1.5	318.04	6FE	-1.4	-3.8	-1.8	0.5
341.92	-0.79	1.5	328.02	16FE	-1.1	-3.7	-1.9	0.5
349.51	-0.70	1.5	337.94	26FE	0.1	-3.6	-2.0	0.5
357.02	-0.60	1.5	347.80	8MR	1.8	-3.5	-2.0	0.5
4.45	-0.50	1.5	357.61	18MR	3.4	-3.5	-2.0	0.4
11.80	-0.40	1.5	7.35	28MR	1.9	-3.4	-2.0	0.4
19.06	-0.30	1.5	17.05	7AP	1.1	-3.4	-2.0	0.3
26.24	-0.20	1.6	26.70	17AP	0.6	-3.4	-2.0	0.3
33.33	-0.09	1.6	36.30	27AP	0.0	-3.3	-2.0	0.3
40.34	0.01	1.7	45.88	7MY	-0.8	-3.3	-2.0	0.4
47.27	0.11	1.7	55.43	17MY	-1.7	-3.3	-1.9	0.4
54.11	0.21	1.8	64.95	27MY	-1.2	-3.3	-1.8	0.4
60.89	0.31	1.8	74.48	6JN	-0.3	-3.3	-1.8	0.4
67.58	0.41	1.8	84.00	16JN	0.3	-3.3	-1.7	0.4
74.21	0.51	1.9	93.53	26JN	0.7	-3.3	-1.7	0.4
80.77	0.61	1.9	103.08	6JL	1.2	-3.4	-1.6	0.4
87.27	0.70	1.9	112.66	16JL	2.2	-3.4	-1.5	0.3
93.70	0.80	1.9	122.28	26JL	3.0	-3.4	-1.5	0.3
100.07	0.89	1.9	131.94	5AU	0.8	-3.5	-1.4	0.2
106.37	0.98	1.9	141.65	15AU	-0.7	-3.5	-1.4	0.2
112.62	1.07	1.9	151.41	25AU	-1.2	-3.5	-1.3	0.2
118.80	1.17	1.9	161.23	4SE	-1.2	-3.4	-1.3	0.1
124.91	1.26	1.9	171.11	14SE	-0.7	-3.4	-1.3	0.0
130.95	1.35	1.8	181.05	24SE	-0.4	-3.4	-1.3	-0.0
136.90	1.44	1.8	191.04	4OC	-0.2	-3.4	-1.3	-0.1
142.76	1.54	1.7	201.09	14OC	-0.2	-3.3	-1.3	-0.2
148.52	1.64	1.6	211.18	24OC	0.0	-3.3	-1.3	-0.2
154.16	1.74	1.5	221.32	3NO	1.2	-3.3	-1.3	-0.2
159.64	1.84	1.4	231.49	13NO	2.1	-3.3	-1.3	-0.1
164.96	1.95	1.3	241.68	23NO	-0.3	-3.3	-1.3	-0.0
170.07	2.06	1.2	251.89	3DE	-0.2	-3.4	-1.3	0.0
174.92	2.18	1.0	262.10	13DE	-0.2	-3.4	-1.4	0.1
179.46	2.30	0.9	272.29	23DE	-0.3	-3.4	-1.4	0.2

263

♂ LONG	LAT	MAG	☉ LONG	16.00UT	☿	♀	♃	♄
183.62	2.43	0.7	282.47	2JA	-0.6	-3.4	-1.5	0.2
187.29	2.55	0.5	292.62	12JA	-1.0	-3.5	-1.5	0.3
190.37	2.68	0.2	302.72	22JA	-1.2	-3.5	-1.6	0.3
192.70	2.80	-0.0	312.79	1FE	-1.0	-3.6	-1.7	0.3
194.11	2.91	-0.3	322.79	11FE	0.1	-3.6	-1.7	0.4
194.43	2.97	-0.6	332.74	21FE	2.2	-3.7	-1.8	0.4
193.49	2.98	-0.9	342.64	3MR	2.7	-3.8	-1.9	0.4
191.31	2.89	-1.2	352.47	13MR	1.4	-3.9	-1.9	0.4
188.11	2.68	-1.4	2.25	23MR	0.8	-3.9	-2.0	0.3
184.42	2.34	-1.5	11.97	2AP	0.4	-4.0	-2.0	0.3
180.98	1.90	-1.3	21.64	12AP	-0.1	-4.1	-2.1	0.3
178.42	1.41	-1.2	31.27	22AP	-0.8	-4.2	-2.1	0.2
177.12	0.93	-1.0	40.86	2MY	-1.8	-4.2	-2.1	0.2
177.17	0.48	-0.8	50.42	12MY	-1.2	-4.1	-2.1	0.2
178.47	0.09	-0.5	59.96	22MY	-0.3	-3.7	-2.1	0.3
180.85	-0.25	-0.4	69.48	1JN	0.4	-2.9	-2.1	0.3
184.15	-0.54	-0.2	79.00	11JN	0.9	-3.3	-2.0	0.3
188.19	-0.78	-0.0	88.53	21JN	1.6	-3.9	-2.0	0.3
192.85	-0.99	0.1	98.07	1JL	2.8	-4.2	-1.9	0.3
198.03	-1.17	0.2	107.63	11JL	2.5	-4.2	-1.8	0.3
203.63	-1.32	0.3	117.23	21JL	0.7	-4.2	-1.8	0.3
209.61	-1.44	0.4	126.87	31JL	-0.7	-4.1	-1.7	0.3
215.90	-1.53	0.5	136.55	10AU	-1.3	-4.0	-1.7	0.3
222.46	-1.61	0.6	146.28	20AU	-1.2	-3.9	-1.6	0.2
229.25	-1.66	0.7	156.08	30AU	-0.6	-3.8	-1.5	0.2
236.24	-1.69	0.7	165.92	9SE	-0.3	-3.8	-1.5	0.1
243.41	-1.70	0.8	175.83	19SE	-0.1	-3.7	-1.5	0.1
250.72	-1.69	0.9	185.80	29SE	0.0	-3.6	-1.4	-0.0
258.16	-1.67	0.9	195.81	9OC	0.2	-3.6	-1.4	-0.1
265.69	-1.63	1.0	205.89	19OC	1.5	-3.5	-1.4	-0.2
273.31	-1.57	1.0	216.01	29OC	1.9	-3.4	-1.3	-0.3
280.97	-1.50	1.1	226.16	8NO	-0.0	-3.5	-1.3	-0.3
288.68	-1.42	1.1	236.34	18NO	-0.4	-3.4	-1.3	-0.3
296.40	-1.32	1.2	246.54	28NO	-0.4	-3.4	-1.3	-0.2
304.12	-1.22	1.2	256.75	8DE	-0.4	-3.4	-1.4	-0.1
311.83	-1.12	1.3	266.95	18DE	-0.7	-3.4	-1.4	-0.0
319.50	-1.00	1.4	277.14	28DE	-1.0	-3.3	-1.4	0.0

Left table (264/265):

♂ LONG	LAT	MAG	☉ LONG	16.00UT 264	☿	♀	♃	♄ MAGNITUDES
327.14	-0.89	1.4	287.30	7JA	-1.1	-3.3	-1.4	0.1
334.72	-0.77	1.5	297.43	17JA	-0.9	-3.3	-1.5	0.2
342.24	-0.65	1.5	307.52	27JA	0.2	-3.3	-1.5	0.2
349.70	-0.53	1.5	317.55	6FE	2.5	-3.3	-1.6	0.2
357.08	-0.41	1.6	327.54	16FE	2.1	-3.4	-1.7	0.3
4.38	-0.29	1.6	337.46	26FE	1.0	-3.4	-1.7	0.3
11.61	-0.18	1.7	347.32	7MR	0.6	-3.4	-1.8	0.3
18.76	-0.07	1.7	357.13	17MR	0.3	-3.5	-1.9	0.3
25.83	0.04	1.7	6.88	27MR	-0.2	-3.5	-1.9	0.3
32.82	0.15	1.8	16.58	6AP	-0.9	-3.5	-2.0	0.3
39.75	0.25	1.8	26.23	16AP	-1.8	-3.4	-2.1	0.3
46.59	0.34	1.8	35.84	26AP	-1.2	-3.4	-2.1	0.2
53.38	0.43	1.8	45.41	6MY	-0.2	-3.4	-2.2	0.2
60.09	0.52	1.8	54.96	16MY	0.6	-3.4	-2.2	0.2
66.75	0.60	1.8	64.49	26MY	1.2	-3.3	-2.2	0.2
73.36	0.68	1.8	74.01	5JN	2.1	-3.3	-2.2	0.2
79.91	0.76	1.9	83.53	15JN	3.4	-3.3	-2.2	0.3
86.42	0.83	1.9	93.07	25JN	2.0	-3.3	-2.2	0.3
92.89	0.90	1.9	102.62	5JL	0.5	-3.3	-2.2	0.3
99.32	0.96	2.0	112.20	15JL	-0.7	-3.4	-2.1	0.3
105.72	1.02	2.0	121.81	25JL	-1.4	-3.4	-2.1	0.3
112.08	1.08	2.0	131.47	4AU	-1.2	-3.4	-2.0	0.3
118.42	1.13	2.0	141.18	14AU	-0.6	-3.5	-1.9	0.3
124.73	1.18	2.0	150.93	24AU	-0.2	-3.5	-1.9	0.3
131.02	1.23	2.0	160.75	3SE	0.0	-3.6	-1.8	0.2
137.28	1.28	2.0	170.63	13SE	-0.2	-3.6	-1.8	0.2
143.52	1.32	2.0	180.56	23SE	0.5	-3.7	-1.7	0.1
149.73	1.36	1.9	190.55	3OC	1.8	-3.8	-1.6	0.0
155.92	1.39	1.9	200.60	13OC	1.7	-3.9	-1.6	-0.0
162.06	1.43	1.8	210.69	23OC	-0.2	-4.0	-1.6	-0.1
168.18	1.46	1.8	220.83	2NO	-0.5	-4.1	-1.5	-0.2
174.25	1.48	1.7	231.00	12NO	-0.5	-4.2	-1.5	-0.2
180.26	1.50	1.6	241.19	22NO	-0.6	-4.3	-1.5	-0.3
186.21	1.52	1.5	251.39	2DE	-0.7	-4.4	-1.4	-0.3
192.08	1.53	1.4	261.60	12DE	-0.8	-4.4	-1.4	-0.2
197.86	1.53	1.3	271.80	22DE	-0.9	-4.2	-1.4	-0.1
				265				
203.52	1.53	1.2	281.97	1JA	-0.7	-3.7	-1.4	-0.1
209.04	1.51	1.0	292.12	11JA	0.3	-3.2	-1.5	0.0
214.40	1.47	0.9	302.23	21JA	2.9	-3.8	-1.5	0.1
219.54	1.42	0.7	312.29	31JA	1.6	-4.2	-1.5	0.1
224.42	1.35	0.5	322.31	10FE	0.7	-4.3	-1.6	0.2
228.97	1.25	0.3	332.26	20FE	0.4	-4.3	-1.6	0.2
233.11	1.11	0.0	342.16	2MR	0.2	-4.2	-1.6	0.3
236.73	0.93	-0.2	352.00	12MR	-0.2	-4.1	-1.7	0.3
239.69	0.69	-0.5	1.78	22MR	-0.9	-4.0	-1.8	0.3
241.84	0.38	-0.8	11.50	1AP	-1.8	-3.9	-1.8	0.3
242.97	-0.02	-1.2	21.18	11AP	-1.2	-3.8	-1.9	0.3
242.95	-0.51	-1.5	30.81	21AP	-0.2	-3.7	-2.0	0.3
241.67	-1.09	-1.8	40.40	1MY	0.7	-3.6	-2.0	0.3
239.28	-1.72	-2.1	49.96	11MY	1.6	-3.5	-2.1	0.2
236.24	-2.35	-2.2	59.50	21MY	2.8	-3.5	-2.2	0.2
233.20	-2.89	-2.1	69.02	31MY	3.1	-3.4	-2.2	0.2
230.89	-3.30	-2.0	78.55	10JN	1.6	-3.4	-2.3	0.2
229.80	-3.56	-1.7	88.07	20JN	0.4	-3.4	-2.3	0.3
230.08	-3.68	-1.5	97.61	30JN	-0.8	-3.4	-2.3	0.3
231.70	-3.71	-1.3	107.18	10JL	-1.5	-3.3	-2.4	0.3
234.48	-3.67	-1.1	116.77	20JL	-1.2	-3.3	-2.4	0.3
238.22	-3.57	-0.9	126.41	30JL	-0.5	-3.3	-2.3	0.4
242.76	-3.44	-0.7	136.09	9AU	-0.1	-3.3	-2.3	0.4
247.93	-3.28	-0.5	145.82	19AU	0.2	-3.4	-2.2	0.3
253.60	-3.10	-0.3	155.60	29AU	0.3	-3.4	-2.2	0.3
259.67	-2.91	-0.2	165.45	8SE	0.7	-3.4	-2.1	0.3
266.07	-2.70	-0.0	175.35	18SE	2.2	-3.4	-2.1	0.3
272.71	-2.49	0.1	185.31	28SE	1.5	-3.4	-2.0	0.2
279.55	-2.27	0.3	195.33	8OC	-0.3	-3.5	-1.9	0.1
286.53	-2.05	0.4	205.39	18OC	-0.7	-3.5	-1.9	0.1
293.62	-1.84	0.5	215.51	28OC	-0.7	-3.5	-1.8	0.0
300.79	-1.62	0.6	225.66	7NO	-0.7	-3.5	-1.7	-0.1
308.00	-1.41	0.8	235.84	17NO	-0.8	-3.4	-1.7	-0.1
315.23	-1.21	0.9	246.04	27NO	-0.7	-3.4	-1.6	-0.2
322.47	-1.01	1.0	256.24	7DE	-0.7	-3.4	-1.6	-0.3
329.70	-0.83	1.1	266.45	17DE	-0.6	-3.4	-1.6	-0.2
336.91	-0.65	1.2	276.64	27DE	0.4	-3.3	-1.6	-0.2

Right table (266/267):

♂ LONG	LAT	MAG	☉ LONG	16.00UT 266	☿	♀	♃	♄ MAGNITUDES
344.08	-0.48	1.3	286.80	6JA	3.0	-3.3	-1.5	-0.1
351.20	-0.32	1.4	296.93	16JA	1.2	-3.3	-1.5	-0.0
358.28	-0.18	1.4	307.02	26JA	0.5	-3.3	-1.5	0.0
5.30	-0.04	1.5	317.06	5FE	0.3	-3.4	-1.5	0.1
12.27	0.09	1.6	327.05	15FE	0.0	-3.4	-1.6	0.2
19.18	0.21	1.7	336.98	25FE	-0.3	-3.4	-1.6	0.2
26.03	0.32	1.7	346.85	7MR	-0.9	-3.4	-1.6	0.3
32.82	0.42	1.8	356.66	17MR	-1.7	-3.4	-1.6	0.3
39.56	0.51	1.8	6.41	27MR	-1.2	-3.5	-1.7	0.3
46.25	0.60	1.9	16.11	6AP	-0.1	-3.5	-1.7	0.3
52.88	0.68	1.9	25.76	16AP	1.0	-3.6	-1.8	0.4
59.47	0.75	1.9	35.38	26AP	2.1	-3.6	-1.8	0.4
66.01	0.81	1.9	44.95	6MY	3.7	-3.7	-1.9	0.3
72.52	0.87	2.0	54.50	16MY	2.4	-3.8	-2.0	0.3
78.99	0.93	2.0	64.04	26MY	1.3	-3.9	-2.0	0.3
85.44	0.97	2.0	73.56	5JN	0.3	-4.0	-2.1	0.3
91.85	1.02	2.0	83.08	15JN	-0.8	-4.0	-2.2	0.2
98.26	1.05	2.0	92.61	25JN	-1.6	-4.1	-2.2	0.3
104.64	1.09	2.0	102.16	5JL	-1.2	-4.2	-2.3	0.3
111.02	1.11	1.9	111.74	15JL	-0.5	-4.2	-2.4	0.4
117.39	1.14	2.0	121.35	25JL	0.0	-4.0	-2.4	0.4
123.76	1.16	2.0	131.01	4AU	0.3	-3.6	-2.4	0.4
130.13	1.17	2.0	140.71	14AU	0.5	-3.2	-2.4	0.4
136.50	1.18	2.0	150.47	24AU	1.0	-3.7	-2.4	0.4
142.89	1.19	2.0	160.28	3SE	2.6	-4.1	-2.4	0.4
149.28	1.19	2.0	170.15	13SE	1.3	-4.3	-2.4	0.4
155.69	1.19	2.0	180.08	23SE	-0.4	-4.3	-2.3	0.4
162.11	1.18	2.0	190.07	3OC	-0.8	-4.3	-2.3	0.3
168.54	1.17	2.0	200.11	13OC	-0.8	-4.2	-2.2	0.3
175.00	1.15	1.9	210.20	23OC	-0.8	-4.1	-2.1	0.2
181.46	1.12	1.9	220.33	2NO	-0.6	-4.0	-2.1	0.2
187.94	1.09	1.8	230.50	12NO	-0.6	-3.9	-2.0	0.1
194.43	1.05	1.8	240.69	22NO	-0.6	-3.8	-1.9	0.0
200.94	1.00	1.7	250.89	2DE	-0.4	-3.7	-1.9	-0.0
207.45	0.94	1.6	261.10	12DE	0.6	-3.7	-1.8	-0.1
213.97	0.87	1.5	271.30	22DE	2.9	-3.6	-1.7	-0.2
				267				
220.48	0.79	1.5	281.48	1JA	0.8	-3.5	-1.7	-0.1
227.00	0.69	1.4	291.63	11JA	0.3	-3.5	-1.7	-0.1
233.51	0.58	1.2	301.74	21JA	0.1	-3.4	-1.6	-0.0
240.00	0.45	1.1	311.81	31JA	-0.1	-3.4	-1.6	0.1
246.48	0.30	1.0	321.82	10FE	-0.4	-3.4	-1.6	0.1
252.91	0.13	0.8	331.78	20FE	-0.9	-3.3	-1.6	0.2
259.30	-0.07	0.7	341.68	2MR	-1.6	-3.3	-1.6	0.2
265.63	-0.30	0.5	351.52	12MR	-1.2	-3.3	-1.6	0.3
271.87	-0.56	0.4	1.31	22MR	-0.1	-3.3	-1.6	0.3
277.99	-0.85	0.2	11.03	1AP	1.2	-3.3	-1.6	0.4
283.97	-1.19	0.0	20.71	11AP	2.8	-3.3	-1.6	0.4
289.74	-1.56	-0.2	30.35	21AP	3.1	-3.3	-1.7	0.4
295.25	-1.99	-0.4	39.94	1MY	1.8	-3.4	-1.7	0.4
300.41	-2.46	-0.6	49.50	11MY	1.0	-3.4	-1.7	0.4
305.11	-2.99	-0.9	59.04	21MY	0.2	-3.4	-1.8	0.4
309.22	-3.58	-1.1	68.56	31MY	-0.8	-3.5	-1.8	0.4
312.56	-4.22	-1.4	78.09	10JN	-1.7	-3.5	-1.9	0.4
314.94	-4.89	-1.7	87.61	20JN	-1.2	-3.5	-2.0	0.4
316.17	-5.57	-1.9	97.15	30JN	-0.4	-3.4	-2.0	0.4
316.08	-6.18	-2.2	106.71	10JL	0.1	-3.4	-2.1	0.4
314.70	-6.64	-2.4	116.31	20JL	0.4	-3.4	-2.2	0.5
312.37	-6.85	-2.6	125.93	30JL	0.7	-3.3	-2.2	0.5
309.68	-6.73	-2.6	135.61	9AU	1.4	-3.3	-2.3	0.5
307.42	-6.30	-2.4	145.34	19AU	3.0	-3.3	-2.4	0.6
306.17	-5.67	-2.1	155.12	29AU	1.2	-3.3	-2.4	0.6
306.19	-4.94	-1.8	164.97	8SE	-0.5	-3.3	-2.4	0.6
307.51	-4.20	-1.5	174.87	18SE	-1.0	-3.3	-2.4	0.6
309.95	-3.51	-1.2	184.82	28SE	-1.0	-3.4	-2.4	0.6
313.32	-2.88	-0.9	194.84	8OC	-0.8	-3.4	-2.4	0.5
317.42	-2.32	-0.6	204.91	18OC	-0.5	-3.4	-2.4	0.5
322.10	-1.84	-0.4	215.02	28OC	-0.4	-3.5	-2.3	0.4
327.21	-1.41	-0.1	225.17	7NO	-0.4	-3.5	-2.3	0.4
332.66	-1.05	0.1	235.35	17NO	-0.3	-3.6	-2.2	0.3
338.36	-0.73	0.3	245.54	27NO	0.8	-3.7	-2.1	0.3
344.25	-0.45	0.5	255.75	7DE	2.6	-3.7	-2.1	0.2
350.29	-0.21	0.7	265.95	17DE	0.5	-3.8	-2.0	0.1
356.44	0.00	0.8	276.14	27DE	0.1	-3.9	-1.9	0.0

♂ LONG	♂ LAT	♂ MAG	☉ LONG	16.00UT	☿ MAG	♀ MAG	♃ MAG	♄ MAG
				268				
2.66	0.18	1.0	286.31	6JA	-0.0	-4.0	-1.9	-0.0
8.93	0.34	1.1	296.44	16JA	-0.2	-4.1	-1.8	0.0
15.24	0.48	1.2	306.53	26JA	-0.4	-4.2	-1.7	0.1
21.58	0.60	1.3	316.58	5FE	-0.9	-4.3	-1.7	0.1
27.92	0.70	1.5	326.56	15FE	-1.4	-4.3	-1.6	0.2
34.26	0.79	1.5	336.49	25FE	-1.1	-4.3	-1.6	0.3
40.61	0.87	1.6	346.37	6MR	-0.0	-4.0	-1.6	0.3
46.94	0.94	1.7	356.18	16MR	1.5	-3.5	-1.6	0.4
53.27	0.99	1.8	5.94	26MR	3.5	-3.2	-1.5	0.4
59.59	1.04	1.8	15.64	5AP	2.3	-3.8	-1.5	0.5
65.89	1.08	1.9	25.30	15AP	1.3	-4.1	-1.5	0.5
72.19	1.11	1.9	34.91	25AP	0.7	-4.2	-1.6	0.5
78.48	1.14	2.0	44.49	5MY	0.1	-4.2	-1.6	0.6
84.77	1.16	2.0	54.04	15MY	-0.8	-4.1	-1.6	0.6
91.05	1.17	2.0	63.57	25MY	-1.7	-4.0	-1.6	0.6
97.34	1.18	2.0	73.09	4JN	-1.2	-3.9	-1.6	0.6
103.64	1.19	2.0	82.61	14JN	-0.4	-3.8	-1.7	0.6
109.94	1.19	2.0	92.15	24JN	0.2	-3.7	-1.7	0.6
116.26	1.18	2.0	101.70	4JL	0.6	-3.7	-1.8	0.6
122.59	1.17	2.0	111.27	14JL	1.0	-3.6	-1.8	0.6
128.95	1.15	2.0	120.88	24JL	1.8	-3.5	-1.9	0.6
135.34	1.14	2.0	130.54	3AU	3.2	-3.5	-2.0	0.7
141.75	1.11	1.9	140.24	13AU	1.0	-3.5	-2.0	0.7
148.20	1.08	1.9	149.99	23AU	-0.6	-3.4	-2.1	0.7
154.69	1.05	1.9	159.81	2SE	-1.1	-3.4	-2.2	0.8
161.21	1.01	1.9	169.67	12SE	-1.1	-3.4	-2.2	0.8
167.78	0.97	1.9	179.60	22SE	-0.7	-3.4	-2.3	0.8
174.40	0.92	1.9	189.59	2OC	-0.4	-3.4	-2.3	0.8
181.06	0.86	1.9	199.63	12OC	-0.3	-3.4	-2.4	0.8
187.77	0.80	1.9	209.72	22OC	-0.3	-3.4	-2.4	0.7
194.53	0.74	1.9	219.85	1NO	-0.1	-3.4	-2.4	0.7
201.34	0.66	1.8	230.01	11NO	1.0	-3.4	-2.4	0.6
208.21	0.58	1.8	240.20	21NO	2.4	-3.4	-2.3	0.6
215.12	0.49	1.8	250.40	1DE	0.3	-3.4	-2.3	0.5
222.09	0.39	1.7	260.61	11DE	-0.1	-3.4	-2.2	0.5
229.11	0.29	1.6	270.81	21DE	-0.2	-3.4	-2.1	0.4
236.17	0.17	1.6	280.99	31DE	-0.3	-3.5	-2.1	0.3
				269				
243.29	0.05	1.5	291.13	10JA	-0.5	-3.5	-2.0	0.2
250.45	-0.09	1.4	301.25	20JA	-1.0	-3.5	-1.9	0.2
257.64	-0.23	1.4	311.32	30JA	-1.3	-3.4	-1.9	0.2
264.88	-0.39	1.3	321.33	9FE	-1.0	-3.4	-1.8	0.3
272.15	-0.55	1.2	331.29	19FE	0.0	-3.4	-1.7	0.3
279.44	-0.72	1.1	341.20	1MR	1.9	-3.4	-1.7	0.4
286.75	-0.90	1.0	351.04	11MR	3.2	-3.3	-1.6	0.5
294.07	-1.09	0.9	0.83	21MR	1.7	-3.3	-1.6	0.5
301.38	-1.28	0.8	10.56	31MR	1.0	-3.3	-1.5	0.6
308.66	-1.48	0.7	20.24	10AP	0.5	-3.3	-1.5	0.6
315.91	-1.68	0.6	29.87	20AP	-0.0	-3.3	-1.5	0.7
323.08	-1.87	0.5	39.47	30AP	-0.8	-3.4	-1.5	0.7
330.17	-2.06	0.4	49.03	10MY	-1.7	-3.4	-1.5	0.7
337.14	-2.25	0.3	58.57	20MY	-1.2	-3.4	-1.5	0.8
343.95	-2.43	0.2	68.10	30MY	-0.3	-3.4	-1.5	0.8
350.56	-2.60	0.1	77.62	9JN	0.3	-3.5	-1.5	0.8
356.92	-2.75	-0.0	87.15	19JN	0.8	-3.5	-1.5	0.8
2.96	-2.90	-0.2	96.69	29JN	1.3	-3.6	-1.5	0.8
8.61	-3.02	-0.3	106.25	9JL	2.4	-3.7	-1.6	0.8
13.77	-3.12	-0.5	115.84	19JL	2.9	-3.7	-1.6	0.8
18.32	-3.20	-0.6	125.47	29JL	0.8	-3.8	-1.6	0.8
22.10	-3.25	-0.8	135.15	8AU	-0.6	-3.9	-1.7	0.9
24.93	-3.26	-1.0	144.87	18AU	-1.2	-4.0	-1.7	0.9
26.59	-3.20	-1.3	154.65	28AU	-1.2	-4.1	-1.8	0.9
26.89	-3.07	-1.5	164.49	7SE	-0.7	-4.2	-1.9	1.0
25.71	-2.82	-1.7	174.39	17SE	-0.3	-4.3	-1.9	1.0
23.19	-2.45	-1.9	184.35	27SE	-0.2	-4.3	-2.0	1.0
19.78	-1.96	-2.0	194.35	7OC	-0.1	-4.2	-2.1	1.0
16.24	-1.39	-1.9	204.42	17OC	0.1	-3.7	-2.1	1.0
13.38	-0.82	-1.6	214.53	27OC	1.2	-3.1	-2.2	1.0
11.71	-0.31	-1.3	224.67	6NO	2.1	-3.8	-2.2	1.0
11.39	0.12	-0.9	234.85	16NO	0.1	-4.2	-2.2	0.9
12.36	0.47	-0.6	245.05	26NO	-0.3	-4.4	-2.3	0.9
14.39	0.74	-0.3	255.25	6DE	-0.3	-4.4	-2.3	0.8
17.31	0.95	-0.0	265.46	16DE	-0.4	-4.3	-2.2	0.8
20.91	1.10	0.2	275.65	26DE	-0.6	-4.2	-2.2	0.7

♂ LONG	♂ LAT	♂ MAG	☉ LONG	16.00UT	☿ MAG	♀ MAG	♃ MAG	♄ MAG
				270				
25.04	1.22	0.4	285.82	5JA	-1.0	-4.1	-2.2	0.6
29.59	1.31	0.7	295.95	15JA	-1.1	-4.0	-2.1	0.6
34.46	1.38	0.8	306.05	25JA	-1.0	-3.9	-2.0	0.5
39.57	1.42	1.0	316.09	4FE	0.1	-3.8	-2.0	0.5
44.89	1.45	1.2	326.08	14FE	2.3	-3.7	-1.9	0.5
50.36	1.47	1.3	336.02	24FE	2.5	-3.6	-1.8	0.6
55.95	1.48	1.4	345.89	6MR	1.3	-3.5	-1.7	0.6
61.64	1.49	1.5	355.71	16MR	0.7	-3.5	-1.7	0.7
67.42	1.48	1.6	5.47	26MR	0.4	-3.4	-1.6	0.7
73.26	1.47	1.7	15.17	5AP	-0.1	-3.4	-1.6	0.8
79.16	1.45	1.8	24.83	15AP	-0.8	-3.4	-1.5	0.8
85.11	1.43	1.8	34.45	25AP	-1.7	-3.3	-1.5	0.9
91.10	1.41	1.9	44.03	5MY	-1.3	-3.3	-1.4	0.9
97.14	1.38	1.9	53.58	15MY	-0.3	-3.3	-1.4	1.0
103.23	1.34	2.0	63.11	25MY	0.4	-3.3	-1.4	1.0
109.35	1.31	2.0	72.63	4JN	1.0	-3.3	-1.4	1.0
115.52	1.27	2.0	82.16	14JN	1.8	-3.3	-1.4	1.1
121.73	1.22	2.0	91.69	24JN	3.0	-3.3	-1.4	1.1
127.99	1.18	2.0	101.24	4JL	2.4	-3.4	-1.4	1.1
134.29	1.12	2.0	110.81	14JL	0.7	-3.4	-1.4	1.1
140.65	1.07	2.0	120.42	24JL	-0.7	-3.4	-1.4	1.1
147.07	1.01	2.0	130.07	3AU	-1.4	-3.5	-1.4	1.1
153.54	0.95	2.0	139.77	13AU	-1.2	-3.5	-1.5	1.1
160.07	0.89	1.9	149.52	23AU	-0.6	-3.5	-1.5	1.1
166.67	0.82	1.9	159.33	2SE	-0.2	-3.4	-1.5	1.2
173.34	0.74	1.8	169.19	12SE	-0.0	-3.4	-1.6	1.2
180.07	0.67	1.8	179.12	22SE	0.1	-3.4	-1.6	1.3
186.87	0.59	1.8	189.10	2OC	0.3	-3.4	-1.7	1.3
193.75	0.50	1.8	199.14	12OC	1.5	-3.3	-1.8	1.3
200.70	0.41	1.8	209.22	22OC	1.9	-3.3	-1.8	1.3
207.73	0.32	1.7	219.35	1NO	-0.1	-3.3	-1.9	1.3
214.83	0.22	1.7	229.51	11NO	-0.4	-3.3	-2.0	1.3
222.00	0.11	1.7	239.70	21NO	-0.4	-3.3	-2.0	1.3
229.25	0.01	1.7	249.90	1DE	-0.5	-3.4	-2.1	1.2
236.57	-0.11	1.6	260.11	11DE	-0.7	-3.4	-2.1	1.2
243.95	-0.22	1.6	270.31	21DE	-0.9	-3.4	-2.1	1.1
251.39	-0.34	1.6	280.49	31DE	-1.0	-3.4	-2.1	1.1
				271				
258.90	-0.46	1.5	290.64	10JA	-0.8	-3.5	-2.1	1.0
266.45	-0.58	1.5	300.76	20JA	0.2	-3.5	-2.1	0.9
274.05	-0.70	1.5	310.83	30JA	2.6	-3.6	-2.1	0.9
281.69	-0.82	1.4	320.85	9FE	1.9	-3.6	-2.0	0.8
289.35	-0.94	1.4	330.81	19FE	0.9	-3.7	-2.0	0.8
297.03	-1.05	1.3	340.72	1MR	0.5	-3.8	-1.9	0.8
304.72	-1.16	1.3	350.57	11MR	0.2	-3.9	-1.8	0.9
312.40	-1.26	1.2	0.36	21MR	-0.2	-3.9	-1.8	0.9
320.06	-1.35	1.2	10.09	31MR	-0.8	-4.0	-1.7	1.0
327.69	-1.43	1.2	19.77	10AP	-1.7	-4.1	-1.6	1.0
335.27	-1.50	1.1	29.41	20AP	-1.3	-4.2	-1.6	1.1
342.79	-1.55	1.1	39.01	30AP	-0.3	-4.2	-1.5	1.1
350.22	-1.59	1.0	48.57	10MY	0.6	-4.1	-1.5	1.2
357.56	-1.62	1.0	58.11	20MY	1.4	-3.7	-1.4	1.2
4.79	-1.62	0.9	67.64	30MY	2.4	-2.9	-1.4	1.3
11.88	-1.61	0.9	77.16	9JN	3.4	-3.3	-1.3	1.3
18.83	-1.59	0.8	86.68	19JN	1.9	-3.9	-1.3	1.3
25.60	-1.54	0.8	96.23	29JN	0.5	-4.2	-1.3	1.4
32.16	-1.48	0.7	105.78	9JL	-0.7	-4.2	-1.3	1.4
38.50	-1.40	0.6	115.37	19JL	-1.5	-4.2	-1.3	1.4
44.57	-1.29	0.5	125.00	29JL	-1.2	-4.1	-1.3	1.4
50.33	-1.17	0.4	134.68	8AU	-0.6	-4.0	-1.3	1.4
55.72	-1.02	0.3	144.40	18AU	-0.1	-3.9	-1.3	1.3
60.66	-0.85	0.2	154.18	28AU	0.1	-3.8	-1.3	1.3
65.06	-0.64	0.1	164.01	7SE	0.2	-3.8	-1.3	1.3
68.80	-0.40	-0.1	173.91	17SE	0.6	-3.7	-1.4	1.3
71.72	-0.12	-0.3	183.86	27SE	1.9	-3.6	-1.4	1.3
73.64	0.21	-0.5	193.87	7OC	1.7	-3.6	-1.5	1.3
74.36	0.59	-0.7	203.93	17OC	-0.2	-3.5	-1.5	1.3
73.70	1.03	-0.9	214.04	27OC	-0.6	-3.5	-1.6	1.2
71.62	1.49	-1.1	224.18	6NO	-0.6	-3.5	-1.6	1.2
68.35	1.94	-1.3	234.36	16NO	-0.6	-3.4	-1.7	1.2
64.44	2.33	-1.4	244.56	26NO	-0.8	-3.4	-1.8	1.1
60.69	2.60	-1.2	254.76	6DE	-0.8	-3.4	-1.8	1.1
57.80	2.75	-0.9	264.97	16DE	-0.8	-3.4	-1.9	1.1
56.16	2.80	-0.6	275.16	26DE	-0.7	-3.3	-1.9	1.0

Left table (272 / 273):

♂ LONG	LAT	MAG	☉ LONG	16.00UT	☿	♀	♃	♄
55.86	2.77	-0.3	285.33	5JA	0.3	-3.3	-2.0	1.0
56.78	2.70	-0.0	295.46	15JA	2.9	-3.3	-2.0	0.9
58.72	2.61	0.2	305.56	25JA	1.4	-3.3	-2.0	0.9
61.48	2.50	0.5	315.60	4FE	0.6	-3.4	-2.1	0.8
64.89	2.39	0.7	325.60	14FE	0.4	-3.4	-2.0	0.8
68.83	2.29	0.9	335.54	24FE	0.1	-3.4	-2.0	0.8
73.17	2.18	1.0	345.41	5MR	-0.2	-3.4	-2.0	0.9
77.83	2.08	1.2	355.23	15MR	-0.9	-3.5	-1.9	0.9
82.76	1.98	1.3	5.00	25MR	-1.7	-3.5	-1.9	0.9
87.91	1.88	1.4	14.70	4AP	-1.2	-3.5	-1.8	1.0
93.23	1.79	1.5	24.36	14AP	-0.2	-3.4	-1.7	1.1
98.70	1.69	1.6	33.98	24AP	0.8	-3.4	-1.7	1.1
104.31	1.60	1.7	43.56	4MY	1.8	-3.4	-1.6	1.2
110.03	1.51	1.7	53.11	14MY	3.1	-3.4	-1.5	1.2
115.86	1.42	1.8	62.64	24MY	2.8	-3.3	-1.5	1.3
121.79	1.33	1.8	72.16	3JN	1.5	-3.3	-1.4	1.3
127.80	1.24	1.9	81.69	13JN	0.4	-3.3	-1.4	1.3
133.91	1.15	1.9	91.22	23JN	-0.7	-3.3	-1.3	1.3
140.10	1.06	1.9	100.76	3JL	-1.5	-3.3	-1.3	1.3
146.38	0.97	1.9	110.34	13JL	-1.2	-3.4	-1.3	1.3
152.75	0.87	1.9	119.95	23JL	-0.5	-3.4	-1.2	1.3
159.20	0.78	1.9	129.59	2AU	-0.1	-3.4	-1.2	1.3
165.75	0.68	1.9	139.29	12AU	0.2	-3.5	-1.2	1.2
172.38	0.59	1.9	149.04	22AU	0.4	-3.5	-1.2	1.2
179.10	0.49	1.8	158.85	1SE	0.8	-3.6	-1.2	1.1
185.91	0.39	1.8	168.71	11SE	2.3	-3.6	-1.2	1.1
192.82	0.29	1.8	178.64	21SE	1.5	-3.7	-1.3	1.1
199.82	0.18	1.7	188.61	1OC	-0.3	-3.8	-1.3	1.1
206.91	0.08	1.7	198.65	11OC	-0.7	-3.9	-1.3	1.1
214.09	-0.03	1.6	208.73	21OC	-0.7	-4.0	-1.3	1.1
221.36	-0.13	1.6	218.86	31OC	-0.7	-4.1	-1.4	1.1
228.71	-0.24	1.5	229.02	10NO	-0.7	-4.2	-1.4	1.1
236.15	-0.34	1.5	239.21	20NO	-0.6	-4.3	-1.5	1.1
243.66	-0.45	1.5	249.41	30NO	-0.7	-4.4	-1.6	1.0
251.25	-0.55	1.5	259.62	10DE	-0.6	-4.4	-1.6	1.0
258.90	-0.65	1.5	269.82	20DE	0.4	-4.2	-1.7	1.0
266.61	-0.74	1.5	280.00	30DE	3.0	-3.7	-1.8	0.9

273

♂ LONG	LAT	MAG	☉ LONG	16.00UT	☿	♀	♃	♄
274.36	-0.83	1.5	290.15	9JA	1.0	-3.2	-1.8	0.9
282.15	-0.91	1.4	300.27	19JA	0.4	-3.8	-1.9	0.8
289.97	-0.98	1.4	310.34	29JA	0.2	-4.2	-1.9	0.8
297.80	-1.05	1.4	320.36	8FE	0.0	-4.3	-2.0	0.7
305.63	-1.10	1.4	330.33	18FE	-0.3	-4.3	-2.0	0.7
313.45	-1.15	1.4	340.24	28FE	-0.9	-4.2	-2.0	0.6
321.25	-1.18	1.4	350.09	10MR	-1.6	-4.1	-2.0	0.6
329.02	-1.20	1.4	359.88	20MR	-1.2	-4.0	-2.0	0.6
336.73	-1.21	1.4	9.62	30MR	-0.2	-3.9	-2.0	0.7
344.40	-1.20	1.4	19.31	9AP	1.0	-3.8	-1.9	0.8
351.98	-1.18	1.4	28.94	19AP	2.3	-3.7	-1.9	0.9
359.50	-1.15	1.4	38.54	29AP	3.7	-3.6	-1.8	0.9
6.92	-1.11	1.4	48.11	9MY	2.2	-3.5	-1.8	1.0
14.24	-1.05	1.4	57.65	19MY	1.2	-3.5	-1.7	1.0
21.46	-0.98	1.3	67.18	29MY	0.3	-3.4	-1.6	1.1
28.57	-0.90	1.3	76.70	8JN	-0.7	-3.4	-1.6	1.1
35.55	-0.81	1.3	86.23	18JN	-1.6	-3.4	-1.5	1.1
42.41	-0.70	1.3	95.76	28JN	-1.2	-3.4	-1.5	1.2
49.12	-0.59	1.3	105.32	8JL	-0.5	-3.3	-1.4	1.2
55.68	-0.46	1.2	114.91	18JL	0.0	-3.3	-1.4	1.2
62.08	-0.32	1.2	124.54	28JL	0.3	-3.3	-1.3	1.2
68.29	-0.18	1.2	134.21	7AU	0.6	-3.3	-1.3	1.1
74.30	-0.01	1.1	143.93	17AU	1.1	-3.4	-1.3	1.1
80.08	0.16	1.0	153.70	27AU	2.7	-3.4	-1.2	1.1
85.60	0.35	1.0	163.54	6SE	1.3	-3.4	-1.2	1.0
90.81	0.57	0.9	173.43	16SE	-0.4	-3.4	-1.2	1.0
95.65	0.80	0.7	183.38	26SE	-0.9	-3.4	-1.2	1.0
100.04	1.06	0.6	193.38	6OC	-0.9	-3.5	-1.2	1.0
103.90	1.35	0.5	203.44	16OC	-0.9	-3.5	-1.2	1.0
107.08	1.68	0.3	213.55	26OC	-0.6	-3.5	-1.3	1.0
109.44	2.04	0.1	223.69	5NO	-0.5	-3.5	-1.3	1.0
110.79	2.45	-0.1	233.86	15NO	-0.5	-3.4	-1.3	1.0
110.95	2.88	-0.4	244.06	25NO	-0.4	-3.4	-1.4	1.0
109.76	3.33	-0.6	254.27	5DE	0.6	-3.4	-1.4	0.9
107.25	3.74	-0.8	264.47	15DE	2.9	-3.4	-1.5	0.9
103.71	4.06	-1.0	274.66	25DE	0.7	-3.3	-1.5	0.9

Right table (274 / 275):

♂ LONG	LAT	MAG	☉ LONG	16.00UT	☿	♀	♃	♄
99.73	4.22	-1.0	284.83	4JA	0.2	-3.3	-1.6	0.8
96.07	4.22	-0.8	294.97	14JA	0.1	-3.3	-1.7	0.8
93.33	4.07	-0.6	305.07	24JA	-0.1	-3.4	-1.7	0.7
91.86	3.83	-0.3	315.12	3FE	-0.4	-3.4	-1.8	0.7
91.68	3.56	-0.1	325.11	13FE	-0.9	-3.4	-1.9	0.6
92.67	3.27	0.2	335.05	23FE	-1.5	-3.4	-1.9	0.6
94.65	3.00	0.4	344.94	5MR	-1.2	-3.4	-2.0	0.5
97.43	2.74	0.6	354.75	15MR	-0.1	-3.4	-2.0	0.5
100.85	2.50	0.8	4.52	25MR	1.3	-3.5	-2.0	0.5
104.79	2.28	0.9	14.24	4AP	3.0	-3.5	-2.0	0.5
109.15	2.07	1.1	23.89	14AP	2.8	-3.6	-2.0	0.6
113.85	1.88	1.2	33.52	24AP	1.6	-3.6	-2.0	0.7
118.85	1.70	1.3	43.10	4MY	0.9	-3.7	-2.0	0.7
124.08	1.54	1.4	52.65	14MY	0.2	-3.8	-1.9	0.8
129.53	1.38	1.5	62.19	24MY	-0.8	-3.9	-1.9	0.9
135.16	1.23	1.5	71.71	3JN	-1.7	-4.0	-1.8	0.9
140.95	1.08	1.6	81.23	13JN	-1.3	-4.1	-1.8	1.0
146.91	0.94	1.6	90.76	23JN	-0.4	-4.1	-1.7	1.0
153.00	0.80	1.6	100.31	3JL	0.1	-4.2	-1.6	1.0
159.23	0.67	1.7	109.88	13JL	0.5	-4.2	-1.6	1.0
165.59	0.54	1.7	119.49	23JL	0.8	-4.0	-1.5	1.0
172.07	0.42	1.7	129.13	2AU	1.5	-3.6	-1.5	1.0
178.68	0.29	1.7	138.82	12AU	3.1	-3.2	-1.4	1.0
185.41	0.17	1.7	148.57	22AU	1.2	-3.7	-1.4	1.0
192.26	0.05	1.7	158.37	1SE	-0.5	-4.1	-1.3	1.0
199.22	-0.07	1.7	168.23	11SE	-1.0	-4.3	-1.3	0.9
206.29	-0.18	1.6	178.15	21SE	-1.0	-4.3	-1.3	0.9
213.47	-0.29	1.6	188.13	1OC	-0.8	-4.2	-1.3	0.8
220.76	-0.39	1.6	198.16	11OC	-0.5	-4.2	-1.3	0.9
228.15	-0.49	1.6	208.24	21OC	-0.4	-4.1	-1.3	0.9
235.64	-0.59	1.5	218.37	31OC	-0.3	-4.0	-1.3	0.9
243.21	-0.68	1.5	228.53	10NO	-0.2	-3.9	-1.3	0.9
250.86	-0.76	1.5	238.72	20NO	0.8	-3.8	-1.3	0.9
258.58	-0.84	1.4	248.92	30NO	2.6	-3.7	-1.3	0.9
266.36	-0.90	1.4	259.12	10DE	0.5	-3.7	-1.4	0.9
274.19	-0.96	1.4	269.33	20DE	0.0	-3.6	-1.4	0.9
282.05	-1.01	1.3	279.51	30DE	-0.1	-3.5	-1.5	0.8

275

♂ LONG	LAT	MAG	☉ LONG	16.00UT	☿	♀	♃	♄
289.93	-1.05	1.3	289.66	9JA	-0.2	-3.5	-1.5	0.8
297.83	-1.07	1.3	299.79	19JA	-0.5	-3.4	-1.6	0.8
305.72	-1.09	1.3	309.86	29JA	-0.9	-3.4	-1.6	0.7
313.59	-1.09	1.3	319.88	8FE	-1.4	-3.4	-1.7	0.7
321.44	-1.08	1.4	329.86	18FE	-1.1	-3.3	-1.8	0.6
329.25	-1.06	1.4	339.77	28FE	-0.0	-3.3	-1.8	0.5
337.00	-1.03	1.4	349.62	10MR	1.6	-3.3	-1.9	0.5
344.69	-0.98	1.4	359.49	20MR	3.5	-3.3	-2.0	0.4
352.31	-0.93	1.5	9.15	30MR	2.1	-3.3	-2.0	0.4
359.86	-0.86	1.5	18.84	9AP	1.2	-3.3	-2.1	0.4
7.32	-0.79	1.5	28.48	19AP	0.7	-3.3	-2.1	0.5
14.69	-0.71	1.5	38.08	29AP	0.1	-3.4	-2.1	0.5
21.97	-0.62	1.6	47.65	9MY	-0.8	-3.4	-2.1	0.6
29.16	-0.53	1.6	57.19	19MY	-1.7	-3.4	-2.1	0.7
36.25	-0.43	1.6	66.72	29MY	-1.3	-3.5	-2.1	0.7
43.23	-0.32	1.6	76.24	8JN	-0.4	-3.5	-2.1	0.8
50.12	-0.21	1.6	85.77	18JN	0.2	-3.4	-2.0	0.8
56.90	-0.09	1.6	95.30	28JN	0.7	-3.4	-2.0	0.9
63.58	0.03	1.6	104.86	8JL	1.1	-3.4	-1.9	0.9
70.14	0.15	1.6	114.44	18JL	2.0	-3.4	-1.8	0.9
76.60	0.28	1.6	124.06	28JL	3.1	-3.3	-1.8	0.9
82.94	0.42	1.6	133.73	7AU	1.0	-3.3	-1.7	0.9
89.15	0.56	1.6	143.45	17AU	-0.6	-3.3	-1.7	0.9
95.22	0.70	1.5	153.22	27AU	-1.2	-3.3	-1.6	0.9
101.15	0.86	1.5	163.05	6SE	-1.2	-3.3	-1.5	0.9
106.90	1.02	1.4	172.94	16SE	-0.7	-3.3	-1.5	0.9
112.47	1.19	1.3	182.89	26SE	-0.4	-3.4	-1.5	0.8
117.82	1.37	1.3	192.89	6OC	-0.2	-3.4	-1.4	0.8
122.90	1.57	1.1	202.95	16OC	-0.2	-3.4	-1.4	0.8
127.67	1.79	1.0	213.05	26OC	-0.0	-3.5	-1.4	0.8
132.06	2.03	0.9	223.19	5NO	1.0	-3.5	-1.4	0.8
135.99	2.29	0.7	233.37	15NO	2.4	-3.6	-1.3	0.8
139.34	2.57	0.5	243.56	25NO	0.3	-3.7	-1.3	0.8
141.99	2.89	0.3	253.77	5DE	-0.2	-3.7	-1.4	0.8
143.74	3.23	0.1	263.98	15DE	-0.2	-3.8	-1.4	0.8
144.45	3.59	-0.1	274.17	25DE	-0.3	-3.9	-1.4	0.8

♂ LONG	♂ LAT	♂ MAG	⊙ LONG	16.00UT	☿	♀	♃	♄
				276	MAGNITUDES			
143.92	3.95	-0.4	284.34	4JA	-0.5	-4.0	-1.4	0.8
142.08	4.27	-0.7	294.48	14JA	-1.0	-4.1	-1.5	0.8
139.06	4.50	-0.9	304.58	24JA	-1.2	-4.2	-1.5	0.7
135.28	4.57	-1.0	314.64	3FE	-1.0	-4.3	-1.6	0.7
131.40	4.46	-0.9	324.64	13FE	0.0	-4.3	-1.6	0.6
128.14	4.19	-0.7	334.57	23FE	2.0	-4.3	-1.7	0.6
125.98	3.83	-0.5	344.46	4MR	2.9	-4.0	-1.7	0.5
125.12	3.42	-0.3	354.29	14MR	1.5	-3.5	-1.8	0.5
125.51	3.01	-0.1	4.05	24MR	0.9	-3.2	-1.9	0.4
127.00	2.63	0.1	13.77	3AP	0.5	-3.8	-2.0	0.4
129.42	2.27	0.3	23.43	13AP	-0.0	-4.1	-2.0	0.3
132.59	1.95	0.5	33.05	23AP	-0.8	-4.2	-2.1	0.4
136.38	1.66	0.7	42.64	3MY	-1.7	-4.2	-2.1	0.4
140.68	1.39	0.8	52.19	13MY	-1.3	-4.1	-2.2	0.5
145.40	1.15	0.9	61.72	23MY	-0.3	-4.0	-2.2	0.5
150.47	0.92	1.0	71.25	2JN	0.4	-3.9	-2.2	0.6
155.84	0.72	1.1	80.77	12JN	0.9	-3.8	-2.2	0.7
161.48	0.53	1.2	90.30	22JN	1.5	-3.7	-2.2	0.7
167.36	0.35	1.2	99.85	2JL	2.6	-3.7	-2.2	0.8
173.44	0.18	1.3	109.42	12JL	2.7	-3.6	-2.2	0.8
179.73	0.02	1.3	119.02	22JL	0.8	-3.5	-2.1	0.8
186.19	-0.13	1.3	128.66	1AU	-0.6	-3.5	-2.1	0.9
192.82	-0.27	1.4	138.35	11AU	-1.3	-3.5	-2.0	0.9
199.61	-0.40	1.4	148.09	21AU	-1.2	-3.4	-1.9	0.9
206.55	-0.52	1.4	157.90	31AU	-0.7	-3.4	-1.9	0.9
213.63	-0.63	1.4	167.75	10SE	-0.3	-3.4	-1.8	0.9
220.84	-0.73	1.4	177.67	20SE	-0.1	-3.4	-1.8	0.8
228.18	-0.82	1.4	187.64	30SE	-0.0	-3.4	-1.7	0.8
235.62	-0.90	1.4	197.67	10OC	0.2	-3.4	-1.6	0.8
243.17	-0.97	1.4	207.75	20OC	1.3	-3.4	-1.6	0.7
250.81	-1.03	1.4	217.87	30OC	2.2	-3.4	-1.6	0.7
258.53	-1.08	1.4	228.03	9NO	0.1	-3.4	-1.5	0.8
266.31	-1.11	1.4	238.22	19NO	-0.3	-3.4	-1.5	0.8
274.14	-1.14	1.4	248.42	29NO	-0.4	-3.4	-1.5	0.8
282.01	-1.15	1.4	258.62	9DE	-0.4	-3.4	-1.5	0.8
289.90	-1.14	1.4	268.82	19DE	-0.6	-3.4	-1.4	0.8
297.80	-1.13	1.4	279.01	29DE	-1.0	-3.5	-1.5	0.8
				277				
305.69	-1.10	1.4	289.17	8JA	-1.1	-3.5	-1.5	0.8
313.55	-1.06	1.4	299.29	18JA	-0.9	-3.5	-1.5	0.8
321.39	-1.01	1.4	309.37	28JA	0.1	-3.4	-1.5	0.7
329.17	-0.96	1.4	319.40	7FE	2.3	-3.4	-1.5	0.7
336.90	-0.89	1.4	329.37	17FE	2.3	-3.4	-1.6	0.7
344.57	-0.81	1.4	339.29	27FE	1.1	-3.4	-1.6	0.6
352.16	-0.73	1.4	349.14	9MR	0.7	-3.3	-1.7	0.5
359.68	-0.65	1.4	358.94	19MR	0.3	-3.3	-1.7	0.5
7.11	-0.55	1.5	8.68	29MR	-0.1	-3.3	-1.8	0.4
14.46	-0.46	1.5	18.37	8AP	-0.8	-3.3	-1.9	0.4
21.72	-0.36	1.6	28.01	18AP	-1.7	-3.3	-1.9	0.3
28.89	-0.26	1.6	37.62	28AP	-1.3	-3.4	-2.0	0.3
35.97	-0.15	1.7	47.18	8MY	-0.3	-3.4	-2.1	0.3
42.97	-0.05	1.7	56.73	18MY	0.5	-3.4	-2.1	0.4
49.88	0.06	1.7	66.26	28MY	1.1	-3.4	-2.2	0.5
56.71	0.16	1.8	75.78	7JN	2.0	-3.5	-2.3	0.5
63.46	0.27	1.8	85.30	17JN	3.2	-3.5	-2.3	0.6
70.13	0.37	1.8	94.84	27JN	2.2	-3.6	-2.3	0.6
76.72	0.48	1.8	104.39	7JL	0.7	-3.7	-2.4	0.7
83.23	0.58	1.9	113.98	17JL	-0.7	-3.7	-2.4	0.7
89.67	0.69	1.9	123.60	27JL	-1.4	-3.8	-2.4	0.8
96.03	0.79	1.9	133.27	6AU	-1.2	-3.9	-2.3	0.8
102.32	0.90	1.8	142.98	16AU	-0.6	-4.0	-2.3	0.8
108.52	1.01	1.8	152.75	26AU	-0.2	-4.1	-2.2	0.8
114.65	1.12	1.8	162.58	5SE	0.0	-4.2	-2.2	0.8
120.68	1.23	1.8	172.46	15SE	0.1	-4.3	-2.1	0.8
126.61	1.34	1.7	182.41	25SE	0.4	-4.3	-2.1	0.8
132.42	1.46	1.7	192.40	5OC	1.6	-4.1	-2.0	0.8
138.12	1.58	1.6	202.46	15OC	1.9	-3.6	-1.9	0.8
143.66	1.71	1.5	212.56	25OC	-0.1	-3.1	-1.9	0.7
149.03	1.84	1.4	222.70	4NO	-0.5	-3.8	-1.8	0.7
154.19	1.99	1.3	232.87	14NO	-0.5	-4.2	-1.7	0.7
159.10	2.14	1.1	243.07	24NO	-0.5	-4.4	-1.7	0.7
163.70	2.30	1.0	253.27	4DE	-0.7	-4.4	-1.6	0.8
167.93	2.47	0.8	263.48	14DE	-0.9	-4.3	-1.6	0.8
171.70	2.65	0.6	273.67	24DE	-0.9	-4.2	-1.6	0.8

♂ LONG	♂ LAT	♂ MAG	⊙ LONG	16.00UT	☿	♀	♃	♄
				278	MAGNITUDES			
174.88	2.85	0.4	283.84	3JA	-0.8	-4.1	-1.6	0.8
177.35	3.05	0.2	293.99	13JA	0.2	-4.0	-1.5	0.8
178.92	3.26	-0.1	304.09	23JA	2.7	-3.9	-1.5	0.8
179.43	3.45	-0.4	314.14	2FE	1.7	-3.8	-1.5	0.7
178.70	3.60	-0.7	324.15	12FE	0.8	-3.7	-1.5	0.7
176.70	3.67	-0.9	334.10	22FE	0.5	-3.6	-1.6	0.7
173.60	3.62	-1.2	343.98	4MR	0.2	-3.5	-1.6	0.6
169.87	3.42	-1.3	353.81	14MR	-0.2	-3.5	-1.6	0.6
166.20	3.08	-1.2	3.59	24MR	-0.8	-3.4	-1.6	0.5
163.29	2.64	-1.0	13.30	3AP	-1.7	-3.4	-1.7	0.5
161.54	2.17	-0.8	22.97	13AP	-1.3	-3.4	-1.7	0.4
161.14	1.70	-0.6	32.59	23AP	-0.3	-3.3	-1.8	0.3
161.99	1.26	-0.4	42.18	3MY	0.7	-3.3	-1.9	0.3
163.95	0.87	-0.2	51.74	13MY	1.5	-3.3	-1.9	0.3
166.85	0.53	-0.0	61.27	23MY	2.6	-3.3	-2.0	0.3
170.51	0.22	0.1	70.79	2JN	3.3	-3.3	-2.1	0.4
174.80	-0.04	0.3	80.32	12JN	1.8	-3.3	-2.1	0.4
179.62	-0.28	0.4	89.84	22JN	0.5	-3.3	-2.2	0.5
184.87	-0.49	0.5	99.38	2JL	-0.7	-3.4	-2.3	0.6
190.50	-0.67	0.6	108.95	12JL	-1.5	-3.4	-2.3	0.6
196.46	-0.83	0.7	118.55	22JL	-1.3	-3.4	-2.4	0.7
202.70	-0.97	0.8	128.19	1AU	-0.6	-3.5	-2.4	0.7
209.20	-1.09	0.8	137.88	11AU	-0.1	-3.5	-2.4	0.7
215.91	-1.20	0.9	147.62	21AU	0.1	-3.5	-2.4	0.8
222.82	-1.28	0.9	157.42	31AU	0.3	-3.4	-2.4	0.8
229.91	-1.35	1.0	167.27	10SE	0.6	-3.4	-2.4	0.8
237.15	-1.40	1.0	177.18	20SE	2.0	-3.4	-2.4	0.8
244.52	-1.43	1.1	187.15	30SE	1.7	-3.4	-2.3	0.8
252.02	-1.44	1.1	197.18	10OC	-0.2	-3.3	-2.3	0.8
259.61	-1.44	1.1	207.26	20OC	-0.6	-3.3	-2.2	0.8
267.27	-1.43	1.2	217.38	30OC	-0.6	-3.3	-2.1	0.7
275.01	-1.40	1.2	227.53	9NO	-0.7	-3.3	-2.1	0.7
282.78	-1.35	1.2	237.72	19NO	-0.8	-3.3	-2.0	0.7
290.58	-1.30	1.3	247.92	29NO	-0.7	-3.4	-1.9	0.7
298.38	-1.23	1.3	258.12	9DE	-0.8	-3.4	-1.8	0.7
306.18	-1.15	1.3	268.32	19DE	-0.7	-3.4	-1.8	0.8
313.96	-1.07	1.4	278.51	29DE	0.3	-3.4	-1.7	0.8
				279				
321.70	-0.97	1.4	288.67	8JA	2.9	-3.5	-1.7	0.8
329.39	-0.88	1.5	298.79	18JA	1.3	-3.5	-1.7	0.8
337.03	-0.77	1.5	308.88	28JA	0.5	-3.6	-1.6	0.8
344.60	-0.67	1.5	318.91	7FE	0.3	-3.6	-1.6	0.8
352.10	-0.56	1.6	328.88	17FE	0.1	-3.7	-1.6	0.7
359.52	-0.45	1.6	338.80	27FE	-0.3	-3.8	-1.6	0.7
6.87	-0.34	1.6	348.67	9MR	-0.8	-3.9	-1.6	0.7
14.13	-0.23	1.6	358.46	19MR	-1.7	-4.0	-1.6	0.6
21.31	-0.12	1.7	8.21	29MR	-1.3	-4.0	-1.6	0.6
28.40	-0.02	1.7	17.91	8AP	-0.2	-4.1	-1.6	0.5
35.42	0.09	1.7	27.55	18AP	0.9	-4.2	-1.6	0.4
42.36	0.19	1.7	37.15	28AP	2.0	-4.2	-1.7	0.4
49.23	0.28	1.7	46.73	8MY	3.5	-4.1	-1.7	0.3
56.02	0.38	1.7	56.27	18MY	2.6	-3.7	-1.7	0.2
62.74	0.47	1.8	65.80	28MY	1.4	-2.8	-1.8	0.3
69.41	0.56	1.8	75.32	7JN	0.4	-3.3	-1.9	0.3
76.01	0.64	1.9	84.84	17JN	-0.7	-3.9	-1.9	0.4
82.56	0.73	1.9	94.38	27JN	-1.6	-4.2	-2.0	0.4
89.06	0.80	1.9	103.93	7JL	-1.3	-4.2	-2.1	0.5
95.51	0.88	2.0	113.51	17JL	-0.5	-4.2	-2.1	0.6
101.91	0.96	2.0	123.13	27JL	-0.0	-4.1	-2.2	0.6
108.28	1.03	2.0	132.80	6AU	0.3	-4.0	-2.3	0.7
114.60	1.10	2.0	142.51	16AU	0.5	-3.9	-2.3	0.7
120.88	1.16	2.0	152.27	26AU	0.9	-3.8	-2.4	0.7
127.13	1.23	2.0	162.10	5SE	2.4	-3.8	-2.4	0.8
133.33	1.29	2.0	171.98	15SE	1.5	-3.7	-2.4	0.8
139.49	1.35	1.9	181.92	25SE	-0.4	-3.6	-2.4	0.8
145.61	1.41	1.9	191.92	5OC	-0.8	-3.6	-2.4	0.8
151.67	1.47	1.8	201.97	15OC	-0.8	-3.5	-2.4	0.8
157.68	1.52	1.8	212.07	25OC	-0.8	-3.5	-2.4	0.8
163.62	1.58	1.7	222.21	4NO	-0.7	-3.5	-2.3	0.7
169.48	1.63	1.6	232.38	14NO	-0.6	-3.4	-2.3	0.7
175.25	1.68	1.5	242.57	24NO	-0.6	-3.4	-2.2	0.7
180.90	1.73	1.4	252.78	4DE	-0.5	-3.4	-2.1	0.7
186.43	1.78	1.3	262.98	14DE	0.4	-3.4	-2.0	0.7
191.79	1.82	1.1	273.18	24DE	3.0	-3.3	-2.0	0.8

♂			☉	16.00UT	☿	♀	♃	♄
LONG	LAT	MAG	LONG	280	MAGNITUDES			
196.95	1.86	1.0	283.35	3JA	0.9	-3.3	-1.9	0.8
201.87	1.88	0.8	293.49	13JA	0.3	-3.3	-1.8	0.8
206.49	1.90	0.6	303.60	23JA	0.1	-3.3	-1.8	0.8
210.73	1.91	0.4	313.66	2FE	-0.0	-3.4	-1.7	0.8
214.51	1.89	0.2	323.66	12FE	-0.3	-3.4	-1.7	0.8
217.69	1.85	-0.1	333.61	22FE	-0.9	-3.4	-1.6	0.8
220.14	1.76	-0.4	343.50	3MR	-1.6	-3.4	-1.6	0.7
221.69	1.63	-0.7	353.33	13MR	-1.2	-3.5	-1.6	0.7
222.13	1.43	-1.0	3.11	23MR	-0.2	-3.5	-1.6	0.7
221.35	1.14	-1.3	12.82	2AP	1.1	-3.5	-1.5	0.6
219.33	0.75	-1.6	22.49	12AP	2.6	-3.4	-1.5	0.5
216.33	0.27	-1.9	32.12	22AP	3.4	-3.4	-1.5	0.5
212.92	-0.26	-1.9	41.71	2MY	1.9	-3.4	-1.5	0.4
209.80	-0.78	-1.8	51.26	12MY	1.1	-3.4	-1.6	0.4
207.63	-1.25	-1.6	60.80	22MY	0.2	-3.3	-1.6	0.3
206.77	-1.63	-1.4	70.33	1JN	-0.7	-3.3	-1.6	0.2
207.29	-1.94	-1.1	79.85	11JN	-1.6	-3.3	-1.6	0.3
209.07	-2.16	-0.9	89.38	21JN	-1.3	-3.3	-1.7	0.3
211.96	-2.32	-0.7	98.92	1JL	-0.5	-3.3	-1.7	0.4
215.75	-2.42	-0.6	108.48	11JL	0.1	-3.4	-1.8	0.5
220.28	-2.49	-0.4	118.09	21JL	0.4	-3.4	-1.9	0.5
225.43	-2.52	-0.2	127.72	31JL	0.7	-3.4	-1.9	0.6
231.07	-2.51	-0.1	137.41	10AU	1.3	-3.5	-2.0	0.6
237.12	-2.48	0.0	147.15	20AU	2.8	-3.5	-2.0	0.7
243.51	-2.42	0.1	156.94	30AU	1.3	-3.6	-2.1	0.7
250.17	-2.35	0.3	166.79	9SE	-0.4	-3.6	-2.2	0.7
257.06	-2.25	0.4	176.70	19SE	-0.9	-3.7	-2.2	0.8
264.12	-2.14	0.5	186.67	29SE	-0.9	-3.8	-2.3	0.8
271.33	-2.02	0.6	196.69	9OC	-0.8	-3.9	-2.3	0.8
278.65	-1.88	0.7	206.77	19OC	-0.6	-4.0	-2.4	0.8
286.05	-1.73	0.8	216.88	29OC	-0.4	-4.1	-2.4	0.8
293.50	-1.58	0.9	227.04	8NO	-0.4	-4.2	-2.4	0.8
300.98	-1.43	0.9	237.22	18NO	-0.3	-4.3	-2.3	0.7
308.48	-1.27	1.0	247.42	28NO	0.6	-4.4	-2.3	0.7
315.96	-1.11	1.1	257.63	8DE	2.9	-4.4	-2.2	0.7
323.43	-0.95	1.2	267.83	18DE	0.7	-4.2	-2.2	0.7
330.86	-0.80	1.3	278.02	28DE	0.1	-3.7	-2.1	0.8
				281				
338.24	-0.65	1.3	288.18	7JA	-0.0	-3.2	-2.0	0.8
345.58	-0.51	1.4	298.31	17JA	-0.1	-3.8	-2.0	0.8
352.85	-0.37	1.5	308.39	27JA	-0.4	-4.2	-1.9	0.8
0.06	-0.23	1.5	318.42	6FE	-0.9	-4.3	-1.8	0.8
7.20	-0.10	1.6	328.40	16FE	-1.5	-4.3	-1.8	0.8
14.28	0.02	1.7	338.32	26FE	-1.2	-4.2	-1.7	0.8
21.28	0.13	1.7	348.18	8MR	-0.1	-4.1	-1.6	0.8
28.22	0.24	1.8	357.99	18MR	1.4	-4.0	-1.6	0.8
35.09	0.34	1.8	7.73	28MR	3.3	-3.9	-1.6	0.7
41.90	0.43	1.8	17.43	7AP	2.5	-3.8	-1.5	0.7
48.64	0.52	1.9	27.08	17AP	1.5	-3.7	-1.5	0.6
55.33	0.60	1.9	36.68	27AP	0.8	-3.6	-1.5	0.5
61.96	0.68	1.9	46.26	7MY	0.1	-3.5	-1.5	0.5
68.55	0.75	1.9	55.81	17MY	-0.7	-3.5	-1.5	0.4
75.09	0.82	1.9	65.34	27MY	-1.7	-3.4	-1.5	0.4
81.60	0.88	1.9	74.86	6JN	-1.3	-3.4	-1.5	0.3
88.07	0.93	1.9	84.39	16JN	-0.4	-3.4	-1.5	0.3
94.50	0.98	1.9	93.92	26JN	0.2	-3.4	-1.5	0.3
100.92	1.03	1.9	103.47	6JL	0.5	-3.3	-1.5	0.4
107.31	1.07	2.0	113.05	16JL	0.9	-3.3	-1.6	0.4
113.69	1.11	2.0	122.67	26JL	1.7	-3.3	-1.6	0.5
120.05	1.14	2.0	132.33	5AU	3.2	-3.3	-1.6	0.5
126.41	1.17	2.0	142.04	15AU	1.2	-3.4	-1.7	0.6
132.76	1.20	2.0	151.80	25AU	-0.5	-3.4	-1.7	0.6
139.10	1.22	2.0	161.62	4SE	-1.1	-3.4	-1.8	0.7
145.44	1.24	2.0	171.50	14SE	-1.1	-3.4	-1.9	0.7
151.79	1.25	2.0	181.44	24SE	-0.8	-3.4	-1.9	0.8
158.13	1.26	2.0	191.43	4OC	-0.5	-3.5	-2.0	0.8
164.47	1.26	1.9	201.48	14OC	-0.3	-3.5	-2.1	0.8
170.81	1.26	1.9	211.57	24OC	-0.3	-3.5	-2.1	0.8
177.14	1.26	1.9	221.71	3NO	-0.1	-3.5	-2.2	0.8
183.47	1.24	1.8	231.88	13NO	0.8	-3.4	-2.2	0.8
189.79	1.22	1.7	242.07	23NO	2.6	-3.4	-2.2	0.8
196.10	1.20	1.6	252.28	3DE	0.4	-3.4	-2.2	0.7
202.39	1.16	1.6	262.48	13DE	-0.1	-3.4	-2.2	0.7
208.65	1.11	1.5	272.68	23DE	-0.1	-3.3	-2.2	0.7

♂			☉	16.00UT	☿	♀	♃	♄
LONG	LAT	MAG	LONG	282	MAGNITUDES			
214.88	1.05	1.4	282.86	2JA	-0.2	-3.3	-2.2	0.8
221.07	0.98	1.2	293.00	12JA	-0.5	-3.3	-2.1	0.8
227.20	0.89	1.1	303.10	22JA	-0.9	-3.3	-2.1	0.8
233.27	0.78	1.0	313.17	1FE	-1.3	-3.4	-2.0	0.9
239.25	0.64	0.8	323.17	11FE	-1.1	-3.4	-1.9	0.9
245.12	0.48	0.6	333.13	21FE	-0.1	-3.4	-1.9	0.9
250.86	0.29	0.4	343.02	3MR	1.7	-3.4	-1.8	0.9
256.42	0.06	0.3	352.85	13MR	3.4	-3.4	-1.7	0.8
261.77	-0.22	0.0	2.63	23MR	1.9	-3.5	-1.7	0.8
266.83	-0.54	-0.2	12.36	2AP	1.1	-3.5	-1.6	0.8
271.54	-0.93	-0.4	22.03	12AP	0.6	-3.6	-1.5	0.7
275.78	-1.39	-0.7	31.65	22AP	0.0	-3.6	-1.5	0.7
279.42	-1.92	-1.0	41.25	2MY	-0.8	-3.7	-1.5	0.6
282.30	-2.54	-1.3	50.80	12MY	-1.7	-3.8	-1.4	0.6
284.23	-3.25	-1.6	60.34	22MY	-1.3	-3.9	-1.4	0.5
285.02	-4.02	-1.9	69.87	1JN	-0.4	-4.0	-1.4	0.4
284.57	-4.81	-2.2	79.39	11JN	0.3	-4.1	-1.4	0.4
282.94	-5.53	-2.4	88.92	21JN	0.7	-4.1	-1.3	0.3
280.48	-6.07	-2.6	98.46	1JL	1.2	-4.2	-1.3	0.3
277.86	-6.34	-2.5	108.03	11JL	2.2	-4.2	-1.4	0.4
275.78	-6.32	-2.4	117.62	21JL	3.0	-4.0	-1.4	0.4
274.80	-6.04	-2.1	127.26	31JL	1.0	-3.6	-1.4	0.5
275.15	-5.61	-1.8	136.94	10AU	-0.6	-3.2	-1.4	0.5
276.78	-5.10	-1.6	146.68	20AU	-1.2	-3.8	-1.5	0.6
279.56	-4.56	-1.3	156.47	30AU	-1.2	-4.1	-1.5	0.6
283.28	-4.03	-1.0	166.31	9SE	-0.7	-4.3	-1.5	0.7
287.73	-3.53	-0.8	176.22	19SE	-0.4	-4.3	-1.6	0.7
292.77	-3.06	-0.6	186.19	29SE	-0.2	-4.2	-1.6	0.8
298.25	-2.62	-0.4	196.21	9OC	-0.1	-4.2	-1.7	0.8
304.07	-2.22	-0.1	206.28	19OC	0.0	-4.1	-1.8	0.8
310.16	-1.86	0.0	216.40	29OC	1.1	-4.0	-1.8	0.8
316.44	-1.52	0.2	226.55	8NO	2.4	-3.9	-1.9	0.8
322.86	-1.22	0.4	236.73	18NO	0.2	-3.8	-2.0	0.8
329.39	-0.95	0.6	246.93	28NO	-0.2	-3.7	-2.0	0.8
335.99	-0.70	0.7	257.13	8DE	-0.3	-3.7	-2.1	0.8
342.63	-0.47	0.9	267.34	18DE	-0.4	-3.6	-2.1	0.8
349.30	-0.27	1.0	277.53	28DE	-0.6	-3.5	-2.1	0.8
				283				
355.97	-0.09	1.1	287.69	7JA	-1.0	-3.5	-2.1	0.8
2.65	0.07	1.2	297.82	17JA	-1.2	-3.4	-2.1	0.8
9.31	0.22	1.3	307.90	27JA	-1.0	-3.4	-2.1	0.9
15.95	0.35	1.4	317.94	6FE	0.0	-3.4	-2.1	0.9
22.57	0.46	1.5	327.92	16FE	2.0	-3.3	-2.0	0.9
29.16	0.57	1.6	337.85	26FE	2.7	-3.3	-1.9	0.9
35.72	0.66	1.7	347.71	8MR	1.4	-3.3	-1.9	0.9
42.25	0.74	1.8	357.52	18MR	0.8	-3.3	-1.8	0.9
48.74	0.81	1.8	7.27	28MR	0.4	-3.3	-1.7	0.9
55.21	0.88	1.9	16.96	7AP	-0.0	-3.3	-1.7	0.9
61.65	0.93	1.9	26.62	17AP	-0.8	-3.3	-1.6	0.8
68.06	0.98	1.9	36.22	27AP	-1.7	-3.4	-1.5	0.8
74.45	1.02	2.0	45.80	7MY	-1.3	-3.4	-1.5	0.7
80.83	1.06	2.0	55.35	17MY	-0.4	-3.4	-1.4	0.7
87.18	1.09	2.0	64.88	27MY	0.4	-3.5	-1.4	0.6
93.53	1.11	2.0	74.40	6JN	1.0	-3.5	-1.4	0.6
99.88	1.13	2.0	83.92	16JN	1.6	-3.4	-1.3	0.5
106.22	1.15	2.0	93.45	26JN	2.8	-3.4	-1.3	0.4
112.56	1.16	2.0	103.00	6JL	2.6	-3.4	-1.3	0.4
118.91	1.16	2.0	112.59	16JL	0.8	-3.4	-1.3	0.4
125.27	1.16	2.0	122.20	26JL	-0.6	-3.3	-1.3	0.4
131.65	1.15	1.9	131.86	5AU	-1.3	-3.3	-1.3	0.5
138.05	1.15	2.0	141.57	15AU	-1.3	-3.3	-1.3	0.5
144.47	1.13	2.0	151.33	25AU	-0.6	-3.3	-1.3	0.6
150.91	1.11	2.0	161.14	4SE	-0.3	-3.3	-1.3	0.6
157.39	1.09	2.0	171.02	14SE	-0.1	-3.3	-1.3	0.7
163.89	1.06	2.0	180.95	24SE	0.0	-3.4	-1.4	0.7
170.44	1.02	2.0	190.95	4OC	0.3	-3.4	-1.4	0.8
177.01	0.98	1.9	200.99	14OC	1.3	-3.4	-1.5	0.8
183.63	0.94	1.9	211.09	24OC	2.2	-3.5	-1.5	0.8
190.28	0.88	1.9	221.22	3NO	0.0	-3.5	-1.6	0.9
196.97	0.82	1.8	231.39	13NO	-0.4	-3.6	-1.7	0.9
203.70	0.75	1.8	241.58	23NO	-0.4	-3.7	-1.7	0.9
210.47	0.68	1.7	251.78	3DE	-0.5	-3.7	-1.8	0.9
217.28	0.59	1.7	261.99	13DE	-0.7	-3.8	-1.8	0.9
224.12	0.49	1.6	272.19	23DE	-0.9	-3.9	-1.9	0.9

♂ LONG	LAT	MAG	☉ LONG	16.00UT 284	☿	♀	♃	♄
231.00	0.39	1.5	282.36	2JA	-1.0	-4.0	-2.0	0.9
237.91	0.27	1.5	292.51	12JA	-0.9	-4.1	-2.0	0.8
244.85	0.13	1.4	302.62	22JA	0.1	-4.2	-2.0	0.9
251.83	-0.01	1.3	312.68	1FE	2.4	-4.3	-2.0	0.9
258.82	-0.17	1.2	322.70	11FE	2.1	-4.3	-2.0	1.0
265.84	-0.35	1.1	332.65	21FE	1.0	-4.3	-2.0	1.0
272.87	-0.54	1.0	342.54	2MR	0.6	-4.0	-2.0	1.0
279.90	-0.74	0.9	352.38	12MR	0.3	-3.5	-2.0	1.0
286.92	-0.96	0.7	2.16	22MR	-0.1	-3.2	-1.9	1.0
293.92	-1.20	0.6	11.89	1AP	-0.8	-3.8	-1.8	1.0
300.88	-1.45	0.5	21.56	11AP	-1.7	-4.1	-1.8	1.0
307.77	-1.71	0.4	31.19	21AP	-1.3	-4.2	-1.7	1.0
314.57	-1.99	0.2	40.78	1MY	-0.3	-4.2	-1.7	0.9
321.24	-2.27	0.1	50.34	11MY	0.5	-4.1	-1.6	0.9
327.73	-2.56	-0.1	59.88	21MY	1.3	-4.0	-1.5	0.8
333.99	-2.86	-0.2	69.41	31MY	2.2	-3.9	-1.5	0.8
339.95	-3.17	-0.4	78.93	10JN	3.4	-3.8	-1.4	0.7
345.52	-3.47	-0.6	88.45	20JN	2.1	-3.7	-1.4	0.6
350.60	-3.77	-0.8	98.00	30JN	0.6	-3.7	-1.3	0.6
355.03	-4.07	-1.0	107.56	10JL	-0.6	-3.6	-1.3	0.5
358.66	-4.34	-1.2	117.16	20JL	-1.4	-3.5	-1.3	0.4
1.28	-4.59	-1.4	126.79	30JL	-1.3	-3.5	-1.2	0.5
2.66	-4.77	-1.6	136.47	9AU	-0.6	-3.5	-1.2	0.5
2.66	-4.85	-1.9	146.20	19AU	-0.2	-3.4	-1.2	0.6
1.21	-4.78	-2.1	155.99	29AU	0.1	-3.4	-1.2	0.6
358.56	-4.50	-2.2	165.84	8SE	0.2	-3.4	-1.2	0.7
355.33	-4.01	-2.3	175.74	18SE	0.5	-3.4	-1.2	0.7
352.30	-3.37	-2.1	185.71	28SE	1.7	-3.4	-1.3	0.8
350.19	-2.66	-1.8	195.73	8OC	2.0	-3.4	-1.3	0.8
349.37	-1.98	-1.4	205.79	18OC	-0.1	-3.4	-1.3	0.9
349.86	-1.37	-1.1	215.91	28OC	-0.6	-3.4	-1.4	0.9
351.55	-0.86	-0.8	226.06	7NO	-0.6	-3.4	-1.4	0.9
354.22	-0.43	-0.5	236.24	17NO	-0.6	-3.4	-1.5	0.9
357.67	-0.08	-0.2	246.44	27NO	-0.7	-3.4	-1.5	1.0
1.74	0.21	0.0	256.64	7DE	-0.8	-3.4	-1.6	1.0
6.28	0.44	0.3	266.84	17DE	-0.8	-3.5	-1.6	1.0
11.18	0.63	0.5	277.03	27DE	-0.7	-3.5	-1.7	1.0
				285				
16.36	0.78	0.7	287.20	6JA	0.2	-3.5	-1.8	1.0
21.76	0.90	0.9	297.32	16JA	2.7	-3.5	-1.8	1.0
27.33	1.00	1.0	307.41	26JA	1.6	-3.4	-1.9	1.0
33.02	1.09	1.2	317.45	5FE	0.7	-3.4	-1.9	1.0
38.81	1.15	1.3	327.43	15FE	0.4	-3.4	-2.0	1.1
44.69	1.20	1.4	337.36	25FE	0.2	-3.4	-2.0	1.1
50.62	1.24	1.5	347.23	7MR	-0.2	-3.3	-2.0	1.1
56.59	1.27	1.6	357.04	17MR	-0.8	-3.3	-2.0	1.1
62.61	1.29	1.7	6.80	27MR	-1.7	-3.3	-2.0	1.2
68.66	1.31	1.8	16.49	6AP	-1.3	-3.3	-2.0	1.1
74.73	1.31	1.8	26.15	16AP	-0.3	-3.3	-1.9	1.1
80.83	1.31	1.9	35.76	26AP	0.7	-3.4	-1.9	1.1
86.95	1.31	1.9	45.33	6MY	1.7	-3.4	-1.8	1.1
93.10	1.30	2.0	54.88	16MY	2.9	-3.4	-1.7	1.0
99.27	1.29	2.0	64.42	26MY	3.0	-3.4	-1.7	1.0
105.46	1.27	2.0	73.94	5JN	1.6	-3.5	-1.6	0.9
111.68	1.24	2.0	83.46	15JN	0.5	-3.5	-1.6	0.9
117.93	1.22	2.0	92.99	25JN	-0.7	-3.6	-1.5	0.8
124.21	1.19	2.0	102.54	5JL	-1.5	-3.7	-1.4	0.7
130.53	1.15	2.0	112.12	15JL	-1.3	-3.7	-1.4	0.7
136.90	1.11	2.0	121.74	25JL	-0.6	-3.8	-1.3	0.6
143.30	1.07	2.0	131.39	4AU	-0.1	-3.9	-1.3	0.6
149.75	1.02	2.0	141.09	14AU	0.2	-4.0	-1.3	0.6
156.26	0.97	1.9	150.86	24AU	0.4	-4.1	-1.3	0.6
162.82	0.91	1.9	160.67	3SE	0.7	-4.2	-1.2	0.7
169.43	0.85	1.8	170.54	13SE	2.1	-4.3	-1.2	0.7
176.10	0.79	1.8	180.48	23SE	1.8	-4.3	-1.2	0.8
182.84	0.72	1.8	190.46	3OC	-0.3	-4.1	-1.2	0.8
189.64	0.64	1.8	200.51	13OC	-0.7	-3.6	-1.2	0.9
196.51	0.56	1.8	210.60	23OC	-0.7	-3.2	-1.3	0.9
203.43	0.47	1.8	220.73	2NO	-0.7	-3.9	-1.3	1.0
210.43	0.38	1.8	230.90	12NO	-0.7	-4.3	-1.3	1.0
217.50	0.28	1.7	241.09	22NO	-0.7	-4.4	-1.3	1.0
224.62	0.18	1.7	251.29	2DE	-0.7	-4.4	-1.4	1.1
231.82	0.07	1.7	261.50	12DE	-0.6	-4.3	-1.4	1.1
239.08	-0.05	1.6	271.70	22DE	0.3	-4.2	-1.5	1.1

♂ LONG	LAT	MAG	☉ LONG	16.00UT 286	☿	♀	♃	♄
246.39	-0.17	1.6	281.87	1JA	3.0	-4.1	-1.6	1.1
253.76	-0.29	1.5	292.02	11JA	1.2	-4.0	-1.6	1.1
261.19	-0.42	1.5	302.13	21JA	0.5	-3.9	-1.7	1.1
268.66	-0.56	1.4	312.20	31JA	0.2	-3.8	-1.8	1.1
276.17	-0.69	1.4	322.21	10FE	0.0	-3.7	-1.8	1.1
283.71	-0.83	1.3	332.17	20FE	-0.3	-3.6	-1.9	1.2
291.28	-0.97	1.3	342.07	2MR	-0.8	-3.5	-1.9	1.2
298.86	-1.11	1.2	351.91	12MR	-1.6	-3.5	-2.0	1.3
306.44	-1.24	1.1	1.70	22MR	-1.3	-3.4	-2.0	1.3
314.01	-1.37	1.1	11.42	1AP	-0.2	-3.4	-2.0	1.3
321.55	-1.49	1.0	21.10	11AP	0.9	-3.4	-2.1	1.3
329.05	-1.60	0.9	30.73	21AP	2.2	-3.3	-2.0	1.3
336.49	-1.70	0.9	40.32	1MY	3.8	-3.3	-2.0	1.3
343.85	-1.79	0.8	49.89	11MY	2.3	-3.3	-2.0	1.3
351.11	-1.86	0.8	59.43	21MY	1.3	-3.3	-1.9	1.2
358.25	-1.92	0.7	68.95	31MY	0.3	-3.3	-1.9	1.2
5.24	-1.95	0.6	78.47	10JN	-0.7	-3.3	-1.8	1.2
12.05	-1.98	0.5	88.00	20JN	-1.6	-3.3	-1.8	1.1
18.65	-1.98	0.5	97.54	30JN	-1.3	-3.4	-1.7	1.0
25.01	-1.96	0.4	107.10	10JL	-0.5	-3.4	-1.6	1.0
31.07	-1.92	0.3	116.69	20JL	-0.0	-3.4	-1.6	0.9
36.77	-1.86	0.2	126.32	30JL	0.3	-3.5	-1.5	0.8
42.05	-1.77	0.0	136.00	9AU	0.6	-3.5	-1.5	0.8
46.81	-1.65	-0.1	145.73	19AU	1.0	-3.5	-1.4	0.7
50.94	-1.50	-0.3	155.51	29AU	2.5	-3.4	-1.4	0.8
54.28	-1.30	-0.4	165.36	8SE	1.5	-3.4	-1.3	0.8
56.65	-1.06	-0.6	175.26	18SE	-0.4	-3.4	-1.3	0.8
57.85	-0.76	-0.9	185.21	28SE	-0.9	-3.4	-1.3	0.9
57.68	-0.40	-1.1	195.23	8OC	-0.9	-3.3	-1.3	0.9
56.06	0.02	-1.3	205.30	18OC	-0.9	-3.3	-1.3	1.0
53.14	0.49	-1.5	215.41	28OC	-0.6	-3.3	-1.3	1.0
49.40	0.95	-1.6	225.56	7NO	-0.5	-3.3	-1.3	1.1
45.62	1.36	-1.4	235.74	17NO	-0.5	-3.3	-1.3	1.1
42.56	1.68	-1.2	245.94	27NO	-0.4	-3.4	-1.3	1.2
40.71	1.89	-0.9	256.15	7DE	0.5	-3.4	-1.3	1.2
40.23	2.02	-0.6	266.35	17DE	3.0	-3.4	-1.4	1.2
41.00	2.09	-0.3	276.54	27DE	0.8	-3.5	-1.4	1.3
				287				
42.84	2.11	0.0	286.71	6JA	0.2	-3.5	-1.5	1.3
45.55	2.11	0.3	296.84	16JA	0.1	-3.5	-1.5	1.3
48.94	2.08	0.5	306.93	26JA	-0.1	-3.6	-1.6	1.3
52.86	2.04	0.7	316.97	5FE	-0.4	-3.7	-1.7	1.3
57.21	2.00	0.9	326.96	15FE	-0.9	-3.7	-1.7	1.3
61.89	1.95	1.1	336.89	25FE	-1.5	-3.8	-1.8	1.3
66.83	1.89	1.2	346.76	7MR	-1.2	-3.9	-1.9	1.4
71.98	1.84	1.3	356.57	17MR	-0.2	-4.0	-1.9	1.4
77.30	1.78	1.5	6.33	27MR	1.2	-4.0	-2.0	1.3
82.77	1.72	1.6	16.03	6AP	2.8	-4.1	-2.0	1.3
88.36	1.66	1.6	25.68	16AP	3.1	-4.2	-2.1	1.3
94.05	1.59	1.7	35.30	26AP	1.8	-4.2	-2.1	1.3
99.84	1.53	1.8	44.88	6MY	1.0	-4.1	-2.1	1.2
105.70	1.46	1.8	54.42	16MY	0.2	-3.7	-2.1	1.2
111.65	1.40	1.9	63.96	26MY	-0.7	-2.8	-2.1	1.2
117.66	1.33	1.9	73.48	5JN	-1.6	-3.3	-2.1	1.1
123.75	1.26	1.9	83.00	15JN	-1.3	-3.9	-2.1	1.1
129.91	1.18	1.9	92.53	25JN	-0.5	-4.2	-2.1	1.0
136.14	1.11	2.0	102.08	5JL	0.1	-4.2	-2.0	1.0
142.43	1.03	2.0	111.66	15JL	0.4	-4.2	-1.9	0.9
148.80	0.96	1.9	121.27	25JL	0.8	-4.1	-1.8	0.9
155.25	0.88	1.9	130.92	4AU	1.4	-4.0	-1.8	0.8
161.77	0.79	1.9	140.62	14AU	3.0	-3.9	-1.7	0.8
168.37	0.71	1.9	150.38	24AU	1.3	-3.8	-1.7	0.8
175.05	0.62	1.9	160.19	3SE	-0.4	-3.8	-1.6	0.8
181.81	0.53	1.8	170.06	13SE	-1.0	-3.7	-1.5	0.9
188.66	0.43	1.8	179.99	23SE	-1.0	-3.6	-1.5	0.9
195.59	0.34	1.7	189.98	3OC	-0.8	-3.6	-1.5	1.0
202.60	0.24	1.7	200.01	13OC	-0.5	-3.5	-1.4	1.1
209.70	0.14	1.6	210.10	23OC	-0.4	-3.5	-1.4	1.1
216.89	0.03	1.6	220.24	2NO	-0.4	-3.5	-1.4	1.2
224.16	-0.07	1.6	230.40	12NO	-0.3	-3.4	-1.4	1.2
231.50	-0.18	1.6	240.59	22NO	0.6	-3.4	-1.4	1.3
238.93	-0.29	1.6	250.80	2DE	2.9	-3.4	-1.4	1.3
246.42	-0.40	1.5	261.00	12DE	0.6	-3.4	-1.4	1.3
253.99	-0.50	1.5	271.20	22DE	0.1	-3.3	-1.4	1.3

♂			☉	16.00UT	☿	♀	♃	♄
LONG	LAT	MAG	LONG	288	MAGNITUDES			
261.61	-0.61	1.5	281.38	1JA	-0.1	-3.3	-1.4	1.3
269.28	-0.71	1.5	291.53	11JA	-0.2	-3.3	-1.4	1.3
277.00	-0.81	1.5	301.64	21JA	-0.4	-3.3	-1.5	1.2
284.75	-0.90	1.5	311.71	31JA	-0.9	-3.4	-1.5	1.2
292.53	-0.99	1.4	321.73	10FE	-1.4	-3.4	-1.6	1.2
300.32	-1.07	1.4	331.69	20FE	-1.2	-3.4	-1.6	1.1
308.10	-1.14	1.4	341.59	1MR	-0.1	-3.4	-1.7	1.0
315.88	-1.20	1.4	351.43	11MR	1.5	-3.5	-1.8	1.1
323.63	-1.25	1.3	1.22	21MR	3.4	-3.5	-1.8	1.1
331.34	-1.29	1.3	10.95	31MR	2.3	-3.5	-1.9	1.1
339.00	-1.31	1.3	20.63	10AP	1.3	-3.4	-2.0	1.0
346.61	-1.32	1.3	30.26	20AP	0.7	-3.4	-2.1	1.0
354.13	-1.31	1.3	39.86	30AP	0.1	-3.4	-2.1	1.0
1.57	-1.29	1.2	49.42	10MY	-0.7	-3.4	-2.2	1.0
8.91	-1.26	1.2	58.96	20MY	-1.6	-3.3	-2.2	1.0
16.15	-1.21	1.2	68.49	30MY	-1.3	-3.3	-2.2	0.9
23.26	-1.15	1.2	78.00	9JN	-0.4	-3.3	-2.3	0.9
30.24	-1.07	1.2	87.53	19JN	0.2	-3.3	-2.3	0.8
37.07	-0.98	1.1	97.07	29JN	0.6	-3.3	-2.3	0.8
43.75	-0.88	1.1	106.63	9JL	1.0	-3.4	-2.2	0.7
50.25	-0.76	1.0	116.23	19JL	1.8	-3.4	-2.2	0.7
56.55	-0.63	1.0	125.85	29JL	3.2	-3.4	-2.1	0.6
62.63	-0.48	0.9	135.53	8AU	1.1	-3.5	-2.1	0.6
68.45	-0.31	0.8	145.26	18AU	-0.5	-3.5	-2.0	0.5
73.99	-0.13	0.8	155.04	28AU	-1.1	-3.6	-1.9	0.5
79.17	0.07	0.7	164.87	7SE	-1.2	-3.6	-1.9	0.5
83.95	0.30	0.5	174.77	17SE	-0.8	-3.7	-1.8	0.5
88.22	0.56	0.4	184.73	27SE	-0.4	-3.8	-1.8	0.6
91.88	0.86	0.2	194.74	7OC	-0.3	-3.9	-1.7	0.6
94.79	1.19	0.1	204.81	17OC	-0.2	-4.0	-1.6	0.7
96.76	1.57	-0.2	214.92	27OC	-0.1	-4.1	-1.6	0.8
97.62	1.99	-0.4	225.06	6NO	0.9	-4.2	-1.6	0.8
97.16	2.44	-0.6	235.24	16NO	2.7	-4.3	-1.5	0.9
95.33	2.90	-0.8	245.44	26NO	0.3	-4.4	-1.5	0.9
92.26	3.31	-1.0	255.64	6DE	-0.1	-4.4	-1.5	1.0
88.40	3.62	-1.1	265.85	16DE	-0.2	-4.2	-1.5	1.0
84.49	3.78	-1.0	276.04	26DE	-0.3	-3.6	-1.5	1.0
				289				
81.26	3.79	-0.8	286.21	5JA	-0.5	-3.2	-1.5	1.0
79.17	3.68	-0.5	296.34	15JA	-0.9	-3.9	-1.5	1.0
78.41	3.50	-0.2	306.43	25JA	-1.2	-4.2	-1.5	1.0
78.90	3.28	0.0	316.48	4FE	-1.1	-4.3	-1.5	0.9
80.46	3.06	0.3	326.47	14FE	-0.1	-4.3	-1.6	0.9
82.91	2.85	0.5	336.40	24FE	1.8	-4.2	-1.6	0.9
86.06	2.64	0.7	346.28	6MR	3.2	-4.1	-1.6	0.8
89.78	2.45	0.9	356.10	16MR	1.7	-4.0	-1.7	0.8
93.95	2.27	1.0	5.85	26MR	1.0	-3.9	-1.8	0.8
98.48	2.11	1.2	15.56	5AP	0.5	-3.8	-1.8	0.8
103.31	1.95	1.3	25.22	15AP	0.0	-3.7	-1.9	0.8
108.39	1.80	1.4	34.83	25AP	-0.7	-3.6	-2.0	0.8
113.68	1.66	1.5	44.41	5MY	-1.7	-3.5	-2.0	0.8
119.16	1.53	1.5	53.97	15MY	-1.3	-3.5	-2.1	0.8
124.79	1.40	1.6	63.50	25MY	-0.4	-3.4	-2.2	0.8
130.57	1.27	1.7	73.02	4JN	0.3	-3.4	-2.2	0.7
136.48	1.15	1.7	82.54	14JN	0.8	-3.4	-2.3	0.7
142.51	1.03	1.7	92.07	24JN	1.4	-3.3	-2.3	0.7
148.66	0.91	1.8	101.62	4JL	2.4	-3.3	-2.4	0.6
154.93	0.79	1.8	111.20	14JL	2.9	-3.3	-2.4	0.6
161.30	0.68	1.8	120.81	24JL	1.0	-3.3	-2.4	0.5
167.79	0.56	1.8	130.46	3AU	-0.6	-3.3	-2.4	0.5
174.38	0.45	1.8	140.16	13AU	-1.2	-3.4	-2.3	0.4
181.08	0.33	1.8	149.91	23AU	-1.3	-3.4	-2.3	0.3
187.89	0.22	1.7	159.72	2SE	-0.7	-3.4	-2.2	0.3
194.80	0.11	1.7	169.58	12SE	-0.3	-3.4	-2.2	0.2
201.82	0.00	1.7	179.50	22SE	-0.1	-3.4	-2.1	0.2
208.94	-0.11	1.7	189.49	2OC	-0.1	-3.5	-2.1	0.2
216.16	-0.22	1.6	199.53	12OC	0.1	-3.5	-2.0	0.3
223.48	-0.32	1.6	209.61	22OC	1.1	-3.5	-1.9	0.4
230.89	-0.42	1.5	219.74	1NO	2.4	-3.5	-1.9	0.5
238.40	-0.52	1.5	229.90	11NO	0.1	-3.4	-1.8	0.5
245.98	-0.61	1.5	240.09	21NO	-0.3	-3.4	-1.7	0.6
253.63	-0.70	1.4	250.29	1DE	-0.3	-3.4	-1.7	0.6
261.35	-0.79	1.4	260.50	11DE	-0.4	-3.4	-1.6	0.7
269.13	-0.86	1.4	270.69	21DE	-0.6	-3.3	-1.6	0.7
276.95	-0.93	1.4	280.88	31DE	-1.0	-3.3	-1.6	0.7

♂			☉	16.00UT	☿	♀	♃	♄
LONG	LAT	MAG	LONG	290	MAGNITUDES			
284.80	-0.99	1.4	291.03	10JA	-1.1	-3.3	-1.6	0.8
292.67	-1.03	1.4	301.14	20JA	-1.0	-3.3	-1.5	0.8
300.55	-1.07	1.4	311.22	30JA	-0.0	-3.4	-1.5	0.7
308.42	-1.10	1.4	321.24	9FE	2.1	-3.4	-1.5	0.7
316.28	-1.11	1.4	331.20	19FE	2.5	-3.4	-1.6	0.7
324.11	-1.11	1.4	341.11	1MR	1.2	-3.4	-1.6	0.7
331.89	-1.10	1.4	350.96	11MR	0.7	-3.4	-1.6	0.6
339.62	-1.08	1.4	0.74	21MR	0.4	-3.5	-1.6	0.6
347.29	-1.05	1.4	10.48	31MR	-0.1	-3.5	-1.7	0.5
354.88	-1.00	1.5	20.16	10AP	-0.7	-3.6	-1.7	0.6
2.40	-0.94	1.5	29.80	20AP	-1.7	-3.6	-1.8	0.6
9.83	-0.88	1.5	39.40	30AP	-1.3	-3.7	-1.8	0.6
17.17	-0.80	1.5	48.96	10MY	-0.4	-3.8	-1.9	0.6
24.41	-0.72	1.5	58.50	20MY	0.4	-3.9	-1.9	0.6
31.54	-0.62	1.5	68.03	30MY	1.1	-4.0	-2.0	0.6
38.57	-0.52	1.5	77.55	9JN	1.8	-4.1	-2.1	0.6
45.50	-0.41	1.5	87.07	19JN	3.1	-4.1	-2.2	0.5
52.30	-0.30	1.5	96.62	29JN	2.4	-4.2	-2.2	0.5
58.99	-0.18	1.5	106.17	9JL	0.8	-4.2	-2.3	0.5
65.56	-0.05	1.5	115.76	19JL	-0.6	-4.0	-2.3	0.4
71.99	0.09	1.5	125.39	29JL	-1.4	-3.5	-2.4	0.4
78.28	0.23	1.4	135.06	8AU	-1.3	-3.2	-2.4	0.3
84.42	0.39	1.4	144.79	18AU	-0.6	-3.8	-2.4	0.3
90.40	0.55	1.4	154.57	28AU	-0.2	-4.1	-2.4	0.2
96.18	0.72	1.3	164.40	7SE	-0.0	-4.3	-2.4	0.1
101.76	0.90	1.2	174.29	17SE	0.1	-4.3	-2.4	0.1
107.08	1.10	1.1	184.25	27SE	0.3	-4.2	-2.4	0.0
112.11	1.31	1.0	194.26	7OC	1.4	-4.1	-2.3	-0.0
116.79	1.55	0.9	204.32	17OC	2.2	-4.1	-2.3	0.0
121.05	1.81	0.8	214.43	27OC	-0.0	-4.0	-2.2	0.1
124.78	2.09	0.6	224.57	6NO	-0.5	-3.9	-2.1	0.2
127.87	2.41	0.4	234.75	16NO	-0.5	-3.8	-2.0	0.2
130.16	2.76	0.2	244.95	26NO	-0.5	-3.7	-2.0	0.3
131.47	3.15	-0.0	255.15	6DE	-0.7	-3.6	-1.9	0.4
131.63	3.55	-0.3	265.35	16DE	-0.9	-3.6	-1.8	0.4
130.48	3.95	-0.5	275.55	26DE	-0.9	-3.5	-1.8	0.5
				291				
128.04	4.29	-0.8	285.71	5JA	-0.8	-3.5	-1.7	0.5
124.56	4.52	-0.9	295.85	15JA	0.1	-3.4	-1.7	0.5
120.61	4.58	-1.0	305.95	25JA	2.4	-3.4	-1.6	0.6
116.92	4.46	-0.8	315.99	4FE	1.9	-3.4	-1.6	0.6
114.12	4.20	-0.6	325.99	14FE	0.9	-3.3	-1.6	0.6
112.56	3.86	-0.4	335.92	24FE	0.5	-3.3	-1.6	0.5
112.29	3.49	-0.1	345.80	6MR	0.2	-3.3	-1.6	0.5
113.20	3.13	0.1	355.62	16MR	-0.1	-3.3	-1.6	0.5
115.12	2.78	0.3	5.39	26MR	-0.8	-3.3	-1.6	0.4
117.87	2.47	0.5	15.09	5AP	-1.7	-3.3	-1.6	0.4
121.28	2.18	0.7	24.75	15AP	-1.3	-3.3	-1.6	0.4
125.25	1.92	0.8	34.37	25AP	-0.3	-3.4	-1.6	0.4
129.66	1.67	1.0	43.95	5MY	0.6	-3.4	-1.7	0.4
134.44	1.45	1.1	53.50	15MY	1.4	-3.4	-1.7	0.4
139.54	1.24	1.2	63.04	25MY	2.4	-3.5	-1.8	0.4
144.90	1.05	1.2	72.56	4JN	3.4	-3.5	-1.8	0.4
150.50	0.87	1.3	82.08	14JN	1.9	-3.4	-1.9	0.4
156.32	0.70	1.4	91.61	24JN	0.6	-3.4	-1.9	0.4
162.32	0.54	1.4	101.16	4JL	-0.6	-3.4	-2.0	0.4
168.49	0.38	1.4	110.73	14JL	-1.4	-3.4	-2.1	0.4
174.84	0.24	1.5	120.34	24JL	-1.3	-3.3	-2.1	0.3
181.33	0.09	1.5	129.98	3AU	-0.6	-3.3	-2.2	0.3
187.98	-0.04	1.5	139.68	13AU	-0.1	-3.3	-2.3	0.3
194.76	-0.17	1.5	149.43	23AU	0.1	-3.3	-2.3	0.2
201.68	-0.29	1.5	159.23	2SE	0.3	-3.3	-2.4	0.1
208.74	-0.41	1.5	169.10	12SE	0.6	-3.3	-2.4	0.1
215.91	-0.52	1.5	179.02	22SE	1.8	-3.4	-2.4	0.0
223.20	-0.62	1.5	189.00	2OC	2.0	-3.4	-2.4	-0.1
230.60	-0.71	1.5	199.04	12OC	-0.2	-3.4	-2.4	-0.1
238.11	-0.80	1.5	209.12	22OC	-0.6	-3.5	-2.4	-0.2
245.70	-0.87	1.5	219.25	1NO	-0.6	-3.5	-2.4	-0.1
253.38	-0.94	1.4	229.41	11NO	-0.6	-3.6	-2.3	-0.0
261.13	-0.99	1.4	239.60	21NO	-0.8	-3.7	-2.2	0.0
268.93	-1.04	1.4	249.80	1DE	-0.7	-3.7	-2.2	0.1
276.79	-1.07	1.4	260.01	11DE	-0.8	-3.8	-2.1	0.2
284.67	-1.09	1.4	270.20	21DE	-0.7	-3.9	-2.0	0.2
292.57	-1.10	1.4	280.38	31DE	0.2	-4.0	-2.0	0.3

♂ LONG LAT MAG	☉ LONG	16.00UT 292	☿	♀	♃	♄
300.47 -1.10 1.4	290.54	10JA	2.8	-4.1	-1.9	0.3
308.36 -1.09 1.4	300.66	20JA	1.4	-4.2	-1.8	0.4
316.23 -1.06 1.3	310.73	30JA	0.6	-4.3	-1.8	0.4
324.07 -1.02 1.3	320.75	9FE	0.3	-4.3	-1.7	0.4
331.85 -0.98 1.3	330.72	19FE	0.1	-4.3	-1.7	0.4
339.58 -0.92 1.4	340.62	29FE	-0.2	-4.0	-1.6	0.4
347.24 -0.85 1.4	350.48	10MR	-0.8	-3.5	-1.6	0.4
354.83 -0.78 1.4	0.27	20MR	-1.6	-3.3	-1.6	0.4
2.34 -0.70 1.5	10.00	30MR	-1.3	-3.8	-1.5	0.4
9.77 -0.61 1.5	19.69	9AP	-0.3	-4.1	-1.5	0.3
17.10 -0.52 1.6	29.33	19AP	0.8	-4.2	-1.5	0.3
24.35 -0.42 1.6	38.93	29AP	1.8	-4.2	-1.5	0.3
31.51 -0.32 1.6	48.50	9MY	3.2	-4.1	-1.5	0.3
38.58 -0.22 1.7	58.04	19MY	2.8	-4.0	-1.6	0.3
45.55 -0.11 1.7	67.56	29MY	1.5	-3.9	-1.6	0.3
52.44 -0.00 1.7	77.09	8JN	0.5	-3.8	-1.6	0.4
59.23 0.11 1.7	86.61	18JN	-0.6	-3.7	-1.6	0.4
65.94 0.22 1.8	96.15	28JN	-1.5	-3.7	-1.7	0.4
72.56 0.33 1.8	105.71	8JL	-1.3	-3.6	-1.7	0.4
79.09 0.45 1.8	115.30	18JL	-0.6	-3.5	-1.8	0.3
85.53 0.56 1.8	124.92	28JL	-0.1	-3.5	-1.9	0.3
91.88 0.68 1.8	134.60	7AU	0.2	-3.5	-1.9	0.3
98.13 0.80 1.8	144.32	17AU	0.4	-3.4	-2.0	0.2
104.29 0.92 1.7	154.09	27AU	0.8	-3.4	-2.0	0.2
110.34 1.05 1.7	163.93	6SE	2.2	-3.4	-2.1	0.1
116.26 1.18 1.6	173.81	16SE	1.8	-3.4	-2.2	0.1
122.06 1.32 1.6	183.77	26SE	-0.3	-3.4	-2.2	-0.0
127.71 1.46 1.5	193.77	6OC	-0.8	-3.4	-2.3	-0.1
133.19 1.61 1.4	203.83	16OC	-0.8	-3.4	-2.3	-0.2
138.46 1.77 1.3	213.94	26OC	-0.8	-3.4	-2.3	-0.2
143.49 1.94 1.2	224.08	5NO	-0.7	-3.4	-2.4	-0.3
148.23 2.13 1.1	234.25	15NO	-0.6	-3.4	-2.3	-0.2
152.62 2.33 0.9	244.45	25NO	-0.6	-3.4	-2.3	-0.1
156.56 2.55 0.8	254.66	5DE	-0.5	-3.4	-2.3	-0.1
159.97 2.78 0.6	264.86	15DE	0.3	-3.5	-2.2	0.0
162.70 3.04 0.3	275.05	25DE	3.0	-3.5	-2.2	0.1
		293				
164.58 3.31 0.1	285.22	4JA	1.1	-3.5	-2.1	0.1
165.46 3.59 -0.2	295.36	14JA	0.4	-3.5	-2.0	0.2
165.15 3.85 -0.4	305.45	24JA	0.2	-3.4	-1.9	0.3
163.55 4.06 -0.7	315.50	3FE	-0.0	-3.4	-1.9	0.3
160.75 4.17 -1.0	325.50	13FE	-0.3	-3.4	-1.8	0.3
157.11 4.13 -1.1	335.44	23FE	-0.8	-3.4	-1.7	0.3
153.26 3.92 -1.1	345.32	5MR	-1.6	-3.3	-1.7	0.3
149.94 3.58 -0.9	355.14	15MR	-1.3	-3.3	-1.6	0.3
147.66 3.14 -0.7	4.90	25MR	-0.3	-3.3	-1.6	0.3
146.67 2.68 -0.5	14.62	4AP	1.0	-3.3	-1.5	0.3
146.98 2.24 -0.3	24.28	14AP	2.4	-3.3	-1.5	0.3
148.43 1.82 -0.1	33.90	24AP	3.7	-3.4	-1.5	0.2
150.86 1.45 0.1	43.48	4MY	2.1	-3.4	-1.5	0.2
154.09 1.11 0.3	53.03	14MY	1.1	-3.4	-1.4	0.2
157.98 0.81 0.4	62.57	24MY	0.3	-3.4	-1.4	0.2
162.42 0.54 0.6	72.09	3JN	-0.7	-3.5	-1.4	0.3
167.32 0.30 0.7	81.61	13JN	-1.6	-3.5	-1.4	0.3
172.60 0.08 0.8	91.15	23JN	-1.3	-3.6	-1.5	0.3
178.22 -0.12 0.8	100.69	3JL	-0.5	-3.7	-1.5	0.3
184.12 -0.30 0.9	110.26	13JL	0.0	-3.7	-1.5	0.3
190.28 -0.46 1.0	119.87	23JL	0.4	-3.8	-1.5	0.3
196.68 -0.61 1.0	129.52	2AU	-0.3	-3.9	-1.6	0.3
203.28 -0.74 1.1	139.21	12AU	1.2	-4.0	-1.6	0.3
210.07 -0.86 1.1	148.96	22AU	2.6	-4.1	-1.7	0.2
217.03 -0.96 1.1	158.76	1SE	1.5	-4.2	-1.7	0.2
224.15 -1.05 1.2	168.62	11SE	-0.4	-4.3	-1.8	0.1
231.42 -1.13 1.2	178.54	21SE	-0.9	-4.3	-1.9	0.1
238.81 -1.19 1.2	188.52	1OC	-0.9	-4.1	-1.9	0.0
246.32 -1.23 1.2	198.55	11OC	-0.9	-3.6	-2.0	-0.1
253.92 -1.27 1.3	208.63	21OC	-0.6	-3.2	-2.1	-0.1
261.61 -1.28 1.3	218.76	31OC	-0.5	-3.9	-2.1	-0.2
269.36 -1.29 1.3	228.91	10NO	-0.5	-4.3	-2.2	-0.3
277.17 -1.28 1.3	239.10	20NO	-0.4	-4.4	-2.2	-0.3
285.00 -1.25 1.3	249.30	30NO	0.5	-4.3	-2.2	-0.2
292.86 -1.21 1.4	259.51	10DE	3.0	-4.3	-2.2	-0.2
300.71 -1.17 1.4	269.71	20DE	0.8	-4.2	-2.2	-0.1
308.56 -1.11 1.4	279.89	30DE	0.2	-4.1	-2.2	-0.0

♂ LONG LAT MAG	☉ LONG	16.00UT 294	☿	♀	♃	♄
316.37 -1.04 1.4	290.05	9JA	0.0	-4.0	-2.1	0.0
324.15 -0.96 1.4	300.17	19JA	-0.1	-3.9	-2.1	0.1
331.87 -0.88 1.5	310.24	29JA	-0.4	-3.8	-2.0	0.2
339.54 -0.79 1.5	320.27	8FE	-0.9	-3.7	-2.0	0.2
347.14 -0.69 1.5	330.24	18FE	-1.5	-3.6	-1.9	0.3
354.66 -0.59 1.5	340.15	28FE	-1.2	-3.5	-1.8	0.3
2.10 -0.49 1.6	350.00	10MR	-0.2	-3.5	-1.8	0.3
9.46 -0.39 1.6	359.80	20MR	1.2	-3.4	-1.7	0.3
16.74 -0.28 1.6	9.54	30MR	3.0	-3.4	-1.6	0.3
23.93 -0.18 1.6	19.22	9AP	2.8	-3.4	-1.6	0.3
31.04 -0.08 1.6	28.87	19AP	1.6	-3.3	-1.5	0.3
38.06 0.03 1.6	38.46	29AP	0.9	-3.3	-1.5	0.3
45.01 0.13 1.7	48.03	9MY	0.2	-3.3	-1.4	0.2
51.88 0.23 1.7	57.58	19MY	-0.7	-3.3	-1.4	0.2
58.67 0.33 1.8	67.10	29MY	-1.6	-3.3	-1.4	0.2
65.39 0.42 1.8	76.63	8JN	-1.3	-3.3	-1.3	0.2
72.04 0.52 1.9	86.15	18JN	-0.5	-3.3	-1.3	0.3
78.63 0.61 1.9	95.69	28JN	0.1	-3.4	-1.3	0.3
85.16 0.70 1.9	105.24	8JL	0.5	-3.4	-1.3	0.3
91.63 0.79 1.9	114.83	18JL	0.9	-3.4	-1.3	0.3
98.04 0.88 1.9	124.46	28JL	1.5	-3.4	-1.4	0.3
104.40 0.96 1.9	134.12	7AU	3.1	-3.5	-1.4	0.3
110.71 1.04 1.9	143.84	17AU	1.3	-3.5	-1.4	0.3
116.96 1.13 1.9	153.61	27AU	-0.4	-3.4	-1.4	0.3
123.15 1.21 1.9	163.44	6SE	-1.0	-3.4	-1.5	0.3
129.29 1.29 1.9	173.33	16SE	-1.1	-3.4	-1.5	0.2
135.36 1.37 1.9	183.28	26SE	-0.8	-3.4	-1.6	0.2
141.37 1.45 1.8	193.28	6OC	-0.5	-3.3	-1.6	0.1
147.29 1.53 1.7	203.34	16OC	-0.3	-3.3	-1.7	0.0
153.13 1.61 1.7	213.44	26OC	-0.3	-3.3	-1.8	-0.0
158.86 1.70 1.6	223.58	5NO	-0.2	-3.3	-1.8	-0.1
164.48 1.78 1.5	233.76	15NO	0.7	-3.3	-1.9	-0.2
169.94 1.87 1.4	243.95	25NO	2.9	-3.4	-2.0	-0.3
175.24 1.96 1.3	254.16	5DE	0.5	-3.4	-2.0	-0.3
180.32 2.04 1.1	264.36	15DE	-0.0	-3.4	-2.1	-0.2
185.14 2.13 1.0	274.55	25DE	-0.1	-3.5	-2.1	-0.1
		295				
189.64 2.22 0.8	284.73	4JA	-0.2	-3.5	-2.1	-0.1
193.75 2.31 0.6	294.87	14JA	-0.5	-3.5	-2.1	-0.0
197.37 2.40 0.4	304.96	24JA	-0.9	-3.6	-2.1	0.1
200.37 2.47 0.1	315.02	3FE	-1.3	-3.7	-2.1	0.1
202.60 2.52 -0.2	325.02	13FE	-1.1	-3.7	-2.0	0.2
203.89 2.55 -0.5	334.96	23FE	-0.2	-3.8	-2.0	0.2
204.06 2.52 -0.8	344.85	5MR	1.5	-3.9	-1.9	0.3
202.97 2.42 -1.1	354.67	15MR	3.5	-4.0	-1.8	0.3
200.66 2.22 -1.4	4.44	25MR	2.0	-4.1	-1.8	0.3
197.40 1.90 -1.6	14.15	4AP	1.2	-4.1	-1.7	0.3
193.78 1.48 -1.6	23.82	14AP	0.6	-4.2	-1.6	0.3
190.53 0.99 -1.5	33.43	24AP	0.1	-4.2	-1.6	0.3
188.25 0.50 -1.3	43.02	4MY	-0.7	-4.1	-1.5	0.3
187.29 0.04 -1.1	52.57	14MY	-1.6	-3.6	-1.5	0.3
187.68 -0.36 -0.9	62.11	24MY	-1.4	-2.8	-1.4	0.3
189.30 -0.70 -0.6	71.63	3JN	-0.4	-3.4	-1.4	0.2
192.00 -0.99 -0.5	81.15	13JN	0.2	-3.9	-1.3	0.2
195.57 -1.22 -0.3	90.68	23JN	0.7	-4.2	-1.3	0.3
199.88 -1.42 -0.1	100.23	3JL	1.2	-4.2	-1.3	0.3
204.79 -1.57 -0.0	109.80	13JL	2.0	-4.2	-1.3	0.3
210.20 -1.69 0.1	119.40	23JL	3.2	-4.1	-1.3	0.4
216.03 -1.78 0.2	129.05	2AU	1.1	-4.0	-1.2	0.4
222.21 -1.85 0.3	138.74	12AU	-0.5	-3.9	-1.3	0.4
228.69 -1.89 0.4	148.48	22AU	-1.2	-3.8	-1.3	0.4
235.42 -1.91 0.5	158.28	1SE	-1.2	-3.7	-1.3	0.4
242.37 -1.91 0.6	168.14	11SE	-0.7	-3.7	-1.3	0.4
249.50 -1.88 0.7	178.06	21SE	-0.4	-3.6	-1.3	0.3
256.78 -1.84 0.7	188.03	1OC	-0.2	-3.6	-1.4	0.3
264.18 -1.78 0.8	198.06	11OC	-0.1	-3.5	-1.4	0.2
271.68 -1.71 0.9	208.14	21OC	-0.0	-3.4	-1.5	0.2
279.26 -1.62 1.0	218.27	31OC	0.9	-3.5	-1.5	0.1
286.88 -1.52 1.0	228.42	10NO	2.7	-3.4	-1.6	0.0
294.53 -1.41 1.1	238.61	20NO	0.3	-3.4	-1.7	-0.0
302.19 -1.30 1.1	248.81	30NO	-0.2	-3.4	-1.7	-0.1
309.85 -1.17 1.2	259.01	10DE	-0.3	-3.4	-1.8	-0.2
317.48 -1.05 1.3	269.21	20DE	-0.3	-3.3	-1.9	-0.2
325.08 -0.92 1.3	279.40	30DE	-0.5	-3.3	-1.9	-0.1

♂ LONG	♂ LAT	♂ MAG	☉ LONG	16.00UT	☿	♀	♃	♄
				296	MAGNITUDES			
332.63	-0.79	1.4	289.56	9JA	-0.9	-3.3	-2.0	-0.1
340.13	-0.66	1.5	299.68	19JA	-1.2	-3.3	-2.0	-0.0
347.56	-0.53	1.5	309.76	29JA	-1.0	-3.4	-2.0	0.1
354.93	-0.40	1.6	319.78	8FE	-0.1	-3.4	-2.0	0.1
2.22	-0.28	1.6	329.76	18FE	1.8	-3.4	-2.0	0.2
9.44	-0.16	1.7	339.67	28FE	2.9	-3.4	-2.0	0.2
16.58	-0.05	1.7	349.52	9MR	1.5	-3.5	-2.0	0.3
23.64	0.06	1.7	359.32	19MR	0.9	-3.5	-1.9	0.3
30.64	0.17	1.8	9.07	29MR	0.5	-3.5	-1.9	0.4
37.55	0.27	1.8	18.75	8AP	0.0	-3.4	-1.8	0.4
44.40	0.36	1.8	28.40	18AP	-0.7	-3.4	-1.8	0.4
51.18	0.45	1.8	38.00	28AP	-1.6	-3.4	-1.7	0.4
57.90	0.54	1.8	47.57	8MY	-1.4	-3.4	-1.6	0.4
64.57	0.62	1.9	57.11	18MY	-0.4	-3.3	-1.6	0.4
71.18	0.70	1.9	66.64	28MY	0.3	-3.3	-1.5	0.4
77.73	0.77	1.9	76.16	7JN	0.9	-3.3	-1.4	0.4
84.25	0.84	1.9	85.68	17JN	1.5	-3.3	-1.4	0.3
90.73	0.90	1.9	95.22	27JN	2.6	-3.3	-1.4	0.3
97.17	0.96	1.9	104.77	7JL	2.8	-3.4	-1.3	0.4
103.58	1.02	2.0	114.36	17JL	0.9	-3.4	-1.3	0.4
109.97	1.07	2.0	123.99	27JL	-0.5	-3.4	-1.3	0.5
116.33	1.11	2.0	133.65	6AU	-1.3	-3.5	-1.2	0.5
122.67	1.16	2.0	143.37	16AU	-1.3	-3.5	-1.2	0.5
128.99	1.20	2.0	153.14	26AU	-0.7	-3.6	-1.2	0.5
135.29	1.24	2.0	162.96	5SE	-0.3	-3.6	-1.2	0.5
141.58	1.27	2.0	172.85	15SE	-0.1	-3.7	-1.2	0.5
147.86	1.30	2.0	182.80	25SE	0.0	-3.8	-1.2	0.5
154.11	1.33	2.0	192.79	5OC	0.2	-3.9	-1.3	0.4
160.35	1.35	1.9	202.85	15OC	1.2	-4.0	-1.3	0.4
166.57	1.37	1.9	212.95	25OC	2.4	-4.1	-1.3	0.4
172.76	1.39	1.8	223.09	4NO	0.1	-4.2	-1.4	0.3
178.92	1.40	1.7	233.27	14NO	-0.4	-4.3	-1.4	0.2
185.04	1.40	1.7	243.46	24NO	-0.4	-4.4	-1.5	0.2
191.12	1.40	1.6	253.66	4DE	-0.4	-4.4	-1.5	0.1
197.14	1.39	1.5	263.87	14DE	-0.6	-4.2	-1.6	0.0
203.10	1.37	1.3	274.06	24DE	-0.9	-3.6	-1.7	-0.1
				297				
208.97	1.34	1.2	284.23	3JA	-1.0	-3.2	-1.7	-0.1
214.74	1.30	1.1	294.38	13JA	-0.9	-3.9	-1.8	-0.0
220.38	1.24	0.9	304.48	23JA	-0.0	-4.2	-1.9	0.0
225.87	1.16	0.8	314.53	2FE	2.2	-4.3	-1.9	0.1
231.17	1.05	0.6	324.53	12FE	2.3	-4.3	-2.0	0.2
236.23	0.92	0.4	334.48	22FE	1.1	-4.2	-2.0	0.2
240.99	0.75	0.1	344.36	4MR	0.6	-4.1	-2.0	0.3
245.38	0.53	-0.1	354.20	14MR	0.3	-4.0	-2.0	0.3
249.28	0.26	-0.4	3.97	24MR	-0.1	-3.9	-2.0	0.4
252.58	-0.08	-0.6	13.68	3AP	-0.7	-3.8	-2.0	0.4
255.11	-0.50	-0.9	23.35	13AP	-1.6	-3.7	-2.0	0.5
256.70	-1.01	-1.2	32.97	23AP	-1.4	-3.6	-1.9	0.5
257.17	-1.61	-1.6	42.56	3MY	-0.4	-3.5	-1.9	0.5
256.39	-2.29	-1.9	52.12	13MY	0.5	-3.5	-1.8	0.5
254.45	-3.01	-2.2	61.65	23MY	1.2	-3.4	-1.7	0.5
251.71	-3.68	-2.4	71.17	2JN	2.0	-3.4	-1.7	0.5
248.77	-4.21	-2.3	80.70	12JN	3.3	-3.4	-1.6	0.5
246.41	-4.55	-2.2	90.23	22JN	2.2	-3.3	-1.5	0.5
245.16	-4.69	-1.9	99.77	2JL	0.7	-3.3	-1.5	0.5
245.26	-4.66	-1.7	109.34	12JL	-0.6	-3.3	-1.4	0.5
246.72	-4.53	-1.5	118.94	22JL	-1.4	-3.3	-1.4	0.5
249.36	-4.33	-1.2	128.58	1AU	-1.3	-3.3	-1.3	0.6
253.00	-4.08	-1.0	138.27	11AU	-0.6	-3.3	-1.3	0.6
257.46	-3.80	-0.8	148.01	21AU	-0.2	-3.4	-1.3	0.7
262.55	-3.52	-0.6	157.81	31AU	0.0	-3.4	-1.2	0.7
268.15	-3.22	-0.4	167.66	10SE	0.2	-3.4	-1.2	0.7
274.14	-2.93	-0.2	177.57	20SE	0.4	-3.4	-1.2	0.7
280.44	-2.64	-0.1	187.55	30SE	1.5	-3.5	-1.2	0.7
286.98	-2.36	0.1	197.57	10OC	2.2	-3.5	-1.2	0.7
293.70	-2.08	0.2	207.65	20OC	-0.0	-3.5	-1.2	0.6
300.55	-1.82	0.4	217.77	30OC	-0.5	-3.5	-1.3	0.6
307.50	-1.56	0.5	227.93	9NO	-0.5	-3.4	-1.3	0.5
314.50	-1.32	0.7	238.11	19NO	-0.6	-3.4	-1.3	0.5
321.54	-1.09	0.8	248.31	29NO	-0.7	-3.4	-1.4	0.4
328.60	-0.88	0.9	258.52	9DE	-0.8	-3.4	-1.4	0.3
335.66	-0.68	1.0	268.72	19DE	-0.9	-3.3	-1.5	0.3
342.70	-0.50	1.1	278.90	29DE	-0.8	-3.3	-1.5	0.2

♂ LONG	♂ LAT	♂ MAG	☉ LONG	16.00UT	☿	♀	♃	♄
				298	MAGNITUDES			
349.71	-0.32	1.2	289.06	8JA	0.1	-3.3	-1.6	0.1
356.69	-0.16	1.3	299.19	18JA	2.5	-3.3	-1.7	0.1
3.63	-0.02	1.4	309.27	28JA	1.7	-3.4	-1.7	0.2
10.53	0.12	1.5	319.30	7FE	0.8	-3.4	-1.8	0.2
17.37	0.24	1.6	329.27	17FE	0.4	-3.4	-1.9	0.3
24.17	0.35	1.7	339.19	27FE	0.2	-3.4	-1.9	0.3
30.92	0.46	1.7	349.05	9MR	-0.2	-3.4	-2.0	0.4
37.61	0.55	1.8	358.85	19MR	-0.8	-3.5	-2.0	0.5
44.26	0.64	1.8	8.60	29MR	-1.6	-3.5	-2.0	0.5
50.87	0.71	1.9	18.29	8AP	-1.4	-3.6	-2.1	0.6
57.43	0.78	1.9	27.93	18AP	-0.4	-3.6	-2.1	0.6
63.95	0.85	1.9	37.54	28AP	0.6	-3.7	-2.0	0.6
70.44	0.90	2.0	47.11	8MY	1.5	-3.8	-2.0	0.7
76.89	0.95	2.0	56.65	18MY	2.7	-3.9	-2.0	0.7
83.32	1.00	2.0	66.19	28MY	3.2	-4.0	-1.9	0.7
89.73	1.04	2.0	75.71	7JN	1.8	-4.1	-1.9	0.7
96.12	1.07	2.0	85.23	17JN	0.6	-4.1	-1.8	0.7
102.49	1.10	2.0	94.77	27JN	-0.6	-4.2	-1.8	0.7
108.86	1.12	2.0	104.32	7JL	-1.5	-4.2	-1.7	0.7
115.23	1.14	2.0	113.90	17JL	-1.3	-4.0	-1.6	0.7
121.60	1.15	2.0	123.52	27JL	-0.6	-3.5	-1.6	0.7
127.97	1.16	2.0	133.19	6AU	-0.1	-3.2	-1.5	0.8
134.35	1.17	2.0	142.89	16AU	0.1	-3.8	-1.5	0.8
140.74	1.17	2.0	152.66	26AU	0.3	-4.2	-1.4	0.8
147.15	1.16	2.0	162.49	5SE	0.7	-4.3	-1.4	0.9
153.57	1.16	2.0	172.37	15SE	1.9	-4.3	-1.3	0.9
160.02	1.14	2.0	182.31	25SE	2.0	-4.2	-1.3	0.9
166.48	1.12	2.0	192.31	5OC	-0.2	-4.1	-1.3	0.9
172.97	1.10	2.0	202.36	15OC	-0.7	-4.0	-1.3	0.9
179.49	1.07	1.9	212.46	25OC	-0.7	-4.0	-1.3	0.9
186.02	1.03	1.9	222.60	4NO	-0.7	-3.9	-1.3	0.9
192.58	0.98	1.8	232.77	14NO	-0.8	-3.8	-1.3	0.8
199.17	0.93	1.8	242.97	24NO	-0.7	-3.7	-1.3	0.8
205.77	0.87	1.7	253.17	4DE	-0.7	-3.6	-1.3	0.7
212.40	0.80	1.6	263.37	14DE	-0.6	-3.6	-1.4	0.6
219.04	0.71	1.6	273.57	24DE	0.2	-3.5	-1.4	0.6
				299				
225.70	0.62	1.5	283.75	3JA	2.8	-3.5	-1.5	0.5
232.37	0.51	1.4	293.89	13JA	1.3	-3.4	-1.5	0.4
239.05	0.39	1.3	304.00	23JA	0.5	-3.4	-1.6	0.4
245.74	0.24	1.2	314.05	2FE	0.3	-3.4	-1.6	0.4
252.42	0.08	1.0	324.05	12FE	0.1	-3.3	-1.7	0.4
259.09	-0.10	0.9	334.01	22FE	-0.2	-3.3	-1.8	0.5
265.75	-0.30	0.8	343.89	4MR	-0.8	-3.3	-1.8	0.5
272.37	-0.52	0.6	353.72	14MR	-1.6	-3.3	-1.9	0.6
278.94	-0.78	0.5	3.50	24MR	-1.3	-3.3	-2.0	0.6
285.44	-1.06	0.3	13.22	3AP	-0.3	-3.3	-2.0	0.7
291.84	-1.37	0.2	22.88	13AP	0.8	-3.3	-2.1	0.7
298.11	-1.71	-0.0	32.51	23AP	2.0	-3.4	-2.1	0.8
304.19	-2.08	-0.2	42.10	3MY	3.6	-3.4	-2.1	0.8
310.04	-2.48	-0.4	51.65	13MY	2.5	-3.4	-2.2	0.9
315.55	-2.93	-0.6	61.19	23MY	1.4	-3.5	-2.2	0.9
320.65	-3.40	-0.8	70.72	2JN	0.4	-3.5	-2.1	0.9
325.20	-3.91	-1.0	80.24	12JN	-0.6	-3.4	-2.1	0.9
329.03	-4.44	-1.3	89.77	22JN	-1.5	-3.4	-2.1	1.0
331.96	-4.98	-1.5	99.31	2JL	-1.3	-3.4	-2.0	1.0
333.78	-5.51	-1.8	108.87	12JL	-0.6	-3.4	-2.0	1.0
334.29	-5.98	-2.0	118.47	22JL	-0.0	-3.3	-1.9	1.0
333.43	-6.30	-2.3	128.11	1AU	0.3	-3.3	-1.8	1.0
331.35	-6.40	-2.5	137.80	11AU	0.5	-3.3	-1.8	1.0
328.57	-6.21	-2.6	147.54	21AU	0.9	-3.3	-1.7	1.0
325.85	-5.72	-2.4	157.33	31AU	2.3	-3.3	-1.7	1.1
323.88	-5.04	-2.1	167.18	10SE	1.8	-3.3	-1.6	1.1
323.12	-4.26	-1.8	177.09	20SE	-0.3	-3.4	-1.5	1.2
323.69	-3.50	-1.5	187.06	30SE	-0.8	-3.4	-1.5	1.2
325.45	-2.80	-1.2	197.08	10OC	-0.8	-3.4	-1.5	1.2
328.24	-2.18	-0.9	207.16	20OC	-0.8	-3.5	-1.4	1.2
331.85	-1.65	-0.6	217.27	30OC	-0.7	-3.5	-1.4	1.2
336.09	-1.19	-0.3	227.43	9NO	-0.5	-3.6	-1.4	1.2
340.83	-0.80	-0.1	237.62	19NO	-0.5	-3.7	-1.4	1.1
345.95	-0.48	0.2	247.81	29NO	-0.5	-3.7	-1.4	1.1
351.35	-0.20	0.4	258.02	9DE	0.3	-3.8	-1.4	1.0
356.98	0.04	0.6	268.23	19DE	3.0	-3.9	-1.4	1.0
2.78	0.24	0.7	278.41	29DE	1.0	-4.0	-1.4	0.9

♂ LONG	♂ LAT	♂ MAG	☉ LONG	16.00UT 300	☿	♀	♃	♄
8.70	0.41	0.9	288.57	8JA	0.3	-4.1	-1.4	0.8
14.71	0.56	1.1	298.70	18JA	0.1	-4.2	-1.5	0.7
20.80	0.68	1.2	308.78	28JA	-0.0	-4.3	-1.5	0.7
26.93	0.79	1.3	318.82	7FE	-0.3	-4.3	-1.6	0.6
33.10	0.88	1.4	328.80	17FE	-0.8	-4.3	-1.6	0.6
39.30	0.95	1.5	338.71	27FE	-1.5	-4.0	-1.7	0.7
45.51	1.01	1.6	348.58	8MR	-1.3	-3.4	-1.7	0.7
51.73	1.07	1.7	358.38	18MR	-0.3	-3.3	-1.8	0.8
57.95	1.11	1.8	8.12	28MR	1.0	-3.8	-1.9	0.8
64.17	1.15	1.8	17.82	7AP	2.6	-4.1	-2.0	0.9
70.40	1.17	1.9	27.47	17AP	3.3	-4.2	-2.0	0.9
76.63	1.19	1.9	37.07	27AP	1.9	-4.2	-2.1	1.0
82.87	1.21	2.0	46.65	7MY	1.0	-4.1	-2.1	1.0
89.11	1.22	2.0	56.19	17MY	0.3	-4.0	-2.2	1.1
95.35	1.22	2.0	65.72	27MY	-0.6	-3.9	-2.2	1.1
101.61	1.22	2.0	75.25	6JN	-1.6	-3.8	-2.3	1.2
107.88	1.21	2.0	84.77	16JN	-1.4	-3.7	-2.3	1.2
114.16	1.20	2.0	94.30	26JN	-0.5	-3.6	-2.3	1.2
120.47	1.18	2.0	103.86	6JL	0.0	-3.6	-2.3	1.2
126.81	1.16	2.0	113.44	16JL	0.4	-3.5	-2.2	1.3
133.17	1.14	2.0	123.06	26JL	0.7	-3.5	-2.2	1.3
139.57	1.11	2.0	132.72	5AU	1.3	-3.4	-2.1	1.3
146.00	1.08	1.9	142.43	15AU	2.8	-3.4	-2.1	1.3
152.47	1.04	1.9	152.19	25AU	1.5	-3.4	-2.1	1.3
158.99	1.00	1.9	162.01	4SE	-0.4	-3.4	-2.0	1.3
165.55	0.95	1.9	171.89	14SE	-1.0	-3.4	-1.9	1.4
172.16	0.90	1.9	181.83	24SE	-1.0	-3.4	-1.8	1.4
178.83	0.84	1.9	191.83	4OC	-0.9	-3.4	-1.8	1.4
185.55	0.77	1.9	201.87	14OC	-0.6	-3.4	-1.7	1.3
192.32	0.70	1.9	211.97	24OC	-0.4	-3.4	-1.7	1.3
199.15	0.63	1.8	222.11	3NO	-0.4	-3.4	-1.6	1.3
206.04	0.55	1.8	232.27	13NO	-0.3	-3.4	-1.6	1.2
212.98	0.46	1.8	242.46	23NO	0.5	-3.4	-1.5	1.2
219.98	0.36	1.7	252.67	3DE	3.1	-3.4	-1.5	1.2
227.04	0.26	1.7	262.87	13DE	0.7	-3.5	-1.5	1.1
234.16	0.14	1.6	273.07	23DE	0.1	-3.5	-1.5	1.1

301

♂ LONG	♂ LAT	♂ MAG	☉ LONG	16.00UT	☿	♀	♃	♄
241.32	0.02	1.6	283.25	2JA	-0.0	-3.5	-1.5	1.0
248.54	-0.11	1.5	293.39	12JA	-0.1	-3.5	-1.5	1.0
255.81	-0.24	1.5	303.50	22JA	-0.4	-3.4	-1.5	0.9
263.12	-0.39	1.4	313.56	1FE	-0.9	-3.4	-1.5	0.9
270.47	-0.54	1.3	323.57	11FE	-1.4	-3.4	-1.5	0.9
277.86	-0.69	1.2	333.52	21FE	-1.2	-3.4	-1.6	0.8
285.27	-0.86	1.2	343.41	3MR	-0.2	-3.3	-1.6	0.9
292.70	-1.02	1.1	353.25	13MR	1.3	-3.3	-1.7	1.0
300.13	-1.19	1.0	3.02	23MR	3.2	-3.3	-1.7	1.0
307.55	-1.36	0.9	12.75	2AP	2.5	-3.3	-1.8	1.1
314.95	-1.52	0.8	22.42	12AP	1.4	-3.3	-1.8	1.1
322.31	-1.69	0.7	32.04	22AP	0.8	-3.4	-1.9	1.2
329.60	-1.84	0.7	41.63	2MY	0.2	-3.4	-2.0	1.2
336.81	-1.99	0.6	51.19	12MY	-0.7	-3.4	-2.1	1.3
343.90	-2.13	0.5	60.73	22MY	-1.6	-3.4	-2.1	1.4
350.85	-2.25	0.4	70.26	1JN	-1.4	-3.5	-2.2	1.4
357.62	-2.36	0.3	79.78	11JN	-0.5	-3.5	-2.3	1.4
4.17	-2.46	0.2	89.30	21JN	0.1	-3.6	-2.3	1.4
10.44	-2.53	0.1	98.85	1JL	0.6	-3.7	-2.4	1.4
16.38	-2.59	-0.1	108.41	11JL	1.0	-3.7	-2.4	1.4
21.91	-2.62	-0.2	118.01	21JL	1.7	-3.8	-2.4	1.4
26.93	-2.63	-0.3	127.65	31JL	3.2	-3.9	-2.4	1.3
31.32	-2.61	-0.5	137.33	10AU	1.3	-4.0	-2.4	1.3
34.94	-2.55	-0.7	147.07	20AU	-0.4	-4.1	-2.4	1.2
37.59	-2.44	-0.9	156.86	30AU	-1.1	-4.2	-2.3	1.2
39.07	-2.28	-1.1	166.70	9SE	-1.1	-4.3	-2.3	1.2
39.18	-2.04	-1.3	176.61	19SE	-0.8	-4.3	-2.2	1.2
37.82	-1.70	-1.5	186.58	29SE	-0.5	-4.1	-2.1	1.2
35.12	-1.27	-1.7	196.60	9OC	-0.3	-3.5	-2.1	1.2
31.56	-0.77	-1.8	206.67	19OC	-0.2	-3.2	-2.0	1.2
27.91	-0.26	-1.7	216.79	29OC	-0.1	-3.9	-1.9	1.2
24.96	0.22	-1.4	226.94	8NO	0.7	-4.2	-1.9	1.1
23.21	0.62	-1.1	237.12	18NO	2.9	-4.4	-1.8	1.1
22.82	0.94	-0.8	247.32	28NO	0.4	-4.4	-1.7	1.1
23.71	1.17	-0.5	257.52	8DE	-0.1	-4.3	-1.7	1.0
25.67	1.34	-0.2	267.73	18DE	-0.2	-4.2	-1.6	1.0
28.50	1.46	0.1	277.91	28DE	-0.3	-4.1	-1.6	1.0

♂ LONG	♂ LAT	♂ MAG	☉ LONG	16.00UT 302	☿	♀	♃	♄
32.02	1.54	0.3	288.07	7JA	-0.5	-4.0	-1.6	0.9
36.06	1.59	0.6	298.21	17JA	-0.9	-3.9	-1.6	0.9
40.53	1.63	0.8	308.29	27JA	-1.3	-3.8	-1.6	0.8
45.32	1.64	0.9	318.33	6FE	-1.1	-3.7	-1.6	0.8
50.36	1.65	1.1	328.31	16FE	-0.2	-3.6	-1.6	0.7
55.60	1.64	1.2	338.24	26FE	1.6	-3.5	-1.6	0.7
61.00	1.63	1.4	348.10	8MR	3.3	-3.5	-1.6	0.7
66.54	1.61	1.5	357.91	18MR	1.8	-3.4	-1.6	0.7
72.18	1.59	1.6	7.66	28MR	1.0	-3.4	-1.6	0.8
77.91	1.56	1.7	17.36	7AP	0.6	-3.4	-1.7	0.9
83.72	1.52	1.7	27.01	17AP	0.1	-3.3	-1.7	0.9
89.59	1.49	1.8	36.62	27AP	-0.7	-3.3	-1.8	1.0
95.52	1.45	1.9	46.19	7MY	-1.6	-3.3	-1.8	1.1
101.51	1.41	1.9	55.74	17MY	-1.4	-3.3	-1.9	1.1
107.56	1.36	1.9	65.27	27MY	-0.5	-3.3	-2.0	1.2
113.65	1.31	2.0	74.79	6JN	0.2	-3.3	-2.0	1.2
119.80	1.26	2.0	84.31	16JN	0.7	-3.3	-2.1	1.2
126.00	1.20	2.0	93.85	26JN	1.3	-3.4	-2.2	1.2
132.25	1.15	2.0	103.40	6JL	2.2	-3.4	-2.2	1.2
138.57	1.09	2.0	112.98	16JL	3.1	-3.4	-2.3	1.2
144.94	1.02	2.0	122.59	26JL	1.1	-3.5	-2.4	1.2
151.37	0.96	2.0	132.25	5AU	-0.5	-3.5	-2.4	1.2
157.87	0.89	1.9	141.96	15AU	-1.2	-3.5	-2.4	1.2
164.44	0.81	1.9	151.72	25AU	-1.3	-3.4	-2.5	1.1
171.07	0.73	1.9	161.53	4SE	-0.7	-3.4	-2.5	1.1
177.79	0.65	1.8	171.41	14SE	-0.4	-3.4	-2.4	1.0
184.57	0.57	1.8	181.34	24SE	-0.2	-3.4	-2.4	1.0
191.43	0.48	1.7	191.33	4OC	-0.1	-3.3	-2.4	1.0
198.37	0.39	1.7	201.38	14OC	0.1	-3.3	-2.3	1.0
205.38	0.29	1.7	211.47	24OC	1.0	-3.3	-2.2	1.0
212.48	0.20	1.7	221.61	3NO	2.7	-3.3	-2.2	1.0
219.65	0.09	1.7	231.78	13NO	0.2	-3.3	-2.1	1.0
226.90	-0.01	1.7	241.97	23NO	-0.3	-3.4	-2.0	1.0
234.22	-0.12	1.6	252.17	3DE	-0.3	-3.4	-2.0	1.0
241.61	-0.23	1.6	262.38	13DE	-0.4	-3.4	-1.9	0.9
249.07	-0.35	1.6	272.57	23DE	-0.6	-3.5	-1.8	0.9

303

♂ LONG	♂ LAT	♂ MAG	☉ LONG	16.00UT	☿	♀	♃	♄
256.59	-0.46	1.6	282.75	2JA	-0.9	-3.5	-1.8	0.9
264.17	-0.58	1.5	292.90	12JA	-1.1	-3.5	-1.7	0.8
271.79	-0.69	1.5	303.00	22JA	-1.0	-3.6	-1.7	0.8
279.46	-0.80	1.5	313.07	1FE	-0.1	-3.7	-1.6	0.7
287.15	-0.91	1.4	323.08	11FE	1.9	-3.7	-1.6	0.7
294.87	-1.01	1.4	333.03	21FE	2.7	-3.8	-1.6	0.6
302.60	-1.11	1.3	342.93	3MR	1.3	-3.9	-1.6	0.6
310.32	-1.20	1.3	352.77	13MR	0.8	-4.0	-1.6	0.5
318.03	-1.28	1.3	2.55	23MR	0.4	-4.1	-1.6	0.5
325.72	-1.35	1.2	12.28	2AP	-0.0	-4.1	-1.6	0.6
333.36	-1.40	1.2	21.95	12AP	-0.7	-4.2	-1.6	0.7
340.95	-1.45	1.2	31.58	22AP	-1.6	-4.2	-1.6	0.7
348.47	-1.47	1.1	41.17	2MY	-1.4	-4.2	-1.6	0.8
355.90	-1.49	1.1	50.73	12MY	-0.4	-3.6	-1.7	0.9
3.23	-1.48	1.1	60.27	22MY	0.4	-2.7	-1.7	0.9
10.46	-1.47	1.0	69.80	1JN	1.0	-3.4	-1.8	1.0
17.55	-1.43	1.0	79.32	11JN	1.7	-3.9	-1.8	1.0
24.49	-1.38	0.9	88.84	21JN	2.9	-4.2	-1.9	1.0
31.28	-1.31	0.9	98.38	1JL	2.6	-4.2	-1.9	1.1
37.87	-1.23	0.8	107.95	11JL	0.9	-4.2	-2.0	1.1
44.25	-1.12	0.8	117.54	21JL	-0.5	-4.1	-2.1	1.1
50.40	-1.00	0.7	127.18	31JL	-1.3	-4.0	-2.1	1.1
56.25	-0.86	0.6	136.86	10AU	-1.3	-3.9	-2.2	1.1
61.79	-0.70	0.5	146.59	20AU	-0.7	-3.8	-2.3	1.0
66.93	-0.52	0.4	156.38	30AU	-0.3	-3.7	-2.3	1.0
71.60	-0.30	0.3	166.23	9SE	-0.0	-3.7	-2.4	1.0
75.71	-0.06	0.1	176.13	19SE	0.1	-3.6	-2.4	0.9
79.11	0.23	-0.0	186.09	29SE	0.3	-3.6	-2.4	0.9
81.64	0.55	-0.2	196.11	9OC	1.2	-3.5	-2.4	0.9
83.13	0.93	-0.4	206.18	19OC	2.5	-3.5	-2.4	0.9
83.35	1.35	-0.7	216.30	29OC	2.9	-3.4	-2.4	0.9
82.18	1.81	-0.9	226.45	8NO	-0.4	-3.4	-2.3	0.9
79.65	2.27	-1.1	236.62	18NO	-0.5	-3.4	-2.3	0.9
76.07	2.69	-1.2	246.83	28NO	-0.5	-3.4	-2.2	0.9
72.11	2.99	-1.2	257.03	8DE	-0.7	-3.4	-2.2	0.9
68.54	3.16	-1.0	267.23	18DE	-0.9	-3.3	-2.1	0.9
65.97	3.20	-0.7	277.42	28DE	-1.0	-3.3	-2.0	0.8

Left table

♂ LONG	♂ LAT	♂ MAG	☉ LONG	16.00UT 304	☿	♀	♃	♄
64.71	3.15	-0.5	287.58	7JA	-0.9	-3.3	-1.9	0.8
64.76	3.04	-0.2	297.71	17JA	-0.0	-3.3	-1.9	0.8
65.96	2.90	0.1	307.80	27JA	2.2	-3.4	-1.8	0.7
68.13	2.76	0.3	317.84	6FE	2.1	-3.4	-1.7	0.7
71.06	2.61	0.6	327.82	16FE	1.0	-3.4	-1.7	0.6
74.61	2.46	0.8	337.75	26FE	0.5	-3.4	-1.6	0.6
78.64	2.32	0.9	347.62	7MR	0.3	-3.5	-1.6	0.5
83.06	2.19	1.1	357.42	17MR	-0.1	-3.5	-1.6	0.5
87.79	2.07	1.2	7.18	27MR	-0.7	-3.5	-1.5	0.4
92.77	1.95	1.3	16.88	6AP	-1.6	-3.4	-1.5	0.5
97.96	1.83	1.5	26.53	16AP	-1.4	-3.4	-1.5	0.5
103.33	1.72	1.5	36.14	26AP	-0.4	-3.4	-1.5	0.6
108.86	1.61	1.6	45.72	6MY	0.5	-3.4	-1.5	0.6
114.52	1.51	1.7	55.27	16MY	1.3	-3.4	-1.5	0.7
120.30	1.40	1.7	64.80	26MY	2.3	-3.3	-1.5	0.8
126.19	1.30	1.8	74.32	5JN	3.4	-3.3	-1.6	0.8
132.19	1.20	1.8	83.84	15JN	2.1	-3.3	-1.6	0.9
138.29	1.10	1.8	93.38	25JN	0.7	-3.3	-1.6	0.9
144.48	0.99	1.9	102.93	5JL	-0.5	-3.4	-1.7	0.9
150.77	0.89	1.9	112.51	15JL	-1.4	-3.4	-1.7	1.0
157.16	0.79	1.9	122.12	25JL	-1.3	-3.4	-1.8	1.0
163.64	0.69	1.9	131.78	4AU	-0.6	-3.5	-1.8	1.0
170.21	0.59	1.8	141.48	14AU	-0.2	-3.5	-1.9	1.0
176.88	0.48	1.8	151.24	24AU	0.1	-3.6	-2.0	1.0
183.65	0.38	1.8	161.05	3SE	0.2	-3.6	-2.0	0.9
190.51	0.27	1.8	170.93	13SE	0.5	-3.7	-2.1	0.9
197.47	0.17	1.7	180.86	23SE	1.6	-3.8	-2.2	0.9
204.53	0.06	1.7	190.85	30C	2.2	-3.9	-2.2	0.8
211.68	-0.05	1.6	200.89	130C	-0.1	-4.0	-2.3	0.8
218.93	-0.15	1.6	210.99	230C	-0.6	-4.1	-2.3	0.8
226.27	-0.26	1.5	221.12	2NO	-0.6	-4.2	-2.3	0.8
233.69	-0.36	1.5	231.28	12NO	-0.6	-4.3	-2.3	0.9
241.19	-0.46	1.4	241.48	22NO	-0.7	-4.4	-2.3	0.9
248.77	-0.56	1.4	251.68	2DE	-0.8	-4.4	-2.3	0.9
256.42	-0.65	1.4	261.88	12DE	-0.8	-4.1	-2.2	0.9
264.13	-0.74	1.4	272.08	22DE	-0.7	-3.6	-2.2	0.8
				305				
271.89	-0.83	1.4	282.26	1JA	0.1	-3.3	-2.1	0.8
279.69	-0.90	1.4	292.40	11JA	2.5	-3.9	-2.1	0.8
287.52	-0.97	1.4	302.52	21JA	1.6	-4.2	-2.0	0.7
295.37	-1.03	1.4	312.58	31JA	0.7	-4.3	-1.9	0.7
303.22	-1.08	1.4	322.59	10FE	0.4	-4.3	-1.8	0.6
311.06	-1.12	1.4	332.55	20FE	0.1	-4.2	-1.8	0.6
318.89	-1.15	1.4	342.45	2MR	-0.2	-4.1	-1.7	0.5
326.68	-1.16	1.4	352.29	12MR	-0.7	-4.0	-1.7	0.5
334.43	-1.16	1.4	2.07	22MR	-1.6	-3.9	-1.6	0.4
342.13	-1.15	1.4	11.80	1AP	-1.4	-3.8	-1.6	0.4
349.76	-1.13	1.4	21.48	11AP	-0.4	-3.7	-1.5	0.3
357.32	-1.09	1.4	31.11	21AP	0.7	-3.6	-1.5	0.4
4.79	-1.05	1.4	40.70	1MY	1.7	-3.5	-1.5	0.5
12.17	-0.99	1.4	50.27	11MY	3.0	-3.5	-1.4	0.5
19.45	-0.91	1.4	59.81	21MY	3.0	-3.4	-1.4	0.6
26.63	-0.83	1.4	69.33	31MY	1.6	-3.4	-1.4	0.6
33.69	-0.74	1.4	78.86	10JN	0.5	-3.4	-1.4	0.7
40.64	-0.64	1.4	88.38	20JN	-0.6	-3.3	-1.4	0.8
47.46	-0.53	1.4	97.92	30JN	-1.5	-3.3	-1.4	0.8
54.15	-0.41	1.4	107.49	10JL	-1.4	-3.3	-1.5	0.8
60.69	-0.28	1.3	117.08	20JL	-0.6	-3.3	-1.5	0.9
67.08	-0.14	1.3	126.71	30JL	-0.1	-3.3	-1.5	0.9
73.31	0.01	1.3	136.39	9AU	0.2	-3.3	-1.6	0.9
79.35	0.17	1.2	146.12	19AU	0.4	-3.4	-1.6	0.9
85.18	0.35	1.2	155.90	29AU	0.8	-3.4	-1.7	0.9
90.77	0.54	1.1	165.75	8SE	2.0	-3.4	-1.7	0.9
96.08	0.74	1.0	175.65	18SE	2.0	-3.4	-1.8	0.9
101.08	0.97	0.9	185.61	28SE	-0.2	-3.5	-1.9	0.8
105.67	1.21	0.8	195.62	80C	-0.7	-3.5	-1.9	0.8
109.79	1.49	0.6	205.69	180C	-0.8	-3.5	-2.0	0.7
113.33	1.79	0.4	215.80	280C	-0.7	-3.5	-2.1	0.8
116.14	2.14	0.3	225.95	7NO	-0.7	-3.4	-2.1	0.8
118.06	2.52	0.0	236.13	17NO	-0.6	-3.4	-2.2	0.8
118.91	2.93	-0.2	246.32	27NO	-0.6	-3.4	-2.2	0.8
118.79	3.37	-0.4	256.53	7DE	-0.6	-3.4	-2.2	0.8
116.74	3.79	-0.7	266.73	17DE	0.2	-3.3	-2.2	0.8
113.76	4.14	-0.9	276.92	27DE	2.8	-3.3	-2.2	0.8

Right table

♂ LONG	♂ LAT	♂ MAG	☉ LONG	16.00UT 306	☿	♀	♃	♄
109.95	4.36	-1.0	287.09	6JA	1.2	-3.3	-2.2	0.8
106.02	4.41	-0.9	297.22	16JA	0.4	-3.3	-2.1	0.8
102.68	4.30	-0.7	307.31	26JA	0.2	-3.4	-2.1	0.7
100.43	4.06	-0.5	317.35	5FE	0.0	-3.4	-2.0	0.7
99.48	3.76	-0.2	327.34	15FE	-0.3	-3.4	-1.9	0.7
99.79	3.44	0.0	337.27	25FE	-0.8	-3.4	-1.9	0.6
101.19	3.13	0.3	347.14	7MR	-1.6	-3.4	-1.8	0.5
103.51	2.83	0.5	356.95	17MR	-1.3	-3.5	-1.7	0.5
106.57	2.56	0.7	6.71	27MR	-0.3	-3.5	-1.7	0.4
110.22	2.31	0.8	16.41	6AP	0.9	-3.6	-1.6	0.4
114.36	2.08	1.0	26.07	16AP	2.2	-3.6	-1.5	0.3
118.88	1.87	1.1	35.68	26AP	3.9	-3.7	-1.5	0.3
123.74	1.67	1.2	45.26	6MY	2.3	-3.8	-1.4	0.4
128.87	1.48	1.3	54.81	16MY	1.2	-3.9	-1.4	0.4
134.24	1.31	1.4	64.34	26MY	0.4	-4.0	-1.4	0.5
139.82	1.15	1.4	73.87	5JN	-0.6	-4.1	-1.3	0.6
145.58	0.99	1.5	83.39	15JN	-1.5	-4.1	-1.3	0.6
151.51	0.84	1.5	92.92	25JN	-1.4	-4.2	-1.3	0.7
157.60	0.69	1.6	102.47	5JL	-0.6	-4.2	-1.3	0.7
163.83	0.55	1.6	112.04	15JL	-0.0	-3.9	-1.3	0.8
170.21	0.42	1.6	121.66	25JL	0.3	-3.5	-1.3	0.8
176.72	0.28	1.6	131.31	4AU	0.6	-3.2	-1.3	0.8
183.36	0.16	1.6	141.01	14AU	1.1	-3.8	-1.4	0.8
190.13	0.03	1.6	150.77	24AU	2.4	-4.2	-1.4	0.9
197.02	-0.09	1.6	160.58	3SE	1.8	-4.3	-1.4	0.9
204.03	-0.20	1.6	170.45	13SE	-0.3	-4.3	-1.5	0.8
211.16	-0.32	1.6	180.38	23SE	-0.9	-4.2	-1.5	0.8
218.40	-0.42	1.6	190.37	30C	-0.9	-4.1	-1.6	0.8
225.74	-0.52	1.5	200.40	130C	-0.9	-4.0	-1.6	0.8
233.19	-0.62	1.5	210.49	230C	-0.6	-4.0	-1.7	0.7
240.73	-0.71	1.5	220.63	2NO	-0.5	-3.9	-1.8	0.7
248.36	-0.79	1.5	230.79	12NO	-0.5	-3.8	-1.8	0.7
256.06	-0.86	1.4	240.98	22NO	-0.4	-3.7	-1.9	0.8
263.83	-0.93	1.4	251.18	2DE	0.3	-3.6	-2.0	0.8
271.64	-0.98	1.4	261.39	12DE	3.0	-3.5	-2.0	0.8
279.50	-1.02	1.4	271.59	22DE	0.9	-3.5	-2.1	0.8
				307				
287.39	-1.06	1.3	281.77	1JA	0.2	-3.5	-2.1	0.8
295.29	-1.08	1.3	291.91	11JA	0.1	-3.4	-2.1	0.8
303.20	-1.09	1.3	302.03	21JA	-0.1	-3.4	-2.1	0.8
311.08	-1.08	1.3	312.10	31JA	-0.3	-3.4	-2.1	0.7
318.95	-1.07	1.3	322.11	10FE	-0.8	-3.3	-2.0	0.7
326.77	-1.04	1.4	332.07	20FE	-1.5	-3.3	-2.0	0.7
334.55	-1.01	1.4	341.98	2MR	-1.3	-3.3	-1.9	0.6
342.27	-0.96	1.4	351.82	12MR	-0.3	-3.3	-1.9	0.6
349.92	-0.90	1.4	1.60	22MR	1.1	-3.3	-1.8	0.5
357.49	-0.84	1.5	11.33	1AP	2.8	-3.3	-1.7	0.4
4.99	-0.76	1.5	21.01	11AP	3.0	-3.3	-1.7	0.4
12.40	-0.68	1.5	30.65	21AP	1.7	-3.4	-1.6	0.3
19.71	-0.59	1.6	40.24	1MY	0.9	-3.4	-1.6	0.3
26.94	-0.50	1.6	49.80	11MY	0.3	-3.4	-1.5	0.3
34.07	-0.40	1.6	59.34	21MY	-0.6	-3.5	-1.4	0.3
41.10	-0.29	1.6	68.87	31MY	-1.6	-3.5	-1.4	0.4
48.04	-0.18	1.7	78.39	10JN	-1.4	-3.4	-1.4	0.5
54.88	-0.07	1.7	87.92	20JN	-0.5	-3.4	-1.3	0.5
61.62	0.05	1.7	97.46	30JN	0.1	-3.4	-1.3	0.6
68.26	0.17	1.7	107.02	10JL	0.5	-3.4	-1.3	0.7
74.80	0.29	1.7	116.61	20JL	0.8	-3.3	-1.2	0.7
81.24	0.41	1.7	126.24	30JL	1.4	-3.3	-1.2	0.7
87.56	0.54	1.7	135.92	9AU	2.9	-3.3	-1.2	0.8
93.77	0.68	1.6	145.65	19AU	1.5	-3.3	-1.2	0.8
99.86	0.82	1.6	155.43	29AU	-0.4	-3.3	-1.3	0.8
105.81	0.96	1.6	165.26	8SE	-1.0	-3.4	-1.3	0.8
111.62	1.11	1.5	175.17	18SE	-1.1	-3.4	-1.3	0.8
117.25	1.27	1.4	185.12	28SE	-0.8	-3.4	-1.3	0.8
122.68	1.45	1.3	195.13	80C	-0.5	-3.4	-1.4	0.8
127.89	1.63	1.2	205.20	180C	-0.4	-3.5	-1.4	0.8
132.81	1.83	1.1	215.31	280C	-0.3	-3.5	-1.5	0.7
137.41	2.04	1.0	225.46	7NO	-0.2	-3.6	-1.5	0.7
141.60	2.27	0.8	235.64	17NO	0.5	-3.7	-1.6	0.7
145.29	2.53	0.7	245.83	27NO	3.1	-3.7	-1.7	0.7
148.36	2.82	0.5	256.04	7DE	0.6	-3.8	-1.7	0.8
150.65	3.12	0.2	266.24	17DE	0.0	-3.9	-1.8	0.8
152.01	3.45	0.0	276.43	27DE	-0.1	-4.0	-1.9	0.8

♂ LONG	♂ LAT	♂ MAG	⊙ LONG	16.00UT 308	☿	♀	♃	♄
					MAGNITUDES			
152.24	3.79	-0.3	286.60	6JA	-0.2	-4.1	-1.9	0.8
151.20	4.10	-0.5	296.73	16JA	-0.4	-4.2	-2.0	0.8
148.88	4.35	-0.8	306.82	26JA	-0.9	-4.3	-2.0	0.8
145.51	4.47	-1.0	316.87	5FE	-1.3	-4.3	-2.0	0.8
141.62	4.42	-1.0	326.86	15FE	-1.2	-4.3	-2.0	0.7
137.94	4.21	-0.9	336.79	25FE	-0.3	-4.0	-2.0	0.7
135.08	3.86	-0.7	346.66	6MR	1.4	-3.4	-2.0	0.7
133.45	3.44	-0.5	356.48	16MR	3.4	-3.3	-2.0	0.6
133.12	3.01	-0.2	6.24	26MR	2.2	-3.9	-1.9	0.5
133.99	2.60	-0.0	15.94	5AP	1.3	-4.1	-1.9	0.5
135.90	2.21	0.2	25.60	15AP	0.7	-4.2	-1.8	0.4
138.68	1.86	0.3	35.21	25AP	0.2	-4.2	-1.7	0.4
142.16	1.55	0.5	44.79	5MY	-0.6	-4.1	-1.7	0.3
146.23	1.26	0.6	54.35	15MY	-1.6	-4.0	-1.6	0.2
150.77	1.00	0.8	63.88	25MY	-1.4	-3.9	-1.5	0.3
155.70	0.77	0.9	73.40	4JN	-0.5	-3.8	-1.5	0.4
160.99	0.55	1.0	82.92	14JN	0.2	-3.7	-1.4	0.4
166.56	0.35	1.0	92.45	24JN	0.6	-3.6	-1.4	0.5
172.40	0.16	1.1	102.00	4JL	1.1	-3.6	-1.3	0.5
178.47	-0.01	1.2	111.58	14JL	1.9	-3.5	-1.3	0.6
184.75	-0.17	1.2	121.19	24JL	3.2	-3.5	-1.3	0.6
191.22	-0.31	1.2	130.84	3AU	1.3	-3.4	-1.2	0.7
197.88	-0.45	1.3	140.54	13AU	-0.4	-3.4	-1.2	0.7
204.69	-0.57	1.3	150.29	23AU	-1.1	-3.4	-1.2	0.8
211.67	-0.69	1.3	160.10	2SE	-1.2	-3.4	-1.2	0.8
218.79	-0.79	1.3	169.97	12SE	-0.8	-3.4	-1.2	0.8
226.04	-0.88	1.3	179.90	22SE	-0.4	-3.4	-1.2	0.8
233.42	-0.97	1.3	189.88	2OC	-0.2	-3.4	-1.2	0.8
240.91	-1.03	1.3	199.92	12OC	-0.2	-3.4	-1.3	0.8
248.49	-1.09	1.4	210.01	22OC	-0.1	-3.4	-1.3	0.8
256.17	-1.13	1.4	220.14	1NO	0.8	-3.4	-1.3	0.7
263.91	-1.16	1.4	230.30	11NO	3.0	-3.4	-1.4	0.7
271.71	-1.18	1.4	240.49	21NO	0.4	-3.4	-1.4	0.7
279.56	-1.18	1.4	250.69	1DE	-0.2	-3.4	-1.5	0.7
287.44	-1.18	1.4	260.90	11DE	-0.2	-3.5	-1.5	0.7
295.32	-1.16	1.4	271.09	21DE	-0.3	-3.5	-1.6	0.8
303.20	-1.12	1.4	281.27	31DE	-0.5	-3.5	-1.7	0.8
				309				
311.07	-1.08	1.4	291.42	10JA	-0.9	-3.5	-1.7	0.8
318.91	-1.02	1.4	301.54	20JA	-1.2	-3.4	-1.8	0.8
326.71	-0.96	1.4	311.61	30JA	-1.1	-3.4	-1.9	0.8
334.45	-0.89	1.4	321.63	9FE	-0.2	-3.4	-1.9	0.8
342.13	-0.81	1.5	331.59	19FE	1.7	-3.4	-2.0	0.8
349.74	-0.72	1.5	341.49	1MR	3.1	-3.3	-2.0	0.7
357.27	-0.63	1.5	351.34	11MR	1.6	-3.3	-2.0	0.7
4.72	-0.54	1.5	1.13	21MR	0.9	-3.3	-2.0	0.6
12.09	-0.44	1.5	10.86	31MR	0.5	-3.3	-2.0	0.6
19.38	-0.34	1.5	20.54	10AP	0.1	-3.3	-2.0	0.5
26.57	-0.24	1.6	30.18	20AP	-0.7	-3.4	-1.9	0.5
33.68	-0.13	1.6	39.78	30AP	-1.6	-3.4	-1.9	0.4
40.71	-0.03	1.7	49.34	10MY	-1.4	-3.4	-1.8	0.3
47.65	0.07	1.7	58.88	20MY	-0.5	-3.4	-1.8	0.3
54.51	0.18	1.8	68.41	30MY	0.3	-3.5	-1.7	0.3
61.29	0.28	1.8	77.93	9JN	0.8	-3.5	-1.7	0.3
67.99	0.38	1.8	87.45	19JN	1.4	-3.6	-1.6	0.4
74.62	0.48	1.9	97.00	29JN	2.5	-3.7	-1.5	0.4
81.18	0.58	1.9	106.56	9JL	2.9	-3.7	-1.5	0.5
87.67	0.68	1.9	116.14	19JL	1.1	-3.8	-1.4	0.5
94.09	0.78	1.9	125.78	29JL	-0.5	-3.9	-1.4	0.6
100.44	0.88	1.9	135.45	8AU	-1.2	-4.0	-1.3	0.7
106.72	0.98	1.9	145.17	18AU	-1.3	-4.1	-1.3	0.7
112.93	1.08	1.9	154.95	28AU	-0.7	-4.2	-1.3	0.7
119.07	1.17	1.8	164.79	7SE	-0.3	-4.3	-1.3	0.8
125.13	1.27	1.8	174.68	17SE	-0.1	-4.3	-1.2	0.8
131.10	1.38	1.8	184.64	27SE	-0.0	-4.1	-1.2	0.8
136.97	1.48	1.7	194.65	7OC	0.1	-3.5	-1.2	0.8
142.74	1.59	1.6	204.71	17OC	1.0	-3.3	-1.2	0.8
148.37	1.70	1.6	214.82	27OC	2.7	-3.9	-1.3	0.8
153.86	1.81	1.5	224.96	6NO	0.2	-4.3	-1.3	0.8
159.17	1.93	1.3	235.14	16NO	-0.3	-4.4	-1.3	0.7
164.26	2.06	1.2	245.34	26NO	-0.4	-4.4	-1.3	0.7
169.10	2.19	1.1	255.54	6DE	-0.4	-4.3	-1.4	0.7
173.61	2.34	0.9	265.75	16DE	-0.6	-4.2	-1.4	0.7
177.74	2.48	0.7	275.94	26DE	-0.9	-4.1	-1.5	0.8

♂ LONG	♂ LAT	♂ MAG	⊙ LONG	16.00UT 310	☿	♀	♃	♄
					MAGNITUDES			
181.37	2.64	0.5	286.11	5JA	-1.0	-4.0	-1.5	0.8
184.40	2.80	0.3	296.24	15JA	-1.0	-3.9	-1.6	0.8
186.67	2.96	0.0	306.34	25JA	-0.1	-3.8	-1.7	0.8
188.01	3.10	-0.2	316.38	4FE	2.0	-3.7	-1.7	0.8
188.24	3.22	-0.5	326.38	14FE	2.5	-3.6	-1.8	0.8
187.21	3.29	-0.8	336.32	24FE	1.2	-3.5	-1.9	0.8
184.94	3.26	-1.1	346.19	6MR	0.7	-3.5	-1.9	0.8
181.66	3.10	-1.3	356.01	16MR	0.4	-3.4	-2.0	0.7
177.92	2.80	-1.4	5.77	26MR	-0.0	-3.4	-2.0	0.7
174.45	2.39	-1.2	15.48	5AP	-0.7	-3.4	-2.1	0.6
171.87	1.91	-1.1	25.14	15AP	-1.6	-3.3	-2.1	0.6
170.55	1.43	-0.9	34.76	25AP	-1.4	-3.3	-2.1	0.5
170.57	0.97	-0.6	44.33	5MY	-0.4	-3.3	-2.1	0.5
171.83	0.57	-0.4	53.89	15MY	0.4	-3.3	-2.0	0.4
174.16	0.21	-0.3	63.42	25MY	1.1	-3.3	-2.0	0.3
177.38	-0.10	-0.1	72.94	4JN	1.9	-3.3	-1.9	0.3
181.34	-0.37	0.1	82.47	14JN	3.1	-3.3	-1.9	0.3
185.91	-0.60	0.2	92.00	24JN	2.4	-3.4	-1.8	0.3
190.98	-0.81	0.3	101.54	4JL	0.9	-3.4	-1.8	0.4
196.48	-0.98	0.4	111.12	14JL	-0.5	-3.5	-1.7	0.4
202.35	-1.13	0.5	120.72	24JL	-1.3	-3.5	-1.6	0.5
208.53	-1.25	0.6	130.37	3AU	-1.4	-3.5	-1.6	0.6
214.99	-1.35	0.7	140.07	13AU	-0.7	-3.5	-1.5	0.6
221.68	-1.44	0.7	149.82	23AU	-0.2	-3.4	-1.5	0.7
228.58	-1.50	0.8	159.62	2SE	-0.0	-3.4	-1.4	0.7
235.66	-1.54	0.9	169.49	12SE	0.1	-3.4	-1.4	0.7
242.91	-1.57	0.9	179.41	22SE	0.4	-3.4	-1.4	0.8
250.28	-1.57	1.0	189.39	2OC	1.3	-3.3	-1.3	0.8
257.78	-1.56	1.0	199.43	12OC	2.5	-3.3	-1.3	0.8
265.36	-1.54	1.1	209.51	22OC	0.0	-3.3	-1.3	0.8
273.02	-1.50	1.1	219.64	1NO	-0.5	-3.3	-1.3	0.8
280.73	-1.44	1.2	229.80	11NO	-0.5	-3.3	-1.3	0.8
288.48	-1.37	1.2	239.99	21NO	-0.5	-3.4	-1.3	0.8
296.24	-1.29	1.2	250.19	1DE	-0.7	-3.4	-1.3	0.7
304.01	-1.21	1.3	260.40	11DE	-0.8	-3.4	-1.4	0.7
311.75	-1.11	1.3	270.60	21DE	-0.9	-3.5	-1.4	0.7
319.47	-1.01	1.4	280.78	31DE	-0.8	-3.5	-1.4	0.8
				311				
327.15	-0.90	1.4	290.94	10JA	-0.0	-3.5	-1.5	0.8
334.77	-0.79	1.5	301.05	20JA	2.3	-3.6	-1.5	0.8
342.33	-0.67	1.5	311.12	30JA	1.9	-3.7	-1.6	0.8
349.82	-0.56	1.5	321.15	9FE	0.8	-3.7	-1.7	0.9
357.24	-0.44	1.6	331.11	19FE	0.5	-3.8	-1.7	0.9
4.59	-0.33	1.6	341.02	1MR	0.2	-3.9	-1.8	0.8
11.85	-0.22	1.7	350.87	11MR	-0.1	-4.0	-1.9	0.8
19.03	-0.11	1.7	0.66	21MR	-0.7	-4.1	-1.9	0.8
26.13	0.00	1.7	10.40	31MR	-1.6	-4.1	-2.0	0.8
33.16	0.10	1.7	20.08	10AP	-1.4	-4.2	-2.1	0.7
40.11	0.20	1.7	29.72	20AP	-0.4	-4.2	-2.1	0.7
46.98	0.30	1.8	39.31	30AP	0.5	-4.0	-2.1	0.6
53.78	0.40	1.8	48.88	10MY	1.4	-3.6	-2.2	0.5
60.52	0.49	1.8	58.42	20MY	2.5	-2.7	-2.2	0.5
67.20	0.57	1.8	67.95	30MY	3.4	-3.4	-2.2	0.4
73.82	0.65	1.8	77.47	9JN	1.9	-4.0	-2.2	0.3
80.38	0.73	1.9	87.00	19JN	0.7	-4.2	-2.1	0.3
86.90	0.81	1.9	96.53	29JN	-0.5	-4.2	-2.1	0.3
93.37	0.88	1.9	106.09	9JL	-1.4	-4.1	-2.0	0.4
99.80	0.95	2.0	115.68	19JL	-1.4	-4.1	-2.0	0.4
106.19	1.01	2.0	125.31	29JL	-0.6	-4.0	-1.9	0.5
112.55	1.07	2.0	134.98	8AU	-0.2	-3.9	-1.8	0.5
118.87	1.13	2.0	144.70	18AU	0.1	-3.8	-1.8	0.6
125.17	1.19	2.0	154.47	28AU	0.3	-3.7	-1.7	0.6
131.43	1.24	2.0	164.31	7SE	0.6	-3.7	-1.7	0.7
137.66	1.30	2.0	174.20	17SE	1.7	-3.6	-1.6	0.7
143.86	1.35	1.9	184.15	27SE	2.2	-3.6	-1.6	0.8
150.02	1.39	1.9	194.16	7OC	-0.1	-3.5	-1.5	0.8
156.14	1.44	1.9	204.22	17OC	-0.6	-3.5	-1.5	0.8
162.22	1.48	1.8	214.32	27OC	-0.7	-3.4	-1.4	0.8
168.24	1.52	1.7	224.47	6NO	-0.7	-3.4	-1.4	0.8
174.21	1.55	1.7	234.64	16NO	-0.8	-3.4	-1.4	0.8
180.10	1.59	1.6	244.84	26NO	-0.7	-3.4	-1.4	0.8
185.90	1.61	1.5	255.05	6DE	-0.7	-3.4	-1.4	0.8
191.60	1.64	1.3	265.25	16DE	-0.7	-3.4	-1.4	0.8
197.17	1.65	1.2	275.45	26DE	0.1	-3.3	-1.4	0.7

♂ LONG	♂ LAT	♂ MAG	☉ LONG	16.00UT	☿	♀	♃	♄
LONG	LAT	MAG	LONG	312	MAGNITUDES			
202.59	1.66	1.1	285.62	5JA	2.6	-3.3	-1.4	0.8
207.81	1.66	0.9	295.75	15JA	1.5	-3.3	-1.5	0.8
212.81	1.64	0.7	305.85	25JA	0.6	-3.4	-1.5	0.9
217.51	1.61	0.5	315.90	4FE	0.3	-3.4	-1.5	0.9
221.85	1.55	0.3	325.89	14FE	0.1	-3.4	-1.6	0.9
225.74	1.46	0.1	335.83	24FE	-0.2	-3.4	-1.6	0.9
229.06	1.33	-0.2	345.71	5MR	-0.7	-3.5	-1.7	0.9
231.66	1.15	-0.5	355.53	15MR	-1.5	-3.5	-1.8	0.9
233.36	0.91	-0.8	5.30	25MR	-1.4	-3.5	-1.8	0.9
233.99	0.59	-1.1	15.01	4AP	-0.4	-3.4	-1.9	0.8
233.42	0.17	-1.4	24.67	14AP	0.7	-3.4	-2.0	0.8
231.60	-0.34	-1.8	34.29	24AP	1.8	-3.4	-2.1	0.8
228.78	-0.91	-2.0	43.87	4MY	3.3	-3.4	-2.1	0.7
225.52	-1.49	-2.1	53.42	14MY	2.7	-3.3	-2.2	0.6
222.51	-2.01	-1.9	62.96	24MY	1.5	-3.3	-2.2	0.6
220.44	-2.43	-1.7	72.48	3JN	0.5	-3.3	-2.3	0.5
219.67	-2.74	-1.5	82.00	13JN	-0.5	-3.3	-2.3	0.4
220.27	-2.93	-1.3	91.53	23JN	-1.5	-3.3	-2.3	0.4
222.17	-3.05	-1.1	101.08	3JL	-1.4	-3.4	-2.3	0.3
225.17	-3.09	-0.9	110.65	13JL	-0.6	-3.4	-2.3	0.4
229.08	-3.09	-0.7	120.26	23JL	-0.1	-3.4	-2.2	0.4
233.74	-3.05	-0.5	129.90	2AU	0.2	-3.5	-2.2	0.5
239.00	-2.97	-0.4	139.59	12AU	0.5	-3.5	-2.1	0.5
244.74	-2.87	-0.2	149.34	22AU	0.9	-3.6	-2.1	0.6
250.89	-2.75	-0.1	159.14	1SE	2.1	-3.6	-2.0	0.6
257.34	-2.61	0.1	169.00	11SE	2.0	-3.7	-2.0	0.7
264.06	-2.45	0.2	178.93	21SE	-0.2	-3.8	-1.9	0.7
270.97	-2.29	0.3	188.90	1OC	-0.8	-3.9	-1.8	0.8
278.03	-2.11	0.4	198.93	11OC	-0.8	-4.0	-1.8	0.8
285.22	-1.93	0.5	209.02	21OC	-0.8	-4.1	-1.7	0.8
292.49	-1.74	0.7	219.14	31OC	-0.7	-4.2	-1.7	0.9
299.81	-1.56	0.8	229.30	10NO	-0.6	-4.3	-1.6	0.9
307.17	-1.37	0.9	239.49	20NO	-0.6	-4.4	-1.6	0.9
314.53	-1.19	1.0	249.69	30NO	-0.5	-4.4	-1.5	0.9
321.89	-1.01	1.1	259.90	10DE	0.2	-4.1	-1.5	0.9
329.23	-0.84	1.2	270.10	20DE	2.8	-3.5	-1.5	0.8
336.54	-0.67	1.2	280.28	30DE	1.1	-3.3	-1.5	0.8
				313				
343.80	-0.52	1.3	290.44	9JA	0.3	-3.9	-1.5	0.8
351.02	-0.36	1.4	300.56	19JA	0.1	-4.3	-1.5	0.9
358.18	-0.22	1.5	310.63	29JA	-0.0	-4.3	-1.5	0.9
5.27	-0.09	1.5	320.66	8FE	-0.3	-4.3	-1.5	0.9
12.31	0.04	1.6	330.63	18FE	-0.8	-4.2	-1.6	1.0
19.28	0.16	1.7	340.54	28FE	-1.5	-4.1	-1.6	1.0
26.19	0.27	1.7	350.39	10MR	-1.3	-4.0	-1.6	1.0
33.03	0.37	1.8	0.19	20MR	-0.4	-3.9	-1.7	1.0
39.82	0.46	1.8	9.93	30MR	0.9	-3.8	-1.7	1.0
46.54	0.55	1.9	19.61	9AP	2.4	-3.7	-1.8	0.9
53.22	0.63	1.9	29.26	19AP	3.6	-3.6	-1.9	0.9
59.84	0.71	1.9	38.85	29AP	2.0	-3.5	-1.9	0.9
66.41	0.78	1.9	48.42	9MY	1.1	-3.5	-2.0	0.8
72.95	0.84	1.9	57.97	19MY	0.4	-3.4	-2.1	0.8
79.44	0.89	2.0	67.49	29MY	-0.6	-3.4	-2.2	0.7
85.91	0.95	2.0	77.02	8JN	-1.5	-3.4	-2.2	0.6
92.35	0.99	1.9	86.54	18JN	-1.4	-3.3	-2.3	0.6
98.76	1.04	1.9	96.08	28JN	-0.6	-3.3	-2.3	0.5
105.15	1.07	1.9	105.63	8JL	0.0	-3.3	-2.4	0.4
111.53	1.11	2.0	115.22	18JL	0.4	-3.3	-2.4	0.4
117.90	1.13	2.0	124.84	28JL	0.7	-3.3	-2.4	0.5
124.27	1.16	2.0	134.51	7AU	1.2	-3.3	-2.4	0.5
130.64	1.18	2.0	144.23	17AU	2.6	-3.4	-2.4	0.6
137.00	1.19	2.0	154.00	27AU	1.7	-3.4	-2.4	0.6
143.37	1.20	2.0	163.83	6SE	-0.3	-3.4	-2.3	0.7
149.74	1.21	2.0	173.72	16SE	-0.9	-3.4	-2.3	0.7
156.12	1.21	2.0	183.67	26SE	-1.0	-3.5	-2.2	0.8
162.51	1.21	2.0	193.67	6OC	-0.9	-3.5	-2.1	0.8
168.90	1.20	1.9	203.73	16OC	-0.6	-3.5	-2.1	0.8
175.31	1.19	1.9	213.83	26OC	-0.4	-3.5	-2.0	0.9
181.72	1.17	1.9	223.97	5NO	-0.4	-3.4	-1.9	0.9
188.13	1.14	1.8	234.15	15NO	-0.3	-3.4	-1.9	0.9
194.55	1.11	1.7	244.34	25NO	0.4	-3.4	-1.8	0.9
200.97	1.07	1.7	254.54	5DE	3.1	-3.4	-1.7	0.9
207.39	1.01	1.6	264.75	15DE	0.8	-3.3	-1.7	0.9
213.79	0.95	1.5	274.94	25DE	0.1	-3.3	-1.7	0.9

♂ LONG	♂ LAT	♂ MAG	☉ LONG	16.00UT	☿	♀	♃	♄
LONG	LAT	MAG	LONG	314	MAGNITUDES			
220.19	0.87	1.4	285.11	4JA	-0.0	-3.3	-1.6	0.9
226.56	0.78	1.3	295.25	14JA	-0.1	-3.3	-1.6	0.9
232.90	0.67	1.2	305.35	24JA	-0.4	-3.4	-1.6	0.9
239.21	0.54	1.0	315.41	3FE	-0.8	-3.4	-1.6	1.0
245.47	0.39	0.9	325.41	13FE	-1.4	-3.4	-1.6	1.0
251.66	0.22	0.7	335.35	23FE	-1.3	-3.4	-1.6	1.1
257.77	0.01	0.6	345.23	5MR	-0.3	-3.4	-1.6	1.1
263.76	-0.23	0.4	355.06	15MR	1.2	-3.5	-1.6	1.1
269.62	-0.51	0.2	4.83	25MR	3.0	-3.5	-1.6	1.1
275.29	-0.83	-0.0	14.54	4AP	2.7	-3.6	-1.7	1.1
280.72	-1.20	-0.2	24.21	14AP	1.5	-3.6	-1.7	1.1
285.84	-1.63	-0.5	33.83	24AP	0.9	-3.7	-1.7	1.1
290.55	-2.12	-0.7	43.41	4MY	0.2	-3.8	-1.8	1.0
294.73	-2.68	-1.0	52.97	14MY	-0.6	-3.9	-1.8	1.0
298.23	-3.30	-1.2	62.50	24MY	-1.5	-4.0	-1.9	0.9
300.86	-4.00	-1.5	72.03	3JN	-1.4	-4.1	-2.0	0.9
302.43	-4.75	-1.8	81.55	13JN	-0.5	-4.1	-2.0	0.8
302.78	-5.49	-2.1	91.07	23JN	0.1	-4.2	-2.1	0.7
301.83	-6.17	-2.4	100.62	3JL	0.5	-4.2	-2.2	0.7
299.81	-6.65	-2.6	110.19	13JL	0.9	-3.9	-2.3	0.6
297.23	-6.84	-2.6	119.79	23JL	1.6	-3.4	-2.3	0.5
294.80	-6.71	-2.5	129.44	2AU	3.0	-3.2	-2.4	0.5
293.22	-6.30	-2.3	139.13	12AU	1.5	-3.8	-2.4	0.6
292.86	-5.71	-2.0	148.87	22AU	-0.3	-4.2	-2.4	0.6
293.81	-5.05	-1.7	158.67	1SE	-1.1	-4.3	-2.5	0.7
295.98	-4.38	-1.4	168.53	11SE	-1.1	-4.3	-2.5	0.7
299.15	-3.75	-1.1	178.44	21SE	-0.8	-4.2	-2.4	0.8
303.14	-3.17	-0.8	188.42	1OC	-0.5	-4.1	-2.4	0.8
307.77	-2.64	-0.6	198.45	11OC	-0.3	-4.0	-2.4	0.9
312.88	-2.17	-0.4	208.53	21OC	-0.3	-3.9	-2.3	0.9
318.37	-1.75	-0.1	218.65	31OC	-0.2	-3.9	-2.2	0.9
324.15	-1.38	0.1	228.81	10NO	0.6	-3.8	-2.2	1.0
330.14	-1.05	0.3	238.99	20NO	3.2	-3.7	-2.1	1.0
336.29	-0.76	0.5	249.20	30NO	0.6	-3.6	-2.0	1.0
342.57	-0.50	0.6	259.40	10DE	-0.1	-3.6	-1.9	1.0
348.92	-0.27	0.8	269.60	20DE	-0.1	-3.5	-1.9	1.1
355.34	-0.07	0.9	279.79	30DE	-0.2	-3.5	-1.8	1.1
				315				
1.79	0.11	1.1	289.94	9JA	-0.4	-3.4	-1.8	1.1
8.26	0.26	1.2	300.06	19JA	-0.9	-3.4	-1.7	1.0
14.74	0.40	1.3	310.14	29JA	-1.3	-3.4	-1.7	1.0
21.22	0.52	1.4	320.17	8FE	-1.2	-3.3	-1.6	1.1
27.69	0.63	1.5	330.14	18FE	-0.3	-3.3	-1.6	1.1
34.15	0.72	1.6	340.06	28FE	1.4	-3.3	-1.6	1.2
40.59	0.80	1.7	349.91	10MR	3.4	-3.3	-1.6	1.2
47.01	0.87	1.8	359.71	20MR	2.0	-3.3	-1.6	1.2
53.41	0.93	1.8	9.46	30MR	1.1	-3.3	-1.6	1.2
59.79	0.99	1.9	19.14	9AP	0.6	-3.3	-1.6	1.2
66.15	1.03	1.9	28.79	19AP	0.1	-3.4	-1.6	1.2
72.50	1.07	2.0	38.39	29AP	-0.6	-3.4	-1.6	1.2
78.84	1.10	2.0	47.96	9MY	-1.5	-3.4	-1.6	1.2
85.16	1.12	2.0	57.50	19MY	-1.5	-3.5	-1.7	1.2
91.48	1.14	2.0	67.03	29MY	-0.5	-3.5	-1.7	1.1
97.79	1.16	2.0	76.55	8JN	0.2	-3.4	-1.8	1.1
104.11	1.17	2.0	86.08	18JN	0.7	-3.4	-1.8	1.0
110.43	1.17	2.0	95.61	28JN	1.2	-3.4	-1.9	0.9
116.76	1.17	2.0	105.17	8JL	2.1	-3.4	-1.9	0.9
123.11	1.16	2.0	114.75	18JL	3.2	-3.3	-2.0	0.8
129.47	1.15	2.0	124.38	28JL	1.3	-3.3	-2.1	0.7
135.86	1.14	2.0	134.04	7AU	-0.4	-3.3	-2.1	0.7
142.27	1.12	1.9	143.76	17AU	-1.2	-3.3	-2.2	0.7
148.72	1.10	2.0	153.53	27AU	-1.3	-3.3	-2.3	0.7
155.19	1.07	2.0	163.35	6SE	-0.8	-3.4	-2.3	0.7
161.70	1.03	2.0	173.24	16SE	-0.4	-3.4	-2.4	0.8
168.25	1.00	2.0	183.18	26SE	-0.2	-3.4	-2.4	0.8
174.85	0.95	1.9	193.18	6OC	-0.1	-3.4	-2.4	0.9
181.48	0.90	1.9	203.24	16OC	0.0	-3.5	-2.4	0.9
188.16	0.84	1.9	213.34	26OC	0.8	-3.5	-2.4	1.0
194.88	0.78	1.9	223.48	5NO	3.0	-3.6	-2.4	1.0
201.65	0.71	1.8	233.65	15NO	0.3	-3.7	-2.3	1.1
208.47	0.63	1.8	243.85	25NO	-0.2	-3.8	-2.3	1.1
215.33	0.54	1.7	254.05	5DE	-0.3	-3.8	-2.2	1.1
222.24	0.45	1.7	264.26	15DE	-0.4	-3.9	-2.1	1.2
229.19	0.34	1.6	274.45	25DE	-0.5	-4.0	-2.1	1.2

♂			⊙	16.00UT	☿	♀	♃	♄
LONG	LAT	MAG	LONG	316	MAGNITUDES			
236.19	0.23	1.5	284.62	4JA	-0.9	-4.1	-2.0	1.2
243.23	0.10	1.5	294.76	14JA	-1.1	-4.2	-1.9	1.2
250.30	-0.04	1.4	304.86	24JA	-1.0	-4.3	-1.8	1.2
257.42	-0.18	1.3	314.92	3FE	-0.2	-4.3	-1.8	1.2
264.56	-0.35	1.2	324.92	13FE	1.7	-4.3	-1.7	1.2
271.73	-0.52	1.1	334.87	23FE	2.9	-4.0	-1.7	1.3
278.92	-0.70	1.0	344.75	4MR	1.5	-3.4	-1.6	1.3
286.12	-0.90	0.9	354.58	14MR	0.8	-3.3	-1.6	1.4
293.32	-1.10	0.8	4.35	24MR	0.5	-3.9	-1.6	1.4
300.50	-1.31	0.7	14.07	3AP	0.0	-4.2	-1.5	1.4
307.66	-1.53	0.6	23.74	13AP	-0.6	-4.2	-1.5	1.4
314.75	-1.76	0.5	33.36	23AP	-1.5	-4.2	-1.5	1.4
321.77	-1.98	0.4	42.94	3MY	-1.5	-4.1	-1.5	1.4
328.67	-2.21	0.3	52.50	13MY	-0.5	-4.0	-1.5	1.3
335.43	-2.44	0.1	62.04	23MY	0.3	-3.9	-1.5	1.3
341.99	-2.66	-0.0	71.56	2JN	0.9	-3.8	-1.5	1.3
348.31	-2.88	-0.1	81.09	12JN	1.6	-3.7	-1.6	1.2
354.31	-3.09	-0.3	90.61	22JN	2.7	-3.6	-1.6	1.2
359.91	-3.29	-0.4	100.16	2JL	2.8	-3.6	-1.6	1.1
5.00	-3.47	-0.6	109.73	12JL	1.0	-3.5	-1.7	1.1
9.46	-3.64	-0.8	119.33	22JL	-0.4	-3.5	-1.7	1.0
13.12	-3.77	-1.0	128.97	1AU	-1.3	-3.4	-1.8	1.0
15.78	-3.86	-1.2	138.66	11AU	-1.4	-3.4	-1.8	0.9
17.22	-3.89	-1.4	148.40	21AU	-0.7	-3.4	-1.9	0.8
17.27	-3.83	-1.7	158.20	31AU	-0.3	-3.4	-2.0	0.8
15.85	-3.63	-1.9	168.06	10SE	-0.1	-3.4	-2.0	0.9
13.16	-3.28	-2.1	177.97	20SE	0.0	-3.4	-2.1	0.9
9.77	-2.77	-2.1	187.94	30SE	0.2	-3.4	-2.2	1.0
6.47	-2.16	-2.0	197.97	10OC	1.1	-3.4	-2.2	1.0
4.01	-1.53	-1.7	208.04	20OC	2.8	-3.4	-2.3	1.1
2.80	-0.95	-1.3	218.16	30OC	0.2	-3.4	-2.3	1.1
2.94	-0.44	-1.0	228.32	9NO	-0.4	-3.4	-2.3	1.2
4.31	-0.03	-0.7	238.50	19NO	-0.4	-3.4	-2.3	1.2
6.70	0.30	-0.4	248.70	29NO	-0.5	-3.4	-2.3	1.3
9.91	0.56	-0.1	258.91	9DE	-0.6	-3.5	-2.3	1.3
13.76	0.77	0.2	269.11	19DE	-0.9	-3.5	-2.2	1.3
18.10	0.93	0.4	279.29	29DE	-1.0	-3.5	-2.2	1.4
				317				
22.83	1.06	0.6	289.45	8JA	-0.9	-3.5	-2.1	1.4
27.85	1.15	0.8	299.57	18JA	-0.1	-3.4	-2.0	1.4
33.10	1.23	1.0	309.65	28JA	2.0	-3.4	-2.0	1.3
38.54	1.29	1.1	319.68	7FE	2.3	-3.4	-1.9	1.3
44.11	1.33	1.3	329.66	17FE	1.1	-3.4	-1.8	1.2
49.79	1.36	1.4	339.57	27FE	0.6	-3.3	-1.7	1.2
55.56	1.39	1.5	349.43	9MR	0.3	-3.3	-1.7	1.2
61.41	1.40	1.6	359.23	19MR	-0.1	-3.3	-1.6	1.2
67.31	1.40	1.7	8.98	29MR	-0.7	-3.3	-1.6	1.2
73.26	1.40	1.8	18.67	8AP	-1.5	-3.3	-1.5	1.2
79.24	1.39	1.8	28.31	18AP	-1.5	-3.4	-1.5	1.2
85.27	1.38	1.9	37.92	28AP	-0.5	-3.4	-1.5	1.1
91.34	1.36	1.9	47.49	8MY	0.4	-3.4	-1.4	1.1
97.43	1.34	2.0	57.04	18MY	1.2	-3.4	-1.4	1.1
103.56	1.31	2.0	66.57	28MY	2.1	-3.5	-1.4	1.0
109.72	1.28	2.0	76.09	7JN	3.4	-3.5	-1.4	1.0
115.92	1.25	2.0	85.61	17JN	2.3	-3.6	-1.4	1.0
122.16	1.21	2.0	95.15	27JN	0.8	-3.7	-1.4	0.9
128.44	1.17	2.0	104.71	7JL	-0.5	-3.7	-1.4	0.9
134.76	1.12	2.0	114.29	17JL	-1.4	-3.8	-1.5	0.8
141.14	1.07	2.0	123.91	27JL	-1.4	-3.9	-1.5	0.8
147.56	1.02	2.0	133.58	6AU	-0.7	-4.0	-1.5	0.7
154.03	0.97	1.9	143.29	16AU	-0.2	-4.1	-1.6	0.7
160.57	0.90	1.9	153.06	26AU	0.0	-4.2	-1.6	0.6
167.16	0.84	1.9	162.88	5SE	0.2	-4.3	-1.7	0.6
173.82	0.77	1.8	172.76	15SE	0.4	-4.3	-1.7	0.7
180.54	0.70	1.8	182.71	25SE	1.4	-4.0	-1.8	0.7
187.33	0.62	1.8	192.70	5OC	2.5	-3.5	-1.9	0.8
194.19	0.54	1.8	202.75	15OC	0.0	-3.3	-1.9	0.9
201.12	0.45	1.8	212.85	25OC	-0.6	-4.0	-2.0	0.9
208.12	0.36	1.8	222.99	4NO	-0.6	-4.3	-2.1	1.0
215.19	0.26	1.7	233.16	14NO	-0.6	-4.4	-2.1	1.0
222.33	0.16	1.7	243.35	24NO	-0.7	-4.4	-2.1	1.1
229.54	0.05	1.7	253.56	4DE	-0.8	-4.3	-2.2	1.1
236.82	-0.06	1.7	263.76	14DE	-0.8	-4.2	-2.2	1.1
244.16	-0.18	1.6	273.96	24DE	-0.8	-4.1	-2.2	1.1

♂			⊙	16.00UT	☿	♀	♃	♄
LONG	LAT	MAG	LONG	318	MAGNITUDES			
251.56	-0.30	1.6	284.13	3JA	-0.0	-4.0	-2.2	1.1
259.02	-0.42	1.5	294.27	13JA	2.3	-3.9	-2.1	1.1
266.53	-0.55	1.5	304.38	23JA	1.8	-3.8	-2.1	1.1
274.09	-0.68	1.4	314.43	2FE	0.7	-3.7	-2.0	1.1
281.68	-0.80	1.4	324.44	12FE	0.4	-3.6	-2.0	1.0
289.30	-0.93	1.3	334.39	22FE	0.2	-3.5	-1.9	1.0
296.94	-1.05	1.3	344.28	4MR	-0.1	-3.5	-1.8	0.9
304.59	-1.17	1.2	354.11	14MR	-0.7	-3.4	-1.8	0.9
312.23	-1.28	1.2	3.88	24MR	-1.5	-3.4	-1.7	0.9
319.85	-1.39	1.1	13.60	3AP	-1.4	-3.4	-1.6	0.9
327.44	-1.48	1.1	23.27	13AP	-0.4	-3.3	-1.6	0.9
334.98	-1.56	1.0	32.90	23AP	0.6	-3.3	-1.5	0.9
342.45	-1.63	1.0	42.48	3MY	1.5	-3.3	-1.5	0.9
349.84	-1.69	0.9	52.04	13MY	2.8	-3.3	-1.4	0.9
357.13	-1.73	0.9	61.58	23MY	3.2	-3.3	-1.4	0.8
4.30	-1.75	0.8	71.10	2JN	1.8	-3.3	-1.4	0.8
11.32	-1.75	0.8	80.62	12JN	0.6	-3.3	-1.3	0.8
18.18	-1.74	0.7	90.15	22JN	-0.5	-3.4	-1.3	0.7
24.84	-1.71	0.6	99.70	2JL	-1.4	-3.4	-1.3	0.7
31.28	-1.65	0.5	109.26	12JL	-1.4	-3.5	-1.3	0.6
37.45	-1.58	0.5	118.87	22JL	-0.6	-3.5	-1.3	0.6
43.31	-1.49	0.4	128.50	1AU	-0.1	-3.5	-1.3	0.5
48.80	-1.37	0.2	138.19	11AU	0.2	-3.5	-1.3	0.5
53.84	-1.22	0.1	147.93	21AU	0.4	-3.4	-1.4	0.4
58.34	-1.05	-0.0	157.72	31AU	0.7	-3.4	-1.4	0.4
62.18	-0.83	-0.2	167.57	10SE	1.8	-3.4	-1.4	0.3
65.19	-0.58	-0.4	177.49	20SE	2.2	-3.4	-1.5	0.3
67.20	-0.28	-0.6	187.45	30SE	-0.1	-3.3	-1.5	0.4
68.00	0.08	-0.8	197.48	10OC	-0.7	-3.3	-1.6	0.5
67.41	0.50	-1.0	207.55	20OC	-0.7	-3.3	-1.7	0.5
65.39	0.96	-1.2	217.67	30OC	-0.7	-3.3	-1.7	0.6
62.16	1.42	-1.4	227.82	9NO	-0.8	-3.3	-1.8	0.7
58.28	1.84	-1.5	238.01	19NO	-0.6	-3.4	-1.8	0.7
54.56	2.16	-1.3	248.20	29NO	-0.7	-3.4	-1.9	0.8
51.69	2.37	-1.0	258.41	9DE	-0.6	-3.4	-2.0	0.8
50.09	2.47	-0.7	268.61	19DE	0.1	-3.5	-2.0	0.8
49.83	2.50	-0.4	278.80	29DE	2.6	-3.5	-2.0	0.9
				319				
50.79	2.48	-0.1	288.96	8JA	1.4	-3.6	-2.1	0.9
52.77	2.43	0.2	299.09	18JA	0.5	-3.6	-2.1	0.9
55.58	2.36	0.4	309.17	28JA	0.2	-3.7	-2.1	0.9
59.03	2.28	0.6	319.20	7FE	0.1	-3.7	-2.1	0.8
63.01	2.20	0.8	329.18	17FE	-0.2	-3.8	-2.0	0.8
67.38	2.12	1.0	339.10	27FE	-0.7	-3.9	-2.0	0.8
72.07	2.04	1.1	348.96	9MR	-1.5	-4.0	-1.9	0.7
77.02	1.95	1.3	358.77	19MR	-1.4	-4.1	-1.9	0.7
82.18	1.87	1.4	8.51	29MR	-0.4	-4.1	-1.8	0.7
87.52	1.79	1.5	18.21	8AP	0.8	-4.2	-1.7	0.7
93.00	1.71	1.6	27.85	18AP	2.0	-4.2	-1.7	0.7
98.60	1.63	1.7	37.46	28AP	3.7	-4.0	-1.6	0.7
104.31	1.55	1.7	47.03	8MY	2.5	-3.6	-1.5	0.7
110.12	1.47	1.8	56.58	18MY	1.4	-2.7	-1.5	0.7
116.03	1.39	1.8	66.10	28MY	0.5	-3.4	-1.4	0.7
122.01	1.31	1.9	75.63	7JN	-0.5	-4.0	-1.4	0.6
128.08	1.23	1.9	85.15	17JN	-1.5	-4.2	-1.3	0.6
134.23	1.14	1.9	94.68	27JN	-1.4	-4.2	-1.3	0.6
140.46	1.06	1.9	104.24	7JL	-0.6	-4.1	-1.3	0.5
146.76	0.97	1.9	113.82	17JL	-0.1	-4.1	-1.3	0.5
153.15	0.89	1.9	123.44	27JL	0.3	-4.0	-1.2	0.4
159.62	0.80	1.9	133.10	6AU	0.5	-3.9	-1.2	0.4
166.17	0.71	1.9	142.81	16AU	1.0	-3.8	-1.2	0.3
172.80	0.61	1.9	152.58	26AU	2.2	-3.7	-1.2	0.3
179.53	0.52	1.8	162.40	5SE	2.0	-3.7	-1.3	0.2
186.34	0.42	1.8	172.28	15SE	-0.2	-3.6	-1.3	0.1
193.24	0.32	1.8	182.22	25SE	-0.8	-3.6	-1.3	0.1
200.22	0.22	1.7	192.22	5OC	-0.9	-3.5	-1.3	0.1
207.30	0.12	1.7	202.26	15OC	-0.9	-3.5	-1.4	0.2
214.46	0.01	1.6	212.36	25OC	-0.6	-3.4	-1.4	0.2
221.72	-0.09	1.6	222.50	4NO	-0.5	-3.4	-1.5	0.3
229.05	-0.20	1.6	232.67	14NO	-0.5	-3.4	-1.6	0.4
236.47	-0.30	1.5	242.86	24NO	-0.5	-3.4	-1.6	0.4
243.96	-0.41	1.5	253.07	4DE	0.2	-3.4	-1.7	0.5
251.52	-0.51	1.5	263.27	14DE	2.9	-3.4	-1.7	0.5
259.15	-0.61	1.5	273.46	24DE	1.0	-3.3	-1.8	0.6

♂ LONG	♂ LAT	♂ MAG	☉ LONG	16.00UT	☿	♀	♃	♄
				320	MAGNITUDES			
266.83	-0.71	1.5	283.64	3JA	0.3	-3.3	-1.9	0.6
274.56	-0.80	1.5	293.78	13JA	0.1	-3.4	-1.9	0.6
282.33	-0.89	1.5	303.89	23JA	-0.1	-3.4	-2.0	0.6
290.13	-0.97	1.4	313.95	2FE	-0.3	-3.4	-2.0	0.6
297.94	-1.05	1.4	323.95	12FE	-0.8	-3.4	-2.0	0.6
305.75	-1.11	1.4	333.90	22FE	-1.5	-3.4	-2.0	0.6
313.56	-1.16	1.4	343.80	3MR	-1.3	-3.5	-2.0	0.6
321.34	-1.20	1.4	353.63	13MR	-0.4	-3.5	-2.0	0.5
329.10	-1.23	1.4	3.41	23MR	1.0	-3.5	-1.9	0.5
336.81	-1.25	1.4	13.13	2AP	2.6	-3.4	-1.9	0.4
344.46	-1.25	1.3	22.80	12AP	3.2	-3.4	-1.8	0.5
352.04	-1.24	1.3	32.42	22AP	1.8	-3.4	-1.8	0.5
359.54	-1.21	1.3	42.02	2MY	1.0	-3.4	-1.7	0.5
6.95	-1.18	1.3	51.57	12MY	0.3	-3.3	-1.7	0.5
14.27	-1.12	1.3	61.11	22MY	-0.5	-3.3	-1.6	0.5
21.47	-1.06	1.3	70.64	1JN	-1.5	-3.3	-1.5	0.5
28.55	-0.98	1.3	80.16	11JN	-1.5	-3.3	-1.5	0.5
35.51	-0.89	1.2	89.68	21JN	-0.6	-3.3	-1.4	0.5
42.33	-0.79	1.2	99.23	1JL	0.0	-3.4	-1.4	0.4
48.99	-0.67	1.2	108.79	11JL	0.4	-3.4	-1.3	0.4
55.50	-0.55	1.1	118.39	21JL	0.7	-3.4	-1.3	0.4
61.81	-0.41	1.1	128.03	31JL	1.3	-3.5	-1.3	0.3
67.93	-0.25	1.0	137.71	10AU	2.7	-3.5	-1.2	0.3
73.81	-0.09	1.0	147.45	20AU	1.7	-3.6	-1.2	0.2
79.43	0.10	0.9	157.24	30AU	-0.3	-3.6	-1.2	0.2
84.75	0.30	0.8	167.09	9SE	-1.0	-3.7	-1.2	0.1
89.70	0.53	0.7	177.00	19SE	-1.0	-3.8	-1.2	0.0
94.20	0.78	0.6	186.97	29SE	-0.9	-3.9	-1.2	-0.0
98.18	1.06	0.4	196.99	9OC	-0.5	-4.0	-1.2	-0.1
101.50	1.38	0.2	207.06	19OC	-0.4	-4.1	-1.3	-0.1
104.00	1.74	0.0	217.18	29OC	-0.4	-4.2	-1.3	-0.0
105.51	2.14	-0.2	227.33	8NO	-0.3	-4.3	-1.3	0.0
105.83	2.58	-0.4	237.51	18NO	0.4	-4.4	-1.4	0.1
104.81	3.03	-0.6	247.71	28NO	3.1	-4.4	-1.4	0.2
102.45	3.47	-0.8	257.92	8DE	0.7	-4.1	-1.5	0.3
99.00	3.82	-1.0	268.12	18DE	0.1	-3.5	-1.6	0.3
95.03	4.03	-1.1	278.31	28DE	-0.1	-3.3	-1.6	0.4
				321				
91.30	4.08	-0.9	288.46	7JA	-0.2	-4.0	-1.7	0.4
88.45	3.98	-0.6	298.59	17JA	-0.4	-4.3	-1.8	0.4
86.85	3.78	-0.4	308.68	27JA	-0.8	-4.3	-1.8	0.5
86.55	3.54	-0.1	318.71	6FE	-1.4	-4.3	-1.9	0.5
87.44	3.28	0.1	328.69	16FE	-1.2	-4.2	-1.9	0.5
89.34	3.02	0.4	338.62	26FE	-0.3	-4.1	-2.0	0.5
92.05	2.78	0.6	348.48	8MR	1.2	-4.0	-2.0	0.4
95.41	2.56	0.8	358.29	18MR	3.2	-3.9	-2.0	0.4
99.31	2.35	0.9	8.04	28MR	2.4	-3.8	-2.0	0.4
103.62	2.15	1.1	17.74	7AP	1.4	-3.7	-2.0	0.3
108.28	1.97	1.2	27.39	17AP	0.8	-3.6	-2.0	0.3
113.22	1.80	1.3	37.00	27AP	0.2	-3.5	-1.9	0.3
118.41	1.65	1.4	46.57	7MY	-0.6	-3.5	-1.9	0.3
123.80	1.49	1.5	56.12	17MY	-1.5	-3.4	-1.8	0.4
129.38	1.35	1.5	65.65	27MY	-1.5	-3.4	-1.8	0.4
135.11	1.21	1.6	75.17	6JN	-0.6	-3.4	-1.7	0.4
140.99	1.08	1.6	84.70	16JN	0.1	-3.3	-1.6	0.4
147.01	0.94	1.7	94.23	26JN	0.6	-3.3	-1.6	0.4
153.16	0.82	1.7	103.78	6JL	1.0	-3.3	-1.5	0.4
159.44	0.69	1.7	113.36	16JL	1.7	-3.3	-1.5	0.3
165.83	0.57	1.7	122.98	26JL	3.2	-3.3	-1.4	0.3
172.34	0.45	1.7	132.63	5AU	1.5	-3.3	-1.4	0.3
178.97	0.33	1.7	142.35	15AU	-0.3	-3.4	-1.3	0.2
185.71	0.21	1.7	152.10	25AU	-1.1	-3.4	-1.3	0.2
192.57	0.09	1.7	161.92	4SE	-1.2	-3.4	-1.3	0.1
199.53	-0.02	1.7	171.80	14SE	-0.8	-3.4	-1.3	0.0
206.61	-0.13	1.7	181.73	24SE	-0.4	-3.5	-1.2	-0.0
213.79	-0.24	1.6	191.73	4OC	-0.3	-3.5	-1.2	-0.1
221.07	-0.34	1.6	201.78	14OC	-0.2	-3.5	-1.2	-0.2
228.45	-0.45	1.6	211.87	24OC	-0.1	-3.5	-1.2	-0.2
235.92	-0.54	1.5	222.00	3NO	0.6	-3.4	-1.3	-0.2
243.48	-0.64	1.5	232.17	13NO	3.2	-3.4	-1.3	-0.1
251.12	-0.72	1.4	242.36	23NO	0.5	-3.4	-1.3	-0.1
258.83	-0.80	1.4	252.56	3DE	-0.1	-3.4	-1.4	0.0
266.60	-0.87	1.4	262.77	13DE	-0.2	-3.3	-1.4	0.1
274.41	-0.94	1.3	272.97	23DE	-0.3	-3.3	-1.5	0.1

♂ LONG	♂ LAT	♂ MAG	☉ LONG	16.00UT	☿	♀	♃	♄
				322	MAGNITUDES			
282.26	-0.99	1.3	283.14	2JA	-0.5	-3.3	-1.5	0.2
290.14	-1.03	1.3	293.29	12JA	-0.9	-3.3	-1.6	0.3
298.03	-1.07	1.3	303.40	22JA	-1.2	-3.4	-1.6	0.3
305.91	-1.09	1.4	313.46	1FE	-1.1	-3.4	-1.7	0.3
313.79	-1.10	1.4	323.47	11FE	-0.3	-3.4	-1.8	0.4
321.63	-1.09	1.4	333.42	21FE	1.5	-3.4	-1.8	0.4
329.44	-1.08	1.4	343.32	3MR	3.3	-3.5	-1.9	0.4
337.20	-1.05	1.4	353.16	13MR	1.8	-3.5	-2.0	0.4
344.89	-1.01	1.4	2.93	23MR	1.0	-3.5	-2.0	0.3
352.52	-0.97	1.5	12.66	2AP	0.6	-3.6	-2.1	0.3
0.07	-0.91	1.5	22.34	12AP	0.1	-3.6	-2.1	0.3
7.54	-0.84	1.5	31.96	22AP	-0.6	-3.7	-2.1	0.2
14.92	-0.76	1.5	41.56	2MY	-1.5	-3.8	-2.1	0.2
22.21	-0.68	1.5	51.12	12MY	-1.5	-3.9	-2.1	0.2
29.39	-0.58	1.6	60.65	22MY	-0.5	-4.0	-2.0	0.3
36.48	-0.48	1.6	70.18	1JN	0.2	-4.1	-2.0	0.3
43.47	-0.38	1.6	79.70	11JN	0.8	-4.1	-1.9	0.3
50.35	-0.26	1.6	89.23	21JN	1.3	-4.2	-1.9	0.3
57.12	-0.15	1.6	98.77	1JL	2.3	-4.1	-1.8	0.3
63.78	-0.02	1.6	108.34	11JL	3.1	-3.9	-1.8	0.3
70.32	0.11	1.6	117.93	21JL	1.2	-3.4	-1.7	0.3
76.74	0.24	1.6	127.57	31JL	-0.4	-3.2	-1.6	0.3
83.03	0.38	1.5	137.25	10AU	-1.2	-3.8	-1.6	0.3
89.18	0.53	1.5	146.98	20AU	-1.3	-4.2	-1.5	0.2
95.18	0.68	1.4	156.77	30AU	-0.8	-4.3	-1.5	0.2
101.01	0.85	1.4	166.61	9SE	-0.4	-4.3	-1.4	0.1
106.64	1.02	1.3	176.51	19SE	-0.1	-4.2	-1.4	0.1
112.06	1.21	1.2	186.48	29SE	-0.0	-4.1	-1.4	-0.0
117.21	1.41	1.1	196.50	9OC	0.1	-4.0	-1.3	-0.1
122.06	1.63	1.0	206.57	19OC	0.9	-3.9	-1.3	-0.2
126.53	1.87	0.9	216.68	29OC	3.0	-3.9	-1.3	-0.2
130.54	2.13	0.7	226.84	8NO	0.3	-3.8	-1.3	-0.3
133.99	2.42	0.5	237.02	18NO	-0.3	-3.7	-1.3	-0.3
136.73	2.74	0.4	247.22	28NO	-0.4	-3.6	-1.3	-0.2
138.62	3.09	0.1	257.42	8DE	-0.4	-3.6	-1.3	-0.1
139.46	3.47	-0.1	267.62	18DE	-0.6	-3.5	-1.4	-0.1
139.07	3.85	-0.4	277.82	28DE	-0.9	-3.5	-1.4	0.0
				323				
137.38	4.20	-0.6	287.98	7JA	-1.1	-3.4	-1.5	0.1
134.47	4.47	-0.8	298.11	17JA	-1.0	-3.4	-1.5	0.1
130.72	4.59	-1.0	308.20	27JA	-0.2	-3.4	-1.6	0.2
126.81	4.53	-0.9	318.23	6FE	1.7	-3.3	-1.6	0.2
123.44	4.31	-0.7	328.22	16FE	2.7	-3.3	-1.7	0.3
121.13	3.97	-0.5	338.14	26FE	1.3	-3.3	-1.8	0.3
120.11	3.59	-0.3	348.01	8MR	0.7	-3.3	-1.8	0.3
120.35	3.19	-0.1	357.82	18MR	0.4	-3.3	-1.9	0.3
121.71	2.82	0.2	7.57	28MR	0.0	-3.3	-2.0	0.3
124.02	2.47	0.4	17.27	7AP	-0.6	-3.3	-2.0	0.3
127.08	2.15	0.5	26.92	17AP	-1.5	-3.4	-2.1	0.3
130.77	1.86	0.7	36.53	27AP	-1.5	-3.4	-2.1	0.2
134.97	1.60	0.8	46.11	7MY	-0.5	-3.4	-2.2	0.2
139.59	1.36	0.9	55.65	17MY	0.3	-3.5	-2.2	0.2
144.57	1.13	1.0	65.19	27MY	1.0	-3.5	-2.2	0.2
149.85	0.93	1.1	74.71	6JN	1.7	-3.4	-2.2	0.2
155.39	0.74	1.2	84.23	16JN	3.0	-3.4	-2.2	0.3
161.16	0.56	1.3	93.77	26JN	2.6	-3.4	-2.1	0.3
167.15	0.39	1.3	103.31	6JL	1.0	-3.4	-2.1	0.3
173.32	0.23	1.4	112.89	16JL	-0.4	-3.3	-2.0	0.3
179.67	0.08	1.4	122.51	26JL	-1.3	-3.3	-2.0	0.3
186.19	-0.07	1.4	132.16	5AU	-1.4	-3.3	-1.9	0.3
192.86	-0.20	1.4	141.87	15AU	-0.7	-3.3	-1.8	0.3
199.69	-0.33	1.4	151.63	25AU	-0.3	-3.3	-1.8	0.3
206.65	-0.45	1.4	161.44	4SE	-0.0	-3.4	-1.7	0.2
213.75	-0.56	1.5	171.31	14SE	0.1	-3.4	-1.7	0.2
220.97	-0.66	1.4	181.25	24SE	0.3	-3.4	-1.6	0.1
228.32	-0.76	1.4	191.24	4OC	1.2	-3.4	-1.6	0.1
235.77	-0.84	1.4	201.28	14OC	2.8	-3.5	-1.5	-0.0
243.32	-0.92	1.4	211.37	24OC	0.1	-3.5	-1.5	-0.1
250.96	-0.98	1.4	221.50	3NO	-0.5	-3.6	-1.5	-0.2
258.68	-1.03	1.4	231.67	13NO	-0.5	-3.7	-1.4	-0.2
266.47	-1.07	1.4	241.87	23NO	-0.5	-3.8	-1.4	-0.3
274.30	-1.10	1.4	252.07	3DE	-0.7	-3.8	-1.4	-0.3
282.17	-1.12	1.4	262.27	13DE	-0.9	-3.9	-1.4	-0.2
290.06	-1.12	1.4	272.47	23DE	-0.9	-4.0	-1.4	-0.1

Left Table

♂ LONG	LAT	MAG	☉ LONG	16.00UT 324	☿	♀	♃	♄
297.97	-1.12	1.4	282.65	2JA	-0.9	-4.1	-1.4	-0.1
305.86	-1.10	1.4	292.80	12JA	-0.1	-4.2	-1.5	0.0
313.74	-1.07	1.4	302.91	22JA	2.0	-4.3	-1.5	0.1
321.58	-1.02	1.4	312.97	1FE	2.1	-4.3	-1.5	0.1
329.38	-0.97	1.4	322.99	11FE	0.9	-4.3	-1.6	0.2
337.13	-0.91	1.4	332.95	21FE	0.5	-4.0	-1.6	0.2
344.81	-0.84	1.4	342.84	2MR	0.3	-3.4	-1.7	0.3
352.42	-0.76	1.4	352.68	12MR	-0.1	-3.4	-1.7	0.3
359.95	-0.68	1.4	2.47	22MR	-0.7	-3.9	-1.8	0.3
7.39	-0.59	1.5	12.19	1AP	-1.5	-4.2	-1.9	0.3
14.76	-0.50	1.5	21.87	11AP	-1.5	-4.2	-1.9	0.3
22.03	-0.40	1.6	31.50	21AP	-0.5	-4.2	-2.0	0.3
29.22	-0.30	1.6	41.09	1MY	0.5	-4.1	-2.1	0.3
36.32	-0.20	1.7	50.65	11MY	1.3	-4.0	-2.1	0.3
43.33	-0.09	1.7	60.19	21MY	2.3	-3.9	-2.2	0.2
50.25	0.01	1.7	69.72	31MY	3.5	-3.8	-2.3	0.2
57.08	0.12	1.8	79.24	10JN	2.1	-3.7	-2.3	0.3
63.83	0.23	1.8	88.77	20JN	0.8	-3.6	-2.3	0.3
70.50	0.34	1.8	98.31	30JN	-0.4	-3.6	-2.3	0.3
77.09	0.45	1.8	107.87	10JL	-1.4	-3.5	-2.3	0.3
83.59	0.56	1.8	117.47	20JL	-1.4	-3.5	-2.3	0.3
90.01	0.67	1.8	127.10	30JL	-0.7	-3.4	-2.3	0.4
96.35	0.78	1.8	136.78	9AU	-0.2	-3.4	-2.2	0.3
102.60	0.89	1.8	146.51	19AU	0.1	-3.4	-2.2	0.3
108.77	1.01	1.8	156.29	29AU	0.3	-3.4	-2.1	0.3
114.84	1.12	1.7	166.14	8SE	0.5	-3.4	-2.0	0.3
120.80	1.24	1.7	176.04	18SE	1.5	-3.4	-2.0	0.3
126.65	1.37	1.6	186.00	28SE	2.5	-3.4	-1.9	0.2
132.36	1.50	1.6	196.01	8OC	0.0	-3.4	-1.8	0.2
137.93	1.63	1.5	206.08	18OC	-0.6	-3.4	-1.8	0.1
143.32	1.78	1.4	216.19	28OC	-0.6	-3.4	-1.7	0.0
148.49	1.93	1.3	226.34	7NO	-0.6	-3.4	-1.7	-0.0
153.42	2.09	1.2	236.52	17NO	-0.8	-3.4	-1.6	-0.1
158.04	2.26	1.0	246.71	27NO	-0.7	-3.4	-1.6	-0.2
162.28	2.45	0.9	256.92	7DE	-0.7	-3.5	-1.6	-0.3
166.05	2.65	0.7	267.12	17DE	-0.7	-3.5	-1.5	-0.2
169.25	2.87	0.5	277.31	27DE	-0.0	-3.5	-1.5	-0.2
				325				
171.72	3.10	0.2	287.48	6JA	2.3	-3.5	-1.5	-0.1
173.31	3.33	-0.0	297.61	16JA	1.6	-3.4	-1.5	-0.0
173.83	3.56	-0.3	307.70	26JA	0.6	-3.4	-1.5	0.0
173.11	3.76	-0.6	317.74	5FE	0.3	-3.4	-1.5	0.1
171.12	3.88	-0.9	327.73	15FE	0.1	-3.4	-1.6	0.2
168.02	3.88	-1.1	337.66	25FE	-0.2	-3.3	-1.6	0.2
164.26	3.73	-1.2	347.53	7MR	-0.7	-3.3	-1.6	0.3
160.55	3.43	-1.1	357.34	17MR	-1.5	-3.3	-1.7	0.3
157.56	3.02	-0.9	7.10	27MR	-1.4	-3.3	-1.7	0.3
155.74	2.55	-0.7	16.80	6AP	-0.5	-3.3	-1.8	0.3
155.25	2.09	-0.5	26.45	16AP	0.6	-3.4	-1.8	0.4
156.01	1.65	-0.3	36.07	26AP	1.7	-3.4	-1.9	0.4
157.87	1.25	-0.1	45.65	6MY	3.1	-3.4	-2.0	0.4
160.67	0.89	0.1	55.19	16MY	2.9	-3.4	-2.0	0.3
164.22	0.58	0.2	64.73	26MY	1.6	-3.5	-2.1	0.3
168.40	0.30	0.4	74.25	5JN	0.6	-3.5	-2.2	0.3
173.11	0.05	0.5	83.77	15JN	-0.5	-3.6	-2.2	0.3
178.25	-0.17	0.6	93.30	25JN	-1.4	-3.7	-2.3	0.3
183.77	-0.37	0.7	102.86	5JL	-1.5	-3.7	-2.3	0.3
189.61	-0.54	0.8	112.43	15JL	-0.6	-3.8	-2.4	0.4
195.73	-0.70	0.8	122.04	25JL	-0.1	-3.9	-2.4	0.4
202.11	-0.84	0.9	131.70	4AU	0.2	-4.0	-2.4	0.4
208.70	-0.96	0.9	141.40	14AU	0.4	-4.1	-2.4	0.4
215.50	-1.07	1.0	151.16	24AU	0.8	-4.2	-2.4	0.4
222.48	-1.16	1.0	160.97	3SE	1.9	-4.3	-2.4	0.4
229.62	-1.23	1.1	170.84	13SE	2.2	-4.3	-2.3	0.4
236.91	-1.29	1.1	180.77	23SE	-0.1	-4.0	-2.2	0.4
244.33	-1.33	1.1	190.76	3OC	-0.8	-3.5	-2.2	0.4
251.85	-1.36	1.2	200.79	13OC	-0.8	-3.4	-2.1	0.3
259.47	-1.37	1.2	210.88	23OC	-0.8	-4.0	-2.0	0.3
267.17	-1.36	1.2	221.02	2NO	-0.7	-4.3	-2.0	0.2
274.93	-1.34	1.3	231.18	12NO	-0.6	-4.4	-1.9	0.1
282.73	-1.31	1.3	241.37	22NO	-0.6	-4.3	-1.8	0.1
290.55	-1.27	1.3	251.57	2DE	-0.6	-4.3	-1.8	-0.0
298.38	-1.21	1.3	261.77	12DE	0.1	-4.2	-1.7	-0.1
306.21	-1.14	1.4	271.97	22DE	2.6	-4.1	-1.7	-0.2

Right Table

♂ LONG	LAT	MAG	☉ LONG	16.00UT 326	☿	♀	♃	♄
314.01	-1.07	1.4	282.15	1JA	1.3	-3.9	-1.6	-0.1
321.78	-0.98	1.4	292.30	11JA	0.4	-3.8	-1.6	-0.0
329.51	-0.89	1.5	302.42	21JA	0.2	-3.8	-1.6	-0.0
337.17	-0.79	1.5	312.49	31JA	0.0	-3.7	-1.6	0.1
344.77	-0.69	1.5	322.50	10FE	-0.2	-3.6	-1.6	0.1
352.30	-0.59	1.5	332.46	20FE	-0.7	-3.5	-1.6	0.2
359.76	-0.48	1.6	342.37	2MR	-1.5	-3.5	-1.6	0.2
7.13	-0.38	1.6	352.21	12MR	-1.4	-3.4	-1.6	0.3
14.41	-0.27	1.6	2.00	22MR	-0.4	-3.4	-1.6	0.3
21.62	-0.16	1.6	11.73	1AP	0.8	-3.3	-1.6	0.4
28.74	-0.06	1.7	21.40	11AP	2.2	-3.3	-1.7	0.4
35.78	0.05	1.7	31.04	21AP	3.8	-3.3	-1.7	0.4
42.74	0.15	1.7	40.63	1MY	2.2	-3.3	-1.7	0.5
49.63	0.25	1.7	50.19	11MY	1.2	-3.3	-1.8	0.5
56.44	0.34	1.7	59.74	21MY	0.4	-3.3	-1.9	0.5
63.18	0.44	1.8	69.26	31MY	-0.5	-3.3	-1.9	0.5
69.85	0.53	1.8	78.78	10JN	-1.5	-3.3	-2.0	0.4
76.47	0.62	1.9	88.31	20JN	-1.5	-3.4	-2.1	0.4
83.02	0.70	1.9	97.85	30JN	-0.6	-3.4	-2.1	0.4
89.52	0.78	1.9	107.41	10JL	-0.0	-3.5	-2.2	0.4
95.97	0.87	2.0	117.00	20JL	0.3	-3.5	-2.3	0.5
102.36	0.94	2.0	126.63	30JL	0.6	-3.5	-2.3	0.5
108.72	1.02	2.0	136.31	9AU	1.1	-3.5	-2.4	0.5
115.02	1.10	2.0	146.04	19AU	2.4	-3.4	-2.4	0.6
121.28	1.17	2.0	155.82	29AU	1.9	-3.4	-2.4	0.6
127.49	1.24	1.9	165.65	8SE	-0.2	-3.4	-2.5	0.6
133.66	1.31	1.9	175.55	18SE	-0.9	-3.4	-2.4	0.6
139.77	1.38	1.9	185.51	28SE	-1.0	-3.3	-2.4	0.6
145.82	1.45	1.8	195.52	8OC	-0.9	-3.3	-2.4	0.6
151.81	1.51	1.8	205.59	18OC	-0.6	-3.3	-2.3	0.5
157.73	1.58	1.7	215.70	28OC	-0.5	-3.3	-2.3	0.5
163.56	1.65	1.6	225.84	7NO	-0.4	-3.3	-2.2	0.4
169.29	1.71	1.5	236.02	17NO	-0.4	-3.4	-2.1	0.3
174.90	1.78	1.4	246.22	27NO	0.3	-3.4	-2.1	0.3
180.37	1.84	1.3	256.42	7DE	2.9	-3.4	-2.0	0.2
185.67	1.90	1.2	266.63	17DE	0.9	-3.5	-1.9	0.1
190.75	1.96	1.0	276.82	27DE	0.2	-3.5	-1.9	0.1
				327				
195.58	2.02	0.9	286.98	6JA	0.0	-3.6	-1.8	0.0
200.09	2.07	0.7	297.12	16JA	-0.1	-3.6	-1.7	0.0
204.20	2.11	0.5	307.21	26JA	-0.3	-3.7	-1.7	0.1
207.82	2.14	0.2	317.25	5FE	-0.8	-3.7	-1.7	0.2
210.81	2.14	-0.0	327.25	15FE	-1.4	-3.8	-1.6	0.2
213.03	2.12	-0.3	337.18	25FE	-1.3	-3.9	-1.6	0.3
214.29	2.05	-0.6	347.05	7MR	-0.4	-4.0	-1.6	0.3
214.42	1.92	-0.9	356.87	17MR	1.0	-4.1	-1.6	0.4
213.30	1.70	-1.2	6.63	27MR	2.8	-4.2	-1.6	0.4
210.97	1.38	-1.5	16.33	6AP	2.9	-4.2	-1.6	0.5
207.74	0.96	-1.8	25.99	16AP	1.6	-4.2	-1.6	0.5
204.23	0.47	-1.7	35.60	26AP	0.9	-4.0	-1.6	0.6
201.15	-0.04	-1.6	45.18	6MY	0.3	-3.6	-1.6	0.6
199.11	-0.52	-1.4	54.74	16MY	-0.5	-2.7	-1.6	0.6
198.41	-0.93	-1.2	64.26	26MY	-1.5	-3.5	-1.7	0.6
199.06	-1.27	-1.0	73.79	5JN	-1.5	-4.0	-1.7	0.6
200.94	-1.55	-0.8	83.31	15JN	-0.6	-4.2	-1.8	0.6
203.88	-1.76	-0.6	92.84	25JN	0.0	-4.2	-1.8	0.6
207.69	-1.92	-0.4	102.39	5JL	0.5	-4.1	-1.9	0.6
212.23	-2.04	-0.3	111.97	15JL	0.8	-4.1	-1.9	0.6
217.35	-2.13	-0.1	121.57	25JL	1.4	-4.0	-2.0	0.6
222.96	-2.18	0.0	131.23	4AU	2.9	-3.9	-2.1	0.7
228.92	-2.20	0.1	140.93	14AU	1.7	-3.8	-2.1	0.7
235.32	-2.20	0.2	150.68	24AU	-0.3	-3.7	-2.2	0.7
241.95	-2.18	0.3	160.49	3SE	-1.0	-3.7	-2.3	0.8
248.82	-2.13	0.4	170.36	13SE	-1.1	-3.6	-2.3	0.8
255.87	-2.06	0.5	180.28	23SE	-0.9	-3.6	-2.4	0.8
263.08	-1.98	0.6	190.27	3OC	-0.5	-3.5	-2.4	0.8
270.42	-1.88	0.7	200.31	13OC	-0.3	-3.5	-2.4	0.8
277.85	-1.77	0.8	210.39	23OC	-0.3	-3.4	-2.4	0.8
285.36	-1.65	0.8	220.52	2NO	-0.2	-3.4	-2.4	0.7
292.90	-1.52	0.9	230.69	12NO	0.5	-3.4	-2.3	0.7
300.47	-1.39	1.0	240.87	22NO	3.2	-3.4	-2.3	0.6
308.05	-1.25	1.1	251.08	2DE	-0.7	-3.4	-2.2	0.6
315.62	-1.10	1.2	261.28	12DE	-0.0	-3.3	-2.2	0.5
323.16	-0.96	1.2	271.48	22DE	-0.1	-3.3	-2.1	0.4

♂ LONG	♂ LAT	♂ MAG	☉ LONG	16.00UT	☿	♀	♃	♄
				328	MAGNITUDES			
330.66	-0.82	1.3	281.66	1JA	-0.2	-3.3	-2.0	0.3
338.12	-0.67	1.4	291.81	11JA	-0.4	-3.4	-2.0	0.3
345.51	-0.54	1.4	301.92	21JA	-0.8	-3.4	-1.9	0.2
352.85	-0.40	1.5	312.00	31JA	-1.3	-3.4	-1.8	0.3
0.12	-0.27	1.6	322.01	10FE	-1.2	-3.4	-1.8	0.3
7.31	-0.15	1.6	331.97	20FE	-0.4	-3.4	-1.7	0.4
14.44	-0.03	1.7	341.88	1MR	1.3	-3.5	-1.6	0.4
21.49	0.08	1.7	351.73	11MR	3.3	-3.5	-1.6	0.5
28.47	0.19	1.8	1.51	21MR	2.2	-3.5	-1.6	0.5
35.38	0.29	1.8	11.25	31MR	1.2	-3.4	-1.5	0.6
42.22	0.39	1.8	20.93	10AP	0.7	-3.4	-1.5	0.6
49.00	0.48	1.8	30.56	20AP	0.2	-3.4	-1.5	0.7
55.72	0.56	1.9	40.16	30AP	-0.6	-3.4	-1.5	0.7
62.38	0.64	1.9	49.73	10MY	-1.5	-3.3	-1.5	0.8
68.99	0.72	1.9	59.27	20MY	-1.5	-3.3	-1.5	0.8
75.55	0.79	1.9	68.80	30MY	-0.6	-3.3	-1.5	0.8
82.07	0.85	1.9	78.32	9JN	0.1	-3.3	-1.5	0.8
88.55	0.91	1.9	87.84	19JN	0.6	-3.3	-1.6	0.8
95.00	0.96	1.9	97.38	29JN	1.1	-3.4	-1.6	0.8
101.42	1.01	2.0	106.94	9JL	1.9	-3.4	-1.6	0.8
107.82	1.06	2.0	116.53	19JL	3.2	-3.4	-1.7	0.8
114.19	1.10	2.0	126.16	29JL	1.4	-3.5	-1.7	0.8
120.55	1.14	2.0	135.84	8AU	-0.3	-3.5	-1.8	0.9
126.90	1.18	2.0	145.56	18AU	-1.1	-3.6	-1.8	0.9
133.23	1.21	2.0	155.34	28AU	-1.2	-3.6	-1.9	1.0
139.56	1.23	2.0	165.18	7SE	-0.8	-3.7	-2.0	1.0
145.88	1.26	2.0	175.07	17SE	-0.4	-3.8	-2.0	1.0
152.19	1.28	2.0	185.03	27SE	-0.2	-3.9	-2.1	1.0
158.49	1.29	2.0	195.03	7OC	-0.1	-4.0	-2.2	1.1
164.78	1.30	1.9	205.10	17OC	-0.0	-4.1	-2.2	1.1
171.06	1.31	1.9	215.21	27OC	0.7	-4.2	-2.3	1.0
177.33	1.31	1.8	225.35	6NO	3.3	-4.3	-2.3	1.0
183.59	1.31	1.8	235.53	16NO	0.5	-4.4	-2.3	1.0
189.82	1.29	1.7	245.73	26NO	-0.2	-4.3	-2.3	0.9
196.02	1.27	1.6	255.93	6DE	-0.3	-4.1	-2.3	0.9
202.19	1.24	1.5	266.13	16DE	-0.3	-3.4	-2.2	0.8
208.31	1.21	1.4	276.33	26DE	-0.5	-3.3	-2.2	0.7
				329				
214.37	1.15	1.3	286.49	5JA	-0.9	-4.0	-2.1	0.7
220.37	1.09	1.1	296.63	15JA	-1.2	-4.3	-2.1	0.6
226.28	1.00	1.0	306.72	25JA	-1.1	-4.3	-2.0	0.5
232.08	0.90	0.8	316.76	4FE	-0.3	-4.3	-1.9	0.5
237.76	0.77	0.7	326.76	14FE	1.5	-4.2	-1.8	0.5
243.26	0.61	0.5	336.70	24FE	3.1	-4.1	-1.8	0.6
248.56	0.41	0.3	346.57	6MR	1.6	-4.0	-1.7	0.6
253.60	0.17	0.1	356.39	16MR	0.9	-3.9	-1.7	0.7
258.30	-0.12	-0.2	6.15	26MR	0.5	-3.8	-1.6	0.7
262.57	-0.47	-0.4	15.86	5AP	0.1	-3.7	-1.5	0.8
266.31	-0.89	-0.7	25.52	15AP	-0.6	-3.6	-1.5	0.8
269.34	-1.40	-1.0	35.14	25AP	-1.5	-3.5	-1.5	0.9
271.51	-2.00	-1.3	44.72	5MY	-1.5	-3.5	-1.4	0.9
272.62	-2.69	-1.6	54.27	15MY	-0.6	-3.4	-1.4	1.0
272.52	-3.45	-1.9	63.81	25MY	0.2	-3.4	-1.4	1.0
271.21	-4.22	-2.2	73.33	4JN	0.8	-3.4	-1.4	1.0
268.90	-4.91	-2.5	82.85	14JN	1.5	-3.3	-1.4	1.1
266.14	-5.42	-2.5	92.38	24JN	2.5	-3.3	-1.4	1.1
263.66	-5.68	-2.4	101.93	4JL	3.0	-3.3	-1.4	1.1
262.10	-5.69	-2.2	111.51	14JL	1.2	-3.3	-1.4	1.1
261.83	-5.50	-1.9	121.12	24JL	-0.3	-3.3	-1.5	1.1
262.92	-5.18	-1.7	130.76	3AU	-1.2	-3.3	-1.5	1.1
265.24	-4.80	-1.4	140.46	13AU	-1.4	-3.4	-1.5	1.1
268.60	-4.38	-1.2	150.21	23AU	-0.8	-3.4	-1.6	1.2
272.81	-3.96	-0.9	160.01	2SE	-0.3	-3.4	-1.6	1.2
277.68	-3.55	-0.7	169.88	12SE	-0.1	-3.4	-1.7	1.3
283.09	-3.15	-0.5	179.80	22SE	0.0	-3.5	-1.7	1.3
288.89	-2.77	-0.3	189.78	2OC	0.2	-3.5	-1.8	1.3
294.99	-2.41	-0.1	199.82	12OC	0.9	-3.5	-1.9	1.3
301.34	-2.07	0.1	209.90	22OC	3.1	-3.5	-1.9	1.4
307.86	-1.76	0.2	220.02	1NO	0.3	-3.4	-2.0	1.4
314.51	-1.47	0.4	230.19	11NO	-0.4	-3.4	-2.1	1.3
321.24	-1.20	0.6	240.37	21NO	-0.4	-3.4	-2.1	1.3
328.04	-0.95	0.7	250.57	1DE	-0.4	-3.4	-2.1	1.2
334.87	-0.72	0.8	260.78	11DE	-0.6	-3.3	-2.2	1.2
341.72	-0.51	1.0	270.98	21DE	-0.9	-3.3	-2.2	1.1
348.57	-0.32	1.1	281.16	31DE	-1.0	-3.3	-2.2	1.1

♂ LONG	♂ LAT	♂ MAG	☉ LONG	16.00UT	☿	♀	♃	♄
				330	MAGNITUDES			
355.40	-0.15	1.2	291.32	10JA	-1.0	-3.3	-2.1	1.0
2.22	0.01	1.3	301.43	20JA	-0.2	-3.4	-2.1	1.0
9.00	0.15	1.4	311.50	30JA	1.8	-3.4	-2.1	0.9
15.76	0.28	1.5	321.53	9FE	2.5	-3.4	-2.0	0.8
22.47	0.40	1.6	331.49	19FE	1.2	-3.4	-1.9	0.8
29.15	0.50	1.7	341.40	1MR	0.6	-3.5	-1.9	0.8
35.79	0.60	1.7	351.25	11MR	0.4	-3.5	-1.8	0.9
42.38	0.68	1.8	1.04	21MR	-0.0	-3.5	-1.7	0.9
48.94	0.76	1.8	10.78	31MR	-0.6	-3.6	-1.7	1.0
55.46	0.83	1.9	20.46	10AP	-1.5	-3.7	-1.6	1.0
61.95	0.89	1.9	30.10	20AP	-1.5	-3.7	-1.5	1.1
68.40	0.94	2.0	39.70	30AP	-0.5	-3.8	-1.5	1.1
74.83	0.99	2.0	49.27	10MY	0.4	-3.9	-1.4	1.2
81.24	1.03	2.0	58.81	20MY	1.1	-4.0	-1.4	1.2
87.62	1.06	2.0	68.34	30MY	1.9	-4.1	-1.4	1.3
94.00	1.09	2.0	77.86	9JN	3.2	-4.1	-1.3	1.3
100.36	1.11	2.0	87.38	19JN	2.4	-4.2	-1.3	1.3
106.71	1.13	2.0	96.92	29JN	0.9	-4.1	-1.3	1.4
113.07	1.15	2.0	106.48	9JL	-0.4	-3.9	-1.3	1.4
119.43	1.15	2.0	116.07	19JL	-1.3	-3.4	-1.3	1.4
125.79	1.16	2.0	125.70	29JL	-1.4	-3.2	-1.3	1.4
132.17	1.16	2.0	135.37	8AU	-0.7	-3.8	-1.3	1.4
138.56	1.15	2.0	145.09	18AU	-0.3	-4.2	-1.3	1.3
144.97	1.14	2.0	154.87	28AU	0.0	-4.3	-1.4	1.3
151.41	1.13	2.0	164.70	7SE	0.2	-4.3	-1.4	1.3
157.87	1.11	2.0	174.59	17SE	0.4	-4.2	-1.4	1.3
164.35	1.09	2.0	184.54	27SE	1.2	-4.1	-1.5	1.3
170.87	1.06	2.0	194.55	7OC	2.8	-4.0	-1.5	1.3
177.41	1.02	1.9	204.61	17OC	0.1	-3.9	-1.6	1.2
183.99	0.98	1.9	214.71	27OC	-0.5	-3.9	-1.7	1.2
190.60	0.93	1.9	224.86	6NO	-0.6	-3.8	-1.7	1.2
197.24	0.87	1.8	235.03	16NO	-0.6	-3.7	-1.8	1.2
203.91	0.81	1.8	245.23	26NO	-0.7	-3.6	-1.9	1.1
210.61	0.74	1.7	255.44	6DE	-0.8	-3.6	-1.9	1.1
217.35	0.65	1.7	265.64	16DE	-0.8	-3.5	-2.0	1.1
224.11	0.56	1.6	275.83	26DE	-0.8	-3.5	-2.0	1.0
				331				
230.90	0.45	1.5	286.00	5JA	-0.1	-3.4	-2.0	1.0
237.72	0.33	1.4	296.14	15JA	2.1	-3.4	-2.1	0.9
244.56	0.20	1.3	306.24	25JA	2.0	-3.4	-2.1	0.9
251.41	0.05	1.2	316.29	4FE	0.8	-3.3	-2.1	0.8
258.28	-0.11	1.1	326.28	14FE	0.4	-3.3	-2.0	0.8
265.16	-0.30	1.0	336.22	24FE	0.2	-3.3	-2.0	0.7
272.03	-0.50	0.9	346.10	6MR	-0.1	-3.3	-1.9	0.8
278.90	-0.72	0.7	355.92	16MR	-0.6	-3.3	-1.9	0.8
285.74	-0.96	0.6	5.68	26MR	-1.5	-3.3	-1.8	0.9
292.53	-1.22	0.5	15.40	5AP	-1.5	-3.3	-1.8	1.0
299.26	-1.50	0.3	25.05	15AP	-0.5	-3.4	-1.7	1.0
305.90	-1.80	0.2	34.67	25AP	0.5	-3.4	-1.6	1.1
312.40	-2.12	0.0	44.26	5MY	1.4	-3.5	-1.6	1.2
318.72	-2.46	-0.2	53.81	15MY	2.6	-3.5	-1.5	1.2
324.81	-2.81	-0.3	63.34	25MY	3.4	-3.5	-1.5	1.3
330.57	-3.18	-0.5	72.87	4JN	1.9	-3.4	-1.4	1.3
335.94	-3.57	-0.7	82.39	14JN	0.7	-3.4	-1.4	1.3
340.77	-3.96	-0.9	91.92	24JN	-0.4	-3.4	-1.3	1.3
344.92	-4.36	-1.1	101.47	4JL	-1.4	-3.4	-1.3	1.3
348.21	-4.75	-1.4	111.04	14JL	-1.5	-3.3	-1.3	1.3
350.41	-5.11	-1.6	120.64	24JL	-0.7	-3.3	-1.2	1.3
351.33	-5.41	-1.8	130.29	3AU	-0.2	-3.3	-1.2	1.3
350.84	-5.58	-2.1	139.98	13AU	0.1	-3.3	-1.2	1.2
348.98	-5.55	-2.3	149.73	23AU	0.3	-3.3	-1.2	1.2
346.18	-5.28	-2.4	159.54	2SE	0.6	-3.4	-1.2	1.1
343.12	-4.76	-2.4	169.40	12SE	1.6	-3.4	-1.3	1.1
340.58	-4.07	-2.1	179.32	22SE	2.5	-3.4	-1.3	1.1
339.16	-3.32	-1.8	189.30	2OC	0.0	-3.4	-1.3	1.1
339.05	-2.59	-1.4	199.33	12OC	-0.7	-3.5	-1.4	1.1
340.21	-1.94	-1.1	209.41	22OC	-0.7	-3.5	-1.4	1.1
342.48	-1.37	-0.8	219.54	1NO	-0.7	-3.6	-1.4	1.1
345.63	-0.90	-0.5	229.70	11NO	-0.8	-3.7	-1.5	1.1
349.49	-0.51	-0.2	239.89	21NO	-0.7	-3.8	-1.6	1.0
353.90	-0.18	0.0	250.08	1DE	-0.7	-3.8	-1.6	1.0
358.72	0.09	0.2	260.29	11DE	-0.7	-3.9	-1.7	1.0
3.87	0.31	0.5	270.49	21DE	-0.0	-4.0	-1.8	0.9
9.26	0.50	0.7	280.67	31DE	2.3	-4.1	-1.8	0.9

♂			☉	16.00UT	☿	♀	♃	♄
LONG	LAT	MAG	LONG	332	MAGNITUDES			
14.84	0.66	0.8	290.83	10JA	1.5	-4.2	-1.9	0.9
20.57	0.78	1.0	300.95	20JA	0.5	-4.3	-1.9	0.8
26.41	0.89	1.1	311.02	30JA	0.3	-4.3	-2.0	0.8
32.33	0.98	1.3	321.04	9FE	0.1	-4.3	-2.0	0.7
38.31	1.05	1.4	331.01	19FE	-0.2	-4.0	-2.0	0.7
44.35	1.11	1.5	340.92	29FE	-0.7	-3.4	-2.0	0.6
50.41	1.16	1.6	350.77	10MR	-1.5	-3.4	-2.0	0.6
56.51	1.19	1.7	0.57	20MR	-1.4	-3.9	-2.0	0.6
62.63	1.22	1.8	10.31	30MR	-0.5	-4.2	-1.9	0.7
68.76	1.24	1.8	19.99	9AP	0.7	-4.2	-1.9	0.8
74.91	1.26	1.9	29.64	19AP	1.9	-4.2	-1.8	0.8
81.07	1.27	1.9	39.23	29AP	3.4	-4.1	-1.8	0.9
87.24	1.27	2.0	48.80	9MY	2.7	-4.0	-1.7	1.0
93.44	1.27	2.0	58.35	19MY	1.5	-3.9	-1.6	1.0
99.64	1.26	2.0	67.87	29MY	0.6	-3.8	-1.6	1.1
105.87	1.24	2.0	77.40	8JN	-0.4	-3.7	-1.5	1.1
112.12	1.23	2.0	86.92	18JN	-1.4	-3.6	-1.5	1.1
118.39	1.20	2.0	96.46	28JN	-1.5	-3.6	-1.4	1.1
124.69	1.18	2.0	106.02	8JL	-0.7	-3.5	-1.4	1.2
131.03	1.15	2.0	115.61	18JL	-0.1	-3.5	-1.3	1.1
137.40	1.11	2.0	125.23	28JL	0.2	-3.4	-1.3	1.1
143.81	1.08	2.0	134.90	7AU	0.5	-3.4	-1.3	1.1
150.26	1.03	1.9	144.62	17AU	0.9	-3.4	-1.2	1.1
156.76	0.99	1.9	154.39	27AU	2.0	-3.4	-1.2	1.1
163.31	0.93	1.9	164.22	6SE	2.2	-3.4	-1.2	1.0
169.91	0.88	1.9	174.12	16SE	-0.1	-3.4	-1.2	1.0
176.57	0.82	1.9	184.06	26SE	-0.8	-3.4	-1.2	0.9
183.29	0.75	1.9	194.07	6OC	-0.9	-3.4	-1.2	1.0
190.07	0.68	1.9	204.13	16OC	-0.8	-3.4	-1.3	1.0
196.90	0.60	1.8	214.23	26OC	-0.7	-3.4	-1.3	1.0
203.81	0.52	1.8	224.37	5NO	-0.5	-3.4	-1.3	1.0
210.77	0.43	1.8	234.55	15NO	-0.5	-3.4	-1.4	1.0
217.79	0.33	1.8	244.74	25NO	-0.5	-3.4	-1.4	1.0
224.88	0.23	1.7	254.94	5DE	0.1	-3.5	-1.5	0.9
232.03	0.12	1.7	265.15	15DE	2.6	-3.5	-1.5	0.9
239.23	0.00	1.6	275.34	25DE	1.2	-3.5	-1.6	0.9
				333				
246.50	-0.12	1.6	285.51	4JA	0.3	-3.5	-1.7	0.8
253.82	-0.25	1.5	295.65	14JA	0.1	-3.4	-1.7	0.8
261.18	-0.39	1.5	305.74	24JA	-0.0	-3.4	-1.8	0.7
268.60	-0.52	1.4	315.80	3FE	-0.3	-3.4	-1.9	0.7
276.05	-0.67	1.3	325.79	13FE	-0.7	-3.4	-1.9	0.6
283.53	-0.82	1.3	335.73	23FE	-1.4	-3.3	-2.0	0.6
291.04	-0.97	1.2	345.62	5MR	-1.4	-3.3	-2.0	0.5
298.56	-1.11	1.1	355.44	15MR	-0.5	-3.3	-2.0	0.5
306.08	-1.26	1.1	5.21	25MR	0.9	-3.3	-2.0	0.5
313.59	-1.41	1.0	14.92	4AP	2.4	-3.3	-2.0	0.5
321.07	-1.54	0.9	24.59	14AP	3.5	-3.4	-2.0	0.6
328.50	-1.67	0.8	34.21	24AP	2.0	-3.4	-2.0	0.7
335.88	-1.79	0.8	43.79	4MY	1.1	-3.4	-1.9	0.7
343.16	-1.90	0.7	53.35	14MY	0.4	-3.4	-1.9	0.8
350.33	-2.00	0.6	62.88	24MY	-0.5	-3.5	-1.8	0.8
357.37	-2.08	0.5	72.41	3JN	-1.4	-3.5	-1.8	0.9
4.24	-2.14	0.5	81.93	13JN	-1.5	-3.6	-1.7	0.9
10.91	-2.18	0.4	91.46	23JN	-0.6	-3.7	-1.6	1.0
17.33	-2.20	0.3	101.01	3JL	-0.0	-3.7	-1.6	1.0
23.46	-2.21	0.2	110.58	13JL	0.4	-3.8	-1.5	1.0
29.24	-2.19	0.0	120.18	23JL	0.7	-3.9	-1.5	1.0
34.59	-2.14	-0.1	129.83	2AU	1.2	-4.0	-1.4	1.0
39.42	-2.07	-0.2	139.52	12AU	2.5	-4.1	-1.4	1.0
43.60	-1.96	-0.4	149.26	22AU	1.9	-4.2	-1.3	1.0
46.98	-1.81	-0.6	159.06	1SE	-0.2	-4.3	-1.3	1.0
49.37	-1.62	-0.8	168.92	11SE	-0.9	-4.2	-1.3	0.9
50.57	-1.36	-1.0	178.84	21SE	-1.0	-4.0	-1.3	0.9
50.39	-1.03	-1.2	188.82	1OC	-0.9	-3.4	-1.2	0.8
48.74	-0.63	-1.4	198.84	11OC	-0.6	-3.4	-1.2	0.9
45.80	-0.17	-1.6	208.92	21OC	-0.4	-4.0	-1.2	0.9
42.09	0.32	-1.7	219.05	31OC	-0.4	-4.3	-1.3	0.9
38.40	0.77	-1.5	229.20	10NO	-0.3	-4.4	-1.3	0.9
35.47	1.15	-1.3	239.39	20NO	0.3	-4.3	-1.3	0.9
33.79	1.43	-0.9	249.59	30NO	2.9	-4.3	-1.3	0.9
33.47	1.63	-0.6	259.80	10DE	0.9	-4.2	-1.4	0.9
34.39	1.75	-0.3	270.00	20DE	0.1	-4.0	-1.4	0.9
36.37	1.83	-0.0	280.18	30DE	-0.0	-3.9	-1.5	0.8

♂			☉	16.00UT	☿	♀	♃	♄
LONG	LAT	MAG	LONG	334	MAGNITUDES			
39.20	1.87	0.2	290.34	9JA	-0.1	-3.8	-1.5	0.8
42.69	1.88	0.5	300.46	19JA	-0.4	-3.8	-1.6	0.8
46.71	1.88	0.7	310.54	29JA	-0.8	-3.7	-1.7	0.7
51.14	1.86	0.9	320.56	8FE	-1.4	-3.6	-1.7	0.7
55.89	1.84	1.0	330.53	18FE	-1.3	-3.5	-1.8	0.6
60.89	1.81	1.2	340.45	28FE	-0.4	-3.5	-1.9	0.6
66.09	1.77	1.3	350.30	10MR	1.1	-3.4	-1.9	0.5
71.46	1.73	1.4	0.10	20MR	3.0	-3.4	-2.0	0.4
76.96	1.68	1.5	9.84	30MR	2.6	-3.3	-2.0	0.4
82.58	1.64	1.6	19.53	9AP	1.5	-3.3	-2.1	0.4
88.29	1.59	1.7	29.17	19AP	0.8	-3.3	-2.1	0.4
94.08	1.54	1.8	38.78	29AP	0.3	-3.3	-2.1	0.5
99.95	1.48	1.8	48.34	9MY	-0.5	-3.3	-2.1	0.6
105.89	1.42	1.9	57.89	19MY	-1.5	-3.3	-2.1	0.6
111.90	1.37	1.9	67.42	29MY	-1.5	-3.3	-2.0	0.7
117.96	1.31	1.9	76.94	8JN	-0.6	-3.3	-2.0	0.8
124.09	1.24	2.0	86.46	18JN	0.1	-3.4	-1.9	0.8
130.28	1.18	2.0	96.00	28JN	0.5	-3.4	-1.9	0.9
136.53	1.11	2.0	105.55	8JL	0.9	-3.5	-1.8	0.9
142.85	1.04	2.0	115.14	18JL	1.6	-3.5	-1.7	0.9
149.24	0.97	2.0	124.76	28JL	3.0	-3.5	-1.7	0.9
155.69	0.89	1.9	134.43	7AU	1.6	-3.5	-1.6	0.9
162.22	0.81	1.9	144.14	17AU	-0.2	-3.4	-1.6	0.9
168.82	0.73	1.9	153.92	27AU	-1.1	-3.4	-1.5	0.9
175.50	0.65	1.9	163.74	6SE	-1.2	-3.4	-1.5	0.9
182.26	0.56	1.8	173.63	16SE	-0.9	-3.3	-1.4	0.9
189.09	0.47	1.8	183.57	26SE	-0.5	-3.3	-1.4	0.8
196.01	0.37	1.7	193.57	6OC	-0.3	-3.3	-1.4	0.8
203.01	0.28	1.7	203.63	16OC	-0.2	-3.3	-1.3	0.8
210.09	0.18	1.7	213.73	26OC	-0.1	-3.3	-1.3	0.8
217.26	0.07	1.7	223.87	5NO	0.5	-3.3	-1.3	0.8
224.50	-0.03	1.6	234.05	15NO	3.2	-3.4	-1.3	0.8
231.82	-0.14	1.6	244.24	25NO	0.6	-3.4	-1.3	0.8
239.22	-0.25	1.6	254.44	5DE	-0.1	-3.4	-1.3	0.8
246.69	-0.36	1.6	264.65	15DE	-0.2	-3.5	-1.4	0.8
254.22	-0.47	1.6	274.84	25DE	-0.3	-3.5	-1.4	0.8
				335				
261.81	-0.58	1.5	285.01	4JA	-0.4	-3.6	-1.4	0.8
269.45	-0.68	1.5	295.16	14JA	-0.8	-3.6	-1.5	0.8
277.14	-0.79	1.5	305.26	24JA	-1.2	-3.7	-1.5	0.7
284.86	-0.89	1.4	315.31	3FE	-1.2	-3.7	-1.6	0.7
292.61	-0.98	1.4	325.32	13FE	-0.4	-3.8	-1.6	0.6
300.37	-1.07	1.4	335.26	23FE	1.3	-3.9	-1.7	0.6
308.14	-1.15	1.4	345.14	5MR	3.4	-4.0	-1.8	0.5
315.89	-1.22	1.3	354.97	15MR	1.9	-4.1	-1.9	0.5
323.62	-1.28	1.3	4.74	25MR	1.1	-4.2	-1.9	0.4
331.31	-1.33	1.3	14.46	4AP	0.6	-4.2	-2.0	0.4
338.96	-1.36	1.3	24.13	14AP	0.2	-4.2	-2.1	0.3
346.54	-1.38	1.2	33.75	24AP	-0.5	-4.0	-2.1	0.3
354.05	-1.38	1.2	43.33	4MY	-1.5	-3.6	-2.2	0.4
1.47	-1.37	1.2	52.89	14MY	-1.5	-2.7	-2.2	0.5
8.79	-1.35	1.2	62.42	24MY	-0.6	-3.5	-2.2	0.5
16.00	-1.31	1.1	71.95	3JN	0.2	-4.0	-2.2	0.6
23.07	-1.25	1.1	81.47	13JN	0.7	-4.2	-2.2	0.7
30.01	-1.18	1.1	91.00	23JN	1.2	-4.2	-2.2	0.7
36.79	-1.09	1.0	100.54	3JL	2.1	-4.1	-2.1	0.8
43.40	-0.99	1.0	110.11	13JL	3.2	-4.1	-2.1	0.8
49.80	-0.87	0.9	119.71	23JL	1.4	-4.0	-2.0	0.8
55.99	-0.74	0.9	129.35	2AU	-0.3	-3.9	-2.0	0.9
61.93	-0.59	0.8	139.05	12AU	-1.2	-3.8	-1.9	0.9
67.57	-0.42	0.7	148.78	22AU	-1.3	-3.7	-1.8	0.9
72.87	-0.23	0.6	158.58	1SE	-0.8	-3.7	-1.8	0.9
77.76	-0.01	0.5	168.44	11SE	-0.4	-3.6	-1.7	0.9
82.15	0.24	0.3	178.35	21SE	-0.2	-3.6	-1.7	0.8
85.93	0.51	0.2	188.32	1OC	-0.1	-3.5	-1.6	0.8
88.97	0.83	-0.0	198.35	11OC	0.0	-3.5	-1.6	0.8
91.08	1.20	-0.2	208.43	21OC	0.7	-3.4	-1.5	0.7
92.08	1.61	-0.4	218.55	31OC	3.3	-3.4	-1.5	0.7
91.77	2.06	-0.6	228.71	10NO	0.4	-3.4	-1.5	0.8
90.08	2.52	-0.9	238.89	20NO	-0.3	-3.4	-1.4	0.8
87.11	2.96	-1.1	249.09	30NO	-0.3	-3.4	-1.4	0.8
83.29	3.30	-1.2	259.30	10DE	-0.4	-3.4	-1.4	0.8
79.37	3.51	-1.1	269.50	20DE	-0.5	-3.3	-1.4	0.8
76.07	3.57	-0.9	279.69	30DE	-0.9	-3.3	-1.4	0.8

Left table (336 / 337):

♂ LONG	LAT	MAG	☉ LONG	16.00UT	☿	♀	♃	♄
				336		MAGNITUDES		
73.90	3.51	-0.6	289.85	9JA	-1.1	-3.4	-1.4	0.8
73.06	3.37	-0.3	299.97	19JA	-1.0	-3.4	-1.5	0.8
73.48	3.20	-0.0	310.05	29JA	-0.3	-3.4	-1.5	0.7
75.00	3.01	0.2	320.08	8FE	1.6	-3.4	-1.5	0.7
77.41	2.82	0.5	330.05	18FE	2.9	-3.4	-1.6	0.7
80.53	2.63	0.7	339.97	28FE	1.4	-3.5	-1.6	0.7
84.23	2.46	0.8	349.83	9MR	0.8	-3.5	-1.7	0.6
88.38	2.30	1.0	359.62	19MR	0.4	-3.5	-1.8	0.5
92.89	2.15	1.1	9.37	29MR	0.0	-3.4	-1.8	0.4
97.70	2.00	1.3	19.06	8AP	-0.6	-3.4	-1.9	0.4
102.76	1.86	1.4	28.70	18AP	-1.5	-3.4	-2.0	0.3
108.02	1.73	1.5	38.31	28AP	-1.5	-3.3	-2.0	0.3
113.46	1.61	1.6	47.88	8MY	-0.6	-3.3	-2.1	0.3
119.06	1.49	1.6	57.42	18MY	0.3	-3.3	-2.2	0.4
124.79	1.37	1.7	66.95	28MY	0.9	-3.3	-2.2	0.4
130.65	1.25	1.7	76.48	7JN	1.6	-3.3	-2.3	0.5
136.63	1.14	1.8	86.00	17JN	2.8	-3.3	-2.3	0.6
142.71	1.03	1.8	95.54	27JN	2.8	-3.4	-2.3	0.6
148.91	0.92	1.8	105.09	7JL	1.1	-3.4	-2.3	0.7
155.21	0.81	1.8	114.67	17JL	-0.3	-3.4	-2.3	0.7
161.61	0.70	1.8	124.30	27JL	-1.3	-3.5	-2.3	0.8
168.12	0.59	1.8	133.96	6AU	-1.4	-3.5	-2.3	0.8
174.72	0.48	1.8	143.67	16AU	-0.8	-3.6	-2.2	0.8
181.44	0.37	1.8	153.44	26AU	-0.3	-3.6	-2.2	0.8
188.25	0.26	1.8	163.26	5SE	-0.1	-3.7	-2.1	0.8
195.16	0.15	1.7	173.15	15SE	0.1	-3.8	-2.0	0.8
202.18	0.04	1.7	183.09	25SE	0.2	-3.9	-2.0	0.8
209.29	-0.06	1.7	193.09	5OC	1.0	-4.0	-1.9	0.8
216.50	-0.17	1.6	203.14	15OC	3.1	-4.1	-1.8	0.8
223.81	-0.28	1.6	213.24	25OC	0.2	-4.2	-1.8	0.7
231.21	-0.38	1.5	223.38	4NO	-0.4	-4.3	-1.7	0.7
238.70	-0.48	1.5	233.55	14NO	-0.5	-4.4	-1.7	0.7
246.26	-0.57	1.4	243.74	24NO	-0.5	-4.3	-1.6	0.7
253.90	-0.67	1.4	253.94	4DE	-0.6	-4.1	-1.6	0.8
261.60	-0.75	1.4	264.15	14DE	-0.9	-3.4	-1.6	0.8
269.37	-0.83	1.4	274.35	24DE	-0.9	-3.3	-1.5	0.8
				337				
277.17	-0.91	1.4	284.52	3JA	-0.9	-4.0	-1.5	0.8
285.01	-0.97	1.4	294.66	13JA	-0.2	-4.3	-1.5	0.8
292.87	-1.02	1.4	304.77	23JA	1.8	-4.3	-1.5	0.8
300.73	-1.07	1.4	314.82	2FE	2.3	-4.3	-1.5	0.8
308.60	-1.10	1.4	324.82	12FE	1.0	-4.2	-1.5	0.7
316.45	-1.12	1.4	334.78	22FE	0.6	-4.1	-1.6	0.7
324.27	-1.13	1.4	344.66	4MR	0.3	-4.0	-1.6	0.6
332.05	-1.13	1.4	354.49	14MR	-0.0	-3.9	-1.6	0.6
339.78	-1.11	1.4	4.27	24MR	-0.6	-3.8	-1.7	0.5
347.45	-1.08	1.4	13.99	3AP	-1.4	-3.7	-1.7	0.5
355.05	-1.05	1.4	23.66	13AP	-1.5	-3.6	-1.8	0.4
2.57	-0.99	1.5	33.28	23AP	-0.6	-3.5	-1.8	0.3
9.99	-0.93	1.5	42.87	3MY	0.4	-3.5	-1.9	0.3
17.33	-0.86	1.5	52.43	13MY	1.2	-3.4	-2.0	0.3
24.57	-0.78	1.5	61.97	23MY	2.2	-3.4	-2.1	0.3
31.70	-0.69	1.5	71.49	2JN	3.4	-3.4	-2.1	0.4
38.72	-0.59	1.5	81.01	12JN	2.2	-3.3	-2.2	0.4
45.63	-0.48	1.5	90.54	22JN	0.9	-3.3	-2.3	0.5
52.42	-0.36	1.5	100.08	2JL	-0.4	-3.3	-2.3	0.6
59.08	-0.24	1.4	109.65	12JL	-1.3	-3.3	-2.4	0.6
65.61	-0.10	1.4	119.25	22JL	-1.5	-3.4	-2.4	0.7
71.99	0.04	1.4	128.89	1AU	-0.7	-3.3	-2.4	0.7
78.23	0.19	1.4	138.58	11AU	-0.2	-3.4	-2.4	0.7
84.29	0.35	1.3	148.32	21AU	0.1	-3.4	-2.4	0.8
90.16	0.52	1.3	158.11	31AU	0.2	-3.4	-2.4	0.8
95.82	0.70	1.2	167.96	10SE	0.5	-3.4	-2.4	0.8
101.22	0.89	1.1	177.87	20SE	1.3	-3.5	-2.3	0.8
106.34	1.11	1.0	187.84	30SE	2.8	-3.5	-2.2	0.8
111.12	1.34	0.9	197.86	10OC	0.1	-3.5	-2.2	0.8
115.47	1.60	0.7	207.94	20OC	-0.6	-3.5	-2.1	0.8
119.31	1.88	0.6	218.05	30OC	-0.6	-3.4	-2.0	0.7
122.51	2.20	0.4	228.21	9NO	-0.6	-3.4	-2.0	0.7
124.93	2.56	0.2	238.39	19NO	-0.7	-3.4	-1.9	0.7
126.40	2.95	-0.0	248.59	29NO	-0.7	-3.4	-1.8	0.7
126.72	3.36	-0.3	258.80	9DE	-0.8	-3.3	-1.8	0.7
125.73	3.77	-0.5	269.00	19DE	-0.7	-3.3	-1.7	0.8
123.44	4.15	-0.7	279.18	29DE	-0.1	-3.3	-1.7	0.8

Right table (338 / 339):

♂ LONG	LAT	MAG	☉ LONG	16.00UT	☿	♀	♃	♄
				338		MAGNITUDES		
120.05	4.42	-0.9	289.34	8JA	2.1	-3.3	-1.6	0.8
116.11	4.53	-1.0	299.47	18JA	1.8	-3.4	-1.6	0.8
112.35	4.46	-0.8	309.55	28JA	0.7	-3.4	-1.6	0.8
109.42	4.25	-0.6	319.59	7FE	0.4	-3.4	-1.6	0.8
107.70	3.94	-0.4	329.56	17FE	0.2	-3.4	-1.6	0.7
107.28	3.60	-0.1	339.48	27FE	-0.1	-3.5	-1.6	0.7
108.05	3.25	0.1	349.35	9MR	-0.6	-3.5	-1.6	0.7
109.86	2.91	0.3	359.15	19MR	-1.4	-3.5	-1.6	0.6
112.50	2.61	0.5	8.90	29MR	-1.5	-3.6	-1.6	0.6
115.83	2.33	0.7	18.59	8AP	-0.5	-3.7	-1.6	0.5
119.71	2.07	0.8	28.24	18AP	0.5	-3.7	-1.7	0.4
124.05	1.83	1.0	37.85	28AP	1.6	-3.8	-1.7	0.4
128.75	1.62	1.1	47.42	8MY	2.9	-3.9	-1.8	0.3
133.77	1.41	1.2	56.97	18MY	3.1	-4.0	-1.8	0.3
139.05	1.23	1.3	66.50	28MY	1.7	-4.1	-1.9	0.3
144.57	1.05	1.3	76.02	7JN	0.7	-4.1	-1.9	0.3
150.30	0.88	1.4	85.54	17JN	-0.4	-4.2	-2.0	0.4
156.21	0.72	1.5	95.08	27JN	-1.4	-4.1	-2.1	0.4
162.29	0.57	1.5	104.63	7JL	-1.5	-3.9	-2.1	0.5
168.53	0.42	1.5	114.21	17JL	-0.7	-3.3	-2.2	0.6
174.92	0.28	1.5	123.83	27JL	-0.2	-3.2	-2.3	0.6
181.46	0.14	1.6	133.49	6AU	0.2	-3.8	-2.3	0.7
188.14	0.01	1.6	143.20	16AU	0.4	-4.2	-2.4	0.7
194.94	-0.11	1.6	152.97	26AU	0.7	-4.3	-2.4	0.7
201.88	-0.23	1.6	162.79	5SE	1.7	-4.3	-2.4	0.8
208.94	-0.35	1.6	172.67	15SE	2.5	-4.2	-2.5	0.8
216.12	-0.46	1.5	182.61	25SE	0.0	-4.1	-2.4	0.8
223.42	-0.56	1.5	192.60	5OC	-0.7	-4.0	-2.4	0.8
230.82	-0.65	1.5	202.65	15OC	-0.8	-3.9	-2.4	0.8
238.32	-0.74	1.5	212.75	25OC	-0.8	-3.8	-2.3	0.8
245.91	-0.82	1.5	222.89	4NO	-0.7	-3.8	-2.3	0.7
253.58	-0.89	1.5	233.05	14NO	-0.6	-3.7	-2.2	0.7
261.33	-0.95	1.4	243.25	24NO	-0.6	-3.6	-2.1	0.7
269.13	-1.00	1.4	253.45	4DE	-0.6	-3.6	-2.1	0.7
276.98	-1.04	1.4	263.65	14DE	-0.0	-3.5	-2.0	0.7
284.85	-1.07	1.4	273.85	24DE	2.4	-3.5	-1.9	0.8
				339				
292.76	-1.09	1.4	284.03	3JA	1.4	-3.4	-1.8	0.8
300.66	-1.09	1.3	294.17	13JA	0.5	-3.4	-1.8	0.8
308.55	-1.08	1.3	304.27	23JA	0.2	-3.4	-1.7	0.8
316.43	-1.07	1.3	314.34	2FE	0.0	-3.3	-1.7	0.8
324.27	-1.04	1.3	324.34	12FE	-0.2	-3.3	-1.6	0.8
332.06	-1.00	1.3	334.29	22FE	-0.7	-3.3	-1.6	0.8
339.80	-0.94	1.4	344.19	4MR	-1.4	-3.3	-1.6	0.7
347.47	-0.88	1.4	354.02	14MR	-1.4	-3.3	-1.6	0.7
355.08	-0.81	1.5	3.79	24MR	-0.5	-3.3	-1.6	0.7
2.60	-0.74	1.5	13.52	3AP	0.7	-3.3	-1.6	0.6
10.04	-0.65	1.5	23.19	13AP	2.0	-3.4	-1.6	0.6
17.39	-0.56	1.6	32.81	23AP	3.8	-3.4	-1.6	0.5
24.65	-0.47	1.6	42.41	3MY	2.4	-3.5	-1.6	0.4
31.81	-0.37	1.6	51.96	13MY	1.3	-3.5	-1.6	0.4
38.89	-0.27	1.7	61.50	23MY	0.5	-3.5	-1.6	0.3
45.87	-0.16	1.7	71.03	2JN	-0.4	-3.4	-1.7	0.2
52.76	-0.05	1.7	80.55	12JN	-1.4	-3.4	-1.7	0.3
59.56	0.06	1.7	90.07	22JN	-1.5	-3.4	-1.8	0.3
66.26	0.18	1.7	99.62	2JL	-0.7	-3.4	-1.8	0.4
72.86	0.29	1.7	109.18	12JL	-0.1	-3.3	-1.9	0.5
79.38	0.41	1.7	118.78	22JL	0.3	-3.3	-1.9	0.5
85.80	0.54	1.7	128.42	1AU	0.6	-3.3	-2.0	0.6
92.11	0.66	1.7	138.10	11AU	1.0	-3.3	-2.1	0.6
98.33	0.79	1.7	147.84	21AU	2.2	-3.3	-2.1	0.7
104.43	0.92	1.7	157.63	31AU	2.2	-3.4	-2.2	0.7
110.41	1.05	1.6	167.48	10SE	-0.1	-3.4	-2.3	0.7
116.25	1.20	1.6	177.39	20SE	-0.9	-3.4	-2.3	0.8
121.95	1.34	1.5	187.35	30SE	-0.9	-3.4	-2.4	0.8
127.46	1.50	1.4	197.37	10OC	-0.9	-3.5	-2.4	0.8
132.78	1.67	1.3	207.45	20OC	-0.6	-3.5	-2.4	0.8
137.85	1.85	1.2	217.57	30OC	-0.5	-3.6	-2.4	0.8
142.63	2.04	1.1	227.72	9NO	-0.5	-3.7	-2.4	0.8
147.06	2.25	0.9	237.90	19NO	-0.4	-3.8	-2.3	0.7
151.05	2.48	0.8	248.10	29NO	0.2	-3.8	-2.3	0.7
154.50	2.73	0.6	258.30	9DE	2.7	-3.9	-2.2	0.7
157.29	3.00	0.4	268.51	19DE	1.1	-4.0	-2.1	0.7
159.24	3.29	0.1	278.69	29DE	0.2	-4.1	-2.1	0.8

♂			⊙	16.00UT	☿	♀	♃	♄
LONG	LAT	MAG	LONG	340	MAGNITUDES			
160.19	3.59	-0.1	288.85	8JA	0.1	-4.2	-2.0	0.8
159.96	3.88	-0.4	298.98	18JA	-0.1	-4.3	-1.9	0.8
158.44	4.13	-0.7	309.07	28JA	-0.3	-4.3	-1.9	0.8
155.70	4.29	-0.9	319.10	7FE	-0.7	-4.2	-1.8	0.8
152.07	4.30	-1.1	329.08	17FE	-1.4	-3.9	-1.7	0.8
148.20	4.14	-1.0	339.00	27FE	-1.3	-3.4	-1.7	0.8
144.79	3.83	-0.9	348.86	8MR	-0.5	-3.4	-1.6	0.8
142.39	3.42	-0.7	358.67	18MR	0.9	-3.9	-1.6	0.8
141.29	2.98	-0.5	8.42	28MR	2.6	-4.2	-1.5	0.7
141.46	2.54	-0.2	18.12	7AP	3.1	-4.2	-1.5	0.7
142.79	2.12	-0.0	27.77	17AP	1.8	-4.2	-1.5	0.6
145.11	1.75	0.2	37.38	27AP	1.0	-4.1	-1.5	0.6
148.23	1.41	0.3	46.95	7MY	0.4	-4.0	-1.5	0.5
152.01	1.10	0.5	56.50	17MY	-0.5	-3.9	-1.5	0.4
156.34	0.83	0.6	66.03	27MY	-1.4	-3.8	-1.5	0.4
161.13	0.58	0.7	75.55	6JN	-1.5	-3.7	-1.5	0.3
166.30	0.36	0.8	85.08	16JN	-0.6	-3.6	-1.5	0.3
171.80	0.15	0.9	94.61	26JN	0.0	-3.6	-1.6	0.3
177.59	-0.04	1.0	104.17	6JL	0.4	-3.5	-1.6	0.4
183.63	-0.21	1.0	113.75	16JL	0.8	-3.5	-1.6	0.4
189.91	-0.37	1.1	123.37	26JL	1.3	-3.4	-1.7	0.5
196.39	-0.51	1.1	133.02	5AU	2.7	-3.4	-1.7	0.5
203.06	-0.64	1.2	142.73	15AU	1.9	-3.4	-1.8	0.6
209.91	-0.76	1.2	152.49	25AU	-0.2	-3.4	-1.8	0.6
216.91	-0.87	1.2	162.31	4SE	-1.0	-3.4	-1.9	0.7
224.07	-0.96	1.2	172.19	14SE	-1.1	-3.4	-2.0	0.7
231.36	-1.04	1.3	182.13	24SE	-0.9	-3.4	-2.0	0.8
238.78	-1.11	1.3	192.12	4OC	-0.5	-3.4	-2.1	0.8
246.30	-1.16	1.3	202.17	14OC	-0.4	-3.4	-2.2	0.8
253.92	-1.20	1.3	212.26	24OC	-0.3	-3.4	-2.2	0.8
261.62	-1.22	1.3	222.39	3NO	-0.2	-3.4	-2.3	0.8
269.39	-1.24	1.3	232.56	13NO	0.3	-3.4	-2.3	0.8
277.21	-1.23	1.3	242.75	23NO	3.0	-3.4	-2.3	0.8
285.06	-1.22	1.4	252.95	3DE	0.8	-3.5	-2.3	0.8
292.93	-1.19	1.4	263.16	13DE	0.0	-3.5	-2.3	0.7
300.80	-1.15	1.4	273.35	23DE	-0.1	-3.5	-2.2	0.7
				341				
308.66	-1.10	1.4	283.53	2JA	-0.2	-3.5	-2.2	0.8
316.49	-1.04	1.4	293.68	12JA	-0.4	-3.4	-2.1	0.8
324.29	-0.97	1.4	303.78	22JA	-0.8	-3.4	-2.0	0.8
332.04	-0.89	1.5	313.84	1FE	-1.3	-3.4	-2.0	0.9
339.72	-0.81	1.5	323.86	11FE	-1.2	-3.4	-1.9	0.9
347.34	-0.72	1.5	333.81	21FE	-0.5	-3.3	-1.8	0.9
354.89	-0.63	1.5	343.70	3MR	1.1	-3.3	-1.8	0.9
2.35	-0.53	1.5	353.54	13MR	3.2	-3.3	-1.7	0.9
9.74	-0.43	1.5	3.32	23MR	2.3	-3.3	-1.6	0.8
17.04	-0.32	1.6	13.04	2AP	1.3	-3.3	-1.6	0.8
24.25	-0.22	1.6	22.72	12AP	0.8	-3.4	-1.5	0.8
31.38	-0.12	1.6	32.34	22AP	0.2	-3.4	-1.5	0.7
38.42	-0.01	1.6	41.94	2MY	-0.5	-3.4	-1.4	0.6
45.39	0.09	1.7	51.50	12MY	-1.4	-3.4	-1.4	0.6
52.27	0.19	1.7	61.03	22MY	-1.6	-3.5	-1.4	0.5
59.07	0.29	1.8	70.56	1JN	-0.6	-3.5	-1.4	0.5
65.80	0.39	1.8	80.09	11JN	0.1	-3.6	-1.4	0.4
72.46	0.49	1.9	89.61	21JN	0.6	-3.7	-1.4	0.3
79.06	0.58	1.9	99.16	1JL	1.0	-3.8	-1.4	0.3
85.59	0.68	1.9	108.72	11JL	1.8	-3.8	-1.4	0.4
92.05	0.77	1.9	118.32	21JL	3.2	-3.9	-1.4	0.4
98.45	0.86	1.9	127.96	31JL	1.6	-4.0	-1.4	0.5
104.80	0.95	1.9	137.64	10AU	-0.2	-4.2	-1.5	0.5
111.08	1.04	1.9	147.37	20AU	-1.1	-4.2	-1.5	0.6
117.30	1.13	1.9	157.16	30AU	-1.2	-4.3	-1.6	0.6
123.45	1.22	1.9	167.01	9SE	-0.9	-4.2	-1.6	0.7
129.54	1.31	1.8	176.91	19SE	-0.5	-4.0	-1.7	0.7
135.55	1.40	1.8	186.87	29SE	-0.2	-3.4	-1.7	0.8
141.48	1.49	1.7	196.89	9OC	-0.2	-3.4	-1.8	0.8
147.31	1.58	1.7	206.96	19OC	-0.1	-4.0	-1.9	0.8
153.04	1.68	1.6	217.07	29OC	0.6	-4.3	-1.9	0.8
158.64	1.77	1.5	227.23	8NO	3.3	-4.4	-2.0	0.8
164.09	1.87	1.4	237.40	18NO	0.6	-4.3	-2.1	0.8
169.35	1.98	1.3	247.60	28NO	-0.1	-4.3	-2.1	0.8
174.40	2.08	1.1	257.81	8DE	-0.2	-4.2	-2.1	0.8
179.18	2.19	1.0	268.01	18DE	-0.3	-4.0	-2.2	0.8
183.63	2.31	0.8	278.20	28DE	-0.5	-3.9	-2.2	0.8

♂			⊙	16.00UT	☿	♀	♃	♄
LONG	LAT	MAG	LONG	342	MAGNITUDES			
187.68	2.42	0.6	288.36	7JA	-0.8	-3.8	-2.1	0.8
191.22	2.54	0.4	298.49	17JA	-1.2	-3.8	-2.1	0.8
194.12	2.65	0.2	308.58	27JA	-1.1	-3.7	-2.1	0.9
196.24	2.75	-0.1	318.62	6FE	-0.4	-3.6	-2.0	0.9
197.39	2.83	-0.4	328.60	16FE	1.3	-3.5	-2.0	0.9
197.40	2.86	-0.7	338.53	26FE	3.3	-3.5	-1.9	0.9
196.15	2.82	-1.0	348.40	8MR	1.7	-3.4	-1.8	0.9
193.68	2.68	-1.3	358.20	18MR	1.0	-3.4	-1.8	0.9
190.30	2.42	-1.5	7.96	28MR	0.6	-3.3	-1.7	0.9
186.62	2.03	-1.5	17.66	7AP	0.1	-3.3	-1.6	0.9
183.37	1.57	-1.3	27.31	17AP	-0.5	-3.3	-1.6	0.9
181.12	1.08	-1.1	36.92	27AP	-1.4	-3.3	-1.5	0.8
180.19	0.61	-0.9	46.49	7MY	-1.6	-3.3	-1.5	0.8
180.59	0.19	-0.7	56.04	17MY	-0.6	-3.3	-1.4	0.7
182.21	-0.18	-0.5	65.57	27MY	0.2	-3.3	-1.4	0.6
184.87	-0.49	-0.3	75.10	6JN	0.8	-3.3	-1.3	0.6
188.40	-0.76	-0.2	84.62	16JN	1.4	-3.4	-1.3	0.5
192.64	-0.98	-0.0	94.15	26JN	2.3	-3.4	-1.3	0.4
197.47	-1.17	0.1	103.70	6JL	3.1	-3.5	-1.3	0.4
202.79	-1.33	0.2	113.28	16JL	1.3	-3.5	-1.3	0.4
208.53	-1.46	0.3	122.90	26JL	-0.3	-3.5	-1.3	0.4
214.62	-1.56	0.4	132.55	5AU	-1.2	-3.5	-1.3	0.5
221.01	-1.64	0.5	142.26	15AU	-1.4	-3.4	-1.3	0.5
227.66	-1.69	0.6	152.02	25AU	-0.8	-3.4	-1.3	0.6
234.53	-1.73	0.7	161.83	4SE	-0.4	-3.4	-1.4	0.6
241.59	-1.74	0.7	171.71	14SE	-0.1	-3.3	-1.4	0.7
248.82	-1.74	0.8	181.64	24SE	-0.0	-3.3	-1.4	0.7
256.19	-1.71	0.9	191.63	4OC	0.1	-3.3	-1.5	0.8
263.66	-1.67	0.9	201.67	14OC	0.8	-3.3	-1.5	0.8
271.23	-1.62	1.0	211.77	24OC	3.4	-3.3	-1.6	0.8
278.86	-1.55	1.0	221.90	3NO	0.4	-3.3	-1.7	0.9
286.54	-1.46	1.1	232.06	13NO	-0.3	-3.4	-1.7	0.9
294.25	-1.37	1.2	242.26	23NO	-0.4	-3.4	-1.8	0.9
301.97	-1.27	1.2	252.46	3DE	-0.4	-3.4	-1.9	0.9
309.67	-1.16	1.3	262.66	13DE	-0.6	-3.5	-1.9	0.9
317.36	-1.05	1.3	272.86	23DE	-0.9	-3.5	-2.0	0.9
				343				
325.01	-0.93	1.4	283.04	2JA	-1.0	-3.6	-2.0	0.9
332.61	-0.80	1.4	293.19	12JA	-1.0	-3.6	-2.0	0.8
340.15	-0.68	1.5	303.30	22JA	-0.3	-3.7	-2.1	0.9
347.63	-0.56	1.5	313.36	1FE	1.6	-3.7	-2.1	0.9
355.04	-0.44	1.6	323.37	11FE	2.7	-3.8	-2.0	1.0
2.38	-0.32	1.6	333.33	21FE	1.3	-3.9	-2.0	1.0
9.63	-0.20	1.6	343.23	3MR	0.7	-4.0	-2.0	1.0
16.81	-0.09	1.7	353.07	13MR	0.4	-4.1	-1.9	1.0
23.92	0.02	1.7	2.85	23MR	0.0	-4.2	-1.9	1.0
30.94	0.12	1.7	12.58	2AP	-0.6	-4.2	-1.8	1.0
37.89	0.23	1.8	22.25	12AP	-1.4	-4.2	-1.7	1.0
44.77	0.32	1.8	31.88	22AP	-1.6	-4.0	-1.7	1.0
51.57	0.42	1.8	41.47	2MY	-0.6	-3.5	-1.6	0.9
58.32	0.50	1.8	51.04	12MY	0.3	-2.7	-1.5	0.9
65.00	0.59	1.8	60.58	22MY	1.0	-3.5	-1.5	0.8
71.63	0.67	1.8	70.10	1JN	1.8	-4.0	-1.4	0.8
78.20	0.74	1.8	79.62	11JN	3.0	-4.2	-1.4	0.7
84.73	0.81	1.9	89.15	21JN	2.6	-4.2	-1.3	0.6
91.21	0.88	1.9	98.69	1JL	1.1	-4.1	-1.3	0.6
97.66	0.94	2.0	108.26	11JL	-0.3	-4.0	-1.3	0.5
104.07	1.00	2.0	117.85	21JL	-1.3	-4.0	-1.3	0.5
110.45	1.06	2.0	127.48	31JL	-1.5	-3.9	-1.2	0.5
116.81	1.11	2.0	137.16	10AU	-0.8	-3.8	-1.2	0.5
123.13	1.16	2.0	146.89	20AU	-0.3	-3.7	-1.2	0.6
129.44	1.21	2.0	156.68	30AU	-0.0	-3.7	-1.2	0.6
135.72	1.25	2.0	166.53	9SE	0.1	-3.6	-1.2	0.7
141.98	1.29	2.0	176.43	19SE	0.3	-3.6	-1.3	0.7
148.21	1.33	2.0	186.39	29SE	1.1	-3.5	-1.3	0.8
154.42	1.37	1.9	196.41	9OC	3.1	-3.5	-1.3	0.8
160.60	1.40	1.9	206.47	19OC	0.2	-3.4	-1.4	0.9
166.75	1.42	1.8	216.58	29OC	-0.5	-3.4	-1.4	0.9
172.86	1.45	1.8	226.74	8NO	-0.5	-3.4	-1.5	0.9
178.92	1.47	1.7	236.91	18NO	-0.6	-3.4	-1.5	1.0
184.94	1.48	1.6	247.11	28NO	-0.7	-3.4	-1.6	1.0
190.88	1.49	1.5	257.32	8DE	-0.8	-3.4	-1.6	1.0
196.75	1.49	1.4	267.52	18DE	-0.9	-3.3	-1.7	1.0
202.52	1.48	1.3	277.71	28DE	-0.8	-3.3	-1.8	1.0

♂ LONG	♂ LAT	♂ MAG	☉ LONG	16.00UT	☿	♀	♃	♄
				344	MAGNITUDES			
208.17	1.46	1.1	287.87	7JA	-0.2	-3.4	-1.8	1.0
213.68	1.43	1.0	298.00	17JA	1.8	-3.4	-1.9	1.0
219.01	1.38	0.8	308.09	27JA	2.2	-3.4	-1.9	1.0
224.12	1.31	0.6	318.13	6FE	0.9	-3.4	-2.0	1.0
228.97	1.22	0.4	328.12	16FE	0.5	-3.4	-2.0	1.1
233.47	1.09	0.2	338.04	26FE	0.3	-3.5	-2.0	1.1
237.55	0.93	-0.1	347.92	7MR	-0.1	-3.5	-2.0	1.1
241.08	0.71	-0.3	357.73	17MR	-0.6	-3.5	-2.0	1.2
243.94	0.43	-0.6	7.48	27MR	-1.4	-3.4	-2.0	1.2
245.95	0.08	-0.9	17.19	6AP	-1.5	-3.4	-1.9	1.2
246.92	-0.37	-1.2	26.84	16AP	-0.6	-3.4	-1.9	1.2
246.70	-0.91	-1.6	36.45	26AP	0.4	-3.3	-1.8	1.1
245.23	-1.53	-1.9	46.03	6MY	1.3	-3.3	-1.7	1.1
242.73	-2.18	-2.2	55.57	16MY	2.4	-3.3	-1.7	1.1
239.66	-2.80	-2.2	65.11	26MY	3.5	-3.3	-1.6	1.0
236.73	-3.31	-2.1	74.63	5JN	2.1	-3.3	-1.6	1.0
234.66	-3.66	-1.9	84.15	15JN	0.8	-3.3	-1.5	0.9
233.83	-3.86	-1.7	93.68	25JN	-0.3	-3.4	-1.4	0.8
234.40	-3.92	-1.5	103.24	5JL	-1.3	-3.4	-1.4	0.8
236.28	-3.90	-1.3	112.81	15JL	-1.5	-3.4	-1.3	0.7
239.27	-3.80	-1.1	122.43	25JL	-0.7	-3.5	-1.3	0.6
243.20	-3.67	-0.9	132.08	4AU	-0.2	-3.5	-1.3	0.6
247.88	-3.50	-0.7	141.78	14AU	0.1	-3.6	-1.2	0.6
253.16	-3.30	-0.5	151.54	24AU	0.3	-3.6	-1.2	0.6
258.93	-3.09	-0.3	161.36	3SE	0.5	-3.7	-1.2	0.7
265.07	-2.87	-0.2	171.22	13SE	1.4	-3.8	-1.2	0.7
271.51	-2.65	-0.0	181.16	23SE	2.8	-3.9	-1.2	0.8
278.19	-2.42	0.1	191.15	3OC	0.1	-4.0	-1.2	0.8
285.05	-2.18	0.3	201.18	13OC	-0.6	-4.1	-1.2	0.9
292.03	-1.95	0.4	211.28	23OC	-0.7	-4.2	-1.3	0.9
299.12	-1.73	0.5	221.41	2NO	-0.7	-4.3	-1.3	1.0
306.28	-1.51	0.7	231.57	12NO	-0.8	-4.4	-1.3	1.0
313.47	-1.29	0.8	241.76	22NO	-0.7	-4.3	-1.4	1.1
320.68	-1.09	0.9	251.97	2DE	-0.7	-4.0	-1.4	1.1
327.88	-0.89	1.0	262.17	12DE	-0.7	-3.4	-1.5	1.1
335.07	-0.71	1.1	272.37	22DE	-0.1	-3.4	-1.5	1.1
				345				
342.24	-0.53	1.2	282.55	1JA	2.1	-4.0	-1.6	1.1
349.37	-0.37	1.3	292.69	11JA	1.7	-4.3	-1.7	1.1
356.45	-0.21	1.4	302.81	21JA	0.6	-4.3	-1.7	1.1
3.48	-0.07	1.5	312.87	31JA	0.3	-4.3	-1.8	1.1
10.46	0.06	1.5	322.89	10FE	0.1	-4.2	-1.9	1.1
17.38	0.19	1.6	332.85	20FE	-0.2	-4.1	-1.9	1.2
24.25	0.30	1.7	342.75	2MR	-0.6	-4.0	-2.0	1.2
31.06	0.40	1.7	352.59	12MR	-1.4	-3.9	-2.0	1.3
37.81	0.50	1.8	2.38	22MR	-1.5	-3.8	-2.0	1.3
44.51	0.59	1.8	12.11	1AP	-0.6	-3.7	-2.0	1.3
51.16	0.67	1.9	21.79	11AP	0.6	-3.6	-2.0	1.3
57.76	0.74	1.9	31.42	21AP	1.7	-3.5	-2.0	1.3
64.32	0.81	1.9	41.02	1MY	3.2	-3.5	-2.0	1.3
70.84	0.87	2.0	50.58	11MY	2.9	-3.4	-1.9	1.3
77.32	0.92	2.0	60.12	21MY	1.6	-3.4	-1.9	1.3
83.78	0.97	2.0	69.65	31MY	0.6	-3.4	-1.8	1.2
90.20	1.01	2.0	79.17	10JN	-0.4	-3.3	-1.8	1.2
96.61	1.05	2.0	88.70	20JN	-1.4	-3.3	-1.7	1.1
103.00	1.08	2.0	98.23	30JN	-1.5	-3.3	-1.6	1.1
109.38	1.11	1.9	107.79	10JL	-0.7	-3.3	-1.6	1.0
115.75	1.13	2.0	117.39	20JL	-0.1	-3.3	-1.5	0.9
122.12	1.15	2.0	127.02	30JL	0.2	-3.3	-1.5	0.8
128.48	1.17	2.0	136.69	9AU	0.5	-3.4	-1.4	0.8
134.86	1.18	2.0	146.42	19AU	0.8	-3.4	-1.4	0.7
141.24	1.18	2.0	156.20	29AU	1.8	-3.4	-1.3	0.8
147.63	1.18	2.0	166.04	8SE	2.4	-3.4	-1.3	0.8
154.04	1.18	2.0	175.95	18SE	0.0	-3.5	-1.3	0.9
160.46	1.17	2.0	185.90	28SE	-0.8	-3.5	-1.3	0.9
166.89	1.16	2.0	195.91	8OC	-0.8	-3.5	-1.3	1.0
173.34	1.14	1.9	205.98	18OC	-0.8	-3.5	-1.3	1.0
179.81	1.11	1.9	216.09	28OC	-0.7	-3.4	-1.3	1.1
186.30	1.08	1.9	226.24	7NO	-0.6	-3.4	-1.3	1.1
192.80	1.04	1.8	236.42	17NO	-0.5	-3.4	-1.3	1.2
199.31	0.99	1.8	246.61	27NO	-0.5	-3.4	-1.3	1.2
205.84	0.94	1.7	256.82	7DE	0.0	-3.3	-1.4	1.2
212.38	0.87	1.6	267.02	17DE	2.4	-3.3	-1.4	1.3
218.92	0.79	1.5	277.21	27DE	1.3	-3.3	-1.4	1.3

♂ LONG	♂ LAT	♂ MAG	☉ LONG	16.00UT	☿	♀	♃	♄
				346	MAGNITUDES			
225.47	0.70	1.4	287.38	6JA	0.4	-3.3	-1.5	1.3
232.01	0.59	1.3	297.51	16JA	0.1	-3.4	-1.6	1.3
238.55	0.47	1.2	307.60	26JA	0.0	-3.4	-1.6	1.3
245.08	0.33	1.1	317.64	5FE	-0.2	-3.4	-1.7	1.3
251.58	0.16	0.9	327.64	15FE	-0.7	-3.4	-1.8	1.3
258.05	-0.02	0.8	337.57	25FE	-1.4	-3.5	-1.8	1.3
264.47	-0.24	0.7	347.44	7MR	-1.4	-3.5	-1.9	1.3
270.83	-0.48	0.5	357.26	17MR	-0.5	-3.5	-2.0	1.3
277.10	-0.75	0.3	7.01	27MR	0.7	-3.6	-2.0	1.3
283.25	-1.06	0.1	16.72	6AP	2.2	-3.7	-2.1	1.3
289.25	-1.40	-0.0	26.38	16AP	3.8	-3.7	-2.1	1.3
295.03	-1.79	-0.2	35.99	26AP	2.2	-3.8	-2.1	1.3
300.54	-2.23	-0.5	45.57	6MY	1.2	-3.9	-2.1	1.2
305.69	-2.71	-0.7	55.12	16MY	0.5	-4.0	-2.1	1.2
310.36	-3.24	-0.9	64.65	26MY	-0.4	-4.1	-2.1	1.2
314.42	-3.82	-1.2	74.18	5JN	-1.4	-4.2	-2.0	1.1
317.68	-4.45	-1.4	83.70	15JN	-1.6	-4.2	-2.0	1.1
319.95	-5.10	-1.7	93.23	25JN	-0.7	-4.1	-1.9	1.0
321.02	-5.74	-2.0	102.78	5JL	-0.1	-3.9	-1.9	1.0
320.76	-6.30	-2.2	112.36	15JL	0.3	-3.3	-1.8	0.9
319.24	-6.68	-2.5	121.96	25JL	0.6	-3.3	-1.7	0.9
316.79	-6.79	-2.6	131.62	4AU	1.1	-3.9	-1.7	0.8
314.10	-6.57	-2.6	141.32	14AU	2.3	-4.2	-1.6	0.8
311.91	-6.07	-2.3	151.07	24AU	2.1	-4.3	-1.6	0.7
310.77	-5.39	-2.1	160.88	3SE	-0.1	-4.3	-1.5	0.7
310.94	-4.63	-1.7	170.75	13SE	-0.9	-4.2	-1.5	0.8
312.37	-3.89	-1.4	180.67	23SE	-1.0	-4.1	-1.4	0.9
314.90	-3.20	-1.1	190.66	3OC	-0.9	-4.0	-1.4	1.0
318.34	-2.59	-0.8	200.70	13OC	-0.6	-3.9	-1.4	1.0
322.49	-2.05	-0.6	210.78	23OC	-0.4	-3.8	-1.3	1.1
327.18	-1.59	-0.3	220.91	2NO	-0.4	-3.8	-1.3	1.1
332.31	-1.18	-0.1	231.08	12NO	-0.4	-3.7	-1.3	1.2
337.75	-0.83	0.1	241.26	22NO	0.2	-3.6	-1.3	1.2
343.45	-0.53	0.3	251.47	2DE	2.7	-3.6	-1.3	1.3
349.33	-0.27	0.5	261.68	12DE	1.0	-3.5	-1.4	1.3
355.34	-0.04	0.7	271.87	22DE	0.2	-3.5	-1.4	1.3
				347				
1.46	0.15	0.9	282.06	1JA	-0.0	-3.4	-1.4	1.3
7.66	0.32	1.0	292.21	11JA	-0.1	-3.4	-1.5	1.2
13.91	0.46	1.1	302.32	21JA	-0.3	-3.4	-1.5	1.2
20.19	0.59	1.3	312.39	31JA	-0.7	-3.3	-1.6	1.2
26.49	0.70	1.4	322.41	10FE	-1.4	-3.3	-1.6	1.1
32.81	0.79	1.5	332.37	20FE	-1.3	-3.3	-1.7	1.1
39.13	0.87	1.6	342.28	2MR	-0.5	-3.3	-1.7	1.0
45.44	0.94	1.7	352.12	12MR	0.9	-3.3	-1.8	1.0
51.75	1.00	1.7	1.91	22MR	2.8	-3.3	-1.9	1.0
58.06	1.05	1.8	11.64	1AP	2.8	-3.3	-2.0	1.0
64.36	1.09	1.9	21.32	11AP	1.6	-3.4	-2.0	1.0
70.64	1.12	1.9	30.95	21AP	0.9	-3.4	-2.1	1.0
76.93	1.15	2.0	40.55	1MY	0.3	-3.5	-2.1	1.0
83.21	1.17	2.0	50.12	11MY	-0.4	-3.5	-2.2	1.0
89.48	1.18	2.0	59.66	21MY	-1.4	-3.5	-2.2	0.9
95.76	1.19	2.0	69.19	31MY	-1.6	-3.4	-2.2	0.9
102.05	1.19	2.0	78.71	10JN	-0.7	-3.4	-2.2	0.9
108.34	1.19	2.0	88.23	20JN	0.0	-3.4	-2.2	0.8
114.65	1.19	2.0	97.77	30JN	0.5	-3.4	-2.2	0.8
120.97	1.18	2.0	107.33	10JL	0.9	-3.3	-2.1	0.7
127.32	1.16	2.0	116.92	20JL	1.5	-3.3	-2.1	0.7
133.69	1.14	2.0	126.55	30JL	2.9	-3.3	-2.0	0.6
140.08	1.12	2.0	136.22	9AU	1.8	-3.3	-2.0	0.6
146.51	1.09	1.9	145.95	19AU	-0.1	-3.3	-1.9	0.5
152.98	1.05	1.9	155.73	29AU	-1.0	-3.4	-1.8	0.5
159.49	1.02	1.9	165.56	8SE	-1.1	-3.4	-1.8	0.4
166.04	0.97	1.9	175.46	18SE	-0.9	-3.4	-1.7	0.5
172.64	0.92	1.9	185.42	28SE	-0.5	-3.4	-1.7	0.5
179.28	0.87	1.9	195.42	8OC	-0.3	-3.5	-1.6	0.6
185.97	0.81	1.9	205.48	18OC	-0.2	-3.5	-1.6	0.7
192.71	0.74	1.9	215.60	28OC	-0.2	-3.6	-1.5	0.7
199.51	0.67	1.9	225.74	7NO	0.4	-3.7	-1.5	0.8
206.35	0.59	1.8	235.92	17NO	3.0	-3.8	-1.5	0.9
213.25	0.51	1.8	246.12	27NO	0.7	-3.9	-1.4	0.9
220.21	0.41	1.7	256.32	7DE	-0.0	-3.9	-1.4	0.9
227.21	0.31	1.7	266.52	17DE	-0.2	-4.0	-1.4	1.0
234.27	0.20	1.6	276.72	27DE	-0.2	-4.1	-1.4	1.0

Left table

♂ LONG	LAT	MAG	☉ LONG	16.00UT	☿	♀	♃	♄
				348		MAGN	TUDES	
241.37	0.07	1.6	286.88	6JA	-0.4	-4.2	-1.4	1.0
248.53	-0.06	1.5	297.02	16JA	-0.8	-4.3	-1.5	1.0
255.72	-0.20	1.4	307.12	26JA	-1.2	-4.3	-1.5	1.0
262.96	-0.34	1.3	317.16	5FE	-1.2	-4.2	-1.5	0.9
270.23	-0.50	1.3	327.15	15FE	-0.5	-3.9	-1.6	0.9
277.54	-0.67	1.2	337.09	25FE	1.1	-3.4	-1.6	0.9
284.86	-0.84	1.1	346.96	6MR	3.3	-3.5	-1.7	0.8
292.20	-1.02	1.0	356.78	16MR	2.1	-4.0	-1.7	0.7
299.54	-1.21	0.9	6.55	26MR	1.2	-4.2	-1.8	0.8
306.86	-1.39	0.8	16.25	5AP	0.7	-4.2	-1.9	0.8
314.16	-1.58	0.7	25.91	15AP	0.2	-4.2	-1.9	0.8
321.40	-1.77	0.6	35.53	25AP	-0.5	-4.1	-2.0	0.8
328.58	-1.95	0.5	45.10	5MY	-1.4	-4.0	-2.1	0.8
335.65	-2.13	0.4	54.66	15MY	-1.6	-3.9	-2.1	0.8
342.59	-2.30	0.3	64.19	25MY	-0.7	-3.8	-2.2	0.7
349.36	-2.46	0.2	73.71	4JN	0.1	-3.7	-2.3	0.7
355.91	-2.61	0.1	83.24	14JN	0.6	-3.6	-2.3	0.7
2.19	-2.74	-0.0	92.77	24JN	1.1	-3.6	-2.3	0.6
8.15	-2.86	-0.2	102.32	4JL	2.0	-3.5	-2.3	0.6
13.68	-2.96	-0.3	111.89	14JL	3.2	-3.5	-2.3	0.6
18.70	-3.03	-0.5	121.50	24JL	1.5	-3.4	-2.3	0.5
23.07	-3.08	-0.7	131.15	3AU	-0.2	-3.4	-2.3	0.4
26.64	-3.09	-0.8	140.85	13AU	-1.1	-3.4	-2.3	0.4
29.20	-3.05	-1.1	150.60	23AU	-1.3	-3.4	-2.2	0.3
30.57	-2.95	-1.3	160.40	2SE	-0.9	-3.4	-2.2	0.3
30.52	-2.76	-1.5	170.27	12SE	-0.4	-3.4	-2.1	0.2
29.00	-2.47	-1.7	180.19	22SE	-0.2	-3.4	-2.0	0.2
26.21	-2.05	-1.9	190.17	2OC	-0.1	-3.4	-2.0	0.2
22.66	-1.54	-2.0	200.21	12OC	0.0	-3.4	-1.9	0.3
19.19	-0.97	-1.8	210.29	22OC	0.6	-3.4	-1.8	0.3
16.53	-0.43	-1.5	220.42	1NO	3.3	-3.4	-1.8	0.4
15.12	0.04	-1.2	230.58	11NO	0.5	-3.4	-1.7	0.5
15.07	0.42	-0.8	240.77	21NO	-0.2	-3.4	-1.7	0.6
16.26	0.72	-0.5	250.97	1DE	-0.3	-3.5	-1.6	0.6
18.48	0.95	-0.2	261.18	11DE	-0.4	-3.5	-1.6	0.7
21.54	1.12	0.0	271.37	21DE	-0.5	-3.5	-1.6	0.7
25.25	1.25	0.3	281.55	31DE	-0.8	-3.5	-1.5	0.7
				349				
29.46	1.35	0.5	291.71	10JA	-1.1	-3.4	-1.5	0.7
34.06	1.42	0.7	301.82	20JA	-1.1	-3.4	-1.5	0.7
38.97	1.46	0.9	311.89	30JA	-0.4	-3.4	-1.5	0.7
44.11	1.50	1.1	321.92	9FE	1.4	-3.4	-1.5	0.7
49.45	1.52	1.2	331.88	19FE	3.1	-3.3	-1.6	0.7
54.93	1.52	1.3	341.79	1MR	1.5	-3.3	-1.6	0.7
60.54	1.52	1.5	351.64	11MR	0.9	-3.3	-1.6	0.6
66.24	1.52	1.6	1.43	21MR	0.5	-3.3	-1.7	0.6
72.02	1.50	1.6	11.17	31MR	0.1	-3.4	-1.7	0.5
77.86	1.49	1.7	20.85	10AP	-0.5	-3.4	-1.7	0.5
83.77	1.46	1.8	30.49	20AP	-1.4	-3.4	-1.8	0.6
89.72	1.44	1.9	40.09	30AP	-1.6	-3.4	-1.9	0.6
95.73	1.40	1.9	49.66	10MY	-0.6	-3.4	-1.9	0.6
101.78	1.37	1.9	59.20	20MY	0.2	-3.5	-2.0	0.6
107.87	1.33	2.0	68.72	30MY	0.8	-3.6	-2.1	0.6
114.01	1.29	2.0	78.25	9JN	1.5	-3.6	-2.1	0.6
120.19	1.24	2.0	87.77	19JN	2.6	-3.7	-2.2	0.5
126.42	1.20	2.0	97.31	29JN	3.0	-3.8	-2.3	0.5
132.70	1.14	2.0	106.87	9JL	1.3	-3.8	-2.3	0.5
139.02	1.09	2.0	116.46	19JL	-0.2	-3.9	-2.4	0.4
145.41	1.03	2.0	126.09	29JL	-1.2	-4.0	-2.4	0.4
151.85	0.97	2.0	135.76	8AU	-1.4	-4.2	-2.4	0.3
158.35	0.90	1.9	145.48	18AU	-0.8	-4.2	-2.4	0.3
164.92	0.83	1.9	155.26	28AU	-0.3	-4.3	-2.4	0.2
171.55	0.76	1.9	165.09	7SE	-0.1	-4.2	-2.4	0.1
178.25	0.68	1.8	174.98	17SE	0.0	-3.9	-2.4	0.1
185.03	0.60	1.8	184.93	27SE	0.2	-3.4	-2.3	0.0
191.87	0.52	1.8	194.94	7OC	0.9	-3.5	-2.2	-0.0
198.79	0.43	1.8	205.00	17OC	3.4	-4.0	-2.2	0.0
205.79	0.33	1.7	215.11	27OC	0.4	-4.3	-2.1	0.1
212.86	0.24	1.7	225.25	6NO	-0.4	-4.4	-2.0	0.2
220.00	0.14	1.7	235.42	16NO	-0.5	-4.3	-2.0	0.2
227.22	0.03	1.7	245.62	26NO	-0.5	-4.2	-1.9	0.3
234.51	-0.08	1.7	255.83	6DE	-0.6	-4.1	-1.8	0.4
241.86	-0.19	1.6	266.03	16DE	-0.9	-4.0	-1.8	0.4
249.28	-0.31	1.6	276.22	26DE	-0.9	-3.9	-1.7	0.5

Right table

♂ LONG	LAT	MAG	☉ LONG	16.00UT	☿	♀	♃	♄
				350		MAGN	TUDES	
256.77	-0.43	1.6	286.39	5JA	-0.9	-3.8	-1.7	0.5
264.30	-0.55	1.5	296.52	15JA	-0.3	-3.8	-1.6	0.5
271.89	-0.66	1.5	306.62	25JA	1.6	-3.7	-1.6	0.5
279.52	-0.78	1.4	316.67	4FE	2.5	-3.6	-1.6	0.5
287.18	-0.90	1.4	326.67	14FE	1.1	-3.5	-1.6	0.5
294.86	-1.01	1.4	336.61	24FE	0.6	-3.5	-1.6	0.5
302.55	-1.12	1.3	346.49	6MR	0.3	-3.4	-1.6	0.5
310.24	-1.22	1.3	356.31	16MR	-0.0	-3.4	-1.6	0.5
317.93	-1.31	1.2	6.07	26MR	-0.6	-3.3	-1.6	0.4
325.58	-1.39	1.2	15.79	5AP	-1.4	-3.3	-1.6	0.4
333.19	-1.46	1.1	25.44	15AP	-1.6	-3.3	-1.6	0.4
340.75	-1.51	1.1	35.07	25AP	-0.6	-3.3	-1.7	0.4
348.23	-1.55	1.1	44.65	5MY	0.3	-3.3	-1.7	0.4
355.63	-1.58	1.0	54.20	15MY	1.1	-3.3	-1.8	0.4
2.93	-1.59	1.0	63.74	25MY	2.0	-3.3	-1.8	0.4
10.11	-1.58	0.9	73.26	4JN	3.3	-3.3	-1.9	0.4
17.14	-1.56	0.9	82.78	14JN	2.4	-3.4	-1.9	0.4
24.02	-1.51	0.8	92.31	24JN	1.0	-3.4	-2.0	0.4
30.72	-1.45	0.8	101.86	4JL	-0.3	-3.5	-2.1	0.4
37.20	-1.38	0.7	111.43	14JL	-1.3	-3.5	-2.1	0.4
43.45	-1.28	0.6	121.04	24JL	-1.5	-3.5	-2.2	0.3
49.42	-1.16	0.5	130.68	3AU	-0.8	-3.5	-2.3	0.3
55.06	-1.02	0.4	140.37	13AU	-0.3	-3.4	-2.3	0.2
60.31	-0.86	0.3	150.12	23AU	0.0	-3.4	-2.4	0.2
65.09	-0.67	0.2	159.93	2SE	0.2	-3.4	-2.4	0.1
69.31	-0.44	0.1	169.79	12SE	0.4	-3.3	-2.4	0.1
72.82	-0.18	-0.1	179.71	22SE	1.2	-3.3	-2.4	-0.0
75.47	0.13	-0.3	189.69	2OC	3.1	-3.3	-2.4	-0.1
77.07	0.48	-0.5	199.72	12OC	0.2	-3.3	-2.4	-0.1
77.40	0.89	-0.7	209.80	22OC	-0.5	-3.3	-2.4	-0.2
76.33	1.34	-1.0	219.93	1NO	-0.6	-3.3	-2.3	-0.1
73.89	1.81	-1.2	230.08	11NO	-0.6	-3.4	-2.2	-0.1
70.36	2.24	-1.3	240.27	21NO	-0.7	-3.4	-2.2	0.0
66.42	2.59	-1.3	250.47	1DE	-0.8	-3.4	-2.1	0.1
62.84	2.81	-1.1	260.68	11DE	-0.8	-3.5	-2.0	0.2
60.25	2.91	-0.8	270.88	21DE	-0.8	-3.5	-2.0	0.2
58.98	2.91	-0.5	281.06	31DE	-0.2	-3.6	-1.9	0.3
				351				
59.02	2.85	-0.2	291.21	10JA	1.9	-3.6	-1.8	0.3
60.22	2.75	0.0	301.33	20JA	2.0	-3.7	-1.8	0.4
62.40	2.64	0.3	311.41	30JA	0.8	-3.8	-1.7	0.4
65.34	2.52	0.5	321.43	9FE	0.4	-3.8	-1.7	0.4
68.90	2.40	0.7	331.40	19FE	0.2	-3.9	-1.6	0.4
72.95	2.28	0.9	341.31	1MR	-0.1	-4.0	-1.6	0.4
77.37	2.17	1.1	351.16	11MR	-0.6	-4.1	-1.6	0.4
82.11	2.06	1.2	0.96	21MR	-1.4	-4.2	-1.6	0.4
87.10	1.96	1.3	10.70	31MR	-1.5	-4.2	-1.5	0.3
92.29	1.85	1.4	20.38	10AP	-0.6	-4.2	-1.5	0.3
97.66	1.75	1.5	30.02	20AP	0.4	-4.0	-1.6	0.3
103.17	1.66	1.6	39.62	30AP	1.4	-3.5	-1.6	0.2
108.81	1.56	1.7	49.19	10MY	2.7	-2.7	-1.6	0.3
114.57	1.47	1.7	58.74	20MY	3.3	-3.5	-1.6	0.3
120.43	1.37	1.8	68.26	30MY	1.9	-4.0	-1.6	0.3
126.39	1.28	1.8	77.79	9JN	0.8	-4.2	-1.7	0.3
132.44	1.19	1.9	87.31	19JN	-0.3	-4.2	-1.7	0.4
138.58	1.09	1.9	96.85	29JN	-1.3	-4.1	-1.8	0.4
144.81	1.00	1.9	106.41	9JL	-1.5	-4.0	-1.8	0.3
151.13	0.90	1.9	115.99	19JL	-0.7	-4.0	-1.9	0.3
157.54	0.81	1.9	125.62	29JL	-0.2	-3.9	-1.9	0.3
164.03	0.71	1.9	135.29	8AU	0.1	-3.8	-2.0	0.3
170.62	0.61	1.9	145.01	18AU	0.3	-3.7	-2.1	0.2
177.29	0.51	1.8	154.78	28AU	0.6	-3.7	-2.2	0.2
184.06	0.41	1.8	164.61	7SE	1.6	-3.6	-2.2	0.1
190.92	0.31	1.8	174.50	17SE	2.7	-3.5	-2.3	0.1
197.88	0.21	1.7	184.45	27SE	0.1	-3.5	-2.3	-0.0
204.93	0.10	1.7	194.45	7OC	-0.7	-3.5	-2.4	-0.1
212.07	-0.00	1.6	204.51	17OC	-0.8	-3.4	-2.4	-0.2
219.30	-0.11	1.6	214.61	27OC	-0.7	-3.4	-2.4	-0.2
226.62	-0.21	1.5	224.76	6NO	-0.8	-3.4	-2.4	-0.3
234.02	-0.32	1.5	234.93	16NO	-0.6	-3.4	-2.3	-0.2
241.50	-0.42	1.5	245.12	26NO	-0.6	-3.4	-2.3	-0.1
249.06	-0.52	1.5	255.33	6DE	-0.6	-3.4	-2.2	-0.1
256.69	-0.62	1.5	265.54	16DE	-0.1	-3.4	-2.2	-0.0
264.37	-0.71	1.5	275.73	26DE	2.1	-3.4	-2.1	0.1

♂ LONG	♂ LAT	♂ MAG	⊙ LONG	16.00UT 352	☿	♀	♃	♄
272.11	-0.80	1.5	285.90	5JA	1.6	-3.4	-2.0	0.1
279.89	-0.88	1.4	296.04	15JA	0.5	-3.4	-2.0	0.2
287.70	-0.96	1.4	306.13	25JA	0.2	-3.4	-1.9	0.2
295.53	-1.03	1.4	316.18	4FE	0.1	-3.4	-1.8	0.3
303.37	-1.08	1.4	326.18	14FE	-0.2	-3.4	-1.8	0.3
311.20	-1.13	1.4	336.12	24FE	-0.6	-3.5	-1.7	0.3
319.01	-1.16	1.4	346.00	5MR	-1.4	-3.5	-1.7	0.3
326.80	-1.19	1.4	355.83	15MR	-1.5	-3.4	-1.6	0.3
334.54	-1.20	1.4	5.59	25MR	-0.6	-3.4	-1.6	0.3
342.23	-1.19	1.4	15.31	4AP	0.6	-3.4	-1.5	0.3
349.86	-1.18	1.4	24.97	14AP	1.9	-3.4	-1.5	0.3
357.41	-1.15	1.4	34.59	24AP	3.5	-3.3	-1.5	0.2
4.88	-1.11	1.4	44.18	4MY	2.6	-3.3	-1.5	0.2
12.25	-1.05	1.4	53.73	14MY	1.5	-3.3	-1.5	0.2
19.52	-0.98	1.4	63.26	24MY	0.6	-3.3	-1.5	0.2
26.69	-0.91	1.4	72.79	3JN	-0.4	-3.3	-1.5	0.3
33.74	-0.82	1.3	82.31	13JN	-1.4	-3.3	-1.5	0.3
40.66	-0.72	1.3	91.84	23JN	-1.6	-3.4	-1.5	0.3
47.45	-0.60	1.3	101.39	3JL	-0.7	-3.4	-1.5	0.3
54.10	-0.48	1.3	110.96	13JL	-0.1	-3.4	-1.6	0.3
60.59	-0.35	1.3	120.57	23JL	0.3	-3.5	-1.6	0.3
66.92	-0.21	1.2	130.21	2AU	0.5	-3.5	-1.7	0.3
73.05	-0.05	1.2	139.90	12AU	0.9	-3.6	-1.7	0.3
78.99	0.12	1.1	149.65	22AU	2.0	-3.6	-1.8	0.2
84.68	0.30	1.0	159.45	1SE	2.4	-3.7	-1.8	0.2
90.09	0.50	0.9	169.31	11SE	0.0	-3.8	-1.9	0.2
95.18	0.72	0.8	179.22	21SE	-0.8	-3.9	-2.0	0.1
99.89	0.96	0.7	189.20	1OC	-0.9	-4.0	-2.0	0.0
104.12	1.23	0.6	199.23	11OC	-0.9	-4.1	-2.1	-0.0
107.78	1.53	0.4	209.31	21OC	-0.7	-4.2	-2.2	-0.1
110.72	1.87	0.2	219.44	31OC	-0.5	-4.3	-2.2	-0.2
112.79	2.25	0.0	229.59	10NO	-0.5	-4.4	-2.2	-0.3
113.80	2.67	-0.2	239.78	20NO	-0.5	-4.3	-2.3	-0.3
113.56	3.11	-0.4	249.98	30NO	0.1	-4.0	-2.3	-0.2
111.98	3.55	-0.7	260.18	10DE	2.4	-3.3	-2.3	-0.2
109.12	3.93	-0.9	270.38	20DE	1.2	-3.4	-2.2	-0.1
105.37	4.19	-1.0	280.57	30DE	0.3	-4.0	-2.2	-0.0
				353				
101.41	4.30	-1.0	290.72	9JA	0.1	-4.3	-2.1	0.0
97.97	4.23	-0.8	300.84	19JA	-0.0	-4.4	-2.1	0.1
95.59	4.04	-0.5	310.92	29JA	-0.3	-4.3	-2.0	0.2
94.50	3.77	-0.3	320.94	8FE	-0.7	-4.2	-1.9	0.2
94.68	3.48	-0.0	330.92	18FE	-1.4	-4.1	-1.9	0.2
95.98	3.18	0.2	340.83	28FE	-1.4	-4.0	-1.8	0.3
98.21	2.90	0.4	350.68	10MR	-0.6	-3.9	-1.7	0.3
101.19	2.65	0.6	0.48	20MR	0.8	-3.8	-1.7	0.3
104.77	2.40	0.8	10.22	30MR	2.4	-3.7	-1.6	0.3
108.85	2.18	1.0	19.91	9AP	3.4	-3.6	-1.5	0.3
113.32	1.98	1.1	29.55	19AP	1.9	-3.5	-1.5	0.3
118.12	1.79	1.2	39.16	29AP	1.1	-3.5	-1.5	0.3
123.19	1.61	1.3	48.73	9MY	0.4	-3.4	-1.4	0.2
128.49	1.45	1.4	58.27	19MY	-0.4	-3.4	-1.4	0.2
134.01	1.29	1.5	67.80	29MY	-1.4	-3.4	-1.4	0.2
139.70	1.13	1.5	77.32	8JN	-1.6	-3.3	-1.4	0.2
145.55	0.99	1.6	86.85	18JN	-0.7	-3.3	-1.4	0.3
151.56	0.85	1.6	96.39	28JN	-0.0	-3.3	-1.4	0.3
157.71	0.71	1.6	105.94	8JL	0.4	-3.3	-1.4	0.3
163.99	0.58	1.7	115.53	18JL	0.7	-3.3	-1.4	0.3
170.41	0.45	1.7	125.16	28JL	1.2	-3.3	-1.4	0.3
176.95	0.32	1.7	134.82	7AU	2.5	-3.4	-1.4	0.3
183.61	0.20	1.7	144.54	17AU	2.1	-3.4	-1.5	0.3
190.40	0.08	1.7	154.31	27AU	-0.1	-3.4	-1.5	0.3
197.30	-0.04	1.7	164.13	6SE	-1.0	-3.4	-1.6	0.3
204.32	-0.15	1.6	174.02	16SE	-1.1	-3.5	-1.6	0.2
211.45	-0.26	1.6	183.97	26SE	-0.9	-3.5	-1.7	0.2
218.68	-0.37	1.6	193.97	6OC	-0.6	-3.5	-1.7	0.1
226.03	-0.47	1.6	204.02	16OC	-0.4	-3.5	-1.8	0.1
233.47	-0.57	1.5	214.12	26OC	-0.3	-3.4	-1.9	-0.0
241.00	-0.66	1.5	224.26	5NO	-0.3	-3.4	-1.9	-0.1
248.62	-0.75	1.5	234.43	15NO	0.2	-3.4	-2.0	-0.2
256.31	-0.82	1.4	244.63	25NO	2.7	-3.4	-2.0	-0.2
264.06	-0.89	1.4	254.83	5DE	0.9	-3.3	-2.1	-0.3
271.87	-0.95	1.4	265.04	15DE	0.1	-3.3	-2.1	-0.2
279.72	-1.00	1.3	275.23	25DE	-0.1	-3.3	-2.1	-0.2

♂ LONG	♂ LAT	♂ MAG	⊙ LONG	16.00UT 354	☿	♀	♃	♄
287.60	-1.04	1.3	285.40	4JA	-0.2	-3.3	-2.1	-0.1
295.50	-1.07	1.3	295.54	14JA	-0.4	-3.4	-2.1	-0.0
303.40	-1.08	1.3	305.64	24JA	-0.7	-3.4	-2.1	0.0
311.29	-1.09	1.3	315.69	3FE	-1.3	-3.4	-2.1	0.1
319.15	-1.08	1.3	325.70	13FE	-1.3	-3.4	-2.0	0.2
326.98	-1.06	1.4	335.64	23FE	-0.5	-3.5	-1.9	0.2
334.76	-1.03	1.4	345.52	5MR	1.0	-3.5	-1.9	0.3
342.49	-0.99	1.4	355.35	15MR	3.0	-3.5	-1.8	0.3
350.14	-0.94	1.5	5.12	25MR	2.5	-3.6	-1.7	0.3
357.73	-0.88	1.5	14.84	4AP	1.4	-3.7	-1.7	0.3
5.23	-0.80	1.5	24.51	14AP	0.8	-3.7	-1.6	0.3
12.65	-0.73	1.5	34.13	24AP	0.3	-3.8	-1.5	0.3
19.97	-0.64	1.6	43.71	4MY	-0.4	-3.9	-1.5	0.3
27.20	-0.55	1.6	53.27	14MY	-1.4	-4.0	-1.4	0.3
34.34	-0.45	1.6	62.81	24MY	-1.6	-4.1	-1.4	0.3
41.37	-0.34	1.6	72.33	3JN	-0.7	-4.2	-1.4	0.2
48.31	-0.23	1.6	81.86	13JN	0.0	-4.2	-1.3	0.2
55.14	-0.12	1.6	91.38	23JN	0.5	-4.1	-1.3	0.3
61.88	0.00	1.6	100.93	3JL	1.0	-3.8	-1.3	0.3
68.50	0.12	1.6	110.50	13JL	1.6	-3.3	-1.3	0.4
75.02	0.25	1.6	120.10	23JL	3.0	-3.3	-1.3	0.4
81.42	0.38	1.6	129.74	2AU	1.8	-3.9	-1.3	0.4
87.71	0.52	1.6	139.44	12AU	-0.1	-4.2	-1.3	0.4
93.86	0.66	1.6	149.18	22AU	-1.1	-4.3	-1.3	0.4
99.88	0.81	1.5	158.97	1SE	-1.2	-4.2	-1.3	0.4
105.75	0.96	1.5	168.83	11SE	-0.9	-4.2	-1.4	0.4
111.44	1.12	1.4	178.74	21SE	-0.5	-4.1	-1.4	0.3
116.94	1.30	1.3	188.72	1OC	-0.3	-4.0	-1.4	0.3
122.20	1.48	1.2	198.75	11OC	-0.2	-3.9	-1.5	0.3
127.19	1.69	1.1	208.82	21OC	-0.1	-3.8	-1.5	0.2
131.85	1.90	1.0	218.94	31OC	0.4	-3.8	-1.6	0.1
136.10	2.14	0.8	229.10	10NO	3.0	-3.7	-1.7	0.1
139.87	2.41	0.7	239.28	20NO	0.7	-3.6	-1.7	-0.0
143.02	2.70	0.5	249.48	30NO	-0.1	-3.6	-1.8	-0.1
145.42	3.02	0.3	259.69	10DE	-0.2	-3.5	-1.9	-0.2
146.89	3.36	0.0	269.89	20DE	-0.3	-3.5	-1.9	-0.2
147.24	3.72	-0.2	280.07	30DE	-0.4	-3.4	-2.0	-0.1
				355				
146.33	4.06	-0.5	290.23	9JA	-0.8	-3.4	-2.0	-0.1
144.12	4.35	-0.7	300.35	19JA	-1.2	-3.4	-2.0	-0.0
140.83	4.52	-0.9	310.43	29JA	-1.2	-3.3	-2.1	0.0
136.94	4.52	-1.0	320.47	8FE	-0.5	-3.3	-2.0	0.1
133.19	4.35	-0.9	330.44	18FE	1.2	-3.3	-2.0	0.2
130.21	4.04	-0.7	340.35	28FE	3.3	-3.3	-2.0	0.2
128.43	3.65	-0.5	350.21	10MR	1.9	-3.3	-2.0	0.3
127.95	3.23	-0.2	0.01	20MR	1.1	-3.3	-1.9	0.3
128.68	2.82	-0.0	9.75	30MR	0.6	-3.3	-1.8	0.4
130.47	2.44	0.2	19.45	9AP	0.2	-3.4	-1.8	0.4
133.13	2.10	0.4	29.09	19AP	-0.5	-3.4	-1.7	0.4
136.50	1.78	0.5	38.69	29AP	-1.4	-3.5	-1.6	0.4
140.47	1.50	0.7	48.26	9MY	-1.6	-3.5	-1.6	0.4
144.90	1.24	0.8	57.81	19MY	-0.7	-3.5	-1.5	0.4
149.74	1.00	0.9	67.34	29MY	0.1	-3.4	-1.5	0.4
154.92	0.78	1.0	76.86	8JN	0.7	-3.4	-1.4	0.4
160.39	0.58	1.1	86.38	18JN	1.3	-3.4	-1.4	0.4
166.12	0.39	1.2	95.92	28JN	2.2	-3.4	-1.3	0.3
172.08	0.22	1.2	105.48	8JL	3.2	-3.3	-1.3	0.4
178.25	0.06	1.3	115.06	18JL	1.5	-3.3	-1.3	0.4
184.60	-0.10	1.3	124.68	28JL	-0.2	-3.3	-1.2	0.5
191.14	-0.24	1.3	134.35	7AU	-1.1	-3.3	-1.2	0.5
197.84	-0.37	1.3	144.06	17AU	-1.3	-3.3	-1.2	0.5
204.70	-0.50	1.3	153.83	27AU	-0.8	-3.4	-1.2	0.5
211.70	-0.61	1.4	163.65	6SE	-0.4	-3.4	-1.2	0.5
218.84	-0.72	1.4	173.54	16SE	-0.2	-3.4	-1.2	0.5
226.12	-0.81	1.4	183.48	26SE	-0.0	-3.4	-1.3	0.5
233.51	-0.90	1.4	193.48	6OC	0.1	-3.5	-1.3	0.5
241.00	-0.97	1.4	203.53	16OC	0.7	-3.6	-1.3	0.4
248.60	-1.03	1.4	213.63	26OC	3.4	-3.6	-1.4	0.4
256.28	-1.08	1.4	223.77	5NO	0.5	-3.7	-1.4	0.3
264.03	-1.12	1.4	233.94	15NO	-0.3	-3.8	-1.5	0.3
271.84	-1.14	1.4	244.14	25NO	-0.4	-3.9	-1.5	0.2
279.69	-1.15	1.4	254.34	5DE	-0.4	-3.9	-1.6	0.1
287.57	-1.15	1.4	264.54	15DE	-0.5	-4.0	-1.7	0.0
295.46	-1.14	1.4	274.74	25DE	-0.8	-4.2	-1.7	-0.0

♂ LONG	♂ LAT	♂ MAG	⊙ LONG	16.00UT	☿	♀	♃	♄
				356	MAGNITUDES			
303.36	-1.11	1.4	284.91	4JA	-1.0	-4.2	-1.8	-0.1
311.24	-1.08	1.4	295.05	14JA	-1.0	-4.3	-1.9	-0.0
319.09	-1.03	1.4	305.16	24JA	-0.4	-4.3	-1.9	0.0
326.90	-0.97	1.4	315.21	3FE	1.4	-4.2	-2.0	0.1
334.66	-0.91	1.4	325.21	13FE	2.9	-3.9	-2.0	0.2
342.35	-0.83	1.4	335.16	23FE	1.4	-3.3	-2.0	0.2
349.98	-0.75	1.4	345.05	4MR	0.8	-3.5	-2.0	0.3
357.53	-0.67	1.4	354.88	14MR	0.4	-4.0	-2.0	0.3
5.01	-0.58	1.4	4.65	24MR	0.1	-4.2	-2.0	0.4
12.39	-0.48	1.5	14.37	3AP	-0.5	-4.2	-1.9	0.4
19.69	-0.38	1.5	24.04	13AP	-1.4	-4.2	-1.9	0.5
26.91	-0.28	1.6	33.66	23AP	-1.6	-4.1	-1.8	0.5
34.03	-0.18	1.6	43.25	3MY	-0.7	-4.0	-1.8	0.5
41.07	-0.07	1.7	52.81	13MY	0.2	-3.9	-1.7	0.5
48.03	0.03	1.7	62.35	23MY	0.9	-3.8	-1.7	0.5
54.89	0.14	1.8	71.87	2JN	1.7	-3.7	-1.6	0.5
61.68	0.24	1.8	81.39	12JN	2.8	-3.6	-1.5	0.5
68.39	0.35	1.8	90.92	22JN	2.8	-3.6	-1.5	0.5
75.02	0.45	1.8	100.46	2JL	1.2	-3.5	-1.4	0.5
81.57	0.56	1.9	110.03	12JL	-0.2	-3.5	-1.4	0.5
88.05	0.66	1.9	119.64	22JL	-1.2	-3.4	-1.3	0.6
94.45	0.76	1.9	129.27	1AU	-1.5	-3.4	-1.3	0.6
100.78	0.87	1.9	138.96	11AU	-0.8	-3.4	-1.3	0.6
107.04	0.97	1.8	148.71	21AU	-0.3	-3.4	-1.2	0.7
113.21	1.08	1.8	158.50	31AU	-0.0	-3.4	-1.2	0.7
119.30	1.18	1.8	168.35	10SE	0.1	-3.4	-1.2	0.7
125.30	1.29	1.8	178.27	20SE	0.3	-3.4	-1.2	0.7
131.19	1.41	1.7	188.23	30SE	1.0	-3.4	-1.2	0.7
136.97	1.52	1.6	198.26	10OC	3.4	-3.4	-1.2	0.7
142.62	1.64	1.6	208.34	20OC	0.4	-3.4	-1.2	0.6
148.11	1.77	1.5	218.45	30OC	-0.4	-3.4	-1.3	0.6
153.43	1.90	1.4	228.61	9NO	-0.5	-3.4	-1.3	0.6
158.52	2.04	1.2	238.79	19NO	-0.5	-3.4	-1.3	0.5
163.35	2.18	1.1	248.99	29NO	-0.6	-3.5	-1.4	0.4
167.86	2.34	0.9	259.20	9DE	-0.8	-3.5	-1.4	0.4
171.97	2.51	0.8	269.40	19DE	-0.9	-3.5	-1.5	0.3
175.60	2.69	0.6	279.58	29DE	-0.9	-3.5	-1.6	0.2
				357				
178.61	2.87	0.3	289.74	8JA	-0.3	-3.4	-1.6	0.2
180.85	3.07	0.1	299.86	18JA	1.6	-3.4	-1.7	0.1
182.16	3.25	-0.2	309.94	28JA	2.4	-3.4	-1.8	0.2
182.35	3.42	-0.5	319.98	7FE	1.0	-3.4	-1.8	0.2
181.28	3.53	-0.7	329.95	17FE	0.5	-3.3	-1.9	0.3
178.97	3.56	-1.0	339.87	27FE	0.3	-3.3	-1.9	0.3
175.64	3.45	-1.3	349.73	9MR	-0.0	-3.3	-2.0	0.4
171.86	3.20	-1.3	359.54	19MR	-0.6	-3.3	-2.0	0.5
168.35	2.81	-1.2	9.28	29MR	-1.4	-3.4	-2.0	0.5
165.73	2.36	-1.0	18.98	8AP	-1.6	-3.4	-2.0	0.6
164.36	1.88	-0.8	28.62	18AP	-0.7	-3.4	-2.0	0.6
164.32	1.42	-0.5	38.23	28AP	0.3	-3.4	-2.0	0.6
165.50	1.00	-0.3	47.80	8MY	1.2	-3.5	-2.0	0.7
167.76	0.63	-0.2	57.35	18MY	2.2	-3.5	-1.9	0.7
170.89	0.30	0.0	66.88	28MY	3.5	-3.5	-1.9	0.7
174.75	0.01	0.2	76.40	7JN	2.2	-3.6	-1.8	0.7
179.22	-0.24	0.3	85.93	17JN	0.9	-3.7	-1.7	0.7
184.18	-0.46	0.4	95.46	27JN	-0.3	-3.8	-1.7	0.7
189.57	-0.65	0.5	105.02	7JL	-1.3	-3.8	-1.6	0.7
195.33	-0.82	0.6	114.60	17JL	-1.5	-3.9	-1.6	0.7
201.39	-0.97	0.7	124.21	27JL	-0.8	-4.0	-1.5	0.7
207.73	-1.10	0.8	133.88	6AU	-0.2	-4.2	-1.4	0.8
214.32	-1.20	0.8	143.59	16AU	0.1	-4.2	-1.4	0.8
221.11	-1.29	0.9	153.35	26AU	0.3	-4.3	-1.4	0.9
228.09	-1.36	0.9	163.18	5SE	0.5	-4.2	-1.3	0.9
235.25	-1.42	1.0	173.06	15SE	1.3	-3.9	-1.3	0.9
242.55	-1.45	1.0	183.00	25SE	3.0	-3.4	-1.3	0.9
249.98	-1.47	1.1	193.00	5OC	0.2	-3.5	-1.3	0.9
257.51	-1.47	1.1	203.04	15OC	-0.6	-4.1	-1.3	0.9
265.14	-1.46	1.1	213.14	25OC	-0.7	-4.3	-1.3	0.9
272.83	-1.43	1.2	223.28	4NO	-0.7	-4.4	-1.3	0.9
280.58	-1.39	1.2	233.45	14NO	-0.7	-4.3	-1.3	0.8
288.36	-1.33	1.2	243.64	24NO	-0.7	-4.2	-1.3	0.8
296.16	-1.27	1.3	253.85	4DE	-0.7	-4.1	-1.3	0.7
303.95	-1.19	1.3	264.05	14DE	-0.7	-4.0	-1.4	0.7
311.73	-1.10	1.4	274.24	24DE	-0.2	-3.9	-1.4	0.6

♂ LONG	♂ LAT	♂ MAG	⊙ LONG	16.00UT	☿	♀	♃	♄
				358	MAGNITUDES			
319.49	-1.01	1.4	284.42	3JA	1.9	-3.8	-1.5	0.5
327.20	-0.91	1.4	294.56	13JA	1.9	-3.7	-1.5	0.5
334.86	-0.80	1.5	304.67	23JA	0.7	-3.7	-1.6	0.4
342.45	-0.70	1.5	314.73	2FE	0.3	-3.6	-1.7	0.4
349.98	-0.59	1.5	324.73	12FE	0.2	-3.5	-1.7	0.4
357.44	-0.48	1.6	334.68	22FE	-0.1	-3.5	-1.8	0.5
4.81	-0.37	1.6	344.58	4MR	-0.6	-3.4	-1.9	0.5
12.11	-0.26	1.6	354.41	14MR	-1.4	-3.4	-1.9	0.6
19.32	-0.15	1.7	4.19	24MR	-1.5	-3.3	-2.0	0.6
26.45	-0.04	1.7	13.91	3AP	-0.6	-3.3	-2.0	0.7
33.50	0.06	1.7	23.58	13AP	0.5	-3.3	-2.1	0.7
40.48	0.16	1.7	33.20	23AP	1.6	-3.3	-2.1	0.8
47.37	0.26	1.7	42.79	3MY	2.9	-3.3	-2.1	0.8
54.20	0.36	1.7	52.35	13MY	3.1	-3.3	-2.1	0.9
60.95	0.45	1.8	61.89	23MY	1.7	-3.3	-2.1	0.9
67.65	0.54	1.8	71.41	2JN	0.7	-3.3	-2.1	0.9
74.28	0.62	1.8	80.93	12JN	-0.3	-3.4	-2.0	1.0
80.85	0.71	1.9	90.46	22JN	-1.3	-3.4	-2.0	1.0
87.37	0.79	1.9	100.01	2JL	-1.6	-3.5	-1.9	1.0
93.85	0.86	1.9	109.57	12JL	-0.8	-3.5	-1.9	1.0
100.28	0.93	2.0	119.17	22JL	-0.2	-3.5	-1.8	1.0
106.66	1.00	2.0	128.81	1AU	0.2	-3.5	-1.7	1.0
113.01	1.07	2.0	138.49	11AU	0.4	-3.4	-1.7	1.0
119.31	1.14	2.0	148.23	21AU	0.7	-3.4	-1.6	1.1
125.58	1.20	2.0	158.02	31AU	1.7	-3.4	-1.6	1.1
131.81	1.26	2.0	167.86	10SE	2.7	-3.3	-1.5	1.1
138.01	1.32	1.9	177.78	20SE	0.1	-3.3	-1.5	1.2
144.16	1.38	1.9	187.74	30SE	-0.7	-3.3	-1.4	1.2
150.26	1.43	1.9	197.76	10OC	-0.8	-3.3	-1.4	1.2
156.32	1.48	1.8	207.84	20OC	-0.8	-3.3	-1.4	1.2
162.31	1.54	1.8	217.96	30OC	-0.7	-3.3	-1.4	1.2
168.23	1.58	1.7	228.11	9NO	-0.6	-3.4	-1.3	1.2
174.08	1.63	1.6	238.29	19NO	-0.6	-3.4	-1.3	1.2
179.83	1.68	1.5	248.49	29NO	-0.6	-3.4	-1.4	1.1
185.46	1.72	1.4	258.70	9DE	-0.1	-3.5	-1.4	1.1
190.96	1.75	1.2	268.90	19DE	2.1	-3.5	-1.4	1.0
196.29	1.79	1.1	279.09	29DE	1.5	-3.6	-1.4	0.9
				359				
201.41	1.81	0.9	289.25	8JA	0.4	-3.6	-1.4	0.9
206.28	1.82	0.8	299.38	18JA	0.2	-3.7	-1.5	0.8
210.83	1.83	0.6	309.46	28JA	0.0	-3.8	-1.5	0.7
214.99	1.81	0.3	319.49	7FE	-0.2	-3.8	-1.6	0.7
218.67	1.77	0.1	329.48	17FE	-0.6	-3.9	-1.6	0.7
221.72	1.70	-0.2	339.40	27FE	-1.3	-4.0	-1.7	0.7
224.01	1.58	-0.5	349.26	9MR	-1.4	-4.1	-1.8	0.8
225.34	1.41	-0.8	359.07	19MR	-0.6	-4.2	-1.8	0.8
225.55	1.16	-1.1	8.82	29MR	0.6	-4.2	-1.9	0.9
224.52	0.82	-1.4	18.51	8AP	2.0	-4.2	-2.0	0.9
222.27	0.38	-1.7	28.16	18AP	3.9	-4.0	-2.1	1.0
219.15	-0.13	-1.9	37.77	28AP	2.3	-3.5	-2.1	1.0
215.77	-0.67	-1.9	47.34	8MY	1.3	-2.7	-2.2	1.1
212.83	-1.18	-1.7	56.89	18MY	0.5	-3.5	-2.2	1.1
210.96	-1.62	-1.6	66.42	28MY	-0.3	-4.0	-2.2	1.2
210.43	-1.97	-1.3	75.94	7JN	-1.3	-4.2	-2.2	1.2
211.26	-2.23	-1.1	85.47	17JN	-1.6	-4.2	-2.2	1.2
213.34	-2.41	-0.9	95.00	27JN	-0.7	-4.1	-2.2	1.2
216.46	-2.53	-0.7	104.55	7JL	-0.1	-4.0	-2.2	1.3
220.46	-2.60	-0.5	114.13	17JL	0.3	-3.9	-2.2	1.3
225.17	-2.63	-0.4	123.75	27JL	0.6	-3.9	-2.1	1.3
230.46	-2.63	-0.2	133.41	6AU	1.0	-3.8	-2.0	1.3
236.23	-2.59	-0.1	143.12	16AU	2.1	-3.7	-2.0	1.3
242.38	-2.54	0.0	152.88	26AU	2.4	-3.6	-1.9	1.3
248.85	-2.46	0.2	162.70	5SE	0.0	-3.6	-1.8	1.3
255.58	-2.36	0.3	172.58	15SE	-0.9	-3.5	-1.8	1.3
262.53	-2.24	0.4	182.51	25SE	-1.0	-3.5	-1.7	1.3
269.63	-2.11	0.5	192.50	5OC	-0.9	-3.5	-1.7	1.3
276.87	-1.97	0.6	202.55	15OC	-0.6	-3.4	-1.6	1.3
284.20	-1.82	0.7	212.64	25OC	-0.5	-3.4	-1.6	1.3
291.61	-1.66	0.8	222.78	4NO	-0.4	-3.4	-1.5	1.3
299.05	-1.50	0.9	232.95	14NO	-0.4	-3.4	-1.5	1.2
306.52	-1.34	1.0	243.14	24NO	0.1	-3.4	-1.5	1.2
313.99	-1.17	1.0	253.35	4DE	2.4	-3.4	-1.5	1.2
321.46	-1.01	1.1	263.55	14DE	1.2	-3.4	-1.5	1.1
328.89	-0.85	1.2	273.75	24DE	0.2	-3.4	-1.5	1.1

Left Table

♂ LONG	LAT	MAG	☉ LONG	16.00UT 360	☿	♀	♃	♄
336.28	-0.70	1.3	283.92	3JA	0.0	-3.4	-1.5	1.0
343.62	-0.55	1.4	294.07	13JA	-0.1	-3.4	-1.5	1.0
350.91	-0.40	1.4	304.18	23JA	-0.3	-3.4	-1.5	0.9
358.14	-0.27	1.5	314.24	2FE	-0.7	-3.4	-1.5	0.9
5.30	-0.13	1.6	324.25	12FE	-1.3	-3.4	-1.6	0.8
12.39	-0.01	1.6	334.20	22FE	-1.4	-3.5	-1.6	0.8
19.42	0.11	1.7	344.10	3MR	-0.6	-3.5	-1.6	0.9
26.38	0.22	1.7	353.93	13MR	0.8	-3.4	-1.7	0.9
33.27	0.32	1.8	3.71	23MR	2.6	-3.4	-1.8	1.0
40.10	0.42	1.8	13.43	2AP	3.1	-3.4	-1.8	1.1
46.87	0.51	1.8	23.11	12AP	1.7	-3.4	-1.9	1.1
53.57	0.59	1.9	32.73	22AP	1.0	-3.3	-2.0	1.2
60.23	0.67	1.9	42.33	2MY	0.4	-3.3	-2.0	1.3
66.83	0.74	1.9	51.89	12MY	-0.4	-3.3	-2.1	1.3
73.39	0.80	1.9	61.42	22MY	-1.3	-3.3	-2.2	1.3
79.90	0.87	1.9	70.95	1JN	-1.6	-3.3	-2.2	1.4
86.39	0.92	1.9	80.47	11JN	-0.7	-3.3	-2.3	1.4
92.84	0.97	1.9	90.00	21JN	-0.0	-3.4	-2.3	1.4
99.26	1.02	1.9	99.54	1JL	0.4	-3.4	-2.4	1.4
105.66	1.06	1.9	109.11	11JL	0.8	-3.4	-2.4	1.4
112.05	1.10	2.0	118.70	21JL	1.4	-3.5	-2.4	1.3
118.42	1.13	2.0	128.34	31JL	2.7	-3.5	-2.4	1.3
124.78	1.16	2.0	138.02	10AU	-0.0	-3.6	-2.3	1.3
131.13	1.18	2.0	147.75	20AU	-0.0	-3.6	-2.3	1.2
137.48	1.21	2.0	157.54	30AU	-1.0	-3.7	-2.2	1.2
143.83	1.22	2.0	167.39	9SE	-1.1	-3.8	-2.2	1.2
150.18	1.24	2.0	177.29	19SE	-0.9	-3.9	-2.1	1.2
156.53	1.24	2.0	187.26	29SE	-0.5	-4.0	-2.0	1.2
162.88	1.25	2.0	197.28	9OC	-0.3	-4.1	-2.0	1.2
169.23	1.24	1.9	207.34	19OC	-0.3	-4.3	-1.9	1.2
175.59	1.24	1.9	217.46	29OC	-0.2	-4.3	-1.8	1.1
181.93	1.22	1.8	227.61	8NO	0.3	-4.4	-1.8	1.1
188.28	1.21	1.8	237.79	18NO	2.7	-4.3	-1.7	1.1
194.61	1.18	1.7	247.99	28NO	0.9	-4.0	-1.7	1.1
200.93	1.14	1.6	258.20	8DE	0.0	-3.3	-1.6	1.0
207.24	1.10	1.5	268.40	18DE	-0.1	-3.4	-1.6	1.0
213.52	1.04	1.4	278.59	28DE	-0.2	-4.1	-1.6	0.9
				361				
219.76	0.97	1.3	288.75	7JA	-0.4	-4.3	-1.6	0.9
225.97	0.88	1.2	298.88	17JA	-0.7	-4.4	-1.5	0.8
232.12	0.78	1.1	308.97	27JA	-1.3	-4.3	-1.5	0.8
238.20	0.65	0.9	319.01	6FE	-1.2	-4.2	-1.5	0.8
244.20	0.50	0.7	328.99	16FE	-0.6	-4.1	-1.6	0.7
250.09	0.32	0.6	338.92	26FE	1.0	-4.0	-1.6	0.7
255.83	0.11	0.4	348.78	8MR	3.1	-3.9	-1.6	0.6
261.41	-0.14	0.2	358.59	18MR	2.3	-3.8	-1.6	0.7
266.76	-0.44	-0.0	8.35	28MR	1.3	-3.7	-1.7	0.8
271.82	-0.79	-0.2	18.05	7AP	0.7	-3.6	-1.7	0.8
276.51	-1.21	-0.5	27.70	17AP	0.3	-3.5	-1.8	0.9
280.72	-1.69	-0.8	37.31	27AP	-0.4	-3.4	-1.8	1.0
284.31	-2.25	-1.0	46.88	7MY	-1.3	-3.4	-1.9	1.0
287.12	-2.90	-1.3	56.43	17MY	-1.6	-3.4	-2.0	1.1
288.95	-3.63	-1.6	65.97	27MY	-0.7	-3.4	-2.0	1.1
289.63	-4.41	-1.9	75.49	6JN	0.1	-3.3	-2.1	1.2
289.05	-5.18	-2.2	85.01	16JN	0.6	-3.3	-2.2	1.2
287.30	-5.86	-2.5	94.54	26JN	1.1	-3.3	-2.2	1.2
284.81	-6.34	-2.6	104.09	6JL	1.8	-3.3	-2.3	1.2
282.22	-6.51	-2.5	113.67	16JL	3.2	-3.3	-2.4	1.2
280.27	-6.39	-2.3	123.29	26JL	1.7	-3.3	-2.4	1.2
279.46	-6.04	-2.1	132.94	5AU	-0.1	-3.4	-2.4	1.2
279.98	-5.54	-1.8	142.65	15AU	-1.1	-3.4	-2.4	1.1
281.77	-4.98	-1.5	152.41	25AU	-1.3	-3.4	-2.4	1.1
284.68	-4.41	-1.3	162.22	4SE	-0.9	-3.4	-2.4	1.1
288.48	-3.86	-1.0	172.10	14SE	-0.5	-3.5	-2.4	1.0
293.00	-3.34	-0.8	182.03	24SE	-0.2	-3.5	-2.4	1.0
298.08	-2.87	-0.5	192.02	4OC	-0.1	-3.5	-2.3	1.0
303.58	-2.43	-0.3	202.06	14OC	-0.0	-3.4	-2.2	1.0
309.42	-2.03	-0.1	212.15	24OC	0.5	-3.4	-2.2	1.0
315.50	-1.67	0.1	222.28	3NO	3.0	-3.4	-2.1	1.0
321.76	-1.34	0.3	232.45	13NO	0.7	-3.4	-2.0	1.0
328.16	-1.04	0.4	242.64	23NO	-0.2	-3.4	-2.0	1.0
334.66	-0.78	0.6	252.84	3DE	-0.3	-3.3	-1.9	1.0
341.22	-0.54	0.8	263.05	13DE	-0.3	-3.3	-1.8	0.9
347.83	-0.33	0.9	273.25	23DE	-0.5	-3.3	-1.8	0.9

Right Table

♂ LONG	LAT	MAG	☉ LONG	16.00UT 362	☿	♀	♃	♄
354.47	-0.14	1.0	283.42	2JA	-0.8	-3.3	-1.7	0.9
1.10	0.04	1.2	293.57	12JA	-1.1	-3.4	-1.7	0.8
7.74	0.19	1.3	303.68	22JA	-1.1	-3.4	-1.6	0.8
14.36	0.33	1.4	313.74	1FE	-0.5	-3.4	-1.6	0.7
20.97	0.45	1.5	323.76	11FE	1.2	-3.4	-1.6	0.7
27.55	0.55	1.6	333.72	21FE	3.2	-3.5	-1.6	0.6
34.10	0.65	1.6	343.61	3MR	1.7	-3.5	-1.6	0.6
40.63	0.74	1.7	353.46	13MR	0.9	-3.5	-1.6	0.5
47.13	0.81	1.8	3.24	23MR	0.5	-3.6	-1.6	0.5
53.59	0.88	1.8	12.96	2AP	0.1	-3.7	-1.6	0.6
60.03	0.93	1.9	22.64	12AP	-0.5	-3.7	-1.6	0.6
66.45	0.98	1.9	32.27	22AP	-1.3	-3.8	-1.7	0.7
72.84	1.03	2.0	41.87	2MY	-1.6	-3.9	-1.7	0.8
79.21	1.06	2.0	51.43	12MY	-0.7	-4.0	-1.7	0.8
85.57	1.09	2.0	60.97	22MY	0.1	-4.1	-1.8	0.9
91.92	1.12	2.0	70.49	1JN	0.8	-4.2	-1.8	1.0
98.26	1.14	2.0	80.02	11JN	1.4	-4.2	-1.9	1.0
104.59	1.15	2.0	89.54	21JN	2.4	-4.1	-2.0	1.0
110.93	1.16	2.0	99.08	1JL	3.1	-3.8	-2.0	1.1
117.27	1.16	2.0	108.65	11JL	1.4	-3.2	-2.1	1.1
123.63	1.16	2.0	118.24	21JL	-0.1	-3.3	-2.2	1.1
130.00	1.16	2.0	127.87	31JL	-1.2	-3.9	-2.2	1.1
136.38	1.15	1.9	137.56	10AU	-1.4	-4.2	-2.3	1.1
142.79	1.13	2.0	147.28	20AU	-0.9	-4.3	-2.3	1.0
149.23	1.11	2.0	157.07	30AU	-0.4	-4.2	-2.4	1.0
155.69	1.09	2.0	166.91	9SE	-0.1	-4.2	-2.4	1.0
162.19	1.06	2.0	176.82	19SE	0.0	-4.1	-2.4	0.9
168.72	1.03	2.0	186.77	29SE	0.2	-4.0	-2.4	0.9
175.28	0.99	2.0	196.79	9OC	0.8	-3.9	-2.4	0.9
181.89	0.94	1.9	206.86	19OC	3.4	-3.8	-2.4	0.9
188.53	0.89	1.9	216.97	29OC	0.5	-3.8	-2.4	0.9
195.21	0.83	1.9	227.12	8NO	-0.3	-3.7	-2.3	0.9
201.94	0.76	1.8	237.30	18NO	-0.4	-3.6	-2.2	0.9
208.70	0.69	1.8	247.50	28NO	-0.5	-3.6	-2.2	0.9
215.50	0.60	1.7	257.71	8DE	-0.6	-3.5	-2.1	0.9
222.34	0.51	1.7	267.91	18DE	-0.8	-3.5	-2.0	0.9
229.23	0.40	1.6	278.09	28DE	-1.0	-3.4	-1.9	0.8
				363				
236.14	0.29	1.5	288.26	7JA	-1.0	-3.4	-1.9	0.8
243.09	0.16	1.4	298.39	17JA	-0.4	-3.4	-1.8	0.8
250.08	0.02	1.3	308.48	27JA	1.4	-3.3	-1.8	0.7
257.09	-0.13	1.2	318.52	6FE	2.7	-3.3	-1.7	0.7
264.12	-0.30	1.1	328.50	16FE	1.2	-3.3	-1.7	0.6
271.18	-0.48	1.0	338.43	26FE	0.7	-3.3	-1.6	0.6
278.24	-0.68	0.9	348.31	8MR	0.4	-3.3	-1.6	0.5
285.31	-0.88	0.8	358.12	18MR	0.0	-3.3	-1.6	0.5
292.36	-1.11	0.7	7.87	28MR	-0.5	-3.3	-1.6	0.4
299.39	-1.34	0.6	17.58	7AP	-1.3	-3.4	-1.5	0.4
306.36	-1.59	0.5	27.23	17AP	-1.6	-3.4	-1.5	0.5
313.27	-1.85	0.3	36.84	27AP	-0.7	-3.5	-1.5	0.6
320.06	-2.12	0.2	46.42	7MY	0.2	-3.5	-1.6	0.6
326.72	-2.39	0.1	55.97	17MY	1.0	-3.5	-1.6	0.7
333.19	-2.67	-0.1	65.50	27MY	1.9	-3.4	-1.6	0.8
339.40	-2.95	-0.2	75.02	6JN	3.1	-3.4	-1.6	0.8
345.30	-3.23	-0.4	84.54	16JN	2.6	-3.4	-1.7	0.9
350.79	-3.51	-0.6	94.08	26JN	1.1	-3.4	-1.7	0.9
355.75	-3.78	-0.8	103.63	6JL	-0.2	-3.3	-1.8	0.9
0.04	-4.04	-1.0	113.20	16JL	-1.2	-3.3	-1.8	1.0
3.48	-4.28	-1.2	122.82	26JL	-1.5	-3.3	-1.9	1.0
5.86	-4.47	-1.4	132.47	5AU	-0.8	-3.3	-2.0	1.0
6.99	-4.59	-1.6	142.17	15AU	-0.3	-3.3	-2.0	1.0
6.68	-4.61	-1.9	151.93	25AU	-0.0	-3.4	-2.1	1.0
4.95	-4.46	-2.1	161.75	4SE	0.2	-3.4	-2.2	0.9
2.11	-4.12	-2.2	171.61	14SE	0.4	-3.4	-2.2	0.9
358.81	-3.58	-2.2	181.55	24SE	1.1	-3.4	-2.3	0.9
355.87	-2.93	-2.0	191.53	4OC	3.3	-3.5	-2.3	0.8
353.94	-2.23	-1.7	201.57	14OC	0.3	-3.6	-2.3	0.8
353.31	-1.59	-1.4	211.66	24OC	-0.5	-3.6	-2.4	0.8
354.00	-1.02	-1.0	221.80	3NO	-0.6	-3.7	-2.4	0.8
355.86	-0.55	-0.7	231.96	13NO	-0.6	-3.8	-2.4	0.9
358.65	-0.16	-0.4	242.15	23NO	-0.7	-3.9	-2.3	0.9
2.20	0.15	-0.1	252.35	3DE	-0.8	-4.0	-2.3	0.9
6.33	0.41	0.1	262.56	13DE	-0.8	-4.1	-2.2	0.8
10.92	0.61	0.4	272.76	23DE	-0.8	-4.2	-2.2	0.8

LONG	LAT	MAG	LONG		☿	♀	♃	♄
15.85	0.78	0.6	282.93	2JA	-0.3	-4.3	-2.1	0.8
21.05	0.91	0.8	293.08	12JA	1.6	-4.3	-2.0	0.8
26.46	1.02	0.9	303.19	22JA	2.2	-4.3	-1.9	0.7
32.03	1.10	1.1	313.26	1FE	0.9	-4.2	-1.9	0.7
37.72	1.17	1.2	323.27	11FE	0.5	-3.9	-1.8	0.7
43.51	1.22	1.3	333.23	21FE	0.2	-3.3	-1.7	0.6
49.38	1.26	1.5	343.13	2MR	-0.1	-3.5	-1.7	0.5
55.30	1.29	1.6	352.97	12MR	-0.6	-4.0	-1.6	0.5
61.27	1.32	1.7	2.76	22MR	-1.3	-4.2	-1.6	0.4
67.28	1.33	1.7	12.49	1AP	-1.6	-4.2	-1.5	0.4
73.32	1.34	1.8	22.17	11AP	-0.7	-4.2	-1.5	0.3
79.39	1.34	1.9	31.80	21AP	0.4	-4.1	-1.5	0.4
85.49	1.33	1.9	41.40	1MY	1.3	-4.0	-1.5	0.4
91.61	1.32	2.0	50.96	11MY	2.5	-3.9	-1.5	0.5
97.75	1.30	2.0	60.50	21MY	3.5	-3.8	-1.5	0.6
103.92	1.28	2.0	70.03	31MY	2.0	-3.7	-1.5	0.6
110.12	1.26	2.0	79.55	10JN	0.9	-3.6	-1.5	0.7
116.35	1.23	2.0	89.08	20JN	-0.2	-3.6	-1.5	0.7
122.61	1.20	2.0	98.62	30JN	-1.3	-3.5	-1.5	0.8
128.91	1.16	2.0	108.18	10JL	-1.6	-3.5	-1.5	0.8
135.25	1.12	2.0	117.78	20JL	-0.8	-3.4	-1.6	0.9
141.63	1.08	2.0	127.41	30JL	-0.2	-3.4	-1.6	0.9
148.05	1.03	2.0	137.09	9AU	0.1	-3.4	-1.7	0.9
154.53	0.98	1.9	146.82	19AU	0.3	-3.4	-1.7	0.9
161.06	0.92	1.9	156.60	29AU	0.6	-3.4	-1.8	0.9
167.65	0.86	1.9	166.44	8SE	1.4	-3.4	-1.8	0.9
174.30	0.80	1.8	176.34	18SE	3.0	-3.4	-1.9	0.9
181.00	0.73	1.8	186.29	28SE	0.2	-3.4	-2.0	0.8
187.78	0.65	1.8	196.31	8OC	-0.7	-3.4	-2.0	0.8
194.61	0.57	1.8	206.37	18OC	-0.7	-3.4	-2.1	0.7
201.52	0.49	1.8	216.48	28OC	-0.7	-3.4	-2.2	0.7
208.49	0.40	1.8	226.63	7NO	-0.8	-3.4	-2.2	0.8
215.53	0.30	1.8	236.81	17NO	-0.7	-3.4	-2.2	0.8
222.63	0.20	1.7	247.00	27NO	-0.7	-3.5	-2.2	0.8
229.81	0.09	1.7	257.21	7DE	-0.7	-3.5	-2.2	0.8
237.04	-0.02	1.7	267.41	17DE	-0.2	-3.5	-2.2	0.8
244.34	-0.14	1.6	277.60	27DE	1.9	-3.5	-2.2	0.8
				365				
251.70	-0.26	1.6	287.76	6JA	1.8	-3.4	-2.2	0.8
259.11	-0.39	1.5	297.90	16JA	0.6	-3.4	-2.1	0.8
266.57	-0.52	1.5	307.99	26JA	0.3	-3.4	-2.0	0.7
274.08	-0.65	1.4	318.03	5FE	0.1	-3.4	-2.0	0.7
281.62	-0.79	1.4	328.02	15FE	-0.1	-3.3	-1.9	0.7
289.19	-0.92	1.3	337.95	25FE	-0.6	-3.3	-1.8	0.6
296.78	-1.06	1.2	347.82	7MR	-1.3	-3.3	-1.8	0.5
304.38	-1.19	1.2	357.64	17MR	-1.5	-3.3	-1.7	0.5
311.97	-1.31	1.1	7.39	27MR	-0.7	-3.4	-1.6	0.4
319.55	-1.43	1.1	17.10	6AP	0.5	-3.4	-1.6	0.4
327.09	-1.54	1.0	26.76	16AP	1.7	-3.4	-1.5	0.3
334.57	-1.64	0.9	36.37	26AP	3.3	-3.4	-1.5	0.3
341.99	-1.72	0.9	45.95	6MY	2.8	-3.5	-1.4	0.4
349.32	-1.80	0.8	55.50	16MY	1.6	-3.5	-1.4	0.4
356.54	-1.85	0.8	65.03	26MY	0.7	-3.5	-1.4	0.5
3.64	-1.89	0.7	74.56	5JN	-0.3	-3.6	-1.4	0.5
10.56	-1.91	0.6	84.09	15JN	-1.3	-3.7	-1.4	0.6
17.31	-1.92	0.5	93.62	25JN	-1.6	-3.8	-1.3	0.7
23.83	-1.90	0.5	103.17	5JL	-0.8	-3.8	-1.3	0.7
30.09	-1.86	0.4	112.75	15JL	-0.2	-3.9	-1.4	0.8
36.04	-1.80	0.3	122.35	25JL	0.2	-4.1	-1.4	0.8
41.62	-1.72	0.1	132.01	4AU	0.5	-4.2	-1.4	0.8
46.75	-1.61	0.0	141.71	14AU	0.8	-4.2	-1.4	0.8
51.34	-1.46	-0.1	151.46	24AU	1.8	-4.3	-1.5	0.8
55.26	-1.29	-0.3	161.27	3SE	2.7	-4.2	-1.5	0.8
58.34	-1.07	-0.5	171.14	13SE	0.1	-3.9	-1.6	0.8
60.41	-0.80	-0.7	181.07	23SE	-0.8	-3.3	-1.6	0.8
61.26	-0.47	-0.9	191.05	3OC	-0.9	-3.5	-1.7	0.8
60.70	-0.07	-1.1	201.09	13OC	-0.9	-4.1	-1.7	0.8
58.70	0.37	-1.3	211.17	23OC	-0.7	-4.3	-1.8	0.7
55.49	0.84	-1.5	221.31	2NO	-0.5	-4.4	-1.9	0.7
51.65	1.29	-1.6	231.47	12NO	-0.5	-4.3	-1.9	0.7
47.99	1.66	-1.4	241.65	22NO	-0.5	-4.2	-2.0	0.8
45.19	1.93	-1.1	251.86	2DE	-0.0	-4.1	-2.0	0.8
43.68	2.09	-0.8	262.06	12DE	2.1	-4.0	-2.1	0.8
43.51	2.18	-0.5	272.26	22DE	1.4	-3.9	-2.1	0.8

LONG	LAT	MAG	LONG		☿	♀	♃	♄
44.55	2.21	-0.2	282.44	1JA	0.3	-3.8	-2.1	0.8
46.62	2.21	0.1	292.59	11JA	0.1	-3.7	-2.1	0.8
49.50	2.18	0.4	302.70	21JA	-0.0	-3.7	-2.1	0.8
53.02	2.14	0.6	312.78	31JA	-0.2	-3.6	-2.1	0.8
57.06	2.09	0.8	322.79	10FE	-0.6	-3.5	-2.0	0.7
61.48	2.03	1.0	332.75	20FE	-1.3	-3.5	-2.0	0.7
66.22	1.97	1.1	342.66	2MR	-1.4	-3.4	-1.9	0.6
71.21	1.90	1.3	352.53	12MR	-0.7	-3.4	-1.8	0.6
76.40	1.84	1.4	2.29	22MR	0.7	-3.3	-1.8	0.5
81.76	1.77	1.5	12.03	1AP	2.2	-3.3	-1.7	0.5
87.25	1.71	1.6	21.71	11AP	3.7	-3.3	-1.6	0.4
92.86	1.64	1.7	31.34	21AP	2.1	-3.3	-1.6	0.3
98.58	1.57	1.7	40.94	1MY	1.2	-3.3	-1.5	0.3
104.39	1.50	1.8	50.50	11MY	0.5	-3.3	-1.5	0.3
110.27	1.43	1.8	60.04	21MY	-0.3	-3.3	-1.4	0.3
116.24	1.36	1.9	69.57	31MY	-1.3	-3.3	-1.4	0.4
122.28	1.29	1.9	79.09	10JN	-1.7	-3.4	-1.3	0.5
128.39	1.21	1.9	88.62	20JN	-0.8	-3.4	-1.3	0.5
134.58	1.14	1.9	98.16	30JN	-0.1	-3.5	-1.3	0.6
140.83	1.06	2.0	107.72	10JL	0.3	-3.5	-1.3	0.6
147.16	0.98	1.9	117.31	20JL	0.7	-3.5	-1.3	0.7
153.57	0.90	1.9	126.94	30JL	1.1	-3.5	-1.3	0.7
160.05	0.81	1.9	136.61	9AU	2.3	-3.4	-1.3	0.8
166.60	0.73	1.9	146.34	19AU	2.3	-3.4	-1.3	0.8
173.24	0.64	1.9	156.12	29AU	0.1	-3.4	-1.3	0.8
179.96	0.55	1.8	165.96	8SE	-0.9	-3.3	-1.3	0.8
186.77	0.45	1.8	175.85	18SE	-1.0	-3.3	-1.4	0.8
193.66	0.36	1.7	185.81	28SE	-1.0	-3.3	-1.4	0.8
200.63	0.26	1.7	195.82	8OC	-0.6	-3.3	-1.4	0.8
207.69	0.16	1.6	205.88	18OC	-0.4	-3.3	-1.5	0.8
214.84	0.06	1.6	215.99	28OC	-0.4	-3.4	-1.6	0.7
222.07	-0.05	1.6	226.13	7NO	-0.3	-3.4	-1.6	0.7
229.39	-0.15	1.6	236.31	17NO	0.1	-3.4	-1.7	0.7
236.78	-0.26	1.6	246.51	27NO	2.4	-3.4	-1.7	0.7
244.24	-0.37	1.6	256.71	7DE	1.1	-3.5	-1.8	0.8
251.78	-0.47	1.5	266.91	17DE	0.1	-3.5	-1.9	0.8
259.38	-0.58	1.5	277.11	27DE	-0.0	-3.6	-1.9	0.8
				367				
267.04	-0.68	1.5	287.27	6JA	-0.1	-3.6	-2.0	0.8
274.74	-0.78	1.5	297.41	16JA	-0.3	-3.7	-2.0	0.8
282.48	-0.87	1.5	307.50	26JA	-0.7	-3.8	-2.0	0.8
290.25	-0.96	1.4	317.55	5FE	-1.3	-3.8	-2.0	0.8
298.04	-1.04	1.4	327.54	15FE	-1.3	-3.9	-2.0	0.7
305.84	-1.11	1.4	337.48	25FE	-0.6	-4.0	-2.0	0.7
313.63	-1.18	1.4	347.35	7MR	0.8	-4.1	-2.0	0.7
321.40	-1.23	1.4	357.17	17MR	2.7	-4.2	-1.9	0.6
329.14	-1.26	1.3	6.93	27MR	2.8	-4.2	-1.9	0.6
336.83	-1.29	1.3	16.63	6AP	1.5	-4.2	-1.8	0.5
344.47	-1.30	1.3	26.29	16AP	0.9	-4.0	-1.7	0.4
352.04	-1.30	1.3	35.91	26AP	0.3	-3.5	-1.7	0.4
359.53	-1.28	1.3	45.49	6MY	-0.4	-2.8	-1.6	0.3
6.93	-1.25	1.3	55.04	16MY	-1.3	-3.6	-1.6	0.2
14.23	-1.20	1.2	64.58	26MY	-1.7	-4.0	-1.5	0.3
21.41	-1.14	1.2	74.10	5JN	-0.7	-4.2	-1.4	0.3
28.47	-1.07	1.2	83.62	15JN	-0.0	-4.2	-1.4	0.4
35.39	-0.99	1.2	93.15	25JN	0.5	-4.1	-1.3	0.5
42.16	-0.88	1.1	102.70	5JL	0.9	-4.0	-1.3	0.5
48.77	-0.77	1.1	112.28	15JL	1.5	-3.9	-1.3	0.6
55.19	-0.64	1.0	121.89	25JL	2.8	-3.9	-1.2	0.6
61.41	-0.50	1.0	131.53	4AU	2.0	-3.8	-1.2	0.7
67.40	-0.34	0.9	141.23	14AU	-0.0	-3.7	-1.2	0.7
73.13	-0.17	0.8	150.99	24AU	-1.0	-3.6	-1.2	0.8
78.55	0.03	0.7	160.79	3SE	-1.2	-3.6	-1.2	0.8
83.61	0.24	0.6	170.66	13SE	-0.9	-3.5	-1.2	0.8
88.23	0.49	0.5	180.58	23SE	-0.5	-3.5	-1.2	0.8
92.33	0.76	0.4	190.56	3OC	-0.3	-3.5	-1.3	0.8
95.77	1.07	0.2	200.60	13OC	-0.2	-3.4	-1.3	0.8
98.41	1.42	0.0	210.69	23OC	-0.1	-3.4	-1.3	0.8
100.07	1.81	-0.2	220.81	2NO	0.3	-3.4	-1.4	0.7
100.55	2.24	-0.4	230.98	12NO	2.7	-3.4	-1.4	0.7
99.70	2.70	-0.7	241.16	22NO	0.8	-3.4	-1.5	0.7
97.48	3.15	-0.9	251.36	2DE	-0.1	-3.4	-1.5	0.7
94.11	3.53	-1.1	261.57	12DE	-0.2	-3.4	-1.6	0.7
90.16	3.79	-1.1	271.77	22DE	-0.3	-3.4	-1.7	0.8

♂ LONG	♂ LAT	♂ MAG	⊙ LONG	16.00UT	☿	♀	♃	♄
				368	MAGNITUDES			
86.38	3.89	-0.9	281.95	1JA	-0.4	-3.4	-1.7	0.8
83.43	3.84	-0.7	292.10	11JA	-0.8	-3.4	-1.8	0.8
81.72	3.69	-0.4	302.22	21JA	-1.2	-3.4	-1.9	0.8
81.33	3.48	-0.2	312.29	31JA	-1.2	-3.4	-1.9	0.8
82.13	3.25	0.1	322.31	10FE	-0.6	-3.4	-2.0	0.8
83.96	3.02	0.3	332.27	20FE	1.0	-3.5	-2.0	0.8
86.62	2.80	0.5	342.18	1MR	3.2	-3.5	-2.0	0.7
89.94	2.59	0.7	352.03	11MR	2.1	-3.4	-2.0	0.7
93.80	2.39	0.9	1.82	21MR	1.1	-3.4	-2.0	0.7
98.07	2.21	1.1	11.55	31MR	0.7	-3.4	-2.0	0.6
102.70	2.04	1.2	21.24	10AP	0.2	-3.4	-1.9	0.5
107.61	1.89	1.3	30.87	20AP	-0.4	-3.3	-1.9	0.5
112.75	1.74	1.4	40.47	30AP	-1.3	-3.3	-1.8	0.4
118.11	1.59	1.5	50.04	10MY	-1.7	-3.3	-1.8	0.3
123.63	1.46	1.6	59.58	20MY	-0.7	-3.3	-1.7	0.3
129.31	1.32	1.6	69.10	30MY	0.1	-3.3	-1.6	0.2
135.14	1.20	1.7	78.63	9JN	0.6	-3.3	-1.6	0.3
141.10	1.07	1.7	88.15	19JN	1.2	-3.4	-1.5	0.4
147.17	0.95	1.7	97.69	29JN	2.0	-3.4	-1.5	0.4
153.37	0.83	1.8	107.25	9JL	3.2	-3.4	-1.4	0.5
159.69	0.71	1.8	116.84	19JL	1.6	-3.5	-1.4	0.5
166.11	0.59	1.8	126.47	29JL	-0.1	-3.5	-1.3	0.6
172.65	0.48	1.8	136.14	8AU	-1.1	-3.6	-1.3	0.6
179.29	0.36	1.8	145.86	18AU	-1.3	-3.6	-1.3	0.7
186.04	0.25	1.7	155.64	28AU	-0.9	-3.7	-1.2	0.7
192.90	0.14	1.7	165.48	7SE	-0.4	-3.8	-1.2	0.8
199.87	0.03	1.7	175.37	17SE	-0.2	-3.9	-1.2	0.8
206.94	-0.08	1.7	185.32	27SE	-0.1	-4.0	-1.2	0.8
214.12	-0.19	1.6	195.33	7OC	0.0	-4.1	-1.2	0.8
221.39	-0.30	1.6	205.39	17OC	0.6	-4.3	-1.2	0.8
228.76	-0.40	1.6	215.50	27OC	3.0	-4.3	-1.3	0.8
236.22	-0.50	1.5	225.65	6NO	0.6	-4.4	-1.3	0.8
243.76	-0.59	1.5	235.82	16NO	-0.2	-4.3	-1.3	0.7
251.39	-0.68	1.4	246.01	26NO	-0.3	-4.0	-1.4	0.7
259.08	-0.77	1.4	256.22	6DE	-0.4	-3.2	-1.4	0.7
266.83	-0.84	1.4	266.42	16DE	-0.5	-3.5	-1.5	0.7
274.64	-0.91	1.3	276.61	26DE	-0.8	-4.1	-1.5	0.8
				369				
282.48	-0.97	1.4	286.78	5JA	-1.1	-4.3	-1.6	0.8
290.34	-1.02	1.4	296.92	15JA	-1.1	-4.4	-1.7	0.8
298.22	-1.06	1.4	307.01	25JA	-0.5	-4.3	-1.7	0.8
306.10	-1.09	1.4	317.06	4FE	1.2	-4.2	-1.8	0.8
313.97	-1.10	1.4	327.05	14FE	3.1	-4.1	-1.9	0.8
321.81	-1.11	1.4	336.99	24FE	1.5	-4.0	-1.9	0.8
329.62	-1.10	1.4	346.87	6MR	0.8	-3.9	-2.0	0.8
337.37	-1.08	1.4	356.69	16MR	0.5	-3.8	-2.0	0.8
345.07	-1.05	1.4	6.46	26MR	0.1	-3.7	-2.0	0.7
352.71	-1.01	1.5	16.17	5AP	-0.5	-3.6	-2.0	0.7
0.26	-0.95	1.5	25.83	15AP	-1.3	-3.5	-2.0	0.6
7.74	-0.89	1.5	35.45	25AP	-1.7	-3.5	-2.0	0.5
15.12	-0.81	1.5	45.03	5MY	-0.7	-3.4	-2.0	0.5
22.41	-0.73	1.5	54.58	15MY	0.2	-3.4	-2.0	0.4
29.60	-0.64	1.5	64.12	25MY	0.8	-3.4	-1.9	0.3
36.69	-0.54	1.5	73.64	4JN	1.5	-3.3	-1.9	0.3
43.66	-0.43	1.5	83.16	14JN	2.6	-3.3	-1.8	0.3
50.53	-0.32	1.5	92.70	24JN	3.0	-3.3	-1.7	0.3
57.29	-0.20	1.5	102.24	4JL	1.3	-3.3	-1.7	0.4
63.92	-0.07	1.5	111.81	14JL	-0.1	-3.3	-1.6	0.4
70.43	0.06	1.5	121.42	24JL	-1.2	-3.3	-1.5	0.5
76.81	0.20	1.5	131.07	3AU	-1.5	-3.4	-1.5	0.6
83.04	0.34	1.4	140.76	13AU	-0.9	-3.4	-1.4	0.6
89.12	0.50	1.4	150.51	23AU	-0.4	-3.4	-1.4	0.7
95.03	0.66	1.4	160.31	2SE	-0.1	-3.5	-1.4	0.7
100.74	0.84	1.3	170.18	12SE	0.1	-3.5	-1.3	0.7
106.24	1.02	1.2	180.10	22SE	0.2	-3.5	-1.3	0.8
111.47	1.23	1.1	190.08	2OC	0.8	-3.5	-1.3	0.8
116.39	1.44	1.0	200.11	12OC	3.4	-3.4	-1.3	0.8
120.95	1.68	0.9	210.19	22OC	0.5	-3.4	-1.3	0.8
125.05	1.95	0.7	220.32	1NO	-0.4	-3.4	-1.3	0.8
128.60	2.24	0.6	230.48	11NO	-0.5	-3.4	-1.3	0.8
131.46	2.57	0.4	240.67	21NO	-0.5	-3.4	-1.3	0.8
133.48	2.93	0.1	250.87	1DE	-0.6	-3.3	-1.3	0.7
134.46	3.31	-0.1	261.07	11DE	-0.8	-3.3	-1.4	0.7
134.24	3.71	-0.3	271.27	21DE	-0.9	-3.3	-1.4	0.7
132.70	4.09	-0.6	281.45	31DE	-0.9	-3.3	-1.4	0.8

♂ LONG	♂ LAT	♂ MAG	⊙ LONG	16.00UT	☿	♀	♃	♄
				370	MAGNITUDES			
129.91	4.40	-0.8	291.61	10JA	-0.4	-3.4	-1.5	0.8
126.22	4.57	-1.0	301.73	20JA	1.4	-3.4	-1.6	0.8
122.28	4.56	-0.9	311.80	30JA	2.6	-3.4	-1.6	0.9
118.80	4.38	-0.8	321.82	9FE	1.1	-3.4	-1.7	0.9
116.34	4.08	-0.5	331.79	19FE	0.6	-3.5	-1.8	0.9
115.17	3.72	-0.3	341.70	1MR	0.3	-3.5	-1.8	0.9
115.26	3.35	-0.1	351.55	11MR	0.0	-3.6	-1.9	0.8
116.49	2.98	0.2	1.35	21MR	-0.5	-3.6	-2.0	0.8
118.67	2.64	0.4	11.08	31MR	-1.3	-3.7	-2.0	0.8
121.63	2.33	0.6	20.77	10AP	-1.6	-3.7	-2.1	0.7
125.23	2.04	0.7	30.41	20AP	-0.7	-3.8	-2.1	0.7
129.34	1.78	0.9	40.01	30AP	0.3	-3.9	-2.1	0.6
133.88	1.55	1.0	49.58	10MY	1.1	-4.0	-2.1	0.6
138.77	1.33	1.1	59.12	20MY	2.1	-4.1	-2.1	0.5
143.96	1.12	1.2	68.65	30MY	3.4	-4.2	-2.1	0.4
149.41	0.93	1.2	78.17	9JN	2.4	-4.2	-2.1	0.4
155.09	0.75	1.3	87.70	19JN	1.1	-4.1	-2.0	0.3
160.97	0.59	1.4	97.23	29JN	-0.2	-3.8	-2.0	0.3
167.04	0.43	1.4	106.79	9JL	-1.2	-3.2	-1.9	0.4
173.29	0.27	1.4	116.38	19JL	-1.6	-3.3	-1.9	0.4
179.70	0.13	1.5	126.00	29JL	-0.8	-3.9	-1.8	0.5
186.26	-0.01	1.5	135.67	8AU	-0.3	-4.2	-1.7	0.5
192.97	-0.14	1.5	145.39	18AU	0.0	-4.3	-1.7	0.6
199.82	-0.27	1.5	155.16	28AU	0.2	-4.2	-1.6	0.6
206.80	-0.38	1.5	165.00	7SE	0.4	-4.2	-1.6	0.7
213.91	-0.50	1.5	174.89	17SE	1.2	-4.1	-1.5	0.7
221.15	-0.60	1.5	184.83	27SE	3.3	-4.0	-1.5	0.8
228.49	-0.70	1.5	194.84	7OC	0.3	-3.9	-1.4	0.8
235.95	-0.78	1.5	204.90	17OC	-0.6	-3.8	-1.4	0.8
243.50	-0.86	1.5	215.00	27OC	-0.7	-3.8	-1.4	0.8
251.14	-0.93	1.4	225.15	6NO	-0.6	-3.7	-1.4	0.8
258.86	-0.99	1.4	235.32	16NO	-0.7	-3.6	-1.4	0.8
266.64	-1.03	1.4	245.52	26NO	-0.7	-3.6	-1.4	0.8
274.47	-1.07	1.4	255.72	6DE	-0.7	-3.5	-1.4	0.8
282.34	-1.09	1.4	265.93	16DE	-0.7	-3.5	-1.4	0.8
290.24	-1.10	1.4	276.12	26DE	-0.3	-3.4	-1.4	0.7
				371				
298.14	-1.10	1.4	286.29	5JA	1.6	-3.4	-1.4	0.8
306.04	-1.09	1.4	296.43	15JA	2.1	-3.4	-1.5	0.8
313.92	-1.07	1.4	306.53	25JA	0.8	-3.3	-1.5	0.9
321.78	-1.03	1.3	316.58	4FE	0.4	-3.3	-1.6	0.9
329.59	-0.99	1.3	326.58	14FE	0.2	-3.3	-1.6	0.9
337.34	-0.93	1.3	336.52	24FE	-0.1	-3.3	-1.7	0.9
345.03	-0.87	1.4	346.40	6MR	-0.6	-3.3	-1.7	0.9
352.66	-0.80	1.4	356.22	16MR	-1.3	-3.3	-1.8	0.9
0.20	-0.72	1.5	5.99	26MR	-1.6	-3.4	-1.9	0.9
7.67	-0.63	1.5	15.70	5AP	-0.7	-3.4	-2.0	0.9
15.05	-0.54	1.5	25.36	15AP	0.4	-3.4	-2.0	0.8
22.33	-0.44	1.6	34.98	25AP	1.4	-3.5	-2.1	0.8
29.53	-0.34	1.6	44.57	5MY	2.7	-3.5	-2.1	0.7
36.64	-0.24	1.7	54.12	15MY	3.3	-3.5	-2.2	0.7
43.66	-0.14	1.7	63.65	25MY	1.9	-3.4	-2.2	0.6
50.59	-0.03	1.7	73.18	4JN	0.8	-3.4	-2.3	0.5
57.43	0.08	1.7	82.70	14JN	-0.2	-3.4	-2.3	0.5
64.18	0.19	1.8	92.23	24JN	-1.3	-3.4	-2.3	0.4
70.84	0.30	1.8	101.78	4JL	-1.6	-3.3	-2.2	0.3
77.42	0.42	1.8	111.35	14JL	-0.8	-3.3	-2.2	0.4
83.91	0.53	1.8	120.95	24JL	-0.2	-3.3	-2.2	0.4
90.31	0.65	1.8	130.60	3AU	0.1	-3.3	-2.1	0.5
96.62	0.76	1.8	140.29	13AU	0.4	-3.3	-2.0	0.5
102.84	0.88	1.8	150.03	23AU	0.7	-3.4	-2.0	0.6
108.95	1.01	1.7	159.84	2SE	1.5	-3.4	-1.9	0.6
114.96	1.13	1.7	169.69	12SE	3.0	-3.4	-1.9	0.7
120.84	1.26	1.6	179.61	22SE	0.2	-3.5	-1.8	0.7
126.60	1.40	1.6	189.59	2OC	-0.7	-3.5	-1.7	0.8
132.19	1.54	1.5	199.62	12OC	-0.8	-3.6	-1.7	0.8
137.61	1.69	1.4	209.70	22OC	-0.8	-3.6	-1.6	0.8
142.81	1.85	1.3	219.82	1NO	-0.8	-3.7	-1.6	0.9
147.76	2.02	1.2	229.98	11NO	-0.6	-3.8	-1.5	0.9
152.41	2.21	1.0	240.17	21NO	-0.6	-3.9	-1.5	0.9
156.68	2.41	0.9	250.37	1DE	-0.6	-4.0	-1.5	0.9
160.48	2.62	0.7	260.57	11DE	-0.2	-4.1	-1.5	0.9
163.71	2.86	0.5	270.77	21DE	1.9	-4.2	-1.5	0.9
166.22	3.11	0.3	280.96	31DE	1.7	-4.3	-1.5	0.8

Left table (372 / 373):

♂ LONG	LAT	MAG	☉ LONG	16.00UT	☿	♀	♃	♄
				372		MAGNITUDES		
167.84	3.37	0.0	291.11	10JA	0.5	-4.3	-1.5	0.8
168.40	3.63	-0.3	301.24	20JA	0.2	-4.3	-1.5	0.9
167.73	3.87	-0.5	311.31	30JA	0.1	-4.2	-1.5	0.9
165.78	4.04	-0.8	321.34	9FE	-0.2	-3.9	-1.5	1.0
162.69	4.09	-1.0	331.31	19FE	-0.6	-3.3	-1.6	1.0
158.93	3.99	-1.2	341.22	29FE	-1.3	-3.5	-1.6	1.0
155.17	3.73	-1.1	351.08	10MR	-1.5	-4.0	-1.7	1.0
152.10	3.34	-0.9	0.87	20MR	-0.7	-4.2	-1.7	1.0
150.18	2.89	-0.7	10.62	30MR	0.5	-4.2	-1.8	1.0
149.58	2.43	-0.4	20.30	9AP	1.9	-4.2	-1.9	1.0
150.23	1.99	-0.2	29.94	19AP	3.6	-4.1	-1.9	0.9
151.99	1.59	-0.0	39.55	29AP	2.5	-4.0	-2.0	0.9
154.68	1.23	0.1	49.11	9MY	1.4	-3.9	-2.1	0.8
158.12	0.91	0.3	58.66	19MY	0.6	-3.8	-2.1	0.8
162.20	0.62	0.4	68.19	29MY	-0.3	-3.7	-2.2	0.7
166.79	0.36	0.6	77.71	8JN	-1.3	-3.6	-2.3	0.7
171.82	0.13	0.7	87.24	18JN	-1.7	-3.6	-2.3	0.6
177.23	-0.08	0.8	96.77	28JN	-0.8	-3.5	-2.3	0.5
182.95	-0.26	0.8	106.33	8JL	-0.1	-3.5	-2.4	0.5
188.95	-0.44	0.9	115.92	18JL	0.3	-3.4	-2.4	0.4
195.20	-0.59	1.0	125.54	28JL	0.5	-3.4	-2.4	0.5
201.68	-0.73	1.0	135.21	7AU	0.9	-3.4	-2.4	0.5
208.36	-0.85	1.1	144.93	17AU	2.0	-3.4	-2.3	0.6
215.23	-0.96	1.1	154.70	27AU	2.6	-3.4	-2.3	0.6
222.26	-1.05	1.1	164.52	6SE	0.2	-3.4	-2.2	0.7
229.44	-1.13	1.2	174.41	16SE	-0.8	-3.4	-2.2	0.7
236.77	-1.19	1.2	184.36	26SE	-1.0	-3.4	-2.1	0.8
244.21	-1.24	1.2	194.36	6OC	-0.9	-3.4	-2.0	0.8
251.77	-1.28	1.2	204.41	16OC	-0.7	-3.4	-2.0	0.8
259.41	-1.30	1.3	214.51	26OC	-0.5	-3.4	-1.9	0.9
267.13	-1.30	1.3	224.65	5NO	-0.4	-3.4	-1.8	0.9
274.90	-1.30	1.3	234.83	15NO	-0.4	-3.5	-1.8	0.9
282.72	-1.27	1.3	245.02	25NO	0.0	-3.4	-1.7	0.9
290.57	-1.24	1.3	255.22	5DE	2.1	-3.5	-1.7	1.0
298.42	-1.19	1.4	265.43	15DE	1.3	-3.5	-1.6	1.0
306.27	-1.13	1.4	275.62	25DE	0.3	-3.4	-1.6	1.0
				373				
314.09	-1.06	1.4	285.79	4JA	0.0	-3.4	-1.6	0.9
321.89	-0.99	1.4	295.93	14JA	-0.1	-3.4	-1.6	0.9
329.64	-0.90	1.5	306.03	24JA	-0.3	-3.4	-1.5	0.9
337.33	-0.81	1.5	316.08	3FE	-0.7	-3.4	-1.5	1.0
344.96	-0.72	1.5	326.09	13FE	-1.3	-3.3	-1.6	1.0
352.51	-0.62	1.5	336.03	23FE	-1.4	-3.3	-1.6	1.1
359.99	-0.52	1.6	345.91	5MR	-0.7	-3.3	-1.6	1.1
7.39	-0.41	1.6	355.74	15MR	0.7	-3.3	-1.6	1.1
14.70	-0.31	1.6	5.51	25MR	2.4	-3.4	-1.6	1.1
21.93	-0.20	1.6	15.23	4AP	3.3	-3.4	-1.7	1.1
29.08	-0.10	1.6	24.90	14AP	1.9	-3.4	-1.7	1.1
36.14	0.00	1.6	34.52	24AP	1.1	-3.4	-1.8	1.1
43.12	0.11	1.7	44.10	4MY	0.5	-3.5	-1.8	1.0
50.02	0.21	1.7	53.66	14MY	-0.3	-3.5	-1.9	1.0
56.85	0.31	1.8	63.20	24MY	-1.3	-3.5	-2.0	0.9
63.60	0.40	1.8	72.72	3JN	-1.7	-3.6	-2.0	0.9
70.29	0.50	1.8	82.24	13JN	-0.8	-3.7	-2.1	0.8
76.91	0.59	1.9	91.77	23JN	-0.1	-3.8	-2.2	0.8
83.46	0.68	1.9	101.32	3JL	0.4	-3.8	-2.3	0.7
89.96	0.76	1.9	110.89	13JL	0.7	-3.9	-2.3	0.6
96.40	0.85	1.9	120.49	23JL	1.3	-4.1	-2.4	0.6
102.79	0.93	2.0	130.13	2AU	2.5	-4.2	-2.4	0.5
109.13	1.02	2.0	139.83	12AU	2.2	-4.2	-2.4	0.6
115.41	1.10	1.9	149.57	22AU	0.1	-4.3	-2.5	0.6
121.64	1.18	1.9	159.36	1SE	-0.9	-4.2	-2.4	0.7
127.82	1.25	1.9	169.22	11SE	-1.1	-3.9	-2.4	0.7
133.94	1.33	1.9	179.13	21SE	-1.0	-3.3	-2.4	0.8
139.99	1.41	1.8	189.11	1OC	-0.6	-3.6	-2.3	0.8
145.98	1.49	1.8	199.14	11OC	-0.4	-4.1	-2.3	0.9
151.88	1.56	1.7	209.21	21OC	-0.3	-4.3	-2.2	0.9
157.70	1.64	1.7	219.33	31OC	-0.2	-4.4	-2.2	1.0
163.40	1.72	1.6	229.49	10NO	0.2	-4.3	-2.1	1.0
168.98	1.80	1.5	239.67	20NO	2.4	-4.2	-2.0	1.0
174.41	1.88	1.3	249.87	30NO	1.0	-4.1	-1.9	1.0
179.66	1.96	1.2	260.08	10DE	0.1	-4.0	-1.9	1.1
184.69	2.04	1.1	270.28	20DE	-0.1	-3.9	-1.8	1.1
189.45	2.12	0.9	280.46	30DE	-0.2	-3.8	-1.8	1.1

Right table (374 / 375):

♂ LONG	LAT	MAG	☉ LONG	16.00UT	☿	♀	♃	♄
				374		MAGNITUDES		
193.87	2.20	0.7	290.62	9JA	-0.4	-3.7	-1.7	1.1
197.88	2.28	0.5	300.74	19JA	-0.7	-3.7	-1.7	1.1
201.38	2.34	0.3	310.82	29JA	-1.2	-3.6	-1.6	1.1
204.21	2.39	0.0	320.85	8FE	-1.3	-3.5	-1.6	1.1
206.25	2.42	-0.3	330.82	18FE	-0.6	-3.5	-1.6	1.1
207.30	2.42	-0.5	340.74	28FE	0.8	-3.4	-1.6	1.2
207.31	2.35	-0.9	350.60	10MR	2.9	-3.4	-1.6	1.2
205.81	2.20	-1.2	0.40	20MR	2.5	-3.3	-1.6	1.2
203.25	1.94	-1.4	10.15	30MR	1.4	-3.3	-1.6	1.3
199.86	1.57	-1.7	19.84	9AP	0.8	-3.3	-1.6	1.3
196.28	1.12	-1.6	29.48	19AP	0.3	-3.3	-1.6	1.3
193.23	0.63	-1.4	39.09	29AP	-0.4	-3.3	-1.7	1.2
191.28	0.14	-1.2	48.66	9MY	-1.3	-3.3	-1.7	1.2
190.67	-0.29	-1.0	58.20	19MY	-1.7	-3.3	-1.7	1.2
191.39	-0.66	-0.8	67.73	29MY	-0.8	-3.3	-1.8	1.1
193.32	-0.98	-0.6	77.25	8JN	0.0	-3.4	-1.8	1.1
196.26	-1.23	-0.4	86.77	18JN	0.5	-3.4	-1.9	1.0
200.06	-1.44	-0.3	96.31	28JN	1.0	-3.5	-2.0	1.0
204.55	-1.60	-0.1	105.87	8JL	1.7	-3.5	-2.0	0.9
209.62	-1.73	0.0	115.45	18JL	3.0	-3.5	-2.1	0.8
215.16	-1.83	0.1	125.07	28JL	1.9	-3.5	-2.2	0.8
221.11	-1.90	0.2	134.74	7AU	0.0	-3.4	-2.2	0.7
227.39	-1.95	0.3	144.45	17AU	-1.0	-3.4	-2.3	0.7
233.96	-1.97	0.4	154.22	27AU	-1.3	-3.4	-2.3	0.7
240.77	-1.97	0.5	164.04	6SE	-0.9	-3.3	-2.4	0.8
247.78	-1.94	0.6	173.92	16SE	-0.5	-3.3	-2.4	0.8
254.97	-1.90	0.7	183.87	26SE	-0.3	-3.3	-2.4	0.9
262.29	-1.84	0.8	193.87	6OC	-0.2	-3.3	-2.4	0.9
269.72	-1.77	0.8	203.92	16OC	-0.1	-3.3	-2.4	1.0
277.24	-1.68	0.9	214.02	26OC	0.4	-3.4	-2.4	1.0
284.83	-1.58	1.0	224.16	5NO	2.7	-3.4	-2.3	1.1
292.45	-1.47	1.0	234.33	15NO	0.8	-3.4	-2.3	1.1
300.09	-1.35	1.1	244.52	25NO	-0.1	-3.4	-2.2	1.1
307.74	-1.23	1.2	254.73	5DE	-0.3	-3.5	-2.1	1.2
315.37	-1.10	1.2	264.93	15DE	-0.3	-3.5	-2.1	1.2
322.97	-0.96	1.3	275.13	25DE	-0.4	-3.6	-2.0	1.2
				375				
330.54	-0.83	1.4	285.30	4JA	-0.8	-3.6	-1.9	1.2
338.05	-0.70	1.4	295.44	14JA	-1.1	-3.7	-1.9	1.2
345.50	-0.57	1.5	305.54	24JA	-1.1	-3.8	-1.8	1.2
352.89	-0.44	1.5	315.60	3FE	-0.6	-3.8	-1.7	1.2
0.21	-0.31	1.6	325.60	13FE	1.0	-3.9	-1.7	1.2
7.45	-0.19	1.6	335.55	23FE	3.2	-4.0	-1.6	1.3
14.62	-0.07	1.7	345.44	5MR	1.8	-4.1	-1.6	1.3
21.72	0.04	1.7	355.27	15MR	1.0	-4.2	-1.6	1.4
28.74	0.15	1.7	5.04	25MR	0.6	-4.2	-1.6	1.4
35.68	0.25	1.8	14.76	4AP	0.2	-4.2	-1.5	1.4
42.56	0.35	1.8	24.43	14AP	-0.4	-4.0	-1.5	1.4
49.37	0.44	1.8	34.05	24AP	-1.3	-3.5	-1.5	1.4
56.11	0.52	1.8	43.64	4MY	-1.7	-2.8	-1.5	1.4
62.80	0.61	1.9	53.20	14MY	-0.8	-3.6	-1.5	1.3
69.43	0.68	1.9	62.74	24MY	0.1	-4.0	-1.6	1.3
76.01	0.76	1.9	72.26	3JN	0.7	-4.2	-1.6	1.2
82.54	0.82	1.9	81.78	13JN	1.3	-4.2	-1.6	1.2
89.04	0.89	1.9	91.31	23JN	2.2	-4.1	-1.7	1.1
95.49	0.95	1.9	100.85	3JL	3.2	-4.0	-1.7	1.1
101.92	1.00	2.0	110.42	13JL	1.6	-3.9	-1.8	1.0
108.32	1.05	2.0	120.02	23JL	-0.0	-3.9	-1.8	1.0
114.69	1.10	2.0	129.66	2AU	-1.1	-3.8	-1.9	1.0
121.04	1.14	2.0	139.35	12AU	-1.4	-3.7	-1.9	0.9
127.38	1.18	2.0	149.09	22AU	-0.9	-3.6	-2.0	0.9
133.69	1.22	2.0	158.88	1SE	-0.4	-3.6	-2.1	0.9
140.00	1.25	2.0	168.74	11SE	-0.1	-3.5	-2.1	0.9
146.28	1.28	2.0	178.65	21SE	-0.0	-3.4	-2.2	0.9
152.56	1.31	2.0	188.62	1OC	0.1	-3.5	-2.3	1.0
158.82	1.33	1.9	198.65	11OC	0.6	-3.4	-2.3	1.0
165.06	1.35	1.9	208.72	21OC	3.1	-3.4	-2.3	1.1
171.28	1.36	1.8	218.84	31OC	0.6	-3.4	-2.3	1.1
177.47	1.37	1.8	229.00	10NO	-0.3	-3.4	-2.3	1.2
183.64	1.37	1.7	239.18	20NO	-0.4	-3.4	-2.3	1.2
189.76	1.37	1.6	249.38	30NO	-0.4	-3.4	-2.3	1.3
195.85	1.36	1.5	259.58	10DE	-0.5	-3.4	-2.2	1.3
201.87	1.34	1.4	269.79	20DE	-0.8	-3.4	-2.2	1.4
207.83	1.31	1.3	279.97	30DE	-1.0	-3.4	-2.1	1.4

Left

♂ LONG	LAT	MAG	☉ LONG	16.00UT 376	☿	♀	♃	♄
213.70	1.27	1.2	290.13	9JA	-1.0	-3.4	-2.1	1.4
219.47	1.21	1.0	300.25	19JA	-0.5	-3.4	-2.0	1.4
225.11	1.13	0.9	310.33	29JA	1.2	-3.4	-1.9	1.3
230.60	1.04	0.7	320.36	8FE	2.9	-3.4	-1.8	1.3
235.89	0.91	0.5	330.34	18FE	1.4	-3.5	-1.8	1.2
240.93	0.75	0.3	340.26	28FE	0.7	-3.5	-1.7	1.2
245.67	0.56	0.1	350.12	9MR	0.4	-3.4	-1.7	1.2
250.02	0.31	-0.2	359.92	19MR	0.1	-3.4	-1.6	1.2
253.88	0.01	-0.4	9.67	29MR	-0.5	-3.4	-1.6	1.2
257.11	-0.37	-0.7	19.36	8AP	-1.3	-3.4	-1.5	1.2
259.55	-0.83	-1.0	29.01	18AP	-1.7	-3.3	-1.5	1.1
261.02	-1.39	-1.3	38.61	28AP	-0.8	-3.3	-1.5	1.1
261.34	-2.04	-1.7	48.19	8MY	0.2	-3.3	-1.4	1.1
260.42	-2.75	-2.0	57.73	18MY	0.9	-3.3	-1.4	1.1
258.37	-3.48	-2.2	67.26	28MY	1.7	-3.3	-1.4	1.0
255.58	-4.12	-2.4	76.79	7JN	2.9	-3.3	-1.4	1.0
252.74	-4.60	-2.3	86.31	17JN	2.8	-3.4	-1.5	0.9
250.55	-4.87	-2.2	95.84	27JN	1.3	-3.4	-1.5	0.9
249.52	-4.94	-1.9	105.40	7JL	-0.1	-3.4	-1.5	0.8
249.88	-4.84	-1.7	114.98	17JL	-1.2	-3.5	-1.5	0.8
251.55	-4.65	-1.5	124.60	27JL	-1.5	-3.5	-1.6	0.7
254.38	-4.39	-1.2	134.27	6AU	-0.9	-3.6	-1.6	0.7
258.18	-4.10	-1.0	143.98	16AU	-0.3	-3.7	-1.7	0.6
262.74	-3.79	-0.8	153.74	26AU	-0.0	-3.7	-1.7	0.6
267.93	-3.47	-0.6	163.57	5SE	0.1	-3.8	-1.8	0.6
273.59	-3.15	-0.4	173.45	15SE	0.3	-3.9	-1.8	0.6
279.63	-2.83	-0.2	183.38	25SE	0.9	-4.0	-1.9	0.7
285.96	-2.53	-0.0	193.38	5OC	3.4	-4.1	-2.0	0.8
292.51	-2.23	0.1	203.43	15OC	0.5	-4.3	-2.0	0.8
299.23	-1.95	0.3	213.53	25OC	-0.5	-4.3	-2.1	0.9
306.06	-1.68	0.4	223.67	4NO	-0.6	-4.4	-2.1	1.0
312.99	-1.42	0.6	233.83	14NO	-0.6	-4.3	-2.2	1.0
319.96	-1.18	0.7	244.03	24NO	-0.7	-3.9	-2.2	1.1
326.97	-0.96	0.8	254.23	4DE	-0.8	-3.2	-2.2	1.1
333.99	-0.75	0.9	264.43	14DE	-0.8	-3.5	-2.2	1.1
341.00	-0.55	1.1	274.63	24DE	-0.8	-4.1	-2.2	1.1

377

♂ LONG	LAT	MAG	☉ LONG	16.00UT 377	☿	♀	♃	♄
348.00	-0.37	1.2	284.81	3JA	-0.4	-4.3	-2.2	1.1
354.97	-0.20	1.3	294.95	13JA	1.4	-4.4	-2.1	1.1
1.90	-0.05	1.4	305.05	23JA	2.4	-4.3	-2.1	1.1
8.79	0.09	1.5	315.11	2FE	1.0	-4.2	-2.0	1.0
15.64	0.22	1.5	325.12	12FE	0.5	-4.1	-1.9	1.0
22.44	0.34	1.6	335.07	22FE	0.3	-4.0	-1.9	1.0
29.20	0.44	1.7	344.96	4MR	-0.0	-3.9	-1.8	0.9
35.90	0.54	1.7	354.79	14MR	-0.5	-3.8	-1.7	0.9
42.56	0.63	1.8	4.57	24MR	-1.3	-3.7	-1.7	0.9
49.17	0.71	1.8	14.29	3AP	-1.6	-3.6	-1.6	0.9
55.74	0.78	1.9	23.96	13AP	-0.8	-3.5	-1.5	0.9
62.27	0.84	1.9	33.59	23AP	-0.3	-3.5	-1.5	0.9
68.77	0.90	2.0	43.18	3MY	1.2	-3.4	-1.4	0.9
75.23	0.95	2.0	52.73	13MY	2.3	-3.4	-1.4	0.9
81.66	0.99	2.0	62.27	23MY	3.6	-3.4	-1.4	0.8
88.08	1.03	2.0	71.80	2JN	2.2	-3.3	-1.4	0.8
94.47	1.07	2.0	81.32	12JN	1.0	-3.3	-1.3	0.8
100.85	1.09	2.0	90.85	22JN	-0.2	-3.3	-1.3	0.7
107.22	1.12	2.0	100.40	2JL	-1.2	-3.3	-1.3	0.7
113.58	1.14	2.0	109.96	12JL	-1.6	-3.3	-1.3	0.6
119.95	1.15	2.0	119.56	22JL	-0.8	-3.3	-1.3	0.6
126.31	1.16	2.0	129.20	1AU	-0.3	-3.4	-1.4	0.5
132.69	1.16	2.0	138.88	11AU	0.1	-3.4	-1.4	0.5
139.08	1.16	2.0	148.62	21AU	0.3	-3.4	-1.4	0.4
145.48	1.16	2.0	158.41	31AU	0.5	-3.5	-1.5	0.4
151.90	1.15	2.0	168.26	10SE	1.3	-3.5	-1.5	0.3
158.34	1.14	2.0	178.17	20SE	3.2	-3.5	-1.6	0.3
164.80	1.12	2.0	188.14	30SE	0.4	-3.5	-1.6	0.4
171.28	1.09	2.0	198.16	10OC	-0.6	-3.4	-1.7	0.4
177.79	1.06	1.9	208.23	20OC	-0.7	-3.4	-1.7	0.5
184.32	1.02	1.9	218.35	30OC	-0.7	-3.4	-1.8	0.6
190.88	0.98	1.9	228.50	9NO	-0.8	-3.4	-1.9	0.6
197.47	0.93	1.8	238.68	19NO	-0.7	-3.3	-1.9	0.7
204.08	0.87	1.8	248.88	29NO	-0.7	-3.3	-2.0	0.7
210.71	0.80	1.7	259.08	9DE	-0.7	-3.4	-2.0	0.8
217.36	0.72	1.6	269.29	19DE	-0.3	-3.3	-2.1	0.8
224.04	0.63	1.5	279.47	29DE	1.6	-3.3	-2.1	0.8

Right

♂ LONG	LAT	MAG	☉ LONG	16.00UT 378	☿	♀	♃	♄
230.73	0.52	1.4	289.63	8JA	2.0	-3.4	-2.1	0.8
237.43	0.41	1.3	299.76	18JA	0.7	-3.4	-2.1	0.8
244.14	0.27	1.2	309.84	28JA	0.3	-3.4	-2.1	0.8
250.86	0.12	1.1	319.88	7FE	0.1	-3.4	-2.0	0.8
257.58	-0.05	1.0	329.86	17FE	-0.1	-3.5	-2.0	0.8
264.29	-0.24	0.9	339.78	27FE	-0.6	-3.5	-1.9	0.7
270.98	-0.45	0.7	349.64	9MR	-1.3	-3.6	-1.9	0.7
277.64	-0.69	0.6	359.45	19MR	-1.6	-3.6	-1.8	0.6
284.24	-0.95	0.5	9.20	29MR	-0.7	-3.7	-1.7	0.6
290.77	-1.24	0.3	18.90	8AP	0.4	-3.7	-1.7	0.6
297.20	-1.55	0.1	28.55	18AP	1.6	-3.8	-1.6	0.7
303.48	-1.90	-0.0	38.15	28AP	3.0	-3.9	-1.5	0.7
309.57	-2.27	-0.2	47.73	8MY	3.0	-4.0	-1.5	0.7
315.41	-2.68	-0.4	57.28	18MY	1.7	-4.1	-1.4	0.7
320.90	-3.11	-0.6	66.81	28MY	0.8	-4.2	-1.4	0.6
325.96	-3.58	-0.8	76.33	7JN	-0.2	-4.2	-1.4	0.6
330.43	-4.06	-1.1	85.86	17JN	-1.2	-4.1	-1.3	0.6
334.17	-4.57	-1.3	95.39	27JN	-1.7	-3.8	-1.3	0.6
336.97	-5.08	-1.5	104.94	7JL	-0.8	-3.2	-1.3	0.5
338.60	-5.56	-1.8	114.52	17JL	-0.2	-3.3	-1.3	0.5
338.91	-5.96	-2.1	124.14	27JL	0.2	-3.9	-1.3	0.4
337.84	-6.21	-2.3	133.80	6AU	0.4	-4.2	-1.3	0.4
335.59	-6.22	-2.5	143.51	16AU	0.8	-4.3	-1.3	0.3
332.74	-5.93	-2.5	153.27	26AU	1.6	-4.2	-1.3	0.3
330.04	-5.38	-2.3	163.09	5SE	2.9	-4.2	-1.3	0.2
328.18	-4.66	-2.1	172.97	15SE	0.3	-4.1	-1.3	0.1
327.58	-3.88	-1.7	182.90	25SE	-0.7	-4.0	-1.4	0.1
328.28	-3.13	-1.4	192.90	5OC	-0.9	-3.9	-1.4	0.1
330.17	-2.45	-1.1	202.95	15OC	-0.8	-3.8	-1.4	0.1
333.05	-1.86	-0.8	213.04	25OC	-0.7	-3.7	-1.5	0.2
336.72	-1.36	-0.5	223.17	4NO	-0.6	-3.7	-1.6	0.3
341.01	-0.93	-0.3	233.34	14NO	-0.5	-3.6	-1.6	0.3
345.78	-0.57	-0.0	243.53	24NO	-0.5	-3.6	-1.7	0.4
350.90	-0.27	0.2	253.74	4DE	-0.1	-3.5	-1.8	0.5
356.31	-0.01	0.4	263.94	14DE	1.9	-3.5	-1.8	0.5
1.94	0.21	0.6	274.14	24DE	1.6	-3.4	-1.9	0.6

379

♂ LONG	LAT	MAG	☉ LONG	16.00UT 379	☿	♀	♃	♄
7.72	0.39	0.8	284.31	3JA	0.4	-3.4	-1.9	0.6
13.62	0.54	1.0	294.46	13JA	0.1	-3.4	-2.0	0.6
19.62	0.68	1.1	304.56	23JA	0.0	-3.3	-2.0	0.6
25.68	0.79	1.2	314.63	2FE	-0.2	-3.3	-2.0	0.6
31.80	0.88	1.4	324.64	12FE	-0.6	-3.3	-2.0	0.6
37.94	0.96	1.5	334.59	22FE	-1.3	-3.3	-2.0	0.6
44.11	1.02	1.6	344.48	4MR	-1.5	-3.3	-2.0	0.6
50.30	1.08	1.6	354.32	14MR	-0.7	-3.3	-2.0	0.5
56.50	1.12	1.7	4.10	24MR	0.5	-3.4	-1.9	0.5
62.70	1.16	1.8	13.82	3AP	2.0	-3.4	-1.9	0.4
68.91	1.19	1.9	23.49	13AP	3.9	-3.4	-1.8	0.4
75.13	1.21	1.9	33.12	23AP	2.3	-3.5	-1.7	0.5
81.35	1.22	1.9	42.71	3MY	1.3	-3.5	-1.7	0.5
87.57	1.23	2.0	52.27	13MY	0.6	-3.5	-1.6	0.5
93.80	1.23	2.0	61.81	23MY	-0.3	-3.4	-1.5	0.5
100.04	1.23	2.0	71.34	2JN	-1.3	-3.4	-1.5	0.5
106.30	1.22	2.0	80.86	12JN	-1.7	-3.4	-1.4	0.5
112.57	1.21	2.0	90.38	22JN	-0.8	-3.4	-1.4	0.5
118.86	1.19	2.0	99.93	2JL	-0.1	-3.3	-1.3	0.4
125.18	1.17	2.0	109.49	12JL	0.3	-3.3	-1.3	0.4
131.53	1.15	2.0	119.09	22JL	0.6	-3.3	-1.3	0.4
137.90	1.12	2.0	128.73	1AU	1.0	-3.3	-1.2	0.3
144.32	1.08	2.0	138.41	11AU	2.1	-3.3	-1.2	0.3
150.77	1.05	1.9	148.14	21AU	2.5	-3.4	-1.2	0.2
157.26	1.00	1.9	157.93	31AU	0.2	-3.4	-1.2	0.2
163.81	0.96	1.9	167.78	10SE	-0.9	-3.4	-1.2	0.1
170.40	0.90	1.9	177.69	20SE	-1.0	-3.5	-1.2	0.0
177.04	0.85	1.9	187.65	30SE	-1.0	-3.5	-1.2	-0.0
183.74	0.78	1.9	197.67	10OC	-0.6	-3.6	-1.3	-0.1
190.49	0.71	1.9	207.74	20OC	-0.4	-3.6	-1.3	-0.1
197.30	0.64	1.9	217.86	30OC	-0.4	-3.7	-1.3	-0.0
204.16	0.56	1.8	228.01	9NO	-0.4	-3.8	-1.4	0.0
211.09	0.47	1.8	238.19	19NO	0.0	-3.9	-1.4	0.1
218.07	0.38	1.8	248.39	29NO	2.1	-4.0	-1.5	0.2
225.11	0.28	1.7	258.59	9DE	1.3	-4.1	-1.6	0.2
232.21	0.17	1.7	268.79	19DE	0.2	-4.2	-1.6	0.3
239.36	0.05	1.6	278.98	29DE	-0.0	-4.3	-1.7	0.4

♂			☉	16.00UT	☿	♀	♃	♄
LONG	LAT	MAG	LONG	380	MAGNITUDES			
246.57	-0.07	1.6	289.14	8JA	-0.1	-4.3	-1.8	0.4
253.83	-0.21	1.5	299.27	18JA	-0.3	-4.4	-1.8	0.4
261.14	-0.35	1.4	309.36	28JA	-0.7	-4.2	-1.9	0.5
268.48	-0.49	1.4	319.39	7FE	-1.3	-3.9	-1.9	0.5
275.87	-0.64	1.3	329.37	17FE	-1.4	-3.3	-2.0	0.5
283.28	-0.80	1.2	339.30	27FE	-0.7	-3.6	-2.0	0.5
290.72	-0.96	1.1	349.17	8MR	0.7	-4.0	-2.0	0.4
298.17	-1.12	1.1	358.97	18MR	2.5	-4.2	-2.0	0.4
305.62	-1.29	1.0	8.73	28MR	3.0	-4.2	-2.0	0.4
313.05	-1.45	0.9	18.43	7AP	1.7	-4.2	-2.0	0.3
320.45	-1.61	0.8	28.08	17AP	1.0	-4.1	-1.9	0.3
327.80	-1.76	0.7	37.69	27AP	0.4	-4.0	-1.9	0.3
335.08	-1.90	0.6	47.26	7MY	-0.3	-3.9	-1.8	0.3
342.26	-2.03	0.6	56.81	17MY	-1.3	-3.8	-1.8	0.3
349.32	-2.16	0.5	66.35	27MY	-1.7	-3.7	-1.7	0.4
356.22	-2.26	0.4	75.87	6JN	-0.8	-3.6	-1.6	0.4
2.93	-2.35	0.3	85.39	16JN	-0.1	-3.6	-1.6	0.4
9.40	-2.42	0.2	94.93	26JN	0.4	-3.5	-1.5	0.4
15.58	-2.48	0.1	104.48	6JL	0.8	-3.5	-1.5	0.4
21.42	-2.51	-0.1	114.06	16JL	1.4	-3.4	-1.4	0.3
26.81	-2.51	-0.2	123.68	26JL	2.7	-3.4	-1.4	0.3
31.68	-2.49	-0.4	133.33	5AU	2.2	-3.4	-1.3	0.3
35.88	-2.44	-0.5	143.04	15AU	0.1	-3.4	-1.3	0.2
39.26	-2.35	-0.7	152.80	25AU	-1.0	-3.4	-1.3	0.2
41.64	-2.21	-0.9	162.61	4SE	-1.2	-3.4	-1.2	0.1
42.79	-2.00	-1.1	172.49	14SE	-1.0	-3.4	-1.2	0.0
42.53	-1.71	-1.4	182.43	24SE	-0.5	-3.4	-1.2	-0.0
40.81	-1.34	-1.6	192.41	4OC	-0.3	-3.4	-1.2	-0.1
37.83	-0.88	-1.7	202.46	14OC	-0.2	-3.4	-1.2	-0.2
34.15	-0.37	-1.8	212.55	24OC	-0.2	-3.4	-1.2	-0.2
30.59	0.12	-1.6	222.68	3NO	0.2	-3.4	-1.3	-0.2
27.86	0.56	-1.3	232.85	13NO	2.4	-3.5	-1.3	-0.2
26.41	0.92	-1.0	243.04	23NO	1.0	-3.5	-1.3	-0.1
26.32	1.18	-0.7	253.24	3DE	-0.0	-3.5	-1.4	-0.0
27.46	1.37	-0.4	263.45	13DE	-0.2	-3.5	-1.4	0.1
29.63	1.51	-0.1	273.65	23DE	-0.2	-3.4	-1.5	0.1
				381				
32.62	1.60	0.2	283.82	2JA	-0.4	-3.4	-1.6	0.2
36.25	1.65	0.4	293.97	12JA	-0.7	-3.4	-1.6	0.2
40.39	1.69	0.6	304.08	22JA	-1.2	-3.4	-1.7	0.3
44.92	1.70	0.8	314.14	1FE	-1.2	-3.4	-1.8	0.3
49.75	1.70	1.0	324.15	11FE	-0.6	-3.3	-1.8	0.3
54.84	1.70	1.1	334.11	21FE	0.9	-3.3	-1.9	0.4
60.11	1.68	1.3	344.00	3MR	3.0	-3.3	-1.9	0.4
65.53	1.66	1.4	353.84	13MR	2.2	-3.3	-2.0	0.4
71.08	1.63	1.5	3.62	23MR	1.2	-3.4	-2.0	0.3
76.74	1.60	1.6	13.35	2AP	0.7	-3.4	-2.1	0.3
82.48	1.56	1.7	23.03	12AP	0.3	-3.4	-2.1	0.3
88.30	1.52	1.8	32.66	22AP	-0.4	-3.4	-2.1	0.2
94.18	1.48	1.8	42.25	2MY	-1.3	-3.5	-2.0	0.2
100.13	1.44	1.9	51.81	12MY	-1.7	-3.5	-2.0	0.2
106.13	1.39	1.9	61.35	22MY	-0.8	-3.6	-2.0	0.2
112.18	1.34	1.9	70.88	1JN	0.0	-3.6	-1.9	0.3
118.30	1.28	2.0	80.40	11JN	0.6	-3.7	-1.9	0.3
124.46	1.23	2.0	89.93	21JN	1.1	-3.8	-1.8	0.3
130.68	1.17	2.0	99.47	1JL	1.8	-3.9	-1.7	0.3
136.96	1.11	2.0	109.03	11JL	3.2	-4.0	-1.7	0.3
143.30	1.04	2.0	118.63	21JL	1.8	-4.1	-1.6	0.3
149.70	0.98	2.0	128.26	31JL	0.0	-4.2	-1.5	0.3
156.16	0.90	2.0	137.94	10AU	-1.1	-4.2	-1.5	0.3
162.69	0.83	1.9	147.67	20AU	-1.3	-4.3	-1.4	0.2
169.29	0.75	1.9	157.46	30AU	-0.9	-4.2	-1.4	0.2
175.97	0.67	1.8	167.31	9SE	-0.5	-3.8	-1.4	0.1
182.72	0.59	1.8	177.21	19SE	-0.2	-3.3	-1.3	0.1
189.54	0.50	1.8	187.17	29SE	-0.1	-3.6	-1.3	-0.0
196.45	0.41	1.7	197.19	9OC	-0.0	-4.1	-1.3	-0.1
203.43	0.32	1.7	207.25	19OC	0.5	-4.3	-1.3	-0.1
210.49	0.22	1.7	217.36	29OC	2.7	-4.4	-1.3	-0.2
217.63	0.12	1.7	227.52	8NO	0.8	-4.3	-1.3	-0.3
224.85	0.01	1.7	237.69	18NO	-0.2	-4.2	-1.3	-0.3
232.14	-0.10	1.6	247.89	28NO	-0.3	-4.1	-1.3	-0.2
239.51	-0.21	1.6	258.10	8DE	-0.4	-4.0	-1.3	-0.1
246.94	-0.32	1.6	268.30	18DE	-0.5	-3.9	-1.4	-0.1
254.44	-0.43	1.6	278.49	28DE	-0.8	-3.8	-1.4	0.0

♂			☉	16.00UT	☿	♀	♃	♄
LONG	LAT	MAG	LONG	382	MAGNITUDES			
262.00	-0.54	1.5	288.66	7JA	-1.1	-3.7	-1.5	0.1
269.61	-0.66	1.5	298.78	17JA	-1.1	-3.7	-1.5	0.1
277.26	-0.77	1.5	308.87	27JA	-0.6	-3.6	-1.6	0.2
284.95	-0.88	1.4	318.91	6FE	1.0	-3.5	-1.7	0.2
292.67	-0.98	1.4	328.90	16FE	3.2	-3.5	-1.7	0.3
300.40	-1.07	1.4	338.83	26FE	1.7	-3.4	-1.8	0.3
308.13	-1.16	1.3	348.70	8MR	0.9	-3.4	-1.9	0.3
315.86	-1.24	1.3	358.51	18MR	0.5	-3.3	-1.9	0.3
323.57	-1.31	1.3	8.26	28MR	0.2	-3.3	-2.0	0.3
331.24	-1.37	1.2	17.97	7AP	-0.4	-3.3	-2.0	0.3
338.86	-1.41	1.2	27.62	17AP	-1.3	-3.3	-2.1	0.3
346.42	-1.44	1.2	37.23	27AP	-1.7	-3.3	-2.1	0.2
353.91	-1.46	1.1	46.81	7MY	-0.8	-3.3	-2.2	0.2
1.30	-1.46	1.1	56.35	17MY	0.1	-3.3	-2.2	0.2
8.59	-1.44	1.1	65.89	27MY	0.8	-3.3	-2.2	0.2
15.76	-1.41	1.0	75.41	6JN	1.4	-3.4	-2.1	0.2
22.79	-1.36	1.0	84.93	16JN	2.4	-3.4	-2.1	0.3
29.67	-1.30	0.9	94.47	26JN	3.1	-3.5	-2.1	0.3
36.38	-1.22	0.9	104.02	6JL	1.5	-3.5	-2.0	0.3
42.90	-1.12	0.8	113.59	16JL	-0.0	-3.5	-1.9	0.3
49.19	-1.00	0.8	123.21	26JL	-1.1	-3.5	-1.9	0.3
55.24	-0.87	0.7	132.86	5AU	-1.4	-3.4	-1.8	0.3
60.99	-0.72	0.6	142.56	15AU	-0.9	-3.4	-1.7	0.3
66.40	-0.54	0.5	152.32	25AU	-0.4	-3.4	-1.7	0.3
71.40	-0.34	0.4	162.13	4SE	-0.1	-3.3	-1.6	0.2
75.91	-0.11	0.3	172.00	14SE	0.0	-3.3	-1.6	0.2
79.81	0.15	0.1	181.93	24SE	0.2	-3.3	-1.5	0.1
82.97	0.45	-0.1	191.92	4OC	0.7	-3.3	-1.5	0.1
85.21	0.80	-0.3	201.96	14OC	3.1	-3.3	-1.4	-0.0
86.35	1.20	-0.5	212.05	24OC	0.6	-3.4	-1.4	-0.1
86.18	1.64	-0.7	222.19	3NO	-0.4	-3.4	-1.4	-0.2
84.61	2.11	-0.9	232.35	13NO	-0.5	-3.4	-1.4	-0.2
81.73	2.56	-1.1	242.54	23NO	-0.5	-3.4	-1.4	-0.3
77.96	2.94	-1.3	252.75	3DE	-0.6	-3.5	-1.4	-0.3
74.03	3.19	-1.2	262.95	13DE	-0.8	-3.5	-1.4	-0.2
70.68	3.31	-0.9	273.15	23DE	-0.9	-3.6	-1.4	-0.2
				383				
68.46	3.30	-0.6	283.33	2JA	-0.9	-3.6	-1.4	-0.1
67.57	3.21	-0.4	293.48	12JA	-0.5	-3.7	-1.5	-0.0
67.95	3.07	-0.1	303.59	22JA	1.2	-3.8	-1.5	0.1
69.43	2.91	0.2	313.66	1FE	2.8	-3.9	-1.5	0.1
71.82	2.75	0.4	323.67	11FE	1.2	-3.9	-1.6	0.2
74.94	2.59	0.6	333.63	21FE	0.6	-4.0	-1.6	0.2
78.63	2.44	0.8	343.53	3MR	0.4	-4.1	-1.7	0.3
82.77	2.30	1.0	353.37	13MR	0.0	-4.2	-1.8	0.3
87.27	2.16	1.1	3.16	23MR	-0.5	-4.2	-1.8	0.3
92.08	2.03	1.3	12.88	2AP	-1.2	-4.2	-1.9	0.3
97.12	1.91	1.4	22.56	12AP	-1.7	-4.0	-2.0	0.3
102.37	1.79	1.5	32.19	22AP	-0.8	-3.5	-2.1	0.3
107.79	1.67	1.6	41.79	2MY	0.2	-2.8	-2.1	0.3
113.35	1.56	1.6	51.35	12MY	1.0	-3.6	-2.2	0.3
119.05	1.45	1.7	60.89	22MY	1.9	-4.0	-2.2	0.2
124.87	1.34	1.7	70.42	1JN	3.2	-4.2	-2.3	0.2
130.80	1.24	1.8	79.94	11JN	2.6	-4.2	-2.3	0.2
136.84	1.13	1.8	89.47	21JN	1.2	-4.1	-2.3	0.3
142.97	1.03	1.8	99.01	1JL	-0.1	-4.0	-2.3	0.3
149.20	0.93	1.8	108.57	11JL	-1.2	-3.9	-2.3	0.3
155.54	0.82	1.9	118.16	21JL	-1.6	-3.8	-2.2	0.3
161.96	0.72	1.8	127.79	31JL	-0.9	-3.8	-2.2	0.4
168.48	0.61	1.8	137.47	10AU	-0.3	-3.7	-2.1	0.4
175.10	0.51	1.8	147.20	20AU	0.0	-3.6	-2.1	0.4
181.82	0.40	1.8	156.98	30AU	0.2	-3.6	-2.0	0.3
188.63	0.30	1.8	166.82	9SE	0.4	-3.5	-1.9	0.3
195.55	0.19	1.7	176.73	19SE	1.0	-3.5	-1.9	0.3
202.56	0.08	1.7	186.68	29SE	3.3	-3.5	-1.8	0.2
209.66	-0.02	1.7	196.69	9OC	0.5	-3.4	-1.7	0.2
216.87	-0.13	1.6	206.76	19OC	-0.5	-3.4	-1.7	0.1
224.16	-0.23	1.6	216.87	29OC	-0.6	-3.4	-1.6	0.0
231.54	-0.34	1.5	227.02	8NO	-0.6	-3.4	-1.6	-0.0
239.01	-0.44	1.5	237.20	18NO	-0.7	-3.4	-1.5	-0.1
246.56	-0.54	1.4	247.39	28NO	-0.8	-3.4	-1.5	-0.2
254.18	-0.63	1.4	257.60	8DE	-0.8	-3.4	-1.5	-0.2
261.87	-0.72	1.4	267.80	18DE	-0.8	-3.4	-1.5	-0.2
269.60	-0.81	1.4	277.99	28DE	-0.4	-3.4	-1.5	-0.2

♂			☉	16.00UT	☿	♀	♃	♄
LONG	LAT	MAG	LONG	384	MAGNITUDES			
277.39	-0.88	1.4	288.16	7JA	1.4	-3.4	-1.5	-0.1
285.21	-0.95	1.4	298.29	17JA	2.3	-3.4	-1.5	-0.0
293.06	-1.01	1.4	308.38	27JA	0.9	-3.4	-1.5	0.0
300.91	-1.06	1.4	318.42	6FE	0.4	-3.4	-1.5	0.1
308.77	-1.11	1.4	328.41	16FE	0.2	-3.5	-1.6	0.2
316.61	-1.13	1.4	338.34	26FE	-0.0	-3.5	-1.6	0.2
324.42	-1.15	1.4	348.22	7MR	-0.5	-3.4	-1.6	0.3
332.20	-1.16	1.4	358.03	17MR	-1.2	-3.4	-1.7	0.3
339.92	-1.15	1.4	7.79	27MR	-1.6	-3.4	-1.7	0.3
347.59	-1.13	1.4	17.49	6AP	-0.8	-3.4	-1.8	0.3
355.19	-1.09	1.4	27.15	16AP	0.3	-3.3	-1.9	0.4
2.70	-1.05	1.4	36.76	26AP	1.3	-3.3	-1.9	0.4
10.13	-0.99	1.4	46.34	6MY	2.5	-3.3	-2.0	0.4
17.46	-0.92	1.4	55.89	16MY	3.5	-3.3	-2.1	0.3
24.70	-0.84	1.4	65.42	26MY	2.0	-3.3	-2.2	0.3
31.82	-0.75	1.4	74.95	5JN	0.9	-3.3	-2.2	0.3
38.83	-0.65	1.4	84.47	15JN	-0.1	-3.4	-2.3	0.3
45.72	-0.54	1.4	94.00	25JN	-1.2	-3.4	-2.3	0.3
52.48	-0.43	1.4	103.55	5JL	-1.6	-3.4	-2.4	0.3
59.11	-0.30	1.4	113.13	15JL	-0.9	-3.5	-2.4	0.4
65.59	-0.17	1.3	122.74	25JL	-0.3	-3.5	-2.4	0.4
71.92	-0.02	1.3	132.39	4AU	0.1	-3.6	-2.4	0.4
78.07	0.13	1.3	142.09	14AU	0.3	-3.7	-2.4	0.5
84.03	0.30	1.2	151.85	24AU	0.6	-3.7	-2.3	0.5
89.78	0.48	1.1	161.66	3SE	1.4	-3.8	-2.3	0.5
95.28	0.67	1.1	171.53	13SE	3.2	-3.9	-2.2	0.4
100.50	0.89	1.0	181.45	23SE	0.4	-4.0	-2.2	0.4
105.37	1.12	0.9	191.44	3OC	-0.7	-4.1	-2.1	0.4
109.82	1.37	0.7	201.48	13OC	-0.8	-4.3	-2.0	0.3
113.77	1.66	0.6	211.56	23OC	-0.8	-4.3	-1.9	0.3
117.10	1.97	0.4	221.69	2NO	-0.8	-4.4	-1.9	0.2
119.66	2.33	0.2	231.86	12NO	-0.6	-4.3	-1.8	0.1
121.27	2.72	-0.0	242.04	22NO	-0.6	-3.9	-1.8	0.1
121.76	3.14	-0.3	252.25	2DE	-0.6	-3.1	-1.7	-0.0
120.95	3.57	-0.5	262.45	12DE	-0.2	-3.6	-1.7	-0.1
118.81	3.97	-0.7	272.65	22DE	1.6	-4.1	-1.6	-0.1
				385				
115.52	4.28	-0.9	282.83	1JA	1.9	-4.3	-1.6	-0.1
111.61	4.44	-1.0	292.98	11JA	0.6	-4.4	-1.6	-0.1
107.78	4.42	-0.9	303.09	21JA	0.2	-4.3	-1.6	-0.0
104.72	4.25	-0.6	313.16	31JA	0.1	-4.2	-1.6	0.1
102.85	3.98	-0.4	323.18	10FE	-0.1	-4.1	-1.6	0.1
102.28	3.66	-0.1	333.14	20FE	-0.6	-4.0	-1.6	0.2
102.92	3.33	0.1	343.05	2MR	-1.2	-3.9	-1.6	0.2
104.62	3.02	0.3	352.89	12MR	-1.5	-3.8	-1.6	0.3
107.17	2.72	0.5	2.68	22MR	-0.8	-3.7	-1.6	0.3
110.42	2.45	0.7	12.41	1AP	0.4	-3.6	-1.7	0.4
114.23	2.20	0.9	22.09	11AP	1.7	-3.5	-1.7	0.4
118.49	1.97	1.0	31.73	21AP	3.3	-3.5	-1.7	0.4
123.12	1.76	1.1	41.33	1MY	2.7	-3.4	-1.8	0.5
128.07	1.56	1.2	50.89	11MY	1.5	-3.4	-1.9	0.5
133.29	1.38	1.3	60.43	21MY	0.7	-3.4	-1.9	0.5
138.73	1.21	1.4	69.96	31MY	-0.2	-3.3	-2.0	0.5
144.38	1.04	1.4	79.48	10JN	-1.2	-3.3	-2.1	0.5
150.20	0.89	1.5	89.01	20JN	-1.7	-3.3	-2.1	0.4
156.20	0.74	1.5	98.55	30JN	-0.8	-3.3	-2.2	0.4
162.34	0.59	1.6	108.11	10JL	-0.2	-3.3	-2.3	0.4
168.64	0.45	1.6	117.70	20JL	0.2	-3.3	-2.3	0.5
175.07	0.32	1.6	127.33	30JL	0.5	-3.4	-2.4	0.5
181.64	0.19	1.6	137.00	9AU	0.9	-3.4	-2.4	0.6
188.34	0.06	1.6	146.73	19AU	1.8	-3.4	-2.4	0.6
195.17	-0.06	1.6	156.51	29AU	2.8	-3.5	-2.5	0.6
202.12	-0.18	1.6	166.35	8SE	0.3	-3.5	-2.4	0.6
209.19	-0.29	1.6	176.24	18SE	-0.8	-3.5	-2.4	0.6
216.37	-0.40	1.6	186.20	28SE	-0.9	-3.5	-2.4	0.6
223.67	-0.50	1.6	196.21	8OC	-0.9	-3.4	-2.3	0.6
231.06	-0.60	1.5	206.27	18OC	-0.7	-3.4	-2.3	0.5
238.56	-0.69	1.5	216.38	28OC	-0.5	-3.4	-2.2	0.5
246.14	-0.77	1.5	226.52	7NO	-0.5	-3.4	-2.1	0.4
253.81	-0.85	1.5	236.70	17NO	-0.4	-3.3	-2.1	0.4
261.55	-0.91	1.4	246.90	27NO	-0.1	-3.3	-2.0	0.3
269.34	-0.97	1.4	257.10	7DE	1.9	-3.3	-1.9	0.2
277.18	-1.02	1.4	267.30	17DE	1.5	-3.3	-1.9	0.2
285.06	-1.05	1.4	277.49	27DE	0.3	-3.3	-1.8	0.1

♂			☉	16.00UT	☿	♀	♃	♄
LONG	LAT	MAG	LONG	386	MAGNITUDES			
292.95	-1.07	1.3	287.66	6JA	0.1	-3.4	-1.7	0.0
300.86	-1.09	1.3	297.79	16JA	-0.0	-3.4	-1.7	0.1
308.75	-1.09	1.3	307.89	26JA	-0.2	-3.4	-1.7	0.1
316.63	-1.07	1.3	317.93	5FE	-0.6	-3.4	-1.6	0.2
324.47	-1.05	1.3	327.92	15FE	-1.2	-3.5	-1.6	0.2
332.28	-1.02	1.4	337.86	25FE	-1.4	-3.5	-1.6	0.3
340.02	-0.97	1.4	347.73	7MR	-0.7	-3.6	-1.6	0.3
347.71	-0.91	1.4	357.55	17MR	0.6	-3.6	-1.6	0.4
355.32	-0.85	1.5	7.32	27MR	2.2	-3.7	-1.6	0.4
2.85	-0.78	1.5	17.02	6AP	3.6	-3.7	-1.6	0.5
10.30	-0.70	1.5	26.68	16AP	2.0	-3.8	-1.6	0.5
17.66	-0.61	1.6	36.30	26AP	1.2	-3.9	-1.6	0.6
24.93	-0.52	1.6	45.88	6MY	0.5	-4.0	-1.7	0.6
32.11	-0.42	1.6	55.43	16MY	-0.3	-4.1	-1.7	0.6
39.19	-0.31	1.6	64.97	26MY	-1.2	-4.2	-1.7	0.6
46.18	-0.21	1.7	74.49	5JN	-1.7	-4.2	-1.8	0.6
53.07	-0.09	1.7	84.01	15JN	-0.8	-4.1	-1.8	0.6
59.86	0.02	1.7	93.54	25JN	-0.1	-3.8	-1.9	0.6
66.55	0.14	1.7	103.09	5JL	0.3	-3.1	-2.0	0.6
73.15	0.26	1.7	112.67	15JL	0.7	-3.3	-2.0	0.6
79.64	0.38	1.7	122.28	25JL	1.2	-3.9	-2.1	0.6
86.03	0.51	1.7	131.92	4AU	2.3	-4.2	-2.2	0.7
92.31	0.64	1.7	141.62	14AU	2.5	-4.3	-2.2	0.7
98.47	0.77	1.6	151.37	24AU	0.2	-4.2	-2.3	0.8
104.51	0.91	1.6	161.18	3SE	-0.9	-4.2	-2.3	0.8
110.41	1.06	1.5	171.05	13SE	-1.1	-4.1	-2.4	0.8
116.15	1.21	1.5	180.97	23SE	-1.0	-4.0	-2.4	0.8
121.72	1.37	1.4	190.95	3OC	-0.6	-3.9	-2.4	0.8
127.09	1.54	1.3	200.99	13OC	-0.4	-3.8	-2.4	0.8
132.21	1.73	1.2	211.07	23OC	-0.3	-3.7	-2.4	0.8
137.04	1.93	1.1	221.20	2NO	-0.3	-3.7	-2.4	0.7
141.52	2.14	1.0	231.37	12NO	0.1	-3.6	-2.3	0.7
145.57	2.38	0.8	241.55	22NO	2.1	-3.6	-2.3	0.6
149.09	2.64	0.6	251.75	2DE	1.2	-3.5	-2.2	0.6
151.94	2.92	0.4	261.96	12DE	0.1	-3.5	-2.1	0.5
153.98	3.23	0.2	272.16	22DE	-0.1	-3.4	-2.0	0.4
				387				
155.03	3.56	-0.1	282.33	1JA	-0.2	-3.4	-2.0	0.4
154.90	3.88	-0.3	292.49	11JA	-0.3	-3.4	-1.9	0.3
153.48	4.17	-0.6	302.60	21JA	-0.7	-3.3	-1.8	0.2
150.83	4.37	-0.8	312.67	31JA	-1.2	-3.3	-1.8	0.3
147.24	4.43	-1.0	322.70	10FE	-1.3	-3.3	-1.7	0.3
143.34	4.32	-1.0	332.66	20FE	-0.7	-3.3	-1.7	0.4
139.84	4.05	-0.9	342.56	2MR	0.7	-3.3	-1.6	0.4
137.32	3.67	-0.6	352.42	12MR	2.7	-3.3	-1.6	0.5
136.07	3.24	-0.4	2.20	22MR	2.7	-3.4	-1.6	0.5
136.11	2.80	-0.2	11.94	1AP	1.5	-3.4	-1.5	0.6
137.31	2.39	0.0	21.62	11AP	0.9	-3.4	-1.5	0.6
139.50	2.02	0.2	31.26	21AP	0.4	-3.5	-1.5	0.7
142.51	1.68	0.4	40.86	1MY	-0.3	-3.5	-1.5	0.7
146.19	1.37	0.5	50.43	11MY	-1.2	-3.5	-1.5	0.8
150.41	1.09	0.7	59.97	21MY	-1.7	-3.4	-1.5	0.8
155.09	0.84	0.8	69.49	31MY	-0.8	-3.4	-1.6	0.8
160.15	0.61	0.9	79.02	10JN	-0.0	-3.4	-1.6	0.8
165.54	0.40	1.0	88.54	20JN	0.5	-3.4	-1.6	0.9
171.22	0.21	1.0	98.08	30JN	0.9	-3.3	-1.7	0.9
177.15	0.03	1.1	107.64	10JL	1.5	-3.3	-1.7	0.9
183.30	-0.13	1.2	117.23	20JL	2.8	-3.3	-1.8	0.9
189.66	-0.29	1.2	126.86	30JL	2.1	-3.3	-1.8	0.9
196.22	-0.43	1.2	136.53	9AU	0.1	-3.3	-1.9	0.9
202.94	-0.56	1.3	146.25	19AU	-1.0	-3.4	-1.9	0.9
209.84	-0.68	1.3	156.03	29AU	-1.2	-3.4	-2.0	1.0
216.88	-0.78	1.3	165.87	8SE	-1.0	-3.4	-2.1	1.0
224.06	-0.88	1.3	175.76	18SE	-0.5	-3.5	-2.1	1.0
231.38	-0.96	1.3	185.71	28SE	-0.3	-3.5	-2.2	1.1
238.81	-1.03	1.3	195.72	8OC	-0.2	-3.6	-2.3	1.1
246.35	-1.09	1.3	205.78	18OC	-0.1	-3.6	-2.3	1.1
253.98	-1.14	1.3	215.89	28OC	0.3	-3.7	-2.3	1.1
261.69	-1.17	1.4	226.03	7NO	2.4	-3.8	-2.3	1.0
269.46	-1.19	1.4	236.20	17NO	1.0	-3.9	-2.3	1.0
277.29	-1.20	1.4	246.40	27NO	-0.1	-4.0	-2.3	1.0
285.15	-1.19	1.4	256.61	7DE	-0.2	-4.1	-2.3	0.9
293.03	-1.17	1.4	266.81	17DE	-0.3	-4.2	-2.2	0.8
300.92	-1.14	1.4	277.00	27DE	-0.4	-4.3	-2.2	0.8

♂			☉	16.00UT	☿	♀	♃	♄
LONG LAT		MAG	LONG	388		MAGNITUDES		
308.79	-1.10	1.4	287.17	6JA	-0.7	-4.3	-2.1	0.7
316.64	-1.04	1.4	297.30	16JA	-1.1	-4.4	-2.0	0.6
324.45	-0.98	1.4	307.40	26JA	-1.2	-4.2	-2.0	0.6
332.22	-0.91	1.4	317.45	5FE	-0.7	-3.9	-1.9	0.5
339.92	-0.83	1.5	327.44	15FE	0.9	-3.3	-1.8	0.5
347.56	-0.75	1.5	337.38	25FE	3.1	-3.6	-1.7	0.6
355.13	-0.66	1.5	347.26	6MR	2.0	-4.1	-1.7	0.6
2.62	-0.56	1.5	357.07	16MR	1.1	-4.2	-1.6	0.7
10.02	-0.46	1.5	6.84	26MR	0.6	-4.2	-1.6	0.7
17.34	-0.36	1.5	16.55	5AP	0.2	-4.2	-1.5	0.8
24.58	-0.26	1.6	26.21	15AP	-0.4	-4.1	-1.5	0.9
31.72	-0.16	1.6	35.83	25AP	-1.2	-4.0	-1.5	0.9
38.79	-0.05	1.7	45.41	5MY	-1.7	-3.9	-1.4	1.0
45.76	0.05	1.7	54.96	15MY	-0.8	-3.8	-1.4	1.0
52.66	0.15	1.7	64.50	25MY	0.0	-3.7	-1.4	1.0
59.48	0.25	1.8	74.03	4JN	0.6	-3.6	-1.4	1.1
66.21	0.36	1.8	83.55	14JN	1.2	-3.6	-1.4	1.1
72.88	0.46	1.8	93.08	24JN	2.0	-3.5	-1.4	1.1
79.47	0.56	1.9	102.63	4JL	3.3	-3.5	-1.5	1.1
86.00	0.65	1.9	112.20	14JL	1.7	-3.4	-1.5	1.1
92.45	0.75	1.9	121.81	24JL	0.1	-3.4	-1.5	1.1
98.84	0.85	1.9	131.46	3AU	-1.1	-3.4	-1.6	1.1
105.17	0.94	1.9	141.15	13AU	-1.4	-3.4	-1.6	1.1
111.42	1.04	1.9	150.90	23AU	-0.9	-3.4	-1.7	1.2
117.60	1.14	1.9	160.71	2SE	-0.4	-3.4	-1.7	1.2
123.71	1.23	1.8	170.57	12SE	-0.2	-3.4	-1.8	1.3
129.74	1.33	1.8	180.49	22SE	-0.0	-3.4	-1.8	1.3
135.68	1.43	1.7	190.47	2OC	0.1	-3.4	-1.9	1.4
141.53	1.53	1.7	200.50	12OC	0.5	-3.4	-2.0	1.4
147.26	1.64	1.6	210.59	22OC	2.8	-3.4	-2.0	1.4
152.85	1.74	1.5	220.71	1NO	0.8	-3.4	-2.1	1.3
158.29	1.85	1.4	230.87	11NO	-0.3	-3.5	-2.1	1.3
163.55	1.97	1.3	241.06	21NO	-0.4	-3.5	-2.2	1.3
168.57	2.09	1.2	251.25	1DE	-0.4	-3.5	-2.2	1.2
173.33	2.22	1.0	261.46	11DE	-0.5	-3.5	-2.2	1.2
177.75	2.36	0.9	271.66	21DE	-0.8	-3.4	-2.2	1.1
181.76	2.50	0.7	281.84	31DE	-1.0	-3.4	-2.2	1.1
				389				
185.25	2.64	0.5	291.99	10JA	-1.0	-3.4	-2.1	1.0
188.10	2.79	0.2	302.11	20JA	-0.6	-3.4	-2.1	1.0
190.14	2.93	-0.0	312.18	30JA	1.0	-3.4	-2.0	0.9
191.21	3.06	-0.3	322.21	9FE	3.1	-3.3	-2.0	0.9
191.12	3.14	-0.6	332.18	19FE	1.5	-3.3	-1.9	0.8
189.77	3.16	-0.9	342.08	1MR	0.8	-3.3	-1.8	0.9
187.20	3.08	-1.2	351.93	11MR	0.5	-3.3	-1.8	0.9
183.74	2.87	-1.4	1.73	21MR	0.1	-3.4	-1.7	0.9
180.02	2.52	-1.4	11.46	31MR	-0.4	-3.4	-1.6	1.0
176.74	2.08	-1.2	21.15	10AP	-1.2	-3.4	-1.6	1.0
174.48	1.60	-1.0	30.79	20AP	-1.7	-3.4	-1.5	1.1
173.54	1.12	-0.8	40.39	30AP	-0.8	-3.5	-1.5	1.2
173.91	0.69	-0.6	49.96	10MY	0.1	-3.5	-1.4	1.2
175.49	0.30	-0.4	59.50	20MY	0.8	-3.6	-1.4	1.3
178.10	-0.04	-0.2	69.03	30MY	1.6	-3.6	-1.4	1.3
181.55	-0.33	-0.1	78.55	9JN	2.7	-3.7	-1.3	1.3
185.71	-0.58	0.1	88.08	19JN	3.0	-3.8	-1.3	1.4
190.45	-0.79	0.2	97.62	29JN	1.4	-3.9	-1.3	1.4
195.67	-0.97	0.3	107.18	9JL	-0.0	-4.0	-1.3	1.4
201.30	-1.13	0.4	116.77	19JL	-1.1	-4.1	-1.3	1.4
207.29	-1.26	0.5	126.39	29JL	-1.5	-4.2	-1.3	1.4
213.57	-1.37	0.6	136.07	8AU	-0.9	-4.2	-1.4	1.3
220.12	-1.46	0.7	145.79	18AU	-0.4	-4.3	-1.4	1.3
226.90	-1.52	0.7	155.56	28AU	-0.1	-4.2	-1.4	1.2
233.87	-1.57	0.8	165.39	7SE	0.1	-3.8	-1.5	1.2
241.03	-1.60	0.9	175.28	17SE	0.3	-3.3	-1.5	1.2
248.33	-1.61	0.9	185.23	27SE	0.8	-3.6	-1.6	1.2
255.76	-1.60	1.0	195.24	7OC	3.1	-4.1	-1.6	1.2
263.29	-1.58	1.0	205.29	17OC	0.6	-4.3	-1.7	1.2
270.91	-1.54	1.1	215.39	27OC	-0.4	-4.4	-1.7	1.2
278.59	-1.48	1.1	225.54	6NO	-0.5	-4.3	-1.8	1.2
286.32	-1.41	1.2	235.71	16NO	-0.5	-4.2	-1.9	1.2
294.06	-1.33	1.2	245.90	26NO	-0.6	-4.1	-1.9	1.1
301.82	-1.25	1.3	256.11	6DE	-0.8	-4.0	-2.0	1.1
309.57	-1.15	1.3	266.32	16DE	-0.9	-3.9	-2.0	1.0
317.30	-1.04	1.4	276.51	26DE	-0.9	-3.8	-2.1	1.0

♂			☉	16.00UT	☿	♀	♃	♄
LONG LAT		MAG	LONG	390		MAGNITUDES		
324.99	-0.93	1.4	286.68	5JA	-0.5	-3.7	-2.1	0.9
332.64	-0.82	1.4	296.82	15JA	1.2	-3.7	-2.1	0.9
340.22	-0.71	1.5	306.91	25JA	2.6	-3.6	-2.1	0.9
347.74	-0.59	1.5	316.97	4FE	1.1	-3.5	-2.1	0.8
355.19	-0.47	1.6	326.96	14FE	0.6	-3.5	-2.0	0.8
2.56	-0.36	1.6	336.90	24FE	0.3	-3.4	-2.0	0.7
9.86	-0.24	1.6	346.78	6MR	0.0	-3.4	-1.9	0.7
17.07	-0.13	1.7	356.61	16MR	-0.5	-3.3	-1.9	0.8
24.20	-0.02	1.7	6.37	26MR	-1.2	-3.3	-1.8	0.9
31.26	0.08	1.7	16.09	5AP	-1.7	-3.3	-1.7	0.9
38.24	0.18	1.7	25.75	15AP	-0.8	-3.3	-1.7	1.0
45.14	0.28	1.8	35.37	25AP	0.2	-3.3	-1.6	1.1
51.97	0.38	1.8	44.95	5MY	1.1	-3.3	-1.5	1.1
58.74	0.47	1.8	54.51	15MY	2.1	-3.3	-1.5	1.2
65.44	0.55	1.8	64.04	25MY	3.5	-3.3	-1.4	1.2
72.08	0.64	1.8	73.57	4JN	2.4	-3.4	-1.4	1.3
78.67	0.71	1.9	83.09	14JN	1.1	-3.5	-1.3	1.3
85.21	0.79	1.9	92.62	24JN	-0.1	-3.5	-1.3	1.3
91.70	0.86	1.9	102.17	4JL	-1.2	-3.5	-1.3	1.3
98.15	0.93	2.0	111.73	14JL	-1.6	-3.5	-1.3	1.3
104.56	0.99	2.0	121.34	24JL	-0.9	-3.5	-1.2	1.3
110.93	1.05	2.0	130.99	3AU	-0.3	-3.4	-1.2	1.2
117.27	1.11	2.0	140.68	13AU	0.0	-3.4	-1.2	1.2
123.59	1.17	2.0	150.42	23AU	0.2	-3.4	-1.3	1.2
129.87	1.22	2.0	160.23	2SE	0.5	-3.3	-1.3	1.1
136.12	1.27	2.0	170.09	12SE	1.1	-3.3	-1.3	1.1
142.34	1.32	2.0	180.00	22SE	3.3	-3.3	-1.3	1.1
148.53	1.36	1.9	189.98	2OC	0.5	-3.3	-1.4	1.1
154.69	1.40	1.9	200.01	12OC	-0.6	-3.3	-1.4	1.1
160.81	1.44	1.8	210.09	22OC	-0.7	-3.4	-1.5	1.1
166.88	1.48	1.8	220.22	1NO	-0.7	-3.4	-1.5	1.1
172.89	1.51	1.7	230.37	11NO	-0.7	-3.4	-1.6	1.1
178.85	1.54	1.6	240.56	21NO	-0.7	-3.4	-1.6	1.0
184.73	1.57	1.5	250.76	1DE	-0.7	-3.5	-1.7	1.0
190.52	1.59	1.4	260.96	11DE	-0.7	-3.5	-1.8	1.0
196.21	1.60	1.3	271.17	21DE	-0.3	-3.6	-1.8	0.9
201.76	1.61	1.2	281.35	31DE	1.4	-3.6	-1.9	0.9
				391				
207.15	1.60	1.0	291.50	10JA	2.2	-3.7	-1.9	0.9
212.35	1.58	0.8	301.62	20JA	0.8	-3.8	-2.0	0.8
217.30	1.55	0.7	311.70	30JA	0.4	-3.9	-2.0	0.8
221.95	1.49	0.4	321.72	9FE	0.2	-3.9	-2.0	0.7
226.23	1.41	0.2	331.70	19FE	-0.1	-4.0	-2.0	0.7
230.03	1.30	-0.0	341.61	1MR	-0.5	-4.1	-2.0	0.6
233.23	1.14	-0.3	351.46	11MR	-1.2	-4.2	-2.0	0.6
235.69	0.92	-0.6	1.26	21MR	-1.6	-4.2	-1.9	0.6
237.22	0.63	-0.9	11.00	31MR	-0.8	-4.2	-1.9	0.7
237.65	0.26	-1.2	20.69	10AP	0.3	-4.0	-1.8	0.7
236.84	-0.21	-1.5	30.33	20AP	1.4	-3.4	-1.8	0.8
234.83	-0.76	-1.8	39.93	30AP	2.8	-2.9	-1.7	0.9
231.91	-1.35	-2.1	49.50	10MY	3.2	-3.6	-1.6	0.9
228.66	-1.93	-2.1	59.04	20MY	1.8	-4.0	-1.6	1.0
225.82	-2.42	-1.9	68.57	30MY	0.9	-4.2	-1.5	1.0
224.02	-2.80	-1.7	78.09	9JN	-0.1	-4.2	-1.5	1.1
223.55	-3.05	-1.5	87.62	19JN	-1.2	-4.1	-1.4	1.1
224.47	-3.19	-1.3	97.16	29JN	-1.7	-4.0	-1.4	1.1
226.63	-3.26	-1.1	106.71	9JL	-0.9	-3.9	-1.3	1.1
229.85	-3.26	-0.9	116.30	19JL	-0.2	-3.8	-1.3	1.1
233.96	-3.22	-0.7	125.92	29JL	0.1	-3.8	-1.3	1.1
238.77	-3.14	-0.5	135.59	8AU	0.4	-3.7	-1.2	1.1
244.16	-3.03	-0.4	145.31	18AU	0.7	-3.6	-1.2	1.1
250.01	-2.90	-0.2	155.08	28AU	1.5	-3.6	-1.2	1.1
256.24	-2.75	-0.1	164.91	7SE	3.1	-3.5	-1.2	1.0
262.76	-2.59	0.1	174.80	17SE	0.4	-3.5	-1.2	1.0
269.52	-2.41	0.2	184.75	27SE	-0.7	-3.5	-1.2	0.9
276.46	-2.23	0.3	194.75	7OC	-0.8	-3.4	-1.3	1.0
283.55	-2.04	0.5	204.81	17OC	-0.8	-3.4	-1.3	1.0
290.75	-1.84	0.6	214.91	27OC	-0.8	-3.4	-1.3	1.0
298.01	-1.65	0.7	225.05	6NO	-0.6	-3.4	-1.4	1.0
305.33	-1.46	0.8	235.22	16NO	-0.5	-3.4	-1.4	1.0
312.66	-1.26	0.9	245.42	26NO	-0.5	-3.4	-1.5	0.9
320.00	-1.08	1.0	255.62	6DE	-0.2	-3.4	-1.5	0.9
327.33	-0.90	1.1	265.82	16DE	1.6	-3.4	-1.6	0.9
334.64	-0.73	1.2	276.02	26DE	1.8	-3.4	-1.6	0.9

♂ LONG	♂ LAT	♂ MAG	⊙ LONG	16.00UT 392	☿	♀	♃	♄
					MAGNITUDES			
341.90	-0.56	1.3	286.18	5JA	0.5	-3.4	-1.7	0.8
349.12	-0.41	1.4	296.33	15JA	0.2	-3.4	-1.8	0.8
356.29	-0.26	1.4	306.42	25JA	0.0	-3.4	-1.8	0.7
3.41	-0.12	1.5	316.48	4FE	-0.2	-3.5	-1.9	0.7
10.46	0.01	1.6	326.48	14FE	-0.6	-3.5	-1.9	0.6
17.45	0.13	1.6	336.42	24FE	-1.2	-3.5	-2.0	0.6
24.37	0.25	1.7	346.30	5MR	-1.5	-3.4	-2.0	0.5
31.24	0.35	1.7	356.13	15MR	-0.8	-3.4	-2.0	0.5
38.04	0.45	1.8	5.90	25MR	0.4	-3.4	-2.0	0.5
44.79	0.54	1.8	15.61	4AP	1.8	-3.4	-2.0	0.5
51.48	0.62	1.9	25.28	14AP	3.6	-3.3	-2.0	0.6
58.12	0.70	1.9	34.90	24AP	2.4	-3.3	-1.9	0.6
64.71	0.77	1.9	44.48	4MY	1.4	-3.3	-1.9	0.7
71.26	0.83	1.9	54.04	14MY	0.6	-3.3	-1.8	0.8
77.76	0.89	2.0	63.58	24MY	-0.2	-3.3	-1.8	0.8
84.24	0.94	2.0	73.10	3JN	-1.2	-3.3	-1.7	0.9
90.68	0.99	2.0	82.62	13JN	-1.7	-3.4	-1.6	0.9
97.10	1.03	2.0	92.15	23JN	-0.9	-3.4	-1.6	1.0
103.50	1.07	1.9	101.70	3JL	-0.2	-3.4	-1.5	1.0
109.89	1.10	1.9	111.27	13JL	0.3	-3.5	-1.5	1.0
116.26	1.13	2.0	120.87	23JL	0.6	-3.5	-1.4	1.0
122.63	1.15	2.0	130.52	2AU	1.0	-3.6	-1.4	1.0
129.00	1.17	2.0	140.21	12AU	1.9	-3.7	-1.3	1.0
135.36	1.18	2.0	149.95	22AU	2.8	-3.7	-1.3	1.0
141.73	1.19	2.0	159.75	1SE	0.3	-3.8	-1.3	1.0
148.11	1.20	2.0	169.61	11SE	-0.8	-3.9	-1.2	0.9
154.49	1.20	2.0	179.52	21SE	-1.0	-4.0	-1.2	0.9
160.88	1.20	2.0	189.50	1OC	-1.0	-4.2	-1.2	0.8
167.28	1.19	2.0	199.53	11OC	-0.7	-4.3	-1.2	0.8
173.69	1.18	1.9	209.60	21OC	-0.5	-4.3	-1.2	0.9
180.11	1.16	1.9	219.73	31OC	-0.4	-4.4	-1.3	0.9
186.54	1.13	1.8	229.88	10NO	-0.4	-4.3	-1.3	0.9
192.97	1.10	1.8	240.07	20NO	-0.0	-3.9	-1.3	0.9
199.41	1.06	1.7	250.27	30NO	1.9	-3.0	-1.4	0.9
205.85	1.01	1.6	260.47	10DE	1.4	-3.6	-1.4	0.9
212.28	0.94	1.6	270.67	20DE	0.2	-4.1	-1.5	0.9
218.71	0.87	1.5	280.86	30DE	0.0	-4.3	-1.5	0.8
				393				
225.13	0.78	1.4	291.02	9JA	-0.1	-4.4	-1.6	0.8
231.52	0.68	1.2	301.13	19JA	-0.3	-4.3	-1.6	0.8
237.90	0.56	1.1	311.21	29JA	-0.6	-4.2	-1.7	0.7
244.23	0.42	1.0	321.24	8FE	-1.2	-4.1	-1.8	0.7
250.51	0.25	0.8	331.21	18FE	-1.4	-4.0	-1.8	0.6
256.73	0.06	0.7	341.13	28FE	-0.8	-3.9	-1.9	0.6
262.86	-0.16	0.5	350.99	10MR	0.6	-3.8	-2.0	0.5
268.89	-0.42	0.3	0.79	20MR	2.3	-3.7	-2.0	0.4
274.76	-0.72	0.1	10.53	30MR	3.2	-3.6	-2.0	0.4
280.45	-1.06	-0.1	20.22	9AP	1.8	-3.5	-2.1	0.4
285.89	-1.45	-0.3	29.87	19AP	1.0	-3.5	-2.1	0.4
291.01	-1.89	-0.5	39.47	29AP	0.5	-3.4	-2.1	0.5
295.70	-2.40	-0.8	49.04	9MY	-0.3	-3.4	-2.0	0.6
299.86	-2.97	-1.0	58.59	19MY	-1.2	-3.4	-2.0	0.6
303.30	-3.61	-1.3	68.12	29MY	-1.8	-3.3	-2.0	0.7
305.85	-4.31	-1.6	77.64	8JN	-0.9	-3.3	-1.9	0.7
307.32	-5.04	-1.9	87.16	18JN	-0.1	-3.3	-1.9	0.8
307.52	-5.76	-2.1	96.70	28JN	0.4	-3.3	-1.8	0.8
306.44	-6.37	-2.4	106.26	8JL	0.8	-3.3	-1.7	0.9
304.32	-6.78	-2.6	115.84	18JL	1.3	-3.3	-1.7	0.9
301.70	-6.88	-2.6	125.46	28JL	2.5	-3.4	-1.6	0.9
299.34	-6.65	-2.5	135.13	7AU	2.4	-3.4	-1.5	0.9
297.87	-6.16	-2.2	144.84	17AU	0.2	-3.4	-1.5	0.9
297.65	-5.52	-1.9	154.61	27AU	-0.9	-3.5	-1.4	0.9
298.75	-4.82	-1.6	164.43	6SE	-1.1	-3.5	-1.4	0.9
301.02	-4.14	-1.3	174.31	16SE	-1.0	-3.5	-1.4	0.9
304.27	-3.50	-1.1	184.26	26SE	-0.6	-3.5	-1.3	0.8
308.32	-2.92	-0.8	194.26	6OC	-0.3	-3.4	-1.3	0.8
312.97	-2.40	-0.5	204.31	16OC	-0.3	-3.4	-1.3	0.8
318.10	-1.93	-0.3	214.41	26OC	-0.2	-3.4	-1.3	0.8
323.60	-1.53	-0.1	224.55	5NO	0.2	-3.4	-1.3	0.8
329.36	-1.17	0.1	234.72	15NO	2.1	-3.3	-1.3	0.8
335.34	-0.85	0.3	244.92	25NO	1.2	-3.3	-1.3	0.8
341.47	-0.58	0.5	255.12	5DE	0.0	-3.3	-1.3	0.8
347.72	-0.33	0.7	265.32	15DE	-0.1	-3.3	-1.4	0.8
354.04	-0.12	0.8	275.52	25DE	-0.2	-3.3	-1.4	0.8

♂ LONG	♂ LAT	♂ MAG	⊙ LONG	16.00UT 394	☿	♀	♃	♄
					MAGNITUDES			
0.43	0.07	1.0	285.69	4JA	-0.4	-3.4	-1.4	0.8
6.84	0.23	1.1	295.83	14JA	-0.7	-3.4	-1.5	0.8
13.28	0.38	1.2	305.94	24JA	-1.2	-3.4	-1.6	0.7
19.73	0.51	1.3	315.99	3FE	-1.3	-3.4	-1.6	0.7
26.17	0.62	1.5	325.99	13FE	-0.7	-3.5	-1.7	0.7
32.61	0.71	1.5	335.94	23FE	0.7	-3.5	-1.7	0.6
39.04	0.80	1.6	345.83	5MR	2.8	-3.6	-1.8	0.5
45.45	0.87	1.7	355.66	15MR	2.4	-3.6	-1.9	0.5
51.84	0.93	1.8	5.43	25MR	1.3	-3.7	-2.0	0.4
58.22	0.99	1.8	15.15	4AP	0.8	-3.8	-2.0	0.4
64.58	1.04	1.9	24.82	14AP	0.3	-3.8	-2.1	0.3
70.92	1.07	1.9	34.44	24AP	-0.3	-3.9	-2.1	0.3
77.25	1.11	2.0	44.03	4MY	-1.2	-4.0	-2.2	0.4
83.57	1.13	2.0	53.59	14MY	-1.8	-4.1	-2.2	0.5
89.89	1.15	2.0	63.12	24MY	-0.9	-4.2	-2.2	0.5
96.20	1.16	2.0	72.65	3JN	-0.0	-4.2	-2.2	0.6
102.51	1.17	2.0	82.17	13JN	0.5	-4.1	-2.1	0.6
108.82	1.18	2.0	91.70	23JN	1.0	-3.7	-2.1	0.7
115.14	1.18	2.0	101.24	3JL	1.7	-3.1	-2.1	0.7
121.48	1.17	2.0	110.81	13JL	3.0	-3.3	-2.0	0.8
127.83	1.16	2.0	120.41	23JL	2.0	-3.9	-1.9	0.8
134.21	1.14	2.0	130.05	2AU	0.1	-4.2	-1.9	0.8
140.60	1.12	1.9	139.74	12AU	-1.0	-4.2	-1.8	0.9
147.03	1.10	1.9	149.48	22AU	-1.3	-4.2	-1.7	0.9
153.49	1.07	2.0	159.27	1SE	-1.0	-4.1	-1.7	0.9
159.99	1.04	2.0	169.13	11SE	-0.5	-4.1	-1.6	0.9
166.52	1.00	2.0	179.04	21SE	-0.2	-4.0	-1.6	0.8
173.10	0.95	1.9	189.01	1OC	-0.1	-3.9	-1.5	0.8
179.72	0.91	1.9	199.03	11OC	-0.0	-3.8	-1.5	0.8
186.38	0.85	1.9	209.11	21OC	0.4	-3.7	-1.5	0.7
193.09	0.79	1.9	219.23	31OC	2.5	-3.7	-1.4	0.7
199.84	0.72	1.9	229.39	10NO	1.0	-3.6	-1.4	0.7
206.65	0.64	1.8	239.57	20NO	-0.1	-3.6	-1.4	0.8
213.50	0.56	1.8	249.77	30NO	-0.3	-3.5	-1.4	0.8
220.40	0.46	1.7	259.98	10DE	-0.3	-3.5	-1.4	0.8
227.34	0.36	1.7	270.18	20DE	-0.5	-3.4	-1.4	0.8
234.33	0.25	1.6	280.36	30DE	-0.7	-3.4	-1.4	0.8
				395				
241.36	0.13	1.5	290.52	9JA	-1.1	-3.4	-1.4	0.8
248.44	-0.00	1.4	300.65	19JA	-1.1	-3.3	-1.5	0.8
255.56	-0.15	1.4	310.73	29JA	-0.7	-3.3	-1.5	0.7
262.71	-0.30	1.3	320.76	8FE	0.9	-3.3	-1.6	0.7
269.89	-0.47	1.2	330.74	18FE	3.2	-3.3	-1.6	0.7
277.10	-0.64	1.1	340.65	28FE	1.8	-3.3	-1.7	0.6
284.32	-0.83	1.0	350.52	10MR	1.0	-3.3	-1.7	0.6
291.55	-1.02	0.9	0.32	20MR	0.6	-3.4	-1.8	0.5
298.77	-1.22	0.8	10.06	30MR	0.2	-3.4	-1.9	0.4
305.98	-1.43	0.7	19.76	9AP	-0.4	-3.4	-1.9	0.4
313.14	-1.65	0.6	29.40	19AP	-1.2	-3.5	-2.0	0.3
320.24	-1.86	0.5	39.00	29AP	-1.8	-3.5	-2.1	0.3
327.26	-2.08	0.4	48.58	9MY	-0.9	-3.5	-2.1	0.3
334.14	-2.30	0.2	58.13	19MY	0.0	-3.4	-2.2	0.4
340.87	-2.51	0.1	67.65	29MY	0.7	-3.4	-2.2	0.4
347.39	-2.71	-0.0	77.18	8JN	1.3	-3.4	-2.3	0.5
353.65	-2.91	-0.2	86.70	18JN	2.3	-3.4	-2.3	0.6
359.57	-3.09	-0.3	96.23	28JN	3.3	-3.3	-2.3	0.6
5.06	-3.26	-0.5	105.79	8JL	1.6	-3.3	-2.3	0.7
10.02	-3.41	-0.6	115.37	18JL	0.1	-3.3	-2.3	0.7
14.32	-3.54	-0.8	124.99	28JL	-1.1	-3.3	-2.2	0.8
17.77	-3.64	-1.0	134.66	7AU	-1.4	-3.3	-2.2	0.8
20.17	-3.68	-1.2	144.37	17AU	-0.9	-3.4	-2.1	0.8
21.33	-3.66	-1.5	154.13	27AU	-0.4	-3.4	-2.0	0.8
21.05	-3.54	-1.7	163.95	6SE	-0.1	-3.4	-2.0	0.8
19.32	-3.28	-1.9	173.83	16SE	0.0	-3.5	-1.9	0.8
16.40	-2.88	-2.1	183.77	26SE	0.1	-3.5	-1.9	0.8
12.92	-2.34	-2.1	193.77	6OC	0.6	-3.6	-1.8	0.8
9.71	-1.73	-1.9	203.82	16OC	2.8	-3.6	-1.7	0.8
7.44	-1.12	-1.6	213.92	26OC	0.8	-3.7	-1.7	0.7
6.48	-0.58	-1.2	224.06	5NO	-0.3	-3.8	-1.6	0.7
6.86	-0.12	-0.9	234.22	15NO	-0.5	-3.9	-1.6	0.7
8.42	0.24	-0.6	244.42	25NO	-0.5	-4.0	-1.6	0.7
10.96	0.53	-0.3	254.62	5DE	-0.6	-4.1	-1.5	0.8
14.29	0.76	-0.0	264.83	15DE	-0.8	-4.2	-1.5	0.8
18.22	0.94	0.2	275.02	25DE	-0.9	-4.3	-1.5	0.8

Left table (396/397):

♂ LONG	LAT	MAG	☉ LONG	16.00UT	☿	♀	♃	♄
						MAGNITUDES		
22.62	1.07	0.5	285.20	4JA	-1.0	-4.3	-1.5	0.8
27.40	1.18	0.7	295.34	14JA	-0.6	-4.4	-1.5	0.8
32.44	1.26	0.9	305.44	24JA	1.1	-4.2	-1.5	0.8
37.72	1.32	1.0	315.50	3FE	3.0	-3.8	-1.5	0.8
43.16	1.37	1.2	325.51	13FE	1.3	-3.3	-1.5	0.7
48.74	1.40	1.3	335.46	23FE	0.7	-3.6	-1.6	0.7
54.43	1.42	1.4	345.35	4MR	0.4	-4.1	-1.6	0.6
60.20	1.43	1.5	355.18	14MR	0.1	-4.2	-1.7	0.6
66.04	1.43	1.6	4.96	24MR	-0.4	-4.2	-1.7	0.5
71.94	1.43	1.7	14.68	3AP	-1.2	-4.2	-1.8	0.5
77.89	1.42	1.8	24.35	13AP	-1.7	-4.1	-1.8	0.4
83.88	1.41	1.8	33.98	23AP	-0.9	-4.0	-1.9	0.3
89.91	1.39	1.9	43.57	3MY	0.1	-3.9	-2.0	0.3
95.98	1.36	1.9	53.12	13MY	0.9	-3.8	-2.0	0.2
102.08	1.33	2.0	62.66	23MY	1.8	-3.7	-2.1	0.2
108.22	1.30	2.0	72.19	2JN	3.0	-3.6	-2.2	0.4
114.39	1.27	2.0	81.71	12JN	2.8	-3.6	-2.2	0.4
120.60	1.23	2.0	91.24	22JN	1.3	-3.5	-2.3	0.5
126.86	1.19	2.0	100.78	2JL	-0.0	-3.5	-2.3	0.5
133.15	1.14	2.0	110.35	12JL	-1.1	-3.4	-2.4	0.6
139.50	1.09	2.0	119.95	22JL	-1.5	-3.4	-2.4	0.7
145.89	1.04	2.0	129.59	1AU	-0.9	-3.4	-2.4	0.7
152.34	0.98	2.0	139.27	11AU	-0.4	-3.4	-2.4	0.7
158.84	0.92	1.9	149.01	21AU	-0.0	-3.4	-2.4	0.8
165.41	0.85	1.9	158.80	31AU	0.2	-3.4	-2.3	0.8
172.03	0.79	1.8	168.65	10SE	0.3	-3.4	-2.3	0.8
178.73	0.71	1.8	178.56	20SE	0.9	-3.4	-2.2	0.8
185.48	0.64	1.8	188.53	30SE	3.1	-3.4	-2.1	0.8
192.31	0.55	1.8	198.55	10OC	0.6	-3.4	-2.1	0.8
199.21	0.47	1.8	208.62	20OC	-0.5	-3.4	-2.0	0.8
206.18	0.38	1.8	218.74	30OC	-0.6	-3.4	-1.9	0.7
213.22	0.28	1.8	228.89	9NO	-0.6	-3.5	-1.9	0.7
220.34	0.18	1.7	239.07	19NO	-0.7	-3.5	-1.8	0.7
227.52	0.07	1.7	249.27	29NO	-0.8	-3.5	-1.8	0.7
234.77	-0.04	1.7	259.47	9DE	-0.8	-3.5	-1.7	0.7
242.09	-0.15	1.6	269.68	19DE	-0.8	-3.4	-1.7	0.8
249.47	-0.27	1.6	279.86	29DE	-0.5	-3.4	-1.6	0.8

397

♂ LONG	LAT	MAG	☉ LONG	16.00UT	☿	♀	♃	♄
256.91	-0.39	1.6	290.02	8JA	1.2	-3.4	-1.6	0.8
264.41	-0.51	1.5	300.15	18JA	2.5	-3.4	-1.6	0.8
271.95	-0.64	1.5	310.23	28JA	1.0	-3.4	-1.6	0.8
279.53	-0.76	1.4	320.26	7FE	0.5	-3.3	-1.6	0.8
287.15	-0.89	1.4	330.25	17FE	0.3	-3.3	-1.6	0.8
294.79	-1.01	1.3	340.17	27FE	-0.0	-3.3	-1.6	0.7
302.45	-1.13	1.3	350.03	9MR	-0.5	-3.3	-1.6	0.7
310.10	-1.24	1.2	359.84	19MR	-1.2	-3.4	-1.6	0.6
317.74	-1.34	1.2	9.59	29MR	-1.6	-3.4	-1.6	0.6
325.36	-1.43	1.1	19.28	8AP	-0.9	-3.4	-1.7	0.5
332.94	-1.52	1.1	28.94	18AP	0.2	-3.4	-1.7	0.5
340.45	-1.58	1.0	38.54	28AP	1.2	-3.5	-1.8	0.4
347.90	-1.64	1.0	48.11	8MY	2.3	-3.5	-1.8	0.3
355.26	-1.68	0.9	57.67	18MY	3.6	-3.6	-1.9	0.3
2.50	-1.70	0.9	67.20	28MY	2.2	-3.6	-1.9	0.3
9.62	-1.71	0.8	76.72	7JN	1.0	-3.7	-2.0	0.3
16.58	-1.70	0.8	86.24	17JN	-0.1	-3.8	-2.1	0.4
23.37	-1.67	0.7	95.78	27JN	-1.1	-3.9	-2.1	0.4
29.95	-1.62	0.6	105.33	7JL	-1.6	-4.0	-2.2	0.5
36.30	-1.55	0.5	114.91	17JL	-0.9	-4.1	-2.3	0.6
42.37	-1.46	0.5	124.53	27JL	-0.3	-4.2	-2.3	0.6
48.12	-1.35	0.4	134.19	6AU	0.1	-4.2	-2.4	0.7
53.47	-1.21	0.2	143.90	16AU	0.3	-4.3	-2.4	0.7
58.35	-1.04	0.1	153.66	26AU	0.5	-4.1	-2.4	0.7
62.67	-0.85	-0.0	163.48	5SE	1.2	-3.8	-2.5	0.8
66.28	-0.61	-0.2	173.36	15SE	3.3	-3.6	-2.4	0.8
69.03	-0.33	-0.4	183.29	25SE	0.5	-3.6	-2.4	0.8
70.71	-0.00	-0.6	193.29	5OC	-0.6	-4.1	-2.4	0.8
71.13	0.39	-0.8	203.34	15OC	-0.8	-4.3	-2.3	0.8
70.14	0.82	-1.0	213.43	25OC	-0.7	-4.4	-2.3	0.8
67.75	1.29	-1.2	223.56	4NO	-0.8	-4.3	-2.2	0.7
64.27	1.74	-1.4	233.73	14NO	-0.7	-4.2	-2.1	0.7
60.35	2.12	-1.4	243.92	24NO	-0.6	-4.1	-2.1	0.7
56.79	2.40	-1.2	254.13	4DE	-0.6	-4.0	-2.0	0.7
54.22	2.55	-0.9	264.33	14DE	-0.3	-3.9	-1.9	0.7
52.98	2.61	-0.6	274.52	24DE	1.4	-3.8	-1.8	0.8

Right table (398/399):

♂ LONG	LAT	MAG	☉ LONG	16.00UT	☿	♀	♃	♄
						MAGNITUDES		
53.05	2.61	-0.3	284.70	3JA	2.1	-3.7	-1.8	0.8
54.29	2.56	-0.0	294.85	13JA	0.6	-3.7	-1.7	0.8
56.50	2.48	0.2	304.95	23JA	0.3	-3.6	-1.7	0.8
59.48	2.40	0.5	315.01	2FE	0.1	-3.5	-1.7	0.8
63.08	2.31	0.7	325.02	12FE	-0.1	-3.5	-1.6	0.8
67.16	2.21	0.9	334.97	22FE	-0.5	-3.4	-1.6	0.8
71.61	2.12	1.0	344.87	4MR	-1.2	-3.4	-1.6	0.8
76.38	2.03	1.2	354.71	14MR	-1.6	-3.3	-1.6	0.7
81.38	1.94	1.3	4.48	24MR	-0.8	-3.3	-1.6	0.7
86.58	1.85	1.4	14.21	3AP	0.3	-3.3	-1.6	0.6
91.96	1.77	1.5	23.88	13AP	1.6	-3.3	-1.6	0.6
97.47	1.68	1.6	33.51	23AP	3.1	-3.3	-1.6	0.5
103.10	1.60	1.7	43.10	3MY	2.9	-3.3	-1.6	0.4
108.85	1.51	1.8	52.66	13MY	1.7	-3.3	-1.7	0.4
114.69	1.43	1.8	62.20	23MY	0.8	-3.3	-1.7	0.3
120.62	1.35	1.8	71.73	2JN	-0.1	-3.4	-1.7	0.3
126.63	1.26	1.9	81.25	12JN	-1.2	-3.4	-1.8	0.3
132.73	1.18	1.9	90.78	22JN	-1.7	-3.5	-1.8	0.3
138.91	1.09	1.9	100.32	2JL	-0.9	-3.5	-1.9	0.4
145.17	1.00	1.9	109.89	12JL	-0.2	-3.5	-2.0	0.4
151.51	0.91	1.9	119.48	22JL	0.2	-3.5	-2.0	0.5
157.94	0.82	1.9	129.12	1AU	0.5	-3.4	-2.1	0.6
164.44	0.73	1.9	138.80	11AU	0.8	-3.4	-2.2	0.6
171.04	0.64	1.9	148.53	21AU	1.6	-3.4	-2.2	0.7
177.72	0.54	1.8	158.32	31AU	3.1	-3.3	-2.3	0.7
184.48	0.44	1.8	168.17	10SE	0.4	-3.3	-2.3	0.7
191.34	0.34	1.8	178.07	20SE	-0.7	-3.3	-2.4	0.8
198.29	0.24	1.7	188.04	30SE	-0.9	-3.3	-2.4	0.8
205.32	0.14	1.7	198.06	10OC	-0.9	-3.3	-2.4	0.8
212.45	0.04	1.6	208.13	20OC	-0.7	-3.4	-2.4	0.8
219.66	-0.07	1.6	218.25	30OC	-0.5	-3.4	-2.4	0.8
226.96	-0.17	1.5	228.40	9NO	-0.5	-3.4	-2.3	0.8
234.34	-0.28	1.5	238.58	19NO	-0.5	-3.4	-2.3	0.7
241.81	-0.38	1.5	248.78	29NO	-0.2	-3.5	-2.2	0.7
249.34	-0.49	1.5	258.98	9DE	1.6	-3.5	-2.2	0.7
256.94	-0.59	1.5	269.18	19DE	1.7	-3.6	-2.1	0.7
264.61	-0.68	1.5	279.37	29DE	0.4	-3.6	-2.0	0.8

399

♂ LONG	LAT	MAG	☉ LONG	16.00UT	☿	♀	♃	♄
272.32	-0.78	1.5	289.53	8JA	0.1	-3.7	-1.9	0.8
280.07	-0.87	1.5	299.66	18JA	-0.0	-3.8	-1.9	0.8
287.86	-0.95	1.4	309.75	28JA	-0.2	-3.9	-1.8	0.8
295.67	-1.02	1.4	319.78	7FE	-0.6	-3.9	-1.7	0.8
303.49	-1.08	1.4	329.76	17FE	-1.2	-4.0	-1.7	0.8
311.31	-1.14	1.4	339.69	27FE	-1.5	-4.1	-1.6	0.8
319.11	-1.18	1.4	349.55	9MR	-0.8	-4.2	-1.6	0.8
326.88	-1.21	1.4	359.36	19MR	0.5	-4.2	-1.6	0.8
334.61	-1.23	1.4	9.12	29MR	2.0	-4.2	-1.5	0.7
342.30	-1.24	1.4	18.81	8AP	3.8	-4.0	-1.5	0.7
349.91	-1.23	1.3	28.46	18AP	2.2	-3.4	-1.5	0.6
357.46	-1.21	1.3	38.08	28AP	1.3	-2.9	-1.5	0.6
4.92	-1.17	1.3	47.65	8MY	0.6	-3.7	-1.5	0.5
12.28	-1.12	1.3	57.20	18MY	-0.2	-4.1	-1.5	0.4
19.55	-1.06	1.3	66.73	28MY	-1.2	-4.2	-1.5	0.4
26.69	-0.98	1.3	76.25	7JN	-1.8	-4.2	-1.6	0.3
33.72	-0.90	1.3	85.78	17JN	-0.9	-4.1	-1.6	0.3
40.62	-0.80	1.3	95.31	27JN	-0.2	-4.0	-1.6	0.3
47.37	-0.69	1.2	104.86	7JL	0.3	-3.9	-1.7	0.4
53.97	-0.57	1.2	114.44	17JL	0.6	-3.8	-1.7	0.4
60.39	-0.43	1.2	124.06	27JL	1.1	-3.8	-1.8	0.5
66.63	-0.28	1.1	133.72	6AU	2.1	-3.7	-1.8	0.5
72.67	-0.12	1.0	143.43	16AU	2.7	-3.6	-1.9	0.6
78.46	0.05	1.0	153.19	26AU	0.3	-3.6	-1.9	0.6
83.98	0.25	0.9	163.00	5SE	-0.8	-3.5	-2.0	0.7
89.17	0.46	0.8	172.88	15SE	-1.1	-3.5	-2.1	0.7
93.98	0.70	0.7	182.81	25SE	-1.0	-3.5	-2.1	0.8
98.33	0.96	0.5	192.80	5OC	-0.6	-3.4	-2.2	0.8
102.11	1.26	0.4	202.85	15OC	-0.4	-3.4	-2.3	0.8
105.19	1.59	0.2	212.94	25OC	-0.3	-3.4	-2.3	0.8
107.41	1.96	0.0	223.07	4NO	-0.3	-3.4	-2.3	0.8
108.58	2.37	-0.2	233.24	14NO	0.0	-3.4	-2.3	0.8
108.51	2.82	-0.4	243.43	24NO	1.9	-3.4	-2.3	0.8
107.09	3.27	-0.7	253.63	4DE	1.4	-3.4	-2.3	0.8
104.36	3.68	-0.9	263.84	14DE	0.2	-3.4	-2.2	0.7
100.68	3.98	-1.1	274.03	24DE	-0.1	-3.4	-2.2	0.7

Left table

♂ LONG	♂ LAT	♂ MAG	☉ LONG	16.00UT 400	☿	♀	♃	♄
96.71	4.14	-1.0	284.21	3JA	-0.1	-3.4	-2.1	0.8
93.17	4.12	-0.8	294.36	13JA	-0.3	-3.4	-2.1	0.8
90.67	3.97	-0.6	304.47	23JA	-0.6	-3.4	-2.0	0.8
89.45	3.75	-0.3	314.52	2FE	-1.2	-3.5	-1.9	0.9
89.52	3.48	-0.0	324.54	12FE	-1.3	-3.5	-1.9	0.9
90.72	3.21	0.2	334.49	22FE	-0.8	-3.5	-1.8	0.9
92.87	2.95	0.4	344.39	3MR	0.6	-3.4	-1.7	0.9
95.79	2.70	0.6	354.23	13MR	2.5	-3.4	-1.7	0.9
99.32	2.48	0.8	4.01	23MR	2.9	-3.4	-1.6	0.8
103.34	2.27	1.0	13.73	2AP	1.6	-3.4	-1.6	0.8
107.76	2.07	1.1	23.41	12AP	0.9	-3.3	-1.5	0.8
112.52	1.89	1.2	33.04	22AP	0.4	-3.3	-1.5	0.7
117.54	1.72	1.3	42.63	2MY	-0.3	-3.3	-1.5	0.7
122.79	1.56	1.4	52.19	12MY	-1.2	-3.3	-1.4	0.6
128.25	1.41	1.5	61.73	22MY	-1.8	-3.3	-1.4	0.5
133.88	1.27	1.6	71.26	1JN	-0.9	-3.3	-1.4	0.5
139.67	1.13	1.6	80.78	11JN	-0.1	-3.4	-1.4	0.4
145.61	0.99	1.6	90.31	21JN	0.4	-3.4	-1.4	0.3
151.68	0.86	1.7	99.85	1JL	0.8	-3.4	-1.4	0.3
157.88	0.73	1.7	109.42	11JL	1.4	-3.5	-1.5	0.4
164.21	0.60	1.7	119.01	21JL	2.7	-3.5	-1.5	0.4
170.65	0.48	1.7	128.65	31JL	2.3	-3.6	-1.5	0.5
177.22	0.36	1.7	138.33	10AU	0.2	-3.7	-1.6	0.5
183.90	0.24	1.7	148.06	20AU	-0.9	-3.7	-1.6	0.6
190.70	0.12	1.7	157.85	30AU	-1.2	-3.8	-1.7	0.6
197.61	0.01	1.7	167.69	9SE	-1.0	-3.9	-1.7	0.7
204.63	-0.11	1.7	177.60	19SE	-0.6	-4.0	-1.8	0.7
211.76	-0.22	1.6	187.56	29SE	-0.3	-4.2	-1.8	0.8
218.99	-0.32	1.6	197.58	9OC	-0.2	-4.3	-1.9	0.8
226.33	-0.42	1.6	207.64	19OC	-0.1	-4.3	-2.0	0.8
233.76	-0.52	1.5	217.75	29OC	0.2	-4.4	-2.0	0.8
241.28	-0.62	1.5	227.91	8NO	2.2	-4.3	-2.1	0.8
248.89	-0.70	1.5	238.08	18NO	1.1	-3.9	-2.1	0.8
256.56	-0.78	1.4	248.28	28NO	-0.0	-3.0	-2.2	0.8
264.31	-0.86	1.4	258.49	8DE	-0.2	-3.6	-2.2	0.8
272.11	-0.92	1.4	268.69	18DE	-0.3	-4.2	-2.2	0.8
279.95	-0.98	1.3	278.87	28DE	-0.4	-4.4	-2.2	0.8
				401				
287.82	-1.02	1.3	289.04	7JA	-0.7	-4.4	-2.2	0.8
295.71	-1.06	1.3	299.17	17JA	-1.1	-4.3	-2.1	0.8
303.60	-1.08	1.3	309.26	27JA	-1.2	-4.2	-2.1	0.9
311.48	-1.09	1.4	319.30	6FE	-0.7	-4.1	-2.0	0.9
319.34	-1.09	1.4	329.28	16FE	0.7	-4.0	-1.9	0.9
327.17	-1.08	1.4	339.21	26FE	3.0	-3.9	-1.9	0.9
334.96	-1.05	1.4	349.08	8MR	2.2	-3.8	-1.8	1.0
342.69	-1.02	1.4	358.89	18MR	1.2	-3.7	-1.7	0.9
350.35	-0.97	1.5	8.64	28MR	0.7	-3.6	-1.7	0.9
357.94	-0.92	1.5	18.35	7AP	0.3	-3.5	-1.6	0.9
5.45	-0.85	1.5	28.00	17AP	-0.3	-3.5	-1.5	0.9
12.87	-0.77	1.5	37.61	27AP	-1.2	-3.4	-1.5	0.8
20.20	-0.69	1.5	47.19	7MY	-1.8	-3.4	-1.4	0.8
27.44	-0.60	1.6	56.74	17MY	-0.9	-3.4	-1.4	0.7
34.58	-0.50	1.6	66.27	27MY	-0.0	-3.3	-1.4	0.7
41.61	-0.40	1.6	75.80	6JN	0.6	-3.3	-1.3	0.6
48.54	-0.29	1.6	85.32	16JN	1.1	-3.3	-1.3	0.5
55.37	-0.17	1.6	94.85	26JN	1.9	-3.3	-1.3	0.5
62.09	-0.05	1.6	104.41	6JL	3.2	-3.3	-1.3	0.4
68.69	0.08	1.6	113.98	16JL	1.9	-3.3	-1.3	0.4
75.18	0.21	1.6	123.60	26JL	0.2	-3.4	-1.3	0.4
81.55	0.34	1.6	133.25	5AU	-1.0	-3.4	-1.3	0.5
87.78	0.49	1.5	142.95	15AU	-1.3	-3.4	-1.4	0.5
93.87	0.64	1.5	152.71	25AU	-1.0	-3.5	-1.4	0.6
99.81	0.79	1.4	162.53	4SE	-0.5	-3.5	-1.4	0.6
105.57	0.96	1.4	172.40	14SE	-0.2	-3.5	-1.5	0.7
111.13	1.13	1.3	182.33	24SE	-0.1	-3.5	-1.5	0.7
116.46	1.32	1.2	192.32	4OC	0.0	-3.4	-1.6	0.8
121.52	1.53	1.1	202.36	14OC	0.5	-3.4	-1.6	0.8
126.25	1.75	1.0	212.45	24OC	2.5	-3.4	-1.7	0.8
130.59	1.99	0.8	222.58	3NO	0.9	-3.4	-1.8	0.9
134.44	2.26	0.7	232.74	13NO	-0.2	-3.3	-1.8	0.9
137.69	2.56	0.5	242.93	23NO	-0.4	-3.3	-1.9	0.9
140.19	2.89	0.3	253.13	3DE	-0.4	-3.3	-1.9	0.9
141.78	3.24	0.1	263.34	13DE	-0.5	-3.3	-2.0	0.9
142.27	3.61	-0.2	273.54	23DE	-0.8	-3.3	-2.0	0.9

Right table

♂ LONG	♂ LAT	♂ MAG	☉ LONG	16.00UT 402	☿	♀	♃	♄
141.50	3.98	-0.4	283.71	2JA	-1.0	-3.4	-2.1	0.9
139.43	4.30	-0.7	293.86	12JA	-1.1	-3.4	-2.1	0.9
136.22	4.52	-0.9	303.97	22JA	-0.6	-3.4	-2.1	0.9
132.36	4.58	-1.0	314.04	1FE	0.9	-3.4	-2.1	0.9
128.54	4.45	-0.9	324.05	11FE	3.2	-3.5	-2.0	1.0
125.44	4.18	-0.7	334.01	21FE	1.6	-3.5	-2.0	1.0
123.51	3.82	-0.5	343.91	3MR	0.9	-3.6	-2.0	1.0
122.87	3.42	-0.2	353.75	13MR	0.5	-3.6	-1.9	1.0
123.46	3.02	0.0	3.54	23MR	0.2	-3.7	-1.8	1.0
125.12	2.65	0.2	13.27	2AP	-0.4	-3.8	-1.8	1.0
127.67	2.31	0.4	22.94	12AP	-1.2	-3.8	-1.7	1.0
130.94	1.99	0.6	32.58	22AP	-1.8	-3.9	-1.6	1.0
134.80	1.71	0.7	42.17	2MY	-0.9	-4.0	-1.6	1.0
139.14	1.45	0.9	51.73	12MY	0.1	-4.1	-1.5	0.9
143.89	1.22	1.0	61.28	22MY	0.8	-4.2	-1.5	0.9
148.97	1.00	1.1	70.80	1JN	1.5	-4.2	-1.4	0.8
154.34	0.80	1.1	80.32	11JN	2.5	-4.1	-1.4	0.7
159.97	0.61	1.2	89.85	21JN	3.2	-3.7	-1.3	0.7
165.83	0.43	1.3	99.39	1JL	1.5	-3.1	-1.3	0.6
171.88	0.27	1.3	108.95	11JL	0.1	-3.4	-1.3	0.5
178.13	0.11	1.4	118.55	21JL	-1.1	-3.9	-1.3	0.5
184.55	-0.03	1.4	128.18	31JL	-1.5	-4.2	-1.2	0.5
191.14	-0.17	1.4	137.86	10AU	-1.0	-4.2	-1.2	0.5
197.88	-0.30	1.4	147.59	20AU	-0.4	-4.2	-1.2	0.6
204.77	-0.43	1.4	157.37	30AU	-0.1	-4.1	-1.3	0.6
211.80	-0.54	1.4	167.21	9SE	0.1	-4.1	-1.3	0.7
218.96	-0.65	1.4	177.12	19SE	0.2	-4.0	-1.3	0.7
226.24	-0.74	1.4	187.07	29SE	0.7	-3.9	-1.3	0.8
233.64	-0.83	1.4	197.09	9OC	2.8	-3.8	-1.4	0.8
241.14	-0.91	1.4	207.15	19OC	0.8	-3.7	-1.4	0.9
248.74	-0.97	1.4	217.26	29OC	-0.4	-3.7	-1.5	0.9
256.42	-1.03	1.4	227.41	8NO	-0.5	-3.6	-1.5	0.9
264.17	-1.07	1.4	237.59	18NO	-0.5	-3.6	-1.6	1.0
271.98	-1.10	1.4	247.79	28NO	-0.6	-3.5	-1.6	1.0
279.84	-1.12	1.4	257.99	8DE	-0.8	-3.5	-1.7	1.0
287.72	-1.13	1.4	268.20	18DE	-0.9	-3.4	-1.8	1.0
295.62	-1.12	1.4	278.38	28DE	-0.9	-3.4	-1.8	1.0
				403				
303.52	-1.11	1.4	288.55	7JA	-0.5	-3.4	-1.9	1.0
311.41	-1.08	1.4	298.68	17JA	1.0	-3.3	-2.0	1.0
319.27	-1.04	1.4	308.77	27JA	2.8	-3.3	-2.0	1.0
327.09	-0.99	1.4	318.81	6FE	1.2	-3.3	-2.0	1.0
334.87	-0.93	1.4	328.80	16FE	0.6	-3.3	-2.0	1.1
342.58	-0.86	1.4	338.73	26FE	0.3	-3.3	-2.0	1.1
350.22	-0.78	1.4	348.60	8MR	0.1	-3.3	-2.0	1.1
357.79	-0.70	1.4	358.42	18MR	-0.4	-3.4	-2.0	1.2
5.28	-0.61	1.5	8.17	28MR	-1.2	-3.4	-1.9	1.2
12.69	-0.52	1.5	17.88	7AP	-1.7	-3.4	-1.9	1.2
20.00	-0.42	1.6	27.53	17AP	-0.9	-3.5	-1.8	1.2
27.23	-0.32	1.6	37.14	27AP	0.1	-3.5	-1.8	1.2
34.37	-0.22	1.6	46.72	7MY	1.0	-3.5	-1.7	1.1
41.42	-0.12	1.7	56.28	17MY	2.0	-3.4	-1.6	1.1
48.38	-0.01	1.7	65.81	27MY	3.3	-3.4	-1.6	1.0
55.26	0.10	1.7	75.33	6JN	2.6	-3.4	-1.5	1.0
62.05	0.20	1.8	84.85	16JN	1.2	-3.4	-1.4	0.9
68.76	0.31	1.8	94.38	26JN	0.0	-3.3	-1.4	0.9
75.39	0.42	1.8	103.94	6JL	-1.1	-3.3	-1.3	0.8
81.94	0.53	1.8	113.51	16JL	-1.6	-3.3	-1.3	0.7
88.40	0.64	1.8	123.12	26JL	-0.9	-3.3	-1.3	0.7
94.79	0.75	1.8	132.78	5AU	-0.3	-3.3	-1.3	0.6
101.09	0.86	1.8	142.48	15AU	0.0	-3.4	-1.2	0.6
107.31	0.97	1.8	152.23	25AU	0.2	-3.4	-1.2	0.7
113.44	1.08	1.8	162.05	4SE	0.4	-3.4	-1.2	0.7
119.47	1.20	1.7	171.92	14SE	1.0	-3.5	-1.2	0.7
125.39	1.31	1.7	181.84	24SE	3.2	-3.5	-1.2	0.8
131.20	1.44	1.6	191.83	4OC	0.6	-3.6	-1.2	0.9
136.87	1.57	1.6	201.87	14OC	-0.5	-3.6	-1.3	0.9
142.38	1.70	1.5	211.95	24OC	-0.7	-3.7	-1.3	0.9
147.71	1.84	1.4	222.09	3NO	-0.7	-3.8	-1.3	1.0
152.82	1.99	1.3	232.25	13NO	-0.7	-3.9	-1.4	1.0
157.66	2.15	1.1	242.44	23NO	-0.7	-4.0	-1.4	1.1
162.18	2.32	1.0	252.64	3DE	-0.7	-4.1	-1.5	1.1
166.30	2.51	0.8	262.85	13DE	-0.7	-4.2	-1.5	1.1
169.93	2.70	0.6	273.04	23DE	-0.4	-4.3	-1.6	1.1

Left table (404 / 405)

♂ LONG	LAT	MAG	☉ LONG	16.00UT 404	☿	♀	♃	♄
172.94	2.91	0.4	283.22	2JA	1.2	-4.4	-1.7	1.1
175.18	3.14	0.1	293.37	12JA	2.4	-4.4	-1.7	1.1
176.50	3.36	-0.1	303.48	22JA	0.8	-4.2	-1.8	1.1
176.69	3.56	-0.4	313.55	1FE	0.4	-3.8	-1.9	1.1
175.62	3.73	-0.7	323.57	11FE	0.2	-3.3	-1.9	1.2
173.30	3.80	-0.9	333.53	21FE	-0.0	-3.7	-2.0	1.2
169.96	3.75	-1.2	343.43	2MR	-0.5	-4.1	-2.0	1.3
166.15	3.54	-1.2	353.28	12MR	-1.2	-4.2	-2.0	1.3
162.59	3.19	-1.1	3.06	22MR	-1.6	-4.2	-2.0	1.3
159.90	2.75	-0.9	12.80	1AP	-0.9	-4.2	-2.0	1.3
158.46	2.28	-0.7	22.48	11AP	0.2	-4.1	-2.0	1.4
158.33	1.82	-0.5	32.11	21AP	1.3	-4.0	-2.0	1.4
159.43	1.39	-0.3	41.71	1MY	2.6	-3.9	-1.9	1.3
161.59	1.01	-0.1	51.27	11MY	3.5	-3.8	-1.9	1.3
164.62	0.67	0.1	60.81	21MY	2.0	-3.7	-1.8	1.3
168.38	0.38	0.2	70.34	31MY	0.9	-3.6	-1.7	1.3
172.74	0.11	0.4	79.86	10JN	-0.1	-3.5	-1.7	1.2
177.60	-0.13	0.5	89.39	20JN	-1.1	-3.5	-1.6	1.1
182.87	-0.33	0.6	98.93	30JN	-1.7	-3.5	-1.6	1.1
188.51	-0.52	0.7	108.49	10JL	-0.9	-3.4	-1.5	1.0
194.46	-0.69	0.8	118.08	20JL	-0.3	-3.4	-1.4	0.9
200.68	-0.83	0.8	127.72	30JL	0.1	-3.4	-1.4	0.9
207.14	-0.96	0.9	137.39	9AU	0.4	-3.4	-1.4	0.8
213.82	-1.07	0.9	147.12	19AU	0.6	-3.4	-1.3	0.8
220.70	-1.16	1.0	156.90	29AU	1.4	-3.4	-1.3	0.8
227.75	-1.24	1.0	166.74	8SE	3.3	-3.4	-1.3	0.8
234.96	-1.30	1.1	176.63	18SE	0.5	-3.4	-1.2	0.9
242.30	-1.35	1.1	186.59	28SE	-0.7	-3.4	-1.2	0.9
249.77	-1.38	1.1	196.60	8OC	-0.8	-3.4	-1.2	1.0
257.34	-1.39	1.2	206.66	18OC	-0.8	-3.4	-1.2	1.0
264.99	-1.39	1.2	216.77	28OC	-0.8	-3.4	-1.3	1.1
272.72	-1.37	1.2	226.92	7NO	-0.6	-3.5	-1.3	1.1
280.49	-1.34	1.3	237.10	17NO	-0.6	-3.5	-1.3	1.2
288.30	-1.30	1.3	247.29	27NO	-0.6	-3.5	-1.3	1.2
296.12	-1.24	1.3	257.50	7DE	-0.3	-3.5	-1.4	1.2
303.95	-1.17	1.4	267.70	17DE	1.4	-3.4	-1.4	1.3
311.76	-1.10	1.4	277.89	27DE	2.0	-3.4	-1.5	1.3

405

♂ LONG	LAT	MAG	☉ LONG	16.00UT 405	☿	♀	♃	♄
319.54	-1.01	1.4	288.05	6JA	0.6	-3.4	-1.5	1.3
327.28	-0.92	1.4	298.19	16JA	0.2	-3.4	-1.6	1.3
334.97	-0.82	1.5	308.28	26JA	0.1	-3.4	-1.7	1.3
342.60	-0.72	1.5	318.32	5FE	-0.1	-3.3	-1.7	1.3
350.15	-0.62	1.5	328.31	15FE	-0.5	-3.3	-1.8	1.3
357.64	-0.51	1.6	338.25	25FE	-1.2	-3.3	-1.9	1.3
5.04	-0.40	1.6	348.12	7MR	-1.5	-3.3	-1.9	1.3
12.37	-0.30	1.6	357.94	17MR	-0.9	-3.4	-2.0	1.3
19.61	-0.19	1.6	7.70	27MR	0.3	-3.4	-2.0	1.3
26.77	-0.08	1.7	17.41	6AP	1.7	-3.4	-2.1	1.3
33.85	0.02	1.7	27.07	16AP	3.4	-3.4	-2.1	1.3
40.84	0.12	1.7	36.68	26AP	2.6	-3.5	-2.1	1.2
47.76	0.22	1.7	46.26	6MY	1.5	-3.5	-2.1	1.2
54.60	0.32	1.7	55.82	16MY	0.7	-3.6	-2.0	1.2
61.38	0.42	1.8	65.35	26MY	-0.1	-3.6	-2.0	1.1
68.08	0.51	1.8	74.87	5JN	-1.1	-3.7	-2.0	1.1
74.73	0.60	1.9	84.40	15JN	-1.8	-3.8	-1.9	1.0
81.31	0.68	1.9	93.93	25JN	-0.9	-3.9	-1.8	1.0
87.83	0.76	1.9	103.48	5JL	-0.2	-4.0	-1.8	0.9
94.31	0.84	1.9	113.05	15JL	0.2	-4.1	-1.7	0.9
100.73	0.92	2.0	122.66	25JL	0.5	-4.2	-1.7	0.8
107.11	1.00	2.0	132.31	4AU	0.9	-4.2	-1.6	0.8
113.44	1.07	2.0	142.01	14AU	1.8	-4.2	-1.5	0.8
119.73	1.14	2.0	151.76	24AU	3.0	-4.1	-1.5	0.7
125.97	1.21	2.0	161.57	3SE	0.4	-3.8	-1.4	0.7
132.17	1.28	1.9	171.44	13SE	-0.8	-3.3	-1.4	0.8
138.32	1.34	1.9	181.36	23SE	-1.0	-3.7	-1.4	0.9
144.41	1.41	1.9	191.34	3OC	-0.9	-4.1	-1.3	0.9
150.45	1.47	1.8	201.38	13OC	-0.7	-4.3	-1.3	1.0
156.42	1.53	1.8	211.47	23OC	-0.5	-4.4	-1.3	1.1
162.32	1.60	1.7	221.59	2NO	-0.4	-4.3	-1.3	1.1
168.14	1.66	1.6	231.76	12NO	-0.4	-4.2	-1.3	1.2
173.84	1.72	1.5	241.94	22NO	-0.1	-4.1	-1.3	1.2
179.43	1.77	1.4	252.14	2DE	1.6	-4.0	-1.3	1.2
184.86	1.83	1.3	262.35	12DE	1.6	-3.9	-1.4	1.2
190.12	1.88	1.1	272.55	22DE	0.3	-3.8	-1.4	1.2

Right table (406 / 407)

♂ LONG	LAT	MAG	☉ LONG	16.00UT 406	☿	♀	♃	♄
195.16	1.93	1.0	282.73	1JA	0.0	-3.7	-1.4	1.2
199.92	1.98	0.8	292.89	11JA	-0.1	-3.7	-1.5	1.2
204.35	2.01	0.6	303.00	21JA	-0.2	-3.6	-1.5	1.2
208.37	2.04	0.4	313.07	31JA	-0.6	-3.5	-1.6	1.2
211.87	2.04	0.1	323.09	10FE	-1.2	-3.5	-1.6	1.1
214.71	2.02	-0.1	333.05	20FE	-1.4	-3.4	-1.7	1.1
216.74	1.97	-0.4	342.96	2MR	-0.8	-3.4	-1.8	1.0
217.76	1.86	-0.7	352.81	12MR	0.5	-3.3	-1.8	1.0
217.63	1.68	-1.0	2.60	22MR	2.1	-3.3	-1.9	1.0
216.24	1.41	-1.3	12.33	1AP	3.5	-3.3	-2.0	1.0
213.68	1.04	-1.6	22.02	11AP	2.0	-3.3	-2.0	1.0
210.34	0.58	-1.8	31.65	21AP	1.1	-3.3	-2.1	1.0
206.89	0.08	-1.7	41.25	1MY	0.5	-3.3	-2.1	1.0
204.03	-0.43	-1.6	50.82	11MY	-0.2	-3.3	-2.2	1.0
202.31	-0.88	-1.4	60.36	21MY	-1.1	-3.3	-2.2	0.9
201.94	-1.26	-1.2	69.88	31MY	-1.8	-3.4	-2.2	0.9
202.92	-1.56	-1.0	79.41	10JN	-0.9	-3.4	-2.2	0.9
205.10	-1.80	-0.8	88.93	20JN	-0.2	-3.5	-2.1	0.8
208.28	-1.98	-0.6	98.47	30JN	0.3	-3.5	-2.1	0.8
212.31	-2.11	-0.4	108.03	10JL	0.7	-3.5	-2.1	0.7
217.02	-2.20	-0.3	117.62	20JL	1.2	-3.4	-2.0	0.7
222.29	-2.26	-0.1	127.25	30JL	2.3	-3.4	-1.9	0.6
228.03	-2.29	0.0	136.92	9AU	2.6	-3.4	-1.9	0.6
234.16	-2.29	0.1	146.64	19AU	0.3	-3.4	-1.8	0.5
240.60	-2.26	0.2	156.42	29AU	-0.9	-3.3	-1.7	0.5
247.31	-2.21	0.3	166.25	8SE	-1.1	-3.3	-1.7	0.4
254.24	-2.15	0.4	176.14	18SE	-1.0	-3.3	-1.6	0.4
261.35	-2.06	0.5	186.10	28SE	-0.6	-3.3	-1.6	0.5
268.60	-1.96	0.6	196.11	8OC	-0.4	-3.3	-1.5	0.6
275.97	-1.85	0.7	206.17	18OC	-0.3	-3.4	-1.5	0.6
283.41	-1.73	0.8	216.27	28OC	-0.2	-3.4	-1.5	0.7
290.92	-1.59	0.9	226.42	7NO	0.1	-3.4	-1.4	0.8
298.47	-1.45	1.0	236.59	17NO	1.9	-3.4	-1.4	0.8
306.03	-1.31	1.0	246.79	27NO	1.4	-3.5	-1.4	0.9
313.58	-1.16	1.1	257.00	7DE	0.1	-3.5	-1.4	0.9
321.12	-1.01	1.2	267.20	17DE	-0.1	-3.6	-1.4	0.9
328.63	-0.86	1.3	277.39	27DE	-0.2	-3.6	-1.4	1.0

407

♂ LONG	LAT	MAG	☉ LONG	16.00UT 407	☿	♀	♃	♄
336.10	-0.72	1.3	287.56	6JA	-0.3	-3.7	-1.4	1.0
343.51	-0.58	1.4	297.70	16JA	-0.6	-3.8	-1.5	1.0
350.86	-0.44	1.5	307.80	26JA	-1.2	-3.9	-1.5	0.9
358.15	-0.31	1.5	317.84	5FE	-1.3	-4.0	-1.5	0.9
5.37	-0.18	1.6	327.83	15FE	-0.8	-4.0	-1.6	0.9
12.52	-0.06	1.6	337.77	25FE	0.6	-4.1	-1.6	0.8
19.59	0.06	1.7	347.65	7MR	2.6	-4.2	-1.7	0.8
26.60	0.17	1.7	357.47	17MR	2.6	-4.2	-1.8	0.7
33.53	0.27	1.8	7.24	27MR	1.4	-4.2	-1.8	0.7
40.40	0.37	1.8	16.94	6AP	0.8	-4.0	-1.9	0.8
47.20	0.46	1.8	26.60	16AP	0.4	-3.4	-2.0	0.8
53.94	0.55	1.9	36.22	26AP	-0.3	-2.9	-2.0	0.8
60.62	0.63	1.9	45.80	6MY	-1.1	-3.7	-2.1	0.8
67.25	0.70	1.9	55.35	16MY	-1.8	-4.1	-2.2	0.7
73.83	0.77	1.9	64.89	26MY	-0.9	-4.2	-2.2	0.7
80.37	0.84	1.9	74.41	5JN	-0.1	-4.2	-2.3	0.7
86.86	0.90	1.9	83.93	15JN	0.5	-4.1	-2.3	0.7
93.33	0.95	1.9	93.47	25JN	0.9	-4.0	-2.3	0.6
99.76	1.00	1.9	103.01	5JL	1.6	-3.9	-2.3	0.6
106.17	1.05	2.0	112.59	15JL	2.9	-3.8	-2.3	0.5
112.55	1.09	2.0	122.20	25JL	2.2	-3.8	-2.3	0.5
118.92	1.13	2.0	131.84	4AU	0.2	-3.7	-2.2	0.4
125.28	1.16	2.0	141.54	14AU	-0.9	-3.6	-2.2	0.4
131.62	1.19	2.0	151.29	24AU	-1.3	-3.6	-2.1	0.3
137.95	1.22	2.0	161.09	3SE	-1.0	-3.5	-2.1	0.3
144.28	1.24	2.0	170.95	13SE	-0.5	-3.5	-2.0	0.2
150.60	1.26	2.0	180.88	23SE	-0.3	-3.5	-1.9	0.2
156.92	1.27	2.0	190.86	3OC	-0.1	-3.4	-1.9	0.2
163.23	1.28	1.9	200.89	13OC	-0.1	-3.4	-1.8	0.3
169.53	1.29	1.9	210.98	23OC	0.3	-3.4	-1.7	0.3
175.82	1.29	1.9	221.10	2NO	2.2	-3.4	-1.7	0.4
182.10	1.28	1.8	231.26	12NO	1.1	-3.4	-1.6	0.5
188.36	1.27	1.7	241.45	22NO	-0.1	-3.4	-1.6	0.5
194.60	1.25	1.7	251.65	2DE	-0.3	-3.4	-1.6	0.6
200.81	1.22	1.6	261.85	12DE	-0.3	-3.4	-1.5	0.6
206.99	1.18	1.5	272.05	22DE	-0.4	-3.4	-1.5	0.7

♂ LONG	♂ LAT	♂ MAG	⊙ LONG	16.00UT	☿	♀	♃	♄
				408	MAGNITUDES			
213.12	1.13	1.4	282.23	1JA	-0.7	-3.4	-1.5	0.7
219.20	1.07	1.2	292.39	11JA	-1.1	-3.4	-1.5	0.7
225.20	0.99	1.1	302.51	21JA	-1.1	-3.4	-1.5	0.7
231.12	0.89	0.9	312.58	31JA	-0.7	-3.5	-1.5	0.7
236.93	0.77	0.8	322.60	10FE	0.7	-3.5	-1.5	0.7
242.61	0.62	0.6	332.57	20FE	3.0	-3.5	-1.6	0.7
248.11	0.44	0.4	342.48	1MR	2.0	-3.4	-1.6	0.6
253.41	0.22	0.2	352.33	11MR	1.1	-3.4	-1.6	0.6
258.44	-0.04	-0.0	2.12	21MR	0.6	-3.4	-1.7	0.6
263.12	-0.36	-0.2	11.86	31MR	0.2	-3.4	-1.7	0.5
267.37	-0.75	-0.5	21.54	10AP	-0.3	-3.3	-1.8	0.5
271.04	-1.21	-0.8	31.18	20AP	-1.1	-3.3	-1.9	0.5
274.01	-1.75	-1.1	40.78	30AP	-1.8	-3.3	-1.9	0.6
276.09	-2.38	-1.4	50.35	10MY	-0.9	-3.3	-2.0	0.6
277.07	-3.10	-1.7	59.89	20MY	-0.0	-3.3	-2.1	0.6
276.84	-3.88	-2.0	69.42	30MY	0.6	-3.3	-2.1	0.6
275.40	-4.64	-2.3	78.94	9JN	1.2	-3.4	-2.2	0.5
273.01	-5.30	-2.5	88.47	19JN	2.1	-3.4	-2.3	0.5
270.29	-5.74	-2.5	98.00	29JN	3.3	-3.4	-2.3	0.5
267.92	-5.92	-2.4	107.56	9JL	1.8	-3.5	-2.4	0.5
266.55	-5.84	-2.1	117.16	19JL	0.2	-3.5	-2.4	0.4
266.50	-5.57	-1.9	126.78	29JL	-1.0	-3.6	-2.4	0.4
267.77	-5.19	-1.6	136.45	8AU	-1.4	-3.7	-2.4	0.3
270.26	-4.76	-1.4	146.17	18AU	-1.0	-3.7	-2.4	0.3
273.76	-4.30	-1.1	155.94	28AU	-0.5	-3.8	-2.4	0.2
278.06	-3.85	-0.9	165.78	7SE	-0.2	-3.9	-2.3	0.1
283.01	-3.42	-0.7	175.67	17SE	-0.0	-4.0	-2.3	0.1
288.46	-3.00	-0.5	185.62	27SE	0.1	-4.2	-2.2	0.0
294.29	-2.61	-0.3	195.62	7OC	0.5	-4.3	-2.1	-0.0
300.41	-2.25	-0.1	205.68	17OC	2.5	-4.3	-2.1	-0.0
306.75	-1.91	0.1	215.78	27OC	0.9	-4.4	-2.0	0.1
313.26	-1.60	0.3	225.93	6NO	-0.3	-4.2	-1.9	0.1
319.89	-1.31	0.4	236.10	16NO	-0.4	-3.8	-1.9	0.2
326.60	-1.04	0.6	246.29	26NO	-0.4	-2.9	-1.8	0.3
333.37	-0.80	0.7	256.50	6DE	-0.5	-3.7	-1.8	0.3
340.16	-0.58	0.9	266.70	16DE	-0.8	-4.2	-1.7	0.4
346.97	-0.38	1.0	276.89	26DE	-1.0	-4.4	-1.7	0.4
				409				
353.78	-0.19	1.1	287.06	5JA	-1.0	-4.4	-1.6	0.5
0.58	-0.03	1.2	297.20	15JA	-0.6	-4.3	-1.6	0.5
7.35	0.12	1.3	307.30	25JA	0.9	-4.2	-1.6	0.5
14.09	0.26	1.4	317.35	4FE	3.1	-4.1	-1.6	0.5
20.81	0.38	1.5	327.35	14FE	1.5	-4.0	-1.6	0.5
27.48	0.49	1.6	337.29	24FE	0.8	-3.9	-1.6	0.5
34.12	0.59	1.7	347.17	6MR	0.4	-3.8	-1.6	0.5
40.72	0.67	1.7	357.00	16MR	0.1	-3.7	-1.6	0.5
47.28	0.75	1.8	6.76	26MR	-0.4	-3.6	-1.6	0.4
53.81	0.82	1.9	16.48	5AP	-1.1	-3.5	-1.6	0.4
60.30	0.88	1.9	26.14	15AP	-1.8	-3.5	-1.7	0.3
66.76	0.94	1.9	35.76	25AP	-0.9	-3.4	-1.7	0.4
73.20	0.98	2.0	45.35	5MY	0.1	-3.4	-1.8	0.4
79.61	1.03	2.0	54.90	15MY	0.8	-3.4	-1.8	0.4
85.99	1.06	2.0	64.43	25MY	1.6	-3.3	-1.9	0.4
92.37	1.09	2.0	73.96	4JN	2.8	-3.3	-2.0	0.4
98.73	1.11	2.0	83.48	14JN	3.0	-3.3	-2.0	0.4
105.08	1.13	2.0	93.01	24JN	1.4	-3.3	-2.1	0.4
111.44	1.14	2.0	102.56	4JL	0.1	-3.3	-2.2	0.4
117.79	1.15	2.0	112.13	14JL	-1.1	-3.3	-2.2	0.4
124.15	1.16	2.0	121.73	24JL	-1.5	-3.4	-2.3	0.3
130.52	1.16	2.0	131.38	3AU	-1.0	-3.4	-2.3	0.3
136.90	1.15	2.0	141.07	13AU	-0.4	-3.4	-2.4	0.2
143.31	1.14	2.0	150.82	23AU	-0.1	-3.5	-2.4	0.2
149.74	1.13	2.0	160.62	2SE	0.1	-3.5	-2.5	0.1
156.18	1.11	2.0	170.48	12SE	0.3	-3.5	-2.5	0.1
162.66	1.09	2.0	180.39	22SE	0.8	-3.5	-2.4	-0.0
169.17	1.06	2.0	190.37	2OC	2.9	-3.4	-2.4	-0.1
175.71	1.02	2.0	200.40	12OC	0.8	-3.4	-2.4	-0.1
182.28	0.98	1.9	210.48	22OC	-0.4	-3.4	-2.3	-0.2
188.88	0.93	1.9	220.61	1NO	-0.6	-3.4	-2.3	-0.1
195.51	0.88	1.9	230.76	11NO	-0.6	-3.3	-2.2	-0.1
202.19	0.81	1.8	240.95	21NO	-0.6	-3.3	-2.1	-0.0
208.89	0.74	1.8	251.15	1DE	-0.8	-3.3	-2.0	0.1
215.63	0.66	1.7	261.35	11DE	-0.8	-3.3	-2.0	0.1
222.39	0.57	1.6	271.55	21DE	-0.8	-3.3	-1.9	0.2
229.19	0.47	1.5	281.74	31DE	-0.5	-3.4	-1.8	0.3

♂ LONG	♂ LAT	♂ MAG	⊙ LONG	16.00UT	☿	♀	♃	♄
				410	MAGNITUDES			
236.02	0.35	1.5	291.89	10JA	1.0	-3.4	-1.8	0.3
242.87	0.23	1.4	302.01	20JA	2.7	-3.4	-1.7	0.3
249.75	0.08	1.3	312.09	30JA	1.1	-3.4	-1.7	0.4
256.64	-0.07	1.2	322.11	9FE	0.5	-3.5	-1.6	0.4
263.55	-0.25	1.1	332.08	19FE	0.3	-3.5	-1.6	0.4
270.46	-0.44	1.0	342.00	1MR	0.0	-3.6	-1.6	0.4
277.37	-0.65	0.8	351.85	11MR	-0.4	-3.6	-1.6	0.4
284.28	-0.87	0.7	1.65	21MR	-1.1	-3.7	-1.6	0.4
291.14	-1.12	0.6	11.39	31MR	-1.7	-3.8	-1.6	0.3
297.96	-1.38	0.4	21.08	10AP	-0.9	-3.8	-1.6	0.3
304.71	-1.66	0.3	30.72	20AP	0.1	-3.9	-1.6	0.3
311.35	-1.96	0.1	40.32	30AP	1.1	-4.0	-1.6	0.2
317.86	-2.27	-0.0	49.89	10MY	2.2	-4.1	-1.6	0.3
324.17	-2.60	-0.2	59.44	20MY	3.6	-4.2	-1.7	0.3
330.22	-2.94	-0.4	68.97	30MY	2.4	-4.2	-1.7	0.3
335.94	-3.30	-0.5	78.49	9JN	1.1	-4.1	-1.7	0.3
341.24	-3.66	-0.7	88.01	19JN	0.0	-3.7	-1.8	0.3
345.96	-4.03	-0.9	97.55	29JN	-1.1	-3.0	-1.9	0.3
349.98	-4.39	-1.1	107.10	9JL	-1.6	-3.4	-1.9	0.3
353.09	-4.74	-1.4	116.69	19JL	-1.0	-3.9	-2.0	0.3
355.08	-5.05	-1.6	126.32	29JL	-0.3	-4.2	-2.0	0.3
355.75	-5.28	-1.9	135.98	8AU	0.0	-4.2	-2.1	0.3
354.98	-5.38	-2.1	145.70	18AU	0.3	-4.2	-2.2	0.2
352.90	-5.27	-2.3	155.47	28AU	0.5	-4.1	-2.2	0.2
349.96	-4.92	-2.4	165.30	7SE	1.1	-4.0	-2.3	0.1
346.88	-4.35	-2.3	175.19	17SE	3.2	-4.0	-2.3	0.1
344.46	-3.64	-2.0	185.14	27SE	0.7	-3.9	-2.4	-0.0
343.21	-2.90	-1.7	195.14	7OC	-0.6	-3.8	-2.4	-0.1
343.27	-2.20	-1.4	205.19	17OC	-0.7	-3.7	-2.4	-0.2
344.60	-1.58	-1.0	215.30	27OC	-0.7	-3.7	-2.4	-0.2
347.00	-1.06	-0.7	225.43	6NO	-0.8	-3.6	-2.4	-0.3
350.25	-0.62	-0.4	235.61	16NO	-0.7	-3.6	-2.3	-0.2
354.18	-0.26	-0.2	245.80	26NO	-0.7	-3.5	-2.3	-0.2
358.63	0.03	0.1	256.00	6DE	-0.7	-3.5	-2.2	-0.1
3.48	0.28	0.3	266.21	16DE	-0.4	-3.4	-2.1	-0.0
8.64	0.48	0.5	276.40	26DE	1.2	-3.4	-2.1	0.1
				411				
14.04	0.65	0.7	286.57	5JA	2.3	-3.4	-2.0	0.1
19.63	0.78	0.9	296.71	15JA	0.7	-3.3	-1.9	0.2
25.35	0.90	1.0	306.81	25JA	0.3	-3.3	-1.9	0.2
31.18	0.99	1.2	316.86	4FE	0.2	-3.3	-1.8	0.3
37.09	1.06	1.3	326.86	14FE	-0.1	-3.3	-1.7	0.3
43.06	1.12	1.4	336.81	24FE	-0.5	-3.3	-1.7	0.3
49.08	1.17	1.5	346.69	6MR	-1.1	-3.3	-1.6	0.3
55.14	1.21	1.6	356.52	16MR	-1.6	-3.4	-1.6	0.3
61.22	1.24	1.7	6.29	26MR	-0.9	-3.4	-1.6	0.3
67.32	1.26	1.8	16.00	5AP	0.2	-3.4	-1.5	0.3
73.45	1.28	1.8	25.67	15AP	1.4	-3.5	-1.5	0.3
79.59	1.28	1.9	35.29	25AP	2.9	-3.5	-1.5	0.2
85.74	1.28	1.9	44.87	5MY	3.2	-3.5	-1.5	0.2
91.92	1.28	2.0	54.43	15MY	1.8	-3.4	-1.5	0.2
98.10	1.27	2.0	63.97	25MY	0.9	-3.4	-1.5	0.2
104.31	1.26	2.0	73.49	4JN	-0.1	-3.4	-1.5	0.2
110.54	1.24	2.0	83.02	14JN	-1.1	-3.4	-1.5	0.3
116.79	1.22	2.0	92.54	24JN	-1.7	-3.3	-1.6	0.3
123.08	1.19	2.0	102.09	4JL	-0.9	-3.3	-1.6	0.3
129.39	1.16	2.0	111.66	14JL	-0.3	-3.3	-1.7	0.3
135.74	1.13	2.0	121.26	24JL	0.1	-3.3	-1.7	0.3
142.13	1.09	2.0	130.91	3AU	0.4	-3.3	-1.8	0.3
148.56	1.04	2.0	140.60	13AU	0.7	-3.4	-1.8	0.3
155.04	1.00	1.9	150.34	23AU	1.5	-3.4	-1.9	0.3
161.56	0.94	1.9	160.14	2SE	3.2	-3.4	-1.9	0.2
168.14	0.89	1.9	170.00	12SE	0.5	-3.5	-2.0	0.2
174.78	0.83	1.9	179.91	22SE	-0.7	-3.5	-2.1	0.1
181.47	0.76	1.9	189.88	2OC	-0.9	-3.6	-2.1	0.0
188.22	0.69	1.9	199.92	12OC	-0.9	-3.6	-2.2	-0.0
195.03	0.61	1.8	209.99	22OC	-0.8	-3.7	-2.2	-0.1
201.91	0.53	1.8	220.11	1NO	-0.6	-3.8	-2.3	-0.2
208.85	0.44	1.8	230.27	11NO	-0.5	-3.9	-2.3	-0.3
215.85	0.35	1.8	240.45	21NO	-0.5	-4.0	-2.3	-0.3
222.92	0.25	1.7	250.65	1DE	-0.2	-4.1	-2.3	-0.3
230.05	0.14	1.7	260.86	11DE	1.4	-4.2	-2.3	-0.2
237.24	0.03	1.7	271.06	21DE	1.9	-4.3	-2.2	-0.1
244.49	-0.09	1.6	281.24	31DE	0.5	-4.4	-2.2	-0.0

♂ LONG	♂ LAT	♂ MAG	⊙ LONG	16.00UT	☿	♀	♃	♄
				412	MAGNITUDES			
251.80	-0.22	1.6	291.40	10JA	0.1	-4.4	-2.1	0.0
259.16	-0.35	1.5	301.52	20JA	0.0	-4.2	-2.0	0.1
266.57	-0.48	1.4	311.60	30JA	-0.2	-3.8	-2.0	0.1
274.02	-0.62	1.4	321.63	9FE	-0.5	-3.3	-1.9	0.2
281.50	-0.77	1.3	331.60	19FE	-1.1	-3.7	-1.8	0.2
289.02	-0.91	1.2	341.51	29FE	-1.5	-4.1	-1.8	0.3
296.55	-1.06	1.2	351.37	10MR	-0.9	-4.3	-1.7	0.3
304.09	-1.20	1.1	1.17	20MR	0.3	-4.3	-1.6	0.3
311.62	-1.34	1.0	10.91	30MR	1.8	-4.2	-1.6	0.3
319.13	-1.48	1.0	20.60	9AP	3.7	-4.1	-1.5	0.3
326.61	-1.60	0.9	30.25	19AP	2.4	-4.0	-1.5	0.3
334.04	-1.72	0.8	39.85	29AP	1.4	-3.9	-1.5	0.3
341.39	-1.83	0.8	49.42	9MY	0.7	-3.8	-1.4	0.2
348.64	-1.92	0.7	58.97	19MY	-0.1	-3.7	-1.4	0.2
355.77	-2.00	0.6	68.50	29MY	-1.1	-3.6	-1.4	0.2
2.76	-2.06	0.5	78.02	8JN	-1.8	-3.5	-1.4	0.2
9.57	-2.10	0.5	87.55	18JN	-0.9	-3.5	-1.4	0.3
16.16	-2.12	0.4	97.09	28JN	-0.2	-3.4	-1.4	0.3
22.49	-2.13	0.3	106.64	8JL	0.3	-3.4	-1.4	0.3
28.52	-2.11	0.2	116.23	18JL	0.6	-3.4	-1.4	0.3
34.18	-2.07	0.0	125.85	28JL	1.0	-3.4	-1.5	0.3
39.38	-2.00	-0.1	135.52	7AU	1.9	-3.4	-1.5	0.3
44.04	-1.90	-0.2	145.23	17AU	2.9	-3.4	-1.6	0.3
48.01	-1.76	-0.4	155.00	27AU	0.4	-3.3	-1.6	0.3
51.14	-1.58	-0.6	164.83	6SE	-0.8	-3.4	-1.7	0.3
53.24	-1.35	-0.8	174.71	16SE	-1.0	-3.4	-1.7	0.2
54.08	-1.06	-1.0	184.66	26SE	-1.0	-3.4	-1.8	0.2
53.52	-0.70	-1.2	194.65	6OC	-0.7	-3.4	-1.8	0.1
51.51	-0.27	-1.4	204.70	16OC	-0.4	-3.4	-1.9	0.1
48.29	0.21	-1.6	214.81	26OC	-0.4	-3.4	-2.0	-0.0
44.50	0.68	-1.7	224.94	5NO	-0.3	-3.5	-2.0	-0.1
40.92	1.10	-1.4	235.11	15NO	-0.1	-3.5	-2.1	-0.2
38.25	1.43	-1.2	245.31	25NO	1.7	-3.5	-2.1	-0.2
36.89	1.66	-0.8	255.51	5DE	1.6	-3.5	-2.2	-0.3
36.86	1.82	-0.5	265.71	15DE	0.2	-3.4	-2.2	-0.2
38.05	1.91	-0.2	275.91	25DE	-0.0	-3.4	-2.2	-0.2
				413				
40.24	1.95	0.1	286.08	4JA	-0.1	-3.4	-2.2	-0.1
43.23	1.97	0.3	296.22	14JA	-0.3	-3.4	-2.1	-0.0
46.85	1.96	0.5	306.32	24JA	-0.6	-3.4	-2.1	0.0
50.96	1.94	0.7	316.37	3FE	-1.2	-3.3	-2.0	0.1
55.46	1.91	0.9	326.38	13FE	-1.4	-3.3	-2.0	0.2
60.26	1.87	1.1	336.32	23FE	-0.8	-3.3	-1.9	0.2
65.31	1.83	1.2	346.21	5MR	0.5	-3.3	-1.8	0.3
70.54	1.79	1.4	356.04	15MR	2.3	-3.4	-1.8	0.3
75.94	1.74	1.5	5.81	25MR	3.2	-3.4	-1.7	0.3
81.46	1.69	1.6	15.53	4AP	1.8	-3.4	-1.6	0.3
87.10	1.63	1.7	25.20	14AP	1.0	-3.4	-1.6	0.3
92.83	1.58	1.7	34.82	24AP	0.5	-3.5	-1.5	0.3
98.64	1.52	1.8	44.41	4MY	-0.2	-3.5	-1.5	0.3
104.53	1.46	1.8	53.97	14MY	-1.1	-3.6	-1.4	0.3
110.48	1.40	1.9	63.51	24MY	-1.8	-3.6	-1.4	0.3
116.51	1.34	1.9	73.03	3JN	-0.9	-3.7	-1.4	0.3
122.59	1.27	1.9	82.55	13JN	-0.1	-3.8	-1.3	0.2
128.75	1.20	2.0	92.09	23JN	0.4	-3.9	-1.3	0.3
134.96	1.13	2.0	101.63	3JL	0.8	-4.0	-1.3	0.3
141.24	1.06	2.0	111.20	13JL	1.3	-4.1	-1.3	0.4
147.59	0.99	2.0	120.80	23JL	2.5	-4.2	-1.3	0.4
154.01	0.91	2.0	130.44	2AU	2.5	-4.2	-1.3	0.4
160.50	0.83	1.9	140.13	12AU	0.4	-4.2	-1.3	0.4
167.06	0.75	1.9	149.87	22AU	-0.9	-4.1	-1.4	0.4
173.70	0.67	1.9	159.67	1SE	-1.2	-3.7	-1.4	0.4
180.41	0.58	1.8	169.52	11SE	-1.1	-3.3	-1.4	0.4
187.21	0.49	1.8	179.44	21SE	-0.6	-3.7	-1.5	0.4
194.09	0.39	1.7	189.40	1OC	-0.3	-4.1	-1.5	0.3
201.05	0.30	1.7	199.43	11OC	-0.2	-4.3	-1.6	0.3
208.10	0.20	1.7	209.51	21OC	-0.2	-4.3	-1.6	0.2
215.22	0.10	1.6	219.62	31OC	0.1	-4.3	-1.7	0.1
222.43	-0.01	1.6	229.78	10NO	1.9	-4.2	-1.8	0.1
229.72	-0.11	1.6	239.96	20NO	1.3	-4.1	-1.8	-0.0
237.09	-0.22	1.6	250.16	30NO	0.0	-4.0	-1.9	-0.1
244.53	-0.33	1.6	260.37	10DE	-0.2	-3.9	-1.9	-0.1
252.03	-0.44	1.6	270.57	20DE	-0.2	-3.8	-2.0	-0.2
259.60	-0.55	1.5	280.75	30DE	-0.4	-3.7	-2.0	-0.1

♂ LONG	♂ LAT	♂ MAG	⊙ LONG	16.00UT	☿	♀	♃	♄
				414	MAGNITUDES			
267.23	-0.65	1.5	290.91	9JA	-0.7	-3.7	-2.1	-0.1
274.91	-0.76	1.5	301.03	19JA	-1.1	-3.6	-2.1	-0.0
282.62	-0.86	1.5	311.11	29JA	-1.2	-3.5	-2.1	0.0
290.37	-0.95	1.4	321.15	8FE	-0.8	-3.5	-2.1	0.1
298.13	-1.04	1.4	331.12	18FE	0.6	-3.4	-2.0	0.2
305.90	-1.12	1.4	341.04	28FE	2.7	-3.4	-2.0	0.2
313.67	-1.19	1.4	350.90	10MR	2.4	-3.4	-1.9	0.3
321.42	-1.25	1.3	0.70	20MR	1.3	-3.3	-1.9	0.3
329.14	-1.30	1.3	10.44	30MR	0.8	-3.3	-1.8	0.4
336.82	-1.33	1.3	20.14	9AP	0.3	-3.3	-1.7	0.4
344.44	-1.35	1.3	29.79	19AP	-0.3	-3.3	-1.7	0.4
352.00	-1.36	1.2	39.39	29AP	-1.1	-3.3	-1.6	0.4
359.47	-1.35	1.2	48.96	9MY	-1.8	-3.3	-1.5	0.4
6.85	-1.33	1.2	58.51	19MY	-0.9	-3.3	-1.5	0.4
14.12	-1.29	1.2	68.04	29MY	-0.1	-3.4	-1.4	0.4
21.28	-1.24	1.1	77.56	8JN	0.5	-3.4	-1.4	0.4
28.30	-1.17	1.1	87.09	18JN	1.0	-3.5	-1.3	0.4
35.17	-1.09	1.1	96.62	28JN	1.8	-3.5	-1.3	0.3
41.88	-0.99	1.0	106.18	8JL	3.1	-3.5	-1.3	0.4
48.41	-0.88	1.0	115.76	18JL	2.1	-3.4	-1.3	0.4
54.74	-0.75	0.9	125.38	28JL	0.3	-3.4	-1.2	0.5
60.84	-0.61	0.8	135.05	7AU	-1.0	-3.4	-1.2	0.5
66.68	-0.44	0.8	144.76	17AU	-1.3	-3.4	-1.2	0.5
72.21	-0.26	0.7	154.52	27AU	-1.0	-3.3	-1.2	0.5
77.38	-0.06	0.6	164.35	6SE	-0.5	-3.3	-1.2	0.5
82.12	0.17	0.5	174.23	16SE	-0.2	-3.3	-1.3	0.5
86.33	0.43	0.3	184.17	26SE	-0.1	-3.3	-1.3	0.5
89.90	0.73	0.1	194.17	6OC	-0.0	-3.3	-1.3	0.5
92.68	1.07	-0.0	204.21	16OC	0.4	-3.4	-1.4	0.4
94.48	1.45	-0.2	214.31	26OC	2.2	-3.4	-1.4	0.4
95.12	1.87	-0.5	224.45	5NO	1.1	-3.4	-1.5	0.3
94.41	2.33	-0.7	234.62	15NO	-0.2	-3.4	-1.5	0.3
92.33	2.79	-0.9	244.81	25NO	-0.3	-3.5	-1.6	0.2
89.05	3.20	-1.1	255.02	5DE	-0.4	-3.5	-1.7	0.1
85.13	3.50	-1.2	265.22	15DE	-0.5	-3.6	-1.7	0.1
81.30	3.65	-1.0	275.41	25DE	-0.7	-3.6	-1.8	-0.0
				415				
78.28	3.66	-0.8	285.59	4JA	-1.0	-3.7	-1.9	-0.1
76.48	3.55	-0.5	295.73	14JA	-1.1	-3.8	-1.9	-0.0
76.00	3.38	-0.2	305.83	24JA	-0.7	-3.9	-2.0	0.0
76.74	3.19	0.1	315.89	3FE	0.7	-4.0	-2.0	0.1
78.52	2.98	0.3	325.90	13FE	3.1	-4.0	-2.0	0.2
81.14	2.78	0.5	335.85	23FE	1.8	-4.1	-2.0	0.2
84.43	2.59	0.7	345.74	5MR	0.9	-4.2	-2.0	0.3
88.26	2.42	0.9	355.57	15MR	0.6	-4.3	-2.0	0.3
92.51	2.25	1.0	5.34	25MR	0.2	-4.2	-2.0	0.4
97.12	2.09	1.2	15.06	4AP	-0.3	-4.0	-1.9	0.4
102.00	1.95	1.3	24.73	14AP	-1.1	-3.4	-1.9	0.5
107.12	1.81	1.4	34.36	24AP	-1.8	-3.0	-1.8	0.5
112.44	1.67	1.5	43.95	4MY	-0.9	-3.7	-1.7	0.5
117.93	1.55	1.6	53.50	14MY	-0.0	-4.1	-1.7	0.5
123.57	1.42	1.6	63.04	24MY	0.7	-4.2	-1.6	0.6
129.35	1.30	1.7	72.57	3JN	1.4	-4.2	-1.5	0.6
135.25	1.18	1.7	82.09	13JN	2.3	-4.1	-1.5	0.6
141.27	1.07	1.8	91.62	23JN	3.3	-4.0	-1.4	0.5
147.40	0.95	1.8	101.16	3JL	1.7	-3.9	-1.4	0.5
153.64	0.84	1.8	110.73	13JL	0.2	-3.8	-1.3	0.5
159.98	0.73	1.8	120.33	23JL	-1.0	-3.8	-1.3	0.6
166.43	0.62	1.8	129.97	2AU	-1.5	-3.7	-1.3	0.6
172.98	0.51	1.8	139.66	12AU	-1.0	-3.6	-1.2	0.6
179.64	0.40	1.8	149.40	22AU	-0.4	-3.6	-1.2	0.7
186.40	0.29	1.8	159.19	1SE	-0.1	-3.5	-1.2	0.7
193.26	0.18	1.7	169.04	11SE	0.0	-3.5	-1.2	0.7
200.23	0.07	1.7	178.95	21SE	0.2	-3.5	-1.2	0.7
207.30	-0.04	1.7	188.92	1OC	0.6	-3.4	-1.2	0.7
214.46	-0.15	1.6	198.94	11OC	2.5	-3.4	-1.2	0.7
221.73	-0.25	1.6	209.02	21OC	0.9	-3.4	-1.3	0.7
229.08	-0.36	1.6	219.14	31OC	-0.3	-3.4	-1.3	0.6
236.53	-0.46	1.5	229.29	10NO	-0.5	-3.4	-1.3	0.6
244.06	-0.55	1.5	239.47	20NO	-0.5	-3.4	-1.4	0.5
251.67	-0.65	1.4	249.67	30NO	-0.6	-3.4	-1.4	0.5
259.34	-0.73	1.4	259.87	10DE	-0.8	-3.4	-1.5	0.4
267.08	-0.81	1.4	270.07	20DE	-0.9	-3.4	-1.6	0.3
274.86	-0.89	1.4	280.26	30DE	-0.9	-3.4	-1.6	0.2

♂ LONG	♂ LAT	♂ MAG	⊙ LONG	16.00UT	☿	♀	♃	♄
				416	MAGNITUDES			
282.69	-0.95	1.4	290.42	9JA	-0.6	-3.4	-1.7	0.2
290.54	-1.01	1.4	300.54	19JA	0.9	-3.4	-1.8	0.1
298.41	-1.05	1.4	310.63	29JA	3.0	-3.5	-1.8	0.2
306.28	-1.09	1.4	320.66	8FE	1.3	-3.5	-1.9	0.2
314.14	-1.11	1.4	330.64	18FE	0.7	-3.5	-1.9	0.3
321.98	-1.12	1.4	340.56	28FE	0.4	-3.4	-2.0	0.4
329.78	-1.12	1.4	350.42	9MR	0.1	-3.4	-2.0	0.4
337.54	-1.11	1.4	0.22	19MR	-0.4	-3.4	-2.0	0.5
345.24	-1.08	1.4	9.97	29MR	-1.1	-3.4	-2.0	0.5
352.87	-1.05	1.4	19.67	8AP	-1.7	-3.3	-2.0	0.6
0.43	-1.00	1.5	29.32	18AP	-1.0	-3.3	-2.0	0.6
7.91	-0.94	1.5	38.92	28AP	0.1	-3.3	-1.9	0.7
15.29	-0.87	1.5	48.49	8MY	0.9	-3.3	-1.9	0.7
22.58	-0.79	1.5	58.05	18MY	1.8	-3.3	-1.8	0.7
29.77	-0.70	1.5	67.57	28MY	3.1	-3.4	-1.8	0.7
36.85	-0.60	1.5	77.10	7JN	2.8	-3.4	-1.7	0.8
43.82	-0.50	1.5	86.62	17JN	1.3	-3.4	-1.7	0.8
50.68	-0.38	1.5	96.16	27JN	0.1	-3.4	-1.6	0.8
57.41	-0.26	1.5	105.71	7JL	-1.0	-3.5	-1.5	0.8
64.02	-0.13	1.5	115.29	17JL	-1.6	-3.5	-1.5	0.7
70.48	0.01	1.4	124.91	27JL	-1.0	-3.6	-1.4	0.7
76.81	0.15	1.4	134.57	6AU	-0.4	-3.7	-1.4	0.8
82.98	0.30	1.4	144.28	16AU	-0.0	-3.7	-1.3	0.8
88.97	0.47	1.3	154.05	26AU	0.2	-3.8	-1.3	0.9
94.77	0.64	1.2	163.87	5SE	0.4	-3.9	-1.3	0.9
100.34	0.83	1.2	173.75	15SE	0.9	-4.0	-1.3	0.9
105.66	1.03	1.1	183.68	25SE	2.9	-4.2	-1.2	1.0
110.67	1.25	1.0	193.68	5OC	0.8	-4.3	-1.2	1.0
115.32	1.49	0.9	203.73	15OC	-0.5	-4.3	-1.2	1.0
119.52	1.75	0.7	213.82	25OC	-0.6	-4.4	-1.2	0.9
123.18	2.05	0.5	223.96	4NO	-0.6	-4.2	-1.3	0.9
126.16	2.37	0.4	234.13	14NO	-0.7	-3.8	-1.3	0.9
128.31	2.74	0.1	244.32	24NO	-0.8	-2.9	-1.3	0.8
129.45	3.13	-0.1	254.52	4DE	-0.7	-3.7	-1.4	0.8
129.39	3.54	-0.3	264.73	14DE	-0.8	-4.2	-1.4	0.7
128.01	3.94	-0.6	274.92	24DE	-0.5	-4.4	-1.4	0.6
				417				
125.35	4.28	-0.8	285.09	3JA	1.0	-4.4	-1.5	0.6
121.74	4.50	-1.0	295.24	13JA	2.6	-4.3	-1.6	0.5
117.78	4.55	-1.0	305.34	23JA	0.9	-4.2	-1.6	0.4
114.20	4.42	-0.8	315.41	2FE	0.4	-4.1	-1.7	0.4
111.60	4.16	-0.5	325.42	12FE	0.2	-4.0	-1.8	0.4
110.27	3.82	-0.3	335.36	22FE	-0.0	-3.9	-1.8	0.5
110.21	3.47	-0.1	345.26	4MR	-0.4	-3.8	-1.9	0.5
111.31	3.11	0.2	355.10	14MR	-1.1	-3.7	-2.0	0.6
113.38	2.78	0.4	4.87	24MR	-1.7	-3.6	-2.0	0.6
116.24	2.48	0.6	14.60	3AP	-0.9	-3.5	-2.1	0.7
119.75	2.20	0.7	24.27	13AP	0.2	-3.5	-2.1	0.8
123.78	1.95	0.9	33.90	23AP	1.2	-3.4	-2.1	0.8
128.23	1.71	1.0	43.49	3MY	2.4	-3.4	-2.1	0.8
133.05	1.50	1.1	53.05	13MY	3.6	-3.3	-2.1	0.9
138.16	1.30	1.2	62.59	23MY	2.2	-3.3	-2.0	0.9
143.52	1.11	1.3	72.11	2JN	1.0	-3.3	-2.0	1.0
149.12	0.94	1.4	81.64	12JN	0.0	-3.3	-2.0	1.0
154.91	0.77	1.4	91.16	22JN	-1.1	-3.3	-1.9	1.0
160.89	0.61	1.5	100.71	2JL	-1.7	-3.3	-1.8	1.0
167.04	0.46	1.5	110.27	12JL	-1.0	-3.3	-1.8	1.0
173.34	0.31	1.5	119.87	22JL	-0.3	-3.4	-1.7	1.0
179.80	0.18	1.5	129.51	1AU	0.1	-3.4	-1.7	1.0
186.40	0.04	1.5	139.19	11AU	0.3	-3.4	-1.6	1.0
193.13	-0.09	1.6	148.92	21AU	0.6	-3.5	-1.5	1.1
200.00	-0.21	1.6	158.71	31AU	1.2	-3.5	-1.5	1.1
207.00	-0.33	1.5	168.56	10SE	3.2	-3.5	-1.4	1.2
214.12	-0.44	1.5	178.46	20SE	0.7	-3.5	-1.4	1.2
221.36	-0.54	1.5	188.43	30SE	-0.6	-3.4	-1.4	1.2
228.71	-0.64	1.5	198.45	10OC	-0.8	-3.4	-1.3	1.2
236.16	-0.73	1.5	208.52	20OC	-0.8	-3.4	-1.3	1.2
243.71	-0.81	1.5	218.64	30OC	-0.8	-3.4	-1.3	1.2
251.35	-0.88	1.5	228.79	9NO	-0.6	-3.3	-1.3	1.2
259.06	-0.94	1.4	238.97	19NO	-0.6	-3.3	-1.3	1.2
266.84	-1.00	1.4	249.17	29NO	-0.6	-3.3	-1.3	1.2
274.66	-1.04	1.4	259.37	9DE	-0.3	-3.3	-1.4	1.1
282.53	-1.07	1.4	269.57	19DE	1.2	-3.3	-1.4	1.0
290.43	-1.09	1.4	279.76	29DE	2.2	-3.4	-1.4	1.0

♂ LONG	♂ LAT	♂ MAG	⊙ LONG	16.00UT	☿	♀	♃	♄
				418	MAGNITUDES			
298.33	-1.09	1.4	289.93	8JA	0.6	-3.4	-1.5	0.9
306.23	-1.09	1.3	300.05	18JA	0.3	-3.4	-1.5	0.8
314.12	-1.07	1.3	310.14	28JA	0.1	-3.4	-1.6	0.8
321.97	-1.04	1.3	320.17	7FE	-0.1	-3.5	-1.6	0.7
329.79	-1.01	1.3	330.16	17FE	-0.5	-3.5	-1.7	0.7
337.56	-0.96	1.4	340.08	27FE	-1.1	-3.6	-1.7	0.7
345.26	-0.90	1.4	349.95	9MR	-1.6	-3.6	-1.8	0.8
352.90	-0.83	1.4	359.75	19MR	-0.9	-3.7	-1.9	0.8
0.46	-0.75	1.5	9.51	29MR	0.2	-3.8	-2.0	0.9
7.93	-0.67	1.5	19.20	8AP	1.5	-3.8	-2.0	0.9
15.33	-0.58	1.5	28.85	18AP	3.1	-3.9	-2.1	1.0
22.63	-0.49	1.6	38.47	28AP	2.9	-4.0	-2.1	1.0
29.84	-0.39	1.6	48.04	8MY	1.6	-4.1	-2.2	1.1
36.96	-0.29	1.6	57.59	18MY	0.8	-4.2	-2.2	1.1
43.99	-0.18	1.7	67.12	28MY	-0.1	-4.2	-2.2	1.2
50.92	-0.07	1.7	76.64	7JN	-1.1	-4.1	-2.2	1.2
57.76	0.04	1.7	86.17	17JN	-1.8	-3.7	-2.2	1.2
64.51	0.15	1.7	95.70	27JN	-1.0	-3.0	-2.2	1.3
71.17	0.27	1.7	105.25	7JL	-0.3	-3.4	-2.1	1.3
77.73	0.38	1.7	114.83	17JL	0.2	-3.9	-2.1	1.3
84.20	0.50	1.7	124.45	27JL	0.5	-4.2	-2.0	1.3
90.58	0.62	1.7	134.10	6AU	0.8	-4.2	-1.9	1.3
96.86	0.75	1.7	143.81	16AU	1.6	-4.2	-1.9	1.3
103.03	0.87	1.7	153.57	26AU	3.2	-4.1	-1.8	1.3
109.08	1.00	1.7	163.39	5SE	0.6	-4.0	-1.7	1.3
115.02	1.14	1.6	173.26	15SE	-0.7	-4.0	-1.7	1.3
120.81	1.28	1.6	183.20	25SE	-1.0	-3.9	-1.6	1.3
126.45	1.43	1.5	193.19	5OC	-0.9	-3.8	-1.6	1.3
131.91	1.59	1.4	203.23	15OC	-0.7	-3.7	-1.5	1.3
137.15	1.75	1.3	213.33	25OC	-0.5	-3.7	-1.5	1.3
142.14	1.93	1.2	223.46	4NO	-0.4	-3.6	-1.5	1.3
146.82	2.13	1.0	233.63	14NO	-0.4	-3.5	-1.4	1.2
151.13	2.34	0.9	243.82	24NO	-0.2	-3.5	-1.4	1.2
154.97	2.56	0.7	254.02	4DE	1.4	-3.5	-1.4	1.1
158.25	2.81	0.5	264.23	14DE	1.8	-3.4	-1.4	1.1
160.81	3.08	0.3	274.43	24DE	0.4	-3.4	-1.4	1.1
				419				
162.50	3.37	0.1	284.60	3JA	0.1	-3.4	-1.4	1.0
163.13	3.66	-0.2	294.75	13JA	-0.0	-3.3	-1.5	1.0
162.53	3.93	-0.5	304.86	23JA	-0.2	-3.3	-1.5	0.9
160.65	4.15	-0.7	314.92	2FE	-0.5	-3.3	-1.5	0.9
157.61	4.25	-1.0	324.93	12FE	-1.1	-3.3	-1.6	0.8
153.85	4.20	-1.1	334.89	22FE	-1.4	-3.3	-1.6	0.8
150.05	3.98	-1.0	344.78	4MR	-0.9	-3.3	-1.7	0.8
146.88	3.63	-0.8	354.62	14MR	0.4	-3.4	-1.7	0.9
144.85	3.20	-0.6	4.40	24MR	1.9	-3.4	-1.8	1.0
144.12	2.74	-0.4	14.13	3AP	3.7	-3.4	-1.9	1.0
144.65	2.31	-0.2	23.80	13AP	2.1	-3.5	-1.9	1.1
146.29	1.90	0.0	33.43	23AP	1.2	-3.5	-2.0	1.2
148.86	1.54	0.2	43.02	3MY	0.6	-3.5	-2.1	1.2
152.20	1.21	0.4	52.58	13MY	-0.1	-3.4	-2.1	1.3
156.16	0.92	0.5	62.12	23MY	-1.1	-3.4	-2.2	1.3
160.65	0.66	0.6	71.65	2JN	-1.8	-3.4	-2.3	1.3
165.57	0.42	0.7	81.17	12JN	-1.0	-3.4	-2.3	1.4
170.86	0.20	0.8	90.70	22JN	-0.2	-3.3	-2.3	1.4
176.46	0.00	0.9	100.24	2JL	0.3	-3.3	-2.3	1.4
182.35	-0.18	1.0	109.81	12JL	0.7	-3.3	-2.3	1.4
188.48	-0.34	1.0	119.40	22JL	1.1	-3.3	-2.3	1.3
194.84	-0.49	1.1	129.03	1AU	2.1	-3.3	-2.3	1.3
201.40	-0.63	1.1	138.72	11AU	2.8	-3.4	-2.2	1.3
208.15	-0.75	1.2	148.45	21AU	0.5	-3.4	-2.2	1.2
215.06	-0.86	1.2	158.23	31AU	-0.8	-3.4	-2.1	1.2
222.14	-0.96	1.2	168.08	10SE	-1.1	-3.5	-2.1	1.1
229.36	-1.04	1.2	177.98	20SE	-1.0	-3.5	-2.0	1.1
236.71	-1.11	1.3	187.94	30SE	-0.6	-3.6	-1.9	1.1
244.18	-1.16	1.3	197.96	10OC	-0.4	-3.6	-1.9	1.1
251.75	-1.21	1.3	208.03	20OC	-0.3	-3.7	-1.8	1.1
259.41	-1.23	1.3	218.14	30OC	-0.3	-3.8	-1.7	1.1
267.14	-1.25	1.3	228.29	9NO	0.0	-3.9	-1.7	1.1
274.93	-1.25	1.3	238.47	19NO	1.7	-4.0	-1.6	1.1
282.77	-1.24	1.3	248.67	29NO	1.6	-4.1	-1.6	1.1
290.63	-1.21	1.4	258.88	9DE	0.2	-4.2	-1.6	1.0
298.50	-1.17	1.4	269.08	19DE	-0.1	-4.3	-1.5	1.0
306.36	-1.12	1.4	279.26	29DE	-0.2	-4.4	-1.5	0.9

♂ LONG	♂ LAT	♂ MAG	☉ LONG	16.00UT	☿	♀	♃	♄
				420	MAGNITUDES			
314.21	-1.06	1.4	289.43	8JA	-0.3	-4.4	-1.5	0.9
322.02	-1.00	1.4	299.56	18JA	-0.6	-4.2	-1.5	0.8
329.79	-0.92	1.5	309.65	28JA	-1.1	-3.8	-1.5	0.8
337.51	-0.83	1.5	319.69	7FE	-1.3	-3.3	-1.5	0.7
345.16	-0.74	1.5	329.67	17FE	-0.9	-3.7	-1.6	0.7
352.74	-0.65	1.5	339.60	27FE	0.5	-4.1	-1.6	0.7
0.24	-0.55	1.5	349.47	8MR	2.4	-4.3	-1.6	0.6
7.66	-0.45	1.5	359.28	18MR	2.9	-4.3	-1.7	0.7
15.00	-0.35	1.6	9.03	28MR	1.6	-4.2	-1.7	0.7
22.25	-0.24	1.6	18.74	7AP	0.9	-4.1	-1.8	0.8
29.41	-0.14	1.6	28.39	17AP	0.4	-4.0	-1.8	0.9
36.50	-0.04	1.6	38.00	27AP	-0.2	-3.9	-1.9	0.9
43.50	0.07	1.7	47.58	7MY	-1.1	-3.8	-2.0	1.0
50.41	0.17	1.7	57.13	17MY	-1.9	-3.7	-2.0	1.1
57.25	0.27	1.8	66.66	27MY	-1.0	-3.6	-2.1	1.1
64.02	0.37	1.8	76.18	6JN	-0.1	-3.5	-2.2	1.1
70.71	0.46	1.8	85.70	16JN	0.4	-3.5	-2.2	1.2
77.34	0.56	1.9	95.24	26JN	0.9	-3.4	-2.3	1.2
83.90	0.65	1.9	104.79	6JL	1.5	-3.4	-2.3	1.2
90.39	0.74	1.9	114.37	16JL	2.7	-3.4	-2.4	1.2
96.83	0.83	1.9	123.98	26JL	2.4	-3.4	-2.4	1.2
103.21	0.92	1.9	133.64	5AU	0.4	-3.4	-2.4	1.2
109.52	1.01	1.9	143.34	15AU	-0.9	-3.3	-2.4	1.1
115.78	1.10	1.9	153.10	25AU	-1.2	-3.3	-2.4	1.1
121.98	1.18	1.9	162.92	4SE	-1.1	-3.4	-2.4	1.0
128.11	1.27	1.9	172.79	14SE	-0.6	-3.4	-2.3	1.0
134.18	1.36	1.8	182.72	24SE	-0.3	-3.4	-2.3	1.0
140.16	1.44	1.8	192.71	4OC	-0.2	-3.4	-2.2	1.0
146.07	1.53	1.7	202.75	14OC	-0.1	-3.4	-2.1	1.0
151.88	1.62	1.7	212.84	24OC	0.2	-3.4	-2.1	1.0
157.57	1.71	1.6	222.97	3NO	1.9	-3.5	-2.0	1.0
163.13	1.80	1.5	233.13	13NO	1.3	-3.5	-1.9	1.0
168.54	1.90	1.4	243.32	23NO	-0.0	-3.5	-1.9	1.0
173.76	1.99	1.2	253.52	3DE	-0.2	-3.5	-1.8	1.0
178.75	2.09	1.1	263.73	13DE	-0.3	-3.4	-1.8	0.9
183.46	2.20	0.9	273.92	23DE	-0.4	-3.4	-1.7	0.9
				421				
187.82	2.30	0.8	284.10	2JA	-0.7	-3.4	-1.7	0.9
191.76	2.41	0.6	294.25	12JA	-1.1	-3.4	-1.6	0.8
195.16	2.51	0.3	304.36	22JA	-1.2	-3.4	-1.6	0.8
197.89	2.60	0.1	314.43	1FE	-0.8	-3.3	-1.6	0.7
199.80	2.68	-0.2	324.44	11FE	0.6	-3.3	-1.6	0.7
200.69	2.73	-0.5	334.40	21FE	2.9	-3.3	-1.6	0.6
200.40	2.72	-0.8	344.30	3MR	2.2	-3.3	-1.6	0.6
198.85	2.64	-1.1	354.14	13MR	1.1	-3.4	-1.6	0.5
196.12	2.44	-1.4	3.93	23MR	0.7	-3.4	-1.6	0.5
192.62	2.12	-1.6	13.65	2AP	0.3	-3.4	-1.6	0.6
189.00	1.70	-1.5	23.33	12AP	-0.3	-3.4	-1.7	0.6
185.96	1.22	-1.3	32.97	22AP	-1.1	-3.5	-1.7	0.7
184.05	0.74	-1.1	42.56	2MY	-1.8	-3.5	-1.7	0.8
183.47	0.29	-0.9	52.12	12MY	-1.0	-3.6	-1.8	0.8
184.21	-0.11	-0.7	61.67	22MY	-0.1	-3.6	-1.8	0.9
186.13	-0.45	-0.5	71.19	1JN	0.6	-3.7	-1.9	0.9
189.05	-0.74	-0.3	80.71	11JN	1.1	-3.8	-2.0	1.0
192.79	-0.98	-0.2	90.24	21JN	1.9	-3.9	-2.0	1.0
197.22	-1.18	-0.0	99.78	1JL	3.2	-4.0	-2.1	1.0
202.21	-1.35	0.1	109.34	11JL	2.0	-4.1	-2.2	1.1
207.67	-1.48	0.2	118.94	21JL	0.3	-4.2	-2.2	1.1
213.53	-1.59	0.3	128.57	31JL	-1.0	-4.2	-2.3	1.1
219.72	-1.67	0.4	138.25	10AU	-1.4	-4.2	-2.4	1.0
226.21	-1.73	0.5	147.98	20AU	-1.0	-4.1	-2.4	1.0
232.94	-1.77	0.6	157.76	30AU	-0.5	-3.7	-2.4	1.0
239.88	-1.79	0.7	167.60	9SE	-0.2	-3.3	-2.5	1.0
247.01	-1.78	0.7	177.51	19SE	-0.0	-3.7	-2.5	0.9
254.29	-1.76	0.8	187.46	29SE	0.1	-4.2	-2.4	0.9
261.69	-1.72	0.9	197.48	9OC	0.4	-4.3	-2.4	0.9
269.20	-1.67	0.9	207.54	19OC	2.2	-4.3	-2.4	0.9
276.79	-1.60	1.0	217.65	29OC	1.1	-4.3	-2.3	0.9
284.43	-1.51	1.0	227.80	8NO	-0.2	-4.2	-2.2	0.9
292.12	-1.42	1.1	237.98	18NO	-0.4	-4.1	-2.2	0.9
299.82	-1.32	1.2	248.17	28NO	-0.4	-4.0	-2.1	0.9
307.52	-1.21	1.2	258.38	8DE	-0.5	-3.9	-2.0	0.9
315.20	-1.09	1.3	268.58	18DE	-0.7	-3.8	-2.0	0.9
322.86	-0.97	1.3	278.77	28DE	-1.0	-3.7	-1.9	0.8

♂ LONG	♂ LAT	♂ MAG	☉ LONG	16.00UT	☿	♀	♃	♄
				422	MAGNITUDES			
330.47	-0.85	1.4	288.94	7JA	-1.0	-3.7	-1.8	0.8
338.04	-0.72	1.4	299.07	17JA	-0.7	-3.6	-1.8	0.8
345.54	-0.60	1.5	309.16	27JA	0.7	-3.5	-1.7	0.7
352.98	-0.47	1.5	319.20	6FE	3.1	-3.5	-1.7	0.7
0.34	-0.35	1.6	329.19	16FE	1.6	-3.4	-1.6	0.6
7.63	-0.23	1.6	339.12	26FE	0.8	-3.4	-1.6	0.6
14.84	-0.12	1.7	348.99	8MR	0.5	-3.4	-1.6	0.5
21.97	-0.01	1.7	358.81	18MR	0.2	-3.3	-1.6	0.5
29.03	0.10	1.7	8.56	28MR	-0.3	-3.3	-1.6	0.4
36.01	0.20	1.8	18.27	7AP	-1.1	-3.3	-1.6	0.4
42.91	0.30	1.8	27.92	17AP	-1.8	-3.3	-1.6	0.5
49.75	0.40	1.8	37.54	27AP	-1.0	-3.3	-1.6	0.5
56.52	0.49	1.8	47.12	7MY	-0.0	-3.3	-1.6	0.6
63.23	0.57	1.8	56.67	17MY	0.8	-3.3	-1.6	0.7
69.88	0.65	1.8	66.20	27MY	1.5	-3.4	-1.7	0.7
76.47	0.73	1.8	75.72	6JN	2.6	-3.4	-1.7	0.8
83.02	0.80	1.9	85.25	16JN	3.1	-3.5	-1.7	0.8
89.52	0.86	1.9	94.78	26JN	1.6	-3.5	-1.8	0.9
95.99	0.93	1.9	104.33	6JL	0.2	-3.5	-1.9	0.9
102.42	0.99	2.0	113.91	16JL	-1.0	-3.4	-1.9	0.9
108.81	1.04	2.0	123.51	26JL	-1.5	-3.4	-2.0	1.0
115.18	1.10	2.0	133.17	5AU	-1.0	-3.4	-2.1	1.0
121.52	1.14	2.0	142.87	15AU	-0.4	-3.4	-2.1	1.0
127.84	1.19	2.0	152.62	25AU	-0.1	-3.3	-2.2	0.9
134.14	1.23	2.0	162.44	4SE	0.1	-3.3	-2.2	0.9
140.41	1.27	2.0	172.30	14SE	0.2	-3.3	-2.3	0.9
146.67	1.31	2.0	182.23	24SE	0.7	-3.3	-2.3	0.9
152.90	1.34	2.0	192.22	4OC	2.6	-3.3	-2.4	0.8
159.11	1.37	1.9	202.26	14OC	1.0	-3.4	-2.4	0.8
165.29	1.40	1.9	212.34	24OC	-0.4	-3.4	-2.4	0.8
171.44	1.42	1.8	222.48	3NO	-0.6	-3.4	-2.4	0.8
177.55	1.43	1.7	232.64	13NO	-0.6	-3.4	-2.4	0.8
183.61	1.45	1.7	242.83	23NO	-0.6	-3.5	-2.3	0.9
189.62	1.45	1.6	253.03	3DE	-0.8	-3.5	-2.3	0.9
195.56	1.45	1.5	263.23	13DE	-0.8	-3.6	-2.2	0.8
201.43	1.44	1.3	273.43	23DE	-0.9	-3.6	-2.1	0.8
				423				
207.19	1.42	1.2	283.61	2JA	-0.6	-3.7	-2.0	0.8
212.83	1.39	1.1	293.76	12JA	0.9	-3.8	-2.0	0.8
218.33	1.35	0.9	303.87	22JA	2.9	-3.9	-1.9	0.7
223.65	1.28	0.7	313.94	1FE	1.2	-4.0	-1.8	0.7
228.74	1.19	0.5	323.96	11FE	0.6	-4.1	-1.8	0.7
233.55	1.08	0.3	333.92	21FE	0.3	-4.1	-1.7	0.6
238.02	0.92	0.1	343.82	3MR	0.1	-4.2	-1.7	0.5
242.04	0.73	-0.1	353.66	13MR	-0.4	-4.3	-1.6	0.5
245.50	0.48	-0.4	3.45	23MR	-1.1	-4.2	-1.6	0.4
248.25	0.16	-0.7	13.18	2AP	-1.7	-4.0	-1.5	0.4
250.13	-0.25	-1.0	22.86	12AP	-1.0	-3.4	-1.5	0.3
250.95	-0.74	-1.3	32.50	22AP	0.1	-3.0	-1.5	0.4
250.54	-1.33	-1.7	42.10	2MY	1.0	-3.7	-1.5	0.4
248.93	-1.98	-2.0	51.66	12MY	2.0	-4.1	-1.5	0.5
246.32	-2.65	-2.2	61.20	22MY	3.4	-4.2	-1.5	0.6
243.25	-3.25	-2.3	70.73	1JN	2.5	-4.2	-1.5	0.6
240.47	-3.71	-2.1	80.25	11JN	1.2	-4.1	-1.5	0.7
238.62	-4.01	-1.9	89.78	21JN	0.1	-4.0	-1.5	0.7
238.07	-4.14	-1.7	99.32	1JL	-1.0	-3.9	-1.6	0.8
238.91	-4.14	-1.5	108.88	11JL	-1.6	-3.8	-1.6	0.8
241.01	-4.06	-1.3	118.47	21JL	-1.0	-3.7	-1.7	0.9
244.21	-3.92	-1.0	128.10	31JL	-0.4	-3.7	-1.7	0.9
248.30	-3.74	-0.8	137.78	10AU	0.0	-3.6	-1.8	0.9
253.11	-3.53	-0.6	147.50	20AU	0.2	-3.6	-1.8	0.9
258.50	-3.30	-0.5	157.29	30AU	0.4	-3.5	-1.9	0.9
264.35	-3.07	-0.3	167.12	9SE	1.0	-3.5	-1.9	0.9
270.55	-2.82	-0.1	177.02	19SE	3.0	-3.4	-2.0	0.8
277.03	-2.57	0.0	186.98	29SE	0.8	-3.4	-2.1	0.8
283.73	-2.33	0.2	196.99	9OC	-0.5	-3.4	-2.1	0.8
290.60	-2.08	0.3	207.05	19OC	-0.7	-3.4	-2.2	0.7
297.59	-1.84	0.4	217.16	29OC	-0.7	-3.4	-2.2	0.7
304.66	-1.61	0.6	227.31	8NO	-0.7	-3.4	-2.3	0.8
311.79	-1.38	0.7	237.49	18NO	-0.7	-3.4	-2.3	0.8
318.96	-1.17	0.8	247.68	28NO	-0.7	-3.4	-2.3	0.8
326.13	-0.96	0.9	257.88	8DE	-0.7	-3.4	-2.3	0.8
333.30	-0.77	1.0	268.09	18DE	-0.4	-3.4	-2.2	0.8
340.45	-0.58	1.1	278.28	28DE	1.0	-3.4	-2.2	0.8

♂ LONG	♂ LAT	♂ MAG	☉ LONG	16.00UT	☿	♀	♃	♄
				424	MAGNITUDES			
347.57	-0.41	1.2	288.44	7JA	2.5	-3.4	-2.1	0.8
354.65	-0.25	1.3	298.58	17JA	0.8	-3.4	-2.1	0.8
1.68	-0.10	1.4	308.67	27JA	0.4	-3.5	-2.0	0.7
8.66	0.03	1.5	318.71	6FE	0.2	-3.5	-1.9	0.7
15.60	0.16	1.6	328.70	16FE	-0.0	-3.5	-1.9	0.7
22.47	0.28	1.6	338.64	26FE	-0.4	-3.4	-1.8	0.6
29.29	0.39	1.7	348.51	7MR	-1.1	-3.4	-1.7	0.6
36.06	0.48	1.8	358.33	17MR	-1.6	-3.4	-1.7	0.5
42.77	0.57	1.8	8.09	27MR	-1.0	-3.4	-1.6	0.4
49.43	0.66	1.9	17.79	6AP	0.2	-3.3	-1.6	0.4
56.05	0.73	1.9	27.45	16AP	1.3	-3.3	-1.5	0.3
62.62	0.80	1.9	37.07	26AP	2.6	-3.3	-1.5	0.3
69.15	0.86	1.9	46.64	6MY	3.4	-3.3	-1.4	0.3
75.64	0.91	2.0	56.20	16MY	2.0	-3.3	-1.4	0.4
82.10	0.96	2.0	65.73	26MY	1.0	-3.4	-1.4	0.5
88.54	1.01	2.0	75.26	5JN	0.0	-3.4	-1.4	0.5
94.95	1.04	2.0	84.78	15JN	-1.0	-3.4	-1.4	0.6
101.34	1.08	2.0	94.31	25JN	-1.7	-3.4	-1.4	0.6
107.72	1.10	2.0	103.86	5JL	-1.0	-3.5	-1.4	0.7
114.10	1.13	1.9	113.44	15JL	-0.3	-3.5	-1.4	0.7
120.47	1.15	2.0	123.05	25JL	0.1	-3.6	-1.4	0.8
126.83	1.16	2.0	132.70	4AU	0.4	-3.7	-1.5	0.8
133.21	1.17	2.0	142.40	14AU	0.7	-3.7	-1.5	0.8
139.58	1.17	2.0	152.15	24AU	1.3	-3.8	-1.5	0.8
145.97	1.17	2.0	161.96	3SE	3.2	-3.9	-1.6	0.8
152.38	1.17	2.0	171.83	13SE	0.7	-4.0	-1.6	0.8
158.79	1.16	2.0	181.75	23SE	-0.6	-4.2	-1.7	0.8
165.23	1.15	2.0	191.73	3OC	-0.9	-4.3	-1.8	0.8
171.68	1.13	2.0	201.77	13OC	-0.8	-4.3	-1.8	0.8
178.15	1.10	1.9	211.86	23OC	-0.8	-4.4	-1.9	0.7
184.63	1.07	1.9	221.98	2NO	-0.6	-4.2	-2.0	0.7
191.14	1.03	1.8	232.15	12NO	-0.5	-3.8	-2.0	0.7
197.66	0.99	1.8	242.33	22NO	-0.5	-2.8	-2.1	0.8
204.20	0.93	1.7	252.53	2DE	-0.3	-3.7	-2.1	0.8
210.75	0.87	1.7	262.74	12DE	1.2	-4.2	-2.1	0.8
217.31	0.79	1.6	272.94	22DE	2.1	-4.4	-2.2	0.8
				425				
223.88	0.70	1.5	283.12	1JA	0.5	-4.4	-2.2	0.8
230.46	0.60	1.4	293.27	11JA	0.2	-4.3	-2.1	0.8
237.03	0.48	1.3	303.38	21JA	0.0	-4.2	-2.1	0.8
243.60	0.35	1.2	313.45	31JA	-0.1	-4.1	-2.1	0.8
250.16	0.20	1.0	323.48	10FE	-0.5	-4.0	-2.0	0.7
256.69	0.02	0.9	333.44	20FE	-1.1	-3.9	-1.9	0.7
263.20	-0.18	0.8	343.34	2MR	-1.5	-3.8	-1.9	0.6
269.65	-0.40	0.6	353.19	12MR	-1.0	-3.7	-1.8	0.6
276.04	-0.66	0.4	2.98	22MR	0.2	-3.6	-1.7	0.5
282.35	-0.94	0.3	12.71	1AP	1.7	-3.5	-1.7	0.5
288.52	-1.26	0.1	22.40	11AP	3.4	-3.5	-1.6	0.4
294.54	-1.62	-0.1	32.03	21AP	2.6	-3.4	-1.5	0.3
300.34	-2.02	-0.3	41.63	1MY	1.5	-3.4	-1.5	0.3
305.85	-2.46	-0.5	51.20	11MY	0.7	-3.3	-1.4	0.3
310.98	-2.94	-0.7	60.74	21MY	-0.1	-3.3	-1.4	0.3
315.63	-3.47	-1.0	70.27	31MY	-1.0	-3.3	-1.4	0.4
319.62	-4.05	-1.2	79.80	10JN	-1.8	-3.3	-1.3	0.5
322.79	-4.66	-1.5	89.32	20JN	-1.0	-3.3	-1.3	0.5
324.94	-5.28	-1.7	98.86	30JN	-0.3	-3.3	-1.3	0.6
325.85	-5.87	-2.0	108.42	10JL	0.2	-3.3	-1.3	0.6
325.43	-6.37	-2.3	118.01	20JL	0.5	-3.4	-1.3	0.7
323.75	-6.67	-2.5	127.64	30JL	0.9	-3.4	-1.3	0.7
321.21	-6.68	-2.6	137.31	9AU	1.8	-3.4	-1.3	0.8
318.51	-6.37	-2.5	147.03	19AU	3.1	-3.5	-1.3	0.8
316.39	-5.80	-2.3	156.81	29AU	0.6	-3.5	-1.3	0.8
315.38	-5.07	-2.0	166.65	8SE	-0.7	-3.5	-1.4	0.8
315.69	-4.30	-1.7	176.54	18SE	-1.0	-3.5	-1.4	0.8
317.23	-3.56	-1.4	186.49	28SE	-1.0	-3.4	-1.5	0.8
319.86	-2.89	-1.1	196.50	8OC	-0.7	-3.4	-1.5	0.8
323.37	-2.30	-0.8	206.56	18OC	-0.5	-3.4	-1.6	0.8
327.55	-1.78	-0.5	216.67	28OC	-0.4	-3.4	-1.6	0.7
332.28	-1.33	-0.3	226.81	7NO	-0.4	-3.3	-1.7	0.7
337.41	-0.95	-0.0	236.99	17NO	-0.1	-3.3	-1.8	0.7
342.85	-0.62	0.2	247.18	27NO	1.4	-3.3	-1.8	0.7
348.54	-0.34	0.4	257.39	7DE	1.8	-3.3	-1.9	0.8
354.40	-0.10	0.6	267.59	17DE	0.3	-3.3	-1.9	0.8
0.39	0.11	0.8	277.78	27DE	0.0	-3.4	-2.0	0.8

♂ LONG	♂ LAT	♂ MAG	☉ LONG	16.00UT	☿	♀	♃	♄
				426	MAGNITUDES			
6.49	0.29	0.9	287.95	6JA	-0.1	-3.4	-2.0	0.8
12.66	0.45	1.1	298.08	16JA	-0.2	-3.4	-2.1	0.8
18.88	0.58	1.2	308.18	26JA	-0.6	-3.4	-2.1	0.8
25.13	0.69	1.3	318.23	5FE	-1.1	-3.5	-2.1	0.8
31.40	0.79	1.4	328.22	15FE	-1.4	-3.5	-2.0	0.7
37.69	0.87	1.5	338.16	25FE	-0.9	-3.6	-2.0	0.7
43.98	0.95	1.6	348.04	7MR	0.4	-3.6	-2.0	0.7
50.27	1.01	1.7	357.85	17MR	2.1	-3.7	-1.9	0.6
56.56	1.06	1.8	7.62	27MR	3.4	-3.8	-1.8	0.6
62.84	1.10	1.8	17.33	6AP	1.9	-3.8	-1.8	0.5
69.12	1.13	1.9	26.99	16AP	1.1	-3.9	-1.7	0.4
75.39	1.16	1.9	36.61	26AP	0.5	-4.0	-1.6	0.4
81.66	1.18	2.0	46.19	6MY	-0.1	-4.1	-1.6	0.3
87.92	1.19	2.0	55.74	16MY	-1.0	-4.2	-1.5	0.3
94.19	1.20	2.0	65.28	26MY	-1.8	-4.2	-1.5	0.3
100.47	1.20	2.0	74.80	5JN	-1.0	-4.0	-1.4	0.3
106.75	1.20	2.0	84.32	15JN	-0.2	-3.7	-1.4	0.4
113.04	1.19	2.0	93.86	25JN	0.3	-3.0	-1.3	0.5
119.35	1.18	2.0	103.40	5JL	0.7	-3.4	-1.3	0.5
125.68	1.17	2.0	112.97	15JL	1.2	-4.0	-1.3	0.6
132.04	1.15	2.0	122.59	25JL	2.3	-4.2	-1.2	0.6
138.42	1.12	2.0	132.23	4AU	2.7	-4.2	-1.2	0.7
144.83	1.09	1.9	141.93	14AU	0.5	-4.2	-1.2	0.7
151.28	1.06	1.9	151.68	24AU	-0.8	-4.1	-1.2	0.8
157.77	1.02	1.9	161.48	3SE	-1.2	-4.0	-1.2	0.8
164.30	0.98	1.9	171.35	13SE	-1.1	-3.9	-1.2	0.8
170.88	0.93	1.9	181.27	23SE	-0.6	-3.9	-1.3	0.8
177.50	0.88	1.9	191.25	3OC	-0.4	-3.8	-1.3	0.8
184.17	0.82	1.9	201.28	13OC	-0.2	-3.7	-1.3	0.8
190.90	0.75	1.9	211.37	23OC	-0.2	-3.7	-1.4	0.8
197.67	0.68	1.9	221.49	2NO	0.1	-3.6	-1.4	0.7
204.50	0.60	1.8	231.65	12NO	1.7	-3.5	-1.5	0.7
211.39	0.52	1.8	241.84	22NO	1.5	-3.5	-1.5	0.7
218.32	0.43	1.8	252.04	2DE	0.1	-3.5	-1.6	0.7
225.31	0.33	1.7	262.25	12DE	-0.2	-3.4	-1.7	0.7
232.36	0.22	1.7	272.45	22DE	-0.2	-3.4	-1.7	0.8
				427				
239.45	0.10	1.6	282.62	1JA	-0.3	-3.4	-1.8	0.8
246.59	-0.02	1.5	292.78	11JA	-0.6	-3.3	-1.9	0.8
253.78	-0.16	1.5	302.90	21JA	-1.1	-3.3	-1.9	0.8
261.02	-0.30	1.4	312.97	31JA	-1.3	-3.3	-2.0	0.8
268.29	-0.46	1.3	322.99	10FE	-0.9	-3.3	-2.0	0.8
275.60	-0.62	1.2	332.96	20FE	0.5	-3.3	-2.0	0.8
282.93	-0.78	1.1	342.86	2MR	2.5	-3.3	-2.0	0.7
290.29	-0.96	1.1	352.72	12MR	2.6	-3.4	-2.0	0.7
297.65	-1.13	1.0	2.51	22MR	1.4	-3.4	-2.0	0.7
305.01	-1.31	0.9	12.24	1AP	0.8	-3.4	-1.9	0.6
312.35	-1.49	0.8	21.93	11AP	0.4	-3.5	-1.9	0.5
319.65	-1.68	0.7	31.57	21AP	-0.2	-3.5	-1.8	0.5
326.89	-1.85	0.6	41.17	1MY	-1.0	-3.5	-1.8	0.4
334.05	-2.02	0.5	50.73	11MY	-1.9	-3.4	-1.7	0.4
341.10	-2.19	0.4	60.28	21MY	-1.0	-3.4	-1.6	0.3
348.00	-2.34	0.3	69.80	31MY	-0.1	-3.4	-1.6	0.2
354.72	-2.48	0.2	79.33	10JN	0.5	-3.3	-1.5	0.3
1.21	-2.61	0.1	88.85	20JN	1.0	-3.3	-1.5	0.4
7.42	-2.72	-0.1	98.39	30JN	1.6	-3.3	-1.4	0.4
13.27	-2.81	-0.2	107.95	10JL	2.9	-3.3	-1.4	0.5
18.68	-2.88	-0.3	117.54	20JL	2.3	-3.3	-1.3	0.5
23.55	-2.92	-0.5	127.16	30JL	0.4	-3.3	-1.3	0.6
27.74	-2.93	-0.7	136.84	9AU	-0.9	-3.4	-1.3	0.6
31.07	-2.90	-0.9	146.56	19AU	-1.3	-3.4	-1.2	0.7
33.37	-2.83	-1.1	156.33	29AU	-1.1	-3.4	-1.2	0.7
34.41	-2.68	-1.3	166.17	8SE	-0.5	-3.5	-1.2	0.8
34.01	-2.44	-1.5	176.06	18SE	-0.3	-3.5	-1.2	0.8
32.16	-2.10	-1.7	186.01	28SE	-0.1	-3.6	-1.2	0.8
29.11	-1.65	-1.9	196.02	8OC	-0.0	-3.6	-1.2	0.8
25.49	-1.12	-2.0	206.08	18OC	0.3	-3.7	-1.2	0.8
22.11	-0.57	-1.7	216.18	28OC	2.0	-3.8	-1.3	0.8
19.68	-0.06	-1.4	226.32	7NO	1.3	-3.9	-1.3	0.8
18.55	0.36	-1.1	236.50	17NO	-0.1	-4.0	-1.4	0.7
18.77	0.70	-0.7	246.69	27NO	-0.3	-4.1	-1.4	0.7
20.18	0.95	-0.4	256.90	7DE	-0.3	-4.2	-1.5	0.7
22.59	1.15	-0.1	267.10	17DE	-0.4	-4.3	-1.5	0.7
25.78	1.29	0.1	277.29	27DE	-0.7	-4.4	-1.6	0.8

Left Table

♂ LONG	♂ LAT	♂ MAG	☉ LONG	16.00UT 428	☿	♀	♃	♄
29.58	1.39	0.4	287.46	6JA	-1.0	-4.4	-1.6	0.8
33.87	1.46	0.6	297.60	16JA	-1.1	-4.2	-1.7	0.8
38.53	1.51	0.8	307.69	26JA	-0.8	-3.8	-1.8	0.8
43.48	1.54	1.0	317.74	5FE	0.6	-3.3	-1.8	0.8
48.66	1.56	1.1	327.74	15FE	2.9	-3.7	-1.9	0.8
54.01	1.57	1.3	337.67	25FE	2.0	-4.1	-1.9	0.8
59.51	1.57	1.4	347.56	6MR	1.0	-4.3	-2.0	0.8
65.12	1.56	1.5	357.38	16MR	0.6	-4.3	-2.0	0.8
70.83	1.54	1.6	7.14	26MR	0.2	-4.2	-2.0	0.7
76.62	1.52	1.7	16.86	5AP	-0.3	-4.1	-2.0	0.7
82.47	1.50	1.8	26.52	15AP	-1.0	-4.0	-2.0	0.6
88.38	1.47	1.8	36.14	25AP	-1.8	-3.9	-2.0	0.6
94.34	1.43	1.9	45.73	5MY	-1.0	-3.8	-1.9	0.5
100.35	1.40	1.9	55.28	15MY	-0.1	-3.7	-1.9	0.4
106.41	1.36	2.0	64.81	25MY	0.6	-3.6	-1.8	0.4
112.51	1.31	2.0	74.34	4JN	1.3	-3.5	-1.8	0.3
118.66	1.27	2.0	83.86	14JN	2.2	-3.4	-1.7	0.3
124.86	1.22	2.0	93.39	24JN	3.3	-3.4	-1.7	0.3
131.11	1.16	2.0	102.94	4JL	1.8	-3.4	-1.6	0.4
137.41	1.11	2.0	112.51	14JL	0.3	-3.4	-1.5	0.4
143.76	1.05	2.0	122.12	24JL	-0.9	-3.4	-1.5	0.5
150.17	0.99	2.0	131.77	3AU	-1.4	-3.4	-1.4	0.6
156.64	0.92	2.0	141.46	13AU	-1.0	-3.3	-1.4	0.6
163.17	0.85	1.9	151.21	23AU	-0.5	-3.4	-1.3	0.7
169.77	0.78	1.9	161.01	2SE	-0.2	-3.4	-1.3	0.7
176.44	0.70	1.8	170.87	12SE	0.0	-3.4	-1.3	0.7
183.18	0.62	1.8	180.79	22SE	0.1	-3.4	-1.3	0.8
189.99	0.54	1.7	190.77	2OC	0.5	-3.4	-1.3	0.8
196.88	0.45	1.7	200.80	12OC	2.3	-3.4	-1.2	0.8
203.84	0.35	1.7	210.88	22OC	1.1	-3.4	-1.3	0.8
210.88	0.26	1.7	221.00	1NO	-0.3	-3.5	-1.3	0.8
217.99	0.16	1.7	231.16	11NO	-0.5	-3.5	-1.3	0.8
225.18	0.05	1.7	241.34	21NO	-0.5	-3.5	-1.3	0.8
232.44	-0.05	1.7	251.55	1DE	-0.6	-3.5	-1.3	0.8
239.77	-0.17	1.6	261.75	11DE	-0.8	-3.4	-1.4	0.7
247.17	-0.28	1.6	271.95	21DE	-0.9	-3.4	-1.4	0.7
254.64	-0.39	1.6	282.13	31DE	-1.0	-3.4	-1.5	0.8

429

♂ LONG	♂ LAT	♂ MAG	☉ LONG	16.00UT	☿	♀	♃	♄
262.15	-0.51	1.5	292.28	10JA	-0.7	-3.4	-1.5	0.8
269.73	-0.63	1.5	302.40	20JA	0.7	-3.3	-1.6	0.8
277.34	-0.75	1.5	312.48	30JA	3.1	-3.3	-1.7	0.9
285.00	-0.86	1.4	322.50	9FE	1.4	-3.3	-1.7	0.9
292.68	-0.97	1.4	332.47	19FE	0.7	-3.3	-1.8	0.9
300.38	-1.08	1.3	342.39	1MR	0.4	-3.3	-1.9	0.9
308.08	-1.18	1.3	352.24	11MR	0.1	-3.4	-1.9	0.8
315.77	-1.27	1.3	2.03	21MR	-0.3	-3.4	-2.0	0.8
323.45	-1.35	1.2	11.77	31MR	-1.1	-3.4	-2.0	0.8
331.09	-1.42	1.2	21.46	10AP	-1.8	-3.4	-2.1	0.7
338.69	-1.47	1.1	31.10	20AP	-1.0	-3.5	-2.1	0.7
346.22	-1.52	1.1	40.71	30AP	-0.0	-3.5	-2.1	0.6
353.67	-1.54	1.1	50.27	10MY	0.8	-3.6	-2.1	0.6
1.03	-1.55	1.0	59.82	20MY	1.7	-3.6	-2.1	0.5
8.28	-1.55	1.0	69.35	30MY	2.9	-3.7	-2.1	0.4
15.40	-1.53	0.9	78.87	9JN	3.0	-3.8	-2.0	0.4
22.38	-1.49	0.9	88.40	19JN	1.5	-3.9	-2.0	0.3
29.19	-1.43	0.8	97.94	29JN	0.2	-4.0	-1.9	0.3
35.81	-1.36	0.8	107.49	9JL	-1.0	-4.1	-1.8	0.4
42.21	-1.27	0.7	117.08	19JL	-1.5	-4.2	-1.8	0.4
48.36	-1.15	0.6	126.70	29JL	-1.0	-4.2	-1.7	0.5
54.22	-1.02	0.5	136.37	8AU	-0.4	-4.2	-1.7	0.5
59.74	-0.87	0.4	146.09	18AU	-0.1	-4.1	-1.6	0.6
64.85	-0.69	0.3	155.86	28AU	0.2	-3.7	-1.5	0.6
69.47	-0.48	0.2	165.69	7SE	0.3	-3.3	-1.5	0.7
73.48	-0.23	0.0	175.58	17SE	0.8	-3.7	-1.4	0.7
76.76	0.05	-0.1	185.53	27SE	2.7	-4.2	-1.4	0.8
79.12	0.38	-0.3	195.53	7OC	1.0	-4.3	-1.4	0.8
80.37	0.76	-0.5	205.59	17OC	-0.4	-4.3	-1.4	0.8
80.31	1.19	-0.8	215.69	27OC	-0.6	-4.3	-1.3	0.8
78.85	1.65	-1.0	225.83	6NO	-0.6	-4.2	-1.3	0.8
76.04	2.11	-1.2	236.00	16NO	-0.7	-4.1	-1.3	0.8
72.32	2.52	-1.3	246.20	26NO	-0.8	-4.0	-1.3	0.8
68.40	2.82	-1.2	256.40	6DE	-0.8	-3.9	-1.3	0.8
65.03	2.99	-1.0	266.61	16DE	-0.8	-3.8	-1.4	0.8
62.79	3.03	-0.7	276.80	26DE	-0.5	-3.7	-1.4	0.8

Right Table

♂ LONG	♂ LAT	♂ MAG	☉ LONG	16.00UT 430	☿	♀	♃	♄
61.87	2.99	-0.4	286.97	5JA	0.9	-3.7	-1.4	0.8
62.24	2.90	-0.1	297.11	15JA	2.8	-3.6	-1.5	0.8
63.73	2.78	0.1	307.21	25JA	1.0	-3.5	-1.5	0.9
66.12	2.66	0.4	317.26	4FE	0.5	-3.5	-1.6	0.9
69.24	2.52	0.6	327.26	14FE	0.3	-3.4	-1.6	0.9
72.94	2.40	0.8	337.20	24FE	0.0	-3.4	-1.7	0.9
77.09	2.27	1.0	347.08	6MR	-0.4	-3.4	-1.8	0.9
81.61	2.15	1.1	356.91	16MR	-1.1	-3.3	-1.8	0.9
86.42	2.04	1.3	6.68	26MR	-1.7	-3.3	-1.9	0.9
91.46	1.93	1.4	16.39	5AP	-1.0	-3.3	-2.0	0.9
96.70	1.82	1.5	26.06	15AP	0.1	-3.3	-2.1	0.8
102.11	1.72	1.6	35.68	25AP	1.1	-3.3	-2.1	0.8
107.66	1.61	1.6	45.26	5MY	2.2	-3.3	-2.2	0.7
113.33	1.51	1.7	54.82	15MY	3.6	-3.3	-2.2	0.7
119.12	1.42	1.8	64.36	25MY	2.3	-3.4	-2.2	0.6
125.02	1.32	1.8	73.88	4JN	1.2	-3.4	-2.2	0.5
131.01	1.22	1.8	83.41	14JN	0.1	-3.5	-2.2	0.5
137.09	1.13	1.9	92.93	24JN	-1.0	-3.5	-2.2	0.4
143.27	1.03	1.9	102.48	4JL	-1.6	-3.5	-2.2	0.4
149.54	0.93	1.9	112.05	14JL	-1.0	-3.4	-2.1	0.4
155.89	0.84	1.9	121.65	24JL	-0.4	-3.4	-2.1	0.4
162.34	0.74	1.9	131.30	3AU	0.0	-3.4	-2.0	0.5
168.87	0.64	1.9	140.99	13AU	0.3	-3.4	-1.9	0.5
175.50	0.54	1.8	150.73	23AU	0.5	-3.3	-1.9	0.6
182.22	0.44	1.8	160.53	2SE	1.1	-3.3	-1.8	0.6
189.04	0.33	1.8	170.39	12SE	3.0	-3.3	-1.8	0.7
195.94	0.23	1.8	180.30	22SE	0.8	-3.3	-1.7	0.7
202.95	0.13	1.7	190.27	2OC	-0.5	-3.3	-1.6	0.8
210.04	0.02	1.7	200.30	12OC	-0.8	-3.4	-1.6	0.8
217.23	-0.08	1.6	210.38	22OC	-0.8	-3.4	-1.5	0.8
224.51	-0.19	1.6	220.50	1NO	-0.8	-3.4	-1.5	0.9
231.88	-0.29	1.5	230.66	11NO	-0.7	-3.4	-1.5	0.9
239.24	-0.40	1.5	240.84	21NO	-0.6	-3.5	-1.4	0.9
246.85	-0.50	1.5	251.05	1DE	-0.6	-3.5	-1.4	0.9
254.45	-0.60	1.5	261.25	11DE	-0.4	-3.6	-1.4	0.9
262.12	-0.69	1.5	271.45	21DE	1.0	-3.7	-1.4	0.9
269.84	-0.78	1.5	281.64	31DE	2.4	-3.7	-1.4	0.8

431

♂ LONG	♂ LAT	♂ MAG	☉ LONG	16.00UT	☿	♀	♃	♄
277.60	-0.86	1.4	291.80	10JA	0.7	-3.8	-1.5	0.8
285.40	-0.94	1.4	301.91	20JA	0.3	-3.9	-1.5	0.9
293.23	-1.00	1.4	311.99	30JA	0.1	-4.0	-1.5	0.9
301.07	-1.06	1.4	322.02	9FE	-0.1	-4.1	-1.5	1.0
308.91	-1.11	1.4	331.99	19FE	-0.4	-4.2	-1.6	1.0
316.74	-1.15	1.4	341.91	1MR	-1.1	-4.2	-1.6	1.0
324.55	-1.17	1.4	351.77	11MR	-1.6	-4.3	-1.7	1.0
332.31	-1.18	1.4	1.56	21MR	-1.0	-4.2	-1.8	1.0
340.04	-1.18	1.4	11.31	31MR	0.2	-4.0	-1.8	1.0
347.70	-1.17	1.4	21.00	10AP	1.4	-3.4	-1.9	1.0
355.29	-1.14	1.4	30.64	20AP	2.9	-3.4	-2.0	0.9
2.80	-1.10	1.4	40.24	30AP	3.1	-3.7	-2.0	0.9
10.23	-1.05	1.4	49.81	10MY	1.8	-4.1	-2.1	0.9
17.55	-0.99	1.4	59.36	20MY	0.9	-4.2	-2.2	0.8
24.74	-0.91	1.4	68.89	30MY	0.0	-4.2	-2.2	0.7
31.89	-0.82	1.4	78.41	9JN	-1.0	-4.1	-2.3	0.7
38.88	-0.73	1.4	87.93	19JN	-1.7	-4.0	-2.3	0.6
45.75	-0.62	1.3	97.47	29JN	-1.0	-3.9	-2.3	0.5
52.47	-0.50	1.3	107.03	9JL	-0.3	-3.8	-2.4	0.5
59.06	-0.37	1.3	116.61	19JL	0.1	-3.7	-2.3	0.4
65.49	-0.23	1.3	126.24	29JL	0.4	-3.7	-2.3	0.5
71.74	-0.08	1.2	135.90	8AU	0.7	-3.6	-2.3	0.5
77.80	0.08	1.2	145.62	18AU	1.5	-3.6	-2.2	0.6
83.64	0.25	1.1	155.39	28AU	3.2	-3.5	-2.2	0.6
89.24	0.44	1.0	165.21	7SE	0.7	-3.5	-2.1	0.7
94.55	0.65	0.9	175.10	17SE	-0.7	-3.4	-2.1	0.7
99.52	0.88	0.8	185.04	27SE	-0.9	-3.4	-2.0	0.8
104.08	1.13	0.7	195.04	7OC	-0.9	-3.4	-1.9	0.8
108.14	1.41	0.5	205.09	17OC	-0.8	-3.4	-1.9	0.9
111.59	1.72	0.4	215.20	27OC	-0.5	-3.4	-1.8	0.9
114.29	2.07	0.2	225.33	6NO	-0.5	-3.4	-1.7	0.9
116.06	2.46	-0.0	235.50	16NO	-0.5	-3.4	-1.7	0.9
116.71	2.89	-0.3	245.70	26NO	-0.2	-3.4	-1.6	1.0
116.07	3.33	-0.5	255.90	6DE	1.2	-3.4	-1.6	1.0
114.09	3.75	-0.7	266.10	16DE	2.1	-3.4	-1.6	1.0
110.92	4.10	-0.9	276.30	26DE	0.5	-3.4	-1.6	1.0

432 / 433

♂ LONG	LAT	MAG	☉ LONG	16.00UT	☿	♀	♃	♄
107.03	4.30	-1.0	286.47	5JA	0.1	-3.4	-1.5	1.0
103.15	4.34	-0.9	296.61	15JA	-0.0	-3.4	-1.5	0.9
99.98	4.22	-0.7	306.72	25JA	-0.2	-3.5	-1.5	1.0
97.97	3.99	-0.4	316.77	4FE	-0.5	-3.5	-1.5	1.0
97.26	3.70	-0.2	326.77	14FE	-1.1	-3.5	-1.5	1.0
97.78	3.39	0.1	336.72	24FE	-1.5	-3.4	-1.6	1.1
99.37	3.09	0.3	346.60	5MR	-1.0	-3.4	-1.6	1.1
101.84	2.81	0.5	356.43	15MR	0.2	-3.4	-1.6	1.1
105.01	2.55	0.7	6.21	25MR	1.8	-3.4	-1.7	1.1
108.75	2.31	0.9	15.92	4AP	3.7	-3.3	-1.7	1.1
112.95	2.09	1.0	25.59	14AP	2.3	-3.3	-1.8	1.1
117.53	1.89	1.1	35.21	24AP	1.3	-3.3	-1.8	1.1
122.42	1.70	1.2	44.80	4MY	0.7	-3.3	-1.9	1.1
127.57	1.52	1.3	54.36	14MY	-0.1	-3.3	-2.0	1.0
132.95	1.35	1.4	63.89	24MY	-1.0	-3.4	-2.0	1.0
138.53	1.19	1.5	73.42	3JN	-1.8	-3.4	-2.1	0.9
144.28	1.04	1.5	82.94	13JN	-1.0	-3.4	-2.2	0.8
150.19	0.89	1.6	92.47	23JN	-0.2	-3.4	-2.3	0.8
156.26	0.75	1.6	102.01	3JL	0.3	-3.5	-2.3	0.7
162.46	0.62	1.6	111.58	13JL	0.6	-3.5	-2.4	0.6
168.81	0.48	1.7	121.19	23JL	1.0	-3.6	-2.4	0.6
175.28	0.36	1.7	130.83	2AU	1.9	-3.7	-2.4	0.5
181.87	0.23	1.7	140.52	12AU	3.0	-3.7	-2.4	0.6
188.59	0.11	1.7	150.26	22AU	0.6	-3.8	-2.4	0.6
195.43	-0.01	1.7	160.05	1SE	-0.8	-3.9	-2.4	0.7
202.39	-0.13	1.6	169.91	11SE	-1.1	-4.0	-2.4	0.7
209.46	-0.24	1.6	179.82	21SE	-1.0	-4.2	-2.3	0.8
216.65	-0.35	1.6	189.79	1OC	-0.7	-4.3	-2.3	0.8
223.94	-0.45	1.6	199.82	11OC	-0.4	-4.3	-2.2	0.9
231.33	-0.55	1.5	209.90	21OC	-0.3	-4.4	-2.1	0.9
238.82	-0.64	1.5	220.01	31OC	-0.3	-4.2	-2.1	1.0
246.40	-0.73	1.5	230.17	10NO	-0.1	-3.7	-2.0	1.0
254.05	-0.81	1.5	240.35	20NO	1.4	-2.8	-1.9	1.0
261.78	-0.88	1.4	250.55	30NO	1.8	-3.8	-1.9	1.1
269.56	-0.94	1.4	260.75	10DE	0.2	-4.4	-1.8	1.1
277.40	-0.99	1.4	270.95	20DE	-0.1	-4.4	-1.8	1.1
285.26	-1.03	1.3	281.14	30DE	-0.1	-4.4	-1.7	1.1
				433				
293.16	-1.06	1.3	291.30	9JA	-0.3	-4.3	-1.7	1.1
301.06	-1.08	1.3	301.42	19JA	-0.6	-4.2	-1.6	1.1
308.95	-1.09	1.3	311.50	29JA	-1.1	-4.1	-1.6	1.1
316.83	-1.08	1.3	321.53	8FE	-1.3	-4.0	-1.6	1.1
324.68	-1.06	1.4	331.51	18FE	-0.9	-3.9	-1.6	1.2
332.48	-1.04	1.4	341.42	28FE	0.3	-3.8	-1.6	1.2
340.23	-1.00	1.4	351.29	10MR	2.2	-3.7	-1.6	1.2
347.93	-0.95	1.4	1.09	20MR	3.1	-3.6	-1.6	1.3
355.55	-0.89	1.5	10.83	30MR	1.7	-3.5	-1.6	1.3
3.09	-0.82	1.5	20.53	9AP	1.0	-3.5	-1.6	1.3
10.55	-0.74	1.5	30.18	19AP	0.5	-3.4	-1.7	1.3
17.92	-0.66	1.5	39.78	29AP	-0.1	-3.4	-1.7	1.3
25.20	-0.56	1.6	49.36	9MY	-1.0	-3.3	-1.7	1.2
32.38	-0.47	1.6	58.90	19MY	-1.9	-3.3	-1.8	1.2
39.46	-0.36	1.6	68.43	29MY	-1.0	-3.3	-1.9	1.2
46.45	-0.26	1.6	77.96	8JN	-0.2	-3.3	-1.9	1.1
53.34	-0.14	1.6	87.48	18JN	0.4	-3.3	-2.0	1.1
60.12	-0.03	1.7	97.01	28JN	0.8	-3.3	-2.0	1.0
66.81	0.09	1.7	106.57	8JL	1.4	-3.3	-2.1	0.9
73.38	0.22	1.7	116.15	18JL	2.5	-3.4	-2.2	0.8
79.85	0.35	1.6	125.77	28JL	2.6	-3.4	-2.3	0.8
86.21	0.48	1.6	135.44	7AU	0.5	-3.4	-2.3	0.7
92.44	0.61	1.6	145.15	17AU	-0.8	-3.5	-2.4	0.7
98.54	0.76	1.6	154.91	27AU	-1.2	-3.5	-2.4	0.7
104.51	0.91	1.5	164.74	6SE	-1.1	-3.5	-2.4	0.8
110.31	1.06	1.5	174.62	16SE	-0.6	-3.5	-2.5	0.8
115.94	1.23	1.4	184.56	26SE	-0.3	-3.4	-2.5	0.9
121.36	1.40	1.3	194.56	6OC	-0.2	-3.4	-2.4	0.9
126.54	1.59	1.2	204.60	16OC	-0.1	-3.4	-2.4	1.0
131.43	1.80	1.1	214.70	26OC	-0.3	-3.4	-2.3	1.0
135.97	2.02	1.0	224.84	5NO	1.7	-3.3	-2.3	1.1
140.09	2.26	0.8	235.01	15NO	1.5	-3.3	-2.2	1.1
143.69	2.53	0.6	245.20	25NO	0.0	-3.3	-2.2	1.1
146.63	2.82	0.4	255.40	5DE	-0.2	-3.3	-2.1	1.2
148.77	3.14	0.2	265.60	15DE	-0.3	-3.3	-2.0	1.2
149.93	3.49	-0.0	275.80	25DE	-0.4	-3.4	-1.9	1.2

434 / 435

♂ LONG	LAT	MAG	☉ LONG	16.00UT	☿	♀	♃	♄
149.92	3.83	-0.3	285.97	4JA	-0.6	-3.4	-1.9	1.2
148.63	4.15	-0.6	296.11	14JA	-1.1	-3.4	-1.8	1.2
146.07	4.40	-0.8	306.22	24JA	-1.2	-3.4	-1.7	1.3
142.53	4.51	-1.0	316.28	3FE	-0.8	-3.5	-1.7	1.2
138.62	4.45	-1.0	326.28	13FE	0.5	-3.5	-1.7	1.2
135.03	4.22	-0.8	336.23	23FE	2.6	-3.6	-1.6	1.3
132.38	3.87	-0.6	346.12	5MR	2.4	-3.6	-1.6	1.3
130.99	3.46	-0.4	355.95	15MR	1.2	-3.7	-1.6	1.4
130.88	3.04	-0.2	5.73	25MR	0.7	-3.8	-1.6	1.4
131.95	2.63	0.0	15.45	4AP	0.3	-3.9	-1.5	1.4
134.02	2.26	0.2	25.12	14AP	-0.2	-3.9	-1.5	1.4
136.91	1.92	0.4	34.75	24AP	-1.0	-4.0	-1.6	1.4
140.48	1.62	0.6	44.34	4MY	-1.9	-4.1	-1.6	1.3
144.60	1.34	0.7	53.90	14MY	-1.0	-4.2	-1.6	1.3
149.17	1.09	0.8	63.43	24MY	-0.1	-4.2	-1.6	1.3
154.14	0.85	0.9	72.96	3JN	0.5	-4.0	-1.7	1.2
159.42	0.64	1.0	82.48	13JN	1.1	-3.6	-1.7	1.2
164.99	0.45	1.1	92.01	23JN	1.8	-2.9	-1.7	1.1
170.81	0.26	1.2	101.56	3JL	3.1	-3.4	-1.8	1.1
176.85	0.09	1.2	111.12	13JL	2.1	-4.0	-1.9	1.0
183.10	-0.06	1.3	120.72	23JL	0.4	-4.2	-1.9	1.0
189.53	-0.21	1.3	130.36	2AU	-0.9	-4.2	-2.0	0.9
196.14	-0.35	1.3	140.05	12AU	-1.4	-4.2	-2.0	0.9
202.92	-0.48	1.3	149.79	22AU	-1.1	-4.1	-2.1	0.9
209.84	-0.60	1.4	159.58	1SE	-0.5	-4.0	-2.2	0.9
216.91	-0.70	1.4	169.43	11SE	-0.2	-3.9	-2.2	0.9
224.12	-0.80	1.4	179.34	21SE	-0.1	-3.9	-2.3	1.0
231.45	-0.89	1.4	189.31	1OC	0.0	-3.8	-2.3	1.0
238.89	-0.96	1.4	199.33	11OC	0.4	-3.7	-2.4	1.1
246.44	-1.03	1.4	209.40	21OC	2.0	-3.7	-2.4	1.1
254.08	-1.08	1.4	219.52	31OC	1.3	-3.6	-2.4	1.2
261.79	-1.12	1.4	229.67	10NO	-0.2	-3.5	-2.4	1.2
269.57	-1.15	1.4	239.86	20NO	-0.4	-3.5	-2.3	1.3
277.40	-1.16	1.4	250.06	30NO	-0.4	-3.3	-2.3	1.3
285.27	-1.16	1.4	260.26	10DE	-0.5	-3.4	-2.2	1.3
293.16	-1.15	1.4	270.46	20DE	-0.7	-3.4	-2.2	1.4
301.05	-1.13	1.4	280.65	30DE	-1.0	-3.4	-2.1	1.4
				435				
308.94	-1.09	1.4	290.81	9JA	-1.0	-3.3	-2.0	1.4
316.80	-1.05	1.4	300.93	19JA	-0.8	-3.3	-1.9	1.3
324.63	-0.99	1.4	311.01	29JA	0.6	-3.3	-1.9	1.3
332.41	-0.93	1.4	321.04	8FE	3.0	-3.3	-1.8	1.3
340.13	-0.86	1.4	331.03	18FE	1.8	-3.3	-1.7	1.2
347.79	-0.78	1.4	340.95	28FE	0.9	-3.3	-1.7	1.1
355.37	-0.69	1.4	350.81	10MR	0.5	-3.4	-1.6	1.2
2.88	-0.60	1.5	0.61	20MR	0.2	-3.4	-1.6	1.2
10.30	-0.50	1.5	10.36	30MR	-0.3	-3.4	-1.5	1.2
17.64	-0.40	1.5	20.05	9AP	-1.0	-3.5	-1.5	1.1
24.90	-0.30	1.6	29.70	19AP	-1.8	-3.5	-1.5	1.1
32.06	-0.20	1.6	39.31	29AP	-1.0	-3.5	-1.5	1.1
39.14	-0.10	1.7	48.88	9MY	-0.1	-3.4	-1.5	1.1
46.13	0.01	1.7	58.43	19MY	0.7	-3.4	-1.5	1.1
53.04	0.11	1.7	67.96	29MY	1.4	-3.4	-1.5	1.0
59.86	0.22	1.8	77.49	8JN	2.4	-3.3	-1.5	1.0
66.61	0.32	1.8	87.01	18JN	3.3	-3.3	-1.5	0.9
73.27	0.42	1.8	96.55	28JN	1.7	-3.3	-1.5	0.9
79.87	0.53	1.9	106.10	8JL	0.3	-3.3	-1.6	0.8
86.39	0.63	1.9	115.68	18JL	-0.9	-3.3	-1.6	0.8
92.83	0.73	1.9	125.30	28JL	-1.5	-3.3	-1.6	0.7
99.20	0.83	1.9	134.96	7AU	-1.1	-3.4	-1.7	0.7
105.50	0.94	1.9	144.67	17AU	-0.5	-3.4	-1.8	0.6
111.72	1.04	1.8	154.43	27AU	-0.1	-3.4	-1.8	0.6
117.86	1.14	1.8	164.26	6SE	0.1	-3.5	-1.9	0.6
123.92	1.25	1.8	174.14	16SE	0.2	-3.5	-1.9	0.6
129.88	1.35	1.7	184.07	26SE	0.6	-3.6	-2.0	0.7
135.74	1.46	1.7	194.07	6OC	2.3	-3.6	-2.1	0.7
141.48	1.58	1.6	204.12	16OC	1.1	-3.7	-2.1	0.8
147.09	1.69	1.5	214.21	26OC	-0.3	-3.8	-2.2	0.9
152.53	1.82	1.4	224.35	5NO	-0.5	-3.9	-2.2	0.9
157.79	1.95	1.3	234.52	15NO	-0.5	-4.0	-2.3	1.0
162.81	2.08	1.2	244.70	25NO	-0.6	-4.1	-2.3	1.0
167.56	2.23	1.1	254.91	5DE	-0.8	-4.2	-2.3	1.1
171.97	2.38	0.9	265.11	15DE	-0.8	-4.3	-2.2	1.1
175.96	2.54	0.7	275.31	25DE	-0.9	-4.4	-2.2	1.1

Left table (436 / 437)

♂ LONG	♂ LAT	♂ MAG	☉ LONG	16.00UT	☿	♀	♃	♄
				436				
179.43	2.71	0.5	285.48	4JA	-0.6	-4.4	-2.2	1.1
182.25	2.89	0.3	295.63	14JA	0.7	-4.2	-2.1	1.1
184.26	3.07	0.0	305.73	24JA	3.0	-3.7	-2.0	1.1
185.28	3.24	-0.3	315.79	3FE	1.3	-3.3	-2.0	1.0
185.14	3.37	-0.6	325.80	13FE	0.6	-3.8	-1.9	1.0
183.74	3.45	-0.8	335.75	23FE	0.4	-4.1	-1.8	1.0
181.11	3.42	-1.1	345.64	4MR	0.1	-4.3	-1.8	0.9
177.61	3.26	-1.3	355.48	14MR	-0.3	-4.3	-1.7	0.9
173.84	2.95	-1.3	5.25	24MR	-1.0	-4.2	-1.6	0.9
170.52	2.53	-1.1	14.98	3AP	-1.8	-4.1	-1.6	0.9
168.23	2.06	-0.9	24.65	13AP	-1.0	-4.0	-1.5	0.9
167.24	1.59	-0.7	34.28	23AP	-0.0	-3.9	-1.5	0.9
167.56	1.14	-0.5	43.87	3MY	0.9	-3.8	-1.4	0.9
169.08	0.74	-0.3	53.43	13MY	1.8	-3.7	-1.4	0.8
171.61	0.39	-0.1	62.97	23MY	3.2	-3.6	-1.4	0.8
174.97	0.08	0.0	72.50	2JN	2.7	-3.5	-1.4	0.8
179.04	-0.19	0.2	82.02	12JN	1.4	-3.5	-1.4	0.8
183.67	-0.42	0.3	91.55	22JN	0.2	-3.4	-1.4	0.7
188.79	-0.63	0.4	101.09	2JL	-1.0	-3.4	-1.4	0.7
194.31	-0.81	0.5	110.66	12JL	-1.6	-3.4	-1.4	0.6
200.18	-0.96	0.6	120.26	22JL	-1.1	-3.4	-1.4	0.6
206.35	-1.10	0.7	129.90	1AU	-0.4	-3.4	-1.4	0.5
212.79	-1.21	0.8	139.58	11AU	-0.0	-3.3	-1.5	0.5
219.46	-1.30	0.8	149.31	21AU	0.2	-3.3	-1.5	0.4
226.34	-1.38	0.9	159.11	31AU	0.4	-3.4	-1.5	0.4
233.40	-1.44	0.9	168.95	10SE	0.9	-3.4	-1.6	0.3
240.61	-1.47	1.0	178.86	20SE	2.7	-3.4	-1.6	0.3
247.97	-1.50	1.0	188.83	30SE	1.0	-3.4	-1.7	0.3
255.45	-1.50	1.1	198.84	10OC	-0.5	-3.4	-1.8	0.4
263.03	-1.49	1.1	208.91	20OC	-0.7	-3.4	-1.8	0.5
270.68	-1.46	1.1	219.03	30OC	-0.7	-3.5	-1.9	0.5
278.40	-1.42	1.2	229.18	9NO	-0.7	-3.5	-2.0	0.6
286.16	-1.37	1.2	239.36	19NO	-0.7	-3.5	-2.0	0.7
293.95	-1.30	1.3	249.56	29NO	-0.7	-3.5	-2.1	0.7
301.74	-1.22	1.3	259.76	9DE	-0.7	-3.4	-2.1	0.8
309.53	-1.14	1.3	269.96	19DE	-0.5	-3.4	-2.1	0.8
317.29	-1.04	1.4	280.15	29DE	0.9	-3.4	-2.1	0.8
				437				
325.02	-0.94	1.4	290.31	8JA	2.7	-3.4	-2.1	0.8
332.69	-0.84	1.5	300.44	18JA	0.9	-3.3	-2.1	0.8
340.31	-0.73	1.5	310.52	28JA	0.4	-3.3	-2.1	0.8
347.87	-0.62	1.5	320.56	7FE	0.2	-3.3	-2.0	0.8
355.35	-0.51	1.6	330.54	17FE	-0.0	-3.3	-2.0	0.8
2.76	-0.39	1.6	340.47	27FE	-0.4	-3.3	-1.9	0.7
10.09	-0.28	1.6	350.33	9MR	-1.0	-3.4	-1.8	0.7
17.33	-0.17	1.7	0.14	19MR	-1.7	-3.4	-1.8	0.6
24.50	-0.07	1.7	9.89	29MR	-1.0	-3.4	-1.7	0.6
31.58	0.04	1.7	19.58	8AP	0.1	-3.4	-1.6	0.6
38.59	0.14	1.7	29.24	18AP	1.2	-3.5	-1.6	0.6
45.51	0.24	1.7	38.85	28AP	2.4	-3.5	-1.5	0.7
52.37	0.34	1.7	48.42	8MY	3.6	-3.6	-1.5	0.7
59.15	0.43	1.7	57.97	18MY	2.1	-3.6	-1.4	0.6
65.87	0.52	1.8	67.50	28MY	1.1	-3.7	-1.4	0.6
72.53	0.61	1.8	77.03	7JN	0.1	-3.8	-1.3	0.6
79.13	0.69	1.9	86.55	17JN	-1.0	-3.9	-1.3	0.6
85.67	0.77	1.9	96.09	27JN	-1.7	-4.0	-1.3	0.6
92.17	0.84	1.9	105.64	7JL	-1.0	-4.1	-1.3	0.5
98.62	0.91	2.0	115.22	17JL	-0.4	-4.2	-1.3	0.5
105.03	0.98	2.0	124.84	27JL	0.1	-4.2	-1.3	0.4
111.39	1.05	2.0	134.50	6AU	0.3	-4.2	-1.3	0.4
117.72	1.11	2.0	144.20	16AU	0.6	-4.1	-1.3	0.3
124.02	1.17	2.0	153.97	26AU	1.2	-3.7	-1.3	0.3
130.27	1.23	2.0	163.78	5SE	3.1	-3.3	-1.3	0.2
136.49	1.29	2.0	173.66	15SE	0.9	-3.8	-1.4	0.1
142.67	1.34	1.9	183.59	25SE	-0.6	-4.2	-1.4	0.1
148.81	1.40	1.9	193.56	5OC	-0.8	-4.3	-1.5	0.0
154.91	1.45	1.9	203.63	15OC	-0.8	-4.3	-1.5	0.1
160.95	1.49	1.8	213.72	25OC	-0.8	-4.3	-1.6	0.2
166.93	1.54	1.7	223.85	4NO	-0.6	-4.2	-1.6	0.3
172.85	1.58	1.6	234.02	14NO	-0.6	-4.1	-1.7	0.3
178.67	1.62	1.6	244.21	24NO	-0.6	-4.0	-1.8	0.4
184.40	1.66	1.5	254.41	4DE	-0.4	-3.9	-1.8	0.4
190.02	1.69	1.3	264.62	14DE	1.0	-3.8	-1.9	0.5
195.49	1.72	1.2	274.82	24DE	2.4	-3.7	-1.9	0.5

Right table (438 / 439)

♂ LONG	♂ LAT	♂ MAG	☉ LONG	16.00UT	☿	♀	♃	♄
				438				
200.78	1.74	1.0	284.99	3JA	0.6	-3.7	-2.0	0.6
205.86	1.75	0.9	295.14	13JA	0.2	-3.6	-2.0	0.6
210.67	1.75	0.7	305.25	23JA	0.1	-3.5	-2.0	0.6
215.16	1.74	0.5	315.31	2FE	-0.1	-3.5	-2.1	0.6
219.24	1.70	0.3	325.32	12FE	-0.5	-3.4	-2.0	0.6
222.81	1.64	0.0	335.27	22FE	-1.1	-3.4	-2.0	0.6
225.73	1.53	-0.3	345.17	4MR	-1.5	-3.4	-2.0	0.6
227.84	1.38	-0.5	355.01	14MR	-1.0	-3.3	-1.9	0.5
228.96	1.16	-0.9	4.79	24MR	0.2	-3.3	-1.9	0.5
228.94	0.86	-1.2	14.51	3AP	1.5	-3.3	-1.8	0.4
227.65	0.47	-1.5	24.19	13AP	3.2	-3.3	-1.7	0.4
225.20	-0.01	-1.8	33.82	23AP	2.8	-3.3	-1.7	0.4
222.00	-0.55	-2.0	43.41	3MY	1.6	-3.3	-1.6	0.5
218.66	-1.09	-1.9	52.97	13MY	0.8	-3.3	-1.5	0.5
215.95	-1.58	-1.7	62.51	23MY	0.0	-3.4	-1.5	0.5
214.38	-1.98	-1.5	72.03	2JN	-1.0	-3.4	-1.4	0.5
214.17	-2.28	-1.3	81.56	12JN	-1.8	-3.5	-1.4	0.5
215.33	-2.50	-1.1	91.09	22JN	-1.0	-3.5	-1.3	0.5
217.67	-2.64	-0.9	100.63	2JL	-0.3	-3.5	-1.3	0.4
221.03	-2.72	-0.7	110.19	12JL	0.2	-3.4	-1.3	0.4
225.23	-2.75	-0.5	119.79	22JL	0.5	-3.4	-1.2	0.4
230.10	-2.75	-0.4	129.42	1AU	0.8	-3.4	-1.2	0.3
235.53	-2.72	-0.2	139.11	11AU	1.6	-3.4	-1.2	0.3
241.42	-2.66	-0.1	148.84	21AU	3.2	-3.3	-1.2	0.2
247.66	-2.57	0.0	158.62	31AU	0.7	-3.3	-1.2	0.2
254.21	-2.47	0.2	168.47	10SE	-0.7	-3.3	-1.2	0.1
261.01	-2.35	0.3	178.37	20SE	-1.0	-3.3	-1.2	0.0
267.99	-2.21	0.4	188.34	30SE	-1.0	-3.3	-1.3	-0.0
275.14	-2.07	0.5	198.36	10OC	-0.7	-3.4	-1.3	-0.1
282.40	-1.91	0.6	208.42	20OC	-0.5	-3.4	-1.3	-0.1
289.74	-1.75	0.7	218.54	30OC	-0.4	-3.4	-1.4	-0.1
297.15	-1.58	0.8	228.69	9NO	-0.4	-3.4	-1.4	0.0
304.59	-1.41	0.9	238.87	19NO	-0.2	-3.5	-1.5	0.1
312.04	-1.24	1.0	249.06	29NO	1.2	-3.5	-1.6	0.2
319.49	-1.07	1.1	259.27	9DE	2.0	-3.6	-1.6	0.2
326.91	-0.91	1.2	269.47	19DE	0.4	-3.7	-1.7	0.3
334.31	-0.75	1.2	279.66	29DE	0.0	-3.7	-1.8	0.3
				439				
341.66	-0.59	1.3	289.82	8JA	-0.1	-3.8	-1.8	0.4
348.96	-0.44	1.4	299.95	18JA	-0.2	-3.9	-1.9	0.4
356.21	-0.30	1.5	310.04	28JA	-0.5	-4.0	-1.9	0.4
3.39	-0.17	1.5	320.08	7FE	-1.1	-4.1	-2.0	0.5
10.51	-0.04	1.6	330.06	17FE	-1.4	-4.2	-2.0	0.5
17.55	0.08	1.7	339.99	27FE	-1.0	-4.2	-2.0	0.5
24.54	0.20	1.7	349.86	9MR	0.2	-4.3	-2.0	0.4
31.45	0.30	1.8	359.67	19MR	1.9	-4.2	-2.0	0.4
38.30	0.40	1.8	9.42	29MR	3.6	-3.9	-2.0	0.4
45.09	0.49	1.8	19.12	8AP	2.1	-3.4	-1.9	0.3
51.81	0.58	1.9	28.77	18AP	1.2	-3.1	-1.9	0.3
58.49	0.66	1.9	38.38	28AP	0.6	-3.8	-1.8	0.3
65.11	0.73	1.9	47.96	8MY	-0.1	-4.1	-1.8	0.3
71.68	0.79	1.9	57.51	18MY	-1.0	-4.2	-1.7	0.3
78.21	0.86	1.9	67.04	28MY	-1.8	-4.2	-1.6	0.4
84.71	0.91	1.9	76.57	7JN	-1.1	-4.1	-1.6	0.4
91.17	0.96	1.9	86.09	17JN	-0.2	-4.0	-1.5	0.4
97.60	1.01	1.9	95.62	27JN	0.3	-3.9	-1.5	0.4
104.01	1.05	1.9	105.18	7JL	0.7	-3.8	-1.4	0.4
110.40	1.09	2.0	114.75	17JL	1.1	-3.7	-1.4	0.3
116.78	1.12	2.0	124.37	27JL	2.1	-3.7	-1.3	0.3
123.14	1.15	2.0	134.03	6AU	2.9	-3.6	-1.3	0.3
129.50	1.17	2.0	143.73	16AU	0.6	-3.6	-1.3	0.2
135.86	1.19	2.0	153.49	26AU	-0.8	-3.5	-1.2	0.2
142.21	1.21	2.0	163.30	5SE	-1.1	-3.5	-1.2	0.1
148.56	1.22	2.0	173.18	15SE	-1.1	-3.4	-1.2	0.1
154.92	1.23	2.0	183.11	25SE	-0.7	-3.4	-1.2	-0.0
161.28	1.23	2.0	193.10	5OC	-0.4	-3.4	-1.2	-0.1
167.64	1.23	2.0	203.14	15OC	-0.3	-3.4	-1.2	-0.2
174.00	1.22	1.9	213.23	25OC	-0.2	-3.4	-1.3	-0.2
180.36	1.21	1.9	223.37	4NO	0.0	-3.4	-1.3	-0.2
186.73	1.19	1.8	233.53	14NO	1.5	-3.4	-1.3	-0.1
193.08	1.16	1.7	243.72	24NO	1.7	-3.4	-1.4	-0.1
199.43	1.13	1.7	253.92	4DE	0.2	-3.4	-1.4	-0.0
205.77	1.08	1.6	264.12	14DE	-0.1	-3.4	-1.5	0.0
212.10	1.03	1.5	274.32	24DE	-0.2	-3.4	-1.5	0.1

♂ LONG	♂ LAT	♂ MAG	⊙ LONG	16.00UT 440	☿	♀	♃	♄
					MAGNITUDES			
218.39	0.96	1.4	284.50	3JA	-0.3	-3.4	-1.6	0.2
224.66	0.88	1.3	294.65	13JA	-0.6	-3.4	-1.7	0.2
230.88	0.78	1.1	304.76	23JA	-1.1	-3.5	-1.7	0.3
237.05	0.66	1.0	314.82	2FE	-1.3	-3.5	-1.8	0.3
243.15	0.52	0.9	324.83	12FE	-0.9	-3.5	-1.9	0.3
249.17	0.35	0.7	334.79	22FE	0.3	-3.4	-1.9	0.4
255.07	0.15	0.5	344.69	3MR	2.3	-3.4	-2.0	0.4
260.84	-0.08	0.3	354.53	13MR	2.8	-3.4	-2.0	0.4
266.43	-0.35	0.1	4.31	23MR	1.5	-3.4	-2.0	0.3
271.79	-0.67	-0.1	14.04	2AP	0.9	-3.3	-2.0	0.3
276.85	-1.05	-0.3	23.72	12AP	0.4	-3.3	-2.0	0.3
281.53	-1.49	-0.6	33.35	22AP	-0.1	-3.3	-2.0	0.2
285.71	-2.00	-0.8	42.94	2MY	-1.0	-3.3	-2.0	0.2
289.26	-2.59	-1.1	52.50	12MY	-1.9	-3.3	-1.9	0.2
292.00	-3.25	-1.4	62.05	22MY	-1.1	-3.4	-1.9	0.2
293.73	-3.99	-1.7	71.57	1JN	-0.2	-3.4	-1.8	0.3
294.30	-4.77	-2.0	81.09	11JN	0.4	-3.4	-1.8	0.3
293.58	-5.53	-2.3	90.62	21JN	0.9	-3.4	-1.7	0.3
291.74	-6.16	-2.5	100.16	1JL	1.5	-3.5	-1.6	0.3
289.21	-6.55	-2.6	109.73	11JL	2.7	-3.5	-1.6	0.3
286.67	-6.63	-2.5	119.32	21JL	2.5	-3.6	-1.5	0.3
284.85	-6.42	-2.3	128.96	31JL	0.5	-3.7	-1.5	0.3
284.20	-5.98	-2.1	138.63	10AU	-0.8	-3.7	-1.4	0.3
284.87	-5.43	-1.8	148.37	20AU	-1.3	-3.8	-1.4	0.2
286.81	-4.83	-1.5	158.15	30AU	-1.1	-3.9	-1.3	0.2
289.82	-4.23	-1.2	167.99	9SE	-0.6	-4.1	-1.3	0.1
293.70	-3.67	-1.0	177.89	19SE	-0.3	-4.2	-1.3	0.1
298.28	-3.14	-0.7	187.85	29SE	-0.1	-4.3	-1.3	0.0
303.39	-2.66	-0.5	197.87	9OC	-0.1	-4.3	-1.3	-0.1
308.91	-2.22	-0.3	207.94	19OC	0.2	-4.3	-1.3	-0.1
314.74	-1.83	-0.1	218.04	29OC	1.7	-4.2	-1.3	-0.2
320.81	-1.47	0.1	228.19	8NO	1.5	-3.7	-1.3	-0.3
327.06	-1.16	0.3	238.37	18NO	-0.0	-2.9	-1.3	-0.3
333.44	-0.87	0.5	248.57	28NO	-0.3	-3.8	-1.3	-0.2
339.90	-0.61	0.7	258.77	8DE	-0.3	-4.2	-1.4	-0.2
346.44	-0.39	0.8	268.98	18DE	-0.4	-4.4	-1.4	-0.1
353.01	-0.18	0.9	279.16	28DE	-0.7	-4.4	-1.4	-0.0
				441				
359.60	-0.00	1.1	289.33	7JA	-1.0	-4.3	-1.5	0.1
6.21	0.16	1.2	299.46	17JA	-1.1	-4.2	-1.6	0.1
12.80	0.30	1.3	309.55	27JA	-0.8	-4.1	-1.6	0.2
19.39	0.43	1.4	319.59	6FE	0.4	-4.0	-1.7	0.2
25.96	0.54	1.5	329.58	16FE	2.7	-3.8	-1.8	0.3
32.50	0.64	1.6	339.51	26FE	2.1	-3.8	-1.8	0.3
39.02	0.73	1.7	349.38	8MR	1.1	-3.7	-1.9	0.3
45.52	0.81	1.7	359.20	18MR	0.7	-3.6	-2.0	0.3
51.98	0.87	1.8	8.95	28MR	0.3	-3.5	-2.0	0.3
58.42	0.93	1.9	18.66	7AP	-0.2	-3.5	-2.1	0.3
64.84	0.98	1.9	28.31	17AP	-1.0	-3.4	-2.1	0.3
71.23	1.03	1.9	37.92	27AP	-1.9	-3.4	-2.1	0.2
77.61	1.06	2.0	47.50	7MY	-1.1	-3.3	-2.1	0.2
83.96	1.09	2.0	57.06	17MY	-0.1	-3.3	-2.1	0.2
90.31	1.12	2.0	66.59	27MY	0.6	-3.3	-2.1	0.2
96.65	1.14	2.0	76.11	6JN	1.2	-3.3	-2.1	0.2
102.98	1.15	2.0	85.64	16JN	2.0	-3.3	-2.0	0.2
109.31	1.16	2.0	95.17	26JN	3.3	-3.3	-2.0	0.3
115.65	1.16	2.0	104.72	6JL	2.0	-3.3	-1.9	0.3
121.99	1.16	2.0	114.29	16JL	0.4	-3.4	-1.8	0.3
128.35	1.16	2.0	123.90	26JL	-0.9	-3.4	-1.8	0.3
134.73	1.15	2.0	133.56	5AU	-1.4	-3.4	-1.7	0.3
141.13	1.13	2.0	143.26	15AU	-1.1	-3.5	-1.7	0.3
147.55	1.11	2.0	153.01	25AU	-0.5	-3.5	-1.6	0.3
154.00	1.09	2.0	162.83	4SE	-0.2	-3.5	-1.5	0.2
160.48	1.06	2.0	172.70	14SE	-0.0	-3.5	-1.5	0.2
167.00	1.03	2.0	182.62	24SE	0.1	-3.4	-1.5	0.1
173.55	0.99	2.0	192.61	4OC	0.4	-3.4	-1.4	0.1
180.14	0.94	1.9	202.65	14OC	2.0	-3.4	-1.4	0.0
186.77	0.89	1.9	212.73	24OC	1.3	-3.4	-1.4	-0.1
193.44	0.83	1.9	222.87	3NO	-0.2	-3.3	-1.4	-0.1
200.15	0.77	1.8	233.03	13NO	-0.4	-3.3	-1.3	-0.2
206.91	0.69	1.8	243.22	23NO	-0.5	-3.3	-1.3	-0.3
213.71	0.61	1.8	253.42	3DE	-0.5	-3.3	-1.4	-0.3
220.54	0.52	1.7	263.63	13DE	-0.7	-3.3	-1.4	-0.2
227.42	0.42	1.6	273.82	23DE	-0.9	-3.4	-1.4	-0.2

♂ LONG	♂ LAT	♂ MAG	⊙ LONG	16.00UT 442	☿	♀	♃	♄
					MAGNITUDES			
234.34	0.31	1.6	284.00	2JA	-1.0	-3.4	-1.4	-0.1
241.29	0.19	1.5	294.15	12JA	-0.7	-3.4	-1.5	-0.0
248.28	0.05	1.4	304.27	22JA	0.6	-3.4	-1.5	0.1
255.30	-0.09	1.3	314.34	1FE	3.0	-3.5	-1.6	0.1
262.35	-0.25	1.2	324.35	11FE	1.6	-3.5	-1.6	0.2
269.43	-0.43	1.1	334.31	21FE	0.8	-3.6	-1.7	0.2
276.52	-0.61	1.0	344.22	3MR	0.5	-3.6	-1.7	0.3
283.62	-0.81	0.9	354.06	13MR	0.2	-3.7	-1.8	0.3
290.72	-1.02	0.8	3.84	23MR	-0.3	-3.8	-1.9	0.3
297.80	-1.25	0.7	13.58	2AP	-1.0	-3.9	-1.9	0.3
304.84	-1.48	0.6	23.26	12AP	-1.8	-3.9	-2.0	0.3
311.84	-1.72	0.4	32.89	22AP	-1.1	-4.0	-2.1	0.3
318.75	-1.98	0.3	42.49	2MY	-0.1	-4.1	-2.1	0.3
325.54	-2.23	0.2	52.05	12MY	0.8	-4.2	-2.2	0.3
332.18	-2.49	0.0	61.59	22MY	1.5	-4.2	-2.2	0.2
338.62	-2.76	-0.1	71.12	1JN	2.7	-4.0	-2.2	0.2
344.79	-3.02	-0.3	80.64	11JN	3.1	-3.6	-2.2	0.2
350.62	-3.28	-0.4	90.17	21JN	1.6	-2.9	-2.2	0.3
356.01	-3.53	-0.6	99.71	1JL	0.3	-3.4	-2.2	0.3
0.85	-3.77	-0.8	109.27	11JL	-0.9	-4.0	-2.2	0.3
4.98	-3.99	-1.0	118.86	21JL	-1.5	-4.2	-2.1	0.3
8.23	-4.18	-1.2	128.49	31JL	-1.1	-4.2	-2.1	0.4
10.37	-4.32	-1.4	138.17	10AU	-0.5	-4.2	-2.0	0.4
11.22	-4.39	-1.7	147.89	20AU	-0.1	-4.1	-1.9	0.4
10.60	-4.34	-1.9	157.68	30AU	0.1	-4.0	-1.9	0.3
8.60	-4.12	-2.1	167.51	9SE	0.3	-3.9	-1.8	0.3
5.58	-3.72	-2.2	177.41	19SE	0.7	-3.9	-1.8	0.3
2.24	-3.15	-2.2	187.37	29SE	2.4	-3.8	-1.7	0.2
359.42	-2.48	-1.9	197.38	9OC	1.2	-3.7	-1.6	0.2
357.67	-1.81	-1.6	207.44	19OC	-0.4	-3.6	-1.6	0.1
357.26	-1.20	-1.3	217.55	29OC	-0.6	-3.6	-1.6	0.1
358.16	-0.68	-0.9	227.70	8NO	-0.6	-3.5	-1.5	-0.0
0.16	-0.25	-0.6	237.88	18NO	-0.6	-3.5	-1.5	-0.1
3.09	0.09	-0.3	248.07	28NO	-0.8	-3.5	-1.5	-0.2
6.73	0.37	-0.1	258.28	8DE	-0.8	-3.4	-1.5	-0.2
10.92	0.60	0.2	268.48	18DE	-0.8	-3.4	-1.4	-0.2
15.55	0.78	0.4	278.67	28DE	-0.6	-3.4	-1.5	-0.2
				443				
20.52	0.92	0.6	288.84	7JA	0.7	-3.3	-1.5	-0.1
25.73	1.03	0.8	298.97	17JA	3.0	-3.3	-1.5	-0.0
31.15	1.12	1.0	309.06	27JA	1.2	-3.3	-1.5	0.0
36.72	1.19	1.1	319.11	6FE	0.5	-3.3	-1.5	0.1
42.42	1.25	1.3	329.10	16FE	0.3	-3.3	-1.6	0.2
48.20	1.29	1.4	339.03	26FE	0.1	-3.3	-1.6	0.2
54.06	1.32	1.5	348.90	8MR	-0.3	-3.4	-1.7	0.3
59.98	1.34	1.6	358.72	18MR	-1.0	-3.4	-1.7	0.3
65.95	1.35	1.7	8.48	28MR	-1.7	-3.4	-1.8	0.3
71.95	1.36	1.8	18.18	7AP	-1.1	-3.5	-1.9	0.3
77.99	1.36	1.8	27.84	17AP	-0.0	-3.5	-1.9	0.4
84.05	1.35	1.9	37.46	27AP	1.0	-3.5	-2.0	0.4
90.15	1.34	1.9	47.04	7MY	2.0	-3.4	-2.1	0.4
96.27	1.32	2.0	56.59	17MY	3.5	-3.4	-2.1	0.4
102.41	1.30	2.0	66.12	27MY	2.5	-3.4	-2.2	0.3
108.59	1.28	2.0	75.65	6JN	1.3	-3.3	-2.3	0.3
114.80	1.25	2.0	85.17	16JN	0.2	-3.3	-2.3	0.3
121.04	1.22	2.0	94.70	26JN	-0.9	-3.3	-2.3	0.3
127.31	1.18	2.0	104.25	6JL	-1.6	-3.3	-2.4	0.3
133.62	1.14	2.0	113.83	16JL	-1.1	-3.3	-2.4	0.4
139.98	1.09	2.0	123.44	26JL	-0.4	-3.3	-2.4	0.4
146.38	1.05	2.0	133.09	5AU	0.0	-3.4	-2.3	0.4
152.83	0.99	2.0	142.79	15AU	0.3	-3.4	-2.3	0.5
159.34	0.94	1.9	152.54	25AU	0.5	-3.4	-2.2	0.5
165.90	0.88	1.9	162.35	4SE	1.0	-3.5	-2.2	0.5
172.52	0.81	1.8	172.22	14SE	2.8	-3.5	-2.1	0.4
179.20	0.74	1.8	182.14	24SE	1.0	-3.6	-2.1	0.4
185.94	0.67	1.8	192.12	4OC	-0.5	-3.6	-2.0	0.4
192.75	0.59	1.8	202.16	14OC	-0.8	-3.7	-1.9	0.3
199.62	0.51	1.8	212.24	24OC	-0.7	-3.8	-1.9	0.3
206.57	0.42	1.8	222.37	3NO	-0.8	-3.9	-1.8	0.2
213.58	0.32	1.8	232.54	13NO	-0.7	-4.0	-1.7	0.2
220.66	0.22	1.7	242.72	23NO	-0.6	-4.1	-1.7	0.1
227.81	0.12	1.7	252.92	3DE	-0.7	-4.2	-1.6	0.0
235.02	0.01	1.7	263.13	13DE	-0.5	-4.3	-1.6	-0.1
242.29	-0.11	1.6	273.33	23DE	0.8	-4.4	-1.6	-0.1

Left section:

♂ LONG	LAT	MAG	☉ LONG	16.00UT 444	☿	♀	♃	♄
249.63	-0.23	1.6	283.50	2JA	2.6	-4.4	-1.6	-0.1
257.03	-0.35	1.5	293.66	12JA	0.8	-4.2	-1.5	-0.1
264.48	-0.48	1.5	303.77	22JA	0.3	-3.7	-1.5	-0.0
271.97	-0.61	1.4	313.84	1FE	0.2	-3.3	-1.5	0.1
279.51	-0.74	1.4	323.86	11FE	-0.0	-3.8	-1.5	0.1
287.08	-0.87	1.3	333.82	21FE	-0.4	-4.2	-1.6	0.2
294.67	-1.01	1.3	343.73	2MR	-1.0	-4.3	-1.6	0.2
302.28	-1.13	1.2	353.58	12MR	-1.6	-4.3	-1.6	0.3
309.89	-1.26	1.2	3.37	22MR	-1.1	-4.2	-1.6	0.3
317.49	-1.37	1.1	13.10	1AP	0.1	-4.1	-1.7	0.4
325.06	-1.48	1.0	22.79	11AP	1.3	-4.0	-1.7	0.4
332.58	-1.58	1.0	32.42	21AP	2.7	-3.9	-1.8	0.4
340.06	-1.67	0.9	42.02	1MY	3.3	-3.8	-1.9	0.5
347.45	-1.74	0.9	51.59	11MY	1.9	-3.7	-1.9	0.5
354.74	-1.79	0.8	61.13	21MY	1.0	-3.6	-2.0	0.5
1.92	-1.83	0.7	70.66	31MY	0.1	-3.5	-2.1	0.5
8.96	-1.86	0.7	80.18	10JN	-0.9	-3.5	-2.1	0.5
15.83	-1.86	0.6	89.71	20JN	-1.7	-3.4	-2.2	0.4
22.50	-1.85	0.5	99.25	30JN	-1.1	-3.4	-2.3	0.4
28.93	-1.81	0.5	108.81	10JL	-0.3	-3.4	-2.3	0.4
35.10	-1.75	0.4	118.40	20JL	0.1	-3.4	-2.4	0.5
40.94	-1.67	0.3	128.03	30JL	0.4	-3.3	-2.4	0.5
46.39	-1.57	0.1	137.70	9AU	0.7	-3.3	-2.4	0.6
51.36	-1.44	0.0	147.42	19AU	1.3	-3.3	-2.4	0.6
55.77	-1.27	-0.1	157.20	29AU	3.2	-3.4	-2.4	0.6
59.46	-1.07	-0.3	167.04	8SE	0.9	-3.4	-2.4	0.6
62.28	-0.82	-0.5	176.93	18SE	-0.6	-3.4	-2.4	0.6
64.03	-0.52	-0.7	186.89	28SE	-0.9	-3.4	-2.3	0.6
64.50	-0.16	-0.9	196.90	8OC	-0.9	-3.4	-2.3	0.6
63.55	0.26	-1.1	206.95	18OC	-0.8	-3.4	-2.2	0.6
61.19	0.72	-1.3	217.06	28OC	-0.6	-3.5	-2.1	0.5
57.74	1.19	-1.5	227.21	7NO	-0.5	-3.5	-2.1	0.5
53.86	1.61	-1.5	237.38	17NO	-0.5	-3.5	-2.0	0.4
50.34	1.93	-1.3	247.57	27NO	-0.3	-3.5	-1.9	0.3
47.85	2.15	-1.0	257.78	7DE	1.0	-3.4	-1.9	0.3
46.68	2.27	-0.7	267.98	17DE	2.3	-3.4	-1.8	0.2
46.82	2.32	-0.4	278.17	27DE	0.5	-3.4	-1.7	0.1
				445				
48.14	2.32	-0.1	288.34	6JA	0.1	-3.4	-1.7	0.0
50.42	2.29	0.2	298.47	16JA	0.0	-3.3	-1.7	0.1
53.47	2.24	0.4	308.57	26JA	-0.1	-3.3	-1.6	0.1
57.13	2.18	0.6	318.61	5FE	-0.5	-3.3	-1.6	0.2
61.26	2.12	0.8	328.60	15FE	-1.0	-3.3	-1.6	0.2
65.76	2.05	1.0	338.54	25FE	-1.5	-3.3	-1.6	0.3
70.56	1.98	1.2	348.42	7MR	-1.0	-3.4	-1.6	0.3
75.60	1.91	1.3	358.24	17MR	0.1	-3.4	-1.6	0.4
80.83	1.84	1.4	8.01	27MR	1.6	-3.4	-1.6	0.4
86.22	1.76	1.5	17.71	6AP	3.5	-3.4	-1.6	0.5
91.74	1.69	1.6	27.37	16AP	2.5	-3.5	-1.6	0.5
97.38	1.62	1.7	36.99	26AP	1.4	-3.5	-1.7	0.6
103.12	1.55	1.8	46.58	6MY	0.7	-3.6	-1.7	0.6
108.95	1.47	1.8	56.15	16MY	-0.0	-3.6	-1.8	0.6
114.86	1.40	1.9	65.67	26MY	-1.0	-3.7	-1.8	0.6
120.85	1.32	1.9	75.19	5JN	-1.8	-3.8	-1.9	0.6
126.92	1.25	1.9	84.71	15JN	-1.1	-3.9	-1.9	0.7
133.06	1.17	1.9	94.25	25JN	-0.3	-4.0	-2.0	0.6
139.27	1.09	1.9	103.79	5JL	0.2	-4.1	-2.1	0.6
145.56	1.01	1.9	113.37	15JL	0.6	-4.2	-2.1	0.6
151.92	0.92	1.9	122.98	25JL	0.9	-4.2	-2.2	0.6
158.36	0.84	1.9	132.62	4AU	1.8	-4.2	-2.3	0.7
164.88	0.75	1.9	142.32	14AU	3.1	-4.0	-2.3	0.7
171.48	0.66	1.9	152.07	24AU	0.7	-3.6	-2.4	0.8
178.16	0.57	1.9	161.87	3SE	-0.7	-3.3	-2.4	0.8
184.92	0.48	1.8	171.74	13SE	-1.1	-3.8	-2.4	0.8
191.77	0.38	1.8	181.66	23SE	-1.0	-4.2	-2.4	0.8
198.71	0.28	1.7	191.64	3OC	-0.7	-4.3	-2.4	0.8
205.73	0.18	1.7	201.67	13OC	-0.5	-4.3	-2.4	0.8
212.84	0.08	1.6	211.76	23OC	-0.4	-4.3	-2.4	0.8
220.03	-0.02	1.6	221.88	2NO	-0.3	-4.2	-2.3	0.8
227.31	-0.13	1.6	232.04	12NO	-0.1	-4.1	-2.3	0.7
234.67	-0.24	1.6	242.23	22NO	1.2	-4.0	-2.2	0.7
242.11	-0.34	1.6	252.43	2DE	2.0	-3.9	-2.1	0.6
249.62	-0.45	1.5	262.63	12DE	0.3	-3.8	-2.1	0.5
257.19	-0.55	1.5	272.83	22DE	-0.0	-3.7	-2.0	0.5

Right section:

♂ LONG	LAT	MAG	☉ LONG	16.00UT 446	☿	♀	♃	♄
264.83	-0.65	1.5	283.01	1JA	-0.1	-3.6	-1.9	0.4
272.51	-0.75	1.5	293.17	11JA	-0.2	-3.6	-1.8	0.3
280.24	-0.85	1.5	303.28	21JA	-0.5	-3.5	-1.8	0.3
288.01	-0.93	1.5	313.35	31JA	-1.1	-3.5	-1.7	0.3
295.79	-1.02	1.4	323.38	10FE	-1.4	-3.4	-1.7	0.3
303.59	-1.09	1.4	333.35	20FE	-1.0	-3.4	-1.6	0.4
311.39	-1.15	1.4	343.25	2MR	0.2	-3.4	-1.6	0.4
319.17	-1.20	1.4	353.10	12MR	2.0	-3.3	-1.6	0.5
326.93	-1.24	1.4	2.90	22MR	3.3	-3.3	-1.6	0.5
334.65	-1.27	1.3	12.63	1AP	1.8	-3.3	-1.5	0.6
342.32	-1.28	1.3	22.32	11AP	1.1	-3.3	-1.5	0.7
349.93	-1.28	1.3	31.96	21AP	0.5	-3.3	-1.5	0.7
357.47	-1.27	1.3	41.56	1MY	-0.1	-3.3	-1.5	0.7
4.91	-1.24	1.3	51.12	11MY	-1.0	-3.3	-1.6	0.8
12.27	-1.20	1.3	60.67	21MY	-1.8	-3.4	-1.6	0.8
19.51	-1.14	1.2	70.20	31MY	-1.1	-3.4	-1.6	0.8
26.64	-1.07	1.2	79.72	10JN	-0.2	-3.5	-1.7	0.9
33.64	-0.99	1.2	89.25	20JN	0.3	-3.5	-1.7	0.9
40.49	-0.89	1.2	98.78	30JN	0.7	-3.5	-1.7	0.9
47.20	-0.78	1.1	108.34	10JL	1.3	-3.4	-1.8	0.9
53.73	-0.66	1.1	117.93	20JL	2.3	-3.4	-1.9	0.9
60.08	-0.52	1.0	127.56	30JL	2.8	-3.4	-1.9	0.9
66.22	-0.37	1.0	137.23	9AU	0.6	-3.4	-2.0	0.9
72.11	-0.20	0.9	146.95	19AU	-0.8	-3.3	-2.0	0.9
77.74	-0.02	0.8	156.72	29AU	-1.2	-3.3	-2.1	1.0
83.04	0.19	0.7	166.56	8SE	-1.1	-3.3	-2.2	1.0
87.96	0.41	0.6	176.45	18SE	-0.6	-3.3	-2.2	1.1
92.43	0.67	0.5	186.40	28SE	-0.3	-3.3	-2.3	1.1
96.33	0.95	0.3	196.41	8OC	-0.2	-3.4	-2.3	1.1
99.54	1.28	0.2	206.46	18OC	-0.2	-3.4	-2.4	1.1
101.91	1.64	-0.0	216.57	28OC	0.1	-3.4	-2.4	1.1
103.23	2.05	-0.2	226.71	7NO	1.5	-3.4	-2.4	1.1
103.33	2.50	-0.5	236.89	17NO	1.7	-3.5	-2.3	1.0
102.07	2.95	-0.7	247.08	27NO	0.1	-3.5	-2.3	1.0
99.47	3.39	-0.9	257.28	7DE	-0.2	-3.6	-2.3	0.9
95.86	3.73	-1.1	267.49	17DE	-0.2	-3.7	-2.2	0.9
91.88	3.93	-1.1	277.68	27DE	-0.3	-3.7	-2.1	0.8
				447				
88.28	3.97	-0.9	287.85	6JA	-0.6	-3.8	-2.1	0.7
85.67	3.87	-0.6	297.98	16JA	-1.1	-3.9	-2.0	0.7
84.34	3.68	-0.3	308.08	26JA	-1.2	-4.0	-1.9	0.6
84.30	3.45	-0.1	318.13	5FE	-0.9	-4.1	-1.8	0.5
85.42	3.20	0.2	328.12	15FE	0.3	-4.2	-1.8	0.6
87.50	2.96	0.4	338.06	25FE	2.4	-4.2	-1.7	0.6
90.36	2.74	0.6	347.95	7MR	2.6	-4.3	-1.7	0.6
93.85	2.52	0.8	357.77	17MR	1.4	-4.2	-1.6	0.7
97.84	2.33	1.0	7.53	27MR	0.8	-3.9	-1.6	0.8
102.22	2.15	1.1	17.24	6AP	0.4	-3.3	-1.5	0.8
106.94	1.97	1.2	26.90	16AP	-0.2	-3.1	-1.5	0.9
111.92	1.81	1.3	36.52	26AP	-1.0	-3.8	-1.5	0.9
117.13	1.66	1.4	46.11	6MY	-1.9	-4.1	-1.5	1.0
122.54	1.52	1.5	55.66	16MY	-1.1	-4.2	-1.5	1.0
128.12	1.38	1.6	65.20	26MY	-0.2	-4.2	-1.5	1.0
133.85	1.25	1.6	74.73	5JN	0.5	-4.1	-1.5	1.1
139.73	1.12	1.7	84.25	15JN	1.0	-4.0	-1.5	1.1
145.73	0.99	1.7	93.78	25JN	1.7	-3.9	-1.5	1.1
151.86	0.87	1.7	103.33	5JL	2.9	-3.8	-1.5	1.2
158.11	0.75	1.7	112.90	15JL	2.3	-3.7	-1.6	1.2
164.47	0.63	1.8	122.51	25JL	0.5	-3.7	-1.6	1.2
170.94	0.51	1.8	132.16	4AU	-0.8	-3.6	-1.6	1.2
177.53	0.39	1.8	141.85	14AU	-1.3	-3.6	-1.7	1.2
184.23	0.28	1.7	151.60	24AU	-1.1	-3.5	-1.7	1.2
191.03	0.16	1.7	161.40	3SE	-0.6	-3.5	-1.8	1.2
197.95	0.05	1.7	171.26	13SE	-0.2	-3.4	-1.9	1.3
204.97	-0.06	1.7	181.18	23SE	-0.1	-3.4	-1.9	1.3
212.09	-0.17	1.6	191.16	3OC	0.0	-3.4	-2.0	1.4
219.32	-0.27	1.6	201.19	13OC	0.3	-3.4	-2.1	1.4
226.65	-0.38	1.6	211.27	23OC	1.8	-3.4	-2.1	1.4
234.07	-0.48	1.5	221.39	2NO	1.5	-3.4	-2.2	1.3
241.58	-0.57	1.5	231.55	12NO	-0.1	-3.4	-2.2	1.3
249.17	-0.66	1.5	241.73	22NO	-0.4	-3.4	-2.2	1.3
256.83	-0.75	1.4	251.94	2DE	-0.4	-3.4	-2.2	1.2
264.56	-0.83	1.4	262.14	12DE	-0.5	-3.4	-2.2	1.2
272.35	-0.89	1.3	272.34	22DE	-0.7	-3.4	-2.2	1.1

448 / 449

♂ LONG	LAT	MAG	☉ LONG	16.00UT	☿	♀	♃	♄
280.17	-0.96	1.3	282.52	1JA	-1.0	-3.4	-2.2	1.1
288.03	-1.01	1.3	292.67	11JA	-1.1	-3.4	-2.1	1.0
295.91	-1.05	1.4	302.79	21JA	-0.8	-3.5	-2.1	1.0
303.80	-1.08	1.4	312.87	31JA	0.4	-3.5	-2.0	0.9
311.67	-1.10	1.4	322.89	10FE	2.8	-3.5	-1.9	0.9
319.53	-1.10	1.4	332.86	20FE	1.9	-3.4	-1.9	0.8
327.36	-1.10	1.4	342.77	1MR	1.0	-3.4	-1.8	0.9
335.14	-1.08	1.4	352.62	11MR	0.6	-3.4	-1.7	0.9
342.87	-1.05	1.4	2.42	21MR	0.2	-3.4	-1.7	1.0
350.54	-1.01	1.4	12.16	31MR	-0.2	-3.3	-1.6	1.0
358.13	-0.96	1.5	21.84	10AP	-1.0	-3.3	-1.5	1.1
5.65	-0.90	1.5	31.48	20AP	-1.9	-3.3	-1.5	1.1
13.07	-0.82	1.5	41.09	30AP	-1.1	-3.3	-1.5	1.2
20.41	-0.74	1.5	50.65	10MY	-0.1	-3.3	-1.4	1.2
27.65	-0.65	1.5	60.20	20MY	0.6	-3.4	-1.4	1.3
34.79	-0.56	1.5	69.73	30MY	1.3	-3.4	-1.4	1.3
41.82	-0.45	1.5	79.25	9JN	2.2	-3.4	-1.4	1.4
48.74	-0.34	1.5	88.78	19JN	3.3	-3.4	-1.4	1.4
55.56	-0.23	1.5	98.32	29JN	1.9	-3.5	-1.3	1.4
62.26	-0.10	1.5	107.87	9JL	0.4	-3.5	-1.4	1.4
68.84	0.03	1.5	117.47	19JL	-0.9	-3.6	-1.4	1.4
75.29	0.16	1.5	127.09	29JL	-1.4	-3.7	-1.4	1.4
81.61	0.30	1.5	136.76	8AU	-1.1	-3.7	-1.4	1.3
87.78	0.45	1.4	146.48	18AU	-0.5	-3.8	-1.5	1.3
93.80	0.61	1.4	156.25	28AU	-0.1	-3.9	-1.5	1.2
99.63	0.78	1.3	166.08	7SE	0.0	-4.1	-1.5	1.2
105.27	0.96	1.3	175.97	17SE	0.2	-4.2	-1.6	1.2
110.68	1.15	1.2	185.92	27SE	0.5	-4.3	-1.6	1.2
115.81	1.35	1.1	195.92	7OC	2.1	-4.3	-1.7	1.2
120.62	1.58	1.0	205.98	17OC	1.3	-4.3	-1.8	1.2
125.04	1.82	0.8	216.08	27OC	-0.3	-4.2	-1.8	1.2
128.99	2.09	0.7	226.22	6NO	-0.5	-3.7	-1.9	1.2
132.34	2.39	0.5	236.39	16NO	-0.5	-2.9	-2.0	1.1
134.96	2.73	0.3	246.58	26NO	-0.6	-3.8	-2.0	1.1
136.69	3.09	0.1	256.79	6DE	-0.8	-4.2	-2.1	1.1
137.32	3.48	-0.2	266.99	16DE	-0.9	-4.4	-2.1	1.0
136.71	3.87	-0.4	277.18	26DE	-0.9	-4.4	-2.1	1.0
				449				
134.78	4.22	-0.7	287.35	5JA	-0.7	-4.3	-2.1	0.9
131.68	4.48	-0.9	297.49	15JA	0.6	-4.2	-2.1	0.9
127.84	4.58	-1.0	307.59	25JA	3.0	-4.1	-2.1	0.8
123.97	4.51	-0.9	317.64	4FE	1.4	-4.0	-2.1	0.8
120.75	4.28	-0.7	327.64	14FE	0.7	-3.8	-2.0	0.8
118.66	3.95	-0.5	337.58	24FE	0.4	-3.8	-1.9	0.7
117.87	3.57	-0.2	347.47	6MR	0.1	-3.7	-1.9	0.7
118.32	3.19	0.0	357.30	16MR	-0.3	-3.6	-1.8	0.8
119.85	2.82	0.2	7.06	26MR	-1.0	-3.5	-1.7	0.8
122.29	2.49	0.4	16.78	5AP	-1.8	-3.5	-1.7	0.9
125.45	2.18	0.6	26.44	15AP	-1.1	-3.4	-1.6	1.0
129.22	1.90	0.8	36.06	25AP	-0.1	-3.4	-1.6	1.1
133.47	1.65	0.9	45.65	5MY	0.8	-3.3	-1.5	1.1
138.13	1.41	1.0	55.21	15MY	1.7	-3.3	-1.4	1.2
143.12	1.20	1.1	64.74	25MY	3.0	-3.3	-1.4	1.2
148.40	1.00	1.2	74.27	4JN	2.9	-3.3	-1.4	1.2
153.94	0.81	1.3	83.79	14JN	1.5	-3.3	-1.3	1.3
159.70	0.64	1.3	93.32	24JN	0.3	-3.3	-1.3	1.3
165.65	0.47	1.4	102.87	4JL	-0.9	-3.3	-1.3	1.3
171.80	0.31	1.4	112.44	14JL	-1.6	-3.4	-1.3	1.3
178.11	0.16	1.4	122.04	24JL	-1.1	-3.4	-1.3	1.3
184.59	0.02	1.5	131.69	3AU	-0.4	-3.4	-1.3	1.2
191.22	-0.11	1.5	141.38	13AU	-0.1	-3.5	-1.3	1.2
197.99	-0.24	1.5	151.12	23AU	0.2	-3.5	-1.3	1.2
204.90	-0.36	1.5	160.92	2SE	0.4	-3.5	-1.3	1.1
211.95	-0.48	1.5	170.78	12SE	0.8	-3.5	-1.3	1.1
219.12	-0.58	1.5	180.69	22SE	2.5	-3.4	-1.4	1.1
226.41	-0.68	1.5	190.67	2OC	1.2	-3.4	-1.4	1.1
233.81	-0.77	1.5	200.70	12OC	-0.4	-3.4	-1.5	1.1
241.31	-0.85	1.5	210.77	22OC	-0.7	-3.4	-1.5	1.1
248.91	-0.92	1.4	220.90	1NO	-0.7	-3.3	-1.6	1.1
256.59	-0.98	1.4	231.06	11NO	-0.7	-3.3	-1.6	1.0
264.34	-1.03	1.4	241.24	21NO	-0.7	-3.3	-1.7	1.0
272.16	-1.07	1.4	251.44	1DE	-0.7	-3.3	-1.8	1.0
280.01	-1.09	1.4	261.64	11DE	-0.7	-3.4	-1.8	1.0
287.90	-1.11	1.4	271.84	21DE	-0.6	-3.4	-1.9	0.9
295.80	-1.11	1.4	282.02	31DE	0.7	-3.4	-2.0	0.9

450 / 451

♂ LONG	LAT	MAG	☉ LONG	16.00UT	☿	♀	♃	♄
303.71	-1.10	1.4	292.18	10JA	2.9	-3.4	-2.0	0.8
311.60	-1.08	1.4	302.30	20JA	1.0	-3.4	-2.0	0.8
319.47	-1.05	1.4	312.38	30JA	0.5	-3.5	-2.0	0.7
327.30	-1.00	1.4	322.41	9FE	0.2	-3.5	-2.0	0.7
335.08	-0.95	1.4	332.38	19FE	0.0	-3.6	-2.0	0.6
342.81	-0.89	1.4	342.29	1MR	-0.4	-3.6	-2.0	0.6
350.47	-0.81	1.4	352.15	11MR	-1.0	-3.7	-2.0	0.6
358.05	-0.74	1.4	1.95	21MR	-1.7	-3.8	-1.9	0.6
5.56	-0.65	1.5	11.69	31MR	-1.1	-3.9	-1.9	0.6
12.98	-0.56	1.5	21.38	10AP	-0.0	-3.9	-1.8	0.7
20.31	-0.46	1.6	31.02	20AP	1.1	-4.0	-1.7	0.8
27.55	-0.37	1.6	40.63	30AP	2.3	-4.1	-1.7	0.8
34.70	-0.26	1.6	50.20	10MY	3.7	-4.2	-1.6	0.9
41.76	-0.16	1.7	59.74	20MY	2.3	-4.2	-1.5	1.0
48.74	-0.05	1.7	69.27	30MY	1.2	-4.0	-1.5	1.0
55.62	0.06	1.7	78.80	9JN	0.2	-3.6	-1.4	1.1
62.41	0.17	1.8	88.32	19JN	-0.9	-2.9	-1.4	1.1
69.12	0.28	1.8	97.86	29JN	-1.6	-3.5	-1.3	1.1
75.74	0.39	1.8	107.41	9JL	-1.1	-4.0	-1.3	1.1
82.28	0.50	1.8	117.00	19JL	-0.4	-4.2	-1.3	1.1
88.73	0.61	1.8	126.62	29JL	0.0	-4.2	-1.2	1.1
95.09	0.73	1.8	136.29	8AU	0.3	-4.2	-1.2	1.1
101.36	0.84	1.8	146.00	18AU	0.5	-4.1	-1.2	1.1
107.54	0.96	1.8	155.78	28AU	1.1	-4.0	-1.2	1.0
113.61	1.08	1.7	165.60	7SE	2.9	-3.9	-1.2	1.0
119.57	1.21	1.7	175.49	17SE	1.0	-3.8	-1.2	1.0
125.42	1.34	1.6	185.43	27SE	-0.5	-3.8	-1.3	0.9
131.12	1.47	1.6	195.43	7OC	-0.8	-3.7	-1.3	0.9
136.66	1.61	1.5	205.49	17OC	-0.8	-3.6	-1.3	0.9
142.02	1.76	1.4	215.59	27OC	-0.8	-3.6	-1.4	1.0
147.15	1.92	1.3	225.73	6NO	-0.6	-3.5	-1.4	1.0
152.01	2.09	1.1	235.90	16NO	-0.6	-3.5	-1.5	1.0
156.56	2.28	1.0	246.09	26NO	-0.6	-3.5	-1.5	0.9
160.70	2.47	0.8	256.29	6DE	-0.6	-3.4	-1.6	0.9
164.35	2.69	0.6	266.50	16DE	0.8	-3.4	-1.6	0.9
167.39	2.92	0.4	276.69	26DE	2.6	-3.4	-1.7	0.9
				451				
169.67	3.17	0.2	286.86	5JA	0.7	-3.4	-1.8	0.8
171.01	3.42	-0.1	297.00	15JA	0.3	-3.3	-1.8	0.8
171.24	3.66	-0.3	307.11	25JA	0.1	-3.3	-1.9	0.7
170.21	3.87	-0.6	317.16	4FE	-0.1	-3.3	-1.9	0.7
167.92	3.99	-0.9	327.16	14FE	-0.4	-3.3	-2.0	0.6
164.59	3.99	-1.1	337.11	24FE	-1.0	-3.3	-2.0	0.6
160.76	3.83	-1.2	346.99	6MR	-1.6	-3.4	-2.0	0.5
157.14	3.52	-1.0	356.82	16MR	-1.1	-3.4	-2.0	0.5
154.37	3.10	-0.8	6.59	26MR	0.1	-3.4	-2.0	0.4
152.83	2.64	-0.6	16.30	5AP	1.4	-3.5	-2.0	0.5
152.59	2.18	-0.4	25.97	15AP	3.0	-3.5	-1.9	0.5
153.59	1.75	-0.2	35.60	25AP	3.0	-3.5	-1.9	0.6
155.64	1.37	-0.0	45.18	5MY	1.7	-3.4	-1.8	0.7
158.56	1.02	0.2	54.74	15MY	0.9	-3.4	-1.7	0.7
162.22	0.71	0.3	64.28	25MY	0.1	-3.4	-1.7	0.8
166.47	0.44	0.5	73.80	4JN	-0.9	-3.3	-1.6	0.9
171.21	0.19	0.6	83.32	14JN	-1.7	-3.4	-1.6	0.9
176.37	-0.03	0.7	92.85	24JN	-1.1	-3.3	-1.5	0.9
181.89	-0.23	0.8	102.40	4JL	-0.3	-3.3	-1.4	1.0
187.72	-0.41	0.8	111.97	14JL	0.1	-3.3	-1.4	1.0
193.82	-0.57	0.9	121.57	24JL	0.5	-3.3	-1.3	1.0
200.17	-0.71	1.0	131.21	3AU	0.8	-3.4	-1.3	1.0
206.73	-0.84	1.0	140.90	13AU	1.5	-3.4	-1.3	1.0
213.49	-0.95	1.1	150.64	23AU	3.2	-3.4	-1.3	1.0
220.43	-1.05	1.1	160.44	2SE	0.9	-3.5	-1.2	1.0
227.53	-1.13	1.1	170.30	12SE	-0.6	-3.5	-1.2	0.9
234.78	-1.20	1.2	180.21	22SE	-1.0	-3.6	-1.2	0.9
242.16	-1.25	1.2	190.18	2OC	-0.9	-3.6	-1.2	0.8
249.65	-1.29	1.2	200.21	12OC	-0.8	-3.7	-1.2	0.8
257.25	-1.31	1.2	210.29	22OC	-0.5	-3.8	-1.2	0.9
264.92	-1.32	1.3	220.40	1NO	-0.4	-3.9	-1.3	0.9
272.67	-1.32	1.3	230.56	11NO	-0.4	-4.0	-1.3	0.9
280.47	-1.30	1.3	240.75	21NO	-0.2	-4.1	-1.3	0.9
288.29	-1.26	1.3	250.94	1DE	1.0	-4.2	-1.4	0.9
296.14	-1.22	1.4	261.15	11DE	2.3	-4.3	-1.4	0.9
303.99	-1.16	1.4	271.35	21DE	0.5	-4.4	-1.5	0.8
311.82	-1.09	1.4	281.53	31DE	0.1	-4.4	-1.6	0.8

Left Table

♂ LONG	LAT	MAG	☉ LONG	16.00UT	☿	♀	♃	♄
				452				
319.63	-1.02	1.4	291.69	10JA	-0.0	-4.2	-1.6	0.8
327.40	-0.93	1.4	301.81	20JA	-0.2	-3.7	-1.7	0.8
335.11	-0.84	1.5	311.89	30JA	-0.5	-3.3	-1.8	0.7
342.77	-0.75	1.5	321.92	9FE	-1.0	-3.8	-1.8	0.7
350.35	-0.65	1.5	331.90	19FE	-1.4	-4.2	-1.9	0.6
357.86	-0.54	1.5	341.81	29FE	-1.0	-4.3	-1.9	0.6
5.30	-0.44	1.6	351.67	10MR	0.1	-4.3	-2.0	0.5
12.65	-0.33	1.6	1.47	20MR	1.7	-4.2	-2.0	0.4
19.91	-0.23	1.6	11.22	30MR	3.6	-4.1	-2.0	0.4
27.10	-0.12	1.6	20.91	9AP	2.2	-4.0	-2.1	0.4
34.20	-0.02	1.6	30.56	19AP	1.3	-3.9	-2.0	0.4
41.21	0.08	1.6	40.16	29AP	0.7	-3.8	-2.0	0.5
48.15	0.18	1.7	49.74	9MY	-0.0	-3.7	-2.0	0.5
55.01	0.28	1.7	59.28	19MY	-0.9	-3.6	-1.9	0.6
61.80	0.38	1.8	68.81	29MY	-1.8	-3.5	-1.9	0.7
68.52	0.47	1.8	78.34	8JN	-1.1	-3.5	-1.8	0.7
75.17	0.57	1.9	87.86	18JN	-0.3	-3.4	-1.8	0.8
81.76	0.66	1.9	97.39	28JN	0.2	-3.4	-1.7	0.8
88.29	0.74	1.9	106.95	8JL	0.6	-3.4	-1.6	0.9
94.76	0.83	1.9	116.54	18JL	1.0	-3.4	-1.6	0.9
101.18	0.91	2.0	126.16	28JL	1.9	-3.3	-1.5	0.9
107.54	0.99	2.0	135.82	7AU	3.1	-3.3	-1.5	0.9
113.86	1.07	2.0	145.53	17AU	0.8	-3.3	-1.4	0.9
120.12	1.14	1.9	155.30	27AU	-0.7	-3.3	-1.4	0.9
126.33	1.22	1.9	165.13	6SE	-1.1	-3.4	-1.3	0.9
132.49	1.30	1.9	175.01	16SE	-1.1	-3.4	-1.3	0.9
138.59	1.37	1.9	184.95	26SE	-0.7	-3.4	-1.3	0.8
144.62	1.44	1.8	194.95	6OC	-0.4	-3.4	-1.3	0.8
150.58	1.52	1.8	205.00	16OC	-0.3	-3.4	-1.3	0.8
156.47	1.59	1.7	215.09	26OC	-0.3	-3.5	-1.3	0.8
162.26	1.66	1.6	225.24	5NO	-0.1	-3.5	-1.3	0.8
167.94	1.74	1.5	235.40	15NO	1.3	-3.5	-1.3	0.8
173.49	1.81	1.4	245.59	25NO	2.0	-3.5	-1.3	0.8
178.88	1.88	1.3	255.80	5DE	0.2	-3.4	-1.3	0.8
184.08	1.96	1.2	266.00	15DE	-0.1	-3.4	-1.4	0.8
189.05	2.03	1.0	276.20	25DE	-0.2	-3.4	-1.4	0.8
				453				
193.74	2.10	0.8	286.37	4JA	-0.3	-3.4	-1.5	0.8
198.08	2.17	0.7	296.51	14JA	-0.6	-3.3	-1.5	0.8
201.98	2.23	0.4	306.61	24JA	-1.0	-3.3	-1.6	0.7
205.34	2.28	0.2	316.67	3FE	-1.3	-3.3	-1.7	0.7
208.01	2.31	-0.1	326.67	13FE	-1.0	-3.3	-1.7	0.7
209.84	2.31	-0.3	336.62	23FE	0.2	-3.4	-1.8	0.6
210.63	2.26	-0.6	346.51	5MR	2.1	-3.4	-1.9	0.5
210.24	2.15	-1.0	356.34	15MR	3.0	-3.4	-1.9	0.5
208.58	1.95	-1.3	6.12	25MR	1.6	-3.4	-2.0	0.4
205.78	1.64	-1.5	15.84	4AP	1.0	-3.4	-2.0	0.4
202.30	1.23	-1.7	25.51	14AP	0.5	-3.5	-2.1	0.3
198.80	0.75	-1.6	35.13	24AP	-0.1	-3.5	-2.1	0.3
196.00	0.26	-1.4	44.72	4MY	-0.9	-3.6	-2.1	0.4
194.38	-0.20	-1.2	54.28	14MY	-1.8	-3.6	-2.1	0.5
194.11	-0.61	-1.0	63.82	24MY	-1.1	-3.7	-2.1	0.5
195.17	-0.95	-0.8	73.35	3JN	-0.2	-3.8	-2.1	0.6
197.39	-1.23	-0.6	82.87	13JN	0.4	-3.9	-2.1	0.6
200.58	-1.46	-0.4	92.40	23JN	0.8	-4.0	-2.0	0.7
204.59	-1.64	-0.3	101.94	3JL	1.4	-4.1	-2.0	0.7
209.26	-1.78	-0.1	111.51	13JL	2.5	-4.2	-1.9	0.8
214.48	-1.89	0.0	121.11	23JL	2.6	-4.2	-1.8	0.8
220.16	-1.96	0.1	130.75	2AU	0.6	-4.2	-1.8	0.8
226.22	-2.01	0.2	140.43	12AU	-0.8	-4.0	-1.7	0.9
232.60	-2.03	0.3	150.17	22AU	-1.2	-3.8	-1.7	0.9
239.26	-2.03	0.4	159.97	1SE	-1.1	-3.5	-1.6	0.9
246.14	-2.01	0.5	169.82	11SE	-0.6	-3.8	-1.5	0.9
253.22	-1.97	0.6	179.73	21SE	-0.3	-4.2	-1.5	0.8
260.46	-1.91	0.7	189.70	1OC	-0.2	-4.3	-1.5	0.8
267.81	-1.83	0.8	199.72	11OC	-0.1	-4.3	-1.4	0.8
275.28	-1.74	0.8	209.79	21OC	0.1	-4.3	-1.4	0.7
282.82	-1.64	0.9	219.91	31OC	1.5	-4.2	-1.4	0.7
290.40	-1.53	1.0	230.06	10NO	1.7	-4.1	-1.4	0.7
298.02	-1.41	1.1	240.25	20NO	0.0	-4.0	-1.4	0.8
305.66	-1.28	1.1	250.45	30NO	-0.3	-3.9	-1.4	0.8
313.28	-1.15	1.2	260.65	10DE	-0.3	-3.8	-1.4	0.8
320.89	-1.01	1.3	270.86	20DE	-0.4	-3.7	-1.4	0.8
328.46	-0.88	1.3	281.04	30DE	-0.6	-3.6	-1.4	0.8

Right Table

♂ LONG	LAT	MAG	☉ LONG	16.00UT	☿	♀	♃	♄
				454				
335.98	-0.74	1.4	291.20	9JA	-1.0	-3.6	-1.4	0.8
343.46	-0.61	1.4	301.33	19JA	-1.2	-3.5	-1.5	0.8
350.87	-0.47	1.5	311.41	29JA	-0.9	-3.5	-1.5	0.7
358.21	-0.35	1.6	321.44	8FE	0.3	-3.4	-1.6	0.7
5.48	-0.22	1.6	331.42	18FE	2.5	-3.4	-1.6	0.7
12.67	-0.10	1.7	341.34	28FE	2.3	-3.4	-1.7	0.6
19.80	0.01	1.7	351.20	10MR	1.2	-3.3	-1.8	0.6
26.84	0.12	1.7	1.01	20MR	0.7	-3.3	-1.8	0.5
33.82	0.23	1.8	10.75	30MR	0.3	-3.3	-1.9	0.4
40.72	0.33	1.8	20.45	9AP	-0.2	-3.3	-2.0	0.4
47.55	0.42	1.8	30.10	19AP	-0.9	-3.3	-2.0	0.3
54.32	0.51	1.8	39.70	29AP	-1.8	-3.3	-2.1	0.3
61.03	0.59	1.9	49.27	9MY	-1.1	-3.3	-2.2	0.3
67.69	0.67	1.9	58.83	19MY	-0.2	-3.4	-2.2	0.4
74.28	0.74	1.9	68.35	29MY	0.5	-3.4	-2.2	0.4
80.84	0.81	1.9	77.88	8JN	1.1	-3.5	-2.3	0.5
87.35	0.87	1.9	87.40	18JN	1.9	-3.5	-2.3	0.5
93.82	0.93	1.9	96.93	28JN	3.1	-3.5	-2.2	0.6
100.26	0.99	1.9	106.49	8JL	2.2	-3.4	-2.2	0.7
106.67	1.04	2.0	116.07	18JL	0.5	-3.4	-2.2	0.7
113.06	1.08	2.0	125.69	28JL	-0.8	-3.4	-2.1	0.8
119.42	1.13	2.0	135.35	7AU	-1.4	-3.4	-2.1	0.8
125.77	1.17	2.0	145.06	17AU	-1.1	-3.3	-2.0	0.8
132.10	1.20	2.0	154.82	27AU	-0.6	-3.3	-1.9	0.8
138.41	1.23	2.0	164.64	6SE	-0.2	-3.3	-1.9	0.8
144.71	1.26	2.0	174.52	16SE	-0.0	-3.3	-1.8	0.8
151.00	1.29	2.0	184.46	26SE	0.1	-3.3	-1.8	0.8
157.28	1.31	2.0	194.45	6OC	0.4	-3.4	-1.7	0.8
163.54	1.32	1.9	204.50	16OC	1.8	-3.4	-1.6	0.8
169.78	1.34	1.9	214.60	26OC	1.5	-3.4	-1.6	0.7
176.01	1.34	1.9	224.73	5NO	-0.1	-3.4	-1.6	0.7
182.21	1.35	1.8	234.91	15NO	-0.4	-3.5	-1.5	0.7
188.38	1.34	1.7	245.09	25NO	-0.4	-3.5	-1.5	0.7
194.51	1.33	1.6	255.30	5DE	-0.5	-3.6	-1.5	0.8
200.59	1.31	1.5	265.51	15DE	-0.7	-3.7	-1.5	0.8
206.62	1.28	1.4	275.70	25DE	-0.9	-3.7	-1.5	0.8
				455				
212.58	1.24	1.3	285.88	4JA	-1.0	-3.8	-1.5	0.8
218.46	1.18	1.1	296.02	14JA	-0.8	-3.9	-1.5	0.8
224.22	1.11	0.9	306.13	24JA	0.4	-4.0	-1.5	0.8
229.87	1.02	0.8	316.19	3FE	2.9	-4.1	-1.5	0.8
235.34	0.90	0.6	326.20	13FE	1.8	-4.2	-1.5	0.7
240.62	0.76	0.4	336.14	23FE	0.9	-4.2	-1.6	0.7
245.65	0.58	0.2	346.04	5MR	0.5	-4.3	-1.6	0.6
250.36	0.35	0.0	355.87	15MR	0.2	-4.2	-1.7	0.6
254.67	0.08	-0.2	5.65	25MR	-0.2	-3.9	-1.7	0.5
258.47	-0.26	-0.5	15.37	4AP	-0.9	-3.3	-1.8	0.5
261.62	-0.68	-0.8	25.04	14AP	-1.8	-3.2	-1.9	0.4
263.96	-1.18	-1.1	34.67	24AP	-1.1	-3.8	-1.9	0.4
265.29	-1.78	-1.4	44.26	4MY	-0.1	-4.1	-2.0	0.3
265.45	-2.47	-1.7	53.82	14MY	0.7	-4.2	-2.1	0.2
264.39	-3.20	-2.0	63.36	24MY	1.4	-4.2	-2.2	0.3
262.22	-3.93	-2.3	72.88	3JN	2.5	-4.1	-2.2	0.4
259.64	-4.54	-2.4	82.41	13JN	3.3	-4.0	-2.3	0.4
256.67	-4.96	-2.3	91.93	23JN	1.8	-3.9	-2.3	0.5
254.66	-5.15	-2.2	101.48	3JL	0.4	-3.8	-2.4	0.5
253.88	-5.14	-1.9	111.04	13JL	-0.8	-3.7	-2.4	0.6
254.47	-4.98	-1.7	120.64	23JL	-1.5	-3.7	-2.4	0.7
256.35	-4.72	-1.4	130.28	2AU	-1.1	-3.6	-2.4	0.7
259.37	-4.42	-1.2	139.97	12AU	-0.5	-3.5	-2.3	0.7
263.31	-4.08	-1.0	149.70	22AU	-0.1	-3.5	-2.3	0.8
267.99	-3.74	-0.8	159.49	1SE	0.1	-3.5	-2.2	0.8
273.27	-3.39	-0.6	169.34	11SE	0.2	-3.4	-2.2	0.8
278.99	-3.05	-0.4	179.25	21SE	0.6	-3.4	-2.1	0.8
285.07	-2.72	-0.2	189.21	1OC	2.2	-3.4	-2.1	0.8
291.43	-2.40	-0.0	199.23	11OC	1.4	-3.4	-2.0	0.8
297.98	-2.10	0.1	209.30	21OC	-0.3	-3.4	-1.9	0.8
304.70	-1.81	0.3	219.42	31OC	-0.6	-3.4	-1.9	0.7
311.52	-1.53	0.5	229.57	10NO	-0.6	-3.4	-1.8	0.7
318.42	-1.28	0.6	239.75	20NO	-0.6	-3.4	-1.7	0.7
325.37	-1.04	0.7	249.95	30NO	-0.8	-3.4	-1.7	0.7
332.35	-0.82	0.9	260.15	10DE	-0.8	-3.4	-1.6	0.7
339.33	-0.61	1.0	270.35	20DE	-0.8	-3.4	-1.6	0.8
346.30	-0.42	1.1	280.54	30DE	-0.7	-3.4	-1.6	0.8

♂ LONG	♂ LAT	♂ MAG	⊙ LONG	16.00UT	☿	♀	♃	♄
				456	MAGNITUDES			
353.26	-0.25	1.2	290.70	9JA	0.5	-3.4	-1.6	0.8
0.18	-0.09	1.3	300.83	19JA	3.0	-3.5	-1.5	0.8
7.07	0.06	1.4	310.92	29JA	1.3	-3.5	-1.5	0.8
13.92	0.19	1.5	320.95	8FE	0.6	-3.5	-1.5	0.8
20.72	0.32	1.6	330.93	18FE	0.3	-3.4	-1.6	0.8
27.48	0.43	1.6	340.86	28FE	0.1	-3.4	-1.6	0.7
34.20	0.53	1.7	350.72	9MR	-0.3	-3.4	-1.6	0.7
40.86	0.62	1.8	0.53	19MR	-1.0	-3.4	-1.6	0.6
47.49	0.70	1.8	10.28	29MR	-1.8	-3.3	-1.7	0.6
54.07	0.77	1.9	19.98	8AP	-1.1	-3.3	-1.7	0.5
60.60	0.84	1.9	29.63	18AP	-0.1	-3.3	-1.8	0.5
67.11	0.89	1.9	39.24	28AP	0.9	-3.3	-1.8	0.4
73.58	0.95	2.0	48.81	8MY	1.9	-3.3	-1.9	0.3
80.02	0.99	2.0	58.36	18MY	3.3	-3.4	-1.9	0.3
86.43	1.03	2.0	67.89	28MY	2.7	-3.4	-2.0	0.3
92.83	1.06	2.0	77.41	7JN	1.4	-3.4	-2.1	0.3
99.21	1.09	2.0	86.94	17JN	0.3	-3.4	-2.1	0.4
105.58	1.12	2.0	96.48	27JN	-0.9	-3.5	-2.2	0.4
111.94	1.13	2.0	106.03	7JL	-1.6	-3.5	-2.3	0.5
118.31	1.15	2.0	115.61	17JL	-1.1	-3.6	-2.3	0.5
124.67	1.16	2.0	125.23	27JL	-0.4	-3.7	-2.4	0.6
131.04	1.16	2.0	134.88	6AU	-0.0	-3.8	-2.4	0.7
137.43	1.16	2.0	144.59	16AU	0.2	-3.8	-2.4	0.7
143.82	1.16	2.0	154.35	26AU	0.4	-3.9	-2.4	0.7
150.24	1.15	2.0	164.17	5SE	0.9	-4.1	-2.4	0.8
156.67	1.13	2.0	174.05	15SE	2.6	-4.2	-2.4	0.8
163.13	1.11	2.0	183.98	25SE	1.2	-4.3	-2.4	0.8
169.60	1.09	2.0	193.97	5OC	-0.4	-4.3	-2.3	0.8
176.11	1.06	2.0	204.02	15OC	-0.7	-4.3	-2.3	0.8
182.64	1.02	1.9	214.11	25OC	-0.7	-4.1	-2.2	0.8
189.20	0.98	1.9	224.24	4NO	-0.7	-3.6	-2.1	0.8
195.79	0.93	1.8	234.41	14NO	-0.7	-3.0	-2.0	0.7
202.40	0.87	1.8	244.60	24NO	-0.7	-3.8	-2.0	0.7
209.04	0.80	1.7	254.80	4DE	-0.7	-4.3	-1.9	0.7
215.70	0.72	1.7	265.01	14DE	-0.5	-4.4	-1.8	0.7
222.38	0.64	1.6	275.20	24DE	0.7	-4.4	-1.8	0.8
				457				
229.09	0.54	1.5	285.38	3JA	2.8	-4.3	-1.7	0.8
235.81	0.42	1.4	295.52	13JA	0.9	-4.2	-1.7	0.8
242.55	0.29	1.3	305.63	23JA	0.4	-4.1	-1.6	0.8
249.30	0.15	1.2	315.69	2FE	0.2	-3.9	-1.6	0.8
256.06	-0.01	1.1	325.71	12FE	-0.0	-3.8	-1.6	0.8
262.81	-0.19	1.0	335.66	22FE	-0.4	-3.8	-1.6	0.8
269.55	-0.39	0.8	345.56	4MR	-1.0	-3.7	-1.6	0.8
276.28	-0.61	0.7	355.40	14MR	-1.7	-3.6	-1.6	0.7
282.96	-0.86	0.6	5.18	24MR	-1.1	-3.5	-1.6	0.7
289.60	-1.13	0.4	14.90	3AP	-0.0	-3.5	-1.6	0.6
296.15	-1.42	0.3	24.58	13AP	1.2	-3.4	-1.6	0.6
302.59	-1.74	0.1	34.21	23AP	2.5	-3.4	-1.6	0.5
308.89	-2.08	-0.1	43.80	3MY	3.6	-3.3	-1.7	0.5
314.98	-2.46	-0.3	53.37	13MY	2.1	-3.3	-1.7	0.4
320.79	-2.86	-0.4	62.90	23MY	1.1	-3.3	-1.8	0.3
326.26	-3.28	-0.6	72.43	2JN	0.2	-3.3	-1.8	0.3
331.25	-3.73	-0.9	81.95	12JN	-0.9	-3.3	-1.9	0.3
335.64	-4.19	-1.1	91.48	22JN	-1.7	-3.3	-1.9	0.3
339.26	-4.67	-1.3	101.02	2JL	-1.1	-3.3	-2.0	0.4
341.89	-5.14	-1.6	110.59	12JL	-0.4	-3.4	-2.1	0.4
343.33	-5.57	-1.8	120.18	22JL	0.1	-3.4	-2.1	0.5
343.42	-5.90	-2.1	129.82	1AU	0.4	-3.4	-2.2	0.6
342.11	-6.07	-2.3	139.50	11AU	0.6	-3.5	-2.3	0.6
339.70	-5.98	-2.5	149.23	21AU	1.2	-3.5	-2.3	0.7
336.78	-5.62	-2.5	159.02	31AU	3.0	-3.5	-2.4	0.7
334.12	-5.01	-2.3	168.86	10SE	1.0	-3.5	-2.4	0.7
332.40	-4.26	-2.0	178.77	20SE	-0.5	-3.4	-2.4	0.8
331.95	-3.49	-1.7	188.73	30SE	-0.9	-3.4	-2.4	0.8
332.80	-2.76	-1.3	198.75	10OC	-0.9	-3.4	-2.4	0.8
334.82	-2.11	-1.0	208.81	20OC	-0.8	-3.4	-2.4	0.8
337.79	-1.55	-0.7	218.92	30OC	-0.6	-3.3	-2.4	0.8
341.53	-1.08	-0.4	229.08	9NO	-0.5	-3.3	-2.3	0.8
345.87	-0.68	-0.2	239.25	19NO	-0.5	-3.3	-2.2	0.8
350.65	-0.35	0.1	249.45	29NO	-0.4	-3.3	-2.2	0.7
355.80	-0.07	0.3	259.66	9DE	0.9	-3.3	-2.1	0.7
1.21	0.17	0.5	269.86	19DE	2.5	-3.4	-2.0	0.7
6.83	0.36	0.7	280.04	29DE	0.6	-3.4	-2.0	0.8

♂ LONG	♂ LAT	♂ MAG	⊙ LONG	16.00UT	☿	♀	♃	♄
				458	MAGNITUDES			
12.60	0.53	0.8	290.21	8JA	0.2	-3.4	-1.9	0.8
18.50	0.67	1.0	300.34	18JA	0.1	-3.5	-1.8	0.8
24.47	0.79	1.1	310.42	28JA	-0.1	-3.5	-1.8	0.8
30.52	0.88	1.3	320.46	7FE	-0.4	-3.5	-1.7	0.8
36.62	0.97	1.4	330.45	17FE	-1.0	-3.6	-1.7	0.8
42.74	1.03	1.5	340.37	27FE	-1.5	-3.6	-1.6	0.8
48.90	1.09	1.6	350.24	9MR	-1.1	-3.7	-1.6	0.8
55.06	1.14	1.7	0.05	19MR	0.1	-3.8	-1.6	0.8
61.24	1.17	1.8	9.81	29MR	1.5	-3.9	-1.5	0.7
67.43	1.20	1.8	19.51	8AP	3.2	-4.0	-1.5	0.7
73.63	1.22	1.9	29.16	18AP	2.7	-4.0	-1.5	0.6
79.83	1.23	1.9	38.77	28AP	1.6	-4.1	-1.5	0.6
86.04	1.24	2.0	48.35	8MY	0.8	-4.2	-1.5	0.5
92.26	1.24	2.0	57.90	18MY	0.1	-4.2	-1.6	0.5
98.49	1.24	2.0	67.43	28MY	-0.9	-4.0	-1.6	0.4
104.73	1.23	2.0	76.96	7JN	-1.7	-3.6	-1.6	0.3
110.98	1.22	2.0	86.48	17JN	-1.1	-2.8	-1.6	0.3
117.26	1.20	2.0	96.02	27JN	-0.3	-3.5	-1.7	0.3
123.56	1.18	2.0	105.57	7JL	0.2	-4.0	-1.7	0.4
129.89	1.16	2.0	115.14	17JL	0.5	-4.2	-1.8	0.4
136.25	1.13	2.0	124.76	27JL	0.9	-4.2	-1.9	0.5
142.64	1.09	2.0	134.42	6AU	1.6	-4.2	-1.9	0.5
149.07	1.05	1.9	144.12	16AU	3.2	-4.1	-2.0	0.6
155.55	1.01	1.9	153.88	26AU	0.9	-4.0	-2.1	0.6
162.07	0.96	1.9	163.69	5SE	-0.6	-3.9	-2.1	0.7
168.64	0.91	1.9	173.56	15SE	-1.0	-3.8	-2.2	0.7
175.26	0.86	1.9	183.50	25SE	-1.0	-3.8	-2.2	0.8
181.93	0.79	1.9	193.49	5OC	-0.8	-3.7	-2.3	0.8
188.66	0.73	1.9	203.53	15OC	-0.5	-3.6	-2.3	0.8
195.45	0.65	1.9	213.62	25OC	-0.4	-3.6	-2.3	0.8
202.29	0.57	1.8	223.75	4NO	-0.4	-3.5	-2.4	0.8
209.20	0.49	1.8	233.92	14NO	-0.2	-3.5	-2.3	0.8
216.16	0.39	1.8	244.11	24NO	1.0	-3.5	-2.3	0.8
223.19	0.30	1.7	254.31	4DE	2.2	-3.4	-2.3	0.8
230.27	0.19	1.7	264.51	14DE	0.4	-3.4	-2.2	0.7
237.41	0.08	1.6	274.71	24DE	0.0	-3.4	-2.2	0.7
				459				
244.61	-0.04	1.6	284.89	3JA	-0.1	-3.4	-2.1	0.8
251.86	-0.17	1.5	295.03	13JA	-0.2	-3.3	-2.0	0.8
259.16	-0.31	1.5	305.15	23JA	-0.5	-3.3	-2.0	0.8
266.50	-0.45	1.4	315.21	2FE	-1.0	-3.3	-1.9	0.9
273.89	-0.60	1.3	325.22	12FE	-1.4	-3.3	-1.8	0.9
281.31	-0.75	1.3	335.18	22FE	-1.0	-3.3	-1.7	0.9
288.75	-0.90	1.2	345.08	4MR	0.1	-3.4	-1.7	0.9
296.22	-1.06	1.1	354.91	14MR	1.8	-3.4	-1.6	0.9
303.69	-1.22	1.0	4.70	24MR	3.5	-3.4	-1.6	0.9
311.15	-1.38	1.0	14.43	3AP	2.0	-3.5	-1.5	0.8
318.59	-1.53	0.9	24.10	13AP	1.2	-3.5	-1.5	0.8
325.99	-1.68	0.8	33.74	23AP	0.6	-3.5	-1.5	0.7
333.33	-1.82	0.7	43.33	3MY	-0.0	-3.4	-1.5	0.7
340.59	-1.94	0.6	52.89	13MY	-0.9	-3.4	-1.4	0.6
347.74	-2.06	0.6	62.44	23MY	-1.8	-3.4	-1.4	0.6
354.76	-2.17	0.5	71.96	2JN	-1.1	-3.3	-1.4	0.5
1.61	-2.25	0.4	81.48	12JN	-0.3	-3.3	-1.4	0.4
8.25	-2.32	0.3	91.01	22JN	0.3	-3.3	-1.5	0.4
14.64	-2.37	0.2	100.55	2JL	0.7	-3.3	-1.5	0.3
20.73	-2.40	0.1	110.12	12JL	1.2	-3.3	-1.5	0.4
26.44	-2.41	-0.1	119.72	22JL	2.1	-3.3	-1.5	0.4
31.69	-2.39	-0.2	129.35	1AU	3.0	-3.4	-1.6	0.5
36.39	-2.34	-0.4	139.03	11AU	0.8	-3.4	-1.6	0.5
40.39	-2.25	-0.5	148.76	21AU	-0.7	-3.4	-1.7	0.6
43.52	-2.13	-0.7	158.54	31AU	-1.2	-3.5	-1.7	0.6
45.61	-1.95	-0.9	168.38	10SE	-1.1	-3.5	-1.8	0.7
46.41	-1.70	-1.2	178.29	20SE	-0.7	-3.6	-1.9	0.7
45.78	-1.37	-1.4	188.24	30SE	-0.4	-3.6	-1.9	0.8
43.71	-0.96	-1.6	198.26	10OC	-0.2	-3.7	-2.0	0.8
40.48	-0.49	-1.7	208.33	20OC	-0.2	-3.8	-2.1	0.8
36.73	0.02	-1.8	218.43	30OC	0.0	-3.9	-2.1	0.8
33.29	0.49	-1.5	228.58	9NO	1.3	-4.0	-2.2	0.8
30.82	0.88	-1.2	238.76	19NO	2.0	-4.1	-2.2	0.8
29.68	1.19	-0.9	248.96	29NO	0.2	-4.2	-2.2	0.8
29.86	1.40	-0.6	259.16	9DE	-0.2	-4.3	-2.2	0.8
31.24	1.56	-0.3	269.37	19DE	-0.2	-4.4	-2.2	0.8
33.60	1.66	-0.0	279.55	29DE	-0.3	-4.4	-2.2	0.8

♂ LONG	♂ LAT	♂ MAG	⊙ LONG	16.00UT	☿	♀	♃	♄
				460	MAGNITUDES			
36.74	1.72	0.3	289.72	8JA	-0.6	-4.2	-2.2	0.8
40.48	1.75	0.5	299.85	18JA	-1.0	-3.7	-2.1	0.9
44.71	1.77	0.7	309.93	28JA	-1.2	-3.3	-2.0	0.9
49.30	1.77	0.9	319.98	7FE	-1.0	-3.8	-2.0	0.9
54.18	1.76	1.0	329.96	17FE	0.2	-4.2	-1.9	0.9
59.30	1.74	1.2	339.89	27FE	2.2	-4.3	-1.8	1.0
64.60	1.71	1.3	349.76	8MR	2.8	-4.3	-1.8	1.0
70.05	1.68	1.4	359.58	18MR	1.5	-4.2	-1.7	1.0
75.61	1.64	1.5	9.33	28MR	0.9	-4.1	-1.6	0.9
81.28	1.61	1.6	19.04	7AP	0.4	-4.0	-1.6	0.9
87.04	1.56	1.7	28.69	17AP	-0.1	-3.9	-1.5	0.9
92.87	1.52	1.8	38.30	27AP	-0.9	-3.8	-1.5	0.8
98.76	1.47	1.8	47.88	7MY	-1.8	-3.7	-1.4	0.8
104.72	1.42	1.9	57.44	17MY	-1.2	-3.6	-1.4	0.7
110.74	1.37	1.9	66.97	27MY	-0.2	-3.5	-1.4	0.7
116.81	1.31	2.0	76.50	6JN	0.4	-3.5	-1.4	0.6
122.94	1.25	2.0	86.02	16JN	0.9	-3.4	-1.3	0.5
129.12	1.19	2.0	95.55	26JN	1.5	-3.4	-1.3	0.5
135.37	1.13	2.0	105.11	6JL	2.7	-3.4	-1.3	0.4
141.67	1.06	2.0	114.68	16JL	2.5	-3.4	-1.3	0.4
148.03	1.00	2.0	124.29	26JL	0.6	-3.3	-1.4	0.4
154.46	0.92	2.0	133.95	5AU	-0.8	-3.3	-1.4	0.5
160.96	0.85	1.9	143.65	15AU	-1.3	-3.3	-1.4	0.5
167.52	0.77	1.9	153.41	25AU	-1.2	-3.3	-1.4	0.6
174.16	0.69	1.9	163.22	4SE	-0.6	-3.4	-1.5	0.6
180.87	0.61	1.8	173.09	14SE	-0.3	-3.4	-1.5	0.7
187.66	0.52	1.8	183.02	24SE	-0.1	-3.4	-1.6	0.7
194.53	0.43	1.7	193.01	4OC	-0.0	-3.4	-1.7	0.8
201.48	0.34	1.7	203.04	14OC	0.2	-3.4	-1.7	0.8
208.50	0.24	1.7	213.13	24OC	1.6	-3.5	-1.8	0.9
215.61	0.14	1.7	223.26	3NO	1.7	-3.5	-1.8	0.9
222.79	0.04	1.7	233.42	13NO	-0.0	-3.5	-1.9	0.9
230.05	-0.07	1.7	243.61	23NO	-0.3	-3.5	-2.0	0.9
237.39	-0.18	1.6	253.81	3DE	-0.4	-3.4	-2.0	0.9
244.80	-0.29	1.6	264.02	13DE	-0.4	-3.4	-2.1	0.9
252.27	-0.40	1.6	274.21	23DE	-0.7	-3.4	-2.1	0.9
				461				
259.81	-0.51	1.6	284.39	2JA	-1.0	-3.4	-2.1	0.9
267.40	-0.62	1.5	294.54	12JA	-1.1	-3.3	-2.1	0.9
275.05	-0.73	1.5	304.65	22JA	-0.9	-3.3	-2.1	0.9
282.73	-0.84	1.5	314.72	1FE	0.3	-3.3	-2.1	1.0
290.44	-0.94	1.4	324.73	11FE	2.6	-3.3	-2.0	1.0
298.18	-1.04	1.4	334.69	21FE	2.1	-3.4	-2.0	1.0
305.92	-1.13	1.4	344.60	3MR	1.1	-3.4	-1.9	1.0
313.66	-1.21	1.3	354.44	13MR	0.6	-3.4	-1.9	1.1
321.39	-1.28	1.3	4.22	23MR	0.3	-3.4	-1.8	1.1
329.09	-1.34	1.3	13.96	2AP	-0.2	-3.4	-1.7	1.1
336.75	-1.38	1.2	23.64	12AP	-0.9	-3.5	-1.7	1.0
344.35	-1.41	1.2	33.27	22AP	-1.8	-3.5	-1.6	1.0
351.88	-1.43	1.2	42.87	2MY	-1.2	-3.6	-1.5	1.0
359.33	-1.43	1.1	52.43	12MY	-0.2	-3.6	-1.5	0.9
6.69	-1.42	1.1	61.97	22MY	0.6	-3.7	-1.4	0.9
13.93	-1.39	1.1	71.50	1JN	1.2	-3.8	-1.4	0.8
21.04	-1.34	1.0	81.02	11JN	2.1	-3.9	-1.3	0.8
28.02	-1.28	1.0	90.55	21JN	3.3	-4.0	-1.3	0.7
34.83	-1.21	1.0	100.10	1JL	2.1	-4.1	-1.3	0.6
41.47	-1.11	0.9	109.66	11JL	0.5	-4.2	-1.3	0.6
47.91	-1.00	0.8	119.25	21JL	-0.8	-4.2	-1.3	0.5
54.11	-0.88	0.8	128.88	31JL	-1.4	-4.2	-1.3	0.5
60.06	-0.73	0.7	138.56	10AU	-1.2	-4.0	-1.3	0.5
65.70	-0.56	0.6	148.29	20AU	-0.5	-3.6	-1.3	0.6
70.98	-0.37	0.5	158.07	30AU	-0.2	-3.3	-1.3	0.6
75.83	-0.16	0.4	167.91	9SE	0.0	-3.8	-1.3	0.7
80.16	0.09	0.2	177.81	19SE	0.1	-4.2	-1.3	0.7
83.85	0.37	0.1	187.76	29SE	0.4	-4.3	-1.4	0.8
86.76	0.69	-0.1	197.77	9OC	1.9	-4.3	-1.4	0.8
88.70	1.06	-0.3	207.84	19OC	1.5	-4.3	-1.5	0.9
89.46	1.47	-0.5	217.95	29OC	-0.2	-4.2	-1.5	0.9
88.90	1.93	-0.7	228.09	8NO	-0.5	-4.1	-1.6	0.9
86.93	2.40	-1.0	238.27	18NO	-0.5	-4.0	-1.7	1.0
83.74	2.83	-1.1	248.47	28NO	-0.5	-3.9	-1.7	1.0
79.85	3.17	-1.3	258.67	8DE	-0.7	-3.8	-1.8	1.0
76.00	3.37	-1.1	268.87	18DE	-0.9	-3.7	-1.9	1.0
72.93	3.42	-0.8	279.06	28DE	-0.9	-3.6	-1.9	1.0

♂ LONG	♂ LAT	♂ MAG	⊙ LONG	16.00UT	☿	♀	♃	♄
				462	MAGNITUDES			
71.07	3.37	-0.6	289.22	7JA	-0.7	-3.6	-2.0	1.0
70.53	3.24	-0.3	299.36	17JA	0.4	-3.5	-2.0	1.0
71.23	3.09	0.0	309.45	27JA	2.9	-3.5	-2.0	1.0
72.97	2.91	0.3	319.49	6FE	1.6	-3.4	-2.0	1.0
75.57	2.74	0.5	329.48	16FE	0.8	-3.4	-2.0	1.1
78.85	2.57	0.7	339.42	26FE	0.4	-3.4	-2.0	1.1
82.67	2.41	0.9	349.29	8MR	0.2	-3.3	-2.0	1.2
86.91	2.26	1.0	359.11	18MR	-0.2	-3.3	-1.9	1.2
91.51	2.12	1.2	8.87	28MR	-0.9	-3.3	-1.9	1.2
96.38	1.99	1.3	18.57	7AP	-1.8	-3.3	-1.8	1.2
101.48	1.86	1.4	28.23	17AP	-1.2	-3.3	-1.8	1.2
106.78	1.74	1.5	37.84	27AP	-0.1	-3.3	-1.7	1.2
112.24	1.62	1.6	47.42	7MY	0.7	-3.3	-1.6	1.1
117.85	1.50	1.7	56.98	17MY	1.6	-3.4	-1.6	1.1
123.59	1.39	1.7	66.51	27MY	2.7	-3.4	-1.5	1.1
129.45	1.28	1.8	76.03	6JN	3.1	-3.5	-1.5	1.0
135.41	1.17	1.8	85.56	16JN	1.6	-3.5	-1.4	0.9
141.49	1.07	1.8	95.09	26JN	0.4	-3.5	-1.4	0.9
147.66	0.96	1.8	104.64	6JL	-0.8	-3.4	-1.3	0.8
153.94	0.85	1.8	114.22	16JL	-1.5	-3.4	-1.3	0.7
160.31	0.75	1.8	123.82	26JL	-1.2	-3.4	-1.3	0.7
166.78	0.64	1.8	133.47	5AU	-0.5	-3.4	-1.2	0.6
173.35	0.54	1.8	143.18	15AU	-0.1	-3.3	-1.2	0.6
180.01	0.43	1.8	152.93	25AU	0.1	-3.3	-1.2	0.7
186.78	0.32	1.8	162.74	4SE	0.3	-3.3	-1.2	0.7
193.64	0.22	1.8	172.61	14SE	0.7	-3.3	-1.2	0.8
200.60	0.11	1.7	182.53	24SE	2.2	-3.3	-1.2	0.8
207.66	0.00	1.7	192.51	4OC	1.4	-3.4	-1.3	0.9
214.82	-0.10	1.6	202.56	14OC	-0.3	-3.4	-1.3	0.9
222.07	-0.21	1.6	212.64	24OC	-0.6	-3.4	-1.3	1.0
229.41	-0.31	1.6	222.77	3NO	-0.6	-3.4	-1.4	1.0
236.84	-0.41	1.5	232.93	13NO	-0.7	-3.5	-1.4	1.0
244.36	-0.51	1.4	243.12	23NO	-0.8	-3.5	-1.5	1.1
251.94	-0.61	1.4	253.32	3DE	-0.7	-3.6	-1.5	1.1
259.60	-0.70	1.4	263.53	13DE	-0.8	-3.7	-1.6	1.1
267.32	-0.78	1.4	273.72	23DE	-0.6	-3.7	-1.7	1.1
				463				
275.09	-0.86	1.4	283.90	2JA	0.5	-3.8	-1.7	1.2
282.90	-0.93	1.4	294.05	12JA	3.0	-3.9	-1.8	1.2
290.73	-1.00	1.4	304.16	22JA	1.2	-4.0	-1.9	1.2
298.59	-1.05	1.4	314.23	1FE	0.5	-4.1	-1.9	1.2
306.45	-1.09	1.4	324.25	11FE	0.3	-4.2	-2.0	1.2
314.30	-1.12	1.4	334.21	21FE	0.1	-4.2	-2.0	1.2
322.13	-1.14	1.4	344.12	3MR	-0.3	-4.3	-2.0	1.3
329.93	-1.15	1.4	353.97	13MR	-0.9	-4.2	-2.0	1.3
337.68	-1.14	1.4	3.75	23MR	-1.7	-3.9	-2.0	1.3
345.38	-1.12	1.4	13.49	2AP	-1.2	-3.3	-2.0	1.4
353.02	-1.09	1.4	23.17	12AP	-0.1	-3.2	-2.0	1.4
0.58	-1.05	1.4	32.80	22AP	1.0	-3.8	-1.9	1.4
8.05	-0.99	1.4	42.40	2MY	2.1	-4.1	-1.9	1.4
15.44	-0.93	1.4	51.97	12MY	3.6	-4.2	-1.8	1.3
22.72	-0.85	1.4	61.51	22MY	2.5	-4.2	-1.7	1.3
29.91	-0.76	1.4	71.04	1JN	1.3	-4.1	-1.7	1.3
36.98	-0.67	1.4	80.56	11JN	0.3	-4.0	-1.6	1.2
43.93	-0.56	1.4	90.09	21JN	-0.8	-3.9	-1.5	1.2
50.77	-0.45	1.4	99.63	1JL	-1.6	-3.8	-1.5	1.1
57.47	-0.32	1.4	109.19	11JL	-1.2	-3.7	-1.4	1.0
64.04	-0.19	1.4	118.78	21JL	-0.4	-3.7	-1.4	1.0
70.46	-0.05	1.3	128.41	31JL	0.0	-3.6	-1.3	0.9
76.72	0.10	1.3	138.09	10AU	0.3	-3.5	-1.3	0.8
82.80	0.26	1.3	147.81	20AU	0.5	-3.5	-1.3	0.8
88.69	0.43	1.2	157.59	30AU	1.0	-3.5	-1.3	0.8
94.36	0.61	1.1	167.43	9SE	2.7	-3.4	-1.2	0.8
99.76	0.81	1.0	177.32	19SE	1.2	-3.4	-1.2	0.9
104.87	1.03	0.9	187.28	29SE	-0.5	-3.4	-1.2	0.9
109.61	1.27	0.8	197.29	9OC	-0.8	-3.4	-1.2	1.0
113.91	1.53	0.7	207.35	19OC	-0.8	-3.4	-1.2	1.0
117.68	1.83	0.5	217.46	29OC	-0.8	-3.4	-1.3	1.1
120.79	2.15	0.3	227.60	8NO	-0.7	-3.4	-1.3	1.1
123.08	2.52	0.1	237.78	18NO	-0.6	-3.4	-1.3	1.2
124.38	2.92	-0.1	247.97	28NO	-0.6	-3.4	-1.4	1.2
124.48	3.34	-0.3	258.18	8DE	-0.5	-3.4	-1.4	1.3
123.28	3.76	-0.6	268.38	18DE	0.7	-3.4	-1.5	1.3
120.76	4.13	-0.8	278.57	28DE	2.8	-3.4	-1.5	1.3

♂ LONG	LAT	MAG	☉ LONG	16.00UT 464	☿	♀	♃	♄
117.22	4.39	-1.0	288.74	7JA	0.8	-3.4	-1.6	1.3
113.26	4.49	-1.0	298.87	17JA	0.3	-3.5	-1.6	1.3
109.59	4.41	-0.8	308.96	27JA	0.1	-3.5	-1.7	1.3
106.86	4.19	-0.6	319.01	6FE	-0.0	-3.5	-1.8	1.3
105.37	3.89	-0.3	329.00	16FE	-0.4	-3.4	-1.8	1.3
105.17	3.55	-0.1	338.93	26FE	-1.0	-3.4	-1.9	1.3
106.15	3.22	0.2	348.81	7MR	-1.6	-3.4	-2.0	1.3
108.11	2.90	0.4	358.63	17MR	-1.1	-3.4	-2.0	1.3
110.88	2.61	0.6	8.39	27MR	-0.0	-3.3	-2.0	1.3
114.31	2.34	0.7	18.10	6AP	1.2	-3.3	-2.1	1.3
118.26	2.09	0.9	27.76	16AP	2.7	-3.3	-2.1	1.2
122.64	1.86	1.0	37.38	26AP	3.2	-3.3	-2.0	1.2
127.38	1.65	1.1	46.96	6MY	1.9	-3.3	-2.0	1.2
132.42	1.46	1.2	56.51	16MY	1.0	-3.4	-2.0	1.2
137.71	1.28	1.3	66.05	26MY	0.1	-3.4	-1.9	1.1
143.23	1.10	1.4	75.57	5JN	-0.8	-3.4	-1.9	1.1
148.94	0.94	1.4	85.09	15JN	-1.7	-3.4	-1.8	1.0
154.83	0.78	1.5	94.63	25JN	-1.2	-3.5	-1.8	1.0
160.89	0.64	1.5	104.17	5JL	-0.4	-3.5	-1.7	0.9
167.10	0.49	1.6	113.75	15JL	0.1	-3.6	-1.6	0.9
173.46	0.35	1.6	123.36	25JL	0.4	-3.7	-1.6	0.8
179.95	0.22	1.6	133.01	4AU	0.7	-3.8	-1.5	0.8
186.58	0.09	1.6	142.70	14AU	1.4	-3.8	-1.5	0.7
193.34	-0.03	1.6	152.46	24AU	3.1	-3.9	-1.4	0.7
200.23	-0.15	1.6	162.26	3SE	1.1	-4.1	-1.4	0.7
207.24	-0.27	1.6	172.13	13SE	-0.6	-4.2	-1.3	0.8
214.36	-0.38	1.6	182.05	23SE	-0.9	-4.3	-1.3	0.8
221.60	-0.48	1.6	192.03	3OC	-0.9	-4.3	-1.3	0.9
228.95	-0.58	1.5	202.06	13OC	-0.8	-4.3	-1.3	1.0
236.40	-0.68	1.5	212.15	23OC	-0.5	-4.1	-1.3	1.0
243.95	-0.76	1.5	222.27	2NO	-0.5	-3.6	-1.3	1.1
251.57	-0.84	1.5	232.44	12NO	-0.4	-3.0	-1.3	1.1
259.28	-0.90	1.4	242.62	22NO	-0.3	-3.9	-1.3	1.2
267.05	-0.96	1.4	252.82	2DE	0.9	-4.3	-1.3	1.2
274.87	-1.01	1.4	263.03	12DE	2.5	-4.4	-1.4	1.2
282.73	-1.05	1.4	273.23	22DE	0.6	-4.4	-1.4	1.2

465

♂ LONG	LAT	MAG	☉ LONG	16.00UT	☿	♀	♃	♄
290.62	-1.07	1.3	283.41	1JA	0.1	-4.3	-1.4	1.2
298.52	-1.08	1.3	293.56	11JA	-0.0	-4.2	-1.5	1.2
306.43	-1.09	1.3	303.68	21JA	-0.2	-4.1	-1.6	1.2
314.31	-1.08	1.3	313.75	31JA	-0.4	-3.9	-1.6	1.1
322.17	-1.06	1.3	323.77	10FE	-1.0	-3.8	-1.7	1.1
330.00	-1.02	1.3	333.74	20FE	-1.5	-3.7	-1.7	1.1
337.77	-0.98	1.4	343.64	2MR	-1.1	-3.7	-1.8	1.0
345.48	-0.93	1.4	353.49	12MR	0.0	-3.6	-1.9	1.0
353.13	-0.86	1.4	3.29	22MR	1.6	-3.5	-2.0	1.0
0.70	-0.79	1.5	13.02	1AP	3.5	-3.5	-2.0	1.0
8.19	-0.71	1.5	22.71	11AP	2.4	-3.4	-2.1	1.0
15.59	-0.63	1.5	32.35	21AP	1.4	-3.4	-2.1	1.0
22.90	-0.53	1.6	41.94	1MY	0.7	-3.3	-2.1	1.0
30.13	-0.44	1.6	51.51	11MY	0.1	-3.3	-2.2	0.9
37.25	-0.34	1.6	61.06	21MY	-0.9	-3.3	-2.2	0.9
44.29	-0.23	1.7	70.58	31MY	-1.7	-3.3	-2.1	0.9
51.22	-0.12	1.7	80.11	10JN	-1.2	-3.3	-2.1	0.8
58.07	-0.01	1.7	89.63	20JN	-0.3	-3.3	-2.1	0.8
64.81	0.11	1.7	99.17	30JN	0.2	-3.3	-2.0	0.8
71.46	0.23	1.7	108.73	10JL	0.6	-3.4	-2.0	0.7
78.01	0.35	1.7	118.32	20JL	1.0	-3.4	-1.9	0.7
84.46	0.47	1.7	127.94	30JL	1.8	-3.4	-1.8	0.6
90.80	0.60	1.7	137.62	9AU	3.2	-3.5	-1.8	0.6
97.03	0.73	1.7	147.34	19AU	0.9	-3.5	-1.7	0.5
103.15	0.86	1.6	157.11	29AU	-0.6	-3.5	-1.7	0.5
109.14	1.00	1.6	166.95	8SE	-1.1	-3.5	-1.6	0.4
114.99	1.15	1.5	176.84	18SE	-1.1	-3.4	-1.5	0.4
120.67	1.30	1.5	186.79	28SE	-0.7	-3.4	-1.5	0.5
126.18	1.47	1.4	196.79	8OC	-0.4	-3.4	-1.5	0.5
131.47	1.64	1.3	206.85	18OC	-0.3	-3.4	-1.4	0.6
136.51	1.82	1.2	216.96	28OC	-0.3	-3.3	-1.4	0.7
141.24	2.02	1.1	227.10	7NO	-0.1	-3.3	-1.4	0.7
145.59	2.24	0.9	237.28	17NO	1.1	-3.3	-1.4	0.8
149.50	2.48	0.7	247.47	27NO	2.2	-3.3	-1.4	0.9
152.84	2.74	0.5	257.68	7DE	0.3	-3.4	-1.4	0.9
155.47	3.03	0.3	267.88	17DE	-0.1	-3.4	-1.4	0.9
157.24	3.33	0.1	278.07	27DE	-0.1	-3.4	-1.4	0.9

♂ LONG	LAT	MAG	☉ LONG	16.00UT 466	☿	♀	♃	♄
157.96	3.65	-0.2	288.24	6JA	-0.3	-3.4	-1.4	0.9
157.47	3.95	-0.4	298.38	16JA	-0.5	-3.5	-1.5	0.9
155.68	4.21	-0.7	308.47	26JA	-1.0	-3.5	-1.5	0.9
152.70	4.36	-0.9	318.52	5FE	-1.3	-3.5	-1.6	0.9
148.96	4.36	-1.1	328.52	15FE	-1.0	-3.6	-1.6	0.9
145.11	4.19	-1.0	338.46	25FE	0.1	-3.6	-1.7	0.8
141.85	3.87	-0.8	348.34	7MR	1.9	-3.7	-1.7	0.8
139.69	3.46	-0.6	358.16	17MR	3.3	-3.8	-1.8	0.7
138.82	3.02	-0.4	7.92	27MR	1.8	-3.9	-1.9	0.7
139.22	2.59	-0.1	17.64	6AP	1.0	-4.0	-1.9	0.7
140.74	2.19	0.1	27.30	16AP	0.5	-4.0	-2.0	0.7
143.19	1.82	0.2	36.92	26AP	-0.0	-4.1	-2.1	0.7
146.41	1.49	0.4	46.50	6MY	-0.9	-4.2	-2.1	0.7
150.27	1.20	0.6	56.05	16MY	-1.8	-4.2	-2.2	0.7
154.65	0.93	0.7	65.59	26MY	-1.2	-4.0	-2.2	0.7
159.46	0.68	0.8	75.12	5JN	-0.3	-3.6	-2.3	0.7
164.64	0.46	0.9	84.64	15JN	0.3	-2.8	-2.3	0.7
170.13	0.26	1.0	94.17	25JN	0.8	-3.5	-2.3	0.6
175.91	0.07	1.0	103.72	5JL	1.3	-4.0	-2.3	0.6
181.92	-0.10	1.1	113.29	15JL	2.3	-4.2	-2.2	0.5
188.16	-0.26	1.1	122.89	25JL	2.8	-4.2	-2.2	0.5
194.60	-0.40	1.2	132.54	4AU	0.8	-4.2	-2.1	0.4
201.23	-0.54	1.2	142.23	14AU	-0.7	-4.1	-2.1	0.4
208.03	-0.66	1.3	151.98	24AU	-1.2	-4.0	-2.0	0.3
214.99	-0.77	1.3	161.79	3SE	-1.2	-3.9	-1.9	0.3
222.10	-0.87	1.3	171.64	13SE	-0.7	-3.8	-1.9	0.2
229.35	-0.96	1.3	181.56	23SE	-0.3	-3.8	-1.8	0.2
236.72	-1.03	1.3	191.54	3OC	-0.2	-3.7	-1.8	0.2
244.20	-1.09	1.3	201.57	13OC	-0.1	-3.6	-1.7	0.2
251.79	-1.14	1.3	211.65	23OC	0.1	-3.6	-1.6	0.3
259.46	-1.18	1.3	221.78	2NO	1.3	-3.5	-1.6	0.4
267.20	-1.20	1.4	231.94	12NO	2.0	-3.5	-1.6	0.4
275.01	-1.21	1.4	242.12	22NO	0.1	-3.5	-1.5	0.5
282.85	-1.20	1.4	252.33	2DE	-0.2	-3.4	-1.5	0.6
290.72	-1.19	1.4	262.53	12DE	-0.3	-3.4	-1.5	0.6
298.60	-1.16	1.4	272.73	22DE	-0.4	-3.4	-1.5	0.7

467

♂ LONG	LAT	MAG	☉ LONG	16.00UT	☿	♀	♃	♄
306.48	-1.12	1.4	282.91	1JA	-0.6	-3.4	-1.5	0.7
314.34	-1.06	1.4	293.07	11JA	-1.0	-3.3	-1.5	0.7
322.18	-1.00	1.4	303.19	21JA	-1.2	-3.3	-1.5	0.7
329.96	-0.93	1.4	313.26	31JA	-0.9	-3.3	-1.5	0.7
337.70	-0.86	1.5	323.28	10FE	0.2	-3.3	-1.5	0.7
345.37	-0.77	1.5	333.25	20FE	2.3	-3.3	-1.6	0.7
352.97	-0.68	1.5	343.17	2MR	2.5	-3.4	-1.6	0.6
0.49	-0.59	1.5	353.02	12MR	1.3	-3.4	-1.7	0.6
7.93	-0.49	1.5	2.81	22MR	0.8	-3.4	-1.7	0.5
15.29	-0.39	1.5	12.55	1AP	0.4	-3.5	-1.8	0.5
22.56	-0.28	1.5	22.24	11AP	-0.1	-3.5	-1.8	0.5
29.75	-0.18	1.6	31.88	21AP	-0.9	-3.5	-1.9	0.5
36.85	-0.08	1.6	41.48	1MY	-1.8	-3.4	-2.0	0.5
43.86	0.03	1.7	51.05	11MY	-1.2	-3.4	-2.0	0.5
50.80	0.13	1.7	60.59	21MY	-0.3	-3.4	-2.1	0.6
57.65	0.23	1.8	70.12	31MY	0.5	-3.3	-2.2	0.5
64.42	0.33	1.8	79.64	10JN	1.0	-3.3	-2.2	0.5
71.13	0.43	1.8	89.17	20JN	1.7	-3.3	-2.3	0.5
77.75	0.53	1.9	98.71	30JN	3.0	-3.3	-2.3	0.5
84.31	0.63	1.9	108.26	10JL	2.4	-3.3	-2.4	0.5
90.80	0.72	1.9	117.85	20JL	0.6	-3.3	-2.4	0.4
97.23	0.82	1.9	127.48	30JL	-0.7	-3.4	-2.4	0.4
103.59	0.91	1.9	137.14	9AU	-1.3	-3.4	-2.4	0.3
109.88	1.01	1.9	146.87	19AU	-1.2	-3.4	-2.3	0.3
116.11	1.10	1.9	156.64	29AU	-0.6	-3.5	-2.3	0.2
122.27	1.19	1.9	166.47	8SE	-0.2	-3.5	-2.2	0.1
128.35	1.29	1.8	176.36	18SE	-0.1	-3.6	-2.2	0.1
134.35	1.38	1.8	186.30	28SE	0.0	-3.6	-2.1	0.0
140.27	1.48	1.7	196.31	8OC	0.3	-3.7	-2.0	-0.0
146.08	1.58	1.7	206.36	18OC	1.6	-3.8	-2.0	-0.0
151.77	1.68	1.6	216.47	28OC	1.8	-3.9	-1.9	0.0
157.33	1.78	1.5	226.61	7NO	-0.1	-4.0	-1.8	0.1
162.72	1.89	1.4	236.78	17NO	-0.4	-4.1	-1.8	0.2
167.92	2.00	1.3	246.97	27NO	-0.4	-4.2	-1.7	0.3
172.89	2.12	1.1	257.18	7DE	-0.5	-4.3	-1.7	0.3
177.57	2.24	1.0	267.38	17DE	-0.7	-4.4	-1.6	0.4
181.89	2.37	0.8	277.57	27DE	-1.0	-4.4	-1.6	0.4

Left table (468 / 469):

♂ LONG	LAT	MAG	☉ LONG	16.00UT	☿	♀	♃	♄
				468		MAGNITUDES		
185.78	2.50	0.6	287.74	6JA	-1.0	-4.1	-1.6	0.5
189.12	2.64	0.4	297.88	16JA	-0.8	-3.6	-1.6	0.5
191.78	2.77	0.1	307.98	26JA	0.3	-3.9	-1.6	0.5
193.60	2.89	-0.1	318.03	5FE	2.7	-3.9	-1.5	0.5
194.39	2.99	-0.4	328.03	15FE	1.9	-4.2	-1.6	0.5
193.99	3.04	-0.7	337.97	25FE	0.9	-4.3	-1.6	0.5
192.32	3.02	-1.0	347.85	6MR	0.6	-4.3	-1.6	0.5
189.48	2.88	-1.3	357.68	16MR	0.2	-4.2	-1.6	0.5
185.90	2.61	-1.5	7.45	26MR	-0.2	-4.1	-1.6	0.4
182.24	2.22	-1.4	17.16	5AP	-0.9	-4.0	-1.7	0.4
179.19	1.76	-1.2	26.83	15AP	-1.8	-3.9	-1.7	0.3
177.28	1.28	-1.0	36.45	25AP	-1.2	-3.8	-1.8	0.4
176.70	0.82	-0.8	46.04	5MY	-0.2	-3.7	-1.8	0.4
177.42	0.40	-0.6	55.59	15MY	0.6	-3.6	-1.9	0.4
179.31	0.04	-0.4	65.13	25MY	1.3	-3.5	-2.0	0.4
182.17	-0.28	-0.2	74.65	4JN	2.3	-3.5	-2.0	0.4
185.84	-0.54	-0.0	84.18	14JN	3.4	-3.4	-2.1	0.4
190.19	-0.77	0.1	93.71	24JN	1.9	-3.4	-2.2	0.4
195.08	-0.97	0.2	103.25	4JL	0.5	-3.4	-2.2	0.4
200.45	-1.14	0.3	112.83	14JL	-0.8	-3.4	-2.3	0.4
206.21	-1.28	0.4	122.43	24JL	-1.5	-3.3	-2.4	0.3
212.30	-1.39	0.5	132.08	3AU	-1.2	-3.3	-2.4	0.3
218.68	-1.48	0.6	141.77	13AU	-0.5	-3.3	-2.4	0.2
225.32	-1.55	0.7	151.51	23AU	-0.1	-3.3	-2.4	0.2
232.17	-1.60	0.7	161.31	2SE	0.1	-3.4	-2.5	0.1
239.22	-1.63	0.8	171.17	12SE	0.2	-3.4	-2.4	0.1
246.43	-1.65	0.9	181.08	22SE	0.5	-3.4	-2.4	-0.0
253.78	-1.64	0.9	191.06	2OC	1.9	-3.4	-2.4	-0.1
261.25	-1.62	1.0	201.09	12OC	1.6	-3.4	-2.3	-0.1
268.82	-1.58	1.0	211.16	22OC	-0.2	-3.5	-2.2	-0.2
276.46	-1.52	1.1	221.29	1NO	-0.5	-3.5	-2.2	-0.2
284.15	-1.46	1.1	231.45	11NO	-0.6	-3.5	-2.1	-0.1
291.88	-1.38	1.2	241.63	21NO	-0.6	-3.5	-2.0	-0.0
299.63	-1.29	1.2	251.83	1DE	-0.8	-3.4	-2.0	0.0
307.38	-1.19	1.3	262.03	11DE	-0.8	-3.4	-1.9	0.1
315.10	-1.08	1.3	272.23	21DE	-0.9	-3.4	-1.8	0.2
322.81	-0.97	1.4	282.41	31DE	-0.7	-3.4	-1.8	0.2
				469				
330.47	-0.86	1.4	292.57	10JA	0.4	-3.3	-1.7	0.3
338.07	-0.74	1.5	302.69	20JA	3.0	-3.3	-1.7	0.3
345.61	-0.62	1.5	312.77	30JA	1.4	-3.3	-1.6	0.4
353.09	-0.51	1.5	322.79	9FE	0.7	-3.3	-1.6	0.4
0.49	-0.39	1.6	332.76	19FE	0.4	-3.4	-1.6	0.4
7.82	-0.27	1.6	342.68	1MR	0.1	-3.4	-1.6	0.4
15.07	-0.16	1.7	352.54	11MR	-0.3	-3.4	-1.6	0.4
22.24	-0.05	1.7	2.33	21MR	-0.9	-3.4	-1.6	0.4
29.32	0.06	1.7	12.08	31MR	-1.8	-3.4	-1.6	0.3
36.33	0.16	1.7	21.77	10AP	-1.2	-3.5	-1.6	0.3
43.27	0.26	1.8	31.41	20AP	-0.2	-3.5	-1.6	0.3
50.13	0.36	1.8	41.02	30AP	0.8	-3.6	-1.6	0.2
56.93	0.45	1.8	50.59	10MY	1.7	-3.6	-1.7	0.3
63.65	0.54	1.8	60.13	20MY	3.0	-3.7	-1.7	0.3
70.32	0.62	1.8	69.66	30MY	2.9	-3.8	-1.8	0.3
76.93	0.70	1.8	79.19	9JN	1.5	-3.9	-1.8	0.3
83.49	0.77	1.9	88.71	19JN	0.4	-4.0	-1.9	0.3
90.01	0.84	1.9	98.25	29JN	-0.8	-4.1	-1.9	0.3
96.47	0.91	1.9	107.81	9JL	-1.5	-4.2	-2.0	0.3
102.90	0.97	2.0	117.39	19JL	-1.2	-4.2	-2.1	0.3
109.30	1.04	2.0	127.02	29JL	-0.5	-4.2	-2.1	0.3
115.66	1.09	2.0	136.68	8AU	-0.1	-4.0	-2.2	0.3
121.99	1.15	2.0	146.39	18AU	0.2	-3.5	-2.3	0.2
128.29	1.20	2.0	156.17	28AU	0.4	-3.3	-2.3	0.2
134.56	1.25	2.0	165.99	7SE	0.8	-3.8	-2.4	0.1
140.81	1.29	2.0	175.88	17SE	2.3	-4.2	-2.4	0.1
147.02	1.34	2.0	185.82	27SE	1.4	-4.3	-2.4	-0.0
153.21	1.38	1.9	195.82	7OC	-0.4	-4.3	-2.4	-0.1
159.36	1.41	1.9	205.87	17OC	-0.7	-4.2	-2.4	-0.1
165.47	1.45	1.8	215.98	27OC	-0.7	-4.2	-2.4	-0.2
171.53	1.48	1.8	226.12	6NO	-0.7	-4.1	-2.3	-0.3
177.55	1.50	1.7	236.29	16NO	-0.7	-4.0	-2.3	-0.2
183.50	1.53	1.6	246.48	26NO	-0.7	-3.9	-2.2	-0.2
189.37	1.54	1.5	256.68	6DE	-0.7	-3.8	-2.2	-0.1
195.15	1.55	1.4	266.89	16DE	-0.6	-3.7	-2.1	-0.0
200.82	1.56	1.3	277.08	26DE	0.5	-3.6	-2.0	0.0

Right table (470 / 471):

♂ LONG	LAT	MAG	☉ LONG	16.00UT	☿	♀	♃	♄
				470		MAGNITUDES		
206.36	1.55	1.1	287.25	5JA	3.0	-3.6	-1.9	0.1
211.73	1.53	1.0	297.39	15JA	1.1	-3.5	-1.9	0.2
216.90	1.50	0.8	307.49	25JA	0.4	-3.5	-1.8	0.2
221.81	1.45	0.6	317.54	4FE	0.2	-3.4	-1.7	0.3
226.42	1.37	0.4	327.55	14FE	0.0	-3.4	-1.7	0.3
230.63	1.27	0.1	337.49	24FE	-0.3	-3.4	-1.6	0.3
234.36	1.12	-0.1	347.38	6MR	-0.9	-3.3	-1.6	0.3
237.46	0.93	-0.4	357.21	16MR	-1.7	-3.3	-1.6	0.3
239.78	0.67	-0.7	6.98	26MR	-1.2	-3.3	-1.6	0.3
241.15	0.34	-1.0	16.69	5AP	-0.1	-3.3	-1.5	0.3
241.38	-0.09	-1.3	26.36	15AP	1.0	-3.3	-1.5	0.3
240.37	-0.60	-1.6	35.99	25AP	2.3	-3.3	-1.5	0.2
238.19	-1.19	-1.9	45.57	5MY	3.7	-3.3	-1.5	0.2
235.18	-1.80	-2.2	55.13	15MY	2.2	-3.4	-1.5	0.2
231.98	-2.37	-2.1	64.67	25MY	1.2	-3.4	-1.6	0.2
229.35	-2.82	-1.9	74.19	4JN	0.2	-3.5	-1.6	0.2
227.81	-3.15	-1.7	83.72	14JN	-0.8	-3.5	-1.6	0.3
227.64	-3.34	-1.5	93.25	24JN	-1.6	-3.5	-1.6	0.3
228.84	-3.44	-1.3	102.79	4JL	-1.2	-3.4	-1.7	0.3
231.25	-3.45	-1.1	112.36	14JL	-0.4	-3.4	-1.7	0.3
234.69	-3.41	-0.9	121.96	24JL	0.0	-3.4	-1.8	0.3
238.97	-3.33	-0.7	131.60	3AU	0.3	-3.4	-1.9	0.3
243.92	-3.21	-0.5	141.29	13AU	0.6	-3.3	-1.9	0.3
249.43	-3.07	-0.3	151.04	23AU	1.1	-3.3	-2.0	0.3
255.38	-2.91	-0.0	160.83	2SE	2.8	-3.3	-2.1	0.2
261.67	-2.74	-0.0	170.69	12SE	1.2	-3.3	-2.1	0.2
268.25	-2.55	0.1	180.60	22SE	-0.5	-3.3	-2.2	0.1
275.05	-2.36	0.2	190.57	2OC	-0.9	-3.4	-2.2	0.0
282.02	-2.16	0.4	200.60	12OC	-0.8	-3.4	-2.3	-0.0
289.12	-1.95	0.5	210.67	22OC	-0.8	-3.4	-2.3	-0.1
296.31	-1.75	0.6	220.79	1NO	-0.6	-3.4	-2.3	-0.2
303.56	-1.55	0.7	230.95	11NO	-0.5	-3.5	-2.3	-0.2
310.86	-1.35	0.8	241.13	21NO	-0.5	-3.5	-2.3	-0.3
318.17	-1.15	0.9	251.33	1DE	-0.4	-3.6	-2.3	-0.3
325.47	-0.96	1.0	261.54	11DE	0.7	-3.7	-2.3	-0.2
332.76	-0.78	1.1	271.74	21DE	2.8	-3.7	-2.2	-0.1
340.02	-0.61	1.2	281.92	31DE	0.7	-3.8	-2.1	-0.1
				471				
347.25	-0.45	1.3	292.08	10JA	0.2	-3.9	-2.1	0.0
354.42	-0.30	1.4	302.20	20JA	0.1	-4.0	-2.0	0.1
1.54	-0.15	1.5	312.28	30JA	-0.1	-4.1	-1.9	0.1
8.60	-0.02	1.5	322.31	9FE	-0.4	-4.2	-1.9	0.2
15.61	0.11	1.6	332.28	19FE	-0.9	-4.3	-1.8	0.2
22.55	0.22	1.7	342.20	1MR	-1.5	-4.3	-1.7	0.3
29.43	0.33	1.7	352.06	11MR	-1.1	-4.2	-1.7	0.3
36.25	0.43	1.8	1.86	21MR	-0.0	-3.9	-1.6	0.3
43.02	0.52	1.8	11.60	31MR	1.3	-3.3	-1.6	0.3
49.72	0.61	1.9	21.30	10AP	3.0	-3.2	-1.5	0.3
56.38	0.69	1.9	30.94	20AP	2.9	-3.8	-1.5	0.3
62.98	0.76	1.9	40.55	30AP	1.7	-4.1	-1.5	0.3
69.55	0.82	1.9	50.12	10MY	0.9	-4.2	-1.4	0.3
76.07	0.88	2.0	59.67	20MY	0.1	-4.2	-1.4	0.2
82.55	0.93	2.0	69.20	30MY	-0.8	-4.1	-1.4	0.2
89.01	0.98	2.0	78.72	9JN	-1.7	-4.0	-1.4	0.2
95.44	1.02	2.0	88.25	19JN	-1.2	-3.9	-1.4	0.3
101.84	1.06	2.0	97.78	29JN	-0.4	-3.8	-1.5	0.3
108.23	1.09	1.9	107.34	9JL	0.1	-3.7	-1.5	0.3
114.61	1.12	2.0	116.92	19JL	0.5	-3.7	-1.5	0.3
120.98	1.14	2.0	126.55	29JL	0.8	-3.6	-1.5	0.3
127.35	1.16	2.0	136.21	8AU	1.5	-3.5	-1.6	0.3
133.71	1.18	2.0	145.92	18AU	3.1	-3.5	-1.6	0.3
140.08	1.19	2.0	155.69	28AU	-1.1	-3.5	-1.7	0.3
146.46	1.19	2.0	165.52	7SE	-0.6	-3.4	-1.7	0.3
152.84	1.19	2.0	175.40	17SE	-1.0	-3.4	-1.8	0.2
159.23	1.19	2.0	185.34	27SE	-1.0	-3.4	-1.9	0.2
165.63	1.18	2.0	195.34	7OC	-0.8	-3.4	-1.9	0.1
172.05	1.17	2.0	205.39	17OC	-0.5	-3.4	-2.0	0.1
178.47	1.15	1.9	215.49	27OC	-0.4	-3.4	-2.1	0.0
184.91	1.12	1.9	225.63	6NO	-0.4	-3.4	-2.1	-0.1
191.35	1.09	1.8	235.79	16NO	-0.2	-3.4	-2.2	-0.1
197.81	1.05	1.8	245.99	26NO	0.9	-3.4	-2.2	-0.2
204.26	1.00	1.7	256.19	6DE	2.5	-3.4	-2.2	-0.3
210.72	0.94	1.6	266.39	16DE	0.5	-3.4	-2.2	-0.2
217.18	0.87	1.5	276.59	26DE	0.0	-3.4	-2.2	-0.2

Left table (472 / 473):

♂ LONG	♂ LAT	♂ MAG	☉ LONG	16.00UT	☿	♀	♃	♄
				472				
223.64	0.78	1.4	286.76	5JA	-0.1	-3.4	-2.2	-0.1
230.08	0.69	1.3	296.90	15JA	-0.2	-3.5	-2.1	-0.0
236.50	0.57	1.2	307.00	25JA	-0.5	-3.5	-2.1	0.0
242.90	0.44	1.1	317.06	4FE	-1.0	-3.5	-2.0	0.1
249.26	0.28	0.9	327.06	14FE	-1.4	-3.4	-1.9	0.2
255.57	0.10	0.8	337.01	24FE	-1.1	-3.4	-1.9	0.2
261.82	-0.10	0.6	346.90	5MR	0.0	-3.4	-1.8	0.3
267.98	-0.34	0.5	356.73	15MR	1.7	-3.4	-1.7	0.3
274.03	-0.62	0.3	6.50	25MR	3.6	-3.3	-1.7	0.3
279.93	-0.93	0.1	16.22	4AP	2.2	-3.3	-1.6	0.3
285.64	-1.29	-0.1	25.89	14AP	1.3	-3.3	-1.5	0.3
291.10	-1.69	-0.3	35.52	24AP	0.7	-3.3	-1.5	0.3
296.21	-2.15	-0.6	45.11	4MY	0.0	-3.3	-1.5	0.3
300.90	-2.67	-0.8	54.66	14MY	-0.8	-3.4	-1.4	0.3
305.02	-3.25	-1.1	64.20	24MY	-1.7	-3.4	-1.4	0.3
308.41	-3.89	-1.3	73.73	3JN	-1.2	-3.4	-1.4	0.3
310.88	-4.59	-1.6	83.25	13JN	-0.3	-3.4	-1.3	0.2
312.23	-5.30	-1.9	92.78	23JN	0.2	-3.5	-1.3	0.3
312.29	-5.98	-2.2	102.33	3JL	0.6	-3.5	-1.3	0.3
311.09	-6.53	-2.4	111.89	13JL	1.1	-3.6	-1.3	0.4
308.87	-6.85	-2.6	121.50	23JL	2.0	-3.7	-1.3	0.4
306.24	-6.85	-2.6	131.14	2AU	3.1	-3.8	-1.4	0.4
303.94	-6.53	-2.4	140.82	12AU	0.9	-3.8	-1.4	0.4
302.58	-5.97	-2.2	150.57	22AU	-0.6	-4.0	-1.4	0.4
302.50	-5.28	-1.9	160.36	1SE	-1.1	-4.1	-1.4	0.4
303.72	-4.56	-1.6	170.21	11SE	-1.1	-4.2	-1.5	0.4
306.08	-3.87	-1.3	180.12	21SE	-0.7	-4.3	-1.5	0.4
309.42	-3.23	-1.0	190.09	1OC	-0.4	-4.3	-1.6	0.3
313.51	-2.65	-0.7	200.11	11OC	-0.3	-4.3	-1.7	0.3
318.19	-2.15	-0.5	210.19	21OC	-0.2	-4.1	-1.7	0.2
323.34	-1.70	-0.2	220.30	31OC	-0.0	-3.5	-1.8	0.2
328.83	-1.30	-0.0	230.46	10NO	1.1	-3.1	-1.8	0.1
334.59	-0.96	0.2	240.64	20NO	2.2	-3.9	-1.9	0.0
340.55	-0.66	0.4	250.84	30NO	0.2	-4.3	-2.0	-0.1
346.65	-0.40	0.6	261.04	10DE	-0.1	-4.4	-2.0	-0.1
352.87	-0.17	0.7	271.24	20DE	-0.2	-4.4	-2.1	-0.2
359.16	0.03	0.9	281.43	30DE	-0.3	-4.3	-2.1	-0.1
				473				
5.51	0.21	1.0	291.58	9JA	-0.5	-4.2	-2.1	-0.1
11.89	0.36	1.2	301.71	19JA	-1.0	-4.1	-2.1	-0.0
18.30	0.49	1.3	311.79	29JA	-1.3	-3.9	-2.1	0.0
24.71	0.61	1.4	321.82	8FE	-1.0	-3.8	-2.0	0.1
31.12	0.71	1.5	331.80	18FE	0.1	-3.7	-2.0	0.2
37.53	0.80	1.6	341.72	28FE	2.0	-3.7	-1.9	0.2
43.92	0.87	1.7	351.58	10MR	3.0	-3.6	-1.9	0.3
50.30	0.94	1.7	1.39	20MR	1.6	-3.5	-1.8	0.3
56.67	0.99	1.8	11.14	30MR	0.9	-3.5	-1.8	0.4
63.02	1.04	1.9	20.83	9AP	0.5	-3.4	-1.7	0.4
69.36	1.08	1.9	30.48	19AP	-0.0	-3.4	-1.6	0.4
75.68	1.11	1.9	40.09	29AP	-0.8	-3.3	-1.6	0.4
81.99	1.14	2.0	49.66	9MY	-1.8	-3.3	-1.5	0.4
88.30	1.16	2.0	59.21	19MY	-1.2	-3.3	-1.4	0.4
94.60	1.17	2.0	68.74	29MY	-0.3	-3.3	-1.4	0.4
100.91	1.18	2.0	78.26	8JN	0.4	-3.3	-1.4	0.4
107.21	1.18	2.0	87.79	18JN	0.8	-3.3	-1.3	0.4
113.53	1.18	2.0	97.32	28JN	1.4	-3.3	-1.3	0.4
119.85	1.17	2.0	106.88	8JL	2.5	-3.4	-1.3	0.4
126.19	1.16	2.0	116.46	18JL	2.7	-3.4	-1.3	0.4
132.55	1.15	2.0	126.08	28JL	0.8	-3.4	-1.2	0.5
138.94	1.13	2.0	135.74	7AU	-0.7	-3.5	-1.2	0.5
145.35	1.10	1.9	145.46	17AU	-1.3	-3.5	-1.2	0.5
151.79	1.08	1.9	155.22	27AU	-1.2	-3.5	-1.3	0.5
158.27	1.04	1.9	165.04	6SE	-0.6	-3.5	-1.3	0.5
164.79	1.00	2.0	174.92	16SE	-0.3	-3.4	-1.3	0.5
171.35	0.96	1.9	184.85	26SE	-0.1	-3.4	-1.3	0.5
177.95	0.91	1.9	194.85	6OC	-0.1	-3.4	-1.4	0.5
184.60	0.86	1.9	204.90	16OC	0.2	-3.4	-1.4	0.5
191.29	0.79	1.9	214.99	26OC	1.4	-3.3	-1.5	0.4
198.03	0.73	1.9	225.13	5NO	2.0	-3.3	-1.5	0.4
204.82	0.65	1.8	235.30	15NO	0.1	-3.3	-1.6	0.3
211.65	0.57	1.8	245.49	25NO	-0.3	-3.3	-1.7	0.2
218.54	0.48	1.7	255.69	5DE	-0.3	-3.4	-1.7	0.1
225.47	0.38	1.7	265.90	15DE	-0.4	-3.4	-1.8	0.1
232.45	0.27	1.6	276.09	25DE	-0.6	-3.4	-1.9	0.0

Right table (474 / 475):

♂ LONG	♂ LAT	♂ MAG	☉ LONG	16.00UT	☿	♀	♃	♄
				474				
239.48	0.16	1.6	286.26	4JA	-1.0	-3.4	-1.9	-0.1
246.56	0.03	1.5	296.41	14JA	-1.1	-3.5	-2.0	-0.0
253.67	-0.11	1.4	306.51	24JA	-0.9	-3.5	-2.0	0.1
260.82	-0.26	1.3	316.57	3FE	0.2	-3.5	-2.0	0.1
268.01	-0.42	1.3	326.58	13FE	2.4	-3.6	-2.0	0.2
275.23	-0.58	1.2	336.53	23FE	2.3	-3.7	-2.0	0.2
282.47	-0.76	1.1	346.42	5MR	1.2	-3.7	-2.0	0.3
289.72	-0.95	1.0	356.26	15MR	0.7	-3.8	-2.0	0.3
296.98	-1.14	0.9	6.03	25MR	0.3	-3.9	-1.9	0.4
304.23	-1.34	0.8	15.76	4AP	-0.1	-4.0	-1.9	0.4
311.44	-1.55	0.7	25.43	14AP	-0.9	-4.0	-1.8	0.5
318.62	-1.76	0.6	35.05	24AP	-1.8	-4.1	-1.7	0.5
325.72	-1.96	0.5	44.65	4MY	-1.2	-4.2	-1.7	0.5
332.72	-2.17	0.3	54.21	14MY	-0.3	-4.2	-1.6	0.6
339.59	-2.37	0.2	63.74	24MY	0.5	-4.0	-1.5	0.6
346.28	-2.56	0.1	73.27	3JN	1.1	-3.5	-1.5	0.6
352.75	-2.75	-0.0	82.79	13JN	1.9	-2.8	-1.4	0.6
358.93	-2.92	-0.2	92.32	23JN	3.2	-3.5	-1.4	0.6
4.77	-3.08	-0.3	101.87	3JL	2.2	-4.0	-1.3	0.5
10.14	-3.22	-0.5	111.43	13JL	0.6	-4.2	-1.3	0.5
14.97	-3.34	-0.6	121.03	23JL	-0.7	-4.2	-1.3	0.6
19.08	-3.43	-0.8	130.67	2AU	-1.4	-4.2	-1.2	0.6
22.31	-3.48	-1.0	140.35	12AU	-1.2	-4.1	-1.2	0.7
24.46	-3.48	-1.3	150.09	22AU	-0.6	-4.0	-1.2	0.7
25.29	-3.40	-1.5	159.88	1SE	-0.2	-3.9	-1.2	0.7
24.68	-3.22	-1.7	169.73	11SE	-0.0	-3.8	-1.2	0.7
22.64	-2.91	-1.9	179.64	21SE	0.1	-3.8	-1.2	0.7
19.51	-2.46	-2.0	189.61	1OC	0.4	-3.7	-1.2	0.7
16.00	-1.90	-2.0	199.63	11OC	1.7	-3.6	-1.3	0.7
12.90	-1.30	-1.8	209.70	21OC	1.8	-3.6	-1.3	0.7
10.85	-0.72	-1.5	219.82	31OC	-0.1	-3.5	-1.3	0.6
10.15	-0.23	-1.1	229.97	10NO	-0.5	-3.5	-1.4	0.6
10.76	0.18	-0.8	240.15	20NO	-0.5	-3.5	-1.4	0.5
12.52	0.50	-0.5	250.35	30NO	-0.5	-3.4	-1.5	0.5
15.22	0.75	-0.2	260.55	10DE	-0.7	-3.4	-1.5	0.4
18.66	0.95	0.1	270.75	20DE	-0.9	-3.4	-1.6	0.3
22.67	1.10	0.3	280.94	30DE	-1.0	-3.4	-1.7	0.3
				475				
27.14	1.21	0.5	291.10	9JA	-0.8	-3.3	-1.7	0.2
31.95	1.29	0.7	301.22	19JA	0.3	-3.3	-1.8	0.2
37.04	1.36	0.9	311.31	29JA	2.8	-3.3	-1.9	0.2
42.33	1.40	1.1	321.34	8FE	1.7	-3.3	-1.9	0.2
47.78	1.43	1.2	331.32	18FE	0.8	-3.3	-2.0	0.3
53.37	1.45	1.3	341.25	28FE	0.5	-3.4	-2.0	0.4
59.06	1.46	1.5	351.11	10MR	0.2	-3.4	-2.0	0.4
64.84	1.47	1.6	0.91	20MR	-0.2	-3.4	-2.0	0.5
70.68	1.46	1.7	10.67	30MR	-0.9	-3.5	-2.0	0.5
76.58	1.45	1.7	20.36	9AP	-1.8	-3.5	-2.0	0.6
82.53	1.43	1.8	30.01	19AP	-1.2	-3.5	-1.9	0.6
88.52	1.41	1.9	39.62	29AP	-0.2	-3.4	-1.9	0.7
94.55	1.39	1.9	49.19	9MY	0.7	-3.4	-1.8	0.7
100.62	1.36	1.9	58.74	19MY	1.5	-3.4	-1.8	0.7
106.73	1.33	2.0	68.28	29MY	2.5	-3.3	-1.7	0.7
112.87	1.29	2.0	77.80	8JN	3.3	-3.3	-1.7	0.8
119.05	1.25	2.0	87.32	18JN	1.8	-3.3	-1.6	0.8
125.28	1.21	2.0	96.86	28JN	0.5	-3.3	-1.5	0.8
131.55	1.16	2.0	106.41	8JL	-0.8	-3.3	-1.5	0.8
137.87	1.11	2.0	115.99	18JL	-1.5	-3.4	-1.4	0.8
144.23	1.05	2.0	125.61	28JL	-1.2	-3.4	-1.4	0.8
150.65	1.00	2.0	135.27	7AU	-0.5	-3.4	-1.3	0.8
157.12	0.94	1.9	144.98	17AU	-0.1	-3.4	-1.3	0.8
163.66	0.87	1.9	154.74	27AU	0.1	-3.5	-1.3	0.9
170.25	0.80	1.9	164.56	6SE	0.3	-3.5	-1.2	0.9
176.92	0.73	1.8	174.44	16SE	0.6	-3.6	-1.2	1.0
183.64	0.65	1.8	184.37	26SE	2.0	-3.6	-1.2	1.0
190.44	0.57	1.8	194.36	6OC	1.6	-3.7	-1.2	1.0
197.31	0.48	1.8	204.41	16OC	-0.3	-3.8	-1.2	1.0
204.25	0.39	1.8	214.51	26OC	-0.6	-3.9	-1.2	1.0
211.26	0.30	1.8	224.64	5NO	-0.6	-4.0	-1.3	0.9
218.35	0.20	1.7	234.81	15NO	-0.6	-4.1	-1.3	0.9
225.50	0.10	1.7	245.00	25NO	-0.8	-4.2	-1.3	0.8
232.73	-0.01	1.7	255.20	5DE	-0.8	-4.3	-1.4	0.8
240.03	-0.12	1.7	265.40	15DE	-0.8	-4.4	-1.4	0.7
247.39	-0.24	1.6	275.60	25DE	-0.7	-4.4	-1.5	0.7

♂ LONG	LAT	MAG	☉ LONG	16.00UT 476	☿	♀	♃	♄
254.81	-0.36	1.6	285.77	4JA	0.4	-4.1	-1.5	0.6
262.29	-0.48	1.5	295.92	14JA	3.0	-3.6	-1.6	0.5
269.82	-0.60	1.5	306.03	24JA	1.3	-3.3	-1.7	0.4
277.39	-0.72	1.4	316.08	3FE	0.6	-3.9	-1.7	0.4
285.00	-0.85	1.4	326.10	13FE	0.3	-4.2	-1.8	0.4
292.64	-0.97	1.4	336.05	23FE	0.1	-4.3	-1.9	0.5
300.30	-1.08	1.3	345.94	4MR	-0.3	-4.3	-1.9	0.5
307.97	-1.19	1.3	355.78	14MR	-0.9	-4.2	-2.0	0.6
315.63	-1.29	1.2	5.56	24MR	-1.7	-4.1	-2.0	0.7
323.27	-1.39	1.2	15.28	3AP	-1.2	-4.0	-2.1	0.7
330.87	-1.47	1.1	24.96	13AP	-0.2	-3.9	-2.1	0.8
338.43	-1.54	1.1	34.59	23AP	0.9	-3.8	-2.1	0.8
345.93	-1.59	1.0	44.18	3MY	1.9	-3.7	-2.0	0.9
353.34	-1.64	1.0	53.75	13MY	3.4	-3.6	-2.0	0.9
0.65	-1.66	0.9	63.28	23MY	2.6	-3.5	-2.0	0.9
7.85	-1.67	0.9	72.81	2JN	1.4	-3.5	-1.9	1.0
14.91	-1.66	0.8	82.34	12JN	0.3	-3.4	-1.9	1.0
21.81	-1.63	0.8	91.86	22JN	-0.8	-3.4	-1.8	1.0
28.53	-1.59	0.7	101.40	2JL	-1.6	-3.4	-1.8	1.0
35.03	-1.52	0.6	110.97	12JL	-1.2	-3.4	-1.7	1.0
41.28	-1.44	0.5	120.57	22JL	-0.5	-3.3	-1.6	1.0
47.25	-1.33	0.4	130.20	1AU	-0.0	-3.3	-1.6	1.0
52.87	-1.20	0.3	139.89	11AU	0.2	-3.3	-1.5	1.0
58.09	-1.04	0.2	149.62	21AU	0.5	-3.3	-1.5	1.1
62.81	-0.86	0.1	159.41	31AU	0.9	-3.4	-1.4	1.1
66.93	-0.64	-0.1	169.25	10SE	2.4	-3.4	-1.4	1.2
70.30	-0.38	-0.2	179.16	20SE	1.4	-3.4	-1.3	1.2
72.77	-0.08	-0.4	189.12	30SE	-0.4	-3.4	-1.3	1.2
74.11	0.28	-0.6	199.14	10OC	-0.8	-3.4	-1.3	1.3
74.14	0.69	-0.9	209.21	20OC	-0.8	-3.5	-1.3	1.3
72.75	1.15	-1.1	219.32	30OC	-0.8	-3.5	-1.3	1.3
70.01	1.61	-1.3	229.47	9NO	-0.7	-3.5	-1.3	1.2
66.33	2.05	-1.4	239.65	19NO	-0.6	-3.5	-1.3	1.2
62.43	2.39	-1.3	249.85	29NO	-0.6	-3.4	-1.3	1.2
59.07	2.61	-1.1	260.05	9DE	-0.5	-3.4	-1.3	1.1
56.84	2.72	-0.8	270.25	19DE	0.5	-3.4	-1.4	1.1
55.94	2.73	-0.5	280.44	29DE	3.0	-3.4	-1.4	1.0

♂ LONG	LAT	MAG	☉ LONG	16.00UT 477	☿	♀	♃	♄
56.34	2.69	-0.2	290.60	8JA	0.9	-3.3	-1.5	0.9
57.85	2.62	0.1	300.73	18JA	0.3	-3.3	-1.5	0.9
60.27	2.52	0.3	310.82	28JA	0.2	-3.3	-1.6	0.8
63.43	2.42	0.5	320.86	7FE	-0.0	-3.3	-1.6	0.7
67.16	2.32	0.7	330.84	17FE	-0.3	-3.4	-1.7	0.7
71.33	2.22	0.9	340.76	27FE	-0.9	-3.4	-1.8	0.7
75.87	2.12	1.1	350.63	9MR	-1.6	-3.4	-1.8	0.8
80.70	2.02	1.2	0.44	19MR	-1.2	-3.4	-1.9	0.8
85.75	1.93	1.4	10.19	29MR	-0.1	-3.4	-2.0	0.9
91.00	1.83	1.5	19.90	8AP	1.1	-3.5	-2.0	0.9
96.41	1.74	1.6	29.55	18AP	2.5	-3.5	-2.1	1.0
101.96	1.65	1.6	39.16	28AP	3.5	-3.6	-2.1	1.0
107.62	1.56	1.7	48.74	8MY	2.0	-3.6	-2.2	1.1
113.39	1.47	1.8	58.29	18MY	1.1	-3.7	-2.2	1.1
119.26	1.39	1.8	67.82	28MY	0.2	-3.8	-2.2	1.2
125.22	1.30	1.9	77.34	7JN	-0.8	-3.9	-2.1	1.2
131.27	1.21	1.9	86.87	17JN	-1.6	-4.0	-2.1	1.3
137.40	1.12	1.9	96.40	27JN	-1.2	-4.1	-2.1	1.3
143.61	1.03	1.9	105.96	7JL	-0.4	-4.2	-2.0	1.3
149.90	0.94	1.9	115.53	17JL	0.1	-4.2	-2.0	1.3
156.28	0.85	1.9	125.15	27JL	0.4	-4.2	-1.9	1.3
162.74	0.76	1.9	134.81	6AU	0.7	-4.0	-1.8	1.3
169.29	0.66	1.9	144.51	16AU	1.2	-3.5	-1.8	1.3
175.92	0.57	1.9	154.27	26AU	2.9	-3.3	-1.7	1.3
182.65	0.47	1.8	164.08	5SE	1.2	-3.9	-1.7	1.3
189.46	0.37	1.8	173.95	15SE	-0.5	-4.2	-1.6	1.3
196.36	0.27	1.8	183.89	25SE	-0.9	-4.3	-1.5	1.3
203.35	0.17	1.7	193.88	5OC	-0.9	-4.3	-1.5	1.3
210.44	0.06	1.7	203.92	15OC	-0.8	-4.2	-1.5	1.3
217.61	-0.04	1.6	214.01	25OC	-0.6	-4.2	-1.4	1.3
224.87	-0.15	1.5	224.14	4NO	-0.5	-4.1	-1.4	1.2
232.22	-0.25	1.5	234.31	14NO	-0.5	-4.0	-1.4	1.2
239.65	-0.36	1.5	244.50	24NO	-0.3	-3.9	-1.4	1.2
247.15	-0.46	1.5	254.70	4DE	0.7	-3.8	-1.4	1.1
254.73	-0.56	1.5	264.91	14DE	2.7	-3.7	-1.4	1.1
262.37	-0.66	1.5	275.10	24DE	0.6	-3.6	-1.4	1.0

♂ LONG	LAT	MAG	☉ LONG	16.00UT 478	☿	♀	♃	♄
270.06	-0.75	1.5	285.28	3JA	0.2	-3.6	-1.4	1.0
277.80	-0.84	1.5	295.43	13JA	0.0	-3.5	-1.5	0.9
285.59	-0.92	1.5	305.54	23JA	-0.1	-3.5	-1.5	0.9
293.39	-1.00	1.4	315.60	2FE	-0.4	-3.4	-1.5	0.9
301.21	-1.06	1.4	325.61	12FE	-0.9	-3.4	-1.6	0.8
309.04	-1.12	1.4	335.57	22FE	-1.5	-3.4	-1.6	0.8
316.85	-1.16	1.4	345.47	4MR	-1.1	-3.3	-1.7	0.8
324.64	-1.20	1.4	355.31	14MR	-0.1	-3.3	-1.8	0.9
332.40	-1.22	1.4	5.09	24MR	1.4	-3.3	-1.8	0.9
340.11	-1.22	1.4	14.82	3AP	3.2	-3.3	-1.9	1.0
347.77	-1.22	1.4	24.50	13AP	2.6	-3.3	-2.0	1.1
355.36	-1.20	1.4	34.13	23AP	1.5	-3.3	-2.0	1.2
2.86	-1.16	1.3	43.72	3MY	0.8	-3.3	-2.1	1.2
10.28	-1.12	1.3	53.28	13MY	0.1	-3.4	-2.2	1.3
17.60	-1.06	1.3	62.83	23MY	-0.8	-3.4	-2.2	1.3
24.80	-0.99	1.3	72.35	2JN	-1.7	-3.5	-2.3	1.3
31.90	-0.90	1.3	81.87	12JN	-1.2	-3.5	-2.3	1.3
38.87	-0.81	1.3	91.40	22JN	-0.4	-3.5	-2.3	1.4
45.70	-0.70	1.3	100.94	2JL	0.2	-3.4	-2.3	1.4
52.39	-0.58	1.2	110.50	12JL	0.5	-3.4	-2.3	1.3
58.92	-0.45	1.2	120.10	22JL	0.9	-3.4	-2.2	1.3
65.27	-0.31	1.2	129.73	1AU	1.6	-3.4	-2.2	1.3
71.44	-0.16	1.1	139.41	11AU	3.2	-3.3	-2.1	1.3
77.38	0.01	1.0	149.15	21AU	1.1	-3.3	-2.1	1.2
83.08	0.20	1.0	158.93	31AU	-0.6	-3.3	-2.0	1.2
88.49	0.40	0.9	168.77	10SE	-1.1	-3.3	-2.0	1.1
93.57	0.62	0.8	178.67	20SE	-1.1	-3.3	-1.9	1.1
98.24	0.87	0.6	188.63	30SE	-0.8	-3.4	-1.8	1.1
102.42	1.14	0.5	198.65	10OC	-0.5	-3.4	-1.8	1.1
106.00	1.45	0.3	208.71	20OC	-0.3	-3.5	-1.7	1.1
108.83	1.79	0.2	218.82	30OC	-0.3	-3.4	-1.7	1.1
110.75	2.18	-0.0	228.97	9NO	-0.2	-3.5	-1.6	1.1
111.57	2.60	-0.3	239.15	19NO	0.9	-3.5	-1.6	1.1
111.10	3.05	-0.3	249.35	29NO	2.5	-3.6	-1.5	1.0
109.28	3.49	-0.7	259.55	9DE	0.4	-3.7	-1.5	1.0
106.22	3.87	-0.9	269.76	19DE	-0.0	-3.7	-1.5	1.0
102.38	4.12	-1.1	279.94	29DE	-0.1	-3.8	-1.5	0.9

♂ LONG	LAT	MAG	☉ LONG	16.00UT 479	☿	♀	♃	♄
98.46	4.21	-1.0	290.11	8JA	-0.2	-3.9	-1.5	0.9
95.18	4.14	-0.7	300.24	18JA	-0.5	-4.0	-1.5	0.8
93.04	3.95	-0.5	310.33	28JA	-1.0	-4.1	-1.5	0.8
92.20	3.69	-0.2	320.37	7FE	-1.3	-4.2	-1.5	0.7
92.61	3.41	0.0	330.36	17FE	-1.1	-4.3	-1.6	0.7
94.11	3.13	0.3	340.28	27FE	0.0	-4.3	-1.6	0.6
96.50	2.87	0.5	350.16	9MR	1.7	-4.2	-1.6	0.6
99.61	2.62	0.7	359.97	19MR	3.5	-3.9	-1.7	0.7
103.29	2.39	0.9	9.72	29MR	1.9	-3.3	-1.7	0.7
107.44	2.18	1.0	19.43	8AP	1.1	-3.3	-1.8	0.8
111.97	1.99	1.1	29.08	18AP	0.6	-3.9	-1.9	0.9
116.81	1.81	1.2	38.69	28AP	0.0	-4.1	-1.9	0.9
121.90	1.64	1.3	48.27	8MY	-0.8	-4.2	-2.0	1.0
127.23	1.48	1.4	57.82	18MY	-1.7	-4.2	-2.1	1.0
132.74	1.32	1.5	67.35	28MY	-1.2	-4.1	-2.1	1.1
138.43	1.18	1.6	76.88	7JN	-0.4	-4.0	-2.2	1.1
144.28	1.04	1.6	86.41	17JN	0.3	-3.9	-2.3	1.2
150.27	0.90	1.6	95.94	27JN	0.7	-3.8	-2.3	1.2
156.39	0.77	1.7	105.49	7JL	1.2	-3.7	-2.4	1.2
162.65	0.64	1.7	115.07	17JL	2.1	-3.6	-2.4	1.2
169.02	0.51	1.7	124.68	27JL	3.0	-3.6	-2.4	1.2
175.53	0.39	1.7	134.34	6AU	0.9	-3.5	-2.4	1.1
182.15	0.27	1.7	144.04	16AU	-0.6	-3.5	-2.4	1.1
188.88	0.15	1.7	153.79	26AU	-1.2	-3.4	-2.4	1.1
195.73	0.03	1.7	163.61	5SE	-1.2	-3.4	-2.3	1.0
202.70	-0.08	1.7	173.48	15SE	-0.7	-3.4	-2.3	1.0
209.77	-0.19	1.6	183.41	25SE	-0.4	-3.4	-2.2	1.0
216.95	-0.30	1.6	193.40	5OC	-0.2	-3.4	-2.1	1.0
224.24	-0.40	1.6	203.43	15OC	-0.2	-3.4	-2.1	1.0
231.62	-0.50	1.6	213.52	25OC	0.0	-3.4	-2.0	1.0
239.10	-0.60	1.5	223.65	4NO	1.1	-3.4	-1.9	1.0
246.67	-0.69	1.5	233.82	14NO	2.2	-3.4	-1.9	1.0
254.31	-0.77	1.4	244.00	24NO	-0.2	-3.4	-1.8	1.0
262.03	-0.84	1.4	254.21	4DE	-0.2	-3.4	-1.8	0.9
269.80	-0.91	1.4	264.41	14DE	-0.2	-3.4	-1.7	0.9
277.62	-0.97	1.3	274.60	24DE	-0.3	-3.4	-1.7	0.9

♂			☉	16.00UT	☿	♀	♃	♄
LONG	LAT	MAG	LONG	480	MAGNITUDES			
285.48	-1.01	1.3	284.78	3JA	-0.6	-3.4	-1.6	0.8
293.36	-1.05	1.3	294.93	13JA	-1.0	-3.5	-1.6	0.8
301.26	-1.07	1.3	305.04	23JA	-1.2	-3.5	-1.6	0.8
309.15	-1.09	1.3	315.11	2FE	-1.0	-3.5	-1.6	0.7
317.02	-1.09	1.4	325.12	12FE	0.1	-3.4	-1.6	0.7
324.87	-1.08	1.4	335.08	22FE	2.1	-3.4	-1.6	0.6
332.68	-1.06	1.4	344.99	3MR	2.7	-3.4	-1.6	0.6
340.44	-1.02	1.4	354.83	13MR	1.4	-3.4	-1.6	0.5
348.13	-0.98	1.4	4.61	23MR	0.8	-3.3	-1.6	0.5
355.76	-0.93	1.5	14.35	2AP	0.4	-3.3	-1.7	0.5
3.31	-0.86	1.5	24.02	12AP	-0.1	-3.3	-1.7	0.6
10.78	-0.79	1.5	33.66	22AP	-0.8	-3.3	-1.7	0.7
18.15	-0.71	1.5	43.26	2MY	-1.7	-3.3	-1.8	0.7
25.44	-0.62	1.6	52.82	12MY	-1.3	-3.4	-1.8	0.8
32.63	-0.52	1.6	62.36	22MY	-0.3	-3.4	-1.9	0.9
39.71	-0.42	1.6	71.89	1JN	0.4	-3.4	-2.0	0.9
46.70	-0.31	1.6	81.41	11JN	0.9	-3.4	-2.0	1.0
53.58	-0.19	1.6	90.94	21JN	1.6	-3.5	-2.1	1.0
60.36	-0.08	1.6	100.48	1JL	2.8	-3.5	-2.2	1.0
67.02	0.05	1.6	110.04	11JL	2.6	-3.6	-2.3	1.0
73.58	0.18	1.6	119.64	21JL	0.7	-3.7	-2.3	1.1
80.01	0.31	1.6	129.27	31JL	-0.7	-3.8	-2.4	1.0
86.32	0.45	1.6	138.94	10AU	-1.3	-3.9	-2.4	1.0
92.51	0.59	1.5	148.67	20AU	-1.2	-4.0	-2.4	1.0
98.54	0.74	1.5	158.46	30AU	-0.6	-4.1	-2.5	1.0
104.41	0.90	1.4	168.29	9SE	-0.3	-4.2	-2.5	1.0
110.11	1.07	1.4	178.19	19SE	-0.1	-4.3	-2.4	0.9
115.60	1.25	1.3	188.15	29SE	0.0	-4.3	-2.4	0.9
120.84	1.44	1.2	198.16	9OC	0.2	-4.3	-2.4	0.9
125.80	1.65	1.1	208.23	19OC	1.4	-4.1	-2.3	0.9
130.41	1.87	1.0	218.34	29OC	2.0	-3.5	-2.2	0.9
134.61	2.12	0.8	228.48	8NO	0.0	-3.2	-2.2	0.9
138.29	2.39	0.6	238.66	18NO	-0.4	-3.9	-2.1	0.9
141.33	2.70	0.4	248.85	28NO	-0.4	-4.3	-2.0	0.9
143.58	3.03	0.2	259.05	8DE	-0.4	-4.4	-2.0	0.9
144.87	3.38	-0.0	269.26	18DE	-0.6	-4.4	-1.9	0.9
145.00	3.75	-0.3	279.45	28DE	-1.0	-4.3	-1.8	0.8
				481				
143.84	4.10	-0.5	289.61	7JA	-1.0	-4.2	-1.8	0.8
141.40	4.38	-0.8	299.75	17JA	-0.9	-4.0	-1.7	0.8
137.94	4.54	-1.0	309.84	27JA	0.2	-3.9	-1.7	0.7
134.02	4.53	-1.0	319.88	6FE	2.5	-3.8	-1.6	0.7
130.35	4.35	-0.8	329.87	16FE	2.1	-3.7	-1.6	0.6
127.57	4.04	-0.6	339.80	26FE	1.0	-3.7	-1.6	0.6
126.01	3.65	-0.4	349.68	8MR	0.6	-3.6	-1.6	0.5
125.75	3.24	-0.2	359.49	18MR	0.3	-3.5	-1.6	0.5
126.68	2.84	0.1	9.25	28MR	-0.1	-3.5	-1.6	0.4
128.63	2.48	0.3	18.96	7AP	-0.8	-3.4	-1.6	0.4
131.41	2.14	0.4	28.62	17AP	-1.7	-3.4	-1.6	0.5
134.87	1.84	0.6	38.23	27AP	-1.3	-3.3	-1.6	0.5
138.89	1.56	0.8	47.81	7MY	-0.3	-3.3	-1.6	0.6
143.37	1.31	0.9	57.37	17MY	0.5	-3.3	-1.7	0.7
148.24	1.08	1.0	66.90	27MY	1.2	-3.3	-1.7	0.7
153.42	0.86	1.1	76.42	6JN	2.1	-3.3	-1.8	0.8
158.89	0.67	1.1	85.95	16JN	3.3	-3.3	-1.8	0.8
164.61	0.48	1.2	95.48	26JN	2.1	-3.3	-1.9	0.9
170.54	0.31	1.3	105.03	6JL	0.6	-3.4	-1.9	0.9
176.68	0.15	1.3	114.61	16JL	-0.7	-3.4	-2.0	0.9
183.00	-0.00	1.3	124.22	26JL	-1.4	-3.4	-2.1	0.9
189.50	-0.15	1.4	133.87	5AU	-1.2	-3.5	-2.1	1.0
196.15	-0.28	1.4	143.57	15AU	-0.6	-3.5	-2.2	1.0
202.96	-0.41	1.4	153.32	25AU	-0.2	-3.5	-2.3	0.9
209.92	-0.52	1.4	163.13	4SE	0.0	-3.5	-2.3	0.9
217.01	-0.63	1.4	173.00	14SE	0.2	-3.4	-2.4	0.9
224.23	-0.73	1.4	182.92	24SE	0.5	-3.4	-2.4	0.9
231.57	-0.82	1.4	192.90	4OC	1.7	-3.4	-2.4	0.8
239.02	-0.90	1.4	202.94	14OC	1.8	-3.4	-2.4	0.8
246.57	-0.97	1.4	213.03	24OC	-0.2	-3.3	-2.4	0.8
254.21	-1.03	1.4	223.15	3NO	-0.5	-3.3	-2.4	0.8
261.93	-1.07	1.4	233.32	13NO	-0.5	-3.3	-2.3	0.8
269.71	-1.10	1.4	243.50	23NO	-0.6	-3.3	-2.3	0.8
277.54	-1.13	1.4	253.70	3DE	-0.7	-3.4	-2.2	0.8
285.41	-1.14	1.4	263.91	13DE	-0.8	-3.4	-2.1	0.8
293.31	-1.13	1.4	274.11	23DE	-0.9	-3.4	-2.1	0.8

♂			☉	16.00UT	☿	♀	♃	♄
LONG	LAT	MAG	LONG	482	MAGNITUDES			
301.21	-1.12	1.4	284.29	2JA	-0.8	-3.4	-2.0	0.8
309.10	-1.09	1.4	294.44	12JA	0.3	-3.5	-1.9	0.8
316.97	-1.05	1.4	304.55	22JA	2.8	-3.5	-1.9	0.7
324.81	-1.00	1.4	314.62	1FE	1.6	-3.5	-1.8	0.7
332.61	-0.95	1.4	324.64	11FE	0.7	-3.6	-1.7	0.7
340.34	-0.88	1.4	334.60	21FE	0.4	-3.7	-1.7	0.6
348.02	-0.80	1.4	344.51	3MR	0.2	-3.7	-1.6	0.5
355.62	-0.72	1.4	354.35	13MR	-0.2	-3.8	-1.6	0.5
3.15	-0.63	1.4	4.14	23MR	-0.8	-3.9	-1.6	0.4
10.59	-0.54	1.5	13.87	2AP	-1.7	-4.0	-1.5	0.4
17.95	-0.45	1.5	23.56	12AP	-1.3	-4.1	-1.5	0.3
25.21	-0.35	1.6	33.19	22AP	-0.2	-4.1	-1.5	0.4
32.39	-0.24	1.6	42.79	2MY	0.7	-4.2	-1.5	0.4
39.49	-0.14	1.7	52.36	12MY	1.6	-4.2	-1.5	0.5
46.49	-0.03	1.7	61.90	22MY	2.8	-4.0	-1.5	0.5
53.41	0.07	1.7	71.43	1JN	3.1	-3.5	-1.5	0.6
60.24	0.18	1.8	80.95	11JN	1.6	-2.7	-1.6	0.7
66.99	0.29	1.8	90.48	21JN	0.4	-3.5	-1.6	0.7
73.66	0.39	1.8	100.02	1JL	-0.7	-4.0	-1.6	0.8
80.25	0.50	1.8	109.58	11JL	-1.5	-4.2	-1.7	0.8
86.75	0.61	1.8	119.17	21JL	-1.2	-4.2	-1.7	0.8
93.19	0.71	1.8	128.80	31JL	-0.5	-4.2	-1.8	0.9
99.54	0.82	1.8	138.48	10AU	-0.1	-4.1	-1.9	0.9
105.80	0.93	1.8	148.20	20AU	0.2	-4.0	-1.9	0.9
111.99	1.04	1.8	157.98	30AU	0.3	-3.9	-2.0	0.9
118.08	1.15	1.8	167.82	9SE	0.7	-3.8	-2.1	0.9
124.07	1.26	1.7	177.71	19SE	2.1	-3.8	-2.1	0.8
129.96	1.38	1.7	187.66	29SE	1.6	-3.7	-2.2	0.8
135.72	1.50	1.6	197.67	9OC	-0.3	-3.6	-2.2	0.8
141.35	1.63	1.5	207.73	19OC	-0.7	-3.6	-2.3	0.7
146.81	1.76	1.4	217.84	29OC	-0.7	-3.5	-2.3	0.7
152.07	1.90	1.3	227.99	8NO	-0.7	-3.5	-2.3	0.8
157.11	2.05	1.2	238.16	18NO	-0.8	-3.5	-2.3	0.8
161.87	2.20	1.1	248.36	28NO	-0.7	-3.4	-2.3	0.8
166.28	2.37	0.9	258.56	8DE	-0.7	-3.4	-2.3	0.8
170.28	2.55	0.7	268.76	18DE	-0.6	-3.4	-2.2	0.8
173.75	2.75	0.5	278.96	28DE	0.4	-3.4	-2.2	0.8
				483				
176.56	2.95	0.3	289.12	7JA	3.0	-3.3	-2.1	0.8
178.57	3.16	0.1	299.26	17JA	1.2	-3.3	-2.0	0.8
179.59	3.37	-0.2	309.35	27JA	0.5	-3.3	-2.0	0.7
179.44	3.55	-0.5	319.40	6FE	0.3	-3.3	-1.9	0.7
178.03	3.67	-0.8	329.38	16FE	0.1	-3.4	-1.8	0.7
175.39	3.70	-1.0	339.32	26FE	-0.3	-3.4	-1.8	0.6
171.86	3.59	-1.2	349.20	8MR	-0.9	-3.4	-1.7	0.6
168.06	3.32	-1.2	359.01	18MR	-1.7	-3.4	-1.6	0.5
164.69	2.93	-1.1	8.78	28MR	-1.2	-3.5	-1.6	0.4
162.34	2.47	-0.9	18.49	7AP	-0.2	-3.5	-1.5	0.4
161.27	2.00	-0.6	28.14	17AP	0.9	-3.5	-1.5	0.3
161.51	1.55	-0.4	37.76	27AP	2.1	-3.4	-1.5	0.3
162.94	1.14	-0.2	47.34	7MY	3.7	-3.4	-1.4	0.3
165.37	0.78	-0.0	56.89	17MY	2.4	-3.4	-1.4	0.4
168.64	0.46	0.1	66.43	27MY	1.3	-3.3	-1.4	0.5
172.60	0.17	0.3	75.96	6JN	0.3	-3.3	-1.4	0.5
177.13	-0.08	0.4	85.48	16JN	-0.7	-3.3	-1.4	0.6
182.13	-0.30	0.5	95.01	26JN	-1.6	-3.3	-1.4	0.6
187.54	-0.49	0.6	104.56	6JL	-1.2	-3.3	-1.4	0.7
193.29	-0.67	0.7	114.13	16JL	-0.5	-3.3	-1.5	0.7
199.35	-0.82	0.8	123.75	26JL	-0.0	-3.4	-1.5	0.8
205.67	-0.96	0.8	133.39	5AU	0.3	-3.4	-1.5	0.8
212.22	-1.07	0.9	143.09	15AU	0.5	-3.4	-1.6	0.8
218.99	-1.17	0.9	152.85	25AU	1.0	-3.5	-1.6	0.8
225.94	-1.25	1.0	162.65	4SE	2.5	-3.5	-1.7	0.8
233.06	-1.32	1.0	172.51	14SE	1.4	-3.6	-1.7	0.8
240.33	-1.36	1.1	182.44	24SE	-0.4	-3.6	-1.8	0.8
247.73	-1.40	1.1	192.42	4OC	-0.8	-3.7	-1.9	0.8
255.25	-1.41	1.1	202.45	14OC	-0.8	-3.8	-1.9	0.8
262.86	-1.41	1.2	212.54	24OC	-0.8	-3.9	-2.0	0.7
270.55	-1.40	1.2	222.66	3NO	-0.6	-4.0	-2.1	0.7
278.29	-1.37	1.2	232.82	13NO	-0.6	-4.1	-2.1	0.7
286.08	-1.33	1.3	243.01	23NO	-0.6	-4.2	-2.2	0.7
293.90	-1.27	1.3	253.21	3DE	-0.5	-4.3	-2.2	0.8
301.72	-1.20	1.3	263.41	13DE	0.5	-4.4	-2.2	0.8
309.53	-1.13	1.4	273.62	23DE	2.9	-4.3	-2.2	0.8

Left Table

♂ LONG	LAT	MAG	☉ LONG	16.00UT	☿	♀	♃	♄
				484		MAGNITUDES		
317.32	-1.04	1.4	283.79	2JA	0.8	-4.1	-2.2	0.8
325.08	-0.95	1.4	293.95	12JA	0.3	-3.6	-2.1	0.8
332.79	-0.85	1.5	304.06	22JA	0.1	-3.3	-2.1	0.8
340.44	-0.75	1.5	314.13	1FE	-0.1	-3.9	-2.0	0.8
348.03	-0.65	1.5	324.15	11FE	-0.4	-4.2	-2.0	0.7
355.54	-0.54	1.6	334.12	21FE	-0.9	-4.3	-1.9	0.7
2.98	-0.43	1.6	344.02	2MR	-1.6	-4.3	-1.8	0.6
10.34	-0.32	1.6	353.87	12MR	-1.2	-4.2	-1.8	0.6
17.61	-0.21	1.6	3.67	22MR	-0.1	-4.1	-1.7	0.5
24.80	-0.11	1.7	13.40	1AP	1.2	-4.0	-1.6	0.5
31.91	-0.00	1.7	23.09	11AP	2.8	-3.9	-1.6	0.4
38.94	0.10	1.7	32.73	21AP	3.2	-3.8	-1.5	0.3
45.89	0.20	1.7	42.32	1MY	1.8	-3.7	-1.5	0.3
52.77	0.30	1.7	51.89	11MY	1.0	-3.6	-1.4	0.3
59.57	0.39	1.7	61.44	21MY	0.2	-3.5	-1.4	0.3
66.30	0.49	1.8	70.96	31MY	-0.8	-3.5	-1.4	0.4
72.97	0.57	1.8	80.49	10JN	-1.6	-3.4	-1.3	0.4
79.58	0.66	1.9	90.02	20JN	-1.2	-3.4	-1.3	0.5
86.13	0.74	1.9	99.55	30JN	-0.4	-3.4	-1.3	0.6
92.63	0.82	1.9	109.12	10JL	0.1	-3.4	-1.3	0.6
99.08	0.90	2.0	118.71	20JL	0.4	-3.3	-1.3	0.7
105.48	0.97	2.0	128.33	30JL	0.7	-3.3	-1.3	0.7
111.84	1.04	2.0	138.01	9AU	1.4	-3.3	-1.3	0.8
118.15	1.11	2.0	147.73	19AU	3.0	-3.3	-1.4	0.8
124.42	1.18	2.0	157.51	29AU	1.2	-3.4	-1.4	0.8
130.65	1.25	2.0	167.34	8SE	-0.5	-3.4	-1.4	0.8
136.83	1.31	1.9	177.23	18SE	-1.0	-3.4	-1.5	0.8
142.97	1.37	1.9	187.18	28SE	-1.0	-3.4	-1.5	0.8
149.05	1.43	1.9	197.19	8OC	-0.6	-3.4	-1.6	0.8
155.07	1.49	1.8	207.25	18OC	-0.5	-3.5	-1.7	0.8
161.03	1.55	1.7	217.35	28OC	-0.4	-3.5	-1.7	0.7
166.92	1.61	1.7	227.50	7NO	-0.4	-3.5	-1.8	0.7
172.71	1.66	1.6	237.67	17NO	-0.3	-3.5	-1.8	0.7
178.39	1.71	1.5	247.86	27NO	0.7	-3.4	-1.9	0.7
183.95	1.77	1.4	258.07	7DE	2.7	-3.4	-2.0	0.7
189.35	1.81	1.2	268.27	17DE	0.6	-3.4	-2.0	0.8
194.56	1.86	1.1	278.46	27DE	0.1	-3.4	-2.1	0.8
				485				
199.55	1.90	0.9	288.63	6JA	-0.0	-3.3	-2.1	0.8
204.25	1.93	0.7	298.76	16JA	-0.2	-3.3	-2.1	0.8
208.61	1.95	0.5	308.86	26JA	-0.4	-3.3	-2.1	0.8
212.53	1.95	0.3	318.91	5FE	-0.9	-3.3	-2.1	0.8
215.89	1.93	0.1	328.90	15FE	-1.4	-3.4	-2.0	0.8
218.58	1.89	-0.2	338.84	25FE	-1.1	-3.4	-2.0	0.7
220.41	1.80	-0.5	348.72	7MR	-0.1	-3.4	-1.9	0.7
221.20	1.65	-0.8	358.54	17MR	1.5	-3.4	-1.9	0.6
220.81	1.42	-1.1	8.30	27MR	3.4	-3.4	-1.8	0.6
219.14	1.10	-1.4	18.02	6AP	2.4	-3.5	-1.7	0.5
216.38	0.69	-1.7	27.68	16AP	1.4	-3.5	-1.7	0.4
212.98	0.20	-1.8	37.30	26AP	0.7	-3.6	-1.6	0.4
209.61	-0.32	-1.7	46.88	6MY	0.1	-3.6	-1.5	0.3
207.01	-0.81	-1.6	56.44	16MY	-0.8	-3.7	-1.5	0.3
205.61	-1.23	-1.4	65.97	26MY	-1.7	-3.8	-1.4	0.3
205.58	-1.57	-1.1	75.50	5JN	-1.3	-3.9	-1.4	0.3
206.89	-1.84	-0.9	85.02	15JN	-0.4	-4.0	-1.3	0.4
209.34	-2.04	-0.7	94.55	25JN	0.2	-4.1	-1.3	0.5
212.76	-2.19	-0.6	104.10	5JL	0.6	-4.2	-1.3	0.5
216.99	-2.29	-0.4	113.67	15JL	1.0	-4.2	-1.3	0.6
221.86	-2.35	-0.2	123.28	25JL	1.8	-4.2	-1.2	0.6
227.28	-2.38	-0.1	132.93	4AU	3.2	-4.0	-1.2	0.7
233.15	-2.38	0.0	142.62	14AU	1.1	-3.5	-1.2	0.7
239.37	-2.35	0.1	152.37	24AU	-0.6	-3.3	-1.2	0.8
245.91	-2.31	0.3	162.18	3SE	-1.1	-3.9	-1.3	0.8
252.69	-2.24	0.4	172.04	13SE	-1.1	-4.2	-1.3	0.8
259.68	-2.15	0.5	181.96	23SE	-0.7	-4.3	-1.3	0.8
266.83	-2.05	0.6	191.94	3OC	-0.4	-4.3	-1.3	0.8
274.12	-1.93	0.6	201.97	13OC	-0.3	-4.2	-1.4	0.8
281.51	-1.80	0.7	212.05	23OC	-0.2	-4.1	-1.4	0.8
288.97	-1.67	0.8	222.17	2NO	-0.1	-4.1	-1.5	0.7
296.48	-1.52	0.9	232.33	12NO	0.9	-4.0	-1.5	0.7
304.02	-1.37	1.0	242.52	22NO	2.5	-3.9	-1.6	0.7
311.56	-1.22	1.1	252.72	2DE	0.3	-3.8	-1.7	0.7
319.10	-1.07	1.1	262.92	12DE	-0.1	-3.7	-1.7	0.7
326.61	-0.92	1.2	273.12	22DE	-0.2	-3.6	-1.8	0.8

Right Table

♂ LONG	LAT	MAG	☉ LONG	16.00UT	☿	♀	♃	♄
				486		MAGNITUDES		
334.08	-0.77	1.3	283.30	1JA	-0.3	-3.6	-1.9	0.8
341.51	-0.62	1.4	293.46	11JA	-0.5	-3.5	-1.9	0.8
348.88	-0.48	1.4	303.58	21JA	-1.0	-3.5	-2.0	0.8
356.19	-0.34	1.5	313.65	31JA	-1.3	-3.4	-2.0	0.8
3.43	-0.21	1.6	323.67	10FE	-1.1	-3.4	-2.0	0.8
10.60	-0.08	1.6	333.64	20FE	-0.0	-3.4	-2.0	0.8
17.70	0.03	1.7	343.55	2MR	1.8	-3.3	-2.0	0.8
24.73	0.15	1.7	353.40	12MR	3.2	-3.3	-2.0	0.7
31.69	0.25	1.8	3.20	22MR	1.7	-3.3	-1.9	0.7
38.58	0.35	1.8	12.94	1AP	1.0	-3.3	-1.9	0.6
45.41	0.45	1.8	22.62	11AP	0.5	-3.3	-1.8	0.6
52.17	0.53	1.9	32.27	21AP	0.0	-3.3	-1.8	0.5
58.87	0.62	1.9	41.87	1MY	-0.8	-3.3	-1.7	0.4
65.52	0.69	1.9	51.43	11MY	-1.7	-3.4	-1.7	0.4
72.12	0.76	1.9	60.98	21MY	-1.3	-3.4	-1.6	0.3
78.67	0.83	1.9	70.51	31MY	-0.4	-3.5	-1.5	0.3
85.18	0.89	1.9	80.03	10JN	0.3	-3.5	-1.5	0.3
91.66	0.94	1.9	89.55	20JN	0.8	-3.5	-1.4	0.3
98.10	0.99	1.9	99.09	30JN	1.3	-3.4	-1.4	0.4
104.52	1.04	1.9	108.65	10JL	2.4	-3.4	-1.3	0.5
110.91	1.08	2.0	118.24	20JL	2.9	-3.4	-1.3	0.5
117.29	1.12	2.0	127.86	30JL	0.9	-3.4	-1.3	0.6
123.65	1.15	2.0	137.53	9AU	-0.6	-3.3	-1.2	0.6
130.00	1.18	2.0	147.25	19AU	-1.2	-3.3	-1.2	0.7
136.34	1.21	2.0	157.03	29AU	-1.2	-3.3	-1.2	0.7
142.68	1.23	2.0	166.86	8SE	-0.7	-3.3	-1.2	0.8
149.01	1.24	2.0	176.75	18SE	-0.3	-3.3	-1.2	0.8
155.33	1.26	2.0	186.70	28SE	-0.2	-3.4	-1.2	0.8
161.65	1.27	2.0	196.70	8OC	-0.1	-3.4	-1.2	0.8
167.97	1.27	1.9	206.76	18OC	0.1	-3.4	-1.3	0.8
174.28	1.27	1.9	216.86	28OC	1.2	-3.4	-1.3	0.8
180.58	1.26	1.8	227.00	7NO	2.2	-3.5	-1.3	0.8
186.87	1.25	1.8	237.18	17NO	0.1	-3.5	-1.4	0.7
193.14	1.23	1.7	247.37	27NO	-0.3	-3.6	-1.4	0.7
199.39	1.20	1.6	257.57	7DE	-0.3	-3.7	-1.5	0.7
205.61	1.16	1.5	267.78	17DE	-0.4	-3.7	-1.6	0.7
211.80	1.12	1.4	277.97	27DE	-0.6	-3.8	-1.6	0.8
				487				
217.95	1.05	1.3	288.14	6JA	-1.0	-3.9	-1.7	0.8
224.03	0.98	1.2	298.28	16JA	-1.1	-4.0	-1.8	0.8
230.05	0.88	1.0	308.37	26JA	-1.0	-4.1	-1.8	0.8
235.98	0.77	0.9	318.42	5FE	0.1	-4.2	-1.9	0.8
241.80	0.63	0.7	328.42	15FE	2.2	-4.3	-1.9	0.8
247.48	0.46	0.6	338.36	25FE	2.5	-4.3	-2.0	0.8
253.00	0.26	0.4	348.24	7MR	1.3	-4.2	-2.0	0.8
258.29	0.02	0.1	358.07	17MR	0.7	-3.9	-2.0	0.8
263.31	-0.27	-0.1	7.84	27MR	0.4	-3.3	-2.0	0.7
267.98	-0.62	-0.3	17.55	6AP	-0.1	-3.3	-2.0	0.7
272.19	-1.03	-0.6	27.22	16AP	-0.8	-3.9	-2.0	0.6
275.83	-1.53	-0.9	36.83	26AP	-1.7	-4.1	-1.9	0.6
278.73	-2.10	-1.1	46.42	6MY	-1.3	-4.2	-1.9	0.5
280.70	-2.77	-1.5	55.98	16MY	-0.3	-4.2	-1.8	0.4
281.57	-3.51	-1.8	65.51	26MY	0.4	-4.1	-1.8	0.4
281.21	-4.30	-2.1	75.04	5JN	1.0	-4.0	-1.7	0.3
279.64	-5.05	-2.3	84.56	15JN	1.8	-3.9	-1.6	0.3
277.21	-5.66	-2.5	94.09	25JN	3.0	-3.8	-1.6	0.3
274.51	-6.02	-2.5	103.64	5JL	2.4	-3.7	-1.5	0.4
272.27	-6.11	-2.4	113.21	15JL	0.7	-3.6	-1.5	0.4
271.09	-5.94	-2.1	122.81	25JL	-0.7	-3.6	-1.4	0.5
271.22	-5.60	-1.9	132.46	4AU	-1.4	-3.5	-1.4	0.6
272.69	-5.16	-1.6	142.15	14AU	-1.2	-3.5	-1.3	0.6
275.33	-4.68	-1.3	151.90	24AU	-0.6	-3.5	-1.3	0.7
278.94	-4.19	-1.1	161.70	3SE	-0.2	-3.4	-1.3	0.7
283.34	-3.72	-0.9	171.56	13SE	-0.0	-3.4	-1.2	0.7
288.35	-3.27	-0.6	181.47	23SE	0.1	-3.4	-1.2	0.8
293.83	-2.84	-0.4	191.45	3OC	0.3	-3.4	-1.2	0.8
299.69	-2.44	-0.2	201.48	13OC	1.5	-3.4	-1.2	0.8
305.82	-2.08	-0.0	211.56	23OC	2.0	-3.4	-1.2	0.8
312.15	-1.74	0.2	221.68	2NO	-0.0	-3.4	-1.3	0.8
318.65	-1.43	0.3	231.84	12NO	-0.4	-3.4	-1.3	0.8
325.25	-1.14	0.5	242.02	22NO	-0.4	-3.4	-1.3	0.8
331.93	-0.88	0.6	252.22	2DE	-0.5	-3.4	-1.4	0.8
338.67	-0.65	0.8	262.43	12DE	-0.7	-3.4	-1.4	0.7
345.43	-0.43	0.9	272.63	22DE	-0.9	-3.4	-1.4	0.7

♂ LONG	♂ LAT	♂ MAG	☉ LONG	16.00UT 488	☿	♀	♃	♄
352.20	-0.24	1.0	282.81	1JA	-1.0	-3.5	-1.5	0.8
358.97	-0.07	1.2	292.97	11JA	-0.8	-3.5	-1.6	0.8
5.72	0.09	1.3	303.08	21JA	0.2	-3.5	-1.6	0.8
12.45	0.23	1.4	313.16	31JA	2.5	-3.5	-1.7	0.9
19.16	0.36	1.5	323.19	10FE	1.9	-3.4	-1.8	0.9
25.83	0.47	1.6	333.16	20FE	0.9	-3.4	-1.8	0.9
32.47	0.58	1.6	343.07	1MR	0.5	-3.4	-1.9	0.9
39.07	0.67	1.7	352.93	11MR	0.2	-3.4	-2.0	0.9
45.64	0.75	1.8	2.72	21MR	-0.2	-3.3	-2.0	0.8
52.17	0.82	1.8	12.46	31MR	-0.8	-3.3	-2.0	0.8
58.67	0.88	1.9	22.15	10AP	-1.7	-3.3	-2.1	0.8
65.13	0.94	1.9	31.79	20AP	-1.3	-3.3	-2.1	0.7
71.57	0.98	2.0	41.40	30AP	-0.3	-3.3	-2.1	0.6
77.98	1.03	2.0	50.97	10MY	0.6	-3.4	-2.1	0.6
84.37	1.06	2.0	60.51	20MY	1.4	-3.4	-2.0	0.5
90.75	1.09	2.0	70.05	30MY	2.4	-3.4	-2.0	0.5
97.11	1.11	2.0	79.57	9JN	3.4	-3.4	-1.9	0.4
103.46	1.13	2.0	89.09	19JN	1.9	-3.5	-1.9	0.3
109.81	1.15	2.0	98.63	29JN	0.6	-3.5	-1.8	0.3
116.16	1.15	2.0	108.19	9JL	-0.7	-3.6	-1.7	0.4
122.51	1.16	2.0	117.77	19JL	-1.4	-3.7	-1.7	0.4
128.88	1.16	2.0	127.40	29JL	-1.3	-3.8	-1.6	0.5
135.25	1.15	2.0	137.07	8AU	-0.6	-3.9	-1.6	0.5
141.65	1.14	2.0	146.78	18AU	-0.2	-4.0	-1.5	0.6
148.06	1.13	2.0	156.55	28AU	0.1	-4.1	-1.5	0.6
154.50	1.11	2.0	166.38	7SE	0.2	-4.2	-1.4	0.7
160.97	1.08	2.0	176.27	17SE	0.5	-4.3	-1.4	0.7
167.47	1.05	2.0	186.21	27SE	1.8	-4.3	-1.4	0.8
173.99	1.02	2.0	196.21	7OC	1.8	-4.3	-1.3	0.8
180.55	0.98	1.9	206.27	17OC	-0.2	-4.1	-1.3	0.8
187.15	0.93	1.9	216.37	27OC	-0.6	-3.5	-1.3	0.8
193.78	0.88	1.9	226.51	6NO	-0.6	-3.2	-1.3	0.8
200.44	0.82	1.8	236.68	16NO	-0.6	-4.0	-1.3	0.8
207.14	0.75	1.8	246.87	26NO	-0.8	-4.3	-1.3	0.8
213.88	0.67	1.7	257.08	6DE	-0.8	-4.4	-1.3	0.8
220.65	0.58	1.7	267.28	16DE	-0.8	-4.4	-1.4	0.8
227.45	0.48	1.6	277.48	26DE	-0.7	-4.3	-1.4	0.8

♂ LONG	♂ LAT	♂ MAG	☉ LONG	489	☿	♀	♃	♄
234.28	0.37	1.5	287.65	5JA	0.3	-4.2	-1.4	0.8
241.14	0.25	1.4	297.79	15JA	2.8	-4.0	-1.5	0.8
248.03	0.11	1.3	307.89	25JA	1.4	-3.9	-1.5	0.9
254.95	-0.04	1.2	317.94	4FE	0.6	-3.8	-1.6	0.9
261.88	-0.20	1.1	327.94	14FE	0.4	-3.7	-1.7	0.9
268.83	-0.38	1.0	337.89	24FE	0.1	-3.7	-1.7	0.9
275.78	-0.58	0.9	347.77	6MR	-0.2	-3.6	-1.8	0.9
282.73	-0.79	0.8	357.60	16MR	-0.8	-3.5	-1.9	0.9
289.66	-1.02	0.7	7.37	26MR	-1.7	-3.5	-1.9	0.9
296.55	-1.27	0.5	17.09	5AP	-1.3	-3.4	-2.0	0.9
303.40	-1.53	0.4	26.75	15AP	-0.2	-3.4	-2.1	0.8
310.16	-1.81	0.3	36.38	25AP	0.8	-3.3	-2.1	0.8
316.82	-2.11	0.1	45.96	5MY	1.8	-3.3	-2.2	0.7
323.32	-2.41	-0.0	55.52	15MY	3.1	-3.3	-2.2	0.7
329.61	-2.73	-0.2	65.06	25MY	2.8	-3.3	-2.2	0.6
335.63	-3.06	-0.4	74.58	4JN	1.5	-3.3	-2.2	0.6
341.30	-3.39	-0.6	84.11	14JN	0.4	-3.3	-2.2	0.5
346.51	-3.73	-0.7	93.64	24JN	-0.7	-3.3	-2.1	0.4
351.13	-4.07	-0.9	103.18	4JL	-1.5	-3.4	-2.1	0.4
355.01	-4.40	-1.2	112.75	14JL	-1.3	-3.4	-2.0	0.4
357.93	-4.70	-1.4	122.35	24JL	-0.5	-3.4	-2.0	0.4
359.70	-4.95	-1.6	131.99	3AU	-0.1	-3.5	-1.9	0.5
0.11	-5.12	-1.9	141.68	13AU	0.2	-3.5	-1.8	0.5
359.07	-5.14	-2.1	151.43	23AU	0.4	-3.5	-1.8	0.6
356.78	-4.95	-2.3	161.22	2SE	0.8	-3.5	-1.7	0.6
353.70	-4.53	-2.4	171.08	12SE	2.2	-3.4	-1.7	0.7
350.65	-3.92	-2.2	180.99	22SE	1.6	-3.4	-1.6	0.7
348.36	-3.21	-1.9	190.96	2OC	-0.3	-3.4	-1.6	0.8
347.27	-2.48	-1.6	200.99	12OC	-0.7	-3.4	-1.5	0.8
347.53	-1.82	-1.3	211.06	22OC	-0.7	-3.3	-1.5	0.8
349.01	-1.24	-0.9	221.18	1NO	-0.7	-3.3	-1.4	0.9
351.52	-0.76	-0.6	231.34	11NO	-0.7	-3.3	-1.4	0.9
354.87	-0.36	-0.4	241.52	21NO	-0.6	-3.3	-1.4	0.9
358.87	-0.03	-0.1	251.72	1DE	-0.7	-3.4	-1.4	0.9
3.36	0.24	0.2	261.93	11DE	-0.6	-3.4	-1.4	0.9
8.24	0.46	0.4	272.13	21DE	0.4	-3.4	-1.4	0.9
13.42	0.64	0.6	282.31	31DE	3.0	-3.4	-1.4	0.9

♂ LONG	♂ LAT	♂ MAG	☉ LONG	16.00UT 490	☿	♀	♃	♄
18.83	0.78	0.8	292.47	10JA	1.1	-3.5	-1.4	0.8
24.41	0.90	0.9	302.60	20JA	0.4	-3.5	-1.5	0.9
30.13	1.00	1.1	312.67	30JA	0.2	-3.5	-1.5	0.9
35.95	1.08	1.2	322.71	9FE	0.0	-3.6	-1.6	1.0
41.85	1.14	1.4	332.68	19FE	-0.3	-3.7	-1.6	1.0
47.81	1.19	1.5	342.59	1MR	-0.9	-3.7	-1.7	1.0
53.82	1.23	1.6	352.46	11MR	-1.6	-3.8	-1.7	1.0
59.86	1.26	1.7	2.26	21MR	-1.2	-3.9	-1.8	1.0
65.93	1.28	1.7	12.00	31MR	-0.2	-4.0	-1.9	1.0
72.03	1.30	1.8	21.69	10AP	1.0	-4.1	-1.9	1.0
78.14	1.30	1.9	31.34	20AP	2.3	-4.1	-2.0	1.0
84.27	1.30	1.9	40.94	30AP	3.7	-4.2	-2.1	0.9
90.43	1.30	2.0	50.51	10MY	2.2	-4.2	-2.1	0.9
96.59	1.29	2.0	60.06	20MY	1.2	-4.0	-2.2	0.8
102.78	1.27	2.0	69.59	30MY	0.3	-3.5	-2.2	0.8
108.99	1.26	2.0	79.11	9JN	-0.7	-2.7	-2.3	0.7
115.23	1.23	2.0	88.64	19JN	-1.6	-3.5	-2.3	0.6
121.49	1.20	2.0	98.17	29JN	-1.3	-4.0	-2.3	0.6
127.79	1.17	2.0	107.73	9JL	-0.5	-4.2	-2.3	0.5
134.11	1.14	2.0	117.31	19JL	0.0	-4.2	-2.3	0.4
140.48	1.10	2.0	126.93	29JL	0.3	-4.2	-2.3	0.5
146.89	1.05	2.0	136.60	8AU	0.6	-4.1	-2.2	0.5
153.34	1.01	1.9	146.31	18AU	1.1	-4.0	-2.1	0.6
159.84	0.96	1.9	156.08	28AU	2.7	-3.9	-2.1	0.6
166.40	0.90	1.9	165.91	7SE	1.4	-3.8	-2.0	0.7
173.00	0.84	1.9	175.79	17SE	-0.4	-3.7	-2.0	0.7
179.67	0.77	1.9	185.73	27SE	-0.9	-3.7	-1.9	0.8
186.40	0.70	1.9	195.73	7OC	-0.9	-3.6	-1.8	0.8
193.18	0.63	1.8	205.78	17OC	-0.9	-3.6	-1.8	0.9
200.03	0.55	1.8	215.87	27OC	-0.6	-3.5	-1.7	0.9
206.95	0.46	1.8	226.01	6NO	-0.5	-3.5	-1.7	0.9
213.93	0.37	1.8	236.18	16NO	-0.5	-3.4	-1.6	0.9
220.97	0.27	1.8	246.37	26NO	-0.4	-3.4	-1.6	1.0
228.08	0.16	1.7	256.58	6DE	0.5	-3.4	-1.5	1.0
235.25	0.05	1.7	266.78	16DE	2.9	-3.4	-1.5	1.0
242.48	-0.06	1.6	276.98	26DE	0.8	-3.4	-1.5	1.0

♂ LONG	♂ LAT	♂ MAG	☉ LONG	491	☿	♀	♃	♄
249.77	-0.19	1.6	287.15	5JA	0.2	-3.3	-1.5	1.0
257.11	-0.31	1.5	297.29	15JA	0.1	-3.3	-1.5	1.0
264.51	-0.44	1.5	307.40	25JA	-0.1	-3.3	-1.5	1.0
271.95	-0.58	1.4	317.45	4FE	-0.4	-3.3	-1.5	1.0
279.43	-0.72	1.4	327.45	14FE	-0.9	-3.4	-1.5	1.1
286.95	-0.86	1.3	337.40	24FE	-1.5	-3.4	-1.6	1.1
294.49	-1.00	1.2	347.29	6MR	-1.2	-3.4	-1.6	1.1
302.04	-1.14	1.2	357.12	16MR	-0.1	-3.4	-1.7	1.1
309.59	-1.28	1.1	6.90	26MR	1.3	-3.5	-1.7	1.1
317.13	-1.41	1.0	16.62	5AP	3.0	-3.5	-1.8	1.1
324.65	-1.54	1.0	26.28	15AP	2.8	-3.5	-1.8	1.1
332.12	-1.65	0.9	35.91	25AP	1.6	-3.4	-1.9	1.1
339.52	-1.76	0.8	45.50	5MY	0.9	-3.4	-2.0	1.1
346.85	-1.85	0.8	55.05	15MY	0.2	-3.4	-2.0	1.0
354.07	-1.93	0.7	64.59	25MY	-0.7	-3.3	-2.1	1.0
1.16	-1.99	0.6	74.12	4JN	-1.6	-3.3	-2.2	0.9
8.09	-2.03	0.5	83.64	14JN	-1.3	-3.3	-2.2	0.9
14.83	-2.05	0.4	93.17	24JN	-0.4	-3.3	-2.3	0.8
21.34	-2.06	0.4	102.72	4JL	0.1	-3.3	-2.3	0.7
27.59	-2.04	0.3	112.28	14JL	0.5	-3.3	-2.4	0.7
33.51	-2.00	0.2	121.88	24JL	0.8	-3.4	-2.4	0.6
39.03	-1.93	0.0	131.53	3AU	1.5	-3.4	-2.4	0.6
44.09	-1.84	-0.1	141.21	13AU	3.1	-3.4	-2.4	0.6
48.56	-1.71	-0.3	150.95	23AU	1.2	-3.5	-2.4	0.6
52.32	-1.55	-0.4	160.75	2SE	-0.5	-3.5	-2.4	0.7
55.19	-1.34	-0.6	170.6[illegible]	12SE	-1.0	-3.6	-2.3	0.7
56.97	-1.08	-0.8	180.51	22SE	-1.0	-3.7	-2.3	0.8
57.45	-0.75	-1.0	190.48	2OC	-0.8	-3.7	-2.2	0.8
56.51	-0.36	-1.3	200.50	12OC	-0.5	-3.8	-2.1	0.9
54.14	0.10	-1.5	210.58	22OC	-0.4	-3.9	-2.0	0.9
50.70	0.58	-1.6	220.69	1NO	-0.3	-4.0	-2.0	1.0
46.86	1.03	-1.6	230.84	11NO	-0.2	-4.1	-1.9	1.0
43.44	1.41	-1.3	241.03	21NO	0.7	-4.2	-1.8	1.0
41.07	1.69	-1.0	251.23	1DE	2.7	-4.3	-1.8	1.1
40.02	1.88	-0.7	261.43	11DE	0.5	-4.4	-1.7	1.1
40.30	1.99	-0.4	271.63	21DE	0.0	-4.3	-1.7	1.1
41.74	2.04	-0.1	281.81	31DE	-0.1	-4.1	-1.7	1.1

♂			⊙	16.00UT	☿	♀	♃	♄
LONG	LAT	MAG	LONG	492	MAGNITUDES			
44.13	2.06	0.1	291.97	10JA	-0.2	-3.5	-1.6	1.1
47.28	2.05	0.4	302.10	20JA	-0.5	-3.4	-1.6	1.1
51.02	2.03	0.6	312.18	30JA	-0.9	-3.9	-1.6	1.1
55.22	1.99	0.8	322.21	9FE	-1.4	-4.2	-1.6	1.1
59.79	1.95	1.0	332.19	19FE	-1.1	-4.3	-1.6	1.2
64.65	1.90	1.1	342.11	29FE	-0.1	-4.3	-1.6	1.2
69.73	1.85	1.3	351.97	10MR	1.6	-4.2	-1.6	1.2
75.00	1.80	1.4	1.78	20MR	3.5	-4.1	-1.6	1.3
80.43	1.74	1.5	11.52	30MR	2.1	-4.0	-1.6	1.3
85.97	1.68	1.6	21.22	9AP	1.2	-3.9	-1.7	1.3
91.63	1.62	1.7	30.87	19AP	0.7	-3.8	-1.7	1.3
97.38	1.56	1.7	40.48	29AP	0.1	-3.7	-1.8	1.3
103.21	1.50	1.8	50.05	9MY	-0.7	-3.6	-1.8	1.3
109.11	1.43	1.9	59.60	19MY	-1.7	-3.5	-1.9	1.2
115.09	1.37	1.9	69.13	29MY	-1.3	-3.5	-1.9	1.2
121.13	1.30	1.9	78.65	8JN	-0.4	-3.4	-2.0	1.1
127.25	1.23	1.9	88.18	18JN	0.2	-3.4	-2.1	1.1
133.42	1.16	2.0	97.71	28JN	0.7	-3.4	-2.1	1.0
139.66	1.09	2.0	107.27	8JL	1.1	-3.4	-2.2	0.9
145.97	1.01	2.0	116.85	18JL	2.0	-3.3	-2.3	0.9
152.35	0.93	2.0	126.47	28JL	3.2	-3.3	-2.3	0.8
158.80	0.85	1.9	136.13	7AU	1.1	-3.3	-2.4	0.7
165.33	0.77	1.9	145.85	17AU	-0.6	-3.3	-2.4	0.7
171.93	0.69	1.9	155.61	27AU	-1.2	-3.4	-2.4	0.7
178.61	0.60	1.9	165.43	6SE	-1.2	-3.4	-2.5	0.8
185.37	0.51	1.8	175.31	16SE	-0.7	-3.4	-2.5	0.8
192.21	0.42	1.8	185.25	26SE	-0.4	-3.4	-2.4	0.9
199.13	0.32	1.7	195.24	6OC	-0.2	-3.4	-2.4	0.9
206.14	0.22	1.7	205.29	16OC	-0.2	-3.5	-2.4	1.0
213.23	0.12	1.6	215.38	26OC	-0.0	-3.5	-2.3	1.0
220.40	0.02	1.6	225.52	5NO	1.0	-3.5	-2.2	1.1
227.66	-0.09	1.6	235.69	15NO	2.5	-3.5	-2.2	1.1
234.99	-0.20	1.6	245.88	25NO	0.3	-3.4	-2.1	1.2
242.40	-0.30	1.6	256.08	5DE	-0.2	-3.4	-2.0	1.2
249.88	-0.41	1.6	266.28	15DE	-0.2	-3.4	-1.9	1.2
257.43	-0.52	1.6	276.47	25DE	-0.3	-3.4	-1.9	1.2
				493				
265.03	-0.62	1.5	286.65	4JA	-0.5	-3.3	-1.8	1.3
272.69	-0.73	1.5	296.79	14JA	-1.0	-3.3	-1.8	1.3
280.39	-0.83	1.5	306.90	24JA	-1.2	-3.3	-1.7	1.3
288.13	-0.92	1.5	316.96	3FE	-1.0	-3.3	-1.7	1.3
295.89	-1.01	1.4	326.96	13FE	-0.0	-3.4	-1.6	1.3
303.66	-1.09	1.4	336.91	23FE	1.9	-3.4	-1.6	1.3
311.44	-1.16	1.4	346.81	5MR	3.0	-3.4	-1.6	1.4
319.20	-1.22	1.3	356.64	15MR	1.5	-3.4	-1.6	1.4
326.94	-1.27	1.3	6.42	25MR	0.9	-3.4	-1.6	1.4
334.65	-1.31	1.3	16.14	4AP	0.5	-3.5	-1.6	1.4
342.30	-1.33	1.3	25.82	14AP	-0.0	-3.5	-1.6	1.4
349.90	-1.34	1.3	35.44	24AP	-0.8	-3.6	-1.6	1.4
357.42	-1.34	1.2	45.03	4MY	-1.7	-3.6	-1.6	1.3
4.85	-1.32	1.2	54.60	14MY	-1.3	-3.7	-1.6	1.3
12.18	-1.28	1.2	64.13	24MY	-0.4	-3.8	-1.7	1.3
19.40	-1.23	1.2	73.66	3JN	0.3	-3.9	-1.7	1.2
26.50	-1.17	1.1	83.19	13JN	0.9	-4.0	-1.8	1.2
33.46	-1.09	1.1	92.71	23JN	1.5	-4.1	-1.8	1.1
40.27	-0.99	1.1	102.26	3JL	2.6	-4.2	-1.9	1.1
46.91	-0.89	1.0	111.82	13JL	2.8	-4.2	-1.9	1.0
53.36	-0.76	1.0	121.42	23JL	0.9	-4.2	-2.0	1.0
59.61	-0.62	0.9	131.06	2AU	-0.6	-3.9	-2.1	0.9
65.62	-0.47	0.8	140.75	12AU	-1.3	-3.5	-2.1	0.9
71.35	-0.30	0.8	150.48	22AU	-1.3	-3.3	-2.2	0.9
76.77	-0.10	0.7	160.27	1SE	-0.7	-3.9	-2.3	0.9
81.80	0.11	0.6	170.12	11SE	-0.3	-4.2	-2.3	0.9
86.38	0.36	0.4	180.03	21SE	-0.1	-4.3	-2.4	1.0
90.41	0.63	0.3	190.00	1OC	-0.0	-4.3	-2.4	1.0
93.75	0.94	0.1	200.02	11OC	0.2	-4.2	-2.4	1.1
96.25	1.30	-0.1	210.09	21OC	1.2	-4.1	-2.4	1.1
97.72	1.70	-0.3	220.20	31OC	2.3	-4.0	-2.4	1.2
97.97	2.14	-0.5	230.35	10NO	0.1	-4.0	-2.4	1.2
96.86	2.60	-0.7	240.53	20NO	-0.3	-3.9	-2.3	1.3
94.39	3.05	-1.0	250.73	30NO	-0.4	-3.8	-2.2	1.3
90.86	3.43	-1.1	260.94	10DE	-0.4	-3.7	-2.2	1.4
86.88	3.67	-1.1	271.14	20DE	-0.6	-3.6	-2.1	1.4
83.22	3.76	-0.9	281.33	30DE	-1.0	-3.6	-2.0	1.4

♂			⊙	16.00UT	☿	♀	♃	♄
LONG	LAT	MAG	LONG	494	MAGNITUDES			
80.53	3.71	-0.7	291.48	9JA	-1.1	-3.5	-2.0	1.4
79.10	3.57	-0.4	301.61	19JA	-0.9	-3.5	-1.9	1.3
78.98	3.38	-0.1	311.70	29JA	0.0	-3.4	-1.8	1.3
80.04	3.16	0.1	321.73	8FE	2.3	-3.4	-1.8	1.2
82.07	2.95	0.4	331.71	18FE	2.3	-3.4	-1.7	1.2
84.89	2.74	0.6	341.63	28FE	1.1	-3.3	-1.7	1.1
88.35	2.55	0.8	351.49	10MR	0.7	-3.3	-1.6	1.1
92.30	2.36	0.9	1.30	20MR	0.3	-3.3	-1.6	1.1
96.66	2.20	1.1	11.05	30MR	-0.1	-3.3	-1.5	1.1
101.35	2.04	1.2	20.75	9AP	-0.8	-3.3	-1.5	1.1
106.31	1.89	1.3	30.40	19AP	-1.7	-3.3	-1.5	1.1
111.49	1.74	1.4	40.01	29AP	-1.3	-3.3	-1.5	1.1
116.86	1.61	1.5	49.58	9MY	-0.3	-3.4	-1.5	1.1
122.40	1.48	1.6	59.13	19MY	0.5	-3.4	-1.5	1.0
128.09	1.35	1.6	68.67	29MY	1.1	-3.5	-1.5	1.0
133.91	1.23	1.7	78.19	8JN	2.0	-3.5	-1.5	1.0
139.85	1.11	1.7	87.71	18JN	3.2	-3.5	-1.6	0.9
145.92	0.99	1.8	97.25	28JN	2.3	-3.4	-1.6	0.9
152.09	0.88	1.8	106.80	8JL	0.7	-3.4	-1.6	0.8
158.38	0.76	1.8	116.38	18JL	-0.6	-3.4	-1.7	0.8
164.77	0.65	1.8	126.00	28JL	-1.4	-3.4	-1.7	0.7
171.27	0.54	1.8	135.66	7AU	-1.3	-3.3	-1.8	0.7
177.87	0.43	1.8	145.37	17AU	-0.6	-3.3	-1.8	0.6
184.58	0.31	1.8	155.13	27AU	-0.2	-3.3	-1.9	0.6
191.39	0.20	1.8	164.95	6SE	0.0	-3.3	-2.0	0.5
198.30	0.09	1.7	174.83	16SE	0.1	-3.3	-2.0	0.6
205.32	-0.02	1.7	184.76	26SE	0.4	-3.4	-2.1	0.6
212.44	-0.12	1.7	194.75	6OC	1.5	-3.4	-2.2	0.7
219.66	-0.23	1.6	204.80	16OC	2.0	-3.4	-2.2	0.8
226.98	-0.33	1.6	214.89	26OC	-0.1	-3.5	-2.3	0.9
234.39	-0.43	1.5	225.02	5NO	-0.5	-3.5	-2.3	0.9
241.88	-0.53	1.5	235.19	15NO	-0.5	-3.5	-2.3	1.0
249.46	-0.62	1.4	245.38	25NO	-0.5	-3.6	-2.3	1.0
257.10	-0.71	1.4	255.58	5DE	-0.7	-3.7	-2.3	1.0
264.82	-0.79	1.4	265.79	15DE	-0.9	-3.8	-2.2	1.1
272.59	-0.87	1.4	275.99	25DE	-0.9	-3.8	-2.2	1.1
				495				
280.40	-0.93	1.4	286.16	4JA	-0.8	-3.9	-2.1	1.1
288.24	-0.99	1.4	296.31	14JA	0.1	-4.0	-2.1	1.1
296.11	-1.04	1.4	306.41	24JA	2.6	-4.1	-2.0	1.1
303.98	-1.08	1.4	316.47	3FE	1.8	-4.2	-1.9	1.0
311.85	-1.10	1.4	326.48	13FE	0.8	-4.3	-1.9	1.0
319.71	-1.12	1.4	336.44	23FE	0.5	-4.3	-1.8	0.9
327.53	-1.12	1.4	346.33	5MR	0.2	-4.2	-1.7	0.9
335.31	-1.11	1.4	356.17	15MR	-0.2	-3.9	-1.7	0.8
343.04	-1.08	1.4	5.95	25MR	-0.8	-3.3	-1.6	0.9
350.71	-1.05	1.4	15.67	4AP	-1.7	-3.3	-1.6	0.9
358.31	-1.00	1.4	25.34	14AP	-1.3	-3.9	-1.5	0.9
5.82	-0.95	1.5	34.97	24AP	-0.3	-4.1	-1.5	0.9
13.25	-0.88	1.5	44.56	4MY	0.6	-4.2	-1.4	0.9
20.59	-0.80	1.5	54.13	14MY	1.5	-4.2	-1.4	0.8
27.83	-0.71	1.5	63.67	24MY	2.6	-4.1	-1.4	0.8
34.96	-0.62	1.5	73.19	3JN	3.3	-4.0	-1.4	0.8
41.99	-0.51	1.5	82.72	13JN	1.8	-3.9	-1.4	0.8
48.90	-0.40	1.5	92.25	23JN	0.5	-3.8	-1.4	0.7
55.70	-0.28	1.5	101.79	3JL	-0.7	-3.7	-1.4	0.7
62.38	-0.16	1.5	111.36	13JL	-1.5	-3.6	-1.4	0.6
68.92	-0.03	1.5	120.96	23JL	-1.3	-3.6	-1.5	0.6
75.33	0.11	1.4	130.59	2AU	-0.6	-3.5	-1.5	0.5
81.59	0.26	1.4	140.28	12AU	-0.1	-3.5	-1.5	0.5
87.69	0.42	1.4	150.01	22AU	0.1	-3.5	-1.6	0.4
93.61	0.59	1.3	159.80	1SE	0.3	-3.4	-1.6	0.4
99.33	0.76	1.2	169.65	11SE	0.6	-3.4	-1.7	0.3
104.82	0.95	1.2	179.55	21SE	1.9	-3.4	-1.7	0.3
110.04	1.16	1.1	189.51	1OC	1.8	-3.4	-1.8	0.3
114.94	1.38	1.0	199.53	11OC	-0.2	-3.4	-1.9	0.4
119.45	1.63	0.8	209.60	21OC	-0.6	-3.4	-1.9	0.5
123.49	1.90	0.7	219.71	31OC	-0.7	-3.4	-2.0	0.5
126.95	2.21	0.5	229.86	10NO	-0.7	-3.4	-2.1	0.6
129.69	2.54	0.3	240.04	20NO	-0.8	-3.4	-2.1	0.6
131.56	2.91	0.1	250.24	30NO	-0.7	-3.4	-2.1	0.7
132.35	3.31	-0.1	260.44	10DE	-0.7	-3.4	-2.2	0.7
131.90	3.71	-0.4	270.64	20DE	-0.7	-3.4	-2.2	0.8
130.13	4.10	-0.6	280.83	30DE	0.3	-3.5	-2.2	0.8

♂ LONG	♂ LAT	♂ MAG	☉ LONG	16.00UT	☿	♀	♃	♄
				496	MAGNITUDES			
127.14	4.39	-0.8	290.99	9JA	2.9	-3.5	-2.2	0.8
123.35	4.55	-1.0	301.12	19JA	1.3	-3.5	-2.1	0.8
119.43	4.53	-0.9	311.20	29JA	0.5	-3.5	-2.1	0.8
116.10	4.34	-0.7	321.24	8FE	0.3	-3.4	-2.0	0.8
113.86	4.05	-0.5	331.22	18FE	0.1	-3.4	-1.9	0.8
112.91	3.69	-0.2	341.15	28FE	-0.2	-3.4	-1.9	0.7
113.22	3.33	0.0	351.02	9MR	-0.8	-3.4	-1.8	0.7
114.63	2.97	0.2	0.82	19MR	-1.7	-3.3	-1.7	0.6
116.95	2.64	0.4	10.58	29MR	-1.3	-3.3	-1.7	0.6
120.02	2.34	0.6	20.28	8AP	-0.3	-3.3	-1.6	0.6
123.70	2.07	0.8	29.93	18AP	0.8	-3.3	-1.5	0.6
127.87	1.82	0.9	39.54	28AP	2.0	-3.3	-1.5	0.6
132.45	1.59	1.0	49.12	8MY	3.5	-3.4	-1.4	0.6
137.36	1.38	1.1	58.67	18MY	2.6	-3.4	-1.4	0.6
142.56	1.18	1.2	68.20	28MY	1.4	-3.4	-1.4	0.6
148.01	1.00	1.3	77.73	7JN	0.4	-3.4	-1.3	0.6
153.68	0.82	1.4	87.25	17JN	-0.7	-3.5	-1.3	0.6
159.54	0.66	1.4	96.78	27JN	-1.5	-3.5	-1.3	0.6
165.59	0.50	1.5	106.34	7JL	-1.3	-3.6	-1.3	0.5
171.80	0.35	1.5	115.92	17JL	-0.5	-3.7	-1.3	0.5
178.17	0.21	1.5	125.54	27JL	-0.0	-3.8	-1.3	0.4
184.69	0.07	1.5	135.19	6AU	0.3	-3.9	-1.3	0.4
191.35	-0.06	1.5	144.90	16AU	0.5	-4.0	-1.3	0.3
198.15	-0.18	1.5	154.66	26AU	0.9	-4.1	-1.4	0.2
205.08	-0.30	1.5	164.48	5SE	2.3	-4.2	-1.4	0.2
212.14	-0.41	1.5	174.35	15SE	1.6	-4.3	-1.4	0.1
219.32	-0.52	1.5	184.28	25SE	-0.3	-4.3	-1.5	0.1
226.62	-0.62	1.5	194.27	5OC	-0.8	-4.3	-1.5	0.0
234.02	-0.71	1.5	204.31	15OC	-0.8	-4.0	-1.6	0.1
241.52	-0.80	1.5	214.41	25OC	-0.8	-3.4	-1.7	0.2
249.12	-0.87	1.5	224.54	4NO	-0.7	-3.3	-1.7	0.2
256.79	-0.93	1.4	234.70	14NO	-0.6	-4.0	-1.8	0.3
264.54	-0.99	1.4	244.89	24NO	-0.6	-4.3	-1.9	0.4
272.35	-1.03	1.4	255.09	4DE	-0.5	-4.4	-1.9	0.4
280.20	-1.07	1.4	265.29	14DE	0.4	-4.3	-2.0	0.5
288.09	-1.09	1.4	275.49	24DE	3.0	-4.3	-2.0	0.5
				497				
295.99	-1.10	1.4	285.67	3JA	1.0	-4.2	-2.0	0.6
303.90	-1.09	1.3	295.81	13JA	0.3	-4.0	-2.1	0.6
311.80	-1.08	1.3	305.92	23JA	0.1	-3.9	-2.1	0.6
319.67	-1.05	1.3	315.99	2FE	-0.0	-3.8	-2.1	0.6
327.51	-1.02	1.3	326.00	12FE	-0.3	-3.7	-2.0	0.6
335.30	-0.97	1.3	335.96	22FE	-0.8	-3.7	-2.0	0.6
343.04	-0.91	1.4	345.85	4MR	-1.6	-3.6	-2.0	0.6
350.71	-0.85	1.4	355.69	14MR	-1.3	-3.5	-1.9	0.5
358.30	-0.77	1.4	5.48	24MR	-0.2	-3.5	-1.8	0.5
5.82	-0.69	1.5	15.20	3AP	1.1	-3.4	-1.8	0.4
13.25	-0.60	1.5	24.88	13AP	2.5	-3.4	-1.7	0.4
20.60	-0.51	1.6	34.51	23AP	3.4	-3.3	-1.6	0.4
27.85	-0.41	1.6	44.11	3MY	2.0	-3.3	-1.6	0.5
35.02	-0.31	1.6	53.67	13MY	1.1	-3.3	-1.5	0.5
42.09	-0.20	1.7	63.21	23MY	0.3	-3.3	-1.5	0.5
49.07	-0.10	1.7	72.74	2JN	-0.7	-3.3	-1.4	0.5
55.95	0.01	1.7	82.26	12JN	-1.6	-3.3	-1.4	0.5
62.75	0.12	1.7	91.79	22JN	-1.3	-3.3	-1.3	0.5
69.45	0.24	1.7	101.33	2JL	-0.5	-3.4	-1.3	0.4
76.07	0.35	1.8	110.89	12JL	0.1	-3.4	-1.3	0.4
82.59	0.47	1.8	120.49	22JL	0.4	-3.4	-1.2	0.4
89.01	0.59	1.8	130.12	1AU	0.7	-3.5	-1.2	0.3
95.35	0.71	1.7	139.80	11AU	1.2	-3.5	-1.2	0.3
101.58	0.83	1.7	149.54	21AU	2.8	-3.5	-1.2	0.2
107.71	0.96	1.7	159.32	31AU	1.4	-3.5	-1.2	0.2
113.72	1.09	1.7	169.16	10SE	-0.4	-3.4	-1.3	0.1
119.60	1.22	1.6	179.07	20SE	-0.9	-3.4	-1.3	0.0
125.34	1.36	1.5	189.02	30SE	-1.0	-3.4	-1.3	-0.0
130.93	1.51	1.5	199.04	10OC	-0.9	-3.4	-1.3	-0.1
136.32	1.67	1.4	209.11	20OC	-0.6	-3.3	-1.4	-0.1
141.48	1.84	1.3	219.22	30OC	-0.4	-3.3	-1.4	-0.1
146.39	2.01	1.1	229.37	9NO	-0.4	-3.3	-1.5	-0.0
150.96	2.21	1.0	239.55	19NO	-0.3	-3.3	-1.6	0.1
155.14	2.42	0.8	249.74	29NO	0.6	-3.4	-1.6	0.1
158.84	2.65	0.7	259.95	9DE	2.9	-3.4	-1.7	0.2
161.92	2.90	0.5	270.15	19DE	0.7	-3.4	-1.8	0.3
164.25	3.16	0.2	280.33	29DE	0.1	-3.4	-1.8	0.3

♂ LONG	♂ LAT	♂ MAG	☉ LONG	16.00UT	☿	♀	♃	♄
				498	MAGNITUDES			
165.65	3.44	-0.0	290.50	8JA	-0.0	-3.5	-1.9	0.4
165.94	3.72	-0.3	300.63	18JA	-0.1	-3.5	-1.9	0.4
164.99	3.96	-0.6	310.72	28JA	-0.4	-3.5	-2.0	0.4
162.75	4.13	-0.8	320.76	7FE	-0.9	-3.6	-2.0	0.4
159.45	4.18	-1.0	330.74	17FE	-1.5	-3.7	-2.0	0.5
155.61	4.07	-1.1	340.67	27FE	-1.2	-3.7	-2.0	0.4
151.92	3.79	-1.0	350.54	9MR	-0.2	-3.8	-2.0	0.4
149.06	3.40	-0.8	0.36	19MR	1.3	-3.9	-2.0	0.4
147.40	2.96	-0.6	10.11	29MR	3.2	-4.0	-1.9	0.4
147.05	2.51	-0.4	19.81	8AP	2.6	-4.1	-1.9	0.3
147.92	2.08	-0.1	29.47	18AP	1.5	-4.1	-1.8	0.3
149.86	1.69	0.1	39.08	28AP	0.8	-4.2	-1.8	0.3
152.67	1.34	0.2	48.66	8MY	0.2	-4.2	-1.7	0.3
156.22	1.02	0.4	58.21	18MY	-0.7	-4.0	-1.6	0.3
160.36	0.74	0.5	67.74	28MY	-1.6	-3.5	-1.6	0.4
164.99	0.49	0.6	77.27	7JN	-1.3	-2.7	-1.5	0.4
170.05	0.26	0.7	86.79	17JN	-0.5	-3.6	-1.5	0.4
175.45	0.05	0.8	96.32	27JN	0.1	-4.0	-1.4	0.4
181.16	-0.14	0.9	105.87	7JL	0.5	-4.2	-1.4	0.4
187.14	-0.31	1.0	115.45	17JL	0.9	-4.2	-1.3	0.3
193.37	-0.47	1.0	125.06	27JL	1.7	-4.1	-1.3	0.3
199.81	-0.61	1.1	134.72	6AU	3.2	-4.1	-1.3	0.3
206.45	-0.74	1.1	144.42	16AU	1.2	-4.0	-1.2	0.2
213.27	-0.85	1.2	154.18	26AU	-0.5	-3.9	-1.2	0.2
220.26	-0.95	1.2	164.00	5SE	-1.1	-3.8	-1.2	0.1
227.40	-1.04	1.2	173.87	15SE	-1.1	-3.7	-1.2	0.1
234.68	-1.11	1.2	183.79	25SE	-0.8	-3.7	-1.2	-0.0
242.09	-1.17	1.2	193.79	5OC	-0.5	-3.6	-1.2	-0.1
249.61	-1.21	1.3	203.82	15OC	-0.3	-3.6	-1.3	-0.1
257.22	-1.25	1.3	213.91	25OC	-0.3	-3.5	-1.3	-0.2
264.91	-1.26	1.3	224.05	4NO	-0.1	-3.5	-1.3	-0.2
272.68	-1.26	1.3	234.21	14NO	0.8	-3.4	-1.4	-0.2
280.49	-1.25	1.3	244.40	24NO	2.7	-3.4	-1.4	-0.1
288.33	-1.23	1.3	254.60	4DE	0.4	-3.4	-1.5	-0.0
296.20	-1.19	1.4	264.80	14DE	-0.1	-3.4	-1.5	0.0
304.06	-1.15	1.4	275.00	24DE	-0.1	-3.4	-1.6	0.1
				499				
311.92	-1.09	1.4	285.18	3JA	-0.2	-3.3	-1.6	0.2
319.74	-1.02	1.4	295.33	13JA	-0.5	-3.3	-1.7	0.2
327.53	-0.94	1.4	305.44	23JA	-0.9	-3.3	-1.8	0.3
335.27	-0.86	1.5	315.50	2FE	-1.3	-3.3	-1.8	0.3
342.95	-0.77	1.5	325.52	12FE	-1.1	-3.4	-1.9	0.3
350.56	-0.68	1.5	335.47	22FE	-0.1	-3.4	-2.0	0.3
358.10	-0.58	1.5	345.38	4MR	1.6	-3.4	-2.0	0.4
5.56	-0.48	1.5	355.22	14MR	3.4	-3.4	-2.0	0.4
12.93	-0.37	1.6	5.00	24MR	1.9	-3.5	-2.0	0.3
20.22	-0.27	1.6	14.73	3AP	1.1	-3.5	-2.0	0.3
27.43	-0.16	1.6	24.41	13AP	0.6	-3.5	-2.0	0.3
34.55	-0.06	1.6	34.04	23AP	0.1	-3.4	-2.0	0.2
41.58	0.04	1.6	43.64	3MY	-0.7	-3.4	-1.9	0.2
48.54	0.15	1.7	53.20	13MY	-1.7	-3.4	-1.9	0.2
55.41	0.25	1.7	62.74	23MY	-1.3	-3.3	-1.8	0.2
62.22	0.34	1.8	72.27	2JN	-0.4	-3.3	-1.8	0.3
68.94	0.44	1.8	81.79	12JN	0.3	-3.3	-1.7	0.3
75.60	0.54	1.9	91.32	22JN	0.7	-3.3	-1.6	0.3
82.19	0.63	1.9	100.86	2JL	1.2	-3.3	-1.6	0.3
88.72	0.72	1.9	110.42	12JL	2.2	-3.3	-1.5	0.3
95.19	0.81	1.9	120.02	22JL	3.1	-3.4	-1.5	0.3
101.60	0.90	1.9	129.65	1AU	1.0	-3.4	-1.4	0.3
107.95	0.98	1.9	139.33	11AU	-0.5	-3.4	-1.4	0.3
114.24	1.07	1.9	149.06	21AU	-1.2	-3.5	-1.3	0.2
120.47	1.15	1.9	158.84	31AU	-1.2	-3.5	-1.3	0.2
126.65	1.23	1.9	168.68	10SE	-0.7	-3.6	-1.3	0.1
132.76	1.32	1.9	178.58	20SE	-0.4	-3.7	-1.3	0.1
138.80	1.40	1.8	188.54	30SE	-0.2	-3.7	-1.2	0.0
144.76	1.48	1.8	198.55	10OC	-0.1	-3.8	-1.2	-0.1
150.64	1.56	1.7	208.62	20OC	0.0	-3.9	-1.2	-0.1
156.42	1.65	1.6	218.73	30OC	1.0	-4.0	-1.3	-0.2
162.08	1.74	1.6	228.87	9NO	2.5	-4.1	-1.3	-0.3
167.61	1.82	1.5	239.05	19NO	0.2	-4.2	-1.3	-0.3
172.97	1.91	1.3	249.25	29NO	-0.2	-4.3	-1.3	-0.2
178.14	2.00	1.2	259.45	9DE	-0.3	-4.4	-1.4	-0.2
183.06	2.10	1.1	269.65	19DE	-0.4	-4.3	-1.4	-0.1
187.69	2.19	0.9	279.84	29DE	-0.6	-4.1	-1.5	-0.0

♂ LONG	♂ LAT	♂ MAG	☉ LONG	16.00UT	☿	♀	♃	♄
LONG	LAT	MAG	LONG	500	MAGNITUDES			
191.96	2.29	0.7	290.01	8JA	-0.9	-3.5	-1.5	0.1
195.78	2.38	0.5	300.14	18JA	-1.2	-3.4	-1.6	0.1
199.03	2.47	0.3	310.23	28JA	-1.0	-4.0	-1.7	0.2
201.58	2.54	-0.0	320.27	7FE	-0.0	-4.2	-1.7	0.2
203.26	2.59	-0.3	330.26	17FE	2.0	-4.3	-1.8	0.3
203.88	2.61	-0.6	340.19	27FE	2.7	-4.3	-1.9	0.3
203.29	2.56	-0.9	350.06	8MR	1.4	-4.2	-1.9	0.3
201.43	2.43	-1.2	359.88	18MR	0.8	-4.1	-2.0	0.3
198.47	2.18	-1.4	9.64	28MR	0.4	-4.0	-2.0	0.3
194.89	1.81	-1.6	19.34	7AP	-0.0	-3.8	-2.1	0.3
191.36	1.36	-1.5	29.00	17AP	-0.7	-3.8	-2.1	0.3
188.59	0.88	-1.3	38.62	27AP	-1.7	-3.7	-2.1	0.3
187.02	0.40	-1.1	48.19	7MY	-1.3	-3.6	-2.1	0.2
186.80	-0.02	-0.9	57.75	17MY	-0.4	-3.5	-2.1	0.2
187.88	-0.39	-0.7	67.28	27MY	0.4	-3.5	-2.0	0.2
190.10	-0.71	-0.5	76.81	6JN	1.0	-3.4	-2.0	0.2
193.26	-0.97	-0.3	86.33	16JN	1.6	-3.4	-1.9	0.3
197.22	-1.19	-0.1	95.86	26JN	2.8	-3.4	-1.9	0.3
201.83	-1.37	-0.0	105.41	6JL	2.6	-3.3	-1.8	0.3
206.98	-1.51	0.1	114.99	16JL	0.8	-3.3	-1.7	0.3
212.58	-1.63	0.2	124.60	26JL	-0.6	-3.3	-1.7	0.3
218.55	-1.71	0.3	134.25	5AU	-1.3	-3.3	-1.6	0.3
224.85	-1.78	0.4	143.96	15AU	-1.3	-3.3	-1.6	0.3
231.43	-1.82	0.5	153.71	25AU	-0.7	-3.4	-1.5	0.3
238.24	-1.84	0.6	163.52	4SE	-0.3	-3.4	-1.5	0.2
245.26	-1.84	0.7	173.39	14SE	-0.1	-3.4	-1.4	0.2
252.44	-1.82	0.7	183.31	24SE	0.0	-3.4	-1.4	0.1
259.77	-1.78	0.8	193.29	4OC	0.3	-3.4	-1.4	0.1
267.22	-1.72	0.9	203.34	14OC	1.3	-3.5	-1.3	0.0
274.75	-1.65	0.9	213.42	24OC	2.3	-3.5	-1.3	-0.1
282.36	-1.57	1.0	223.55	3NO	0.0	-3.5	-1.3	-0.1
290.01	-1.47	1.1	233.71	13NO	-0.4	-3.5	-1.3	-0.2
297.69	-1.37	1.1	243.90	23NO	-0.4	-3.4	-1.3	-0.3
305.38	-1.26	1.2	254.10	3DE	-0.5	-3.4	-1.3	-0.3
313.07	-1.14	1.2	264.31	13DE	-0.6	-3.4	-1.4	-0.2
320.73	-1.01	1.3	274.50	23DE	-0.9	-3.4	-1.4	-0.2
				501				
328.35	-0.89	1.4	284.68	2JA	-1.0	-3.3	-1.4	-0.1
335.93	-0.76	1.4	294.83	12JA	-0.9	-3.3	-1.5	-0.0
343.45	-0.63	1.5	304.94	22JA	0.0	-3.3	-1.5	0.0
350.91	-0.51	1.5	315.01	1FE	2.3	-3.3	-1.6	0.1
358.30	-0.38	1.6	325.03	11FE	2.1	-3.4	-1.6	0.2
5.62	-0.26	1.6	334.99	21FE	1.0	-3.4	-1.7	0.2
12.86	-0.14	1.7	344.90	3MR	0.6	-3.4	-1.8	0.3
20.02	-0.03	1.7	354.74	13MR	0.3	-3.4	-1.8	0.3
27.10	0.08	1.7	4.53	23MR	-0.1	-3.4	-1.9	0.3
34.11	0.18	1.8	14.26	2AP	-0.8	-3.5	-2.0	0.3
41.05	0.28	1.8	23.95	12AP	-1.7	-3.5	-2.0	0.3
47.91	0.38	1.8	33.58	22AP	-1.3	-3.6	-2.1	0.3
54.71	0.47	1.8	43.18	2MY	-0.4	-3.7	-2.1	0.3
61.44	0.55	1.8	52.75	12MY	0.5	-3.7	-2.2	0.3
68.12	0.63	1.8	62.29	22MY	1.3	-3.8	-2.2	0.2
74.74	0.71	1.8	71.82	1JN	2.2	-3.9	-2.2	0.2
81.31	0.78	1.8	81.34	11JN	3.4	-4.0	-2.2	0.2
87.83	0.85	1.9	90.86	21JN	2.1	-4.1	-2.2	0.3
94.31	0.91	1.9	100.41	1JL	0.7	-4.2	-2.1	0.3
100.76	0.97	2.0	109.97	11JL	-0.6	-4.2	-2.1	0.3
107.17	1.03	2.0	119.56	21JL	-1.4	-4.2	-2.0	0.3
113.55	1.08	2.0	129.19	31JL	-1.3	-3.9	-2.0	0.4
119.91	1.13	2.0	138.86	10AU	-0.6	-3.4	-1.9	0.4
126.24	1.17	2.0	148.59	20AU	-0.2	-3.3	-1.8	0.4
132.55	1.21	2.0	158.37	30AU	0.1	-3.9	-1.8	0.4
138.85	1.25	2.0	168.20	9SE	0.2	-4.2	-1.7	0.3
145.12	1.28	2.0	178.10	19SE	0.5	-4.3	-1.7	0.3
151.37	1.32	2.0	188.05	29SE	1.6	-4.3	-1.6	0.3
157.60	1.34	1.9	198.06	9OC	2.1	-4.2	-1.6	0.2
163.81	1.37	1.9	208.12	19OC	-0.1	-4.1	-1.5	0.1
169.99	1.39	1.8	218.23	29OC	-0.6	-4.0	-1.5	0.1
176.14	1.40	1.8	228.38	8NO	-0.6	-3.9	-1.5	-0.0
182.25	1.41	1.7	238.55	18NO	-0.6	-3.9	-1.4	-0.1
188.31	1.42	1.6	248.75	28NO	-0.7	-3.8	-1.4	-0.2
194.32	1.42	1.5	258.96	8DE	-0.8	-3.7	-1.4	-0.2
200.26	1.41	1.4	269.16	18DE	-0.8	-3.6	-1.4	-0.2
206.12	1.39	1.3	279.35	28DE	-0.8	-3.6	-1.4	-0.2

♂ LONG	♂ LAT	♂ MAG	☉ LONG	16.00UT	☿	♀	♃	♄
LONG	LAT	MAG	LONG	502	MAGNITUDES			
211.87	1.36	1.2	289.52	7JA	0.1	-3.5	-1.4	-0.1
217.50	1.31	1.0	299.65	17JA	2.6	-3.5	-1.5	-0.0
222.99	1.25	0.9	309.75	27JA	1.6	-3.4	-1.5	0.0
228.28	1.17	0.7	319.79	6FE	0.7	-3.4	-1.5	0.1
233.35	1.06	0.5	329.78	16FE	0.4	-3.4	-1.6	0.2
238.13	0.92	0.3	339.72	26FE	0.2	-3.3	-1.6	0.2
242.54	0.74	0.0	349.59	8MR	-0.2	-3.3	-1.7	0.3
246.50	0.51	-0.2	359.41	18MR	-0.8	-3.3	-1.8	0.3
249.88	0.22	-0.5	9.17	28MR	-1.6	-3.3	-1.8	0.3
252.52	-0.14	-0.8	18.88	7AP	-1.3	-3.3	-1.9	0.3
254.25	-0.59	-1.1	28.54	17AP	-0.3	-3.3	-2.0	0.4
254.89	-1.13	-1.4	38.16	27AP	0.7	-3.3	-2.0	0.4
254.31	-1.76	-1.7	47.74	7MY	1.6	-3.4	-2.1	0.4
252.53	-2.44	-2.0	57.29	17MY	2.9	-3.4	-2.2	0.4
249.82	-3.11	-2.3	66.82	27MY	3.0	-3.5	-2.2	0.3
246.80	-3.68	-2.3	76.35	6JN	1.6	-3.5	-2.3	0.3
244.17	-4.09	-2.1	85.87	16JN	0.5	-3.5	-2.3	0.3
242.56	-4.32	-1.9	95.40	26JN	-0.6	-3.4	-2.3	0.3
242.29	-4.38	-1.7	104.95	6JL	-1.5	-3.4	-2.3	0.3
243.39	-4.32	-1.5	114.52	16JL	-1.3	-3.4	-2.3	0.4
245.73	-4.19	-1.2	124.14	26JL	-0.6	-3.4	-2.3	0.4
249.13	-4.00	-1.0	133.78	5AU	-0.1	-3.3	-2.3	0.4
253.38	-3.77	-0.8	143.48	15AU	0.2	-3.3	-2.2	0.5
258.32	-3.53	-0.6	153.23	25AU	0.4	-3.3	-2.2	0.5
263.81	-3.28	-0.4	163.04	4SE	0.7	-3.3	-2.1	0.5
269.73	-3.01	-0.3	172.90	14SE	2.0	-3.3	-2.0	0.5
275.99	-2.75	-0.1	182.83	24SE	1.8	-3.4	-2.0	0.4
282.51	-2.48	0.0	192.80	4OC	-0.2	-3.4	-1.9	0.4
289.23	-2.22	0.2	202.84	14OC	-0.7	-3.4	-1.8	0.4
296.11	-1.97	0.3	212.93	24OC	-0.7	-3.5	-1.8	0.3
303.09	-1.72	0.5	223.05	3NO	-0.7	-3.5	-1.7	0.2
310.15	-1.48	0.6	233.21	13NO	-0.7	-3.6	-1.7	0.2
317.27	-1.25	0.7	243.40	23NO	-0.7	-3.6	-1.6	0.1
324.40	-1.04	0.8	253.60	3DE	-0.7	-3.7	-1.6	0.0
331.54	-0.83	1.0	263.80	13DE	-0.6	-3.8	-1.6	-0.0
338.67	-0.64	1.1	274.01	23DE	0.3	-3.8	-1.5	-0.1
				503				
345.78	-0.46	1.2	284.18	2JA	2.9	-3.9	-1.5	-0.1
352.85	-0.29	1.3	294.34	12JA	1.2	-4.0	-1.5	-0.1
359.89	-0.14	1.4	304.46	22JA	0.5	-4.1	-1.5	-0.0
6.88	0.00	1.4	314.52	1FE	0.2	-4.2	-1.5	0.1
13.82	0.14	1.5	324.55	11FE	0.0	-4.3	-1.5	0.1
20.71	0.26	1.6	334.51	21FE	-0.3	-4.3	-1.6	0.2
27.54	0.37	1.7	344.42	3MR	-0.8	-4.2	-1.6	0.2
34.32	0.47	1.7	354.27	13MR	-1.6	-3.9	-1.6	0.3
41.04	0.56	1.8	4.06	23MR	-1.3	-3.3	-1.7	0.4
47.72	0.65	1.8	13.79	2AP	-0.3	-3.4	-1.7	0.4
54.34	0.72	1.9	23.48	12AP	0.9	-3.9	-1.8	0.4
60.93	0.79	1.9	33.12	22AP	2.1	-4.2	-1.8	0.5
67.47	0.85	1.9	42.71	2MY	3.8	-4.2	-1.9	0.5
73.97	0.91	2.0	52.28	12MY	2.3	-4.2	-2.0	0.5
80.44	0.96	2.0	61.83	22MY	1.3	-4.1	-2.1	0.5
86.88	1.00	2.0	71.35	1JN	0.4	-4.0	-2.1	0.5
93.30	1.04	2.0	80.88	11JN	-0.6	-3.9	-2.2	0.5
99.70	1.07	2.0	90.41	21JN	-1.6	-3.8	-2.3	0.5
106.08	1.10	2.0	99.94	1JL	-1.3	-3.7	-2.3	0.4
112.46	1.12	2.0	109.50	11JL	-0.5	-3.6	-2.4	0.4
118.82	1.14	1.9	119.10	21JL	-0.0	-3.6	-2.4	0.5
125.19	1.15	2.0	128.72	31JL	0.3	-3.5	-2.4	0.5
131.56	1.16	2.0	138.40	10AU	0.6	-3.5	-2.4	0.6
137.94	1.17	2.0	148.12	20AU	1.0	-3.5	-2.4	0.6
144.32	1.17	2.0	157.89	30AU	2.5	-3.4	-2.4	0.6
150.73	1.16	2.0	167.73	9SE	1.6	-3.4	-2.4	0.6
157.14	1.16	2.0	177.62	19SE	-0.3	-3.4	-2.3	0.6
163.57	1.14	2.0	187.57	29SE	-0.9	-3.4	-2.2	0.6
170.02	1.12	2.0	197.58	9OC	-0.9	-3.4	-2.2	0.6
176.49	1.10	2.0	207.64	19OC	-0.9	-3.4	-2.1	0.6
182.98	1.06	1.9	217.74	29OC	-0.6	-3.4	-2.0	0.5
189.49	1.03	1.9	227.89	8NO	-0.5	-3.4	-2.0	0.5
196.02	0.98	1.8	238.06	18NO	-0.5	-3.4	-1.9	0.4
202.56	0.93	1.8	248.25	28NO	-0.4	-3.4	-1.8	0.3
209.13	0.86	1.7	258.46	8DE	0.4	-3.4	-1.8	0.3
215.71	0.79	1.6	268.66	18DE	3.0	-3.4	-1.7	0.2
222.30	0.71	1.5	278.85	28DE	0.9	-3.5	-1.7	0.1

♂			⊙	16.00UT	☿	♀	♃	♄
LONG	LAT	MAG	LONG	504	MAGNITUDES			
228.90	0.61	1.4	289.02	7JA	0.2	-3.5	-1.6	0.1
235.50	0.50	1.3	299.15	17JA	0.1	-3.5	-1.6	0.1
242.11	0.37	1.2	309.25	27JA	-0.1	-3.5	-1.6	0.1
248.71	0.23	1.1	319.30	6FE	-0.3	-3.4	-1.6	0.2
255.30	0.06	1.0	329.29	16FE	-0.8	-3.4	-1.6	0.2
261.87	-0.13	0.9	339.23	26FE	-1.5	-3.4	-1.6	0.3
268.40	-0.34	0.7	349.11	7MR	-1.3	-3.4	-1.6	0.3
274.89	-0.57	0.6	358.93	17MR	-0.2	-3.3	-1.6	0.4
281.31	-0.84	0.4	8.69	27MR	1.1	-3.3	-1.6	0.5
287.64	-1.14	0.2	18.41	6AP	2.8	-3.3	-1.6	0.5
293.83	-1.47	0.1	28.07	16AP	3.1	-3.3	-1.7	0.5
299.86	-1.83	-0.1	37.69	26AP	1.8	-3.3	-1.7	0.6
305.66	-2.23	-0.3	47.27	6MY	1.0	-3.4	-1.8	0.6
311.16	-2.68	-0.5	56.83	16MY	0.2	-3.4	-1.8	0.6
316.27	-3.16	-0.8	66.36	26MY	-0.7	-3.4	-1.9	0.6
320.86	-3.69	-1.0	75.89	5JN	-1.6	-3.4	-1.9	0.7
324.78	-4.25	-1.2	85.41	15JN	-1.3	-3.5	-2.0	0.7
327.85	-4.83	-1.5	94.94	25JN	-0.5	-3.5	-2.1	0.7
329.84	-5.42	-1.8	104.49	5JL	0.1	-3.6	-2.1	0.7
330.58	-5.96	-2.0	114.06	15JL	0.4	-3.7	-2.2	0.6
329.97	-6.38	-2.3	123.67	25JL	0.8	-3.8	-2.3	0.6
328.11	-6.60	-2.5	133.32	4AU	1.4	-3.9	-2.3	0.7
325.48	-6.51	-2.6	143.01	14AU	2.9	-4.0	-2.4	0.7
322.78	-6.12	-2.5	152.76	24AU	1.4	-4.1	-2.4	0.8
320.75	-5.49	-2.2	162.57	3SE	-0.4	-4.2	-2.4	0.8
319.89	-4.73	-1.9	172.43	13SE	-1.0	-4.3	-2.5	0.8
320.32	-3.96	-1.6	182.35	23SE	-1.0	-4.3	-2.4	0.8
322.00	-3.23	-1.3	192.33	3OC	-0.8	-4.3	-2.4	0.8
324.72	-2.57	-1.0	202.36	13OC	-0.5	-4.0	-2.4	0.8
328.29	-2.00	-0.7	212.44	23OC	-0.4	-3.4	-2.3	0.8
332.53	-1.51	-0.4	222.56	2NO	-0.4	-3.3	-2.3	0.8
337.28	-1.08	-0.2	232.72	12NO	-0.3	-4.0	-2.2	0.7
342.42	-0.72	0.0	242.91	22NO	0.6	-4.3	-2.1	0.7
347.87	-0.42	0.3	253.11	2DE	3.0	-4.4	-2.1	0.6
353.55	-0.15	0.5	263.31	12DE	0.6	-4.3	-2.0	0.6
359.39	0.07	0.6	273.51	22DE	0.0	-4.3	-1.9	0.5
				505				
5.38	0.26	0.8	283.69	1JA	-0.1	-4.1	-1.9	0.4
11.45	0.43	1.0	293.84	11JA	-0.2	-4.0	-1.8	0.4
17.59	0.57	1.1	303.96	21JA	-0.4	-3.9	-1.7	0.3
23.79	0.69	1.2	314.04	31JA	-0.9	-3.8	-1.7	0.3
30.02	0.79	1.4	324.06	10FE	-1.4	-3.7	-1.7	0.3
36.26	0.88	1.5	334.03	20FE	-1.2	-3.7	-1.6	0.4
42.53	0.95	1.6	343.94	2MR	-0.2	-3.6	-1.6	0.4
48.79	1.01	1.6	353.79	12MR	1.4	-3.5	-1.6	0.5
55.06	1.06	1.7	3.59	22MR	3.4	-3.5	-1.6	0.6
61.33	1.11	1.8	13.33	1AP	2.3	-3.4	-1.6	0.6
67.59	1.14	1.9	23.01	11AP	1.3	-3.4	-1.6	0.7
73.85	1.17	1.9	32.65	21AP	0.7	-3.3	-1.6	0.7
80.11	1.19	1.9	42.26	1MY	0.1	-3.3	-1.6	0.8
86.37	1.20	2.0	51.82	11MY	-0.7	-3.3	-1.6	0.8
92.63	1.21	2.0	61.37	21MY	-1.6	-3.3	-1.6	0.8
98.89	1.21	2.0	70.90	31MY	-1.4	-3.3	-1.7	0.9
105.16	1.21	2.0	80.42	10JN	-0.5	-3.3	-1.7	0.9
111.44	1.20	2.0	89.95	20JN	0.2	-3.3	-1.8	0.9
117.74	1.19	2.0	99.49	30JN	0.6	-3.4	-1.8	0.9
124.06	1.18	2.0	109.04	10JL	1.0	-3.4	-1.9	0.9
130.40	1.16	2.0	118.63	20JL	1.8	-3.4	-1.9	0.9
136.76	1.13	2.0	128.26	30JL	3.2	-3.5	-2.0	0.9
143.16	1.10	2.0	137.93	9AU	1.2	-3.5	-2.1	0.9
149.59	1.07	1.9	147.65	19AU	-0.5	-3.5	-2.1	1.0
156.06	1.03	1.9	157.42	29AU	-1.1	-3.5	-2.2	1.0
162.57	0.99	1.9	167.25	8SE	-1.2	-3.4	-2.3	1.0
169.13	0.94	1.9	177.14	18SE	-0.8	-3.4	-2.3	1.1
175.73	0.89	1.9	187.09	28SE	-0.4	-3.4	-2.4	1.1
182.38	0.83	1.9	197.09	8OC	-0.3	-3.4	-2.4	1.1
189.09	0.76	1.9	207.15	18OC	-0.2	-3.3	-2.4	1.1
195.85	0.69	1.9	217.25	28OC	-0.1	-3.3	-2.4	1.1
202.66	0.62	1.8	227.39	7NO	0.8	-3.3	-2.4	1.1
209.53	0.53	1.8	237.56	17NO	2.8	-3.3	-2.3	1.1
216.45	0.44	1.8	247.76	27NO	0.4	-3.4	-2.3	1.0
223.42	0.35	1.7	257.96	7DE	-0.1	-3.4	-2.2	1.0
230.45	0.24	1.7	268.16	17DE	-0.2	-3.4	-2.2	0.9
237.54	0.13	1.6	278.35	27DE	-0.3	-3.4	-2.1	0.8

♂			⊙	16.00UT	☿	♀	♃	♄
LONG	LAT	MAG	LONG	506	MAGNITUDES			
244.67	0.00	1.6	288.52	6JA	-0.5	-3.5	-2.0	0.8
251.86	-0.13	1.5	298.66	16JA	-0.9	-3.5	-1.9	0.7
259.09	-0.26	1.4	308.76	26JA	-1.2	-3.6	-1.9	0.6
266.37	-0.41	1.4	318.81	5FE	-1.1	-3.6	-1.8	0.6
273.68	-0.57	1.3	328.81	15FE	-0.1	-3.7	-1.7	0.6
281.02	-0.73	1.2	338.75	25FE	1.7	-3.7	-1.7	0.6
288.39	-0.89	1.1	348.63	7MR	3.2	-3.8	-1.6	0.7
295.77	-1.06	1.0	358.46	17MR	1.7	-3.9	-1.6	0.7
303.16	-1.24	1.0	8.22	27MR	1.0	-4.0	-1.6	0.8
310.53	-1.41	0.9	17.94	6AP	0.5	-4.1	-1.5	0.8
317.88	-1.59	0.8	27.60	16AP	0.0	-4.1	-1.5	0.9
325.18	-1.76	0.7	37.22	26AP	-0.7	-4.2	-1.5	0.9
332.41	-1.92	0.6	46.81	6MY	-1.6	-4.2	-1.5	1.0
339.55	-2.08	0.5	56.37	16MY	-1.4	-4.0	-1.5	1.0
346.57	-2.23	0.4	65.90	26MY	-0.4	-3.5	-1.5	1.1
353.43	-2.36	0.3	75.43	5JN	0.3	-2.7	-1.5	1.1
0.09	-2.48	0.2	84.95	15JN	0.8	-3.6	-1.5	1.1
6.51	-2.59	0.1	94.48	25JN	1.4	-4.0	-1.6	1.1
12.62	-2.67	-0.1	104.03	5JL	2.4	-4.2	-1.6	1.2
18.36	-2.73	-0.2	113.60	15JL	2.9	-4.2	-1.6	1.2
23.64	-2.77	-0.3	123.20	25JL	1.0	-4.1	-1.7	1.2
28.34	-2.79	-0.5	132.85	4AU	-0.5	-4.1	-1.7	1.2
32.32	-2.77	-0.7	142.55	14AU	-1.2	-4.0	-1.8	1.2
35.42	-2.70	-0.9	152.29	24AU	-1.3	-3.9	-1.8	1.2
37.43	-2.58	-1.1	162.09	3SE	-0.7	-3.8	-1.9	1.3
38.12	-2.39	-1.3	171.95	13SE	-0.3	-3.7	-2.0	1.3
37.37	-2.11	-1.5	181.86	23SE	-0.1	-3.7	-2.0	1.4
35.18	-1.72	-1.7	191.84	3OC	-0.1	-3.6	-2.1	1.4
31.91	-1.24	-1.9	201.87	13OC	0.1	-3.6	-2.2	1.4
28.26	-0.70	-1.9	211.95	23OC	1.1	-3.5	-2.2	1.3
25.01	-0.18	-1.6	222.07	2NO	2.5	-3.5	-2.3	1.3
22.83	0.29	-1.3	232.23	12NO	0.2	-3.4	-2.3	1.3
21.99	0.66	-1.0	242.41	22NO	-0.3	-3.4	-2.3	1.2
22.46	0.95	-0.6	252.61	2DE	-0.3	-3.4	-2.3	1.2
24.10	1.17	-0.3	262.82	12DE	-0.4	-3.4	-2.3	1.2
26.67	1.32	-0.1	273.01	22DE	-0.6	-3.4	-2.2	1.1
				507				
29.99	1.43	0.2	283.20	1JA	-0.9	-3.3	-2.2	1.1
33.90	1.51	0.4	293.35	11JA	-1.1	-3.3	-2.1	1.0
38.26	1.56	0.6	303.47	21JA	-1.0	-3.3	-2.0	1.0
42.97	1.59	0.8	313.55	31JA	-0.1	-3.3	-2.0	0.9
47.96	1.61	1.0	323.57	10FE	2.0	-3.4	-1.9	0.9
53.17	1.62	1.2	333.54	20FE	2.5	-3.4	-1.8	0.9
58.54	1.61	1.3	343.46	2MR	1.2	-3.4	-1.8	0.9
64.06	1.60	1.4	353.31	12MR	0.7	-3.4	-1.7	0.9
69.68	1.58	1.5	3.11	22MR	0.4	-3.5	-1.6	1.0
75.40	1.56	1.6	12.85	1AP	-0.1	-3.5	-1.6	1.0
81.19	1.53	1.7	22.54	11AP	-0.7	-3.5	-1.5	1.1
87.05	1.50	1.8	32.18	21AP	-1.6	-3.4	-1.5	1.1
92.97	1.46	1.8	41.78	1MY	-1.4	-3.4	-1.5	1.2
98.94	1.43	1.9	51.35	11MY	-0.4	-3.4	-1.4	1.2
104.96	1.38	1.9	60.90	21MY	0.4	-3.3	-1.4	1.3
111.03	1.34	2.0	70.43	31MY	1.1	-3.3	-1.4	1.3
117.15	1.29	2.0	79.95	10JN	1.8	-3.3	-1.4	1.4
123.31	1.24	2.0	89.48	20JN	3.1	-3.3	-1.4	1.4
129.53	1.18	2.0	99.02	30JN	2.4	-3.3	-1.4	1.4
135.80	1.13	2.0	108.58	10JL	0.8	-3.3	-1.4	1.4
142.12	1.07	2.0	118.16	20JL	-0.6	-3.4	-1.4	1.4
148.49	1.00	2.0	127.79	30JL	-1.3	-3.4	-1.4	1.4
154.93	0.94	2.0	137.45	9AU	-1.3	-3.4	-1.5	1.3
161.43	0.87	1.9	147.17	19AU	-0.7	-3.5	-1.5	1.3
168.00	0.79	1.9	156.95	29AU	-0.2	-3.5	-1.6	1.2
174.63	0.72	1.9	166.77	8SE	-0.0	-3.6	-1.6	1.2
181.34	0.64	1.8	176.66	18SE	0.1	-3.7	-1.7	1.2
188.12	0.55	1.8	186.61	28SE	0.3	-3.7	-1.7	1.2
194.97	0.47	1.7	196.60	8OC	1.4	-3.8	-1.8	1.2
201.90	0.37	1.7	206.66	18OC	2.3	-3.9	-1.9	1.2
208.90	0.28	1.7	216.76	28OC	0.0	-4.0	-1.9	1.2
215.98	0.18	1.7	226.90	7NO	-0.5	-4.1	-2.0	1.2
223.14	0.08	1.7	237.07	17NO	-0.5	-4.3	-2.1	1.1
230.37	-0.03	1.7	247.26	27NO	-0.5	-4.3	-2.1	1.1
237.68	-0.14	1.7	257.46	7DE	-0.7	-4.4	-2.1	1.1
245.05	-0.25	1.6	267.67	17DE	-0.9	-4.3	-2.2	1.0
252.50	-0.36	1.6	277.86	27DE	-0.9	-4.1	-2.2	1.0

♂ LONG	♂ LAT	♂ MAG	⊙ LONG	16.00UT	☿	♀	♃	♄
				508	MAGNITUDES			
260.00	-0.48	1.6	288.03	6JA	-0.8	-3.5	-2.2	0.9
267.55	-0.60	1.5	298.17	16JA	0.0	-3.4	-2.1	0.9
275.16	-0.71	1.5	308.27	26JA	2.4	-4.0	-2.1	0.8
282.81	-0.82	1.4	318.32	5FE	1.9	-4.3	-2.0	0.8
290.49	-0.93	1.4	328.32	15FE	0.9	-4.3	-2.0	0.7
298.19	-1.04	1.4	338.27	25FE	0.5	-4.3	-1.9	0.7
305.90	-1.14	1.3	348.15	6MR	0.2	-4.2	-1.8	0.7
313.61	-1.23	1.3	357.98	16MR	-0.1	-4.1	-1.8	0.8
321.31	-1.31	1.3	7.75	26MR	-0.7	-4.0	-1.7	0.8
328.98	-1.38	1.2	17.46	5AP	-1.6	-3.8	-1.6	0.9
336.61	-1.44	1.2	27.13	15AP	-1.4	-3.8	-1.6	1.0
344.19	-1.48	1.1	36.76	25AP	-0.4	-3.7	-1.5	1.0
351.69	-1.51	1.1	46.34	5MY	0.6	-3.6	-1.5	1.1
359.11	-1.52	1.1	55.90	15MY	1.4	-3.5	-1.4	1.1
6.43	-1.52	1.0	65.44	25MY	2.4	-3.5	-1.4	1.2
13.63	-1.50	1.0	74.96	4JN	3.4	-3.4	-1.4	1.2
20.69	-1.46	0.9	84.49	14JN	1.9	-3.4	-1.3	1.2
27.61	-1.41	0.9	94.02	24JN	0.6	-3.4	-1.3	1.3
34.34	-1.34	0.8	103.56	4JL	-0.6	-3.3	-1.3	1.3
40.89	-1.25	0.8	113.14	14JL	-1.4	-3.3	-1.3	1.3
47.20	-1.14	0.7	122.74	24JL	-1.3	-3.3	-1.3	1.2
53.25	-1.02	0.6	132.38	3AU	-0.6	-3.3	-1.3	1.2
59.01	-0.87	0.5	142.08	13AU	-0.2	-3.3	-1.3	1.2
64.39	-0.70	0.4	151.82	23AU	0.1	-3.4	-1.3	1.1
69.36	-0.51	0.3	161.61	2SE	0.3	-3.4	-1.4	1.1
73.80	-0.28	0.2	171.47	12SE	0.6	-3.4	-1.4	1.0
77.60	-0.02	0.0	181.39	22SE	1.7	-3.4	-1.4	1.0
80.63	0.29	-0.2	191.36	2OC	2.1	-3.4	-1.5	1.1
82.69	0.64	-0.4	201.38	12OC	-0.1	-3.5	-1.5	1.1
83.58	1.04	-0.6	211.46	22OC	-0.6	-3.5	-1.6	1.1
83.13	1.48	-0.8	221.58	1NO	-0.6	-3.5	-1.7	1.1
81.26	1.95	-1.0	231.74	11NO	-0.6	-3.5	-1.7	1.0
78.15	2.40	-1.2	241.92	21NO	-0.8	-3.4	-1.8	1.0
74.29	2.78	-1.3	252.11	1DE	-0.7	-3.4	-1.9	1.0
70.44	3.03	-1.2	262.32	11DE	-0.8	-3.4	-1.9	1.0
67.33	3.14	-0.9	272.52	21DE	-0.7	-3.4	-2.0	0.9
65.44	3.14	-0.6	282.70	31DE	0.1	-3.3	-2.0	0.9
				509				
64.87	3.06	-0.3	292.86	10JA	2.7	-3.3	-2.0	0.8
65.56	2.94	-0.1	302.98	20JA	1.5	-3.3	-2.1	0.8
67.30	2.81	0.2	313.06	30JA	0.6	-3.3	-2.1	0.7
69.90	2.66	0.4	323.09	9FE	0.3	-3.4	-2.0	0.7
73.18	2.52	0.7	333.06	19FE	0.1	-3.4	-2.0	0.6
77.00	2.38	0.8	342.97	1MR	-0.2	-3.4	-2.0	0.6
81.26	2.25	1.0	352.84	11MR	-0.8	-3.4	-1.9	0.6
85.85	2.13	1.2	2.63	21MR	-1.6	-3.5	-1.9	0.6
90.72	2.01	1.3	12.38	31MR	-1.3	-3.5	-1.8	0.6
95.82	1.89	1.4	22.07	10AP	-0.3	-3.5	-1.7	0.7
101.11	1.78	1.5	31.72	20AP	0.7	-3.6	-1.7	0.8
106.56	1.67	1.6	41.32	30AP	1.8	-3.7	-1.6	0.8
112.15	1.57	1.7	50.90	10MY	3.2	-3.7	-1.5	0.9
117.86	1.46	1.7	60.44	20MY	2.8	-3.8	-1.5	0.9
123.68	1.36	1.8	69.97	30MY	1.5	-3.9	-1.4	1.0
129.61	1.26	1.8	79.50	9JN	0.5	-4.0	-1.4	1.0
135.64	1.16	1.8	89.02	19JN	-0.6	-4.1	-1.3	1.1
141.76	1.06	1.9	98.56	29JN	-1.5	-4.2	-1.3	1.1
147.97	0.97	1.9	108.12	9JL	-1.3	-4.2	-1.3	1.1
154.28	0.87	1.9	117.70	19JL	-0.6	-4.1	-1.3	1.1
160.67	0.77	1.9	127.32	29JL	-0.1	-3.9	-1.2	1.1
167.16	0.67	1.9	136.99	8AU	0.2	-3.4	-1.2	1.1
173.74	0.56	1.9	146.70	18AU	0.4	-3.4	-1.2	1.1
180.41	0.46	1.8	156.47	28AU	0.8	-3.9	-1.2	1.0
187.18	0.36	1.8	166.30	7SE	2.1	-4.2	-1.2	1.0
194.04	0.25	1.8	176.18	17SE	1.9	-4.3	-1.3	0.9
200.99	0.15	1.7	186.12	27SE	-0.2	-4.3	-1.3	0.9
208.05	0.04	1.7	196.12	7OC	-0.8	-4.2	-1.3	0.9
215.19	-0.06	1.6	206.17	17OC	-0.8	-4.1	-1.4	0.9
222.43	-0.17	1.6	216.27	27OC	-0.8	-4.0	-1.4	0.9
229.76	-0.27	1.5	226.41	6NO	-0.7	-3.9	-1.5	0.9
237.17	-0.37	1.5	236.58	16NO	-0.6	-3.9	-1.5	0.9
244.66	-0.47	1.5	246.77	26NO	-0.6	-3.8	-1.6	0.9
252.23	-0.57	1.5	256.97	6DE	-0.5	-3.7	-1.6	0.9
259.86	-0.67	1.5	267.17	16DE	0.3	-3.6	-1.7	0.9
267.56	-0.75	1.5	277.37	26DE	2.9	-3.6	-1.8	0.9

♂ LONG	♂ LAT	♂ MAG	⊙ LONG	16.00UT	☿	♀	♃	♄
				510	MAGNITUDES			
275.31	-0.84	1.4	287.54	5JA	1.1	-3.5	-1.8	0.8
283.10	-0.91	1.4	297.68	15JA	0.4	-3.5	-1.9	0.8
290.92	-0.98	1.4	307.78	25JA	0.2	-3.4	-1.9	0.7
298.76	-1.04	1.4	317.84	4FE	0.0	-3.4	-2.0	0.7
306.60	-1.09	1.4	327.84	14FE	-0.3	-3.4	-2.0	0.6
314.44	-1.13	1.4	337.79	24FE	-0.8	-3.3	-2.0	0.6
322.27	-1.16	1.4	347.68	6MR	-1.6	-3.3	-2.0	0.5
330.06	-1.17	1.4	357.51	16MR	-1.3	-3.3	-2.0	0.5
337.81	-1.17	1.4	7.28	26MR	-0.3	-3.3	-2.0	0.4
345.50	-1.16	1.4	17.00	5AP	0.9	-3.3	-1.9	0.5
353.13	-1.14	1.4	26.67	15AP	2.3	-3.3	-1.9	0.5
0.69	-1.10	1.4	36.29	25AP	3.7	-3.4	-1.8	0.6
8.16	-1.05	1.4	45.88	5MY	2.1	-3.4	-1.7	0.7
15.54	-0.99	1.4	55.44	15MY	1.2	-3.4	-1.7	0.7
22.82	-0.92	1.4	64.98	25MY	0.3	-3.5	-1.6	0.8
30.00	-0.83	1.4	74.50	4JN	-0.6	-3.5	-1.6	0.8
37.06	-0.74	1.4	84.02	14JN	-1.5	-3.5	-1.5	0.9
43.99	-0.63	1.4	93.56	24JN	-1.4	-3.4	-1.4	0.9
50.80	-0.52	1.3	103.10	4JL	-0.5	-3.4	-1.4	1.0
57.47	-0.40	1.3	112.67	14JL	0.0	-3.4	-1.3	1.0
63.99	-0.26	1.3	122.27	24JL	0.4	-3.4	-1.3	1.0
70.35	-0.12	1.3	131.91	3AU	0.6	-3.3	-1.3	1.0
76.53	0.04	1.2	141.60	13AU	1.1	-3.3	-1.2	1.0
82.51	0.21	1.2	151.34	23AU	2.6	-3.3	-1.2	1.0
88.27	0.39	1.1	161.13	2SE	1.6	-3.3	-1.2	1.0
93.77	0.58	1.0	170.99	12SE	-0.3	-3.3	-1.2	0.9
98.98	0.80	0.9	180.90	22SE	-0.9	-3.4	-1.2	0.9
103.82	1.03	0.8	190.87	2OC	-0.9	-3.4	-1.2	0.8
108.23	1.30	0.7	200.89	12OC	-0.9	-3.4	-1.2	0.8
112.12	1.59	0.5	210.97	22OC	-0.6	-3.5	-1.3	0.8
115.35	1.91	0.3	221.09	1NO	-0.5	-3.5	-1.3	0.9
117.79	2.27	0.1	231.24	11NO	-0.5	-3.6	-1.3	0.9
119.24	2.67	-0.1	241.43	21NO	-0.4	-3.6	-1.4	0.9
119.51	3.10	-0.3	251.62	1DE	0.4	-3.7	-1.4	0.9
118.48	3.54	-0.6	261.83	11DE	3.1	-3.8	-1.5	0.9
116.11	3.94	-0.8	272.03	21DE	0.8	-3.8	-1.5	0.8
112.67	4.24	-1.0	282.21	31DE	0.2	-3.9	-1.6	0.8
				511				
108.71	4.39	-1.0	292.37	10JA	0.0	-4.0	-1.7	0.8
104.96	4.36	-0.8	302.49	20JA	-0.1	-4.1	-1.7	0.8
102.10	4.18	-0.6	312.57	30JA	-0.4	-4.2	-1.8	0.7
100.47	3.92	-0.4	322.60	9FE	-0.8	-4.3	-1.9	0.7
100.13	3.61	-0.1	332.58	19FE	-1.5	-4.3	-1.9	0.6
100.99	3.29	0.1	342.50	1MR	-1.2	-4.2	-2.0	0.6
102.84	2.99	0.4	352.36	11MR	-0.3	-3.9	-2.0	0.5
105.53	2.71	0.6	2.16	21MR	1.2	-3.2	-2.0	0.4
108.89	2.45	0.7	11.91	31MR	3.0	-3.4	-2.0	0.4
112.77	2.21	0.9	21.60	10AP	2.8	-3.9	-2.0	0.4
117.09	1.99	1.0	31.25	20AP	1.6	-4.2	-2.0	0.4
121.77	1.79	1.2	40.86	30AP	0.9	-4.2	-2.0	0.5
126.75	1.60	1.3	50.43	10MY	0.2	-4.2	-1.9	0.5
131.97	1.42	1.3	59.98	20MY	-0.6	-4.1	-1.9	0.6
137.42	1.25	1.4	69.51	30MY	-1.6	-4.0	-1.8	0.7
143.06	1.10	1.5	79.03	9JN	-1.4	-3.9	-1.8	0.7
148.88	0.94	1.5	88.56	19JN	-0.5	-3.8	-1.7	0.8
154.85	0.80	1.6	98.09	29JN	0.1	-3.7	-1.6	0.8
160.97	0.66	1.6	107.65	9JL	0.5	-3.6	-1.6	0.9
167.24	0.52	1.6	117.24	19JL	0.9	-3.6	-1.5	0.9
173.63	0.39	1.6	126.85	29JL	1.5	-3.5	-1.5	0.9
180.16	0.26	1.7	136.52	8AU	3.0	-3.5	-1.4	0.9
186.82	0.14	1.7	146.23	18AU	1.4	-3.4	-1.4	0.9
193.59	0.01	1.6	155.99	28AU	-0.4	-3.4	-1.3	0.9
200.49	-0.10	1.6	165.82	7SE	-1.0	-3.4	-1.3	0.9
207.51	-0.22	1.6	175.70	17SE	-1.1	-3.4	-1.3	0.9
214.64	-0.33	1.6	185.64	27SE	-0.8	-3.4	-1.3	0.8
221.88	-0.43	1.6	195.63	7OC	-0.5	-3.4	-1.2	0.8
229.22	-0.53	1.6	205.69	17OC	-0.3	-3.4	-1.2	0.8
236.66	-0.63	1.5	215.78	27OC	-0.3	-3.4	-1.3	0.8
244.20	-0.71	1.5	225.92	6NO	-0.2	-3.4	-1.3	0.8
251.82	-0.79	1.5	236.09	16NO	0.6	-3.4	-1.3	0.8
259.52	-0.86	1.4	246.27	26NO	3.0	-3.4	-1.3	0.8
267.28	-0.93	1.4	256.48	6DE	0.5	-3.4	-1.4	0.8
275.09	-0.98	1.4	266.68	16DE	-0.0	-3.4	-1.4	0.8
282.94	-1.02	1.4	276.87	26DE	-0.1	-3.5	-1.4	0.8

♂ LONG	♂ LAT	♂ MAG	☉ LONG	16.00UT	☿	♀	♃	♄
				512	MAGNITUDES			
290.82	-1.06	1.3	287.05	5JA	-0.2	-3.5	-1.5	0.8
298.72	-1.08	1.3	297.19	15JA	-0.4	-3.5	-1.6	0.8
306.62	-1.08	1.3	307.29	25JA	-0.9	-3.5	-1.6	0.7
314.51	-1.08	1.3	317.35	4FE	-1.3	-3.4	-1.7	0.7
322.37	-1.07	1.3	327.36	14FE	-1.2	-3.4	-1.8	0.7
330.20	-1.04	1.4	337.30	24FE	-0.2	-3.4	-1.8	0.6
337.98	-1.00	1.4	347.20	5MR	1.5	-3.4	-1.9	0.6
345.70	-0.96	1.4	357.03	15MR	3.5	-3.3	-2.0	0.5
353.35	-0.90	1.4	6.80	25MR	2.1	-3.3	-2.0	0.4
0.93	-0.83	1.5	16.53	4AP	1.2	-3.3	-2.1	0.4
8.43	-0.76	1.5	26.20	14AP	0.7	-3.3	-2.1	0.3
15.84	-0.67	1.5	35.82	24AP	0.1	-3.3	-2.1	0.3
23.17	-0.58	1.6	45.42	4MY	-0.7	-3.4	-2.1	0.4
30.39	-0.49	1.6	54.98	14MY	-1.6	-3.4	-2.1	0.4
37.53	-0.38	1.6	64.51	24MY	-1.4	-3.4	-2.1	0.5
44.57	-0.28	1.6	74.04	3JN	-0.5	-3.4	-2.0	0.6
51.50	-0.17	1.6	83.57	13JN	0.2	-3.5	-2.0	0.6
58.34	-0.05	1.7	93.09	23JN	0.7	-3.5	-1.9	0.7
65.08	0.07	1.7	102.64	3JL	1.1	-3.6	-1.9	0.7
71.71	0.19	1.7	112.21	13JL	2.0	-3.7	-1.8	0.8
78.24	0.31	1.7	121.80	23JL	3.2	-3.8	-1.7	0.8
84.67	0.44	1.6	131.45	2AU	1.2	-3.9	-1.7	0.8
90.97	0.57	1.6	141.13	12AU	-0.5	-4.0	-1.6	0.9
97.16	0.71	1.6	150.87	22AU	-1.2	-4.1	-1.6	0.9
103.21	0.86	1.6	160.66	1SE	-1.2	-4.2	-1.5	0.9
109.12	1.00	1.5	170.51	11SE	-0.8	-4.3	-1.5	0.9
114.86	1.16	1.4	180.42	21SE	-0.4	-4.3	-1.4	0.8
120.43	1.33	1.4	190.39	1OC	-0.2	-4.3	-1.4	0.8
125.77	1.51	1.3	200.41	11OC	-0.1	-4.0	-1.4	0.8
130.86	1.70	1.2	210.48	21OC	-0.0	-3.4	-1.3	0.7
135.65	1.90	1.1	220.60	31OC	0.9	-3.4	-1.3	0.7
140.07	2.13	0.9	230.75	10NO	2.8	-4.0	-1.3	0.7
144.05	2.37	0.8	240.93	20NO	0.3	-4.3	-1.3	0.8
147.46	2.64	0.6	251.13	30NO	-0.2	-4.4	-1.3	0.8
150.18	2.94	0.4	261.33	10DE	-0.3	-4.3	-1.4	0.8
152.05	3.26	0.1	271.53	20DE	-0.3	-4.3	-1.4	0.8
152.88	3.60	-0.1	281.72	30DE	-0.5	-4.1	-1.4	0.8
				513				
152.51	3.93	-0.4	291.88	9JA	-0.9	-4.0	-1.5	0.8
150.83	4.22	-0.6	302.00	19JA	-1.2	-3.9	-1.5	0.8
147.94	4.42	-0.9	312.09	29JA	-1.1	-3.8	-1.5	0.8
144.24	4.47	-1.0	322.12	8FE	-0.1	-3.7	-1.6	0.7
140.34	4.35	-1.0	332.10	18FE	1.8	-3.7	-1.7	0.7
136.98	4.07	-0.8	342.03	28FE	2.9	-3.6	-1.7	0.6
134.69	3.68	-0.6	351.89	10MR	1.5	-3.5	-1.8	0.6
133.67	3.26	-0.3	1.69	20MR	0.9	-3.5	-1.9	0.5
133.93	2.84	-0.1	11.45	30MR	0.5	-3.4	-1.9	0.5
135.32	2.44	0.1	21.14	9AP	0.0	-3.4	-2.0	0.4
137.65	2.07	0.3	30.79	19AP	-0.7	-3.3	-2.1	0.3
140.76	1.75	0.5	40.40	29AP	-1.6	-3.3	-2.1	0.3
144.51	1.45	0.6	49.97	9MY	-1.4	-3.3	-2.2	0.3
148.78	1.18	0.7	59.52	19MY	-0.4	-3.3	-2.2	0.4
153.49	0.93	0.9	69.06	29MY	0.3	-3.3	-2.2	0.4
158.57	0.71	0.9	78.58	8JN	0.9	-3.3	-2.2	0.5
163.95	0.50	1.0	88.10	18JN	1.5	-3.3	-2.2	0.5
169.62	0.31	1.1	97.64	28JN	2.6	-3.4	-2.2	0.6
175.52	0.13	1.2	107.19	8JL	2.8	-3.4	-2.1	0.7
181.65	-0.03	1.2	116.77	18JL	1.0	-3.4	-2.1	0.7
187.97	-0.18	1.3	126.39	28JL	-0.5	-3.5	-2.0	0.7
194.48	-0.32	1.3	136.05	7AU	-1.3	-3.5	-2.0	0.8
201.16	-0.46	1.3	145.76	17AU	-1.3	-3.5	-1.9	0.8
208.01	-0.58	1.3	155.52	27AU	-0.7	-3.5	-1.8	0.8
215.00	-0.69	1.3	165.33	6SE	-0.3	-3.4	-1.8	0.8
222.13	-0.79	1.4	175.21	16SE	-0.1	-3.4	-1.7	0.8
229.40	-0.88	1.4	185.15	26SE	0.0	-3.4	-1.7	0.8
236.79	-0.96	1.4	195.14	6OC	0.2	-3.3	-1.6	0.8
244.28	-1.02	1.4	205.19	16OC	1.1	-3.3	-1.6	0.8
251.88	-1.08	1.4	215.28	26OC	2.6	-3.3	-1.5	0.7
259.55	-1.12	1.4	225.41	5NO	0.1	-3.3	-1.5	0.7
267.30	-1.15	1.4	235.58	15NO	-0.4	-3.3	-1.5	0.7
275.11	-1.17	1.4	245.77	25NO	-0.4	-3.4	-1.4	0.7
282.96	-1.17	1.4	255.97	5DE	-0.4	-3.4	-1.4	0.8
290.84	-1.16	1.4	266.18	15DE	-0.6	-3.4	-1.4	0.8
298.74	-1.14	1.4	276.38	25DE	-0.9	-3.4	-1.4	0.8

♂ LONG	♂ LAT	♂ MAG	☉ LONG	16.00UT	☿	♀	♃	♄
				514	MAGNITUDES			
306.62	-1.11	1.4	286.55	4JA	-1.0	-3.5	-1.4	0.8
314.50	-1.07	1.4	296.70	14JA	-0.9	-3.5	-1.5	0.8
322.34	-1.01	1.4	306.81	24JA	-0.1	-3.6	-1.5	0.8
330.15	-0.95	1.4	316.87	3FE	2.1	-3.6	-1.5	0.8
337.90	-0.88	1.4	326.88	13FE	2.3	-3.7	-1.6	0.7
345.59	-0.80	1.4	336.83	23FE	1.1	-3.7	-1.6	0.7
353.20	-0.71	1.4	346.72	5MR	0.6	-3.8	-1.7	0.7
0.75	-0.62	1.5	356.56	15MR	0.3	-3.9	-1.7	0.6
8.21	-0.53	1.5	6.34	25MR	-0.1	-4.0	-1.8	0.5
15.59	-0.43	1.5	16.06	4AP	-0.7	-4.1	-1.9	0.5
22.88	-0.33	1.5	25.74	14AP	-1.6	-4.1	-1.9	0.4
30.08	-0.22	1.6	35.37	24AP	-1.4	-4.2	-2.0	0.4
37.20	-0.12	1.6	44.96	4MY	-0.4	-4.2	-2.1	0.3
44.23	-0.02	1.7	54.52	14MY	0.4	-4.0	-2.1	0.2
51.18	0.09	1.7	64.06	24MY	1.2	-3.4	-2.2	0.3
58.04	0.19	1.8	73.58	3JN	2.0	-2.7	-2.2	0.4
64.82	0.30	1.8	83.11	13JN	3.3	-3.6	-2.3	0.4
71.53	0.40	1.8	92.63	23JN	2.3	-4.0	-2.3	0.5
78.16	0.50	1.8	102.18	3JL	0.8	-4.2	-2.3	0.5
84.71	0.60	1.9	111.74	13JL	-0.5	-4.2	-2.3	0.6
91.20	0.70	1.9	121.34	23JL	-1.4	-4.1	-2.3	0.6
97.61	0.80	1.9	130.98	2AU	-1.3	-4.1	-2.3	0.7
103.95	0.90	1.9	140.66	12AU	-0.7	-4.0	-2.3	0.7
110.21	1.00	1.9	150.39	22AU	-0.2	-3.9	-2.2	0.8
116.41	1.10	1.8	160.18	1SE	0.0	-3.8	-2.2	0.8
122.52	1.20	1.8	170.03	11SE	0.2	-3.7	-2.1	0.8
128.54	1.31	1.8	179.93	21SE	0.4	-3.7	-2.0	0.8
134.48	1.41	1.7	189.90	1OC	1.4	-3.6	-2.0	0.8
140.30	1.52	1.7	199.92	11OC	2.3	-3.6	-1.9	0.8
146.00	1.63	1.6	209.98	21OC	-0.0	-3.5	-1.8	0.8
151.57	1.74	1.5	220.10	31OC	-0.5	-3.5	-1.8	0.7
156.96	1.86	1.4	230.25	10NO	-0.6	-3.4	-1.7	0.7
162.16	1.99	1.3	240.43	20NO	-0.6	-3.4	-1.7	0.7
167.12	2.12	1.2	250.63	30NO	-0.7	-3.4	-1.6	0.7
171.78	2.26	1.0	260.83	10DE	-0.8	-3.4	-1.6	0.7
176.09	2.41	0.8	271.03	20DE	-0.9	-3.4	-1.6	0.8
179.96	2.56	0.6	281.22	30DE	-0.8	-3.3	-1.5	0.8
				515				
183.27	2.73	0.4	291.38	9JA	0.0	-3.3	-1.5	0.8
185.89	2.89	0.2	301.51	19JA	2.4	-3.3	-1.5	0.8
187.66	3.05	-0.1	311.60	29JA	1.8	-3.3	-1.5	0.8
188.39	3.20	-0.4	321.64	8FE	0.8	-3.4	-1.5	0.8
187.93	3.30	-0.6	331.62	18FE	0.4	-3.4	-1.6	0.8
186.19	3.33	-0.9	341.54	28FE	0.2	-3.4	-1.6	0.7
183.29	3.25	-1.2	351.41	10MR	-0.2	-3.4	-1.6	0.7
179.67	3.03	-1.4	1.22	20MR	-0.7	-3.5	-1.6	0.6
175.96	2.68	-1.3	10.97	30MR	-1.6	-3.5	-1.7	0.6
172.88	2.23	-1.1	20.67	9AP	-1.4	-3.4	-1.7	0.5
170.94	1.76	-0.9	30.32	19AP	-0.4	-3.4	-1.8	0.5
170.31	1.29	-0.7	39.93	29AP	0.6	-3.4	-1.9	0.4
170.99	0.86	-0.5	49.51	9MY	1.5	-3.4	-1.9	0.3
172.81	0.48	-0.3	59.06	19MY	2.7	-3.3	-2.0	0.3
175.59	0.15	-0.1	68.59	29MY	3.2	-3.3	-2.1	0.3
179.18	-0.14	0.1	78.11	8JN	1.8	-3.3	-2.1	0.3
183.43	-0.39	0.2	87.63	18JN	0.6	-3.3	-2.2	0.4
188.22	-0.61	0.3	97.17	28JN	-0.6	-3.3	-2.3	0.4
193.48	-0.80	0.4	106.72	8JL	-1.4	-3.3	-2.3	0.5
199.13	-0.96	0.5	116.30	18JL	-1.4	-3.4	-2.4	0.5
205.11	-1.10	0.6	125.92	28JL	-0.6	-3.4	-2.4	0.6
211.38	-1.22	0.7	135.58	7AU	-0.1	-3.4	-2.4	0.7
217.91	-1.32	0.8	145.28	17AU	0.1	-3.5	-2.4	0.7
224.66	-1.40	0.8	155.04	27AU	0.3	-3.5	-2.4	0.7
231.61	-1.46	0.9	164.86	6SE	0.7	-3.6	-2.4	0.8
238.73	-1.50	0.9	174.73	16SE	1.8	-3.7	-2.4	0.8
246.01	-1.52	1.0	184.67	26SE	2.1	-3.7	-2.3	0.8
253.42	-1.53	1.0	194.66	6OC	-0.1	-3.8	-2.2	0.8
260.94	-1.52	1.1	204.70	16OC	-0.7	-3.9	-2.2	0.8
268.55	-1.50	1.1	214.79	26OC	-0.7	-4.0	-2.1	0.8
276.23	-1.46	1.1	224.92	5NO	-0.7	-4.1	-2.0	0.8
283.96	-1.40	1.2	235.09	15NO	-0.8	-4.3	-2.0	0.7
291.73	-1.34	1.2	245.28	25NO	-0.7	-4.3	-1.9	0.7
299.52	-1.26	1.3	255.48	5DE	-0.7	-4.4	-1.8	0.7
307.30	-1.17	1.3	265.68	15DE	-0.6	-4.3	-1.8	0.7
315.07	-1.08	1.4	275.88	25DE	0.1	-4.0	-1.7	0.8

♂ LONG	♂ LAT	♂ MAG	⊙ LONG	16.00UT	☿	♀	♃	♄
				516	MAGNITUDES			
322.81	-0.98	1.4	286.05	4JA	2.7	-3.4	-1.7	0.8
330.50	-0.87	1.4	296.20	14JA	1.3	-3.4	-1.6	0.8
338.15	-0.76	1.5	306.31	24JA	0.5	-4.0	-1.6	0.8
345.73	-0.65	1.5	316.37	3FE	0.3	-4.3	-1.6	0.8
353.24	-0.54	1.5	326.38	13FE	0.1	-4.3	-1.6	0.8
0.68	-0.42	1.6	336.34	23FE	-0.2	-4.3	-1.6	0.8
8.04	-0.31	1.6	346.24	4MR	-0.8	-4.2	-1.6	0.8
15.32	-0.20	1.6	356.08	14MR	-1.6	-4.1	-1.6	0.7
22.52	-0.09	1.7	5.86	24MR	-1.4	-4.0	-1.6	0.7
29.64	0.02	1.7	15.59	3AP	-0.4	-3.8	-1.6	0.6
36.68	0.12	1.7	25.27	13AP	0.8	-3.7	-1.6	0.6
43.64	0.22	1.7	34.90	23AP	2.0	-3.7	-1.7	0.5
50.52	0.32	1.7	44.49	3MY	3.6	-3.6	-1.7	0.5
57.34	0.41	1.8	54.06	13MY	2.5	-3.5	-1.8	0.4
64.09	0.50	1.8	63.60	23MY	1.4	-3.5	-1.8	0.3
70.77	0.59	1.8	73.12	2JN	0.5	-3.4	-1.9	0.3
77.39	0.67	1.8	82.65	12JN	-0.6	-3.4	-1.9	0.3
83.96	0.75	1.9	92.18	22JN	-1.5	-3.4	-2.0	0.3
90.48	0.82	1.9	101.72	2JL	-1.4	-3.3	-2.1	0.4
96.95	0.89	1.9	111.28	12JL	-0.6	-3.3	-2.1	0.4
103.38	0.96	2.0	120.88	22JL	-0.0	-3.3	-2.2	0.5
109.77	1.03	2.0	130.51	1AU	0.3	-3.3	-2.3	0.6
116.12	1.09	2.0	140.19	11AU	0.5	-3.3	-2.3	0.6
122.44	1.15	2.0	149.93	21AU	0.9	-3.4	-2.4	0.7
128.72	1.21	2.0	159.71	31AU	2.2	-3.4	-2.4	0.7
134.96	1.26	2.0	169.55	10SE	1.8	-3.4	-2.4	0.7
141.17	1.31	2.0	179.46	20SE	-0.2	-3.4	-2.5	0.8
147.34	1.36	1.9	189.41	30SE	-0.8	-3.5	-2.4	0.8
153.48	1.41	1.9	199.43	10OC	-0.8	-3.5	-2.4	0.8
159.56	1.46	1.8	209.50	20OC	-0.8	-3.5	-2.4	0.8
165.59	1.50	1.8	219.61	30OC	-0.7	-3.5	-2.3	0.8
171.57	1.54	1.7	229.76	9NO	-0.5	-3.5	-2.3	0.8
177.47	1.58	1.6	239.94	19NO	-0.5	-3.4	-2.2	0.8
183.28	1.61	1.5	250.13	29NO	-0.5	-3.4	-2.1	0.7
189.00	1.64	1.4	260.33	9DE	0.3	-3.4	-2.0	0.7
194.59	1.66	1.3	270.54	19DE	3.0	-3.4	-2.0	0.7
200.04	1.68	1.2	280.72	29DE	1.0	-3.3	-1.9	0.8
				517				
205.30	1.69	1.0	290.88	8JA	0.3	-3.3	-1.8	0.8
210.34	1.69	0.8	301.01	18JA	0.1	-3.3	-1.8	0.8
215.11	1.67	0.6	311.10	28JA	-0.0	-3.3	-1.7	0.8
219.54	1.64	0.4	321.14	7FE	-0.3	-3.4	-1.7	0.8
223.54	1.58	0.2	331.13	17FE	-0.8	-3.4	-1.6	0.8
227.01	1.48	-0.1	341.05	27FE	-1.5	-3.4	-1.6	0.8
229.80	1.35	-0.3	350.93	9MR	-1.3	-3.4	-1.6	0.8
231.74	1.16	-0.6	0.74	19MR	-0.3	-3.5	-1.6	0.8
232.68	0.89	-0.9	10.49	29MR	1.0	-3.5	-1.6	0.8
232.42	0.54	-1.3	20.20	8AP	2.6	-3.5	-1.5	0.7
230.91	0.10	-1.6	29.86	18AP	3.3	-3.6	-1.6	0.7
228.30	-0.42	-1.9	39.47	28AP	1.9	-3.7	-1.6	0.6
225.03	-0.98	-2.0	49.05	8MY	1.1	-3.7	-1.6	0.5
221.79	-1.52	-1.9	58.60	18MY	0.3	-3.8	-1.6	0.5
219.31	-1.98	-1.7	68.13	28MY	-0.6	-3.9	-1.6	0.4
218.04	-2.33	-1.5	77.66	7JN	-1.5	-4.0	-1.7	0.3
218.15	-2.59	-1.3	87.18	17JN	-1.4	-4.1	-1.7	0.3
219.59	-2.76	-1.1	96.71	27JN	-0.5	-4.2	-1.8	0.3
222.19	-2.85	-0.9	106.27	7JL	0.0	-4.2	-1.8	0.4
225.78	-2.89	-0.7	115.84	17JL	0.4	-4.1	-1.9	0.4
230.15	-2.89	-0.5	125.46	27JL	0.7	-3.9	-1.9	0.5
235.17	-2.86	-0.4	135.11	6AU	1.3	-3.4	-2.0	0.5
240.73	-2.79	-0.2	144.82	16AU	2.7	-3.4	-2.1	0.6
246.71	-2.71	-0.1	154.57	26AU	1.6	-3.9	-2.2	0.6
253.05	-2.60	0.1	164.39	5SE	-0.3	-4.2	-2.2	0.7
259.67	-2.47	0.2	174.25	15SE	-1.0	-4.3	-2.3	0.7
266.51	-2.33	0.3	184.18	25SE	-1.0	-4.3	-2.3	0.8
273.53	-2.17	0.4	194.17	5OC	-0.9	-4.2	-2.4	0.8
280.70	-2.01	0.5	204.21	15OC	-0.6	-4.1	-2.4	0.8
287.97	-1.84	0.6	214.30	25OC	-0.4	-4.0	-2.4	0.8
295.31	-1.67	0.7	224.43	4NO	-0.4	-3.9	-2.4	0.8
302.71	-1.49	0.8	234.60	14NO	-0.3	-3.9	-2.4	0.8
310.13	-1.31	0.9	244.78	24NO	0.5	-3.8	-2.3	0.8
317.55	-1.14	1.0	254.99	4DE	3.1	-3.7	-2.3	0.8
324.97	-0.97	1.1	265.19	14DE	0.7	-3.6	-2.2	0.8
332.36	-0.80	1.2	275.39	24DE	0.1	-3.6	-2.1	0.7

♂ LONG	♂ LAT	♂ MAG	⊙ LONG	16.00UT	☿	♀	♃	♄
				518	MAGNITUDES			
339.72	-0.64	1.3	285.57	3JA	-0.0	-3.5	-2.1	0.8
347.02	-0.49	1.4	295.71	13JA	-0.1	-3.5	-2.0	0.8
354.28	-0.34	1.4	305.82	23JA	-0.4	-3.4	-1.9	0.9
1.48	-0.20	1.5	315.89	2FE	-0.8	-3.4	-1.8	0.9
8.61	-0.07	1.6	325.90	12FE	-1.4	-3.4	-1.8	0.9
15.68	0.06	1.6	335.86	22FE	-1.2	-3.3	-1.7	0.9
22.68	0.17	1.7	345.76	4MR	-0.3	-3.3	-1.7	0.9
29.61	0.28	1.7	355.60	14MR	1.3	-3.3	-1.6	0.9
36.48	0.38	1.8	5.39	24MR	3.2	-3.3	-1.6	0.9
43.29	0.48	1.8	15.12	3AP	2.5	-3.3	-1.5	0.8
50.04	0.56	1.9	24.80	13AP	1.4	-3.3	-1.5	0.8
56.73	0.64	1.9	34.43	23AP	0.8	-3.4	-1.5	0.7
63.37	0.72	1.9	44.03	3MY	0.2	-3.4	-1.5	0.7
69.96	0.78	1.9	53.59	13MY	-0.6	-3.4	-1.5	0.6
76.51	0.85	1.9	63.13	23MY	-1.6	-3.5	-1.5	0.6
83.02	0.90	1.9	72.66	2JN	-1.4	-3.5	-1.5	0.5
89.49	0.95	1.9	82.18	12JN	-0.5	-3.5	-1.5	0.4
95.93	1.00	1.9	91.71	22JN	0.1	-3.4	-1.5	0.4
102.35	1.04	1.9	101.26	2JL	0.6	-3.4	-1.5	0.3
108.75	1.08	1.9	110.82	12JL	1.0	-3.4	-1.6	0.4
115.13	1.11	2.0	120.41	22JL	1.7	-3.4	-1.6	0.4
121.50	1.14	2.0	130.04	1AU	3.2	-3.3	-1.7	0.5
127.86	1.16	2.0	139.72	11AU	1.4	-3.3	-1.7	0.5
134.22	1.18	2.0	149.45	21AU	-0.4	-3.3	-1.8	0.6
140.58	1.20	2.0	159.23	31AU	-1.1	-3.3	-1.8	0.6
146.93	1.21	2.0	169.07	10SE	-1.1	-3.3	-1.9	0.7
153.29	1.22	2.0	178.97	20SE	-0.8	-3.4	-2.0	0.7
159.66	1.22	2.0	188.93	30SE	-0.5	-3.4	-2.0	0.8
166.02	1.22	2.0	198.94	10OC	-0.3	-3.4	-2.1	0.8
172.39	1.21	1.9	209.00	20OC	-0.2	-3.5	-2.2	0.8
178.77	1.19	1.9	219.12	30OC	-0.1	-3.5	-2.2	0.8
185.14	1.17	1.8	229.26	9NO	0.7	-3.6	-2.2	0.9
191.52	1.15	1.8	239.44	19NO	3.0	-3.6	-2.3	0.9
197.90	1.11	1.7	249.64	29NO	0.5	-3.7	-2.3	0.9
204.26	1.07	1.6	259.84	9DE	-0.1	-3.8	-2.3	0.8
210.62	1.02	1.6	270.04	19DE	-0.2	-3.8	-2.2	0.8
216.96	0.95	1.5	280.23	29DE	-0.3	-3.9	-2.2	0.8
				519				
223.28	0.87	1.3	290.39	8JA	-0.5	-4.0	-2.1	0.8
229.56	0.78	1.2	300.53	18JA	-0.9	-4.1	-2.1	0.9
235.81	0.67	1.1	310.62	28JA	-1.3	-4.2	-2.0	0.9
242.00	0.53	1.0	320.66	7FE	-1.1	-4.3	-1.9	0.9
248.13	0.38	0.8	330.65	17FE	-0.2	-4.3	-1.9	1.0
254.16	0.19	0.6	340.58	27FE	1.5	-4.2	-1.8	1.0
260.09	-0.02	0.5	350.45	9MR	3.4	-3.9	-1.7	1.0
265.88	-0.27	0.3	0.27	19MR	1.8	-3.2	-1.7	1.0
271.49	-0.57	0.1	10.02	29MR	1.0	-3.4	-1.6	1.0
276.86	-0.91	-0.1	19.73	8AP	0.6	-3.9	-1.6	0.9
281.93	-1.31	-0.4	29.39	18AP	0.1	-4.2	-1.5	0.9
286.60	-1.77	-0.6	39.00	28AP	-0.6	-4.2	-1.5	0.9
290.77	-2.30	-0.9	48.58	8MY	-1.6	-4.2	-1.4	0.8
294.27	-2.91	-1.2	58.13	18MY	-1.4	-4.1	-1.4	0.7
296.95	-3.60	-1.4	67.67	28MY	-0.5	-4.0	-1.4	0.7
298.60	-4.34	-1.7	77.19	7JN	0.2	-3.9	-1.4	0.6
299.04	-5.11	-2.0	86.72	17JN	0.7	-3.8	-1.4	0.6
298.21	-5.83	-2.3	96.25	27JN	1.3	-3.7	-1.4	0.5
296.29	-6.40	-2.5	105.80	7JL	2.2	-3.6	-1.4	0.4
293.72	-6.71	-2.6	115.38	17JL	3.1	-3.6	-1.4	0.4
291.25	-6.70	-2.5	124.99	27JL	1.2	-3.5	-1.4	0.4
289.55	-6.40	-2.3	134.64	6AU	-0.4	-3.5	-1.4	0.5
289.04	-5.89	-2.0	144.35	16AU	-1.2	-3.4	-1.5	0.5
289.87	-5.28	-1.7	154.10	26AU	-1.3	-3.4	-1.5	0.6
291.93	-4.65	-1.5	163.91	5SE	-0.7	-3.4	-1.6	0.6
295.03	-4.03	-1.2	173.78	15SE	-0.4	-3.4	-1.6	0.7
298.99	-3.45	-0.9	183.71	25SE	-0.2	-3.4	-1.7	0.7
303.61	-2.92	-0.7	193.69	5OC	-0.1	-3.4	-1.7	0.8
308.74	-2.44	-0.4	203.73	15OC	0.1	-3.4	-1.8	0.8
314.27	-2.01	-0.2	213.81	25OC	0.9	-3.4	-1.9	0.9
320.10	-1.62	-0.0	223.94	4NO	2.8	-3.4	-1.9	0.9
326.15	-1.28	0.2	234.10	14NO	0.3	-3.4	-2.0	0.9
332.38	-0.97	0.4	244.29	24NO	-0.3	-3.4	-2.1	0.9
338.72	-0.69	0.5	254.49	4DE	-0.3	-3.4	-2.1	0.9
345.16	-0.45	0.7	264.70	14DE	-0.4	-3.4	-2.1	0.9
351.66	-0.23	0.8	274.89	24DE	-0.6	-3.5	-2.1	0.9

Left columns (sections 520, 521):

♂ LONG	LAT	MAG	☉ LONG	16.00UT	☿	♀	♃	♄
358.19	-0.04	1.0	285.07	3JA	-0.9	-3.5	-2.1	0.9
4.74	0.13	1.1	295.22	13JA	-1.1	-3.5	-2.1	0.9
11.31	0.28	1.2	305.33	23JA	-1.0	-3.5	-2.1	0.9
17.86	0.41	1.3	315.40	2FE	-0.2	-3.4	-2.1	1.0
24.41	0.53	1.4	325.42	12FE	1.8	-3.4	-2.0	1.0
30.94	0.63	1.5	335.38	22FE	2.7	-3.4	-2.0	1.0
37.45	0.73	1.6	345.28	3MR	1.3	-3.4	-1.9	1.1
43.93	0.81	1.7	355.13	13MR	0.8	-3.3	-1.8	1.1
50.39	0.87	1.8	4.91	23MR	0.4	-3.3	-1.7	1.1
56.83	0.94	1.8	14.65	2AP	-0.0	-3.3	-1.7	1.1
63.24	0.99	1.9	24.33	12AP	-0.7	-3.3	-1.6	1.1
69.63	1.03	1.9	33.96	22AP	-1.6	-3.3	-1.6	1.0
76.00	1.07	2.0	43.56	2MY	-1.4	-3.4	-1.5	1.0
82.36	1.10	2.0	53.13	12MY	-0.5	-3.4	-1.4	0.9
88.70	1.12	2.0	62.67	22MY	0.3	-3.4	-1.4	0.9
95.03	1.14	2.0	72.20	1JN	1.0	-3.4	-1.4	0.8
101.36	1.16	2.0	81.72	11JN	1.7	-3.5	-1.3	0.8
107.69	1.16	2.0	91.25	21JN	2.9	-3.5	-1.3	0.7
114.02	1.17	2.0	100.79	1JL	2.6	-3.6	-1.3	0.6
120.36	1.17	2.0	110.36	11JL	0.9	-3.7	-1.3	0.6
126.71	1.16	2.0	119.95	21JL	-0.5	-3.8	-1.3	0.5
133.08	1.15	2.0	129.58	31JL	-1.3	-3.9	-1.3	0.5
139.46	1.14	1.9	139.26	10AU	-1.4	-4.0	-1.3	0.5
145.87	1.12	2.0	148.98	20AU	-0.7	-4.1	-1.3	0.6
152.31	1.09	2.0	158.76	30AU	-0.3	-4.2	-1.3	0.6
158.78	1.06	2.0	168.60	9SE	-0.0	-4.3	-1.4	0.7
165.28	1.03	2.0	178.49	19SE	0.1	-4.3	-1.4	0.7
171.82	0.99	2.0	188.45	29SE	0.3	-4.3	-1.4	0.8
178.40	0.94	2.0	198.46	9OC	1.2	-4.0	-1.5	0.8
185.01	0.89	1.9	208.52	19OC	2.6	-3.3	-1.5	0.9
191.67	0.84	1.9	218.63	29OC	0.1	-3.4	-1.6	0.9
198.37	0.77	1.9	228.77	8NO	-0.4	-4.0	-1.7	1.0
205.11	0.70	1.8	238.94	18NO	-0.5	-4.3	-1.7	1.0
211.90	0.62	1.8	249.14	28NO	-0.5	-4.4	-1.8	1.0
218.73	0.53	1.7	259.35	8DE	-0.7	-4.3	-1.9	1.0
225.60	0.44	1.7	269.54	18DE	-0.9	-4.2	-1.9	1.0
232.51	0.33	1.6	279.74	28DE	-0.9	-4.1	-2.0	1.0

521

♂ LONG	LAT	MAG	☉ LONG	16.00UT	☿	♀	♃	♄
239.47	0.21	1.5	289.90	7JA	-0.9	-4.0	-2.0	1.0
246.46	0.08	1.5	300.03	17JA	-0.1	-3.9	-2.0	1.0
253.49	-0.06	1.4	310.13	27JA	2.1	-3.8	-2.1	1.0
260.54	-0.21	1.3	320.17	6FE	2.1	-3.7	-2.1	1.1
267.63	-0.37	1.2	330.16	16FE	1.0	-3.7	-2.0	1.1
274.74	-0.55	1.1	340.10	26FE	0.5	-3.6	-2.0	1.1
281.87	-0.74	1.0	349.98	8MR	0.3	-3.5	-2.0	1.2
289.00	-0.94	0.9	359.79	18MR	-0.1	-3.5	-1.9	1.2
296.13	-1.16	0.8	9.56	28MR	-0.7	-3.4	-1.8	1.2
303.23	-1.38	0.6	19.27	7AP	-1.6	-3.4	-1.8	1.2
310.30	-1.61	0.5	28.92	17AP	-1.4	-3.3	-1.7	1.2
317.30	-1.85	0.4	38.54	27AP	-0.4	-3.3	-1.7	1.2
324.21	-2.09	0.3	48.12	7MY	0.5	-3.3	-1.6	1.2
331.00	-2.34	0.1	57.67	17MY	1.3	-3.3	-1.5	1.1
337.62	-2.59	0.0	67.21	27MY	2.3	-3.3	-1.5	1.1
344.02	-2.83	-0.1	76.74	6JN	3.4	-3.3	-1.4	1.0
350.14	-3.07	-0.3	86.26	16JN	2.1	-3.3	-1.4	1.0
355.89	-3.30	-0.4	95.79	26JN	0.8	-3.4	-1.3	0.9
1.18	-3.52	-0.6	105.34	6JL	-0.5	-3.4	-1.3	0.8
5.89	-3.73	-0.8	114.91	16JL	-1.4	-3.4	-1.3	0.8
9.85	-3.91	-1.0	124.53	26JL	-1.4	-3.5	-1.2	0.7
12.89	-4.06	-1.2	134.18	5AU	-0.7	-3.5	-1.2	0.6
14.79	-4.15	-1.4	143.87	15AU	-0.2	-3.5	-1.2	0.6
15.33	-4.16	-1.7	153.63	25AU	0.1	-3.5	-1.2	0.7
14.42	-4.04	-1.9	163.43	4SE	0.2	-3.4	-1.2	0.7
12.15	-3.76	-2.1	173.30	14SE	0.5	-3.4	-1.2	0.8
8.97	-3.30	-2.2	183.22	24SE	1.5	-3.4	-1.3	0.8
5.64	-2.70	-2.1	193.20	4OC	2.3	-3.3	-1.3	0.9
2.94	-2.04	-1.8	203.24	14OC	-0.0	-3.3	-1.3	0.9
1.41	-1.40	-1.5	213.32	24OC	-0.6	-3.3	-1.4	1.0
1.23	-0.83	-1.2	223.45	3NO	-0.6	-3.3	-1.4	1.0
2.31	-0.36	-0.8	233.61	13NO	-0.6	-3.3	-1.5	1.1
4.48	0.03	-0.5	243.80	23NO	-0.7	-3.4	-1.5	1.1
7.53	0.34	-0.2	253.99	3DE	-0.8	-3.4	-1.6	1.1
11.25	0.58	0.0	264.20	13DE	-0.8	-3.4	-1.7	1.1
15.52	0.78	0.3	274.40	23DE	-0.7	-3.4	-1.7	1.2

Right columns (sections 522, 523):

♂ LONG	LAT	MAG	☉ LONG	16.00UT	☿	♀	♃	♄
20.19	0.93	0.5	284.57	2JA	0.0	-3.5	-1.8	1.2
25.18	1.05	0.7	294.73	12JA	2.4	-3.5	-1.8	1.2
30.42	1.15	0.9	304.84	22JA	1.6	-3.6	-1.9	1.2
35.85	1.22	1.0	314.91	1FE	0.7	-3.6	-2.0	1.2
41.42	1.28	1.2	324.93	11FE	0.4	-3.7	-2.0	1.2
47.11	1.32	1.3	334.90	21FE	0.2	-3.7	-2.0	1.2
52.90	1.35	1.4	344.80	3MR	-0.2	-3.8	-2.0	1.3
58.75	1.37	1.5	354.66	13MR	-0.7	-3.9	-2.0	1.3
64.66	1.38	1.6	4.45	23MR	-1.6	-4.0	-2.0	1.3
70.62	1.39	1.7	14.18	2AP	-1.4	-4.1	-2.0	1.4
76.62	1.38	1.8	23.87	12AP	-0.4	-4.2	-1.9	1.4
82.65	1.38	1.8	33.50	22AP	0.6	-4.2	-1.9	1.4
88.71	1.36	1.9	43.10	2MY	1.7	-4.2	-1.8	1.4
94.81	1.35	1.9	52.67	12MY	3.0	-4.0	-1.7	1.4
100.93	1.32	2.0	62.21	22MY	3.0	-3.4	-1.7	1.3
107.08	1.30	2.0	71.74	1JN	1.6	-2.7	-1.6	1.3
113.26	1.27	2.0	81.26	11JN	0.6	-3.6	-1.5	1.2
119.48	1.23	2.0	90.79	21JN	-0.5	-4.0	-1.5	1.2
125.73	1.20	2.0	100.33	1JL	-1.5	-4.2	-1.4	1.1
132.02	1.15	2.0	109.89	11JL	-1.4	-4.2	-1.4	1.1
138.35	1.11	2.0	119.48	21JL	-0.6	-4.1	-1.3	1.0
144.72	1.06	2.0	129.11	31JL	-0.1	-4.1	-1.3	0.9
151.15	1.01	2.0	138.78	10AU	0.2	-4.0	-1.3	0.8
157.62	0.95	1.9	148.51	20AU	0.4	-3.9	-1.2	0.8
164.16	0.89	1.9	158.28	30AU	0.7	-3.8	-1.2	0.8
170.75	0.83	1.9	168.12	9SE	1.9	-3.7	-1.2	0.8
177.40	0.76	1.8	178.01	19SE	2.1	-3.7	-1.2	0.9
184.11	0.68	1.8	187.96	29SE	-0.2	-3.6	-1.2	0.9
190.89	0.60	1.8	197.97	9OC	-0.7	-3.6	-1.2	1.0
197.74	0.52	1.8	208.03	19OC	-0.8	-3.5	-1.2	1.0
204.65	0.43	1.8	218.13	29OC	-0.8	-3.5	-1.3	1.1
211.64	0.34	1.8	228.28	8NO	-0.7	-3.4	-1.3	1.1
218.69	0.24	1.8	238.46	18NO	-0.6	-3.4	-1.3	1.2
225.81	0.14	1.7	248.65	28NO	-0.6	-3.4	-1.4	1.2
233.01	0.03	1.7	258.85	8DE	-0.6	-3.4	-1.4	1.3
240.26	-0.08	1.7	269.06	18DE	0.1	-3.4	-1.5	1.3
247.58	-0.20	1.6	279.24	28DE	2.7	-3.3	-1.6	1.3

523

♂ LONG	LAT	MAG	☉ LONG	16.00UT	☿	♀	♃	♄
254.96	-0.32	1.6	289.41	7JA	1.2	-3.3	-1.6	1.3
262.39	-0.44	1.5	299.55	17JA	0.4	-3.3	-1.7	1.4
269.87	-0.57	1.5	309.64	27JA	0.2	-3.3	-1.8	1.4
277.40	-0.70	1.4	319.69	6FE	0.0	-3.4	-1.8	1.4
284.97	-0.83	1.4	329.68	16FE	-0.3	-3.4	-1.9	1.3
292.56	-0.96	1.3	339.62	26FE	-0.8	-3.4	-1.9	1.3
300.17	-1.09	1.3	349.50	8MR	-1.5	-3.4	-2.0	1.3
307.79	-1.21	1.2	359.32	18MR	-1.4	-3.5	-2.0	1.3
315.41	-1.32	1.2	9.08	28MR	-0.4	-3.5	-2.0	1.3
323.00	-1.43	1.1	18.79	7AP	0.8	-3.4	-2.0	1.3
330.57	-1.53	1.0	28.45	17AP	2.2	-3.4	-2.0	1.2
338.08	-1.61	1.0	38.07	27AP	3.9	-3.4	-2.0	1.2
345.53	-1.68	0.9	47.66	7MY	2.3	-3.4	-2.0	1.2
352.89	-1.74	0.9	57.21	17MY	1.3	-3.3	-1.9	1.2
0.14	-1.78	0.8	66.74	27MY	0.4	-3.3	-1.9	1.1
7.27	-1.81	0.7	76.27	6JN	-0.6	-3.3	-1.8	1.1
14.25	-1.81	0.7	85.79	16JN	-1.5	-3.3	-1.7	1.0
21.05	-1.80	0.6	95.32	26JN	-1.4	-3.3	-1.7	1.0
27.64	-1.76	0.5	104.87	6JL	-0.6	-3.3	-1.6	0.9
33.99	-1.71	0.4	114.45	16JL	-0.0	-3.4	-1.6	0.9
40.06	-1.64	0.4	124.05	26JL	0.3	-3.4	-1.5	0.8
45.78	-1.54	0.2	133.70	5AU	0.6	-3.4	-1.4	0.8
51.09	-1.41	0.1	143.40	15AU	1.0	-3.5	-1.4	0.7
55.91	-1.26	-0.0	153.15	25AU	2.4	-3.5	-1.4	0.7
60.11	-1.07	-0.2	162.95	4SE	1.8	-3.6	-1.3	0.7
63.57	-0.84	-0.3	172.81	14SE	-0.2	-3.7	-1.3	0.7
66.12	-0.57	-0.5	182.74	24SE	-0.9	-3.7	-1.3	0.8
67.53	-0.24	-0.7	192.72	4OC	-0.9	-3.8	-1.3	0.9
67.63	0.15	-1.0	202.75	14OC	-0.9	-3.9	-1.3	0.9
66.28	0.59	-1.2	212.83	24OC	-0.6	-4.0	-1.3	1.0
63.57	1.06	-1.4	222.96	3NO	-0.5	-4.1	-1.3	1.1
59.93	1.52	-1.5	233.11	13NO	-0.5	-4.3	-1.3	1.1
56.05	1.90	-1.4	243.30	23NO	-0.4	-4.3	-1.3	1.2
52.75	2.18	-1.2	253.50	3DE	0.3	-4.4	-1.3	1.2
50.57	2.35	-0.9	263.70	13DE	3.0	-4.3	-1.4	1.2
49.74	2.42	-0.6	273.90	23DE	0.9	-4.0	-1.4	1.2

♂ LONG	♂ LAT	♂ MAG	☉ LONG	16.00UT	☿	♀	♃	♄
				524	MAGNITUDES			
50.20	2.44	-0.3	284.09	2JA	0.2	-3.4	-1.5	1.2
51.78	2.41	0.0	294.24	12JA	0.1	-3.5	-1.5	1.2
54.26	2.36	0.3	304.35	22JA	-0.1	-4.0	-1.6	1.2
57.47	2.29	0.5	314.43	1FE	-0.3	-4.3	-1.6	1.1
61.25	2.22	0.7	324.45	11FE	-0.8	-4.3	-1.7	1.1
65.47	2.14	0.9	334.42	21FE	-1.5	-4.3	-1.8	1.0
70.05	2.06	1.1	344.33	2MR	-1.3	-4.2	-1.9	1.0
74.91	1.98	1.2	354.18	12MR	-0.3	-4.1	-1.9	1.0
79.99	1.90	1.3	3.97	22MR	1.1	-3.9	-2.0	1.0
85.27	1.83	1.4	13.71	1AP	2.8	-3.8	-2.0	1.0
90.69	1.75	1.5	23.39	11AP	3.0	-3.7	-2.1	1.0
96.24	1.67	1.6	33.04	21AP	1.7	-3.7	-2.1	1.0
101.91	1.59	1.7	42.64	1MY	1.0	-3.6	-2.1	1.0
107.67	1.52	1.8	52.21	11MY	0.3	-3.5	-2.1	0.9
113.52	1.44	1.8	61.75	21MY	-0.6	-3.5	-2.1	0.9
119.46	1.36	1.9	71.28	31MY	-1.5	-3.4	-2.1	0.9
125.47	1.28	1.9	80.80	10JN	-1.4	-3.4	-2.0	0.8
131.57	1.20	1.9	90.33	20JN	-0.6	-3.4	-2.0	0.8
137.74	1.12	1.9	99.87	30JN	0.1	-3.3	-1.9	0.8
143.98	1.03	1.9	109.43	10JL	0.5	-3.3	-1.9	0.7
150.30	0.95	1.9	119.02	20JL	0.8	-3.3	-1.8	0.7
156.69	0.86	1.9	128.65	30JL	1.4	-3.3	-1.7	0.6
163.17	0.78	1.9	138.31	9AU	2.9	-3.3	-1.7	0.5
169.72	0.69	1.9	148.04	19AU	1.6	-3.4	-1.6	0.5
176.36	0.59	1.9	157.81	29AU	-0.3	-3.4	-1.6	0.4
183.08	0.50	1.8	167.64	8SE	-1.0	-3.4	-1.5	0.4
189.89	0.40	1.8	177.53	18SE	-1.1	-3.4	-1.5	0.4
196.78	0.30	1.8	187.48	28SE	-0.9	-3.5	-1.4	0.5
203.77	0.20	1.7	197.48	8OC	-0.5	-3.5	-1.4	0.5
210.84	0.10	1.6	207.54	18OC	-0.4	-3.5	-1.4	0.6
217.99	-0.00	1.6	217.64	28OC	-0.3	-3.5	-1.4	0.7
225.23	-0.11	1.6	227.78	7NO	-0.2	-3.5	-1.3	0.7
232.56	-0.21	1.6	237.96	17NO	0.5	-3.4	-1.3	0.8
239.96	-0.32	1.6	248.15	27NO	3.2	-3.4	-1.3	0.8
247.44	-0.42	1.5	258.35	7DE	0.6	-3.4	-1.4	0.9
254.99	-0.53	1.5	268.56	17DE	0.0	-3.4	-1.4	0.9
262.60	-0.63	1.5	278.75	27DE	-0.1	-3.3	-1.4	0.9
				525				
270.27	-0.72	1.5	288.92	6JA	-0.2	-3.3	-1.4	0.9
277.99	-0.82	1.5	299.06	16JA	-0.4	-3.3	-1.5	0.9
285.75	-0.91	1.5	309.15	26JA	-0.8	-3.3	-1.5	0.9
293.53	-0.99	1.4	319.20	5FE	-1.3	-3.4	-1.6	0.9
301.33	-1.06	1.4	329.20	15FE	-1.2	-3.4	-1.6	0.9
309.14	-1.12	1.4	339.14	25FE	-0.3	-3.4	-1.7	0.8
316.93	-1.18	1.4	349.02	7MR	1.3	-3.4	-1.8	0.8
324.71	-1.22	1.4	358.85	17MR	3.4	-3.5	-1.8	0.7
332.46	-1.25	1.4	8.61	27MR	2.2	-3.5	-1.9	0.7
340.16	-1.26	1.3	18.33	6AP	1.3	-3.5	-2.0	0.7
347.80	-1.27	1.3	27.99	16AP	0.7	-3.6	-2.0	0.7
355.38	-1.26	1.3	37.61	26AP	0.2	-3.7	-2.1	0.7
2.87	-1.23	1.3	47.20	6MY	-0.6	-3.7	-2.2	0.7
10.28	-1.19	1.3	56.75	16MY	-1.5	-3.8	-2.2	0.7
17.58	-1.14	1.3	66.29	26MY	-1.4	-3.9	-2.2	0.7
24.77	-1.07	1.2	75.81	5JN	-0.5	-4.0	-2.2	0.7
31.84	-0.99	1.2	85.34	15JN	0.2	-4.1	-2.2	0.7
38.78	-0.90	1.2	94.87	25JN	0.6	-4.2	-2.2	0.6
45.58	-0.79	1.2	104.42	5JL	1.1	-4.2	-2.2	0.6
52.21	-0.67	1.1	113.99	15JL	1.9	-4.1	-2.1	0.5
58.67	-0.54	1.1	123.59	25JL	3.2	-3.9	-2.1	0.5
64.94	-0.40	1.0	133.24	4AU	1.3	-3.3	-2.0	0.4
70.99	-0.24	1.0	142.93	14AU	-0.4	-3.4	-2.0	0.4
76.79	-0.06	0.9	152.68	24AU	-1.1	-3.9	-1.9	0.3
82.31	0.13	0.8	162.48	3SE	-1.2	-4.2	-1.8	0.3
87.49	0.35	0.7	172.34	13SE	-0.8	-4.3	-1.8	0.2
92.28	0.59	0.6	182.25	23SE	-0.4	-4.3	-1.7	0.2
96.58	0.85	0.5	192.23	3OC	-0.2	-4.2	-1.7	0.1
100.28	1.15	0.3	202.26	13OC	-0.2	-4.1	-1.6	0.2
103.25	1.49	0.1	212.34	23OC	-0.1	-4.0	-1.6	0.3
105.32	1.87	-0.1	222.46	2NO	0.7	-3.9	-1.5	0.4
106.29	2.29	-0.3	232.62	12NO	3.1	-3.8	-1.5	0.4
106.00	2.74	-0.5	242.80	22NO	0.4	-3.8	-1.5	0.5
104.33	3.20	-0.8	253.00	2DE	-0.2	-3.7	-1.5	0.5
101.39	3.60	-1.0	263.21	12DE	-0.2	-3.6	-1.4	0.6
97.60	3.90	-1.1	273.41	22DE	-0.3	-3.6	-1.4	0.6

♂ LONG	♂ LAT	♂ MAG	☉ LONG	16.00UT	☿	♀	♃	♄
				526	MAGNITUDES			
93.65	4.04	-1.0	283.59	1JA	-0.5	-3.5	-1.4	0.7
90.28	4.02	-0.8	293.75	11JA	-0.9	-3.5	-1.5	0.7
88.02	3.87	-0.5	303.86	21JA	-1.2	-3.4	-1.5	0.7
87.07	3.65	-0.3	313.94	31JA	-1.1	-3.4	-1.5	0.7
87.39	3.40	-0.0	323.97	10FE	-0.2	-3.4	-1.5	0.7
88.80	3.15	0.2	333.94	20FE	1.6	-3.3	-1.6	0.7
91.12	2.90	0.5	343.85	2MR	3.2	-3.3	-1.6	0.6
94.18	2.67	0.7	353.71	12MR	1.6	-3.3	-1.7	0.6
97.82	2.46	0.8	3.50	22MR	0.9	-3.3	-1.7	0.5
101.92	2.26	1.0	13.24	1AP	0.5	-3.3	-1.8	0.5
106.41	2.07	1.1	22.93	11AP	0.1	-3.3	-1.9	0.5
111.21	1.90	1.3	32.57	21AP	-0.6	-3.4	-1.9	0.5
116.27	1.74	1.4	42.18	1MY	-1.5	-3.4	-2.0	0.5
121.55	1.59	1.4	51.75	11MY	-1.5	-3.4	-2.1	0.5
127.01	1.44	1.5	61.29	21MY	-0.5	-3.5	-2.2	0.5
132.64	1.30	1.6	70.82	31MY	0.3	-3.5	-2.2	0.5
138.43	1.17	1.6	80.34	10JN	0.8	-3.5	-2.3	0.5
144.35	1.04	1.7	89.87	20JN	1.4	-3.4	-2.3	0.5
150.41	0.91	1.7	99.41	30JN	2.4	-3.4	-2.3	0.5
156.58	0.78	1.7	108.96	10JL	3.0	-3.4	-2.4	0.4
162.88	0.66	1.7	118.55	20JL	1.1	-3.4	-2.4	0.4
169.29	0.54	1.8	128.18	30JL	-0.4	-3.3	-2.3	0.4
175.81	0.42	1.8	137.84	9AU	-1.2	-3.3	-2.3	0.3
182.45	0.31	1.7	147.56	19AU	-1.3	-3.3	-2.3	0.3
189.20	0.19	1.7	157.33	29AU	-0.7	-3.3	-2.2	0.2
196.06	0.08	1.7	167.16	8SE	-0.3	-3.3	-2.2	0.1
203.02	-0.03	1.7	177.04	18SE	-0.1	-3.4	-2.1	0.1
210.10	-0.14	1.7	186.99	28SE	-0.0	-3.4	-2.0	0.0
217.27	-0.25	1.6	196.99	8OC	0.1	-3.4	-2.0	-0.0
224.55	-0.36	1.6	207.04	18OC	1.0	-3.5	-1.9	-0.0
231.93	-0.46	1.6	217.15	28OC	2.9	-3.5	-1.8	0.0
239.40	-0.55	1.5	227.29	7NO	0.2	-3.6	-1.8	0.1
246.95	-0.64	1.5	237.46	17NO	-0.3	-3.6	-1.7	0.2
254.58	-0.73	1.4	247.65	27NO	-0.4	-3.7	-1.7	0.2
262.28	-0.81	1.4	257.85	7DE	-0.4	-3.8	-1.6	0.3
270.04	-0.88	1.4	268.06	17DE	-0.6	-3.9	-1.6	0.4
277.85	-0.94	1.3	278.25	27DE	-0.9	-3.9	-1.6	0.4
				527				
285.70	-0.99	1.3	288.42	6JA	-1.0	-4.0	-1.5	0.5
293.57	-1.04	1.3	298.56	16JA	-1.0	-4.1	-1.5	0.5
301.46	-1.07	1.4	308.66	26JA	-0.2	-4.2	-1.5	0.5
309.34	-1.09	1.4	318.71	5FE	1.9	-4.3	-1.5	0.5
317.22	-1.10	1.4	328.71	15FE	2.5	-4.3	-1.5	0.5
325.06	-1.10	1.4	338.66	25FE	1.2	-4.2	-1.6	0.5
332.87	-1.08	1.4	348.54	7MR	0.7	-3.8	-1.6	0.5
340.63	-1.05	1.4	358.37	17MR	0.4	-3.2	-1.6	0.5
348.33	-1.02	1.4	8.14	27MR	-0.0	-3.5	-1.7	0.4
355.96	-0.97	1.5	17.85	6AP	-0.7	-4.0	-1.7	0.4
3.52	-0.91	1.5	27.52	16AP	-1.5	-4.2	-1.8	0.3
10.99	-0.84	1.5	37.15	26AP	-1.5	-4.2	-1.8	0.3
18.37	-0.76	1.5	46.73	6MY	-0.5	-4.2	-1.9	0.4
25.66	-0.67	1.5	56.29	16MY	0.4	-4.1	-2.0	0.4
32.85	-0.57	1.5	65.83	26MY	1.1	-4.0	-2.0	0.4
39.93	-0.47	1.5	75.35	5JN	1.9	-3.9	-2.1	0.4
46.92	-0.36	1.6	84.88	15JN	3.1	-3.8	-2.2	0.4
53.79	-0.25	1.6	94.41	25JN	2.5	-3.7	-2.2	0.4
60.55	-0.13	1.6	103.95	5JL	0.9	-3.6	-2.3	0.4
67.20	-0.00	1.6	113.52	15JL	-0.4	-3.6	-2.3	0.4
73.72	0.13	1.5	123.13	25JL	-1.3	-3.5	-2.4	0.3
80.12	0.27	1.5	132.77	4AU	-1.4	-3.5	-2.4	0.3
86.38	0.41	1.5	142.46	14AU	-0.7	-3.4	-2.4	0.2
92.49	0.56	1.4	152.21	24AU	-0.2	-3.4	-2.4	0.2
98.44	0.72	1.4	162.00	3SE	-0.0	-3.4	-2.4	0.1
104.21	0.89	1.3	171.86	13SE	0.1	-3.4	-2.4	0.1
109.77	1.07	1.3	181.78	23SE	0.4	-3.4	-2.4	0.0
115.09	1.27	1.2	191.75	3OC	1.3	-3.4	-2.3	-0.1
120.13	1.48	1.1	201.78	13OC	2.6	-3.4	-2.2	-0.1
124.82	1.71	0.9	211.85	23OC	0.1	-3.4	-2.2	-0.2
129.10	1.96	0.8	221.97	2NO	-0.5	-3.4	-2.1	-0.2
132.87	2.24	0.6	232.13	12NO	-0.5	-3.4	-2.0	-0.1
136.02	2.54	0.4	242.31	22NO	-0.5	-3.4	-2.0	-0.0
138.39	2.88	0.2	252.50	2DE	-0.7	-3.4	-1.9	0.0
139.81	3.25	0.0	262.71	12DE	-0.8	-3.4	-1.8	0.1
140.09	3.63	-0.2	272.91	22DE	-0.9	-3.5	-1.8	0.2

♂ LONG	♂ LAT	♂ MAG	☉ LONG	16.00UT	☿	♀	♃	♄
				528	MAGNITUDES			
139.09	4.01	-0.5	283.09	1JA	-0.8	-3.5	-1.7	0.2
136.78	4.33	-0.7	293.25	11JA	-0.1	-3.5	-1.7	0.3
133.40	4.53	-0.9	303.37	21JA	2.2	-3.5	-1.6	0.3
129.49	4.58	-1.0	313.44	31JA	2.0	-3.4	-1.6	0.4
125.74	4.44	-0.8	323.48	10FE	0.9	-3.4	-1.6	0.4
122.82	4.16	-0.6	333.45	20FE	0.5	-3.4	-1.6	0.4
121.11	3.80	-0.4	343.36	1MR	0.2	-3.4	-1.6	0.4
120.70	3.41	-0.2	353.22	11MR	-0.1	-3.3	-1.6	0.4
121.49	3.03	0.1	3.02	21MR	-0.7	-3.3	-1.6	0.4
123.31	2.67	0.3	12.76	31MR	-1.5	-3.3	-1.6	0.3
125.98	2.33	0.5	22.46	10AP	-1.4	-3.3	-1.6	0.3
129.35	2.03	0.6	32.10	20AP	-0.5	-3.3	-1.6	0.3
133.27	1.76	0.8	41.71	30AP	0.5	-3.4	-1.7	0.2
137.66	1.51	0.9	51.28	10MY	1.4	-3.4	-1.7	0.2
142.44	1.28	1.0	60.83	20MY	2.5	-3.4	-1.8	0.3
147.53	1.07	1.1	70.36	30MY	3.4	-3.5	-1.8	0.3
152.91	0.87	1.2	79.88	9JN	1.9	-3.5	-1.9	0.3
158.52	0.69	1.3	89.41	19JN	0.7	-3.5	-2.0	0.3
164.36	0.52	1.3	98.94	29JN	-0.5	-3.6	-2.0	0.3
170.39	0.35	1.4	108.50	9JL	-1.4	-3.7	-2.1	0.3
176.61	0.20	1.4	118.09	19JL	-1.4	-3.8	-2.2	0.3
182.99	0.05	1.4	127.71	29JL	-0.7	-3.9	-2.2	0.3
189.54	-0.08	1.5	137.38	8AU	-0.2	-4.0	-2.3	0.3
196.23	-0.22	1.5	147.09	18AU	0.1	-4.1	-2.3	0.2
203.07	-0.34	1.5	156.86	28AU	0.3	-4.2	-2.4	0.2
210.05	-0.46	1.5	166.69	7SE	0.6	-4.3	-2.4	0.1
217.15	-0.56	1.5	176.57	17SE	1.6	-4.3	-2.4	0.1
224.38	-0.66	1.5	186.51	27SE	2.3	-4.3	-2.4	-0.0
231.73	-0.76	1.5	196.51	7OC	-0.1	-4.0	-2.4	-0.1
239.18	-0.84	1.5	206.56	17OC	-0.6	-3.3	-2.4	-0.1
246.73	-0.91	1.4	216.66	27OC	-0.7	-3.4	-2.4	-0.2
254.38	-0.97	1.4	226.80	6NO	-0.7	-4.1	-2.3	-0.3
262.09	-1.03	1.4	236.96	16NO	-0.8	-4.3	-2.2	-0.3
269.87	-1.07	1.4	247.16	26NO	-0.7	-4.4	-2.2	-0.2
277.71	-1.09	1.4	257.36	6DE	-0.7	-4.3	-2.1	-0.1
285.58	-1.11	1.4	267.56	16DE	-0.7	-4.2	-2.0	-0.0
293.47	-1.12	1.4	277.76	26DE	0.0	-4.1	-2.0	0.0
				529				
301.38	-1.11	1.4	287.93	5JA	2.5	-4.0	-1.9	0.1
309.27	-1.09	1.4	298.07	15JA	1.5	-3.9	-1.8	0.2
317.16	-1.06	1.4	308.17	25JA	0.6	-3.8	-1.8	0.2
325.01	-1.02	1.4	318.23	4FE	0.3	-3.7	-1.7	0.3
332.81	-0.96	1.4	328.23	14FE	0.1	-3.7	-1.7	0.3
340.56	-0.90	1.4	338.18	24FE	-0.2	-3.6	-1.6	0.3
348.25	-0.83	1.4	348.07	6MR	-0.7	-3.5	-1.6	0.3
355.87	-0.75	1.4	357.89	16MR	-1.5	-3.5	-1.6	0.3
3.41	-0.67	1.4	7.67	26MR	-1.4	-3.4	-1.6	0.3
10.87	-0.58	1.5	17.39	5AP	-0.4	-3.4	-1.5	0.3
18.24	-0.49	1.5	27.06	15AP	0.7	-3.3	-1.5	0.3
25.52	-0.39	1.6	36.68	25AP	1.8	-3.3	-1.5	0.2
32.72	-0.29	1.6	46.27	5MY	3.3	-3.3	-1.6	0.2
39.82	-0.18	1.7	55.83	15MY	2.7	-3.3	-1.6	0.2
46.84	-0.08	1.7	65.37	25MY	1.5	-3.3	-1.6	0.2
53.76	0.03	1.7	74.90	4JN	0.5	-3.3	-1.6	0.2
60.60	0.14	1.7	84.42	14JN	-0.5	-3.3	-1.7	0.3
67.35	0.25	1.8	93.95	24JN	-1.5	-3.4	-1.7	0.3
74.02	0.36	1.8	103.49	4JL	-1.4	-3.4	-1.8	0.3
80.60	0.47	1.8	113.06	14JL	-0.6	-3.4	-1.8	0.3
87.10	0.58	1.8	122.66	24JL	-0.1	-3.5	-1.9	0.3
93.51	0.69	1.8	132.30	3AU	0.2	-3.5	-2.0	0.3
99.83	0.81	1.8	141.99	13AU	0.5	-3.5	-2.0	0.3
106.06	0.92	1.8	151.73	23AU	0.8	-3.5	-2.1	0.3
112.20	1.04	1.7	161.53	2SE	2.0	-3.4	-2.2	0.2
118.23	1.16	1.7	171.38	12SE	2.1	-3.4	-2.2	0.2
124.16	1.28	1.7	181.29	22SE	-0.2	-3.4	-2.3	0.1
129.95	1.41	1.6	191.26	2OC	-0.8	-3.3	-2.3	0.1
135.60	1.54	1.5	201.28	12OC	-0.8	-3.3	-2.4	-0.0
141.09	1.69	1.4	211.36	22OC	-0.8	-3.3	-2.4	-0.1
146.38	1.83	1.3	221.47	1NO	-0.7	-3.3	-2.4	-0.2
151.44	1.99	1.2	231.63	11NO	-0.6	-3.3	-2.4	-0.2
156.22	2.16	1.1	241.81	21NO	-0.6	-3.4	-2.3	-0.3
160.65	2.34	0.9	252.01	1DE	-0.5	-3.4	-2.3	-0.3
164.67	2.54	0.8	262.21	11DE	0.2	-3.4	-2.2	-0.2
168.16	2.75	0.6	272.41	21DE	2.8	-3.4	-2.2	-0.1
170.99	2.98	0.4	282.60	31DE	1.1	-3.5	-2.1	-0.1

♂ LONG	♂ LAT	♂ MAG	☉ LONG	16.00UT	☿	♀	♃	♄
				530	MAGNITUDES			
173.03	3.22	0.1	292.75	10JA	0.3	-3.5	-2.0	0.0
174.07	3.45	-0.1	302.88	20JA	0.1	-3.6	-2.0	0.1
173.96	3.68	-0.4	312.96	30JA	-0.0	-3.6	-1.9	0.1
172.57	3.85	-0.7	322.99	9FE	-0.3	-3.7	-1.8	0.2
169.95	3.93	-1.0	332.97	19FE	-0.8	-3.7	-1.7	0.2
166.42	3.86	-1.2	342.89	1MR	-1.5	-3.8	-1.7	0.3
162.58	3.64	-1.1	352.75	11MR	-1.3	-3.9	-1.6	0.3
159.15	3.28	-1.0	2.55	21MR	-0.4	-4.0	-1.6	0.3
156.71	2.84	-0.8	12.30	31MR	0.9	-4.1	-1.6	0.3
155.55	2.38	-0.6	21.99	10AP	2.4	-4.2	-1.5	0.3
155.69	1.93	-0.4	31.64	20AP	3.6	-4.2	-1.5	0.3
157.02	1.51	-0.2	41.24	30AP	2.1	-4.2	-1.5	0.3
159.34	1.14	0.0	50.82	10MY	1.1	-3.9	-1.5	0.2
162.51	0.81	0.2	60.37	20MY	0.4	-3.4	-1.5	0.2
166.36	0.51	0.3	69.90	30MY	-0.5	-2.8	-1.5	0.2
170.78	0.25	0.5	79.42	9JN	-1.5	-3.6	-1.5	0.2
175.67	0.02	0.6	88.95	19JN	-1.4	-4.0	-1.5	0.3
180.97	-0.19	0.7	98.48	29JN	-0.6	-4.2	-1.5	0.3
186.60	-0.38	0.8	108.04	9JL	-0.0	-4.2	-1.5	0.3
192.54	-0.55	0.8	117.62	19JL	0.4	-4.1	-1.6	0.3
198.73	-0.70	0.9	127.24	29JL	0.7	-4.0	-1.6	0.3
205.17	-0.83	1.0	136.90	8AU	1.2	-4.0	-1.7	0.3
211.82	-0.95	1.0	146.62	18AU	2.5	-3.9	-1.7	0.3
218.66	-1.05	1.0	156.38	28AU	1.8	-3.8	-1.8	0.3
225.67	-1.14	1.1	166.21	7SE	-0.2	-3.7	-1.8	0.3
232.84	-1.21	1.1	176.09	17SE	-0.9	-3.7	-1.9	0.3
240.15	-1.26	1.1	186.02	27SE	-1.0	-3.6	-2.0	0.2
247.59	-1.31	1.2	196.02	7OC	-0.9	-3.6	-2.0	0.1
255.13	-1.33	1.2	206.07	17OC	-0.6	-3.5	-2.1	0.1
262.77	-1.34	1.2	216.17	27OC	-0.4	-3.5	-2.2	0.0
270.48	-1.34	1.3	226.30	6NO	-0.4	-3.4	-2.2	-0.1
278.25	-1.32	1.3	236.47	16NO	-0.3	-3.4	-2.2	-0.1
286.06	-1.29	1.3	246.66	26NO	0.3	-3.4	-2.3	-0.2
293.89	-1.24	1.3	256.87	6DE	3.0	-3.4	-2.3	-0.3
301.74	-1.19	1.4	267.07	16DE	0.8	-3.4	-2.2	-0.2
309.58	-1.12	1.4	277.26	26DE	0.1	-3.3	-2.2	-0.2
				531				
317.39	-1.04	1.4	287.44	5JA	-0.0	-3.3	-2.2	-0.1
325.18	-0.96	1.4	297.58	15JA	-0.1	-3.3	-2.1	-0.0
332.91	-0.87	1.5	307.68	25JA	-0.4	-3.3	-2.0	0.0
340.59	-0.77	1.5	317.74	4FE	-0.8	-3.4	-2.0	0.1
348.20	-0.67	1.5	327.75	14FE	-1.4	-3.4	-1.9	0.2
355.74	-0.57	1.5	337.69	24FE	-1.3	-3.4	-1.8	0.2
3.21	-0.47	1.6	347.58	6MR	-0.4	-3.4	-1.8	0.3
10.60	-0.36	1.6	357.42	16MR	1.1	-3.5	-1.7	0.3
17.90	-0.25	1.6	7.19	26MR	3.0	-3.5	-1.6	0.3
25.11	-0.15	1.6	16.91	5AP	2.7	-3.4	-1.6	0.3
32.25	-0.04	1.6	26.58	15AP	1.5	-3.4	-1.5	0.3
39.30	0.06	1.6	36.21	25AP	0.9	-3.4	-1.5	0.3
46.27	0.16	1.7	45.80	5MY	0.3	-3.4	-1.4	0.3
53.16	0.26	1.7	55.36	15MY	-0.6	-3.3	-1.4	0.3
59.98	0.36	1.8	64.90	25MY	-1.5	-3.3	-1.4	0.3
66.73	0.45	1.8	74.43	4JN	-1.5	-3.3	-1.4	0.3
73.41	0.54	1.8	83.95	14JN	-0.6	-3.3	-1.4	0.2
80.03	0.63	1.9	93.48	24JN	0.1	-3.3	-1.4	0.3
86.58	0.72	1.9	103.02	4JL	0.5	-3.3	-1.4	0.3
93.08	0.80	1.9	112.59	14JL	0.9	-3.4	-1.4	0.4
99.53	0.88	1.9	122.19	24JL	1.6	-3.4	-1.4	0.4
105.92	0.96	2.0	131.83	3AU	3.0	-3.4	-1.4	0.4
112.26	1.04	2.0	141.52	13AU	1.6	-3.5	-1.4	0.4
118.56	1.12	2.0	151.25	23AU	-0.3	-3.5	-1.5	0.4
124.80	1.19	1.9	161.05	2SE	-1.1	-3.6	-1.5	0.4
130.99	1.26	1.9	170.90	12SE	-1.1	-3.7	-1.6	0.4
137.13	1.33	1.9	180.81	22SE	-0.8	-3.7	-1.6	0.4
143.21	1.40	1.9	190.78	2OC	-0.5	-3.8	-1.7	0.3
149.23	1.47	1.8	200.80	12OC	-0.3	-3.9	-1.7	0.3
155.17	1.54	1.8	210.87	22OC	-0.3	-4.0	-1.8	0.2
161.04	1.61	1.7	220.99	1NO	-0.2	-4.2	-1.9	0.2
166.80	1.68	1.6	231.14	11NO	0.5	-4.3	-1.9	0.1
172.45	1.75	1.5	241.32	21NO	3.2	-4.4	-2.0	0.0
177.97	1.81	1.4	251.52	1DE	0.6	-4.4	-2.0	-0.1
183.32	1.88	1.3	261.72	11DE	-0.1	-4.3	-2.1	-0.1
188.47	1.95	1.1	271.92	21DE	-0.2	-4.0	-2.1	-0.2
193.38	2.01	1.0	282.10	31DE	-0.2	-3.4	-2.1	-0.1

♂			☉	16.00UT	☿	♀	♃	♄
LONG	LAT	MAG	LONG	532	MAGNITUDES			
198.00	2.07	0.8	292.26	10JA	-0.4	-3.5	-2.1	-0.1
202.25	2.12	0.6	302.39	20JA	-0.8	-4.0	-2.1	-0.0
206.04	2.17	0.4	312.47	30JA	-1.3	-4.3	-2.1	0.0
209.25	2.19	0.1	322.50	9FE	-1.2	-4.3	-2.0	0.1
211.75	2.20	-0.1	332.48	19FE	-0.3	-4.3	-2.0	0.2
213.35	2.17	-0.4	342.41	29FE	1.4	-4.2	-1.9	0.2
213.88	2.08	-0.7	352.27	10MR	3.4	-4.1	-1.9	0.3
213.20	1.93	-1.1	2.07	20MR	2.0	-3.9	-1.8	0.3
211.25	1.68	-1.4	11.83	30MR	1.1	-3.8	-1.7	0.4
208.26	1.32	-1.6	21.52	9AP	0.6	-3.7	-1.6	0.4
204.72	0.87	-1.7	31.17	19AP	0.1	-3.7	-1.6	0.4
201.35	0.38	-1.6	40.78	29AP	-0.6	-3.6	-1.5	0.4
198.82	-0.11	-1.4	50.35	9MY	-1.5	-3.5	-1.5	0.4
197.54	-0.55	-1.2	59.90	19MY	-1.5	-3.5	-1.4	0.4
197.63	-0.92	-1.0	69.44	29MY	-0.5	-3.4	-1.4	0.4
199.01	-1.23	-0.8	78.96	8JN	0.2	-3.4	-1.3	0.4
201.51	-1.48	-0.6	88.49	18JN	0.7	-3.4	-1.3	0.4
204.95	-1.68	-0.4	98.02	28JN	1.2	-3.3	-1.3	0.4
209.16	-1.83	-0.3	107.57	8JL	2.1	-3.3	-1.3	0.4
214.00	-1.94	-0.1	117.16	18JL	3.2	-3.3	-1.3	0.4
219.37	-2.02	0.0	126.78	28JL	1.3	-3.3	-1.3	0.5
225.18	-2.07	0.1	136.44	7AU	-0.4	-3.3	-1.3	0.5
231.35	-2.10	0.2	146.15	17AU	-1.2	-3.4	-1.3	0.5
237.84	-2.10	0.3	155.91	27AU	-1.3	-3.4	-1.3	0.5
244.57	-2.08	0.4	165.73	6SE	-0.8	-3.4	-1.3	0.6
251.53	-2.04	0.5	175.61	16SE	-0.4	-3.4	-1.4	0.5
258.66	-1.98	0.6	185.55	26SE	-0.2	-3.5	-1.4	0.5
265.94	-1.90	0.7	195.54	6OC	-0.1	-3.5	-1.4	0.5
273.34	-1.81	0.8	205.58	16OC	0.0	-3.5	-1.5	0.5
280.83	-1.71	0.8	215.68	26OC	0.8	-3.5	-1.6	0.4
288.38	-1.59	0.9	225.81	5NO	3.1	-3.5	-1.6	0.4
295.97	-1.47	1.0	235.98	15NO	0.4	-3.4	-1.7	0.3
303.59	-1.34	1.1	246.17	25NO	-0.2	-3.4	-1.7	0.2
311.20	-1.20	1.1	256.36	5DE	-0.3	-3.4	-1.8	0.2
318.81	-1.06	1.2	266.57	15DE	-0.4	-3.4	-1.9	0.1
326.39	-0.92	1.3	276.77	25DE	-0.5	-3.3	-1.9	0.0
				533				
333.92	-0.78	1.3	286.94	4JA	-0.9	-3.3	-2.0	-0.0
341.41	-0.65	1.4	297.08	14JA	-1.1	-3.3	-2.0	-0.0
348.84	-0.51	1.5	307.19	24JA	-1.1	-3.3	-2.0	0.1
356.21	-0.38	1.5	317.25	3FE	-0.3	-3.4	-2.0	0.1
3.50	-0.25	1.6	327.26	13FE	1.6	-3.4	-2.0	0.2
10.73	-0.13	1.6	337.21	23FE	2.9	-3.4	-2.0	0.2
17.87	-0.01	1.7	347.10	5MR	1.5	-3.4	-2.0	0.3
24.95	0.10	1.7	356.94	15MR	0.8	-3.5	-1.9	0.4
31.95	0.21	1.8	6.72	25MR	0.5	-3.5	-1.9	0.4
38.88	0.31	1.8	16.44	4AP	0.0	-3.5	-1.8	0.4
45.74	0.40	1.8	26.12	14AP	-0.6	-3.6	-1.8	0.5
52.53	0.49	1.8	35.75	24AP	-1.5	-3.7	-1.7	0.5
59.26	0.58	1.9	45.34	4MY	-1.5	-3.7	-1.6	0.5
65.94	0.65	1.9	54.90	14MY	-0.5	-3.8	-1.6	0.6
72.56	0.73	1.9	64.44	24MY	0.3	-3.9	-1.5	0.6
79.13	0.80	1.9	73.97	3JN	0.9	-4.0	-1.4	0.6
85.66	0.86	1.9	83.49	13JN	1.6	-4.1	-1.4	0.6
92.14	0.92	1.9	93.02	23JN	2.7	-4.2	-1.3	0.6
98.60	0.97	1.9	102.56	3JL	2.8	-4.2	-1.3	0.6
105.02	1.02	2.0	112.13	13JL	1.1	-4.1	-1.3	0.5
111.42	1.07	2.0	121.73	23JL	-0.4	-3.8	-1.3	0.6
117.79	1.11	2.0	131.37	2AU	-1.3	-3.3	-1.2	0.6
124.15	1.15	2.0	141.05	12AU	-1.4	-3.4	-1.2	0.7
130.48	1.19	2.0	150.78	22AU	-0.7	-3.9	-1.2	0.7
136.81	1.22	2.0	160.57	1SE	-0.3	-4.2	-1.2	0.7
143.12	1.24	2.0	170.42	11SE	-0.1	-4.3	-1.2	0.7
149.42	1.27	2.0	180.33	21SE	0.0	-4.3	-1.2	0.7
155.71	1.29	2.0	190.29	1OC	0.2	-4.2	-1.3	0.7
161.99	1.30	2.0	200.31	11OC	1.0	-4.1	-1.3	0.7
168.26	1.31	1.9	210.38	21OC	2.9	-4.0	-1.3	0.7
174.51	1.32	1.9	220.49	31OC	0.2	-3.9	-1.4	0.7
180.74	1.32	1.8	230.65	10NO	-0.4	-3.8	-1.4	0.6
186.94	1.32	1.7	240.82	20NO	-0.4	-3.8	-1.5	0.6
193.12	1.30	1.7	251.02	30NO	-0.5	-3.7	-1.5	0.5
199.25	1.28	1.6	261.23	10DE	-0.6	-3.6	-1.6	0.4
205.34	1.25	1.5	271.43	20DE	-0.9	-3.6	-1.7	0.4
211.38	1.21	1.3	281.61	30DE	-1.0	-3.5	-1.7	0.3

♂			☉	16.00UT	☿	♀	♃	♄
LONG	LAT	MAG	LONG	534	MAGNITUDES			
217.34	1.16	1.2	291.78	9JA	-0.9	-3.5	-1.8	0.2
223.22	1.09	1.1	301.90	19JA	-0.2	-3.4	-1.9	0.2
228.99	1.01	0.9	311.99	29JA	1.9	-3.4	-1.9	0.2
234.63	0.90	0.8	322.02	8FE	2.3	-3.4	-2.0	0.3
240.10	0.76	0.6	332.00	18FE	1.1	-3.3	-2.0	0.3
245.37	0.60	0.4	341.93	28FE	0.6	-3.3	-2.0	0.4
250.38	0.39	0.2	351.80	10MR	0.3	-3.3	-2.0	0.4
255.06	0.14	-0.1	1.60	20MR	-0.1	-3.3	-2.0	0.5
259.34	-0.17	-0.3	11.36	30MR	-0.6	-3.3	-2.0	0.5
263.08	-0.55	-0.6	21.06	9AP	-1.5	-3.3	-1.9	0.6
266.15	-1.00	-0.9	30.71	19AP	-1.5	-3.4	-1.9	0.6
268.39	-1.55	-1.2	40.32	29AP	-0.5	-3.4	-1.8	0.7
269.58	-2.18	-1.5	49.90	9MY	0.4	-3.4	-1.8	0.7
269.61	-2.90	-1.8	59.44	19MY	1.2	-3.5	-1.7	0.7
268.39	-3.65	-2.1	68.98	29MY	2.1	-3.5	-1.6	0.8
266.12	-4.37	-2.4	78.50	8JN	3.4	-3.5	-1.6	0.8
263.32	-4.94	-2.5	88.02	18JN	2.3	-3.4	-1.5	0.8
260.66	-5.29	-2.3	97.56	28JN	0.9	-3.4	-1.5	0.8
258.86	-5.40	-2.1	107.11	8JL	-0.4	-3.4	-1.4	0.8
258.31	-5.31	-1.9	116.69	18JL	-1.3	-3.4	-1.4	0.8
259.12	-5.08	-1.7	126.31	28JL	-1.4	-3.3	-1.3	0.8
261.22	-4.76	-1.4	135.96	7AU	-0.7	-3.3	-1.3	0.8
264.41	-4.41	-1.2	145.67	17AU	-0.2	-3.3	-1.3	0.9
268.48	-4.04	-0.9	155.43	27AU	0.0	-3.3	-1.2	0.9
273.27	-3.66	-0.7	165.25	6SE	0.2	-3.3	-1.2	0.9
278.61	-3.29	-0.5	175.12	16SE	0.4	-3.4	-1.2	1.0
284.39	-2.93	-0.3	185.06	26SE	1.4	-3.4	-1.2	1.0
290.50	-2.59	-0.2	195.05	6OC	2.6	-3.4	-1.2	1.0
296.87	-2.26	0.0	205.09	16OC	0.1	-3.5	-1.2	1.0
303.43	-1.95	0.2	215.19	26OC	-0.6	-3.5	-1.3	1.0
310.14	-1.66	0.3	225.32	5NO	-0.6	-3.6	-1.3	0.9
316.95	-1.38	0.5	235.48	15NO	-0.6	-3.6	-1.3	0.9
323.83	-1.13	0.6	245.67	25NO	-0.7	-3.7	-1.4	0.9
330.75	-0.89	0.8	255.88	5DE	-0.8	-3.8	-1.4	0.8
337.69	-0.68	0.9	266.08	15DE	-0.8	-3.9	-1.5	0.7
344.63	-0.48	1.0	276.28	25DE	-0.8	-3.9	-1.5	0.7
				535				
351.56	-0.29	1.1	286.45	4JA	-0.1	-4.0	-1.6	0.6
358.47	-0.12	1.2	296.59	14JA	2.2	-4.1	-1.6	0.5
5.36	0.03	1.3	306.71	24JA	1.8	-4.2	-1.7	0.5
12.21	0.17	1.4	316.77	3FE	0.7	-4.3	-1.8	0.4
19.01	0.30	1.5	326.78	13FE	0.4	-4.3	-1.8	0.5
25.78	0.41	1.6	336.73	23FE	0.2	-4.2	-1.9	0.5
32.50	0.51	1.7	346.63	5MR	-0.1	-3.8	-2.0	0.6
39.17	0.61	1.7	356.47	15MR	-0.7	-3.2	-2.0	0.6
45.80	0.69	1.8	6.25	25MR	-1.5	-3.5	-2.0	0.7
52.39	0.77	1.8	15.98	4AP	-1.5	-4.0	-2.0	0.7
58.94	0.83	1.9	25.65	14AP	-0.5	-4.2	-2.0	0.8
65.45	0.89	1.9	35.29	24AP	0.6	-4.2	-2.0	0.8
71.93	0.94	2.0	44.88	4MY	1.5	-4.2	-2.0	0.9
78.37	0.99	2.0	54.44	14MY	2.8	-4.1	-2.0	0.9
84.79	1.03	2.0	63.98	24MY	3.2	-4.0	-1.9	0.9
91.19	1.06	2.0	73.51	3JN	1.8	-3.9	-1.9	1.0
97.58	1.09	2.0	83.03	13JN	0.7	-3.8	-1.8	1.0
103.95	1.11	2.0	92.56	23JN	-0.5	-3.7	-1.7	1.0
110.31	1.13	2.0	102.10	3JL	-1.4	-3.6	-1.7	1.0
116.67	1.15	2.0	111.67	13JL	-1.4	-3.6	-1.6	1.0
123.03	1.15	2.0	121.26	23JL	-0.7	-3.5	-1.5	1.1
129.40	1.16	2.0	130.90	2AU	-0.1	-3.5	-1.5	1.0
135.77	1.16	2.0	140.58	12AU	0.2	-3.4	-1.4	1.0
142.16	1.15	2.0	150.31	22AU	0.4	-3.4	-1.4	1.1
148.57	1.14	2.0	160.10	1SE	0.7	-3.4	-1.4	1.1
155.00	1.13	2.0	169.94	11SE	1.7	-3.4	-1.3	1.2
161.44	1.11	2.0	179.85	21SE	2.3	-3.4	-1.3	1.2
167.92	1.08	2.0	189.81	1OC	-0.1	-3.4	-1.3	1.3
174.42	1.05	2.0	199.82	11OC	-0.7	-3.4	-1.3	1.3
180.94	1.02	1.9	209.89	21OC	-0.7	-3.4	-1.3	1.3
187.50	0.98	1.9	220.00	31OC	-0.7	-3.4	-1.3	1.3
194.08	0.93	1.9	230.15	10NO	-0.8	-3.4	-1.3	1.3
200.69	0.87	1.8	240.33	20NO	-0.6	-3.4	-1.3	1.2
207.33	0.80	1.8	250.53	30NO	-0.7	-3.4	-1.3	1.2
214.00	0.73	1.7	260.73	10DE	-0.6	-3.4	-1.4	1.1
220.69	0.64	1.6	270.93	20DE	0.0	-3.5	-1.4	1.1
227.40	0.55	1.6	281.12	30DE	2.5	-3.5	-1.4	1.0

♂			⊙	16.00UT	☿	♀	♃	♄
LONG	LAT	MAG	LONG	536	MAGNITUDES			
234.15	0.44	1.5	291.28	9JA	1.4	-3.5	-1.5	1.0
240.90	0.32	1.4	301.41	19JA	0.5	-3.4	-1.5	0.9
247.68	0.18	1.3	311.50	29JA	0.2	-3.4	-1.6	0.8
254.46	0.03	1.2	321.54	8FE	0.1	-3.4	-1.7	0.7
261.25	-0.14	1.1	331.52	18FE	-0.2	-3.4	-1.7	0.7
268.05	-0.33	0.9	341.45	28FE	-0.7	-3.4	-1.8	0.8
274.83	-0.54	0.8	351.32	9MR	-1.5	-3.3	-1.9	0.8
281.58	-0.77	0.7	1.13	19MR	-1.4	-3.3	-1.9	0.8
288.31	-1.02	0.5	10.88	29MR	-0.5	-3.3	-2.0	0.9
294.97	-1.30	0.4	20.59	8AP	0.7	-3.3	-2.1	0.9
301.55	-1.59	0.2	30.24	18AP	2.0	-3.3	-2.1	1.0
308.01	-1.92	0.1	39.85	28AP	3.7	-3.4	-2.1	1.1
314.31	-2.26	-0.1	49.43	8MY	2.5	-3.4	-2.1	1.1
320.39	-2.63	-0.3	58.98	18MY	1.4	-3.4	-2.1	1.2
326.18	-3.02	-0.5	68.51	28MY	0.5	-3.5	-2.1	1.2
331.60	-3.43	-0.7	78.04	7JN	-0.5	-3.5	-2.1	1.2
336.53	-3.85	-0.9	87.57	17JN	-1.5	-3.6	-2.0	1.3
340.83	-4.29	-1.1	97.10	27JN	-1.5	-3.6	-2.0	1.3
344.31	-4.74	-1.3	106.65	7JL	-0.6	-3.7	-1.9	1.3
346.77	-5.16	-1.6	116.23	17JL	-0.1	-3.8	-1.9	1.3
348.01	-5.53	-1.8	125.84	27JL	0.3	-3.9	-1.8	1.3
347.86	-5.80	-2.1	135.50	6AU	0.5	-4.0	-1.7	1.4
346.34	-5.87	-2.3	145.20	16AU	1.0	-4.1	-1.7	1.3
343.77	-5.70	-2.5	154.96	26AU	2.2	-4.2	-1.6	1.3
340.78	-5.26	-2.5	164.78	5SE	2.1	-4.3	-1.6	1.3
338.19	-4.61	-2.2	174.65	15SE	-0.2	-4.3	-1.5	1.3
336.60	-3.85	-1.9	184.57	25SE	-0.8	-4.2	-1.5	1.3
336.31	-3.09	-1.6	194.56	5OC	-0.9	-3.9	-1.4	1.3
337.31	-2.38	-1.3	204.60	15OC	-0.9	-3.3	-1.4	1.3
339.45	-1.77	-0.9	214.69	25OC	-0.6	-3.5	-1.4	1.3
342.52	-1.25	-0.6	224.83	4NO	-0.5	-4.1	-1.4	1.2
346.32	-0.81	-0.4	234.99	14NO	-0.5	-4.3	-1.4	1.2
350.70	-0.44	-0.1	245.18	24NO	-0.4	-4.4	-1.4	1.2
355.52	-0.13	0.1	255.38	4DE	0.2	-4.3	-1.4	1.1
0.68	0.12	0.3	265.58	14DE	2.8	-4.2	-1.4	1.1
6.09	0.34	0.5	275.78	24DE	1.0	-4.1	-1.4	1.0
				537				
11.70	0.51	0.7	285.96	3JA	0.3	-4.0	-1.4	1.0
17.47	0.66	0.9	296.11	13JA	0.1	-3.9	-1.5	0.9
23.35	0.79	1.0	306.22	23JA	-0.1	-3.8	-1.5	0.9
29.31	0.89	1.2	316.28	2FE	-0.3	-3.7	-1.5	0.8
35.34	0.97	1.3	326.30	12FE	-0.8	-3.7	-1.6	0.8
41.42	1.05	1.4	336.25	22FE	-1.4	-3.6	-1.7	0.8
47.53	1.10	1.5	346.16	4MR	-1.3	-3.5	-1.7	0.8
53.66	1.15	1.6	356.00	14MR	-0.4	-3.5	-1.8	0.9
59.81	1.19	1.7	5.78	24MR	0.9	-3.4	-1.9	0.9
65.98	1.21	1.8	15.51	3AP	2.6	-3.4	-1.9	1.0
72.16	1.24	1.8	25.19	13AP	3.2	-3.3	-2.0	1.1
78.34	1.25	1.9	34.82	23AP	1.8	-3.3	-2.1	1.1
84.53	1.26	1.9	44.42	3MY	1.0	-3.3	-2.1	1.2
90.74	1.26	2.0	53.98	13MY	0.4	-3.3	-2.2	1.2
96.95	1.25	2.0	63.53	23MY	-0.5	-3.3	-2.2	1.3
103.18	1.25	2.0	73.05	2JN	-1.5	-3.3	-2.2	1.3
109.42	1.23	2.0	82.57	12JN	-1.5	-3.3	-2.3	1.3
115.68	1.22	2.0	92.10	22JN	-0.6	-3.4	-2.3	1.3
121.96	1.19	2.0	101.65	2JL	0.0	-3.4	-2.2	1.3
128.27	1.17	2.0	111.21	12JL	0.4	-3.4	-2.2	1.3
134.61	1.14	2.0	120.80	22JL	0.7	-3.5	-2.2	1.3
140.99	1.10	2.0	130.44	1AU	1.3	-3.5	-2.1	1.3
147.40	1.06	2.0	140.11	11AU	2.7	-3.5	-2.0	1.2
153.85	1.02	1.9	149.84	21AU	1.8	-3.5	-2.0	1.2
160.35	0.97	1.9	159.62	31AU	-0.2	-3.4	-1.9	1.1
166.89	0.92	1.9	169.46	10SE	-1.0	-3.4	-1.8	1.1
173.49	0.87	1.9	179.36	20SE	-1.0	-3.4	-1.8	1.1
180.14	0.80	1.9	189.32	30SE	-0.9	-3.3	-1.7	1.1
186.84	0.74	1.9	199.33	10OC	-0.5	-3.3	-1.7	1.1
193.61	0.66	1.9	209.39	20OC	-0.4	-3.3	-1.6	1.1
200.43	0.59	1.9	219.51	30OC	-0.3	-3.3	-1.6	1.1
207.31	0.50	1.8	229.65	9NO	-0.3	-3.3	-1.5	1.1
214.25	0.41	1.8	239.83	19NO	0.4	-3.4	-1.5	1.1
221.26	0.32	1.8	250.03	29NO	3.1	-3.4	-1.5	1.0
228.32	0.21	1.7	260.23	9DE	0.8	-3.4	-1.5	1.0
235.44	0.10	1.7	270.43	19DE	0.1	-3.4	-1.5	1.0
242.62	-0.02	1.6	280.62	29DE	-0.1	-3.5	-1.5	0.9

♂			⊙	16.00UT	☿	♀	♃	♄
LONG	LAT	MAG	LONG	538	MAGNITUDES			
249.86	-0.14	1.6	290.78	8JA	-0.2	-3.5	-1.5	0.9
257.15	-0.27	1.5	300.92	18JA	-0.4	-3.6	-1.5	0.8
264.48	-0.41	1.4	311.01	28JA	-0.8	-3.6	-1.5	0.8
271.86	-0.55	1.4	321.05	7FE	-1.4	-3.7	-1.5	0.7
279.29	-0.70	1.3	331.04	17FE	-1.2	-3.8	-1.6	0.7
286.74	-0.85	1.2	340.97	27FE	-0.4	-3.8	-1.6	0.6
294.21	-1.00	1.2	350.84	9MR	1.2	-3.9	-1.7	0.6
301.70	-1.16	1.1	0.66	19MR	3.2	-4.0	-1.7	0.6
309.18	-1.31	1.0	10.42	29MR	2.4	-4.1	-1.8	0.7
316.65	-1.46	0.9	20.12	8AP	1.4	-4.2	-1.8	0.8
324.10	-1.60	0.9	29.78	18AP	0.8	-4.2	-1.9	0.8
331.49	-1.74	0.8	39.39	28AP	0.2	-4.2	-2.0	0.9
338.82	-1.86	0.7	48.97	8MY	-0.5	-3.9	-2.1	1.0
346.05	-1.98	0.6	58.53	18MY	-1.5	-3.4	-2.1	1.0
353.17	-2.08	0.5	68.06	28MY	-1.5	-2.8	-2.2	1.1
0.14	-2.16	0.5	77.58	7JN	-0.6	-3.6	-2.3	1.1
6.93	-2.23	0.4	87.11	17JN	0.1	-4.1	-2.3	1.1
13.50	-2.28	0.3	96.64	27JN	0.6	-4.2	-2.3	1.2
19.81	-2.31	0.2	106.19	7JL	1.0	-4.2	-2.4	1.2
25.79	-2.31	0.0	115.77	17JL	1.7	-4.1	-2.4	1.2
31.38	-2.29	-0.1	125.38	27JL	3.1	-4.0	-2.4	1.2
36.49	-2.25	-0.2	135.03	6AU	1.5	-4.0	-2.4	1.1
41.01	-2.17	-0.4	144.73	16AU	-0.3	-3.9	-2.3	1.1
44.79	-2.05	-0.6	154.49	26AU	-1.1	-3.8	-2.3	1.1
47.67	-1.89	-0.8	164.30	5SE	-1.2	-3.7	-2.2	1.0
49.44	-1.68	-1.0	174.17	15SE	-0.8	-3.7	-2.2	1.0
49.89	-1.39	-1.2	184.09	25SE	-0.5	-3.6	-2.1	1.0
48.88	-1.03	-1.4	194.08	5OC	-0.3	-3.6	-2.0	1.0
46.47	-0.59	-1.6	204.12	15OC	-0.2	-3.5	-2.0	1.0
43.03	-0.10	-1.7	214.20	25OC	-0.1	-3.5	-1.9	1.0
39.26	0.39	-1.7	224.33	4NO	0.6	-3.4	-1.8	1.0
35.98	0.83	-1.4	234.49	14NO	3.3	-3.4	-1.8	1.0
33.80	1.18	-1.1	244.68	24NO	0.5	-3.4	-1.7	1.0
32.96	1.43	-0.8	254.88	4DE	-0.1	-3.4	-1.7	0.9
33.43	1.61	-0.5	265.09	14DE	-0.2	-3.4	-1.6	0.9
35.05	1.72	-0.2	275.28	24DE	-0.3	-3.3	-1.6	0.9
				539				
37.59	1.79	0.1	285.46	3JA	-0.5	-3.3	-1.6	0.8
40.87	1.83	0.3	295.61	13JA	-0.8	-3.3	-1.6	0.8
44.73	1.84	0.6	305.72	23JA	-1.2	-3.4	-1.5	0.7
49.03	1.84	0.8	315.79	2FE	-1.1	-3.4	-1.5	0.7
53.68	1.82	0.9	325.81	12FE	-0.3	-3.4	-1.6	0.6
58.62	1.80	1.1	335.77	22FE	1.4	-3.4	-1.6	0.6
63.76	1.77	1.2	345.67	4MR	3.3	-3.4	-1.6	0.6
69.09	1.73	1.4	355.52	14MR	1.8	-3.5	-1.6	0.5
74.56	1.69	1.5	5.31	24MR	1.0	-3.5	-1.6	0.5
80.14	1.65	1.6	15.04	3AP	0.6	-3.4	-1.7	0.5
85.83	1.61	1.7	24.72	13AP	0.1	-3.4	-1.7	0.6
91.60	1.56	1.7	34.35	23AP	-0.6	-3.4	-1.8	0.7
97.44	1.51	1.8	43.95	3MY	-1.5	-3.4	-1.8	0.7
103.35	1.45	1.9	53.52	13MY	-1.5	-3.3	-1.9	0.8
109.33	1.40	1.9	63.06	23MY	-0.6	-3.3	-2.0	0.8
115.36	1.34	1.9	72.59	2JN	0.2	-3.3	-2.0	0.9
121.45	1.28	2.0	82.11	12JN	0.7	-3.3	-2.1	0.9
127.60	1.22	2.0	91.64	22JN	1.3	-3.3	-2.2	1.0
133.81	1.15	2.0	101.18	2JL	2.3	-3.3	-2.3	1.0
140.08	1.09	2.0	110.74	12JL	3.1	-3.4	-2.3	1.0
146.40	1.02	2.0	120.33	22JL	1.3	-3.4	-2.4	1.0
152.80	0.95	2.0	129.96	1AU	-0.3	-3.4	-2.4	1.0
159.26	0.87	1.9	139.64	11AU	-1.2	-3.5	-2.4	1.0
165.79	0.79	1.9	149.36	21AU	-1.3	-3.5	-2.5	1.0
172.39	0.71	1.9	159.15	31AU	-0.8	-3.6	-2.5	1.0
179.07	0.63	1.9	168.99	10SE	-0.4	-3.7	-2.4	0.9
185.82	0.54	1.8	178.88	20SE	-0.1	-3.7	-2.4	0.9
192.65	0.45	1.8	188.84	30SE	-0.0	-3.8	-2.4	0.9
199.56	0.36	1.7	198.85	10OC	0.1	-3.9	-2.3	0.9
206.55	0.26	1.7	208.91	20OC	0.8	-4.0	-2.2	0.9
213.62	0.16	1.7	219.02	30OC	3.2	-4.2	-2.2	0.9
220.77	0.06	1.7	229.16	9NO	0.3	-4.3	-2.1	0.9
228.00	-0.05	1.7	239.33	19NO	-0.3	-4.4	-2.0	0.9
235.31	-0.15	1.6	249.53	29NO	-0.4	-4.4	-1.9	0.9
242.69	-0.26	1.6	259.73	9DE	-0.4	-4.3	-1.9	0.9
250.13	-0.37	1.6	269.93	19DE	-0.6	-4.0	-1.8	0.9
257.65	-0.48	1.6	280.12	29DE	-0.9	-3.3	-1.8	0.8

♂			⊙	16.00UT	☿	♀	♃	♄
LONG	LAT	MAG	LONG	540	MAGNITUDES			
265.22	-0.59	1.5	290.29	8JA	-1.1	-3.5	-1.7	0.8
272.85	-0.70	1.5	300.42	18JA	-1.0	-4.1	-1.7	0.8
280.52	-0.81	1.5	310.51	28JA	-0.3	-4.3	-1.6	0.7
288.22	-0.91	1.4	320.56	7FE	1.7	-4.3	-1.6	0.7
295.95	-1.01	1.4	330.55	17FE	2.7	-4.3	-1.6	0.6
303.70	-1.09	1.4	340.49	27FE	1.3	-4.2	-1.6	0.6
311.45	-1.18	1.3	350.36	8MR	0.7	-4.1	-1.6	0.5
319.19	-1.25	1.3	0.18	18MR	0.4	-3.9	-1.6	0.5
326.91	-1.31	1.3	9.94	28MR	0.0	-3.8	-1.6	0.4
334.60	-1.35	1.3	19.65	7AP	-0.6	-3.7	-1.6	0.4
342.23	-1.39	1.2	29.31	17AP	-1.5	-3.7	-1.6	0.4
349.81	-1.41	1.2	38.93	27AP	-1.5	-3.6	-1.7	0.5
357.30	-1.41	1.2	48.51	7MY	-0.5	-3.5	-1.7	0.6
4.71	-1.40	1.1	58.06	17MY	0.3	-3.5	-1.7	0.6
12.02	-1.37	1.1	67.60	27MY	1.0	-3.4	-1.8	0.7
19.21	-1.33	1.1	77.12	6JN	1.7	-3.4	-1.8	0.8
26.27	-1.27	1.0	86.65	16JN	2.9	-3.4	-1.9	0.8
33.18	-1.20	1.0	96.18	26JN	2.6	-3.3	-2.0	0.9
39.92	-1.11	1.0	105.73	6JL	1.0	-3.3	-2.0	0.9
46.49	-1.00	0.9	115.30	16JL	-0.4	-3.3	-2.1	0.9
52.84	-0.88	0.8	124.92	26JL	-1.3	-3.3	-2.2	0.9
58.95	-0.74	0.8	134.57	5AU	-1.4	-3.3	-2.2	0.9
64.80	-0.58	0.7	144.26	15AU	-0.7	-3.4	-2.3	0.9
70.32	-0.41	0.6	154.02	25AU	-0.3	-3.4	-2.4	0.9
75.47	-0.20	0.5	163.82	4SE	-0.0	-3.4	-2.4	0.9
80.17	0.03	0.4	173.69	14SE	0.1	-3.4	-2.4	0.9
84.31	0.29	0.2	183.61	24SE	0.3	-3.5	-2.4	0.8
87.78	0.59	0.1	193.59	4OC	1.1	-3.5	-2.4	0.8
90.41	0.93	-0.1	203.63	14OC	2.9	-3.5	-2.4	0.8
92.02	1.32	-0.3	213.71	24OC	0.2	-3.5	-2.4	0.8
92.42	1.75	-0.6	223.84	3NO	-0.5	-3.4	-2.3	0.8
91.44	2.21	-0.8	234.00	13NO	-0.5	-3.4	-2.3	0.8
89.08	2.67	-1.0	244.18	23NO	-0.5	-3.4	-2.2	0.8
85.63	3.08	-1.2	254.38	3DE	-0.7	-3.4	-2.2	0.8
81.67	3.37	-1.2	264.59	13DE	-0.8	-3.4	-2.1	0.8
77.97	3.51	-1.0	274.78	23DE	-0.9	-3.3	-2.0	0.8
				541				
75.21	3.51	-0.7	284.96	2JA	-0.9	-3.3	-1.9	0.8
73.71	3.42	-0.5	295.11	12JA	-0.2	-3.3	-1.9	0.8
73.53	3.26	-0.2	305.23	22JA	2.0	-3.3	-1.8	0.7
74.54	3.08	0.1	315.30	1FE	2.2	-3.4	-1.7	0.7
76.54	2.90	0.3	325.32	11FE	0.9	-3.4	-1.7	0.7
79.34	2.72	0.5	335.28	21FE	0.5	-3.4	-1.7	0.6
82.78	2.54	0.7	345.19	3MR	0.3	-3.4	-1.6	0.5
86.72	2.38	0.9	355.04	13MR	-0.1	-3.5	-1.6	0.5
91.07	2.22	1.1	4.83	23MR	-0.6	-3.5	-1.6	0.4
95.75	2.08	1.2	14.56	2AP	-1.5	-3.5	-1.5	0.4
100.68	1.94	1.3	24.25	12AP	-1.5	-3.6	-1.5	0.3
105.85	1.81	1.4	33.89	22AP	-0.5	-3.7	-1.5	0.3
111.19	1.68	1.5	43.49	2MY	0.4	-3.7	-1.5	0.4
116.70	1.56	1.6	53.06	12MY	1.3	-3.8	-1.6	0.5
122.36	1.44	1.7	62.60	22MY	2.3	-3.9	-1.6	0.5
128.14	1.33	1.7	72.13	1JN	3.5	-4.0	-1.6	0.6
134.03	1.22	1.8	81.65	11JN	2.1	-4.1	-1.6	0.7
140.04	1.11	1.8	91.18	21JN	0.8	-4.2	-1.7	0.7
146.16	1.00	1.8	100.72	1JL	-0.4	-4.2	-1.7	0.8
152.37	0.89	1.8	110.28	11JL	-1.4	-4.1	-1.8	0.8
158.69	0.78	1.8	119.87	21JL	-1.5	-3.8	-1.8	0.8
165.11	0.67	1.8	129.50	31JL	-0.7	-3.3	-1.9	0.9
171.62	0.56	1.8	139.17	10AU	-0.2	-3.4	-2.0	0.9
178.24	0.46	1.8	148.90	20AU	0.1	-4.0	-2.0	0.9
184.95	0.35	1.8	158.67	30AU	0.3	-4.2	-2.1	0.9
191.77	0.24	1.8	168.51	9SE	0.5	-4.3	-2.2	0.9
198.68	0.13	1.7	178.40	19SE	1.5	-4.3	-2.2	0.8
205.70	0.03	1.7	188.35	29SE	2.6	-4.2	-2.3	0.8
212.81	-0.08	1.7	198.36	9OC	0.0	-4.1	-2.3	0.8
220.02	-0.18	1.6	208.42	19OC	-0.6	-4.0	-2.3	0.7
227.32	-0.29	1.6	218.52	29OC	-0.7	-3.9	-2.4	0.7
234.71	-0.39	1.5	228.67	8NO	-0.7	-3.8	-2.4	0.8
242.19	-0.49	1.5	238.84	18NO	-0.8	-3.8	-2.3	0.8
249.75	-0.59	1.4	249.04	28NO	-0.7	-3.7	-2.3	0.8
257.38	-0.68	1.4	259.24	8DE	-0.7	-3.6	-2.3	0.8
265.07	-0.76	1.4	269.44	18DE	-0.7	-3.6	-2.2	0.8
272.82	-0.84	1.4	279.63	28DE	-0.1	-3.5	-2.1	0.8

♂			⊙	16.00UT	☿	♀	♃	♄
LONG	LAT	MAG	LONG	542	MAGNITUDES			
280.62	-0.91	1.4	289.80	7JA	2.2	-3.5	-2.1	0.8
288.45	-0.98	1.4	299.93	17JA	1.7	-3.4	-2.0	0.8
296.30	-1.03	1.4	310.03	27JA	0.6	-3.4	-1.9	0.7
304.16	-1.08	1.4	320.08	6FE	0.3	-3.4	-1.9	0.7
312.02	-1.11	1.4	330.07	16FE	0.1	-3.3	-1.8	0.7
319.86	-1.13	1.4	340.00	26FE	-0.2	-3.3	-1.7	0.6
327.68	-1.14	1.4	349.89	8MR	-0.7	-3.3	-1.7	0.6
335.46	-1.14	1.4	359.70	18MR	-1.5	-3.3	-1.6	0.5
343.19	-1.12	1.4	9.47	28MR	-1.5	-3.3	-1.6	0.4
350.85	-1.09	1.4	19.18	7AP	-0.5	-3.3	-1.5	0.4
358.45	-1.05	1.4	28.84	17AP	0.6	-3.4	-1.5	0.3
5.97	-1.00	1.4	38.46	27AP	1.7	-3.4	-1.5	0.3
13.40	-0.93	1.4	48.04	7MY	3.1	-3.4	-1.5	0.3
20.73	-0.86	1.4	57.60	17MY	2.9	-3.5	-1.4	0.4
27.97	-0.78	1.4	67.13	27MY	1.6	-3.5	-1.4	0.4
35.10	-0.68	1.4	76.66	6JN	0.6	-3.5	-1.4	0.5
42.11	-0.58	1.4	86.18	16JN	-0.4	-3.4	-1.5	0.6
49.01	-0.47	1.4	95.71	26JN	-1.4	-3.4	-1.5	0.6
55.79	-0.35	1.4	105.26	6JL	-1.5	-3.4	-1.5	0.7
62.43	-0.22	1.4	114.84	16JL	-0.7	-3.3	-1.5	0.7
68.94	-0.08	1.4	124.44	26JL	-0.1	-3.3	-1.6	0.8
75.29	0.06	1.4	134.10	5AU	0.2	-3.3	-1.6	0.8
81.48	0.22	1.3	143.79	15AU	0.4	-3.3	-1.7	0.8
87.49	0.38	1.3	153.54	25AU	0.8	-3.3	-1.7	0.8
93.30	0.56	1.2	163.34	4SE	1.9	-3.3	-1.8	0.8
98.88	0.75	1.1	173.20	14SE	2.3	-3.4	-1.8	0.8
104.19	0.95	1.0	183.13	24SE	-0.1	-3.4	-1.9	0.8
109.18	1.17	0.9	193.11	4OC	-0.8	-3.4	-2.0	0.8
113.78	1.42	0.8	203.14	14OC	-0.8	-3.5	-2.0	0.8
117.93	1.69	0.7	213.22	24OC	-0.8	-3.5	-2.1	0.7
121.51	1.99	0.5	223.35	3NO	-0.7	-3.6	-2.2	0.7
124.38	2.33	0.3	233.50	13NO	-0.6	-3.6	-2.2	0.7
126.39	2.70	0.1	243.69	23NO	-0.6	-3.7	-2.2	0.7
127.34	3.11	-0.1	253.89	3DE	-0.6	-3.8	-2.2	0.8
127.06	3.53	-0.4	264.09	13DE	0.1	-3.9	-2.2	0.8
125.45	3.93	-0.6	274.29	23DE	2.5	-4.0	-2.2	0.8
				543				
122.58	4.27	-0.8	284.47	2JA	1.3	-4.1	-2.2	0.8
118.85	4.47	-1.0	294.62	12JA	0.4	-4.1	-2.1	0.8
114.90	4.50	-0.9	304.74	22JA	0.2	-4.2	-2.1	0.8
111.46	4.37	-0.7	314.82	1FE	0.0	-4.3	-2.0	0.8
109.08	4.10	-0.5	324.83	11FE	-0.2	-4.3	-1.9	0.7
107.97	3.78	-0.3	334.80	21FE	-0.7	-4.2	-1.9	0.7
108.14	3.43	-0.0	344.71	3MR	-1.4	-3.8	-1.8	0.6
109.43	3.09	0.2	354.56	13MR	-1.4	-3.2	-1.7	0.6
111.65	2.78	0.4	4.36	23MR	-0.5	-3.5	-1.7	0.5
114.63	2.48	0.6	14.09	2AP	0.8	-4.0	-1.6	0.5
118.23	2.22	0.8	23.78	12AP	2.2	-4.2	-1.6	0.4
122.32	1.97	0.9	33.42	22AP	3.9	-4.2	-1.5	0.3
126.82	1.75	1.1	43.02	2MY	2.2	-4.2	-1.5	0.3
131.66	1.54	1.2	52.59	12MY	1.2	-4.1	-1.4	0.3
136.79	1.35	1.3	62.13	22MY	0.5	-4.0	-1.4	0.3
142.16	1.17	1.3	71.66	1JN	-0.5	-3.9	-1.4	0.4
147.75	1.00	1.4	81.19	11JN	-1.4	-3.8	-1.4	0.4
153.53	0.84	1.5	90.72	21JN	-1.5	-3.7	-1.3	0.5
159.49	0.68	1.5	100.25	1JL	-0.6	-3.6	-1.3	0.6
165.61	0.53	1.5	109.81	11JL	-0.0	-3.6	-1.3	0.6
171.88	0.39	1.6	119.41	21JL	0.3	-3.5	-1.4	0.7
178.30	0.25	1.6	129.03	31JL	0.6	-3.5	-1.4	0.7
184.85	0.12	1.6	138.70	10AU	1.1	-3.4	-1.4	0.8
191.54	-0.01	1.6	148.43	20AU	2.3	-3.4	-1.4	0.8
198.36	-0.13	1.6	158.20	30AU	2.0	-3.4	-1.5	0.8
205.31	-0.25	1.6	168.03	9SE	-0.1	-3.4	-1.5	0.8
212.37	-0.36	1.6	177.92	19SE	-0.9	-3.4	-1.6	0.8
219.56	-0.46	1.6	187.87	29SE	-1.0	-3.4	-1.6	0.8
226.85	-0.56	1.5	197.87	9OC	-0.9	-3.4	-1.7	0.8
234.26	-0.66	1.5	207.93	19OC	-0.6	-3.4	-1.7	0.8
241.76	-0.74	1.5	218.03	29OC	-0.5	-3.4	-1.8	0.7
249.35	-0.82	1.5	228.18	8NO	-0.4	-3.4	-1.9	0.7
257.02	-0.89	1.5	238.35	18NO	-0.4	-3.4	-1.9	0.7
264.76	-0.95	1.4	248.54	28NO	0.2	-3.4	-2.0	0.7
272.56	-1.00	1.4	258.74	8DE	2.8	-3.4	-2.0	0.7
280.41	-1.04	1.4	268.95	18DE	1.0	-3.5	-2.1	0.8
288.29	-1.07	1.4	279.14	28DE	0.2	-3.5	-2.1	0.8

♂ LONG	♂ LAT	♂ MAG	☉ LONG	16.00UT	☿	♀	♃	♄
				544	MAGNITUDES			
296.19	-1.08	1.3	289.30	7JA	0.0	-3.5	-2.1	0.8
304.10	-1.09	1.3	299.44	17JA	-0.1	-3.4	-2.1	0.8
311.99	-1.08	1.3	309.54	27JA	-0.3	-3.4	-2.1	0.8
319.87	-1.06	1.3	319.59	6FE	-0.8	-3.4	-2.1	0.8
327.72	-1.03	1.3	329.58	16FE	-1.4	-3.4	-2.0	0.8
335.51	-0.99	1.4	339.52	26FE	-1.3	-3.4	-2.0	0.7
343.26	-0.94	1.4	349.40	7MR	-0.5	-3.3	-1.9	0.7
350.94	-0.88	1.4	359.23	17MR	1.0	-3.3	-1.8	0.6
358.54	-0.81	1.5	8.99	27MR	2.8	-3.3	-1.8	0.6
6.07	-0.73	1.5	18.71	6AP	2.9	-3.3	-1.7	0.5
13.52	-0.64	1.5	28.37	16AP	1.7	-3.4	-1.6	0.4
20.87	-0.55	1.6	37.99	26AP	0.9	-3.4	-1.6	0.4
28.14	-0.46	1.6	47.58	6MY	0.3	-3.4	-1.5	0.3
35.31	-0.36	1.6	57.13	16MY	-0.5	-3.4	-1.4	0.3
42.39	-0.25	1.6	66.67	26MY	-1.5	-3.5	-1.4	0.3
49.37	-0.14	1.7	76.19	5JN	-1.5	-3.5	-1.4	0.3
56.26	-0.03	1.7	85.72	15JN	-0.6	-3.6	-1.3	0.4
63.06	0.08	1.7	95.25	25JN	0.0	-3.6	-1.3	0.4
69.75	0.20	1.7	104.80	5JL	0.5	-3.7	-1.3	0.5
76.36	0.32	1.7	114.37	15JL	0.8	-3.8	-1.3	0.6
82.86	0.44	1.7	123.98	25JL	1.4	-3.9	-1.3	0.6
89.26	0.56	1.7	133.63	4AU	2.8	-4.0	-1.3	0.7
95.56	0.69	1.7	143.32	14AU	1.7	-4.1	-1.3	0.7
101.75	0.82	1.7	153.07	24AU	-0.2	-4.2	-1.3	0.7
107.82	0.96	1.6	162.87	3SE	-1.0	-4.3	-1.3	0.8
113.75	1.09	1.6	172.73	13SE	-1.1	-4.3	-1.3	0.8
119.55	1.24	1.5	182.65	23SE	-0.9	-4.2	-1.4	0.8
125.17	1.39	1.5	192.62	3OC	-0.5	-3.9	-1.4	0.8
130.61	1.56	1.4	202.65	13OC	-0.3	-3.3	-1.4	0.8
135.82	1.73	1.3	212.73	23OC	-0.3	-3.5	-1.5	0.8
140.77	1.92	1.2	222.86	2NO	-0.2	-4.1	-1.6	0.7
145.39	2.12	1.0	233.01	12NO	0.4	-4.3	-1.6	0.7
149.63	2.34	0.9	243.19	22NO	3.1	-4.4	-1.7	0.7
153.37	2.58	0.7	253.40	2DE	0.7	-4.3	-1.7	0.7
156.52	2.84	0.5	263.60	12DE	-0.0	-4.2	-1.8	0.7
158.92	3.12	0.3	273.80	22DE	-0.1	-4.1	-1.9	0.8
				545				
160.40	3.42	0.0	283.98	1JA	-0.2	-4.0	-1.9	0.8
160.78	3.73	-0.2	294.13	11JA	-0.4	-3.9	-2.0	0.8
159.92	4.01	-0.5	304.25	21JA	-0.8	-3.8	-2.0	0.8
157.76	4.23	-0.8	314.33	31JA	-1.3	-3.7	-2.0	0.8
154.52	4.32	-1.0	324.35	10FE	-1.2	-3.7	-2.0	0.8
150.67	4.26	-1.1	334.32	20FE	-0.4	-3.6	-2.0	0.8
146.93	4.03	-0.9	344.24	2MR	1.2	-3.5	-2.0	0.8
143.96	3.67	-0.8	354.09	12MR	3.3	-3.5	-2.0	0.7
142.17	3.24	-0.5	3.89	22MR	2.2	-3.4	-1.9	0.7
141.68	2.80	-0.3	13.63	1AP	1.2	-3.4	-1.9	0.6
142.43	2.37	-0.1	23.32	11AP	0.7	-3.3	-1.8	0.6
144.24	1.98	0.1	32.96	21AP	0.2	-3.3	-1.7	0.5
146.94	1.62	0.3	42.56	1MY	-0.5	-3.3	-1.7	0.4
150.38	1.31	0.4	52.13	11MY	-1.5	-3.3	-1.6	0.4
154.41	1.02	0.6	61.68	21MY	-1.5	-3.3	-1.5	0.3
158.94	0.76	0.7	71.21	31MY	-0.6	-3.3	-1.5	0.3
163.88	0.53	0.8	80.73	10JN	0.1	-3.3	-1.4	0.3
169.17	0.31	0.9	90.25	20JN	0.6	-3.4	-1.4	0.3
174.77	0.12	1.0	99.79	30JN	1.1	-3.4	-1.3	0.4
180.63	-0.06	1.0	109.35	10JL	1.9	-3.4	-1.3	0.5
186.74	-0.23	1.1	118.94	20JL	3.2	-3.5	-1.3	0.5
193.06	-0.38	1.1	128.56	30JL	1.5	-3.5	-1.2	0.6
199.58	-0.52	1.2	138.23	9AU	-0.3	-3.5	-1.2	0.6
206.28	-0.64	1.2	147.95	19AU	-1.1	-3.5	-1.2	0.7
213.16	-0.76	1.2	157.72	29AU	-1.3	-3.4	-1.2	0.7
220.18	-0.86	1.3	167.55	8SE	-0.8	-3.4	-1.2	0.8
227.36	-0.95	1.3	177.44	18SE	-0.4	-3.4	-1.2	0.8
234.66	-1.03	1.3	187.39	28SE	-0.2	-3.3	-1.2	0.8
242.09	-1.09	1.3	197.38	8OC	-0.1	-3.3	-1.3	0.8
249.62	-1.14	1.3	207.44	18OC	-0.0	-3.3	-1.3	0.8
257.25	-1.18	1.3	217.54	28OC	0.6	-3.3	-1.3	0.8
264.96	-1.21	1.3	227.68	7NO	3.3	-3.3	-1.4	0.8
272.73	-1.22	1.3	237.85	17NO	0.5	-3.4	-1.4	0.7
280.55	-1.22	1.4	248.05	27NO	-0.2	-3.4	-1.5	0.7
288.41	-1.20	1.4	258.25	7DE	-0.3	-3.4	-1.6	0.7
296.29	-1.17	1.4	268.45	17DE	-0.3	-3.4	-1.6	0.7
304.17	-1.14	1.4	278.65	27DE	-0.5	-3.5	-1.7	0.8

♂ LONG	♂ LAT	♂ MAG	☉ LONG	16.00UT	☿	♀	♃	♄
				546	MAGNITUDES			
312.04	-1.09	1.4	288.81	6JA	-0.8	-3.5	-1.7	0.8
319.88	-1.03	1.4	298.95	16JA	-1.1	-3.6	-1.8	0.8
327.69	-0.96	1.4	309.05	26JA	-1.1	-3.6	-1.9	0.8
335.45	-0.88	1.5	319.10	5FE	-0.4	-3.7	-1.9	0.8
343.15	-0.79	1.5	329.10	15FE	1.5	-3.8	-2.0	0.8
350.78	-0.71	1.5	339.05	25FE	3.1	-3.8	-2.0	0.8
358.34	-0.61	1.5	348.93	7MR	1.6	-3.9	-2.0	0.8
5.82	-0.51	1.5	358.76	17MR	0.9	-4.0	-2.0	0.8
13.22	-0.41	1.5	8.53	27MR	0.5	-4.1	-2.0	0.7
20.53	-0.31	1.5	18.24	6AP	0.1	-4.2	-2.0	0.7
27.76	-0.21	1.5	27.91	16AP	-0.6	-4.2	-1.9	0.6
34.90	-0.10	1.6	37.53	26AP	-1.5	-4.2	-1.9	0.6
41.95	0.00	1.6	47.12	6MY	-1.5	-3.9	-1.8	0.5
48.92	0.11	1.7	56.67	16MY	-0.6	-3.3	-1.8	0.4
55.81	0.21	1.7	66.21	26MY	0.2	-2.8	-1.7	0.4
62.62	0.31	1.8	75.73	5JN	0.8	-3.7	-1.6	0.3
69.36	0.41	1.8	85.26	15JN	1.5	-4.1	-1.6	0.3
76.02	0.51	1.8	94.79	25JN	2.5	-4.2	-1.5	0.3
82.62	0.60	1.9	104.33	5JL	3.0	-4.2	-1.5	0.4
89.14	0.70	1.9	113.91	15JL	1.2	-4.1	-1.4	0.4
95.60	0.79	1.9	123.51	25JL	-0.3	-4.0	-1.4	0.5
102.00	0.88	1.9	133.15	4AU	-1.2	-3.9	-1.3	0.5
108.33	0.98	1.9	142.85	14AU	-1.4	-3.9	-1.3	0.6
114.60	1.07	1.9	152.59	24AU	-0.8	-3.8	-1.3	0.7
120.80	1.16	1.9	162.39	3SE	-0.3	-3.7	-1.2	0.7
126.93	1.25	1.9	172.25	13SE	-0.1	-3.7	-1.2	0.7
132.99	1.34	1.8	182.16	23SE	0.0	-3.6	-1.2	0.8
138.96	1.43	1.8	192.13	3OC	0.2	-3.5	-1.2	0.8
144.85	1.52	1.7	202.16	13OC	0.9	-3.5	-1.2	0.8
150.62	1.62	1.6	212.24	23OC	3.2	-3.5	-1.2	0.8
156.28	1.72	1.6	222.36	2NO	0.3	-3.4	-1.3	0.8
161.80	1.82	1.5	232.52	12NO	-0.4	-3.4	-1.3	0.8
167.14	1.92	1.4	242.70	22NO	-0.4	-3.4	-1.3	0.8
172.29	2.03	1.2	252.90	2DE	-0.4	-3.4	-1.4	0.8
177.19	2.14	1.1	263.11	12DE	-0.6	-3.4	-1.4	0.7
181.78	2.25	0.9	273.31	22DE	-0.9	-3.3	-1.5	0.7
				547				
186.00	2.37	0.7	283.49	1JA	-1.0	-3.3	-1.5	0.8
189.77	2.50	0.5	293.65	11JA	-1.0	-3.3	-1.6	0.8
192.95	2.62	0.3	303.76	21JA	-0.3	-3.4	-1.7	0.8
195.41	2.73	0.1	313.84	31JA	1.7	-3.4	-1.7	0.9
196.99	2.83	-0.2	323.87	10FE	2.5	-3.4	-1.8	0.9
197.49	2.90	-0.5	333.84	20FE	1.2	-3.4	-1.9	0.9
196.77	2.92	-0.8	343.75	2MR	0.6	-3.4	-1.9	0.9
194.78	2.84	-1.1	353.61	12MR	0.4	-3.5	-2.0	0.9
191.72	2.65	-1.4	3.41	22MR	-0.0	-3.5	-2.0	0.8
188.06	2.33	-1.5	13.15	1AP	-0.6	-3.4	-2.0	0.8
184.50	1.91	-1.3	22.85	11AP	-1.5	-3.4	-2.1	0.8
181.73	1.43	-1.2	32.49	21AP	-1.5	-3.4	-2.0	0.7
180.17	0.96	-1.0	42.09	1MY	-0.6	-3.4	-2.0	0.7
179.95	0.51	-0.8	51.67	11MY	0.3	-3.3	-2.0	0.6
181.02	0.12	-0.6	61.21	21MY	1.1	-3.3	-2.0	0.5
183.19	-0.22	-0.4	70.74	31MY	1.9	-3.3	-1.9	0.5
186.30	-0.51	-0.2	80.27	10JN	3.2	-3.3	-1.8	0.4
190.19	-0.76	-0.0	89.79	20JN	2.5	-3.3	-1.8	0.3
194.71	-0.96	0.1	99.33	30JN	1.0	-3.3	-1.7	0.3
199.77	-1.14	0.2	108.89	10JL	-0.3	-3.4	-1.7	0.4
205.27	-1.29	0.3	118.47	20JL	-1.3	-3.4	-1.6	0.4
211.15	-1.41	0.4	128.09	30JL	-1.5	-3.4	-1.5	0.5
217.35	-1.51	0.5	137.76	9AU	-0.7	-3.5	-1.5	0.5
223.83	-1.58	0.6	147.47	19AU	-0.3	-3.5	-1.4	0.6
230.55	-1.64	0.7	157.24	29AU	0.0	-3.6	-1.4	0.6
237.48	-1.67	0.7	167.07	8SE	0.2	-3.7	-1.4	0.7
244.60	-1.69	0.8	176.95	18SE	0.4	-3.7	-1.3	0.7
251.86	-1.68	0.9	186.90	28SE	1.2	-3.8	-1.3	0.8
259.27	-1.66	0.9	196.90	8OC	2.9	-3.9	-1.3	0.8
266.77	-1.62	1.0	206.95	18OC	0.2	-4.0	-1.3	0.8
274.37	-1.57	1.0	217.05	28OC	-0.5	-4.2	-1.3	0.8
282.03	-1.50	1.1	227.19	7NO	-0.6	-4.3	-1.3	0.8
289.73	-1.42	1.1	237.36	17NO	-0.6	-4.4	-1.3	0.8
297.46	-1.33	1.2	247.55	27NO	-0.7	-4.4	-1.3	0.8
305.20	-1.23	1.2	257.76	7DE	-0.8	-4.3	-1.3	0.8
312.93	-1.13	1.3	267.96	17DE	-0.8	-4.0	-1.4	0.8
320.64	-1.01	1.3	278.15	27DE	-0.8	-3.3	-1.4	0.8

Left table

♂ LONG	♂ LAT	♂ MAG	☉ LONG	16.00UT 548	☿	♀	♃	♄
328.31	-0.90	1.4	288.32	6JA	-0.2	-3.5	-1.5	0.8
335.93	-0.78	1.4	298.46	16JA	2.0	-4.1	-1.5	0.8
343.50	-0.66	1.5	308.56	26JA	2.0	-4.3	-1.6	0.9
351.00	-0.54	1.5	318.62	5FE	0.8	-4.3	-1.6	0.9
358.43	-0.42	1.6	328.62	15FE	0.4	-4.3	-1.7	0.9
5.79	-0.30	1.6	338.57	25FE	0.2	-4.2	-1.8	0.9
13.07	-0.19	1.6	348.45	6MR	-0.1	-4.1	-1.8	0.9
20.26	-0.07	1.7	358.28	16MR	-0.6	-3.9	-1.9	0.9
27.38	0.04	1.7	8.06	26MR	-1.4	-3.8	-2.0	0.9
34.43	0.14	1.7	17.77	5AP	-1.5	-3.7	-2.0	0.9
41.39	0.24	1.8	27.44	15AP	-0.6	-3.7	-2.1	0.9
48.28	0.34	1.8	37.07	25AP	0.5	-3.6	-2.1	0.8
55.11	0.43	1.8	46.66	5MY	1.4	-3.5	-2.1	0.8
61.86	0.52	1.8	56.21	15MY	2.6	-3.5	-2.1	0.7
68.56	0.60	1.8	65.75	25MY	3.4	-3.4	-2.1	0.6
75.19	0.68	1.8	75.28	4JN	1.9	-3.4	-2.1	0.6
81.78	0.76	1.9	84.80	14JN	0.8	-3.4	-2.1	0.5
88.31	0.83	1.9	94.33	24JN	-0.4	-3.3	-2.0	0.4
94.80	0.89	1.9	103.88	4JL	-1.4	-3.3	-2.0	0.4
101.25	0.96	2.0	113.45	14JL	-1.5	-3.3	-1.9	0.4
107.66	1.02	2.0	123.05	24JL	-0.7	-3.3	-1.9	0.4
114.04	1.07	2.0	132.69	3AU	-0.2	-3.3	-1.8	0.5
120.39	1.13	2.0	142.38	13AU	0.1	-3.4	-1.7	0.5
126.71	1.18	2.0	152.12	23AU	0.3	-3.4	-1.7	0.6
133.00	1.22	2.0	161.91	2SE	0.6	-3.4	-1.6	0.6
139.26	1.27	2.0	171.77	12SE	1.6	-3.4	-1.6	0.7
145.50	1.31	2.0	181.68	22SE	2.6	-3.5	-1.5	0.7
151.71	1.35	1.9	191.65	2OC	0.0	-3.5	-1.5	0.8
157.89	1.38	1.9	201.67	12OC	-0.7	-3.5	-1.4	0.8
164.04	1.41	1.9	211.75	22OC	-0.7	-3.4	-1.4	0.9
170.15	1.44	1.8	221.87	1NO	-0.7	-3.4	-1.4	0.9
176.21	1.47	1.7	232.02	11NO	-0.8	-3.4	-1.4	0.9
182.22	1.49	1.7	242.20	21NO	-0.7	-3.4	-1.4	0.9
188.16	1.50	1.6	252.40	1DE	-0.7	-3.4	-1.4	0.9
194.02	1.51	1.5	262.60	11DE	-0.7	-3.4	-1.4	0.9
199.79	1.51	1.3	272.81	21DE	-0.1	-3.3	-1.4	0.9
205.45	1.50	1.2	282.99	31DE	2.3	-3.3	-1.4	0.9

549

♂ LONG	♂ LAT	♂ MAG	☉ LONG	16.00UT	☿	♀	♃	♄
210.97	1.48	1.1	293.15	10JA	1.6	-3.3	-1.4	0.8
216.31	1.45	0.9	303.27	20JA	0.5	-3.3	-1.5	0.9
221.45	1.40	0.7	313.35	30JA	0.3	-3.4	-1.5	0.9
226.33	1.33	0.5	323.38	9FE	0.1	-3.4	-1.6	1.0
230.88	1.23	0.3	333.36	19FE	-0.2	-3.4	-1.6	1.0
235.03	1.10	0.1	343.28	1MR	-0.7	-3.4	-1.7	1.0
238.67	0.93	-0.2	353.14	11MR	-1.4	-3.5	-1.8	1.0
241.65	0.70	-0.5	2.94	21MR	-1.5	-3.5	-1.8	1.0
243.84	0.40	-0.8	12.69	31MR	-0.5	-3.6	-1.9	1.0
245.03	0.02	-1.1	22.38	10AP	0.6	-3.6	-2.0	1.0
245.05	-0.46	-1.4	32.03	20AP	1.8	-3.7	-2.0	1.0
243.84	-1.02	-1.7	41.63	30AP	3.4	-3.7	-2.1	0.9
241.49	-1.63	-2.0	51.21	10MY	2.7	-3.8	-2.2	0.9
238.42	-2.25	-2.2	60.76	20MY	1.5	-3.9	-2.2	0.8
235.32	-2.79	-2.1	70.29	30MY	0.6	-4.0	-2.2	0.8
232.88	-3.21	-1.9	79.81	9JN	-0.4	-4.1	-2.3	0.7
231.63	-3.47	-1.7	89.34	19JN	-1.4	-4.2	-2.3	0.6
231.77	-3.61	-1.5	98.87	29JN	-1.5	-4.2	-2.3	0.6
233.24	-3.65	-1.3	108.43	9JL	-0.7	-4.1	-2.2	0.5
235.89	-3.62	-1.1	118.01	19JL	-0.1	-3.8	-2.2	0.5
239.54	-3.53	-0.9	127.63	29JL	0.2	-3.2	-2.2	0.5
243.98	-3.41	-0.7	137.29	8AU	0.5	-3.4	-2.1	0.5
249.08	-3.26	-0.5	147.01	18AU	0.9	-4.0	-2.0	0.6
254.70	-3.09	-0.3	156.77	28AU	2.0	-4.2	-2.0	0.6
260.73	-2.90	-0.2	166.59	7SE	2.3	-4.3	-1.9	0.7
267.10	-2.71	-0.0	176.48	17SE	-0.1	-4.3	-1.9	0.7
273.73	-2.50	0.1	186.41	27SE	-0.8	-4.2	-1.8	0.8
280.56	-2.29	0.2	196.41	7OC	-0.9	-4.1	-1.7	0.8
287.55	-2.07	0.4	206.46	17OC	-0.8	-4.0	-1.7	0.9
294.65	-1.86	0.5	216.55	27OC	-0.7	-3.9	-1.6	0.9
301.84	-1.64	0.6	226.69	6NO	-0.5	-3.8	-1.5	0.9
309.09	-1.43	0.7	236.84	16NO	-0.5	-3.8	-1.5	1.0
316.36	-1.23	0.8	247.05	26NO	-0.5	-3.7	-1.5	1.0
323.64	-1.03	0.9	257.26	6DE	0.1	-3.6	-1.5	1.0
330.92	-0.85	1.1	267.46	16DE	2.5	-3.6	-1.5	1.0
338.17	-0.67	1.1	277.65	26DE	1.2	-3.5	-1.5	1.0

Right table

♂ LONG	♂ LAT	♂ MAG	☉ LONG	16.00UT 550	☿	♀	♃	♄
345.39	-0.50	1.2	287.83	5JA	0.3	-3.5	-1.5	1.0
352.57	-0.34	1.3	297.97	15JA	0.1	-3.4	-1.5	1.0
359.69	-0.19	1.4	308.07	25JA	-0.0	-3.4	-1.5	1.0
6.77	-0.05	1.5	318.13	4FE	-0.3	-3.4	-1.5	1.0
13.79	0.08	1.6	328.14	14FE	-0.7	-3.3	-1.6	1.1
20.74	0.20	1.6	338.09	24FE	-1.4	-3.3	-1.6	1.1
27.64	0.31	1.7	347.98	6MR	-1.4	-3.3	-1.6	1.1
34.48	0.42	1.7	357.81	16MR	-0.5	-3.3	-1.7	1.1
41.26	0.51	1.8	7.59	26MR	0.8	-3.3	-1.7	1.2
47.98	0.60	1.8	17.31	5AP	2.4	-3.3	-1.8	1.2
54.65	0.68	1.9	26.98	15AP	3.5	-3.4	-1.9	1.1
61.27	0.75	1.9	36.60	25AP	2.0	-3.4	-1.9	1.1
67.85	0.81	1.9	46.20	5MY	1.1	-3.4	-2.0	1.1
74.38	0.87	1.9	55.75	15MY	0.4	-3.5	-2.1	1.1
80.88	0.93	2.0	65.29	25MY	-0.4	-3.5	-2.2	1.0
87.34	0.97	2.0	74.82	4JN	-1.4	-3.5	-2.2	0.9
93.78	1.02	2.0	84.34	14JN	-1.5	-3.4	-2.3	0.9
100.19	1.05	2.0	93.87	24JN	-0.7	-3.4	-2.3	0.8
106.59	1.09	2.0	103.41	4JL	-0.0	-3.4	-2.4	0.8
112.97	1.11	1.9	112.98	14JL	0.4	-3.3	-2.4	0.7
119.34	1.14	2.0	122.58	24JL	0.7	-3.3	-2.4	0.6
125.71	1.15	2.0	132.22	3AU	1.2	-3.3	-2.4	0.6
132.08	1.17	2.0	141.91	13AU	2.5	-3.3	-2.4	0.6
138.45	1.18	2.0	151.64	23AU	2.0	-3.3	-2.3	0.6
144.82	1.18	2.0	161.44	2SE	-0.1	-3.3	-2.3	0.7
151.21	1.18	2.0	171.29	12SE	-0.9	-3.4	-2.2	0.7
157.60	1.18	2.0	181.19	22SE	-1.0	-3.4	-2.2	0.8
164.00	1.17	2.0	191.16	2OC	-0.9	-3.4	-2.1	0.8
170.42	1.16	2.0	201.18	12OC	-0.6	-3.5	-2.0	0.9
176.85	1.14	1.9	211.25	22OC	-0.4	-3.5	-2.0	0.9
183.29	1.11	1.9	221.37	1NO	-0.4	-3.6	-1.9	1.0
189.74	1.08	1.9	231.52	11NO	-0.3	-3.6	-1.8	1.0
196.21	1.04	1.8	241.70	21NO	0.3	-3.7	-1.8	1.0
202.68	0.99	1.7	251.90	1DE	2.8	-3.8	-1.7	1.1
209.16	0.93	1.7	262.11	11DE	0.9	-3.9	-1.7	1.1
215.65	0.86	1.6	272.31	21DE	0.1	-4.0	-1.6	1.1
222.13	0.78	1.5	282.50	31DE	-0.0	-4.1	-1.6	1.1

551

♂ LONG	♂ LAT	♂ MAG	☉ LONG	16.00UT	☿	♀	♃	♄
228.61	0.69	1.4	292.65	10JA	-0.1	-4.2	-1.6	1.1
235.08	0.58	1.3	302.78	20JA	-0.3	-4.2	-1.6	1.1
241.53	0.45	1.2	312.86	30JA	-0.8	-4.3	-1.6	1.1
247.95	0.31	1.0	322.90	9FE	-1.4	-4.3	-1.6	1.1
254.34	0.14	0.9	332.87	19FE	-1.3	-4.2	-1.6	1.2
260.68	-0.05	0.7	342.80	1MR	-0.5	-3.8	-1.6	1.2
266.96	-0.28	0.6	352.66	11MR	1.0	-3.2	-1.6	1.3
273.15	-0.53	0.4	2.47	21MR	3.0	-3.5	-1.6	1.3
279.22	-0.82	0.2	12.22	31MR	2.6	-4.0	-1.7	1.3
285.14	-1.15	0.0	21.91	10AP	1.5	-4.2	-1.7	1.3
290.86	-1.52	-0.2	31.56	20AP	0.8	-4.2	-1.7	1.3
296.32	-1.94	-0.4	41.17	30AP	0.3	-4.2	-1.8	1.3
301.43	-2.41	-0.6	50.74	10MY	-0.5	-4.1	-1.9	1.3
306.09	-2.94	-0.9	60.29	20MY	-1.4	-4.0	-1.9	1.2
310.16	-3.52	-1.1	69.83	30MY	-1.6	-3.9	-2.0	1.2
313.48	-4.16	-1.4	79.35	9JN	-0.6	-3.8	-2.1	1.2
315.84	-4.84	-1.7	88.87	19JN	0.1	-3.7	-2.1	1.1
317.05	-5.53	-1.9	98.41	29JN	0.5	-3.6	-2.2	1.0
316.97	-6.16	-2.2	107.96	9JL	0.9	-3.6	-2.3	1.0
315.61	-6.64	-2.5	117.55	19JL	1.6	-3.5	-2.3	0.9
313.29	-6.87	-2.6	127.17	29JL	3.0	-3.5	-2.4	0.8
310.65	-6.77	-2.6	136.83	8AU	1.7	-3.4	-2.4	0.8
308.41	-6.36	-2.4	146.54	18AU	-0.2	-3.4	-2.4	0.7
307.18	-5.74	-2.1	156.30	28AU	-1.1	-3.4	-2.5	0.8
307.25	-5.01	-1.8	166.12	7SE	-1.2	-3.4	-2.5	0.8
308.59	-4.28	-1.5	176.00	17SE	-0.9	-3.4	-2.4	0.8
311.07	-3.58	-1.2	185.94	27SE	-0.5	-3.4	-2.4	0.9
314.47	-2.95	-0.9	195.93	7OC	-0.3	-3.4	-2.3	0.9
318.61	-2.39	-0.7	205.98	17OC	-0.2	-3.4	-2.3	1.0
323.34	-1.89	-0.4	216.07	27OC	-0.1	-3.4	-2.2	1.0
328.49	-1.46	-0.2	226.20	6NO	0.5	-3.4	-2.1	1.1
333.98	-1.08	-0.0	236.37	16NO	3.1	-3.4	-2.1	1.1
339.74	-0.76	0.2	246.56	26NO	0.7	-3.4	-2.0	1.2
345.68	-0.47	0.4	256.76	6DE	-0.1	-3.4	-1.9	1.2
351.76	-0.23	0.6	266.96	16DE	-0.2	-3.5	-1.9	1.2
357.96	-0.01	0.8	277.16	26DE	-0.3	-3.5	-1.8	1.3

♂			☉	16.00UT	☿	♀	♃	♄	♂			☉	16.00UT	☿	♀	♃	♄
LONG	LAT	MAG	LONG	552	MAGNITUDES				LONG	LAT	MAG	LONG	554	MAGNITUDES			
4.22	0.17	0.9	287.33	5JA	-0.4	-3.5	-1.8	1.3	27.10	1.24	0.4	286.84	4JA	-0.9	-3.5	-2.1	1.1
10.54	0.34	1.1	297.47	15JA	-0.8	-3.4	-1.7	1.3	31.63	1.33	0.6	296.98	14JA	-0.3	-3.6	-2.0	1.1
16.89	0.48	1.2	307.58	25JA	-1.2	-3.4	-1.7	1.3	36.48	1.39	0.8	307.09	24JA	1.7	-3.6	-2.0	1.0
23.26	0.60	1.3	317.64	4FE	-1.2	-3.4	-1.6	1.3	41.60	1.44	1.0	317.15	3FE	2.4	-3.7	-1.9	1.0
29.65	0.70	1.4	327.65	14FE	-0.4	-3.4	-1.6	1.3	46.91	1.47	1.1	327.16	13FE	1.0	-3.8	-1.8	1.0
36.03	0.80	1.5	337.60	24FE	1.2	-3.4	-1.6	1.3	52.38	1.49	1.3	337.12	23FE	0.6	-3.8	-1.8	0.9
42.40	0.87	1.6	347.49	5MR	3.4	-3.3	-1.6	1.4	57.97	1.50	1.4	347.02	5MR	0.3	-3.9	-1.7	0.9
48.77	0.94	1.7	357.33	15MR	1.9	-3.3	-1.6	1.4	63.67	1.50	1.5	356.85	15MR	-0.0	-4.0	-1.6	0.8
55.13	1.00	1.8	7.11	25MR	1.1	-3.3	-1.6	1.4	69.45	1.50	1.6	6.64	25MR	-0.6	-4.1	-1.6	0.8
61.47	1.05	1.8	16.83	4AP	0.6	-3.3	-1.6	1.4	75.29	1.48	1.7	16.36	4AP	-1.4	-4.2	-1.5	0.9
67.80	1.09	1.9	26.51	14AP	0.2	-3.4	-1.6	1.4	81.19	1.46	1.8	26.04	14AP	-1.5	-4.2	-1.5	0.9
74.12	1.12	1.9	36.14	24AP	-0.5	-3.4	-1.6	1.4	87.14	1.44	1.8	35.67	24AP	-0.6	-4.2	-1.5	0.9
80.42	1.14	2.0	45.73	4MY	-1.4	-3.4	-1.7	1.3	93.14	1.42	1.9	45.26	4MY	0.4	-3.9	-1.5	0.8
86.73	1.16	2.0	55.29	14MY	-1.6	-3.4	-1.7	1.3	99.17	1.38	1.9	54.83	14MY	1.2	-3.3	-1.4	0.8
93.02	1.18	2.0	64.83	24MY	-0.6	-3.5	-1.7	1.2	105.25	1.35	2.0	64.37	24MY	2.1	-2.8	-1.4	0.8
99.32	1.19	2.0	74.36	3JN	0.1	-3.5	-1.8	1.2	111.31	1.31	2.0	73.89	3JN	3.4	-3.7	-1.4	0.8
105.61	1.19	2.0	83.88	13JN	0.7	-3.6	-1.8	1.2	117.52	1.27	2.0	83.42	13JN	2.3	-4.1	-1.4	0.7
111.92	1.19	2.0	93.41	23JN	1.2	-3.6	-1.9	1.1	123.72	1.23	2.0	92.95	23JN	0.9	-4.2	-1.4	0.7
118.23	1.18	2.0	102.95	3JL	2.1	-3.7	-2.0	1.1	129.96	1.18	2.0	102.49	3JL	-0.3	-4.2	-1.5	0.7
124.56	1.17	2.0	112.52	13JL	3.2	-3.8	-2.0	1.0	136.25	1.13	2.0	112.06	13JL	-1.3	-4.1	-1.5	0.6
130.91	1.15	2.0	122.12	23JL	1.4	-3.9	-2.1	1.0	142.59	1.07	2.0	121.65	23JL	-1.5	-4.0	-1.5	0.6
137.28	1.13	2.0	131.76	2AU	-0.2	-4.0	-2.2	0.9	148.97	1.01	2.0	131.29	2AU	-0.7	-3.9	-1.6	0.5
143.68	1.11	1.9	141.44	12AU	-1.2	-4.1	-2.2	0.9	155.42	0.95	2.0	140.97	12AU	-0.2	-3.9	-1.6	0.5
150.11	1.08	1.9	151.17	22AU	-1.3	-4.2	-2.3	0.8	161.92	0.89	1.9	150.70	22AU	0.1	-3.8	-1.7	0.4
156.58	1.05	1.9	160.96	1SE	-0.8	-4.3	-2.4	0.9	168.49	0.82	1.9	160.49	1SE	0.2	-3.7	-1.7	0.4
163.08	1.01	1.9	170.81	11SE	-0.4	-4.3	-2.4	0.9	175.12	0.74	1.9	170.33	11SE	0.5	-3.6	-1.8	0.3
169.62	0.96	1.9	180.72	21SE	-0.2	-4.2	-2.4	1.0	181.81	0.67	1.8	180.24	21SE	1.3	-3.6	-1.8	0.3
176.20	0.92	1.9	190.68	1OC	-0.1	-3.9	-2.4	1.0	188.58	0.59	1.8	190.20	1OC	2.9	-3.5	-1.9	0.3
182.83	0.86	1.9	200.70	11OC	0.1	-3.2	-2.4	1.1	195.41	0.50	1.8	200.21	11OC	0.2	-3.5	-2.0	0.4
189.51	0.80	1.9	210.77	21OC	0.7	-3.6	-2.4	1.1	202.32	0.41	1.8	210.28	21OC	-0.6	-3.5	-2.0	0.4
196.24	0.73	1.9	220.88	31OC	3.4	-4.1	-2.4	1.2	209.31	0.32	1.8	220.39	31OC	-0.6	-3.4	-2.1	0.5
203.01	0.66	1.9	231.04	10NO	0.5	-4.3	-2.3	1.2	216.36	0.22	1.7	230.54	10NO	-0.6	-3.4	-2.1	0.6
209.83	0.58	1.8	241.21	20NO	-0.3	-4.4	-2.3	1.3	223.49	0.12	1.7	240.72	20NO	-0.7	-3.4	-2.2	0.6
216.71	0.49	1.8	251.41	30NO	-0.3	-4.4	-2.2	1.3	230.69	0.01	1.7	250.91	30NO	-0.7	-3.4	-2.2	0.7
223.63	0.40	1.7	261.61	10DE	-0.4	-4.2	-2.1	1.4	237.96	-0.10	1.7	261.12	10DE	-0.8	-3.4	-2.2	0.7
230.60	0.29	1.7	271.81	20DE	-0.5	-4.1	-2.1	1.4	245.30	-0.21	1.6	271.32	20DE	-0.7	-3.3	-2.2	0.8
237.63	0.18	1.6	282.00	30DE	-0.9	-4.0	-2.0	1.4	252.70	-0.33	1.6	281.51	30DE	-0.2	-3.3	-2.2	0.8
				553									555				
244.70	0.06	1.5	292.16	9JA	-1.1	-3.9	-1.9	1.3	260.17	-0.45	1.6	291.67	9JA	2.0	-3.3	-2.2	0.8
251.81	-0.08	1.5	302.29	19JA	-1.0	-3.8	-1.8	1.3	267.68	-0.57	1.5	301.80	19JA	1.9	-3.4	-2.1	0.8
258.96	-0.22	1.4	312.37	29JA	-0.4	-3.7	-1.8	1.3	275.25	-0.69	1.5	311.88	29JA	0.7	-3.4	-2.0	0.8
266.16	-0.37	1.3	322.41	8FE	1.5	-3.6	-1.7	1.2	282.86	-0.81	1.4	321.92	8FE	0.4	-3.4	-2.0	0.8
273.38	-0.53	1.2	332.39	18FE	2.9	-3.6	-1.7	1.2	290.50	-0.92	1.4	331.91	18FE	0.2	-3.4	-1.9	0.8
280.64	-0.70	1.1	342.32	28FE	1.4	-3.5	-1.6	1.1	298.16	-1.04	1.3	341.83	28FE	-0.1	-3.4	-1.8	0.7
287.91	-0.88	1.0	352.19	10MR	0.8	-3.5	-1.6	1.1	305.83	-1.15	1.3	351.70	10MR	-0.6	-3.5	-1.8	0.7
295.19	-1.07	0.9	1.99	20MR	0.5	-3.4	-1.6	1.1	313.51	-1.25	1.2	1.52	20MR	-1.4	-3.5	-1.7	0.6
302.48	-1.26	0.9	11.75	30MR	0.1	-3.4	-1.5	1.1	321.17	-1.34	1.2	11.27	30MR	-1.5	-3.4	-1.6	0.6
309.74	-1.46	0.8	21.45	9AP	-0.5	-3.3	-1.5	1.1	328.81	-1.42	1.2	20.97	9AP	-0.6	-3.4	-1.6	0.6
316.97	-1.66	0.6	31.10	19AP	-1.4	-3.3	-1.5	1.1	336.40	-1.49	1.1	30.62	19AP	0.5	-3.4	-1.5	0.6
324.14	-1.85	0.5	40.71	29AP	-1.6	-3.3	-1.5	1.1	343.94	-1.55	1.1	40.23	29AP	1.5	-3.4	-1.5	0.6
331.23	-2.05	0.4	50.29	9MY	-0.6	-3.3	-1.5	1.1	351.41	-1.59	1.0	49.81	9MY	2.9	-3.3	-1.4	0.6
338.21	-2.24	0.3	59.83	19MY	0.2	-3.3	-1.5	1.0	358.79	-1.62	1.0	59.37	19MY	3.2	-3.3	-1.4	0.6
345.05	-2.42	0.2	69.37	29MY	0.9	-3.3	-1.6	1.0	6.06	-1.63	0.9	68.90	29MY	1.8	-3.3	-1.4	0.6
351.70	-2.60	0.1	78.89	8JN	1.6	-3.3	-1.6	1.0	13.21	-1.62	0.9	78.42	8JN	0.7	-3.3	-1.3	0.6
358.10	-2.76	-0.0	88.42	18JN	2.7	-3.4	-1.6	0.9	20.21	-1.60	0.8	87.95	18JN	-0.4	-3.3	-1.3	0.6
4.21	-2.91	-0.2	97.95	28JN	2.8	-3.4	-1.7	0.9	27.04	-1.55	0.8	97.48	28JN	-1.4	-3.3	-1.3	0.6
9.93	-3.04	-0.3	107.51	8JL	1.2	-3.4	-1.7	0.8	33.68	-1.49	0.7	107.04	8JL	-1.5	-3.4	-1.3	0.5
15.19	-3.15	-0.5	117.08	18JL	-0.3	-3.5	-1.8	0.8	40.09	-1.41	0.6	116.62	18JL	-0.7	-3.4	-1.3	0.5
19.85	-3.23	-0.7	126.70	28JL	-1.2	-3.5	-1.8	0.7	46.25	-1.31	0.5	126.23	28JL	-0.2	-3.4	-1.3	0.4
23.77	-3.29	-0.8	136.36	7AU	-1.4	-3.5	-1.9	0.7	52.11	-1.19	0.4	135.89	7AU	0.2	-3.5	-1.4	0.4
26.77	-3.30	-1.1	146.06	17AU	-0.8	-3.4	-2.0	0.6	57.60	-1.04	0.3	145.59	17AU	0.4	-3.5	-1.4	0.3
28.62	-3.25	-1.3	155.83	27AU	-0.3	-3.4	-2.0	0.6	62.68	-0.87	0.2	155.35	27AU	0.7	-3.6	-1.4	0.2
29.14	-3.12	-1.5	165.64	6SE	-0.1	-3.4	-2.1	0.5	67.22	-0.66	0.1	165.17	6SE	1.7	-3.7	-1.5	0.2
28.19	-2.89	-1.7	175.51	16SE	0.1	-3.4	-2.1	0.6	71.13	-0.42	-0.1	175.04	16SE	2.6	-3.7	-1.5	0.1
25.84	-2.52	-1.9	185.45	26SE	0.2	-3.3	-2.2	0.6	74.26	-0.14	-0.3	184.97	26SE	0.0	-3.8	-1.6	0.1
22.54	-2.04	-2.0	195.44	6OC	1.0	-3.3	-2.3	0.7	76.42	0.19	-0.5	194.96	6OC	-0.7	-3.9	-1.6	0.0
19.02	-1.46	-2.0	205.48	16OC	3.2	-3.3	-2.3	0.8	77.41	0.57	-0.7	205.00	16OC	-0.8	-4.1	-1.7	0.1
16.07	-0.88	-1.7	215.57	26OC	0.3	-3.3	-2.3	0.8	77.06	1.00	-0.9	215.08	26OC	-0.8	-4.2	-1.7	0.1
14.27	-0.35	-1.4	225.71	5NO	-0.4	-3.3	-2.3	0.9	75.27	1.47	-1.1	225.22	5NO	-0.7	-4.4	-1.8	0.2
13.82	0.10	-1.0	235.87	15NO	-0.5	-3.4	-2.3	1.0	72.22	1.93	-1.3	235.38	15NO	-0.6	-4.4	-1.9	0.3
14.66	0.46	-0.7	246.06	25NO	-0.5	-3.4	-2.3	1.0	68.39	2.33	-1.4	245.57	25NO	-0.6	-4.4	-1.9	0.4
16.61	0.74	-0.4	256.26	5DE	-0.6	-3.4	-2.3	1.0	64.55	2.63	-1.3	255.77	5DE	-0.6	-4.3	-2.0	0.4
19.45	0.95	-0.1	266.46	15DE	-0.9	-3.4	-2.2	1.1	61.45	2.80	-1.0	265.97	15DE	-0.0	-3.9	-2.0	0.5
23.00	1.12	0.1	276.66	25DE	-0.9	-3.5	-2.2	1.1	59.56	2.86	-0.7	276.16	25DE	2.3	-3.2	-2.1	0.5

♂ LONG	♂ LAT	♂ MAG	⊙ LONG	16.00UT	☿	♀	♃	♄
				556	MAGNITUDES			
59.01	2.83	-0.4	286.34	4JA	1.5	-3.6	-2.1	0.6
59.71	2.76	-0.1	296.49	14JA	0.5	-4.1	-2.1	0.6
61.47	2.66	0.2	306.60	24JA	0.2	-4.3	-2.1	0.6
64.09	2.55	0.4	316.67	3FE	0.0	-4.3	-2.1	0.6
67.41	2.44	0.6	326.68	13FE	-0.2	-4.3	-2.0	0.6
71.25	2.33	0.8	336.64	23FE	-0.7	-4.2	-2.0	0.6
75.53	2.22	1.0	346.54	4MR	-1.4	-4.1	-1.9	0.6
80.14	2.11	1.1	356.38	14MR	-1.4	-3.9	-1.9	0.5
85.02	2.00	1.3	6.16	24MR	-0.6	-3.8	-1.8	0.5
90.13	1.90	1.4	15.89	3AP	0.7	-3.7	-1.7	0.4
95.42	1.80	1.5	25.57	13AP	2.0	-3.7	-1.7	0.4
100.87	1.71	1.6	35.20	23AP	3.8	-3.6	-1.6	0.4
106.45	1.61	1.7	44.80	3MY	2.4	-3.5	-1.5	0.4
112.14	1.52	1.7	54.36	13MY	1.3	-3.5	-1.5	0.5
117.94	1.43	1.8	63.91	23MY	0.5	-3.4	-1.4	0.5
123.84	1.34	1.8	73.44	2JN	-0.4	-3.4	-1.4	0.5
129.83	1.25	1.9	82.96	12JN	-1.4	-3.4	-1.3	0.5
135.91	1.16	1.9	92.49	22JN	-1.6	-3.3	-1.3	0.5
142.07	1.06	1.9	102.03	2JL	-0.7	-3.3	-1.3	0.4
148.31	0.97	1.9	111.59	12JL	-0.1	-3.3	-1.3	0.4
154.64	0.88	1.9	121.19	22JL	0.3	-3.3	-1.2	0.4
161.06	0.78	1.9	130.83	1AU	0.6	-3.3	-1.2	0.3
167.56	0.69	1.9	140.50	11AU	1.0	-3.4	-1.2	0.3
174.15	0.59	1.9	150.23	21AU	2.1	-3.4	-1.3	0.2
180.83	0.49	1.8	160.02	31AU	2.3	-3.4	-1.3	0.2
187.59	0.39	1.8	169.86	10SE	-0.0	-3.4	-1.3	0.1
194.45	0.29	1.8	179.76	20SE	-0.9	-3.5	-1.3	0.0
201.40	0.19	1.7	189.72	30SE	-0.9	-3.5	-1.4	-0.0
208.44	0.09	1.7	199.73	10OC	-0.9	-3.5	-1.4	-0.1
215.57	-0.02	1.6	209.79	20OC	-0.6	-3.5	-1.5	-0.1
222.79	-0.12	1.6	219.90	30OC	-0.5	-3.4	-1.5	-0.1
230.10	-0.23	1.5	230.05	9NO	-0.5	-3.4	-1.6	-0.0
237.49	-0.33	1.5	240.22	19NO	-0.4	-3.4	-1.6	0.1
244.96	-0.43	1.5	250.42	29NO	0.1	-3.4	-1.7	0.1
252.51	-0.53	1.5	260.62	9DE	2.5	-3.4	-1.8	0.2
260.12	-0.63	1.5	270.82	19DE	1.1	-3.3	-1.8	0.3
267.80	-0.73	1.5	281.01	29DE	0.2	-3.3	-1.9	0.3
				557				
275.52	-0.81	1.5	291.17	8JA	0.0	-3.3	-1.9	0.4
283.29	-0.90	1.5	301.30	18JA	-0.1	-3.3	-2.0	0.4
291.09	-0.97	1.4	311.39	28JA	-0.3	-3.4	-2.0	0.4
298.91	-1.04	1.4	321.43	7FE	-0.7	-3.4	-2.0	0.4
306.74	-1.10	1.4	331.42	17FE	-1.4	-3.4	-2.0	0.4
314.57	-1.14	1.4	341.35	27FE	-1.4	-3.4	-2.0	0.4
322.38	-1.18	1.4	351.22	9MR	-0.5	-3.5	-2.0	0.4
330.16	-1.20	1.4	1.04	19MR	0.9	-3.5	-1.9	0.4
337.90	-1.21	1.4	10.80	29MR	2.6	-3.6	-1.9	0.4
345.59	-1.21	1.4	20.50	8AP	3.2	-3.6	-1.8	0.3
353.22	-1.19	1.4	30.16	18AP	1.8	-3.7	-1.8	0.3
0.77	-1.16	1.4	39.78	28AP	1.0	-3.7	-1.7	0.3
8.23	-1.12	1.4	49.35	8MY	0.4	-3.8	-1.6	0.3
15.61	-1.06	1.3	58.91	18MY	-0.4	-3.9	-1.6	0.3
22.88	-0.99	1.3	68.44	28MY	-1.4	-4.0	-1.5	0.4
30.04	-0.91	1.3	77.96	7JN	-1.6	-4.1	-1.5	0.4
37.08	-0.82	1.3	87.49	17JN	-0.7	-4.2	-1.4	0.4
43.99	-0.71	1.3	97.02	27JN	-0.0	-4.2	-1.4	0.4
50.76	-0.60	1.3	106.57	7JL	0.4	-4.1	-1.3	0.4
57.39	-0.47	1.2	116.15	17JL	0.8	-3.8	-1.3	0.3
63.85	-0.34	1.2	125.77	27JL	1.3	-3.2	-1.3	0.3
70.13	-0.19	1.2	135.42	6AU	2.6	-3.4	-1.2	0.3
76.21	-0.03	1.1	145.12	16AU	1.9	-4.0	-1.2	0.2
82.07	0.15	1.0	154.88	26AU	-0.1	-4.2	-1.2	0.2
87.68	0.34	1.0	164.69	5SE	-1.0	-4.3	-1.2	0.1
92.98	0.55	0.9	174.56	15SE	-1.1	-4.3	-1.2	0.1
97.93	0.78	0.7	184.48	25SE	-0.9	-4.2	-1.2	-0.0
102.45	1.04	0.6	194.47	5OC	-0.5	-4.1	-1.2	-0.1
106.45	1.32	0.5	204.51	15OC	-0.4	-4.0	-1.3	-0.1
109.81	1.64	0.3	214.60	25OC	-0.3	-3.9	-1.3	-0.2
112.39	2.00	0.1	224.72	4NO	-0.2	-3.8	-1.3	-0.3
114.00	2.40	-0.1	234.89	14NO	0.3	-3.8	-1.4	-0.2
114.45	2.83	-0.3	245.07	24NO	2.8	-3.7	-1.4	-0.1
113.58	3.28	-0.6	255.27	4DE	0.8	-3.6	-1.5	-0.0
111.36	3.70	-0.8	265.48	14DE	0.0	-3.6	-1.6	0.0
108.02	4.04	-1.0	275.67	24DE	-0.1	-3.5	-1.6	0.1

♂ LONG	♂ LAT	♂ MAG	⊙ LONG	16.00UT	☿	♀	♃	♄
				558	MAGNITUDES			
104.08	4.24	-1.1	285.85	3JA	-0.2	-3.5	-1.7	0.2
100.27	4.26	-0.9	296.00	13JA	-0.4	-3.4	-1.8	0.2
97.29	4.13	-0.7	306.11	23JA	-0.8	-3.4	-1.8	0.3
95.52	3.91	-0.4	316.18	2FE	-1.3	-3.4	-1.9	0.3
95.05	3.63	-0.1	326.20	12FE	-1.2	-3.3	-1.9	0.3
95.80	3.33	0.1	336.16	22FE	-0.5	-3.3	-2.0	0.3
97.57	3.05	0.3	346.06	4MR	1.1	-3.3	-2.0	0.4
100.18	2.78	0.5	355.91	14MR	3.1	-3.3	-2.0	0.3
103.47	2.53	0.7	5.69	24MR	2.4	-3.3	-2.0	0.3
107.29	2.31	0.9	15.42	3AP	1.3	-3.3	-2.0	0.3
111.56	2.10	1.0	25.11	13AP	0.8	-3.4	-2.0	0.3
116.19	1.90	1.2	34.74	23AP	0.2	-3.4	-1.9	0.3
121.11	1.72	1.3	44.34	3MY	-0.5	-3.4	-1.9	0.2
126.28	1.55	1.4	53.90	13MY	-1.4	-3.5	-1.8	0.2
131.67	1.39	1.4	63.44	23MY	-1.6	-3.5	-1.7	0.2
137.25	1.24	1.5	72.97	2JN	-0.7	-3.5	-1.7	0.3
143.00	1.09	1.6	82.50	12JN	0.1	-3.4	-1.6	0.3
148.89	0.95	1.6	92.02	22JN	0.6	-3.4	-1.6	0.3
154.94	0.81	1.6	101.56	2JL	1.0	-3.4	-1.5	0.3
161.12	0.68	1.7	111.13	12JL	1.8	-3.3	-1.4	0.3
167.42	0.55	1.7	120.72	22JL	3.1	-3.3	-1.4	0.3
173.86	0.42	1.7	130.35	1AU	1.6	-3.3	-1.4	0.3
180.41	0.30	1.7	140.03	11AU	-0.2	-3.3	-1.3	0.3
187.08	0.18	1.7	149.75	21AU	-1.1	-3.3	-1.3	0.2
193.88	0.06	1.7	159.53	31AU	-1.2	-3.3	-1.3	0.2
200.78	-0.06	1.7	169.37	10SE	-0.9	-3.4	-1.2	0.1
207.80	-0.17	1.6	179.27	20SE	-0.5	-3.4	-1.2	0.1
214.93	-0.28	1.6	189.23	30SE	-0.2	-3.4	-1.2	0.0
222.17	-0.38	1.6	199.24	10OC	-0.2	-3.5	-1.2	-0.1
229.51	-0.48	1.6	209.30	20OC	-0.1	-3.5	-1.2	-0.1
236.94	-0.58	1.5	219.41	30OC	0.5	-3.6	-1.3	-0.2
244.47	-0.67	1.5	229.56	9NO	3.2	-3.6	-1.3	-0.3
252.08	-0.75	1.5	239.73	19NO	0.6	-3.7	-1.3	-0.3
259.76	-0.83	1.4	249.93	29NO	-0.1	-3.8	-1.4	-0.2
267.51	-0.89	1.4	260.13	9DE	-0.3	-3.9	-1.4	-0.2
275.31	-0.95	1.4	270.33	19DE	-0.3	-4.0	-1.4	-0.1
283.16	-1.00	1.3	280.52	29DE	-0.5	-4.1	-1.5	-0.0
				559				
291.03	-1.04	1.3	290.69	8JA	-0.8	-4.2	-1.6	0.0
298.92	-1.07	1.3	300.82	18JA	-1.2	-4.3	-1.6	0.1
306.82	-1.08	1.3	310.91	28JA	-1.1	-4.3	-1.7	0.2
314.70	-1.09	1.3	320.95	7FE	-0.4	-4.3	-1.8	0.2
322.57	-1.08	1.3	330.94	17FE	1.3	-4.2	-1.8	0.3
330.40	-1.06	1.4	340.88	27FE	3.3	-3.8	-1.9	0.3
338.18	-1.03	1.4	350.75	9MR	1.7	-3.2	-2.0	0.3
345.90	-0.99	1.4	0.57	19MR	1.0	-3.6	-2.0	0.3
353.57	-0.94	1.5	10.33	29MR	0.6	-4.0	-2.0	0.3
1.15	-0.87	1.5	20.03	8AP	0.1	-4.2	-2.1	0.3
8.66	-0.80	1.5	29.69	18AP	-0.5	-4.2	-2.1	0.3
16.08	-0.72	1.5	39.31	28AP	-1.4	-4.2	-2.1	0.3
23.41	-0.63	1.5	48.89	8MY	-1.6	-4.1	-2.0	0.2
30.65	-0.54	1.6	58.44	18MY	-0.6	-4.0	-2.0	0.2
37.78	-0.44	1.6	67.98	28MY	0.2	-3.9	-2.0	0.2
44.82	-0.33	1.6	77.50	7JN	0.8	-3.8	-1.9	0.2
51.76	-0.22	1.6	87.03	17JN	1.3	-3.7	-1.8	0.3
58.59	-0.10	1.6	96.56	27JN	2.3	-3.6	-1.8	0.3
65.32	0.02	1.6	106.11	7JL	3.1	-3.6	-1.7	0.3
71.93	0.14	1.6	115.69	17JL	1.4	-3.5	-1.7	0.3
78.44	0.27	1.6	125.30	27JL	-0.2	-3.5	-1.6	0.3
84.82	0.41	1.6	134.95	6AU	-1.2	-3.4	-1.5	0.3
91.08	0.55	1.6	144.65	16AU	-1.4	-3.4	-1.5	0.3
97.21	0.69	1.5	154.40	26AU	-0.8	-3.4	-1.4	0.3
103.18	0.85	1.5	164.21	5SE	-0.4	-3.4	-1.4	0.2
109.00	1.01	1.4	174.08	15SE	-0.1	-3.4	-1.4	0.2
114.62	1.18	1.4	184.00	25SE	-0.0	-3.4	-1.3	0.2
120.03	1.36	1.3	193.98	5OC	0.1	-3.4	-1.3	0.1
125.19	1.55	1.2	204.02	15OC	0.8	-3.4	-1.3	0.0
130.05	1.76	1.1	214.11	25OC	3.5	-3.4	-1.3	-0.1
134.54	1.99	0.9	224.23	4NO	0.4	-3.4	-1.3	-0.1
138.59	2.24	0.8	234.39	14NO	-0.3	-3.4	-1.3	-0.2
142.09	2.52	0.6	244.58	24NO	-0.4	-3.4	-1.3	-0.3
144.91	2.83	0.4	254.78	4DE	-0.4	-3.4	-1.3	-0.3
146.89	3.16	0.2	264.98	14DE	-0.6	-3.5	-1.4	-0.2
147.84	3.51	-0.1	275.18	24DE	-0.9	-3.5	-1.4	-0.2

Left (560 / 561)

♂ LONG	LAT	MAG	☉ LONG	16.00UT	☿	♀	♃	♄
				560				
147.61	3.87	-0.3	285.36	3JA	-1.0	-3.5	-1.4	-0.1
146.07	4.20	-0.6	295.51	13JA	-1.0	-3.4	-1.5	-0.0
143.28	4.44	-0.8	305.63	23JA	-0.4	-3.4	-1.5	0.0
139.62	4.54	-1.0	315.69	2FE	1.5	-3.4	-1.6	0.1
135.70	4.46	-1.0	325.71	12FE	2.7	-3.4	-1.7	0.2
132.24	4.22	-0.8	335.68	22FE	1.3	-3.4	-1.7	0.2
129.80	3.87	-0.6	345.58	3MR	0.7	-3.3	-1.8	0.2
128.63	3.47	-0.3	355.43	13MR	0.4	-3.3	-1.9	0.3
128.74	3.05	-0.1	5.22	23MR	0.0	-3.3	-1.9	0.3
129.99	2.66	0.1	14.95	2AP	-0.5	-3.3	-2.0	0.3
132.20	2.30	0.3	24.64	12AP	-1.4	-3.4	-2.1	0.3
135.21	1.97	0.5	34.27	22AP	-1.6	-3.4	-2.1	0.3
138.85	1.68	0.6	43.87	2MY	-0.6	-3.4	-2.1	0.3
143.03	1.41	0.8	53.44	12MY	0.3	-3.4	-2.2	0.3
147.64	1.16	0.9	62.98	22MY	1.0	-3.5	-2.2	0.2
152.61	0.94	1.0	72.51	1JN	1.8	-3.5	-2.2	0.2
157.90	0.73	1.1	82.04	11JN	3.0	-3.6	-2.1	0.2
163.46	0.54	1.2	91.56	21JN	2.6	-3.6	-2.1	0.3
169.26	0.36	1.2	101.10	1JL	1.1	-3.7	-2.1	0.3
175.28	0.19	1.3	110.67	11JL	-0.3	-3.8	-2.0	0.3
181.49	0.03	1.3	120.26	21JL	-1.3	-3.9	-1.9	0.3
187.88	-0.12	1.3	129.88	31JL	-1.5	-4.0	-1.9	0.4
194.45	-0.25	1.4	139.56	10AU	-0.8	-4.1	-1.8	0.4
201.18	-0.38	1.4	149.28	20AU	-0.3	-4.2	-1.7	0.4
208.05	-0.50	1.4	159.06	30AU	-0.0	-4.3	-1.7	0.4
215.07	-0.61	1.4	168.90	9SE	0.1	-4.3	-1.6	0.3
222.23	-0.72	1.4	178.79	19SE	0.3	-4.2	-1.6	0.3
229.50	-0.81	1.4	188.74	29SE	1.1	-3.9	-1.5	0.3
236.90	-0.89	1.4	198.75	9OC	3.2	-3.2	-1.5	0.2
244.40	-0.96	1.4	208.81	19OC	-0.3	-3.6	-1.5	0.1
252.00	-1.02	1.4	218.91	29OC	-0.5	-4.1	-1.4	0.1
259.68	-1.07	1.4	229.06	8NO	-0.5	-4.3	-1.4	0.0
267.44	-1.11	1.4	239.23	18NO	-0.6	-4.4	-1.4	-0.1
275.25	-1.13	1.4	249.43	28NO	-0.7	-4.3	-1.4	-0.1
283.10	-1.14	1.4	259.63	8DE	-0.8	-4.2	-1.4	-0.2
290.99	-1.14	1.4	269.84	18DE	-0.8	-4.1	-1.4	-0.2
298.89	-1.13	1.4	280.02	28DE	-0.8	-4.0	-1.4	-0.2
				561				
306.78	-1.10	1.4	290.20	7JA	-0.3	-3.9	-1.4	-0.1
314.67	-1.07	1.4	300.33	17JA	1.8	-3.8	-1.5	-0.0
322.52	-1.02	1.4	310.42	27JA	2.2	-3.7	-1.5	0.0
330.34	-0.96	1.4	320.47	6FE	0.9	-3.6	-1.6	0.1
338.10	-0.90	1.4	330.46	16FE	0.5	-3.6	-1.6	0.1
345.81	-0.82	1.4	340.40	26FE	0.3	-3.5	-1.7	0.2
353.44	-0.74	1.4	350.28	8MR	-0.1	-3.5	-1.7	0.3
1.00	-0.66	1.4	0.10	18MR	-0.6	-3.4	-1.8	0.3
8.48	-0.56	1.5	9.86	28MR	-1.4	-3.4	-1.9	0.3
15.88	-0.47	1.5	19.57	7AP	-1.5	-3.3	-1.9	0.4
23.19	-0.37	1.5	29.23	17AP	-0.6	-3.3	-2.0	0.4
30.41	-0.27	1.6	38.85	27AP	0.4	-3.3	-2.1	0.4
37.54	-0.16	1.6	48.44	7MY	1.3	-3.3	-2.1	0.4
44.58	-0.06	1.7	57.99	17MY	2.4	-3.3	-2.2	0.4
51.54	0.05	1.7	67.52	27MY	3.5	-3.3	-2.2	0.4
58.42	0.15	1.8	77.05	6JN	2.1	-3.3	-2.3	0.3
65.20	0.26	1.8	86.57	16JN	0.9	-3.4	-2.3	0.3
71.91	0.37	1.8	96.10	26JN	-0.3	-3.4	-2.3	0.3
78.54	0.47	1.8	105.65	6JL	-1.3	-3.4	-2.3	0.3
85.09	0.58	1.8	115.22	16JL	-1.5	-3.5	-2.3	0.4
91.56	0.68	1.8	124.83	26JL	-0.8	-3.5	-2.2	0.4
97.96	0.79	1.8	134.48	5AU	-0.2	-3.5	-2.2	0.4
104.27	0.89	1.8	144.18	15AU	0.1	-3.4	-2.1	0.5
110.51	1.00	1.8	153.93	25AU	0.3	-3.4	-2.1	0.5
116.66	1.11	1.8	163.73	4SE	0.5	-3.4	-2.0	0.5
122.71	1.22	1.8	173.59	14SE	1.4	-3.4	-1.9	0.5
128.68	1.33	1.7	183.51	24SE	2.9	-3.3	-1.9	0.4
134.52	1.44	1.7	193.49	4OC	0.2	-3.3	-1.8	0.4
140.24	1.56	1.6	203.52	14OC	-0.6	-3.3	-1.7	0.4
145.82	1.69	1.5	213.61	24OC	-0.7	-3.3	-1.7	0.3
151.23	1.82	1.4	223.73	3NO	-0.7	-3.3	-1.6	0.3
156.44	1.95	1.3	233.89	13NO	-0.8	-3.4	-1.6	0.2
161.40	2.10	1.2	244.08	23NO	-0.7	-3.4	-1.6	0.1
166.07	2.25	1.0	254.28	3DE	-0.7	-3.4	-1.5	0.0
170.38	2.42	0.9	264.48	13DE	-0.7	-3.4	-1.5	-0.0
174.24	2.59	0.7	274.68	23DE	-0.2	-3.5	-1.5	-0.1

Right (562 / 563)

♂ LONG	LAT	MAG	☉ LONG	16.00UT	☿	♀	♃	♄
				562				
177.54	2.78	0.5	284.86	2JA	2.0	-3.5	-1.5	-0.1
180.16	2.98	0.2	295.01	12JA	1.7	-3.6	-1.5	-0.1
181.92	3.17	-0.0	305.13	22JA	0.6	-3.6	-1.5	-0.0
182.63	3.36	-0.3	315.21	1FE	0.3	-3.7	-1.5	0.1
182.16	3.51	-0.6	325.23	11FE	0.1	-3.8	-1.5	0.1
180.39	3.59	-0.9	335.20	21FE	-0.1	-3.8	-1.6	0.2
177.47	3.57	-1.1	345.11	3MR	-0.6	-3.9	-1.6	0.2
173.81	3.39	-1.3	354.96	13MR	-1.4	-4.0	-1.7	0.3
170.06	3.08	-1.2	4.75	23MR	-1.5	-4.1	-1.7	0.4
166.93	2.66	-1.0	14.49	2AP	-0.6	-4.2	-1.8	0.4
164.93	2.19	-0.8	24.17	12AP	0.5	-4.2	-1.8	0.4
164.23	1.72	-0.6	33.81	22AP	1.7	-4.2	-1.9	0.5
164.83	1.29	-0.4	43.41	2MY	3.2	-3.9	-2.0	0.5
166.57	0.90	-0.2	52.98	12MY	2.9	-3.3	-2.0	0.5
169.26	0.55	-0.0	62.53	22MY	1.6	-2.9	-2.1	0.5
172.76	0.24	0.1	72.05	1JN	0.7	-3.7	-2.2	0.5
176.90	-0.02	0.3	81.57	11JN	-0.3	-4.1	-2.2	0.5
181.59	-0.26	0.4	91.10	21JN	-1.3	-4.2	-2.3	0.5
186.74	-0.47	0.5	100.64	1JL	-1.6	-4.2	-2.3	0.4
192.26	-0.65	0.6	110.20	11JL	-0.7	-4.1	-2.4	0.4
198.13	-0.81	0.7	119.79	21JL	-0.1	-4.0	-2.4	0.5
204.29	-0.95	0.8	129.42	31JL	0.2	-3.9	-2.4	0.5
210.70	-1.07	0.8	139.09	10AU	0.5	-3.9	-2.4	0.6
217.34	-1.18	0.9	148.81	20AU	0.8	-3.8	-2.4	0.6
224.18	-1.26	0.9	158.58	30AU	1.8	-3.7	-2.3	0.6
231.20	-1.33	1.0	168.42	9SE	2.5	-3.6	-2.3	0.6
238.39	-1.38	1.0	178.31	19SE	0.1	-3.6	-2.2	0.6
245.72	-1.42	1.1	188.26	29SE	-0.8	-3.5	-2.2	0.6
253.17	-1.43	1.1	198.26	9OC	-0.9	-3.5	-2.1	0.6
260.73	-1.44	1.1	208.32	19OC	-0.8	-3.5	-2.0	0.6
268.37	-1.42	1.2	218.42	29OC	-0.7	-3.4	-1.9	0.5
276.09	-1.40	1.2	228.56	8NO	-0.6	-3.4	-1.9	0.5
283.85	-1.36	1.2	238.74	18NO	-0.5	-3.4	-1.8	0.4
291.65	-1.30	1.3	248.93	28NO	-0.5	-3.4	-1.8	0.4
299.46	-1.24	1.3	259.13	8DE	-0.0	-3.4	-1.7	0.3
307.28	-1.16	1.3	269.34	18DE	2.3	-3.3	-1.7	0.2
315.08	-1.08	1.4	279.53	28DE	1.4	-3.3	-1.6	0.1
				563				
322.85	-0.98	1.4	289.69	7JA	0.4	-3.3	-1.6	0.1
330.58	-0.89	1.4	299.83	17JA	0.1	-3.4	-1.6	0.1
338.25	-0.78	1.5	309.93	27JA	0.0	-3.4	-1.6	0.1
345.87	-0.68	1.5	319.98	6FE	-0.2	-3.4	-1.6	0.2
353.41	-0.57	1.5	329.98	16FE	-0.7	-3.4	-1.6	0.2
0.88	-0.46	1.6	339.91	26FE	-1.4	-3.4	-1.6	0.3
8.27	-0.35	1.6	349.80	8MR	-1.4	-3.5	-1.6	0.3
15.59	-0.24	1.6	359.62	18MR	-0.6	-3.5	-1.6	0.4
22.81	-0.13	1.7	9.38	28MR	0.7	-3.4	-1.6	0.5
29.96	-0.02	1.7	19.10	7AP	2.2	-3.4	-1.7	0.5
37.02	0.08	1.7	28.76	17AP	3.8	-3.4	-1.7	0.5
44.00	0.18	1.7	38.38	27AP	2.2	-3.4	-1.8	0.6
50.91	0.28	1.7	47.97	7MY	1.2	-3.3	-1.8	0.6
57.75	0.37	1.7	57.52	17MY	0.5	-3.3	-1.9	0.6
64.51	0.47	1.8	67.06	27MY	-0.4	-3.3	-1.9	0.7
71.21	0.56	1.8	76.58	6JN	-1.4	-3.3	-2.0	0.7
77.84	0.64	1.9	86.11	16JN	-1.6	-3.3	-2.1	0.7
84.42	0.72	1.9	95.64	26JN	-0.7	-3.3	-2.1	0.7
90.95	0.80	1.9	105.18	6JL	-0.1	-3.4	-2.2	0.7
97.42	0.88	1.9	114.76	16JL	0.3	-3.4	-2.3	0.7
103.84	0.95	2.0	124.36	26JL	0.6	-3.4	-2.3	0.7
110.23	1.02	2.0	134.01	5AU	1.1	-3.5	-2.4	0.7
116.56	1.09	2.0	143.71	15AU	2.3	-3.5	-2.4	0.8
122.86	1.16	2.0	153.45	25AU	2.2	-3.6	-2.5	0.8
129.11	1.22	2.0	163.25	4SE	-0.0	-3.7	-2.5	0.8
135.33	1.28	2.0	173.12	14SE	-0.9	-3.8	-2.5	0.8
141.50	1.34	1.9	183.03	24SE	-1.0	-3.8	-2.4	0.9
147.62	1.40	1.9	193.01	4OC	-0.9	-3.9	-2.4	0.9
153.69	1.45	1.8	203.04	14OC	-0.6	-4.1	-2.3	0.8
159.70	1.51	1.8	213.12	24OC	-0.4	-4.2	-2.3	0.8
165.64	1.56	1.7	223.24	3NO	-0.4	-4.3	-2.2	0.8
171.51	1.61	1.6	233.40	13NO	-0.3	-4.4	-2.1	0.8
177.28	1.66	1.5	243.58	23NO	0.2	-4.4	-2.1	0.7
182.95	1.71	1.4	253.78	3DE	2.6	-4.3	-2.0	0.6
188.48	1.75	1.3	263.99	13DE	1.0	-3.9	-1.9	0.6
193.85	1.79	1.2	274.18	23DE	0.2	-3.2	-1.9	0.5

Left table (564 / 565)

♂ LONG	LAT	MAG	☉ LONG	16.00UT	☿	♀	♃	♄
				564		MAGNITUDES		
199.03	1.82	1.0	284.37	2JA	-0.0	-3.6	-1.8	0.4
203.96	1.85	0.9	294.52	12JA	-0.1	-4.1	-1.7	0.4
208.61	1.86	0.7	304.64	22JA	-0.3	-4.3	-1.7	0.3
212.89	1.87	0.5	314.71	1FE	-0.7	-4.3	-1.7	0.3
216.70	1.85	0.2	324.74	11FE	-1.3	-4.3	-1.6	0.3
219.95	1.81	-0.0	334.71	21FE	-1.3	-4.2	-1.6	0.4
222.48	1.73	-0.3	344.62	2MR	-0.6	-4.1	-1.6	0.4
224.12	1.60	-0.6	354.47	12MR	0.9	-3.9	-1.6	0.5
224.68	1.41	-0.9	4.27	22MR	2.7	-3.8	-1.6	0.6
224.03	1.14	-1.2	14.01	1AP	2.8	-3.7	-1.6	0.6
222.12	0.76	-1.5	23.70	11AP	1.6	-3.7	-1.6	0.7
219.20	0.31	-1.8	33.34	21AP	0.9	-3.6	-1.6	0.7
215.75	-0.21	-1.9	42.95	1MY	0.3	-3.5	-1.6	0.8
212.53	-0.72	-1.7	52.52	11MY	-0.4	-3.5	-1.7	0.8
210.19	-1.19	-1.6	62.06	21MY	-1.4	-3.4	-1.7	0.8
209.11	-1.58	-1.3	71.60	31MY	-1.6	-3.4	-1.7	0.9
209.42	-1.88	-1.1	81.12	10JN	-0.7	-3.4	-1.8	0.9
211.02	-2.11	-0.9	90.64	20JN	0.0	-3.3	-1.8	0.9
213.73	-2.27	-0.7	100.19	30JN	0.5	-3.3	-1.9	0.9
217.38	-2.38	-0.5	109.74	10JL	0.9	-3.3	-2.0	0.9
221.79	-2.45	-0.4	119.33	20JL	1.5	-3.3	-2.0	0.9
226.83	-2.48	-0.2	128.96	30JL	2.8	-3.3	-2.1	0.9
232.38	-2.48	-0.1	138.62	9AU	1.9	-3.4	-2.2	0.9
238.35	-2.46	0.0	148.34	19AU	-0.1	-3.4	-2.2	1.0
244.68	-2.41	0.2	158.11	29AU	-1.0	-3.4	-2.3	1.0
251.29	-2.34	0.3	167.94	8SE	-1.2	-3.4	-2.4	1.1
258.13	-2.25	0.4	177.83	18SE	-0.9	-3.5	-2.4	1.1
265.17	-2.14	0.5	187.78	28SE	-0.5	-3.5	-2.4	1.1
272.36	-2.02	0.6	197.77	8OC	-0.3	-3.5	-2.4	1.1
279.67	-1.89	0.7	207.83	18OC	-0.2	-3.5	-2.4	1.1
287.07	-1.75	0.7	217.93	28OC	-0.2	-3.4	-2.4	1.1
294.54	-1.60	0.8	228.07	7NO	0.4	-3.4	-2.4	1.1
302.04	-1.44	0.9	238.24	17NO	2.9	-3.4	-2.3	1.1
309.56	-1.29	1.0	248.43	27NO	0.8	-3.4	-2.2	1.0
317.08	-1.13	1.1	258.63	7DE	-0.0	-3.4	-2.2	1.0
324.58	-0.97	1.2	268.84	17DE	-0.2	-3.3	-2.1	0.9
332.06	-0.82	1.2	279.03	27DE	-0.2	-3.3	-2.0	0.8
				565				
339.49	-0.67	1.3	289.20	6JA	-0.4	-3.3	-2.0	0.8
346.87	-0.52	1.4	299.34	16JA	-0.8	-3.3	-1.9	0.7
354.20	-0.38	1.5	309.44	26JA	-1.2	-3.4	-1.8	0.6
1.46	-0.24	1.5	319.49	5FE	-1.2	-3.4	-1.8	0.6
8.66	-0.11	1.6	329.49	15FE	-0.5	-3.4	-1.7	0.6
15.78	0.01	1.6	339.43	25FE	1.1	-3.4	-1.7	0.6
22.84	0.12	1.7	349.31	7MR	3.3	-3.5	-1.6	0.7
29.82	0.23	1.7	359.14	17MR	2.1	-3.5	-1.6	0.7
36.74	0.33	1.8	8.91	27MR	1.2	-3.6	-1.6	0.8
43.58	0.43	1.8	18.63	6AP	0.7	-3.6	-1.5	0.8
50.37	0.52	1.8	28.29	16AP	0.2	-3.7	-1.5	0.9
57.10	0.60	1.9	37.92	26AP	-0.5	-3.7	-1.5	0.9
63.76	0.68	1.9	47.50	6MY	-1.4	-3.8	-1.5	1.0
70.38	0.75	1.9	57.06	16MY	-1.6	-3.9	-1.5	1.0
76.95	0.82	1.9	66.60	26MY	-0.7	-4.0	-1.5	1.1
83.48	0.88	1.9	76.12	5JN	0.1	-4.1	-1.6	1.1
89.97	0.93	1.9	85.65	15JN	0.6	-4.2	-1.6	1.1
96.42	0.98	1.9	95.18	25JN	1.1	-4.2	-1.6	1.2
102.85	1.03	1.9	104.73	5JL	1.9	-4.1	-1.7	1.2
109.26	1.07	2.0	114.30	15JL	3.2	-3.7	-1.7	1.2
115.64	1.11	2.0	123.90	25JL	1.6	-3.2	-1.8	1.2
122.01	1.14	2.0	133.55	4AU	-0.1	-3.5	-1.8	1.2
128.37	1.17	2.0	143.24	14AU	-1.1	-4.0	-1.9	1.2
134.71	1.19	2.0	152.98	24AU	-1.3	-4.2	-1.9	1.2
141.05	1.21	2.0	162.78	3SE	-0.9	-4.3	-2.0	1.3
147.39	1.23	2.0	172.64	13SE	-0.4	-4.2	-2.1	1.3
153.72	1.24	2.0	182.55	23SE	-0.2	-4.2	-2.1	1.4
160.05	1.25	2.0	192.52	3OC	-0.1	-4.1	-2.2	1.4
166.38	1.25	2.0	202.55	13OC	0.0	-4.0	-2.3	1.4
172.70	1.25	1.9	212.63	23OC	0.6	-3.9	-2.3	1.3
179.02	1.24	1.9	222.75	2NO	3.2	-3.8	-2.3	1.3
185.34	1.23	1.8	232.91	12NO	0.6	-3.7	-2.3	1.3
191.63	1.21	1.8	243.09	22NO	-0.2	-3.7	-2.3	1.2
197.92	1.18	1.7	253.29	2DE	-0.3	-3.6	-2.3	1.2
204.18	1.15	1.6	263.49	12DE	-0.4	-3.6	-2.3	1.1
210.42	1.10	1.5	273.69	22DE	-0.5	-3.5	-2.2	1.1

Right table (566 / 567)

♂ LONG	LAT	MAG	☉ LONG	16.00UT	☿	♀	♃	♄
				566		MAGNITUDES		
216.62	1.04	1.4	283.87	1JA	-0.8	-3.5	-2.1	1.0
222.78	0.97	1.3	294.03	11JA	-1.1	-3.4	-2.1	1.0
228.88	0.88	1.1	304.15	21JA	-1.1	-3.4	-2.0	1.0
234.92	0.77	1.0	314.23	31JA	-0.5	-3.4	-1.9	0.9
240.86	0.64	0.8	324.26	10FE	1.3	-3.3	-1.9	0.9
246.69	0.49	0.7	334.23	20FE	3.1	-3.3	-1.8	0.9
252.39	0.30	0.5	344.14	2MR	1.6	-3.3	-1.7	0.9
257.91	0.08	0.3	354.00	12MR	0.9	-3.3	-1.7	0.9
263.21	-0.19	0.1	3.80	22MR	0.5	-3.3	-1.6	1.0
268.24	-0.51	-0.1	13.54	1AP	0.1	-3.3	-1.6	1.0
272.89	-0.88	-0.4	23.23	11AP	-0.5	-3.4	-1.5	1.1
277.08	-1.33	-0.6	32.88	21AP	-1.4	-3.4	-1.5	1.1
280.68	-1.85	-0.9	42.48	1MY	-1.6	-3.4	-1.5	1.2
283.51	-2.46	-1.2	52.05	11MY	-0.7	-3.5	-1.4	1.2
285.40	-3.15	-1.5	61.60	21MY	0.2	-3.5	-1.4	1.3
286.17	-3.92	-1.8	71.13	31MY	0.8	-3.5	-1.4	1.3
285.68	-4.70	-2.1	80.66	10JN	1.5	-3.4	-1.4	1.4
284.03	-5.42	-2.4	90.18	20JN	2.6	-3.4	-1.4	1.4
281.55	-5.98	-2.6	99.72	30JN	3.0	-3.4	-1.4	1.4
278.89	-6.26	-2.5	109.28	10JL	1.3	-3.3	-1.5	1.4
276.79	-6.26	-2.3	118.86	20JL	-0.2	-3.3	-1.5	1.4
275.77	-6.01	-2.1	128.48	30JL	-1.2	-3.3	-1.5	1.4
276.10	-5.59	-1.8	138.15	9AU	-1.4	-3.3	-1.6	1.3
277.73	-5.10	-1.6	147.86	19AU	-0.8	-3.3	-1.6	1.3
280.50	-4.58	-1.3	157.63	29AU	-0.3	-3.3	-1.7	1.2
284.22	-4.06	-1.1	167.46	8SE	-0.1	-3.4	-1.7	1.2
288.69	-3.56	-0.8	177.34	18SE	0.0	-3.4	-1.8	1.2
293.75	-3.10	-0.6	187.29	28SE	0.2	-3.4	-1.8	1.2
299.27	-2.66	-0.4	197.29	8OC	0.9	-3.5	-1.9	1.2
305.13	-2.26	-0.2	207.34	18OC	3.5	-3.5	-2.0	1.2
311.25	-1.89	0.0	217.44	28OC	-0.4	-3.6	-2.0	1.2
317.58	-1.56	0.2	227.58	7NO	-0.4	-3.6	-2.1	1.1
324.05	-1.25	0.4	237.74	17NO	-0.5	-3.7	-2.1	1.1
330.62	-0.97	0.5	247.94	27NO	-0.5	-3.8	-2.2	1.1
337.27	-0.72	0.7	258.14	7DE	-0.6	-3.9	-2.2	1.1
343.97	-0.49	0.8	268.34	17DE	-0.9	-4.0	-2.2	1.0
350.69	-0.29	1.0	278.54	27DE	-0.9	-4.1	-2.2	1.0
				567				
357.42	-0.11	1.1	288.71	6JA	-0.9	-4.2	-2.2	0.9
4.14	0.06	1.2	298.85	16JA	-0.4	-4.3	-2.1	0.9
10.85	0.21	1.3	308.95	26JA	1.5	-4.3	-2.0	0.8
17.54	0.34	1.4	319.00	5FE	2.6	-4.3	-2.0	0.8
24.20	0.46	1.5	329.00	15FE	1.1	-4.2	-1.9	0.7
30.83	0.56	1.6	338.95	25FE	0.6	-3.8	-1.9	0.7
37.43	0.66	1.7	348.84	7MR	0.3	-3.8	-1.8	0.7
44.00	0.74	1.7	358.66	17MR	0.0	-3.6	-1.7	0.7
50.53	0.82	1.8	8.44	27MR	-0.5	-4.0	-1.7	0.8
57.03	0.88	1.8	18.16	6AP	-1.4	-4.2	-1.6	0.9
63.50	0.94	1.9	27.82	16AP	-1.6	-4.2	-1.6	0.9
69.94	0.98	1.9	37.45	26AP	-0.7	-4.2	-1.5	1.0
76.35	1.03	2.0	47.04	6MY	0.3	-4.1	-1.5	1.1
82.74	1.06	2.0	56.59	16MY	1.1	-4.0	-1.4	1.1
89.12	1.09	2.0	66.14	26MY	2.0	-3.9	-1.4	1.2
95.48	1.12	2.0	75.66	5JN	3.3	-3.8	-1.4	1.2
101.83	1.13	2.0	85.18	15JN	2.5	-3.7	-1.3	1.2
108.17	1.15	2.0	94.72	25JN	1.0	-3.6	-1.3	1.2
114.52	1.16	2.0	104.26	5JL	-0.2	-3.6	-1.3	1.2
120.87	1.16	2.0	113.83	15JL	-1.3	-3.5	-1.3	1.2
127.23	1.16	2.0	123.44	25JL	-1.5	-3.5	-1.3	1.2
133.60	1.15	2.0	133.08	4AU	-0.8	-3.4	-1.3	1.2
139.98	1.14	2.0	142.77	14AU	-0.3	-3.4	-1.4	1.2
146.39	1.13	2.0	152.51	24AU	0.0	-3.4	-1.4	1.1
152.82	1.11	2.0	162.31	3SE	0.2	-3.4	-1.4	1.1
159.27	1.08	2.0	172.16	13SE	0.4	-3.4	-1.5	1.0
165.76	1.05	2.0	182.08	23SE	1.2	-3.4	-1.5	1.0
172.27	1.02	2.0	192.04	3OC	3.2	-3.4	-1.6	1.0
178.82	0.98	2.0	202.07	13OC	0.3	-3.4	-1.6	1.0
185.41	0.93	1.9	212.14	23OC	-0.5	-3.4	-1.7	1.0
192.03	0.88	1.9	222.26	2NO	-0.6	-3.4	-1.7	1.0
198.68	0.82	1.9	232.41	12NO	-0.6	-3.4	-1.8	1.0
205.38	0.75	1.8	242.60	22NO	-0.7	-3.4	-1.9	1.0
212.10	0.68	1.8	252.79	2DE	-0.8	-3.4	-1.9	1.0
218.87	0.59	1.7	263.00	12DE	-0.8	-3.5	-2.0	1.0
225.67	0.50	1.6	273.20	22DE	-0.8	-3.5	-2.0	0.9

♂ LONG	♂ LAT	♂ MAG	⊙ LONG	16.00UT 568	☿	♀	♃	♄
					MAGNITUDES			
232.51	0.39	1.6	283.38	1JA	-0.3	-3.5	-2.1	0.9
239.38	0.27	1.5	293.53	11JA	1.8	-3.4	-2.1	0.8
246.28	0.14	1.4	303.66	21JA	2.1	-3.4	-2.1	0.8
253.20	0.00	1.3	313.74	31JA	0.8	-3.4	-2.1	0.7
260.15	-0.16	1.2	323.77	10FE	0.4	-3.4	-2.0	0.7
267.12	-0.33	1.1	333.74	20FE	0.2	-3.4	-2.0	0.6
274.11	-0.52	1.0	343.66	1MR	-0.1	-3.3	-2.0	0.6
281.09	-0.72	0.9	353.52	11MR	-0.6	-3.3	-1.9	0.6
288.08	-0.94	0.8	3.32	21MR	-1.4	-3.3	-1.8	0.5
295.04	-1.17	0.6	13.07	31MR	-1.5	-3.3	-1.8	0.6
301.96	-1.42	0.5	22.76	10AP	-0.6	-3.4	-1.7	0.7
308.83	-1.68	0.4	32.41	20AP	0.4	-3.4	-1.6	0.7
315.60	-1.96	0.2	42.01	30AP	1.4	-3.4	-1.6	0.8
322.26	-2.24	0.1	51.59	10MY	2.6	-3.4	-1.5	0.9
328.75	-2.54	-0.1	61.14	20MY	3.4	-3.5	-1.5	0.9
335.02	-2.84	-0.2	70.67	30MY	1.9	-3.5	-1.4	1.0
341.00	-3.15	-0.4	80.19	9JN	0.8	-3.6	-1.4	1.0
346.61	-3.47	-0.6	89.72	19JN	-0.3	-3.6	-1.3	1.1
351.73	-3.78	-0.8	99.26	29JN	-1.3	-3.7	-1.3	1.1
356.24	-4.08	-1.0	108.81	9JL	-1.6	-3.8	-1.3	1.1
359.95	-4.37	-1.2	118.40	19JL	-0.8	-3.9	-1.3	1.1
2.69	-4.62	-1.4	128.02	29JL	-0.2	-4.0	-1.2	1.1
4.22	-4.82	-1.7	137.68	8AU	0.1	-4.1	-1.2	1.1
4.35	-4.92	-1.9	147.40	18AU	0.3	-4.2	-1.2	1.1
3.05	-4.86	-2.1	157.16	28AU	0.6	-4.3	-1.3	1.0
0.53	-4.60	-2.3	166.99	7SE	1.5	-4.3	-1.3	1.0
357.36	-4.12	-2.3	176.87	17SE	2.8	-4.2	-1.3	0.9
354.36	-3.48	-2.1	186.81	27SE	0.2	-3.8	-1.3	0.9
352.20	-2.76	-1.8	196.81	7OC	-0.7	-3.2	-1.4	0.9
351.31	-2.07	-1.5	206.86	17OC	-0.8	-3.6	-1.4	0.9
351.76	-1.44	-1.2	216.95	27OC	-0.7	-4.1	-1.5	0.9
353.40	-0.91	-0.9	227.09	6NO	-0.8	-4.3	-1.5	0.9
356.04	-0.46	-0.6	237.26	16NO	-0.6	-4.4	-1.6	0.9
359.48	-0.10	-0.3	247.44	26NO	-0.6	-4.3	-1.6	0.9
3.54	0.19	-0.0	257.65	6DE	-0.6	-4.2	-1.7	0.9
8.08	0.43	0.2	267.85	16DE	-0.1	-4.1	-1.8	0.9
12.99	0.63	0.4	278.04	26DE	2.0	-4.0	-1.8	0.9

♂ LONG	♂ LAT	♂ MAG	⊙ LONG	569	☿	♀	♃	♄
18.18	0.78	0.6	288.22	5JA	1.6	-3.9	-1.9	0.8
23.60	0.91	0.8	298.36	15JA	0.5	-3.8	-1.9	0.8
29.18	1.01	1.0	308.46	25JA	0.2	-3.7	-2.0	0.7
34.89	1.10	1.1	318.52	4FE	0.1	-3.6	-2.0	0.7
40.71	1.16	1.3	328.53	14FE	-0.2	-3.6	-2.0	0.6
46.59	1.21	1.4	338.47	24FE	-0.6	-3.5	-2.0	0.6
52.54	1.25	1.5	348.37	6MR	-1.3	-3.5	-2.0	0.5
58.54	1.28	1.6	358.20	16MR	-1.5	-3.4	-2.0	0.5
64.57	1.30	1.7	7.97	26MR	-0.6	-3.4	-1.9	0.4
70.63	1.32	1.8	17.69	5AP	0.6	-3.3	-1.9	0.5
76.72	1.32	1.8	27.36	15AP	1.8	-3.3	-1.8	0.5
82.82	1.32	1.9	36.99	25AP	3.5	-3.3	-1.8	0.6
88.95	1.32	1.9	46.58	5MY	2.6	-3.3	-1.7	0.6
95.10	1.31	2.0	56.14	15MY	1.5	-3.3	-1.6	0.7
101.26	1.29	2.0	65.68	25MY	0.6	-3.3	-1.6	0.8
107.46	1.27	2.0	75.21	4JN	-0.3	-3.3	-1.5	0.8
113.67	1.25	2.0	84.73	14JN	-1.3	-3.4	-1.4	0.9
119.91	1.22	2.0	94.26	24JN	-1.6	-3.4	-1.4	0.9
126.19	1.19	2.0	103.80	4JL	-0.7	-3.4	-1.4	0.9
132.49	1.15	2.0	113.37	14JL	-0.1	-3.5	-1.3	1.0
138.83	1.11	2.0	122.97	24JL	0.2	-3.5	-1.3	1.0
145.22	1.07	2.0	132.61	3AU	0.5	-3.5	-1.2	1.0
151.65	1.02	2.0	142.30	13AU	0.9	-3.4	-1.2	1.0
158.12	0.97	1.9	152.03	23AU	1.9	-3.4	-1.2	1.0
164.65	0.91	1.9	161.83	2SE	2.5	-3.4	-1.2	0.9
171.24	0.85	1.9	171.68	12SE	0.1	-3.4	-1.2	0.9
177.87	0.79	1.9	181.59	22SE	-0.8	-3.3	-1.2	0.9
184.57	0.72	1.9	191.56	2OC	-0.9	-3.3	-1.2	0.8
191.33	0.64	1.8	201.58	12OC	-0.9	-3.3	-1.3	0.8
198.16	0.56	1.8	211.65	22OC	-0.7	-3.3	-1.3	0.8
205.05	0.48	1.8	221.77	1NO	-0.5	-3.3	-1.3	0.8
212.00	0.38	1.8	231.92	11NO	-0.5	-3.4	-1.4	0.9
219.02	0.29	1.8	242.10	21NO	-0.5	-3.4	-1.4	0.9
226.10	0.19	1.7	252.30	1DE	0.0	-3.4	-1.5	0.9
233.25	0.08	1.7	262.50	11DE	2.3	-3.4	-1.5	0.9
240.46	-0.04	1.7	272.70	21DE	1.3	-3.5	-1.6	0.8
247.73	-0.15	1.6	282.89	31DE	0.3	-3.5	-1.7	0.8

♂ LONG	♂ LAT	♂ MAG	⊙ LONG	16.00UT 570	☿	♀	♃	♄
					MAGNITUDES			
255.06	-0.28	1.6	293.05	10JA	0.1	-3.6	-1.7	0.8
262.45	-0.41	1.5	303.17	20JA	-0.0	-3.6	-1.8	0.7
269.88	-0.54	1.5	313.25	30JA	-0.3	-3.7	-1.9	0.7
277.35	-0.68	1.4	323.28	9FE	-0.7	-3.8	-1.9	0.7
284.87	-0.81	1.3	333.26	19FE	-1.3	-3.8	-2.0	0.6
292.41	-0.95	1.3	343.19	1MR	-1.4	-3.9	-2.0	0.6
299.96	-1.09	1.2	353.05	11MR	-0.6	-4.0	-2.0	0.5
307.53	-1.22	1.1	2.85	21MR	0.7	-4.1	-2.0	0.4
315.10	-1.35	1.1	12.60	31MR	2.3	-4.2	-2.0	0.4
322.64	-1.48	1.0	22.30	10AP	3.4	-4.2	-2.0	0.3
330.15	-1.59	1.0	31.95	20AP	1.9	-4.2	-2.0	0.4
337.60	-1.69	0.9	41.56	30AP	1.1	-3.9	-1.9	0.5
344.99	-1.78	0.8	51.13	10MY	0.4	-3.3	-1.9	0.5
352.28	-1.86	0.8	60.68	20MY	-0.4	-2.9	-1.8	0.6
359.46	-1.92	0.7	70.21	30MY	-1.3	-3.7	-1.7	0.6
6.50	-1.96	0.6	79.73	9JN	-1.6	-4.1	-1.7	0.7
13.37	-1.99	0.5	89.26	19JN	-0.7	-4.2	-1.6	0.8
20.04	-1.99	0.4	98.79	29JN	-0.1	-4.2	-1.6	0.8
26.47	-1.98	0.4	108.35	9JL	0.4	-4.1	-1.5	0.8
32.62	-1.94	0.3	117.93	19JL	0.7	-4.0	-1.4	0.9
38.42	-1.87	0.1	127.55	29JL	1.2	-3.9	-1.4	0.9
43.82	-1.79	0.0	137.21	8AU	2.5	-3.8	-1.4	0.9
48.72	-1.67	-0.1	146.92	18AU	2.2	-3.8	-1.3	0.9
52.99	-1.52	-0.3	156.69	28AU	-0.0	-3.7	-1.3	0.9
56.52	-1.33	-0.4	166.51	7SE	-0.9	-3.6	-1.3	0.9
59.11	-1.09	-0.6	176.39	17SE	-1.1	-3.6	-1.2	0.9
60.56	-0.79	-0.9	186.33	27SE	-1.0	-3.5	-1.2	0.8
60.67	-0.43	-1.1	196.32	7OC	-0.6	-3.5	-1.2	0.8
59.33	-0.01	-1.3	206.37	17OC	-0.4	-3.5	-1.2	0.8
56.63	0.46	-1.5	216.46	27OC	-0.3	-3.4	-1.2	0.8
53.00	0.93	-1.6	226.59	6NO	-0.3	-3.4	-1.3	0.8
49.18	1.36	-1.5	236.77	16NO	0.2	-3.4	-1.3	0.8
45.97	1.69	-1.3	246.95	26NO	2.6	-3.4	-1.3	0.8
43.91	1.92	-0.9	257.15	6DE	1.0	-3.4	-1.4	0.8
43.19	2.06	-0.6	267.36	16DE	0.1	-3.4	-1.4	0.8
43.78	2.13	-0.3	277.55	26DE	-0.1	-3.3	-1.5	0.8

♂ LONG	♂ LAT	♂ MAG	⊙ LONG	571	☿	♀	♃	♄
45.46	2.16	-0.0	287.73	5JA	-0.2	-3.3	-1.5	0.8
48.04	2.15	0.2	297.87	15JA	-0.3	-3.4	-1.6	0.8
51.34	2.12	0.5	307.98	25JA	-0.7	-3.4	-1.7	0.7
55.20	2.08	0.7	318.03	4FE	-1.3	-3.4	-1.7	0.7
59.49	2.03	0.9	328.04	14FE	-1.3	-3.4	-1.8	0.7
64.12	1.98	1.0	337.99	24FE	-0.6	-3.4	-1.9	0.6
69.03	1.92	1.2	347.88	6MR	0.9	-3.5	-1.9	0.6
74.16	1.86	1.3	357.72	16MR	2.9	-3.5	-2.0	0.5
79.46	1.80	1.4	7.50	26MR	2.6	-3.4	-2.0	0.4
84.91	1.74	1.5	17.22	5AP	1.4	-3.4	-2.1	0.4
90.48	1.67	1.6	26.89	15AP	0.8	-3.4	-2.1	0.3
96.16	1.61	1.7	36.52	25AP	0.3	-3.4	-2.1	0.3
101.93	1.54	1.8	46.11	5MY	-0.4	-3.3	-2.1	0.4
107.78	1.47	1.8	55.67	15MY	-1.3	-3.3	-2.0	0.4
113.70	1.40	1.9	65.21	25MY	-1.6	-3.3	-2.0	0.5
119.70	1.33	1.9	74.74	4JN	-0.7	-3.3	-2.0	0.5
125.77	1.26	1.9	84.26	14JN	0.0	-3.3	-1.9	0.6
131.90	1.19	2.0	93.79	24JN	0.5	-3.3	-1.8	0.7
138.10	1.11	2.0	103.33	4JL	0.9	-3.4	-1.8	0.7
144.37	1.04	2.0	112.90	14JL	1.6	-3.4	-1.7	0.8
150.71	0.96	2.0	122.50	24JL	3.0	-3.4	-1.7	0.8
157.12	0.88	1.9	132.14	3AU	1.8	-3.5	-1.6	0.8
163.61	0.79	1.9	141.83	13AU	-0.1	-3.5	-1.5	0.8
170.17	0.71	1.9	151.56	23AU	-1.0	-3.6	-1.5	0.9
176.81	0.62	1.9	161.35	2SE	-1.2	-3.7	-1.4	0.9
183.53	0.53	1.8	171.20	12SE	-0.9	-3.8	-1.4	0.8
190.33	0.44	1.8	181.10	22SE	-0.5	-3.8	-1.4	0.8
197.21	0.34	1.7	191.07	2OC	-0.3	-3.9	-1.3	0.8
204.18	0.24	1.7	201.09	12OC	-0.2	-4.1	-1.3	0.8
211.24	0.14	1.6	211.16	22OC	-0.1	-4.2	-1.3	0.7
218.37	0.04	1.6	221.27	1NO	0.4	-4.3	-1.3	0.7
225.59	-0.06	1.6	231.43	11NO	2.9	-4.4	-1.3	0.7
232.89	-0.17	1.6	241.60	21NO	0.8	-4.4	-1.3	0.8
240.27	-0.28	1.6	251.80	1DE	-0.1	-4.3	-1.3	0.8
247.73	-0.38	1.6	262.01	11DE	-0.2	-3.9	-1.4	0.8
255.25	-0.49	1.6	272.21	21DE	-0.3	-3.1	-1.4	0.8
262.83	-0.60	1.5	282.39	31DE	-0.4	-3.6	-1.4	0.8

♂			☉	16.00UT	☿	♀	♃	♄
LONG	LAT	MAG	LONG	572	MAGNITUDES			
270.47	-0.70	1.5	292.56	10JA	-0.8	-4.1	-1.5	0.8
278.16	-0.80	1.5	302.68	20JA	-1.2	-4.3	-1.5	0.8
285.89	-0.89	1.5	312.76	30JA	-1.2	-4.3	-1.6	0.8
293.65	-0.98	1.4	322.80	9FE	-0.5	-4.3	-1.6	0.7
301.42	-1.06	1.4	332.78	19FE	1.1	-4.2	-1.7	0.7
309.20	-1.13	1.4	342.71	29FE	3.3	-4.1	-1.8	0.6
316.98	-1.20	1.4	352.57	10MR	1.9	-3.9	-1.8	0.6
324.74	-1.25	1.3	2.38	20MR	1.1	-3.8	-1.9	0.5
332.47	-1.28	1.3	12.13	30MR	0.6	-3.7	-2.0	0.5
340.16	-1.31	1.3	21.83	9AP	0.2	-3.7	-2.0	0.4
347.78	-1.32	1.3	31.48	19AP	-0.4	-3.6	-2.1	0.3
355.35	-1.32	1.3	41.09	29AP	-1.3	-3.5	-2.1	0.3
2.83	-1.30	1.2	50.67	9MY	-1.6	-3.5	-2.2	0.3
10.21	-1.27	1.2	60.22	19MY	-0.7	-3.4	-2.2	0.3
17.50	-1.22	1.2	69.75	29MY	0.1	-3.4	-2.2	0.4
24.67	-1.16	1.2	79.28	8JN	0.7	-3.4	-2.2	0.5
31.70	-1.09	1.1	88.80	18JN	1.3	-3.3	-2.1	0.5
38.60	-1.00	1.1	98.34	28JN	2.1	-3.3	-2.1	0.6
45.34	-0.89	1.1	107.89	8JL	3.2	-3.3	-2.1	0.6
51.91	-0.77	1.0	117.47	18JL	1.5	-3.3	-2.0	0.7
58.29	-0.64	1.0	127.09	28JL	-0.1	-3.3	-1.9	0.7
64.44	-0.49	0.9	136.75	7AU	-1.1	-3.4	-1.9	0.8
70.36	-0.33	0.8	146.45	17AU	-1.4	-3.4	-1.8	0.8
75.99	-0.14	0.8	156.21	27AU	-0.9	-3.4	-1.7	0.8
81.28	0.06	0.7	166.03	6SE	-0.4	-3.4	-1.7	0.8
86.17	0.29	0.5	175.90	16SE	-0.2	-3.5	-1.6	0.8
90.59	0.54	0.4	185.84	26SE	-0.0	-3.5	-1.6	0.8
94.41	0.83	0.3	195.83	6OC	0.1	-3.5	-1.5	0.8
97.52	1.16	0.1	205.87	16OC	0.7	-3.5	-1.5	0.8
99.73	1.53	-0.1	215.97	26OC	3.2	-3.4	-1.5	0.7
100.86	1.95	-0.3	226.10	5NO	0.6	-3.4	-1.4	0.7
100.73	2.40	-0.6	236.26	15NO	-0.3	-3.4	-1.4	0.7
99.20	2.86	-0.8	246.45	25NO	-0.4	-3.4	-1.4	0.7
96.38	3.29	-1.0	256.65	5DE	-0.4	-3.3	-1.4	0.8
92.65	3.63	-1.1	266.86	15DE	-0.5	-3.3	-1.4	0.8
88.68	3.82	-1.1	277.05	25DE	-0.8	-3.3	-1.4	0.8
				573				
85.24	3.85	-0.9	287.23	4JA	-1.0	-3.3	-1.4	0.8
82.89	3.75	-0.6	297.37	14JA	-1.0	-3.3	-1.5	0.8
81.84	3.57	-0.3	307.48	24JA	-0.5	-3.4	-1.5	0.8
82.07	3.35	-0.0	317.54	3FE	1.3	-3.4	-1.5	0.8
83.42	3.13	0.2	327.55	13FE	2.9	-3.4	-1.6	0.7
85.69	2.90	0.4	337.51	23FE	1.4	-3.4	-1.6	0.7
88.71	2.69	0.6	347.41	5MR	0.8	-3.5	-1.7	0.7
92.31	2.49	0.8	357.24	15MR	0.4	-3.5	-1.8	0.6
96.39	2.31	1.0	7.03	25MR	0.1	-3.6	-1.8	0.6
100.85	2.13	1.1	16.75	4AP	-0.5	-3.6	-1.9	0.5
105.62	1.97	1.3	26.43	14AP	-1.3	-3.7	-2.0	0.4
110.64	1.82	1.4	36.06	24AP	-1.6	-3.8	-2.0	0.4
115.89	1.68	1.5	45.65	4MY	-0.7	-3.8	-2.1	0.3
121.31	1.54	1.5	55.22	14MY	0.2	-3.9	-2.2	0.3
126.90	1.41	1.6	64.76	24MY	0.9	-4.0	-2.2	0.3
132.64	1.28	1.7	74.28	3JN	1.7	-4.1	-2.3	0.3
138.50	1.16	1.7	83.81	13JN	2.8	-4.2	-2.3	0.4
144.50	1.03	1.7	93.34	23JN	2.8	-4.2	-2.3	0.5
150.60	0.92	1.8	102.88	3JL	1.2	-4.1	-2.3	0.5
156.82	0.80	1.8	112.44	13JL	-0.2	-3.7	-2.3	0.6
163.15	0.68	1.8	122.04	23JL	-1.2	-3.2	-2.3	0.6
169.59	0.57	1.8	131.68	2AU	-1.5	-3.5	-2.2	0.7
176.14	0.45	1.8	141.36	12AU	-0.8	-4.0	-2.2	0.7
182.79	0.34	1.8	151.09	22AU	-0.3	-4.2	-2.1	0.8
189.54	0.23	1.8	160.87	1SE	-0.0	-4.3	-2.1	0.8
196.41	0.12	1.7	170.72	11SE	0.1	-4.2	-2.0	0.8
203.37	0.01	1.7	180.62	21SE	0.3	-4.2	-1.9	0.8
210.44	-0.10	1.7	190.58	1OC	0.9	-4.1	-1.9	0.8
217.61	-0.21	1.6	200.60	11OC	3.5	-4.0	-1.8	0.8
224.88	-0.31	1.6	210.67	21OC	0.4	-3.9	-1.7	0.8
232.25	-0.41	1.6	220.78	31OC	-0.4	-3.8	-1.7	0.7
239.70	-0.51	1.5	230.93	10NO	-0.5	-3.7	-1.6	0.7
247.24	-0.60	1.5	241.11	20NO	-0.5	-3.7	-1.6	0.7
254.86	-0.69	1.4	251.30	30NO	-0.6	-3.6	-1.6	0.7
262.54	-0.77	1.4	261.51	10DE	-0.8	-3.6	-1.5	0.7
270.29	-0.85	1.4	271.71	20DE	-0.9	-3.5	-1.5	0.8
278.08	-0.92	1.4	281.90	30DE	-0.9	-3.5	-1.5	0.8

♂			☉	16.00UT	☿	♀	♃	♄
LONG	LAT	MAG	LONG	574	MAGNITUDES			
285.92	-0.98	1.4	292.06	9JA	-0.4	-3.4	-1.5	0.8
293.78	-1.03	1.4	302.19	19JA	1.5	-3.4	-1.5	0.8
301.66	-1.06	1.4	312.28	29JA	2.4	-3.4	-1.5	0.8
309.53	-1.09	1.4	322.32	8FE	1.0	-3.3	-1.5	0.8
317.40	-1.11	1.4	332.30	18FE	0.5	-3.3	-1.6	0.8
325.24	-1.11	1.4	342.23	28FE	0.3	-3.3	-1.6	0.7
333.05	-1.10	1.4	352.10	10MR	-0.0	-3.3	-1.6	0.7
340.81	-1.08	1.4	1.91	20MR	-0.5	-3.3	-1.7	0.7
348.51	-1.05	1.4	11.66	30MR	-1.3	-3.3	-1.7	0.6
356.14	-1.01	1.4	21.37	9AP	-1.6	-3.4	-1.8	0.5
3.70	-0.95	1.5	31.02	19AP	-0.7	-3.4	-1.8	0.5
11.17	-0.89	1.5	40.63	29AP	0.3	-3.4	-1.9	0.4
18.56	-0.81	1.5	50.21	9MY	1.2	-3.5	-2.0	0.3
25.84	-0.73	1.5	59.76	19MY	2.2	-3.5	-2.1	0.3
33.03	-0.63	1.5	69.29	29MY	3.5	-3.5	-2.1	0.2
40.12	-0.53	1.5	78.82	8JN	2.3	-3.4	-2.2	0.3
47.09	-0.42	1.5	88.34	18JN	1.0	-3.4	-2.3	0.4
53.95	-0.31	1.5	97.87	28JN	-0.2	-3.4	-2.3	0.4
60.69	-0.18	1.5	107.43	8JL	-1.3	-3.3	-2.4	0.5
67.31	-0.06	1.5	117.00	18JL	-1.6	-3.3	-2.4	0.5
73.80	0.08	1.5	126.62	28JL	-0.8	-3.3	-2.4	0.6
80.15	0.22	1.4	136.28	7AU	-0.3	-3.3	-2.4	0.6
86.35	0.38	1.4	145.98	17AU	0.1	-3.3	-2.4	0.7
92.38	0.54	1.4	155.74	27AU	0.3	-3.3	-2.4	0.7
98.23	0.70	1.3	165.55	6SE	0.5	-3.4	-2.3	0.8
103.87	0.89	1.2	175.42	16SE	1.3	-3.4	-2.3	0.8
109.27	1.08	1.1	185.35	26SE	3.1	-3.4	-2.2	0.8
114.39	1.29	1.0	195.34	6OC	0.3	-3.5	-2.1	0.8
119.17	1.52	0.9	205.38	16OC	-0.6	-3.5	-2.1	0.8
123.54	1.77	0.8	215.47	26OC	-0.7	-3.6	-2.0	0.8
127.42	2.05	0.6	225.60	5NO	-0.7	-3.6	-1.9	0.8
130.67	2.36	0.4	235.77	15NO	-0.7	-3.7	-1.9	0.7
133.17	2.71	0.2	245.95	25NO	-0.7	-3.8	-1.8	0.7
134.73	3.08	0.0	256.16	5DE	-0.7	-3.9	-1.8	0.7
135.17	3.48	-0.2	266.36	15DE	-0.7	-4.0	-1.7	0.7
134.33	3.88	-0.5	276.55	25DE	-0.2	-4.1	-1.7	0.8
				575				
132.17	4.23	-0.7	286.73	4JA	1.8	-4.2	-1.6	0.8
128.89	4.48	-0.9	296.88	14JA	1.9	-4.3	-1.6	0.8
124.99	4.57	-1.0	306.99	24JA	0.7	-4.3	-1.6	0.8
121.17	4.49	-0.9	317.06	3FE	0.3	-4.3	-1.6	0.8
118.13	4.25	-0.6	327.07	13FE	0.2	-4.2	-1.6	0.8
116.26	3.92	-0.4	337.03	23FE	-0.1	-3.8	-1.6	0.8
115.69	3.55	-0.2	346.93	5MR	-0.6	-3.2	-1.6	0.8
116.35	3.18	0.1	356.77	15MR	-1.3	-3.6	-1.6	0.7
118.04	2.83	0.3	6.55	25MR	-1.5	-4.1	-1.6	0.7
120.61	2.50	0.5	16.28	4AP	-0.7	-4.2	-1.6	0.7
123.88	2.21	0.7	25.96	14AP	0.4	-4.2	-1.7	0.6
127.72	1.94	0.8	35.59	24AP	1.5	-4.2	-1.7	0.5
132.02	1.69	0.9	45.19	4MY	2.9	-4.1	-1.8	0.5
136.71	1.47	1.1	54.75	14MY	3.1	-4.0	-1.8	0.4
141.73	1.26	1.2	64.29	24MY	1.7	-3.9	-1.9	0.3
147.02	1.06	1.2	73.82	3JN	0.8	-3.8	-2.0	0.3
152.55	0.88	1.3	83.34	13JN	-0.3	-3.7	-2.0	0.3
158.29	0.71	1.4	92.87	23JN	-1.3	-3.6	-2.1	0.3
164.23	0.55	1.4	102.42	3JL	-1.6	-3.6	-2.2	0.4
170.34	0.39	1.5	111.98	13JL	-0.8	-3.5	-2.2	0.4
176.62	0.25	1.5	121.58	23JL	-0.2	-3.5	-2.3	0.5
183.06	0.10	1.5	131.21	2AU	0.2	-3.4	-2.3	0.6
189.64	-0.03	1.5	140.89	12AU	0.4	-3.4	-2.4	0.6
196.37	-0.16	1.5	150.62	22AU	0.7	-3.4	-2.4	0.7
203.23	-0.28	1.5	160.41	1SE	1.6	-3.4	-2.5	0.7
210.22	-0.39	1.5	170.24	11SE	2.8	-3.4	-2.5	0.7
217.34	-0.50	1.5	180.15	21SE	0.2	-3.4	-2.4	0.8
224.58	-0.60	1.5	190.11	1OC	-0.7	-3.4	-2.4	0.8
231.92	-0.70	1.5	200.12	11OC	-0.8	-3.4	-2.4	0.8
239.38	-0.78	1.5	210.18	21OC	-0.8	-3.4	-2.3	0.8
246.93	-0.86	1.5	220.29	31OC	-0.7	-3.4	-2.3	0.8
254.57	-0.93	1.5	230.44	10NO	-0.6	-3.4	-2.2	0.8
262.28	-0.98	1.4	240.62	20NO	-0.6	-3.4	-2.1	0.8
270.06	-1.03	1.4	250.81	30NO	-0.6	-3.5	-2.0	0.7
277.89	-1.06	1.4	261.01	10DE	-0.1	-3.5	-2.0	0.7
285.76	-1.09	1.4	271.21	20DE	2.0	-3.5	-1.9	0.7
293.65	-1.10	1.4	281.40	30DE	1.5	-3.5	-1.8	0.8

♂ LONG	♂ LAT	♂ MAG	⊙ LONG	16.00UT 576	☿	♀	♃	♄
301.56	-1.10	1.4	291.56	9JA	0.4	-3.4	-1.8	0.8
309.46	-1.09	1.3	301.69	19JA	0.2	-3.4	-1.7	0.8
317.35	-1.06	1.3	311.78	29JA	0.0	-3.4	-1.7	0.8
325.20	-1.03	1.3	321.82	8FE	-0.2	-3.4	-1.6	0.9
333.02	-0.98	1.3	331.81	18FE	-0.6	-3.4	-1.6	0.9
340.78	-0.93	1.3	341.74	28FE	-1.3	-3.3	-1.6	0.8
348.48	-0.86	1.4	351.61	9MR	-1.5	-3.3	-1.6	0.8
356.11	-0.79	1.4	1.43	19MR	-0.7	-3.3	-1.6	0.8
3.67	-0.71	1.5	11.18	29MR	0.6	-3.3	-1.6	0.8
11.14	-0.62	1.5	20.89	8AP	2.0	-3.4	-1.6	0.7
18.52	-0.53	1.5	30.55	18AP	3.8	-3.4	-1.6	0.7
25.82	-0.43	1.6	40.16	28AP	2.3	-3.4	-1.6	0.6
33.03	-0.33	1.6	49.74	8MY	1.3	-3.4	-1.6	0.5
40.15	-0.23	1.6	59.30	18MY	0.6	-3.5	-1.7	0.5
47.17	-0.12	1.7	68.83	28MY	-0.3	-3.5	-1.7	0.4
54.10	-0.01	1.7	78.35	7JN	-1.3	-3.6	-1.7	0.4
60.94	0.10	1.7	87.88	17JN	-1.6	-3.6	-1.8	0.3
67.69	0.21	1.7	97.41	27JN	-0.8	-3.7	-1.9	0.3
74.35	0.32	1.8	106.96	7JL	-0.1	-3.8	-1.9	0.4
80.92	0.44	1.8	116.54	17JL	0.3	-3.9	-2.0	0.4
87.40	0.56	1.8	126.15	27JL	0.6	-4.0	-2.0	0.5
93.79	0.67	1.8	135.81	6AU	1.0	-4.1	-2.1	0.5
100.08	0.79	1.7	145.51	16AU	2.1	-4.2	-2.2	0.6
106.27	0.92	1.7	155.27	26AU	2.4	-4.3	-2.2	0.6
112.36	1.04	1.7	165.08	5SE	0.1	-4.3	-2.3	0.7
118.33	1.17	1.6	174.95	15SE	-0.9	-4.2	-2.4	0.7
124.16	1.30	1.6	184.87	25SE	-1.0	-3.8	-2.4	0.8
129.85	1.44	1.5	194.86	5OC	-0.9	-3.2	-2.4	0.8
135.38	1.59	1.4	204.90	15OC	-0.6	-3.7	-2.4	0.8
140.70	1.75	1.4	214.98	25OC	-0.5	-4.2	-2.4	0.8
145.79	1.91	1.2	225.11	4NO	-0.4	-4.3	-2.4	0.8
150.60	2.09	1.1	235.28	14NO	-0.4	-4.4	-2.3	0.8
155.07	2.29	1.0	245.46	24NO	0.1	-4.3	-2.3	0.8
159.12	2.50	0.8	255.66	4DE	2.3	-4.2	-2.2	0.8
162.65	2.72	0.6	265.87	14DE	1.2	-4.1	-2.1	0.8
165.53	2.97	0.4	276.06	24DE	0.2	-4.0	-2.1	0.7
				577				
167.62	3.23	0.2	286.24	3JA	0.0	-3.9	-2.0	0.8
168.72	3.50	-0.1	296.39	13JA	-0.1	-3.8	-1.9	0.8
168.67	3.76	-0.4	306.50	23JA	-0.3	-3.7	-1.9	0.9
167.34	3.97	-0.6	316.57	2FE	-0.7	-3.6	-1.8	0.9
164.76	4.10	-0.9	326.59	12FE	-1.3	-3.6	-1.7	0.9
161.25	4.09	-1.1	336.54	22FE	-1.4	-3.5	-1.7	0.9
157.38	3.91	-1.1	346.45	4MR	-0.6	-3.5	-1.6	0.9
153.89	3.59	-0.9	356.29	14MR	0.8	-3.4	-1.6	0.9
151.34	3.17	-0.7	6.08	24MR	2.5	-3.4	-1.6	0.9
150.06	2.71	-0.5	15.81	3AP	3.1	-3.3	-1.5	0.8
150.09	2.27	-0.3	25.49	13AP	1.7	-3.3	-1.5	0.8
151.29	1.85	-0.1	35.13	23AP	1.0	-3.3	-1.5	0.8
153.51	1.47	0.1	44.73	3MY	0.4	-3.3	-1.5	0.7
156.57	1.13	0.3	54.29	13MY	-0.4	-3.3	-1.5	0.6
160.31	0.83	0.4	63.83	23MY	-1.3	-3.3	-1.5	0.6
164.61	0.56	0.5	73.36	2JN	-1.7	-3.3	-1.5	0.5
169.40	0.32	0.7	82.89	12JN	-0.7	-3.4	-1.6	0.4
174.57	0.10	0.8	92.41	22JN	-0.0	-3.4	-1.6	0.4
180.09	-0.10	0.8	101.96	2JL	0.4	-3.4	-1.6	0.3
185.91	-0.28	0.9	111.52	12JL	0.8	-3.5	-1.7	0.4
191.99	-0.44	1.0	121.11	22JL	1.3	-3.5	-1.7	0.4
198.30	-0.59	1.0	130.74	1AU	2.6	-3.5	-1.8	0.5
204.83	-0.72	1.1	140.42	11AU	2.1	-3.4	-1.8	0.5
211.55	-0.84	1.1	150.14	21AU	0.0	-3.4	-1.9	0.6
218.45	-0.95	1.1	159.93	31AU	-1.0	-3.4	-1.9	0.6
225.51	-1.04	1.2	169.76	10SE	-1.1	-3.4	-2.0	0.7
232.72	-1.11	1.2	179.66	20SE	-1.0	-3.3	-2.1	0.7
240.06	-1.18	1.2	189.62	30SE	-0.5	-3.3	-2.1	0.8
247.52	-1.22	1.2	199.63	10OC	-0.3	-3.3	-2.2	0.8
255.08	-1.26	1.3	209.69	20OC	-0.3	-3.3	-2.2	0.8
262.74	-1.28	1.3	219.80	30OC	-0.2	-3.3	-2.3	0.8
270.47	-1.28	1.3	229.94	9NO	0.3	-3.4	-2.3	0.9
278.25	-1.27	1.3	240.12	19NO	2.6	-3.4	-2.3	0.9
286.08	-1.25	1.3	250.31	29NO	0.9	-3.4	-2.3	0.9
293.93	-1.22	1.4	260.52	9DE	0.0	-3.4	-2.3	0.9
301.79	-1.17	1.4	270.72	19DE	-0.1	-3.5	-2.2	0.8
309.65	-1.11	1.4	280.91	29DE	-0.2	-3.5	-2.2	0.8

♂ LONG	♂ LAT	♂ MAG	⊙ LONG	16.00UT 578	☿	♀	♃	♄
317.49	-1.05	1.4	291.07	8JA	-0.4	-3.6	-2.1	0.8
325.29	-0.97	1.4	301.20	18JA	-0.7	-3.6	-2.0	0.9
333.05	-0.89	1.5	311.30	28JA	-1.2	-3.7	-2.0	0.9
340.76	-0.80	1.5	321.34	7FE	-1.2	-3.8	-1.9	0.9
348.39	-0.70	1.5	331.33	17FE	-0.6	-3.9	-1.8	1.0
355.96	-0.60	1.5	341.26	27FE	0.9	-3.9	-1.8	1.0
3.45	-0.50	1.5	351.14	9MR	3.1	-4.0	-1.7	1.0
10.86	-0.40	1.6	0.95	19MR	2.3	-4.1	-1.6	1.0
18.19	-0.29	1.6	10.72	29MR	1.3	-4.2	-1.6	1.0
25.43	-0.19	1.6	20.42	8AP	0.7	-4.2	-1.5	0.9
32.58	-0.08	1.6	30.08	18AP	0.3	-4.2	-1.5	0.9
39.66	0.02	1.6	39.70	28AP	-0.4	-3.9	-1.5	0.9
46.65	0.12	1.7	49.28	8MY	-1.3	-3.3	-1.4	0.8
53.56	0.22	1.7	58.83	18MY	-1.7	-2.9	-1.4	0.8
60.39	0.32	1.8	68.37	28MY	-0.7	-3.7	-1.4	0.7
67.15	0.42	1.8	77.89	7JN	0.0	-4.1	-1.4	0.6
73.84	0.51	1.8	87.41	17JN	0.6	-4.2	-1.4	0.6
80.47	0.61	1.9	96.95	27JN	1.1	-4.2	-1.4	0.5
87.02	0.70	1.9	106.50	7JL	1.8	-4.1	-1.4	0.4
93.52	0.78	1.9	116.07	17JL	3.1	-4.0	-1.4	0.4
99.96	0.87	1.9	125.69	27JL	1.8	-3.9	-1.5	0.4
106.34	0.95	1.9	135.34	6AU	-0.0	-3.8	-1.5	0.5
112.67	1.04	1.9	145.04	16AU	-1.1	-3.8	-1.6	0.5
118.94	1.12	1.9	154.79	26AU	-1.3	-3.7	-1.6	0.6
125.15	1.20	1.9	164.60	5SE	-0.9	-3.6	-1.7	0.6
131.30	1.28	1.9	174.47	15SE	-0.5	-3.6	-1.7	0.7
137.39	1.36	1.9	184.39	25SE	-0.2	-3.5	-1.8	0.7
143.41	1.44	1.8	194.37	5OC	-0.1	-3.5	-1.8	0.8
149.35	1.52	1.8	204.41	15OC	-0.0	-3.5	-1.9	0.8
155.20	1.60	1.7	214.50	25OC	0.5	-3.4	-2.0	0.9
160.95	1.68	1.6	224.62	4NO	2.9	-3.4	-2.0	0.9
166.58	1.76	1.5	234.78	14NO	0.7	-3.4	-2.1	0.9
172.07	1.84	1.4	244.97	24NO	-0.2	-3.4	-2.1	0.9
177.39	1.92	1.3	255.17	4DE	-0.3	-3.4	-2.2	0.9
182.50	2.01	1.2	265.37	14DE	-0.3	-3.4	-2.2	0.9
187.36	2.09	1.0	275.57	24DE	-0.5	-3.3	-2.2	0.9
				579				
191.91	2.18	0.8	285.75	3JA	-0.8	-3.4	-2.2	0.9
196.08	2.26	0.6	295.90	13JA	-1.1	-3.4	-2.1	0.9
199.77	2.34	0.4	306.01	23JA	-1.1	-3.4	-2.1	0.9
202.87	2.41	0.2	316.08	2FE	-0.5	-3.4	-2.0	1.0
205.22	2.46	-0.1	326.10	12FE	1.1	-3.4	-2.0	1.0
206.66	2.49	-0.4	336.06	22FE	3.2	-3.4	-1.9	1.0
207.00	2.47	-0.7	345.97	4MR	1.7	-3.5	-1.8	1.1
206.10	2.38	-1.0	355.81	14MR	0.9	-3.5	-1.8	1.1
203.95	2.19	-1.3	5.60	24MR	0.5	-3.4	-1.7	1.1
200.80	1.89	-1.5	15.34	3AP	0.1	-3.4	-1.6	1.1
197.17	1.49	-1.6	25.02	13AP	-0.4	-3.4	-1.6	1.1
193.79	1.01	-1.4	34.66	23AP	-1.3	-3.4	-1.5	1.0
191.31	0.53	-1.3	44.25	3MY	-1.7	-3.3	-1.5	1.0
190.09	0.07	-1.1	53.82	13MY	-0.7	-3.3	-1.4	1.0
190.23	-0.33	-0.8	63.37	23MY	0.1	-3.3	-1.4	0.9
191.64	-0.67	-0.6	72.89	2JN	0.8	-3.3	-1.4	0.9
194.13	-0.96	-0.5	82.42	12JN	1.4	-3.3	-1.3	0.8
197.55	-1.19	-0.3	91.95	22JN	2.4	-3.3	-1.3	0.7
201.71	-1.38	-0.1	101.48	2JL	3.2	-3.4	-1.3	0.7
206.49	-1.54	0.0	111.05	12JL	1.4	-3.4	-1.3	0.6
211.79	-1.66	0.1	120.64	22JL	-0.1	-3.4	-1.3	0.5
217.52	-1.76	0.2	130.27	1AU	-1.1	-3.5	-1.3	0.5
223.61	-1.82	0.3	139.95	11AU	-1.4	-3.5	-1.3	0.5
230.02	-1.87	0.4	149.67	21AU	-0.9	-3.6	-1.4	0.6
236.68	-1.89	0.5	159.45	31AU	-0.4	-3.7	-1.4	0.6
243.57	-1.89	0.6	169.29	10SE	-0.1	-3.8	-1.4	0.7
250.66	-1.87	0.7	179.18	20SE	0.0	-3.8	-1.5	0.7
257.90	-1.83	0.7	189.13	30SE	0.2	-4.0	-1.5	0.8
265.27	-1.78	0.8	199.14	10OC	0.7	-4.1	-1.6	0.8
272.75	-1.71	0.9	209.20	20OC	3.2	-4.2	-1.6	0.9
280.31	-1.62	0.9	219.30	30OC	0.5	-4.3	-1.7	0.9
287.93	-1.53	1.0	229.45	9NO	-0.3	-4.4	-1.8	1.0
295.59	-1.42	1.1	239.62	19NO	-0.4	-4.4	-1.8	1.0
303.27	-1.31	1.1	249.82	29NO	-0.5	-4.3	-1.9	1.0
310.95	-1.19	1.2	260.02	9DE	-0.6	-3.8	-1.9	1.0
318.61	-1.06	1.3	270.22	19DE	-0.8	-3.1	-2.0	1.0
326.25	-0.93	1.3	280.41	29DE	-1.0	-3.7	-2.0	1.0

♂ LONG	LAT	MAG	☉ LONG	16.00UT 580	☿	♀	♃	♄ MAGNITUDES
333.84	-0.80	1.4	290.58	8JA	-1.0	-4.2	-2.1	1.0
341.38	-0.67	1.4	300.71	18JA	-0.5	-4.3	-2.1	1.0
348.86	-0.54	1.5	310.80	28JA	1.3	-4.3	-2.1	1.0
356.28	-0.42	1.5	320.85	7FE	2.8	-4.3	-2.1	1.1
3.62	-0.29	1.6	330.84	17FE	1.2	-4.2	-2.0	1.1
10.89	-0.17	1.6	340.78	27FE	0.7	-4.0	-2.0	1.2
18.08	-0.06	1.7	350.66	8MR	0.4	-3.9	-1.9	1.2
25.19	0.06	1.7	0.48	18MR	0.0	-3.8	-1.9	1.2
32.23	0.16	1.7	10.24	28MR	-0.5	-3.7	-1.8	1.2
39.19	0.26	1.8	19.96	7AP	-1.3	-3.7	-1.7	1.2
46.08	0.36	1.8	29.61	17AP	-1.6	-3.6	-1.7	1.2
52.91	0.45	1.8	39.23	27AP	-0.7	-3.5	-1.6	1.2
59.67	0.54	1.8	48.82	7MY	0.2	-3.5	-1.5	1.2
66.36	0.62	1.8	58.37	17MY	1.0	-3.4	-1.5	1.1
73.00	0.70	1.8	67.91	27MY	1.8	-3.4	-1.4	1.1
79.59	0.77	1.8	77.43	6JN	3.1	-3.4	-1.4	1.0
86.13	0.84	1.9	86.95	16JN	2.6	-3.3	-1.3	1.0
92.63	0.90	1.9	96.49	26JN	1.2	-3.3	-1.3	0.9
99.09	0.96	1.9	106.04	6JL	-0.2	-3.3	-1.3	0.9
105.52	1.01	2.0	115.61	16JL	-1.2	-3.3	-1.3	0.8
111.92	1.06	2.0	125.22	26JL	-1.5	-3.3	-1.2	0.7
118.29	1.11	2.0	134.87	5AU	-0.8	-3.4	-1.2	0.6
124.63	1.15	2.0	144.57	15AU	-0.3	-3.4	-1.2	0.6
130.96	1.19	2.0	154.32	25AU	-0.0	-3.4	-1.2	0.7
137.26	1.23	2.0	164.13	4SE	0.2	-3.4	-1.3	0.7
143.55	1.26	2.0	173.99	14SE	0.4	-3.5	-1.3	0.8
149.82	1.29	2.0	183.91	24SE	1.0	-3.5	-1.3	0.8
156.07	1.32	2.0	193.89	4OC	3.4	-3.5	-1.3	0.9
162.30	1.34	1.9	203.92	14OC	0.4	-3.5	-1.4	0.9
168.51	1.36	1.9	214.00	24OC	-0.5	-3.4	-1.4	1.0
174.69	1.37	1.8	224.13	3NO	-0.6	-3.4	-1.5	1.0
180.84	1.38	1.8	234.28	13NO	-0.6	-3.4	-1.5	1.1
186.95	1.39	1.7	244.47	23NO	-0.7	-3.4	-1.6	1.1
193.02	1.38	1.6	254.67	3DE	-0.8	-3.3	-1.7	1.1
199.02	1.37	1.5	264.87	13DE	-0.8	-3.3	-1.7	1.2
204.96	1.35	1.4	275.07	23DE	-0.8	-3.3	-1.8	1.2
				581				
210.81	1.32	1.3	285.25	2JA	-0.3	-3.3	-1.8	1.2
216.57	1.28	1.1	295.40	12JA	1.6	-3.3	-1.9	1.2
222.19	1.22	1.0	305.52	22JA	2.3	-3.4	-2.0	1.2
227.66	1.14	0.8	315.59	1FE	0.9	-3.4	-2.0	1.2
232.94	1.04	0.6	325.61	11FE	0.5	-3.4	-2.0	1.2
237.99	0.91	0.4	335.58	21FE	0.2	-3.4	-2.0	1.2
242.73	0.75	0.2	345.49	3MR	-0.0	-3.5	-2.0	1.3
247.10	0.54	-0.0	355.34	13MR	-0.5	-3.5	-2.0	1.3
251.00	0.28	-0.3	5.13	23MR	-1.3	-3.6	-2.0	1.4
254.28	-0.05	-0.6	14.87	2AP	-1.6	-3.6	-1.9	1.4
256.82	-0.45	-0.9	24.55	12AP	-0.7	-3.7	-1.9	1.4
258.41	-0.94	-1.2	34.20	22AP	0.3	-3.8	-1.8	1.4
258.89	-1.53	-1.5	43.80	2MY	1.3	-3.8	-1.7	1.4
258.13	-2.19	-1.8	53.36	12MY	2.4	-3.9	-1.7	1.4
256.19	-2.90	-2.1	62.91	22MY	3.5	-4.0	-1.6	1.4
253.42	-3.56	-2.3	72.44	1JN	2.1	-4.1	-1.5	1.3
250.45	-4.09	-2.3	81.96	11JN	0.9	-4.2	-1.5	1.3
247.99	-4.44	-2.1	91.49	21JN	-0.2	-4.2	-1.4	1.2
246.63	-4.59	-1.9	101.03	1JL	-1.2	-4.1	-1.4	1.2
246.63	-4.59	-1.7	110.59	11JL	-1.6	-3.7	-1.3	1.1
247.97	-4.47	-1.5	120.18	21JL	-0.8	-3.1	-1.3	1.0
250.54	-4.28	-1.2	129.81	31JL	-0.2	-3.5	-1.3	0.9
254.11	-4.05	-1.0	139.48	10AU	0.1	-4.0	-1.2	0.9
258.51	-3.79	-0.8	149.20	20AU	0.3	-4.2	-1.2	0.8
263.57	-3.51	-0.6	158.97	30AU	0.6	-4.3	-1.2	0.8
269.14	-3.23	-0.4	168.81	9SE	1.4	-4.2	-1.2	0.9
275.13	-2.94	-0.2	178.70	19SE	3.1	-4.2	-1.2	0.9
281.43	-2.66	-0.1	188.65	29SE	0.3	-4.1	-1.2	0.9
287.98	-2.38	0.1	198.65	9OC	-0.6	-4.0	-1.2	1.0
294.72	-2.10	0.2	208.71	19OC	-0.7	-3.9	-1.3	1.0
301.60	-1.84	0.4	218.81	29OC	-0.7	-3.8	-1.3	1.1
308.58	-1.59	0.5	228.96	8NO	-0.8	-3.7	-1.3	1.1
315.62	-1.35	0.6	239.13	18NO	-0.7	-3.7	-1.4	1.2
322.71	-1.12	0.8	249.32	28NO	-0.6	-3.6	-1.4	1.2
329.81	-0.90	0.9	259.53	8DE	-0.6	-3.6	-1.5	1.3
336.92	-0.70	1.0	269.73	18DE	-0.2	-3.5	-1.5	1.3
344.01	-0.51	1.1	279.92	28DE	1.8	-3.5	-1.6	1.3

♂ LONG	LAT	MAG	☉ LONG	16.00UT 582	☿	♀	♃	♄ MAGNITUDES
351.08	-0.34	1.2	290.09	7JA	1.8	-3.4	-1.7	1.4
358.11	-0.18	1.3	300.23	17JA	0.6	-3.4	-1.7	1.4
5.10	-0.03	1.4	310.32	27JA	0.3	-3.4	-1.8	1.4
12.05	0.11	1.5	320.37	6FE	0.1	-3.3	-1.9	1.3
18.94	0.24	1.6	330.37	16FE	-0.1	-3.3	-1.9	1.3
25.79	0.35	1.6	340.30	26FE	-0.6	-3.3	-2.0	1.3
32.58	0.45	1.7	350.19	8MR	-1.3	-3.3	-2.0	1.3
39.32	0.55	1.8	0.01	18MR	-1.5	-3.3	-2.0	1.3
46.01	0.64	1.8	9.77	28MR	-0.7	-3.3	-2.0	1.2
52.65	0.71	1.9	19.49	7AP	0.5	-3.4	-2.0	1.2
59.24	0.78	1.9	29.15	17AP	1.7	-3.4	-2.0	1.2
65.79	0.85	1.9	38.77	27AP	3.2	-3.4	-2.0	1.2
72.30	0.90	2.0	48.35	7MY	2.8	-3.5	-1.9	1.2
78.78	0.95	2.0	57.91	17MY	1.6	-3.5	-1.9	1.1
85.23	1.00	2.0	67.44	27MY	0.7	-3.5	-1.8	1.1
91.66	1.04	2.0	76.97	6JN	-0.3	-3.4	-1.7	1.1
98.06	1.07	2.0	86.49	16JN	-1.3	-3.4	-1.7	1.0
104.44	1.10	2.0	96.02	26JN	-1.7	-3.4	-1.6	1.0
110.82	1.12	2.0	105.57	6JL	-0.8	-3.3	-1.5	0.9
117.19	1.14	2.0	115.15	16JL	-0.2	-3.3	-1.5	0.9
123.55	1.15	2.0	124.75	26JL	0.2	-3.3	-1.4	0.8
129.92	1.16	2.0	134.40	5AU	0.5	-3.3	-1.4	0.8
136.29	1.16	2.0	144.09	15AU	0.8	-3.3	-1.3	0.7
142.68	1.16	2.0	153.84	25AU	1.8	-3.3	-1.3	0.7
149.07	1.16	2.0	163.64	4SE	2.7	-3.4	-1.3	0.7
155.48	1.15	2.0	173.50	14SE	0.2	-3.4	-1.3	0.7
161.91	1.14	2.0	183.42	24SE	-0.8	-3.4	-1.2	0.8
168.36	1.12	2.0	193.40	4OC	-0.9	-3.5	-1.2	0.9
174.82	1.09	2.0	203.43	14OC	-0.9	-3.5	-1.2	0.9
181.31	1.06	1.9	213.51	24OC	-0.7	-3.6	-1.2	1.0
187.82	1.02	1.9	223.64	3NO	-0.5	-3.6	-1.3	1.0
194.35	0.98	1.9	233.79	13NO	-0.5	-3.7	-1.3	1.1
200.90	0.93	1.8	243.98	23NO	-0.5	-3.8	-1.3	1.1
207.48	0.86	1.7	254.18	3DE	-0.1	-3.9	-1.3	1.2
214.07	0.79	1.7	264.38	13DE	2.0	-4.0	-1.4	1.2
220.67	0.71	1.6	274.58	23DE	1.4	-4.1	-1.4	1.2
				583				
227.29	0.62	1.5	284.76	2JA	0.4	-4.2	-1.5	1.2
233.90	0.51	1.4	294.92	12JA	0.1	-4.3	-1.6	1.2
240.56	0.39	1.3	305.03	22JA	-0.0	-4.3	-1.6	1.2
247.20	0.25	1.2	315.11	1FE	-0.2	-4.3	-1.7	1.1
253.83	0.10	1.1	325.13	11FE	-0.6	-4.2	-1.8	1.1
260.46	-0.08	1.0	335.10	21FE	-1.3	-3.7	-1.8	1.0
267.06	-0.28	0.8	345.01	3MR	-1.4	-3.7	-1.9	1.0
273.63	-0.50	0.7	354.87	13MR	-0.7	-3.7	-1.9	1.0
280.16	-0.75	0.5	4.66	23MR	0.6	-4.1	-2.0	1.0
286.61	-1.02	0.4	14.40	2AP	2.2	-4.2	-2.0	1.0
292.96	-1.33	0.2	24.09	12AP	3.7	-4.2	-2.1	1.0
299.18	-1.66	0.0	33.73	22AP	2.1	-4.2	-2.1	1.0
305.22	-2.04	-0.2	43.34	2MY	1.2	-4.1	-2.1	0.9
311.02	-2.44	-0.4	52.90	12MY	0.5	-4.0	-2.1	0.9
316.52	-2.88	-0.6	62.45	22MY	-0.3	-3.9	-2.0	0.9
321.59	-3.36	-0.8	71.98	1JN	-1.3	-3.8	-2.0	0.9
326.12	-3.87	-1.0	81.50	11JN	-1.7	-3.7	-2.0	0.8
329.97	-4.42	-1.3	91.03	21JN	-0.8	-3.6	-1.9	0.8
332.91	-4.97	-1.5	100.57	1JL	-0.1	-3.5	-1.8	0.7
334.75	-5.52	-1.8	110.13	11JL	0.3	-3.5	-1.8	0.7
335.32	-6.00	-2.1	119.71	21JL	0.7	-3.5	-1.7	0.6
334.50	-6.35	-2.3	129.34	31JL	1.1	-3.4	-1.7	0.6
332.49	-6.47	-2.5	139.01	10AU	2.3	-3.4	-1.6	0.5
329.76	-6.29	-2.6	148.73	20AU	2.4	-3.4	-1.5	0.5
327.07	-5.82	-2.4	158.50	30AU	0.1	-3.4	-1.5	0.4
325.15	-5.14	-2.2	168.33	9SE	-0.9	-3.4	-1.4	0.4
324.41	-4.37	-1.9	178.22	19SE	-1.0	-3.4	-1.4	0.4
324.98	-3.59	-1.5	188.17	29SE	-1.0	-3.4	-1.4	0.4
326.78	-2.88	-1.2	198.17	9OC	-0.6	-3.4	-1.3	0.5
329.59	-2.25	-0.9	208.22	19OC	-0.4	-3.4	-1.3	0.6
333.22	-1.70	-0.6	218.33	29OC	-0.4	-3.4	-1.3	0.6
337.50	-1.24	-0.1	228.46	8NO	-0.3	-3.4	-1.3	0.7
342.27	-0.84	-0.1	238.64	18NO	0.1	-3.4	-1.3	0.8
347.43	-0.50	0.1	248.83	28NO	2.3	-3.5	-1.3	0.8
352.88	-0.22	0.3	259.03	8DE	1.1	-3.5	-1.3	0.9
358.54	0.02	0.5	269.23	18DE	0.1	-3.5	-1.4	0.9
4.38	0.23	0.7	279.43	28DE	-0.0	-3.5	-1.4	0.9

♂ LONG	♂ LAT	♂ MAG	☉ LONG	16.00UT	☿	♀	♃	♄
LONG	LAT	MAG	LONG	584	MAGNITUDES			
10.34	0.41	0.9	289.59	7JA	-0.1	-3.4	-1.4	0.9
16.39	0.56	1.0	299.73	17JA	-0.3	-3.4	-1.5	0.9
22.51	0.68	1.1	309.83	27JA	-0.7	-3.4	-1.5	0.9
28.68	0.79	1.3	319.88	6FE	-1.3	-3.4	-1.6	0.9
34.89	0.88	1.4	329.88	16FE	-1.3	-3.4	-1.7	0.9
41.11	0.96	1.5	339.82	26FE	-0.7	-3.3	-1.7	0.8
47.35	1.02	1.6	349.71	7MR	0.8	-3.3	-1.8	0.8
53.60	1.07	1.7	359.53	17MR	2.7	-3.3	-1.9	0.7
59.84	1.12	1.7	9.30	27MR	2.8	-3.3	-1.9	0.7
66.09	1.15	1.8	19.01	6AP	1.6	-3.4	-2.0	0.7
72.34	1.18	1.9	28.68	16AP	0.9	-3.4	-2.1	0.7
78.59	1.20	1.9	38.31	26AP	0.4	-3.4	-2.1	0.7
84.83	1.21	2.0	47.89	6MY	-0.3	-3.4	-2.2	0.7
91.08	1.22	2.0	57.45	16MY	-1.3	-3.5	-2.2	0.7
97.33	1.22	2.0	66.99	26MY	-1.7	-3.5	-2.2	0.7
103.59	1.22	2.0	76.51	5JN	-0.8	-3.6	-2.2	0.7
109.86	1.21	2.0	86.04	15JN	-0.0	-3.6	-2.2	0.6
116.15	1.20	2.0	95.57	25JN	0.5	-3.7	-2.2	0.6
122.45	1.18	2.0	105.11	5JL	0.9	-3.8	-2.1	0.6
128.77	1.16	2.0	114.69	15JL	1.5	-3.9	-2.1	0.5
135.12	1.14	2.0	124.29	25JL	2.8	-4.0	-2.0	0.5
141.50	1.11	2.0	133.93	4AU	2.0	-4.1	-1.9	0.4
147.92	1.07	2.0	143.63	14AU	0.0	-4.2	-1.9	0.4
154.37	1.04	1.9	153.37	24AU	-1.0	-4.3	-1.8	0.3
160.86	0.99	1.9	163.17	3SE	-1.2	-4.3	-1.7	0.3
167.39	0.95	1.9	173.03	13SE	-1.0	-4.2	-1.7	0.2
173.98	0.89	1.9	182.94	23SE	-0.5	-3.8	-1.6	0.2
180.61	0.84	1.9	192.91	3OC	-0.3	-3.2	-1.6	0.1
187.29	0.77	1.9	202.94	13OC	-0.2	-3.7	-1.5	0.2
194.03	0.70	1.9	213.02	23OC	-0.1	-4.2	-1.5	0.3
200.82	0.63	1.9	223.14	2NO	0.3	-4.4	-1.5	0.3
207.67	0.55	1.8	233.30	12NO	2.6	-4.4	-1.4	0.4
214.57	0.46	1.8	243.48	22NO	0.9	-4.3	-1.4	0.5
221.53	0.36	1.8	253.68	2DE	-0.1	-4.2	-1.4	0.5
228.54	0.26	1.7	263.89	12DE	-0.2	-4.1	-1.4	0.6
235.61	0.15	1.7	274.08	22DE	-0.3	-4.0	-1.4	0.6
				585				
242.73	0.03	1.6	284.27	1JA	-0.4	-3.9	-1.4	0.7
249.91	-0.09	1.5	294.43	11JA	-0.7	-3.8	-1.5	0.7
257.13	-0.23	1.5	304.55	21JA	-1.2	-3.7	-1.5	0.7
264.41	-0.37	1.4	314.62	31JA	-1.2	-3.6	-1.5	0.7
271.72	-0.52	1.3	324.65	10FE	-0.6	-3.6	-1.6	0.7
279.06	-0.67	1.3	334.62	20FE	1.0	-3.5	-1.6	0.7
286.44	-0.84	1.2	344.54	2MR	3.2	-3.5	-1.7	0.6
293.84	-1.00	1.1	354.40	12MR	2.1	-3.4	-1.7	0.6
301.24	-1.17	1.0	4.19	22MR	1.1	-3.4	-1.8	0.5
308.65	-1.34	0.9	13.94	1AP	0.7	-3.3	-1.9	0.5
316.04	-1.51	0.8	23.63	11AP	0.2	-3.3	-1.9	0.5
323.39	-1.67	0.8	33.27	21AP	-0.4	-3.3	-2.0	0.5
330.69	-1.83	0.7	42.87	1MY	-1.3	-3.3	-2.1	0.5
337.91	-1.98	0.6	52.45	11MY	-1.7	-3.3	-2.1	0.5
345.02	-2.12	0.5	61.99	21MY	-0.8	-3.3	-2.2	0.5
352.01	-2.25	0.4	71.52	31MY	0.1	-3.3	-2.2	0.5
358.82	-2.37	0.3	81.05	10JN	0.6	-3.4	-2.3	0.5
5.42	-2.47	0.2	90.57	20JN	1.2	-3.4	-2.3	0.5
11.76	-2.55	0.1	100.11	30JN	2.0	-3.4	-2.3	0.5
17.78	-2.61	-0.1	109.67	10JL	3.2	-3.5	-2.3	0.4
23.40	-2.64	-0.2	119.25	20JL	1.7	-3.5	-2.3	0.4
28.54	-2.66	-0.4	128.87	30JL	-0.0	-3.5	-2.3	0.4
33.07	-2.64	-0.5	138.54	9AU	-1.1	-3.4	-2.2	0.3
36.84	-2.58	-0.7	148.26	19AU	-1.3	-3.4	-2.2	0.3
39.69	-2.48	-0.9	158.03	29AU	-0.9	-3.4	-2.1	0.2
41.39	-2.32	-1.1	167.85	8SE	-0.4	-3.4	-2.1	0.1
41.74	-2.08	-1.3	177.73	18SE	-0.2	-3.3	-2.0	0.1
40.63	-1.75	-1.6	187.68	28SE	-0.1	-3.3	-1.9	0.0
38.13	-1.33	-1.7	197.68	8OC	0.0	-3.3	-1.9	-0.0
34.68	-0.82	-1.9	207.73	18OC	0.5	-3.3	-1.8	-0.1
31.02	-0.30	-1.8	217.83	28OC	2.9	-3.3	-1.7	0.0
27.95	0.20	-1.5	227.97	7NO	0.7	-3.4	-1.7	0.1
26.04	0.61	-1.2	238.13	17NO	-0.2	-3.4	-1.6	0.2
25.48	0.94	-0.9	248.33	27NO	-0.4	-3.4	-1.6	0.2
26.21	1.18	-0.6	258.53	7DE	-0.4	-3.4	-1.6	0.3
28.06	1.36	-0.3	268.73	17DE	-0.5	-3.5	-1.5	0.4
30.79	1.48	0.0	278.93	27DE	-0.8	-3.5	-1.5	0.4

♂ LONG	♂ LAT	♂ MAG	☉ LONG	16.00UT	☿	♀	♃	♄
LONG	LAT	MAG	LONG	586	MAGNITUDES			
34.24	1.56	0.3	289.10	6JA	-1.0	-3.6	-1.5	0.4
38.24	1.62	0.5	299.24	16JA	-1.1	-3.6	-1.5	0.5
42.67	1.65	0.7	309.34	26JA	-0.5	-3.7	-1.5	0.5
47.43	1.67	0.9	319.40	5FE	1.1	-3.8	-1.5	0.5
52.46	1.67	1.1	329.40	15FE	3.1	-3.9	-1.5	0.5
57.69	1.66	1.2	339.35	25FE	1.5	-3.9	-1.6	0.5
63.09	1.65	1.3	349.23	7MR	0.8	-4.0	-1.6	0.5
68.62	1.63	1.5	359.06	17MR	0.5	-4.1	-1.6	0.5
74.25	1.60	1.6	8.83	27MR	0.1	-4.2	-1.7	0.4
79.98	1.57	1.6	18.55	6AP	-0.4	-4.2	-1.7	0.4
85.78	1.54	1.7	28.22	16AP	-1.3	-4.2	-1.8	0.3
91.65	1.50	1.8	37.84	26AP	-1.7	-3.9	-1.9	0.3
97.58	1.46	1.9	47.43	6MY	-0.8	-3.2	-1.9	0.4
103.56	1.41	1.9	56.99	16MY	0.1	-3.0	-2.0	0.4
109.59	1.37	1.9	66.53	26MY	0.8	-3.7	-2.1	0.4
115.67	1.32	2.0	76.05	5JN	1.5	-4.1	-2.2	0.4
121.81	1.26	2.0	85.57	15JN	2.6	-4.2	-2.2	0.4
127.99	1.21	2.0	95.11	25JN	3.0	-4.2	-2.3	0.4
134.22	1.15	2.0	104.65	5JL	1.4	-4.1	-2.3	0.4
140.51	1.09	2.0	114.22	15JL	-0.1	-4.0	-2.4	0.4
146.86	1.02	2.0	123.83	25JL	-1.2	-3.9	-2.4	0.3
153.27	0.96	2.0	133.47	4AU	-1.5	-3.8	-2.4	0.3
159.73	0.89	2.0	143.16	14AU	-0.9	-3.8	-2.4	0.2
166.27	0.81	1.9	152.90	24AU	-0.4	-3.7	-2.4	0.2
172.87	0.74	1.9	162.69	3SE	-0.1	-3.6	-2.4	0.1
179.54	0.66	1.8	172.55	13SE	0.1	-3.6	-2.3	0.1
186.28	0.57	1.8	182.46	23SE	0.2	-3.5	-2.3	0.0
193.10	0.48	1.7	192.43	3OC	0.8	-3.5	-2.2	-0.1
199.99	0.39	1.7	202.46	13OC	3.3	-3.5	-2.1	-0.1
206.97	0.30	1.7	212.53	23OC	0.5	-3.4	-2.1	-0.2
214.01	0.20	1.7	222.65	2NO	-0.4	-3.4	-2.0	-0.2
221.14	0.10	1.7	232.80	12NO	-0.5	-3.4	-1.9	-0.1
228.34	-0.00	1.7	242.99	22NO	-0.5	-3.4	-1.9	-0.0
235.62	-0.11	1.7	253.18	2DE	-0.6	-3.4	-1.8	0.0
242.96	-0.22	1.6	263.39	12DE	-0.8	-3.4	-1.8	0.1
250.38	-0.34	1.6	273.59	22DE	-0.9	-3.3	-1.7	0.2
				587				
257.86	-0.45	1.6	283.77	1JA	-0.9	-3.4	-1.7	0.2
265.40	-0.56	1.5	293.93	11JA	-0.4	-3.4	-1.6	0.3
272.99	-0.68	1.5	304.05	21JA	1.3	-3.4	-1.6	0.3
280.62	-0.79	1.5	314.13	31JA	2.6	-3.4	-1.6	0.4
288.29	-0.90	1.4	324.16	10FE	1.1	-3.4	-1.6	0.4
295.99	-1.00	1.4	334.14	20FE	0.6	-3.5	-1.6	0.4
303.71	-1.10	1.4	344.05	2MR	0.3	-3.5	-1.6	0.4
311.43	-1.19	1.3	353.91	12MR	0.0	-3.5	-1.6	0.4
319.14	-1.27	1.3	3.71	22MR	-0.5	-3.4	-1.6	0.4
326.83	-1.34	1.2	13.46	1AP	-1.3	-3.4	-1.6	0.3
334.49	-1.40	1.2	23.15	11AP	-1.6	-3.4	-1.7	0.3
342.10	-1.44	1.2	32.80	21AP	-0.8	-3.4	-1.7	0.3
349.65	-1.48	1.1	42.40	1MY	0.2	-3.3	-1.7	0.2
357.12	-1.49	1.1	51.98	11MY	1.1	-3.3	-1.8	0.2
4.50	-1.49	1.1	61.53	21MY	2.0	-3.3	-1.8	0.3
11.77	-1.47	1.0	71.06	31MY	3.4	-3.3	-1.9	0.3
18.92	-1.44	1.0	80.58	10JN	2.4	-3.3	-2.0	0.3
25.92	-1.39	0.9	90.11	20JN	1.1	-3.3	-2.0	0.3
32.77	-1.32	0.9	99.64	30JN	-0.1	-3.4	-2.1	0.3
39.44	-1.24	0.8	109.20	10JL	-1.2	-3.4	-2.2	0.3
45.90	-1.14	0.8	118.79	20JL	-1.6	-3.4	-2.2	0.3
52.12	-1.02	0.7	128.40	30JL	-0.8	-3.5	-2.3	0.3
58.07	-0.88	0.6	138.07	9AU	-0.3	-3.5	-2.4	0.3
63.71	-0.72	0.5	147.79	19AU	0.0	-3.6	-2.4	0.2
68.97	-0.53	0.4	157.55	29AU	0.2	-3.7	-2.4	0.2
73.78	-0.32	0.3	167.37	8SE	0.4	-3.8	-2.5	0.1
78.03	-0.08	0.1	177.26	18SE	1.1	-3.9	-2.5	0.1
81.62	0.21	-0.0	187.20	28SE	3.4	-4.0	-2.4	0.0
84.37	0.53	-0.2	197.19	8OC	0.4	-4.1	-2.4	-0.1
86.11	0.90	-0.4	207.24	18OC	-0.6	-4.2	-2.4	-0.1
86.63	1.32	-0.6	217.34	28OC	-0.7	-4.3	-2.3	-0.2
85.78	1.78	-0.9	227.48	7NO	-0.6	-4.4	-2.2	-0.3
83.52	2.25	-1.1	237.65	17NO	-0.7	-4.4	-2.2	-0.3
80.13	2.68	-1.2	247.83	27NO	-0.7	-4.2	-2.1	-0.2
76.19	3.01	-1.3	258.04	7DE	-0.7	-3.8	-2.0	-0.1
72.48	3.21	-1.1	268.24	17DE	-0.7	-3.1	-2.0	-0.0
69.67	3.26	-0.8	278.43	27DE	-0.3	-3.7	-1.9	0.0

♂ LONG	♂ LAT	♂ MAG	⊙ LONG	16.00UT	☿	♀	♃	♄
				588	MAGNITUDES			
68.14	3.22	-0.5	288.60	6JA	1.6	-4.2	-1.8	0.1
67.93	3.11	-0.2	298.75	16JA	2.1	-4.3	-1.8	0.2
68.92	2.97	0.0	308.84	26JA	0.8	-4.3	-1.7	0.2
70.90	2.81	0.3	318.90	5FE	0.4	-4.3	-1.7	0.3
73.70	2.66	0.5	328.91	15FE	0.2	-4.2	-1.6	0.3
77.14	2.51	0.7	338.85	25FE	-0.1	-4.0	-1.6	0.3
81.08	2.36	0.9	348.75	6MR	-0.5	-3.9	-1.6	0.3
85.44	2.23	1.1	358.58	16MR	-1.3	-3.8	-1.6	0.3
90.11	2.09	1.2	8.35	26MR	-1.6	-3.7	-1.6	0.3
95.04	1.97	1.3	18.08	5AP	-0.7	-3.7	-1.6	0.3
100.20	1.85	1.4	27.75	15AP	0.4	-3.6	-1.6	0.3
105.53	1.74	1.5	37.37	25AP	1.4	-3.5	-1.6	0.3
111.02	1.62	1.6	46.97	5MY	2.7	-3.5	-1.6	0.2
116.65	1.52	1.7	56.53	15MY	3.3	-3.4	-1.6	0.2
122.40	1.41	1.7	66.07	25MY	1.9	-3.4	-1.7	0.2
128.26	1.31	1.8	75.60	4JN	0.8	-3.3	-1.7	0.2
134.22	1.20	1.8	85.12	14JN	-0.2	-3.3	-1.7	0.3
140.28	1.10	1.8	94.65	24JN	-1.2	-3.3	-1.8	0.3
146.44	1.00	1.9	104.19	4JL	-1.6	-3.3	-1.9	0.3
152.69	0.90	1.9	113.76	14JL	-0.8	-3.3	-1.9	0.3
159.04	0.79	1.9	123.36	24JL	-0.2	-3.3	-2.0	0.3
165.48	0.69	1.9	133.00	3AU	0.1	-3.4	-2.1	0.3
172.01	0.59	1.9	142.69	13AU	0.4	-3.4	-2.1	0.3
178.63	0.49	1.8	152.43	23AU	0.7	-3.4	-2.2	0.3
185.35	0.38	1.8	162.22	2SE	1.5	-3.4	-2.2	0.2
192.17	0.28	1.8	172.07	12SE	3.0	-3.5	-2.3	0.2
199.08	0.17	1.7	181.98	22SE	0.3	-3.5	-2.3	0.1
206.09	0.07	1.7	191.95	2OC	-0.7	-3.5	-2.4	0.1
213.19	-0.04	1.7	201.97	12OC	-0.8	-3.5	-2.4	-0.0
220.38	-0.14	1.6	212.04	22OC	-0.8	-3.4	-2.4	-0.1
227.67	-0.25	1.6	222.16	1NO	-0.8	-3.4	-2.4	-0.2
235.05	-0.35	1.5	232.31	11NO	-0.6	-3.4	-2.4	-0.2
242.51	-0.45	1.5	242.49	21NO	-0.6	-3.4	-2.3	-0.3
250.04	-0.55	1.5	252.69	1DE	-0.6	-3.3	-2.3	-0.3
257.65	-0.64	1.5	262.89	11DE	-0.2	-3.3	-2.2	-0.2
265.33	-0.73	1.4	273.09	21DE	1.8	-3.3	-2.1	-0.1
273.06	-0.82	1.4	283.27	31DE	1.7	-3.3	-2.1	-0.1
				589				
280.83	-0.89	1.4	293.43	10JA	0.5	-3.3	-2.0	0.0
288.64	-0.96	1.4	303.55	20JA	0.2	-3.4	-1.9	0.1
296.48	-1.02	1.4	313.64	30JA	0.1	-3.4	-1.8	0.1
304.32	-1.08	1.4	323.67	9FE	-0.2	-3.4	-1.8	0.2
312.17	-1.12	1.4	333.65	19FE	-0.6	-3.4	-1.7	0.2
320.00	-1.14	1.4	343.57	1MR	-1.3	-3.5	-1.7	0.3
327.81	-1.16	1.4	353.43	11MR	-1.5	-3.5	-1.6	0.3
335.59	-1.17	1.4	3.24	21MR	-0.7	-3.6	-1.6	0.3
343.31	-1.16	1.4	12.99	31MR	0.5	-3.6	-1.5	0.3
350.97	-1.13	1.4	22.68	10AP	1.8	-3.7	-1.5	0.3
358.57	-1.10	1.4	32.33	20AP	3.6	-3.8	-1.5	0.3
6.08	-1.05	1.4	41.94	30AP	2.5	-3.8	-1.5	0.3
13.51	-1.00	1.4	51.52	10MY	1.4	-3.9	-1.5	0.3
20.84	-0.92	1.4	61.07	20MY	0.6	-4.0	-1.5	0.2
28.07	-0.84	1.4	70.60	30MY	-0.2	-4.1	-1.5	0.2
35.19	-0.75	1.4	80.12	9JN	-1.2	-4.2	-1.5	0.2
42.19	-0.65	1.4	89.65	19JN	-1.7	-4.2	-1.5	0.2
49.07	-0.54	1.4	99.19	29JN	-0.8	-4.1	-1.6	0.3
55.82	-0.42	1.4	108.74	9JL	-0.2	-3.7	-1.6	0.3
62.42	-0.29	1.3	118.32	19JL	0.3	-3.1	-1.7	0.3
68.88	-0.15	1.3	127.94	29JL	0.5	-3.5	-1.7	0.3
75.16	0.00	1.3	137.60	8AU	0.9	-4.0	-1.8	0.3
81.27	0.16	1.2	147.31	18AU	1.9	-4.2	-1.8	0.3
87.17	0.34	1.1	157.08	28AU	2.7	-4.3	-1.9	0.3
92.84	0.52	1.1	166.90	7SE	0.2	-4.2	-1.9	0.3
98.25	0.73	1.0	176.78	17SE	-0.8	-4.2	-2.0	0.3
103.33	0.95	0.9	186.71	27SE	-1.0	-4.1	-2.1	0.2
108.04	1.19	0.8	196.71	7OC	-0.9	-4.0	-2.1	0.2
112.30	1.46	0.6	206.75	17OC	-0.7	-3.9	-2.2	0.1
115.99	1.76	0.5	216.85	27OC	-0.5	-3.8	-2.2	0.0
119.00	2.10	0.3	226.98	6NO	-0.4	-3.7	-2.3	-0.1
121.15	2.47	0.1	237.15	16NO	-0.4	-3.7	-2.3	-0.1
122.26	2.88	-0.1	247.34	26NO	-0.0	-3.6	-2.3	-0.2
122.16	3.31	-0.4	257.54	6DE	2.0	-3.5	-2.3	-0.3
120.71	3.73	-0.6	267.75	16DE	1.4	-3.5	-2.2	-0.2
117.98	4.10	-0.8	277.94	26DE	0.3	-3.5	-2.2	-0.2

♂ LONG	♂ LAT	♂ MAG	⊙ LONG	16.00UT	☿	♀	♃	♄
				590	MAGNITUDES			
114.32	4.35	-1.0	288.11	5JA	0.0	-3.4	-2.1	-0.1
110.34	4.43	-1.0	298.26	15JA	-0.1	-3.4	-2.1	-0.0
106.81	4.35	-0.8	308.36	25JA	-0.3	-3.4	-2.0	0.0
104.29	4.12	-0.5	318.42	4FE	-0.6	-3.3	-1.9	0.1
103.03	3.83	-0.3	328.43	14FE	-1.3	-3.3	-1.9	0.2
103.07	3.50	-0.0	338.38	24FE	-1.4	-3.3	-1.8	0.2
104.24	3.18	0.2	348.27	6MR	-0.7	-3.3	-1.7	0.3
106.36	2.88	0.4	358.11	16MR	0.6	-3.3	-1.7	0.3
109.26	2.60	0.6	7.88	26MR	2.3	-3.3	-1.6	0.3
112.78	2.34	0.8	17.60	5AP	3.3	-3.4	-1.6	0.3
116.81	2.10	0.9	27.28	15AP	1.9	-3.4	-1.5	0.3
121.25	1.89	1.1	36.91	25AP	1.1	-3.4	-1.5	0.3
126.02	1.68	1.2	46.50	5MY	0.5	-3.5	-1.4	0.3
131.08	1.50	1.3	56.06	15MY	-0.3	-3.5	-1.4	0.3
136.39	1.32	1.4	65.60	25MY	-1.3	-3.5	-1.4	0.3
141.90	1.15	1.4	75.13	4JN	-1.7	-3.4	-1.4	0.3
147.61	1.00	1.5	84.65	14JN	-0.8	-3.4	-1.4	0.2
153.49	0.85	1.5	94.18	24JN	-0.1	-3.4	-1.4	0.3
159.52	0.70	1.6	103.72	4JL	0.4	-3.3	-1.4	0.3
165.70	0.56	1.6	113.29	14JL	0.7	-3.3	-1.4	0.4
172.02	0.42	1.6	122.89	24JL	1.2	-3.3	-1.4	0.4
178.47	0.29	1.6	132.53	3AU	2.4	-3.3	-1.5	0.4
185.06	0.17	1.6	142.22	13AU	2.3	-3.3	-1.5	0.4
191.77	0.04	1.6	151.95	23AU	0.1	-3.3	-1.5	0.4
198.61	-0.08	1.6	161.74	2SE	-0.9	-3.4	-1.6	0.4
205.57	-0.19	1.6	171.59	12SE	-1.1	-3.4	-1.7	0.4
212.64	-0.30	1.6	181.49	22SE	-1.0	-3.4	-1.7	0.4
219.82	-0.41	1.6	191.46	2OC	-0.6	-3.5	-1.8	0.3
227.12	-0.51	1.6	201.48	12OC	-0.4	-3.5	-1.8	0.3
234.52	-0.61	1.5	211.55	22OC	-0.3	-3.6	-1.9	0.2
242.01	-0.70	1.5	221.66	1NO	-0.2	-3.6	-2.0	0.2
249.60	-0.78	1.5	231.82	11NO	0.2	-3.7	-2.0	0.1
257.26	-0.85	1.5	241.99	21NO	2.3	-3.8	-2.1	0.0
264.99	-0.92	1.4	252.19	1DE	1.1	-3.9	-2.1	-0.0
272.79	-0.97	1.4	262.40	11DE	0.1	-4.0	-2.2	-0.1
280.63	-1.02	1.4	272.59	21DE	-0.1	-4.1	-2.2	-0.2
288.50	-1.05	1.3	282.78	31DE	-0.2	-4.2	-2.2	-0.1
				591				
296.40	-1.07	1.3	292.94	10JA	-0.3	-4.3	-2.1	-0.1
304.30	-1.08	1.3	303.07	20JA	-0.7	-4.3	-2.1	-0.0
312.20	-1.08	1.3	313.15	30JA	-1.2	-4.3	-2.1	0.0
320.08	-1.07	1.3	323.19	9FE	-1.3	-4.2	-2.0	0.1
327.92	-1.05	1.3	333.16	19FE	-0.7	-3.7	-2.0	0.2
335.73	-1.01	1.4	343.09	1MR	0.8	-3.2	-1.9	0.2
343.48	-0.97	1.4	352.96	11MR	2.9	-3.7	-1.8	0.3
351.16	-0.91	1.4	2.76	21MR	2.5	-4.1	-1.7	0.3
358.78	-0.84	1.5	12.51	31MR	1.4	-4.2	-1.7	0.4
6.32	-0.77	1.5	22.21	10AP	0.8	-4.2	-1.6	0.4
13.77	-0.69	1.5	31.86	20AP	0.3	-4.2	-1.6	0.4
21.14	-0.60	1.6	41.47	30AP	-0.3	-4.1	-1.5	0.4
28.41	-0.51	1.6	51.05	10MY	-1.3	-3.9	-1.5	0.4
35.59	-0.41	1.6	60.60	20MY	-1.7	-3.8	-1.4	0.4
42.68	-0.30	1.6	70.13	30MY	-0.8	-3.8	-1.4	0.4
49.66	-0.19	1.6	79.66	9JN	-0.0	-3.7	-1.3	0.4
56.55	-0.08	1.7	89.18	19JN	0.5	-3.6	-1.3	0.4
63.34	0.04	1.7	98.72	29JN	1.0	-3.5	-1.3	0.4
70.03	0.16	1.7	108.28	9JL	1.7	-3.5	-1.3	0.4
76.61	0.28	1.7	117.86	19JL	3.0	-3.5	-1.3	0.4
83.09	0.41	1.7	127.48	29JL	1.9	-3.4	-1.3	0.5
89.46	0.54	1.7	137.14	8AU	0.1	-3.4	-1.3	0.5
95.72	0.67	1.6	146.84	18AU	-1.0	-3.4	-1.3	0.5
101.85	0.81	1.6	156.61	28AU	-1.3	-3.4	-1.3	0.6
107.85	0.95	1.6	166.42	7SE	-1.0	-3.4	-1.4	0.6
113.70	1.10	1.5	176.30	17SE	-0.5	-3.4	-1.4	0.6
119.39	1.26	1.4	186.24	27SE	-0.3	-3.4	-1.5	0.5
124.88	1.43	1.4	196.23	7OC	-0.1	-3.4	-1.5	0.5
130.14	1.61	1.3	206.27	17OC	-0.1	-3.4	-1.6	0.5
135.15	1.80	1.2	216.36	27OC	0.4	-3.4	-1.6	0.4
139.83	2.01	1.0	226.49	6NO	2.6	-3.4	-1.7	0.4
144.12	2.23	0.9	236.66	16NO	0.9	-3.4	-1.8	0.3
147.93	2.48	0.7	246.85	26NO	-0.1	-3.5	-1.8	0.2
151.15	2.76	0.5	257.05	6DE	-0.3	-3.5	-1.9	0.2
153.64	3.05	0.3	267.25	16DE	-0.3	-3.5	-1.9	0.1
155.22	3.37	0.1	277.45	26DE	-0.4	-3.5	-2.0	0.0

♂ LONG	♂ LAT	♂ MAG	☉ LONG	16.00UT	☿	♀	♃	♄
				592	MAGNITUDES			
155.72	3.70	-0.2	287.62	5JA	-0.8	-3.4	-2.0	-0.0
154.97	4.01	-0.5	297.76	15JA	-1.1	-3.4	-2.1	0.0
152.92	4.27	-0.7	307.87	25JA	-1.1	-3.4	-2.1	0.1
149.74	4.42	-0.9	317.93	4FE	-0.6	-3.4	-2.1	0.1
145.91	4.40	-1.0	327.94	14FE	1.0	-3.4	-2.0	0.2
142.11	4.22	-0.9	337.90	24FE	3.2	-3.3	-2.0	0.2
139.02	3.89	-0.7	347.79	5MR	1.9	-3.3	-2.0	0.3
137.09	3.48	-0.5	357.63	15MR	1.0	-3.3	-1.9	0.4
136.46	3.05	-0.3	7.41	25MR	0.6	-3.3	-1.8	0.4
137.07	2.63	-0.1	17.13	4AP	0.2	-3.4	-1.8	0.5
138.76	2.24	0.1	26.81	14AP	-0.4	-3.4	-1.7	0.5
141.34	1.89	0.3	36.44	24AP	-1.3	-3.4	-1.6	0.5
144.67	1.57	0.5	46.03	4MY	-1.7	-3.4	-1.6	0.6
148.59	1.28	0.6	55.60	14MY	-0.8	-3.5	-1.5	0.6
153.01	1.02	0.8	65.14	24MY	0.1	-3.5	-1.5	0.6
157.85	0.78	0.9	74.66	3JN	0.7	-3.6	-1.4	0.6
163.03	0.56	1.0	84.19	13JN	1.3	-3.6	-1.4	0.6
168.52	0.36	1.0	93.72	23JN	2.2	-3.7	-1.3	0.6
174.28	0.18	1.1	103.26	3JL	3.3	-3.8	-1.3	0.6
180.27	0.01	1.2	112.83	13JL	1.6	-3.9	-1.3	0.5
186.48	-0.15	1.2	122.43	23JL	-0.0	-4.0	-1.2	0.6
192.88	-0.30	1.2	132.06	2AU	-1.1	-4.1	-1.2	0.6
199.46	-0.43	1.3	141.75	12AU	-1.4	-4.2	-1.2	0.7
206.22	-0.56	1.3	151.48	22AU	-0.9	-4.3	-1.2	0.7
213.13	-0.67	1.3	161.27	1SE	-0.4	-4.3	-1.2	0.7
220.19	-0.78	1.3	171.11	11SE	-0.1	-4.1	-1.2	0.7
227.38	-0.87	1.3	181.02	21SE	-0.0	-3.7	-1.3	0.8
234.71	-0.95	1.4	190.98	1OC	0.1	-3.2	-1.3	0.8
242.14	-1.02	1.4	201.00	11OC	0.6	-3.7	-1.3	0.7
249.69	-1.08	1.4	211.07	21OC	2.9	-4.2	-1.4	0.7
257.33	-1.12	1.4	221.17	31OC	0.7	-4.4	-1.4	0.7
265.04	-1.16	1.4	231.33	10NO	-0.3	-4.4	-1.5	0.6
272.82	-1.17	1.4	241.50	20NO	-0.4	-4.3	-1.5	0.6
280.65	-1.18	1.4	251.70	30NO	-0.4	-4.2	-1.6	0.5
288.52	-1.17	1.4	261.90	10DE	-0.5	-4.1	-1.7	0.4
296.41	-1.16	1.4	272.10	20DE	-0.8	-4.0	-1.7	0.4
304.30	-1.13	1.4	282.29	30DE	-1.0	-3.9	-1.8	0.3
				593				
312.18	-1.08	1.4	292.45	9JA	-1.0	-3.8	-1.9	0.2
320.04	-1.03	1.4	302.58	19JA	-0.5	-3.7	-1.9	0.2
327.86	-0.97	1.4	312.66	29JA	1.2	-3.6	-2.0	0.2
335.64	-0.90	1.4	322.70	8FE	3.0	-3.6	-2.0	0.3
343.36	-0.82	1.4	332.69	18FE	1.4	-3.5	-2.0	0.3
351.01	-0.73	1.5	342.61	28FE	0.7	-3.5	-2.0	0.4
358.59	-0.64	1.5	352.48	10MR	0.4	-3.4	-2.0	0.4
6.09	-0.55	1.5	2.29	20MR	0.1	-3.4	-2.0	0.5
13.51	-0.45	1.5	12.05	30MR	-0.4	-3.3	-1.9	0.5
20.84	-0.35	1.5	21.75	9AP	-1.2	-3.3	-1.9	0.6
28.09	-0.25	1.6	31.40	19AP	-1.7	-3.3	-1.8	0.6
35.24	-0.14	1.6	41.01	29AP	-0.8	-3.3	-1.8	0.7
42.31	-0.04	1.7	50.59	9MY	0.2	-3.3	-1.7	0.7
49.30	0.07	1.7	60.14	19MY	0.9	-3.3	-1.7	0.7
56.20	0.17	1.7	69.67	29MY	1.7	-3.3	-1.6	0.8
63.02	0.27	1.8	79.20	8JN	2.9	-3.4	-1.5	0.8
69.77	0.38	1.8	88.72	18JN	2.8	-3.4	-1.5	0.8
76.43	0.48	1.8	98.26	28JN	1.3	-3.4	-1.4	0.8
83.03	0.58	1.9	107.81	8JL	-0.1	-3.5	-1.4	0.8
89.55	0.68	1.9	117.39	18JL	-1.2	-3.5	-1.3	0.8
96.00	0.77	1.9	127.00	28JL	-1.5	-3.5	-1.3	0.8
102.38	0.87	1.9	136.66	7AU	-0.9	-3.4	-1.3	0.8
108.68	0.97	1.9	146.37	17AU	-0.3	-3.4	-1.2	0.9
114.92	1.07	1.9	156.12	27AU	-0.0	-3.4	-1.2	0.9
121.08	1.17	1.8	165.94	6SE	0.1	-3.4	-1.2	0.9
127.16	1.26	1.8	175.81	16SE	0.3	-3.3	-1.2	1.0
133.16	1.36	1.8	185.74	26SE	0.9	-3.3	-1.2	1.0
139.06	1.47	1.7	195.74	6OC	3.3	-3.3	-1.2	1.0
144.85	1.57	1.7	205.78	16OC	0.5	-3.3	-1.2	1.0
150.51	1.68	1.6	215.87	26OC	-0.5	-3.3	-1.3	1.0
156.03	1.79	1.5	226.00	5NO	-0.6	-3.4	-1.3	1.0
161.37	1.90	1.4	236.16	15NO	-0.6	-3.4	-1.3	0.9
166.51	2.03	1.3	246.35	25NO	-0.7	-3.4	-1.4	0.9
171.39	2.15	1.1	256.55	5DE	-0.8	-3.4	-1.4	0.8
175.96	2.29	1.0	266.76	15DE	-0.8	-3.5	-1.5	0.8
180.17	2.43	0.8	276.95	25DE	-0.8	-3.5	-1.6	0.7

♂ LONG	♂ LAT	♂ MAG	☉ LONG	16.00UT	☿	♀	♃	♄
				594	MAGNITUDES			
183.89	2.58	0.6	287.13	4JA	-0.4	-3.6	-1.6	0.6
187.04	2.73	0.4	297.27	14JA	1.4	-3.6	-1.7	0.6
189.46	2.88	0.1	307.38	24JA	2.5	-3.7	-1.8	0.5
190.97	3.03	-0.2	317.45	3FE	1.0	-3.8	-1.8	0.4
191.40	3.14	-0.4	327.46	13FE	0.5	-3.9	-1.9	0.5
190.61	3.21	-0.7	337.42	23FE	0.3	-4.0	-1.9	0.5
188.54	3.19	-1.0	347.32	5MR	-0.0	-4.0	-2.0	0.6
185.41	3.06	-1.3	357.16	15MR	-0.5	-4.1	-2.0	0.6
181.69	2.78	-1.4	6.94	25MR	-1.2	-4.2	-2.0	0.7
178.10	2.39	-1.2	16.67	4AP	-1.6	-4.2	-2.0	0.7
175.31	1.93	-1.1	26.34	14AP	-0.8	-4.2	-2.0	0.8
173.72	1.45	-0.9	35.98	24AP	0.3	-3.9	-2.0	0.8
173.47	1.00	-0.7	45.57	4MY	1.2	-3.2	-1.9	0.9
174.48	0.59	-0.4	55.14	14MY	2.3	-3.0	-1.9	0.9
176.59	0.23	-0.3	64.68	24MY	3.6	-3.8	-1.8	1.0
179.63	-0.08	-0.1	74.21	3JN	2.2	-4.1	-1.8	1.0
183.44	-0.35	0.1	83.73	13JN	1.0	-4.2	-1.7	1.0
187.86	-0.58	0.2	93.26	23JN	-0.1	-4.2	-1.7	1.0
192.82	-0.78	0.3	102.80	3JL	-1.2	-4.1	-1.6	1.1
198.21	-0.96	0.4	112.36	13JL	-1.6	-4.0	-1.5	1.1
203.98	-1.10	0.5	121.96	23JL	-0.9	-3.9	-1.5	1.1
210.07	-1.23	0.6	131.59	2AU	-0.3	-3.8	-1.4	1.1
216.44	-1.33	0.7	141.27	12AU	0.1	-3.8	-1.4	1.1
223.06	-1.42	0.8	151.00	22AU	0.3	-3.7	-1.3	1.1
229.89	-1.48	0.8	160.79	1SE	0.5	-3.6	-1.3	1.2
236.92	-1.53	0.9	170.63	11SE	1.2	-3.6	-1.3	1.2
244.10	-1.55	0.9	180.53	21SE	3.3	-3.5	-1.3	1.2
251.44	-1.56	1.0	190.49	1OC	0.4	-3.5	-1.2	1.3
258.89	-1.56	1.0	200.51	11OC	-0.6	-3.5	-1.2	1.3
266.45	-1.53	1.1	210.57	21OC	-0.7	-3.4	-1.2	1.3
274.09	-1.49	1.1	220.69	31OC	-0.7	-3.4	-1.3	1.3
281.79	-1.44	1.1	230.83	10NO	-0.8	-3.4	-1.3	1.3
289.54	-1.38	1.2	241.01	20NO	-0.7	-3.4	-1.3	1.3
297.31	-1.30	1.2	251.21	30NO	-0.7	-3.4	-1.3	1.2
305.09	-1.21	1.3	261.41	10DE	-0.7	-3.4	-1.4	1.2
312.86	-1.12	1.3	271.61	20DE	-0.3	-3.4	-1.4	1.1
320.60	-1.02	1.4	281.80	30DE	1.6	-3.4	-1.5	1.0
				595				
328.31	-0.91	1.4	291.96	9JA	2.0	-3.4	-1.5	1.0
335.98	-0.80	1.4	302.09	19JA	0.7	-3.4	-1.6	0.9
343.58	-0.68	1.5	312.18	29JA	0.3	-3.4	-1.6	0.8
351.12	-0.57	1.5	322.22	8FE	0.1	-3.4	-1.7	0.8
358.59	-0.45	1.6	332.21	18FE	-0.1	-3.5	-1.8	0.7
5.98	-0.34	1.6	342.13	28FE	-0.5	-3.5	-1.8	0.8
13.29	-0.23	1.6	352.00	10MR	-1.2	-3.5	-1.9	0.8
20.53	-0.12	1.7	1.82	20MR	-1.6	-3.4	-2.0	0.9
27.68	-0.01	1.7	11.57	30MR	-0.8	-3.4	-2.0	0.9
34.75	0.10	1.7	21.28	9AP	0.4	-3.4	-2.1	1.0
41.74	0.20	1.7	30.93	19AP	1.5	-3.4	-2.1	1.0
48.66	0.30	1.7	40.55	29AP	3.0	-3.3	-2.1	1.1
55.50	0.39	1.8	50.12	9MY	3.0	-3.3	-2.1	1.1
62.28	0.48	1.8	59.68	19MY	1.7	-3.3	-2.1	1.2
68.99	0.57	1.8	69.21	29MY	0.8	-3.3	-2.0	1.2
75.64	0.65	1.8	78.73	8JN	-0.2	-3.3	-2.0	1.2
82.24	0.73	1.9	88.26	18JN	-1.2	-3.4	-2.0	1.3
88.78	0.80	1.9	97.79	28JN	-1.7	-3.4	-1.9	1.3
95.28	0.88	1.9	107.34	8JL	-0.8	-3.4	-1.8	1.3
101.73	0.94	2.0	116.92	18JL	-0.2	-3.4	-1.8	1.4
108.14	1.01	2.0	126.54	28JL	0.2	-3.5	-1.7	1.4
114.51	1.07	2.0	136.19	7AU	0.4	-3.5	-1.6	1.4
120.84	1.13	2.0	145.90	17AU	0.8	-3.6	-1.6	1.3
127.14	1.18	2.0	155.65	27AU	1.6	-3.7	-1.5	1.3
133.41	1.24	2.0	165.46	6SE	3.0	-3.8	-1.5	1.3
139.65	1.29	2.0	175.33	16SE	0.3	-3.9	-1.4	1.3
145.85	1.34	2.0	185.26	26SE	-0.7	-4.0	-1.4	1.3
152.01	1.38	1.9	195.25	6OC	-0.9	-4.1	-1.4	1.3
158.14	1.42	1.9	205.29	16OC	-0.8	-4.2	-1.3	1.3
164.21	1.46	1.8	215.37	26OC	-0.7	-4.3	-1.3	1.2
170.24	1.50	1.8	225.50	5NO	-0.6	-4.4	-1.3	1.2
176.20	1.54	1.7	235.67	15NO	-0.5	-4.4	-1.3	1.2
182.09	1.57	1.6	245.85	25NO	-0.5	-4.2	-1.3	1.2
187.89	1.59	1.5	256.06	5DE	-0.1	-3.8	-1.3	1.1
193.59	1.61	1.4	266.26	15DE	1.8	-3.0	-1.4	1.1
199.16	1.63	1.2	276.46	25DE	1.6	-3.7	-1.4	1.0

♂			⊙	16.00UT	☿	♀	♃	♄
LONG	LAT	MAG	LONG	596	MAGNITUDES			
204.58	1.63	1.1	286.63	4JA	0.4	-4.2	-1.4	1.0
209.81	1.63	0.9	296.78	14JA	0.1	-4.3	-1.5	0.9
214.81	1.61	0.8	306.89	24JA	0.0	-4.3	-1.5	0.9
219.52	1.58	0.6	316.96	3FE	-0.2	-4.3	-1.6	0.8
223.88	1.52	0.4	326.98	13FE	-0.6	-4.2	-1.6	0.8
227.80	1.44	0.1	336.93	23FE	-1.2	-4.0	-1.7	0.8
231.15	1.32	-0.1	346.84	4MR	-1.5	-3.9	-1.8	0.8
233.80	1.14	-0.4	356.68	14MR	-0.8	-3.8	-1.8	0.8
235.58	0.91	-0.7	6.47	24MR	0.5	-3.7	-1.9	0.9
236.29	0.60	-1.0	16.20	3AP	2.0	-3.6	-2.0	1.0
235.80	0.20	-1.4	25.88	13AP	3.9	-3.6	-2.0	1.0
234.07	-0.29	-1.7	35.51	23AP	2.3	-3.5	-2.1	1.1
231.29	-0.84	-2.0	45.11	3MY	1.3	-3.5	-2.1	1.2
228.00	-1.41	-2.0	54.68	13MY	0.6	-3.4	-2.2	1.2
224.90	-1.93	-1.9	64.22	23MY	-0.2	-3.4	-2.2	1.3
222.66	-2.35	-1.7	73.75	2JN	-1.2	-3.3	-2.2	1.3
221.71	-2.66	-1.5	83.27	12JN	-1.7	-3.3	-2.2	1.3
222.14	-2.87	-1.3	92.80	22JN	-0.8	-3.3	-2.2	1.3
223.87	-2.99	-1.1	102.34	2JL	-0.1	-3.3	-2.2	1.3
226.73	-3.04	-0.9	111.91	12JL	0.3	-3.3	-2.1	1.3
230.52	-3.05	-0.7	121.50	22JL	0.6	-3.3	-2.1	1.3
235.07	-3.01	-0.5	131.13	1AU	1.0	-3.4	-2.0	1.3
240.25	-2.94	-0.4	140.81	11AU	2.1	-3.4	-1.9	1.2
245.92	-2.85	-0.2	150.53	21AU	2.6	-3.4	-1.9	1.2
252.01	-2.73	-0.1	160.31	31AU	0.2	-3.4	-1.8	1.1
258.42	-2.60	0.1	170.15	10SE	-0.9	-3.5	-1.7	1.1
265.10	-2.45	0.2	180.05	20SE	-1.0	-3.5	-1.7	1.1
271.99	-2.29	0.3	190.01	30SE	-1.0	-3.5	-1.6	1.1
279.05	-2.12	0.4	200.02	10OC	-0.6	-3.5	-1.6	1.1
286.24	-1.94	0.5	210.08	20OC	-0.4	-3.4	-1.5	1.1
293.52	-1.76	0.6	220.19	30OC	-0.4	-3.4	-1.5	1.1
300.86	-1.58	0.7	230.33	9NO	-0.3	-3.4	-1.5	1.1
308.25	-1.39	0.8	240.51	19NO	0.0	-3.4	-1.5	1.1
315.65	-1.21	0.9	250.70	29NO	2.0	-3.3	-1.4	1.0
323.05	-1.03	1.0	260.91	9DE	1.3	-3.3	-1.4	1.0
330.43	-0.86	1.1	271.11	19DE	0.2	-3.3	-1.4	1.0
337.78	-0.69	1.2	281.30	29DE	-0.0	-3.3	-1.4	0.9
				597				
345.10	-0.53	1.3	291.46	8JA	-0.1	-3.3	-1.4	0.9
352.36	-0.38	1.4	301.59	18JA	-0.3	-3.4	-1.5	0.8
359.57	-0.23	1.5	311.69	28JA	-0.6	-3.4	-1.5	0.8
6.72	-0.10	1.5	321.73	7FE	-1.2	-3.4	-1.5	0.7
13.80	0.03	1.6	331.72	17FE	-1.4	-3.4	-1.6	0.7
20.83	0.15	1.6	341.66	27FE	-0.7	-3.5	-1.6	0.6
27.78	0.26	1.7	351.53	9MR	0.7	-3.5	-1.7	0.6
34.67	0.36	1.8	1.35	19MR	2.5	-3.6	-1.7	0.6
41.50	0.46	1.8	11.11	29MR	3.0	-3.6	-1.8	0.7
48.27	0.55	1.8	20.81	8AP	1.7	-3.7	-1.9	0.8
54.98	0.63	1.9	30.47	18AP	1.0	-3.8	-2.0	0.8
61.64	0.71	1.9	40.09	28AP	0.4	-3.8	-2.0	0.9
68.24	0.78	1.9	49.67	8MY	-0.3	-3.9	-2.1	1.0
74.80	0.84	1.9	59.22	18MY	-1.2	-4.0	-2.2	1.0
81.33	0.90	1.9	68.76	28MY	-1.7	-4.1	-2.2	1.1
87.81	0.95	1.9	78.28	7JN	-0.8	-4.2	-2.3	1.1
94.26	0.99	1.9	87.80	17JN	-0.1	-4.2	-2.3	1.1
100.69	1.03	1.9	97.34	27JN	0.4	-4.0	-2.3	1.1
107.10	1.07	1.9	106.89	7JL	0.8	-3.7	-2.3	1.2
113.49	1.10	2.0	116.46	17JL	1.4	-3.1	-2.3	1.2
119.86	1.13	2.0	126.08	27JL	2.6	-3.5	-2.3	1.1
126.23	1.15	2.0	135.73	6AU	2.2	-4.0	-2.3	1.1
132.59	1.17	2.0	145.43	16AU	0.1	-4.2	-2.2	1.1
138.95	1.19	2.0	155.18	26AU	-1.0	-4.3	-2.2	1.1
145.31	1.20	2.0	164.99	5SE	-1.2	-4.2	-2.1	1.0
151.67	1.21	2.0	174.85	15SE	-1.0	-4.1	-2.1	1.0
158.04	1.21	2.0	184.78	25SE	-0.6	-4.1	-2.0	0.9
164.41	1.20	2.0	194.76	5OC	-0.3	-4.0	-1.9	1.0
170.79	1.19	2.0	204.80	15OC	-0.2	-3.9	-1.9	1.0
177.18	1.18	1.9	214.88	25OC	-0.2	-3.8	-1.8	1.0
183.56	1.16	1.9	225.01	4NO	0.2	-3.7	-1.7	1.0
189.96	1.13	1.8	235.17	14NO	2.3	-3.7	-1.7	1.0
196.35	1.10	1.8	245.36	24NO	1.1	-3.6	-1.6	1.0
202.74	1.06	1.7	255.56	4DE	-0.0	-3.5	-1.6	0.9
209.13	1.01	1.6	265.76	14DE	-0.2	-3.5	-1.6	0.9
215.50	0.94	1.5	275.96	24DE	-0.2	-3.5	-1.5	0.9

♂			⊙	16.00UT	☿	♀	♃	♄
LONG	LAT	MAG	LONG	598	MAGNITUDES			
221.86	0.87	1.4	286.14	3JA	-0.4	-3.4	-1.5	0.8
228.20	0.78	1.3	296.29	13JA	-0.7	-3.4	-1.5	0.8
234.50	0.67	1.2	306.41	23JA	-1.2	-3.4	-1.5	0.7
240.77	0.55	1.1	316.47	2FE	-1.2	-3.3	-1.5	0.7
246.99	0.40	0.9	326.49	12FE	-0.7	-3.3	-1.5	0.6
253.13	0.23	0.8	336.46	22FE	0.8	-3.3	-1.6	0.6
259.19	0.03	0.6	346.36	4MR	3.0	-3.3	-1.6	0.5
265.14	-0.20	0.4	356.21	14MR	2.3	-3.3	-1.6	0.5
270.94	-0.48	0.2	6.00	24MR	1.2	-3.3	-1.7	0.5
276.56	-0.79	0.0	15.73	3AP	0.7	-3.4	-1.7	0.5
281.94	-1.15	-0.2	25.42	13AP	0.3	-3.4	-1.8	0.6
287.00	-1.57	-0.4	35.05	23AP	-0.3	-3.4	-1.8	0.6
291.66	-2.05	-0.7	44.65	3MY	-1.2	-3.5	-1.9	0.7
295.79	-2.61	-0.9	54.22	13MY	-1.7	-3.5	-2.0	0.8
299.24	-3.23	-1.2	63.76	23MY	-0.8	-3.5	-2.0	0.8
301.84	-3.92	-1.5	73.29	2JN	-0.0	-3.4	-2.1	0.9
303.37	-4.67	-1.8	82.81	12JN	0.6	-3.4	-2.2	0.9
303.68	-5.42	-2.1	92.34	22JN	1.1	-3.4	-2.2	1.0
302.73	-6.10	-2.4	101.88	2JL	1.8	-3.3	-2.3	1.0
300.69	-6.60	-2.6	111.44	12JL	3.2	-3.3	-2.4	1.0
298.11	-6.82	-2.6	121.03	22JL	1.9	-3.3	-2.4	1.0
295.70	-6.71	-2.5	130.66	1AU	0.1	-3.3	-2.4	1.0
294.12	-6.32	-2.3	140.34	11AU	-1.0	-3.3	-2.4	1.0
293.78	-5.75	-2.0	150.06	21AU	-1.3	-3.3	-2.4	1.0
294.76	-5.10	-1.7	159.84	31AU	-1.0	-3.4	-2.4	1.0
296.95	-4.44	-1.4	169.67	10SE	-0.5	-3.4	-2.4	0.9
300.16	-3.80	-1.1	179.57	20SE	-0.2	-3.4	-2.3	0.9
304.18	-3.22	-0.9	189.52	30SE	-0.1	-3.5	-2.3	0.9
308.84	-2.69	-0.6	199.53	10OC	0.0	-3.5	-2.2	0.8
314.00	-2.22	-0.4	209.59	20OC	0.5	-3.6	-2.1	0.9
319.54	-1.79	-0.2	219.69	30OC	2.6	-3.6	-2.1	0.9
325.36	-1.42	0.0	229.84	9NO	0.9	-3.7	-2.0	0.9
331.41	-1.08	0.2	240.01	19NO	-0.2	-3.8	-1.9	0.9
337.61	-0.78	0.4	250.20	29NO	-0.3	-3.9	-1.9	0.9
343.93	-0.52	0.6	260.41	9DE	-0.4	-4.0	-1.8	0.9
350.34	-0.29	0.7	270.61	19DE	-0.5	-4.1	-1.8	0.9
356.80	-0.09	0.9	280.80	29DE	-0.8	-4.2	-1.7	0.8
				599				
3.30	0.10	1.0	290.97	8JA	-1.1	-4.3	-1.7	0.8
9.82	0.25	1.2	301.10	18JA	-1.1	-4.3	-1.6	0.8
16.35	0.39	1.3	311.19	28JA	-0.6	-4.3	-1.6	0.7
22.87	0.52	1.4	321.24	7FE	1.0	-4.2	-1.6	0.7
29.38	0.63	1.5	331.23	17FE	3.2	-3.7	-1.6	0.6
35.88	0.72	1.6	341.17	27FE	1.7	-3.2	-1.6	0.6
42.36	0.80	1.7	351.05	9MR	0.9	-3.7	-1.6	0.5
48.81	0.87	1.7	0.87	19MR	0.5	-4.1	-1.6	0.4
55.24	0.94	1.8	10.63	29MR	0.2	-4.2	-1.6	0.4
61.65	0.99	1.8	20.34	8AP	-0.4	-4.2	-1.6	0.4
68.04	1.04	1.9	30.00	18AP	-1.2	-4.2	-1.7	0.4
74.41	1.07	1.9	39.62	28AP	-1.7	-4.1	-1.7	0.5
80.76	1.10	2.0	49.21	8MY	-0.8	-3.9	-1.7	0.6
87.10	1.13	2.0	58.76	18MY	0.1	-3.8	-1.8	0.6
93.43	1.15	2.0	68.30	28MY	0.8	-3.8	-1.9	0.7
99.76	1.16	2.0	77.82	7JN	1.4	-3.7	-1.9	0.8
106.08	1.17	2.0	87.34	17JN	2.4	-3.6	-2.0	0.8
112.40	1.17	2.0	96.88	27JN	3.2	-3.5	-2.1	0.8
118.73	1.17	2.0	106.43	7JL	1.5	-3.5	-2.1	0.9
125.07	1.16	2.0	116.00	17JL	0.0	-3.5	-2.2	0.9
131.43	1.15	2.0	125.61	27JL	-1.1	-3.4	-2.3	0.9
137.80	1.14	2.0	135.26	6AU	-1.4	-3.4	-2.3	0.9
144.20	1.12	1.9	144.96	16AU	-0.9	-3.4	-2.4	0.9
150.63	1.09	2.0	154.71	26AU	-0.4	-3.4	-2.4	0.9
157.08	1.07	2.0	164.52	5SE	-0.1	-3.4	-2.4	0.9
163.57	1.03	2.0	174.38	15SE	0.0	-3.4	-2.5	0.9
170.10	0.99	2.0	184.30	25SE	0.2	-3.4	-2.5	0.8
176.66	0.95	2.0	194.28	5OC	0.7	-3.4	-2.4	0.8
183.26	0.90	1.9	204.31	15OC	3.0	-3.4	-2.4	0.8
189.91	0.84	1.9	214.40	25OC	0.7	-3.4	-2.4	0.8
196.60	0.78	1.9	224.52	4NO	-0.4	-3.4	-2.3	0.8
203.33	0.71	1.9	234.68	14NO	-0.5	-3.4	-2.2	0.8
210.11	0.63	1.8	244.86	24NO	-0.5	-3.5	-2.2	0.8
216.93	0.55	1.8	255.06	4DE	-0.6	-3.5	-2.1	0.8
223.80	0.45	1.7	265.26	14DE	-0.8	-3.5	-2.0	0.8
230.71	0.35	1.6	275.46	24DE	-0.9	-3.5	-1.9	0.8

♂ LONG	♂ LAT	♂ MAG	☉ LONG	16.00UT	☿	♀	♃	♄
				600	MAGNITUDES			
237.66	0.23	1.6	285.64	3JA	-0.9	-3.4	-1.9	0.8
244.65	0.11	1.5	295.79	13JA	-0.5	-3.4	-1.8	0.8
251.68	-0.02	1.4	305.91	23JA	1.2	-3.4	-1.8	0.7
258.75	-0.17	1.3	315.98	2FE	2.8	-3.4	-1.7	0.7
265.84	-0.33	1.2	326.00	12FE	1.2	-3.4	-1.7	0.7
272.97	-0.50	1.2	335.97	22FE	0.6	-3.3	-1.6	0.6
280.12	-0.68	1.1	345.88	3MR	0.4	-3.3	-1.6	0.5
287.28	-0.87	1.0	355.72	13MR	0.1	-3.3	-1.6	0.5
294.44	-1.07	0.8	5.52	23MR	-0.4	-3.3	-1.6	0.4
301.59	-1.29	0.7	15.26	2AP	-1.2	-3.4	-1.6	0.4
308.72	-1.51	0.6	24.94	12AP	-1.7	-3.4	-1.6	0.3
315.79	-1.73	0.5	34.58	22AP	-0.8	-3.4	-1.6	0.3
322.80	-1.96	0.4	44.18	2MY	0.2	-3.4	-1.6	0.4
329.70	-2.19	0.3	53.75	12MY	1.0	-3.5	-1.6	0.5
336.47	-2.43	0.1	63.30	22MY	1.9	-3.5	-1.6	0.5
343.05	-2.66	-0.0	72.83	1JN	3.2	-3.6	-1.7	0.6
349.40	-2.88	-0.1	82.35	11JN	2.6	-3.6	-1.7	0.6
355.45	-3.10	-0.3	91.88	21JN	1.2	-3.7	-1.8	0.7
1.11	-3.30	-0.5	101.42	1JL	-0.1	-3.8	-1.8	0.7
6.29	-3.49	-0.6	110.98	11JL	-1.1	-3.9	-1.9	0.8
10.84	-3.66	-0.8	120.57	21JL	-1.6	-4.0	-1.9	0.8
14.62	-3.81	-1.0	130.20	31JL	-0.9	-4.1	-2.0	0.9
17.43	-3.91	-1.2	139.87	10AU	-0.3	-4.2	-2.1	0.9
19.04	-3.95	-1.5	149.59	20AU	0.0	-4.3	-2.1	0.9
19.28	-3.90	-1.7	159.37	30AU	0.2	-4.3	-2.2	0.9
18.05	-3.71	-1.9	169.20	9SE	0.4	-4.1	-2.3	0.9
15.51	-3.37	-2.1	179.09	19SE	1.0	-3.7	-2.3	0.8
12.21	-2.87	-2.2	189.04	29SE	3.3	-3.2	-2.3	0.8
8.90	-2.25	-2.0	199.04	9OC	0.5	-3.7	-2.4	0.8
6.37	-1.61	-1.7	209.10	19OC	-0.5	-4.2	-2.4	0.7
5.07	-1.01	-1.4	219.21	29OC	-0.6	-4.4	-2.4	0.7
5.11	-0.49	-1.1	229.35	8NO	-0.6	-4.4	-2.4	0.7
6.40	-0.06	-0.7	239.52	18NO	-0.7	-4.3	-2.3	0.8
8.73	0.28	-0.4	249.71	28NO	-0.8	-4.2	-2.3	0.8
11.90	0.56	-0.2	259.92	8DE	-0.8	-4.1	-2.2	0.8
15.72	0.77	0.1	270.12	18DE	-0.8	-4.0	-2.2	0.8
20.05	0.94	0.3	280.31	28DE	-0.4	-3.9	-2.1	0.8
				601				
24.77	1.07	0.6	290.48	7JA	1.4	-3.8	-2.0	0.8
29.79	1.17	0.7	300.61	17JA	2.3	-3.7	-2.0	0.8
35.05	1.25	0.9	310.71	27JA	0.9	-3.6	-1.9	0.7
40.49	1.30	1.1	320.76	6FE	0.4	-3.6	-1.8	0.7
46.07	1.35	1.2	330.75	16FE	0.2	-3.5	-1.8	0.7
51.76	1.38	1.4	340.69	26FE	-0.0	-3.5	-1.7	0.6
57.54	1.40	1.5	350.57	8MR	-0.5	-3.4	-1.6	0.6
63.40	1.41	1.6	0.40	18MR	-1.2	-3.4	-1.6	0.5
69.31	1.41	1.7	10.16	28MR	-1.6	-3.3	-1.6	0.4
75.26	1.41	1.7	19.87	7AP	-0.8	-3.3	-1.5	0.4
81.26	1.40	1.8	29.54	17AP	0.3	-3.3	-1.5	0.3
87.29	1.39	1.9	39.16	27AP	1.3	-3.3	-1.5	0.3
93.35	1.37	1.9	48.74	7MY	2.5	-3.3	-1.5	0.3
99.45	1.35	2.0	58.30	17MY	3.5	-3.3	-1.5	0.4
105.57	1.32	2.0	67.83	27MY	2.0	-3.3	-1.5	0.4
111.73	1.29	2.0	77.36	6JN	0.9	-3.4	-1.5	0.5
117.92	1.25	2.0	86.89	16JN	-0.1	-3.4	-1.5	0.6
124.14	1.21	2.0	96.42	26JN	-1.2	-3.4	-1.5	0.6
130.41	1.17	2.0	105.97	6JL	-1.7	-3.5	-1.6	0.7
136.72	1.12	2.0	115.54	16JL	-0.9	-3.5	-1.6	0.7
143.06	1.08	2.0	125.14	26JL	-0.3	-3.5	-1.7	0.8
149.46	1.02	2.0	134.79	5AU	0.1	-3.4	-1.7	0.8
155.91	0.97	2.0	144.49	15AU	0.3	-3.4	-1.8	0.8
162.42	0.90	1.9	154.23	25AU	0.6	-3.4	-1.8	0.8
168.98	0.84	1.9	164.04	4SE	1.3	-3.4	-1.9	0.8
175.60	0.77	1.8	173.90	14SE	3.2	-3.3	-1.9	0.8
182.28	0.70	1.8	183.81	24SE	0.4	-3.3	-2.0	0.8
189.04	0.62	1.8	193.79	4OC	-0.6	-3.3	-2.1	0.8
195.85	0.54	1.8	203.82	14OC	-0.8	-3.3	-2.1	0.8
202.74	0.45	1.8	213.90	24OC	-0.8	-3.3	-2.2	0.7
209.70	0.36	1.8	224.03	3NO	-0.8	-3.4	-2.2	0.7
216.73	0.26	1.8	234.18	13NO	-0.6	-3.4	-2.3	0.7
223.82	0.16	1.7	244.36	23NO	-0.6	-3.4	-2.3	0.7
230.99	0.06	1.7	254.57	3DE	-0.6	-3.4	-2.3	0.8
238.22	-0.05	1.7	264.77	13DE	-0.3	-3.5	-2.3	0.8
245.52	-0.17	1.6	274.97	23DE	1.6	-3.5	-2.2	0.8

♂ LONG	♂ LAT	♂ MAG	☉ LONG	16.00UT	☿	♀	♃	♄
				602	MAGNITUDES			
252.88	-0.29	1.6	285.15	2JA	1.9	-3.6	-2.2	0.8
260.30	-0.41	1.6	295.30	12JA	0.6	-3.6	-2.1	0.8
267.77	-0.53	1.5	305.42	22JA	0.2	-3.7	-2.1	0.8
275.29	-0.66	1.5	315.50	1FE	0.1	-3.8	-2.0	0.8
282.85	-0.79	1.4	325.52	11FE	-0.1	-3.9	-1.9	0.7
290.45	-0.91	1.4	335.49	21FE	-0.5	-4.0	-1.8	0.7
298.07	-1.04	1.3	345.40	3MR	-1.2	-4.0	-1.8	0.6
305.70	-1.16	1.2	355.25	13MR	-1.5	-4.1	-1.7	0.6
313.33	-1.27	1.2	5.04	23MR	-0.8	-4.2	-1.6	0.5
320.95	-1.38	1.1	14.79	2AP	0.4	-4.2	-1.6	0.5
328.55	-1.47	1.1	24.47	12AP	1.7	-4.2	-1.5	0.4
336.10	-1.56	1.0	34.11	22AP	3.3	-3.9	-1.5	0.3
343.60	-1.63	1.0	43.72	2MY	2.7	-3.2	-1.5	0.3
351.01	-1.69	0.9	53.29	12MY	1.6	-3.0	-1.4	0.2
358.34	-1.73	0.9	62.83	22MY	0.7	-3.8	-1.4	0.3
5.55	-1.76	0.8	72.37	1JN	-0.2	-4.1	-1.4	0.4
12.63	-1.76	0.7	81.89	11JN	-1.2	-4.2	-1.4	0.4
19.54	-1.75	0.7	91.41	21JN	-1.7	-4.2	-1.4	0.5
26.27	-1.72	0.6	100.96	1JL	-0.9	-4.1	-1.4	0.6
32.78	-1.67	0.5	110.51	11JL	-0.2	-4.0	-1.4	0.6
39.04	-1.60	0.4	120.10	21JL	0.2	-3.9	-1.4	0.7
44.99	-1.50	0.3	129.73	31JL	0.5	-3.8	-1.4	0.7
50.58	-1.39	0.2	139.40	10AU	0.8	-3.8	-1.5	0.7
55.75	-1.24	0.1	149.12	20AU	1.8	-3.7	-1.5	0.8
60.39	-1.07	-0.0	158.89	30AU	2.9	-3.6	-1.5	0.8
64.40	-0.86	-0.2	168.72	9SE	0.3	-3.6	-1.6	0.8
67.61	-0.60	-0.4	178.61	19SE	-0.8	-3.5	-1.7	0.8
69.85	-0.30	-0.6	188.56	29SE	-0.9	-3.5	-1.7	0.8
70.92	0.06	-0.8	198.56	9OC	-0.9	-3.5	-1.8	0.8
70.62	0.47	-1.0	208.61	19OC	-0.7	-3.4	-1.8	0.8
68.88	0.93	-1.2	218.72	29OC	-0.5	-3.4	-1.9	0.7
65.87	1.40	-1.4	228.86	8NO	-0.5	-3.4	-2.0	0.7
62.07	1.84	-1.5	239.03	18NO	-0.4	-3.4	-2.0	0.7
58.26	2.18	-1.3	249.22	28NO	-0.1	-3.4	-2.1	0.7
55.21	2.40	-1.1	259.42	8DE	1.8	-3.4	-2.1	0.7
53.36	2.52	-0.8	269.62	18DE	1.6	-3.4	-2.1	0.8
52.87	2.55	-0.5	279.82	28DE	0.3	-3.4	-2.2	0.8
				603				
53.63	2.53	-0.2	289.98	7JA	0.1	-3.4	-2.1	0.8
55.44	2.48	0.1	300.12	17JA	-0.0	-3.4	-2.1	0.8
58.13	2.41	0.3	310.22	27JA	-0.2	-3.4	-2.1	0.8
61.49	2.33	0.6	320.27	6FE	-0.6	-3.4	-2.0	0.8
65.38	2.24	0.8	330.27	16FE	-1.2	-3.5	-2.0	0.8
69.69	2.15	0.9	340.21	26FE	-1.4	-3.5	-1.9	0.7
74.34	2.06	1.1	350.09	8MR	-0.8	-3.5	-1.9	0.7
79.25	1.98	1.2	359.92	18MR	0.5	-3.4	-1.8	0.6
84.39	1.89	1.4	9.69	28MR	2.1	-3.4	-1.7	0.6
89.69	1.81	1.5	19.40	7AP	3.6	-3.4	-1.7	0.5
95.15	1.73	1.6	29.06	17AP	2.0	-3.4	-1.6	0.5
100.73	1.64	1.7	38.69	27AP	1.2	-3.3	-1.5	0.4
106.42	1.56	1.7	48.27	7MY	0.5	-3.3	-1.5	0.3
112.21	1.48	1.8	57.83	17MY	-0.2	-3.3	-1.4	0.3
118.09	1.40	1.8	67.37	27MY	-1.2	-3.3	-1.4	0.3
124.05	1.31	1.9	76.89	6JN	-1.8	-3.3	-1.4	0.3
130.09	1.23	1.9	86.42	16JN	-0.9	-3.4	-1.3	0.4
136.21	1.15	1.9	95.95	26JN	-0.1	-3.4	-1.3	0.4
142.41	1.06	1.9	105.50	6JL	0.3	-3.4	-1.3	0.5
148.68	0.98	1.9	115.07	16JL	0.7	-3.4	-1.3	0.6
155.04	0.89	1.9	124.68	26JL	1.1	-3.5	-1.3	0.6
161.47	0.80	1.9	134.32	5AU	2.2	-3.5	-1.3	0.7
167.98	0.71	1.9	144.02	15AU	2.5	-3.6	-1.3	0.7
174.58	0.62	1.9	153.76	25AU	0.2	-3.7	-1.3	0.7
181.25	0.52	1.9	163.56	4SE	-0.9	-3.8	-1.3	0.8
188.02	0.43	1.8	173.42	14SE	-1.1	-3.9	-1.4	0.8
194.87	0.33	1.8	183.34	24SE	-1.0	-4.0	-1.4	0.8
201.81	0.23	1.7	193.31	4OC	-0.6	-4.1	-1.5	0.8
208.84	0.13	1.7	203.34	14OC	-0.4	-4.2	-1.5	0.8
215.96	0.02	1.6	213.41	24OC	-0.3	-4.3	-1.6	0.8
223.16	-0.08	1.6	223.53	3NO	-0.3	-4.4	-1.6	0.8
230.45	-0.19	1.6	233.69	13NO	0.1	-4.4	-1.7	0.7
237.82	-0.29	1.6	243.87	23NO	2.0	-4.2	-1.8	0.7
245.27	-0.40	1.5	254.07	3DE	1.3	-3.7	-1.8	0.7
252.79	-0.50	1.5	264.28	13DE	0.1	-3.0	-1.9	0.7
260.38	-0.60	1.5	274.47	23DE	-0.1	-3.8	-1.9	0.8

604

♂ LONG	LAT	MAG	☉ LONG	16.00UT	☿	♀	♃	♄
268.03	-0.70	1.5	284.65	2JA	-0.2	-4.2	-2.0	0.8
275.73	-0.79	1.5	294.81	12JA	-0.3	-4.4	-2.0	0.8
283.47	-0.88	1.5	304.93	22JA	-0.7	-4.3	-2.0	0.8
291.25	-0.96	1.4	315.00	1FE	-1.2	-4.3	-2.1	0.8
299.05	-1.04	1.4	325.03	11FE	-1.3	-4.2	-2.0	0.8
306.86	-1.10	1.4	335.00	21FE	-0.7	-4.0	-2.0	0.8
314.67	-1.15	1.4	344.92	2MR	0.7	-3.9	-2.0	0.8
322.46	-1.20	1.4	354.78	12MR	2.6	-3.8	-1.9	0.7
330.23	-1.23	1.4	4.57	22MR	2.7	-3.7	-1.9	0.7
337.96	-1.25	1.4	14.31	1AP	1.5	-3.6	-1.8	0.6
345.64	-1.25	1.3	24.01	11AP	0.9	-3.6	-1.8	0.6
353.26	-1.24	1.3	33.65	21AP	0.4	-3.5	-1.7	0.5
0.80	-1.22	1.3	43.26	1MY	-0.3	-3.5	-1.6	0.4
8.26	-1.18	1.3	52.83	11MY	-1.2	-3.4	-1.6	0.4
15.62	-1.13	1.3	62.37	21MY	-1.8	-3.4	-1.5	0.3
22.87	-1.07	1.3	71.90	31MY	-0.9	-3.3	-1.4	0.3
30.01	-0.99	1.3	81.43	10JN	-0.1	-3.3	-1.4	0.3
37.03	-0.90	1.2	90.96	20JN	0.5	-3.3	-1.3	0.3
43.91	-0.80	1.2	100.49	30JN	0.9	-3.3	-1.3	0.4
50.63	-0.69	1.2	110.05	10JL	1.5	-3.3	-1.3	0.5
57.20	-0.56	1.1	119.64	20JL	2.8	-3.3	-1.3	0.5
63.59	-0.42	1.1	129.26	30JL	2.1	-3.4	-1.2	0.6
69.78	-0.27	1.0	138.93	9AU	0.2	-3.4	-1.2	0.6
75.74	-0.10	1.0	148.65	19AU	-1.0	-3.4	-1.2	0.7
81.45	0.09	0.9	158.42	29AU	-1.2	-3.4	-1.2	0.7
86.86	0.29	0.8	168.25	8SE	-1.0	-3.5	-1.2	0.7
91.91	0.51	0.7	178.13	18SE	-0.5	-3.5	-1.2	0.8
96.55	0.76	0.6	188.07	28SE	-0.3	-3.5	-1.3	0.8
100.67	1.04	0.4	198.07	8OC	-0.2	-3.5	-1.3	0.8
104.16	1.35	0.3	208.12	18OC	-0.1	-3.4	-1.3	0.8
106.88	1.71	0.1	218.22	28OC	0.3	-3.4	-1.4	0.8
108.64	2.10	-0.1	228.36	7NO	2.3	-3.4	-1.4	0.8
109.25	2.53	-0.4	238.53	17NO	1.0	-3.4	-1.5	0.8
108.55	2.99	-0.6	248.72	27NO	-0.1	-3.3	-1.5	0.7
106.49	3.43	-0.8	258.93	7DE	-0.2	-3.3	-1.6	0.7
103.25	3.80	-1.0	269.13	17DE	-0.3	-3.3	-1.7	0.7
99.33	4.04	-1.1	279.32	27DE	-0.4	-3.3	-1.7	0.8

605

♂ LONG	LAT	MAG	☉ LONG	16.00UT	☿	♀	♃	♄
95.47	4.12	-0.9	289.49	6JA	-0.7	-3.4	-1.8	0.8
92.39	4.04	-0.7	299.63	16JA	-1.1	-3.4	-1.9	0.8
90.49	3.86	-0.4	309.73	26JA	-1.2	-3.4	-1.9	0.8
89.92	3.61	-0.2	319.78	5FE	-0.7	-3.4	-2.0	0.8
90.56	3.34	0.1	329.78	15FE	0.8	-3.4	-2.0	0.8
92.25	3.08	0.3	339.73	25FE	3.1	-3.5	-2.0	0.8
94.80	2.83	0.5	349.62	7MR	2.0	-3.5	-2.0	0.8
98.03	2.60	0.7	359.44	17MR	1.1	-3.6	-2.0	0.8
101.81	2.38	0.9	9.22	27MR	0.6	-3.6	-2.0	0.7
106.04	2.18	1.0	18.93	6AP	0.2	-3.7	-1.9	0.7
110.62	2.00	1.2	28.60	16AP	-0.3	-3.8	-1.9	0.6
115.50	1.82	1.3	38.23	26AP	-1.2	-3.8	-1.8	0.6
120.63	1.66	1.4	47.82	6MY	-1.8	-3.9	-1.8	0.5
125.97	1.51	1.5	57.37	16MY	-0.9	-4.0	-1.7	0.5
131.49	1.36	1.5	66.91	26MY	0.0	-4.1	-1.6	0.4
137.18	1.22	1.6	76.44	5JN	0.6	-4.2	-1.5	0.3
143.01	1.08	1.6	85.96	15JN	1.2	-4.2	-1.5	0.3
148.99	0.95	1.7	95.49	25JN	2.0	-4.0	-1.5	0.3
155.09	0.82	1.7	105.04	5JL	3.3	-3.6	-1.4	0.4
161.32	0.70	1.7	114.61	15JL	1.8	-3.1	-1.4	0.4
167.66	0.57	1.7	124.21	25JL	0.1	-3.5	-1.3	0.5
174.12	0.45	1.7	133.85	4AU	-1.0	-4.0	-1.3	0.5
180.70	0.34	1.7	143.54	14AU	-1.4	-4.2	-1.3	0.6
187.39	0.22	1.7	153.29	24AU	-1.0	-4.3	-1.2	0.6
194.19	0.10	1.7	163.08	3SE	-0.5	-4.2	-1.2	0.7
201.10	-0.01	1.7	172.94	13SE	-0.2	-4.1	-1.2	0.7
208.12	-0.12	1.7	182.85	23SE	-0.0	-4.0	-1.2	0.8
215.25	-0.23	1.6	192.82	3OC	0.1	-4.0	-1.2	0.8
222.48	-0.33	1.6	202.85	13OC	0.5	-3.9	-1.3	0.8
229.81	-0.43	1.6	212.93	23OC	2.7	-3.8	-1.3	0.8
237.24	-0.53	1.5	223.04	2NO	0.8	-3.7	-1.3	0.8
244.75	-0.62	1.5	233.20	12NO	-0.3	-3.7	-1.3	0.8
252.35	-0.71	1.5	243.38	22NO	-0.4	-3.6	-1.4	0.8
260.02	-0.79	1.4	253.58	2DE	-0.4	-3.5	-1.4	0.8
267.75	-0.86	1.4	263.78	12DE	-0.5	-3.5	-1.5	0.7
275.55	-0.93	1.3	273.98	22DE	-0.8	-3.5	-1.5	0.7

606

♂ LONG	LAT	MAG	☉ LONG	16.00UT	☿	♀	♃	♄
283.38	-0.98	1.3	284.16	1JA	-1.0	-3.4	-1.6	0.8
291.24	-1.03	1.3	294.32	11JA	-1.0	-3.4	-1.7	0.8
299.12	-1.06	1.3	304.45	21JA	-0.6	-3.4	-1.7	0.8
307.01	-1.08	1.4	314.52	31JA	1.0	-3.3	-1.8	0.9
314.89	-1.09	1.4	324.55	10FE	3.1	-3.3	-1.8	0.9
322.75	-1.09	1.4	334.53	20FE	1.5	-3.3	-1.9	0.9
330.58	-1.08	1.4	344.44	2MR	0.8	-3.3	-2.0	0.9
338.37	-1.06	1.4	354.30	12MR	0.5	-3.3	-2.0	0.9
346.10	-1.02	1.4	4.10	22MR	0.1	-3.3	-2.0	0.8
353.76	-0.97	1.4	13.85	1AP	-0.4	-3.4	-2.0	0.8
1.36	-0.92	1.5	23.54	11AP	-1.2	-3.4	-2.0	0.8
8.87	-0.85	1.5	33.19	21AP	-1.7	-3.4	-2.0	0.7
16.30	-0.77	1.5	42.79	1MY	-0.9	-3.5	-2.0	0.7
23.63	-0.69	1.5	52.36	11MY	0.1	-3.5	-1.9	0.6
30.87	-0.59	1.5	61.91	21MY	0.8	-3.5	-1.9	0.6
38.01	-0.49	1.6	71.44	31MY	1.6	-3.4	-1.8	0.5
45.05	-0.39	1.6	80.97	10JN	2.7	-3.4	-1.8	0.4
51.98	-0.27	1.6	90.49	20JN	3.0	-3.4	-1.7	0.3
58.80	-0.15	1.6	100.03	30JN	1.4	-3.3	-1.6	0.3
65.51	-0.03	1.6	109.58	10JL	0.0	-3.3	-1.6	0.4
72.10	0.10	1.6	119.17	20JL	-1.1	-3.3	-1.5	0.4
78.57	0.23	1.5	128.79	30JL	-1.5	-3.3	-1.5	0.5
84.92	0.37	1.5	138.46	9AU	-0.9	-3.3	-1.4	0.5
91.12	0.52	1.5	148.17	19AU	-0.4	-3.3	-1.4	0.6
97.17	0.67	1.4	157.93	29AU	-0.1	-3.4	-1.3	0.6
103.06	0.84	1.4	167.76	8SE	0.1	-3.4	-1.3	0.7
108.76	1.01	1.3	177.65	18SE	0.3	-3.4	-1.3	0.7
114.24	1.19	1.2	187.58	28SE	0.8	-3.5	-1.3	0.8
119.47	1.39	1.1	197.58	8OC	3.0	-3.5	-1.3	0.8
124.40	1.60	1.0	207.63	18OC	0.7	-3.6	-1.2	0.8
128.97	1.83	0.9	217.73	28OC	-0.4	-3.6	-1.3	0.8
133.11	2.09	0.8	227.87	7NO	-0.5	-3.7	-1.3	0.8
136.70	2.37	0.6	238.04	17NO	-0.5	-3.8	-1.3	0.8
139.63	2.69	0.4	248.23	27NO	-0.6	-3.9	-1.3	0.8
141.73	3.03	0.2	258.43	7DE	-0.8	-4.0	-1.3	0.8
142.83	3.40	-0.1	268.64	17DE	-0.8	-4.1	-1.4	0.8
142.74	3.77	-0.3	278.83	27DE	-0.9	-4.2	-1.4	0.8

607

♂ LONG	LAT	MAG	☉ LONG	16.00UT	☿	♀	♃	♄
141.34	4.13	-0.6	289.00	6JA	-0.5	-4.3	-1.5	0.8
138.68	4.41	-0.8	299.14	16JA	1.2	-4.3	-1.5	0.8
135.08	4.56	-1.0	309.24	26JA	2.7	-4.3	-1.6	0.9
131.14	4.54	-1.0	319.30	5FE	1.1	-4.1	-1.7	0.9
127.58	4.34	-0.8	329.30	15FE	0.6	-3.7	-1.7	0.9
125.00	4.02	-0.6	339.25	25FE	0.3	-3.3	-1.8	0.9
123.67	3.64	-0.3	349.14	7MR	0.0	-3.7	-1.9	0.9
123.64	3.24	-0.1	358.97	17MR	-0.4	-4.1	-1.9	0.9
124.75	2.86	0.1	8.74	27MR	-1.2	-4.2	-2.0	0.9
126.84	2.50	0.3	18.46	6AP	-1.7	-4.2	-2.0	0.9
129.74	2.18	0.5	28.13	16AP	-0.9	-4.1	-2.1	0.9
133.28	1.88	0.7	37.76	26AP	0.2	-4.0	-2.1	0.8
137.36	1.62	0.8	47.35	6MY	1.1	-3.9	-2.1	0.8
141.88	1.37	0.9	56.91	16MY	2.1	-3.8	-2.1	0.7
146.77	1.15	1.0	66.45	26MY	3.5	-3.8	-2.1	0.7
151.96	0.94	1.1	75.98	5JN	2.4	-3.7	-2.1	0.6
157.43	0.75	1.2	85.50	15JN	1.1	-3.6	-2.0	0.5
163.13	0.57	1.3	95.03	25JN	-0.0	-3.5	-2.0	0.5
169.04	0.40	1.3	104.58	5JL	-1.1	-3.5	-1.9	0.4
175.15	0.24	1.4	114.14	15JL	-1.6	-3.4	-1.8	0.4
181.43	0.09	1.4	123.75	25JL	-0.9	-3.4	-1.8	0.4
187.89	-0.05	1.4	133.39	4AU	-0.3	-3.4	-1.7	0.5
194.50	-0.19	1.5	143.07	14AU	0.0	-3.4	-1.6	0.5
201.26	-0.32	1.5	152.81	24AU	0.2	-3.4	-1.6	0.6
208.16	-0.44	1.5	162.61	3SE	0.5	-3.4	-1.5	0.6
215.20	-0.55	1.5	172.46	13SE	1.1	-3.4	-1.5	0.7
222.37	-0.65	1.5	182.37	23SE	3.3	-3.4	-1.4	0.7
229.66	-0.74	1.5	192.34	3OC	0.6	-3.4	-1.4	0.8
237.06	-0.83	1.5	202.36	13OC	-0.6	-3.4	-1.4	0.8
244.56	-0.90	1.4	212.43	23OC	-0.7	-3.4	-1.3	0.9
252.16	-0.97	1.4	222.55	2NO	-0.7	-3.4	-1.3	0.9
259.84	-1.02	1.4	232.70	12NO	-0.7	-3.4	-1.3	0.9
267.60	-1.06	1.4	242.88	22NO	-0.7	-3.5	-1.3	0.9
275.41	-1.10	1.4	253.08	2DE	-0.7	-3.5	-1.3	0.9
283.26	-1.11	1.4	263.28	12DE	-0.7	-3.5	-1.4	0.9
291.15	-1.12	1.4	273.48	22DE	-0.4	-3.5	-1.4	0.9

♂ LONG	♂ LAT	♂ MAG	☉ LONG	16.00UT	☿	♀	♃	♄
				608	MAGNITUDES			
299.05	-1.11	1.4	283.67	1JA	1.3	-3.4	-1.4	0.9
306.95	-1.10	1.4	293.83	11JA	2.2	-3.4	-1.4	0.9
314.84	-1.07	1.4	303.95	21JA	0.8	-3.4	-1.5	0.9
322.71	-1.03	1.4	314.03	31JA	0.4	-3.4	-1.5	0.9
330.53	-0.98	1.4	324.06	10FE	0.2	-3.4	-1.6	1.0
338.31	-0.92	1.4	334.04	20FE	-0.1	-3.3	-1.7	1.0
346.03	-0.85	1.4	343.96	1MR	-0.5	-3.3	-1.7	1.0
353.68	-0.77	1.4	353.82	11MR	-1.2	-3.3	-1.8	1.0
1.25	-0.69	1.4	3.63	21MR	-1.6	-3.3	-1.9	1.0
8.75	-0.60	1.5	13.38	31MR	-0.8	-3.4	-1.9	1.0
16.16	-0.51	1.5	23.07	10AP	0.3	-3.4	-2.0	1.0
23.49	-0.41	1.6	32.72	20AP	1.4	-3.4	-2.1	1.0
30.72	-0.31	1.6	42.33	30AP	2.8	-3.4	-2.1	1.0
37.87	-0.21	1.6	51.90	10MY	3.2	-3.5	-2.2	0.9
44.93	-0.10	1.7	61.45	20MY	1.9	-3.5	-2.2	0.9
51.89	0.01	1.7	70.99	30MY	0.9	-3.6	-2.2	0.8
58.77	0.11	1.7	80.51	9JN	-0.1	-3.6	-2.2	0.7
65.57	0.22	1.8	90.03	19JN	-1.2	-3.7	-2.2	0.7
72.28	0.33	1.8	99.57	29JN	-1.7	-3.8	-2.2	0.6
78.90	0.44	1.8	109.12	9JL	-0.9	-3.9	-2.2	0.5
85.44	0.55	1.8	118.71	19JL	-0.3	-4.0	-2.1	0.5
91.90	0.66	1.8	128.33	29JL	0.1	-4.1	-2.1	0.5
98.27	0.77	1.8	137.99	8AU	0.4	-4.2	-2.0	0.5
104.56	0.88	1.8	147.70	18AU	0.7	-4.3	-1.9	0.6
110.76	1.00	1.8	157.47	28AU	1.5	-4.3	-1.9	0.6
116.86	1.11	1.7	167.29	7SE	3.2	-4.1	-1.8	0.7
122.86	1.23	1.7	177.17	17SE	0.4	-3.7	-1.7	0.7
128.74	1.35	1.7	187.10	27SE	-0.7	-3.2	-1.7	0.8
134.49	1.48	1.6	197.10	7OC	-0.9	-3.8	-1.6	0.8
140.09	1.61	1.5	207.14	17OC	-0.8	-4.2	-1.6	0.9
145.52	1.75	1.4	217.24	27OC	-0.8	-4.4	-1.5	0.9
150.75	1.90	1.3	227.37	6NO	-0.6	-4.4	-1.5	0.9
155.73	2.05	1.2	237.54	16NO	-0.5	-4.3	-1.5	1.0
160.42	2.22	1.1	247.73	26NO	-0.5	-4.2	-1.5	1.0
164.74	2.40	0.9	257.93	6DE	-0.2	-4.1	-1.4	1.0
168.62	2.60	0.7	268.14	16DE	1.6	-4.0	-1.4	1.0
171.94	2.80	0.5	278.33	26DE	1.8	-3.9	-1.4	1.0
				609				
174.57	3.02	0.3	288.50	5JA	0.5	-3.8	-1.5	1.0
176.35	3.25	0.0	298.65	15JA	0.2	-3.7	-1.5	1.0
177.09	3.47	-0.2	308.76	25JA	0.0	-3.6	-1.5	1.0
176.63	3.67	-0.5	318.81	4FE	-0.2	-3.6	-1.5	1.0
174.88	3.80	-0.8	328.82	14FE	-0.5	-3.5	-1.6	1.1
171.97	3.83	-1.0	338.77	24FE	-1.2	-3.5	-1.6	1.1
168.29	3.70	-1.2	348.67	6MR	-1.5	-3.4	-1.7	1.1
164.50	3.43	-1.1	358.50	16MR	-0.8	-3.4	-1.7	1.2
161.31	3.03	-1.0	8.28	26MR	0.4	-3.3	-1.8	1.2
159.22	2.58	-0.7	18.00	5AP	1.8	-3.3	-1.8	1.2
158.43	2.11	-0.5	27.68	15AP	3.6	-3.3	-1.9	1.2
158.94	1.67	-0.3	37.30	25AP	2.5	-3.3	-2.0	1.1
160.57	1.27	-0.1	46.89	5MY	1.4	-3.3	-2.1	1.1
163.16	0.92	0.1	56.45	15MY	0.7	-3.3	-2.1	1.1
166.55	0.60	0.2	65.99	25MY	-0.2	-3.3	-2.2	1.0
170.59	0.32	0.4	75.52	4JN	-1.2	-3.4	-2.2	1.0
175.17	0.07	0.5	85.04	14JN	-1.8	-3.4	-2.3	0.9
180.20	-0.15	0.6	94.57	24JN	-0.9	-3.4	-2.3	0.8
185.61	-0.35	0.7	104.11	4JL	-0.2	-3.5	-2.4	0.8
191.36	-0.53	0.8	113.68	14JL	0.3	-3.5	-2.4	0.7
197.40	-0.68	0.8	123.28	24JL	0.6	-3.5	-2.4	0.6
203.69	-0.82	0.9	132.92	3AU	1.0	-3.4	-2.3	0.6
210.21	-0.95	1.0	142.61	13AU	1.9	-3.4	-2.3	0.6
216.94	-1.05	1.0	152.34	23AU	2.8	-3.4	-2.2	0.7
223.85	-1.14	1.0	162.13	2SE	0.4	-3.4	-2.2	0.7
230.93	-1.22	1.1	171.98	12SE	-0.8	-3.3	-2.1	0.7
238.17	-1.28	1.1	181.88	22SE	-1.0	-3.3	-2.1	0.8
245.53	-1.32	1.1	191.85	2OC	-1.0	-3.3	-2.0	0.8
253.02	-1.35	1.2	201.87	12OC	-0.7	-3.3	-1.9	0.9
260.60	-1.36	1.2	211.94	22OC	-0.5	-3.3	-1.9	0.9
268.28	-1.36	1.2	222.05	1NO	-0.4	-3.4	-1.8	1.0
276.02	-1.34	1.3	232.21	11NO	-0.4	-3.4	-1.7	1.0
283.80	-1.31	1.3	242.38	21NO	-0.0	-3.4	-1.7	1.1
291.62	-1.27	1.3	252.58	1DE	1.8	-3.4	-1.6	1.1
299.47	-1.21	1.3	262.79	11DE	1.5	-3.5	-1.6	1.1
307.30	-1.15	1.4	272.98	21DE	0.3	-3.5	-1.6	1.1
315.13	-1.07	1.4	283.17	31DE	0.0	-3.6	-1.6	1.1

♂ LONG	♂ LAT	♂ MAG	☉ LONG	16.00UT	☿	♀	♃	♄
				610	MAGNITUDES			
322.93	-0.99	1.4	293.33	10JA	-0.1	-3.7	-1.5	1.1
330.68	-0.90	1.4	303.46	20JA	-0.3	-3.7	-1.5	1.1
338.38	-0.80	1.5	313.54	30JA	-0.6	-3.8	-1.5	1.1
346.03	-0.70	1.5	323.58	9FE	-1.2	-3.9	-1.5	1.1
353.60	-0.60	1.5	333.56	19FE	-1.4	-4.0	-1.6	1.2
1.10	-0.49	1.6	343.48	1MR	-0.8	-4.1	-1.6	1.2
8.52	-0.39	1.6	353.35	11MR	0.5	-4.1	-1.6	1.3
15.86	-0.28	1.6	3.16	21MR	2.3	-4.2	-1.6	1.3
23.11	-0.17	1.6	12.91	31MR	3.3	-4.2	-1.7	1.3
30.28	-0.07	1.6	22.61	10AP	1.8	-4.2	-1.7	1.3
37.37	0.04	1.7	32.26	20AP	1.1	-3.9	-1.8	1.3
44.38	0.14	1.7	41.87	30AP	0.5	-3.2	-1.9	1.3
51.30	0.24	1.7	51.44	10MY	-0.2	-3.1	-1.9	1.3
58.15	0.34	1.7	60.99	20MY	-1.2	-3.8	-2.0	1.3
64.93	0.43	1.8	70.52	30MY	-1.8	-4.1	-2.1	1.2
71.64	0.52	1.8	80.05	9JN	-0.9	-4.2	-2.1	1.2
78.29	0.61	1.9	89.57	19JN	-0.1	-4.2	-2.2	1.1
84.88	0.70	1.9	99.11	29JN	0.4	-4.1	-2.3	1.1
91.40	0.78	1.9	108.66	9JL	0.8	-4.0	-2.3	1.0
97.87	0.86	1.9	118.24	19JL	1.3	-3.9	-2.4	0.9
104.30	0.94	2.0	127.86	29JL	2.4	-3.8	-2.4	0.8
110.67	1.02	2.0	137.52	8AU	2.4	-3.7	-2.4	0.8
116.99	1.09	2.0	147.23	18AU	0.3	-3.7	-2.5	0.7
123.26	1.16	2.0	156.99	28AU	-0.9	-3.6	-2.4	0.8
129.49	1.23	1.9	166.81	7SE	-1.2	-3.6	-2.4	0.8
135.66	1.30	1.9	176.68	17SE	-1.0	-3.5	-2.4	0.9
141.79	1.37	1.9	186.62	27SE	-0.6	-3.5	-2.3	0.9
147.85	1.43	1.8	196.61	7OC	-0.3	-3.5	-2.3	0.9
153.85	1.50	1.8	206.66	17OC	-0.3	-3.4	-2.2	1.0
159.78	1.56	1.7	216.75	27OC	-0.2	-3.4	-2.1	1.1
165.62	1.63	1.7	226.88	6NO	0.1	-3.4	-2.1	1.1
171.36	1.69	1.6	237.05	16NO	2.1	-3.4	-2.0	1.1
176.99	1.75	1.5	247.24	26NO	1.2	-3.4	-1.9	1.2
182.47	1.81	1.4	257.44	6DE	0.0	-3.4	-1.9	1.2
187.78	1.87	1.2	267.64	16DE	-0.1	-3.4	-1.8	1.2
192.89	1.93	1.1	277.84	26DE	-0.2	-3.4	-1.8	1.3
				611				
197.75	1.98	0.9	288.01	5JA	-0.4	-3.4	-1.7	1.3
202.29	2.03	0.7	298.15	15JA	-0.7	-3.4	-1.7	1.3
206.45	2.06	0.5	308.26	25JA	-1.2	-3.4	-1.6	1.3
210.13	2.09	0.3	318.32	4FE	-1.3	-3.4	-1.6	1.3
213.21	2.09	0.0	328.33	14FE	-0.7	-3.5	-1.6	1.3
215.52	2.07	-0.2	338.29	24FE	0.7	-3.5	-1.6	1.3
216.91	2.01	-0.5	348.18	6MR	2.8	-3.5	-1.6	1.4
217.19	1.88	-0.8	358.02	16MR	2.5	-3.4	-1.6	1.4
216.22	1.68	-1.1	7.80	26MR	1.3	-3.4	-1.6	1.4
214.03	1.37	-1.4	17.52	5AP	0.8	-3.4	-1.6	1.4
210.88	0.97	-1.7	27.20	15AP	0.3	-3.4	-1.6	1.4
207.32	0.50	-1.7	36.83	25AP	-0.3	-3.3	-1.7	1.3
204.12	-0.00	-1.6	46.42	5MY	-1.2	-3.3	-1.7	1.3
201.88	-0.47	-1.4	55.99	15MY	-1.8	-3.3	-1.8	1.3
200.94	-0.89	-1.2	65.53	25MY	-0.9	-3.3	-1.8	1.2
201.37	-1.23	-1.0	75.05	4JN	-0.1	-3.3	-1.9	1.2
203.04	-1.50	-0.8	84.58	14JN	0.5	-3.4	-1.9	1.2
205.81	-1.72	-0.6	94.11	24JN	1.0	-3.4	-2.0	1.1
209.47	-1.88	-0.4	103.65	4JL	1.7	-3.4	-2.1	1.1
213.86	-2.01	-0.2	113.22	14JL	3.0	-3.4	-2.1	1.0
218.87	-2.09	-0.1	122.82	24JL	2.0	-3.5	-2.2	1.0
224.38	-2.15	0.0	132.45	3AU	0.2	-3.5	-2.3	0.9
230.30	-2.18	0.1	142.13	13AU	-1.0	-3.6	-2.3	0.9
236.58	-2.18	0.2	151.87	23AU	-1.3	-3.7	-2.4	0.8
243.15	-2.16	0.3	161.65	2SE	-1.0	-3.8	-2.4	0.8
249.96	-2.12	0.4	171.50	12SE	-0.5	-3.9	-2.4	0.9
256.97	-2.06	0.5	181.41	22SE	-0.2	-4.0	-2.5	1.0
264.15	-1.98	0.6	191.36	2OC	-0.1	-4.1	-2.5	1.0
271.47	-1.89	0.7	201.38	12OC	-0.0	-4.2	-2.4	1.1
278.89	-1.78	0.8	211.45	22OC	0.4	-4.3	-2.4	1.1
286.39	-1.66	0.9	221.56	1NO	2.3	-4.4	-2.3	1.2
293.95	-1.53	0.9	231.71	11NO	1.0	-4.4	-2.3	1.2
301.54	-1.40	1.0	241.89	21NO	-0.1	-4.2	-2.2	1.3
309.14	-1.26	1.1	252.08	1DE	-0.3	-3.7	-2.1	1.3
316.73	-1.12	1.2	262.29	11DE	-0.3	-3.0	-2.1	1.4
324.31	-0.97	1.2	272.49	21DE	-0.5	-3.8	-2.0	1.4
331.85	-0.83	1.3	282.67	31DE	-0.7	-4.2	-1.9	1.3

♂ LONG	♂ LAT	♂ MAG	⊙ LONG	16.00UT	☿	♀	♃	♄
				612	MAGNITUDES			
339.35	-0.69	1.4	292.84	10JA	-1.1	-4.4	-1.9	1.3
346.80	-0.55	1.4	302.97	20JA	-1.1	-4.3	-1.8	1.3
354.18	-0.42	1.5	313.05	30JA	-0.7	-4.3	-1.7	1.3
1.50	-0.28	1.5	323.09	9FE	0.8	-4.2	-1.7	1.2
8.75	-0.16	1.6	333.07	19FE	3.2	-4.0	-1.7	1.2
15.93	-0.04	1.6	343.00	29FE	1.8	-3.9	-1.6	1.1
23.03	0.08	1.7	352.87	10MR	1.0	-3.8	-1.6	1.1
30.06	0.19	1.7	2.68	20MR	0.6	-3.7	-1.6	1.1
37.01	0.29	1.8	12.43	30MR	0.2	-3.6	-1.6	1.1
43.90	0.38	1.8	22.14	9AP	-0.3	-3.6	-1.5	1.1
50.72	0.48	1.8	31.79	19AP	-1.2	-3.5	-1.5	1.1
57.47	0.56	1.9	41.40	29AP	-1.8	-3.5	-1.6	1.1
64.17	0.64	1.9	50.98	9MY	-0.9	-3.4	-1.6	1.1
70.81	0.71	1.9	60.53	19MY	0.0	-3.4	-1.6	1.0
77.40	0.78	1.9	70.06	29MY	0.7	-3.3	-1.6	1.0
83.95	0.85	1.9	79.59	8JN	1.3	-3.3	-1.7	1.0
90.45	0.91	1.9	89.12	18JN	2.3	-3.3	-1.7	0.9
96.92	0.96	1.9	98.65	28JN	3.3	-3.3	-1.7	0.9
103.36	1.01	1.9	108.21	8JL	1.7	-3.3	-1.8	0.8
109.76	1.06	2.0	117.78	18JL	0.1	-3.3	-1.9	0.8
116.15	1.10	2.0	127.40	28JL	-1.0	-3.3	-1.9	0.7
122.52	1.14	2.0	137.06	7AU	-1.4	-3.4	-2.0	0.7
128.86	1.17	2.0	146.76	17AU	-1.0	-3.4	-2.1	0.6
135.20	1.20	2.0	156.52	27AU	-0.4	-3.4	-2.1	0.6
141.52	1.23	2.0	166.33	6SE	-0.1	-3.5	-2.2	0.5
147.83	1.25	2.0	176.21	16SE	0.0	-3.5	-2.2	0.5
154.14	1.27	2.0	186.14	26SE	0.1	-3.5	-2.3	0.6
160.43	1.28	2.0	196.13	6OC	0.6	-3.5	-2.3	0.7
166.71	1.29	1.9	206.17	16OC	2.7	-3.4	-2.4	0.7
172.98	1.30	1.9	216.25	26OC	0.8	-3.4	-2.4	0.8
179.24	1.30	1.8	226.39	5NO	-0.3	-3.4	-2.4	0.9
185.47	1.29	1.8	236.55	15NO	-0.5	-3.4	-2.4	0.9
191.69	1.28	1.7	246.74	25NO	-0.5	-3.3	-2.3	1.0
197.87	1.26	1.6	256.94	5DE	-0.6	-3.3	-2.3	1.0
204.02	1.23	1.5	267.14	15DE	-0.8	-3.3	-2.2	1.0
210.12	1.19	1.4	277.34	25DE	-0.9	-3.3	-2.1	1.1
				613				
216.16	1.14	1.3	287.51	4JA	-1.0	-3.4	-2.1	1.1
222.13	1.08	1.2	297.66	14JA	-0.6	-3.4	-2.0	1.1
228.02	0.99	1.0	307.77	24JA	1.0	-3.4	-1.9	1.0
233.79	0.89	0.9	317.83	3FE	3.0	-3.4	-1.9	1.0
239.43	0.76	0.7	327.84	13FE	1.3	-3.4	-1.8	1.0
244.91	0.61	0.5	337.80	23FE	0.7	-3.5	-1.7	0.9
250.17	0.42	0.3	347.70	5MR	0.4	-3.5	-1.7	0.9
255.18	0.19	0.1	357.54	15MR	0.1	-3.6	-1.6	0.8
259.85	-0.09	-0.1	7.32	25MR	-0.4	-3.6	-1.6	0.8
264.09	-0.43	-0.4	17.06	4AP	-1.2	-3.7	-1.5	0.8
267.79	-0.84	-0.7	26.73	14AP	-1.7	-3.8	-1.5	0.8
270.80	-1.33	-1.0	36.37	24AP	-0.9	-3.9	-1.5	0.8
272.94	-1.92	-1.3	45.96	4MY	0.1	-3.9	-1.5	0.8
274.03	-2.59	-1.6	55.52	14MY	0.9	-4.0	-1.5	0.8
273.92	-3.34	-1.9	65.07	24MY	1.7	-4.1	-1.5	0.8
272.57	-4.10	-2.2	74.60	3JN	3.0	-4.2	-1.5	0.8
270.25	-4.79	-2.4	84.12	13JN	2.8	-4.2	-1.5	0.7
267.44	-5.31	-2.5	93.65	23JN	1.3	-4.0	-1.5	0.7
264.91	-5.58	-2.3	103.19	3JL	0.0	-3.6	-1.5	0.7
263.29	-5.61	-2.1	112.75	13JL	-1.1	-3.0	-1.6	0.6
262.95	-5.44	-1.9	122.35	23JL	-1.5	-3.6	-1.6	0.6
263.98	-5.14	-1.6	131.99	2AU	-0.9	-4.0	-1.6	0.5
266.26	-4.78	-1.4	141.66	12AU	-0.4	-4.2	-1.7	0.5
269.59	-4.38	-1.2	151.40	22AU	-0.0	-4.3	-1.8	0.4
273.78	-3.97	-0.9	161.18	1SE	0.2	-4.2	-1.8	0.3
278.65	-3.56	-0.7	171.02	11SE	0.3	-4.1	-1.9	0.3
284.05	-3.17	-0.5	180.92	21SE	0.9	-4.0	-1.9	0.3
289.87	-2.80	-0.3	190.88	1OC	3.0	-4.0	-2.0	0.3
296.00	-2.44	-0.1	200.89	11OC	0.7	-3.9	-2.1	0.3
302.37	-2.11	0.1	210.96	21OC	-0.5	-3.8	-2.1	0.4
308.93	-1.79	0.2	221.07	31OC	-0.6	-3.7	-2.2	0.5
315.62	-1.50	0.4	231.22	10NO	-0.6	-3.7	-2.2	0.6
322.40	-1.23	0.5	241.40	20NO	-0.7	-3.6	-2.2	0.6
329.25	-0.98	0.7	251.59	30NO	-0.8	-3.5	-2.3	0.7
336.13	-0.75	0.8	261.79	10DE	-0.8	-3.5	-2.3	0.7
343.03	-0.53	0.9	272.00	20DE	-0.8	-3.5	-2.2	0.8
349.93	-0.34	1.1	282.19	30DE	-0.5	-3.4	-2.2	0.8

♂ LONG	♂ LAT	♂ MAG	⊙ LONG	16.00UT	☿	♀	♃	♄
				614	MAGNITUDES			
356.82	-0.16	1.2	292.35	9JA	1.2	-3.4	-2.1	0.8
3.68	-0.00	1.3	302.48	19JA	2.6	-3.4	-2.1	0.8
10.52	0.14	1.4	312.57	29JA	1.0	-3.3	-2.0	0.8
17.32	0.27	1.5	322.60	8FE	0.5	-3.3	-1.9	0.8
24.08	0.39	1.5	332.59	18FE	0.3	-3.3	-1.9	0.8
30.81	0.50	1.6	342.52	28FE	-0.0	-3.3	-1.8	0.7
37.48	0.60	1.7	352.39	10MR	-0.4	-3.3	-1.7	0.7
44.12	0.68	1.8	2.21	20MR	-1.2	-3.3	-1.7	0.6
50.71	0.76	1.8	11.96	30MR	-1.7	-3.4	-1.6	0.6
57.27	0.83	1.9	21.66	9AP	-0.9	-3.4	-1.6	0.6
63.79	0.89	1.9	31.32	19AP	0.2	-3.4	-1.5	0.6
70.27	0.94	1.9	40.93	29AP	1.2	-3.5	-1.5	0.6
76.72	0.99	2.0	50.51	9MY	2.3	-3.5	-1.4	0.6
83.15	1.03	2.0	60.07	19MY	3.6	-3.5	-1.4	0.6
89.55	1.06	2.0	69.60	29MY	2.2	-3.4	-1.4	0.6
95.94	1.09	2.0	79.12	8JN	1.0	-3.4	-1.4	0.6
102.31	1.11	2.0	88.65	18JN	-0.0	-3.4	-1.4	0.6
108.67	1.13	2.0	98.18	28JN	-1.1	-3.3	-1.4	0.5
115.03	1.14	2.0	107.73	8JL	-1.7	-3.3	-1.4	0.5
121.39	1.15	2.0	117.32	18JL	-0.9	-3.3	-1.4	0.5
127.75	1.16	2.0	126.93	28JL	-0.3	-3.3	-1.4	0.4
134.12	1.16	2.0	136.58	7AU	0.1	-3.3	-1.4	0.4
140.50	1.15	2.0	146.29	17AU	0.3	-3.3	-1.5	0.3
146.90	1.14	2.0	156.04	27AU	0.5	-3.4	-1.5	0.2
153.32	1.13	2.0	165.85	6SE	1.2	-3.4	-1.5	0.2
159.76	1.11	2.0	175.73	16SE	3.3	-3.4	-1.6	0.1
166.22	1.08	2.0	185.65	26SE	0.6	-3.5	-1.6	0.1
172.71	1.05	2.0	195.64	6OC	-0.6	-3.5	-1.7	0.0
179.23	1.02	2.0	205.68	16OC	-0.8	-3.6	-1.8	0.1
185.78	0.97	1.9	215.76	26OC	-0.7	-3.6	-1.8	0.1
192.36	0.93	1.9	225.89	5NO	-0.8	-3.7	-1.9	0.2
198.97	0.87	1.9	236.06	15NO	-0.7	-3.8	-2.0	0.3
205.61	0.81	1.8	246.24	25NO	-0.6	-3.9	-2.0	0.3
212.27	0.73	1.7	256.44	5DE	-0.6	-4.0	-2.1	0.4
218.97	0.65	1.7	266.65	15DE	-0.3	-4.1	-2.1	0.5
225.69	0.56	1.6	276.84	25DE	1.3	-4.2	-2.1	0.5
				615				
232.44	0.46	1.5	287.02	4JA	2.1	-4.3	-2.1	0.5
239.22	0.34	1.4	297.17	14JA	0.7	-4.4	-2.1	0.6
246.01	0.21	1.3	307.28	24JA	0.3	-4.3	-2.1	0.6
252.82	0.06	1.2	317.35	3FE	0.1	-4.1	-2.1	0.6
259.64	-0.10	1.1	327.36	13FE	-0.1	-3.7	-2.0	0.6
266.47	-0.28	1.0	337.32	23FE	-0.5	-3.3	-2.0	0.6
273.30	-0.48	0.9	347.22	5MR	-1.2	-3.8	-1.9	0.6
280.12	-0.69	0.8	357.07	15MR	-1.6	-4.1	-1.8	0.5
286.91	-0.93	0.6	6.85	25MR	-0.9	-4.2	-1.8	0.5
293.67	-1.19	0.5	16.58	4AP	0.3	-4.2	-1.7	0.4
300.36	-1.46	0.3	26.27	14AP	1.5	-4.1	-1.6	0.4
306.96	-1.76	0.2	35.90	24AP	3.1	-4.0	-1.6	0.4
313.43	-2.08	0.0	45.50	4MY	3.0	-3.9	-1.5	0.4
319.73	-2.42	-0.1	55.06	14MY	1.7	-3.8	-1.4	0.5
325.81	-2.78	-0.3	64.60	24MY	0.8	-3.8	-1.4	0.5
331.58	-3.16	-0.5	74.13	3JN	-0.1	-3.7	-1.4	0.5
336.95	-3.55	-0.7	83.66	13JN	-1.1	-3.6	-1.3	0.5
341.81	-3.96	-0.9	93.18	23JN	-1.7	-3.5	-1.3	0.5
346.01	-4.37	-1.1	102.73	3JL	-0.9	-3.5	-1.3	0.4
349.35	-4.77	-1.4	112.29	13JL	-0.2	-3.4	-1.3	0.4
351.64	-5.14	-1.6	121.89	23JL	0.2	-3.4	-1.3	0.4
352.66	-5.46	-1.9	131.52	2AU	0.5	-3.4	-1.3	0.3
352.27	-5.64	-2.1	141.20	12AU	0.8	-3.4	-1.3	0.3
350.54	-5.64	-2.3	150.92	22AU	1.6	-3.4	-1.3	0.2
347.81	-5.39	-2.4	160.71	1SE	3.1	-3.4	-1.3	0.2
344.80	-4.88	-2.4	170.55	11SE	0.5	-3.4	-1.3	0.1
342.28	-4.19	-2.1	180.45	21SE	-0.7	-3.4	-1.4	0.0
340.83	-3.43	-1.8	190.40	1OC	-0.9	-3.4	-1.4	-0.0
340.70	-2.68	-1.5	200.41	11OC	-0.9	-3.4	-1.5	-0.1
341.85	-2.02	-1.2	210.47	21OC	-0.7	-3.4	-1.5	-0.1
344.10	-1.44	-0.9	220.59	31OC	-0.5	-3.4	-1.6	-0.1
347.26	-0.95	-0.6	230.73	10NO	-0.5	-3.4	-1.6	-0.0
351.13	-0.54	-0.3	240.90	20NO	-0.5	-3.5	-1.7	0.0
355.55	-0.21	-0.0	251.10	30NO	-0.2	-3.5	-1.8	0.1
0.39	0.07	0.2	261.30	10DE	1.6	-3.5	-1.8	0.2
5.56	0.30	0.4	271.50	20DE	1.8	-3.5	-1.9	0.3
10.97	0.50	0.6	281.69	30DE	0.4	-3.4	-1.9	0.3

♂ LONG	♂ LAT	♂ MAG	⊙ LONG	16.00UT	☿	♀	♃	♄
				616	MAGNITUDES			
16.58	0.66	0.8	291.85	9JA	0.1	-3.4	-2.0	0.4
22.34	0.79	0.9	301.98	19JA	-0.0	-3.4	-2.0	0.4
28.20	0.90	1.1	312.08	29JA	-0.2	-3.4	-2.0	0.4
34.15	0.99	1.2	322.12	8FE	-0.6	-3.3	-2.0	0.4
40.16	1.06	1.4	332.10	18FE	-1.2	-3.3	-2.0	0.4
46.22	1.12	1.5	342.04	28FE	-1.5	-3.3	-2.0	0.4
52.31	1.17	1.6	351.91	9MR	-0.9	-3.3	-2.0	0.4
58.42	1.20	1.7	1.73	19MR	0.4	-3.3	-1.9	0.4
64.56	1.23	1.7	11.49	29MR	2.0	-3.4	-1.9	0.4
70.71	1.25	1.8	21.19	8AP	3.8	-3.4	-1.8	0.3
76.87	1.27	1.9	30.85	18AP	2.2	-3.4	-1.7	0.3
83.05	1.27	1.9	40.47	28AP	1.3	-3.4	-1.7	0.3
89.23	1.27	2.0	50.05	8MY	0.6	-3.5	-1.6	0.3
95.42	1.27	2.0	59.60	18MY	-0.2	-3.5	-1.5	0.3
101.63	1.26	2.0	69.14	28MY	-1.1	-3.6	-1.5	0.3
107.86	1.25	2.0	78.66	7JN	-1.8	-3.6	-1.4	0.4
114.10	1.23	2.0	88.19	17JN	-0.9	-3.7	-1.4	0.4
120.37	1.21	2.0	97.73	27JN	-0.2	-3.8	-1.3	0.4
126.66	1.18	2.0	107.27	7JL	0.3	-3.9	-1.3	0.4
132.98	1.15	2.0	116.85	17JL	0.6	-4.0	-1.3	0.3
139.34	1.11	2.0	126.47	27JL	1.1	-4.1	-1.2	0.3
145.73	1.07	2.0	136.12	6AU	2.1	-4.2	-1.2	0.3
152.16	1.03	2.0	145.82	16AU	2.7	-4.3	-1.2	0.2
158.64	0.98	1.9	155.57	26AU	0.4	-4.3	-1.2	0.2
165.16	0.93	1.9	165.38	5SE	-0.8	-4.1	-1.2	0.1
171.73	0.88	1.9	175.25	15SE	-1.1	-3.7	-1.2	0.1
178.36	0.81	1.9	185.18	25SE	-1.0	-3.2	-1.2	0.0
185.04	0.75	1.9	195.16	5OC	-0.6	-3.8	-1.3	-0.1
191.78	0.68	1.9	205.19	15OC	-0.4	-4.2	-1.3	-0.1
198.58	0.60	1.9	215.28	25OC	-0.3	-4.4	-1.3	-0.2
205.43	0.52	1.8	225.40	4NO	-0.3	-4.4	-1.4	-0.3
212.35	0.43	1.8	235.57	14NO	0.0	-4.3	-1.4	-0.2
219.34	0.33	1.8	245.75	24NO	1.8	-4.2	-1.5	-0.1
226.38	0.23	1.7	255.95	4DE	1.5	-4.1	-1.6	-0.1
233.48	0.13	1.7	266.16	14DE	0.2	-4.0	-1.6	0.0
240.64	0.01	1.6	276.35	24DE	-0.1	-3.9	-1.7	0.1
				617				
247.86	-0.11	1.6	286.53	3JA	-0.1	-3.8	-1.8	0.2
255.14	-0.24	1.5	296.68	13JA	-0.3	-3.7	-1.8	0.2
262.46	-0.37	1.5	306.80	23JA	-0.6	-3.6	-1.9	0.3
269.84	-0.51	1.4	316.86	2FE	-1.2	-3.6	-1.9	0.3
277.25	-0.65	1.4	326.88	12FE	-1.3	-3.5	-2.0	0.3
284.70	-0.80	1.3	336.84	22FE	-0.8	-3.5	-2.0	0.3
292.18	-0.95	1.2	346.75	4MR	0.5	-3.4	-2.0	0.4
299.68	-1.10	1.1	356.60	14MR	2.4	-3.4	-2.0	0.4
307.18	-1.24	1.1	6.39	24MR	2.9	-3.3	-2.0	0.3
314.68	-1.39	1.0	16.12	3AP	1.6	-3.3	-2.0	0.3
322.16	-1.53	0.9	25.80	13AP	0.9	-3.3	-1.9	0.3
329.60	-1.66	0.9	35.44	23AP	0.4	-3.3	-1.9	0.3
336.98	-1.79	0.8	45.04	3MY	-0.2	-3.3	-1.8	0.2
344.29	-1.90	0.7	54.60	13MY	-1.1	-3.3	-1.7	0.2
351.49	-2.00	0.6	64.15	23MY	-1.8	-3.3	-1.7	0.2
358.57	-2.08	0.5	73.67	2JN	-0.9	-3.4	-1.6	0.3
5.49	-2.14	0.4	83.20	12JN	-0.1	-3.4	-1.6	0.3
12.22	-2.19	0.4	92.73	22JN	0.4	-3.4	-1.5	0.3
18.71	-2.22	0.3	102.26	2JL	0.8	-3.5	-1.4	0.3
24.93	-2.23	0.2	111.83	12JL	1.4	-3.5	-1.4	0.3
30.80	-2.21	0.0	121.42	22JL	2.6	-3.5	-1.3	0.3
36.26	-2.16	-0.1	131.05	1AU	2.3	-3.4	-1.3	0.3
41.22	-2.09	-0.2	140.73	11AU	0.3	-3.4	-1.3	0.3
45.55	-1.99	-0.4	150.45	21AU	-0.9	-3.4	-1.2	0.2
49.11	-1.84	-0.6	160.23	31AU	-1.2	-3.4	-1.2	0.2
51.72	-1.65	-0.8	170.07	10SE	-1.0	-3.3	-1.2	0.2
53.16	-1.39	-1.0	179.96	20SE	-0.6	-3.3	-1.2	0.1
53.25	-1.07	-1.2	189.91	30SE	-0.3	-3.3	-1.2	0.0
51.86	-0.67	-1.4	199.92	10OC	-0.2	-3.3	-1.2	-0.0
49.14	-0.20	-1.6	209.98	20OC	-0.1	-3.3	-1.2	-0.1
45.53	0.29	-1.7	220.09	30OC	0.2	-3.4	-1.3	-0.2
41.79	0.76	-1.6	230.24	9NO	2.1	-3.4	-1.3	-0.3
38.72	1.15	-1.3	240.41	19NO	1.2	-3.4	-1.3	-0.3
36.83	1.45	-1.0	250.60	29NO	-0.0	-3.4	-1.4	-0.2
36.30	1.65	-0.7	260.81	9DE	-0.2	-3.5	-1.4	-0.2
37.06	1.79	-0.4	271.01	19DE	-0.3	-3.5	-1.5	-0.1
38.89	1.87	-0.1	281.20	29DE	-0.4	-3.6	-1.5	-0.0

♂ LONG	♂ LAT	♂ MAG	⊙ LONG	16.00UT	☿	♀	♃	♄
				618	MAGNITUDES			
41.61	1.91	0.2	291.37	8JA	-0.7	-3.7	-1.6	0.0
45.03	1.92	0.4	301.50	18JA	-1.1	-3.7	-1.7	0.1
48.98	1.91	0.6	311.59	28JA	-1.2	-3.8	-1.7	0.2
53.37	1.90	0.8	321.64	7FE	-0.8	-3.9	-1.8	0.2
58.08	1.87	1.0	331.63	17FE	0.7	-4.0	-1.9	0.2
63.05	1.83	1.1	341.56	27FE	2.9	-4.1	-1.9	0.3
68.24	1.79	1.3	351.44	9MR	2.2	-4.1	-2.0	0.3
73.59	1.75	1.4	1.26	19MR	1.2	-4.2	-2.0	0.3
79.08	1.70	1.5	11.02	29MR	0.7	-4.2	-2.0	0.3
84.69	1.65	1.6	20.73	8AP	0.3	-4.2	-2.0	0.3
90.39	1.60	1.7	30.39	18AP	-0.3	-3.8	-2.0	0.3
96.17	1.55	1.8	40.01	28AP	-1.1	-3.2	-2.0	0.3
102.03	1.49	1.8	49.59	8MY	-1.8	-3.1	-2.0	0.2
107.95	1.43	1.9	59.14	18MY	-0.9	-3.8	-1.9	0.2
113.94	1.37	1.9	68.68	28MY	-0.0	-4.1	-1.9	0.2
119.99	1.31	1.9	78.21	7JN	0.6	-4.2	-1.8	0.2
126.10	1.25	2.0	87.73	17JN	1.1	-4.2	-1.8	0.3
132.27	1.18	2.0	97.26	27JN	1.9	-4.1	-1.7	0.3
138.50	1.11	2.0	106.81	7JL	3.2	-4.0	-1.6	0.3
144.79	1.04	2.0	116.38	17JL	1.9	-3.9	-1.6	0.3
151.15	0.97	2.0	126.00	27JL	0.2	-3.8	-1.5	0.3
157.57	0.89	2.0	135.65	6AU	-1.0	-3.7	-1.5	0.3
164.07	0.81	1.9	145.34	16AU	-1.4	-3.7	-1.4	0.3
170.63	0.73	1.9	155.10	26AU	-1.0	-3.6	-1.4	0.3
177.27	0.65	1.9	164.90	5SE	-0.5	-3.6	-1.3	0.3
183.98	0.56	1.8	174.76	15SE	-0.2	-3.5	-1.3	0.2
190.78	0.47	1.8	184.69	25SE	-0.1	-3.5	-1.3	0.2
197.65	0.38	1.7	194.67	5OC	0.0	-3.5	-1.3	0.1
204.61	0.28	1.7	204.70	15OC	0.4	-3.4	-1.3	0.0
211.64	0.18	1.7	214.79	25OC	2.4	-3.4	-1.3	-0.0
218.76	0.08	1.7	224.92	4NO	1.0	-3.4	-1.3	-0.1
225.95	-0.02	1.7	235.07	14NO	-0.2	-3.4	-1.3	-0.2
233.23	-0.13	1.6	245.26	24NO	-0.4	-3.4	-1.3	-0.3
240.58	-0.24	1.6	255.46	4DE	-0.4	-3.4	-1.3	-0.3
248.00	-0.35	1.6	265.66	14DE	-0.5	-3.4	-1.4	-0.2
255.49	-0.45	1.6	275.86	24DE	-0.8	-3.4	-1.4	-0.2
				619				
263.05	-0.56	1.5	286.04	3JA	-1.0	-3.4	-1.5	-0.1
270.66	-0.67	1.5	296.19	13JA	-1.0	-3.4	-1.5	-0.0
278.31	-0.78	1.5	306.31	23JA	-0.7	-3.4	-1.6	0.0
286.01	-0.88	1.5	316.38	2FE	0.8	-3.4	-1.6	0.1
293.74	-0.97	1.4	326.39	12FE	3.2	-3.5	-1.7	0.2
301.49	-1.06	1.4	336.36	22FE	1.6	-3.5	-1.8	0.2
309.24	-1.14	1.4	346.27	4MR	0.9	-3.5	-1.8	0.2
316.99	-1.21	1.3	356.11	14MR	0.5	-3.4	-1.9	0.3
324.73	-1.27	1.3	5.91	24MR	0.2	-3.4	-2.0	0.3
332.44	-1.32	1.3	15.64	3AP	-0.3	-3.4	-2.0	0.3
340.11	-1.36	1.3	25.33	13AP	-1.1	-3.4	-2.1	0.3
347.72	-1.38	1.2	34.97	23AP	-1.8	-3.3	-2.1	0.3
355.26	-1.39	1.2	44.57	3MY	-0.9	-3.3	-2.1	0.3
2.72	-1.38	1.2	54.13	13MY	0.0	-3.3	-2.1	0.3
10.09	-1.36	1.1	63.68	23MY	0.8	-3.3	-2.1	0.3
17.34	-1.32	1.1	73.21	2JN	1.5	-3.3	-2.1	0.2
24.48	-1.26	1.1	82.73	12JN	2.5	-3.4	-2.1	0.2
31.47	-1.19	1.0	92.26	22JN	3.2	-3.4	-2.0	0.3
38.32	-1.11	1.0	101.80	2JL	1.6	-3.4	-2.0	0.3
44.99	-1.01	1.0	111.36	12JL	0.1	-3.4	-1.9	0.3
51.48	-0.89	0.9	120.96	22JL	-1.0	-3.5	-1.8	0.4
57.74	-0.75	0.8	130.58	1AU	-1.5	-3.5	-1.8	0.4
63.77	-0.60	0.8	140.25	11AU	-1.0	-3.6	-1.7	0.4
69.50	-0.43	0.7	149.98	21AU	-0.4	-3.7	-1.7	0.4
74.91	-0.24	0.6	159.75	31AU	-0.1	-3.8	-1.6	0.4
79.92	-0.03	0.5	169.58	10SE	0.1	-3.9	-1.5	0.3
84.45	0.22	0.3	179.48	20SE	0.2	-4.0	-1.5	0.3
88.40	0.49	0.2	189.43	30SE	0.7	-4.1	-1.5	0.3
91.63	0.81	0.0	199.43	10OC	2.7	-4.2	-1.4	0.2
93.98	1.17	-0.2	209.49	20OC	0.8	-4.3	-1.4	0.2
95.25	1.58	-0.4	219.60	30OC	-0.4	-4.4	-1.4	0.1
95.26	2.02	-0.6	229.74	9NO	-0.5	-4.4	-1.4	0.0
93.87	2.49	-0.8	239.92	19NO	-0.5	-4.2	-1.4	-0.1
91.16	2.94	-1.0	250.11	29NO	-0.6	-3.7	-1.4	-0.1
87.48	3.31	-1.2	260.31	9DE	-0.8	-3.0	-1.4	-0.2
83.51	3.55	-1.1	270.51	19DE	-0.9	-3.8	-1.4	-0.2
80.03	3.63	-0.9	280.70	29DE	-0.9	-4.2	-1.4	-0.2

♂			⊙	16.00UT	☿	♀	♃	♄
LONG	LAT	MAG	LONG	620	MAGNITUDES			
77.60	3.58	-0.7	290.87	8JA	-0.6	-4.4	-1.4	-0.1
76.47	3.45	-0.4	301.01	18JA	1.0	-4.3	-1.5	-0.0
76.65	3.27	-0.1	311.10	28JA	2.9	-4.3	-1.5	0.0
77.94	3.07	0.2	321.15	7FE	1.2	-4.2	-1.6	0.1
80.18	2.87	0.4	331.15	17FE	0.6	-4.0	-1.6	0.1
83.17	2.68	0.6	341.08	27FE	0.3	-3.9	-1.7	0.2
86.76	2.50	0.8	350.96	8MR	0.1	-3.8	-1.8	0.3
90.82	2.33	1.0	0.79	18MR	-0.4	-3.7	-1.8	0.3
95.27	2.17	1.1	10.55	28MR	-1.1	-3.6	-1.9	0.3
100.02	2.03	1.2	20.26	7AP	-1.7	-3.6	-2.0	0.4
105.02	1.88	1.4	29.93	17AP	-0.9	-3.5	-2.0	0.4
110.24	1.75	1.5	39.54	27AP	0.1	-3.5	-2.1	0.4
115.64	1.62	1.5	49.13	7MY	1.0	-3.4	-2.2	0.4
121.19	1.50	1.6	58.69	17MY	1.9	-3.4	-2.2	0.4
126.89	1.38	1.7	68.22	27MY	3.3	-3.3	-2.2	0.4
132.71	1.26	1.7	77.75	6JN	2.6	-3.3	-2.3	0.3
138.65	1.15	1.8	87.27	16JN	1.2	-3.3	-2.3	0.3
144.70	1.03	1.8	96.80	26JN	0.0	-3.3	-2.2	0.3
150.85	0.92	1.8	106.35	6JL	-1.1	-3.3	-2.2	0.3
157.11	0.81	1.8	115.93	16JL	-1.6	-3.3	-2.2	0.4
163.47	0.70	1.8	125.53	26JL	-1.0	-3.3	-2.1	0.4
169.93	0.59	1.8	135.18	5AU	-0.4	-3.4	-2.1	0.5
176.49	0.48	1.8	144.88	15AU	0.0	-3.4	-2.0	0.5
183.15	0.38	1.8	154.62	25AU	0.2	-3.4	-1.9	0.5
189.92	0.27	1.8	164.43	4SE	0.4	-3.5	-1.9	0.5
196.78	0.16	1.8	174.29	14SE	1.0	-3.5	-1.8	0.5
203.74	0.05	1.7	184.20	24SE	3.1	-3.5	-1.8	0.5
210.81	-0.06	1.7	194.18	4OC	0.7	-3.5	-1.7	0.4
217.97	-0.16	1.6	204.21	14OC	-0.5	-3.4	-1.6	0.4
225.23	-0.27	1.6	214.29	24OC	-0.7	-3.4	-1.6	0.3
232.58	-0.37	1.6	224.42	3NO	-0.7	-3.4	-1.6	0.3
240.02	-0.47	1.5	234.57	13NO	-0.7	-3.4	-1.5	0.2
247.54	-0.56	1.5	244.75	23NO	-0.7	-3.3	-1.5	0.1
255.14	-0.65	1.4	254.96	3DE	-0.7	-3.3	-1.5	0.1
262.81	-0.74	1.4	265.16	13DE	-0.7	-3.3	-1.5	-0.0
270.54	-0.82	1.4	275.36	23DE	-0.4	-3.3	-1.5	-0.1
				621				
278.32	-0.89	1.4	285.54	2JA	1.2	-3.4	-1.5	-0.1
286.13	-0.96	1.4	295.69	12JA	2.4	-3.4	-1.5	-0.1
293.98	-1.02	1.4	305.81	22JA	0.9	-3.4	-1.5	0.0
301.84	-1.06	1.4	315.89	1FE	0.4	-3.4	-1.5	0.1
309.71	-1.10	1.4	325.91	11FE	0.2	-3.4	-1.5	0.1
317.57	-1.12	1.4	335.88	21FE	-0.0	-3.5	-1.6	0.2
325.40	-1.13	1.4	345.79	3MR	-0.5	-3.5	-1.6	0.2
333.20	-1.13	1.4	355.64	13MR	-1.1	-3.6	-1.7	0.3
340.96	-1.12	1.4	5.44	23MR	-1.6	-3.6	-1.7	0.4
348.66	-1.09	1.4	15.18	2AP	-0.9	-3.7	-1.8	0.4
356.30	-1.05	1.4	24.87	12AP	0.2	-3.8	-1.9	0.4
3.85	-1.00	1.4	34.51	22AP	1.3	-3.9	-1.9	0.5
11.33	-0.94	1.4	44.11	2MY	2.6	-3.9	-2.0	0.5
18.71	-0.87	1.5	53.68	12MY	3.5	-4.0	-2.1	0.5
26.00	-0.79	1.5	63.22	22MY	2.0	-4.1	-2.2	0.5
33.19	-0.70	1.5	72.76	1JN	1.0	-4.2	-2.2	0.5
40.26	-0.60	1.5	82.28	11JN	-0.0	-4.2	-2.3	0.5
47.22	-0.49	1.5	91.80	21JN	-1.1	-4.0	-2.3	0.5
54.07	-0.37	1.4	101.35	1JL	-1.7	-3.6	-2.4	0.5
60.78	-0.25	1.4	110.90	11JL	-0.9	-3.0	-2.4	0.4
67.37	-0.11	1.4	120.49	21JL	-0.3	-3.6	-2.4	0.5
73.81	0.03	1.4	130.12	31JL	0.1	-4.0	-2.4	0.5
80.10	0.18	1.4	139.79	10AU	0.4	-4.2	-2.3	0.6
86.23	0.33	1.3	149.51	20AU	0.6	-4.2	-2.3	0.6
92.16	0.50	1.3	159.28	30AU	1.3	-4.2	-2.2	0.6
97.89	0.68	1.2	169.11	9SE	3.3	-4.1	-2.2	0.6
103.38	0.88	1.1	179.00	19SE	0.6	-4.0	-2.1	0.6
108.58	1.09	1.0	188.95	29SE	-0.6	-3.9	-2.1	0.6
113.46	1.32	0.9	198.95	9OC	-0.8	-3.9	-2.0	0.6
117.93	1.57	0.8	209.00	19OC	-0.8	-3.8	-1.9	0.6
121.90	1.85	0.6	219.11	29OC	-0.8	-3.7	-1.9	0.5
125.28	2.16	0.4	229.24	8NO	-0.6	-3.7	-1.8	0.5
127.91	2.51	0.2	239.42	18NO	-0.6	-3.6	-1.7	0.4
129.61	2.89	0.0	249.61	28NO	-0.6	-3.5	-1.7	0.4
130.22	3.29	-0.2	259.81	8DE	-0.3	-3.5	-1.6	0.3
129.55	3.71	-0.4	270.01	18DE	1.3	-3.5	-1.6	0.2
127.54	4.09	-0.7	280.21	28DE	2.0	-3.4	-1.6	0.2

♂			⊙	16.00UT	☿	♀	♃	♄
LONG	LAT	MAG	LONG	622	MAGNITUDES			
124.37	4.38	-0.9	290.37	7JA	0.6	-3.4	-1.6	0.1
120.50	4.53	-1.0	300.51	17JA	0.2	-3.4	-1.6	0.1
116.62	4.49	-0.9	310.61	27JA	0.1	-3.3	-1.5	0.1
113.46	4.30	-0.7	320.66	6FE	-0.1	-3.3	-1.5	0.2
111.44	4.00	-0.4	330.66	16FE	-0.5	-3.3	-1.6	0.2
110.72	3.66	-0.2	340.60	26FE	-1.1	-3.3	-1.6	0.3
111.24	3.30	0.1	350.48	8MR	-1.5	-3.3	-1.6	0.3
112.81	2.96	0.3	0.31	18MR	-0.9	-3.3	-1.6	0.4
115.28	2.65	0.5	10.08	28MR	0.3	-3.4	-1.7	0.5
118.46	2.36	0.7	19.79	7AP	1.7	-3.4	-1.7	0.5
122.22	2.10	0.8	29.46	17AP	3.4	-3.4	-1.8	0.6
126.45	1.85	1.0	39.08	27AP	2.7	-3.5	-1.8	0.6
131.06	1.63	1.1	48.66	7MY	1.5	-3.5	-1.9	0.6
136.00	1.43	1.2	58.22	17MY	0.7	-3.5	-1.9	0.6
141.21	1.24	1.3	67.76	27MY	-0.1	-3.4	-2.0	0.7
146.66	1.06	1.3	77.28	6JN	-1.1	-3.4	-2.1	0.7
152.32	0.89	1.4	86.81	16JN	-1.8	-3.4	-2.1	0.7
158.17	0.73	1.5	96.34	26JN	-0.9	-3.3	-2.2	0.7
164.19	0.57	1.5	105.88	6JL	-0.2	-3.3	-2.3	0.7
170.37	0.43	1.5	115.46	16JL	0.2	-3.3	-2.3	0.7
176.71	0.29	1.6	125.06	26JL	0.5	-3.3	-2.4	0.7
183.18	0.15	1.6	134.71	5AU	0.9	-3.3	-2.4	0.7
189.80	0.02	1.6	144.40	15AU	1.7	-3.3	-2.4	0.8
196.55	-0.10	1.6	154.15	25AU	3.0	-3.4	-2.5	0.8
203.43	-0.22	1.6	163.95	4SE	0.5	-3.4	-2.4	0.8
210.43	-0.34	1.6	173.81	14SE	-0.8	-3.4	-2.4	0.9
217.56	-0.44	1.6	183.72	24SE	-1.0	-3.5	-2.4	0.9
224.80	-0.55	1.5	193.69	4OC	-0.9	-3.5	-2.3	0.9
232.15	-0.64	1.5	203.73	14OC	-0.7	-3.6	-2.3	0.9
239.60	-0.73	1.5	213.80	24OC	-0.5	-3.6	-2.2	0.8
247.15	-0.81	1.5	223.92	3NO	-0.4	-3.7	-2.1	0.8
254.78	-0.88	1.5	234.08	13NO	-0.4	-3.8	-2.1	0.8
262.49	-0.94	1.4	244.26	23NO	-0.1	-3.9	-2.0	0.7
270.26	-0.99	1.4	254.46	3DE	1.6	-4.0	-1.9	0.7
278.09	-1.04	1.4	264.67	13DE	1.7	-4.1	-1.9	0.6
285.96	-1.07	1.4	274.86	23DE	0.3	-4.2	-1.8	0.5
				623				
293.85	-1.08	1.4	285.04	2JA	0.0	-4.3	-1.7	0.5
301.75	-1.09	1.3	295.20	12JA	-0.1	-4.4	-1.7	0.4
309.66	-1.09	1.3	305.32	22JA	-0.2	-4.3	-1.7	0.3
317.54	-1.07	1.3	315.39	1FE	-0.6	-4.1	-1.6	0.3
325.41	-1.04	1.3	325.42	11FE	-1.2	-3.7	-1.6	0.3
333.23	-1.00	1.3	335.39	21FE	-1.4	-3.3	-1.6	0.4
341.00	-0.95	1.4	345.30	3MR	-0.9	-3.8	-1.6	0.5
348.71	-0.89	1.4	355.16	13MR	0.4	-4.1	-1.6	0.5
356.35	-0.82	1.4	4.96	23MR	2.1	-4.3	-1.6	0.6
3.92	-0.75	1.5	14.70	2AP	3.5	-4.2	-1.6	0.6
11.40	-0.66	1.5	24.40	12AP	2.0	-4.1	-1.6	0.7
18.80	-0.57	1.5	34.04	22AP	1.1	-4.0	-1.6	0.7
26.11	-0.48	1.6	43.64	2MY	0.5	-3.9	-1.7	0.8
33.33	-0.38	1.6	53.22	12MY	-0.2	-3.8	-1.7	0.8
40.45	-0.27	1.6	62.76	22MY	-1.1	-3.7	-1.8	0.8
47.48	-0.17	1.7	72.29	1JN	-1.8	-3.7	-1.8	0.9
54.42	-0.06	1.7	81.82	11JN	-0.9	-3.6	-1.9	0.9
61.26	0.06	1.7	91.34	21JN	-0.2	-3.5	-1.9	0.9
68.01	0.17	1.7	100.88	1JL	0.3	-3.5	-2.0	0.9
74.66	0.29	1.7	110.44	11JL	0.7	-3.4	-2.1	0.9
81.22	0.41	1.7	120.03	21JL	1.2	-3.4	-2.1	0.9
87.68	0.53	1.7	129.65	31JL	2.2	-3.4	-2.2	0.9
94.04	0.65	1.7	139.32	10AU	2.6	-3.4	-2.3	0.9
100.29	0.78	1.7	149.04	20AU	0.4	-3.4	-2.3	1.0
106.43	0.91	1.7	158.81	30AU	-0.9	-3.4	-2.4	1.0
112.45	1.04	1.6	168.64	9SE	-1.1	-3.4	-2.4	1.1
118.34	1.18	1.6	178.52	19SE	-1.1	-3.4	-2.4	1.1
124.08	1.33	1.5	188.46	29SE	-0.6	-3.4	-2.4	1.1
129.64	1.48	1.4	198.47	9OC	-0.4	-3.4	-2.4	1.1
135.01	1.64	1.3	208.52	19OC	-0.3	-3.4	-2.4	1.1
140.15	1.82	1.2	218.62	29OC	-0.2	-3.4	-2.4	1.1
145.00	2.01	1.1	228.76	8NO	0.1	-3.4	-2.3	1.1
149.52	2.21	1.0	238.92	18NO	1.8	-3.5	-2.3	1.1
153.61	2.43	0.8	249.11	28NO	1.4	-3.5	-2.2	1.0
157.20	2.67	0.6	259.32	8DE	0.1	-3.5	-2.1	1.0
160.14	2.93	0.4	269.52	18DE	-0.1	-3.5	-2.0	0.9
162.29	3.21	0.2	279.71	28DE	-0.2	-3.4	-2.0	0.9

♂ LONG	♂ LAT	♂ MAG	⊙ LONG	16.00UT	☿	♀	♃	♄
				624	MAGNITUDES			
163.48	3.50	-0.1	289.88	7JA	-0.3	-3.4	-1.9	0.8
163.51	3.79	-0.3	300.01	17JA	-0.6	-3.4	-1.8	0.7
162.27	4.05	-0.6	310.12	27JA	-1.1	-3.4	-1.8	0.7
159.77	4.22	-0.8	320.17	6FE	-1.3	-3.3	-1.7	0.6
156.30	4.26	-1.0	330.17	16FE	-0.8	-3.3	-1.7	0.6
152.41	4.13	-1.1	340.11	26FE	0.5	-3.3	-1.6	0.6
148.83	3.85	-0.9	350.00	7MR	2.6	-3.3	-1.6	0.7
146.18	3.45	-0.7	359.83	17MR	2.7	-3.3	-1.6	0.7
144.77	3.01	-0.5	9.60	27MR	1.5	-3.4	-1.6	0.8
144.66	2.57	-0.3	19.32	6AP	0.9	-3.4	-1.5	0.8
145.74	2.15	-0.1	28.98	16AP	0.4	-3.4	-1.5	0.9
147.84	1.77	0.1	38.61	26AP	-0.2	-3.4	-1.5	0.9
150.78	1.43	0.3	48.20	6MY	-1.1	-3.5	-1.5	1.0
154.41	1.12	0.5	57.76	16MY	-1.8	-3.5	-1.6	1.0
158.61	0.85	0.6	67.30	26MY	-0.9	-3.6	-1.6	1.1
163.28	0.60	0.7	76.83	5JN	-0.1	-3.6	-1.6	1.1
168.35	0.37	0.8	86.35	15JN	0.5	-3.7	-1.7	1.1
173.75	0.16	0.9	95.88	25JN	0.9	-3.8	-1.7	1.2
179.45	-0.02	1.0	105.43	5JL	1.6	-3.9	-1.7	1.2
185.41	-0.20	1.0	115.00	15JL	2.8	-4.0	-1.8	1.2
191.61	-0.35	1.1	124.60	25JL	2.2	-4.1	-1.9	1.2
198.02	-0.50	1.1	134.25	4AU	0.3	-4.2	-1.9	1.2
204.61	-0.63	1.2	143.93	14AU	-0.9	-4.3	-2.0	1.2
211.39	-0.75	1.2	153.68	24AU	-1.3	-4.2	-2.1	1.2
218.34	-0.85	1.2	163.47	3SE	-1.0	-4.1	-2.1	1.3
225.43	-0.95	1.3	173.33	13SE	-0.5	-3.6	-2.2	1.3
232.67	-1.03	1.3	183.24	23SE	-0.3	-3.3	-2.2	1.4
240.03	-1.09	1.3	193.21	3OC	-0.1	-3.8	-2.3	1.4
247.51	-1.15	1.3	203.24	13OC	-0.1	-4.2	-2.3	1.3
255.09	-1.19	1.3	213.32	23OC	0.3	-4.4	-2.4	1.3
262.76	-1.22	1.3	223.43	2NO	2.1	-4.3	-2.4	1.3
270.50	-1.23	1.3	233.59	12NO	1.2	-4.3	-2.4	1.3
278.30	-1.23	1.3	243.77	22NO	-0.1	-4.2	-2.3	1.2
286.14	-1.22	1.4	253.97	2DE	-0.3	-4.1	-2.3	1.2
294.00	-1.19	1.4	264.17	12DE	-0.3	-4.0	-2.2	1.1
301.88	-1.16	1.4	274.37	22DE	-0.4	-3.9	-2.2	1.1
				625				
309.75	-1.11	1.4	284.55	1JA	-0.7	-3.8	-2.1	1.0
317.61	-1.05	1.4	294.71	11JA	-1.1	-3.7	-2.0	1.0
325.43	-0.98	1.4	304.83	21JA	-1.1	-3.6	-2.0	0.9
333.21	-0.90	1.4	314.91	31JA	-0.7	-3.6	-1.9	0.9
340.93	-0.82	1.5	324.94	10FE	0.7	-3.5	-1.8	0.9
348.60	-0.73	1.5	334.91	20FE	3.0	-3.5	-1.8	0.8
356.19	-0.64	1.5	344.83	2MR	2.0	-3.4	-1.7	0.9
3.70	-0.54	1.5	354.69	12MR	1.1	-3.4	-1.6	0.9
11.13	-0.44	1.5	4.49	22MR	0.6	-3.3	-1.6	1.0
18.48	-0.33	1.5	14.23	1AP	0.3	-3.3	-1.6	1.0
25.74	-0.23	1.5	23.93	11AP	-0.3	-3.3	-1.5	1.1
32.92	-0.13	1.6	33.58	21AP	-1.1	-3.3	-1.5	1.1
40.01	-0.02	1.6	43.18	1MY	-1.8	-3.3	-1.5	1.2
47.02	0.08	1.7	52.75	11MY	-0.9	-3.3	-1.5	1.2
53.95	0.18	1.7	62.30	21MY	-0.0	-3.3	-1.4	1.3
60.79	0.29	1.8	71.83	31MY	0.6	-3.4	-1.4	1.3
67.57	0.39	1.8	81.36	10JN	1.2	-3.4	-1.5	1.4
74.26	0.48	1.8	90.88	20JN	2.1	-3.4	-1.5	1.4
80.89	0.58	1.9	100.42	30JN	3.3	-3.5	-1.5	1.4
87.45	0.67	1.9	109.98	10JL	1.8	-3.5	-1.5	1.4
93.94	0.77	1.9	119.56	20JL	0.2	-3.5	-1.6	1.4
100.37	0.86	1.9	129.18	30JL	-1.0	-3.4	-1.6	1.3
106.74	0.95	1.9	138.85	9AU	-1.4	-3.4	-1.6	1.3
113.05	1.03	1.9	148.56	19AU	-1.0	-3.4	-1.7	1.3
119.29	1.12	1.9	158.33	29AU	-0.5	-3.4	-1.7	1.2
125.46	1.21	1.9	168.15	8SE	-0.2	-3.3	-1.8	1.2
131.57	1.30	1.9	178.04	18SE	-0.0	-3.3	-1.9	1.2
137.60	1.39	1.8	187.97	28SE	0.1	-3.3	-1.9	1.2
143.55	1.47	1.8	197.97	8OC	0.5	-3.3	-2.0	1.2
149.40	1.56	1.7	208.02	18OC	2.4	-3.3	-2.1	1.2
155.15	1.65	1.6	218.12	28OC	1.0	-3.4	-2.1	1.2
160.77	1.75	1.5	228.26	7NO	-0.3	-3.4	-2.2	1.1
166.24	1.84	1.4	238.43	17NO	-0.4	-3.4	-2.2	1.1
171.54	1.94	1.3	248.62	27NO	-0.4	-3.4	-2.2	1.1
176.62	2.04	1.2	258.82	7DE	-0.5	-3.5	-2.2	1.0
181.44	2.15	1.0	269.02	17DE	-0.8	-3.5	-2.2	1.0
185.95	2.26	0.9	279.21	27DE	-0.9	-3.6	-2.2	1.0

♂ LONG	♂ LAT	♂ MAG	⊙ LONG	16.00UT	☿	♀	♃	♄
				626	MAGNITUDES			
190.07	2.37	0.7	289.39	6JA	-1.0	-3.7	-2.2	0.9
193.69	2.48	0.5	299.53	16JA	-0.7	-3.7	-2.1	0.9
196.71	2.59	0.2	309.63	26JA	0.8	-3.8	-2.1	0.8
198.97	2.68	-0.0	319.69	5FE	3.1	-3.9	-2.0	0.8
200.28	2.76	-0.3	329.69	15FE	1.5	-4.0	-1.9	0.7
200.49	2.80	-0.6	339.64	25FE	0.8	-4.1	-1.8	0.7
199.44	2.77	-0.9	349.53	7MR	0.4	-4.1	-1.8	0.7
197.16	2.64	-1.2	359.36	17MR	0.1	-4.2	-1.7	0.7
193.89	2.39	-1.4	9.13	27MR	-0.3	-4.2	-1.6	0.8
190.19	2.03	-1.5	18.85	6AP	-1.1	-4.2	-1.6	0.9
186.80	1.58	-1.3	28.52	16AP	-1.8	-3.8	-1.5	0.9
184.33	1.10	-1.1	38.14	26AP	-1.0	-3.1	-1.5	1.0
183.13	0.64	-0.9	47.74	6MY	0.0	-3.1	-1.4	1.1
183.28	0.22	-0.7	57.29	16MY	0.8	-3.8	-1.4	1.1
184.67	-0.15	-0.5	66.83	26MY	1.6	-4.1	-1.4	1.2
187.13	-0.47	-0.3	76.36	5JN	2.8	-4.2	-1.4	1.2
190.49	-0.73	-0.2	85.88	15JN	3.0	-4.2	-1.3	1.2
194.58	-0.96	-0.0	95.41	25JN	1.5	-4.1	-1.3	1.2
199.28	-1.15	0.1	104.96	5JL	0.1	-4.0	-1.3	1.2
204.49	-1.30	0.2	114.53	15JL	-1.0	-3.9	-1.3	1.2
210.12	-1.43	0.3	124.13	25JL	-1.5	-3.8	-1.4	1.2
216.12	-1.54	0.4	133.78	4AU	-1.0	-3.7	-1.4	1.2
222.43	-1.62	0.5	143.46	14AU	-0.4	-3.7	-1.4	1.2
229.00	-1.67	0.6	153.20	24AU	-0.1	-3.6	-1.4	1.1
235.81	-1.71	0.7	163.00	3SE	0.1	-3.6	-1.5	1.1
242.82	-1.73	0.7	172.85	13SE	0.3	-3.5	-1.5	1.0
249.99	-1.73	0.8	182.76	23SE	0.8	-3.5	-1.6	1.0
257.32	-1.71	0.9	192.73	3OC	2.8	-3.4	-1.6	1.0
264.77	-1.67	0.9	202.75	13OC	0.9	-3.4	-1.7	1.0
272.31	-1.62	1.0	212.82	23OC	-0.4	-3.4	-1.8	1.0
279.93	-1.55	1.0	222.94	2NO	-0.6	-3.4	-1.8	1.0
287.61	-1.47	1.1	233.09	12NO	-0.6	-3.4	-1.9	1.0
295.32	-1.38	1.1	243.27	22NO	-0.6	-3.4	-2.0	1.0
303.05	-1.28	1.2	253.47	2DE	-0.8	-3.4	-2.0	1.0
310.78	-1.17	1.2	263.68	12DE	-0.8	-3.4	-2.1	0.9
318.49	-1.06	1.3	273.87	22DE	-0.8	-3.4	-2.1	0.9
				627				
326.17	-0.94	1.4	284.06	1JA	-0.5	-3.4	-2.1	0.9
333.81	-0.82	1.4	294.22	11JA	1.0	-3.4	-2.1	0.8
341.40	-0.70	1.5	304.34	21JA	2.8	-3.4	-2.1	0.8
348.92	-0.57	1.5	314.42	31JA	1.1	-3.4	-2.1	0.7
356.38	-0.45	1.5	324.45	10FE	0.5	-3.5	-2.0	0.7
3.76	-0.33	1.6	334.43	20FE	0.3	-3.5	-2.0	0.6
11.07	-0.21	1.6	344.35	2MR	0.0	-3.5	-1.9	0.6
18.30	-0.10	1.7	354.21	12MR	-0.4	-3.4	-1.9	0.5
25.45	0.01	1.7	4.01	22MR	-1.1	-3.4	-1.8	0.5
32.52	0.12	1.7	13.76	1AP	-1.7	-3.4	-1.7	0.6
39.51	0.22	1.8	23.45	11AP	-1.0	-3.4	-1.7	0.7
46.44	0.32	1.8	33.10	21AP	0.1	-3.3	-1.6	0.7
53.29	0.41	1.8	42.71	1MY	1.1	-3.3	-1.5	0.8
60.07	0.50	1.8	52.28	11MY	2.1	-3.3	-1.5	0.9
66.79	0.58	1.8	61.83	21MY	3.5	-3.3	-1.4	0.9
73.44	0.66	1.8	71.37	31MY	2.4	-3.3	-1.4	1.0
80.05	0.74	1.8	80.89	10JN	1.2	-3.4	-1.3	1.0
86.60	0.81	1.9	90.42	20JN	0.0	-3.4	-1.3	1.0
93.11	0.88	1.9	99.95	30JN	-1.1	-3.4	-1.3	1.1
99.58	0.94	1.9	109.51	10JL	-1.6	-3.4	-1.3	1.1
106.01	1.00	2.0	119.09	20JL	-1.0	-3.5	-1.3	1.1
112.40	1.06	2.0	128.71	30JL	-0.3	-3.5	-1.3	1.1
118.76	1.11	2.0	138.38	9AU	0.0	-3.6	-1.3	1.1
125.10	1.16	2.0	148.09	19AU	0.3	-3.7	-1.3	1.0
131.41	1.20	2.0	157.86	29AU	0.5	-3.8	-1.3	1.0
137.69	1.25	2.0	167.68	8SE	1.1	-3.9	-1.3	1.0
143.95	1.29	2.0	177.56	18SE	3.1	-4.0	-1.3	0.9
150.19	1.32	2.0	187.50	28SE	0.7	-4.1	-1.4	0.9
156.39	1.35	1.9	197.49	8OC	-0.6	-4.2	-1.4	0.9
162.57	1.38	1.9	207.54	18OC	-0.7	-4.3	-1.5	0.9
168.71	1.41	1.8	217.63	28OC	-0.7	-4.4	-1.5	0.9
174.82	1.43	1.8	227.77	7NO	-0.8	-4.4	-1.6	0.9
180.88	1.45	1.7	237.94	17NO	-0.7	-4.2	-1.7	0.9
186.88	1.46	1.6	248.12	27NO	-0.6	-3.6	-1.7	0.9
192.81	1.47	1.5	258.32	7DE	-0.7	-3.0	-1.8	0.9
198.67	1.47	1.4	268.53	17DE	-0.4	-3.8	-1.8	0.9
204.43	1.46	1.3	278.72	27DE	1.1	-4.2	-1.9	0.8

Left table — 628 / 629

♂ LONG	♂ LAT	♂ MAG	☉ LONG	16.00UT 628	☿	♀	♃	♄
210.07	1.44	1.2	288.89	6JA	2.3	-4.4	-2.0	0.8
215.57	1.41	1.0	299.04	16JA	0.7	-4.3	-2.0	0.8
220.89	1.36	0.8	309.14	26JA	0.3	-4.2	-2.0	0.7
226.00	1.30	0.7	319.20	5FE	0.2	-4.2	-2.0	0.7
230.84	1.21	0.5	329.21	15FE	-0.1	-4.0	-2.0	0.6
235.34	1.08	0.2	339.16	25FE	-0.5	-3.9	-2.0	0.6
239.42	0.92	-0.0	349.05	6MR	-1.1	-3.8	-2.0	0.5
242.97	0.72	-0.3	358.88	16MR	-1.6	-3.7	-1.9	0.5
245.84	0.45	-0.6	8.66	26MR	-0.9	-3.6	-1.9	0.4
247.88	0.11	-0.9	18.38	5AP	0.2	-3.6	-1.8	0.4
248.88	-0.32	-1.2	28.06	15AP	1.4	-3.5	-1.8	0.5
248.71	-0.84	-1.5	37.68	25AP	2.8	-3.5	-1.7	0.6
247.29	-1.44	-1.8	47.27	5MY	3.2	-3.4	-1.6	0.6
244.79	-2.08	-2.1	56.84	15MY	1.8	-3.4	-1.6	0.7
241.70	-2.70	-2.2	66.38	25MY	0.9	-3.3	-1.5	0.8
238.70	-3.21	-2.1	75.90	4JN	-0.0	-3.3	-1.5	0.8
236.49	-3.56	-1.9	85.43	14JN	-1.1	-3.3	-1.4	0.9
235.54	-3.77	-1.7	94.96	24JN	-1.7	-3.3	-1.4	0.9
235.96	-3.85	-1.5	104.50	4JL	-1.0	-3.3	-1.3	0.9
237.70	-3.84	-1.3	114.07	14JL	-0.3	-3.3	-1.3	1.0
240.59	-3.76	-1.0	123.67	24JL	0.1	-3.3	-1.3	1.0
244.42	-3.63	-0.8	133.31	3AU	0.4	-3.4	-1.2	1.0
249.03	-3.47	-0.7	143.00	13AU	0.7	-3.4	-1.2	1.0
254.26	-3.28	-0.5	152.73	23AU	1.5	-3.4	-1.2	1.0
259.98	-3.09	-0.3	162.52	2SE	3.2	-3.5	-1.2	0.9
266.09	-2.87	-0.2	172.37	12SE	0.6	-3.5	-1.2	0.9
272.52	-2.65	-0.0	182.28	22SE	-0.7	-3.5	-1.2	0.9
279.19	-2.43	0.1	192.24	2OC	-0.9	-3.5	-1.3	0.8
286.05	-2.20	0.3	202.26	12OC	-0.9	-3.4	-1.3	0.8
293.06	-1.97	0.4	212.33	22OC	-0.8	-3.4	-1.3	0.8
300.17	-1.75	0.5	222.45	1NO	-0.6	-3.4	-1.4	0.9
307.35	-1.53	0.6	232.60	11NO	-0.5	-3.4	-1.4	0.9
314.58	-1.31	0.8	242.78	21NO	-0.5	-3.3	-1.5	0.9
321.83	-1.11	0.9	252.97	1DE	-0.2	-3.3	-1.5	0.9
329.08	-0.91	1.0	263.18	11DE	1.3	-3.3	-1.6	0.9
336.32	-0.72	1.1	273.38	21DE	2.0	-3.3	-1.6	0.8
343.53	-0.55	1.2	283.56	31DE	0.5	-3.4	-1.7	0.8

629

♂ LONG	♂ LAT	♂ MAG	☉ LONG	16.00UT	☿	♀	♃	♄
350.71	-0.38	1.3	293.72	10JA	0.1	-3.4	-1.8	0.8
357.85	-0.23	1.4	303.85	20JA	0.0	-3.4	-1.8	0.7
4.93	-0.08	1.4	313.93	30JA	-0.2	-3.4	-1.9	0.7
11.96	0.05	1.5	323.97	9FE	-0.5	-3.4	-1.9	0.7
18.93	0.18	1.6	333.94	19FE	-1.1	-3.5	-2.0	0.6
25.85	0.29	1.7	343.87	1MR	-1.5	-3.5	-2.0	0.5
32.70	0.40	1.7	353.74	11MR	-0.9	-3.6	-2.0	0.5
39.50	0.50	1.8	3.54	21MR	0.3	-3.6	-2.0	0.4
46.24	0.59	1.8	13.29	31MR	1.8	-3.7	-2.0	0.4
52.93	0.67	1.9	22.99	10AP	3.7	-3.8	-2.0	0.3
59.56	0.74	1.9	32.64	20AP	2.4	-3.9	-1.9	0.4
66.15	0.81	1.9	42.25	30AP	1.4	-3.9	-1.9	0.4
72.70	0.87	1.9	51.83	10MY	0.7	-4.0	-1.8	0.5
79.20	0.92	2.0	61.38	20MY	-0.1	-4.1	-1.7	0.5
85.68	0.97	2.0	70.91	30MY	-1.1	-4.2	-1.7	0.6
92.12	1.01	2.0	80.44	9JN	-1.8	-4.2	-1.6	0.7
98.54	1.05	2.0	89.96	19JN	-1.0	-4.0	-1.5	0.7
104.94	1.08	2.0	99.50	29JN	-0.2	-3.6	-1.5	0.8
111.33	1.11	2.0	109.05	9JL	0.3	-3.0	-1.4	0.8
117.70	1.13	1.9	118.63	19JL	0.6	-3.6	-1.4	0.9
124.07	1.15	2.0	128.25	29JL	1.0	-4.0	-1.3	0.9
130.44	1.16	2.0	137.91	8AU	1.9	-4.2	-1.3	0.9
136.81	1.17	2.0	147.62	18AU	2.9	-4.2	-1.3	0.9
143.18	1.18	2.0	157.38	28AU	0.5	-4.2	-1.2	0.9
149.56	1.18	2.0	167.20	7SE	-0.8	-4.1	-1.2	0.9
155.95	1.17	2.0	177.07	17SE	-1.0	-4.0	-1.2	0.9
162.36	1.16	2.0	187.01	27SE	-1.0	-3.9	-1.2	0.8
168.77	1.15	2.0	197.00	7OC	-0.7	-3.9	-1.2	0.8
175.20	1.13	2.0	207.05	17OC	-0.4	-3.8	-1.2	0.7
181.65	1.10	1.9	217.14	27OC	-0.4	-3.7	-1.3	0.7
188.11	1.07	1.8	227.28	6NO	-0.3	-3.6	-1.3	0.8
194.58	1.03	1.8	237.44	16NO	-0.1	-3.6	-1.3	0.8
201.07	0.98	1.8	247.63	26NO	1.6	-3.5	-1.4	0.8
207.57	0.93	1.7	257.83	6DE	1.7	-3.5	-1.4	0.8
214.07	0.86	1.6	268.04	16DE	0.2	-3.5	-1.4	0.8
220.58	0.79	1.5	278.23	26DE	-0.0	-3.4	-1.5	0.8

Right table — 630 / 631

♂ LONG	♂ LAT	♂ MAG	☉ LONG	16.00UT 630	☿	♀	♃	♄
227.09	0.70	1.4	288.41	5JA	-0.1	-3.4	-1.6	0.8
233.59	0.59	1.3	298.55	15JA	-0.3	-3.4	-1.6	0.8
240.09	0.47	1.2	308.66	25JA	-0.6	-3.3	-1.7	0.7
246.57	0.33	1.1	318.72	4FE	-1.1	-3.3	-1.8	0.7
253.02	0.17	1.0	328.72	14FE	-1.4	-3.3	-1.8	0.7
259.44	-0.01	0.8	338.68	24FE	-0.9	-3.3	-1.9	0.6
265.81	-0.21	0.7	348.57	6MR	0.4	-3.3	-2.0	0.6
272.12	-0.45	0.5	358.41	16MR	2.2	-3.3	-2.0	0.5
278.34	-0.72	0.4	8.19	26MR	3.2	-3.4	-2.0	0.4
284.44	-1.02	0.2	17.91	5AP	1.8	-3.4	-2.0	0.4
290.38	-1.36	-0.0	27.59	15AP	1.0	-3.4	-2.1	0.3
296.11	-1.75	-0.2	37.22	25AP	0.5	-3.5	-2.0	0.3
301.57	-2.18	-0.4	46.81	5MY	-0.2	-3.5	-2.0	0.3
306.68	-2.66	-0.7	56.37	15MY	-1.1	-3.5	-2.0	0.4
311.31	-3.19	-0.9	65.92	25MY	-1.8	-3.4	-1.9	0.5
315.33	-3.77	-1.2	75.44	4JN	-1.0	-3.4	-1.9	0.5
318.57	-4.40	-1.4	84.96	14JN	-0.2	-3.4	-1.8	0.6
320.82	-5.06	-1.7	94.49	24JN	0.4	-3.3	-1.8	0.7
321.89	-5.71	-2.0	104.03	4JL	0.8	-3.3	-1.7	0.7
321.65	-6.29	-2.2	113.60	14JL	1.3	-3.3	-1.6	0.8
320.13	-6.69	-2.5	123.20	24JL	2.4	-3.3	-1.6	0.8
317.73	-6.82	-2.6	132.84	3AU	2.5	-3.3	-1.5	0.8
315.06	-6.63	-2.6	142.52	13AU	0.4	-3.3	-1.5	0.8
312.96	-6.14	-2.4	152.25	23AU	-0.9	-3.4	-1.4	0.9
311.80	-5.47	-2.1	162.04	2SE	-1.2	-3.4	-1.4	0.9
312.00	-4.72	-1.8	171.89	12SE	-1.1	-3.4	-1.3	0.8
313.47	-3.97	-1.5	181.79	22SE	-0.6	-3.5	-1.3	0.8
316.04	-3.28	-1.2	191.75	2OC	-0.3	-3.5	-1.3	0.8
319.51	-2.66	-0.9	201.77	12OC	-0.2	-3.6	-1.3	0.8
323.71	-2.11	-0.6	211.84	22OC	-0.2	-3.7	-1.3	0.7
328.45	-1.64	-0.4	221.95	1NO	0.1	-3.7	-1.3	0.7
333.62	-1.22	-0.1	232.11	11NO	1.8	-3.8	-1.3	0.7
339.11	-0.87	0.1	242.29	21NO	1.4	-3.9	-1.3	0.8
344.85	-0.56	0.3	252.48	1DE	0.0	-4.0	-1.3	0.8
350.78	-0.29	0.5	262.69	11DE	-0.2	-4.1	-1.4	0.8
356.84	-0.06	0.7	272.89	21DE	-0.2	-4.2	-1.4	0.8
3.01	0.14	0.8	283.07	31DE	-0.4	-4.3	-1.4	0.8

631

♂ LONG	♂ LAT	♂ MAG	☉ LONG	16.00UT	☿	♀	♃	♄
9.25	0.31	1.0	293.23	10JA	-0.7	-4.4	-1.5	0.8
15.54	0.46	1.1	303.36	20JA	-1.1	-4.3	-1.5	0.8
21.86	0.59	1.2	313.44	30JA	-1.2	-4.1	-1.6	0.8
28.20	0.70	1.4	323.48	9FE	-0.8	-3.6	-1.7	0.7
34.56	0.79	1.5	333.47	19FE	0.5	-3.3	-1.7	0.7
40.91	0.88	1.6	343.39	1MR	2.7	-3.8	-1.8	0.6
47.26	0.95	1.6	353.26	11MR	2.4	-4.2	-1.9	0.6
53.60	1.00	1.7	3.07	21MR	1.3	-4.3	-1.9	0.5
59.93	1.05	1.8	12.82	31MR	0.8	-4.2	-2.0	0.5
66.25	1.09	1.8	22.52	10AP	0.3	-4.1	-2.1	0.4
72.56	1.13	1.9	32.18	20AP	-0.2	-4.0	-2.1	0.3
78.86	1.15	1.9	41.79	30AP	-1.1	-3.9	-2.1	0.3
85.16	1.17	2.0	51.37	10MY	-1.9	-3.8	-2.1	0.3
91.45	1.19	2.0	60.92	20MY	-1.0	-3.7	-2.1	0.3
97.73	1.19	2.0	70.45	30MY	-0.1	-3.7	-2.1	0.4
104.02	1.20	2.0	79.98	9JN	0.5	-3.6	-2.1	0.5
110.32	1.19	2.0	89.50	19JN	1.0	-3.5	-2.1	0.5
116.62	1.19	2.0	99.03	29JN	1.7	-3.5	-2.0	0.6
122.94	1.18	2.0	108.59	9JL	3.0	-3.4	-2.0	0.6
129.28	1.16	2.0	118.17	19JL	2.1	-3.4	-1.9	0.7
135.64	1.14	2.0	127.78	29JL	-0.9	-3.4	-1.8	0.7
142.02	1.12	2.0	137.45	8AU	-0.9	-3.4	-1.8	0.8
148.43	1.09	1.9	147.15	18AU	-1.3	-3.4	-1.7	0.8
154.88	1.05	1.9	156.91	28AU	-1.0	-3.4	-1.7	0.8
161.36	1.01	1.9	166.73	7SE	-0.5	-3.4	-1.6	0.8
167.89	0.97	1.9	176.60	17SE	-0.2	-3.4	-1.5	0.8
174.45	0.92	1.9	186.53	27SE	-0.1	-3.4	-1.5	0.8
181.06	0.87	1.9	196.52	7OC	0.0	-3.4	-1.5	0.8
187.72	0.81	1.9	206.56	17OC	0.4	-3.4	-1.4	0.8
194.43	0.74	1.9	216.65	27OC	2.1	-3.4	-1.4	0.7
201.19	0.67	1.9	226.78	6NO	1.2	-3.4	-1.4	0.7
207.99	0.59	1.8	236.95	16NO	-0.1	-3.5	-1.4	0.7
214.85	0.51	1.8	247.13	26NO	-0.3	-3.5	-1.4	0.7
221.76	0.41	1.8	257.34	6DE	-0.4	-3.5	-1.4	0.7
228.72	0.31	1.7	267.54	16DE	-0.5	-3.5	-1.4	0.8
235.74	0.20	1.6	277.73	26DE	-0.7	-3.4	-1.4	0.8

Left half — year 632/633:

♂ LONG	♂ LAT	♂ MAG	☉ LONG	16.00UT 632	☿	♀	♃	♄
242.80	0.08	1.6	287.91	5JA	-1.0	-3.4	-1.4	0.8
249.90	-0.04	1.5	298.05	15JA	-1.1	-3.4	-1.5	0.8
257.06	-0.18	1.4	308.16	25JA	-0.7	-3.4	-1.5	0.8
264.25	-0.33	1.4	318.23	4FE	0.7	-3.3	-1.5	0.8
271.48	-0.48	1.3	328.24	14FE	3.1	-3.3	-1.6	0.7
278.75	-0.65	1.2	338.19	24FE	1.8	-3.3	-1.7	0.7
286.03	-0.82	1.1	348.10	5MR	0.9	-3.3	-1.7	0.7
293.34	-1.00	1.0	357.93	15MR	0.6	-3.3	-1.8	0.6
300.65	-1.18	0.9	7.72	25MR	0.2	-3.4	-1.9	0.6
307.95	-1.37	0.8	17.45	4AP	-0.3	-3.4	-1.9	0.5
315.23	-1.56	0.7	27.12	14AP	-1.1	-3.4	-2.0	0.4
322.47	-1.75	0.6	36.75	24AP	-1.8	-3.4	-2.1	0.4
329.64	-1.94	0.5	46.35	4MY	-1.0	-3.5	-2.1	0.3
336.72	-2.12	0.4	55.91	14MY	-0.0	-3.5	-2.2	0.3
343.68	-2.30	0.3	65.46	24MY	0.7	-3.6	-2.2	0.3
350.48	-2.46	0.2	74.99	3JN	1.4	-3.6	-2.3	0.3
357.07	-2.61	0.1	84.51	13JN	2.3	-3.7	-2.3	0.4
3.42	-2.75	-0.1	94.04	23JN	3.3	-3.8	-2.3	0.5
9.43	-2.88	-0.2	103.58	3JL	1.7	-3.9	-2.3	0.5
15.05	-2.98	-0.3	113.14	13JL	0.2	-4.0	-2.2	0.6
20.17	-3.06	-0.5	122.74	23JL	-1.0	-4.1	-2.2	0.6
24.67	-3.11	-0.7	132.38	2AU	-1.5	-4.2	-2.1	0.7
28.38	-3.12	-0.9	142.05	12AU	-1.0	-4.2	-2.1	0.7
31.13	-3.09	-1.1	151.79	22AU	-0.5	-4.2	-2.0	0.8
32.68	-3.00	-1.3	161.57	1SE	-0.1	-4.1	-1.9	0.8
32.87	-2.82	-1.5	171.41	11SE	0.0	-3.6	-1.9	0.8
31.57	-2.53	-1.7	181.32	21SE	0.2	-3.3	-1.8	0.8
28.95	-2.12	-1.9	191.27	10C	0.6	-3.8	-1.8	0.8
25.50	-1.61	-2.0	201.29	110C	2.5	-4.2	-1.7	0.8
22.00	-1.04	-1.9	211.35	210C	1.0	-4.4	-1.6	0.8
19.24	-0.48	-1.6	221.46	310C	-0.3	-4.3	-1.6	0.7
17.69	0.01	-1.3	231.61	10NO	-0.5	-4.3	-1.6	0.7
17.50	0.41	-0.9	241.79	20NO	-0.5	-4.2	-1.5	0.7
18.57	0.72	-0.6	251.98	30NO	-0.6	-4.1	-1.5	0.7
20.71	0.96	-0.3	262.19	10DE	-0.8	-4.0	-1.5	0.7
23.69	1.14	-0.0	272.39	20DE	-0.9	-3.9	-1.5	0.8
27.35	1.27	0.2	282.58	30DE	-0.9	-3.8	-1.5	0.8

633

♂ LONG	♂ LAT	♂ MAG	☉ LONG	16.00UT	☿	♀	♃	♄
31.53	1.37	0.5	292.74	9JA	-0.6	-3.7	-1.5	0.8
36.11	1.44	0.7	302.87	19JA	0.8	-3.6	-1.5	0.8
41.01	1.48	0.8	312.96	29JA	3.0	-3.6	-1.5	0.8
46.15	1.52	1.0	323.00	8FE	1.3	-3.5	-1.5	0.8
51.48	1.53	1.2	332.99	18FE	0.7	-3.5	-1.6	0.8
56.97	1.54	1.3	342.92	28FE	0.4	-3.4	-1.6	0.7
62.57	1.54	1.4	352.79	10MR	0.1	-3.4	-1.7	0.7
68.27	1.53	1.5	2.60	20MR	-0.4	-3.3	-1.7	0.7
74.06	1.52	1.6	12.36	30MR	-1.1	-3.3	-1.8	0.6
79.90	1.50	1.7	22.06	9AP	-1.8	-3.3	-1.8	0.5
85.81	1.47	1.8	31.72	19AP	-1.0	-3.3	-1.9	0.5
91.76	1.44	1.8	41.33	29AP	0.0	-3.3	-2.0	0.4
97.76	1.41	1.9	50.91	9MY	0.9	-3.3	-2.0	0.4
103.81	1.38	1.9	60.46	19MY	1.8	-3.3	-2.1	0.3
109.89	1.34	2.0	69.99	29MY	3.1	-3.4	-2.2	0.2
116.02	1.29	2.0	79.52	8JN	2.8	-3.4	-2.2	0.3
122.19	1.25	2.0	89.04	18JN	1.4	-3.4	-2.3	0.4
128.40	1.20	2.0	98.58	28JN	0.1	-3.5	-2.3	0.4
134.66	1.15	2.0	108.13	8JL	-1.0	-3.5	-2.4	0.5
140.97	1.09	2.0	117.71	18JL	-1.6	-3.5	-2.4	0.5
147.33	1.03	2.0	127.32	28JL	-1.0	-3.4	-2.4	0.6
153.74	0.97	2.0	136.97	7AU	-0.4	-3.4	-2.4	0.6
160.22	0.90	2.0	146.68	17AU	-0.0	-3.4	-2.3	0.7
166.75	0.83	1.9	156.43	27AU	0.2	-3.4	-2.3	0.7
173.35	0.76	1.9	166.24	6SE	0.4	-3.3	-2.2	0.8
180.01	0.68	1.8	176.12	16SE	0.9	-3.3	-2.2	0.8
186.74	0.60	1.8	186.04	26SE	2.8	-3.3	-2.1	0.8
193.55	0.52	1.8	196.03	60C	0.9	-3.3	-2.1	0.8
200.42	0.43	1.8	206.07	160C	-0.5	-3.3	-2.0	0.8
207.37	0.34	1.8	216.15	260C	-0.7	-3.4	-1.9	0.8
214.40	0.24	1.7	226.28	5NO	-0.6	-3.4	-1.9	0.8
221.49	0.14	1.7	236.45	15NO	-0.7	-3.4	-1.8	0.7
228.66	0.04	1.7	246.63	25NO	-0.7	-3.4	-1.7	0.7
235.91	-0.07	1.7	256.83	5DE	-0.7	-3.5	-1.7	0.7
243.22	-0.18	1.6	267.04	15DE	-0.8	-3.5	-1.6	0.7
250.60	-0.30	1.6	277.23	25DE	-0.5	-3.6	-1.6	0.8

Right half — year 634/635:

♂ LONG	♂ LAT	♂ MAG	☉ LONG	16.00UT 634	☿	♀	♃	♄
258.04	-0.41	1.6	287.41	4JA	1.0	-3.7	-1.6	0.8
265.54	-0.53	1.5	297.56	14JA	2.7	-3.7	-1.6	0.8
273.09	-0.65	1.5	307.67	24JA	1.0	-3.8	-1.6	0.8
280.69	-0.77	1.5	317.74	3FE	0.4	-3.9	-1.5	0.8
288.32	-0.88	1.4	327.76	13FE	0.2	-4.0	-1.6	0.8
295.98	-1.00	1.4	337.71	23FE	-0.0	-4.1	-1.6	0.8
303.66	-1.10	1.3	347.62	5MR	-0.4	-4.1	-1.6	0.8
311.35	-1.21	1.3	357.46	15MR	-1.1	-4.2	-1.6	0.8
319.03	-1.30	1.2	7.24	25MR	-1.7	-4.2	-1.6	0.7
326.69	-1.38	1.2	16.98	4AP	-1.0	-4.2	-1.7	0.7
334.31	-1.45	1.1	26.66	14AP	0.1	-3.8	-1.7	0.6
341.89	-1.51	1.1	36.29	24AP	1.2	-3.1	-1.8	0.5
349.41	-1.55	1.1	45.89	4MY	2.4	-3.2	-1.8	0.5
356.84	-1.58	1.0	55.45	14MY	3.6	-3.8	-1.9	0.4
4.18	-1.59	1.0	64.99	24MY	2.2	-4.1	-2.0	0.4
11.41	-1.59	0.9	74.52	3JN	1.1	-4.2	-2.0	0.3
18.50	-1.57	0.9	84.05	13JN	0.0	-4.2	-2.1	0.3
25.44	-1.53	0.8	93.57	23JN	-1.0	-4.1	-2.2	0.3
32.20	-1.47	0.8	103.12	3JL	-1.7	-4.0	-2.2	0.4
38.76	-1.39	0.7	112.68	13JL	-1.0	-3.9	-2.3	0.4
45.09	-1.29	0.6	122.27	23JL	-0.3	-3.8	-2.4	0.5
51.15	-1.18	0.5	131.91	2AU	0.1	-3.7	-2.4	0.6
56.89	-1.04	0.4	141.59	12AU	0.3	-3.7	-2.4	0.6
62.26	-0.88	0.3	151.31	22AU	0.6	-3.6	-2.5	0.7
67.17	-0.68	0.2	161.10	1SE	1.2	-3.6	-2.5	0.7
71.53	-0.46	0.1	170.94	11SE	3.2	-3.5	-2.4	0.7
75.23	-0.20	-0.1	180.83	21SE	0.7	-3.5	-2.4	0.8
78.09	0.10	-0.3	190.79	10C	-0.6	-3.4	-2.4	0.8
79.93	0.46	-0.5	200.80	110C	-0.8	-3.4	-2.3	0.8
80.56	0.86	-0.7	210.86	210C	-0.8	-3.4	-2.3	0.8
79.80	1.31	-0.9	220.98	310C	-0.8	-3.4	-2.2	0.8
77.62	1.78	-1.1	231.12	10NO	-0.6	-3.4	-2.1	0.8
74.29	2.23	-1.3	241.29	20NO	-0.6	-3.4	-2.0	0.8
70.37	2.60	-1.4	251.49	30NO	-0.6	-3.4	-2.0	0.7
66.66	2.84	-1.2	261.69	10DE	-0.4	-3.4	-1.9	0.7
63.85	2.96	-0.9	271.89	20DE	1.1	-3.4	-1.8	0.7
62.31	2.97	-0.6	282.08	30DE	2.3	-3.4	-1.8	0.8

635

♂ LONG	♂ LAT	♂ MAG	☉ LONG	16.00UT	☿	♀	♃	♄
62.11	2.91	-0.3	292.24	9JA	0.6	-3.4	-1.7	0.8
63.11	2.81	-0.0	302.37	19JA	0.3	-3.4	-1.7	0.8
65.11	2.69	0.2	312.47	29JA	0.1	-3.5	-1.6	0.8
67.93	2.57	0.5	322.51	8FE	-0.1	-3.5	-1.6	0.9
71.39	2.44	0.7	332.49	18FE	-0.5	-3.5	-1.6	0.9
75.36	2.32	0.9	342.43	28FE	-1.1	-3.5	-1.6	0.9
79.72	2.20	1.0	352.30	10MR	-1.6	-3.4	-1.6	0.8
84.41	2.09	1.2	2.12	20MR	-1.0	-3.4	-1.6	0.8
89.35	1.98	1.3	11.88	30MR	0.2	-3.4	-1.6	0.8
94.51	1.87	1.4	21.58	9AP	1.5	-3.4	-1.6	0.7
99.84	1.77	1.5	31.24	19AP	3.1	-3.3	-1.6	0.7
105.33	1.67	1.6	40.86	29AP	2.9	-3.3	-1.6	0.6
110.94	1.57	1.7	50.44	9MY	1.6	-3.3	-1.7	0.6
116.67	1.48	1.7	59.99	19MY	0.8	-3.3	-1.7	0.5
122.50	1.38	1.8	69.53	29MY	-0.0	-3.3	-1.8	0.4
128.43	1.29	1.8	79.05	8JN	-1.0	-3.4	-1.8	0.4
134.45	1.19	1.9	88.58	18JN	-1.8	-3.4	-1.9	0.3
140.56	1.10	1.9	98.11	28JN	-1.0	-3.4	-1.9	0.3
146.76	1.00	1.9	107.66	8JL	-0.3	-3.4	-2.0	0.4
153.04	0.91	1.9	117.24	18JL	0.2	-3.5	-2.1	0.4
159.41	0.81	1.9	126.85	28JL	0.5	-3.5	-2.1	0.5
165.87	0.71	1.9	136.50	7AU	0.8	-3.6	-2.2	0.5
172.41	0.62	1.9	146.21	17AU	1.6	-3.7	-2.3	0.6
179.04	0.52	1.9	155.96	27AU	3.2	-3.8	-2.3	0.6
185.69	0.42	1.8	165.77	6SE	0.6	-3.9	-2.4	0.7
192.58	0.31	1.8	175.64	16SE	-0.7	-4.0	-2.4	0.7
199.49	0.21	1.8	185.56	26SE	-1.0	-4.1	-2.4	0.8
206.48	0.11	1.7	195.54	60C	-0.9	-4.2	-2.4	0.8
213.57	0.00	1.7	205.58	160C	-0.7	-4.3	-2.4	0.8
220.76	-0.10	1.6	215.67	260C	-0.5	-4.4	-2.4	0.8
228.02	-0.20	1.6	225.79	5NO	-0.4	-4.4	-2.4	0.8
235.38	-0.31	1.5	235.96	15NO	-0.4	-4.2	-2.3	0.8
242.82	-0.41	1.5	246.14	25NO	-0.2	-3.6	-2.2	0.8
250.33	-0.51	1.5	256.34	5DE	1.3	-3.0	-2.2	0.8
257.92	-0.61	1.5	266.54	15DE	1.9	-3.9	-2.1	0.8
265.57	-0.70	1.5	276.74	25DE	0.4	-4.3	-2.0	0.7

636

♂ LONG	LAT	MAG	☉ LONG	16.00UT	☿	♀	♃	♄
273.28	-0.79	1.5	286.91	4JA	0.1	-4.4	-1.9	0.8
281.03	-0.87	1.5	297.07	14JA	-0.0	-4.3	-1.9	0.8
288.82	-0.95	1.4	307.18	24JA	-0.2	-4.3	-1.8	0.9
296.64	-1.02	1.4	317.24	3FE	-0.5	-4.1	-1.8	0.9
304.47	-1.08	1.4	327.26	13FE	-1.1	-4.0	-1.7	0.9
312.30	-1.12	1.4	337.23	23FE	-1.4	-3.9	-1.7	0.9
320.12	-1.16	1.4	347.13	4MR	-0.9	-3.8	-1.6	0.9
327.92	-1.18	1.4	356.98	14MR	0.3	-3.7	-1.6	0.9
335.68	-1.20	1.4	6.77	24MR	1.9	-3.6	-1.6	0.9
343.40	-1.20	1.4	16.50	3AP	3.7	-3.6	-1.5	0.9
351.06	-1.18	1.4	26.19	13AP	2.1	-3.5	-1.5	0.8
358.65	-1.15	1.4	35.83	23AP	1.2	-3.5	-1.5	0.8
6.16	-1.11	1.4	45.42	3MY	0.6	-3.4	-1.5	0.7
13.58	-1.06	1.4	54.99	13MY	-0.1	-3.4	-1.5	0.7
20.91	-0.99	1.4	64.54	23MY	-1.0	-3.3	-1.6	0.6
28.12	-0.92	1.3	74.06	2JN	-1.8	-3.3	-1.6	0.5
35.23	-0.83	1.3	83.59	12JN	-1.0	-3.3	-1.6	0.5
42.21	-0.73	1.3	93.12	22JN	-0.2	-3.3	-1.6	0.4
49.06	-0.62	1.3	102.66	2JL	0.3	-3.3	-1.7	0.3
55.77	-0.49	1.3	112.22	12JL	0.7	-3.3	-1.7	0.4
62.32	-0.36	1.2	121.81	22JL	0.3	-3.3	-1.8	0.4
68.72	-0.22	1.2	131.44	1AU	2.1	-3.4	-1.9	0.5
74.92	-0.06	1.2	141.12	11AU	2.9	-3.4	-1.9	0.5
80.92	0.11	1.1	150.84	21AU	0.5	-3.4	-2.0	0.6
86.70	0.29	1.0	160.62	31AU	-0.8	-3.5	-2.1	0.6
92.20	0.49	0.9	170.46	10SE	-1.1	-3.5	-2.1	0.7
97.39	0.70	0.8	180.35	20SE	-1.1	-3.5	-2.2	0.7
102.21	0.94	0.7	190.30	30SE	-0.7	-3.5	-2.2	0.8
106.57	1.21	0.6	200.31	10OC	-0.4	-3.4	-2.3	0.8
110.38	1.51	0.4	210.37	20OC	-0.3	-3.4	-2.3	0.8
113.52	1.84	0.3	220.48	30OC	-0.3	-3.4	-2.3	0.8
115.82	2.21	0.1	230.62	9NO	-0.0	-3.4	-2.3	0.9
117.09	2.62	-0.2	240.80	19NO	1.6	-3.3	-2.3	0.9
117.16	3.06	-0.4	250.99	29NO	1.6	-3.3	-2.3	0.9
115.88	3.50	-0.6	261.20	9DE	0.2	-3.3	-2.3	0.9
113.30	3.89	-0.8	271.39	19DE	-0.1	-3.3	-2.2	0.8
109.71	4.19	-1.0	281.58	29DE	-0.2	-3.4	-2.1	0.8

637

♂ LONG	LAT	MAG	☉ LONG	16.00UT	☿	♀	♃	♄
105.72	4.32	-1.0	291.75	8JA	-0.3	-3.4	-2.1	0.8
102.11	4.28	-0.8	301.88	18JA	-0.6	-3.4	-2.0	0.9
99.45	4.10	-0.6	311.97	28JA	-1.1	-3.4	-1.9	0.9
98.06	3.84	-0.3	322.02	7FE	-1.3	-3.4	-1.9	0.9
97.97	3.55	-0.1	332.01	17FE	-0.9	-3.5	-1.8	1.0
99.03	3.24	0.2	341.95	27FE	0.4	-3.5	-1.7	1.0
101.07	2.96	0.4	351.83	9MR	2.4	-3.6	-1.7	1.0
103.89	2.69	0.6	1.64	19MR	2.9	-3.6	-1.6	1.0
107.35	2.44	0.8	11.40	29MR	1.6	-3.7	-1.6	1.0
111.32	2.21	0.9	21.12	8AP	0.9	-3.8	-1.5	1.0
115.70	2.00	1.1	30.77	18AP	0.4	-3.9	-1.5	0.9
120.42	1.81	1.2	40.39	28AP	-0.2	-4.0	-1.5	0.9
125.43	1.63	1.3	49.98	8MY	-1.0	-4.0	-1.5	0.8
130.67	1.46	1.4	59.53	18MY	-1.9	-4.1	-1.4	0.8
136.13	1.30	1.5	69.07	28MY	-1.0	-4.2	-1.4	0.7
141.77	1.14	1.5	78.60	7JN	-0.2	-4.2	-1.4	0.7
147.57	1.00	1.6	88.12	17JN	0.4	-4.0	-1.4	0.6
153.52	0.86	1.6	97.65	27JN	0.9	-3.5	-1.5	0.5
159.62	0.72	1.6	107.20	7JL	1.5	-3.0	-1.5	0.5
165.86	0.59	1.7	116.78	17JL	2.6	-3.6	-1.5	0.4
172.22	0.46	1.7	126.39	27JL	2.4	-4.1	-1.5	0.5
178.71	0.33	1.7	136.04	6AU	0.4	-4.2	-1.6	0.5
185.32	0.21	1.7	145.73	16AU	-0.9	-4.2	-1.6	0.5
192.05	0.09	1.7	155.49	26AU	-1.2	-4.2	-1.7	0.6
198.90	-0.03	1.7	165.29	5SE	-1.1	-4.1	-1.7	0.6
205.86	-0.14	1.7	175.16	15SE	-0.6	-4.0	-1.8	0.7
212.93	-0.25	1.6	185.08	25SE	-0.3	-3.9	-1.9	0.7
220.12	-0.36	1.6	195.06	5OC	-0.2	-3.9	-1.9	0.8
227.41	-0.46	1.6	205.09	15OC	-0.1	-3.8	-2.0	0.8
234.80	-0.56	1.6	215.18	25OC	0.2	-3.7	-2.1	0.9
242.29	-0.65	1.5	225.30	4NO	1.8	-3.6	-2.1	0.9
249.86	-0.73	1.5	235.46	14NO	1.4	-3.6	-2.2	0.9
257.52	-0.81	1.5	245.65	24NO	-0.0	-3.5	-2.2	0.9
265.24	-0.88	1.4	255.85	4DE	-0.3	-3.5	-2.2	0.9
273.02	-0.94	1.4	266.05	14DE	-0.3	-3.4	-2.2	0.9
280.85	-0.99	1.4	276.25	24DE	-0.4	-3.4	-2.2	0.9

638

♂ LONG	LAT	MAG	☉ LONG	16.00UT	☿	♀	♃	♄
288.72	-1.03	1.3	286.43	3JA	-0.7	-3.4	-2.2	0.9
296.60	-1.06	1.3	296.58	13JA	-1.1	-3.4	-2.1	0.9
304.50	-1.08	1.3	306.69	23JA	-1.2	-3.3	-2.1	0.9
312.40	-1.09	1.3	316.76	2FE	-0.8	-3.3	-2.0	1.0
320.27	-1.08	1.3	326.78	12FE	0.5	-3.3	-2.0	1.0
328.12	-1.06	1.4	336.75	22FE	2.8	-3.3	-1.9	1.0
335.93	-1.03	1.4	346.66	4MR	2.2	-3.3	-1.8	1.1
343.68	-0.99	1.4	356.50	14MR	1.2	-3.3	-1.7	1.1
351.38	-0.94	1.4	6.30	24MR	0.7	-3.4	-1.7	1.1
359.00	-0.88	1.5	16.03	3AP	0.3	-3.4	-1.6	1.1
6.55	-0.81	1.5	25.71	13AP	-0.2	-3.4	-1.6	1.1
14.01	-0.74	1.5	35.36	23AP	-1.0	-3.5	-1.5	1.1
21.38	-0.65	1.5	44.95	3MY	-1.9	-3.5	-1.5	1.0
28.66	-0.56	1.6	54.52	13MY	-1.0	-3.5	-1.4	1.0
35.85	-0.46	1.6	64.07	23MY	-0.1	-3.4	-1.4	0.9
42.93	-0.35	1.6	73.60	2JN	0.6	-3.4	-1.4	0.9
49.92	-0.24	1.6	83.12	12JN	1.1	-3.4	-1.3	0.8
56.80	-0.13	1.6	92.65	22JN	1.9	-3.3	-1.3	0.7
63.58	-0.01	1.6	102.19	2JL	3.2	-3.3	-1.3	0.7
70.26	0.11	1.6	111.75	12JL	2.0	-3.3	-1.3	0.6
76.82	0.24	1.6	121.34	22JL	0.3	-3.3	-1.3	0.5
83.27	0.37	1.6	130.97	1AU	-0.9	-3.3	-1.4	0.5
89.61	0.51	1.6	140.64	11AU	-1.4	-3.3	-1.4	0.6
95.81	0.65	1.6	150.37	21AU	-1.1	-3.4	-1.4	0.6
101.88	0.79	1.5	160.14	31AU	-0.5	-3.4	-1.4	0.6
107.80	0.95	1.5	169.98	10SE	-0.2	-3.5	-1.5	0.7
113.55	1.11	1.4	179.87	20SE	-0.0	-3.5	-1.5	0.7
119.10	1.28	1.3	189.82	30SE	0.1	-3.5	-1.6	0.8
124.43	1.46	1.2	199.83	10OC	0.4	-3.6	-1.7	0.8
129.49	1.66	1.1	209.89	20OC	2.2	-3.7	-1.7	0.9
134.24	1.87	1.0	219.99	30OC	1.2	-3.7	-1.8	0.9
138.61	2.11	0.9	230.13	9NO	-0.2	-3.8	-1.8	1.0
142.56	2.36	0.7	240.31	19NO	-0.4	-3.9	-1.9	1.0
145.80	2.64	0.5	250.50	29NO	-0.4	-4.0	-2.0	1.0
148.39	2.95	0.3	260.70	9DE	-0.5	-4.1	-2.0	1.0
150.08	3.29	0.1	270.90	19DE	-0.7	-4.2	-2.1	1.0
150.70	3.64	-0.2	281.09	29DE	-1.0	-4.3	-2.1	1.1

639

♂ LONG	LAT	MAG	☉ LONG	16.00UT	☿	♀	♃	♄
150.09	3.98	-0.4	291.26	8JA	-1.0	-4.4	-2.1	1.1
148.16	4.27	-0.7	301.39	18JA	-0.7	-4.3	-2.1	1.0
145.07	4.47	-0.9	311.49	28JA	-0.1	-4.1	-2.1	1.0
141.26	4.50	-1.0	321.53	7FE	3.1	-3.6	-2.1	1.1
137.40	4.36	-0.9	331.53	17FE	1.6	-3.3	-2.0	1.1
134.21	4.07	-0.7	341.46	27FE	0.8	-3.8	-2.0	1.2
132.13	3.69	-0.5	351.35	9MR	0.5	-4.2	-1.9	1.2
131.35	3.28	-0.3	1.17	19MR	0.2	-4.1	-1.8	1.2
131.82	2.86	-0.0	10.93	29MR	-0.3	-4.2	-1.8	1.2
133.37	2.48	0.2	20.64	8AP	-1.0	-4.1	-1.7	1.2
135.84	2.13	0.4	30.31	18AP	-1.8	-4.0	-1.6	1.2
139.06	1.81	0.5	39.92	28AP	-1.0	-3.9	-1.6	1.2
142.88	1.52	0.7	49.51	8MY	-0.0	-3.8	-1.5	1.2
147.20	1.25	0.8	59.07	18MY	0.8	-3.7	-1.5	1.2
151.94	1.02	0.9	68.60	28MY	1.5	-3.7	-1.4	1.1
157.02	0.80	1.0	78.13	7JN	2.6	-3.6	-1.4	1.1
162.41	0.59	1.1	87.66	17JN	3.2	-3.5	-1.3	1.0
168.06	0.41	1.2	97.19	27JN	1.6	-3.5	-1.3	0.9
173.95	0.23	1.2	106.74	7JL	0.2	-3.4	-1.3	0.9
180.04	0.07	1.3	116.31	17JL	-1.0	-3.4	-1.3	0.8
186.33	-0.08	1.3	125.92	27JL	-1.5	-3.4	-1.2	0.7
192.80	-0.23	1.3	135.57	6AU	-0.4	-3.4	-1.2	0.7
199.44	-0.36	1.4	145.27	16AU	-0.4	-3.4	-1.3	0.7
206.23	-0.48	1.4	155.01	26AU	-0.1	-3.4	-1.3	0.7
213.17	-0.60	1.4	164.82	5SE	0.1	-3.4	-1.3	0.7
220.25	-0.70	1.4	174.68	15SE	0.3	-3.4	-1.3	0.8
227.47	-0.80	1.4	184.60	25SE	0.7	-3.4	-1.3	0.8
234.80	-0.88	1.4	194.58	5OC	2.5	-3.4	-1.4	0.9
242.25	-0.96	1.4	204.61	15OC	1.0	-3.4	-1.4	0.9
249.80	-1.02	1.4	214.69	25OC	-0.4	-3.4	-1.5	1.0
257.45	-1.07	1.4	224.81	4NO	-0.6	-3.4	-1.5	1.0
265.21	-1.11	1.4	234.97	14NO	-0.6	-3.5	-1.6	1.1
272.95	-1.13	1.4	245.15	24NO	-0.6	-3.5	-1.7	1.1
280.79	-1.15	1.4	255.35	4DE	-0.8	-3.5	-1.7	1.1
288.66	-1.15	1.4	265.55	14DE	-0.8	-3.5	-1.8	1.2
296.55	-1.14	1.4	275.75	24DE	-0.9	-3.4	-1.9	1.2

♂ LONG	♂ LAT	♂ MAG	☉ LONG	16.00UT	☿	♀	♃	♄
				640	MAGNITUDES			
304.45	-1.12	1.4	285.93	3JA	-0.6	-3.4	-1.9	1.2
312.34	-1.08	1.4	296.08	13JA	0.8	-3.4	-2.0	1.2
320.21	-1.04	1.4	306.20	23JA	2.9	-3.4	-2.0	1.2
328.05	-0.98	1.4	316.27	2FE	1.2	-3.3	-2.0	1.2
335.84	-0.92	1.4	326.30	12FE	0.6	-3.3	-2.0	1.2
343.57	-0.84	1.4	336.26	22FE	0.3	-3.3	-2.0	1.2
351.24	-0.76	1.4	346.18	3MR	0.1	-3.3	-2.0	1.3
358.84	-0.68	1.4	356.03	13MR	-0.4	-3.3	-2.0	1.3
6.36	-0.59	1.4	5.82	23MR	-1.1	-3.4	-1.9	1.4
13.79	-0.49	1.5	15.56	2AP	-1.7	-3.4	-1.9	1.4
21.14	-0.39	1.5	25.25	12AP	-1.0	-3.4	-1.8	1.4
28.40	-0.29	1.6	34.89	22AP	0.0	-3.4	-1.7	1.4
35.58	-0.19	1.6	44.49	2MY	1.0	-3.5	-1.7	1.4
42.66	-0.08	1.7	54.06	12MY	2.0	-3.5	-1.6	1.4
49.66	0.02	1.7	63.61	22MY	3.4	-3.6	-1.6	1.4
56.58	0.13	1.7	73.14	1JN	2.6	-3.6	-1.5	1.3
63.41	0.24	1.8	82.66	11JN	1.3	-3.7	-1.4	1.3
70.15	0.34	1.8	92.19	21JN	0.1	-3.8	-1.4	1.2
76.82	0.45	1.8	101.73	1JL	-1.0	-3.9	-1.3	1.2
83.41	0.55	1.8	111.29	11JL	-1.6	-4.0	-1.3	1.1
89.93	0.65	1.9	120.88	21JL	-1.0	-4.1	-1.3	1.0
96.36	0.76	1.9	130.51	31JL	-0.4	-4.2	-1.2	1.0
102.72	0.86	1.9	140.18	10AU	0.0	-4.2	-1.2	0.9
109.00	0.96	1.8	149.90	20AU	0.2	-4.2	-1.2	0.8
115.21	1.07	1.8	159.67	30AU	0.4	-4.0	-1.2	0.8
121.32	1.17	1.8	169.50	9SE	1.0	-3.6	-1.2	0.9
127.35	1.28	1.8	179.39	19SE	2.9	-3.3	-1.2	0.9
133.27	1.39	1.7	189.34	29SE	0.9	-3.9	-1.2	1.0
139.08	1.50	1.6	199.34	9OC	-0.5	-4.2	-1.3	1.0
144.76	1.62	1.6	209.40	19OC	-0.7	-4.3	-1.3	1.1
150.28	1.74	1.5	219.50	29OC	-0.7	-4.3	-1.3	1.1
155.64	1.87	1.4	229.64	8NO	-0.7	-4.3	-1.4	1.2
160.78	2.00	1.3	239.81	18NO	-0.7	-4.2	-1.4	1.2
165.66	2.15	1.1	250.01	28NO	-0.7	-4.1	-1.5	1.3
170.24	2.30	1.0	260.21	8DE	-0.7	-4.0	-1.5	1.3
174.44	2.46	0.8	270.41	18DE	-0.5	-3.9	-1.6	1.3
178.15	2.63	0.6	280.60	28DE	1.0	-3.8	-1.7	1.3
				641				
181.29	2.81	0.4	290.77	7JA	2.6	-3.7	-1.7	1.4
183.69	2.99	0.2	300.91	17JA	0.8	-3.6	-1.8	1.4
185.18	3.17	-0.1	311.00	27JA	0.4	-3.6	-1.9	1.4
185.59	3.33	-0.4	321.05	6FE	0.2	-3.5	-1.9	1.3
184.77	3.45	-0.7	331.05	16FE	-0.0	-3.5	-2.0	1.3
182.67	3.49	-0.9	340.99	26FE	-0.4	-3.4	-2.0	1.2
179.50	3.41	-1.2	350.87	8MR	-1.1	-3.4	-2.0	1.2
175.75	3.18	-1.3	0.70	18MR	-1.6	-3.3	-2.0	1.2
172.11	2.82	-1.2	10.47	28MR	-1.0	-3.3	-2.0	1.2
169.27	2.37	-1.0	20.18	7AP	0.1	-3.3	-2.0	1.2
167.62	1.90	-0.8	29.85	17AP	1.3	-3.3	-1.9	1.2
167.31	1.44	-0.6	39.47	27AP	2.6	-3.3	-1.9	1.2
168.25	1.02	-0.4	49.05	7MY	3.4	-3.3	-1.8	1.2
170.28	0.65	-0.2	58.61	17MY	2.0	-3.3	-1.8	1.1
173.23	0.32	0.0	68.15	27MY	1.0	-3.4	-1.7	1.1
176.94	0.04	0.2	77.67	6JN	0.0	-3.4	-1.7	1.1
181.26	-0.22	0.3	87.20	16JN	-1.0	-3.4	-1.6	1.0
186.11	-0.44	0.4	96.73	26JN	-1.7	-3.5	-1.5	1.0
191.39	-0.63	0.5	106.27	6JL	-1.0	-3.5	-1.5	0.9
197.04	-0.80	0.6	115.85	16JL	-0.3	-3.5	-1.4	0.9
203.02	-0.95	0.7	125.45	26JL	0.1	-3.4	-1.4	0.8
209.27	-1.07	0.8	135.10	5AU	0.4	-3.4	-1.3	0.8
215.77	-1.18	0.8	144.79	15AU	0.7	-3.4	-1.3	0.7
222.50	-1.27	0.9	154.54	25AU	1.3	-3.4	-1.3	0.7
229.42	-1.35	0.9	164.33	4SE	3.2	-3.3	-1.2	0.6
236.51	-1.40	1.0	174.20	14SE	0.8	-3.3	-1.2	0.7
243.76	-1.44	1.0	184.11	24SE	-0.6	-3.3	-1.2	0.8
251.14	-1.46	1.1	194.09	4OC	-0.9	-3.3	-1.2	0.8
258.64	-1.46	1.1	204.12	14OC	-0.8	-3.3	-1.2	0.9
266.24	-1.45	1.1	214.20	24OC	-0.8	-3.4	-1.2	1.0
273.91	-1.43	1.2	224.32	3NO	-0.6	-3.4	-1.3	1.0
281.65	-1.39	1.2	234.48	13NO	-0.5	-3.4	-1.3	1.1
289.42	-1.34	1.2	244.66	23NO	-0.5	-3.5	-1.3	1.1
297.22	-1.27	1.3	254.86	3DE	-0.3	-3.5	-1.4	1.2
305.03	-1.20	1.3	265.06	13DE	1.1	-3.5	-1.4	1.2
312.84	-1.11	1.4	275.26	23DE	2.2	-3.6	-1.5	1.2

♂ LONG	♂ LAT	♂ MAG	☉ LONG	16.00UT	☿	♀	♃	♄
				642	MAGNITUDES			
320.61	-1.02	1.4	285.44	2JA	0.6	-3.7	-1.5	1.2
328.36	-0.92	1.4	295.60	12JA	0.2	-3.7	-1.6	1.2
336.06	-0.82	1.5	305.71	22JA	0.0	-3.8	-1.7	1.1
343.69	-0.71	1.5	315.79	1FE	-0.1	-3.9	-1.7	1.1
351.27	-0.60	1.5	325.82	11FE	-0.5	-4.0	-1.8	1.1
358.77	-0.49	1.6	335.79	21FE	-1.1	-4.1	-1.9	1.0
6.19	-0.38	1.6	345.70	3MR	-1.5	-4.2	-1.9	1.0
13.54	-0.27	1.6	355.56	13MR	-1.0	-4.2	-2.0	0.9
20.80	-0.16	1.6	5.35	23MR	0.2	-4.2	-2.0	1.0
27.98	-0.05	1.7	15.09	2AP	1.6	-4.1	-2.0	1.0
35.08	0.06	1.7	24.78	12AP	3.4	-3.8	-2.1	1.0
42.10	0.16	1.7	34.43	22AP	2.6	-3.1	-2.1	0.9
49.04	0.26	1.7	44.03	2MY	1.5	-3.2	-2.0	0.9
55.91	0.35	1.7	53.60	12MY	0.7	-3.9	-2.0	0.9
62.70	0.45	1.7	63.15	22MY	-0.0	-4.1	-2.0	0.9
69.43	0.54	1.8	72.68	1JN	-1.0	-4.2	-1.9	0.9
76.10	0.62	1.8	82.20	11JN	-1.8	-4.2	-1.9	0.8
82.70	0.70	1.9	91.73	21JN	-1.0	-4.1	-1.8	0.8
89.25	0.78	1.9	101.26	1JL	-0.3	-4.0	-1.8	0.7
95.75	0.86	1.9	110.83	11JL	0.2	-3.9	-1.7	0.7
102.20	0.93	2.0	120.41	21JL	0.5	-3.8	-1.6	0.6
108.60	1.00	2.0	130.04	31JL	0.9	-3.7	-1.6	0.6
114.97	1.07	2.0	139.70	10AU	1.7	-3.7	-1.5	0.5
121.28	1.13	2.0	149.42	20AU	3.1	-3.6	-1.5	0.5
127.57	1.19	2.0	159.19	30AU	0.6	-3.6	-1.4	0.4
133.81	1.25	2.0	169.02	9SE	-0.7	-3.5	-1.4	0.4
140.01	1.31	1.9	178.91	19SE	-1.0	-3.5	-1.3	0.4
146.16	1.36	1.9	188.85	29SE	-1.0	-3.4	-1.3	0.4
152.27	1.42	1.9	198.86	9OC	-0.7	-3.4	-1.3	0.5
158.33	1.47	1.8	208.91	19OC	-0.5	-3.4	-1.3	0.6
164.33	1.52	1.8	219.01	29OC	-0.4	-3.4	-1.3	0.6
170.26	1.57	1.7	229.15	8NO	-0.4	-3.4	-1.3	0.7
176.11	1.61	1.6	239.32	18NO	-0.1	-3.4	-1.3	0.8
181.86	1.65	1.5	249.51	28NO	1.3	-3.4	-1.3	0.8
187.50	1.69	1.4	259.71	8DE	1.9	-3.4	-1.3	0.8
193.00	1.73	1.3	269.91	18DE	0.3	-3.4	-1.4	0.9
198.34	1.76	1.1	280.11	28DE	0.0	-3.4	-1.4	0.9
				643				
203.47	1.78	1.0	290.28	7JA	-0.1	-3.4	-1.5	0.9
208.36	1.79	0.8	300.41	17JA	-0.2	-3.4	-1.5	0.9
212.94	1.79	0.6	310.52	27JA	-0.5	-3.4	-1.6	0.9
217.14	1.78	0.4	320.57	6FE	-1.1	-3.5	-1.6	0.9
220.86	1.74	0.2	330.56	16FE	-1.4	-3.5	-1.7	0.8
223.97	1.67	-0.1	340.51	26FE	-0.9	-3.5	-1.8	0.8
226.33	1.56	-0.4	350.40	8MR	0.3	-3.4	-1.8	0.8
227.77	1.39	-0.7	0.22	18MR	2.0	-3.4	-1.9	0.7
228.10	1.15	-1.0	9.99	28MR	3.4	-3.4	-2.0	0.7
227.18	0.83	-1.3	19.71	7AP	1.9	-3.4	-2.0	0.7
225.05	0.41	-1.6	29.37	17AP	1.1	-3.3	-2.1	0.7
221.98	-0.09	-1.9	39.00	27AP	0.6	-3.3	-2.1	0.7
218.55	-0.62	-1.9	48.59	7MY	-0.1	-3.3	-2.1	0.7
215.49	-1.12	-1.7	58.14	17MY	-1.0	-3.3	-2.2	0.7
213.42	-1.56	-1.5	67.68	27MY	-1.8	-3.3	-2.2	0.7
212.69	-1.91	-1.3	77.21	6JN	-1.0	-3.4	-2.1	0.7
213.32	-2.17	-1.1	86.73	16JN	-0.2	-3.4	-2.1	0.6
215.21	-2.36	-0.9	96.26	26JN	0.3	-3.4	-2.1	0.6
218.18	-2.48	-0.7	105.81	6JL	0.7	-3.4	-2.0	0.6
222.04	-2.56	-0.5	115.38	16JL	1.2	-3.5	-2.0	0.5
226.63	-2.59	-0.4	124.99	26JL	2.3	-3.5	-1.9	0.5
231.83	-2.59	-0.2	134.63	5AU	2.7	-3.6	-1.8	0.4
237.50	-2.57	-0.1	144.32	15AU	0.5	-3.7	-1.8	0.4
243.59	-2.52	0.0	154.06	25AU	-0.8	-3.8	-1.7	0.3
250.00	-2.44	0.2	163.86	4SE	-1.2	-3.9	-1.7	0.2
256.69	-2.35	0.3	173.71	14SE	-1.1	-4.0	-1.6	0.2
263.59	-2.24	0.4	183.63	24SE	-0.6	-4.1	-1.5	0.1
270.68	-2.11	0.5	193.60	4OC	-0.4	-4.2	-1.5	0.1
277.90	-1.98	0.6	203.63	14OC	-0.2	-4.3	-1.5	0.2
285.23	-1.83	0.7	213.71	24OC	-0.2	-4.4	-1.4	0.3
292.64	-1.68	0.8	223.82	3NO	0.1	-4.3	-1.4	0.3
300.10	-1.52	0.9	233.98	13NO	1.6	-4.1	-1.4	0.4
307.60	-1.35	0.9	244.16	23NO	1.6	-3.6	-1.4	0.5
315.10	-1.19	1.0	254.36	3DE	0.1	-3.1	-1.4	0.5
322.60	-1.03	1.1	264.56	13DE	-0.2	-3.9	-1.4	0.6
330.07	-0.87	1.2	274.76	23DE	-0.2	-4.3	-1.4	0.6

♂ LONG	♂ LAT	♂ MAG	⊙ LONG	16.00UT	☿	♀	♃	♄
				644	MAGNITUDES			
337.51	-0.71	1.3	284.94	2JA	-0.3	-4.4	-1.4	0.6
344.90	-0.56	1.3	295.10	12JA	-0.6	-4.3	-1.4	0.7
352.24	-0.42	1.4	305.23	22JA	-1.1	-4.3	-1.5	0.7
359.51	-0.28	1.5	315.30	1FE	-1.3	-4.1	-1.5	0.7
6.73	-0.14	1.5	325.33	11FE	-0.9	-4.0	-1.6	0.7
13.87	-0.02	1.6	335.31	21FE	0.4	-3.9	-1.6	0.6
20.95	0.10	1.7	345.22	2MR	2.5	-3.8	-1.7	0.6
27.96	0.21	1.7	355.08	12MR	2.6	-3.7	-1.8	0.6
34.90	0.32	1.8	4.88	22MR	1.4	-3.6	-1.8	0.5
41.77	0.41	1.8	14.62	1AP	0.8	-3.6	-1.9	0.5
48.58	0.50	1.8	24.32	11AP	0.4	-3.5	-2.0	0.5
55.32	0.59	1.9	33.97	21AP	-0.2	-3.4	-2.0	0.5
62.01	0.67	1.9	43.57	1MY	-1.0	-3.4	-2.1	0.5
68.65	0.74	1.9	53.14	11MY	-1.9	-3.4	-2.2	0.5
75.24	0.80	1.9	62.69	21MY	-1.0	-3.3	-2.2	0.5
81.78	0.87	1.9	72.22	31MY	-0.2	-3.3	-2.2	0.5
88.28	0.92	1.9	81.75	10JN	0.5	-3.3	-2.3	0.5
94.75	0.97	1.9	91.27	20JN	1.0	-3.3	-2.3	0.5
101.19	1.02	1.9	100.81	30JN	1.6	-3.3	-2.3	0.5
107.61	1.06	1.9	110.37	10JL	2.9	-3.3	-2.3	0.4
114.00	1.10	2.0	119.95	20JL	2.3	-3.3	-2.2	0.4
120.38	1.13	2.0	129.57	30JL	0.4	-3.4	-2.2	0.4
126.74	1.16	2.0	139.24	9AU	-0.9	-3.4	-2.1	0.3
133.09	1.18	2.0	148.95	19AU	-1.3	-3.4	-2.1	0.3
139.44	1.20	2.0	158.72	29AU	-1.1	-3.5	-2.0	0.2
145.78	1.22	2.0	168.54	8SE	-0.6	-3.5	-1.9	0.1
152.12	1.23	2.0	178.43	18SE	-0.3	-3.5	-1.9	0.1
158.46	1.23	2.0	188.36	28SE	-0.1	-3.5	-1.8	0.0
164.80	1.24	2.0	198.36	8OC	-0.0	-3.4	-1.8	-0.0
171.14	1.24	1.9	208.41	18OC	0.3	-3.4	-1.7	-0.1
177.47	1.23	1.9	218.51	28OC	1.9	-3.4	-1.6	0.0
183.80	1.21	1.9	228.65	7NO	1.4	-3.4	-1.6	0.1
190.12	1.19	1.8	238.82	17NO	-0.1	-3.3	-1.6	0.1
196.43	1.17	1.7	249.00	27NO	-0.3	-3.3	-1.5	0.2
202.73	1.13	1.6	259.21	7DE	-0.3	-3.3	-1.5	0.3
209.01	1.08	1.6	269.41	17DE	-0.4	-3.3	-1.5	0.3
215.26	1.03	1.5	279.60	27DE	-0.7	-3.4	-1.5	0.4
				645				
221.48	0.96	1.3	289.78	6JA	-1.0	-3.4	-1.5	0.4
227.65	0.87	1.2	299.92	16JA	-1.1	-3.4	-1.5	0.5
233.77	0.77	1.1	310.02	26JA	-0.8	-3.4	-1.5	0.5
239.81	0.65	0.9	320.08	5FE	0.5	-3.5	-1.5	0.5
245.77	0.50	0.8	330.08	15FE	2.9	-3.5	-1.6	0.5
251.62	0.33	0.6	340.03	25FE	2.0	-3.5	-1.6	0.5
257.32	0.12	0.4	349.92	7MR	1.0	-3.6	-1.6	0.5
262.85	-0.12	0.2	359.75	17MR	0.6	-3.6	-1.7	0.5
268.16	-0.41	0.0	9.52	27MR	0.3	-3.7	-1.7	0.4
273.17	-0.75	-0.2	19.25	6AP	-0.2	-3.8	-1.8	0.4
277.81	-1.15	-0.4	28.91	16AP	-1.0	-3.9	-1.9	0.3
281.97	-1.63	-0.7	38.54	26AP	-1.9	-4.0	-1.9	0.3
285.52	-2.18	-1.0	48.13	6MY	-1.0	-4.0	-2.0	0.4
288.28	-2.82	-1.3	57.69	16MY	-0.1	-4.1	-2.1	0.4
290.07	-3.53	-1.6	67.23	26MY	0.6	-4.2	-2.1	0.4
290.70	-4.31	-1.9	76.76	5JN	1.3	-4.2	-2.2	0.4
290.09	-5.08	-2.2	86.28	15JN	2.1	-4.0	-2.3	0.4
288.33	-5.77	-2.5	95.81	25JN	3.3	-3.5	-2.3	0.4
285.81	-6.25	-2.6	105.35	5JL	1.9	-3.0	-2.4	0.4
283.21	-6.45	-2.5	114.92	15JL	0.3	-3.6	-2.4	0.4
281.24	-6.35	-2.3	124.52	25JL	-0.9	-4.1	-2.4	0.3
280.40	-6.02	-2.1	134.17	4AU	-1.4	-4.2	-2.4	0.3
280.92	-5.54	-1.8	143.85	14AU	-1.1	-4.2	-2.4	0.3
282.71	-5.00	-1.5	153.59	24AU	-0.5	-4.2	-2.4	0.2
285.61	-4.44	-1.3	163.39	3SE	-0.2	-4.1	-2.3	0.1
289.44	-3.90	-1.0	173.24	13SE	0.0	-4.0	-2.2	0.1
293.98	-3.38	-0.8	183.15	23SE	0.1	-3.9	-2.2	0.0
299.09	-2.91	-0.6	193.12	3OC	0.5	-3.8	-2.1	-0.1
304.63	-2.47	-0.3	203.14	13OC	2.2	-3.8	-2.0	-0.1
310.50	-2.07	-0.1	213.21	23OC	1.2	-3.7	-2.0	-0.2
316.63	-1.71	0.1	223.33	2NO	-0.3	-3.6	-1.9	-0.2
322.94	-1.37	0.2	233.48	12NO	-0.5	-3.6	-1.8	-0.1
329.39	-1.08	0.4	243.66	22NO	-0.5	-3.5	-1.8	-0.1
335.94	-0.81	0.6	253.86	2DE	-0.6	-3.5	-1.7	0.0
342.56	-0.56	0.7	264.06	12DE	-0.8	-3.4	-1.7	0.1
349.21	-0.35	0.9	274.26	22DE	-0.9	-3.4	-1.6	0.2

♂ LONG	♂ LAT	♂ MAG	⊙ LONG	16.00UT	☿	♀	♃	♄
				646	MAGNITUDES			
355.90	-0.15	1.0	284.45	1JA	-0.9	-3.4	-1.6	0.2
2.59	0.02	1.1	294.61	11JA	-0.7	-3.4	-1.6	0.3
9.27	0.18	1.2	304.73	21JA	0.7	-3.3	-1.6	0.3
15.94	0.32	1.3	314.81	31JA	3.1	-3.3	-1.6	0.3
22.59	0.44	1.4	324.84	10FE	1.5	-3.3	-1.6	0.4
29.22	0.55	1.5	334.82	20FE	0.7	-3.3	-1.6	0.4
35.81	0.65	1.6	344.74	2MR	0.4	-3.3	-1.6	0.4
42.38	0.74	1.7	354.60	12MR	0.1	-3.3	-1.6	0.4
48.91	0.81	1.8	4.41	22MR	-0.3	-3.4	-1.6	0.4
55.41	0.88	1.8	14.16	1AP	-1.0	-3.4	-1.6	0.3
61.88	0.94	1.9	23.85	11AP	-1.8	-3.4	-1.7	0.3
68.32	0.99	1.9	33.50	21AP	-1.0	-3.5	-1.7	0.3
74.73	1.03	1.9	43.11	1MY	-0.0	-3.5	-1.8	0.2
81.13	1.06	2.0	52.68	11MY	0.8	-3.5	-1.8	0.2
87.50	1.09	2.0	62.23	21MY	1.7	-3.4	-1.9	0.3
93.86	1.12	2.0	71.76	31MY	2.9	-3.4	-2.0	0.3
100.21	1.14	2.0	81.28	10JN	3.0	-3.4	-2.0	0.3
106.55	1.15	2.0	90.81	20JN	1.5	-3.3	-2.1	0.3
112.90	1.16	2.0	100.35	30JN	0.2	-3.3	-2.2	0.3
119.24	1.16	2.0	109.90	10JL	-0.9	-3.3	-2.2	0.3
125.59	1.16	2.0	119.49	20JL	-1.5	-3.3	-2.3	0.3
131.95	1.15	2.0	129.11	30JL	-1.1	-3.3	-2.4	0.3
138.33	1.14	1.9	138.77	9AU	-0.4	-3.3	-2.4	0.3
144.73	1.13	2.0	148.48	19AU	-0.1	-3.4	-2.4	0.2
151.14	1.11	2.0	158.24	29AU	0.2	-3.4	-2.5	0.2
157.59	1.08	2.0	168.06	8SE	0.3	-3.4	-2.5	0.1
164.06	1.05	2.0	177.95	18SE	0.8	-3.5	-2.4	0.1
170.57	1.02	2.0	187.88	28SE	2.6	-3.5	-2.4	0.0
177.11	0.98	2.0	197.88	8OC	1.0	-3.6	-2.4	-0.1
183.68	0.93	1.9	207.93	18OC	-0.4	-3.7	-2.3	-0.1
190.29	0.88	1.9	218.02	28OC	-0.6	-3.7	-2.2	-0.2
196.94	0.82	1.9	228.15	7NO	-0.6	-3.8	-2.2	-0.3
203.63	0.76	1.8	238.32	17NO	-0.7	-3.9	-2.1	-0.3
210.36	0.68	1.8	248.51	27NO	-0.8	-4.0	-2.0	-0.2
217.12	0.60	1.7	258.71	7DE	-0.8	-4.1	-2.0	-0.1
223.92	0.51	1.7	268.92	17DE	-0.8	-4.2	-1.9	-0.1
230.76	0.41	1.6	279.11	27DE	-0.6	-4.3	-1.8	0.0
				647				
237.63	0.29	1.5	289.28	6JA	0.8	-4.4	-1.8	0.1
244.54	0.17	1.5	299.43	16JA	2.9	-4.3	-1.7	0.1
251.47	0.03	1.4	309.53	26JA	1.1	-4.1	-1.7	0.2
258.44	-0.12	1.3	319.58	5FE	0.5	-3.6	-1.6	0.2
265.43	-0.28	1.2	329.59	15FE	0.3	-3.3	-1.6	0.3
272.44	-0.46	1.1	339.54	25FE	0.0	-3.9	-1.6	0.3
279.46	-0.65	1.0	349.43	7MR	-0.4	-4.2	-1.6	0.3
286.48	-0.86	0.8	359.27	17MR	-1.0	-4.3	-1.6	0.3
293.49	-1.08	0.7	9.05	27MR	-1.7	-4.2	-1.6	0.3
300.48	-1.32	0.6	18.77	6AP	-1.0	-4.1	-1.6	0.3
307.42	-1.56	0.5	28.44	16AP	0.0	-4.0	-1.6	0.3
314.30	-1.82	0.3	38.07	26AP	1.1	-3.9	-1.6	0.3
321.09	-2.09	0.2	47.66	6MY	2.2	-3.8	-1.6	0.2
327.73	-2.37	0.1	57.23	16MY	3.6	-3.7	-1.7	0.2
334.20	-2.65	-0.1	66.77	26MY	2.3	-3.7	-1.7	0.2
340.44	-2.94	-0.3	76.29	5JN	1.2	-3.6	-1.8	0.2
346.36	-3.23	-0.4	85.82	15JN	0.1	-3.5	-1.8	0.3
351.89	-3.51	-0.6	95.35	25JN	-1.0	-3.5	-1.9	0.3
356.91	-3.79	-0.8	104.89	5JL	-1.6	-3.4	-1.9	0.3
1.28	-4.06	-1.0	114.46	15JL	-1.1	-3.4	-2.0	0.3
4.82	-4.31	-1.2	124.06	25JL	-0.4	-3.4	-2.1	0.3
7.33	-4.52	-1.4	133.70	4AU	0.0	-3.4	-2.1	0.3
8.59	-4.66	-1.7	143.39	14AU	0.3	-3.4	-2.2	0.3
8.45	-4.68	-1.9	153.12	24AU	0.5	-3.3	-2.3	0.3
6.87	-4.55	-2.1	162.91	3SE	1.1	-3.3	-2.3	0.2
4.14	-4.22	-2.3	172.77	13SE	3.0	-3.4	-2.4	0.2
0.91	-3.70	-2.3	182.67	23SE	0.9	-3.4	-2.4	0.1
357.96	-3.03	-2.1	192.64	3OC	-0.5	-3.4	-2.4	0.1
355.97	-2.33	-1.8	202.66	13OC	-0.8	-3.4	-2.4	-0.0
355.30	-1.66	-1.4	212.73	23OC	-0.8	-3.4	-2.4	-0.1
355.93	-1.08	-1.1	222.84	2NO	-0.8	-3.4	-2.4	-0.2
357.73	-0.59	-0.8	232.99	12NO	-0.7	-3.5	-2.3	-0.2
0.50	-0.19	-0.5	243.17	22NO	-0.6	-3.5	-2.3	-0.3
4.03	0.14	-0.2	253.37	2DE	-0.6	-3.5	-2.2	-0.3
8.16	0.40	0.1	263.57	12DE	-0.4	-3.5	-2.1	-0.2
12.75	0.61	0.3	273.77	22DE	1.0	-3.4	-2.1	-0.1

♂ LONG	♂ LAT	♂ MAG	☉ LONG	16.00UT 648	☿	♀	♃	♄
					MAGNITUDES			
17.68	0.78	0.5	283.95	1JA	2.5	-3.4	-2.0	-0.1
22.90	0.92	0.7	294.11	11JA	0.7	-3.4	-1.9	-0.0
28.32	1.03	0.9	304.23	21JA	0.3	-3.4	-1.9	0.1
33.91	1.11	1.0	314.32	31JA	0.1	-3.3	-1.8	0.1
39.62	1.18	1.2	324.35	10FE	-0.1	-3.3	-1.7	0.2
45.42	1.24	1.3	334.33	20FE	-0.4	-3.3	-1.7	0.2
51.30	1.28	1.4	344.25	1MR	-1.0	-3.3	-1.6	0.3
57.25	1.31	1.5	354.12	11MR	-1.6	-3.4	-1.6	0.3
63.23	1.33	1.6	3.92	21MR	-1.0	-3.4	-1.6	0.3
69.25	1.34	1.7	13.68	31MR	0.1	-3.4	-1.5	0.3
75.31	1.35	1.8	23.38	10AP	1.4	-3.4	-1.5	0.3
81.39	1.34	1.8	33.03	20AP	2.9	-3.4	-1.5	0.3
87.49	1.34	1.9	42.64	30AP	3.1	-3.5	-1.5	0.3
93.61	1.33	1.9	52.22	10MY	1.8	-3.5	-1.5	0.3
99.76	1.31	2.0	61.76	20MY	0.9	-3.6	-1.5	0.2
105.93	1.29	2.0	71.30	30MY	0.0	-3.6	-1.5	0.2
112.12	1.26	2.0	80.83	9JN	-1.0	-3.7	-1.6	0.2
118.34	1.24	2.0	90.35	19JN	-1.7	-3.8	-1.6	0.2
124.60	1.20	2.0	99.89	29JN	-1.0	-3.9	-1.6	0.3
130.88	1.17	2.0	109.44	9JL	-0.3	-4.0	-1.7	0.3
137.20	1.12	2.0	119.02	19JL	0.1	-4.1	-1.7	0.3
143.56	1.08	2.0	128.64	29JL	0.4	-4.2	-1.8	0.3
149.97	1.03	2.0	138.30	8AU	0.7	-4.2	-1.9	0.4
156.42	0.98	2.0	148.01	18AU	1.5	-4.2	-1.9	0.3
162.92	0.92	1.9	157.78	28AU	3.2	-4.0	-2.0	0.3
169.48	0.86	1.9	167.59	7SE	0.8	-3.5	-2.1	0.3
176.09	0.80	1.8	177.47	17SE	-0.6	-3.3	-2.1	0.3
182.77	0.73	1.8	187.41	27SE	-0.9	-3.9	-2.2	0.2
189.50	0.65	1.8	197.40	7OC	-0.9	-4.2	-2.2	0.2
196.30	0.58	1.8	207.44	17OC	-0.8	-4.3	-2.3	0.1
203.16	0.49	1.8	217.53	27OC	-0.5	-4.3	-2.3	0.0
210.09	0.40	1.8	227.67	6NO	-0.5	-4.3	-2.3	-0.0
217.08	0.31	1.8	237.83	16NO	-0.5	-4.2	-2.3	-0.1
224.15	0.21	1.7	248.02	26NO	-0.3	-4.1	-2.3	-0.2
231.27	0.10	1.7	258.22	6DE	1.1	-4.0	-2.3	-0.3
238.46	-0.01	1.7	268.42	16DE	2.1	-3.9	-2.2	-0.2
245.72	-0.13	1.6	278.62	26DE	0.5	-3.8	-2.2	-0.2

♂ LONG	♂ LAT	♂ MAG	☉ LONG	16.00UT 649	☿	♀	♃	♄
253.03	-0.25	1.6	288.79	5JA	0.1	-3.7	-2.1	-0.1
260.40	-0.37	1.5	298.94	15JA	-0.0	-3.6	-2.1	-0.0
267.83	-0.50	1.5	309.04	25JA	-0.2	-3.6	-2.0	0.0
275.29	-0.63	1.4	319.10	4FE	-0.5	-3.5	-1.9	0.1
282.81	-0.77	1.4	329.11	14FE	-1.1	-3.5	-1.8	0.2
290.35	-0.90	1.3	339.07	24FE	-1.5	-3.4	-1.8	0.2
297.91	-1.04	1.3	348.96	6MR	-1.0	-3.4	-1.7	0.3
305.50	-1.17	1.2	358.80	16MR	0.2	-3.3	-1.6	0.3
313.08	-1.30	1.1	8.58	26MR	1.7	-3.3	-1.6	0.3
320.64	-1.42	1.1	18.30	5AP	3.7	-3.3	-1.5	0.3
328.19	-1.53	1.0	27.98	15AP	2.3	-3.3	-1.5	0.3
335.69	-1.63	0.9	37.61	25AP	1.3	-3.3	-1.5	0.3
343.13	-1.72	0.9	47.20	5MY	0.7	-3.3	-1.5	0.3
350.49	-1.80	0.8	56.76	15MY	-0.0	-3.3	-1.4	0.3
357.74	-1.86	0.7	66.31	25MY	-1.0	-3.4	-1.4	0.3
4.88	-1.90	0.7	75.83	4JN	-1.8	-3.4	-1.4	0.3
11.86	-1.92	0.6	85.36	14JN	-1.1	-3.4	-1.4	0.3
18.66	-1.93	0.5	94.89	24JN	-0.3	-3.5	-1.4	0.3
25.26	-1.91	0.4	104.43	4JL	0.2	-3.5	-1.4	0.3
31.60	-1.88	0.4	114.00	14JL	0.6	-3.5	-1.5	0.4
37.64	-1.82	0.3	123.60	24JL	1.0	-3.4	-1.5	0.4
43.32	-1.74	0.1	133.23	3AU	1.9	-3.4	-1.5	0.4
48.58	-1.63	0.0	142.91	13AU	3.0	-3.4	-1.6	0.4
53.30	-1.49	-0.1	152.65	23AU	0.7	-3.4	-1.6	0.4
57.38	-1.31	-0.3	162.43	2SE	-0.7	-3.3	-1.7	0.4
60.66	-1.09	-0.5	172.28	12SE	-1.1	-3.3	-1.7	0.4
62.95	-0.82	-0.7	182.19	22SE	-1.0	-3.3	-1.8	0.4
64.06	-0.50	-0.9	192.15	2OC	-0.7	-3.3	-1.9	0.4
63.79	-0.11	-1.1	202.17	12OC	-0.4	-3.3	-1.9	0.3
62.06	0.34	-1.3	212.24	22OC	-0.3	-3.4	-2.0	0.3
59.06	0.82	-1.5	222.35	1NO	-0.3	-3.4	-2.1	0.2
55.29	1.28	-1.6	232.50	11NO	-0.1	-3.4	-2.1	0.1
51.55	1.67	-1.4	242.68	21NO	1.4	-3.5	-2.2	0.0
48.58	1.95	-1.2	252.87	1DE	1.8	-3.5	-2.2	-0.0
46.83	2.13	-0.8	263.07	11DE	0.2	-3.5	-2.2	-0.1
46.44	2.23	-0.5	273.28	21DE	-0.1	-3.6	-2.2	-0.2
47.31	2.26	-0.2	283.46	31DE	-0.1	-3.7	-2.2	-0.1

♂ LONG	♂ LAT	♂ MAG	☉ LONG	16.00UT 650	☿	♀	♃	♄
					MAGNITUDES			
49.22	2.26	0.0	293.62	10JA	-0.3	-3.7	-2.2	-0.1
51.99	2.22	0.3	303.75	20JA	-0.6	-3.8	-2.1	-0.0
55.42	2.18	0.5	313.83	30JA	-1.1	-3.9	-2.1	0.0
59.38	2.12	0.7	323.87	9FE	-1.3	-4.0	-2.0	0.1
63.76	2.06	0.9	333.85	19FE	-0.9	-4.1	-1.9	0.2
68.46	2.00	1.1	343.78	1MR	0.3	-4.2	-1.9	0.2
73.41	1.93	1.2	353.65	11MR	2.2	-4.2	-1.8	0.3
78.58	1.86	1.4	3.46	21MR	3.1	-4.2	-1.7	0.3
83.92	1.79	1.5	13.21	31MR	1.7	-4.1	-1.6	0.4
89.39	1.72	1.6	22.91	10AP	1.0	-3.8	-1.6	0.4
94.99	1.65	1.6	32.56	20AP	0.5	-3.1	-1.5	0.4
100.69	1.58	1.7	42.17	30AP	-0.1	-3.2	-1.5	0.4
106.48	1.51	1.8	51.75	10MY	-1.0	-3.9	-1.4	0.4
112.35	1.44	1.8	61.30	20MY	-1.8	-4.1	-1.4	0.4
118.29	1.37	1.9	70.83	30MY	-1.1	-4.2	-1.4	0.4
124.31	1.29	1.9	80.36	9JN	-0.2	-4.2	-1.3	0.4
130.40	1.22	1.9	89.88	19JN	0.4	-4.1	-1.3	0.4
136.56	1.14	2.0	99.42	29JN	0.8	-4.0	-1.3	0.4
142.79	1.06	2.0	108.98	9JL	1.3	-3.9	-1.3	0.4
149.09	0.98	2.0	118.56	19JL	2.5	-3.8	-1.3	0.4
155.46	0.90	2.0	128.17	29JL	2.6	-3.7	-1.3	0.5
161.90	0.82	1.9	137.83	8AU	0.5	-3.7	-1.3	0.5
168.42	0.73	1.9	147.54	18AU	-0.8	-3.6	-1.4	0.5
175.02	0.64	1.9	157.30	28AU	-1.2	-3.5	-1.4	0.6
181.70	0.55	1.9	167.12	7SE	-1.1	-3.5	-1.4	0.6
188.46	0.46	1.8	176.99	17SE	-0.6	-3.5	-1.5	0.6
195.31	0.36	1.8	186.92	27SE	-0.3	-3.4	-1.5	0.6
202.24	0.27	1.7	196.91	7OC	-0.2	-3.4	-1.6	0.5
209.25	0.17	1.7	206.95	17OC	-0.1	-3.4	-1.7	0.5
216.35	0.06	1.6	217.04	27OC	0.1	-3.4	-1.7	0.4
223.54	-0.04	1.6	227.18	6NO	1.6	-3.4	-1.8	0.4
230.80	-0.14	1.6	237.34	16NO	1.6	-3.4	-1.8	0.3
238.15	-0.25	1.6	247.53	26NO	0.0	-3.4	-1.9	0.3
245.57	-0.36	1.6	257.73	6DE	-0.2	-3.4	-2.0	0.2
253.06	-0.46	1.6	267.93	16DE	-0.3	-3.4	-2.0	0.1
260.62	-0.57	1.5	278.12	26DE	-0.4	-3.4	-2.1	0.0

♂ LONG	♂ LAT	♂ MAG	☉ LONG	16.00UT 651	☿	♀	♃	♄
268.25	-0.67	1.5	288.30	5JA	-0.6	-3.4	-2.1	-0.0
275.92	-0.77	1.5	298.44	15JA	-1.1	-3.4	-2.1	0.0
283.64	-0.86	1.5	308.55	25JA	-1.2	-3.4	-2.1	0.1
291.39	-0.95	1.5	318.62	4FE	-0.9	-3.5	-2.1	0.1
299.17	-1.03	1.4	328.62	14FE	0.4	-3.5	-2.0	0.2
306.95	-1.10	1.4	338.58	24FE	2.6	-3.5	-2.0	0.2
314.74	-1.17	1.4	348.48	6MR	2.4	-3.4	-1.9	0.3
322.52	-1.22	1.4	358.32	16MR	1.3	-3.4	-1.9	0.4
330.27	-1.26	1.3	8.10	26MR	0.7	-3.4	-1.8	0.4
337.99	-1.29	1.3	17.83	5AP	0.3	-3.4	-1.7	0.5
345.65	-1.30	1.3	27.50	15AP	-0.2	-3.3	-1.7	0.5
353.26	-1.30	1.3	37.14	25AP	-1.0	-3.3	-1.6	0.5
0.79	-1.29	1.3	46.73	5MY	-1.9	-3.3	-1.5	0.6
8.23	-1.26	1.2	56.29	15MY	-1.1	-3.3	-1.5	0.6
15.57	-1.21	1.2	65.84	25MY	-0.2	-3.3	-1.4	0.6
22.81	-1.16	1.2	75.37	4JN	0.5	-3.4	-1.4	0.6
29.92	-1.08	1.2	84.89	14JN	1.1	-3.4	-1.3	0.6
36.90	-1.00	1.1	94.42	24JN	1.8	-3.4	-1.3	0.6
43.73	-0.90	1.1	103.96	4JL	3.1	-3.4	-1.3	0.6
50.40	-0.78	1.1	113.53	14JL	2.2	-3.5	-1.3	0.6
56.90	-0.66	1.0	123.13	24JL	0.4	-3.5	-1.2	0.6
63.19	-0.51	1.0	132.76	3AU	-0.9	-3.6	-1.2	0.6
69.27	-0.36	0.9	142.44	13AU	-1.4	-3.7	-1.2	0.7
75.08	-0.18	0.8	152.18	23AU	-1.1	-3.8	-1.2	0.7
80.60	0.01	0.7	161.96	2SE	-0.5	-3.9	-1.3	0.7
85.77	0.23	0.6	171.80	12SE	-0.2	-4.0	-1.3	0.8
90.52	0.47	0.5	181.71	22SE	-0.1	-4.1	-1.3	0.8
94.75	0.74	0.4	191.67	2OC	0.0	-4.2	-1.3	0.8
98.38	1.04	0.2	201.68	12OC	0.3	-4.3	-1.4	0.7
101.23	1.39	0.0	211.75	22OC	1.9	-4.4	-1.4	0.7
103.13	1.77	-0.2	221.86	1NO	1.4	-4.3	-1.5	0.7
103.91	2.20	-0.4	232.00	11NO	-0.1	-4.1	-1.5	0.6
103.37	2.66	-0.6	242.18	21NO	-0.4	-3.5	-1.6	0.6
101.45	3.12	-0.8	252.38	1DE	-0.4	-3.1	-1.7	0.5
98.31	3.52	-1.0	262.58	11DE	-0.5	-3.9	-1.7	0.5
94.42	3.80	-1.1	272.78	21DE	-0.7	-4.3	-1.8	0.4
90.54	3.93	-1.0	282.97	31DE	-1.0	-4.4	-1.9	0.3

♂ LONG	♂ LAT	♂ MAG	☉ LONG	16.00UT 652	☿	♀	♃	♄
					MAGNITUDES			
87.37	3.90	-0.8	293.13	10JA	-1.0	-4.3	-1.9	0.2
85.37	3.76	-0.5	303.26	20JA	-0.8	-4.3	-2.0	0.2
84.69	3.55	-0.2	313.34	30JA	0.5	-4.1	-2.0	0.2
85.26	3.32	0.0	323.38	9FE	3.0	-4.0	-2.0	0.3
86.88	3.08	0.3	333.37	19FE	1.8	-3.9	-2.0	0.3
89.37	2.85	0.5	343.30	29FE	0.9	-3.8	-2.0	0.4
92.56	2.63	0.7	353.17	10MR	0.5	-3.7	-2.0	0.4
96.31	2.43	0.9	2.98	20MR	0.2	-3.6	-2.0	0.5
100.51	2.24	1.0	12.74	30MR	-0.3	-3.6	-1.9	0.5
105.06	2.07	1.2	22.44	9AP	-1.0	-3.5	-1.9	0.6
109.90	1.91	1.3	32.10	19AP	-1.8	-3.4	-1.8	0.6
115.00	1.75	1.4	41.71	29AP	-1.1	-3.4	-1.7	0.7
120.30	1.61	1.5	51.29	9MY	-0.1	-3.4	-1.7	0.7
125.78	1.47	1.6	60.84	19MY	0.7	-3.3	-1.6	0.7
131.42	1.33	1.6	70.38	29MY	1.4	-3.3	-1.5	0.8
137.20	1.20	1.7	79.90	8JN	2.4	-3.3	-1.5	0.8
143.11	1.08	1.7	89.43	18JN	3.3	-3.3	-1.4	0.8
149.15	0.96	1.7	98.96	28JN	1.8	-3.3	-1.4	0.8
155.30	0.83	1.8	108.51	8JL	0.3	-3.3	-1.3	0.8
161.57	0.72	1.8	118.10	18JL	-0.9	-3.3	-1.3	0.8
167.95	0.60	1.8	127.71	28JL	-1.5	-3.4	-1.3	0.8
174.43	0.48	1.8	137.36	7AU	-1.1	-3.4	-1.2	0.8
181.03	0.37	1.8	147.07	17AU	-0.5	-3.4	-1.2	0.9
187.73	0.26	1.8	156.82	27AU	-0.1	-3.5	-1.2	0.9
194.54	0.14	1.7	166.64	6SE	0.1	-3.5	-1.2	1.0
201.45	0.03	1.7	176.51	16SE	0.2	-3.5	-1.2	1.0
208.47	-0.07	1.7	186.44	26SE	0.6	-3.5	-1.2	1.0
215.59	-0.18	1.7	196.42	6OC	2.3	-3.4	-1.2	1.0
222.81	-0.29	1.6	206.47	16OC	1.2	-3.4	-1.3	1.0
230.13	-0.39	1.6	216.55	26OC	-0.3	-3.4	-1.3	1.0
237.54	-0.49	1.5	226.68	5NO	-0.5	-3.4	-1.3	1.0
245.05	-0.58	1.5	236.84	15NO	-0.5	-3.3	-1.4	0.9
252.63	-0.67	1.5	247.03	25NO	-0.6	-3.3	-1.4	0.9
260.28	-0.75	1.4	257.23	5DE	-0.8	-3.3	-1.5	0.8
268.00	-0.83	1.4	267.43	15DE	-0.8	-3.3	-1.6	0.8
275.78	-0.90	1.4	277.63	25DE	-0.9	-3.4	-1.6	0.7
				653				
283.60	-0.96	1.4	287.80	4JA	-0.7	-3.4	-1.7	0.6
291.45	-1.01	1.4	297.95	14JA	0.6	-3.4	-1.7	0.6
299.32	-1.05	1.4	308.06	24JA	3.0	-3.4	-1.8	0.5
307.20	-1.08	1.4	318.13	3FE	1.3	-3.5	-1.9	0.4
315.08	-1.10	1.4	328.14	13FE	0.6	-3.5	-1.9	0.5
322.93	-1.11	1.4	338.10	23FE	0.4	-3.5	-2.0	0.5
330.76	-1.10	1.4	348.00	5MR	0.1	-3.6	-2.0	0.6
338.54	-1.08	1.4	357.85	15MR	-0.3	-3.6	-2.0	0.6
346.28	-1.05	1.4	7.63	25MR	-1.0	-3.7	-2.0	0.7
353.95	-1.01	1.4	17.36	4AP	-1.8	-3.8	-2.0	0.7
1.54	-0.96	1.5	27.04	14AP	-1.1	-3.9	-2.0	0.8
9.06	-0.90	1.5	36.67	24AP	-0.0	-4.0	-1.9	0.8
16.49	-0.82	1.5	46.27	4MY	0.9	-4.0	-1.9	0.9
23.83	-0.74	1.5	55.84	14MY	1.8	-4.1	-1.8	0.9
31.07	-0.65	1.5	65.38	24MY	3.2	-4.2	-1.8	1.0
38.21	-0.55	1.5	74.91	3JN	2.8	-4.2	-1.7	1.0
45.24	-0.44	1.5	84.43	13JN	1.4	-4.0	-1.6	1.0
52.16	-0.33	1.5	93.96	23JN	0.2	-3.5	-1.6	1.0
58.97	-0.21	1.5	103.50	3JL	-0.9	-3.0	-1.5	1.1
65.66	-0.08	1.5	113.07	13JL	-1.6	-3.6	-1.5	1.1
72.22	0.05	1.5	122.66	23JL	-1.1	-4.1	-1.4	1.1
78.66	0.19	1.5	132.29	2AU	-0.4	-4.2	-1.4	1.1
84.95	0.33	1.4	141.97	12AU	-0.0	-4.2	-1.3	1.1
91.08	0.49	1.4	151.70	22AU	0.2	-4.2	-1.3	1.1
97.05	0.65	1.4	161.48	1SE	0.4	-4.1	-1.3	1.2
102.83	0.82	1.3	171.32	11SE	0.9	-4.0	-1.2	1.2
108.39	1.01	1.2	181.22	21SE	2.6	-3.9	-1.2	1.2
113.70	1.21	1.1	191.18	1OC	1.1	-3.8	-1.2	1.3
118.71	1.42	1.0	201.19	11OC	-0.4	-3.8	-1.2	1.3
123.37	1.66	0.9	211.26	21OC	-0.7	-3.7	-1.2	1.3
127.60	1.92	0.7	221.37	31OC	-0.7	-3.6	-1.2	1.3
131.29	2.20	0.6	231.51	10NO	-0.7	-3.6	-1.3	1.3
134.33	2.52	0.4	241.69	20NO	-0.7	-3.5	-1.3	1.3
136.56	2.87	0.2	251.89	30NO	-0.7	-3.5	-1.3	1.2
137.80	3.25	-0.0	262.09	10DE	-0.7	-3.4	-1.4	1.2
137.88	3.64	-0.3	272.29	20DE	-0.5	-3.4	-1.4	1.1
136.64	4.02	-0.5	282.48	30DE	0.8	-3.4	-1.5	1.1

♂ LONG	♂ LAT	♂ MAG	☉ LONG	16.00UT 654	☿	♀	♃	♄
					MAGNITUDES			
134.11	4.34	-0.8	292.64	9JA	2.8	-3.4	-1.6	1.0
130.59	4.54	-0.9	302.77	19JA	0.9	-3.3	-1.6	0.9
126.63	4.57	-1.0	312.86	29JA	0.4	-3.3	-1.7	0.9
122.99	4.42	-0.8	322.90	8FE	0.2	-3.3	-1.8	0.8
120.26	4.14	-0.6	332.89	18FE	-0.0	-3.3	-1.8	0.7
118.78	3.78	-0.3	342.82	28FE	-0.4	-3.3	-1.9	0.8
118.59	3.40	-0.1	352.69	10MR	-1.0	-3.3	-1.9	0.8
119.57	3.03	0.1	2.51	20MR	-1.7	-3.4	-2.0	0.9
121.54	2.68	0.3	12.27	30MR	-1.1	-3.4	-2.0	0.9
124.33	2.36	0.5	21.97	9AP	0.0	-3.4	-2.1	1.0
127.78	2.07	0.7	31.63	19AP	1.2	-3.5	-2.1	1.0
131.78	1.80	0.8	41.25	29AP	2.4	-3.5	-2.1	1.1
136.21	1.56	1.0	50.82	9MY	3.6	-3.5	-2.0	1.1
141.01	1.34	1.1	60.38	19MY	2.1	-3.4	-2.0	1.2
146.12	1.13	1.2	69.91	29MY	1.1	-3.4	-2.0	1.2
151.49	0.94	1.2	79.44	8JN	0.1	-3.4	-1.9	1.3
157.10	0.76	1.3	88.96	18JN	-0.9	-3.3	-1.9	1.3
162.92	0.59	1.4	98.50	28JN	-1.7	-3.3	-1.8	1.3
168.93	0.44	1.4	108.05	8JL	-1.1	-3.3	-1.7	1.3
175.11	0.28	1.4	117.62	18JL	-0.4	-3.3	-1.7	1.4
181.46	0.14	1.5	127.24	28JL	0.1	-3.3	-1.6	1.4
187.96	0.00	1.5	136.89	7AU	0.3	-3.3	-1.6	1.4
194.60	-0.13	1.5	146.59	17AU	0.6	-3.4	-1.5	1.3
201.39	-0.25	1.5	156.35	27AU	1.2	-3.4	-1.5	1.3
208.32	-0.37	1.5	166.15	6SE	3.0	-3.4	-1.4	1.3
215.37	-0.48	1.5	176.02	16SE	0.9	-3.5	-1.4	1.3
222.55	-0.59	1.5	185.95	26SE	-0.6	-3.5	-1.3	1.3
229.84	-0.68	1.5	195.93	6OC	-0.8	-3.6	-1.3	1.3
237.25	-0.77	1.5	205.97	16OC	-0.8	-3.7	-1.3	1.2
244.75	-0.85	1.5	216.06	26OC	-0.8	-3.7	-1.3	1.2
252.35	-0.92	1.5	226.19	5NO	-0.6	-3.8	-1.3	1.2
260.03	-0.98	1.4	236.35	15NO	-0.5	-3.9	-1.3	1.2
267.78	-1.02	1.4	246.54	25NO	-0.6	-4.0	-1.3	1.1
275.59	-1.06	1.4	256.73	5DE	-0.4	-4.1	-1.3	1.1
283.44	-1.09	1.4	266.94	15DE	1.0	-4.2	-1.4	1.1
291.33	-1.10	1.4	277.14	25DE	2.4	-4.3	-1.4	1.0
				655				
299.23	-1.10	1.4	287.31	4JA	0.6	-4.4	-1.4	1.0
307.14	-1.09	1.4	297.46	14JA	0.2	-4.3	-1.5	0.9
315.03	-1.07	1.3	307.58	24JA	0.1	-4.1	-1.5	0.9
322.90	-1.04	1.3	317.64	3FE	-0.1	-3.6	-1.6	0.8
330.74	-0.99	1.3	327.66	13FE	-0.4	-3.4	-1.7	0.8
338.52	-0.94	1.3	337.62	23FE	-1.0	-3.9	-1.7	0.7
346.25	-0.88	1.4	347.52	5MR	-1.5	-4.2	-1.8	0.7
353.92	-0.81	1.4	357.37	15MR	-1.0	-4.3	-1.9	0.8
1.51	-0.73	1.4	7.16	25MR	0.1	-4.2	-1.9	0.9
9.02	-0.64	1.5	16.89	4AP	1.5	-4.1	-2.0	1.0
16.44	-0.55	1.5	26.57	14AP	3.2	-4.0	-2.1	1.0
23.78	-0.45	1.6	36.21	24AP	2.8	-3.9	-2.1	1.1
31.03	-0.35	1.6	45.81	4MY	1.6	-3.8	-2.1	1.1
38.19	-0.25	1.6	55.38	14MY	0.8	-3.7	-2.2	1.2
45.26	-0.14	1.7	64.92	24MY	0.0	-3.7	-2.2	1.2
52.23	-0.04	1.7	74.45	3JN	-0.9	-3.6	-2.2	1.3
59.12	0.07	1.7	83.97	13JN	-1.7	-3.5	-2.2	1.3
65.91	0.18	1.7	93.50	23JN	-1.1	-3.5	-2.1	1.3
72.62	0.30	1.8	103.04	3JL	-0.3	-3.4	-2.1	1.3
79.24	0.41	1.8	112.61	13JL	0.2	-3.4	-2.0	1.3
85.77	0.52	1.8	122.20	23JL	0.5	-3.4	-2.0	1.3
92.21	0.64	1.8	131.83	2AU	0.8	-3.4	-1.9	1.3
98.56	0.76	1.8	141.51	12AU	1.6	-3.4	-1.8	1.2
104.81	0.87	1.7	151.23	22AU	3.2	-3.3	-1.8	1.2
110.96	1.00	1.7	161.01	1SE	0.8	-3.3	-1.7	1.1
117.00	1.12	1.7	170.85	11SE	-0.7	-3.4	-1.6	1.1
122.92	1.25	1.6	180.74	21SE	-1.0	-3.4	-1.6	1.1
128.71	1.38	1.6	190.70	1OC	-1.0	-3.4	-1.5	1.1
134.35	1.52	1.5	200.71	11OC	-0.7	-3.4	-1.5	1.1
139.82	1.67	1.4	210.77	21OC	-0.5	-3.4	-1.5	1.1
145.08	1.82	1.3	220.87	31OC	-0.4	-3.4	-1.4	1.1
150.09	1.99	1.2	231.02	10NO	-0.4	-3.5	-1.4	1.1
154.80	2.17	1.1	241.19	20NO	-0.2	-3.5	-1.4	1.0
159.16	2.36	0.9	251.38	30NO	1.2	-3.5	-1.4	1.0
163.07	2.57	0.7	261.59	10DE	2.1	-3.5	-1.4	1.0
166.43	2.79	0.5	271.79	20DE	0.4	-3.4	-1.4	0.9
169.10	3.03	0.3	281.98	30DE	0.0	-3.4	-1.4	0.9

♂ LONG	LAT	MAG	☉ LONG	16.00UT 656	☿	♀	♃	♄
170.92	3.29	0.1	292.15	9JA	-0.1	-3.4	-1.4	0.9
171.72	3.54	-0.2	302.28	19JA	-0.2	-3.4	-1.5	0.8
171.31	3.78	-0.5	312.37	29JA	-0.5	-3.3	-1.5	0.8
169.62	3.96	-0.7	322.42	8FE	-1.0	-3.3	-1.5	0.7
166.74	4.03	-1.0	332.40	18FE	-1.4	-3.3	-1.6	0.7
163.06	3.96	-1.1	342.34	28FE	-1.0	-3.3	-1.7	0.6
159.23	3.72	-1.1	352.22	9MR	0.2	-3.4	-1.7	0.6
155.96	3.36	-0.9	2.03	19MR	1.9	-3.4	-1.8	0.6
153.77	2.92	-0.7	11.79	29MR	3.6	-3.4	-1.9	0.7
152.87	2.46	-0.5	21.51	8AP	2.1	-3.4	-1.9	0.7
153.26	2.02	-0.3	31.16	18AP	1.2	-3.4	-2.0	0.8
154.77	1.62	-0.1	40.78	28AP	0.6	-3.5	-2.1	0.9
157.26	1.25	0.1	50.37	8MY	-0.0	-3.5	-2.1	0.9
160.54	0.93	0.3	59.92	18MY	-1.0	-3.6	-2.2	1.0
164.46	0.64	0.4	69.45	28MY	-1.8	-3.6	-2.2	1.0
168.93	0.38	0.6	78.98	7JN	-1.1	-3.7	-2.3	1.1
173.85	0.15	0.7	88.50	17JN	-0.3	-3.8	-2.3	1.1
179.15	-0.06	0.8	98.04	27JN	0.3	-3.9	-2.3	1.1
184.78	-0.25	0.8	107.59	7JL	0.7	-4.0	-2.3	1.1
190.70	-0.42	0.9	117.16	17JL	1.1	-4.1	-2.3	1.1
196.87	-0.57	1.0	126.78	27JL	2.1	-4.2	-2.2	1.1
203.27	-0.71	1.0	136.43	6AU	2.9	-4.2	-2.2	1.1
209.88	-0.83	1.1	146.12	16AU	0.7	-4.2	-2.1	1.1
216.67	-0.94	1.1	155.88	26AU	-0.7	-4.0	-2.1	1.1
223.64	-1.04	1.1	165.68	5SE	-1.1	-3.5	-2.0	1.0
230.77	-1.12	1.2	175.55	15SE	-1.1	-3.3	-1.9	1.0
238.04	-1.18	1.2	185.47	25SE	-0.7	-3.9	-1.9	0.9
245.44	-1.23	1.2	195.45	5OC	-0.4	-4.2	-1.8	1.0
252.94	-1.27	1.2	205.48	15OC	-0.3	-4.3	-1.8	1.0
260.55	-1.29	1.3	215.57	25OC	-0.2	-4.3	-1.7	1.0
268.25	-1.30	1.3	225.69	4NO	-0.0	-4.3	-1.7	1.0
276.00	-1.29	1.3	235.85	14NO	1.4	-4.2	-1.6	1.0
283.81	-1.27	1.3	246.04	24NO	1.8	-4.1	-1.6	0.9
291.65	-1.24	1.3	256.24	4DE	0.2	-4.0	-1.5	0.9
299.51	-1.19	1.4	266.44	14DE	-0.1	-3.9	-1.5	0.9
307.37	-1.14	1.4	276.64	24DE	-0.2	-3.8	-1.5	0.9

♂ LONG	LAT	MAG	☉ LONG	16.00UT 657	☿	♀	♃	♄
315.21	-1.07	1.4	286.82	3JA	-0.3	-3.7	-1.5	0.8
323.05	-1.00	1.4	296.97	13JA	-0.6	-3.6	-1.5	0.8
330.81	-0.91	1.4	307.09	23JA	-1.1	-3.6	-1.5	0.7
338.54	-0.82	1.5	317.16	2FE	-1.3	-3.5	-1.5	0.7
346.20	-0.73	1.5	327.18	12FE	-0.9	-3.5	-1.5	0.6
353.81	-0.63	1.5	337.14	22FE	0.3	-3.4	-1.6	0.6
1.33	-0.53	1.5	347.05	4MR	2.3	-3.4	-1.6	0.5
8.77	-0.42	1.6	356.90	14MR	2.8	-3.3	-1.7	0.5
16.14	-0.32	1.6	6.69	24MR	1.5	-3.3	-1.7	0.5
23.42	-0.21	1.6	16.43	3AP	0.9	-3.3	-1.8	0.5
30.61	-0.11	1.6	26.11	13AP	0.4	-3.3	-1.8	0.6
37.72	-0.00	1.6	35.75	23AP	-0.1	-3.3	-1.9	0.6
44.75	0.10	1.6	45.35	3MY	-1.0	-3.3	-2.0	0.7
51.69	0.20	1.7	54.92	13MY	-1.8	-3.3	-2.0	0.8
58.56	0.30	1.7	64.46	23MY	-1.1	-3.4	-2.1	0.8
65.35	0.40	1.8	73.99	2JN	-0.2	-3.4	-2.2	0.9
72.07	0.49	1.8	83.51	12JN	0.4	-3.4	-2.2	0.9
78.73	0.58	1.9	93.04	22JN	0.9	-3.5	-2.3	1.0
85.32	0.67	1.9	102.58	2JL	1.5	-3.5	-2.3	1.0
91.85	0.76	1.9	112.14	12JL	2.7	-3.5	-2.4	1.0
98.32	0.85	1.9	121.73	22JL	2.5	-3.4	-2.4	1.0
104.73	0.93	1.9	131.36	1AU	0.6	-3.4	-2.4	1.0
111.08	1.01	1.9	141.03	11AU	-0.8	-3.4	-2.4	1.0
117.39	1.09	1.9	150.76	21AU	-1.3	-3.4	-2.4	1.0
123.63	1.17	1.9	160.53	31AU	-1.1	-3.3	-2.4	1.0
129.83	1.24	1.9	170.37	10SE	-0.6	-3.3	-2.3	0.9
135.96	1.32	1.9	180.26	20SE	-0.3	-3.3	-2.3	0.9
142.03	1.40	1.8	190.21	30SE	-0.1	-3.3	-2.2	0.8
148.03	1.47	1.8	200.21	10OC	-0.1	-3.3	-2.1	0.8
153.94	1.55	1.7	210.28	20OC	0.2	-3.4	-2.0	0.9
159.77	1.62	1.7	220.38	30OC	1.7	-3.4	-2.0	0.9
165.50	1.70	1.6	230.52	9NO	1.6	-3.4	-1.9	0.9
171.09	1.77	1.5	240.69	19NO	-0.0	-3.5	-1.8	0.9
176.54	1.85	1.4	250.89	29NO	-0.3	-3.5	-1.8	0.9
181.82	1.93	1.3	261.09	9DE	-0.3	-3.5	-1.7	0.9
186.88	2.00	1.1	271.29	19DE	-0.4	-3.6	-1.7	0.8
191.67	2.08	0.9	281.48	29DE	-0.7	-3.7	-1.7	0.8

♂ LONG	LAT	MAG	☉ LONG	16.00UT 658	☿	♀	♃	♄
196.14	2.16	0.8	291.65	8JA	-1.0	-3.7	-1.6	0.8
200.21	2.23	0.6	301.79	18JA	-1.1	-3.8	-1.6	0.7
203.78	2.29	0.3	311.88	28JA	-0.9	-3.9	-1.6	0.7
206.72	2.34	0.1	321.93	7FE	0.4	-4.0	-1.6	0.7
208.87	2.37	-0.2	331.92	17FE	2.7	-4.1	-1.6	0.6
210.07	2.36	-0.5	341.86	27FE	2.2	-4.2	-1.6	0.5
210.13	2.30	-0.8	351.74	9MR	1.1	-4.2	-1.6	0.5
208.93	2.16	-1.1	1.56	19MR	0.7	-4.3	-1.6	0.4
206.53	1.92	-1.4	11.32	29MR	0.3	-4.1	-1.6	0.4
203.21	1.57	-1.6	21.04	8AP	-0.2	-3.8	-1.7	0.4
199.58	1.14	-1.6	30.70	18AP	-1.0	-3.1	-1.7	0.4
196.39	0.65	-1.4	40.32	28AP	-1.9	-3.3	-1.8	0.5
194.21	0.18	-1.2	49.90	8MY	-1.1	-3.9	-1.8	0.5
193.35	-0.25	-1.0	59.46	18MY	-0.2	-4.1	-1.9	0.6
193.84	-0.63	-0.8	68.99	28MY	0.6	-4.2	-1.9	0.7
195.55	-0.94	-0.6	78.52	7JN	1.2	-4.1	-2.0	0.7
198.32	-1.20	-0.3	88.04	17JN	2.0	-4.1	-2.1	0.8
201.95	-1.40	-0.3	97.57	27JN	3.2	-4.0	-2.1	0.8
206.30	-1.57	-0.1	107.13	7JL	2.1	-3.9	-2.2	0.9
211.25	-1.70	0.0	116.70	17JL	0.4	-3.8	-2.3	0.9
216.69	-1.80	0.1	126.31	27JL	-0.8	-3.7	-2.3	0.9
222.54	-1.88	0.2	135.96	6AU	-1.4	-3.7	-2.4	0.9
228.75	-1.92	0.3	145.65	16AU	-1.1	-3.6	-2.4	0.9
235.24	-1.95	0.4	155.40	26AU	-0.5	-3.5	-2.5	0.9
241.99	-1.95	0.5	165.21	5SE	-0.2	-3.5	-2.5	0.9
248.95	-1.93	0.6	175.07	15SE	-0.0	-3.5	-2.5	0.9
256.09	-1.89	0.7	184.99	25SE	0.1	-3.4	-2.4	0.8
263.38	-1.84	0.8	194.97	5OC	0.4	-3.4	-2.4	0.8
270.79	-1.77	0.8	205.00	15OC	2.0	-3.4	-2.4	0.8
278.29	-1.68	0.9	215.08	25OC	1.4	-3.4	-2.3	0.8
285.87	-1.59	1.0	225.20	4NO	-0.2	-3.4	-2.2	0.8
293.50	-1.48	1.0	235.36	14NO	-0.4	-3.4	-2.2	0.8
301.16	-1.36	1.1	245.54	24NO	-0.5	-3.4	-2.1	0.8
308.82	-1.24	1.2	255.74	4DE	-0.5	-3.4	-2.0	0.8
316.48	-1.11	1.2	265.95	14DE	-0.7	-3.4	-1.9	0.8
324.12	-0.98	1.3	276.14	24DE	-0.9	-3.4	-1.9	0.8

♂ LONG	LAT	MAG	☉ LONG	16.00UT 659	☿	♀	♃	♄
331.72	-0.85	1.3	286.32	3JA	-1.0	-3.4	-1.8	0.8
339.28	-0.71	1.4	296.47	13JA	-0.7	-3.4	-1.8	0.8
346.78	-0.58	1.5	306.59	23JA	0.5	-3.4	-1.7	0.7
354.22	-0.45	1.5	316.66	2FE	3.0	-3.5	-1.7	0.7
1.59	-0.32	1.6	326.69	12FE	1.6	-3.5	-1.6	0.6
8.88	-0.20	1.6	336.65	22FE	0.8	-3.5	-1.6	0.6
16.10	-0.08	1.7	346.56	4MR	0.5	-3.4	-1.6	0.5
23.25	0.03	1.7	356.41	14MR	0.2	-3.4	-1.6	0.5
30.31	0.14	1.7	6.21	24MR	-0.3	-3.4	-1.6	0.4
37.31	0.24	1.8	15.95	3AP	-1.0	-3.4	-1.6	0.4
44.23	0.34	1.8	25.64	13AP	-1.8	-3.3	-1.6	0.3
51.07	0.43	1.8	35.28	23AP	-1.1	-3.3	-1.6	0.3
57.86	0.52	1.8	44.88	3MY	-0.1	-3.3	-1.6	0.4
64.58	0.60	1.8	54.45	13MY	0.7	-3.3	-1.6	0.4
71.24	0.68	1.9	63.99	23MY	1.5	-3.3	-1.7	0.5
77.86	0.75	1.9	73.53	2JN	2.6	-3.4	-1.7	0.6
84.42	0.82	1.9	83.05	12JN	3.2	-3.4	-1.8	0.6
90.93	0.88	1.9	92.58	22JN	1.6	-3.4	-1.8	0.7
97.41	0.94	1.9	102.12	2JL	0.3	-3.4	-1.9	0.7
103.85	1.00	2.0	111.68	12JL	-0.9	-3.5	-2.0	0.8
110.26	1.05	2.0	121.27	22JL	-1.5	-3.5	-2.0	0.8
116.65	1.10	2.0	130.89	1AU	-1.1	-3.6	-2.1	0.8
123.01	1.14	2.0	140.56	11AU	-0.5	-3.7	-2.2	0.9
129.35	1.18	2.0	150.28	21AU	-0.1	-3.8	-2.2	0.9
135.67	1.21	2.0	160.06	31AU	0.1	-3.9	-2.3	0.9
141.97	1.25	2.0	169.89	10SE	0.3	-4.0	-2.3	0.9
148.25	1.27	2.0	179.78	20SE	0.7	-4.1	-2.4	0.8
154.52	1.30	2.0	189.73	30SE	2.3	-4.2	-2.4	0.8
160.77	1.32	2.0	199.73	10OC	1.2	-4.3	-2.4	0.8
167.01	1.34	1.9	209.79	20OC	-0.3	-4.4	-2.4	0.7
173.22	1.35	1.9	219.89	30OC	-0.6	-4.3	-2.4	0.7
179.40	1.36	1.8	230.03	9NO	-0.6	-4.1	-2.4	0.7
185.55	1.36	1.7	240.20	19NO	-0.6	-3.5	-2.3	0.8
191.66	1.35	1.7	250.39	29NO	-0.8	-3.2	-2.3	0.8
197.73	1.34	1.6	260.59	9DE	-0.8	-3.9	-2.2	0.8
203.74	1.32	1.5	270.80	19DE	-0.8	-4.3	-2.1	0.8
209.68	1.29	1.3	280.99	29DE	-0.6	-4.4	-2.1	0.8

Left table (660 / 661)

♂ LONG	LAT	MAG	☉ LONG	16.00UT	☿	♀	♃	♄
				660				
215.53	1.25	1.2	291.15	8JA	0.6	-4.3	-2.0	0.8
221.29	1.19	1.1	301.29	18JA	3.0	-4.2	-1.9	0.8
226.91	1.12	0.9	311.39	28JA	1.2	-4.1	-1.8	0.7
232.37	1.03	0.7	321.44	7FE	0.5	-4.0	-1.8	0.7
237.64	0.90	0.6	331.43	17FE	0.3	-3.9	-1.7	0.7
242.66	0.75	0.4	341.38	27FE	0.1	-3.8	-1.7	0.6
247.38	0.56	0.1	351.26	8MR	-0.3	-3.7	-1.6	0.6
251.72	0.33	-0.1	1.08	18MR	-1.0	-3.6	-1.6	0.5
255.55	0.03	-0.4	10.85	28MR	-1.7	-3.6	-1.6	0.4
258.77	-0.33	-0.7	20.56	7AP	-1.1	-3.5	-1.5	0.4
261.21	-0.78	-1.0	30.23	17AP	-0.0	-3.4	-1.5	0.3
262.67	-1.32	-1.3	39.86	27AP	1.0	-3.4	-1.5	0.3
263.00	-1.94	-1.6	49.44	7MY	2.0	-3.4	-1.5	0.3
262.08	-2.64	-1.9	59.00	17MY	3.5	-3.3	-1.5	0.4
260.02	-3.36	-2.2	68.54	27MY	2.5	-3.3	-1.5	0.4
257.22	-4.00	-2.4	78.06	6JN	1.3	-3.3	-1.5	0.5
254.30	-4.49	-2.3	87.59	16JN	0.2	-3.3	-1.6	0.5
252.03	-4.77	-2.1	97.12	26JN	-0.9	-3.3	-1.6	0.6
250.91	-4.85	-1.9	106.67	6JL	-1.6	-3.3	-1.6	0.7
251.15	-4.77	-1.7	116.24	16JL	-1.1	-3.3	-1.7	0.7
252.74	-4.60	-1.4	125.85	26JL	-0.4	-3.4	-1.7	0.8
255.50	-4.35	-1.2	135.49	5AU	0.0	-3.4	-1.8	0.8
259.24	-4.07	-1.0	145.19	15AU	0.3	-3.4	-1.9	0.8
263.77	-3.77	-0.8	154.93	25AU	0.5	-3.5	-1.9	0.8
268.92	-3.47	-0.6	164.73	4SE	1.0	-3.5	-2.0	0.8
274.57	-3.16	-0.4	174.59	14SE	2.7	-3.5	-2.1	0.8
280.61	-2.85	-0.2	184.50	24SE	1.1	-3.5	-2.1	0.8
286.94	-2.55	-0.1	194.48	4OC	-0.5	-3.4	-2.2	0.8
293.51	-2.26	0.1	204.51	14OC	-0.8	-3.4	-2.2	0.8
300.25	-1.98	0.3	214.58	24OC	-0.7	-3.4	-2.3	0.7
307.12	-1.71	0.4	224.70	3NO	-0.8	-3.4	-2.3	0.7
314.08	-1.45	0.5	234.86	13NO	-0.7	-3.3	-2.3	0.7
321.10	-1.21	0.7	245.04	23NO	-0.6	-3.3	-2.3	0.7
328.15	-0.98	0.8	255.24	3DE	-0.6	-3.4	-2.3	0.8
335.22	-0.77	0.9	265.45	13DE	-0.5	-3.3	-2.3	0.8
342.29	-0.57	1.0	275.64	23DE	0.8	-3.4	-2.2	0.8
				661				
349.33	-0.39	1.1	285.82	2JA	2.7	-3.4	-2.2	0.8
356.36	-0.22	1.2	295.98	12JA	0.8	-3.4	-2.1	0.8
3.34	-0.06	1.3	306.10	22JA	0.3	-3.4	-2.0	0.8
10.28	0.08	1.4	316.17	1FE	0.2	-3.5	-1.9	0.8
17.18	0.21	1.5	326.20	11FE	-0.0	-3.5	-1.9	0.7
24.03	0.33	1.6	336.17	21FE	-0.4	-3.5	-1.8	0.7
30.83	0.44	1.7	346.08	3MR	-1.0	-3.6	-1.7	0.6
37.58	0.54	1.7	355.94	13MR	-1.6	-3.7	-1.7	0.6
44.28	0.63	1.8	5.74	23MR	-1.1	-3.7	-1.6	0.5
50.93	0.71	1.8	15.48	2AP	0.0	-3.8	-1.6	0.5
57.54	0.78	1.9	25.17	12AP	1.2	-3.9	-1.5	0.4
64.10	0.84	1.9	34.81	22AP	2.7	-4.0	-1.5	0.3
70.62	0.90	1.9	44.42	2MY	3.3	-4.1	-1.5	0.3
77.11	0.95	2.0	53.99	12MY	1.9	-4.1	-1.4	0.2
83.57	0.99	2.0	63.54	22MY	1.0	-4.2	-1.4	0.3
90.00	1.03	2.0	73.07	1JN	0.1	-4.2	-1.4	0.4
96.40	1.07	2.0	82.59	11JN	-0.9	-4.0	-1.4	0.4
102.79	1.09	2.0	92.12	21JN	-1.7	-3.4	-1.4	0.5
109.17	1.12	2.0	101.66	1JL	-1.1	-2.9	-1.4	0.5
115.54	1.13	2.0	111.22	11JL	-0.4	-3.6	-1.4	0.6
121.90	1.15	2.0	120.80	21JL	0.1	-4.1	-1.5	0.7
128.27	1.16	2.0	130.43	31JL	0.4	-4.2	-1.5	0.7
134.64	1.16	2.0	140.10	10AU	0.7	-4.2	-1.5	0.7
141.02	1.16	2.0	149.81	20AU	1.3	-4.2	-1.6	0.8
147.41	1.15	2.0	159.58	30AU	3.1	-4.1	-1.6	0.8
153.81	1.14	2.0	169.41	9SE	0.9	-4.0	-1.7	0.8
160.23	1.13	2.0	179.30	19SE	-0.6	-3.9	-1.7	0.8
166.68	1.11	2.0	189.24	29SE	-0.9	-3.8	-1.8	0.8
173.14	1.09	2.0	199.25	9OC	-0.9	-3.8	-1.9	0.8
179.62	1.05	2.0	209.30	19OC	-0.8	-3.7	-1.9	0.8
186.13	1.02	1.9	219.40	29OC	-0.6	-3.6	-2.0	0.7
192.66	0.97	1.9	229.54	8NO	-0.5	-3.6	-2.1	0.7
199.22	0.92	1.8	239.70	18NO	-0.5	-3.5	-2.1	0.7
205.80	0.86	1.8	249.90	28NO	-0.3	-3.5	-2.2	0.7
212.39	0.80	1.7	260.10	8DE	1.0	-3.4	-2.2	0.7
219.01	0.72	1.6	270.30	18DE	2.4	-3.4	-2.2	0.8
225.65	0.63	1.6	280.50	28DE	0.6	-3.4	-2.2	0.8

Right table (662 / 663)

♂ LONG	LAT	MAG	☉ LONG	16.00UT	☿	♀	♃	♄
				662				
232.29	0.53	1.5	290.67	7JA	0.1	-3.4	-2.2	0.8
238.96	0.41	1.4	300.80	17JA	0.0	-3.3	-2.1	0.8
245.63	0.28	1.3	310.90	27JA	-0.1	-3.3	-2.1	0.8
252.30	0.13	1.2	320.96	6FE	-0.5	-3.3	-2.0	0.8
258.97	-0.04	1.0	330.95	16FE	-1.0	-3.3	-2.0	0.8
265.63	-0.22	0.9	340.90	26FE	-1.5	-3.3	-1.9	0.7
272.27	-0.43	0.8	350.78	8MR	-1.1	-3.3	-1.8	0.7
278.88	-0.66	0.6	0.61	18MR	0.1	-3.4	-1.8	0.6
285.44	-0.92	0.5	10.38	28MR	1.6	-3.4	-1.7	0.6
291.92	-1.20	0.3	20.10	7AP	3.4	-3.4	-1.6	0.5
298.30	-1.52	0.2	29.76	17AP	2.5	-3.5	-1.6	0.5
304.54	-1.86	-0.0	39.39	27AP	1.4	-3.5	-1.5	0.4
310.59	-2.23	-0.2	48.97	7MY	0.7	-3.5	-1.5	0.3
316.40	-2.64	-0.4	58.53	17MY	0.0	-3.4	-1.4	0.3
321.88	-3.07	-0.6	68.07	27MY	-0.9	-3.4	-1.4	0.3
326.92	-3.54	-0.8	77.60	6JN	-1.8	-3.4	-1.4	0.3
331.40	-4.04	-1.1	87.12	16JN	-1.1	-3.3	-1.3	0.4
335.15	-4.56	-1.3	96.65	26JN	-0.3	-3.3	-1.3	0.4
337.97	-5.08	-1.6	106.20	6JL	0.2	-3.3	-1.3	0.5
339.66	-5.58	-1.8	115.77	16JL	0.6	-3.3	-1.3	0.5
340.03	-6.00	-2.1	125.38	26JL	0.9	-3.3	-1.3	0.6
339.02	-6.27	-2.3	135.02	5AU	1.7	-3.3	-1.3	0.7
336.86	-6.30	-2.5	144.71	15AU	3.2	-3.4	-1.3	0.7
334.05	-6.03	-2.6	154.46	25AU	0.8	-3.4	-1.4	0.7
331.39	-5.49	-2.4	164.25	4SE	-0.7	-3.4	-1.4	0.8
329.56	-4.77	-2.1	174.11	14SE	-1.1	-3.5	-1.4	0.8
328.95	-3.99	-1.8	184.02	24SE	-1.0	-3.5	-1.5	0.8
329.67	-3.22	-1.5	193.99	4OC	-0.7	-3.6	-1.5	0.8
331.57	-2.53	-1.2	204.02	14OC	-0.5	-3.7	-1.6	0.8
334.47	-1.93	-0.9	214.10	24OC	-0.3	-3.7	-1.7	0.8
338.17	-1.41	-0.6	224.22	3NO	-0.3	-3.8	-1.7	0.8
342.48	-0.97	-0.3	234.37	13NO	-0.1	-3.9	-1.8	0.7
347.28	-0.60	-0.1	244.55	23NO	1.2	-4.0	-1.8	0.7
352.44	-0.29	0.2	254.75	3DE	2.1	-4.1	-1.9	0.7
357.88	-0.03	0.4	264.95	13DE	0.3	-4.2	-2.0	0.7
3.54	0.20	0.6	275.15	23DE	-0.0	-4.3	-2.0	0.8
				663				
9.36	0.38	0.7	285.33	2JA	-0.1	-4.4	-2.0	0.8
15.30	0.54	0.9	295.49	12JA	-0.2	-4.3	-2.1	0.8
21.33	0.68	1.1	305.61	22JA	-0.5	-4.1	-2.1	0.8
27.43	0.79	1.2	315.69	1FE	-1.0	-3.5	-2.1	0.8
33.57	0.88	1.3	325.71	11FE	-1.4	-3.4	-2.0	0.8
39.75	0.96	1.4	335.69	21FE	-1.0	-3.9	-2.0	0.8
45.95	1.03	1.5	345.60	3MR	0.2	-4.2	-2.0	0.8
52.16	1.09	1.6	355.46	13MR	2.0	-4.3	-1.9	0.7
58.39	1.13	1.7	5.26	23MR	3.4	-4.2	-1.8	0.7
64.61	1.17	1.8	15.01	2AP	1.9	-4.1	-1.8	0.6
70.84	1.19	1.8	24.70	12AP	1.1	-4.0	-1.7	0.6
77.07	1.21	1.9	34.35	22AP	0.6	-3.9	-1.6	0.5
83.30	1.23	1.9	43.95	2MY	-0.1	-3.8	-1.6	0.4
89.54	1.23	2.0	53.52	12MY	-0.9	-3.7	-1.5	0.4
95.78	1.24	2.0	63.07	22MY	-1.8	-3.7	-1.5	0.3
102.02	1.23	2.0	72.60	1JN	-1.1	-3.6	-1.4	0.3
108.28	1.22	2.0	82.13	11JN	-0.3	-3.5	-1.4	0.3
114.55	1.21	2.0	91.66	21JN	0.3	-3.5	-1.3	0.3
120.84	1.19	2.0	101.19	1JL	0.7	-3.4	-1.3	0.4
127.15	1.17	2.0	110.75	11JL	1.2	-3.4	-1.3	0.4
133.48	1.15	2.0	120.34	21JL	2.3	-3.4	-1.2	0.5
139.84	1.12	2.0	129.96	31JL	2.8	-3.4	-1.2	0.6
146.24	1.08	2.0	139.63	10AU	0.7	-3.4	-1.2	0.6
152.67	1.04	1.9	149.35	20AU	-0.7	-3.3	-1.2	0.7
159.14	1.00	1.9	159.11	30AU	-1.2	-3.3	-1.2	0.7
165.66	0.95	1.9	168.94	9SE	-1.1	-3.4	-1.3	0.7
172.22	0.90	1.9	178.82	19SE	-0.6	-3.4	-1.3	0.8
178.83	0.84	1.9	188.76	29SE	-0.3	-3.4	-1.3	0.8
185.49	0.78	1.9	198.76	9OC	-0.2	-3.4	-1.3	0.8
192.21	0.71	1.9	208.81	19OC	-0.2	-3.4	-1.4	0.8
198.98	0.64	1.9	218.91	29OC	0.1	-3.4	-1.4	0.8
205.80	0.56	1.8	229.05	8NO	1.4	-3.5	-1.5	0.8
212.69	0.47	1.8	239.22	18NO	1.8	-3.5	-1.6	0.8
219.62	0.38	1.7	249.40	28NO	0.1	-3.5	-1.6	0.7
226.62	0.28	1.7	259.61	8DE	-0.2	-3.5	-1.7	0.7
233.68	0.17	1.7	269.81	18DE	-0.2	-3.4	-1.7	0.7
240.78	0.06	1.6	280.00	28DE	-0.3	-3.4	-1.8	0.8

Left table (664 / 665):

♂ LONG	LAT	MAG	☉ LONG	16.00UT	☿	♀	♃	♄
247.95	-0.06	1.6	290.17	7JA	-0.6	-3.4	-1.9	0.8
255.16	-0.19	1.5	300.31	17JA	-1.0	-3.4	-1.9	0.8
262.42	-0.33	1.5	310.41	27JA	-1.2	-3.3	-2.0	0.8
269.73	-0.47	1.4	320.47	6FE	-0.9	-3.3	-2.0	0.8
277.08	-0.62	1.3	330.47	16FE	0.3	-3.3	-2.0	0.8
284.46	-0.78	1.2	340.41	26FE	2.4	-3.3	-2.0	0.8
291.86	-0.94	1.2	350.30	7MR	2.6	-3.4	-2.0	0.8
299.28	-1.10	1.1	0.13	17MR	1.4	-3.4	-2.0	0.8
306.71	-1.27	1.0	9.90	27MR	0.8	-3.4	-1.9	0.7
314.13	-1.43	0.9	19.63	6AP	0.4	-3.4	-1.9	0.7
321.52	-1.59	0.8	29.30	16AP	-0.1	-3.4	-1.8	0.6
328.88	-1.74	0.7	38.92	26AP	-0.9	-3.5	-1.8	0.6
336.17	-1.89	0.6	48.51	6MY	-1.8	-3.5	-1.7	0.5
343.37	-2.03	0.6	58.07	16MY	-1.1	-3.6	-1.6	0.5
350.46	-2.15	0.5	67.61	26MY	-0.2	-3.6	-1.6	0.4
357.40	-2.26	0.4	77.14	5JN	0.5	-3.7	-1.5	0.3
4.16	-2.36	0.3	86.66	15JN	1.0	-3.8	-1.5	0.3
10.69	-2.44	0.2	96.19	25JN	1.7	-3.9	-1.4	0.3
16.95	-2.49	0.0	105.74	5JL	2.9	-4.0	-1.4	0.4
22.87	-2.53	-0.1	115.31	15JL	2.4	-4.1	-1.3	0.4
28.37	-2.54	-0.2	124.91	25JL	0.6	-4.2	-1.3	0.5
33.35	-2.52	-0.4	134.56	4AU	-0.8	-4.2	-1.3	0.5
37.70	-2.47	-0.5	144.24	14AU	-1.3	-4.2	-1.2	0.6
41.26	-2.38	-0.7	153.98	24AU	-1.1	-4.0	-1.2	0.6
43.83	-2.24	-0.9	163.78	3SE	-0.6	-3.5	-1.2	0.7
45.21	-2.04	-1.1	173.63	13SE	-0.2	-3.4	-1.2	0.7
45.21	-1.76	-1.4	183.54	23SE	-0.1	-3.9	-1.2	0.8
43.73	-1.39	-1.6	193.52	3OC	0.0	-4.2	-1.2	0.8
40.94	-0.93	-1.7	203.54	13OC	0.3	-4.3	-1.2	0.8
37.35	-0.42	-1.9	213.61	23OC	1.7	-4.3	-1.3	0.8
33.73	0.09	-1.7	223.73	2NO	1.6	-4.3	-1.3	0.8
30.88	0.55	-1.4	233.88	12NO	-0.1	-4.2	-1.3	0.8
29.25	0.92	-1.1	244.06	22NO	-0.4	-4.1	-1.4	0.8
28.98	1.19	-0.8	254.26	2DE	-0.4	-4.0	-1.4	0.8
29.97	1.39	-0.5	264.46	12DE	-0.5	-3.9	-1.5	0.7
32.02	1.53	-0.2	274.66	22DE	-0.7	-3.8	-1.6	0.7

665

♂ LONG	LAT	MAG	☉ LONG	16.00UT	☿	♀	♃	♄
34.92	1.62	0.1	284.85	1JA	-1.0	-3.7	-1.6	0.8
38.49	1.68	0.4	295.00	11JA	-1.1	-3.6	-1.7	0.8
42.58	1.71	0.6	305.12	21JA	-0.8	-3.6	-1.8	0.8
47.08	1.73	0.8	315.21	31JA	0.4	-3.5	-1.8	0.9
51.89	1.73	0.9	325.23	10FE	2.8	-3.5	-1.9	0.9
56.95	1.72	1.1	335.21	20FE	2.0	-3.4	-1.9	0.9
62.21	1.70	1.3	345.13	2MR	1.0	-3.4	-2.0	0.9
67.63	1.68	1.4	354.99	12MR	0.6	-3.3	-2.0	0.9
73.17	1.65	1.5	4.79	22MR	0.3	-3.3	-2.0	0.8
78.82	1.62	1.6	14.54	1AP	-0.2	-3.3	-2.0	0.8
84.56	1.58	1.7	24.24	11AP	-0.9	-3.3	-2.0	0.8
90.37	1.54	1.7	33.89	21AP	-1.8	-3.3	-2.0	0.7
96.25	1.49	1.8	43.49	1MY	-1.1	-3.3	-1.9	0.7
102.18	1.45	1.9	53.06	11MY	-0.2	-3.3	-1.9	0.6
108.17	1.40	1.9	62.62	21MY	0.6	-3.4	-1.8	0.5
114.22	1.34	1.9	72.15	31MY	1.3	-3.4	-1.8	0.5
120.31	1.29	2.0	81.67	10JN	2.2	-3.4	-1.7	0.4
126.46	1.23	2.0	91.20	20JN	3.3	-3.5	-1.6	0.4
132.66	1.17	2.0	100.73	30JN	1.9	-3.5	-1.6	0.3
138.92	1.11	2.0	110.29	10JL	0.4	-3.5	-1.5	0.4
145.23	1.04	2.0	119.87	20JL	-0.8	-3.4	-1.5	0.4
151.60	0.98	2.0	129.49	30JL	-1.4	-3.4	-1.4	0.5
158.04	0.91	2.0	139.15	9AU	-1.1	-3.4	-1.4	0.5
164.54	0.83	1.9	148.87	19AU	-0.5	-3.4	-1.3	0.6
171.10	0.75	1.9	158.63	29AU	-0.2	-3.3	-1.3	0.6
177.74	0.67	1.9	168.45	8SE	0.0	-3.3	-1.3	0.7
184.45	0.59	1.8	178.34	18SE	0.2	-3.3	-1.2	0.7
191.23	0.50	1.8	188.28	28SE	0.5	-3.3	-1.2	0.8
198.09	0.41	1.7	198.27	8OC	2.0	-3.3	-1.2	0.8
205.03	0.32	1.7	208.32	18OC	1.4	-3.4	-1.2	0.8
212.04	0.22	1.7	218.42	28OC	-0.2	-3.4	-1.2	0.8
219.13	0.12	1.7	228.55	7NO	-0.5	-3.4	-1.3	0.8
226.31	0.02	1.7	238.72	17NO	-0.5	-3.5	-1.3	0.8
233.55	-0.09	1.6	248.91	27NO	-0.6	-3.5	-1.3	0.8
240.87	-0.20	1.6	259.11	7DE	-0.8	-3.6	-1.4	0.8
248.26	-0.31	1.6	269.32	17DE	-0.9	-3.6	-1.4	0.8
255.72	-0.42	1.6	279.51	27DE	-0.9	-3.7	-1.5	0.8

Right table (666 / 667):

♂ LONG	LAT	MAG	☉ LONG	16.00UT	☿	♀	♃	♄
263.24	-0.53	1.6	289.68	6JA	-0.7	-3.7	-1.5	0.8
270.81	-0.64	1.5	299.82	16JA	0.5	-3.8	-1.6	0.8
278.43	-0.75	1.5	309.93	26JA	3.0	-3.9	-1.6	0.9
286.10	-0.86	1.5	319.98	5FE	1.5	-4.0	-1.7	0.9
293.79	-0.96	1.4	329.99	15FE	0.7	-4.1	-1.8	0.9
301.51	-1.06	1.4	339.94	25FE	0.4	-4.2	-1.8	1.0
309.23	-1.15	1.3	349.83	7MR	0.1	-4.2	-1.9	1.0
316.96	-1.23	1.3	359.66	17MR	-0.3	-4.3	-2.0	1.0
324.67	-1.31	1.3	9.44	27MR	-0.9	-4.1	-2.0	0.9
332.35	-1.37	1.2	19.16	6AP	-1.8	-3.8	-2.1	0.9
340.00	-1.41	1.2	28.83	16AP	-1.1	-3.1	-2.1	0.9
347.58	-1.44	1.2	38.46	26AP	-0.1	-3.9	-2.1	0.8
355.10	-1.46	1.1	48.05	6MY	0.8	-3.9	-2.1	0.8
2.54	-1.46	1.1	57.61	16MY	1.7	-4.1	-2.1	0.7
9.87	-1.45	1.1	67.15	26MY	2.9	-4.2	-2.0	0.7
17.09	-1.42	1.0	76.68	5JN	2.9	-4.1	-2.0	0.6
24.18	-1.37	1.0	86.20	15JN	1.5	-4.1	-1.9	0.5
31.12	-1.31	0.9	95.73	25JN	0.3	-4.0	-1.9	0.5
37.90	-1.23	0.9	105.27	5JL	-0.9	-3.9	-1.8	0.4
44.49	-1.13	0.8	114.85	15JL	-1.5	-3.8	-1.7	0.4
50.86	-1.02	0.8	124.44	25JL	-1.1	-3.7	-1.7	0.4
57.00	-0.89	0.7	134.08	4AU	-0.5	-3.6	-1.6	0.5
62.84	-0.73	0.6	143.77	14AU	-0.1	-3.6	-1.6	0.5
68.36	-0.56	0.5	153.51	24AU	0.2	-3.5	-1.5	0.6
73.48	-0.36	0.4	163.30	3SE	0.4	-3.5	-1.5	0.6
78.12	-0.13	0.3	173.15	13SE	0.8	-3.5	-1.4	0.7
82.19	0.13	0.1	183.06	23SE	2.4	-3.4	-1.4	0.7
85.54	0.43	-0.1	193.03	3OC	1.3	-3.4	-1.4	0.8
88.01	0.78	-0.2	203.05	13OC	-0.4	-3.4	-1.3	0.8
89.41	1.17	-0.5	213.12	23OC	-0.7	-3.4	-1.3	0.9
89.55	1.61	-0.7	223.24	2NO	-0.7	-3.4	-1.3	0.9
88.28	2.08	-0.9	233.39	12NO	-0.7	-3.4	-1.3	0.9
85.67	2.54	-1.1	243.57	22NO	-0.7	-3.4	-1.3	0.9
82.04	2.94	-1.3	253.76	2DE	-0.7	-3.4	-1.3	0.9
78.08	3.22	-1.2	263.97	12DE	-0.7	-3.4	-1.3	0.9
74.56	3.36	-1.0	274.17	22DE	-0.6	-3.4	-1.4	0.9

667

♂ LONG	LAT	MAG	☉ LONG	16.00UT	☿	♀	♃	♄
72.08	3.36	-0.7	284.35	1JA	0.6	-3.4	-1.4	0.9
70.92	3.28	-0.4	294.51	11JA	2.9	-3.4	-1.5	0.9
71.06	3.14	-0.2	304.63	21JA	1.1	-3.4	-1.5	0.9
72.33	2.98	0.1	314.72	31JA	0.5	-3.5	-1.6	0.9
74.56	2.81	0.4	324.75	10FE	0.2	-3.5	-1.6	1.0
77.55	2.64	0.6	334.73	20FE	0.0	-3.5	-1.7	1.0
81.13	2.48	0.8	344.65	2MR	-0.3	-3.4	-1.8	1.0
85.19	2.33	0.9	354.51	12MR	-1.0	-3.4	-1.8	1.0
89.63	2.19	1.1	4.32	22MR	-1.7	-3.4	-1.9	1.1
94.38	2.06	1.2	14.07	1AP	-1.1	-3.4	-2.0	1.0
99.38	1.93	1.4	23.77	11AP	-0.0	-3.3	-2.0	1.0
104.58	1.80	1.5	33.41	21AP	1.0	-3.3	-2.1	1.0
109.96	1.69	1.5	43.03	1MY	2.2	-3.3	-2.1	1.0
115.50	1.57	1.6	52.60	11MY	3.7	-3.3	-2.2	0.9
121.16	1.46	1.7	62.15	21MY	2.3	-3.3	-2.2	0.9
126.95	1.35	1.7	71.68	31MY	1.2	-3.4	-2.2	0.8
132.85	1.24	1.8	81.21	10JN	0.2	-3.4	-2.2	0.7
138.84	1.14	1.8	90.73	20JN	-0.9	-3.4	-2.2	0.7
144.94	1.03	1.8	100.27	30JN	-1.6	-3.4	-2.1	0.6
151.14	0.93	1.9	109.83	10JL	-1.1	-3.5	-2.1	0.5
157.43	0.83	1.9	119.41	20JL	-0.4	-3.5	-2.0	0.5
163.82	0.72	1.9	129.03	30JL	0.0	-3.6	-2.0	0.5
170.30	0.62	1.9	138.69	9AU	0.3	-3.7	-1.9	0.5
176.87	0.51	1.8	148.40	19AU	0.5	-3.8	-1.8	0.6
183.54	0.41	1.8	158.16	29AU	1.1	-3.9	-1.8	0.6
190.31	0.30	1.8	167.98	8SE	2.8	-4.0	-1.7	0.7
197.17	0.20	1.8	177.86	18SE	1.1	-4.1	-1.7	0.7
204.13	0.09	1.7	187.79	28SE	-0.5	-4.2	-1.6	0.8
211.19	-0.01	1.7	197.78	8OC	-0.8	-4.3	-1.6	0.8
218.34	-0.12	1.6	207.83	18OC	-0.8	-4.4	-1.5	0.9
225.58	-0.22	1.6	217.92	28OC	-0.8	-4.3	-1.5	0.9
232.92	-0.32	1.5	228.06	7NO	-0.6	-4.1	-1.4	0.9
240.34	-0.43	1.5	238.22	17NO	-0.6	-3.4	-1.4	1.0
247.85	-0.52	1.5	248.41	27NO	-0.6	-3.2	-1.4	1.0
255.43	-0.62	1.4	258.61	7DE	-0.4	-4.0	-1.4	1.0
263.07	-0.71	1.4	268.82	17DE	0.8	-4.3	-1.4	1.0
270.78	-0.79	1.4	279.01	27DE	2.7	-4.4	-1.4	1.0

♂ / ☉ / Magnitudes (668–669)

♂ LONG	♂ LAT	♂ MAG	☉ LONG	16.00UT	☿	♀	♃	♄
				668				
278.54	-0.87	1.4	289.18	6JA	0.7	-4.3	-1.4	1.0
286.34	-0.94	1.4	299.33	16JA	0.3	-4.2	-1.5	1.0
294.17	-1.01	1.4	309.44	26JA	0.1	-4.1	-1.5	1.0
302.02	-1.06	1.4	319.49	5FE	-0.1	-4.0	-1.5	1.0
309.87	-1.10	1.4	329.50	15FE	-0.4	-3.9	-1.6	1.1
317.72	-1.13	1.4	339.46	25FE	-1.0	-3.8	-1.6	1.1
325.54	-1.15	1.4	349.35	6MR	-1.6	-3.7	-1.7	1.1
333.34	-1.16	1.4	359.19	16MR	-1.1	-3.6	-1.7	1.2
341.09	-1.15	1.4	8.97	26MR	0.0	-3.6	-1.8	1.2
348.79	-1.13	1.4	18.69	5AP	1.3	-3.5	-1.9	1.2
356.42	-1.10	1.4	28.37	15AP	2.9	-3.4	-2.0	1.2
3.98	-1.06	1.4	38.00	25AP	3.0	-3.4	-2.0	1.2
11.45	-1.00	1.4	47.59	5MY	1.7	-3.4	-2.1	1.1
18.83	-0.93	1.4	57.15	15MY	0.9	-3.3	-2.2	1.1
26.12	-0.85	1.4	66.70	25MY	0.1	-3.3	-2.2	1.0
33.29	-0.76	1.4	76.22	4JN	-0.9	-3.3	-2.3	1.0
40.36	-0.66	1.4	85.75	14JN	-1.7	-3.3	-2.3	0.9
47.30	-0.56	1.4	95.28	24JN	-1.1	-3.3	-2.3	0.9
54.12	-0.44	1.4	104.82	4JL	-0.4	-3.3	-2.3	0.8
60.81	-0.31	1.3	114.39	14JL	0.1	-3.3	-2.3	0.7
67.35	-0.18	1.3	123.99	24JL	0.5	-3.4	-2.3	0.7
73.74	-0.03	1.3	133.62	3AU	0.8	-3.4	-2.3	0.6
79.96	0.12	1.3	143.30	13AU	1.5	-3.4	-2.2	0.6
85.99	0.29	1.2	153.04	23AU	3.2	-3.5	-2.1	0.7
91.81	0.47	1.1	162.82	2SE	1.0	-3.5	-2.1	0.7
97.39	0.66	1.1	172.67	12SE	-0.6	-3.5	-2.0	0.8
102.69	0.87	1.0	182.58	22SE	-1.0	-3.5	-2.0	0.8
107.66	1.10	0.9	192.54	2OC	-1.0	-3.4	-1.9	0.9
112.23	1.35	0.7	202.56	12OC	-0.8	-3.4	-1.8	0.9
116.32	1.63	0.6	212.62	22OC	-0.5	-3.4	-1.8	0.9
119.81	1.94	0.4	222.73	1NO	-0.4	-3.4	-1.7	1.0
122.58	2.28	0.2	232.89	11NO	-0.4	-3.3	-1.7	1.0
124.43	2.67	0.0	243.06	21NO	-0.2	-3.3	-1.6	1.1
125.20	3.08	-0.2	253.26	1DE	1.0	-3.3	-1.6	1.1
124.70	3.51	-0.4	263.46	11DE	2.3	-3.3	-1.6	1.1
122.86	3.92	-0.7	273.66	21DE	0.5	-3.4	-1.5	1.1
119.80	4.25	-0.9	283.85	31DE	0.1	-3.4	-1.5	1.1
				669				
115.97	4.44	-1.0	294.01	10JA	-0.0	-3.4	-1.5	1.1
112.05	4.46	-0.9	304.14	20JA	-0.2	-3.4	-1.5	1.1
108.77	4.31	-0.7	314.22	30JA	-0.5	-3.5	-1.5	1.1
106.61	4.05	-0.5	324.26	9FE	-1.0	-3.5	-1.5	1.1
105.74	3.73	-0.2	334.25	19FE	-1.4	-3.5	-1.6	1.2
106.13	3.39	0.0	344.17	1MR	-1.1	-3.6	-1.6	1.2
107.60	3.07	0.3	354.04	11MR	0.1	-3.7	-1.6	1.3
109.97	2.76	0.5	3.85	21MR	1.7	-3.7	-1.7	1.3
113.06	2.49	0.7	13.60	31MR	3.6	-3.8	-1.7	1.3
116.75	2.23	0.8	23.30	10AP	2.2	-3.9	-1.8	1.3
120.91	2.00	1.0	32.96	20AP	1.3	-4.0	-1.8	1.3
125.46	1.78	1.1	42.56	30AP	0.7	-4.1	-1.9	1.3
130.33	1.58	1.2	52.14	10MY	0.0	-4.1	-2.0	1.3
135.48	1.39	1.3	61.69	20MY	-0.9	-4.2	-2.0	1.3
140.85	1.22	1.4	71.23	30MY	-1.8	-4.1	-2.1	1.2
146.44	1.05	1.4	80.75	9JN	-1.1	-3.9	-2.2	1.2
152.21	0.90	1.5	90.28	19JN	-0.3	-3.4	-2.3	1.1
158.14	0.74	1.5	99.81	29JN	0.2	-2.9	-2.3	1.1
164.24	0.60	1.6	109.37	9JL	0.6	-3.7	-2.4	1.0
170.48	0.46	1.6	118.95	19JL	1.0	-4.1	-2.4	0.9
176.85	0.33	1.6	128.56	29JL	1.9	-4.2	-2.4	0.9
183.37	0.20	1.6	138.22	8AU	3.1	-4.2	-2.4	0.8
190.01	0.07	1.6	147.93	18AU	0.8	-4.2	-2.4	0.7
196.78	-0.05	1.6	157.69	28AU	-0.4	-4.1	-2.4	0.8
203.68	-0.17	1.6	167.50	7SE	-1.1	-4.0	-2.4	0.8
210.69	-0.28	1.6	177.38	17SE	-1.1	-3.9	-2.3	0.9
217.82	-0.39	1.6	187.31	27SE	-0.7	-3.8	-2.3	0.9
225.06	-0.49	1.6	197.30	7OC	-0.4	-3.8	-2.2	1.0
232.40	-0.59	1.5	207.34	17OC	-0.3	-3.7	-2.1	1.0
239.85	-0.68	1.5	217.43	27OC	-0.3	-3.6	-2.0	1.1
247.39	-0.76	1.5	227.57	6NO	-0.1	-3.6	-2.0	1.1
255.02	-0.84	1.5	237.73	16NO	1.2	-3.5	-1.9	1.2
262.72	-0.90	1.4	247.91	26NO	2.1	-3.5	-1.8	1.2
270.49	-0.96	1.4	258.12	6DE	0.2	-3.4	-1.8	1.2
278.30	-1.01	1.4	268.32	16DE	-0.1	-3.4	-1.7	1.3
286.16	-1.04	1.4	278.51	26DE	-0.2	-3.4	-1.7	1.3

♂ / ☉ / Magnitudes (670–671)

♂ LONG	♂ LAT	♂ MAG	☉ LONG	16.00UT	☿	♀	♃	♄
				670				
294.05	-1.07	1.3	288.69	5JA	-0.3	-3.4	-1.7	1.3
301.95	-1.08	1.3	298.83	15JA	-0.5	-3.3	-1.6	1.3
309.85	-1.08	1.3	308.94	25JA	-1.0	-3.3	-1.6	1.3
317.75	-1.07	1.3	319.01	4FE	-1.3	-3.3	-1.6	1.3
325.61	-1.05	1.3	329.02	14FE	-1.0	-3.3	-1.6	1.3
333.44	-1.02	1.4	338.97	24FE	0.2	-3.3	-1.6	1.3
341.21	-0.98	1.4	348.87	6MR	2.1	-3.3	-1.6	1.4
348.93	-0.92	1.4	358.71	16MR	3.1	-3.4	-1.6	1.4
356.58	-0.86	1.4	8.49	26MR	1.7	-3.4	-1.6	1.4
4.16	-0.79	1.5	18.22	5AP	1.0	-3.4	-1.6	1.4
11.66	-0.71	1.5	27.90	15AP	0.5	-3.5	-1.7	1.4
19.07	-0.62	1.5	37.53	25AP	-0.1	-3.5	-1.7	1.3
26.38	-0.53	1.6	47.13	5MY	-0.9	-3.5	-1.8	1.3
33.61	-0.43	1.6	56.69	15MY	-1.8	-3.4	-1.8	1.3
40.74	-0.32	1.6	66.23	25MY	-1.2	-3.4	-1.9	1.2
47.78	-0.21	1.6	75.76	4JN	-0.3	-3.4	-1.9	1.2
54.71	-0.10	1.7	85.28	14JN	0.4	-3.3	-2.0	1.1
61.55	0.01	1.7	94.81	24JN	0.8	-3.3	-2.1	1.1
68.29	0.13	1.7	104.35	4JL	1.4	-3.3	-2.1	1.0
74.93	0.25	1.7	113.92	14JL	2.5	-3.3	-2.2	1.0
81.47	0.37	1.7	123.51	24JL	2.7	-3.3	-2.3	0.9
87.91	0.50	1.7	133.15	3AU	0.7	-3.3	-2.3	0.9
94.23	0.63	1.7	142.83	13AU	-0.7	-3.4	-2.4	0.9
100.44	0.76	1.6	152.56	23AU	-1.2	-3.4	-2.4	0.8
106.52	0.90	1.6	162.35	2SE	-1.2	-3.4	-2.5	0.8
112.46	1.05	1.5	172.19	12SE	-0.6	-3.5	-2.5	0.9
118.26	1.20	1.5	182.10	22SE	-0.3	-3.5	-2.5	0.9
123.88	1.35	1.4	192.06	2OC	-0.2	-3.6	-2.4	1.0
129.30	1.52	1.3	202.07	12OC	-0.1	-3.7	-2.4	1.1
134.49	1.70	1.2	212.14	22OC	0.1	-3.7	-2.3	1.1
139.39	1.90	1.1	222.25	1NO	1.4	-3.8	-2.3	1.2
143.96	2.11	1.0	232.39	11NO	1.8	-3.9	-2.2	1.3
148.12	2.34	0.8	242.57	21NO	0.0	-4.0	-2.1	1.3
151.77	2.59	0.7	252.77	1DE	-0.3	-4.1	-2.1	1.3
154.79	2.86	0.5	262.97	11DE	-0.3	-4.2	-2.0	1.3
157.03	3.16	0.2	273.17	21DE	-0.4	-4.3	-1.9	1.3
158.31	3.47	-0.0	283.35	31DE	-0.6	-4.4	-1.9	1.3
				671				
158.46	3.79	-0.3	293.51	10JA	-1.0	-4.3	-1.8	1.3
157.32	4.08	-0.5	303.65	20JA	-1.1	-4.1	-1.8	1.3
154.92	4.30	-0.8	313.73	30JA	-0.9	-3.5	-1.7	1.3
151.50	4.39	-1.0	323.77	9FE	0.3	-3.4	-1.7	1.2
147.60	4.31	-1.0	333.76	19FE	2.5	-3.9	-1.6	1.2
143.95	4.06	-0.9	343.68	1MR	2.4	-4.2	-1.6	1.1
141.17	3.70	-0.7	353.55	11MR	1.2	-4.1	-1.6	1.1
139.62	3.27	-0.5	3.37	21MR	0.7	-4.2	-1.6	1.1
139.37	2.84	-0.2	13.12	31MR	0.3	-4.1	-1.6	1.1
140.32	2.43	-0.0	22.83	10AP	-0.1	-4.0	-1.6	1.1
142.29	2.04	0.2	32.48	20AP	-0.9	-3.9	-1.6	1.1
145.12	1.70	0.4	42.10	30AP	-1.8	-3.8	-1.6	1.1
148.64	1.39	0.5	51.68	10MY	-1.2	-3.7	-1.6	1.0
152.74	1.11	0.7	61.23	20MY	-0.2	-3.7	-1.6	1.0
157.30	0.86	0.8	70.76	30MY	0.5	-3.6	-1.7	1.0
162.26	0.63	0.9	80.29	9JN	1.1	-3.5	-1.7	0.9
167.56	0.42	1.0	89.82	19JN	1.8	-3.5	-1.8	0.9
173.15	0.22	1.0	99.35	29JN	3.1	-3.4	-1.8	0.9
179.00	0.05	1.1	108.90	9JL	2.2	-3.4	-1.9	0.8
185.08	-0.12	1.2	118.49	19JL	0.5	-3.4	-2.0	0.8
191.37	-0.27	1.2	128.10	29JL	-0.8	-3.4	-2.0	0.7
197.84	-0.41	1.2	137.75	8AU	-1.4	-3.3	-2.1	0.7
204.51	-0.54	1.3	147.46	18AU	-1.2	-3.3	-2.2	0.6
211.33	-0.66	1.3	157.21	28AU	-0.6	-3.3	-2.2	0.6
218.31	-0.77	1.3	167.03	7SE	-0.2	-3.4	-2.3	0.5
225.44	-0.86	1.3	176.90	17SE	-0.0	-3.4	-2.3	0.5
232.69	-0.95	1.3	186.83	27SE	0.1	-3.4	-2.4	0.6
240.07	-1.02	1.3	196.82	7OC	0.4	-3.4	-2.4	0.7
247.57	-1.08	1.4	206.86	17OC	1.7	-3.4	-2.4	0.7
255.16	-1.13	1.4	216.94	27OC	1.6	-3.4	-2.4	0.8
262.83	-1.16	1.4	227.07	6NO	-0.1	-3.5	-2.4	0.9
270.58	-1.18	1.4	237.24	16NO	-0.4	-3.5	-2.3	0.9
278.38	-1.19	1.4	247.42	26NO	-0.4	-3.5	-2.3	1.0
286.23	-1.19	1.4	257.62	6DE	-0.5	-3.5	-2.2	1.0
294.11	-1.17	1.4	267.82	16DE	-0.7	-3.4	-2.2	1.0
301.99	-1.14	1.4	278.02	26DE	-0.9	-3.4	-2.1	1.0

Left table

♂ LONG	LAT	MAG	☉ LONG	16.00UT 672	☿	♀	♃	♄
309.88	-1.10	1.4	288.19	5JA	-1.0	-3.4	-2.0	1.0
317.75	-1.05	1.4	298.34	15JA	-0.8	-3.4	-2.0	1.0
325.59	-0.99	1.4	308.45	25JA	0.4	-3.3	-1.9	1.0
333.38	-0.92	1.4	318.51	4FE	2.8	-3.3	-1.8	1.0
341.13	-0.84	1.4	328.53	14FE	1.8	-3.3	-1.8	1.0
348.80	-0.76	1.5	338.48	24FE	0.9	-3.3	-1.7	0.9
356.42	-0.67	1.5	348.39	5MR	0.5	-3.4	-1.6	0.9
3.95	-0.57	1.5	358.23	15MR	0.2	-3.4	-1.6	0.8
11.40	-0.47	1.5	8.01	25MR	-0.2	-3.4	-1.6	0.8
18.77	-0.37	1.5	17.75	4AP	-0.9	-3.4	-1.5	0.8
26.06	-0.27	1.5	27.43	14AP	-1.8	-3.4	-1.5	0.8
33.25	-0.17	1.6	37.06	24AP	-1.2	-3.5	-1.5	0.8
40.36	-0.06	1.6	46.66	4MY	-0.2	-3.5	-1.5	0.8
47.39	0.04	1.7	56.22	14MY	0.7	-3.6	-1.5	0.8
54.33	0.15	1.7	65.77	24MY	1.4	-3.7	-1.5	0.8
61.19	0.25	1.8	75.30	3JN	2.5	-3.7	-1.5	0.8
67.97	0.35	1.8	84.82	13JN	3.3	-3.8	-1.5	0.7
74.67	0.45	1.8	94.35	23JN	1.8	-3.9	-1.6	0.7
81.30	0.55	1.9	103.89	3JL	0.4	-4.0	-1.6	0.7
87.86	0.65	1.9	113.46	13JL	-0.8	-4.1	-1.6	0.6
94.35	0.75	1.9	123.05	23JL	-1.5	-4.2	-1.7	0.6
100.76	0.84	1.9	132.69	2AU	-1.2	-4.2	-1.7	0.5
107.11	0.94	1.9	142.37	12AU	-0.5	-4.2	-1.8	0.5
113.39	1.03	1.9	152.09	22AU	-0.1	-4.0	-1.9	0.4
119.60	1.13	1.8	161.88	1SE	0.1	-3.5	-1.9	0.3
125.73	1.22	1.8	171.72	11SE	0.2	-3.4	-2.0	0.3
131.79	1.32	1.8	181.61	21SE	0.6	-3.9	-2.0	0.3
137.75	1.42	1.8	191.57	10C	2.1	-4.2	-2.1	0.3
143.61	1.51	1.7	201.58	110C	1.4	-4.3	-2.2	0.3
149.37	1.62	1.6	211.65	210C	-0.3	-4.3	-2.2	0.4
154.99	1.72	1.5	221.76	310C	-0.6	-4.2	-2.3	0.5
160.46	1.83	1.5	231.90	10NO	-0.6	-4.2	-2.3	0.5
165.75	1.94	1.3	242.08	20NO	-0.6	-4.1	-2.3	0.6
170.82	2.06	1.2	252.27	30NO	-0.8	-4.0	-2.3	0.7
175.63	2.18	1.1	262.48	10DE	-0.8	-3.9	-2.3	0.7
180.12	2.31	0.9	272.68	20DE	-0.8	-3.8	-2.2	0.7
184.20	2.44	0.7	282.87	30DE	-0.7	-3.7	-2.2	0.8
				673				
187.78	2.58	0.5	293.03	9JA	0.5	-3.6	-2.1	0.8
190.75	2.72	0.3	303.16	19JA	3.0	-3.6	-2.1	0.8
192.95	2.86	0.0	313.25	29JA	1.3	-3.5	-2.0	0.8
194.20	2.98	-0.3	323.29	8FE	0.6	-3.5	-1.9	0.8
194.32	3.07	-0.5	333.28	18FE	0.3	-3.4	-1.8	0.7
193.19	3.10	-0.8	343.21	28FE	0.1	-3.4	-1.8	0.7
190.82	3.03	-1.1	353.08	10MR	-0.3	-3.3	-1.7	0.7
187.47	2.84	-1.3	2.89	20MR	-0.9	-3.3	-1.6	0.6
183.72	2.51	-1.4	12.66	30MR	-1.8	-3.3	-1.6	0.6
180.30	2.09	-1.2	22.36	9AP	-1.2	-3.3	-1.5	0.6
177.81	1.62	-1.0	32.02	19AP	-0.1	-3.3	-1.5	0.6
176.59	1.15	-0.8	41.63	29AP	0.9	-3.3	-1.5	0.6
176.72	0.71	-0.6	51.21	9MY	1.9	-3.3	-1.4	0.6
178.06	0.33	-0.4	60.77	19MY	3.2	-3.4	-1.4	0.6
180.46	-0.01	-0.2	70.30	29MY	2.7	-3.4	-1.4	0.6
183.74	-0.30	-0.1	79.83	8JN	1.4	-3.4	-1.4	0.6
187.75	-0.55	0.1	89.35	18JN	0.3	-3.5	-1.4	0.6
192.35	-0.77	0.2	98.89	28JN	-0.8	-3.5	-1.4	0.5
197.46	-0.95	0.3	108.44	8JL	-1.6	-3.5	-1.4	0.5
202.99	-1.11	0.4	118.02	18JL	-1.2	-3.4	-1.4	0.5
208.88	-1.24	0.5	127.63	28JL	-0.5	-3.4	-1.5	0.4
215.08	-1.35	0.6	137.28	7AU	-0.0	-3.4	-1.5	0.4
221.54	-1.44	0.7	146.98	17AU	0.2	-3.4	-1.5	0.3
228.25	-1.51	0.7	156.74	27AU	0.4	-3.3	-1.6	0.2
235.17	-1.56	0.8	166.55	6SE	0.9	-3.3	-1.6	0.2
242.26	-1.59	0.9	176.42	16SE	2.5	-3.3	-1.7	0.1
249.51	-1.60	0.9	186.34	26SE	1.3	-3.3	-1.7	0.1
256.90	-1.59	1.0	196.32	60C	-0.4	-3.3	-1.8	0.0
264.40	-1.57	1.0	206.36	160C	-0.7	-3.4	-1.9	0.1
272.00	-1.53	1.1	216.45	260C	-0.7	-3.4	-1.9	0.1
279.66	-1.48	1.1	226.58	5NO	-0.7	-3.4	-2.0	0.2
287.38	-1.42	1.2	236.74	15NO	-0.7	-3.5	-2.1	0.3
295.14	-1.34	1.2	246.93	25NO	-0.6	-3.5	-2.1	0.3
302.90	-1.25	1.2	257.12	5DE	-0.7	-3.6	-2.1	0.4
310.67	-1.16	1.3	267.33	15DE	-0.5	-3.6	-2.2	0.5
318.43	-1.06	1.3	277.53	25DE	0.6	-3.7	-2.2	0.5

Right table

♂ LONG	LAT	MAG	☉ LONG	16.00UT 674	☿	♀	♃	♄
326.15	-0.95	1.4	287.70	4JA	2.9	-3.8	-2.2	0.5
333.83	-0.83	1.4	297.85	14JA	1.0	-3.8	-2.1	0.6
341.45	-0.72	1.5	307.96	24JA	0.4	-3.9	-2.1	0.6
349.01	-0.60	1.5	318.03	3FE	0.2	-4.0	-2.1	0.6
356.51	-0.48	1.5	328.05	13FE	-0.0	-4.1	-2.0	0.6
3.93	-0.37	1.6	338.01	23FE	-0.4	-4.2	-1.9	0.6
11.27	-0.25	1.6	347.91	5MR	-0.9	-4.2	-1.9	0.5
18.54	-0.14	1.7	357.76	15MR	-1.6	-4.3	-1.8	0.5
25.72	-0.03	1.7	7.54	25MR	-1.1	-4.1	-1.7	0.5
32.82	0.08	1.7	17.27	4AP	-0.1	-3.8	-1.7	0.4
39.85	0.18	1.7	26.96	14AP	1.1	-3.1	-1.6	0.4
46.80	0.28	1.7	36.60	24AP	2.5	-3.3	-1.5	0.4
53.67	0.37	1.8	46.19	4MY	3.6	-3.9	-1.5	0.4
60.48	0.46	1.8	55.76	14MY	2.1	-4.1	-1.4	0.5
67.21	0.55	1.8	65.30	24MY	1.1	-4.2	-1.4	0.5
73.89	0.63	1.8	74.83	3JN	0.2	-4.1	-1.4	0.5
80.51	0.71	1.8	84.36	13JN	-0.8	-4.1	-1.3	0.5
87.07	0.79	1.9	93.88	23JN	-1.7	-4.0	-1.3	0.5
93.59	0.86	1.9	103.42	3JL	-1.2	-3.9	-1.3	0.4
100.06	0.93	1.9	112.99	13JL	-0.4	-3.8	-1.3	0.4
106.49	0.99	2.0	122.58	23JL	0.1	-3.7	-1.3	0.4
112.88	1.05	2.0	132.21	2AU	0.4	-3.6	-1.3	0.3
119.23	1.11	2.0	141.89	12AU	0.6	-3.6	-1.3	0.3
125.56	1.16	2.0	151.62	22AU	1.2	-3.5	-1.3	0.2
131.85	1.21	2.0	161.40	1SE	2.9	-3.5	-1.4	0.2
138.10	1.26	2.0	171.24	11SE	1.1	-3.5	-1.4	0.1
144.33	1.31	2.0	181.14	21SE	-0.5	-3.4	-1.4	0.0
150.52	1.35	1.9	191.09	10C	-0.9	-3.4	-1.5	-0.0
156.68	1.39	1.9	201.10	110C	-0.9	-3.4	-1.5	-0.1
162.79	1.43	1.9	211.16	210C	-0.8	-3.4	-1.6	-0.1
168.86	1.46	1.8	221.27	310C	-0.6	-3.4	-1.7	-0.1
174.88	1.50	1.7	231.41	10NO	-0.5	-3.4	-1.8	-0.0
180.83	1.52	1.6	241.59	20NO	-0.5	-3.4	-1.8	0.0
186.71	1.55	1.6	251.78	30NO	-0.4	-3.4	-1.9	0.1
192.50	1.56	1.5	261.98	10DE	0.8	-3.4	-1.9	0.2
198.18	1.58	1.3	272.18	20DE	2.6	-3.4	-2.0	0.2
203.73	1.58	1.2	282.37	30DE	0.7	-3.4	-2.0	0.3
				675				
209.12	1.57	1.1	292.54	9JA	0.2	-3.4	-2.0	0.3
214.32	1.56	0.9	302.67	19JA	0.0	-3.4	-2.1	0.4
219.28	1.52	0.7	312.76	29JA	-0.1	-3.5	-2.1	0.4
223.94	1.47	0.5	322.80	8FE	-0.4	-3.5	-2.1	0.4
228.23	1.39	0.3	332.79	18FE	-1.0	-3.5	-2.0	0.4
232.06	1.28	0.0	342.72	28FE	-1.5	-3.4	-2.0	0.4
235.29	1.13	-0.2	352.60	10MR	-1.1	-3.4	-1.9	0.4
237.80	0.92	-0.5	2.42	20MR	0.0	-3.4	-1.9	0.4
239.40	0.65	-0.8	12.18	30MR	1.4	-3.4	-1.8	0.4
239.90	0.29	-1.1	21.89	9AP	3.2	-3.3	-1.7	0.3
239.18	-0.16	-1.5	31.55	19AP	2.7	-3.3	-1.7	0.3
237.23	-0.69	-1.8	41.16	29AP	1.6	-3.3	-1.6	0.3
234.51	-1.27	-2.0	50.75	9MY	0.8	-3.3	-1.6	0.3
231.04	-1.84	-2.0	60.30	19MY	0.1	-3.3	-1.5	0.3
228.10	-2.34	-1.9	69.84	29MY	-0.9	-3.4	-1.4	0.3
226.13	-2.72	-1.7	79.36	8JN	-1.7	-3.4	-1.4	0.4
225.50	-2.98	-1.5	88.89	18JN	-1.2	-3.4	-1.3	0.4
226.23	-3.13	-1.3	98.42	28JN	-0.4	-3.4	-1.3	0.4
228.24	-3.20	-1.1	107.97	8JL	0.2	-3.5	-1.3	0.4
231.34	-3.21	-0.9	117.55	18JL	0.5	-3.6	-1.3	0.3
235.33	-3.18	-0.7	127.16	28JL	0.9	-3.6	-1.3	0.3
240.05	-3.11	-0.5	136.81	7AU	1.6	-3.7	-1.2	0.3
245.36	-3.01	-0.3	146.51	17AU	3.2	-3.8	-1.2	0.2
251.15	-2.88	-0.2	156.27	27AU	1.0	-3.9	-1.2	0.2
257.33	-2.74	-0.1	166.07	6SE	-0.6	-4.0	-1.2	0.1
263.82	-2.58	0.1	175.94	16SE	-1.0	-4.1	-1.3	0.1
270.55	-2.42	0.2	185.86	26SE	-1.0	-4.2	-1.3	0.0
277.49	-2.24	0.3	195.85	60C	-0.8	-4.3	-1.3	-0.1
284.57	-2.05	0.4	205.88	160C	-0.5	-4.4	-1.4	-0.1
291.78	-1.86	0.6	215.96	260C	-0.4	-4.3	-1.4	-0.2
299.06	-1.67	0.7	226.09	5NO	-0.3	-4.1	-1.4	-0.3
306.40	-1.47	0.8	236.25	15NO	-0.2	-3.4	-1.5	-0.2
313.77	-1.28	0.9	246.43	25NO	1.0	-3.2	-1.6	-0.1
321.15	-1.10	1.0	256.63	5DE	2.3	-4.0	-1.6	-0.1
328.52	-0.92	1.1	266.83	15DE	0.4	-4.3	-1.7	0.0
335.87	-0.74	1.2	277.03	25DE	0.0	-4.4	-1.8	0.1

Left table (676 / 677)

♂ LONG	LAT	MAG	☉ LONG	16.00UT	☿	♀	♃	♄
				676				
343.18	-0.58	1.2	287.21	4JA	-0.1	-4.3	-1.8	0.1
350.45	-0.42	1.3	297.36	14JA	-0.2	-4.2	-1.9	0.2
357.67	-0.27	1.4	307.47	24JA	-0.5	-4.1	-1.9	0.3
4.84	-0.13	1.5	317.54	3FE	-1.0	-4.0	-2.0	0.3
11.94	0.00	1.5	327.56	13FE	-1.4	-3.9	-2.0	0.3
18.98	0.13	1.6	337.53	23FE	-1.1	-3.8	-2.0	0.3
25.96	0.24	1.7	347.43	4MR	0.1	-3.7	-2.0	0.4
32.87	0.35	1.7	357.28	14MR	1.8	-3.6	-2.0	0.4
39.72	0.45	1.8	7.08	24MR	3.5	-3.6	-2.0	0.3
46.50	0.54	1.8	16.81	3AP	2.0	-3.5	-1.9	0.3
53.23	0.62	1.9	26.49	13AP	1.2	-3.4	-1.9	0.3
59.91	0.70	1.9	36.14	23AP	0.6	-3.4	-1.8	0.3
66.53	0.77	1.9	45.73	3MY	-0.0	-3.4	-1.8	0.2
73.10	0.83	1.9	55.30	13MY	-0.9	-3.3	-1.7	0.2
79.64	0.89	1.9	64.85	23MY	-1.8	-3.3	-1.6	0.2
86.14	0.94	2.0	74.38	2JN	-1.2	-3.3	-1.6	0.3
92.60	0.99	2.0	83.90	12JN	-0.3	-3.3	-1.5	0.3
99.04	1.03	1.9	93.43	22JN	0.3	-3.3	-1.4	0.3
105.45	1.06	1.9	102.97	2JL	0.7	-3.3	-1.4	0.3
111.84	1.10	1.9	112.53	12JL	1.2	-3.3	-1.3	0.3
118.22	1.12	2.0	122.12	22JL	2.1	-3.4	-1.3	0.3
124.59	1.15	2.0	131.75	1AU	3.0	-3.4	-1.3	0.3
130.96	1.17	2.0	141.42	11AU	0.8	-3.4	-1.2	0.3
137.32	1.18	2.0	151.15	21AU	-0.7	-3.5	-1.2	0.3
143.68	1.19	2.0	160.92	31AU	-1.2	-3.5	-1.2	0.2
150.04	1.20	2.0	170.76	10SE	-1.1	-3.5	-1.2	0.2
156.41	1.20	2.0	180.65	20SE	-0.7	-3.5	-1.2	0.1
162.79	1.19	2.0	190.60	30SE	-0.4	-3.4	-1.2	0.0
169.17	1.18	2.0	200.61	10OC	-0.2	-3.4	-1.2	-0.0
175.57	1.17	1.9	210.67	20OC	-0.2	-3.4	-1.3	-0.1
181.96	1.15	1.9	220.77	30OC	0.0	-3.4	-1.3	-0.2
188.37	1.12	1.9	230.92	9NO	1.2	-3.3	-1.3	-0.3
194.78	1.09	1.8	241.09	19NO	2.1	-3.3	-1.4	-0.3
201.19	1.05	1.7	251.28	29NO	0.2	-3.3	-1.4	-0.2
207.60	1.00	1.7	261.48	9DE	-0.2	-3.3	-1.5	-0.2
214.00	0.94	1.6	271.69	19DE	-0.2	-3.4	-1.5	-0.1
220.40	0.86	1.5	281.87	29DE	-0.3	-3.4	-1.6	-0.0
				677				
226.78	0.78	1.4	292.04	8JA	-0.6	-3.4	-1.7	0.0
233.14	0.68	1.3	302.18	18JA	-1.0	-3.4	-1.7	0.1
239.47	0.56	1.1	312.27	28JA	-1.2	-3.5	-1.8	0.2
245.76	0.42	1.0	322.32	7FE	-1.0	-3.5	-1.9	0.2
252.00	0.26	0.9	332.31	17FE	0.2	-3.5	-1.9	0.2
258.17	0.07	0.7	342.25	27FE	2.2	-3.6	-2.0	0.3
264.25	-0.14	0.5	352.13	9MR	2.8	-3.7	-2.0	0.3
270.22	-0.39	0.4	1.95	19MR	1.5	-3.7	-2.0	0.3
276.04	-0.68	0.2	11.71	29MR	0.9	-3.8	-2.0	0.3
281.68	-1.01	-0.0	21.43	8AP	0.4	-3.9	-2.0	0.3
287.06	-1.40	-0.2	31.09	18AP	-0.1	-4.0	-2.0	0.3
292.12	-1.83	-0.5	40.71	28AP	-0.9	-4.1	-2.0	0.3
296.77	-2.34	-0.7	50.29	8MY	-1.8	-4.1	-1.9	0.2
300.86	-2.90	-1.0	59.85	18MY	-1.2	-4.2	-1.9	0.2
304.27	-3.54	-1.3	69.38	28MY	-0.3	-4.1	-1.8	0.2
306.78	-4.24	-1.6	78.91	7JN	0.4	-3.9	-1.7	0.2
308.19	-4.97	-1.8	88.43	17JN	0.9	-3.4	-1.7	0.3
308.38	-5.70	-2.1	97.96	27JN	1.5	-2.9	-1.6	0.3
307.29	-6.33	-2.4	107.51	7JL	2.7	-3.7	-1.6	0.3
305.16	-6.75	-2.6	117.09	17JL	2.6	-4.1	-1.5	0.3
302.56	-6.87	-2.6	126.69	27JL	0.7	-4.2	-1.5	0.3
300.20	-6.66	-2.5	136.35	6AU	-0.7	-4.2	-1.4	0.3
298.76	-6.19	-2.2	146.04	16AU	-1.3	-4.2	-1.4	0.3
298.57	-5.56	-1.9	155.79	26AU	-1.2	-4.1	-1.3	0.3
299.68	-4.88	-1.7	165.59	5SE	-0.6	-4.0	-1.3	0.3
301.99	-4.20	-1.4	175.46	15SE	-0.3	-3.9	-1.3	0.2
305.29	-3.56	-1.1	185.38	25SE	-0.1	-3.8	-1.3	0.2
309.37	-2.97	-0.8	195.36	5OC	-0.0	-3.8	-1.2	0.1
314.07	-2.45	-0.6	205.39	15OC	0.2	-3.7	-1.2	0.0
319.25	-1.98	-0.3	215.47	25OC	1.5	-3.6	-1.2	-0.0
324.79	-1.57	-0.1	225.60	4NO	1.8	-3.6	-1.2	-0.1
330.61	-1.21	0.1	235.75	14NO	-0.0	-3.5	-1.3	-0.2
336.64	-0.88	0.3	245.94	24NO	-0.3	-3.5	-1.3	-0.2
342.82	-0.60	0.5	256.14	4DE	-0.4	-3.4	-1.3	-0.3
349.12	-0.35	0.6	266.34	14DE	-0.4	-3.4	-1.4	-0.2
355.50	-0.13	0.8	276.54	24DE	-0.6	-3.4	-1.4	-0.2

Right table (678 / 679)

♂ LONG	LAT	MAG	☉ LONG	16.00UT	☿	♀	♃	♄
				678				
1.93	0.06	0.9	286.72	3JA	-1.0	-3.4	-1.5	-0.1
8.40	0.23	1.1	296.87	13JA	-1.1	-3.3	-1.5	-0.0
14.88	0.37	1.2	306.99	23JA	-0.9	-3.3	-1.6	0.0
21.37	0.50	1.3	317.06	2FE	0.3	-3.3	-1.7	0.1
27.86	0.62	1.4	327.08	12FE	2.6	-3.3	-1.7	0.2
34.34	0.72	1.5	337.05	22FE	2.1	-3.4	-1.8	0.2
40.80	0.80	1.6	346.96	4MR	1.1	-3.3	-1.9	0.2
47.25	0.88	1.7	356.81	14MR	0.6	-3.4	-1.9	0.3
53.67	0.94	1.8	6.60	24MR	0.3	-3.4	-2.0	0.3
60.08	0.99	1.8	16.34	3AP	-0.2	-3.4	-2.0	0.3
66.46	1.04	1.9	26.03	13AP	-0.9	-3.5	-2.1	0.3
72.83	1.08	1.9	35.67	23AP	-1.8	-3.5	-2.1	0.3
79.18	1.11	2.0	45.27	3MY	-1.2	-3.4	-2.1	0.3
85.51	1.13	2.0	54.84	13MY	-0.2	-3.4	-2.1	0.3
91.84	1.15	2.0	64.38	23MY	0.5	-3.4	-2.1	0.3
98.16	1.17	2.0	73.92	2JN	1.2	-3.4	-2.0	0.2
104.48	1.17	2.0	83.44	12JN	2.0	-3.3	-2.0	0.2
110.79	1.18	2.0	92.96	22JN	3.3	-3.3	-1.9	0.3
117.12	1.18	2.0	102.51	2JL	2.1	-3.3	-1.9	0.3
123.45	1.17	2.0	112.06	12JL	0.5	-3.3	-1.8	0.3
129.79	1.16	2.0	121.65	22JL	-0.8	-3.3	-1.7	0.4
136.16	1.14	2.0	131.28	1AU	-1.4	-3.3	-1.7	0.4
142.54	1.12	1.9	140.95	11AU	-1.2	-3.4	-1.6	0.4
148.95	1.10	1.9	150.67	21AU	-0.6	-3.4	-1.6	0.4
155.39	1.07	1.9	160.45	31AU	-0.2	-3.4	-1.5	0.4
161.87	1.03	2.0	170.28	10SE	0.0	-3.5	-1.5	0.4
168.38	1.00	2.0	180.17	20SE	0.1	-3.5	-1.4	0.3
174.93	0.95	2.0	190.12	30SE	0.4	-3.6	-1.4	0.3
181.51	0.90	1.9	200.12	10OC	1.8	-3.7	-1.4	0.2
188.14	0.85	1.9	210.18	20OC	1.6	-3.7	-1.3	0.2
194.82	0.78	1.9	220.28	30OC	-0.2	-3.8	-1.3	0.1
201.54	0.72	1.9	230.42	9NO	-0.5	-3.9	-1.3	0.0
208.30	0.64	1.8	240.59	19NO	-0.5	-4.0	-1.3	-0.0
215.11	0.56	1.8	250.79	29NO	-0.5	-4.1	-1.3	-0.1
221.97	0.47	1.7	260.99	9DE	-0.7	-4.3	-1.3	-0.2
228.87	0.37	1.7	271.19	19DE	-0.9	-4.3	-1.4	-0.2
235.82	0.26	1.6	281.38	29DE	-0.9	-4.4	-1.4	-0.2
				679				
242.81	0.14	1.5	291.55	8JA	-0.8	-4.3	-1.4	-0.1
249.84	0.01	1.5	301.69	18JA	0.4	-4.0	-1.5	-0.0
256.91	-0.13	1.4	311.78	28JA	2.9	-3.5	-1.5	0.0
264.02	-0.28	1.3	321.83	7FE	1.6	-3.4	-1.6	0.1
271.16	-0.45	1.2	331.83	17FE	0.8	-4.2	-1.7	0.1
278.32	-0.62	1.1	341.77	27FE	0.4	-4.2	-1.7	0.2
285.50	-0.80	1.0	351.65	9MR	0.2	-4.3	-1.8	0.3
292.70	-1.00	0.9	1.47	19MR	-0.2	-4.2	-1.9	0.3
299.89	-1.20	0.8	11.24	29MR	-0.9	-4.1	-1.9	0.3
307.06	-1.41	0.7	20.95	8AP	-1.8	-4.0	-2.0	0.4
314.21	-1.63	0.6	30.62	18AP	-1.2	-3.9	-2.1	0.4
321.29	-1.84	0.5	40.24	28AP	-0.2	-3.8	-2.1	0.4
328.30	-2.06	0.4	49.83	8MY	0.7	-3.7	-2.2	0.4
335.19	-2.28	0.2	59.39	18MY	1.6	-3.7	-2.2	0.4
341.94	-2.50	0.1	68.92	28MY	2.7	-3.6	-2.2	0.4
348.84	-2.71	-0.0	78.45	7JN	3.1	-3.5	-2.2	0.3
354.78	-2.91	-0.2	87.98	17JN	1.7	-3.5	-2.2	0.3
0.75	-3.10	-0.3	97.51	27JN	0.4	-3.4	-2.2	0.3
6.32	-3.28	-0.5	107.05	7JL	-0.8	-3.4	-2.1	0.3
11.37	-3.44	-0.6	116.63	17JL	-1.5	-3.4	-2.1	0.4
15.76	-3.57	-0.8	126.24	27JL	-1.2	-3.4	-2.0	0.4
19.35	-3.67	-1.0	135.88	6AU	-0.5	-3.3	-2.0	0.5
21.91	-3.73	-1.2	145.58	16AU	-0.1	-3.3	-1.9	0.5
23.24	-3.72	-1.5	155.32	26AU	0.1	-3.3	-1.8	0.5
23.17	-3.60	-1.7	165.12	5SE	0.3	-3.4	-1.8	0.5
21.62	-3.36	-1.9	174.98	15SE	0.7	-3.4	-1.7	0.5
18.85	-2.96	-2.1	184.90	25SE	2.2	-3.4	-1.7	0.5
15.44	-2.42	-2.2	194.87	5OC	1.5	-3.4	-1.6	0.4
12.20	-1.81	-1.9	204.90	15OC	-0.3	-3.4	-1.6	0.4
9.85	-1.19	-1.6	214.98	25OC	-0.6	-3.4	-1.5	0.3
8.78	-0.63	-1.3	225.10	4NO	-0.6	-3.5	-1.5	0.3
9.05	-0.16	-1.0	235.26	14NO	-0.7	-3.5	-1.5	0.2
10.54	0.23	-0.6	245.44	24NO	-0.8	-3.5	-1.4	0.1
13.02	0.53	-0.4	255.64	4DE	-0.7	-3.5	-1.4	0.1
16.30	0.76	-0.1	265.84	14DE	-0.8	-3.4	-1.4	-0.0
20.21	0.95	0.2	276.04	24DE	-0.6	-3.4	-1.4	-0.1

♂			⊙	16.00UT	☿	♀	♃	♄
LONG	LAT	MAG	LONG	680	MAGNITUDES			
24.59	1.09	0.4	286.22	3JA	0.5	-3.4	-1.4	-0.1
29.36	1.19	0.6	296.38	13JA	3.0	-3.4	-1.5	-0.1
34.41	1.28	0.8	306.49	23JA	1.2	-3.3	-1.5	0.0
39.68	1.34	1.0	316.57	2FE	0.5	-3.3	-1.5	0.1
45.13	1.38	1.1	326.59	12FE	0.3	-3.3	-1.6	0.1
50.72	1.41	1.3	336.56	22FE	0.1	-3.3	-1.6	0.2
56.42	1.43	1.4	346.48	3MR	-0.3	-3.4	-1.7	0.2
62.20	1.44	1.5	356.33	13MR	-0.9	-3.4	-1.7	0.3
68.05	1.45	1.6	6.13	23MR	-1.7	-3.4	-1.8	0.4
73.96	1.44	1.7	15.87	2AP	-1.2	-3.4	-1.8	0.4
79.91	1.43	1.8	25.56	12AP	-0.1	-3.4	-1.9	0.4
85.91	1.42	1.8	35.20	22AP	0.9	-3.5	-2.0	0.5
91.94	1.39	1.9	44.81	2MY	2.1	-3.5	-2.1	0.5
98.00	1.37	1.9	54.38	12MY	3.6	-3.6	-2.1	0.5
104.10	1.34	2.0	63.92	22MY	2.5	-3.7	-2.2	0.5
110.23	1.31	2.0	73.46	1JN	1.3	-3.7	-2.2	0.5
116.40	1.27	2.0	82.98	11JN	0.3	-3.8	-2.3	0.5
122.60	1.23	2.0	92.51	21JN	-0.8	-3.9	-2.3	0.5
128.84	1.19	2.0	102.05	1JL	-1.6	-4.0	-2.3	0.5
135.12	1.14	2.0	111.61	11JL	-1.2	-4.1	-2.3	0.5
141.44	1.09	2.0	121.19	21JL	-0.5	-4.2	-2.3	0.5
147.81	1.04	2.0	130.82	31JL	0.0	-4.2	-2.3	0.5
154.24	0.98	2.0	140.49	10AU	0.3	-4.2	-2.3	0.6
160.71	0.92	1.9	150.20	20AU	0.5	-3.9	-2.2	0.6
167.24	0.85	1.9	159.98	30AU	1.0	-3.4	-2.2	0.6
173.84	0.79	1.9	169.81	9SE	2.6	-3.4	-2.1	0.7
180.49	0.71	1.8	179.69	19SE	1.3	-4.0	-2.0	0.7
187.21	0.64	1.8	189.64	29SE	-0.4	-4.3	-2.0	0.6
194.00	0.55	1.8	199.64	9OC	-0.8	-4.3	-1.9	0.6
200.85	0.47	1.8	209.69	19OC	-0.8	-4.3	-1.8	0.6
207.78	0.38	1.8	219.79	29OC	-0.8	-4.2	-1.8	0.6
214.78	0.28	1.8	229.93	8NO	-0.7	-4.1	-1.7	0.5
221.85	0.19	1.7	240.10	18NO	-0.6	-4.0	-1.7	0.5
228.99	0.08	1.7	250.29	28NO	-0.6	-4.0	-1.6	0.4
236.19	-0.03	1.7	260.49	8DE	-0.5	-3.9	-1.6	0.3
243.47	-0.14	1.7	270.69	18DE	0.6	-3.8	-1.6	0.2
250.81	-0.26	1.6	280.89	28DE	2.9	-3.7	-1.5	0.2
				681				
258.21	-0.38	1.6	291.06	7JA	0.9	-3.6	-1.5	0.1
265.66	-0.50	1.5	301.19	17JA	0.3	-3.6	-1.5	0.1
273.17	-0.62	1.5	311.29	27JA	0.1	-3.5	-1.5	0.1
280.72	-0.75	1.4	321.35	6FE	-0.0	-3.4	-1.5	0.2
288.31	-0.87	1.4	331.34	16FE	-0.4	-3.4	-1.6	0.2
295.93	-0.99	1.3	341.29	26FE	-0.9	-3.4	-1.6	0.3
303.57	-1.11	1.3	351.18	8MR	-1.6	-3.3	-1.6	0.4
311.21	-1.22	1.2	1.00	18MR	-1.2	-3.3	-1.6	0.4
318.85	-1.33	1.2	10.78	28MR	-0.1	-3.3	-1.7	0.5
326.47	-1.42	1.1	20.49	7AP	1.2	-3.3	-1.7	0.5
334.06	-1.51	1.1	30.16	17AP	2.7	-3.3	-1.8	0.6
341.60	-1.58	1.0	39.78	27AP	3.3	-3.3	-1.9	0.6
349.07	-1.64	1.0	49.37	7MY	1.9	-3.3	-1.9	0.6
356.46	-1.68	0.9	58.92	17MY	1.0	-3.4	-2.0	0.7
3.75	-1.71	0.9	68.46	27MY	0.2	-3.4	-2.1	0.7
10.91	-1.72	0.8	77.99	6JN	-0.8	-3.4	-2.1	0.7
17.93	-1.71	0.7	87.51	16JN	-1.7	-3.5	-2.2	0.7
24.78	-1.68	0.7	97.05	26JN	-1.2	-3.5	-2.3	0.7
31.43	-1.63	0.6	106.59	6JL	-0.4	-3.5	-2.3	0.7
37.86	-1.57	0.5	116.16	16JL	0.1	-3.4	-2.4	0.7
44.02	-1.48	0.4	125.77	26JL	0.4	-3.4	-2.4	0.7
49.86	-1.37	0.3	135.41	5AU	0.7	-3.4	-2.4	0.7
55.33	-1.23	0.2	145.10	15AU	1.3	-3.3	-2.4	0.8
60.34	-1.06	0.1	154.85	25AU	3.0	-3.3	-2.4	0.8
64.81	-0.87	-0.0	164.64	4SE	1.1	-3.3	-2.4	0.8
68.60	-0.63	-0.2	174.50	14SE	-0.5	-3.3	-2.4	0.9
71.55	-0.36	-0.4	184.42	24SE	-0.9	-3.3	-2.3	0.9
73.48	-0.03	-0.6	194.38	4OC	-0.9	-3.3	-2.2	0.9
74.19	0.36	-0.8	204.41	14OC	-0.8	-3.4	-2.2	0.9
73.49	0.79	-1.0	214.49	24OC	-0.5	-3.4	-2.1	0.9
71.37	1.27	-1.2	224.61	3NO	-0.4	-3.4	-2.0	0.8
68.08	1.73	-1.4	234.76	13NO	-0.4	-3.5	-2.0	0.8
64.19	2.13	-1.5	244.94	23NO	-0.3	-3.5	-1.9	0.7
60.51	2.43	-1.3	255.14	3DE	0.8	-3.6	-1.8	0.7
57.73	2.60	-1.0	265.34	13DE	2.6	-3.6	-1.8	0.6
56.23	2.67	-0.7	275.54	23DE	0.6	-3.7	-1.7	0.5

♂			⊙	16.00UT	☿	♀	♃	♄
LONG	LAT	MAG	LONG	682	MAGNITUDES			
56.07	2.66	-0.4	285.72	2JA	0.1	-3.8	-1.7	0.5
57.11	2.61	-0.1	295.88	12JA	-0.0	-3.8	-1.7	0.4
59.17	2.54	0.2	306.00	22JA	-0.1	-3.9	-1.6	0.3
62.03	2.45	0.4	316.08	1FE	-0.4	-4.0	-1.6	0.3
65.53	2.35	0.6	326.11	11FE	-1.0	-4.1	-1.6	0.4
69.54	2.25	0.8	336.08	21FE	-1.5	-4.2	-1.6	0.4
73.94	2.16	1.0	345.99	3MR	-1.1	-4.3	-1.6	0.5
78.65	2.06	1.2	355.85	13MR	-0.0	-4.3	-1.6	0.5
83.62	1.97	1.3	5.66	23MR	1.5	-4.1	-1.6	0.6
88.79	1.88	1.4	15.40	2AP	3.4	-3.8	-1.6	0.6
94.14	1.79	1.5	25.09	12AP	2.4	-3.1	-1.6	0.7
99.63	1.70	1.6	34.74	22AP	1.4	-3.4	-1.7	0.7
105.24	1.61	1.7	44.34	2MY	0.7	-3.9	-1.7	0.8
110.95	1.52	1.7	53.92	12MY	0.1	-4.2	-1.8	0.8
116.77	1.44	1.8	63.46	22MY	-0.8	-4.2	-1.8	0.9
122.68	1.35	1.8	72.99	1JN	-1.7	-4.1	-1.9	0.9
128.67	1.27	1.9	82.52	11JN	-1.2	-4.1	-1.9	0.9
134.74	1.18	1.9	92.05	21JN	-0.4	-4.0	-2.0	0.9
140.89	1.09	1.9	101.58	1JL	0.2	-3.9	-2.1	0.9
147.12	1.01	1.9	111.14	11JL	0.6	-3.8	-2.2	0.9
153.43	0.92	1.9	120.73	21JL	1.0	-3.7	-2.2	0.9
159.81	0.83	1.9	130.35	31JL	1.8	-3.6	-2.3	0.9
166.28	0.73	1.9	140.02	10AU	3.2	-3.6	-2.3	0.9
172.84	0.64	1.9	149.73	20AU	1.0	-3.5	-2.4	1.0
179.47	0.55	1.9	159.50	30AU	-0.6	-3.5	-2.4	1.0
186.20	0.45	1.8	169.33	9SE	-1.1	-3.5	-2.5	1.1
193.01	0.35	1.8	179.21	19SE	-1.1	-3.4	-2.5	1.1
199.91	0.25	1.8	189.15	29SE	-0.7	-3.4	-2.5	1.1
206.89	0.15	1.7	199.15	9OC	-0.4	-3.4	-2.4	1.1
213.97	0.05	1.7	209.20	19OC	-0.3	-3.4	-2.4	1.2
221.13	-0.06	1.6	219.30	29OC	-0.3	-3.4	-2.3	1.1
228.39	-0.16	1.6	229.44	8NO	-0.1	-3.4	-2.3	1.1
235.72	-0.27	1.6	239.61	18NO	1.0	-3.4	-2.2	1.1
243.14	-0.37	1.5	249.79	28NO	2.3	-3.4	-2.1	1.1
250.63	-0.47	1.5	260.00	8DE	0.3	-3.4	-2.1	1.0
258.19	-0.57	1.5	270.20	18DE	-0.1	-3.4	-2.0	1.0
265.81	-0.67	1.5	280.39	28DE	-0.1	-3.4	-1.9	0.9
				683				
273.50	-0.77	1.5	290.56	7JA	-0.3	-3.4	-1.8	0.8
281.22	-0.85	1.5	300.70	17JA	-0.5	-3.4	-1.8	0.7
288.99	-0.94	1.5	310.80	27JA	-1.0	-3.5	-1.7	0.7
296.79	-1.01	1.4	320.85	6FE	-1.3	-3.5	-1.7	0.6
304.60	-1.08	1.4	330.85	16FE	-1.0	-3.5	-1.6	0.6
312.41	-1.13	1.4	340.80	26FE	0.1	-3.4	-1.6	0.6
320.22	-1.18	1.4	350.69	8MR	1.9	-3.4	-1.6	0.7
328.00	-1.21	1.4	0.52	18MR	3.3	-3.4	-1.6	0.7
335.75	-1.23	1.4	10.29	28MR	1.8	-3.4	-1.6	0.8
343.46	-1.24	1.4	20.01	7AP	1.0	-3.3	-1.6	0.8
351.11	-1.23	1.3	29.68	17AP	0.6	-3.3	-1.6	0.9
358.69	-1.21	1.3	39.31	27AP	-0.0	-3.3	-1.6	0.9
6.20	-1.18	1.3	48.90	7MY	-0.8	-3.3	-1.6	1.0
13.61	-1.13	1.3	58.46	17MY	-1.7	-3.3	-1.6	1.0
20.92	-1.07	1.3	68.00	27MY	-1.2	-3.4	-1.6	1.1
28.12	-1.00	1.3	77.53	6JN	-0.3	-3.4	-1.7	1.1
35.21	-0.91	1.3	87.05	16JN	0.3	-3.4	-1.7	1.2
42.16	-0.81	1.2	96.58	26JN	0.8	-3.5	-1.8	1.2
48.98	-0.70	1.2	106.13	6JL	1.3	-3.5	-1.8	1.2
55.64	-0.58	1.2	115.70	16JL	2.3	-3.6	-1.9	1.2
62.13	-0.44	1.1	125.30	26JL	2.9	-3.6	-2.0	1.2
68.45	-0.30	1.1	134.95	5AU	0.8	-3.7	-2.0	1.2
74.55	-0.13	1.0	144.63	15AU	-0.7	-3.8	-2.1	1.2
80.43	0.04	1.0	154.37	25AU	-1.2	-3.9	-2.2	1.2
86.03	0.23	0.9	164.17	4SE	-1.2	-4.0	-2.2	1.3
91.33	0.45	0.8	174.02	14SE	-0.7	-4.1	-2.3	1.3
96.26	0.68	0.7	183.93	24SE	-0.3	-4.2	-2.3	1.4
100.73	0.94	0.5	193.90	4OC	-0.2	-4.3	-2.4	1.3
104.67	1.23	0.4	203.93	14OC	-0.1	-4.4	-2.4	1.3
107.94	1.56	0.2	214.00	24OC	0.1	-4.3	-2.4	1.3
110.38	1.92	0.0	224.12	3NO	1.3	-4.0	-2.4	1.3
111.81	2.33	-0.2	234.27	13NO	2.1	-3.4	-2.4	1.2
112.05	2.77	-0.4	244.45	23NO	0.1	-3.3	-2.3	1.2
110.94	3.22	-0.6	254.65	3DE	-0.2	-4.0	-2.3	1.2
108.49	3.65	-0.9	264.85	13DE	-0.3	-4.3	-2.2	1.1
104.99	3.98	-1.0	275.05	23DE	-0.4	-4.4	-2.1	1.1

Left Table

♂ LONG	♂ LAT	♂ MAG	☉ LONG	16.00UT / 684	☿	♀	♃	♄
101.00	4.16	-1.1	285.23	2JA	-0.6	-4.3	-2.1	1.0
97.32	4.17	-0.9	295.39	12JA	-1.0	-4.2	-2.0	1.0
94.55	4.04	-0.6	305.51	22JA	-1.2	-4.1	-1.9	0.9
93.04	3.82	-0.4	315.59	1FE	-1.0	-4.0	-1.9	0.9
92.83	3.55	-0.1	325.62	11FE	0.1	-3.9	-1.8	0.9
93.79	3.27	0.2	335.60	21FE	2.3	-3.8	-1.7	0.8
95.75	3.00	0.4	345.52	2MR	2.6	-3.7	-1.7	0.9
98.51	2.75	0.6	355.37	12MR	1.3	-3.6	-1.6	0.9
101.91	2.52	0.8	5.18	22MR	0.8	-3.6	-1.6	1.0
105.83	2.30	0.9	14.93	1AP	0.4	-3.5	-1.5	1.0
110.17	2.10	1.1	24.62	11AP	-0.1	-3.4	-1.5	1.1
114.85	1.91	1.2	34.27	21AP	-0.8	-3.4	-1.5	1.2
119.81	1.74	1.3	43.88	1MY	-1.8	-3.4	-1.5	1.2
125.00	1.58	1.4	53.45	11MY	-1.2	-3.3	-1.5	1.3
130.41	1.42	1.5	63.00	21MY	-0.3	-3.3	-1.5	1.3
135.99	1.27	1.5	72.54	31MY	0.4	-3.3	-1.5	1.4
141.73	1.13	1.6	82.06	10JN	1.0	-3.3	-1.5	1.4
147.62	1.00	1.6	91.59	20JN	1.7	-3.3	-1.5	1.4
153.64	0.86	1.7	101.13	30JN	2.9	-3.3	-1.6	1.4
159.79	0.74	1.7	110.68	10JL	2.4	-3.3	-1.6	1.4
166.07	0.61	1.7	120.27	20JL	0.7	-3.4	-1.6	1.4
172.47	0.49	1.7	129.89	30JL	-0.7	-3.4	-1.7	1.3
178.98	0.37	1.7	139.55	9AU	-1.3	-3.4	-1.7	1.3
185.61	0.25	1.7	149.26	19AU	-1.2	-3.5	-1.8	1.2
192.36	0.13	1.7	159.03	29AU	-0.6	-3.5	-1.8	1.2
199.21	0.02	1.7	168.85	8SE	-0.2	-3.5	-1.9	1.2
206.18	-0.10	1.7	178.73	18SE	-0.1	-3.5	-2.0	1.2
213.26	-0.21	1.7	188.67	28SE	0.0	-3.4	-2.0	1.2
220.44	-0.31	1.6	198.66	8OC	0.3	-3.4	-2.1	1.2
227.72	-0.41	1.6	208.71	18OC	1.5	-3.4	-2.2	1.2
235.11	-0.51	1.6	218.81	28OC	1.9	-3.4	-2.2	1.1
242.58	-0.60	1.5	228.94	7NO	-0.1	-3.3	-2.2	1.1
250.14	-0.69	1.5	239.11	17NO	-0.4	-3.3	-2.3	1.1
257.78	-0.77	1.4	249.30	27NO	-0.4	-3.3	-2.3	1.1
265.49	-0.85	1.4	259.50	7DE	-0.5	-3.3	-2.3	1.0
273.26	-0.91	1.4	269.70	17DE	-0.7	-3.4	-2.2	1.0
281.08	-0.97	1.3	279.89	27DE	-0.9	-3.4	-2.2	1.0
				685				
288.93	-1.02	1.3	290.06	6JA	-1.0	-3.4	-2.2	0.9
296.81	-1.05	1.3	300.21	16JA	-0.8	-3.4	-2.1	0.9
304.70	-1.08	1.3	310.31	26JA	0.2	-3.5	-2.0	0.8
312.59	-1.09	1.4	320.37	5FE	2.6	-3.5	-2.0	0.8
320.47	-1.09	1.4	330.37	15FE	1.9	-3.5	-1.9	0.7
328.31	-1.08	1.4	340.32	25FE	1.0	-3.6	-1.8	0.7
336.12	-1.06	1.4	350.21	7MR	0.6	-3.7	-1.7	0.6
343.88	-1.02	1.4	0.05	17MR	0.3	-3.7	-1.7	0.7
351.58	-0.98	1.4	9.82	27MR	-0.2	-3.8	-1.6	0.8
359.21	-0.92	1.5	19.54	6AP	-0.9	-3.9	-1.6	0.8
6.76	-0.86	1.5	29.22	16AP	-1.8	-4.0	-1.5	0.9
14.23	-0.78	1.5	38.84	26AP	-1.2	-4.1	-1.5	1.0
21.61	-0.70	1.5	48.43	6MY	-0.2	-4.1	-1.4	1.0
28.89	-0.61	1.5	58.00	16MY	0.6	-4.2	-1.4	1.1
36.08	-0.51	1.6	67.54	26MY	1.3	-4.1	-1.4	1.1
43.17	-0.41	1.6	77.06	5JN	2.3	-3.9	-1.4	1.2
50.15	-0.30	1.6	86.59	15JN	3.4	-3.4	-1.4	1.2
57.03	-0.18	1.6	96.12	25JN	1.9	-2.9	-1.4	1.2
63.80	-0.06	1.6	105.66	5JL	0.5	-3.7	-1.4	1.2
70.45	0.07	1.6	115.24	15JL	-0.7	-4.1	-1.4	1.2
76.99	0.20	1.6	124.83	25JL	-1.4	-4.2	-1.4	1.2
83.41	0.34	1.5	134.48	4AU	-1.2	-4.2	-1.4	1.2
89.69	0.48	1.5	144.16	14AU	-0.6	-4.2	-1.5	1.2
95.84	0.63	1.5	153.90	24AU	-0.2	-4.1	-1.5	1.1
101.83	0.78	1.4	163.69	3SE	0.1	-4.0	-1.6	1.1
107.65	0.94	1.4	173.54	13SE	0.2	-3.9	-1.6	1.0
113.27	1.12	1.3	183.45	23SE	0.5	-3.8	-1.7	1.0
118.67	1.30	1.2	193.42	3OC	1.9	-3.7	-1.7	1.0
123.81	1.51	1.1	203.44	13OC	1.7	-3.7	-1.8	1.0
128.63	1.72	1.0	213.51	23OC	-0.2	-3.6	-1.9	1.0
133.07	1.96	0.9	223.62	2NO	-0.6	-3.6	-1.9	1.0
137.05	2.22	0.7	233.78	12NO	-0.6	-3.5	-2.0	1.0
140.45	2.51	0.5	243.95	22NO	-0.6	-3.5	-2.0	1.0
143.15	2.83	0.3	254.18	2DE	-0.8	-3.4	-2.1	1.0
144.96	3.17	0.1	264.36	12DE	-0.8	-3.4	-2.1	0.9
145.72	3.54	-0.1	274.55	22DE	-0.9	-3.4	-2.1	0.9

Right Table

♂ LONG	♂ LAT	♂ MAG	☉ LONG	16.00UT / 686	☿	♀	♃	♄
145.25	3.90	-0.4	284.74	1JA	-0.7	-3.4	-2.2	0.9
143.46	4.23	-0.6	294.90	11JA	0.3	-3.3	-2.1	0.8
140.48	4.47	-0.9	305.02	21JA	2.9	-3.3	-2.1	0.8
136.70	4.56	-1.0	315.10	31JA	1.5	-3.3	-2.1	0.7
132.81	4.47	-0.9	325.14	10FE	0.7	-3.3	-2.0	0.7
129.50	4.22	-0.7	335.11	20FE	0.4	-3.3	-2.0	0.6
127.27	3.87	-0.5	345.04	2MR	0.1	-3.4	-1.9	0.6
126.34	3.47	-0.3	354.90	12MR	-0.2	-3.4	-1.8	0.5
126.66	3.07	-0.0	4.70	22MR	-0.9	-3.4	-1.8	0.5
128.08	2.69	0.2	14.45	1AP	-1.7	-3.4	-1.7	0.6
130.44	2.34	0.4	24.15	11AP	-1.2	-3.5	-1.6	0.6
133.54	2.02	0.6	33.80	21AP	-0.2	-3.5	-1.6	0.7
137.27	1.73	0.7	43.41	1MY	0.8	-3.4	-1.5	0.8
141.49	1.47	0.8	52.99	11MY	1.7	-3.4	-1.5	0.8
146.14	1.23	1.0	62.53	21MY	3.0	-3.4	-1.4	0.9
151.31	1.01	1.1	72.07	31MY	2.9	-3.4	-1.4	0.9
156.42	0.81	1.1	81.59	10JN	1.5	-3.3	-1.3	1.0
161.97	0.62	1.2	91.12	20JN	0.4	-3.3	-1.3	1.0
167.75	0.45	1.3	100.66	30JN	-0.8	-3.3	-1.3	1.1
173.74	0.28	1.3	110.21	10JL	-1.5	-3.3	-1.3	1.1
179.92	0.12	1.4	119.79	20JL	-1.2	-3.3	-1.3	1.1
186.28	-0.02	1.4	129.42	30JL	-0.5	-3.3	-1.3	1.1
192.80	-0.16	1.4	139.08	9AU	-0.1	-3.4	-1.3	1.1
199.48	-0.29	1.4	148.78	19AU	0.2	-3.4	-1.3	1.0
206.31	-0.41	1.4	158.55	29AU	0.4	-3.4	-1.3	1.0
213.28	-0.53	1.5	168.37	8SE	0.8	-3.5	-1.4	1.0
220.38	-0.63	1.5	178.25	18SE	2.3	-3.5	-1.4	0.9
227.60	-0.73	1.5	188.19	28SE	1.5	-3.6	-1.4	0.9
234.95	-0.82	1.4	198.18	8OC	-0.3	-3.7	-1.5	0.9
242.40	-0.89	1.4	208.22	18OC	-0.7	-3.7	-1.5	0.9
249.96	-0.96	1.4	218.32	28OC	-0.7	-3.8	-1.6	0.9
257.60	-1.02	1.4	228.45	7NO	-0.7	-3.9	-1.7	0.9
265.32	-1.06	1.4	238.61	17NO	-0.7	-4.0	-1.7	0.9
273.11	-1.10	1.4	248.81	27NO	-0.7	-4.1	-1.8	0.9
280.95	-1.12	1.4	259.00	7DE	-0.7	-4.3	-1.9	0.9
288.82	-1.13	1.4	269.21	17DE	-0.6	-4.3	-1.9	0.9
296.72	-1.12	1.4	279.40	27DE	0.5	-4.4	-2.0	0.8
				687				
304.62	-1.11	1.4	289.57	6JA	3.0	-4.3	-2.0	0.8
312.52	-1.08	1.4	299.72	16JA	1.1	-4.0	-2.0	0.8
320.40	-1.04	1.4	309.82	26JA	0.4	-3.5	-2.1	0.7
328.25	-0.99	1.4	319.88	5FE	0.2	-3.5	-2.1	0.7
336.05	-0.94	1.4	329.89	15FE	0.0	-4.0	-2.0	0.6
343.80	-0.87	1.4	339.84	25FE	-0.3	-4.2	-2.0	0.6
351.48	-0.79	1.4	349.73	7MR	-0.9	-4.3	-2.0	0.5
359.09	-0.71	1.4	359.57	17MR	-1.7	-4.2	-1.9	0.5
6.63	-0.62	1.4	9.35	27MR	-1.2	-4.1	-1.9	0.4
14.08	-0.53	1.5	19.07	6AP	-0.1	-4.0	-1.8	0.4
21.44	-0.43	1.5	28.75	16AP	1.0	-3.9	-1.7	0.5
28.72	-0.33	1.6	38.38	26AP	2.3	-3.8	-1.7	0.5
35.91	-0.23	1.6	47.97	6MY	3.8	-3.7	-1.6	0.6
43.01	-0.12	1.7	57.53	16MY	2.2	-3.6	-1.5	0.7
50.02	-0.02	1.7	67.08	26MY	1.2	-3.6	-1.5	0.7
56.94	0.09	1.7	76.60	5JN	0.3	-3.5	-1.4	0.8
63.78	0.20	1.8	86.13	15JN	-0.8	-3.5	-1.4	0.8
70.53	0.31	1.8	95.66	25JN	-1.6	-3.4	-1.3	0.9
77.19	0.41	1.8	105.20	5JL	-1.2	-3.4	-1.3	0.9
83.78	0.52	1.8	114.77	15JL	-0.5	-3.4	-1.3	0.9
90.28	0.63	1.8	124.37	25JL	0.0	-3.4	-1.2	1.0
96.70	0.74	1.8	134.01	4AU	0.3	-3.3	-1.2	1.0
103.04	0.85	1.8	143.69	14AU	0.6	-3.3	-1.2	1.0
109.29	0.96	1.8	153.43	24AU	1.1	-3.3	-1.2	1.0
115.44	1.07	1.8	163.22	3SE	2.7	-3.4	-1.2	0.9
121.51	1.19	1.7	173.07	13SE	1.3	-3.4	-1.2	0.9
127.46	1.30	1.7	182.97	23SE	-0.4	-3.4	-1.3	0.9
133.30	1.42	1.6	192.93	3OC	-0.9	-3.4	-1.3	0.8
139.01	1.55	1.6	202.95	13OC	-0.8	-3.4	-1.3	0.8
144.56	1.68	1.5	213.02	23OC	-0.8	-3.4	-1.4	0.8
149.93	1.81	1.4	223.13	2NO	-0.6	-3.5	-1.4	0.8
155.09	1.96	1.3	233.29	12NO	-0.5	-3.5	-1.5	0.8
159.99	2.11	1.2	243.46	22NO	-0.5	-3.5	-1.5	0.9
164.58	2.28	1.0	253.66	2DE	-0.4	-3.5	-1.6	0.9
168.79	2.45	0.8	263.86	12DE	0.6	-3.4	-1.6	0.8
172.52	2.64	0.7	274.06	22DE	2.8	-3.4	-1.7	0.8

♂			☉	16.00UT	☿	♀	♃	♄
LONG	LAT	MAG	LONG	688	MAGNITUDES			
175.66	2.85	0.4	284.24	1JA	0.8	-3.4	-1.8	0.8
178.08	3.06	0.2	294.40	11JA	0.2	-3.4	-1.8	0.8
179.58	3.27	-0.0	304.53	21JA	0.1	-3.3	-1.9	0.7
180.01	3.48	-0.3	314.61	31JA	-0.1	-3.3	-1.9	0.7
179.20	3.64	-0.6	324.65	10FE	-0.4	-3.3	-2.0	0.7
177.11	3.73	-0.9	334.63	20FE	-0.9	-3.3	-2.0	0.6
173.94	3.70	-1.1	344.55	1MR	-1.5	-3.4	-2.0	0.5
170.16	3.52	-1.2	354.42	11MR	-1.2	-3.4	-2.0	0.5
166.48	3.19	-1.1	4.23	21MR	-0.1	-3.4	-2.0	0.4
163.57	2.77	-0.9	13.98	31MR	1.3	-3.4	-2.0	0.4
161.84	2.30	-0.7	23.68	10AP	3.0	-3.4	-1.9	0.3
161.43	1.84	-0.5	33.34	20AP	2.9	-3.5	-1.9	0.4
162.28	1.42	-0.3	42.95	30AP	1.7	-3.5	-1.8	0.4
164.21	1.04	-0.1	52.53	10MY	0.9	-3.6	-1.7	0.5
167.06	0.70	0.1	62.08	20MY	0.2	-3.7	-1.7	0.6
170.66	0.39	0.2	71.61	30MY	-0.8	-3.7	-1.6	0.6
174.87	0.13	0.4	81.14	9JN	-1.7	-3.8	-1.5	0.7
179.61	-0.11	0.5	90.66	19JN	-1.2	-3.9	-1.5	0.7
184.77	-0.32	0.6	100.20	29JN	-0.4	-4.0	-1.4	0.8
190.31	-0.50	0.7	109.75	9JL	0.1	-4.1	-1.4	0.8
196.17	-0.67	0.8	119.33	19JL	0.5	-4.2	-1.3	0.9
202.30	-0.81	0.8	128.95	29JL	0.8	-4.2	-1.3	0.9
208.69	-0.94	0.9	138.61	8AU	1.5	-4.2	-1.3	0.9
215.29	-1.05	1.0	148.32	18AU	3.1	-3.9	-1.2	0.9
222.10	-1.15	1.0	158.08	28AU	1.1	-3.4	-1.2	0.9
229.09	-1.22	1.0	167.90	7SE	-0.5	-3.4	-1.2	0.9
236.23	-1.29	1.1	177.77	17SE	-1.0	-4.0	-1.2	0.9
243.53	-1.33	1.1	187.70	27SE	-1.0	-4.3	-1.2	0.8
250.95	-1.36	1.1	197.69	7OC	-0.8	-4.3	-1.2	0.8
258.48	-1.38	1.2	207.74	17OC	-0.5	-4.3	-1.2	0.7
266.10	-1.38	1.2	217.83	27OC	-0.4	-4.2	-1.3	0.7
273.81	-1.37	1.2	227.96	6NO	-0.4	-4.1	-1.3	0.8
281.57	-1.34	1.3	238.12	16NO	-0.2	-4.0	-1.3	0.8
289.37	-1.30	1.3	248.31	26NO	0.8	-3.9	-1.4	0.8
297.20	-1.24	1.3	258.51	6DE	2.6	-3.9	-1.4	0.8
305.03	-1.18	1.3	268.71	16DE	0.5	-3.8	-1.5	0.8
312.86	-1.10	1.4	278.91	26DE	0.0	-3.7	-1.6	0.8
				689				
320.67	-1.02	1.4	289.09	5JA	-0.1	-3.6	-1.6	0.8
328.44	-0.93	1.4	299.23	15JA	-0.2	-3.6	-1.7	0.8
336.17	-0.83	1.5	309.34	25JA	-0.5	-3.5	-1.8	0.7
343.84	-0.73	1.5	319.40	4FE	-1.0	-3.4	-1.8	0.7
351.44	-0.63	1.5	329.41	14FE	-1.4	-3.4	-1.9	0.7
358.97	-0.52	1.5	339.36	24FE	-1.1	-3.4	-1.9	0.6
6.42	-0.41	1.6	349.26	6MR	-0.0	-3.3	-2.0	0.6
13.80	-0.31	1.6	359.10	16MR	1.6	-3.3	-2.0	0.5
21.09	-0.20	1.6	8.88	26MR	3.6	-3.3	-2.0	0.4
28.30	-0.09	1.6	18.61	5AP	2.2	-3.3	-2.0	0.4
35.42	0.01	1.7	28.28	15AP	1.3	-3.3	-2.0	0.3
42.46	0.12	1.7	37.92	25AP	0.7	-3.3	-2.0	0.3
49.43	0.22	1.7	47.51	5MY	0.1	-3.3	-2.0	0.3
56.31	0.32	1.7	57.07	15MY	-0.8	-3.4	-1.9	0.4
63.12	0.41	1.7	66.62	25MY	-1.7	-3.4	-1.9	0.5
69.87	0.50	1.8	76.14	4JN	-1.2	-3.4	-1.8	0.5
76.54	0.59	1.8	85.67	14JN	-0.4	-3.5	-1.7	0.6
83.16	0.68	1.9	95.20	24JN	0.2	-3.5	-1.7	0.6
89.71	0.76	1.9	104.74	4JL	0.6	-3.5	-1.6	0.7
96.21	0.84	1.9	114.30	14JL	1.1	-3.4	-1.6	0.7
102.66	0.92	2.0	123.90	24JL	1.9	-3.4	-1.5	0.8
109.05	0.99	2.0	133.54	3AU	3.1	-3.4	-1.4	0.8
115.40	1.06	2.0	143.22	13AU	1.0	-3.3	-1.4	0.8
121.71	1.13	2.0	152.95	23AU	-0.6	-3.3	-1.4	0.8
127.96	1.20	2.0	162.74	2SE	-1.1	-3.3	-1.3	0.8
134.17	1.27	1.9	172.58	12SE	-1.1	-3.3	-1.3	0.8
140.33	1.33	1.9	182.49	22SE	-0.7	-3.3	-1.3	0.8
146.43	1.40	1.9	192.45	2OC	-0.4	-3.3	-1.3	0.8
152.48	1.46	1.8	202.46	12OC	-0.3	-3.4	-1.3	0.8
158.47	1.52	1.8	212.53	22OC	-0.2	-3.4	-1.3	0.7
164.37	1.58	1.7	222.64	1NO	-0.0	-3.4	-1.3	0.7
170.19	1.64	1.6	232.79	11NO	1.0	-3.5	-1.3	0.7
175.91	1.69	1.5	242.97	21NO	2.3	-3.5	-1.3	0.7
181.50	1.75	1.4	253.16	1DE	0.3	-3.6	-1.3	0.8
186.96	1.80	1.3	263.36	11DE	-0.1	-3.6	-1.4	0.8
192.23	1.85	1.2	273.57	21DE	-0.2	-3.7	-1.4	0.8
197.29	1.90	1.0	283.75	31DE	-0.3	-3.8	-1.5	0.8

♂			☉	16.00UT	☿	♀	♃	♄
LONG	LAT	MAG	LONG	690	MAGNITUDES			
202.08	1.94	0.8	293.91	10JA	-0.5	-3.8	-1.5	0.8
206.55	1.97	0.7	304.04	20JA	-1.0	-3.9	-1.6	0.8
210.61	1.99	0.4	314.13	30JA	-1.3	-4.0	-1.6	0.8
214.17	2.00	0.2	324.17	9FE	-1.0	-4.1	-1.7	0.7
217.10	1.98	-0.0	334.15	19FE	0.0	-4.2	-1.8	0.7
219.22	1.93	-0.3	344.08	1MR	2.0	-4.3	-1.8	0.6
220.39	1.83	-0.6	353.95	11MR	3.0	-4.3	-1.9	0.6
220.40	1.66	-0.9	3.76	21MR	1.6	-4.1	-2.0	0.5
219.15	1.40	-1.2	13.51	31MR	0.9	-3.8	-2.0	0.5
216.72	1.05	-1.5	23.22	10AP	0.5	-3.1	-2.1	0.4
213.42	0.61	-1.8	32.87	20AP	-0.0	-3.4	-2.1	0.3
209.92	0.11	-1.7	42.48	30AP	-0.8	-3.9	-2.1	0.3
206.91	-0.38	-1.6	52.06	10MY	-1.7	-4.2	-2.1	0.3
204.97	-0.83	-1.4	61.62	20MY	-1.2	-4.2	-2.1	0.3
204.39	-1.21	-1.2	71.15	30MY	-0.3	-4.1	-2.1	0.4
205.14	-1.52	-0.9	80.67	9JN	0.3	-4.1	-2.0	0.4
207.13	-1.76	-0.7	90.20	19JN	0.8	-4.0	-2.0	0.5
210.15	-1.94	-0.6	99.73	29JN	1.4	-3.9	-1.9	0.6
214.02	-2.07	-0.4	109.29	9JL	2.5	-3.8	-1.9	0.6
218.61	-2.17	-0.2	118.87	19JL	2.8	-3.7	-1.8	0.7
223.77	-2.23	-0.1	128.48	29JL	0.8	-3.6	-1.7	0.7
229.41	-2.26	0.0	138.14	8AU	-0.7	-3.6	-1.7	0.8
235.46	-2.26	0.1	147.84	18AU	-1.3	-3.5	-1.6	0.8
241.83	-2.24	0.3	157.60	28AU	-1.2	-3.5	-1.6	0.8
248.48	-2.20	0.4	167.41	7SE	-0.7	-3.5	-1.5	0.8
255.36	-2.14	0.5	177.29	17SE	-0.3	-3.4	-1.5	0.8
262.44	-2.06	0.5	187.22	27SE	-0.1	-3.4	-1.4	0.8
269.66	-1.96	0.6	197.20	7OC	-0.1	-3.4	-1.4	0.8
277.01	-1.85	0.7	207.25	17OC	0.2	-3.4	-1.4	0.8
284.45	-1.73	0.8	217.33	27OC	1.3	-3.4	-1.4	0.7
291.97	-1.60	0.9	227.47	6NO	2.1	-3.4	-1.3	0.7
299.52	-1.46	1.0	237.63	16NO	0.1	-3.4	-1.3	0.7
307.10	-1.32	1.0	247.81	26NO	-0.3	-3.4	-1.3	0.7
314.69	-1.18	1.1	258.02	6DE	-0.3	-3.4	-1.3	0.7
322.26	-1.03	1.2	268.22	16DE	-0.4	-3.4	-1.4	0.8
329.81	-0.88	1.3	278.41	26DE	-0.6	-3.4	-1.4	0.8
				691				
337.31	-0.73	1.3	288.59	5JA	-1.0	-3.4	-1.4	0.8
344.77	-0.59	1.4	298.74	15JA	-1.1	-3.5	-1.5	0.8
352.17	-0.45	1.5	308.85	25JA	-0.9	-3.5	-1.5	0.8
359.51	-0.32	1.5	318.91	4FE	0.1	-3.5	-1.6	0.8
6.78	-0.19	1.6	328.93	14FE	2.3	-3.5	-1.6	0.7
13.98	-0.07	1.6	338.88	24FE	2.3	-3.4	-1.7	0.7
21.11	0.05	1.7	348.78	6MR	1.2	-3.4	-1.8	0.7
28.16	0.16	1.7	358.62	16MR	0.7	-3.4	-1.8	0.6
35.14	0.27	1.8	8.41	26MR	0.3	-3.4	-1.9	0.6
42.06	0.37	1.8	18.14	5AP	-0.1	-3.3	-2.0	0.5
48.90	0.46	1.8	27.82	15AP	-0.8	-3.3	-2.0	0.4
55.68	0.55	1.8	37.45	25AP	-1.7	-3.3	-2.1	0.4
62.40	0.63	1.9	47.05	5MY	-1.3	-3.3	-2.1	0.3
69.06	0.70	1.9	56.61	15MY	-0.3	-3.3	-2.2	0.3
75.67	0.77	1.9	66.15	25MY	0.5	-3.4	-2.2	0.3
82.24	0.84	1.9	75.68	4JN	1.1	-3.4	-2.2	0.3
88.76	0.90	1.9	85.21	14JN	1.9	-3.4	-2.2	0.4
95.24	0.95	1.9	94.73	24JN	3.2	-3.5	-2.2	0.5
101.69	1.00	1.9	104.28	4JL	2.3	-3.5	-2.2	0.5
108.11	1.05	2.0	113.84	14JL	0.6	-3.6	-2.1	0.6
114.51	1.09	2.0	123.44	24JL	-0.7	-3.6	-2.1	0.6
120.88	1.13	2.0	133.07	3AU	-1.4	-3.7	-2.0	0.7
127.24	1.16	2.0	142.75	13AU	-1.2	-3.8	-2.0	0.7
133.58	1.19	2.0	152.48	23AU	-0.6	-3.9	-1.9	0.8
139.92	1.21	2.0	162.26	2SE	-0.2	-4.0	-1.8	0.8
146.24	1.23	2.0	172.10	12SE	-0.0	-4.1	-1.8	0.8
152.56	1.25	2.0	182.00	22SE	0.1	-4.2	-1.7	0.8
158.86	1.27	2.0	191.96	2OC	0.4	-4.3	-1.7	0.8
165.16	1.27	2.0	201.97	12OC	1.6	-4.4	-1.6	0.8
171.45	1.28	1.9	212.04	22OC	1.9	-4.3	-1.6	0.8
177.73	1.28	1.9	222.15	1NO	-0.1	-4.0	-1.5	0.7
183.99	1.27	1.8	232.29	11NO	-0.5	-3.3	-1.5	0.7
190.23	1.26	1.8	242.47	21NO	-0.5	-3.3	-1.5	0.7
196.45	1.24	1.7	252.67	1DE	-0.5	-4.0	-1.5	0.7
202.64	1.21	1.6	262.87	11DE	-0.7	-4.3	-1.4	0.7
208.80	1.17	1.5	273.07	21DE	-0.9	-4.4	-1.4	0.8
214.90	1.12	1.4	283.26	31DE	-0.9	-4.3	-1.4	0.8

♂			☉	16.00UT	☿	♀	♃	♄
LONG	LAT	MAG	LONG	692	MAGNITUDES			
220.95	1.06	1.3	293.42	10JA	-0.8	-4.2	-1.5	0.8
226.93	0.98	1.1	303.55	20JA	0.2	-4.1	-1.5	0.8
232.82	0.88	1.0	313.64	30JA	2.7	-4.0	-1.5	0.8
238.60	0.77	0.8	323.68	9FE	1.8	-3.9	-1.5	0.8
244.24	0.62	0.7	333.67	19FE	0.8	-3.8	-1.6	0.8
249.72	0.45	0.5	343.60	29FE	0.5	-3.7	-1.6	0.7
254.97	0.24	0.3	353.47	10MR	0.2	-3.6	-1.7	0.7
259.96	-0.02	0.0	3.29	20MR	-0.2	-3.6	-1.7	0.7
264.61	-0.33	-0.2	13.05	30MR	-0.8	-3.5	-1.8	0.6
268.81	-0.70	-0.5	22.75	9AP	-1.7	-3.4	-1.9	0.5
272.46	-1.15	-0.7	32.41	19AP	-1.3	-3.4	-1.9	0.5
275.39	-1.68	-1.0	42.03	29AP	-0.2	-3.4	-2.0	0.4
277.43	-2.30	-1.3	51.61	9MY	0.6	-3.3	-2.1	0.4
278.39	-3.00	-1.7	61.16	19MY	1.5	-3.3	-2.2	0.3
278.12	-3.77	-2.0	70.70	29MY	2.5	-3.3	-2.2	0.2
276.66	-4.52	-2.3	80.22	8JN	3.3	-3.3	-2.3	0.3
274.25	-5.18	-2.5	89.74	18JN	1.8	-3.3	-2.3	0.3
271.47	-5.64	-2.5	99.28	28JN	0.5	-3.3	-2.3	0.4
269.07	-5.83	-2.3	108.83	8JL	-0.7	-3.3	-2.4	0.5
267.65	-5.77	-2.1	118.41	18JL	-1.5	-3.4	-2.4	0.5
267.54	-5.52	-1.9	128.02	28JL	-1.2	-3.4	-2.3	0.6
268.78	-5.16	-1.6	137.67	7AU	-0.5	-3.4	-2.3	0.6
271.24	-4.75	-1.4	147.37	17AU	-0.1	-3.5	-2.3	0.7
274.72	-4.31	-1.1	157.13	27AU	0.1	-3.5	-2.2	0.7
279.02	-3.87	-0.9	166.94	6SE	0.3	-3.5	-2.2	0.8
283.97	-3.44	-0.7	176.81	16SE	0.6	-3.5	-2.1	0.8
289.43	-3.03	-0.5	186.73	26SE	2.0	-3.4	-2.0	0.8
295.29	-2.65	-0.3	196.71	6OC	1.7	-3.4	-2.0	0.8
301.43	-2.28	-0.1	206.75	16OC	-0.2	-3.4	-1.9	0.8
307.82	-1.94	0.1	216.84	26OC	-0.6	-3.4	-1.8	0.8
314.37	-1.63	0.3	226.96	5NO	-0.6	-3.3	-1.8	0.8
321.04	-1.34	0.4	237.13	15NO	-0.6	-3.3	-1.7	0.7
327.80	-1.07	0.6	247.31	25NO	-0.8	-3.3	-1.7	0.7
334.61	-0.82	0.7	257.51	5DE	-0.7	-3.3	-1.6	0.7
341.46	-0.60	0.8	267.72	15DE	-0.8	-3.4	-1.6	0.7
348.32	-0.39	1.0	277.91	25DE	-0.7	-3.4	-1.6	0.8
				693				
355.18	-0.21	1.1	288.09	4JA	0.3	-3.4	-1.5	0.8
2.03	-0.04	1.2	298.24	14JA	2.9	-3.4	-1.5	0.8
8.85	0.11	1.3	308.35	24JA	1.3	-3.5	-1.5	0.8
15.65	0.25	1.4	318.42	3FE	0.6	-3.5	-1.5	0.8
22.41	0.37	1.5	328.44	13FE	0.3	-3.6	-1.5	0.8
29.13	0.49	1.6	338.40	23FE	0.1	-3.6	-1.6	0.8
35.81	0.59	1.7	348.30	5MR	-0.3	-3.7	-1.6	0.8
42.45	0.67	1.7	358.15	15MR	-0.9	-3.7	-1.6	0.8
49.05	0.75	1.8	7.94	25MR	-1.7	-3.8	-1.7	0.7
55.61	0.82	1.8	17.67	4AP	-1.2	-3.9	-1.7	0.7
62.13	0.89	1.9	27.36	14AP	-0.2	-4.0	-1.8	0.6
68.62	0.94	1.9	36.99	24AP	0.8	-4.1	-1.8	0.5
75.08	0.99	2.0	46.59	4MY	1.9	-4.1	-1.9	0.5
81.51	1.03	2.0	56.16	14MY	3.3	-4.2	-1.9	0.4
87.92	1.06	2.0	65.70	24MY	2.7	-4.1	-2.0	0.4
94.31	1.09	2.0	75.23	3JN	1.4	-3.9	-2.1	0.3
100.68	1.11	2.0	84.75	13JN	0.4	-3.3	-2.2	0.3
107.04	1.13	2.0	94.28	23JN	-0.7	-3.0	-2.2	0.3
113.40	1.14	2.0	103.82	3JL	-1.6	-3.7	-2.3	0.4
119.75	1.15	2.0	113.38	13JL	-1.2	-4.1	-2.3	0.4
126.11	1.16	2.0	122.97	23JL	-0.5	-4.2	-2.4	0.5
132.47	1.16	2.0	132.60	2AU	-0.0	-4.2	-2.4	0.5
138.85	1.15	2.0	142.28	12AU	0.2	-4.1	-2.4	0.6
145.24	1.14	2.0	152.01	22AU	0.5	-4.1	-2.4	0.7
151.65	1.12	2.0	161.79	1SE	0.9	-4.0	-2.4	0.7
158.09	1.10	2.0	171.63	11SE	2.4	-3.9	-2.4	0.7
164.54	1.08	2.0	181.52	21SE	1.5	-3.8	-2.4	0.8
171.03	1.05	2.0	191.48	1OC	-0.4	-3.7	-2.3	0.8
177.54	1.01	2.0	201.49	11OC	-0.8	-3.7	-2.2	0.8
184.08	0.97	1.9	211.55	21OC	-0.8	-3.6	-2.2	0.8
190.66	0.93	1.9	221.65	31OC	-0.8	-3.6	-2.1	0.8
197.26	0.87	1.9	231.80	10NO	-0.7	-3.5	-2.0	0.8
203.90	0.81	1.8	241.97	20NO	-0.6	-3.5	-2.0	0.8
210.56	0.74	1.8	252.17	30NO	-0.6	-3.4	-1.9	0.7
217.26	0.66	1.7	262.37	10DE	-0.5	-3.4	-1.8	0.7
223.99	0.57	1.7	272.57	20DE	0.5	-3.4	-1.8	0.7
230.75	0.47	1.6	282.76	30DE	3.0	-3.4	-1.7	0.8

♂			☉	16.00UT	☿	♀	♃	♄
LONG	LAT	MAG	LONG	694	MAGNITUDES			
237.53	0.36	1.5	292.93	9JA	1.0	-3.3	-1.7	0.8
244.34	0.23	1.4	303.06	19JA	0.3	-3.3	-1.6	0.8
251.17	0.09	1.3	313.15	29JA	0.2	-3.3	-1.6	0.8
258.01	-0.06	1.2	323.19	8FE	-0.0	-3.3	-1.6	0.9
264.87	-0.23	1.1	333.18	18FE	-0.3	-3.3	-1.6	0.9
271.74	-0.42	1.0	343.12	28FE	-0.9	-3.4	-1.6	0.9
278.60	-0.62	0.9	353.00	10MR	-1.6	-3.4	-1.6	0.8
285.45	-0.84	0.7	2.81	20MR	-1.2	-3.4	-1.6	0.8
292.28	-1.09	0.6	12.57	30MR	-0.2	-3.4	-1.6	0.8
299.06	-1.35	0.5	22.28	9AP	1.1	-3.5	-1.6	0.7
305.77	-1.63	0.3	31.94	19AP	2.5	-3.5	-1.6	0.7
312.39	-1.93	0.2	41.56	29AP	3.5	-3.4	-1.7	0.6
318.86	-2.24	-0.0	51.14	9MY	2.0	-3.4	-1.7	0.6
325.16	-2.57	-0.2	60.69	19MY	1.1	-3.4	-1.8	0.5
331.21	-2.92	-0.3	70.23	29MY	0.2	-3.4	-1.8	0.4
336.94	-3.28	-0.5	79.76	8JN	-0.7	-3.3	-1.9	0.4
342.25	-3.65	-0.7	89.28	18JN	-1.6	-3.3	-2.0	0.3
347.02	-4.03	-0.9	98.81	28JN	-1.2	-3.3	-2.0	0.3
351.08	-4.41	-1.2	108.37	8JL	-0.5	-3.3	-2.1	0.4
354.27	-4.77	-1.4	117.94	18JL	0.1	-3.3	-2.2	0.4
356.35	-5.09	-1.6	127.55	28JL	0.4	-3.3	-2.2	0.5
357.13	-5.34	-1.9	137.20	7AU	0.7	-3.4	-2.3	0.5
356.49	-5.45	-2.1	146.90	17AU	1.2	-3.4	-2.3	0.6
354.52	-5.37	-2.3	156.65	27AU	2.8	-3.4	-2.4	0.6
351.65	-5.03	-2.4	166.46	6SE	1.3	-3.5	-2.4	0.7
348.63	-4.47	-2.3	176.33	16SE	-0.5	-3.5	-2.4	0.7
346.21	-3.76	-2.1	186.25	26SE	-0.9	-3.6	-2.5	0.8
344.93	-3.00	-1.8	196.23	6OC	-0.9	-3.7	-2.4	0.8
344.98	-2.29	-1.4	206.27	16OC	-0.9	-3.8	-2.4	0.8
346.28	-1.65	-1.1	216.35	26OC	-0.6	-3.8	-2.4	0.8
348.66	-1.11	-0.8	226.48	5NO	-0.5	-3.9	-2.3	0.8
351.91	-0.66	-0.5	236.63	15NO	-0.5	-4.0	-2.2	0.8
355.85	-0.29	-0.2	246.82	25NO	-0.3	-4.2	-2.2	0.8
0.31	0.01	0.0	257.02	5DE	0.6	-4.3	-2.1	0.8
5.18	0.27	0.3	267.22	15DE	2.8	-4.3	-2.0	0.8
10.37	0.47	0.5	277.42	25DE	0.7	-4.4	-2.0	0.8
				695				
15.79	0.65	0.7	287.60	4JA	0.2	-4.3	-1.9	0.8
21.40	0.79	0.8	297.74	14JA	0.0	-4.0	-1.8	0.8
27.15	0.90	1.0	307.86	24JA	-0.1	-3.4	-1.8	0.9
33.00	1.00	1.1	317.93	3FE	-0.4	-3.5	-1.7	0.9
38.94	1.07	1.3	327.95	13FE	-0.9	-4.0	-1.7	0.9
44.93	1.13	1.4	337.91	23FE	-1.5	-4.2	-1.6	0.9
50.98	1.18	1.5	347.82	5MR	-1.2	-4.3	-1.6	0.9
57.05	1.22	1.6	357.67	15MR	-0.1	-4.2	-1.6	0.9
63.15	1.25	1.7	7.46	25MR	1.4	-4.1	-1.6	0.9
69.28	1.27	1.8	17.20	4AP	3.2	-4.0	-1.5	0.9
75.41	1.28	1.8	26.88	14AP	2.6	-3.9	-1.5	0.8
81.57	1.29	1.9	36.52	24AP	1.5	-3.8	-1.6	0.8
87.73	1.29	1.9	46.12	4MY	0.8	-3.7	-1.6	0.7
93.91	1.29	2.0	55.69	14MY	0.1	-3.6	-1.6	0.7
100.10	1.28	2.0	65.24	24MY	-0.8	-3.6	-1.6	0.6
106.31	1.26	2.0	74.77	3JN	-1.7	-3.5	-1.6	0.5
112.54	1.24	2.0	84.29	13JN	-1.3	-3.5	-1.7	0.5
118.79	1.22	2.0	93.82	23JN	-0.4	-3.4	-1.7	0.4
125.06	1.19	2.0	103.36	3JL	0.2	-3.4	-1.8	0.3
131.36	1.16	2.0	112.92	13JL	0.5	-3.4	-1.8	0.4
137.70	1.13	2.0	122.52	23JL	0.9	-3.4	-1.9	0.4
144.07	1.09	2.0	132.14	2AU	1.6	-3.3	-2.0	0.5
150.48	1.04	2.0	141.82	12AU	3.2	-3.3	-2.0	0.5
156.93	0.99	1.9	151.54	22AU	1.1	-3.3	-2.1	0.6
163.43	0.94	1.9	161.32	1SE	-0.5	-3.4	-2.2	0.6
169.98	0.89	1.9	171.15	11SE	-1.1	-3.4	-2.2	0.7
176.58	0.83	1.9	181.05	21SE	-1.1	-3.4	-2.3	0.7
183.24	0.76	1.9	191.00	1OC	-0.8	-3.4	-2.3	0.8
189.95	0.69	1.9	201.00	11OC	-0.5	-3.4	-2.4	0.8
196.73	0.61	1.9	211.06	21OC	-0.3	-3.5	-2.4	0.8
203.56	0.53	1.8	221.16	31OC	-0.3	-3.5	-2.4	0.9
210.46	0.45	1.8	231.31	10NO	-0.2	-3.5	-2.4	0.9
217.42	0.35	1.8	241.48	20NO	0.8	-3.5	-2.3	0.9
224.44	0.25	1.8	251.67	30NO	2.6	-3.5	-2.3	0.9
231.53	0.15	1.7	261.87	10DE	0.4	-3.4	-2.2	0.9
238.68	0.04	1.7	272.08	20DE	-0.0	-3.4	-2.2	0.9
245.88	-0.08	1.6	282.26	30DE	-0.1	-3.4	-2.1	0.8

♂ LONG LAT MAG	☉ LONG	16.00UT 696	☿ ♀ ♃ ♄ MAGNITUDES
253.14-0.20 1.6	292.43	9JA	-0.2 -3.4 -2.0 0.8
260.46-0.33 1.5	302.56	19JA	-0.5 -3.3 -2.0 0.9
267.82-0.47 1.5	312.65	29JA	-0.9 -3.3 -1.9 0.9
275.24-0.61 1.4	322.70	8FE	-1.3 -3.3 -1.8 0.9
282.69-0.75 1.3	332.70	18FE	-1.1 -3.3 -1.8 1.0
290.17-0.89 1.3	342.63	28FE	-0.0 -3.4 -1.7 1.0
297.68-1.04 1.2	352.51	9MR	1.7 -3.4 -1.6 1.0
305.19-1.18 1.1	2.33	19MR	3.5 -3.4 -1.6 1.0
312.71-1.33 1.1	12.09	29MR	2.0 -3.4 -1.6 1.0
320.22-1.46 1.0	21.81	8AP	1.1 -3.5 -1.5 1.0
327.70-1.59 0.9	31.47	18AP	0.6 -3.5 -1.5 0.9
335.13-1.71 0.8	41.09	28AP	0.0 -3.5 -1.5 0.9
342.50-1.82 0.8	50.68	8MY	-0.8 -3.6 -1.5 0.8
349.78-1.92 0.7	60.23	18MY	-1.7 -3.7 -1.5 0.8
356.95-2.00 0.6	69.77	28MY	-1.3 -3.7 -1.5 0.7
3.98-2.06 0.5	79.30	7JN	-0.4 -3.8 -1.5 0.7
10.84-2.11 0.4	88.82	17JN	0.3 -3.9 -1.5 0.6
17.49-2.14 0.4	98.35	27JN	0.7 -4.0 -1.5 0.5
23.90-2.14 0.3	107.91	7JL	1.2 -4.1 -1.5 0.5
30.02-2.13 0.1	117.48	17JL	2.1 -4.2 -1.6 0.4
35.77-2.09 0.0	127.09	27JL	3.1 -4.2 -1.6 0.5
41.09-2.02 -0.1	136.74	6AU	1.0 -4.2 -1.7 0.5
45.88-1.92 -0.2	146.44	16AU	-0.6 -3.9 -1.7 0.5
50.01-1.79 -0.4	156.18	26AU	-1.2 -3.4 -1.8 0.6
53.33-1.61 -0.6	165.99	5SE	-1.2 -3.4 -1.8 0.7
55.64-1.38 -0.8	175.85	15SE	-0.7 -4.0 -1.9 0.7
56.74-1.09 -1.0	185.77	25SE	-0.4 -4.3 -2.0 0.8
56.45-0.73 -1.2	195.75	5OC	-0.2 -4.3 -2.0 0.8
54.69-0.31 -1.4	205.78	15OC	-0.1 -4.3 -2.1 0.8
51.67 0.17 -1.6	215.86	25OC	0.0 -4.2 -2.2 0.9
47.93 0.66 -1.7	225.99	4NO	1.1 -4.1 -2.2 0.9
44.28 1.09 -1.5	236.14	14NO	2.3 -4.0 -2.2 0.9
41.45 1.44 -1.2	246.33	24NO	0.2 -3.9 -2.3 0.9
39.87 1.69 -0.9	256.53	4DE	-0.2 -3.9 -2.3 0.9
39.66 1.85 -0.6	266.73	14DE	-0.3 -3.8 -2.2 1.0
40.68 1.94 -0.3	276.93	24DE	-0.3 -3.7 -2.2 0.9
		697	
42.73 1.99 -0.0	287.11	3JA	-0.6 -3.6 -2.2 0.9
45.63 2.00 0.2	297.26	13JA	-1.0 -3.6 -2.1 0.9
49.17 2.00 0.5	307.37	23JA	-1.2 -3.5 -2.1 0.9
53.22 1.97 0.7	317.45	2FE	-1.0 -3.4 -2.0 1.0
57.68 1.94 0.9	327.47	12FE	0.0 -3.4 -1.9 1.0
62.45 1.90 1.0	337.43	22FE	2.0 -3.4 -1.8 1.1
67.47 1.86 1.2	347.35	4MR	2.8 -3.3 -1.8 1.1
72.69 1.81 1.3	357.19	14MR	1.4 -3.3 -1.7 1.1
78.07 1.76 1.4	6.99	24MR	0.8 -3.3 -1.6 1.1
83.58 1.70 1.5	16.73	3AP	0.4 -3.3 -1.6 1.1
89.21 1.65 1.6	26.41	13AP	-0.0 -3.3 -1.5 1.1
94.92 1.59 1.7	36.05	23AP	-0.8 -3.3 -1.5 1.1
100.72 1.53 1.8	45.66	3MY	-1.7 -3.3 -1.5 1.0
106.60 1.47 1.8	55.23	13MY	-1.3 -3.4 -1.4 1.0
112.54 1.41 1.9	64.77	23MY	-0.3 -3.4 -1.4 0.9
118.54 1.34 1.9	74.30	2JN	0.4 -3.4 -1.4 0.9
124.61 1.28 1.9	83.83	12JN	0.9 -3.5 -1.4 0.8
130.74 1.21 2.0	93.35	22JN	1.6 -3.5 -1.4 0.8
136.94 1.14 2.0	102.90	2JL	2.7 -3.5 -1.4 0.7
143.19 1.06 2.0	112.45	12JL	2.6 -3.4 -1.4 0.6
149.51 0.99 2.0	122.05	22JL	0.8 -3.4 -1.4 0.6
155.90 0.91 2.0	131.67	1AU	-0.6 -3.4 -1.4 0.5
162.35 0.83 1.9	141.34	11AU	-1.3 -3.3 -1.4 0.6
168.88 0.75 1.9	151.07	21AU	-1.2 -3.3 -1.5 0.6
175.48 0.67 1.9	160.84	31AU	-0.6 -3.3 -1.5 0.7
182.16 0.58 1.9	170.67	10SE	-0.3 -3.3 -1.6 0.7
188.91 0.49 1.8	180.56	20SE	-0.1 -3.3 -1.6 0.8
195.75 0.40 1.8	190.51	30SE	0.0 -3.3 -1.7 0.8
202.66 0.30 1.7	200.51	10OC	0.2 -3.4 -1.7 0.9
209.66 0.21 1.7	210.57	20OC	1.4 -3.4 -1.8 0.9
216.74 0.10 1.7	220.67	30OC	2.1 -3.4 -1.9 0.9
223.90 0.00 1.7	230.81	9NO	0.0 -3.5 -1.9 1.0
231.15-0.10 1.6	240.98	19NO	-0.4 -3.5 -2.0 1.0
238.47-0.21 1.6	251.18	29NO	-0.4 -3.6 -2.0 1.0
245.86-0.32 1.6	261.38	9DE	-0.5 -3.6 -2.1 1.0
253.33-0.43 1.6	271.58	19DE	-0.6 -3.7 -2.1 1.1
260.86-0.53 1.6	281.77	29DE	-0.9 -3.8 -2.1 1.1

♂ LONG LAT MAG	☉ LONG	16.00UT 698	☿ ♀ ♃ ♄ MAGNITUDES
268.45-0.64 1.5	291.94	8JA	-1.0 -3.8 -2.1 1.1
276.09-0.74 1.5	302.07	18JA	-0.9 -3.9 -2.1 1.1
283.78-0.84 1.5	312.17	28JA	0.1 -4.0 -2.1 1.0
291.50-0.94 1.4	322.22	7FE	2.4 -4.1 -2.0 1.1
299.25-1.03 1.4	332.22	17FE	2.1 -4.2 -2.0 1.1
307.02-1.11 1.4	342.16	27FE	1.0 -4.3 -1.9 1.2
314.78-1.18 1.4	352.03	9MR	0.6 -4.3 -1.9 1.2
322.54-1.24 1.3	1.86	19MR	0.3 -4.1 -1.8 1.2
330.27-1.29 1.3	11.63	29MR	-0.1 -3.7 -1.7 1.2
337.96-1.33 1.3	21.34	8AP	-0.8 -3.1 -1.7 1.3
345.61-1.36 1.3	31.00	18AP	-1.7 -3.4 -1.6 1.2
353.20-1.36 1.2	40.63	28AP	-1.3 -4.0 -1.5 1.2
0.71-1.36 1.2	50.21	8MY	-0.3 -4.2 -1.5 1.2
8.13-1.34 1.2	59.77	18MY	0.5 -4.2 -1.4 1.2
15.45-1.30 1.1	69.31	28MY	1.2 -4.1 -1.4 1.1
22.66-1.25 1.1	78.83	7JN	2.1 -4.1 -1.4 1.1
29.73-1.18 1.1	88.36	17JN	3.3 -4.0 -1.3 1.0
36.67-1.10 1.1	97.89	27JN	2.1 -3.9 -1.3 1.0
43.45-1.01 1.0	107.44	7JL	0.6 -3.8 -1.3 0.9
50.05-0.89 1.0	117.01	17JL	-0.7 -3.7 -1.3 0.8
56.45-0.77 0.9	126.62	27JL	-1.4 -3.6 -1.3 0.8
62.63-0.62 0.8	136.27	6AU	-1.3 -3.6 -1.3 0.7
68.55-0.46 0.8	145.97	16AU	-0.6 -3.5 -1.3 0.7
74.18-0.28 0.7	155.71	26AU	-0.2 -3.5 -1.3 0.7
79.46-0.07 0.6	165.51	5SE	0.0 -3.5 -1.3 0.7
84.32 0.16 0.5	175.38	15SE	0.2 -3.4 -1.4 0.8
88.68 0.41 0.3	185.29	25SE	0.5 -3.4 -1.4 0.8
92.42 0.71 0.2	195.27	5OC	1.7 -3.4 -1.4 0.9
95.40 1.04 -0.0	205.30	15OC	1.9 -3.4 -1.5 0.9
97.45 1.42 -0.2	215.38	25OC	-0.1 -3.4 -1.6 1.0
98.37 1.84 -0.4	225.50	4NO	-0.5 -3.4 -1.6 1.0
97.97 2.30 -0.7	235.66	14NO	-0.5 -3.4 -1.7 1.1
96.19 2.76 -0.9	245.83	24NO	-0.6 -3.4 -1.7 1.1
93.15 3.19 -1.1	256.03	4DE	-0.7 -3.4 -1.8 1.2
89.30 3.52 -1.2	266.24	14DE	-0.8 -3.4 -1.9 1.2
85.40 3.69 -1.1	276.43	24DE	-0.9 -3.4 -1.9 1.2
		699	
82.16 3.72 -0.8	286.61	3JA	-0.8 -3.4 -2.0 1.2
80.08 3.62 -0.6	296.77	13JA	0.2 -3.5 -2.0 1.2
79.33 3.46 -0.3	306.88	23JA	2.7 -3.5 -2.0 1.2
79.83 3.25 -0.0	316.96	2FE	1.6 -3.5 -2.0 1.2
81.41 3.04 0.2	326.98	12FE	0.7 -3.5 -2.0 1.2
83.87 2.84 0.5	336.95	22FE	0.4 -3.4 -2.0 1.3
87.03 2.64 0.7	346.86	4MR	0.2 -3.4 -2.0 1.3
90.76 2.45 0.9	356.72	14MR	-0.2 -3.4 -1.9 1.4
94.94 2.28 1.0	6.51	24MR	-0.8 -3.4 -1.9 1.4
99.47 2.12 1.2	16.25	3AP	-1.7 -3.3 -1.8 1.4
104.30 1.97 1.3	25.94	13AP	-1.3 -3.3 -1.8 1.4
109.36 1.82 1.4	35.58	23AP	-0.3 -3.3 -1.7 1.4
114.64 1.69 1.5	45.19	3MY	0.7 -3.3 -1.6 1.4
120.09 1.56 1.6	54.76	13MY	1.6 -3.3 -1.6 1.4
125.69 1.43 1.6	64.30	23MY	2.8 -3.4 -1.5 1.3
131.42 1.31 1.7	73.84	2JN	3.1 -3.4 -1.5 1.3
137.29 1.19 1.7	83.36	12JN	1.7 -3.4 -1.4 1.3
143.27 1.07 1.8	92.89	22JN	0.5 -3.5 -1.4 1.2
149.36 0.96 1.8	102.43	2JL	-0.7 -3.5 -1.3 1.2
155.56 0.85 1.8	111.99	12JL	-1.5 -3.6 -1.3 1.1
161.86 0.73 1.8	121.58	22JL	-1.3 -3.6 -1.3 1.1
168.26 0.62 1.8	131.20	1AU	-0.5 -3.7 -1.2 1.0
174.77 0.51 1.8	140.88	11AU	-0.1 -3.8 -1.2 0.9
181.38 0.40 1.8	150.59	21AU	0.2 -3.9 -1.2 0.9
188.09 0.29 1.8	160.37	31AU	0.3 -4.0 -1.2 0.8
194.90 0.18 1.8	170.20	10SE	0.7 -4.1 -1.2 0.9
201.81 0.08 1.7	180.08	20SE	2.1 -4.2 -1.2 0.9
208.83-0.03 1.7	190.03	30SE	1.7 -4.3 -1.3 1.0
215.95-0.14 1.7	200.03	10OC	-0.3 -4.4 -1.3 1.0
223.16-0.24 1.6	210.08	20OC	-0.7 -4.3 -1.3 1.1
230.47-0.34 1.6	220.19	30OC	-0.7 -4.0 -1.4 1.1
237.86-0.44 1.5	230.33	9NO	-0.7 -3.3 -1.4 1.2
245.35-0.54 1.5	240.49	19NO	-0.8 -3.4 -1.5 1.2
252.91-0.63 1.4	250.69	29NO	-0.7 -4.1 -1.5 1.3
260.55-0.72 1.4	260.89	9DE	-0.7 -4.3 -1.6 1.3
268.25-0.80 1.4	271.09	19DE	-0.6 -4.4 -1.7 1.3
276.01-0.88 1.4	281.28	29DE	0.3 -4.3 -1.7 1.4

♂			☉	16.00UT	☿	♀	♃	♄
LONG	LAT	MAG	LONG	700	MAGNITUDES			
283.82	-0.94	1.4	291.45	8JA	3.0	-4.2	-1.8	1.4
291.65	-1.00	1.4	301.58	18JA	1.2	-4.1	-1.9	1.4
299.51	-1.05	1.4	311.68	28JA	0.5	-4.0	-1.9	1.4
307.38	-1.08	1.4	321.73	7FE	0.3	-3.9	-2.0	1.3
315.25	-1.11	1.4	331.73	17FE	0.1	-3.8	-2.0	1.3
323.10	-1.12	1.4	341.68	27FE	-0.3	-3.7	-2.0	1.2
330.92	-1.12	1.4	351.56	8MR	-0.8	-3.6	-2.0	1.2
338.70	-1.11	1.4	1.39	18MR	-1.7	-3.6	-2.0	1.2
346.44	-1.09	1.4	11.16	28MR	-1.3	-3.5	-2.0	1.2
354.11	-1.05	1.4	20.87	7AP	-0.2	-3.4	-1.9	1.2
1.70	-1.01	1.4	30.54	17AP	0.9	-3.4	-1.9	1.2
9.23	-0.95	1.4	40.17	27AP	2.1	-3.4	-1.8	1.2
16.66	-0.88	1.5	49.75	7MY	3.7	-3.3	-1.8	1.2
24.00	-0.80	1.5	59.31	17MY	2.4	-3.3	-1.7	1.1
31.24	-0.71	1.5	68.85	27MY	1.3	-3.3	-1.7	1.1
38.37	-0.61	1.5	78.38	6JN	0.3	-3.3	-1.6	1.0
45.39	-0.51	1.5	87.90	16JN	-0.7	-3.3	-1.5	1.0
52.30	-0.39	1.5	97.43	26JN	-1.6	-3.3	-1.5	1.0
59.09	-0.27	1.5	106.98	6JL	-1.3	-3.3	-1.4	0.9
65.75	-0.14	1.4	116.55	16JL	-0.5	-3.4	-1.4	0.9
72.28	-0.00	1.4	126.16	26JL	-0.0	-3.4	-1.3	0.8
78.66	0.14	1.4	135.80	5AU	0.3	-3.4	-1.3	0.8
84.89	0.29	1.4	145.49	15AU	0.5	-3.5	-1.3	0.7
90.94	0.45	1.3	155.24	25AU	1.0	-3.5	-1.2	0.7
96.80	0.63	1.2	165.03	4SE	2.5	-3.5	-1.2	0.6
102.45	0.81	1.2	174.89	14SE	1.5	-3.5	-1.2	0.7
107.85	1.01	1.1	184.81	24SE	-0.4	-3.4	-1.2	0.7
112.95	1.23	1.0	194.78	4OC	-0.8	-3.4	-1.2	0.8
117.70	1.46	0.9	204.80	14OC	-0.8	-3.4	-1.2	0.9
122.02	1.72	0.7	214.88	24OC	-0.8	-3.3	-1.2	0.9
125.82	2.01	0.6	225.00	3NO	-0.6	-3.3	-1.3	1.0
128.98	2.33	0.4	235.16	13NO	-0.6	-3.3	-1.3	1.1
131.35	2.68	0.2	245.34	23NO	-0.6	-3.3	-1.4	1.1
132.74	3.07	-0.0	255.53	3DE	-0.5	-3.3	-1.4	1.1
132.98	3.47	-0.3	265.74	13DE	0.5	-3.4	-1.5	1.2
131.90	3.88	-0.5	275.94	23DE	3.0	-3.4	-1.5	1.2
				701				
129.52	4.23	-0.7	286.12	2JA	0.9	-3.4	-1.6	1.2
126.08	4.47	-0.9	296.27	12JA	0.3	-3.4	-1.6	1.2
122.13	4.55	-1.0	306.40	22JA	0.1	-3.5	-1.7	1.1
118.41	4.45	-0.8	316.47	1FE	-0.1	-3.5	-1.8	1.1
115.54	4.21	-0.6	326.50	11FE	-0.4	-3.6	-1.8	1.1
113.91	3.88	-0.4	336.47	21FE	-0.9	-3.6	-1.9	1.0
113.57	3.52	-0.1	346.38	3MR	-1.6	-3.7	-2.0	1.0
114.41	3.16	0.1	356.24	13MR	-1.2	-3.7	-2.0	0.9
116.28	2.83	0.3	6.04	23MR	-0.2	-3.8	-2.0	0.9
118.97	2.51	0.5	15.79	2AP	1.2	-3.9	-2.0	0.9
122.33	2.23	0.7	25.48	12AP	2.7	-4.0	-2.0	0.9
126.24	1.97	0.9	35.13	22AP	3.2	-4.1	-2.0	0.9
130.60	1.73	1.0	44.73	2MY	1.8	-4.1	-2.0	0.9
135.32	1.51	1.1	54.30	12MY	1.0	-4.2	-2.0	0.9
140.35	1.31	1.2	63.85	22MY	0.2	-4.1	-1.9	0.9
145.65	1.12	1.3	73.38	1JN	-0.7	-3.9	-1.9	0.9
151.17	0.95	1.4	82.91	11JN	-1.6	-3.3	-1.8	0.8
156.91	0.78	1.4	92.43	21JN	-1.3	-3.0	-1.7	0.8
162.82	0.62	1.5	101.97	1JL	-0.5	-3.7	-1.7	0.7
168.91	0.47	1.5	111.53	11JL	0.1	-4.1	-1.6	0.7
175.16	0.32	1.5	121.11	21JL	0.4	-4.2	-1.5	0.6
181.55	0.19	1.5	130.73	31JL	0.7	-4.2	-1.5	0.6
188.09	0.05	1.6	140.40	10AU	1.3	-4.1	-1.4	0.5
194.77	-0.08	1.6	150.12	20AU	2.9	-4.1	-1.4	0.5
201.58	-0.20	1.6	159.89	30AU	1.3	-4.0	-1.4	0.4
208.52	-0.31	1.6	169.72	9SE	-0.5	-3.9	-1.3	0.4
215.59	-0.42	1.6	179.60	19SE	-1.0	-3.8	-1.3	0.3
222.77	-0.53	1.5	189.54	29SE	-1.0	-3.7	-1.3	0.4
230.07	-0.63	1.5	199.54	9OC	-0.8	-3.7	-1.3	0.5
237.47	-0.71	1.5	209.59	19OC	-0.5	-3.6	-1.3	0.5
244.97	-0.80	1.5	219.69	29OC	-0.4	-3.6	-1.3	0.6
252.57	-0.87	1.5	229.83	8NO	-0.4	-3.5	-1.3	0.7
260.24	-0.93	1.5	240.00	18NO	-0.3	-3.5	-1.3	0.7
267.98	-0.99	1.4	250.19	28NO	0.7	-3.4	-1.3	0.8
275.79	-1.03	1.4	260.39	8DE	2.8	-3.4	-1.3	0.8
283.64	-1.06	1.4	270.60	18DE	0.6	-3.4	-1.4	0.9
291.52	-1.08	1.4	280.79	28DE	0.1	-3.4	-1.4	0.9

♂			☉	16.00UT	☿	♀	♃	♄
LONG	LAT	MAG	LONG	702	MAGNITUDES			
299.42	-1.09	1.4	290.96	7JA	-0.0	-3.3	-1.5	0.9
307.32	-1.09	1.3	301.10	17JA	-0.2	-3.3	-1.5	0.9
315.22	-1.07	1.3	311.20	27JA	-0.4	-3.3	-1.6	0.9
323.10	-1.05	1.3	321.25	6FE	-0.9	-3.3	-1.7	0.9
330.94	-1.01	1.3	331.25	16FE	-1.4	-3.3	-1.7	0.8
338.73	-0.96	1.3	341.20	26FE	-1.2	-3.4	-1.8	0.8
346.47	-0.91	1.4	351.09	8MR	-0.1	-3.4	-1.9	0.8
354.15	-0.84	1.4	0.91	18MR	1.5	-3.4	-1.9	0.7
1.75	-0.76	1.5	10.68	28MR	3.4	-3.4	-2.0	0.7
9.27	-0.68	1.5	20.40	7AP	2.4	-3.5	-2.0	0.7
16.71	-0.59	1.5	30.07	17AP	1.4	-3.5	-2.1	0.7
24.06	-0.50	1.6	39.70	27AP	0.7	-3.4	-2.1	0.7
31.32	-0.40	1.6	49.29	7MY	0.1	-3.4	-2.1	0.7
38.49	-0.30	1.6	58.84	17MY	-0.7	-3.4	-2.1	0.7
45.57	-0.19	1.7	68.38	27MY	-1.7	-3.4	-2.1	0.7
52.55	-0.08	1.7	77.91	6JN	-1.3	-3.3	-2.1	0.7
59.44	0.03	1.7	87.43	16JN	-0.4	-3.3	-2.0	0.6
66.23	0.14	1.7	96.96	26JN	0.2	-3.3	-2.0	0.6
72.94	0.26	1.7	106.51	6JL	0.6	-3.3	-1.9	0.6
79.54	0.38	1.7	116.08	16JL	1.0	-3.3	-1.9	0.5
86.06	0.49	1.7	125.68	26JL	1.8	-3.3	-1.8	0.5
92.48	0.62	1.7	135.33	5AU	3.2	-3.4	-1.7	0.4
98.79	0.74	1.7	145.02	15AU	1.1	-3.4	-1.7	0.4
105.00	0.87	1.7	154.76	25AU	-0.5	-3.4	-1.6	0.3
111.10	0.99	1.7	164.56	4SE	-1.1	-3.5	-1.6	0.2
117.07	1.13	1.6	174.41	14SE	-1.1	-3.5	-1.5	0.2
122.91	1.27	1.6	184.32	24SE	-0.8	-3.6	-1.5	0.1
128.59	1.41	1.5	194.29	4OC	-0.4	-3.7	-1.4	0.1
134.10	1.57	1.4	204.31	14OC	-0.3	-3.8	-1.4	0.2
139.40	1.73	1.3	214.39	24OC	-0.2	-3.8	-1.4	0.2
144.45	1.90	1.2	224.51	3NO	-0.1	-3.9	-1.4	0.3
149.21	2.09	1.1	234.66	13NO	0.9	-4.0	-1.3	0.4
153.61	2.29	0.9	244.84	23NO	2.6	-4.2	-1.3	0.4
157.56	2.51	0.8	255.04	3DE	0.4	-4.3	-1.4	0.5
160.97	2.75	0.6	265.24	13DE	-0.1	-4.3	-1.4	0.6
163.70	3.01	0.4	275.44	23DE	-0.2	-4.4	-1.4	0.6
				703				
165.59	3.29	0.1	285.63	2JA	-0.3	-4.3	-1.4	0.6
166.46	3.57	-0.1	295.78	12JA	-0.5	-4.0	-1.4	0.7
166.14	3.84	-0.4	305.90	22JA	-0.9	-3.4	-1.5	0.7
164.52	4.06	-0.7	315.99	1FE	-1.3	-3.5	-1.5	0.7
161.71	4.19	-0.9	326.01	11FE	-1.1	-4.0	-1.6	0.7
158.04	4.17	-1.1	335.99	21FE	-0.1	-4.3	-1.7	0.6
154.18	3.98	-1.0	345.91	3MR	1.8	-4.3	-1.7	0.6
150.82	3.64	-0.9	355.76	13MR	3.2	-4.2	-1.8	0.6
148.51	3.23	-0.7	5.57	23MR	1.7	-4.1	-1.9	0.5
147.49	2.78	-0.4	15.32	2AP	1.0	-4.0	-1.9	0.5
147.75	2.34	-0.2	25.01	12AP	0.5	-3.9	-2.0	0.5
149.14	1.93	-0.0	34.66	22AP	0.0	-3.8	-2.1	0.5
151.51	1.56	0.2	44.27	2MY	-0.8	-3.7	-2.1	0.5
154.68	1.23	0.3	53.84	12MY	-1.7	-3.6	-2.2	0.5
158.50	0.94	0.5	63.39	22MY	-1.3	-3.6	-2.2	0.5
162.86	0.67	0.6	72.92	1JN	-0.4	-3.5	-2.2	0.5
167.66	0.43	0.7	82.44	11JN	0.3	-3.5	-2.2	0.5
172.85	0.22	0.8	91.97	21JN	0.8	-3.4	-2.2	0.5
178.37	0.02	0.9	101.51	1JL	1.3	-3.4	-2.2	0.5
184.17	-0.16	1.0	111.07	11JL	2.3	-3.4	-2.2	0.4
190.22	-0.33	1.0	120.65	21JL	2.9	-3.4	-2.2	0.4
196.50	-0.48	1.1	130.27	31JL	0.9	-3.3	-2.1	0.4
202.99	-0.61	1.1	139.94	10AU	-0.6	-3.3	-2.0	0.3
209.67	-0.73	1.2	149.65	20AU	-1.2	-3.3	-2.0	0.3
216.52	-0.84	1.2	159.42	30AU	-1.2	-3.4	-1.9	0.2
223.53	-0.94	1.2	169.24	9SE	-0.7	-3.4	-1.8	0.1
230.69	-1.02	1.2	179.12	19SE	-0.3	-3.4	-1.8	0.1
237.99	-1.09	1.3	189.06	29SE	-0.2	-3.4	-1.7	0.0
245.41	-1.15	1.3	199.05	9OC	-0.1	-3.4	-1.7	-0.0
252.94	-1.20	1.3	209.10	19OC	0.1	-3.5	-1.6	-0.1
260.56	-1.23	1.3	219.20	29OC	1.1	-3.5	-1.6	-0.0
268.27	-1.24	1.3	229.33	8NO	2.4	-3.5	-1.5	0.1
276.04	-1.25	1.3	239.50	18NO	0.2	-3.5	-1.5	0.1
283.86	-1.23	1.3	249.69	28NO	-0.3	-3.5	-1.5	0.2
291.71	-1.21	1.4	259.89	8DE	-0.3	-3.4	-1.5	0.3
299.58	-1.18	1.4	270.10	18DE	-0.4	-3.4	-1.5	0.3
307.46	-1.13	1.4	280.29	28DE	-0.6	-3.4	-1.5	0.4

Left table

♂ LONG	♂ LAT	♂ MAG	☉ LONG	16.00UT 704	☿	♀	♃	♄
315.32	-1.07	1.4	290.46	7JA	-1.0	-3.4	-1.5	0.4
323.16	-1.00	1.4	300.60	17JA	-1.1	-3.3	-1.5	0.5
330.96	-0.93	1.4	310.70	27JA	-1.0	-3.3	-1.5	0.5
338.70	-0.84	1.5	320.76	6FE	0.0	-3.3	-1.5	0.5
346.39	-0.76	1.5	330.77	16FE	2.1	-3.3	-1.6	0.5
354.02	-0.66	1.5	340.71	26FE	2.5	-3.4	-1.6	0.5
1.56	-0.56	1.5	350.60	7MR	1.3	-3.4	-1.6	0.5
9.03	-0.46	1.5	0.44	17MR	0.7	-3.4	-1.7	0.4
16.42	-0.36	1.5	10.21	27MR	0.4	-3.4	-1.8	0.4
23.72	-0.25	1.6	19.93	6AP	-0.1	-3.5	-1.8	0.4
30.93	-0.15	1.6	29.61	16AP	-0.8	-3.5	-1.9	0.3
38.06	-0.04	1.6	39.23	26AP	-1.7	-3.5	-2.0	0.3
45.11	0.06	1.6	48.82	6MY	-1.3	-3.6	-2.0	0.4
52.07	0.16	1.7	58.39	16MY	-0.4	-3.7	-2.1	0.4
58.95	0.26	1.7	67.93	26MY	0.4	-3.7	-2.2	0.4
65.76	0.36	1.8	77.45	5JN	1.0	-3.8	-2.2	0.4
72.49	0.46	1.8	86.98	15JN	1.8	-3.9	-2.3	0.4
79.15	0.56	1.9	96.51	25JN	3.0	-4.0	-2.3	0.4
85.75	0.65	1.9	106.05	5JL	2.5	-4.1	-2.4	0.4
92.27	0.74	1.9	115.63	15JL	0.8	-4.2	-2.4	0.4
98.74	0.83	1.9	125.22	25JL	-0.6	-4.2	-2.4	0.3
105.14	0.92	1.9	134.87	4AU	-1.3	-4.1	-2.4	0.3
111.48	1.00	1.9	144.55	14AU	-1.3	-3.9	-2.3	0.3
117.76	1.09	1.9	154.29	24AU	-0.6	-3.4	-2.3	0.2
123.98	1.18	1.9	164.08	3SE	-0.2	-3.5	-2.2	0.1
130.13	1.26	1.9	173.93	13SE	-0.0	-4.0	-2.2	0.1
136.21	1.34	1.8	183.84	23SE	0.1	-4.3	-2.1	0.0
142.22	1.43	1.8	193.81	3OC	0.3	-4.3	-2.0	-0.1
148.14	1.51	1.7	203.83	13OC	1.4	-4.3	-2.0	-0.1
153.96	1.60	1.7	213.90	23OC	2.1	-4.2	-1.9	-0.2
159.68	1.69	1.6	224.01	2NO	-0.0	-4.1	-1.8	-0.2
165.26	1.78	1.5	234.17	12NO	-0.4	-4.0	-1.8	-0.1
170.69	1.87	1.4	244.34	22NO	-0.5	-3.9	-1.7	-0.1
175.94	1.96	1.3	254.54	2DE	-0.5	-3.8	-1.7	0.0
180.97	2.06	1.1	264.75	12DE	-0.7	-3.8	-1.6	0.1
185.72	2.15	1.0	274.94	22DE	-0.9	-3.7	-1.6	0.1

705

♂ LONG	♂ LAT	♂ MAG	☉ LONG	16.00UT 705	☿	♀	♃	♄
190.14	2.25	0.8	285.13	1JA	-1.0	-3.6	-1.6	0.2
194.14	2.35	0.6	295.29	11JA	-0.9	-3.6	-1.6	0.3
197.63	2.45	0.4	305.41	21JA	0.1	-3.5	-1.5	0.3
200.48	2.54	0.1	315.50	31JA	2.5	-3.4	-1.5	0.3
202.52	2.62	-0.1	325.53	10FE	2.0	-3.4	-1.5	0.4
203.58	2.66	-0.4	335.51	20FE	0.9	-3.4	-1.6	0.4
203.49	2.66	-0.7	345.43	2MR	0.5	-3.3	-1.6	0.4
202.13	2.59	-1.0	355.29	12MR	0.2	-3.3	-1.6	0.4
199.57	2.41	-1.3	5.10	22MR	-0.1	-3.3	-1.6	0.4
196.13	2.11	-1.5	14.85	1AP	-0.8	-3.3	-1.7	0.3
192.46	1.71	-1.5	24.55	11AP	-1.7	-3.3	-1.7	0.3
189.27	1.24	-1.3	34.19	21AP	-1.3	-3.3	-1.8	0.3
187.11	0.77	-1.1	43.81	1MY	-0.3	-3.3	-1.8	0.2
186.28	0.32	-0.9	53.38	11MY	0.6	-3.4	-1.9	0.2
186.78	-0.08	-0.7	62.93	21MY	1.3	-3.4	-2.0	0.3
188.48	-0.42	-0.5	72.46	31MY	2.3	-3.4	-2.0	0.3
191.22	-0.71	-0.3	81.99	10JN	3.4	-3.5	-2.1	0.3
194.80	-0.95	-0.2	91.51	20JN	2.0	-3.5	-2.2	0.3
199.08	-1.15	-0.0	101.05	30JN	0.6	-3.5	-2.2	0.3
203.95	-1.32	0.1	110.60	10JL	-0.6	-3.4	-2.3	0.3
209.30	-1.46	0.2	120.18	20JL	-1.4	-3.4	-2.4	0.3
215.06	-1.57	0.3	129.81	30JL	-1.3	-3.4	-2.4	0.3
221.17	-1.65	0.4	139.47	9AU	-0.6	-3.3	-2.4	0.3
227.57	-1.71	0.5	149.17	19AU	-0.2	-3.3	-2.4	0.2
234.23	-1.75	0.6	158.94	29AU	0.1	-3.3	-2.4	0.2
241.12	-1.77	0.7	168.76	8SE	0.2	-3.3	-2.4	0.1
248.19	-1.77	0.7	178.64	18SE	0.5	-3.3	-2.4	0.1
255.43	-1.75	0.8	188.58	28SE	1.8	-3.3	-2.4	0.0
262.80	-1.72	0.9	198.57	8OC	1.9	-3.4	-2.3	-0.1
270.28	-1.67	0.9	208.61	18OC	-0.2	-3.4	-2.2	-0.2
277.85	-1.60	1.0	218.71	28OC	-0.6	-3.4	-2.2	-0.2
285.49	-1.52	1.0	228.84	7NO	-0.6	-3.5	-2.1	-0.3
293.17	-1.43	1.1	239.00	17NO	-0.6	-3.5	-2.0	-0.3
300.89	-1.33	1.2	249.20	27NO	-0.8	-3.6	-1.9	-0.1
308.61	-1.22	1.2	259.39	7DE	-0.8	-3.6	-1.9	-0.1
316.32	-1.10	1.3	269.60	17DE	-0.8	-3.7	-1.8	-0.1
324.01	-0.98	1.3	279.79	27DE	-0.7	-3.8	-1.8	0.0

Right table

♂ LONG	♂ LAT	♂ MAG	☉ LONG	16.00UT 706	☿	♀	♃	♄
331.66	-0.86	1.4	289.96	6JA	0.2	-3.9	-1.7	0.1
339.26	-0.73	1.4	300.11	16JA	2.8	-3.9	-1.7	0.1
346.81	-0.61	1.5	310.21	26JA	1.5	-4.0	-1.6	0.2
354.29	-0.48	1.5	320.27	5FE	0.6	-4.1	-1.6	0.2
1.70	-0.36	1.6	330.28	15FE	0.4	-4.2	-1.6	0.3
9.04	-0.24	1.6	340.23	25FE	0.1	-4.3	-1.6	0.3
16.30	-0.13	1.6	350.12	7MR	-0.2	-4.3	-1.6	0.3
23.48	-0.01	1.7	359.96	17MR	-0.8	-4.1	-1.6	0.3
30.59	0.10	1.7	9.74	27MR	-1.7	-3.7	-1.6	0.3
37.61	0.20	1.7	19.46	6AP	-1.3	-3.1	-1.6	0.3
44.57	0.30	1.8	29.14	16AP	-0.3	-3.5	-1.6	0.3
51.44	0.39	1.8	38.77	26AP	0.8	-4.0	-1.7	0.3
58.25	0.48	1.8	48.36	6MY	1.8	-4.2	-1.7	0.2
65.00	0.57	1.8	57.92	16MY	3.1	-4.2	-1.7	0.2
71.68	0.65	1.8	67.47	26MY	2.9	-4.1	-1.8	0.2
78.31	0.72	1.8	76.99	5JN	1.5	-4.1	-1.8	0.2
84.89	0.80	1.8	86.52	15JN	0.5	-4.0	-1.9	0.3
91.41	0.86	1.9	96.05	25JN	-0.7	-3.9	-2.0	0.3
97.90	0.93	1.9	105.59	5JL	-1.5	-3.8	-2.0	0.3
104.35	0.98	2.0	115.16	15JL	-1.3	-3.7	-2.1	0.3
110.76	1.04	2.0	124.76	25JL	-0.5	-3.6	-2.2	0.3
117.14	1.09	2.0	134.40	4AU	-0.1	-3.6	-2.2	0.3
123.49	1.14	2.0	144.08	14AU	0.2	-3.5	-2.3	0.3
129.81	1.18	2.0	153.82	24AU	0.4	-3.5	-2.4	0.3
136.11	1.23	2.0	163.61	3SE	0.8	-3.5	-2.4	0.2
142.39	1.26	2.0	173.46	13SE	2.2	-3.4	-2.4	0.2
148.65	1.30	2.0	183.36	23SE	1.7	-3.4	-2.4	0.1
154.88	1.33	2.0	193.32	3OC	-0.3	-3.4	-2.4	0.1
161.08	1.36	1.9	203.34	13OC	-0.7	-3.4	-2.4	0.0
167.26	1.38	1.9	213.41	23OC	-0.7	-3.4	-2.4	-0.1
173.40	1.40	1.8	223.52	2NO	-0.7	-3.4	-2.4	-0.1
179.50	1.42	1.8	233.68	12NO	-0.7	-3.4	-2.3	-0.2
185.56	1.43	1.7	243.85	22NO	-0.6	-3.4	-2.2	-0.3
191.56	1.43	1.6	254.05	2DE	-0.6	-3.4	-2.2	-0.3
197.49	1.43	1.5	264.25	12DE	-0.6	-3.4	-2.1	-0.2
203.34	1.42	1.4	274.45	22DE	0.3	-3.4	-2.0	-0.2

707

♂ LONG	♂ LAT	♂ MAG	☉ LONG	16.00UT 707	☿	♀	♃	♄
209.09	1.40	1.3	284.63	1JA	3.0	-3.4	-1.9	-0.1
214.73	1.37	1.1	294.79	11JA	1.1	-3.5	-1.9	-0.0
220.21	1.33	1.0	304.92	21JA	0.4	-3.5	-1.8	0.1
225.52	1.26	0.8	315.00	31JA	0.2	-3.5	-1.8	0.1
230.60	1.18	0.6	325.04	10FE	0.0	-3.4	-1.7	0.2
235.41	1.06	0.4	335.02	20FE	-0.3	-3.4	-1.7	0.2
239.87	0.92	0.2	344.94	2MR	-0.8	-3.4	-1.6	0.3
243.89	0.73	-0.1	354.81	12MR	-1.6	-3.4	-1.6	0.3
247.35	0.49	-0.3	4.62	22MR	-1.3	-3.4	-1.6	0.3
250.13	0.18	-0.6	14.37	1AP	-0.2	-3.3	-1.6	0.3
252.02	-0.21	-0.9	24.07	11AP	1.0	-3.3	-1.5	0.3
252.87	-0.68	-1.3	33.72	21AP	2.3	-3.3	-1.5	0.3
252.52	-1.25	-1.6	43.33	1MY	3.7	-3.3	-1.5	0.3
250.92	-1.89	-1.9	52.91	11MY	2.2	-3.3	-1.6	0.3
248.32	-2.54	-2.2	62.46	21MY	1.2	-3.4	-1.6	0.2
245.22	-3.14	-2.2	72.00	31MY	0.3	-3.4	-1.6	0.2
242.35	-3.61	-2.1	81.52	10JN	-0.7	-3.4	-1.6	0.2
240.38	-3.91	-1.9	91.05	20JN	-1.6	-3.5	-1.7	0.2
239.69	-4.05	-1.7	100.58	30JN	-1.3	-3.5	-1.7	0.3
240.38	-4.07	-1.5	110.14	10JL	-0.5	-3.6	-1.8	0.3
242.38	-4.00	-1.2	119.72	20JL	0.0	-3.6	-1.8	0.3
245.47	-3.87	-1.0	129.34	30JL	0.3	-3.7	-1.9	0.4
249.48	-3.70	-0.8	139.00	9AU	0.6	-3.8	-2.0	0.4
254.23	-3.51	-0.6	148.71	19AU	1.1	-3.9	-2.0	0.4
259.56	-3.29	-0.5	158.47	29AU	2.6	-4.0	-2.1	0.3
265.37	-3.06	-0.3	168.29	8SE	1.5	-4.1	-2.2	0.3
271.55	-2.83	-0.1	178.16	18SE	-0.4	-4.2	-2.2	0.3
278.02	-2.58	0.0	188.09	28SE	-0.9	-4.3	-2.3	0.2
284.72	-2.34	0.2	198.08	8OC	-0.9	-4.3	-2.3	0.2
291.60	-2.10	0.3	208.12	18OC	-0.9	-4.3	-2.3	0.1
298.61	-1.86	0.4	218.22	28OC	-0.6	-4.0	-2.4	0.0
305.71	-1.63	0.6	228.35	7NO	-0.5	-3.2	-2.4	-0.0
312.88	-1.41	0.7	238.51	17NO	-0.5	-3.4	-2.4	-0.1
320.08	-1.19	0.8	248.70	27NO	-0.4	-4.1	-2.3	-0.2
327.30	-0.98	0.9	258.90	7DE	0.5	-4.3	-2.3	-0.2
334.51	-0.79	1.0	269.10	17DE	3.0	-4.4	-2.2	-0.2
341.71	-0.60	1.1	279.30	27DE	0.8	-4.3	-2.2	-0.2

Table 708 / 709

♂ LONG	LAT	MAG	☉ LONG	16.00UT	☿	♀	♃	♄
348.88	-0.43	1.2	289.47	6JA	0.2	-4.2	-2.1	-0.1
356.01	-0.27	1.3	299.61	16JA	0.0	-4.1	-2.0	-0.0
3.10	-0.11	1.4	309.72	26JA	-0.1	-4.0	-1.9	0.0
10.14	0.02	1.5	319.79	5FE	-0.4	-3.9	-1.9	0.1
17.12	0.15	1.5	329.79	15FE	-0.9	-3.8	-1.8	0.1
24.05	0.27	1.6	339.75	25FE	-1.5	-3.7	-1.7	0.2
30.91	0.38	1.7	349.65	6MR	-1.2	-3.6	-1.7	0.2
37.73	0.48	1.7	359.48	16MR	-0.2	-3.6	-1.6	0.3
44.48	0.57	1.8	9.27	26MR	1.2	-3.5	-1.6	0.3
51.18	0.66	1.8	19.00	5AP	3.0	-3.4	-1.5	0.3
57.83	0.73	1.9	28.67	15AP	2.9	-3.4	-1.5	0.3
64.44	0.80	1.9	38.31	25AP	1.6	-3.4	-1.5	0.4
71.00	0.86	1.9	47.90	5MY	0.9	-3.3	-1.5	0.3
77.52	0.91	2.0	57.46	15MY	0.2	-3.3	-1.5	0.3
84.00	0.96	2.0	67.01	25MY	-0.7	-3.3	-1.5	0.3
90.45	1.01	2.0	76.54	4JN	-1.6	-3.3	-1.5	0.3
96.88	1.04	2.0	86.06	14JN	-1.3	-3.3	-1.5	0.3
103.29	1.08	2.0	95.59	24JN	-0.5	-3.3	-1.5	0.3
109.68	1.10	2.0	105.13	4JL	0.1	-3.3	-1.5	0.3
116.05	1.13	1.9	114.70	14JL	0.5	-3.4	-1.5	0.4
122.43	1.14	2.0	124.30	24JL	0.8	-3.4	-1.6	0.4
128.79	1.16	2.0	133.93	3AU	1.5	-3.4	-1.6	0.4
135.16	1.17	2.0	143.61	13AU	3.1	-3.5	-1.7	0.4
141.53	1.17	2.0	153.35	23AU	1.3	-3.5	-1.7	0.4
147.91	1.17	2.0	163.13	2SE	-0.5	-3.5	-1.8	0.4
154.30	1.17	2.0	172.97	12SE	-1.0	-3.5	-1.8	0.4
160.70	1.16	2.0	182.88	22SE	-1.0	-3.4	-1.9	0.4
167.12	1.14	2.0	192.84	2OC	-0.8	-3.4	-2.0	0.4
173.55	1.12	2.0	202.85	12OC	-0.5	-3.4	-2.0	0.3
180.00	1.10	1.9	212.92	22OC	-0.4	-3.3	-2.1	0.3
186.46	1.06	1.9	223.03	1NO	-0.3	-3.3	-2.2	0.2
192.94	1.03	1.9	233.18	11NO	-0.2	-3.3	-2.2	0.1
199.43	0.98	1.8	243.36	21NO	0.7	-3.3	-2.2	0.1
205.94	0.92	1.7	253.55	1DE	2.8	-3.3	-2.2	-0.0
212.46	0.86	1.7	263.75	11DE	0.5	-3.4	-2.2	-0.1
218.99	0.79	1.6	273.95	21DE	0.0	-3.4	-2.2	-0.2
225.52	0.70	1.5	284.14	31DE	-0.1	-3.4	-2.2	-0.1
232.06	0.60	1.4	294.30	10JA	-0.2	-3.4	-2.1	-0.1
238.59	0.49	1.3	304.43	20JA	-0.4	-3.5	-2.1	-0.0
245.12	0.36	1.2	314.52	30JA	-0.9	-3.5	-2.0	0.0
251.63	0.21	1.1	324.55	9FE	-1.4	-3.6	-2.0	0.1
258.12	0.04	0.9	334.54	19FE	-1.1	-3.6	-1.9	0.2
264.57	-0.16	0.8	344.47	1MR	-0.1	-3.7	-1.8	0.2
270.98	-0.38	0.6	354.33	11MR	1.5	-3.7	-1.7	0.3
277.32	-0.63	0.5	4.15	21MR	3.5	-3.8	-1.7	0.3
283.56	-0.91	0.3	13.90	31MR	2.1	-3.9	-1.6	0.4
289.69	-1.22	0.1	23.60	10AP	1.2	-4.0	-1.6	0.4
295.66	-1.58	-0.1	33.26	20AP	0.7	-4.1	-1.5	0.4
301.41	-1.97	-0.3	42.87	30AP	0.1	-4.2	-1.5	0.4
306.88	-2.41	-0.5	52.45	10MY	-0.7	-4.2	-1.4	0.4
311.97	-2.89	-0.7	62.01	20MY	-1.6	-4.1	-1.4	0.4
316.58	-3.42	-0.9	71.54	30MY	-1.3	-3.9	-1.4	0.4
320.56	-4.00	-1.2	81.06	9JN	-0.4	-3.3	-1.4	0.4
323.71	-4.62	-1.5	90.59	19JN	0.2	-3.0	-1.3	0.4
325.86	-5.25	-1.7	100.12	29JN	0.7	-3.7	-1.3	0.4
326.79	-5.86	-2.0	109.68	9JL	1.1	-4.1	-1.3	0.4
326.38	-6.37	-2.3	119.26	19JL	2.0	-4.2	-1.4	0.5
324.74	-6.69	-2.5	128.87	29JL	3.2	-4.2	-1.4	0.5
322.24	-6.73	-2.6	138.53	8AU	1.1	-4.1	-1.4	0.5
319.57	-6.44	-2.5	148.24	18AU	-0.5	-4.1	-1.4	0.6
317.49	-5.88	-2.3	157.99	28AU	-1.2	-4.0	-1.5	0.6
316.50	-5.16	-2.0	167.81	7SE	-1.2	-3.9	-1.5	0.6
316.83	-4.39	-1.7	177.68	17SE	-0.7	-3.8	-1.6	0.6
318.41	-3.65	-1.4	187.61	27SE	-0.4	-3.7	-1.6	0.6
321.07	-2.97	-1.1	197.60	7OC	-0.2	-3.7	-1.7	0.5
324.61	-2.36	-0.8	207.64	17OC	-0.2	-3.6	-1.7	0.5
328.84	-1.84	-0.6	217.72	27OC	-0.0	-3.6	-1.8	0.5
333.60	-1.38	-0.3	227.86	6NO	0.9	-3.5	-1.9	0.4
338.77	-0.99	-0.1	238.02	16NO	2.6	-3.5	-1.9	0.3
344.26	-0.65	0.2	248.20	26NO	0.3	-3.4	-2.0	0.3
349.99	-0.36	0.4	258.40	6DE	-0.2	-3.4	-2.0	0.2
355.89	-0.11	0.5	268.61	16DE	-0.2	-3.4	-2.1	0.1
1.93	0.10	0.7	278.80	26DE	-0.3	-3.4	-2.1	0.1

Table 710 / 711

♂ LONG	LAT	MAG	☉ LONG	16.00UT	☿	♀	♃	♄
8.07	0.28	0.9	288.98	5JA	-0.5	-3.3	-2.1	-0.0
14.28	0.44	1.0	299.13	15JA	-0.9	-3.3	-2.1	0.0
20.54	0.58	1.2	309.23	25JA	-1.2	-3.3	-2.1	0.1
26.83	0.69	1.3	319.30	4FE	-1.1	-3.3	-2.1	0.1
33.14	0.79	1.4	329.31	14FE	-0.1	-3.3	-2.0	0.2
39.46	0.88	1.5	339.27	24FE	1.9	-3.4	-2.0	0.2
45.78	0.95	1.6	349.17	6MR	3.0	-3.4	-1.9	0.3
52.11	1.01	1.7	359.01	16MR	1.6	-3.4	-1.8	0.4
58.42	1.06	1.7	8.79	26MR	0.9	-3.4	-1.8	0.4
64.73	1.10	1.8	18.52	5AP	0.5	-3.5	-1.7	0.5
71.03	1.14	1.9	28.20	15AP	0.0	-3.5	-1.6	0.5
77.32	1.16	1.9	37.83	25AP	-0.7	-3.5	-1.6	0.5
83.60	1.18	2.0	47.43	5MY	-1.7	-3.4	-1.5	0.6
89.88	1.20	2.0	57.00	15MY	-1.3	-3.4	-1.5	0.6
96.16	1.20	2.0	66.54	25MY	-0.4	-3.4	-1.4	0.6
102.44	1.21	2.0	76.07	4JN	0.3	-3.3	-1.4	0.6
108.73	1.20	2.0	85.59	14JN	0.9	-3.3	-1.3	0.6
115.02	1.20	2.0	95.12	24JN	1.5	-3.3	-1.3	0.6
121.33	1.18	2.0	104.67	4JL	2.6	-3.3	-1.3	0.6
127.65	1.17	2.0	114.23	14JL	2.8	-3.3	-1.3	0.6
133.99	1.15	2.0	123.82	24JL	0.9	-3.3	-1.3	0.6
140.36	1.12	2.0	133.46	3AU	-0.6	-3.4	-1.3	0.6
146.76	1.09	2.0	143.14	13AU	-1.3	-3.4	-1.3	0.7
153.19	1.06	1.9	152.87	23AU	-1.3	-3.4	-1.3	0.7
159.66	1.02	1.9	162.66	2SE	-0.7	-3.5	-1.3	0.7
166.16	0.98	1.9	172.50	12SE	-0.3	-3.5	-1.3	0.8
172.71	0.93	1.9	182.40	22SE	-0.1	-3.6	-1.4	0.8
179.30	0.87	1.9	192.36	2OC	-0.0	-3.7	-1.4	0.8
185.94	0.82	1.9	202.37	12OC	0.2	-3.8	-1.4	0.8
192.63	0.75	1.9	212.43	22OC	1.2	-3.8	-1.5	0.7
199.37	0.68	1.9	222.54	1NO	2.4	-3.9	-1.6	0.7
206.16	0.60	1.9	232.69	11NO	0.1	-4.1	-1.6	0.7
213.00	0.52	1.8	242.86	21NO	-0.3	-4.2	-1.7	0.6
219.89	0.43	1.8	253.06	1DE	-0.4	-4.3	-1.7	0.5
226.84	0.33	1.7	263.26	11DE	-0.4	-4.4	-1.8	0.5
233.84	0.22	1.7	273.46	21DE	-0.6	-4.4	-1.9	0.4
240.89	0.11	1.6	283.65	31DE	-0.9	-4.3	-1.9	0.3
247.99	-0.01	1.6	293.81	10JA	-1.1	-4.0	-2.0	0.3
255.13	-0.15	1.5	303.94	20JA	-0.9	-3.4	-2.0	0.2
262.33	-0.29	1.4	314.03	30JA	0.0	-3.5	-2.0	0.2
269.56	-0.44	1.3	324.07	9FE	2.2	-4.0	-2.0	0.3
276.82	-0.60	1.3	334.05	19FE	2.3	-4.3	-2.0	0.3
284.12	-0.76	1.2	343.99	1MR	1.1	-4.3	-2.0	0.4
291.44	-0.93	1.1	353.86	11MR	0.7	-4.2	-2.0	0.4
298.77	-1.11	1.0	3.67	21MR	0.3	-4.1	-1.9	0.5
306.11	-1.29	0.9	13.43	31MR	-0.1	-4.0	-1.9	0.5
313.43	-1.47	0.8	23.13	10AP	-0.8	-3.9	-1.8	0.6
320.72	-1.66	0.7	32.79	20AP	-1.7	-3.8	-1.7	0.6
327.96	-1.84	0.5	42.41	30AP	-1.3	-3.7	-1.7	0.7
335.12	-2.01	0.5	51.99	10MY	-0.4	-3.6	-1.6	0.7
342.19	-2.18	0.4	61.54	20MY	0.5	-3.6	-1.5	0.8
349.12	-2.34	0.3	71.08	30MY	1.1	-3.5	-1.5	0.8
355.88	-2.48	0.2	80.60	9JN	2.0	-3.5	-1.4	0.8
2.42	-2.62	0.1	90.13	19JN	3.2	-3.4	-1.4	0.8
8.69	-2.73	-0.1	99.67	29JN	2.3	-3.4	-1.3	0.8
14.61	-2.83	-0.2	109.22	9JL	0.7	-3.4	-1.3	0.8
20.12	-2.90	-0.3	118.80	19JL	-0.6	-3.3	-1.3	0.8
25.10	-2.95	-0.5	128.41	29JL	-1.4	-3.3	-1.2	0.8
29.42	-2.96	-0.7	138.06	8AU	-1.3	-3.3	-1.2	0.8
32.92	-2.94	-0.9	147.77	18AU	-0.6	-3.3	-1.2	0.9
35.40	-2.87	-1.1	157.52	28AU	-0.2	-3.4	-1.2	0.9
36.65	-2.73	-1.3	167.33	7SE	0.0	-3.4	-1.2	1.0
36.49	-2.50	-1.5	177.20	17SE	0.1	-3.4	-1.2	1.0
34.86	-2.16	-1.7	187.13	27SE	0.4	-3.4	-1.2	1.0
31.98	-1.71	-1.9	197.11	7OC	1.5	-3.4	-1.3	1.0
28.42	-1.18	-2.0	207.15	17OC	2.2	-3.5	-1.3	1.0
24.99	-0.62	-1.8	217.24	27OC	-0.1	-3.5	-1.3	1.0
22.44	-0.10	-1.5	227.36	6NO	-0.5	-3.5	-1.4	1.0
21.16	0.34	-1.2	237.53	16NO	-0.5	-3.5	-1.4	1.0
21.23	0.69	-0.8	247.71	26NO	-0.5	-3.5	-1.5	0.9
22.53	0.96	-0.5	257.91	6DE	-0.7	-3.4	-1.5	0.9
24.83	1.16	-0.2	268.11	16DE	-0.8	-3.4	-1.6	0.8
27.96	1.31	0.1	278.31	26DE	-0.9	-3.4	-1.7	0.7

♂ LONG	♂ LAT	♂ MAG	☉ LONG	16.00UT	☿ MAG	♀ MAG	♃ MAG	♄ MAG
				712				
31.72	1.41	0.3	288.48	5JA	-0.8	-3.4	-1.7	0.7
35.97	1.48	0.5	298.63	15JA	0.1	-3.3	-1.8	0.6
40.61	1.53	0.7	308.74	25JA	2.5	-3.3	-1.9	0.5
45.54	1.57	0.9	318.81	4FE	1.8	-3.3	-1.9	0.5
50.71	1.58	1.1	328.83	14FE	0.8	-3.3	-2.0	0.5
56.07	1.59	1.2	338.79	24FE	0.5	-3.4	-2.0	0.5
61.56	1.59	1.3	348.69	5MR	0.2	-3.4	-2.0	0.6
67.17	1.57	1.5	358.53	15MR	-0.2	-3.4	-2.0	0.6
72.88	1.56	1.6	8.32	25MR	-0.8	-3.4	-2.0	0.7
78.67	1.54	1.7	18.05	4AP	-1.6	-3.5	-2.0	0.7
84.52	1.51	1.7	27.74	14AP	-1.3	-3.5	-1.9	0.8
90.43	1.48	1.8	37.37	24AP	-0.3	-3.5	-1.9	0.8
96.39	1.44	1.9	46.97	4MY	0.6	-3.6	-1.8	0.9
102.39	1.40	1.9	56.54	14MY	1.5	-3.7	-1.8	0.9
108.44	1.36	1.9	66.08	24MY	2.6	-3.7	-1.7	1.0
114.54	1.32	2.0	75.61	3JN	3.3	-3.8	-1.6	1.0
120.68	1.27	2.0	85.14	13JN	1.8	-3.9	-1.6	1.0
126.86	1.22	2.0	94.66	23JN	0.6	-4.0	-1.5	1.1
133.09	1.17	2.0	104.20	3JL	-0.6	-4.1	-1.5	1.1
139.37	1.11	2.0	113.77	13JL	-1.5	-4.2	-1.4	1.1
145.69	1.05	2.0	123.36	23JL	-1.3	-4.2	-1.4	1.1
152.08	0.99	2.0	132.99	2AU	-0.6	-4.1	-1.3	1.1
158.52	0.92	2.0	142.67	12AU	-0.1	-3.8	-1.3	1.1
165.02	0.85	1.9	152.40	22AU	0.1	-3.3	-1.3	1.1
171.59	0.78	1.9	162.18	1SE	0.3	-3.5	-1.2	1.2
178.22	0.70	1.9	172.02	11SE	0.6	-4.0	-1.2	1.2
184.92	0.62	1.8	181.92	21SE	1.9	-4.3	-1.2	1.3
191.69	0.54	1.8	191.87	1OC	1.9	-4.3	-1.2	1.3
198.54	0.45	1.8	201.88	11OC	-0.2	-4.3	-1.2	1.3
205.45	0.36	1.8	211.94	21OC	-0.7	-4.2	-1.2	1.3
212.45	0.26	1.7	222.05	31OC	-0.7	-4.1	-1.3	1.3
219.51	0.16	1.7	232.20	10NO	-0.7	-4.0	-1.3	1.3
226.65	0.06	1.7	242.37	20NO	-0.8	-3.9	-1.3	1.3
233.87	-0.05	1.7	252.56	30NO	-0.7	-3.8	-1.4	1.2
241.16	-0.15	1.7	262.77	10DE	-0.7	-3.8	-1.4	1.2
248.51	-0.27	1.6	272.97	20DE	-0.7	-3.7	-1.5	1.1
255.93	-0.38	1.6	283.16	30DE	0.2	-3.6	-1.5	1.1
				713				
263.41	-0.50	1.6	293.32	9JA	2.8	-3.6	-1.6	1.0
270.94	-0.62	1.5	303.45	19JA	1.3	-3.5	-1.7	0.9
278.53	-0.73	1.5	313.54	29JA	0.5	-3.4	-1.7	0.9
286.15	-0.85	1.4	323.59	8FE	0.3	-3.4	-1.8	0.8
293.81	-0.96	1.4	333.57	18FE	0.1	-3.4	-1.9	0.7
301.49	-1.06	1.4	343.51	28FE	-0.2	-3.3	-1.9	0.8
309.18	-1.16	1.3	353.39	10MR	-0.8	-3.3	-2.0	0.8
316.87	-1.26	1.3	3.20	20MR	-1.6	-3.3	-2.0	0.9
324.56	-1.34	1.2	12.96	30MR	-1.3	-3.3	-2.0	0.9
332.21	-1.41	1.2	22.67	9AP	-0.3	-3.3	-2.0	1.0
339.82	-1.47	1.1	32.33	19AP	0.8	-3.3	-2.0	1.0
347.38	-1.52	1.1	41.95	29AP	1.9	-3.3	-2.0	1.1
354.86	-1.55	1.1	51.53	9MY	3.4	-3.4	-2.0	1.1
2.26	-1.56	1.0	61.08	19MY	2.6	-3.4	-2.0	1.2
9.56	-1.56	1.0	70.62	29MY	1.4	-3.4	-1.9	1.2
16.73	-1.54	0.9	80.14	8JN	0.4	-3.5	-1.8	1.3
23.76	-1.50	0.9	89.67	18JN	-0.6	-3.5	-1.8	1.3
30.64	-1.45	0.8	99.20	28JN	-1.5	-3.5	-1.7	1.3
37.33	-1.37	0.8	108.75	8JL	-1.3	-3.4	-1.7	1.4
43.81	-1.28	0.7	118.33	18JL	-0.5	-3.4	-1.6	1.4
50.05	-1.17	0.6	127.94	28JL	-0.0	-3.4	-1.5	1.4
56.00	-1.04	0.5	137.59	7AU	0.3	-3.3	-1.5	1.4
61.62	-0.88	0.4	147.29	17AU	0.5	-3.3	-1.4	1.3
66.85	-0.70	0.3	157.04	27AU	0.9	-3.3	-1.4	1.3
71.61	-0.49	0.2	166.85	6SE	2.3	-3.3	-1.4	1.2
75.78	-0.25	0.0	176.72	16SE	1.7	-3.3	-1.3	1.3
79.25	0.03	-0.1	186.64	26SE	-0.3	-3.3	-1.3	1.3
81.83	0.35	-0.3	196.62	6OC	-0.8	-3.4	-1.3	1.2
83.35	0.73	-0.5	206.66	16OC	-0.8	-3.4	-1.3	1.2
83.59	1.16	-0.8	216.75	26OC	-0.8	-3.4	-1.3	1.2
82.42	1.62	-1.0	226.87	5NO	-0.7	-3.5	-1.3	1.2
79.89	2.09	-1.2	237.03	15NO	-0.6	-3.5	-1.3	1.2
76.31	2.52	-1.3	247.22	25NO	-0.6	-3.6	-1.3	1.1
72.37	2.84	-1.3	257.42	5DE	-0.5	-3.6	-1.3	1.1
68.84	3.03	-1.1	267.62	15DE	0.3	-3.7	-1.4	1.1
66.35	3.09	-0.8	277.82	25DE	3.0	-3.8	-1.4	1.0

♂ LONG	♂ LAT	♂ MAG	☉ LONG	16.00UT	☿ MAG	♀ MAG	♃ MAG	♄ MAG
				714				
65.17	3.06	-0.5	287.99	4JA	1.0	-3.9	-1.4	1.0
65.31	2.97	-0.2	298.14	14JA	0.3	-3.9	-1.5	0.9
66.59	2.85	0.1	308.26	24JA	0.1	-4.0	-1.6	0.9
68.83	2.71	0.3	318.33	3FE	-0.0	-4.1	-1.6	0.8
71.82	2.57	0.5	328.34	13FE	-0.3	-4.2	-1.7	0.8
75.42	2.44	0.7	338.31	23FE	-0.8	-4.3	-1.8	0.7
79.50	2.31	0.9	348.21	5MR	-1.6	-4.3	-1.8	0.7
83.95	2.18	1.1	358.06	15MR	-1.3	-4.1	-1.9	0.8
88.71	2.06	1.2	7.85	25MR	-0.3	-3.7	-2.0	0.9
93.71	1.95	1.3	17.59	4AP	1.0	-3.1	-2.0	0.9
98.92	1.84	1.5	27.27	14AP	2.5	-3.5	-2.1	1.0
104.29	1.73	1.5	36.91	24AP	3.4	-4.0	-2.1	1.1
109.81	1.63	1.6	46.51	4MY	2.0	-4.2	-2.1	1.1
115.46	1.53	1.7	56.07	14MY	1.1	-4.2	-2.1	1.2
121.22	1.43	1.8	65.62	24MY	0.3	-4.1	-2.1	1.2
127.08	1.33	1.8	75.15	3JN	-0.7	-4.0	-2.1	1.3
133.04	1.23	1.8	84.67	13JN	-1.6	-4.0	-2.1	1.3
139.10	1.13	1.9	94.20	23JN	-1.3	-3.9	-2.0	1.3
145.24	1.03	1.9	103.74	3JL	-0.5	-3.8	-2.0	1.3
151.47	0.94	1.9	113.30	13JL	0.0	-3.7	-1.9	1.3
157.79	0.84	1.9	122.90	23JL	0.4	-3.6	-1.9	1.3
164.20	0.74	1.9	132.53	2AU	0.7	-3.6	-1.8	1.2
170.69	0.64	1.9	142.20	12AU	1.2	-3.5	-1.7	1.2
177.28	0.54	1.9	151.93	22AU	2.7	-3.5	-1.7	1.2
183.95	0.44	1.8	161.70	1SE	1.5	-3.4	-1.6	1.1
190.72	0.34	1.8	171.54	11SE	-0.4	-3.4	-1.6	1.1
197.58	0.24	1.8	181.44	21SE	-0.9	-3.4	-1.5	1.1
204.53	0.13	1.7	191.39	1OC	-1.0	-3.4	-1.5	1.1
211.58	0.03	1.7	201.39	11OC	-0.9	-3.4	-1.4	1.1
218.72	-0.08	1.6	211.46	21OC	-0.6	-3.4	-1.4	1.1
225.95	-0.18	1.6	221.56	31OC	-0.4	-3.4	-1.4	1.1
233.27	-0.28	1.5	231.70	10NO	-0.4	-3.4	-1.4	1.1
240.67	-0.39	1.5	241.88	20NO	-0.3	-3.4	-1.4	1.0
248.15	-0.49	1.5	252.07	30NO	0.5	-3.4	-1.4	1.0
255.71	-0.58	1.5	262.27	10DE	3.0	-3.4	-1.4	1.0
263.34	-0.68	1.5	272.47	20DE	0.7	-3.4	-1.4	0.9
271.02	-0.77	1.5	282.66	30DE	0.1	-3.4	-1.4	0.9
				715				
278.76	-0.85	1.5	292.83	9JA	-0.0	-3.5	-1.4	0.9
286.54	-0.93	1.4	302.96	19JA	-0.1	-3.5	-1.5	0.8
294.35	-1.00	1.4	313.05	29JA	-0.4	-3.5	-1.5	0.8
302.18	-1.06	1.4	323.10	8FE	-0.9	-3.4	-1.6	0.7
310.02	-1.10	1.4	333.09	18FE	-1.4	-3.4	-1.6	0.7
317.85	-1.14	1.4	343.03	28FE	-1.2	-3.4	-1.7	0.6
325.67	-1.17	1.4	352.90	10MR	-0.2	-3.4	-1.7	0.6
333.45	-1.18	1.4	2.73	20MR	1.3	-3.4	-1.8	0.6
341.20	-1.19	1.4	12.49	30MR	3.2	-3.3	-1.9	0.7
348.89	-1.17	1.4	22.20	9AP	2.6	-3.3	-2.0	0.7
356.52	-1.15	1.4	31.86	19AP	1.5	-3.3	-2.0	0.8
4.07	-1.11	1.4	41.48	29AP	0.8	-3.3	-2.1	0.9
11.54	-1.06	1.4	51.06	9MY	0.2	-3.3	-2.1	0.9
18.91	-1.00	1.4	60.62	19MY	-0.7	-3.4	-2.2	1.0
26.19	-0.92	1.4	70.15	29MY	-1.6	-3.4	-2.2	1.0
33.35	-0.84	1.4	79.68	8JN	-1.3	-3.4	-2.3	1.1
40.40	-0.74	1.3	89.21	18JN	-0.5	-3.5	-2.3	1.1
47.32	-0.63	1.3	98.74	28JN	0.1	-3.5	-2.3	1.1
54.11	-0.51	1.3	108.29	8JL	0.5	-3.6	-2.2	1.1
60.76	-0.38	1.3	117.87	18JL	0.9	-3.6	-2.2	1.1
67.25	-0.25	1.2	127.47	28JL	1.6	-3.7	-2.2	1.1
73.57	-0.10	1.2	137.13	7AU	3.2	-3.8	-2.1	1.1
79.70	0.07	1.2	146.82	17AU	1.3	-3.9	-2.0	1.1
85.62	0.24	1.1	156.57	27AU	-0.4	-4.0	-2.0	1.1
91.30	0.43	1.0	166.38	6SE	-1.1	-4.1	-1.9	1.0
96.70	0.63	0.9	176.24	16SE	-1.1	-4.2	-1.8	1.0
101.77	0.86	0.8	186.16	26SE	-0.8	-4.3	-1.8	0.9
106.45	1.11	0.7	196.14	6OC	-0.5	-4.3	-1.7	0.9
110.65	1.38	0.6	206.17	16OC	-0.3	-4.3	-1.7	1.0
114.27	1.69	0.4	216.25	26OC	-0.3	-3.9	-1.6	1.0
117.16	2.03	0.2	226.38	5NO	-0.1	-3.2	-1.6	1.0
119.17	2.42	0.0	236.53	15NO	0.7	-3.5	-1.5	1.0
120.11	2.83	-0.2	246.72	25NO	2.9	-4.1	-1.5	0.9
119.78	3.27	-0.5	256.92	5DE	0.5	-4.3	-1.5	0.9
118.10	3.70	-0.7	267.12	15DE	-0.1	-4.4	-1.5	0.9
115.17	4.07	-0.9	277.32	25DE	-0.1	-4.3	-1.5	0.9

♂			☉	16.00UT	☿	♀	♃	♄
LONG	LAT	MAG	LONG	716	MAGNITUDES			
111.39	4.30	-1.0	287.50	4JA	-0.2	-4.2	-1.5	0.8
107.44	4.37	-1.0	297.65	14JA	-0.5	-4.1	-1.5	0.8
104.05	4.28	-0.7	307.76	24JA	-0.9	-4.0	-1.5	0.7
101.75	4.05	-0.5	317.84	3FE	-1.3	-3.9	-1.5	0.7
100.75	3.77	-0.2	327.86	13FE	-1.1	-3.8	-1.5	0.6
101.01	3.45	0.0	337.83	23FE	-0.2	-3.7	-1.6	0.6
102.38	3.15	0.3	347.74	4MR	1.6	-3.6	-1.6	0.5
104.66	2.86	0.5	357.59	14MR	3.4	-3.6	-1.7	0.5
107.68	2.59	0.7	7.38	24MR	1.9	-3.5	-1.7	0.4
111.30	2.34	0.8	17.12	3AP	1.1	-3.4	-1.8	0.5
115.40	2.12	1.0	26.81	13AP	0.6	-3.4	-1.9	0.5
119.89	1.91	1.1	36.45	23AP	0.1	-3.4	-1.9	0.6
124.71	1.71	1.2	46.05	3MY	-0.7	-3.3	-2.0	0.7
129.80	1.53	1.3	55.62	13MY	-1.6	-3.3	-2.1	0.7
135.11	1.36	1.4	65.16	23MY	-1.4	-3.3	-2.1	0.8
140.63	1.20	1.5	74.70	2JN	-0.4	-3.3	-2.2	0.9
146.33	1.05	1.5	84.22	12JN	0.2	-3.3	-2.3	0.9
152.19	0.90	1.6	93.74	22JN	0.7	-3.3	-2.3	0.9
158.20	0.76	1.6	103.29	2JL	1.2	-3.3	-2.4	1.0
164.36	0.62	1.6	112.84	12JL	2.2	-3.4	-2.4	1.0
170.64	0.49	1.7	122.44	22JL	3.1	-3.4	-2.4	1.0
177.06	0.36	1.7	132.07	1AU	1.1	-3.4	-2.4	1.0
183.60	0.24	1.7	141.73	11AU	-0.5	-3.5	-2.4	1.0
190.27	0.12	1.7	151.46	21AU	-1.2	-3.5	-2.3	1.0
197.05	-0.00	1.7	161.23	31AU	-1.2	-3.5	-2.3	1.0
203.96	-0.12	1.7	171.06	10SE	-0.7	-3.4	-2.2	0.9
210.97	-0.23	1.6	180.95	20SE	-0.4	-3.4	-2.2	0.9
218.11	-0.34	1.6	190.90	30SE	-0.2	-3.4	-2.1	0.8
225.34	-0.44	1.6	200.90	10OC	-0.1	-3.4	-2.0	0.8
232.69	-0.54	1.6	210.96	20OC	0.0	-3.3	-1.9	0.8
240.13	-0.63	1.5	221.06	30OC	1.0	-3.3	-1.9	0.9
247.66	-0.72	1.5	231.20	9NO	2.6	-3.3	-1.8	0.9
255.27	-0.80	1.5	241.37	19NO	0.2	-3.3	-1.8	0.9
262.97	-0.87	1.4	251.57	29NO	-0.2	-3.3	-1.7	0.9
270.72	-0.93	1.4	261.77	9DE	-0.3	-3.4	-1.7	0.9
278.53	-0.98	1.4	271.97	19DE	-0.4	-3.4	-1.6	0.8
286.38	-1.02	1.3	282.16	29DE	-0.6	-3.4	-1.6	0.8
				717				
294.26	-1.06	1.3	292.33	8JA	-0.9	-3.4	-1.6	0.8
302.16	-1.08	1.3	302.46	18JA	-1.1	-3.5	-1.6	0.7
310.06	-1.08	1.3	312.56	28JA	-1.0	-3.5	-1.6	0.7
317.95	-1.08	1.3	322.61	7FE	-0.1	-3.6	-1.6	0.7
325.81	-1.07	1.4	332.61	17FE	1.9	-3.6	-1.6	0.6
333.64	-1.04	1.4	342.55	27FE	2.8	-3.7	-1.6	0.5
341.42	-1.00	1.4	352.43	9MR	1.4	-3.7	-1.6	0.5
349.15	-0.95	1.4	2.25	19MR	0.8	-3.8	-1.6	0.4
356.81	-0.90	1.4	12.02	29MR	0.4	-3.9	-1.7	0.4
4.39	-0.83	1.5	21.73	8AP	-0.0	-4.0	-1.7	0.3
11.90	-0.75	1.5	31.40	18AP	-0.7	-4.1	-1.7	0.4
19.32	-0.67	1.5	41.02	28AP	-1.6	-4.2	-1.8	0.5
26.64	-0.57	1.6	50.60	8MY	-1.4	-4.2	-1.9	0.5
33.88	-0.48	1.6	60.16	18MY	-0.4	-4.1	-1.9	0.6
41.01	-0.37	1.6	69.70	28MY	0.4	-3.9	-2.0	0.7
48.05	-0.26	1.6	79.22	7JN	0.9	-3.3	-2.1	0.7
54.99	-0.15	1.6	88.75	17JN	1.6	-3.0	-2.1	0.8
61.82	-0.04	1.6	98.28	27JN	2.8	-3.7	-2.2	0.8
68.55	0.09	1.6	107.83	7JL	2.6	-4.1	-2.3	0.9
75.17	0.21	1.6	117.40	17JL	0.9	-4.2	-2.3	0.9
81.69	0.34	1.6	127.01	27JL	-0.5	-4.2	-2.4	0.9
88.09	0.47	1.6	136.65	6AU	-1.3	-4.1	-2.4	0.9
94.37	0.61	1.6	146.35	16AU	-1.3	-4.0	-2.4	0.9
100.52	0.75	1.6	156.10	26AU	-0.7	-4.0	-2.5	0.9
106.54	0.90	1.5	165.90	5SE	-0.3	-3.9	-2.5	0.9
112.39	1.05	1.5	175.76	15SE	-0.1	-3.8	-2.4	0.9
118.08	1.21	1.4	185.68	25SE	0.0	-3.7	-2.4	0.8
123.56	1.38	1.3	195.65	5OC	0.2	-3.7	-2.3	0.8
128.81	1.57	1.2	205.69	15OC	1.2	-3.6	-2.3	0.7
133.77	1.77	1.1	215.76	25OC	2.4	-3.6	-2.2	0.8
138.41	1.98	1.0	225.88	4NO	0.1	-3.5	-2.2	0.8
142.64	2.22	0.8	236.04	14NO	-0.4	-3.5	-2.1	0.8
146.36	2.48	0.7	246.22	24NO	-0.4	-3.4	-2.0	0.8
149.47	2.76	0.5	256.42	4DE	-0.5	-3.4	-1.9	0.8
151.81	3.08	0.3	266.63	14DE	-0.6	-3.4	-1.9	0.8
153.20	3.41	0.0	276.82	24DE	-0.9	-3.4	-1.8	0.8

♂			☉	16.00UT	☿	♀	♃	♄
LONG	LAT	MAG	LONG	718	MAGNITUDES			
153.48	3.75	-0.2	287.00	3JA	-1.0	-3.3	-1.8	0.8
152.47	4.07	-0.5	297.16	13JA	-0.9	-3.3	-1.7	0.8
150.18	4.33	-0.7	307.27	23JA	-0.0	-3.3	-1.7	0.7
146.83	4.47	-0.9	317.35	2FE	2.2	-3.3	-1.6	0.7
142.93	4.44	-1.0	327.37	12FE	2.1	-3.3	-1.6	0.6
139.21	4.24	-0.9	337.34	22FE	1.0	-3.4	-1.6	0.6
136.30	3.91	-0.7	347.25	4MR	0.6	-3.4	-1.6	0.5
134.60	3.50	-0.4	357.11	14MR	0.3	-3.4	-1.6	0.5
134.21	3.08	-0.2	6.90	24MR	-0.1	-3.4	-1.6	0.4
135.02	2.67	0.0	16.64	3AP	-0.7	-3.5	-1.6	0.4
136.87	2.29	0.2	26.33	13AP	-1.6	-3.5	-1.6	0.3
139.58	1.95	0.4	35.97	23AP	-1.4	-3.4	-1.6	0.3
142.99	1.64	0.6	45.58	3MY	-0.4	-3.4	-1.7	0.4
146.98	1.36	0.7	55.15	13MY	0.5	-3.4	-1.7	0.4
151.44	1.10	0.8	64.70	23MY	1.2	-3.4	-1.7	0.5
156.30	0.87	0.9	74.23	2JN	2.2	-3.3	-1.8	0.6
161.50	0.66	1.0	83.75	12JN	3.4	-3.3	-1.9	0.6
166.98	0.46	1.1	93.28	22JN	2.1	-3.3	-1.9	0.7
172.72	0.28	1.2	102.82	2JL	0.7	-3.3	-2.0	0.7
178.69	0.11	1.2	112.38	12JL	-0.6	-3.3	-2.0	0.8
184.87	-0.05	1.3	121.96	22JL	-1.4	-3.3	-2.1	0.8
191.23	-0.20	1.3	131.59	1AU	-1.3	-3.4	-2.2	0.8
197.77	-0.34	1.3	141.26	11AU	-0.6	-3.4	-2.2	0.8
204.48	-0.46	1.4	150.98	21AU	-0.2	-3.4	-2.3	0.9
211.34	-0.58	1.4	160.75	31AU	0.1	-3.5	-2.4	0.9
218.35	-0.69	1.4	170.58	10SE	0.2	-3.5	-2.4	0.8
225.50	-0.79	1.4	180.47	20SE	0.5	-3.6	-2.4	0.8
232.77	-0.87	1.4	190.42	30SE	1.6	-3.7	-2.4	0.8
240.17	-0.95	1.4	200.42	10OC	2.2	-3.8	-2.4	0.8
247.66	-1.02	1.4	210.47	20OC	-0.1	-3.9	-2.4	0.7
255.26	-1.07	1.4	220.57	30OC	-0.6	-4.0	-2.4	0.7
262.94	-1.11	1.4	230.71	9NO	-0.6	-4.1	-2.3	0.7
270.69	-1.14	1.4	240.88	19NO	-0.6	-4.2	-2.3	0.8
278.51	-1.16	1.4	251.07	29NO	-0.7	-4.3	-2.2	0.8
286.36	-1.16	1.4	261.28	9DE	-0.8	-4.4	-2.1	0.8
294.24	-1.15	1.4	271.47	19DE	-0.8	-4.4	-2.1	0.8
302.14	-1.13	1.4	281.67	29DE	-0.8	-4.3	-2.0	0.8
				719				
310.03	-1.10	1.4	291.83	8JA	0.1	-4.0	-1.9	0.8
317.91	-1.05	1.4	301.97	18JA	2.6	-3.4	-1.9	0.8
325.76	-1.00	1.4	312.07	28JA	1.6	-3.6	-1.8	0.7
333.57	-0.94	1.4	322.12	7FE	0.7	-4.1	-1.7	0.7
341.33	-0.86	1.4	332.12	17FE	0.4	-4.3	-1.7	0.7
349.03	-0.78	1.4	342.06	27FE	0.2	-4.3	-1.6	0.6
356.66	-0.70	1.4	351.94	9MR	-0.2	-4.2	-1.6	0.6
4.21	-0.61	1.4	1.77	19MR	-0.8	-4.1	-1.6	0.5
11.68	-0.51	1.4	11.54	29MR	-1.6	-4.0	-1.6	0.4
19.07	-0.41	1.5	21.26	8AP	-1.4	-3.9	-1.5	0.4
26.37	-0.31	1.5	30.92	18AP	-0.3	-3.8	-1.5	0.3
33.59	-0.21	1.6	40.55	28AP	0.7	-3.7	-1.5	0.3
40.71	-0.10	1.6	50.14	8MY	1.6	-3.6	-1.5	0.3
47.75	0.00	1.7	59.69	18MY	2.9	-3.6	-1.6	0.3
54.70	0.11	1.7	69.24	28MY	3.1	-3.5	-1.6	0.4
61.57	0.21	1.8	78.76	7JN	1.7	-3.5	-1.6	0.5
68.36	0.32	1.8	88.29	17JN	0.5	-3.4	-1.6	0.5
75.07	0.42	1.8	97.82	27JN	-0.6	-3.4	-1.7	0.6
81.70	0.52	1.8	107.37	7JL	-1.5	-3.4	-1.7	0.6
88.25	0.63	1.9	116.94	17JL	-1.3	-3.3	-1.8	0.7
94.73	0.73	1.9	126.55	27JL	-0.6	-3.3	-1.8	0.7
101.13	0.83	1.9	136.19	6AU	-0.1	-3.3	-1.9	0.8
107.46	0.93	1.9	145.88	16AU	0.2	-3.3	-2.0	0.8
113.71	1.03	1.8	155.63	26AU	0.4	-3.3	-2.0	0.8
119.88	1.13	1.8	165.43	5SE	0.7	-3.4	-2.1	0.8
125.96	1.24	1.8	175.28	15SE	2.0	-3.4	-2.2	0.8
131.95	1.34	1.7	185.20	25SE	1.9	-3.4	-2.2	0.8
137.83	1.45	1.7	195.17	5OC	-0.2	-3.4	-2.3	0.8
143.61	1.56	1.6	205.20	15OC	-0.7	-3.5	-2.3	0.8
149.24	1.67	1.6	215.27	25OC	-0.7	-3.5	-2.3	0.7
154.72	1.79	1.5	225.39	4NO	-0.7	-3.5	-2.4	0.7
160.02	1.92	1.4	235.55	14NO	-0.7	-3.5	-2.4	0.7
165.09	2.05	1.2	245.73	24NO	-0.7	-3.4	-2.3	0.7
169.89	2.19	1.1	255.92	4DE	-0.7	-3.4	-2.3	0.7
174.38	2.33	0.9	266.13	14DE	-0.6	-3.4	-2.3	0.8
178.45	2.49	0.8	276.33	24DE	0.2	-3.4	-2.2	0.8

♂

LONG	LAT	MAG	☉ LONG	16.00UT 720	☿	♀	♃	♄
182.02	2.65	0.6	286.50	3JA	2.8	-3.4	-2.1	0.8
184.97	2.82	0.3	296.66	13JA	1.2	-3.3	-2.1	0.8
187.14	2.99	0.1	306.78	23JA	0.5	-3.3	-2.0	0.8
188.37	3.15	-0.2	316.85	2FE	0.2	-3.3	-1.9	0.8
188.46	3.29	-0.5	326.88	12FE	0.0	-3.3	-1.8	0.7
187.28	3.37	-0.8	336.86	22FE	-0.3	-3.4	-1.8	0.7
184.87	3.36	-1.0	346.77	3MR	-0.8	-3.4	-1.7	0.6
181.49	3.22	-1.3	356.63	13MR	-1.6	-3.4	-1.7	0.6
177.70	2.94	-1.3	6.43	23MR	-1.3	-3.4	-1.6	0.5
174.24	2.54	-1.1	16.17	2AP	-0.3	-3.5	-1.6	0.5
171.70	2.08	-0.9	25.86	12AP	0.9	-3.5	-1.5	0.4
170.43	1.61	-0.7	35.51	22AP	2.1	-3.5	-1.5	0.3
170.49	1.17	-0.5	45.11	2MY	3.8	-3.6	-1.5	0.3
171.75	0.76	-0.3	54.69	12MY	2.4	-3.7	-1.5	0.2
174.08	0.41	-0.1	64.24	22MY	1.3	-3.7	-1.4	0.3
177.27	0.10	0.0	73.77	1JN	0.4	-3.8	-1.4	0.3
181.17	-0.17	0.2	83.29	11JN	-0.6	-3.9	-1.4	0.4
185.68	-0.40	0.3	92.82	21JN	-1.5	-4.0	-1.5	0.5
190.67	-0.61	0.4	102.36	1JL	-1.4	-4.1	-1.5	0.5
196.09	-0.79	0.5	111.92	11JL	-0.5	-4.2	-1.5	0.6
201.86	-0.94	0.6	121.51	21JL	-0.0	-4.2	-1.5	0.6
207.95	-1.08	0.7	131.13	31JL	0.3	-4.1	-1.6	0.7
214.30	-1.19	0.8	140.80	10AU	0.6	-3.8	-1.6	0.7
220.90	-1.29	0.8	150.51	20AU	1.0	-3.3	-1.7	0.8
227.70	-1.36	0.9	160.28	30AU	2.4	-3.5	-1.7	0.8
234.70	-1.42	0.9	170.11	9SE	1.7	-4.0	-1.8	0.8
241.86	-1.46	1.0	179.99	19SE	-0.3	-4.3	-1.8	0.8
249.17	-1.48	1.0	189.93	29SE	-0.9	-4.3	-1.9	0.8
256.61	-1.49	1.1	199.93	9OC	-0.9	-4.3	-2.0	0.8
264.15	-1.48	1.1	209.98	19OC	-0.9	-4.2	-2.0	0.8
271.78	-1.46	1.1	220.08	29OC	-0.6	-4.1	-2.1	0.7
279.49	-1.42	1.2	230.22	8NO	-0.5	-4.0	-2.1	0.7
287.24	-1.37	1.2	240.39	18NO	-0.5	-3.9	-2.2	0.7
295.02	-1.31	1.3	250.58	28NO	-0.4	-3.8	-2.2	0.7
302.83	-1.23	1.3	260.78	8DE	0.4	-3.8	-2.2	0.7
310.63	-1.15	1.3	270.98	18DE	3.0	-3.7	-2.2	0.8
318.41	-1.05	1.4	281.17	28DE	0.9	-3.6	-2.2	0.8
				721				
326.17	-0.95	1.4	291.35	7JA	0.2	-3.6	-2.2	0.8
333.88	-0.85	1.4	301.48	17JA	0.1	-3.5	-2.1	0.8
341.54	-0.74	1.5	311.58	27JA	-0.1	-3.4	-2.1	0.8
349.14	-0.63	1.5	321.64	6FE	-0.3	-3.4	-2.0	0.8
356.67	-0.52	1.5	331.64	16FE	-0.8	-3.4	-1.9	0.8
4.12	-0.41	1.6	341.58	26FE	-1.5	-3.3	-1.9	0.7
11.50	-0.29	1.6	351.47	8MR	-1.3	-3.3	-1.8	0.7
18.80	-0.18	1.6	1.30	18MR	-0.3	-3.3	-1.7	0.6
26.01	-0.07	1.7	11.07	28MR	1.1	-3.3	-1.7	0.6
33.14	0.03	1.7	20.79	7AP	2.7	-3.3	-1.6	0.5
40.19	0.14	1.7	30.46	17AP	3.1	-3.3	-1.5	0.5
47.16	0.24	1.7	40.08	27AP	1.8	-3.3	-1.5	0.4
54.06	0.33	1.7	49.67	7MY	1.0	-3.4	-1.4	0.3
60.89	0.43	1.7	59.23	17MY	0.3	-3.4	-1.4	0.3
67.64	0.52	1.8	68.77	27MY	-0.6	-3.4	-1.4	0.2
74.33	0.60	1.8	78.30	6JN	-1.6	-3.5	-1.4	0.3
80.96	0.68	1.8	87.82	16JN	-1.4	-3.5	-1.3	0.4
87.54	0.76	1.9	97.35	26JN	-0.5	-3.5	-1.3	0.4
94.06	0.84	1.9	106.90	6JL	0.1	-3.4	-1.3	0.5
100.53	0.91	1.9	116.47	16JL	0.4	-3.4	-1.3	0.5
106.96	0.98	2.0	126.08	26JL	0.8	-3.4	-1.3	0.6
113.34	1.04	2.0	135.72	5AU	1.4	-3.3	-1.4	0.6
119.69	1.11	2.0	145.41	15AU	2.9	-3.3	-1.4	0.7
125.99	1.17	2.0	155.15	25AU	1.5	-3.3	-1.4	0.7
132.26	1.23	2.0	164.95	4SE	-0.4	-3.3	-1.5	0.8
138.48	1.28	2.0	174.80	14SE	-1.0	-3.3	-1.5	0.8
144.67	1.33	1.9	184.71	24SE	-1.0	-3.3	-1.6	0.8
150.82	1.38	1.9	194.68	4OC	-0.8	-3.4	-1.6	0.8
156.91	1.43	1.9	204.71	14OC	-0.5	-3.4	-1.7	0.8
162.96	1.48	1.8	214.78	24OC	-0.4	-3.4	-1.7	0.8
168.95	1.52	1.7	224.90	3NO	-0.4	-3.5	-1.8	0.7
174.86	1.56	1.7	235.05	13NO	-0.3	-3.5	-1.9	0.7
180.69	1.60	1.6	245.23	23NO	0.5	-3.6	-1.9	0.7
186.43	1.64	1.5	255.43	3DE	3.0	-3.6	-2.0	0.7
192.04	1.67	1.4	265.63	13DE	0.6	-3.7	-2.0	0.7
197.52	1.69	1.2	275.83	23DE	0.0	-3.8	-2.1	0.7

♂

LONG	LAT	MAG	☉ LONG	16.00UT 722	☿	♀	♃	♄
202.82	1.71	1.1	286.02	2JA	-0.1	-3.9	-2.1	0.8
207.91	1.72	0.9	296.17	12JA	-0.2	-3.9	-2.1	0.8
212.75	1.72	0.7	306.29	22JA	-0.4	-4.0	-2.1	0.8
217.25	1.71	0.5	316.37	1FE	-0.9	-4.1	-2.1	0.8
221.37	1.67	0.3	326.40	11FE	-1.4	-4.2	-2.0	0.8
224.98	1.61	0.1	336.38	21FE	-1.2	-4.3	-2.0	0.8
227.95	1.51	-0.2	346.30	3MR	-0.2	-4.3	-1.9	0.8
230.15	1.36	-0.5	356.15	13MR	1.4	-4.1	-1.9	0.7
231.37	1.15	-0.8	5.96	23MR	3.4	-3.7	-1.8	0.7
231.44	0.87	-1.1	15.70	2AP	2.3	-3.1	-1.7	0.6
230.28	0.49	-1.4	25.39	12AP	1.3	-3.5	-1.7	0.6
227.92	0.03	-1.7	35.04	22AP	0.7	-4.0	-1.6	0.5
224.73	-0.49	-1.9	44.65	2MY	0.2	-4.2	-1.5	0.5
221.35	-1.03	-1.9	54.22	12MY	-0.7	-4.2	-1.5	0.4
218.48	-1.52	-1.7	63.77	22MY	-1.6	-4.1	-1.4	0.3
216.73	-1.92	-1.5	73.30	1JN	-1.4	-4.0	-1.4	0.3
216.33	-2.22	-1.3	82.83	11JN	-0.5	-3.9	-1.3	0.3
217.28	-2.44	-1.1	92.36	21JN	0.2	-3.9	-1.3	0.3
219.46	-2.59	-0.9	101.89	1JL	0.6	-3.8	-1.3	0.4
222.68	-2.67	-0.7	111.45	11JL	1.0	-3.7	-1.3	0.4
226.74	-2.71	-0.5	121.04	21JL	1.8	-3.6	-1.3	0.5
231.51	-2.72	-0.4	130.66	31JL	3.2	-3.6	-1.2	0.6
236.85	-2.69	-0.2	140.32	10AU	1.3	-3.5	-1.2	0.6
242.65	-2.63	-0.1	150.04	20AU	-0.4	-3.5	-1.3	0.7
248.84	-2.56	0.0	159.81	30AU	-1.1	-3.4	-1.3	0.7
255.34	-2.46	0.2	169.63	9SE	-1.2	-3.4	-1.3	0.7
262.09	-2.35	0.3	179.51	19SE	-0.8	-3.4	-1.3	0.8
269.06	-2.22	0.4	189.45	29SE	-0.4	-3.4	-1.4	0.8
276.18	-2.07	0.5	199.45	9OC	-0.3	-3.4	-1.4	0.8
283.43	-1.92	0.6	209.50	19OC	-0.2	-3.4	-1.4	0.8
290.78	-1.76	0.7	219.59	29OC	-0.1	-3.4	-1.5	0.8
298.20	-1.60	0.8	229.73	8NO	0.8	-3.4	-1.6	0.8
305.66	-1.43	0.9	239.90	18NO	2.9	-3.4	-1.6	0.8
313.14	-1.26	1.0	250.09	28NO	0.4	-3.4	-1.7	0.7
320.62	-1.09	1.1	260.28	8DE	-0.1	-3.4	-1.8	0.7
328.09	-0.93	1.1	270.49	18DE	-0.2	-3.4	-1.8	0.7
335.53	-0.76	1.2	280.68	28DE	-0.3	-3.4	-1.9	0.8
				723				
342.93	-0.61	1.3	290.85	7JA	-0.5	-3.5	-1.9	0.8
350.28	-0.46	1.4	300.99	17JA	-0.9	-3.5	-2.0	0.8
357.57	-0.31	1.4	311.09	27JA	-1.2	-3.5	-2.0	0.8
4.81	-0.18	1.5	321.15	6FE	-1.1	-3.4	-2.0	0.8
11.97	-0.05	1.6	331.15	16FE	-0.2	-3.4	-2.0	0.8
19.07	0.07	1.6	341.10	26FE	1.7	-3.4	-2.0	0.8
26.10	0.19	1.7	350.99	8MR	3.2	-3.4	-2.0	0.8
33.06	0.30	1.7	0.82	18MR	1.7	-3.4	-2.0	0.8
39.96	0.40	1.8	10.60	28MR	1.0	-3.3	-1.9	0.7
46.79	0.49	1.8	20.32	7AP	0.5	-3.3	-1.8	0.7
53.55	0.57	1.9	29.99	17AP	0.1	-3.3	-1.8	0.7
60.26	0.65	1.9	39.62	27AP	-0.7	-3.3	-1.7	0.6
66.92	0.73	1.9	49.21	7MY	-1.6	-3.3	-1.7	0.5
73.52	0.79	1.9	58.77	17MY	-1.4	-3.4	-1.6	0.5
80.08	0.86	1.9	68.31	27MY	-0.5	-3.4	-1.5	0.5
86.60	0.91	1.9	77.84	6JN	0.3	-3.5	-1.5	0.3
93.08	0.96	1.9	87.36	16JN	0.8	-3.5	-1.4	0.3
99.53	1.01	1.9	96.89	26JN	1.4	-3.5	-1.4	0.3
105.95	1.05	1.9	106.44	6JL	2.4	-3.6	-1.3	0.4
112.35	1.09	2.0	116.01	16JL	3.0	-3.6	-1.3	0.4
118.74	1.12	2.0	125.61	26JL	1.1	-3.7	-1.3	0.5
125.11	1.15	2.0	135.25	5AU	-0.5	-3.8	-1.2	0.5
131.47	1.17	2.0	144.94	15AU	-1.2	-3.9	-1.2	0.6
137.82	1.19	2.0	154.68	25AU	-1.3	-4.0	-1.2	0.6
144.17	1.20	2.0	164.47	4SE	-0.7	-4.1	-1.2	0.7
150.51	1.22	2.0	174.33	14SE	-0.3	-4.2	-1.2	0.7
156.86	1.22	2.0	184.23	24SE	-0.1	-4.3	-1.2	0.8
163.20	1.22	2.0	194.20	4OC	-0.0	-4.3	-1.2	0.8
169.55	1.22	2.0	204.22	14OC	0.1	-4.3	-1.3	0.8
175.89	1.21	1.9	214.29	24OC	1.0	-3.9	-1.3	0.8
182.24	1.20	1.9	224.41	3NO	2.7	-3.2	-1.3	0.8
188.58	1.18	1.8	234.56	13NO	0.2	-3.5	-1.4	0.8
194.91	1.15	1.8	244.74	23NO	-0.3	-4.1	-1.4	0.8
201.24	1.12	1.7	254.94	3DE	-0.3	-4.4	-1.5	0.8
207.55	1.07	1.6	265.14	13DE	-0.4	-4.4	-1.6	0.8
213.84	1.02	1.5	275.34	23DE	-0.6	-4.3	-1.6	0.7

♂ LONG	LAT	MAG	☉ LONG	16.00UT 724	☿	♀	♃	♄
220.11	0.95	1.4	285.52	2JA	-0.9	-4.2	-1.7	0.8
226.34	0.87	1.3	295.68	12JA	-1.1	-4.1	-1.8	0.8
232.53	0.77	1.2	305.80	22JA	-1.0	-4.0	-1.8	0.8
238.67	0.66	1.0	315.89	1FE	-0.1	-3.9	-1.9	0.9
244.73	0.52	0.9	325.92	11FE	2.0	-3.8	-1.9	0.9
250.70	0.36	0.7	335.89	21FE	2.5	-3.7	-2.0	0.9
256.56	0.17	0.6	345.82	2MR	1.2	-3.6	-2.0	0.9
262.28	-0.06	0.4	355.68	12MR	0.7	-3.6	-2.0	0.9
267.82	-0.32	0.2	5.48	22MR	0.4	-3.5	-2.0	0.9
273.13	-0.64	-0.0	15.24	1AP	-0.0	-3.4	-2.0	0.8
278.13	-1.00	-0.3	24.93	11AP	-0.7	-3.4	-2.0	0.8
282.76	-1.43	-0.5	34.58	21AP	-1.6	-3.4	-1.9	0.7
286.89	-1.93	-0.8	44.19	1MY	-1.4	-3.3	-1.9	0.7
290.38	-2.51	-1.1	53.77	11MY	-0.4	-3.3	-1.8	0.6
293.07	-3.17	-1.4	63.31	21MY	0.4	-3.3	-1.7	0.6
294.76	-3.90	-1.7	72.85	31MY	1.0	-3.3	-1.7	0.5
295.27	-4.68	-2.0	82.37	10JN	1.8	-3.3	-1.6	0.4
294.53	-5.43	-2.2	91.90	20JN	3.0	-3.3	-1.6	0.4
292.65	-6.07	-2.5	101.44	30JN	2.5	-3.3	-1.5	0.3
290.10	-6.48	-2.6	110.99	10JL	0.9	-3.4	-1.4	0.4
287.56	-6.58	-2.5	120.57	20JL	-0.5	-3.4	-1.4	0.4
285.72	-6.39	-2.3	130.20	30JL	-1.3	-3.4	-1.4	0.5
285.08	-5.98	-2.1	139.86	9AU	-1.3	-3.5	-1.3	0.5
285.76	-5.44	-1.8	149.56	19AU	-0.7	-3.5	-1.3	0.6
287.70	-4.86	-1.5	159.33	29AU	-0.2	-3.5	-1.3	0.6
290.74	-4.27	-1.2	169.15	8SE	-0.0	-3.4	-1.2	0.7
294.65	-3.71	-1.0	179.03	18SE	0.1	-3.4	-1.2	0.7
299.25	-3.19	-0.7	188.97	28SE	0.3	-3.4	-1.2	0.8
304.41	-2.71	-0.5	198.96	8OC	1.3	-3.4	-1.2	0.8
309.97	-2.27	-0.3	209.00	18OC	2.4	-3.3	-1.2	0.8
315.85	-1.87	-0.1	219.10	28OC	0.0	-3.3	-1.2	0.8
321.97	-1.51	0.1	229.23	7NO	-0.5	-3.3	-1.3	0.8
328.26	-1.19	0.3	239.40	17NO	-0.5	-3.3	-1.3	0.9
334.69	-0.90	0.5	249.59	27NO	-0.5	-3.3	-1.3	0.8
341.22	-0.64	0.6	259.79	7DE	-0.7	-3.4	-1.4	0.8
347.80	-0.41	0.8	269.99	17DE	-0.9	-3.4	-1.4	0.8
354.43	-0.20	0.9	280.19	27DE	-0.9	-3.4	-1.5	0.8

725

♂ LONG	LAT	MAG	☉ LONG	16.00UT 725	☿	♀	♃	♄
1.07	-0.01	1.0	290.36	6JA	-0.8	-3.4	-1.6	0.8
7.72	0.15	1.2	300.50	16JA	-0.0	-3.5	-1.6	0.8
14.37	0.30	1.3	310.61	26JA	2.3	-3.5	-1.7	0.9
21.00	0.43	1.4	320.66	5FE	2.0	-3.6	-1.8	0.9
27.62	0.54	1.5	330.67	15FE	0.9	-3.6	-1.8	0.9
34.20	0.64	1.6	340.62	25FE	0.5	-3.7	-1.9	1.0
40.76	0.73	1.6	350.52	7MR	0.2	-3.8	-1.9	1.0
47.29	0.81	1.7	0.35	17MR	-0.1	-3.8	-2.0	1.0
53.79	0.88	1.8	10.13	27MR	-0.7	-3.9	-2.0	0.9
60.26	0.94	1.8	19.85	6AP	-1.6	-4.0	-2.0	0.9
66.71	0.99	1.9	29.53	16AP	-1.4	-4.1	-2.1	0.9
73.13	1.03	1.9	39.16	26AP	-0.4	-4.2	-2.0	0.8
79.52	1.07	2.0	48.75	6MY	0.5	-4.2	-2.0	0.8
85.89	1.10	2.0	58.31	16MY	1.4	-4.1	-2.0	0.7
92.25	1.12	2.0	67.85	26MY	2.4	-3.9	-2.0	0.7
98.60	1.14	2.0	77.38	5JN	3.4	-3.2	-1.9	0.6
104.94	1.15	2.0	86.90	15JN	2.0	-3.0	-1.8	0.5
111.28	1.16	2.0	96.44	25JN	0.7	-3.8	-1.8	0.5
117.62	1.16	2.0	105.98	5JL	-0.5	-4.1	-1.7	0.4
123.96	1.16	2.0	115.55	15JL	-1.4	-4.2	-1.7	0.4
130.31	1.16	2.0	125.15	25JL	-1.4	-4.2	-1.6	0.4
136.68	1.15	2.0	134.78	4AU	-0.6	-4.1	-1.5	0.5
143.07	1.13	2.0	144.47	14AU	-0.2	-4.0	-1.5	0.5
149.47	1.11	2.0	154.20	24AU	0.1	-4.0	-1.4	0.6
155.91	1.09	2.0	163.99	3SE	0.3	-3.9	-1.4	0.6
162.37	1.06	2.0	173.84	13SE	0.6	-3.8	-1.4	0.7
168.86	1.02	2.0	183.75	23SE	1.7	-3.7	-1.3	0.7
175.39	0.98	2.0	193.71	3OC	2.2	-3.7	-1.3	0.8
181.95	0.94	2.0	203.73	13OC	-0.1	-3.6	-1.3	0.8
188.55	0.89	1.9	213.80	23OC	-0.6	-3.6	-1.3	0.9
195.19	0.83	1.9	223.92	2NO	-0.6	-3.5	-1.3	0.9
201.87	0.76	1.9	234.07	12NO	-0.6	-3.5	-1.3	0.9
208.58	0.69	1.8	244.25	22NO	-0.8	-3.4	-1.3	0.9
215.34	0.61	1.8	254.44	2DE	-0.7	-3.4	-1.3	0.9
222.14	0.52	1.7	264.65	12DE	-0.8	-3.4	-1.3	0.9
228.97	0.42	1.7	274.85	22DE	-0.7	-3.4	-1.4	0.9

♂ LONG	LAT	MAG	☉ LONG	16.00UT 726	☿	♀	♃	♄
235.85	0.31	1.6	285.03	1JA	0.1	-3.3	-1.4	0.9
242.76	0.19	1.5	295.19	11JA	2.6	-3.3	-1.5	0.9
249.70	0.06	1.4	305.32	21JA	1.5	-3.3	-1.5	0.9
256.68	-0.08	1.3	315.40	31JA	0.6	-3.3	-1.6	1.0
263.68	-0.24	1.2	325.44	10FE	0.3	-3.3	-1.7	1.0
270.71	-0.41	1.1	335.42	20FE	0.1	-3.4	-1.7	1.0
277.76	-0.59	1.0	345.34	2MR	-0.2	-3.4	-1.8	1.0
284.81	-0.79	0.9	355.21	12MR	-0.8	-3.4	-1.9	1.1
291.87	-1.00	0.8	5.01	22MR	-1.6	-3.4	-1.9	1.1
298.91	-1.22	0.7	14.76	1AP	-1.4	-3.5	-2.0	1.1
305.93	-1.45	0.6	24.46	11AP	-0.4	-3.5	-2.0	1.0
312.89	-1.70	0.4	34.11	21AP	0.7	-3.4	-2.1	1.0
319.78	-1.95	0.3	43.72	1MY	1.8	-3.4	-2.1	1.0
326.57	-2.21	0.2	53.30	11MY	3.2	-3.4	-2.1	0.9
333.20	-2.48	0.0	62.85	21MY	2.8	-3.4	-2.2	0.9
339.65	-2.74	-0.1	72.38	31MY	1.5	-3.3	-2.1	0.8
345.94	-3.01	-0.3	81.91	10JN	0.5	-3.3	-2.1	0.8
351.71	-3.28	-0.4	91.44	20JN	-0.6	-3.3	-2.1	0.7
357.15	-3.54	-0.6	100.97	30JN	-1.5	-3.3	-2.0	0.6
2.06	-3.79	-0.8	110.53	10JL	-1.4	-3.3	-2.0	0.6
6.28	-4.02	-1.0	120.11	20JL	-0.6	-3.3	-1.9	0.5
9.64	-4.22	-1.2	129.72	30JL	-0.1	-3.4	-1.9	0.5
11.91	-4.38	-1.4	139.39	9AU	0.2	-3.4	-1.8	0.5
12.91	-4.45	-1.7	149.06	19AU	0.4	-3.4	-1.7	0.6
12.47	-4.42	-1.9	158.85	29AU	0.8	-3.5	-1.7	0.6
10.61	-4.22	-2.1	168.67	8SE	2.1	-3.5	-1.6	0.7
7.71	-3.82	-2.3	178.55	18SE	1.9	-3.6	-1.6	0.7
4.42	-3.25	-2.2	188.48	28SE	-0.2	-3.7	-1.5	0.8
1.56	-2.58	-2.0	198.47	8OC	-0.8	-3.8	-1.5	0.8
359.76	-1.90	-1.7	208.51	18OC	-0.8	-3.9	-1.4	0.9
359.28	-1.27	-1.3	218.61	28OC	-0.8	-4.0	-1.4	0.9
0.11	-0.73	-1.0	228.74	7NO	-0.7	-4.1	-1.4	0.9
2.08	-0.29	-0.7	238.90	17NO	-0.6	-4.2	-1.4	1.0
4.96	0.07	-0.4	249.09	27NO	-0.6	-4.3	-1.4	1.0
8.58	0.36	-0.1	259.30	7DE	-0.5	-4.4	-1.4	1.0
12.78	0.59	0.1	269.49	17DE	0.2	-4.4	-1.4	1.0
17.40	0.78	0.4	279.69	27DE	2.9	-4.3	-1.4	1.0

727

♂ LONG	LAT	MAG	☉ LONG	16.00UT 727	☿	♀	♃	♄
22.37	0.93	0.6	289.87	6JA	1.1	-3.9	-1.4	1.0
27.60	1.04	0.8	300.01	16JA	0.4	-3.3	-1.5	1.0
33.03	1.13	0.9	310.12	26JA	0.2	-3.6	-1.5	1.0
38.62	1.21	1.1	320.18	5FE	0.0	-4.1	-1.5	1.0
44.33	1.26	1.2	330.19	15FE	-0.3	-4.3	-1.6	1.1
50.13	1.30	1.4	340.14	25FE	-0.8	-4.3	-1.6	1.1
56.01	1.33	1.5	350.04	7MR	-1.5	-4.2	-1.7	1.2
61.94	1.35	1.6	359.87	17MR	-1.3	-4.1	-1.8	1.2
67.92	1.37	1.7	9.66	27MR	-0.3	-4.0	-1.8	1.2
73.94	1.37	1.7	19.39	6AP	0.9	-3.9	-1.9	1.2
79.98	1.37	1.8	29.06	16AP	2.3	-3.8	-2.0	1.2
86.06	1.36	1.9	38.69	26AP	3.7	-3.7	-2.1	1.2
92.16	1.35	1.9	48.29	6MY	2.1	-3.6	-2.1	1.1
98.28	1.33	2.0	57.85	16MY	1.2	-3.6	-2.2	1.1
104.43	1.31	2.0	67.39	26MY	0.4	-3.5	-2.2	1.1
110.60	1.28	2.0	76.92	5JN	-0.6	-3.5	-2.3	1.0
116.80	1.25	2.0	86.45	15JN	-1.5	-3.4	-2.3	0.9
123.03	1.22	2.0	95.98	25JN	-1.4	-3.4	-2.3	0.9
129.29	1.18	2.0	105.52	5JL	-0.6	-3.4	-2.3	0.8
135.59	1.14	2.0	115.09	15JL	0.0	-3.3	-2.3	0.7
141.93	1.09	2.0	124.69	25JL	0.3	-3.3	-2.2	0.7
148.31	1.05	2.0	134.32	4AU	0.6	-3.3	-2.2	0.6
154.73	0.99	2.0	144.00	14AU	1.1	-3.3	-2.1	0.6
161.21	0.94	1.9	153.74	24AU	2.5	-3.3	-2.0	0.7
167.74	0.88	1.9	163.52	3SE	1.7	-3.4	-2.0	0.7
174.33	0.81	1.9	173.37	13SE	-0.3	-3.4	-1.9	0.8
180.97	0.74	1.8	183.27	23SE	-0.9	-3.4	-1.8	0.8
187.68	0.67	1.8	193.23	3OC	-0.9	-3.4	-1.8	0.9
194.45	0.59	1.8	203.24	13OC	-0.9	-3.5	-1.7	0.9
201.28	0.51	1.8	213.31	23OC	-0.6	-3.5	-1.7	1.0
208.18	0.42	1.8	223.42	2NO	-0.5	-3.5	-1.6	1.0
215.15	0.33	1.8	233.57	12NO	-0.4	-3.5	-1.6	1.0
222.19	0.23	1.8	243.75	22NO	-0.4	-3.4	-1.5	1.1
229.29	0.13	1.7	253.94	2DE	0.4	-3.4	-1.5	1.1
236.45	0.02	1.7	264.14	12DE	3.1	-3.4	-1.5	1.1
243.69	-0.10	1.7	274.35	22DE	0.8	-3.4	-1.5	1.1

♂ LONG	♂ LAT	♂ MAG	☉ LONG	16.00UT	☿	♀	♃	♄
				728	MAGNITUDES			
250.98	-0.22	1.6	284.53	1JA	0.2	-3.4	-1.5	1.2
258.33	-0.34	1.6	294.69	11JA	0.0	-3.3	-1.5	1.2
265.74	-0.47	1.5	304.82	21JA	-0.1	-3.3	-1.5	1.2
273.20	-0.59	1.5	314.91	31JA	-0.4	-3.3	-1.5	1.1
280.70	-0.73	1.4	324.94	10FE	-0.8	-3.3	-1.5	1.2
288.25	-0.86	1.4	334.93	20FE	-1.5	-3.4	-1.6	1.2
295.81	-0.99	1.3	344.86	1MR	-1.3	-3.4	-1.6	1.3
303.40	-1.12	1.2	354.72	11MR	-0.3	-3.4	-1.6	1.3
311.00	-1.24	1.2	4.54	21MR	1.2	-3.4	-1.7	1.3
318.59	-1.36	1.1	14.29	31MR	3.0	-3.5	-1.8	1.3
326.16	-1.47	1.1	23.99	10AP	2.8	-3.5	-1.8	1.4
333.70	-1.57	1.0	33.65	20AP	1.6	-3.5	-1.9	1.4
341.18	-1.66	0.9	43.26	30AP	0.9	-3.6	-2.0	1.4
348.60	-1.74	0.9	52.84	10MY	0.2	-3.7	-2.0	1.3
355.93	-1.80	0.8	62.40	20MY	-0.6	-3.7	-2.1	1.3
3.15	-1.84	0.7	71.93	30MY	-1.6	-3.8	-2.2	1.3
10.23	-1.87	0.7	81.45	9JN	-1.4	-3.9	-2.2	1.2
17.16	-1.87	0.6	90.98	19JN	-0.5	-4.0	-2.3	1.2
23.89	-1.86	0.5	100.51	29JN	0.1	-4.1	-2.3	1.1
30.41	-1.83	0.4	110.07	9JL	0.5	-4.2	-2.4	1.0
36.65	-1.77	0.3	119.65	19JL	0.9	-4.2	-2.4	1.0
42.59	-1.69	0.2	129.26	29JL	1.5	-4.1	-2.4	0.9
48.15	-1.59	0.1	138.92	8AU	3.0	-3.8	-2.4	0.8
53.25	-1.46	0.0	148.63	18AU	1.5	-3.3	-2.4	0.8
57.80	-1.29	-0.1	158.38	28AU	-0.4	-3.5	-2.3	0.8
61.66	-1.09	-0.3	168.20	7SE	-1.0	-4.0	-2.3	0.8
64.68	-0.85	-0.5	178.07	17SE	-1.1	-4.3	-2.2	0.9
66.67	-0.55	-0.7	188.00	27SE	-0.8	-4.3	-2.2	0.9
67.42	-0.19	-0.9	197.99	7OC	-0.5	-4.3	-2.1	1.0
66.75	0.23	-1.1	208.03	17OC	-0.3	-4.2	-2.0	1.0
64.66	0.69	-1.3	218.11	27OC	-0.3	-4.1	-1.9	1.1
61.38	1.17	-1.5	228.25	6NO	-0.2	-4.0	-1.9	1.1
57.52	1.61	-1.6	238.41	16NO	0.6	-3.9	-1.8	1.2
53.91	1.95	-1.3	248.59	26NO	3.1	-3.8	-1.8	1.2
51.20	2.18	-1.1	258.80	6DE	0.6	-3.8	-1.7	1.2
49.80	2.32	-0.7	269.00	16DE	-0.0	-3.7	-1.7	1.3
49.74	2.37	-0.4	279.19	26DE	-0.1	-3.6	-1.6	1.3
				729				
50.87	2.37	-0.1	289.37	5JA	-0.2	-3.5	-1.6	1.3
53.01	2.34	0.1	299.52	15JA	-0.4	-3.5	-1.6	1.3
55.95	2.29	0.4	309.62	25JA	-0.9	-3.4	-1.6	1.3
59.52	2.22	0.6	319.69	4FE	-1.3	-3.4	-1.6	1.3
63.58	2.15	0.8	329.70	14FE	-1.2	-3.4	-1.6	1.3
68.04	2.08	1.0	339.66	24FE	-0.3	-3.3	-1.6	1.4
72.80	2.01	1.1	349.56	6MR	1.4	-3.3	-1.6	1.4
77.81	1.93	1.3	359.40	16MR	3.5	-3.3	-1.6	1.4
83.01	1.86	1.4	9.19	26MR	2.1	-3.3	-1.6	1.4
88.38	1.78	1.5	18.92	5AP	1.2	-3.3	-1.7	1.4
93.89	1.71	1.6	28.60	15AP	0.7	-3.3	-1.7	1.3
99.50	1.63	1.7	38.23	25AP	0.1	-3.3	-1.8	1.3
105.23	1.56	1.7	47.83	5MY	-0.6	-3.4	-1.8	1.3
111.04	1.48	1.8	57.39	15MY	-1.6	-3.4	-1.9	1.3
116.93	1.41	1.8	66.93	25MY	-1.4	-3.4	-1.9	1.2
122.90	1.33	1.9	76.46	4JN	-0.5	-3.5	-2.0	1.2
128.94	1.25	1.9	85.99	14JN	0.2	-3.5	-2.1	1.1
135.06	1.17	1.9	95.51	24JN	0.7	-3.5	-2.1	1.1
141.24	1.09	1.9	105.06	4JL	1.1	-3.4	-2.2	1.0
147.50	1.01	2.0	114.62	14JL	2.0	-3.4	-2.3	1.0
153.83	0.93	2.0	124.22	24JL	3.2	-3.4	-2.3	0.9
160.24	0.84	1.9	133.85	3AU	1.2	-3.3	-2.4	0.9
166.72	0.75	1.9	143.53	13AU	-0.4	-3.3	-2.4	0.8
173.28	0.66	1.9	153.26	23AU	-1.2	-3.3	-2.5	0.8
179.92	0.57	1.9	163.05	2SE	-1.2	-3.3	-2.5	0.8
186.64	0.48	1.8	172.89	12SE	-0.8	-3.3	-2.5	0.9
193.44	0.38	1.8	182.78	22SE	-0.4	-3.3	-2.4	0.9
200.33	0.29	1.7	192.75	2OC	-0.2	-3.4	-2.4	1.0
207.31	0.19	1.7	202.76	12OC	-0.1	-3.4	-2.3	1.1
214.37	0.09	1.6	212.82	22OC	-0.0	-3.4	-2.3	1.1
221.51	-0.02	1.6	222.93	1NO	0.8	-3.5	-2.2	1.2
228.74	-0.12	1.6	233.08	11NO	2.9	-3.5	-2.1	1.2
236.05	-0.23	1.6	243.25	21NO	0.3	-3.6	-2.1	1.3
243.45	-0.33	1.6	253.45	1DE	-0.2	-3.6	-2.0	1.3
250.91	-0.44	1.6	263.65	11DE	-0.3	-3.7	-1.9	1.3
258.44	-0.54	1.5	273.85	21DE	-0.3	-3.8	-1.9	1.3
266.04	-0.64	1.5	284.04	31DE	-0.5	-3.9	-1.8	1.3

♂ LONG	♂ LAT	♂ MAG	☉ LONG	16.00UT	☿	♀	♃	♄
				730	MAGNITUDES			
273.70	-0.74	1.5	294.20	10JA	-0.9	-4.0	-1.8	1.3
281.40	-0.83	1.5	304.33	20JA	-1.2	-4.0	-1.7	1.3
289.14	-0.92	1.5	314.42	30JA	-1.1	-4.1	-1.7	1.2
296.91	-1.00	1.4	324.46	9FE	-0.2	-4.2	-1.6	1.2
304.70	-1.08	1.4	334.44	19FE	1.7	-4.3	-1.6	1.1
312.49	-1.14	1.4	344.38	1MR	3.0	-4.3	-1.6	1.1
320.28	-1.20	1.4	354.24	11MR	1.5	-4.1	-1.6	1.1
328.05	-1.24	1.4	4.06	21MR	0.9	-3.7	-1.6	1.1
335.79	-1.27	1.3	13.82	31MR	0.5	-3.1	-1.6	1.1
343.48	-1.28	1.3	23.52	10AP	0.0	-3.6	-1.6	1.1
351.12	-1.29	1.3	33.18	20AP	-0.7	-4.0	-1.6	1.1
358.69	-1.27	1.3	42.80	30AP	-1.6	-4.2	-1.6	1.1
6.18	-1.25	1.3	52.38	10MY	-1.4	-4.2	-1.7	1.0
13.58	-1.21	1.2	61.93	20MY	-0.5	-4.1	-1.7	1.0
20.88	-1.15	1.2	71.47	30MY	0.3	-4.0	-1.7	1.0
28.06	-1.08	1.2	80.99	9JN	0.9	-3.9	-1.8	0.9
35.11	-1.00	1.2	90.52	19JN	1.5	-3.8	-1.9	0.9
42.03	-0.90	1.2	100.05	29JN	2.6	-3.8	-1.9	0.8
48.80	-0.79	1.1	109.60	9JL	2.8	-3.7	-2.0	0.8
55.40	-0.67	1.1	119.18	19JL	1.0	-3.6	-2.0	0.7
61.82	-0.54	1.0	128.80	29JL	-0.5	-3.6	-2.1	0.7
68.03	-0.38	1.0	138.45	8AU	-1.3	-3.5	-2.2	0.6
74.01	-0.22	0.9	148.16	18AU	-1.3	-3.5	-2.3	0.6
79.73	-0.03	0.8	157.91	28AU	-0.7	-3.4	-2.3	0.5
85.13	0.17	0.7	167.72	7SE	-0.3	-3.4	-2.4	0.5
90.17	0.40	0.6	177.59	17SE	-0.1	-3.4	-2.4	0.5
94.76	0.65	0.5	187.52	27SE	0.0	-3.4	-2.4	0.6
98.82	0.93	0.4	197.50	7OC	0.2	-3.4	-2.4	0.6
102.22	1.25	0.2	207.54	17OC	1.1	-3.4	-2.4	0.7
104.80	1.61	-0.0	217.63	27OC	2.7	-3.4	-2.4	0.8
106.38	2.01	-0.2	227.76	6NO	0.2	-3.4	-2.4	0.8
106.78	2.45	-0.4	237.92	16NO	-0.4	-3.4	-2.3	0.9
105.83	2.91	-0.7	248.10	26NO	-0.4	-3.4	-2.3	0.9
103.52	3.36	-0.9	258.30	6DE	-0.4	-3.4	-2.2	1.0
100.09	3.72	-1.1	268.51	16DE	-0.6	-3.4	-2.1	1.0
96.13	3.95	-1.1	278.70	26DE	-0.9	-3.4	-2.0	1.0
				731				
92.39	4.02	-0.9	288.87	5JA	-1.0	-3.5	-2.0	1.0
89.52	3.94	-0.7	299.02	15JA	-0.9	-3.5	-1.9	1.0
87.90	3.75	-0.4	309.13	25JA	-0.1	-3.5	-1.8	1.0
87.59	3.52	-0.1	319.20	4FE	2.0	-3.4	-1.8	1.0
88.48	3.27	0.1	329.21	14FE	2.3	-3.4	-1.7	1.0
90.37	3.02	0.3	339.17	24FE	1.1	-3.4	-1.7	0.9
93.08	2.79	0.6	349.07	6MR	0.6	-3.4	-1.6	0.9
96.44	2.57	0.8	358.92	16MR	0.3	-3.4	-1.6	0.8
100.33	2.36	0.9	8.71	26MR	-0.1	-3.3	-1.6	0.8
104.63	2.17	1.1	18.44	5AP	-0.7	-3.3	-1.5	0.8
109.27	2.00	1.2	28.12	15AP	-1.6	-3.3	-1.5	0.8
114.20	1.83	1.3	37.76	25AP	-1.4	-3.3	-1.5	0.8
119.35	1.68	1.4	47.35	5MY	-0.4	-3.3	-1.5	0.8
124.72	1.53	1.5	56.92	15MY	0.4	-3.4	-1.5	0.8
130.25	1.39	1.6	66.47	25MY	1.1	-3.4	-1.5	0.8
135.94	1.26	1.6	76.00	4JN	2.0	-3.4	-1.6	0.8
141.77	1.12	1.7	85.52	14JN	3.3	-3.5	-1.6	0.7
147.73	1.00	1.7	95.05	24JN	2.3	-3.5	-1.6	0.7
153.81	0.87	1.7	104.59	4JL	0.8	-3.6	-1.7	0.7
160.02	0.75	1.8	114.16	14JL	-0.5	-3.6	-1.7	0.6
166.33	0.63	1.8	123.75	24JL	-1.4	-3.7	-1.8	0.6
172.76	0.52	1.8	133.38	3AU	-1.4	-3.8	-1.8	0.5
179.30	0.40	1.8	143.06	13AU	-0.7	-3.9	-1.9	0.4
185.94	0.28	1.8	152.79	23AU	-0.2	-4.0	-2.0	0.4
192.70	0.17	1.7	162.57	2SE	0.0	-4.1	-2.0	0.3
199.56	0.06	1.7	172.41	12SE	0.2	-4.2	-2.1	0.3
206.53	-0.05	1.7	182.31	22SE	0.4	-4.3	-2.2	0.2
213.60	-0.16	1.7	192.26	2OC	1.4	-4.3	-2.2	0.3
220.78	-0.26	1.6	202.27	12OC	2.4	-4.2	-2.3	0.3
228.05	-0.37	1.6	212.33	22OC	0.0	-3.9	-2.3	0.4
235.42	-0.47	1.6	222.44	1NO	-0.5	-3.1	-2.3	0.5
242.89	-0.56	1.5	232.59	11NO	-0.6	-3.6	-2.3	0.5
250.43	-0.65	1.5	242.76	21NO	-0.6	-4.1	-2.3	0.6
258.06	-0.74	1.4	252.95	1DE	-0.7	-4.4	-2.3	0.6
265.75	-0.81	1.4	263.16	11DE	-0.8	-4.4	-2.3	0.7
273.51	-0.88	1.3	273.35	21DE	-0.8	-4.3	-2.2	0.7
281.31	-0.95	1.3	283.54	31DE	-0.8	-4.2	-2.2	0.8

♂ LONG	LAT	MAG	☉ LONG	16.00UT 732	☿	♀	♃	♄
289.15	-1.00	1.4	293.71	10JA	-0.0	-4.1	-2.1	0.8
297.02	-1.04	1.4	303.84	20JA	2.3	-4.0	-2.0	0.8
304.90	-1.07	1.4	313.93	30JA	1.8	-3.9	-1.9	0.8
312.78	-1.09	1.4	323.97	9FE	0.8	-3.8	-1.9	0.8
320.65	-1.10	1.4	333.96	19FE	0.4	-3.7	-1.8	0.7
328.49	-1.10	1.4	343.89	29FE	0.2	-3.6	-1.7	0.7
336.30	-1.08	1.4	353.77	10MR	-0.1	-3.6	-1.7	0.7
344.06	-1.06	1.4	3.58	20MR	-0.7	-3.5	-1.6	0.6
351.76	-1.02	1.4	13.35	30MR	-1.6	-3.4	-1.6	0.6
359.39	-0.97	1.5	23.06	9AP	-1.4	-3.4	-1.5	0.6
6.95	-0.91	1.5	32.71	19AP	-0.4	-3.4	-1.5	0.6
14.42	-0.83	1.5	42.33	29AP	0.6	-3.3	-1.5	0.6
21.80	-0.75	1.5	51.92	9MY	1.5	-3.3	-1.4	0.6
29.09	-0.66	1.5	61.47	19MY	2.7	-3.3	-1.4	0.6
36.28	-0.57	1.5	71.01	29MY	3.3	-3.3	-1.4	0.6
43.37	-0.46	1.5	80.53	8JN	1.8	-3.3	-1.4	0.6
50.34	-0.35	1.5	90.06	18JN	0.6	-3.3	-1.4	0.6
57.21	-0.24	1.5	99.59	28JN	-0.5	-3.3	-1.4	0.5
63.96	-0.11	1.5	109.14	8JL	-1.4	-3.4	-1.5	0.5
70.59	0.02	1.5	118.72	18JL	-1.4	-3.4	-1.5	0.5
77.10	0.15	1.5	128.33	28JL	-0.6	-3.4	-1.5	0.4
83.47	0.30	1.5	137.99	7AU	-0.1	-3.5	-1.6	0.4
89.70	0.44	1.4	147.68	17AU	0.1	-3.5	-1.6	0.3
95.77	0.60	1.4	157.44	27AU	0.3	-3.5	-1.7	0.2
101.67	0.77	1.3	167.25	6SE	0.6	-3.4	-1.7	0.2
107.37	0.94	1.3	177.11	16SE	1.8	-3.4	-1.8	0.1
112.85	1.13	1.2	187.04	26SE	2.2	-3.4	-1.8	0.1
118.06	1.33	1.1	197.02	6OC	-0.1	-3.4	-1.9	0.0
122.97	1.55	1.0	207.05	16OC	-0.7	-3.3	-2.0	0.0
127.50	1.79	0.9	217.14	26OC	-0.7	-3.3	-2.0	0.1
131.56	2.06	0.7	227.26	5NO	-0.7	-3.3	-2.1	0.2
135.07	2.35	0.5	237.42	15NO	-0.8	-3.3	-2.1	0.3
137.88	2.67	0.3	247.61	25NO	-0.7	-3.3	-2.2	0.3
139.83	3.03	0.1	257.80	5DE	-0.7	-3.4	-2.2	0.4
140.74	3.41	-0.1	268.00	15DE	-0.6	-3.4	-2.2	0.4
140.42	3.79	-0.4	278.20	25DE	0.1	-3.4	-2.2	0.5

733

♂ LONG	LAT	MAG	☉ LONG	16.00UT	☿	♀	♃	♄
138.79	4.15	-0.6	288.38	4JA	2.6	-3.4	-2.2	0.5
135.92	4.43	-0.8	298.53	14JA	1.4	-3.5	-2.1	0.6
132.19	4.57	-1.0	308.64	24JA	0.5	-3.6	-2.1	0.6
128.26	4.53	-0.9	318.71	3FE	0.3	-3.6	-2.0	0.6
124.84	4.32	-0.7	328.73	13FE	0.1	-3.6	-2.0	0.6
122.47	4.00	-0.5	338.70	23FE	-0.2	-3.7	-1.9	0.6
121.38	3.63	-0.3	348.60	5MR	-0.7	-3.8	-1.8	0.5
121.55	3.24	-0.0	358.44	15MR	-1.5	-3.8	-1.8	0.5
122.85	2.87	0.2	8.24	25MR	-1.4	-3.9	-1.7	0.5
125.09	2.52	0.4	17.97	4AP	-0.4	-4.0	-1.6	0.4
128.09	2.21	0.6	27.65	14AP	0.8	-4.1	-1.6	0.4
131.72	1.93	0.7	37.30	24AP	2.0	-4.2	-1.5	0.4
135.86	1.67	0.9	46.89	4MY	3.5	-4.2	-1.5	0.4
140.42	1.43	1.0	56.46	14MY	2.5	-4.1	-1.4	0.4
145.33	1.21	1.1	66.01	24MY	1.4	-3.8	-1.4	0.5
150.53	1.01	1.2	75.54	3JN	0.5	-3.2	-1.4	0.5
155.99	0.82	1.3	85.06	13JN	-0.5	-3.0	-1.3	0.5
161.68	0.65	1.3	94.59	23JN	-1.5	-3.8	-1.3	0.5
167.57	0.48	1.4	104.13	3JL	-1.4	-4.1	-1.3	0.4
173.65	0.32	1.4	113.69	13JL	-0.6	-4.2	-1.3	0.4
179.90	0.17	1.4	123.29	23JL	-0.1	-4.2	-1.3	0.4
186.31	0.03	1.5	132.92	2AU	0.3	-4.1	-1.3	0.3
192.88	-0.10	1.5	142.59	12AU	0.5	-4.0	-1.4	0.3
199.59	-0.23	1.5	152.32	22AU	0.9	-4.0	-1.4	0.2
206.45	-0.35	1.5	162.09	1SE	2.2	-3.9	-1.4	0.2
213.43	-0.46	1.5	171.93	11SE	1.9	-3.8	-1.5	0.1
220.55	-0.57	1.5	181.83	21SE	-0.2	-3.7	-1.5	0.0
227.78	-0.67	1.5	191.78	1OC	-0.8	-3.7	-1.6	-0.0
235.13	-0.76	1.5	201.79	11OC	-0.9	-3.6	-1.6	-0.1
242.59	-0.84	1.5	211.85	21OC	-0.8	-3.6	-1.7	-0.1
250.15	-0.91	1.5	221.95	31OC	-0.7	-3.5	-1.7	-0.1
257.79	-0.97	1.5	232.09	10NO	-0.5	-3.5	-1.8	-0.0
265.51	-1.02	1.4	242.27	20NO	-0.5	-3.4	-1.9	0.0
273.29	-1.06	1.4	252.46	30NO	-0.5	-3.4	-1.9	0.1
281.13	-1.09	1.4	262.66	10DE	0.2	-3.4	-2.0	0.2
289.00	-1.10	1.4	272.86	20DE	2.9	-3.4	-2.0	0.2
296.90	-1.11	1.4	283.05	30DE	1.0	-3.3	-2.1	0.3

♂ LONG	LAT	MAG	☉ LONG	16.00UT 734	☿	♀	♃	♄
304.81	-1.10	1.4	293.22	9JA	0.3	-3.3	-2.1	0.3
312.71	-1.08	1.4	303.35	19JA	0.1	-3.3	-2.1	0.4
320.59	-1.05	1.4	313.44	29JA	-0.0	-3.3	-2.1	0.4
328.45	-1.01	1.3	323.49	8FE	-0.3	-3.3	-2.1	0.4
336.26	-0.95	1.3	333.48	18FE	-0.8	-3.4	-2.0	0.4
344.02	-0.89	1.3	343.41	28FE	-1.5	-3.4	-2.0	0.4
351.72	-0.82	1.4	353.29	10MR	-1.3	-3.4	-1.9	0.4
359.34	-0.75	1.4	3.11	20MR	-0.4	-3.4	-1.8	0.4
6.89	-0.66	1.5	12.87	30MR	1.0	-3.5	-1.8	0.4
14.36	-0.57	1.5	22.58	9AP	2.5	-3.5	-1.7	0.3
21.74	-0.47	1.5	32.25	19AP	3.3	-3.4	-1.6	0.3
29.03	-0.38	1.6	41.86	29AP	1.9	-3.4	-1.6	0.3
36.23	-0.27	1.6	51.45	9MY	1.1	-3.4	-1.5	0.3
43.34	-0.17	1.7	61.00	19MY	0.3	-3.4	-1.5	0.3
50.35	-0.06	1.7	70.54	29MY	-0.6	-3.3	-1.4	0.3
57.28	0.05	1.7	80.06	8JN	-1.5	-3.3	-1.4	0.4
64.12	0.16	1.7	89.59	18JN	-1.4	-3.3	-1.3	0.4
70.87	0.27	1.8	99.12	28JN	-0.6	-3.3	-1.3	0.4
77.54	0.38	1.8	108.67	8JL	0.0	-3.3	-1.3	0.4
84.11	0.49	1.8	118.25	18JL	0.4	-3.3	-1.3	0.4
90.60	0.61	1.8	127.86	28JL	0.7	-3.4	-1.2	0.3
97.00	0.72	1.8	137.51	7AU	1.3	-3.4	-1.2	0.3
103.30	0.84	1.8	147.21	17AU	2.7	-3.4	-1.2	0.3
109.52	0.95	1.7	156.96	27AU	1.7	-3.5	-1.2	0.2
115.63	1.07	1.7	166.77	6SE	-0.3	-3.5	-1.3	0.1
121.62	1.20	1.7	176.63	16SE	-1.0	-3.6	-1.3	0.1
127.50	1.32	1.6	186.55	26SE	-1.0	-3.7	-1.3	0.0
133.24	1.46	1.6	196.53	6OC	-0.9	-3.8	-1.4	-0.1
138.83	1.59	1.5	206.56	16OC	-0.6	-3.9	-1.4	-0.1
144.23	1.74	1.4	216.64	26OC	-0.4	-4.0	-1.5	-0.2
149.42	1.89	1.3	226.77	5NO	-0.4	-4.1	-1.5	-0.3
154.34	2.06	1.2	236.93	15NO	-0.3	-4.2	-1.6	-0.2
158.96	2.24	1.0	247.11	25NO	0.4	-4.3	-1.6	-0.1
163.20	2.43	0.9	257.31	5DE	3.1	-4.4	-1.7	-0.1
166.96	2.63	0.7	267.51	15DE	0.7	-4.3	-1.8	0.0
170.13	2.85	0.5	277.71	25DE	0.1	-4.3	-1.8	0.1

735

♂ LONG	LAT	MAG	☉ LONG	16.00UT	☿	♀	♃	♄
172.58	3.09	0.3	287.89	4JA	-0.0	-3.9	-1.9	0.1
174.13	3.33	0.0	298.04	14JA	-0.1	-3.3	-1.9	0.2
174.60	3.57	-0.3	308.15	24JA	-0.4	-3.6	-2.0	0.2
173.83	3.78	-0.5	318.23	3FE	-0.8	-4.1	-2.0	0.3
171.79	3.92	-0.8	328.24	13FE	-1.4	-4.3	-2.0	0.3
168.64	3.94	-1.0	338.21	23FE	-1.2	-4.3	-2.0	0.4
164.84	3.80	-1.2	348.12	5MR	-0.3	-4.2	-2.0	0.4
161.11	3.52	-1.0	357.97	15MR	1.2	-4.1	-2.0	0.4
158.12	3.12	-0.9	7.76	25MR	3.2	-4.0	-1.9	0.3
156.29	2.66	-0.6	17.50	4AP	2.5	-3.9	-1.9	0.3
155.78	2.21	-0.4	27.19	14AP	1.4	-3.8	-1.8	0.3
156.51	1.78	-0.2	36.83	24AP	0.8	-3.7	-1.8	0.3
158.33	1.39	-0.0	46.43	4MY	0.2	-3.6	-1.7	0.2
161.07	1.04	0.2	56.00	14MY	-0.6	-3.6	-1.6	0.2
164.56	0.73	0.3	65.54	24MY	-1.5	-3.5	-1.6	0.2
168.67	0.45	0.5	75.08	3JN	-1.4	-3.5	-1.5	0.3
173.30	0.21	0.6	84.60	13JN	-0.5	-3.4	-1.4	0.3
178.35	-0.01	0.7	94.13	23JN	0.1	-3.4	-1.4	0.3
183.77	-0.21	0.8	103.67	3JL	0.5	-3.4	-1.4	0.3
189.51	-0.39	0.9	113.23	13JL	1.0	-3.3	-1.3	0.3
195.52	-0.55	0.9	122.82	23JL	1.7	-3.3	-1.3	0.3
201.79	-0.70	1.0	132.45	2AU	3.1	-3.3	-1.2	0.3
208.27	-0.82	1.0	142.12	12AU	1.4	-3.3	-1.2	0.3
214.96	-0.94	1.1	151.85	22AU	-0.3	-3.3	-1.2	0.3
221.83	-1.03	1.1	161.62	1SE	-1.1	-3.4	-1.2	0.2
228.87	-1.12	1.1	171.45	11SE	-1.2	-3.4	-1.2	0.2
236.06	-1.19	1.2	181.35	21SE	-0.8	-3.4	-1.2	0.1
243.39	-1.24	1.2	191.30	1OC	-0.5	-3.4	-1.2	0.0
250.84	-1.28	1.2	201.30	11OC	-0.3	-3.5	-1.2	-0.0
258.40	-1.30	1.2	211.36	21OC	-0.2	-3.5	-1.3	-0.1
266.05	-1.32	1.3	221.46	31OC	0.6	-3.5	-1.4	-0.2
273.77	-1.31	1.3	231.60	10NO	0.6	-3.5	-1.4	-0.2
281.55	-1.29	1.3	241.77	20NO	3.1	-3.4	-1.4	-0.3
289.37	-1.26	1.3	251.96	30NO	0.5	-3.4	-1.5	-0.3
297.22	-1.22	1.3	262.16	10DE	-0.1	-3.4	-1.5	-0.2
305.07	-1.16	1.4	272.37	20DE	-0.2	-3.4	-1.6	-0.1
312.92	-1.10	1.4	282.55	30DE	-0.3	-3.4	-1.6	-0.0

♂ LONG	♂ LAT	♂ MAG	☉ LONG	16.00UT	☿	♀	♃	♄
				736	MAGNITUDES			
320.75	-1.02	1.4	292.72	9JA	-0.5	-3.3	-1.7	0.0
328.55	-0.94	1.4	302.86	19JA	-0.9	-3.3	-1.8	0.1
336.30	-0.85	1.5	312.95	29JA	-1.2	-3.3	-1.8	0.2
343.99	-0.76	1.5	323.00	8FE	-1.1	-3.3	-1.9	0.2
351.62	-0.66	1.5	332.99	18FE	-0.3	-3.4	-1.9	0.2
359.18	-0.56	1.5	342.93	28FE	1.5	-3.4	-2.0	0.3
6.66	-0.45	1.6	352.81	9MR	3.4	-3.4	-2.0	0.3
14.06	-0.34	1.6	2.64	19MR	1.8	-3.4	-2.0	0.3
21.38	-0.24	1.6	12.40	29MR	1.1	-3.5	-2.0	0.3
28.61	-0.13	1.6	22.11	8AP	0.6	-3.5	-2.0	0.3
35.76	-0.03	1.6	31.78	18AP	0.1	-3.6	-2.0	0.3
42.82	0.08	1.6	41.40	28AP	-0.6	-3.6	-1.9	0.3
49.80	0.18	1.7	50.98	8MY	-1.5	-3.7	-1.9	0.2
56.71	0.28	1.7	60.55	18MY	-1.5	-3.7	-1.8	0.2
63.54	0.38	1.8	70.08	28MY	-0.5	-3.8	-1.7	0.2
70.29	0.47	1.8	79.61	7JN	0.2	-3.9	-1.7	0.2
76.98	0.56	1.8	89.13	17JN	0.7	-4.0	-1.6	0.3
83.60	0.65	1.9	98.67	27JN	1.3	-4.1	-1.6	0.3
90.16	0.74	1.9	108.21	7JL	2.2	-4.2	-1.5	0.3
96.66	0.82	1.9	117.79	17JL	3.1	-4.2	-1.4	0.3
103.10	0.90	1.9	127.40	27JL	1.2	-4.1	-1.4	0.3
109.49	0.98	2.0	137.04	6AU	-0.4	-3.8	-1.3	0.3
115.82	1.06	2.0	146.74	16AU	-1.2	-3.3	-1.3	0.3
122.10	1.14	1.9	156.49	26AU	-1.3	-3.5	-1.3	0.3
128.33	1.21	1.9	166.29	5SE	-0.8	-4.0	-1.3	0.3
134.50	1.29	1.9	176.15	15SE	-0.4	-4.3	-1.2	0.2
140.61	1.36	1.9	186.07	25SE	-0.2	-4.3	-1.2	0.2
146.66	1.43	1.8	196.04	5OC	-0.1	-4.3	-1.2	0.1
152.63	1.50	1.8	206.08	15OC	0.1	-4.2	-1.2	0.1
158.53	1.57	1.7	216.16	25OC	0.9	-4.1	-1.2	-0.0
164.33	1.64	1.6	226.28	4NO	3.0	-4.0	-1.3	-0.1
170.03	1.71	1.6	236.44	14NO	0.3	-3.9	-1.3	-0.2
175.59	1.78	1.5	246.62	24NO	-0.3	-3.8	-1.3	-0.2
181.00	1.85	1.3	256.82	4DE	-0.3	-3.7	-1.4	-0.3
186.23	1.92	1.2	267.02	14DE	-0.4	-3.7	-1.4	-0.2
191.23	1.99	1.1	277.22	24DE	-0.6	-3.6	-1.5	-0.2
				737				
195.96	2.06	0.9	287.40	3JA	-0.9	-3.5	-1.5	-0.1
200.34	2.12	0.7	297.55	13JA	-1.1	-3.5	-1.6	-0.0
204.30	2.18	0.5	307.67	23JA	-1.0	-3.4	-1.6	0.0
207.74	2.22	0.3	317.74	2FE	-0.2	-3.4	-1.7	0.1
210.51	2.25	0.0	327.77	12FE	1.8	-3.4	-1.8	0.2
212.46	2.25	-0.3	337.73	22FE	2.8	-3.3	-1.8	0.2
213.41	2.21	-0.6	347.64	4MR	1.3	-3.3	-1.9	0.2
213.17	2.11	-0.9	357.50	14MR	0.8	-3.3	-2.0	0.3
211.68	1.92	-1.2	7.29	24MR	0.4	-3.3	-2.0	0.3
209.02	1.63	-1.5	17.03	3AP	0.0	-3.3	-2.0	0.3
205.58	1.24	-1.7	26.72	13AP	-0.6	-3.3	-2.1	0.3
202.02	0.78	-1.6	36.36	23AP	-1.5	-3.3	-2.1	0.3
199.04	0.29	-1.4	45.97	3MY	-1.5	-3.4	-2.1	0.3
197.19	-0.17	-1.2	55.54	13MY	-0.5	-3.4	-2.0	0.3
196.69	-0.57	-1.0	65.08	23MY	0.3	-3.4	-2.0	0.3
197.52	-0.92	-0.8	74.61	2JN	1.0	-3.5	-2.0	0.2
199.54	-1.20	-0.6	84.14	12JN	1.7	-3.5	-1.9	0.2
202.56	-1.43	-0.4	93.66	22JN	2.9	-3.5	-1.8	0.3
206.41	-1.61	-0.3	103.20	2JL	2.7	-3.4	-1.8	0.3
210.95	-1.75	-0.1	112.77	12JL	1.0	-3.4	-1.7	0.3
216.06	-1.86	0.0	122.35	22JL	-0.4	-3.4	-1.7	0.4
221.64	-1.93	0.1	131.98	1AU	-1.3	-3.3	-1.6	0.4
227.61	-1.98	0.2	141.65	11AU	-1.4	-3.3	-1.5	0.4
233.92	-2.01	0.3	151.37	21AU	-0.7	-3.3	-1.5	0.4
240.50	-2.01	0.4	161.14	31AU	-0.3	-3.3	-1.4	0.4
247.33	-2.00	0.5	170.97	10SE	-0.0	-3.3	-1.4	0.4
254.36	-1.96	0.6	180.86	20SE	0.1	-3.3	-1.4	0.3
261.55	-1.90	0.7	190.81	30SE	0.3	-3.4	-1.3	0.3
268.89	-1.83	0.8	200.81	10OC	1.2	-3.4	-1.3	0.2
276.33	-1.75	0.8	210.86	20OC	2.7	-3.4	-1.3	0.2
283.86	-1.65	0.9	220.96	30OC	0.1	-3.5	-1.3	0.1
291.45	-1.54	1.0	231.11	9NO	-0.4	-3.5	-1.3	0.0
299.08	-1.42	1.0	241.27	19NO	-0.5	-3.6	-1.3	-0.0
306.73	-1.29	1.1	251.47	29NO	-0.5	-3.6	-1.3	-0.1
314.38	-1.16	1.2	261.67	9DE	-0.6	-3.7	-1.3	-0.2
322.02	-1.03	1.2	271.87	19DE	-0.9	-3.8	-1.4	-0.2
329.63	-0.89	1.3	282.06	29DE	-0.9	-3.9	-1.4	-0.2

♂ LONG	♂ LAT	♂ MAG	☉ LONG	16.00UT	☿	♀	♃	♄
				738	MAGNITUDES			
337.19	-0.75	1.4	292.23	8JA	-0.9	-4.0	-1.5	-0.1
344.71	-0.62	1.4	302.37	18JA	-0.1	-4.1	-1.5	-0.0
352.17	-0.49	1.5	312.47	28JA	2.1	-4.1	-1.6	0.0
359.56	-0.36	1.5	322.52	7FE	2.2	-4.2	-1.6	0.1
6.88	-0.23	1.6	332.51	17FE	1.0	-4.3	-1.7	0.1
14.13	-0.11	1.6	342.46	27FE	0.5	-4.3	-1.8	0.2
21.30	0.01	1.7	352.34	9MR	0.3	-4.1	-1.8	0.3
28.39	0.12	1.7	2.16	19MR	-0.1	-3.7	-1.9	0.3
35.42	0.22	1.7	11.93	29MR	-0.7	-3.1	-2.0	0.3
42.36	0.32	1.8	21.65	8AP	-1.5	-3.6	-2.0	0.4
49.24	0.42	1.8	31.31	18AP	-1.4	-4.0	-2.1	0.4
56.05	0.51	1.8	40.94	28AP	-0.5	-4.2	-2.1	0.4
62.80	0.59	1.8	50.52	8MY	0.5	-4.2	-2.2	0.4
69.48	0.67	1.9	60.08	18MY	1.3	-4.1	-2.2	0.4
76.12	0.74	1.9	69.62	28MY	2.2	-4.0	-2.2	0.4
82.70	0.81	1.9	79.15	7JN	3.4	-3.9	-2.2	0.4
89.23	0.87	1.9	88.67	17JN	2.1	-3.8	-2.1	0.3
95.73	0.93	1.9	98.21	27JN	0.8	-3.8	-2.1	0.3
102.19	0.99	1.9	107.75	7JL	-0.5	-3.7	-2.0	0.4
108.61	1.04	2.0	117.32	17JL	-1.4	-3.6	-2.0	0.4
115.01	1.08	2.0	126.93	27JL	-1.4	-3.6	-1.9	0.4
121.38	1.12	2.0	136.58	6AU	-0.7	-3.5	-1.9	0.5
127.73	1.16	2.0	146.27	16AU	-0.2	-3.5	-1.8	0.5
134.07	1.20	2.0	156.02	26AU	0.1	-3.4	-1.7	0.5
140.38	1.23	2.0	165.81	5SE	0.2	-3.4	-1.7	0.5
146.68	1.25	2.0	175.67	15SE	0.5	-3.4	-1.6	0.5
152.97	1.28	2.0	185.59	25SE	1.5	-3.4	-1.6	0.5
159.23	1.30	2.0	195.56	5OC	2.4	-3.4	-1.5	0.4
165.49	1.31	1.9	205.59	15OC	-0.0	-3.4	-1.5	0.4
171.72	1.32	1.9	215.67	25OC	-0.6	-3.4	-1.5	0.4
177.93	1.33	1.8	225.78	4NO	-0.6	-3.4	-1.4	0.3
184.12	1.33	1.8	235.94	14NO	-0.6	-3.4	-1.4	0.2
190.28	1.33	1.7	246.12	24NO	-0.7	-3.4	-1.4	0.2
196.39	1.31	1.6	256.32	4DE	-0.8	-3.4	-1.4	0.1
202.46	1.29	1.5	266.52	14DE	-0.8	-3.4	-1.4	0.0
208.47	1.26	1.4	276.72	24DE	-0.7	-3.4	-1.4	-0.1
				739				
214.41	1.22	1.3	286.90	3JA	-0.0	-3.5	-1.4	-0.1
220.27	1.17	1.2	297.06	13JA	2.4	-3.5	-1.4	-0.1
226.01	1.10	1.0	307.18	23JA	1.7	-3.5	-1.5	0.0
231.63	1.01	0.9	317.25	2FE	0.7	-3.4	-1.5	0.1
237.08	0.90	0.7	327.28	12FE	0.4	-3.4	-1.6	0.1
242.34	0.76	0.5	337.25	22FE	0.2	-3.4	-1.6	0.2
247.34	0.58	0.3	347.16	4MR	-0.2	-3.4	-1.7	0.3
252.03	0.37	0.1	357.02	14MR	-0.7	-3.4	-1.7	0.3
256.31	0.10	-0.2	6.82	24MR	-1.5	-3.3	-1.8	0.4
260.09	-0.23	-0.5	16.56	3AP	-1.4	-3.3	-1.9	0.4
263.23	-0.63	-0.7	26.25	13AP	-0.4	-3.3	-2.0	0.4
265.55	-1.12	-1.0	35.90	23AP	0.6	-3.3	-2.0	0.5
266.88	-1.70	-1.4	45.50	3MY	1.6	-3.4	-2.1	0.5
267.05	-2.37	-1.7	55.07	13MY	3.0	-3.4	-2.2	0.5
265.97	-3.09	-2.0	64.62	23MY	3.0	-3.4	-2.2	0.5
263.80	-3.81	-2.3	74.15	2JN	1.7	-3.4	-2.3	0.5
260.96	-4.42	-2.4	83.68	12JN	0.6	-3.5	-2.3	0.5
258.14	-4.85	-2.3	93.21	22JN	-0.5	-3.5	-2.3	0.5
256.06	-5.06	-2.1	102.74	2JL	-1.4	-3.6	-2.3	0.5
255.18	-5.06	-1.9	112.30	12JL	-1.4	-3.6	-2.3	0.5
255.69	-4.92	-1.7	121.89	22JL	-0.6	-3.7	-2.3	0.5
257.50	-4.68	-1.4	131.51	1AU	-0.1	-3.8	-2.2	0.6
260.45	-4.39	-1.2	141.18	11AU	0.2	-3.9	-2.2	0.6
264.35	-4.07	-1.0	150.90	21AU	0.4	-4.0	-2.1	0.6
269.00	-3.73	-0.8	160.67	31AU	0.7	-4.1	-2.1	0.7
274.25	-3.40	-0.6	170.50	10SE	1.9	-4.2	-2.0	0.7
279.97	-3.06	-0.4	180.38	20SE	2.2	-4.3	-1.9	0.7
286.05	-2.74	-0.2	190.32	30SE	-0.1	-4.3	-1.9	0.7
292.42	-2.43	-0.0	200.32	10OC	-0.7	-4.2	-1.8	0.6
299.00	-2.12	0.1	210.37	20OC	-0.8	-3.9	-1.7	0.6
305.74	-1.84	0.3	220.47	30OC	-0.8	-3.1	-1.7	0.6
312.61	-1.56	0.4	230.61	9NO	-0.7	-3.6	-1.6	0.5
319.55	-1.31	0.6	240.78	19NO	-0.6	-4.1	-1.6	0.5
326.54	-1.06	0.7	250.97	29NO	-0.6	-4.4	-1.6	0.4
333.57	-0.84	0.8	261.17	9DE	-0.6	-4.4	-1.5	0.3
340.60	-0.63	1.0	271.37	19DE	0.1	-4.3	-1.5	0.3
347.62	-0.44	1.1	281.56	29DE	2.7	-4.2	-1.5	0.2

Left table:

♂ LONG	♂ LAT	♂ MAG	☉ LONG	16.00UT	☿	♀	♃	♄
				740				
354.63	-0.26	1.2	291.74	8JA	1.3	-4.1	-1.5	0.1
1.61	-0.10	1.3	301.87	18JA	0.4	-4.0	-1.5	0.1
8.55	0.05	1.4	311.97	28JA	0.2	-3.9	-1.5	0.1
15.45	0.19	1.5	322.03	7FE	0.0	-3.8	-1.5	0.2
22.30	0.31	1.6	332.03	17FE	-0.2	-3.7	-1.6	0.2
29.11	0.42	1.6	341.98	27FE	-0.7	-3.6	-1.6	0.3
35.86	0.52	1.7	351.87	8MR	-1.5	-3.6	-1.6	0.4
42.57	0.62	1.7	1.69	18MR	-1.4	-3.5	-1.7	0.4
49.23	0.70	1.8	11.47	28MR	-0.4	-3.4	-1.7	0.5
55.85	0.77	1.8	21.19	7AP	0.8	-3.4	-1.8	0.5
62.42	0.84	1.9	30.85	17AP	2.1	-3.4	-1.8	0.6
68.95	0.90	1.9	40.48	27AP	3.9	-3.3	-1.9	0.6
75.45	0.95	2.0	50.07	7MY	2.3	-3.3	-2.0	0.6
81.91	0.99	2.0	59.63	17MY	1.3	-3.3	-2.1	0.7
88.35	1.03	2.0	69.16	27MY	0.4	-3.3	-2.1	0.7
94.76	1.06	2.0	78.69	6JN	-0.5	-3.3	-2.2	0.7
101.16	1.09	2.0	88.22	16JN	-1.5	-3.3	-2.3	0.7
107.53	1.12	2.0	97.75	26JN	-1.4	-3.3	-2.3	0.7
113.90	1.13	2.0	107.30	6JL	-0.6	-3.4	-2.4	0.7
120.27	1.15	2.0	116.86	16JL	-0.0	-3.4	-2.4	0.7
126.63	1.15	2.0	126.47	26JL	0.3	-3.4	-2.4	0.7
133.00	1.16	2.0	136.11	5AU	0.6	-3.5	-2.4	0.7
139.37	1.16	2.0	145.80	15AU	1.0	-3.5	-2.4	0.8
145.76	1.15	2.0	155.54	25AU	2.3	-3.5	-2.4	0.8
152.16	1.14	2.0	165.34	4SE	1.9	-3.4	-2.3	0.8
158.58	1.13	2.0	175.19	14SE	-0.2	-3.4	-2.3	0.9
165.02	1.11	2.0	185.10	24SE	-0.9	-3.4	-2.2	0.9
171.47	1.08	2.0	195.07	4OC	-0.9	-3.4	-2.2	0.9
177.95	1.05	2.0	205.10	14OC	-0.9	-3.3	-2.1	0.9
184.46	1.01	1.9	215.17	24OC	-0.6	-3.3	-2.0	0.9
190.99	0.97	1.9	225.29	3NO	-0.5	-3.3	-1.9	0.8
197.54	0.92	1.9	235.44	13NO	-0.5	-3.3	-1.9	0.8
204.13	0.86	1.8	245.62	23NO	-0.4	-3.3	-1.8	0.8
210.73	0.80	1.8	255.82	3DE	0.3	-3.4	-1.8	0.7
217.35	0.72	1.7	266.02	13DE	2.9	-3.4	-1.7	0.6
224.00	0.63	1.6	276.22	23DE	0.9	-3.4	-1.7	0.6
				741				
230.66	0.54	1.5	286.40	2JA	0.2	-3.4	-1.6	0.5
237.34	0.43	1.4	296.56	12JA	0.0	-3.5	-1.6	0.4
244.04	0.30	1.3	306.68	22JA	-0.1	-3.5	-1.6	0.4
250.74	0.16	1.2	316.76	1FE	-0.3	-3.6	-1.6	0.3
257.45	0.00	1.1	326.79	11FE	-0.8	-3.6	-1.6	0.4
264.16	-0.17	1.0	336.77	21FE	-1.4	-3.7	-1.6	0.4
270.85	-0.37	0.9	346.69	3MR	-1.3	-3.8	-1.6	0.5
277.52	-0.59	0.7	356.54	13MR	-0.4	-3.8	-1.6	0.5
284.17	-0.83	0.6	6.35	23MR	1.0	-3.9	-1.6	0.6
290.75	-1.09	0.4	16.10	2AP	2.7	-4.0	-1.6	0.6
297.26	-1.38	0.3	25.79	12AP	3.0	-4.1	-1.7	0.7
303.66	-1.70	0.1	35.44	22AP	1.7	-4.2	-1.7	0.7
309.91	-2.05	-0.1	45.05	2MY	1.0	-4.2	-1.8	0.8
315.97	-2.42	-0.2	54.62	12MY	0.3	-4.1	-1.9	0.8
321.76	-2.82	-0.4	64.17	22MY	-0.5	-3.8	-1.9	0.9
327.21	-3.25	-0.6	73.70	1JN	-1.5	-3.2	-2.0	0.9
332.20	-3.70	-0.9	83.22	11JN	-1.5	-3.0	-2.0	0.9
336.60	-4.18	-1.1	92.75	21JN	-0.6	-3.8	-2.1	0.9
340.24	-4.67	-1.3	102.29	1JL	0.0	-4.1	-2.2	0.9
342.92	-5.15	-1.6	111.84	11JL	0.4	-4.2	-2.3	0.9
344.41	-5.60	-1.8	121.43	21JL	0.8	-4.2	-2.3	0.9
344.56	-5.95	-2.1	131.05	31JL	1.4	-4.1	-2.4	0.9
343.35	-6.14	-2.3	140.71	10AU	2.8	-4.0	-2.4	0.9
341.01	-6.07	-2.5	150.43	20AU	1.7	-3.9	-2.4	1.0
338.14	-5.72	-2.5	160.20	30AU	-0.3	-3.9	-2.5	1.0
335.53	-5.12	-2.3	170.02	9SE	-1.0	-3.8	-2.5	1.1
333.81	-4.38	-2.0	179.90	19SE	-1.1	-3.7	-2.5	1.1
333.37	-3.59	-1.7	189.84	29SE	-0.9	-3.7	-2.4	1.1
334.23	-2.85	-1.4	199.84	9OC	-0.5	-3.6	-2.4	1.2
336.26	-2.19	-1.1	209.89	19OC	-0.4	-3.5	-2.3	1.2
339.25	-1.61	-0.8	219.98	29OC	-0.3	-3.5	-2.3	1.2
343.01	-1.13	-0.5	230.12	8NO	-0.2	-3.5	-2.2	1.1
347.37	-0.72	-0.2	240.29	18NO	0.5	-3.4	-2.1	1.1
352.20	-0.37	0.0	250.47	28NO	3.2	-3.4	-2.1	1.1
357.37	-0.09	0.2	260.67	8DE	0.7	-3.4	-2.0	1.0
2.82	0.15	0.4	270.88	18DE	0.0	-3.4	-1.9	1.0
8.47	0.36	0.6	281.07	28DE	-0.1	-3.3	-1.9	0.9

Right table:

♂ LONG	♂ LAT	♂ MAG	☉ LONG	16.00UT	☿	♀	♃	♄
				742				
14.28	0.53	0.8	291.24	7JA	-0.2	-3.3	-1.8	0.8
20.20	0.67	1.0	301.38	17JA	-0.4	-3.3	-1.7	0.8
26.22	0.79	1.1	311.48	27JA	-0.8	-3.3	-1.7	0.7
32.29	0.89	1.2	321.54	6FE	-1.3	-3.3	-1.7	0.6
38.42	0.97	1.4	331.54	16FE	-1.2	-3.4	-1.6	0.6
44.58	1.04	1.5	341.49	26FE	-0.3	-3.4	-1.6	0.6
50.75	1.10	1.6	351.38	8MR	1.3	-3.4	-1.6	0.7
56.95	1.14	1.7	1.21	18MR	3.3	-3.4	-1.6	0.7
63.15	1.18	1.7	10.99	28MR	2.2	-3.5	-1.6	0.8
69.36	1.21	1.8	20.71	7AP	1.3	-3.5	-1.6	0.8
75.58	1.23	1.9	30.38	17AP	0.7	-3.4	-1.6	0.9
81.79	1.24	1.9	40.01	27AP	0.2	-3.4	-1.6	1.0
88.01	1.25	2.0	49.60	7MY	-0.6	-3.4	-1.6	1.0
94.24	1.25	2.0	59.16	17MY	-1.5	-3.4	-1.7	1.1
100.47	1.25	2.0	68.70	27MY	-1.5	-3.3	-1.7	1.1
106.72	1.24	2.0	78.23	6JN	-0.6	-3.3	-1.7	1.1
112.97	1.22	2.0	87.75	16JN	0.1	-3.3	-1.8	1.2
119.25	1.21	2.0	97.28	26JN	0.6	-3.3	-1.9	1.2
125.54	1.18	2.0	106.83	6JL	1.1	-3.3	-1.9	1.2
131.86	1.16	2.0	116.40	16JL	1.8	-3.3	-2.0	1.2
138.20	1.13	2.0	126.00	26JL	3.2	-3.4	-2.0	1.2
144.58	1.09	2.0	135.64	5AU	1.4	-3.4	-2.1	1.2
150.99	1.05	2.0	145.33	15AU	-0.3	-3.4	-2.2	1.2
157.44	1.01	1.9	155.07	25AU	-1.1	-3.5	-2.2	1.2
163.94	0.96	1.9	164.86	4SE	-1.2	-3.5	-2.3	1.3
170.48	0.91	1.9	174.71	14SE	-0.8	-3.6	-2.4	1.3
177.07	0.85	1.9	184.62	24SE	-0.4	-3.7	-2.4	1.3
183.71	0.79	1.9	194.59	4OC	-0.2	-3.8	-2.4	1.3
190.40	0.72	1.9	204.61	14OC	-0.2	-3.9	-2.4	1.3
197.15	0.65	1.9	214.68	24OC	-0.1	-4.0	-2.4	1.3
203.96	0.57	1.9	224.80	3NO	0.7	-4.1	-2.4	1.3
210.82	0.49	1.8	234.95	13NO	3.2	-4.2	-2.3	1.2
217.74	0.40	1.8	245.13	23NO	0.5	-4.3	-2.3	1.2
224.73	0.30	1.8	255.33	3DE	-0.2	-4.4	-2.2	1.2
231.76	0.20	1.7	265.53	13DE	-0.2	-4.4	-2.2	1.1
238.86	0.08	1.7	275.72	23DE	-0.3	-4.3	-2.1	1.1
				743				
246.01	-0.03	1.6	285.91	2JA	-0.5	-3.9	-2.0	1.0
253.21	-0.16	1.6	296.07	12JA	-0.9	-3.3	-1.9	1.0
260.47	-0.29	1.5	306.19	22JA	-1.2	-3.6	-1.9	0.9
267.77	-0.43	1.4	316.27	1FE	-1.1	-4.1	-1.8	0.9
275.12	-0.58	1.4	326.30	11FE	-0.3	-4.3	-1.7	0.8
282.50	-0.73	1.3	336.28	21FE	1.5	-4.3	-1.7	0.8
289.91	-0.88	1.2	346.20	3MR	3.2	-4.2	-1.6	0.8
297.35	-1.04	1.1	356.06	13MR	1.6	-4.1	-1.6	0.9
304.79	-1.20	1.1	5.87	23MR	0.9	-4.0	-1.6	1.0
312.24	-1.36	1.0	15.62	2AP	0.5	-3.9	-1.5	1.1
319.67	-1.51	0.9	25.32	12AP	0.1	-3.8	-1.5	1.1
327.06	-1.66	0.8	34.96	22AP	-0.6	-3.7	-1.5	1.2
334.41	-1.81	0.7	44.58	2MY	-1.5	-3.6	-1.5	1.2
341.68	-1.94	0.6	54.15	12MY	-1.5	-3.6	-1.5	1.3
348.86	-2.06	0.5	63.70	22MY	-0.5	-3.5	-1.5	1.3
355.91	-2.17	0.5	73.24	1JN	0.2	-3.5	-1.5	1.4
2.80	-2.26	0.4	82.76	11JN	0.8	-3.4	-1.6	1.4
9.50	-2.33	0.3	92.29	21JN	1.4	-3.4	-1.6	1.4
15.96	-2.39	0.2	101.83	1JL	2.4	-3.4	-1.6	1.4
22.12	-2.42	0.0	111.38	11JL	3.0	-3.3	-1.7	1.4
27.92	-2.43	-0.1	120.97	21JL	1.2	-3.3	-1.7	1.4
33.29	-2.41	-0.2	130.59	31JL	-0.4	-3.3	-1.8	1.3
38.11	-2.37	-0.4	140.25	10AU	-1.2	-3.3	-1.8	1.3
42.26	-2.29	-0.5	149.96	20AU	-1.3	-3.3	-1.9	1.2
45.58	-2.16	-0.7	159.73	30AU	-0.8	-3.4	-2.0	1.2
47.86	-1.98	-0.9	169.55	9SE	-0.3	-3.4	-2.0	1.1
48.91	-1.74	-1.2	179.42	19SE	-0.1	-3.4	-2.1	1.2
48.53	-1.42	-1.4	189.36	29SE	-0.0	-3.4	-2.1	1.2
46.70	-1.01	-1.6	199.35	9OC	0.1	-3.5	-2.2	1.2
43.64	-0.53	-1.8	209.40	19OC	0.9	-3.5	-2.3	1.2
39.94	-0.02	-1.8	219.49	29OC	3.0	-3.5	-2.3	1.1
36.43	0.46	-1.6	229.62	8NO	0.3	-3.5	-2.3	1.1
33.81	0.88	-1.3	239.79	18NO	-0.3	-3.4	-2.3	1.1
32.48	1.19	-1.0	249.98	28NO	-0.4	-3.4	-2.3	1.1
32.50	1.42	-0.7	260.18	8DE	-0.4	-3.4	-2.3	1.0
33.74	1.58	-0.4	270.38	18DE	-0.6	-3.4	-2.2	1.0
35.98	1.68	-0.1	280.57	28DE	-0.9	-3.4	-2.2	0.9

♂			☉	16.00UT	☿	♀	♃	♄
LONG	LAT	MAG	LONG	744	MAGNITUDES			
39.04	1.75	0.2	290.74	7JA	-1.0	-3.3	-2.1	0.9
42.72	1.78	0.4	300.89	17JA	-1.0	-3.3	-2.1	0.8
46.90	1.80	0.6	310.99	27JA	-0.2	-3.3	-2.0	0.8
51.46	1.79	0.8	321.05	6FE	1.8	-3.3	-1.9	0.8
56.32	1.78	1.0	331.05	16FE	2.6	-3.4	-1.8	0.7
61.42	1.76	1.2	341.01	26FE	1.2	-3.4	-1.8	0.7
66.71	1.73	1.3	350.90	7MR	0.7	-3.4	-1.7	0.6
72.14	1.70	1.4	0.73	17MR	0.4	-3.4	-1.7	0.7
77.71	1.66	1.5	10.51	27MR	-0.0	-3.5	-1.6	0.7
83.37	1.62	1.6	20.23	6AP	-0.6	-3.5	-1.6	0.8
89.12	1.58	1.7	29.91	16AP	-1.5	-3.6	-1.5	0.9
94.94	1.53	1.8	39.54	26AP	-1.5	-3.6	-1.5	1.0
100.83	1.48	1.8	49.13	6MY	-0.5	-3.7	-1.4	1.0
106.78	1.43	1.9	58.70	16MY	0.4	-3.8	-1.4	1.1
112.78	1.37	1.9	68.24	26MY	1.1	-3.8	-1.4	1.1
118.84	1.32	2.0	77.77	5JN	1.9	-3.9	-1.4	1.2
124.95	1.26	2.0	87.29	15JN	3.1	-4.0	-1.4	1.2
131.12	1.20	2.0	96.83	25JN	2.5	-4.1	-1.4	1.2
137.34	1.13	2.0	106.37	5JL	0.9	-4.2	-1.4	1.2
143.62	1.07	2.0	115.94	15JL	-0.4	-4.2	-1.5	1.2
149.96	1.00	2.0	125.54	25JL	-1.3	-4.1	-1.5	1.2
156.36	0.93	2.0	135.18	4AU	-1.4	-3.8	-1.5	1.2
162.82	0.85	2.0	144.86	14AU	-0.7	-3.2	-1.6	1.1
169.35	0.77	1.9	154.60	24AU	-0.3	-3.6	-1.6	1.1
175.95	0.69	1.9	164.39	3SE	-0.0	-4.0	-1.7	1.1
182.63	0.61	1.8	174.24	13SE	0.1	-4.3	-1.7	1.0
189.37	0.52	1.8	184.14	23SE	0.3	-4.3	-1.8	1.0
196.20	0.43	1.7	194.10	3OC	1.2	-4.3	-1.8	1.0
203.10	0.34	1.7	204.13	13OC	2.7	-4.2	-1.9	1.0
210.08	0.24	1.7	214.20	23OC	0.1	-4.1	-2.0	1.0
217.14	0.15	1.7	224.31	2NO	-0.5	-4.0	-2.0	1.0
224.28	0.04	1.7	234.46	12NO	-0.5	-3.9	-2.1	1.0
231.49	-0.06	1.7	244.64	22NO	-0.6	-3.8	-2.1	1.0
238.78	-0.17	1.6	254.83	2DE	-0.7	-3.7	-2.2	1.0
246.15	-0.28	1.6	265.04	12DE	-0.8	-3.7	-2.2	0.9
253.58	-0.39	1.6	275.23	22DE	-0.9	-3.6	-2.2	0.9
				745				
261.08	-0.50	1.6	285.42	1JA	-0.8	-3.5	-2.2	0.9
268.64	-0.61	1.5	295.58	11JA	-0.1	-3.5	-2.1	0.8
276.25	-0.72	1.5	305.70	21JA	2.1	-3.4	-2.1	0.8
283.90	-0.83	1.5	315.78	31JA	2.0	-3.4	-2.1	0.7
291.59	-0.93	1.4	325.82	10FE	0.9	-3.4	-2.0	0.7
299.31	-1.03	1.4	335.80	20FE	0.5	-3.3	-1.9	0.6
307.04	-1.12	1.4	345.72	2MR	0.2	-3.3	-1.9	0.6
314.78	-1.20	1.3	355.59	12MR	-0.1	-3.3	-1.8	0.5
322.51	-1.27	1.3	5.40	22MR	-0.7	-3.3	-1.7	0.5
330.22	-1.33	1.3	15.15	1AP	-1.5	-3.3	-1.7	0.6
337.89	-1.38	1.2	24.85	11AP	-1.5	-3.3	-1.6	0.6
345.51	-1.41	1.2	34.50	21AP	-0.5	-3.3	-1.5	0.7
353.08	-1.43	1.2	44.11	1MY	0.5	-3.4	-1.5	0.8
0.56	-1.44	1.1	53.69	11MY	1.4	-3.4	-1.4	0.8
7.96	-1.43	1.1	63.24	21MY	2.5	-3.4	-1.4	0.9
15.25	-1.40	1.1	72.77	31MY	3.4	-3.5	-1.4	0.9
22.42	-1.36	1.0	82.30	10JN	2.0	-3.5	-1.3	1.0
29.45	-1.30	1.0	91.82	20JN	0.7	-3.5	-1.3	1.0
36.32	-1.22	0.9	101.36	30JN	-0.4	-3.4	-1.3	1.0
43.03	-1.13	0.9	110.92	10JL	-1.4	-3.4	-1.3	1.1
49.54	-1.02	0.8	120.50	20JL	-1.4	-3.4	-1.3	1.1
55.83	-0.89	0.8	130.12	30JL	-0.7	-3.3	-1.3	1.1
61.86	-0.74	0.7	139.78	9AU	-0.2	-3.3	-1.3	1.0
67.60	-0.58	0.6	149.48	19AU	0.1	-3.3	-1.4	1.0
72.99	-0.39	0.5	159.25	29AU	0.3	-3.3	-1.4	1.0
77.96	-0.18	0.4	169.07	8SE	0.6	-3.3	-1.4	1.0
82.44	0.07	0.2	178.94	18SE	1.6	-3.3	-1.5	0.9
86.30	0.35	0.1	188.87	28SE	2.4	-3.4	-1.5	0.9
89.40	0.67	-0.1	198.87	8OC	-0.0	-3.4	-1.6	0.9
91.58	1.03	-0.3	208.91	18OC	-0.6	-3.4	-1.6	0.9
92.64	1.44	-0.5	219.00	28OC	-0.7	-3.5	-1.7	0.9
92.37	1.90	-0.7	229.13	7NO	-0.7	-3.5	-1.7	0.9
90.71	2.37	-0.9	239.30	17NO	-0.8	-3.6	-1.8	0.9
87.75	2.82	-1.1	249.48	27NO	-0.7	-3.6	-1.9	0.9
83.95	3.18	-1.3	259.69	7DE	-0.7	-3.7	-1.9	0.9
80.04	3.40	-1.2	269.89	17DE	-0.7	-3.8	-2.0	0.9
76.75	3.48	-0.9	280.08	27DE	-0.0	-3.9	-2.0	0.8

♂			☉	16.00UT	☿	♀	♃	♄
LONG	LAT	MAG	LONG	746	MAGNITUDES			
74.62	3.44	-0.6	290.26	6JA	2.4	-4.0	-2.1	0.8
73.82	3.32	-0.3	300.40	16JA	1.5	-4.1	-2.1	0.8
74.28	3.15	-0.1	310.50	26JA	0.6	-4.2	-2.1	0.7
75.84	2.97	0.2	320.57	5FE	0.3	-4.2	-2.1	0.7
78.28	2.79	0.4	330.57	15FE	0.1	-4.3	-2.0	0.6
81.43	2.62	0.6	340.53	25FE	-0.2	-4.3	-2.0	0.6
85.16	2.45	0.8	350.43	7MR	-0.7	-4.1	-1.9	0.5
89.32	2.30	1.0	0.26	17MR	-1.5	-3.7	-1.9	0.5
93.85	2.15	1.1	10.04	27MR	-1.4	-3.1	-1.8	0.4
98.67	2.01	1.3	19.77	6AP	-0.5	-3.6	-1.7	0.4
103.73	1.88	1.4	29.44	16AP	0.7	-4.0	-1.7	0.5
108.98	1.75	1.5	39.07	26AP	1.8	-4.2	-1.6	0.5
114.41	1.63	1.6	48.67	6MY	3.3	-4.2	-1.6	0.6
119.98	1.51	1.6	58.23	16MY	2.7	-4.1	-1.5	0.7
125.69	1.40	1.7	67.77	26MY	1.5	-4.0	-1.4	0.7
131.51	1.29	1.8	77.30	5JN	0.6	-3.9	-1.4	0.8
137.45	1.18	1.8	86.83	15JN	-0.5	-3.8	-1.3	0.8
143.48	1.07	1.8	96.36	25JN	-1.4	-3.8	-1.3	0.9
149.62	0.96	1.8	105.90	5JL	-1.5	-3.7	-1.3	0.9
155.86	0.86	1.9	115.47	15JL	-0.6	-3.6	-1.3	0.9
162.19	0.75	1.9	125.07	25JL	-0.1	-3.6	-1.2	1.0
168.62	0.65	1.9	134.71	4AU	0.2	-3.5	-1.2	1.0
175.14	0.54	1.8	144.39	14AU	0.5	-3.5	-1.2	1.0
181.76	0.43	1.8	154.13	24AU	0.8	-3.4	-1.2	0.9
188.48	0.33	1.8	163.91	3SE	2.0	-3.4	-1.2	0.9
195.29	0.22	1.8	173.76	13SE	2.2	-3.4	-1.3	0.9
202.20	0.12	1.7	183.67	23SE	-0.1	-3.4	-1.3	0.9
209.21	0.01	1.7	193.63	3OC	-0.8	-3.4	-1.3	0.8
216.32	-0.09	1.7	203.64	13OC	-0.8	-3.4	-1.4	0.8
223.52	-0.20	1.6	213.71	23OC	-0.8	-3.4	-1.4	0.8
230.81	-0.30	1.6	223.82	2NO	-0.7	-3.4	-1.5	0.8
238.19	-0.40	1.5	233.97	12NO	-0.6	-3.4	-1.5	0.8
245.66	-0.50	1.5	244.15	22NO	-0.6	-3.4	-1.6	0.8
253.21	-0.60	1.4	254.34	2DE	-0.5	-3.4	-1.6	0.8
260.83	-0.69	1.4	264.54	12DE	0.1	-3.4	-1.7	0.8
268.51	-0.77	1.4	274.74	22DE	2.7	-3.4	-1.8	0.8
				747				
276.25	-0.85	1.4	284.92	1JA	1.2	-3.5	-1.8	0.8
284.04	-0.92	1.4	295.08	11JA	0.3	-3.5	-1.9	0.8
291.86	-0.99	1.4	305.21	21JA	0.1	-3.5	-1.9	0.7
299.70	-1.04	1.4	315.30	31JA	-0.0	-3.4	-2.0	0.7
307.56	-1.08	1.4	325.33	10FE	-0.3	-3.4	-2.0	0.6
315.41	-1.12	1.4	335.32	20FE	-0.7	-3.4	-2.0	0.6
323.25	-1.14	1.4	345.24	2MR	-1.5	-3.4	-2.0	0.5
331.07	-1.15	1.4	355.11	12MR	-1.4	-3.4	-2.0	0.5
338.85	-1.14	1.4	4.92	22MR	-0.4	-3.3	-2.0	0.4
346.57	-1.13	1.4	14.67	1AP	0.9	-3.3	-1.9	0.4
354.24	-1.10	1.4	24.37	11AP	2.3	-3.3	-1.9	0.3
1.84	-1.06	1.4	34.03	21AP	3.6	-3.3	-1.8	0.4
9.36	-1.00	1.4	43.64	1MY	2.1	-3.4	-1.7	0.4
16.79	-0.94	1.4	53.22	11MY	1.2	-3.4	-1.7	0.5
24.13	-0.86	1.4	62.78	21MY	0.4	-3.4	-1.6	0.5
31.36	-0.78	1.4	72.31	31MY	-0.5	-3.4	-1.5	0.6
38.49	-0.68	1.4	81.83	10JN	-1.5	-3.5	-1.5	0.7
45.50	-0.57	1.4	91.36	20JN	-1.5	-3.5	-1.4	0.7
52.39	-0.46	1.4	100.90	30JN	-0.6	-3.6	-1.4	0.8
59.15	-0.34	1.4	110.45	10JL	-0.0	-3.6	-1.3	0.8
65.78	-0.20	1.4	120.03	20JL	0.4	-3.7	-1.3	0.8
72.26	-0.06	1.3	129.65	30JL	0.7	-3.8	-1.3	0.9
78.58	0.09	1.3	139.31	9AU	1.2	-3.9	-1.2	0.9
84.73	0.25	1.3	149.02	19AU	2.5	-4.0	-1.2	0.9
90.69	0.42	1.2	158.77	29AU	1.9	-4.1	-1.2	0.9
96.43	0.60	1.1	168.59	8SE	-0.2	-4.2	-1.2	0.9
101.91	0.80	1.1	178.46	18SE	-0.9	-4.3	-1.2	0.8
107.11	1.01	1.0	188.39	28SE	-1.0	-4.3	-1.2	0.8
111.95	1.25	0.8	198.38	8OC	-0.9	-4.2	-1.2	0.8
116.38	1.51	0.7	208.43	18OC	-0.6	-3.8	-1.3	0.7
120.29	1.79	0.6	218.51	28OC	-0.4	-3.1	-1.3	0.7
123.58	2.11	0.4	228.64	7NO	-0.4	-3.6	-1.3	0.8
126.08	2.47	0.2	238.81	17NO	-0.3	-4.2	-1.4	0.8
127.63	2.86	-0.0	248.99	27NO	0.3	-4.4	-1.4	0.8
128.03	3.28	-0.3	259.19	7DE	3.0	-4.4	-1.5	0.8
127.13	3.70	-0.5	269.40	17DE	0.9	-4.3	-1.5	0.8
124.91	4.08	-0.7	279.59	27DE	0.1	-4.2	-1.6	0.8

♂			⊙	16.00UT	☿	♀	♃	♄
LONG	LAT	MAG	LONG	748	MAGNITUDES			
121.56	4.37	-0.9	289.76	6JA	-0.0	-4.1	-1.7	0.8
117.63	4.50	-1.0	299.91	16JA	-0.1	-4.0	-1.7	0.8
113.83	4.45	-0.9	310.02	26JA	-0.3	-3.9	-1.8	0.7
110.84	4.25	-0.6	320.08	5FE	-0.8	-3.8	-1.9	0.7
109.05	3.95	-0.4	330.09	15FE	-1.4	-3.7	-1.9	0.7
108.57	3.62	-0.1	340.05	25FE	-1.3	-3.6	-2.0	0.6
109.28	3.27	0.1	349.95	6MR	-0.4	-3.6	-2.0	0.6
111.04	2.95	0.3	359.79	16MR	1.1	-3.5	-2.0	0.5
113.64	2.65	0.5	9.57	26MR	2.9	-3.4	-2.0	0.4
116.92	2.37	0.7	19.30	5AP	2.7	-3.4	-2.0	0.4
120.76	2.12	0.9	28.98	15AP	1.5	-3.4	-2.0	0.3
125.04	1.88	1.0	38.61	25AP	0.9	-3.3	-2.0	0.3
129.70	1.67	1.1	48.21	5MY	0.3	-3.3	-1.9	0.3
134.66	1.47	1.2	57.78	15MY	-0.5	-3.3	-1.9	0.4
139.88	1.29	1.3	67.32	25MY	-1.5	-3.3	-1.8	0.4
145.34	1.11	1.4	76.85	4JN	-1.5	-3.3	-1.7	0.5
150.99	0.95	1.5	86.37	14JN	-0.6	-3.3	-1.7	0.6
156.82	0.79	1.5	95.90	24JN	0.1	-3.3	-1.6	0.6
162.82	0.64	1.5	105.44	4JL	0.5	-3.4	-1.5	0.7
168.97	0.50	1.6	115.01	14JL	0.9	-3.4	-1.5	0.7
175.27	0.36	1.6	124.60	24JL	1.5	-3.4	-1.4	0.8
181.71	0.23	1.6	134.24	3AU	3.0	-3.5	-1.4	0.8
188.28	0.10	1.6	143.92	13AU	1.6	-3.5	-1.3	0.8
194.98	-0.02	1.6	153.65	23AU	-0.3	-3.5	-1.3	0.8
201.81	-0.14	1.6	163.44	2SE	-1.1	-3.4	-1.3	0.8
208.77	-0.26	1.6	173.28	12SE	-1.1	-3.4	-1.3	0.8
215.84	-0.37	1.6	183.18	22SE	-0.9	-3.4	-1.2	0.8
223.02	-0.47	1.6	193.14	2OC	-0.5	-3.4	-1.2	0.8
230.32	-0.57	1.6	203.15	12OC	-0.3	-3.3	-1.2	0.8
237.72	-0.66	1.5	213.21	22OC	-0.3	-3.3	-1.2	0.7
245.22	-0.75	1.5	223.33	1NO	-0.2	-3.3	-1.3	0.7
252.80	-0.82	1.5	233.47	11NO	0.5	-3.3	-1.3	0.7
260.47	-0.89	1.5	243.65	21NO	3.2	-3.3	-1.3	0.7
268.21	-0.95	1.4	253.84	1DE	0.6	-3.4	-1.3	0.8
276.00	-1.00	1.4	264.04	11DE	-0.1	-3.4	-1.4	0.8
283.84	-1.04	1.4	274.24	21DE	-0.2	-3.4	-1.4	0.8
291.72	-1.07	1.4	284.43	31DE	-0.2	-3.4	-1.5	0.8
				749				
299.62	-1.08	1.3	294.59	10JA	-0.4	-3.5	-1.5	0.8
307.52	-1.09	1.3	304.72	20JA	-0.8	-3.5	-1.6	0.8
315.42	-1.08	1.3	314.81	30JA	-1.3	-3.6	-1.7	0.7
323.30	-1.06	1.3	324.85	9FE	-1.2	-3.6	-1.7	0.7
331.14	-1.03	1.3	334.84	19FE	-0.4	-3.7	-1.8	0.7
338.94	-0.99	1.4	344.77	1MR	1.3	-3.8	-1.9	0.6
346.69	-0.93	1.4	354.64	11MR	3.4	-3.8	-1.9	0.6
354.38	-0.87	1.4	4.45	21MR	2.0	-3.9	-2.0	0.5
1.99	-0.80	1.5	14.21	31MR	1.1	-4.0	-2.0	0.5
9.52	-0.72	1.5	23.91	10AP	0.6	-4.1	-2.1	0.4
16.97	-0.64	1.5	33.57	20AP	0.2	-4.2	-2.1	0.3
24.34	-0.54	1.6	43.19	30AP	-0.6	-4.2	-2.1	0.3
31.61	-0.45	1.6	52.76	10MY	-1.5	-4.1	-2.1	0.3
38.78	-0.34	1.6	62.32	20MY	-1.5	-3.8	-2.0	0.3
45.87	-0.24	1.6	71.85	30MY	-0.6	-3.1	-2.0	0.4
52.85	-0.13	1.7	81.38	9JN	0.2	-3.1	-2.0	0.4
59.74	-0.01	1.7	90.90	19JN	0.7	-3.8	-1.9	0.5
66.54	0.10	1.7	100.44	29JN	1.2	-4.1	-1.8	0.6
73.23	0.22	1.7	109.99	9JL	2.0	-4.2	-1.8	0.6
79.82	0.34	1.7	119.57	19JL	3.2	-4.2	-1.7	0.7
86.32	0.46	1.7	129.18	29JL	1.4	-4.1	-1.7	0.7
92.70	0.59	1.7	138.83	8AU	-0.3	-4.0	-1.6	0.8
98.98	0.72	1.7	148.54	18AU	-1.2	-3.9	-1.5	0.8
105.14	0.86	1.6	158.30	28AU	-1.3	-3.9	-1.5	0.8
111.17	0.99	1.6	168.11	7SE	-0.8	-3.8	-1.4	0.8
117.07	1.14	1.5	177.98	17SE	-0.4	-3.7	-1.4	0.8
122.80	1.29	1.5	187.91	27SE	-0.2	-3.6	-1.4	0.8
128.36	1.45	1.4	197.89	7OC	-0.1	-3.6	-1.3	0.8
133.71	1.62	1.3	207.93	17OC	0.0	-3.5	-1.3	0.8
138.81	1.80	1.2	218.02	27OC	0.7	-3.5	-1.3	0.7
143.62	1.99	1.1	228.15	6NO	3.2	-3.5	-1.3	0.7
148.08	2.20	0.9	238.31	16NO	0.4	-3.4	-1.3	0.7
152.09	2.43	0.8	248.50	26NO	-0.2	-3.4	-1.3	0.7
155.57	2.69	0.6	258.70	6DE	-0.3	-3.4	-1.3	0.7
158.37	2.96	0.4	268.90	16DE	-0.4	-3.4	-1.4	0.8
160.35	3.26	0.2	279.10	26DE	-0.5	-3.4	-1.4	0.8

♂			⊙	16.00UT	☿	♀	♃	♄
LONG	LAT	MAG	LONG	750	MAGNITUDES			
161.32	3.56	-0.1	289.27	5JA	-0.9	-3.3	-1.4	0.8
161.10	3.86	-0.4	299.42	15JA	-1.1	-3.3	-1.5	0.8
159.58	4.13	-0.6	309.53	25JA	-1.1	-3.3	-1.5	0.8
156.85	4.30	-0.9	319.59	4FE	-0.3	-3.3	-1.6	0.8
153.22	4.32	-1.0	329.61	14FE	1.6	-3.4	-1.7	0.7
149.32	4.18	-1.0	339.57	24FE	3.0	-3.4	-1.7	0.7
145.88	3.89	-0.8	349.47	6MR	1.5	-3.4	-1.8	0.7
143.44	3.49	-0.6	359.31	16MR	0.8	-3.4	-1.9	0.6
142.28	3.06	-0.4	9.10	26MR	0.5	-3.5	-1.9	0.6
142.39	2.62	-0.2	18.83	5AP	0.1	-3.5	-2.0	0.5
143.66	2.22	0.0	28.51	15AP	-0.6	-3.4	-2.1	0.4
145.91	1.85	0.2	38.15	25AP	-1.5	-3.4	-2.1	0.4
148.96	1.51	0.4	47.74	5MY	-1.5	-3.4	-2.1	0.3
152.67	1.21	0.5	57.31	15MY	-0.6	-3.4	-2.2	0.3
156.93	0.94	0.7	66.85	25MY	0.3	-3.3	-2.2	0.3
161.63	0.70	0.8	76.38	4JN	0.9	-3.3	-2.2	0.3
166.71	0.48	0.9	85.91	14JN	1.6	-3.3	-2.2	0.4
172.11	0.27	1.0	95.43	24JN	2.7	-3.3	-2.1	0.4
177.80	0.09	1.0	104.98	4JL	2.8	-3.3	-2.1	0.5
183.74	-0.08	1.1	114.54	14JL	1.1	-3.3	-2.1	0.6
189.90	-0.24	1.2	124.14	24JL	-0.4	-3.4	-2.0	0.6
196.27	-0.39	1.2	133.77	3AU	-1.3	-3.4	-1.9	0.7
202.83	-0.52	1.2	143.45	13AU	-1.4	-3.4	-1.9	0.7
209.56	-0.64	1.3	153.17	23AU	-0.8	-3.5	-1.8	0.7
216.46	-0.76	1.3	162.95	2SE	-0.3	-3.5	-1.7	0.8
223.51	-0.85	1.3	172.80	12SE	-0.1	-3.6	-1.7	0.8
230.69	-0.94	1.3	182.69	22SE	0.0	-3.7	-1.6	0.8
238.01	-1.02	1.3	192.65	2OC	0.2	-3.8	-1.6	0.8
245.45	-1.08	1.3	202.66	12OC	1.0	-3.9	-1.5	0.8
252.99	-1.13	1.3	212.72	22OC	3.0	-4.0	-1.5	0.8
260.62	-1.17	1.4	222.83	1NO	0.2	-4.1	-1.5	0.7
268.34	-1.19	1.4	232.98	11NO	-0.4	-4.2	-1.4	0.7
276.12	-1.20	1.4	243.15	21NO	-0.5	-4.3	-1.4	0.7
283.94	-1.20	1.4	253.34	1DE	-0.5	-4.4	-1.4	0.7
291.81	-1.19	1.4	263.55	11DE	-0.6	-4.4	-1.4	0.7
299.69	-1.16	1.4	273.75	21DE	-0.9	-4.3	-1.4	0.8
307.58	-1.12	1.4	283.93	31DE	-1.0	-3.9	-1.4	0.8
				751				
315.46	-1.07	1.4	294.10	10JA	-0.9	-3.2	-1.4	0.8
323.31	-1.01	1.4	304.23	20JA	-0.2	-3.7	-1.5	0.8
331.12	-0.94	1.4	314.32	30JA	1.8	-4.1	-1.5	0.8
338.89	-0.86	1.4	324.37	9FE	2.4	-4.3	-1.5	0.8
346.60	-0.78	1.5	334.35	19FE	1.1	-4.3	-1.6	0.8
354.24	-0.69	1.5	344.29	1MR	0.6	-4.2	-1.6	0.7
1.81	-0.60	1.5	354.16	11MR	0.3	-4.1	-1.7	0.7
9.30	-0.50	1.5	3.98	21MR	-0.0	-4.0	-1.8	0.7
16.70	-0.40	1.5	13.74	31MR	-0.6	-3.9	-1.8	0.6
24.03	-0.29	1.5	23.45	10AP	-1.5	-3.8	-1.9	0.6
31.26	-0.19	1.6	33.10	20AP	-1.5	-3.7	-2.0	0.5
38.41	-0.09	1.6	42.72	30AP	-0.5	-3.6	-2.1	0.4
45.47	0.02	1.7	52.30	10MY	0.4	-3.6	-2.1	0.4
52.45	0.12	1.7	61.85	20MY	1.2	-3.5	-2.2	0.3
59.35	0.23	1.7	71.39	30MY	2.1	-3.5	-2.2	0.2
66.16	0.33	1.8	80.92	9JN	3.4	-3.4	-2.3	0.3
72.90	0.43	1.8	90.44	19JN	2.3	-3.4	-2.3	0.3
79.57	0.53	1.8	99.98	29JN	0.9	-3.4	-2.3	0.4
86.16	0.62	1.9	109.53	9JL	-0.4	-3.3	-2.3	0.5
92.69	0.72	1.9	119.11	19JL	-1.3	-3.3	-2.3	0.5
99.14	0.81	1.9	128.72	29JL	-1.4	-3.3	-2.3	0.6
105.53	0.91	1.9	138.37	8AU	-0.7	-3.3	-2.2	0.6
111.85	1.00	1.9	148.07	18AU	-0.2	-3.3	-2.2	0.7
118.10	1.09	1.9	157.83	28AU	0.0	-3.4	-2.1	0.7
124.28	1.18	1.9	167.64	7SE	0.2	-3.4	-2.1	0.7
130.39	1.28	1.8	177.50	17SE	0.4	-3.4	-2.0	0.8
136.41	1.37	1.8	187.43	27SE	1.3	-3.4	-1.9	0.8
142.34	1.46	1.7	197.41	7OC	2.7	-3.5	-1.9	0.8
148.18	1.56	1.7	207.44	17OC	0.1	-3.5	-1.8	0.8
153.90	1.66	1.6	217.53	27OC	-0.6	-3.5	-1.7	0.8
159.48	1.76	1.5	227.65	6NO	-0.6	-3.5	-1.7	0.8
164.91	1.86	1.4	237.81	16NO	-0.6	-3.4	-1.6	0.7
170.14	1.97	1.3	248.00	26NO	-0.7	-3.4	-1.6	0.7
175.15	2.08	1.2	258.19	6DE	-0.8	-3.4	-1.6	0.7
179.88	2.20	1.0	268.39	16DE	-0.8	-3.4	-1.5	0.7
184.27	2.32	0.8	278.59	26DE	-0.8	-3.4	-1.5	0.8

♂			☉	16.00UT	☿	♀	♃	♄
LONG	LAT	MAG	LONG	752	MAGNITUDES			
188.24	2.45	0.7	288.77	5JA	-0.1	-3.3	-1.5	0.8
191.68	2.57	0.4	298.92	15JA	2.1	-3.3	-1.5	0.8
194.46	2.70	0.2	309.03	25JA	1.8	-3.3	-1.5	0.8
196.43	2.82	-0.1	319.10	4FE	0.7	-3.3	-1.5	0.8
197.41	2.92	-0.3	329.12	14FE	0.4	-3.4	-1.5	0.8
197.22	2.97	-0.6	339.09	24FE	0.2	-3.4	-1.6	0.8
195.76	2.95	-0.9	348.99	5MR	-0.1	-3.4	-1.6	0.8
193.11	2.83	-1.2	358.84	15MR	-0.7	-3.4	-1.6	0.8
189.60	2.59	-1.4	8.63	25MR	-1.5	-3.5	-1.7	0.7
185.88	2.22	-1.4	18.36	4AP	-1.5	-3.5	-1.7	0.7
182.66	1.77	-1.2	28.04	14AP	-0.5	-3.6	-1.8	0.6
180.50	1.30	-1.0	37.69	24AP	0.5	-3.6	-1.9	0.6
179.65	0.84	-0.8	47.28	4MY	1.5	-3.7	-1.9	0.5
180.12	0.43	-0.6	56.85	14MY	2.8	-3.8	-2.0	0.4
181.78	0.06	-0.4	66.40	24MY	3.2	-3.8	-2.1	0.4
184.45	-0.25	-0.2	75.93	3JN	1.8	-3.9	-2.1	0.3
187.96	-0.52	-0.0	85.45	13JN	0.7	-4.0	-2.2	0.3
192.16	-0.75	0.1	94.98	23JN	-0.4	-4.1	-2.3	0.3
196.93	-0.95	0.2	104.52	3JL	-1.4	-4.2	-2.3	0.4
202.18	-1.11	0.3	114.08	13JL	-1.5	-4.2	-2.4	0.4
207.83	-1.25	0.4	123.68	23JL	-0.7	-4.1	-2.4	0.5
213.84	-1.37	0.5	133.31	2AU	-0.2	-3.7	-2.4	0.5
220.13	-1.46	0.6	142.98	12AU	0.2	-3.2	-2.4	0.6
226.70	-1.53	0.7	152.71	22AU	0.4	-3.6	-2.4	0.6
233.48	-1.59	0.7	162.48	1SE	0.7	-4.1	-2.4	0.7
240.47	-1.62	0.8	172.32	11SE	1.7	-4.3	-2.3	0.7
247.63	-1.63	0.9	182.22	21SE	2.4	-4.3	-2.3	0.8
254.94	-1.63	0.9	192.17	1OC	-0.0	-4.3	-2.2	0.8
262.37	-1.61	1.0	202.17	11OC	-0.7	-4.2	-2.1	0.8
269.91	-1.57	1.0	212.24	21OC	-0.7	-4.1	-2.1	0.8
277.53	-1.52	1.1	222.34	31OC	-0.7	-4.0	-2.0	0.8
285.22	-1.46	1.1	232.48	10NO	-0.8	-3.9	-1.9	0.8
292.95	-1.38	1.2	242.66	20NO	-0.6	-3.8	-1.9	0.8
300.71	-1.30	1.2	252.85	30NO	-0.6	-3.7	-1.8	0.8
308.47	-1.20	1.3	263.05	10DE	-0.6	-3.7	-1.8	0.7
316.22	-1.10	1.3	273.25	20DE	0.0	-3.6	-1.7	0.7
323.95	-0.99	1.4	283.44	30DE	2.4	-3.5	-1.7	0.8
				753				
331.65	-0.87	1.4	293.60	9JA	1.4	-3.5	-1.6	0.8
339.29	-0.76	1.4	303.74	19JA	0.5	-3.4	-1.6	0.8
346.88	-0.64	1.5	313.83	29JA	0.2	-3.4	-1.6	0.8
354.40	-0.52	1.5	323.88	8FE	0.1	-3.4	-1.6	0.9
1.86	-0.40	1.6	333.87	18FE	-0.2	-3.3	-1.6	0.9
9.23	-0.28	1.6	343.80	28FE	-0.7	-3.3	-1.6	0.9
16.53	-0.17	1.6	353.68	10MR	-1.5	-3.3	-1.6	0.8
23.75	-0.06	1.7	3.51	20MR	-1.4	-3.3	-1.6	0.8
30.88	0.05	1.7	13.27	30MR	-0.5	-3.3	-1.6	0.8
37.94	0.16	1.7	22.98	9AP	0.7	-3.3	-1.7	0.7
44.92	0.26	1.7	32.64	19AP	2.0	-3.3	-1.7	0.7
51.82	0.35	1.8	42.26	29AP	3.6	-3.4	-1.7	0.6
58.66	0.44	1.8	51.84	9MY	2.5	-3.4	-1.8	0.6
65.42	0.53	1.8	61.40	19MY	1.4	-3.4	-1.8	0.5
72.12	0.62	1.8	70.93	29MY	0.5	-3.5	-1.9	0.4
78.77	0.70	1.8	80.46	8JN	-0.5	-3.5	-2.0	0.4
85.35	0.77	1.9	89.98	18JN	-1.4	-3.5	-2.0	0.3
91.89	0.84	1.9	99.51	28JN	-1.5	-3.4	-2.1	0.3
98.38	0.91	1.9	109.07	8JL	-0.7	-3.4	-2.2	0.4
104.83	0.97	2.0	118.64	18JL	-0.1	-3.4	-2.2	0.4
111.24	1.03	2.0	128.25	28JL	0.3	-3.3	-2.3	0.5
117.62	1.09	2.0	137.90	7AU	0.5	-3.3	-2.4	0.5
123.96	1.14	2.0	147.60	17AU	0.9	-3.3	-2.4	0.6
130.27	1.19	2.0	157.35	27AU	2.1	-3.3	-2.4	0.6
136.55	1.24	2.0	167.16	6SE	2.1	-3.3	-2.5	0.7
142.79	1.28	2.0	177.02	16SE	-0.1	-3.3	-2.5	0.7
149.01	1.33	2.0	186.94	26SE	-0.8	-3.4	-2.4	0.8
155.20	1.36	1.9	196.92	6OC	-0.9	-3.4	-2.4	0.8
161.35	1.40	1.9	206.95	16OC	-0.9	-3.4	-2.4	0.8
167.46	1.43	1.8	217.03	26OC	-0.7	-3.5	-2.3	0.8
173.52	1.46	1.8	227.16	5NO	-0.5	-3.5	-2.3	0.8
179.53	1.48	1.7	237.32	15NO	-0.5	-3.6	-2.2	0.8
185.48	1.51	1.6	247.50	25NO	-0.4	-3.6	-2.1	0.8
191.35	1.52	1.5	257.70	5DE	0.2	-3.7	-2.0	0.8
197.12	1.53	1.4	267.90	15DE	2.7	-3.8	-2.0	0.8
202.79	1.53	1.3	278.10	25DE	1.1	-3.9	-1.9	0.8

♂			☉	16.00UT	☿	♀	♃	♄
LONG	LAT	MAG	LONG	754	MAGNITUDES			
208.32	1.52	1.2	288.28	4JA	0.3	-4.0	-1.8	0.8
213.69	1.51	1.0	298.43	14JA	0.1	-4.1	-1.8	0.8
218.86	1.47	0.8	308.54	24JA	-0.1	-4.2	-1.7	0.9
223.78	1.42	0.6	318.62	3FE	-0.3	-4.2	-1.7	0.9
228.39	1.35	0.4	328.63	13FE	-0.7	-4.3	-1.6	0.9
232.62	1.25	0.2	338.60	23FE	-1.4	-4.3	-1.6	0.9
236.36	1.11	-0.0	348.51	5MR	-1.3	-4.1	-1.6	0.9
239.50	0.92	-0.3	358.36	15MR	-0.5	-3.7	-1.6	0.9
241.87	0.68	-0.6	8.15	25MR	0.9	-3.1	-1.6	0.9
243.29	0.36	-0.9	17.89	4AP	2.5	-3.6	-1.6	0.9
243.59	-0.05	-1.2	27.57	14AP	3.3	-4.1	-1.6	0.8
242.66	-0.55	-1.6	37.21	24AP	1.9	-4.2	-1.6	0.8
240.52	-1.12	-1.9	46.82	4MY	1.0	-4.2	-1.6	0.7
237.53	-1.72	-2.1	56.39	14MY	0.4	-4.1	-1.6	0.7
234.28	-2.28	-2.1	65.93	24MY	-0.5	-4.0	-1.7	0.6
231.51	-2.74	-1.9	75.47	3JN	-1.4	-3.9	-1.7	0.5
229.83	-3.07	-1.7	84.99	13JN	-1.5	-3.8	-1.7	0.5
229.49	-3.27	-1.5	94.52	23JN	-0.6	-3.8	-1.8	0.4
230.52	-3.37	-1.3	104.06	3JL	0.0	-3.7	-1.9	0.4
232.79	-3.40	-1.1	113.62	13JL	0.4	-3.6	-1.9	0.4
236.10	-3.36	-0.9	123.21	23JL	0.7	-3.6	-2.0	0.4
240.28	-3.29	-0.7	132.84	2AU	1.3	-3.5	-2.1	0.5
245.15	-3.18	-0.5	142.51	12AU	2.6	-3.5	-2.1	0.5
250.58	-3.05	-0.3	152.23	22AU	1.9	-3.4	-2.2	0.6
256.48	-2.90	-0.2	162.01	1SE	-0.2	-3.4	-2.2	0.6
262.73	-2.73	-0.0	171.84	11SE	-1.0	-3.4	-2.3	0.7
269.28	-2.55	0.1	181.74	21SE	-1.1	-3.4	-2.4	0.7
276.06	-2.36	0.2	191.69	1OC	-0.9	-3.4	-2.4	0.8
283.03	-2.17	0.3	201.69	11OC	-0.6	-3.4	-2.4	0.8
290.13	-1.97	0.5	211.75	21OC	-0.4	-3.4	-2.4	0.8
297.34	-1.77	0.6	221.85	31OC	-0.3	-3.4	-2.4	0.9
304.61	-1.56	0.7	231.99	10NO	-0.3	-3.4	-2.4	0.9
311.94	-1.37	0.8	242.16	20NO	0.3	-3.4	-2.3	0.9
319.28	-1.17	0.9	252.36	30NO	3.0	-3.4	-2.3	0.9
326.63	-0.98	1.0	262.55	10DE	0.8	-3.4	-2.2	0.9
333.97	-0.80	1.1	272.76	20DE	0.1	-3.4	-2.1	0.9
341.28	-0.63	1.2	282.95	30DE	-0.1	-3.5	-2.1	0.8
				755				
348.55	-0.47	1.3	293.11	9JA	-0.2	-3.5	-2.0	0.8
355.77	-0.31	1.4	303.25	19JA	-0.4	-3.5	-1.9	0.9
2.95	-0.16	1.4	313.34	29JA	-0.8	-3.4	-1.8	0.9
10.06	-0.03	1.5	323.38	8FE	-1.3	-3.4	-1.8	0.9
17.12	0.10	1.6	333.38	18FE	-1.3	-3.4	-1.7	1.0
24.11	0.22	1.6	343.32	28FE	-0.4	-3.4	-1.7	1.0
31.04	0.33	1.7	353.20	10MR	1.1	-3.4	-1.6	1.0
37.91	0.43	1.8	3.02	20MR	3.1	-3.3	-1.6	1.0
44.71	0.52	1.8	12.79	30MR	2.4	-3.3	-1.6	1.0
51.46	0.61	1.8	22.50	9AP	1.4	-3.3	-1.5	1.0
58.15	0.69	1.9	32.17	19AP	0.8	-3.3	-1.5	0.9
64.79	0.76	1.9	41.79	29AP	0.2	-3.4	-1.5	0.9
71.38	0.82	1.9	51.37	9MY	-0.5	-3.4	-1.5	0.9
77.93	0.88	1.9	60.93	19MY	-1.5	-3.4	-1.5	0.8
84.44	0.93	2.0	70.47	29MY	-1.5	-3.4	-1.5	0.7
90.92	0.98	2.0	79.99	8JN	-0.6	-3.5	-1.5	0.7
97.36	1.02	2.0	89.52	18JN	0.1	-3.5	-1.5	0.6
103.78	1.06	1.9	99.05	28JN	0.6	-3.6	-1.6	0.5
110.18	1.09	1.9	108.60	8JL	1.0	-3.6	-1.6	0.5
116.57	1.12	1.9	118.18	18JL	1.7	-3.7	-1.7	0.4
122.94	1.14	2.0	127.79	28JL	3.1	-3.8	-1.7	0.5
129.31	1.16	2.0	137.43	7AU	1.6	-3.9	-1.8	0.5
135.68	1.17	2.0	147.13	17AU	-0.2	-4.0	-1.8	0.5
142.04	1.18	2.0	156.88	27AU	-1.1	-4.1	-1.9	0.6
148.40	1.19	2.0	166.68	6SE	-1.2	-4.2	-1.9	0.7
154.78	1.19	2.0	176.54	16SE	-0.9	-4.3	-2.0	0.7
161.15	1.18	2.0	186.46	26SE	-0.5	-4.3	-2.1	0.8
167.54	1.17	2.0	196.44	6OC	-0.3	-4.2	-2.1	0.8
173.94	1.16	2.0	206.47	16OC	-0.2	-3.8	-2.2	0.8
180.34	1.14	1.9	216.55	26OC	-0.1	-3.1	-2.2	0.9
186.75	1.11	1.9	226.67	5NO	0.6	-3.7	-2.3	0.9
193.17	1.08	1.8	236.83	15NO	3.3	-4.2	-2.3	0.9
199.60	1.04	1.8	247.01	25NO	0.6	-4.4	-2.3	0.9
206.03	0.99	1.7	257.20	5DE	-0.1	-4.4	-2.3	1.0
212.46	0.93	1.6	267.41	15DE	-0.2	-4.3	-2.3	1.0
218.89	0.86	1.5	277.60	25DE	-0.3	-4.2	-2.2	1.0

♂			⊙	16.00UT	☿	♀	♃	♄
LONG	LAT	MAG	LONG	756	MAGNITUDES			
225.31	0.78	1.5	287.78	4JA	-0.5	-4.1	-2.2	0.9
231.71	0.68	1.3	297.94	14JA	-0.8	-4.0	-2.1	0.9
238.09	0.57	1.2	308.05	24JA	-1.2	-3.9	-2.0	0.9
244.45	0.44	1.1	318.13	3FE	-1.1	-3.8	-2.0	1.0
250.77	0.29	1.0	328.15	13FE	-0.4	-3.7	-1.9	1.0
257.03	0.12	0.8	338.12	23FE	1.4	-3.6	-1.8	1.1
263.24	-0.09	0.7	348.03	4MR	3.3	-3.6	-1.7	1.1
269.35	-0.32	0.5	357.89	14MR	1.8	-3.5	-1.7	1.1
275.34	-0.59	0.3	7.68	24MR	1.0	-3.4	-1.6	1.1
281.20	-0.89	0.1	17.42	3AP	0.6	-3.4	-1.6	1.1
286.85	-1.24	-0.1	27.11	13AP	0.1	-3.4	-1.5	1.1
292.24	-1.64	-0.3	36.75	23AP	-0.5	-3.3	-1.5	1.1
297.31	-2.10	-0.5	46.36	3MY	-1.5	-3.3	-1.5	1.1
301.94	-2.61	-0.8	55.93	13MY	-1.5	-3.3	-1.4	1.0
306.02	-3.19	-1.0	65.47	23MY	-0.6	-3.3	-1.4	1.0
309.37	-3.83	-1.3	75.01	2JN	0.2	-3.3	-1.4	0.9
311.79	-4.52	-1.6	84.53	12JN	0.7	-3.3	-1.4	0.8
313.12	-5.24	-1.9	94.06	22JN	1.3	-3.3	-1.4	0.8
313.17	-5.93	-2.2	103.60	2JL	2.2	-3.4	-1.4	0.7
311.95	-6.50	-2.4	113.16	12JL	3.2	-3.4	-1.4	0.6
309.75	-6.84	-2.6	122.75	22JL	1.3	-3.4	-1.4	0.6
307.12	-6.86	-2.6	132.37	1AU	-0.3	-3.5	-1.5	0.5
304.84	-6.56	-2.4	142.04	11AU	-1.2	-3.5	-1.5	0.6
303.52	-6.02	-2.2	151.76	21AU	-1.3	-3.5	-1.6	0.6
303.46	-5.34	-1.9	161.54	31AU	-0.8	-3.4	-1.6	0.7
304.71	-4.63	-1.6	171.37	10SE	-0.4	-3.4	-1.7	0.7
307.11	-3.93	-1.3	181.25	20SE	-0.1	-3.4	-1.7	0.8
310.48	-3.29	-1.0	191.20	30SE	-0.0	-3.4	-1.8	0.8
314.61	-2.71	-0.8	201.20	10OC	0.1	-3.3	-1.8	0.9
319.34	-2.20	-0.5	211.25	20OC	0.8	-3.3	-1.9	0.9
324.53	-1.75	-0.3	221.35	30OC	3.3	-3.3	-2.0	0.9
330.07	-1.35	-0.1	231.49	9NO	0.4	-3.3	-2.0	1.0
335.88	-1.00	0.1	241.66	19NO	-0.3	-3.3	-2.1	1.0
341.88	-0.69	0.3	251.86	29NO	-0.4	-3.4	-2.1	1.0
348.04	-0.42	0.5	262.06	9DE	-0.4	-3.4	-2.2	1.1
354.31	-0.19	0.7	272.26	19DE	-0.6	-3.4	-2.2	1.1
0.65	0.02	0.8	282.45	29DE	-0.9	-3.4	-2.2	1.1
				757				
7.05	0.20	1.0	292.62	8JA	-1.0	-3.5	-2.2	1.1
13.47	0.35	1.1	302.75	18JA	-1.0	-3.5	-2.1	1.1
19.92	0.49	1.2	312.85	28JA	-0.3	-3.6	-2.1	1.1
26.38	0.61	1.4	322.90	7FE	1.6	-3.6	-2.0	1.1
32.83	0.71	1.5	332.90	17FE	2.8	-3.7	-2.0	1.1
39.27	0.80	1.6	342.84	27FE	1.3	-3.8	-1.9	1.2
45.70	0.88	1.6	352.73	9MR	0.7	-3.8	-1.8	1.2
52.12	0.94	1.7	2.55	19MR	0.4	-3.9	-1.8	1.2
58.51	1.00	1.8	12.32	29MR	0.0	-4.0	-1.7	1.3
64.89	1.05	1.8	22.04	8AP	-0.6	-4.1	-1.6	1.3
71.25	1.08	1.9	31.70	18AP	-1.5	-4.2	-1.6	1.3
77.60	1.12	1.9	41.33	28AP	-1.5	-4.2	-1.5	1.3
83.93	1.14	2.0	50.91	8MY	-0.6	-4.1	-1.5	1.2
90.25	1.16	2.0	60.47	18MY	0.3	-3.8	-1.4	1.2
96.56	1.17	2.0	70.01	28MY	1.0	-3.1	-1.4	1.2
102.87	1.18	2.0	79.53	7JN	1.7	-3.1	-1.3	1.1
109.18	1.18	2.0	89.06	17JN	2.9	-3.8	-1.3	1.0
115.50	1.18	2.0	98.59	27JN	2.7	-4.1	-1.3	1.0
121.82	1.17	2.0	108.14	7JL	1.1	-4.2	-1.3	0.9
128.16	1.16	2.0	117.71	17JL	-0.3	-4.2	-1.3	0.8
134.51	1.15	2.0	127.32	27JL	-1.3	-4.1	-1.3	0.8
140.88	1.13	2.0	136.97	6AU	-1.4	-4.0	-1.3	0.7
147.28	1.10	1.9	146.66	16AU	-0.8	-3.9	-1.3	0.7
153.71	1.07	1.9	156.41	26AU	-0.3	-3.9	-1.3	0.7
160.16	1.04	2.0	166.20	5SE	-0.0	-3.8	-1.4	0.8
166.66	1.00	2.0	176.06	15SE	0.1	-3.7	-1.4	0.8
173.19	0.96	2.0	185.98	25SE	0.3	-3.6	-1.5	0.8
179.76	0.91	2.0	195.95	5OC	1.1	-3.6	-1.5	0.9
186.38	0.85	1.9	205.98	15OC	3.0	-3.5	-1.6	1.0
193.04	0.79	1.9	216.06	25OC	0.2	-3.5	-1.6	1.0
199.74	0.72	1.9	226.18	4NO	-0.5	-3.5	-1.7	1.0
206.49	0.65	1.9	236.33	14NO	-0.5	-3.4	-1.8	1.1
213.29	0.57	1.8	246.52	24NO	-0.5	-3.4	-1.8	1.1
220.13	0.48	1.8	256.71	4DE	-0.7	-3.4	-1.9	1.2
227.02	0.38	1.7	266.91	14DE	-0.8	-3.4	-1.9	1.2
233.96	0.28	1.7	277.11	24DE	-0.9	-3.4	-2.0	1.2

♂			⊙	16.00UT	☿	♀	♃	♄
LONG	LAT	MAG	LONG	758	MAGNITUDES			
240.95	0.16	1.6	287.29	3JA	-0.9	-3.3	-2.0	1.2
247.97	0.04	1.5	297.45	13JA	-0.2	-3.3	-2.1	1.2
255.04	-0.10	1.4	307.57	23JA	1.9	-3.3	-2.1	1.2
262.15	-0.24	1.4	317.64	2FE	2.2	-3.4	-2.1	1.2
269.30	-0.40	1.3	327.67	12FE	0.9	-3.4	-2.0	1.2
276.47	-0.57	1.2	337.64	22FE	0.5	-3.4	-2.0	1.3
283.67	-0.74	1.1	347.55	4MR	0.3	-3.4	-2.0	1.3
290.89	-0.93	1.0	357.41	14MR	-0.1	-3.5	-1.9	1.4
298.11	-1.12	0.9	7.21	24MR	-0.6	-3.5	-1.9	1.4
305.33	-1.32	0.8	16.95	3AP	-1.4	-3.5	-1.8	1.4
312.53	-1.53	0.7	26.64	13AP	-1.5	-3.4	-1.7	1.4
319.68	-1.73	0.6	36.28	23AP	-0.6	-3.4	-1.7	1.4
326.78	-1.94	0.5	45.89	3MY	0.4	-3.4	-1.6	1.4
333.78	-2.15	0.3	55.46	13MY	1.3	-3.4	-1.5	1.4
340.66	-2.36	0.2	65.01	23MY	2.3	-3.3	-1.5	1.3
347.37	-2.56	0.1	74.54	2JN	3.5	-3.3	-1.4	1.3
353.88	-2.75	-0.0	84.06	12JN	2.1	-3.3	-1.4	1.2
0.11	-2.93	-0.2	93.59	22JN	0.8	-3.3	-1.3	1.2
6.01	-3.09	-0.3	103.13	2JL	-0.4	-3.3	-1.3	1.1
11.47	-3.24	-0.5	112.69	12JL	-1.3	-3.4	-1.3	1.1
16.38	-3.36	-0.7	122.28	22JL	-1.5	-3.4	-1.2	1.0
20.62	-3.46	-0.8	131.90	1AU	-0.7	-3.4	-1.2	1.0
24.00	-3.52	-1.0	141.57	11AU	-0.2	-3.4	-1.2	0.9
26.31	-3.53	-1.3	151.29	21AU	0.1	-3.5	-1.2	0.9
27.35	-3.46	-1.5	161.06	31AU	0.3	-3.5	-1.2	0.9
26.94	-3.29	-1.7	170.89	10SE	0.5	-3.6	-1.2	0.9
25.09	-2.99	-1.9	180.77	20SE	1.4	-3.7	-1.3	0.9
22.11	-2.54	-2.1	190.72	30SE	2.7	-3.8	-1.3	1.0
18.62	-1.98	-2.1	200.72	10OC	0.1	-3.9	-1.3	1.0
15.48	-1.37	-1.8	210.77	20OC	-0.6	-4.0	-1.4	1.1
13.34	-0.78	-1.5	220.87	30OC	-0.7	-4.1	-1.4	1.1
12.51	-0.27	-1.2	231.01	9NO	-0.7	-4.2	-1.5	1.2
13.01	0.16	-0.9	241.17	19NO	-0.8	-4.3	-1.5	1.2
14.68	0.49	-0.6	251.36	29NO	-0.7	-4.4	-1.6	1.3
17.31	0.75	-0.3	261.57	9DE	-0.7	-4.4	-1.7	1.3
20.71	0.95	0.0	271.77	19DE	-0.7	-4.3	-1.7	1.3
24.69	1.11	0.3	281.95	29DE	-0.1	-3.8	-1.8	1.4
				759				
29.14	1.22	0.5	292.13	8JA	2.1	-3.2	-1.9	1.4
33.95	1.31	0.7	302.26	18JA	1.7	-3.7	-1.9	1.4
39.02	1.37	0.9	312.36	28JA	0.6	-4.2	-2.0	1.4
44.32	1.42	1.0	322.42	7FE	0.3	-4.3	-2.0	1.3
49.78	1.45	1.2	332.41	17FE	0.1	-4.3	-2.0	1.3
55.37	1.47	1.3	342.36	27FE	-0.1	-4.2	-2.0	1.2
61.07	1.48	1.4	352.25	9MR	-0.7	-4.1	-2.0	1.2
66.85	1.48	1.5	2.07	19MR	-1.4	-4.0	-2.0	1.2
72.70	1.47	1.6	11.85	29MR	-1.5	-3.9	-1.9	1.2
78.60	1.46	1.7	21.57	8AP	-0.5	-3.8	-1.9	1.2
84.55	1.44	1.8	31.23	18AP	0.6	-3.7	-1.8	1.2
90.55	1.42	1.8	40.86	28AP	1.7	-3.6	-1.8	1.2
96.58	1.40	1.9	50.45	8MY	3.1	-3.6	-1.7	1.1
102.65	1.37	1.9	60.01	18MY	3.0	-3.5	-1.7	1.1
108.75	1.33	2.0	69.55	28MY	1.6	-3.5	-1.6	1.1
114.89	1.29	2.0	79.08	7JN	0.6	-3.4	-1.5	1.0
121.06	1.25	2.0	88.60	17JN	-0.4	-3.4	-1.5	1.0
127.27	1.21	2.0	98.13	27JN	-1.4	-3.4	-1.4	0.9
133.53	1.16	2.0	107.68	7JL	-1.5	-3.3	-1.4	0.9
139.83	1.11	2.0	117.25	17JL	-0.7	-3.3	-1.3	0.8
146.17	1.05	2.0	126.86	27JL	-0.1	-3.3	-1.3	0.8
152.56	1.00	2.0	136.50	6AU	0.2	-3.3	-1.3	0.7
159.01	0.93	2.0	146.19	16AU	0.4	-3.3	-1.2	0.7
165.51	0.87	1.9	155.93	26AU	0.8	-3.4	-1.2	0.7
172.08	0.80	1.9	165.73	5SE	1.8	-3.4	-1.2	0.6
178.70	0.73	1.8	175.58	15SE	2.4	-3.4	-1.2	0.7
185.39	0.65	1.8	185.50	25SE	-0.0	-3.4	-1.2	0.7
192.15	0.57	1.8	195.47	5OC	-0.8	-3.5	-1.2	0.8
198.98	0.49	1.8	205.49	15OC	-0.8	-3.5	-1.2	0.9
205.87	0.40	1.8	215.57	25OC	-0.8	-3.5	-1.3	0.9
212.84	0.30	1.8	225.69	4NO	-0.7	-3.5	-1.3	1.0
219.88	0.21	1.8	235.84	14NO	-0.6	-3.4	-1.3	1.0
226.99	0.10	1.7	246.02	24NO	-0.6	-3.4	-1.4	1.1
234.17	-0.00	1.7	256.22	4DE	-0.5	-3.4	-1.4	1.1
241.42	-0.11	1.7	266.42	14DE	0.0	-3.4	-1.5	1.1
248.74	-0.23	1.6	276.61	24DE	2.4	-3.4	-1.6	1.2

Left section:

♂ LONG	LAT	MAG	☉ LONG	16.00UT 760	☿	♀	♃	♄
256.11	-0.34	1.6	286.80	3JA	1.3	-3.3	-1.6	1.2
263.55	-0.46	1.6	296.95	13JA	0.4	-3.3	-1.7	1.1
271.04	-0.59	1.5	307.07	23JA	0.2	-3.3	-1.8	1.1
278.58	-0.71	1.5	317.15	2FE	0.0	-3.3	-1.8	1.1
286.17	-0.83	1.4	327.18	12FE	-0.2	-3.4	-1.9	1.1
293.78	-0.95	1.4	337.15	22FE	-0.7	-3.4	-1.9	1.0
301.42	-1.07	1.3	347.07	3MR	-1.4	-3.4	-2.0	1.0
309.07	-1.18	1.3	356.93	13MR	-1.4	-3.4	-2.0	0.9
316.73	-1.28	1.2	6.73	23MR	-0.5	-3.5	-2.0	0.9
324.37	-1.38	1.2	16.48	2AP	0.7	-3.5	-2.0	0.9
331.98	-1.46	1.1	26.17	12AP	2.1	-3.6	-2.0	0.9
339.56	-1.53	1.1	35.82	22AP	3.9	-3.6	-2.0	0.9
347.07	-1.59	1.0	45.43	2MY	2.2	-3.7	-1.9	0.9
354.52	-1.64	1.0	55.00	12MY	1.3	-3.8	-1.9	0.9
1.87	-1.67	0.9	64.55	22MY	0.5	-3.8	-1.8	0.9
9.11	-1.68	0.9	74.08	1JN	-0.4	-3.9	-1.8	0.8
16.23	-1.67	0.8	83.61	11JN	-1.4	-4.0	-1.7	0.8
23.19	-1.65	0.7	93.14	21JN	-1.5	-4.1	-1.7	0.8
29.97	-1.60	0.7	102.67	1JL	-0.7	-4.2	-1.6	0.7
36.55	-1.54	0.6	112.23	11JL	-0.1	-4.2	-1.5	0.7
42.88	-1.45	0.5	121.82	21JL	0.3	-4.1	-1.5	0.6
48.94	-1.35	0.4	131.44	31JL	0.6	-3.7	-1.4	0.6
54.67	-1.22	0.3	141.10	10AU	1.1	-3.2	-1.4	0.5
60.00	-1.06	0.2	150.82	20AU	2.3	-3.6	-1.3	0.5
64.85	-0.88	0.1	160.59	30AU	2.1	-4.1	-1.3	0.4
69.13	-0.66	-0.1	170.41	9SE	-0.1	-4.3	-1.3	0.4
72.69	-0.40	-0.2	180.29	19SE	-0.9	-4.3	-1.3	0.3
75.37	-0.10	-0.4	190.23	29SE	-1.0	-4.3	-1.2	0.4
76.98	0.25	-0.6	200.23	9OC	-0.9	-4.2	-1.2	0.4
77.30	0.66	-0.8	210.28	19OC	-0.6	-4.1	-1.2	0.5
76.21	1.12	-1.1	220.38	29OC	-0.5	-4.0	-1.2	0.6
73.73	1.59	-1.3	230.51	8NO	-0.4	-3.9	-1.3	0.7
70.19	2.04	-1.4	240.68	18NO	-0.4	-3.8	-1.3	0.7
66.27	2.41	-1.4	250.87	28NO	0.2	-3.7	-1.3	0.8
62.77	2.65	-1.2	261.07	8DE	2.7	-3.7	-1.4	0.8
60.30	2.77	-0.9	271.28	18DE	1.0	-3.6	-1.4	0.8
59.16	2.79	-0.6	281.47	28DE	0.2	-3.5	-1.5	0.9

761

♂ LONG	LAT	MAG	☉ LONG	16.00UT	☿	♀	♃	♄
59.32	2.75	-0.3	291.64	7JA	0.0	-3.5	-1.5	0.9
60.64	2.67	0.0	301.78	17JA	-0.1	-3.4	-1.6	0.9
62.92	2.58	0.3	311.88	27JA	-0.3	-3.4	-1.6	0.9
65.96	2.47	0.5	321.93	6FE	-0.7	-3.4	-1.7	0.9
69.60	2.36	0.7	331.94	16FE	-1.4	-3.3	-1.8	0.8
73.71	2.26	0.9	341.88	26FE	-1.3	-3.3	-1.8	0.8
78.19	2.15	1.1	351.77	8MR	-0.5	-3.3	-1.9	0.7
82.97	2.05	1.2	1.61	18MR	0.9	-3.3	-2.0	0.7
87.99	1.95	1.3	11.38	28MR	2.7	-3.3	-2.0	0.7
93.21	1.85	1.4	21.10	7AP	2.9	-3.3	-2.1	0.7
98.59	1.76	1.5	30.77	17AP	1.7	-3.3	-2.1	0.7
104.11	1.67	1.6	40.40	27AP	0.9	-3.4	-2.1	0.7
109.75	1.57	1.7	49.99	7MY	0.3	-3.4	-2.1	0.7
115.50	1.48	1.8	59.55	17MY	-0.5	-3.4	-2.1	0.7
121.34	1.39	1.8	69.09	27MY	-1.4	-3.5	-2.0	0.7
127.27	1.30	1.8	78.61	6JN	-1.6	-3.5	-2.0	0.7
133.29	1.22	1.9	88.14	16JN	-0.6	-3.5	-2.0	0.6
139.39	1.13	1.9	97.67	26JN	0.0	-3.4	-1.9	0.6
145.57	1.04	1.9	107.22	6JL	0.5	-3.4	-1.8	0.6
151.83	0.94	1.9	116.79	16JL	0.8	-3.4	-1.8	0.5
158.17	0.85	1.9	126.39	26JL	1.4	-3.3	-1.7	0.5
164.60	0.76	1.9	136.03	5AU	2.8	-3.3	-1.6	0.4
171.11	0.66	1.9	145.72	15AU	1.8	-3.3	-1.6	0.4
177.70	0.57	1.9	155.45	25AU	-0.2	-3.3	-1.5	0.3
184.38	0.47	1.8	165.25	4SE	-1.0	-3.3	-1.5	0.2
191.14	0.37	1.8	175.10	14SE	-1.1	-3.3	-1.4	0.2
198.00	0.27	1.8	185.01	24SE	-0.9	-3.4	-1.4	0.1
204.95	0.17	1.7	194.98	4OC	-0.5	-3.4	-1.4	0.1
211.98	0.07	1.7	205.00	14OC	-0.3	-3.4	-1.3	0.2
219.11	-0.03	1.6	215.07	24OC	-0.3	-3.5	-1.3	0.2
226.32	-0.14	1.6	225.19	3NO	-0.2	-3.5	-1.3	0.3
233.62	-0.24	1.5	235.35	13NO	0.4	-3.6	-1.3	0.4
241.00	-0.35	1.5	245.52	23NO	3.0	-3.6	-1.3	0.4
248.46	-0.45	1.5	255.72	3DE	0.7	-3.7	-1.3	0.5
255.99	-0.55	1.5	265.93	13DE	-0.0	-3.8	-1.4	0.5
263.59	-0.65	1.5	276.12	23DE	-0.1	-3.9	-1.4	0.6

Right section:

♂ LONG	LAT	MAG	☉ LONG	16.00UT 762	☿	♀	♃	♄
271.26	-0.74	1.5	286.31	2JA	-0.2	-4.0	-1.4	0.6
278.97	-0.83	1.5	296.46	12JA	-0.4	-4.1	-1.5	0.6
286.73	-0.91	1.5	306.59	22JA	-0.8	-4.2	-1.5	0.7
294.52	-0.99	1.4	316.67	1FE	-1.3	-4.3	-1.6	0.7
302.32	-1.05	1.4	326.70	11FE	-1.2	-4.3	-1.6	0.7
310.14	-1.11	1.4	336.67	21FE	-0.4	-4.3	-1.7	0.6
317.96	-1.16	1.4	346.60	3MR	1.2	-4.1	-1.7	0.6
325.76	-1.19	1.4	356.46	13MR	3.3	-3.6	-1.8	0.6
333.53	-1.21	1.4	6.26	23MR	2.2	-3.1	-1.9	0.5
341.27	-1.22	1.4	16.01	2AP	1.2	-3.7	-2.0	0.5
348.95	-1.22	1.4	25.70	12AP	0.7	-4.1	-2.0	0.5
356.57	-1.20	1.3	35.35	22AP	0.2	-4.2	-2.1	0.5
4.12	-1.17	1.3	44.96	2MY	-0.5	-4.2	-2.1	0.5
11.58	-1.13	1.3	54.54	12MY	-1.4	-4.1	-2.2	0.5
18.94	-1.07	1.3	64.09	22MY	-1.6	-4.0	-2.2	0.5
26.21	-1.00	1.3	73.62	1JN	-0.6	-3.9	-2.2	0.5
33.35	-0.92	1.3	83.15	11JN	0.1	-3.8	-2.2	0.5
40.38	-0.82	1.3	92.67	21JN	0.6	-3.8	-2.2	0.5
47.27	-0.71	1.2	102.21	1JL	1.1	-3.7	-2.2	0.5
54.02	-0.60	1.2	111.77	11JL	1.9	-3.6	-2.1	0.4
60.62	-0.47	1.2	121.35	21JL	3.2	-3.5	-2.1	0.4
67.04	-0.32	1.1	130.97	31JL	1.5	-3.5	-2.0	0.4
73.27	-0.17	1.1	140.64	10AU	-0.2	-3.5	-1.9	0.3
79.30	0.00	1.0	150.35	20AU	-1.1	-3.4	-1.9	0.3
85.08	0.18	1.0	160.11	30AU	-1.3	-3.4	-1.8	0.3
90.58	0.38	0.9	169.93	9SE	-0.8	-3.4	-1.7	0.1
95.76	0.60	0.8	179.81	19SE	-0.4	-3.4	-1.7	0.1
100.55	0.85	0.7	189.75	29SE	-0.2	-3.4	-1.6	0.0
104.86	1.12	0.5	199.74	9OC	-0.1	-3.4	-1.6	-0.0
108.61	1.42	0.4	209.79	19OC	-0.0	-3.4	-1.5	-0.1
111.64	1.76	0.2	219.89	29OC	0.6	-3.4	-1.5	-0.0
113.79	2.14	-0.0	230.02	8NO	3.3	-3.4	-1.5	0.0
114.89	2.56	-0.2	240.18	18NO	0.5	-3.4	-1.4	0.1
114.73	3.00	-0.5	250.38	28NO	-0.2	-3.4	-1.4	0.2
113.21	3.45	-0.7	260.57	8DE	-0.3	-3.4	-1.4	0.3
110.41	3.84	-0.9	270.78	18DE	-0.3	-3.4	-1.4	0.3
106.69	4.13	-1.1	280.97	28DE	-0.5	-3.5	-1.4	0.4

763

♂ LONG	LAT	MAG	☉ LONG	16.00UT	☿	♀	♃	♄
102.72	4.25	-1.0	291.14	7JA	-0.8	-3.5	-1.4	0.4
99.24	4.20	-0.8	301.28	17JA	-1.1	-3.5	-1.5	0.5
96.82	4.02	-0.5	311.39	27JA	-1.1	-3.4	-1.5	0.5
95.70	3.77	-0.3	321.44	6FE	-0.4	-3.4	-1.5	0.5
95.84	3.48	-0.0	331.45	16FE	1.4	-3.4	-1.6	0.5
97.12	3.19	0.2	341.40	26FE	3.2	-3.4	-1.6	0.5
99.32	2.92	0.4	351.29	8MR	1.6	-3.4	-1.7	0.5
102.28	2.66	0.6	1.13	18MR	0.9	-3.3	-1.7	0.4
105.85	2.43	0.8	10.91	28MR	0.5	-3.3	-1.8	0.4
109.90	2.21	1.0	20.63	7AP	0.1	-3.3	-1.9	0.4
114.34	2.01	1.1	30.30	17AP	-0.5	-3.3	-1.9	0.3
119.11	1.83	1.2	39.93	27AP	-1.4	-3.4	-2.0	0.3
124.15	1.65	1.3	49.52	7MY	-1.6	-3.4	-2.1	0.3
129.41	1.49	1.4	59.08	17MY	-0.6	-3.4	-2.1	0.4
134.88	1.33	1.5	68.63	27MY	0.2	-3.4	-2.2	0.4
140.51	1.19	1.6	78.15	6JN	0.8	-3.5	-2.3	0.4
146.31	1.05	1.6	87.68	16JN	1.4	-3.5	-2.3	0.4
152.25	0.91	1.6	97.21	26JN	2.5	-3.6	-2.3	0.4
158.33	0.78	1.7	106.75	6JL	3.0	-3.6	-2.3	0.4
164.53	0.65	1.7	116.32	16JL	1.3	-3.7	-2.3	0.4
170.86	0.52	1.7	125.93	26JL	-0.3	-3.8	-2.3	0.3
177.31	0.40	1.7	135.56	5AU	-1.2	-3.9	-2.3	0.3
183.88	0.28	1.7	145.25	15AU	-1.4	-4.0	-2.2	0.3
190.56	0.16	1.7	154.99	25AU	-0.8	-4.1	-2.2	0.2
197.36	0.04	1.7	164.78	4SE	-0.3	-4.2	-2.1	0.1
204.27	-0.07	1.7	174.63	14SE	-0.1	-4.3	-2.1	0.1
211.29	-0.18	1.7	184.53	24SE	0.0	-4.3	-2.0	0.0
218.42	-0.29	1.6	194.50	4OC	0.2	-4.2	-1.9	-0.1
225.65	-0.39	1.6	204.52	14OC	0.9	-3.8	-1.9	-0.1
232.99	-0.49	1.6	214.59	24OC	3.3	-3.1	-1.8	-0.2
240.42	-0.59	1.5	224.70	3NO	0.4	-3.7	-1.7	-0.2
247.94	-0.67	1.5	234.85	13NO	-0.4	-4.2	-1.7	-0.1
255.54	-0.76	1.5	245.03	23NO	-0.4	-4.4	-1.6	-0.1
263.22	-0.83	1.4	255.22	3DE	-0.5	-4.4	-1.6	-0.0
270.97	-0.90	1.4	265.43	13DE	-0.6	-4.3	-1.6	0.1
278.76	-0.96	1.4	275.62	23DE	-0.9	-4.2	-1.5	0.1

♂ LONG	LAT	MAG	☉ LONG	16.00UT 764	☿	♀	♃	♄
286.60	-1.00	1.3	285.81	2JA	-1.0	-4.1	-1.5	0.2
294.48	-1.04	1.3	295.97	12JA	-1.0	-4.0	-1.5	0.3
302.37	-1.07	1.3	306.09	22JA	-0.3	-3.9	-1.5	0.3
310.26	-1.09	1.3	316.18	1FE	1.6	-3.8	-1.5	0.3
318.14	-1.09	1.4	326.21	11FE	2.6	-3.7	-1.5	0.4
326.01	-1.08	1.4	336.19	21FE	1.2	-3.6	-1.6	0.4
333.84	-1.06	1.4	346.12	2MR	0.6	-3.6	-1.6	0.4
341.62	-1.03	1.4	355.98	12MR	0.4	-3.5	-1.6	0.4
349.35	-0.99	1.4	5.79	22MR	0.0	-3.4	-1.7	0.4
357.02	-0.93	1.5	15.54	1AP	-0.6	-3.4	-1.7	0.3
4.61	-0.87	1.5	25.24	11AP	-1.4	-3.4	-1.8	0.3
12.12	-0.80	1.5	34.89	21AP	-1.6	-3.3	-1.8	0.3
19.54	-0.72	1.5	44.50	1MY	-0.6	-3.3	-1.9	0.2
26.88	-0.63	1.5	54.08	11MY	0.3	-3.3	-2.0	0.2
34.11	-0.53	1.6	63.63	21MY	1.1	-3.3	-2.0	0.3
41.25	-0.43	1.6	73.16	31MY	1.9	-3.3	-2.1	0.3
48.29	-0.32	1.6	82.69	10JN	3.2	-3.3	-2.2	0.3
55.22	-0.20	1.6	92.22	20JN	2.5	-3.3	-2.2	0.3
62.05	-0.09	1.6	101.75	30JN	1.0	-3.4	-2.3	0.3
68.76	0.04	1.6	111.31	10JL	-0.3	-3.4	-2.3	0.3
75.36	0.17	1.6	120.89	20JL	-1.3	-3.4	-2.4	0.3
81.85	0.30	1.6	130.51	30JL	-1.5	-3.5	-2.4	0.3
88.21	0.44	1.6	140.17	9AU	-0.8	-3.5	-2.4	0.3
94.44	0.58	1.5	149.87	19AU	-0.3	-3.5	-2.4	0.3
100.53	0.73	1.5	159.64	29AU	0.0	-3.4	-2.4	0.2
106.46	0.89	1.4	169.46	8SE	0.2	-3.4	-2.4	0.2
112.21	1.05	1.4	179.33	18SE	0.4	-3.4	-2.3	0.1
117.76	1.23	1.3	189.26	28SE	1.2	-3.4	-2.3	0.0
123.08	1.42	1.2	199.26	8OC	3.0	-3.3	-2.2	-0.0
128.12	1.62	1.1	209.30	18OC	0.2	-3.3	-2.1	-0.1
132.83	1.84	1.0	219.39	28OC	-0.5	-3.3	-2.1	-0.2
137.13	2.08	0.8	229.52	7NO	-0.6	-3.3	-2.0	-0.3
140.94	2.35	0.7	239.68	17NO	-0.6	-3.3	-1.9	-0.3
144.15	2.64	0.5	249.87	27NO	-0.7	-3.4	-1.9	-0.2
146.59	2.96	0.3	260.07	7DE	-0.8	-3.4	-1.8	-0.1
148.12	3.31	0.0	270.27	17DE	-0.8	-3.4	-1.8	-0.1
148.53	3.67	-0.2	280.47	27DE	-0.8	-3.4	-1.7	0.0

765

♂ LONG	LAT	MAG	☉ LONG	16.00UT 765	☿	♀	♃	♄
147.66	4.02	-0.5	290.64	6JA	-0.2	-3.5	-1.7	0.1
145.51	4.32	-0.7	300.78	16JA	1.9	-3.5	-1.6	0.1
142.23	4.50	-0.9	310.89	26JA	2.0	-3.6	-1.6	0.2
138.35	4.52	-1.0	320.96	5FE	0.8	-3.6	-1.6	0.2
134.56	4.37	-0.9	330.96	15FE	0.4	-3.7	-1.6	0.3
131.53	4.08	-0.7	340.92	25FE	0.2	-3.8	-1.6	0.3
129.68	3.69	-0.4	350.82	7MR	-0.1	-3.9	-1.6	0.3
129.13	3.29	-0.2	0.65	17MR	-0.6	-3.9	-1.6	0.3
129.79	2.89	0.0	10.43	27MR	-1.4	-4.0	-1.6	0.3
131.52	2.51	0.2	20.16	6AP	-1.5	-4.1	-1.6	0.3
134.12	2.17	0.4	29.83	16AP	-0.6	-4.2	-1.7	0.3
137.42	1.86	0.6	39.47	26AP	0.5	-4.2	-1.7	0.3
141.32	1.58	0.7	49.06	6MY	1.4	-4.1	-1.7	0.2
145.69	1.32	0.9	58.62	16MY	2.6	-3.8	-1.8	0.2
150.45	1.09	1.0	68.17	26MY	3.4	-3.1	-1.8	0.2
155.55	0.88	1.1	77.70	5JN	1.9	-3.1	-1.9	0.2
160.94	0.68	1.2	87.22	15JN	0.8	-3.8	-2.0	0.3
166.58	0.49	1.2	96.75	25JN	-0.3	-4.1	-2.0	0.3
172.44	0.32	1.3	106.29	5JL	-1.3	-4.2	-2.1	0.3
178.51	0.16	1.3	115.86	15JL	-1.5	-4.2	-2.2	0.3
184.76	0.01	1.4	125.46	25JL	-0.7	-4.1	-2.3	0.3
191.19	-0.13	1.4	135.10	4AU	-0.2	-4.0	-2.3	0.3
197.78	-0.27	1.4	144.78	14AU	0.1	-3.9	-2.4	0.3
204.53	-0.39	1.4	154.51	24AU	0.3	-3.8	-2.4	0.3
211.42	-0.51	1.4	164.30	3SE	0.6	-3.8	-2.4	0.2
218.45	-0.62	1.4	174.14	13SE	1.5	-3.7	-2.5	0.2
225.61	-0.72	1.4	184.05	23SE	2.7	-3.6	-2.5	0.1
232.90	-0.81	1.4	194.01	3OC	0.1	-3.6	-2.4	0.1
240.30	-0.89	1.4	204.03	13OC	-0.7	-3.5	-2.4	0.0
247.80	-0.96	1.4	214.10	23OC	-0.7	-3.5	-2.4	-0.1
255.40	-1.01	1.4	224.21	2NO	-0.7	-3.5	-2.3	-0.1
263.09	-1.06	1.4	234.36	12NO	-0.8	-3.4	-2.2	-0.2
270.84	-1.10	1.4	244.54	22NO	-0.7	-3.4	-2.2	-0.3
278.65	-1.12	1.4	254.73	2DE	-0.7	-3.4	-2.1	-0.3
286.51	-1.13	1.4	264.93	12DE	-0.6	-3.4	-2.0	-0.2
294.39	-1.13	1.4	275.13	22DE	-0.1	-3.4	-1.9	-0.2

♂ LONG	LAT	MAG	☉ LONG	16.00UT 766	☿	♀	♃	♄
302.29	-1.12	1.4	285.31	1JA	2.2	-3.3	-1.9	-0.1
310.19	-1.09	1.4	295.47	11JA	1.6	-3.3	-1.8	-0.0
318.08	-1.06	1.4	305.60	21JA	0.6	-3.3	-1.8	0.1
325.94	-1.01	1.4	315.68	31JA	0.3	-3.4	-1.7	0.1
333.77	-0.95	1.4	325.72	10FE	0.1	-3.4	-1.7	0.2
341.54	-0.89	1.4	335.71	20FE	-0.2	-3.4	-1.6	0.2
349.25	-0.81	1.4	345.63	2MR	-0.6	-3.4	-1.6	0.3
356.90	-0.73	1.4	355.50	12MR	-1.4	-3.5	-1.6	0.3
4.47	-0.64	1.4	5.31	22MR	-1.5	-3.5	-1.6	0.3
11.95	-0.55	1.5	15.06	1AP	-0.6	-3.5	-1.6	0.3
19.36	-0.45	1.5	24.76	11AP	0.6	-3.4	-1.6	0.3
26.68	-0.35	1.6	34.42	21AP	1.8	-3.4	-1.6	0.3
33.91	-0.25	1.6	44.03	1MY	3.4	-3.4	-1.6	0.3
41.05	-0.15	1.6	53.61	11MY	2.7	-3.4	-1.6	0.3
48.10	-0.04	1.7	63.16	21MY	1.5	-3.3	-1.6	0.2
55.06	0.07	1.7	72.70	31MY	0.6	-3.3	-1.7	0.2
61.94	0.17	1.7	82.22	10JN	-0.4	-3.3	-1.7	0.2
68.73	0.28	1.8	91.75	20JN	-1.4	-3.3	-1.7	0.2
75.44	0.39	1.8	101.28	30JN	-1.5	-3.3	-1.8	0.3
82.07	0.49	1.8	110.84	10JL	-0.7	-3.4	-1.9	0.3
88.61	0.60	1.8	120.42	20JL	-0.1	-3.4	-1.9	0.3
95.08	0.71	1.8	130.04	30JL	0.2	-3.4	-2.0	0.4
101.46	0.81	1.8	139.69	9AU	0.5	-3.4	-2.1	0.4
107.76	0.92	1.8	149.40	19AU	0.9	-3.5	-2.1	0.4
113.97	1.03	1.8	159.16	29AU	2.0	-3.5	-2.2	0.3
120.10	1.14	1.8	168.98	8SE	2.4	-3.6	-2.3	0.3
126.12	1.25	1.7	178.85	18SE	-0.0	-3.7	-2.3	0.3
132.04	1.37	1.7	188.78	28SE	-0.8	-3.8	-2.4	0.2
137.84	1.49	1.6	198.77	8OC	-0.9	-3.9	-2.4	0.2
143.50	1.61	1.6	208.81	18OC	-0.9	-4.0	-2.4	0.1
148.99	1.74	1.5	218.90	28OC	-0.7	-4.1	-2.4	0.1
154.31	1.87	1.4	229.03	7NO	-0.5	-4.2	-2.4	-0.0
159.39	2.01	1.2	239.19	17NO	-0.5	-4.3	-2.4	-0.1
164.21	2.17	1.1	249.38	27NO	-0.5	-4.4	-2.3	-0.2
168.70	2.33	1.0	259.58	7DE	0.1	-4.4	-2.2	-0.2
172.78	2.50	0.8	269.78	17DE	2.4	-4.2	-2.2	-0.2
176.37	2.69	0.6	279.97	27DE	1.2	-3.8	-2.1	-0.2

767

♂ LONG	LAT	MAG	☉ LONG	16.00UT 767	☿	♀	♃	♄
179.32	2.88	0.4	290.15	6JA	0.3	-3.2	-2.0	-0.1
181.50	3.08	0.1	300.29	16JA	0.1	-3.7	-2.0	-0.0
182.73	3.28	-0.1	310.40	26JA	-0.0	-4.2	-1.9	0.0
182.82	3.46	-0.4	320.46	5FE	-0.3	-4.3	-1.8	0.1
181.65	3.59	-0.7	330.48	15FE	-0.7	-4.3	-1.8	0.1
179.24	3.63	-1.0	340.43	25FE	-1.4	-4.2	-1.7	0.2
175.84	3.55	-1.2	350.33	7MR	-1.4	-4.1	-1.7	0.2
172.02	3.31	-1.2	0.17	17MR	-0.5	-4.0	-1.6	0.3
168.51	2.94	-1.1	9.95	27MR	0.8	-3.9	-1.6	0.3
165.90	2.49	-0.9	19.69	6AP	2.3	-3.8	-1.5	0.3
164.55	2.03	-0.7	29.36	16AP	3.5	-3.7	-1.5	0.4
164.52	1.58	-0.4	39.00	26AP	2.0	-3.6	-1.5	0.4
165.70	1.17	-0.2	48.60	6MY	1.1	-3.6	-1.5	0.4
167.93	0.80	-0.1	58.16	16MY	0.4	-3.5	-1.5	0.3
171.01	0.48	0.1	67.70	26MY	-0.4	-3.4	-1.5	0.3
174.82	0.19	0.3	77.24	5JN	-1.4	-3.4	-1.5	0.3
179.22	-0.06	0.4	86.76	15JN	-1.6	-3.4	-1.5	0.3
184.10	-0.28	0.5	96.29	25JN	-0.7	-3.4	-1.5	0.3
189.40	-0.48	0.6	105.83	5JL	-0.0	-3.3	-1.6	0.3
195.06	-0.65	0.7	115.40	15JL	0.4	-3.3	-1.6	0.4
201.02	-0.80	0.8	125.00	25JL	0.7	-3.3	-1.7	0.4
207.26	-0.94	0.8	134.63	4AU	1.2	-3.3	-1.7	0.4
213.73	-1.05	0.9	144.31	14AU	2.4	-3.3	-1.8	0.4
220.42	-1.15	1.0	154.04	24AU	2.1	-3.4	-1.8	0.4
227.31	-1.23	1.0	163.83	3SE	-0.1	-3.4	-1.9	0.4
234.37	-1.30	1.0	173.67	13SE	-0.9	-3.4	-2.0	0.4
241.59	-1.35	1.1	183.57	23SE	-1.0	-3.4	-2.0	0.4
248.94	-1.38	1.1	193.53	3OC	-0.9	-3.5	-2.1	0.4
256.41	-1.40	1.1	203.54	13OC	-0.6	-3.5	-2.1	0.3
263.99	-1.41	1.2	213.60	23OC	-0.4	-3.5	-2.2	0.3
271.65	-1.39	1.2	223.72	2NO	-0.4	-3.5	-2.2	0.2
279.38	-1.37	1.2	233.86	12NO	-0.3	-3.4	-2.3	0.1
287.16	-1.33	1.3	244.03	22NO	0.2	-3.4	-2.3	0.1
294.97	-1.27	1.3	254.23	2DE	2.7	-3.4	-2.3	-0.0
302.80	-1.21	1.3	264.43	12DE	0.9	-3.4	-2.3	-0.0
310.63	-1.14	1.4	274.63	22DE	0.1	-3.3	-2.2	-0.1

Left table (768 / 769):

♂ LONG	LAT	MAG	☉ LONG	16.00UT	☿	♀	♃	♄
318.44	-1.05	1.4	284.82	1JA	-0.0	-3.3	-2.2	-0.1
326.22	-0.96	1.4	294.98	11JA	-0.1	-3.3	-2.1	-0.1
333.96	-0.87	1.4	305.11	21JA	-0.3	-3.3	-2.1	-0.0
341.66	-0.76	1.5	315.20	31JA	-0.7	-3.3	-2.0	0.0
349.28	-0.66	1.5	325.23	10FE	-1.3	-3.4	-1.9	0.1
356.84	-0.55	1.5	335.22	20FE	-1.3	-3.4	-1.9	0.2
4.33	-0.44	1.6	345.15	1MR	-0.5	-3.4	-1.8	0.2
11.73	-0.33	1.6	355.02	11MR	1.0	-3.4	-1.7	0.3
19.06	-0.22	1.6	4.83	21MR	2.9	-3.5	-1.7	0.3
26.30	-0.12	1.6	14.59	31MR	2.6	-3.5	-1.6	0.4
33.46	-0.01	1.7	24.29	10AP	1.5	-3.6	-1.5	0.4
40.53	0.10	1.7	33.95	20AP	0.9	-3.6	-1.5	0.4
47.53	0.20	1.7	43.57	30AP	0.3	-3.7	-1.5	0.4
54.45	0.29	1.7	53.15	10MY	-0.5	-3.8	-1.4	0.5
61.29	0.39	1.7	62.70	20MY	-1.4	-3.8	-1.4	0.5
68.07	0.48	1.8	72.24	30MY	-1.6	-3.9	-1.4	0.5
74.77	0.57	1.8	81.76	9JN	-0.7	-4.0	-1.4	0.4
81.41	0.66	1.9	91.29	19JN	0.0	-4.1	-1.4	0.4
87.99	0.74	1.9	100.83	29JN	0.5	-4.2	-1.4	0.4
94.52	0.82	1.9	110.38	9JL	0.9	-4.2	-1.4	0.4
100.99	0.90	1.9	119.96	19JL	1.6	-4.1	-1.4	0.5
107.42	0.97	2.0	129.58	29JL	3.0	-3.7	-1.4	0.5
113.79	1.04	2.0	139.23	8AU	1.8	-3.2	-1.5	0.5
120.12	1.11	2.0	148.93	18AU	-0.1	-3.6	-1.5	0.5
126.40	1.17	2.0	158.69	28AU	-1.1	-4.1	-1.5	0.6
132.64	1.24	2.0	168.50	7SE	-1.2	-4.3	-1.6	0.6
138.84	1.30	1.9	178.37	17SE	-0.9	-4.3	-1.7	0.6
144.98	1.36	1.9	188.30	27SE	-0.5	-4.2	-1.7	0.6
151.07	1.42	1.9	198.28	7OC	-0.3	-4.2	-1.8	0.6
157.10	1.48	1.8	208.32	17OC	-0.2	-4.1	-1.8	0.5
163.06	1.53	1.8	218.41	27OC	0.1	-4.0	-1.9	0.5
168.95	1.59	1.7	228.54	6NO	0.4	-3.9	-2.0	0.4
174.75	1.64	1.6	238.70	16NO	3.0	-3.8	-2.0	0.4
180.44	1.69	1.5	248.89	26NO	0.7	-3.7	-2.1	0.3
186.01	1.74	1.4	259.08	6DE	-0.1	-3.7	-2.1	0.2
191.43	1.78	1.3	269.29	16DE	-0.2	-3.6	-2.1	0.2
196.66	1.83	1.1	279.48	26DE	-0.3	-3.5	-2.2	0.1

769

♂ LONG	LAT	MAG	☉ LONG	16.00UT	☿	♀	♃	♄
201.66	1.86	1.0	289.66	5JA	-0.4	-3.5	-2.2	0.0
206.40	1.89	0.8	299.81	15JA	-0.8	-3.4	-2.1	0.0
210.78	1.91	0.6	309.92	25JA	-1.2	-3.4	-2.1	0.1
214.75	1.91	0.4	319.98	4FE	-1.2	-3.4	-2.1	0.1
218.18	1.90	0.1	330.00	14FE	-0.5	-3.3	-2.0	0.2
220.94	1.85	-0.1	339.96	24FE	1.2	-3.3	-1.9	0.2
222.88	1.76	-0.4	349.86	6MR	3.3	-3.3	-1.9	0.3
223.80	1.62	-0.7	359.70	16MR	2.0	-3.3	-1.8	0.4
223.53	1.41	-1.0	9.49	26MR	1.1	-3.3	-1.7	0.4
222.02	1.10	-1.3	19.22	5AP	0.6	-3.3	-1.7	0.5
219.35	0.70	-1.6	28.90	15AP	0.2	-3.3	-1.6	0.5
215.96	0.23	-1.8	38.53	25AP	-0.5	-3.4	-1.5	0.5
212.53	-0.28	-1.7	48.13	5MY	-1.4	-3.4	-1.5	0.6
209.75	-0.76	-1.5	57.70	15MY	-1.6	-3.4	-1.4	0.6
208.15	-1.18	-1.4	67.24	25MY	-0.7	-3.5	-1.4	0.6
207.90	-1.53	-1.1	76.77	4JN	0.1	-3.5	-1.4	0.6
208.99	-1.79	-0.9	86.30	14JN	0.7	-3.5	-1.3	0.6
211.27	-2.00	-0.7	95.82	24JN	1.2	-3.4	-1.3	0.6
214.54	-2.15	-0.5	105.36	4JL	2.1	-3.4	-1.3	0.6
218.62	-2.25	-0.4	114.93	14JL	3.2	-3.4	-1.3	0.6
223.39	-2.31	-0.2	124.52	24JL	1.5	-3.3	-1.3	0.6
228.70	-2.35	-0.1	134.16	3AU	-0.2	-3.3	-1.3	0.6
234.48	-2.35	0.0	143.84	13AU	-1.1	-3.3	-1.3	0.7
240.63	-2.33	0.2	153.56	23AU	-1.3	-3.3	-1.3	0.7
247.10	-2.29	0.3	163.35	2SE	-0.8	-3.3	-1.3	0.8
253.83	-2.23	0.4	173.19	12SE	-0.4	-3.3	-1.4	0.8
260.78	-2.14	0.5	183.08	22SE	-0.2	-3.4	-1.4	0.8
267.91	-2.05	0.6	193.04	2OC	-0.1	-3.4	-1.5	0.8
275.18	-1.93	0.6	203.05	12OC	0.1	-3.4	-1.5	0.8
282.56	-1.81	0.7	213.11	22OC	0.7	-3.5	-1.6	0.8
290.02	-1.68	0.8	223.22	1NO	3.4	-3.5	-1.6	0.7
297.54	-1.53	0.9	233.37	11NO	0.5	-3.6	-1.7	0.6
305.09	-1.39	1.0	243.54	21NO	-0.3	-3.6	-1.8	0.6
312.66	-1.24	1.1	253.74	1DE	-0.3	-3.7	-1.8	0.6
320.23	-1.08	1.1	263.94	11DE	-0.4	-3.8	-1.9	0.5
327.77	-0.93	1.2	274.14	21DE	-0.5	-3.9	-1.9	0.5
335.29	-0.78	1.3	284.32	31DE	-0.8	-4.0	-2.0	0.4

Right table (770 / 771):

♂ LONG	LAT	MAG	☉ LONG	16.00UT	☿	♀	♃	♄
342.76	-0.64	1.3	294.49	10JA	-1.1	-4.1	-2.0	0.3
350.18	-0.49	1.4	304.62	20JA	-1.0	-4.2	-2.1	0.2
357.53	-0.35	1.5	314.71	30JA	-0.4	-4.3	-2.1	0.2
4.83	-0.22	1.5	324.75	9FE	1.4	-4.3	-2.1	0.3
12.05	-0.09	1.6	334.74	19FE	3.0	-4.3	-2.0	0.3
19.20	0.03	1.6	344.67	1MR	1.4	-4.1	-2.0	0.4
26.28	0.14	1.7	354.55	11MR	0.8	-3.6	-1.9	0.4
33.29	0.25	1.7	4.36	21MR	0.5	-3.2	-1.9	0.5
40.22	0.35	1.8	14.12	31MR	0.1	-3.7	-1.8	0.6
47.09	0.44	1.8	23.83	10AP	-0.5	-4.1	-1.8	0.6
53.89	0.53	1.8	33.48	20AP	-1.4	-4.2	-1.7	0.7
60.63	0.61	1.9	43.10	30AP	-1.6	-4.2	-1.6	0.7
67.32	0.69	1.9	52.68	10MY	-0.6	-4.1	-1.6	0.7
73.95	0.76	1.9	62.24	20MY	0.2	-4.0	-1.5	0.8
80.53	0.83	1.9	71.77	30MY	0.9	-3.9	-1.4	0.8
87.06	0.89	1.9	81.30	9JN	1.6	-3.8	-1.4	0.8
93.56	0.94	1.9	90.82	19JN	2.7	-3.7	-1.4	0.8
100.02	0.99	1.9	100.36	29JN	2.9	-3.7	-1.3	0.8
106.45	1.04	1.9	109.91	9JL	1.2	-3.6	-1.3	0.8
112.86	1.08	2.0	119.49	19JL	-0.2	-3.5	-1.3	0.8
119.24	1.11	2.0	129.11	29JL	-1.2	-3.5	-1.2	0.8
125.61	1.15	2.0	138.76	8AU	-1.4	-3.5	-1.2	0.8
131.96	1.18	2.0	148.46	18AU	-0.8	-3.4	-1.2	0.9
138.31	1.20	2.0	158.22	28AU	-0.3	-3.4	-1.2	0.9
144.64	1.22	2.0	168.03	7SE	-0.1	-3.4	-1.2	1.0
150.96	1.24	2.0	177.89	17SE	0.1	-3.4	-1.2	1.0
157.28	1.25	2.0	187.82	27SE	0.2	-3.4	-1.3	1.0
163.59	1.26	2.0	197.80	7OC	0.9	-3.4	-1.3	1.0
169.89	1.26	1.9	207.84	17OC	3.3	-3.4	-1.3	1.0
176.19	1.26	1.9	217.92	27OC	-0.3	-3.4	-1.4	1.0
182.47	1.25	1.9	228.05	6NO	-0.4	-3.4	-1.4	1.0
188.74	1.24	1.8	238.21	16NO	-0.5	-3.4	-1.5	1.0
194.99	1.22	1.7	248.39	26NO	-0.5	-3.4	-1.5	0.9
201.22	1.19	1.6	258.59	6DE	-0.6	-3.4	-1.6	0.9
207.42	1.15	1.6	268.79	16DE	-0.8	-3.4	-1.7	0.8
213.59	1.10	1.5	278.99	26DE	-0.9	-3.5	-1.7	0.8

771

♂ LONG	LAT	MAG	☉ LONG	16.00UT	☿	♀	♃	♄
219.70	1.04	1.3	289.16	5JA	-0.9	-3.5	-1.8	0.7
225.76	0.97	1.2	299.31	15JA	-0.3	-3.5	-1.9	0.6
231.75	0.88	1.1	309.43	25JA	1.7	-3.4	-1.9	0.5
237.64	0.77	0.9	319.49	4FE	2.4	-3.4	-2.0	0.5
243.43	0.63	0.8	329.51	14FE	1.0	-3.4	-2.0	0.5
249.08	0.47	0.6	339.47	24FE	0.6	-3.4	-2.0	0.5
254.55	0.27	0.4	349.37	6MR	0.3	-3.4	-2.0	0.6
259.81	0.04	0.2	359.22	16MR	-0.0	-3.3	-2.0	0.6
264.78	-0.24	-0.0	9.01	26MR	-0.6	-3.3	-2.0	0.7
269.40	-0.58	-0.3	18.74	5AP	-1.4	-3.3	-1.9	0.7
273.58	-0.98	-0.5	28.43	15AP	-1.6	-3.3	-1.9	0.8
277.17	-1.46	-0.8	38.07	25AP	-0.6	-3.4	-1.8	0.9
280.02	-2.03	-1.1	47.66	5MY	0.3	-3.4	-1.8	0.9
281.96	-2.68	-1.4	57.23	15MY	1.2	-3.4	-1.7	0.9
282.79	-3.41	-1.7	66.78	25MY	2.1	-3.4	-1.6	1.0
282.39	-4.18	-2.0	76.31	4JN	3.4	-3.5	-1.6	1.0
280.80	-4.93	-2.3	85.83	14JN	2.3	-3.5	-1.5	1.1
278.32	-5.54	-2.5	95.36	24JN	1.0	-3.6	-1.5	1.1
275.59	-5.92	-2.5	104.90	4JL	-0.3	-3.6	-1.4	1.1
273.32	-6.03	-2.3	114.47	14JL	-1.3	-3.7	-1.4	1.1
272.09	-5.88	-2.1	124.06	24JL	-1.5	-3.8	-1.3	1.1
272.20	-5.57	-1.9	133.69	3AU	-0.8	-3.9	-1.3	1.1
273.63	-5.15	-1.6	143.37	13AU	-0.2	-4.0	-1.3	1.1
276.24	-4.68	-1.3	153.10	23AU	0.1	-4.1	-1.2	1.1
279.87	-4.21	-1.1	162.87	2SE	0.2	-4.2	-1.2	1.2
284.27	-3.74	-0.9	172.71	12SE	0.5	-4.3	-1.2	1.2
289.30	-3.29	-0.6	182.61	22SE	1.3	-4.3	-1.2	1.3
294.81	-2.87	-0.4	192.56	2OC	3.0	-4.2	-1.2	1.3
300.68	-2.48	-0.2	202.57	12OC	0.2	-3.7	-1.2	1.3
306.85	-2.11	-0.0	212.63	22OC	-0.6	-3.1	-1.2	1.3
313.23	-1.77	0.1	222.73	1NO	-0.6	-3.7	-1.3	1.3
319.77	-1.46	0.3	232.88	11NO	-0.6	-4.2	-1.3	1.3
326.43	-1.17	0.5	243.05	21NO	-0.7	-4.4	-1.3	1.3
333.16	-0.91	0.6	253.24	1DE	-0.7	-4.4	-1.4	1.2
339.94	-0.67	0.8	263.44	11DE	-0.8	-4.3	-1.5	1.2
346.76	-0.45	0.9	273.65	21DE	-0.7	-4.2	-1.5	1.1
353.58	-0.26	1.0	283.83	31DE	-0.2	-4.1	-1.6	1.1

♂ LONG	♂ LAT	♂ MAG	⊙ LONG	16.00UT	☿	♀	♃	♄
				772	MAGNITUDES			
0.40	-0.08	1.1	294.00	10JA	1.9	-4.0	-1.6	1.0
7.21	0.08	1.2	304.13	20JA	1.9	-3.9	-1.7	1.0
13.99	0.23	1.3	314.22	30JA	0.7	-3.8	-1.8	0.9
20.74	0.36	1.4	324.27	9FE	0.4	-3.7	-1.8	0.8
27.47	0.47	1.5	334.26	19FE	0.2	-3.6	-1.9	0.8
34.15	0.57	1.6	344.19	29FE	-0.1	-3.6	-1.9	0.8
40.79	0.67	1.7	354.07	10MR	-0.6	-3.5	-2.0	0.8
47.40	0.75	1.7	3.89	20MR	-1.4	-3.4	-2.0	0.9
53.96	0.82	1.8	13.65	30MR	-1.5	-3.4	-2.0	0.9
60.49	0.88	1.9	23.37	9AP	-0.6	-3.4	-2.0	1.0
66.99	0.94	1.9	33.03	19AP	0.5	-3.3	-2.0	1.0
73.45	0.99	1.9	42.64	29AP	1.5	-3.3	-2.0	1.1
79.88	1.03	2.0	52.23	9MY	2.8	-3.3	-1.9	1.1
86.29	1.06	2.0	61.78	19MY	3.2	-3.3	-1.9	1.2
92.68	1.09	2.0	71.32	29MY	1.8	-3.3	-1.8	1.2
99.06	1.12	2.0	80.85	8JN	0.7	-3.3	-1.8	1.3
105.42	1.13	2.0	90.37	18JN	-0.3	-3.3	-1.7	1.3
111.77	1.15	2.0	99.90	28JN	-1.3	-3.4	-1.6	1.3
118.12	1.15	2.0	109.46	8JL	-1.6	-3.4	-1.6	1.4
124.48	1.16	2.0	119.03	18JL	-0.7	-3.4	-1.5	1.4
130.84	1.16	2.0	128.64	28JL	-0.2	-3.5	-1.5	1.4
137.21	1.15	2.0	138.29	7AU	0.2	-3.5	-1.4	1.4
143.59	1.14	2.0	147.99	17AU	0.4	-3.5	-1.4	1.3
149.99	1.12	2.0	157.74	27AU	0.7	-3.4	-1.3	1.3
156.42	1.10	2.0	167.55	6SE	1.6	-3.4	-1.3	1.2
162.86	1.08	2.0	177.41	16SE	2.7	-3.4	-1.3	1.2
169.34	1.05	2.0	187.33	26SE	0.1	-3.4	-1.3	1.2
175.84	1.01	2.0	197.31	6OC	-0.7	-3.3	-1.2	1.2
182.37	0.97	2.0	207.34	16OC	-0.8	-3.3	-1.2	1.2
188.94	0.93	1.9	217.43	26OC	-0.8	-3.3	-1.2	1.2
195.54	0.87	1.9	227.55	5NO	-0.7	-3.3	-1.3	1.2
202.17	0.81	1.9	237.71	15NO	-0.6	-3.3	-1.3	1.2
208.83	0.74	1.8	247.89	25NO	-0.6	-3.4	-1.3	1.1
215.53	0.67	1.8	258.10	5DE	-0.6	-3.4	-1.3	1.1
222.26	0.58	1.7	268.30	15DE	-0.1	-3.4	-1.4	1.0
229.02	0.48	1.6	278.49	25DE	2.2	-3.4	-1.4	1.0
				773				
235.81	0.38	1.5	288.67	4JA	1.5	-3.5	-1.5	1.0
242.63	0.26	1.5	298.82	14JA	0.5	-3.5	-1.5	0.9
249.47	0.12	1.4	308.94	24JA	0.2	-3.6	-1.6	0.9
256.34	-0.02	1.3	319.01	3FE	0.1	-3.6	-1.7	0.8
263.22	-0.18	1.2	329.03	13FE	-0.2	-3.7	-1.7	0.8
270.12	-0.36	1.1	338.99	23FE	-0.6	-3.8	-1.8	0.7
277.03	-0.56	0.9	348.90	5MR	-1.4	-3.9	-1.9	0.7
283.93	-0.77	0.8	358.75	15MR	-1.5	-3.9	-1.9	0.8
290.81	-1.00	0.7	8.54	25MR	-0.6	-4.0	-2.0	0.8
297.67	-1.24	0.6	18.28	4AP	0.6	-4.1	-2.0	0.9
304.48	-1.50	0.4	27.97	14AP	2.0	-4.2	-2.1	1.0
311.21	-1.78	0.3	37.61	24AP	3.7	-4.2	-2.1	1.1
317.83	-2.08	0.1	47.21	4MY	2.4	-4.1	-2.1	1.1
324.31	-2.39	-0.0	56.78	14MY	1.4	-3.8	-2.1	1.2
330.60	-2.71	-0.2	66.32	24MY	0.6	-3.0	-2.1	1.2
336.62	-3.04	-0.4	75.85	3JN	-0.4	-3.1	-2.0	1.2
342.30	-3.38	-0.6	85.37	13JN	-1.4	-3.8	-2.0	1.3
347.55	-3.73	-0.8	94.90	23JN	-1.6	-4.1	-1.9	1.3
352.21	-4.08	-1.0	104.45	3JL	-0.7	-4.2	-1.9	1.3
356.14	-4.42	-1.2	114.00	13JL	-0.1	-4.2	-1.8	1.3
359.16	-4.73	-1.4	123.60	23JL	0.3	-4.1	-1.8	1.3
1.02	-5.00	-1.7	133.23	2AU	0.6	-4.0	-1.7	1.2
1.55	-5.19	-1.9	142.90	12AU	1.0	-3.9	-1.6	1.2
0.65	-5.22	-2.1	152.62	22AU	2.1	-3.8	-1.6	1.2
358.45	-5.05	-2.3	162.40	1SE	2.3	-3.8	-1.5	1.1
355.47	-4.65	-2.4	172.23	11SE	0.0	-3.7	-1.5	1.1
352.44	-4.04	-2.3	182.12	21SE	-0.9	-3.6	-1.4	1.1
350.14	-3.32	-2.0	192.07	1OC	-0.9	-3.6	-1.4	1.1
349.03	-2.58	-1.7	202.08	11OC	-0.9	-3.5	-1.4	1.1
349.25	-1.90	-1.3	212.14	21OC	-0.6	-3.5	-1.4	1.1
350.71	-1.30	-1.0	222.24	31OC	-0.5	-3.5	-1.3	1.1
353.22	-0.80	-0.7	232.38	10NO	-0.4	-3.4	-1.3	1.0
356.56	-0.39	-0.4	242.56	20NO	-0.4	-3.4	-1.3	1.0
0.56	-0.05	-0.1	252.75	30NO	0.1	-3.4	-1.3	1.0
5.07	0.22	0.1	262.95	10DE	2.4	-3.4	-1.3	1.0
9.97	0.45	0.3	273.15	20DE	1.2	-3.4	-1.4	0.9
15.17	0.63	0.5	283.34	30DE	0.2	-3.3	-1.4	0.9

♂ LONG	♂ LAT	♂ MAG	⊙ LONG	16.00UT	☿	♀	♃	♄
				774	MAGNITUDES			
20.60	0.79	0.7	293.51	9JA	0.0	-3.3	-1.4	0.8
26.20	0.91	0.9	303.64	19JA	-0.1	-3.3	-1.5	0.8
31.95	1.01	1.0	313.74	29JA	-0.3	-3.4	-1.5	0.7
37.79	1.09	1.2	323.78	8FE	-0.7	-3.4	-1.6	0.7
43.71	1.15	1.3	333.78	18FE	-1.4	-3.4	-1.6	0.7
49.70	1.20	1.4	343.72	28FE	-1.4	-3.4	-1.7	0.6
55.72	1.24	1.5	353.59	10MR	-0.6	-3.5	-1.8	0.6
61.79	1.27	1.6	3.42	20MR	0.8	-3.5	-1.8	0.6
67.88	1.29	1.7	13.18	30MR	2.5	-3.5	-1.9	0.6
73.99	1.30	1.8	22.89	9AP	3.2	-3.4	-2.0	0.7
80.11	1.31	1.8	32.56	19AP	1.8	-3.4	-2.1	0.8
86.26	1.31	1.9	42.18	29AP	1.0	-3.4	-2.1	0.8
92.42	1.30	1.9	51.76	9MY	0.4	-3.4	-2.2	0.9
98.59	1.29	2.0	61.32	19MY	-0.4	-3.3	-2.2	1.0
104.78	1.28	2.0	70.86	29MY	-1.4	-3.3	-2.2	1.0
110.99	1.26	2.0	80.38	8JN	-1.6	-3.3	-2.2	1.0
117.22	1.23	2.0	89.91	18JN	-0.7	-3.3	-2.2	1.1
123.48	1.21	2.0	99.44	28JN	-0.0	-3.3	-2.2	1.1
129.76	1.17	2.0	108.99	8JL	0.4	-3.4	-2.2	1.1
136.08	1.14	2.0	118.57	18JL	0.8	-3.4	-2.1	1.1
142.42	1.10	2.0	128.17	28JL	1.3	-3.4	-2.1	1.1
148.81	1.05	2.0	137.82	7AU	2.6	-3.4	-2.0	1.1
155.24	1.01	2.0	147.52	17AU	2.0	-3.5	-1.9	1.1
161.72	0.95	1.9	157.26	27AU	-0.1	-3.5	-1.9	1.0
168.24	0.90	1.9	167.07	6SE	-1.0	-3.6	-1.8	1.0
174.82	0.84	1.9	176.93	16SE	-1.1	-3.7	-1.7	1.0
181.45	0.77	1.9	186.85	26SE	-0.9	-3.8	-1.7	0.9
188.14	0.70	1.9	196.82	6OC	-0.6	-3.9	-1.6	0.9
194.89	0.63	1.9	206.86	16OC	-0.4	-4.0	-1.6	0.9
201.70	0.55	1.8	216.93	26OC	-0.3	-4.1	-1.5	0.9
208.57	0.46	1.8	227.06	5NO	-0.2	-4.2	-1.5	0.9
215.51	0.37	1.8	237.22	15NO	0.3	-4.3	-1.5	0.9
222.51	0.27	1.8	247.40	25NO	2.7	-4.4	-1.5	0.9
229.57	0.17	1.7	257.60	5DE	0.9	-4.4	-1.4	0.9
236.70	0.06	1.7	267.80	15DE	0.0	-4.2	-1.4	0.9
243.88	-0.05	1.7	278.00	25DE	-0.1	-3.8	-1.4	0.9
				775				
251.13	-0.17	1.6	288.18	4JA	-0.2	-3.2	-1.4	0.8
258.43	-0.30	1.6	298.33	14JA	-0.4	-3.8	-1.5	0.8
265.79	-0.43	1.5	308.44	24JA	-0.7	-4.2	-1.5	0.7
273.19	-0.57	1.4	318.52	3FE	-1.3	-4.3	-1.5	0.7
280.64	-0.70	1.4	328.54	13FE	-1.3	-4.3	-1.6	0.6
288.12	-0.84	1.3	338.51	23FE	-0.5	-4.2	-1.6	0.6
295.63	-0.99	1.2	348.42	5MR	1.0	-4.1	-1.6	0.5
303.16	-1.13	1.2	358.28	15MR	3.1	-4.0	-1.7	0.5
310.69	-1.27	1.1	8.07	25MR	2.4	-3.9	-1.8	0.4
318.23	-1.40	1.0	17.81	4AP	1.3	-3.8	-1.8	0.5
325.74	-1.53	1.0	27.50	14AP	0.8	-3.7	-1.9	0.5
333.21	-1.64	0.9	37.14	24AP	0.3	-3.6	-2.0	0.6
340.64	-1.75	0.8	46.74	4MY	-0.4	-3.6	-2.0	0.7
347.98	-1.85	0.8	56.32	14MY	-1.4	-3.5	-2.1	0.7
355.23	-1.93	0.7	65.86	24MY	-1.6	-3.4	-2.2	0.8
2.36	-1.99	0.6	75.39	3JN	-0.7	-3.4	-2.2	0.8
9.34	-2.04	0.5	84.92	13JN	0.1	-3.4	-2.3	0.9
16.14	-2.06	0.4	94.45	23JN	0.6	-3.4	-2.3	0.9
22.72	-2.07	0.3	103.99	3JL	1.0	-3.3	-2.3	1.0
29.04	-2.06	0.2	113.55	13JL	1.7	-3.3	-2.4	1.0
35.05	-2.02	0.1	123.14	23JL	3.1	-3.3	-2.3	1.0
40.69	-1.95	0.0	132.76	2AU	1.7	-3.3	-2.3	1.0
45.86	-1.86	-0.1	142.44	12AU	-0.1	-3.3	-2.3	1.0
50.48	-1.74	-0.3	152.15	22AU	-1.1	-3.4	-2.2	1.0
54.40	-1.58	-0.4	161.93	1SE	-1.2	-3.4	-2.2	1.0
57.46	-1.37	-0.6	171.76	11SE	-0.9	-3.4	-2.1	0.9
59.47	-1.11	-0.8	181.64	21SE	-0.5	-3.4	-2.1	0.9
60.22	-0.78	-1.0	191.59	1OC	-0.2	-3.5	-2.0	0.8
59.54	-0.39	-1.3	201.59	11OC	-0.2	-3.5	-1.9	0.8
57.43	0.06	-1.5	211.64	21OC	-0.1	-3.5	-1.9	0.8
54.15	0.55	-1.6	221.75	31OC	0.5	-3.5	-1.8	0.8
50.34	1.02	-1.7	231.89	10NO	3.1	-3.4	-1.7	0.9
46.82	1.41	-1.4	242.05	20NO	0.7	-3.4	-1.7	0.9
44.25	1.71	-1.1	252.25	30NO	-0.1	-3.4	-1.6	0.9
43.00	1.91	-0.8	262.45	10DE	-0.3	-3.4	-1.6	0.9
43.09	2.02	-0.5	272.65	20DE	-0.3	-3.3	-1.6	0.8
44.36	2.08	-0.2	282.84	30DE	-0.5	-3.3	-1.6	0.8

♂ LONG	♂ LAT	♂ MAG	☉ LONG	16.00UT	☿	♀	♃	♄
				776	MAGNITUDES			
46.63	2.10	0.1	293.01	9JA	-0.8	-3.3	-1.5	0.8
49.68	2.09	0.3	303.14	19JA	-1.1	-3.3	-1.5	0.7
53.34	2.06	0.5	313.24	29JA	-1.1	-3.4	-1.5	0.7
57.49	2.03	0.7	323.29	8FE	-0.5	-3.4	-1.5	0.6
62.02	1.98	0.9	333.29	18FE	1.2	-3.4	-1.6	0.6
66.84	1.93	1.1	343.23	28FE	3.3	-3.4	-1.6	0.5
71.91	1.87	1.2	353.12	9MR	1.8	-3.4	-1.6	0.5
77.16	1.82	1.4	2.94	19MR	1.0	-3.5	-1.6	0.4
82.56	1.76	1.5	12.71	29MR	0.6	-3.5	-1.7	0.4
88.10	1.70	1.6	22.43	8AP	0.1	-3.6	-1.7	0.3
93.74	1.64	1.7	32.09	18AP	-0.5	-3.6	-1.8	0.4
99.48	1.57	1.7	41.72	28AP	-1.4	-3.7	-1.8	0.5
105.29	1.51	1.8	51.30	8MY	-1.6	-3.8	-1.9	0.5
111.18	1.44	1.8	60.86	18MY	-0.7	-3.8	-2.0	0.6
117.15	1.38	1.9	70.40	28MY	0.1	-3.9	-2.1	0.6
123.17	1.31	1.9	79.93	7JN	0.7	-4.0	-2.1	0.7
129.26	1.24	1.9	89.45	17JN	1.3	-4.1	-2.2	0.8
135.42	1.16	2.0	98.99	27JN	2.3	-4.2	-2.3	0.8
141.63	1.09	2.0	108.53	7JL	3.2	-4.2	-2.3	0.8
147.91	1.01	2.0	118.10	17JL	1.4	-4.0	-2.4	0.9
154.26	0.94	2.0	127.71	27JL	-0.2	-3.6	-2.4	0.9
160.68	0.86	2.0	137.36	6AU	-1.2	-3.2	-2.4	0.9
167.17	0.77	1.9	147.05	16AU	-1.4	-3.6	-2.4	0.9
173.74	0.69	1.9	156.80	26AU	-0.8	-4.1	-2.4	0.9
180.37	0.60	1.9	166.59	5SE	-0.4	-4.3	-2.4	0.9
187.09	0.51	1.8	176.45	15SE	-0.1	-4.3	-2.4	0.9
193.89	0.42	1.8	186.37	25SE	-0.0	-4.2	-2.3	0.8
200.77	0.32	1.7	196.34	5OC	0.1	-4.2	-2.3	0.8
207.73	0.23	1.7	206.37	15OC	0.7	-4.1	-2.2	0.7
214.77	0.13	1.7	216.45	25OC	3.4	-4.0	-2.1	0.8
221.90	0.02	1.6	226.57	4NO	0.5	-3.9	-2.1	0.8
229.10	-0.08	1.6	236.72	14NO	-0.3	-3.8	-2.0	0.8
236.39	-0.19	1.6	246.90	24NO	-0.4	-3.7	-1.9	0.8
243.75	-0.29	1.6	257.10	4DE	-0.4	-3.7	-1.9	0.8
251.19	-0.40	1.6	267.30	14DE	-0.6	-3.6	-1.8	0.8
258.70	-0.51	1.6	277.50	24DE	-0.8	-3.5	-1.8	0.8
				777				
266.26	-0.61	1.5	287.68	3JA	-1.0	-3.5	-1.7	0.8
273.89	-0.71	1.5	297.83	13JA	-1.0	-3.4	-1.7	0.8
281.56	-0.81	1.5	307.96	23JA	-0.4	-3.4	-1.6	0.7
289.27	-0.91	1.5	318.03	2FE	1.4	-3.4	-1.6	0.7
297.02	-1.00	1.4	328.06	12FE	2.8	-3.3	-1.6	0.6
304.78	-1.08	1.4	338.03	22FE	1.3	-3.3	-1.6	0.6
312.55	-1.15	1.4	347.94	4MR	0.7	-3.3	-1.6	0.5
320.32	-1.22	1.4	357.80	14MR	0.4	-3.3	-1.6	0.5
328.07	-1.27	1.3	7.60	24MR	0.0	-3.3	-1.6	0.4
335.78	-1.31	1.3	17.34	3AP	-0.5	-3.3	-1.6	0.4
343.46	-1.33	1.3	27.03	13AP	-1.4	-3.3	-1.6	0.3
351.09	-1.34	1.3	36.68	23AP	-1.6	-3.4	-1.7	0.3
358.64	-1.34	1.2	46.28	3MY	-0.7	-3.4	-1.7	0.3
6.11	-1.32	1.2	55.85	13MY	0.3	-3.4	-1.8	0.4
13.49	-1.29	1.2	65.40	23MY	1.0	-3.5	-1.8	0.5
20.76	-1.24	1.2	74.93	2JN	1.8	-3.5	-1.9	0.5
27.92	-1.18	1.1	84.46	12JN	3.0	-3.5	-1.9	0.6
34.93	-1.10	1.1	93.98	22JN	2.7	-3.4	-2.0	0.7
41.81	-1.01	1.1	103.52	2JL	1.1	-3.4	-2.1	0.7
48.51	-0.90	1.0	113.08	12JL	-0.2	-3.4	-2.1	0.8
55.04	-0.78	1.0	122.67	22JL	-1.2	-3.3	-2.2	0.8
61.36	-0.64	0.9	132.29	1AU	-1.5	-3.3	-2.3	0.8
67.45	-0.48	0.8	141.96	11AU	-0.8	-3.3	-2.3	0.8
73.27	-0.31	0.8	151.68	21AU	-0.3	-3.3	-2.4	0.8
78.79	-0.12	0.7	161.45	31AU	-0.0	-3.3	-2.4	0.8
83.94	0.10	0.6	171.28	10SE	0.1	-3.3	-2.4	0.8
88.65	0.34	0.4	181.16	20SE	0.3	-3.4	-2.5	0.8
92.83	0.61	0.3	191.10	30SE	1.0	-3.4	-2.5	0.8
96.35	0.92	0.1	201.11	10OC	3.3	-3.4	-2.4	0.8
99.07	1.27	-0.1	211.16	20OC	0.3	-3.5	-2.4	0.7
100.80	1.67	-0.3	221.25	30OC	-0.5	-3.5	-2.4	0.7
101.35	2.10	-0.5	231.40	9NO	-0.6	-3.6	-2.3	0.7
100.54	2.57	-0.7	241.56	19NO	-0.6	-3.6	-2.2	0.7
98.37	3.03	-0.9	251.75	29NO	-0.7	-3.7	-2.2	0.8
95.03	3.42	-1.1	261.96	9DE	-0.8	-3.8	-2.1	0.8
91.09	3.70	-1.2	272.16	19DE	-0.8	-3.9	-2.0	0.8
87.30	3.81	-1.0	282.34	29DE	-0.8	-4.0	-1.9	0.8

♂ LONG	♂ LAT	♂ MAG	☉ LONG	16.00UT	☿	♀	♃	♄
				778	MAGNITUDES			
84.35	3.78	-0.7	292.52	8JA	-0.3	-4.1	-1.9	0.8
82.65	3.64	-0.5	302.65	18JA	1.7	-4.2	-1.8	0.8
82.26	3.45	-0.2	312.75	28JA	2.2	-4.3	-1.8	0.7
83.08	3.23	0.1	322.80	7FE	0.9	-4.3	-1.7	0.7
84.93	3.01	0.3	332.80	17FE	0.5	-4.3	-1.7	0.7
87.59	2.79	0.5	342.75	27FE	0.3	-4.1	-1.6	0.6
90.93	2.59	0.7	352.64	9MR	-0.1	-3.6	-1.6	0.6
94.79	2.40	0.9	2.46	19MR	-0.6	-3.2	-1.6	0.5
99.07	2.23	1.1	12.23	29MR	-1.4	-3.7	-1.6	0.4
103.69	2.06	1.2	21.95	8AP	-1.6	-4.1	-1.6	0.4
108.59	1.91	1.3	31.62	18AP	-0.7	-4.2	-1.6	0.3
113.72	1.76	1.4	41.24	28AP	0.4	-4.2	-1.6	0.3
119.05	1.62	1.5	50.84	8MY	1.3	-4.1	-1.6	0.3
124.55	1.49	1.6	60.39	18MY	2.4	-4.0	-1.6	0.3
130.20	1.36	1.6	69.93	28MY	3.5	-3.9	-1.6	0.4
135.98	1.24	1.7	79.46	7JN	2.1	-3.8	-1.7	0.5
141.89	1.12	1.7	88.99	17JN	0.9	-3.7	-1.7	0.5
147.91	1.00	1.8	98.52	27JN	-0.3	-3.7	-1.8	0.6
154.05	0.88	1.8	108.07	7JL	-1.3	-3.6	-1.8	0.6
160.29	0.77	1.8	117.64	17JL	-1.6	-3.5	-1.9	0.7
166.63	0.65	1.8	127.25	27JL	-0.8	-3.5	-1.9	0.7
173.09	0.54	1.8	136.89	6AU	-0.2	-3.5	-2.0	0.8
179.64	0.43	1.8	146.58	16AU	0.1	-3.4	-2.1	0.8
186.30	0.32	1.8	156.32	26AU	0.3	-3.4	-2.1	0.8
193.06	0.21	1.8	166.12	5SE	0.5	-3.4	-2.2	0.8
199.93	0.10	1.7	175.97	15SE	1.4	-3.4	-2.3	0.8
206.89	-0.01	1.7	185.89	25SE	3.0	-3.4	-2.3	0.8
213.96	-0.11	1.7	195.86	5OC	0.2	-3.4	-2.4	0.8
221.13	-0.22	1.6	205.88	15OC	-0.6	-3.4	-2.4	0.8
228.40	-0.32	1.6	215.96	25OC	-0.7	-3.4	-2.4	0.7
235.75	-0.42	1.6	226.08	4NO	-0.7	-3.4	-2.4	0.7
243.20	-0.52	1.5	236.23	14NO	-0.8	-3.4	-2.4	0.7
250.73	-0.61	1.5	246.41	24NO	-0.7	-3.4	-2.3	0.7
258.34	-0.70	1.4	256.61	4DE	-0.7	-3.4	-2.3	0.7
266.02	-0.78	1.4	266.81	14DE	-0.7	-3.5	-2.2	0.8
273.75	-0.86	1.4	277.01	24DE	-0.2	-3.5	-2.2	0.8
				779				
281.54	-0.92	1.4	287.19	3JA	1.9	-3.5	-2.1	0.8
289.37	-0.98	1.4	297.34	13JA	1.8	-3.5	-2.0	0.8
297.22	-1.03	1.4	307.46	23JA	0.6	-3.4	-1.9	0.8
305.09	-1.07	1.4	317.54	2FE	0.3	-3.4	-1.9	0.8
312.96	-1.10	1.4	327.56	12FE	0.1	-3.4	-1.8	0.7
320.82	-1.11	1.4	337.54	22FE	-0.1	-3.4	-1.7	0.7
328.66	-1.12	1.4	347.46	4MR	-0.6	-3.3	-1.7	0.6
336.46	-1.11	1.4	357.31	14MR	-1.3	-3.3	-1.6	0.6
344.22	-1.09	1.4	7.12	24MR	-1.5	-3.3	-1.6	0.5
351.92	-1.06	1.4	16.86	3AP	-0.6	-3.3	-1.6	0.5
359.55	-1.01	1.4	26.55	13AP	0.5	-3.3	-1.5	0.4
7.11	-0.95	1.4	36.20	23AP	1.7	-3.4	-1.5	0.4
14.59	-0.89	1.5	45.81	3MY	3.1	-3.4	-1.5	0.3
21.97	-0.81	1.5	55.38	13MY	2.9	-3.4	-1.5	0.2
29.26	-0.72	1.5	64.93	23MY	1.6	-3.4	-1.5	0.3
36.45	-0.63	1.5	74.47	2JN	0.7	-3.5	-1.5	0.3
43.53	-0.53	1.5	83.99	12JN	-0.3	-3.5	-1.5	0.4
50.50	-0.41	1.5	93.52	22JN	-1.3	-3.6	-1.5	0.5
57.35	-0.29	1.5	103.06	2JL	-1.6	-3.6	-1.5	0.5
64.08	-0.17	1.5	112.62	12JL	-0.8	-3.7	-1.6	0.6
70.68	-0.04	1.5	122.21	22JL	-0.2	-3.8	-1.6	0.6
77.15	0.10	1.4	131.83	1AU	0.2	-3.9	-1.7	0.7
83.47	0.25	1.4	141.49	11AU	0.5	-4.0	-1.7	0.7
89.63	0.41	1.4	151.21	21AU	0.8	-4.1	-1.8	0.8
95.61	0.57	1.3	160.98	31AU	1.8	-4.2	-1.8	0.8
101.40	0.75	1.2	170.80	10SE	2.6	-4.3	-1.9	0.8
106.96	0.94	1.2	180.69	20SE	0.1	-4.3	-2.0	0.8
112.26	1.14	1.1	190.62	30SE	-0.8	-4.1	-2.0	0.8
117.25	1.36	1.0	200.62	10OC	-0.9	-3.7	-2.1	0.8
121.87	1.60	0.8	210.67	20OC	-0.8	-3.1	-2.1	0.8
126.04	1.87	0.7	220.77	30OC	-0.7	-3.8	-2.2	0.7
129.66	2.17	0.5	230.90	9NO	-0.6	-4.2	-2.2	0.7
132.59	2.49	0.3	241.07	19NO	-0.5	-4.4	-2.3	0.7
134.67	2.85	0.1	251.26	29NO	-0.5	-4.4	-2.3	0.7
135.74	3.24	-0.1	261.46	9DE	-0.0	-4.3	-2.3	0.7
135.58	3.65	-0.3	271.66	19DE	2.2	-4.2	-2.2	0.7
134.11	4.03	-0.6	281.85	29DE	1.4	-4.1	-2.2	0.8

♂			☉	16.00UT	☿	♀	♃	♄
LONG	LAT	MAG	LONG	780	MAGNITUDES			
131.38	4.35	-0.8	292.02	8JA	0.4	-4.0	-2.2	0.8
127.71	4.53	-1.0	302.17	18JA	0.1	-3.9	-2.1	0.8
123.76	4.55	-0.9	312.26	28JA	0.0	-3.8	-2.0	0.8
120.24	4.39	-0.8	322.32	7FE	-0.2	-3.7	-2.0	0.8
117.71	4.10	-0.5	332.32	17FE	-0.7	-3.6	-1.9	0.8
116.47	3.75	-0.3	342.27	27FE	-1.3	-3.6	-1.8	0.7
116.50	3.38	-0.0	352.16	8MR	-1.4	-3.5	-1.8	0.7
117.66	3.02	0.2	1.99	18MR	-0.6	-3.4	-1.7	0.6
119.79	2.68	0.4	11.76	28MR	0.7	-3.4	-1.6	0.6
122.70	2.38	0.6	21.48	7AP	2.1	-3.4	-1.6	0.5
126.24	2.10	0.7	31.16	17AP	3.8	-3.3	-1.5	0.5
130.30	1.84	0.9	40.78	27AP	2.2	-3.3	-1.5	0.4
134.78	1.61	1.0	50.37	7MY	1.2	-3.3	-1.4	0.3
139.61	1.39	1.1	59.94	17MY	0.5	-3.3	-1.4	0.3
144.73	1.19	1.2	69.47	27MY	-0.3	-3.3	-1.4	0.2
150.11	1.01	1.3	79.00	6JN	-1.3	-3.3	-1.4	0.3
155.71	0.83	1.4	88.53	16JN	-1.6	-3.3	-1.4	0.3
161.51	0.67	1.4	98.06	26JN	-0.7	-3.4	-1.4	0.4
167.50	0.51	1.5	107.61	6JL	-0.1	-3.4	-1.4	0.5
173.65	0.36	1.5	117.18	16JL	0.3	-3.4	-1.4	0.5
179.96	0.22	1.5	126.78	26JL	0.6	-3.5	-1.4	0.6
186.42	0.08	1.5	136.42	5AU	1.1	-3.5	-1.4	0.6
193.02	-0.05	1.6	146.11	15AU	2.2	-3.5	-1.5	0.7
199.76	-0.17	1.6	155.85	25AU	2.3	-3.4	-1.5	0.7
206.64	-0.29	1.6	165.64	4SE	0.0	-3.4	-1.5	0.7
213.64	-0.40	1.6	175.50	14SE	-0.9	-3.4	-1.6	0.8
220.76	-0.51	1.5	185.40	24SE	-1.0	-3.4	-1.7	0.8
228.00	-0.61	1.5	195.37	4OC	-1.0	-3.3	-1.7	0.8
235.36	-0.70	1.5	205.39	14OC	-0.6	-3.3	-1.8	0.8
242.81	-0.78	1.5	215.46	24OC	-0.4	-3.3	-1.8	0.8
250.36	-0.86	1.5	225.58	3NO	-0.4	-3.3	-1.9	0.7
258.00	-0.92	1.5	235.73	13NO	-0.3	-3.3	-2.0	0.7
265.72	-0.98	1.4	245.91	23NO	0.1	-3.4	-2.0	0.7
273.50	-1.03	1.4	256.11	3DE	2.5	-3.4	-2.1	0.7
281.33	-1.06	1.4	266.31	13DE	1.1	-3.4	-2.1	0.7
289.20	-1.08	1.4	276.51	23DE	0.2	-3.4	-2.1	0.7
				781				
297.09	-1.09	1.4	286.69	2JA	-0.0	-3.5	-2.1	0.8
305.00	-1.09	1.3	296.85	12JA	-0.1	-3.5	-2.1	0.8
312.91	-1.08	1.3	306.97	22JA	-0.3	-3.6	-2.1	0.8
320.80	-1.06	1.3	317.05	1FE	-0.7	-3.6	-2.1	0.8
328.66	-1.02	1.3	327.08	11FE	-1.3	-3.7	-2.0	0.8
336.47	-0.97	1.3	337.06	21FE	-1.3	-3.8	-2.0	0.8
344.24	-0.92	1.4	346.98	3MR	-0.6	-3.9	-1.9	0.8
351.95	-0.85	1.4	356.84	13MR	0.8	-3.9	-1.8	0.7
359.59	-0.78	1.4	6.64	23MR	2.7	-4.0	-1.8	0.7
7.15	-0.70	1.5	16.40	2AP	2.9	-4.1	-1.7	0.6
14.63	-0.61	1.5	26.09	12AP	1.6	-4.2	-1.6	0.6
22.02	-0.52	1.5	35.74	22AP	0.9	-4.2	-1.6	0.5
29.32	-0.42	1.6	45.35	2MY	0.4	-4.1	-1.5	0.5
36.53	-0.32	1.6	54.92	12MY	-0.4	-3.8	-1.5	0.4
43.65	-0.21	1.6	64.47	22MY	-1.3	-3.0	-1.4	0.3
50.68	-0.11	1.7	74.01	1JN	-1.6	-3.2	-1.4	0.3
57.61	0.01	1.7	83.53	11JN	-0.7	-3.8	-1.3	0.3
64.45	0.12	1.7	93.06	21JN	-0.0	-4.1	-1.3	0.3
71.20	0.23	1.7	102.60	1JL	0.5	-4.2	-1.3	0.4
77.86	0.35	1.7	112.15	11JL	0.8	-4.2	-1.3	0.4
84.42	0.46	1.7	121.74	21JL	1.5	-4.1	-1.3	0.5
90.89	0.58	1.7	131.36	31JL	2.8	-4.0	-1.3	0.5
97.26	0.70	1.7	141.02	10AU	2.0	-3.9	-1.3	0.6
103.53	0.82	1.7	150.73	20AU	-0.0	-3.8	-1.3	0.7
109.70	0.95	1.7	160.50	30AU	-1.0	-3.8	-1.3	0.7
115.75	1.08	1.7	170.32	9SE	-1.2	-3.7	-1.3	0.7
121.68	1.21	1.6	180.20	19SE	-0.9	-3.6	-1.4	0.8
127.46	1.35	1.5	190.14	29SE	-0.5	-3.6	-1.4	0.8
133.09	1.49	1.5	200.13	9OC	-0.3	-3.5	-1.5	0.8
138.53	1.65	1.4	210.18	19OC	-0.2	-3.5	-1.5	0.8
143.75	1.81	1.3	220.28	29OC	-0.2	-3.5	-1.6	0.8
148.72	1.98	1.2	230.41	8NO	0.3	-3.4	-1.6	0.8
153.38	2.17	1.0	240.58	18NO	2.8	-3.4	-1.7	0.8
157.65	2.37	0.9	250.77	28NO	0.8	-3.4	-1.8	0.7
161.46	2.59	0.7	260.96	8DE	-0.0	-3.4	-1.8	0.7
164.69	2.83	0.5	271.17	18DE	-0.2	-3.4	-1.9	0.7
167.19	3.09	0.3	281.36	28DE	-0.2	-3.3	-1.9	0.7

♂			☉	16.00UT	☿	♀	♃	♄
LONG	LAT	MAG	LONG	782	MAGNITUDES			
168.80	3.36	0.0	291.53	7JA	-0.4	-3.3	-2.0	0.8
169.35	3.63	-0.2	301.67	17JA	-0.8	-3.3	-2.0	0.8
168.66	3.87	-0.5	311.78	27JA	-1.2	-3.4	-2.0	0.8
166.68	4.06	-0.7	321.83	6FE	-1.2	-3.4	-2.0	0.8
163.57	4.13	-1.0	331.84	16FE	-0.5	-3.4	-2.0	0.8
159.78	4.04	-1.1	341.79	26FE	1.0	-3.4	-2.0	0.8
156.00	3.80	-1.0	351.68	8MR	3.2	-3.5	-2.0	0.8
152.91	3.42	-0.8	1.51	18MR	2.1	-3.5	-1.9	0.8
150.96	2.99	-0.6	11.29	28MR	1.2	-3.5	-1.9	0.8
150.32	2.54	-0.4	21.01	7AP	0.7	-3.4	-1.8	0.7
150.93	2.11	-0.2	30.68	17AP	0.2	-3.4	-1.7	0.7
152.64	1.71	0.0	40.31	27AP	-0.4	-3.4	-1.7	0.6
155.27	1.36	0.2	49.90	7MY	-1.3	-3.4	-1.6	0.5
158.64	1.04	0.4	59.47	17MY	-1.6	-3.3	-1.5	0.5
162.65	0.76	0.5	69.01	27MY	-0.7	-3.3	-1.5	0.4
167.16	0.50	0.6	78.53	6JN	0.1	-3.3	-1.4	0.3
172.10	0.27	0.7	88.06	16JN	0.6	-3.3	-1.4	0.3
177.40	0.06	0.8	97.59	26JN	1.1	-3.3	-1.3	0.3
183.02	-0.12	0.9	107.13	6JL	1.9	-3.4	-1.3	0.4
188.92	-0.30	1.0	116.71	16JL	3.2	-3.4	-1.3	0.4
195.07	-0.45	1.0	126.31	26JL	1.6	-3.4	-1.2	0.5
201.43	-0.59	1.1	135.95	5AU	-0.1	-3.4	-1.2	0.5
208.00	-0.72	1.1	145.64	15AU	-1.1	-3.5	-1.2	0.6
214.75	-0.83	1.2	155.37	25AU	-1.3	-3.5	-1.2	0.6
221.68	-0.94	1.2	165.16	4SE	-0.9	-3.6	-1.2	0.7
228.76	-1.02	1.2	175.02	14SE	-0.4	-3.7	-1.2	0.7
235.98	-1.10	1.2	184.92	24SE	-0.2	-3.8	-1.2	0.8
243.34	-1.16	1.3	194.89	4OC	-0.1	-3.9	-1.3	0.8
250.81	-1.20	1.3	204.91	14OC	0.0	-4.0	-1.3	0.8
258.38	-1.24	1.3	214.98	24OC	0.6	-4.1	-1.3	0.8
266.05	-1.26	1.3	225.09	3NO	3.1	-4.2	-1.4	0.8
273.78	-1.26	1.3	235.24	13NO	0.6	-4.3	-1.4	0.8
281.58	-1.25	1.3	245.42	23NO	-0.2	-4.4	-1.5	0.8
289.42	-1.23	1.3	255.61	3DE	-0.3	-4.4	-1.6	0.8
297.28	-1.20	1.4	265.82	13DE	-0.4	-4.2	-1.6	0.8
305.15	-1.15	1.4	276.02	23DE	-0.5	-3.7	-1.7	0.7
				783				
313.02	-1.09	1.4	286.20	2JA	-0.8	-3.2	-1.8	0.8
320.87	-1.03	1.4	296.36	12JA	-1.1	-3.8	-1.8	0.8
328.69	-0.95	1.4	306.48	22JA	-1.1	-4.2	-1.9	0.8
336.46	-0.87	1.5	316.56	1FE	-0.5	-4.3	-1.9	0.9
344.18	-0.78	1.5	326.60	11FE	1.2	-4.3	-2.0	0.9
351.83	-0.69	1.5	336.58	21FE	3.1	-4.2	-2.0	0.9
359.41	-0.59	1.5	346.50	3MR	1.6	-4.1	-2.0	0.9
6.92	-0.49	1.5	356.37	13MR	0.9	-4.0	-2.0	0.9
14.34	-0.38	1.5	6.17	23MR	0.5	-3.9	-2.0	0.9
21.68	-0.28	1.6	15.92	2AP	0.1	-3.8	-2.0	0.8
28.93	-0.17	1.6	25.62	12AP	-0.5	-3.7	-1.9	0.8
36.10	-0.07	1.6	35.27	22AP	-1.3	-3.6	-1.9	0.7
43.19	0.04	1.6	44.88	2MY	-1.6	-3.6	-1.8	0.7
50.19	0.14	1.7	54.46	12MY	-0.7	-3.5	-1.7	0.6
57.11	0.24	1.7	64.01	22MY	0.2	-3.4	-1.7	0.6
63.95	0.34	1.8	73.55	1JN	0.8	-3.4	-1.6	0.5
70.72	0.44	1.8	83.07	11JN	1.5	-3.4	-1.6	0.4
77.41	0.53	1.8	92.60	21JN	2.5	-3.4	-1.5	0.4
84.04	0.63	1.9	102.14	1JL	3.0	-3.3	-1.4	0.3
90.60	0.72	1.9	111.69	11JL	1.3	-3.3	-1.4	0.4
97.09	0.80	1.9	121.27	21JL	-0.2	-3.3	-1.3	0.4
103.52	0.89	1.9	130.89	31JL	-1.2	-3.3	-1.3	0.5
109.90	0.98	1.9	140.56	10AU	-1.4	-3.3	-1.3	0.5
116.21	1.06	1.9	150.26	20AU	-0.8	-3.4	-1.2	0.6
122.47	1.14	1.9	160.03	30AU	-0.4	-3.4	-1.2	0.6
128.66	1.22	1.9	169.85	9SE	-0.1	-3.4	-1.2	0.7
134.79	1.31	1.9	179.72	19SE	0.1	-3.4	-1.2	0.7
140.85	1.39	1.8	189.66	29SE	0.2	-3.5	-1.2	0.8
146.83	1.47	1.8	199.65	9OC	0.8	-3.5	-1.2	0.8
152.72	1.55	1.7	209.69	19OC	3.4	-3.5	-1.2	0.8
158.52	1.63	1.7	219.78	29OC	0.5	-3.5	-1.3	0.8
164.20	1.71	1.6	229.92	8NO	-0.4	-3.4	-1.3	0.8
169.75	1.80	1.5	240.08	18NO	-0.5	-3.4	-1.3	0.9
175.14	1.88	1.4	250.27	28NO	-0.5	-3.4	-1.4	0.9
180.33	1.97	1.2	260.47	8DE	-0.6	-3.4	-1.4	0.8
185.29	2.06	1.1	270.67	18DE	-0.8	-3.3	-1.5	0.8
189.97	2.15	0.9	280.86	28DE	-0.9	-3.3	-1.5	0.8

♂			☉	16.00UT	☿	♀	♃	♄
LONG	LAT	MAG	LONG	784	MAGNITUDES			
194.30	2.24	0.7	291.04	7JA	-0.9	-3.3	-1.6	0.8
198.19	2.33	0.5	301.18	17JA	-0.4	-3.3	-1.7	0.8
201.53	2.41	0.3	311.28	27JA	1.5	-3.4	-1.7	0.9
204.19	2.48	0.1	321.35	6FE	2.6	-3.4	-1.8	0.9
206.02	2.53	-0.2	331.35	16FE	1.1	-3.4	-1.9	0.9
206.81	2.55	-0.5	341.31	26FE	0.6	-3.4	-1.9	1.0
206.41	2.51	-0.8	351.20	7MR	0.3	-3.4	-2.0	1.0
204.75	2.38	-1.1	1.04	17MR	0.0	-3.5	-2.0	1.0
201.93	2.15	-1.4	10.82	27MR	-0.5	-3.5	-2.0	0.9
198.38	1.81	-1.5	20.55	6AP	-1.3	-3.6	-2.0	0.9
194.78	1.38	-1.4	30.22	16AP	-1.6	-3.6	-2.0	0.9
191.82	0.90	-1.3	39.85	26AP	-0.7	-3.7	-2.0	0.9
190.00	0.43	-1.1	49.45	6MY	0.3	-3.8	-2.0	0.8
189.54	0.01	-0.9	59.01	16MY	1.1	-3.9	-1.9	0.7
190.37	-0.36	-0.7	68.55	26MY	2.0	-3.9	-1.9	0.7
192.38	-0.68	-0.5	78.08	5JN	3.3	-4.0	-1.8	0.6
195.37	-0.94	-0.3	87.60	15JN	2.5	-4.1	-1.8	0.6
199.17	-1.16	-0.1	97.14	25JN	1.1	-4.2	-1.7	0.5
203.64	-1.34	0.0	106.68	5JL	-0.2	-4.2	-1.6	0.4
208.67	-1.48	0.1	116.25	15JL	-1.2	-4.0	-1.6	0.4
214.16	-1.60	0.2	125.85	25JL	-1.5	-3.6	-1.5	0.4
220.04	-1.69	0.4	135.48	4AU	-0.8	-3.2	-1.5	0.5
226.26	-1.76	0.4	145.16	14AU	-0.3	-3.7	-1.4	0.5
232.76	-1.80	0.5	154.90	24AU	0.0	-4.1	-1.4	0.6
239.51	-1.82	0.6	164.69	3SE	0.2	-4.3	-1.3	0.6
246.46	-1.82	0.7	174.53	13SE	0.4	-4.3	-1.3	0.7
253.60	-1.80	0.7	184.44	23SE	1.1	-4.2	-1.3	0.7
260.89	-1.77	0.8	194.40	3OC	3.3	-4.2	-1.3	0.8
268.30	-1.72	0.9	204.42	13OC	0.3	-4.1	-1.3	0.8
275.82	-1.65	0.9	214.49	23OC	-0.5	-4.0	-1.3	0.9
283.42	-1.57	1.0	224.60	2NO	-0.6	-3.9	-1.3	0.9
291.07	-1.48	1.1	234.75	12NO	-0.6	-3.8	-1.3	0.9
298.76	-1.38	1.1	244.93	22NO	-0.7	-3.7	-1.3	0.9
306.46	-1.27	1.2	255.12	2DE	-0.8	-3.7	-1.3	0.9
314.17	-1.15	1.2	265.32	12DE	-0.8	-3.6	-1.4	0.9
321.86	-1.03	1.3	275.53	22DE	-0.8	-3.5	-1.4	0.9
				785				
329.52	-0.90	1.3	285.71	1JA	-0.3	-3.5	-1.4	0.9
337.14	-0.77	1.4	295.87	11JA	1.7	-3.4	-1.5	0.9
344.70	-0.65	1.4	306.00	21JA	2.1	-3.4	-1.6	0.9
352.21	-0.52	1.5	316.08	31JA	0.8	-3.4	-1.6	1.0
359.65	-0.39	1.5	326.12	10FE	0.4	-3.3	-1.7	1.0
7.01	-0.27	1.6	336.10	20FE	0.2	-3.3	-1.8	1.0
14.30	-0.15	1.6	346.03	2MR	-0.1	-3.3	-1.8	1.0
21.51	-0.04	1.7	355.89	12MR	-0.6	-3.3	-1.9	1.1
28.65	0.07	1.7	5.70	22MR	-1.3	-3.3	-2.0	1.1
35.71	0.18	1.7	15.46	1AP	-1.6	-3.3	-2.0	1.1
42.69	0.28	1.8	25.16	11AP	-0.7	-3.3	-2.1	1.0
49.60	0.37	1.8	34.81	21AP	0.4	-3.4	-2.1	1.0
56.43	0.47	1.8	44.42	1MY	1.4	-3.4	-2.1	1.0
63.20	0.55	1.8	54.00	11MY	2.6	-3.4	-2.1	0.9
69.92	0.63	1.8	63.55	21MY	3.4	-3.5	-2.1	0.9
76.57	0.71	1.8	73.09	31MY	1.9	-3.5	-2.1	0.8
83.16	0.78	1.8	82.61	10JN	0.8	-3.5	-2.1	0.8
89.71	0.85	1.9	92.14	20JN	-0.2	-3.4	-2.0	0.7
96.22	0.91	1.9	101.67	30JN	-1.3	-3.4	-2.0	0.6
102.68	0.97	1.9	111.23	10JL	-1.6	-3.4	-1.9	0.6
109.11	1.02	2.0	120.81	20JL	-0.8	-3.3	-1.8	0.5
115.51	1.08	2.0	130.42	30JL	-0.2	-3.3	-1.8	0.5
121.87	1.12	2.0	140.08	9AU	0.1	-3.3	-1.7	0.5
128.21	1.17	2.0	149.79	19AU	0.3	-3.3	-1.6	0.6
134.53	1.21	2.0	159.55	29AU	0.6	-3.3	-1.6	0.6
140.83	1.24	2.0	169.36	8SE	1.5	-3.3	-1.5	0.7
147.10	1.28	2.0	179.24	18SE	2.9	-3.4	-1.5	0.7
153.35	1.31	2.0	189.17	28SE	0.2	-3.4	-1.4	0.8
159.58	1.33	1.9	199.16	8OC	-0.7	-3.4	-1.4	0.8
165.78	1.36	1.9	209.20	18OC	-0.8	-3.5	-1.4	0.9
171.95	1.37	1.9	219.29	28OC	-0.8	-3.5	-1.4	0.9
178.09	1.39	1.8	229.42	7NO	-0.8	-3.6	-1.3	1.0
184.19	1.40	1.7	239.59	17NO	-0.6	-3.7	-1.3	1.0
190.24	1.40	1.7	249.77	27NO	-0.6	-3.7	-1.3	1.0
196.24	1.40	1.6	259.97	7DE	-0.6	-3.8	-1.4	1.0
202.17	1.39	1.5	270.18	17DE	-0.2	-3.9	-1.4	1.0
208.01	1.37	1.3	280.37	27DE	1.9	-4.0	-1.4	1.0

♂			☉	16.00UT	☿	♀	♃	♄
LONG	LAT	MAG	LONG	786	MAGNITUDES			
213.76	1.34	1.2	290.55	6JA	1.7	-4.1	-1.4	1.0
219.38	1.29	1.1	300.69	16JA	0.5	-4.2	-1.5	1.0
224.85	1.23	0.9	310.80	26JA	0.2	-4.3	-1.5	1.0
230.13	1.15	0.7	320.86	5FE	0.1	-4.3	-1.6	1.0
235.19	1.05	0.5	330.87	15FE	-0.2	-4.3	-1.6	1.1
239.96	0.91	0.3	340.83	25FE	-0.6	-4.1	-1.7	1.1
244.37	0.74	0.1	350.73	7MR	-1.3	-3.6	-1.7	1.2
248.32	0.52	-0.2	0.57	17MR	-1.5	-3.2	-1.8	1.2
251.70	0.24	-0.4	10.35	27MR	-0.7	-3.7	-1.9	1.2
254.36	-0.11	-0.7	20.08	6AP	0.5	-4.1	-2.0	1.2
256.11	-0.54	-1.0	29.75	16AP	1.8	-4.2	-2.0	1.2
256.79	-1.06	-1.4	39.39	26AP	3.5	-4.2	-2.1	1.2
256.23	-1.67	-1.7	48.98	6MY	2.6	-4.1	-2.1	1.2
254.47	-2.34	-2.0	58.55	16MY	1.5	-4.0	-2.2	1.1
251.77	-3.00	-2.2	68.09	26MY	0.6	-3.9	-2.2	1.1
248.69	-3.57	-2.2	77.62	5JN	-0.3	-3.8	-2.2	1.0
245.97	-3.98	-2.1	87.14	15JN	-1.3	-3.7	-2.2	1.0
244.24	-4.22	-1.9	96.67	25JN	-1.6	-3.7	-2.2	0.9
243.84	-4.30	-1.7	106.22	5JL	-0.8	-3.6	-2.2	0.8
244.82	-4.26	-1.5	115.78	15JL	-0.1	-3.5	-2.2	0.8
247.05	-4.14	-1.2	125.38	25JL	0.2	-3.5	-2.1	0.7
250.35	-3.96	-1.0	135.02	4AU	0.5	-3.5	-2.1	0.6
254.53	-3.75	-0.8	144.70	14AU	0.9	-3.4	-2.0	0.6
259.41	-3.52	-0.6	154.43	24AU	1.9	-3.4	-1.9	0.7
264.86	-3.27	-0.4	164.22	3SE	2.6	-3.4	-1.9	0.7
270.75	-3.02	-0.3	174.06	13SE	0.1	-3.4	-1.8	0.8
276.99	-2.76	-0.1	183.96	23SE	-0.8	-3.4	-1.7	0.8
283.51	-2.50	0.0	193.92	3OC	-0.9	-3.4	-1.7	0.9
290.24	-2.24	0.2	203.93	13OC	-0.9	-3.4	-1.6	0.9
297.13	-1.99	0.3	214.00	23OC	-0.7	-3.4	-1.6	1.0
304.14	-1.74	0.5	224.11	2NO	-0.5	-3.4	-1.5	1.0
311.23	-1.50	0.6	234.25	12NO	-0.5	-3.4	-1.5	1.0
318.38	-1.28	0.7	244.43	22NO	-0.4	-3.4	-1.5	1.1
325.56	-1.06	0.8	254.63	2DE	-0.0	-3.4	-1.5	1.1
332.74	-0.85	0.9	264.82	12DE	2.2	-3.5	-1.5	1.1
339.92	-0.66	1.0	275.02	22DE	1.3	-3.5	-1.4	1.2
				787				
347.08	-0.48	1.1	285.21	1JA	0.3	-3.5	-1.5	1.2
354.21	-0.31	1.2	295.37	11JA	0.1	-3.5	-1.5	1.2
1.29	-0.15	1.3	305.50	21JA	-0.0	-3.4	-1.5	1.2
8.34	-0.01	1.4	315.59	31JA	-0.3	-3.4	-1.5	1.2
15.33	0.13	1.5	325.63	10FE	-0.7	-3.4	-1.5	1.2
22.26	0.25	1.6	335.61	20FE	-1.3	-3.4	-1.6	1.2
29.14	0.36	1.6	345.54	2MR	-1.4	-3.3	-1.6	1.3
35.97	0.47	1.7	355.41	12MR	-0.6	-3.3	-1.7	1.3
42.73	0.56	1.8	5.23	22MR	0.7	-3.3	-1.7	1.3
49.45	0.65	1.8	14.99	1AP	2.3	-3.3	-1.8	1.4
56.12	0.72	1.9	24.69	11AP	3.4	-3.3	-1.9	1.4
62.73	0.79	1.9	34.34	21AP	1.9	-3.4	-1.9	1.4
69.30	0.85	1.9	43.96	1MY	1.1	-3.4	-2.0	1.4
75.83	0.91	1.9	53.54	11MY	0.5	-3.4	-2.1	1.4
82.33	0.96	2.0	63.09	21MY	-0.3	-3.4	-2.1	1.3
88.79	1.01	2.0	72.63	31MY	-1.3	-3.5	-2.2	1.3
95.22	1.04	2.0	82.15	10JN	-1.7	-3.5	-2.3	1.3
101.63	1.08	2.0	91.68	20JN	-0.8	-3.6	-2.3	1.3
108.03	1.10	2.0	101.21	30JN	-0.1	-3.6	-2.4	1.3
114.41	1.12	2.0	110.76	10JL	0.4	-3.7	-2.4	1.3
120.78	1.14	1.9	120.35	20JL	0.7	-3.8	-2.4	1.2
127.15	1.15	2.0	129.96	30JL	1.2	-3.9	-2.5	1.2
133.52	1.16	2.0	139.62	9AU	2.4	-4.0	-2.4	1.1
139.89	1.17	2.0	149.32	19AU	2.2	-4.1	-2.4	1.1
146.27	1.17	2.0	159.08	29AU	0.0	-4.2	-2.4	1.0
152.65	1.16	2.0	168.89	8SE	-0.9	-4.3	-2.3	1.0
159.05	1.15	2.0	178.76	18SE	-1.1	-4.3	-2.2	0.9
165.47	1.14	2.0	188.69	28SE	-1.0	-4.1	-2.1	0.9
171.90	1.11	2.0	198.67	8OC	-0.6	-3.7	-2.0	1.0
178.35	1.09	2.0	208.71	18OC	-0.4	-3.1	-2.0	1.0
184.81	1.06	1.9	218.80	28OC	-0.3	-3.8	-1.9	1.1
191.30	1.02	1.9	228.92	7NO	-0.3	-4.2	-1.9	1.0
197.80	0.97	1.8	239.09	17NO	0.2	-4.4	-1.8	1.1
204.31	0.92	1.8	249.27	27NO	2.5	-4.4	-1.7	1.2
210.85	0.86	1.7	259.47	7DE	1.0	-4.3	-1.6	1.3
217.39	0.79	1.6	269.68	17DE	0.1	-4.2	-1.6	1.3
223.94	0.70	1.6	279.87	27DE	-0.1	-4.1	-1.6	1.3

♂ LONG	♂ LAT	♂ MAG	⊙ LONG	16.00UT	☿	♀	♃	♄
				788	MAGNITUDES			
230.50	0.61	1.5	290.05	6JA	-0.2	-4.0	-1.6	1.3
237.07	0.50	1.4	300.20	16JA	-0.3	-3.9	-1.5	1.3
243.63	0.38	1.3	310.31	26JA	-0.7	-3.8	-1.5	1.3
250.19	0.23	1.2	320.37	5FE	-1.3	-3.7	-1.5	1.4
256.73	0.07	1.0	330.39	15FE	-1.3	-3.6	-1.6	1.3
263.25	-0.11	0.9	340.35	25FE	-0.6	-3.6	-1.6	1.4
269.73	-0.32	0.7	350.25	6MR	0.9	-3.5	-1.6	1.4
276.17	-0.55	0.6	0.10	16MR	2.9	-3.4	-1.6	1.4
282.54	-0.81	0.4	9.88	26MR	2.6	-3.4	-1.7	1.3
288.81	-1.10	0.3	19.61	5AP	1.4	-3.4	-1.7	1.3
294.96	-1.43	0.1	29.30	15AP	0.8	-3.3	-1.8	1.3
300.94	-1.79	-0.1	38.93	25AP	0.3	-3.3	-1.8	1.3
306.70	-2.19	-0.3	48.52	5MY	-0.4	-3.3	-1.9	1.3
312.16	-2.63	-0.5	58.09	15MY	-1.3	-3.3	-1.9	1.2
317.23	-3.12	-0.7	67.64	25MY	-1.7	-3.3	-2.0	1.2
321.80	-3.64	-1.0	77.16	4JN	-0.7	-3.3	-2.1	1.2
325.70	-4.21	-1.2	86.69	14JN	0.0	-3.3	-2.1	1.1
328.76	-4.81	-1.5	96.22	24JN	0.5	-3.4	-2.2	1.1
330.77	-5.41	-1.8	105.76	4JL	0.9	-3.4	-2.3	1.0
331.52	-5.97	-2.0	115.33	14JL	1.6	-3.4	-2.3	1.0
330.94	-6.41	-2.3	124.92	24JL	3.0	-3.5	-2.4	0.9
329.14	-6.64	-2.5	134.55	3AU	1.9	-3.5	-2.4	0.9
326.54	-6.58	-2.6	144.23	13AU	-0.0	-3.5	-2.4	0.8
323.89	-6.20	-2.5	153.96	23AU	-1.0	-3.4	-2.5	0.8
321.90	-5.58	-2.3	163.74	2SE	-1.2	-3.4	-2.4	0.8
321.05	-4.83	-2.0	173.58	12SE	-0.9	-3.4	-2.4	0.8
321.52	-4.05	-1.7	183.48	22SE	-0.5	-3.4	-2.4	0.9
323.22	-3.31	-1.3	193.43	2OC	-0.3	-3.3	-2.3	1.0
325.98	-2.65	-1.0	203.45	12OC	-0.2	-3.3	-2.3	1.0
329.59	-2.06	-0.8	213.50	22OC	-0.1	-3.3	-2.2	1.1
333.86	-1.56	-0.5	223.61	1NO	0.4	-3.3	-2.1	1.2
338.65	-1.13	-0.2	233.76	11NO	2.8	-3.3	-2.1	1.2
343.84	-0.76	-0.0	243.93	21NO	0.8	-3.4	-2.0	1.2
349.32	-0.44	0.2	254.12	1DE	-0.1	-3.4	-1.9	1.3
355.04	-0.17	0.4	264.33	11DE	-0.2	-3.4	-1.9	1.3
0.94	0.06	0.6	274.52	21DE	-0.3	-3.4	-1.8	1.3
6.95	0.25	0.8	284.71	31DE	-0.4	-3.5	-1.7	1.3
				789				
13.07	0.42	0.9	294.88	10JA	-0.8	-3.5	-1.7	1.3
19.26	0.57	1.1	305.01	20JA	-1.2	-3.6	-1.7	1.3
25.49	0.69	1.2	315.10	30JA	-1.2	-3.6	-1.6	1.2
31.75	0.79	1.3	325.14	9FE	-0.6	-3.7	-1.6	1.2
38.04	0.88	1.4	335.13	19FE	1.1	-3.8	-1.6	1.1
44.33	0.96	1.5	345.06	1MR	3.3	-3.9	-1.6	1.1
50.63	1.02	1.6	354.94	11MR	1.9	-4.0	-1.6	1.1
56.92	1.07	1.7	4.75	21MR	1.1	-4.0	-1.6	1.1
63.22	1.11	1.8	14.51	31MR	0.6	-4.1	-1.6	1.1
69.50	1.15	1.8	24.22	10AP	0.2	-4.2	-1.6	1.1
75.78	1.17	1.9	33.88	20AP	-0.4	-4.2	-1.6	1.0
82.06	1.19	1.9	43.50	30AP	-1.3	-4.1	-1.7	1.0
88.33	1.21	2.0	53.08	10MY	-1.7	-3.8	-1.7	1.0
94.60	1.21	2.0	62.63	20MY	-0.7	-3.0	-1.8	1.0
100.87	1.22	2.0	72.17	30MY	0.1	-3.2	-1.8	1.0
107.14	1.21	2.0	81.70	9JN	0.7	-3.9	-1.9	0.9
113.43	1.21	2.0	91.22	19JN	1.2	-4.1	-1.9	0.9
119.72	1.19	2.0	100.75	29JN	2.1	-4.2	-2.0	0.8
126.03	1.18	2.0	110.31	9JL	3.2	-4.2	-2.1	0.8
132.36	1.15	2.0	119.88	19JL	1.6	-4.1	-2.1	0.7
138.72	1.13	2.0	129.50	29JL	-0.1	-4.0	-2.2	0.7
145.10	1.10	2.0	139.15	8AU	-1.1	-3.9	-2.3	0.6
151.51	1.07	1.9	148.85	18AU	-1.4	-3.8	-2.3	0.6
157.96	1.03	1.9	158.61	28AU	-0.9	-3.8	-2.4	0.5
164.45	0.98	1.9	168.41	7SE	-0.4	-3.7	-2.4	0.5
170.98	0.94	1.9	178.28	17SE	-0.2	-3.6	-2.4	0.5
177.55	0.88	1.9	188.21	27SE	-0.0	-3.6	-2.5	0.5
184.17	0.82	1.9	198.19	7OC	0.1	-3.5	-2.4	0.6
190.84	0.76	1.9	208.23	17OC	0.6	-3.5	-2.4	0.7
197.57	0.69	1.9	218.31	27OC	3.1	-3.5	-2.4	0.7
204.34	0.62	1.9	228.44	6NO	0.6	-3.4	-2.3	0.8
211.17	0.53	1.8	238.60	16NO	-0.3	-3.4	-2.3	0.9
218.05	0.45	1.8	248.78	26NO	-0.4	-3.4	-2.2	0.9
224.98	0.35	1.8	258.98	6DE	-0.4	-3.4	-2.1	1.0
231.97	0.25	1.7	269.18	16DE	-0.5	-3.4	-2.1	1.0
239.01	0.13	1.6	279.38	26DE	-0.8	-3.3	-2.0	1.0

♂ LONG	♂ LAT	♂ MAG	⊙ LONG	16.00UT	☿	♀	♃	♄
				790	MAGNITUDES			
246.10	0.01	1.6	289.55	5JA	-1.0	-3.3	-1.9	1.0
253.24	-0.11	1.5	299.70	15JA	-1.0	-3.3	-1.9	1.0
260.42	-0.25	1.5	309.82	25JA	-0.5	-3.4	-1.8	1.0
267.65	-0.40	1.4	319.88	4FE	1.3	-3.4	-1.7	1.0
274.92	-0.55	1.3	329.90	14FE	3.0	-3.4	-1.7	0.9
282.23	-0.71	1.2	339.86	24FE	1.4	-3.4	-1.6	0.9
289.56	-0.87	1.1	349.76	6MR	0.8	-3.5	-1.6	0.9
296.91	-1.04	1.1	359.61	16MR	0.4	-3.5	-1.6	0.8
304.26	-1.22	1.0	9.40	26MR	0.1	-3.5	-1.6	0.8
311.62	-1.39	0.9	19.13	5AP	-0.5	-3.4	-1.6	0.8
318.95	-1.57	0.8	28.82	15AP	-1.3	-3.4	-1.5	0.8
326.24	-1.74	0.7	38.46	25AP	-1.7	-3.4	-1.5	0.8
333.48	-1.91	0.6	48.05	5MY	-0.7	-3.4	-1.6	0.8
340.63	-2.07	0.5	57.62	15MY	0.2	-3.3	-1.6	0.8
347.67	-2.22	0.4	67.17	25MY	0.9	-3.3	-1.6	0.8
354.56	-2.36	0.3	76.70	4JN	1.6	-3.3	-1.6	0.8
1.27	-2.49	0.2	86.22	14JN	2.8	-3.3	-1.7	0.7
7.74	-2.60	0.1	95.75	24JN	2.9	-3.3	-1.7	0.7
13.92	-2.69	-0.1	105.29	4JL	1.3	-3.4	-1.8	0.6
19.74	-2.76	-0.2	114.86	14JL	-0.1	-3.4	-1.8	0.6
25.12	-2.80	-0.4	124.45	24JL	-1.2	-3.4	-1.9	0.6
29.94	-2.82	-0.5	134.08	3AU	-1.5	-3.4	-1.9	0.5
34.07	-2.80	-0.7	143.76	13AU	-0.8	-3.5	-2.0	0.4
37.33	-2.74	-0.9	153.49	23AU	-0.3	-3.6	-2.1	0.4
39.53	-2.63	-1.1	163.26	2SE	-0.0	-3.6	-2.1	0.3
40.46	-2.44	-1.3	173.10	12SE	0.1	-3.7	-2.2	0.3
39.93	-2.16	-1.5	183.00	22SE	0.3	-3.8	-2.3	0.2
37.97	-1.78	-1.7	192.95	2OC	0.9	-3.9	-2.3	0.2
34.85	-1.30	-1.9	202.96	12OC	3.4	-4.0	-2.3	0.3
31.22	-0.76	-1.9	213.02	22OC	0.5	-4.1	-2.4	0.4
27.91	-0.22	-1.7	223.12	1NO	-0.4	-4.2	-2.4	0.4
25.59	0.26	-1.4	233.27	11NO	-0.5	-4.3	-2.4	0.5
24.59	0.65	-1.1	243.44	21NO	-0.5	-4.4	-2.3	0.6
24.93	0.95	-0.7	253.63	1DE	-0.6	-4.4	-2.3	0.6
26.44	1.18	-0.4	263.83	11DE	-0.8	-4.2	-2.3	0.7
28.93	1.34	-0.1	274.03	21DE	-0.9	-3.7	-2.2	0.7
32.19	1.45	0.1	284.22	31DE	-0.9	-3.2	-2.1	0.7
				791				
36.04	1.53	0.4	294.38	10JA	-0.4	-3.8	-2.1	0.8
40.38	1.59	0.6	304.52	20JA	1.5	-4.2	-2.0	0.8
45.07	1.62	0.8	314.61	30JA	2.5	-4.3	-1.9	0.8
50.04	1.63	1.0	324.65	9FE	1.0	-4.3	-1.8	0.8
55.24	1.64	1.1	334.64	19FE	0.5	-4.2	-1.8	0.7
60.61	1.63	1.3	344.58	1MR	0.3	-4.1	-1.7	0.7
66.12	1.62	1.4	354.45	11MR	-0.0	-4.0	-1.7	0.7
71.75	1.60	1.5	4.27	21MR	-0.5	-3.9	-1.6	0.6
77.46	1.58	1.6	14.03	31MR	-1.3	-3.8	-1.6	0.6
83.25	1.55	1.7	23.75	10AP	-1.6	-3.7	-1.5	0.6
89.11	1.51	1.8	33.41	20AP	-0.7	-3.6	-1.5	0.6
95.03	1.47	1.8	43.03	30AP	0.3	-3.6	-1.5	0.6
100.99	1.43	1.9	52.61	10MY	1.2	-3.5	-1.5	0.6
107.01	1.39	1.9	62.17	20MY	2.2	-3.4	-1.5	0.6
113.07	1.34	2.0	71.70	30MY	3.5	-3.4	-1.5	0.6
119.17	1.29	2.0	81.24	9JN	2.3	-3.4	-1.5	0.6
125.32	1.24	2.0	90.76	19JN	1.0	-3.3	-1.5	0.6
131.52	1.19	2.0	100.29	29JN	-0.2	-3.3	-1.5	0.5
137.77	1.13	2.0	109.85	9JL	-1.2	-3.3	-1.5	0.5
144.07	1.07	2.0	119.42	19JL	-1.6	-3.3	-1.6	0.5
150.42	1.01	2.0	129.03	29JL	-0.8	-3.3	-1.6	0.4
156.83	0.94	2.0	138.69	8AU	-0.3	-3.3	-1.6	0.4
163.30	0.87	2.0	148.38	18AU	0.1	-3.4	-1.7	0.3
169.83	0.80	1.9	158.13	28AU	0.3	-3.4	-1.8	0.2
176.43	0.72	1.9	167.94	7SE	0.5	-3.4	-1.8	0.2
183.10	0.64	1.8	177.80	17SE	1.2	-3.4	-1.9	0.1
189.84	0.56	1.8	187.73	27SE	3.2	-3.5	-1.9	0.1
196.65	0.47	1.8	197.71	7OC	0.3	-3.5	-2.0	0.0
203.54	0.38	1.7	207.74	17OC	-0.6	-3.5	-2.1	0.0
210.50	0.28	1.7	217.82	27OC	-0.7	-3.5	-2.1	0.1
217.53	0.19	1.7	227.95	6NO	-0.7	-3.4	-2.2	0.2
224.64	0.08	1.7	238.10	16NO	-0.7	-3.4	-2.2	0.2
231.83	-0.02	1.7	248.28	26NO	-0.7	-3.4	-2.2	0.3
239.09	-0.13	1.7	258.48	6DE	-0.7	-3.4	-2.3	0.4
246.42	-0.24	1.6	268.68	16DE	-0.7	-3.3	-2.2	0.4
253.82	-0.35	1.6	278.88	26DE	-0.3	-3.3	-2.2	0.5

♂ LONG	LAT	MAG	☉ LONG	16.00UT 792	☿	♀	♃	♄
261.28	-0.47	1.6	289.06	5JA	1.7	-3.3	-2.2	0.5
268.80	-0.58	1.5	299.21	15JA	2.0	-3.3	-2.1	0.5
276.37	-0.70	1.5	309.32	25JA	0.7	-3.4	-2.1	0.6
283.99	-0.81	1.5	319.39	4FE	0.3	-3.4	-2.0	0.6
291.64	-0.92	1.4	329.41	14FE	0.2	-3.4	-1.9	0.6
299.32	-1.03	1.4	339.38	24FE	-0.1	-3.4	-1.9	0.6
307.02	-1.12	1.3	349.29	5MR	-0.6	-3.4	-1.8	0.5
314.73	-1.22	1.3	359.13	15MR	-1.3	-3.5	-1.7	0.5
322.42	-1.30	1.3	8.92	25MR	-1.6	-3.5	-1.7	0.5
330.10	-1.37	1.2	18.66	4AP	-0.7	-3.6	-1.6	0.4
337.75	-1.43	1.2	28.35	14AP	0.4	-3.6	-1.5	0.4
345.34	-1.48	1.1	37.99	24AP	1.5	-3.7	-1.5	0.4
352.87	-1.51	1.1	47.59	4MY	2.9	-3.8	-1.5	0.4
0.33	-1.53	1.1	57.16	14MY	3.1	-3.9	-1.4	0.4
7.69	-1.53	1.0	66.71	24MY	1.8	-3.9	-1.4	0.5
14.94	-1.51	1.0	76.24	3JN	0.8	-4.0	-1.4	0.5
22.05	-1.48	0.9	85.76	13JN	-0.2	-4.1	-1.4	0.5
29.03	-1.42	0.9	95.29	23JN	-1.3	-4.2	-1.3	0.5
35.83	-1.35	0.8	104.84	3JL	-1.6	-4.2	-1.3	0.4
42.44	-1.27	0.8	114.40	13JL	-0.8	-4.0	-1.4	0.4
48.84	-1.16	0.7	123.99	23JL	-0.2	-3.6	-1.4	0.4
54.98	-1.04	0.6	133.62	2AU	0.2	-3.1	-1.4	0.3
60.82	-0.89	0.5	143.29	12AU	0.4	-3.7	-1.4	0.3
66.32	-0.72	0.4	153.01	22AU	0.7	-4.0	-1.5	0.2
71.40	-0.52	0.3	162.79	1SE	1.6	-4.3	-1.5	0.2
75.99	-0.30	0.2	172.62	11SE	2.9	-4.3	-1.5	0.1
79.96	-0.04	0.0	182.52	21SE	0.2	-4.2	-1.6	0.0
83.18	0.26	-0.2	192.47	1OC	-0.7	-4.1	-1.6	-0.0
85.47	0.61	-0.4	202.47	11OC	-0.8	-4.1	-1.7	-0.1
86.64	1.01	-0.6	212.53	21OC	-0.8	-4.0	-1.8	-0.1
86.49	1.45	-0.8	222.63	31OC	-0.7	-3.9	-1.8	-0.1
84.92	1.93	-1.0	232.77	10NO	-0.6	-3.8	-1.9	-0.1
82.04	2.39	-1.2	242.95	20NO	-0.6	-3.7	-2.0	0.0
78.28	2.78	-1.3	253.14	30NO	-0.5	-3.7	-2.0	0.0
74.38	3.06	-1.2	263.34	10DE	-0.1	-3.6	-2.1	0.2
71.07	3.19	-1.0	273.54	20DE	1.9	-3.5	-2.1	0.2
68.92	3.20	-0.7	283.73	30DE	1.6	-3.5	-2.1	0.3

793

♂ LONG	LAT	MAG	☉ LONG	16.00UT	☿	♀	♃	♄
68.11	3.13	-0.4	293.89	9JA	0.4	-3.4	-2.1	0.3
68.56	3.01	-0.1	304.03	19JA	0.2	-3.4	-2.1	0.4
70.11	2.87	0.1	314.12	29JA	0.0	-3.4	-2.1	0.4
72.57	2.72	0.4	324.17	8FE	-0.2	-3.3	-2.0	0.4
75.73	2.57	0.6	334.16	18FE	-0.6	-3.3	-2.0	0.4
79.47	2.42	0.8	344.10	28FE	-1.3	-3.3	-1.9	0.4
83.64	2.29	1.0	353.98	10MR	-1.5	-3.3	-1.9	0.4
88.18	2.16	1.1	3.80	20MR	-0.7	-3.3	-1.8	0.4
93.01	2.03	1.3	13.57	30MR	0.6	-3.3	-1.7	0.4
98.06	1.91	1.4	23.28	9AP	2.0	-3.3	-1.7	0.3
103.31	1.80	1.5	32.94	19AP	3.8	-3.4	-1.6	0.3
108.73	1.69	1.6	42.56	29AP	2.4	-3.4	-1.5	0.3
114.29	1.58	1.6	52.15	9MY	1.3	-3.4	-1.5	0.3
119.97	1.47	1.7	61.71	19MY	0.6	-3.5	-1.4	0.3
125.76	1.37	1.8	71.24	29MY	-0.3	-3.5	-1.4	0.3
131.66	1.27	1.8	80.77	8JN	-1.3	-3.5	-1.4	0.4
137.66	1.17	1.8	90.30	18JN	-1.7	-3.4	-1.3	0.4
143.74	1.07	1.9	99.83	28JN	-0.8	-3.4	-1.3	0.4
149.92	0.97	1.9	109.38	8JL	-0.1	-3.4	-1.3	0.4
156.19	0.87	1.9	118.95	18JL	0.3	-3.3	-1.3	0.4
162.55	0.77	1.9	128.56	28JL	0.6	-3.3	-1.3	0.3
168.99	0.67	1.9	138.21	7AU	1.0	-3.3	-1.3	0.3
175.53	0.57	1.9	147.91	17AU	2.1	-3.3	-1.3	0.3
182.16	0.47	1.8	157.66	27AU	2.5	-3.3	-1.3	0.2
188.88	0.36	1.8	167.46	6SE	0.1	-3.3	-1.3	0.2
195.69	0.26	1.8	177.32	16SE	-0.9	-3.4	-1.3	0.1
202.60	0.16	1.8	187.24	26SE	-1.0	-3.4	-1.4	0.0
209.60	0.05	1.7	197.22	6OC	-0.9	-3.4	-1.4	-0.0
216.70	-0.05	1.7	207.25	16OC	-0.6	-3.5	-1.5	-0.1
223.88	-0.16	1.6	217.33	26OC	-0.5	-3.5	-1.5	-0.2
231.16	-0.26	1.6	227.45	5NO	-0.4	-3.6	-1.6	-0.2
238.53	-0.36	1.5	237.61	15NO	-0.4	-3.7	-1.6	-0.2
245.98	-0.46	1.5	247.79	25NO	0.0	-3.7	-1.7	-0.2
253.50	-0.56	1.5	257.99	5DE	2.2	-3.8	-1.8	-0.1
261.10	-0.65	1.5	268.19	15DE	1.3	-3.9	-1.9	-0.0
268.76	-0.74	1.5	278.39	25DE	0.2	-4.0	-1.9	0.1

♂ LONG	LAT	MAG	☉ LONG	16.00UT 794	☿	♀	♃	♄
276.48	-0.83	1.5	288.57	4JA	0.0	-4.1	-1.9	0.1
284.25	-0.90	1.4	298.72	14JA	-0.1	-4.2	-2.0	0.2
292.05	-0.97	1.4	308.83	24JA	-0.3	-4.3	-2.0	0.2
299.88	-1.04	1.4	318.91	3FE	-0.7	-4.3	-2.0	0.3
307.72	-1.09	1.4	328.93	13FE	-1.3	-4.3	-2.0	0.3
315.56	-1.13	1.4	338.90	23FE	-1.4	-4.1	-2.0	0.3
323.39	-1.16	1.4	348.81	5MR	-0.7	-3.6	-2.0	0.3
331.20	-1.17	1.4	358.66	15MR	0.7	-3.2	-1.9	0.4
338.96	-1.18	1.4	8.45	25MR	2.5	-3.8	-1.9	0.3
346.69	-1.17	1.4	18.19	4AP	3.1	-4.1	-1.8	0.3
354.35	-1.14	1.4	27.88	14AP	1.7	-4.2	-1.8	0.3
1.95	-1.11	1.4	37.52	24AP	1.0	-4.2	-1.7	0.3
9.46	-1.06	1.4	47.12	4MY	0.4	-4.1	-1.6	0.2
16.89	-1.00	1.4	56.70	14MY	-0.3	-4.0	-1.6	0.2
24.22	-0.93	1.4	66.24	24MY	-1.3	-3.9	-1.5	0.2
31.45	-0.85	1.4	75.77	3JN	-1.7	-3.8	-1.5	0.3
38.56	-0.75	1.4	85.30	13JN	-0.8	-3.7	-1.4	0.3
45.55	-0.65	1.3	94.83	23JN	-0.1	-3.7	-1.4	0.3
52.42	-0.53	1.3	104.37	3JL	0.4	-3.6	-1.3	0.3
59.15	-0.41	1.3	113.93	13JL	0.8	-3.5	-1.3	0.3
65.73	-0.27	1.3	123.52	23JL	1.3	-3.5	-1.3	0.3
72.15	-0.13	1.2	133.15	2AU	2.6	-3.5	-1.2	0.3
78.40	0.03	1.2	142.82	12AU	2.2	-3.4	-1.2	0.3
84.45	0.20	1.2	152.54	22AU	0.1	-3.4	-1.2	0.3
90.29	0.38	1.1	162.32	1SE	-1.0	-3.4	-1.2	0.2
95.87	0.57	1.0	172.15	11SE	-1.1	-3.4	-1.2	0.2
101.16	0.78	0.9	182.04	21SE	-1.0	-3.4	-1.2	0.1
106.11	1.02	0.8	191.99	1OC	-0.6	-3.4	-1.2	0.1
110.64	1.27	0.7	201.99	11OC	-0.3	-3.4	-1.3	-0.0
114.67	1.56	0.5	212.04	21OC	-0.3	-3.4	-1.3	-0.1
118.08	1.88	0.4	222.15	31OC	-0.2	-3.4	-1.4	-0.2
120.72	2.23	0.2	232.29	10NO	0.2	-3.4	-1.4	-0.2
122.43	2.62	-0.0	242.45	20NO	2.5	-3.4	-1.5	-0.3
123.00	3.05	-0.3	252.65	30NO	1.0	-3.4	-1.5	-0.3
122.27	3.48	-0.5	262.85	10DE	0.0	-3.5	-1.6	-0.2
120.20	3.89	-0.7	273.04	20DE	-0.1	-3.5	-1.6	-0.1
116.96	4.22	-0.9	283.24	30DE	-0.2	-3.5	-1.7	-0.0

795

♂ LONG	LAT	MAG	☉ LONG	16.00UT	☿	♀	♃	♄
113.05	4.39	-1.0	293.40	9JA	-0.4	-3.5	-1.8	0.0
109.20	4.40	-0.9	303.53	19JA	-0.7	-3.4	-1.8	0.1
106.09	4.24	-0.7	313.63	29JA	-1.2	-3.4	-1.9	0.1
104.16	3.98	-0.4	323.68	8FE	-1.2	-3.4	-1.9	0.2
103.54	3.67	-0.2	333.67	18FE	-0.6	-3.4	-2.0	0.2
104.13	3.35	0.1	343.62	28FE	0.9	-3.3	-2.0	0.3
105.79	3.04	0.3	353.50	10MR	3.0	-3.3	-2.0	0.3
108.30	2.75	0.5	3.32	20MR	2.3	-3.3	-2.0	0.3
111.51	2.48	0.7	13.09	30MR	1.3	-3.3	-2.0	0.3
115.29	2.24	0.9	22.81	9AP	0.7	-3.3	-2.0	0.3
119.51	2.01	1.0	32.47	19AP	0.3	-3.4	-1.9	0.3
124.11	1.81	1.1	42.10	29AP	-0.4	-3.4	-1.9	0.3
129.01	1.61	1.2	51.68	9MY	-1.3	-3.4	-1.8	0.2
134.17	1.43	1.3	61.24	19MY	-1.7	-3.4	-1.7	0.2
139.56	1.26	1.4	70.78	29MY	-0.8	-3.5	-1.7	0.2
145.14	1.10	1.5	80.31	8JN	0.0	-3.5	-1.6	0.2
150.90	0.95	1.5	89.83	18JN	0.6	-3.6	-1.5	0.3
156.82	0.81	1.6	99.37	28JN	1.0	-3.6	-1.5	0.3
162.89	0.66	1.6	108.91	8JL	1.8	-3.7	-1.4	0.3
169.10	0.53	1.6	118.49	18JL	3.1	-3.8	-1.4	0.3
175.44	0.40	1.7	128.10	28JL	1.8	-3.9	-1.3	0.3
181.92	0.27	1.7	137.74	7AU	0.0	-4.0	-1.3	0.3
188.52	0.14	1.7	147.44	17AU	-1.1	-4.1	-1.3	0.3
195.24	0.02	1.7	157.19	27AU	-1.3	-4.2	-1.3	0.3
202.08	-0.09	1.7	166.98	6SE	-0.9	-4.3	-1.2	0.3
209.04	-0.21	1.6	176.84	16SE	-0.5	-4.3	-1.2	0.2
216.12	-0.32	1.6	186.76	26SE	-0.2	-4.1	-1.2	0.2
223.30	-0.42	1.6	196.74	6OC	-0.1	-3.6	-1.2	0.1
230.59	-0.52	1.6	206.76	16OC	-0.0	-3.2	-1.2	0.1
237.99	-0.61	1.5	216.84	26OC	0.5	-3.8	-1.2	-0.0
245.48	-0.70	1.5	226.96	5NO	2.8	-4.2	-1.3	-0.1
253.05	-0.78	1.5	237.12	15NO	0.8	-4.4	-1.3	-0.1
260.71	-0.85	1.5	247.30	25NO	-0.2	-4.4	-1.3	-0.2
268.44	-0.92	1.4	257.50	5DE	-0.3	-4.3	-1.4	-0.3
276.23	-0.97	1.4	267.70	15DE	-0.3	-4.2	-1.4	-0.2
284.06	-1.02	1.4	277.90	25DE	-0.5	-4.1	-1.5	-0.2

Left table (796 / 797)

♂ LONG	LAT	MAG	☉ LONG	16.00UT	☿	♀	♃	♄
291.93	-1.05	1.3	288.07	4JA	-0.8	-4.0	-1.6	-0.1
299.82	-1.07	1.3	298.23	14JA	-1.1	-3.9	-1.6	-0.0
307.72	-1.08	1.3	308.35	24JA	-1.1	-3.8	-1.7	0.0
315.61	-1.08	1.3	318.42	3FE	-0.6	-3.7	-1.8	0.1
323.49	-1.07	1.3	328.45	13FE	1.1	-3.6	-1.8	0.2
331.34	-1.05	1.4	338.42	23FE	3.2	-3.6	-1.9	0.2
339.15	-1.01	1.4	348.33	4MR	1.7	-3.5	-1.9	0.2
346.90	-0.96	1.4	358.19	14MR	0.9	-3.4	-2.0	0.3
354.59	-0.91	1.4	7.99	24MR	0.5	-3.4	-2.0	0.3
2.21	-0.84	1.5	17.73	3AP	0.2	-3.4	-2.0	0.3
9.76	-0.77	1.5	27.42	13AP	-0.4	-3.3	-2.0	0.3
17.22	-0.68	1.5	37.06	23AP	-1.3	-3.3	-2.0	0.3
24.59	-0.59	1.5	46.67	3MY	-1.7	-3.3	-2.0	0.3
31.87	-0.50	1.6	56.24	13MY	-0.8	-3.3	-2.0	0.3
39.05	-0.39	1.6	65.79	23MY	0.1	-3.3	-1.9	0.3
46.14	-0.29	1.6	75.32	2JN	0.8	-3.3	-1.9	0.3
53.13	-0.18	1.6	84.84	12JN	1.4	-3.3	-1.8	0.2
60.02	-0.06	1.6	94.37	22JN	2.4	-3.4	-1.8	0.3
66.80	0.06	1.6	103.91	2JL	3.2	-3.4	-1.7	0.3
73.49	0.18	1.7	113.47	12JL	1.5	-3.4	-1.6	0.3
80.06	0.30	1.6	123.06	22JL	-0.1	-3.5	-1.6	0.4
86.53	0.43	1.6	132.68	1AU	-1.1	-3.5	-1.5	0.4
92.88	0.57	1.6	142.35	11AU	-1.4	-3.5	-1.5	0.4
99.11	0.70	1.6	152.07	21AU	-0.9	-3.4	-1.4	0.4
105.21	0.85	1.6	161.84	31AU	-0.4	-3.4	-1.4	0.4
111.17	0.99	1.5	171.67	10SE	-0.1	-3.4	-1.3	0.4
116.96	1.15	1.5	181.55	20SE	0.0	-3.4	-1.3	0.3
122.58	1.31	1.4	191.49	30SE	0.2	-3.3	-1.3	0.3
127.99	1.49	1.3	201.50	10OC	0.7	-3.3	-1.3	0.3
133.15	1.67	1.2	211.55	20OC	3.1	-3.3	-1.3	0.2
138.02	1.87	1.1	221.65	30OC	0.6	-3.3	-1.3	0.1
142.54	2.09	0.9	231.79	9NO	-0.3	-3.3	-1.3	0.1
146.62	2.33	0.8	241.96	19NO	-0.4	-3.4	-1.3	-0.0
150.18	2.59	0.6	252.15	29NO	-0.5	-3.4	-1.3	-0.1
153.07	2.88	0.4	262.35	9DE	-0.6	-3.4	-1.3	-0.2
155.14	3.19	0.2	272.55	19DE	-0.8	-3.4	-1.4	-0.2
156.23	3.52	-0.1	282.74	29DE	-0.9	-3.5	-1.4	-0.2

797

♂ LONG	LAT	MAG	☉ LONG	16.00UT	☿	♀	♃	♄
156.13	3.85	-0.3	292.91	8JA	-1.0	-3.5	-1.5	-0.1
154.74	4.14	-0.6	303.05	18JA	-0.5	-3.6	-1.5	-0.0
152.10	4.36	-0.8	313.15	28JA	1.3	-3.7	-1.6	0.0
148.52	4.44	-1.0	323.20	7FE	2.8	-3.7	-1.7	0.1
144.60	4.34	-1.0	333.20	17FE	1.3	-3.8	-1.7	0.1
141.07	4.09	-0.8	343.14	27FE	0.7	-3.9	-1.8	0.2
138.49	3.72	-0.6	353.03	9MR	0.4	-4.0	-1.9	0.3
137.19	3.30	-0.4	2.85	19MR	0.1	-4.0	-1.9	0.3
137.16	2.88	-0.2	12.62	29MR	-0.5	-4.1	-2.0	0.3
138.30	2.47	0.1	22.34	8AP	-1.3	-4.2	-2.0	0.4
140.42	2.10	0.3	32.01	18AP	-1.7	-4.2	-2.1	0.4
143.36	1.77	0.4	41.63	28AP	-0.8	-4.1	-2.1	0.4
146.97	1.47	0.6	51.22	8MY	0.2	-3.7	-2.1	0.4
151.12	1.19	0.7	60.78	18MY	1.0	-3.0	-2.1	0.4
155.71	0.95	0.8	70.32	28MY	1.8	-3.2	-2.1	0.4
160.69	0.72	0.9	79.85	7JN	3.1	-3.9	-2.1	0.4
165.99	0.51	1.0	89.37	17JN	2.7	-4.1	-2.1	0.3
171.57	0.32	1.1	98.90	27JN	1.2	-4.2	-2.0	0.3
177.40	0.15	1.2	108.45	7JL	-0.1	-4.2	-2.0	0.4
183.45	-0.02	1.2	118.02	17JL	-1.2	-4.1	-1.9	0.4
189.70	-0.17	1.3	127.63	27JL	-1.5	-4.0	-1.8	0.4
196.15	-0.31	1.3	137.28	6AU	-0.9	-3.9	-1.8	0.5
202.76	-0.44	1.3	146.96	16AU	-0.3	-3.8	-1.7	0.5
209.54	-0.56	1.3	156.71	26AU	-0.0	-3.7	-1.6	0.5
216.47	-0.68	1.4	166.51	5SE	0.2	-3.7	-1.6	0.5
223.54	-0.78	1.4	176.36	15SE	0.4	-3.6	-1.5	0.5
230.75	-0.87	1.4	186.28	25SE	1.0	-3.6	-1.5	0.5
238.09	-0.94	1.4	196.25	5OC	3.4	-3.5	-1.5	0.5
245.53	-1.01	1.4	206.27	15OC	0.5	-3.5	-1.4	0.4
253.08	-1.07	1.4	216.35	25OC	-0.5	-3.5	-1.4	0.4
260.72	-1.11	1.4	226.47	4NO	-0.6	-3.4	-1.4	0.3
268.44	-1.14	1.4	236.62	14NO	-0.6	-3.4	-1.4	0.3
276.23	-1.16	1.4	246.80	24NO	-0.7	-3.4	-1.4	0.2
284.06	-1.17	1.4	257.00	4DE	-0.8	-3.4	-1.4	0.1
291.93	-1.16	1.4	267.20	14DE	-0.8	-3.4	-1.4	0.0
299.82	-1.14	1.4	277.40	24DE	-0.8	-3.3	-1.4	-0.0

Right table (798 / 799)

♂ LONG	LAT	MAG	☉ LONG	16.00UT	☿	♀	♃	♄
307.72	-1.11	1.4	287.58	3JA	-0.4	-3.3	-1.4	-0.1
315.60	-1.07	1.4	297.74	13JA	1.5	-3.3	-1.4	-0.0
323.47	-1.02	1.4	307.86	23JA	2.3	-3.4	-1.5	0.0
331.30	-0.96	1.4	317.94	2FE	0.9	-3.4	-1.5	0.1
339.08	-0.88	1.4	327.96	12FE	0.5	-3.4	-1.6	0.1
346.81	-0.81	1.4	337.94	22FE	0.2	-3.4	-1.6	0.2
354.47	-0.72	1.4	347.85	4MR	-0.0	-3.5	-1.7	0.3
2.05	-0.63	1.4	357.71	14MR	-0.5	-3.5	-1.8	0.3
9.56	-0.54	1.5	7.51	24MR	-1.3	-3.5	-1.8	0.4
16.99	-0.44	1.5	17.25	3AP	-1.6	-3.4	-1.9	0.4
24.33	-0.34	1.5	26.94	13AP	-0.8	-3.4	-2.0	0.4
31.58	-0.23	1.6	36.59	23AP	0.3	-3.4	-2.1	0.5
38.75	-0.13	1.6	46.20	3MY	1.3	-3.4	-2.1	0.5
45.82	-0.02	1.7	55.77	13MY	2.4	-3.3	-2.2	0.5
52.82	0.08	1.7	65.32	23MY	3.5	-3.3	-2.2	0.5
59.72	0.19	1.7	74.85	2JN	2.1	-3.3	-2.2	0.5
66.55	0.29	1.8	84.38	12JN	0.9	-3.3	-2.3	0.5
73.30	0.39	1.8	93.91	22JN	-0.2	-3.3	-2.3	0.5
79.96	0.50	1.8	103.44	2JL	-1.2	-3.4	-2.2	0.5
86.55	0.60	1.8	113.00	12JL	-1.6	-3.4	-2.2	0.5
93.07	0.70	1.9	122.59	22JL	-0.8	-3.4	-2.2	0.5
99.52	0.80	1.9	132.21	1AU	-0.2	-3.4	-2.1	0.6
105.89	0.90	1.9	141.88	11AU	0.1	-3.5	-2.1	0.6
112.18	1.00	1.9	151.59	21AU	0.3	-3.6	-2.0	0.6
118.40	1.10	1.8	161.36	31AU	0.6	-3.6	-1.9	0.6
124.54	1.20	1.8	171.19	10SE	1.3	-3.7	-1.9	0.7
130.59	1.30	1.8	181.07	20SE	3.2	-3.8	-1.8	0.7
136.55	1.40	1.7	191.01	30SE	0.3	-3.9	-1.8	0.7
142.40	1.50	1.7	201.01	10OC	-0.6	-4.0	-1.7	0.7
148.13	1.61	1.6	211.06	20OC	-0.8	-4.1	-1.6	0.6
153.72	1.72	1.5	221.15	30OC	-0.7	-4.2	-1.6	0.6
159.15	1.84	1.4	231.29	9NO	-0.8	-4.3	-1.6	0.6
164.39	1.96	1.3	241.46	19NO	-0.7	-4.4	-1.5	0.5
169.40	2.08	1.2	251.65	29NO	-0.6	-4.4	-1.5	0.4
174.12	2.22	1.0	261.85	9DE	-0.4	-4.2	-1.5	0.4
178.50	2.36	0.9	272.05	19DE	-0.2	-3.7	-1.5	0.3
182.45	2.51	0.7	282.24	29DE	1.7	-3.2	-1.5	0.2

799

♂ LONG	LAT	MAG	☉ LONG	16.00UT	☿	♀	♃	♄
185.87	2.66	0.5	292.41	8JA	1.9	-3.8	-1.5	0.1
188.63	2.82	0.2	302.56	18JA	0.6	-4.2	-1.5	0.1
190.56	2.98	-0.0	312.65	28JA	0.3	-4.3	-1.5	0.1
191.50	3.12	-0.3	322.71	7FE	0.1	-4.3	-1.5	0.2
191.26	3.23	-0.6	332.71	17FE	-0.1	-4.2	-1.6	0.2
189.75	3.26	-0.9	342.66	27FE	-0.6	-4.1	-1.6	0.3
187.04	3.20	-1.1	352.55	9MR	-1.3	-4.0	-1.6	0.4
183.49	3.00	-1.3	2.38	19MR	-1.5	-3.9	-1.7	0.4
179.73	2.67	-1.3	12.15	29MR	-0.7	-3.8	-1.8	0.5
176.47	2.25	-1.1	21.87	8AP	0.4	-3.7	-1.8	0.5
174.27	1.78	-0.9	31.55	18AP	1.7	-3.6	-1.9	0.6
173.38	1.32	-0.7	41.17	28AP	3.2	-3.6	-2.0	0.6
173.79	0.89	-0.5	50.76	8MY	2.8	-3.5	-2.0	0.6
175.38	0.51	-0.3	60.32	18MY	1.6	-3.4	-2.1	0.7
177.97	0.18	-0.1	69.86	28MY	0.7	-3.4	-2.2	0.7
181.39	-0.11	0.1	79.39	7JN	-0.2	-3.4	-2.2	0.7
185.49	-0.37	0.2	88.92	17JN	-1.2	-3.3	-2.3	0.7
190.16	-0.58	0.3	98.45	27JN	-1.7	-3.3	-2.3	0.7
195.30	-0.77	0.4	107.99	7JL	-0.8	-3.3	-2.4	0.7
200.84	-0.94	0.5	117.57	17JL	-0.2	-3.3	-2.4	0.7
206.73	-1.08	0.6	127.17	27JL	0.2	-3.3	-2.4	0.7
212.92	-1.20	0.7	136.81	6AU	0.5	-3.3	-2.4	0.7
219.37	-1.30	0.8	146.50	16AU	0.8	-3.4	-2.3	0.8
226.05	-1.38	0.8	156.24	26AU	1.7	-3.4	-2.3	0.8
232.93	-1.44	0.9	166.03	5SE	2.8	-3.4	-2.2	0.9
240.00	-1.48	0.9	175.89	15SE	0.2	-3.4	-2.2	0.9
247.22	-1.51	1.0	185.79	25SE	-0.8	-3.5	-2.1	0.9
254.59	-1.52	1.0	195.76	5OC	-0.9	-3.5	-2.1	0.9
262.07	-1.52	1.1	205.79	15OC	-0.9	-3.5	-2.0	0.9
269.65	-1.49	1.1	215.85	25OC	-0.7	-3.5	-1.9	0.9
277.31	-1.46	1.1	225.97	4NO	-0.5	-3.4	-1.9	0.9
285.04	-1.41	1.2	236.12	14NO	-0.5	-3.4	-1.8	0.8
292.80	-1.34	1.2	246.30	24NO	-0.5	-3.4	-1.7	0.8
300.59	-1.27	1.3	256.50	4DE	-0.1	-3.4	-1.7	0.7
308.39	-1.18	1.3	266.70	14DE	1.9	-3.3	-1.7	0.7
316.18	-1.09	1.3	276.90	24DE	1.5	-3.3	-1.6	0.6

♂ LONG	♂ LAT	♂ MAG	☉ LONG	16.00UT	☿	♀	♃	♄
				800	MAGNITUDES			
323.95	-0.99	1.4	287.08	3JA	0.4	-3.3	-1.6	0.5
331.68	-0.89	1.4	297.24	13JA	0.1	-3.3	-1.6	0.4
339.36	-0.78	1.5	307.36	23JA	-0.0	-3.4	-1.6	0.4
346.98	-0.66	1.5	317.44	2FE	-0.2	-3.4	-1.6	0.3
354.54	-0.55	1.5	327.47	12FE	-0.6	-3.4	-1.6	0.4
2.02	-0.44	1.6	337.45	22FE	-1.3	-3.4	-1.6	0.4
9.44	-0.32	1.6	347.37	3MR	-1.4	-3.4	-1.6	0.5
16.76	-0.21	1.6	357.23	13MR	-0.7	-3.5	-1.6	0.5
24.01	-0.10	1.7	7.03	23MR	0.6	-3.5	-1.6	0.6
31.18	0.01	1.7	16.79	2AP	2.1	-3.6	-1.7	0.6
38.26	0.11	1.7	26.48	12AP	3.7	-3.6	-1.7	0.7
45.27	0.22	1.7	36.13	22AP	2.1	-3.7	-1.8	0.7
52.20	0.31	1.7	45.74	2MY	1.2	-3.8	-1.8	0.8
59.06	0.41	1.7	55.32	12MY	0.5	-3.9	-1.9	0.8
65.84	0.50	1.7	64.86	22MY	-0.3	-3.9	-2.0	0.9
72.56	0.58	1.8	74.40	1JN	-1.3	-4.0	-2.0	0.9
79.22	0.67	1.8	83.92	11JN	-1.7	-4.1	-2.1	0.9
85.82	0.74	1.9	93.45	21JN	-0.8	-4.2	-2.2	0.9
92.36	0.82	1.9	102.99	1JL	-0.1	-4.2	-2.2	1.0
98.86	0.89	1.9	112.54	11JL	0.3	-4.0	-2.3	1.0
105.31	0.96	2.0	122.13	21JL	0.7	-3.6	-2.4	1.0
111.71	1.02	2.0	131.75	31JL	1.1	-3.1	-2.4	1.0
118.08	1.09	2.0	141.41	10AU	2.2	-3.7	-2.4	1.0
124.41	1.15	2.0	151.12	20AU	2.5	-4.1	-2.5	1.0
130.70	1.20	2.0	160.89	30AU	0.2	-4.3	-2.5	1.1
136.95	1.25	2.0	170.71	9SE	-0.9	-4.3	-2.4	1.1
143.17	1.31	2.0	180.59	19SE	-1.1	-4.2	-2.4	1.1
149.34	1.35	1.9	190.53	29SE	-1.0	-4.1	-2.4	1.2
155.48	1.40	1.9	200.52	9OC	-0.6	-4.1	-2.3	1.2
161.57	1.44	1.9	210.57	19OC	-0.4	-4.0	-2.3	1.2
167.60	1.48	1.8	220.67	29OC	-0.3	-3.9	-2.2	1.2
173.57	1.52	1.7	230.80	8NO	-0.3	-3.8	-2.1	1.2
179.48	1.56	1.6	240.96	18NO	0.1	-3.7	-2.0	1.1
185.30	1.59	1.6	251.16	28NO	2.2	-3.7	-2.0	1.1
191.01	1.62	1.4	261.35	8DE	1.2	-3.6	-1.9	1.1
196.61	1.64	1.3	271.55	18DE	0.1	-3.5	-1.8	1.0
202.06	1.65	1.2	281.75	28DE	-0.1	-3.5	-1.8	0.9
				801				
207.33	1.66	1.0	291.92	7JA	-0.1	-3.4	-1.7	0.9
212.38	1.66	0.9	302.06	17JA	-0.3	-3.4	-1.7	0.8
217.16	1.64	0.7	312.17	27JA	-0.7	-3.4	-1.7	0.7
221.61	1.61	0.5	322.22	6FE	-1.3	-3.3	-1.6	0.7
225.64	1.55	0.3	332.23	16FE	-1.3	-3.3	-1.6	0.6
229.15	1.46	0.0	342.18	26FE	-0.7	-3.3	-1.6	0.7
231.99	1.33	-0.3	352.07	8MR	0.7	-3.3	-1.6	0.7
234.02	1.15	-0.6	1.90	18MR	2.7	-3.3	-1.6	0.8
235.03	0.89	-0.9	11.68	28MR	2.8	-3.3	-1.6	0.8
234.88	0.56	-1.2	21.40	7AP	1.6	-3.3	-1.6	0.9
233.47	0.14	-1.5	31.08	17AP	0.9	-3.4	-1.6	0.9
230.92	-0.37	-1.8	40.71	27AP	0.4	-3.4	-1.6	1.0
227.66	-0.91	-2.0	50.30	7MY	-0.3	-3.4	-1.7	1.0
224.35	-1.45	-1.9	59.86	17MY	-1.3	-3.5	-1.7	1.1
221.71	-1.91	-1.7	69.40	27MY	-1.7	-3.5	-1.8	1.1
220.26	-2.27	-1.5	78.93	6JN	-0.8	-3.5	-1.8	1.1
220.18	-2.53	-1.3	88.45	16JN	-0.0	-3.4	-1.9	1.2
221.44	-2.70	-1.1	97.98	26JN	0.5	-3.4	-1.9	1.2
223.89	-2.80	-0.9	107.53	6JL	0.9	-3.4	-2.0	1.2
227.32	-2.85	-0.7	117.10	16JL	1.5	-3.3	-2.1	1.2
231.58	-2.85	-0.5	126.70	26JL	2.8	-3.3	-2.1	1.3
236.51	-2.83	-0.4	136.34	5AU	2.1	-3.3	-2.2	1.3
241.98	-2.77	-0.2	146.02	15AU	0.1	-3.3	-2.3	1.3
247.90	-2.69	-0.1	155.76	25AU	-1.0	-3.3	-2.3	1.3
254.17	-2.58	0.1	165.55	4SE	-1.2	-3.3	-2.4	1.3
260.74	-2.46	0.2	175.40	14SE	-1.0	-3.4	-2.4	1.3
267.56	-2.33	0.3	185.31	24SE	-0.5	-3.4	-2.4	1.3
274.56	-2.18	0.4	195.27	4OC	-0.3	-3.4	-2.4	1.3
281.71	-2.02	0.5	205.30	14OC	-0.2	-3.5	-2.4	1.3
288.99	-1.85	0.6	215.37	24OC	-0.1	-3.5	-2.4	1.3
296.34	-1.68	0.7	225.48	3NO	0.3	-3.6	-2.4	1.3
303.76	-1.51	0.8	235.63	13NO	2.5	-3.7	-2.3	1.2
311.20	-1.33	0.9	245.81	23NO	1.0	-3.7	-2.3	1.2
318.66	-1.16	1.0	256.00	3DE	-0.1	-3.8	-2.2	1.2
326.12	-0.99	1.1	266.21	13DE	-0.2	-3.9	-2.1	1.1
333.55	-0.82	1.2	276.41	23DE	-0.3	-4.0	-2.0	1.1

♂ LONG	♂ LAT	♂ MAG	☉ LONG	16.00UT	☿	♀	♃	♄
				802	MAGNITUDES			
340.95	-0.66	1.2	286.59	2JA	-0.4	-4.1	-2.0	1.0
348.31	-0.50	1.3	296.75	12JA	-0.7	-4.2	-1.9	1.0
355.62	-0.35	1.4	306.87	22JA	-1.2	-4.3	-1.8	0.9
2.86	-0.21	1.5	316.95	1FE	-1.2	-4.3	-1.8	0.9
10.05	-0.08	1.5	326.99	11FE	-0.6	-4.3	-1.7	0.8
17.17	0.05	1.6	336.96	21FE	0.9	-4.1	-1.7	0.8
24.22	0.17	1.7	346.89	3MR	3.1	-3.6	-1.6	0.8
31.20	0.28	1.7	356.75	13MR	2.1	-3.2	-1.6	0.9
38.12	0.38	1.8	6.56	23MR	1.1	-3.8	-1.6	1.0
44.97	0.47	1.8	16.31	2AP	0.7	-4.1	-1.6	1.0
51.76	0.56	1.8	26.01	12AP	0.2	-4.2	-1.5	1.1
58.49	0.64	1.9	35.66	22AP	-0.4	-4.2	-1.5	1.2
65.16	0.72	1.9	45.27	2MY	-1.3	-4.1	-1.5	1.2
71.79	0.78	1.9	54.85	12MY	-1.7	-4.0	-1.5	1.3
78.36	0.85	1.9	64.40	22MY	-0.8	-3.9	-1.6	1.3
84.89	0.90	1.9	73.93	1JN	0.0	-3.8	-1.6	1.3
91.39	0.95	1.9	83.46	11JN	0.6	-3.7	-1.6	1.4
97.85	1.00	1.9	92.99	21JN	1.2	-3.7	-1.7	1.4
104.28	1.04	1.9	102.52	1JL	2.0	-3.6	-1.7	1.4
110.69	1.08	1.9	112.08	11JL	3.2	-3.5	-1.7	1.4
117.08	1.11	2.0	121.66	21JL	1.7	-3.5	-1.8	1.3
123.46	1.14	2.0	131.28	31JL	0.0	-3.4	-1.9	1.3
129.83	1.16	2.0	140.95	10AU	-1.1	-3.4	-1.9	1.3
136.18	1.18	2.0	150.65	20AU	-1.3	-3.4	-2.0	1.2
142.53	1.19	2.0	160.42	30AU	-0.9	-3.4	-2.1	1.2
148.89	1.20	2.0	170.24	9SE	-0.4	-3.4	-2.1	1.1
155.23	1.21	2.0	180.11	19SE	-0.2	-3.4	-2.2	1.1
161.59	1.21	2.0	190.05	29SE	-0.1	-3.4	-2.2	1.1
167.94	1.21	2.0	200.04	9OC	0.0	-3.4	-2.3	1.1
174.29	1.20	2.0	210.08	19OC	0.5	-3.4	-2.3	1.1
180.65	1.18	1.9	220.18	29OC	2.8	-3.4	-2.4	1.1
187.01	1.16	1.9	230.31	8NO	0.8	-3.4	-2.4	1.1
193.36	1.14	1.8	240.47	18NO	-0.2	-3.4	-2.3	1.1
199.72	1.10	1.7	250.66	28NO	-0.4	-3.4	-2.3	1.1
206.06	1.06	1.7	260.86	8DE	-0.4	-3.5	-2.3	1.0
212.39	1.01	1.6	271.06	18DE	-0.5	-3.5	-2.2	1.0
218.70	0.94	1.5	281.25	28DE	-0.8	-3.5	-2.2	0.9
				803				
224.98	0.87	1.4	291.43	7JA	-1.0	-3.5	-2.1	0.9
231.23	0.77	1.3	301.57	17JA	-1.1	-3.4	-2.0	0.8
237.44	0.66	1.1	311.67	27JA	-0.6	-3.4	-1.9	0.8
243.60	0.54	1.0	321.73	6FE	1.1	-3.4	-1.9	0.7
249.68	0.38	0.8	331.74	16FE	3.1	-3.4	-1.8	0.7
255.68	0.21	0.7	341.69	26FE	1.5	-3.3	-1.7	0.7
261.56	-0.00	0.5	351.59	8MR	0.8	-3.3	-1.7	0.6
267.29	-0.25	0.3	1.42	18MR	0.5	-3.3	-1.6	0.7
272.85	-0.54	0.1	11.20	28MR	0.1	-3.3	-1.6	0.7
278.17	-0.87	-0.1	20.93	7AP	-0.4	-3.3	-1.5	0.8
283.18	-1.26	-0.3	30.60	17AP	-1.2	-3.4	-1.5	0.9
287.80	-1.71	-0.6	40.24	27AP	-1.7	-3.4	-1.5	0.9
291.91	-2.24	-0.8	49.83	7MY	-0.8	-3.4	-1.5	1.0
295.37	-2.84	-1.1	59.39	17MY	0.1	-3.4	-1.4	1.1
297.99	-3.51	-1.4	68.94	27MY	0.8	-3.5	-1.4	1.1
299.59	-4.25	-1.7	78.47	6JN	1.5	-3.5	-1.4	1.1
300.00	-5.02	-2.0	87.99	16JN	2.6	-3.6	-1.4	1.2
299.14	-5.75	-2.3	97.52	26JN	3.0	-3.6	-1.5	1.2
297.18	-6.33	-2.5	107.07	6JL	1.4	-3.7	-1.5	1.2
294.61	-6.66	-2.6	116.64	16JL	-0.0	-3.8	-1.5	1.2
292.12	-6.67	-2.5	126.24	26JL	-1.1	-3.9	-1.5	1.2
290.42	-6.38	-2.3	135.88	5AU	-1.5	-4.0	-1.6	1.2
289.92	-5.90	-2.0	145.56	15AU	-0.9	-4.1	-1.6	1.1
290.76	-5.31	-1.7	155.30	25AU	-0.4	-4.2	-1.7	1.1
292.84	-4.69	-1.5	165.08	4SE	-0.1	-4.3	-1.7	1.1
295.97	-4.08	-1.2	174.93	14SE	0.1	-4.3	-1.8	1.0
299.95	-3.50	-0.9	184.83	24SE	0.2	-4.1	-1.9	1.0
304.61	-2.97	-0.7	194.79	4OC	0.8	-3.6	-1.9	1.0
309.78	-2.49	-0.5	204.81	14OC	3.2	-3.2	-2.0	1.0
315.35	-2.06	-0.2	214.88	24OC	0.6	-3.8	-2.1	1.0
321.23	-1.66	-0.0	224.99	3NO	-0.4	-4.2	-2.1	1.0
327.33	-1.31	0.2	235.14	13NO	-0.5	-4.4	-2.2	1.0
333.61	-1.00	0.3	245.32	23NO	-0.5	-4.4	-2.2	1.0
340.01	-0.72	0.5	255.51	3DE	-0.6	-4.3	-2.2	1.0
346.49	-0.47	0.7	265.71	13DE	-0.8	-4.2	-2.2	0.9
353.04	-0.25	0.8	275.91	23DE	-0.9	-4.1	-2.2	0.9

♂			☉	16.00UT	☿ ♀ ♃ ♄					♂			☉	16.00UT	☿ ♀ ♃ ♄			
LONG	LAT	MAG	LONG	804	MAGNITUDES					LONG	LAT	MAG	LONG	806	MAGNITUDES			
359.63	-0.06	1.0	286.10	2JA	-0.9	-4.0	-2.2	0.9		22.07	0.93	0.4	285.61	1JA	1.8	-3.3	-1.9	0.8
6.23	0.12	1.1	296.25	12JA	-0.5	-3.9	-2.1	0.8		27.07	1.06	0.6	295.76	11JA	0.5	-3.4	-2.0	0.8
12.85	0.27	1.2	306.38	22JA	1.3	-3.8	-2.1	0.8		32.31	1.16	0.8	305.89	21JA	0.2	-3.4	-2.0	0.7
19.45	0.41	1.3	316.47	1FE	2.7	-3.7	-2.0	0.7		37.75	1.23	1.0	315.98	31JA	0.1	-3.4	-2.0	0.7
26.04	0.53	1.4	326.50	11FE	1.1	-3.6	-2.0	0.7		43.34	1.29	1.1	326.02	10FE	-0.2	-3.4	-2.0	0.6
32.62	0.63	1.5	336.49	21FE	0.6	-3.6	-1.9	0.6		49.04	1.33	1.3	336.00	20FE	-0.6	-3.4	-2.0	0.6
39.17	0.73	1.6	346.41	2MR	0.3	-3.5	-1.8	0.6		54.84	1.36	1.4	345.93	2MR	-1.2	-3.5	-2.0	0.5
45.69	0.81	1.7	356.28	12MR	0.0	-3.4	-1.8	0.5		60.71	1.38	1.5	355.80	12MR	-1.5	-3.5	-2.0	0.5
52.19	0.88	1.7	6.09	22MR	-0.5	-3.4	-1.7	0.5		66.63	1.39	1.6	5.61	22MR	-0.8	-3.5	-1.9	0.4
58.65	0.94	1.8	15.84	1AP	-1.2	-3.4	-1.6	0.5		72.61	1.40	1.7	15.37	1AP	0.5	-3.4	-1.9	0.4
65.10	0.99	1.9	25.54	11AP	-1.7	-3.3	-1.6	0.6		78.61	1.39	1.8	25.07	11AP	1.8	-3.4	-1.8	0.3
71.51	1.03	1.9	35.20	21AP	-0.8	-3.3	-1.5	0.7		84.65	1.39	1.8	34.73	21AP	3.5	-3.4	-1.8	0.3
77.91	1.07	1.9	44.81	1MY	0.2	-3.3	-1.5	0.7		90.72	1.37	1.9	44.34	1MY	2.5	-3.4	-1.7	0.4
84.28	1.10	2.0	54.39	11MY	1.1	-3.3	-1.4	0.8		96.82	1.35	1.9	53.92	11MY	1.4	-3.3	-1.6	0.5
90.64	1.13	2.0	63.94	21MY	2.0	-3.3	-1.4	0.9		102.94	1.33	2.0	63.47	21MY	0.7	-3.3	-1.6	0.5
96.99	1.14	2.0	73.47	31MY	2.4	-3.3	-1.4	0.9		109.09	1.30	2.0	73.01	31MY	-0.2	-3.3	-1.5	0.6
103.32	1.16	2.0	83.00	10JN	2.5	-3.3	-1.4	1.0		115.26	1.27	2.0	82.53	10JN	-1.2	-3.3	-1.4	0.7
109.66	1.17	2.0	92.53	20JN	1.1	-3.4	-1.3	1.0		121.47	1.24	2.0	92.06	20JN	-1.7	-3.4	-1.4	0.7
115.99	1.17	2.0	102.06	30JN	-0.1	-3.4	-1.3	1.0		127.71	1.20	2.0	101.60	30JN	-0.8	-3.4	-1.3	0.8
122.33	1.17	2.0	111.62	10JL	-1.2	-3.4	-1.3	1.0		133.99	1.16	2.0	111.15	10JL	-0.2	-3.4	-1.3	0.8
128.67	1.16	2.0	121.20	20JL	-1.6	-3.5	-1.3	1.1		140.30	1.11	2.0	120.73	20JL	0.2	-3.4	-1.3	0.8
135.03	1.15	2.0	130.82	30JL	-0.9	-3.5	-1.4	1.1		146.66	1.06	2.0	130.35	30JL	0.5	-3.4	-1.2	0.9
141.41	1.13	2.0	140.48	9AU	-0.3	-3.5	-1.4	1.0		153.06	1.01	2.0	140.00	9AU	0.9	-3.5	-1.2	0.9
147.80	1.11	2.0	150.18	19AU	0.0	-3.4	-1.4	1.0		159.51	0.95	2.0	149.71	19AU	1.9	-3.6	-1.2	0.9
154.22	1.09	2.0	159.94	29AU	0.2	-3.4	-1.4	1.0		166.01	0.89	1.9	159.47	29AU	2.8	-3.6	-1.2	0.9
160.67	1.06	2.0	169.76	8SE	0.4	-3.4	-1.5	1.0		172.57	0.82	1.9	169.28	8SE	0.3	-3.7	-1.2	0.9
167.15	1.02	2.0	179.63	18SE	1.1	-3.4	-1.5	0.9		179.19	0.76	1.8	179.15	18SE	-0.8	-3.8	-1.2	0.8
173.66	0.98	2.0	189.56	28SE	3.4	-3.3	-1.6	0.9		185.87	0.68	1.8	189.08	28SE	-1.0	-3.9	-1.2	0.8
180.21	0.94	2.0	199.55	8OC	0.5	-3.3	-1.6	0.9		192.61	0.61	1.8	199.07	8OC	-0.9	-4.0	-1.3	0.8
186.80	0.89	1.9	209.59	18OC	-0.5	-3.3	-1.7	0.9		199.41	0.52	1.8	209.11	18OC	-0.7	-4.1	-1.3	0.7
193.43	0.83	1.9	219.68	28OC	-0.7	-3.3	-1.8	0.9		206.29	0.44	1.8	219.20	28OC	-0.5	-4.2	-1.3	0.7
200.09	0.77	1.9	229.81	7NO	-0.7	-3.3	-1.8	0.9		213.23	0.35	1.8	229.32	7NO	-0.4	-4.3	-1.4	0.7
206.80	0.70	1.8	239.98	17NO	-0.7	-3.4	-1.9	0.9		220.24	0.25	1.8	239.49	17NO	-0.4	-4.4	-1.4	0.8
213.55	0.62	1.8	250.16	27NO	-0.7	-3.4	-2.0	0.9		227.31	0.15	1.7	249.67	27NO	-0.4	-4.4	-1.5	0.8
220.34	0.53	1.8	260.36	7DE	-0.7	-3.4	-2.0	0.9		234.46	0.04	1.7	259.87	7DE	1.9	-4.2	-1.5	0.8
227.17	0.44	1.7	270.57	17DE	-0.7	-3.5	-2.1	0.9		241.67	-0.07	1.7	270.07	17DE	1.4	-3.6	-1.6	0.8
234.04	0.33	1.6	280.76	27DE	-0.3	-3.5	-2.1	0.8		248.94	-0.19	1.6	280.27	27DE	0.3	-3.2	-1.7	0.8
				805										**807**				
240.95	0.22	1.6	290.93	6JA	1.5	-3.5	-2.1	0.8		256.27	-0.31	1.6	290.44	6JA	0.0	-3.9	-1.7	0.8
247.89	0.09	1.5	301.08	16JA	2.2	-3.6	-2.1	0.8		263.67	-0.43	1.5	300.59	16JA	-0.1	-4.2	-1.8	0.8
254.88	-0.05	1.4	311.18	26JA	0.8	-3.7	-2.1	0.7		271.11	-0.56	1.5	310.70	26JA	-0.3	-4.3	-1.8	0.7
261.89	-0.20	1.3	321.25	5FE	0.4	-3.7	-2.1	0.7		278.60	-0.69	1.4	320.76	5FE	-0.6	-4.3	-1.9	0.7
268.93	-0.36	1.2	331.26	15FE	0.2	-3.8	-2.0	0.6		286.14	-0.81	1.4	330.77	15FE	-1.2	-4.2	-1.9	0.7
276.00	-0.53	1.1	341.21	25FE	-0.1	-3.9	-2.0	0.6		293.70	-0.94	1.3	340.73	25FE	-1.4	-4.1	-2.0	0.6
283.08	-0.72	1.0	351.11	7MR	-0.5	-4.0	-1.9	0.5		301.29	-1.07	1.3	350.63	7MR	-0.7	-4.0	-2.0	0.6
290.17	-0.92	0.9	0.95	17MR	-1.2	-4.1	-1.8	0.5		308.90	-1.19	1.2	0.48	17MR	0.6	-3.9	-2.0	0.5
297.26	-1.13	0.8	10.73	27MR	-1.6	-4.1	-1.8	0.4		316.50	-1.31	1.2	10.26	27MR	2.3	-3.8	-2.0	0.4
304.34	-1.35	0.7	20.46	6AP	-0.8	-4.2	-1.7	0.4		324.10	-1.42	1.1	19.99	6AP	3.4	-3.7	-2.0	0.4
311.37	-1.58	0.5	30.14	16AP	0.3	-4.2	-1.6	0.5		331.67	-1.52	1.0	29.68	16AP	1.9	-3.6	-1.9	0.3
318.36	-1.82	0.4	39.77	26AP	1.4	-4.1	-1.6	0.5		339.19	-1.61	1.0	39.31	26AP	1.1	-3.6	-1.9	0.3
325.26	-2.07	0.3	49.37	6MY	2.7	-3.7	-1.5	0.6		346.66	-1.68	0.9	48.91	6MY	0.5	-3.5	-1.8	0.4
332.04	-2.32	0.2	58.93	16MY	3.3	-2.9	-1.5	0.6		354.06	-1.74	0.9	58.48	16MY	-0.3	-3.4	-1.8	0.4
338.66	-2.57	0.0	68.47	26MY	1.9	-3.2	-1.4	0.7		1.35	-1.79	0.8	68.02	26MY	-1.2	-3.4	-1.7	0.4
345.08	-2.82	-0.1	78.01	5JN	0.9	-3.9	-1.4	0.8		8.52	-1.81	0.7	77.55	5JN	-1.7	-3.4	-1.7	0.5
351.23	-3.07	-0.3	87.53	15JN	-0.2	-4.1	-1.3	0.8		15.56	-1.82	0.7	87.07	15JN	-0.8	-3.3	-1.6	0.6
357.03	-3.31	-0.5	97.06	25JN	-1.2	-4.2	-1.3	0.9		22.42	-1.81	0.6	96.60	25JN	-0.1	-3.3	-1.5	0.6
2.38	-3.54	-0.6	106.61	5JL	-1.7	-4.2	-1.3	0.9		29.08	-1.78	0.5	106.14	5JL	0.4	-3.3	-1.5	0.7
7.16	-3.75	-0.8	116.17	15JL	-0.9	-4.1	-1.3	0.9		35.51	-1.73	0.4	115.71	15JL	0.7	-3.3	-1.4	0.7
11.23	-3.94	-1.0	125.77	25JL	-0.2	-4.0	-1.3	0.9		41.66	-1.65	0.3	125.31	25JL	1.2	-3.3	-1.4	0.8
14.39	-4.10	-1.2	135.41	4AU	0.1	-3.9	-1.2	0.9		47.48	-1.56	0.2	134.94	4AU	2.4	-3.3	-1.3	0.8
16.43	-4.20	-1.5	145.09	14AU	0.4	-3.8	-1.3	0.9		52.91	-1.43	0.1	144.62	14AU	2.4	-3.4	-1.3	0.8
17.15	-4.22	-1.7	154.82	24AU	0.7	-3.7	-1.3	0.9		57.85	-1.28	-0.0	154.35	24AU	0.2	-3.4	-1.3	0.8
16.41	-4.12	-1.9	164.61	3SE	1.5	-3.7	-1.3	0.9		62.22	-1.09	-0.1	164.13	3SE	-0.9	-3.4	-1.2	0.8
14.29	-3.85	-2.1	174.45	13SE	3.1	-3.6	-1.3	0.9		65.86	-0.86	-0.3	173.97	13SE	-1.1	-3.4	-1.2	0.8
11.22	-3.40	-2.2	184.35	23SE	0.3	-3.6	-1.3	0.9		68.61	-0.59	-0.5	183.87	23SE	-1.0	-3.5	-1.2	0.8
7.91	-2.80	-2.2	194.31	3OC	-0.7	-3.5	-1.4	0.8		70.28	-0.26	-0.7	193.83	3OC	-0.6	-3.5	-1.2	0.8
5.18	-2.13	-1.9	204.33	13OC	-0.8	-3.5	-1.4	0.8		70.65	0.12	-0.9	203.84	13OC	-0.4	-3.5	-1.2	0.8
3.57	-1.48	-1.6	214.39	23OC	-0.8	-3.5	-1.5	0.8		69.60	0.56	-1.2	213.90	23OC	-0.3	-3.5	-1.2	0.7
3.30	-0.89	-1.2	224.50	2NO	-0.8	-3.4	-1.5	0.8		67.15	1.04	-1.4	224.01	2NO	-0.2	-3.4	-1.3	0.7
4.32	-0.40	-0.9	234.65	12NO	-0.6	-3.4	-1.6	0.8		63.63	1.51	-1.5	234.16	12NO	0.1	-3.4	-1.3	0.7
6.43	0.00	-0.6	244.82	22NO	-0.6	-3.4	-1.7	0.8		59.76	1.91	-1.5	244.33	22NO	2.2	-3.4	-1.3	0.7
9.44	0.32	-0.3	255.02	2DE	-0.6	-3.4	-1.7	0.8		56.31	2.21	-1.2	254.52	2DE	1.2	-3.4	-1.4	0.8
13.15	0.58	-0.0	265.22	12DE	-0.2	-3.4	-1.8	0.8		53.91	2.39	-1.0	264.72	12DE	0.1	-3.3	-1.4	0.8
17.40	0.78	0.2	275.42	22DE	1.7	-3.3	-1.8	0.8		52.84	2.48	-0.6	274.92	22DE	-0.1	-3.3	-1.5	0.8

Left Table (808 / 809)

♂ LONG	♂ LAT	♂ MAG	☉ LONG	16.00UT	☿	♀	♃	♄
				808				
53.09	2.49	-0.3	285.11	1JA	-0.2	-3.3	-1.5	0.8
54.49	2.46	-0.1	295.27	11JA	-0.3	-3.3	-1.6	0.8
56.84	2.41	0.2	305.40	21JA	-0.7	-3.4	-1.6	0.8
59.94	2.34	0.4	315.49	31JA	-1.2	-3.4	-1.7	0.7
63.64	2.26	0.7	325.53	10FE	-1.3	-3.4	-1.8	0.7
67.81	2.18	0.8	335.52	20FE	-0.7	-3.4	-1.8	0.7
72.33	2.09	1.0	345.45	1MR	0.7	-3.4	-1.9	0.6
77.16	2.01	1.2	355.32	11MR	2.8	-3.5	-2.0	0.6
82.21	1.93	1.3	5.14	21MR	2.5	-3.5	-2.0	0.5
87.45	1.85	1.4	14.90	31MR	1.4	-3.6	-2.0	0.5
92.85	1.77	1.5	24.61	10AP	0.8	-3.6	-2.1	0.4
98.38	1.69	1.6	34.26	20AP	0.3	-3.7	-2.1	0.3
104.03	1.61	1.7	43.88	30AP	-0.3	-3.8	-2.1	0.3
109.77	1.53	1.8	53.46	10MY	-1.2	-3.9	-2.0	0.2
115.61	1.45	1.8	63.02	20MY	-1.7	-4.0	-2.0	0.3
121.52	1.37	1.9	72.55	30MY	-0.8	-4.0	-1.9	0.4
127.52	1.28	1.9	82.08	9JN	-0.0	-4.1	-1.9	0.4
133.58	1.20	1.9	91.60	19JN	0.5	-4.2	-1.8	0.5
139.72	1.12	1.9	101.14	29JN	1.0	-4.2	-1.8	0.5
145.94	1.04	1.9	110.69	9JL	1.6	-4.0	-1.7	0.6
152.22	0.95	1.9	120.27	19JL	3.0	-3.5	-1.6	0.7
158.59	0.87	1.9	129.88	29JL	2.0	-3.1	-1.6	0.7
165.03	0.78	1.9	139.54	8AU	0.1	-3.7	-1.5	0.7
171.54	0.69	1.9	149.24	18AU	-1.0	-4.1	-1.5	0.8
178.14	0.60	1.9	158.99	28AU	-1.3	-4.3	-1.4	0.8
184.82	0.50	1.9	168.80	7SE	-1.0	-4.3	-1.4	0.8
191.58	0.41	1.8	178.67	17SE	-0.5	-4.2	-1.3	0.8
198.43	0.31	1.8	188.60	27SE	-0.3	-4.1	-1.3	0.8
205.37	0.21	1.7	198.58	7OC	-0.1	-4.0	-1.3	0.8
212.39	0.11	1.7	208.62	17OC	-0.1	-4.0	-1.3	0.8
219.50	0.01	1.6	218.70	27OC	0.4	-3.9	-1.3	0.7
226.69	-0.10	1.6	228.83	6NO	2.5	-3.8	-1.3	0.7
233.97	-0.20	1.6	238.99	16NO	0.9	-3.7	-1.3	0.7
241.33	-0.31	1.6	249.18	26NO	-0.3	-3.7	-1.3	0.7
248.76	-0.41	1.6	259.38	6DE	-0.3	-3.6	-1.3	0.7
256.27	-0.51	1.5	269.58	16DE	-0.3	-3.5	-1.4	0.8
263.85	-0.61	1.5	279.78	26DE	-0.4	-3.5	-1.4	0.8
				809				
271.48	-0.71	1.5	289.95	5JA	-0.7	-3.4	-1.4	0.8
279.17	-0.81	1.5	300.10	15JA	-1.1	-3.4	-1.5	0.8
286.90	-0.90	1.5	310.21	25JA	-1.1	-3.4	-1.6	0.8
294.66	-0.98	1.5	320.28	4FE	-0.6	-3.3	-1.6	0.8
302.45	-1.05	1.4	330.29	14FE	0.9	-3.3	-1.7	0.7
310.25	-1.12	1.4	340.26	24FE	3.2	-3.3	-1.8	0.7
318.04	-1.17	1.4	350.16	6MR	1.9	-3.3	-1.8	0.7
325.83	-1.22	1.4	0.00	16MR	1.0	-3.3	-1.9	0.6
333.59	-1.25	1.4	9.79	26MR	0.6	-3.3	-2.0	0.6
341.31	-1.27	1.3	19.52	5AP	0.2	-3.3	-2.0	0.5
348.98	-1.27	1.3	29.21	15AP	-0.0	-3.4	-2.1	0.4
356.59	-1.26	1.3	38.85	25AP	-1.2	-3.4	-2.1	0.4
4.12	-1.24	1.3	48.44	5MY	-1.7	-3.4	-2.1	0.3
11.57	-1.20	1.3	58.01	15MY	-0.8	-3.5	-2.2	0.3
18.92	-1.15	1.3	67.56	25MY	0.0	-3.5	-2.2	0.3
26.17	-1.08	1.2	77.08	4JN	0.7	-3.5	-2.1	0.3
33.29	-1.00	1.2	86.61	14JN	1.3	-3.4	-2.1	0.4
40.29	-0.91	1.2	96.14	24JN	2.2	-3.4	-2.1	0.4
47.14	-0.81	1.2	105.68	4JL	3.3	-3.4	-2.0	0.5
53.84	-0.69	1.1	115.24	14JL	1.6	-3.3	-2.0	0.6
60.37	-0.55	1.1	124.84	24JL	0.0	-3.3	-1.9	0.6
66.71	-0.41	1.0	134.47	3AU	-1.1	-3.3	-1.8	0.7
72.84	-0.25	1.0	144.14	13AU	-1.4	-3.3	-1.8	0.7
78.73	-0.07	0.9	153.87	23AU	-0.9	-3.3	-1.7	0.7
84.34	0.12	0.8	163.65	2SE	-0.4	-3.3	-1.6	0.8
89.63	0.33	0.7	173.49	12SE	-0.1	-3.4	-1.6	0.8
94.53	0.57	0.6	183.39	22SE	-0.0	-3.4	-1.5	0.8
98.96	0.83	0.5	193.34	2OC	0.1	-3.4	-1.5	0.8
102.83	1.13	0.3	203.35	12OC	0.6	-3.5	-1.5	0.8
105.99	1.46	0.1	213.41	22OC	2.8	-3.5	-1.4	0.8
108.29	1.84	-0.1	223.51	1NO	0.7	-3.6	-1.4	0.7
109.55	2.25	-0.3	233.66	11NO	-0.3	-3.7	-1.4	0.7
109.55	2.70	-0.5	243.83	21NO	-0.7	-3.7	-1.4	0.7
108.20	3.16	-0.7	254.02	1DE	-0.4	-3.8	-1.4	0.7
105.52	3.58	-0.9	264.23	11DE	-0.5	-3.9	-1.4	0.7
101.86	3.90	-1.1	274.43	21DE	-0.8	-4.0	-1.4	0.7
97.89	4.07	-1.1	284.61	31DE	-1.0	-4.1	-1.4	0.8

Right Table (810 / 811)

♂ LONG	♂ LAT	♂ MAG	☉ LONG	16.00UT	☿	♀	♃	♄
				810				
94.34	4.07	-0.9	294.78	10JA	-1.0	-4.2	-1.4	0.8
91.81	3.94	-0.6	304.91	20JA	-0.6	-4.3	-1.5	0.8
90.57	3.73	-0.3	315.00	30JA	1.1	-4.3	-1.5	0.8
90.62	3.47	-0.1	325.05	9FE	3.0	-4.3	-1.6	0.8
91.81	3.21	0.2	335.04	19FE	1.4	-4.1	-1.6	0.8
93.95	2.95	0.4	344.97	1MR	0.7	-3.5	-1.7	0.7
96.86	2.72	0.6	354.85	11MR	0.4	-3.3	-1.7	0.7
100.38	2.49	0.8	4.67	21MR	0.1	-3.8	-1.8	0.7
104.39	2.29	1.0	14.43	31MR	-0.4	-4.1	-1.9	0.6
108.80	2.10	1.1	24.14	10AP	-1.2	-4.2	-1.9	0.6
113.53	1.92	1.2	33.80	20AP	-1.7	-4.2	-2.0	0.5
118.53	1.76	1.3	43.41	30AP	-0.8	-4.1	-2.1	0.4
123.75	1.60	1.4	53.00	10MY	0.1	-4.0	-2.1	0.4
129.17	1.45	1.5	62.55	20MY	0.9	-3.9	-2.2	0.3
134.76	1.31	1.6	72.09	30MY	1.7	-3.8	-2.2	0.2
140.50	1.17	1.6	81.62	9JN	2.9	-3.7	-2.3	0.3
146.38	1.04	1.7	91.14	19JN	2.8	-3.7	-2.3	0.3
152.38	0.91	1.7	100.68	29JN	1.3	-3.6	-2.3	0.4
158.52	0.79	1.7	110.23	9JL	-0.0	-3.5	-2.3	0.4
164.76	0.67	1.8	119.81	19JL	-1.1	-3.5	-2.2	0.5
171.13	0.55	1.8	129.42	29JL	-1.5	-3.4	-2.2	0.6
177.60	0.43	1.8	139.07	8AU	-0.9	-3.4	-2.1	0.6
184.19	0.31	1.8	148.77	18AU	-0.4	-3.4	-2.1	0.7
190.89	0.20	1.7	158.52	28AU	-0.0	-3.4	-2.0	0.7
197.69	0.08	1.7	168.33	7SE	0.1	-3.4	-1.9	0.7
204.61	-0.03	1.7	178.19	17SE	0.3	-3.4	-1.9	0.8
211.63	-0.13	1.7	188.12	27SE	0.9	-3.4	-1.8	0.8
218.75	-0.24	1.7	198.10	7OC	3.2	-3.4	-1.8	0.8
225.96	-0.34	1.6	208.13	17OC	0.6	-3.4	-1.7	0.8
233.31	-0.44	1.6	218.21	27OC	-0.5	-3.4	-1.6	0.8
240.73	-0.54	1.5	228.34	6NO	-0.6	-3.4	-1.6	0.8
248.24	-0.63	1.5	238.49	16NO	-0.6	-3.4	-1.6	0.7
255.83	-0.72	1.5	248.68	26NO	-0.7	-3.4	-1.5	0.7
263.49	-0.80	1.4	258.88	6DE	-0.8	-3.5	-1.5	0.7
271.22	-0.87	1.4	269.08	16DE	-0.8	-3.5	-1.5	0.7
279.00	-0.93	1.3	279.27	26DE	-0.8	-3.5	-1.5	0.7
				811				
286.83	-0.99	1.3	289.45	5JA	-0.4	-3.5	-1.5	0.8
294.69	-1.03	1.3	299.60	15JA	1.3	-3.4	-1.5	0.8
302.57	-1.06	1.4	309.71	25JA	2.5	-3.4	-1.5	0.8
310.46	-1.09	1.4	319.79	4FE	1.0	-3.4	-1.5	0.8
318.33	-1.10	1.4	329.80	14FE	0.5	-3.4	-1.5	0.8
326.19	-1.10	1.4	339.77	24FE	0.3	-3.3	-1.6	0.8
334.02	-1.08	1.4	349.68	6MR	-0.0	-3.3	-1.6	0.8
341.81	-1.06	1.4	359.52	16MR	-0.5	-3.3	-1.7	0.8
349.54	-1.02	1.4	9.32	26MR	-1.2	-3.3	-1.7	0.7
357.21	-0.97	1.4	19.05	5AP	-1.6	-3.3	-1.8	0.7
4.80	-0.91	1.5	28.74	15AP	-0.8	-3.4	-1.8	0.6
12.32	-0.85	1.5	38.38	25AP	0.2	-3.4	-1.9	0.6
19.75	-0.77	1.5	47.98	5MY	1.2	-3.4	-2.0	0.5
27.08	-0.68	1.5	57.55	15MY	2.3	-3.4	-2.1	0.4
34.32	-0.59	1.5	67.09	25MY	3.6	-3.5	-2.1	0.4
41.46	-0.48	1.5	76.63	4JN	2.3	-3.5	-2.2	0.3
48.49	-0.37	1.5	86.15	14JN	1.0	-3.6	-2.3	0.3
55.42	-0.26	1.5	95.68	24JN	-0.1	-3.6	-2.3	0.3
62.23	-0.14	1.5	105.22	4JL	-1.2	-3.7	-2.4	0.4
68.93	-0.01	1.5	114.78	14JL	-1.6	-3.8	-2.4	0.4
75.51	0.12	1.5	124.38	24JL	-0.9	-3.9	-2.4	0.5
81.96	0.26	1.5	134.01	3AU	-0.3	-4.0	-2.4	0.5
88.27	0.40	1.5	143.68	13AU	0.1	-4.1	-2.4	0.6
94.44	0.55	1.4	153.40	23AU	0.3	-4.2	-2.3	0.6
100.45	0.71	1.4	163.18	2SE	0.5	-4.3	-2.3	0.7
106.28	0.88	1.3	173.01	12SE	1.2	-4.3	-2.2	0.7
111.91	1.06	1.3	182.91	22SE	3.3	-4.1	-2.2	0.8
117.30	1.25	1.2	192.86	2OC	0.5	-3.6	-2.1	0.8
122.41	1.46	1.1	202.86	12OC	-0.6	-3.2	-2.1	0.8
127.21	1.68	1.0	212.92	22OC	-0.7	-3.9	-2.0	0.8
131.60	1.92	0.8	223.02	1NO	-0.7	-4.3	-1.9	0.8
135.50	2.19	0.7	233.16	11NO	-0.8	-4.4	-1.9	0.8
138.82	2.49	0.5	243.34	21NO	-0.7	-4.4	-1.8	0.8
141.38	2.82	0.3	253.53	1DE	-0.4	-4.3	-1.7	0.8
143.04	3.18	0.1	263.73	11DE	-0.7	-4.2	-1.7	0.8
143.60	3.56	-0.2	273.93	21DE	-0.3	-4.1	-1.6	0.7
142.89	3.93	-0.4	284.12	31DE	1.5	-4.0	-1.6	0.8

♂			☉	16.00UT	☿	♀	♃	♄
LONG	LAT	MAG	LONG	812	MAGNITUDES			
140.88	4.26	-0.7	294.28	10JA	2.1	-3.9	-1.6	0.8
137.70	4.49	-0.9	304.42	20JA	0.7	-3.8	-1.6	0.8
133.84	4.57	-1.0	314.51	30JA	0.3	-3.7	-1.6	0.8
130.00	4.46	-0.9	324.56	9FE	0.1	-3.6	-1.6	0.9
126.84	4.21	-0.7	334.56	19FE	-0.1	-3.5	-1.6	0.9
124.84	3.85	-0.4	344.49	29FE	-0.5	-3.5	-1.6	0.9
124.13	3.46	-0.2	354.37	10MR	-1.2	-3.4	-1.6	0.8
124.65	3.07	0.0	4.20	20MR	-1.6	-3.4	-1.6	0.8
126.25	2.71	0.2	13.96	30MR	-0.8	-3.4	-1.6	0.8
128.74	2.37	0.4	23.67	9AP	0.3	-3.3	-1.7	0.8
131.95	2.06	0.6	33.34	19AP	1.5	-3.3	-1.7	0.7
135.75	1.78	0.8	42.96	29AP	3.0	-3.3	-1.8	0.6
140.02	1.53	0.9	52.54	9MY	3.0	-3.3	-1.8	0.6
144.70	1.30	1.0	62.10	19MY	1.7	-3.3	-1.9	0.5
149.71	1.08	1.1	71.63	29MY	0.8	-3.3	-2.0	0.5
155.00	0.88	1.2	81.16	8JN	-0.1	-3.3	-2.0	0.4
160.55	0.70	1.3	90.69	18JN	-1.2	-3.4	-2.1	0.3
166.32	0.53	1.3	100.22	28JN	-1.7	-3.4	-2.2	0.3
172.28	0.36	1.4	109.77	8JL	-0.9	-3.4	-2.2	0.4
178.43	0.21	1.4	119.35	18JL	-0.2	-3.5	-2.3	0.4
184.75	0.06	1.4	128.95	28JL	0.2	-3.5	-2.4	0.5
191.23	-0.07	1.5	138.60	7AU	0.4	-3.5	-2.4	0.5
197.86	-0.20	1.5	148.30	17AU	0.7	-3.4	-2.4	0.6
204.64	-0.33	1.5	158.05	27AU	1.6	-3.4	-2.5	0.6
211.56	-0.44	1.5	167.85	6SE	3.1	-3.4	-2.5	0.7
218.60	-0.55	1.5	177.71	16SE	0.4	-3.4	-2.4	0.7
225.78	-0.65	1.5	187.63	26SE	-0.7	-3.3	-2.4	0.8
233.07	-0.74	1.5	197.61	6OC	-0.9	-3.3	-2.4	0.8
240.47	-0.83	1.5	207.64	16OC	-0.9	-3.3	-2.3	0.8
247.98	-0.90	1.5	217.72	26OC	-0.7	-3.3	-2.2	0.8
255.58	-0.96	1.4	227.84	5NO	-0.6	-3.3	-2.2	0.8
263.26	-1.02	1.4	238.00	15NO	-0.5	-3.4	-2.1	0.8
271.01	-1.06	1.4	248.18	25NO	-0.5	-3.4	-2.0	0.8
278.82	-1.09	1.4	258.38	5DE	-0.2	-3.4	-2.0	0.8
286.68	-1.11	1.4	268.58	15DE	1.7	-3.5	-1.9	0.8
294.57	-1.11	1.4	278.77	25DE	1.7	-3.5	-1.8	0.8
				813				
302.47	-1.11	1.4	288.95	4JA	0.4	-3.5	-1.8	0.8
310.38	-1.09	1.4	299.11	14JA	0.1	-3.6	-1.7	0.8
318.27	-1.06	1.4	309.22	24JA	0.0	-3.7	-1.7	0.9
326.14	-1.02	1.4	319.30	3FE	-0.2	-3.7	-1.6	0.9
333.97	-0.97	1.4	329.32	13FE	-0.6	-3.8	-1.6	0.9
341.75	-0.91	1.3	339.28	23FE	-1.2	-3.9	-1.6	0.9
349.48	-0.84	1.4	349.20	5MR	-1.5	-4.0	-1.6	0.9
357.14	-0.76	1.4	359.05	15MR	-0.8	-4.1	-1.6	0.9
4.72	-0.68	1.4	8.84	25MR	0.5	-4.1	-1.6	0.9
12.23	-0.59	1.5	18.58	4AP	1.9	-4.2	-1.6	0.9
19.65	-0.50	1.5	28.27	14AP	3.8	-4.2	-1.6	0.8
26.98	-0.40	1.6	37.91	24AP	2.3	-4.1	-1.6	0.8
34.23	-0.30	1.6	47.52	4MY	1.3	-3.7	-1.6	0.7
41.38	-0.19	1.6	57.09	14MY	0.6	-2.9	-1.7	0.7
48.44	-0.08	1.7	66.63	24MY	-0.2	-3.3	-1.7	0.6
55.42	0.02	1.7	76.17	3JN	-1.2	-3.9	-1.8	0.6
62.30	0.13	1.7	85.69	13JN	-1.7	-4.1	-1.8	0.5
69.09	0.24	1.8	95.22	23JN	-0.9	-4.2	-1.9	0.4
75.80	0.35	1.8	104.76	3JL	-0.2	-4.2	-1.9	0.4
82.42	0.46	1.8	114.32	13JL	0.3	-4.1	-2.0	0.4
88.96	0.57	1.8	123.91	23JL	0.6	-4.0	-2.1	0.4
95.41	0.69	1.8	133.54	2AU	1.0	-3.9	-2.1	0.5
101.76	0.80	1.8	143.21	12AU	2.0	-3.8	-2.2	0.5
108.03	0.91	1.8	152.93	22AU	2.7	-3.7	-2.3	0.6
114.20	1.03	1.7	162.70	1SE	0.3	-3.7	-2.3	0.6
120.27	1.15	1.7	172.54	11SE	-0.8	-3.6	-2.4	0.7
126.23	1.27	1.7	182.42	21SE	-1.0	-3.6	-2.4	0.7
132.06	1.40	1.6	192.37	1OC	-1.0	-3.5	-2.4	0.8
137.75	1.53	1.5	202.38	11OC	-0.6	-3.5	-2.4	0.8
143.28	1.66	1.5	212.43	21OC	-0.4	-3.5	-2.4	0.8
148.62	1.81	1.4	222.53	31OC	-0.4	-3.4	-2.4	0.9
153.73	1.96	1.3	232.67	10NO	-0.3	-3.4	-2.3	0.9
158.58	2.12	1.1	242.84	20NO	0.0	-3.4	-2.3	0.9
163.09	2.30	1.0	253.04	30NO	1.9	-3.4	-2.2	0.9
167.20	2.49	0.8	263.24	10DE	1.4	-3.4	-2.2	0.9
170.81	2.69	0.6	273.43	20DE	0.2	-3.3	-2.1	0.9
173.79	2.91	0.4	283.63	30DE	-0.0	-3.3	-2.0	0.9

♂			☉	16.00UT	☿	♀	♃	♄
LONG	LAT	MAG	LONG	814	MAGNITUDES			
176.00	3.14	0.2	293.79	9JA	-0.1	-3.4	-1.9	0.8
177.27	3.37	-0.1	303.92	19JA	-0.3	-3.4	-1.9	0.9
177.41	3.59	-0.4	314.02	29JA	-0.6	-3.4	-1.8	0.9
176.28	3.76	-0.6	324.07	8FE	-1.2	-3.4	-1.7	1.0
173.90	3.86	-0.9	334.06	18FE	-1.4	-3.4	-1.7	1.0
170.51	3.82	-1.1	344.01	28FE	-0.8	-3.5	-1.6	1.0
166.67	3.62	-1.2	353.89	10MR	0.6	-3.5	-1.6	1.0
163.09	3.29	-1.0	3.71	20MR	2.4	-3.5	-1.6	1.0
160.39	2.86	-0.8	13.48	30MR	3.0	-3.4	-1.6	1.0
158.95	2.40	-0.6	23.19	9AP	1.7	-3.4	-1.5	1.0
158.81	1.95	-0.4	32.86	19AP	1.0	-3.4	-1.5	1.0
159.88	1.54	-0.2	42.48	29AP	0.4	-3.4	-1.5	0.9
161.99	1.16	0.0	52.07	9MY	-0.3	-3.3	-1.5	0.9
164.97	0.83	0.2	61.63	19MY	-1.2	-3.3	-1.5	0.8
168.67	0.53	0.3	71.17	29MY	-1.8	-3.3	-1.6	0.8
172.96	0.27	0.5	80.69	8JN	-0.9	-3.3	-1.6	0.7
177.73	0.03	0.6	90.22	18JN	-0.1	-3.3	-1.6	0.6
182.91	-0.18	0.7	99.75	28JN	0.4	-3.4	-1.6	0.6
188.45	-0.36	0.8	109.30	8JL	0.8	-3.4	-1.7	0.5
194.29	-0.53	0.9	118.88	18JL	1.4	-3.4	-1.7	0.4
200.41	-0.68	0.9	128.49	28JL	2.6	-3.4	-1.8	0.5
206.77	-0.82	1.0	138.13	7AU	2.3	-3.5	-1.9	0.5
213.34	-0.93	1.0	147.83	17AU	0.2	-3.6	-1.9	0.5
220.11	-1.04	1.1	157.57	27AU	-0.9	-3.6	-2.0	0.6
227.05	-1.12	1.1	167.37	6SE	-1.2	-3.7	-2.1	0.7
234.16	-1.19	1.1	177.23	16SE	-1.0	-3.8	-2.1	0.7
241.42	-1.25	1.2	187.15	26SE	-0.6	-3.9	-2.2	0.8
248.81	-1.29	1.2	197.12	6OC	-0.3	-4.0	-2.2	0.8
256.31	-1.32	1.2	207.15	16OC	-0.2	-4.1	-2.3	0.8
263.91	-1.33	1.2	217.23	26OC	-0.2	-4.2	-2.3	0.9
271.59	-1.33	1.3	227.35	5NO	0.2	-4.3	-2.3	0.9
279.34	-1.32	1.3	237.51	15NO	2.2	-4.4	-2.3	0.9
287.14	-1.29	1.3	247.69	25NO	1.1	-4.4	-2.3	1.0
294.97	-1.25	1.3	257.88	5DE	-0.0	-4.2	-2.3	1.0
302.82	-1.19	1.3	268.09	15DE	-0.2	-3.6	-2.3	1.0
310.67	-1.13	1.4	278.28	25DE	-0.2	-3.2	-2.2	1.0
				815				
318.50	-1.05	1.4	288.46	4JA	-0.4	-3.9	-2.1	1.0
326.31	-0.97	1.4	298.62	14JA	-0.7	-4.2	-2.1	0.9
334.08	-0.88	1.4	308.73	24JA	-1.2	-4.3	-2.0	0.9
341.80	-0.79	1.5	318.80	3FE	-1.2	-4.3	-1.9	1.0
349.45	-0.69	1.5	328.83	13FE	-0.7	-4.2	-1.8	1.0
357.04	-0.58	1.5	338.80	23FE	0.8	-4.1	-1.8	1.1
4.55	-0.48	1.5	348.71	5MR	2.9	-4.0	-1.7	1.1
11.98	-0.37	1.6	358.57	15MR	2.3	-3.9	-1.7	1.1
19.34	-0.26	1.6	8.37	25MR	1.2	-3.8	-1.6	1.1
26.60	-0.16	1.6	18.11	4AP	0.7	-3.7	-1.6	1.1
33.79	-0.05	1.6	27.80	14AP	0.3	-3.6	-1.5	1.1
40.89	0.05	1.6	37.45	24AP	-0.3	-3.6	-1.5	1.1
47.90	0.16	1.6	47.05	4MY	-1.2	-3.5	-1.5	1.1
54.84	0.26	1.7	56.62	14MY	-1.8	-3.4	-1.4	1.0
61.70	0.35	1.7	66.17	24MY	-0.9	-3.4	-1.4	1.0
68.49	0.45	1.8	75.70	3JN	-0.0	-3.4	-1.4	0.9
75.21	0.54	1.8	85.23	13JN	0.6	-3.3	-1.4	0.9
81.86	0.63	1.9	94.76	23JN	1.1	-3.3	-1.4	0.8
88.44	0.72	1.9	104.30	3JL	1.8	-3.3	-1.5	0.7
94.97	0.80	1.9	113.86	13JL	3.1	-3.3	-1.5	0.7
101.44	0.88	1.9	123.45	23JL	1.9	-3.3	-1.5	0.6
107.86	0.96	2.0	133.07	2AU	0.1	-3.3	-1.5	0.5
114.22	1.04	2.0	142.74	12AU	-1.0	-3.4	-1.6	0.6
120.53	1.11	2.0	152.46	22AU	-1.3	-3.4	-1.6	0.6
126.79	1.18	1.9	162.23	1SE	-1.0	-3.4	-1.7	0.7
133.00	1.25	1.9	172.06	11SE	-0.5	-3.4	-1.7	0.7
139.15	1.32	1.9	181.95	21SE	-0.2	-3.5	-1.8	0.8
145.25	1.39	1.9	191.89	1OC	-0.1	-3.5	-1.9	0.8
151.27	1.46	1.8	201.89	11OC	0.0	-3.5	-1.9	0.9
157.23	1.52	1.8	211.94	21OC	0.4	-3.5	-2.0	0.9
163.10	1.59	1.7	222.04	31OC	2.5	-3.4	-2.1	1.0
168.88	1.66	1.6	232.18	10NO	0.9	-3.4	-2.1	1.0
174.54	1.72	1.5	242.35	20NO	-0.2	-3.4	-2.2	1.0
180.08	1.79	1.4	252.53	30NO	-0.3	-3.4	-2.2	1.0
185.45	1.85	1.3	262.74	10DE	-0.4	-3.3	-2.2	1.1
190.62	1.91	1.2	272.94	20DE	-0.5	-3.3	-2.2	1.1
195.57	1.97	1.0	283.12	30DE	-0.8	-3.3	-2.2	1.1

Left table

♂ LONG	LAT	MAG	☉ LONG	16.00UT	☿	♀	♃	♄
				816				
200.22	2.03	0.8	293.30	9JA	-1.0	-3.3	-2.2	1.1
204.52	2.08	0.6	303.43	19JA	-1.1	-3.4	-2.1	1.1
208.37	2.12	0.4	313.53	29JA	-0.6	-3.4	-2.1	1.1
211.66	2.14	0.2	323.58	8FE	0.9	-3.4	-2.0	1.1
214.26	2.15	-0.1	333.58	18FE	3.2	-3.4	-1.9	1.1
215.99	2.12	-0.4	343.52	28FE	1.7	-3.4	-1.9	1.2
216.68	2.04	-0.7	353.41	9MR	0.9	-3.5	-1.8	1.2
216.16	1.90	-1.0	3.24	19MR	0.5	-3.5	-1.7	1.3
214.38	1.66	-1.3	13.01	29MR	0.2	-3.6	-1.7	1.3
211.49	1.32	-1.5	22.73	8AP	-0.4	-3.6	-1.6	1.3
207.97	0.89	-1.7	32.39	18AP	-1.2	-3.7	-1.5	1.3
204.50	0.41	-1.6	42.02	28AP	-1.7	-3.8	-1.5	1.3
201.78	-0.07	-1.4	51.61	8MY	-0.9	-3.9	-1.4	1.2
200.28	-0.51	-1.2	61.17	18MY	0.1	-4.0	-1.4	1.2
200.13	-0.88	-1.0	70.71	28MY	0.8	-4.0	-1.4	1.2
201.29	-1.19	-0.8	80.24	7JN	1.4	-4.1	-1.4	1.1
203.60	-1.44	-0.6	89.76	17JN	2.4	-4.2	-1.3	1.1
206.86	-1.64	-0.4	99.29	27JN	3.2	-4.2	-1.3	1.0
210.93	-1.79	-0.2	108.84	7JL	1.5	-4.0	-1.3	0.9
215.65	-1.91	-0.1	118.41	17JL	0.1	-3.5	-1.3	0.9
220.91	-1.99	0.0	128.02	27JL	-1.1	-3.1	-1.3	0.8
226.62	-2.05	0.1	137.67	6AU	-1.5	-3.7	-1.4	0.7
232.71	-2.08	0.3	147.36	16AU	-0.9	-4.1	-1.4	0.7
239.12	-2.08	0.4	157.10	26AU	-0.4	-4.3	-1.4	0.7
245.80	-2.07	0.4	166.90	5SE	-0.1	-4.3	-1.4	0.8
252.70	-2.03	0.5	176.75	15SE	0.1	-4.2	-1.5	0.8
259.79	-1.97	0.6	186.67	25SE	0.2	-4.1	-1.5	0.9
267.04	-1.90	0.7	196.64	5OC	0.7	-4.0	-1.6	0.9
274.41	-1.81	0.8	206.66	15OC	2.9	-3.9	-1.7	1.0
281.89	-1.71	0.8	216.74	25OC	0.7	-3.9	-1.7	1.0
289.44	-1.60	0.9	226.86	4NO	-0.4	-3.8	-1.8	1.1
297.04	-1.48	1.0	237.01	14NO	-0.5	-3.7	-1.8	1.1
304.66	-1.35	1.1	247.19	24NO	-0.5	-3.6	-1.9	1.1
312.31	-1.22	1.1	257.39	4DE	-0.6	-3.6	-2.0	1.2
319.94	-1.08	1.2	267.59	14DE	-0.8	-3.5	-2.0	1.2
327.55	-0.94	1.3	277.79	24DE	-0.9	-3.5	-2.1	1.2
				817				
335.13	-0.80	1.3	287.97	3JA	-0.9	-3.4	-2.1	1.2
342.66	-0.66	1.4	298.12	13JA	-0.5	-3.4	-2.1	1.2
350.14	-0.52	1.4	308.25	23JA	1.1	-3.4	-2.1	1.3
357.55	-0.39	1.5	318.32	2FE	2.9	-3.3	-2.1	1.3
4.89	-0.26	1.6	328.35	12FE	1.2	-3.3	-2.0	1.2
12.17	-0.14	1.6	338.32	22FE	0.6	-3.3	-2.0	1.3
19.36	-0.02	1.7	348.24	4MR	0.4	-3.3	-1.9	1.3
26.49	0.09	1.7	358.10	14MR	0.1	-3.3	-1.9	1.4
33.54	0.20	1.7	7.90	24MR	-0.4	-3.3	-1.8	1.4
40.51	0.30	1.8	17.64	3AP	-1.2	-3.3	-1.7	1.4
47.41	0.40	1.8	27.33	13AP	-1.7	-3.4	-1.7	1.4
54.25	0.49	1.8	36.98	23AP	-0.9	-3.4	-1.6	1.4
61.02	0.57	1.8	46.59	3MY	0.1	-3.4	-1.6	1.4
67.73	0.65	1.9	56.16	13MY	1.0	-3.5	-1.5	1.4
74.38	0.73	1.9	65.71	23MY	1.9	-3.5	-1.4	1.3
80.98	0.80	1.9	75.24	2JN	3.2	-3.5	-1.4	1.3
87.53	0.86	1.9	84.76	12JN	2.6	-3.4	-1.4	1.2
94.04	0.92	1.9	94.29	22JN	1.2	-3.4	-1.3	1.2
100.52	0.97	1.9	103.83	2JL	-0.0	-3.4	-1.3	1.1
106.95	1.02	1.9	113.39	12JL	-1.1	-3.3	-1.3	1.0
113.36	1.07	2.0	122.98	22JL	-1.6	-3.3	-1.3	1.0
119.75	1.11	2.0	132.60	1AU	-0.9	-3.3	-1.2	1.0
126.11	1.15	2.0	142.27	11AU	-0.3	-3.3	-1.2	0.9
132.45	1.18	2.0	151.98	21AU	-0.0	-3.3	-1.2	0.9
138.78	1.21	2.0	161.75	31AU	0.2	-3.3	-1.3	0.9
145.09	1.24	2.0	171.58	10SE	0.4	-3.4	-1.3	0.9
151.39	1.26	2.0	181.46	20SE	1.0	-3.4	-1.3	0.9
157.67	1.28	2.0	191.40	30SE	3.2	-3.4	-1.3	1.0
163.94	1.29	2.0	201.40	10OC	0.6	-3.5	-1.4	1.0
170.20	1.30	1.9	211.45	20OC	-0.5	-3.5	-1.4	1.1
176.44	1.31	1.9	221.55	30OC	-0.6	-3.6	-1.5	1.1
182.65	1.31	1.8	231.68	9NO	-0.6	-3.7	-1.5	1.2
188.84	1.30	1.8	241.86	19NO	-0.7	-3.7	-1.6	1.2
195.00	1.29	1.7	252.04	29NO	-0.7	-3.8	-1.7	1.3
201.12	1.27	1.6	262.24	9DE	-0.7	-3.9	-1.7	1.3
207.20	1.24	1.5	272.45	19DE	-0.8	-4.0	-1.8	1.4
213.21	1.20	1.4	282.63	29DE	-0.4	-4.1	-1.9	1.4

Right table

♂ LONG	LAT	MAG	☉ LONG	16.00UT	☿	♀	♃	♄
				818				
219.15	1.15	1.3	292.80	8JA	1.3	-4.2	-1.9	1.4
225.01	1.08	1.1	302.95	18JA	2.4	-4.3	-2.0	1.4
230.75	1.00	1.0	313.04	28JA	0.9	-4.3	-2.0	1.3
236.36	0.89	0.8	323.10	7FE	0.4	-4.3	-2.0	1.3
241.81	0.76	0.6	333.10	17FE	0.2	-4.1	-2.0	1.2
247.05	0.60	0.4	343.04	27FE	-0.0	-3.5	-2.0	1.2
252.04	0.40	0.2	352.93	9MR	-0.5	-3.3	-2.0	1.2
256.69	0.16	-0.0	2.77	19MR	-1.2	-3.8	-2.0	1.2
260.94	-0.14	-0.3	12.54	29MR	-1.6	-4.1	-1.9	1.2
264.66	-0.51	-0.5	22.26	8AP	-0.9	-4.2	-1.9	1.2
267.70	-0.95	-0.8	31.93	18AP	0.2	-4.2	-1.8	1.2
269.92	-1.48	-1.1	41.55	28AP	1.3	-4.1	-1.7	1.1
271.11	-2.10	-1.4	51.14	8MY	2.5	-4.0	-1.7	1.1
271.11	-2.80	-1.8	60.71	18MY	3.5	-3.9	-1.6	1.1
269.89	-3.54	-2.1	70.24	28MY	2.1	-3.8	-1.5	1.1
267.61	-4.24	-2.3	79.77	7JN	1.0	-3.7	-1.5	1.0
264.75	-4.82	-2.4	89.30	17JN	-0.1	-3.7	-1.4	1.0
262.05	-5.18	-2.3	98.83	27JN	-1.1	-3.6	-1.4	0.9
260.16	-5.31	-2.1	108.38	7JL	-1.7	-3.5	-1.3	0.9
259.53	-5.23	-1.9	117.95	17JL	-0.9	-3.5	-1.3	0.8
260.28	-5.02	-1.6	127.55	27JL	-0.3	-3.4	-1.3	0.8
262.30	-4.73	-1.4	137.20	6AU	0.1	-3.4	-1.2	0.7
265.44	-4.39	-1.2	146.89	16AU	0.3	-3.4	-1.2	0.7
269.48	-4.03	-0.9	156.63	26AU	0.6	-3.4	-1.2	0.6
274.25	-3.67	-0.7	166.42	5SE	1.3	-3.4	-1.2	0.6
279.58	-3.30	-0.5	176.28	15SE	3.3	-3.4	-1.2	0.6
285.36	-2.95	-0.3	186.19	25SE	0.5	-3.4	-1.2	0.7
291.49	-2.61	-0.2	196.16	5OC	-0.6	-3.4	-1.2	0.8
297.88	-2.29	0.0	206.18	15OC	-0.8	-3.4	-1.3	0.8
304.47	-1.98	0.2	216.25	25OC	-0.8	-3.4	-1.3	0.9
311.21	-1.69	0.3	226.37	4NO	-0.8	-3.4	-1.3	1.0
318.06	-1.41	0.5	236.52	14NO	-0.6	-3.4	-1.4	1.0
324.98	-1.16	0.6	246.70	24NO	-0.6	-3.4	-1.4	1.1
331.95	-0.92	0.7	256.90	4DE	-0.6	-3.5	-1.5	1.1
338.94	-0.70	0.9	267.10	14DE	-0.3	-3.5	-1.5	1.1
345.94	-0.49	1.0	277.29	24DE	1.5	-3.5	-1.6	1.1
				819				
352.92	-0.31	1.1	287.48	3JA	2.0	-3.5	-1.7	1.1
359.89	-0.14	1.2	297.63	13JA	0.6	-3.4	-1.7	1.1
6.82	0.02	1.3	307.75	23JA	0.2	-3.4	-1.8	1.1
13.72	0.16	1.4	317.83	2FE	0.1	-3.4	-1.9	1.1
20.58	0.29	1.5	327.86	12FE	-0.1	-3.4	-1.9	1.0
27.39	0.41	1.6	337.84	22FE	-0.5	-3.3	-2.0	1.0
34.15	0.51	1.7	347.76	4MR	-1.2	-3.3	-2.0	0.9
40.87	0.61	1.7	357.62	14MR	-1.5	-3.3	-2.0	0.9
47.54	0.69	1.8	7.42	24MR	-0.8	-3.3	-2.0	0.9
54.17	0.77	1.8	17.17	3AP	0.4	-3.3	-2.0	0.9
60.75	0.83	1.9	26.86	13AP	1.7	-3.4	-2.0	0.9
67.29	0.89	1.9	36.51	23AP	3.3	-3.4	-1.9	0.9
73.80	0.95	1.9	46.12	3MY	2.7	-3.4	-1.9	0.9
80.27	0.99	2.0	55.70	13MY	1.6	-3.4	-1.8	0.9
86.71	1.03	2.0	65.25	23MY	0.7	-3.5	-1.8	0.9
93.13	1.06	2.0	74.78	2JN	-0.1	-3.5	-1.7	0.8
99.52	1.09	2.0	84.31	12JN	-1.2	-3.6	-1.6	0.8
105.90	1.11	2.0	93.83	22JN	-1.7	-3.7	-1.6	0.8
112.27	1.13	2.0	103.37	2JL	-0.9	-3.7	-1.5	0.7
118.64	1.14	2.0	112.93	12JL	-0.2	-3.8	-1.5	0.7
125.00	1.15	2.0	122.51	22JL	0.2	-3.9	-1.4	0.6
131.36	1.16	2.0	132.14	1AU	0.5	-4.0	-1.4	0.6
137.73	1.16	2.0	141.80	11AU	0.8	-4.1	-1.3	0.5
144.11	1.15	2.0	151.51	21AU	1.7	-4.2	-1.3	0.5
150.51	1.14	2.0	161.28	31AU	3.0	-4.3	-1.3	0.4
156.92	1.12	2.0	171.10	10SE	0.4	-4.3	-1.2	0.4
163.35	1.10	2.0	180.98	20SE	-0.8	-4.1	-1.2	0.3
169.80	1.08	2.0	190.92	30SE	-0.9	-3.6	-1.2	0.4
176.28	1.05	2.0	200.92	10OC	-0.9	-3.3	-1.2	0.4
182.78	1.01	2.0	210.96	20OC	-0.7	-3.9	-1.2	0.5
189.30	0.97	1.9	221.06	30OC	-0.5	-4.3	-1.2	0.6
195.86	0.92	1.9	231.19	9NO	-0.5	-4.4	-1.3	0.6
202.44	0.86	1.8	241.36	19NO	-0.4	-4.4	-1.3	0.7
209.04	0.80	1.8	251.55	29NO	-0.1	-4.3	-1.3	0.7
215.67	0.73	1.7	261.75	9DE	1.7	-4.2	-1.4	0.8
222.33	0.64	1.7	271.95	19DE	1.6	-4.1	-1.4	0.8
229.00	0.55	1.6	282.15	29DE	0.3	-4.0	-1.5	0.9

Left table (820–821)

♂ LONG	♂ LAT	♂ MAG	☉ LONG	16.00UT 820	☿	♀	♃	♄
235.70	0.44	1.5	292.31	8JA	0.1	-3.9	-1.5	0.9
242.42	0.32	1.4	302.46	18JA	-0.0	-3.8	-1.6	0.9
249.14	0.19	1.3	312.56	28JA	-0.2	-3.7	-1.7	0.9
255.88	0.04	1.2	322.61	7FE	-0.6	-3.6	-1.7	0.8
262.63	-0.13	1.1	332.62	17FE	-1.2	-3.5	-1.8	0.8
269.37	-0.31	1.0	342.57	27FE	-1.4	-3.5	-1.9	0.8
276.10	-0.52	0.8	352.46	8MR	-0.8	-3.4	-1.9	0.7
282.81	-0.74	0.7	2.30	18MR	0.5	-3.4	-2.0	0.7
289.48	-0.99	0.6	12.07	28MR	2.1	-3.4	-2.0	0.6
296.10	-1.26	0.4	21.79	7AP	3.6	-3.3	-2.1	0.7
302.63	-1.56	0.2	31.47	17AP	2.0	-3.3	-2.1	0.7
309.05	-1.88	0.1	41.10	27AP	1.2	-3.3	-2.1	0.7
315.32	-2.22	-0.1	50.69	7MY	0.5	-3.3	-2.0	0.7
321.37	-2.59	-0.3	60.25	17MY	-0.2	-3.3	-2.0	0.7
327.15	-2.99	-0.5	69.79	27MY	-1.2	-3.3	-2.0	0.7
332.56	-3.40	-0.7	79.32	6JN	-1.8	-3.3	-1.9	0.7
337.50	-3.84	-0.9	88.84	16JN	-0.9	-3.4	-1.9	0.6
341.81	-4.29	-1.1	98.37	26JN	-0.1	-3.4	-1.8	0.6
345.33	-4.74	-1.4	107.92	6JL	0.3	-3.4	-1.7	0.6
347.84	-5.18	-1.6	117.49	16JL	0.7	-3.5	-1.7	0.5
349.14	-5.57	-1.9	127.09	26JL	1.1	-3.5	-1.6	0.5
349.08	-5.86	-2.1	136.73	5AU	2.2	-3.5	-1.6	0.4
347.64	-5.96	-2.3	146.42	15AU	2.6	-3.4	-1.5	0.4
345.14	-5.80	-2.5	156.15	25AU	0.3	-3.4	-1.5	0.3
342.23	-5.38	-2.5	165.94	4SE	-0.9	-3.4	-1.4	0.2
339.66	-4.73	-2.3	175.79	14SE	-1.1	-3.4	-1.4	0.2
338.08	-3.96	-2.0	185.70	24SE	-1.0	-3.3	-1.3	0.1
337.79	-3.19	-1.6	195.67	4OC	-0.6	-3.3	-1.3	0.1
338.79	-2.47	-1.3	205.69	14OC	-0.4	-3.3	-1.3	0.1
340.94	-1.84	-1.0	215.76	24OC	-0.3	-3.3	-1.3	0.2
344.02	-1.30	-0.7	225.87	3NO	-0.3	-3.3	-1.3	0.3
347.85	-0.85	-0.4	236.03	13NO	0.1	-3.4	-1.3	0.3
352.25	-0.47	-0.2	246.20	23NO	1.9	-3.4	-1.3	0.4
357.10	-0.16	0.1	256.40	3DE	1.3	-3.4	-1.3	0.5
2.28	0.11	0.3	266.60	13DE	0.1	-3.5	-1.3	0.5
7.73	0.33	0.5	276.80	23DE	-0.1	-3.5	-1.4	0.6
				821				
13.38	0.51	0.7	286.98	2JA	-0.2	-3.5	-1.4	0.6
19.18	0.66	0.9	297.15	12JA	-0.3	-3.6	-1.5	0.6
25.09	0.79	1.0	307.27	22JA	-0.6	-3.7	-1.5	0.6
31.08	0.89	1.2	317.35	1FE	-1.2	-3.7	-1.6	0.7
37.14	0.98	1.3	327.38	11FE	-1.3	-3.8	-1.6	0.7
43.25	1.05	1.4	337.36	21FE	-0.8	-3.9	-1.7	0.6
49.38	1.11	1.5	347.28	3MR	0.6	-4.0	-1.8	0.6
55.55	1.16	1.6	357.15	13MR	2.6	-4.1	-1.9	0.6
61.72	1.19	1.7	6.95	23MR	2.7	-4.1	-1.9	0.5
67.91	1.22	1.8	16.70	2AP	1.5	-4.2	-2.0	0.5
74.10	1.24	1.8	26.40	12AP	0.9	-4.2	-2.0	0.4
80.30	1.26	1.9	36.05	22AP	0.4	-4.1	-2.1	0.5
86.51	1.26	1.9	45.66	2MY	-0.3	-3.7	-2.1	0.5
92.72	1.26	2.0	55.24	12MY	-1.2	-2.9	-2.2	0.5
98.94	1.26	2.0	64.79	22MY	-1.8	-3.3	-2.2	0.5
105.17	1.25	2.0	74.32	1JN	-0.9	-3.9	-2.2	0.5
111.41	1.24	2.0	83.85	11JN	-0.1	-4.1	-2.1	0.5
117.67	1.22	2.0	93.37	21JN	0.5	-4.2	-2.1	0.5
123.95	1.20	2.0	102.91	1JL	0.9	-4.1	-2.1	0.5
130.25	1.17	2.0	112.47	11JL	1.5	-4.1	-2.0	0.4
136.57	1.14	2.0	122.05	21JL	2.8	-4.0	-2.0	0.4
142.93	1.10	2.0	131.67	31JL	2.2	-3.9	-1.9	0.4
149.33	1.06	2.0	141.33	10AU	0.2	-3.8	-1.8	0.3
155.76	1.02	1.9	151.04	20AU	-1.0	-3.7	-1.8	0.3
162.23	0.97	1.9	160.80	30AU	-1.2	-3.7	-1.7	0.2
168.75	0.92	1.9	170.63	9SE	-1.0	-3.6	-1.6	0.1
175.31	0.86	1.9	180.50	19SE	-0.5	-3.6	-1.6	0.1
181.93	0.80	1.9	190.44	29SE	-0.3	-3.5	-1.5	0.0
188.60	0.74	1.9	200.43	9OC	-0.2	-3.5	-1.5	-0.1
195.33	0.66	1.9	210.47	19OC	-0.1	-3.4	-1.5	-0.1
202.11	0.59	1.9	220.57	29OC	0.3	-3.4	-1.4	-0.0
208.96	0.50	1.8	230.70	8NO	2.2	-3.4	-1.4	0.0
215.86	0.41	1.8	240.86	18NO	1.1	-3.4	-1.4	0.1
222.82	0.32	1.8	251.05	28NO	-0.1	-3.4	-1.4	0.2
229.84	0.22	1.7	261.26	8DE	-0.2	-3.4	-1.4	0.2
236.91	0.11	1.7	271.45	18DE	-0.3	-3.4	-1.4	0.3
244.05	-0.01	1.6	281.65	28DE	-0.4	-3.3	-1.4	0.4

Right table (822–823)

♂ LONG	♂ LAT	♂ MAG	☉ LONG	16.00UT 822	☿	♀	♃	♄
251.24	-0.13	1.6	291.82	7JA	-0.7	-3.4	-1.4	0.4
258.48	-0.26	1.5	301.96	17JA	-1.1	-3.4	-1.5	0.4
265.78	-0.39	1.5	312.07	27JA	-1.2	-3.4	-1.5	0.5
273.12	-0.53	1.4	322.13	6FE	-0.7	-3.4	-1.5	0.5
280.50	-0.68	1.3	332.13	16FE	0.8	-3.4	-1.6	0.5
287.92	-0.83	1.3	342.09	26FE	3.1	-3.5	-1.6	0.5
295.36	-0.98	1.2	351.98	8MR	2.1	-3.5	-1.7	0.5
302.82	-1.14	1.1	1.82	18MR	1.1	-3.5	-1.8	0.4
310.29	-1.29	1.0	11.60	28MR	0.7	-3.4	-1.8	0.4
317.75	-1.44	1.0	21.32	7AP	0.3	-3.4	-1.9	0.4
325.18	-1.59	0.9	30.99	17AP	-0.3	-3.4	-2.0	0.3
332.58	-1.73	0.8	40.63	27AP	-1.2	-3.4	-2.0	0.3
339.92	-1.85	0.7	50.22	7MY	-1.8	-3.3	-2.1	0.3
347.17	-1.97	0.6	59.78	17MY	-0.9	-3.3	-2.2	0.4
354.32	-2.08	0.5	69.33	27MY	-0.0	-3.3	-2.2	0.4
1.33	-2.17	0.4	78.85	6JN	0.6	-3.3	-2.3	0.4
8.17	-2.24	0.4	88.38	16JN	1.2	-3.3	-2.3	0.4
14.80	-2.29	0.3	97.91	26JN	2.0	-3.4	-2.3	0.4
21.17	-2.32	0.1	107.46	6JL	3.3	-3.4	-2.3	0.4
27.24	-2.33	0.0	117.02	16JL	1.8	-3.4	-2.3	0.4
32.93	-2.32	-0.1	126.62	26JL	0.1	-3.5	-2.2	0.3
38.15	-2.27	-0.2	136.26	5AU	-1.0	-3.5	-2.2	0.3
42.81	-2.20	-0.4	145.94	15AU	-1.4	-3.6	-2.1	0.3
46.75	-2.09	-0.6	155.68	25AU	-1.0	-3.6	-2.1	0.2
49.81	-1.93	-0.8	165.47	4SE	-0.5	-3.7	-2.0	0.2
51.81	-1.71	-1.0	175.31	14SE	-0.2	-3.8	-1.9	0.1
52.50	-1.43	-1.2	185.22	24SE	-0.0	-3.9	-1.9	0.0
51.76	-1.07	-1.4	195.18	4OC	0.1	-4.0	-1.8	-0.0
49.58	-0.63	-1.6	205.20	14OC	0.5	-4.1	-1.8	-0.1
46.28	-0.13	-1.8	215.27	24OC	2.5	-4.2	-1.7	-0.2
42.54	0.37	-1.8	225.38	3NO	0.9	-4.3	-1.6	-0.2
39.16	0.82	-1.5	235.53	13NO	-0.2	-4.4	-1.6	-0.2
36.80	1.18	-1.2	245.71	23NO	-0.4	-4.4	-1.6	-0.1
35.77	1.45	-0.9	255.90	3DE	-0.4	-4.1	-1.5	-0.0
36.07	1.63	-0.6	266.10	13DE	-0.5	-3.6	-1.5	0.1
37.54	1.75	-0.3	276.30	23DE	-0.8	-3.2	-1.5	0.1
				823				
39.98	1.82	0.0	286.49	2JA	-1.0	-3.9	-1.5	0.2
43.18	1.86	0.3	296.65	12JA	-1.0	-4.3	-1.5	0.2
46.97	1.87	0.5	306.77	22JA	-0.6	-4.4	-1.5	0.3
51.23	1.87	0.7	316.86	1FE	0.9	-4.3	-1.5	0.3
55.85	1.85	0.9	326.89	11FE	3.1	-4.2	-1.5	0.4
60.76	1.82	1.1	336.88	21FE	1.5	-4.1	-1.6	0.4
65.90	1.79	1.2	346.80	3MR	0.8	-4.0	-1.6	0.4
71.21	1.76	1.3	356.67	13MR	0.5	-3.9	-1.6	0.4
76.67	1.71	1.5	6.48	23MR	0.1	-3.8	-1.7	0.4
82.25	1.67	1.6	16.23	2AP	-0.4	-3.7	-1.7	0.3
87.92	1.62	1.6	25.93	12AP	-1.2	-3.6	-1.8	0.3
93.69	1.57	1.7	35.59	22AP	-1.7	-3.6	-1.9	0.3
99.52	1.52	1.8	45.20	2MY	-0.9	-3.5	-1.9	0.2
105.42	1.46	1.8	54.78	12MY	0.1	-3.4	-2.0	0.2
111.39	1.41	1.9	64.33	22MY	0.8	-3.4	-2.1	0.3
117.40	1.35	1.9	73.86	1JN	1.6	-3.4	-2.2	0.3
123.48	1.29	2.0	83.39	11JN	2.7	-3.3	-2.2	0.3
129.61	1.22	2.0	92.92	21JN	3.0	-3.3	-2.3	0.3
135.80	1.16	2.0	102.45	1JL	1.5	-3.3	-2.3	0.3
142.04	1.09	2.0	112.01	11JL	0.1	-3.3	-2.4	0.3
148.35	1.02	2.0	121.59	21JL	-1.1	-3.3	-2.4	0.3
154.71	0.95	2.0	131.21	31JL	-1.5	-3.3	-2.4	0.3
161.14	0.87	2.0	140.87	10AU	-1.0	-3.4	-2.4	0.3
167.64	0.79	1.9	150.58	20AU	-0.4	-3.4	-2.4	0.3
174.20	0.71	1.9	160.33	30AU	-0.1	-3.4	-2.4	0.2
180.84	0.63	1.9	170.15	9SE	0.1	-3.4	-2.3	0.2
187.55	0.54	1.8	180.03	19SE	0.3	-3.5	-2.2	0.1
194.34	0.45	1.8	189.95	29SE	0.8	-3.5	-2.2	0.0
201.20	0.36	1.7	199.94	9OC	2.9	-3.5	-2.1	-0.0
208.15	0.27	1.7	209.99	19OC	0.7	-3.5	-2.0	-0.1
215.17	0.17	1.7	220.07	29OC	-0.4	-3.4	-2.0	-0.2
222.28	0.07	1.7	230.20	8NO	-0.6	-3.4	-1.9	-0.2
229.46	-0.04	1.7	240.37	18NO	-0.5	-3.4	-1.8	-0.3
236.72	-0.14	1.7	250.55	28NO	-0.6	-3.4	-1.8	-0.2
244.05	-0.25	1.6	260.75	8DE	-0.8	-3.3	-1.7	-0.2
251.46	-0.36	1.6	270.95	18DE	-0.8	-3.3	-1.7	-0.1
258.93	-0.47	1.6	281.14	28DE	-0.9	-3.3	-1.6	-0.0

♂ / ☉ / 16.00UT 824 / Magnitudes (☿ ♀ ♃ ♄)

♂ LONG	LAT	MAG	☉ LONG	16.00UT 824	☿	♀	♃	♄
266.47	-0.58	1.6	291.32	7JA	-0.5	-3.3	-1.6	0.1
274.06	-0.69	1.5	301.47	17JA	1.1	-3.4	-1.6	0.1
281.70	-0.79	1.5	311.57	27JA	2.7	-3.4	-1.6	0.2
289.38	-0.90	1.5	321.64	6FE	1.1	-3.4	-1.6	0.2
297.09	-0.99	1.4	331.65	16FE	0.6	-3.4	-1.6	0.3
304.82	-1.08	1.4	341.60	26FE	0.3	-3.4	-1.6	0.3
312.57	-1.17	1.4	351.50	7MR	0.0	-3.5	-1.6	0.3
320.31	-1.24	1.3	1.34	17MR	-0.4	-3.5	-1.6	0.3
328.03	-1.30	1.3	11.12	27MR	-1.2	-3.6	-1.6	0.3
335.73	-1.35	1.3	20.85	6AP	-1.7	-3.6	-1.7	0.3
343.39	-1.39	1.2	30.53	16AP	-0.9	-3.7	-1.7	0.3
350.99	-1.41	1.2	40.16	26AP	0.2	-3.8	-1.7	0.3
358.52	-1.41	1.2	49.76	6MY	1.1	-3.9	-1.8	0.2
5.97	-1.41	1.1	59.33	16MY	2.1	-4.0	-1.9	0.2
13.32	-1.38	1.1	68.87	26MY	3.4	-4.1	-1.9	0.2
20.56	-1.34	1.1	78.40	5JN	2.4	-4.1	-2.0	0.2
27.68	-1.28	1.0	87.92	15JN	1.2	-4.2	-2.1	0.3
34.65	-1.21	1.0	97.45	25JN	-0.0	-4.2	-2.1	0.3
41.46	-1.12	0.9	107.00	5JL	-1.1	-4.0	-2.2	0.3
48.09	-1.02	0.9	116.56	15JL	-1.6	-3.5	-2.3	0.3
54.52	-0.90	0.8	126.16	25JL	-0.9	-3.1	-2.3	0.3
60.72	-0.76	0.8	135.80	4AU	-0.3	-3.7	-2.4	0.3
66.65	-0.60	0.7	145.48	14AU	0.0	-4.1	-2.4	0.3
72.28	-0.42	0.6	155.21	24AU	0.2	-4.3	-2.4	0.3
77.54	-0.22	0.5	165.00	3SE	0.5	-4.3	-2.5	0.2
82.36	0.01	0.4	174.84	13SE	1.1	-4.2	-2.5	0.2
86.66	0.27	0.2	184.74	23SE	3.2	-4.1	-2.4	0.2
90.30	0.57	0.1	194.70	3OC	0.6	-4.0	-2.4	0.1
93.15	0.90	-0.1	204.71	13OC	-0.6	-3.9	-2.4	0.0
95.01	1.29	-0.3	214.78	23OC	-0.7	-3.9	-2.3	-0.0
95.69	1.72	-0.5	224.89	2NO	-0.7	-3.8	-2.2	-0.1
95.03	2.18	-0.8	235.04	12NO	-0.7	-3.7	-2.2	-0.2
92.97	2.65	-1.0	245.21	22NO	-0.7	-3.6	-2.1	-0.3
89.70	3.07	-1.2	255.41	2DE	-0.7	-3.6	-2.0	-0.3
85.79	3.39	-1.3	265.61	12DE	-0.7	-3.5	-1.9	-0.2
81.98	3.56	-1.1	275.81	22DE	-0.4	-3.5	-1.9	-0.2

825

♂ LONG	LAT	MAG	☉ LONG	16.00UT 825	☿	♀	♃	♄
78.98	3.58	-0.8	285.99	1JA	1.3	-3.4	-1.8	-0.1
77.22	3.49	-0.5	296.15	11JA	2.3	-3.4	-1.8	-0.0
76.77	3.34	-0.3	306.28	21JA	0.8	-3.4	-1.7	0.0
77.56	3.15	0.0	316.37	31JA	0.4	-3.3	-1.7	0.1
79.37	2.96	0.3	326.40	10FE	0.2	-3.3	-1.6	0.2
82.02	2.77	0.5	336.39	20FE	-0.1	-3.3	-1.6	0.2
85.35	2.59	0.7	346.32	2MR	-0.5	-3.3	-1.6	0.3
89.20	2.41	0.9	356.19	12MR	-1.2	-3.3	-1.6	0.3
93.47	2.25	1.0	6.00	22MR	-1.6	-3.3	-1.6	0.3
98.08	2.10	1.2	15.76	1AP	-0.9	-3.3	-1.6	0.3
102.97	1.96	1.3	25.46	11AP	0.3	-3.4	-1.6	0.3
108.09	1.82	1.4	35.12	21AP	1.4	-3.4	-1.6	0.3
113.39	1.70	1.5	44.73	1MY	2.8	-3.4	-1.6	0.3
118.87	1.57	1.6	54.31	11MY	3.3	-3.5	-1.6	0.3
124.48	1.45	1.7	63.87	21MY	1.9	-3.5	-1.7	0.2
130.23	1.34	1.7	73.40	31MY	0.9	-3.4	-1.7	0.2
136.09	1.22	1.8	82.93	10JN	-0.1	-3.4	-1.8	0.2
142.06	1.11	1.8	92.45	20JN	-1.1	-3.4	-1.8	0.2
148.14	1.00	1.8	101.99	30JN	-1.7	-3.4	-1.9	0.3
154.32	0.89	1.8	111.54	10JL	-0.9	-3.3	-2.0	0.3
160.59	0.78	1.8	121.12	20JL	-0.3	-3.3	-2.0	0.3
166.97	0.68	1.8	130.74	30JL	0.1	-3.3	-2.1	0.4
173.44	0.57	1.8	140.39	9AU	0.4	-3.3	-2.2	0.4
180.01	0.46	1.8	150.10	19AU	0.7	-3.3	-2.2	0.4
186.68	0.35	1.8	159.86	29AU	1.4	-3.3	-2.3	0.4
193.44	0.25	1.8	169.67	8SE	3.2	-3.4	-2.3	0.3
200.31	0.14	1.8	179.54	18SE	0.5	-3.4	-2.4	0.3
207.27	0.03	1.7	189.47	28SE	-0.7	-3.4	-2.4	0.2
214.33	-0.07	1.7	199.46	8OC	-0.9	-3.5	-2.4	0.2
221.49	-0.18	1.6	209.50	18OC	-0.8	-3.5	-2.4	0.1
228.74	-0.28	1.6	219.58	28OC	-0.8	-3.6	-2.4	0.1
236.09	-0.38	1.5	229.71	7NO	-0.6	-3.7	-2.4	-0.0
243.52	-0.48	1.5	239.88	17NO	-0.5	-3.7	-2.3	-0.1
251.03	-0.57	1.4	250.06	27NO	-0.5	-3.8	-2.3	-0.2
258.62	-0.66	1.4	260.26	7DE	-0.2	-3.9	-2.2	-0.2
266.28	-0.75	1.4	270.46	17DE	1.5	-4.0	-2.1	-0.2
274.00	-0.83	1.4	280.65	27DE	1.9	-4.1	-2.1	-0.2

♂ / ☉ / 16.00UT 826 / Magnitudes (☿ ♀ ♃ ♄)

♂ LONG	LAT	MAG	☉ LONG	16.00UT 826	☿	♀	♃	♄
281.76	-0.90	1.4	290.83	6JA	0.5	-4.2	-2.0	-0.1
289.57	-0.97	1.4	300.98	16JA	0.2	-4.3	-1.9	-0.1
297.41	-1.02	1.4	311.08	26JA	0.0	-4.3	-1.8	0.0
305.26	-1.07	1.4	321.15	5FE	-0.2	-4.3	-1.8	0.1
313.12	-1.10	1.4	331.16	15FE	-0.5	-4.0	-1.7	0.1
320.97	-1.13	1.4	341.12	25FE	-1.2	-3.5	-1.7	0.2
328.80	-1.14	1.4	351.02	7MR	-1.5	-3.3	-1.6	0.2
336.60	-1.14	1.4	0.86	17MR	-0.9	-3.9	-1.6	0.3
344.36	-1.12	1.4	10.65	27MR	0.4	-4.2	-1.6	0.3
352.06	-1.10	1.4	20.38	6AP	1.8	-4.2	-1.5	0.3
359.69	-1.06	1.4	30.06	16AP	3.6	-4.2	-1.5	0.4
7.25	-1.01	1.4	39.69	26AP	2.5	-4.1	-1.5	0.4
14.72	-0.94	1.4	49.29	6MY	1.4	-4.0	-1.5	0.4
22.11	-0.87	1.4	58.86	16MY	0.7	-3.9	-1.5	0.4
29.39	-0.79	1.4	68.40	26MY	-0.1	-3.8	-1.5	0.3
36.58	-0.69	1.4	77.93	5JN	-1.1	-3.7	-1.5	0.3
43.65	-0.59	1.4	87.46	15JN	-1.8	-3.7	-1.6	0.3
50.60	-0.48	1.4	96.99	25JN	-0.9	-3.6	-1.6	0.3
57.43	-0.36	1.4	106.53	5JL	-0.2	-3.5	-1.6	0.3
64.14	-0.23	1.4	116.10	15JL	0.2	-3.5	-1.7	0.4
70.70	-0.09	1.4	125.70	25JL	0.6	-3.4	-1.7	0.4
77.12	0.05	1.3	135.33	4AU	0.9	-3.4	-1.8	0.4
83.37	0.20	1.3	145.01	14AU	1.9	-3.4	-1.9	0.4
89.44	0.37	1.3	154.74	24AU	2.9	-3.4	-1.9	0.5
95.32	0.54	1.2	164.52	3SE	0.4	-3.4	-2.0	0.5
100.98	0.73	1.1	174.37	13SE	-0.8	-3.4	-2.1	0.4
106.37	0.93	1.0	184.26	23SE	-1.0	-3.4	-2.1	0.4
111.45	1.16	0.9	194.22	3OC	-1.0	-3.4	-2.2	0.4
116.17	1.40	0.8	204.23	13OC	-0.7	-3.4	-2.2	0.3
120.44	1.66	0.7	214.29	23OC	-0.5	-3.4	-2.3	0.3
124.17	1.96	0.5	224.40	2NO	-0.4	-3.4	-2.3	0.2
127.23	2.29	0.3	234.55	12NO	-0.4	-3.4	-2.3	0.2
129.46	2.65	0.1	244.72	22NO	-0.1	-3.4	-2.3	0.1
130.68	3.05	-0.1	254.91	2DE	1.7	-3.5	-2.3	0.0
130.69	3.46	-0.3	265.12	12DE	1.6	-3.5	-2.3	-0.1
129.39	3.87	-0.6	275.31	22DE	0.3	-3.5	-2.2	-0.1

827

♂ LONG	LAT	MAG	☉ LONG	16.00UT 827	☿	♀	♃	♄
126.80	4.23	-0.8	285.50	1JA	0.0	-3.5	-2.2	-0.1
123.21	4.46	-1.0	295.66	11JA	-0.1	-3.4	-2.1	-0.1
119.24	4.52	-1.0	305.79	21JA	-0.3	-3.4	-2.0	-0.0
115.62	4.41	-0.8	315.88	31JA	-0.6	-3.4	-2.0	0.0
112.96	4.16	-0.6	325.92	10FE	-1.2	-3.4	-1.9	0.1
111.57	3.84	-0.3	335.90	20FE	-1.4	-3.3	-1.8	0.2
111.45	3.49	-0.1	345.83	2MR	-0.8	-3.3	-1.8	0.2
112.50	3.15	0.2	355.71	12MR	0.5	-3.3	-1.7	0.3
114.52	2.82	0.4	5.52	22MR	2.2	-3.3	-1.6	0.3
117.33	2.52	0.6	15.28	1AP	3.3	-3.3	-1.6	0.4
120.80	2.25	0.8	24.99	11AP	1.8	-3.4	-1.5	0.4
124.78	2.00	0.9	34.64	21AP	1.1	-3.4	-1.5	0.4
129.19	1.77	1.0	44.26	1MY	0.5	-3.4	-1.5	0.5
133.95	1.56	1.1	53.85	11MY	-0.2	-3.4	-1.4	0.5
139.00	1.36	1.2	63.40	21MY	-1.1	-3.5	-1.4	0.5
144.30	1.18	1.3	72.94	31MY	-1.8	-3.5	-1.4	0.5
149.83	1.01	1.4	82.47	10JN	-0.9	-3.6	-1.4	0.5
155.55	0.84	1.5	91.99	20JN	-0.1	-3.7	-1.4	0.4
161.45	0.69	1.5	101.53	30JN	0.4	-3.7	-1.4	0.4
167.51	0.54	1.5	111.08	10JL	0.8	-3.8	-1.4	0.4
173.72	0.40	1.6	120.66	20JL	1.3	-3.9	-1.5	0.5
180.08	0.26	1.6	130.28	30JL	2.4	-4.0	-1.5	0.5
186.58	0.13	1.6	139.93	9AU	2.5	-4.1	-1.5	0.5
193.21	0.00	1.6	149.63	19AU	0.3	-4.2	-1.6	0.6
199.98	-0.12	1.6	159.39	29AU	-0.9	-4.3	-1.6	0.6
206.87	-0.24	1.6	169.20	8SE	-1.2	-4.2	-1.7	0.6
213.88	-0.35	1.6	179.06	18SE	-1.0	-4.0	-1.7	0.6
221.01	-0.45	1.6	188.99	28SE	-0.6	-3.5	-1.8	0.6
228.25	-0.55	1.5	198.97	8OC	-0.3	-3.3	-1.9	0.6
235.60	-0.65	1.5	209.01	18OC	-0.2	-3.9	-1.9	0.5
243.06	-0.73	1.5	219.10	28OC	-0.2	-4.3	-2.0	0.5
250.60	-0.81	1.5	229.22	7NO	0.1	-4.4	-2.1	0.4
258.23	-0.88	1.5	239.38	17NO	2.0	-4.3	-2.1	0.4
265.94	-0.94	1.4	249.57	27NO	1.3	-4.3	-2.2	0.3
273.71	-0.99	1.4	259.76	7DE	0.1	-4.2	-2.2	0.2
281.54	-1.03	1.4	269.96	17DE	-0.2	-4.1	-2.2	0.2
289.40	-1.06	1.4	280.16	27DE	-0.2	-4.0	-2.2	0.1

♂ LONG	♂ LAT	♂ MAG	⊙ LONG	16.00UT	☿	♀	♃	♄
				828	MAGNITUDES			
297.30	-1.08	1.3	290.34	6JA	-0.4	-3.9	-2.2	0.0
305.20	-1.09	1.3	300.48	16JA	-0.7	-3.8	-2.1	0.0
313.11	-1.08	1.3	310.60	26JA	-1.2	-3.7	-2.1	0.1
320.99	-1.06	1.3	320.66	5FE	-1.3	-3.6	-2.0	0.1
328.86	-1.04	1.3	330.68	15FE	-0.8	-3.5	-2.0	0.2
336.68	-1.00	1.3	340.64	25FE	0.6	-3.5	-1.9	0.2
344.46	-0.95	1.4	350.54	6MR	2.7	-3.4	-1.8	0.3
352.17	-0.89	1.4	0.39	16MR	2.5	-3.4	-1.8	0.4
359.82	-0.82	1.4	10.18	26MR	1.3	-3.4	-1.7	0.4
7.39	-0.74	1.5	19.91	5AP	0.8	-3.3	-1.6	0.5
14.88	-0.65	1.5	29.59	15AP	0.3	-3.3	-1.6	0.5
22.29	-0.56	1.5	39.23	25AP	-0.3	-3.3	-1.5	0.5
29.60	-0.47	1.6	48.83	5MY	-1.1	-3.3	-1.5	0.6
36.82	-0.37	1.6	58.40	15MY	-1.8	-3.3	-1.4	0.6
43.95	-0.26	1.6	67.94	25MY	-0.9	-3.3	-1.4	0.6
50.98	-0.15	1.7	77.47	4JN	-0.1	-3.3	-1.4	0.6
57.92	-0.04	1.7	87.00	14JN	0.5	-3.4	-1.3	0.6
64.76	0.07	1.7	96.53	24JN	1.0	-3.4	-1.3	0.6
71.50	0.19	1.7	106.07	4JL	1.7	-3.4	-1.3	0.6
78.15	0.31	1.7	115.63	14JL	3.0	-3.5	-1.3	0.6
84.70	0.43	1.7	125.23	24JL	2.1	-3.5	-1.3	0.6
91.14	0.56	1.7	134.86	3AU	0.2	-3.5	-1.3	0.7
97.48	0.68	1.7	144.54	13AU	-1.0	-3.4	-1.3	0.7
103.71	0.81	1.7	154.27	23AU	-1.3	-3.4	-1.4	0.7
109.82	0.94	1.6	164.04	2SE	-1.0	-3.4	-1.4	0.8
115.80	1.08	1.6	173.88	12SE	-0.5	-3.4	-1.4	0.8
121.64	1.23	1.5	183.78	22SE	-0.2	-3.3	-1.5	0.8
127.31	1.38	1.5	193.73	2OC	-0.1	-3.3	-1.5	0.8
132.81	1.54	1.4	203.74	12OC	-0.0	-3.3	-1.6	0.8
138.08	1.71	1.3	213.80	22OC	0.3	-3.3	-1.7	0.8
143.09	1.89	1.2	223.90	1NO	2.2	-3.3	-1.7	0.7
147.80	2.08	1.0	234.05	11NO	1.1	-3.4	-1.8	0.7
152.13	2.29	0.9	244.22	21NO	-0.1	-3.4	-1.9	0.7
155.99	2.53	0.7	254.41	1DE	-0.3	-3.4	-1.9	0.6
159.28	2.78	0.5	264.62	11DE	-0.3	-3.5	-2.0	0.5
161.86	3.05	0.3	274.82	21DE	-0.5	-3.5	-2.0	0.5
163.56	3.34	0.1	285.00	31DE	-0.7	-3.5	-2.1	0.4
				829				
164.20	3.64	-0.2	295.17	10JA	-1.1	-3.6	-2.1	0.3
163.60	3.92	-0.4	305.30	20JA	-1.1	-3.7	-2.1	0.2
161.72	4.15	-0.7	315.39	30JA	-0.7	-3.7	-2.1	0.2
158.68	4.27	-0.9	325.44	9FE	0.8	-3.8	-2.1	0.3
154.90	4.23	-1.1	335.43	19FE	3.1	-3.9	-2.0	0.3
151.07	4.03	-1.0	345.36	1MR	1.8	-4.0	-2.0	0.4
147.87	3.69	-0.8	355.24	11MR	1.0	-4.1	-1.9	0.5
145.80	3.27	-0.6	5.05	21MR	0.6	-4.2	-1.9	0.5
145.03	2.83	-0.3	14.81	31MR	0.2	-4.2	-1.8	0.6
145.50	2.40	-0.1	24.52	10AP	-0.3	-4.2	-1.7	0.6
147.09	2.01	0.1	34.18	20AP	-1.1	-4.1	-1.7	0.7
149.59	1.65	0.3	43.80	30AP	-1.8	-3.7	-1.6	0.7
152.86	1.33	0.4	53.38	10MY	-0.9	-2.9	-1.5	0.7
156.76	1.04	0.6	62.94	20MY	-0.0	-3.3	-1.5	0.8
161.16	0.78	0.7	72.47	30MY	0.7	-3.9	-1.4	0.8
165.99	0.54	0.8	82.00	9JN	1.3	-4.1	-1.4	0.8
171.19	0.33	0.9	91.53	19JN	2.2	-4.2	-1.3	0.8
176.70	0.13	1.0	101.06	29JN	3.3	-4.1	-1.3	0.9
182.48	-0.05	1.1	110.62	9JL	1.7	-4.1	-1.3	0.9
188.51	-0.21	1.1	120.19	19JL	0.2	-4.0	-1.3	0.8
194.75	-0.36	1.2	129.80	29JL	-1.0	-3.9	-1.2	0.8
201.20	-0.50	1.2	139.46	8AU	-1.4	-3.8	-1.2	0.9
207.84	-0.63	1.2	149.16	18AU	-1.0	-3.7	-1.2	0.9
214.64	-0.74	1.3	158.91	28AU	-0.4	-3.7	-1.2	1.0
221.61	-0.85	1.3	168.72	7SE	-0.1	-3.6	-1.3	1.0
228.72	-0.94	1.3	178.58	17SE	0.0	-3.6	-1.3	1.0
235.97	-1.01	1.3	188.51	27SE	0.1	-3.5	-1.3	1.0
243.35	-1.08	1.3	198.49	7OC	0.6	-3.5	-1.3	1.1
250.84	-1.13	1.3	208.52	17OC	2.6	-3.4	-1.4	1.1
258.42	-1.17	1.3	218.61	27OC	0.9	-3.4	-1.4	1.1
266.10	-1.20	1.3	228.73	6NO	-0.3	-3.4	-1.5	1.0
273.85	-1.21	1.4	238.89	16NO	-0.5	-3.4	-1.6	1.0
281.65	-1.21	1.4	249.07	26NO	-0.5	-3.4	-1.6	1.0
289.50	-1.20	1.4	259.27	6DE	-0.6	-3.4	-1.7	0.9
297.37	-1.18	1.4	269.47	16DE	-0.8	-3.4	-1.7	0.8
305.26	-1.14	1.4	279.67	26DE	-0.9	-3.4	-1.8	0.8

♂ LONG	♂ LAT	♂ MAG	⊙ LONG	16.00UT	☿	♀	♃	♄
				830	MAGNITUDES			
313.14	-1.09	1.4	289.85	5JA	-1.0	-3.4	-1.9	0.7
321.01	-1.03	1.4	299.99	15JA	-0.6	-3.4	-1.9	0.6
328.84	-0.96	1.4	310.11	25JA	0.9	-3.4	-2.0	0.6
336.63	-0.89	1.4	320.18	4FE	3.0	-3.4	-2.0	0.5
344.37	-0.80	1.5	330.19	14FE	1.4	-3.4	-2.0	0.5
352.04	-0.72	1.5	340.16	24FE	0.7	-3.5	-2.0	0.6
359.65	-0.62	1.5	350.07	6MR	0.4	-3.5	-2.0	0.6
7.17	-0.52	1.5	359.91	16MR	0.1	-3.5	-2.0	0.6
14.62	-0.42	1.5	9.70	26MR	-0.4	-3.4	-1.9	0.7
21.98	-0.32	1.5	19.44	5AP	-1.1	-3.4	-1.9	0.8
29.25	-0.21	1.5	29.12	15AP	-1.7	-3.4	-1.8	0.8
36.44	-0.11	1.6	38.76	25AP	-0.9	-3.4	-1.8	0.9
43.54	-0.00	1.6	48.36	5MY	0.1	-3.3	-1.7	0.9
50.56	0.10	1.7	57.93	15MY	0.9	-3.3	-1.6	1.0
57.49	0.20	1.7	67.48	25MY	1.7	-3.3	-1.6	1.0
64.35	0.30	1.8	77.01	4JN	3.0	-3.3	-1.5	1.0
71.12	0.40	1.8	86.53	14JN	2.8	-3.3	-1.5	1.1
77.82	0.50	1.8	96.06	24JN	1.4	-3.4	-1.4	1.1
84.45	0.60	1.9	105.60	4JL	0.1	-3.4	-1.4	1.1
91.01	0.69	1.9	115.16	14JL	-1.1	-3.4	-1.3	1.1
97.50	0.79	1.9	124.76	24JL	-1.6	-3.5	-1.3	1.1
103.92	0.88	1.9	134.39	3AU	-1.0	-3.5	-1.3	1.1
110.28	0.97	1.9	144.06	13AU	-0.4	-3.6	-1.2	1.1
116.57	1.06	1.9	153.79	23AU	-0.0	-3.6	-1.2	1.1
122.80	1.15	1.9	163.57	2SE	0.2	-3.7	-1.2	1.2
128.95	1.24	1.9	173.40	12SE	0.3	-3.8	-1.2	1.2
135.03	1.33	1.8	183.30	22SE	0.9	-3.9	-1.2	1.3
141.02	1.42	1.8	193.25	2OC	2.9	-4.0	-1.2	1.3
146.93	1.51	1.7	203.25	12OC	0.8	-4.1	-1.2	1.3
152.73	1.60	1.7	213.31	22OC	-0.5	-4.2	-1.3	1.4
158.41	1.69	1.6	223.42	1NO	-0.6	-4.3	-1.3	1.3
163.95	1.79	1.5	233.56	11NO	-0.6	-4.4	-1.3	1.3
169.33	1.89	1.4	243.73	21NO	-0.7	-4.4	-1.4	1.3
174.51	1.99	1.3	253.92	1DE	-0.8	-4.1	-1.4	1.2
179.45	2.10	1.1	264.12	11DE	-0.8	-3.5	-1.5	1.2
184.10	2.21	1.0	274.32	21DE	-0.8	-3.2	-1.6	1.1
188.38	2.32	0.8	284.51	31DE	-0.5	-3.9	-1.6	1.1
				831				
192.23	2.44	0.6	294.67	10JA	1.1	-4.3	-1.7	1.0
195.51	2.56	0.4	304.81	20JA	2.6	-4.4	-1.8	1.0
198.10	2.67	0.1	314.90	30JA	1.0	-4.3	-1.8	0.9
199.84	2.76	-0.1	324.95	9FE	0.5	-4.2	-1.9	0.9
200.53	2.83	-0.4	334.94	19FE	0.3	-4.1	-1.9	0.8
200.02	2.85	-0.7	344.88	1MR	0.0	-4.0	-2.0	0.8
198.24	2.79	-1.0	354.75	11MR	-0.4	-3.9	-2.0	0.9
195.32	2.61	-1.3	4.58	21MR	-1.1	-3.8	-2.0	0.9
191.71	2.32	-1.4	14.34	31MR	-1.7	-3.7	-2.0	0.9
188.06	1.91	-1.3	24.05	10AP	-0.9	-3.6	-2.0	1.0
185.09	1.45	-1.2	33.72	20AP	0.2	-3.6	-2.0	1.1
183.28	0.98	-1.0	43.34	30AP	1.2	-3.5	-1.9	1.1
182.80	0.54	-0.8	52.92	10MY	2.3	-3.4	-1.9	1.2
183.62	0.15	-0.6	62.48	20MY	3.6	-3.4	-1.8	1.2
185.58	-0.19	-0.4	72.02	30MY	2.2	-3.4	-1.8	1.3
188.51	-0.48	-0.2	81.54	9JN	1.1	-3.3	-1.7	1.3
192.24	-0.73	-0.0	91.07	19JN	-0.0	-3.3	-1.6	1.3
196.62	-0.94	0.1	100.60	29JN	-1.1	-3.3	-1.6	1.4
201.55	-1.12	0.2	110.15	9JL	-1.7	-3.3	-1.5	1.4
206.94	-1.26	0.3	119.73	19JL	-1.0	-3.3	-1.5	1.4
212.73	-1.39	0.4	129.34	29JL	-0.3	-3.3	-1.4	1.4
218.83	-1.49	0.5	138.99	8AU	0.1	-3.4	-1.4	1.3
225.24	-1.56	0.6	148.69	18AU	0.3	-3.4	-1.3	1.3
231.89	-1.62	0.7	158.44	28AU	0.5	-3.4	-1.3	1.2
238.75	-1.65	0.7	168.24	7SE	1.2	-3.4	-1.3	1.2
245.81	-1.67	0.8	178.10	17SE	3.2	-3.5	-1.2	1.2
253.03	-1.67	0.9	188.02	27SE	0.6	-3.5	-1.2	1.2
260.39	-1.65	0.9	198.00	7OC	-0.6	-3.5	-1.2	1.2
267.87	-1.62	1.0	208.03	17OC	-0.8	-3.5	-1.2	1.2
275.44	-1.57	1.0	218.11	27OC	-0.7	-3.4	-1.2	1.2
283.09	-1.50	1.1	228.24	6NO	-0.8	-3.4	-1.3	1.2
290.79	-1.43	1.1	238.40	16NO	-0.7	-3.4	-1.3	1.1
298.52	-1.34	1.2	248.57	26NO	-0.6	-3.4	-1.3	1.1
306.27	-1.24	1.2	258.77	6DE	-0.6	-3.3	-1.3	1.1
314.03	-1.14	1.3	268.98	16DE	-0.3	-3.3	-1.4	1.0
321.76	-1.03	1.3	279.17	26DE	1.3	-3.3	-1.4	1.0

♂			☉	16.00UT	☿	♀	♃	♄
LONG	LAT	MAG	LONG	832	MAGNITUDES			
329.46	-0.91	1.4	289.35	5JA	2.2	-3.3	-1.5	0.9
337.13	-0.79	1.4	299.50	15JA	0.7	-3.4	-1.6	0.9
344.74	-0.67	1.5	309.61	25JA	0.3	-3.4	-1.6	0.8
352.28	-0.55	1.5	319.69	4FE	0.1	-3.4	-1.7	0.8
359.76	-0.43	1.5	329.71	14FE	-0.1	-3.4	-1.8	0.8
7.17	-0.31	1.6	339.67	24FE	-0.5	-3.5	-1.8	0.7
14.50	-0.20	1.6	349.59	5MR	-1.1	-3.5	-1.9	0.7
21.75	-0.08	1.7	359.44	15MR	-1.6	-3.5	-2.0	0.8
28.92	0.03	1.7	9.23	25MR	-0.9	-3.6	-2.0	0.8
36.01	0.13	1.7	18.97	4AP	0.3	-3.6	-2.0	0.9
43.02	0.24	1.7	28.66	14AP	1.5	-3.7	-2.1	1.0
49.95	0.33	1.8	38.30	24AP	3.0	-3.8	-2.1	1.0
56.82	0.43	1.8	47.90	4MY	3.0	-3.9	-2.1	1.1
63.61	0.51	1.8	57.48	14MY	1.7	-4.0	-2.0	1.1
70.34	0.60	1.8	67.02	24MY	0.8	-4.1	-2.0	1.2
77.01	0.68	1.8	76.55	3JN	-0.1	-4.1	-2.0	1.2
83.63	0.75	1.8	86.08	13JN	-1.1	-4.2	-1.9	1.3
90.19	0.82	1.9	95.60	23JN	-1.7	-4.1	-1.9	1.3
96.70	0.89	1.9	105.15	3JL	-0.9	-3.9	-1.8	1.3
103.17	0.96	1.9	114.71	13JL	-0.3	-3.4	-1.7	1.3
109.60	1.01	2.0	124.29	23JL	0.2	-3.1	-1.7	1.2
115.99	1.07	2.0	133.93	2AU	0.5	-3.7	-1.6	1.2
122.35	1.12	2.0	143.60	12AU	0.8	-4.1	-1.6	1.2
128.68	1.17	2.0	153.31	22AU	1.6	-4.3	-1.5	1.2
134.98	1.22	2.0	163.09	1SE	3.2	-4.3	-1.5	1.1
141.25	1.26	2.0	172.92	11SE	0.5	-4.2	-1.4	1.1
147.49	1.30	2.0	182.81	21SE	-0.7	-4.1	-1.4	1.0
153.70	1.34	2.0	192.76	1OC	-0.9	-4.0	-1.3	1.1
159.88	1.37	1.9	202.77	11OC	-0.9	-3.9	-1.3	1.1
166.02	1.40	1.9	212.82	21OC	-0.7	-3.9	-1.3	1.1
172.13	1.43	1.8	222.92	31OC	-0.5	-3.8	-1.3	1.1
178.18	1.45	1.8	233.07	10NO	-0.5	-3.7	-1.3	1.0
184.18	1.47	1.7	243.23	20NO	-0.5	-3.6	-1.3	1.0
190.12	1.48	1.6	253.43	30NO	-0.2	-3.6	-1.3	1.0
195.98	1.49	1.5	263.63	10DE	1.5	-3.5	-1.3	1.0
201.74	1.49	1.4	273.83	20DE	1.8	-3.5	-1.4	0.9
207.40	1.48	1.2	284.02	30DE	0.4	-3.4	-1.4	0.9
				833				
212.91	1.46	1.1	294.19	9JA	0.1	-3.4	-1.4	0.8
218.25	1.43	0.9	304.32	19JA	-0.0	-3.4	-1.5	0.8
223.39	1.38	0.8	314.42	29JA	-0.2	-3.3	-1.5	0.7
228.26	1.31	0.6	324.47	8FE	-0.5	-3.3	-1.6	0.7
232.83	1.22	0.4	334.46	18FE	-1.1	-3.3	-1.7	0.6
236.98	1.09	0.1	344.40	28FE	-1.5	-3.3	-1.7	0.6
240.64	0.92	-0.1	354.28	10MR	-0.9	-3.3	-1.8	0.6
243.66	0.70	-0.4	4.11	20MR	0.4	-3.3	-1.9	0.6
245.88	0.41	-0.7	13.88	30MR	1.9	-3.3	-1.9	0.6
247.12	0.05	-1.0	23.59	9AP	3.8	-3.4	-2.0	0.7
247.22	-0.41	-1.3	33.25	19AP	2.2	-3.4	-2.1	0.7
246.06	-0.95	-1.7	42.88	29AP	1.3	-3.4	-2.1	0.8
243.76	-1.55	-1.9	52.46	9MY	0.6	-3.5	-2.2	0.9
240.69	-2.16	-2.1	62.02	19MY	-0.1	-3.5	-2.2	0.9
237.50	-2.70	-2.1	71.56	29MY	-1.1	-3.4	-2.2	1.0
234.95	-3.12	-1.9	81.08	8JN	-1.8	-3.4	-2.2	1.0
233.55	-3.39	-1.7	90.61	18JN	-0.9	-3.4	-2.2	1.1
233.51	-3.54	-1.5	100.14	28JN	-0.2	-3.4	-2.1	1.1
234.84	-3.59	-1.3	109.69	8JL	0.3	-3.3	-2.1	1.1
237.36	-3.57	-1.0	119.26	18JL	0.6	-3.3	-2.0	1.1
240.89	-3.49	-0.8	128.87	28JL	1.1	-3.3	-2.0	1.1
245.25	-3.38	-0.7	138.52	7AU	2.0	-3.3	-1.9	1.1
250.26	-3.24	-0.5	148.21	17AU	2.8	-3.3	-1.8	1.1
255.82	-3.07	-0.3	157.96	27AU	0.4	-3.3	-1.8	1.0
261.81	-2.90	-0.2	167.76	6SE	-0.8	-3.4	-1.7	1.0
268.13	-2.70	-0.0	177.62	16SE	-1.1	-3.4	-1.6	1.0
274.74	-2.50	0.1	187.54	26SE	-1.0	-3.4	-1.6	0.9
281.57	-2.30	0.2	197.51	6OC	-0.7	-3.5	-1.5	0.9
288.56	-2.09	0.4	207.54	16OC	-0.4	-3.5	-1.5	0.9
295.68	-1.87	0.5	217.62	26OC	-0.3	-3.6	-1.5	0.9
302.88	-1.66	0.6	227.74	5NO	-0.3	-3.7	-1.4	0.9
310.15	-1.45	0.7	237.90	15NO	-0.0	-3.7	-1.4	0.9
317.46	-1.25	0.8	248.08	25NO	1.7	-3.8	-1.4	0.9
324.78	-1.05	0.9	258.28	5DE	1.5	-3.9	-1.4	0.9
332.10	-0.87	1.0	268.48	15DE	0.2	-4.0	-1.4	0.9
339.40	-0.68	1.1	278.68	25DE	-0.1	-4.1	-1.4	0.9

♂			☉	16.00UT	☿	♀	♃	♄
LONG	LAT	MAG	LONG	834	MAGNITUDES			
346.67	-0.51	1.2	288.86	4JA	-0.1	-4.2	-1.4	0.8
353.89	-0.35	1.3	299.01	14JA	-0.3	-4.3	-1.4	0.8
1.08	-0.20	1.4	309.13	24JA	-0.6	-4.3	-1.5	0.7
8.20	-0.06	1.5	319.20	3FE	-1.2	-4.3	-1.5	0.7
15.27	0.07	1.5	329.23	13FE	-1.3	-4.0	-1.6	0.6
22.28	0.19	1.6	339.20	23FE	-0.8	-3.5	-1.6	0.6
29.22	0.31	1.7	349.11	5MR	0.5	-3.4	-1.7	0.5
36.11	0.41	1.7	358.96	15MR	2.4	-3.9	-1.7	0.5
42.93	0.51	1.8	8.76	25MR	3.0	-4.2	-1.8	0.4
49.70	0.60	1.8	18.50	4AP	1.6	-4.2	-1.9	0.4
56.41	0.68	1.9	28.19	14AP	1.0	-4.2	-1.9	0.5
63.06	0.75	1.9	37.84	24AP	0.4	-4.1	-2.0	0.6
69.67	0.81	1.9	47.44	4MY	-0.2	-4.0	-2.1	0.6
76.23	0.87	1.9	57.01	14MY	-1.1	-3.9	-2.1	0.7
82.76	0.93	2.0	66.56	24MY	-1.8	-3.8	-2.2	0.8
89.24	0.97	2.0	76.09	3JN	-0.9	-3.7	-2.2	0.8
95.70	1.02	2.0	85.62	13JN	-0.1	-3.6	-2.3	0.9
102.13	1.05	2.0	95.15	23JN	0.4	-3.6	-2.3	0.9
108.54	1.08	2.0	104.68	3JL	0.8	-3.5	-2.3	0.9
114.93	1.11	1.9	114.25	13JL	1.4	-3.5	-2.3	1.0
121.30	1.13	2.0	123.84	23JL	2.6	-3.4	-2.3	1.0
127.67	1.15	2.0	133.46	2AU	2.4	-3.4	-2.2	1.0
134.04	1.16	2.0	143.13	12AU	0.3	-3.4	-2.2	1.0
140.41	1.17	2.0	152.85	22AU	-0.9	-3.4	-2.1	1.0
146.77	1.18	2.0	162.62	1SE	-1.2	-3.4	-2.1	0.9
153.15	1.18	2.0	172.45	11SE	-1.0	-3.4	-2.0	0.9
159.52	1.17	2.0	182.34	21SE	-0.6	-3.4	-1.9	0.9
165.91	1.16	2.0	192.28	1OC	-0.3	-3.4	-1.9	0.8
172.31	1.15	2.0	202.28	11OC	-0.2	-3.4	-1.8	0.8
178.72	1.13	2.0	212.33	21OC	-0.1	-3.4	-1.8	0.8
185.14	1.10	1.9	222.43	31OC	0.2	-3.4	-1.7	0.8
191.57	1.07	1.9	232.57	10NO	2.0	-3.4	-1.7	0.9
198.01	1.03	1.8	242.74	20NO	1.3	-3.4	-1.6	0.9
204.46	0.98	1.8	252.93	30NO	-0.0	-3.5	-1.6	0.9
210.91	0.93	1.7	263.13	10DE	-0.2	-3.5	-1.5	0.8
217.35	0.86	1.6	273.33	20DE	-0.3	-3.5	-1.5	0.8
223.80	0.78	1.5	283.52	30DE	-0.4	-3.5	-1.5	0.8
				835				
230.25	0.69	1.4	293.69	9JA	-0.7	-3.4	-1.5	0.8
236.67	0.58	1.3	303.83	19JA	-1.1	-3.4	-1.5	0.7
243.08	0.46	1.2	313.92	29JA	-1.2	-3.4	-1.5	0.7
249.47	0.32	1.1	323.98	8FE	-0.8	-3.4	-1.5	0.6
255.81	0.15	0.9	333.97	18FE	0.6	-3.3	-1.6	0.6
262.10	-0.04	0.8	343.92	28FE	2.9	-3.3	-1.6	0.5
268.33	-0.25	0.6	353.81	10MR	2.2	-3.3	-1.6	0.5
274.46	-0.50	0.4	3.63	20MR	1.2	-3.3	-1.7	0.4
280.49	-0.78	0.3	13.40	30MR	0.7	-3.3	-1.7	0.4
286.35	-1.11	0.1	23.12	9AP	0.3	-3.4	-1.8	0.3
292.01	-1.47	-0.1	32.79	19AP	-0.3	-3.4	-1.8	0.4
297.42	-1.88	-0.4	42.41	29AP	-1.1	-3.4	-1.9	0.4
302.48	-2.35	-0.6	52.00	9MY	-1.8	-3.4	-2.0	0.5
307.09	-2.88	-0.8	61.56	19MY	-0.9	-3.5	-2.0	0.6
311.12	-3.46	-1.1	71.10	29MY	-0.1	-3.5	-2.1	0.6
314.39	-4.10	-1.4	80.63	8JN	0.6	-3.6	-2.2	0.7
316.73	-4.79	-1.6	90.15	18JN	1.1	-3.7	-2.2	0.7
317.92	-5.48	-1.9	99.68	28JN	1.9	-3.7	-2.3	0.8
317.82	-6.13	-2.2	109.23	8JL	3.2	-3.8	-2.4	0.8
316.47	-6.63	-2.5	118.80	18JL	2.0	-3.9	-2.4	0.9
314.16	-6.87	-2.6	128.41	28JL	0.2	-4.0	-2.4	0.9
311.53	-6.80	-2.6	138.06	7AU	-1.0	-4.1	-2.4	0.9
309.33	-6.41	-2.4	147.75	17AU	-1.4	-4.2	-2.4	0.9
308.13	-5.80	-2.1	157.49	27AU	-1.0	-4.3	-2.4	0.9
308.22	-5.08	-1.9	167.29	6SE	-0.5	-4.2	-2.4	0.9
309.61	-4.35	-1.6	177.14	16SE	-0.2	-4.0	-2.3	0.9
312.12	-3.65	-1.3	187.06	26SE	-0.1	-3.5	-2.2	0.8
315.57	-3.01	-1.0	197.03	6OC	0.0	-3.3	-2.2	0.8
319.76	-2.45	-0.7	207.05	16OC	0.4	-3.9	-2.1	0.7
324.52	-1.95	-0.5	217.13	26OC	2.3	-4.3	-2.0	0.7
329.73	-1.51	-0.2	227.25	5NO	1.1	-4.4	-2.0	0.8
335.27	-1.12	-0.0	237.40	15NO	-0.2	-4.3	-1.9	0.8
341.06	-0.79	0.2	247.58	25NO	-0.4	-4.3	-1.8	0.8
347.06	-0.50	0.4	257.78	5DE	-0.4	-4.2	-1.8	0.8
353.19	-0.25	0.6	267.98	15DE	-0.5	-4.1	-1.7	0.8
359.43	-0.03	0.7	278.18	25DE	-0.7	-4.0	-1.7	0.8

♂			☉	16.00UT	☿	♀	♃	♄
LONG	LAT	MAG	LONG	836		MAGNITUDES		
5.75	0.16	0.9	288.36	4JA	-1.0	-3.9	-1.7	0.8
12.11	0.33	1.0	298.51	14JA	-1.0	-3.8	-1.6	0.8
18.51	0.47	1.2	308.64	24JA	-0.7	-3.7	-1.6	0.7
24.92	0.60	1.3	318.71	3FE	0.8	-3.6	-1.6	0.7
31.35	0.71	1.4	328.74	13FE	3.1	-3.5	-1.6	0.6
37.77	0.80	1.5	338.72	23FE	1.7	-3.5	-1.6	0.6
44.18	0.88	1.6	348.63	4MR	0.9	-3.4	-1.6	0.5
50.58	0.95	1.7	358.49	14MR	0.5	-3.4	-1.6	0.5
56.96	1.00	1.7	8.29	24MR	0.2	-3.4	-1.6	0.4
63.34	1.05	1.8	18.04	3AP	-0.3	-3.3	-1.6	0.4
69.69	1.09	1.9	27.73	13AP	-1.1	-3.3	-1.7	0.3
76.03	1.12	1.9	37.38	23AP	-1.8	-3.3	-1.7	0.3
82.36	1.15	1.9	46.98	3MY	-1.0	-3.3	-1.8	0.3
88.67	1.17	2.0	56.55	13MY	0.0	-3.3	-1.8	0.4
94.98	1.18	2.0	66.10	23MY	0.7	-3.3	-1.9	0.5
101.29	1.19	2.0	75.64	2JN	1.5	-3.3	-1.9	0.5
107.59	1.19	2.0	85.16	12JN	2.5	-3.4	-2.0	0.6
113.90	1.19	2.0	94.69	22JN	3.2	-3.4	-2.1	0.6
120.21	1.18	2.0	104.23	2JL	1.6	-3.4	-2.1	0.7
126.53	1.17	2.0	113.78	12JL	0.2	-3.5	-2.2	0.7
132.88	1.15	2.0	123.37	22JL	-1.0	-3.5	-2.3	0.8
139.24	1.13	2.0	132.99	1AU	-1.5	-3.5	-2.3	0.8
145.62	1.11	2.0	142.66	11AU	-1.0	-3.4	-2.4	0.8
152.03	1.08	1.9	152.38	21AU	-0.4	-3.4	-2.4	0.8
158.48	1.04	1.9	162.14	31AU	-0.1	-3.4	-2.5	0.8
164.96	1.00	2.0	171.97	10SE	0.1	-3.4	-2.5	0.8
171.47	0.96	2.0	181.86	20SE	0.2	-3.3	-2.5	0.8
178.03	0.91	2.0	191.79	30SE	0.7	-3.3	-2.4	0.8
184.63	0.86	1.9	201.79	10OC	2.6	-3.3	-2.4	0.8
191.27	0.80	1.9	211.84	20OC	0.9	-3.3	-2.4	0.7
197.96	0.73	1.9	221.94	30OC	-0.4	-3.3	-2.3	0.7
204.70	0.66	1.9	232.07	9NO	-0.5	-3.4	-2.2	0.7
211.48	0.58	1.8	242.24	19NO	-0.5	-3.4	-2.2	0.7
218.32	0.49	1.8	252.43	29NO	-0.6	-3.4	-2.1	0.8
225.20	0.40	1.7	262.63	9DE	-0.8	-3.5	-2.0	0.8
232.13	0.30	1.7	272.83	19DE	-0.9	-3.5	-1.9	0.8
239.11	0.19	1.6	283.02	29DE	-0.9	-3.6	-1.9	0.8
				837				
246.13	0.06	1.6	293.19	8JA	-0.6	-3.6	-1.8	0.8
253.20	-0.07	1.5	303.33	18JA	0.9	-3.7	-1.8	0.8
260.30	-0.21	1.4	313.43	28JA	2.9	-3.7	-1.7	0.7
267.45	-0.36	1.3	323.49	7FE	1.2	-3.8	-1.7	0.7
274.64	-0.52	1.2	333.49	17FE	0.6	-3.9	-1.6	0.7
281.85	-0.68	1.2	343.43	27FE	0.3	-4.0	-1.6	0.6
289.08	-0.86	1.1	353.32	9MR	0.1	-4.1	-1.6	0.6
296.33	-1.05	1.0	3.16	19MR	-0.4	-4.2	-1.6	0.5
303.58	-1.24	0.9	12.93	29MR	-1.1	-4.2	-1.6	0.4
310.82	-1.43	0.8	22.65	8AP	-1.7	-4.2	-1.6	0.4
318.03	-1.63	0.7	32.32	18AP	-1.0	-4.1	-1.6	0.3
325.19	-1.83	0.6	41.95	28AP	0.1	-3.7	-1.6	0.3
332.28	-2.03	0.4	51.54	8MY	1.0	-2.9	-1.6	0.3
339.27	-2.23	0.3	61.10	18MY	1.9	-3.4	-1.7	0.3
346.12	-2.42	0.2	70.64	28MY	3.3	-3.9	-1.7	0.4
352.80	-2.60	0.1	80.17	7JN	2.6	-4.2	-1.7	0.4
359.24	-2.77	-0.1	89.69	17JN	1.3	-4.2	-1.8	0.5
5.41	-2.92	-0.2	99.22	27JN	0.1	-4.1	-1.8	0.6
11.20	-3.06	-0.3	108.77	7JL	-1.0	-4.1	-1.9	0.6
16.54	-3.17	-0.5	118.34	17JL	-1.6	-4.0	-2.0	0.7
21.31	-3.26	-0.7	127.94	27JL	-1.0	-3.9	-2.0	0.7
25.36	-3.32	-0.9	137.59	6AU	-0.4	-3.8	-2.1	0.8
28.50	-3.34	-1.1	147.28	16AU	0.0	-3.7	-2.2	0.8
30.54	-3.30	-1.3	157.02	26AU	0.2	-3.7	-2.2	0.8
31.26	-3.18	-1.5	166.81	5SE	0.4	-3.6	-2.3	0.8
30.51	-2.96	-1.7	176.67	15SE	1.0	-3.6	-2.3	0.8
28.36	-2.60	-1.9	186.58	25SE	3.0	-3.5	-2.4	0.8
25.17	-2.12	-2.1	196.55	5OC	0.8	-3.5	-2.4	0.8
21.67	-1.54	-2.0	206.57	15OC	-0.5	-3.4	-2.4	0.8
18.66	-0.94	-1.8	216.64	25OC	-0.7	-3.4	-2.4	0.7
16.74	-0.39	-1.4	226.76	4NO	-0.7	-3.4	-2.4	0.7
16.17	0.07	-1.1	236.91	14NO	-0.7	-3.4	-2.4	0.7
16.91	0.45	-0.8	247.09	24NO	-0.7	-3.4	-2.3	0.7
18.77	0.74	-0.5	257.29	4DE	-0.7	-3.4	-2.3	0.7
21.56	0.96	-0.2	267.49	14DE	-0.7	-3.4	-2.2	0.8
25.06	1.13	0.1	277.68	24DE	-0.5	-3.4	-2.1	0.8

♂			☉	16.00UT	☿	♀	♃	♄
LONG	LAT	MAG	LONG	838		MAGNITUDES		
29.13	1.25	0.3	287.87	3JA	1.1	-3.4	-2.0	0.8
33.64	1.35	0.5	298.02	13JA	2.5	-3.4	-2.0	0.8
38.49	1.41	0.7	308.14	23JA	0.9	-3.4	-1.9	0.8
43.59	1.46	0.9	318.22	2FE	0.4	-3.4	-1.8	0.8
48.91	1.49	1.1	328.25	12FE	0.2	-3.4	-1.8	0.7
54.38	1.51	1.2	338.22	22FE	-0.0	-3.5	-1.7	0.7
59.98	1.52	1.4	348.15	4MR	-0.4	-3.5	-1.7	0.6
65.69	1.52	1.5	358.00	14MR	-1.1	-3.5	-1.6	0.6
71.47	1.51	1.6	7.81	24MR	-1.6	-3.4	-1.6	0.5
77.32	1.50	1.7	17.56	3AP	-0.9	-3.4	-1.6	0.5
83.22	1.48	1.7	27.25	13AP	0.2	-3.4	-1.5	0.4
89.17	1.45	1.8	36.90	23AP	1.3	-3.4	-1.5	0.4
95.17	1.42	1.9	46.51	3MY	2.6	-3.3	-1.5	0.3
101.21	1.39	1.9	56.08	13MY	3.5	-3.3	-1.5	0.2
107.28	1.36	2.0	65.63	23MY	2.0	-3.3	-1.5	0.3
113.39	1.32	2.0	75.17	2JN	1.0	-3.3	-1.5	0.3
119.53	1.27	2.0	84.69	12JN	0.0	-3.3	-1.5	0.3
125.72	1.23	2.0	94.22	22JN	-1.1	-3.4	-1.6	0.4
131.95	1.18	2.0	103.76	2JL	-1.7	-3.4	-1.6	0.5
138.22	1.13	2.0	113.32	12JL	-1.0	-3.4	-1.6	0.6
144.53	1.07	2.0	122.90	22JL	-0.3	-3.5	-1.7	0.6
150.90	1.01	2.0	132.53	1AU	0.1	-3.5	-1.7	0.7
157.32	0.95	2.0	142.19	11AU	0.4	-3.6	-1.8	0.7
163.79	0.89	1.9	151.90	21AU	0.6	-3.6	-1.9	0.8
170.33	0.82	1.9	161.67	31AU	1.3	-3.7	-1.9	0.8
176.92	0.74	1.9	171.49	10SE	3.3	-3.8	-2.0	0.8
183.58	0.67	1.8	181.37	20SE	0.7	-3.9	-2.1	0.8
190.31	0.59	1.8	191.31	30SE	-0.6	-4.0	-2.1	0.8
197.10	0.50	1.8	201.31	10OC	-0.8	-4.1	-2.2	0.8
203.97	0.42	1.8	211.35	20OC	-0.8	-4.2	-2.2	0.8
210.91	0.32	1.8	221.45	30OC	-0.8	-4.3	-2.3	0.7
217.92	0.23	1.8	231.58	9NO	-0.6	-4.4	-2.3	0.7
225.00	0.13	1.7	241.75	19NO	-0.6	-4.3	-2.3	0.7
232.16	0.02	1.7	251.94	29NO	-0.6	-4.1	-2.3	0.7
239.38	-0.09	1.7	262.14	9DE	-0.3	-3.5	-2.3	0.7
246.68	-0.20	1.7	272.34	19DE	1.3	-3.2	-2.3	0.7
254.04	-0.31	1.6	282.53	29DE	2.1	-4.0	-2.2	0.8
				839				
261.46	-0.43	1.6	292.70	8JA	0.6	-4.3	-2.1	0.8
268.94	-0.55	1.5	302.84	18JA	0.2	-4.4	-2.1	0.8
276.47	-0.67	1.5	312.94	28JA	0.1	-4.3	-2.0	0.8
284.04	-0.79	1.4	323.00	7FE	-0.1	-4.2	-1.9	0.8
291.65	-0.91	1.4	333.00	17FE	-0.5	-4.1	-1.9	0.8
299.29	-1.02	1.3	342.95	27FE	-1.1	-4.0	-1.8	0.7
306.95	-1.13	1.3	352.84	9MR	-1.5	-3.9	-1.7	0.7
314.62	-1.24	1.3	2.68	19MR	-0.9	-3.8	-1.7	0.6
322.28	-1.33	1.2	12.45	29MR	0.3	-3.7	-1.6	0.6
329.92	-1.42	1.2	22.17	8AP	1.6	-3.6	-1.6	0.5
337.53	-1.49	1.1	31.85	18AP	3.3	-3.5	-1.5	0.5
345.09	-1.55	1.1	41.48	28AP	2.7	-3.5	-1.5	0.4
352.58	-1.59	1.0	51.07	8MY	1.5	-3.4	-1.4	0.3
359.99	-1.62	1.0	60.63	18MY	0.7	-3.4	-1.4	0.3
7.30	-1.64	0.9	70.18	28MY	-0.1	-3.4	-1.4	0.2
14.50	-1.63	0.9	79.70	7JN	-1.1	-3.3	-1.4	0.3
21.56	-1.61	0.8	89.23	17JN	-1.8	-3.3	-1.4	0.3
28.45	-1.57	0.7	98.76	27JN	-1.0	-3.3	-1.4	0.4
35.16	-1.51	0.7	108.31	7JL	-0.2	-3.3	-1.4	0.5
41.65	-1.43	0.6	117.88	17JL	0.2	-3.3	-1.4	0.5
47.89	-1.33	0.5	127.48	27JL	0.5	-3.3	-1.5	0.6
53.84	-1.21	0.4	137.12	6AU	0.9	-3.4	-1.5	0.6
59.45	-1.06	0.3	146.81	16AU	1.7	-3.4	-1.5	0.7
64.63	-0.89	0.2	156.55	26AU	3.1	-3.4	-1.6	0.7
69.32	-0.68	0.1	166.34	5SE	0.5	-3.4	-1.6	0.7
73.40	-0.45	-0.1	176.19	15SE	-0.7	-3.5	-1.7	0.8
76.71	-0.17	-0.3	186.10	25SE	-1.0	-3.5	-1.7	0.8
79.11	0.16	-0.4	196.06	5OC	-1.0	-3.5	-1.8	0.8
80.37	0.54	-0.7	206.08	15OC	-0.7	-3.5	-1.9	0.8
80.30	0.97	-0.9	216.15	25OC	-0.5	-3.4	-1.9	0.8
78.81	1.44	-1.1	226.26	4NO	-0.4	-3.4	-2.0	0.7
75.99	1.91	-1.3	236.42	14NO	-0.4	-3.4	-2.1	0.7
72.27	2.34	-1.4	246.59	24NO	-0.1	-3.4	-2.1	0.7
68.38	2.66	-1.3	256.79	4DE	1.5	-3.3	-2.1	0.7
65.09	2.84	-1.1	266.99	14DE	1.8	-3.3	-2.2	0.7
62.96	2.91	-0.8	277.19	24DE	0.3	-3.3	-2.2	0.7

♂			⊙	16.00UT	☿	♀	♃	♄
LONG	LAT	MAG	LONG	840	MAGNITUDES			
62.16	2.89	-0.5	287.37	3JA	0.0	-3.3	-2.2	0.8
62.64	2.82	-0.2	297.53	13JA	-0.1	-3.4	-2.2	0.8
64.23	2.72	0.1	307.65	23JA	-0.2	-3.4	-2.1	0.8
66.71	2.61	0.3	317.73	2FE	-0.6	-3.4	-2.1	0.8
69.91	2.49	0.6	327.77	12FE	-1.1	-3.4	-2.0	0.8
73.68	2.37	0.8	337.74	22FE	-1.4	-3.5	-1.9	0.8
77.88	2.25	0.9	347.66	3MR	-0.9	-3.5	-1.9	0.8
82.45	2.14	1.1	357.53	13MR	0.4	-3.5	-1.8	0.7
87.29	2.03	1.2	7.34	23MR	2.1	-3.6	-1.7	0.7
92.36	1.92	1.4	17.08	2AP	3.5	-3.6	-1.7	0.6
97.62	1.82	1.5	26.79	12AP	2.0	-3.7	-1.6	0.6
103.04	1.72	1.6	36.43	22AP	1.1	-3.8	-1.5	0.5
108.59	1.63	1.6	46.05	2MY	0.6	-3.9	-1.5	0.5
114.26	1.53	1.7	55.62	12MY	-0.1	-4.0	-1.4	0.4
120.04	1.44	1.8	65.17	22MY	-1.1	-4.1	-1.4	0.3
125.91	1.34	1.8	74.71	1JN	-1.8	-4.1	-1.4	0.3
131.87	1.25	1.9	84.24	11JN	-1.0	-4.2	-1.3	0.3
137.92	1.16	1.9	93.76	21JN	-0.2	-4.1	-1.3	0.3
144.05	1.07	1.9	103.30	1JL	0.3	-3.9	-1.3	0.4
150.26	0.97	1.9	112.86	11JL	0.7	-3.4	-1.3	0.4
156.56	0.88	1.9	122.44	21JL	1.2	-3.1	-1.3	0.5
162.94	0.79	1.9	132.06	31JL	2.2	-3.8	-1.3	0.5
169.40	0.69	1.9	141.73	10AU	2.7	-4.1	-1.3	0.6
175.94	0.59	1.9	151.43	20AU	0.4	-4.3	-1.3	0.6
182.58	0.50	1.9	161.20	30AU	-0.8	-4.2	-1.4	0.7
189.30	0.40	1.8	171.02	9SE	-1.1	-4.2	-1.4	0.7
196.11	0.30	1.8	180.89	19SE	-1.1	-4.1	-1.4	0.8
203.01	0.19	1.8	190.83	29SE	-0.6	-4.0	-1.5	0.8
210.01	0.09	1.7	200.82	9OC	-0.4	-3.9	-1.5	0.8
217.09	-0.01	1.7	210.87	19OC	-0.3	-3.8	-1.6	0.8
224.26	-0.11	1.6	220.96	29OC	-0.2	-3.8	-1.7	0.8
231.52	-0.22	1.5	231.09	8NO	0.1	-3.7	-1.7	0.8
238.87	-0.32	1.5	241.26	18NO	1.7	-3.6	-1.8	0.8
246.29	-0.42	1.5	251.44	28NO	1.5	-3.6	-1.9	0.7
253.80	-0.52	1.5	261.65	8DE	0.1	-3.5	-1.9	0.7
261.37	-0.62	1.5	271.84	18DE	-0.1	-3.5	-2.0	0.7
269.01	-0.71	1.5	282.04	28DE	-0.2	-3.4	-2.0	0.7
				841				
276.71	-0.80	1.5	292.21	7JA	-0.3	-3.4	-2.0	0.8
284.45	-0.89	1.5	302.35	17JA	-0.6	-3.4	-2.1	0.8
292.23	-0.96	1.5	312.46	27JA	-1.1	-3.3	-2.1	0.8
300.04	-1.03	1.4	322.52	6FE	-1.3	-3.3	-2.1	0.8
307.86	-1.09	1.4	332.52	16FE	-0.8	-3.3	-2.0	0.8
315.69	-1.14	1.4	342.48	26FE	0.5	-3.3	-2.0	0.8
323.50	-1.17	1.4	352.37	8MR	2.5	-3.3	-1.9	0.8
331.30	-1.20	1.4	2.20	18MR	2.7	-3.3	-1.9	0.8
339.06	-1.21	1.4	11.98	28MR	1.5	-3.3	-1.8	0.8
346.77	-1.21	1.4	21.71	7AP	0.9	-3.4	-1.8	0.7
354.43	-1.19	1.4	31.38	17AP	0.4	-3.4	-1.7	0.7
2.02	-1.17	1.3	41.01	27AP	-0.2	-3.4	-1.6	0.6
9.53	-1.12	1.3	50.61	7MY	-1.1	-3.5	-1.6	0.5
16.95	-1.07	1.3	60.17	17MY	-1.9	-3.5	-1.5	0.5
24.27	-1.00	1.3	69.71	27MY	-1.0	-3.4	-1.4	0.4
31.48	-0.92	1.3	79.24	6JN	-0.1	-3.4	-1.4	0.3
38.57	-0.83	1.3	88.76	16JN	0.5	-3.4	-1.4	0.3
45.54	-0.73	1.3	98.30	26JN	0.9	-3.4	-1.3	0.3
52.37	-0.61	1.3	107.84	6JL	1.6	-3.3	-1.3	0.3
59.06	-0.49	1.2	117.41	16JL	2.8	-3.3	-1.3	0.4
65.59	-0.35	1.2	127.01	26JL	2.3	-3.3	-1.2	0.5
71.93	-0.20	1.1	136.65	5AU	0.3	-3.3	-1.2	0.5
78.09	-0.04	1.1	146.33	15AU	-0.9	-3.3	-1.2	0.6
84.03	0.14	1.0	156.07	25AU	-1.3	-3.3	-1.2	0.6
89.71	0.33	1.0	165.86	4SE	-1.1	-3.4	-1.2	0.7
95.11	0.54	0.9	175.71	14SE	-0.6	-3.4	-1.3	0.7
100.17	0.77	0.8	185.62	24SE	-0.3	-3.4	-1.3	0.7
104.81	1.02	0.6	195.58	4OC	-0.1	-3.5	-1.3	0.8
108.96	1.30	0.5	205.59	14OC	-0.1	-3.5	-1.3	0.8
112.49	1.61	0.3	215.67	24OC	0.3	-3.6	-1.4	0.8
115.27	1.96	0.1	225.78	3NO	2.0	-3.7	-1.4	0.8
117.13	2.36	-0.1	235.92	13NO	1.3	-3.7	-1.5	0.8
117.86	2.78	-0.3	246.10	23NO	-0.1	-3.8	-1.6	0.8
117.31	3.23	-0.5	256.30	3DE	-0.3	-3.9	-1.6	0.8
115.40	3.66	-0.8	266.50	13DE	-0.3	-4.0	-1.7	0.8
112.27	4.02	-0.9	276.70	23DE	-0.4	-4.1	-1.8	0.7

♂			⊙	16.00UT	☿	♀	♃	♄
LONG	LAT	MAG	LONG	842	MAGNITUDES			
108.40	4.25	-1.1	286.88	2JA	-0.7	-4.2	-1.8	0.8
104.50	4.30	-0.9	297.04	12JA	-1.1	-4.3	-1.9	0.8
101.29	4.20	-0.7	307.17	22JA	-1.1	-4.4	-1.9	0.8
99.23	3.98	-0.5	317.25	1FE	-0.8	-4.3	-2.0	0.9
98.48	3.70	-0.2	327.28	11FE	0.6	-4.0	-2.0	0.9
98.96	3.40	0.1	337.26	21FE	3.0	-3.5	-2.0	0.9
100.53	3.11	0.3	347.19	3MR	2.0	-3.4	-2.0	0.9
102.96	2.83	0.5	357.05	13MR	1.1	-3.9	-2.0	0.9
106.11	2.57	0.7	6.86	23MR	0.6	-4.2	-2.0	0.9
109.83	2.34	0.9	16.61	2AP	0.3	-4.2	-1.9	0.8
114.00	2.12	1.0	26.31	12AP	-0.3	-4.2	-1.9	0.8
118.55	1.92	1.1	35.97	22AP	-1.1	-4.1	-1.8	0.8
123.40	1.74	1.3	45.58	2MY	-1.8	-4.0	-1.8	0.7
128.51	1.56	1.4	55.16	12MY	-1.0	-3.9	-1.7	0.6
133.84	1.40	1.4	64.71	22MY	-0.1	-3.8	-1.6	0.6
139.37	1.24	1.5	74.24	1JN	0.6	-3.7	-1.6	0.5
145.06	1.10	1.6	83.77	11JN	1.2	-3.6	-1.5	0.4
150.91	0.95	1.6	93.30	21JN	2.1	-3.6	-1.4	0.4
156.90	0.82	1.6	102.83	1JL	3.3	-3.5	-1.4	0.3
163.03	0.68	1.7	112.39	11JL	1.9	-3.5	-1.3	0.4
169.29	0.56	1.7	121.98	21JL	0.3	-3.4	-1.3	0.4
175.67	0.43	1.7	131.59	31JL	-1.0	-3.4	-1.3	0.5
182.17	0.31	1.7	141.25	10AU	-1.4	-3.4	-1.2	0.5
188.80	0.19	1.7	150.96	20AU	-1.0	-3.4	-1.2	0.6
195.53	0.07	1.7	160.72	30AU	-0.5	-3.4	-1.2	0.6
202.38	-0.05	1.7	170.54	9SE	-0.2	-3.4	-1.2	0.7
209.35	-0.16	1.7	180.42	19SE	-0.0	-3.4	-1.2	0.7
216.42	-0.27	1.6	190.35	29SE	0.1	-3.4	-1.2	0.8
223.61	-0.37	1.6	200.34	9OC	0.5	-3.4	-1.2	0.8
230.90	-0.47	1.6	210.38	19OC	2.3	-3.4	-1.2	0.8
238.28	-0.57	1.6	220.47	29OC	1.1	-3.4	-1.3	0.8
245.76	-0.66	1.5	230.60	8NO	-0.2	-3.4	-1.3	0.8
253.33	-0.74	1.5	240.77	18NO	-0.4	-3.4	-1.4	0.9
260.97	-0.82	1.4	250.95	28NO	-0.4	-3.5	-1.4	0.9
268.69	-0.88	1.4	261.15	8DE	-0.5	-3.5	-1.5	0.8
276.46	-0.94	1.4	271.35	18DE	-0.8	-3.5	-1.5	0.8
284.28	-0.99	1.3	281.54	28DE	-0.9	-3.5	-1.6	0.8
				843				
292.14	-1.03	1.3	291.72	7JA	-1.0	-3.4	-1.6	0.8
300.03	-1.06	1.3	301.86	17JA	-0.7	-3.4	-1.7	0.8
307.92	-1.08	1.3	311.96	27JA	0.8	-3.4	-1.8	0.9
315.81	-1.09	1.3	322.03	6FE	3.1	-3.4	-1.8	0.9
323.69	-1.08	1.4	332.04	16FE	1.5	-3.3	-1.9	0.9
331.53	-1.06	1.4	341.99	26FE	0.8	-3.3	-1.9	1.0
339.34	-1.03	1.4	351.89	8MR	0.4	-3.3	-2.0	1.0
347.10	-0.99	1.4	1.73	18MR	0.1	-3.3	-2.0	1.0
354.80	-0.94	1.4	11.51	28MR	-0.3	-3.3	-2.0	1.0
2.43	-0.88	1.5	21.24	7AP	-1.1	-3.4	-2.0	0.9
9.98	-0.81	1.5	30.91	17AP	-1.8	-3.4	-2.0	0.9
17.45	-0.73	1.5	40.55	27AP	-1.0	-3.4	-2.0	0.9
24.83	-0.64	1.5	50.14	7MY	0.0	-3.4	-1.9	0.8
32.11	-0.55	1.5	59.71	17MY	0.8	-3.5	-1.9	0.8
39.30	-0.45	1.6	69.25	27MY	1.6	-3.5	-1.8	0.7
46.39	-0.34	1.6	78.78	6JN	2.8	-3.5	-1.7	0.6
53.38	-0.23	1.6	88.31	16JN	3.0	-3.7	-1.7	0.6
60.26	-0.11	1.6	97.83	26JN	1.5	-3.7	-1.6	0.5
67.04	0.01	1.6	107.38	6JL	0.2	-3.8	-1.6	0.4
73.70	0.14	1.6	116.95	16JL	-1.0	-3.9	-1.5	0.4
80.26	0.27	1.6	126.55	26JL	-1.5	-4.0	-1.4	0.4
86.69	0.40	1.6	136.19	5AU	-1.0	-4.1	-1.4	0.5
93.00	0.54	1.6	145.87	15AU	-0.4	-4.2	-1.4	0.5
99.18	0.68	1.5	155.60	25AU	-0.1	-4.3	-1.3	0.6
105.21	0.83	1.5	165.39	4SE	0.1	-4.2	-1.3	0.6
111.07	0.99	1.4	175.23	14SE	0.3	-4.0	-1.3	0.7
116.76	1.16	1.4	185.13	24SE	0.8	-3.5	-1.2	0.7
122.23	1.34	1.3	195.10	4OC	2.7	-3.4	-1.2	0.8
127.46	1.53	1.2	205.11	14OC	0.9	-4.0	-1.2	0.8
132.40	1.73	1.1	215.17	24OC	-0.4	-4.3	-1.2	0.9
136.99	1.96	0.9	225.29	3NO	-0.6	-4.4	-1.2	0.9
141.15	2.20	0.8	235.43	13NO	-0.6	-4.3	-1.3	0.9
144.79	2.47	0.6	245.61	23NO	-0.6	-4.3	-1.3	0.9
147.78	2.77	0.4	255.80	3DE	-0.8	-4.2	-1.3	0.9
149.97	3.09	0.2	266.00	13DE	-0.8	-4.1	-1.4	0.9
151.18	3.44	-0.0	276.20	23DE	-0.8	-4.0	-1.4	0.9

Left table:

♂ LONG	LAT	MAG	⊙ LONG	16.00UT 844	☿	♀	♃	♄
151.23	3.79	-0.3	286.39	2JA	-0.6	-3.9	-1.5	0.9
149.98	4.12	-0.5	296.55	12JA	0.9	-3.8	-1.5	0.9
147.46	4.38	-0.8	306.68	22JA	2.8	-3.7	-1.6	0.9
143.94	4.50	-1.0	316.77	1FE	1.1	-3.6	-1.7	1.0
140.01	4.46	-1.0	326.80	11FE	0.5	-3.5	-1.7	1.0
136.39	4.25	-0.8	336.78	21FE	0.3	-3.5	-1.8	1.0
133.68	3.91	-0.6	346.71	2MR	0.0	-3.4	-1.9	1.1
132.22	3.51	-0.4	356.58	12MR	-0.4	-3.4	-1.9	1.1
132.03	3.10	-0.1	6.39	22MR	-1.1	-3.4	-2.0	1.1
133.04	2.70	0.1	16.15	1AP	-1.7	-3.3	-2.0	1.1
135.04	2.33	0.3	25.85	11AP	-1.0	-3.3	-2.1	1.1
137.86	2.00	0.5	35.51	21AP	0.1	-3.3	-2.1	1.0
141.37	1.70	0.6	45.12	1MY	1.1	-3.3	-2.1	1.0
145.42	1.43	0.8	54.70	11MY	2.1	-3.3	-2.1	1.0
149.92	1.18	0.9	64.25	21MY	3.5	-3.3	-2.1	0.9
154.80	0.95	1.0	73.79	31MY	2.4	-3.3	-2.0	0.9
160.00	0.74	1.1	83.31	10JN	1.2	-3.4	-2.0	0.8
165.48	0.55	1.2	92.84	20JN	0.1	-3.4	-1.9	0.7
171.20	0.37	1.2	102.38	30JN	-1.0	-3.4	-1.9	0.7
177.15	0.20	1.3	111.93	10JL	-1.6	-3.5	-1.8	0.6
183.29	0.04	1.3	121.51	20JL	-1.0	-3.5	-1.7	0.5
189.62	-0.10	1.4	131.13	30JL	-0.4	-3.5	-1.7	0.5
196.12	-0.24	1.4	140.78	9AU	0.0	-3.4	-1.6	0.5
202.78	-0.37	1.4	150.49	19AU	0.3	-3.4	-1.6	0.6
209.59	-0.49	1.4	160.25	29AU	0.5	-3.4	-1.5	0.6
216.55	-0.60	1.4	170.06	8SE	1.1	-3.3	-1.5	0.7
223.65	-0.70	1.4	179.93	18SE	3.1	-3.3	-1.4	0.7
230.87	-0.80	1.4	189.86	28SE	0.8	-3.3	-1.4	0.8
238.21	-0.88	1.4	199.85	8OC	-0.5	-3.3	-1.4	0.8
245.67	-0.95	1.4	209.89	18OC	-0.8	-3.3	-1.3	0.9
253.22	-1.01	1.4	219.98	28OC	-0.7	-3.3	-1.3	0.9
260.86	-1.06	1.4	230.10	7NO	-0.8	-3.4	-1.3	1.0
268.59	-1.10	1.4	240.27	17NO	-0.7	-3.4	-1.3	1.0
276.37	-1.12	1.4	250.45	27NO	-0.6	-3.4	-1.3	1.0
284.21	-1.14	1.4	260.65	7DE	-0.7	-3.5	-1.3	1.0
292.08	-1.14	1.4	270.86	17DE	-0.4	-3.5	-1.4	1.0
299.98	-1.13	1.4	281.05	27DE	1.1	-3.6	-1.4	1.0
				845				
307.88	-1.10	1.4	291.22	6JA	2.4	-3.6	-1.4	1.0
315.77	-1.07	1.4	301.37	16JA	0.8	-3.7	-1.5	1.0
323.65	-1.03	1.4	311.48	26JA	0.3	-3.7	-1.5	1.0
331.49	-0.97	1.4	321.54	5FE	0.1	-3.8	-1.6	1.0
339.28	-0.91	1.4	331.56	15FE	-0.1	-3.9	-1.6	1.1
347.02	-0.83	1.4	341.52	25FE	-0.4	-4.0	-1.7	1.1
354.70	-0.75	1.4	351.41	7MR	-1.1	-4.1	-1.8	1.2
2.30	-0.67	1.4	1.26	17MR	-1.6	-4.2	-1.8	1.2
9.83	-0.57	1.4	11.04	27MR	-1.0	-4.2	-1.9	1.2
17.27	-0.48	1.5	20.77	6AP	0.2	-4.2	-2.0	1.2
24.63	-0.38	1.5	30.45	16AP	1.4	-4.1	-2.0	1.2
31.90	-0.28	1.6	40.08	26AP	2.8	-3.7	-2.1	1.2
39.08	-0.17	1.6	49.68	6MY	3.2	-2.8	-2.1	1.2
46.17	-0.07	1.7	59.25	16MY	1.8	-3.4	-2.2	1.1
53.18	0.04	1.7	68.79	26MY	0.9	-3.9	-2.2	1.1
60.10	0.15	1.7	78.32	5JN	0.0	-4.2	-2.2	1.0
66.93	0.25	1.8	87.85	15JN	-1.0	-4.2	-2.2	1.0
73.68	0.36	1.8	97.37	25JN	-1.7	-4.1	-2.2	0.9
80.35	0.47	1.8	106.92	5JL	-1.0	-4.1	-2.1	0.8
86.94	0.57	1.8	116.49	15JL	-0.3	-4.0	-2.1	0.8
93.44	0.68	1.8	126.08	25JL	0.1	-3.9	-2.0	0.7
99.87	0.78	1.8	135.71	4AU	0.4	-3.8	-2.0	0.6
106.22	0.89	1.8	145.40	14AU	0.7	-3.7	-1.9	0.6
112.48	0.99	1.8	155.12	24AU	1.4	-3.7	-1.8	0.7
118.67	1.10	1.8	164.91	3SE	3.2	-3.6	-1.8	0.7
124.75	1.21	1.8	174.75	13SE	0.7	-3.6	-1.7	0.8
130.74	1.32	1.7	184.65	23SE	-0.7	-3.5	-1.6	0.8
136.62	1.43	1.7	194.61	3OC	-0.9	-3.5	-1.6	0.9
142.38	1.55	1.6	204.62	13OC	-0.9	-3.4	-1.6	0.9
147.99	1.67	1.5	214.68	23OC	-0.8	-3.4	-1.5	1.0
153.44	1.79	1.4	224.79	2NO	-0.6	-3.4	-1.5	1.0
158.69	1.92	1.3	234.94	12NO	-0.5	-3.4	-1.4	1.1
163.70	2.06	1.2	245.11	22NO	-0.5	-3.4	-1.4	1.1
168.44	2.21	1.1	255.31	2DE	-0.3	-3.4	-1.4	1.1
172.82	2.37	0.9	265.51	12DE	1.3	-3.4	-1.4	1.1
176.77	2.54	0.7	275.71	22DE	2.0	-3.4	-1.4	1.2

Right table:

♂ LONG	LAT	MAG	⊙ LONG	16.00UT 846	☿	♀	♃	♄
180.20	2.72	0.5	285.89	1JA	0.5	-3.4	-1.4	1.2
182.95	2.90	0.3	296.06	11JA	0.1	-3.4	-1.4	1.2
184.89	3.09	0.0	306.18	21JA	0.0	-3.4	-1.5	1.2
185.83	3.28	-0.2	316.27	31JA	-0.2	-3.4	-1.5	1.2
185.58	3.43	-0.5	326.32	10FE	-0.5	-3.4	-1.5	1.2
184.07	3.52	-0.8	336.30	20FE	-1.1	-3.5	-1.6	1.2
181.35	3.51	-1.0	346.23	2MR	-1.5	-3.5	-1.6	1.3
177.77	3.36	-1.2	356.10	12MR	-0.9	-3.5	-1.7	1.3
173.98	3.07	-1.2	5.92	22MR	0.3	-3.4	-1.8	1.3
170.66	2.67	-1.0	15.68	1AP	1.7	-3.4	-1.8	1.4
168.40	2.21	-0.8	25.38	11AP	3.6	-3.4	-1.9	1.4
167.43	1.75	-0.6	35.04	21AP	2.4	-3.4	-2.0	1.4
167.75	1.31	-0.4	44.65	1MY	1.4	-3.3	-2.0	1.4
169.26	0.92	-0.2	54.24	11MY	0.7	-3.3	-2.1	1.4
171.75	0.57	-0.0	63.79	21MY	-0.1	-3.3	-2.2	1.3
175.07	0.26	0.1	73.33	31MY	-1.0	-3.3	-2.2	1.3
179.07	-0.00	0.3	82.85	10JN	-1.8	-3.3	-2.3	1.3
183.62	-0.24	0.4	92.38	20JN	-1.0	-3.4	-2.3	1.2
188.65	-0.45	0.5	101.91	30JN	-0.2	-3.4	-2.3	1.1
194.08	-0.63	0.6	111.47	10JL	0.2	-3.4	-2.3	1.1
199.85	-0.79	0.7	121.04	20JL	0.6	-3.5	-2.3	1.0
205.92	-0.93	0.8	130.66	30JL	1.0	-3.5	-2.3	0.9
212.25	-1.05	0.8	140.31	9AU	1.9	-3.6	-2.3	0.9
218.81	-1.16	0.9	150.02	19AU	3.0	-3.6	-2.2	0.8
225.58	-1.24	0.9	159.77	29AU	0.6	-3.7	-2.1	0.8
232.55	-1.31	1.0	169.58	8SE	-0.8	-3.8	-2.1	0.8
239.67	-1.37	1.0	179.45	18SE	-1.1	-3.9	-2.0	0.9
246.95	-1.40	1.1	189.38	28SE	-1.0	-4.0	-1.9	0.9
254.36	-1.43	1.1	199.36	8OC	-0.7	-4.1	-1.9	1.0
261.88	-1.43	1.1	209.40	18OC	-0.4	-4.2	-1.8	1.0
269.49	-1.42	1.2	219.49	28OC	-0.4	-4.3	-1.8	1.1
277.19	-1.40	1.2	229.61	7NO	-0.3	-4.4	-1.7	1.1
284.94	-1.36	1.2	239.77	17NO	-0.1	-4.3	-1.7	1.2
292.73	-1.31	1.3	249.96	27NO	1.5	-4.1	-1.6	1.2
300.55	-1.24	1.3	260.15	7DE	1.7	-3.4	-1.6	1.3
308.37	-1.17	1.3	270.35	17DE	0.3	-3.3	-1.6	1.3
316.19	-1.09	1.4	280.55	27DE	-0.0	-4.0	-1.5	1.3
				847				
323.99	-1.00	1.4	290.73	6JA	-0.1	-4.3	-1.5	1.3
331.75	-0.90	1.4	300.87	16JA	-0.3	-4.4	-1.5	1.4
339.46	-0.80	1.5	310.99	26JA	-0.6	-4.3	-1.5	1.4
347.11	-0.69	1.5	321.05	5FE	-1.1	-4.2	-1.5	1.4
354.70	-0.58	1.5	331.07	15FE	-1.4	-4.1	-1.5	1.4
2.22	-0.47	1.6	341.03	25FE	-0.9	-4.0	-1.6	1.3
9.66	-0.36	1.6	350.93	7MR	0.4	-3.9	-1.6	1.3
17.02	-0.25	1.6	0.78	17MR	2.2	-3.8	-1.6	1.3
24.30	-0.14	1.6	10.57	27MR	3.2	-3.7	-1.7	1.3
31.49	-0.03	1.7	20.30	6AP	1.8	-3.6	-1.7	1.3
38.60	0.07	1.7	29.98	16AP	1.0	-3.5	-1.8	1.3
45.63	0.17	1.7	39.62	26AP	0.5	-3.5	-1.9	1.3
52.58	0.27	1.7	49.22	6MY	-0.1	-3.4	-1.9	1.2
59.46	0.37	1.7	58.79	16MY	-1.0	-3.4	-2.0	1.2
66.26	0.46	1.7	68.34	26MY	-1.8	-3.4	-2.1	1.2
73.00	0.55	1.8	77.86	5JN	-1.0	-3.3	-2.1	1.1
79.67	0.64	1.8	87.39	15JN	-0.2	-3.3	-2.2	1.1
86.28	0.72	1.9	96.92	25JN	0.4	-3.3	-2.3	1.0
92.83	0.80	1.9	106.46	5JL	0.8	-3.3	-2.3	1.0
99.33	0.87	1.9	116.00	15JL	1.3	-3.3	-2.4	0.9
105.78	0.95	2.0	125.62	25JL	2.4	-3.3	-2.4	0.9
112.17	1.02	2.0	135.25	4AU	2.6	-3.4	-2.4	0.8
118.53	1.08	2.0	144.93	14AU	0.5	-3.4	-2.4	0.8
124.84	1.15	2.0	154.66	24AU	-0.8	-3.4	-2.4	0.8
131.11	1.21	2.0	164.43	3SE	-1.2	-3.4	-2.4	0.7
137.33	1.27	2.0	174.27	13SE	-1.1	-3.5	-2.4	0.8
143.51	1.33	1.9	184.17	23SE	-0.6	-3.5	-2.3	0.9
149.64	1.39	1.9	194.12	3OC	-0.3	-3.5	-2.2	0.9
155.72	1.44	1.9	204.13	13OC	-0.2	-3.5	-2.2	1.0
161.73	1.49	1.8	214.19	23OC	-0.2	-3.4	-2.1	1.1
167.68	1.54	1.7	224.29	2NO	0.1	-3.4	-2.0	1.1
173.56	1.59	1.7	234.44	12NO	1.7	-3.4	-2.0	1.2
179.33	1.64	1.6	244.61	22NO	1.5	-3.4	-1.9	1.2
185.00	1.68	1.5	254.80	2DE	0.0	-3.3	-1.8	1.2
190.55	1.72	1.4	265.00	12DE	-0.2	-3.3	-1.8	1.3
195.93	1.76	1.2	275.21	22DE	-0.2	-3.3	-1.7	1.3

♂ LONG	♂ LAT	♂ MAG	☉ LONG	16.00UT	☿	♀	♃	♄
				848	MAGNITUDES			
201.12	1.79	1.1	285.39	1JA	-0.4	-3.3	-1.7	1.3
206.08	1.81	0.9	295.56	11JA	-0.6	-3.4	-1.7	1.3
210.75	1.83	0.7	305.69	21JA	-1.1	-3.4	-1.6	1.2
215.06	1.83	0.5	315.78	31JA	-1.2	-3.4	-1.6	1.2
218.93	1.81	0.3	325.82	10FE	-0.8	-3.4	-1.6	1.2
222.24	1.77	0.1	335.81	20FE	0.5	-3.5	-1.6	1.1
224.85	1.70	-0.2	345.75	1MR	2.6	-3.5	-1.6	1.1
226.59	1.58	-0.5	355.62	11MR	2.4	-3.5	-1.6	1.0
227.27	1.39	-0.8	5.44	21MR	1.3	-3.6	-1.6	1.0
226.75	1.13	-1.1	15.20	31MR	0.8	-3.7	-1.6	1.0
224.98	0.78	-1.4	24.92	10AP	0.3	-3.7	-1.6	1.0
222.12	0.34	-1.7	34.58	20AP	-0.2	-3.8	-1.7	1.0
218.69	-0.17	-1.8	44.19	30AP	-1.0	-3.9	-1.7	1.0
215.36	-0.67	-1.7	53.78	10MY	-1.9	-4.0	-1.8	1.0
212.84	-1.14	-1.5	63.33	20MY	-1.0	-4.1	-1.8	1.0
211.55	-1.53	-1.3	72.87	30MY	-0.1	-4.1	-1.9	0.9
211.64	-1.83	-1.1	82.40	9JN	0.5	-4.2	-1.9	0.9
213.05	-2.06	-0.9	91.92	19JN	1.0	-4.1	-2.0	0.9
215.59	-2.23	-0.7	101.45	29JN	1.7	-3.9	-2.1	0.8
219.08	-2.34	-0.5	111.01	9JL	3.0	-3.4	-2.2	0.8
223.37	-2.41	-0.4	120.58	19JL	2.2	-3.1	-2.2	0.7
228.29	-2.45	-0.2	130.19	29JL	0.4	-3.8	-2.3	0.7
233.75	-2.45	-0.1	139.85	8AU	-0.9	-4.1	-2.3	0.6
239.64	-2.43	0.0	149.55	18AU	-1.3	-4.2	-2.4	0.6
245.89	-2.39	0.2	159.30	28AU	-1.1	-4.2	-2.4	0.5
252.44	-2.32	0.3	169.11	7SE	-0.5	-4.2	-2.5	0.5
259.25	-2.24	0.4	178.97	17SE	-0.2	-4.1	-2.5	0.4
266.25	-2.14	0.5	188.90	27SE	-0.1	-4.0	-2.5	0.5
273.42	-2.02	0.6	198.88	7OC	0.0	-3.9	-2.4	0.6
280.72	-1.89	0.7	208.91	17OC	0.3	-3.8	-2.4	0.6
288.11	-1.75	0.7	218.99	27OC	2.0	-3.8	-2.3	0.7
295.58	-1.61	0.8	229.12	6NO	1.3	-3.7	-2.3	0.8
303.10	-1.46	0.9	239.28	16NO	-0.1	-3.6	-2.2	0.8
310.64	-1.30	1.0	249.46	26NO	-0.3	-3.6	-2.1	0.9
318.20	-1.14	1.1	259.66	6DE	-0.4	-3.5	-2.1	0.9
325.74	-0.99	1.2	269.86	16DE	-0.5	-3.5	-2.0	1.0
333.25	-0.83	1.2	280.06	26DE	-0.7	-3.4	-1.9	1.0
				849				
340.73	-0.68	1.3	290.24	5JA	-1.0	-3.4	-1.9	1.0
348.16	-0.53	1.4	300.38	15JA	-1.1	-3.4	-1.8	1.0
355.53	-0.39	1.4	310.49	25JA	-0.8	-3.4	-1.7	1.0
2.85	-0.25	1.5	320.57	4FE	0.6	-3.3	-1.7	1.0
10.09	-0.12	1.6	330.58	14FE	3.0	-3.3	-1.7	0.9
17.26	-0.00	1.6	340.55	24FE	1.8	-3.3	-1.6	0.9
24.37	0.12	1.7	350.46	6MR	0.9	-3.3	-1.6	0.8
31.40	0.23	1.7	0.30	16MR	0.6	-3.3	-1.6	0.8
38.36	0.33	1.8	10.09	26MR	0.2	-3.3	-1.6	0.8
45.26	0.43	1.8	19.83	5AP	-0.3	-3.4	-1.6	0.8
52.08	0.52	1.8	29.51	15AP	-1.0	-3.4	-1.6	0.8
58.85	0.60	1.9	39.15	25AP	-1.8	-3.4	-1.6	0.8
65.55	0.68	1.9	48.76	5MY	-1.0	-3.5	-1.6	0.8
72.20	0.75	1.9	58.32	15MY	-0.1	-3.5	-1.6	0.8
78.80	0.81	1.9	67.87	25MY	0.7	-3.4	-1.7	0.8
85.36	0.88	1.9	77.40	4JN	1.3	-3.4	-1.7	0.7
91.87	0.93	1.9	86.92	14JN	2.3	-3.4	-1.7	0.7
98.34	0.98	1.9	96.45	24JN	3.3	-3.4	-1.8	0.7
104.79	1.03	1.9	106.00	4JL	1.8	-3.3	-1.8	0.6
111.20	1.07	1.9	115.56	14JL	0.3	-3.3	-1.9	0.6
117.60	1.10	2.0	125.15	24JL	-0.9	-3.3	-2.0	0.5
123.97	1.14	2.0	134.78	3AU	-1.5	-3.3	-2.0	0.5
130.33	1.16	2.0	144.45	13AU	-1.0	-3.3	-2.1	0.4
136.68	1.19	2.0	154.18	23AU	-0.5	-3.3	-2.2	0.4
143.02	1.21	2.0	163.96	2SE	-0.1	-3.4	-2.2	0.3
149.35	1.22	2.0	173.79	12SE	0.0	-3.4	-2.3	0.3
155.68	1.23	2.0	183.69	22SE	0.2	-3.4	-2.3	0.2
162.00	1.24	2.0	193.64	2OC	0.6	-3.5	-2.4	0.2
168.32	1.24	2.0	203.64	12OC	2.4	-3.5	-2.4	0.3
174.63	1.24	1.9	213.70	22OC	1.1	-3.6	-2.4	0.3
180.93	1.23	1.9	223.81	1NO	-0.3	-3.7	-2.4	0.4
187.22	1.22	1.8	233.95	11NO	-0.5	-3.8	-2.4	0.5
193.51	1.20	1.8	244.12	21NO	-0.5	-3.8	-2.3	0.5
199.77	1.17	1.7	254.31	1DE	-0.6	-3.9	-2.3	0.6
206.01	1.13	1.6	264.51	11DE	-0.8	-4.0	-2.2	0.6
212.23	1.09	1.5	274.71	21DE	-0.9	-4.1	-2.2	0.7
218.40	1.03	1.4	284.90	31DE	-0.9	-4.2	-2.1	0.7

♂ LONG	♂ LAT	♂ MAG	☉ LONG	16.00UT	☿	♀	♃	♄
				850	MAGNITUDES			
224.53	0.96	1.3	295.06	10JA	-0.6	-4.3	-2.0	0.7
230.60	0.87	1.2	305.20	20JA	0.7	-4.4	-1.9	0.8
236.60	0.77	1.0	315.29	30JA	3.1	-4.3	-1.9	0.8
242.51	0.64	0.9	325.33	9FE	1.3	-4.0	-1.8	0.7
248.31	0.49	0.7	335.33	19FE	0.7	-3.5	-1.7	0.7
253.97	0.31	0.5	345.26	1MR	0.4	-3.4	-1.7	0.7
259.45	0.09	0.3	355.14	11MR	0.1	-3.9	-1.6	0.7
264.71	-0.17	0.1	4.96	21MR	-0.3	-4.2	-1.6	0.6
269.68	-0.47	-0.1	14.73	31MR	-1.1	-4.2	-1.6	0.6
274.30	-0.84	-0.3	24.44	10AP	-1.8	-4.2	-1.5	0.6
278.44	-1.27	-0.6	34.10	20AP	-1.0	-4.1	-1.5	0.6
281.99	-1.78	-0.9	43.72	30AP	0.0	-4.0	-1.5	0.6
284.79	-2.38	-1.2	53.31	10MY	0.9	-3.9	-1.5	0.6
286.64	-3.06	-1.5	62.87	20MY	1.8	-3.8	-1.5	0.6
287.36	-3.81	-1.8	72.40	30MY	3.0	-3.7	-1.5	0.6
286.85	-4.59	-2.1	81.93	9JN	2.8	-3.6	-1.5	0.6
285.15	-5.31	-2.4	91.46	19JN	1.4	-3.6	-1.5	0.6
282.65	-5.87	-2.6	100.99	29JN	0.2	-3.5	-1.6	0.5
279.96	-6.18	-2.5	110.54	9JL	-1.0	-3.5	-1.6	0.5
277.81	-6.19	-2.3	120.12	19JL	-1.6	-3.4	-1.6	0.5
276.77	-5.96	-2.1	129.73	29JL	-1.0	-3.4	-1.7	0.4
277.06	-5.57	-1.8	139.38	8AU	-0.4	-3.4	-1.7	0.4
278.67	-5.10	-1.6	149.08	18AU	-0.0	-3.4	-1.8	0.3
281.44	-4.59	-1.3	158.83	28AU	0.2	-3.4	-1.9	0.2
285.15	-4.08	-1.1	168.63	7SE	0.4	-3.4	-1.9	0.2
289.64	-3.59	-0.8	178.50	17SE	0.9	-3.4	-2.0	0.1
294.72	-3.13	-0.6	188.42	27SE	2.7	-3.4	-2.1	0.1
300.25	-2.70	-0.4	198.39	7OC	0.9	-3.4	-2.1	0.0
306.15	-2.30	-0.2	208.43	17OC	-0.4	-3.4	-2.2	0.0
312.32	-1.93	-0.0	218.50	27OC	-0.7	-3.4	-2.2	0.1
318.68	-1.59	0.2	228.63	6NO	-0.6	-3.4	-2.3	0.1
325.20	-1.28	0.3	238.79	16NO	-0.7	-3.4	-2.3	0.2
331.83	-1.00	0.5	248.97	26NO	-0.7	-3.5	-2.3	0.3
338.53	-0.75	0.6	259.16	6DE	-0.7	-3.5	-2.3	0.3
345.27	-0.52	0.8	269.37	16DE	-0.8	-3.5	-2.3	0.4
352.05	-0.31	0.9	279.56	26DE	-0.5	-3.5	-2.2	0.5
				851				
358.83	-0.12	1.1	289.74	5JA	0.9	-3.4	-2.2	0.5
5.60	0.05	1.2	299.89	15JA	2.7	-3.4	-2.1	0.5
12.37	0.20	1.3	310.00	25JA	1.0	-3.4	-2.0	0.6
19.10	0.33	1.4	320.07	4FE	0.4	-3.4	-2.0	0.6
25.81	0.46	1.5	330.10	14FE	0.2	-3.3	-1.9	0.6
32.49	0.56	1.6	340.06	24FE	0.0	-3.3	-1.8	0.6
39.13	0.66	1.6	349.97	6MR	-0.4	-3.3	-1.8	0.5
45.74	0.74	1.7	359.82	16MR	-1.1	-3.3	-1.7	0.5
52.31	0.82	1.8	9.61	26MR	-1.7	-3.4	-1.6	0.5
58.84	0.88	1.8	19.35	5AP	-1.0	-3.4	-1.6	0.4
65.34	0.94	1.9	29.04	15AP	0.1	-3.4	-1.5	0.4
71.81	0.99	1.9	38.68	25AP	1.2	-3.4	-1.5	0.4
78.24	1.03	2.0	48.29	5MY	2.4	-3.4	-1.5	0.4
84.66	1.07	2.0	57.86	15MY	3.6	-3.5	-1.4	0.4
91.05	1.09	2.0	67.40	25MY	2.2	-3.5	-1.4	0.4
97.43	1.12	2.0	76.94	4JN	1.1	-3.6	-1.4	0.5
103.79	1.14	2.0	86.47	14JN	0.1	-3.7	-1.4	0.5
110.14	1.15	2.0	95.99	24JN	-1.0	-3.7	-1.4	0.4
116.49	1.16	2.0	105.54	4JL	-1.7	-3.8	-1.4	0.4
122.84	1.16	2.0	115.10	14JL	-1.0	-3.9	-1.4	0.4
129.19	1.16	2.0	124.69	24JL	-0.3	-4.0	-1.4	0.4
135.56	1.15	2.0	134.32	3AU	0.1	-4.1	-1.5	0.3
141.93	1.14	2.0	143.99	13AU	0.3	-4.2	-1.5	0.3
148.32	1.12	2.0	153.71	23AU	0.6	-4.3	-1.5	0.2
154.74	1.10	2.0	163.49	2SE	1.2	-4.2	-1.6	0.2
161.18	1.08	2.0	173.32	12SE	3.1	-4.0	-1.6	0.1
167.64	1.05	2.0	183.21	22SE	0.8	-3.4	-1.7	0.0
174.13	1.01	2.0	193.16	2OC	-0.6	-3.4	-1.7	-0.0
180.65	0.97	2.0	203.16	12OC	-0.8	-4.0	-1.8	-0.1
187.21	0.93	1.9	213.21	22OC	-0.8	-4.3	-1.9	-0.1
193.80	0.88	1.9	223.32	1NO	-0.8	-4.4	-1.9	-0.1
200.42	0.82	1.9	233.46	11NO	-0.6	-4.3	-2.0	-0.1
207.08	0.75	1.8	243.62	21NO	-0.6	-4.3	-2.1	-0.0
213.78	0.67	1.8	253.82	1DE	-0.6	-4.2	-2.1	0.1
220.50	0.59	1.7	264.02	11DE	-0.4	-4.1	-2.1	0.1
227.27	0.50	1.7	274.22	21DE	1.1	-4.0	-2.2	0.2
234.06	0.39	1.6	284.41	31DE	2.3	-3.9	-2.2	0.3

♂ LONG	♂ LAT	♂ MAG	⊙ LONG	16.00UT	☿	♀	♃	♄
				852	MAGNITUDES			
240.89	0.28	1.5	294.57	10JA	0.7	-3.8	-2.2	0.3
247.74	0.15	1.4	304.71	20JA	0.2	-3.7	-2.1	0.4
254.62	0.01	1.3	314.81	30JA	0.1	-3.6	-2.1	0.4
261.53	-0.14	1.2	324.85	9FE	-0.1	-3.5	-2.0	0.4
268.45	-0.31	1.1	334.85	19FE	-0.5	-3.5	-2.0	0.4
275.39	-0.50	1.0	344.79	29FE	-1.1	-3.4	-1.9	0.4
282.33	-0.70	0.9	354.67	10MR	-1.6	-3.4	-1.8	0.4
289.27	-0.91	0.8	4.49	20MR	-1.0	-3.4	-1.8	0.4
296.19	-1.14	0.7	14.26	30MR	0.2	-3.3	-1.7	0.4
303.08	-1.39	0.5	23.98	9AP	1.5	-3.3	-1.6	0.3
309.91	-1.65	0.4	33.64	19AP	3.1	-3.3	-1.6	0.3
316.66	-1.93	0.2	43.26	29AP	2.9	-3.3	-1.5	0.3
323.30	-2.22	0.1	52.85	9MY	1.7	-3.3	-1.5	0.3
329.78	-2.52	-0.1	62.41	19MY	0.8	-3.3	-1.4	0.3
336.05	-2.83	-0.2	71.95	29MY	-0.0	-3.3	-1.4	0.3
342.04	-3.14	-0.4	81.47	8JN	-1.0	-3.4	-1.3	0.4
347.68	-3.46	-0.6	91.00	18JN	-1.8	-3.4	-1.3	0.4
352.84	-3.78	-0.8	100.53	28JN	-1.0	-3.4	-1.3	0.4
357.40	-4.10	-1.0	110.08	8JL	-0.3	-3.5	-1.3	0.4
1.20	-4.39	-1.2	119.66	18JL	0.2	-3.5	-1.3	0.4
4.02	-4.66	-1.4	129.27	28JL	0.5	-3.5	-1.3	0.3
5.67	-4.88	-1.7	138.91	7AU	0.8	-3.4	-1.3	0.3
5.95	-4.99	-1.9	148.61	17AU	1.6	-3.4	-1.3	0.3
4.78	-4.95	-2.1	158.36	27AU	3.2	-3.4	-1.3	0.2
2.39	-4.70	-2.3	168.15	6SE	0.7	-3.3	-1.4	0.2
359.30	-4.24	-2.4	178.02	16SE	-0.7	-3.3	-1.4	0.1
356.30	-3.59	-2.2	187.93	26SE	-1.0	-3.3	-1.4	0.0
354.13	-2.87	-1.9	197.91	6OC	-0.9	-3.3	-1.5	-0.0
353.19	-2.15	-1.6	207.94	16OC	-0.7	-3.3	-1.5	-0.1
353.58	-1.51	-1.2	218.02	26OC	-0.5	-3.3	-1.6	-0.2
355.19	-0.96	-0.9	228.13	5NO	-0.4	-3.4	-1.7	-0.2
357.81	-0.50	-0.6	238.29	15NO	-0.4	-3.4	-1.7	-0.2
1.24	-0.13	-0.3	248.47	25NO	-0.2	-3.4	-1.8	-0.2
5.30	0.18	-0.1	258.67	5DE	1.3	-3.5	-1.9	-0.1
9.84	0.42	0.2	268.87	15DE	2.0	-3.5	-1.9	-0.0
14.77	0.62	0.4	279.07	25DE	0.4	-3.6	-2.0	0.1
				853				
19.98	0.79	0.6	289.24	4JA	0.1	-3.6	-2.0	0.1
25.41	0.92	0.8	299.40	14JA	-0.0	-3.7	-2.0	0.2
31.01	1.02	0.9	309.52	24JA	-0.2	-3.7	-2.1	0.2
36.74	1.11	1.1	319.59	3FE	-0.5	-3.8	-2.1	0.3
42.58	1.17	1.2	329.61	13FE	-1.1	-3.9	-2.0	0.3
48.49	1.22	1.4	339.59	23FE	-1.4	-4.0	-2.0	0.3
54.46	1.26	1.5	349.49	5MR	-1.0	-4.1	-2.0	0.3
60.47	1.29	1.6	359.35	15MR	0.3	-4.2	-1.9	0.4
66.52	1.31	1.7	9.15	25MR	1.9	-4.2	-1.9	0.3
72.59	1.33	1.7	18.89	4AP	3.7	-4.2	-1.8	0.3
78.69	1.33	1.8	28.58	14AP	2.1	-4.1	-1.7	0.3
84.81	1.33	1.9	38.22	24AP	1.2	-3.7	-1.7	0.3
90.94	1.32	1.9	47.82	4MY	0.6	-2.8	-1.6	0.2
97.10	1.31	2.0	57.40	14MY	-0.1	-3.4	-1.5	0.2
103.27	1.30	2.0	66.94	24MY	-1.0	-3.9	-1.5	0.2
109.46	1.28	2.0	76.47	3JN	-1.8	-4.2	-1.4	0.3
115.67	1.25	2.0	86.00	13JN	-1.0	-4.2	-1.4	0.3
121.91	1.22	2.0	95.53	23JN	-0.2	-4.1	-1.3	0.3
128.17	1.19	2.0	105.07	3JL	0.3	-4.1	-1.3	0.3
134.46	1.15	2.0	114.63	13JL	0.6	-4.0	-1.3	0.3
140.79	1.11	2.0	124.22	23JL	1.1	-3.9	-1.2	0.3
147.16	1.07	2.0	133.85	2AU	2.0	-3.8	-1.2	0.3
153.56	1.02	2.0	143.52	12AU	2.9	-3.7	-1.2	0.3
160.02	0.97	1.9	153.24	22AU	0.6	-3.7	-1.2	0.3
166.52	0.91	1.9	163.01	1SE	-0.8	-3.6	-1.2	0.2
173.07	0.85	1.9	172.84	11SE	-1.1	-3.5	-1.2	0.2
179.68	0.78	1.9	182.73	21SE	-1.1	-3.5	-1.3	0.1
186.34	0.71	1.9	192.67	1OC	-0.7	-3.5	-1.3	0.1
193.06	0.64	1.9	202.68	11OC	-0.4	-3.4	-1.3	-0.0
199.85	0.56	1.9	212.73	21OC	-0.3	-3.4	-1.4	-0.1
206.69	0.48	1.8	222.83	31OC	-0.3	-3.4	-1.4	-0.1
213.61	0.39	1.8	232.97	10NO	-0.0	-3.4	-1.5	-0.2
220.58	0.29	1.8	243.14	20NO	1.5	-3.4	-1.5	-0.3
227.62	0.19	1.8	253.32	30NO	1.7	-3.4	-1.6	-0.3
234.72	0.09	1.7	263.53	10DE	0.2	-3.4	-1.6	-0.2
241.89	-0.03	1.7	273.73	20DE	-0.1	-3.4	-1.7	-0.1
249.12	-0.14	1.6	283.91	30DE	-0.2	-3.4	-1.8	-0.1

♂ LONG	♂ LAT	♂ MAG	⊙ LONG	16.00UT	☿	♀	♃	♄
				854	MAGNITUDES			
256.40	-0.27	1.6	294.08	9JA	-0.3	-3.4	-1.8	0.0
263.74	-0.39	1.5	304.22	19JA	-0.6	-3.4	-1.9	0.1
271.13	-0.53	1.5	314.31	29JA	-1.1	-3.4	-1.9	0.1
278.57	-0.66	1.4	324.37	8FE	-1.3	-3.4	-2.0	0.2
286.05	-0.80	1.4	334.36	18FE	-0.9	-3.5	-2.0	0.2
293.56	-0.94	1.3	344.30	28FE	0.4	-3.5	-2.0	0.3
301.10	-1.07	1.2	354.19	10MR	2.3	-3.5	-2.0	0.3
308.64	-1.21	1.2	4.02	20MR	2.9	-3.4	-2.0	0.3
316.20	-1.34	1.1	13.79	30MR	1.6	-3.4	-2.0	0.3
323.74	-1.46	1.0	23.50	9AP	0.9	-3.4	-1.9	0.3
331.24	-1.58	1.0	33.17	19AP	0.4	-3.4	-1.9	0.3
338.71	-1.69	0.9	42.79	29AP	-0.2	-3.3	-1.8	0.3
346.12	-1.78	0.8	52.38	9MY	-1.0	-3.3	-1.7	0.3
353.44	-1.86	0.7	61.94	19MY	-1.9	-3.3	-1.7	0.2
0.66	-1.92	0.7	71.48	29MY	-1.0	-3.3	-1.6	0.2
7.74	-1.97	0.6	81.01	8JN	-0.2	-3.3	-1.5	0.2
14.66	-2.00	0.5	90.53	18JN	0.4	-3.4	-1.5	0.3
21.40	-2.01	0.4	100.07	28JN	0.9	-3.4	-1.4	0.3
27.90	-1.99	0.3	109.62	8JL	1.4	-3.4	-1.4	0.3
34.13	-1.96	0.2	119.19	18JL	2.6	-3.5	-1.3	0.3
40.04	-1.90	0.1	128.80	28JL	2.5	-3.5	-1.3	0.3
45.54	-1.81	0.0	138.44	7AU	0.5	-3.6	-1.3	0.3
50.57	-1.69	-0.1	148.13	17AU	-0.8	-3.6	-1.2	0.3
55.00	-1.54	-0.3	157.88	27AU	-1.3	-3.7	-1.2	0.3
58.70	-1.35	-0.5	167.68	6SE	-1.1	-3.8	-1.2	0.3
61.50	-1.12	-0.6	177.53	16SE	-0.6	-3.9	-1.2	0.3
63.19	-0.82	-0.8	187.45	26SE	-0.3	-4.0	-1.2	0.2
63.57	-0.46	-1.1	197.43	6OC	-0.2	-4.1	-1.2	0.1
62.52	-0.04	-1.3	207.45	16OC	-0.1	-4.2	-1.2	0.1
60.05	0.43	-1.5	217.53	26OC	0.2	-4.3	-1.3	0.0
56.55	0.91	-1.6	227.65	5NO	1.8	-4.4	-1.3	-0.1
52.73	1.36	-1.6	237.80	15NO	1.5	-4.3	-1.3	-0.1
49.37	1.71	-1.3	247.98	25NO	-0.0	-4.1	-1.4	-0.2
47.11	1.95	-1.0	258.18	5DE	-0.3	-3.4	-1.4	-0.3
46.18	2.10	-0.7	268.38	15DE	-0.3	-3.3	-1.5	-0.2
46.56	2.18	-0.4	278.57	25DE	-0.4	-4.0	-1.5	-0.2
				855				
48.09	2.20	-0.1	288.75	4JA	-0.7	-4.3	-1.6	-0.1
50.55	2.19	0.1	298.91	14JA	-1.1	-4.4	-1.7	-0.0
53.75	2.16	0.4	309.03	24JA	-1.2	-4.3	-1.7	0.0
57.53	2.12	0.6	319.10	3FE	-0.8	-4.2	-1.8	0.1
61.77	2.07	0.8	329.13	13FE	0.5	-4.1	-1.9	0.2
66.36	2.01	1.0	339.10	23FE	2.7	-4.0	-1.9	0.2
71.24	1.95	1.1	349.02	5MR	2.2	-3.9	-2.0	0.2
76.34	1.88	1.3	358.87	15MR	1.2	-3.8	-2.0	0.3
81.63	1.82	1.4	8.68	25MR	0.7	-3.7	-2.0	0.3
87.06	1.75	1.5	18.42	4AP	0.3	-3.6	-2.0	0.3
92.62	1.69	1.6	28.11	14AP	-0.2	-3.5	-2.0	0.3
98.28	1.62	1.7	37.76	24AP	-1.0	-3.5	-2.0	0.3
104.03	1.55	1.8	47.36	4MY	-1.9	-3.4	-2.0	0.3
109.86	1.48	1.8	56.94	14MY	-1.0	-3.4	-1.9	0.3
115.77	1.41	1.9	66.49	24MY	-0.1	-3.4	-1.9	0.3
121.75	1.34	1.9	76.02	3JN	0.6	-3.3	-1.8	0.3
127.80	1.27	1.9	85.54	13JN	1.1	-3.3	-1.7	0.2
133.91	1.19	2.0	95.08	23JN	1.9	-3.3	-1.7	0.3
140.09	1.12	2.0	104.61	3JL	3.2	-3.3	-1.6	0.3
146.33	1.04	2.0	114.17	13JL	2.0	-3.3	-1.6	0.3
152.64	0.96	2.0	123.76	23JL	0.4	-3.3	-1.5	0.4
159.02	0.88	2.0	133.38	2AU	-0.9	-3.4	-1.4	0.4
165.47	0.80	1.9	143.05	12AU	-1.4	-3.4	-1.4	0.4
172.00	0.71	1.9	152.77	22AU	-1.1	-3.4	-1.4	0.4
178.60	0.62	1.9	162.54	1SE	-0.5	-3.4	-1.3	0.4
185.27	0.53	1.9	172.36	11SE	-0.2	-3.5	-1.3	0.4
192.03	0.44	1.8	182.25	21SE	-0.0	-3.5	-1.3	0.4
198.87	0.35	1.8	192.19	1OC	0.1	-3.5	-1.3	0.3
205.79	0.25	1.7	202.19	11OC	0.4	-3.5	-1.2	0.3
212.80	0.15	1.7	212.24	21OC	2.1	-3.4	-1.2	0.2
219.89	0.05	1.6	222.33	31OC	1.3	-3.4	-1.2	0.2
227.06	-0.06	1.6	232.47	10NO	-0.2	-3.4	-1.3	0.1
234.32	-0.16	1.6	242.64	20NO	-0.4	-3.4	-1.3	0.0
241.65	-0.27	1.6	252.83	30NO	-0.4	-3.3	-1.3	-0.1
249.06	-0.37	1.6	263.03	10DE	-0.5	-3.3	-1.3	-0.1
256.54	-0.48	1.6	273.23	20DE	-0.7	-3.3	-1.4	-0.2
264.08	-0.58	1.6	283.42	30DE	-1.0	-3.3	-1.4	-0.2

Left table (856 / 857):

♂ LONG	♂ LAT	♂ MAG	☉ LONG	Date 16.00UT	☿	♀	♃	♄
271.69	-0.68	1.5	293.59	9JA	-1.0	-3.4	-1.5	-0.1
279.34	-0.78	1.5	303.73	19JA	-0.7	-3.4	-1.6	-0.0
287.04	-0.88	1.5	313.82	29JA	0.6	-3.4	-1.6	0.0
294.78	-0.97	1.4	323.88	8FE	3.1	-3.4	-1.7	0.1
302.54	-1.05	1.4	333.88	18FE	1.6	-3.5	-1.8	0.2
310.31	-1.12	1.4	343.82	28FE	0.8	-3.5	-1.8	0.2
318.09	-1.19	1.4	353.71	9MR	0.5	-3.5	-1.9	0.3
325.86	-1.24	1.3	3.55	19MR	0.2	-3.6	-2.0	0.3
333.60	-1.28	1.3	13.32	29MR	-0.3	-3.7	-2.0	0.3
341.30	-1.31	1.3	23.04	8AP	-1.0	-3.7	-2.1	0.4
348.96	-1.32	1.3	32.71	18AP	-1.8	-3.8	-2.1	0.4
356.55	-1.32	1.3	42.33	28AP	-1.0	-3.9	-2.1	0.4
4.07	-1.31	1.2	51.92	8MY	-0.1	-4.0	-2.1	0.4
11.50	-1.28	1.2	61.49	18MY	0.7	-4.1	-2.1	0.4
18.83	-1.23	1.2	71.02	28MY	1.5	-4.1	-2.1	0.4
26.06	-1.17	1.2	80.55	7JN	2.6	-4.2	-2.0	0.4
33.15	-1.10	1.1	90.08	17JN	3.2	-4.1	-2.0	0.4
40.11	-1.01	1.1	99.61	27JN	1.6	-3.9	-1.9	0.3
46.91	-0.91	1.1	109.16	7JL	0.3	-3.4	-1.9	0.4
53.55	-0.79	1.0	118.73	17JL	-0.9	-3.1	-1.8	0.4
60.00	-0.66	1.0	128.33	27JL	-1.5	-3.8	-1.7	0.5
66.24	-0.51	0.9	137.98	6AU	-1.1	-4.1	-1.7	0.5
72.23	-0.34	0.8	147.67	16AU	-0.5	-4.2	-1.6	0.5
77.95	-0.16	0.8	157.40	26AU	-0.1	-4.2	-1.6	0.5
83.35	0.04	0.7	167.20	5SE	0.1	-4.2	-1.5	0.5
88.36	0.27	0.5	177.06	15SE	0.3	-4.1	-1.5	0.5
92.91	0.53	0.4	186.97	25SE	0.7	-4.0	-1.4	0.5
96.90	0.81	0.3	196.94	5OC	2.4	-3.9	-1.4	0.5
100.19	1.14	0.1	206.96	15OC	1.1	-3.8	-1.4	0.4
102.64	1.50	-0.1	217.03	25OC	-0.3	-3.8	-1.3	0.4
104.03	1.91	-0.3	227.15	4NO	-0.6	-3.7	-1.3	0.3
104.19	2.36	-0.5	237.31	14NO	-0.6	-3.6	-1.3	0.3
102.99	2.83	-0.8	247.48	24NO	-0.6	-3.6	-1.3	0.2
100.43	3.27	-1.0	257.68	4DE	-0.8	-3.5	-1.3	0.1
96.85	3.63	-1.1	267.88	14DE	-0.8	-3.5	-1.4	0.1
92.87	3.85	-1.1	278.08	24DE	-0.8	-3.4	-1.4	-0.0

857

♂ LONG	♂ LAT	♂ MAG	☉ LONG	Date 16.00UT	☿	♀	♃	♄
89.26	3.90	-0.9	288.26	3JA	-0.6	-3.4	-1.4	-0.1
86.65	3.82	-0.7	298.42	13JA	0.7	-3.4	-1.5	-0.0
85.32	3.65	-0.4	308.54	23JA	3.0	-3.4	-1.5	0.0
85.28	3.43	-0.1	318.62	2FE	1.2	-3.3	-1.6	0.1
86.42	3.19	0.1	328.65	12FE	0.6	-3.3	-1.6	0.1
88.51	2.96	0.4	338.62	22FE	0.3	-3.3	-1.7	0.2
91.38	2.74	0.6	348.54	4MR	-0.1	-3.3	-1.7	0.3
94.87	2.53	0.8	358.40	14MR	-0.3	-3.3	-1.8	0.3
98.86	2.34	1.0	8.20	24MR	-1.0	-3.3	-1.9	0.4
103.24	2.16	1.1	17.95	3AP	-1.7	-3.4	-2.0	0.4
107.95	2.00	1.2	27.64	13AP	-1.0	-3.4	-2.0	0.5
112.91	1.84	1.3	37.29	23AP	0.0	-3.4	-2.1	0.5
118.11	1.69	1.4	46.90	3MY	1.0	-3.5	-2.1	0.5
123.49	1.55	1.5	56.47	13MY	2.0	-3.5	-2.2	0.5
129.04	1.42	1.6	66.02	23MY	3.4	-3.4	-2.2	0.5
134.73	1.29	1.7	75.56	2JN	2.6	-3.4	-2.2	0.5
140.56	1.16	1.7	85.08	12JN	1.3	-3.4	-2.2	0.5
146.50	1.04	1.7	94.61	22JN	0.2	-3.4	-2.2	0.5
152.57	0.92	1.8	104.15	2JL	-1.0	-3.3	-2.2	0.5
158.75	0.80	1.8	113.70	12JL	-1.6	-3.3	-2.1	0.5
165.03	0.69	1.8	123.29	22JL	-1.1	-3.3	-2.1	0.5
171.43	0.57	1.8	132.91	1AU	-0.4	-3.3	-2.0	0.6
177.93	0.46	1.8	142.58	11AU	0.0	-3.3	-2.0	0.6
184.53	0.35	1.8	152.29	21AU	0.2	-3.3	-1.9	0.7
191.24	0.24	1.8	162.06	31AU	0.4	-3.4	-1.8	0.7
198.05	0.13	1.8	171.88	10SE	1.0	-3.4	-1.8	0.7
204.96	0.02	1.7	181.76	20SE	2.8	-3.4	-1.7	0.7
211.98	-0.09	1.7	191.70	30SE	1.0	-3.5	-1.7	0.7
219.10	-0.20	1.7	201.69	10OC	-0.5	-3.5	-1.6	0.7
226.32	-0.30	1.6	211.74	20OC	-0.7	-3.6	-1.6	0.7
233.64	-0.40	1.6	221.84	30OC	-0.7	-3.7	-1.5	0.6
241.04	-0.50	1.5	231.97	9NO	-0.7	-3.8	-1.5	0.6
248.54	-0.59	1.5	242.14	19NO	-0.7	-3.8	-1.5	0.5
256.11	-0.68	1.4	252.33	29NO	-0.7	-3.9	-1.4	0.5
263.76	-0.76	1.4	262.53	9DE	-0.7	-4.0	-1.4	0.4
271.47	-0.84	1.4	272.73	19DE	-0.5	-4.1	-1.4	0.3
279.24	-0.91	1.4	282.93	29DE	0.9	-4.2	-1.4	0.2

Right table (858 / 859):

♂ LONG	♂ LAT	♂ MAG	☉ LONG	Date 16.00UT	☿	♀	♃	♄
287.05	-0.97	1.4	293.09	8JA	2.6	-4.3	-1.4	0.2
294.90	-1.02	1.4	303.24	18JA	0.9	-4.4	-1.5	0.1
302.77	-1.06	1.4	313.34	28JA	0.4	-4.3	-1.5	0.2
310.64	-1.09	1.4	323.39	7FE	0.2	-4.0	-1.5	0.2
318.51	-1.11	1.4	333.40	17FE	-0.0	-3.4	-1.6	0.3
326.37	-1.11	1.4	343.35	27FE	-0.4	-3.4	-1.6	0.3
334.19	-1.11	1.4	353.24	9MR	-1.0	-4.0	-1.7	0.4
341.98	-1.09	1.4	3.07	19MR	-1.6	-4.2	-1.7	0.4
349.71	-1.06	1.4	12.85	29MR	-1.0	-4.3	-1.8	0.5
357.38	-1.01	1.4	22.56	8AP	0.1	-4.2	-1.9	0.5
4.97	-0.96	1.4	32.24	18AP	1.3	-4.1	-1.9	0.6
12.49	-0.90	1.5	41.87	28AP	2.6	-4.0	-2.0	0.6
19.92	-0.82	1.5	51.46	8MY	3.4	-3.9	-2.1	0.7
27.26	-0.74	1.5	61.02	18MY	2.0	-3.8	-2.1	0.7
34.50	-0.64	1.5	70.56	28MY	1.0	-3.7	-2.2	0.7
41.63	-0.54	1.5	80.09	7JN	0.1	-3.6	-2.3	0.7
48.66	-0.43	1.5	89.62	17JN	-1.0	-3.6	-2.3	0.7
55.58	-0.32	1.5	99.15	27JN	-1.7	-3.5	-2.3	0.7
62.37	-0.20	1.5	108.69	7JL	-1.1	-3.5	-2.4	0.7
69.05	-0.07	1.5	118.27	17JL	-0.3	-3.4	-2.4	0.7
75.59	0.07	1.5	127.87	27JL	0.1	-3.4	-2.3	0.7
82.00	0.21	1.4	137.51	6AU	0.4	-3.4	-2.3	0.7
88.25	0.37	1.4	147.20	16AU	0.7	-3.4	-2.3	0.8
94.35	0.52	1.4	156.94	26AU	1.3	-3.4	-2.2	0.8
100.26	0.69	1.3	166.73	5SE	3.2	-3.4	-2.2	0.9
105.97	0.87	1.2	176.58	15SE	0.8	-3.4	-2.1	0.9
111.44	1.06	1.1	186.49	25SE	-0.6	-3.4	-2.0	0.9
116.65	1.27	1.1	196.45	5OC	-0.9	-3.4	-2.0	0.9
121.52	1.50	0.9	206.48	15OC	-0.9	-3.4	-1.9	0.9
126.01	1.75	0.8	216.54	25OC	-0.8	-3.4	-1.8	0.9
130.02	2.02	0.7	226.66	4NO	-0.6	-3.4	-1.8	0.9
133.43	2.32	0.5	236.81	14NO	-0.5	-3.4	-1.7	0.9
136.12	2.65	0.3	246.99	24NO	-0.5	-3.5	-1.7	0.8
137.92	3.02	0.1	257.18	4DE	-0.3	-3.5	-1.6	0.8
138.64	3.41	-0.2	267.38	14DE	1.1	-3.5	-1.6	0.7
138.10	3.80	-0.4	277.58	24DE	2.3	-3.5	-1.6	0.6

859

♂ LONG	♂ LAT	♂ MAG	☉ LONG	Date 16.00UT	☿	♀	♃	♄
136.24	4.17	-0.6	287.76	3JA	0.6	-3.4	-1.5	0.5
133.18	4.44	-0.9	297.92	13JA	0.2	-3.4	-1.5	0.5
129.36	4.57	-1.0	308.04	23JA	0.0	-3.4	-1.5	0.4
125.46	4.51	-0.9	318.12	2FE	-0.1	-3.4	-1.5	0.4
122.19	4.30	-0.7	328.16	12FE	-0.5	-3.3	-1.5	0.4
120.03	3.98	-0.5	338.13	22FE	-1.1	-3.3	-1.6	0.4
119.17	3.61	-0.2	348.06	4MR	-1.5	-3.3	-1.6	0.5
119.55	3.23	0.0	357.92	14MR	-1.0	-3.3	-1.6	0.5
121.03	2.87	0.2	7.73	24MR	0.2	-3.4	-1.7	0.6
123.40	2.54	0.4	17.47	3AP	1.6	-3.4	-1.7	0.7
126.52	2.24	0.6	27.18	13AP	3.4	-3.4	-1.8	0.7
130.23	1.96	0.8	36.82	23AP	2.6	-3.4	-1.8	0.8
134.42	1.71	0.9	46.43	3MY	1.5	-3.4	-1.9	0.8
139.02	1.48	1.0	56.01	13MY	0.8	-3.5	-1.9	0.9
143.95	1.27	1.1	65.56	23MY	-0.0	-3.5	-2.0	0.9
149.16	1.07	1.2	75.10	2JN	-1.0	-3.6	-2.1	0.9
154.62	0.89	1.3	84.63	12JN	-1.8	-3.7	-2.2	0.9
160.30	0.72	1.4	94.15	22JN	-1.1	-3.7	-2.2	1.0
166.17	0.56	1.4	103.69	2JL	-0.3	-3.8	-2.3	1.0
172.22	0.40	1.5	113.25	12JL	0.2	-3.9	-2.3	1.0
178.44	0.25	1.5	122.83	22JL	0.5	-4.0	-2.4	1.0
184.81	0.11	1.5	132.45	1AU	0.9	-4.1	-2.4	1.0
191.34	-0.02	1.5	142.12	11AU	1.7	-4.2	-2.4	1.0
198.00	-0.15	1.5	151.82	21AU	3.2	-4.3	-2.4	1.0
204.80	-0.27	1.5	161.59	31AU	0.7	-4.2	-2.4	1.1
211.74	-0.38	1.5	171.41	10SE	-0.7	-4.0	-2.4	1.1
218.80	-0.49	1.5	181.29	20SE	-1.0	-3.4	-2.4	1.2
225.98	-0.59	1.5	191.22	30SE	-1.0	-3.4	-2.3	1.2
233.28	-0.68	1.5	201.22	10OC	-0.7	-4.0	-2.2	1.2
240.68	-0.77	1.5	211.26	20OC	-0.5	-4.3	-2.2	1.2
248.18	-0.85	1.5	221.35	30OC	-0.4	-4.4	-2.1	1.2
255.78	-0.92	1.5	231.48	9NO	-0.4	-4.3	-2.0	1.2
263.46	-0.97	1.4	241.65	19NO	-0.1	-4.2	-2.0	1.2
271.21	-1.02	1.4	251.83	29NO	1.3	-4.2	-1.9	1.1
279.02	-1.06	1.4	262.04	9DE	2.0	-4.0	-1.8	1.1
286.87	-1.08	1.4	272.23	19DE	0.3	-3.9	-1.8	1.0
294.75	-1.10	1.4	282.43	29DE	0.0	-3.9	-1.7	1.0

Left table (860/861):

♂ LONG	LAT	MAG	☉ LONG	16.00UT 860	☿	♀	♃	♄
302.66	-1.10	1.4	292.60	8JA	-0.1	-3.8	-1.7	0.9
310.56	-1.09	1.3	302.74	18JA	-0.2	-3.7	-1.6	0.8
318.46	-1.06	1.3	312.85	28JA	-0.5	-3.6	-1.6	0.8
326.34	-1.03	1.3	322.91	7FE	-1.1	-3.5	-1.6	0.7
334.18	-0.99	1.3	332.91	17FE	-1.4	-3.5	-1.6	0.6
341.97	-0.93	1.3	342.87	27FE	-1.0	-3.4	-1.6	0.7
349.71	-0.87	1.4	352.76	8MR	0.3	-3.4	-1.6	0.7
357.38	-0.80	1.4	2.59	18MR	2.0	-3.4	-1.6	0.8
4.98	-0.72	1.4	12.38	28MR	3.4	-3.3	-1.6	0.8
12.50	-0.63	1.5	22.10	7AP	1.9	-3.3	-1.6	0.9
19.93	-0.54	1.5	31.77	17AP	1.1	-3.3	-1.7	0.9
27.28	-0.44	1.6	41.41	27AP	0.6	-3.3	-1.7	1.0
34.53	-0.34	1.6	51.00	7MY	-0.1	-3.3	-1.7	1.0
41.70	-0.24	1.6	60.56	17MY	-1.0	-3.3	-1.8	1.1
48.77	-0.13	1.7	70.11	27MY	-1.8	-3.3	-1.8	1.1
55.75	-0.02	1.7	79.63	6JN	-1.1	-3.4	-1.9	1.2
62.63	0.09	1.7	89.16	16JN	-0.2	-3.4	-2.0	1.2
69.43	0.20	1.7	98.69	26JN	0.3	-3.4	-2.0	1.2
76.13	0.32	1.7	108.23	6JL	0.7	-3.5	-2.1	1.3
82.75	0.43	1.8	117.80	16JL	1.2	-3.5	-2.2	1.3
89.27	0.55	1.8	127.40	26JL	2.2	-3.5	-2.2	1.3
95.69	0.67	1.8	137.04	5AU	2.8	-3.4	-2.3	1.3
102.02	0.79	1.7	146.72	15AU	0.6	-3.4	-2.4	1.3
108.25	0.91	1.7	156.46	25AU	-0.8	-3.4	-2.4	1.3
114.37	1.03	1.7	166.25	4SE	-1.2	-3.3	-2.4	1.3
120.38	1.16	1.6	176.10	14SE	-1.1	-3.3	-2.5	1.3
126.26	1.29	1.6	186.01	24SE	-0.6	-3.3	-2.5	1.3
131.99	1.43	1.5	195.97	4OC	-0.4	-3.3	-2.4	1.3
137.56	1.57	1.5	205.98	14OC	-0.2	-3.3	-2.4	1.3
142.93	1.72	1.4	216.05	24OC	-0.2	-3.3	-2.4	1.3
148.08	1.89	1.3	226.16	3NO	0.1	-3.4	-2.3	1.2
152.96	2.06	1.1	236.31	13NO	1.5	-3.4	-2.3	1.2
157.51	2.25	1.0	246.49	23NO	1.7	-3.4	-2.2	1.2
161.66	2.45	0.8	256.68	3DE	0.1	-3.5	-2.1	1.1
165.31	2.67	0.7	266.89	13DE	-0.2	-3.5	-2.0	1.1
168.34	2.90	0.4	277.09	23DE	-0.2	-3.6	-2.0	1.1

861

♂ LONG	LAT	MAG	☉ LONG	16.00UT	☿	♀	♃	♄
170.61	3.15	0.2	287.27	2JA	-0.3	-3.6	-1.9	1.0
171.94	3.41	-0.0	297.43	12JA	-0.6	-3.7	-1.8	1.0
172.14	3.67	-0.3	307.55	22JA	-1.1	-3.8	-1.8	0.9
171.08	3.89	-0.6	317.63	1FE	-1.2	-3.8	-1.7	0.9
168.76	4.03	-0.8	327.67	11FE	-0.9	-3.9	-1.7	0.8
165.39	4.04	-1.0	337.65	21FE	-0.4	-4.0	-1.6	0.8
161.54	3.89	-1.1	347.57	3MR	2.4	-4.1	-1.6	0.8
157.90	3.59	-1.0	357.44	13MR	2.6	-4.2	-1.6	0.9
155.10	3.19	-0.8	7.25	23MR	1.4	-4.2	-1.6	0.9
153.54	2.74	-0.6	17.00	2AP	0.8	-4.2	-1.6	1.0
153.27	2.29	-0.3	26.70	12AP	0.4	-4.1	-1.6	1.1
154.23	1.87	-0.1	36.36	22AP	-0.2	-3.6	-1.6	1.1
156.23	1.49	0.1	45.97	2MY	-1.0	-2.8	-1.6	1.2
159.10	1.15	0.2	55.55	12MY	-1.9	-3.4	-1.6	1.3
162.68	0.85	0.4	65.10	22MY	-1.1	-4.0	-1.6	1.3
166.86	0.57	0.5	74.63	1JN	-0.2	-4.2	-1.6	1.3
171.52	0.33	0.7	84.16	11JN	0.5	-4.2	-1.7	1.3
176.59	0.11	0.8	93.69	21JN	1.0	-4.1	-1.7	1.4
182.01	-0.09	0.8	103.22	1JL	1.6	-4.1	-1.8	1.4
187.74	-0.27	0.9	112.78	11JL	2.8	-4.0	-1.8	1.3
193.74	-0.43	1.0	122.37	21JL	2.4	-3.9	-1.9	1.3
199.97	-0.58	1.0	131.98	31JL	0.5	-3.8	-2.0	1.3
206.42	-0.71	1.1	141.64	10AU	-0.8	-3.7	-2.0	1.3
213.07	-0.83	1.1	151.35	20AU	-1.3	-3.6	-2.1	1.2
219.90	-0.93	1.2	161.11	30AU	-1.1	-3.6	-2.2	1.2
226.90	-1.02	1.2	170.93	9SE	-0.6	-3.5	-2.2	1.1
234.05	-1.10	1.2	180.81	19SE	-0.3	-3.5	-2.3	1.1
241.33	-1.16	1.2	190.74	29SE	-0.1	-3.5	-2.3	1.1
248.74	-1.21	1.3	200.73	9OC	-0.0	-3.4	-2.4	1.1
256.27	-1.25	1.3	210.77	19OC	0.3	-3.4	-2.4	1.1
263.88	-1.27	1.3	220.86	29OC	1.8	-3.4	-2.4	1.1
271.58	-1.28	1.3	230.99	8NO	1.5	-3.4	-2.4	1.1
279.35	-1.27	1.3	241.16	18NO	-0.1	-3.4	-2.4	1.1
287.16	-1.25	1.3	251.34	28NO	-0.3	-3.4	-2.3	1.0
295.01	-1.22	1.4	261.54	8DE	-0.3	-3.4	-2.3	1.0
302.87	-1.17	1.4	271.74	18DE	-0.4	-3.4	-2.2	1.0
310.74	-1.12	1.4	281.93	28DE	-0.7	-3.4	-2.1	0.9

Right table (862/863):

♂ LONG	LAT	MAG	☉ LONG	16.00UT 862	☿	♀	♃	♄
318.60	-1.05	1.4	292.11	7JA	-1.0	-3.4	-2.1	0.9
326.43	-0.98	1.4	302.25	17JA	-1.1	-3.4	-2.0	0.8
334.21	-0.90	1.4	312.35	27JA	-0.8	-3.4	-1.9	0.8
341.95	-0.81	1.5	322.42	6FE	0.5	-3.4	-1.8	0.7
349.63	-0.71	1.5	332.43	16FE	2.8	-3.5	-1.8	0.7
357.24	-0.61	1.5	342.38	26FE	2.0	-3.5	-1.7	0.7
4.78	-0.51	1.5	352.28	8MR	1.0	-3.5	-1.7	0.6
12.24	-0.41	1.5	2.11	18MR	0.6	-3.4	-1.6	0.6
19.61	-0.30	1.6	11.89	28MR	0.3	-3.4	-1.6	0.7
26.90	-0.20	1.6	21.62	7AP	-0.2	-3.4	-1.5	0.8
34.11	-0.09	1.6	31.30	17AP	-1.0	-3.4	-1.5	0.8
41.23	0.01	1.6	40.93	27AP	-1.9	-3.3	-1.5	0.9
48.27	0.12	1.6	50.53	7MY	-1.1	-3.3	-1.5	1.0
55.22	0.22	1.7	60.09	17MY	-0.1	-3.3	-1.5	1.0
62.10	0.32	1.7	69.63	27MY	0.6	-3.3	-1.5	1.1
68.90	0.41	1.8	79.17	6JN	1.3	-3.3	-1.5	1.1
75.63	0.51	1.8	88.69	16JN	2.1	-3.4	-1.5	1.1
82.29	0.60	1.9	98.22	26JN	3.3	-3.4	-1.5	1.2
88.88	0.69	1.9	107.77	6JL	1.9	-3.4	-1.6	1.2
95.41	0.78	1.9	117.33	16JL	0.4	-3.5	-1.6	1.2
101.87	0.87	1.9	126.93	26JL	-0.9	-3.5	-1.6	1.2
108.28	0.95	1.9	136.57	5AU	-1.4	-3.6	-1.7	1.1
114.63	1.03	1.9	146.25	15AU	-1.1	-3.6	-1.7	1.1
120.92	1.11	1.9	155.99	25AU	-0.5	-3.7	-1.8	1.1
127.15	1.19	1.9	165.78	4SE	-0.2	-3.8	-1.9	1.0
133.32	1.27	1.9	175.62	14SE	0.0	-3.9	-1.9	1.0
139.42	1.35	1.9	185.52	24SE	0.1	-4.0	-2.0	1.0
145.46	1.42	1.8	195.48	4OC	0.5	-4.1	-2.0	1.0
151.41	1.50	1.8	205.50	14OC	2.1	-4.2	-2.1	1.0
157.28	1.58	1.7	215.56	24OC	1.3	-4.3	-2.2	1.0
163.05	1.65	1.6	225.68	3NO	-0.2	-4.4	-2.2	1.0
168.70	1.73	1.5	235.82	13NO	-0.5	-4.3	-2.2	1.0
174.21	1.81	1.4	246.00	23NO	-0.5	-4.0	-2.3	1.0
179.55	1.89	1.3	256.19	3DE	-0.6	-3.3	-2.3	0.9
184.69	1.97	1.2	266.39	13DE	-0.8	-3.3	-2.3	0.9
189.59	2.05	1.0	276.59	23DE	-0.9	-4.0	-2.2	0.9

863

♂ LONG	LAT	MAG	☉ LONG	16.00UT	☿	♀	♃	♄
194.19	2.13	0.9	286.78	2JA	-0.9	-4.3	-2.2	0.8
198.41	2.21	0.7	296.93	12JA	-0.7	-4.4	-2.1	0.8
202.18	2.29	0.5	307.06	22JA	0.6	-4.3	-2.1	0.8
205.37	2.35	0.2	317.15	1FE	3.1	-4.2	-2.0	0.7
207.84	2.41	-0.0	327.18	11FE	1.5	-4.1	-1.9	0.7
209.43	2.43	-0.3	337.17	21FE	0.7	-4.0	-1.9	0.6
209.94	2.41	-0.6	347.10	3MR	0.4	-3.9	-1.8	0.5
209.24	2.33	-0.9	356.96	13MR	0.1	-3.8	-1.7	0.5
207.27	2.16	-1.2	6.78	23MR	-0.3	-3.7	-1.7	0.5
204.22	1.88	-1.5	16.53	2AP	-1.0	-3.6	-1.6	0.5
200.61	1.49	-1.6	26.23	12AP	-1.8	-3.5	-1.6	0.6
197.11	1.03	-1.4	35.89	22AP	-1.1	-3.5	-1.5	0.7
194.43	0.56	-1.3	45.51	2MY	-0.1	-3.4	-1.5	0.7
192.97	0.10	-1.1	55.08	12MY	0.8	-3.4	-1.4	0.8
192.86	-0.30	-0.8	64.64	22MY	1.6	-3.4	-1.4	0.8
194.04	-0.64	-0.6	74.18	1JN	2.8	-3.3	-1.4	0.9
196.34	-0.92	-0.5	83.70	11JN	3.0	-3.3	-1.4	0.9
199.58	-1.16	-0.3	93.23	21JN	1.5	-3.3	-1.4	1.0
203.59	-1.35	-0.1	102.77	1JL	0.3	-3.3	-1.4	1.0
208.24	-1.51	0.0	112.32	11JL	-0.9	-3.3	-1.4	1.0
213.43	-1.63	0.1	121.90	21JL	-1.5	-3.3	-1.4	1.0
219.06	-1.73	0.3	131.52	31JL	-1.1	-3.4	-1.4	1.0
225.06	-1.80	0.4	141.17	10AU	-0.4	-3.4	-1.4	1.0
231.38	-1.85	0.4	150.88	20AU	-0.1	-3.4	-1.5	1.0
237.98	-1.87	0.5	160.64	30AU	0.2	-3.4	-1.5	1.0
244.81	-1.88	0.6	170.45	9SE	0.3	-3.5	-1.6	1.0
251.84	-1.86	0.7	180.33	19SE	0.8	-3.5	-1.6	0.9
259.04	-1.83	0.7	190.26	29SE	2.5	-3.5	-1.7	0.9
266.38	-1.77	0.8	200.24	9OC	1.1	-3.5	-1.7	0.9
273.83	-1.71	0.9	210.28	19OC	-0.4	-3.4	-1.8	0.9
281.38	-1.63	0.9	220.37	29OC	-0.6	-3.4	-1.9	0.9
288.99	-1.53	1.0	230.49	8NO	-0.6	-3.4	-1.9	0.9
296.66	-1.43	1.1	240.66	18NO	-0.7	-3.4	-2.0	0.9
304.35	-1.32	1.1	250.84	28NO	-0.8	-3.3	-2.0	0.9
312.04	-1.20	1.2	261.04	8DE	-0.7	-3.3	-2.1	0.9
319.73	-1.07	1.2	271.24	18DE	-0.8	-3.3	-2.1	0.9
327.40	-0.95	1.3	281.44	28DE	-0.6	-3.3	-2.1	0.8

♂ LONG	♂ LAT	♂ MAG	☉ LONG	16.00UT 864	☿	♀	♃	♄
					MAGNITUDES			
335.03	-0.82	1.4	291.61	7JA	0.7	-3.4	-2.1	0.8
342.61	-0.69	1.4	301.76	17JA	2.9	-3.4	-2.1	0.8
350.14	-0.56	1.5	311.87	27JA	1.1	-3.4	-2.1	0.7
357.60	-0.43	1.5	321.93	6FE	0.5	-3.4	-2.1	0.7
4.99	-0.30	1.6	331.94	16FE	0.3	-3.5	-2.0	0.6
12.31	-0.18	1.6	341.90	26FE	0.0	-3.5	-1.9	0.6
19.55	-0.06	1.7	351.80	7MR	-0.4	-3.5	-1.9	0.5
26.71	0.05	1.7	1.64	17MR	-1.0	-3.6	-1.8	0.5
33.80	0.16	1.7	11.43	27MR	-1.7	-3.7	-1.7	0.4
40.81	0.26	1.8	21.15	6AP	-1.1	-3.7	-1.7	0.4
47.74	0.36	1.8	30.84	16AP	0.0	-3.8	-1.6	0.4
54.61	0.45	1.8	40.47	26AP	1.0	-3.9	-1.5	0.5
61.41	0.54	1.8	50.07	6MY	2.2	-4.0	-1.5	0.6
68.14	0.62	1.8	59.64	16MY	3.6	-4.1	-1.4	0.6
74.81	0.69	1.8	69.18	26MY	2.4	-4.1	-1.4	0.7
81.43	0.77	1.8	78.71	5JN	1.2	-4.2	-1.4	0.8
88.00	0.83	1.8	88.23	15JN	0.2	-4.1	-1.3	0.8
94.52	0.90	1.9	97.76	25JN	-0.9	-3.9	-1.3	0.8
101.00	0.96	1.9	107.31	5JL	-1.6	-3.3	-1.3	0.9
107.45	1.01	2.0	116.87	15JL	-1.1	-3.1	-1.3	0.9
113.86	1.06	2.0	126.47	25JL	-0.4	-3.8	-1.3	0.9
120.24	1.11	2.0	136.10	4AU	0.0	-4.1	-1.3	0.9
126.60	1.15	2.0	145.79	14AU	0.3	-4.2	-1.3	0.9
132.93	1.19	2.0	155.51	24AU	0.5	-4.2	-1.3	0.9
139.24	1.22	2.0	165.30	3SE	1.1	-4.2	-1.3	0.9
145.53	1.26	2.0	175.14	13SE	2.9	-4.1	-1.4	0.9
151.79	1.28	2.0	185.04	23SE	1.0	-4.0	-1.4	0.8
158.04	1.31	2.0	195.00	3OC	-0.5	-3.9	-1.4	0.8
164.27	1.33	1.9	205.01	13OC	-0.8	-3.8	-1.5	0.8
170.47	1.35	1.9	215.07	23OC	-0.8	-3.8	-1.5	0.8
176.64	1.36	1.8	225.18	2NO	-0.8	-3.7	-1.6	0.8
182.78	1.37	1.8	235.33	12NO	-0.7	-3.6	-1.7	0.8
188.88	1.37	1.7	245.50	22NO	-0.6	-3.6	-1.7	0.8
194.93	1.37	1.6	255.70	2DE	-0.6	-3.5	-1.8	0.8
200.92	1.35	1.5	265.90	12DE	-0.4	-3.5	-1.9	0.8
206.85	1.33	1.4	276.10	22DE	0.9	-3.4	-1.9	0.8
				865				
212.69	1.30	1.3	286.28	1JA	2.6	-3.4	-2.0	0.8
218.42	1.26	1.2	296.45	11JA	0.8	-3.4	-2.0	0.8
224.03	1.20	1.0	306.57	21JA	0.3	-3.4	-2.0	0.7
229.49	1.13	0.8	316.66	31JA	0.1	-3.3	-2.0	0.7
234.75	1.03	0.7	326.70	10FE	-0.1	-3.3	-2.0	0.6
239.78	0.91	0.5	336.69	20FE	-0.4	-3.3	-2.0	0.6
244.51	0.75	0.3	346.62	2MR	-1.0	-3.3	-2.0	0.5
248.87	0.55	0.0	356.49	12MR	-1.6	-3.3	-1.9	0.5
252.76	0.29	-0.2	6.30	22MR	-1.1	-3.3	-1.9	0.4
256.05	-0.02	-0.5	16.06	1AP	0.1	-3.4	-1.8	0.4
258.58	-0.41	-0.8	25.77	11AP	1.3	-3.4	-1.8	0.3
260.19	-0.88	-1.1	35.42	21AP	2.9	-3.4	-1.7	0.3
260.69	-1.45	-1.5	45.04	1MY	3.1	-3.5	-1.6	0.4
259.95	-2.10	-1.8	54.62	11MY	1.8	-3.5	-1.6	0.4
258.03	-2.79	-2.1	64.17	21MY	0.9	-3.4	-1.5	0.5
255.24	-3.44	-2.3	73.71	31MY	0.1	-3.4	-1.5	0.6
252.20	-3.98	-2.3	83.24	10JN	-0.9	-3.4	-1.4	0.6
249.66	-4.34	-2.1	92.76	20JN	-1.7	-3.4	-1.4	0.7
248.18	-4.50	-1.9	102.30	30JN	-1.1	-3.3	-1.3	0.7
248.06	-4.51	-1.7	111.85	10JL	-0.3	-3.3	-1.3	0.8
249.30	-4.41	-1.4	121.43	20JL	0.1	-3.3	-1.3	0.8
251.77	-4.24	-1.2	131.05	30JL	0.4	-3.3	-1.2	0.8
255.27	-4.02	-1.0	140.70	9AU	0.7	-3.3	-1.2	0.9
259.61	-3.77	-0.8	150.40	19AU	1.4	-3.3	-1.2	0.9
264.61	-3.50	-0.6	160.16	29AU	3.2	-3.4	-1.2	0.9
270.16	-3.23	-0.4	169.97	8SE	0.8	-3.4	-1.2	0.9
276.13	-2.95	-0.3	179.84	18SE	-0.6	-3.4	-1.2	0.8
282.42	-2.67	-0.1	189.77	28SE	-0.9	-3.5	-1.3	0.8
288.98	-2.40	0.1	199.75	8OC	-0.9	-3.5	-1.3	0.8
295.73	-2.13	0.2	209.79	18OC	-0.8	-3.6	-1.3	0.7
302.63	-1.87	0.3	219.88	28OC	-0.5	-3.7	-1.4	0.7
309.64	-1.61	0.5	230.01	7NO	-0.5	-3.8	-1.4	0.7
316.72	-1.37	0.6	240.16	17NO	-0.5	-3.8	-1.5	0.8
323.85	-1.14	0.7	250.35	27NO	-0.3	-3.9	-1.5	0.8
331.00	-0.93	0.9	260.55	7DE	1.1	-4.0	-1.6	0.8
338.15	-0.72	1.0	270.75	17DE	2.2	-4.1	-1.7	0.8
345.30	-0.53	1.1	280.95	27DE	0.5	-4.2	-1.7	0.8

♂ LONG	♂ LAT	♂ MAG	☉ LONG	16.00UT 866	☿	♀	♃	♄
					MAGNITUDES			
352.42	-0.35	1.2	291.12	6JA	0.1	-4.3	-1.8	0.8
359.50	-0.19	1.3	301.27	16JA	-0.0	-4.4	-1.8	0.8
6.54	-0.04	1.4	311.38	26JA	-0.2	-4.3	-1.9	0.7
13.54	0.10	1.5	321.44	5FE	-0.5	-4.0	-1.9	0.7
20.48	0.23	1.5	331.46	15FE	-1.0	-3.4	-2.0	0.7
27.38	0.35	1.6	341.42	25FE	-1.5	-3.5	-2.0	0.6
34.21	0.45	1.7	351.32	7MR	-1.0	-4.0	-2.0	0.6
40.99	0.55	1.7	1.16	17MR	0.2	-4.2	-2.0	0.5
47.73	0.64	1.8	10.95	27MR	1.7	-4.3	-2.0	0.4
54.40	0.71	1.8	20.68	6AP	3.6	-4.2	-1.9	0.4
61.03	0.79	1.9	30.37	16AP	2.3	-4.1	-1.9	0.3
67.62	0.85	1.9	40.01	26AP	1.3	-4.0	-1.8	0.3
74.16	0.91	1.9	49.60	6MY	0.7	-3.9	-1.8	0.3
80.66	0.96	2.0	59.17	16MY	-0.0	-3.8	-1.7	0.4
87.13	1.00	2.0	68.72	26MY	-0.9	-3.7	-1.7	0.4
93.58	1.04	2.0	78.25	5JN	-1.8	-3.6	-1.6	0.5
99.99	1.07	2.0	87.77	15JN	-1.1	-3.6	-1.5	0.5
106.39	1.10	2.0	97.30	25JN	-0.3	-3.5	-1.5	0.6
112.78	1.12	2.0	106.84	5JL	0.2	-3.5	-1.4	0.7
119.15	1.14	2.0	116.41	15JL	0.6	-3.4	-1.4	0.7
125.52	1.15	2.0	126.01	25JL	1.0	-3.4	-1.3	0.8
131.88	1.16	2.0	135.64	4AU	1.9	-3.4	-1.3	0.8
138.25	1.16	2.0	145.32	14AU	3.1	-3.4	-1.3	0.8
144.63	1.16	2.0	155.05	24AU	0.7	-3.4	-1.2	0.8
151.01	1.15	2.0	164.83	3SE	-0.7	-3.4	-1.2	0.8
157.41	1.14	2.0	174.67	13SE	-1.1	-3.4	-1.2	0.8
163.82	1.13	2.0	184.56	23SE	-1.1	-3.4	-1.2	0.8
170.25	1.11	2.0	194.52	3OC	-0.7	-3.4	-1.2	0.8
176.69	1.08	2.0	204.53	13OC	-0.4	-3.4	-1.2	0.8
183.16	1.05	2.0	214.59	23OC	-0.3	-3.4	-1.2	0.7
189.64	1.02	1.9	224.69	2NO	-0.3	-3.4	-1.3	0.7
196.15	0.97	1.9	234.84	12NO	-0.1	-3.4	-1.3	0.7
202.67	0.92	1.8	245.01	22NO	1.3	-3.5	-1.3	0.7
209.21	0.86	1.8	255.20	2DE	1.9	-3.5	-1.4	0.7
215.76	0.79	1.7	265.41	12DE	0.3	-3.5	-1.4	0.8
222.33	0.71	1.6	275.60	22DE	-0.1	-3.5	-1.5	0.8
				867				
228.91	0.62	1.5	285.79	1JA	-0.1	-3.4	-1.6	0.8
235.50	0.51	1.4	295.95	11JA	-0.3	-3.4	-1.6	0.8
242.10	0.39	1.3	306.08	21JA	-0.6	-3.4	-1.7	0.8
248.69	0.26	1.2	316.17	31JA	-1.1	-3.4	-1.8	0.7
255.28	0.11	1.1	326.22	10FE	-1.3	-3.3	-1.8	0.7
261.85	-0.06	1.0	336.20	20FE	-1.0	-3.3	-1.9	0.7
268.41	-0.26	0.9	346.13	2MR	0.3	-3.3	-1.9	0.6
274.93	-0.48	0.7	356.01	12MR	2.1	-3.3	-2.0	0.6
281.40	-0.72	0.6	5.83	22MR	3.1	-3.4	-2.0	0.5
287.79	-0.99	0.4	15.59	1AP	1.7	-3.4	-2.0	0.5
294.10	-1.29	0.2	25.30	11AP	1.0	-3.4	-2.0	0.4
300.27	-1.62	0.0	34.96	21AP	0.5	-3.4	-2.0	0.3
306.26	-1.99	-0.1	44.58	1MY	-0.1	-3.4	-2.0	0.3
312.02	-2.40	-0.3	54.16	11MY	-1.0	-3.5	-2.0	0.2
317.47	-2.84	-0.6	63.72	21MY	-1.8	-3.5	-1.9	0.3
322.52	-3.32	-0.8	73.25	31MY	-1.1	-3.6	-1.9	0.3
327.04	-3.84	-1.0	82.78	10JN	-0.2	-3.7	-1.8	0.4
330.86	-4.39	-1.3	92.31	20JN	0.4	-3.7	-1.7	0.5
333.81	-4.96	-1.5	101.84	30JN	0.8	-3.8	-1.7	0.5
335.66	-5.52	-1.8	111.40	10JL	1.3	-3.9	-1.6	0.6
336.25	-6.02	-2.1	120.97	20JL	2.4	-4.0	-1.5	0.6
335.48	-6.39	-2.3	130.58	30JL	2.7	-4.1	-1.5	0.7
333.51	-6.53	-2.5	140.24	9AU	0.6	-4.2	-1.4	0.7
330.84	-6.38	-2.6	149.94	19AU	-0.8	-4.3	-1.4	0.8
328.19	-5.92	-2.5	159.69	29AU	-1.2	-4.2	-1.4	0.8
326.29	-5.24	-2.2	169.50	8SE	-1.1	-3.9	-1.3	0.8
325.59	-4.47	-1.9	179.36	18SE	-0.6	-3.4	-1.3	0.8
326.20	-3.69	-1.6	189.29	28SE	-0.3	-3.4	-1.3	0.8
328.02	-2.96	-1.3	199.27	8OC	-0.2	-4.0	-1.3	0.8
330.87	-2.32	-1.0	209.31	18OC	-0.1	-4.3	-1.3	0.8
334.54	-1.77	-0.7	219.39	28OC	0.1	-4.4	-1.3	0.7
338.86	-1.29	-0.4	229.52	7NO	1.5	-4.3	-1.3	0.7
343.68	-0.88	-0.2	239.67	17NO	1.7	-4.2	-1.3	0.7
348.87	-0.54	0.1	249.86	27NO	0.1	-4.1	-1.3	0.7
354.36	-0.24	0.3	260.06	7DE	-0.2	-4.0	-1.3	0.7
0.07	0.01	0.5	270.26	17DE	-0.3	-3.9	-1.4	0.8
5.94	0.22	0.6	280.45	27DE	-0.4	-3.8	-1.4	0.8

♂ LONG	♂ LAT	♂ MAG	☉ LONG	16.00UT 868	☿	♀	♃	♄
					MAGNITUDES			
11.95	0.40	0.8	290.63	6JA	-0.6	-3.8	-1.5	0.8
18.04	0.55	1.0	300.78	16JA	-1.1	-3.7	-1.5	0.8
24.20	0.68	1.1	310.89	26JA	-1.2	-3.6	-1.6	0.8
30.41	0.79	1.2	320.96	5FE	-0.9	-3.5	-1.7	0.8
36.65	0.89	1.4	330.98	15FE	0.4	-3.5	-1.7	0.7
42.91	0.96	1.5	340.94	25FE	2.5	-3.4	-1.8	0.7
49.18	1.03	1.6	350.85	6MR	2.4	-3.4	-1.9	0.7
55.45	1.08	1.6	0.69	16MR	1.3	-3.4	-1.9	0.6
61.72	1.12	1.7	10.49	26MR	0.7	-3.3	-2.0	0.6
68.00	1.16	1.8	20.22	5AP	0.4	-3.3	-2.0	0.5
74.26	1.19	1.9	29.90	15AP	-0.2	-3.3	-2.1	0.4
80.53	1.21	1.9	39.54	25AP	-1.0	-3.3	-2.1	0.4
86.79	1.22	1.9	49.15	5MY	-1.9	-3.3	-2.1	0.3
93.05	1.23	2.0	58.71	15MY	-1.1	-3.3	-2.1	0.3
99.31	1.23	2.0	68.26	25MY	-0.2	-3.3	-2.1	0.2
105.58	1.22	2.0	77.79	4JN	0.5	-3.4	-2.1	0.3
111.85	1.22	2.0	87.31	14JN	1.1	-3.4	-2.0	0.4
118.13	1.20	2.0	96.84	24JN	1.8	-3.4	-2.0	0.4
124.43	1.19	2.0	106.39	4JL	3.0	-3.5	-1.9	0.5
130.74	1.16	2.0	115.95	14JL	2.2	-3.5	-1.9	0.5
137.08	1.14	2.0	125.54	24JL	0.5	-3.5	-1.8	0.6
143.45	1.11	2.0	135.17	3AU	-0.8	-3.4	-1.7	0.6
149.84	1.07	2.0	144.84	13AU	-1.4	-3.4	-1.7	0.7
156.27	1.03	1.9	154.57	23AU	-1.1	-3.4	-1.6	0.7
162.74	0.99	1.9	164.35	2SE	-0.6	-3.3	-1.6	0.8
169.25	0.94	1.9	174.18	12SE	-0.2	-3.3	-1.5	0.8
175.81	0.89	1.9	184.08	22SE	-0.1	-3.3	-1.5	0.8
182.41	0.83	1.9	194.03	2OC	0.0	-3.3	-1.4	0.8
189.06	0.77	1.9	204.03	12OC	0.3	-3.3	-1.4	0.8
195.76	0.70	1.9	214.09	22OC	1.8	-3.3	-1.4	0.8
202.51	0.63	1.9	224.20	1NO	1.5	-3.4	-1.4	0.7
209.32	0.55	1.9	234.34	11NO	-0.1	-3.4	-1.3	0.7
216.18	0.46	1.8	244.51	21NO	-0.4	-3.4	-1.3	0.7
223.10	0.37	1.8	254.71	1DE	-0.4	-3.5	-1.3	0.7
230.07	0.27	1.7	264.91	11DE	-0.5	-3.5	-1.4	0.7
237.10	0.16	1.7	275.11	21DE	-0.7	-3.6	-1.4	0.7
244.17	0.04	1.6	285.30	31DE	-1.0	-3.6	-1.4	0.8
				869				
251.31	-0.08	1.6	295.46	10JA	-1.0	-3.7	-1.4	0.8
258.49	-0.22	1.5	305.60	20JA	-0.8	-3.8	-1.5	0.8
265.71	-0.36	1.4	315.69	30JA	0.5	-3.8	-1.5	0.8
272.98	-0.50	1.4	325.73	9FE	2.9	-3.9	-1.6	0.8
280.29	-0.66	1.3	335.73	19FE	1.8	-4.0	-1.6	0.8
287.63	-0.82	1.2	345.66	1MR	0.9	-4.1	-1.7	0.7
294.99	-0.98	1.1	355.54	11MR	0.5	-4.2	-1.8	0.7
302.37	-1.15	1.0	5.36	21MR	0.2	-4.2	-1.8	0.7
309.75	-1.32	0.9	15.12	31MR	-0.2	-4.2	-1.9	0.6
317.12	-1.49	0.9	24.83	10AP	-1.0	-4.1	-2.0	0.6
324.46	-1.65	0.8	34.50	20AP	-1.9	-3.6	-2.0	0.5
331.76	-1.82	0.7	44.12	30AP	-1.1	-2.8	-2.1	0.4
338.99	-1.97	0.6	53.70	10MY	-0.1	-3.5	-2.2	0.4
346.12	-2.12	0.5	63.26	20MY	0.7	-4.0	-2.2	0.3
353.13	-2.25	0.4	72.79	30MY	1.4	-4.2	-2.2	0.3
359.98	-2.37	0.3	82.32	9JN	2.4	-4.2	-2.2	0.3
6.63	-2.48	0.2	91.85	19JN	3.3	-4.1	-2.2	0.3
13.03	-2.56	0.0	101.38	29JN	1.8	-4.0	-2.2	0.4
19.12	-2.63	-0.1	110.93	9JL	0.4	-4.0	-2.2	0.4
24.83	-2.67	-0.2	120.51	19JL	-0.9	-3.9	-2.1	0.5
30.07	-2.68	-0.4	130.12	29JL	-1.5	-3.8	-2.1	0.6
34.73	-2.67	-0.5	139.77	8AU	-1.1	-3.7	-2.0	0.6
38.65	-2.62	-0.7	149.47	18AU	-0.5	-3.6	-2.0	0.7
41.67	-2.52	-0.9	159.22	28AU	-0.1	-3.6	-1.9	0.7
43.58	-2.36	-1.1	169.02	7SE	0.1	-3.5	-1.8	0.7
44.16	-2.13	-1.3	178.89	17SE	0.2	-3.5	-1.8	0.8
43.29	-1.80	-1.6	188.81	27SE	0.6	-3.5	-1.7	0.8
41.00	-1.38	-1.8	198.78	7OC	2.2	-3.4	-1.7	0.8
37.67	-0.88	-1.9	208.82	17OC	1.3	-3.4	-1.6	0.8
34.02	-0.34	-1.9	218.90	27OC	-0.3	-3.4	-1.6	0.8
30.86	0.16	-1.6	229.02	6NO	-0.5	-3.4	-1.5	0.8
28.79	0.60	-1.3	239.18	16NO	-0.5	-3.4	-1.5	0.7
28.08	0.94	-0.9	249.36	26NO	-0.6	-3.4	-1.5	0.7
28.66	1.19	-0.6	259.56	6DE	-0.8	-3.4	-1.5	0.7
30.39	1.38	-0.3	269.76	16DE	-0.8	-3.4	-1.4	0.7
33.05	1.50	-0.0	279.96	26DE	-0.9	-3.4	-1.4	0.7

♂ LONG	♂ LAT	♂ MAG	☉ LONG	16.00UT 870	☿	♀	♃	♄
					MAGNITUDES			
36.43	1.59	0.2	290.14	5JA	-0.7	-3.4	-1.5	0.8
40.39	1.64	0.4	300.29	15JA	0.6	-3.4	-1.5	0.8
44.79	1.68	0.7	310.40	25JA	3.0	-3.4	-1.5	0.8
49.53	1.69	0.8	320.47	4FE	1.3	-3.4	-1.5	0.8
54.54	1.69	1.0	330.49	14FE	0.6	-3.5	-1.6	0.8
59.77	1.69	1.2	340.46	24FE	0.4	-3.5	-1.6	0.8
65.16	1.67	1.3	350.37	6MR	0.1	-3.5	-1.6	0.8
70.69	1.65	1.4	0.22	16MR	-0.3	-3.4	-1.7	0.8
76.32	1.62	1.5	10.01	26MR	-1.0	-3.4	-1.8	0.7
82.04	1.59	1.6	19.75	5AP	-1.8	-3.4	-1.8	0.7
87.85	1.55	1.7	29.44	15AP	-1.1	-3.4	-1.9	0.6
93.71	1.51	1.8	39.07	25AP	-0.1	-3.3	-2.0	0.6
99.63	1.47	1.8	48.68	5MY	0.9	-3.3	-2.0	0.5
105.61	1.42	1.9	58.25	15MY	1.8	-3.3	-2.1	0.4
111.63	1.37	1.9	67.79	25MY	3.1	-3.3	-2.2	0.4
117.70	1.32	2.0	77.33	4JN	2.8	-3.3	-2.2	0.3
123.82	1.27	2.0	86.85	14JN	1.4	-3.4	-2.3	0.3
129.99	1.21	2.0	96.38	24JN	0.3	-3.4	-2.3	0.3
136.21	1.15	2.0	105.92	4JL	-0.9	-3.4	-2.4	0.4
142.48	1.09	2.0	115.48	14JL	-1.6	-3.5	-2.4	0.4
148.80	1.02	2.0	125.07	24JL	-1.1	-3.5	-2.4	0.5
155.18	0.96	2.0	134.71	3AU	-0.4	-3.6	-2.3	0.5
161.62	0.89	2.0	144.38	13AU	-0.0	-3.6	-2.3	0.6
168.11	0.81	1.9	154.10	23AU	0.2	-3.7	-2.3	0.6
174.68	0.74	1.9	163.88	2SE	0.4	-3.8	-2.2	0.7
181.32	0.66	1.9	173.71	12SE	0.9	-3.9	-2.2	0.7
188.02	0.57	1.8	183.60	22SE	2.6	-4.0	-2.1	0.8
194.80	0.49	1.8	193.55	2OC	1.1	-4.1	-2.0	0.8
201.65	0.40	1.7	203.55	12OC	-0.4	-4.2	-1.9	0.8
208.57	0.30	1.7	213.60	22OC	-0.7	-4.3	-1.9	0.8
215.58	0.21	1.7	223.71	1NO	-0.7	-4.4	-1.8	0.8
222.66	0.11	1.7	233.85	11NO	-0.7	-4.3	-1.8	0.8
229.81	0.00	1.7	244.01	21NO	-0.7	-4.0	-1.7	0.8
237.04	-0.10	1.7	254.21	1DE	-0.7	-3.3	-1.7	0.8
244.34	-0.21	1.7	264.41	11DE	-0.7	-3.4	-1.6	0.7
251.71	-0.32	1.6	274.61	21DE	-0.5	-4.0	-1.6	0.7
259.15	-0.44	1.6	284.80	31DE	0.7	-4.3	-1.6	0.8
				871				
266.65	-0.55	1.6	294.96	10JA	2.8	-4.4	-1.5	0.8
274.20	-0.66	1.5	305.10	20JA	1.0	-4.3	-1.5	0.8
281.81	-0.77	1.5	315.20	30JA	0.4	-4.2	-1.5	0.8
289.45	-0.88	1.4	325.24	9FE	0.2	-4.1	-1.5	0.9
297.13	-0.99	1.4	335.24	19FE	-0.0	-4.0	-1.6	0.9
304.83	-1.09	1.4	345.18	1MR	-0.4	-3.9	-1.6	0.9
312.54	-1.18	1.3	355.06	11MR	-1.0	-3.8	-1.6	0.9
320.25	-1.26	1.3	4.88	21MR	-1.7	-3.7	-1.6	0.8
327.95	-1.33	1.2	14.65	31MR	-1.1	-3.6	-1.7	0.8
335.62	-1.40	1.2	24.37	10AP	0.0	-3.5	-1.7	0.8
343.25	-1.44	1.2	34.03	20AP	1.1	-3.5	-1.8	0.7
350.83	-1.48	1.1	43.66	30AP	2.4	-3.4	-1.8	0.7
358.33	-1.50	1.1	53.24	10MY	3.6	-3.4	-1.9	0.6
5.75	-1.50	1.1	62.80	20MY	2.1	-3.4	-2.0	0.5
13.06	-1.48	1.0	72.34	30MY	1.1	-3.3	-2.0	0.5
20.26	-1.45	1.0	81.86	9JN	0.2	-3.3	-2.1	0.4
27.33	-1.40	0.9	91.39	19JN	-0.9	-3.3	-2.2	0.3
34.23	-1.34	0.9	100.93	29JN	-1.7	-3.3	-2.2	0.3
40.97	-1.26	0.8	110.47	9JL	-1.1	-3.3	-2.3	0.3
47.50	-1.16	0.8	120.05	19JL	-0.4	-3.3	-2.4	0.4
53.81	-1.04	0.7	129.66	29JL	0.1	-3.4	-2.4	0.5
59.85	-0.90	0.6	139.30	8AU	0.3	-3.4	-2.4	0.5
65.58	-0.73	0.5	149.00	18AU	0.6	-3.4	-2.4	0.6
70.95	-0.55	0.4	158.75	28AU	1.2	-3.4	-2.4	0.6
75.89	-0.34	0.3	168.55	7SE	3.0	-3.5	-2.4	0.7
80.30	-0.10	0.1	178.41	17SE	1.0	-3.5	-2.4	0.7
84.05	0.18	-0.0	188.33	27SE	-0.5	-3.5	-2.4	0.7
87.02	0.51	-0.2	198.30	7OC	-0.9	-3.5	-2.3	0.8
89.00	0.88	-0.4	208.33	17OC	-0.8	-3.4	-2.2	0.8
89.81	1.29	-0.6	218.41	27OC	-0.8	-3.4	-2.2	0.8
89.26	1.75	-0.8	228.52	6NO	-0.6	-3.4	-2.1	0.8
87.30	2.23	-1.0	238.68	16NO	-0.5	-3.4	-2.0	0.8
84.11	2.67	-1.2	248.86	26NO	-0.5	-3.3	-2.0	0.8
80.23	3.03	-1.3	259.06	6DE	-0.4	-3.3	-1.9	0.8
76.41	3.25	-1.2	269.26	16DE	0.9	-3.3	-1.8	0.8
73.38	3.32	-0.9	279.46	26DE	2.5	-3.3	-1.8	0.8

Left table

♂ LONG	♂ LAT	♂ MAG	☉ LONG	16.00UT 872	☿	♀	♃	♄
71.59	3.29	-0.6	289.63	5JA	0.7	-3.4	-1.7	0.8
71.12	3.18	-0.3	299.79	15JA	0.2	-3.4	-1.7	0.8
71.89	3.03	-0.0	309.90	25JA	0.1	-3.4	-1.6	0.9
73.70	2.87	0.2	319.98	4FE	-0.1	-3.4	-1.6	0.9
76.35	2.71	0.5	330.00	14FE	-0.4	-3.5	-1.6	0.9
79.68	2.55	0.7	339.97	24FE	-1.0	-3.5	-1.6	0.9
83.54	2.40	0.9	349.88	5MR	-1.5	-3.5	-1.6	0.9
87.82	2.26	1.0	359.74	15MR	-1.1	-3.6	-1.6	0.9
92.43	2.12	1.2	9.54	25MR	0.1	-3.7	-1.6	0.9
97.32	1.99	1.3	19.28	4AP	1.4	-3.7	-1.6	0.9
102.43	1.87	1.4	28.97	14AP	3.1	-3.8	-1.6	0.9
107.73	1.75	1.5	38.61	24AP	2.8	-3.9	-1.7	0.9
113.19	1.64	1.6	48.22	4MY	1.6	-4.0	-1.7	0.8
118.79	1.53	1.7	57.79	14MY	0.8	-4.1	-1.7	0.7
124.50	1.42	1.7	67.34	24MY	0.1	-4.1	-1.8	0.6
130.33	1.31	1.8	76.87	3JN	-0.9	-4.2	-1.8	0.6
136.26	1.21	1.8	86.40	13JN	-1.7	-4.1	-1.9	0.5
142.29	1.11	1.8	95.92	23JN	-1.1	-3.9	-2.0	0.4
148.42	1.00	1.9	105.46	3JL	-0.3	-3.3	-2.0	0.4
154.63	0.90	1.9	115.03	13JL	0.2	-3.2	-2.1	0.4
160.94	0.80	1.9	124.61	23JL	0.5	-3.8	-2.2	0.4
167.34	0.70	1.9	134.24	2AU	0.8	-4.1	-2.2	0.5
173.83	0.59	1.9	143.91	12AU	1.6	-4.2	-2.3	0.5
180.41	0.49	1.9	153.63	22AU	3.2	-4.2	-2.4	0.6
187.08	0.39	1.8	163.40	1SE	0.9	-4.2	-2.4	0.6
193.85	0.28	1.8	173.23	11SE	-0.6	-4.1	-2.4	0.7
200.71	0.18	1.8	183.12	21SE	-1.0	-4.0	-2.5	0.7
207.67	0.08	1.7	193.06	1OC	-1.0	-3.9	-2.5	0.8
214.72	-0.03	1.7	203.07	11OC	-0.8	-3.8	-2.4	0.8
221.87	-0.13	1.6	213.12	21OC	-0.5	-3.8	-2.4	0.8
229.10	-0.24	1.6	223.22	31OC	-0.4	-3.7	-2.4	0.9
236.43	-0.34	1.5	233.36	10NO	-0.4	-3.6	-2.3	0.9
243.84	-0.44	1.5	243.52	20NO	-0.2	-3.6	-2.2	0.9
251.34	-0.54	1.5	253.71	30NO	1.1	-3.5	-2.2	0.9
258.91	-0.63	1.5	263.92	10DE	2.2	-3.5	-2.1	0.9
266.54	-0.72	1.5	274.11	20DE	0.4	-3.4	-2.0	0.9
274.24	-0.80	1.5	284.30	30DE	0.0	-3.4	-1.9	0.9
				873				
281.99	-0.88	1.4	294.47	9JA	-0.1	-3.4	-1.9	0.8
289.77	-0.95	1.4	304.61	19JA	-0.2	-3.4	-1.8	0.9
297.59	-1.02	1.4	314.70	29JA	-0.5	-3.3	-1.8	0.9
305.43	-1.07	1.4	324.76	8FE	-1.0	-3.3	-1.7	1.0
313.28	-1.11	1.4	334.75	18FE	-1.4	-3.3	-1.7	1.0
321.12	-1.14	1.4	344.69	28FE	-1.0	-3.3	-1.6	1.0
328.94	-1.16	1.4	354.58	10MR	0.2	-3.3	-1.6	1.0
336.73	-1.17	1.4	4.40	20MR	1.8	-3.3	-1.6	1.0
344.48	-1.16	1.4	14.17	30MR	3.6	-3.4	-1.6	1.0
352.17	-1.14	1.4	23.89	9AP	2.1	-3.4	-1.5	1.0
359.80	-1.11	1.4	33.56	19AP	1.2	-3.4	-1.5	1.0
7.36	-1.06	1.4	43.18	29AP	0.6	-3.5	-1.6	0.9
14.83	-1.01	1.4	52.77	9MY	-0.0	-3.5	-1.6	0.9
22.21	-0.94	1.4	62.33	19MY	-0.9	-3.4	-1.6	0.8
29.49	-0.86	1.4	71.87	29MY	-1.8	-3.4	-1.6	0.8
36.66	-0.76	1.4	81.40	8JN	-1.1	-3.4	-1.6	0.7
43.72	-0.66	1.4	90.92	18JN	-0.3	-3.4	-1.7	0.6
50.66	-0.55	1.4	100.46	28JN	0.3	-3.3	-1.7	0.6
57.46	-0.43	1.3	110.01	8JL	0.7	-3.3	-1.8	0.5
64.13	-0.30	1.3	119.58	18JL	1.1	-3.3	-1.8	0.5
70.64	-0.16	1.3	129.19	28JL	2.1	-3.3	-1.9	0.5
77.00	-0.01	1.3	138.83	7AU	3.0	-3.3	-2.0	0.5
83.17	0.15	1.2	148.52	17AU	0.7	-3.3	-2.0	0.5
89.15	0.32	1.1	158.27	27AU	-0.7	-3.4	-2.1	0.6
94.89	0.51	1.1	168.07	6SE	-1.1	-3.4	-2.2	0.7
100.38	0.71	1.0	177.92	16SE	-1.1	-3.4	-2.2	0.7
105.56	0.93	0.9	187.84	26SE	-0.7	-3.5	-2.3	0.8
110.38	1.17	0.8	197.81	6OC	-0.4	-3.5	-2.3	0.8
114.76	1.44	0.6	207.84	16OC	-0.3	-3.6	-2.4	0.8
118.61	1.73	0.5	217.92	26OC	-0.2	-3.7	-2.4	0.9
121.80	2.06	0.3	228.04	5NO	-0.0	-3.8	-2.4	0.9
124.17	2.42	0.1	238.19	15NO	1.3	-3.9	-2.4	0.9
125.55	2.82	-0.1	248.37	25NO	1.9	-3.9	-2.3	1.0
125.74	3.25	-0.3	258.57	5DE	0.2	-4.0	-2.3	1.0
124.61	3.68	-0.6	268.77	15DE	-0.1	-4.2	-2.2	1.0
122.16	4.06	-0.8	278.96	25DE	-0.2	-4.3	-2.2	1.0

Right table

♂ LONG	♂ LAT	♂ MAG	☉ LONG	16.00UT 874	☿	♀	♃	♄
118.66	4.34	-1.0	289.14	4JA	-0.3	-4.3	-2.1	1.0
114.69	4.45	-1.0	299.29	14JA	-0.6	-4.4	-2.0	1.0
110.99	4.39	-0.8	309.42	24JA	-1.0	-4.3	-1.9	0.9
108.20	4.19	-0.6	319.49	3FE	-1.3	-4.0	-1.9	1.0
106.66	3.90	-0.3	329.51	13FE	-1.0	-3.4	-1.8	1.0
106.40	3.57	-0.1	339.49	23FE	0.2	-3.5	-1.7	1.1
107.34	3.24	0.2	349.40	5MR	2.2	-4.0	-1.7	1.1
109.26	2.93	0.4	359.26	15MR	2.9	-4.2	-1.6	1.1
111.99	2.64	0.6	9.06	25MR	1.5	-4.3	-1.6	1.1
115.38	2.37	0.8	18.80	4AP	0.9	-4.2	-1.5	1.1
119.30	2.13	0.9	28.49	14AP	0.4	-4.1	-1.5	1.1
123.64	1.91	1.0	38.14	24AP	-0.1	-4.0	-1.5	1.1
128.34	1.70	1.2	47.75	4MY	-0.9	-3.9	-1.5	1.1
133.33	1.51	1.3	57.32	14MY	-1.8	-3.8	-1.5	1.0
138.57	1.33	1.4	66.87	24MY	-1.1	-3.7	-1.5	1.0
144.03	1.16	1.4	76.41	3JN	-0.2	-3.6	-1.5	0.9
149.67	1.01	1.5	85.93	13JN	0.4	-3.6	-1.5	0.9
155.50	0.85	1.5	95.46	23JN	0.9	-3.5	-1.5	0.8
161.48	0.71	1.6	105.00	3JL	1.5	-3.5	-1.5	0.8
167.60	0.57	1.6	114.56	13JL	2.6	-3.4	-1.5	0.7
173.87	0.43	1.6	124.15	23JL	2.5	-3.4	-1.6	0.6
180.27	0.30	1.6	133.77	2AU	0.6	-3.4	-1.6	0.6
186.80	0.17	1.7	143.44	12AU	-0.8	-3.4	-1.7	0.6
193.46	0.05	1.7	153.16	22AU	-1.3	-3.4	-1.7	0.6
200.24	-0.07	1.6	162.93	1SE	-1.1	-3.4	-1.8	0.7
207.14	-0.18	1.6	172.75	11SE	-0.6	-3.4	-1.8	0.7
214.16	-0.29	1.6	182.64	21SE	-0.3	-3.4	-1.9	0.8
221.29	-0.40	1.6	192.58	1OC	-0.1	-3.4	-2.0	0.8
228.53	-0.50	1.6	202.58	11OC	-0.1	-3.4	-2.0	0.9
235.88	-0.60	1.6	212.63	21OC	0.2	-3.4	-2.1	0.9
243.33	-0.68	1.5	222.73	31OC	1.6	-3.4	-2.2	1.0
250.86	-0.77	1.5	232.86	10NO	1.7	-3.4	-2.2	1.0
258.49	-0.84	1.5	243.03	20NO	-0.0	-3.5	-2.2	1.0
266.18	-0.91	1.4	253.22	30NO	-0.3	-3.5	-2.3	1.1
273.95	-0.96	1.4	263.42	10DE	-0.3	-3.5	-2.3	1.1
281.76	-1.01	1.4	273.62	20DE	-0.4	-3.5	-2.2	1.1
289.62	-1.04	1.3	283.81	30DE	-0.6	-3.4	-2.2	1.1
				875				
297.50	-1.07	1.3	293.98	9JA	-1.0	-3.4	-2.2	1.1
305.41	-1.08	1.3	304.12	19JA	-1.1	-3.4	-2.1	1.1
313.31	-1.08	1.3	314.21	29JA	-0.9	-3.4	-2.0	1.1
321.19	-1.07	1.3	324.26	8FE	0.3	-3.3	-2.0	1.1
329.06	-1.05	1.3	334.27	18FE	2.6	-3.3	-1.9	1.2
336.89	-1.02	1.4	344.21	28FE	2.2	-3.3	-1.8	1.2
344.67	-0.97	1.4	354.10	10MR	1.1	-3.3	-1.8	1.2
352.39	-0.92	1.4	3.93	20MR	0.7	-3.4	-1.7	1.3
0.04	-0.85	1.4	13.70	30MR	0.3	-3.4	-1.6	1.3
7.63	-0.78	1.5	23.42	9AP	-0.2	-3.4	-1.6	1.3
15.13	-0.70	1.5	33.09	19AP	-0.9	-3.4	-1.5	1.3
22.54	-0.61	1.5	42.71	29AP	-1.8	-3.5	-1.5	1.3
29.86	-0.52	1.6	52.31	9MY	-1.1	-3.5	-1.4	1.3
37.09	-0.42	1.6	61.87	19MY	-0.2	-3.5	-1.4	1.2
44.22	-0.31	1.6	71.41	29MY	0.5	-3.6	-1.4	1.2
51.26	-0.20	1.6	80.94	8JN	1.2	-3.7	-1.4	1.1
58.20	-0.09	1.6	90.47	18JN	2.0	-3.7	-1.4	1.1
65.03	0.03	1.7	100.00	28JN	3.2	-3.8	-1.4	1.0
71.77	0.15	1.7	109.55	8JL	2.1	-3.9	-1.4	1.0
78.40	0.27	1.7	119.12	18JL	0.5	-4.0	-1.4	0.9
84.93	0.40	1.7	128.72	28JL	-0.8	-4.1	-1.4	0.8
91.35	0.53	1.6	138.37	7AU	-1.4	-4.2	-1.4	0.8
97.65	0.66	1.6	148.06	17AU	-1.1	-4.3	-1.4	0.7
103.83	0.80	1.6	157.80	27AU	-0.5	-4.2	-1.5	0.7
109.87	0.94	1.6	167.60	6SE	-0.2	-3.9	-1.5	0.8
115.77	1.09	1.5	177.45	16SE	-0.0	-3.4	-1.6	0.8
121.51	1.24	1.4	187.36	26SE	0.1	-3.5	-1.6	0.9
127.06	1.41	1.4	197.33	6OC	0.4	-4.0	-1.7	0.9
132.39	1.58	1.3	207.36	16OC	1.9	-4.3	-1.7	1.0
137.46	1.77	1.2	217.43	26OC	1.5	-4.4	-1.8	1.0
142.23	1.97	1.0	227.55	5NO	-0.2	-4.3	-1.9	1.1
146.62	2.19	0.9	237.70	15NO	-0.5	-4.2	-1.9	1.1
150.55	2.43	0.7	247.87	25NO	-0.5	-4.1	-2.0	1.1
153.92	2.70	0.6	258.07	5DE	-0.5	-4.0	-2.0	1.2
156.58	2.99	0.3	268.27	15DE	-0.7	-3.9	-2.1	1.2
158.38	3.30	0.1	278.47	25DE	-0.9	-3.8	-2.1	1.2

876 / 877

♂ LONG	LAT	MAG	☉ LONG	16.00UT	☿	♀	♃	♄
159.13	3.62	-0.1	288.65	4JA	-1.0	-3.8	-2.1	1.3
158.66	3.93	-0.4	298.81	14JA	-0.8	-3.7	-2.1	1.3
156.90	4.19	-0.6	308.93	24JA	0.4	-3.6	-2.1	1.3
153.94	4.36	-0.9	319.01	3FE	3.0	-3.5	-2.1	1.3
150.19	4.38	-1.0	329.03	13FE	1.6	-3.5	-2.0	1.3
146.31	4.22	-1.0	339.01	23FE	0.8	-3.4	-2.0	1.3
143.01	3.92	-0.8	348.93	4MR	0.5	-3.4	-1.9	1.3
140.80	3.52	-0.6	358.79	14MR	0.2	-3.4	-1.8	1.4
139.88	3.09	-0.3	8.59	24MR	-0.2	-3.3	-1.8	1.4
140.21	2.67	-0.1	18.34	3AP	-0.9	-3.3	-1.7	1.4
141.67	2.27	0.1	28.03	13AP	-1.8	-3.3	-1.6	1.4
144.06	1.91	0.3	37.68	23AP	-1.1	-3.3	-1.6	1.4
147.21	1.59	0.5	47.29	3MY	-0.1	-3.3	-1.5	1.4
151.00	1.30	0.6	56.86	13MY	0.7	-3.3	-1.5	1.3
155.30	1.03	0.8	66.41	23MY	1.5	-3.3	-1.4	1.3
160.03	0.79	0.9	75.95	2JN	2.6	-3.4	-1.3	1.2
165.12	0.58	1.0	85.47	12JN	3.2	-3.4	-1.3	1.2
170.52	0.37	1.0	95.00	22JN	1.7	-3.4	-1.3	1.2
176.20	0.19	1.1	104.54	2JL	0.4	-3.5	-1.3	1.1
182.11	0.02	1.2	114.09	12JL	-0.8	-3.5	-1.3	1.1
188.24	-0.14	1.2	123.68	22JL	-1.5	-3.5	-1.3	1.0
194.58	-0.29	1.3	133.31	1AU	-1.1	-3.4	-1.3	1.0
201.09	-0.42	1.3	142.97	11AU	-0.5	-3.4	-1.3	0.9
207.78	-0.55	1.3	152.68	21AU	-0.1	-3.4	-1.3	0.9
214.63	-0.66	1.3	162.45	31AU	0.1	-3.3	-1.3	0.9
221.62	-0.76	1.3	172.27	10SE	0.3	-3.3	-1.3	0.9
228.76	-0.86	1.4	182.16	20SE	0.7	-3.3	-1.4	1.0
236.03	-0.94	1.4	192.10	30SE	2.3	-3.3	-1.4	1.0
243.42	-1.01	1.4	202.09	10OC	1.3	-3.3	-1.4	1.1
250.92	-1.07	1.4	212.14	20OC	-0.3	-3.3	-1.5	1.1
258.52	-1.11	1.4	222.24	30OC	-0.6	-3.4	-1.6	1.2
266.20	-1.15	1.4	232.37	9NO	-0.6	-3.4	-1.6	1.2
273.95	-1.17	1.4	242.54	19NO	-0.6	-3.4	-1.7	1.3
281.77	-1.18	1.4	252.73	29NO	-0.8	-3.5	-1.7	1.3
289.62	-1.17	1.4	262.92	9DE	-0.8	-3.5	-1.8	1.3
297.50	-1.16	1.4	273.12	19DE	-0.8	-3.6	-1.9	1.4
305.40	-1.13	1.4	283.32	29DE	-0.6	-3.6	-1.9	1.4

877

♂ LONG	LAT	MAG	☉ LONG	16.00UT	☿	♀	♃	♄
313.29	-1.09	1.4	293.48	8JA	0.6	-3.7	-2.0	1.4
321.17	-1.04	1.4	303.63	18JA	3.0	-3.8	-2.0	1.4
329.02	-0.98	1.4	313.73	28JA	1.2	-3.8	-2.0	1.3
336.82	-0.91	1.4	323.78	7FE	0.5	-3.9	-2.0	1.3
344.58	-0.83	1.4	333.79	17FE	0.3	-4.0	-2.0	1.2
352.27	-0.74	1.4	343.74	27FE	0.1	-4.1	-2.0	1.2
359.89	-0.65	1.4	353.62	9MR	-0.3	-4.2	-2.0	1.2
7.43	-0.56	1.5	3.46	19MR	-1.0	-4.2	-1.9	1.2
14.90	-0.46	1.5	13.23	29MR	-1.7	-4.2	-1.9	1.2
22.28	-0.36	1.5	22.95	8AP	-1.1	-4.1	-1.8	1.2
29.57	-0.26	1.5	32.63	18AP	-0.1	-3.6	-1.7	1.1
36.78	-0.15	1.6	42.25	28AP	0.9	-2.8	-1.7	1.1
43.90	-0.05	1.6	51.84	8MY	2.0	-3.5	-1.6	1.1
50.93	0.06	1.7	61.41	18MY	3.5	-4.0	-1.6	1.1
57.88	0.16	1.7	70.95	28MY	2.5	-4.2	-1.5	1.0
64.74	0.27	1.8	80.47	7JN	1.3	-4.2	-1.4	1.0
71.52	0.37	1.8	90.00	17JN	0.2	-4.1	-1.4	1.0
78.23	0.47	1.8	99.53	27JN	-0.9	-4.0	-1.3	0.9
84.86	0.57	1.8	109.08	7JL	-1.6	-4.0	-1.3	0.9
91.41	0.67	1.9	118.65	17JL	-1.1	-3.9	-1.3	0.8
97.89	0.77	1.9	128.25	27JL	-0.4	-3.8	-1.2	0.8
104.30	0.87	1.9	137.89	6AU	-0.0	-3.7	-1.2	0.7
110.64	0.96	1.9	147.59	16AU	0.3	-3.6	-1.2	0.7
116.90	1.06	1.9	157.32	26AU	0.5	-3.6	-1.2	0.6
123.09	1.16	1.8	167.12	5SE	1.0	-3.5	-1.2	0.6
129.20	1.25	1.8	176.97	15SE	2.7	-3.5	-1.2	0.6
135.22	1.35	1.8	186.88	25SE	1.2	-3.5	-1.2	0.7
141.14	1.45	1.7	196.85	5OC	-0.5	-3.4	-1.3	0.7
146.96	1.55	1.7	206.87	15OC	-0.8	-3.4	-1.3	0.8
152.65	1.66	1.6	216.94	25OC	-0.7	-3.4	-1.3	0.9
158.20	1.76	1.5	227.05	4NO	-0.8	-3.4	-1.4	0.9
163.58	1.88	1.4	237.21	14NO	-0.7	-3.4	-1.4	1.0
168.75	1.99	1.3	247.38	24NO	-0.6	-3.4	-1.5	1.0
173.69	2.12	1.1	257.58	4DE	-0.6	-3.4	-1.5	1.1
178.32	2.24	1.0	267.78	14DE	-0.5	-3.4	-1.6	1.1
182.59	2.38	0.8	277.98	24DE	0.7	-3.4	-1.7	1.1

878 / 879

♂ LONG	LAT	MAG	☉ LONG	16.00UT	☿	♀	♃	♄
186.42	2.52	0.6	288.16	3JA	2.8	-3.4	-1.7	1.1
189.67	2.67	0.4	298.32	13JA	0.9	-3.4	-1.8	1.1
192.23	2.81	0.2	308.44	23JA	0.3	-3.4	-1.9	1.1
193.92	2.95	-0.1	318.52	2FE	0.2	-3.4	-1.9	1.1
194.56	3.07	-0.4	328.55	12FE	-0.0	-3.5	-2.0	1.0
193.99	3.14	-0.7	338.52	22FE	-0.4	-3.5	-2.0	1.0
192.16	3.13	-0.9	348.44	4MR	-1.0	-3.4	-2.0	0.9
189.18	3.01	-1.2	358.31	14MR	-1.6	-3.4	-2.0	0.9
185.51	2.76	-1.4	8.11	24MR	-1.1	-3.4	-2.0	0.9
181.83	2.39	-1.2	17.86	3AP	-0.0	-3.4	-2.0	0.9
178.82	1.94	-1.1	27.56	13AP	1.2	-3.4	-1.9	0.9
176.97	1.47	-0.9	37.21	23AP	2.6	-3.3	-1.9	0.9
176.45	1.02	-0.7	46.82	3MY	3.4	-3.3	-1.8	0.9
177.21	0.62	-0.5	56.40	13MY	1.9	-3.3	-1.8	0.9
179.11	0.26	-0.3	65.94	23MY	1.0	-3.3	-1.7	0.8
181.95	-0.06	-0.1	75.48	2JN	0.1	-3.3	-1.6	0.8
185.59	-0.33	0.1	85.01	12JN	-0.9	-3.4	-1.6	0.8
189.88	-0.56	0.2	94.53	22JN	-1.7	-3.4	-1.5	0.8
194.71	-0.76	0.3	104.07	2JL	-1.1	-3.4	-1.5	0.7
199.99	-0.93	0.4	113.63	12JL	-0.4	-3.5	-1.4	0.7
205.66	-1.08	0.5	123.21	22JL	0.1	-3.5	-1.4	0.6
211.65	-1.21	0.6	132.83	1AU	0.4	-3.6	-1.3	0.6
217.94	-1.31	0.7	142.50	11AU	0.7	-3.6	-1.3	0.5
224.48	-1.40	0.8	152.21	21AU	1.3	-3.7	-1.3	0.4
231.25	-1.46	0.8	161.97	31AU	3.1	-3.8	-1.2	0.4
238.21	-1.51	0.9	171.80	10SE	1.0	-3.9	-1.2	0.3
245.34	-1.54	0.9	181.67	20SE	-0.6	-4.0	-1.2	0.3
252.62	-1.55	1.0	191.61	30SE	-0.9	-4.1	-1.2	0.3
260.04	-1.55	1.0	201.61	10OC	-0.9	-4.2	-1.2	0.4
267.57	-1.53	1.1	211.65	20OC	-0.8	-4.3	-1.2	0.5
275.18	-1.49	1.1	221.75	30OC	-0.6	-4.4	-1.3	0.5
282.87	-1.44	1.1	231.88	9NO	-0.5	-4.3	-1.3	0.6
290.61	-1.38	1.2	242.04	19NO	-0.4	-4.0	-1.3	0.7
298.38	-1.31	1.2	252.23	29NO	-0.3	-3.2	-1.4	0.7
306.17	-1.22	1.3	262.43	9DE	0.9	-3.4	-1.4	0.8
313.96	-1.13	1.3	272.63	19DE	2.5	-4.1	-1.5	0.8
321.73	-1.03	1.4	282.82	29DE	0.6	-4.3	-1.5	0.8

879

♂ LONG	LAT	MAG	☉ LONG	16.00UT	☿	♀	♃	♄
329.47	-0.92	1.4	293.00	8JA	0.1	-4.4	-1.6	0.8
337.17	-0.81	1.4	303.13	18JA	0.0	-4.3	-1.7	0.8
344.81	-0.70	1.5	313.24	28JA	-0.1	-4.2	-1.7	0.8
352.40	-0.58	1.5	323.30	7FE	-0.4	-4.1	-1.8	0.8
359.91	-0.47	1.5	333.30	17FE	-1.0	-4.0	-1.9	0.8
7.36	-0.35	1.6	343.25	27FE	-1.5	-3.9	-1.9	0.8
14.72	-0.24	1.6	353.15	9MR	-1.1	-3.8	-2.0	0.7
22.00	-0.12	1.6	2.98	19MR	0.1	-3.7	-2.0	0.7
29.20	-0.02	1.7	12.76	29MR	1.5	-3.6	-2.0	0.6
36.32	0.09	1.7	22.49	8AP	3.4	-3.5	-2.0	0.6
43.36	0.19	1.7	32.16	18AP	2.5	-3.5	-2.0	0.7
50.33	0.29	1.7	41.79	28AP	1.4	-3.4	-2.0	0.7
57.21	0.39	1.7	51.39	8MY	0.8	-3.4	-2.0	0.7
64.03	0.48	1.8	60.95	18MY	0.0	-3.4	-2.0	0.7
70.78	0.57	1.8	70.49	28MY	-0.9	-3.3	-1.9	0.7
77.46	0.65	1.8	80.02	7JN	-1.7	-3.3	-1.8	0.6
84.09	0.73	1.8	89.54	17JN	-1.1	-3.3	-1.8	0.6
90.66	0.80	1.9	99.08	27JN	-0.3	-3.3	-1.7	0.6
97.18	0.87	1.9	108.62	7JL	0.2	-3.3	-1.7	0.6
103.65	0.94	2.0	118.19	17JL	0.6	-3.3	-1.6	0.5
110.08	1.01	2.0	127.79	27JL	0.9	-3.4	-1.5	0.5
116.47	1.07	2.0	137.43	6AU	1.7	-3.4	-1.5	0.4
122.82	1.12	2.0	147.11	16AU	3.2	-3.4	-1.4	0.3
129.13	1.18	2.0	156.85	26AU	0.9	-3.4	-1.4	0.3
135.41	1.23	2.0	166.64	5SE	-0.6	-3.5	-1.4	0.2
141.65	1.28	2.0	176.49	15SE	-1.1	-3.5	-1.3	0.2
147.85	1.33	2.0	186.40	25SE	-1.0	-3.5	-1.3	0.1
154.02	1.37	1.9	196.36	5OC	-0.7	-3.5	-1.3	0.1
160.14	1.41	1.9	206.38	15OC	-0.5	-3.4	-1.3	0.1
166.22	1.45	1.8	216.45	25OC	-0.3	-3.4	-1.3	0.2
172.25	1.48	1.8	226.56	4NO	-0.3	-3.4	-1.3	0.2
178.21	1.52	1.7	236.71	14NO	-0.1	-3.4	-1.3	0.3
184.10	1.54	1.6	246.89	24NO	1.1	-3.3	-1.3	0.4
189.90	1.57	1.4	257.08	4DE	2.2	-3.3	-1.4	0.4
195.60	1.59	1.4	267.28	14DE	0.3	-3.3	-1.4	0.5
201.17	1.60	1.3	277.48	24DE	-0.0	-3.3	-1.4	0.5

Left table

♂ LONG	♂ LAT	♂ MAG	☉ LONG	16.00UT	☿	♀	♃	♄
				880				
206.59	1.61	1.1	287.66	3JA	-0.1	-3.4	-1.4	0.6
211.83	1.60	1.0	297.82	13JA	-0.2	-3.4	-1.5	0.6
216.84	1.58	0.8	307.95	23JA	-0.5	-3.4	-1.6	0.6
221.57	1.55	0.6	318.03	2FE	-1.0	-3.4	-1.6	0.6
225.95	1.50	0.4	328.06	12FE	-1.4	-3.5	-1.7	0.6
229.89	1.41	0.2	338.05	22FE	-1.0	-3.5	-1.8	0.6
233.29	1.30	-0.1	347.97	3MR	0.1	-3.6	-1.8	0.6
236.00	1.13	-0.4	357.83	13MR	1.9	-3.6	-1.9	0.6
237.84	0.91	-0.7	7.64	23MR	3.4	-3.7	-2.0	0.5
238.65	0.61	-1.0	17.39	2AP	1.9	-3.7	-2.0	0.5
238.26	0.23	-1.3	27.10	12AP	1.1	-3.8	-2.1	0.4
236.61	-0.24	-1.6	36.75	22AP	0.6	-3.9	-2.1	0.4
233.90	-0.78	-1.9	46.36	2MY	-0.0	-4.0	-2.1	0.5
230.59	-1.34	-2.0	55.94	12MY	-0.9	-4.1	-2.1	0.5
227.40	-1.86	-1.9	65.49	22MY	-1.8	-4.2	-2.1	0.5
225.01	-2.28	-1.7	75.02	1JN	-1.2	-4.2	-2.1	0.5
223.86	-2.60	-1.5	84.55	11JN	-0.3	-4.1	-2.1	0.5
224.11	-2.81	-1.3	94.08	21JN	0.3	-3.9	-2.0	0.5
225.67	-2.94	-1.1	103.61	1JL	0.7	-3.3	-2.0	0.5
228.37	-3.00	-0.9	113.17	11JL	1.2	-3.2	-1.9	0.4
232.03	-3.00	-0.7	122.75	21JL	2.2	-3.8	-1.9	0.4
236.47	-2.98	-0.5	132.37	31JL	2.9	-4.1	-1.8	0.4
241.55	-2.91	-0.3	142.03	10AU	0.7	-4.2	-1.7	0.3
247.15	-2.83	-0.2	151.74	20AU	-0.7	-4.2	-1.7	0.3
253.17	-2.72	-0.1	161.50	30AU	-1.2	-4.2	-1.6	0.2
259.53	-2.59	0.1	171.32	9SE	-1.2	-4.1	-1.6	0.1
266.18	-2.45	0.2	181.19	19SE	-0.7	-4.0	-1.5	0.1
273.04	-2.29	0.3	191.12	29SE	-0.3	-3.9	-1.5	0.0
280.08	-2.13	0.4	201.12	9OC	-0.2	-3.8	-1.4	-0.1
287.27	-1.95	0.5	211.16	19OC	-0.2	-3.7	-1.4	-0.1
294.55	-1.77	0.6	221.25	29OC	0.1	-3.7	-1.4	-0.1
301.91	-1.59	0.7	231.38	8NO	1.3	-3.6	-1.4	0.0
309.32	-1.41	0.8	241.55	18NO	1.9	-3.6	-1.4	0.1
316.75	-1.23	0.9	251.73	28NO	0.1	-3.5	-1.4	0.2
324.18	-1.05	1.0	261.94	8DE	-0.2	-3.5	-1.4	0.2
331.61	-0.88	1.1	272.14	18DE	-0.2	-3.4	-1.4	0.3
339.01	-0.71	1.2	282.33	28DE	-0.3	-3.4	-1.4	0.3
				881				
346.37	-0.55	1.3	292.50	7JA	-0.6	-3.4	-1.4	0.4
353.68	-0.39	1.4	302.65	17JA	-1.0	-3.4	-1.5	0.4
0.94	-0.25	1.4	312.75	27JA	-1.2	-3.3	-1.5	0.5
8.14	-0.11	1.5	322.82	6FE	-0.9	-3.3	-1.6	0.5
15.28	0.02	1.6	332.82	16FE	0.2	-3.3	-1.6	0.5
22.35	0.14	1.6	342.78	26FE	2.3	-3.3	-1.7	0.5
29.35	0.25	1.7	352.67	8MR	2.6	-3.3	-1.7	0.5
36.29	0.36	1.7	2.51	18MR	1.4	-3.4	-1.8	0.4
43.17	0.46	1.8	12.29	28MR	0.8	-3.4	-1.9	0.4
49.98	0.55	1.8	22.02	7AP	0.4	-3.4	-1.9	0.4
56.73	0.63	1.9	31.69	17AP	-0.1	-3.4	-2.0	0.3
63.42	0.71	1.9	41.32	27AP	-0.9	-3.5	-2.1	0.3
70.06	0.77	1.9	50.92	7MY	-1.8	-3.5	-2.1	0.3
76.65	0.84	1.9	60.48	17MY	-1.2	-3.4	-2.2	0.4
83.20	0.90	1.9	70.03	27MY	-0.2	-3.4	-2.2	0.4
89.71	0.95	1.9	79.56	6JN	0.4	-3.4	-2.2	0.4
96.18	0.99	1.9	89.08	16JN	1.0	-3.4	-2.3	0.4
102.62	1.03	1.9	98.61	26JN	1.7	-3.3	-2.3	0.4
109.04	1.07	1.9	108.16	6JL	2.9	-3.3	-2.2	0.4
115.44	1.10	1.9	117.72	16JL	2.4	-3.3	-2.2	0.4
121.82	1.13	2.0	127.32	26JL	0.6	-3.3	-2.2	0.3
128.19	1.15	2.0	136.96	5AU	-0.8	-3.3	-2.1	0.3
134.55	1.17	2.0	146.64	15AU	-1.3	-3.3	-2.0	0.3
140.91	1.18	2.0	156.37	25AU	-1.2	-3.4	-2.0	0.2
147.27	1.19	2.0	166.16	4SE	-0.6	-3.4	-1.9	0.2
153.62	1.20	2.0	176.01	14SE	-0.3	-3.4	-1.8	0.1
159.98	1.20	2.0	185.91	24SE	-0.1	-3.5	-1.8	0.0
166.34	1.20	2.0	195.87	4OC	0.0	-3.5	-1.7	-0.0
172.70	1.19	2.0	205.88	14OC	0.3	-3.6	-1.7	-0.1
179.07	1.17	1.9	215.95	24OC	1.6	-3.7	-1.6	-0.2
185.44	1.15	1.9	226.07	3NO	1.7	-3.8	-1.6	-0.2
191.81	1.12	1.8	236.21	13NO	-0.1	-3.9	-1.5	-0.2
198.18	1.09	1.8	246.39	23NO	-0.4	-4.0	-1.5	-0.1
204.54	1.05	1.7	256.58	3DE	-0.4	-4.1	-1.5	-0.0
210.90	1.00	1.6	266.78	13DE	-0.5	-4.2	-1.5	0.0
217.25	0.94	1.6	276.98	23DE	-0.7	-4.3	-1.5	0.1

Right table

♂ LONG	♂ LAT	♂ MAG	☉ LONG	16.00UT	☿	♀	♃	♄
				882				
223.57	0.86	1.5	287.17	2JA	-1.0	-4.3	-1.5	0.2
229.88	0.77	1.3	297.33	12JA	-1.1	-4.4	-1.5	0.2
236.15	0.67	1.2	307.46	22JA	-0.8	-4.3	-1.5	0.3
242.37	0.55	1.1	317.54	1FE	0.3	-4.0	-1.5	0.3
248.55	0.41	1.0	327.58	11FE	2.7	-3.4	-1.5	0.3
254.65	0.24	0.8	337.56	21FE	2.0	-3.5	-1.6	0.4
260.67	0.04	0.6	347.49	3MR	1.0	-4.0	-1.6	0.4
266.56	-0.18	0.5	357.36	13MR	0.6	-4.2	-1.7	0.4
272.32	-0.45	0.3	7.17	23MR	0.3	-4.3	-1.7	0.4
277.88	-0.75	0.1	16.93	2AP	-0.2	-4.2	-1.8	0.3
283.21	-1.11	-0.2	26.63	12AP	-0.9	-4.1	-1.8	0.3
288.22	-1.52	-0.4	36.28	22AP	-1.8	-4.0	-1.9	0.3
292.82	-1.99	-0.6	45.90	2MY	-1.2	-3.9	-2.0	0.2
296.90	-2.54	-0.9	55.47	12MY	-0.2	-3.8	-2.1	0.2
300.31	-3.16	-1.2	65.03	22MY	0.6	-3.7	-2.1	0.2
302.85	-3.84	-1.5	74.57	1JN	1.3	-3.6	-2.2	0.3
304.35	-4.59	-1.8	84.09	11JN	2.2	-3.6	-2.3	0.3
304.62	-5.34	-2.1	93.62	21JN	3.3	-3.5	-2.3	0.3
303.64	-6.03	-2.3	103.16	1JL	2.0	-3.5	-2.3	0.3
301.60	-6.55	-2.6	112.71	11JL	0.5	-3.4	-2.4	0.3
299.00	-6.78	-2.6	122.29	21JL	-0.8	-3.4	-2.4	0.3
296.58	-6.69	-2.5	131.91	31JL	-1.4	-3.4	-2.4	0.3
295.02	-6.32	-2.3	141.56	10AU	-1.2	-3.4	-2.4	0.3
294.68	-5.77	-2.0	151.27	20AU	-0.5	-3.4	-2.3	0.3
295.68	-5.13	-1.7	161.03	30AU	-0.2	-3.3	-2.3	0.2
297.90	-4.48	-1.4	170.84	9SE	0.0	-3.4	-2.2	0.2
301.13	-3.86	-1.2	180.72	19SE	0.2	-3.4	-2.2	0.1
305.19	-3.27	-0.9	190.65	29SE	0.5	-3.4	-2.1	0.0
309.89	-2.74	-0.6	200.63	9OC	2.0	-3.4	-2.0	-0.0
315.09	-2.27	-0.4	210.67	19OC	1.5	-3.4	-1.9	-0.1
320.68	-1.84	-0.2	220.76	29OC	-0.2	-3.4	-1.9	-0.2
326.55	-1.46	0.0	230.89	8NO	-0.5	-3.5	-1.8	-0.2
332.64	-1.12	0.2	241.05	18NO	-0.5	-3.5	-1.8	-0.3
338.90	-0.81	0.4	251.23	28NO	-0.6	-3.5	-1.7	-0.2
345.27	-0.55	0.6	261.43	8DE	-0.8	-3.5	-1.7	-0.2
351.73	-0.31	0.7	271.64	18DE	-0.8	-3.5	-1.6	-0.1
358.24	-0.10	0.9	281.83	28DE	-0.9	-3.4	-1.6	-0.0
				883				
4.79	0.08	1.0	292.00	7JA	-0.7	-3.4	-1.6	0.1
11.36	0.25	1.1	302.15	17JA	0.4	-3.4	-1.6	0.1
17.93	0.39	1.2	312.26	27JA	3.0	-3.4	-1.5	0.2
24.50	0.52	1.4	322.32	6FE	1.5	-3.3	-1.5	0.2
31.06	0.63	1.5	332.33	16FE	0.7	-3.3	-1.6	0.3
37.59	0.72	1.5	342.29	26FE	0.4	-3.3	-1.6	0.3
44.11	0.81	1.6	352.19	8MR	0.1	-3.3	-1.6	0.3
50.60	0.88	1.7	2.03	18MR	-0.3	-3.4	-1.6	0.3
57.06	0.94	1.8	11.82	28MR	-0.9	-3.4	-1.6	0.3
63.50	0.99	1.8	21.55	7AP	-1.8	-3.4	-1.7	0.3
69.92	1.04	1.9	31.23	17AP	-1.2	-3.4	-1.7	0.3
76.31	1.08	1.9	40.86	27AP	-0.1	-3.5	-1.8	0.3
82.68	1.11	2.0	50.46	7MY	0.8	-3.5	-1.9	0.2
89.04	1.13	2.0	60.03	17MY	1.7	-3.5	-1.9	0.2
95.39	1.15	2.0	69.57	27MY	2.9	-3.6	-2.0	0.2
101.72	1.16	2.0	79.10	6JN	3.0	-3.7	-2.1	0.3
108.05	1.17	2.0	88.63	16JN	1.5	-3.7	-2.1	0.3
114.38	1.17	2.0	98.15	26JN	0.3	-3.8	-2.2	0.3
120.71	1.17	2.0	107.70	6JL	-0.8	-3.9	-2.3	0.3
127.05	1.16	2.0	117.27	16JL	-1.5	-4.0	-2.3	0.3
133.40	1.15	2.0	126.86	26JL	-1.2	-4.1	-2.4	0.3
139.76	1.14	2.0	136.50	5AU	-0.5	-4.2	-2.4	0.3
146.15	1.12	1.9	146.18	15AU	-0.1	-4.2	-2.4	0.3
152.56	1.09	2.0	155.91	25AU	0.2	-4.2	-2.5	0.3
158.99	1.06	2.0	165.69	4SE	0.4	-3.9	-2.5	0.3
165.46	1.03	2.0	175.54	14SE	0.8	-3.3	-2.4	0.2
171.96	0.99	2.0	185.43	24SE	2.3	-3.5	-2.4	0.2
178.50	0.94	2.0	195.39	4OC	1.3	-4.0	-2.3	0.1
185.07	0.89	2.0	205.41	14OC	-0.4	-4.3	-2.3	0.0
191.69	0.84	1.9	215.47	24OC	-0.7	-4.4	-2.2	-0.0
198.34	0.78	1.9	225.57	3NO	-0.7	-4.3	-2.2	-0.1
205.04	0.71	1.9	235.72	13NO	-0.7	-4.2	-2.1	-0.2
211.78	0.63	1.8	245.89	23NO	-0.7	-4.1	-2.0	-0.3
218.56	0.55	1.8	256.09	3DE	-0.7	-4.0	-1.9	-0.3
225.39	0.45	1.7	266.29	13DE	-0.7	-3.9	-1.9	-0.2
232.25	0.35	1.7	276.48	23DE	-0.6	-3.8	-1.8	-0.2

♂			☉	16.00UT	☿	♀	♃	♄
LONG	LAT	MAG	LONG	884	MAGNITUDES			
239.16	0.24	1.6	286.67	2JA	0.6	-3.8	-1.8	-0.1
246.11	0.12	1.5	296.84	12JA	3.0	-3.7	-1.7	-0.0
253.09	-0.01	1.4	306.96	22JA	1.1	-3.6	-1.7	0.0
260.11	-0.16	1.4	317.05	1FE	0.5	-3.5	-1.6	0.1
267.16	-0.31	1.3	327.09	11FE	0.2	-3.5	-1.6	0.2
274.24	-0.48	1.2	337.08	21FE	0.0	-3.4	-1.6	0.2
281.35	-0.66	1.1	347.01	2MR	-0.3	-3.4	-1.6	0.3
288.47	-0.85	1.0	356.88	12MR	-0.9	-3.4	-1.6	0.3
295.59	-1.05	0.9	6.70	22MR	-1.7	-3.3	-1.6	0.3
302.71	-1.26	0.8	16.46	1AP	-1.2	-3.3	-1.6	0.3
309.80	-1.48	0.6	26.16	11AP	-0.1	-3.3	-1.6	0.3
316.86	-1.71	0.5	35.82	21AP	1.0	-3.3	-1.6	0.3
323.85	-1.94	0.4	45.44	1MY	2.2	-3.3	-1.7	0.3
330.74	-2.18	0.3	55.02	11MY	3.7	-3.3	-1.7	0.3
337.51	-2.41	0.1	64.57	21MY	2.3	-3.3	-1.7	0.3
344.11	-2.65	-0.0	74.11	31MY	1.2	-3.4	-1.8	0.2
350.48	-2.88	-0.2	83.63	10JN	0.2	-3.4	-1.8	0.2
356.57	-3.10	-0.3	93.16	20JN	-0.8	-3.4	-1.9	0.2
2.29	-3.31	-0.5	102.70	30JN	-1.6	-3.5	-2.0	0.3
7.53	-3.51	-0.6	112.25	10JL	-1.2	-3.5	-2.0	0.3
12.18	-3.69	-0.8	121.83	20JL	-0.4	-3.5	-2.1	0.3
16.07	-3.84	-1.0	131.44	30JL	0.0	-3.4	-2.2	0.4
19.01	-3.96	-1.2	141.10	9AU	0.3	-3.4	-2.2	0.4
20.79	-4.01	-1.5	150.80	19AU	0.5	-3.4	-2.3	0.4
21.20	-3.96	-1.7	160.56	29AU	1.1	-3.3	-2.4	0.4
20.15	-3.79	-1.9	170.37	8SE	2.8	-3.3	-2.4	0.3
17.78	-3.46	-2.1	180.23	18SE	1.2	-3.3	-2.4	0.3
14.56	-2.96	-2.2	190.16	28SE	-0.5	-3.3	-2.4	0.3
11.27	-2.35	-2.1	200.15	8OC	-0.8	-3.3	-2.4	0.2
8.68	-1.69	-1.8	210.18	18OC	-0.8	-3.3	-2.4	0.2
7.28	-1.07	-1.5	220.27	28OC	-0.8	-3.4	-2.4	0.1
7.25	-0.53	-1.1	230.40	7NO	-0.6	-3.4	-2.3	0.0
8.46	-0.09	-0.8	240.56	17NO	-0.6	-3.4	-2.3	-0.1
10.73	0.27	-0.5	250.74	27NO	-0.6	-3.5	-2.2	-0.1
13.86	0.55	-0.2	260.94	7DE	-0.4	-3.5	-2.1	-0.2
17.66	0.77	0.0	271.14	17DE	0.7	-3.6	-2.1	-0.2
21.97	0.94	0.3	281.34	27DE	2.7	-3.6	-2.0	-0.2
				885				
26.69	1.08	0.5	291.51	6JA	0.8	-3.7	-1.9	-0.1
31.71	1.18	0.7	301.66	16JA	0.3	-3.8	-1.9	-0.1
36.98	1.26	0.9	311.77	26JA	0.1	-3.8	-1.8	0.0
42.43	1.32	1.0	321.83	5FE	-0.1	-3.9	-1.7	0.1
48.02	1.36	1.2	331.85	15FE	-0.4	-4.0	-1.7	0.1
53.72	1.39	1.3	341.81	25FE	-1.0	-4.1	-1.6	0.2
59.52	1.41	1.4	351.71	7MR	-1.6	-4.2	-1.6	0.2
65.38	1.42	1.5	1.55	17MR	-1.1	-4.2	-1.6	0.3
71.30	1.43	1.6	11.34	27MR	-0.0	-4.2	-1.6	0.3
77.27	1.42	1.7	21.07	6AP	1.3	-4.1	-1.5	0.3
83.27	1.41	1.8	30.76	16AP	2.9	-3.6	-1.5	0.4
89.31	1.40	1.8	40.40	26AP	3.0	-2.8	-1.5	0.4
95.37	1.38	1.9	49.99	6MY	1.7	-3.5	-1.5	0.4
101.47	1.35	1.9	59.56	16MY	0.9	-4.0	-1.6	0.4
107.59	1.32	2.0	69.11	26MY	0.1	-4.2	-1.6	0.3
113.75	1.29	2.0	78.64	5JN	-0.8	-4.2	-1.6	0.3
119.93	1.26	2.0	88.16	15JN	-1.7	-4.1	-1.6	0.3
126.15	1.22	2.0	97.69	25JN	-1.2	-4.0	-1.7	0.3
132.40	1.17	2.0	107.23	5JL	-0.4	-3.9	-1.7	0.3
138.69	1.13	2.0	116.80	15JL	0.1	-3.9	-1.8	0.4
145.02	1.08	2.0	126.40	25JL	0.5	-3.8	-1.8	0.4
151.39	1.02	2.0	136.03	4AU	0.8	-3.7	-1.9	0.4
157.82	0.96	2.0	145.71	14AU	1.4	-3.6	-2.0	0.5
164.30	0.90	1.9	155.44	24AU	3.1	-3.6	-2.0	0.5
170.83	0.84	1.9	165.22	3SE	1.0	-3.5	-2.1	0.5
177.42	0.77	1.9	175.06	13SE	-0.6	-3.5	-2.2	0.5
184.07	0.70	1.8	184.96	23SE	-1.0	-3.5	-2.2	0.4
190.78	0.62	1.8	194.91	3OC	-1.0	-3.4	-2.3	0.4
197.56	0.54	1.8	204.92	13OC	-0.8	-3.4	-2.3	0.4
204.41	0.45	1.8	214.98	23OC	-0.5	-3.4	-2.3	0.3
211.32	0.36	1.8	225.08	2NO	-0.4	-3.4	-2.4	0.3
218.30	0.27	1.8	235.23	12NO	-0.4	-3.4	-2.4	0.2
225.35	0.17	1.8	245.40	22NO	-0.3	-3.4	-2.3	0.1
232.47	0.06	1.7	255.59	2DE	0.9	-3.4	-2.3	0.0
239.66	-0.04	1.7	265.80	12DE	2.4	-3.4	-2.3	-0.0
246.92	-0.16	1.7	276.00	22DE	0.5	-3.4	-2.2	-0.1

♂			☉	16.00UT	☿	♀	♃	♄
LONG	LAT	MAG	LONG	886	MAGNITUDES			
254.23	-0.28	1.6	286.18	1JA	0.1	-3.4	-2.1	-0.1
261.61	-0.40	1.6	296.34	11JA	-0.0	-3.4	-2.1	-0.1
269.04	-0.52	1.5	306.47	21JA	-0.2	-3.4	-2.0	-0.0
276.52	-0.65	1.5	316.56	31JA	-0.5	-3.4	-1.9	0.0
284.05	-0.77	1.4	326.60	10FE	-1.0	-3.5	-1.8	0.1
291.62	-0.90	1.4	336.59	20FE	-1.4	-3.5	-1.8	0.2
299.21	-1.02	1.3	346.52	2MR	-1.1	-3.4	-1.7	0.2
306.82	-1.14	1.3	356.40	12MR	0.1	-3.4	-1.7	0.3
314.44	-1.26	1.2	6.21	22MR	1.6	-3.4	-1.6	0.3
322.06	-1.37	1.1	15.97	1AP	3.6	-3.4	-1.6	0.4
329.65	-1.47	1.1	25.68	11AP	2.3	-3.4	-1.5	0.4
337.21	-1.55	1.0	35.34	21AP	1.3	-3.3	-1.5	0.4
344.73	-1.63	1.0	44.96	1MY	0.7	-3.3	-1.5	0.5
352.17	-1.69	0.9	54.55	11MY	0.0	-3.3	-1.5	0.5
359.53	-1.73	0.9	64.10	21MY	-0.9	-3.3	-1.4	0.5
6.78	-1.76	0.8	73.64	31MY	-1.7	-3.3	-1.4	0.5
13.91	-1.77	0.7	83.17	10JN	-1.2	-3.4	-1.5	0.5
20.88	-1.76	0.7	92.69	20JN	-0.3	-3.4	-1.5	0.5
27.67	-1.74	0.6	102.23	30JN	0.2	-3.4	-1.5	0.4
34.26	-1.69	0.5	111.78	10JL	0.6	-3.5	-1.5	0.4
40.59	-1.62	0.4	121.36	20JL	1.0	-3.5	-1.5	0.5
46.64	-1.52	0.3	130.97	30JL	1.9	-3.6	-1.6	0.5
52.34	-1.41	0.2	140.63	9AU	3.1	-3.6	-1.6	0.6
57.62	-1.26	0.1	150.33	19AU	0.9	-3.7	-1.7	0.6
62.40	-1.09	-0.0	160.08	29AU	-0.7	-3.8	-1.7	0.6
66.56	-0.88	-0.2	169.89	8SE	-1.1	-3.9	-1.8	0.6
69.97	-0.63	-0.4	179.76	18SE	-1.1	-4.0	-1.8	0.6
72.44	-0.33	-0.5	189.68	28SE	-0.7	-4.1	-1.9	0.6
73.76	0.03	-0.8	199.67	8OC	-0.4	-4.2	-2.0	0.6
73.75	0.44	-1.0	209.70	18OC	-0.3	-4.3	-2.0	0.6
72.30	0.90	-1.2	219.78	28OC	-0.2	-4.4	-2.1	0.5
69.51	1.38	-1.4	229.91	7NO	-0.1	-4.3	-2.2	0.5
65.82	1.83	-1.5	240.06	17NO	1.1	-4.0	-2.2	0.4
61.97	2.19	-1.4	250.25	27NO	2.2	-3.2	-2.2	0.3
58.73	2.44	-1.1	260.45	7DE	0.3	-3.5	-2.2	0.3
56.66	2.57	-0.9	270.65	17DE	-0.1	-4.1	-2.2	0.2
55.93	2.61	-0.5	280.84	27DE	-0.2	-4.3	-2.2	0.1
				887				
56.48	2.59	-0.2	291.02	6JA	-0.3	-4.4	-2.2	0.1
58.14	2.53	0.0	301.16	16JA	-0.5	-4.3	-2.1	0.0
60.69	2.46	0.3	311.28	26JA	-1.0	-4.2	-2.1	0.1
63.95	2.37	0.5	321.35	5FE	-1.3	-4.1	-2.0	0.1
67.77	2.28	0.7	331.36	15FE	-1.0	-4.0	-1.9	0.2
72.02	2.19	0.9	341.32	25FE	0.1	-3.9	-1.9	0.3
76.62	2.10	1.1	351.23	7MR	2.0	-3.8	-1.8	0.3
81.50	2.01	1.2	1.08	17MR	3.1	-3.7	-1.7	0.4
86.60	1.92	1.3	10.87	27MR	1.7	-3.6	-1.7	0.4
91.88	1.83	1.5	20.61	6AP	1.0	-3.5	-1.6	0.5
97.31	1.74	1.6	30.29	16AP	0.5	-3.5	-1.5	0.5
102.87	1.66	1.6	39.93	26AP	-0.1	-3.4	-1.5	0.6
108.54	1.57	1.7	49.53	6MY	-0.9	-3.4	-1.5	0.6
114.31	1.49	1.8	59.10	16MY	-1.8	-3.4	-1.4	0.6
120.17	1.40	1.8	68.65	26MY	-1.2	-3.3	-1.4	0.6
126.11	1.32	1.9	78.18	5JN	-0.3	-3.3	-1.4	0.6
132.13	1.24	1.9	87.70	15JN	0.3	-3.3	-1.4	0.6
138.22	1.15	1.9	97.23	25JN	0.8	-3.3	-1.3	0.6
144.39	1.07	1.9	106.78	5JL	1.4	-3.3	-1.3	0.6
150.63	0.98	1.9	116.34	15JL	2.5	-3.3	-1.3	0.6
156.95	0.89	1.9	125.93	25JL	2.7	-3.4	-1.4	0.6
163.35	0.80	1.9	135.57	4AU	0.7	-3.4	-1.4	0.7
169.82	0.71	1.9	145.24	14AU	-0.7	-3.4	-1.4	0.7
176.38	0.62	1.9	154.97	24AU	-1.2	-3.4	-1.4	0.8
183.02	0.53	1.9	164.74	3SE	-1.2	-3.5	-1.5	0.8
189.74	0.43	1.8	174.58	13SE	-0.6	-3.5	-1.5	0.8
196.55	0.33	1.8	184.47	23SE	-0.3	-3.5	-1.6	0.8
203.44	0.23	1.7	194.42	3OC	-0.2	-3.4	-1.6	0.8
210.42	0.13	1.7	204.43	13OC	-0.1	-3.4	-1.7	0.8
217.49	0.03	1.6	214.49	23OC	0.1	-3.4	-1.7	0.8
224.64	-0.07	1.6	224.59	2NO	1.4	-3.4	-1.8	0.8
231.88	-0.18	1.6	234.73	12NO	1.9	-3.3	-1.9	0.7
239.21	-0.28	1.6	244.91	22NO	0.1	-3.3	-1.9	0.7
246.61	-0.38	1.6	255.10	2DE	-0.3	-3.3	-2.0	0.6
254.09	-0.49	1.6	265.29	12DE	-0.3	-3.3	-2.0	0.6
261.64	-0.59	1.5	275.50	22DE	-0.4	-3.3	-2.1	0.5

888

♂ LONG	LAT	MAG	☉ LONG	16.00UT 888	☿	♀	♃	♄
269.25	-0.68	1.5	285.68	1JA	-0.6	-3.4	-2.1	0.4
276.92	-0.78	1.5	295.85	11JA	-1.0	-3.4	-2.1	0.3
284.64	-0.87	1.5	305.98	21JA	-1.1	-3.4	-2.1	0.3
292.40	-0.95	1.5	316.07	31JA	-0.9	-3.4	-2.1	0.3
300.18	-1.03	1.4	326.11	10FE	0.2	-3.5	-2.1	0.3
307.98	-1.09	1.4	336.11	20FE	2.4	-3.5	-2.0	0.4
315.79	-1.15	1.4	346.05	1MR	2.4	-3.6	-1.9	0.4
323.59	-1.19	1.4	355.92	11MR	1.2	-3.6	-1.9	0.5
331.37	-1.23	1.4	5.74	21MR	0.7	-3.7	-1.8	0.5
339.12	-1.25	1.3	15.51	31MR	0.3	-3.7	-1.7	0.6
346.82	-1.25	1.3	25.22	10AP	-0.1	-3.8	-1.7	0.6
354.47	-1.25	1.3	34.88	20AP	-0.9	-3.9	-1.6	0.7
2.04	-1.23	1.3	44.50	30AP	-1.8	-4.0	-1.6	0.7
9.54	-1.19	1.3	54.08	10MY	-1.2	-4.1	-1.5	0.8
16.95	-1.14	1.3	63.64	20MY	-0.2	-4.2	-1.4	0.8
24.25	-1.08	1.3	73.18	30MY	0.5	-4.2	-1.4	0.8
31.45	-1.00	1.2	82.71	9JN	1.1	-4.1	-1.4	0.8
38.52	-0.92	1.2	92.24	19JN	1.8	-3.8	-1.3	0.9
45.45	-0.81	1.2	101.77	29JN	3.1	-3.2	-1.3	0.9
52.25	-0.70	1.2	111.32	9JL	2.3	-3.2	-1.3	0.9
58.88	-0.57	1.1	120.90	19JL	0.6	-3.8	-1.3	0.9
65.33	-0.43	1.1	130.51	29JL	-0.8	-4.1	-1.3	0.9
71.59	-0.28	1.0	140.16	8AU	-1.4	-4.2	-1.3	0.9
77.64	-0.11	1.0	149.86	18AU	-1.2	-4.2	-1.3	0.9
83.43	0.07	0.9	159.61	28AU	-0.6	-4.2	-1.3	1.0
88.93	0.27	0.8	169.41	7SE	-0.2	-4.1	-1.3	1.0
94.09	0.50	0.7	179.28	17SE	-0.0	-4.0	-1.3	1.0
98.85	0.74	0.6	189.20	27SE	0.1	-3.9	-1.4	1.1
103.12	1.02	0.4	199.18	7OC	0.3	-3.8	-1.4	1.1
106.77	1.33	0.3	209.21	17OC	1.7	-3.7	-1.5	1.1
109.69	1.67	0.1	219.29	27OC	1.7	-3.7	-1.5	1.1
111.70	2.06	-0.1	229.41	6NO	-0.1	-3.6	-1.6	1.1
112.59	2.49	-0.3	239.57	16NO	-0.4	-3.6	-1.6	1.0
112.21	2.94	-0.5	249.75	26NO	-0.4	-3.5	-1.7	1.0
110.45	3.39	-0.8	259.95	6DE	-0.5	-3.5	-1.8	0.9
107.43	3.78	-1.0	270.15	16DE	-0.7	-3.4	-1.8	0.9
103.61	4.05	-1.1	280.35	26DE	-0.9	-3.4	-1.9	0.8

889

♂ LONG	LAT	MAG	☉ LONG	16.00UT 889	☿	♀	♃	♄
99.68	4.16	-1.0	290.53	5JA	-1.0	-3.4	-1.9	0.7
96.37	4.10	-0.8	300.68	15JA	-0.8	-3.4	-2.0	0.7
94.20	3.93	-0.5	310.79	25JA	0.3	-3.3	-2.0	0.6
93.33	3.68	-0.2	320.86	4FE	2.8	-3.3	-2.0	0.5
93.73	3.41	0.0	330.88	14FE	1.8	-3.3	-2.0	0.5
95.21	3.14	0.3	340.84	24FE	0.9	-3.3	-2.0	0.6
97.58	2.88	0.5	350.75	6MR	0.5	-3.4	-2.0	0.6
100.68	2.64	0.7	0.61	16MR	0.2	-3.4	-2.0	0.7
104.35	2.42	0.9	10.39	26MR	-0.2	-3.4	-1.9	0.7
108.48	2.21	1.0	20.13	5AP	-0.9	-3.4	-1.9	0.8
112.99	2.02	1.1	29.82	15AP	-1.8	-3.4	-1.8	0.8
117.80	1.84	1.3	39.46	25AP	-1.2	-3.5	-1.7	0.9
122.87	1.68	1.4	49.06	5MY	-0.2	-3.5	-1.7	0.9
128.16	1.52	1.5	58.63	15MY	0.6	-3.4	-1.6	1.0
133.63	1.37	1.5	68.18	25MY	1.4	-3.4	-1.5	1.0
139.27	1.23	1.6	77.71	4JN	2.4	-3.4	-1.5	1.1
145.06	1.09	1.6	87.24	14JN	3.3	-3.4	-1.4	1.1
150.99	0.96	1.7	96.76	24JN	1.8	-3.3	-1.4	1.1
157.05	0.83	1.7	106.31	4JL	0.5	-3.3	-1.3	1.1
163.23	0.70	1.7	115.87	14JL	-0.4	-3.3	-1.3	1.1
169.52	0.58	1.7	125.46	24JL	-1.5	-3.3	-1.3	1.2
175.94	0.46	1.8	135.09	3AU	-1.2	-3.3	-1.2	1.2
182.46	0.34	1.8	144.76	13AU	-0.5	-3.3	-1.2	1.1
189.10	0.23	1.7	154.49	23AU	-0.1	-3.4	-1.2	1.2
195.85	0.11	1.7	164.26	2SE	0.1	-3.4	-1.2	1.2
202.71	-0.00	1.7	174.10	12SE	0.2	-3.4	-1.2	1.3
209.68	-0.11	1.7	183.99	22SE	0.6	-3.5	-1.2	1.3
216.75	-0.22	1.7	193.94	2OC	2.0	-3.5	-1.2	1.3
223.93	-0.32	1.6	203.94	12OC	1.5	-3.6	-1.3	1.4
231.21	-0.42	1.6	214.00	22OC	-0.3	-3.7	-1.3	1.4
238.58	-0.52	1.6	224.10	1NO	-0.6	-3.8	-1.3	1.3
246.05	-0.61	1.5	234.24	11NO	-0.6	-3.9	-1.4	1.3
253.60	-0.70	1.5	244.41	21NO	-0.6	-4.0	-1.4	1.3
261.24	-0.78	1.4	254.61	1DE	-0.8	-4.1	-1.5	1.2
268.94	-0.85	1.4	264.80	11DE	-0.8	-4.2	-1.6	1.2
276.70	-0.92	1.4	275.00	21DE	-0.8	-4.2	-1.6	1.1
284.51	-0.97	1.3	285.19	31DE	-0.7	-4.3	-1.7	1.1

890

♂ LONG	LAT	MAG	☉ LONG	16.00UT 890	☿	♀	♃	♄
292.36	-1.02	1.3	295.36	10JA	0.4	-4.4	-1.8	1.0
300.23	-1.05	1.3	305.49	20JA	3.0	-4.3	-1.8	1.0
308.12	-1.08	1.4	315.59	30JA	1.3	-3.9	-1.9	0.9
316.00	-1.09	1.4	325.63	9FE	0.6	-3.4	-1.9	0.9
323.87	-1.09	1.4	335.62	19FE	0.3	-3.5	-2.0	0.8
331.72	-1.08	1.4	345.57	1MR	0.1	-4.0	-2.0	0.8
339.53	-1.06	1.4	355.44	11MR	-0.3	-4.2	-2.0	0.9
347.29	-1.03	1.4	5.27	21MR	-0.9	-4.3	-2.0	0.9
354.99	-0.98	1.4	15.04	31MR	-1.8	-4.2	-2.0	1.0
2.63	-0.92	1.5	24.75	10AP	-1.2	-4.1	-2.0	1.0
10.18	-0.86	1.5	34.41	20AP	-0.1	-4.0	-1.9	1.1
17.66	-0.78	1.5	44.04	30AP	0.9	-3.9	-1.9	1.1
25.04	-0.70	1.5	53.62	10MY	1.9	-3.8	-1.8	1.2
32.33	-0.60	1.5	63.18	20MY	3.2	-3.7	-1.8	1.2
39.52	-0.50	1.5	72.72	30MY	2.7	-3.6	-1.7	1.3
46.61	-0.40	1.5	82.24	9JN	1.4	-3.6	-1.6	1.3
53.59	-0.28	1.6	91.77	19JN	0.3	-3.5	-1.6	1.4
60.47	-0.16	1.6	101.31	29JN	-0.8	-3.5	-1.5	1.4
67.23	-0.04	1.6	110.86	9JL	-1.6	-3.4	-1.4	1.4
73.88	0.09	1.5	120.43	19JL	-1.2	-3.4	-1.4	1.4
80.40	0.22	1.5	130.04	29JL	-0.5	-3.4	-1.4	1.4
86.79	0.36	1.5	139.69	8AU	-0.0	-3.4	-1.3	1.3
93.05	0.51	1.5	149.39	18AU	0.2	-3.4	-1.3	1.3
99.16	0.66	1.4	159.14	28AU	0.4	-3.3	-1.3	1.2
105.10	0.82	1.4	168.94	7SE	0.9	-3.4	-1.2	1.2
110.86	0.99	1.3	178.80	17SE	2.4	-3.4	-1.2	1.2
116.41	1.17	1.3	188.72	27SE	1.4	-3.4	-1.2	1.2
121.71	1.37	1.2	198.69	7OC	-0.4	-3.4	-1.2	1.2
126.73	1.58	1.1	208.72	17OC	-0.7	-3.4	-1.2	1.2
131.40	1.81	0.9	218.80	27OC	-0.7	-3.4	-1.2	1.2
135.65	2.06	0.8	228.92	6NO	-0.7	-3.5	-1.3	1.2
139.39	2.33	0.6	239.08	16NO	-0.7	-3.5	-1.3	1.1
142.48	2.63	0.4	249.26	26NO	-0.6	-3.5	-1.3	1.1
144.79	2.97	0.2	259.46	6DE	-0.7	-3.5	-1.4	1.1
146.14	3.32	-0.0	269.66	16DE	-0.5	-3.4	-1.4	1.0
146.33	3.70	-0.2	279.85	26DE	0.6	-3.4	-1.5	1.0

891

♂ LONG	LAT	MAG	☉ LONG	16.00UT 891	☿	♀	♃	♄
145.24	4.05	-0.5	290.03	5JA	2.9	-3.4	-1.5	0.9
142.85	4.35	-0.7	300.18	15JA	1.0	-3.4	-1.6	0.9
139.42	4.53	-0.9	310.30	25JA	0.4	-3.4	-1.7	0.8
135.49	4.54	-1.0	320.37	4FE	0.2	-3.3	-1.7	0.8
131.78	4.37	-0.8	330.39	14FE	-0.0	-3.3	-1.8	0.7
128.94	4.07	-0.6	340.36	24FE	-0.3	-3.3	-1.9	0.7
127.31	3.69	-0.4	350.27	6MR	-0.9	-3.3	-1.9	0.7
126.98	3.29	-0.1	0.13	16MR	-1.6	-3.4	-2.0	0.7
127.84	2.90	0.1	9.92	26MR	-1.2	-3.4	-2.0	0.8
129.72	2.54	0.3	19.66	5AP	-0.1	-3.4	-2.0	0.9
132.44	2.21	0.5	29.35	15AP	1.1	-3.4	-2.0	0.9
135.84	1.91	0.7	39.00	25AP	2.4	-3.5	-2.0	1.0
139.80	1.64	0.8	48.60	5MY	3.6	-3.5	-2.0	1.1
144.21	1.39	0.9	58.17	15MY	2.1	-3.5	-2.0	1.1
149.00	1.16	1.0	67.72	25MY	1.1	-3.6	-1.9	1.2
154.11	0.95	1.1	77.25	4JN	0.2	-3.7	-1.9	1.2
159.50	0.76	1.2	86.78	14JN	-0.8	-3.7	-1.8	1.2
165.13	0.58	1.3	96.31	24JN	-1.6	-3.8	-1.8	1.3
170.97	0.41	1.3	105.85	4JL	-1.2	-3.9	-1.7	1.3
177.01	0.25	1.4	115.41	14JL	-0.4	-4.0	-1.7	1.3
183.23	0.10	1.4	125.00	24JL	0.1	-4.1	-1.6	1.2
189.62	-0.04	1.4	134.63	3AU	0.4	-4.2	-1.5	1.2
196.16	-0.18	1.5	144.30	13AU	0.6	-4.2	-1.5	1.2
202.86	-0.30	1.5	154.02	23AU	1.2	-4.2	-1.4	1.1
209.70	-0.42	1.5	163.79	2SE	2.9	-3.9	-1.4	1.1
216.68	-0.53	1.5	173.62	12SE	1.2	-3.3	-1.3	1.0
223.79	-0.64	1.5	183.51	22SE	-0.5	-3.5	-1.3	1.0
231.03	-0.73	1.5	193.45	2OC	-0.9	-4.1	-1.3	1.0
238.38	-0.82	1.5	203.46	12OC	-0.9	-4.3	-1.3	1.0
245.83	-0.89	1.5	213.51	22OC	-0.8	-4.4	-1.3	1.0
253.39	-0.96	1.4	223.61	1NO	-0.6	-4.3	-1.3	1.0
261.03	-1.01	1.4	233.75	11NO	-0.5	-4.2	-1.3	1.0
268.75	-1.06	1.4	243.92	21NO	-0.5	-4.1	-1.3	1.0
276.54	-1.09	1.4	254.11	1DE	-0.4	-4.0	-1.3	1.0
284.37	-1.11	1.4	264.31	11DE	0.7	-3.9	-1.3	1.0
292.25	-1.12	1.4	274.51	21DE	2.7	-3.8	-1.4	0.9
300.14	-1.11	1.4	284.70	31DE	0.7	-3.8	-1.4	0.9

♂ LONG	♂ LAT	♂ MAG	☉ LONG	16.00UT	☿	♀	♃	♄
				892	MAGNITUDES			
308.05	-1.10	1.4	294.87	10JA	0.2	-3.7	-1.5	0.8
315.95	-1.07	1.4	305.00	20JA	0.0	-3.6	-1.5	0.8
323.83	-1.03	1.4	315.10	30JA	-0.1	-3.5	-1.6	0.7
331.68	-0.98	1.4	325.15	9FE	-0.4	-3.5	-1.6	0.7
339.49	-0.93	1.4	335.15	19FE	-0.9	-3.4	-1.7	0.6
347.24	-0.86	1.4	345.09	29FE	-1.5	-3.4	-1.8	0.6
354.93	-0.78	1.4	354.98	10MR	-1.1	-3.4	-1.8	0.6
2.55	-0.70	1.4	4.80	20MR	-0.0	-3.3	-1.9	0.5
10.09	-0.61	1.4	14.57	30MR	1.4	-3.3	-2.0	0.6
17.55	-0.52	1.5	24.29	9AP	3.2	-3.3	-2.0	0.7
24.92	-0.42	1.5	33.95	19AP	2.7	-3.3	-2.1	0.7
32.21	-0.32	1.6	43.57	29AP	1.6	-3.3	-2.1	0.8
39.40	-0.22	1.6	53.16	9MY	0.8	-3.3	-2.1	0.9
46.51	-0.11	1.7	62.72	19MY	0.1	-3.3	-2.1	0.9
53.52	-0.00	1.7	72.26	29MY	-0.8	-3.4	-2.1	1.0
60.45	0.11	1.7	81.79	8JN	-1.7	-3.4	-2.1	1.0
67.29	0.22	1.7	91.31	18JN	-1.2	-3.4	-2.1	1.1
74.04	0.33	1.8	100.85	28JN	-0.4	-3.5	-2.0	1.1
80.71	0.43	1.8	110.40	8JL	0.2	-3.5	-2.0	1.1
87.29	0.54	1.8	119.97	18JL	0.5	-3.5	-1.9	1.1
93.78	0.65	1.8	129.57	28JL	0.9	-3.4	-1.9	1.1
100.19	0.77	1.8	139.22	7AU	1.6	-3.4	-1.8	1.1
106.51	0.88	1.8	148.91	17AU	3.2	-3.4	-1.7	1.1
112.74	0.99	1.8	158.66	27AU	1.0	-3.3	-1.7	1.0
118.88	1.10	1.7	168.46	6SE	-0.6	-3.3	-1.6	1.0
124.91	1.22	1.7	178.31	16SE	-1.0	-3.3	-1.6	0.9
130.82	1.34	1.7	188.23	26SE	-1.0	-3.3	-1.5	0.9
136.61	1.46	1.6	198.20	6OC	-0.8	-3.3	-1.5	0.9
142.25	1.59	1.5	208.23	16OC	-0.5	-3.3	-1.4	0.9
147.73	1.73	1.4	218.31	26OC	-0.4	-3.4	-1.4	0.9
153.00	1.87	1.3	228.43	5NO	-0.3	-3.4	-1.4	0.9
158.04	2.02	1.2	238.58	15NO	-0.2	-3.4	-1.4	0.9
162.79	2.18	1.1	248.76	25NO	0.9	-3.5	-1.4	0.9
167.20	2.36	0.9	258.96	5DE	2.4	-3.5	-1.4	0.9
171.17	2.54	0.8	269.16	15DE	0.4	-3.6	-1.4	0.9
174.62	2.74	0.6	279.36	25DE	0.0	-3.6	-1.4	0.8
				893				
177.40	2.95	0.3	289.54	4JA	-0.1	-3.7	-1.4	0.8
179.36	3.17	0.1	299.69	14JA	-0.2	-3.8	-1.4	0.8
180.33	3.39	-0.2	309.81	24JA	-0.5	-3.9	-1.5	0.7
180.12	3.58	-0.4	319.89	3FE	-1.0	-3.9	-1.5	0.7
178.63	3.72	-0.7	329.91	13FE	-1.4	-4.0	-1.6	0.6
175.93	3.76	-1.0	339.89	23FE	-1.1	-4.1	-1.6	0.6
172.35	3.67	-1.2	349.80	5MR	0.0	-4.2	-1.7	0.5
168.52	3.42	-1.1	359.65	15MR	1.7	-4.3	-1.8	0.5
165.14	3.04	-1.0	9.45	25MR	3.6	-4.2	-1.8	0.4
162.79	2.60	-0.8	19.20	4AP	2.0	-4.1	-1.9	0.4
161.72	2.14	-0.6	28.89	14AP	1.2	-3.6	-2.0	0.5
161.94	1.70	-0.3	38.53	24AP	0.6	-2.9	-2.0	0.6
163.34	1.30	-0.1	48.14	4MY	0.0	-3.5	-2.1	0.6
165.73	0.94	0.1	57.71	14MY	-0.8	-4.0	-2.2	0.7
168.94	0.62	0.2	67.26	24MY	-1.7	-4.2	-2.2	0.8
172.83	0.34	0.4	76.79	3JN	-1.2	-4.2	-2.2	0.8
177.28	0.09	0.5	86.32	13JN	-0.3	-4.1	-2.3	0.9
182.20	-0.13	0.6	95.85	23JN	0.3	-4.0	-2.3	0.9
187.51	-0.33	0.7	105.39	3JL	0.7	-3.9	-2.3	0.9
193.16	-0.51	0.8	114.94	13JL	1.2	-3.9	-2.2	1.0
199.11	-0.67	0.9	124.54	23JL	2.1	-3.8	-2.2	1.0
205.32	-0.81	0.9	134.16	2AU	3.0	-3.7	-2.2	1.0
211.76	-0.93	1.0	143.83	12AU	0.9	-3.6	-2.1	1.0
218.42	-1.03	1.0	153.55	22AU	-0.6	-3.6	-2.0	1.0
225.27	-1.13	1.1	163.31	1SE	-1.2	-3.5	-2.0	0.9
232.28	-1.20	1.1	173.14	11SE	-1.2	-3.5	-1.9	0.9
239.46	-1.26	1.1	183.03	21SE	-0.7	-3.5	-1.8	0.9
246.78	-1.31	1.2	192.97	1OC	-0.4	-3.4	-1.8	0.8
254.21	-1.34	1.2	202.97	11OC	-0.2	-3.4	-1.7	0.8
261.76	-1.35	1.2	213.02	21OC	-0.2	-3.4	-1.7	0.8
269.40	-1.35	1.2	223.12	31OC	0.0	-3.4	-1.6	0.8
277.12	-1.34	1.3	233.25	10NO	1.2	-3.4	-1.6	0.8
284.89	-1.31	1.3	243.42	20NO	2.2	-3.4	-1.5	0.8
292.71	-1.27	1.3	253.61	30NO	0.2	-3.4	-1.5	0.8
300.55	-1.22	1.3	263.81	10DE	-0.2	-3.4	-1.5	0.8
308.40	-1.16	1.4	274.02	20DE	-0.2	-3.4	-1.5	0.8
316.24	-1.08	1.4	284.20	30DE	-0.3	-3.4	-1.5	0.8

♂ LONG	♂ LAT	♂ MAG	☉ LONG	16.00UT	☿	♀	♃	♄
				894	MAGNITUDES			
324.06	-1.00	1.4	294.37	9JA	-0.6	-3.4	-1.5	0.8
331.84	-0.91	1.4	304.51	19JA	-1.0	-3.4	-1.5	0.7
339.58	-0.82	1.5	314.61	29JA	-1.2	-3.5	-1.5	0.7
347.26	-0.71	1.5	324.66	8FE	-1.0	-3.5	-1.5	0.6
354.88	-0.61	1.5	334.66	18FE	0.1	-3.5	-1.6	0.6
2.42	-0.50	1.5	344.61	28FE	2.1	-3.4	-1.6	0.5
9.89	-0.40	1.6	354.49	10MR	2.8	-3.4	-1.6	0.5
17.28	-0.29	1.6	4.32	20MR	1.5	-3.4	-1.7	0.4
24.58	-0.18	1.6	14.09	30MR	0.9	-3.4	-1.7	0.4
31.80	-0.07	1.6	23.81	9AP	0.4	-3.4	-1.8	0.3
38.93	0.03	1.6	33.48	19AP	-0.1	-3.3	-1.9	0.4
45.99	0.13	1.6	43.10	29AP	-0.8	-3.3	-1.9	0.4
52.96	0.23	1.7	52.70	9MY	-1.8	-3.3	-2.0	0.5
59.85	0.33	1.7	62.26	19MY	-1.2	-3.3	-2.1	0.6
66.67	0.43	1.8	71.79	29MY	-0.3	-3.3	-2.2	0.6
73.42	0.52	1.8	81.32	8JN	0.4	-3.4	-2.2	0.7
80.10	0.61	1.8	90.85	18JN	0.9	-3.4	-2.3	0.7
86.72	0.69	1.9	100.38	28JN	1.5	-3.4	-2.3	0.8
93.28	0.78	1.9	109.93	8JL	2.7	-3.5	-2.4	0.8
99.78	0.86	1.9	119.50	18JL	2.6	-3.5	-2.4	0.9
106.22	0.94	1.9	129.11	28JL	0.7	-3.6	-2.4	0.9
112.61	1.01	2.0	138.75	7AU	-0.7	-3.6	-2.4	0.9
118.95	1.08	2.0	148.45	17AU	-1.3	-3.7	-2.4	0.9
125.24	1.15	2.0	158.18	27AU	-1.2	-3.8	-2.3	0.9
131.49	1.22	1.9	167.98	6SE	-0.6	-3.9	-2.3	0.9
137.67	1.29	1.9	177.84	16SE	-0.3	-4.0	-2.2	0.9
143.81	1.36	1.9	187.75	26SE	-0.1	-4.1	-2.1	0.8
149.89	1.42	1.9	197.72	6OC	-0.0	-4.2	-2.1	0.8
155.89	1.48	1.8	207.74	16OC	0.2	-4.3	-2.0	0.7
161.83	1.54	1.8	217.82	26OC	1.4	-4.4	-1.9	0.7
167.68	1.61	1.7	227.94	5NO	1.9	-4.3	-1.9	0.8
173.44	1.67	1.6	238.09	15NO	0.0	-3.9	-1.8	0.8
179.07	1.72	1.5	248.26	25NO	-0.3	-3.1	-1.8	0.8
184.57	1.78	1.4	258.46	5DE	-0.4	-3.5	-1.7	0.8
189.90	1.84	1.3	268.66	15DE	-0.4	-4.1	-1.7	0.8
195.04	1.89	1.1	278.86	25DE	-0.6	-4.3	-1.6	0.8
				895				
199.92	1.94	1.0	289.04	4JA	-1.0	-4.4	-1.6	0.8
204.50	1.98	0.8	299.20	14JA	-1.1	-4.3	-1.6	0.8
208.71	2.02	0.6	309.31	24JA	-0.9	-4.2	-1.6	0.7
212.45	2.04	0.3	319.40	3FE	0.2	-4.1	-1.6	0.7
215.60	2.05	0.1	329.42	13FE	2.5	-4.0	-1.6	0.6
218.02	2.03	-0.2	339.40	23FE	2.2	-3.9	-1.6	0.6
219.53	1.97	-0.5	349.32	5MR	1.1	-3.8	-1.6	0.5
219.96	1.85	-0.8	359.18	15MR	0.6	-3.7	-1.6	0.5
219.16	1.65	-1.1	8.98	25MR	0.3	-3.6	-1.6	0.4
217.10	1.37	-1.4	18.73	4AP	-0.1	-3.5	-1.7	0.4
214.03	0.98	-1.6	28.42	14AP	-0.9	-3.5	-1.7	0.3
210.47	0.53	-1.7	38.07	24AP	-1.8	-3.4	-1.8	0.3
207.14	0.04	-1.6	47.68	4MY	-1.2	-3.4	-1.8	0.3
204.70	-0.43	-1.4	57.25	14MY	-0.3	-3.4	-1.9	0.4
203.53	-0.84	-1.2	66.80	24MY	0.5	-3.3	-1.9	0.4
203.73	-1.19	-1.0	76.34	3JN	1.2	-3.3	-2.0	0.5
205.21	-1.46	-0.8	85.86	13JN	2.0	-3.3	-2.1	0.6
207.79	-1.68	-0.6	95.39	23JN	3.3	-3.3	-2.1	0.6
211.29	-1.85	-0.4	104.93	3JL	2.1	-3.3	-2.2	0.7
215.56	-1.97	-0.2	114.49	13JL	0.6	-3.3	-2.3	0.7
220.44	-2.06	-0.1	124.07	23JL	-0.7	-3.4	-2.3	0.8
225.85	-2.12	0.0	133.70	2AU	-1.4	-3.4	-2.4	0.8
231.69	-2.15	0.2	143.36	12AU	-1.2	-3.4	-2.4	0.8
237.88	-2.16	0.3	153.07	22AU	-0.6	-3.4	-2.5	0.8
244.38	-2.14	0.4	162.84	1SE	-0.2	-3.5	-2.5	0.8
251.13	-2.10	0.5	172.66	11SE	0.0	-3.5	-2.5	0.8
258.10	-2.05	0.5	182.55	21SE	0.1	-3.5	-2.4	0.8
265.25	-1.97	0.6	192.49	1OC	0.4	-3.4	-2.4	0.8
272.54	-1.89	0.7	202.48	11OC	1.7	-3.4	-2.3	0.8
279.94	-1.78	0.8	212.53	21OC	1.7	-3.4	-2.3	0.7
287.44	-1.67	0.9	222.63	31OC	-0.2	-3.4	-2.2	0.7
295.00	-1.54	0.9	232.76	10NO	-0.5	-3.3	-2.1	0.7
302.60	-1.41	1.0	242.92	20NO	-0.5	-3.3	-2.1	0.7
310.22	-1.27	1.1	253.11	30NO	-0.5	-3.3	-2.0	0.8
317.84	-1.13	1.1	263.31	10DE	-0.7	-3.3	-1.9	0.8
325.45	-0.99	1.2	273.51	20DE	-0.9	-3.3	-1.9	0.8
333.04	-0.85	1.3	283.70	30DE	-0.9	-3.4	-1.8	0.8

♂ LONG	♂ LAT	♂ MAG	☉ LONG	16.00UT	☿	♀	♃	♄
				896	MAGNITUDES			
340.58	-0.70	1.3	293.87	9JA	-0.8	-3.4	-1.8	0.8
348.07	-0.56	1.4	304.01	19JA	0.3	-3.4	-1.7	0.8
355.50	-0.43	1.5	314.12	29JA	2.8	-3.4	-1.7	0.7
2.87	-0.30	1.5	324.17	8FE	1.6	-3.5	-1.6	0.7
10.17	-0.17	1.6	334.17	18FE	0.8	-3.5	-1.6	0.7
17.40	-0.05	1.6	344.12	28FE	0.4	-3.6	-1.6	0.6
24.55	0.07	1.7	354.01	9MR	0.2	-3.6	-1.6	0.6
31.62	0.18	1.7	3.85	19MR	-0.2	-3.7	-1.6	0.5
38.63	0.28	1.8	13.62	29MR	-0.9	-3.7	-1.6	0.4
45.56	0.38	1.8	23.34	8AP	-1.8	-3.8	-1.6	0.4
52.42	0.47	1.8	33.02	18AP	-1.2	-3.9	-1.6	0.3
59.22	0.56	1.8	42.65	28AP	-0.2	-4.0	-1.6	0.3
65.95	0.64	1.9	52.24	8MY	0.7	-4.1	-1.7	0.2
72.62	0.71	1.9	61.80	18MY	1.6	-4.2	-1.7	0.3
79.24	0.78	1.9	71.34	28MY	2.7	-4.2	-1.8	0.4
85.82	0.85	1.9	80.87	7JN	3.2	-4.1	-1.8	0.4
92.34	0.91	1.9	90.40	17JN	1.7	-3.8	-1.9	0.5
98.83	0.96	1.9	99.93	27JN	0.4	-3.2	-1.9	0.6
105.29	1.01	1.9	109.47	7JL	-0.8	-3.2	-2.0	0.6
111.71	1.06	2.0	119.04	17JL	-1.5	-3.8	-2.1	0.7
118.10	1.10	2.0	128.64	27JL	-1.2	-4.1	-2.1	0.7
124.48	1.13	2.0	138.28	6AU	-0.5	-4.2	-2.2	0.7
130.83	1.17	2.0	147.98	16AU	-0.1	-4.2	-2.3	0.8
137.17	1.20	2.0	157.71	26AU	0.1	-4.1	-2.3	0.8
143.49	1.22	2.0	167.50	5SE	0.3	-4.1	-2.4	0.8
149.80	1.24	2.0	177.36	15SE	0.7	-4.0	-2.4	0.8
156.10	1.26	2.0	187.27	25SE	2.1	-3.9	-2.4	0.8
162.39	1.27	2.0	197.23	5OC	1.6	-3.8	-2.4	0.8
168.66	1.28	2.0	207.26	15OC	-0.3	-3.7	-2.4	0.8
174.92	1.29	1.9	217.33	25OC	-0.7	-3.7	-2.4	0.7
181.16	1.29	1.9	227.44	4NO	-0.6	-3.6	-2.4	0.7
187.39	1.28	1.8	237.60	14NO	-0.7	-3.6	-2.3	0.7
193.58	1.27	1.7	247.77	24NO	-0.8	-3.5	-2.3	0.7
199.75	1.25	1.6	257.97	4DE	-0.7	-3.5	-2.2	0.7
205.88	1.22	1.6	268.17	14DE	-0.8	-3.4	-2.1	0.7
211.95	1.18	1.5	278.37	24DE	-0.6	-3.4	-2.1	0.8
				897				
217.98	1.13	1.3	288.55	3JA	0.4	-3.4	-2.0	0.8
223.92	1.06	1.2	298.71	13JA	3.0	-3.4	-1.9	0.8
229.78	0.98	1.1	308.83	23JA	1.2	-3.3	-1.8	0.8
235.53	0.88	0.9	318.91	2FE	0.5	-3.3	-1.8	0.7
241.15	0.76	0.8	328.94	12FE	0.3	-3.3	-1.7	0.7
246.59	0.61	0.6	338.91	22FE	0.1	-3.3	-1.7	0.7
251.83	0.43	0.4	348.83	4MR	-0.3	-3.3	-1.6	0.6
256.80	0.20	0.2	358.70	14MR	-0.9	-3.4	-1.6	0.6
261.43	-0.07	-0.1	8.50	24MR	-1.7	-3.4	-1.6	0.5
265.65	-0.40	-0.3	18.25	3AP	-1.2	-3.4	-1.6	0.5
269.31	-0.80	-0.6	27.95	13AP	-0.2	-3.4	-1.5	0.4
272.29	-1.27	-0.9	37.60	23AP	0.9	-3.5	-1.5	0.4
274.42	-1.84	-1.2	47.21	3MY	2.1	-3.5	-1.5	0.3
275.48	-2.50	-1.5	56.79	13MY	3.6	-3.4	-1.5	0.2
275.35	-3.23	-1.8	66.34	23MY	2.5	-3.4	-1.6	0.2
274.00	-3.98	-2.1	75.87	2JN	1.3	-3.4	-1.6	0.3
271.63	-4.67	-2.4	85.40	12JN	0.3	-3.4	-1.6	0.4
268.80	-5.19	-2.4	94.92	22JN	-0.8	-3.3	-1.6	0.4
266.20	-5.48	-2.3	104.46	2JL	-1.6	-3.3	-1.7	0.5
264.51	-5.52	-2.1	114.02	12JL	-1.2	-3.3	-1.7	0.6
264.11	-5.38	-1.9	123.60	22JL	-0.5	-3.3	-1.8	0.6
265.07	-5.10	-1.6	133.23	1AU	-0.0	-3.3	-1.8	0.7
267.29	-4.75	-1.4	142.89	11AU	0.3	-3.4	-1.9	0.7
270.59	-4.37	-1.1	152.60	21AU	0.5	-3.4	-2.0	0.7
274.75	-3.97	-0.9	162.36	31AU	1.0	-3.4	-2.0	0.8
279.61	-3.58	-0.7	172.19	10SE	2.5	-3.4	-2.1	0.8
285.01	-3.19	-0.5	182.06	20SE	1.4	-3.5	-2.2	0.8
290.83	-2.82	-0.3	192.00	30SE	-0.4	-3.5	-2.2	0.8
296.98	-2.47	-0.1	202.00	10OC	-0.8	-3.6	-2.3	0.8
303.39	-2.14	0.0	212.04	20OC	-0.8	-3.7	-2.3	0.8
309.97	-1.82	0.2	222.13	30OC	-0.8	-3.8	-2.3	0.7
316.70	-1.53	0.4	232.27	9NO	-0.7	-3.9	-2.4	0.7
323.53	-1.26	0.5	242.43	19NO	-0.6	-4.0	-2.3	0.7
330.42	-1.00	0.6	252.62	29NO	-0.6	-4.1	-2.3	0.7
337.35	-0.77	0.8	262.82	9DE	-0.5	-4.2	-2.3	0.7
344.30	-0.56	0.9	273.02	19DE	0.6	-4.3	-2.2	0.7
351.26	-0.36	1.0	283.21	29DE	2.9	-4.3	-2.2	0.8

♂ LONG	♂ LAT	♂ MAG	☉ LONG	16.00UT	☿	♀	♃	♄
				898	MAGNITUDES			
358.20	-0.18	1.1	293.38	8JA	0.9	-4.4	-2.1	0.8
5.12	-0.02	1.2	303.52	18JA	0.3	-4.3	-2.0	0.8
12.01	0.13	1.3	313.63	28JA	0.1	-3.9	-2.0	0.8
18.86	0.27	1.4	323.68	7FE	-0.0	-3.4	-1.9	0.8
25.67	0.39	1.5	333.69	17FE	-0.4	-3.6	-1.8	0.8
32.44	0.50	1.6	343.64	27FE	-0.9	-4.0	-1.8	0.7
39.16	0.59	1.7	353.53	9MR	-1.6	-4.2	-1.7	0.7
45.84	0.68	1.7	3.37	19MR	-1.2	-4.3	-1.6	0.6
52.47	0.76	1.8	13.15	29MR	-0.1	-4.2	-1.6	0.6
59.06	0.83	1.8	22.87	8AP	1.2	-4.1	-1.5	0.5
65.61	0.89	1.9	32.54	18AP	2.7	-4.0	-1.5	0.5
72.12	0.94	1.9	42.18	28AP	3.3	-3.9	-1.5	0.4
78.60	0.99	2.0	51.77	8MY	1.9	-3.8	-1.5	0.3
85.05	1.03	2.0	61.33	18MY	1.0	-3.7	-1.4	0.3
91.47	1.06	2.0	70.88	28MY	0.2	-3.6	-1.4	0.2
97.88	1.09	2.0	80.41	7JN	-0.8	-3.6	-1.4	0.3
104.26	1.11	2.0	89.93	17JN	-1.7	-3.5	-1.4	0.3
110.63	1.13	2.0	99.47	27JN	-1.2	-3.5	-1.5	0.4
116.99	1.14	2.0	109.01	7JL	-0.4	-3.4	-1.5	0.4
123.35	1.15	2.0	118.58	17JL	0.1	-3.4	-1.5	0.5
129.71	1.16	2.0	128.18	27JL	0.4	-3.4	-1.5	0.6
136.08	1.15	2.0	137.82	6AU	0.7	-3.4	-1.6	0.6
142.45	1.15	2.0	147.51	16AU	1.3	-3.3	-1.6	0.7
148.84	1.14	2.0	157.25	26AU	3.0	-3.3	-1.7	0.7
155.25	1.12	2.0	167.04	5SE	1.2	-3.3	-1.7	0.7
161.67	1.10	2.0	176.88	15SE	-0.5	-3.4	-1.8	0.8
168.12	1.08	2.0	186.79	25SE	-0.9	-3.4	-1.9	0.8
174.58	1.05	2.0	196.75	5OC	-0.9	-3.4	-1.9	0.8
181.08	1.01	2.0	206.77	15OC	-0.8	-3.4	-2.0	0.8
187.60	0.97	1.9	216.84	25OC	-0.5	-3.4	-2.0	0.8
194.15	0.92	1.9	226.95	4NO	-0.4	-3.5	-2.1	0.7
200.73	0.87	1.9	237.10	14NO	-0.4	-3.5	-2.2	0.7
207.33	0.80	1.8	247.28	24NO	-0.3	-3.5	-2.2	0.7
213.96	0.73	1.8	257.47	4DE	0.7	-3.5	-2.2	0.7
220.62	0.65	1.7	267.67	14DE	2.7	-3.4	-2.2	0.7
227.31	0.56	1.6	277.87	24DE	0.6	-3.4	-2.2	0.7
				899				
234.02	0.46	1.6	288.05	3JA	0.1	-3.4	-2.2	0.8
240.75	0.34	1.5	298.21	13JA	-0.0	-3.4	-2.2	0.8
247.50	0.22	1.4	308.34	23JA	-0.1	-3.4	-2.1	0.8
254.26	0.07	1.3	318.42	2FE	-0.4	-3.3	-2.0	0.8
261.04	-0.09	1.2	328.45	12FE	-0.9	-3.3	-2.0	0.8
267.82	-0.26	1.0	338.43	22FE	-1.5	-3.3	-1.9	0.8
274.60	-0.46	0.9	348.35	4MR	-1.1	-3.3	-1.8	0.8
281.37	-0.67	0.8	358.22	14MR	-0.0	-3.4	-1.8	0.7
288.12	-0.90	0.7	8.03	24MR	1.5	-3.4	-1.7	0.7
294.82	-1.16	0.5	17.78	3AP	3.4	-3.4	-1.6	0.6
301.47	-1.43	0.4	27.48	13AP	2.5	-3.4	-1.6	0.6
308.03	-1.73	0.2	37.13	23AP	1.4	-3.5	-1.5	0.5
314.47	-2.05	0.0	46.74	3MY	0.8	-3.5	-1.5	0.5
320.75	-2.39	-0.1	56.32	13MY	0.1	-3.5	-1.4	0.4
326.80	-2.75	-0.3	65.88	23MY	-0.8	-3.6	-1.4	0.3
332.56	-3.13	-0.5	75.41	2JN	-1.7	-3.7	-1.4	0.3
337.94	-3.53	-0.7	84.94	12JN	-1.2	-3.8	-1.4	0.2
342.81	-3.95	-0.9	94.47	22JN	-0.4	-3.8	-1.3	0.3
347.03	-4.37	-1.1	104.00	2JL	0.2	-3.9	-1.3	0.3
350.43	-4.79	-1.4	113.56	12JL	0.6	-4.0	-1.3	0.4
352.77	-5.18	-1.6	123.15	22JL	1.0	-4.1	-1.3	0.5
353.88	-5.50	-1.9	132.76	1AU	1.7	-4.2	-1.3	0.5
353.59	-5.71	-2.1	142.43	11AU	3.2	-4.2	-1.4	0.6
351.94	-5.73	-2.3	152.14	21AU	1.0	-4.1	-1.4	0.6
349.31	-5.49	-2.5	161.89	31AU	-0.6	-3.8	-1.4	0.7
346.34	-5.00	-2.4	171.71	10SE	-1.1	-3.3	-1.5	0.7
343.84	-4.31	-2.2	181.59	20SE	-1.1	-3.6	-1.5	0.7
342.40	-3.54	-1.9	191.52	30SE	-0.7	-4.1	-1.6	0.8
342.25	-2.78	-1.6	201.51	10OC	-0.4	-4.3	-1.6	0.8
343.40	-2.09	-1.2	211.56	20OC	-0.3	-4.3	-1.7	0.8
345.66	-1.50	-0.9	221.64	30OC	-0.3	-4.3	-1.8	0.8
348.82	-1.00	-0.6	231.78	9NO	-0.1	-4.2	-1.8	0.8
352.71	-0.58	-0.4	241.94	19NO	0.9	-4.1	-1.9	0.8
357.14	-0.23	-0.1	252.12	29NO	2.4	-4.0	-1.9	0.7
2.01	0.06	0.1	262.33	9DE	0.3	-3.9	-2.0	0.7
7.21	0.29	0.4	272.53	19DE	-0.1	-3.8	-2.0	0.7
12.65	0.49	0.6	282.72	29DE	-0.1	-3.7	-2.1	0.7

♂ LONG	♂ LAT	♂ MAG	☉ LONG	16.00UT	☿	♀	♃	♄
				900	MAGNITUDES			
18.29	0.65	0.7	292.89	8JA	-0.3	-3.7	-2.1	0.8
24.07	0.79	0.9	303.04	18JA	-0.5	-3.6	-2.1	0.8
29.97	0.90	1.1	313.14	28JA	-1.0	-3.5	-2.1	0.8
35.94	0.99	1.2	323.20	7FE	-1.3	-3.5	-2.1	0.8
41.98	1.07	1.3	333.21	17FE	-1.1	-3.4	-2.0	0.8
48.07	1.13	1.4	343.16	27FE	0.0	-3.4	-2.0	0.8
54.18	1.17	1.5	353.06	8MR	1.8	-3.4	-1.9	0.8
60.32	1.21	1.6	2.90	18MR	3.3	-3.3	-1.9	0.8
66.48	1.24	1.7	12.68	28MR	1.8	-3.3	-1.8	0.8
72.65	1.26	1.8	22.41	7AP	1.0	-3.3	-1.7	0.7
78.83	1.27	1.8	32.08	17AP	0.6	-3.3	-1.7	0.7
85.01	1.28	1.9	41.71	27AP	0.0	-3.3	-1.6	0.6
91.21	1.28	1.9	51.31	7MY	-0.8	-3.3	-1.5	0.5
97.41	1.28	2.0	60.87	17MY	-1.7	-3.3	-1.5	0.5
103.62	1.27	2.0	70.41	27MY	-1.2	-3.4	-1.4	0.4
109.85	1.25	2.0	79.95	6JN	-0.3	-3.4	-1.4	0.4
116.10	1.23	2.0	89.47	16JN	0.3	-3.4	-1.3	0.3
122.36	1.21	2.0	99.00	26JN	0.8	-3.5	-1.3	0.3
128.64	1.18	2.0	108.55	6JL	1.3	-3.5	-1.3	0.3
134.95	1.15	2.0	118.11	16JL	2.3	-3.5	-1.3	0.4
141.29	1.11	2.0	127.71	26JL	2.9	-3.4	-1.2	0.4
147.66	1.07	2.0	137.35	5AU	0.9	-3.4	-1.2	0.5
154.07	1.03	2.0	147.03	15AU	-0.6	-3.4	-1.2	0.6
160.53	0.98	1.9	156.77	25AU	-1.2	-3.3	-1.3	0.6
167.02	0.93	1.9	166.56	4SE	-1.2	-3.3	-1.3	0.7
173.57	0.87	1.9	176.40	14SE	-0.7	-3.3	-1.3	0.7
180.16	0.81	1.9	186.31	24SE	-0.3	-3.3	-1.3	0.7
186.81	0.75	1.9	196.27	4OC	-0.2	-3.3	-1.4	0.8
193.51	0.68	1.9	206.28	14OC	-0.1	-3.3	-1.4	0.8
200.27	0.60	1.9	216.35	24OC	0.1	-3.4	-1.5	0.8
207.09	0.52	1.9	226.46	3NO	1.2	-3.4	-1.5	0.8
213.97	0.43	1.8	236.61	13NO	2.2	-3.4	-1.6	0.8
220.91	0.34	1.8	246.78	23NO	0.1	-3.5	-1.6	0.8
227.91	0.24	1.8	256.98	3DE	-0.2	-3.5	-1.7	0.8
234.96	0.13	1.7	267.18	13DE	-0.3	-3.6	-1.8	0.8
242.08	0.02	1.7	277.38	23DE	-0.4	-3.6	-1.8	0.7
				901				
249.26	-0.10	1.6	287.56	2JA	-0.6	-3.7	-1.9	0.8
256.49	-0.22	1.6	297.72	12JA	-1.0	-3.8	-1.9	0.8
263.77	-0.36	1.5	307.85	22JA	-1.2	-3.9	-2.0	0.8
271.10	-0.49	1.4	317.93	1FE	-1.0	-3.9	-2.0	0.9
278.48	-0.63	1.4	327.97	11FE	0.1	-4.0	-2.0	0.9
285.90	-0.78	1.3	337.95	21FE	2.2	-4.1	-2.0	0.9
293.34	-0.93	1.2	347.88	3MR	2.6	-4.2	-2.0	0.9
300.81	-1.08	1.2	357.74	13MR	1.3	-4.3	-2.0	0.9
308.30	-1.23	1.1	7.56	23MR	0.8	-4.2	-1.9	0.9
315.78	-1.37	1.0	17.31	2AP	0.4	-4.1	-1.9	0.8
323.25	-1.52	0.9	27.01	12AP	-0.1	-3.6	-1.8	0.8
330.69	-1.65	0.9	36.67	22AP	-0.8	-2.9	-1.8	0.8
338.08	-1.78	0.8	46.28	2MY	-1.7	-3.6	-1.7	0.7
345.40	-1.89	0.7	55.86	12MY	-1.3	-4.0	-1.6	0.6
352.63	-2.00	0.6	65.41	22MY	-0.3	-4.2	-1.6	0.6
359.75	-2.08	0.5	74.95	1JN	0.4	-4.2	-1.5	0.5
6.71	-2.15	0.4	84.47	11JN	1.0	-4.1	-1.5	0.5
13.50	-2.20	0.3	94.00	21JN	1.7	-4.0	-1.4	0.4
20.05	-2.23	0.2	103.54	1JL	2.9	-3.9	-1.4	0.3
26.35	-2.24	0.1	113.09	11JL	2.5	-3.8	-1.3	0.3
32.31	-2.23	0.0	122.68	21JL	0.7	-3.8	-1.3	0.4
37.87	-2.19	-0.1	132.30	31JL	-0.7	-3.7	-1.3	0.4
42.95	-2.12	-0.2	141.95	10AU	-1.3	-3.6	-1.2	0.5
47.43	-2.01	-0.4	151.66	20AU	-1.2	-3.6	-1.2	0.6
51.15	-1.87	-0.6	161.42	30AU	-0.6	-3.5	-1.2	0.6
53.96	-1.68	-0.8	171.23	9SE	-0.2	-3.5	-1.2	0.7
55.63	-1.43	-1.0	181.11	19SE	-0.1	-3.5	-1.2	0.7
55.97	-1.11	-1.2	191.04	29SE	0.0	-3.4	-1.2	0.8
54.86	-0.71	-1.4	201.03	9OC	0.3	-3.4	-1.2	0.8
52.35	-0.24	-1.6	211.07	19OC	1.5	-3.4	-1.3	0.8
48.85	0.26	-1.8	221.16	29OC	2.0	-3.4	-1.3	0.8
45.11	0.74	-1.7	231.29	8NO	-0.0	-3.4	-1.3	0.8
41.90	1.15	-1.4	241.45	18NO	-0.4	-3.4	-1.4	0.9
39.83	1.46	-1.1	251.64	28NO	-0.4	-3.4	-1.5	0.9
39.11	1.68	-0.8	261.83	8DE	-0.5	-3.4	-1.5	0.8
39.69	1.82	-0.5	272.04	18DE	-0.7	-3.4	-1.6	0.8
41.39	1.90	-0.2	282.23	28DE	-0.9	-3.4	-1.6	0.8

♂ LONG	♂ LAT	♂ MAG	☉ LONG	16.00UT	☿	♀	♃	♄
				902	MAGNITUDES			
44.01	1.94	0.1	292.40	7JA	-1.0	-3.4	-1.7	0.8
47.34	1.95	0.3	302.55	17JA	-0.9	-3.4	-1.8	0.8
51.24	1.95	0.6	312.65	27JA	0.2	-3.5	-1.8	0.9
55.58	1.93	0.8	322.71	6FE	2.6	-3.5	-1.9	0.9
60.26	1.90	0.9	332.73	16FE	2.0	-3.5	-1.9	0.9
65.21	1.86	1.1	342.68	26FE	1.0	-3.4	-2.0	1.0
70.38	1.82	1.2	352.58	8MR	0.6	-3.4	-2.0	1.0
75.72	1.77	1.4	2.42	18MR	0.3	-3.4	-2.0	1.0
81.20	1.72	1.5	12.20	28MR	-0.2	-3.4	-2.0	1.0
86.79	1.67	1.6	21.93	7AP	-0.8	-3.4	-2.0	0.9
92.48	1.61	1.7	31.61	17AP	-1.7	-3.3	-2.0	0.9
98.26	1.56	1.7	41.24	27AP	-1.3	-3.3	-1.9	0.9
104.10	1.50	1.8	50.84	7MY	-0.3	-3.3	-1.9	0.8
110.02	1.44	1.9	60.41	17MY	0.6	-3.3	-1.8	0.8
115.99	1.38	1.9	69.95	27MY	1.3	-3.3	-1.7	0.7
122.03	1.32	1.9	79.48	6JN	2.3	-3.4	-1.7	0.6
128.12	1.25	2.0	89.01	16JN	3.4	-3.4	-1.6	0.6
134.27	1.18	2.0	98.54	26JN	2.0	-3.4	-1.5	0.5
140.48	1.11	2.0	108.08	6JL	0.6	-3.5	-1.5	0.5
146.75	1.04	2.0	117.65	16JL	-0.7	-3.5	-1.4	0.4
153.08	0.97	2.0	127.25	26JL	-1.4	-3.6	-1.4	0.4
159.47	0.89	2.0	136.88	5AU	-1.2	-3.6	-1.3	0.5
165.93	0.81	1.9	146.57	15AU	-0.6	-3.7	-1.3	0.5
172.46	0.73	1.9	156.29	25AU	-0.2	-3.8	-1.3	0.6
179.06	0.65	1.9	166.08	4SE	0.1	-3.9	-1.2	0.6
185.74	0.56	1.9	175.93	14SE	0.2	-4.0	-1.2	0.7
192.49	0.47	1.8	185.82	24SE	0.5	-4.1	-1.2	0.7
199.32	0.38	1.8	195.78	4OC	1.8	-4.2	-1.2	0.8
206.23	0.29	1.7	205.80	14OC	1.8	-4.3	-1.2	0.8
213.22	0.19	1.7	215.86	24OC	-0.2	-4.4	-1.2	0.9
220.29	0.09	1.7	225.97	3NO	-0.6	-4.3	-1.2	0.9
227.44	-0.01	1.7	236.12	13NO	-0.6	-3.9	-1.3	0.9
234.66	-0.12	1.7	246.29	23NO	-0.6	-3.1	-1.3	0.9
241.97	-0.23	1.6	256.48	3DE	-0.8	-3.5	-1.3	0.9
249.35	-0.33	1.6	266.69	13DE	-0.8	-4.1	-1.4	0.9
256.79	-0.44	1.6	276.88	23DE	-0.8	-4.3	-1.4	0.9
				903				
264.31	-0.55	1.6	287.07	2JA	-0.7	-4.4	-1.5	0.9
271.88	-0.66	1.5	297.23	12JA	0.3	-4.3	-1.6	0.9
279.50	-0.76	1.5	307.36	22JA	2.9	-4.2	-1.6	0.9
287.17	-0.86	1.5	317.44	1FE	1.5	-4.1	-1.7	1.0
294.88	-0.96	1.4	327.48	11FE	0.7	-4.0	-1.8	1.0
302.61	-1.05	1.4	337.47	21FE	0.4	-3.9	-1.8	1.0
310.35	-1.13	1.4	347.40	3MR	0.1	-3.8	-1.9	1.1
318.10	-1.21	1.3	357.27	13MR	-0.2	-3.7	-2.0	1.1
325.84	-1.27	1.3	7.08	23MR	-0.9	-3.6	-2.0	1.1
333.56	-1.32	1.3	16.84	2AP	-1.7	-3.5	-2.0	1.1
341.25	-1.36	1.3	26.55	12AP	-1.3	-3.5	-2.1	1.1
348.88	-1.38	1.2	36.20	22AP	-0.2	-3.4	-2.1	1.0
356.46	-1.39	1.2	45.82	2MY	0.8	-3.4	-2.0	1.0
3.95	-1.39	1.2	55.40	12MY	1.7	-3.4	-2.0	1.0
11.36	-1.36	1.1	64.96	22MY	3.0	-3.3	-2.0	0.9
18.67	-1.33	1.1	74.49	1JN	2.9	-3.3	-1.9	0.9
25.85	-1.27	1.1	84.02	11JN	1.6	-3.3	-1.9	0.8
32.91	-1.21	1.0	93.55	21JN	0.4	-3.3	-1.8	0.7
39.82	-1.12	1.0	103.08	1JL	-0.7	-3.3	-1.8	0.7
46.56	-1.02	0.9	112.64	11JL	-1.5	-3.3	-1.7	0.6
53.12	-0.90	0.9	122.21	21JL	-1.2	-3.4	-1.6	0.5
59.46	-0.77	0.8	131.83	31JL	-0.5	-3.4	-1.6	0.5
65.57	-0.62	0.8	141.49	10AU	-0.1	-3.4	-1.5	0.5
71.40	-0.45	0.7	151.19	20AU	0.2	-3.4	-1.5	0.6
76.91	-0.26	0.6	160.95	30AU	0.4	-3.5	-1.4	0.6
82.03	-0.04	0.5	170.76	9SE	0.8	-3.5	-1.4	0.7
86.70	0.20	0.3	180.63	19SE	2.2	-3.5	-1.4	0.7
90.81	0.47	0.2	190.56	29SE	1.6	-3.4	-1.3	0.8
94.22	0.79	0.0	200.54	9OC	-0.3	-3.4	-1.3	0.8
96.80	1.14	-0.2	210.58	19OC	-0.7	-3.4	-1.3	0.9
98.34	1.54	-0.4	220.66	29OC	-0.7	-3.4	-1.3	0.9
98.64	1.99	-0.6	230.79	8NO	-0.7	-3.3	-1.3	1.0
97.57	2.46	-0.8	240.95	18NO	-0.7	-3.3	-1.3	1.0
95.12	2.92	-1.0	251.14	28NO	-0.7	-3.3	-1.3	1.0
91.61	3.31	-1.2	261.33	8DE	-0.7	-3.3	-1.3	1.0
87.64	3.57	-1.2	271.53	18DE	-0.6	-3.3	-1.4	1.0
83.99	3.68	-1.0	281.73	28DE	0.4	-3.4	-1.4	1.0

♂			⊙	16.00UT	☿	♀	♃	♄
LONG	LAT	MAG	LONG	904	MAGNITUDES			
81.32	3.65	-0.7	291.90	7JA	3.0	-3.4	-1.4	1.0
79.92	3.52	-0.4	302.05	17JA	1.1	-3.4	-1.5	1.0
79.83	3.34	-0.2	312.16	27JA	0.4	-3.4	-1.5	1.0
80.92	3.14	0.1	322.23	6FE	0.2	-3.5	-1.6	1.1
82.98	2.93	0.3	332.24	16FE	0.0	-3.5	-1.7	1.1
85.83	2.73	0.6	342.20	26FE	-0.3	-3.6	-1.7	1.1
89.31	2.54	0.8	352.10	7MR	-0.9	-3.6	-1.8	1.2
93.28	2.37	0.9	1.95	17MR	-1.7	-3.7	-1.9	1.2
97.65	2.20	1.1	11.74	27MR	-1.2	-3.7	-1.9	1.2
102.34	2.05	1.2	21.46	6AP	-0.2	-3.8	-2.0	1.2
107.30	1.90	1.3	31.15	16AP	1.0	-3.9	-2.1	1.2
112.47	1.77	1.4	40.79	26AP	2.3	-4.0	-2.1	1.2
117.83	1.64	1.5	50.38	6MY	3.8	-4.1	-2.1	1.2
123.34	1.51	1.6	59.95	16MY	2.3	-4.2	-2.2	1.2
129.00	1.39	1.7	69.50	26MY	1.2	-4.2	-2.2	1.1
134.79	1.27	1.7	79.02	5JN	0.3	-4.1	-2.1	1.1
140.69	1.15	1.8	88.55	15JN	-0.7	-3.8	-2.1	1.0
146.70	1.04	1.8	98.08	25JN	-1.6	-3.2	-2.1	0.9
152.82	0.93	1.8	107.62	5JL	-1.2	-3.2	-2.0	0.9
159.03	0.82	1.8	117.19	15JL	-0.5	-3.8	-2.0	0.8
165.35	0.71	1.8	126.79	25JL	0.0	-4.1	-1.9	0.7
171.77	0.60	1.8	136.42	4AU	0.3	-4.2	-1.9	0.7
178.29	0.49	1.8	146.09	14AU	0.6	-4.2	-1.8	0.6
184.90	0.38	1.8	155.82	24AU	1.1	-4.1	-1.7	0.7
191.62	0.27	1.8	165.60	3SE	2.7	-4.1	-1.7	0.7
198.43	0.17	1.8	175.44	13SE	1.4	-4.0	-1.6	0.8
205.35	0.06	1.7	185.34	23SE	-0.4	-3.9	-1.6	0.8
212.36	-0.05	1.7	195.29	3OC	-0.9	-3.8	-1.5	0.9
219.47	-0.15	1.7	205.31	13OC	-0.9	-3.7	-1.5	0.9
226.68	-0.26	1.6	215.37	23OC	-0.9	-3.7	-1.4	1.0
233.98	-0.36	1.6	225.47	2NO	-0.6	-3.6	-1.4	1.0
241.37	-0.46	1.5	235.62	12NO	-0.5	-3.6	-1.4	1.1
248.85	-0.55	1.5	245.80	22NO	-0.5	-3.5	-1.4	1.1
256.41	-0.64	1.4	255.99	2DE	-0.4	-3.5	-1.4	1.1
264.04	-0.73	1.4	266.19	12DE	0.6	-3.4	-1.4	1.2
271.73	-0.81	1.4	276.39	22DE	2.9	-3.4	-1.4	1.2
				905				
279.48	-0.88	1.4	286.57	1JA	0.8	-3.4	-1.4	1.2
287.28	-0.95	1.4	296.74	11JA	0.2	-3.4	-1.4	1.2
295.11	-1.01	1.4	306.87	21JA	0.1	-3.3	-1.5	1.2
302.96	-1.05	1.4	316.96	31JA	-0.1	-3.3	-1.5	1.2
310.82	-1.09	1.4	327.00	10FE	-0.4	-3.3	-1.6	1.2
318.68	-1.12	1.4	336.99	20FE	-0.9	-3.3	-1.6	1.2
326.53	-1.13	1.4	346.92	2MR	-1.5	-3.3	-1.7	1.3
334.35	-1.13	1.4	356.80	12MR	-1.2	-3.4	-1.7	1.3
342.13	-1.12	1.4	6.61	22MR	-0.1	-3.4	-1.8	1.4
349.86	-1.10	1.4	16.37	1AP	1.3	-3.4	-1.9	1.4
357.52	-1.06	1.4	26.08	11AP	2.9	-3.5	-1.9	1.4
5.12	-1.01	1.4	35.74	21AP	3.0	-3.5	-2.0	1.4
12.64	-0.95	1.4	45.35	1MY	1.7	-3.5	-2.1	1.4
20.07	-0.88	1.4	54.94	11MY	0.9	-3.4	-2.1	1.4
27.41	-0.80	1.4	64.49	21MY	0.2	-3.4	-2.2	1.4
34.64	-0.71	1.4	74.03	31MY	-0.8	-3.4	-2.2	1.3
41.77	-0.61	1.4	83.56	10JN	-1.7	-3.4	-2.3	1.3
48.79	-0.50	1.4	93.08	20JN	-1.3	-3.3	-2.3	1.2
55.69	-0.38	1.4	102.62	30JN	-0.4	-3.3	-2.3	1.2
62.46	-0.26	1.4	112.17	10JL	0.1	-3.3	-2.3	1.1
69.10	-0.12	1.4	121.75	20JL	0.5	-3.3	-2.2	1.0
75.60	0.02	1.4	131.36	30JL	0.8	-3.3	-2.2	1.0
81.96	0.17	1.3	141.02	9AU	1.5	-3.4	-2.2	0.9
88.14	0.32	1.3	150.71	19AU	3.1	-3.4	-2.1	0.8
94.15	0.49	1.3	160.47	29AU	1.2	-3.4	-2.0	0.8
99.94	0.67	1.2	170.28	8SE	-0.5	-3.4	-2.0	0.9
105.51	0.86	1.1	180.14	18SE	-1.0	-3.5	-1.9	0.9
110.80	1.07	1.0	190.07	28SE	-1.0	-3.6	-1.8	0.9
115.77	1.30	0.9	200.05	8OC	-0.8	-3.6	-1.8	1.0
120.35	1.55	0.8	210.09	18OC	-0.5	-3.7	-1.7	1.0
124.47	1.82	0.6	220.17	28OC	-0.4	-3.8	-1.7	1.1
128.00	2.12	0.5	230.30	7NO	-0.4	-3.9	-1.6	1.1
130.82	2.46	0.3	240.45	17NO	-0.2	-4.0	-1.6	1.2
132.77	2.83	0.1	250.64	27NO	0.8	-4.1	-1.5	1.2
133.64	3.23	-0.2	260.84	7DE	2.7	-4.2	-1.5	1.3
133.28	3.64	-0.4	271.04	17DE	0.5	-4.3	-1.5	1.3
131.58	4.03	-0.6	281.23	27DE	0.0	-4.4	-1.5	1.3

♂			⊙	16.00UT	☿	♀	♃	♄
LONG	LAT	MAG	LONG	906	MAGNITUDES			
128.64	4.35	-0.8	291.41	6JA	-0.1	-4.4	-1.5	1.4
124.87	4.52	-1.0	301.56	16JA	-0.2	-4.3	-1.5	1.4
120.93	4.52	-0.9	311.67	26JA	-0.5	-3.9	-1.5	1.4
117.55	4.35	-0.7	321.74	5FE	-0.9	-3.3	-1.5	1.4
115.25	4.06	-0.5	331.75	15FE	-1.4	-3.6	-1.5	1.3
114.23	3.72	-0.2	341.72	25FE	-1.1	-4.1	-1.6	1.3
114.47	3.36	0.0	351.62	7MR	-0.1	-4.3	-1.6	1.3
115.83	3.01	0.2	1.47	17MR	1.6	-4.3	-1.7	1.3
118.10	2.69	0.4	11.26	27MR	3.6	-4.2	-1.7	1.3
121.13	2.39	0.6	21.00	6AP	2.2	-4.1	-1.8	1.3
124.76	2.12	0.8	30.68	16AP	1.3	-4.0	-1.8	1.3
128.88	1.88	0.9	40.32	26AP	0.7	-3.9	-1.9	1.2
133.40	1.65	1.1	49.92	6MY	0.1	-3.8	-2.0	1.2
138.26	1.44	1.2	59.49	16MY	-0.8	-3.7	-2.0	1.2
143.40	1.25	1.3	69.03	26MY	-1.7	-3.6	-2.1	1.2
148.78	1.07	1.3	78.57	5JN	-1.3	-3.6	-2.2	1.1
154.38	0.90	1.4	88.09	15JN	-0.4	-3.5	-2.2	1.1
160.17	0.74	1.5	97.62	25JN	0.2	-3.5	-2.3	1.0
166.13	0.58	1.5	107.16	5JL	0.6	-3.4	-2.4	1.0
172.25	0.44	1.5	116.73	15JL	1.1	-3.4	-2.4	0.9
178.53	0.30	1.6	126.32	25JL	1.9	-3.4	-2.4	0.9
184.95	0.16	1.6	135.95	4AU	3.2	-3.4	-2.4	0.8
191.51	0.03	1.6	145.63	14AU	1.0	-3.3	-2.4	0.8
198.20	-0.09	1.6	155.36	24AU	-0.6	-3.3	-2.4	0.7
205.02	-0.21	1.6	165.14	3SE	-1.1	-3.3	-2.3	0.7
211.97	-0.32	1.6	174.97	13SE	-1.1	-3.4	-2.3	0.7
219.04	-0.43	1.6	184.87	23SE	-0.7	-3.4	-2.2	0.8
226.22	-0.53	1.6	194.82	3OC	-0.4	-3.4	-2.1	0.9
233.52	-0.63	1.5	204.82	13OC	-0.3	-3.4	-2.1	1.0
240.92	-0.72	1.5	214.88	23OC	-0.2	-3.4	-2.0	1.0
248.42	-0.80	1.5	224.98	2NO	-0.1	-3.5	-1.9	1.1
256.01	-0.87	1.5	235.12	12NO	1.0	-3.5	-1.9	1.1
263.68	-0.93	1.5	245.30	22NO	2.4	-3.5	-1.8	1.2
271.43	-0.99	1.4	255.49	2DE	0.3	-3.5	-1.8	1.2
279.23	-1.03	1.4	265.69	12DE	-0.1	-3.4	-1.7	1.2
287.07	-1.06	1.4	275.89	22DE	-0.2	-3.4	-1.7	1.2
				907				
294.96	-1.08	1.4	286.07	1JA	-0.3	-3.4	-1.6	1.2
302.85	-1.09	1.3	296.24	11JA	-0.5	-3.4	-1.6	1.2
310.76	-1.08	1.3	306.37	21JA	-1.0	-3.4	-1.6	1.2
318.66	-1.07	1.3	316.46	31JA	-1.2	-3.3	-1.6	1.2
326.54	-1.04	1.3	326.50	10FE	-1.0	-3.3	-1.6	1.1
334.39	-1.01	1.3	336.50	20FE	0.0	-3.3	-1.6	1.1
342.19	-0.96	1.4	346.43	2MR	1.9	-3.3	-1.6	1.0
349.93	-0.90	1.4	356.31	12MR	3.1	-3.4	-1.6	1.0
357.61	-0.83	1.4	6.13	22MR	1.6	-3.4	-1.6	1.0
5.22	-0.76	1.5	15.90	1AP	0.9	-3.4	-1.6	1.0
12.75	-0.67	1.5	25.61	11AP	0.5	-3.4	-1.7	1.0
20.20	-0.58	1.5	35.27	21AP	-0.0	-3.5	-1.7	1.0
27.56	-0.49	1.6	44.89	1MY	-0.8	-3.5	-1.8	1.0
34.82	-0.39	1.6	54.47	11MY	-1.7	-3.6	-1.8	1.0
42.00	-0.28	1.6	64.03	21MY	-1.3	-3.6	-1.9	1.0
49.07	-0.18	1.6	73.57	31MY	-0.4	-3.7	-2.0	0.9
56.06	-0.07	1.7	83.10	10JN	0.3	-3.8	-2.0	0.9
62.95	0.05	1.7	92.63	20JN	0.8	-3.8	-2.1	0.8
69.74	0.16	1.7	102.16	30JN	1.4	-3.9	-2.2	0.8
76.44	0.28	1.7	111.71	10JL	2.5	-4.0	-2.2	0.8
83.04	0.40	1.7	121.29	20JL	2.8	-4.1	-2.3	0.7
89.54	0.52	1.7	130.90	30JL	0.9	-4.2	-2.4	0.7
95.94	0.64	1.7	140.55	9AU	-0.6	-4.2	-2.4	0.6
102.24	0.77	1.7	150.25	19AU	-1.3	-4.1	-2.4	0.5
108.42	0.90	1.7	160.00	29AU	-1.2	-3.8	-2.5	0.5
114.48	1.03	1.6	169.81	8SE	-0.7	-3.3	-2.5	0.5
120.41	1.17	1.6	179.67	18SE	-0.3	-3.6	-2.5	0.4
126.20	1.31	1.5	189.59	28SE	-0.1	-4.1	-2.4	0.5
131.81	1.46	1.4	199.57	8OC	-0.0	-4.3	-2.4	0.5
137.24	1.62	1.4	209.60	18OC	0.1	-4.3	-2.3	0.6
142.43	1.79	1.3	219.68	28OC	1.2	-4.3	-2.3	0.7
147.36	1.97	1.1	229.80	7NO	2.2	-4.2	-2.2	0.7
151.96	2.17	1.0	239.96	17NO	0.1	-4.1	-2.1	0.8
156.16	2.38	0.9	250.14	27NO	-0.3	-4.0	-2.1	0.9
159.87	2.61	0.7	260.34	7DE	-0.3	-3.9	-2.0	0.9
162.96	2.87	0.5	270.54	17DE	-0.4	-3.8	-1.9	0.9
165.30	3.14	0.3	280.74	27DE	-0.6	-3.7	-1.9	1.0

♂ LONG	♂ LAT	♂ MAG	☉ LONG	16.00UT	☿	♀	♃	♄
				908	MAGNITUDES			
166.71	3.42	0.0	290.91	6JA	-1.0	-3.7	-1.8	1.0
167.00	3.71	-0.3	301.07	16JA	-1.1	-3.6	-1.7	1.0
166.05	3.96	-0.5	311.18	26JA	-0.9	-3.5	-1.7	1.0
163.80	4.15	-0.8	321.25	5FE	0.1	-3.5	-1.7	0.9
160.49	4.21	-1.0	331.27	15FE	2.3	-3.4	-1.6	0.9
156.63	4.11	-1.1	341.23	25FE	2.4	-3.4	-1.6	0.9
152.92	3.86	-0.9	351.14	6MR	1.2	-3.4	-1.6	0.8
150.02	3.48	-0.7	1.00	16MR	0.7	-3.3	-1.6	0.8
148.32	3.04	-0.5	10.79	26MR	0.3	-3.3	-1.6	0.7
147.92	2.60	-0.3	20.53	5AP	-0.1	-3.3	-1.6	0.8
148.75	2.18	-0.1	30.21	15AP	-0.8	-3.3	-1.6	0.8
150.63	1.80	0.1	39.85	25AP	-1.7	-3.3	-1.6	0.8
153.38	1.45	0.3	49.46	5MY	-1.3	-3.3	-1.6	0.8
156.86	1.14	0.5	59.03	15MY	-0.3	-3.3	-1.7	0.8
160.92	0.86	0.6	68.57	25MY	0.5	-3.4	-1.7	0.7
165.47	0.61	0.7	78.11	4JN	1.1	-3.4	-1.8	0.7
170.44	0.38	0.8	87.63	14JN	1.9	-3.4	-1.8	0.7
175.74	0.18	0.9	97.16	24JN	3.1	-3.5	-1.9	0.7
181.36	-0.01	1.0	106.70	4JL	2.3	-3.5	-1.9	0.6
187.24	-0.18	1.1	116.26	14JL	0.7	-3.5	-2.0	0.6
193.35	-0.34	1.1	125.85	24JL	-0.7	-3.4	-2.1	0.5
199.68	-0.48	1.2	135.48	3AU	-1.4	-3.4	-2.1	0.5
206.21	-0.61	1.2	145.16	13AU	-1.3	-3.4	-2.2	0.4
212.92	-0.73	1.2	154.88	23AU	-0.6	-3.3	-2.3	0.4
219.80	-0.84	1.2	164.66	2SE	-0.2	-3.3	-2.3	0.3
226.84	-0.93	1.3	174.49	12SE	-0.0	-3.3	-2.4	0.3
234.01	-1.01	1.3	184.38	22SE	0.1	-3.3	-2.4	0.2
241.32	-1.08	1.3	194.33	2OC	0.4	-3.3	-2.4	0.2
248.75	-1.14	1.3	204.33	12OC	1.5	-3.3	-2.4	0.2
256.29	-1.18	1.3	214.39	22OC	2.0	-3.4	-2.4	0.3
263.92	-1.21	1.3	224.49	1NO	-0.1	-3.4	-2.4	0.4
271.63	-1.22	1.3	234.63	11NO	-0.5	-3.4	-2.4	0.4
279.40	-1.23	1.4	244.80	21NO	-0.5	-3.5	-2.3	0.5
287.23	-1.22	1.4	255.00	1DE	-0.5	-3.5	-2.2	0.6
295.09	-1.19	1.4	265.19	11DE	-0.7	-3.6	-2.2	0.6
302.97	-1.16	1.4	275.39	21DE	-0.9	-3.6	-2.1	0.7
310.85	-1.11	1.4	285.58	31DE	-0.9	-3.7	-2.0	0.7
				909				
318.72	-1.05	1.4	295.75	10JA	-0.8	-3.8	-2.0	0.7
326.56	-0.99	1.4	305.88	20JA	0.2	-3.9	-1.9	0.7
334.37	-0.91	1.4	315.98	30JA	2.6	-3.9	-1.8	0.7
342.13	-0.83	1.4	326.02	9FE	1.8	-4.0	-1.8	0.7
349.83	-0.74	1.5	336.01	19FE	0.8	-4.1	-1.7	0.7
357.47	-0.65	1.5	345.96	1MR	0.5	-4.2	-1.7	0.7
5.02	-0.55	1.5	355.83	11MR	0.2	-4.3	-1.6	0.6
12.50	-0.45	1.5	5.66	21MR	-0.2	-4.2	-1.6	0.6
19.90	-0.34	1.5	15.42	31MR	-0.8	-4.0	-1.6	0.5
27.21	-0.24	1.5	25.14	10AP	-1.7	-3.6	-1.5	0.5
34.44	-0.13	1.5	34.80	20AP	-1.3	-2.9	-1.5	0.6
41.58	-0.03	1.6	44.42	30AP	-0.3	-3.6	-1.5	0.6
48.64	0.08	1.6	54.01	10MY	0.6	-4.0	-1.5	0.6
55.61	0.18	1.7	63.57	20MY	1.4	-4.2	-1.5	0.6
62.50	0.28	1.7	73.11	30MY	2.5	-4.2	-1.5	0.6
69.31	0.38	1.8	82.63	9JN	3.3	-4.1	-1.6	0.6
76.05	0.48	1.8	92.16	19JN	1.8	-4.0	-1.6	0.6
82.71	0.57	1.8	101.70	29JN	0.5	-3.9	-1.6	0.5
89.30	0.67	1.9	111.24	9JL	-0.7	-3.8	-1.7	0.5
95.83	0.76	1.9	120.82	19JL	-1.5	-3.8	-1.7	0.5
102.29	0.85	1.9	130.43	29JL	-1.3	-3.7	-1.8	0.4
108.68	0.94	1.9	140.08	8AU	-0.6	-3.6	-1.8	0.4
115.01	1.03	1.9	149.78	18AU	-0.1	-3.6	-1.9	0.3
121.28	1.12	1.9	159.53	28AU	0.1	-3.5	-2.0	0.2
127.47	1.20	1.9	169.33	7SE	0.3	-3.5	-2.0	0.2
133.60	1.29	1.9	179.19	17SE	0.6	-3.5	-2.1	0.1
139.65	1.37	1.8	189.11	27SE	1.9	-3.4	-2.2	0.1
145.62	1.46	1.8	199.08	7OC	1.8	-3.4	-2.2	0.0
151.49	1.55	1.7	209.11	17OC	-0.2	-3.4	-2.3	-0.0
157.26	1.63	1.6	219.19	27OC	-0.6	-3.4	-2.3	0.0
162.90	1.73	1.6	229.31	6NO	-0.6	-3.4	-2.3	0.1
168.41	1.82	1.5	239.47	16NO	-0.6	-3.4	-2.3	0.2
173.73	1.91	1.4	249.65	26NO	-0.8	-3.4	-2.3	0.3
178.85	2.01	1.2	259.85	6DE	-0.7	-3.4	-2.3	0.3
183.73	2.11	1.1	270.05	16DE	-0.8	-3.4	-2.3	0.4
188.29	2.21	0.9	280.25	26DE	-0.7	-3.4	-2.2	0.4

♂ LONG	♂ LAT	♂ MAG	☉ LONG	16.00UT	☿	♀	♃	♄
				910	MAGNITUDES			
192.47	2.32	0.7	290.42	5JA	0.3	-3.4	-2.1	0.5
196.18	2.42	0.5	300.57	15JA	2.9	-3.4	-2.1	0.5
199.31	2.53	0.3	310.69	25JA	1.4	-3.5	-2.0	0.5
201.70	2.62	0.0	320.76	4FE	0.6	-3.5	-1.9	0.5
203.19	2.69	-0.2	330.78	14FE	0.3	-3.5	-1.9	0.6
203.58	2.73	-0.5	340.75	24FE	0.1	-3.4	-1.8	0.5
202.76	2.71	-0.8	350.66	6MR	-0.2	-3.4	-1.7	0.5
200.66	2.59	-1.1	0.51	16MR	-0.8	-3.4	-1.7	0.5
197.52	2.37	-1.4	10.31	26MR	-1.7	-3.4	-1.6	0.5
193.84	2.03	-1.4	20.05	5AP	-1.3	-3.3	-1.6	0.4
190.31	1.59	-1.3	29.74	15AP	-0.2	-3.3	-1.5	0.4
187.63	1.12	-1.1	39.38	25AP	0.8	-3.3	-1.5	0.4
186.17	0.66	-0.9	48.98	5MY	1.9	-3.3	-1.5	0.4
186.06	0.24	-0.7	58.56	15MY	3.3	-3.3	-1.4	0.4
187.22	-0.13	-0.5	68.11	25MY	2.7	-3.3	-1.4	0.4
189.47	-0.44	-0.3	77.64	4JN	1.4	-3.4	-1.4	0.4
192.65	-0.71	-0.2	87.17	14JN	0.4	-3.4	-1.4	0.4
196.59	-0.93	-0.0	96.69	24JN	-0.7	-3.4	-1.4	0.4
201.15	-1.12	0.1	106.23	4JL	-1.5	-3.5	-1.4	0.4
206.24	-1.28	0.2	115.80	14JL	-1.3	-3.5	-1.5	0.4
211.77	-1.41	0.4	125.39	24JL	-0.5	-3.6	-1.5	0.4
217.67	-1.51	0.4	135.01	3AU	-0.0	-3.6	-1.5	0.3
223.89	-1.59	0.5	144.69	13AU	0.2	-3.7	-1.6	0.3
230.39	-1.65	0.6	154.41	23AU	0.5	-3.8	-1.6	0.2
237.13	-1.69	0.7	164.18	2SE	0.9	-3.9	-1.7	0.2
244.08	-1.71	0.7	174.01	12SE	2.3	-4.0	-1.7	0.1
251.20	-1.71	0.8	183.90	22SE	1.6	-4.1	-1.8	0.0
258.48	-1.70	0.9	193.85	2OC	-0.3	-4.2	-1.8	-0.0
265.89	-1.66	0.9	203.85	12OC	-0.8	-4.3	-1.9	-0.1
273.41	-1.61	1.0	213.90	22OC	-0.8	-4.4	-2.0	-0.1
281.01	-1.55	1.0	224.00	1NO	-0.8	-4.3	-2.0	-0.2
288.68	-1.47	1.1	234.14	11NO	-0.7	-3.9	-2.1	-0.1
296.39	-1.39	1.1	244.31	21NO	-0.6	-3.0	-2.1	-0.0
304.13	-1.29	1.2	254.50	1DE	-0.6	-3.6	-2.2	0.1
311.87	-1.18	1.2	264.70	11DE	-0.5	-4.1	-2.2	0.1
319.61	-1.07	1.3	274.90	21DE	0.4	-4.4	-2.2	0.2
327.32	-0.95	1.3	285.09	31DE	3.0	-4.4	-2.2	0.3
				911				
335.00	-0.83	1.4	295.26	10JA	1.0	-4.3	-2.2	0.3
342.62	-0.71	1.4	305.39	20JA	0.3	-4.2	-2.1	0.3
350.19	-0.59	1.5	315.49	30JA	0.2	-4.1	-2.1	0.4
357.69	-0.46	1.5	325.54	9FE	-0.0	-4.0	-2.0	0.4
5.13	-0.34	1.6	335.53	19FE	-0.3	-3.9	-1.9	0.4
12.48	-0.22	1.6	345.47	1MR	-0.9	-3.8	-1.9	0.4
19.76	-0.11	1.6	355.36	11MR	-1.6	-3.7	-1.8	0.4
26.96	0.00	1.7	5.18	21MR	-1.2	-3.6	-1.7	0.4
34.08	0.11	1.7	14.95	31MR	-0.2	-3.5	-1.7	0.4
41.12	0.21	1.7	24.67	10AP	1.1	-3.5	-1.6	0.3
48.09	0.31	1.8	34.33	20AP	2.5	-3.4	-1.5	0.3
54.98	0.41	1.8	43.96	30AP	3.5	-3.4	-1.5	0.3
61.80	0.50	1.8	53.55	10MY	2.0	-3.4	-1.5	0.3
68.56	0.58	1.8	63.11	20MY	1.1	-3.3	-1.4	0.3
75.25	0.66	1.8	72.65	30MY	0.3	-3.3	-1.4	0.3
81.89	0.74	1.8	82.18	9JN	-0.7	-3.3	-1.4	0.3
88.47	0.81	1.9	91.70	19JN	-1.6	-3.3	-1.3	0.4
95.00	0.88	1.9	101.24	29JN	-1.3	-3.3	-1.3	0.4
101.49	0.94	1.9	110.79	9JL	-0.5	-3.3	-1.3	0.4
107.94	1.00	2.0	120.36	19JL	0.0	-3.4	-1.3	0.4
114.35	1.05	2.0	129.97	29JL	0.4	-3.4	-1.3	0.3
120.73	1.10	2.0	139.62	8AU	0.7	-3.4	-1.3	0.3
127.07	1.15	2.0	149.31	18AU	1.2	-3.4	-1.4	0.3
133.39	1.20	2.0	159.05	28AU	2.8	-3.5	-1.4	0.2
139.68	1.24	2.0	168.85	7SE	1.4	-3.5	-1.4	0.2
145.94	1.28	2.0	178.71	17SE	-0.4	-3.5	-1.5	0.1
152.18	1.31	2.0	188.63	27SE	-0.9	-3.4	-1.5	0.0
158.38	1.34	1.9	198.60	7OC	-0.9	-3.4	-1.6	-0.0
164.55	1.37	1.9	208.62	17OC	-0.9	-3.4	-1.6	-0.1
170.69	1.40	1.9	218.70	27OC	-0.6	-3.4	-1.7	-0.2
176.79	1.42	1.8	228.82	6NO	-0.5	-3.3	-1.8	-0.2
182.84	1.43	1.7	238.97	16NO	-0.5	-3.3	-1.8	-0.2
188.84	1.44	1.6	249.15	26NO	-0.4	-3.3	-1.9	-0.2
194.77	1.45	1.6	259.35	6DE	0.6	-3.3	-1.9	-0.1
200.62	1.45	1.5	269.55	16DE	2.9	-3.3	-2.0	-0.0
206.37	1.44	1.3	279.75	26DE	0.7	-3.4	-2.0	0.0

912

♂ LONG	LAT	MAG	☉ LONG	16.00UT	☿	♀	♃	♄
212.00	1.42	1.2	289.93	5JA	0.2	-3.4	-2.1	0.1
217.50	1.39	1.1	300.08	15JA	0.0	-3.4	-2.1	0.2
222.82	1.34	0.9	310.20	25JA	-0.1	-3.4	-2.1	0.2
227.92	1.28	0.7	320.27	4FE	-0.4	-3.5	-2.1	0.3
232.76	1.19	0.5	330.30	14FE	-0.9	-3.5	-2.0	0.3
237.26	1.07	0.3	340.27	24FE	-1.5	-3.6	-2.0	0.3
241.35	0.92	0.1	350.18	5MR	-1.2	-3.6	-1.9	0.3
244.92	0.72	-0.2	0.04	15MR	-0.1	-3.7	-1.9	0.3
247.81	0.46	-0.5	9.84	25MR	1.3	-3.8	-1.8	0.3
249.88	0.13	-0.8	19.58	4AP	3.2	-3.8	-1.8	0.3
250.95	-0.28	-1.1	29.27	14AP	2.7	-3.9	-1.7	0.3
250.82	-0.78	-1.4	38.92	24AP	1.5	-4.0	-1.6	0.3
249.46	-1.37	-1.7	48.53	4MY	0.8	-4.1	-1.6	0.2
247.00	-1.99	-2.0	58.10	14MY	0.2	-4.2	-1.5	0.2
243.88	-2.60	-2.2	67.65	24MY	-0.7	-4.2	-1.4	0.2
240.80	-3.11	-2.1	77.18	3JN	-1.6	-4.1	-1.4	0.2
238.47	-3.48	-1.9	86.70	13JN	-1.3	-3.8	-1.4	0.3
237.35	-3.69	-1.7	96.24	23JN	-0.4	-3.1	-1.3	0.3
237.63	-3.78	-1.5	105.77	3JL	0.1	-3.2	-1.3	0.3
239.23	-3.78	-1.2	115.33	13JL	0.5	-3.9	-1.3	0.3
242.00	-3.71	-1.0	124.92	23JL	0.9	-4.1	-1.2	0.3
245.74	-3.59	-0.8	134.55	2AU	1.6	-4.2	-1.2	0.3
250.25	-3.44	-0.6	144.22	12AU	3.2	-4.2	-1.2	0.3
255.41	-3.27	-0.5	153.94	22AU	1.2	-4.1	-1.2	0.3
261.08	-3.08	-0.3	163.70	1SE	-0.5	-4.0	-1.2	0.2
267.15	-2.87	-0.2	173.53	11SE	-1.1	-4.0	-1.3	0.2
273.56	-2.66	-0.0	183.42	21SE	-1.1	-3.9	-1.3	0.1
280.22	-2.44	0.1	193.36	1OC	-0.8	-3.8	-1.3	0.1
287.07	-2.21	0.3	203.36	11OC	-0.5	-3.7	-1.4	0.0
294.09	-1.99	0.4	213.42	21OC	-0.3	-3.7	-1.4	-0.1
301.22	-1.77	0.5	223.51	31OC	-0.3	-3.6	-1.5	-0.1
308.42	-1.55	0.6	233.65	10NO	-0.2	-3.6	-1.5	-0.2
315.68	-1.34	0.7	243.82	20NO	0.8	-3.5	-1.6	-0.3
322.97	-1.13	0.8	254.00	30NO	2.7	-3.5	-1.6	-0.3
330.26	-0.93	1.0	264.21	10DE	0.4	-3.4	-1.7	-0.2
337.55	-0.74	1.1	274.41	20DE	-0.0	-3.4	-1.8	-0.1
344.81	-0.56	1.2	284.59	30DE	-0.1	-3.4	-1.8	-0.1

913

♂ LONG	LAT	MAG	☉ LONG	16.00UT	☿	♀	♃	♄
352.04	-0.40	1.2	294.76	9JA	-0.2	-3.4	-1.9	0.0
359.23	-0.24	1.3	304.90	19JA	-0.5	-3.3	-1.9	0.1
6.36	-0.09	1.4	315.00	29JA	-0.9	-3.3	-2.0	0.1
13.44	0.05	1.5	325.05	8FE	-1.3	-3.3	-2.0	0.2
20.46	0.17	1.6	335.05	18FE	-1.1	-3.3	-2.0	0.2
27.42	0.29	1.6	344.99	28FE	-0.1	-3.3	-2.0	0.3
34.33	0.40	1.7	354.88	10MR	1.6	-3.4	-2.0	0.3
41.17	0.49	1.7	4.71	20MR	3.5	-3.4	-2.0	0.3
47.95	0.58	1.8	14.48	30MR	2.0	-3.4	-1.9	0.3
54.67	0.67	1.8	24.20	9AP	1.1	-3.5	-1.9	0.3
61.35	0.74	1.9	33.87	19AP	0.6	-3.5	-1.8	0.3
67.97	0.81	1.9	43.49	29AP	0.1	-3.5	-1.7	0.3
74.54	0.87	1.9	53.08	9MY	-0.7	-3.4	-1.7	0.3
81.08	0.92	2.0	62.64	19MY	-1.7	-3.4	-1.6	0.2
87.57	0.97	2.0	72.18	29MY	-1.3	-3.4	-1.6	0.2
94.04	1.01	2.0	81.71	8JN	-0.4	-3.4	-1.5	0.2
100.48	1.05	2.0	91.24	18JN	0.2	-3.3	-1.4	0.3
106.89	1.08	2.0	100.77	28JN	0.7	-3.3	-1.4	0.3
113.28	1.11	2.0	110.32	8JL	1.2	-3.3	-1.3	0.3
119.67	1.13	1.9	119.89	18JL	2.1	-3.3	-1.3	0.3
126.04	1.15	2.0	129.49	28JL	3.1	-3.3	-1.3	0.4
132.40	1.16	2.0	139.14	7AU	1.0	-3.4	-1.2	0.4
138.77	1.17	2.0	148.83	17AU	-0.6	-3.4	-1.2	0.3
145.14	1.17	2.0	158.57	27AU	-1.2	-3.4	-1.2	0.3
151.51	1.17	2.0	168.37	6SE	-1.2	-3.4	-1.2	0.3
157.89	1.17	2.0	178.23	16SE	-0.7	-3.5	-1.2	0.3
164.28	1.16	2.0	188.14	26SE	-0.4	-3.6	-1.2	0.2
170.68	1.14	2.0	198.11	6OC	-0.2	-3.6	-1.2	0.2
177.09	1.12	2.0	208.14	16OC	-0.1	-3.7	-1.3	0.1
183.51	1.09	1.9	218.21	26OC	0.0	-3.8	-1.3	0.0
189.95	1.06	1.9	228.33	5NO	1.0	-3.9	-1.3	-0.0
196.40	1.02	1.9	238.48	15NO	2.4	-4.0	-1.4	-0.1
202.86	0.98	1.8	248.66	25NO	0.2	-4.1	-1.4	-0.2
209.32	0.92	1.7	258.86	5DE	-0.2	-4.2	-1.5	-0.3
215.79	0.86	1.7	269.06	15DE	-0.3	-4.3	-1.5	-0.2
222.27	0.78	1.6	279.25	25DE	-0.3	-4.4	-1.6	-0.2

914

♂ LONG	LAT	MAG	☉ LONG	16.00UT	☿	♀	♃	♄
228.74	0.69	1.5	289.44	4JA	-0.6	-4.4	-1.7	-0.1
235.21	0.59	1.4	299.59	14JA	-1.0	-4.3	-1.7	-0.0
241.66	0.47	1.3	309.71	24JA	-1.2	-3.9	-1.8	0.0
248.10	0.34	1.1	319.79	3FE	-1.0	-3.3	-1.9	0.1
254.51	0.18	1.0	329.81	13FE	-0.0	-3.6	-1.9	0.2
260.88	0.01	0.9	339.79	23FE	2.0	-4.1	-2.0	0.2
267.20	-0.20	0.7	349.71	5MR	2.8	-4.3	-2.0	0.3
273.46	-0.43	0.6	359.56	15MR	1.4	-4.3	-2.0	0.3
279.62	-0.69	0.4	9.36	25MR	0.8	-4.2	-2.0	0.3
285.66	-0.98	0.2	19.11	4AP	0.4	-4.1	-2.0	0.3
291.55	-1.32	0.0	28.81	14AP	-0.0	-4.0	-2.0	0.3
297.23	-1.70	-0.2	38.45	24AP	-0.8	-3.9	-1.9	0.3
302.64	-2.13	-0.4	48.06	4MY	-1.7	-3.8	-1.9	0.3
307.69	-2.60	-0.6	57.64	14MY	-1.3	-3.7	-1.8	0.3
312.28	-3.13	-0.9	67.19	24MY	-0.4	-3.6	-1.8	0.3
316.26	-3.72	-1.1	76.72	3JN	0.4	-3.6	-1.7	0.3
319.46	-4.35	-1.4	86.25	13JN	0.9	-3.5	-1.7	0.2
321.69	-5.02	-1.7	95.77	23JN	1.6	-3.5	-1.6	0.3
322.74	-5.69	-2.0	105.32	3JL	2.7	-3.4	-1.5	0.3
322.49	-6.28	-2.2	114.87	13JL	2.6	-3.4	-1.5	0.3
320.99	-6.70	-2.5	124.46	23JL	0.8	-3.4	-1.4	0.4
318.59	-6.85	-2.6	134.09	2AU	-0.6	-3.4	-1.4	0.4
315.97	-6.68	-2.4	143.75	12AU	-1.3	-3.3	-1.3	0.4
313.84	-6.21	-2.4	153.47	22AU	-1.3	-3.3	-1.3	0.4
312.77	-5.54	-2.1	163.24	1SE	-0.7	-3.3	-1.3	0.4
313.01	-4.79	-1.8	173.06	11SE	-0.3	-3.4	-1.3	0.4
314.51	-4.05	-1.5	182.94	21SE	-0.1	-3.4	-1.2	0.4
317.13	-3.35	-1.2	192.88	1OC	0.0	-3.4	-1.2	0.3
320.65	-2.72	-0.9	202.88	11OC	0.2	-3.4	-1.2	0.3
324.88	-2.17	-0.7	212.93	21OC	1.3	-3.4	-1.2	0.2
329.68	-1.69	-0.4	223.02	31OC	2.2	-3.5	-1.2	0.2
334.90	-1.27	-0.2	233.16	10NO	0.0	-3.5	-1.3	0.1
340.44	-0.90	0.1	243.32	20NO	-0.4	-3.5	-1.3	0.0
346.23	-0.59	0.2	253.51	30NO	-0.4	-3.5	-1.3	-0.0
352.20	-0.31	0.4	263.71	10DE	-0.5	-3.4	-1.4	-0.1
358.31	-0.08	0.6	273.91	20DE	-0.6	-3.4	-1.4	-0.2
4.53	0.13	0.8	284.10	30DE	-0.9	-3.4	-1.5	-0.2

915

♂ LONG	LAT	MAG	☉ LONG	16.00UT	☿	♀	♃	♄
10.82	0.30	0.9	294.27	9JA	-1.0	-3.4	-1.5	-0.1
17.15	0.46	1.1	304.41	19JA	-0.9	-3.4	-1.6	-0.0
23.52	0.59	1.2	314.51	29JA	0.1	-3.3	-1.7	0.0
29.90	0.70	1.3	324.56	8FE	2.3	-3.3	-1.7	0.1
36.29	0.80	1.4	334.57	18FE	2.2	-3.3	-1.8	0.2
42.68	0.88	1.5	344.51	28FE	1.0	-3.3	-1.9	0.2
49.06	0.95	1.6	354.40	10MR	0.6	-3.4	-1.9	0.3
55.44	1.01	1.7	4.23	20MR	0.3	-3.4	-2.0	0.3
61.80	1.06	1.8	14.01	30MR	-0.1	-3.4	-2.0	0.3
68.14	1.10	1.8	23.73	9AP	-0.8	-3.4	-2.1	0.4
74.48	1.13	1.9	33.40	19AP	-1.7	-3.5	-2.1	0.4
80.80	1.16	1.9	43.03	29AP	-1.3	-3.5	-2.1	0.4
87.11	1.18	2.0	52.62	9MY	-0.3	-3.6	-2.1	0.4
93.41	1.19	2.0	62.19	19MY	0.5	-3.6	-2.0	0.4
99.71	1.20	2.0	71.73	29MY	1.2	-3.7	-2.0	0.4
106.00	1.20	2.0	81.25	8JN	2.1	-3.8	-1.9	0.4
112.30	1.20	2.0	90.78	18JN	3.3	-3.8	-1.9	0.4
118.61	1.19	2.0	100.31	28JN	2.1	-3.9	-1.8	0.4
124.92	1.18	2.0	109.86	8JL	0.7	-4.0	-1.8	0.4
131.25	1.16	2.0	119.43	18JL	-0.6	-4.1	-1.7	0.4
137.60	1.14	2.0	129.04	28JL	-1.4	-4.2	-1.6	0.5
143.97	1.11	2.0	138.68	7AU	-1.3	-4.2	-1.6	0.5
150.37	1.08	1.9	148.37	17AU	-0.6	-4.1	-1.5	0.5
156.80	1.05	1.9	158.11	27AU	-0.2	-3.8	-1.5	0.5
163.26	1.01	1.9	167.90	6SE	0.0	-3.3	-1.4	0.5
169.76	0.97	1.9	177.75	16SE	0.2	-3.6	-1.4	0.5
176.30	0.92	1.9	187.66	26SE	0.4	-4.1	-1.4	0.5
182.88	0.86	1.9	197.63	6OC	1.6	-4.3	-1.3	0.5
189.51	0.81	1.9	207.65	16OC	2.0	-4.3	-1.3	0.5
196.18	0.74	1.9	217.72	26OC	-0.1	-4.3	-1.3	0.4
202.90	0.67	1.9	227.84	5NO	-0.5	-4.2	-1.3	0.3
209.67	0.59	1.9	237.99	15NO	-0.5	-4.1	-1.3	0.3
216.49	0.51	1.8	248.17	25NO	-0.6	-4.0	-1.3	0.2
223.36	0.42	1.8	258.36	5DE	-0.7	-3.9	-1.3	0.2
230.28	0.32	1.7	268.57	15DE	-0.8	-3.8	-1.3	0.1
237.24	0.21	1.7	278.76	25DE	-0.9	-3.7	-1.4	0.0

916

♂ LONG	LAT	MAG	☉ LONG	16.00UT 916	☿	♀	♃	♄
244.26	0.09	1.6	288.94	4JA	-0.8	-3.7	-1.4	-0.0
251.32	-0.03	1.5	299.10	14JA	0.2	-3.6	-1.5	-0.0
258.43	-0.17	1.5	309.22	24JA	2.7	-3.5	-1.5	0.0
265.58	-0.31	1.4	319.30	3FE	1.6	-3.5	-1.6	0.1
272.77	-0.47	1.3	329.34	13FE	0.7	-3.4	-1.6	0.1
279.99	-0.63	1.2	339.31	23FE	0.4	-3.4	-1.7	0.2
287.24	-0.80	1.1	349.23	4MR	0.2	-3.4	-1.8	0.3
294.50	-0.98	1.0	359.10	14MR	-0.2	-3.3	-1.8	0.3
301.78	-1.16	0.9	8.90	24MR	-0.8	-3.3	-1.9	0.4
309.06	-1.35	0.8	18.65	3AP	-1.7	-3.3	-2.0	0.4
316.31	-1.54	0.7	28.35	13AP	-1.3	-3.3	-2.0	0.5
323.54	-1.73	0.6	37.99	23AP	-0.3	-3.3	-2.1	0.5
330.71	-1.92	0.5	47.60	3MY	0.7	-3.3	-2.1	0.5
337.79	-2.11	0.4	57.18	13MY	1.6	-3.3	-2.2	0.5
344.76	-2.29	0.3	66.73	23MY	2.8	-3.4	-2.2	0.6
351.58	-2.46	0.2	76.26	2JN	3.1	-3.4	-2.2	0.6
358.21	-2.62	0.1	85.79	12JN	1.7	-3.4	-2.2	0.6
4.60	-2.76	-0.1	95.32	22JN	0.5	-3.5	-2.1	0.6
10.69	-2.89	-0.2	104.85	2JL	-0.7	-3.5	-2.1	0.5
16.38	-3.00	-0.3	114.41	12JL	-1.5	-3.5	-2.0	0.5
21.60	-3.08	-0.5	123.99	22JL	-1.3	-3.4	-2.0	0.5
26.21	-3.14	-0.7	133.62	1AU	-0.6	-3.4	-1.9	0.6
30.06	-3.16	-0.9	143.28	11AU	-0.1	-3.4	-1.9	0.6
32.97	-3.14	-1.1	152.99	21AU	0.2	-3.3	-1.8	0.7
34.72	-3.05	-1.3	162.76	31AU	0.3	-3.3	-1.7	0.7
35.11	-2.88	-1.5	172.58	10SE	0.7	-3.3	-1.7	0.7
34.04	-2.60	-1.7	182.45	20SE	2.0	-3.3	-1.6	0.7
31.59	-2.20	-1.9	192.39	30SE	1.8	-3.3	-1.6	0.7
28.24	-1.68	-2.1	202.39	10OC	-0.2	-3.3	-1.5	0.7
24.75	-1.10	-1.9	212.43	20OC	-0.7	-3.4	-1.5	0.7
21.89	-0.53	-1.7	222.53	30OC	-0.7	-3.4	-1.5	0.6
20.22	-0.03	-1.3	232.66	9NO	-0.7	-3.4	-1.4	0.6
19.90	0.39	-1.0	242.82	19NO	-0.7	-3.5	-1.4	0.5
20.85	0.71	-0.7	253.01	29NO	-0.7	-3.5	-1.4	0.5
22.90	0.96	-0.4	263.21	9DE	-0.7	-3.6	-1.4	0.4
25.83	1.15	-0.1	273.41	19DE	-0.6	-3.6	-1.4	0.3
29.44	1.29	0.2	283.61	29DE	0.3	-3.7	-1.4	0.3

917

♂ LONG	LAT	MAG	☉ LONG	16.00UT 917	☿	♀	♃	♄
33.59	1.39	0.4	293.78	8JA	2.9	-3.8	-1.4	0.2
38.15	1.46	0.6	303.92	18JA	1.2	-3.9	-1.5	0.1
43.04	1.51	0.8	314.02	28JA	0.5	-4.0	-1.5	0.2
48.18	1.54	1.0	324.08	7FE	0.3	-4.0	-1.5	0.2
53.51	1.55	1.1	334.08	17FE	0.1	-4.1	-1.6	0.3
58.99	1.56	1.3	344.04	27FE	-0.3	-4.2	-1.6	0.3
64.60	1.56	1.4	353.93	9MR	-0.8	-4.3	-1.7	0.4
70.31	1.55	1.5	3.76	19MR	-1.6	-4.2	-1.8	0.4
76.09	1.53	1.6	13.54	29MR	-1.3	-4.0	-1.8	0.5
81.95	1.51	1.7	23.26	8AP	-0.3	-3.5	-1.9	0.6
87.85	1.48	1.8	32.94	18AP	0.9	-2.9	-2.0	0.6
93.81	1.45	1.8	42.57	28AP	2.1	-3.6	-2.0	0.6
99.81	1.42	1.9	52.16	8MY	3.7	-4.0	-2.1	0.7
105.85	1.38	1.9	61.72	18MY	2.4	-4.2	-2.2	0.7
111.93	1.34	2.0	71.27	28MY	1.3	-4.2	-2.2	0.7
118.04	1.30	2.0	80.79	7JN	0.4	-4.1	-2.3	0.7
124.20	1.25	2.0	90.32	17JN	-0.7	-4.0	-2.3	0.8
130.40	1.20	2.0	99.85	27JN	-1.6	-3.9	-2.3	0.8
136.64	1.15	2.0	109.40	7JL	-1.3	-3.8	-2.3	0.8
142.93	1.09	2.0	118.97	17JL	-0.5	-3.8	-2.3	0.8
149.27	1.03	2.0	128.57	27JL	-0.0	-3.7	-2.3	0.7
155.66	0.97	2.0	138.21	6AU	0.3	-3.6	-2.2	0.8
162.10	0.90	2.0	147.90	16AU	0.5	-3.6	-2.2	0.8
168.61	0.83	1.9	157.63	26AU	1.0	-3.5	-2.1	0.9
175.17	0.76	1.9	167.42	5SE	2.4	-3.5	-2.0	0.9
181.80	0.68	1.9	177.27	15SE	1.6	-3.4	-2.0	0.9
188.49	0.61	1.8	187.18	25SE	-0.3	-3.4	-1.9	0.9
195.26	0.52	1.8	197.14	5OC	-0.8	-3.4	-1.8	1.0
202.09	0.43	1.8	207.16	15OC	-0.8	-3.4	-1.8	1.0
209.00	0.34	1.8	217.23	25OC	-0.8	-3.4	-1.7	0.9
215.98	0.25	1.8	227.34	4NO	-0.6	-3.4	-1.7	0.9
223.03	0.15	1.7	237.49	14NO	-0.5	-3.4	-1.6	0.9
230.16	0.05	1.7	247.67	24NO	-0.5	-3.4	-1.6	0.8
237.35	-0.06	1.7	257.86	4DE	-0.5	-3.4	-1.6	0.8
244.62	-0.17	1.7	268.07	14DE	0.4	-3.4	-1.5	0.7
251.96	-0.29	1.6	278.27	24DE	3.0	-3.4	-1.5	0.7

918

♂ LONG	LAT	MAG	☉ LONG	16.00UT 918	☿	♀	♃	♄
259.35	-0.40	1.6	288.45	3JA	0.9	-3.4	-1.5	0.6
266.81	-0.52	1.6	298.60	13JA	0.3	-3.4	-1.5	0.5
274.33	-0.64	1.5	308.73	23JA	0.1	-3.5	-1.5	0.4
281.89	-0.75	1.5	318.81	2FE	-0.1	-3.5	-1.5	0.4
289.50	-0.87	1.4	328.84	12FE	-0.3	-3.5	-1.5	0.4
297.13	-0.98	1.4	338.83	22FE	-0.9	-3.4	-1.6	0.5
304.79	-1.09	1.3	348.75	4MR	-1.6	-3.4	-1.6	0.5
312.47	-1.19	1.3	358.61	14MR	-1.2	-3.4	-1.6	0.6
320.14	-1.29	1.2	8.42	24MR	-0.2	-3.4	-1.7	0.6
327.80	-1.37	1.2	18.17	3AP	1.1	-3.3	-1.7	0.7
335.44	-1.45	1.1	27.87	13AP	2.7	-3.3	-1.8	0.7
343.04	-1.51	1.1	37.52	23AP	3.2	-3.3	-1.9	0.8
350.57	-1.55	1.1	47.13	3MY	1.8	-3.3	-1.9	0.8
358.04	-1.59	1.0	56.71	13MY	1.0	-3.3	-2.0	0.9
5.42	-1.60	1.0	66.27	23MY	0.2	-3.3	-2.1	0.9
12.69	-1.60	0.9	75.80	2JN	-0.7	-3.4	-2.1	0.9
19.83	-1.58	0.9	85.33	12JN	-1.6	-3.4	-2.2	1.0
26.83	-1.54	0.8	94.85	22JN	-1.3	-3.4	-2.3	1.0
33.66	-1.48	0.7	104.39	2JL	-0.5	-3.5	-2.3	1.0
40.29	-1.41	0.7	113.95	12JL	0.1	-3.5	-2.4	1.0
46.70	-1.31	0.6	123.53	22JL	0.4	-3.6	-2.4	1.0
52.84	-1.20	0.5	133.15	1AU	0.7	-3.6	-2.4	1.0
58.68	-1.06	0.4	142.81	11AU	1.3	-3.7	-2.4	1.0
64.16	-0.89	0.3	152.52	21AU	2.9	-3.8	-2.4	1.0
69.20	-0.70	0.2	162.28	31AU	1.4	-3.9	-2.4	1.1
73.71	-0.48	0.1	172.10	10SE	-0.4	-4.0	-2.3	1.1
77.58	-0.22	-0.1	181.98	20SE	-1.0	-4.1	-2.3	1.2
80.65	0.08	-0.3	191.91	30SE	-1.0	-4.2	-2.2	1.2
82.73	0.43	-0.5	201.91	10OC	-0.8	-4.3	-2.1	1.2
83.63	0.83	-0.7	211.95	20OC	-0.5	-4.4	-2.1	1.2
83.18	1.28	-0.9	222.04	30OC	-0.4	-4.2	-2.0	1.2
81.29	1.76	-1.1	232.17	9NO	-0.4	-3.8	-1.9	1.2
78.16	2.22	-1.3	242.33	19NO	-0.3	-3.0	-1.9	1.2
74.31	2.61	-1.4	252.51	29NO	0.6	-3.6	-1.8	1.2
70.50	2.88	-1.2	262.72	9DE	2.9	-4.2	-1.8	1.1
67.48	3.01	-1.0	272.92	19DE	0.6	-4.4	-1.7	1.1
65.69	3.03	-0.7	283.11	29DE	0.1	-4.4	-1.7	1.0

919

♂ LONG	LAT	MAG	☉ LONG	16.00UT 919	☿	♀	♃	♄
65.24	2.98	-0.4	293.28	8JA	-0.0	-4.3	-1.6	0.9
66.03	2.87	-0.1	303.42	18JA	-0.2	-4.2	-1.6	0.9
67.86	2.75	0.2	313.53	28JA	-0.4	-4.1	-1.6	0.8
70.54	2.62	0.4	323.59	7FE	-0.9	-4.0	-1.6	0.7
73.90	2.49	0.6	333.60	17FE	-1.4	-3.9	-1.6	0.7
77.78	2.36	0.8	343.55	27FE	-1.2	-3.8	-1.6	0.7
82.09	2.24	1.0	353.45	9MR	-0.2	-3.7	-1.6	0.7
86.72	2.12	1.1	3.29	19MR	1.4	-3.6	-1.6	0.8
91.62	2.00	1.3	13.07	29MR	3.4	-3.5	-1.6	0.8
96.75	1.89	1.4	22.80	8AP	2.4	-3.5	-1.7	0.9
102.05	1.79	1.5	32.47	18AP	1.4	-3.4	-1.7	1.0
107.50	1.68	1.6	42.10	28AP	0.8	-3.4	-1.7	1.0
113.09	1.58	1.7	51.70	8MY	0.1	-3.4	-1.8	1.1
118.79	1.49	1.7	61.27	18MY	-0.7	-3.3	-1.8	1.1
124.60	1.39	1.8	70.81	28MY	-1.6	-3.3	-1.9	1.1
130.50	1.29	1.8	80.34	7JN	-1.3	-3.3	-2.0	1.2
136.49	1.20	1.9	89.86	17JN	-0.4	-3.3	-2.0	1.2
142.57	1.10	1.9	99.39	27JN	0.2	-3.3	-2.1	1.3
148.74	1.01	1.9	108.94	7JL	0.6	-3.3	-2.2	1.3
154.98	0.91	1.9	118.51	17JL	1.0	-3.4	-2.2	1.3
161.32	0.81	1.9	128.11	27JL	1.8	-3.4	-2.3	1.3
167.73	0.72	1.9	137.75	6AU	3.2	-3.4	-2.4	1.3
174.23	0.62	1.9	147.43	16AU	1.2	-3.4	-2.4	1.3
180.82	0.52	1.9	157.16	26AU	-0.5	-3.5	-2.4	1.3
187.50	0.42	1.8	166.95	5SE	-1.1	-3.5	-2.5	1.3
194.27	0.32	1.8	176.79	15SE	-1.1	-3.5	-2.5	1.3
201.13	0.22	1.8	186.70	25SE	-0.8	-3.4	-2.5	1.3
208.08	0.12	1.7	196.66	5OC	-0.4	-3.4	-2.4	1.3
215.12	0.01	1.7	206.67	15OC	-0.3	-3.4	-2.4	1.3
222.25	-0.09	1.6	216.74	25OC	-0.2	-3.4	-2.3	1.3
229.47	-0.19	1.6	226.85	4NO	-0.1	-3.3	-2.3	1.2
236.78	-0.30	1.5	237.00	14NO	0.8	-3.3	-2.2	1.2
244.17	-0.40	1.5	247.17	24NO	2.7	-3.3	-2.1	1.2
251.64	-0.50	1.5	257.37	4DE	0.4	-3.3	-2.0	1.1
259.18	-0.60	1.5	267.56	14DE	-0.1	-3.3	-2.0	1.1
266.80	-0.69	1.5	277.76	24DE	-0.2	-3.4	-1.9	1.0

♂ LONG	♂ LAT	♂ MAG	⊙ LONG	16.00UT	☿	♀	♃	♄
				920	MAGNITUDES			
274.47	-0.78	1.5	287.95	3JA	-0.3	-3.4	-1.8	1.0
282.19	-0.86	1.5	298.11	13JA	-0.5	-3.4	-1.8	0.9
289.96	-0.94	1.5	308.23	23JA	-0.9	-3.4	-1.7	0.9
297.76	-1.01	1.4	318.32	2FE	-1.3	-3.5	-1.7	0.9
305.58	-1.07	1.4	328.35	12FE	-1.1	-3.5	-1.7	0.8
313.41	-1.12	1.4	338.34	22FE	-0.1	-3.6	-1.6	0.8
321.23	-1.16	1.4	348.27	3MR	1.7	-3.6	-1.6	0.8
329.04	-1.18	1.4	358.13	13MR	3.3	-3.7	-1.6	0.8
336.83	-1.20	1.4	7.94	23MR	1.8	-3.8	-1.6	0.9
344.57	-1.20	1.4	17.70	2AP	1.0	-3.8	-1.6	1.0
352.25	-1.19	1.4	27.40	12AP	0.6	-3.9	-1.6	1.1
359.88	-1.16	1.4	37.06	22AP	0.0	-4.0	-1.6	1.1
7.43	-1.12	1.4	46.67	2MY	-0.7	-4.1	-1.6	1.2
14.90	-1.07	1.3	56.25	12MY	-1.7	-4.2	-1.6	1.2
22.27	-1.01	1.3	65.81	22MY	-1.3	-4.2	-1.7	1.3
29.54	-0.93	1.3	75.34	1JN	-0.4	-4.1	-1.7	1.3
36.70	-0.84	1.3	84.87	11JN	0.3	-3.8	-1.8	1.3
43.74	-0.74	1.3	94.40	21JN	0.8	-3.1	-1.8	1.3
50.65	-0.63	1.3	103.93	1JL	1.3	-3.2	-1.9	1.3
57.42	-0.51	1.3	113.48	11JL	2.3	-3.9	-1.9	1.3
64.04	-0.37	1.2	123.07	21JL	3.0	-4.2	-2.0	1.3
70.49	-0.23	1.2	132.69	31JL	1.0	-4.2	-2.1	1.3
76.77	-0.07	1.1	142.34	10AU	-0.5	-4.2	-2.1	1.3
82.85	0.09	1.1	152.05	20AU	-1.2	-4.1	-2.2	1.2
88.69	0.28	1.0	161.81	30AU	-1.3	-4.0	-2.3	1.2
94.28	0.47	0.9	171.62	9SE	-0.7	-4.0	-2.3	1.1
99.57	0.69	0.8	181.50	19SE	-0.3	-3.9	-2.4	1.1
104.49	0.93	0.7	191.43	29SE	-0.2	-3.8	-2.4	1.1
108.99	1.19	0.6	201.42	9OC	-0.1	-3.7	-2.4	1.1
112.95	1.48	0.5	211.46	19OC	0.1	-3.7	-2.4	1.1
116.27	1.80	0.3	221.55	29OC	1.1	-3.6	-2.4	1.1
118.79	2.17	0.1	231.67	8NO	2.5	-3.6	-2.4	1.1
120.32	2.57	-0.1	241.84	18NO	0.2	-3.5	-2.3	1.1
120.68	3.00	-0.3	252.02	28NO	-0.3	-3.5	-2.3	1.0
119.72	3.45	-0.6	262.22	8DE	-0.3	-3.4	-2.2	1.0
117.41	3.86	-0.8	272.43	18DE	-0.4	-3.4	-2.1	1.0
114.01	4.17	-1.0	282.62	28DE	-0.6	-3.4	-2.1	0.9
				921				
110.05	4.34	-1.1	292.79	7JA	-1.0	-3.4	-2.0	0.9
106.28	4.33	-0.9	302.93	17JA	-1.1	-3.3	-1.9	0.8
103.38	4.17	-0.6	313.04	27JA	-1.0	-3.3	-1.9	0.8
101.70	3.91	-0.4	323.10	6FE	-0.0	-3.3	-1.8	0.7
101.32	3.61	-0.1	333.11	16FE	2.1	-3.3	-1.7	0.7
102.14	3.31	0.1	343.07	26FE	2.6	-3.3	-1.7	0.6
103.97	3.01	0.4	352.96	8MR	1.3	-3.4	-1.6	0.6
106.64	2.73	0.6	2.81	18MR	0.7	-3.4	-1.6	0.6
109.96	2.48	0.7	12.59	28MR	0.4	-3.4	-1.6	0.7
113.82	2.24	0.9	22.32	7AP	-0.1	-3.5	-1.5	0.8
118.12	2.03	1.1	32.00	17AP	-0.7	-3.5	-1.5	0.8
122.76	1.83	1.2	41.63	27AP	-1.7	-3.5	-1.5	0.9
127.70	1.64	1.3	51.23	7MY	-1.3	-3.4	-1.5	1.0
132.88	1.47	1.4	60.80	17MY	-0.4	-3.4	-1.5	1.0
138.28	1.31	1.5	70.34	27MY	0.4	-3.4	-1.5	1.1
143.87	1.15	1.5	79.87	6JN	1.0	-3.4	-1.5	1.1
149.62	1.00	1.6	89.40	16JN	1.8	-3.3	-1.6	1.1
155.53	0.86	1.6	98.93	26JN	3.0	-3.3	-1.6	1.2
161.57	0.72	1.6	108.47	6JL	2.5	-3.3	-1.6	1.2
167.76	0.59	1.7	118.04	16JL	0.8	-3.3	-1.7	1.2
174.07	0.46	1.7	127.64	26JL	-0.6	-3.3	-1.7	1.2
180.50	0.34	1.7	137.27	5AU	-1.3	-3.4	-1.8	1.1
187.06	0.21	1.7	146.96	15AU	-1.3	-3.4	-1.8	1.1
193.74	0.10	1.7	156.68	25AU	-0.7	-3.4	-1.9	1.1
200.53	-0.02	1.7	166.47	4SE	-0.2	-3.4	-2.0	1.0
207.44	-0.13	1.7	176.32	14SE	-0.0	-3.5	-2.0	1.0
214.46	-0.24	1.7	186.21	24SE	0.1	-3.6	-2.1	0.9
221.59	-0.35	1.6	196.17	4OC	0.3	-3.6	-2.2	1.0
228.83	-0.45	1.6	206.19	14OC	1.4	-3.7	-2.2	1.0
236.17	-0.55	1.6	216.25	24OC	2.2	-3.8	-2.3	1.0
243.61	-0.64	1.5	226.36	3NO	0.0	-3.9	-2.3	1.0
251.14	-0.72	1.5	236.51	13NO	-0.4	-4.0	-2.3	1.0
258.75	-0.80	1.5	246.68	23NO	-0.5	-4.1	-2.3	1.0
266.44	-0.87	1.4	256.87	3DE	-0.5	-4.2	-2.3	0.9
274.19	-0.93	1.4	267.08	13DE	-0.7	-4.3	-2.3	0.9
281.99	-0.98	1.4	277.27	23DE	-0.9	-4.4	-2.2	0.9

♂ LONG	♂ LAT	♂ MAG	⊙ LONG	16.00UT	☿	♀	♃	♄
				922	MAGNITUDES			
289.84	-1.03	1.3	287.46	2JA	-1.0	-4.4	-2.2	0.8
297.72	-1.06	1.3	297.62	12JA	-0.9	-4.2	-2.1	0.8
305.61	-1.08	1.3	307.74	22JA	0.1	-3.9	-2.0	0.7
313.50	-1.09	1.3	317.83	1FE	2.4	-3.3	-2.0	0.7
321.39	-1.08	1.3	327.87	11FE	2.0	-3.7	-1.9	0.6
329.25	-1.07	1.4	337.85	21FE	0.9	-4.1	-1.8	0.6
337.08	-1.04	1.4	347.78	3MR	0.5	-4.3	-1.8	0.6
344.87	-1.00	1.4	357.65	13MR	0.3	-4.3	-1.7	0.5
352.59	-0.95	1.4	7.46	23MR	-0.1	-4.2	-1.6	0.5
0.26	-0.89	1.5	17.22	2AP	-0.8	-4.1	-1.6	0.5
7.84	-0.82	1.5	26.93	12AP	-1.7	-4.0	-1.5	0.6
15.35	-0.75	1.5	36.58	22AP	-1.3	-3.9	-1.5	0.6
22.77	-0.66	1.5	46.20	2MY	-0.3	-3.8	-1.5	0.7
30.10	-0.57	1.5	55.78	12MY	0.6	-3.7	-1.4	0.8
37.34	-0.47	1.6	65.34	22MY	1.3	-3.6	-1.4	0.8
44.48	-0.36	1.6	74.88	1JN	2.3	-3.6	-1.4	0.9
51.51	-0.25	1.6	84.40	11JN	3.4	-3.5	-1.4	0.9
58.45	-0.14	1.6	93.93	21JN	2.0	-3.5	-1.4	1.0
65.28	-0.02	1.6	103.47	1JL	0.6	-3.4	-1.4	1.0
72.00	0.11	1.6	113.02	11JL	-0.6	-3.4	-1.4	1.0
78.62	0.23	1.6	122.60	21JL	-1.4	-3.4	-1.4	1.0
85.12	0.36	1.6	132.22	31JL	-1.3	-3.4	-1.5	1.0
91.50	0.50	1.6	141.88	10AU	-0.6	-3.3	-1.5	1.0
97.75	0.64	1.6	151.58	20AU	-0.2	-3.3	-1.6	1.0
103.87	0.78	1.5	161.34	30AU	0.1	-3.3	-1.6	1.0
109.84	0.94	1.5	171.15	9SE	0.2	-3.4	-1.7	0.9
115.64	1.10	1.4	181.02	19SE	0.5	-3.4	-1.7	0.9
121.26	1.26	1.3	190.95	29SE	1.7	-3.4	-1.8	0.9
126.65	1.44	1.3	200.93	9OC	2.0	-3.4	-1.8	0.8
131.79	1.64	1.2	210.97	19OC	-0.1	-3.4	-1.9	0.9
136.63	1.84	1.0	221.06	29OC	-0.6	-3.5	-2.0	0.9
141.09	2.07	0.9	231.18	8NO	-0.6	-3.5	-2.0	0.9
145.10	2.32	0.7	241.34	18NO	-0.6	-3.5	-2.1	0.9
148.56	2.59	0.6	251.53	28NO	-0.8	-3.5	-2.1	0.9
151.32	2.89	0.4	261.72	8DE	-0.8	-3.4	-2.2	0.9
153.24	3.22	0.1	271.92	18DE	-0.8	-3.4	-2.2	0.8
154.11	3.56	-0.1	282.12	28DE	-0.7	-3.4	-2.2	0.8
				923				
153.78	3.90	-0.4	292.29	7JA	0.2	-3.4	-2.2	0.8
152.15	4.20	-0.6	302.44	17JA	2.7	-3.4	-2.1	0.8
149.29	4.41	-0.8	312.55	27JA	1.5	-3.3	-2.1	0.7
145.59	4.48	-1.0	322.61	6FE	0.6	-3.3	-2.0	0.7
141.67	4.37	-0.9	332.62	16FE	0.4	-3.3	-2.0	0.6
138.27	4.10	-0.8	342.59	26FE	0.1	-3.3	-1.9	0.6
135.91	3.73	-0.5	352.48	8MR	-0.2	-3.4	-1.8	0.5
134.83	3.32	-0.3	2.33	18MR	-0.8	-3.4	-1.8	0.4
135.02	2.91	-0.1	12.12	28MR	-1.6	-3.4	-1.7	0.4
136.35	2.51	0.1	21.85	7AP	-1.3	-3.4	-1.6	0.4
138.61	2.15	0.3	31.53	17AP	-0.3	-3.5	-1.6	0.4
141.66	1.83	0.5	41.17	27AP	0.7	-3.5	-1.5	0.5
145.34	1.54	0.7	50.76	7MY	1.7	-3.6	-1.5	0.5
149.54	1.27	0.8	60.33	17MY	3.1	-3.6	-1.4	0.6
154.18	1.03	0.9	69.88	27MY	2.9	-3.7	-1.4	0.7
159.17	0.81	1.0	79.41	6JN	1.6	-3.8	-1.4	0.7
164.47	0.61	1.1	88.94	16JN	0.5	-3.8	-1.3	0.8
170.05	0.42	1.2	98.47	26JN	-0.6	-3.9	-1.3	0.8
175.85	0.24	1.2	108.01	6JL	-1.5	-4.0	-1.3	0.9
181.88	0.08	1.3	117.58	16JL	-1.3	-4.1	-1.3	0.9
188.10	-0.07	1.3	127.18	26JL	-0.6	-4.2	-1.3	0.9
194.50	-0.21	1.3	136.81	5AU	-0.1	-4.2	-1.3	0.9
201.07	-0.35	1.4	146.49	15AU	0.2	-4.1	-1.3	0.9
207.80	-0.47	1.4	156.22	25AU	0.4	-3.8	-1.4	0.9
214.68	-0.58	1.4	166.00	4SE	0.8	-3.3	-1.4	0.9
221.70	-0.69	1.4	175.84	14SE	2.1	-3.6	-1.4	0.9
228.86	-0.78	1.4	185.74	24SE	1.8	-4.1	-1.5	0.8
236.14	-0.87	1.4	195.69	4OC	-0.3	-4.3	-1.5	0.8
243.54	-0.94	1.4	205.70	14OC	-0.7	-4.3	-1.6	0.8
251.05	-1.01	1.4	215.76	24OC	-0.8	-4.3	-1.6	0.8
258.65	-1.06	1.4	225.87	3NO	-0.8	-4.2	-1.7	0.8
266.33	-1.10	1.4	236.02	13NO	-0.7	-4.1	-1.8	0.8
274.09	-1.13	1.4	246.19	23NO	-0.6	-4.0	-1.8	0.8
281.90	-1.14	1.4	256.38	3DE	-0.6	-3.9	-1.9	0.8
289.76	-1.15	1.4	266.58	13DE	-0.6	-3.8	-1.9	0.8
297.65	-1.14	1.4	276.78	23DE	0.3	-3.7	-2.0	0.8

♂ LONG	♂ LAT	♂ MAG	⊙ LONG	16.00UT	☿	♀	♃	♄
				924	MAGNITUDES			
305.55	-1.12	1.4	286.96	2JA	2.9	-3.7	-2.0	0.8
313.45	-1.08	1.4	297.13	12JA	1.1	-3.6	-2.1	0.8
321.34	-1.04	1.4	307.26	22JA	0.4	-3.5	-2.1	0.7
329.20	-0.99	1.4	317.34	1FE	0.2	-3.5	-2.1	0.7
337.02	-0.92	1.4	327.39	11FE	0.0	-3.4	-2.1	0.6
344.79	-0.85	1.4	337.38	21FE	-0.3	-3.4	-2.0	0.6
352.49	-0.77	1.4	347.31	2MR	-0.8	-3.4	-2.0	0.5
0.13	-0.69	1.4	357.18	12MR	-1.6	-3.3	-1.9	0.5
7.70	-0.60	1.4	7.00	22MR	-1.3	-3.3	-1.9	0.4
15.18	-0.50	1.5	16.76	1AP	-0.3	-3.3	-1.8	0.4
22.58	-0.40	1.5	26.47	11AP	0.9	-3.3	-1.7	0.3
29.88	-0.30	1.5	36.12	21AP	2.3	-3.3	-1.7	0.3
37.11	-0.19	1.6	45.74	1MY	3.8	-3.3	-1.6	0.4
44.24	-0.09	1.6	55.32	11MY	2.2	-3.3	-1.5	0.4
51.29	0.02	1.7	64.88	21MY	1.2	-3.4	-1.5	0.5
58.24	0.12	1.7	74.41	31MY	0.3	-3.4	-1.4	0.6
65.12	0.23	1.8	83.94	10JN	-0.6	-3.4	-1.4	0.6
71.91	0.33	1.8	93.47	20JN	-1.6	-3.5	-1.3	0.7
78.61	0.44	1.8	103.00	30JN	-1.3	-3.5	-1.3	0.7
85.24	0.54	1.8	112.56	10JL	-0.5	-3.5	-1.3	0.8
91.79	0.65	1.8	122.14	20JL	0.0	-3.4	-1.2	0.8
98.26	0.75	1.8	131.75	30JL	0.3	-3.4	-1.2	0.8
104.65	0.85	1.8	141.41	9AU	0.6	-3.4	-1.2	0.9
110.96	0.96	1.8	151.11	19AU	1.1	-3.3	-1.2	0.9
117.20	1.06	1.8	160.86	29AU	2.6	-3.3	-1.2	0.9
123.34	1.16	1.8	170.67	8SE	1.6	-3.3	-1.2	0.9
129.39	1.27	1.8	180.54	18SE	-0.3	-3.3	-1.3	0.8
135.34	1.38	1.7	190.46	28SE	-0.9	-3.3	-1.3	0.8
141.18	1.49	1.7	200.45	8OC	-0.9	-3.3	-1.3	0.8
146.89	1.60	1.6	210.48	18OC	-0.9	-3.4	-1.4	0.7
152.46	1.72	1.5	220.56	28OC	-0.6	-3.4	-1.4	0.7
157.85	1.84	1.4	230.69	7NO	-0.5	-3.4	-1.5	0.7
163.03	1.97	1.3	240.85	17NO	-0.5	-3.5	-1.5	0.8
167.98	2.11	1.2	251.03	27NO	-0.4	-3.5	-1.6	0.8
172.62	2.25	1.0	261.23	7DE	0.4	-3.6	-1.7	0.8
176.89	2.41	0.9	271.43	17DE	3.0	-3.6	-1.7	0.8
180.71	2.57	0.7	281.63	27DE	0.8	-3.7	-1.8	0.8
				925				
183.96	2.74	0.5	291.80	6JA	0.2	-3.8	-1.8	0.8
186.51	2.92	0.2	301.95	16JA	0.0	-3.9	-1.9	0.8
188.19	3.09	-0.0	312.06	26JA	-0.1	-4.0	-1.9	0.7
188.82	3.25	-0.3	322.13	5FE	-0.4	-4.1	-2.0	0.7
188.24	3.37	-0.6	332.14	15FE	-0.9	-4.1	-2.0	0.7
186.38	3.42	-0.9	342.11	25FE	-1.5	-4.2	-2.0	0.6
183.37	3.36	-1.1	352.01	7MR	-1.2	-4.3	-2.0	0.6
179.68	3.15	-1.3	1.86	17MR	-0.2	-4.2	-2.0	0.5
175.95	2.81	-1.2	11.65	27MR	1.2	-4.0	-1.9	0.4
172.89	2.39	-1.0	21.38	6AP	2.9	-3.5	-1.9	0.4
170.98	1.92	-0.8	31.06	16AP	2.9	-3.0	-1.8	0.3
170.39	1.47	-0.6	40.70	26AP	1.6	-3.6	-1.8	0.3
171.07	1.05	-0.4	50.30	6MY	0.9	-4.0	-1.7	0.3
172.88	0.67	-0.2	59.87	16MY	0.2	-4.2	-1.7	0.3
175.64	0.34	0.0	69.42	26MY	-0.7	-4.2	-1.6	0.4
179.18	0.06	0.2	78.95	5JN	-1.6	-4.1	-1.5	0.5
183.36	-0.20	0.3	88.47	15JN	-1.3	-4.0	-1.5	0.5
188.08	-0.42	0.4	98.00	25JN	-0.5	-3.9	-1.4	0.6
193.25	-0.61	0.5	107.55	5JL	0.1	-3.8	-1.4	0.6
198.80	-0.78	0.6	117.11	15JL	0.5	-3.8	-1.3	0.7
204.68	-0.93	0.7	126.71	25JL	0.8	-3.7	-1.3	0.7
210.86	-1.06	0.8	136.34	4AU	1.5	-3.6	-1.3	0.8
217.28	-1.16	0.8	146.01	14AU	3.0	-3.6	-1.2	0.8
223.93	-1.25	0.9	155.74	24AU	1.4	-3.5	-1.2	0.8
230.78	-1.33	0.9	165.52	3SE	-0.4	-3.5	-1.2	0.8
237.81	-1.38	1.0	175.36	13SE	-1.0	-3.4	-1.2	0.8
245.01	-1.42	1.0	185.26	23SE	-1.1	-3.4	-1.2	0.8
252.34	-1.45	1.1	195.21	3OC	-0.8	-3.4	-1.2	0.8
259.80	-1.46	1.1	205.22	13OC	-0.5	-3.4	-1.2	0.8
267.36	-1.45	1.1	215.28	23OC	-0.4	-3.4	-1.3	0.7
275.01	-1.42	1.2	225.38	2NO	-0.3	-3.4	-1.3	0.7
282.73	-1.39	1.2	235.52	12NO	-0.2	-3.4	-1.3	0.7
290.50	-1.34	1.2	245.70	22NO	0.6	-3.4	-1.4	0.7
298.30	-1.28	1.3	255.89	2DE	2.9	-3.4	-1.4	0.7
306.12	-1.20	1.3	266.09	12DE	0.5	-3.4	-1.5	0.8
313.94	-1.12	1.3	276.29	22DE	0.0	-3.4	-1.5	0.8

♂ LONG	♂ LAT	♂ MAG	⊙ LONG	16.00UT	☿	♀	♃	♄
				926	MAGNITUDES			
321.74	-1.03	1.4	286.47	1JA	-0.1	-3.4	-1.6	0.8
329.51	-0.93	1.4	296.63	11JA	-0.2	-3.4	-1.7	0.8
337.24	-0.83	1.4	306.77	21JA	-0.4	-3.5	-1.7	0.8
344.92	-0.72	1.5	316.86	31JA	-0.9	-3.5	-1.8	0.7
352.54	-0.61	1.5	326.90	10FE	-1.4	-3.5	-1.9	0.7
0.09	-0.50	1.5	336.89	20FE	-1.2	-3.4	-1.9	0.7
7.56	-0.39	1.6	346.82	2MR	-0.2	-3.4	-2.0	0.6
14.95	-0.28	1.6	356.70	12MR	1.5	-3.4	-2.0	0.6
22.27	-0.17	1.6	6.52	22MR	3.5	-3.4	-2.0	0.5
29.50	-0.06	1.6	16.28	1AP	2.1	-3.3	-2.0	0.5
36.64	0.05	1.7	25.99	11AP	1.2	-3.3	-2.0	0.4
43.71	0.15	1.7	35.66	21AP	0.7	-3.3	-2.0	0.3
50.69	0.25	1.7	45.27	1MY	0.1	-3.3	-1.9	0.3
57.60	0.35	1.7	54.86	11MY	-0.7	-3.3	-1.9	0.2
64.44	0.44	1.7	64.42	21MY	-1.6	-3.3	-1.8	0.3
71.21	0.53	1.8	73.95	31MY	-1.4	-3.4	-1.8	0.3
77.91	0.62	1.8	83.48	10JN	-0.5	-3.4	-1.7	0.4
84.54	0.70	1.9	93.01	20JN	0.2	-3.4	-1.7	0.5
91.12	0.78	1.9	102.54	30JN	0.6	-3.5	-1.6	0.5
97.65	0.85	1.9	112.09	10JL	1.1	-3.5	-1.5	0.6
104.12	0.93	1.9	121.67	20JL	1.9	-3.6	-1.5	0.6
110.54	1.00	2.0	131.28	30JL	3.2	-3.6	-1.4	0.7
116.92	1.06	2.0	140.93	9AU	1.2	-3.7	-1.4	0.7
123.26	1.13	2.0	150.64	19AU	-0.5	-3.8	-1.3	0.8
129.55	1.19	2.0	160.38	29AU	-1.2	-3.9	-1.3	0.8
135.81	1.24	2.0	170.19	8SE	-1.2	-4.0	-1.3	0.8
142.01	1.30	2.0	180.06	18SE	-0.8	-4.1	-1.3	0.8
148.18	1.35	1.9	189.98	28SE	-0.4	-4.2	-1.2	0.8
154.29	1.40	1.9	199.96	8OC	-0.2	-4.3	-1.2	0.8
160.36	1.45	1.8	210.00	18OC	-0.2	-4.4	-1.2	0.8
166.36	1.50	1.8	220.07	28OC	-0.0	-4.2	-1.2	0.7
172.29	1.55	1.7	230.20	7NO	0.9	-3.8	-1.3	0.7
178.15	1.59	1.6	240.36	17NO	2.7	-2.9	-1.3	0.7
183.90	1.63	1.5	250.54	27NO	0.3	-3.6	-1.3	0.7
189.55	1.67	1.4	260.74	7DE	-0.2	-4.2	-1.3	0.7
195.06	1.70	1.3	270.94	17DE	-0.2	-4.4	-1.4	0.7
200.41	1.73	1.2	281.13	27DE	-0.3	-4.4	-1.4	0.8
				927				
205.56	1.75	1.0	291.31	6JA	-0.5	-4.3	-1.5	0.8
210.47	1.76	0.9	301.46	16JA	-0.9	-4.2	-1.6	0.8
215.07	1.76	0.7	311.57	26JA	-1.2	-4.1	-1.6	0.8
219.30	1.74	0.5	321.64	5FE	-1.1	-4.0	-1.7	0.8
223.06	1.70	0.2	331.66	15FE	-0.1	-3.9	-1.8	0.7
226.24	1.64	-0.0	341.63	25FE	1.8	-3.8	-1.8	0.7
228.69	1.53	-0.3	351.54	7MR	3.0	-3.7	-1.9	0.7
230.22	1.37	-0.6	1.39	17MR	1.6	-3.6	-2.0	0.6
230.67	1.14	-0.9	11.18	27MR	0.9	-3.5	-2.0	0.6
229.88	0.83	-1.2	20.92	6AP	0.5	-3.5	-2.0	0.5
227.85	0.43	-1.5	30.60	16AP	0.0	-3.4	-2.1	0.4
224.84	-0.05	-1.8	40.24	26AP	-0.7	-3.4	-2.1	0.4
221.38	-0.57	-1.8	49.85	6MY	-1.6	-3.4	-2.1	0.3
218.21	-1.07	-1.7	59.42	16MY	-1.4	-3.3	-2.1	0.3
215.97	-1.50	-1.5	68.96	26MY	-0.4	-3.3	-2.0	0.2
215.01	-1.85	-1.3	78.49	5JN	0.3	-3.3	-2.0	0.3
215.45	-2.12	-1.1	88.02	15JN	0.9	-3.3	-1.9	0.3
217.15	-2.31	-0.9	97.55	25JN	1.5	-3.3	-1.9	0.4
219.95	-2.44	-0.7	107.09	5JL	2.5	-3.3	-1.8	0.5
223.68	-2.52	-0.5	116.65	15JL	2.8	-3.4	-1.8	0.5
228.15	-2.56	-0.4	126.24	25JL	1.0	-3.4	-1.7	0.6
233.23	-2.56	-0.2	135.87	4AU	-0.5	-3.4	-1.6	0.6
238.83	-2.54	-0.1	145.55	14AU	-1.3	-3.4	-1.6	0.7
244.83	-2.50	0.0	155.27	24AU	-1.3	-3.5	-1.5	0.7
251.18	-2.43	0.2	165.05	3SE	-0.7	-3.5	-1.5	0.8
257.82	-2.34	0.3	174.88	13SE	-0.3	-3.5	-1.4	0.8
264.68	-2.23	0.4	184.77	23SE	-0.1	-3.4	-1.4	0.8
271.73	-2.11	0.5	194.72	3OC	-0.0	-3.4	-1.4	0.8
278.94	-1.98	0.6	204.72	13OC	0.2	-3.4	-1.3	0.8
286.26	-1.84	0.7	214.78	23OC	1.1	-3.4	-1.3	0.8
293.68	-1.69	0.8	224.88	2NO	2.5	-3.3	-1.3	0.7
301.15	-1.53	0.8	235.02	12NO	0.1	-3.3	-1.3	0.7
308.67	-1.37	0.9	245.19	22NO	-0.3	-3.3	-1.3	0.7
316.20	-1.21	1.0	255.39	2DE	-0.4	-3.3	-1.3	0.7
323.72	-1.05	1.1	265.59	12DE	-0.4	-3.3	-1.3	0.7
331.23	-0.89	1.2	275.79	22DE	-0.6	-3.4	-1.4	0.7

♂			☉	16.00UT	☿	♀	♃	♄
LONG	LAT	MAG	LONG	928	MAGNITUDES			
338.72	-0.73	1.3	285.98	1JA	-0.9	-3.4	-1.4	0.8
346.15	-0.58	1.3	296.14	11JA	-1.0	-3.4	-1.4	0.8
353.54	-0.43	1.4	306.28	21JA	-0.9	-3.4	-1.5	0.8
0.87	-0.29	1.5	316.37	31JA	-0.0	-3.5	-1.5	0.8
8.13	-0.16	1.5	326.42	10FE	2.1	-3.5	-1.6	0.8
15.33	-0.03	1.6	336.41	20FE	2.4	-3.6	-1.7	0.8
22.46	0.09	1.6	346.35	1MR	1.1	-3.6	-1.7	0.8
29.52	0.20	1.7	356.23	11MR	0.7	-3.7	-1.8	0.7
36.50	0.31	1.7	6.05	21MR	0.3	-3.8	-1.9	0.7
43.42	0.41	1.8	15.82	31MR	-0.1	-3.8	-1.9	0.6
50.27	0.50	1.8	25.53	10AP	-0.7	-3.9	-2.0	0.6
57.06	0.59	1.8	35.19	20AP	-1.6	-4.0	-2.1	0.5
63.79	0.66	1.9	44.82	30AP	-1.4	-4.1	-2.1	0.4
70.46	0.74	1.9	54.40	10MY	-0.4	-4.2	-2.2	0.4
77.07	0.80	1.9	63.96	20MY	0.4	-4.2	-2.2	0.3
83.65	0.86	1.9	73.50	30MY	1.1	-4.1	-2.2	0.3
90.17	0.92	1.9	83.02	9JN	1.9	-3.8	-2.2	0.3
96.66	0.97	1.9	92.55	19JN	3.2	-3.1	-2.2	0.3
103.12	1.02	1.9	102.09	29JN	2.3	-3.3	-2.1	0.4
109.55	1.06	1.9	111.63	9JL	0.8	-3.9	-2.1	0.4
115.95	1.09	2.0	121.21	19JL	-0.6	-4.2	-2.1	0.5
122.34	1.13	2.0	130.82	29JL	-1.4	-4.2	-2.0	0.5
128.70	1.15	2.0	140.47	8AU	-1.3	-4.2	-1.9	0.6
135.06	1.18	2.0	150.16	18AU	-0.6	-4.1	-1.9	0.7
141.41	1.20	2.0	159.91	28AU	-0.2	-4.0	-1.8	0.7
147.75	1.21	2.0	169.71	7SE	0.0	-4.0	-1.7	0.7
154.08	1.22	2.0	179.58	17SE	0.1	-3.9	-1.7	0.8
160.41	1.23	2.0	189.50	27SE	0.4	-3.8	-1.6	0.8
166.74	1.23	2.0	199.47	7OC	1.4	-3.7	-1.6	0.8
173.06	1.23	2.0	209.50	17OC	2.3	-3.7	-1.5	0.8
179.38	1.22	1.9	219.58	27OC	-0.0	-3.6	-1.5	0.8
185.69	1.20	1.9	229.70	6NO	-0.5	-3.6	-1.5	0.8
191.99	1.18	1.8	239.86	16NO	-0.5	-3.5	-1.4	0.7
198.28	1.15	1.7	250.04	26NO	-0.5	-3.5	-1.4	0.7
204.56	1.12	1.7	260.24	6DE	-0.7	-3.4	-1.4	0.7
210.81	1.07	1.6	270.44	16DE	-0.8	-3.4	-1.4	0.7
217.04	1.02	1.5	280.64	26DE	-0.9	-3.4	-1.4	0.7
				929				
223.23	0.95	1.4	290.82	5JA	-0.8	-3.4	-1.4	0.8
229.37	0.87	1.3	300.97	15JA	0.0	-3.3	-1.5	0.8
235.45	0.77	1.1	311.09	25JA	2.4	-3.3	-1.5	0.8
241.46	0.65	1.0	321.16	4FE	1.8	-3.3	-1.5	0.8
247.39	0.51	0.8	331.18	14FE	0.8	-3.3	-1.6	0.8
253.19	0.34	0.7	341.15	24FE	0.5	-3.3	-1.6	0.8
258.86	0.14	0.5	351.06	6MR	0.2	-3.4	-1.7	0.8
264.34	-0.10	0.3	0.91	16MR	-0.2	-3.4	-1.7	0.8
269.60	-0.38	0.1	10.71	26MR	-0.8	-3.4	-1.8	0.7
274.57	-0.71	-0.2	20.44	5AP	-1.6	-3.5	-1.9	0.7
279.16	-1.11	-0.4	30.13	15AP	-1.4	-3.5	-1.9	0.6
283.27	-1.57	-0.7	39.78	25AP	-0.4	-3.5	-2.0	0.6
286.77	-2.11	-0.9	49.38	5MY	0.6	-3.4	-2.1	0.5
289.49	-2.73	-1.2	58.95	15MY	1.5	-3.4	-2.1	0.5
291.23	-3.44	-1.6	68.50	25MY	2.6	-3.4	-2.2	0.4
291.84	-4.20	-1.9	78.03	4JN	3.3	-3.4	-2.3	0.3
291.19	-4.98	-2.2	87.56	14JN	1.8	-3.3	-2.3	0.3
289.39	-5.66	-2.4	97.08	24JN	0.6	-3.3	-2.3	0.3
286.85	-6.16	-2.6	106.62	4JL	-0.6	-3.3	-2.3	0.3
284.21	-6.38	-2.5	116.19	14JL	-1.4	-3.3	-2.3	0.4
282.22	-6.30	-2.3	125.78	24JL	-1.3	-3.3	-2.3	0.5
281.37	-5.99	-2.1	135.40	3AU	-0.6	-3.4	-2.3	0.5
281.85	-5.53	-1.8	145.08	13AU	-0.1	-3.4	-2.2	0.6
283.64	-5.00	-1.5	154.79	23AU	0.1	-3.4	-2.2	0.6
286.55	-4.46	-1.3	164.57	2SE	0.3	-3.4	-2.1	0.7
290.38	-3.93	-1.0	174.40	12SE	0.6	-3.5	-2.0	0.7
294.94	-3.42	-0.8	184.29	22SE	1.8	-3.6	-2.0	0.7
300.07	-2.95	-0.6	194.23	2OC	2.0	-3.6	-1.9	0.8
305.64	-2.51	-0.4	204.24	12OC	-0.2	-3.7	-1.8	0.8
311.56	-2.11	-0.2	214.29	22OC	-0.7	-3.8	-1.8	0.8
317.72	-1.74	0.0	224.39	1NO	-0.7	-3.9	-1.7	0.8
324.08	-1.41	0.2	234.53	11NO	-0.7	-4.0	-1.7	0.8
330.58	-1.11	0.4	244.70	21NO	-0.8	-4.1	-1.6	0.8
337.18	-0.83	0.5	254.89	1DE	-0.7	-4.2	-1.6	0.8
343.85	-0.59	0.7	265.09	11DE	-0.7	-4.3	-1.6	0.7
350.56	-0.36	0.8	275.29	21DE	-0.7	-4.4	-1.5	0.7
357.29	-0.17	1.0	285.48	31DE	0.2	-4.4	-1.5	0.8

♂			☉	16.00UT	☿	♀	♃	♄
LONG	LAT	MAG	LONG	930	MAGNITUDES			
4.04	0.01	1.1	295.65	10JA	2.7	-4.2	-1.5	0.8
10.77	0.17	1.2	305.78	20JA	1.4	-3.8	-1.5	0.8
17.49	0.31	1.3	315.88	30JA	0.5	-3.3	-1.5	0.8
24.19	0.44	1.4	325.93	9FE	0.3	-3.7	-1.5	0.9
30.86	0.55	1.5	335.92	19FE	0.1	-4.1	-1.6	0.9
37.50	0.65	1.6	345.87	1MR	-0.2	-4.3	-1.6	0.9
44.10	0.74	1.7	355.75	11MR	-0.8	-4.3	-1.6	0.9
50.67	0.81	1.7	5.57	21MR	-1.6	-4.2	-1.7	0.8
57.21	0.88	1.8	15.34	31MR	-1.3	-4.1	-1.7	0.8
63.71	0.94	1.8	25.06	10AP	-0.3	-4.0	-1.8	0.8
70.18	0.99	1.9	34.73	20AP	0.8	-3.9	-1.8	0.7
76.62	1.03	1.9	44.35	30AP	1.9	-3.8	-1.9	0.7
83.04	1.07	2.0	53.94	10MY	3.4	-3.7	-2.0	0.6
89.43	1.10	2.0	63.50	20MY	2.6	-3.6	-2.0	0.5
95.81	1.12	2.0	73.04	30MY	1.4	-3.6	-2.1	0.5
102.17	1.14	2.0	82.57	9JN	0.5	-3.5	-2.2	0.4
108.52	1.15	2.0	92.09	19JN	-0.6	-3.4	-2.2	0.3
114.87	1.16	2.0	101.63	29JN	-1.5	-3.4	-2.3	0.3
121.21	1.16	2.0	111.18	9JL	-1.3	-3.4	-2.3	0.3
127.56	1.16	2.0	120.75	19JL	-0.6	-3.4	-2.4	0.4
133.92	1.15	2.0	130.36	29JL	-0.1	-3.3	-2.4	0.4
140.29	1.14	2.0	140.01	8AU	0.3	-3.3	-2.4	0.5
146.67	1.13	2.0	149.70	18AU	0.5	-3.3	-2.4	0.6
153.08	1.11	2.0	159.45	28AU	0.9	-3.3	-2.4	0.6
159.51	1.08	2.0	169.25	7SE	2.2	-3.4	-2.4	0.7
165.96	1.05	2.0	179.10	17SE	1.8	-3.4	-2.3	0.7
172.44	1.02	2.0	189.02	27SE	-0.3	-3.4	-2.3	0.7
178.96	0.98	2.0	198.99	7OC	-0.8	-3.4	-2.2	0.8
185.50	0.93	2.0	209.02	17OC	-0.8	-3.4	-2.1	0.8
192.08	0.88	1.9	219.09	27OC	-0.8	-3.5	-2.1	0.8
198.70	0.82	1.9	229.21	6NO	-0.7	-3.5	-2.0	0.8
205.35	0.75	1.9	239.36	16NO	-0.6	-3.5	-1.9	0.8
212.04	0.68	1.8	249.55	26NO	-0.6	-3.5	-1.9	0.8
218.76	0.60	1.8	259.74	6DE	-0.5	-3.4	-1.8	0.8
225.53	0.51	1.7	269.94	16DE	0.3	-3.4	-1.8	0.8
232.32	0.41	1.6	280.14	26DE	3.0	-3.4	-1.7	0.8
				931				
239.15	0.30	1.6	290.32	5JA	1.0	-3.4	-1.7	0.8
246.01	0.18	1.5	300.47	15JA	0.3	-3.4	-1.6	0.8
252.90	0.04	1.4	310.59	25JA	0.1	-3.3	-1.6	0.9
259.82	-0.10	1.3	320.66	4FE	-0.0	-3.3	-1.6	0.9
266.76	-0.27	1.2	330.69	14FE	-0.3	-3.3	-1.6	0.9
273.72	-0.44	1.1	340.66	24FE	-0.8	-3.3	-1.6	0.9
280.70	-0.63	1.0	350.57	6MR	-1.5	-3.4	-1.6	0.9
287.67	-0.84	0.9	0.43	16MR	-1.3	-3.4	-1.6	0.9
294.64	-1.05	0.7	10.23	26MR	-0.3	-3.4	-1.6	0.9
301.59	-1.29	0.6	19.97	5AP	1.0	-3.4	-1.6	0.9
308.50	-1.53	0.5	29.66	15AP	2.5	-3.5	-1.7	0.9
315.35	-1.79	0.4	39.31	25AP	3.4	-3.5	-1.7	0.8
322.11	-2.06	0.2	48.92	5MY	2.0	-3.6	-1.7	0.8
328.75	-2.34	0.1	58.49	15MY	1.1	-3.6	-1.8	0.7
335.22	-2.63	-0.1	68.04	25MY	0.3	-3.7	-1.9	0.7
341.45	-2.93	-0.3	77.57	4JN	-0.6	-3.8	-1.9	0.6
347.40	-3.22	-0.4	87.10	14JN	-1.6	-3.9	-2.0	0.5
352.96	-3.52	-0.6	96.63	24JN	-1.4	-3.9	-2.0	0.5
358.03	-3.81	-0.8	106.17	4JL	-0.5	-4.0	-2.1	0.4
2.47	-4.09	-1.0	115.73	14JL	0.0	-4.1	-2.2	0.4
6.09	-4.35	-1.2	125.32	24JL	0.4	-4.2	-2.3	0.4
8.71	-4.56	-1.4	134.94	3AU	0.7	-4.2	-2.3	0.5
10.12	-4.72	-1.7	144.61	13AU	1.2	-4.1	-2.4	0.5
10.10	-4.76	-1.9	154.33	23AU	2.7	-3.7	-2.4	0.6
8.67	-4.65	-2.1	164.10	2SE	1.6	-3.3	-2.4	0.6
6.07	-4.33	-2.3	173.93	12SE	-0.3	-3.7	-2.5	0.7
2.89	-3.81	-2.4	183.81	22SE	-0.9	-4.1	-2.5	0.7
359.95	-3.14	-2.1	193.75	2OC	-1.0	-4.3	-2.4	0.8
357.93	-2.43	-1.8	203.75	12OC	-0.9	-4.3	-2.4	0.8
357.20	-1.74	-1.5	213.81	22OC	-0.6	-4.3	-2.4	0.8
357.79	-1.14	-1.2	223.90	1NO	-0.4	-4.2	-2.3	0.9
359.55	-0.63	-0.8	234.04	11NO	-0.4	-4.1	-2.2	0.9
2.30	-0.22	-0.5	244.21	21NO	-0.3	-4.0	-2.2	0.9
5.82	0.12	-0.3	254.39	1DE	0.5	-3.9	-2.1	0.9
9.95	0.39	0.0	264.60	11DE	3.1	-3.8	-2.0	0.9
14.54	0.61	0.2	274.80	21DE	0.7	-3.7	-2.0	0.9
19.49	0.78	0.5	284.98	31DE	0.1	-3.7	-1.9	0.9

Left panel (932 / 933)

♂ LONG	♂ LAT	♂ MAG	☉ LONG	16.00UT 932	☿	♀	♃	♄
24.72	0.92	0.7	295.15	10JA	-0.0	-3.6	-1.8	0.8
30.16	1.04	0.8	305.29	20JA	-0.1	-3.5	-1.8	0.9
35.76	1.12	1.0	315.39	30JA	-0.4	-3.5	-1.7	0.9
41.49	1.19	1.1	325.44	9FE	-0.8	-3.4	-1.7	1.0
47.32	1.25	1.3	335.44	19FE	-1.4	-3.4	-1.6	1.0
53.22	1.29	1.4	345.38	29FE	-1.2	-3.4	-1.6	1.0
59.18	1.32	1.5	355.27	10MR	-0.2	-3.3	-1.6	1.0
65.18	1.34	1.6	5.10	20MR	1.3	-3.3	-1.6	1.0
71.22	1.35	1.7	14.87	30MR	3.2	-3.3	-1.6	1.0
77.29	1.36	1.8	24.59	9AP	2.6	-3.3	-1.6	1.0
83.38	1.35	1.8	34.26	19AP	1.5	-3.3	-1.6	1.0
89.49	1.35	1.9	43.89	29AP	0.8	-3.3	-1.6	0.9
95.62	1.33	1.9	53.48	9MY	0.2	-3.3	-1.6	0.9
101.77	1.32	2.0	63.04	19MY	-0.6	-3.4	-1.6	0.9
107.94	1.29	2.0	72.58	29MY	-1.6	-3.4	-1.7	0.8
114.13	1.27	2.0	82.11	8JN	-1.4	-3.4	-1.7	0.7
120.35	1.24	2.0	91.63	18JN	-0.5	-3.5	-1.8	0.7
126.59	1.20	2.0	101.16	28JN	0.1	-3.5	-1.8	0.6
132.86	1.17	2.0	110.71	8JL	0.5	-3.5	-1.9	0.5
139.17	1.13	2.0	120.29	18JL	0.9	-3.4	-1.9	0.5
145.51	1.08	2.0	129.89	28JL	1.6	-3.4	-2.0	0.5
151.90	1.03	2.0	139.54	7AU	3.1	-3.4	-2.1	0.5
158.33	0.98	2.0	149.23	17AU	1.4	-3.3	-2.1	0.6
164.80	0.92	1.9	158.97	27AU	-0.4	-3.3	-2.2	0.6
171.33	0.86	1.9	168.77	6SE	-1.1	-3.3	-2.3	0.7
177.91	0.80	1.9	178.62	16SE	-1.1	-3.3	-2.3	0.7
184.55	0.73	1.9	188.53	26SE	-0.8	-3.3	-2.4	0.8
191.25	0.65	1.9	198.51	6OC	-0.5	-3.3	-2.4	0.8
198.01	0.58	1.9	208.53	16OC	-0.3	-3.4	-2.4	0.8
204.83	0.49	1.8	218.60	26OC	-0.3	-3.4	-2.4	0.9
211.72	0.40	1.8	228.72	5NO	-0.1	-3.4	-2.4	0.9
218.67	0.31	1.8	238.87	15NO	0.7	-3.5	-2.4	0.9
225.69	0.21	1.8	249.05	25NO	3.0	-3.5	-2.3	1.0
232.77	0.11	1.7	259.25	5DE	0.5	-3.6	-2.3	1.0
239.91	-0.00	1.7	269.45	15DE	-0.1	-3.6	-2.2	1.0
247.12	-0.12	1.7	279.64	25DE	-0.2	-3.7	-2.1	1.0

933

♂ LONG	♂ LAT	♂ MAG	☉ LONG	16.00UT 933	☿	♀	♃	♄
254.39	-0.24	1.6	289.83	4JA	-0.2	-3.8	-2.0	1.0
261.72	-0.36	1.6	299.98	14JA	-0.5	-3.9	-2.0	1.0
269.10	-0.49	1.5	310.10	24JA	-0.9	-4.0	-1.9	1.0
276.53	-0.62	1.4	320.18	3FE	-1.3	-4.1	-1.8	1.0
284.01	-0.75	1.4	330.20	13FE	-1.1	-4.1	-1.8	1.0
291.52	-0.89	1.3	340.18	23FE	-0.2	-4.2	-1.7	1.1
299.06	-1.02	1.3	350.09	5MR	1.5	-4.3	-1.7	1.1
306.61	-1.15	1.2	359.95	15MR	3.4	-4.2	-1.6	1.1
314.18	-1.28	1.1	9.75	25MR	1.9	-4.0	-1.6	1.1
321.74	-1.41	1.1	19.50	4AP	1.1	-3.5	-1.5	1.2
329.28	-1.52	1.0	29.19	14AP	0.6	-3.0	-1.5	1.1
336.79	-1.62	0.9	38.84	24AP	0.1	-3.7	-1.5	1.1
344.24	-1.72	0.9	48.45	4MY	-0.7	-4.1	-1.5	1.1
351.62	-1.80	0.8	58.02	14MY	-1.6	-4.2	-1.5	1.1
358.91	-1.86	0.7	67.57	24MY	-1.4	-4.2	-1.5	1.0
6.09	-1.91	0.7	77.11	3JN	-0.5	-4.1	-1.5	1.0
13.12	-1.93	0.6	86.63	13JN	0.2	-4.0	-1.5	0.9
19.98	-1.94	0.5	96.16	23JN	0.7	-3.9	-1.5	0.8
26.64	-1.93	0.4	105.71	3JL	1.2	-3.8	-1.6	0.8
33.05	-1.90	0.3	115.26	13JL	2.1	-3.8	-1.6	0.7
39.19	-1.84	0.2	124.85	23JL	3.1	-3.7	-1.7	0.6
44.97	-1.76	0.1	134.48	2AU	1.1	-3.6	-1.7	0.6
50.33	-1.65	0.0	144.14	12AU	-0.5	-3.6	-1.8	0.6
55.19	-1.51	-0.1	153.86	22AU	-1.2	-3.5	-1.8	0.6
59.42	-1.34	-0.3	163.63	1SE	-1.2	-3.5	-1.9	0.7
62.89	-1.12	-0.5	173.45	11SE	-0.7	-3.4	-2.0	0.7
65.40	-0.85	-0.7	183.33	21SE	-0.4	-3.4	-2.0	0.8
66.76	-0.53	-0.9	193.28	1OC	-0.2	-3.4	-2.1	0.8
66.76	-0.14	-1.1	203.27	11OC	-0.1	-3.4	-2.1	0.9
65.31	0.31	-1.3	213.32	21OC	0.0	-3.4	-2.2	0.9
62.52	0.79	-1.5	223.42	31OC	0.9	-3.4	-2.2	1.0
58.85	1.26	-1.6	233.55	10NO	2.7	-3.4	-2.3	1.0
55.06	1.67	-1.5	243.72	20NO	0.3	-3.4	-2.3	1.0
51.92	1.97	-1.2	253.90	30NO	-0.2	-3.4	-2.3	1.1
49.97	2.17	-0.9	264.10	10DE	-0.3	-3.4	-2.3	1.1
49.37	2.27	-0.6	274.30	20DE	-0.4	-3.4	-2.2	1.1
50.05	2.31	-0.3	284.49	30DE	-0.5	-3.4	-2.2	1.1

Right panel (934 / 935)

♂ LONG	♂ LAT	♂ MAG	☉ LONG	16.00UT 934	☿	♀	♃	♄
51.81	2.30	-0.0	294.66	9JA	-0.9	-3.4	-2.1	1.1
54.46	2.27	0.2	304.80	19JA	-1.1	-3.5	-2.1	1.1
57.81	2.22	0.5	314.90	29JA	-1.0	-3.5	-2.0	1.1
61.71	2.16	0.7	324.95	8FE	-0.1	-3.5	-1.9	1.1
66.03	2.10	0.9	334.95	18FE	1.9	-3.4	-1.9	1.2
70.69	2.03	1.0	344.90	28FE	2.8	-3.4	-1.8	1.2
75.62	1.96	1.2	354.79	10MR	1.4	-3.4	-1.7	1.2
80.76	1.89	1.3	4.62	20MR	0.8	-3.4	-1.7	1.3
86.08	1.81	1.4	14.39	30MR	0.4	-3.3	-1.6	1.3
91.54	1.74	1.5	24.11	9AP	-0.0	-3.3	-1.6	1.3
97.12	1.67	1.6	33.79	19AP	-0.7	-3.3	-1.5	1.3
102.80	1.60	1.7	43.41	29AP	-1.6	-3.3	-1.5	1.3
108.57	1.52	1.8	53.00	9MY	-1.4	-3.3	-1.4	1.3
114.43	1.45	1.8	62.57	19MY	-0.4	-3.4	-1.4	1.3
120.36	1.38	1.9	72.11	29MY	0.3	-3.4	-1.4	1.2
126.36	1.30	1.9	81.64	8JN	0.9	-3.4	-1.4	1.2
132.42	1.22	1.9	91.17	18JN	1.6	-3.4	-1.4	1.1
138.56	1.15	2.0	100.70	28JN	2.8	-3.5	-1.4	1.1
144.76	1.07	2.0	110.25	8JL	2.7	-3.5	-1.4	1.0
151.03	0.99	2.0	119.82	18JL	0.9	-3.6	-1.4	0.9
157.37	0.90	2.0	129.42	28JL	-0.5	-3.6	-1.4	0.8
163.78	0.82	1.9	139.07	7AU	-1.3	-3.7	-1.5	0.8
170.27	0.73	1.9	148.76	17AU	-1.3	-3.8	-1.5	0.7
176.83	0.64	1.9	158.50	27AU	-0.7	-3.9	-1.6	0.7
183.47	0.55	1.9	168.29	6SE	-0.3	-4.0	-1.6	0.8
190.19	0.46	1.8	178.15	16SE	-0.1	-4.1	-1.7	0.8
196.99	0.37	1.8	188.05	26SE	0.0	-4.2	-1.7	0.9
203.87	0.27	1.7	198.02	6OC	0.2	-4.3	-1.8	0.9
210.84	0.17	1.7	208.05	16OC	1.2	-4.3	-1.8	1.0
217.89	0.07	1.6	218.11	26OC	2.5	-4.2	-1.9	1.0
225.03	-0.03	1.6	228.23	5NO	0.1	-3.8	-2.0	1.1
232.25	-0.14	1.6	238.38	15NO	-0.4	-2.9	-2.0	1.1
239.55	-0.24	1.6	248.56	25NO	-0.4	-3.7	-2.1	1.2
246.93	-0.35	1.6	258.75	5DE	-0.5	-4.2	-2.1	1.2
254.38	-0.45	1.6	268.95	15DE	-0.6	-4.4	-2.2	1.2
261.90	-0.55	1.6	279.15	25DE	-0.9	-4.4	-2.2	1.3

935

♂ LONG	♂ LAT	♂ MAG	☉ LONG	16.00UT 935	☿	♀	♃	♄
269.48	-0.66	1.5	289.33	4JA	-1.0	-4.3	-2.2	1.3
277.13	-0.76	1.5	299.49	14JA	-0.9	-4.2	-2.1	1.3
284.81	-0.85	1.5	309.61	24JA	-0.1	-4.1	-2.1	1.3
292.54	-0.94	1.4	319.69	3FE	2.2	-4.0	-2.1	1.3
300.30	-1.02	1.4	329.72	13FE	2.2	-3.9	-2.0	1.3
308.08	-1.10	1.4	339.69	23FE	1.0	-3.8	-1.9	1.3
315.86	-1.16	1.4	349.61	5MR	0.6	-3.7	-1.9	1.4
323.64	-1.22	1.4	359.48	15MR	0.3	-3.6	-1.8	1.4
331.40	-1.26	1.3	9.28	25MR	-0.1	-3.5	-1.7	1.4
339.14	-1.29	1.3	19.03	4AP	-0.7	-3.5	-1.7	1.4
346.83	-1.30	1.3	28.73	14AP	-1.6	-3.4	-1.6	1.4
354.46	-1.31	1.3	38.38	24AP	-1.4	-3.4	-1.5	1.4
2.02	-1.29	1.3	47.99	4MY	-0.4	-3.4	-1.5	1.3
9.50	-1.27	1.2	57.57	14MY	0.5	-3.3	-1.4	1.3
16.89	-1.22	1.2	67.11	24MY	1.2	-3.3	-1.4	1.3
24.18	-1.17	1.2	76.65	3JN	2.2	-3.3	-1.4	1.2
31.34	-1.10	1.2	86.18	13JN	3.4	-3.3	-1.3	1.2
38.38	-1.01	1.1	95.70	23JN	2.2	-3.3	-1.3	1.1
45.28	-0.91	1.1	105.24	3JL	0.7	-3.3	-1.3	1.1
52.01	-0.80	1.1	114.80	13JL	-0.5	-3.4	-1.3	1.0
58.57	-0.67	1.0	124.39	23JL	-1.4	-3.4	-1.3	1.0
64.94	-0.53	1.0	134.01	2AU	-1.4	-3.4	-1.3	0.9
71.09	-0.37	0.9	143.67	12AU	-0.6	-3.4	-1.3	0.9
76.99	-0.20	0.8	153.38	22AU	-0.2	-3.5	-1.3	0.9
82.61	-0.00	0.7	163.15	1SE	0.1	-3.5	-1.3	0.8
87.88	0.21	0.6	172.97	11SE	0.2	-3.5	-1.4	0.9
92.75	0.45	0.5	182.85	21SE	0.5	-3.4	-1.4	1.0
97.13	0.72	0.4	192.79	1OC	1.5	-3.4	-1.5	1.0
100.92	1.02	0.2	202.78	11OC	2.3	-3.4	-1.5	1.1
103.97	1.36	0.1	212.83	21OC	-0.0	-3.4	-1.6	1.1
106.12	1.74	-0.1	222.92	31OC	-0.6	-3.3	-1.6	1.2
107.16	2.16	-0.4	233.05	10NO	-0.6	-3.3	-1.7	1.2
106.94	2.62	-0.6	243.22	20NO	-0.6	-3.4	-1.8	1.3
105.32	3.08	-0.8	253.41	30NO	-0.7	-3.3	-1.8	1.3
102.42	3.50	-1.0	263.60	10DE	-0.8	-3.3	-1.9	1.4
98.65	3.81	-1.1	273.80	20DE	-0.8	-3.4	-1.9	1.4
94.69	3.97	-1.1	284.00	30DE	-0.8	-3.4	-2.0	1.4

Left Table

♂ LONG	♂ LAT	♂ MAG	☉ LONG	16.00UT 936	☿	♀	♃	♄
91.32	3.96	-0.8	294.17	9JA	0.0	-3.4	-2.0	1.4
89.05	3.83	-0.6	304.30	19JA	2.5	-3.4	-2.1	1.3
88.09	3.63	-0.3	314.41	29JA	1.7	-3.5	-2.1	1.3
88.40	3.39	-0.0	324.47	8FE	0.7	-3.5	-2.1	1.3
89.82	3.14	0.2	334.47	18FE	0.4	-3.6	-2.0	1.2
92.14	2.90	0.5	344.42	28FE	0.2	-3.6	-2.0	1.1
95.20	2.68	0.7	354.31	9MR	-0.2	-3.7	-2.0	1.1
98.84	2.47	0.8	4.15	19MR	-0.7	-3.8	-1.9	1.1
102.95	2.27	1.0	13.93	29MR	-1.6	-3.8	-1.8	1.1
107.43	2.09	1.1	23.65	8AP	-1.4	-3.9	-1.8	1.1
112.21	1.93	1.3	33.32	18AP	-0.4	-4.0	-1.7	1.1
117.25	1.77	1.4	42.96	28AP	0.6	-4.1	-1.6	1.1
122.50	1.62	1.5	52.55	8MY	1.6	-4.2	-1.6	1.1
127.94	1.48	1.5	62.11	18MY	2.9	-4.2	-1.5	1.0
133.53	1.34	1.6	71.65	28MY	3.1	-4.1	-1.5	1.0
139.28	1.21	1.7	81.18	7JN	1.7	-3.7	-1.4	1.0
145.15	1.08	1.7	90.71	17JN	0.6	-3.0	-1.4	0.9
151.14	0.96	1.7	100.24	27JN	-0.6	-3.3	-1.3	0.9
157.25	0.84	1.8	109.78	7JL	-1.5	-3.9	-1.3	0.8
163.47	0.72	1.8	119.35	17JL	-1.4	-4.2	-1.3	0.8
169.81	0.60	1.8	128.96	27JL	-0.6	-4.2	-1.2	0.7
176.25	0.49	1.8	138.60	6AU	-0.1	-4.2	-1.2	0.7
182.79	0.38	1.8	148.28	16AU	0.2	-4.1	-1.2	0.6
189.45	0.26	1.8	158.02	26AU	0.4	-4.0	-1.2	0.6
196.20	0.15	1.8	167.81	5SE	0.7	-3.9	-1.2	0.6
203.06	0.04	1.7	177.66	15SE	1.9	-3.9	-1.2	0.6
210.03	-0.07	1.7	187.57	25SE	2.0	-3.8	-1.3	0.6
217.10	-0.17	1.7	197.54	5OC	-0.2	-3.7	-1.3	0.7
224.27	-0.28	1.6	207.56	15OC	-0.7	-3.7	-1.3	0.8
231.54	-0.38	1.6	217.63	25OC	-0.7	-3.6	-1.4	0.8
238.91	-0.48	1.6	227.74	4NO	-0.7	-3.5	-1.4	0.9
246.36	-0.57	1.5	237.89	14NO	-0.7	-3.5	-1.5	0.9
253.90	-0.66	1.5	248.07	24NO	-0.6	-3.5	-1.5	1.0
261.52	-0.74	1.4	258.26	4DE	-0.7	-3.4	-1.6	1.0
269.20	-0.82	1.4	268.46	14DE	-0.6	-3.4	-1.7	1.1
276.95	-0.89	1.4	278.66	24DE	0.2	-3.4	-1.7	1.1
				937				
284.74	-0.95	1.4	288.84	3JA	2.8	-3.4	-1.8	1.1
292.58	-1.00	1.4	299.00	13JA	1.3	-3.3	-1.9	1.1
300.44	-1.05	1.4	309.12	23JA	0.5	-3.3	-1.9	1.1
308.31	-1.08	1.4	319.20	2FE	0.2	-3.3	-2.0	1.0
316.19	-1.10	1.4	329.23	12FE	0.0	-3.3	-2.0	1.0
324.06	-1.11	1.4	339.22	22FE	-0.2	-3.3	-2.0	1.0
331.90	-1.10	1.4	349.13	4MR	-0.8	-3.4	-2.0	0.9
339.71	-1.09	1.4	359.00	14MR	-1.6	-3.4	-2.0	0.9
347.47	-1.06	1.4	8.81	24MR	-1.3	-3.4	-2.0	0.9
355.17	-1.02	1.4	18.56	3AP	-0.4	-3.5	-1.9	0.9
2.81	-0.97	1.4	28.26	13AP	0.8	-3.5	-1.9	0.9
10.37	-0.91	1.5	37.91	23AP	2.1	-3.5	-1.8	0.9
17.84	-0.83	1.5	47.52	3MY	3.8	-3.4	-1.8	0.9
25.23	-0.75	1.5	57.10	13MY	2.4	-3.4	-1.7	0.8
32.52	-0.66	1.5	66.65	23MY	1.3	-3.4	-1.6	0.8
39.71	-0.56	1.5	76.18	2JN	0.4	-3.4	-1.6	0.8
46.80	-0.45	1.5	85.71	12JN	-0.6	-3.3	-1.5	0.8
53.77	-0.34	1.5	95.24	22JN	-1.5	-3.3	-1.5	0.7
60.63	-0.22	1.5	104.77	2JL	-1.4	-3.3	-1.4	0.7
67.38	-0.09	1.5	114.33	12JL	-0.6	-3.3	-1.4	0.6
73.99	0.04	1.5	123.92	22JL	-0.0	-3.3	-1.3	0.6
80.48	0.18	1.5	133.53	1AU	0.3	-3.4	-1.3	0.5
86.83	0.32	1.4	143.20	11AU	0.6	-3.4	-1.3	0.5
93.02	0.48	1.4	152.91	21AU	1.0	-3.4	-1.2	0.4
99.05	0.64	1.3	162.67	31AU	2.4	-3.5	-1.2	0.4
104.89	0.81	1.3	172.49	10SE	1.8	-3.5	-1.2	0.3
110.52	0.99	1.2	182.37	20SE	-0.3	-3.6	-1.2	0.3
115.90	1.19	1.1	192.30	30SE	-0.9	-3.6	-1.2	0.3
121.00	1.40	1.0	202.30	10OC	-0.9	-3.7	-1.2	0.4
125.76	1.63	0.9	212.34	20OC	-0.9	-3.8	-1.2	0.4
130.10	1.88	0.8	222.43	30OC	-0.6	-3.9	-1.3	0.5
133.94	2.16	0.6	232.56	9NO	-0.5	-4.0	-1.3	0.6
137.15	2.47	0.4	242.73	19NO	-0.5	-4.1	-1.3	0.6
139.59	2.81	0.2	252.91	29NO	-0.4	-4.2	-1.4	0.7
141.08	3.18	0.0	263.11	9DE	0.3	-4.3	-1.4	0.7
141.43	3.57	-0.2	273.31	19DE	3.0	-4.4	-1.5	0.8
140.51	3.95	-0.5	283.50	29DE	0.9	-4.4	-1.6	0.8

Right Table

♂ LONG	♂ LAT	♂ MAG	☉ LONG	16.00UT 938	☿	♀	♃	♄
138.26	4.28	-0.7	293.68	8JA	0.2	-4.2	-1.6	0.8
134.92	4.51	-0.9	303.82	18JA	0.1	-3.8	-1.7	0.8
131.00	4.57	-1.0	313.92	28JA	-0.1	-3.3	-1.8	0.8
127.22	4.45	-0.8	323.98	7FE	-0.3	-3.7	-1.8	0.8
124.24	4.19	-0.6	333.99	17FE	-0.8	-4.1	-1.9	0.8
122.46	3.84	-0.4	343.94	27FE	-1.5	-4.3	-1.9	0.7
121.98	3.45	-0.1	353.84	9MR	-1.3	-4.3	-2.0	0.7
122.70	3.07	0.1	3.67	19MR	-0.3	-4.2	-2.0	0.7
124.46	2.72	0.3	13.45	29MR	1.1	-4.1	-2.0	0.6
127.07	2.39	0.5	23.18	8AP	2.7	-4.0	-2.0	0.6
130.38	2.10	0.7	32.85	18AP	3.1	-3.9	-2.0	0.6
134.25	1.83	0.8	42.49	28AP	1.8	-3.8	-2.0	0.6
138.58	1.58	1.0	52.08	8MY	1.0	-3.7	-1.9	0.7
143.29	1.35	1.1	61.65	18MY	0.3	-3.6	-1.9	0.6
148.31	1.15	1.2	71.19	28MY	-0.6	-3.6	-1.8	0.6
153.61	0.95	1.2	80.72	7JN	-1.5	-3.5	-1.8	0.6
159.16	0.77	1.3	90.25	17JN	-1.4	-3.4	-1.7	0.6
164.91	0.60	1.4	99.78	27JN	-0.5	-3.4	-1.6	0.6
170.85	0.44	1.4	109.33	7JL	0.1	-3.4	-1.6	0.5
176.97	0.29	1.5	118.89	17JL	0.4	-3.4	-1.5	0.5
183.25	0.15	1.5	128.49	27JL	0.8	-3.3	-1.5	0.5
189.69	0.01	1.5	138.13	6AU	1.4	-3.3	-1.4	0.4
196.28	-0.12	1.5	147.82	16AU	2.8	-3.3	-1.4	0.3
203.01	-0.24	1.5	157.55	26AU	1.6	-3.3	-1.3	0.3
209.87	-0.36	1.5	167.34	5SE	-0.3	-3.4	-1.3	0.2
216.87	-0.47	1.5	177.19	15SE	-1.0	-3.4	-1.3	0.2
223.99	-0.57	1.5	187.09	25SE	-1.0	-3.4	-1.3	0.1
231.23	-0.67	1.5	197.06	5OC	-0.9	-3.4	-1.2	0.1
238.58	-0.76	1.5	207.07	15OC	-0.5	-3.4	-1.2	0.1
246.04	-0.84	1.5	217.14	25OC	-0.3	-3.5	-1.2	0.1
253.59	-0.91	1.5	227.25	4NO	-0.3	-3.5	-1.3	0.2
261.23	-0.97	1.5	237.39	14NO	-0.3	-3.5	-1.3	0.3
268.95	-1.02	1.4	247.57	24NO	0.5	-3.5	-1.3	0.4
276.75	-1.05	1.4	257.77	4DE	3.1	-3.4	-1.3	0.4
284.56	-1.08	1.4	267.96	14DE	0.7	-3.4	-1.4	0.5
292.43	-1.10	1.4	278.16	24DE	0.0	-3.4	-1.4	0.5
				939				
300.33	-1.10	1.4	288.35	3JA	-0.1	-3.4	-1.5	0.6
308.23	-1.09	1.4	298.50	13JA	-0.2	-3.4	-1.5	0.6
316.14	-1.07	1.3	308.63	23JA	-0.4	-3.3	-1.6	0.6
324.02	-1.04	1.3	318.71	2FE	-0.8	-3.3	-1.7	0.6
331.88	-1.00	1.3	328.75	12FE	-1.4	-3.3	-1.7	0.6
339.69	-0.95	1.3	338.73	22FE	-1.2	-3.3	-1.8	0.6
347.46	-0.89	1.3	348.66	4MR	-0.3	-3.4	-1.8	0.6
355.16	-0.82	1.4	358.52	14MR	1.3	-3.4	-1.9	0.5
2.79	-0.74	1.4	8.33	24MR	3.4	-3.4	-2.0	0.5
10.35	-0.65	1.5	18.09	3AP	2.3	-3.4	-2.0	0.5
17.82	-0.56	1.5	27.79	13AP	1.3	-3.5	-2.1	0.4
25.21	-0.46	1.5	37.44	23AP	0.7	-3.5	-2.1	0.4
32.51	-0.36	1.6	47.06	3MY	0.2	-3.6	-2.1	0.5
39.72	-0.26	1.6	56.64	13MY	-0.6	-3.6	-2.1	0.5
46.83	-0.15	1.6	66.19	23MY	-1.6	-3.7	-2.1	0.5
53.86	-0.04	1.7	75.73	2JN	-1.4	-3.8	-2.0	0.5
60.79	0.07	1.7	85.25	12JN	-0.5	-3.9	-2.0	0.5
67.63	0.18	1.7	94.78	22JN	0.1	-3.9	-1.9	0.5
74.38	0.29	1.7	104.32	2JL	0.6	-4.0	-1.9	0.5
81.04	0.40	1.8	113.87	12JL	1.0	-4.1	-1.8	0.4
87.62	0.52	1.8	123.46	22JL	1.8	-4.2	-1.8	0.4
94.09	0.63	1.8	133.07	1AU	3.2	-4.2	-1.7	0.4
100.48	0.75	1.8	142.73	11AU	1.3	-4.1	-1.6	0.3
106.77	0.87	1.7	152.44	21AU	-0.4	-3.7	-1.6	0.3
112.96	0.99	1.7	162.20	31AU	-1.1	-3.2	-1.5	0.2
119.03	1.11	1.7	172.00	10SE	-1.2	-3.7	-1.5	0.1
125.00	1.24	1.6	181.89	20SE	-0.8	-4.1	-1.4	0.1
130.83	1.37	1.6	191.82	30SE	-0.4	-4.3	-1.4	0.0
136.51	1.50	1.5	201.81	10OC	-0.3	-4.3	-1.4	-0.1
142.02	1.65	1.4	211.85	20OC	-0.2	-4.3	-1.3	-0.1
147.33	1.80	1.3	221.94	30OC	-0.1	-4.2	-1.3	-0.1
152.40	1.96	1.2	232.07	9NO	0.7	-4.1	-1.3	-0.0
157.19	2.13	1.1	242.23	19NO	3.0	-4.0	-1.3	0.1
161.62	2.32	1.0	252.42	29NO	0.4	-3.9	-1.3	0.1
165.64	2.52	0.8	262.62	9DE	-0.1	-3.8	-1.3	0.2
169.12	2.73	0.6	272.82	19DE	-0.2	-3.7	-1.4	0.3
171.95	2.96	0.4	283.01	29DE	-0.3	-3.7	-1.4	0.3

♂ LONG	♂ LAT	♂ MAG	⊙ LONG	16.00UT 940	☿	♀	♃	♄
					MAGNITUDES			
173.97	3.21	0.1	293.18	8JA	-0.5	-3.6	-1.4	0.4
174.99	3.46	-0.1	303.33	18JA	-0.9	-3.5	-1.5	0.4
174.84	3.69	-0.4	313.44	28JA	-1.2	-3.5	-1.5	0.4
173.42	3.88	-0.7	323.50	7FE	-1.1	-3.4	-1.6	0.5
170.76	3.97	-0.9	333.51	17FE	-0.2	-3.4	-1.6	0.5
167.20	3.92	-1.1	343.46	27FE	1.6	-3.4	-1.7	0.5
163.33	3.71	-1.1	353.36	8MR	3.2	-3.3	-1.8	0.5
159.88	3.37	-0.9	3.20	18MR	1.7	-3.3	-1.8	0.4
157.43	2.94	-0.7	12.98	28MR	1.0	-3.3	-1.9	0.4
156.24	2.49	-0.5	22.71	7AP	0.5	-3.3	-2.0	0.4
156.34	2.05	-0.3	32.39	17AP	0.1	-3.3	-2.0	0.3
157.63	1.64	-0.1	42.03	27AP	-0.6	-3.3	-2.1	0.3
159.90	1.27	0.1	51.62	7MY	-1.6	-3.3	-2.1	0.3
163.01	0.95	0.3	61.19	17MY	-1.4	-3.4	-2.2	0.3
166.79	0.66	0.4	70.73	27MY	-0.5	-3.4	-2.2	0.4
171.12	0.40	0.6	80.26	6JN	0.3	-3.4	-2.2	0.4
175.93	0.16	0.7	89.79	16JN	0.8	-3.5	-2.2	0.4
181.13	-0.05	0.8	99.32	26JN	1.4	-3.5	-2.2	0.4
186.66	-0.23	0.9	108.86	6JL	2.4	-3.5	-2.2	0.4
192.49	-0.40	0.9	118.43	16JL	3.0	-3.4	-2.1	0.4
198.58	-0.56	1.0	128.02	26JL	1.1	-3.4	-2.1	0.3
204.90	-0.69	1.0	137.66	5AU	-0.4	-3.4	-2.0	0.3
211.44	-0.82	1.1	147.34	15AU	-1.2	-3.3	-1.9	0.3
218.16	-0.93	1.1	157.07	25AU	-1.3	-3.3	-1.9	0.2
225.07	-1.02	1.2	166.86	4SE	-0.7	-3.3	-1.8	0.2
232.13	-1.10	1.2	176.70	14SE	-0.3	-3.3	-1.7	0.1
239.35	-1.17	1.2	186.60	24SE	-0.1	-3.3	-1.7	0.0
246.69	-1.22	1.2	196.56	4OC	-0.0	-3.3	-1.6	-0.0
254.16	-1.26	1.2	206.58	14OC	0.1	-3.4	-1.6	-0.1
261.73	-1.28	1.3	216.64	24OC	1.0	-3.4	-1.5	-0.2
269.38	-1.29	1.3	226.75	3NO	2.8	-3.4	-1.5	-0.2
277.12	-1.29	1.3	236.90	13NO	0.2	-3.5	-1.5	-0.2
284.91	-1.27	1.3	247.07	23NO	-0.3	-3.5	-1.4	-0.1
292.74	-1.24	1.3	257.27	3DE	-0.4	-3.6	-1.4	-0.0
300.60	-1.20	1.4	267.47	13DE	-0.4	-3.6	-1.4	0.0
308.46	-1.14	1.4	277.67	23DE	-0.6	-3.7	-1.4	0.1
				941				
316.32	-1.08	1.4	287.85	2JA	-0.9	-3.8	-1.4	0.2
324.16	-1.01	1.4	298.01	12JA	-1.1	-3.9	-1.5	0.2
331.97	-0.92	1.4	308.14	22JA	-1.0	-4.0	-1.5	0.3
339.73	-0.83	1.5	318.23	1FE	-0.1	-4.1	-1.5	0.3
347.44	-0.74	1.5	328.27	11FE	1.9	-4.2	-1.5	0.3
355.08	-0.64	1.5	338.25	21FE	2.6	-4.2	-1.6	0.4
2.65	-0.54	1.5	348.18	3MR	1.2	-4.3	-1.6	0.4
10.14	-0.43	1.5	358.05	13MR	0.7	-4.2	-1.7	0.4
17.55	-0.33	1.6	7.86	23MR	0.4	-4.0	-1.8	0.4
24.88	-0.22	1.6	17.62	2AP	-0.0	-3.5	-1.8	0.3
32.12	-0.12	1.6	27.32	12AP	-0.7	-3.0	-1.9	0.3
39.27	-0.01	1.6	36.98	22AP	-1.6	-3.7	-2.0	0.3
46.35	0.09	1.6	46.60	2MY	-1.4	-4.1	-2.0	0.2
53.34	0.19	1.7	56.18	12MY	-0.5	-4.2	-2.1	0.2
60.25	0.29	1.7	65.73	22MY	0.4	-4.2	-2.2	0.2
67.09	0.39	1.8	75.27	1JN	1.0	-4.1	-2.2	0.3
73.85	0.49	1.8	84.79	11JN	1.8	-4.0	-2.3	0.3
80.54	0.58	1.8	94.32	21JN	3.0	-3.9	-2.3	0.3
87.16	0.67	1.9	103.86	1JL	2.5	-3.8	-2.3	0.3
93.72	0.76	1.9	113.41	11JL	0.9	-3.7	-2.3	0.3
100.22	0.84	1.9	122.99	21JL	-0.5	-3.7	-2.3	0.3
106.66	0.92	1.9	132.61	31JL	-1.3	-3.6	-2.3	0.3
113.04	1.00	1.9	142.26	10AU	-1.4	-3.6	-2.3	0.3
119.36	1.08	1.9	151.97	20AU	-0.7	-3.5	-2.2	0.3
125.63	1.16	1.9	161.73	30AU	-0.3	-3.5	-2.2	0.2
131.84	1.24	1.9	171.54	9SE	-0.0	-3.4	-2.1	0.2
137.99	1.31	1.9	181.41	19SE	0.1	-3.4	-2.0	0.1
144.07	1.38	1.9	191.34	29SE	0.3	-3.4	-2.0	0.1
150.08	1.46	1.8	201.32	9OC	1.3	-3.4	-1.9	-0.0
156.02	1.53	1.8	211.36	19OC	2.5	-3.4	-1.8	-0.1
161.86	1.60	1.7	221.45	29OC	0.1	-3.4	-1.8	-0.2
167.60	1.67	1.6	231.57	8NO	-0.5	-3.4	-1.7	-0.2
173.21	1.75	1.5	241.73	18NO	-0.5	-3.4	-1.7	-0.3
178.69	1.82	1.4	251.92	28NO	-0.5	-3.4	-1.6	-0.2
183.98	1.89	1.3	262.12	8DE	-0.7	-3.4	-1.6	-0.2
189.08	1.97	1.1	272.32	18DE	-0.9	-3.4	-1.6	-0.1
193.91	2.04	1.0	282.51	28DE	-0.9	-3.4	-1.5	-0.0

♂ LONG	♂ LAT	♂ MAG	⊙ LONG	16.00UT 942	☿	♀	♃	♄
					MAGNITUDES			
198.42	2.11	0.8	292.69	7JA	-0.8	-3.4	-1.5	0.0
202.56	2.18	0.6	302.83	17JA	-0.1	-3.5	-1.5	0.1
206.20	2.24	0.4	312.94	27JA	2.2	-3.5	-1.5	0.2
209.23	2.28	0.2	323.01	6FE	2.0	-3.5	-1.5	0.2
211.51	2.31	-0.1	333.02	16FE	0.9	-3.4	-1.5	0.3
212.86	2.31	-0.4	342.98	26FE	0.5	-3.4	-1.6	0.3
213.09	2.25	-0.7	352.88	8MR	0.2	-3.4	-1.6	0.3
212.08	2.12	-1.0	2.72	18MR	-0.1	-3.4	-1.6	0.3
209.83	1.90	-1.3	12.51	28MR	-0.7	-3.3	-1.7	0.3
206.61	1.57	-1.5	22.24	7AP	-1.6	-3.3	-1.7	0.3
202.96	1.15	-1.6	31.92	17AP	-1.4	-3.3	-1.8	0.3
199.63	0.68	-1.4	41.56	27AP	-0.4	-3.3	-1.8	0.3
197.24	0.21	-1.2	51.15	7MY	0.5	-3.3	-1.9	0.2
196.13	-0.22	-1.0	60.72	17MY	1.4	-3.4	-2.0	0.2
196.38	-0.59	-0.8	70.27	27MY	2.4	-3.4	-2.0	0.2
197.87	-0.91	-0.6	79.80	6JN	3.4	-3.4	-2.1	0.2
200.44	-1.16	-0.4	89.32	16JN	2.0	-3.4	-2.2	0.3
203.92	-1.37	-0.3	98.86	26JN	0.7	-3.5	-2.3	0.3
208.13	-1.54	-0.1	108.40	6JL	-0.5	-3.5	-2.3	0.3
212.95	-1.67	0.0	117.97	16JL	-1.4	-3.6	-2.4	0.3
218.28	-1.78	0.1	127.56	26JL	-1.4	-3.6	-2.4	0.3
224.03	-1.85	0.3	137.19	5AU	-0.6	-3.7	-2.4	0.3
230.15	-1.90	0.4	146.88	15AU	-0.2	-3.8	-2.4	0.3
236.57	-1.93	0.4	156.61	25AU	0.1	-3.9	-2.4	0.3
243.25	-1.93	0.5	166.39	4SE	0.3	-4.0	-2.4	0.3
250.15	-1.92	0.6	176.23	14SE	0.6	-4.1	-2.4	0.2
257.25	-1.89	0.7	186.13	24SE	1.6	-4.2	-2.3	0.2
264.50	-1.83	0.8	196.08	4OC	2.3	-4.3	-2.3	0.1
271.88	-1.77	0.8	206.09	14OC	-0.1	-4.3	-2.2	0.1
279.37	-1.69	0.9	216.15	24OC	-0.6	-4.2	-2.1	-0.0
286.94	-1.59	1.0	226.26	3NO	-0.6	-3.8	-2.1	-0.1
294.56	-1.49	1.0	236.40	13NO	-0.7	-2.9	-2.0	-0.2
302.23	-1.37	1.1	246.58	23NO	-0.8	-3.7	-1.9	-0.2
309.91	-1.25	1.1	256.77	3DE	-0.7	-4.2	-1.9	-0.3
317.60	-1.12	1.2	266.97	13DE	-0.8	-4.4	-1.8	-0.2
325.27	-0.99	1.3	277.17	23DE	-0.7	-4.4	-1.7	-0.2
				943				
332.90	-0.86	1.3	287.35	2JA	0.0	-4.3	-1.7	-0.1
340.50	-0.73	1.4	297.52	12JA	2.5	-4.2	-1.7	-0.0
348.05	-0.59	1.4	307.65	22JA	1.5	-4.1	-1.6	0.0
355.53	-0.46	1.5	317.73	1FE	0.6	-4.0	-1.6	0.1
2.95	-0.34	1.5	327.78	11FE	0.3	-3.9	-1.6	0.2
10.29	-0.21	1.6	337.77	21FE	0.1	-3.8	-1.6	0.2
17.56	-0.09	1.6	347.70	3MR	-0.2	-3.7	-1.6	0.3
24.76	0.02	1.7	357.57	13MR	-0.7	-3.6	-1.6	0.3
31.87	0.13	1.7	7.39	23MR	-1.5	-3.5	-1.6	0.3
38.91	0.24	1.7	17.15	2AP	-1.4	-3.5	-1.6	0.3
45.88	0.34	1.8	26.86	12AP	-0.4	-3.4	-1.6	0.3
52.77	0.43	1.8	36.52	22AP	0.7	-3.4	-1.7	0.3
59.60	0.52	1.8	46.13	2MY	1.8	-3.4	-1.7	0.3
66.36	0.60	1.8	55.72	12MY	3.2	-3.3	-1.8	0.3
73.05	0.68	1.8	65.27	22MY	2.8	-3.3	-1.8	0.3
79.69	0.75	1.8	74.81	1JN	1.5	-3.3	-1.9	0.2
86.28	0.82	1.8	84.34	11JN	0.5	-3.3	-1.9	0.2
92.83	0.88	1.9	93.86	21JN	-0.5	-3.3	-2.0	0.2
99.32	0.94	1.9	103.40	1JL	-1.5	-3.3	-2.1	0.3
105.79	1.00	1.9	112.95	11JL	-1.4	-3.4	-2.1	0.3
112.21	1.05	2.0	122.53	21JL	-0.6	-3.4	-2.2	0.4
118.61	1.09	2.0	132.14	31JL	-0.1	-3.4	-2.3	0.4
124.98	1.13	2.0	141.80	10AU	0.2	-3.4	-2.3	0.4
131.32	1.17	2.0	151.50	20AU	0.4	-3.5	-2.4	0.4
137.65	1.21	2.0	161.25	30AU	0.8	-3.5	-2.4	0.4
143.95	1.24	2.0	171.06	9SE	2.0	-3.5	-2.4	0.4
150.23	1.27	2.0	180.93	19SE	2.0	-3.4	-2.5	0.3
156.50	1.29	2.0	190.85	29SE	-0.2	-3.4	-2.5	0.3
162.75	1.31	2.0	200.84	9OC	-0.8	-3.4	-2.4	0.2
168.97	1.32	1.9	210.87	19OC	-0.8	-3.4	-2.4	0.2
175.17	1.34	1.9	220.95	29OC	-0.8	-3.3	-2.3	0.1
181.35	1.34	1.8	231.08	8NO	-0.7	-3.3	-2.3	0.0
187.49	1.34	1.8	241.24	18NO	-0.6	-3.3	-2.2	-0.0
193.59	1.34	1.7	251.42	28NO	-0.6	-3.3	-2.2	-0.1
199.64	1.33	1.6	261.62	8DE	-0.5	-3.3	-2.1	-0.2
205.64	1.30	1.5	271.82	18DE	0.2	-3.4	-2.0	-0.2
211.56	1.27	1.4	282.01	28DE	2.8	-3.4	-1.9	-0.2

♂			⊙	16.00UT	☿	♀	♃	♄
LONG	LAT	MAG	LONG	944	MAGNITUDES			
217.40	1.23	1.3	292.19	7JA	1.2	-3.4	-1.9	-0.1
223.13	1.18	1.1	302.34	17JA	0.4	-3.4	-1.8	-0.1
228.74	1.11	1.0	312.45	27JA	0.2	-3.5	-1.8	0.0
234.18	1.01	0.8	322.52	6FE	0.0	-3.5	-1.7	0.1
239.43	0.90	0.6	332.53	16FE	-0.3	-3.6	-1.7	0.1
244.44	0.75	0.4	342.49	26FE	-0.8	-3.6	-1.6	0.2
249.14	0.57	0.2	352.40	7MR	-1.5	-3.7	-1.6	0.2
253.46	0.34	-0.1	2.25	17MR	-1.3	-3.8	-1.6	0.3
257.29	0.05	-0.3	12.04	27MR	-0.4	-3.8	-1.6	0.3
260.50	-0.30	-0.6	21.77	6AP	0.9	-3.9	-1.6	0.4
262.93	-0.73	-0.9	31.45	16AP	2.3	-4.0	-1.6	0.4
264.41	-1.25	-1.2	41.10	26AP	3.7	-4.1	-1.6	0.4
264.74	-1.86	-1.5	50.70	6MY	2.1	-4.2	-1.6	0.4
263.85	-2.54	-1.9	60.26	16MY	1.2	-4.2	-1.6	0.4
261.79	-3.24	-2.1	69.81	26MY	0.4	-4.1	-1.6	0.4
258.95	-3.88	-2.3	79.34	5JN	-0.6	-3.7	-1.7	0.3
255.99	-4.37	-2.3	88.87	15JN	-1.5	-3.0	-1.7	0.3
253.61	-4.66	-2.1	98.40	25JN	-1.4	-3.3	-1.8	0.3
252.38	-4.76	-1.9	107.94	5JL	-0.6	-3.9	-1.8	0.3
252.52	-4.71	-1.7	117.50	15JL	-0.0	-4.2	-1.9	0.4
254.00	-4.54	-1.4	127.10	25JL	0.3	-4.2	-1.9	0.4
256.68	-4.31	-1.2	136.73	4AU	0.6	-4.2	-2.0	0.4
260.35	-4.05	-1.0	146.40	14AU	1.1	-4.1	-2.1	0.5
264.82	-3.76	-0.8	156.13	24AU	2.5	-4.0	-2.1	0.5
269.94	-3.46	-0.6	165.91	3SE	1.8	-3.9	-2.2	0.5
275.57	-3.16	-0.4	175.75	13SE	-0.3	-3.9	-2.3	0.5
281.59	-2.86	-0.2	185.64	23SE	-0.9	-3.8	-2.3	0.5
287.93	-2.57	-0.1	195.60	3OC	-0.9	-3.7	-2.4	0.4
294.51	-2.28	0.1	205.60	13OC	-0.9	-3.6	-2.4	0.4
301.27	-2.00	0.2	215.67	23OC	-0.6	-3.6	-2.4	0.3
308.17	-1.73	0.4	225.77	2NO	-0.5	-3.5	-2.4	0.3
315.16	-1.48	0.5	235.91	12NO	-0.4	-3.5	-2.4	0.2
322.22	-1.23	0.6	246.09	22NO	-0.4	-3.5	-2.3	0.1
329.32	-1.00	0.8	256.28	2DE	0.3	-3.4	-2.3	0.1
336.43	-0.79	0.9	266.47	12DE	3.0	-3.4	-2.2	-0.0
343.55	-0.59	1.0	276.68	22DE	0.8	-3.4	-2.2	-0.1
				945				
350.65	-0.40	1.1	286.86	1JA	0.2	-3.4	-2.1	-0.1
357.72	-0.23	1.2	297.02	11JA	0.0	-3.3	-2.0	-0.1
4.76	-0.07	1.3	307.16	21JA	-0.1	-3.3	-1.9	-0.0
11.76	0.07	1.4	317.25	31JA	-0.4	-3.3	-1.9	0.0
18.71	0.20	1.5	327.29	10FE	-0.8	-3.3	-1.8	0.1
25.61	0.33	1.6	337.28	20FE	-1.4	-3.3	-1.7	0.2
32.45	0.44	1.6	347.21	2MR	-1.3	-3.4	-1.7	0.2
39.25	0.54	1.7	357.09	12MR	-0.3	-3.4	-1.6	0.3
45.99	0.63	1.8	6.91	22MR	1.1	-3.4	-1.6	0.3
52.68	0.71	1.8	16.67	1AP	2.9	-3.5	-1.6	0.4
59.32	0.78	1.9	26.38	11AP	2.8	-3.5	-1.5	0.4
65.92	0.84	1.9	36.05	21AP	1.6	-3.5	-1.5	0.4
72.47	0.90	1.9	45.66	1MY	0.9	-3.4	-1.5	0.5
78.99	0.95	2.0	55.25	11MY	0.3	-3.4	-1.5	0.5
85.47	1.00	2.0	64.81	21MY	-0.6	-3.4	-1.5	0.5
91.92	1.03	2.0	74.34	31MY	-1.5	-3.4	-1.5	0.5
98.34	1.07	2.0	83.87	10JN	-1.4	-3.3	-1.5	0.5
104.74	1.09	2.0	93.40	20JN	-0.5	-3.3	-1.5	0.5
111.13	1.12	2.0	102.93	30JN	0.1	-3.3	-1.5	0.5
117.50	1.13	2.0	112.48	10JL	0.5	-3.3	-1.6	0.4
123.87	1.15	2.0	122.06	20JL	0.9	-3.3	-1.6	0.5
130.24	1.15	2.0	131.67	30JL	1.5	-3.4	-1.7	0.5
136.60	1.16	2.0	141.33	9AU	3.0	-3.4	-1.7	0.6
142.98	1.16	2.0	151.03	19AU	1.5	-3.4	-1.8	0.6
149.36	1.15	2.0	160.78	29AU	-0.3	-3.5	-1.8	0.6
155.75	1.14	2.0	170.58	8SE	-1.0	-3.5	-1.9	0.6
162.16	1.12	2.0	180.45	18SE	-1.1	-3.6	-2.0	0.6
168.58	1.10	2.0	190.37	28SE	-0.8	-3.6	-2.0	0.6
175.02	1.08	2.0	200.35	8OC	-0.5	-3.7	-2.1	0.6
181.49	1.05	2.0	210.39	18OC	-0.3	-3.8	-2.1	0.6
187.97	1.01	1.9	220.46	28OC	-0.3	-3.9	-2.2	0.5
194.47	0.97	1.9	230.59	7NO	-0.2	-4.0	-2.2	0.5
201.00	0.92	1.9	240.75	17NO	0.5	-4.1	-2.3	0.4
207.54	0.86	1.8	250.93	27NO	3.1	-4.2	-2.3	0.4
214.11	0.79	1.7	261.13	7DE	0.6	-4.3	-2.3	0.3
220.69	0.71	1.7	271.33	17DE	-0.0	-4.4	-2.3	0.2
227.29	0.63	1.6	281.52	27DE	-0.1	-4.4	-2.2	0.2

♂			⊙	16.00UT	☿	♀	♃	♄
LONG	LAT	MAG	LONG	946	MAGNITUDES			
233.90	0.53	1.5	291.70	6JA	-0.2	-4.2	-2.2	0.1
240.52	0.41	1.4	301.85	16JA	-0.4	-3.8	-2.1	0.0
247.15	0.28	1.3	311.96	26JA	-0.8	-3.3	-2.0	0.1
253.78	0.14	1.2	322.03	5FE	-1.3	-3.7	-2.0	0.1
260.40	-0.02	1.1	332.05	15FE	-1.2	-4.2	-1.9	0.2
267.01	-0.21	0.9	342.01	25FE	-0.3	-4.3	-1.8	0.3
273.61	-0.41	0.8	351.92	7MR	1.4	-4.3	-1.8	0.3
280.16	-0.64	0.7	1.77	17MR	3.4	-4.2	-1.7	0.4
286.67	-0.89	0.5	11.56	27MR	2.1	-4.1	-1.6	0.4
293.10	-1.17	0.3	21.30	6AP	1.2	-4.0	-1.6	0.5
299.43	-1.48	0.2	30.99	16AP	0.7	-3.9	-1.5	0.5
305.63	-1.82	-0.0	40.62	26AP	0.1	-3.8	-1.5	0.6
311.64	-2.19	-0.2	50.23	6MY	-0.6	-3.7	-1.5	0.6
317.41	-2.60	-0.4	59.80	16MY	-1.5	-3.6	-1.4	0.6
322.86	-3.04	-0.6	69.34	26MY	-1.5	-3.5	-1.4	0.6
327.88	-3.51	-0.8	78.88	5JN	-0.5	-3.5	-1.4	0.7
332.34	-4.01	-1.1	88.41	15JN	0.2	-3.4	-1.4	0.7
336.10	-4.54	-1.3	97.93	25JN	0.7	-3.4	-1.4	0.7
338.93	-5.07	-1.6	107.48	5JL	1.1	-3.4	-1.4	0.7
340.64	-5.59	-1.8	117.04	15JL	2.0	-3.4	-1.4	0.6
341.06	-6.03	-2.1	126.64	25JL	3.2	-3.3	-1.4	0.6
340.11	-6.32	-2.3	136.27	4AU	1.3	-3.3	-1.4	0.7
337.99	-6.37	-2.5	145.94	14AU	-0.4	-3.3	-1.5	0.7
335.25	-6.12	-2.6	155.66	24AU	-1.2	-3.3	-1.5	0.8
332.62	-5.59	-2.4	165.44	3SE	-1.2	-3.4	-1.6	0.8
330.82	-4.88	-2.1	175.28	13SE	-0.8	-3.4	-1.6	0.8
330.24	-4.09	-1.8	185.17	23SE	-0.4	-3.4	-1.7	0.8
330.97	-3.32	-1.5	195.12	3OC	-0.2	-3.4	-1.7	0.8
332.90	-2.61	-1.2	205.12	13OC	-0.1	-3.4	-1.8	0.8
335.82	-2.00	-0.9	215.18	23OC	-0.0	-3.5	-1.8	0.8
339.55	-1.47	-0.6	225.28	2NO	0.8	-3.5	-1.9	0.8
343.90	-1.02	-0.4	235.42	12NO	3.0	-3.5	-2.0	0.8
348.73	-0.64	-0.1	245.59	22NO	0.4	-3.5	-2.0	0.7
353.93	-0.32	0.1	255.78	2DE	-0.2	-3.4	-2.1	0.7
359.41	-0.05	0.3	265.98	12DE	-0.3	-3.4	-2.1	0.6
5.11	0.18	0.5	276.18	22DE	-0.3	-3.4	-2.1	0.5
				947				
10.97	0.38	0.7	286.37	1JA	-0.5	-3.4	-2.2	0.4
16.95	0.54	0.9	296.53	11JA	-0.9	-3.4	-2.1	0.4
23.01	0.68	1.0	306.66	21JA	-1.2	-3.3	-2.1	0.3
29.15	0.79	1.2	316.76	31JA	-1.1	-3.3	-2.1	0.3
35.33	0.89	1.3	326.80	10FE	-0.2	-3.3	-2.0	0.3
41.54	0.97	1.4	336.79	20FE	1.7	-3.3	-2.0	0.4
47.77	1.04	1.5	346.73	2MR	3.0	-3.4	-1.9	0.4
54.01	1.09	1.6	356.61	12MR	1.5	-3.4	-1.8	0.5
60.26	1.14	1.7	6.43	22MR	0.9	-3.4	-1.8	0.5
66.52	1.17	1.8	16.20	1AP	0.5	-3.4	-1.7	0.6
72.77	1.20	1.8	25.91	11AP	0.0	-3.5	-1.6	0.6
79.01	1.22	1.9	35.58	21AP	-0.6	-3.5	-1.6	0.7
85.26	1.23	1.9	45.20	1MY	-1.5	-3.6	-1.5	0.7
91.51	1.24	2.0	54.78	11MY	-1.5	-3.6	-1.5	0.8
97.76	1.24	2.0	64.34	21MY	-0.5	-3.7	-1.4	0.8
104.01	1.24	2.0	73.88	31MY	0.3	-3.8	-1.4	0.8
110.27	1.23	2.0	83.41	10JN	0.9	-3.9	-1.3	0.9
116.54	1.21	2.0	92.94	20JN	1.5	-4.0	-1.3	0.9
122.83	1.20	2.0	102.48	30JN	2.6	-4.0	-1.3	0.9
129.13	1.17	2.0	112.02	10JL	2.9	-4.1	-1.3	0.9
135.45	1.15	2.0	121.60	20JL	1.1	-4.2	-1.3	0.9
141.80	1.12	2.0	131.21	30JL	-0.4	-4.2	-1.3	0.9
148.18	1.08	2.0	140.86	9AU	-1.3	-4.1	-1.3	0.9
154.60	1.04	2.0	150.56	19AU	-1.3	-3.7	-1.3	0.9
161.05	1.00	1.9	160.31	29AU	-0.7	-3.2	-1.3	1.0
167.53	0.95	1.9	170.11	8SE	-0.3	-3.7	-1.3	1.0
174.07	0.90	1.9	179.97	18SE	-0.1	-4.1	-1.4	1.1
180.65	0.84	1.9	189.89	28SE	0.0	-4.3	-1.4	1.1
187.28	0.78	1.9	199.87	8OC	0.2	-4.3	-1.5	1.1
193.96	0.71	1.9	209.90	18OC	1.0	-4.3	-1.5	1.1
200.69	0.64	1.9	219.98	28OC	2.8	-4.2	-1.6	1.1
207.48	0.56	1.9	230.10	7NO	0.2	-4.1	-1.6	1.1
214.32	0.48	1.8	240.26	17NO	-0.4	-4.0	-1.7	1.1
221.22	0.38	1.8	250.44	27NO	-0.4	-3.9	-1.8	1.0
228.17	0.29	1.8	260.63	7DE	-0.5	-3.8	-1.8	1.0
235.18	0.18	1.7	270.83	17DE	-0.6	-3.7	-1.9	0.9
242.25	0.07	1.7	281.03	27DE	-0.9	-3.7	-2.0	0.8

Left table:

♂ LONG	LAT	MAG	⊙ LONG	16.00UT 948	☿	♀	♃	♄
249.36	-0.05	1.6	291.21	6JA	-1.0	-3.6	-2.0	0.8
256.53	-0.18	1.5	301.36	16JA	-0.9	-3.5	-2.0	0.7
263.75	-0.32	1.5	311.47	26JA	-0.2	-3.5	-2.0	0.6
271.02	-0.46	1.4	321.54	5FE	2.0	-3.4	-2.1	0.6
278.32	-0.61	1.3	331.57	15FE	2.4	-3.4	-2.0	0.6
285.67	-0.76	1.3	341.54	25FE	1.1	-3.4	-2.0	0.6
293.04	-0.92	1.2	351.44	6MR	0.6	-3.3	-2.0	0.6
300.43	-1.08	1.1	1.30	16MR	0.3	-3.3	-1.9	0.7
307.83	-1.25	1.0	11.09	26MR	-0.0	-3.3	-1.9	0.7
315.23	-1.41	0.9	20.83	5AP	-0.7	-3.3	-1.8	0.8
322.62	-1.57	0.8	30.52	15AP	-1.5	-3.3	-1.7	0.8
329.97	-1.73	0.7	40.16	25AP	-1.5	-3.3	-1.7	0.9
337.26	-1.88	0.7	49.77	5MY	-0.5	-3.3	-1.6	1.0
344.48	-2.02	0.6	59.34	15MY	0.4	-3.4	-1.6	1.0
351.59	-2.15	0.5	68.89	25MY	1.1	-3.4	-1.5	1.0
358.57	-2.27	0.4	78.42	4JN	2.0	-3.4	-1.4	1.1
5.37	-2.37	0.3	87.95	14JN	3.3	-3.5	-1.4	1.1
11.96	-2.45	0.2	97.47	24JN	2.3	-3.5	-1.3	1.1
18.29	-2.51	0.0	107.01	4JL	0.9	-3.5	-1.3	1.2
24.28	-2.55	-0.1	116.58	14JL	-0.5	-3.4	-1.3	1.2
29.88	-2.56	-0.2	126.17	24JL	-1.3	-3.4	-1.2	1.2
34.98	-2.55	-0.4	135.79	3AU	-1.4	-3.4	-1.2	1.2
39.47	-2.50	-0.5	145.47	13AU	-0.7	-3.3	-1.2	1.2
43.18	-2.41	-0.7	155.19	23AU	-0.2	-3.3	-1.2	1.2
45.95	-2.28	-0.9	164.96	2SE	0.0	-3.3	-1.2	1.2
47.55	-2.08	-1.1	174.80	12SE	0.2	-3.3	-1.2	1.3
47.79	-1.80	-1.4	184.68	22SE	0.4	-3.3	-1.2	1.3
46.56	-1.44	-1.6	194.63	2OC	1.4	-3.3	-1.3	1.4
43.96	-0.98	-1.8	204.64	12OC	2.6	-3.4	-1.3	1.4
40.47	-0.46	-1.9	214.69	22OC	0.0	-3.4	-1.3	1.3
36.84	0.06	-1.8	224.79	1NO	-0.5	-3.4	-1.4	1.3
33.85	0.53	-1.5	234.93	11NO	-0.6	-3.5	-1.4	1.3
32.06	0.91	-1.2	245.10	21NO	-0.6	-3.5	-1.5	1.2
31.62	1.20	-0.8	255.29	1DE	-0.7	-3.6	-1.6	1.2
32.47	1.41	-0.5	265.49	11DE	-0.8	-3.7	-1.6	1.2
34.40	1.56	-0.2	275.68	21DE	-0.8	-3.7	-1.7	1.1
37.21	1.65	0.0	285.87	31DE	-0.8	-3.8	-1.8	1.1
				949				
40.71	1.71	0.3	296.04	10JA	-0.1	-3.9	-1.8	1.0
44.76	1.74	0.5	306.17	20JA	2.2	-4.0	-1.9	1.0
49.22	1.76	0.7	316.27	30JA	1.8	-4.1	-1.9	0.9
54.01	1.76	0.9	326.32	9FE	0.8	-4.2	-2.0	0.9
59.06	1.74	1.1	336.31	19FE	0.4	-4.2	-2.0	0.9
64.31	1.73	1.2	346.25	1MR	0.2	-4.3	-2.0	0.9
69.72	1.70	1.3	356.14	11MR	-0.1	-4.2	-2.0	0.9
75.26	1.67	1.5	5.96	21MR	-0.7	-4.0	-2.0	0.9
80.90	1.63	1.6	15.73	31MR	-1.5	-3.5	-2.0	1.0
86.64	1.59	1.7	25.45	10AP	-1.4	-3.1	-1.9	1.0
92.44	1.55	1.7	35.11	20AP	-0.5	-3.7	-1.9	1.1
98.31	1.50	1.8	44.73	30AP	0.6	-4.1	-1.8	1.1
104.25	1.45	1.9	54.32	10MY	1.5	-4.2	-1.8	1.2
110.23	1.40	1.9	63.88	20MY	2.7	-4.2	-1.7	1.3
116.26	1.35	1.9	73.42	30MY	3.3	-4.1	-1.6	1.3
122.35	1.29	2.0	82.95	9JN	1.8	-4.0	-1.6	1.3
128.48	1.24	2.0	92.47	19JN	0.7	-3.9	-1.5	1.4
134.66	1.17	2.0	102.01	29JN	-0.5	-3.8	-1.4	1.4
140.90	1.11	2.0	111.56	9JL	-1.4	-3.7	-1.4	1.4
147.19	1.05	2.0	121.13	19JL	-1.4	-3.7	-1.3	1.4
153.54	0.98	2.0	130.74	29JL	-0.7	-3.6	-1.3	1.4
159.94	0.91	2.0	140.39	8AU	-0.1	-3.6	-1.3	1.3
166.41	0.83	2.0	150.08	18AU	0.1	-3.5	-1.2	1.3
172.94	0.76	1.9	159.83	28AU	0.3	-3.5	-1.2	1.2
179.54	0.68	1.9	169.64	7SE	0.6	-3.4	-1.2	1.2
186.21	0.59	1.8	179.49	17SE	1.7	-3.4	-1.2	1.2
192.95	0.51	1.8	189.41	27SE	2.3	-3.4	-1.2	1.2
199.77	0.42	1.7	199.39	7OC	-0.1	-3.4	-1.2	1.2
206.66	0.32	1.7	209.41	17OC	-0.7	-3.4	-1.2	1.2
213.63	0.23	1.7	219.49	27OC	-0.7	-3.4	-1.3	1.2
220.68	0.13	1.7	229.61	6NO	-0.7	-3.4	-1.3	1.2
227.80	0.03	1.7	239.76	16NO	-0.8	-3.4	-1.3	1.1
235.00	-0.08	1.7	249.95	26NO	-0.7	-3.4	-1.4	1.1
242.28	-0.19	1.7	260.14	6DE	-0.7	-3.4	-1.4	1.1
249.62	-0.30	1.6	270.34	16DE	-0.6	-3.4	-1.5	1.0
257.03	-0.41	1.6	280.54	26DE	0.1	-3.4	-1.5	1.0

Right table:

♂ LONG	LAT	MAG	⊙ LONG	16.00UT 950	☿	♀	♃	♄
264.51	-0.52	1.6	290.72	5JA	2.5	-3.4	-1.6	0.9
272.05	-0.63	1.5	300.87	15JA	1.4	-3.5	-1.7	0.9
279.64	-0.74	1.5	310.98	25JA	0.5	-3.5	-1.7	0.8
287.27	-0.85	1.5	321.06	4FE	0.3	-3.5	-1.8	0.8
294.94	-0.95	1.4	331.08	14FE	0.1	-3.4	-1.9	0.7
302.64	-1.05	1.4	341.05	24FE	-0.2	-3.4	-1.9	0.7
310.35	-1.14	1.4	350.96	6MR	-0.7	-3.4	-2.0	0.7
318.07	-1.23	1.3	0.82	16MR	-1.5	-3.4	-2.0	0.7
325.78	-1.30	1.3	10.62	26MR	-1.4	-3.3	-2.0	0.8
333.47	-1.36	1.2	20.36	5AP	-0.4	-3.3	-2.0	0.9
341.13	-1.41	1.2	30.05	15AP	0.7	-3.3	-2.0	0.9
348.74	-1.45	1.2	39.70	25AP	1.9	-3.3	-2.0	1.0
356.29	-1.47	1.1	49.30	5MY	3.5	-3.3	-2.0	1.0
3.76	-1.47	1.1	58.87	15MY	2.6	-3.4	-1.9	1.1
11.14	-1.46	1.1	68.42	25MY	1.4	-3.4	-1.9	1.2
18.41	-1.43	1.0	77.95	4JN	0.5	-3.4	-1.8	1.2
25.55	-1.39	1.0	87.48	14JN	-0.5	-3.4	-1.8	1.2
32.56	-1.33	0.9	97.01	24JN	-1.5	-3.5	-1.7	1.2
39.40	-1.25	0.9	106.55	4JL	-1.4	-3.5	-1.6	1.2
46.06	-1.15	0.8	116.11	14JL	-0.6	-3.6	-1.6	1.2
52.51	-1.04	0.8	125.70	24JL	-0.1	-3.7	-1.5	1.2
58.73	-0.90	0.7	135.33	3AU	0.3	-3.7	-1.5	1.2
64.67	-0.75	0.6	145.00	13AU	0.5	-3.8	-1.4	1.2
70.29	-0.57	0.5	154.72	23AU	0.3	-3.9	-1.4	1.1
75.52	-0.38	0.4	164.49	2SE	2.2	-4.0	-1.3	1.1
80.30	-0.15	0.3	174.32	12SE	2.0	-4.1	-1.3	1.0
84.52	0.11	0.1	184.21	22SE	-0.2	-4.2	-1.3	1.0
88.06	0.41	-0.0	194.15	2OC	-0.8	-4.3	-1.3	1.0
90.76	0.75	-0.2	204.15	12OC	-0.9	-4.3	-1.2	1.0
92.42	1.14	-0.4	214.20	22OC	-0.8	-4.2	-1.2	1.0
92.85	1.58	-0.7	224.30	1NO	-0.7	-3.7	-1.3	1.0
91.90	2.05	-0.9	234.44	11NO	-0.5	-2.9	-1.3	1.0
89.55	2.52	-1.1	244.61	21NO	-0.5	-3.7	-1.3	1.0
86.10	2.94	-1.2	254.79	1DE	-0.5	-4.2	-1.3	1.0
82.15	3.25	-1.3	264.99	11DE	0.2	-4.4	-1.3	0.9
78.48	3.41	-1.1	275.19	21DE	2.8	-4.4	-1.4	0.9
75.77	3.43	-0.8	285.38	31DE	1.1	-4.3	-1.4	0.9
				951				
74.33	3.35	-0.5	295.55	10JA	0.3	-4.2	-1.5	0.8
74.21	3.21	-0.2	305.69	20JA	0.1	-4.1	-1.5	0.8
75.28	3.04	0.0	315.78	30JA	-0.0	-4.0	-1.6	0.7
77.33	2.87	0.3	325.83	9FE	-0.3	-3.9	-1.7	0.7
80.18	2.69	0.5	335.84	19FE	-0.8	-3.8	-1.7	0.6
83.66	2.53	0.7	345.78	1MR	-1.5	-3.7	-1.8	0.6
87.64	2.37	0.9	355.78	11MR	-1.3	-3.6	-1.9	0.5
92.01	2.22	1.1	5.49	21MR	-0.4	-3.5	-1.9	0.5
96.71	2.08	1.2	15.26	31MR	0.9	-3.5	-2.0	0.6
101.66	1.95	1.3	24.98	10AP	2.5	-3.4	-2.0	0.6
106.82	1.82	1.4	34.65	20AP	3.4	-3.4	-2.1	0.7
112.17	1.70	1.5	44.27	30AP	1.9	-3.4	-2.1	0.8
117.66	1.58	1.6	53.86	10MY	1.1	-3.3	-2.1	0.8
123.30	1.47	1.7	63.43	20MY	0.4	-3.3	-2.1	0.9
129.05	1.36	1.7	72.96	30MY	-0.5	-3.3	-2.1	1.0
134.91	1.25	1.8	82.49	9JN	-1.5	-3.3	-2.0	1.0
140.88	1.14	1.8	92.02	19JN	-1.5	-3.3	-2.0	1.0
146.94	1.04	1.8	101.55	29JN	-0.6	-3.3	-1.9	1.1
153.10	0.93	1.9	111.10	9JL	0.0	-3.4	-1.9	1.1
159.36	0.83	1.9	120.67	19JL	0.4	-3.4	-1.8	1.1
165.70	0.73	1.9	130.28	29JL	0.7	-3.4	-1.8	1.1
172.14	0.62	1.9	139.93	8AU	1.2	-3.4	-1.7	1.1
178.67	0.52	1.9	149.62	18AU	2.6	-3.5	-1.6	1.0
185.30	0.41	1.8	159.36	28AU	1.8	-3.5	-1.6	1.0
192.02	0.31	1.8	169.16	7SE	-0.3	-3.5	-1.5	1.0
198.83	0.20	1.8	179.01	17SE	-1.0	-3.4	-1.5	0.9
205.74	0.10	1.7	188.92	27SE	-1.0	-3.4	-1.4	0.9
212.75	-0.01	1.7	198.90	7OC	-0.9	-3.4	-1.4	0.9
219.85	-0.11	1.7	208.92	17OC	-0.6	-3.4	-1.4	0.9
227.05	-0.21	1.6	218.99	27OC	-0.4	-3.3	-1.4	0.9
234.33	-0.31	1.6	229.11	6NO	-0.4	-3.3	-1.3	0.9
241.71	-0.41	1.5	239.27	16NO	-0.3	-3.3	-1.3	0.9
249.17	-0.51	1.5	249.44	26NO	0.4	-3.3	-1.3	0.9
256.70	-0.61	1.5	259.64	6DE	3.1	-3.3	-1.3	0.9
264.31	-0.70	1.5	269.84	16DE	0.8	-3.4	-1.4	0.9
271.99	-0.78	1.4	280.04	26DE	0.1	-3.4	-1.4	0.8

♂ LONG	♂ LAT	♂ MAG	☉ LONG	16.00UT	☿	♀	♃	♄
				952	MAGNITUDES			
279.72	-0.86	1.4	290.22	5JA	-0.0	-3.4	-1.4	0.8
287.49	-0.93	1.4	300.37	15JA	-0.1	-3.4	-1.4	0.8
295.30	-1.00	1.4	310.49	25JA	-0.4	-3.5	-1.5	0.7
303.14	-1.05	1.4	320.57	4FE	-0.8	-3.5	-1.5	0.7
310.99	-1.09	1.4	330.60	14FE	-1.4	-3.6	-1.6	0.6
318.83	-1.13	1.4	340.57	24FE	-1.3	-3.6	-1.7	0.6
326.67	-1.15	1.4	350.49	5MR	-0.4	-3.7	-1.7	0.5
334.48	-1.16	1.4	0.35	15MR	1.2	-3.8	-1.8	0.5
342.25	-1.15	1.4	10.15	25MR	3.1	-3.9	-1.9	0.4
349.98	-1.14	1.4	19.89	4AP	2.5	-3.9	-1.9	0.4
357.64	-1.11	1.4	29.59	14AP	1.4	-4.0	-2.0	0.5
5.24	-1.06	1.4	39.23	24AP	0.8	-4.1	-2.1	0.5
12.76	-1.01	1.4	48.84	4MY	0.2	-4.2	-2.1	0.6
20.18	-0.94	1.4	58.41	14MY	-0.6	-4.2	-2.2	0.7
27.52	-0.86	1.4	67.96	24MY	-1.5	-4.1	-2.2	0.7
34.75	-0.78	1.4	77.50	3JN	-1.5	-3.7	-2.2	0.8
41.86	-0.68	1.4	87.02	13JN	-0.6	-3.0	-2.2	0.8
48.87	-0.57	1.4	96.55	23JN	0.1	-3.3	-2.2	0.9
55.74	-0.45	1.4	106.09	3JL	0.5	-3.9	-2.2	0.9
62.49	-0.32	1.3	115.65	13JL	1.0	-4.2	-2.2	0.9
69.09	-0.19	1.3	125.24	23JL	1.7	-4.2	-2.1	1.0
75.54	-0.04	1.3	134.86	2AU	3.1	-4.2	-2.1	1.0
81.82	0.11	1.3	144.53	12AU	1.5	-4.1	-2.0	1.0
87.92	0.28	1.2	154.24	22AU	-0.3	-4.0	-1.9	1.0
93.82	0.46	1.1	164.01	1SE	-1.1	-3.9	-1.9	0.9
99.47	0.65	1.1	173.83	11SE	-1.2	-3.9	-1.8	0.9
104.86	0.85	1.0	183.72	21SE	-0.8	-3.8	-1.7	0.9
109.93	1.08	0.9	193.66	1OC	-0.5	-3.7	-1.7	0.8
114.61	1.33	0.8	203.66	11OC	-0.3	-3.6	-1.6	0.8
118.84	1.60	0.6	213.71	21OC	-0.2	-3.6	-1.6	0.8
122.49	1.90	0.5	223.80	31OC	-0.1	-3.5	-1.5	0.8
125.45	2.24	0.3	233.94	10NO	0.6	-3.5	-1.5	0.8
127.54	2.62	0.1	244.11	20NO	3.2	-3.5	-1.5	0.8
128.57	3.02	-0.2	254.30	30NO	0.5	-3.4	-1.5	0.8
128.38	3.45	-0.4	264.49	10DE	-0.1	-3.4	-1.4	0.8
126.85	3.86	-0.6	274.70	20DE	-0.2	-3.4	-1.4	0.8
124.04	4.21	-0.8	284.89	30DE	-0.3	-3.4	-1.4	0.8
				953				
120.33	4.43	-1.0	295.05	9JA	-0.5	-3.3	-1.4	0.8
116.36	4.48	-1.0	305.20	19JA	-0.8	-3.3	-1.5	0.7
112.88	4.36	-0.8	315.30	29JA	-1.2	-3.3	-1.5	0.7
110.44	4.11	-0.5	325.35	8FE	-1.1	-3.3	-1.5	0.6
109.28	3.79	-0.3	335.35	18FE	-0.3	-3.3	-1.6	0.6
109.40	3.45	-0.0	345.30	28FE	1.4	-3.4	-1.6	0.5
110.63	3.12	0.2	355.18	10MR	3.4	-3.4	-1.7	0.5
112.81	2.81	0.4	5.02	20MR	1.9	-3.4	-1.7	0.4
115.76	2.52	0.6	14.79	30MR	1.1	-3.5	-1.8	0.4
119.31	2.26	0.8	24.51	9AP	0.6	-3.5	-1.8	0.3
123.37	2.02	0.9	34.18	19AP	0.1	-3.5	-1.9	0.3
127.83	1.80	1.1	43.81	29AP	-0.6	-3.4	-2.0	0.4
132.62	1.60	1.2	53.40	9MY	-1.5	-3.4	-2.1	0.5
137.69	1.41	1.3	62.96	19MY	-1.5	-3.4	-2.1	0.5
143.01	1.23	1.4	72.50	29MY	-0.5	-3.4	-2.2	0.6
148.53	1.06	1.4	82.03	8JN	0.2	-3.3	-2.2	0.7
154.25	0.90	1.5	91.56	18JN	0.7	-3.3	-2.3	0.7
160.13	0.75	1.5	101.09	28JN	1.3	-3.3	-2.3	0.8
166.17	0.61	1.6	110.63	8JL	2.2	-3.3	-2.4	0.8
172.35	0.47	1.6	120.21	18JL	3.2	-3.3	-2.4	0.8
178.68	0.33	1.6	129.81	28JL	1.3	-3.4	-2.3	0.9
185.13	0.20	1.6	139.45	7AU	-0.4	-3.4	-2.3	0.9
191.72	0.08	1.6	149.14	17AU	-1.2	-3.4	-2.3	0.9
198.43	-0.04	1.6	158.88	27AU	-1.3	-3.5	-2.2	0.9
205.27	-0.16	1.6	168.68	6SE	-0.8	-3.5	-2.2	0.9
212.23	-0.27	1.6	178.53	16SE	-0.4	-3.6	-2.1	0.8
219.30	-0.38	1.6	188.44	26SE	-0.2	-3.6	-2.0	0.8
226.49	-0.48	1.6	198.41	6OC	-0.1	-3.7	-2.0	0.8
233.78	-0.58	1.6	208.43	16OC	0.1	-3.8	-1.9	0.7
241.18	-0.67	1.5	218.50	26OC	0.8	-3.9	-1.8	0.7
248.68	-0.75	1.5	228.62	5NO	3.1	-4.0	-1.8	0.7
256.26	-0.83	1.5	238.77	15NO	0.3	-4.1	-1.7	0.8
263.92	-0.89	1.5	248.95	25NO	-0.3	-4.2	-1.7	0.8
271.66	-0.95	1.4	259.14	5DE	-0.3	-4.3	-1.6	0.8
279.45	-1.00	1.4	269.35	15DE	-0.4	-4.4	-1.6	0.8
287.29	-1.04	1.4	279.54	25DE	-0.5	-4.4	-1.6	0.8

♂ LONG	♂ LAT	♂ MAG	☉ LONG	16.00UT	☿	♀	♃	♄
				954	MAGNITUDES			
295.17	-1.06	1.3	289.72	4JA	-0.9	-4.2	-1.6	0.8
303.06	-1.08	1.3	299.88	14JA	-1.1	-3.8	-1.5	0.8
310.97	-1.08	1.3	310.00	24JA	-1.0	-3.3	-1.5	0.7
318.86	-1.08	1.3	320.08	3FE	-0.2	-3.8	-1.5	0.7
326.74	-1.06	1.3	330.11	13FE	1.7	-4.2	-1.5	0.6
334.59	-1.02	1.3	340.08	23FE	2.8	-4.3	-1.6	0.6
342.40	-0.98	1.4	350.00	5MR	1.4	-4.3	-1.6	0.5
350.15	-0.93	1.4	359.87	15MR	0.8	-4.2	-1.6	0.5
357.84	-0.87	1.4	9.67	25MR	0.4	-4.1	-1.6	0.4
5.46	-0.80	1.5	19.42	4AP	0.0	-4.0	-1.7	0.4
13.00	-0.72	1.5	29.12	14AP	-0.6	-3.9	-1.7	0.3
20.45	-0.63	1.5	38.77	24AP	-1.5	-3.8	-1.8	0.3
27.82	-0.54	1.5	48.38	4MY	-1.5	-3.7	-1.9	0.3
35.10	-0.44	1.6	57.95	14MY	-0.5	-3.6	-1.9	0.4
42.27	-0.33	1.6	67.50	24MY	0.3	-3.5	-2.0	0.4
49.36	-0.22	1.6	77.04	3JN	1.0	-3.5	-2.1	0.5
56.35	-0.11	1.6	86.57	13JN	1.7	-3.4	-2.1	0.6
63.23	0.00	1.7	96.09	23JN	2.8	-3.4	-2.2	0.6
70.02	0.12	1.7	105.63	3JL	2.7	-3.4	-2.3	0.7
76.71	0.24	1.7	115.19	13JL	1.0	-3.4	-2.3	0.7
83.29	0.37	1.7	124.78	23JL	-0.4	-3.3	-2.4	0.8
89.77	0.49	1.7	134.40	2AU	-1.3	-3.3	-2.4	0.8
96.14	0.62	1.6	144.06	12AU	-1.4	-3.3	-2.4	0.8
102.39	0.75	1.6	153.77	22AU	-0.7	-3.3	-2.4	0.8
108.52	0.89	1.6	163.54	1SE	-0.3	-3.4	-2.4	0.8
114.51	1.03	1.5	173.37	11SE	-0.0	-3.4	-2.4	0.8
120.36	1.18	1.5	183.24	21SE	0.1	-3.4	-2.4	0.8
126.03	1.34	1.4	193.18	1OC	0.3	-3.4	-2.3	0.8
131.51	1.50	1.3	203.18	11OC	1.1	-3.4	-2.3	0.8
136.76	1.68	1.3	213.22	21OC	2.8	-3.5	-2.2	0.7
141.74	1.87	1.1	223.31	31OC	0.2	-3.5	-2.1	0.7
146.40	2.07	1.0	233.45	10NO	-0.4	-3.5	-2.1	0.7
150.66	2.29	0.9	243.61	20NO	-0.5	-3.5	-2.0	0.7
154.44	2.54	0.7	253.80	30NO	-0.5	-3.4	-1.9	0.7
157.61	2.80	0.5	264.00	10DE	-0.6	-3.4	-1.9	0.8
160.04	3.09	0.3	274.19	20DE	-0.9	-3.4	-1.8	0.8
161.55	3.39	0.0	284.39	30DE	-0.9	-3.4	-1.7	0.8
				955				
161.96	3.70	-0.2	294.55	9JA	-0.9	-3.3	-1.7	0.8
161.12	4.00	-0.5	304.69	19JA	-0.2	-3.3	-1.7	0.8
158.98	4.22	-0.7	314.80	29JA	2.0	-3.3	-1.6	0.7
155.74	4.34	-0.9	324.85	8FE	2.2	-3.3	-1.6	0.7
151.89	4.29	-1.0	334.86	18FE	1.0	-3.3	-1.6	0.7
148.11	4.07	-0.9	344.81	28FE	0.5	-3.4	-1.6	0.6
145.10	3.73	-0.7	354.70	10MR	0.3	-3.4	-1.6	0.6
143.25	3.31	-0.5	4.53	20MR	-0.1	-3.4	-1.6	0.5
142.71	2.88	-0.3	14.32	30MR	-0.7	-3.4	-1.6	0.4
143.40	2.46	-0.0	24.04	9AP	-1.5	-3.5	-1.6	0.4
145.15	2.07	0.2	33.71	19AP	-1.5	-3.5	-1.6	0.3
147.78	1.72	0.3	43.34	29AP	-0.5	-3.6	-1.7	0.3
151.15	1.41	0.5	52.94	9MY	0.4	-3.6	-1.7	0.2
155.11	1.13	0.6	62.50	19MY	1.2	-3.7	-1.8	0.3
159.55	0.87	0.8	72.04	29MY	2.2	-3.8	-1.8	0.4
164.41	0.64	0.9	81.57	8JN	3.4	-3.9	-1.9	0.4
169.61	0.43	1.0	91.10	18JN	2.2	-4.0	-1.9	0.5
175.12	0.24	1.0	100.63	28JN	0.8	-4.1	-2.0	0.5
180.89	0.06	1.1	110.18	8JL	-0.4	-4.1	-2.1	0.6
186.89	-0.11	1.2	119.75	18JL	-1.4	-4.2	-2.1	0.7
193.10	-0.26	1.2	129.35	28JL	-1.4	-4.2	-2.2	0.7
199.51	-0.40	1.3	138.99	7AU	-0.7	-4.1	-2.3	0.7
206.10	-0.53	1.3	148.68	17AU	-0.2	-3.7	-2.3	0.8
212.86	-0.65	1.3	158.42	27AU	0.1	-3.2	-2.4	0.8
219.78	-0.75	1.3	168.21	6SE	0.2	-3.7	-2.4	0.8
226.84	-0.85	1.3	178.05	16SE	0.5	-4.1	-2.5	0.8
234.04	-0.93	1.4	187.96	26SE	1.4	-4.3	-2.5	0.8
241.37	-1.01	1.4	197.93	6OC	2.6	-4.3	-2.5	0.8
248.81	-1.07	1.4	207.95	16OC	0.0	-4.3	-2.4	0.7
256.36	-1.12	1.4	218.02	26OC	-0.6	-4.2	-2.4	0.7
264.00	-1.15	1.4	228.13	5NO	-0.6	-4.1	-2.3	0.7
271.71	-1.18	1.4	238.28	15NO	-0.6	-4.0	-2.3	0.6
279.50	-1.19	1.4	248.46	25NO	-0.7	-3.9	-2.2	0.7
287.33	-1.19	1.4	258.65	5DE	-0.7	-3.8	-2.1	0.7
295.20	-1.17	1.4	268.85	15DE	-0.8	-3.7	-2.1	0.7
303.09	-1.14	1.4	279.05	25DE	-0.7	-3.7	-2.0	0.8

♂ LONG	♂ LAT	♂ MAG	⊙ LONG	16.00UT	☿	♀	♃	♄
				956	MAGNITUDES			
310.98	-1.11	1.4	289.23	4JA	-0.1	-3.6	-1.9	0.8
318.86	-1.06	1.4	299.39	14JA	2.3	-3.5	-1.9	0.8
326.72	-1.00	1.4	309.51	24JA	1.7	-3.5	-1.8	0.8
334.54	-0.93	1.4	319.59	3FE	0.7	-3.4	-1.7	0.7
342.32	-0.85	1.4	329.62	13FE	0.4	-3.4	-1.7	0.7
350.04	-0.77	1.4	339.60	23FE	0.2	-3.4	-1.7	0.7
357.69	-0.68	1.4	349.52	4MR	-0.2	-3.3	-1.6	0.6
5.27	-0.58	1.5	359.39	14MR	-0.7	-3.3	-1.6	0.6
12.77	-0.48	1.5	9.20	24MR	-1.5	-3.3	-1.6	0.5
20.19	-0.38	1.5	18.95	3AP	-1.4	-3.3	-1.6	0.5
27.52	-0.28	1.5	28.65	13AP	-0.5	-3.3	-1.6	0.4
34.77	-0.18	1.6	38.30	23AP	0.6	-3.3	-1.6	0.4
41.93	-0.07	1.6	47.91	3MY	1.6	-3.3	-1.6	0.3
49.00	0.03	1.7	57.49	13MY	3.0	-3.4	-1.6	0.2
55.99	0.14	1.7	67.04	23MY	3.0	-3.4	-1.6	0.2
62.89	0.24	1.7	76.57	2JN	1.7	-3.4	-1.6	0.3
69.71	0.34	1.8	86.10	12JN	0.6	-3.5	-1.7	0.4
76.45	0.45	1.8	95.63	22JN	-0.5	-3.5	-1.7	0.4
83.12	0.55	1.8	105.17	2JL	-1.4	-3.4	-1.8	0.5
89.71	0.64	1.9	114.73	12JL	-1.5	-3.4	-1.8	0.5
96.23	0.74	1.9	124.31	22JL	-0.7	-3.4	-1.9	0.6
102.68	0.84	1.9	133.93	1AU	-0.1	-3.4	-1.9	0.6
109.06	0.93	1.9	143.59	11AU	0.2	-3.3	-2.0	0.7
115.36	1.03	1.9	153.30	21AU	0.4	-3.3	-2.1	0.7
121.60	1.12	1.9	163.06	31AU	0.7	-3.3	-2.1	0.8
127.76	1.21	1.8	172.88	10SE	1.8	-3.3	-2.2	0.8
133.83	1.31	1.8	182.76	20SE	2.3	-3.3	-2.3	0.8
139.82	1.40	1.8	192.69	30SE	-0.1	-3.3	-2.3	0.8
145.71	1.50	1.7	202.69	10OC	-0.7	-3.4	-2.4	0.8
151.49	1.60	1.6	212.73	20OC	-0.8	-3.4	-2.4	0.8
157.14	1.70	1.6	222.82	30OC	-0.8	-3.4	-2.4	0.7
162.64	1.80	1.5	232.95	9NO	-0.7	-3.5	-2.4	0.7
167.96	1.91	1.4	243.12	19NO	-0.6	-3.5	-2.4	0.7
173.08	2.02	1.2	253.30	29NO	-0.6	-3.6	-2.3	0.7
177.94	2.14	1.1	263.50	9DE	-0.6	-3.7	-2.3	0.7
182.48	2.26	0.9	273.70	19DE	0.1	-3.7	-2.2	0.7
186.65	2.39	0.8	283.89	29DE	2.6	-3.8	-2.1	0.8
				957				
190.33	2.52	0.6	294.07	8JA	1.3	-3.9	-2.1	0.8
193.41	2.66	0.3	304.21	18JA	0.4	-4.0	-2.0	0.8
195.76	2.79	0.1	314.31	28JA	0.2	-4.1	-1.9	0.8
197.19	2.91	-0.2	324.37	7FE	0.0	-4.2	-1.9	0.8
197.53	2.99	-0.5	334.38	17FE	-0.2	-4.2	-1.8	0.8
196.63	3.03	-0.8	344.33	27FE	-0.7	-4.3	-1.7	0.7
194.46	2.97	-1.0	354.22	9MR	-1.5	-4.2	-1.7	0.7
191.25	2.80	-1.3	4.06	19MR	-1.4	-4.0	-1.6	0.6
187.51	2.50	-1.4	13.84	29MR	-0.5	-3.5	-1.6	0.6
183.96	2.09	-1.2	23.57	8AP	0.8	-3.1	-1.5	0.5
181.25	1.63	-1.0	33.24	18AP	2.1	-3.7	-1.5	0.5
179.76	1.17	-0.8	42.87	28AP	3.9	-4.1	-1.5	0.4
179.61	0.74	-0.6	52.47	8MY	2.3	-4.2	-1.5	0.3
180.72	0.35	-0.4	62.03	18MY	1.3	-4.2	-1.5	0.3
182.90	0.01	-0.2	71.57	28MY	0.5	-4.1	-1.5	0.2
186.00	-0.28	-0.1	81.11	7JN	-0.5	-4.0	-1.5	0.3
189.86	-0.53	0.1	90.63	17JN	-1.5	-3.9	-1.5	0.3
194.32	-0.75	0.2	100.16	27JN	-1.5	-3.8	-1.5	0.4
199.31	-0.93	0.3	109.71	7JL	-0.6	-3.7	-1.5	0.4
204.73	-1.09	0.4	119.28	17JL	-0.0	-3.7	-1.6	0.5
210.52	-1.22	0.5	128.88	27JL	0.3	-3.6	-1.6	0.6
216.63	-1.33	0.6	138.52	6AU	0.6	-3.5	-1.7	0.6
223.02	-1.42	0.7	148.21	16AU	1.0	-3.5	-1.7	0.7
229.65	-1.49	0.8	157.94	26AU	2.3	-3.5	-1.8	0.7
236.50	-1.54	0.8	167.73	5SE	2.0	-3.4	-1.8	0.7
243.53	-1.57	0.9	177.58	15SE	-0.2	-3.4	-1.9	0.8
250.73	-1.59	0.9	187.48	25SE	-0.9	-3.4	-2.0	0.8
258.08	-1.58	1.0	197.45	5OC	-0.9	-3.4	-2.0	0.8
265.54	-1.57	1.0	207.46	15OC	-0.9	-3.4	-2.1	0.8
273.10	-1.53	1.1	217.53	25OC	-0.6	-3.4	-2.1	0.8
280.75	-1.48	1.1	227.64	4NO	-0.5	-3.4	-2.2	0.7
288.46	-1.42	1.2	237.79	14NO	-0.5	-3.4	-2.2	0.7
296.21	-1.35	1.2	247.96	24NO	-0.4	-3.4	-2.3	0.7
303.99	-1.26	1.2	258.16	4DE	0.2	-3.4	-2.3	0.7
311.77	-1.17	1.3	268.35	14DE	2.8	-3.4	-2.3	0.7
319.54	-1.07	1.3	278.55	24DE	1.0	-3.4	-2.2	0.7

♂ LONG	♂ LAT	♂ MAG	⊙ LONG	16.00UT	☿	♀	♃	♄
				958	MAGNITUDES			
327.29	-0.96	1.4	288.74	3JA	0.2	-3.4	-2.2	0.8
335.00	-0.85	1.4	298.89	13JA	0.0	-3.5	-2.1	0.8
342.67	-0.73	1.5	309.02	23JA	-0.1	-3.5	-2.1	0.8
350.27	-0.61	1.5	319.10	2FE	-0.3	-3.5	-2.0	0.8
357.81	-0.50	1.5	329.14	12FE	-0.8	-3.4	-1.9	0.8
5.28	-0.38	1.6	339.12	22FE	-1.4	-3.4	-1.9	0.8
12.68	-0.26	1.6	349.04	4MR	-1.3	-3.4	-1.8	0.8
19.99	-0.15	1.6	358.91	14MR	-0.4	-3.4	-1.7	0.7
27.22	-0.04	1.7	8.72	24MR	1.0	-3.3	-1.7	0.7
34.37	0.07	1.7	18.47	3AP	2.7	-3.3	-1.6	0.7
41.44	0.17	1.7	28.17	13AP	3.0	-3.3	-1.6	0.6
48.44	0.27	1.7	37.83	23AP	1.7	-3.3	-1.5	0.5
55.36	0.37	1.7	47.44	3MY	1.0	-3.3	-1.5	0.5
62.20	0.46	1.8	57.02	13MY	0.3	-3.4	-1.4	0.4
68.98	0.55	1.8	66.57	23MY	-0.5	-3.4	-1.4	0.3
75.69	0.63	1.8	76.11	2JN	-1.5	-3.4	-1.4	0.3
82.34	0.71	1.8	85.64	12JN	-1.5	-3.4	-1.4	0.2
88.93	0.78	1.9	95.17	22JN	-0.6	-3.5	-1.4	0.3
95.48	0.86	1.9	104.70	2JL	0.0	-3.5	-1.4	0.3
101.97	0.92	1.9	114.26	12JL	0.4	-3.6	-1.4	0.4
108.42	0.99	2.0	123.85	22JL	0.8	-3.7	-1.4	0.5
114.83	1.05	2.0	133.46	1AU	1.4	-3.7	-1.4	0.5
121.20	1.10	2.0	143.12	11AU	2.8	-3.8	-1.4	0.6
127.53	1.16	2.0	152.83	21AU	1.7	-3.9	-1.5	0.6
133.83	1.21	2.0	162.59	31AU	-0.2	-4.0	-1.5	0.7
140.10	1.25	2.0	172.41	10SE	-1.0	-4.1	-1.6	0.7
146.33	1.30	2.0	182.28	20SE	-1.1	-4.2	-1.6	0.7
152.52	1.34	1.9	192.21	30SE	-0.9	-4.3	-1.7	0.8
158.68	1.38	1.9	202.20	10OC	-0.5	-4.3	-1.7	0.8
164.80	1.42	1.9	212.25	20OC	-0.4	-4.2	-1.8	0.8
170.86	1.45	1.8	222.33	30OC	-0.3	-3.7	-1.9	0.8
176.88	1.48	1.7	232.46	9NO	-0.2	-3.0	-1.9	0.8
182.83	1.50	1.7	242.62	19NO	0.4	-3.8	-2.0	0.8
188.70	1.53	1.6	252.81	29NO	3.1	-4.2	-2.0	0.7
194.49	1.54	1.5	263.00	9DE	0.7	-4.4	-2.1	0.7
200.17	1.55	1.4	273.21	19DE	0.0	-4.4	-2.1	0.7
205.72	1.56	1.2	283.40	29DE	-0.1	-4.3	-2.1	0.7
				959				
211.11	1.55	1.1	293.57	8JA	-0.2	-4.2	-2.1	0.8
216.32	1.53	0.9	303.72	18JA	-0.4	-4.1	-2.1	0.8
221.28	1.50	0.8	313.82	28JA	-0.8	-4.0	-2.1	0.8
225.96	1.45	0.6	323.88	7FE	-1.3	-3.9	-2.1	0.8
230.27	1.37	0.3	333.90	17FE	-1.2	-3.8	-2.0	0.8
234.12	1.26	0.1	343.85	27FE	-0.4	-3.7	-2.0	0.8
237.40	1.12	-0.2	353.75	9MR	1.2	-3.6	-1.9	0.8
239.95	0.92	-0.4	3.59	19MR	3.3	-3.5	-1.8	0.8
241.61	0.65	-0.8	13.37	29MR	2.3	-3.5	-1.8	0.8
242.20	0.31	-1.1	23.10	8AP	1.3	-3.4	-1.7	0.7
241.56	-0.12	-1.4	32.78	18AP	0.7	-3.4	-1.6	0.7
239.70	-0.64	-1.7	42.41	28AP	0.2	-3.4	-1.6	0.6
236.84	-1.20	-2.0	52.01	8MY	-0.5	-3.3	-1.5	0.6
233.51	-1.76	-2.0	61.58	18MY	-1.5	-3.3	-1.4	0.5
230.47	-2.26	-1.9	71.12	28MY	-1.5	-3.3	-1.4	0.4
228.34	-2.64	-1.7	80.65	7JN	-0.6	-3.3	-1.4	0.4
227.52	-2.91	-1.5	90.18	17JN	0.1	-3.3	-1.3	0.3
228.09	-3.07	-1.3	99.70	27JN	0.6	-3.3	-1.3	0.3
229.93	-3.15	-1.1	109.25	7JL	1.1	-3.4	-1.3	0.3
232.89	-3.17	-0.9	118.82	17JL	1.8	-3.4	-1.3	0.4
236.77	-3.14	-0.7	128.42	27JL	3.2	-3.4	-1.3	0.4
241.38	-3.07	-0.5	138.05	6AU	1.5	-3.4	-1.3	0.5
246.61	-2.98	-0.3	147.74	16AU	-0.3	-3.5	-1.3	0.6
252.34	-2.87	-0.2	157.47	26AU	-1.1	-3.5	-1.3	0.6
258.46	-2.73	-0.0	167.25	5SE	-1.2	-3.5	-1.3	0.7
264.90	-2.58	0.1	177.10	15SE	-0.8	-3.4	-1.3	0.7
271.61	-2.42	0.2	187.00	25SE	-0.4	-3.4	-1.4	0.7
278.53	-2.24	0.3	196.96	5OC	-0.2	-3.4	-1.4	0.8
285.61	-2.06	0.4	206.97	15OC	-0.2	-3.4	-1.5	0.8
292.81	-1.87	0.5	217.03	25OC	-0.1	-3.3	-1.5	0.8
300.11	-1.68	0.7	227.14	4NO	0.6	-3.3	-1.6	0.8
307.47	-1.49	0.8	237.29	14NO	3.3	-3.3	-1.6	0.8
314.86	-1.30	0.9	247.46	24NO	0.5	-3.3	-1.7	0.8
322.28	-1.12	0.9	257.66	4DE	-0.2	-3.3	-1.8	0.8
329.69	-0.94	1.0	267.86	14DE	-0.2	-3.4	-1.8	0.8
337.08	-0.76	1.1	278.05	24DE	-0.3	-3.4	-1.9	0.7

♂			☉	16.00UT	☿	♀	♃	♄
LONG	LAT	MAG	LONG	960	MAGNITUDES			
344.44	-0.60	1.2	288.24	3JA	-0.5	-3.4	-2.0	0.8
351.76	-0.44	1.3	298.40	13JA	-0.9	-3.4	-2.0	0.8
359.03	-0.28	1.4	308.53	23JA	-1.2	-3.5	-2.0	0.8
6.25	-0.14	1.5	318.62	2FE	-1.1	-3.5	-2.0	0.9
13.40	-0.01	1.5	328.65	12FE	-0.3	-3.6	-2.0	0.9
20.49	0.12	1.6	338.64	22FE	1.5	-3.6	-2.0	0.9
27.51	0.23	1.7	348.57	3MR	3.2	-3.7	-2.0	0.9
34.47	0.34	1.7	358.44	13MR	1.7	-3.8	-2.0	0.9
41.37	0.44	1.8	8.25	23MR	0.9	-3.9	-1.9	0.9
48.20	0.53	1.8	18.01	2AP	0.5	-3.9	-1.8	0.8
54.97	0.62	1.8	27.71	12AP	0.1	-4.0	-1.8	0.8
61.68	0.70	1.9	37.37	22AP	-0.6	-4.1	-1.7	0.8
68.33	0.77	1.9	46.98	2MY	-1.5	-4.2	-1.7	0.7
74.94	0.83	1.9	56.56	12MY	-1.5	-4.2	-1.6	0.7
81.50	0.89	1.9	66.11	22MY	-0.6	-4.1	-1.5	0.6
88.02	0.94	1.9	75.65	1JN	0.2	-3.7	-1.5	0.5
94.51	0.99	1.9	85.18	11JN	0.8	-2.9	-1.4	0.5
100.96	1.03	1.9	94.70	21JN	1.4	-3.3	-1.4	0.4
107.39	1.06	1.9	104.24	1JL	2.4	-3.9	-1.3	0.3
113.79	1.10	1.9	113.80	11JL	3.0	-4.2	-1.3	0.3
120.18	1.12	2.0	123.38	21JL	1.2	-4.2	-1.3	0.4
126.56	1.14	2.0	133.00	31JL	-0.3	-4.2	-1.2	0.4
132.92	1.16	2.0	142.65	10AU	-1.2	-4.1	-1.2	0.5
139.28	1.18	2.0	152.36	20AU	-1.4	-4.0	-1.2	0.5
145.64	1.18	2.0	162.12	30AU	-0.8	-3.9	-1.2	0.6
152.00	1.19	2.0	171.93	9SE	-0.3	-3.8	-1.2	0.7
158.36	1.19	2.0	181.80	19SE	-0.1	-3.8	-1.2	0.7
164.72	1.18	2.0	191.73	29SE	-0.0	-3.7	-1.2	0.7
171.09	1.18	2.0	201.72	9OC	0.1	-3.6	-1.3	0.8
177.46	1.16	2.0	211.75	19OC	0.9	-3.6	-1.3	0.8
183.84	1.14	1.9	221.84	29OC	3.1	-3.5	-1.3	0.8
190.22	1.11	1.9	231.97	8NO	0.3	-3.5	-1.4	0.8
196.61	1.08	1.8	242.13	18NO	-0.3	-3.5	-1.4	0.9
202.99	1.04	1.8	252.32	28NO	-0.4	-3.4	-1.5	0.9
209.38	0.99	1.7	262.51	8DE	-0.4	-3.4	-1.6	0.9
215.75	0.93	1.6	272.71	18DE	-0.6	-3.4	-1.6	0.8
222.12	0.86	1.5	282.91	28DE	-0.9	-3.4	-1.7	0.8
				961				
228.46	0.77	1.4	293.08	7JA	-1.0	-3.3	-1.8	0.8
234.79	0.68	1.3	303.23	17JA	-1.0	-3.3	-1.8	0.8
241.08	0.56	1.2	313.34	27JA	-0.3	-3.3	-1.9	0.9
247.32	0.43	1.0	323.40	6FE	1.7	-3.3	-1.9	0.9
253.52	0.27	0.9	333.41	16FE	2.6	-3.3	-2.0	0.9
259.65	0.09	0.7	343.37	26FE	1.2	-3.4	-2.0	1.0
265.68	-0.12	0.6	353.27	8MR	0.7	-3.4	-2.0	1.0
271.60	-0.37	0.4	3.11	18MR	0.4	-3.4	-2.0	1.0
277.37	-0.65	0.2	12.90	28MR	-0.0	-3.5	-2.0	1.0
282.95	-0.98	0.0	22.63	7AP	-0.6	-3.5	-2.0	1.0
288.28	-1.35	-0.2	32.31	17AP	-1.5	-3.5	-1.9	0.9
293.28	-1.78	-0.4	41.95	27AP	-1.5	-3.4	-1.9	0.9
297.87	-2.27	-0.7	51.54	7MY	-0.5	-3.4	-1.8	0.8
301.92	-2.84	-1.0	61.11	17MY	0.3	-3.4	-1.7	0.8
305.26	-3.47	-1.2	70.66	27MY	1.0	-3.4	-1.7	0.7
307.72	-4.16	-1.5	80.18	6JN	1.9	-3.3	-1.6	0.7
309.10	-4.90	-1.8	89.71	16JN	3.1	-3.3	-1.5	0.6
309.24	-5.63	-2.1	99.24	26JN	2.5	-3.3	-1.5	0.5
308.13	-6.27	-2.4	108.78	6JL	1.0	-3.3	-1.4	0.5
305.99	-6.71	-2.6	118.35	16JL	-0.4	-3.3	-1.4	0.4
303.38	-6.85	-2.6	127.95	26JL	-1.3	-3.4	-1.3	0.4
301.05	-6.66	-2.5	137.58	5AU	-1.4	-3.4	-1.3	0.5
299.61	-6.21	-2.2	147.26	15AU	-0.7	-3.4	-1.3	0.5
299.45	-5.60	-2.0	156.99	25AU	-0.3	-3.5	-1.2	0.6
300.60	-4.92	-1.7	166.77	4SE	-0.0	-3.5	-1.2	0.6
302.93	-4.25	-1.4	176.62	14SE	0.1	-3.6	-1.2	0.7
306.26	-3.61	-1.1	186.52	24SE	0.3	-3.6	-1.2	0.7
310.39	-3.03	-0.8	196.47	4OC	1.2	-3.7	-1.2	0.8
315.13	-2.50	-0.6	206.49	14OC	2.8	-3.8	-1.2	0.8
320.36	-2.03	-0.4	216.55	24OC	0.1	-3.9	-1.2	0.9
325.96	-1.61	-0.1	226.65	3NO	-0.5	-4.0	-1.3	0.9
331.82	-1.24	0.1	236.80	13NO	-0.5	-4.1	-1.3	0.9
337.91	-0.92	0.2	246.97	23NO	-0.6	-4.2	-1.3	0.9
344.14	-0.63	0.4	257.16	3DE	-0.7	-4.3	-1.4	0.9
350.49	-0.37	0.6	267.37	13DE	-0.8	-4.4	-1.4	0.9
356.92	-0.15	0.8	277.56	23DE	-0.9	-4.4	-1.5	0.9

♂			☉	16.00UT	☿	♀	♃	♄
LONG	LAT	MAG	LONG	962	MAGNITUDES			
3.40	0.05	0.9	287.75	2JA	-0.8	-4.2	-1.5	0.9
9.92	0.22	1.0	297.91	12JA	-0.2	-3.7	-1.6	0.9
16.45	0.37	1.2	308.04	22JA	2.0	-3.3	-1.7	0.9
22.99	0.50	1.3	318.12	1FE	2.0	-3.8	-1.7	1.0
29.52	0.62	1.4	328.17	11FE	0.9	-4.2	-1.8	1.0
36.04	0.72	1.5	338.15	21FE	0.5	-4.3	-1.9	1.0
42.54	0.80	1.6	348.08	3MR	0.2	-4.3	-1.9	1.1
49.02	0.88	1.7	357.96	13MR	-0.1	-4.2	-2.0	1.1
55.48	0.94	1.7	7.77	23MR	-0.6	-4.1	-2.0	1.1
61.92	1.00	1.8	17.53	2AP	-1.5	-4.0	-2.0	1.1
68.33	1.04	1.9	27.24	12AP	-1.5	-3.9	-2.0	1.1
74.73	1.08	1.9	36.90	22AP	-0.5	-3.8	-2.0	1.1
81.10	1.11	1.9	46.52	2MY	0.5	-3.7	-2.0	1.0
87.45	1.14	2.0	56.10	12MY	1.4	-3.6	-2.0	1.0
93.79	1.16	2.0	65.66	22MY	2.5	-3.5	-1.9	0.9
100.12	1.17	2.0	75.19	1JN	3.4	-3.5	-1.9	0.9
106.45	1.18	2.0	84.72	11JN	2.0	-3.4	-1.8	0.8
112.77	1.18	2.0	94.25	21JN	0.8	-3.4	-1.7	0.8
119.09	1.18	2.0	103.78	1JL	-0.4	-3.4	-1.7	0.7
125.43	1.17	2.0	113.34	11JL	-1.4	-3.4	-1.6	0.6
131.77	1.16	2.0	122.92	21JL	-1.5	-3.3	-1.6	0.6
138.12	1.14	2.0	132.53	31JL	-0.7	-3.3	-1.5	0.5
144.50	1.12	2.0	142.19	10AU	-0.2	-3.3	-1.5	0.5
150.89	1.10	1.9	151.89	20AU	0.1	-3.3	-1.4	0.6
157.32	1.07	2.0	161.64	30AU	0.3	-3.4	-1.4	0.6
163.77	1.03	2.0	171.46	9SE	0.6	-3.4	-1.3	0.7
170.26	0.99	2.0	181.32	19SE	1.5	-3.4	-1.3	0.7
176.78	0.95	2.0	191.25	29SE	2.6	-3.4	-1.3	0.8
183.34	0.90	2.0	201.23	9OC	0.0	-3.4	-1.3	0.8
189.94	0.84	1.9	211.27	19OC	-0.6	-3.5	-1.3	0.9
196.58	0.78	1.9	221.35	29OC	-0.7	-3.5	-1.3	0.9
203.26	0.71	1.9	231.48	8NO	-0.7	-3.5	-1.3	1.0
209.99	0.64	1.9	241.63	18NO	-0.8	-3.5	-1.3	1.0
216.76	0.56	1.8	251.82	28NO	-0.7	-3.4	-1.3	1.0
223.58	0.47	1.8	262.02	8DE	-0.7	-3.4	-1.3	1.0
230.44	0.37	1.7	272.22	18DE	-0.7	-3.4	-1.4	1.0
237.34	0.26	1.6	282.41	28DE	-0.0	-3.4	-1.4	1.1
				963				
244.29	0.14	1.6	292.59	7JA	2.3	-3.3	-1.5	1.1
251.27	0.02	1.5	302.73	17JA	1.6	-3.3	-1.5	1.0
258.30	-0.12	1.4	312.84	27JA	0.6	-3.3	-1.6	1.0
265.36	-0.27	1.3	322.91	6FE	0.3	-3.3	-1.6	1.1
272.45	-0.43	1.2	332.92	16FE	0.1	-3.4	-1.7	1.1
279.57	-0.60	1.1	342.89	26FE	-0.2	-3.4	-1.8	1.1
286.71	-0.78	1.0	352.79	8MR	-0.7	-3.4	-1.8	1.2
293.86	-0.98	0.9	2.64	18MR	-1.5	-3.4	-1.9	1.2
301.02	-1.18	0.8	12.43	28MR	-1.4	-3.4	-2.0	1.2
308.16	-1.39	0.7	22.16	7AP	-0.5	-3.5	-2.0	1.2
315.28	-1.60	0.6	31.84	17AP	0.6	-3.5	-2.1	1.2
322.35	-1.82	0.5	41.48	27AP	1.8	-3.6	-2.1	1.2
329.34	-2.05	0.4	51.08	7MY	3.3	-3.6	-2.1	1.2
336.23	-2.27	0.2	60.65	17MY	2.8	-3.7	-2.1	1.2
342.99	-2.49	0.1	70.20	27MY	1.5	-3.8	-2.1	1.1
349.55	-2.71	-0.0	79.73	6JN	0.6	-3.9	-2.1	1.1
355.88	-2.92	-0.2	89.25	16JN	-0.4	-4.0	-2.0	1.0
1.90	-3.11	-0.3	98.79	26JN	-1.4	-4.1	-2.0	1.0
7.52	-3.30	-0.5	108.33	6JL	-1.5	-4.1	-1.9	0.9
12.65	-3.46	-0.7	117.89	16JL	-0.7	-4.2	-1.9	0.8
17.15	-3.61	-0.8	127.49	26JL	-0.1	-4.2	-1.8	0.8
20.85	-3.72	-1.0	137.12	5AU	0.2	-4.0	-1.8	0.7
23.56	-3.78	-1.3	146.80	15AU	0.5	-3.6	-1.7	0.7
25.06	-3.78	-1.5	156.52	25AU	0.8	-3.2	-1.6	0.7
25.16	-3.68	-1.7	166.30	4SE	2.0	-3.7	-1.6	0.7
23.81	-3.44	-1.9	176.14	14SE	2.3	-4.1	-1.5	0.8
21.18	-3.05	-2.1	186.04	24SE	-0.1	-4.3	-1.5	0.8
17.84	-2.52	-2.2	195.99	4OC	-0.8	-4.3	-1.4	0.9
14.60	-1.89	-2.0	206.00	14OC	-0.8	-4.3	-1.4	0.9
12.17	-1.26	-1.7	216.06	24OC	-0.8	-4.2	-1.4	1.0
11.01	-0.68	-1.4	226.16	3NO	-0.7	-4.1	-1.4	1.0
11.19	-0.19	-1.0	236.30	13NO	-0.6	-4.0	-1.3	1.1
12.60	0.20	-0.7	246.48	23NO	-0.5	-3.9	-1.3	1.1
15.03	0.52	-0.4	256.67	3DE	-0.5	-3.8	-1.3	1.1
18.27	0.76	-0.1	266.87	13DE	0.1	-3.7	-1.4	1.2
22.15	0.95	0.1	277.07	23DE	2.6	-3.7	-1.4	1.2

♂ LONG	♂ LAT	♂ MAG	⊙ LONG	16.00UT 964	☿	♀	♃	♄
					MAGNITUDES			
26.53	1.10	0.4	287.26	2JA	1.2	-3.6	-1.4	1.2
31.29	1.21	0.6	297.42	12JA	0.3	-3.5	-1.4	1.2
36.34	1.29	0.8	307.55	22JA	0.1	-3.5	-1.5	1.2
41.63	1.35	0.9	317.64	1FE	-0.0	-3.4	-1.5	1.2
47.09	1.40	1.1	327.69	11FE	-0.3	-3.4	-1.6	1.2
52.69	1.43	1.2	337.68	21FE	-0.7	-3.4	-1.6	1.2
58.39	1.45	1.4	347.61	2MR	-1.4	-3.3	-1.7	1.3
64.18	1.46	1.5	357.49	12MR	-1.4	-3.3	-1.8	1.3
70.05	1.46	1.6	7.31	22MR	-0.5	-3.3	-1.8	1.4
75.96	1.45	1.7	17.07	1AP	0.8	-3.3	-1.9	1.4
81.92	1.44	1.7	26.78	11AP	2.3	-3.3	-2.0	1.4
87.93	1.42	1.8	36.44	21AP	3.6	-3.3	-2.0	1.4
93.96	1.40	1.9	46.06	1MY	2.1	-3.3	-2.1	1.4
100.03	1.38	1.9	55.64	11MY	1.2	-3.4	-2.2	1.4
106.13	1.35	2.0	65.20	21MY	0.4	-3.4	-2.2	1.4
112.25	1.31	2.0	74.73	31MY	-0.5	-3.4	-2.2	1.3
118.41	1.28	2.0	84.26	10JN	-1.4	-3.5	-2.2	1.3
124.61	1.23	2.0	93.79	20JN	-1.5	-3.5	-2.2	1.2
130.83	1.19	2.0	103.32	30JN	-0.6	-3.4	-2.2	1.2
137.10	1.14	2.0	112.88	10JL	-0.0	-3.4	-2.2	1.1
143.40	1.09	2.0	122.46	20JL	0.4	-3.4	-2.2	1.1
149.75	1.04	2.0	132.06	30JL	0.7	-3.4	-2.1	1.0
156.15	0.98	2.0	141.72	9AU	1.1	-3.3	-2.1	0.9
162.60	0.92	2.0	151.42	19AU	2.4	-3.3	-2.0	0.9
169.10	0.85	1.9	161.17	29AU	2.0	-3.3	-1.9	0.8
175.66	0.79	1.9	170.97	8SE	-0.2	-3.3	-1.9	0.9
182.29	0.71	1.8	180.84	18SE	-0.9	-3.3	-1.8	0.9
188.97	0.64	1.8	190.76	28SE	-1.0	-3.3	-1.7	1.0
195.72	0.56	1.8	200.74	8OC	-0.9	-3.4	-1.7	1.0
202.53	0.47	1.8	210.78	18OC	-0.6	-3.4	-1.6	1.1
209.42	0.38	1.8	220.85	28OC	-0.4	-3.4	-1.6	1.1
216.37	0.29	1.8	230.98	7NO	-0.4	-3.5	-1.5	1.2
223.39	0.19	1.8	241.14	17NO	-0.3	-3.5	-1.5	1.2
230.49	0.09	1.7	251.32	27NO	0.3	-3.6	-1.5	1.2
237.65	-0.02	1.7	261.52	7DE	2.9	-3.7	-1.5	1.3
244.88	-0.13	1.7	271.72	17DE	0.9	-3.7	-1.5	1.3
252.17	-0.25	1.6	281.91	27DE	0.1	-3.8	-1.5	1.4
				965				
259.53	-0.36	1.6	292.09	6JA	-0.0	-3.9	-1.5	1.4
266.95	-0.49	1.5	302.24	16JA	-0.1	-4.0	-1.5	1.4
274.41	-0.61	1.5	312.35	26JA	-0.3	-4.1	-1.5	1.4
281.93	-0.73	1.5	322.43	5FE	-0.8	-4.2	-1.5	1.4
289.49	-0.86	1.4	332.45	15FE	-1.4	-4.3	-1.5	1.3
297.08	-0.98	1.3	342.41	25FE	-1.3	-4.3	-1.6	1.3
304.70	-1.10	1.3	352.32	7MR	-0.4	-4.2	-1.6	1.3
312.33	-1.21	1.2	2.16	17MR	1.0	-4.0	-1.7	1.3
319.96	-1.32	1.2	11.95	27MR	2.9	-3.5	-1.7	1.3
327.58	-1.42	1.1	21.69	6AP	2.7	-3.1	-1.8	1.2
335.17	-1.50	1.1	31.38	16AP	1.5	-3.8	-1.9	1.2
342.73	-1.58	1.0	41.02	26AP	0.9	-4.1	-1.9	1.2
350.22	-1.64	1.0	50.62	6MY	0.3	-4.2	-2.0	1.2
357.64	-1.69	0.9	60.19	16MY	-0.5	-4.2	-2.1	1.2
4.96	-1.72	0.9	69.73	26MY	-1.5	-4.1	-2.2	1.1
12.17	-1.73	0.8	79.27	5JN	-1.5	-4.0	-2.2	1.1
19.24	-1.72	0.7	88.79	15JN	-0.6	-3.9	-2.3	1.0
26.15	-1.70	0.7	98.32	25JN	0.1	-3.8	-2.3	1.0
32.88	-1.65	0.6	107.87	5JL	0.5	-3.7	-2.4	0.9
39.38	-1.58	0.5	117.43	15JL	0.9	-3.7	-2.4	0.9
45.62	-1.50	0.4	127.02	25JL	1.5	-3.6	-2.4	0.8
51.56	-1.39	0.3	136.66	4AU	3.0	-3.5	-2.4	0.8
57.13	-1.25	0.2	146.33	14AU	1.7	-3.5	-2.3	0.8
62.27	-1.09	0.1	156.05	24AU	-0.2	-3.5	-2.3	0.7
66.88	-0.89	-0.0	165.83	3SE	-1.0	-3.4	-2.2	0.7
70.83	-0.66	-0.2	175.67	13SE	-1.1	-3.4	-2.2	0.7
73.99	-0.38	-0.4	185.56	23SE	-0.9	-3.4	-2.1	0.8
76.15	-0.05	-0.6	195.51	3OC	-0.5	-3.4	-2.0	0.8
77.12	0.33	-0.8	205.51	13OC	-0.3	-3.4	-2.0	0.9
76.73	0.76	-1.0	215.57	23OC	-0.2	-3.4	-1.9	1.0
74.88	1.24	-1.2	225.67	2NO	-0.2	-3.4	-1.8	1.0
71.78	1.72	-1.4	235.81	12NO	0.5	-3.4	-1.8	1.1
67.96	2.14	-1.5	245.98	22NO	3.2	-3.4	-1.7	1.1
64.19	2.45	-1.3	256.17	2DE	0.7	-3.4	-1.7	1.2
61.22	2.64	-1.1	266.37	12DE	-0.1	-3.4	-1.6	1.2
59.48	2.72	-0.8	276.57	22DE	-0.2	-3.4	-1.6	1.2

♂ LONG	♂ LAT	♂ MAG	⊙ LONG	16.00UT 966	☿	♀	♃	♄
					MAGNITUDES			
59.09	2.72	-0.4	286.76	1JA	-0.2	-3.4	-1.6	1.2
59.94	2.67	-0.2	296.92	11JA	-0.4	-3.5	-1.6	1.2
61.83	2.59	0.1	307.05	21JA	-0.8	-3.5	-1.5	1.2
64.57	2.50	0.4	317.15	31JA	-1.3	-3.5	-1.5	1.2
67.98	2.39	0.6	327.19	10FE	-1.2	-3.4	-1.5	1.1
71.91	2.29	0.8	337.19	20FE	-0.4	-3.4	-1.6	1.1
76.26	2.19	1.0	347.13	2MR	1.3	-3.4	-1.6	1.0
80.93	2.09	1.1	357.00	12MR	3.4	-3.4	-1.6	1.0
85.86	1.99	1.3	6.83	22MR	2.0	-3.3	-1.6	1.0
91.00	1.90	1.4	16.59	1AP	1.1	-3.3	-1.7	1.0
96.32	1.80	1.5	26.30	11AP	0.7	-3.3	-1.7	1.0
101.79	1.71	1.6	35.97	21AP	0.2	-3.3	-1.8	1.0
107.38	1.62	1.7	45.59	1MY	-0.5	-3.3	-1.8	1.0
113.07	1.54	1.7	55.17	11MY	-1.5	-3.4	-1.9	0.9
118.87	1.45	1.8	64.73	21MY	-1.5	-3.4	-1.9	0.9
124.75	1.36	1.8	74.27	31MY	-0.6	-3.4	-2.0	0.9
130.72	1.27	1.9	83.80	10JN	0.1	-3.4	-2.1	0.9
136.76	1.19	1.9	93.33	20JN	0.7	-3.5	-2.2	0.8
142.89	1.10	1.9	102.86	30JN	1.2	-3.5	-2.2	0.8
149.09	1.01	1.9	112.41	10JL	2.0	-3.6	-2.3	0.7
155.36	0.92	1.9	121.99	20JL	3.2	-3.7	-2.3	0.7
161.72	0.83	1.9	131.60	30JL	1.4	-3.7	-2.4	0.6
168.15	0.74	1.9	141.25	9AU	-0.3	-3.8	-2.4	0.6
174.66	0.64	1.9	150.95	19AU	-1.2	-3.9	-2.5	0.5
181.26	0.55	1.9	160.70	29AU	-1.3	-4.0	-2.5	0.5
187.94	0.45	1.9	170.50	8SE	-0.8	-4.1	-2.4	0.4
194.70	0.35	1.8	180.37	18SE	-0.4	-4.2	-2.4	0.4
201.56	0.26	1.8	190.29	28SE	-0.2	-4.3	-2.4	0.4
208.49	0.15	1.7	200.26	8OC	-0.1	-4.3	-2.3	0.5
215.52	0.05	1.7	210.29	18OC	0.0	-4.1	-2.3	0.6
222.64	-0.05	1.6	220.37	28OC	0.7	-3.7	-2.2	0.6
229.84	-0.15	1.6	230.49	7NO	3.3	-3.0	-2.1	0.7
237.13	-0.26	1.6	240.65	17NO	0.4	-3.8	-2.0	0.8
244.50	-0.36	1.6	250.83	27NO	-0.2	-4.2	-2.0	0.8
251.94	-0.46	1.5	261.02	7DE	-0.3	-4.4	-1.9	0.9
259.46	-0.56	1.5	271.22	17DE	-0.4	-4.4	-1.8	0.9
267.05	-0.66	1.5	281.42	27DE	-0.5	-4.3	-1.8	0.9
				967				
274.70	-0.75	1.5	291.59	6JA	-0.9	-4.2	-1.7	0.9
282.40	-0.84	1.5	301.75	16JA	-1.1	-4.1	-1.7	0.9
290.14	-0.92	1.5	311.86	26JA	-1.1	-4.0	-1.7	0.9
297.92	-1.00	1.4	321.93	5FE	-0.3	-3.9	-1.6	0.9
305.71	-1.07	1.4	331.96	15FE	1.5	-3.8	-1.6	0.9
313.52	-1.13	1.4	341.92	25FE	3.0	-3.7	-1.6	0.8
321.33	-1.17	1.4	351.83	7MR	1.5	-3.6	-1.6	0.8
329.13	-1.21	1.4	1.69	17MR	0.8	-3.5	-1.6	0.8
336.89	-1.23	1.4	11.48	27MR	0.5	-3.5	-1.6	0.7
344.62	-1.24	1.3	21.22	6AP	0.1	-3.4	-1.6	0.7
352.30	-1.24	1.3	30.91	16AP	-0.6	-3.4	-1.6	0.7
359.92	-1.22	1.3	40.56	26AP	-1.5	-3.4	-1.6	0.7
7.46	-1.19	1.3	50.16	6MY	-1.5	-3.3	-1.7	0.7
14.92	-1.14	1.3	59.73	16MY	-0.6	-3.3	-1.7	0.7
22.28	-1.08	1.3	69.28	26MY	0.2	-3.3	-1.8	0.7
29.53	-1.01	1.3	78.81	5JN	0.9	-3.3	-1.8	0.7
36.67	-0.92	1.2	88.34	15JN	1.5	-3.3	-1.9	0.7
43.68	-0.82	1.2	97.87	25JN	2.6	-3.3	-1.9	0.7
50.56	-0.71	1.2	107.41	5JL	2.9	-3.4	-2.0	0.6
57.29	-0.59	1.2	116.97	15JL	1.2	-3.4	-2.1	0.6
63.85	-0.46	1.1	126.56	25JL	-0.3	-3.4	-2.2	0.5
70.23	-0.31	1.1	136.19	4AU	-1.2	-3.4	-2.2	0.5
76.41	-0.15	1.0	145.86	14AU	-1.4	-3.5	-2.3	0.4
82.36	0.03	1.0	155.58	24AU	-0.8	-3.5	-2.3	0.4
88.06	0.22	0.9	165.35	3SE	-0.3	-3.5	-2.4	0.3
93.45	0.43	0.8	175.19	13SE	-0.1	-3.4	-2.4	0.2
98.48	0.66	0.7	185.07	23SE	0.1	-3.4	-2.4	0.2
103.09	0.92	0.6	195.02	3OC	0.2	-3.4	-2.5	0.1
107.18	1.21	0.4	205.03	13OC	1.0	-3.4	-2.4	0.2
110.62	1.53	0.2	215.08	23OC	3.1	-3.3	-2.4	0.3
113.28	1.89	0.1	225.18	2NO	0.3	-3.3	-2.4	0.3
114.97	2.29	-0.1	235.32	12NO	-0.4	-3.3	-2.3	0.4
115.50	2.72	-0.4	245.48	22NO	-0.5	-3.3	-2.3	0.5
114.70	3.18	-0.6	255.67	2DE	-0.5	-3.3	-2.2	0.5
112.54	3.61	-0.8	265.88	12DE	-0.6	-3.4	-2.1	0.6
109.24	3.96	-1.0	276.07	22DE	-0.9	-3.4	-2.0	0.6

♂ LONG	♂ LAT	♂ MAG	☉ LONG	16.00UT 968	☿	♀	♃	♄
105.30	4.18	-1.1	286.26	1JA	-0.9	-3.4	-2.0	0.7
101.48	4.22	-0.9	296.43	11JA	-0.9	-3.4	-1.9	0.7
98.47	4.11	-0.7	306.56	21JA	-0.3	-3.5	-1.8	0.7
96.67	3.89	-0.4	316.66	31JA	1.8	-3.5	-1.8	0.7
96.18	3.62	-0.2	326.71	10FE	2.4	-3.6	-1.7	0.7
96.90	3.34	0.1	336.70	20FE	1.1	-3.6	-1.7	0.7
98.66	3.06	0.3	346.64	1MR	0.6	-3.7	-1.6	0.7
101.26	2.80	0.5	356.53	11MR	0.3	-3.8	-1.6	0.6
104.53	2.56	0.7	6.35	21MR	-0.0	-3.9	-1.6	0.6
108.34	2.33	0.9	16.12	31MR	-0.6	-3.9	-1.6	0.5
112.60	2.13	1.1	25.84	10AP	-1.4	-4.0	-1.5	0.5
117.20	1.93	1.2	35.50	20AP	-1.5	-4.1	-1.5	0.5
122.09	1.76	1.3	45.13	30AP	-0.6	-4.2	-1.5	0.5
127.24	1.59	1.4	54.71	10MY	0.4	-4.2	-1.6	0.6
132.58	1.43	1.5	64.27	20MY	1.1	-4.1	-1.6	0.6
138.11	1.28	1.5	73.81	30MY	2.1	-3.7	-1.6	0.6
143.81	1.14	1.6	83.34	9JN	3.3	-2.9	-1.6	0.6
149.65	1.00	1.6	92.87	19JN	2.3	-3.3	-1.7	0.5
155.63	0.87	1.7	102.40	29JN	0.9	-3.9	-1.7	0.5
161.74	0.74	1.7	111.95	9JL	-0.3	-4.2	-1.8	0.5
167.97	0.62	1.7	121.52	19JL	-1.3	-4.2	-1.8	0.4
174.32	0.49	1.7	131.13	29JL	-1.5	-4.2	-1.9	0.4
180.78	0.37	1.7	140.78	8AU	-0.7	-4.1	-1.9	0.4
187.36	0.25	1.7	150.47	18AU	-0.2	-4.0	-2.0	0.3
194.05	0.14	1.7	160.22	28AU	0.0	-3.9	-2.1	0.2
200.86	0.02	1.7	170.02	7SE	0.2	-3.8	-2.1	0.2
207.77	-0.09	1.7	179.88	17SE	0.4	-3.8	-2.2	0.1
214.79	-0.20	1.7	189.80	27SE	1.3	-3.7	-2.3	0.1
221.92	-0.30	1.6	199.77	7OC	2.8	-3.6	-2.3	-0.0
229.15	-0.40	1.6	209.80	17OC	0.1	-3.6	-2.3	-0.0
236.49	-0.50	1.6	219.88	27OC	-0.6	-3.5	-2.4	0.0
243.92	-0.59	1.5	230.00	6NO	-0.6	-3.5	-2.4	0.1
251.43	-0.68	1.5	240.15	16NO	-0.6	-3.5	-2.4	0.2
259.03	-0.76	1.5	250.34	26NO	-0.7	-3.4	-2.3	0.2
266.70	-0.84	1.4	260.53	6DE	-0.8	-3.4	-2.3	0.3
274.44	-0.90	1.4	270.73	16DE	-0.8	-3.4	-2.2	0.4
282.23	-0.96	1.3	280.93	26DE	-0.8	-3.4	-2.2	0.4

♂ LONG	♂ LAT	♂ MAG	☉ LONG	969	☿	♀	♃	♄
290.07	-1.01	1.3	291.11	5JA	-0.2	-3.3	-2.1	0.5
297.93	-1.05	1.3	301.25	15JA	2.0	-3.3	-2.0	0.5
305.81	-1.07	1.3	311.37	25JA	1.9	-3.3	-2.0	0.5
313.70	-1.09	1.4	321.45	4FE	0.8	-3.3	-1.9	0.5
321.58	-1.09	1.4	331.47	14FE	0.4	-3.4	-1.8	0.5
329.45	-1.08	1.4	341.44	24FE	0.2	-3.4	-1.8	0.5
337.27	-1.06	1.4	351.35	6MR	-0.1	-3.4	-1.7	0.5
345.06	-1.03	1.4	1.20	16MR	-0.6	-3.4	-1.6	0.5
352.79	-0.99	1.4	11.00	26MR	-1.4	-3.5	-1.6	0.5
0.45	-0.93	1.4	20.74	5AP	-1.5	-3.5	-1.6	0.4
8.05	-0.87	1.5	30.43	15AP	-0.5	-3.5	-1.5	0.4
15.56	-0.79	1.5	40.08	25AP	0.5	-3.4	-1.5	0.4
22.99	-0.71	1.5	49.69	5MY	1.5	-3.4	-1.5	0.4
30.32	-0.62	1.5	59.26	15MY	2.7	-3.4	-1.5	0.4
37.56	-0.52	1.5	68.81	25MY	3.2	-3.4	-1.5	0.4
44.70	-0.42	1.5	78.34	4JN	1.8	-3.3	-1.5	0.4
51.73	-0.31	1.6	87.87	14JN	0.7	-3.3	-1.5	0.4
58.66	-0.19	1.6	97.40	24JN	-0.4	-3.3	-1.5	0.4
65.48	-0.07	1.6	106.94	4JL	-1.4	-3.3	-1.5	0.4
72.19	0.06	1.6	116.50	14JL	-1.5	-3.3	-1.5	0.4
78.78	0.19	1.6	126.09	24JL	-0.7	-3.4	-1.6	0.4
85.25	0.33	1.5	135.72	3AU	-0.2	-3.4	-1.6	0.3
91.59	0.47	1.5	145.39	13AU	0.2	-3.4	-1.7	0.3
97.79	0.62	1.5	155.11	23AU	0.4	-3.5	-1.7	0.2
103.83	0.77	1.4	164.88	2SE	0.7	-3.5	-1.8	0.2
109.71	0.93	1.4	174.71	12SE	1.7	-3.6	-1.8	0.1
115.40	1.10	1.3	184.59	22SE	2.5	-3.6	-1.9	0.1
120.86	1.29	1.2	194.54	2OC	0.0	-3.7	-2.0	-0.0
126.08	1.48	1.1	204.54	12OC	-0.7	-3.8	-2.0	-0.1
130.99	1.70	1.0	214.59	22OC	-0.8	-3.9	-2.1	-0.1
135.53	1.93	0.9	224.69	1NO	-0.7	-4.0	-2.1	-0.2
139.63	2.18	0.7	234.82	11NO	-0.8	-4.1	-2.2	-0.1
143.18	2.46	0.6	244.99	21NO	-0.6	-4.2	-2.2	-0.0
146.05	2.77	0.4	255.18	1DE	-0.6	-4.3	-2.2	0.0
148.09	3.10	0.2	265.38	11DE	-0.6	-4.4	-2.2	0.1
149.11	3.46	-0.1	275.58	21DE	-0.0	-4.4	-2.2	0.2
148.93	3.82	-0.3	285.77	31DE	2.3	-4.2	-2.2	0.2

♂ LONG	♂ LAT	♂ MAG	☉ LONG	16.00UT 970	☿	♀	♃	♄
147.44	4.16	-0.6	295.94	10JA	1.5	-3.7	-2.2	0.3
144.70	4.41	-0.8	306.07	20JA	0.5	-3.3	-2.1	0.3
141.05	4.53	-1.0	316.17	30JA	0.2	-3.8	-2.1	0.4
137.11	4.47	-0.9	326.22	9FE	0.1	-4.2	-2.0	0.4
133.61	4.25	-0.8	336.22	19FE	-0.2	-4.3	-1.9	0.4
131.11	3.91	-0.5	346.16	1MR	-0.7	-4.3	-1.8	0.4
129.87	3.52	-0.3	356.05	11MR	-1.4	-4.2	-1.8	0.4
129.92	3.11	-0.1	5.87	21MR	-1.4	-4.1	-1.7	0.4
131.10	2.73	0.1	15.64	31MR	-0.5	-4.0	-1.6	0.4
133.25	2.37	0.3	25.36	10AP	0.7	-3.9	-1.6	0.3
136.19	2.05	0.5	35.03	20AP	1.9	-3.8	-1.5	0.3
139.78	1.75	0.7	44.65	30AP	3.6	-3.7	-1.5	0.3
143.88	1.49	0.8	54.25	10MY	2.5	-3.6	-1.4	0.3
148.43	1.25	0.9	63.81	20MY	1.4	-3.5	-1.4	0.3
153.33	1.03	1.1	73.35	30MY	0.5	-3.5	-1.4	0.3
158.54	0.82	1.1	82.88	9JN	-0.4	-3.4	-1.4	0.3
164.01	0.63	1.2	92.41	19JN	-1.4	-3.4	-1.4	0.4
169.72	0.45	1.3	101.94	29JN	-1.5	-3.4	-1.3	0.4
175.64	0.29	1.3	111.49	9JL	-0.7	-3.4	-1.3	0.4
181.76	0.13	1.4	121.06	19JL	-0.1	-3.3	-1.4	0.4
188.05	-0.01	1.4	130.67	29JL	0.3	-3.3	-1.4	0.3
194.51	-0.15	1.4	140.32	8AU	0.5	-3.3	-1.4	0.3
201.12	-0.28	1.4	150.01	18AU	0.9	-3.3	-1.4	0.3
207.89	-0.40	1.5	159.75	28AU	2.1	-3.4	-1.5	0.2
214.80	-0.52	1.5	169.55	7SE	2.2	-3.4	-1.5	0.2
221.84	-0.62	1.5	179.41	17SE	-0.1	-3.4	-1.5	0.1
229.01	-0.72	1.5	189.32	27SE	-0.8	-3.4	-1.6	0.0
236.30	-0.80	1.5	199.29	7OC	-0.9	-3.4	-1.7	-0.0
243.71	-0.88	1.5	209.32	17OC	-0.9	-3.5	-1.7	-0.1
251.21	-0.95	1.4	219.39	27OC	-0.7	-3.5	-1.8	-0.2
258.82	-1.01	1.4	229.51	6NO	-0.5	-3.5	-1.8	-0.2
266.50	-1.05	1.4	239.66	16NO	-0.5	-3.5	-1.9	-0.3
274.26	-1.09	1.4	249.84	26NO	-0.4	-3.4	-2.0	-0.2
282.08	-1.11	1.4	260.03	6DE	0.1	-3.4	-2.0	-0.1
289.93	-1.12	1.4	270.23	16DE	2.6	-3.4	-2.1	-0.0
297.82	-1.12	1.4	280.43	26DE	1.1	-3.4	-2.1	0.0

♂ LONG	♂ LAT	♂ MAG	☉ LONG	971	☿	♀	♃	♄
305.73	-1.11	1.4	290.61	5JA	0.3	-3.3	-2.1	0.1
313.63	-1.08	1.4	300.76	15JA	0.1	-3.3	-2.1	0.2
321.53	-1.05	1.4	310.88	25JA	-0.1	-3.3	-2.1	0.2
329.39	-1.00	1.4	320.96	4FE	-0.3	-3.3	-2.1	0.3
337.22	-0.94	1.4	330.98	14FE	-0.7	-3.4	-2.0	0.3
345.00	-0.88	1.4	340.95	24FE	-1.4	-3.4	-2.0	0.3
352.72	-0.80	1.4	350.87	6MR	-1.4	-3.4	-1.9	0.3
0.38	-0.72	1.4	0.73	16MR	-0.5	-3.4	-1.9	0.3
7.95	-0.63	1.4	10.53	26MR	0.9	-3.4	-1.8	0.3
15.45	-0.54	1.5	20.28	5AP	2.5	-3.5	-1.7	0.3
22.86	-0.44	1.5	29.97	15AP	3.3	-3.5	-1.7	0.3
30.19	-0.34	1.6	39.62	25AP	1.9	-3.6	-1.6	0.3
37.43	-0.24	1.6	49.23	5MY	1.1	-3.6	-1.5	0.2
44.57	-0.13	1.6	58.80	15MY	0.4	-3.7	-1.5	0.2
51.63	-0.03	1.7	68.35	25MY	-0.5	-3.8	-1.4	0.2
58.60	0.08	1.7	77.88	4JN	-1.4	-3.9	-1.4	0.2
65.48	0.19	1.7	87.41	14JN	-1.5	-4.0	-1.3	0.3
72.27	0.30	1.8	96.94	24JN	-0.7	-4.1	-1.3	0.3
78.98	0.41	1.8	106.48	4JL	-0.0	-4.1	-1.3	0.3
85.60	0.52	1.8	116.04	14JL	0.4	-4.2	-1.3	0.3
92.14	0.62	1.8	125.63	24JL	0.7	-4.2	-1.3	0.3
98.60	0.73	1.8	135.25	3AU	1.3	-4.0	-1.3	0.3
104.97	0.84	1.8	144.92	13AU	2.6	-3.6	-1.3	0.3
111.25	0.95	1.8	154.64	23AU	1.9	-3.2	-1.3	0.3
117.44	1.06	1.8	164.41	2SE	-0.1	-3.8	-1.3	0.3
123.54	1.17	1.7	174.23	12SE	-1.0	-4.1	-1.3	0.2
129.53	1.29	1.7	184.12	22SE	-1.1	-4.3	-1.3	0.2
135.40	1.41	1.6	194.06	2OC	-0.9	-4.3	-1.4	0.1
141.14	1.53	1.6	204.05	12OC	-0.6	-4.2	-1.4	0.0
146.73	1.66	1.5	214.10	22OC	-0.4	-4.2	-1.5	-0.0
152.15	1.79	1.4	224.20	1NO	-0.3	-4.1	-1.5	-0.1
157.35	1.93	1.3	234.33	11NO	-0.3	-4.0	-1.6	-0.2
162.32	2.08	1.2	244.50	21NO	0.3	-3.9	-1.7	-0.3
166.97	2.24	1.0	254.69	1DE	2.9	-3.8	-1.7	-0.3
171.26	2.41	0.9	264.89	11DE	0.8	-3.7	-1.8	-0.2
175.10	2.59	0.7	275.09	21DE	0.1	-3.7	-1.8	-0.1
178.37	2.78	0.5	285.28	31DE	-0.1	-3.6	-1.9	-0.1

♂			⊙	16.00UT	☿	♀	♃	♄
LONG	LAT	MAG	LONG	972	MAGNITUDES			
180.94	2.98	0.3	295.44	10JA	-0.2	-3.5	-2.0	-0.0
182.64	3.19	0.0	305.58	20JA	-0.4	-3.5	-2.0	0.1
183.29	3.39	-0.3	315.68	30JA	-0.8	-3.4	-2.0	0.1
182.74	3.56	-0.5	325.74	9FE	-1.3	-3.4	-2.0	0.2
180.89	3.66	-0.8	335.74	19FE	-1.3	-3.4	-2.0	0.2
177.90	3.64	-1.0	345.69	29FE	-0.5	-3.3	-2.0	0.3
174.19	3.49	-1.2	355.57	10MR	1.1	-3.3	-2.0	0.3
170.41	3.19	-1.1	5.40	20MR	3.1	-3.3	-1.9	0.3
167.29	2.78	-0.9	15.18	30MR	2.5	-3.3	-1.9	0.3
165.29	2.33	-0.7	24.89	9AP	1.4	-3.3	-1.8	0.3
164.60	1.87	-0.5	34.57	19AP	0.8	-3.3	-1.8	0.3
165.19	1.44	-0.3	44.19	29AP	0.3	-3.3	-1.7	0.3
166.90	1.06	-0.1	53.78	9MY	-0.5	-3.4	-1.6	0.3
169.55	0.71	0.1	63.35	19MY	-1.4	-3.4	-1.6	0.2
172.99	0.41	0.2	72.89	29MY	-1.6	-3.4	-1.5	0.2
177.06	0.15	0.4	82.41	8JN	-0.6	-3.5	-1.4	0.2
181.67	-0.09	0.5	91.94	18JN	0.1	-3.5	-1.4	0.3
186.72	-0.30	0.6	101.47	28JN	0.6	-3.4	-1.3	0.3
192.16	-0.48	0.7	111.02	8JL	1.0	-3.4	-1.3	0.3
197.92	-0.65	0.8	120.60	18JL	1.7	-3.4	-1.3	0.3
203.97	-0.80	0.9	130.20	28JL	3.1	-3.4	-1.2	0.4
210.27	-0.92	0.9	139.84	7AU	1.6	-3.3	-1.2	0.4
216.80	-1.03	1.0	149.53	17AU	-0.2	-3.3	-1.2	0.4
223.54	-1.13	1.0	159.27	27AU	-1.1	-3.3	-1.2	0.3
230.46	-1.21	1.1	169.07	6SE	-1.2	-3.3	-1.2	0.3
237.55	-1.27	1.1	178.93	16SE	-0.9	-3.3	-1.2	0.3
244.79	-1.32	1.1	188.83	26SE	-0.5	-3.4	-1.2	0.2
252.16	-1.35	1.2	198.80	6OC	-0.3	-3.4	-1.3	0.2
259.65	-1.37	1.2	208.83	16OC	-0.2	-3.4	-1.3	0.1
267.24	-1.38	1.2	218.90	26OC	-0.1	-3.4	-1.3	0.1
274.92	-1.36	1.2	229.01	5NO	0.5	-3.5	-1.4	-0.0
282.66	-1.34	1.3	239.17	15NO	3.2	-3.5	-1.4	-0.1
290.46	-1.30	1.3	249.34	25NO	0.6	-3.6	-1.5	-0.2
298.28	-1.25	1.3	259.54	5DE	-0.1	-3.7	-1.5	-0.2
306.12	-1.19	1.3	269.74	15DE	-0.2	-3.7	-1.6	-0.2
313.97	-1.11	1.4	279.94	25DE	-0.3	-3.8	-1.7	-0.2
				973				
321.79	-1.03	1.4	290.12	4JA	-0.5	-3.9	-1.7	-0.1
329.59	-0.94	1.4	300.27	14JA	-0.8	-4.0	-1.8	-0.0
337.35	-0.85	1.4	310.39	24JA	-1.2	-4.1	-1.8	0.0
345.06	-0.74	1.5	320.47	3FE	-1.1	-4.2	-1.9	0.1
352.71	-0.64	1.5	330.50	13FE	-0.4	-4.3	-1.9	0.1
0.28	-0.53	1.5	340.47	23FE	1.3	-4.3	-2.0	0.2
7.78	-0.42	1.6	350.39	5MR	3.3	-4.2	-2.0	0.2
15.21	-0.32	1.6	0.26	15MR	1.8	-4.0	-2.0	0.3
22.54	-0.21	1.6	10.06	25MR	1.0	-3.4	-2.0	0.3
29.80	-0.10	1.6	19.80	4AP	0.6	-3.2	-2.0	0.3
36.97	0.01	1.6	29.50	14AP	0.1	-3.8	-1.9	0.4
44.06	0.11	1.6	39.15	24AP	-0.5	-4.1	-1.9	0.4
51.07	0.21	1.7	48.76	4MY	-1.4	-4.2	-1.8	0.3
58.00	0.31	1.7	58.34	14MY	-1.6	-4.2	-1.8	0.3
64.85	0.41	1.7	67.88	24MY	-0.6	-4.1	-1.7	0.3
71.64	0.50	1.8	77.42	3JN	0.2	-4.0	-1.7	0.3
78.35	0.59	1.8	86.95	13JN	0.7	-3.9	-1.6	0.3
84.99	0.67	1.9	96.47	23JN	1.3	-3.8	-1.5	0.3
91.58	0.76	1.9	106.01	3JL	2.2	-3.7	-1.5	0.3
98.11	0.84	1.9	115.57	13JL	3.2	-3.7	-1.4	0.4
104.58	0.91	1.9	125.16	23JL	1.4	-3.6	-1.4	0.4
111.00	0.99	2.0	134.78	2AU	-0.2	-3.5	-1.3	0.4
117.37	1.06	2.0	144.45	12AU	-1.2	-3.5	-1.3	0.4
123.69	1.13	2.0	154.16	22AU	-1.3	-3.5	-1.3	0.4
129.96	1.19	2.0	163.93	1SE	-0.8	-3.4	-1.2	0.4
136.18	1.26	1.9	173.75	11SE	-0.4	-3.4	-1.2	0.4
142.35	1.32	1.9	183.63	21SE	-0.1	-3.4	-1.2	0.4
148.47	1.38	1.9	193.57	1OC	-0.0	-3.4	-1.2	0.4
154.52	1.44	1.8	203.57	11OC	0.1	-3.4	-1.2	0.3
160.51	1.50	1.8	213.61	21OC	0.8	-3.4	-1.2	0.3
166.43	1.56	1.7	223.71	31OC	3.4	-3.4	-1.2	0.2
172.26	1.61	1.7	233.84	10NO	0.4	-3.4	-1.3	0.1
177.99	1.67	1.6	244.01	20NO	-0.3	-3.4	-1.3	0.1
183.60	1.72	1.5	254.20	30NO	-0.4	-3.4	-1.3	-0.0
189.06	1.77	1.3	264.40	10DE	-0.4	-3.4	-1.4	-0.1
194.35	1.82	1.2	274.59	20DE	-0.6	-3.4	-1.5	-0.2
199.44	1.86	1.1	284.79	30DE	-0.9	-3.4	-1.5	-0.2

♂			⊙	16.00UT	☿	♀	♃	♄
LONG	LAT	MAG	LONG	974	MAGNITUDES			
204.26	1.90	0.9	294.95	9JA	-1.0	-3.5	-1.6	-0.1
208.77	1.93	0.7	305.09	19JA	-1.0	-3.5	-1.6	-0.0
212.88	1.95	0.5	315.20	29JA	-0.3	-3.5	-1.7	0.0
216.50	1.96	0.3	325.25	8FE	1.5	-3.4	-1.8	0.1
219.51	1.94	0.0	335.25	18FE	2.8	-3.4	-1.8	0.2
221.74	1.89	-0.3	345.20	28FE	1.3	-3.4	-1.9	0.2
223.03	1.79	-0.5	355.09	10MR	0.7	-3.4	-2.0	0.3
223.19	1.63	-0.9	4.92	20MR	0.4	-3.3	-2.0	0.3
222.10	1.39	-1.2	14.70	30MR	0.0	-3.3	-2.0	0.4
219.80	1.05	-1.5	24.42	9AP	-0.6	-3.3	-2.0	0.4
216.57	0.63	-1.7	34.10	19AP	-1.4	-3.3	-2.0	0.4
213.01	0.15	-1.7	43.73	29AP	-1.6	-3.3	-2.0	0.4
209.87	-0.34	-1.5	53.32	9MY	-0.6	-3.4	-2.0	0.4
207.73	-0.79	-1.4	62.88	19MY	0.3	-3.4	-2.0	0.4
206.90	-1.17	-1.2	72.43	29MY	1.0	-3.4	-1.9	0.4
207.45	-1.48	-0.9	81.95	8JN	1.7	-3.4	-1.9	0.4
209.23	-1.72	-0.7	91.48	18JN	2.9	-3.5	-1.8	0.4
212.07	-1.90	-0.6	101.01	28JN	2.7	-3.5	-1.7	0.4
215.80	-2.04	-0.4	110.56	8JL	1.1	-3.6	-1.7	0.4
220.25	-2.14	-0.2	120.13	18JL	-0.3	-3.7	-1.6	0.4
225.30	-2.20	-0.1	129.74	28JL	-1.3	-3.7	-1.6	0.5
230.85	-2.23	0.0	139.37	7AU	-1.4	-3.8	-1.5	0.5
236.80	-2.24	0.2	149.06	17AU	-0.8	-3.9	-1.5	0.5
243.10	-2.22	0.3	158.80	27AU	-0.3	-4.0	-1.4	0.6
249.69	-2.19	0.4	168.59	6SE	-0.0	-4.1	-1.4	0.6
256.52	-2.13	0.5	178.44	16SE	0.1	-4.2	-1.3	0.6
263.55	-2.05	0.5	188.36	26SE	0.3	-4.3	-1.3	0.5
270.74	-1.96	0.6	198.32	6OC	1.1	-4.3	-1.3	0.5
278.07	-1.86	0.7	208.34	16OC	3.1	-4.1	-1.3	0.5
285.50	-1.74	0.8	218.41	26OC	0.3	-3.6	-1.3	0.4
293.01	-1.61	0.9	228.52	5NO	-0.5	-3.0	-1.3	0.4
300.58	-1.48	0.9	238.67	15NO	-0.5	-3.8	-1.3	0.3
308.17	-1.34	1.0	248.85	25NO	-0.5	-4.3	-1.3	0.3
315.78	-1.19	1.1	259.04	5DE	-0.7	-4.4	-1.3	0.2
323.39	-1.04	1.2	269.25	15DE	-0.8	-4.4	-1.4	0.1
330.97	-0.90	1.2	279.45	25DE	-0.9	-4.3	-1.4	0.0
				975				
338.52	-0.75	1.3	289.62	4JA	-0.9	-4.2	-1.4	-0.0
346.02	-0.61	1.4	299.78	14JA	-0.3	-4.1	-1.5	-0.0
353.47	-0.47	1.4	309.91	24JA	1.8	-4.0	-1.5	0.0
0.86	-0.33	1.5	319.99	3FE	2.2	-3.9	-1.6	0.1
8.18	-0.20	1.5	330.02	13FE	0.9	-3.8	-1.7	0.2
15.43	-0.08	1.6	340.00	23FE	0.5	-3.7	-1.7	0.2
22.61	0.04	1.6	349.92	5MR	0.3	-3.6	-1.8	0.3
29.71	0.16	1.7	359.78	15MR	-0.1	-3.5	-1.9	0.3
36.74	0.26	1.7	9.59	25MR	-0.6	-3.5	-1.9	0.4
43.70	0.36	1.8	19.34	4AP	-1.4	-3.4	-2.0	0.4
50.59	0.46	1.8	29.04	14AP	-1.5	-3.4	-2.1	0.5
57.41	0.54	1.8	38.69	24AP	-0.6	-3.4	-2.1	0.5
64.17	0.63	1.9	48.30	4MY	0.4	-3.3	-2.1	0.5
70.87	0.70	1.9	57.88	14MY	1.3	-3.3	-2.1	0.6
77.51	0.77	1.9	67.43	24MY	2.3	-3.3	-2.1	0.6
84.10	0.84	1.9	76.96	3JN	3.5	-3.3	-2.1	0.6
90.65	0.90	1.9	86.49	13JN	2.1	-3.3	-2.1	0.6
97.15	0.95	1.9	96.02	23JN	0.9	-3.3	-2.1	0.6
103.62	1.00	1.9	105.56	3JL	-0.3	-3.4	-2.0	0.6
110.06	1.04	1.9	115.12	13JL	-1.3	-3.4	-1.9	0.5
116.46	1.09	2.0	124.70	23JL	-1.5	-3.4	-1.9	0.6
122.85	1.12	2.0	134.32	2AU	-0.7	-3.4	-1.8	0.6
129.21	1.15	2.0	143.98	12AU	-0.2	-3.5	-1.8	0.6
135.56	1.18	2.0	153.69	22AU	0.1	-3.5	-1.7	0.7
141.89	1.21	2.0	163.45	1SE	0.3	-3.5	-1.6	0.7
148.22	1.23	2.0	173.27	11SE	0.5	-3.4	-1.6	0.7
154.52	1.24	2.0	183.15	21SE	1.4	-3.4	-1.5	0.7
160.82	1.26	2.0	193.08	1OC	2.8	-3.4	-1.5	0.7
167.11	1.26	2.0	203.08	11OC	0.1	-3.4	-1.4	0.7
173.39	1.27	1.9	213.12	21OC	-0.6	-3.3	-1.4	0.7
179.65	1.27	1.9	223.21	31OC	-0.7	-3.3	-1.4	0.7
185.90	1.26	1.8	233.34	10NO	-0.7	-3.3	-1.4	0.6
192.13	1.24	1.8	243.51	20NO	-0.8	-3.3	-1.4	0.6
198.33	1.22	1.7	253.69	30NO	-0.7	-3.3	-1.4	0.5
204.50	1.20	1.6	263.90	10DE	-0.7	-3.4	-1.4	0.5
210.64	1.16	1.5	274.10	20DE	-0.7	-3.4	-1.4	0.4
216.72	1.11	1.4	284.29	30DE	-0.1	-3.4	-1.4	0.3

Left table (♂ / ☉ / 16.00UT 976–977 / Magnitudes ☿ ♀ ♃ ♄):

♂ LONG	LAT	MAG	☉ LONG	16.00UT 976	☿	♀	♃	♄
222.75	1.05	1.3	294.46	9JA	2.1	-3.5	-1.4	0.2
228.70	0.97	1.2	304.60	19JA	1.8	-3.5	-1.5	0.2
234.56	0.88	1.0	314.71	29JA	0.7	-3.5	-1.5	0.2
240.31	0.76	0.9	324.77	8FE	0.3	-3.6	-1.5	0.2
245.93	0.62	0.7	334.77	18FE	0.1	-3.6	-1.6	0.3
251.37	0.45	0.5	344.72	28FE	-0.1	-3.7	-1.7	0.3
256.59	0.25	0.3	354.62	9MR	-0.6	-3.8	-1.7	0.4
261.55	-0.00	0.1	4.46	19MR	-1.4	-3.9	-1.8	0.5
266.16	-0.30	-0.2	14.23	29MR	-1.5	-4.0	-1.9	0.5
270.33	-0.66	-0.4	23.96	8AP	-0.6	-4.0	-1.9	0.6
273.94	-1.09	-0.7	33.64	18AP	0.5	-4.1	-2.0	0.6
276.85	-1.61	-1.0	43.27	28AP	1.6	-4.2	-2.1	0.7
278.86	-2.21	-1.3	52.86	8MY	3.0	-4.2	-2.1	0.7
279.80	-2.90	-1.6	62.42	18MY	3.0	-4.1	-2.2	0.7
279.53	-3.65	-1.9	71.97	28MY	1.7	-3.6	-2.2	0.7
278.03	-4.40	-2.2	81.50	7JN	0.7	-2.8	-2.3	0.8
275.60	-5.06	-2.4	91.02	17JN	-0.4	-3.4	-2.3	0.8
272.78	-5.53	-2.5	100.55	27JN	-1.4	-3.9	-2.3	0.8
270.31	-5.73	-2.3	110.10	7JL	-1.5	-4.2	-2.3	0.8
268.84	-5.70	-2.1	119.67	17JL	-0.7	-4.2	-2.2	0.8
268.66	-5.47	-1.9	129.27	27JL	-0.1	-4.2	-2.2	0.8
269.84	-5.13	-1.6	138.91	6AU	0.2	-4.1	-2.1	0.8
272.26	-4.73	-1.4	148.59	16AU	0.4	-4.0	-2.1	0.8
275.71	-4.31	-1.1	158.33	26AU	0.8	-3.9	-2.0	0.9
279.99	-3.88	-0.9	168.12	5SE	1.8	-3.8	-1.9	0.9
284.94	-3.46	-0.7	177.96	15SE	2.5	-3.8	-1.9	1.0
290.40	-3.06	-0.5	187.87	25SE	0.0	-3.7	-1.8	1.0
296.28	-2.68	-0.3	197.83	5OC	-0.8	-3.6	-1.7	1.0
302.45	-2.32	-0.1	207.85	15OC	-0.8	-3.6	-1.7	1.0
308.86	-1.98	0.1	217.92	25OC	-0.8	-3.5	-1.6	1.0
315.45	-1.66	0.2	228.03	4NO	-0.7	-3.5	-1.6	1.0
322.16	-1.37	0.4	238.17	14NO	-0.6	-3.5	-1.6	0.9
328.97	-1.10	0.5	248.35	24NO	-0.6	-3.4	-1.5	0.9
335.83	-0.85	0.7	258.55	4DE	-0.5	-3.4	-1.5	0.8
342.73	-0.62	0.8	268.75	14DE	-0.0	-3.4	-1.5	0.8
349.64	-0.41	0.9	278.95	24DE	2.3	-3.4	-1.5	0.7

977

♂ LONG	LAT	MAG	☉ LONG	16.00UT 977	☿	♀	♃	♄
356.56	-0.22	1.1	289.13	3JA	1.4	-3.3	-1.5	0.6
3.46	-0.05	1.2	299.29	13JA	0.4	-3.3	-1.5	0.6
10.33	0.10	1.3	309.41	23JA	0.2	-3.3	-1.5	0.5
17.18	0.24	1.4	319.50	2FE	0.0	-3.3	-1.5	0.4
23.98	0.37	1.5	329.53	12FE	-0.2	-3.4	-1.5	0.4
30.75	0.48	1.6	339.52	22FE	-0.7	-3.4	-1.6	0.5
37.48	0.58	1.6	349.44	4MR	-1.4	-3.4	-1.6	0.5
44.16	0.67	1.7	359.30	14MR	-1.4	-3.4	-1.7	0.6
50.80	0.75	1.8	9.12	24MR	-0.6	-3.5	-1.7	0.6
57.40	0.83	1.8	18.87	3AP	0.7	-3.5	-1.8	0.7
63.95	0.89	1.9	28.57	13AP	2.1	-3.5	-1.8	0.7
70.47	0.94	1.9	38.22	23AP	3.9	-3.4	-1.9	0.8
76.96	0.99	1.9	47.84	3MY	2.2	-3.4	-2.0	0.8
83.41	1.03	2.0	57.41	13MY	1.3	-3.4	-2.0	0.9
89.84	1.06	2.0	66.97	23MY	0.5	-3.4	-2.1	0.9
96.24	1.09	2.0	76.50	2JN	-0.4	-3.3	-2.2	1.0
102.63	1.12	2.0	86.03	12JN	-1.4	-3.3	-2.3	1.0
109.00	1.13	2.0	95.56	22JN	-1.6	-3.3	-2.3	1.0
115.37	1.15	2.0	105.09	2JL	-0.7	-3.3	-2.3	1.0
121.72	1.15	2.0	114.65	12JL	-0.1	-3.3	-2.4	1.0
128.08	1.16	2.0	124.23	22JL	0.3	-3.4	-2.4	1.0
134.44	1.15	2.0	133.85	1AU	0.6	-3.4	-2.4	1.0
140.81	1.15	2.0	143.51	11AU	1.1	-3.4	-2.4	1.0
147.20	1.14	2.0	153.22	21AU	2.2	-3.5	-2.3	1.1
153.59	1.12	2.0	162.98	31AU	2.2	-3.5	-2.3	1.1
160.01	1.10	2.0	172.79	10SE	-0.1	-3.6	-2.2	1.2
166.45	1.08	2.0	182.67	20SE	-0.9	-3.6	-2.2	1.2
172.91	1.04	2.0	192.60	30SE	-1.0	-3.7	-2.1	1.2
179.40	1.01	2.0	202.59	10OC	-0.9	-3.8	-2.0	1.3
185.92	0.97	2.0	212.63	20OC	-0.6	-3.9	-2.0	1.3
192.46	0.92	1.9	222.72	30OC	-0.5	-4.0	-1.9	1.3
199.03	0.87	1.9	232.85	9NO	-0.4	-4.1	-1.8	1.3
205.64	0.80	1.9	243.01	19NO	-0.4	-4.2	-1.8	1.2
212.27	0.74	1.8	253.20	29NO	0.2	-4.3	-1.7	1.2
218.93	0.66	1.7	263.39	9DE	2.6	-4.4	-1.7	1.2
225.62	0.57	1.7	273.60	19DE	1.0	-4.4	-1.6	1.1
232.34	0.47	1.6	283.79	29DE	0.2	-4.2	-1.6	1.0

Right table (♂ / ☉ / 16.00UT 978–979 / Magnitudes ☿ ♀ ♃ ♄):

♂ LONG	LAT	MAG	☉ LONG	16.00UT 978	☿	♀	♃	♄
239.08	0.36	1.5	293.96	8JA	0.0	-3.7	-1.6	1.0
245.84	0.24	1.4	304.11	18JA	-0.1	-3.3	-1.6	0.9
252.62	0.10	1.3	314.21	28JA	-0.3	-3.9	-1.6	0.8
259.42	-0.05	1.2	324.27	7FE	-0.7	-4.2	-1.6	0.8
266.24	-0.21	1.1	334.29	17FE	-1.4	-4.3	-1.6	0.7
273.05	-0.40	1.0	344.24	27FE	-1.3	-4.3	-1.6	0.7
279.87	-0.60	0.9	354.14	9MR	-0.5	-4.2	-1.6	0.8
286.67	-0.82	0.8	3.98	19MR	0.9	-4.1	-1.6	0.8
293.45	-1.06	0.6	13.76	29MR	2.7	-4.0	-1.6	0.9
300.19	-1.32	0.5	23.49	8AP	3.0	-3.9	-1.7	0.9
306.86	-1.60	0.3	33.17	18AP	1.7	-3.8	-1.7	1.0
313.44	-1.89	0.2	42.80	28AP	1.0	-3.7	-1.8	1.0
319.89	-2.21	0.0	52.40	8MY	0.3	-3.6	-1.8	1.1
326.16	-2.54	-0.2	61.97	18MY	-0.4	-3.5	-1.9	1.1
332.20	-2.90	-0.3	71.51	28MY	-1.4	-3.5	-2.0	1.2
337.93	-3.26	-0.5	81.04	7JN	-1.6	-3.4	-2.0	1.2
343.25	-3.64	-0.7	90.57	17JN	-0.7	-3.4	-2.1	1.3
348.04	-4.03	-0.9	100.10	27JN	0.0	-3.4	-2.2	1.3
352.15	-4.42	-1.2	109.64	7JL	0.5	-3.4	-2.2	1.3
355.39	-4.80	-1.4	119.21	17JL	0.8	-3.3	-2.3	1.3
357.56	-5.14	-1.7	128.81	27JL	1.4	-3.3	-2.4	1.3
358.43	-5.40	-1.9	138.45	6AU	2.8	-3.3	-2.4	1.3
357.89	-5.53	-2.1	148.15	16AU	1.9	-3.3	-2.4	1.3
356.03	-5.46	-2.3	157.86	26AU	-0.1	-3.4	-2.5	1.3
353.25	-5.15	-2.5	167.65	5SE	-1.0	-3.4	-2.5	1.3
350.26	-4.59	-2.4	177.49	15SE	-1.1	-3.4	-2.4	1.3
347.86	-3.87	-2.1	187.39	25SE	-0.9	-3.4	-2.4	1.3
346.57	-3.11	-1.8	197.35	5OC	-0.5	-3.4	-2.4	1.3
346.60	-2.38	-1.5	207.37	15OC	-0.3	-3.5	-2.3	1.3
347.90	-1.73	-1.2	217.43	25OC	-0.3	-3.5	-2.2	1.2
350.27	-1.17	-0.8	227.54	4NO	-0.2	-3.5	-2.2	1.2
353.53	-0.71	-0.5	237.68	14NO	0.4	-3.5	-2.1	1.2
357.48	-0.32	-0.3	247.85	24NO	2.9	-3.4	-2.0	1.2
1.96	-0.01	-0.0	258.05	4DE	0.8	-3.4	-2.0	1.1
6.86	0.25	0.2	268.25	14DE	-0.0	-3.4	-1.9	1.1
12.06	0.47	0.4	278.44	24DE	-0.1	-3.4	-1.8	1.0

979

♂ LONG	LAT	MAG	☉ LONG	16.00UT 979	☿	♀	♃	♄
17.51	0.64	0.6	288.63	3JA	-0.2	-3.3	-1.8	1.0
23.15	0.79	0.8	298.79	13JA	-0.4	-3.3	-1.7	0.9
28.92	0.91	1.0	308.91	23JA	-0.8	-3.3	-1.7	0.9
34.80	1.00	1.1	319.00	2FE	-1.3	-3.3	-1.7	0.8
40.77	1.08	1.2	329.04	12FE	-1.2	-3.4	-1.6	0.8
46.79	1.14	1.4	339.02	22FE	-0.5	-3.4	-1.6	0.8
52.86	1.19	1.5	348.95	4MR	1.1	-3.4	-1.6	0.7
58.96	1.23	1.6	358.82	14MR	3.2	-3.4	-1.6	0.8
65.08	1.26	1.7	8.63	24MR	2.2	-3.4	-1.6	0.9
71.22	1.28	1.7	18.40	3AP	1.2	-3.5	-1.6	1.0
77.38	1.29	1.8	28.10	13AP	0.7	-3.5	-1.6	1.0
83.54	1.30	1.9	37.75	23AP	0.2	-3.6	-1.6	1.1
89.72	1.30	1.9	47.37	3MY	-0.5	-3.6	-1.7	1.2
95.91	1.29	2.0	56.95	13MY	-1.4	-3.7	-1.7	1.2
102.10	1.28	2.0	66.51	23MY	-1.6	-3.8	-1.7	1.2
108.32	1.27	2.0	76.05	2JN	-0.7	-3.9	-1.8	1.3
114.54	1.25	2.0	85.57	12JN	0.1	-4.0	-1.8	1.3
120.79	1.22	2.0	95.10	22JN	0.6	-4.1	-1.9	1.3
127.05	1.19	2.0	104.64	2JL	1.1	-4.1	-2.0	1.3
133.35	1.16	2.0	114.19	12JL	1.9	-4.2	-2.0	1.3
139.67	1.13	2.0	123.77	22JL	3.2	-4.2	-2.1	1.3
146.02	1.09	2.0	133.39	1AU	1.6	-4.0	-2.2	1.3
152.41	1.04	2.0	143.05	11AU	-0.2	-3.6	-2.2	1.2
158.84	0.99	2.0	152.75	21AU	-1.1	-3.2	-2.3	1.2
165.32	0.94	1.9	162.51	31AU	-1.3	-3.8	-2.3	1.1
171.84	0.88	1.9	172.32	10SE	-0.9	-4.2	-2.4	1.1
178.41	0.82	1.9	182.19	20SE	-0.4	-4.3	-2.4	1.1
185.04	0.76	1.9	192.12	30SE	-0.2	-4.3	-2.4	1.1
191.72	0.69	1.9	202.11	10OC	-0.1	-4.2	-2.4	1.1
198.45	0.61	1.9	212.15	20OC	-0.0	-4.2	-2.4	1.1
205.25	0.53	1.9	222.24	30OC	0.6	-4.1	-2.4	1.1
212.11	0.45	1.8	232.36	9NO	3.2	-4.0	-2.3	1.1
219.03	0.36	1.8	242.52	19NO	0.6	-3.9	-2.3	1.1
226.01	0.26	1.8	252.71	29NO	-0.2	-3.8	-2.2	1.0
233.04	0.15	1.7	262.90	9DE	-0.3	-3.7	-2.2	1.0
240.15	0.04	1.7	273.10	19DE	-0.3	-3.6	-2.1	1.0
247.31	-0.07	1.6	283.30	29DE	-0.5	-3.6	-2.0	0.9

♂			☉	16.00UT	☿	♀	♃	♄	♂			☉	16.00UT	☿	♀	♃	♄
LONG	LAT	MAG	LONG	980		MAGNITUDES			LONG	LAT	MAG	LONG	982		MAGNITUDES		
254.52	-0.19	1.6	293.47	8JA	-0.8	-3.5	-1.9	0.9	269.70	-0.63	1.5	292.98	7JA	-0.3	-3.5	-2.2	0.8
261.79	-0.32	1.5	303.61	18JA	-1.1	-3.5	-1.9	0.8	277.31	-0.73	1.5	303.12	17JA	1.8	-3.5	-2.1	0.8
269.12	-0.45	1.5	313.73	28JA	-1.1	-3.4	-1.8	0.8	284.97	-0.83	1.5	313.23	27JA	2.1	-3.5	-2.1	0.7
276.49	-0.59	1.4	323.79	7FE	-0.4	-3.4	-1.8	0.7	292.67	-0.93	1.5	323.30	6FE	0.8	-3.4	-2.0	0.7
283.90	-0.73	1.4	333.80	17FE	1.3	-3.4	-1.7	0.7	300.40	-1.02	1.4	333.31	16FE	0.4	-3.4	-1.9	0.6
291.35	-0.88	1.3	343.76	27FE	3.2	-3.3	-1.7	0.6	308.15	-1.10	1.4	343.27	26FE	0.2	-3.4	-1.9	0.6
298.82	-1.02	1.2	353.66	8MR	1.6	-3.3	-1.6	0.6	315.90	-1.17	1.4	353.18	8MR	-0.1	-3.4	-1.8	0.5
306.32	-1.17	1.1	3.50	18MR	0.9	-3.3	-1.6	0.6	323.66	-1.24	1.3	3.02	18MR	-0.6	-3.3	-1.7	0.4
313.82	-1.31	1.1	13.29	28MR	0.5	-3.3	-1.6	0.7	331.40	-1.29	1.3	12.81	28MR	-1.4	-3.3	-1.7	0.4
321.31	-1.45	1.0	23.02	7AP	0.1	-3.3	-1.5	0.7	339.11	-1.33	1.3	22.54	7AP	-1.5	-3.3	-1.6	0.4
328.79	-1.58	0.9	32.70	17AP	-0.5	-3.3	-1.5	0.8	346.78	-1.36	1.3	32.22	17AP	-0.6	-3.3	-1.6	0.4
336.22	-1.70	0.8	42.34	27AP	-1.4	-3.3	-1.5	0.9	354.39	-1.37	1.2	41.86	27AP	0.4	-3.3	-1.5	0.5
343.60	-1.82	0.8	51.93	7MY	-1.6	-3.4	-1.5	0.9	1.93	-1.36	1.2	51.46	7MY	1.4	-3.4	-1.5	0.5
350.91	-1.92	0.7	61.50	17MY	-0.6	-3.5	-1.5	1.0	9.39	-1.35	1.2	61.03	17MY	2.5	-3.4	-1.4	0.6
358.10	-2.00	0.6	71.05	27MY	0.2	-3.5	-1.6	1.0	16.76	-1.31	1.1	70.58	27MY	3.4	-3.4	-1.4	0.7
5.17	-2.07	0.5	80.58	6JN	0.8	-3.5	-1.6	1.1	24.02	-1.26	1.1	80.11	6JN	2.0	-3.4	-1.4	0.7
12.08	-2.12	0.4	90.10	16JN	1.4	-3.5	-1.6	1.1	31.15	-1.20	1.1	89.64	16JN	0.8	-3.5	-1.3	0.8
18.80	-2.15	0.3	99.64	26JN	2.5	-3.4	-1.7	1.1	38.14	-1.12	1.0	99.17	26JN	-0.3	-3.5	-1.3	0.8
25.28	-2.16	0.2	109.18	6JL	3.0	-3.4	-1.7	1.1	44.98	-1.02	1.0	108.72	6JL	-1.3	-3.6	-1.3	0.9
31.48	-2.15	0.1	118.75	16JL	1.3	-3.4	-1.8	1.2	51.65	-0.91	0.9	118.28	16JL	-1.5	-3.7	-1.3	0.9
37.32	-2.11	0.0	128.34	26JL	-0.2	-3.4	-1.8	1.1	58.13	-0.78	0.9	127.88	26JL	-0.7	-3.7	-1.3	0.9
42.76	-2.05	-0.1	137.97	5AU	-1.2	-3.3	-1.9	1.1	64.39	-0.64	0.8	137.51	5AU	-0.2	-3.8	-1.4	0.9
47.67	-1.95	-0.3	147.66	15AU	-1.4	-3.3	-1.9	1.1	70.40	-0.47	0.8	147.19	15AU	0.1	-3.9	-1.4	0.9
51.95	-1.82	-0.4	157.39	25AU	-0.8	-3.3	-2.0	1.1	76.12	-0.29	0.7	156.92	25AU	0.3	-4.0	-1.4	0.9
55.45	-1.64	-0.6	167.17	4SE	-0.4	-3.3	-2.1	1.0	81.50	-0.09	0.6	166.70	4SE	0.6	-4.1	-1.4	0.9
57.97	-1.42	-0.8	177.01	14SE	-0.1	-3.3	-2.1	1.0	86.49	0.14	0.5	176.53	14SE	1.5	-4.2	-1.5	0.9
59.31	-1.13	-1.0	186.91	24SE	0.0	-3.4	-2.2	0.9	90.99	0.39	0.3	186.43	24SE	2.8	-4.3	-1.5	0.8
59.29	-0.77	-1.2	196.86	4OC	0.2	-3.4	-2.3	1.0	94.89	0.68	0.2	196.39	4OC	0.1	-4.3	-1.6	0.8
57.79	-0.34	-1.4	206.88	14OC	0.8	-3.4	-2.3	1.0	98.08	1.01	0.0	206.39	14OC	-0.7	-4.1	-1.7	0.8
54.97	0.14	-1.6	216.94	24OC	3.5	-3.4	-2.3	1.0	100.36	1.39	-0.2	216.45	24OC	-0.7	-3.6	-1.7	0.8
51.33	0.63	-1.7	227.04	3NO	0.4	-3.5	-2.4	1.0	101.55	1.80	-0.4	226.56	3NO	-0.7	-3.1	-1.8	0.8
47.62	1.09	-1.6	237.19	13NO	-0.4	-3.5	-2.4	1.0	101.47	2.26	-0.6	236.70	13NO	-0.8	-3.9	-1.8	0.8
44.64	1.45	-1.3	247.36	23NO	-0.4	-3.6	-2.3	0.9	99.98	2.73	-0.9	246.87	23NO	-0.7	-4.3	-1.9	0.8
42.87	1.71	-1.0	257.55	3DE	-0.5	-3.7	-2.3	0.9	97.19	3.17	-1.1	257.06	3DE	-0.7	-4.4	-2.0	0.8
42.46	1.88	-0.7	267.76	13DE	-0.6	-3.7	-2.3	0.9	93.47	3.52	-1.2	267.26	13DE	-0.6	-4.4	-2.0	0.8
43.31	1.98	-0.4	277.95	23DE	-0.9	-3.8	-2.2	0.9	89.51	3.73	-1.1	277.46	23DE	-0.1	-4.3	-2.1	0.8
				981									983				
45.24	2.03	-0.1	288.14	2JA	-1.0	-3.9	-2.1	0.8	86.08	3.78	-0.9	287.64	2JA	2.1	-4.2	-2.1	0.8
48.02	2.04	0.2	298.30	12JA	-1.0	-4.0	-2.1	0.8	83.74	3.70	-0.6	297.81	12JA	1.6	-4.1	-2.1	0.8
51.49	2.03	0.4	308.43	22JA	-0.4	-4.1	-2.0	0.7	82.71	3.53	-0.4	307.94	22JA	0.6	-4.0	-2.1	0.7
55.49	2.01	0.6	318.51	1FE	1.6	-4.2	-1.9	0.7	82.97	3.33	-0.1	318.03	1FE	0.3	-3.9	-2.1	0.7
59.90	1.97	0.8	328.56	11FE	2.6	-4.3	-1.9	0.6	84.34	3.11	0.2	328.07	11FE	0.1	-3.8	-2.0	0.6
64.65	1.93	1.0	338.54	21FE	1.2	-4.3	-1.8	0.6	86.63	2.89	0.4	338.06	21FE	-0.2	-3.7	-2.0	0.6
69.64	1.88	1.2	348.47	3MR	0.6	-4.2	-1.7	0.5	89.67	2.69	0.6	348.00	3MR	-0.6	-3.6	-2.0	0.5
74.84	1.83	1.3	358.35	13MR	0.4	-4.0	-1.7	0.5	93.29	2.49	0.8	357.87	13MR	-1.4	-3.5	-1.9	0.5
80.21	1.78	1.4	8.16	23MR	0.0	-3.4	-1.6	0.5	97.38	2.32	1.0	7.69	23MR	-1.5	-3.5	-1.8	0.4
85.71	1.72	1.5	17.92	2AP	-0.6	-3.2	-1.6	0.5	101.84	2.15	1.1	17.45	2AP	-0.6	-3.4	-1.8	0.4
91.32	1.66	1.6	27.63	12AP	-1.4	-3.8	-1.5	0.5	106.61	1.99	1.3	27.16	12AP	0.6	-3.4	-1.7	0.3
97.03	1.60	1.7	37.28	22AP	-1.6	-4.1	-1.5	0.6	111.63	1.84	1.4	36.82	22AP	1.8	-3.4	-1.6	0.3
102.82	1.54	1.8	46.90	2MY	-0.6	-4.2	-1.5	0.7	116.86	1.70	1.5	46.44	2MY	3.4	-3.3	-1.6	0.4
108.68	1.48	1.8	56.49	12MY	0.3	-4.2	-1.5	0.7	122.27	1.57	1.6	56.02	12MY	2.7	-3.3	-1.5	0.4
114.61	1.41	1.9	66.04	22MY	-1.4	-4.1	-1.4	0.8	127.83	1.44	1.6	65.58	22MY	1.5	-3.3	-1.4	0.5
120.60	1.35	1.9	75.58	1JN	1.9	-4.0	-1.4	0.9	133.53	1.32	1.7	75.12	1JN	0.6	-3.3	-1.4	0.6
126.65	1.28	1.9	85.11	11JN	3.2	-3.9	-1.4	0.9	139.35	1.20	1.7	84.65	11JN	-0.3	-3.3	-1.4	0.6
132.76	1.21	2.0	94.64	21JN	2.5	-3.8	-1.4	1.0	145.30	1.08	1.8	94.18	21JN	-1.3	-3.3	-1.3	0.7
138.93	1.14	2.0	104.17	1JL	1.0	-3.7	-1.5	1.0	151.35	0.96	1.8	103.71	1JL	-1.6	-3.4	-1.3	0.7
145.17	1.07	2.0	113.73	11JL	-0.3	-3.7	-1.5	1.0	157.50	0.85	1.8	113.26	11JL	-0.7	-3.4	-1.3	0.8
151.46	0.99	2.0	123.31	21JL	-1.3	-3.6	-1.5	1.0	163.77	0.74	1.8	122.84	21JL	-0.1	-3.5	-1.3	0.8
157.82	0.91	2.0	132.92	31JL	-1.5	-3.5	-1.6	1.0	170.13	0.63	1.8	132.45	31JL	0.2	-3.5	-1.2	0.8
164.24	0.84	2.0	142.58	10AU	-0.8	-3.5	-1.6	1.0	176.59	0.52	1.8	142.11	10AU	0.5	-3.5	-1.2	0.8
170.73	0.75	1.9	152.28	20AU	-0.3	-3.5	-1.6	1.0	183.15	0.41	1.8	151.81	20AU	0.9	-3.5	-1.2	0.9
177.30	0.67	1.9	162.04	30AU	0.0	-3.4	-1.7	1.0	189.81	0.30	1.8	161.56	30AU	1.9	-3.5	-1.3	0.9
183.94	0.58	1.9	171.85	9SE	0.2	-3.4	-1.8	0.9	196.58	0.19	1.8	171.37	9SE	2.5	-3.4	-1.3	0.8
190.65	0.49	1.8	181.72	19SE	0.4	-3.4	-1.8	0.9	203.44	0.08	1.8	181.24	19SE	0.0	-3.4	-1.3	0.8
197.44	0.40	1.8	191.64	29SE	1.1	-3.4	-1.9	0.9	210.40	-0.02	1.7	191.16	29SE	-0.8	-3.4	-1.3	0.8
204.32	0.31	1.7	201.63	9OC	3.1	-3.4	-1.9	0.8	217.47	-0.13	1.7	201.14	9OC	-0.8	-3.4	-1.4	0.8
211.27	0.21	1.7	211.66	19OC	0.2	-3.4	-2.0	0.8	224.63	-0.23	1.6	211.17	19OC	-0.9	-3.3	-1.4	0.7
218.30	0.11	1.7	221.74	29OC	-0.5	-3.4	-2.1	0.9	231.89	-0.33	1.6	221.25	29OC	-0.7	-3.3	-1.5	0.7
225.42	0.01	1.7	231.87	8NO	-0.6	-3.4	-2.1	0.9	239.24	-0.43	1.6	231.38	8NO	-0.5	-3.3	-1.5	0.7
232.61	-0.09	1.7	242.03	18NO	-0.6	-3.4	-2.2	0.9	246.68	-0.53	1.5	241.53	18NO	-0.5	-3.4	-1.6	0.7
239.89	-0.20	1.6	252.21	28NO	-0.7	-3.4	-2.2	0.9	254.20	-0.62	1.5	251.71	28NO	-0.5	-3.3	-1.7	0.8
247.24	-0.31	1.6	262.41	8DE	-0.8	-3.4	-2.2	0.9	261.80	-0.71	1.4	261.91	8DE	0.0	-3.4	-1.7	0.8
254.66	-0.41	1.6	272.61	18DE	-0.8	-3.4	-2.2	0.8	269.47	-0.79	1.4	272.11	18DE	2.3	-3.4	-1.8	0.8
262.15	-0.52	1.6	282.80	28DE	-0.8	-3.5	-2.2	0.8	277.19	-0.87	1.4	282.30	28DE	1.3	-3.4	-1.9	0.8

984

♂ LONG	LAT	MAG	☉ LONG	16.00UT	☿	♀	♃	♄
284.97	-0.93	1.4	292.48	7JA	0.3	-3.5	-1.9	0.8
292.79	-0.99	1.4	302.63	17JA	0.1	-3.5	-2.0	0.8
300.64	-1.04	1.4	312.74	27JA	-0.0	-3.5	-2.0	0.7
308.50	-1.08	1.4	322.81	6FE	-0.3	-3.6	-2.0	0.7
316.37	-1.11	1.4	332.83	16FE	-0.7	-3.7	-2.0	0.7
324.23	-1.12	1.4	342.79	26FE	-1.4	-3.7	-2.0	0.6
332.06	-1.12	1.4	352.70	7MR	-1.4	-3.8	-2.0	0.6
339.87	-1.12	1.4	2.55	17MR	-0.6	-3.9	-2.0	0.5
347.63	-1.09	1.4	12.34	27MR	0.7	-4.0	-1.9	0.4
355.33	-1.06	1.4	22.08	6AP	2.3	-4.0	-1.9	0.4
2.97	-1.01	1.4	31.76	16AP	3.5	-4.1	-1.8	0.3
10.53	-0.96	1.4	41.40	26AP	2.0	-4.2	-1.7	0.3
18.01	-0.89	1.4	51.01	6MY	1.1	-4.2	-1.7	0.3
25.39	-0.81	1.4	60.57	16MY	0.5	-4.0	-1.6	0.3
32.68	-0.72	1.5	70.12	26MY	-0.4	-3.6	-1.5	0.4
39.87	-0.62	1.5	79.65	5JN	-1.4	-2.8	-1.5	0.5
46.94	-0.52	1.5	89.18	15JN	-1.6	-3.4	-1.4	0.5
53.91	-0.40	1.5	98.71	25JN	-0.7	-3.9	-1.4	0.6
60.75	-0.28	1.4	108.25	5JL	-0.1	-4.2	-1.3	0.6
67.47	-0.15	1.4	117.81	15JL	0.4	-4.2	-1.3	0.7
74.05	-0.01	1.4	127.41	25JL	0.7	-4.2	-1.3	0.7
80.49	0.13	1.4	137.04	4AU	1.2	-4.1	-1.2	0.8
86.78	0.28	1.3	146.71	14AU	2.4	-4.0	-1.2	0.8
92.89	0.44	1.3	156.44	24AU	2.1	-3.9	-1.2	0.8
98.83	0.62	1.2	166.22	3SE	-0.0	-3.8	-1.2	0.8
104.54	0.80	1.2	176.05	13SE	-0.9	-3.8	-1.2	0.8
110.02	0.99	1.1	185.95	23SE	-1.0	-3.7	-1.2	0.8
115.20	1.21	1.0	195.90	3OC	-1.0	-3.6	-1.2	0.8
120.05	1.44	0.9	205.91	13OC	-0.6	-3.6	-1.3	0.8
124.50	1.69	0.8	215.96	23OC	-0.4	-3.5	-1.3	0.7
128.44	1.97	0.6	226.07	2NO	-0.4	-3.5	-1.3	0.7
131.77	2.29	0.4	236.21	12NO	-0.3	-3.5	-1.4	0.7
134.34	2.63	0.2	246.38	22NO	0.2	-3.4	-1.4	0.7
135.98	3.01	0.0	256.57	2DE	2.6	-3.4	-1.5	0.7
136.50	3.41	-0.2	266.77	12DE	1.0	-3.4	-1.5	0.8
135.74	3.81	-0.5	276.97	22DE	0.1	-3.4	-1.6	0.8

985

♂ LONG	LAT	MAG	☉ LONG	16.00UT	☿	♀	♃	♄
133.64	4.18	-0.7	287.16	1JA	-0.0	-3.3	-1.7	0.8
130.41	4.44	-0.9	297.32	11JA	-0.1	-3.3	-1.7	0.8
126.51	4.55	-1.0	307.45	21JA	-0.3	-3.3	-1.8	0.8
122.66	4.49	-0.7	317.54	31JA	-0.7	-3.3	-1.9	0.7
119.57	4.26	-0.7	327.59	10FE	-1.3	-3.4	-1.9	0.7
117.63	3.94	-0.4	337.58	20FE	-1.3	-3.4	-2.0	0.7
117.00	3.58	-0.2	347.52	2MR	-0.5	-3.4	-2.0	0.6
117.59	3.22	0.1	357.39	12MR	0.9	-3.4	-2.0	0.6
119.23	2.87	0.3	7.21	22MR	2.9	-3.5	-2.0	0.5
121.74	2.55	0.5	16.98	1AP	2.7	-3.5	-2.0	0.5
124.96	2.26	0.7	26.69	11AP	1.5	-3.5	-2.0	0.4
128.75	1.99	0.8	36.35	21AP	0.9	-3.4	-1.9	0.3
133.00	1.75	1.0	45.98	1MY	0.3	-3.4	-1.9	0.3
137.64	1.53	1.1	55.56	11MY	-0.4	-3.4	-1.8	0.2
142.59	1.33	1.2	65.12	21MY	-1.4	-3.4	-1.8	0.3
147.81	1.13	1.3	74.66	31MY	-1.6	-3.3	-1.7	0.3
153.27	0.96	1.4	84.18	10JN	-0.7	-3.3	-1.6	0.4
158.94	0.79	1.4	93.71	20JN	0.0	-3.3	-1.6	0.5
164.80	0.63	1.5	103.25	30JN	0.5	-3.3	-1.5	0.5
170.82	0.48	1.5	112.79	10JL	0.9	-3.3	-1.5	0.6
177.01	0.33	1.5	122.37	20JL	1.6	-3.4	-1.4	0.6
183.35	0.19	1.6	131.98	30JL	2.9	-3.4	-1.4	0.7
189.83	0.06	1.6	141.63	9AU	1.8	-3.4	-1.3	0.7
196.45	-0.07	1.6	151.33	19AU	-0.1	-3.5	-1.3	0.7
203.20	-0.19	1.6	161.08	29AU	-1.0	-3.5	-1.3	0.8
210.08	-0.30	1.6	170.89	8SE	-1.2	-3.6	-1.2	0.8
217.09	-0.41	1.6	180.75	18SE	-0.9	-3.6	-1.2	0.8
224.22	-0.52	1.6	190.67	28SE	-0.5	-3.7	-1.2	0.8
231.46	-0.61	1.5	200.65	8OC	-0.3	-3.8	-1.2	0.8
238.81	-0.70	1.5	210.68	18OC	-0.2	-3.9	-1.2	0.8
246.27	-0.78	1.5	220.76	28OC	-0.1	-4.0	-1.2	0.7
253.81	-0.86	1.5	230.88	7NO	-0.0	-4.1	-1.3	0.7
261.45	-0.92	1.5	241.04	17NO	2.9	-4.2	-1.3	0.7
269.16	-0.98	1.4	251.22	27NO	0.7	-4.3	-1.3	0.7
276.93	-1.02	1.4	261.42	7DE	-0.1	-4.4	-1.4	0.7
284.76	-1.06	1.4	271.62	17DE	-0.2	-4.4	-1.4	0.7
292.63	-1.08	1.4	281.82	27DE	-0.3	-4.2	-1.5	0.8

986

♂ LONG	LAT	MAG	☉ LONG	16.00UT	☿	♀	♃	♄
300.52	-1.09	1.4	291.99	6JA	-0.4	-3.6	-1.5	0.8
308.43	-1.09	1.3	302.14	16JA	-0.8	-3.3	-1.6	0.8
316.33	-1.08	1.3	312.26	26JA	-1.2	-3.9	-1.7	0.8
324.22	-1.05	1.3	322.32	5FE	-1.2	-4.2	-1.7	0.8
332.08	-1.02	1.3	332.35	15FE	-0.5	-4.3	-1.8	0.7
339.90	-0.97	1.3	342.31	25FE	1.1	-4.3	-1.9	0.7
347.68	-0.91	1.4	352.22	7MR	3.3	-4.2	-1.9	0.7
355.39	-0.85	1.4	2.07	17MR	2.0	-4.1	-2.0	0.6
3.03	-0.77	1.4	11.87	27MR	1.1	-4.0	-2.0	0.6
10.60	-0.69	1.5	21.60	6AP	0.6	-3.9	-2.0	0.5
18.09	-0.60	1.5	31.30	16AP	0.2	-3.8	-2.1	0.5
25.49	-0.51	1.5	40.94	26AP	-0.5	-3.7	-2.1	0.4
32.80	-0.41	1.6	50.54	6MY	-1.4	-3.6	-2.0	0.3
40.02	-0.31	1.6	60.11	16MY	-1.6	-3.5	-2.0	0.3
47.14	-0.20	1.6	69.66	26MY	-0.7	-3.5	-2.0	0.2
54.17	-0.09	1.7	79.19	5JN	0.1	-3.4	-1.9	0.3
61.11	0.02	1.7	88.72	15JN	0.7	-3.4	-1.9	0.3
67.95	0.14	1.7	98.25	25JN	1.2	-3.4	-1.8	0.4
74.70	0.25	1.7	107.79	5JL	2.1	-3.4	-1.7	0.5
81.35	0.37	1.7	117.36	15JL	3.2	-3.3	-1.7	0.5
87.91	0.49	1.7	126.95	25JL	1.5	-3.3	-1.6	0.6
94.37	0.61	1.7	136.57	4AU	-0.2	-3.3	-1.6	0.6
100.72	0.73	1.7	146.25	14AU	-1.1	-3.3	-1.5	0.7
106.97	0.86	1.7	155.97	24AU	-1.3	-3.4	-1.5	0.7
113.11	0.99	1.7	165.74	3SE	-0.9	-3.4	-1.4	0.7
119.13	1.12	1.6	175.58	13SE	-0.4	-3.4	-1.4	0.8
125.01	1.25	1.6	185.47	23SE	-0.2	-3.4	-1.3	0.8
130.74	1.40	1.5	195.41	3OC	-0.1	-3.4	-1.3	0.8
136.29	1.55	1.4	205.42	13OC	0.1	-3.5	-1.3	0.8
141.64	1.71	1.3	215.47	23OC	0.6	-3.5	-1.3	0.8
146.76	1.87	1.2	225.57	2NO	3.3	-3.5	-1.3	0.7
151.59	2.06	1.1	235.71	12NO	0.5	-3.5	-1.3	0.7
156.08	2.25	1.0	245.88	22NO	-0.3	-3.4	-1.3	0.7
160.14	2.46	0.8	256.07	2DE	-0.3	-3.4	-1.3	0.7
163.68	2.70	0.6	266.27	12DE	-0.4	-3.4	-1.3	0.7
166.57	2.95	0.4	276.47	22DE	-0.5	-3.4	-1.4	0.7

987

♂ LONG	LAT	MAG	☉ LONG	16.00UT	☿	♀	♃	♄
168.66	3.21	0.2	286.66	1JA	-0.8	-3.3	-1.4	0.8
169.76	3.49	-0.1	296.83	11JA	-1.1	-3.3	-1.5	0.8
169.71	3.75	-0.3	306.96	21JA	-1.0	-3.3	-1.5	0.8
168.37	3.98	-0.6	317.05	31JA	-0.4	-3.3	-1.6	0.8
165.78	4.12	-0.8	327.10	10FE	1.4	-3.4	-1.6	0.8
162.26	4.13	-1.0	337.09	20FE	3.0	-3.4	-1.7	0.8
158.37	3.97	-1.0	347.03	2MR	1.4	-3.4	-1.8	0.8
154.84	3.66	-0.9	356.92	12MR	0.8	-3.4	-1.8	0.7
152.27	3.25	-0.7	6.74	22MR	0.5	-3.4	-1.9	0.7
150.94	2.81	-0.5	16.51	1AP	0.1	-3.5	-2.0	0.6
150.92	2.37	-0.2	26.22	11AP	-0.5	-3.5	-2.0	0.6
152.08	1.96	-0.0	35.89	21AP	-1.4	-3.6	-2.1	0.5
154.23	1.59	0.2	45.51	1MY	-1.6	-3.6	-2.1	0.5
157.22	1.25	0.3	55.10	11MY	-0.7	-3.7	-2.1	0.4
160.89	0.95	0.5	64.66	21MY	0.2	-3.8	-2.2	0.3
165.12	0.69	0.6	74.20	31MY	0.9	-3.9	-2.2	0.3
169.81	0.45	0.7	83.73	10JN	1.6	-4.0	-2.1	0.2
174.90	0.23	0.8	93.25	20JN	2.7	-4.1	-2.1	0.3
180.32	0.03	0.9	102.79	30JN	2.9	-4.1	-2.1	0.4
186.04	-0.15	1.0	112.34	10JL	1.2	-4.2	-2.0	0.4
192.01	-0.31	1.1	121.91	20JL	-0.2	-4.2	-2.0	0.5
198.21	-0.46	1.1	131.52	30JL	-1.2	-4.0	-1.9	0.5
204.62	-0.60	1.2	141.17	9AU	-1.5	-3.5	-1.8	0.6
211.23	-0.72	1.2	150.86	19AU	-0.8	-3.2	-1.8	0.6
218.02	-0.83	1.2	160.61	29AU	-0.3	-3.8	-1.7	0.7
224.97	-0.93	1.2	170.41	8SE	-0.1	-4.2	-1.6	0.7
232.07	-1.01	1.3	180.27	18SE	0.1	-4.3	-1.6	0.8
239.31	-1.08	1.3	190.19	28SE	0.2	-4.3	-1.5	0.8
246.68	-1.14	1.3	200.16	8OC	0.9	-4.2	-1.5	0.8
254.16	-1.19	1.3	210.19	18OC	3.5	-4.1	-1.5	0.8
261.74	-1.22	1.3	220.27	28OC	0.4	-4.1	-1.4	0.8
269.42	-1.24	1.3	230.39	7NO	-0.4	-4.0	-1.4	0.8
277.16	-1.24	1.3	240.54	17NO	-0.5	-3.9	-1.4	0.7
284.96	-1.23	1.3	250.73	27NO	-0.5	-3.8	-1.4	0.7
292.80	-1.21	1.4	260.92	7DE	-0.6	-3.7	-1.4	0.7
300.67	-1.18	1.4	271.12	17DE	-0.8	-3.6	-1.4	0.7
308.55	-1.13	1.4	281.32	27DE	-0.9	-3.6	-1.4	0.7

♂ LONG	LAT	MAG	☉ LONG	16.00UT 988	☿	♀	♃	♄
316.43	-1.08	1.4	291.50	6JA	-0.9	-3.5	-1.4	0.8
324.29	-1.01	1.4	301.65	16JA	-0.3	-3.5	-1.4	0.8
332.11	-0.94	1.4	311.77	26JA	1.6	-3.4	-1.5	0.8
339.89	-0.85	1.4	321.84	5FE	2.4	-3.4	-1.5	0.8
347.62	-0.77	1.5	331.87	15FE	1.0	-3.4	-1.6	0.8
355.28	-0.67	1.5	341.84	25FE	0.6	-3.3	-1.6	0.8
2.87	-0.57	1.5	351.75	6MR	0.3	-3.3	-1.7	0.8
10.39	-0.47	1.5	1.60	16MR	-0.0	-3.3	-1.8	0.8
17.82	-0.37	1.5	11.40	26MR	-0.5	-3.3	-1.8	0.7
25.17	-0.26	1.5	21.14	5AP	-1.4	-3.3	-1.9	0.7
32.44	-0.16	1.5	30.83	15AP	-1.6	-3.3	-2.0	0.6
39.62	-0.05	1.6	40.48	25AP	-0.7	-3.3	-2.0	0.6
46.71	0.05	1.6	50.08	5MY	0.3	-3.4	-2.1	0.5
53.72	0.16	1.7	59.65	15MY	1.2	-3.4	-2.2	0.5
60.64	0.26	1.7	69.20	25MY	2.1	-3.5	-2.2	0.4
67.49	0.36	1.8	78.73	4JN	3.4	-3.5	-2.3	0.3
74.26	0.45	1.8	88.26	14JN	2.3	-3.5	-2.3	0.3
80.96	0.55	1.8	97.79	24JN	1.0	-3.4	-2.3	0.3
87.59	0.64	1.9	107.33	4JL	-0.2	-3.4	-2.3	0.3
94.15	0.74	1.9	116.89	14JL	-1.3	-3.4	-2.3	0.4
100.64	0.83	1.9	126.48	24JL	-1.5	-3.4	-2.2	0.5
107.07	0.91	1.9	136.10	3AU	-0.8	-3.3	-2.2	0.5
113.43	1.00	1.9	145.77	13AU	-0.3	-3.3	-2.1	0.6
119.74	1.08	1.9	155.50	23AU	0.0	-3.3	-2.1	0.6
125.97	1.17	1.9	165.26	2SE	0.2	-3.3	-2.0	0.7
132.15	1.25	1.9	175.09	12SE	0.5	-3.3	-1.9	0.7
138.25	1.33	1.8	184.98	22SE	1.2	-3.4	-1.9	0.7
144.27	1.41	1.8	194.92	2OC	3.1	-3.4	-1.8	0.8
150.21	1.50	1.8	204.93	12OC	0.2	-3.4	-1.7	0.8
156.06	1.58	1.7	214.98	22OC	-0.6	-3.4	-1.7	0.8
161.79	1.66	1.6	225.07	1NO	-0.7	-3.5	-1.6	0.8
167.40	1.75	1.5	235.21	11NO	-0.6	-3.5	-1.6	0.8
172.86	1.84	1.4	245.38	21NO	-0.7	-3.6	-1.6	0.8
178.14	1.93	1.3	255.57	1DE	-0.7	-3.7	-1.5	0.8
183.20	2.02	1.2	265.77	11DE	-0.7	-3.7	-1.5	0.7
188.00	2.11	1.0	275.97	21DE	-0.7	-3.8	-1.5	0.7
192.48	2.21	0.9	286.16	31DE	-0.2	-3.9	-1.5	0.7

♂ LONG	LAT	MAG	☉ LONG	16.00UT 989	☿	♀	♃	♄
196.55	2.30	0.7	296.33	10JA	1.8	-4.0	-1.5	0.8
200.13	2.39	0.4	306.47	20JA	1.9	-4.1	-1.5	0.8
203.08	2.48	0.2	316.56	30JA	0.7	-4.2	-1.5	0.8
205.26	2.55	-0.1	326.62	9FE	0.4	-4.3	-1.5	0.9
206.49	2.60	-0.3	336.61	19FE	0.2	-4.3	-1.6	0.9
206.60	2.60	-0.6	346.55	1MR	-0.1	-4.2	-1.6	0.9
205.44	2.54	-0.9	356.44	11MR	-0.6	-4.0	-1.6	0.9
203.05	2.37	-1.2	6.27	21MR	-1.4	-3.4	-1.7	0.8
199.72	2.10	-1.4	16.04	31MR	-1.5	-3.2	-1.7	0.8
196.02	1.71	-1.4	25.76	10AP	-0.6	-3.8	-1.8	0.8
192.67	1.26	-1.3	35.42	20AP	0.5	-4.1	-1.9	0.7
190.30	0.79	-1.1	45.05	30AP	1.5	-4.2	-1.9	0.7
189.20	0.35	-0.9	54.64	10MY	2.8	-4.2	-2.0	0.6
189.45	-0.05	-0.7	64.20	20MY	3.2	-4.1	-2.1	0.6
190.93	-0.39	-0.5	73.74	30MY	1.8	-4.0	-2.1	0.5
193.46	-0.68	-0.3	83.27	9JN	0.8	-3.9	-2.2	0.4
196.88	-0.92	-0.1	92.79	19JN	-0.3	-3.8	-2.3	0.4
201.01	-1.13	0.0	102.32	29JN	-1.3	-3.7	-2.3	0.3
205.75	-1.29	0.1	111.88	9JL	-1.6	-3.7	-2.4	0.3
210.99	-1.43	0.3	121.45	19JL	-0.8	-3.6	-2.4	0.4
216.65	-1.54	0.4	131.06	29JL	-0.2	-3.5	-2.4	0.4
222.66	-1.63	0.5	140.70	8AU	0.2	-3.5	-2.4	0.5
228.94	-1.69	0.5	150.40	18AU	0.4	-3.5	-2.4	0.6
235.58	-1.73	0.6	160.14	28AU	0.7	-3.4	-2.3	0.6
242.39	-1.76	0.7	169.94	7SE	1.6	-3.4	-2.3	0.7
249.41	-1.76	0.8	179.80	17SE	2.8	-3.4	-2.2	0.7
256.60	-1.74	0.8	189.71	27SE	0.1	-3.4	-2.2	0.7
263.93	-1.71	0.9	199.68	7OC	-0.7	-3.4	-2.1	0.8
271.39	-1.66	0.9	209.71	17OC	-0.8	-3.4	-2.0	0.8
278.94	-1.60	1.0	219.78	27OC	-0.8	-3.4	-2.0	0.8
286.57	-1.52	1.0	229.90	6NO	-0.7	-3.4	-1.9	0.8
294.25	-1.44	1.1	240.05	16NO	-0.6	-3.4	-1.8	0.8
301.96	-1.34	1.1	250.23	26NO	-0.6	-3.4	-1.8	0.8
309.70	-1.23	1.2	260.43	6DE	-0.6	-3.4	-1.7	0.8
317.43	-1.11	1.3	270.62	16DE	-0.1	-3.4	-1.7	0.8
325.15	-1.00	1.3	280.82	26DE	2.1	-3.5	-1.6	0.8

♂ LONG	LAT	MAG	☉ LONG	16.00UT 990	☿	♀	♃	♄
332.83	-0.87	1.4	291.00	5JA	1.5	-3.5	-1.6	0.8
340.48	-0.75	1.4	301.15	15JA	0.5	-3.5	-1.6	0.9
348.06	-0.62	1.5	311.27	25JA	0.2	-3.5	-1.6	0.9
355.59	-0.50	1.5	321.35	4FE	0.1	-3.4	-1.6	0.9
3.05	-0.37	1.5	331.37	14FE	-0.2	-3.4	-1.6	0.9
10.44	-0.25	1.6	341.35	24FE	-0.6	-3.4	-1.6	0.9
17.75	-0.14	1.6	351.26	6MR	-1.3	-3.4	-1.6	0.9
24.98	-0.02	1.7	1.12	16MR	-1.5	-3.3	-1.6	0.9
32.13	0.09	1.7	10.92	26MR	-0.6	-3.3	-1.6	0.9
39.21	0.19	1.7	20.67	5AP	0.6	-3.3	-1.7	0.9
46.20	0.29	1.8	30.36	15AP	1.9	-3.3	-1.7	0.9
53.13	0.39	1.8	40.01	25AP	3.7	-3.3	-1.7	0.8
59.98	0.48	1.8	49.62	5MY	2.4	-3.4	-1.8	0.8
66.76	0.57	1.8	59.19	15MY	1.4	-3.4	-1.9	0.7
73.48	0.65	1.8	68.74	25MY	0.6	-3.4	-1.9	0.7
80.14	0.72	1.8	78.27	4JN	-0.3	-3.4	-2.0	0.6
86.74	0.79	1.8	87.80	14JN	-1.3	-3.5	-2.1	0.5
93.30	0.86	1.9	97.33	24JN	-1.6	-3.5	-2.1	0.5
99.81	0.92	1.9	106.87	4JL	-0.7	-3.6	-2.2	0.4
106.27	0.98	1.9	116.43	14JL	-0.1	-3.7	-2.3	0.4
112.70	1.04	2.0	126.02	24JL	0.3	-3.7	-2.3	0.4
119.10	1.09	2.0	135.64	3AU	0.6	-3.8	-2.4	0.5
125.46	1.14	2.0	145.31	13AU	1.0	-3.9	-2.4	0.5
131.79	1.18	2.0	155.03	23AU	2.1	-4.0	-2.4	0.6
138.10	1.22	2.0	164.80	2SE	2.4	-4.1	-2.5	0.6
144.38	1.26	2.0	174.62	12SE	0.1	-4.2	-2.5	0.7
150.64	1.29	2.0	184.51	22SE	-0.9	-4.3	-2.4	0.7
156.87	1.32	2.0	194.45	2OC	-1.0	-4.3	-2.4	0.8
163.07	1.35	1.9	204.44	12OC	-0.9	-4.1	-2.4	0.8
169.24	1.37	1.9	214.49	22OC	-0.6	-3.6	-2.3	0.8
175.38	1.39	1.8	224.59	1NO	-0.5	-3.1	-2.2	0.9
181.47	1.40	1.8	234.72	11NO	-0.4	-3.9	-2.2	0.9
187.52	1.41	1.7	244.89	21NO	-0.4	-4.3	-2.1	0.9
193.51	1.42	1.6	255.08	1DE	0.1	-4.4	-2.0	0.9
199.44	1.41	1.5	265.27	11DE	2.3	-4.4	-2.0	0.9
205.28	1.40	1.4	275.48	21DE	1.2	-4.3	-1.9	0.9
211.02	1.38	1.3	285.67	31DE	0.2	-4.2	-1.8	0.9

♂ LONG	LAT	MAG	☉ LONG	16.00UT 991	☿	♀	♃	♄
216.64	1.35	1.2	295.83	10JA	0.0	-4.1	-1.8	0.9
222.12	1.31	1.0	305.97	20JA	-0.1	-4.0	-1.7	0.9
227.42	1.24	0.8	316.07	30JA	-0.3	-3.9	-1.7	0.9
232.50	1.16	0.6	326.12	9FE	-0.7	-3.8	-1.6	1.0
237.30	1.05	0.4	336.13	19FE	-1.3	-3.7	-1.6	1.0
241.75	0.91	0.2	346.07	1MR	-1.4	-3.6	-1.6	1.0
245.78	0.73	-0.0	355.96	11MR	-0.6	-3.5	-1.6	1.0
249.26	0.49	-0.3	5.79	21MR	0.8	-3.5	-1.6	1.0
252.04	0.20	-0.6	15.57	31MR	2.5	-3.4	-1.6	1.0
253.97	-0.17	-0.9	25.29	10AP	3.2	-3.4	-1.6	1.0
254.86	-0.63	-1.2	34.96	20AP	1.8	-3.4	-1.6	1.0
254.54	-1.18	-1.5	44.59	30AP	1.0	-3.3	-1.6	1.0
253.00	-1.80	-1.8	54.18	10MY	0.4	-3.3	-1.7	0.9
250.40	-2.44	-2.1	63.74	20MY	-0.4	-3.3	-1.7	0.9
247.27	-3.04	-2.2	73.28	30MY	-1.3	-3.3	-1.7	0.8
244.31	-3.51	-2.1	82.81	9JN	-1.6	-3.3	-1.8	0.8
242.20	-3.82	-1.9	92.34	19JN	-0.7	-3.3	-1.8	0.7
241.38	-3.97	-1.7	101.87	29JN	-0.0	-3.4	-1.9	0.6
241.94	-4.00	-1.4	111.42	9JL	0.4	-3.4	-2.0	0.5
243.80	-3.95	-1.2	120.99	19JL	0.8	-3.4	-2.0	0.5
246.79	-3.83	-1.0	130.59	29JL	1.3	-3.5	-2.1	0.5
250.71	-3.67	-0.8	140.23	8AU	2.6	-3.5	-2.2	0.5
255.38	-3.48	-0.6	149.93	18AU	2.1	-3.5	-2.2	0.6
260.66	-3.28	-0.5	159.67	28AU	-0.0	-3.5	-2.3	0.6
266.42	-3.06	-0.3	169.46	7SE	-1.0	-3.4	-2.3	0.7
272.57	-2.83	-0.1	179.32	17SE	-1.1	-3.4	-2.4	0.7
279.03	-2.59	0.0	189.22	27SE	-0.9	-3.4	-2.4	0.8
285.72	-2.36	0.2	199.19	7OC	-0.6	-3.4	-2.4	0.8
292.61	-2.12	0.3	209.22	17OC	-0.4	-3.3	-2.4	0.8
299.63	-1.88	0.4	219.29	27OC	-0.3	-3.3	-2.4	0.9
306.75	-1.65	0.5	229.40	6NO	-0.2	-3.3	-2.3	0.9
313.95	-1.43	0.7	239.56	16NO	0.3	-3.3	-2.3	1.0
321.19	-1.21	0.8	249.73	26NO	2.6	-3.3	-2.3	1.0
328.45	-1.00	0.9	259.93	6DE	0.9	-3.4	-2.2	1.0
335.71	-0.81	1.0	270.13	16DE	0.0	-3.4	-2.1	1.0
342.96	-0.62	1.1	280.32	26DE	-0.1	-3.4	-2.1	1.0

♂ LONG	♂ LAT	♂ MAG	⊙ LONG	16.00UT 992	☿	♀	♃	♄
					MAGNITUDES			
350.18	-0.44	1.2	290.50	5JA	-0.2	-3.5	-2.0	1.0
357.36	-0.28	1.3	300.66	15JA	-0.4	-3.5	-1.9	1.0
4.50	-0.13	1.4	310.78	25JA	-0.7	-3.5	-1.9	1.0
11.59	0.01	1.4	320.86	4FE	-1.3	-3.6	-1.8	1.0
18.63	0.15	1.5	330.89	14FE	-1.3	-3.7	-1.7	1.1
25.60	0.27	1.6	340.86	24FE	-0.6	-3.7	-1.7	1.1
32.52	0.38	1.7	350.78	5MR	1.0	-3.8	-1.6	1.1
39.38	0.48	1.7	0.65	15MR	3.0	-3.9	-1.6	1.1
46.17	0.57	1.8	10.45	25MR	2.4	-4.0	-1.6	1.2
52.92	0.65	1.8	20.20	4AP	1.3	-4.1	-1.5	1.2
59.61	0.73	1.9	29.89	14AP	0.8	-4.1	-1.5	1.2
66.24	0.80	1.9	39.54	24AP	0.3	-4.2	-1.5	1.1
72.83	0.86	1.9	49.15	4MY	-0.4	-4.2	-1.5	1.1
79.38	0.92	1.9	58.73	14MY	-1.3	-4.0	-1.5	1.1
85.89	0.96	2.0	68.28	24MY	-1.6	-3.6	-1.5	1.0
92.37	1.01	2.0	77.81	3JN	-0.7	-2.8	-1.6	1.0
98.81	1.04	2.0	87.34	13JN	0.0	-3.4	-1.6	0.9
105.23	1.08	2.0	96.87	23JN	0.6	-4.0	-1.6	0.9
111.63	1.10	2.0	106.41	3JL	1.0	-4.2	-1.7	0.8
118.02	1.12	1.9	115.97	13JL	1.7	-4.2	-1.7	0.7
124.39	1.14	1.9	125.55	23JL	3.1	-4.2	-1.7	0.7
130.76	1.15	2.0	135.17	2AU	1.8	-4.1	-1.8	0.6
137.13	1.16	2.0	144.84	12AU	-0.1	-4.0	-1.9	0.6
143.49	1.17	2.0	154.55	22AU	-1.1	-3.9	-1.9	0.6
149.86	1.17	2.0	164.32	1SE	-1.3	-3.8	-2.0	0.7
156.24	1.16	2.0	174.14	11SE	-0.9	-3.8	-2.1	0.7
162.63	1.15	2.0	184.02	21SE	-0.5	-3.7	-2.1	0.8
169.03	1.13	2.0	193.96	1OC	-0.2	-3.6	-2.2	0.8
175.44	1.11	2.0	203.96	11OC	-0.1	-3.6	-2.2	0.9
181.87	1.09	2.0	214.00	21OC	-0.1	-3.5	-2.3	0.9
188.31	1.06	1.9	224.10	31OC	0.5	-3.5	-2.3	1.0
194.77	1.02	1.9	234.23	10NO	3.0	-3.5	-2.3	1.0
201.23	0.97	1.8	244.40	20NO	0.7	-3.4	-2.3	1.0
207.71	0.92	1.8	254.59	30NO	-0.1	-3.4	-2.3	1.1
214.20	0.85	1.7	264.78	10DE	-0.3	-3.4	-2.3	1.1
220.70	0.78	1.6	274.98	20DE	-0.3	-3.4	-2.2	1.1
227.20	0.70	1.5	285.18	30DE	-0.5	-3.3	-2.2	1.1
				993				
233.69	0.60	1.4	295.34	9JA	-0.8	-3.3	-2.1	1.1
240.19	0.49	1.3	305.48	19JA	-1.1	-3.3	-2.0	1.1
246.67	0.36	1.2	315.59	29JA	-1.1	-3.3	-2.0	1.1
253.14	0.21	1.1	325.64	8FE	-0.5	-3.4	-1.9	1.1
259.58	0.05	1.0	335.64	18FE	1.2	-3.4	-1.8	1.2
265.99	-0.14	0.8	345.59	28FE	3.3	-3.4	-1.8	1.2
272.34	-0.36	0.7	355.48	10MR	1.8	-3.4	-1.7	1.3
278.63	-0.60	0.5	5.31	20MR	1.0	-3.5	-1.6	1.3
284.83	-0.88	0.3	15.09	30MR	0.6	-3.5	-1.6	1.3
290.90	-1.19	0.2	24.81	9AP	0.2	-3.5	-1.5	1.3
296.82	-1.53	-0.0	34.48	19AP	-0.5	-3.4	-1.5	1.3
302.51	-1.92	-0.2	44.12	29AP	-1.3	-3.4	-1.5	1.3
307.93	-2.36	-0.5	53.71	9MY	-1.6	-3.4	-1.5	1.3
312.98	-2.84	-0.7	63.27	19MY	-0.7	-3.4	-1.4	1.3
317.55	-3.37	-0.9	72.81	29MY	0.1	-3.3	-1.4	1.2
321.48	-3.95	-1.2	82.34	8JN	0.7	-3.3	-1.4	1.2
324.62	-4.58	-1.5	91.87	18JN	1.3	-3.3	-1.4	1.1
326.74	-5.22	-1.7	101.40	28JN	2.3	-3.3	-1.4	1.1
327.67	-5.84	-2.0	110.95	8JL	3.2	-3.3	-1.5	1.0
327.28	-6.37	-2.3	120.52	18JL	1.4	-3.4	-1.5	0.9
325.64	-6.71	-2.5	130.13	28JL	-0.1	-3.4	-1.5	0.9
323.17	-6.77	-2.6	139.77	7AU	-1.2	-3.4	-1.5	0.8
320.54	-6.50	-2.6	149.45	17AU	-1.4	-3.5	-1.6	0.7
318.48	-5.96	-2.3	159.19	27AU	-0.9	-3.5	-1.6	0.8
317.54	-5.25	-2.1	168.98	6SE	-0.4	-3.6	-1.7	0.8
317.89	-4.48	-1.7	178.84	16SE	-0.1	-3.6	-1.8	0.8
319.51	-3.73	-1.4	188.75	26SE	-0.0	-3.7	-1.8	0.9
322.21	-3.04	-1.1	198.71	6OC	0.1	-3.8	-1.9	0.9
325.78	-2.43	-0.9	208.73	16OC	0.7	-3.9	-1.9	1.0
330.06	-1.89	-0.6	218.80	26OC	3.3	-4.0	-2.0	1.0
334.87	-1.43	-0.3	228.91	5NO	0.5	-4.1	-2.1	1.1
340.08	-1.03	-0.1	239.06	15NO	-0.3	-4.2	-2.1	1.1
345.62	-0.68	0.1	249.24	25NO	-0.4	-4.3	-2.2	1.2
351.39	-0.39	0.3	259.43	5DE	-0.4	-4.4	-2.2	1.2
357.34	-0.13	0.5	269.63	15DE	-0.6	-4.4	-2.2	1.2
3.43	0.09	0.7	279.83	25DE	-0.8	-4.1	-2.2	1.3

♂ LONG	♂ LAT	♂ MAG	⊙ LONG	16.00UT 994	☿	♀	♃	♄
					MAGNITUDES			
9.62	0.28	0.8	290.01	4JA	-1.0	-3.6	-2.2	1.3
15.87	0.44	1.0	300.17	14JA	-1.0	-3.3	-2.1	1.3
22.17	0.58	1.1	310.29	24JA	-0.4	-3.9	-2.1	1.3
28.51	0.69	1.2	320.37	3FE	1.4	-4.2	-2.0	1.3
34.86	0.80	1.4	330.40	13FE	2.8	-4.3	-2.0	1.3
41.22	0.88	1.5	340.38	23FE	1.3	-4.3	-1.9	1.3
47.57	0.96	1.6	350.30	5MR	0.7	-4.2	-1.8	1.4
53.93	1.02	1.6	0.17	15MR	0.4	-4.1	-1.8	1.4
60.27	1.07	1.7	9.97	25MR	0.1	-4.0	-1.7	1.4
66.61	1.11	1.8	19.72	4AP	-0.5	-3.9	-1.6	1.4
72.93	1.14	1.8	29.42	14AP	-1.3	-3.8	-1.6	1.4
79.24	1.17	1.9	39.07	24AP	-1.6	-3.7	-1.5	1.3
85.54	1.19	1.9	48.68	4MY	-0.7	-3.6	-1.5	1.3
91.84	1.20	2.0	58.26	14MY	0.2	-3.5	-1.4	1.3
98.13	1.21	2.0	67.82	24MY	1.0	-3.5	-1.4	1.2
104.42	1.21	2.0	77.35	3JN	1.8	-3.4	-1.4	1.2
110.71	1.21	2.0	86.88	13JN	3.0	-3.4	-1.3	1.2
117.00	1.20	2.0	96.41	23JN	2.7	-3.4	-1.3	1.1
123.31	1.18	2.0	105.95	3JL	1.2	-3.3	-1.3	1.1
129.63	1.17	2.0	115.51	13JL	-0.2	-3.3	-1.3	1.0
135.96	1.15	2.0	125.09	23JL	-1.2	-3.3	-1.3	1.0
142.32	1.12	2.0	134.71	2AU	-1.5	-3.3	-1.3	0.9
148.70	1.09	2.0	144.38	12AU	-0.8	-3.3	-1.4	0.9
155.12	1.06	1.9	154.09	22AU	-0.3	-3.4	-1.4	0.8
161.56	1.02	1.9	163.85	1SE	-0.0	-3.4	-1.4	0.8
168.04	0.97	1.9	173.67	11SE	0.1	-3.4	-1.4	0.9
174.56	0.93	1.9	183.55	21SE	0.3	-3.4	-1.5	0.9
181.13	0.87	1.9	193.48	1OC	1.0	-3.4	-1.5	1.0
187.73	0.81	1.9	203.47	11OC	3.4	-3.5	-1.6	1.1
194.39	0.75	1.9	213.52	21OC	0.4	-3.5	-1.7	1.1
201.09	0.68	1.9	223.61	31OC	-0.5	-3.5	-1.7	1.2
207.85	0.60	1.9	233.74	10NO	-0.6	-3.5	-1.8	1.2
214.65	0.52	1.8	243.90	20NO	-0.6	-3.4	-1.9	1.3
221.50	0.43	1.8	254.09	30NO	-0.7	-3.4	-1.9	1.3
228.41	0.33	1.7	264.29	10DE	-0.8	-3.4	-2.0	1.3
235.36	0.23	1.7	274.48	20DE	-0.8	-3.4	-2.0	1.3
242.37	0.12	1.6	284.67	30DE	-0.8	-3.3	-2.1	1.3
				995				
249.42	-0.01	1.6	294.85	9JA	-0.3	-3.3	-2.1	1.3
256.52	-0.14	1.5	304.99	19JA	1.6	-3.3	-2.1	1.3
263.67	-0.27	1.4	315.09	29JA	2.3	-3.3	-2.1	1.3
270.86	-0.42	1.4	325.15	8FE	0.9	-3.4	-2.1	1.2
278.08	-0.58	1.3	335.16	18FE	0.5	-3.4	-2.0	1.2
285.34	-0.74	1.2	345.11	28FE	0.3	-3.4	-2.0	1.1
292.62	-0.91	1.1	355.00	10MR	-0.0	-3.4	-1.9	1.1
299.92	-1.09	1.0	4.84	20MR	-0.5	-3.4	-1.9	1.1
307.23	-1.27	0.9	14.62	30MR	-1.3	-3.5	-1.8	1.1
314.53	-1.45	0.8	24.35	9AP	-1.6	-3.5	-1.7	1.1
321.80	-1.64	0.7	34.02	19AP	-0.7	-3.6	-1.7	1.1
329.03	-1.82	0.6	43.65	29AP	0.3	-3.6	-1.6	1.1
336.20	-2.00	0.5	53.25	9MY	1.3	-3.7	-1.5	1.1
343.27	-2.17	0.4	62.81	19MY	2.4	-3.8	-1.5	1.0
350.23	-2.33	0.3	72.35	29MY	3.5	-3.9	-1.4	1.0
357.02	-2.49	0.2	81.89	8JN	2.1	-4.0	-1.4	1.0
3.60	-2.62	0.1	91.41	18JN	0.9	-4.1	-1.3	0.9
9.93	-2.74	-0.1	100.94	28JN	-0.2	-4.1	-1.3	0.9
15.92	-2.84	-0.2	110.49	8JL	-1.3	-4.2	-1.3	0.8
21.51	-2.92	-0.4	120.06	18JL	-1.6	-4.2	-1.3	0.8
26.60	-2.98	-0.5	129.66	28JL	-0.8	-4.0	-1.2	0.7
31.04	-3.00	-0.7	139.30	7AU	-0.2	-3.5	-1.2	0.7
34.68	-2.98	-0.9	148.99	17AU	0.1	-3.2	-1.2	0.6
37.35	-2.91	-1.1	158.72	27AU	0.3	-3.8	-1.2	0.6
38.80	-2.78	-1.3	168.52	6SE	0.5	-4.2	-1.3	0.5
38.86	-2.56	-1.5	178.36	16SE	1.3	-4.3	-1.3	0.5
37.45	-2.22	-1.8	188.27	26SE	3.1	-4.3	-1.3	0.6
34.73	-1.78	-1.9	198.23	6OC	0.3	-4.2	-1.3	0.6
31.26	-1.24	-2.0	208.24	16OC	-0.6	-4.1	-1.4	0.7
27.80	-0.67	-1.9	218.31	26OC	-0.7	-4.0	-1.4	0.8
25.13	-0.14	-1.6	228.43	5NO	-0.7	-4.0	-1.5	0.8
23.72	0.32	-1.2	238.57	15NO	-0.8	-3.9	-1.6	0.9
23.65	0.69	-0.9	248.75	25NO	-0.7	-3.8	-1.6	1.0
24.83	0.96	-0.6	258.94	5DE	-0.7	-3.7	-1.7	1.0
27.06	1.17	-0.3	269.14	15DE	-0.7	-3.6	-1.7	1.0
30.11	1.32	-0.0	279.34	25DE	-0.2	-3.6	-1.8	1.0

♂			☉	16.00UT	☿	♀	♃	♄
LONG	LAT	MAG	LONG	996	MAGNITUDES			
33.83	1.43	0.2	289.52	4JA	1.8	-3.5	-1.9	1.0
38.05	1.51	0.5	299.68	14JA	1.8	-3.5	-1.9	1.0
42.67	1.56	0.7	309.80	24JA	0.6	-3.4	-2.0	1.0
47.59	1.59	0.9	319.89	3FE	0.3	-3.4	-2.0	1.0
52.75	1.60	1.0	329.92	13FE	0.1	-3.4	-2.0	1.0
58.10	1.61	1.2	339.90	23FE	-0.1	-3.3	-2.0	0.9
63.60	1.60	1.3	349.83	4MR	-0.6	-3.3	-2.0	0.9
69.21	1.59	1.4	359.69	14MR	-1.3	-3.3	-2.0	0.8
74.92	1.57	1.5	9.50	24MR	-1.5	-3.3	-1.9	0.8
80.71	1.55	1.6	19.26	3AP	-0.7	-3.3	-1.9	0.8
86.57	1.52	1.7	28.95	13AP	0.5	-3.3	-1.8	0.8
92.48	1.49	1.8	38.61	23AP	1.6	-3.3	-1.8	0.8
98.43	1.45	1.8	48.22	3MY	3.1	-3.4	-1.7	0.8
104.44	1.41	1.9	57.80	13MY	2.9	-3.4	-1.7	0.8
110.48	1.37	1.9	67.36	23MY	1.6	-3.5	-1.6	0.8
116.57	1.32	2.0	76.89	2JN	0.7	-3.5	-1.5	0.8
122.69	1.27	2.0	86.41	12JN	-0.3	-3.5	-1.5	0.8
128.87	1.22	2.0	95.95	22JN	-1.3	-3.4	-1.4	0.7
135.08	1.17	2.0	105.48	2JL	-1.6	-3.4	-1.4	0.7
141.34	1.11	2.0	115.04	12JL	-0.8	-3.4	-1.3	0.6
147.65	1.05	2.0	124.62	22JL	-0.2	-3.4	-1.3	0.6
154.01	0.99	2.0	134.24	1AU	0.2	-3.3	-1.3	0.5
160.42	0.92	2.0	143.90	11AU	0.4	-3.3	-1.2	0.5
166.89	0.85	2.0	153.61	21AU	0.8	-3.3	-1.2	0.4
173.43	0.78	1.9	163.37	31AU	1.7	-3.3	-1.2	0.4
180.02	0.70	1.9	173.19	10SE	2.7	-3.3	-1.2	0.3
186.69	0.62	1.8	183.07	20SE	0.2	-3.4	-1.2	0.3
193.42	0.54	1.8	193.00	30SE	-0.8	-3.4	-1.2	0.3
200.22	0.45	1.8	202.98	10OC	-0.9	-3.4	-1.2	0.3
207.10	0.36	1.8	213.03	20OC	-0.8	-3.4	-1.3	0.4
214.04	0.27	1.8	223.12	30OC	-0.7	-3.5	-1.3	0.5
221.07	0.17	1.7	233.25	9NO	-0.6	-3.5	-1.3	0.5
228.16	0.07	1.7	243.41	19NO	-0.5	-3.6	-1.4	0.6
235.33	-0.04	1.7	253.60	29NO	-0.5	-3.7	-1.4	0.7
242.57	-0.15	1.7	263.79	9DE	-0.1	-3.7	-1.5	0.7
249.88	-0.26	1.6	274.00	19DE	2.1	-3.8	-1.6	0.7
257.25	-0.37	1.6	284.18	29DE	1.4	-3.9	-1.6	0.8
				997				
264.69	-0.49	1.6	294.36	8JA	0.4	-4.0	-1.7	0.8
272.19	-0.60	1.5	304.50	18JA	0.1	-4.1	-1.8	0.8
279.74	-0.72	1.5	314.60	28JA	0.0	-4.2	-1.8	0.8
287.33	-0.83	1.5	324.66	7FE	-0.2	-4.3	-1.9	0.8
294.96	-0.94	1.4	334.68	17FE	-0.6	-4.3	-1.9	0.8
302.62	-1.05	1.4	344.63	27FE	-1.3	-4.2	-2.0	0.7
310.30	-1.15	1.3	354.52	9MR	-1.4	-4.0	-2.0	0.7
317.99	-1.25	1.3	4.37	19MR	-0.7	-3.4	-2.0	0.6
325.66	-1.33	1.2	14.15	29MR	0.6	-3.3	-2.0	0.6
333.32	-1.41	1.2	23.87	8AP	2.1	-3.8	-2.0	0.6
340.95	-1.47	1.1	33.55	18AP	3.8	-4.1	-2.0	0.6
348.53	-1.52	1.1	43.18	28AP	2.2	-4.2	-1.9	0.6
356.04	-1.55	1.1	52.78	8MY	1.2	-4.2	-1.9	0.6
3.48	-1.57	1.0	62.35	18MY	0.5	-4.1	-1.8	0.6
10.81	-1.57	1.0	71.89	28MY	-0.3	-4.0	-1.8	0.6
18.04	-1.55	0.9	81.42	7JN	-1.3	-3.9	-1.7	0.6
25.12	-1.51	0.9	90.95	17JN	-1.7	-3.8	-1.6	0.6
32.06	-1.46	0.8	100.48	27JN	-0.8	-3.7	-1.6	0.6
38.82	-1.39	0.7	110.03	7JL	-0.1	-3.6	-1.5	0.5
45.37	-1.30	0.7	119.60	17JL	0.3	-3.6	-1.5	0.5
51.69	-1.19	0.6	129.19	27JL	0.6	-3.5	-1.4	0.4
57.74	-1.06	0.5	138.83	6AU	1.1	-3.5	-1.4	0.4
63.46	-0.90	0.4	148.52	16AU	2.2	-3.5	-1.3	0.3
68.81	-0.72	0.3	158.25	26AU	2.4	-3.4	-1.3	0.3
73.70	-0.51	0.2	168.04	5SE	0.1	-3.4	-1.3	0.2
78.02	-0.27	0.0	177.88	15SE	-0.9	-3.4	-1.2	0.2
81.67	0.01	-0.1	187.78	25SE	-1.0	-3.4	-1.2	0.1
84.47	0.33	-0.3	197.75	5OC	-1.0	-3.4	-1.2	0.0
86.24	0.70	-0.5	207.76	15OC	-0.6	-3.4	-1.2	0.0
86.77	1.13	-0.7	217.82	25OC	-0.4	-3.4	-1.2	0.1
85.91	1.59	-1.0	227.94	4NO	-0.4	-3.4	-1.3	0.2
83.64	2.07	-1.2	238.08	14NO	-0.3	-3.4	-1.3	0.3
80.24	2.52	-1.3	248.25	24NO	0.1	-3.4	-1.3	0.3
76.31	2.86	-1.4	258.45	4DE	2.4	-3.4	-1.3	0.4
72.65	3.08	-1.2	268.65	14DE	1.1	-3.4	-1.4	0.4
69.93	3.15	-0.9	278.84	24DE	0.2	-3.5	-1.4	0.5

♂			☉	16.00UT	☿	♀	♃	♄
LONG	LAT	MAG	LONG	998	MAGNITUDES			
68.49	3.13	-0.6	289.03	3JA	-0.0	-3.5	-1.5	0.5
68.39	3.04	-0.3	299.19	13JA	-0.1	-3.5	-1.6	0.6
69.47	2.91	-0.0	309.31	23JA	-0.3	-3.5	-1.6	0.6
71.54	2.77	0.2	319.40	2FE	-0.7	-3.4	-1.7	0.6
74.41	2.63	0.5	329.43	12FE	-1.3	-3.4	-1.8	0.6
77.91	2.48	0.7	339.42	22FE	-1.3	-3.4	-1.8	0.6
81.91	2.35	0.9	349.34	4MR	-0.6	-3.4	-1.9	0.6
86.31	2.22	1.0	359.21	14MR	0.8	-3.3	-1.9	0.5
91.01	2.09	1.2	9.02	24MR	2.7	-3.3	-2.0	0.5
95.97	1.97	1.3	18.78	3AP	2.9	-3.3	-2.0	0.5
101.14	1.86	1.4	28.48	13AP	1.6	-3.3	-2.1	0.4
106.48	1.75	1.5	38.14	23AP	0.9	-3.3	-2.1	0.4
111.97	1.64	1.6	47.76	3MY	0.4	-3.4	-2.1	0.4
117.60	1.54	1.7	57.34	13MY	-0.4	-3.4	-2.0	0.5
123.33	1.43	1.7	66.89	23MY	-1.3	-3.4	-2.0	0.5
129.16	1.33	1.8	76.43	2JN	-1.7	-3.4	-2.0	0.5
135.10	1.23	1.8	85.96	12JN	-0.7	-3.5	-1.9	0.5
141.12	1.14	1.9	95.48	22JN	-0.0	-3.5	-1.9	0.5
147.23	1.04	1.9	105.02	2JL	0.5	-3.6	-1.8	0.5
153.43	0.94	1.9	114.58	12JL	0.8	-3.7	-1.7	0.4
159.71	0.84	1.9	124.16	22JL	1.4	-3.7	-1.7	0.4
166.08	0.74	1.9	133.78	1AU	2.8	-3.8	-1.6	0.4
172.54	0.64	1.9	143.43	11AU	2.0	-3.9	-1.6	0.3
179.08	0.54	1.9	153.14	21AU	0.0	-4.0	-1.5	0.3
185.71	0.44	1.9	162.90	31AU	-1.0	-4.2	-1.5	0.2
192.43	0.34	1.8	172.71	10SE	-1.2	-4.2	-1.4	0.1
199.25	0.24	1.8	182.58	20SE	-0.9	-4.3	-1.4	0.1
206.16	0.14	1.8	192.52	30SE	-0.5	-4.3	-1.3	0.0
213.15	0.04	1.7	202.50	10OC	-0.3	-4.1	-1.3	-0.1
220.24	-0.07	1.7	212.54	20OC	-0.2	-3.5	-1.3	-0.1
227.42	-0.17	1.6	222.63	30OC	-0.2	-3.2	-1.3	-0.1
234.69	-0.27	1.6	232.75	9NO	0.3	-3.9	-1.3	-0.0
242.05	-0.37	1.5	242.92	19NO	2.6	-4.3	-1.3	0.0
249.49	-0.47	1.5	253.10	29NO	0.9	-4.4	-1.3	0.1
257.00	-0.57	1.5	263.30	9DE	-0.0	-4.4	-1.3	0.2
264.59	-0.67	1.5	273.50	19DE	-0.2	-4.3	-1.4	0.2
272.24	-0.75	1.5	283.69	29DE	-0.2	-4.2	-1.4	0.3
				999				
279.94	-0.84	1.5	293.86	8JA	-0.4	-4.1	-1.4	0.3
287.70	-0.92	1.5	304.01	18JA	-0.7	-4.0	-1.5	0.4
295.49	-0.99	1.4	314.12	28JA	-1.2	-3.9	-1.5	0.4
303.30	-1.05	1.4	324.18	7FE	-1.2	-3.8	-1.6	0.4
311.13	-1.10	1.4	334.19	17FE	-0.6	-3.7	-1.7	0.5
318.97	-1.14	1.4	344.15	27FE	1.0	-3.6	-1.7	0.5
326.79	-1.17	1.4	354.05	9MR	3.2	-3.5	-1.8	0.4
334.59	-1.18	1.4	3.89	19MR	2.1	-3.5	-1.9	0.4
342.35	-1.19	1.4	13.68	29MR	1.2	-3.4	-1.9	0.4
350.07	-1.18	1.4	23.41	8AP	0.7	-3.4	-2.0	0.4
357.73	-1.16	1.4	33.09	18AP	0.2	-3.3	-2.1	0.3
5.32	-1.12	1.4	42.73	28AP	-0.4	-3.3	-2.1	0.3
12.83	-1.07	1.4	52.32	8MY	-1.3	-3.3	-2.1	0.3
20.26	-1.01	1.4	61.89	18MY	-1.7	-3.3	-2.2	0.3
27.58	-0.93	1.3	71.44	28MY	-0.7	-3.3	-2.2	0.4
34.80	-0.85	1.3	80.97	7JN	0.1	-3.3	-2.2	0.4
41.90	-0.75	1.3	90.49	17JN	0.6	-3.3	-2.1	0.4
48.88	-0.64	1.3	100.02	27JN	1.1	-3.4	-2.1	0.4
55.73	-0.53	1.3	109.57	7JL	1.9	-3.4	-2.1	0.4
62.44	-0.40	1.3	119.13	17JL	3.2	-3.4	-2.0	0.4
68.99	-0.26	1.2	128.73	27JL	1.7	-3.5	-2.0	0.3
75.37	-0.11	1.2	138.36	6AU	-0.1	-3.5	-1.9	0.3
81.58	0.05	1.1	148.05	16AU	-1.1	-3.5	-1.8	0.3
87.57	0.23	1.1	157.78	26AU	-1.3	-3.5	-1.8	0.2
93.33	0.42	1.0	167.56	5SE	-0.9	-3.4	-1.7	0.2
98.82	0.62	0.9	177.40	15SE	-0.4	-3.4	-1.6	0.1
103.99	0.84	0.8	187.30	25SE	-0.2	-3.4	-1.6	0.0
108.78	1.09	0.7	197.25	5OC	-0.1	-3.4	-1.5	-0.0
113.12	1.36	0.6	207.27	15OC	0.0	-3.3	-1.5	-0.1
116.89	1.66	0.4	217.33	25OC	0.5	-3.3	-1.5	-0.2
119.98	2.00	0.2	227.43	4NO	3.0	-3.3	-1.4	-0.2
122.21	2.37	0.0	237.58	14NO	0.7	-3.3	-1.4	-0.2
123.41	2.78	-0.2	247.76	24NO	-0.2	-3.3	-1.4	-0.1
123.40	3.22	-0.4	257.95	4DE	-0.3	-3.4	-1.4	-0.1
122.03	3.65	-0.6	268.15	14DE	-0.4	-3.4	-1.4	0.0
119.36	4.03	-0.9	278.35	24DE	-0.5	-3.4	-1.4	0.1

Left table

♂ LONG	♂ LAT	♂ MAG	☉ LONG	16.00UT	☿	♀	♃	♄
				1000				
115.72	4.30	-1.0	288.53	3JA	-0.8	-3.5	-1.4	0.1
111.74	4.40	-1.0	298.70	13JA	-1.1	-3.5	-1.4	0.2
108.17	4.33	-0.8	308.82	23JA	-1.1	-3.5	-1.5	0.3
105.60	4.12	-0.6	318.91	2FE	-0.5	-3.6	-1.5	0.3
104.30	3.84	-0.3	328.95	12FE	1.2	-3.7	-1.6	0.4
104.29	3.52	-0.0	338.94	22FE	3.2	-3.7	-1.6	0.4
105.42	3.20	0.2	348.87	3MR	1.6	-3.8	-1.7	0.4
107.51	2.91	0.4	358.74	13MR	0.9	-3.9	-1.7	0.4
110.39	2.63	0.6	8.56	23MR	0.5	-4.0	-1.8	0.4
113.88	2.37	0.8	18.31	2AP	0.1	-4.1	-1.9	0.3
117.88	2.14	1.0	28.02	12AP	-0.5	-4.1	-1.9	0.3
122.28	1.93	1.1	37.68	22AP	-1.3	-4.2	-2.0	0.3
127.02	1.73	1.2	47.29	2MY	-1.7	-4.2	-2.1	0.2
132.04	1.55	1.3	56.88	12MY	-0.7	-4.0	-2.1	0.2
137.30	1.37	1.4	66.43	22MY	0.1	-3.6	-2.2	0.2
142.76	1.21	1.5	75.97	1JN	0.8	-2.7	-2.2	0.2
148.41	1.06	1.5	85.50	11JN	1.5	-3.4	-2.3	0.3
154.22	0.91	1.6	95.02	21JN	2.5	-4.0	-2.3	0.3
160.18	0.77	1.6	104.56	1JL	3.1	-4.2	-2.3	0.3
166.28	0.63	1.6	114.12	11JL	1.4	-4.2	-2.3	0.3
172.51	0.50	1.7	123.69	21JL	-0.1	-4.2	-2.3	0.3
178.88	0.37	1.7	133.31	31JL	-1.2	-4.1	-2.2	0.3
185.37	0.24	1.7	142.97	10AU	-1.4	-4.0	-2.2	0.3
191.98	0.12	1.7	152.67	20AU	-0.9	-3.9	-2.1	0.3
198.71	0.00	1.7	162.42	30AU	-0.4	-3.8	-2.1	0.2
205.56	-0.11	1.7	172.24	9SE	-0.1	-3.7	-2.0	0.2
212.52	-0.22	1.7	182.10	19SE	0.1	-3.7	-1.9	0.1
219.60	-0.33	1.6	192.03	29SE	0.2	-3.6	-1.9	0.1
226.79	-0.43	1.6	202.01	9OC	0.8	-3.6	-1.8	-0.0
234.08	-0.53	1.6	212.05	19OC	3.3	-3.5	-1.7	-0.1
241.47	-0.62	1.6	222.13	29OC	0.5	-3.5	-1.7	-0.2
248.96	-0.71	1.5	232.26	8NO	-0.4	-3.5	-1.6	-0.2
256.53	-0.78	1.5	242.42	18NO	-0.5	-3.4	-1.6	-0.3
264.18	-0.86	1.5	252.60	28NO	-0.5	-3.4	-1.6	-0.3
271.91	-0.92	1.4	262.80	8DE	-0.6	-3.4	-1.5	-0.2
279.69	-0.97	1.4	273.00	18DE	-0.8	-3.4	-1.5	-0.1
287.52	-1.02	1.3	283.20	28DE	-0.9	-3.3	-1.5	-0.0
				1001				
295.39	-1.05	1.3	293.37	7JA	-0.9	-3.3	-1.5	0.0
303.27	-1.07	1.3	303.52	17JA	-0.4	-3.3	-1.5	0.1
311.17	-1.08	1.3	313.63	27JA	1.4	-3.3	-1.5	0.2
319.07	-1.08	1.3	323.69	6FE	2.7	-3.4	-1.5	0.2
326.95	-1.07	1.3	333.71	16FE	1.2	-3.4	-1.5	0.2
334.80	-1.04	1.4	343.67	26FE	0.6	-3.4	-1.6	0.3
342.60	-1.01	1.4	353.57	8MR	0.3	-3.4	-1.6	0.3
350.36	-0.96	1.4	3.41	18MR	0.0	-3.5	-1.7	0.3
358.06	-0.90	1.4	13.20	28MR	-0.5	-3.5	-1.7	0.3
5.68	-0.84	1.5	22.94	7AP	-1.3	-3.5	-1.8	0.3
13.23	-0.76	1.5	32.62	17AP	-1.6	-3.4	-1.8	0.3
20.69	-0.68	1.5	42.26	27AP	-0.7	-3.4	-1.9	0.3
28.07	-0.58	1.5	51.86	7MY	0.3	-3.4	-2.0	0.3
35.35	-0.49	1.6	61.42	17MY	1.1	-3.3	-2.0	0.2
42.53	-0.38	1.6	70.97	27MY	2.0	-3.3	-2.1	0.2
49.62	-0.27	1.6	80.50	6JN	3.3	-3.3	-2.2	0.2
56.61	-0.16	1.6	90.03	16JN	2.5	-3.3	-2.2	0.3
63.49	-0.04	1.6	99.56	26JN	1.1	-3.3	-2.3	0.3
70.27	0.08	1.6	109.10	6JL	-0.2	-3.3	-2.3	0.3
76.94	0.20	1.6	118.66	16JL	-1.2	-3.4	-2.4	0.3
83.51	0.33	1.6	128.26	26JL	-1.5	-3.4	-2.4	0.3
89.96	0.46	1.6	137.90	5AU	-0.8	-3.4	-2.4	0.3
96.28	0.60	1.6	147.57	15AU	-0.3	-3.5	-2.4	0.3
102.49	0.74	1.6	157.30	25AU	0.0	-3.5	-2.4	0.3
108.55	0.88	1.5	167.08	4SE	0.2	-3.6	-2.3	0.3
114.46	1.04	1.5	176.92	14SE	0.4	-3.6	-2.3	0.2
120.20	1.20	1.4	186.82	24SE	1.1	-3.7	-2.2	0.2
125.74	1.37	1.3	196.77	4OC	3.4	-3.8	-2.2	0.1
131.05	1.55	1.2	206.78	14OC	0.4	-3.9	-2.1	0.1
136.10	1.74	1.1	216.84	24OC	-0.5	-4.0	-2.0	-0.0
140.82	1.95	1.0	226.94	3NO	-0.6	-4.1	-2.0	-0.1
145.16	2.18	0.9	237.09	13NO	-0.6	-4.2	-1.9	-0.1
149.01	2.43	0.7	247.26	23NO	-0.7	-4.3	-1.8	-0.2
152.27	2.71	0.5	257.45	3DE	-0.8	-4.4	-1.8	-0.3
154.81	3.01	0.3	267.65	13DE	-0.8	-4.1	-1.7	-0.2
156.43	3.33	0.1	277.85	23DE	-0.8	-4.1	-1.7	-0.2

Right table

♂ LONG	♂ LAT	♂ MAG	☉ LONG	16.00UT	☿	♀	♃	♄
				1002				
156.97	3.67	-0.2	288.03	2JA	-0.3	-3.6	-1.6	-0.1
156.26	3.99	-0.4	298.20	12JA	1.6	-3.3	-1.6	-0.0
154.24	4.26	-0.7	308.33	22JA	2.1	-3.9	-1.6	0.0
151.09	4.42	-0.9	318.42	1FE	0.8	-4.2	-1.6	0.1
147.25	4.42	-1.0	328.46	11FE	0.4	-4.3	-1.6	0.2
143.42	4.25	-0.9	338.45	21FE	0.2	-4.3	-1.6	0.2
140.29	3.94	-0.7	348.38	3MR	-0.1	-4.2	-1.6	0.2
138.30	3.54	-0.5	358.26	13MR	-0.5	-4.1	-1.6	0.3
137.60	3.12	-0.2	8.08	23MR	-1.3	-4.0	-1.6	0.3
138.15	2.71	-0.0	17.84	2AP	-1.6	-3.9	-1.6	0.3
139.77	2.32	0.2	27.55	12AP	-0.7	-3.8	-1.7	0.3
142.29	1.97	0.4	37.21	22AP	0.4	-3.7	-1.7	0.3
145.55	1.66	0.5	46.83	2MY	1.4	-3.6	-1.8	0.3
149.40	1.37	0.7	56.41	12MY	2.6	-3.5	-1.8	0.3
153.75	1.12	0.8	65.97	22MY	3.4	-3.5	-1.9	0.3
158.51	0.88	0.9	75.51	1JN	1.9	-3.4	-1.9	0.2
163.61	0.67	1.0	85.04	11JN	0.9	-3.4	-2.0	0.2
169.01	0.47	1.1	94.57	21JN	-0.2	-3.4	-2.1	0.2
174.68	0.29	1.2	104.10	1JL	-1.3	-3.3	-2.1	0.3
180.57	0.12	1.2	113.65	11JL	-1.6	-3.3	-2.2	0.3
186.67	-0.04	1.3	123.24	21JL	-0.8	-3.3	-2.3	0.4
192.97	-0.19	1.3	132.84	31JL	-0.2	-3.3	-2.3	0.4
199.44	-0.32	1.3	142.50	10AU	0.1	-3.3	-2.4	0.4
206.08	-0.45	1.4	152.20	20AU	0.3	-3.4	-2.4	0.4
212.88	-0.57	1.4	161.95	30AU	0.6	-3.4	-2.5	0.4
219.83	-0.68	1.4	171.76	9SE	1.5	-3.4	-2.5	0.4
226.92	-0.77	1.4	181.63	19SE	3.0	-3.4	-2.5	0.3
234.13	-0.86	1.4	191.55	29SE	0.3	-3.4	-2.4	0.3
241.47	-0.94	1.4	201.53	9OC	-0.7	-3.5	-2.4	0.2
248.92	-1.00	1.4	211.56	19OC	-0.8	-3.5	-2.3	0.2
256.48	-1.06	1.4	221.64	29OC	-0.8	-3.5	-2.3	0.1
264.12	-1.10	1.4	231.77	8NO	-0.8	-3.5	-2.2	0.1
271.84	-1.13	1.4	241.93	18NO	-0.6	-3.4	-2.1	-0.0
279.63	-1.15	1.4	252.10	28NO	-0.6	-3.4	-2.1	-0.1
287.46	-1.16	1.4	262.30	8DE	-0.6	-3.4	-2.0	-0.2
295.33	-1.15	1.4	272.50	18DE	-0.2	-3.4	-1.9	-0.2
303.23	-1.13	1.4	282.69	28DE	1.8	-3.3	-1.9	-0.2
				1003				
311.13	-1.10	1.4	292.87	7JA	1.7	-3.3	-1.8	-0.1
319.02	-1.06	1.4	303.02	17JA	0.5	-3.3	-1.8	-0.1
326.89	-1.01	1.4	313.13	27JA	0.2	-3.3	-1.7	0.0
334.73	-0.94	1.4	323.20	6FE	0.1	-3.4	-1.7	0.1
342.52	-0.87	1.4	333.22	16FE	-0.2	-3.4	-1.6	0.1
350.26	-0.79	1.4	343.18	26FE	-0.6	-3.4	-1.6	0.2
357.93	-0.71	1.4	353.09	8MR	-1.3	-3.4	-1.6	0.3
5.52	-0.62	1.4	2.94	18MR	-1.5	-3.4	-1.6	0.3
13.04	-0.52	1.4	12.73	28MR	-0.7	-3.5	-1.6	0.3
20.48	-0.42	1.5	22.47	7AP	0.5	-3.5	-1.6	0.4
27.83	-0.32	1.5	32.15	17AP	1.8	-3.6	-1.6	0.4
35.09	-0.22	1.6	41.79	27AP	3.4	-3.6	-1.6	0.4
42.27	-0.11	1.6	51.40	7MY	2.6	-3.7	-1.6	0.4
49.35	-0.01	1.7	60.97	17MY	1.5	-3.8	-1.7	0.4
56.35	0.10	1.7	70.51	27MY	0.6	-3.9	-1.7	0.4
63.27	0.20	1.7	80.05	6JN	-0.3	-4.0	-1.7	0.4
70.10	0.31	1.8	89.57	16JN	-1.3	-4.1	-1.8	0.3
76.85	0.41	1.8	99.10	26JN	-1.7	-4.1	-1.8	0.3
83.51	0.52	1.8	108.65	6JL	-0.8	-4.2	-1.9	0.3
90.10	0.62	1.8	118.21	16JL	-0.2	-4.2	-2.0	0.4
96.61	0.72	1.8	127.80	26JL	0.2	-4.0	-2.0	0.4
103.05	0.82	1.9	137.43	5AU	0.5	-3.5	-2.1	0.5
109.41	0.92	1.9	147.11	15AU	0.9	-3.3	-2.2	0.5
115.69	1.02	1.8	156.83	25AU	1.9	-3.8	-2.2	0.5
121.88	1.13	1.8	166.61	4SE	2.7	-4.2	-2.3	0.5
127.99	1.23	1.8	176.44	14SE	0.2	-4.3	-2.3	0.5
134.01	1.33	1.8	186.34	24SE	-0.8	-4.3	-2.4	0.5
139.93	1.43	1.7	196.29	4OC	-0.9	-4.2	-2.4	0.4
145.72	1.54	1.6	206.29	14OC	-0.9	-4.1	-2.4	0.4
151.39	1.65	1.6	216.35	24OC	-0.7	-4.0	-2.4	0.4
156.91	1.77	1.5	226.46	3NO	-0.5	-3.9	-2.4	0.3
162.24	1.89	1.4	236.60	13NO	-0.5	-3.9	-2.4	0.2
167.37	2.01	1.3	246.77	23NO	-0.4	-3.8	-2.3	0.2
172.23	2.15	1.1	256.96	3DE	-0.0	-3.7	-2.3	0.1
176.77	2.29	1.0	267.16	13DE	2.1	-3.6	-2.2	0.0
180.93	2.43	0.8	277.36	23DE	1.4	-3.6	-2.1	-0.1

δ LONG	LAT	MAG	☉ LONG	16.00UT 1004	☿	♀	♃	♄
184.61	2.59	0.6	287.54	2JA	0.3	-3.5	-2.0	-0.1
187.68	2.75	0.4	297.71	12JA	0.1	-3.5	-2.0	-0.1
190.02	2.92	0.1	307.84	22JA	-0.0	-3.4	-1.9	-0.0
191.43	3.08	-0.1	317.93	1FE	-0.2	-3.4	-1.8	0.1
191.74	3.21	-0.4	327.97	11FE	-0.6	-3.4	-1.8	0.1
190.82	3.30	-0.7	337.97	21FE	-1.3	-3.3	-1.7	0.2
188.62	3.30	-1.0	347.91	2MR	-1.4	-3.3	-1.7	0.2
185.39	3.18	-1.2	357.78	12MR	-0.7	-3.3	-1.6	0.3
181.61	2.92	-1.3	7.60	22MR	0.7	-3.3	-1.6	0.3
178.01	2.54	-1.1	17.37	1AP	2.3	-3.3	-1.6	0.4
175.25	2.09	-1.0	27.08	11AP	3.5	-3.3	-1.5	0.4
173.70	1.63	-0.8	36.74	21AP	2.0	-3.3	-1.5	0.5
173.48	1.19	-0.5	46.37	1MY	1.1	-3.4	-1.5	0.5
174.51	0.79	-0.3	55.95	11MY	0.5	-3.4	-1.5	0.5
176.61	0.43	-0.1	65.51	21MY	-0.3	-3.5	-1.5	0.5
179.62	0.12	0.0	75.05	31MY	-1.3	-3.5	-1.5	0.5
183.37	-0.15	0.2	84.57	10JN	-1.7	-3.5	-1.5	0.5
187.74	-0.38	0.3	94.10	20JN	-0.8	-3.4	-1.6	0.5
192.61	-0.59	0.4	103.64	30JN	-0.1	-3.4	-1.6	0.5
197.92	-0.77	0.5	113.19	10JL	0.4	-3.4	-1.6	0.5
203.59	-0.92	0.6	122.77	20JL	0.7	-3.4	-1.7	0.5
209.59	-1.06	0.7	132.38	30JL	1.2	-3.3	-1.7	0.5
215.86	-1.17	0.8	142.03	9AU	2.4	-3.3	-1.8	0.6
222.37	-1.27	0.8	151.73	19AU	2.3	-3.3	-1.9	0.6
229.11	-1.35	0.9	161.48	29AU	0.1	-3.3	-1.9	0.6
236.04	-1.40	0.9	171.28	8SE	-0.9	-3.3	-2.0	0.6
243.14	-1.45	1.0	181.14	18SE	-1.1	-3.4	-2.1	0.7
250.40	-1.47	1.0	191.06	28SE	-1.0	-3.4	-2.1	0.6
257.79	-1.48	1.1	201.04	8OC	-0.6	-3.4	-2.2	0.6
265.30	-1.48	1.1	211.07	18OC	-0.4	-3.4	-2.2	0.6
272.90	-1.46	1.1	221.15	28OC	-0.3	-3.5	-2.3	0.6
280.58	-1.42	1.2	231.27	7NO	-0.3	-3.5	-2.3	0.5
288.32	-1.37	1.2	241.43	17NO	0.2	-3.6	-2.3	0.5
296.10	-1.31	1.2	251.61	27NO	2.4	-3.7	-2.3	0.4
303.91	-1.24	1.3	261.81	7DE	1.1	-3.7	-2.3	0.3
311.72	-1.16	1.3	272.01	17DE	0.1	-3.8	-2.3	0.3
319.53	-1.06	1.4	282.20	27DE	-0.1	-3.9	-2.2	0.2
				1005				
327.31	-0.97	1.4	292.38	6JA	-0.2	-4.0	-2.1	0.1
335.05	-0.86	1.4	302.53	16JA	-0.3	-4.1	-2.1	0.1
342.75	-0.75	1.5	312.65	26JA	-0.7	-4.2	-2.0	0.1
350.39	-0.64	1.5	322.71	5FE	-1.3	-4.3	-1.9	0.2
357.96	-0.53	1.5	332.74	15FE	-1.3	-4.3	-1.9	0.2
5.46	-0.42	1.6	342.70	25FE	-0.6	-4.2	-1.8	0.3
12.89	-0.30	1.6	352.61	7MR	0.8	-4.0	-1.7	0.3
20.23	-0.19	1.6	2.46	17MR	2.8	-3.4	-1.7	0.4
27.50	-0.08	1.6	12.25	27MR	2.6	-3.3	-1.6	0.4
34.68	0.03	1.7	21.99	6AP	1.4	-3.9	-1.6	0.5
41.78	0.13	1.7	31.68	16AP	0.8	-4.1	-1.5	0.5
48.80	0.23	1.7	41.32	26AP	0.3	-4.2	-1.5	0.6
55.74	0.33	1.7	50.92	6MY	-0.4	-4.2	-1.5	0.6
62.60	0.42	1.7	60.50	16MY	-1.3	-4.1	-1.4	0.6
69.40	0.51	1.7	70.04	26MY	-1.7	-4.0	-1.4	0.7
76.13	0.60	1.8	79.58	5JN	-0.8	-3.9	-1.4	0.7
82.79	0.68	1.8	89.11	15JN	-0.0	-3.8	-1.4	0.7
89.40	0.76	1.9	98.64	25JN	0.5	-3.7	-1.4	0.7
95.95	0.84	1.9	108.18	5JL	0.9	-3.6	-1.4	0.7
102.44	0.91	1.9	117.74	15JL	1.6	-3.6	-1.4	0.7
108.89	0.98	2.0	127.33	25JL	2.9	-3.5	-1.5	0.7
115.30	1.04	2.0	136.96	4AU	1.9	-3.5	-1.5	0.7
121.66	1.10	2.0	146.64	14AU	0.0	-3.4	-1.5	0.7
127.97	1.16	2.0	156.36	24AU	-1.0	-3.4	-1.6	0.8
134.25	1.22	2.0	166.14	3SE	-1.2	-3.4	-1.6	0.8
140.49	1.27	2.0	175.97	13SE	-0.9	-3.4	-1.7	0.9
146.68	1.32	1.9	185.86	23SE	-0.5	-3.4	-1.8	0.9
152.84	1.37	1.9	195.81	3OC	-0.3	-3.4	-1.8	0.9
158.94	1.42	1.9	205.81	13OC	-0.2	-3.4	-1.9	0.9
164.99	1.46	1.8	215.86	23OC	-0.1	-3.4	-1.9	0.9
170.98	1.51	1.8	225.96	2NO	0.4	-3.4	-2.0	0.8
176.89	1.55	1.7	236.11	12NO	2.7	-3.4	-2.1	0.8
182.73	1.58	1.6	246.27	22NO	0.8	-3.4	-2.1	0.7
188.47	1.61	1.5	256.46	2DE	-0.1	-3.4	-2.2	0.7
194.09	1.64	1.4	266.66	12DE	-0.2	-3.4	-2.2	0.6
199.57	1.67	1.3	276.86	22DE	-0.3	-3.5	-2.2	0.6

δ LONG	LAT	MAG	☉ LONG	16.00UT 1006	☿	♀	♃	♄
204.89	1.68	1.1	287.05	1JA	-0.4	-3.5	-2.2	0.5
209.99	1.69	1.0	297.21	11JA	-0.8	-3.5	-2.2	0.4
214.84	1.69	0.8	307.34	21JA	-1.2	-3.5	-2.1	0.3
219.38	1.67	0.6	317.44	31JA	-1.2	-3.4	-2.1	0.3
223.53	1.64	0.4	327.49	10FE	-0.6	-3.4	-2.0	0.3
227.18	1.58	0.1	337.48	20FE	1.0	-3.4	-2.0	0.4
230.23	1.48	-0.1	347.42	2MR	3.2	-3.4	-1.9	0.4
232.84	1.34	-0.4	357.30	12MR	1.9	-3.3	-1.8	0.5
233.82	1.14	-0.7	7.12	22MR	1.1	-3.3	-1.7	0.5
234.02	0.87	-1.0	16.89	1AP	0.6	-3.3	-1.7	0.6
232.97	0.51	-1.3	26.61	11AP	0.2	-3.3	-1.6	0.7
230.72	0.06	-1.6	36.27	21AP	-0.4	-3.3	-1.6	0.7
227.57	-0.45	-1.9	45.89	1MY	-1.3	-3.4	-1.5	0.8
224.13	-0.97	-1.8	55.48	11MY	-1.7	-3.4	-1.5	0.8
221.14	-1.46	-1.7	65.04	21MY	-0.8	-3.4	-1.4	0.8
219.19	-1.86	-1.5	74.58	31MY	0.1	-3.4	-1.4	0.9
218.58	-2.17	-1.3	84.11	10JN	0.7	-3.5	-1.3	0.9
219.34	-2.39	-1.1	93.64	20JN	1.2	-3.5	-1.3	0.9
221.34	-2.54	-0.9	103.17	30JN	2.1	-3.6	-1.3	0.9
224.40	-2.63	-0.7	112.73	10JL	3.3	-3.7	-1.3	0.9
228.33	-2.68	-0.5	122.30	20JL	1.6	-3.7	-1.3	0.9
232.98	-2.68	-0.4	131.91	30JL	-0.0	-3.8	-1.3	0.9
238.22	-2.66	-0.2	141.56	9AU	-1.1	-3.9	-1.3	0.9
243.94	-2.61	-0.1	151.26	19AU	-1.4	-4.0	-1.3	1.0
250.06	-2.54	0.1	161.01	29AU	-0.9	-4.2	-1.4	1.0
256.50	-2.45	0.2	170.81	8SE	-0.2	-4.3	-1.4	1.1
263.21	-2.34	0.3	180.67	18SE	-0.2	-4.3	-1.4	1.1
270.14	-2.22	0.4	190.59	28SE	-0.0	-4.3	-1.5	1.1
277.24	-2.08	0.5	200.56	8OC	0.1	-4.1	-1.5	1.1
284.48	-1.93	0.6	210.59	18OC	0.6	-3.5	-1.6	1.1
291.83	-1.77	0.7	220.67	28OC	3.0	-3.2	-1.7	1.1
299.26	-1.61	0.8	230.78	7NO	0.7	-3.9	-1.7	1.1
306.73	-1.44	0.9	240.94	17NO	-0.3	-4.3	-1.8	1.1
314.24	-1.28	1.0	251.12	27NO	-0.4	-4.4	-1.9	1.1
321.75	-1.11	1.0	261.31	7DE	-0.4	-4.4	-1.9	1.0
329.25	-0.94	1.1	271.51	17DE	-0.5	-4.3	-2.0	1.0
336.73	-0.78	1.2	281.71	27DE	-0.8	-4.2	-2.0	0.9
				1007				
344.18	-0.62	1.3	291.89	6JA	-1.0	-4.1	-2.1	0.8
351.58	-0.47	1.4	302.04	16JA	-1.0	-4.0	-2.1	0.7
358.92	-0.33	1.4	312.16	26JA	-0.5	-3.9	-2.1	0.7
6.20	-0.19	1.5	322.23	5FE	1.2	-3.8	-2.1	0.6
13.42	-0.06	1.6	332.25	15FE	3.0	-3.7	-2.0	0.6
20.57	0.07	1.6	342.22	25FE	1.4	-3.6	-2.0	0.6
27.65	0.18	1.7	352.13	7MR	0.8	-3.5	-2.0	0.7
34.66	0.29	1.7	1.99	17MR	0.4	-3.5	-1.9	0.7
41.60	0.39	1.8	11.79	27MR	0.1	-3.4	-1.8	0.8
48.47	0.49	1.8	21.53	6AP	-0.5	-3.4	-1.8	0.8
55.28	0.57	1.8	31.22	16AP	-1.3	-3.3	-1.7	0.9
62.03	0.65	1.9	40.86	26AP	-1.7	-3.3	-1.6	0.9
68.72	0.73	1.9	50.47	6MY	-0.8	-3.3	-1.6	1.0
75.35	0.79	1.9	60.04	16MY	0.2	-3.3	-1.5	1.0
81.94	0.86	1.9	69.59	26MY	0.9	-3.3	-1.5	1.1
88.48	0.91	1.9	79.12	5JN	1.6	-3.3	-1.4	1.1
94.99	0.96	1.9	88.65	15JN	2.8	-3.3	-1.4	1.1
101.45	1.01	1.9	98.18	25JN	2.9	-3.4	-1.3	1.2
107.89	1.05	1.9	107.72	5JL	1.3	-3.4	-1.3	1.2
114.31	1.08	1.9	117.28	15JL	-0.1	-3.4	-1.3	1.2
120.70	1.12	2.0	126.87	25JL	-1.2	-3.5	-1.2	1.2
127.07	1.14	2.0	136.50	4AU	-1.5	-3.5	-1.2	1.2
133.44	1.17	2.0	146.17	14AU	-0.9	-3.5	-1.2	1.2
139.79	1.18	2.0	155.89	24AU	-0.3	-3.5	-1.2	1.2
146.13	1.20	2.0	165.66	3SE	-0.0	-3.4	-1.2	1.3
152.47	1.21	2.0	175.49	13SE	0.1	-3.4	-1.3	1.3
158.81	1.21	2.0	185.38	23SE	0.3	-3.4	-1.3	1.4
165.14	1.22	2.0	195.32	3OC	0.9	-3.4	-1.3	1.4
171.48	1.21	2.0	205.32	13OC	3.3	-3.3	-1.4	1.3
177.81	1.20	1.9	215.38	23OC	0.5	-3.3	-1.4	1.3
184.13	1.19	1.9	225.47	2NO	-0.4	-3.3	-1.4	1.3
190.46	1.17	1.9	235.61	12NO	-0.5	-3.3	-1.5	1.3
196.77	1.14	1.8	245.78	22NO	-0.5	-3.3	-1.6	1.2
203.07	1.10	1.7	255.96	2DE	-0.6	-3.4	-1.6	1.2
209.36	1.06	1.6	266.17	12DE	-0.8	-3.4	-1.7	1.2
215.63	1.01	1.6	276.37	22DE	-0.9	-3.4	-1.8	1.1

♂ LONG	♂ LAT	♂ MAG	☉ LONG	16.00UT	☿	♀	♃	♄
				1008	MAGNITUDES			
221.87	0.94	1.5	286.55	1JA	-0.9	-3.5	-1.8	1.1
228.07	0.86	1.3	296.72	11JA	-0.4	-3.5	-1.9	1.0
234.22	0.77	1.2	306.86	21JA	1.4	-3.5	-1.9	1.0
240.32	0.66	1.1	316.95	31JA	2.5	-3.6	-2.0	0.9
246.35	0.52	0.9	327.00	10FE	1.0	-3.7	-2.0	0.9
252.28	0.36	0.8	337.00	20FE	0.5	-3.7	-2.0	0.8
258.10	0.18	0.6	346.94	1MR	0.3	-3.8	-2.0	0.9
263.78	-0.04	0.4	356.83	11MR	-0.0	-3.9	-2.0	0.9
269.26	-0.30	0.2	6.66	21MR	-0.5	-4.0	-2.0	1.0
274.52	-0.60	0.0	16.42	31MR	-1.3	-4.1	-1.9	1.0
279.48	-0.96	-0.2	26.14	10AP	-1.6	-4.1	-1.9	1.1
284.05	-1.38	-0.5	35.81	20AP	-0.8	-4.2	-1.8	1.1
288.14	-1.87	-0.7	45.43	30AP	0.3	-4.2	-1.8	1.2
291.58	-2.44	-1.0	55.02	10MY	1.2	-4.0	-1.7	1.2
294.22	-3.08	-1.3	64.58	20MY	2.2	-3.6	-1.6	1.3
295.87	-3.81	-1.6	74.12	30MY	3.5	-2.7	-1.6	1.3
296.34	-4.58	-1.9	83.65	9JN	2.3	-3.4	-1.5	1.4
295.55	-5.33	-2.2	93.18	19JN	1.0	-4.0	-1.5	1.4
293.66	-5.98	-2.5	102.71	29JN	-0.1	-4.2	-1.4	1.4
291.08	-6.40	-2.6	112.26	9JL	-1.2	-4.2	-1.4	1.4
288.51	-6.52	-2.5	121.84	19JL	-1.6	-4.1	-1.3	1.4
286.66	-6.35	-2.3	131.44	29JL	-0.8	-4.1	-1.3	1.4
285.99	-5.96	-2.0	141.09	8AU	-0.3	-4.0	-1.2	1.3
286.68	-5.45	-1.8	150.78	18AU	0.1	-3.9	-1.2	1.3
288.62	-4.88	-1.5	160.53	28AU	0.3	-3.8	-1.2	1.2
291.66	-4.30	-1.2	170.33	7SE	0.5	-3.7	-1.2	1.2
295.60	-3.75	-1.0	180.19	17SE	1.2	-3.7	-1.2	1.2
300.23	-3.23	-0.8	190.10	27SE	3.3	-3.6	-1.2	1.2
305.41	-2.75	-0.5	200.07	7OC	0.4	-3.6	-1.2	1.2
311.01	-2.31	-0.3	210.10	17OC	-0.6	-3.5	-1.2	1.2
316.93	-1.91	-0.1	220.17	27OC	-0.7	-3.5	-1.3	1.2
323.10	-1.55	0.1	230.29	6NO	-0.7	-3.4	-1.3	1.1
329.45	-1.22	0.3	240.45	16NO	-0.7	-3.4	-1.4	1.1
335.93	-0.93	0.4	250.62	26NO	-0.7	-3.4	-1.4	1.1
342.50	-0.66	0.6	260.82	6DE	-0.7	-3.4	-1.5	1.1
349.14	-0.43	0.7	271.02	16DE	-0.7	-3.4	-1.5	1.0
355.82	-0.22	0.9	281.22	26DE	-0.3	-3.3	-1.6	1.0
				1009				
2.51	-0.03	1.0	291.40	5JA	1.6	-3.3	-1.6	0.9
9.22	0.14	1.1	301.55	15JA	2.0	-3.3	-1.7	0.9
15.91	0.29	1.2	311.67	25JA	0.7	-3.3	-1.8	0.8
22.60	0.42	1.4	321.74	4FE	0.3	-3.4	-1.8	0.8
29.25	0.54	1.4	331.77	14FE	0.2	-3.4	-1.9	0.7
35.88	0.64	1.5	341.74	24FE	-0.1	-3.4	-1.9	0.7
42.49	0.73	1.6	351.66	6MR	-0.5	-3.4	-2.0	0.7
49.05	0.81	1.7	1.51	16MR	-1.3	-3.5	-2.0	0.7
55.59	0.88	1.8	11.31	26MR	-1.6	-3.5	-2.0	0.8
62.09	0.94	1.8	21.06	5AP	-0.7	-3.5	-2.0	0.8
68.57	0.99	1.9	30.75	15AP	0.4	-3.4	-2.0	0.9
75.01	1.04	1.9	40.39	25AP	1.5	-3.4	-2.0	1.0
81.43	1.07	2.0	50.00	5MY	2.9	-3.4	-1.9	1.0
87.82	1.10	2.0	59.57	15MY	3.1	-3.3	-1.9	1.1
94.20	1.12	2.0	69.12	25MY	1.8	-3.3	-1.8	1.1
100.56	1.14	2.0	78.66	4JN	0.8	-3.3	-1.7	1.2
106.91	1.16	2.0	88.18	14JN	-0.2	-3.3	-1.7	1.2
113.25	1.16	2.0	97.71	24JN	-1.2	-3.3	-1.6	1.2
119.59	1.16	2.0	107.25	4JL	-1.7	-3.3	-1.6	1.2
125.94	1.16	2.0	116.81	14JL	-0.8	-3.4	-1.5	1.2
132.29	1.16	2.0	126.40	24JL	-0.2	-3.4	-1.4	1.2
138.65	1.14	2.0	136.03	3AU	0.2	-3.4	-1.4	1.2
145.02	1.13	2.0	145.69	13AU	0.4	-3.5	-1.4	1.2
151.42	1.11	2.0	155.41	23AU	0.7	-3.5	-1.3	1.1
157.83	1.08	2.0	165.18	2SE	1.6	-3.6	-1.3	1.1
164.28	1.05	2.0	175.01	12SE	3.0	-3.6	-1.3	1.0
170.75	1.02	2.0	184.89	22SE	0.3	-3.7	-1.2	1.0
177.25	0.98	2.0	194.84	2OC	-0.7	-3.8	-1.2	1.0
183.79	0.93	2.0	204.83	12OC	-0.8	-3.9	-1.2	1.0
190.36	0.88	1.9	214.89	22OC	-0.8	-4.0	-1.2	1.0
196.97	0.82	1.9	224.98	1NO	-0.7	-4.1	-1.2	1.0
203.61	0.76	1.9	235.12	11NO	-0.6	-4.2	-1.3	1.0
210.29	0.69	1.8	245.29	21NO	-0.5	-4.3	-1.3	1.0
217.01	0.61	1.8	255.48	1DE	-0.5	-4.4	-1.3	1.0
223.77	0.52	1.7	265.67	11DE	-0.1	-4.4	-1.4	0.9
230.56	0.43	1.7	275.87	21DE	1.8	-4.1	-1.4	0.9
237.39	0.32	1.6	286.06	31DE	1.6	-3.5	-1.5	0.9
				1010	MAGNITUDES			
244.26	0.20	1.5	296.23	10JA	0.4	-3.3	-1.5	0.8
251.16	0.07	1.5	306.37	20JA	0.2	-3.9	-1.6	0.8
258.09	-0.07	1.4	316.47	30JA	0.0	-4.3	-1.6	0.7
265.04	-0.22	1.3	326.52	9FE	-0.2	-4.3	-1.7	0.7
272.03	-0.39	1.2	336.52	19FE	-0.6	-4.3	-1.8	0.6
279.02	-0.57	1.1	346.46	1MR	-1.3	-4.2	-1.9	0.6
286.03	-0.76	1.0	356.35	11MR	-1.5	-4.1	-1.9	0.5
293.05	-0.97	0.8	6.18	21MR	-0.7	-4.0	-2.0	0.5
300.05	-1.19	0.7	15.95	31MR	0.5	-3.9	-2.0	0.6
307.03	-1.43	0.6	25.67	10AP	1.9	-3.8	-2.1	0.6
313.96	-1.67	0.5	35.35	20AP	3.8	-3.7	-2.1	0.7
320.82	-1.93	0.3	44.97	30AP	2.4	-3.6	-2.1	0.8
327.59	-2.19	0.2	54.56	10MY	1.3	-3.5	-2.1	0.8
334.22	-2.46	0.0	64.13	20MY	0.6	-3.5	-2.0	0.9
340.67	-2.73	-0.1	73.67	30MY	-0.3	-3.4	-2.0	0.9
346.88	-3.01	-0.3	83.19	9JN	-1.3	-3.4	-2.0	1.0
352.77	-3.28	-0.4	92.72	19JN	-1.7	-3.4	-1.9	1.0
358.25	-3.55	-0.6	102.26	29JN	-0.8	-3.3	-1.9	1.0
3.22	-3.81	-0.8	111.80	9JL	-0.1	-3.3	-1.8	1.1
7.52	-4.05	-1.0	121.38	19JL	0.3	-3.3	-1.7	1.1
10.97	-4.26	-1.2	130.98	29JL	0.6	-3.3	-1.7	1.1
13.37	-4.43	-1.5	140.63	8AU	1.0	-3.3	-1.6	1.1
14.51	-4.52	-1.7	150.32	18AU	2.0	-3.3	-1.5	1.0
14.22	-4.50	-1.9	160.06	28AU	2.6	-3.4	-1.5	1.0
12.52	-4.31	-2.1	169.85	7SE	0.2	-3.4	-1.4	1.0
9.71	-3.93	-2.3	179.71	17SE	-0.8	-3.4	-1.4	0.9
6.48	-3.36	-2.3	189.62	27SE	-1.0	-3.4	-1.4	0.9
3.62	-2.68	-2.0	199.59	7OC	-1.0	-3.5	-1.3	0.9
1.77	-1.99	-1.7	209.61	17OC	-0.6	-3.5	-1.3	0.9
1.24	-1.34	-1.4	219.68	27OC	-0.5	-3.5	-1.3	0.9
2.01	-0.78	-1.1	229.80	6NO	-0.4	-3.5	-1.3	0.9
3.93	-0.32	-0.7	239.95	16NO	-0.4	-3.4	-1.3	0.9
6.80	0.05	-0.5	250.13	26NO	0.0	-3.4	-1.3	0.9
10.41	0.35	-0.2	260.32	6DE	2.1	-3.4	-1.3	0.9
14.60	0.59	0.1	270.53	16DE	1.3	-3.4	-1.3	0.9
19.23	0.78	0.3	280.72	26DE	0.2	-3.3	-1.4	0.8
				1011				
24.21	0.93	0.5	290.90	5JA	0.0	-3.3	-1.4	0.8
29.45	1.05	0.7	301.06	15JA	-0.1	-3.3	-1.5	0.8
34.90	1.14	0.9	311.17	25JA	-0.3	-3.3	-1.5	0.7
40.51	1.22	1.0	321.25	4FE	-0.6	-3.4	-1.6	0.7
46.24	1.27	1.2	331.28	14FE	-1.3	-3.4	-1.6	0.6
52.06	1.32	1.3	341.26	24FE	-1.4	-3.4	-1.7	0.6
57.95	1.35	1.4	351.17	6MR	-0.7	-3.4	-1.8	0.5
63.90	1.37	1.5	1.04	16MR	0.7	-3.4	-1.8	0.5
69.89	1.38	1.6	10.84	26MR	2.4	-3.5	-1.9	0.4
75.92	1.38	1.7	20.58	5AP	3.1	-3.5	-2.0	0.4
81.98	1.38	1.8	30.28	15AP	1.7	-3.6	-2.0	0.5
88.06	1.37	1.8	39.93	25AP	1.0	-3.6	-2.1	0.5
94.17	1.36	1.9	49.54	5MY	0.4	-3.7	-2.1	0.6
100.30	1.34	1.9	59.12	15MY	-0.3	-3.8	-2.2	0.7
106.44	1.31	2.0	68.67	25MY	-1.3	-3.9	-2.2	0.7
112.62	1.29	2.0	78.20	4JN	-1.7	-4.0	-2.2	0.8
118.81	1.26	2.0	87.73	14JN	-0.8	-4.1	-2.2	0.8
125.03	1.22	2.0	97.26	24JN	-0.1	-4.1	-2.2	0.9
131.29	1.18	2.0	106.80	4JL	0.4	-4.2	-2.1	0.9
137.57	1.14	2.0	116.36	14JL	0.8	-4.2	-2.1	0.9
143.89	1.09	2.0	125.94	24JL	1.3	-3.9	-2.0	1.0
150.25	1.05	2.0	135.57	3AU	2.6	-3.5	-1.9	1.0
156.66	0.99	2.0	145.23	13AU	2.2	-3.3	-1.9	1.0
163.11	0.94	2.0	154.94	23AU	0.1	-3.8	-1.8	0.9
169.61	0.87	1.9	164.71	2SE	-1.0	-4.2	-1.8	0.9
176.17	0.81	1.9	174.53	12SE	-1.1	-4.3	-1.7	0.9
182.78	0.74	1.8	184.41	22SE	-1.0	-4.3	-1.6	0.9
189.45	0.67	1.9	194.35	2OC	-0.6	-4.2	-1.6	0.8
196.18	0.59	1.8	204.35	12OC	-0.3	-4.1	-1.5	0.8
202.97	0.51	1.8	214.39	22OC	-0.3	-4.0	-1.5	0.8
209.83	0.42	1.8	224.49	1NO	-0.2	-3.9	-1.5	0.8
216.76	0.33	1.8	234.62	11NO	0.2	-3.9	-1.4	0.8
223.75	0.23	1.8	244.79	21NO	2.4	-3.8	-1.4	0.8
230.81	0.13	1.8	254.98	1DE	1.0	-3.7	-1.4	0.8
237.93	0.02	1.7	265.18	11DE	0.0	-3.6	-1.4	0.8
245.12	-0.09	1.7	275.38	21DE	-0.1	-3.6	-1.4	0.8
252.37	-0.21	1.6	285.57	31DE	-0.2	-3.5	-1.4	0.8

♂ LONG	♂ LAT	♂ MAG	☉ LONG	16.00UT	☿	♀	♃	♄
				1012	MAGNITUDES			
259.68	-0.33	1.6	295.74	10JA	-0.4	-3.5	-1.4	0.8
267.05	-0.45	1.5	305.88	20JA	-0.7	-3.4	-1.5	0.7
274.47	-0.58	1.5	315.98	30JA	-1.2	-3.4	-1.5	0.7
281.93	-0.71	1.4	326.04	9FE	-1.2	-3.4	-1.5	0.6
289.44	-0.84	1.4	336.04	19FE	-0.7	-3.3	-1.6	0.6
296.98	-0.97	1.3	345.99	29FE	0.8	-3.3	-1.6	0.5
304.54	-1.10	1.2	355.88	10MR	3.0	-3.3	-1.7	0.5
312.12	-1.23	1.2	5.71	20MR	2.3	-3.3	-1.7	0.4
319.70	-1.35	1.1	15.49	30MR	1.3	-3.3	-1.8	0.4
327.27	-1.46	1.1	25.21	9AP	0.8	-3.3	-1.9	0.3
334.81	-1.57	1.0	34.88	19AP	0.3	-3.3	-2.0	0.3
342.31	-1.66	0.9	44.51	29AP	-0.4	-3.4	-2.0	0.4
349.74	-1.74	0.9	54.10	9MY	-1.3	-3.4	-2.1	0.5
357.10	-1.80	0.8	63.66	19MY	-1.7	-3.5	-2.2	0.5
4.36	-1.85	0.7	73.21	29MY	-0.8	-3.5	-2.2	0.6
11.48	-1.88	0.7	82.73	8JN	0.0	-3.5	-2.3	0.6
18.46	-1.89	0.6	92.26	18JN	0.6	-3.4	-2.3	0.7
25.26	-1.88	0.5	101.79	28JN	1.0	-3.4	-2.3	0.8
31.84	-1.85	0.4	111.34	8JL	1.8	-3.4	-2.3	0.8
38.17	-1.79	0.3	120.91	18JL	3.1	-3.4	-2.3	0.8
44.20	-1.72	0.2	130.52	28JL	1.9	-3.3	-2.3	0.9
49.86	-1.61	0.1	140.16	7AU	0.0	-3.3	-2.2	0.9
55.08	-1.48	-0.0	149.84	17AU	-1.0	-3.3	-2.2	0.9
59.77	-1.32	-0.1	159.58	27AU	-1.3	-3.3	-2.1	0.9
63.80	-1.12	-0.3	169.37	6SE	-0.9	-3.3	-2.1	0.9
67.02	-0.87	-0.5	179.22	16SE	-0.5	-3.4	-2.0	0.8
69.23	-0.58	-0.7	189.14	26SE	-0.2	-3.4	-1.9	0.8
70.25	-0.22	-0.9	199.10	6OC	-0.1	-3.4	-1.9	0.8
69.87	0.20	-1.1	209.12	16OC	-0.0	-3.4	-1.8	0.7
68.03	0.66	-1.3	219.19	26OC	0.4	-3.5	-1.7	0.7
64.95	1.15	-1.5	229.30	5NO	2.7	-3.5	-1.7	0.7
61.16	1.60	-1.6	239.45	15NO	0.8	-3.6	-1.6	0.8
57.45	1.97	-1.4	249.63	25NO	-0.2	-3.7	-1.6	0.8
54.57	2.22	-1.1	259.82	5DE	-0.3	-3.8	-1.6	0.8
52.94	2.36	-0.8	270.03	15DE	-0.3	-3.8	-1.5	0.8
52.66	2.42	-0.5	280.23	25DE	-0.5	-3.9	-1.5	0.8
				1013				
53.61	2.42	-0.2	290.40	4JA	-0.8	-4.0	-1.5	0.8
55.60	2.39	0.1	300.56	14JA	-1.1	-4.1	-1.5	0.8
58.43	2.33	0.3	310.69	24JA	-1.1	-4.2	-1.5	0.7
61.92	2.27	0.5	320.76	3FE	-0.6	-4.3	-1.5	0.7
65.91	2.19	0.7	330.80	13FE	1.0	-4.3	-1.5	0.6
70.32	2.12	0.9	340.78	23FE	3.2	-4.2	-1.6	0.6
75.04	2.04	1.1	350.69	5MR	1.7	-4.0	-1.6	0.5
80.02	1.96	1.2	0.56	15MR	0.9	-3.4	-1.6	0.5
85.20	1.88	1.4	10.37	25MR	0.6	-3.3	-1.7	0.4
90.55	1.80	1.5	20.11	4AP	0.2	-3.9	-1.7	0.4
96.04	1.72	1.6	29.81	14AP	-0.4	-4.2	-1.8	0.3
101.64	1.65	1.7	39.47	24AP	-1.3	-4.2	-1.8	0.3
107.34	1.57	1.7	49.07	4MY	-1.7	-4.2	-1.9	0.3
113.14	1.49	1.8	58.65	14MY	-0.8	-4.1	-2.0	0.4
119.01	1.41	1.8	68.21	24MY	0.1	-4.0	-2.1	0.4
124.96	1.33	1.9	77.74	3JN	0.7	-3.9	-2.1	0.5
130.99	1.26	1.9	87.27	13JN	1.4	-3.8	-2.2	0.5
137.08	1.18	1.9	96.80	23JN	2.3	-3.7	-2.3	0.6
143.24	1.09	2.0	106.33	3JL	3.2	-3.6	-2.3	0.7
149.47	1.01	2.0	115.89	13JL	1.5	-3.6	-2.4	0.7
155.77	0.93	2.0	125.48	23JL	-0.0	-3.5	-2.4	0.8
162.14	0.84	2.0	135.10	2AU	-1.1	-3.5	-2.4	0.8
168.59	0.76	1.9	144.76	12AU	-1.4	-3.4	-2.4	0.8
175.11	0.67	1.9	154.48	22AU	-0.9	-3.4	-2.4	0.8
181.71	0.58	1.9	164.24	1SE	-0.4	-3.4	-2.4	0.8
188.39	0.48	1.9	174.06	11SE	-0.1	-3.4	-2.3	0.8
195.15	0.39	1.8	183.94	21SE	0.0	-3.4	-2.3	0.8
201.99	0.29	1.8	193.87	1OC	0.2	-3.4	-2.2	0.8
208.92	0.19	1.7	203.87	11OC	0.7	-3.4	-2.2	0.8
215.94	0.09	1.7	213.91	21OC	3.0	-3.4	-2.1	0.7
223.03	-0.01	1.6	224.00	31OC	0.6	-3.4	-2.0	0.7
230.22	-0.11	1.6	234.13	10NO	-0.3	-3.4	-2.0	0.7
237.48	-0.22	1.6	244.29	20NO	-0.5	-3.4	-1.9	0.7
244.82	-0.32	1.6	254.48	30NO	-0.5	-3.4	-1.8	0.7
252.24	-0.42	1.6	264.68	10DE	-0.6	-3.4	-1.8	0.8
259.74	-0.53	1.6	274.88	20DE	-0.8	-3.5	-1.7	0.8
267.29	-0.63	1.5	285.07	30DE	-0.9	-3.5	-1.7	0.8

♂ LONG	♂ LAT	♂ MAG	☉ LONG	16.00UT	☿	♀	♃	♄
				1014	MAGNITUDES			
274.92	-0.73	1.5	295.24	9JA	-1.0	-3.5	-1.6	0.8
282.59	-0.82	1.5	305.38	19JA	-0.5	-3.5	-1.6	0.8
290.30	-0.91	1.5	315.48	29JA	1.2	-3.4	-1.6	0.7
298.05	-0.99	1.5	325.54	8FE	2.9	-3.4	-1.6	0.7
305.83	-1.07	1.4	335.55	18FE	1.3	-3.4	-1.6	0.7
313.61	-1.13	1.4	345.50	28FE	0.7	-3.4	-1.6	0.6
321.40	-1.19	1.4	355.39	10MR	0.4	-3.3	-1.6	0.6
329.18	-1.23	1.4	5.23	20MR	0.1	-3.3	-1.6	0.5
336.93	-1.27	1.3	15.01	30MR	-0.5	-3.3	-1.6	0.5
344.65	-1.28	1.3	24.73	9AP	-1.2	-3.3	-1.6	0.4
352.31	-1.29	1.3	34.41	19AP	-1.7	-3.3	-1.7	0.3
359.91	-1.28	1.3	44.04	29AP	-0.8	-3.4	-1.7	0.3
7.44	-1.25	1.3	53.64	9MY	0.2	-3.4	-1.8	0.2
14.88	-1.22	1.2	63.20	19MY	1.0	-3.4	-1.8	0.3
22.23	-1.16	1.2	72.74	29MY	1.8	-3.4	-1.9	0.3
29.46	-1.09	1.2	82.28	8JN	3.1	-3.5	-1.9	0.4
36.57	-1.01	1.2	91.80	18JN	2.7	-3.5	-2.0	0.5
43.55	-0.92	1.1	101.33	28JN	1.2	-3.6	-2.1	0.5
50.38	-0.81	1.1	110.88	8JL	-0.1	-3.7	-2.2	0.6
57.05	-0.69	1.1	120.45	18JL	-1.2	-3.7	-2.2	0.6
63.54	-0.55	1.0	130.05	28JL	-1.5	-3.8	-2.3	0.7
69.83	-0.40	1.0	139.69	7AU	-0.9	-3.9	-2.3	0.7
75.89	-0.23	0.9	149.38	17AU	-0.3	-4.0	-2.4	0.8
81.69	-0.05	0.8	159.11	27AU	-0.0	-4.2	-2.4	0.8
87.19	0.16	0.7	168.91	6SE	0.2	-4.2	-2.5	0.8
92.34	0.38	0.6	178.75	16SE	0.4	-4.3	-2.5	0.8
97.06	0.63	0.5	188.66	26SE	1.0	-4.3	-2.5	0.8
101.27	0.91	0.4	198.62	6OC	3.3	-4.0	-2.4	0.8
104.84	1.22	0.2	208.64	16OC	0.5	-3.5	-2.4	0.7
107.64	1.58	0.0	218.70	26OC	-0.5	-3.3	-2.3	0.7
109.47	1.98	-0.2	228.82	5NO	-0.6	-4.0	-2.3	0.7
110.16	2.41	-0.4	238.96	15NO	-0.6	-4.3	-2.2	0.6
109.53	2.87	-0.6	249.14	25NO	-0.7	-4.4	-2.1	0.7
107.51	3.33	-0.9	259.33	5DE	-0.8	-4.3	-2.1	0.7
104.30	3.71	-1.0	269.53	15DE	-0.8	-4.3	-2.0	0.7
100.39	3.97	-1.1	279.73	25DE	-0.8	-4.2	-1.9	0.8
				1015				
96.53	4.06	-1.0	289.91	4JA	-0.4	-4.1	-1.9	0.8
93.44	4.00	-0.8	300.07	14JA	1.4	-3.9	-1.8	0.8
91.53	3.83	-0.5	310.19	24JA	2.4	-3.8	-1.7	0.8
90.95	3.59	-0.2	320.28	3FE	0.9	-3.8	-1.7	0.8
91.59	3.34	0.1	330.31	13FE	0.5	-3.7	-1.7	0.7
93.28	3.08	0.3	340.29	23FE	0.2	-3.6	-1.6	0.7
95.83	2.84	0.5	350.22	5MR	-0.0	-3.5	-1.6	0.7
99.07	2.61	0.7	0.08	15MR	-0.5	-3.5	-1.6	0.6
102.84	2.40	0.9	9.89	25MR	-1.2	-3.4	-1.6	0.6
107.07	2.20	1.0	19.64	4AP	-1.6	-3.4	-1.6	0.5
111.64	2.02	1.2	29.34	14AP	-0.8	-3.3	-1.6	0.4
116.50	1.85	1.3	39.00	24AP	0.3	-3.3	-1.6	0.4
121.61	1.69	1.4	48.61	4MY	1.3	-3.3	-1.6	0.3
126.92	1.54	1.5	58.19	14MY	2.4	-3.3	-1.6	0.2
132.41	1.40	1.6	67.75	24MY	3.5	-3.3	-1.7	0.2
138.05	1.26	1.6	77.28	3JN	2.1	-3.3	-1.7	0.3
143.84	1.13	1.7	86.81	13JN	1.0	-3.3	-1.7	0.3
149.76	1.00	1.7	96.34	23JN	-0.1	-3.4	-1.8	0.4
155.80	0.88	1.7	105.88	3JL	-1.2	-3.4	-1.8	0.5
161.96	0.76	1.8	115.43	13JL	-1.6	-3.4	-1.9	0.5
168.23	0.64	1.8	125.02	23JL	-0.9	-3.5	-2.0	0.6
174.61	0.52	1.8	134.63	2AU	-0.3	-3.5	-2.0	0.6
181.10	0.40	1.8	144.29	12AU	0.1	-3.5	-2.1	0.7
187.69	0.29	1.8	154.00	22AU	0.3	-3.5	-2.2	0.7
194.40	0.18	1.8	163.76	1SE	0.6	-3.4	-2.2	0.7
201.21	0.07	1.7	173.58	11SE	1.3	-3.4	-2.3	0.8
208.12	-0.04	1.7	183.46	21SE	3.3	-3.4	-2.3	0.8
215.15	-0.15	1.7	193.39	1OC	0.4	-3.4	-2.4	0.8
222.27	-0.25	1.7	203.37	11OC	-0.6	-3.3	-2.4	0.8
229.49	-0.36	1.6	213.42	21OC	-0.8	-3.3	-2.4	0.8
236.81	-0.46	1.6	223.51	31OC	-0.7	-3.3	-2.4	0.7
244.23	-0.55	1.5	233.64	10NO	-0.8	-3.3	-2.4	0.7
251.73	-0.64	1.5	243.80	20NO	-0.7	-3.3	-2.3	0.7
259.31	-0.72	1.4	253.98	30NO	-0.6	-3.4	-2.3	0.7
266.97	-0.80	1.4	264.18	10DE	-0.6	-3.4	-2.2	0.7
274.69	-0.87	1.4	274.38	20DE	-0.3	-3.4	-2.2	0.7
282.47	-0.94	1.4	284.57	30DE	1.6	-3.5	-2.1	0.8

♂			☉	16.00UT	☿	♀	♃	♄
LONG	LAT	MAG	LONG	1016	MAGNITUDES			
290.29	-0.99	1.4	294.74	9JA	1.9	-3.5	-2.0	0.8
298.14	-1.04	1.4	304.89	19JA	0.6	-3.6	-1.9	0.8
306.01	-1.07	1.4	314.99	29JA	0.3	-3.6	-1.9	0.8
313.89	-1.09	1.4	325.05	8FE	0.1	-3.7	-1.8	0.8
321.77	-1.10	1.4	335.06	18FE	-0.1	-3.7	-1.7	0.8
329.62	-1.10	1.4	345.02	28FE	-0.5	-3.8	-1.7	0.7
337.45	-1.09	1.4	354.91	9MR	-1.2	-3.9	-1.6	0.7
345.23	-1.06	1.4	4.75	19MR	-1.5	-4.0	-1.6	0.7
352.96	-1.02	1.4	14.53	29MR	-0.8	-4.1	-1.6	0.6
0.63	-0.97	1.4	24.26	8AP	0.4	-4.1	-1.5	0.5
8.23	-0.91	1.5	33.94	18AP	1.6	-4.2	-1.5	0.5
15.75	-0.84	1.5	43.57	28AP	3.2	-4.2	-1.5	0.4
23.18	-0.76	1.5	53.17	8MY	2.8	-4.0	-1.5	0.4
30.52	-0.68	1.5	62.74	18MY	1.6	-3.6	-1.5	0.3
37.76	-0.58	1.5	72.28	28MY	0.7	-2.7	-1.5	0.2
44.89	-0.47	1.5	81.81	7JN	-0.2	-3.5	-1.5	0.2
51.93	-0.36	1.5	91.34	17JN	-1.2	-4.0	-1.5	0.3
58.84	-0.25	1.5	100.87	27JN	-1.7	-4.2	-1.6	0.4
65.65	-0.12	1.5	110.42	7JL	-0.8	-4.2	-1.6	0.4
72.34	0.01	1.5	119.99	17JL	-0.2	-4.1	-1.6	0.5
78.90	0.14	1.5	129.58	27JL	0.2	-4.1	-1.7	0.5
85.33	0.29	1.5	139.22	6AU	0.5	-4.0	-1.7	0.6
91.61	0.43	1.4	148.91	16AU	0.8	-3.9	-1.8	0.6
97.74	0.59	1.4	158.64	26AU	1.7	-3.8	-1.9	0.7
103.70	0.75	1.3	168.43	5SE	2.9	-3.7	-1.9	0.7
109.47	0.93	1.3	178.27	15SE	0.3	-3.7	-2.0	0.7
115.01	1.11	1.2	188.17	25SE	-0.8	-3.6	-2.1	0.8
120.31	1.31	1.1	198.13	5OC	-0.9	-3.6	-2.1	0.8
125.30	1.53	1.0	208.15	15OC	-0.9	-3.5	-2.2	0.8
129.93	1.76	0.9	218.21	25OC	-0.7	-3.5	-2.2	0.8
134.12	2.02	0.7	228.32	4NO	-0.5	-3.4	-2.3	0.7
137.78	2.31	0.6	238.47	14NO	-0.5	-3.4	-2.3	0.7
140.77	2.62	0.4	248.64	24NO	-0.5	-3.4	-2.3	0.7
142.94	2.96	0.2	258.84	4DE	-0.1	-3.4	-2.3	0.7
144.10	3.34	-0.1	269.04	14DE	1.8	-3.4	-2.3	0.7
144.08	3.72	-0.3	279.23	24DE	1.6	-3.3	-2.2	0.7
				1017				
142.75	4.08	-0.5	289.42	3JA	0.4	-3.3	-2.2	0.8
140.14	4.38	-0.8	299.58	13JA	0.1	-3.3	-2.1	0.8
136.56	4.54	-1.0	309.70	23JA	-0.0	-3.3	-2.1	0.8
132.60	4.54	-1.0	319.79	2FE	-0.2	-3.4	-2.0	0.8
129.01	4.36	-0.8	329.82	12FE	-0.6	-3.4	-1.9	0.8
126.36	4.05	-0.6	339.80	22FE	-1.2	-3.4	-1.8	0.8
124.97	3.68	-0.3	349.73	4MR	-1.4	-3.4	-1.8	0.8
124.87	3.29	-0.1	359.60	14MR	-0.8	-3.5	-1.7	0.7
125.92	2.91	0.1	9.41	24MR	0.5	-3.5	-1.6	0.7
127.95	2.56	0.4	19.17	3AP	2.1	-3.5	-1.6	0.7
130.79	2.24	0.5	28.87	13AP	3.7	-3.4	-1.5	0.6
134.28	1.95	0.7	38.52	23AP	2.1	-3.4	-1.5	0.5
138.30	1.69	0.9	48.14	3MY	1.2	-3.4	-1.5	0.5
142.76	1.45	1.0	57.72	13MY	0.5	-3.3	-1.4	0.4
147.57	1.23	1.1	67.27	23MY	-0.2	-3.3	-1.4	0.4
152.70	1.02	1.2	76.81	2JN	-1.2	-3.3	-1.4	0.3
158.09	0.83	1.3	86.34	12JN	-1.7	-3.3	-1.4	0.2
163.70	0.65	1.3	95.87	22JN	-0.8	-3.3	-1.4	0.3
169.53	0.49	1.4	105.41	2JL	-0.1	-3.3	-1.4	0.3
175.54	0.33	1.4	114.96	12JL	0.3	-3.4	-1.4	0.4
181.73	0.18	1.5	124.54	22JL	0.7	-3.4	-1.4	0.4
188.08	0.04	1.5	134.16	1AU	1.1	-3.4	-1.5	0.5
194.59	-0.09	1.5	143.82	11AU	2.2	-3.5	-1.5	0.6
201.24	-0.22	1.5	153.53	21AU	2.5	-3.5	-1.5	0.6
208.04	-0.34	1.5	163.29	31AU	0.2	-3.6	-1.6	0.7
214.96	-0.45	1.5	173.10	10SE	-0.9	-3.6	-1.6	0.7
222.02	-0.56	1.5	182.97	20SE	-1.1	-3.7	-1.7	0.7
229.20	-0.66	1.5	192.91	30SE	-1.0	-3.8	-1.8	0.8
236.50	-0.74	1.5	202.89	10OC	-0.6	-3.9	-1.8	0.8
243.91	-0.83	1.5	212.93	20OC	-0.4	-4.0	-1.9	0.8
251.42	-0.90	1.5	223.02	30OC	-0.3	-4.1	-2.0	0.8
259.02	-0.96	1.5	233.14	9NO	-0.3	-4.2	-2.0	0.8
266.70	-1.01	1.4	243.31	19NO	0.1	-4.3	-2.1	0.8
274.45	-1.05	1.4	253.49	29NO	2.1	-4.4	-2.1	0.7
282.26	-1.08	1.4	263.69	9DE	1.3	-4.3	-2.2	0.7
290.12	-1.10	1.4	273.89	19DE	0.1	-4.1	-2.2	0.7
298.01	-1.11	1.4	284.08	29DE	-0.1	-3.5	-2.2	0.7

♂			☉	16.00UT	☿	♀	♃	♄
LONG	LAT	MAG	LONG	1018	MAGNITUDES			
305.91	-1.10	1.4	294.25	8JA	-0.1	-3.3	-2.2	0.8
313.82	-1.08	1.4	304.40	18JA	-0.3	-4.0	-2.1	0.8
321.72	-1.05	1.3	314.51	28JA	-0.7	-4.3	-2.1	0.8
329.59	-1.01	1.3	324.56	7FE	-1.2	-4.3	-2.0	0.8
337.43	-0.96	1.3	334.58	17FE	-1.3	-4.3	-2.0	0.8
345.22	-0.90	1.3	344.54	27FE	-0.7	-4.2	-1.9	0.8
352.95	-0.83	1.4	354.43	9MR	0.7	-4.1	-1.9	0.8
0.62	-0.75	1.4	4.28	19MR	2.6	-4.0	-1.8	0.8
8.21	-0.67	1.4	14.06	29MR	2.8	-3.9	-1.7	0.8
15.72	-0.58	1.5	23.79	8AP	1.6	-3.8	-1.6	0.7
23.15	-0.48	1.5	33.47	18AP	0.9	-3.7	-1.6	0.7
30.49	-0.39	1.6	43.11	28AP	0.4	-3.6	-1.5	0.6
37.74	-0.28	1.6	52.70	8MY	-0.3	-3.5	-1.5	0.6
44.90	-0.18	1.6	62.27	18MY	-1.2	-3.5	-1.4	0.5
51.96	-0.07	1.7	71.82	28MY	-1.7	-3.4	-1.4	0.4
58.94	0.04	1.7	81.35	7JN	-0.8	-3.4	-1.4	0.4
65.82	0.15	1.7	90.88	17JN	-0.1	-3.4	-1.3	0.3
72.62	0.26	1.7	100.41	27JN	0.5	-3.3	-1.3	0.3
79.32	0.37	1.8	109.95	7JL	0.9	-3.3	-1.3	0.3
85.94	0.49	1.8	119.52	17JL	1.5	-3.3	-1.3	0.4
92.47	0.60	1.8	129.12	27JL	2.7	-3.3	-1.3	0.4
98.90	0.71	1.8	138.76	6AU	2.1	-3.3	-1.3	0.5
105.25	0.83	1.8	148.44	16AU	0.1	-3.3	-1.3	0.5
111.50	0.95	1.7	158.17	26AU	-1.0	-3.4	-1.3	0.6
117.64	1.06	1.7	167.95	5SE	-1.2	-3.4	-1.4	0.6
123.68	1.19	1.7	177.80	15SE	-1.0	-3.4	-1.4	0.7
129.59	1.31	1.6	187.70	25SE	-0.5	-3.4	-1.4	0.7
135.37	1.44	1.6	197.65	5OC	-0.3	-3.5	-1.5	0.8
141.00	1.57	1.5	207.66	15OC	-0.2	-3.5	-1.6	0.8
146.45	1.72	1.4	217.73	25OC	-0.1	-3.5	-1.6	0.8
151.69	1.87	1.3	227.83	4NO	0.3	-3.5	-1.7	0.8
156.68	2.02	1.2	237.98	14NO	2.4	-3.4	-1.7	0.8
161.37	2.20	1.1	248.15	24NO	1.0	-3.4	-1.8	0.8
165.69	2.38	0.9	258.34	4DE	-0.1	-3.4	-1.9	0.8
169.56	2.58	0.7	268.54	14DE	-0.2	-3.4	-1.9	0.8
172.87	2.79	0.5	278.74	24DE	-0.3	-3.3	-2.0	0.7
				1019				
175.48	3.02	0.3	288.92	3JA	-0.4	-3.3	-2.0	0.8
177.23	3.25	0.1	299.08	13JA	-0.7	-3.3	-2.1	0.8
177.94	3.49	-0.2	309.21	23JA	-1.2	-3.3	-2.1	0.8
177.44	3.69	-0.5	319.29	2FE	-1.2	-3.4	-2.1	0.9
175.64	3.84	-0.7	329.34	12FE	-0.7	-3.4	-2.1	0.9
172.69	3.88	-1.0	339.32	22FE	0.8	-3.4	-2.0	0.9
168.97	3.77	-1.1	349.25	4MR	3.1	-3.4	-2.0	0.9
165.16	3.51	-1.1	359.13	14MR	2.1	-3.5	-1.9	0.9
161.95	3.13	-0.9	8.94	24MR	1.1	-3.5	-1.9	0.9
159.85	2.69	-0.7	18.70	3AP	0.7	-3.5	-1.8	0.9
159.04	2.24	-0.5	28.41	13AP	0.3	-3.6	-1.7	0.8
159.52	1.80	-0.2	38.06	23AP	-0.4	-3.6	-1.7	0.8
161.11	1.41	-0.0	47.68	3MY	-1.2	-3.7	-1.6	0.7
163.65	1.06	0.1	57.26	13MY	-1.7	-3.8	-1.5	0.7
166.98	0.75	0.3	66.82	23MY	-0.8	-3.9	-1.5	0.6
170.94	0.47	0.5	76.35	2JN	0.0	-4.0	-1.4	0.5
175.44	0.22	0.6	85.88	12JN	0.6	-4.1	-1.4	0.5
180.38	0.00	0.7	95.41	22JN	1.2	-4.2	-1.3	0.4
185.70	-0.20	0.8	104.95	2JL	2.0	-4.2	-1.3	0.4
191.34	-0.38	0.9	114.50	12JL	3.2	-4.1	-1.3	0.3
197.27	-0.54	0.9	124.08	22JL	1.8	-3.9	-1.3	0.4
203.45	-0.68	1.0	133.70	1AU	0.1	-3.4	-1.2	0.4
209.86	-0.81	1.0	143.36	11AU	-1.1	-3.3	-1.2	0.5
216.48	-0.92	1.1	153.06	21AU	-1.3	-3.8	-1.2	0.5
223.28	-1.02	1.1	162.81	31AU	-0.9	-4.2	-1.2	0.6
230.26	-1.10	1.1	172.63	10SE	-0.5	-4.3	-1.2	0.7
237.39	-1.17	1.2	182.49	20SE	-0.2	-4.3	-1.3	0.7
244.67	-1.23	1.2	192.42	30SE	-0.1	-4.2	-1.3	0.7
252.07	-1.27	1.2	202.41	10OC	0.0	-4.1	-1.3	0.8
259.58	-1.30	1.2	212.44	20OC	0.5	-4.0	-1.4	0.8
267.20	-1.31	1.3	222.53	30OC	2.7	-3.9	-1.4	0.8
274.89	-1.31	1.3	232.66	9NO	0.8	-3.9	-1.5	0.9
282.65	-1.29	1.3	242.81	19NO	-0.2	-3.8	-1.5	0.9
290.46	-1.26	1.3	253.00	29NO	-0.4	-3.7	-1.6	0.9
298.31	-1.22	1.3	263.20	9DE	-0.4	-3.6	-1.6	0.9
306.17	-1.17	1.4	273.39	19DE	-0.5	-3.6	-1.7	0.8
314.03	-1.11	1.4	283.59	29DE	-0.8	-3.5	-1.8	0.8

♂ LONG	LAT	MAG	☉ LONG	16.00UT	☿	♀	♃	♄
				1020	MAGNITUDES			
321.88	-1.03	1.4	293.76	8JA	-1.0	-3.5	-1.8	0.8
329.70	-0.95	1.4	303.91	18JA	-1.0	-3.4	-1.9	0.8
337.49	-0.86	1.4	314.02	28JA	-0.6	-3.4	-1.9	0.9
345.22	-0.77	1.5	324.08	7FE	1.0	-3.4	-2.0	0.9
352.89	-0.67	1.5	334.10	17FE	3.2	-3.3	-2.0	1.0
0.49	-0.57	1.5	344.06	27FE	1.6	-3.3	-2.0	1.0
8.02	-0.46	1.5	353.96	8MR	0.8	-3.3	-2.0	1.0
15.46	-0.35	1.6	3.80	18MR	0.5	-3.3	-2.0	1.0
22.83	-0.25	1.6	13.59	28MR	0.1	-3.3	-2.0	1.0
30.11	-0.14	1.6	23.33	7AP	-0.4	-3.3	-1.9	1.0
37.31	-0.03	1.6	33.01	17AP	-1.2	-3.3	-1.9	0.9
44.42	0.07	1.6	42.65	27AP	-1.7	-3.4	-1.8	0.9
51.45	0.17	1.6	52.24	7MY	-0.8	-3.4	-1.7	0.9
58.39	0.27	1.7	61.81	17MY	0.1	-3.5	-1.7	0.8
65.26	0.37	1.7	71.36	27MY	0.8	-3.5	-1.6	0.7
72.06	0.47	1.8	80.89	6JN	1.5	-3.5	-1.6	0.7
78.78	0.56	1.8	90.41	16JN	2.6	-3.4	-1.5	0.6
85.44	0.65	1.9	99.95	26JN	3.1	-3.4	-1.4	0.5
92.02	0.73	1.9	109.49	6JL	1.4	-3.4	-1.4	0.5
98.55	0.82	1.9	119.05	16JL	0.0	-3.4	-1.3	0.4
105.02	0.90	1.9	128.65	26JL	-1.1	-3.3	-1.3	0.4
111.43	0.98	1.9	138.28	5AU	-1.5	-3.3	-1.3	0.5
117.78	1.06	1.9	147.96	15AU	-0.9	-3.3	-1.2	0.5
124.09	1.13	1.9	157.69	25AU	-0.4	-3.3	-1.2	0.6
130.33	1.20	1.9	167.47	4SE	-0.1	-3.3	-1.2	0.6
136.52	1.28	1.9	177.31	14SE	0.1	-3.4	-1.2	0.7
142.64	1.35	1.9	187.21	24SE	0.2	-3.4	-1.2	0.7
148.70	1.42	1.8	197.16	4OC	0.8	-3.4	-1.2	0.8
154.69	1.49	1.8	207.17	14OC	3.1	-3.4	-1.2	0.8
160.60	1.55	1.7	217.24	24OC	0.6	-3.5	-1.3	0.9
166.42	1.62	1.7	227.34	3NO	-0.4	-3.5	-1.3	0.9
172.13	1.69	1.6	237.48	13NO	-0.5	-3.6	-1.3	0.9
177.71	1.76	1.5	247.66	23NO	-0.5	-3.7	-1.4	0.9
183.15	1.82	1.4	257.85	3DE	-0.6	-3.8	-1.4	0.9
188.40	1.89	1.2	268.05	13DE	-0.8	-3.8	-1.5	0.9
193.43	1.96	1.1	278.25	23DE	-0.9	-3.9	-1.5	0.9
				1021				
198.19	2.02	0.9	288.43	2JA	-0.9	-4.0	-1.6	0.9
202.63	2.08	0.8	298.59	12JA	-0.5	-4.1	-1.7	0.9
206.65	2.13	0.6	308.72	22JA	1.2	-4.2	-1.7	0.9
210.16	2.17	0.3	318.81	1FE	2.7	-4.3	-1.8	1.0
213.04	2.20	0.1	328.85	11FE	1.1	-4.3	-1.9	1.0
215.11	2.20	-0.2	338.84	21FE	0.6	-4.2	-1.9	1.0
216.21	2.16	-0.5	348.77	3MR	0.3	-3.9	-2.0	1.1
216.16	2.07	-0.8	358.65	13MR	0.0	-3.4	-2.0	1.1
214.84	1.89	-1.1	8.47	23MR	-0.5	-3.4	-2.0	1.1
212.33	1.62	-1.4	18.23	2AP	-1.2	-3.9	-2.0	1.1
208.95	1.25	-1.6	27.93	12AP	-1.7	-4.2	-2.0	1.1
205.33	0.80	-1.6	37.60	22AP	-0.8	-4.2	-2.0	1.1
202.21	0.32	-1.4	47.21	2MY	0.2	-4.2	-2.0	1.0
200.13	-0.14	-1.2	56.80	12MY	1.1	-4.1	-1.9	1.0
199.38	-0.54	-1.0	66.36	22MY	2.0	-4.0	-1.9	1.0
199.98	-0.88	-0.8	75.89	1JN	3.3	-3.9	-1.8	0.9
201.78	-1.17	-0.6	85.42	11JN	2.5	-3.8	-1.7	0.8
204.62	-1.39	-0.4	94.95	21JN	1.1	-3.7	-1.7	0.8
208.31	-1.58	-0.2	104.48	1JL	-0.1	-3.6	-1.6	0.7
212.71	-1.72	-0.1	114.04	11JL	-1.2	-3.6	-1.5	0.6
217.70	-1.83	0.0	123.62	21JL	-1.6	-3.5	-1.5	0.6
223.18	-1.91	0.2	133.23	31JL	-0.9	-3.5	-1.4	0.5
229.05	-1.96	0.3	142.88	10AU	-0.3	-3.4	-1.4	0.5
235.28	-1.99	0.4	152.59	20AU	0.0	-3.4	-1.3	0.6
241.79	-2.00	0.5	162.34	30AU	0.2	-3.4	-1.3	0.6
248.55	-1.98	0.5	172.15	9SE	0.4	-3.4	-1.3	0.7
255.53	-1.95	0.6	182.02	19SE	1.1	-3.4	-1.3	0.7
262.68	-1.90	0.7	191.94	29SE	3.3	-3.4	-1.2	0.8
269.98	-1.83	0.8	201.92	9OC	0.5	-3.4	-1.2	0.8
277.41	-1.75	0.8	211.96	19OC	-0.5	-3.4	-1.2	0.9
284.92	-1.65	0.9	222.04	29OC	-0.7	-3.4	-1.2	0.9
292.51	-1.55	1.0	232.16	8NO	-0.7	-3.4	-1.3	1.0
300.14	-1.43	1.0	242.32	18NO	-0.7	-3.4	-1.3	1.0
307.80	-1.31	1.1	252.50	28NO	-0.7	-3.4	-1.3	1.0
315.48	-1.18	1.2	262.70	8DE	-0.7	-3.4	-1.3	1.0
323.14	-1.04	1.2	272.90	18DE	-0.7	-3.5	-1.4	1.1
330.78	-0.91	1.3	283.09	28DE	-0.4	-3.5	-1.4	1.1

♂ LONG	LAT	MAG	☉ LONG	16.00UT	☿	♀	♃	♄
				1022	MAGNITUDES			
338.39	-0.77	1.3	293.27	7JA	1.4	-3.5	-1.5	1.1
345.95	-0.63	1.4	303.42	17JA	2.2	-3.5	-1.5	1.1
353.45	-0.50	1.5	313.53	27JA	0.8	-3.4	-1.6	1.1
0.89	-0.37	1.5	323.60	6FE	0.4	-3.4	-1.7	1.1
8.27	-0.24	1.6	333.61	16FE	0.2	-3.4	-1.7	1.1
15.56	-0.12	1.6	343.57	26FE	-0.1	-3.4	-1.8	1.2
22.79	-0.00	1.7	353.48	8MR	-0.5	-3.3	-1.9	1.2
29.93	0.11	1.7	3.33	18MR	-1.2	-3.3	-1.9	1.2
37.00	0.22	1.7	13.11	28MR	-1.6	-3.3	-2.0	1.2
44.00	0.32	1.8	22.85	7AP	-0.8	-3.3	-2.0	1.3
50.92	0.41	1.8	32.54	17AP	0.3	-3.3	-2.1	1.3
57.77	0.50	1.8	42.18	27AP	1.4	-3.4	-2.1	1.2
64.56	0.59	1.8	51.78	7MY	2.7	-3.4	-2.1	1.2
71.28	0.67	1.8	61.35	17MY	3.3	-3.4	-2.1	1.2
77.94	0.74	1.9	70.89	27MY	1.9	-3.4	-2.1	1.2
84.56	0.81	1.9	80.43	6JN	0.9	-3.5	-2.0	1.1
91.12	0.87	1.9	89.96	16JN	-0.1	-3.5	-2.0	1.1
97.64	0.93	1.9	99.48	26JN	-1.2	-3.6	-1.9	1.0
104.11	0.98	1.9	109.03	6JL	-1.7	-3.7	-1.9	0.9
110.56	1.03	2.0	118.59	16JL	-0.9	-3.7	-1.8	0.9
116.97	1.08	2.0	128.19	26JL	-0.2	-3.8	-1.7	0.8
123.35	1.12	2.0	137.82	5AU	0.1	-3.9	-1.7	0.7
129.71	1.16	2.0	147.49	15AU	0.4	-4.0	-1.6	0.7
136.04	1.19	2.0	157.22	25AU	0.7	-4.2	-1.5	0.7
142.36	1.22	2.0	167.00	4SE	1.4	-4.2	-1.5	0.7
148.66	1.25	2.0	176.83	14SE	3.2	-4.3	-1.4	0.8
154.94	1.27	2.0	186.73	24SE	0.4	-4.3	-1.4	0.8
161.21	1.29	2.0	196.68	4OC	-0.7	-4.0	-1.4	0.9
167.45	1.30	2.0	206.69	14OC	-0.8	-3.4	-1.3	0.9
173.68	1.31	1.9	216.74	24OC	-0.8	-3.3	-1.3	1.0
179.88	1.32	1.9	226.85	3NO	-0.8	-4.0	-1.3	1.0
186.06	1.32	1.8	236.99	13NO	-0.6	-4.3	-1.3	1.1
192.20	1.31	1.7	247.16	23NO	-0.6	-4.4	-1.3	1.1
198.30	1.30	1.6	257.35	3DE	-0.6	-4.3	-1.3	1.1
204.36	1.28	1.6	267.55	13DE	-0.2	-4.3	-1.3	1.2
210.35	1.25	1.5	277.75	23DE	1.6	-4.2	-1.4	1.2
				1023				
216.27	1.21	1.3	287.94	2JA	1.8	-4.0	-1.4	1.2
222.11	1.15	1.2	298.10	12JA	0.5	-3.9	-1.4	1.2
227.84	1.09	1.1	308.23	22JA	0.2	-3.8	-1.5	1.2
233.43	1.00	0.9	318.33	1FE	0.1	-3.8	-1.5	1.2
238.87	0.89	0.7	328.37	11FE	-0.2	-3.7	-1.6	1.2
244.10	0.75	0.5	338.36	21FE	-0.6	-3.6	-1.7	1.2
249.08	0.58	0.3	348.30	3MR	-1.2	-3.5	-1.7	1.3
253.75	0.37	0.1	358.18	13MR	-1.5	-3.5	-1.8	1.3
258.02	0.12	-0.1	8.00	23MR	-0.8	-3.4	-1.9	1.4
261.78	-0.20	-0.4	17.76	2AP	0.4	-3.4	-1.9	1.4
264.91	-0.59	-0.7	27.47	12AP	1.8	-3.3	-2.0	1.4
267.23	-1.06	-1.0	37.14	22AP	3.5	-3.3	-2.1	1.4
268.56	-1.62	-1.3	46.76	2MY	2.6	-3.3	-2.1	1.4
268.73	-2.27	-1.6	56.34	12MY	1.5	-3.3	-2.2	1.4
267.67	-2.98	-1.9	65.90	22MY	0.7	-3.3	-2.2	1.4
265.48	-3.69	-2.2	75.44	1JN	-0.2	-3.3	-2.2	1.3
262.61	-4.30	-2.4	84.96	11JN	-1.2	-3.3	-2.2	1.3
259.73	-4.74	-2.3	94.49	21JN	-1.7	-3.4	-2.2	1.2
257.55	-4.96	-2.1	104.03	1JL	-0.9	-3.4	-2.2	1.2
256.58	-4.98	-1.9	113.58	11JL	-0.2	-3.4	-2.1	1.1
256.98	-4.86	-1.6	123.16	21JL	0.2	-3.5	-2.1	1.1
258.71	-4.64	-1.4	132.77	31JL	0.5	-3.5	-2.0	1.0
261.59	-4.36	-1.2	142.42	10AU	0.9	-3.5	-2.0	1.0
265.42	-4.05	-1.0	152.12	20AU	1.9	-3.5	-1.9	0.9
270.03	-3.73	-0.8	161.87	30AU	2.8	-3.4	-1.8	0.8
275.26	-3.40	-0.6	171.67	9SE	0.3	-3.4	-1.8	0.9
280.96	-3.08	-0.4	181.53	19SE	-0.8	-3.4	-1.7	0.9
287.04	-2.76	-0.2	191.45	29SE	-1.0	-3.3	-1.6	1.0
293.41	-2.45	-0.0	201.43	9OC	-0.9	-3.3	-1.6	1.0
300.01	-2.15	0.1	211.46	19OC	-0.7	-3.3	-1.5	1.1
306.78	-1.86	0.3	221.54	29OC	-0.5	-3.3	-1.5	1.1
313.67	-1.59	0.4	231.66	8NO	-0.4	-3.3	-1.5	1.2
320.65	-1.33	0.5	241.82	18NO	-0.4	-3.4	-1.4	1.2
327.69	-1.09	0.7	252.00	28NO	-0.1	-3.4	-1.4	1.3
334.76	-0.86	0.8	262.20	8DE	1.8	-3.4	-1.4	1.3
341.84	-0.65	0.9	272.40	18DE	1.5	-3.4	-1.4	1.3
348.92	-0.46	1.0	282.60	28DE	0.3	-3.5	-1.4	1.4

♂			☉	16.00UT	☿	♀	♃	♄
LONG	LAT	MAG	LONG	1024	MAGNITUDES			
355.98	-0.28	1.1	292.77	7JA	0.0	-3.5	-1.4	1.4
3.01	-0.11	1.2	302.93	17JA	-0.1	-3.6	-1.5	1.4
10.00	0.04	1.3	313.04	27JA	-0.2	-3.6	-1.5	1.4
16.95	0.18	1.4	323.11	6FE	-0.6	-3.7	-1.5	1.3
23.85	0.30	1.5	333.13	16FE	-1.2	-3.7	-1.6	1.3
30.71	0.42	1.6	343.10	26FE	-1.4	-3.8	-1.6	1.2
37.51	0.52	1.7	353.00	7MR	-0.8	-3.9	-1.7	1.2
44.26	0.61	1.7	2.86	17MR	0.6	-4.0	-1.7	1.2
50.97	0.70	1.8	12.65	27MR	2.2	-4.1	-1.8	1.2
57.62	0.77	1.8	22.39	6AP	3.4	-4.2	-1.8	1.2
64.23	0.84	1.9	32.08	16AP	1.9	-4.2	-1.9	1.2
70.79	0.90	1.9	41.72	26AP	1.1	-4.2	-2.0	1.2
77.31	0.95	1.9	51.32	6MY	0.5	-4.0	-2.1	1.2
83.80	0.99	2.0	60.89	16MY	-0.2	-3.5	-2.1	1.1
90.26	1.03	2.0	70.44	26MY	-1.2	-2.6	-2.2	1.1
96.69	1.07	2.0	79.97	5JN	-1.8	-3.5	-2.2	1.1
103.10	1.09	2.0	89.50	15JN	-0.9	-4.0	-2.3	1.0
109.49	1.12	2.0	99.03	25JN	-0.1	-4.2	-2.3	1.0
115.87	1.13	2.0	108.57	5JL	0.4	-4.2	-2.3	0.9
122.24	1.14	2.0	118.13	15JL	0.7	-4.1	-2.3	0.9
128.60	1.15	2.0	127.72	25JL	1.2	-4.1	-2.3	0.8
134.97	1.16	2.0	137.35	4AU	2.4	-4.0	-2.3	0.8
141.34	1.15	2.0	147.03	14AU	2.4	-3.9	-2.2	0.7
147.71	1.15	2.0	156.75	24AU	0.2	-3.8	-2.2	0.7
154.10	1.14	2.0	166.52	3SE	-0.9	-3.7	-2.1	0.6
160.51	1.12	2.0	176.36	13SE	-1.1	-3.7	-2.1	0.7
166.93	1.10	2.0	186.25	23SE	-1.0	-3.6	-2.0	0.7
173.37	1.08	2.0	196.19	3OC	-0.6	-3.6	-1.9	0.8
179.83	1.04	2.0	206.20	13OC	-0.4	-3.5	-1.9	0.9
186.31	1.01	2.0	216.25	23OC	-0.3	-3.5	-1.8	0.9
192.81	0.96	1.9	226.35	2NO	-0.2	-3.4	-1.7	1.0
199.34	0.92	1.9	236.50	12NO	0.1	-3.4	-1.7	1.0
205.89	0.86	1.8	246.66	22NO	2.1	-3.4	-1.6	1.1
212.46	0.79	1.8	256.85	2DE	1.2	-3.4	-1.6	1.1
219.05	0.72	1.7	267.06	12DE	0.1	-3.4	-1.6	1.2
225.66	0.63	1.6	277.25	22DE	-0.1	-3.3	-1.6	1.2
				1025				
232.28	0.54	1.6	287.44	1JA	-0.2	-3.3	-1.5	1.2
238.92	0.43	1.5	297.61	11JA	-0.3	-3.3	-1.5	1.2
245.57	0.31	1.4	307.74	21JA	-0.7	-3.4	-1.5	1.1
252.23	0.17	1.3	317.83	31JA	-1.2	-3.4	-1.5	1.1
258.89	0.01	1.2	327.88	10FE	-1.3	-3.4	-1.5	1.1
265.55	-0.16	1.0	337.88	20FE	-0.7	-3.4	-1.6	1.0
272.20	-0.35	0.9	347.82	2MR	0.7	-3.4	-1.6	1.0
278.82	-0.56	0.8	357.70	12MR	2.8	-3.5	-1.6	0.9
285.41	-0.80	0.6	7.52	22MR	2.5	-3.5	-1.7	0.9
291.95	-1.06	0.5	17.29	1AP	1.4	-3.5	-1.7	0.9
298.41	-1.35	0.3	27.00	11AP	0.8	-3.4	-1.7	1.0
304.76	-1.66	0.1	36.67	21AP	0.3	-3.4	-1.8	0.9
310.98	-2.01	-0.0	46.29	1MY	-0.3	-3.4	-1.9	0.9
316.99	-2.38	-0.2	55.88	11MY	-1.2	-3.3	-1.9	0.9
322.76	-2.78	-0.4	65.43	21MY	-1.8	-3.3	-2.0	0.9
328.18	-3.22	-0.6	74.97	31MY	-0.9	-3.3	-2.1	0.9
333.16	-3.68	-0.9	84.50	10JN	-0.0	-3.3	-2.1	0.8
337.56	-4.16	-1.1	94.03	20JN	0.5	-3.3	-2.2	0.8
341.21	-4.66	-1.3	103.56	30JN	1.0	-3.3	-2.3	0.8
343.91	-5.16	-1.6	113.12	10JL	1.6	-3.4	-2.3	0.7
345.46	-5.62	-1.9	122.69	20JL	2.9	-3.4	-2.4	0.7
345.66	-6.00	-2.1	132.30	30JL	2.0	-3.4	-2.4	0.6
344.51	-6.20	-2.3	141.95	9AU	0.2	-3.5	-2.4	0.6
342.25	-6.16	-2.5	151.64	19AU	-1.0	-3.5	-2.4	0.5
339.42	-5.83	-2.6	161.39	29AU	-1.3	-3.6	-2.4	0.5
336.84	-5.23	-2.4	171.20	8SE	-1.0	-3.6	-2.4	0.4
335.16	-4.49	-2.1	181.05	18SE	-0.5	-3.7	-2.3	0.4
334.73	-3.69	-1.8	190.97	28SE	-0.3	-3.8	-2.3	0.4
335.60	-2.94	-1.4	200.95	8OC	-0.1	-3.9	-2.2	0.5
337.65	-2.26	-1.1	210.97	18OC	-0.1	-4.0	-2.2	0.5
340.66	-1.68	-0.8	221.06	28OC	0.3	-4.1	-2.1	0.6
344.46	-1.18	-0.5	231.17	7NO	2.4	-4.2	-2.0	0.7
348.84	-0.76	-0.3	241.33	17NO	1.0	-4.3	-2.0	0.7
353.70	-0.41	-0.0	251.51	27NO	-0.1	-4.4	-1.9	0.8
358.91	-0.11	0.2	261.70	7DE	-0.3	-4.3	-1.8	0.8
4.39	0.14	0.4	271.90	17DE	-0.3	-4.1	-1.8	0.9
10.08	0.35	0.6	282.10	27DE	-0.4	-3.5	-1.7	0.9

♂			☉	16.00UT	☿	♀	♃	♄
LONG	LAT	MAG	LONG	1026	MAGNITUDES			
15.93	0.52	0.8	292.27	6JA	-0.7	-3.4	-1.7	0.9
21.89	0.67	0.9	302.42	16JA	-1.1	-4.0	-1.6	0.9
27.94	0.79	1.1	312.54	26JA	-1.1	-4.3	-1.6	0.9
34.05	0.89	1.2	322.61	5FE	-0.7	-4.3	-1.6	0.9
40.20	0.98	1.3	332.64	15FE	0.9	-4.3	-1.6	0.9
46.39	1.05	1.4	342.61	25FE	3.2	-4.2	-1.6	0.8
52.60	1.11	1.5	352.52	7MR	1.9	-4.1	-1.6	0.8
58.82	1.15	1.6	2.37	17MR	1.0	-4.0	-1.6	0.7
65.05	1.19	1.7	12.17	27MR	0.6	-3.9	-1.6	0.7
71.28	1.21	1.8	21.91	6AP	0.2	-3.8	-1.6	0.7
77.52	1.23	1.8	31.61	16AP	-0.4	-3.7	-1.7	0.7
83.75	1.25	1.9	41.25	26AP	-1.2	-3.6	-1.7	0.7
89.99	1.25	1.9	50.86	6MY	-1.8	-3.5	-1.7	0.7
96.22	1.25	2.0	60.43	16MY	-0.9	-3.5	-1.8	0.7
102.46	1.25	2.0	69.98	26MY	0.0	-3.4	-1.8	0.7
108.71	1.24	2.0	79.51	5JN	0.7	-3.4	-1.9	0.7
114.97	1.23	2.0	89.04	15JN	1.3	-3.4	-2.0	0.7
121.24	1.21	2.0	98.57	25JN	2.2	-3.3	-2.0	0.6
127.53	1.18	2.0	108.11	5JL	3.3	-3.3	-2.1	0.6
133.84	1.16	2.0	117.67	15JL	1.7	-3.3	-2.2	0.6
140.17	1.13	2.0	127.27	25JL	0.1	-3.3	-2.2	0.5
146.53	1.09	2.0	136.89	4AU	-1.1	-3.3	-2.3	0.5
152.93	1.05	2.0	146.56	14AU	-1.4	-3.3	-2.4	0.4
159.36	1.01	1.9	156.28	24AU	-1.0	-3.4	-2.4	0.3
165.83	0.96	1.9	166.05	3SE	-0.4	-3.4	-2.4	0.3
172.35	0.91	1.9	175.88	13SE	-0.1	-3.4	-2.5	0.2
178.91	0.85	1.9	185.77	23SE	-0.0	-3.4	-2.5	0.2
185.51	0.79	1.9	195.71	3OC	0.1	-3.5	-2.5	0.1
192.18	0.72	1.9	205.71	13OC	0.6	-3.5	-2.4	0.2
198.89	0.65	1.9	215.77	23OC	2.7	-3.5	-2.4	0.2
205.66	0.57	1.9	225.86	2NO	0.8	-3.5	-2.3	0.3
212.48	0.49	1.8	236.00	12NO	-0.3	-3.4	-2.3	0.4
219.36	0.40	1.8	246.17	22NO	-0.4	-3.4	-2.2	0.4
226.30	0.30	1.8	256.35	2DE	-0.4	-3.4	-2.1	0.5
233.29	0.20	1.7	266.56	12DE	-0.5	-3.4	-2.0	0.6
240.34	0.09	1.7	276.75	22DE	-0.8	-3.3	-2.0	0.6
				1027				
247.45	-0.03	1.6	286.94	1JA	-1.0	-3.3	-1.9	0.6
254.61	-0.15	1.6	297.11	11JA	-1.0	-3.3	-1.8	0.7
261.82	-0.28	1.5	307.24	21JA	-0.6	-3.3	-1.8	0.7
269.07	-0.42	1.4	317.34	31JA	1.0	-3.4	-1.7	0.7
276.38	-0.56	1.4	327.39	10FE	3.0	-3.4	-1.7	0.7
283.72	-0.71	1.3	337.39	20FE	1.4	-3.4	-1.6	0.7
291.10	-0.87	1.2	347.33	2MR	0.7	-3.4	-1.6	0.6
298.50	-1.02	1.1	357.21	12MR	0.4	-3.5	-1.6	0.6
305.92	-1.18	1.1	7.04	22MR	0.1	-3.5	-1.6	0.6
313.34	-1.34	1.0	16.81	1AP	-0.4	-3.5	-1.6	0.5
320.75	-1.50	0.9	26.53	11AP	-1.2	-3.6	-1.6	0.5
328.14	-1.65	0.8	36.20	21AP	-1.7	-3.7	-1.6	0.5
335.49	-1.79	0.7	45.82	1MY	-0.9	-3.7	-1.6	0.5
342.77	-1.93	0.6	55.42	11MY	0.1	-3.8	-1.6	0.5
349.97	-2.06	0.5	64.98	21MY	0.9	-3.9	-1.6	0.5
357.05	-2.17	0.5	74.52	31MY	1.7	-4.0	-1.7	0.5
3.98	-2.27	0.4	84.05	10JN	2.9	-4.1	-1.7	0.5
10.72	-2.34	0.3	93.57	20JN	2.9	-4.2	-1.7	0.5
17.24	-2.40	0.1	103.11	30JN	1.4	-4.2	-1.8	0.5
23.48	-2.44	0.0	112.66	10JL	0.0	-4.1	-1.8	0.5
29.37	-2.45	-0.1	122.23	20JL	-1.1	-3.9	-1.9	0.4
34.84	-2.44	-0.2	131.84	30JL	-1.5	-3.4	-2.0	0.4
39.78	-2.40	-0.4	141.49	9AU	-0.9	-3.3	-2.0	0.3
44.07	-2.32	-0.6	151.18	19AU	-0.4	-3.9	-2.1	0.3
47.56	-2.20	-0.7	160.92	29AU	-0.0	-4.2	-2.2	0.2
50.05	-2.02	-0.9	170.72	8SE	0.1	-4.3	-2.2	0.2
51.32	-1.78	-1.2	180.58	18SE	0.3	-4.3	-2.3	0.1
51.20	-1.47	-1.4	190.49	28SE	0.9	-4.2	-2.3	0.0
49.60	-1.06	-1.6	200.47	8OC	3.1	-4.1	-2.4	-0.0
46.72	-0.58	-1.8	210.49	18OC	0.7	-4.0	-2.4	-0.1
43.10	-0.06	-1.9	220.56	28OC	-0.4	-3.9	-2.4	-0.0
39.53	0.44	-1.7	230.69	7NO	-0.6	-3.9	-2.4	0.1
36.78	0.87	-1.4	240.84	17NO	-0.6	-3.8	-2.4	0.1
35.27	1.20	-1.1	251.01	27NO	-0.7	-3.7	-2.3	0.2
35.12	1.44	-0.7	261.21	7DE	-0.8	-3.6	-2.3	0.3
36.22	1.61	-0.4	271.41	17DE	-0.8	-3.6	-2.2	0.3
38.35	1.71	-0.1	281.61	27DE	-0.8	-3.5	-2.1	0.4

♂			⊙	16.00UT	☿	♀	♃	♄
LONG	LAT	MAG	LONG	1028	MAGNITUDES			
41.32	1.78	0.1	291.79	6JA	-0.5	-3.5	-2.1	0.4
44.95	1.81	0.4	301.94	16JA	1.2	-3.4	-2.0	0.5
49.08	1.83	0.6	312.05	26JA	2.6	-3.4	-1.9	0.5
53.61	1.82	0.8	322.13	5FE	1.0	-3.4	-1.8	0.5
58.45	1.81	1.0	332.15	15FE	0.5	-3.3	-1.8	0.5
63.54	1.79	1.1	342.13	25FE	0.3	-3.3	-1.7	0.5
68.81	1.75	1.3	352.04	6MR	0.0	-3.3	-1.7	0.5
74.24	1.72	1.4	1.90	16MR	-0.5	-3.3	-1.6	0.5
79.80	1.68	1.5	11.70	26MR	-1.2	-3.3	-1.6	0.4
85.46	1.64	1.6	21.44	5AP	-1.7	-3.3	-1.5	0.4
91.20	1.59	1.7	31.13	15AP	-0.9	-3.3	-1.5	0.4
97.02	1.54	1.8	40.78	25AP	0.2	-3.4	-1.5	0.3
102.90	1.49	1.8	50.39	5MY	1.2	-3.4	-1.5	0.4
108.84	1.44	1.9	59.96	15MY	2.2	-3.5	-1.5	0.4
114.83	1.38	1.9	69.51	25MY	3.6	-3.5	-1.5	0.4
120.88	1.32	1.9	79.05	4JN	2.3	-3.5	-1.5	0.4
126.98	1.26	2.0	88.57	14JN	1.1	-3.4	-1.5	0.4
133.13	1.20	2.0	98.10	24JN	-0.1	-3.4	-1.5	0.4
139.33	1.14	2.0	107.65	4JL	-1.1	-3.4	-1.6	0.4
145.59	1.07	2.0	117.20	14JL	-1.6	-3.4	-1.6	0.4
151.90	1.00	2.0	126.80	24JL	-0.9	-3.3	-1.6	0.4
158.28	0.93	2.0	136.42	3AU	-0.3	-3.3	-1.7	0.3
164.71	0.85	2.0	146.09	13AU	0.1	-3.3	-1.7	0.3
171.21	0.77	1.9	155.81	23AU	0.3	-3.3	-1.8	0.2
177.77	0.69	1.9	165.58	2SE	0.5	-3.3	-1.9	0.2
184.41	0.61	1.9	175.40	12SE	1.2	-3.4	-1.9	0.1
191.12	0.53	1.8	185.29	22SE	3.3	-3.4	-2.0	0.1
197.90	0.44	1.8	195.23	2OC	0.5	-3.4	-2.1	-0.0
204.76	0.34	1.7	205.22	12OC	-0.6	-3.4	-2.1	-0.1
211.70	0.25	1.7	215.28	22OC	-0.7	-3.5	-2.2	-0.1
218.71	0.15	1.7	225.37	1NO	-0.7	-3.5	-2.2	-0.2
225.80	0.05	1.7	235.51	11NO	-0.8	-3.6	-2.3	-0.1
232.97	-0.05	1.7	245.68	21NO	-0.7	-3.7	-2.3	-0.1
240.22	-0.16	1.7	255.86	1DE	-0.7	-3.8	-2.3	0.0
247.54	-0.27	1.6	266.06	11DE	-0.7	-3.8	-2.3	0.1
254.93	-0.38	1.6	276.26	21DE	-0.3	-3.9	-2.2	0.1
262.38	-0.49	1.6	286.45	31DE	1.4	-4.0	-2.2	0.2
				1029				
269.90	-0.60	1.6	296.62	10JA	2.1	-4.1	-2.2	0.3
277.48	-0.71	1.5	306.76	20JA	0.7	-4.2	-2.1	0.3
285.10	-0.81	1.5	316.85	30JA	0.3	-4.3	-2.0	0.4
292.76	-0.92	1.5	326.90	9FE	0.1	-4.3	-1.9	0.4
300.46	-1.02	1.4	336.91	19FE	-0.1	-4.2	-1.9	0.4
308.17	-1.11	1.4	346.85	1MR	-0.5	-3.9	-1.8	0.4
315.90	-1.19	1.3	356.73	11MR	-1.2	-3.4	-1.7	0.4
323.63	-1.26	1.3	6.57	21MR	-1.6	-3.4	-1.7	0.4
331.34	-1.33	1.3	16.34	31MR	-0.8	-3.9	-1.6	0.4
339.02	-1.38	1.2	26.05	10AP	0.3	-4.2	-1.6	0.3
346.67	-1.41	1.2	35.73	20AP	1.5	-4.2	-1.5	0.3
354.25	-1.44	1.2	45.35	30AP	3.0	-4.2	-1.5	0.3
1.77	-1.44	1.1	54.94	10MY	3.1	-4.1	-1.4	0.2
9.21	-1.44	1.1	64.51	20MY	1.7	-4.0	-1.4	0.3
16.54	-1.41	1.1	74.05	30MY	0.8	-3.9	-1.4	0.3
23.76	-1.37	1.0	83.58	9JN	-0.1	-3.8	-1.4	0.3
30.85	-1.31	1.0	93.11	19JN	-1.2	-3.7	-1.4	0.3
37.79	-1.24	0.9	102.64	29JN	-1.7	-3.6	-1.4	0.4
44.56	-1.14	0.9	112.19	9JL	-0.9	-3.6	-1.4	0.4
51.14	-1.03	0.8	121.77	19JL	-0.2	-3.5	-1.4	0.4
57.50	-0.91	0.8	131.37	29JL	0.2	-3.5	-1.4	0.3
63.62	-0.76	0.7	141.02	8AU	0.4	-3.4	-1.5	0.3
69.46	-0.60	0.6	150.71	18AU	0.7	-3.4	-1.5	0.3
74.95	-0.41	0.5	160.45	28AU	1.6	-3.4	-1.5	0.2
80.05	-0.19	0.4	170.25	7SE	3.1	-3.4	-1.6	0.2
84.66	0.05	0.3	180.10	17SE	0.4	-3.4	-1.6	0.1
88.68	0.32	0.1	190.01	27SE	-0.7	-3.4	-1.7	0.1
91.98	0.64	-0.1	199.98	7OC	-0.9	-3.4	-1.8	-0.0
94.39	1.00	-0.3	210.01	17OC	-0.9	-3.4	-1.8	-0.1
95.71	1.41	-0.5	220.08	27OC	-0.7	-3.4	-1.9	-0.2
95.75	1.86	-0.7	230.19	6NO	-0.6	-3.4	-1.9	-0.2
94.38	2.34	-0.9	240.35	16NO	-0.5	-3.4	-2.0	-0.3
91.68	2.80	-1.1	250.52	26NO	-0.5	-3.4	-2.1	-0.2
88.01	3.18	-1.3	260.72	6DE	-0.2	-3.4	-2.1	-0.1
84.06	3.44	-1.2	270.92	16DE	1.6	-3.5	-2.1	-0.1
80.60	3.54	-1.0	281.11	26DE	1.7	-3.5	-2.2	0.0

♂			⊙	16.00UT	☿	♀	♃	♄
LONG	LAT	MAG	LONG	1030	MAGNITUDES			
78.21	3.51	-0.7	291.29	5JA	0.4	-3.5	-2.2	0.1
77.14	3.39	-0.4	301.45	15JA	0.1	-3.5	-2.1	0.1
77.37	3.22	-0.1	311.56	25JA	0.0	-3.4	-2.1	0.2
78.72	3.04	0.1	321.64	4FE	-0.2	-3.4	-2.1	0.2
81.01	2.85	0.4	331.67	14FE	-0.6	-3.4	-2.0	0.3
84.04	2.67	0.6	341.64	24FE	-1.2	-3.4	-2.0	0.3
87.66	2.49	0.8	351.56	6MR	-1.5	-3.3	-1.9	0.3
91.75	2.33	1.0	1.42	16MR	-0.8	-3.3	-1.8	0.3
96.21	2.18	1.1	11.22	26MR	0.4	-3.3	-1.7	0.3
100.98	2.04	1.2	20.97	5AP	1.9	-3.3	-1.7	0.3
105.99	1.90	1.4	30.66	15AP	3.8	-3.3	-1.6	0.3
111.20	1.77	1.5	40.31	25AP	2.3	-3.4	-1.6	0.3
116.59	1.64	1.6	49.92	5MY	1.3	-3.4	-1.5	0.3
122.13	1.52	1.6	59.50	15MY	0.6	-3.4	-1.5	0.2
127.81	1.41	1.7	69.05	25MY	-0.2	-3.4	-1.4	0.2
133.60	1.30	1.7	78.58	4JN	-1.2	-3.5	-1.4	0.2
139.50	1.18	1.8	88.11	14JN	-1.8	-3.5	-1.3	0.3
145.50	1.08	1.8	97.64	24JN	-0.9	-3.6	-1.3	0.3
151.60	0.97	1.8	107.18	4JL	-0.2	-3.7	-1.3	0.3
157.80	0.86	1.9	116.74	14JL	0.3	-3.8	-1.3	0.3
164.09	0.76	1.9	126.33	24JL	0.6	-3.8	-1.3	0.3
170.48	0.65	1.9	135.96	3AU	1.0	-3.9	-1.3	0.3
176.96	0.54	1.9	145.62	13AU	2.0	-4.0	-1.3	0.3
183.53	0.44	1.8	155.33	23AU	2.7	-4.2	-1.3	0.3
190.20	0.33	1.8	165.11	2SE	0.3	-4.2	-1.3	0.3
196.97	0.23	1.8	174.93	12SE	-0.8	-4.3	-1.4	0.2
203.83	0.12	1.8	184.81	22SE	-1.0	-4.2	-1.4	0.2
210.79	0.02	1.7	194.75	2OC	-1.0	-4.0	-1.4	0.1
217.85	-0.09	1.7	204.75	12OC	-0.7	-3.4	-1.5	0.0
225.00	-0.19	1.6	214.79	22OC	-0.4	-3.4	-1.6	-0.0
232.25	-0.29	1.6	224.89	1NO	-0.4	-4.0	-1.6	-0.1
239.58	-0.39	1.5	235.02	11NO	-0.3	-4.3	-1.7	-0.2
247.00	-0.49	1.5	245.18	21NO	-0.0	-4.4	-1.7	-0.2
254.51	-0.58	1.5	255.37	1DE	1.8	-4.3	-1.8	-0.3
262.08	-0.67	1.4	265.57	11DE	1.4	-4.3	-1.9	-0.2
269.73	-0.76	1.4	275.77	21DE	0.2	-4.2	-1.9	-0.2
277.44	-0.84	1.4	285.96	31DE	-0.0	-4.0	-2.0	-0.1
				1031				
285.20	-0.91	1.4	296.13	10JA	-0.1	-3.9	-2.0	-0.0
293.00	-0.98	1.4	306.26	20JA	-0.3	-3.8	-2.0	0.1
300.83	-1.03	1.4	316.37	30JA	-0.6	-3.8	-2.1	0.1
308.68	-1.08	1.4	326.42	9FE	-1.2	-3.7	-2.1	0.2
316.53	-1.11	1.4	336.42	19FE	-1.4	-3.6	-2.0	0.2
324.38	-1.14	1.4	346.37	1MR	-0.8	-3.5	-2.0	0.3
332.21	-1.15	1.4	356.26	11MR	0.6	-3.5	-2.0	0.3
340.01	-1.15	1.4	6.09	21MR	2.4	-3.4	-1.9	0.3
347.76	-1.13	1.4	15.87	31MR	3.1	-3.4	-1.8	0.3
355.46	-1.10	1.4	25.59	10AP	1.7	-3.3	-1.8	0.3
3.10	-1.06	1.4	35.26	20AP	1.0	-3.3	-1.7	0.3
10.66	-1.01	1.4	44.90	30AP	0.4	-3.3	-1.6	0.3
18.13	-0.95	1.4	54.49	10MY	-0.2	-3.3	-1.6	0.3
25.52	-0.87	1.4	64.05	20MY	-1.2	-3.3	-1.5	0.3
32.80	-0.79	1.4	73.59	30MY	-1.8	-3.3	-1.5	0.2
39.98	-0.69	1.4	83.12	9JN	-0.9	-3.3	-1.4	0.2
47.05	-0.59	1.4	92.65	19JN	-0.1	-3.3	-1.4	0.2
53.99	-0.47	1.4	102.18	29JN	0.4	-3.4	-1.3	0.3
60.81	-0.35	1.4	111.73	9JL	0.8	-3.4	-1.3	0.3
67.49	-0.22	1.4	121.30	19JL	1.4	-3.5	-1.3	0.3
74.03	-0.07	1.3	130.91	29JL	2.6	-3.5	-1.2	0.4
80.42	0.08	1.3	140.55	8AU	2.3	-3.5	-1.2	0.4
86.63	0.24	1.3	150.23	18AU	0.3	-3.5	-1.2	0.4
92.65	0.41	1.2	159.98	28AU	-0.9	-3.4	-1.2	0.4
98.46	0.59	1.1	169.77	7SE	-1.2	-3.4	-1.2	0.3
104.03	0.78	1.1	179.62	17SE	-1.0	-3.4	-1.2	0.3
109.32	1.00	1.0	189.53	27SE	-0.6	-3.3	-1.3	0.3
114.26	1.23	0.9	199.49	7OC	-0.3	-3.3	-1.3	0.2
118.81	1.48	0.7	209.52	17OC	-0.2	-3.3	-1.3	0.1
122.86	1.76	0.6	219.59	27OC	-0.2	-3.3	-1.4	0.1
126.31	2.08	0.4	229.70	6NO	0.2	-3.3	-1.4	-0.0
129.02	2.42	0.2	239.85	16NO	2.1	-3.4	-1.5	-0.1
130.81	2.80	0.0	250.03	26NO	1.2	-3.4	-1.5	-0.1
131.49	3.21	-0.2	260.22	6DE	0.0	-3.4	-1.6	-0.2
130.91	3.64	-0.5	270.42	16DE	-0.2	-3.4	-1.7	-0.2
128.97	4.03	-0.7	280.62	26DE	-0.2	-3.5	-1.7	-0.2

♂ LONG	♂ LAT	♂ MAG	☉ LONG	16.00UT	☿	♀	♃	♄
				1032	MAGNITUDES			
125.85	4.34	-0.9	290.80	5JA	-0.4	-3.5	-1.8	-0.1
121.99	4.50	-1.0	300.95	15JA	-0.7	-3.6	-1.8	-0.1
118.09	4.48	-0.9	311.08	25JA	-1.2	-3.6	-1.9	0.0
114.88	4.30	-0.7	321.15	4FE	-1.2	-3.7	-1.9	0.1
112.79	4.02	-0.4	331.19	14FE	-0.7	-3.7	-2.0	0.1
112.02	3.68	-0.2	341.16	24FE	0.7	-3.8	-2.0	0.2
112.48	3.33	0.1	351.08	5MR	2.9	-3.9	-2.0	0.2
114.01	3.00	0.3	0.95	15MR	2.3	-4.0	-2.0	0.3
116.43	2.69	0.5	10.75	25MR	1.2	-4.1	-2.0	0.3
119.57	2.40	0.7	20.50	4AP	0.7	-4.2	-1.9	0.3
123.28	2.14	0.8	30.20	14AP	0.3	-4.2	-1.9	0.4
127.47	1.91	1.0	39.85	24AP	-0.3	-4.2	-1.8	0.4
132.03	1.69	1.1	49.46	4MY	-1.2	-4.0	-1.8	0.4
136.92	1.49	1.2	59.04	14MY	-1.8	-3.5	-1.7	0.3
142.08	1.30	1.3	68.59	24MY	-0.9	-2.6	-1.7	0.3
147.47	1.12	1.4	78.12	3JN	-0.0	-3.5	-1.6	0.3
153.06	0.96	1.5	87.65	13JN	0.6	-4.0	-1.5	0.3
158.84	0.80	1.5	97.18	23JN	1.1	-4.2	-1.5	0.3
164.78	0.65	1.5	106.72	3JL	1.8	-4.2	-1.4	0.3
170.88	0.51	1.6	116.28	13JL	3.1	-4.1	-1.4	0.4
177.12	0.37	1.6	125.86	23JL	1.9	-4.1	-1.3	0.4
183.50	0.24	1.6	135.48	2AU	0.2	-4.0	-1.3	0.4
190.02	0.11	1.6	145.15	12AU	-1.0	-3.9	-1.3	0.4
196.67	-0.02	1.6	154.86	22AU	-1.3	-3.8	-1.2	0.4
203.44	-0.13	1.6	164.62	1SE	-1.0	-3.7	-1.2	0.4
210.33	-0.25	1.6	174.45	11SE	-0.5	-3.7	-1.2	0.4
217.35	-0.36	1.6	184.33	21SE	-0.2	-3.6	-1.2	0.4
224.48	-0.46	1.6	194.26	1OC	-0.1	-3.6	-1.2	0.4
231.72	-0.56	1.6	204.26	11OC	0.0	-3.5	-1.2	0.3
239.07	-0.65	1.5	214.30	21OC	0.4	-3.5	-1.2	0.3
246.52	-0.74	1.5	224.39	31OC	2.4	-3.4	-1.3	0.2
254.06	-0.81	1.5	234.53	10NO	1.0	-3.4	-1.3	0.2
261.69	-0.88	1.5	244.69	20NO	-0.2	-3.4	-1.3	0.1
269.39	-0.94	1.4	254.88	30NO	-0.3	-3.4	-1.4	0.0
277.16	-0.99	1.4	265.08	10DE	-0.4	-3.4	-1.4	-0.1
284.98	-1.03	1.4	275.28	20DE	-0.5	-3.4	-1.5	-0.1
292.84	-1.06	1.4	285.46	30DE	-0.8	-3.3	-1.6	-0.2
				1033				
300.73	-1.08	1.3	295.64	9JA	-1.0	-3.3	-1.6	-0.1
308.63	-1.08	1.3	305.78	19JA	-1.1	-3.4	-1.7	-0.0
316.53	-1.08	1.3	315.88	29JA	-0.7	-3.4	-1.8	0.0
324.42	-1.06	1.3	325.94	8FE	0.9	-3.4	-1.8	0.1
332.28	-1.03	1.3	335.94	18FE	3.2	-3.4	-1.9	0.2
340.11	-0.99	1.4	345.89	28FE	1.7	-3.4	-1.9	0.2
347.89	-0.94	1.4	355.78	10MR	0.9	-3.5	-2.0	0.3
355.61	-0.88	1.4	5.62	20MR	0.5	-3.5	-2.0	0.3
3.26	-0.81	1.4	15.39	30MR	0.2	-3.5	-2.0	0.4
10.84	-0.73	1.5	25.12	9AP	-0.4	-3.4	-2.0	0.4
18.34	-0.65	1.5	34.79	19AP	-1.2	-3.4	-2.0	0.4
25.75	-0.56	1.5	44.42	29AP	-1.8	-3.4	-2.0	0.4
33.07	-0.46	1.6	54.02	9MY	-0.9	-3.3	-2.0	0.4
40.30	-0.35	1.6	63.58	19MY	0.0	-3.3	-1.9	0.5
47.43	-0.25	1.6	73.12	29MY	0.7	-3.3	-1.8	0.4
54.47	-0.14	1.6	82.66	8JN	1.4	-3.3	-1.8	0.4
61.40	-0.02	1.7	92.18	18JN	2.4	-3.3	-1.7	0.4
68.24	0.09	1.7	101.71	28JN	3.2	-3.3	-1.7	0.4
74.99	0.21	1.7	111.26	8JL	1.6	-3.4	-1.6	0.4
81.63	0.33	1.7	120.83	18JL	0.1	-3.4	-1.5	0.4
88.16	0.46	1.7	130.43	28JL	-1.1	-3.4	-1.5	0.5
94.60	0.58	1.7	140.08	7AU	-1.5	-3.5	-1.4	0.5
100.92	0.71	1.7	149.76	17AU	-1.0	-3.5	-1.4	0.5
107.12	0.85	1.6	159.50	27AU	-0.4	-3.6	-1.3	0.6
113.20	0.98	1.6	169.29	6SE	-0.1	-3.6	-1.3	0.6
119.14	1.13	1.5	179.14	16SE	0.1	-3.7	-1.3	0.6
124.92	1.27	1.5	189.04	26SE	0.2	-3.8	-1.3	0.6
130.53	1.43	1.4	199.01	6OC	0.7	-3.9	-1.3	0.5
135.94	1.59	1.3	209.03	16OC	2.8	-4.0	-1.2	0.5
141.11	1.77	1.2	219.10	26OC	0.8	-4.1	-1.2	0.5
146.00	1.96	1.1	229.21	5NO	-0.3	-4.2	-1.3	0.4
150.54	2.16	1.0	239.36	15NO	-0.5	-4.3	-1.3	0.4
154.67	2.39	0.8	249.53	25NO	-0.5	-4.4	-1.3	0.3
158.27	2.63	0.6	259.73	5DE	-0.6	-4.3	-1.3	0.2
161.24	2.90	0.4	269.93	15DE	-0.8	-4.1	-1.4	0.2
163.42	3.18	0.2	280.12	25DE	-0.9	-3.4	-1.4	0.1

♂ LONG	♂ LAT	♂ MAG	☉ LONG	16.00UT	☿	♀	♃	♄
				1034	MAGNITUDES			
164.62	3.48	-0.0	290.31	4JA	-0.9	-3.4	-1.5	0.0
164.68	3.78	-0.3	300.46	14JA	-0.6	-4.0	-1.5	-0.0
163.45	4.04	-0.5	310.59	24JA	1.0	-4.3	-1.6	0.0
160.96	4.23	-0.8	320.67	3FE	2.9	-4.3	-1.6	0.1
157.48	4.28	-1.0	330.70	13FE	1.2	-4.3	-1.7	0.2
153.57	4.17	-1.0	340.68	23FE	0.6	-4.2	-1.8	0.2
149.97	3.90	-0.9	350.60	5MR	0.4	-4.1	-1.8	0.3
147.27	3.52	-0.7	0.47	15MR	0.1	-4.0	-1.9	0.3
145.80	3.09	-0.4	10.28	25MR	-0.4	-3.9	-2.0	0.4
145.64	2.66	-0.2	20.03	4AP	-1.2	-3.8	-2.0	0.4
146.66	2.25	0.0	29.73	14AP	-1.7	-3.7	-2.1	0.5
148.69	1.87	0.2	39.39	24AP	-0.9	-3.6	-2.1	0.5
151.57	1.53	0.4	49.00	4MY	0.1	-3.5	-2.1	0.6
155.13	1.23	0.5	58.58	14MY	1.0	-3.5	-2.1	0.6
159.25	0.96	0.7	68.13	24MY	1.9	-3.4	-2.1	0.6
163.84	0.71	0.8	77.67	3JN	3.1	-3.4	-2.1	0.6
168.82	0.49	0.9	87.19	13JN	2.7	-3.4	-2.0	0.6
174.13	0.29	1.0	96.72	23JN	1.3	-3.3	-2.0	0.6
179.74	0.10	1.1	106.26	3JL	0.0	-3.3	-1.9	0.6
185.59	-0.07	1.1	115.82	13JL	-1.1	-3.3	-1.9	0.6
191.68	-0.23	1.2	125.40	23JL	-1.6	-3.3	-1.8	0.6
197.98	-0.38	1.2	135.02	2AU	-0.9	-3.3	-1.7	0.6
204.46	-0.51	1.2	144.68	12AU	-0.4	-3.3	-1.7	0.7
211.13	-0.63	1.3	154.39	22AU	-0.0	-3.4	-1.6	0.7
217.96	-0.74	1.3	164.15	1SE	0.2	-3.4	-1.5	0.7
224.95	-0.84	1.3	173.97	11SE	0.4	-3.4	-1.5	0.8
232.08	-0.93	1.3	183.85	21SE	1.0	-3.4	-1.5	0.8
239.34	-1.00	1.3	193.78	1OC	3.1	-3.5	-1.4	0.8
246.72	-1.07	1.3	203.77	11OC	0.7	-3.5	-1.4	0.8
254.22	-1.12	1.4	213.81	21OC	-0.5	-3.5	-1.4	0.7
261.81	-1.16	1.4	223.90	31OC	-0.6	-3.5	-1.3	0.7
269.49	-1.18	1.4	234.03	10NO	-0.6	-3.4	-1.3	0.7
277.24	-1.20	1.4	244.19	20NO	-0.7	-3.4	-1.3	0.6
285.05	-1.20	1.4	254.38	30NO	-0.7	-3.4	-1.3	0.6
292.90	-1.18	1.4	264.58	10DE	-0.7	-3.4	-1.3	0.5
300.78	-1.16	1.4	274.78	20DE	-0.8	-3.3	-1.4	0.4
308.67	-1.12	1.4	284.97	30DE	-0.4	-3.3	-1.4	0.3
				1035				
316.56	-1.08	1.4	295.14	9JA	1.2	-3.3	-1.4	0.3
324.43	-1.02	1.4	305.28	19JA	2.5	-3.3	-1.5	0.2
332.27	-0.95	1.4	315.39	29JA	0.9	-3.4	-1.5	0.2
340.07	-0.87	1.4	325.45	8FE	0.4	-3.4	-1.6	0.2
347.81	-0.79	1.4	335.46	18FE	0.2	-3.4	-1.6	0.3
355.50	-0.70	1.5	345.41	28FE	-0.0	-3.4	-1.7	0.4
3.11	-0.61	1.5	355.31	10MR	-0.5	-3.5	-1.8	0.4
10.64	-0.51	1.5	5.15	20MR	-1.2	-3.5	-1.8	0.5
18.10	-0.41	1.5	14.92	30MR	-1.6	-3.5	-1.9	0.5
25.47	-0.30	1.5	24.65	9AP	-0.9	-3.6	-2.0	0.6
32.75	-0.20	1.5	34.33	19AP	0.2	-3.7	-2.0	0.6
39.95	-0.09	1.6	43.96	29AP	1.3	-3.7	-2.1	0.7
47.06	0.01	1.6	53.56	9MY	2.5	-3.8	-2.1	0.7
54.09	0.12	1.7	63.13	19MY	3.5	-3.9	-2.2	0.7
61.03	0.22	1.7	72.67	29MY	2.1	-4.0	-2.2	0.8
67.89	0.32	1.8	82.20	8JN	1.0	-4.1	-2.2	0.8
74.67	0.42	1.8	91.73	18JN	-0.0	-4.2	-2.2	0.8
81.37	0.52	1.8	101.26	28JN	-1.1	-4.2	-2.2	0.8
88.00	0.62	1.9	110.81	8JL	-1.7	-4.1	-2.2	0.8
94.56	0.71	1.9	120.38	18JL	-0.9	-3.9	-2.1	0.8
101.04	0.81	1.9	129.97	28JL	-0.3	-3.4	-2.1	0.8
107.46	0.90	1.9	139.61	7AU	0.1	-3.3	-2.0	0.8
113.81	0.99	1.9	149.30	17AU	0.3	-3.9	-2.0	0.9
120.09	1.09	1.9	159.03	27AU	0.6	-4.2	-1.9	0.9
126.29	1.18	1.9	168.82	6SE	1.3	-4.3	-1.8	0.9
132.42	1.27	1.8	178.66	16SE	3.3	-4.3	-1.8	1.0
138.47	1.36	1.8	188.56	26SE	0.5	-4.2	-1.7	1.0
144.42	1.45	1.8	198.52	6OC	-0.6	-4.1	-1.6	1.0
150.28	1.54	1.7	208.54	16OC	-0.8	-4.0	-1.6	1.0
156.02	1.64	1.6	218.60	26OC	-0.8	-3.9	-1.6	1.0
161.64	1.73	1.5	228.71	5NO	-0.8	-3.8	-1.5	1.0
167.09	1.83	1.4	238.86	15NO	-0.6	-3.8	-1.5	1.0
172.36	1.94	1.3	249.03	25NO	-0.6	-3.7	-1.5	0.9
177.42	2.05	1.2	259.23	5DE	-0.6	-3.6	-1.4	0.9
182.20	2.16	1.1	269.43	15DE	-0.3	-3.6	-1.4	0.8
186.65	2.27	0.9	279.63	25DE	1.4	-3.5	-1.4	0.7

Left table (1036 / 1037):

♂ LONG	LAT	MAG	☉ LONG	16.00UT	☿	♀	♃	♄
190.70	2.39	0.7	289.81	4JA	2.0	-3.5	-1.4	0.7
194.23	2.51	0.5	299.97	14JA	0.6	-3.4	-1.5	0.6
197.14	2.64	0.3	310.10	24JA	0.2	-3.4	-1.5	0.5
199.26	2.75	0.0	320.18	3FE	0.1	-3.4	-1.5	0.5
200.42	2.84	-0.3	330.22	13FE	-0.1	-3.3	-1.5	0.5
200.45	2.90	-0.6	340.20	23FE	-0.5	-3.3	-1.6	0.5
199.21	2.89	-0.9	350.13	4MR	-1.2	-3.3	-1.6	0.6
196.74	2.79	-1.1	360.00	14MR	-1.5	-3.3	-1.7	0.6
193.34	2.56	-1.4	9.81	24MR	-0.9	-3.3	-1.7	0.7
189.59	2.22	-1.3	19.56	3AP	0.3	-3.3	-1.8	0.7
186.22	1.78	-1.2	29.27	13AP	1.6	-3.3	-1.9	0.8
183.81	1.32	-1.0	38.92	23AP	3.3	-3.4	-2.0	0.8
182.70	0.86	-0.8	48.54	3MY	2.8	-3.4	-2.0	0.9
182.91	0.45	-0.6	58.11	13MY	1.6	-3.5	-2.1	0.9
184.34	0.09	-0.4	67.67	23MY	0.8	-3.5	-2.2	1.0
186.81	-0.23	-0.2	77.21	2JN	-0.1	-3.5	-2.2	1.0
190.15	-0.50	-0.0	86.73	12JN	-1.1	-3.4	-2.3	1.0
194.19	-0.73	0.1	96.26	22JN	-1.8	-3.4	-2.3	1.0
198.83	-0.92	0.2	105.80	2JL	-0.9	-3.4	-2.3	1.1
203.97	-1.09	0.4	115.35	12JL	-0.2	-3.4	-2.4	1.1
209.52	-1.23	0.5	124.93	22JL	0.2	-3.3	-2.4	1.1
215.43	-1.35	0.5	134.55	1AU	0.5	-3.3	-2.3	1.1
221.64	-1.44	0.6	144.21	11AU	0.8	-3.3	-2.3	1.1
228.12	-1.51	0.7	153.92	21AU	1.7	-3.3	-2.3	1.1
234.84	-1.57	0.8	163.68	31AU	3.0	-3.3	-2.2	1.1
241.76	-1.60	0.8	173.49	10SE	0.4	-3.4	-2.1	1.2
248.87	-1.62	0.9	183.36	20SE	-0.8	-3.4	-2.1	1.2
256.13	-1.62	0.9	193.29	30SE	-1.0	-3.4	-2.0	1.3
263.52	-1.60	1.0	203.28	10OC	-0.9	-3.4	-1.9	1.3
271.03	-1.57	1.0	213.32	20OC	-0.7	-3.5	-1.9	1.3
278.63	-1.52	1.1	223.41	30OC	-0.5	-3.6	-1.8	1.3
286.30	-1.46	1.1	233.53	9NO	-0.4	-3.6	-1.8	1.3
294.15	-1.39	1.2	243.69	19NO	-0.4	-3.7	-1.7	1.3
301.79	-1.30	1.2	253.88	29NO	-0.1	-3.8	-1.7	1.2
309.56	-1.21	1.2	264.08	9DE	1.6	-3.8	-1.6	1.2
317.33	-1.11	1.3	274.28	19DE	1.7	-3.9	-1.6	1.2
325.09	-1.00	1.3	284.47	29DE	0.3	-4.0	-1.6	1.1

1037

♂ LONG	LAT	MAG	☉ LONG	16.00UT	☿	♀	♃	♄
332.82	-0.88	1.4	294.64	8JA	0.1	-4.1	-1.5	1.0
340.50	-0.77	1.4	304.79	18JA	-0.0	-4.2	-1.5	1.0
348.13	-0.65	1.5	314.90	28JA	-0.2	-4.3	-1.5	0.9
355.69	-0.53	1.5	324.96	7FE	-0.6	-4.3	-1.5	0.8
3.19	-0.41	1.5	334.97	17FE	-1.2	-4.2	-1.5	0.7
10.62	-0.29	1.6	344.93	27FE	-1.4	-3.9	-1.6	0.8
17.96	-0.18	1.6	354.83	9MR	-0.8	-3.3	-1.6	0.8
25.23	-0.07	1.7	4.67	19MR	0.4	-3.4	-1.6	0.8
32.41	0.04	1.7	14.45	29MR	2.1	-3.9	-1.7	0.9
39.52	0.15	1.7	24.18	8AP	3.6	-4.2	-1.7	0.9
46.54	0.25	1.7	33.86	18AP	2.1	-4.2	-1.8	1.0
53.49	0.35	1.7	43.50	28AP	1.2	-4.2	-1.8	1.1
60.37	0.44	1.8	53.09	8MY	0.6	-4.1	-1.9	1.1
67.17	0.53	1.8	62.66	18MY	-0.2	-4.0	-2.0	1.2
73.91	0.61	1.8	72.21	28MY	-1.1	-3.9	-2.0	1.2
80.59	0.69	1.8	81.74	7JN	-1.8	-3.8	-2.1	1.2
87.21	0.77	1.8	91.27	17JN	-0.9	-3.7	-2.2	1.3
93.77	0.84	1.9	100.80	27JN	-0.2	-3.6	-2.2	1.3
100.29	0.91	1.9	110.34	7JL	0.3	-3.6	-2.3	1.3
106.76	0.97	2.0	119.91	17JL	0.7	-3.5	-2.4	1.4
113.19	1.03	2.0	129.51	27JL	1.1	-3.5	-2.4	1.4
119.58	1.08	2.0	139.14	6AU	2.2	-3.4	-2.4	1.4
125.93	1.14	2.0	148.83	16AU	2.7	-3.4	-2.4	1.3
132.25	1.19	2.0	158.56	26AU	0.4	-3.4	-2.4	1.3
138.54	1.23	2.0	168.34	5SE	-0.9	-3.4	-2.4	1.3
144.80	1.28	2.0	178.19	15SE	-1.1	-3.4	-2.4	1.3
151.02	1.32	2.0	188.09	25SE	-1.0	-3.4	-2.4	1.3
157.20	1.35	1.9	198.04	5OC	-0.6	-3.4	-2.3	1.3
163.36	1.39	1.9	208.06	15OC	-0.4	-3.4	-2.2	1.2
169.46	1.42	1.9	218.12	25OC	-0.3	-3.4	-2.2	1.2
175.53	1.44	1.8	228.22	4NO	-0.3	-3.4	-2.1	1.2
181.53	1.47	1.7	238.37	14NO	0.1	-3.4	-2.0	1.2
187.47	1.49	1.6	248.54	24NO	1.9	-3.4	-2.0	1.1
193.34	1.50	1.6	258.73	4DE	1.4	-3.4	-1.9	1.1
199.12	1.51	1.4	268.93	14DE	0.1	-3.5	-1.8	1.1
204.78	1.51	1.3	279.13	24DE	-0.1	-3.5	-1.8	1.0

Right table (1038 / 1039):

♂ LONG	LAT	MAG	☉ LONG	16.00UT	☿	♀	♃	♄
210.31	1.50	1.2	289.31	3JA	-0.2	-3.5	-1.7	1.0
215.68	1.48	1.0	299.47	13JA	-0.3	-3.5	-1.7	0.9
220.85	1.45	0.9	309.60	23JA	-0.6	-3.4	-1.6	0.9
225.78	1.40	0.7	319.68	2FE	-1.2	-3.4	-1.6	0.8
230.40	1.33	0.5	329.73	12FE	-1.3	-3.4	-1.6	0.8
234.64	1.23	0.3	339.71	22FE	-0.8	-3.4	-1.6	0.8
238.41	1.10	0.0	349.64	4MR	0.6	-3.3	-1.6	0.7
241.58	0.92	-0.2	359.51	14MR	2.5	-3.3	-1.6	0.8
244.00	0.68	-0.5	9.33	24MR	2.8	-3.3	-1.6	0.9
245.49	0.37	-0.8	19.09	3AP	1.5	-3.3	-1.6	0.9
245.86	-0.02	-1.2	28.79	13AP	0.9	-3.3	-1.6	1.0
245.01	-0.50	-1.5	38.45	23AP	0.4	-3.4	-1.7	1.1
242.95	-1.05	-1.8	48.07	3MY	-0.2	-3.4	-1.7	1.1
239.96	-1.64	-2.0	57.65	13MY	-1.1	-3.4	-1.7	1.2
236.67	-2.19	-2.0	67.21	23MY	-1.8	-3.4	-1.8	1.2
233.79	-2.65	-1.9	76.74	2JN	-0.9	-3.5	-1.8	1.3
231.93	-2.99	-1.7	86.27	12JN	-0.1	-3.5	-1.9	1.3
231.42	-3.20	-1.5	95.80	22JN	0.5	-3.6	-2.0	1.3
232.28	-3.31	-1.3	105.34	2JL	0.9	-3.7	-2.0	1.3
234.40	-3.34	-1.0	114.89	12JL	1.5	-3.8	-2.1	1.3
237.59	-3.32	-0.8	124.47	22JL	2.8	-3.8	-2.2	1.3
241.65	-3.25	-0.7	134.09	1AU	2.2	-3.9	-2.2	1.3
246.43	-3.15	-0.5	143.75	11AU	0.3	-4.1	-2.3	1.2
251.79	-3.03	-0.3	153.45	21AU	-0.9	-4.2	-2.4	1.2
257.61	-2.89	-0.2	163.20	31AU	-1.2	-4.2	-2.4	1.1
263.82	-2.73	-0.0	173.02	10SE	-1.0	-4.3	-2.4	1.1
270.34	-2.55	0.1	182.89	20SE	-0.5	-4.2	-2.5	1.1
277.09	-2.37	0.2	192.81	30SE	-0.3	-4.0	-2.5	1.1
284.05	-2.18	0.3	202.80	10OC	-0.2	-3.4	-2.4	1.1
291.15	-1.98	0.5	212.83	20OC	-0.1	-3.4	-2.4	1.1
298.36	-1.78	0.6	222.92	30OC	0.3	-4.0	-2.4	1.1
305.66	-1.58	0.7	233.05	9NO	2.1	-4.3	-2.3	1.1
313.01	-1.39	0.8	243.20	19NO	1.2	-4.4	-2.2	1.0
320.39	-1.19	0.9	253.39	29NO	-0.3	-4.3	-2.2	1.0
327.78	-1.00	1.0	263.59	9DE	-0.3	-4.3	-2.1	1.0
335.16	-0.82	1.1	273.78	19DE	-0.3	-4.1	-2.0	1.0
342.51	-0.65	1.2	283.98	29DE	-0.4	-4.0	-2.0	0.9

1039

♂ LONG	LAT	MAG	☉ LONG	16.00UT	☿	♀	♃	♄
349.83	-0.48	1.2	294.15	8JA	-0.7	-3.9	-1.9	0.9
357.11	-0.32	1.3	304.29	18JA	-1.1	-3.8	-1.8	0.8
4.34	-0.18	1.4	314.41	28JA	-1.2	-3.7	-1.8	0.8
11.50	-0.04	1.5	324.47	7FE	-0.7	-3.7	-1.7	0.7
18.61	0.09	1.6	334.48	17FE	0.7	-3.6	-1.7	0.7
25.65	0.21	1.6	344.45	27FE	3.0	-3.5	-1.6	0.6
32.63	0.32	1.7	354.35	9MR	2.1	-3.5	-1.6	0.6
39.55	0.42	1.7	4.19	19MR	1.1	-3.4	-1.6	0.6
46.40	0.52	1.8	13.98	29MR	0.7	-3.4	-1.6	0.6
53.19	0.61	1.8	23.72	8AP	0.3	-3.3	-1.6	0.7
59.92	0.68	1.9	33.40	18AP	-0.3	-3.3	-1.6	0.8
66.59	0.76	1.9	43.04	28AP	-1.1	-3.3	-1.6	0.8
73.22	0.82	1.9	52.64	8MY	-1.8	-3.3	-1.6	0.9
79.79	0.88	1.9	62.20	18MY	-0.9	-3.3	-1.6	1.0
86.33	0.93	1.9	71.75	28MY	-0.0	-3.3	-1.6	1.0
92.83	0.98	2.0	81.28	7JN	0.6	-3.3	-1.7	1.1
99.29	1.02	2.0	90.81	17JN	1.2	-3.3	-1.7	1.1
105.73	1.06	1.9	100.34	27JN	2.0	-3.4	-1.7	1.1
112.14	1.09	1.9	109.88	7JL	3.3	-3.4	-1.8	1.1
118.53	1.12	1.9	119.45	17JL	1.8	-3.5	-1.8	1.1
124.91	1.14	2.0	129.05	27JL	0.2	-3.5	-1.9	1.1
131.28	1.16	2.0	138.68	6AU	-1.0	-3.5	-2.0	1.1
137.65	1.17	2.0	148.35	16AU	-1.4	-3.5	-2.0	1.1
144.01	1.18	2.0	158.08	26AU	-1.0	-3.4	-2.1	1.1
150.37	1.18	2.0	167.86	5SE	-0.5	-3.4	-2.2	1.0
156.73	1.18	2.0	177.70	15SE	-0.2	-3.4	-2.2	1.0
163.10	1.18	2.0	187.60	25SE	-0.0	-3.3	-2.3	0.9
169.47	1.17	2.0	197.55	5OC	0.1	-3.3	-2.3	0.9
175.85	1.15	2.0	207.56	15OC	0.5	-3.3	-2.4	0.9
182.23	1.13	1.9	217.63	25OC	2.5	-3.3	-2.4	1.0
188.63	1.10	1.9	227.73	4NO	1.0	-3.3	-2.4	1.0
195.02	1.07	1.9	237.87	14NO	-0.2	-3.4	-2.4	1.0
201.43	1.03	1.8	248.05	24NO	-0.4	-3.4	-2.3	0.9
207.83	0.98	1.7	258.24	4DE	-0.4	-3.4	-2.3	0.9
214.23	0.92	1.7	268.43	14DE	-0.5	-3.4	-2.2	0.9
220.63	0.85	1.6	278.64	24DE	-0.8	-3.5	-2.2	0.9

♂ LONG	♂ LAT	♂ MAG	☉ LONG	16.00UT	☿	♀	♃	♄
				1040	MAGNITUDES			
227.01	0.77	1.5	288.82	3JA	-1.0	-3.5	-2.1	0.8
233.38	0.68	1.4	298.98	13JA	-1.0	-3.6	-2.0	0.8
239.73	0.57	1.3	309.11	23JA	-0.6	-3.6	-2.0	0.7
246.05	0.44	1.1	319.20	2FE	0.9	-3.7	-1.9	0.7
252.33	0.30	1.0	329.24	12FE	3.1	-3.7	-1.8	0.6
258.55	0.13	0.9	339.23	22FE	1.5	-3.8	-1.8	0.6
264.70	-0.07	0.7	349.16	3MR	0.8	-3.9	-1.7	0.5
270.76	-0.30	0.5	359.04	13MR	0.5	-4.0	-1.6	0.5
276.71	-0.56	0.4	8.86	23MR	0.1	-4.1	-1.6	0.5
282.50	-0.86	0.2	18.62	2AP	-0.4	-4.2	-1.6	0.5
288.10	-1.20	-0.0	28.32	12AP	-1.1	-4.2	-1.5	0.5
293.44	-1.59	-0.3	37.99	22AP	-1.8	-4.2	-1.5	0.6
298.45	-2.04	-0.5	47.60	2MY	-0.9	-4.0	-1.5	0.7
303.04	-2.55	-0.7	57.19	12MY	0.0	-3.5	-1.5	0.7
307.05	-3.12	-1.0	66.75	22MY	0.8	-2.6	-1.5	0.8
310.35	-3.76	-1.3	76.28	1JN	1.6	-3.5	-1.5	0.8
312.75	-4.46	-1.6	85.81	11JN	2.7	-4.0	-1.5	0.9
314.02	-5.18	-1.9	95.34	21JN	3.1	-4.2	-1.5	0.9
314.05	-5.88	-2.2	104.87	1JL	1.5	-4.2	-1.5	1.0
312.82	-6.46	-2.4	114.43	11JL	0.1	-4.1	-1.6	1.0
310.60	-6.82	-2.6	124.01	21JL	-1.0	-4.0	-1.6	1.0
308.00	-6.86	-2.6	133.62	31JL	-1.5	-4.0	-1.6	1.0
305.72	-6.58	-2.4	143.28	10AU	-1.0	-3.9	-1.7	1.0
304.41	-6.06	-2.2	152.98	20AU	-0.4	-3.8	-1.7	1.0
304.38	-5.39	-1.9	162.73	30AU	-0.1	-3.7	-1.8	1.0
305.65	-4.69	-1.6	172.54	9SE	0.1	-3.7	-1.9	0.9
308.09	-4.00	-1.3	182.41	19SE	0.3	-3.6	-1.9	0.9
311.50	-3.35	-1.1	192.33	29SE	0.8	-3.6	-2.0	0.9
315.67	-2.77	-0.8	202.31	9OC	2.8	-3.5	-2.1	0.8
320.45	-2.25	-0.5	212.35	19OC	0.8	-3.5	-2.1	0.8
325.68	-1.79	-0.3	222.43	29OC	-0.4	-3.4	-2.2	0.9
331.27	-1.39	-0.1	232.55	8NO	-0.6	-3.4	-2.2	0.9
337.13	-1.03	0.1	242.71	18NO	-0.6	-3.4	-2.2	0.9
343.19	-0.72	0.3	252.89	28NO	-0.6	-3.4	-2.3	0.9
349.39	-0.45	0.5	263.09	8DE	-0.8	-3.4	-2.3	0.9
355.71	-0.21	0.6	273.29	18DE	-0.8	-3.4	-2.2	0.8
2.10	0.00	0.8	283.48	28DE	-0.9	-3.3	-2.2	0.8
				1041				
8.55	0.19	0.9	293.66	7JA	-0.5	-3.3	-2.2	0.8
15.03	0.35	1.1	303.81	17JA	1.0	-3.4	-2.1	0.8
21.52	0.49	1.2	313.92	27JA	2.8	-3.4	-2.1	0.7
28.02	0.61	1.3	323.98	6FE	1.1	-3.4	-2.0	0.7
34.52	0.71	1.4	334.00	16FE	0.6	-3.4	-1.9	0.6
41.00	0.80	1.5	343.96	26FE	0.3	-3.4	-1.8	0.6
47.47	0.88	1.6	353.87	8MR	0.0	-3.5	-1.8	0.5
53.92	0.95	1.7	3.71	18MR	-0.4	-3.5	-1.7	0.4
60.34	1.00	1.8	13.50	28MR	-1.1	-3.5	-1.6	0.4
66.75	1.05	1.8	23.24	7AP	-1.7	-3.4	-1.6	0.4
73.14	1.09	1.9	32.92	17AP	-0.9	-3.4	-1.5	0.4
79.51	1.12	1.9	42.56	27AP	0.1	-3.4	-1.5	0.4
85.86	1.15	2.0	52.17	7MY	1.1	-3.3	-1.5	0.5
92.20	1.16	2.0	61.73	17MY	2.1	-3.3	-1.4	0.6
98.53	1.18	2.0	71.28	27MY	3.4	-3.3	-1.4	0.6
104.85	1.18	2.0	80.82	6JN	2.4	-3.3	-1.4	0.7
111.16	1.19	2.0	90.34	16JN	1.2	-3.3	-1.4	0.8
117.48	1.18	2.0	99.87	26JN	0.0	-3.3	-1.4	0.8
123.80	1.18	2.0	109.42	6JL	-1.1	-3.4	-1.4	0.8
130.13	1.16	2.0	118.98	16JL	-1.6	-3.4	-1.4	0.9
136.48	1.15	2.0	128.57	26JL	-1.0	-3.4	-1.4	0.9
142.84	1.13	2.0	138.21	5AU	-0.3	-3.5	-1.4	0.9
149.23	1.10	1.9	147.88	15AU	0.0	-3.5	-1.4	0.9
155.64	1.07	1.9	157.61	25AU	0.2	-3.6	-1.5	0.9
162.08	1.04	2.0	167.39	4SE	0.5	-3.6	-1.5	0.9
168.55	1.00	2.0	177.23	14SE	1.1	-3.7	-1.6	0.9
175.06	0.95	2.0	187.12	24SE	3.2	-3.8	-1.6	0.8
181.60	0.90	2.0	197.07	4OC	0.7	-3.9	-1.7	0.8
188.18	0.85	1.9	207.08	14OC	-0.5	-4.0	-1.8	0.8
194.81	0.79	1.9	217.14	24OC	-0.7	-4.1	-1.8	0.7
201.48	0.72	1.9	227.24	3NO	-0.7	-4.2	-1.9	0.8
208.20	0.65	1.9	237.38	13NO	-0.7	-4.3	-1.9	0.8
214.96	0.57	1.8	247.55	23NO	-0.7	-4.4	-2.0	0.8
221.76	0.48	1.8	257.74	3DE	-0.7	-4.3	-2.1	0.8
228.61	0.39	1.7	267.94	13DE	-0.7	-4.0	-2.1	0.8
235.51	0.28	1.7	278.14	23DE	-0.4	-3.4	-2.1	0.8

♂ LONG	♂ LAT	♂ MAG	☉ LONG	16.00UT	☿	♀	♃	♄
				1042	MAGNITUDES			
242.45	0.17	1.6	288.33	2JA	1.2	-3.4	-2.1	0.8
249.43	0.05	1.5	298.49	12JA	2.3	-4.0	-2.1	0.8
256.46	-0.09	1.5	308.62	22JA	0.8	-4.3	-2.1	0.7
263.52	-0.23	1.4	318.71	1FE	0.4	-4.3	-2.1	0.7
270.62	-0.38	1.3	328.75	11FE	0.2	-4.3	-2.0	0.7
277.75	-0.55	1.2	338.75	21FE	-0.1	-4.2	-2.0	0.6
284.91	-0.72	1.1	348.68	3MR	-0.5	-4.1	-1.9	0.5
292.09	-0.91	1.0	358.56	13MR	-1.1	-4.0	-1.9	0.5
299.27	-1.10	0.9	8.38	23MR	-1.6	-3.9	-1.8	0.4
306.46	-1.30	0.8	18.15	2AP	-0.9	-3.8	-1.7	0.4
313.63	-1.50	0.7	27.85	12AP	0.2	-3.7	-1.7	0.3
320.77	-1.71	0.6	37.52	22AP	1.4	-3.6	-1.6	0.3
327.84	-1.93	0.5	47.14	2MY	2.7	-3.5	-1.5	0.3
334.84	-2.14	0.3	56.72	12MY	3.3	-3.5	-1.5	0.4
341.73	-2.35	0.2	66.28	22MY	1.9	-3.4	-1.4	0.5
348.46	-2.55	0.1	75.82	1JN	0.9	-3.4	-1.4	0.5
354.99	-2.75	-0.0	85.35	11JN	-0.0	-3.4	-1.3	0.6
1.27	-2.93	-0.2	94.88	21JN	-1.1	-3.3	-1.3	0.7
7.21	-3.10	-0.3	104.42	1JL	-1.7	-3.3	-1.3	0.7
12.75	-3.26	-0.5	113.97	11JL	-0.9	-3.3	-1.3	0.8
17.76	-3.39	-0.7	123.55	21JL	-0.3	-3.3	-1.3	0.8
22.09	-3.50	-0.9	133.16	31JL	0.1	-3.3	-1.3	0.8
25.61	-3.56	-1.1	142.81	10AU	0.4	-3.3	-1.3	0.8
28.08	-3.58	-1.3	152.51	20AU	0.7	-3.4	-1.3	0.8
29.29	-3.52	-1.5	162.26	30AU	1.4	-3.4	-1.3	0.9
29.09	-3.36	-1.7	172.07	9SE	3.3	-3.4	-1.3	0.8
27.43	-3.07	-1.9	181.93	19SE	0.6	-3.4	-1.4	0.8
24.58	-2.63	-2.1	191.85	29SE	-0.7	-3.5	-1.4	0.8
21.16	-2.07	-2.2	201.83	9OC	-0.9	-3.5	-1.4	0.8
17.98	-1.44	-1.9	211.86	19OC	-0.8	-3.5	-1.5	0.7
15.75	-0.84	-1.6	221.94	29OC	-0.8	-3.5	-1.6	0.7
14.82	-0.31	-1.3	232.06	8NO	-0.6	-3.4	-1.6	0.7
15.21	0.13	-0.9	242.22	18NO	-0.5	-3.4	-1.7	0.7
16.80	0.48	-0.6	252.40	28NO	-0.5	-3.4	-1.7	0.8
19.38	0.75	-0.3	262.59	8DE	-0.3	-3.4	-1.8	0.8
22.72	0.96	-0.1	272.79	18DE	1.4	-3.3	-1.9	0.8
26.69	1.12	0.2	282.99	28DE	2.0	-3.3	-1.9	0.8
				1043				
31.12	1.24	0.4	293.16	7JA	0.5	-3.3	-2.0	0.8
35.91	1.33	0.6	303.31	17JA	0.2	-3.3	-2.0	0.8
40.99	1.39	0.8	313.42	27JA	0.0	-3.4	-2.0	0.7
46.29	1.44	1.0	323.49	6FE	-0.2	-3.4	-2.0	0.7
51.76	1.47	1.1	333.52	16FE	-0.5	-3.4	-2.0	0.7
57.36	1.49	1.3	343.48	26FE	-1.1	-3.4	-2.0	0.6
63.06	1.50	1.4	353.39	8MR	-1.5	-3.5	-2.0	0.6
68.85	1.49	1.5	3.24	18MR	-0.9	-3.5	-1.9	0.5
74.71	1.49	1.6	13.03	28MR	0.3	-3.5	-1.9	0.4
80.62	1.47	1.7	22.77	7AP	1.8	-3.6	-1.8	0.4
86.58	1.46	1.8	32.46	17AP	3.6	-3.7	-1.8	0.3
92.58	1.43	1.8	42.10	27AP	2.5	-3.7	-1.7	0.3
98.61	1.40	1.9	51.70	7MY	1.4	-3.8	-1.6	0.3
104.68	1.37	1.9	61.28	17MY	0.7	-3.9	-1.6	0.3
110.78	1.34	2.0	70.82	27MY	-0.1	-4.0	-1.5	0.4
116.91	1.30	2.0	80.36	6JN	-1.1	-4.1	-1.4	0.4
123.08	1.26	2.0	89.89	16JN	-1.8	-4.2	-1.4	0.5
129.28	1.21	2.0	99.41	26JN	-0.9	-4.2	-1.3	0.6
135.52	1.16	2.0	108.96	6JL	-0.2	-4.1	-1.3	0.6
141.80	1.11	2.0	118.52	16JL	0.2	-3.9	-1.3	0.7
148.13	1.05	2.0	128.11	26JL	0.6	-3.3	-1.2	0.7
154.50	1.00	2.0	137.74	5AU	0.9	-3.3	-1.2	0.8
160.92	0.93	2.0	147.42	15AU	1.9	-3.9	-1.2	0.8
167.39	0.87	1.9	157.14	25AU	3.0	-4.2	-1.2	0.8
173.92	0.80	1.9	166.92	4SE	0.5	-4.3	-1.2	0.8
180.52	0.73	1.9	176.75	14SE	-0.8	-4.3	-1.2	0.8
187.17	0.65	1.8	186.64	24SE	-1.0	-4.2	-1.2	0.8
193.89	0.57	1.8	196.59	4OC	-1.0	-4.1	-1.3	0.8
200.68	0.49	1.8	206.60	14OC	-0.7	-4.0	-1.3	0.8
207.53	0.40	1.8	216.65	24OC	-0.5	-3.9	-1.3	0.7
214.45	0.31	1.8	226.75	3NO	-0.4	-3.8	-1.4	0.7
221.45	0.21	1.8	236.89	13NO	-0.4	-3.8	-1.4	0.7
228.51	0.11	1.7	247.06	23NO	-0.1	-3.7	-1.5	0.7
235.65	0.00	1.7	257.25	3DE	1.6	-3.6	-1.5	0.7
242.85	-0.10	1.7	267.45	13DE	1.6	-3.6	-1.6	0.8
250.12	-0.22	1.7	277.65	23DE	0.3	-3.5	-1.7	0.8

♂ LONG	♂ LAT	♂ MAG	⊙ LONG	16.00UT	☿	♀	♃	♄
				1044	MAGNITUDES			
257.46	-0.33	1.6	287.83	2JA	0.0	-3.5	-1.7	0.8
264.85	-0.45	1.6	298.00	12JA	-0.1	-3.4	-1.8	0.8
272.31	-0.57	1.5	308.13	22JA	-0.3	-3.4	-1.9	0.8
279.81	-0.69	1.5	318.23	1FE	-0.6	-3.4	-1.9	0.8
287.36	-0.82	1.4	328.27	11FE	-1.2	-3.3	-2.0	0.7
294.95	-0.94	1.4	338.26	21FE	-1.4	-3.3	-2.0	0.7
302.57	-1.05	1.3	348.20	2MR	-0.9	-3.3	-2.0	0.6
310.20	-1.16	1.3	358.09	12MR	0.4	-3.3	-2.0	0.6
317.84	-1.27	1.2	7.91	22MR	2.2	-3.3	-2.0	0.5
325.48	-1.37	1.2	17.67	1AP	3.3	-3.3	-2.0	0.5
333.10	-1.46	1.1	27.39	11AP	1.8	-3.3	-1.9	0.4
340.68	-1.53	1.1	37.05	21AP	1.1	-3.4	-1.9	0.4
348.22	-1.59	1.0	46.68	1MY	0.5	-3.4	-1.8	0.3
355.69	-1.64	1.0	56.26	11MY	-0.2	-3.5	-1.8	0.2
3.08	-1.67	0.9	65.82	21MY	-1.1	-3.5	-1.7	0.3
10.37	-1.69	0.9	75.36	31MY	-1.8	-3.5	-1.6	0.3
17.53	-1.68	0.8	84.89	10JN	-0.9	-3.4	-1.6	0.4
24.54	-1.66	0.7	94.41	20JN	-0.2	-3.4	-1.5	0.4
31.39	-1.62	0.7	103.95	30JN	0.4	-3.4	-1.5	0.5
38.04	-1.55	0.6	113.50	10JL	0.8	-3.4	-1.4	0.6
44.46	-1.47	0.5	123.08	20JL	1.3	-3.3	-1.4	0.6
50.60	-1.37	0.4	132.69	30JL	2.4	-3.3	-1.3	0.7
56.43	-1.24	0.3	142.34	9AU	2.6	-3.3	-1.3	0.7
61.88	-1.08	0.2	152.03	19AU	0.4	-3.3	-1.3	0.7
66.86	-0.90	0.1	161.78	29AU	-0.9	-3.3	-1.2	0.8
71.29	-0.68	-0.1	171.59	8SE	-1.2	-3.4	-1.2	0.8
75.03	-0.43	-0.2	181.44	18SE	-1.1	-3.4	-1.2	0.8
77.92	-0.13	-0.4	191.37	28SE	-0.6	-3.4	-1.2	0.8
79.77	0.23	-0.6	201.34	8OC	-0.3	-3.4	-1.2	0.8
80.39	0.63	-0.8	211.37	18OC	-0.2	-3.5	-1.2	0.8
79.59	1.09	-1.0	221.45	28OC	-0.2	-3.6	-1.2	0.7
77.37	1.57	-1.3	231.57	7NO	0.1	-3.6	-1.3	0.7
74.01	2.04	-1.4	241.72	17NO	1.9	-3.7	-1.3	0.7
70.11	2.42	-1.5	251.91	27NO	1.4	-3.8	-1.4	0.7
66.48	2.69	-1.2	262.10	7DE	0.1	-3.9	-1.4	0.7
63.79	2.82	-1.0	272.30	17DE	-0.2	-3.9	-1.5	0.7
62.40	2.85	-0.7	282.50	27DE	-0.2	-4.0	-1.5	0.8
				1045				
62.34	2.81	-0.4	292.67	6JA	-0.3	-4.1	-1.6	0.8
63.47	2.73	-0.1	302.82	16JA	-0.6	-4.2	-1.6	0.8
65.59	2.63	0.2	312.94	26JA	-1.1	-4.3	-1.7	0.8
68.51	2.52	0.4	323.01	5FE	-1.2	-4.3	-1.8	0.8
72.05	2.41	0.6	333.03	15FE	-0.8	-4.2	-1.8	0.8
76.09	2.30	0.8	343.00	25FE	0.6	-3.9	-1.9	0.7
80.52	2.18	1.0	352.91	7MR	2.7	-3.3	-2.0	0.7
85.25	2.08	1.2	2.76	17MR	2.5	-3.5	-2.0	0.6
90.24	1.97	1.3	12.56	27MR	1.3	-4.0	-2.0	0.6
95.43	1.87	1.4	22.30	6AP	0.8	-4.2	-2.0	0.5
100.78	1.78	1.5	31.99	16AP	0.4	-4.2	-2.0	0.5
106.27	1.68	1.6	41.64	26AP	-0.2	-4.2	-2.0	0.4
111.89	1.59	1.7	51.24	6MY	-1.1	-4.1	-2.0	0.3
117.61	1.49	1.7	60.81	16MY	-1.8	-4.0	-1.9	0.3
123.44	1.40	1.8	70.36	26MY	-0.9	-3.9	-1.9	0.2
129.34	1.31	1.8	79.89	5JN	-0.1	-3.8	-1.8	0.3
135.34	1.22	1.9	89.42	15JN	0.5	-3.7	-1.8	0.3
141.41	1.13	1.9	98.95	25JN	1.0	-3.6	-1.7	0.4
147.56	1.04	1.9	108.49	5JL	1.7	-3.6	-1.7	0.4
153.79	0.95	1.9	118.06	15JL	3.0	-3.5	-1.6	0.5
160.10	0.85	1.9	127.65	25JL	2.1	-3.5	-1.5	0.6
166.49	0.76	1.9	137.27	4AU	0.3	-3.4	-1.5	0.6
172.96	0.67	1.9	146.95	14AU	-0.9	-3.4	-1.4	0.7
179.51	0.57	1.9	156.67	24AU	-1.3	-3.4	-1.4	0.7
186.15	0.48	1.9	166.44	3SE	-1.0	-3.4	-1.3	0.7
192.87	0.38	1.8	176.27	13SE	-0.5	-3.4	-1.3	0.8
199.68	0.28	1.8	186.16	23SE	-0.2	-3.4	-1.3	0.8
206.58	0.18	1.8	196.11	3OC	-0.1	-3.4	-1.3	0.8
213.57	0.08	1.7	206.11	13OC	-0.0	-3.4	-1.3	0.8
220.64	-0.03	1.7	216.16	23OC	0.3	-3.4	-1.3	0.8
227.81	-0.13	1.6	226.26	2NO	2.2	-3.4	-1.3	0.7
235.06	-0.23	1.6	236.40	12NO	1.1	-3.4	-1.3	0.7
242.39	-0.33	1.6	246.57	22NO	-0.1	-3.4	-1.3	0.7
249.80	-0.44	1.5	256.75	2DE	-0.3	-3.4	-1.3	0.7
257.30	-0.54	1.5	266.96	12DE	-0.3	-3.5	-1.3	0.7
264.86	-0.63	1.5	277.16	22DE	-0.5	-3.5	-1.4	0.7

♂ LONG	♂ LAT	♂ MAG	⊙ LONG	16.00UT	☿	♀	♃	♄
				1046	MAGNITUDES			
272.48	-0.73	1.5	287.34	1JA	-0.7	-3.5	-1.4	0.8
280.16	-0.82	1.5	297.51	11JA	-1.0	-3.5	-1.5	0.8
287.89	-0.90	1.5	307.64	21JA	-1.1	-3.4	-1.5	0.8
295.66	-0.98	1.5	317.73	31JA	-0.7	-3.4	-1.6	0.8
303.45	-1.04	1.4	327.78	10FE	0.7	-3.4	-1.7	0.8
311.26	-1.10	1.4	337.78	20FE	3.1	-3.4	-1.7	0.8
319.08	-1.15	1.4	347.72	2MR	1.9	-3.3	-1.8	0.8
326.88	-1.19	1.4	357.61	12MR	1.0	-3.3	-1.9	0.7
334.67	-1.21	1.4	7.43	22MR	0.6	-3.3	-1.9	0.7
342.42	-1.23	1.4	17.20	1AP	0.2	-3.3	-2.0	0.6
350.13	-1.22	1.4	26.92	11AP	-0.3	-3.4	-2.0	0.6
357.78	-1.21	1.3	36.58	21AP	-1.1	-3.4	-2.1	0.5
5.36	-1.18	1.3	46.21	1MY	-1.8	-3.4	-2.1	0.5
12.87	-1.14	1.3	55.80	11MY	-1.0	-3.4	-2.1	0.4
20.28	-1.08	1.3	65.36	21MY	-0.0	-3.4	-2.1	0.3
27.59	-1.01	1.3	74.90	31MY	0.7	-3.5	-2.1	0.3
34.79	-0.93	1.3	84.43	10JN	1.3	-3.5	-2.1	0.2
41.88	-0.83	1.3	93.96	20JN	2.2	-3.6	-2.0	0.3
48.83	-0.73	1.2	103.49	30JN	3.3	-3.7	-2.0	0.4
55.64	-0.61	1.2	113.04	10JL	1.7	-3.8	-1.9	0.4
62.30	-0.48	1.2	122.62	20JL	0.2	-3.8	-1.9	0.5
68.79	-0.34	1.1	132.22	30JL	-1.0	-3.9	-1.8	0.5
75.10	-0.18	1.1	141.87	9AU	-1.4	-4.1	-1.7	0.6
81.19	-0.01	1.0	151.57	19AU	-1.0	-4.2	-1.7	0.6
87.06	0.17	1.0	161.31	29AU	-0.5	-4.2	-1.6	0.7
92.65	0.37	0.9	171.11	8SE	-0.1	-4.3	-1.6	0.7
97.93	0.59	0.8	180.97	18SE	0.0	-4.2	-1.5	0.7
102.83	0.83	0.7	190.88	28SE	0.2	-3.9	-1.5	0.8
107.28	1.10	0.5	200.86	8OC	0.6	-3.3	-1.4	0.8
111.18	1.39	0.4	210.88	18OC	2.5	-3.4	-1.4	0.8
114.40	1.73	0.2	220.96	28OC	1.0	-4.0	-1.4	0.8
116.78	2.10	0.0	231.08	7NO	-0.3	-4.3	-1.4	0.8
118.14	2.51	-0.2	241.23	17NO	-0.5	-4.4	-1.3	0.8
118.30	2.96	-0.4	251.41	27NO	-0.5	-4.3	-1.3	0.7
117.09	3.40	-0.7	261.61	7DE	-0.6	-4.2	-1.4	0.7
114.56	3.81	-0.9	271.81	17DE	-0.8	-4.1	-1.4	0.7
111.00	4.12	-1.0	282.00	27DE	-0.9	-4.0	-1.4	0.7
				1047				
107.01	4.27	-1.1	292.18	6JA	-0.9	-3.9	-1.4	0.8
103.38	4.25	-0.9	302.33	16JA	-0.6	-3.8	-1.5	0.8
100.68	4.09	-0.6	312.45	26JA	0.9	-3.7	-1.5	0.8
99.26	3.84	-0.3	322.53	5FE	3.1	-3.7	-1.5	0.8
99.14	3.55	-0.1	332.55	15FE	1.4	-3.6	-1.6	0.8
100.18	3.26	0.2	342.52	25FE	0.7	-3.5	-1.7	0.8
102.19	2.97	0.4	352.44	7MR	0.4	-3.5	-1.7	0.8
105.00	2.71	0.6	2.29	17MR	0.1	-3.4	-1.8	0.8
108.44	2.47	0.8	12.09	27MR	-0.4	-3.4	-1.9	0.8
112.39	2.24	0.9	21.84	6AP	-1.1	-3.3	-1.9	0.7
116.75	2.04	1.1	31.53	16AP	-1.8	-3.3	-2.0	0.7
121.45	1.85	1.2	41.17	26AP	-1.0	-3.3	-2.1	0.6
126.42	1.67	1.3	50.78	6MY	0.1	-3.3	-2.1	0.5
131.63	1.50	1.4	60.35	16MY	0.9	-3.3	-2.2	0.5
137.04	1.35	1.5	69.90	26MY	1.7	-3.3	-2.2	0.4
142.63	1.20	1.6	79.44	5JN	2.9	-3.3	-2.2	0.3
148.38	1.05	1.6	88.96	15JN	2.9	-3.3	-2.2	0.3
154.27	0.91	1.7	98.49	25JN	1.4	-3.4	-2.2	0.3
160.30	0.78	1.7	108.04	5JL	0.1	-3.4	-2.2	0.3
166.45	0.65	1.7	117.59	15JL	-1.0	-3.5	-2.2	0.4
172.73	0.53	1.7	127.18	25JL	-1.6	-3.5	-2.1	0.4
179.13	0.40	1.7	136.81	4AU	-1.0	-3.5	-2.1	0.5
185.65	0.28	1.7	146.48	14AU	-0.4	-3.5	-2.0	0.6
192.28	0.16	1.7	156.19	24AU	-0.0	-3.4	-2.0	0.6
199.03	0.05	1.7	165.97	3SE	0.2	-3.4	-1.9	0.7
205.88	-0.06	1.7	175.79	13SE	0.3	-3.4	-1.8	0.7
212.85	-0.17	1.7	185.68	23SE	0.8	-3.3	-1.8	0.7
219.93	-0.28	1.7	195.62	3OC	2.9	-3.3	-1.7	0.8
227.11	-0.38	1.6	205.61	13OC	0.8	-3.3	-1.7	0.8
234.39	-0.48	1.6	215.67	23OC	-0.4	-3.3	-1.6	0.8
241.77	-0.57	1.6	225.76	2NO	-0.6	-3.3	-1.6	0.8
249.25	-0.66	1.5	235.90	12NO	-0.6	-3.4	-1.5	0.8
256.81	-0.74	1.5	246.07	22NO	-0.7	-3.4	-1.5	0.8
264.45	-0.82	1.4	256.26	2DE	-0.8	-3.4	-1.5	0.8
272.16	-0.89	1.4	266.45	12DE	-0.8	-3.4	-1.5	0.7
279.93	-0.95	1.4	276.66	22DE	-0.8	-3.5	-1.5	0.7

♂ LONG	♂ LAT	♂ MAG	☉ LONG	16.00UT	☿	♀	♃	♄
				1048	MAGNITUDES			
287.75	-1.00	1.3	286.84	1JA	-0.5	-3.5	-1.5	0.7
295.61	-1.04	1.3	297.01	11JA	1.0	-3.6	-1.5	0.8
303.49	-1.07	1.3	307.15	21JA	2.7	-3.6	-1.5	0.8
311.38	-1.08	1.3	317.25	31JA	1.0	-3.7	-1.5	0.8
319.27	-1.09	1.4	327.30	10FE	0.5	-3.8	-1.5	0.9
327.14	-1.08	1.4	337.30	20FE	0.3	-3.8	-1.6	0.9
334.99	-1.06	1.4	347.25	1MR	0.0	-3.9	-1.6	0.9
342.80	-1.03	1.4	357.13	11MR	-0.4	-4.0	-1.7	0.9
350.56	-0.99	1.4	6.96	21MR	-1.1	-4.1	-1.7	0.9
358.26	-0.94	1.4	16.73	31MR	-1.7	-4.2	-1.8	0.8
5.89	-0.88	1.5	26.45	10AP	-1.0	-4.2	-1.8	0.8
13.44	-0.81	1.5	36.12	20AP	0.1	-4.2	-1.9	0.7
20.91	-0.73	1.5	45.75	30AP	1.2	-4.0	-2.0	0.7
28.29	-0.64	1.5	55.34	10MY	2.3	-3.5	-2.1	0.6
35.58	-0.54	1.5	64.90	20MY	3.6	-2.6	-2.1	0.6
42.77	-0.44	1.6	74.44	30MY	2.2	-3.5	-2.2	0.5
49.85	-0.33	1.6	83.97	9JN	1.1	-4.0	-2.2	0.4
56.84	-0.21	1.6	93.50	19JN	0.0	-4.2	-2.3	0.4
63.72	-0.09	1.6	103.03	29JN	-1.1	-4.2	-2.3	0.3
70.48	0.03	1.6	112.58	9JL	-1.7	-4.1	-2.4	0.3
77.14	0.16	1.6	122.15	19JL	-1.0	-4.0	-2.4	0.4
83.67	0.29	1.6	131.76	29JL	-0.3	-4.0	-2.4	0.4
90.09	0.43	1.5	141.40	8AU	0.1	-3.9	-2.3	0.5
96.37	0.57	1.5	151.10	18AU	0.3	-3.8	-2.3	0.5
102.51	0.72	1.5	160.84	28AU	0.5	-3.7	-2.3	0.6
108.49	0.88	1.4	170.63	7SE	1.2	-3.7	-2.2	0.7
114.31	1.04	1.4	180.49	17SE	3.2	-3.6	-2.1	0.7
119.92	1.21	1.3	190.40	27SE	0.7	-3.5	-2.1	0.7
125.30	1.40	1.2	200.37	7OC	-0.6	-3.5	-2.0	0.8
130.42	1.60	1.1	210.40	17OC	-0.8	-3.5	-1.9	0.8
135.22	1.81	1.0	220.47	27OC	-0.8	-3.4	-1.9	0.8
139.64	2.05	0.9	230.58	6NO	-0.8	-3.4	-1.8	0.8
143.58	2.31	0.7	240.74	16NO	-0.6	-3.4	-1.8	0.8
146.94	2.59	0.5	250.91	26NO	-0.6	-3.4	-1.7	0.8
149.58	2.90	0.3	261.11	6DE	-0.6	-3.4	-1.7	0.8
151.33	3.24	0.1	271.31	16DE	-0.4	-3.4	-1.6	0.8
152.01	3.59	-0.1	281.50	26DE	1.2	-3.3	-1.6	0.8
				1049				
151.45	3.94	-0.4	291.68	5JA	2.3	-3.3	-1.6	0.8
149.57	4.25	-0.6	301.84	15JA	0.7	-3.4	-1.6	0.8
146.52	4.45	-0.9	311.96	25JA	0.3	-3.4	-1.5	0.9
142.71	4.51	-1.0	322.03	4FE	0.1	-3.4	-1.5	0.9
138.83	4.39	-0.9	332.06	14FE	-0.1	-3.4	-1.5	0.9
135.58	4.11	-0.7	342.03	24FE	-0.5	-3.4	-1.6	0.9
133.44	3.74	-0.5	351.95	6MR	-1.1	-3.5	-1.6	0.9
132.59	3.33	-0.2	1.81	16MR	-1.6	-3.5	-1.6	0.9
132.99	2.93	-0.0	11.61	26MR	-0.9	-3.4	-1.6	0.9
134.48	2.55	0.2	21.36	5AP	0.2	-3.4	-1.7	0.9
136.89	2.20	0.4	31.06	15AP	1.5	-3.4	-1.7	0.9
140.04	1.88	0.6	40.70	25AP	3.0	-3.4	-1.8	0.8
143.79	1.60	0.7	50.31	5MY	3.0	-3.3	-1.8	0.8
148.05	1.34	0.9	59.89	15MY	1.7	-3.3	-1.9	0.7
152.71	1.10	1.0	69.44	25MY	0.8	-3.3	-2.0	0.7
157.72	0.89	1.1	78.97	4JN	-0.0	-3.3	-2.1	0.6
163.03	0.69	1.2	88.50	14JN	-1.1	-3.3	-2.1	0.6
168.59	0.50	1.2	98.03	24JN	-1.7	-3.3	-2.2	0.5
174.38	0.33	1.3	107.57	4JL	-1.0	-3.4	-2.3	0.4
180.38	0.17	1.3	117.13	14JL	-0.3	-3.4	-2.3	0.4
186.57	0.02	1.4	126.71	24JL	0.2	-3.4	-2.4	0.4
192.93	-0.12	1.4	136.34	3AU	0.5	-3.5	-2.4	0.5
199.46	-0.26	1.4	146.01	13AU	0.8	-3.5	-2.4	0.5
206.14	-0.38	1.4	155.72	23AU	1.6	-3.6	-2.5	0.6
212.97	-0.50	1.4	165.49	2SE	3.2	-3.6	-2.4	0.6
219.94	-0.61	1.5	175.32	12SE	0.6	-3.7	-2.4	0.7
227.04	-0.70	1.5	185.20	22SE	-0.7	-3.8	-2.4	0.7
234.27	-0.79	1.5	195.14	2OC	-0.9	-3.9	-2.3	0.8
241.62	-0.87	1.5	205.13	12OC	-0.9	-4.0	-2.3	0.8
249.07	-0.94	1.4	215.18	22OC	-0.7	-4.1	-2.2	0.8
256.63	-1.00	1.4	225.28	1NO	-0.5	-4.3	-2.2	0.9
264.28	-1.05	1.4	235.41	11NO	-0.5	-4.3	-2.1	0.9
272.00	-1.09	1.4	245.57	21NO	-0.5	-4.4	-2.0	0.9
279.78	-1.11	1.4	255.76	1DE	-0.2	-4.3	-1.9	0.9
287.62	-1.13	1.4	265.96	11DE	1.4	-4.0	-1.9	0.9
295.50	-1.13	1.4	276.15	21DE	1.9	-3.3	-1.8	0.9
303.39	-1.12	1.4	286.35	31DE	0.4	-3.5	-1.8	0.9

♂ LONG	♂ LAT	♂ MAG	☉ LONG	16.00UT	☿	♀	♃	♄
				1050	MAGNITUDES			
311.30	-1.09	1.4	296.51	10JA	0.1	-4.1	-1.7	0.9
319.20	-1.06	1.4	306.65	20JA	-0.0	-4.3	-1.7	0.9
327.08	-1.02	1.4	316.75	30JA	-0.2	-4.3	-1.6	0.9
334.93	-0.96	1.4	326.81	9FE	-0.5	-4.3	-1.6	1.0
342.73	-0.90	1.4	336.81	19FE	-1.1	-4.2	-1.6	1.0
350.48	-0.82	1.4	346.76	1MR	-1.5	-4.1	-1.6	1.0
358.16	-0.74	1.4	356.65	11MR	-0.9	-4.0	-1.6	1.0
5.78	-0.65	1.4	6.48	21MR	0.3	-3.9	-1.6	1.0
13.31	-0.56	1.4	16.26	31MR	1.9	-3.8	-1.6	1.0
20.77	-0.46	1.5	25.98	10AP	3.8	-3.7	-1.6	1.0
28.13	-0.36	1.5	35.65	20AP	2.2	-3.6	-1.6	1.0
35.41	-0.26	1.6	45.29	30AP	1.3	-3.5	-1.7	1.0
42.60	-0.16	1.6	54.88	10MY	0.6	-3.5	-1.7	0.9
49.70	-0.05	1.7	64.44	20MY	-0.1	-3.4	-1.7	0.9
56.71	0.06	1.7	73.98	30MY	-1.1	-3.4	-1.8	0.8
63.63	0.17	1.7	83.51	9JN	-1.8	-3.4	-1.9	0.8
70.47	0.27	1.8	93.04	19JN	-1.0	-3.3	-1.9	0.7
77.22	0.38	1.8	102.57	29JN	-0.2	-3.3	-2.0	0.6
83.89	0.49	1.8	112.12	9JL	0.3	-3.3	-2.0	0.6
90.47	0.59	1.8	121.69	19JL	0.6	-3.3	-2.1	0.5
96.97	0.70	1.8	131.30	29JL	1.0	-3.3	-2.2	0.5
103.39	0.81	1.8	140.94	8AU	2.0	-3.3	-2.2	0.5
109.72	0.92	1.8	150.63	18AU	2.9	-3.4	-2.3	0.6
115.97	1.02	1.8	160.37	28AU	0.5	-3.4	-2.4	0.6
122.12	1.13	1.8	170.16	7SE	-0.8	-3.4	-2.4	0.7
128.18	1.24	1.7	180.01	17SE	-1.1	-3.4	-2.4	0.7
134.13	1.35	1.7	189.92	27SE	-1.0	-3.5	-2.5	0.8
139.96	1.47	1.6	199.88	7OC	-0.7	-3.5	-2.4	0.8
145.65	1.59	1.6	209.91	17OC	-0.4	-3.5	-2.4	0.9
151.19	1.71	1.5	219.98	27OC	-0.3	-3.5	-2.4	0.9
156.55	1.84	1.4	230.09	6NO	-0.3	-3.4	-2.3	0.9
161.69	1.98	1.3	240.24	16NO	-0.0	-3.4	-2.3	1.0
166.57	2.13	1.1	250.42	26NO	1.6	-3.4	-2.2	1.0
171.13	2.28	1.0	260.61	6DE	1.6	-3.4	-2.2	1.0
175.30	2.45	0.8	270.81	16DE	0.2	-3.3	-2.1	1.0
178.99	2.63	0.6	281.01	26DE	-0.1	-3.3	-2.0	1.0
				1051				
182.08	2.81	0.4	291.18	5JA	-0.1	-3.3	-1.9	1.0
184.43	3.01	0.2	301.34	15JA	-0.3	-3.3	-1.9	1.0
185.86	3.20	-0.1	311.46	25JA	-0.6	-3.4	-1.8	1.0
186.19	3.38	-0.3	321.54	4FE	-1.1	-3.4	-1.7	1.0
185.28	3.51	-0.6	331.57	14FE	-1.3	-3.4	-1.7	1.1
183.10	3.57	-0.9	341.55	24FE	-0.9	-3.4	-1.7	1.1
179.86	3.50	-1.1	351.47	6MR	0.4	-3.5	-1.6	1.1
176.06	3.29	-1.2	1.33	16MR	2.3	-3.5	-1.6	1.2
172.41	2.94	-1.1	11.14	26MR	3.0	-3.5	-1.6	1.2
169.58	2.51	-0.9	20.89	5AP	1.6	-3.6	-1.6	1.2
167.94	2.05	-0.7	30.59	15AP	1.0	-3.7	-1.5	1.2
167.63	1.60	-0.5	40.24	25AP	0.5	-3.7	-1.5	1.2
168.56	1.19	-0.3	49.85	5MY	-0.2	-3.8	-1.5	1.1
170.56	0.82	-0.1	59.43	15MY	-1.1	-3.9	-1.6	1.1
173.47	0.50	0.1	68.98	25MY	-1.8	-4.0	-1.6	1.1
177.11	0.21	0.3	78.51	4JN	-1.0	-4.1	-1.6	1.0
181.36	-0.04	0.4	88.05	14JN	-0.2	-4.2	-1.6	1.0
186.13	-0.26	0.5	97.57	24JN	0.4	-4.2	-1.7	0.9
191.32	-0.46	0.6	107.11	4JL	0.8	-4.1	-1.7	0.8
196.88	-0.63	0.7	116.67	14JL	1.4	-3.8	-1.8	0.8
202.75	-0.79	0.8	126.26	24JL	2.6	-3.3	-1.8	0.7
208.90	-0.92	0.9	135.88	3AU	2.4	-3.3	-1.9	0.6
215.29	-1.04	0.9	145.54	13AU	0.4	-3.9	-2.0	0.6
221.91	-1.14	1.0	155.25	23AU	-0.9	-4.2	-2.0	0.6
228.73	-1.22	1.0	165.02	2SE	-1.2	-4.3	-2.1	0.7
235.72	-1.29	1.0	174.84	12SE	-1.1	-4.2	-2.2	0.7
242.88	-1.34	1.1	184.72	22SE	-0.6	-4.2	-2.2	0.8
250.18	-1.37	1.1	194.65	2OC	-0.3	-4.1	-2.3	0.8
257.61	-1.39	1.1	204.65	12OC	-0.2	-4.0	-2.3	0.9
265.15	-1.40	1.2	214.69	22OC	-0.1	-3.9	-2.4	0.9
272.78	-1.39	1.2	224.78	1NO	0.2	-3.8	-2.4	1.0
280.48	-1.37	1.2	234.92	11NO	1.9	-3.8	-2.4	1.0
288.25	-1.33	1.3	245.08	21NO	1.3	-3.7	-2.4	1.1
296.05	-1.28	1.3	255.26	1DE	-0.0	-3.6	-2.3	1.1
303.88	-1.22	1.3	265.47	11DE	-0.2	-3.6	-2.3	1.1
311.72	-1.14	1.3	275.66	21DE	-0.3	-3.5	-2.2	1.1
319.55	-1.06	1.4	285.85	31DE	-0.4	-3.5	-2.1	1.1

♂ LONG	LAT	MAG	⊙ LONG	16.00UT 1052	☿	♀	♃	♄
327.36	-0.97	1.4	296.03	10JA	-0.7	-3.4	-2.1	1.1
335.13	-0.88	1.4	306.16	20JA	-1.1	-3.4	-2.0	1.1
342.86	-0.78	1.5	316.27	30JA	-1.2	-3.4	-1.9	1.1
350.53	-0.67	1.5	326.32	9FE	-0.8	-3.3	-1.9	1.1
358.13	-0.56	1.5	336.33	19FE	0.6	-3.3	-1.8	1.2
5.66	-0.45	1.5	346.28	29FE	2.8	-3.3	-1.7	1.2
13.12	-0.34	1.6	356.17	10MR	2.3	-3.3	-1.7	1.3
20.49	-0.23	1.6	6.01	20MR	1.2	-3.3	-1.6	1.3
27.78	-0.12	1.6	15.78	30MR	0.7	-3.3	-1.6	1.3
34.99	-0.02	1.6	25.51	9AP	0.3	-3.4	-1.5	1.3
42.11	0.09	1.7	35.18	19AP	-0.2	-3.4	-1.5	1.4
49.16	0.19	1.7	44.81	29AP	-1.1	-3.4	-1.5	1.3
56.12	0.29	1.7	54.41	9MY	-1.8	-3.5	-1.5	1.3
63.01	0.39	1.7	63.97	19MY	-1.0	-3.5	-1.5	1.3
69.82	0.48	1.7	73.52	29MY	-0.1	-3.5	-1.5	1.3
76.56	0.57	1.8	83.05	8JN	0.6	-3.4	-1.5	1.2
83.24	0.65	1.8	92.57	18JN	1.1	-3.4	-1.5	1.2
89.85	0.74	1.9	102.11	28JN	1.9	-3.4	-1.5	1.1
96.41	0.82	1.9	111.65	8JL	3.1	-3.4	-1.5	1.0
102.90	0.89	1.9	121.22	18JL	2.0	-3.3	-1.5	1.0
109.35	0.96	2.0	130.83	28JL	0.3	-3.3	-1.6	0.9
115.75	1.04	2.0	140.47	7AU	-0.9	-3.3	-1.6	0.8
122.09	1.10	2.0	150.15	17AU	-1.4	-3.3	-1.7	0.8
128.39	1.17	2.0	159.89	27AU	-1.0	-3.3	-1.7	0.8
134.65	1.23	2.0	169.68	6SE	-0.5	-3.4	-1.8	0.8
140.85	1.29	1.9	179.53	16SE	-0.2	-3.4	-1.9	0.8
147.00	1.35	1.9	189.44	26SE	-0.1	-3.4	-1.9	0.9
153.10	1.41	1.9	199.40	6OC	0.0	-3.5	-2.0	0.9
159.14	1.46	1.8	209.42	16OC	0.4	-3.5	-2.1	1.0
165.12	1.52	1.8	219.49	26OC	2.2	-3.6	-2.1	1.0
171.01	1.57	1.7	229.60	5NO	1.1	-3.6	-2.2	1.1
176.82	1.62	1.6	239.74	15NO	-0.2	-3.7	-2.2	1.1
182.52	1.67	1.5	249.92	25NO	-0.4	-3.8	-2.2	1.2
188.10	1.71	1.4	260.12	5DE	-0.4	-3.9	-2.2	1.2
193.53	1.75	1.3	270.31	15DE	-0.5	-4.0	-2.2	1.3
198.78	1.79	1.2	280.51	25DE	-0.7	-4.0	-2.2	1.3
				1053				
203.81	1.83	1.0	290.70	4JA	-1.0	-4.2	-2.2	1.3
208.58	1.85	0.8	300.85	14JA	-1.0	-4.2	-2.1	1.3
213.01	1.87	0.6	310.98	24JA	-0.7	-4.3	-2.1	1.3
217.02	1.87	0.4	321.06	3FE	0.7	-4.2	-2.0	1.3
220.52	1.85	0.2	331.09	13FE	3.1	-4.2	-2.0	1.3
223.37	1.81	-0.1	341.07	23FE	1.7	-3.9	-1.9	1.3
225.40	1.73	-0.3	350.99	5MR	0.9	-3.3	-1.8	1.4
226.46	1.59	-0.6	0.85	15MR	0.5	-3.5	-1.7	1.4
226.34	1.39	-1.0	10.66	25MR	0.2	-4.0	-1.7	1.4
224.97	1.10	-1.3	20.41	4AP	-0.3	-4.2	-1.6	1.3
222.43	0.72	-1.6	30.11	14AP	-1.1	-4.2	-1.6	1.3
219.08	0.26	-1.8	39.77	24AP	-1.8	-4.2	-1.5	1.3
215.58	-0.24	-1.7	49.38	4MY	-1.0	-4.1	-1.5	1.3
212.65	-0.72	-1.5	58.96	14MY	-0.0	-4.0	-1.4	1.2
210.82	-1.14	-1.3	68.51	24MY	0.7	-3.9	-1.4	1.2
210.35	-1.48	-1.1	78.05	3JN	1.4	-3.8	-1.4	1.2
211.23	-1.75	-0.9	87.58	13JN	2.5	-3.7	-1.4	1.1
213.31	-1.96	-0.7	97.11	23JN	3.2	-3.6	-1.3	1.1
216.41	-2.11	-0.5	106.65	3JL	1.6	-3.6	-1.3	1.0
220.35	-2.22	-0.4	116.20	13JL	0.2	-3.5	-1.4	1.0
224.99	-2.28	-0.2	125.79	23JL	-1.0	-3.5	-1.4	0.9
230.20	-2.32	-0.1	135.41	2AU	-1.5	-3.4	-1.4	0.9
235.88	-2.33	0.0	145.07	12AU	-1.0	-3.4	-1.4	0.8
241.95	-2.31	0.2	154.78	22AU	-0.4	-3.4	-1.4	0.8
248.35	-2.27	0.3	164.55	1SE	-0.1	-3.4	-1.5	0.8
255.25	-2.22	0.4	174.36	11SE	0.1	-3.4	-1.5	0.8
261.93	-2.14	0.5	184.24	21SE	0.2	-3.4	-1.6	0.9
269.02	-2.05	0.6	194.17	1OC	0.7	-3.4	-1.6	1.0
276.26	-1.94	0.6	204.16	11OC	2.5	-3.4	-1.7	1.0
283.62	-1.82	0.7	214.21	21OC	1.0	-3.4	-1.8	1.1
291.08	-1.69	0.8	224.29	31OC	-0.3	-3.4	-1.8	1.1
298.60	-1.55	0.9	234.42	10NO	-0.5	-3.4	-1.9	1.2
306.17	-1.40	1.0	244.59	20NO	-0.5	-3.4	-1.9	1.2
313.76	-1.25	1.0	254.77	30NO	-0.6	-3.4	-2.0	1.3
321.35	-1.10	1.1	264.97	10DE	-0.8	-3.5	-2.1	1.3
328.93	-0.95	1.2	275.17	20DE	-0.8	-3.5	-2.1	1.3
336.48	-0.80	1.3	285.36	30DE	-0.9	-3.5	-2.1	1.3

♂ LONG	LAT	MAG	⊙ LONG	16.00UT 1054	☿	♀	♃	♄
343.99	-0.65	1.3	295.53	9JA	-0.6	-3.5	-2.1	1.3
351.46	-0.51	1.4	305.67	19JA	0.8	-3.4	-2.1	1.3
358.86	-0.37	1.5	315.77	29JA	3.0	-3.4	-2.1	1.2
6.21	-0.23	1.5	325.83	8FE	1.2	-3.4	-2.1	1.2
13.48	-0.10	1.6	335.84	18FE	0.6	-3.4	-2.0	1.2
20.68	0.02	1.6	345.79	28FE	0.3	-3.3	-2.0	1.1
27.81	0.13	1.7	355.69	10MR	0.1	-3.3	-1.9	1.1
34.87	0.24	1.7	5.53	20MR	-0.4	-3.3	-1.8	1.1
41.85	0.34	1.8	15.31	30MR	-1.1	-3.3	-1.8	1.1
48.76	0.44	1.8	25.04	9AP	-1.7	-3.4	-1.7	1.1
55.61	0.53	1.8	34.72	19AP	-1.0	-3.4	-1.6	1.1
62.38	0.61	1.8	44.35	29AP	0.1	-3.4	-1.6	1.0
69.10	0.69	1.9	53.94	9MY	1.0	-3.4	-1.5	1.0
75.77	0.76	1.9	63.51	19MY	1.9	-3.5	-1.4	1.0
82.38	0.82	1.9	73.05	29MY	3.2	-3.5	-1.4	1.0
88.94	0.88	1.9	82.58	8JN	2.6	-3.5	-1.4	0.9
95.46	0.94	1.9	92.11	18JN	1.3	-3.6	-1.3	0.9
101.94	0.99	1.9	101.64	28JN	0.1	-3.7	-1.3	0.8
108.39	1.03	1.9	111.19	8JL	-1.0	-3.8	-1.3	0.8
114.81	1.08	2.0	120.76	18JL	-1.6	-3.8	-1.3	0.7
121.21	1.11	2.0	130.36	28JL	-1.0	-3.9	-1.3	0.7
127.58	1.14	2.0	140.00	7AU	-0.4	-4.1	-1.3	0.6
133.94	1.17	2.0	149.69	17AU	-0.0	-4.2	-1.3	0.6
140.28	1.19	2.0	159.42	27AU	0.2	-4.2	-1.3	0.5
146.61	1.21	2.0	169.21	6SE	0.4	-4.3	-1.3	0.5
152.93	1.23	2.0	179.06	16SE	0.9	-4.3	-1.3	0.5
159.24	1.24	2.0	188.96	26SE	2.9	-3.9	-1.4	0.5
165.54	1.25	2.0	198.92	6OC	0.8	-3.3	-1.4	0.6
171.84	1.25	2.0	208.94	16OC	-0.5	-3.5	-1.5	0.7
178.12	1.25	1.9	219.00	26OC	-0.7	-4.1	-1.5	0.7
184.39	1.24	1.9	229.11	5NO	-0.7	-4.3	-1.6	0.8
190.64	1.23	1.8	239.26	15NO	-0.7	-4.4	-1.6	0.9
196.88	1.20	1.8	249.43	25NO	-0.7	-4.3	-1.7	0.9
203.09	1.18	1.7	259.62	5DE	-0.7	-4.2	-1.8	1.0
209.27	1.14	1.6	269.82	15DE	-0.7	-4.1	-1.8	1.0
215.41	1.09	1.5	280.02	25DE	-0.5	-4.0	-1.9	1.0
				1055				
221.50	1.03	1.4	290.20	4JA	1.0	-3.9	-1.9	1.0
227.53	0.96	1.3	300.36	14JA	2.6	-3.8	-2.0	1.0
233.49	0.87	1.1	310.48	24JA	0.9	-3.7	-2.0	1.0
239.36	0.76	1.0	320.57	3FE	0.4	-3.7	-2.0	1.0
245.11	0.63	0.8	330.61	13FE	0.2	-3.6	-2.0	1.0
250.73	0.47	0.6	340.59	23FE	-0.0	-3.5	-2.0	0.9
256.16	0.28	0.4	350.52	5MR	-0.4	-3.5	-2.0	0.9
261.38	0.06	0.2	0.38	15MR	-1.1	-3.4	-2.0	0.8
266.32	-0.22	0.0	10.19	25MR	-1.6	-3.4	-1.9	0.8
270.90	-0.54	-0.2	19.95	4AP	-1.0	-3.3	-1.9	0.8
275.03	-0.94	-0.5	29.65	14AP	0.1	-3.3	-1.8	0.8
278.59	-1.40	-0.8	39.31	24AP	1.2	-3.3	-1.7	0.8
281.41	-1.95	-1.1	48.92	4MY	2.5	-3.3	-1.7	0.8
283.31	-2.59	-1.4	58.50	14MY	3.5	-3.3	-1.6	0.8
284.12	-3.31	-1.7	68.06	24MY	2.0	-3.3	-1.5	0.8
283.69	-4.07	-2.0	77.59	3JN	1.0	-3.3	-1.5	0.8
282.07	-4.81	-2.3	87.12	13JN	0.0	-3.3	-1.4	0.7
279.58	-5.43	-2.5	96.65	23JN	-1.0	-3.4	-1.4	0.7
276.79	-5.82	-2.5	106.19	3JL	-1.7	-3.4	-1.3	0.7
274.47	-5.95	-2.3	115.74	13JL	-1.0	-3.5	-1.3	0.6
273.19	-5.82	-2.1	125.32	23JL	-0.3	-3.5	-1.3	0.6
273.25	-5.53	-1.8	134.94	2AU	0.1	-3.5	-1.2	0.5
274.47	-5.13	-1.6	144.60	12AU	0.4	-3.5	-1.2	0.5
277.24	-4.68	-1.3	154.31	22AU	0.6	-3.4	-1.2	0.4
280.83	-4.22	-1.1	164.07	1SE	1.3	-3.4	-1.2	0.3
285.24	-3.76	-0.9	173.88	11SE	3.2	-3.4	-1.2	0.3
290.26	-3.32	-0.7	183.76	21SE	0.7	-3.3	-1.2	0.2
295.79	-2.90	-0.4	193.69	1OC	-0.6	-3.3	-1.2	0.2
301.69	-2.51	-0.3	203.67	11OC	-0.8	-3.3	-1.3	0.3
307.89	-2.15	-0.1	213.71	21OC	-0.8	-3.3	-1.3	0.3
314.31	-1.81	0.1	223.80	31OC	-0.8	-3.4	-1.3	0.4
320.89	-1.49	0.3	233.93	10NO	-0.6	-3.4	-1.4	0.5
327.58	-1.20	0.4	244.09	20NO	-0.5	-3.4	-1.4	0.6
334.37	-0.94	0.6	254.28	30NO	-0.6	-3.4	-1.5	0.6
341.20	-0.69	0.7	264.47	10DE	-0.3	-3.4	-1.6	0.7
348.07	-0.47	0.9	274.67	20DE	1.2	-3.5	-1.6	0.7
354.95	-0.27	1.0	284.87	30DE	2.2	-3.5	-1.7	0.7

♂ LONG	LAT	MAG	☉ LONG	16.00UT	☿	♀	♃	♄
				1056	MAGNITUDES			
1.82	-0.09	1.1	295.04	9JA	0.6	-3.6	-1.8	0.8
8.67	0.07	1.2	305.18	19JA	0.2	-3.6	-1.8	0.8
15.51	0.22	1.3	315.29	29JA	0.1	-3.7	-1.9	0.8
22.31	0.35	1.4	325.35	8FE	-0.1	-3.8	-1.9	0.8
29.08	0.47	1.5	335.36	18FE	-0.5	-3.8	-2.0	0.7
35.81	0.57	1.6	345.32	28FE	-1.1	-3.9	-2.0	0.7
42.49	0.67	1.7	355.22	9MR	-1.5	-4.0	-2.0	0.7
49.14	0.75	1.7	5.06	19MR	-1.0	-4.1	-2.0	0.6
55.74	0.82	1.8	14.84	29MR	0.2	-4.2	-2.0	0.6
62.30	0.89	1.8	24.57	8AP	1.6	-4.2	-2.0	0.6
68.83	0.94	1.9	34.25	18AP	3.3	-4.2	-1.9	0.6
75.32	0.99	1.9	43.89	28AP	2.7	-4.0	-1.9	0.6
81.78	1.03	2.0	53.48	8MY	1.5	-3.5	-1.8	0.6
88.21	1.07	2.0	63.05	18MY	0.8	-2.7	-1.8	0.6
94.62	1.09	2.0	72.60	28MY	-0.0	-3.6	-1.7	0.6
101.01	1.12	2.0	82.12	7JN	-1.0	-4.0	-1.6	0.6
107.38	1.13	2.0	91.65	17JN	-1.8	-4.2	-1.6	0.6
113.74	1.15	2.0	101.18	27JN	-1.0	-4.2	-1.5	0.5
120.10	1.15	2.0	110.73	7JL	-0.3	-4.1	-1.4	0.5
126.45	1.16	2.0	120.30	17JL	0.2	-4.0	-1.4	0.5
132.81	1.15	2.0	129.89	27JL	0.5	-3.9	-1.4	0.4
139.17	1.15	2.0	139.53	6AU	0.9	-3.9	-1.3	0.4
145.55	1.14	2.0	149.21	16AU	1.7	-3.8	-1.3	0.3
151.94	1.12	2.0	158.94	26AU	3.1	-3.7	-1.3	0.3
158.35	1.10	2.0	168.73	5SE	0.6	-3.6	-1.2	0.2
164.78	1.07	2.0	178.57	15SE	-0.7	-3.6	-1.2	0.1
171.23	1.04	2.0	188.48	25SE	-1.0	-3.5	-1.2	0.1
177.71	1.01	2.0	198.43	5OC	-1.0	-3.5	-1.2	0.0
184.22	0.97	2.0	208.45	15OC	-0.7	-3.5	-1.2	0.0
190.76	0.92	1.9	218.51	25OC	-0.5	-3.4	-1.2	0.1
197.33	0.87	1.9	228.62	4NO	-0.4	-3.4	-1.3	0.1
203.93	0.81	1.9	238.77	14NO	-0.4	-3.4	-1.3	0.2
210.56	0.74	1.8	248.94	24NO	-0.1	-3.4	-1.3	0.3
217.22	0.66	1.8	259.13	4DE	1.4	-3.4	-1.4	0.4
223.91	0.58	1.7	269.33	14DE	1.8	-3.4	-1.4	0.4
230.63	0.48	1.6	279.53	24DE	0.3	-3.3	-1.5	0.5
				1057				
237.38	0.38	1.6	289.71	3JA	0.0	-3.4	-1.5	0.5
244.16	0.26	1.5	299.87	13JA	-0.1	-3.4	-1.6	0.5
250.95	0.13	1.4	310.00	23JA	-0.2	-3.4	-1.7	0.6
257.77	-0.01	1.3	320.08	2FE	-0.5	-3.4	-1.7	0.6
264.61	-0.17	1.2	330.12	12FE	-1.1	-3.4	-1.8	0.6
271.46	-0.35	1.1	340.11	22FE	-1.4	-3.4	-1.9	0.6
278.32	-0.54	1.0	350.03	4MR	-0.9	-3.5	-1.9	0.6
285.17	-0.74	0.8	359.91	14MR	0.3	-3.5	-2.0	0.5
292.01	-0.97	0.7	9.72	24MR	2.0	-3.4	-2.0	0.5
298.82	-1.21	0.6	19.48	3AP	3.5	-3.4	-2.0	0.4
305.58	-1.47	0.4	29.18	13AP	2.0	-3.4	-2.0	0.4
312.28	-1.75	0.3	38.84	23AP	1.2	-3.4	-2.0	0.4
318.87	-2.05	0.1	48.45	3MY	0.6	-3.3	-2.0	0.4
325.33	-2.36	-0.0	58.04	13MY	-0.1	-3.3	-2.0	0.4
331.59	-2.68	-0.2	67.59	23MY	-1.0	-3.3	-1.9	0.5
337.62	-3.02	-0.4	77.13	2JN	-1.8	-3.3	-1.9	0.5
343.30	-3.37	-0.6	86.66	12JN	-1.0	-3.3	-1.8	0.5
348.56	-3.73	-0.8	96.18	22JN	-0.2	-3.3	-1.8	0.5
353.26	-4.09	-1.0	105.72	2JL	0.3	-3.4	-1.7	0.4
357.24	-4.44	-1.2	115.28	12JL	0.7	-3.4	-1.6	0.4
0.32	-4.77	-1.4	124.86	22JL	1.2	-3.4	-1.6	0.4
2.29	-5.05	-1.7	134.47	1AU	2.2	-3.5	-1.5	0.3
2.91	-5.25	-1.9	144.13	11AU	2.8	-3.5	-1.5	0.3
2.13	-5.31	-2.1	153.83	21AU	0.5	-3.6	-1.4	0.2
0.05	-5.16	-2.3	163.59	31AU	-0.8	-3.7	-1.4	0.2
357.13	-4.77	-2.5	173.41	10SE	-1.1	-3.7	-1.3	0.1
354.15	-4.16	-2.3	183.27	20SE	-1.1	-3.8	-1.3	0.1
351.85	-3.43	-2.1	193.20	30SE	-0.6	-3.9	-1.3	-0.0
350.72	-2.68	-1.7	203.19	10OC	-0.4	-4.0	-1.3	-0.1
350.93	-1.98	-1.4	213.22	20OC	-0.3	-4.1	-1.3	-0.1
352.37	-1.37	-1.1	223.31	30OC	-0.2	-4.3	-1.3	-0.1
354.87	-0.85	-0.8	233.44	9NO	0.0	-4.3	-1.3	-0.1
358.22	-0.43	-0.5	243.60	19NO	1.6	-4.4	-1.3	0.0
2.22	-0.08	-0.2	253.78	29NO	1.6	-4.3	-1.3	0.1
6.75	0.21	0.0	263.98	9DE	0.1	-4.0	-1.3	0.1
11.67	0.44	0.3	274.18	19DE	-0.1	-3.3	-1.4	0.2
16.89	0.63	0.5	284.37	29DE	-0.2	-3.5	-1.4	0.3

♂ LONG	LAT	MAG	☉ LONG	16.00UT	☿	♀	♃	♄
				1058	MAGNITUDES			
22.35	0.79	0.7	294.55	8JA	-0.3	-4.1	-1.5	0.3
27.98	0.91	0.8	304.69	18JA	-0.6	-4.3	-1.5	0.4
33.75	1.01	1.0	314.80	28JA	-1.1	-4.3	-1.6	0.4
39.62	1.10	1.2	324.86	7FE	-1.3	-4.3	-1.6	0.4
45.57	1.16	1.3	334.88	17FE	-0.9	-4.2	-1.7	0.4
51.58	1.21	1.4	344.84	27FE	0.4	-4.1	-1.8	0.4
57.63	1.25	1.5	354.74	9MR	2.5	-4.0	-1.8	0.4
63.71	1.28	1.6	4.58	19MR	2.7	-3.9	-1.9	0.4
69.82	1.30	1.7	14.37	29MR	1.5	-3.8	-2.0	0.4
75.95	1.31	1.8	24.10	8AP	0.9	-3.7	-2.0	0.4
82.09	1.32	1.8	33.78	18AP	0.4	-3.6	-2.1	0.3
88.25	1.32	1.9	43.42	28AP	-0.2	-3.5	-2.1	0.3
94.42	1.31	1.9	53.02	8MY	-1.0	-3.5	-2.1	0.3
100.60	1.30	2.0	62.59	18MY	-1.9	-3.4	-2.1	0.3
106.79	1.28	2.0	72.14	28MY	-1.0	-3.4	-2.1	0.3
113.00	1.26	2.0	81.67	7JN	-0.1	-3.4	-2.1	0.4
119.23	1.24	2.0	91.20	17JN	0.4	-3.3	-2.1	0.4
125.48	1.21	2.0	100.73	27JN	0.9	-3.3	-2.0	0.4
131.75	1.18	2.0	110.27	7JL	1.6	-3.3	-2.0	0.4
138.06	1.14	2.0	119.84	17JL	2.8	-3.3	-1.9	0.4
144.39	1.10	2.0	129.44	27JL	2.3	-3.3	-1.9	0.3
150.76	1.05	2.0	139.07	6AU	0.4	-3.3	-1.8	0.3
157.17	1.00	2.0	148.75	16AU	-0.9	-3.4	-1.7	0.3
163.62	0.95	1.9	158.48	26AU	-1.3	-3.4	-1.7	0.2
170.12	0.90	1.9	168.26	5SE	-1.1	-3.4	-1.6	0.2
176.67	0.84	1.9	178.09	15SE	-0.6	-3.4	-1.6	0.1
183.27	0.77	1.9	187.99	25SE	-0.3	-3.5	-1.5	0.0
189.92	0.70	1.9	197.95	5OC	-0.1	-3.5	-1.5	-0.0
196.63	0.63	1.9	207.96	15OC	-0.1	-3.5	-1.4	-0.1
203.41	0.55	1.9	218.02	25OC	0.3	-3.4	-1.4	-0.2
210.24	0.46	1.8	228.12	4NO	1.9	-3.4	-1.4	-0.2
217.13	0.37	1.8	238.26	14NO	1.3	-3.4	-1.4	-0.2
224.09	0.28	1.8	248.44	24NO	-0.1	-3.4	-1.4	-0.2
231.11	0.18	1.8	258.63	4DE	-0.3	-3.4	-1.4	-0.1
238.19	0.07	1.7	268.83	14DE	-0.3	-3.3	-1.4	-0.0
245.33	-0.04	1.7	279.03	24DE	-0.4	-3.3	-1.4	0.1
				1059				
252.53	-0.16	1.6	289.21	3JA	-0.7	-3.3	-1.4	0.1
259.79	-0.29	1.6	299.37	13JA	-1.1	-3.3	-1.4	0.2
267.10	-0.42	1.5	309.51	23JA	-1.1	-3.4	-1.5	0.2
274.46	-0.55	1.5	319.59	2FE	-0.8	-3.4	-1.5	0.3
281.87	-0.69	1.4	329.63	12FE	0.6	-3.4	-1.6	0.3
289.32	-0.83	1.3	339.62	22FE	2.9	-3.4	-1.6	0.3
296.80	-0.97	1.3	349.55	4MR	2.0	-3.5	-1.7	0.4
304.30	-1.11	1.2	359.43	14MR	1.1	-3.5	-1.8	0.4
311.82	-1.25	1.1	9.25	24MR	0.6	-3.5	-1.8	0.4
319.33	-1.39	1.0	19.01	3AP	0.3	-3.6	-1.9	0.3
326.84	-1.51	1.0	28.72	13AP	-0.3	-3.7	-2.0	0.3
334.32	-1.64	0.9	38.38	23AP	-1.0	-3.7	-2.0	0.3
341.74	-1.75	0.8	47.99	3MY	-1.9	-3.8	-2.1	0.2
349.11	-1.84	0.8	57.58	13MY	-1.0	-3.9	-2.2	0.2
356.39	-1.93	0.7	67.14	23MY	-0.1	-4.0	-2.2	0.2
3.55	-2.00	0.6	76.67	2JN	0.6	-4.1	-2.2	0.3
10.57	-2.05	0.5	86.20	12JN	1.2	-4.2	-2.3	0.3
17.43	-2.08	0.4	95.73	22JN	2.1	-4.2	-2.3	0.3
24.07	-2.09	0.3	105.27	2JL	3.3	-4.1	-2.3	0.3
30.47	-2.08	0.2	114.82	12JL	1.9	-3.8	-2.2	0.3
36.56	-2.04	0.1	124.40	22JL	0.3	-3.3	-2.2	0.3
42.30	-1.98	0.0	134.01	1AU	-0.9	-3.3	-2.1	0.3
47.59	-1.89	-0.1	143.67	11AU	-1.4	-3.9	-2.1	0.3
52.33	-1.77	-0.3	153.37	21AU	-1.1	-4.2	-2.0	0.3
56.42	-1.61	-0.4	163.12	31AU	-0.5	-4.3	-2.0	0.2
59.67	-1.40	-0.6	172.93	10SE	-0.2	-4.2	-1.9	0.2
61.90	-1.14	-0.8	182.80	20SE	-0.0	-4.2	-1.8	0.1
62.90	-0.82	-1.0	192.72	30SE	0.1	-4.1	-1.8	0.1
62.49	-0.43	-1.3	202.70	10OC	0.5	-4.0	-1.7	0.0
60.63	0.03	-1.5	212.74	20OC	2.2	-3.9	-1.7	-0.1
57.53	0.52	-1.6	222.82	30OC	1.2	-3.8	-1.6	-0.1
53.77	1.00	-1.7	232.94	9NO	-0.2	-3.8	-1.6	-0.2
50.16	1.41	-1.5	243.10	19NO	-0.4	-3.7	-1.5	-0.3
47.43	1.73	-1.2	253.28	29NO	-0.5	-3.6	-1.5	-0.3
45.97	1.94	-0.9	263.48	9DE	-0.5	-3.6	-1.5	-0.2
45.87	2.06	-0.6	273.68	19DE	-0.8	-3.5	-1.5	-0.1
46.98	2.12	-0.3	283.87	29DE	-0.9	-3.5	-1.5	-0.1

♂			☉	16.00UT	☿ ♀ ♃ ♄				♂			☉	16.00UT	☿ ♀ ♃ ♄			
LONG	LAT	MAG	LONG	1060	MAGNITUDES				LONG	LAT	MAG	LONG	1062	MAGNITUDES			
49.11	2.14	-0.0	294.05	8JA	-1.0	-3.4	-1.5	0.0	88.23	3.85	-0.8	293.55	7JA	2.5	-3.5	-1.8	-0.1
52.07	2.13	0.3	304.20	18JA	-0.7	-3.4	-1.5	0.1	86.24	3.72	-0.5	303.70	17JA	0.8	-3.4	-1.7	-0.1
55.66	2.10	0.5	314.31	28JA	0.7	-3.4	-1.5	0.1	85.58	3.52	-0.3	313.81	27JA	0.3	-3.4	-1.7	0.0
59.76	2.06	0.7	324.38	7FE	3.1	-3.3	-1.5	0.2	86.17	3.30	0.0	323.88	6FE	0.1	-3.4	-1.6	0.1
64.25	2.01	0.9	334.40	17FE	1.5	-3.3	-1.5	0.2	87.81	3.07	0.3	333.90	16FE	-0.1	-3.4	-1.6	0.1
69.04	1.96	1.1	344.36	27FE	0.8	-3.3	-1.6	0.3	90.33	2.84	0.5	343.87	26FE	-0.4	-3.3	-1.6	0.2
74.08	1.90	1.2	354.26	8MR	0.4	-3.3	-1.6	0.3	93.54	2.63	0.7	353.77	8MR	-1.1	-3.3	-1.6	0.2
79.31	1.84	1.3	4.11	18MR	0.1	-3.3	-1.7	0.3	97.30	2.44	0.9	3.63	18MR	-1.6	-3.3	-1.6	0.3
84.70	1.78	1.4	13.90	28MR	-0.3	-3.3	-1.7	0.3	101.50	2.26	1.0	13.42	28MR	-1.0	-3.3	-1.6	0.3
90.23	1.71	1.6	23.64	7AP	-1.0	-3.4	-1.8	0.3	106.05	2.09	1.2	23.16	7AP	0.1	-3.4	-1.6	0.4
95.86	1.65	1.6	33.32	17AP	-1.8	-3.4	-1.9	0.3	110.90	1.93	1.3	32.85	17AP	1.4	-3.4	-1.6	0.4
101.58	1.58	1.7	42.96	27AP	-0.3	-3.4	-1.9	0.3	115.98	1.78	1.4	42.49	27AP	2.8	-3.4	-1.6	0.4
107.39	1.52	1.8	52.56	7MY	-0.0	-3.5	-2.0	0.3	121.26	1.64	1.5	52.09	7MY	3.2	-3.4	-1.7	0.4
113.27	1.45	1.8	62.13	17MY	0.8	-3.5	-2.1	0.2	126.72	1.50	1.6	61.67	17MY	1.8	-3.5	-1.7	0.4
119.21	1.38	1.9	71.67	27MY	1.6	-3.5	-2.1	0.2	132.33	1.37	1.6	71.21	27MY	0.9	-3.5	-1.8	0.4
125.22	1.31	1.9	81.21	6JN	2.7	-3.4	-2.2	0.2	138.07	1.25	1.7	80.74	6JN	0.0	-3.5	-1.8	0.4
131.29	1.24	1.9	90.73	16JN	3.0	-3.4	-2.3	0.2	143.94	1.12	1.7	90.28	16JN	-1.0	-3.6	-1.9	0.3
137.42	1.17	2.0	100.26	26JN	1.5	-3.4	-2.3	0.3	149.93	1.00	1.8	99.80	26JN	-1.7	-3.7	-1.9	0.3
143.62	1.09	2.0	109.81	6JL	0.2	-3.4	-2.4	0.3	156.02	0.89	1.8	109.35	6JL	-1.0	-3.8	-2.0	0.4
149.88	1.02	2.0	119.37	16JL	-1.0	-3.3	-2.4	0.3	162.22	0.77	1.8	118.91	16JL	-0.3	-3.9	-2.1	0.4
156.20	0.94	2.0	128.97	26JL	-1.5	-3.3	-2.4	0.3	168.53	0.66	1.8	128.50	26JL	0.1	-4.0	-2.1	0.4
162.58	0.86	2.0	138.60	5AU	-1.0	-3.3	-2.4	0.3	174.93	0.55	1.8	138.13	5AU	0.4	-4.1	-2.2	0.5
169.04	0.77	1.9	148.27	15AU	-0.4	-3.3	-2.4	0.3	181.44	0.44	1.8	147.81	15AU	0.7	-4.2	-2.3	0.5
175.57	0.69	1.9	158.00	25AU	-0.1	-3.3	-2.3	0.3	188.05	0.33	1.8	157.53	25AU	1.4	-4.2	-2.3	0.5
182.17	0.60	1.9	167.78	4SE	0.1	-3.4	-2.3	0.3	194.77	0.22	1.8	167.31	4SE	3.2	-4.3	-2.4	0.5
188.84	0.51	1.9	177.62	14SE	0.3	-3.4	-2.2	0.3	201.58	0.11	1.8	177.14	14SE	0.7	-4.2	-2.4	0.5
195.60	0.42	1.8	187.51	24SE	0.7	-3.4	-2.1	0.2	208.50	0.00	1.7	187.03	24SE	-0.6	-3.9	-2.4	0.5
202.43	0.33	1.8	197.46	4OC	2.6	-3.5	-2.1	0.1	215.51	-0.11	1.7	196.98	4OC	-0.9	-3.3	-2.4	0.5
209.35	0.23	1.7	207.47	14OC	1.0	-3.5	-2.0	0.1	222.63	-0.21	1.7	206.99	14OC	-0.9	-3.5	-2.4	0.4
216.35	0.13	1.7	217.53	24OC	-0.4	-3.6	-1.9	0.0	229.84	-0.31	1.6	217.04	24OC	-0.8	-4.1	-2.4	0.4
223.43	0.03	1.7	227.63	3NO	-0.6	-3.6	-1.9	-0.1	237.15	-0.41	1.6	227.14	3NO	-0.6	-4.3	-2.4	0.3
230.58	-0.07	1.7	237.77	13NO	-0.6	-3.7	-1.8	-0.1	244.55	-0.51	1.5	237.28	13NO	-0.5	-4.4	-2.3	0.3
237.82	-0.18	1.6	247.94	23NO	-0.6	-3.8	-1.8	-0.2	252.04	-0.60	1.5	247.45	23NO	-0.5	-4.3	-2.3	0.2
245.14	-0.28	1.6	258.13	3DE	-0.8	-3.9	-1.7	-0.3	259.60	-0.69	1.4	257.64	3DE	-0.3	-4.2	-2.2	0.1
252.53	-0.39	1.6	268.33	13DE	-0.8	-4.0	-1.7	-0.3	267.24	-0.77	1.4	267.84	13DE	1.2	-4.1	-2.1	0.0
260.00	-0.49	1.6	278.53	23DE	-0.8	-4.1	-1.6	-0.2	274.94	-0.85	1.4	278.04	23DE	2.1	-4.0	-2.1	-0.0
				1061									1063				
267.53	-0.60	1.6	288.72	2JA	-0.6	-4.2	-1.6	-0.1	282.70	-0.92	1.4	288.22	2JA	0.5	-3.9	-2.0	-0.1
275.11	-0.70	1.5	298.88	12JA	0.8	-4.3	-1.6	-0.1	290.51	-0.98	1.4	298.39	12JA	0.1	-3.8	-1.9	-0.1
282.76	-0.80	1.5	309.01	22JA	2.9	-4.3	-1.6	0.0	298.34	-1.03	1.4	308.52	22JA	0.0	-3.7	-1.8	-0.0
290.44	-0.90	1.5	319.10	1FE	1.1	-4.3	-1.5	0.1	306.20	-1.07	1.4	318.61	1FE	-0.2	-3.7	-1.8	0.1
298.16	-0.99	1.4	329.14	11FE	0.5	-4.2	-1.5	0.1	314.07	-1.10	1.4	328.66	11FE	-0.5	-3.6	-1.7	0.1
305.91	-1.07	1.4	339.14	21FE	0.3	-3.9	-1.6	0.2	321.93	-1.11	1.4	338.65	21FE	-1.1	-3.5	-1.7	0.2
313.67	-1.14	1.4	349.07	3MR	0.0	-3.3	-1.6	0.2	329.79	-1.12	1.4	348.59	3MR	-1.5	-3.5	-1.6	0.2
321.44	-1.21	1.4	358.95	13MR	-0.4	-3.5	-1.6	0.3	337.61	-1.11	1.4	358.47	13MR	-1.0	-3.4	-1.6	0.3
329.19	-1.26	1.3	8.77	23MR	-1.1	-4.0	-1.6	0.3	345.39	-1.09	1.4	8.29	23MR	0.2	-3.4	-1.6	0.4
336.92	-1.30	1.3	18.54	2AP	-1.7	-4.2	-1.7	0.3	353.12	-1.06	1.4	18.06	2AP	1.7	-3.3	-1.6	0.4
344.62	-1.33	1.3	28.24	12AP	-1.0	-4.2	-1.7	0.3	0.79	-1.02	1.4	27.78	12AP	3.6	-3.3	-1.5	0.4
352.27	-1.35	1.2	37.91	22AP	0.1	-4.2	-1.8	0.3	8.39	-0.96	1.4	37.44	22AP	2.4	-3.3	-1.5	0.5
359.85	-1.35	1.2	47.53	2MY	1.0	-4.1	-1.8	0.3	15.91	-0.90	1.4	47.07	2MY	1.4	-3.3	-1.5	0.5
7.36	-1.33	1.2	57.11	12MY	2.1	-4.0	-1.9	0.3	23.34	-0.82	1.5	56.65	12MY	0.7	-3.3	-1.5	0.5
14.79	-1.30	1.2	66.67	22MY	3.5	-3.9	-1.9	0.3	30.68	-0.74	1.5	66.21	22MY	-0.1	-3.3	-1.6	0.5
22.10	-1.25	1.1	76.21	1JN	2.4	-3.8	-2.0	0.3	37.92	-0.64	1.5	75.75	1JN	-1.0	-3.3	-1.6	0.5
29.31	-1.19	1.1	85.74	11JN	1.2	-3.7	-2.1	0.2	45.05	-0.54	1.5	85.28	11JN	-1.8	-3.3	-1.6	0.5
36.39	-1.11	1.1	95.27	21JN	0.1	-3.6	-2.1	0.2	52.07	-0.43	1.5	94.81	21JN	-1.0	-3.4	-1.6	0.5
43.32	-1.02	1.0	104.81	1JL	-1.0	-3.6	-2.2	0.3	58.98	-0.31	1.5	104.34	1JL	-0.3	-3.4	-1.7	0.5
50.09	-0.91	1.0	114.36	11JL	-1.6	-3.5	-2.3	0.3	65.77	-0.18	1.5	113.90	11JL	0.2	-3.5	-1.7	0.5
56.69	-0.79	1.0	123.94	21JL	-1.0	-3.5	-2.3	0.4	72.42	-0.05	1.4	123.47	21JL	0.6	-3.5	-1.8	0.5
63.08	-0.65	0.9	133.55	31JL	-0.4	-3.4	-2.4	0.4	78.95	0.09	1.4	133.08	31JL	1.0	-3.5	-1.8	0.6
69.25	-0.50	0.8	143.20	10AU	0.0	-3.4	-2.4	0.4	85.33	0.24	1.4	142.73	10AU	1.9	-3.5	-1.9	0.6
75.17	-0.33	0.8	152.90	20AU	0.3	-3.4	-2.4	0.4	91.55	0.40	1.3	152.42	20AU	3.1	-3.4	-2.0	0.6
80.78	-0.13	0.7	162.65	30AU	0.5	-3.4	-2.5	0.4	97.59	0.56	1.3	162.18	30AU	0.6	-3.4	-2.0	0.7
86.04	0.08	0.6	172.46	9SE	1.1	-3.4	-2.4	0.4	103.45	0.74	1.2	171.98	9SE	-0.7	-3.4	-2.1	0.7
90.87	0.32	0.4	182.32	19SE	3.0	-3.4	-2.4	0.4	109.08	0.92	1.2	181.84	19SE	-1.1	-3.3	-2.2	0.7
95.19	0.59	0.3	192.24	29SE	0.9	-3.4	-2.4	0.3	114.46	1.12	1.1	191.76	29SE	-1.0	-3.3	-2.2	0.7
98.90	0.90	0.1	202.22	9OC	-0.5	-3.4	-2.3	0.3	119.54	1.34	1.0	201.73	9OC	-0.7	-3.3	-2.3	0.7
101.82	1.24	-0.0	212.25	19OC	-0.8	-3.4	-2.3	0.2	124.27	1.58	0.9	211.76	19OC	-0.4	-3.3	-2.3	0.6
103.80	1.63	-0.2	222.33	29OC	-0.7	-3.4	-2.2	0.1	128.56	1.84	0.7	221.84	29OC	-0.3	-3.3	-2.3	0.6
104.63	2.07	-0.5	232.45	8NO	-0.8	-3.4	-2.1	0.1	132.32	2.13	0.6	231.96	8NO	-0.0	-3.4	-2.4	0.5
104.14	2.53	-0.7	242.61	18NO	-0.7	-3.4	-2.1	0.0	135.43	2.45	0.4	242.11	18NO	-0.1	-3.4	-2.4	0.5
102.26	3.00	-0.9	252.79	28NO	-0.6	-3.4	-2.0	-0.1	137.74	2.80	0.2	252.30	28NO	1.4	-3.4	-2.3	0.4
99.15	3.41	-1.1	262.98	8DE	-0.7	-3.5	-1.9	-0.1	139.05	3.18	-0.0	262.49	8DE	1.8	-3.4	-2.3	0.4
95.27	3.71	-1.2	273.19	18DE	-0.4	-3.5	-1.9	-0.2	139.20	3.58	-0.3	272.69	18DE	0.3	-3.5	-2.2	0.3
91.39	3.86	-1.1	283.38	28DE	1.0	-3.5	-1.8	-0.2	138.03	3.97	-0.5	282.89	28DE	-0.0	-3.5	-2.2	0.2

417

Left table (1064 / 1065)

♂ LONG	LAT	MAG	☉ LONG	16.00UT	☿	♀	♃	♄
135.57	4.30	-0.8	293.06	7JA	-0.1	-3.6	-2.1	0.2
132.07	4.51	-0.9	303.21	17JA	-0.3	-3.6	-2.0	0.1
128.11	4.56	-1.0	313.33	27JA	-0.6	-3.7	-2.0	0.1
124.43	4.43	-0.8	323.40	6FE	-1.1	-3.8	-1.9	0.2
121.65	4.16	-0.6	333.42	16FE	-1.4	-3.8	-1.8	0.2
120.10	3.81	-0.3	343.39	26FE	-0.9	-3.9	-1.8	0.3
119.84	3.44	-0.1	353.30	7MR	0.3	-4.0	-1.7	0.3
120.76	3.07	0.1	3.15	17MR	2.1	-4.1	-1.6	0.4
122.68	2.73	0.4	12.95	27MR	3.2	-4.2	-1.6	0.5
125.43	2.41	0.5	22.69	6AP	1.8	-4.2	-1.6	0.5
128.83	2.13	0.7	32.38	16AP	1.0	-4.2	-1.5	0.6
132.77	1.86	0.9	42.02	26AP	0.5	-4.0	-1.5	0.6
137.15	1.63	1.0	51.63	6MY	-0.1	-3.5	-1.5	0.6
141.89	1.41	1.1	61.20	16MY	-1.0	-2.7	-1.4	0.7
146.94	1.21	1.2	70.75	26MY	-1.8	-3.6	-1.4	0.7
152.25	1.02	1.3	80.28	5JN	-1.0	-4.0	-1.4	0.7
157.78	0.84	1.4	89.81	15JN	-0.2	-4.2	-1.4	0.7
163.52	0.68	1.4	99.34	25JN	0.4	-4.2	-1.5	0.7
169.44	0.52	1.5	108.88	5JL	0.8	-4.1	-1.5	0.7
175.54	0.37	1.5	118.44	15JL	1.3	-4.0	-1.5	0.7
181.79	0.23	1.5	128.04	25JL	2.4	-3.9	-1.5	0.7
188.19	0.09	1.6	137.66	4AU	2.6	-3.9	-1.6	0.7
194.73	-0.04	1.6	147.34	14AU	0.5	-3.8	-1.6	0.8
201.42	-0.16	1.6	157.06	24AU	-0.8	-3.7	-1.7	0.8
208.23	-0.28	1.6	166.83	3SE	-1.2	-3.6	-1.7	0.8
215.17	-0.39	1.6	176.66	13SE	-1.1	-3.6	-1.8	0.9
222.24	-0.50	1.6	186.55	23SE	-0.6	-3.5	-1.9	0.9
229.43	-0.60	1.5	196.50	3OC	-0.3	-3.5	-1.9	0.9
236.73	-0.69	1.5	206.50	13OC	-0.2	-3.5	-2.0	0.9
244.14	-0.77	1.5	216.55	23OC	-0.2	-3.4	-2.1	0.9
251.64	-0.85	1.5	226.65	2NO	0.1	-3.4	-2.1	0.9
259.24	-0.91	1.5	236.79	12NO	1.7	-3.4	-2.2	0.8
266.92	-0.97	1.4	246.96	22NO	1.6	-3.4	-2.2	0.8
274.66	-1.02	1.4	257.14	2DE	0.1	-3.4	-2.2	0.7
282.47	-1.05	1.4	267.35	12DE	-0.3	-3.4	-2.2	0.7
290.32	-1.08	1.4	277.54	22DE	-0.2	-3.3	-2.2	0.6

1065

♂ LONG	LAT	MAG	☉ LONG	16.00UT	☿	♀	♃	♄
298.21	-1.09	1.4	287.73	1JA	-0.4	-3.4	-2.2	0.5
306.11	-1.09	1.3	297.90	11JA	-0.6	-3.4	-2.2	0.5
314.02	-1.08	1.3	308.03	21JA	-1.1	-3.4	-2.1	0.4
321.92	-1.06	1.3	318.12	31JA	-1.2	-3.4	-2.1	0.3
329.80	-1.03	1.3	328.17	10FE	-0.9	-3.4	-2.0	0.4
337.64	-0.98	1.3	338.17	20FE	0.4	-3.4	-1.9	0.4
345.43	-0.93	1.3	348.11	2MR	2.6	-3.5	-1.8	0.5
353.18	-0.86	1.4	357.99	12MR	2.5	-3.5	-1.8	0.5
0.85	-0.79	1.4	7.82	22MR	1.3	-3.4	-1.7	0.6
8.45	-0.71	1.5	17.58	1AP	0.8	-3.4	-1.6	0.6
15.98	-0.62	1.5	27.30	11AP	0.4	-3.4	-1.6	0.7
23.42	-0.53	1.5	36.97	21AP	-0.2	-3.4	-1.5	0.7
30.77	-0.43	1.6	46.59	1MY	-1.0	-3.3	-1.5	0.8
38.03	-0.33	1.6	56.18	11MY	-1.9	-3.3	-1.4	0.8
45.20	-0.22	1.6	65.74	21MY	-1.0	-3.3	-1.4	0.9
52.27	-0.11	1.7	75.28	31MY	-0.1	-3.3	-1.4	0.9
59.26	-0.00	1.7	84.81	10JN	0.5	-3.3	-1.4	0.9
66.14	0.11	1.7	94.34	20JN	1.0	-3.3	-1.3	0.9
72.94	0.22	1.7	103.87	30JN	1.7	-3.4	-1.3	0.9
79.64	0.34	1.7	113.43	10JL	3.0	-3.4	-1.3	1.0
86.24	0.46	1.7	123.00	20JL	2.2	-3.4	-1.3	1.0
92.75	0.57	1.7	132.61	30JL	0.4	-3.5	-1.4	1.0
99.17	0.69	1.7	142.26	9AU	-0.9	-3.5	-1.4	0.9
105.48	0.82	1.7	151.95	19AU	-1.3	-3.6	-1.4	1.0
111.68	0.94	1.7	161.70	29AU	-1.1	-3.7	-1.4	1.0
117.78	1.07	1.7	171.50	8SE	-0.5	-3.7	-1.5	1.1
123.74	1.20	1.6	181.36	18SE	-0.2	-3.8	-1.5	1.1
129.57	1.33	1.6	191.27	28SE	-0.1	-3.9	-1.6	1.1
135.24	1.48	1.5	201.25	8OC	0.0	-4.0	-1.6	1.2
140.73	1.63	1.4	211.27	18OC	0.3	-4.1	-1.7	1.2
146.02	1.78	1.3	221.35	28OC	1.9	-4.3	-1.8	1.2
151.05	1.95	1.2	231.47	7NO	1.3	-4.3	-1.8	1.2
155.78	2.13	1.1	241.62	17NO	-0.1	-4.4	-1.9	1.1
160.15	2.33	0.9	251.80	27NO	-0.4	-4.3	-1.9	1.1
164.07	2.54	0.8	262.00	7DE	-0.4	-4.0	-2.0	1.1
167.43	2.77	0.6	272.19	17DE	-0.5	-3.2	-2.1	1.0
170.10	3.02	0.3	282.39	27DE	-0.7	-3.5	-2.1	0.9

Right table (1066 / 1067)

♂ LONG	LAT	MAG	☉ LONG	16.00UT	☿	♀	♃	♄
171.92	3.28	0.1	292.57	6JA	-1.0	-4.1	-2.1	0.9
172.71	3.54	-0.1	302.72	16JA	-1.1	-4.3	-2.1	0.8
172.29	3.79	-0.4	312.84	26JA	-0.8	-4.4	-2.1	0.7
170.58	3.98	-0.7	322.91	5FE	0.6	-4.3	-2.1	0.7
167.68	4.07	-0.9	332.93	15FE	3.0	-4.2	-2.0	0.6
163.98	4.01	-1.1	342.90	25FE	1.8	-4.1	-2.0	0.6
160.12	3.79	-1.0	352.82	7MR	0.9	-4.0	-1.9	0.7
156.83	3.44	-0.8	2.67	17MR	0.6	-3.9	-1.9	0.7
154.60	3.01	-0.6	12.47	27MR	0.2	-3.8	-1.8	0.8
153.67	2.57	-0.4	22.22	6AP	-0.3	-3.7	-1.7	0.8
154.02	2.13	-0.2	31.91	16AP	-1.0	-3.6	-1.7	0.9
155.48	1.74	0.0	41.56	26AP	-1.8	-3.5	-1.6	0.9
157.91	1.38	0.2	51.17	6MY	-1.0	-3.5	-1.5	1.0
161.12	1.06	0.4	60.74	16MY	-0.1	-3.4	-1.5	1.0
164.98	0.77	0.5	70.29	26MY	0.7	-3.4	-1.4	1.1
169.37	0.51	0.6	79.82	5JN	1.3	-3.4	-1.4	1.1
174.20	0.28	0.8	89.35	15JN	2.3	-3.3	-1.3	1.2
179.40	0.08	0.8	98.88	25JN	3.3	-3.3	-1.3	1.2
184.94	-0.11	0.9	108.42	5JL	1.8	-3.3	-1.3	1.2
190.75	-0.28	1.0	117.98	15JL	0.3	-3.3	-1.3	1.2
196.81	-0.44	1.1	127.57	25JL	-0.9	-3.3	-1.3	1.2
203.10	-0.58	1.1	137.20	4AU	-1.5	-3.3	-1.2	1.2
209.59	-0.71	1.1	146.87	14AU	-1.1	-3.4	-1.3	1.2
216.28	-0.82	1.2	156.59	24AU	-0.5	-3.4	-1.3	1.2
223.14	-0.92	1.2	166.36	3SE	-0.1	-3.4	-1.3	1.3
230.16	-1.01	1.2	176.18	13SE	0.1	-3.5	-1.3	1.3
237.32	-1.08	1.3	186.07	23SE	0.2	-3.5	-1.3	1.4
244.63	-1.14	1.3	196.02	3OC	0.6	-3.5	-1.4	1.3
252.05	-1.19	1.3	206.01	13OC	2.3	-3.5	-1.4	1.3
259.58	-1.23	1.3	216.06	23OC	1.2	-3.4	-1.5	1.3
267.21	-1.25	1.3	226.16	2NO	-0.3	-3.4	-1.5	1.3
274.92	-1.26	1.3	236.29	12NO	-0.5	-3.4	-1.6	1.3
282.69	-1.25	1.3	246.46	22NO	-0.5	-3.4	-1.6	1.2
290.52	-1.23	1.3	256.65	2DE	-0.6	-3.4	-1.7	1.2
298.37	-1.20	1.4	266.84	12DE	-0.8	-3.3	-1.8	1.1
306.25	-1.15	1.4	277.05	22DE	-0.9	-3.3	-1.8	1.1

1067

♂ LONG	LAT	MAG	☉ LONG	16.00UT	☿	♀	♃	♄
314.13	-1.10	1.4	287.23	1JA	-0.9	-3.3	-1.9	1.0
321.99	-1.04	1.4	297.40	11JA	-0.7	-3.3	-1.9	1.0
329.84	-0.96	1.4	307.54	21JA	0.7	-3.4	-2.0	0.9
337.64	-0.88	1.4	317.63	31JA	3.1	-3.4	-2.0	0.9
345.39	-0.79	1.5	327.68	10FE	1.4	-3.4	-2.0	0.9
353.08	-0.70	1.5	337.69	20FE	0.7	-3.4	-2.0	0.8
0.71	-0.60	1.5	347.63	2MR	0.4	-3.5	-2.0	0.8
8.26	-0.50	1.5	357.51	12MR	0.1	-3.5	-2.0	0.9
15.73	-0.39	1.5	7.35	22MR	-0.3	-3.6	-1.9	1.0
23.12	-0.29	1.5	17.12	1AP	-1.0	-3.6	-1.9	1.0
30.42	-0.18	1.5	26.83	11AP	-1.8	-3.7	-1.8	1.1
37.64	-0.08	1.6	36.51	21AP	-1.0	-3.7	-1.8	1.1
44.77	0.03	1.6	46.13	1MY	-0.0	-3.8	-1.7	1.2
51.82	0.13	1.6	55.72	11MY	0.9	-3.9	-1.6	1.3
58.78	0.23	1.7	65.29	21MY	1.8	-4.0	-1.6	1.3
65.67	0.33	1.7	74.83	31MY	3.0	-4.1	-1.5	1.4
72.47	0.43	1.8	84.35	10JN	2.8	-4.2	-1.5	1.4
79.20	0.53	1.8	93.88	20JN	1.4	-4.2	-1.4	1.4
85.86	0.62	1.9	103.42	30JN	0.2	-4.1	-1.4	1.4
92.46	0.71	1.9	112.97	10JL	-0.9	-3.8	-1.3	1.4
98.98	0.80	1.9	122.54	20JL	-1.6	-3.2	-1.3	1.4
105.44	0.89	1.9	132.15	30JL	-1.1	-3.3	-1.3	1.3
111.84	0.97	1.9	141.79	9AU	-0.4	-3.9	-1.2	1.3
118.18	1.05	1.9	151.49	19AU	-0.0	-4.2	-1.3	1.3
124.46	1.14	1.9	161.23	29AU	0.2	-4.3	-1.2	1.2
130.67	1.22	1.9	171.02	8SE	0.4	-4.2	-1.2	1.1
136.82	1.30	1.9	180.88	18SE	0.8	-4.2	-1.2	1.2
142.89	1.37	1.8	190.79	28SE	2.7	-4.1	-1.2	1.2
148.89	1.45	1.8	200.76	8OC	1.0	-4.0	-1.2	1.2
154.80	1.53	1.7	210.79	18OC	-0.4	-3.9	-1.3	1.2
160.62	1.61	1.7	220.86	28OC	-0.7	-3.8	-1.3	1.1
166.32	1.69	1.6	230.98	7NO	-0.7	-3.8	-1.3	1.1
171.89	1.77	1.5	241.13	17NO	-0.7	-3.7	-1.4	1.1
177.31	1.85	1.4	251.31	27NO	-0.7	-3.6	-1.5	1.1
182.53	1.94	1.3	261.50	7DE	-0.7	-3.6	-1.5	1.0
187.53	2.02	1.1	271.71	17DE	-0.7	-3.5	-1.6	1.0
192.26	2.11	1.0	281.90	27DE	-0.5	-3.5	-1.6	1.0

Left table (1068–1069):

♂ LONG	LAT	MAG	☉ LONG	16.00UT	☿	♀	♃	♄
				1068				
196.64	2.19	0.8	292.08	6JA	0.8	-3.4	-1.7	0.9
200.60	2.27	0.6	302.23	16JA	2.8	-3.4	-1.8	0.9
204.04	2.35	0.4	312.35	26JA	1.0	-3.4	-1.8	0.8
206.82	2.42	0.1	322.43	5FE	0.4	-3.3	-1.9	0.8
208.77	2.47	-0.1	332.45	15FE	0.2	-3.3	-1.9	0.7
209.74	2.49	-0.4	342.43	25FE	0.0	-3.3	-2.0	0.7
209.54	2.45	-0.7	352.34	6MR	-0.4	-3.3	-2.0	0.6
208.07	2.34	-1.0	2.20	16MR	-1.0	-3.3	-2.0	0.7
205.42	2.13	-1.3	12.00	26MR	-1.7	-3.3	-2.0	0.7
201.93	1.80	-1.5	21.75	5AP	-1.0	-3.4	-2.0	0.8
198.26	1.39	-1.4	31.45	15AP	0.1	-3.4	-2.0	0.9
195.14	0.92	-1.3	41.09	25AP	1.1	-3.4	-1.9	0.9
193.08	0.46	-1.1	50.70	5MY	2.3	-3.5	-1.9	1.0
192.36	0.04	-0.9	60.28	15MY	3.6	-3.5	-1.8	1.1
192.96	-0.33	-0.7	69.82	25MY	2.2	-3.5	-1.7	1.1
194.74	-0.65	-0.5	79.36	4JN	1.1	-3.4	-1.7	1.1
197.55	-0.91	-0.3	88.89	14JN	0.1	-3.4	-1.6	1.2
201.18	-1.13	-0.1	98.41	24JN	-1.0	-3.4	-1.6	1.2
205.51	-1.31	0.0	107.96	4JL	-1.7	-3.4	-1.5	1.2
210.42	-1.46	0.1	117.52	14JL	-1.1	-3.3	-1.4	1.2
215.80	-1.57	0.3	127.10	24JL	-0.4	-3.3	-1.4	1.2
221.58	-1.67	0.4	136.73	3AU	0.1	-3.3	-1.3	1.2
227.71	-1.73	0.5	146.39	13AU	0.3	-3.3	-1.3	1.2
234.13	-1.78	0.5	156.11	23AU	0.6	-3.3	-1.3	1.1
240.81	-1.80	0.6	165.88	2SE	1.2	-3.4	-1.2	1.1
247.70	-1.81	0.7	175.70	12SE	3.1	-3.4	-1.2	1.0
254.79	-1.79	0.8	185.58	22SE	0.9	-3.4	-1.2	1.0
262.03	-1.76	0.8	195.53	2OC	-0.6	-3.5	-1.2	1.0
269.42	-1.72	0.9	205.52	12OC	-0.8	-3.5	-1.2	1.0
276.91	-1.65	0.9	215.57	22OC	-0.8	-3.6	-1.2	1.0
284.49	-1.58	1.0	225.67	1NO	-0.3	-3.6	-1.2	1.0
292.13	-1.49	1.0	235.80	11NO	-0.6	-3.7	-1.3	1.0
299.82	-1.39	1.1	245.97	21NO	-0.6	-3.8	-1.3	1.0
307.54	-1.28	1.2	256.16	1DE	-0.9	-3.9	-1.3	1.0
315.26	-1.16	1.2	266.35	11DE	-0.4	-4.0	-1.4	0.9
322.98	-1.04	1.3	276.55	21DE	1.0	-4.1	-1.4	0.9
330.67	-0.92	1.3	286.74	31DE	2.4	-4.2	-1.5	0.9
				1069				
338.33	-0.79	1.4	296.91	10JA	0.7	-4.3	-1.6	0.8
345.93	-0.66	1.4	307.05	20JA	0.2	-4.3	-1.6	0.8
353.49	-0.53	1.5	317.15	30JA	0.1	-4.3	-1.7	0.7
0.97	-0.41	1.5	327.20	9FE	-0.1	-4.2	-1.8	0.7
8.39	-0.28	1.6	337.20	19FE	-0.4	-3.9	-1.8	0.6
15.73	-0.16	1.6	347.15	1MR	-1.0	-3.3	-1.9	0.6
22.99	-0.05	1.6	357.04	11MR	-1.6	-3.5	-1.9	0.5
30.17	0.06	1.7	6.87	21MR	-1.0	-4.0	-2.0	0.5
37.28	0.17	1.7	16.65	31MR	0.1	-4.2	-2.0	0.5
44.31	0.27	1.7	26.37	10AP	1.5	-4.2	-2.0	0.6
51.26	0.37	1.8	36.04	20AP	3.1	-4.2	-2.1	0.7
58.14	0.46	1.8	45.67	30AP	2.9	-4.1	-2.0	0.7
64.95	0.55	1.8	55.26	10MY	1.7	-4.0	-2.0	0.8
71.70	0.63	1.8	64.82	20MY	0.8	-3.9	-2.0	0.9
78.39	0.71	1.8	74.37	30MY	0.0	-3.8	-1.9	0.9
85.01	0.78	1.8	83.89	9JN	-1.0	-3.7	-1.9	1.0
91.59	0.85	1.8	93.42	19JN	-1.7	-3.6	-1.8	1.0
98.12	0.91	1.9	102.96	29JN	-1.1	-3.6	-1.8	1.0
104.61	0.97	1.9	112.50	9JL	-0.3	-3.5	-1.7	1.0
111.05	1.02	2.0	122.08	19JL	0.2	-3.5	-1.6	1.1
117.46	1.07	2.0	131.68	29JL	0.5	-3.4	-1.6	1.1
123.84	1.12	2.0	141.32	8AU	0.8	-3.4	-1.5	1.1
130.21	1.16	2.0	151.01	18AU	1.6	-3.4	-1.5	1.0
136.52	1.20	2.0	160.76	28AU	3.2	-3.4	-1.4	1.0
142.81	1.24	2.0	170.55	7SE	0.8	-3.4	-1.4	1.0
149.09	1.27	2.0	180.40	17SE	-0.7	-3.4	-1.3	0.9
155.34	1.30	2.0	190.31	27SE	-1.0	-3.4	-1.3	0.9
161.56	1.32	2.0	200.28	7OC	-0.9	-3.4	-1.3	0.9
167.76	1.34	1.9	210.30	17OC	-0.8	-3.4	-1.3	0.9
173.93	1.36	1.9	220.37	27OC	-0.5	-3.4	-1.3	0.9
180.06	1.37	1.8	230.48	6NO	-0.4	-3.4	-1.3	0.9
186.16	1.38	1.8	240.64	16NO	-0.4	-3.4	-1.3	0.9
192.20	1.38	1.7	250.82	26NO	-0.2	-3.4	-1.3	0.9
198.19	1.38	1.6	261.01	6DE	1.2	-3.5	-1.3	0.9
204.11	1.37	1.5	271.21	16DE	2.1	-3.5	-1.4	0.9
209.94	1.35	1.4	281.41	26DE	0.4	-3.5	-1.4	0.8

Right table (1070–1071):

♂ LONG	LAT	MAG	☉ LONG	16.00UT	☿	♀	♃	♄
				1070				
215.67	1.32	1.2	291.58	5JA	0.1	-3.4	-1.4	0.8
221.29	1.27	1.1	301.74	15JA	-0.0	-3.4	-1.5	0.8
226.74	1.21	0.9	311.86	25JA	-0.2	-3.4	-1.5	0.7
232.02	1.14	0.8	321.94	4FE	-0.5	-3.4	-1.6	0.7
237.07	1.03	0.6	331.97	14FE	-1.1	-3.4	-1.7	0.6
241.83	0.90	0.4	341.94	24FE	-1.4	-3.3	-1.7	0.6
246.24	0.74	0.1	351.86	6MR	-1.0	-3.3	-1.8	0.5
250.20	0.52	-0.1	1.72	16MR	0.2	-3.3	-1.9	0.5
253.58	0.26	-0.4	11.53	26MR	1.8	-3.3	-1.9	0.4
256.26	-0.08	-0.7	21.28	5AP	3.7	-3.4	-2.0	0.4
258.04	-0.49	-1.0	30.98	15AP	2.2	-3.4	-2.1	0.4
258.75	-1.00	-1.3	40.63	25AP	1.2	-3.4	-2.1	0.5
258.25	-1.59	-1.6	50.23	5MY	0.6	-3.4	-2.1	0.6
256.51	-2.24	-1.9	59.81	15MY	-0.1	-3.5	-2.1	0.6
253.81	-2.89	-2.2	69.37	25MY	-1.0	-3.5	-2.1	0.7
250.69	-3.46	-2.2	78.90	4JN	-1.8	-3.5	-2.1	0.8
247.87	-3.88	-2.1	88.43	14JN	-1.1	-3.6	-2.1	0.8
246.02	-4.13	-1.9	97.96	24JN	-0.3	-3.7	-2.1	0.9
245.48	-4.22	-1.7	107.50	4JL	0.3	-3.8	-2.0	0.9
246.31	-4.20	-1.4	117.06	14JL	0.6	-3.9	-2.0	0.9
248.43	-4.08	-1.2	126.64	24JL	1.1	-4.0	-1.9	0.9
251.63	-3.92	-1.0	136.26	3AU	2.0	-4.1	-1.8	0.9
255.73	-3.72	-0.8	145.93	13AU	3.0	-4.2	-1.9	0.9
260.55	-3.50	-0.6	155.64	23AU	0.6	-4.2	-1.7	0.9
265.94	-3.26	-0.4	165.41	2SE	-0.8	-4.3	-1.7	0.9
271.79	-3.02	-0.3	175.23	12SE	-1.1	-4.2	-1.6	0.9
278.01	-2.77	-0.1	185.11	22SE	-1.1	-3.9	-1.6	0.9
284.51	-2.51	0.0	195.04	2OC	-0.7	-3.3	-1.5	0.8
291.24	-2.26	0.2	205.04	12OC	-0.4	-3.5	-1.5	0.8
298.15	-2.01	0.3	215.08	22OC	-0.3	-4.1	-1.4	0.8
305.17	-1.77	0.4	225.17	1NO	-0.3	-4.3	-1.4	0.8
312.30	-1.53	0.6	235.31	11NO	-0.0	-4.4	-1.4	0.8
319.48	-1.30	0.7	245.47	21NO	1.4	-4.3	-1.4	0.8
326.69	-1.08	0.8	255.66	1DE	1.8	-4.2	-1.4	0.8
333.92	-0.87	0.9	265.86	11DE	0.2	-4.1	-1.4	0.8
341.15	-0.68	1.0	276.06	21DE	-0.1	-4.0	-1.4	0.8
348.36	-0.50	1.1	286.25	31DE	-0.2	-3.9	-1.4	0.8
				1071				
355.54	-0.32	1.2	296.42	10JA	-0.3	-3.8	-1.4	0.8
2.68	-0.16	1.3	306.56	20JA	-0.6	-3.7	-1.5	0.7
9.77	-0.02	1.4	316.66	30JA	-1.1	-3.7	-1.5	0.7
16.81	0.12	1.5	326.72	9FE	-1.3	-3.6	-1.5	0.7
23.80	0.24	1.5	336.72	19FE	-0.9	-3.5	-1.6	0.6
30.73	0.36	1.6	346.67	1MR	0.3	-3.4	-1.7	0.5
37.60	0.46	1.7	356.57	11MR	2.3	-3.4	-1.7	0.5
44.41	0.56	1.7	6.40	21MR	2.9	-3.4	-1.8	0.4
51.17	0.64	1.8	16.18	31MR	1.6	-3.3	-1.9	0.4
57.88	0.72	1.8	25.91	10AP	0.9	-3.3	-1.9	0.3
64.53	0.79	1.9	35.58	20AP	0.5	-3.3	-2.0	0.3
71.13	0.85	1.9	45.21	30AP	-0.1	-3.3	-2.1	0.4
77.69	0.91	1.9	54.80	10MY	-1.0	-3.3	-2.1	0.4
84.21	0.96	2.0	64.37	20MY	-1.9	-3.3	-2.2	0.5
90.70	1.00	2.0	73.91	30MY	-1.1	-3.3	-2.2	0.6
97.16	1.04	2.0	83.44	9JN	-0.2	-3.3	-2.3	0.6
103.58	1.07	2.0	92.96	19JN	0.4	-3.4	-2.3	0.7
109.99	1.10	2.0	102.50	29JN	0.9	-3.4	-2.3	0.7
116.38	1.12	2.0	112.05	9JL	1.4	-3.5	-2.3	0.8
122.76	1.14	1.9	121.61	19JL	2.6	-3.5	-2.2	0.8
129.13	1.15	2.0	131.22	29JL	2.5	-3.5	-2.2	0.8
135.49	1.16	2.0	140.86	8AU	0.5	-3.5	-2.1	0.9
141.86	1.16	2.0	150.54	18AU	-0.8	-3.4	-2.1	0.9
148.23	1.16	2.0	160.28	28AU	-1.3	-3.4	-2.0	0.9
154.61	1.15	2.0	170.07	7SE	-1.1	-3.4	-2.0	0.9
161.00	1.14	2.0	179.92	17SE	-0.6	-3.3	-1.9	0.8
167.40	1.13	2.0	189.83	27SE	-0.3	-3.3	-1.8	0.8
173.81	1.11	2.0	199.79	7OC	-0.2	-3.3	-1.8	0.8
180.24	1.08	2.0	209.81	17OC	-0.1	-3.3	-1.7	0.7
186.68	1.05	1.9	219.88	27OC	0.2	-3.3	-1.7	0.7
193.14	1.01	1.9	229.99	6NO	1.7	-3.4	-1.6	0.7
199.61	0.97	1.9	240.13	16NO	1.5	-3.4	-1.6	0.8
206.10	0.91	1.8	250.31	26NO	-0.0	-3.4	-1.5	0.8
212.60	0.85	1.7	260.51	6DE	-0.3	-3.4	-1.5	0.8
219.11	0.78	1.7	270.71	16DE	-0.3	-3.5	-1.5	0.8
225.63	0.70	1.6	280.91	26DE	-0.4	-3.5	-1.5	0.8

♂ LONG	LAT	MAG	☉ LONG	16.00UT 1072	☿	♀	♃	♄
232.15	0.61	1.5	291.09	5JA	-0.7	-3.6	-1.5	0.8
238.68	0.50	1.4	301.24	15JA	-1.1	-3.6	-1.5	0.8
245.20	0.38	1.3	311.37	25JA	-1.2	-3.7	-1.5	0.7
251.71	0.24	1.2	321.45	4FE	-0.8	-3.8	-1.5	0.7
258.21	0.08	1.1	331.48	14FE	0.4	-3.8	-1.5	0.7
264.68	-0.09	0.9	341.47	24FE	2.7	-3.9	-1.6	0.6
271.12	-0.30	0.8	351.39	5MR	2.2	-4.0	-1.6	0.6
277.50	-0.52	0.6	1.25	15MR	1.2	-4.1	-1.7	0.5
283.82	-0.78	0.5	11.06	25MR	0.7	-4.2	-1.7	0.4
290.04	-1.07	0.3	20.81	4AP	0.3	-4.2	-1.8	0.4
296.14	-1.39	0.1	30.51	14AP	-0.2	-4.2	-1.8	0.3
302.07	-1.74	-0.1	40.17	24AP	-1.0	-4.0	-1.9	0.3
307.78	-2.14	-0.3	49.78	4MY	-1.9	-3.4	-2.0	0.3
313.20	-2.58	-0.5	59.35	14MY	-1.1	-2.7	-2.0	0.3
318.23	-3.07	-0.7	68.91	24MY	-0.1	-3.6	-2.1	0.4
322.76	-3.60	-1.0	78.44	3JN	0.5	-4.0	-2.2	0.5
326.64	-4.17	-1.2	87.97	13JN	1.1	-4.2	-2.2	0.5
329.68	-4.77	-1.5	97.50	23JN	1.9	-4.2	-2.3	0.6
331.68	-5.39	-1.8	107.04	3JL	3.2	-4.1	-2.3	0.7
332.46	-5.96	-2.0	116.59	13JL	2.1	-4.0	-2.4	0.7
331.89	-6.42	-2.3	126.18	23JL	0.4	-3.9	-2.4	0.7
330.12	-6.68	-2.5	135.80	2AU	-0.9	-3.9	-2.4	0.8
327.57	-6.64	-2.6	145.46	12AU	-1.4	-3.8	-2.4	0.8
324.93	-6.28	-2.5	155.17	22AU	-1.1	-3.7	-2.4	0.8
322.98	-5.67	-2.3	164.93	1SE	-0.5	-3.6	-2.3	0.8
322.16	-4.92	-2.0	174.75	11SE	-0.2	-3.6	-2.3	0.8
322.66	-4.14	-1.7	184.63	21SE	-0.0	-3.5	-2.2	0.8
324.40	-3.39	-1.4	194.56	1OC	0.1	-3.5	-2.1	0.8
327.18	-2.72	-1.1	204.55	11OC	0.4	-3.5	-2.1	0.8
330.83	-2.13	-0.8	214.60	21OC	2.0	-3.4	-2.0	0.7
335.15	-1.62	-0.5	224.68	31OC	1.4	-3.4	-1.9	0.7
339.98	-1.17	-0.3	234.81	10NO	-0.2	-3.4	-1.9	0.7
345.21	-0.80	-0.0	244.98	20NO	-0.4	-3.4	-1.8	0.7
350.74	-0.47	0.2	255.16	30NO	-0.4	-3.4	-1.7	0.7
356.50	-0.20	0.4	265.36	10DE	-0.5	-3.4	-1.7	0.8
2.44	0.04	0.6	275.56	20DE	-0.7	-3.4	-1.7	0.8
8.51	0.24	0.7	285.75	30DE	-0.9	-3.4	-1.6	0.8
				1073				
14.66	0.42	0.9	295.92	9JA	-1.0	-3.4	-1.6	0.8
20.89	0.56	1.0	306.07	19JA	-0.7	-3.4	-1.6	0.8
27.16	0.69	1.2	316.17	29JA	0.5	-3.4	-1.6	0.7
33.47	0.80	1.3	326.23	8FE	3.0	-3.4	-1.6	0.7
39.79	0.89	1.4	336.24	18FE	1.7	-3.4	-1.6	0.7
46.12	0.96	1.5	346.19	28FE	0.8	-3.5	-1.6	0.6
52.45	1.02	1.6	356.08	10MR	0.5	-3.5	-1.6	0.6
58.77	1.08	1.7	5.92	20MR	0.2	-3.4	-1.6	0.5
65.09	1.12	1.7	15.70	30MR	-0.3	-3.4	-1.6	0.5
71.40	1.15	1.8	25.43	9AP	-1.0	-3.4	-1.7	0.4
77.71	1.18	1.9	35.11	19AP	-1.8	-3.4	-1.7	0.3
84.00	1.20	1.9	44.74	29AP	-1.1	-3.3	-1.8	0.3
90.29	1.21	2.0	54.34	9MY	-0.1	-3.3	-1.8	0.2
96.57	1.22	2.0	63.90	19MY	0.7	-3.3	-1.9	0.3
102.85	1.22	2.0	73.44	29MY	1.5	-3.3	-1.9	0.3
109.13	1.22	2.0	82.98	8JN	2.5	-3.3	-2.0	0.4
115.42	1.21	2.0	92.50	18JN	3.2	-3.3	-2.1	0.5
121.71	1.19	2.0	102.03	28JN	1.7	-3.4	-2.2	0.5
128.02	1.18	2.0	111.58	8JL	0.3	-3.4	-2.2	0.6
134.34	1.15	2.0	121.15	18JL	-0.9	-3.4	-2.3	0.6
140.69	1.13	2.0	130.75	28JL	-1.5	-3.5	-2.3	0.7
147.06	1.10	2.0	140.39	7AU	-1.1	-3.5	-2.4	0.7
153.45	1.06	2.0	150.07	17AU	-0.5	-3.6	-2.4	0.8
159.88	1.02	1.9	159.81	27AU	-0.1	-3.7	-2.5	0.8
166.35	0.98	1.9	169.60	6SE	0.1	-3.7	-2.5	0.8
172.85	0.93	1.9	179.44	16SE	0.3	-3.8	-2.4	0.8
179.40	0.88	1.9	189.35	26SE	0.7	-3.9	-2.4	0.8
185.99	0.82	1.9	199.31	6OC	2.3	-4.0	-2.4	0.8
192.63	0.76	1.9	209.32	16OC	1.2	-4.2	-2.3	0.8
199.31	0.69	1.9	219.39	26OC	-0.3	-4.3	-2.3	0.7
206.05	0.62	1.9	229.50	5NO	-0.6	-4.3	-2.2	0.7
212.84	0.53	1.9	239.65	15NO	-0.6	-4.4	-2.1	0.6
219.68	0.45	1.8	249.82	25NO	-0.6	-4.3	-2.1	0.7
226.57	0.35	1.7	260.01	5DE	-0.8	-3.9	-2.0	0.7
233.51	0.25	1.7	270.21	15DE	-0.8	-3.2	-1.9	0.8
240.51	0.14	1.7	280.41	25DE	-0.8	-3.5	-1.9	0.8

♂ LONG	LAT	MAG	☉ LONG	16.00UT 1074	☿	♀	♃	♄
247.55	0.02	1.6	290.59	4JA	-0.6	-4.1	-1.8	0.8
254.64	-0.10	1.5	300.75	14JA	0.7	-4.3	-1.7	0.8
261.79	-0.24	1.5	310.87	24JA	3.0	-4.4	-1.7	0.8
268.97	-0.38	1.4	320.96	3FE	1.2	-4.3	-1.7	0.8
276.20	-0.53	1.3	330.99	13FE	0.6	-4.2	-1.6	0.7
283.46	-0.69	1.2	340.97	23FE	0.3	-4.1	-1.6	0.7
290.75	-0.85	1.2	350.90	5MR	0.1	-4.0	-1.6	0.7
298.07	-1.02	1.1	0.77	15MR	-0.3	-3.9	-1.6	0.6
305.39	-1.20	1.0	10.58	25MR	-1.0	-3.8	-1.6	0.6
312.72	-1.37	0.9	20.34	4AP	-1.7	-3.7	-1.6	0.5
320.03	-1.55	0.8	30.04	14AP	-1.1	-3.6	-1.6	0.4
327.32	-1.73	0.7	39.70	24AP	-0.0	-3.5	-1.6	0.4
334.55	-1.90	0.6	49.31	4MY	0.9	-3.5	-1.6	0.3
341.71	-2.06	0.5	58.89	14MY	2.0	-3.4	-1.7	0.3
348.76	-2.22	0.4	68.45	24MY	3.3	-3.4	-1.7	0.2
355.68	-2.36	0.3	77.99	3JN	2.6	-3.4	-1.8	0.3
2.42	-2.49	0.2	87.51	13JN	1.3	-3.3	-1.8	0.3
8.95	-2.61	0.0	97.04	23JN	0.2	-3.3	-1.9	0.4
15.19	-2.70	-0.1	106.58	3JL	-0.9	-3.3	-1.9	0.5
21.09	-2.78	-0.2	116.14	13JL	-1.6	-3.3	-2.0	0.5
26.56	-2.83	-0.4	125.72	23JL	-1.1	-3.3	-2.1	0.6
31.49	-2.85	-0.5	135.34	2AU	-0.4	-3.3	-2.1	0.6
35.75	-2.84	-0.7	145.00	12AU	-0.0	-3.4	-2.2	0.7
39.18	-2.78	-0.9	154.70	22AU	0.2	-3.4	-2.3	0.7
41.56	-2.67	-1.1	164.46	1SE	0.4	-3.4	-2.3	0.7
42.71	-2.49	-1.3	174.28	11SE	0.9	-3.5	-2.4	0.8
42.42	-2.22	-1.6	184.15	21SE	2.7	-3.5	-2.4	0.8
40.66	-1.84	-1.8	194.08	1OC	1.0	-3.5	-2.4	0.8
37.70	-1.36	-1.9	204.07	11OC	-0.5	-3.5	-2.4	0.8
34.13	-0.81	-2.0	214.11	21OC	-0.7	-3.4	-2.4	0.8
30.76	-0.26	-1.8	224.20	31OC	-0.7	-3.4	-2.4	0.7
28.32	0.23	-1.5	234.32	10NO	-0.7	-3.4	-2.4	0.7
27.17	0.64	-1.1	244.48	20NO	-0.7	-3.4	-2.3	0.7
27.37	0.95	-0.8	254.67	30NO	-0.7	-3.4	-2.2	0.6
28.77	1.19	-0.5	264.86	10DE	-0.7	-3.3	-2.2	0.7
31.17	1.36	-0.2	275.06	20DE	-0.5	-3.3	-2.1	0.7
34.37	1.48	0.1	285.25	30DE	0.8	-3.3	-2.0	0.7
				1075				
38.18	1.56	0.3	295.42	9JA	2.7	-3.3	-2.0	0.8
42.48	1.61	0.5	305.57	19JA	0.9	-3.4	-1.9	0.8
47.15	1.64	0.7	315.68	29JA	0.4	-3.4	-1.8	0.8
52.11	1.66	0.9	325.73	8FE	0.2	-3.4	-1.8	0.8
57.30	1.66	1.1	335.75	18FE	-0.0	-3.4	-1.7	0.8
62.67	1.65	1.2	345.70	28FE	-0.4	-3.5	-1.7	0.7
68.18	1.64	1.4	355.60	10MR	-1.0	-3.5	-1.6	0.7
73.80	1.62	1.5	5.44	20MR	-1.6	-3.6	-1.6	0.7
79.52	1.59	1.6	15.23	30MR	-1.1	-3.6	-1.6	0.6
85.32	1.56	1.7	24.96	9AP	0.1	-3.7	-1.5	0.6
91.17	1.52	1.7	34.64	19AP	1.2	-3.7	-1.5	0.5
97.09	1.49	1.8	44.28	29AP	2.6	-3.8	-1.5	0.4
103.05	1.44	1.9	53.87	9MY	3.4	-3.9	-1.5	0.4
109.06	1.40	1.9	63.44	19MY	2.0	-4.0	-1.5	0.3
115.11	1.35	1.9	72.99	29MY	1.0	-4.1	-1.6	0.2
121.21	1.30	2.0	82.52	8JN	0.1	-4.2	-1.6	0.2
127.35	1.25	2.0	92.05	18JN	-0.9	-4.2	-1.6	0.3
133.53	1.19	2.0	101.58	28JN	-1.7	-4.1	-1.6	0.4
139.76	1.13	2.0	111.12	8JL	-1.1	-3.8	-1.7	0.4
146.04	1.07	2.0	120.69	18JL	-0.4	-3.2	-1.7	0.5
152.37	1.01	2.0	130.29	28JL	0.1	-3.4	-1.8	0.5
158.76	0.94	2.0	139.92	7AU	0.4	-3.9	-1.8	0.6
165.20	0.87	2.0	149.61	17AU	0.7	-4.2	-1.9	0.6
171.70	0.80	1.9	159.34	27AU	1.3	-4.3	-2.0	0.7
178.26	0.72	1.9	169.12	6SE	3.1	-4.3	-2.0	0.7
184.90	0.64	1.9	178.97	16SE	0.9	-4.2	-2.1	0.7
191.60	0.56	1.8	188.87	26SE	-0.6	-4.1	-2.2	0.8
198.37	0.47	1.8	198.82	6OC	-0.9	-4.0	-2.2	0.8
205.21	0.38	1.8	208.84	16OC	-0.9	-3.9	-2.3	0.8
212.13	0.29	1.8	218.90	26OC	-0.8	-3.8	-2.3	0.8
219.12	0.19	1.7	229.01	5NO	-0.6	-3.7	-2.3	0.7
226.19	0.09	1.7	239.15	15NO	-0.5	-3.7	-2.3	0.7
233.33	-0.01	1.7	249.33	25NO	-0.5	-3.6	-2.3	0.7
240.54	-0.12	1.7	259.52	5DE	-0.3	-3.6	-2.3	0.7
247.83	-0.23	1.7	269.72	15DE	1.0	-3.5	-2.3	0.7
255.18	-0.34	1.6	279.92	25DE	2.3	-3.5	-2.2	0.7

♂ LONG	♂ LAT	♂ MAG	☉ LONG	16.00UT 1076	☿	♀	♃	♄
					MAGNITUDES			
262.60	-0.45	1.6	290.10	4JA	0.6	-3.4	-2.2	0.8
270.08	-0.57	1.6	300.26	14JA	0.2	-3.4	-2.1	0.8
277.61	-0.68	1.5	310.39	24JA	0.0	-3.4	-2.0	0.8
285.20	-0.80	1.5	320.47	3FE	-0.1	-3.3	-1.9	0.8
292.82	-0.91	1.4	330.51	13FE	-0.5	-3.3	-1.9	0.8
300.48	-1.01	1.4	340.49	23FE	-1.0	-3.3	-1.8	0.8
308.16	-1.11	1.3	350.42	4MR	-1.5	-3.3	-1.7	0.8
315.85	-1.21	1.3	0.29	14MR	-1.0	-3.3	-1.7	0.8
323.55	-1.29	1.3	10.11	24MR	0.1	-3.3	-1.6	0.7
331.23	-1.37	1.2	19.86	3AP	1.6	-3.4	-1.6	0.7
338.88	-1.43	1.2	29.57	13AP	3.3	-3.4	-1.5	0.6
346.49	-1.48	1.1	39.23	23AP	2.6	-3.4	-1.5	0.6
354.05	-1.51	1.1	48.84	3MY	1.5	-3.5	-1.5	0.5
1.53	-1.53	1.0	58.43	13MY	0.8	-3.5	-1.5	0.4
8.93	-1.53	1.0	67.98	23MY	0.0	-3.5	-1.4	0.4
16.22	-1.52	1.0	77.52	2JN	-0.9	-3.4	-1.4	0.3
23.39	-1.49	0.9	87.05	12JN	-1.8	-3.4	-1.4	0.2
30.42	-1.44	0.9	96.57	22JN	-1.1	-3.4	-1.4	0.3
37.29	-1.37	0.8	106.11	2JL	-0.3	-3.4	-1.5	0.3
43.97	-1.28	0.7	115.67	12JL	0.2	-3.3	-1.5	0.4
50.44	-1.18	0.7	125.25	22JL	0.5	-3.3	-1.5	0.4
56.67	-1.05	0.6	134.87	1AU	0.9	-3.3	-1.5	0.5
62.60	-0.91	0.5	144.53	11AU	1.7	-3.3	-1.6	0.6
68.21	-0.74	0.4	154.23	21AU	3.2	-3.3	-1.6	0.6
73.41	-0.54	0.3	163.99	31AU	0.8	-3.4	-1.7	0.7
78.13	-0.32	0.2	173.80	10SE	-0.7	-3.4	-1.7	0.7
82.27	-0.06	0.0	183.67	20SE	-1.0	-3.4	-1.8	0.7
85.68	0.24	-0.2	193.60	30SE	-1.0	-3.5	-1.9	0.8
88.20	0.59	-0.3	203.58	10OC	-0.7	-3.5	-1.9	0.8
89.63	0.98	-0.5	213.62	20OC	-0.5	-3.6	-2.0	0.8
89.78	1.42	-0.8	223.70	30OC	-0.4	-3.6	-2.1	0.8
88.51	1.90	-1.0	233.83	9NO	-0.4	-3.7	-2.1	0.8
85.89	2.37	-1.2	243.99	19NO	-0.1	-3.8	-2.2	0.8
82.26	2.79	-1.3	254.17	29NO	1.2	-3.9	-2.2	0.7
78.32	3.09	-1.3	264.37	9DE	2.0	-4.0	-2.2	0.7
74.86	3.24	-1.1	274.57	19DE	0.3	-4.1	-2.2	0.7
72.46	3.27	-0.8	284.76	29DE	-0.0	-4.2	-2.2	0.7
				1077				
71.38	3.20	-0.5	294.94	8JA	-0.1	-4.3	-2.2	0.8
71.60	3.08	-0.2	305.08	18JA	-0.2	-4.3	-2.1	0.8
72.96	2.93	0.1	315.19	28JA	-0.5	-4.3	-2.1	0.8
75.26	2.77	0.3	325.25	7FE	-1.1	-4.2	-2.0	0.8
78.31	2.62	0.5	335.26	17FE	-1.4	-3.9	-2.0	0.8
81.94	2.47	0.8	345.22	27FE	-1.0	-3.3	-1.9	0.8
86.05	2.32	0.9	355.12	9MR	0.2	-3.6	-1.8	0.8
90.53	2.19	1.1	4.96	19MR	1.9	-4.0	-1.7	0.8
95.30	2.06	1.2	14.75	29MR	3.5	-4.2	-1.7	0.8
100.32	1.93	1.4	24.49	8AP	1.9	-4.2	-1.6	0.7
105.54	1.82	1.5	34.17	18AP	1.1	-4.2	-1.6	0.7
110.92	1.70	1.6	43.81	28AP	0.6	-4.1	-1.5	0.6
116.45	1.59	1.6	53.40	8MY	-0.1	-4.0	-1.5	0.6
122.10	1.48	1.7	62.97	18MY	-1.0	-3.9	-1.4	0.5
127.87	1.38	1.8	72.52	28MY	-1.8	-3.8	-1.4	0.5
133.74	1.28	1.8	82.05	7JN	-1.1	-3.7	-1.4	0.4
139.70	1.17	1.8	91.58	17JN	-0.3	-3.6	-1.3	0.3
145.76	1.07	1.9	101.11	27JN	0.3	-3.6	-1.3	0.3
151.90	0.97	1.9	110.66	7JL	0.7	-3.5	-1.3	0.3
158.13	0.87	1.9	120.22	17JL	1.2	-3.5	-1.3	0.4
164.46	0.77	1.9	129.82	27JL	2.2	-3.4	-1.3	0.4
170.86	0.67	1.9	139.46	6AU	2.8	-3.4	-1.3	0.5
177.36	0.57	1.9	149.14	16AU	0.6	-3.4	-1.4	0.5
183.94	0.47	1.9	158.87	26AU	-0.8	-3.4	-1.4	0.6
190.62	0.37	1.8	168.65	5SE	-1.2	-3.4	-1.4	0.6
197.39	0.27	1.8	178.49	15SE	-1.1	-3.4	-1.5	0.7
204.25	0.16	1.8	188.39	25SE	-0.6	-3.4	-1.5	0.7
211.20	0.06	1.7	198.34	5OC	-0.4	-3.4	-1.6	0.8
218.25	-0.04	1.7	208.35	15OC	-0.2	-3.4	-1.6	0.8
225.39	-0.15	1.6	218.42	25OC	-0.2	-3.4	-1.7	0.8
232.61	-0.25	1.6	228.52	4NO	0.0	-3.4	-1.8	0.8
239.93	-0.35	1.5	238.66	14NO	1.4	-3.4	-1.8	0.8
247.33	-0.45	1.5	248.83	24NO	1.8	-3.5	-1.9	0.8
254.82	-0.55	1.5	259.02	4DE	0.1	-3.5	-2.0	0.8
262.37	-0.64	1.5	269.22	14DE	-0.2	-3.5	-2.0	0.8
270.00	-0.73	1.5	279.42	24DE	-0.2	-3.5	-2.0	0.8

♂ LONG	♂ LAT	♂ MAG	☉ LONG	16.00UT 1078	☿	♀	♃	♄
					MAGNITUDES			
277.69	-0.82	1.5	289.60	3JA	-0.3	-3.4	-2.1	0.8
285.42	-0.89	1.5	299.76	13JA	-0.6	-3.4	-2.1	0.8
293.20	-0.96	1.4	309.89	23JA	-1.1	-3.4	-2.1	0.8
301.01	-1.03	1.4	319.98	2FE	-1.2	-3.4	-2.1	0.9
308.85	-1.08	1.4	330.02	12FE	-0.9	-3.4	-2.1	0.9
316.69	-1.12	1.4	340.01	22FE	0.3	-3.3	-2.0	0.9
324.52	-1.15	1.4	349.94	4MR	2.4	-3.3	-2.0	0.9
332.34	-1.17	1.4	359.81	14MR	2.7	-3.3	-1.9	0.9
340.13	-1.18	1.4	9.63	24MR	1.4	-3.3	-1.8	0.9
347.87	-1.17	1.4	19.39	3AP	0.8	-3.4	-1.8	0.9
355.57	-1.15	1.4	29.10	13AP	0.4	-3.4	-1.7	0.8
3.20	-1.12	1.4	38.76	23AP	-0.1	-3.4	-1.6	0.8
10.75	-1.07	1.4	48.38	3MY	-1.0	-3.4	-1.6	0.7
18.22	-1.01	1.4	57.96	13MY	-1.8	-3.5	-1.5	0.7
25.60	-0.94	1.4	67.52	23MY	-1.1	-3.5	-1.5	0.6
32.88	-0.86	1.4	77.05	2JN	-0.2	-3.6	-1.4	0.6
40.05	-0.76	1.3	86.59	12JN	0.4	-3.6	-1.4	0.5
47.09	-0.66	1.3	96.12	22JN	0.9	-3.7	-1.3	0.4
54.01	-0.54	1.3	105.65	2JL	1.6	-3.8	-1.3	0.4
60.80	-0.42	1.3	115.21	12JL	2.8	-3.9	-1.3	0.3
67.45	-0.28	1.3	124.79	22JL	2.4	-4.0	-1.3	0.4
73.93	-0.14	1.2	134.40	1AU	0.5	-4.1	-1.2	0.4
80.25	0.02	1.2	144.06	11AU	-0.8	-4.2	-1.2	0.5
86.37	0.18	1.1	153.76	21AU	-1.3	-4.2	-1.2	0.5
92.28	0.36	1.1	163.51	31AU	-1.1	-4.3	-1.3	0.6
97.95	0.56	1.0	173.33	10SE	-0.6	-4.2	-1.3	0.6
103.33	0.77	0.9	183.19	20SE	-0.3	-3.8	-1.3	0.7
108.38	1.00	0.8	193.12	30SE	-0.1	-3.3	-1.3	0.7
113.03	1.25	0.7	203.10	10OC	-0.0	-3.6	-1.4	0.8
117.20	1.53	0.5	213.13	20OC	0.3	-4.1	-1.4	0.8
120.78	1.84	0.4	223.21	30OC	1.7	-4.3	-1.5	0.8
123.62	2.19	0.2	233.34	9NO	1.6	-4.4	-1.5	0.9
125.56	2.57	-0.0	243.50	19NO	-0.1	-4.3	-1.6	0.9
126.41	2.99	-0.2	253.68	29NO	-0.3	-4.2	-1.6	0.9
126.00	3.43	-0.5	263.88	9DE	-0.4	-4.1	-1.7	0.9
124.23	3.84	-0.7	274.08	19DE	-0.4	-4.0	-1.8	0.9
121.23	4.19	-0.9	284.27	29DE	-0.7	-3.9	-1.8	0.8
				1079				
117.41	4.40	-1.0	294.44	8JA	-1.0	-3.8	-1.9	0.8
113.48	4.43	-0.9	304.59	18JA	-1.1	-3.7	-1.9	0.8
110.16	4.30	-0.7	314.70	28JA	-0.8	-3.7	-2.0	0.9
107.93	4.05	-0.5	324.77	7FE	0.4	-3.6	-2.0	0.9
107.02	3.74	-0.2	334.78	17FE	2.8	-3.5	-2.0	1.0
107.36	3.41	0.0	344.74	27FE	2.0	-3.5	-2.0	1.0
108.79	3.10	0.3	354.65	9MR	1.0	-3.4	-2.0	1.0
111.12	2.80	0.5	4.50	19MR	0.6	-3.4	-2.0	1.0
114.19	2.52	0.7	14.28	29MR	0.3	-3.3	-1.9	1.0
117.84	2.27	0.8	24.02	8AP	-0.2	-3.3	-1.9	1.0
121.97	2.04	1.0	33.71	18AP	-1.0	-3.3	-1.8	0.9
126.48	1.83	1.1	43.34	28AP	-1.9	-3.3	-1.8	0.9
131.31	1.63	1.2	52.95	8MY	-1.1	-3.3	-1.7	0.9
136.40	1.45	1.3	62.51	18MY	-0.1	-3.3	-1.6	0.8
141.73	1.28	1.4	72.06	28MY	0.6	-3.3	-1.6	0.8
147.26	1.11	1.5	81.59	7JN	1.2	-3.3	-1.5	0.7
152.96	0.96	1.5	91.12	17JN	2.1	-3.4	-1.4	0.6
158.83	0.81	1.6	100.65	27JN	3.3	-3.4	-1.4	0.6
164.85	0.67	1.6	110.20	7JL	2.0	-3.5	-1.3	0.5
171.00	0.53	1.6	119.76	17JL	0.4	-3.5	-1.3	0.4
177.29	0.40	1.7	129.35	27JL	-0.9	-3.5	-1.3	0.4
183.71	0.27	1.7	138.99	6AU	-1.4	-3.5	-1.2	0.5
190.26	0.15	1.7	148.66	16AU	-1.1	-3.4	-1.2	0.5
196.93	0.03	1.7	158.39	26AU	-0.5	-3.4	-1.2	0.6
203.71	-0.09	1.7	168.17	5SE	-0.2	-3.4	-1.2	0.6
210.62	-0.20	1.7	178.01	15SE	0.0	-3.3	-1.2	0.7
217.64	-0.31	1.6	187.90	25SE	0.1	-3.3	-1.2	0.7
224.77	-0.41	1.6	197.86	5OC	0.5	-3.3	-1.2	0.8
232.01	-0.51	1.6	207.86	15OC	2.0	-3.3	-1.3	0.8
239.35	-0.60	1.6	217.92	25OC	1.4	-3.3	-1.3	0.9
246.79	-0.69	1.5	228.03	4NO	-0.2	-3.4	-1.3	0.9
254.33	-0.77	1.5	238.17	14NO	-0.5	-3.4	-1.4	0.9
261.94	-0.84	1.5	248.34	24NO	-0.5	-3.4	-1.4	0.9
269.64	-0.91	1.4	258.53	4DE	-0.6	-3.4	-1.5	0.9
277.39	-0.96	1.4	268.73	14DE	-0.8	-3.5	-1.5	1.0
285.20	-1.01	1.4	278.93	24DE	-0.9	-3.5	-1.6	1.0

♂ / ☉ / 16.00UT — MAGNITUDES (☿ ♀ ♃ ♄)

♂ LONG LAT MAG	☉ LONG	16.00UT 1080	☿	♀	♃	♄
293.05 -1.04 1.3	289.11	3JA	-0.9	-3.6	-1.7	0.9
300.93 -1.07 1.3	299.27	13JA	-0.7	-3.6	-1.7	0.9
308.83 -1.08 1.3	309.40	23JA	0.5	-3.7	-1.8	0.9
316.73 -1.08 1.3	319.50	2FE	3.0	-3.8	-1.8	1.0
324.62 -1.07 1.3	329.54	12FE	1.5	-3.9	-1.9	1.0
332.49 -1.05 1.4	339.53	22FE	0.7	-3.9	-1.9	1.0
340.32 -1.02 1.4	349.47	3MR	0.4	-4.0	-2.0	1.1
348.10 -0.97 1.4	359.34	13MR	0.1	-4.1	-2.0	1.1
355.83 -0.91 1.4	9.16	23MR	-0.3	-4.2	-2.0	1.1
3.49 -0.85 1.4	18.92	2AP	-1.0	-4.2	-2.0	1.1
11.07 -0.78 1.5	28.63	12AP	-1.8	-4.2	-2.0	1.1
18.58 -0.69 1.5	38.30	22AP	-1.1	-4.0	-1.9	1.1
26.00 -0.60 1.5	47.92	2MY	-0.1	-3.4	-1.9	1.1
33.33 -0.51 1.6	57.50	12MY	0.8	-2.8	-1.8	1.0
40.56 -0.40 1.6	67.06	22MY	1.6	-3.6	-1.8	1.0
47.70 -0.30 1.6	76.60	1JN	2.8	-4.0	-1.7	0.9
54.74 -0.19 1.6	86.12	11JN	3.0	-4.2	-1.7	0.9
61.68 -0.07 1.6	95.65	21JN	1.6	-4.2	-1.6	0.8
68.51 0.05 1.6	105.19	1JL	0.3	-4.1	-1.5	0.7
75.24 0.17 1.6	114.74	11JL	-0.9	-4.0	-1.5	0.7
81.87 0.30 1.6	124.32	21JL	-1.5	-3.9	-1.4	0.6
88.38 0.43 1.6	133.93	31JL	-1.1	-3.8	-1.4	0.5
94.78 0.56 1.6	143.58	10AU	-0.5	-3.8	-1.3	0.6
101.06 0.69 1.6	153.29	20AU	-0.1	-3.7	-1.3	0.6
107.21 0.84 1.6	163.04	30AU	0.2	-3.6	-1.3	0.6
113.21 0.98 1.5	172.84	9SE	0.3	-3.6	-1.2	0.7
119.06 1.14 1.5	182.71	19SE	0.7	-3.5	-1.2	0.7
124.74 1.30 1.4	192.63	29SE	2.4	-3.5	-1.2	0.8
130.21 1.47 1.3	202.61	9OC	1.2	-3.5	-1.2	0.8
135.44 1.65 1.2	212.65	19OC	-0.4	-3.4	-1.2	0.9
140.38 1.84 1.1	222.73	29OC	-0.6	-3.4	-1.2	0.9
145.00 2.06 1.0	232.85	8NO	-0.6	-3.4	-1.3	1.0
149.19 2.29 0.8	243.01	18NO	-0.7	-3.4	-1.3	1.0
152.88 2.54 0.7	253.19	28NO	-0.8	-3.4	-1.3	1.0
155.94 2.82 0.5	263.38	8DE	-0.7	-3.4	-1.4	1.1
158.21 3.12 0.2	273.59	18DE	-0.8	-3.4	-1.4	1.1
159.53 3.44 0.0	283.78	28DE	-0.6	-3.4	-1.5	1.1

1081

♂ LONG LAT MAG	☉ LONG	16.00UT	☿	♀	♃	♄
159.72 3.76 -0.2	293.95	7JA	0.7	-3.4	-1.5	1.1
158.62 4.06 -0.5	304.10	17JA	2.9	-3.4	-1.6	1.1
156.25 4.29 -0.8	314.21	27JA	1.1	-3.4	-1.7	1.1
152.83 4.40 -1.0	324.28	6FE	0.5	-3.4	-1.7	1.1
148.92 4.33 -1.0	334.30	16FE	0.3	-3.4	-1.8	1.1
145.24 4.11 -0.9	344.26	26FE	0.0	-3.5	-1.9	1.2
142.41 3.76 -0.7	354.17	8MR	-0.3	-3.5	-1.9	1.2
140.79 3.34 -0.4	4.02	18MR	-1.0	-3.4	-2.0	1.2
140.48 2.92 -0.2	13.81	28MR	-1.7	-3.4	-2.0	1.3
141.36 2.51 0.0	23.55	7AP	-1.1	-3.4	-2.0	1.3
143.27 2.13 0.2	33.24	17AP	-0.0	-3.4	-2.1	1.3
146.03 1.79 0.4	42.87	27AP	1.0	-3.3	-2.1	1.3
149.48 1.49 0.6	52.48	7MY	2.2	-3.3	-2.0	1.2
153.50 1.21 0.7	62.05	17MY	3.6	-3.3	-2.0	1.2
157.99 0.96 0.8	71.60	27MY	2.4	-3.3	-2.0	1.2
162.87 0.73 0.9	81.13	6JN	1.2	-3.3	-1.9	1.1
168.08 0.53 1.0	90.66	16JN	0.2	-3.3	-1.9	1.1
173.58 0.33 1.1	100.19	26JN	-0.9	-3.4	-1.8	1.0
179.33 0.16 1.2	109.73	6JL	-1.6	-3.4	-1.8	0.9
185.30 -0.01 1.2	119.29	16JL	-1.1	-3.4	-1.7	0.9
191.49 -0.16 1.3	128.89	26JL	-0.4	-3.5	-1.6	0.8
197.86 -0.30 1.3	138.52	5AU	0.0	-3.5	-1.6	0.7
204.41 -0.43 1.3	148.19	15AU	0.3	-3.6	-1.5	0.7
211.12 -0.55 1.4	157.92	25AU	0.5	-3.7	-1.5	0.7
217.98 -0.66 1.4	167.69	4SE	1.1	-3.7	-1.4	0.7
225.00 -0.76 1.4	177.53	14SE	2.8	-3.8	-1.4	0.8
232.15 -0.85 1.4	187.42	24SE	1.1	-3.9	-1.3	0.8
239.43 -0.93 1.4	197.37	4OC	-0.5	-4.0	-1.3	0.9
246.82 -1.00 1.4	207.38	14OC	-0.8	-4.2	-1.3	0.9
254.33 -1.06 1.4	217.43	24OC	-0.8	-4.3	-1.3	1.0
261.92 -1.10 1.4	227.53	3NO	-0.8	-4.4	-1.3	1.0
269.61 -1.14 1.4	237.68	13NO	-0.7	-4.4	-1.3	1.1
277.36 -1.16 1.4	247.84	23NO	-0.6	-4.3	-1.3	1.1
285.18 -1.16 1.4	258.03	3DE	-0.6	-3.9	-1.3	1.2
293.04 -1.16 1.4	268.24	13DE	-0.4	-3.1	-1.3	1.2
300.92 -1.14 1.4	278.43	23DE	0.8	-3.6	-1.4	1.2

♂ / ☉ / 16.00UT — MAGNITUDES (☿ ♀ ♃ ♄)

♂ LONG LAT MAG	☉ LONG	16.00UT 1082	☿	♀	♃	♄
308.82 -1.11 1.4	288.62	2JA	2.6	-4.1	-1.4	1.2
316.71 -1.07 1.4	298.78	12JA	0.8	-4.3	-1.5	1.2
324.59 -1.02 1.4	308.91	22JA	0.3	-4.4	-1.5	1.2
332.45 -0.96 1.4	319.01	1FE	0.1	-4.3	-1.6	1.2
340.26 -0.89 1.4	329.05	11FE	-0.1	-4.2	-1.6	1.2
348.02 -0.82 1.4	339.04	21FE	-0.4	-4.1	-1.7	1.3
355.72 -0.73 1.4	348.98	3MR	-1.0	-4.0	-1.8	1.3
3.35 -0.64 1.4	358.87	13MR	-1.6	-3.9	-1.8	1.4
10.90 -0.55 1.4	8.68	23MR	-1.1	-3.8	-1.9	1.4
18.38 -0.45 1.4	18.45	2AP	0.0	-3.7	-2.0	1.4
25.77 -0.35 1.5	28.17	12AP	1.3	-3.6	-2.0	1.4
33.07 -0.24 1.5	37.83	22AP	2.8	-3.5	-2.1	1.4
40.28 -0.14 1.6	47.45	2MY	3.1	-3.5	-2.1	1.4
47.41 -0.03 1.6	57.04	12MY	1.8	-3.4	-2.1	1.4
54.45 0.07 1.7	66.60	22MY	0.9	-3.4	-2.2	1.3
61.40 0.18 1.7	76.14	1JN	0.1	-3.4	-2.2	1.3
68.27 0.28 1.8	85.67	11JN	-0.9	-3.3	-2.1	1.2
75.06 0.39 1.8	95.19	21JN	-1.7	-3.3	-2.1	1.2
81.77 0.49 1.8	104.73	1JL	-1.1	-3.3	-2.1	1.1
88.40 0.59 1.8	114.28	11JL	-0.4	-3.3	-2.0	1.1
94.95 0.69 1.8	123.86	21JL	0.1	-3.3	-2.0	1.0
101.43 0.79 1.9	133.47	31JL	0.4	-3.3	-1.9	1.0
107.83 0.89 1.9	143.12	10AU	0.7	-3.4	-1.8	1.0
114.15 0.99 1.9	152.81	20AU	1.4	-3.4	-1.8	0.9
120.40 1.09 1.8	162.57	30AU	3.2	-3.4	-1.7	0.9
126.57 1.19 1.8	172.37	9SE	0.9	-3.5	-1.7	0.9
132.64 1.28 1.8	182.23	19SE	-0.6	-3.5	-1.6	0.9
138.63 1.39 1.7	192.15	29SE	-0.9	-3.5	-1.6	1.0
144.51 1.49 1.7	202.12	9OC	-0.9	-3.5	-1.5	1.0
150.27 1.59 1.6	212.15	19OC	-0.8	-3.4	-1.5	1.1
155.89 1.70 1.5	222.23	29OC	-0.5	-3.4	-1.4	1.1
161.36 1.81 1.3	232.35	8NO	-0.5	-3.4	-1.4	1.2
166.64 1.93 1.3	242.50	18NO	-0.4	-3.4	-1.4	1.2
171.69 2.05 1.2	252.69	28NO	-0.3	-3.3	-1.4	1.3
176.47 2.18 1.1	262.88	8DE	1.0	-3.3	-1.4	1.3
180.92 2.31 0.9	273.08	18DE	2.3	-3.3	-1.4	1.4
184.96 2.45 0.7	283.28	28DE	0.5	-3.3	-1.4	1.4

1083

♂ LONG LAT MAG	☉ LONG	16.00UT	☿	♀	♃	♄
188.48 2.60 0.5	293.45	7JA	0.1	-3.3	-1.4	1.4
191.37 2.75 0.3	303.60	17JA	-0.0	-3.4	-1.4	1.4
193.47 2.90 0.1	313.72	27JA	-0.2	-3.4	-1.5	1.4
194.61 3.04 -0.2	323.79	6FE	-0.5	-3.4	-1.5	1.3
194.60 3.15 -0.5	333.81	16FE	-1.0	-3.4	-1.6	1.3
193.33 3.19 -0.8	343.78	26FE	-1.5	-3.5	-1.6	1.2
190.82 3.14 -1.0	353.69	8MR	-1.0	-3.5	-1.7	1.2
187.38 2.97 -1.3	3.54	18MR	0.1	-3.6	-1.7	1.2
183.59 2.66 -1.3	13.34	28MR	1.7	-3.6	-1.8	1.2
180.18 2.25 -1.1	23.08	7AP	3.6	-3.7	-1.9	1.2
177.72 1.80 -0.9	32.77	17AP	2.3	-3.7	-2.0	1.2
176.55 1.34 -0.7	42.42	27AP	1.3	-3.8	-2.0	1.2
176.70 0.91 -0.5	52.02	7MY	0.7	-3.9	-2.1	1.1
178.05 0.53 -0.3	61.59	17MY	0.0	-4.0	-2.2	1.1
180.44 0.20 -0.1	71.14	27MY	-0.9	-4.1	-2.2	1.1
183.68 -0.10 0.1	80.67	6JN	-1.8	-4.2	-2.2	1.0
187.63 -0.35 0.2	90.20	16JN	-1.1	-4.2	-2.3	1.0
192.17 -0.57 0.3	99.73	26JN	-0.3	-4.1	-2.3	0.9
197.19 -0.76 0.4	109.27	6JL	0.2	-3.8	-2.3	0.9
202.63 -0.92 0.5	118.83	16JL	0.6	-3.2	-2.3	0.8
208.42 -1.06 0.6	128.43	26JL	1.0	-3.4	-2.2	0.8
214.52 -1.18 0.7	138.05	5AU	1.9	-3.9	-2.2	0.7
220.88 -1.28 0.8	147.73	15AU	3.1	-4.2	-2.2	0.7
227.49 -1.36 0.8	157.45	25AU	0.8	-4.3	-2.1	0.7
234.31 -1.43 0.9	167.22	4SE	-0.7	-4.2	-2.0	0.6
241.31 -1.47 0.9	177.05	14SE	-1.1	-4.2	-2.0	0.6
248.48 -1.50 1.0	186.94	24SE	-1.1	-4.1	-1.9	0.7
255.80 -1.51 1.0	196.88	4OC	-0.7	-4.0	-1.8	0.7
263.24 -1.51 1.1	206.89	14OC	-0.4	-3.9	-1.8	0.8
270.79 -1.49 1.1	216.94	24OC	-0.3	-3.8	-1.7	0.9
278.43 -1.46 1.1	227.04	3NO	-0.3	-3.7	-1.7	0.9
286.13 -1.41 1.2	237.18	13NO	-0.1	-3.7	-1.6	1.0
293.89 -1.35 1.2	247.35	23NO	1.2	-3.6	-1.6	1.0
301.68 -1.27 1.3	257.53	3DE	2.0	-3.6	-1.5	1.1
309.49 -1.19 1.3	267.74	13DE	0.3	-3.5	-1.5	1.1
317.30 -1.10 1.3	277.94	23DE	-0.1	-3.5	-1.5	1.1

♂ LONG	♂ LAT	♂ MAG	⊙ LONG	16.00UT 1084	☿	♀	♃	♄
					MAGNITUDES			
325.08	-1.00	1.4	288.12	2JA	-0.1	-3.4	-1.5	1.1
332.84	-0.90	1.4	298.29	12JA	-0.3	-3.4	-1.5	1.1
340.56	-0.79	1.4	308.43	22JA	-0.5	-3.4	-1.5	1.1
348.22	-0.68	1.5	318.52	1FE	-1.0	-3.3	-1.5	1.1
355.82	-0.56	1.5	328.57	11FE	-1.3	-3.3	-1.5	1.1
3.36	-0.45	1.5	338.57	21FE	-1.0	-3.3	-1.6	1.0
10.81	-0.33	1.6	348.50	2MR	0.2	-3.3	-1.6	1.0
18.19	-0.22	1.6	358.39	12MR	2.1	-3.3	-1.6	0.9
25.49	-0.11	1.6	8.22	22MR	3.2	-3.3	-1.7	0.9
32.70	0.00	1.7	17.98	1AP	1.7	-3.4	-1.7	0.9
39.84	0.11	1.7	27.70	11AP	1.0	-3.4	-1.8	0.9
46.89	0.21	1.7	37.37	21AP	0.5	-3.4	-1.9	0.9
53.86	0.31	1.7	46.99	1MY	-0.1	-3.5	-1.9	0.9
60.76	0.40	1.7	56.58	11MY	-0.9	-3.5	-2.0	0.9
67.59	0.49	1.7	66.14	21MY	-1.8	-3.5	-2.1	0.9
74.34	0.58	1.8	75.67	31MY	-1.1	-3.4	-2.1	0.9
81.04	0.66	1.8	85.21	10JN	-0.3	-3.4	-2.2	0.8
87.67	0.74	1.9	94.73	20JN	0.3	-3.4	-2.3	0.8
94.24	0.82	1.9	104.26	30JN	0.8	-3.4	-2.3	0.7
100.76	0.89	1.9	113.82	10JL	1.3	-3.3	-2.4	0.7
107.23	0.96	2.0	123.39	20JL	2.4	-3.3	-2.4	0.6
113.66	1.02	2.0	133.00	30JL	2.7	-3.3	-2.4	0.6
120.05	1.08	2.0	142.65	9AU	0.7	-3.3	-2.4	0.5
126.39	1.14	2.0	152.34	19AU	-0.8	-3.3	-2.4	0.5
132.69	1.19	2.0	162.09	29AU	-1.2	-3.4	-2.4	0.4
138.95	1.25	2.0	171.89	8SE	-1.1	-3.4	-2.3	0.4
145.18	1.30	2.0	181.75	18SE	-0.6	-3.4	-2.3	0.3
151.36	1.34	1.9	191.66	28SE	-0.3	-3.5	-2.2	0.3
157.50	1.39	1.9	201.64	8OC	-0.2	-3.5	-2.1	0.4
163.59	1.43	1.9	211.66	18OC	-0.1	-3.6	-2.1	0.5
169.63	1.47	1.8	221.74	28OC	0.1	-3.6	-2.0	0.6
175.61	1.50	1.7	231.86	7NO	1.5	-3.7	-1.9	0.6
181.51	1.54	1.7	242.01	17NO	1.8	-3.8	-1.9	0.7
187.33	1.57	1.6	252.19	27NO	0.1	-3.9	-1.8	0.7
193.05	1.59	1.5	262.39	7DE	-0.2	-4.0	-1.7	0.8
198.65	1.61	1.4	272.58	17DE	-0.3	-4.1	-1.7	0.8
204.11	1.63	1.2	282.78	27DE	-0.4	-4.2	-1.7	0.9
				1085				
209.39	1.63	1.1	292.96	6JA	-0.6	-4.3	-1.6	0.9
214.45	1.63	0.9	303.11	16JA	-1.0	-4.3	-1.6	0.9
219.26	1.61	0.7	313.23	26JA	-1.2	-4.3	-1.6	0.9
223.73	1.58	0.5	323.30	5FE	-0.9	-4.2	-1.6	0.9
227.79	1.52	0.3	333.32	15FE	0.3	-3.8	-1.6	0.8
231.34	1.43	0.1	343.30	25FE	2.5	-3.3	-1.6	0.8
234.25	1.31	-0.2	353.21	7MR	2.4	-3.6	-1.6	0.8
236.35	1.13	-0.5	3.06	17MR	1.3	-4.1	-1.6	0.7
237.47	0.89	-0.8	12.86	27MR	0.7	-4.2	-1.6	0.7
237.42	0.57	-1.1	22.61	6AP	0.4	-4.2	-1.6	0.7
236.13	0.16	-1.4	32.30	16AP	-0.2	-4.2	-1.7	0.7
233.66	-0.32	-1.7	41.95	26AP	-0.9	-4.1	-1.7	0.7
230.41	-0.86	-1.9	51.56	6MY	-1.8	-4.0	-1.8	0.7
227.04	-1.38	-1.8	61.13	16MY	-1.1	-3.9	-1.8	0.7
224.26	-1.84	-1.7	70.68	26MY	-0.2	-3.8	-1.9	0.7
222.61	-2.21	-1.5	80.21	5JN	0.5	-3.7	-2.0	0.7
222.33	-2.47	-1.3	89.74	15JN	1.0	-3.6	-2.0	0.6
223.40	-2.65	-1.1	99.27	25JN	1.8	-3.6	-2.1	0.6
225.68	-2.75	-0.9	108.81	5JL	3.0	-3.5	-2.2	0.6
228.97	-2.81	-0.7	118.37	15JL	2.3	-3.5	-2.2	0.5
233.10	-2.82	-0.5	127.96	25JL	0.5	-3.4	-2.3	0.5
237.92	-2.79	-0.3	137.59	4AU	-0.8	-3.4	-2.4	0.4
243.30	-2.74	-0.2	147.26	14AU	-1.4	-3.4	-2.4	0.4
249.13	-2.67	-0.1	156.98	24AU	-1.1	-3.4	-2.4	0.3
255.35	-2.57	0.1	166.75	3SE	-0.6	-3.4	-2.5	0.3
261.87	-2.46	0.2	176.58	13SE	-0.2	-3.4	-2.5	0.2
268.64	-2.33	0.3	186.46	23SE	-0.1	-3.4	-2.4	0.2
275.62	-2.18	0.4	196.41	3OC	0.0	-3.4	-2.4	0.1
282.76	-2.03	0.5	206.40	13OC	0.3	-3.4	-2.4	0.1
290.03	-1.87	0.6	216.46	23OC	1.8	-3.4	-2.3	0.2
297.39	-1.70	0.7	226.55	2NO	1.6	-3.4	-2.3	0.3
304.82	-1.52	0.8	236.69	12NO	-0.1	-3.4	-2.2	0.3
312.28	-1.35	0.9	246.85	22NO	-0.4	-3.5	-2.1	0.4
319.77	-1.18	1.0	257.04	2DE	-0.4	-3.5	-2.0	0.5
327.26	-1.00	1.1	267.24	12DE	-0.5	-3.5	-2.0	0.5
334.74	-0.84	1.1	277.44	22DE	-0.7	-3.5	-1.9	0.6

♂ LONG	♂ LAT	♂ MAG	⊙ LONG	16.00UT 1086	☿	♀	♃	♄
					MAGNITUDES			
342.18	-0.67	1.2	287.62	1JA	-1.0	-3.4	-1.8	0.6
349.59	-0.52	1.3	297.79	11JA	-1.0	-3.4	-1.8	0.6
356.94	-0.37	1.4	307.93	21JA	-0.8	-3.4	-1.7	0.7
4.24	-0.22	1.4	318.02	31JA	0.4	-3.4	-1.7	0.7
11.48	-0.09	1.5	328.07	10FE	2.9	-3.4	-1.7	0.7
18.65	0.04	1.6	338.07	20FE	1.8	-3.3	-1.6	0.6
25.75	0.16	1.6	348.02	2MR	0.9	-3.3	-1.6	0.6
32.78	0.27	1.7	357.90	12MR	0.5	-3.3	-1.6	0.6
39.75	0.37	1.7	7.73	22MR	0.2	-3.3	-1.6	0.6
46.64	0.47	1.8	17.50	1AP	-0.2	-3.4	-1.6	0.5
53.47	0.56	1.8	27.22	11AP	-0.9	-3.4	-1.6	0.5
60.24	0.64	1.9	36.90	21AP	-1.8	-3.4	-1.6	0.5
66.95	0.72	1.9	46.52	1MY	-1.1	-3.4	-1.6	0.5
73.61	0.78	1.9	56.11	11MY	-0.2	-3.5	-1.6	0.5
80.22	0.85	1.9	65.68	21MY	0.7	-3.5	-1.7	0.5
86.78	0.90	1.9	75.22	31MY	1.4	-3.6	-1.7	0.5
93.29	0.95	1.9	84.75	10JN	2.4	-3.6	-1.8	0.5
99.78	1.00	1.9	94.28	20JN	3.3	-3.7	-1.8	0.5
106.23	1.04	1.9	103.81	30JN	1.8	-3.8	-1.9	0.5
112.65	1.08	1.9	113.36	10JL	0.4	-3.9	-1.9	0.5
119.05	1.11	2.0	122.93	20JL	-0.8	-4.0	-2.0	0.4
125.43	1.13	2.0	132.54	30JL	-1.5	-4.1	-2.1	0.4
131.80	1.16	2.0	142.18	9AU	-1.1	-4.2	-2.1	0.3
138.16	1.18	2.0	151.88	19AU	-0.5	-4.2	-2.2	0.3
144.51	1.19	2.0	161.62	29AU	-0.1	-4.3	-2.3	0.2
150.86	1.20	2.0	171.42	8SE	0.1	-4.2	-2.3	0.2
157.20	1.20	2.0	181.27	18SE	0.2	-3.8	-2.4	0.1
163.54	1.20	2.0	191.18	28SE	0.6	-3.2	-2.4	0.0
169.88	1.20	2.0	201.15	8OC	2.1	-3.6	-2.4	-0.0
176.22	1.19	2.0	211.18	18OC	1.4	-4.1	-2.4	-0.1
182.56	1.17	1.9	221.25	28OC	-0.3	-4.3	-2.4	-0.0
188.90	1.15	1.9	231.37	7NO	-0.6	-4.4	-2.4	0.0
195.24	1.13	1.8	241.52	17NO	-0.6	-4.3	-2.3	0.1
201.57	1.09	1.8	251.70	27NO	-0.6	-4.2	-2.3	0.2
207.89	1.05	1.7	261.89	7DE	-0.8	-4.1	-2.2	0.2
214.19	1.00	1.6	272.09	17DE	-0.8	-4.0	-2.2	0.3
220.47	0.93	1.5	282.29	27DE	-0.9	-3.9	-2.1	0.4
				1087				
226.73	0.86	1.4	292.47	6JA	-0.7	-3.8	-2.0	0.4
232.95	0.77	1.3	302.62	16JA	0.5	-3.7	-1.9	0.4
239.12	0.66	1.2	312.74	26JA	3.0	-3.7	-1.9	0.5
245.24	0.54	1.0	322.81	5FE	1.4	-3.6	-1.8	0.5
251.28	0.39	0.9	332.84	15FE	0.6	-3.5	-1.7	0.5
257.24	0.21	0.7	342.81	25FE	0.4	-3.5	-1.7	0.5
263.08	0.01	0.6	352.73	7MR	0.1	-3.4	-1.6	0.5
268.77	-0.23	0.4	2.59	17MR	-0.3	-3.4	-1.6	0.5
274.27	-0.51	0.2	12.39	27MR	-1.0	-3.3	-1.6	0.4
279.54	-0.83	-0.1	22.14	6AP	-1.8	-3.3	-1.5	0.4
284.50	-1.21	-0.3	31.84	16AP	-1.1	-3.3	-1.5	0.4
289.07	-1.66	-0.5	41.48	26AP	-0.1	-3.3	-1.5	0.3
293.12	-2.17	-0.8	51.09	6MY	0.9	-3.3	-1.5	0.3
296.52	-2.76	-1.1	60.67	16MY	1.8	-3.3	-1.5	0.4
299.10	-3.43	-1.4	70.22	26MY	3.1	-3.3	-1.5	0.4
300.66	-4.16	-1.7	79.75	5JN	2.8	-3.3	-1.5	0.4
301.01	-4.93	-2.0	89.28	15JN	1.4	-3.4	-1.6	0.4
300.13	-5.66	-2.3	98.81	25JN	0.3	-3.4	-1.6	0.4
298.14	-6.25	-2.5	108.35	5JL	-0.9	-3.5	-1.6	0.4
295.55	-6.59	-2.6	117.91	15JL	-1.6	-3.5	-1.7	0.4
293.05	-6.62	-2.5	127.50	25JL	-1.1	-3.5	-1.7	0.4
291.33	-6.36	-2.3	137.12	4AU	-0.5	-3.4	-1.8	0.3
290.83	-5.90	-2.0	146.79	14AU	-0.0	-3.4	-1.8	0.3
291.67	-5.33	-1.8	156.50	24AU	0.2	-3.4	-1.9	0.2
293.75	-4.72	-1.5	166.27	3SE	0.4	-3.4	-2.0	0.2
296.91	-4.12	-1.2	176.10	13SE	0.8	-3.3	-2.0	0.1
300.91	-3.55	-1.0	185.98	23SE	2.5	-3.3	-2.1	0.1
305.59	-3.02	-0.7	195.92	3OC	1.2	-3.3	-2.2	-0.0
310.81	-2.54	-0.5	205.91	13OC	-0.4	-3.3	-2.2	-0.1
316.42	-2.10	-0.3	215.96	23OC	-0.7	-3.3	-2.3	-0.1
322.34	-1.71	-0.1	226.06	2NO	-0.7	-3.4	-2.3	-0.2
328.49	-1.35	0.1	236.19	12NO	-0.7	-3.4	-2.3	-0.2
334.82	-1.03	0.3	246.36	22NO	-0.7	-3.4	-2.3	-0.1
341.27	-0.75	0.5	256.55	2DE	-0.7	-3.4	-2.3	-0.0
347.81	-0.50	0.6	266.74	12DE	-0.7	-3.5	-2.3	0.1
354.41	-0.27	0.8	276.94	22DE	-0.5	-3.5	-2.2	0.1

♂ LONG	♂ LAT	♂ MAG	☉ LONG	16.00UT	☿	♀	♃	♄
				1088	MAGNITUDES			
1.05	-0.07	0.9	287.13	1JA	0.7	-3.6	-2.2	0.2
7.71	0.11	1.0	297.30	11JA	2.9	-3.6	-2.1	0.2
14.37	0.26	1.2	307.44	21JA	1.0	-3.7	-2.1	0.3
21.02	0.40	1.3	317.54	31JA	0.4	-3.8	-2.0	0.3
27.66	0.53	1.4	327.59	10FE	0.2	-3.9	-1.9	0.4
34.28	0.63	1.5	337.59	20FE	0.0	-3.9	-1.8	0.4
40.87	0.73	1.6	347.54	1MR	-0.4	-4.0	-1.8	0.4
47.43	0.81	1.6	357.43	11MR	-1.0	-4.1	-1.7	0.4
53.97	0.88	1.7	7.26	21MR	-1.7	-4.2	-1.6	0.4
60.47	0.94	1.8	17.04	31MR	-1.1	-4.2	-1.6	0.4
66.94	0.99	1.8	26.75	10AP	-0.0	-4.2	-1.5	0.3
73.39	1.04	1.9	36.43	20AP	1.1	-4.0	-1.5	0.3
79.81	1.08	1.9	46.06	30AP	2.4	-3.4	-1.5	0.3
86.20	1.11	2.0	55.65	10MY	3.6	-2.8	-1.5	0.2
92.58	1.13	2.0	65.21	20MY	2.1	-3.6	-1.4	0.3
98.94	1.15	2.0	74.75	30MY	1.1	-4.1	-1.4	0.3
105.29	1.16	2.0	84.28	9JN	0.2	-4.2	-1.4	0.3
111.63	1.17	2.0	93.81	19JN	-0.9	-4.2	-1.4	0.3
117.97	1.17	2.0	103.35	29JN	-1.7	-4.1	-1.4	0.4
124.30	1.17	2.0	112.89	9JL	-1.1	-4.0	-1.4	0.4
130.65	1.16	2.0	122.47	19JL	-0.4	-3.9	-1.5	0.4
137.00	1.15	2.0	132.07	29JL	0.1	-3.8	-1.5	0.3
143.37	1.13	2.0	141.71	8AU	0.3	-3.8	-1.5	0.3
149.75	1.11	2.0	151.41	18AU	0.6	-3.7	-1.6	0.3
156.16	1.08	2.0	161.15	28AU	1.2	-3.6	-1.6	0.2
162.59	1.05	2.0	170.94	7SE	2.9	-3.6	-1.7	0.2
169.05	1.02	2.0	180.79	17SE	1.1	-3.5	-1.7	0.1
175.54	0.98	2.0	190.71	27SE	-0.5	-3.5	-1.8	0.1
182.06	0.94	2.0	200.67	7OC	-0.9	-3.5	-1.9	-0.0
188.62	0.89	2.0	210.69	17OC	-0.8	-3.4	-1.9	-0.1
195.22	0.83	1.9	220.76	27OC	-0.8	-3.4	-2.0	-0.1
201.85	0.77	1.9	230.88	6NO	-0.6	-3.4	-2.0	-0.2
208.52	0.70	1.9	241.03	16NO	-0.5	-3.4	-2.1	-0.3
215.24	0.62	1.8	251.21	26NO	-0.5	-3.4	-2.1	-0.2
221.99	0.53	1.8	261.40	6DE	-0.4	-3.4	-2.2	-0.2
228.78	0.44	1.7	271.60	16DE	0.8	-3.4	-2.2	-0.1
235.61	0.34	1.7	281.80	26DE	2.6	-3.4	-2.2	-0.0
				1089				
242.47	0.22	1.6	291.97	5JA	0.7	-3.4	-2.2	0.1
249.38	0.10	1.5	302.13	15JA	0.2	-3.4	-2.2	0.1
256.31	-0.04	1.4	312.25	25JA	0.1	-3.4	-2.1	0.2
263.28	-0.18	1.3	322.32	4FE	-0.1	-3.4	-2.1	0.2
270.28	-0.34	1.2	332.36	14FE	-0.4	-3.5	-2.0	0.3
277.30	-0.51	1.1	342.33	24FE	-1.0	-3.5	-1.9	0.3
284.34	-0.70	1.0	352.25	6MR	-1.5	-3.5	-1.9	0.3
291.39	-0.90	0.9	2.11	16MR	-1.1	-3.4	-1.8	0.3
298.44	-1.11	0.8	11.92	26MR	0.0	-3.4	-1.7	0.3
305.47	-1.33	0.7	21.66	5AP	1.4	-3.4	-1.6	0.3
312.48	-1.56	0.6	31.36	15AP	3.1	-3.4	-1.6	0.3
319.44	-1.80	0.4	41.01	25AP	2.8	-3.3	-1.5	0.3
326.32	-2.05	0.3	50.62	5MY	1.6	-3.3	-1.5	0.3
333.09	-2.30	0.2	60.20	15MY	0.8	-3.3	-1.4	0.2
339.71	-2.55	0.0	69.75	25MY	0.1	-3.3	-1.4	0.2
346.14	-2.81	-0.1	79.28	4JN	-0.9	-3.3	-1.4	0.2
352.31	-3.06	-0.3	88.82	14JN	-1.7	-3.3	-1.3	0.3
358.14	-3.31	-0.5	98.34	24JN	-1.1	-3.4	-1.3	0.3
3.55	-3.55	-0.6	107.88	4JL	-0.4	-3.4	-1.3	0.3
8.41	-3.77	-0.8	117.45	14JL	0.2	-3.4	-1.3	0.3
12.56	-3.98	-1.0	127.03	24JL	0.5	-3.5	-1.3	0.3
15.84	-4.14	-1.2	136.65	3AU	0.8	-3.5	-1.3	0.3
18.01	-4.26	-1.5	146.32	13AU	1.6	-3.6	-1.3	0.3
18.88	-4.29	-1.7	156.03	23AU	3.2	-3.7	-1.4	0.3
18.31	-4.20	-1.9	165.80	2SE	0.9	-3.7	-1.4	0.3
16.34	-3.94	-2.1	175.62	12SE	-0.6	-3.8	-1.4	0.2
13.37	-3.50	-2.3	185.50	22SE	-1.0	-3.9	-1.5	0.2
10.11	-2.90	-2.2	195.44	2OC	-1.0	-4.0	-1.5	0.1
7.34	-2.23	-2.0	205.43	12OC	-0.8	-4.2	-1.6	0.1
5.67	-1.55	-1.6	215.48	22OC	-0.5	-4.3	-1.6	-0.0
5.33	-0.95	-1.3	225.57	1NO	-0.4	-4.4	-1.7	-0.1
6.28	-0.44	-1.0	235.70	11NO	-0.4	-4.4	-1.8	-0.2
8.36	-0.03	-0.7	245.87	21NO	-0.2	-4.3	-1.8	-0.2
11.33	0.30	-0.4	256.05	1DE	1.0	-3.9	-1.9	-0.3
15.02	0.57	-0.1	266.25	11DE	2.3	-3.1	-2.0	-0.2
19.26	0.78	0.2	276.45	21DE	0.4	-3.6	-2.0	-0.2
23.93	0.94	0.4	286.64	31DE	0.0	-4.1	-2.0	-0.1

♂ LONG	♂ LAT	♂ MAG	☉ LONG	16.00UT	☿	♀	♃	♄
				1090	MAGNITUDES			
28.94	1.07	0.6	296.81	10JA	-0.1	-4.3	-2.1	-0.0
34.19	1.17	0.8	306.94	20JA	-0.2	-4.4	-2.1	0.0
39.64	1.24	0.9	317.05	30JA	-0.5	-4.3	-2.1	0.1
45.25	1.30	1.1	327.10	9FE	-1.0	-4.2	-2.1	0.2
50.97	1.35	1.2	337.11	19FE	-1.4	-4.1	-2.0	0.2
56.78	1.38	1.4	347.06	1MR	-1.0	-4.0	-2.0	0.3
62.67	1.40	1.5	356.95	11MR	0.1	-3.9	-1.9	0.3
68.61	1.41	1.6	6.78	21MR	1.8	-3.8	-1.9	0.3
74.59	1.41	1.7	16.56	31MR	3.6	-3.7	-1.8	0.3
80.61	1.40	1.7	26.29	10AP	2.1	-3.6	-1.7	0.3
86.66	1.40	1.8	35.96	20AP	1.2	-3.5	-1.7	0.3
92.73	1.38	1.9	45.59	30AP	0.6	-3.5	-1.6	0.3
98.84	1.36	1.9	55.19	10MY	-0.0	-3.4	-1.5	0.3
104.96	1.34	2.0	64.75	20MY	-0.9	-3.4	-1.5	0.3
111.11	1.31	2.0	74.29	30MY	-1.8	-3.4	-1.4	0.2
117.28	1.28	2.0	83.83	9JN	-1.1	-3.3	-1.4	0.2
123.48	1.24	2.0	93.35	19JN	-0.3	-3.3	-1.3	0.2
129.71	1.20	2.0	102.89	29JN	0.3	-3.3	-1.3	0.3
135.98	1.16	2.0	112.44	9JL	0.7	-3.3	-1.3	0.3
142.28	1.11	2.0	122.00	19JL	1.1	-3.3	-1.3	0.3
148.61	1.06	2.0	131.61	29JL	2.0	-3.3	-1.2	0.4
154.99	1.01	2.0	141.25	8AU	3.0	-3.4	-1.2	0.4
161.42	0.95	2.0	150.94	18AU	0.8	-3.4	-1.2	0.4
167.90	0.89	1.9	160.68	28AU	-0.7	-3.4	-1.2	0.4
174.42	0.82	1.9	170.47	7SE	-1.1	-3.5	-1.3	0.3
181.01	0.76	1.9	180.32	17SE	-1.1	-3.5	-1.3	0.3
187.65	0.68	1.8	190.22	27SE	-0.7	-3.5	-1.3	0.3
194.36	0.61	1.8	200.19	7OC	-0.4	-3.5	-1.3	0.2
201.12	0.52	1.8	210.20	17OC	-0.3	-3.4	-1.4	0.2
207.95	0.44	1.8	220.27	27OC	-0.2	-3.4	-1.4	0.1
214.85	0.35	1.8	230.39	6NO	-0.0	-3.4	-1.5	0.0
221.82	0.25	1.8	240.53	16NO	1.2	-3.4	-1.5	-0.1
228.85	0.15	1.8	250.71	26NO	2.0	-3.3	-1.6	-0.1
235.95	0.05	1.7	260.90	6DE	0.2	-3.3	-1.7	-0.2
243.11	-0.06	1.7	271.10	16DE	-0.1	-3.3	-1.7	-0.2
250.34	-0.18	1.7	281.30	26DE	-0.2	-3.3	-1.8	-0.2
				1091				
257.63	-0.29	1.6	291.48	5JA	-0.3	-3.3	-1.8	-0.1
264.98	-0.42	1.6	301.63	15JA	-0.6	-3.4	-1.9	-0.1
272.38	-0.54	1.5	311.76	25JA	-1.0	-3.4	-2.0	0.0
279.84	-0.67	1.5	321.84	4FE	-1.3	-3.4	-2.0	0.1
287.34	-0.80	1.4	331.87	14FE	-1.0	-3.4	-2.0	0.1
294.88	-0.93	1.3	341.85	24FE	0.2	-3.5	-2.0	0.2
302.44	-1.05	1.3	351.77	6MR	2.2	-3.5	-2.0	0.2
310.03	-1.18	1.2	1.63	16MR	2.9	-3.6	-2.0	0.3
317.62	-1.30	1.2	11.45	26MR	1.5	-3.6	-2.0	0.3
325.21	-1.41	1.1	21.20	5AP	0.9	-3.7	-1.9	0.3
332.78	-1.51	1.0	30.89	15AP	0.5	-3.8	-1.9	0.4
340.32	-1.60	1.0	40.55	25AP	-0.1	-3.8	-1.8	0.4
347.80	-1.68	0.9	50.16	5MY	-0.9	-3.9	-1.7	0.4
355.22	-1.74	0.9	59.74	15MY	-1.8	-4.0	-1.7	0.4
2.55	-1.79	0.8	69.30	25MY	-1.2	-4.1	-1.6	0.3
9.76	-1.82	0.7	78.83	4JN	-0.3	-4.2	-1.5	0.3
16.85	-1.83	0.7	88.36	14JN	0.4	-4.2	-1.5	0.3
23.77	-1.83	0.6	97.89	24JN	0.9	-4.1	-1.4	0.3
30.49	-1.80	0.5	107.42	4JL	1.5	-3.7	-1.4	0.3
37.00	-1.75	0.4	116.98	14JL	2.6	-3.1	-1.3	0.4
43.24	-1.67	0.3	126.57	24JL	2.6	-3.4	-1.3	0.4
49.15	-1.58	0.2	136.19	3AU	0.7	-3.9	-1.3	0.4
54.69	-1.45	0.1	145.85	13AU	-0.7	-4.2	-1.2	0.4
59.76	-1.30	-0.0	155.56	23AU	-1.3	-4.3	-1.2	0.5
64.27	-1.12	-0.2	165.32	2SE	-1.2	-4.2	-1.2	0.5
68.09	-0.89	-0.3	175.14	12SE	-0.6	-4.1	-1.2	0.4
71.05	-0.62	-0.5	185.02	22SE	-0.3	-4.1	-1.2	0.4
72.96	-0.29	-0.7	194.96	2OC	-0.1	-4.0	-1.2	0.4
73.61	0.09	-0.9	204.95	12OC	-0.1	-3.9	-1.2	0.4
72.84	0.53	-1.2	214.99	22OC	0.2	-3.8	-1.3	0.3
70.64	1.02	-1.4	225.08	1NO	1.5	-3.7	-1.3	0.2
67.30	1.50	-1.5	235.21	11NO	1.8	-3.7	-1.3	0.2
63.43	1.92	-1.6	245.38	21NO	0.0	-3.6	-1.4	0.1
59.87	2.23	-1.3	255.56	1DE	-0.3	-3.6	-1.4	0.0
57.26	2.43	-1.0	265.76	11DE	-0.3	-3.5	-1.5	-0.0
55.96	2.53	-0.7	275.96	21DE	-0.4	-3.5	-1.5	-0.1
56.00	2.54	-0.4	286.15	31DE	-0.6	-3.4	-1.6	-0.1

♂ LONG	♂ LAT	♂ MAG	⊙ LONG	16.00UT 1092	☿	♀	♃	♄
					MAGNITUDES			
57.22	2.51	-0.1	296.32	10JA	-1.0	-3.4	-1.7	-0.1
59.43	2.46	0.1	306.46	20JA	-1.1	-3.4	-1.7	-0.0
62.43	2.38	0.4	316.56	30JA	-0.9	-3.3	-1.8	0.0
66.04	2.30	0.6	326.62	9FE	0.3	-3.3	-1.9	0.1
70.14	2.21	0.8	336.63	19FE	2.6	-3.3	-1.9	0.2
74.62	2.13	1.0	346.58	29FE	2.2	-3.3	-2.0	0.2
79.40	2.04	1.1	356.47	10MR	1.1	-3.3	-2.0	0.3
84.43	1.95	1.3	6.31	20MR	0.7	-3.3	-2.0	0.3
89.64	1.87	1.4	16.09	30MR	0.3	-3.4	-2.0	0.4
95.02	1.78	1.5	25.82	9AP	-0.2	-3.4	-2.0	0.4
100.54	1.70	1.6	35.50	19AP	-0.9	-3.4	-2.0	0.4
106.16	1.62	1.7	45.13	29AP	-1.8	-3.5	-1.9	0.4
111.89	1.54	1.7	54.72	9MY	-1.2	-3.5	-1.9	0.5
117.70	1.45	1.8	64.29	19MY	-0.2	-3.5	-1.8	0.5
123.60	1.37	1.9	73.83	29MY	0.5	-3.4	-1.8	0.5
129.57	1.29	1.9	83.36	8JN	1.2	-3.4	-1.7	0.5
135.62	1.21	1.9	92.89	18JN	2.0	-3.4	-1.6	0.4
141.73	1.12	1.9	102.42	28JN	3.2	-3.3	-1.6	0.4
147.92	1.04	2.0	111.97	8JL	2.1	-3.3	-1.5	0.4
154.18	0.95	2.0	121.54	18JL	0.5	-3.3	-1.5	0.5
160.51	0.87	2.0	131.14	28JL	-0.8	-3.3	-1.4	0.5
166.91	0.78	1.9	140.78	7AU	-1.4	-3.3	-1.4	0.5
173.39	0.69	1.9	150.46	17AU	-1.2	-3.3	-1.3	0.6
179.95	0.60	1.9	160.19	27AU	-0.6	-3.4	-1.3	0.6
186.59	0.51	1.9	169.99	6SE	-0.2	-3.4	-1.3	0.6
193.31	0.41	1.8	179.83	16SE	-0.0	-3.4	-1.2	0.6
200.12	0.31	1.8	189.74	26SE	0.1	-3.5	-1.2	0.6
207.01	0.22	1.7	199.70	6OC	0.4	-3.5	-1.2	0.6
213.98	0.12	1.7	209.72	16OC	1.8	-3.6	-1.2	0.5
221.04	0.01	1.6	219.78	26OC	1.6	-3.6	-1.2	0.5
228.19	-0.09	1.6	229.90	5NO	-0.2	-3.7	-1.2	0.5
235.42	-0.19	1.6	240.04	15NO	-0.5	-3.8	-1.3	0.4
242.73	-0.30	1.6	250.21	25NO	-0.5	-3.9	-1.3	0.3
250.12	-0.40	1.6	260.41	5DE	-0.5	-4.0	-1.3	0.3
257.58	-0.50	1.6	270.61	15DE	-0.7	-4.1	-1.4	0.2
265.11	-0.60	1.5	280.80	25DE	-0.9	-4.2	-1.4	0.1
				1093				
272.71	-0.70	1.5	290.99	4JA	-1.0	-4.3	-1.5	0.0
280.36	-0.79	1.5	301.15	14JA	-0.8	-4.3	-1.5	0.0
288.07	-0.88	1.5	311.27	24JA	0.4	-4.3	-1.6	0.1
295.81	-0.97	1.5	321.35	3FE	2.9	-4.2	-1.7	0.1
303.58	-1.04	1.4	331.39	13FE	1.7	-3.8	-1.7	0.2
311.36	-1.11	1.4	341.37	23FE	0.8	-3.3	-1.8	0.2
319.16	-1.17	1.4	351.29	5MR	0.5	-3.6	-1.9	0.3
326.95	-1.21	1.4	1.16	15MR	0.2	-4.1	-1.9	0.4
334.72	-1.25	1.4	10.97	25MR	-0.2	-4.2	-2.0	0.4
342.46	-1.27	1.3	20.73	4AP	-0.9	-4.2	-2.0	0.5
350.15	-1.27	1.3	30.43	14AP	-1.8	-4.2	-2.1	0.5
357.79	-1.27	1.3	40.08	24AP	-1.2	-4.1	-2.1	0.5
5.36	-1.24	1.3	49.70	4MY	-0.2	-4.0	-2.1	0.6
12.85	-1.21	1.3	59.28	14MY	0.7	-3.9	-2.1	0.6
20.25	-1.16	1.2	68.83	24MY	1.5	-3.8	-2.0	0.6
27.54	-1.09	1.2	78.37	3JN	2.6	-3.7	-2.0	0.6
34.72	-1.02	1.2	87.90	13JN	3.2	-3.6	-1.9	0.6
41.78	-0.92	1.2	97.42	23JN	1.7	-3.6	-1.9	0.6
48.69	-0.82	1.1	106.97	3JL	0.4	-3.5	-1.8	0.6
55.46	-0.70	1.1	116.52	13JL	-0.8	-3.5	-1.8	0.6
62.06	-0.57	1.1	126.10	23JL	-1.5	-3.4	-1.7	0.6
68.47	-0.42	1.0	135.72	2AU	-1.2	-3.4	-1.6	0.6
74.67	-0.26	1.0	145.38	12AU	-0.5	-3.4	-1.6	0.7
80.65	-0.09	0.9	155.09	22AU	-0.1	-3.4	-1.5	0.7
86.35	0.10	0.8	164.85	1SE	0.1	-3.4	-1.5	0.8
91.73	0.32	0.7	174.67	11SE	0.3	-3.4	-1.4	0.8
96.75	0.55	0.6	184.54	21SE	0.7	-3.4	-1.4	0.8
101.31	0.81	0.5	194.48	1OC	2.2	-3.4	-1.4	0.8
105.34	1.10	0.3	204.46	11OC	1.4	-3.4	-1.3	0.8
108.69	1.43	0.2	214.50	21OC	-0.3	-3.4	-1.3	0.8
111.21	1.80	-0.0	224.59	31OC	-0.6	-3.4	-1.3	0.7
112.72	2.21	-0.2	234.72	10NO	-0.6	-3.4	-1.3	0.7
113.04	2.66	-0.5	244.88	20NO	-0.6	-3.5	-1.3	0.7
111.99	3.12	-0.7	255.07	30NO	-0.8	-3.5	-1.3	0.6
109.60	3.55	-0.9	265.26	10DE	-0.8	-3.5	-1.3	0.5
106.11	3.90	-1.1	275.46	20DE	-0.8	-3.5	-1.4	0.5
102.14	4.10	-1.1	285.65	30DE	-0.6	-3.4	-1.4	0.4

♂ LONG	♂ LAT	♂ MAG	⊙ LONG	16.00UT 1094	☿	♀	♃	♄
					MAGNITUDES			
98.44	4.13	-0.9	295.82	9JA	0.5	-3.4	-1.4	0.3
95.65	4.01	-0.7	305.97	19JA	3.0	-3.4	-1.5	0.3
94.12	3.80	-0.4	316.07	29JA	1.2	-3.4	-1.5	0.2
93.91	3.54	-0.1	326.13	8FE	0.5	-3.4	-1.6	0.3
94.86	3.27	0.1	336.14	18FE	0.3	-3.3	-1.7	0.3
96.82	3.01	0.4	346.10	28FE	0.1	-3.3	-1.7	0.4
99.57	2.76	0.6	355.99	10MR	-0.3	-3.3	-1.8	0.4
102.97	2.53	0.8	5.83	20MR	-0.9	-3.3	-1.9	0.5
106.89	2.32	0.9	15.62	30MR	-1.7	-3.4	-1.9	0.5
111.22	2.13	1.1	25.34	9AP	-1.2	-3.4	-2.0	0.6
115.88	1.94	1.2	35.03	19AP	-0.1	-3.4	-2.1	0.6
120.82	1.77	1.3	44.66	29AP	0.9	-3.4	-2.1	0.7
125.99	1.61	1.4	54.26	9MY	2.0	-3.5	-2.1	0.7
131.35	1.46	1.5	63.83	19MY	3.4	-3.5	-2.2	0.8
136.90	1.32	1.6	73.37	29MY	2.5	-3.6	-2.2	0.8
142.59	1.18	1.6	82.90	8JN	1.3	-3.6	-2.2	0.8
148.43	1.05	1.7	92.43	18JN	0.3	-3.7	-2.2	0.8
154.39	0.92	1.7	101.96	28JN	-0.8	-3.8	-2.1	0.8
160.48	0.79	1.7	111.51	8JL	-1.6	-3.9	-2.1	0.8
166.68	0.67	1.8	121.08	18JL	-1.2	-4.0	-2.0	0.8
173.00	0.55	1.8	130.68	28JL	-0.4	-4.1	-2.0	0.8
179.42	0.43	1.8	140.31	7AU	-0.0	-4.2	-1.9	0.8
185.96	0.32	1.8	150.00	17AU	0.3	-4.2	-1.9	0.9
192.61	0.20	1.8	159.73	27AU	0.5	-4.3	-1.8	0.9
199.36	0.09	1.7	169.51	6SE	1.0	-4.1	-1.7	1.0
206.23	-0.02	1.7	179.36	16SE	2.6	-3.8	-1.7	1.0
213.20	-0.13	1.7	189.26	26SE	1.2	-3.2	-1.6	1.0
220.27	-0.23	1.7	199.21	6OC	-0.4	-3.6	-1.6	1.0
227.44	-0.33	1.6	209.23	16OC	-0.8	-4.1	-1.5	1.1
234.72	-0.43	1.6	219.29	26OC	-0.8	-4.3	-1.5	1.0
242.09	-0.53	1.6	229.40	5NO	-0.8	-4.4	-1.5	1.0
249.56	-0.62	1.5	239.55	15NO	-0.7	-4.3	-1.4	1.0
257.10	-0.71	1.5	249.72	25NO	-0.6	-4.2	-1.4	1.0
264.73	-0.79	1.4	259.91	5DE	-0.6	-4.1	-1.4	0.9
272.42	-0.86	1.4	270.11	15DE	-0.5	-4.0	-1.4	0.8
280.18	-0.92	1.3	280.31	25DE	0.7	-3.9	-1.4	0.8
				1095				
287.98	-0.98	1.3	290.49	4JA	2.8	-3.8	-1.4	0.7
295.82	-1.02	1.4	300.66	14JA	0.9	-3.7	-1.4	0.6
303.69	-1.06	1.4	310.78	24JA	0.3	-3.7	-1.5	0.6
311.57	-1.08	1.4	320.87	3FE	0.2	-3.6	-1.5	0.5
319.46	-1.10	1.4	330.91	13FE	-0.0	-3.5	-1.6	0.5
327.33	-1.10	1.4	340.89	23FE	-0.4	-3.5	-1.6	0.5
335.17	-1.09	1.4	350.82	5MR	-1.0	-3.4	-1.7	0.6
342.98	-1.06	1.4	0.69	15MR	-1.6	-3.4	-1.7	0.6
350.74	-1.03	1.4	10.50	25MR	-1.1	-3.3	-1.8	0.7
358.44	-0.98	1.4	20.26	4AP	-0.0	-3.3	-1.9	0.7
6.08	-0.92	1.5	29.97	14AP	1.2	-3.3	-1.9	0.8
13.63	-0.86	1.5	39.62	24AP	2.6	-3.3	-2.0	0.8
21.11	-0.78	1.5	49.24	4MY	3.4	-3.3	-2.1	0.9
28.49	-0.69	1.5	58.82	14MY	1.9	-3.3	-2.1	0.9
35.78	-0.60	1.5	68.37	24MY	1.0	-3.3	-2.2	1.0
42.97	-0.49	1.5	77.91	3JN	0.2	-3.3	-2.2	1.0
50.06	-0.39	1.5	87.44	13JN	-0.8	-3.4	-2.3	1.0
57.03	-0.27	1.5	96.97	23JN	-1.7	-3.4	-2.3	1.1
63.90	-0.15	1.5	106.50	3JL	-1.2	-3.5	-2.3	1.1
70.65	-0.02	1.5	116.06	13JL	-0.4	-3.5	-2.3	1.1
77.28	0.11	1.5	125.64	23JL	0.1	-3.5	-2.3	1.1
83.79	0.25	1.5	135.26	2AU	0.4	-3.4	-2.3	1.1
90.15	0.39	1.5	144.91	12AU	0.7	-3.4	-2.2	1.1
96.38	0.54	1.4	154.62	22AU	1.3	-3.4	-2.2	1.1
102.44	0.70	1.4	164.37	1SE	3.0	-3.4	-2.1	1.2
108.33	0.87	1.3	174.19	11SE	1.1	-3.3	-2.0	1.2
114.02	1.04	1.3	184.06	21SE	-0.5	-3.3	-2.0	1.3
119.49	1.23	1.2	193.99	1OC	-0.9	-3.3	-1.9	1.3
124.69	1.43	1.1	203.97	11OC	-0.9	-3.3	-1.8	1.3
129.57	1.65	1.0	214.01	21OC	-0.8	-3.3	-1.8	1.3
134.07	1.89	0.8	224.09	31OC	-0.6	-3.4	-1.7	1.3
138.10	2.16	0.7	234.22	10NO	-0.5	-3.4	-1.7	1.3
141.57	2.45	0.5	244.38	20NO	-0.5	-3.4	-1.6	1.3
144.33	2.76	0.3	254.56	30NO	-0.3	-3.4	-1.6	1.3
146.21	3.11	0.1	264.76	10DE	0.8	-3.5	-1.6	1.2
147.04	3.48	-0.1	274.96	20DE	2.5	-3.5	-1.5	1.2
146.64	3.86	-0.4	285.15	30DE	0.6	-3.6	-1.5	1.1

♂ LONG	♂ LAT	♂ MAG	☉ LONG	16.00UT	☿	♀	♃	♄
				1096	MAGNITUDES			
144.91	4.19	-0.6	295.33	9JA	0.1	-3.6	-1.5	1.1
141.98	4.45	-0.8	305.47	19JA	0.0	-3.7	-1.5	1.0
138.21	4.55	-1.0	315.58	29JA	-0.1	-3.8	-1.5	0.9
134.29	4.48	-0.9	325.65	8FE	-0.4	-3.9	-1.5	0.9
130.93	4.24	-0.7	335.66	18FE	-1.0	-3.9	-1.5	0.8
128.64	3.90	-0.5	345.62	28FE	-1.5	-4.0	-1.6	0.8
127.63	3.51	-0.2	355.52	9MR	-1.1	-4.1	-1.6	0.8
127.89	3.12	-0.0	5.36	19MR	0.0	-4.2	-1.7	0.9
129.25	2.75	0.2	15.15	29MR	1.5	-4.2	-1.7	0.9
131.54	2.40	0.4	24.88	8AP	3.4	-4.2	-1.8	1.0
134.59	2.09	0.6	34.56	18AP	2.5	-4.0	-1.8	1.0
138.25	1.80	0.7	44.20	28AP	1.5	-3.4	-1.9	1.1
142.42	1.55	0.9	53.80	8MY	0.8	-2.9	-1.9	1.1
147.00	1.31	1.0	63.36	18MY	0.1	-3.7	-2.0	1.2
151.92	1.09	1.1	72.91	28MY	-0.9	-4.1	-2.1	1.2
157.14	0.89	1.2	82.44	7JN	-1.7	-4.2	-2.2	1.3
162.61	0.71	1.3	91.97	17JN	-1.2	-4.2	-2.2	1.3
168.31	0.54	1.3	101.50	27JN	-0.4	-4.1	-2.3	1.3
174.21	0.37	1.4	111.05	7JL	0.2	-4.0	-2.3	1.4
180.29	0.22	1.4	120.61	17JL	0.5	-3.9	-2.4	1.4
186.55	0.07	1.5	130.21	27JL	0.9	-3.8	-2.4	1.4
192.97	-0.06	1.5	139.85	6AU	1.7	-3.8	-2.4	1.4
199.54	-0.19	1.5	149.52	16AU	3.2	-3.7	-2.4	1.3
206.25	-0.32	1.5	159.26	26AU	0.9	-3.6	-2.4	1.3
213.11	-0.43	1.5	169.04	5SE	-0.6	-3.6	-2.4	1.2
220.10	-0.54	1.5	178.88	15SE	-1.1	-3.5	-2.3	1.2
227.22	-0.64	1.5	188.78	25SE	-1.1	-3.5	-2.3	1.2
234.45	-0.73	1.5	198.73	5OC	-0.7	-3.5	-2.2	1.2
241.80	-0.81	1.5	208.74	15OC	-0.5	-3.4	-2.1	1.2
249.26	-0.89	1.5	218.81	25OC	-0.3	-3.4	-2.1	1.2
256.82	-0.95	1.5	228.91	4NO	-0.3	-3.4	-2.0	1.2
264.46	-1.01	1.4	239.05	14NO	-0.1	-3.4	-1.9	1.2
272.18	-1.05	1.4	249.23	24NO	1.0	-3.4	-1.9	1.1
279.97	-1.08	1.4	259.41	4DE	2.3	-3.4	-1.8	1.1
287.80	-1.10	1.4	269.61	14DE	0.3	-3.4	-1.8	1.1
295.68	-1.11	1.4	279.81	24DE	-0.0	-3.4	-1.7	1.0
				1097				
303.57	-1.11	1.4	289.99	3JA	-0.1	-3.4	-1.7	1.0
311.48	-1.09	1.4	300.16	13JA	-0.2	-3.4	-1.6	0.9
319.39	-1.06	1.4	310.29	23JA	-0.5	-3.4	-1.6	0.9
327.27	-1.03	1.3	320.37	2FE	-1.0	-3.4	-1.6	0.8
335.13	-0.98	1.3	330.41	12FE	-1.3	-3.5	-1.6	0.8
342.94	-0.92	1.3	340.40	22FE	-1.0	-3.5	-1.6	0.7
350.70	-0.85	1.3	350.33	4MR	0.1	-3.5	-1.6	0.7
358.40	-0.77	1.4	0.20	14MR	1.9	-3.4	-1.6	0.8
6.03	-0.69	1.4	10.02	24MR	3.4	-3.4	-1.6	0.8
13.58	-0.60	1.5	19.78	3AP	1.9	-3.4	-1.6	0.9
21.04	-0.51	1.5	29.49	13AP	1.1	-3.4	-1.7	1.0
28.43	-0.41	1.5	39.15	23AP	0.6	-3.3	-1.7	1.0
35.72	-0.31	1.6	48.77	3MY	-0.0	-3.3	-1.7	1.1
42.92	-0.20	1.6	58.35	13MY	-0.9	-3.3	-1.8	1.1
50.03	-0.09	1.7	67.91	23MY	-1.8	-3.3	-1.9	1.2
57.05	0.02	1.7	77.44	2JN	-1.2	-3.3	-1.9	1.2
63.98	0.12	1.7	86.97	12JN	-0.3	-3.3	-2.0	1.3
70.82	0.24	1.7	96.50	22JN	0.3	-3.4	-2.1	1.3
77.57	0.35	1.8	106.04	2JL	0.7	-3.4	-2.1	1.3
84.23	0.46	1.8	115.59	12JL	1.2	-3.4	-2.2	1.3
90.81	0.57	1.8	125.17	22JL	2.2	-3.5	-2.3	1.3
97.29	0.68	1.8	134.79	1AU	2.9	-3.5	-2.3	1.2
103.69	0.79	1.8	144.44	11AU	0.8	-3.6	-2.4	1.2
109.99	0.91	1.8	154.15	21AU	-0.7	-3.7	-2.4	1.2
116.20	1.02	1.7	163.90	31AU	-1.2	-3.7	-2.4	1.1
122.31	1.14	1.7	173.71	10SE	-1.2	-3.8	-2.5	1.1
128.30	1.26	1.7	183.58	20SE	-0.7	-3.9	-2.5	1.1
134.17	1.38	1.6	193.50	30SE	-0.4	-4.0	-2.4	1.1
139.90	1.51	1.6	203.49	10OC	-0.2	-4.2	-2.4	1.1
145.47	1.64	1.5	213.52	20OC	-0.1	-4.3	-2.4	1.1
150.86	1.78	1.4	223.60	30OC	0.1	-4.4	-2.3	1.1
156.03	1.93	1.3	233.73	9NO	1.3	-4.4	-2.2	1.1
160.93	2.09	1.2	243.89	19NO	2.0	-4.3	-2.2	1.0
165.53	2.26	1.0	254.07	29NO	0.1	-3.8	-2.1	1.0
169.73	2.44	0.9	264.27	9DE	-0.2	-3.0	-2.0	1.0
173.45	2.63	0.7	274.47	19DE	-0.3	-3.6	-2.0	0.9
176.58	2.84	0.5	284.65	29DE	-0.3	-4.2	-1.9	0.9

♂ LONG	♂ LAT	♂ MAG	☉ LONG	16.00UT	☿	♀	♃	♄
				1098	MAGNITUDES			
178.96	3.06	0.2	294.83	8JA	-0.6	-4.3	-1.8	0.9
180.43	3.29	-0.0	304.98	18JA	-1.0	-4.4	-1.8	0.8
180.82	3.50	-0.3	315.08	28JA	-1.2	-4.3	-1.7	0.8
179.95	3.68	-0.6	325.15	7FE	-1.0	-4.2	-1.7	0.7
177.81	3.78	-0.8	335.17	17FE	0.2	-4.1	-1.6	0.7
174.60	3.77	-1.1	345.13	27FE	2.3	-4.0	-1.6	0.6
170.78	3.60	-1.2	355.04	9MR	2.6	-3.9	-1.6	0.6
167.08	3.29	-1.0	4.88	19MR	1.4	-3.8	-1.6	0.5
164.16	2.88	-0.8	14.67	29MR	0.8	-3.7	-1.6	0.6
162.42	2.43	-0.6	24.41	8AP	0.4	-3.6	-1.6	0.7
162.00	1.98	-0.4	34.09	18AP	-0.1	-3.5	-1.6	0.7
162.81	1.56	-0.2	43.73	28AP	-0.9	-3.5	-1.6	0.8
164.70	1.18	0.0	53.34	8MY	-1.8	-3.4	-1.6	0.9
167.50	0.85	0.2	62.90	18MY	-1.2	-3.4	-1.6	0.9
171.03	0.55	0.3	72.45	28MY	-0.3	-3.4	-1.7	1.0
175.18	0.28	0.5	81.99	7JN	0.4	-3.3	-1.7	1.0
179.83	0.05	0.6	91.51	17JN	1.0	-3.3	-1.8	1.1
184.91	-0.16	0.7	101.04	27JN	1.6	-3.3	-1.8	1.1
190.35	-0.35	0.8	110.59	7JL	2.8	-3.3	-1.9	1.1
196.10	-0.52	0.9	120.15	17JL	2.5	-3.3	-1.9	1.1
202.13	-0.67	0.9	129.75	27JL	0.6	-3.3	-2.0	1.1
208.40	-0.80	1.0	139.38	6AU	-0.7	-3.4	-2.1	1.1
214.90	-0.92	1.0	149.06	16AU	-1.3	-3.4	-2.1	1.1
221.60	-1.02	1.1	158.78	26AU	-1.2	-3.4	-2.2	1.1
228.48	-1.11	1.1	168.57	5SE	-0.6	-3.5	-2.3	1.0
235.52	-1.18	1.1	178.40	15SE	-0.3	-3.5	-2.3	1.0
242.72	-1.24	1.2	188.29	25SE	-0.1	-3.5	-2.4	0.9
250.05	-1.28	1.2	198.25	5OC	0.0	-3.5	-2.4	0.9
257.51	-1.31	1.2	208.25	15OC	0.3	-3.4	-2.4	0.9
265.07	-1.33	1.2	218.31	25OC	1.6	-3.4	-2.4	0.9
272.72	-1.33	1.3	228.42	4NO	1.8	-3.4	-2.4	0.9
280.45	-1.31	1.3	238.56	14NO	-0.0	-3.4	-2.4	0.9
288.23	-1.29	1.3	248.73	24NO	-0.4	-3.3	-2.3	0.9
296.05	-1.25	1.3	258.92	4DE	-0.4	-3.3	-2.3	0.9
303.90	-1.20	1.3	269.12	14DE	-0.5	-3.3	-2.2	0.9
311.76	-1.13	1.4	279.31	24DE	-0.7	-3.3	-2.1	0.9
				1099				
319.61	-1.06	1.4	289.50	3JA	-1.0	-3.3	-2.1	0.8
327.44	-0.98	1.4	299.66	13JA	-1.0	-3.4	-2.0	0.8
335.24	-0.89	1.4	309.79	23JA	-0.9	-3.4	-1.9	0.7
342.99	-0.80	1.5	319.88	2FE	0.3	-3.4	-1.8	0.7
350.69	-0.70	1.5	329.92	12FE	2.6	-3.4	-1.8	0.6
358.32	-0.59	1.5	339.91	22FE	2.0	-3.5	-1.7	0.6
5.88	-0.49	1.5	349.85	4MR	1.0	-3.5	-1.7	0.5
13.36	-0.38	1.6	359.72	14MR	0.6	-3.6	-1.6	0.5
20.76	-0.27	1.6	9.55	24MR	0.3	-3.6	-1.6	0.5
28.07	-0.17	1.6	19.31	3AP	-0.2	-3.7	-1.6	0.4
35.31	-0.06	1.6	29.02	13AP	-0.9	-3.8	-1.5	0.5
42.45	0.05	1.6	38.68	23AP	-1.8	-3.8	-1.5	0.6
49.52	0.15	1.6	48.30	3MY	-1.2	-3.9	-1.5	0.6
56.50	0.25	1.7	57.89	13MY	-0.2	-4.0	-1.5	0.7
63.41	0.35	1.7	67.45	23MY	0.6	-4.1	-1.5	0.8
70.24	0.44	1.8	76.99	2JN	1.3	-4.2	-1.5	0.8
76.99	0.54	1.8	86.51	12JN	2.2	-4.2	-1.5	0.9
83.68	0.63	1.8	96.04	22JN	3.3	-4.1	-1.6	0.9
90.30	0.71	1.9	105.58	2JL	2.0	-3.7	-1.6	0.9
96.85	0.80	1.9	115.13	12JL	0.5	-3.1	-1.6	1.0
103.35	0.88	1.9	124.71	22JL	-0.8	-3.4	-1.7	1.0
109.79	0.95	1.9	134.32	1AU	-1.4	-3.9	-1.7	1.0
116.18	1.03	2.0	143.98	11AU	-1.2	-4.2	-1.8	1.0
122.51	1.10	2.0	153.68	21AU	-0.5	-4.2	-1.8	1.0
128.79	1.18	1.9	163.43	31AU	-0.2	-4.2	-1.9	1.0
135.01	1.24	1.9	173.23	10SE	0.0	-4.1	-2.0	0.9
141.18	1.31	1.9	183.10	20SE	0.2	-4.1	-2.0	0.9
147.28	1.38	1.9	193.02	30SE	0.5	-4.0	-2.1	0.8
153.32	1.44	1.8	203.00	10OC	1.9	-3.9	-2.2	0.8
159.29	1.51	1.8	213.04	20OC	1.6	-3.8	-2.2	0.8
165.18	1.57	1.7	223.11	30OC	-0.2	-3.7	-2.3	0.8
170.97	1.64	1.6	233.24	9NO	-0.5	-3.7	-2.3	0.9
176.65	1.70	1.6	243.40	19NO	-0.5	-3.6	-2.3	0.9
182.20	1.76	1.5	253.58	29NO	-0.6	-3.6	-2.3	0.9
187.59	1.82	1.3	263.77	9DE	-0.7	-3.5	-2.3	0.9
192.80	1.88	1.2	273.97	19DE	-0.8	-3.5	-2.3	0.8
197.77	1.93	1.0	284.17	29DE	-0.9	-3.4	-2.2	0.8

♂ LONG	♂ LAT	♂ MAG	⊙ LONG	16.00UT	☿	♀	♃	♄
				1100	MAGNITUDES			
202.47	1.99	0.9	294.34	8JA	-0.7	-3.4	-2.2	0.8
206.81	2.03	0.7	304.49	18JA	0.4	-3.4	-2.1	0.8
210.72	2.07	0.5	314.60	28JA	2.9	-3.3	-2.0	0.7
214.10	2.10	0.2	324.67	7FE	1.5	-3.3	-1.9	0.7
216.80	2.10	-0.0	334.69	17FE	0.7	-3.3	-1.9	0.6
218.66	2.07	-0.3	344.65	27FE	0.4	-3.3	-1.8	0.6
219.50	2.00	-0.6	354.56	8MR	0.1	-3.3	-1.7	0.5
219.16	1.86	-0.9	4.41	18MR	-0.2	-3.3	-1.7	0.4
217.55	1.64	-1.2	14.20	28MR	-0.9	-3.4	-1.6	0.4
214.80	1.32	-1.5	23.93	7AP	-1.8	-3.4	-1.6	0.4
211.30	0.90	-1.7	33.62	17AP	-1.2	-3.4	-1.5	0.4
207.76	0.44	-1.6	43.26	27AP	-0.2	-3.5	-1.5	0.4
204.86	-0.04	-1.4	52.86	7MY	0.8	-3.5	-1.5	0.5
203.12	-0.47	-1.2	62.44	17MY	1.7	-3.5	-1.4	0.6
202.73	-0.85	-1.0	71.98	27MY	2.9	-3.4	-1.4	0.6
203.66	-1.16	-0.8	81.52	6JN	3.0	-3.4	-1.4	0.7
205.77	-1.41	-0.6	91.05	16JN	1.6	-3.4	-1.4	0.7
208.87	-1.61	-0.4	100.57	26JN	0.4	-3.3	-1.4	0.8
212.78	-1.76	-0.2	110.12	6JL	-0.8	-3.3	-1.4	0.8
217.37	-1.88	-0.1	119.69	16JL	-1.5	-3.3	-1.4	0.9
222.52	-1.97	0.0	129.28	26JL	-1.2	-3.3	-1.5	0.9
228.13	-2.02	0.2	138.91	5AU	-0.5	-3.3	-1.5	0.9
234.13	-2.06	0.3	148.59	15AU	-0.1	-3.3	-1.5	0.9
240.46	-2.06	0.4	158.31	25AU	0.2	-3.4	-1.6	0.9
247.06	-2.05	0.5	168.09	4SE	0.4	-3.4	-1.6	0.9
253.91	-2.02	0.5	177.92	14SE	0.8	-3.4	-1.7	0.9
260.95	-1.97	0.6	187.81	24SE	2.3	-3.5	-1.7	0.8
268.16	-1.90	0.7	197.76	4OC	1.4	-3.5	-1.8	0.8
275.51	-1.81	0.8	207.77	14OC	-0.3	-3.6	-1.9	0.8
282.97	-1.72	0.8	217.82	24OC	-0.7	-3.6	-1.9	0.7
290.50	-1.61	0.9	227.92	3NO	-0.7	-3.7	-2.0	0.8
298.11	-1.49	1.0	238.07	13NO	-0.7	-3.8	-2.0	0.8
305.74	-1.36	1.0	248.23	23NO	-0.7	-3.9	-2.1	0.8
313.40	-1.23	1.1	258.42	3DE	-0.7	-4.0	-2.1	0.8
321.06	-1.09	1.2	268.62	13DE	-0.7	-4.1	-2.2	0.8
328.70	-0.95	1.2	278.82	23DE	-0.6	-4.2	-2.2	0.8
				1101				
336.31	-0.81	1.3	289.01	2JA	0.5	-4.3	-2.2	0.8
343.89	-0.68	1.4	299.17	12JA	3.0	-4.3	-2.2	0.8
351.40	-0.54	1.4	309.30	22JA	1.1	-4.3	-2.1	0.7
358.86	-0.40	1.5	319.40	1FE	0.5	-4.2	-2.1	0.7
6.26	-0.27	1.5	329.44	11FE	0.2	-3.8	-2.0	0.7
13.58	-0.15	1.6	339.43	21FE	0.0	-3.3	-2.0	0.6
20.83	-0.03	1.6	349.37	3MR	-0.3	-3.7	-1.9	0.6
28.00	0.09	1.7	359.25	13MR	-0.9	-4.1	-1.8	0.5
35.10	0.19	1.7	9.07	23MR	-1.7	-4.2	-1.8	0.4
42.12	0.30	1.7	18.84	2AP	-1.2	-4.2	-1.7	0.4
49.07	0.39	1.8	28.55	12AP	-0.1	-4.2	-1.6	0.3
55.95	0.49	1.8	38.21	22AP	1.0	-4.1	-1.6	0.3
62.76	0.57	1.8	47.84	2MY	2.2	-4.0	-1.5	0.3
69.50	0.65	1.8	57.42	12MY	3.7	-3.9	-1.5	0.4
76.19	0.73	1.9	66.98	22MY	2.3	-3.8	-1.4	0.5
82.82	0.79	1.9	76.52	1JN	1.2	-3.7	-1.4	0.5
89.40	0.86	1.9	86.05	11JN	0.3	-3.6	-1.3	0.6
95.94	0.92	1.9	95.58	21JN	-0.8	-3.5	-1.3	0.6
102.43	0.97	1.9	105.12	1JL	-1.6	-3.5	-1.3	0.7
108.89	1.02	1.9	114.67	11JL	-1.2	-3.5	-1.3	0.7
115.31	1.07	2.0	124.25	21JL	-0.4	-3.4	-1.3	0.8
121.71	1.11	2.0	133.86	31JL	0.0	-3.4	-1.3	0.8
128.08	1.14	2.0	143.51	10AU	0.3	-3.4	-1.3	0.8
134.43	1.18	2.0	153.21	20AU	0.5	-3.4	-1.3	0.8
140.76	1.21	2.0	162.96	30AU	1.1	-3.4	-1.4	0.8
147.07	1.23	2.0	172.76	9SE	2.7	-3.4	-1.4	0.8
153.37	1.25	2.0	182.62	19SE	1.3	-3.4	-1.4	0.8
159.65	1.27	2.0	192.55	29SE	-0.5	-3.4	-1.5	0.8
165.91	1.28	2.0	202.52	9OC	-0.8	-3.4	-1.5	0.8
172.16	1.29	1.9	212.55	19OC	-0.8	-3.4	-1.6	0.7
178.39	1.30	1.9	222.63	29OC	-0.8	-3.4	-1.6	0.7
184.59	1.29	1.8	232.75	8NO	-0.6	-3.4	-1.7	0.7
190.77	1.29	1.8	242.90	18NO	-0.6	-3.5	-1.8	0.7
196.92	1.27	1.7	253.08	28NO	-0.6	-3.5	-1.8	0.8
203.02	1.25	1.6	263.28	8DE	-0.4	-3.5	-1.9	0.8
209.08	1.22	1.5	273.47	18DE	0.7	-3.5	-2.0	0.8
215.07	1.18	1.4	283.67	28DE	2.8	-3.4	-2.0	0.8

♂ LONG	♂ LAT	♂ MAG	⊙ LONG	16.00UT	☿	♀	♃	♄
				1102	MAGNITUDES			
220.99	1.13	1.3	293.84	7JA	0.8	-3.4	-2.0	0.8
226.83	1.07	1.2	303.99	17JA	0.3	-3.4	-2.1	0.8
232.55	0.98	1.0	314.11	27JA	0.1	-3.4	-2.1	0.7
238.14	0.88	0.9	324.18	6FE	-0.1	-3.4	-2.1	0.7
243.57	0.76	0.7	334.20	16FE	-0.4	-3.3	-2.0	0.7
248.78	0.60	0.5	344.17	26FE	-0.9	-3.3	-2.0	0.6
253.74	0.41	0.3	354.08	8MR	-1.6	-3.3	-2.0	0.6
258.38	0.17	0.0	3.93	18MR	-1.2	-3.3	-1.9	0.5
262.60	-0.12	-0.2	13.73	28MR	-0.1	-3.4	-1.8	0.5
266.30	-0.47	-0.5	23.46	7AP	1.3	-3.4	-1.8	0.4
269.34	-0.90	-0.8	33.15	17AP	2.9	-3.4	-1.7	0.3
271.54	-1.41	-1.1	42.80	27AP	3.1	-3.4	-1.6	0.3
272.73	-2.01	-1.4	52.40	7MY	1.7	-3.5	-1.6	0.2
272.74	-2.69	-1.7	61.97	17MY	0.9	-3.5	-1.5	0.3
271.51	-3.42	-2.0	71.53	27MY	0.1	-3.6	-1.5	0.4
269.22	-4.12	-2.3	81.06	6JN	-0.8	-3.6	-1.4	0.4
266.32	-4.70	-2.4	90.59	16JN	-1.7	-3.7	-1.4	0.5
263.54	-5.07	-2.3	100.12	26JN	-1.2	-3.8	-1.3	0.6
261.58	-5.21	-2.1	109.66	6JL	-0.4	-3.9	-1.3	0.6
260.85	-5.16	-1.9	119.22	16JL	0.1	-4.0	-1.3	0.7
261.51	-4.97	-1.6	128.82	26JL	0.4	-4.1	-1.2	0.7
263.47	-4.69	-1.4	138.44	5AU	0.8	-4.2	-1.2	0.7
266.54	-4.37	-1.2	148.12	15AU	1.4	-4.2	-1.2	0.8
270.53	-4.02	-0.9	157.84	25AU	3.1	-4.3	-1.2	0.8
275.27	-3.67	-0.7	167.61	4SE	1.1	-4.1	-1.2	0.8
280.58	-3.32	-0.5	177.45	14SE	-0.5	-3.8	-1.2	0.8
286.36	-2.97	-0.4	187.34	24SE	-1.0	-3.2	-1.3	0.8
292.49	-2.64	-0.2	197.28	4OC	-1.0	-3.7	-1.3	0.8
298.89	-2.31	-0.0	207.28	14OC	-0.8	-4.1	-1.3	0.8
305.51	-2.01	0.1	217.34	24OC	-0.5	-4.3	-1.4	0.7
312.28	-1.72	0.3	227.43	3NO	-0.4	-4.4	-1.4	0.7
319.16	-1.44	0.4	237.57	13NO	-0.4	-4.3	-1.5	0.7
326.13	-1.18	0.6	247.74	23NO	-0.3	-4.2	-1.5	0.7
333.14	-0.94	0.7	257.93	3DE	0.8	-4.1	-1.6	0.7
340.18	-0.72	0.8	268.13	13DE	2.5	-4.0	-1.7	0.7
347.23	-0.51	1.0	278.33	23DE	0.5	-3.9	-1.7	0.8
				1103				
354.26	-0.32	1.1	288.51	2JA	0.1	-3.8	-1.8	0.8
1.28	-0.15	1.2	298.68	12JA	-0.0	-3.7	-1.9	0.8
8.27	0.01	1.3	308.82	22JA	-0.2	-3.7	-1.9	0.8
15.22	0.15	1.4	318.91	1FE	-0.5	-3.6	-2.0	0.8
22.12	0.28	1.5	328.96	11FE	-1.0	-3.5	-2.0	0.7
28.98	0.40	1.5	338.95	21FE	-1.4	-3.5	-2.0	0.7
35.79	0.51	1.6	348.89	3MR	-1.1	-3.4	-2.0	0.7
42.56	0.60	1.7	358.78	13MR	0.0	-3.4	-2.0	0.6
49.27	0.69	1.7	8.60	23MR	1.6	-3.3	-2.0	0.5
55.93	0.77	1.8	18.37	2AP	3.6	-3.3	-1.9	0.5
62.55	0.84	1.8	28.09	12AP	2.3	-3.3	-1.9	0.4
69.12	0.90	1.9	37.75	22AP	1.3	-3.3	-1.8	0.4
75.66	0.95	1.9	47.38	2MY	0.7	-3.3	-1.8	0.3
82.15	0.99	2.0	56.97	12MY	0.1	-3.3	-1.7	0.2
88.62	1.03	2.0	66.53	22MY	-0.8	-3.3	-1.6	0.2
95.05	1.07	2.0	76.06	1JN	-1.7	-3.3	-1.6	0.3
101.47	1.09	2.0	85.60	11JN	-1.2	-3.4	-1.5	0.4
107.86	1.12	2.0	95.12	21JN	-0.4	-3.4	-1.5	0.4
114.24	1.13	2.0	104.65	1JL	0.2	-3.5	-1.4	0.5
120.61	1.14	2.0	114.21	11JL	0.6	-3.5	-1.4	0.6
126.97	1.15	2.0	123.78	21JL	1.0	-3.5	-1.3	0.6
133.33	1.15	2.0	133.39	31JL	1.9	-3.4	-1.3	0.7
139.70	1.15	2.0	143.04	10AU	3.2	-3.4	-1.3	0.7
146.07	1.15	2.0	152.73	20AU	0.9	-3.4	-1.2	0.7
152.46	1.13	2.0	162.48	30AU	-0.6	-3.4	-1.2	0.8
158.86	1.12	2.0	172.28	9SE	-1.1	-3.3	-1.2	0.8
165.27	1.10	2.0	182.14	19SE	-1.1	-3.3	-1.2	0.8
171.70	1.07	2.0	192.06	29SE	-0.7	-3.3	-1.2	0.8
178.16	1.04	2.0	202.03	9OC	-0.4	-3.3	-1.2	0.8
184.64	1.01	2.0	212.06	19OC	-0.3	-3.3	-1.2	0.8
191.14	0.96	1.9	222.13	29OC	-0.2	-3.4	-1.3	0.7
197.66	0.91	1.9	232.26	8NO	-0.1	-3.4	-1.3	0.7
204.21	0.86	1.9	242.41	18NO	1.1	-3.4	-1.3	0.7
210.79	0.79	1.8	252.59	28NO	2.3	-3.4	-1.4	0.7
217.38	0.72	1.8	262.78	8DE	0.3	-3.5	-1.4	0.7
224.00	0.64	1.7	272.98	18DE	-0.1	-3.5	-1.5	0.7
230.64	0.55	1.6	283.18	28DE	-0.2	-3.6	-1.6	0.8

♂ LONG	♂ LAT	♂ MAG	☉ LONG	16.00UT 1104	☿ MAGNITUDES	♀ MAGNITUDES	♃ MAGNITUDES	♄ MAGNITUDES
237.30	0.44	1.5	293.36	7JA	-0.3	-3.6	-1.6	0.8
243.97	0.33	1.4	303.51	17JA	-0.5	-3.7	-1.7	0.8
250.65	0.20	1.3	313.62	27JA	-1.0	-3.8	-1.8	0.8
257.35	0.05	1.2	323.70	6FE	-1.3	-3.9	-1.8	0.8
264.04	-0.11	1.1	333.72	16FE	-1.0	-4.0	-1.9	0.8
270.74	-0.30	1.0	343.69	26FE	0.1	-4.0	-1.9	0.7
277.42	-0.50	0.9	353.60	7MR	2.0	-4.1	-2.0	0.7
284.07	-0.72	0.7	3.46	17MR	3.1	-4.2	-2.0	0.6
290.70	-0.96	0.6	13.26	27MR	1.7	-4.2	-2.0	0.6
297.26	-1.23	0.4	23.00	6AP	1.0	-4.2	-2.0	0.5
303.75	-1.53	0.3	32.69	16AP	0.5	-4.0	-2.0	0.5
310.13	-1.84	0.1	42.34	26AP	-0.0	-3.4	-2.0	0.4
316.35	-2.19	-0.1	51.94	6MY	-0.8	-2.9	-1.9	0.3
322.38	-2.56	-0.3	61.51	16MY	-1.8	-3.7	-1.9	0.3
328.13	-2.95	-0.5	71.06	26MY	-1.2	-4.1	-1.8	0.2
333.52	-3.38	-0.7	80.60	5JN	-0.3	-4.2	-1.8	0.3
338.45	-3.82	-0.9	90.12	15JN	0.3	-4.2	-1.7	0.3
342.78	-4.28	-1.1	99.66	25JN	0.8	-4.1	-1.6	0.4
346.31	-4.75	-1.4	109.20	5JL	1.4	-4.0	-1.6	0.4
348.87	-5.20	-1.6	118.76	15JL	2.4	-3.9	-1.5	0.5
350.22	-5.61	-1.9	128.35	25JL	2.8	-3.8	-1.5	0.6
350.22	-5.91	-2.1	137.98	4AU	0.8	-3.8	-1.4	0.6
348.87	-6.04	-2.4	147.64	14AU	-0.7	-3.7	-1.4	0.7
346.44	-5.90	-2.5	157.37	24AU	-1.2	-3.6	-1.3	0.7
343.57	-5.49	-2.5	167.14	3SE	-1.2	-3.6	-1.3	0.7
341.04	-4.84	-2.3	176.96	13SE	-0.7	-3.5	-1.3	0.8
339.47	-4.07	-2.0	186.85	23SE	-0.3	-3.5	-1.3	0.8
339.20	-3.29	-1.7	196.80	3OC	-0.2	-3.5	-1.2	0.8
340.21	-2.56	-1.4	206.80	13OC	-0.1	-3.4	-1.2	0.8
342.37	-1.91	-1.1	216.85	23OC	0.1	-3.4	-1.2	0.8
345.48	-1.36	-0.8	226.95	2NO	1.3	-3.4	-1.2	0.7
349.33	-0.90	-0.5	237.08	12NO	2.0	-3.4	-1.3	0.7
353.76	-0.51	-0.2	247.25	22NO	0.1	-3.4	-1.3	0.7
358.64	-0.18	0.0	257.44	2DE	-0.3	-3.4	-1.3	0.7
3.86	0.09	0.2	267.64	12DE	-0.3	-3.4	-1.4	0.7
9.34	0.32	0.4	277.84	22DE	-0.4	-3.4	-1.4	0.7
				1105				
15.03	0.50	0.6	288.02	1JA	-0.6	-3.4	-1.5	0.8
20.86	0.66	0.8	298.19	11JA	-1.0	-3.4	-1.5	0.8
26.81	0.79	1.0	308.33	21JA	-1.1	-3.4	-1.6	0.8
32.84	0.90	1.1	318.42	31JA	-0.9	-3.4	-1.6	0.8
38.93	0.99	1.2	328.47	10FE	0.2	-3.5	-1.7	0.8
45.06	1.06	1.4	338.47	20FE	2.3	-3.5	-1.8	0.8
51.23	1.12	1.5	348.41	2MR	2.4	-3.5	-1.8	0.8
57.42	1.17	1.6	358.30	12MR	1.2	-3.4	-1.9	0.7
63.62	1.20	1.7	8.13	22MR	0.7	-3.4	-2.0	0.7
69.83	1.23	1.7	17.90	1AP	0.4	-3.4	-2.0	0.7
76.04	1.25	1.8	27.61	11AP	-0.1	-3.4	-2.1	0.6
82.26	1.26	1.9	37.28	21AP	-0.8	-3.3	-2.1	0.5
88.48	1.27	1.9	46.91	1MY	-1.8	-3.3	-2.1	0.5
94.70	1.27	2.0	56.50	11MY	-1.2	-3.3	-2.1	0.4
100.93	1.26	2.0	66.06	21MY	-0.3	-3.3	-2.1	0.3
107.17	1.25	2.0	75.60	31MY	0.5	-3.3	-2.0	0.3
113.41	1.24	2.0	85.13	10JN	1.1	-3.3	-2.0	0.2
119.67	1.22	2.0	94.66	20JN	1.8	-3.4	-1.9	0.3
125.94	1.20	2.0	104.19	30JN	3.1	-3.4	-1.9	0.3
132.24	1.17	2.0	113.74	10JL	2.3	-3.4	-1.8	0.4
138.56	1.14	2.0	123.32	20JL	0.6	-3.5	-1.8	0.5
144.90	1.10	2.0	132.92	30JL	-0.7	-3.5	-1.7	0.5
151.28	1.06	2.0	142.57	9AU	-1.4	-3.6	-1.6	0.6
157.69	1.02	2.0	152.26	19AU	-1.2	-3.7	-1.6	0.6
164.14	0.97	1.9	162.01	29AU	-0.6	-3.7	-1.5	0.7
170.63	0.92	1.9	171.81	8SE	-0.2	-3.8	-1.5	0.7
177.17	0.86	1.9	181.66	18SE	-0.0	-3.9	-1.4	0.7
183.76	0.80	1.9	191.57	28SE	0.1	-4.0	-1.4	0.8
190.39	0.73	1.9	201.55	8OC	0.3	-4.2	-1.4	0.8
197.09	0.66	1.9	211.57	18OC	1.6	-4.3	-1.3	0.8
203.83	0.59	1.9	221.64	28OC	1.8	-4.4	-1.3	0.8
210.64	0.50	1.9	231.76	7NO	-0.1	-4.4	-1.3	0.8
217.50	0.42	1.8	241.92	17NO	-0.4	-4.2	-1.3	0.8
224.41	0.32	1.8	252.09	27NO	-0.4	-3.8	-1.3	0.7
231.39	0.22	1.8	262.29	7DE	-0.5	-3.0	-1.3	0.7
238.42	0.11	1.7	272.49	17DE	-0.7	-3.7	-1.4	0.7
245.51	0.00	1.7	282.68	27DE	-0.9	-4.2	-1.4	0.7

♂ LONG	♂ LAT	♂ MAG	☉ LONG	16.00UT 1106	☿ MAGNITUDES	♀ MAGNITUDES	♃ MAGNITUDES	♄ MAGNITUDES
252.66	-0.12	1.6	292.86	6JA	-1.0	-4.4	-1.4	0.8
259.86	-0.25	1.6	303.02	16JA	-0.8	-4.4	-1.5	0.8
267.11	-0.38	1.5	313.13	26JA	0.3	-4.3	-1.5	0.8
274.41	-0.52	1.4	323.21	5FE	2.7	-4.2	-1.6	0.8
281.75	-0.66	1.4	333.24	15FE	1.8	-4.1	-1.6	0.8
289.13	-0.81	1.3	343.21	25FE	0.9	-4.0	-1.7	0.8
296.54	-0.97	1.2	353.12	7MR	0.5	-3.9	-1.8	0.8
303.97	-1.12	1.1	2.98	17MR	0.2	-3.8	-1.8	0.8
311.41	-1.27	1.0	12.78	27MR	-0.2	-3.7	-1.9	0.8
318.85	-1.43	1.0	22.53	6AP	-0.9	-3.6	-2.0	0.7
326.28	-1.57	0.9	32.23	16AP	-1.8	-3.5	-2.0	0.7
333.67	-1.71	0.8	41.87	26AP	-1.2	-3.5	-2.1	0.6
341.02	-1.85	0.7	51.48	6MY	-0.2	-3.4	-2.1	0.6
348.28	-1.97	0.6	61.06	16MY	0.6	-3.4	-2.2	0.5
355.45	-2.08	0.5	70.61	26MY	1.4	-3.4	-2.2	0.4
2.50	-2.17	0.4	80.14	5JN	2.4	-3.3	-2.2	0.4
9.38	-2.25	0.3	89.67	15JN	3.3	-3.3	-2.2	0.3
16.07	-2.30	0.2	99.20	25JN	1.8	-3.3	-2.2	0.3
22.51	-2.34	0.1	108.74	5JL	0.5	-3.3	-2.1	0.3
28.65	-2.35	0.0	118.30	15JL	-0.7	-3.3	-2.1	0.4
34.44	-2.34	-0.1	127.89	25JL	-1.5	-3.3	-2.0	0.4
39.77	-2.30	-0.2	137.52	4AU	-1.2	-3.4	-2.0	0.5
44.55	-2.23	-0.4	147.18	14AU	-0.5	-3.4	-1.9	0.6
48.65	-2.12	-0.6	156.89	24AU	-0.1	-3.4	-1.9	0.6
51.89	-1.96	-0.8	166.67	3SE	0.1	-3.5	-1.8	0.7
54.09	-1.75	-1.0	176.49	13SE	0.2	-3.5	-1.7	0.7
55.04	-1.47	-1.2	186.37	23SE	0.6	-3.5	-1.7	0.7
54.54	-1.11	-1.4	196.31	3OC	2.0	-3.5	-1.6	0.8
52.60	-0.68	-1.6	206.31	13OC	1.6	-3.4	-1.6	0.8
49.47	-0.18	-1.8	216.35	23OC	-0.2	-3.4	-1.5	0.8
45.75	0.33	-1.8	226.45	2NO	-0.6	-3.4	-1.5	0.8
42.30	0.80	-1.6	236.58	12NO	-0.6	-3.4	-1.5	0.8
39.78	1.18	-1.3	246.75	22NO	-0.6	-3.3	-1.4	0.8
38.57	1.46	-1.0	256.94	2DE	-0.8	-3.3	-1.4	0.8
38.71	1.66	-0.6	267.14	12DE	-0.8	-3.3	-1.4	0.8
40.04	1.78	-0.3	277.33	22DE	-0.8	-3.3	-1.4	0.7
				1107				
42.36	1.85	-0.1	287.53	1JA	-0.7	-3.3	-1.4	0.7
45.48	1.89	0.2	297.69	11JA	0.4	-3.4	-1.4	0.8
49.21	1.90	0.4	307.83	21JA	3.0	-3.4	-1.5	0.8
53.43	1.90	0.7	317.93	31JA	1.4	-3.4	-1.5	0.9
58.03	1.88	0.8	327.98	10FE	0.6	-3.4	-1.5	0.9
62.91	1.85	1.0	337.99	20FE	0.3	-3.5	-1.6	0.9
68.03	1.82	1.2	347.93	2MR	0.1	-3.5	-1.6	0.9
73.33	1.78	1.3	357.82	12MR	-0.3	-3.6	-1.7	0.9
78.78	1.73	1.4	7.65	22MR	-0.9	-3.6	-1.7	0.9
84.36	1.69	1.5	17.43	1AP	-1.7	-3.7	-1.8	0.8
90.03	1.64	1.6	27.15	11AP	-1.2	-3.8	-1.9	0.8
95.78	1.58	1.7	36.82	21AP	-0.2	-3.8	-2.0	0.8
101.61	1.53	1.8	46.45	1MY	0.8	-3.9	-2.0	0.7
107.50	1.47	1.8	56.04	11MY	1.8	-4.0	-2.1	0.6
113.45	1.41	1.9	65.60	21MY	3.2	-4.1	-2.2	0.6
119.46	1.35	1.9	75.15	31MY	2.7	-4.2	-2.2	0.5
125.53	1.29	2.0	84.68	10JN	1.4	-4.2	-2.3	0.5
131.64	1.23	2.0	94.20	20JN	0.4	-4.1	-2.3	0.4
137.81	1.16	2.0	103.74	30JN	-0.8	-3.7	-2.3	0.3
144.03	1.09	2.0	113.28	10JL	-1.6	-3.1	-2.3	0.3
150.31	1.02	2.0	122.86	20JL	-1.2	-3.4	-2.3	0.4
156.65	0.95	2.0	132.46	30JL	-0.5	-4.0	-2.3	0.4
163.05	0.87	2.0	142.11	9AU	-0.0	-4.2	-2.3	0.5
169.51	0.79	2.0	151.80	19AU	0.2	-4.2	-2.2	0.5
176.05	0.71	1.9	161.54	29AU	0.4	-4.2	-2.2	0.6
182.64	0.63	1.9	171.33	8SE	0.9	-4.1	-2.1	0.7
189.32	0.54	1.8	181.18	18SE	2.4	-4.1	-2.0	0.7
196.06	0.46	1.8	191.10	28SE	1.4	-4.0	-2.0	0.7
202.89	0.36	1.7	201.06	8OC	-0.4	-3.9	-1.9	0.8
209.79	0.27	1.7	211.08	18OC	-0.7	-3.8	-1.8	0.8
216.77	0.17	1.7	221.15	28OC	-0.7	-3.7	-1.8	0.8
223.82	0.07	1.7	231.27	7NO	-0.7	-3.7	-1.7	0.8
230.96	-0.03	1.7	241.42	17NO	-0.7	-3.6	-1.7	0.8
238.17	-0.14	1.7	251.60	27NO	-0.6	-3.6	-1.6	0.8
245.46	-0.24	1.7	261.79	7DE	-0.7	-3.5	-1.6	0.8
252.82	-0.35	1.6	271.99	17DE	-0.5	-3.5	-1.6	0.8
260.25	-0.46	1.6	282.19	27DE	0.5	-3.4	-1.5	0.8

♂ LONG	♂ LAT	♂ MAG	⊙ LONG	16.00UT 1108	☿	♀	♃	♄
267.75	-0.57	1.6	292.36	6JA	3.0	-3.4	-1.5	0.8
275.30	-0.68	1.5	302.52	16JA	1.0	-3.4	-1.5	0.8
282.91	-0.78	1.5	312.64	26JA	0.4	-3.3	-1.5	0.9
290.56	-0.88	1.5	322.72	5FE	0.2	-3.3	-1.5	0.9
298.25	-0.98	1.4	332.75	15FE	-0.0	-3.3	-1.5	0.9
305.97	-1.07	1.4	342.73	25FE	-0.3	-3.3	-1.6	0.9
313.70	-1.16	1.4	352.64	6MR	-0.9	-3.3	-1.6	1.0
321.44	-1.23	1.3	2.51	16MR	-1.6	-3.3	-1.6	1.0
329.16	-1.29	1.3	12.31	26MR	-1.2	-3.4	-1.7	0.9
336.87	-1.35	1.3	22.06	5AP	-0.1	-3.4	-1.7	0.9
344.55	-1.38	1.2	31.76	15AP	1.1	-3.4	-1.8	0.9
352.17	-1.41	1.2	41.41	25AP	2.4	-3.5	-1.8	0.9
359.73	-1.42	1.2	51.01	5MY	3.6	-3.5	-1.9	0.8
7.22	-1.41	1.1	60.59	15MY	2.1	-3.5	-2.0	0.8
14.61	-1.39	1.1	70.14	25MY	1.1	-3.4	-2.0	0.7
21.90	-1.35	1.1	79.68	4JN	0.2	-3.4	-2.1	0.6
29.07	-1.30	1.0	89.21	14JN	-0.8	-3.4	-2.2	0.6
36.09	-1.23	1.0	98.74	24JN	-1.6	-3.3	-2.2	0.5
42.97	-1.14	0.9	108.27	4JL	-1.2	-3.3	-2.3	0.4
49.67	-1.03	0.9	117.84	14JL	-0.4	-3.3	-2.4	0.4
56.17	-0.91	0.8	127.42	24JL	0.0	-3.3	-2.4	0.4
62.45	-0.77	0.8	137.04	3AU	0.4	-3.3	-2.4	0.5
68.47	-0.62	0.7	146.71	13AU	0.6	-3.3	-2.4	0.5
74.19	-0.44	0.6	156.42	23AU	1.2	-3.4	-2.4	0.6
79.57	-0.24	0.5	166.19	2SE	2.8	-3.4	-2.4	0.6
84.52	-0.01	0.4	176.01	12SE	1.3	-3.4	-2.4	0.7
88.95	0.25	0.2	185.89	22SE	-0.5	-3.5	-2.3	0.7
92.78	0.54	0.1	195.83	2OC	-0.9	-3.5	-2.3	0.8
95.83	0.88	-0.1	205.82	12OC	-0.9	-3.6	-2.2	0.8
97.93	1.26	-0.3	215.87	22OC	-0.9	-3.6	-2.1	0.8
98.90	1.68	-0.5	225.96	1NO	-0.6	-3.7	-2.1	0.9
98.54	2.15	-0.7	236.09	11NO	-0.5	-3.8	-2.0	0.9
96.78	2.62	-1.0	246.26	21NO	-0.5	-3.9	-1.9	0.9
93.75	3.06	-1.1	256.44	1DE	-0.4	-4.0	-1.9	0.9
89.92	3.40	-1.3	266.64	11DE	0.7	-4.1	-1.8	0.9
86.03	3.60	-1.1	276.84	21DE	2.8	-4.2	-1.7	0.9
82.82	3.64	-0.9	287.03	31DE	0.7	-4.3	-1.7	0.9
				1109				
80.78	3.56	-0.6	297.20	10JA	0.2	-4.4	-1.7	0.9
80.07	3.41	-0.3	307.34	20JA	0.0	-4.4	-1.6	0.9
80.62	3.22	-0.0	317.44	30JA	-0.1	-4.2	-1.6	0.9
82.24	3.02	0.2	327.50	9FE	-0.4	-3.8	-1.6	1.0
84.74	2.82	0.4	337.50	19FE	-0.9	-3.3	-1.6	1.0
87.94	2.63	0.7	347.45	1MR	-1.5	-3.7	-1.6	1.0
91.70	2.45	0.8	357.34	11MR	-1.2	-4.1	-1.6	1.1
95.89	2.29	1.0	7.17	21MR	-0.1	-4.2	-1.6	1.1
100.44	2.13	1.2	16.95	31MR	1.4	-4.2	-1.6	1.1
105.28	1.98	1.3	26.68	10AP	3.1	-4.2	-1.6	1.1
110.35	1.84	1.4	36.35	20AP	2.7	-4.1	-1.7	1.0
115.62	1.71	1.5	45.98	30AP	1.6	-4.0	-1.7	1.0
121.05	1.58	1.6	55.58	10MY	0.8	-3.9	-1.8	1.0
126.63	1.46	1.6	65.14	20MY	0.1	-3.8	-1.8	0.9
132.34	1.34	1.7	74.68	30MY	-0.8	-3.7	-1.9	0.9
138.17	1.23	1.8	84.22	9JN	-1.7	-3.6	-1.9	0.8
144.11	1.12	1.8	93.74	19JN	-1.2	-3.5	-2.0	0.7
150.15	1.00	1.8	103.28	29JN	-0.4	-3.5	-2.1	0.7
156.29	0.90	1.8	112.83	9JL	0.1	-3.5	-2.1	0.6
162.53	0.79	1.9	122.39	19JL	0.5	-3.4	-2.2	0.5
168.87	0.68	1.9	132.00	29JL	0.9	-3.4	-2.3	0.5
175.30	0.57	1.9	141.64	8AU	1.6	-3.4	-2.3	0.5
181.82	0.47	1.8	151.33	18AU	3.2	-3.4	-2.4	0.6
188.44	0.36	1.8	161.07	28AU	1.1	-3.4	-2.4	0.6
195.16	0.25	1.8	170.86	7SE	-0.6	-3.4	-2.4	0.7
201.98	0.15	1.8	180.71	17SE	-1.0	-3.4	-2.5	0.7
208.89	0.04	1.7	190.62	27SE	-1.0	-3.4	-2.5	0.8
215.90	-0.06	1.7	200.58	7OC	-0.8	-3.4	-2.4	0.8
223.01	-0.17	1.7	210.60	17OC	-0.5	-3.4	-2.4	0.9
230.21	-0.27	1.6	220.67	27OC	-0.4	-3.4	-2.4	0.9
237.50	-0.37	1.6	230.78	6NO	-0.3	-3.4	-2.3	0.9
244.88	-0.47	1.5	240.93	16NO	-0.2	-3.5	-2.2	1.0
252.35	-0.56	1.5	251.10	26NO	0.9	-3.5	-2.2	1.0
259.90	-0.65	1.4	261.29	6DE	2.5	-3.5	-2.1	1.0
267.52	-0.74	1.4	271.49	16DE	0.4	-3.5	-2.0	1.0
275.20	-0.82	1.4	281.69	26DE	-0.0	-3.4	-1.9	1.0

♂ LONG	♂ LAT	♂ MAG	⊙ LONG	16.00UT 1110	☿	♀	♃	♄
282.94	-0.89	1.4	291.87	5JA	-0.1	-3.4	-1.9	1.0
290.72	-0.96	1.4	302.02	15JA	-0.2	-3.4	-1.8	1.0
298.54	-1.02	1.4	312.15	25JA	-0.5	-3.4	-1.8	1.0
306.39	-1.06	1.4	322.23	4FE	-1.0	-3.4	-1.7	1.0
314.24	-1.10	1.4	332.25	14FE	-1.4	-3.3	-1.7	1.1
322.10	-1.13	1.4	342.24	24FE	-1.1	-3.3	-1.6	1.1
329.94	-1.14	1.4	352.16	6MR	-0.0	-3.3	-1.6	1.1
337.75	-1.14	1.4	2.02	16MR	1.7	-3.3	-1.6	1.2
345.53	-1.13	1.4	11.83	26MR	3.6	-3.4	-1.6	1.2
353.26	-1.10	1.4	21.58	5AP	2.0	-3.4	-1.6	1.2
0.92	-1.07	1.4	31.28	15AP	1.2	-3.4	-1.6	1.2
8.52	-1.02	1.4	40.94	25AP	0.6	-3.4	-1.6	1.2
16.04	-0.95	1.4	50.55	5MY	0.0	-3.5	-1.6	1.2
23.47	-0.88	1.4	60.13	15MY	-0.8	-3.5	-1.6	1.1
30.81	-0.80	1.4	69.68	25MY	-1.7	-3.6	-1.6	1.1
38.04	-0.71	1.4	79.22	4JN	-1.2	-3.6	-1.7	1.0
45.17	-0.60	1.4	88.75	14JN	-0.4	-3.7	-1.7	1.0
52.18	-0.49	1.4	98.28	24JN	0.3	-3.8	-1.8	0.9
59.07	-0.37	1.4	107.82	4JL	0.7	-3.9	-1.8	0.9
65.83	-0.24	1.4	117.37	14JL	1.1	-4.0	-1.9	0.8
72.45	-0.11	1.4	126.96	24JL	2.1	-4.1	-1.9	0.7
78.93	0.04	1.3	136.58	3AU	3.1	-4.2	-2.0	0.6
85.24	0.19	1.3	146.24	13AU	0.9	-4.2	-2.1	0.6
91.39	0.36	1.3	155.96	23AU	-0.6	-4.2	-2.1	0.7
97.33	0.53	1.2	165.72	2SE	-1.2	-4.1	-2.2	0.7
103.06	0.72	1.1	175.54	12SE	-1.2	-3.7	-2.3	0.7
108.53	0.92	1.0	185.42	22SE	-0.7	-3.2	-2.3	0.8
113.71	1.14	0.9	195.35	2OC	-0.4	-3.7	-2.4	0.8
118.53	1.37	0.8	205.34	12OC	-0.2	-4.2	-2.4	0.9
122.93	1.64	0.7	215.38	22OC	-0.2	-4.3	-2.4	0.9
126.80	1.93	0.5	225.47	1NO	-0.0	-4.4	-2.4	1.0
130.04	2.25	0.4	235.60	11NO	1.1	-4.3	-2.4	1.0
132.48	2.60	0.2	245.77	21NO	2.3	-4.2	-2.3	1.1
133.96	2.99	-0.1	255.95	1DE	0.2	-4.1	-2.3	1.1
134.28	3.40	-0.3	266.15	11DE	-0.2	-4.0	-2.2	1.1
133.28	3.81	-0.5	276.35	21DE	-0.2	-3.9	-2.2	1.1
130.97	4.18	-0.8	286.53	31DE	-0.3	-3.8	-2.1	1.2
				1111				
127.56	4.44	-0.9	296.71	10JA	-0.6	-3.7	-2.0	1.2
123.61	4.53	-1.0	306.85	20JA	-1.0	-3.7	-2.0	1.2
119.86	4.45	-0.8	316.95	30JA	-1.2	-3.6	-1.9	1.2
116.94	4.22	-0.6	327.01	9FE	-1.0	-3.5	-1.8	1.2
115.24	3.91	-0.4	337.02	19FE	0.1	-3.5	-1.8	1.2
114.84	3.55	-0.1	346.97	1MR	2.1	-3.4	-1.7	1.2
115.63	3.20	0.1	356.86	11MR	2.9	-3.4	-1.6	1.3
117.45	2.87	0.3	6.70	21MR	1.5	-3.3	-1.6	1.3
120.10	2.56	0.5	16.48	31MR	0.9	-3.3	-1.6	1.3
123.42	2.28	0.7	26.21	10AP	0.5	-3.3	-1.5	1.4
127.29	2.02	0.9	35.89	20AP	-0.1	-3.3	-1.5	1.4
131.60	1.79	1.0	45.51	30AP	-0.8	-3.3	-1.5	1.4
136.27	1.57	1.1	55.11	10MY	-1.7	-3.3	-1.5	1.4
141.25	1.37	1.2	64.68	20MY	-1.3	-3.3	-1.5	1.3
146.49	1.19	1.3	74.22	30MY	-0.3	-3.3	-1.5	1.3
151.95	1.02	1.4	83.75	9JN	0.4	-3.4	-1.5	1.3
157.61	0.85	1.5	93.28	19JN	0.9	-3.4	-1.5	1.2
163.45	0.70	1.5	102.81	29JN	1.5	-3.5	-1.5	1.1
169.46	0.55	1.5	112.36	9JL	2.7	-3.5	-1.6	1.1
175.62	0.40	1.6	121.93	19JL	2.6	-3.5	-1.6	1.0
181.92	0.27	1.6	131.53	29JL	0.8	-3.4	-1.7	0.9
188.36	0.14	1.6	141.17	8AU	-0.7	-3.4	-1.7	0.9
194.94	0.01	1.6	150.86	18AU	-1.3	-3.4	-1.8	0.8
201.64	-0.11	1.6	160.59	28AU	-1.2	-3.4	-1.8	0.8
208.47	-0.23	1.6	170.38	7SE	-0.6	-3.3	-1.9	0.8
215.43	-0.34	1.6	180.23	17SE	-0.3	-3.3	-2.0	0.9
222.50	-0.44	1.6	190.13	27SE	-0.1	-3.3	-2.0	0.9
229.69	-0.54	1.6	200.09	7OC	-0.0	-3.3	-2.1	1.0
236.99	-0.64	1.6	210.11	17OC	0.2	-3.3	-2.2	1.0
244.39	-0.72	1.5	220.17	27OC	1.4	-3.4	-2.2	1.1
251.90	-0.80	1.5	230.29	6NO	2.0	-3.4	-2.2	1.1
259.49	-0.87	1.5	240.43	16NO	0.0	-3.4	-2.3	1.2
267.15	-0.93	1.5	250.60	26NO	-0.3	-3.4	-2.3	1.2
274.90	-0.98	1.4	260.80	6DE	-0.4	-3.5	-2.3	1.2
282.69	-1.03	1.4	271.00	16DE	-0.4	-3.5	-2.3	1.3
290.54	-1.06	1.4	281.19	26DE	-0.6	-3.6	-2.2	1.3

♂ LONG	♂ LAT	♂ MAG	⊙ LONG	16.00UT	☿	♀	♃	♄
				1112	MAGNITUDES			
298.42	-1.08	1.3	291.38	5JA	-1.0	-3.6	-2.2	1.3
306.32	-1.09	1.3	301.53	15JA	-1.1	-3.7	-2.1	1.3
314.22	-1.08	1.3	311.66	25JA	-0.9	-3.8	-2.1	1.3
322.12	-1.07	1.3	321.74	4FE	0.1	-3.9	-2.0	1.4
330.00	-1.04	1.3	331.78	14FE	2.4	-4.0	-1.9	1.4
337.85	-1.00	1.3	341.76	24FE	2.2	-4.0	-1.8	1.3
345.65	-0.95	1.4	351.68	5MR	1.1	-4.1	-1.8	1.3
353.40	-0.89	1.4	1.55	15MR	0.6	-4.2	-1.7	1.3
1.08	-0.83	1.4	11.36	25MR	0.3	-4.2	-1.6	1.3
8.70	-0.75	1.5	21.11	4AP	-0.1	-4.2	-1.6	1.3
16.23	-0.67	1.5	30.81	14AP	-0.8	-3.9	-1.5	1.3
23.68	-0.57	1.5	40.47	24AP	-1.7	-3.3	-1.5	1.3
31.04	-0.48	1.6	50.08	4MY	-1.3	-2.9	-1.5	1.3
38.31	-0.38	1.6	59.66	14MY	-0.3	-3.7	-1.4	1.2
45.49	-0.27	1.6	69.22	24MY	0.5	-4.1	-1.4	1.2
52.57	-0.16	1.6	78.76	3JN	1.2	-4.2	-1.4	1.1
59.56	-0.05	1.7	88.28	13JN	2.0	-4.2	-1.4	1.1
66.45	0.07	1.7	97.81	23JN	3.3	-4.1	-1.4	1.1
73.24	0.18	1.7	107.35	3JL	2.2	-4.0	-1.4	1.0
79.93	0.30	1.7	116.91	13JL	0.6	-3.9	-1.4	1.0
86.52	0.42	1.7	126.49	23JL	-0.7	-3.8	-1.4	0.9
93.01	0.55	1.7	136.11	2AU	-1.4	-3.8	-1.4	0.9
99.39	0.67	1.7	145.77	12AU	-1.2	-3.7	-1.5	0.8
105.67	0.80	1.7	155.48	22AU	-0.6	-3.6	-1.5	0.8
111.82	0.94	1.6	165.24	1SE	-0.2	-3.6	-1.6	0.7
117.85	1.07	1.6	175.06	11SE	0.0	-3.5	-1.6	0.8
123.73	1.21	1.5	184.93	21SE	0.1	-3.5	-1.7	0.8
129.46	1.36	1.5	194.87	1OC	0.4	-3.5	-1.7	0.9
135.00	1.52	1.4	204.85	11OC	1.7	-3.4	-1.8	1.0
140.34	1.68	1.3	214.89	21OC	1.8	-3.4	-1.9	1.0
145.42	1.86	1.2	224.98	31OC	-0.1	-3.4	-1.9	1.1
150.20	2.05	1.1	235.11	10NO	-0.5	-3.4	-2.0	1.2
154.63	2.25	0.9	245.27	20NO	-0.5	-3.4	-2.0	1.2
158.61	2.48	0.8	255.46	30NO	-0.5	-3.4	-2.1	1.2
162.04	2.72	0.6	265.65	10DE	-0.7	-3.4	-2.1	1.3
164.79	2.98	0.4	275.85	20DE	-0.9	-3.4	-2.2	1.3
166.69	3.27	0.1	286.04	30DE	-0.9	-3.4	-2.2	1.3
				1113				
167.58	3.56	-0.1	296.21	9JA	-0.8	-3.4	-2.2	1.3
167.27	3.84	-0.4	306.36	19JA	0.2	-3.4	-2.1	1.2
165.66	4.07	-0.6	316.46	29JA	2.8	-3.4	-2.1	1.2
162.85	4.21	-0.9	326.52	8FE	1.7	-3.5	-2.0	1.2
159.17	4.20	-1.0	336.53	18FE	0.8	-3.5	-2.0	1.1
155.28	4.03	-1.0	346.49	28FE	0.4	-3.5	-1.9	1.1
151.90	3.71	-0.8	356.38	10MR	0.2	-3.4	-1.9	1.0
149.54	3.30	-0.6	6.22	20MR	-0.2	-3.4	-1.8	1.0
148.47	2.86	-0.4	16.00	30MR	-0.8	-3.4	-1.7	1.0
148.67	2.43	-0.2	25.73	9AP	-1.7	-3.4	-1.7	1.0
150.01	2.03	0.1	35.41	19AP	-1.3	-3.3	-1.6	1.0
152.32	1.67	0.2	45.05	29AP	-0.2	-3.3	-1.5	1.0
155.42	1.34	0.4	54.64	9MY	0.7	-3.3	-1.5	1.0
159.16	1.05	0.6	64.21	19MY	1.5	-3.3	-1.4	1.0
163.45	0.79	0.7	73.75	29MY	2.7	-3.3	-1.4	0.9
168.17	0.55	0.8	83.28	8JN	3.2	-3.3	-1.4	0.9
173.27	0.34	0.9	92.81	18JN	1.7	-3.4	-1.3	0.9
178.69	0.14	1.0	102.35	28JN	0.5	-3.4	-1.3	0.8
184.38	-0.04	1.1	111.89	8JL	-0.7	-3.4	-1.3	0.8
190.33	-0.20	1.1	121.46	18JL	-1.5	-3.5	-1.3	0.7
196.51	-0.35	1.2	131.06	28JL	-1.2	-3.5	-1.3	0.7
202.88	-0.49	1.2	140.70	7AU	-0.5	-3.6	-1.3	0.6
209.45	-0.62	1.2	150.39	17AU	-0.1	-3.7	-1.3	0.6
216.19	-0.73	1.3	160.12	27AU	0.1	-3.7	-1.3	0.5
223.09	-0.83	1.3	169.90	6SE	0.3	-3.8	-1.4	0.5
230.14	-0.92	1.3	179.75	16SE	0.7	-3.9	-1.4	0.4
237.33	-1.00	1.3	189.65	26SE	2.1	-4.1	-1.4	0.5
244.65	-1.07	1.3	199.61	6OC	1.6	-4.2	-1.5	0.6
252.10	-1.12	1.3	209.63	16OC	-0.3	-4.3	-1.5	0.6
259.64	-1.16	1.3	219.69	26OC	-0.7	-4.4	-1.6	0.7
267.28	-1.19	1.4	229.79	5NO	-0.7	-4.4	-1.6	0.8
275.00	-1.21	1.4	239.94	15NO	-0.7	-4.2	-1.7	0.8
282.78	-1.21	1.4	250.11	25NO	-0.8	-3.8	-1.8	0.9
290.61	-1.20	1.4	260.30	5DE	-0.7	-2.9	-1.8	0.9
298.48	-1.18	1.4	270.50	15DE	-0.7	-3.7	-1.9	1.0
306.37	-1.14	1.4	280.70	25DE	-0.6	-4.2	-2.0	1.0

♂ LONG	♂ LAT	♂ MAG	⊙ LONG	16.00UT	☿	♀	♃	♄
				1114	MAGNITUDES			
314.26	-1.10	1.4	290.88	4JA	0.4	-4.4	-2.0	1.0
322.14	-1.04	1.4	301.04	14JA	3.0	-4.4	-2.0	1.0
329.99	-0.97	1.4	311.17	24JA	1.2	-4.3	-2.1	1.0
337.81	-0.90	1.4	321.25	3FE	0.5	-4.2	-2.1	1.0
345.58	-0.81	1.4	331.29	13FE	0.3	-4.1	-2.0	0.9
353.29	-0.73	1.5	341.27	23FE	0.1	-4.0	-2.0	0.9
0.94	-0.63	1.5	351.20	5MR	-0.3	-3.8	-2.0	0.8
8.51	-0.53	1.5	1.07	15MR	-0.9	-3.8	-1.9	0.8
16.00	-0.43	1.5	10.88	25MR	-1.7	-3.7	-1.9	0.8
23.41	-0.33	1.5	20.64	4AP	-1.2	-3.6	-1.8	0.8
30.74	-0.22	1.5	30.35	14AP	-0.2	-3.5	-1.8	0.8
37.97	-0.12	1.6	40.00	24AP	0.9	-3.5	-1.7	0.8
45.12	-0.01	1.6	49.62	4MY	2.0	-3.4	-1.6	0.8
52.19	0.09	1.7	59.20	14MY	3.5	-3.4	-1.6	0.8
59.16	0.20	1.7	68.76	24MY	2.5	-3.4	-1.5	0.8
66.06	0.30	1.7	78.30	3JN	1.3	-3.3	-1.4	0.7
72.88	0.40	1.8	87.83	13JN	0.3	-3.3	-1.4	0.7
79.62	0.50	1.8	97.35	23JN	-0.7	-3.3	-1.3	0.7
86.28	0.59	1.8	106.89	3JL	-1.6	-3.3	-1.3	0.6
92.88	0.69	1.9	116.45	13JL	-1.2	-3.3	-1.3	0.6
99.40	0.78	1.9	126.03	23JL	-0.5	-3.3	-1.3	0.6
105.85	0.87	1.9	135.65	2AU	-0.0	-3.4	-1.2	0.5
112.24	0.96	1.9	145.31	12AU	0.3	-3.4	-1.2	0.4
118.55	1.05	1.9	155.01	22AU	0.5	-3.4	-1.2	0.4
124.80	1.14	1.9	164.77	1SE	1.0	-3.5	-1.2	0.3
130.98	1.23	1.9	174.58	11SE	2.5	-3.5	-1.2	0.3
137.07	1.32	1.8	184.45	21SE	1.5	-3.5	-1.2	0.2
143.09	1.40	1.8	194.38	1OC	-0.4	-3.5	-1.3	0.2
149.02	1.49	1.7	204.37	11OC	-0.8	-3.4	-1.3	0.2
154.84	1.58	1.7	214.40	21OC	-0.8	-3.4	-1.3	0.3
160.55	1.67	1.6	224.49	31OC	-0.8	-3.4	-1.4	0.4
166.12	1.77	1.5	234.62	10NO	-0.7	-3.4	-1.4	0.5
171.53	1.86	1.4	244.77	20NO	-0.6	-3.3	-1.5	0.5
176.74	1.96	1.3	254.96	30NO	-0.6	-3.3	-1.6	0.6
181.73	2.06	1.2	265.16	10DE	-0.5	-3.3	-1.6	0.6
186.43	2.17	1.0	275.35	20DE	0.5	-3.3	-1.7	0.7
190.78	2.27	0.8	285.55	30DE	3.0	-3.3	-1.8	0.7
				1115				
194.70	2.39	0.6	295.72	9JA	0.9	-3.4	-1.8	0.7
198.09	2.49	0.4	305.86	19JA	0.3	-3.4	-1.9	0.7
200.82	2.60	0.2	315.97	29JA	0.1	-3.4	-1.9	0.7
202.71	2.69	-0.1	326.03	8FE	-0.0	-3.4	-2.0	0.7
203.59	2.76	-0.4	336.04	18FE	-0.3	-3.5	-2.0	0.7
203.30	2.78	-0.7	346.00	28FE	-0.9	-3.5	-2.0	0.7
201.73	2.73	-0.9	355.91	10MR	-1.6	-3.5	-2.0	0.7
198.99	2.57	-1.2	5.75	20MR	-1.2	-3.6	-2.0	0.6
195.43	2.30	-1.4	15.54	30MR	-0.1	-3.7	-2.0	0.6
191.72	1.91	-1.3	25.27	9AP	1.1	-3.8	-1.9	0.5
188.57	1.46	-1.2	34.95	19AP	2.7	-3.8	-1.9	0.6
186.50	1.00	-1.0	44.59	29AP	3.3	-3.9	-1.8	0.6
185.76	0.56	-0.8	54.18	9MY	1.9	-4.0	-1.8	0.6
186.33	0.17	-0.6	63.75	19MY	1.0	-4.1	-1.7	0.6
188.07	-0.17	-0.4	73.30	29MY	0.2	-4.2	-1.6	0.6
190.80	-0.46	-0.2	82.83	8JN	-0.8	-4.2	-1.6	0.6
194.36	-0.71	-0.0	92.35	18JN	-1.6	-4.1	-1.5	0.6
198.60	-0.92	0.1	101.89	28JN	-1.2	-3.7	-1.4	0.5
203.41	-1.09	0.2	111.43	8JL	-0.4	-3.0	-1.4	0.5
208.69	-1.24	0.4	121.00	18JL	0.1	-3.4	-1.4	0.5
214.37	-1.37	0.5	130.60	28JL	0.4	-4.0	-1.3	0.4
220.39	-1.46	0.5	140.23	7AU	0.7	-4.2	-1.3	0.4
226.70	-1.54	0.6	149.91	17AU	1.3	-4.2	-1.2	0.3
233.28	-1.60	0.7	159.64	27AU	2.9	-4.2	-1.2	0.3
240.08	-1.64	0.8	169.43	6SE	1.3	-4.1	-1.2	0.2
247.07	-1.66	0.8	179.27	16SE	-0.5	-4.0	-1.2	0.1
254.24	-1.66	0.9	189.17	26SE	-1.0	-4.0	-1.2	0.1
261.56	-1.65	0.9	199.12	6OC	-0.9	-3.9	-1.2	0.0
269.00	-1.61	1.0	209.13	16OC	-0.8	-3.8	-1.2	-0.0
276.55	-1.57	1.0	219.20	26OC	-0.5	-3.7	-1.3	0.0
284.18	-1.51	1.1	229.30	5NO	-0.4	-3.7	-1.3	0.1
291.87	-1.43	1.1	239.45	15NO	-0.4	-3.6	-1.3	0.2
299.60	-1.35	1.2	249.62	25NO	-0.3	-3.5	-1.4	0.3
307.36	-1.25	1.2	259.81	5DE	0.7	-3.5	-1.4	0.3
315.13	-1.15	1.3	270.01	15DE	2.8	-3.5	-1.5	0.4
322.89	-1.04	1.3	280.21	25DE	0.6	-3.4	-1.5	0.4

♂ LONG	♂ LAT	♂ MAG	⊙ LONG	16.00UT	☿	♀	♃	♄
				1116	MAGNITUDES			
330.62	-0.92	1.4	290.39	4JA	0.1	-3.4	-1.6	0.5
338.32	-0.81	1.4	300.55	14JA	-0.0	-3.4	-1.7	0.5
345.97	-0.69	1.4	310.68	24JA	-0.1	-3.3	-1.7	0.5
353.56	-0.56	1.5	320.77	3FE	-0.4	-3.3	-1.8	0.6
1.09	-0.44	1.5	330.81	13FE	-0.9	-3.3	-1.9	0.6
8.54	-0.32	1.6	340.80	23FE	-1.5	-3.3	-1.9	0.6
15.92	-0.21	1.6	350.72	4MR	-1.2	-3.3	-2.0	0.5
23.22	-0.09	1.6	0.60	14MR	-0.1	-3.3	-2.0	0.5
30.44	0.02	1.7	10.41	24MR	1.4	-3.4	-2.0	0.5
37.58	0.13	1.7	20.17	3AP	3.4	-3.4	-2.0	0.4
44.63	0.23	1.7	29.88	13AP	2.5	-3.4	-2.0	0.4
51.62	0.33	1.7	39.54	23AP	1.4	-3.5	-2.0	0.4
58.52	0.42	1.8	49.15	3MY	0.8	-3.5	-2.0	0.4
65.36	0.51	1.8	58.74	13MY	0.1	-3.5	-1.9	0.4
72.13	0.60	1.8	68.30	23MY	-0.8	-3.4	-1.9	0.4
78.83	0.68	1.8	77.83	2JN	-1.7	-3.4	-1.8	0.4
85.47	0.75	1.8	87.36	12JN	-1.3	-3.4	-1.8	0.5
92.06	0.82	1.9	96.89	22JN	-0.4	-3.3	-1.7	0.4
98.60	0.89	1.9	106.42	2JL	0.2	-3.3	-1.6	0.4
105.09	0.95	1.9	115.98	12JL	0.6	-3.3	-1.6	0.4
111.54	1.01	2.0	125.56	22JL	1.0	-3.3	-1.5	0.4
117.95	1.07	2.0	135.17	1AU	1.7	-3.3	-1.5	0.3
124.32	1.12	2.0	144.83	11AU	3.2	-3.3	-1.4	0.3
130.66	1.17	2.0	154.53	21AU	1.1	-3.4	-1.4	0.2
136.97	1.21	2.0	164.29	31AU	-0.5	-3.4	-1.3	0.2
143.25	1.25	2.0	174.10	10SE	-1.1	-3.4	-1.3	0.1
149.49	1.29	2.0	183.97	20SE	-1.1	-3.5	-1.3	0.1
155.70	1.33	2.0	193.89	30SE	-0.8	-3.5	-1.3	-0.0
161.89	1.36	1.9	203.88	10OC	-0.4	-3.6	-1.2	-0.1
168.03	1.39	1.9	213.92	20OC	-0.3	-3.6	-1.2	-0.1
174.13	1.41	1.8	224.00	30OC	-0.3	-3.7	-1.2	-0.2
180.18	1.43	1.8	234.13	9NO	-0.1	-3.8	-1.3	-0.1
186.18	1.45	1.7	244.28	19NO	0.9	-3.9	-1.3	-0.0
192.11	1.46	1.6	254.46	29NO	2.5	-4.0	-1.3	0.0
197.97	1.47	1.5	264.66	9DE	0.4	-4.1	-1.3	0.1
203.73	1.47	1.4	274.86	19DE	-0.1	-4.2	-1.4	0.2
209.38	1.46	1.3	285.05	29DE	-0.1	-4.3	-1.4	0.2
				1117				
214.89	1.44	1.1	295.23	8JA	-0.3	-4.4	-1.5	0.3
220.23	1.41	1.0	305.38	18JA	-0.5	-4.4	-1.5	0.3
225.37	1.36	0.8	315.48	28JA	-1.0	-4.2	-1.6	0.4
230.25	1.29	0.6	325.55	7FE	-1.3	-3.8	-1.7	0.4
234.82	1.20	0.4	335.56	17FE	-1.1	-3.3	-1.7	0.4
239.00	1.08	0.2	345.52	27FE	-0.0	-3.7	-1.8	0.4
242.68	0.91	-0.1	355.43	9MR	1.8	-4.1	-1.9	0.4
245.73	0.70	-0.3	5.27	19MR	3.3	-4.3	-1.9	0.4
248.00	0.42	-0.6	15.06	29MR	1.8	-4.2	-2.0	0.4
249.31	0.07	-0.9	24.80	8AP	1.0	-4.2	-2.0	0.4
249.47	-0.37	-1.3	34.48	18AP	0.6	-4.1	-2.1	0.3
248.40	-0.89	-1.6	44.12	28AP	0.0	-4.0	-2.1	0.3
246.14	-1.48	-1.9	53.72	8MY	-0.8	-3.9	-2.1	0.3
243.08	-2.07	-2.1	63.29	18MY	-1.7	-3.8	-2.1	0.3
239.84	-2.61	-2.0	72.83	28MY	-1.3	-3.7	-2.1	0.3
237.15	-3.03	-1.9	82.37	7JN	-0.4	-3.6	-2.0	0.3
235.59	-3.31	-1.7	91.90	17JN	0.3	-3.5	-2.0	0.4
235.38	-3.47	-1.5	101.43	27JN	0.8	-3.5	-1.9	0.4
236.54	-3.53	-1.2	110.97	7JL	1.3	-3.4	-1.9	0.4
238.93	-3.52	-1.0	120.54	17JL	2.3	-3.4	-1.8	0.4
242.33	-3.45	-0.8	130.13	27JL	3.0	-3.4	-1.8	0.3
246.58	-3.34	-0.6	139.77	6AU	0.9	-3.4	-1.7	0.3
251.51	-3.21	-0.5	149.44	16AU	-0.6	-3.4	-1.6	0.3
256.99	-3.06	-0.3	159.17	26AU	-1.2	-3.4	-1.6	0.2
262.93	-2.89	-0.2	168.96	5SE	-1.2	-3.3	-1.5	0.2
269.22	-2.70	-0.0	178.79	15SE	-0.7	-3.4	-1.5	0.1
275.79	-2.51	0.1	188.69	25SE	-0.3	-3.4	-1.4	0.0
282.60	-2.31	0.2	198.64	5OC	-0.2	-3.4	-1.4	-0.0
289.59	-2.10	0.4	208.65	15OC	-0.1	-3.4	-1.4	-0.1
296.71	-1.89	0.5	218.71	25OC	0.1	-3.4	-1.3	-0.2
303.93	-1.68	0.6	228.81	4NO	1.1	-3.4	-1.3	-0.2
311.22	-1.48	0.7	238.95	14NO	2.3	-3.5	-1.3	-0.2
318.56	-1.27	0.8	249.12	24NO	0.2	-3.5	-1.3	-0.2
325.91	-1.08	0.9	259.32	4DE	-0.2	-3.5	-1.3	-0.1
333.27	-0.89	1.0	269.51	14DE	-0.3	-3.5	-1.4	-0.0
340.62	-0.70	1.1	279.71	24DE	-0.4	-3.4	-1.4	0.0

♂ LONG	♂ LAT	♂ MAG	⊙ LONG	16.00UT	☿	♀	♃	♄
				1118	MAGNITUDES			
347.93	-0.53	1.2	289.90	3JA	-0.6	-3.4	-1.4	0.1
355.21	-0.37	1.3	300.06	13JA	-1.0	-3.4	-1.4	0.2
2.44	-0.21	1.4	310.19	23JA	-1.2	-3.4	-1.5	0.2
9.62	-0.07	1.4	320.28	2FE	-1.0	-3.4	-1.5	0.3
16.74	0.06	1.5	330.32	12FE	0.0	-3.3	-1.6	0.3
23.80	0.19	1.6	340.31	22FE	2.1	-3.3	-1.7	0.3
30.80	0.30	1.6	350.24	4MR	2.6	-3.3	-1.7	0.3
37.73	0.41	1.7	0.12	14MR	1.3	-3.3	-1.8	0.3
44.60	0.50	1.8	9.94	24MR	0.8	-3.4	-1.9	0.3
51.41	0.59	1.8	19.70	3AP	0.4	-3.4	-1.9	0.3
58.16	0.67	1.8	29.41	13AP	-0.1	-3.4	-2.0	0.3
64.85	0.75	1.9	39.07	23AP	-0.8	-3.4	-2.1	0.3
71.49	0.81	1.9	48.69	3MY	-1.7	-3.5	-2.1	0.2
78.09	0.87	1.9	58.28	13MY	-1.3	-3.5	-2.2	0.2
84.64	0.93	1.9	67.84	23MY	-0.3	-3.6	-2.2	0.2
91.15	0.97	2.0	77.38	2JN	0.4	-3.6	-2.2	0.2
97.62	1.02	2.0	86.90	12JN	1.0	-3.7	-2.2	0.3
104.07	1.05	2.0	96.43	22JN	1.7	-3.8	-2.2	0.3
110.49	1.08	1.9	105.97	2JL	2.9	-3.9	-2.2	0.3
116.89	1.11	1.9	115.52	12JL	2.5	-4.0	-2.2	0.3
123.28	1.13	1.9	125.10	22JL	0.8	-4.1	-2.1	0.3
129.65	1.15	2.0	134.72	1AU	-0.6	-4.2	-2.0	0.3
136.02	1.16	2.0	144.37	11AU	-1.3	-4.2	-2.0	0.3
142.38	1.17	2.0	154.07	21AU	-1.2	-4.2	-1.9	0.3
148.74	1.17	2.0	163.82	31AU	-0.6	-4.1	-1.9	0.2
155.10	1.17	2.0	173.62	10SE	-0.3	-3.7	-1.8	0.2
161.47	1.17	2.0	183.49	20SE	-0.1	-3.2	-1.7	0.1
167.85	1.16	2.0	193.41	30SE	0.0	-3.7	-1.7	0.1
174.23	1.14	2.0	203.39	10OC	0.3	-4.2	-1.6	0.0
180.62	1.12	2.0	213.43	20OC	1.4	-4.3	-1.6	-0.1
187.02	1.09	1.9	223.51	30OC	2.1	-4.3	-1.5	-0.1
193.43	1.06	1.9	233.63	9NO	-0.0	-4.3	-1.5	-0.2
199.85	1.02	1.8	243.79	19NO	-0.4	-4.2	-1.5	-0.3
206.26	0.97	1.8	253.97	29NO	-0.4	-4.1	-1.4	-0.3
212.68	0.92	1.7	264.16	9DE	-0.5	-4.0	-1.4	-0.2
219.11	0.85	1.6	274.37	19DE	-0.7	-3.9	-1.4	-0.1
225.52	0.77	1.5	284.56	29DE	-0.9	-3.8	-1.4	-0.1
				1119				
231.93	0.68	1.4	294.73	8JA	-1.0	-3.7	-1.4	-0.0
238.32	0.58	1.3	304.89	18JA	-0.9	-3.7	-1.5	0.1
244.69	0.46	1.2	315.00	28JA	0.1	-3.6	-1.5	0.1
251.03	0.32	1.1	325.06	7FE	2.5	-3.5	-1.5	0.2
257.33	0.16	1.0	335.09	17FE	2.0	-3.5	-1.6	0.2
263.58	-0.02	0.8	345.05	27FE	1.0	-3.4	-1.6	0.3
269.76	-0.23	0.6	354.95	9MR	0.6	-3.4	-1.7	0.3
275.84	-0.48	0.5	4.81	19MR	0.3	-3.3	-1.7	0.3
281.81	-0.75	0.3	14.60	29MR	-0.1	-3.3	-1.8	0.3
287.63	-1.07	0.1	24.33	8AP	-0.8	-3.3	-1.8	0.3
293.24	-1.43	-0.1	34.02	18AP	-1.7	-3.3	-1.9	0.3
298.58	-1.84	-0.3	43.66	28AP	-1.3	-3.3	-2.0	0.3
303.60	-2.30	-0.6	53.26	8MY	-0.3	-3.3	-2.1	0.3
308.15	-2.82	-0.8	62.83	18MY	0.6	-3.3	-2.1	0.2
312.14	-3.40	-1.1	72.38	28MY	1.3	-3.3	-2.2	0.2
315.38	-4.04	-1.3	81.91	7JN	2.3	-3.4	-2.2	0.2
317.67	-4.73	-1.6	91.44	17JN	3.4	-3.4	-2.3	0.2
318.83	-5.43	-1.9	100.97	27JN	2.0	-3.5	-2.3	0.3
318.73	-6.09	-2.2	110.51	7JL	0.6	-3.5	-2.3	0.3
317.36	-6.60	-2.5	120.08	17JL	-0.7	-3.5	-2.4	0.3
315.07	-6.87	-2.6	129.67	27JL	-1.4	-3.4	-2.3	0.3
312.45	-6.82	-2.6	139.30	6AU	-1.3	-3.4	-2.3	0.3
310.26	-6.45	-2.4	148.98	16AU	-0.6	-3.4	-2.3	0.3
309.09	-5.85	-2.2	158.70	26AU	-0.2	-3.4	-2.2	0.3
309.21	-5.15	-1.9	168.48	5SE	0.1	-3.3	-2.2	0.3
310.62	-4.42	-1.6	178.31	15SE	0.2	-3.3	-2.1	0.3
313.17	-3.72	-1.3	188.20	25SE	0.5	-3.3	-2.0	0.2
316.65	-3.08	-1.0	198.15	5OC	1.8	-3.3	-2.0	0.2
320.88	-2.51	-0.7	208.16	15OC	1.9	-3.3	-1.9	0.1
325.69	-2.00	-0.5	218.21	25OC	-0.2	-3.4	-1.8	0.0
330.93	-1.55	-0.3	228.31	4NO	-0.6	-3.4	-1.8	-0.0
336.53	-1.16	-0.0	238.46	14NO	-0.6	-3.4	-1.7	-0.1
342.37	-0.82	0.2	248.62	24NO	-0.6	-3.4	-1.7	-0.2
348.41	-0.53	0.4	258.81	4DE	-0.7	-3.5	-1.6	-0.3
354.60	-0.27	0.5	269.01	14DE	-0.8	-3.5	-1.6	-0.3
0.89	-0.04	0.7	279.21	24DE	-0.8	-3.6	-1.6	-0.2

♂ LONG	♂ LAT	♂ MAG	☉ LONG	16.00UT	☿ MAGNITUDES	♀ MAGNITUDES	♃ MAGNITUDES	♄ MAGNITUDES
				1120				
7.25	0.15	0.8	289.40	3JA	-0.7	-3.6	-1.5	-0.1
13.66	0.32	1.0	299.56	13JA	0.2	-3.7	-1.5	-0.1
20.11	0.47	1.1	309.69	23JA	2.8	-3.8	-1.5	0.0
26.56	0.60	1.2	319.79	2FE	1.5	-3.9	-1.5	0.1
33.03	0.71	1.4	329.84	12FE	0.7	-4.0	-1.5	0.1
39.49	0.80	1.5	339.83	22FE	0.4	-4.1	-1.6	0.2
45.94	0.88	1.6	349.77	3MR	0.1	-4.1	-1.6	0.2
52.38	0.95	1.6	359.65	13MR	-0.2	-4.2	-1.6	0.3
58.79	1.01	1.7	9.47	23MR	-0.8	-4.3	-1.6	0.3
65.19	1.06	1.8	19.23	2AP	-1.7	-4.2	-1.7	0.3
71.58	1.10	1.8	28.95	12AP	-1.3	-3.9	-1.7	0.3
77.94	1.13	1.9	38.61	22AP	-0.3	-3.3	-1.8	0.3
84.29	1.15	1.9	48.23	2MY	0.7	-3.0	-1.9	0.3
90.62	1.17	2.0	57.82	12MY	1.7	-3.7	-1.9	0.3
96.94	1.19	2.0	67.37	22MY	3.0	-4.1	-2.0	0.3
103.26	1.19	2.0	76.92	1JN	2.9	-4.2	-2.1	0.3
109.57	1.19	2.0	86.44	11JN	1.6	-4.2	-2.1	0.2
115.88	1.19	2.0	95.97	21JN	0.4	-4.1	-2.2	0.2
122.20	1.18	2.0	105.51	1JL	-0.7	-4.0	-2.3	0.3
128.52	1.17	2.0	115.06	11JL	-1.5	-3.9	-2.3	0.3
134.85	1.15	2.0	124.64	21JL	-1.3	-3.8	-2.4	0.4
141.21	1.13	2.0	134.25	31JL	-0.5	-3.7	-2.4	0.4
147.58	1.11	2.0	143.90	10AU	-0.1	-3.7	-2.4	0.4
153.98	1.08	1.9	153.60	20AU	0.2	-3.6	-2.4	0.4
160.41	1.04	1.9	163.35	30AU	0.4	-3.6	-2.4	0.4
166.86	1.00	2.0	173.15	9SE	0.8	-3.5	-2.4	0.4
173.35	0.96	2.0	183.01	19SE	2.2	-3.5	-2.4	0.4
179.89	0.91	2.0	192.94	29SE	1.7	-3.4	-2.3	0.3
186.45	0.85	2.0	202.91	9OC	-0.3	-3.4	-2.3	0.3
193.07	0.80	1.9	212.94	19OC	-0.7	-3.4	-2.2	0.2
199.72	0.73	1.9	223.02	29OC	-0.7	-3.4	-2.1	0.2
206.42	0.66	1.9	233.14	8NO	-0.7	-3.4	-2.0	0.1
213.17	0.58	1.9	243.29	18NO	-0.7	-3.4	-2.0	0.0
219.97	0.49	1.8	253.48	28NO	-0.7	-3.4	-1.9	-0.0
226.81	0.40	1.8	263.67	8DE	-0.7	-3.4	-1.8	-0.1
233.69	0.30	1.7	273.87	18DE	-0.6	-3.4	-1.8	-0.2
240.63	0.19	1.7	284.06	28DE	0.4	-3.4	-1.7	-0.2
				1121				
247.60	0.07	1.6	294.24	7JA	3.0	-3.4	-1.7	-0.1
254.63	-0.06	1.5	304.39	17JA	1.1	-3.4	-1.7	-0.1
261.69	-0.19	1.4	314.50	27JA	0.4	-3.4	-1.6	0.0
268.79	-0.34	1.4	324.57	6FE	0.2	-3.5	-1.6	0.1
275.93	-0.50	1.3	334.59	16FE	0.0	-3.5	-1.6	0.1
283.10	-0.67	1.2	344.56	26FE	-0.3	-3.5	-1.6	0.2
290.29	-0.84	1.1	354.47	8MR	-0.9	-3.4	-1.6	0.2
297.50	-1.03	1.0	4.32	18MR	-1.6	-3.4	-1.6	0.3
304.72	-1.22	0.9	14.12	28MR	-1.3	-3.4	-1.6	0.3
311.92	-1.41	0.8	23.85	7AP	-0.2	-3.4	-1.6	0.4
319.11	-1.61	0.7	33.54	17AP	1.0	-3.3	-1.6	0.4
326.26	-1.82	0.6	43.19	27AP	2.2	-3.3	-1.7	0.4
333.34	-2.02	0.4	52.79	7MY	3.8	-3.3	-1.7	0.4
340.33	-2.22	0.3	62.37	17MY	2.3	-3.3	-1.8	0.4
347.20	-2.41	0.2	71.92	27MY	1.2	-3.3	-1.8	0.4
353.90	-2.59	0.1	81.45	6JN	0.3	-3.4	-1.9	0.4
0.38	-2.77	-0.1	90.98	16JN	-0.7	-3.4	-1.9	0.4
6.59	-2.93	-0.2	100.51	26JN	-1.6	-3.4	-2.0	0.3
12.45	-3.07	-0.3	110.05	6JL	-1.3	-3.4	-2.1	0.4
17.88	-3.20	-0.5	119.61	16JL	-0.5	-3.5	-2.1	0.4
22.74	-3.30	-0.7	129.21	26JL	0.0	-3.5	-2.2	0.4
26.91	-3.36	-0.9	138.83	5AU	0.3	-3.6	-2.3	0.5
30.20	-3.39	-1.1	148.51	15AU	0.6	-3.7	-2.3	0.5
32.40	-3.35	-1.3	158.23	25AU	1.1	-3.7	-2.4	0.5
33.32	-3.24	-1.5	168.00	4SE	2.6	-3.8	-2.4	0.5
32.78	-3.03	-1.7	177.84	14SE	1.5	-3.9	-2.4	0.5
30.81	-2.68	-1.9	187.73	24SE	-0.4	-4.1	-2.5	0.5
27.76	-2.20	-2.1	197.67	4OC	-0.9	-4.2	-2.5	0.5
24.28	-1.62	-2.1	207.68	14OC	-0.9	-4.3	-2.4	0.5
21.22	-1.01	-1.8	217.73	24OC	-0.9	-4.4	-2.4	0.4
19.21	-0.44	-1.5	227.83	3NO	-0.6	-4.4	-2.3	0.4
18.51	0.04	-1.2	237.97	13NO	-0.5	-4.2	-2.3	0.3
19.15	0.43	-0.9	248.13	23NO	-0.5	-3.7	-2.2	0.2
20.93	0.73	-0.5	258.32	3DE	-0.4	-2.9	-2.1	0.2
23.65	0.96	-0.2	268.52	13DE	0.5	-3.7	-2.1	0.1
27.11	1.14	0.0	278.72	23DE	3.0	-4.2	-2.0	0.0

♂ LONG	♂ LAT	♂ MAG	☉ LONG	16.00UT	☿ MAGNITUDES	♀ MAGNITUDES	♃ MAGNITUDES	♄ MAGNITUDES
				1122				
31.16	1.27	0.3	288.90	2JA	0.8	-4.4	-1.9	-0.1
35.65	1.36	0.5	299.07	12JA	0.2	-4.4	-1.9	-0.1
40.49	1.43	0.7	309.20	22JA	0.1	-4.3	-1.8	0.0
45.59	1.48	0.9	319.29	1FE	-0.1	-4.2	-1.7	0.1
50.91	1.51	1.0	329.34	11FE	-0.4	-4.1	-1.7	0.1
56.39	1.53	1.2	339.34	21FE	-0.9	-4.0	-1.7	0.2
61.99	1.54	1.3	349.28	3MR	-1.5	-3.8	-1.6	0.2
67.70	1.53	1.4	359.16	13MR	-1.2	-3.8	-1.6	0.3
73.49	1.52	1.5	8.99	23MR	-0.2	-3.7	-1.6	0.4
79.35	1.51	1.6	18.76	2AP	1.2	-3.6	-1.6	0.4
85.26	1.49	1.7	28.47	12AP	2.9	-3.5	-1.6	0.4
91.22	1.46	1.8	38.14	22AP	3.0	-3.5	-1.6	0.5
97.22	1.43	1.8	47.76	2MY	1.7	-3.4	-1.6	0.5
103.25	1.40	1.9	57.36	12MY	0.9	-3.4	-1.6	0.5
109.32	1.36	1.9	66.92	22MY	0.2	-3.3	-1.6	0.5
115.43	1.32	2.0	76.46	1JN	-0.7	-3.3	-1.6	0.5
121.57	1.28	2.0	85.99	11JN	-1.6	-3.3	-1.7	0.5
127.74	1.23	2.0	95.52	21JN	-1.3	-3.3	-1.7	0.5
133.96	1.18	2.0	105.05	1JL	-0.5	-3.3	-1.8	0.5
140.21	1.13	2.0	114.60	11JL	0.1	-3.3	-1.8	0.5
146.51	1.07	2.0	124.18	21JL	0.5	-3.3	-1.9	0.5
152.85	1.01	2.0	133.79	31JL	0.8	-3.4	-1.9	0.6
159.25	0.95	2.0	143.44	10AU	1.4	-3.4	-2.0	0.6
165.69	0.89	2.0	153.13	20AU	3.0	-3.4	-2.1	0.6
172.20	0.82	1.9	162.87	30AU	1.3	-3.5	-2.1	0.7
178.76	0.74	1.9	172.68	9SE	-0.5	-3.5	-2.2	0.7
185.38	0.67	1.8	182.53	19SE	-1.0	-3.5	-2.3	0.7
192.07	0.59	1.8	192.45	29SE	-1.0	-3.5	-2.3	0.7
198.83	0.51	1.8	202.43	9OC	-0.8	-3.4	-2.4	0.7
205.66	0.42	1.8	212.45	19OC	-0.5	-3.4	-2.4	0.7
212.55	0.33	1.8	222.53	29OC	-0.4	-3.4	-2.4	0.6
219.52	0.23	1.8	232.65	8NO	-0.4	-3.4	-2.4	0.6
226.56	0.13	1.8	242.80	18NO	-0.2	-3.3	-2.4	0.5
233.67	0.03	1.7	252.98	28NO	0.7	-3.3	-2.3	0.5
240.85	-0.08	1.7	263.17	8DE	2.8	-3.3	-2.3	0.4
248.09	-0.19	1.7	273.37	18DE	0.5	-3.3	-2.2	0.3
255.41	-0.30	1.6	283.56	28DE	0.0	-3.3	-2.1	0.3
				1123				
262.79	-0.42	1.6	293.74	7JA	-0.1	-3.4	-2.1	0.2
270.22	-0.54	1.6	303.89	17JA	-0.2	-3.4	-2.0	0.1
277.72	-0.66	1.5	314.01	27JA	-0.4	-3.4	-1.9	0.1
285.26	-0.78	1.5	324.08	6FE	-0.9	-3.4	-1.9	0.2
292.84	-0.90	1.4	334.10	16FE	-1.4	-3.5	-1.8	0.2
300.45	-1.01	1.4	344.07	26FE	-1.1	-3.5	-1.7	0.3
308.09	-1.12	1.3	353.99	8MR	-0.1	-3.6	-1.7	0.4
315.74	-1.23	1.3	3.84	18MR	1.5	-3.6	-1.6	0.4
323.39	-1.32	1.2	13.64	28MR	3.5	-3.7	-1.6	0.5
331.03	-1.41	1.2	23.39	7AP	2.2	-3.8	-1.5	0.5
338.65	-1.48	1.1	33.08	17AP	1.3	-3.8	-1.5	0.6
346.22	-1.55	1.1	42.72	27AP	0.7	-3.9	-1.5	0.6
353.74	-1.60	1.0	52.33	7MY	0.1	-4.0	-1.5	0.7
1.18	-1.63	1.0	61.90	17MY	-0.7	-4.1	-1.5	0.7
8.53	-1.64	0.9	71.46	27MY	-1.7	-4.2	-1.5	0.7
15.77	-1.64	0.9	80.99	6JN	-1.3	-4.2	-1.5	0.7
22.88	-1.62	0.8	90.52	16JN	-0.4	-4.0	-1.5	0.7
29.83	-1.58	0.7	100.05	26JN	0.2	-3.7	-1.5	0.7
36.61	-1.53	0.7	109.59	6JL	0.6	-3.0	-1.5	0.7
43.17	-1.45	0.6	119.15	16JL	1.1	-3.4	-1.6	0.7
49.50	-1.35	0.5	128.74	26JL	1.9	-4.0	-1.6	0.7
55.54	-1.23	0.4	138.37	5AU	3.2	-4.2	-1.7	0.7
61.25	-1.08	0.3	148.04	15AU	1.1	-4.2	-1.7	0.8
66.55	-0.91	0.2	157.76	25AU	-0.5	-4.2	-1.8	0.8
71.37	-0.71	0.1	167.53	4SE	-1.1	-4.1	-1.8	0.9
75.61	-0.47	-0.1	177.36	14SE	-1.2	-4.0	-1.9	0.9
79.12	-0.19	-0.2	187.25	24SE	-0.7	-4.0	-2.0	0.9
81.73	0.13	-0.4	197.19	4OC	-0.4	-3.9	-2.0	0.9
83.25	0.51	-0.6	207.19	14OC	-0.3	-3.8	-2.1	0.9
83.47	0.94	-0.9	217.24	24OC	-0.2	-3.7	-2.2	0.9
82.27	1.41	-1.1	227.34	3NO	-0.1	-3.7	-2.2	0.9
79.70	1.90	-1.3	237.47	13NO	0.9	-3.6	-2.2	0.9
76.11	2.34	-1.4	247.64	23NO	2.5	-3.5	-2.3	0.8
72.20	2.68	-1.4	257.83	3DE	0.3	-3.5	-2.3	0.8
68.76	2.89	-1.1	268.02	13DE	-0.1	-3.5	-2.3	0.7
66.38	2.97	-0.9	278.23	23DE	-0.2	-3.4	-2.2	0.6

♂ LONG	♂ LAT	♂ MAG	⊙ LONG	16.00UT 1124	☿	♀	♃	♄
					MAGNITUDES			
65.35	2.96	-0.6	288.41	2JA	-0.3	-3.4	-2.2	0.6
65.61	2.89	-0.3	298.58	12JA	-0.5	-3.4	-2.2	0.5
67.01	2.78	0.0	308.72	22JA	-0.9	-3.3	-2.1	0.4
69.35	2.66	0.3	318.81	1FE	-1.2	-3.3	-2.0	0.4
72.44	2.53	0.5	328.86	11FE	-1.1	-3.3	-2.0	0.4
76.12	2.41	0.7	338.86	21FE	-0.0	-3.3	-1.9	0.4
80.26	2.29	0.9	348.80	2MR	1.9	-3.3	-1.8	0.5
84.76	2.17	1.1	358.68	12MR	3.1	-3.3	-1.7	0.5
89.56	2.06	1.2	8.51	22MR	1.6	-3.4	-1.7	0.6
94.60	1.95	1.3	18.28	1AP	0.9	-3.4	-1.6	0.6
99.83	1.84	1.4	28.00	11AP	0.5	-3.4	-1.6	0.7
105.22	1.74	1.5	37.67	21AP	-0.0	-3.5	-1.5	0.7
110.75	1.64	1.6	47.30	1MY	-0.8	-3.5	-1.5	0.8
116.39	1.54	1.7	56.88	11MY	-1.7	-3.5	-1.4	0.8
122.15	1.45	1.8	66.45	21MY	-1.3	-3.4	-1.4	0.9
127.99	1.35	1.8	75.99	31MY	-0.4	-3.4	-1.4	0.9
133.93	1.26	1.9	85.52	10JN	0.3	-3.4	-1.4	0.9
139.95	1.16	1.9	95.05	20JN	0.8	-3.3	-1.4	1.0
146.05	1.07	1.9	104.58	30JN	1.4	-3.3	-1.4	1.0
152.24	0.98	1.9	114.13	10JL	2.5	-3.3	-1.4	1.0
158.50	0.88	1.9	123.71	20JL	2.8	-3.3	-1.4	1.0
164.84	0.79	1.9	133.31	30JL	0.9	-3.3	-1.4	1.0
171.27	0.69	1.9	142.96	9AU	-0.6	-3.3	-1.4	1.0
177.78	0.60	1.9	152.66	19AU	-1.3	-3.4	-1.5	1.0
184.37	0.50	1.9	162.40	29AU	-1.3	-3.4	-1.5	1.1
191.05	0.40	1.8	172.20	8SE	-0.7	-3.4	-1.6	1.1
197.81	0.30	1.8	182.06	18SE	-0.3	-3.5	-1.6	1.1
204.67	0.20	1.8	191.97	28SE	-0.1	-3.5	-1.7	1.2
211.61	0.10	1.7	201.94	8OC	-0.0	-3.6	-1.7	1.2
218.65	-0.00	1.7	211.97	18OC	0.1	-3.6	-1.8	1.2
225.77	-0.11	1.6	222.04	28OC	1.2	-3.7	-1.9	1.2
232.98	-0.21	1.6	232.15	7NO	2.3	-3.8	-1.9	1.2
240.28	-0.31	1.6	242.31	17NO	0.1	-3.9	-2.0	1.2
247.66	-0.41	1.5	252.48	27NO	-0.3	-4.0	-2.0	1.1
255.12	-0.51	1.5	262.68	7DE	-0.3	-4.1	-2.1	1.1
262.66	-0.61	1.5	272.88	17DE	-0.4	-4.2	-2.1	1.0
270.26	-0.70	1.5	283.07	27DE	-0.6	-4.3	-2.1	1.0
				1125				
277.92	-0.79	1.5	293.25	6JA	-1.0	-4.4	-2.1	0.9
285.64	-0.88	1.5	303.40	16JA	-1.1	-4.4	-2.1	0.8
293.39	-0.95	1.5	313.52	26JA	-0.9	-4.2	-2.1	0.8
301.18	-1.02	1.4	323.60	5FE	0.0	-3.7	-2.1	0.7
308.99	-1.08	1.4	333.62	15FE	2.2	-3.3	-2.0	0.6
316.81	-1.13	1.4	343.59	25FE	2.4	-3.7	-2.0	0.7
324.63	-1.17	1.4	353.51	7MR	1.2	-4.1	-1.9	0.7
332.44	-1.20	1.4	3.37	17MR	0.7	-4.3	-1.8	0.8
340.21	-1.21	1.4	13.16	27MR	0.4	-4.2	-1.8	0.8
347.95	-1.21	1.4	22.91	6AP	-0.1	-4.2	-1.7	0.9
355.64	-1.20	1.4	32.61	16AP	-0.8	-4.1	-1.6	0.9
3.26	-1.17	1.3	42.25	26AP	-1.7	-4.0	-1.6	1.0
10.81	-1.13	1.3	51.86	6MY	-1.3	-3.9	-1.5	1.0
18.27	-1.08	1.3	61.44	16MY	-0.3	-3.8	-1.5	1.1
25.64	-1.01	1.3	70.99	26MY	0.4	-3.7	-1.4	1.1
32.90	-0.93	1.3	80.53	5JN	1.1	-3.6	-1.4	1.2
40.05	-0.84	1.3	90.05	15JN	1.9	-3.5	-1.3	1.2
47.07	-0.74	1.3	99.58	25JN	3.1	-3.5	-1.3	1.2
53.97	-0.62	1.2	109.13	5JL	2.3	-3.4	-1.3	1.2
60.71	-0.50	1.2	118.69	15JL	0.7	-3.4	-1.3	1.3
67.30	-0.36	1.2	128.28	25JL	-0.6	-3.4	-1.3	1.3
73.72	-0.21	1.1	137.90	4AU	-1.4	-3.4	-1.3	1.3
79.95	-0.05	1.1	147.57	14AU	-1.3	-3.4	-1.3	1.3
85.96	0.12	1.0	157.29	24AU	-0.6	-3.3	-1.3	1.3
91.73	0.31	1.0	167.06	3SE	-0.2	-3.3	-1.3	1.3
97.22	0.52	0.9	176.89	13SE	-0.0	-3.4	-1.4	1.3
102.38	0.75	0.8	186.77	23SE	0.1	-3.4	-1.4	1.3
107.14	1.00	0.6	196.71	3OC	0.4	-3.4	-1.4	1.3
111.42	1.27	0.5	206.71	13OC	1.5	-3.4	-1.5	1.3
115.12	1.58	0.4	216.75	23OC	2.1	-3.4	-1.5	1.3
118.10	1.93	0.2	226.85	2NO	-0.1	-3.4	-1.6	1.3
120.19	2.31	-0.0	236.98	12NO	-0.5	-3.5	-1.7	1.2
121.21	2.73	-0.3	247.15	22NO	-0.5	-3.5	-1.7	1.2
120.96	3.18	-0.5	257.33	2DE	-0.5	-3.5	-1.8	1.2
119.36	3.62	-0.7	267.53	12DE	-0.7	-3.5	-1.8	1.1
116.48	4.00	-0.9	277.73	22DE	-0.9	-3.4	-1.9	1.1

♂ LONG	♂ LAT	♂ MAG	⊙ LONG	16.00UT 1126	☿	♀	♃	♄
					MAGNITUDES			
112.72	4.25	-1.1	287.92	1JA	-0.9	-3.4	-2.0	1.0
108.76	4.34	-1.0	298.08	11JA	-0.8	-3.4	-2.0	1.0
105.34	4.26	-0.8	308.22	21JA	0.1	-3.4	-2.0	0.9
103.00	4.05	-0.5	318.32	31JA	2.5	-3.4	-2.0	0.9
101.96	3.77	-0.3	328.37	10FE	1.8	-3.3	-2.1	0.8
102.19	3.47	-0.0	338.37	20FE	0.8	-3.3	-2.0	0.8
103.53	3.16	0.2	348.32	2MR	0.5	-3.3	-2.0	0.8
105.79	2.88	0.5	358.20	12MR	0.2	-3.3	-2.0	0.9
108.79	2.61	0.7	8.03	22MR	-0.2	-3.4	-1.9	0.9
112.39	2.37	0.8	17.81	1AP	-0.8	-3.4	-1.9	1.0
116.46	2.15	1.0	27.53	11AP	-1.7	-3.4	-1.8	1.1
120.93	1.94	1.1	37.20	21AP	-1.3	-3.4	-1.7	1.1
125.71	1.75	1.2	46.83	1MY	-0.3	-3.5	-1.7	1.2
130.76	1.58	1.3	56.42	11MY	0.6	-3.5	-1.6	1.3
136.04	1.41	1.4	65.98	21MY	1.4	-3.6	-1.5	1.3
141.51	1.25	1.5	75.53	31MY	2.5	-3.6	-1.5	1.3
147.15	1.10	1.6	85.06	10JN	3.3	-3.7	-1.4	1.4
152.95	0.96	1.6	94.59	20JN	1.9	-3.8	-1.4	1.4
158.90	0.82	1.7	104.12	30JN	0.6	-3.9	-1.3	1.4
164.98	0.69	1.7	113.67	10JL	-0.6	-4.0	-1.3	1.4
171.19	0.56	1.7	123.25	20JL	-1.5	-4.1	-1.3	1.3
177.52	0.43	1.7	132.85	30JL	-1.3	-4.2	-1.2	1.3
183.97	0.31	1.7	142.50	9AU	-0.6	-4.2	-1.2	1.3
190.54	0.19	1.7	152.19	19AU	-0.1	-4.2	-1.2	1.2
197.22	0.08	1.7	161.93	29AU	0.1	-4.1	-1.2	1.2
204.02	-0.04	1.7	171.72	8SE	0.3	-3.7	-1.2	1.1
210.93	-0.15	1.7	181.58	18SE	0.6	-3.2	-1.2	1.1
217.95	-0.26	1.7	191.49	28SE	1.8	-3.7	-1.3	1.1
225.08	-0.36	1.6	201.46	8OC	1.9	-4.2	-1.3	1.1
232.32	-0.46	1.6	211.48	18OC	-0.2	-4.3	-1.3	1.1
239.65	-0.55	1.6	221.55	28OC	-0.6	-4.3	-1.4	1.1
247.09	-0.64	1.5	231.66	7NO	-0.6	-4.3	-1.4	1.1
254.61	-0.73	1.5	241.82	17NO	-0.6	-4.2	-1.5	1.1
262.21	-0.80	1.5	251.99	27NO	-0.8	-4.1	-1.5	1.1
269.89	-0.87	1.4	262.18	7DE	-0.7	-4.0	-1.6	1.0
277.64	-0.93	1.4	272.39	17DE	-0.8	-3.9	-1.6	1.0
285.43	-0.99	1.3	282.58	27DE	-0.7	-3.8	-1.7	1.0
				1127				
293.27	-1.03	1.3	292.76	6JA	0.2	-3.7	-1.8	0.9
301.15	-1.06	1.3	302.91	16JA	2.8	-3.6	-1.8	0.9
309.04	-1.08	1.3	313.03	26JA	1.4	-3.6	-1.9	0.8
316.93	-1.09	1.3	323.11	5FE	0.6	-3.5	-1.9	0.8
324.82	-1.08	1.4	333.14	15FE	0.3	-3.5	-2.0	0.7
332.68	-1.07	1.4	343.12	25FE	0.1	-3.4	-2.0	0.7
340.51	-1.04	1.4	353.03	7MR	-0.2	-3.4	-2.0	0.6
348.30	-1.00	1.4	2.90	17MR	-0.8	-3.3	-2.0	0.6
356.03	-0.95	1.4	12.70	27MR	-1.7	-3.3	-2.0	0.7
3.70	-0.89	1.4	22.44	6AP	-1.3	-3.3	-2.0	0.8
11.29	-0.82	1.5	32.14	16AP	-0.3	-3.3	-1.9	0.8
18.80	-0.74	1.5	41.79	26AP	0.8	-3.3	-1.9	0.9
26.23	-0.65	1.5	51.40	6MY	1.9	-3.3	-1.8	1.0
33.57	-0.56	1.5	60.98	16MY	3.3	-3.3	-1.7	1.0
40.80	-0.46	1.5	70.53	26MY	2.7	-3.4	-1.7	1.1
47.94	-0.35	1.6	80.06	5JN	1.4	-3.4	-1.6	1.1
54.98	-0.24	1.6	89.59	15JN	0.4	-3.4	-1.6	1.2
61.92	-0.12	1.6	99.12	25JN	-0.7	-3.5	-1.5	1.2
68.74	0.00	1.6	108.66	5JL	-1.5	-3.5	-1.4	1.2
75.46	0.13	1.6	118.22	15JL	-1.3	-3.5	-1.4	1.2
82.06	0.26	1.6	127.81	25JL	-0.5	-3.4	-1.3	1.2
88.55	0.39	1.6	137.43	4AU	-0.1	-3.4	-1.3	1.2
94.91	0.53	1.5	147.10	14AU	0.2	-3.4	-1.3	1.1
101.13	0.67	1.5	156.81	24AU	0.5	-3.4	-1.2	1.1
107.21	0.82	1.5	166.58	3SE	0.9	-3.3	-1.2	1.1
113.14	0.98	1.4	176.40	13SE	2.3	-3.3	-1.2	1.0
118.88	1.15	1.4	186.28	23SE	1.7	-3.3	-1.2	1.0
124.42	1.32	1.3	196.22	3OC	-0.3	-3.3	-1.2	1.0
129.72	1.51	1.2	206.22	13OC	-0.8	-3.3	-1.2	1.0
134.74	1.71	1.1	216.26	23OC	-0.8	-3.4	-1.2	1.0
139.42	1.93	1.0	226.35	2NO	-0.8	-3.4	-1.3	1.0
143.70	2.17	0.8	236.49	12NO	-0.7	-3.4	-1.3	1.0
147.47	2.43	0.7	246.65	22NO	-0.6	-3.4	-1.3	1.0
150.63	2.71	0.5	256.84	2DE	-0.6	-3.5	-1.4	1.0
153.02	3.03	0.3	267.04	12DE	-0.5	-3.5	-1.4	0.9
154.46	3.36	0.0	277.24	22DE	0.4	-3.6	-1.5	0.9

♂			⊙	16.00UT	☿	♀	♃	♄
LONG	LAT	MAG	LONG	1128	MAGNITUDES			
154.79	3.71	-0.2	287.42	1JA	3.0	-3.7	-1.5	0.9
153.84	4.04	-0.5	297.60	11JA	1.0	-3.7	-1.6	0.8
151.59	4.31	-0.7	307.73	21JA	0.3	-3.8	-1.7	0.8
148.26	4.46	-0.9	317.83	31JA	0.2	-3.9	-1.7	0.7
144.35	4.45	-1.0	327.89	10FE	-0.0	-4.0	-1.8	0.7
140.60	4.27	-0.9	337.89	20FE	-0.3	-4.1	-1.9	0.6
137.64	3.95	-0.6	347.84	1MR	-0.8	-4.1	-1.9	0.6
135.87	3.56	-0.4	357.73	11MR	-1.6	-4.2	-2.0	0.5
135.41	3.14	-0.2	7.56	21MR	-1.3	-4.3	-2.0	0.5
136.15	2.74	0.0	17.34	31MR	-0.2	-4.2	-2.0	0.5
137.93	2.37	0.3	27.07	10AP	1.0	-3.9	-2.0	0.6
140.58	2.03	0.4	36.74	20AP	2.5	-3.3	-2.0	0.6
143.93	1.72	0.6	46.37	30AP	3.5	-3.0	-2.0	0.7
147.85	1.44	0.8	55.96	10MY	2.0	-3.7	-2.0	0.8
152.24	1.19	0.9	65.52	20MY	1.1	-4.1	-1.9	0.8
157.02	0.96	1.0	75.07	30MY	0.3	-4.2	-1.9	0.9
162.14	0.75	1.1	84.60	9JN	-0.7	-4.2	-1.8	0.9
167.54	0.56	1.2	94.12	19JN	-1.6	-4.1	-1.7	1.0
173.19	0.38	1.2	103.66	29JN	-1.3	-4.0	-1.7	1.0
179.06	0.21	1.3	113.21	9JL	-0.5	-3.9	-1.6	1.0
185.13	0.05	1.3	122.78	19JL	0.0	-3.8	-1.6	1.0
191.39	-0.09	1.4	132.38	29JL	0.4	-3.7	-1.5	1.0
197.83	-0.23	1.4	142.03	8AU	0.7	-3.7	-1.5	1.0
204.42	-0.36	1.4	151.71	18AU	1.2	-3.6	-1.4	1.0
211.17	-0.48	1.4	161.45	28AU	2.7	-3.6	-1.4	1.0
218.07	-0.59	1.4	171.25	7SE	1.5	-3.5	-1.3	1.0
225.11	-0.69	1.4	181.09	17SE	-0.4	-3.5	-1.3	0.9
232.27	-0.78	1.4	191.01	27SE	-0.9	-3.4	-1.3	0.9
239.56	-0.87	1.4	200.97	7OC	-0.9	-3.4	-1.3	0.9
246.96	-0.94	1.4	210.99	17OC	-0.9	-3.4	-1.3	0.9
254.47	-1.00	1.4	221.06	27OC	-0.6	-3.4	-1.3	0.9
262.07	-1.05	1.4	231.18	6NO	-0.5	-3.4	-1.3	0.9
269.76	-1.09	1.4	241.32	16NO	-0.5	-3.4	-1.3	0.9
277.52	-1.12	1.4	251.50	26NO	-0.4	-3.4	-1.3	0.9
285.33	-1.13	1.4	261.69	6DE	0.5	-3.4	-1.3	0.9
293.19	-1.14	1.4	271.89	16DE	3.0	-3.4	-1.4	0.9
301.07	-1.13	1.4	282.09	26DE	0.7	-3.4	-1.4	0.8
				1129				
308.98	-1.11	1.4	292.27	5JA	0.1	-3.4	-1.5	0.8
316.88	-1.07	1.4	302.42	15JA	0.0	-3.4	-1.5	0.8
324.77	-1.03	1.4	312.55	25JA	-0.1	-3.4	-1.6	0.7
332.63	-0.98	1.4	322.62	4FE	-0.4	-3.5	-1.6	0.7
340.45	-0.91	1.4	332.65	14FE	-0.9	-3.5	-1.7	0.6
348.23	-0.84	1.4	342.63	24FE	-1.5	-3.5	-1.8	0.6
355.94	-0.76	1.4	352.55	6MR	-1.2	-3.4	-1.8	0.5
3.59	-0.68	1.4	2.41	16MR	-0.2	-3.4	-1.9	0.5
11.16	-0.58	1.4	12.22	26MR	1.3	-3.4	-2.0	0.4
18.65	-0.49	1.5	21.97	5AP	3.1	-3.4	-2.0	0.4
26.06	-0.39	1.5	31.67	15AP	2.7	-3.3	-2.1	0.4
33.38	-0.29	1.6	41.32	25AP	1.5	-3.3	-2.1	0.5
40.61	-0.18	1.6	50.93	5MY	0.8	-3.3	-2.1	0.6
47.75	-0.07	1.6	60.51	15MY	0.2	-3.3	-2.1	0.6
54.80	0.03	1.7	70.07	25MY	-0.7	-3.3	-2.1	0.7
61.76	0.14	1.7	79.60	4JN	-1.6	-3.4	-2.1	0.7
68.64	0.25	1.7	89.13	14JN	-1.3	-3.4	-2.0	0.8
75.44	0.35	1.8	98.66	24JN	-0.5	-3.4	-2.0	0.8
82.14	0.46	1.8	108.20	4JL	0.1	-3.4	-1.9	0.9
88.77	0.57	1.8	117.76	14JL	0.5	-3.5	-1.9	0.9
95.32	0.67	1.8	127.35	24JL	0.9	-3.5	-1.8	0.9
101.78	0.78	1.8	136.96	3AU	1.6	-3.6	-1.7	0.9
108.16	0.88	1.8	146.63	13AU	3.1	-3.7	-1.7	0.9
114.46	0.99	1.8	156.34	23AU	1.3	-3.7	-1.6	0.9
120.67	1.09	1.8	166.10	2SE	-0.5	-3.8	-1.6	0.9
126.79	1.20	1.8	175.92	12SE	-1.1	-3.9	-1.5	0.9
132.81	1.31	1.7	185.80	22SE	-1.1	-4.1	-1.5	0.9
138.72	1.42	1.7	195.74	2OC	-0.8	-4.2	-1.4	0.8
144.51	1.53	1.6	205.73	12OC	-0.5	-4.3	-1.4	0.8
150.16	1.65	1.5	215.77	22OC	-0.3	-4.4	-1.4	0.8
155.64	1.77	1.5	225.86	1NO	-0.3	-4.4	-1.4	0.8
160.94	1.90	1.4	235.99	11NO	-0.2	-4.2	-1.3	0.8
166.01	2.03	1.2	246.16	21NO	0.7	-3.7	-1.3	0.8
170.80	2.17	1.1	256.34	1DE	2.8	-2.9	-1.3	0.8
175.26	2.32	1.0	266.54	11DE	0.5	-3.8	-1.3	0.8
179.31	2.49	0.8	276.74	21DE	-0.0	-4.2	-1.4	0.8
182.84	2.66	0.6	286.93	31DE	-0.1	-4.4	-1.4	0.8

♂			⊙	16.00UT	☿	♀	♃	♄
LONG	LAT	MAG	LONG	1130	MAGNITUDES			
185.74	2.83	0.4	297.10	10JA	-0.2	-4.4	-1.4	0.8
187.85	3.02	0.1	307.24	20JA	-0.5	-4.3	-1.5	0.7
188.99	3.19	-0.2	317.34	30JA	-0.9	-4.2	-1.5	0.7
189.00	3.34	-0.4	327.40	9FE	-1.3	-4.1	-1.6	0.7
187.72	3.44	-0.7	337.41	19FE	-1.1	-3.9	-1.6	0.6
185.22	3.45	-1.0	347.36	1MR	-0.1	-3.8	-1.7	0.6
181.77	3.32	-1.2	357.26	11MR	1.6	-3.7	-1.7	0.5
177.94	3.06	-1.2	7.09	21MR	3.5	-3.7	-1.8	0.4
174.47	2.67	-1.0	16.87	31MR	2.0	-3.6	-1.9	0.4
171.95	2.23	-0.9	26.60	10AP	1.1	-3.5	-2.0	0.3
170.70	1.77	-0.6	36.28	20AP	0.6	-3.5	-2.0	0.3
170.76	1.33	-0.4	45.91	30AP	0.1	-3.4	-2.1	0.4
172.01	0.94	-0.2	55.50	10MY	-0.7	-3.4	-2.1	0.4
174.30	0.59	-0.0	65.07	20MY	-1.6	-3.3	-2.2	0.5
177.44	0.28	0.1	74.61	30MY	-1.3	-3.3	-2.2	0.6
181.29	0.01	0.3	84.14	9JN	-0.4	-3.3	-2.2	0.6
185.72	-0.22	0.4	93.67	19JN	0.2	-3.3	-2.2	0.7
190.63	-0.43	0.5	103.20	29JN	0.7	-3.3	-2.2	0.7
195.95	-0.61	0.6	112.75	9JL	1.2	-3.3	-2.2	0.8
201.62	-0.77	0.7	122.32	19JL	2.1	-3.3	-2.2	0.8
207.60	-0.92	0.8	131.92	29JL	3.1	-3.4	-2.1	0.8
213.85	-1.04	0.9	141.56	8AU	1.1	-3.4	-2.0	0.9
220.34	-1.14	0.9	151.25	18AU	-0.5	-3.4	-2.0	0.9
227.03	-1.23	1.0	160.98	28AU	-1.2	-3.5	-1.9	0.9
233.93	-1.30	1.0	170.77	7SE	-1.2	-3.5	-1.9	0.9
241.00	-1.35	1.0	180.62	17SE	-0.7	-3.5	-1.8	0.8
248.22	-1.39	1.1	190.52	27SE	-0.4	-3.5	-1.7	0.8
255.58	-1.42	1.1	200.48	7OC	-0.2	-3.4	-1.7	0.8
263.06	-1.42	1.1	210.50	17OC	-0.1	-3.4	-1.6	0.7
270.64	-1.42	1.2	220.56	27OC	0.0	-3.4	-1.6	0.7
278.30	-1.39	1.2	230.68	6NO	1.0	-3.4	-1.5	0.7
286.04	-1.36	1.2	240.82	16NO	2.5	-3.3	-1.5	0.8
293.82	-1.31	1.3	250.99	26NO	0.2	-3.3	-1.5	0.8
301.64	-1.25	1.3	261.19	6DE	-0.2	-3.3	-1.5	0.8
309.47	-1.18	1.3	271.39	16DE	-0.3	-3.3	-1.4	0.8
317.30	-1.09	1.4	281.58	26DE	-0.3	-3.4	-1.4	0.8
				1131				
325.12	-1.01	1.4	291.77	5JA	-0.5	-3.4	-1.4	0.8
332.90	-0.91	1.4	301.93	15JA	-0.9	-3.4	-1.5	0.8
340.65	-0.81	1.4	312.05	25JA	-1.2	-3.4	-1.5	0.7
348.34	-0.70	1.5	322.14	4FE	-1.0	-3.4	-1.5	0.7
355.97	-0.59	1.5	332.17	14FE	-0.1	-3.5	-1.5	0.7
3.54	-0.48	1.5	342.15	24FE	1.9	-3.5	-1.6	0.6
11.02	-0.37	1.6	352.08	6MR	2.8	-3.6	-1.6	0.6
18.43	-0.26	1.6	1.94	16MR	1.4	-3.6	-1.7	0.5
25.76	-0.15	1.6	11.75	26MR	0.8	-3.7	-1.7	0.4
33.00	-0.04	1.6	21.51	5AP	0.4	-3.8	-1.8	0.4
40.16	0.06	1.7	31.21	15AP	-0.0	-3.9	-1.9	0.3
47.24	0.17	1.7	40.86	25AP	-0.7	-3.9	-1.9	0.3
54.24	0.27	1.7	50.48	5MY	-1.7	-4.0	-2.0	0.3
61.15	0.36	1.7	60.06	15MY	-1.3	-4.1	-2.1	0.3
68.00	0.46	1.7	69.61	25MY	-0.4	-4.2	-2.1	0.4
74.77	0.55	1.8	79.15	4JN	0.4	-4.2	-2.2	0.5
81.48	0.63	1.8	88.67	14JN	0.9	-4.0	-2.3	0.5
88.12	0.72	1.9	98.20	24JN	1.6	-3.6	-2.3	0.6
94.70	0.80	1.9	107.74	4JL	2.7	-3.0	-2.3	0.6
101.23	0.87	1.9	117.30	14JL	2.7	-3.5	-2.4	0.7
107.70	0.94	1.9	126.88	24JL	0.9	-4.0	-2.4	0.7
114.12	1.01	2.0	136.50	3AU	-0.6	-4.2	-2.3	0.8
120.49	1.08	2.0	146.16	13AU	-1.3	-4.2	-2.3	0.8
126.82	1.14	2.0	155.87	23AU	-1.3	-4.2	-2.3	0.8
133.11	1.20	2.0	165.63	2SE	-0.7	-4.1	-2.2	0.8
139.34	1.26	2.0	175.45	12SE	-0.3	-4.0	-2.2	0.8
145.53	1.32	1.9	185.32	22SE	-0.1	-3.9	-2.1	0.8
151.67	1.37	1.9	195.26	2OC	0.0	-3.9	-2.0	0.8
157.75	1.43	1.9	205.24	12OC	0.2	-3.8	-2.0	0.8
163.78	1.48	1.8	215.28	22OC	1.2	-3.7	-1.9	0.7
169.73	1.52	1.8	225.37	1NO	2.3	-3.7	-1.8	0.7
175.61	1.57	1.7	235.50	11NO	0.1	-3.6	-1.8	0.7
181.40	1.61	1.6	245.66	21NO	-0.4	-3.5	-1.7	0.7
187.08	1.66	1.5	255.85	1DE	-0.4	-3.5	-1.7	0.7
192.63	1.69	1.4	266.04	11DE	-0.5	-3.5	-1.6	0.8
198.03	1.73	1.3	276.24	21DE	-0.6	-3.4	-1.6	0.8
203.24	1.76	1.1	286.44	31DE	-0.9	-3.4	-1.6	0.8

Left Table

♂ LONG	♂ LAT	♂ MAG	☉ LONG	16.00UT 1132	☿	♀	♃	♄
208.22	1.78	1.0	296.60	10JA	-1.0	-3.4	-1.5	0.8
212.92	1.79	0.8	306.75	20JA	-0.9	-3.3	-1.5	0.8
217.27	1.79	0.6	316.86	30JA	0.0	-3.3	-1.5	0.7
221.19	1.78	0.4	326.91	9FE	2.3	-3.3	-1.5	0.7
224.56	1.74	0.1	336.93	19FE	2.2	-3.3	-1.5	0.7
227.25	1.66	-0.2	346.88	29FE	1.0	-3.3	-1.6	0.6
229.09	1.55	-0.4	356.78	10MR	0.6	-3.3	-1.6	0.6
229.91	1.37	-0.7	6.62	20MR	0.3	-3.4	-1.6	0.5
229.52	1.12	-1.1	16.40	30MR	-0.1	-3.4	-1.7	0.5
227.89	0.78	-1.4	26.13	9AP	-0.8	-3.4	-1.7	0.4
225.13	0.36	-1.6	35.81	19AP	-1.7	-3.5	-1.8	0.3
221.70	-0.13	-1.8	45.44	29AP	-1.3	-3.5	-1.8	0.3
218.29	-0.63	-1.7	55.04	9MY	-0.4	-3.5	-1.9	0.2
215.59	-1.09	-1.5	64.60	19MY	0.5	-3.4	-1.9	0.3
214.10	-1.48	-1.3	74.15	29MY	1.2	-3.4	-2.0	0.3
213.98	-1.79	-1.1	83.68	8JN	2.1	-3.4	-2.1	0.4
215.17	-2.02	-0.9	93.21	18JN	3.3	-3.3	-2.2	0.4
217.53	-2.19	-0.7	102.74	28JN	2.2	-3.3	-2.2	0.5
220.87	-2.30	-0.5	112.28	8JL	0.7	-3.3	-2.3	0.6
225.01	-2.38	-0.4	121.85	18JL	-0.6	-3.3	-2.3	0.6
229.82	-2.42	-0.2	131.45	28JL	-1.4	-3.3	-2.4	0.7
235.18	-2.43	-0.1	141.09	7AU	-1.3	-3.3	-2.4	0.7
240.98	-2.41	0.1	150.77	17AU	-0.6	-3.4	-2.4	0.7
247.16	-2.37	0.2	160.51	27AU	-0.2	-3.4	-2.4	0.8
253.65	-2.31	0.3	170.29	6SE	0.0	-3.4	-2.4	0.8
260.40	-2.23	0.4	180.14	16SE	0.2	-3.5	-2.4	0.8
267.36	-2.13	0.5	190.04	26SE	0.4	-3.5	-2.4	0.8
274.50	-2.02	0.6	200.00	6OC	1.6	-3.6	-2.3	0.8
281.78	-1.90	0.6	210.01	16OC	2.1	-3.6	-2.2	0.8
289.17	-1.76	0.7	220.08	26OC	-0.1	-3.7	-2.2	0.7
296.64	-1.62	0.8	230.18	5NO	-0.5	-3.8	-2.1	0.7
304.17	-1.47	0.9	240.33	15NO	-0.5	-3.9	-2.0	0.7
311.73	-1.32	1.0	250.50	25NO	-0.6	-4.0	-2.0	0.7
319.30	-1.16	1.1	260.69	5DE	-0.7	-4.1	-1.9	0.7
326.88	-1.01	1.1	270.90	15DE	-0.8	-4.2	-1.8	0.7
334.43	-0.85	1.2	281.09	25DE	-0.9	-4.3	-1.8	0.8

1133

♂ LONG	♂ LAT	♂ MAG	☉ LONG	16.00UT	☿	♀	♃	♄
341.95	-0.70	1.3	291.27	4JA	-0.8	-4.4	-1.7	0.8
349.42	-0.55	1.3	301.43	14JA	0.1	-4.4	-1.7	0.8
356.85	-0.41	1.4	311.56	24JA	2.6	-4.2	-1.7	0.8
4.21	-0.27	1.5	321.64	3FE	1.7	-3.7	-1.6	0.8
11.50	-0.14	1.5	331.60	13FE	0.7	-3.3	-1.6	0.7
18.73	-0.01	1.6	341.66	23FE	0.4	-3.8	-1.6	0.7
25.88	0.11	1.6	351.59	5MR	0.2	-4.1	-1.6	0.7
32.97	0.22	1.7	1.46	15MR	-0.2	-4.3	-1.6	0.6
39.98	0.32	1.7	11.27	25MR	-0.8	-4.3	-1.6	0.6
46.91	0.42	1.8	21.03	4AP	-1.6	-4.2	-1.6	0.5
53.79	0.51	1.8	30.74	14AP	-1.3	-4.1	-1.6	0.4
60.59	0.60	1.8	40.39	24AP	-0.3	-4.0	-1.6	0.4
67.33	0.68	1.9	50.01	4MY	0.7	-3.9	-1.7	0.3
74.02	0.75	1.9	59.59	14MY	1.6	-3.8	-1.7	0.3
80.65	0.81	1.9	69.15	24MY	2.8	-3.7	-1.8	0.2
87.23	0.87	1.9	78.69	3JN	3.1	-3.6	-1.8	0.3
93.77	0.93	1.9	88.22	13JN	1.7	-3.5	-1.9	0.3
100.26	0.98	1.9	97.74	23JN	0.5	-3.5	-1.9	0.4
106.72	1.02	1.9	107.28	3JL	-0.6	-3.4	-2.0	0.5
113.16	1.07	1.9	116.84	13JL	-1.5	-3.4	-2.1	0.5
119.56	1.10	2.0	126.42	23JL	-1.3	-3.4	-2.2	0.6
125.95	1.13	2.0	136.04	2AU	-0.6	-3.4	-2.2	0.6
132.31	1.16	2.0	145.70	12AU	-0.1	-3.4	-2.3	0.7
138.66	1.18	2.0	155.40	22AU	0.2	-3.3	-2.3	0.7
145.00	1.20	2.0	165.16	1SE	0.3	-3.3	-2.4	0.7
151.33	1.22	2.0	174.98	11SE	0.7	-3.4	-2.4	0.8
157.65	1.23	2.0	184.84	21SE	1.9	-3.4	-2.4	0.8
163.97	1.23	2.0	194.78	1OC	1.9	-3.4	-2.5	0.8
170.27	1.23	2.0	204.76	11OC	-0.2	-3.4	-2.4	0.8
176.57	1.23	1.9	214.80	21OC	-0.7	-3.4	-2.4	0.8
182.86	1.22	1.9	224.88	31OC	-0.7	-3.4	-2.4	0.7
189.14	1.21	1.9	235.01	10NO	-0.7	-3.5	-2.3	0.7
195.40	1.19	1.8	245.17	20NO	-0.7	-3.5	-2.3	0.7
201.65	1.16	1.7	255.35	30NO	-0.7	-3.5	-2.2	0.6
207.87	1.12	1.6	265.55	10DE	-0.7	-3.5	-2.1	0.7
214.06	1.08	1.6	275.74	20DE	-0.6	-3.4	-2.0	0.7
220.21	1.02	1.5	285.94	30DE	0.2	-3.4	-2.0	0.7

Right Table

♂ LONG	♂ LAT	♂ MAG	☉ LONG	16.00UT 1134	☿	♀	♃	♄
226.31	0.95	1.3	296.11	9JA	2.9	-3.4	-1.9	0.8
232.36	0.86	1.2	306.25	19JA	1.3	-3.4	-1.8	0.8
238.33	0.76	1.1	316.36	29JA	0.5	-3.4	-1.8	0.8
244.21	0.64	0.9	326.42	8FE	0.3	-3.3	-1.7	0.8
249.97	0.49	0.8	336.43	18FE	0.1	-3.3	-1.7	0.8
255.59	0.32	0.6	346.39	28FE	-0.3	-3.3	-1.6	0.7
261.04	0.11	0.4	356.29	10MR	-0.8	-3.3	-1.6	0.7
266.26	-0.15	0.2	6.13	20MR	-1.6	-3.4	-1.6	0.7
271.19	-0.44	-0.0	15.92	30MR	-1.3	-3.4	-1.6	0.6
275.75	-0.80	-0.3	25.65	9AP	-0.3	-3.4	-1.6	0.6
279.86	-1.22	-0.6	35.33	19AP	0.9	-3.4	-1.5	0.5
283.37	-1.72	-0.8	44.97	29AP	2.1	-3.5	-1.6	0.4
286.12	-2.30	-1.1	54.57	9MY	3.6	-3.5	-1.6	0.4
287.93	-2.97	-1.4	64.14	19MY	2.4	-3.6	-1.6	0.3
288.62	-3.71	-1.8	73.69	29MY	1.3	-3.6	-1.6	0.3
288.07	-4.48	-2.1	83.22	8JN	0.4	-3.7	-1.6	0.2
286.36	-5.20	-2.3	92.75	18JN	-0.6	-3.8	-1.7	0.3
283.80	-5.77	-2.5	102.28	28JN	-1.5	-3.9	-1.7	0.3
281.08	-6.08	-2.5	111.83	8JL	-1.3	-4.0	-1.8	0.4
278.89	-6.12	-2.3	121.39	18JL	-0.5	-4.1	-1.8	0.5
277.80	-5.91	-2.1	130.99	28JL	-0.0	-4.2	-1.9	0.5
278.06	-5.54	-1.8	140.63	7AU	0.3	-4.2	-1.9	0.6
279.64	-5.09	-1.6	150.31	17AU	0.5	-4.2	-2.0	0.6
282.38	-4.59	-1.3	160.04	27AU	1.0	-4.1	-2.1	0.7
286.10	-4.10	-1.1	169.82	6SE	2.4	-3.7	-2.1	0.7
290.58	-3.62	-0.8	179.66	16SE	1.7	-3.2	-2.2	0.7
295.67	-3.16	-0.6	189.56	26SE	-0.3	-3.8	-2.3	0.8
301.24	-2.73	-0.4	199.52	6OC	-0.8	-4.2	-2.3	0.8
307.16	-2.34	-0.2	209.53	16OC	-0.8	-4.3	-2.4	0.8
313.36	-1.97	-0.0	219.59	26OC	-0.8	-4.3	-2.4	0.8
319.77	-1.63	0.1	229.69	5NO	-0.6	-4.3	-2.4	0.8
326.33	-1.32	0.3	239.84	15NO	-0.5	-4.2	-2.4	0.7
333.01	-1.03	0.5	250.01	25NO	-0.5	-4.1	-2.3	0.7
339.76	-0.77	0.6	260.20	5DE	-0.5	-4.0	-2.3	0.7
346.56	-0.54	0.8	270.40	15DE	0.4	-3.9	-2.2	0.7
353.38	-0.33	0.9	280.60	25DE	3.0	-3.8	-2.2	0.7

1135

♂ LONG	♂ LAT	♂ MAG	☉ LONG	16.00UT	☿	♀	♃	♄
0.22	-0.14	1.0	290.78	4JA	0.9	-3.7	-2.1	0.8
7.05	0.04	1.1	300.94	14JA	0.3	-3.6	-2.0	0.8
13.86	0.19	1.2	311.07	24JA	0.1	-3.6	-2.0	0.8
20.65	0.33	1.3	321.16	3FE	-0.0	-3.5	-1.9	0.8
27.41	0.45	1.4	331.20	13FE	-0.3	-3.5	-1.8	0.8
34.13	0.56	1.5	341.18	23FE	-0.8	-3.4	-1.8	0.8
40.82	0.66	1.6	351.11	5MR	-1.5	-3.4	-1.7	0.8
47.47	0.74	1.7	0.99	15MR	-1.3	-3.3	-1.7	0.8
54.07	0.82	1.7	10.80	25MR	-0.2	-3.3	-1.6	0.7
60.64	0.88	1.8	20.56	4AP	1.1	-3.3	-1.6	0.7
67.17	0.94	1.9	30.27	14AP	2.7	-3.3	-1.5	0.6
73.67	0.99	1.9	39.93	24AP	3.2	-3.3	-1.5	0.6
80.14	1.03	1.9	49.55	4MY	1.8	-3.3	-1.5	0.5
86.57	1.07	2.0	59.13	14MY	1.0	-3.3	-1.5	0.4
92.98	1.10	2.0	68.69	24MY	0.3	-3.4	-1.5	0.4
99.37	1.12	2.0	78.22	3JN	-0.7	-3.4	-1.5	0.3
105.75	1.14	2.0	87.75	13JN	-1.6	-3.4	-1.5	0.3
112.11	1.15	2.0	97.28	23JN	-1.3	-3.5	-1.5	0.3
118.46	1.16	2.0	106.82	3JL	-0.5	-3.5	-1.5	0.3
124.82	1.16	2.0	116.38	13JL	0.1	-3.5	-1.5	0.4
131.17	1.16	2.0	125.96	23JL	0.4	-3.4	-1.6	0.4
137.53	1.15	2.0	135.57	2AU	0.7	-3.4	-1.6	0.5
143.90	1.14	2.0	145.23	12AU	1.3	-3.4	-1.7	0.5
150.28	1.12	2.0	154.93	22AU	2.9	-3.4	-1.7	0.6
156.68	1.10	2.0	164.68	1SE	1.5	-3.3	-1.8	0.7
163.10	1.08	2.0	174.50	11SE	-0.4	-3.3	-1.8	0.7
169.55	1.05	2.0	184.36	21SE	-1.0	-3.3	-1.9	0.7
176.02	1.01	2.0	194.29	1OC	-1.0	-3.3	-2.0	0.8
182.52	0.97	2.0	204.27	11OC	-0.8	-3.3	-2.0	0.8
189.05	0.92	2.0	214.31	21OC	-0.5	-3.4	-2.1	0.8
195.61	0.87	1.9	224.39	31OC	-0.4	-3.4	-2.2	0.8
202.20	0.81	1.9	234.52	10NO	-0.4	-3.4	-2.2	0.8
208.83	0.75	1.9	244.67	20NO	-0.3	-3.4	-2.2	0.8
215.49	0.67	1.8	254.85	30NO	0.6	-3.5	-2.3	0.7
222.18	0.59	1.8	265.05	10DE	3.0	-3.5	-2.3	0.7
228.90	0.50	1.7	275.25	20DE	0.6	-3.6	-2.2	0.7
235.66	0.40	1.6	285.44	30DE	0.1	-3.7	-2.2	0.7

♂ LONG	♂ LAT	♂ MAG	⊙ LONG	16.00UT	☿	♀	♃	♄
				1136	MAGNITUDES			
242.44	0.28	1.5	295.62	9JA	-0.0	-3.7	-2.2	0.8
249.25	0.16	1.5	305.77	19JA	-0.2	-3.8	-2.1	0.8
256.09	0.02	1.4	315.87	29JA	-0.4	-3.9	-2.1	0.8
262.95	-0.13	1.3	325.94	8FE	-0.9	-4.0	-2.0	0.8
269.82	-0.30	1.2	335.95	18FE	-1.4	-4.1	-1.9	0.9
276.71	-0.48	1.0	345.91	28FE	-1.2	-4.2	-1.9	0.9
283.61	-0.67	0.9	355.82	9MR	-0.2	-4.2	-1.8	0.8
290.50	-0.89	0.8	5.66	19MR	1.4	-4.3	-1.7	0.8
297.38	-1.12	0.7	15.45	29MR	3.4	-4.2	-1.7	0.8
304.23	-1.36	0.5	25.19	8AP	2.4	-3.9	-1.6	0.8
311.02	-1.62	0.4	34.87	18AP	1.4	-3.3	-1.5	0.7
317.74	-1.90	0.3	44.51	28AP	0.8	-3.1	-1.5	0.7
324.35	-2.19	0.1	54.11	8MY	0.2	-3.8	-1.5	0.6
330.81	-2.49	-0.1	63.68	18MY	-0.7	-4.1	-1.4	0.5
337.07	-2.80	-0.2	73.22	28MY	-1.6	-4.2	-1.4	0.5
343.07	-3.13	-0.4	82.76	7JN	-1.4	-4.2	-1.4	0.4
348.71	-3.45	-0.6	92.28	17JN	-0.5	-4.1	-1.4	0.3
353.91	-3.78	-0.8	101.82	27JN	0.2	-4.0	-1.4	0.3
358.51	-4.11	-1.0	111.36	7JL	0.6	-3.9	-1.4	0.3
2.37	-4.42	-1.2	120.93	17JL	1.0	-3.8	-1.4	0.4
5.29	-4.70	-1.4	130.52	27JL	1.8	-3.7	-1.4	0.4
7.03	-4.93	-1.7	140.16	6AU	3.2	-3.7	-1.4	0.5
7.43	-5.06	-1.9	149.84	16AU	1.2	-3.6	-1.4	0.5
6.40	-5.04	-2.2	159.57	26AU	-0.5	-3.6	-1.5	0.6
4.10	-4.81	-2.3	169.35	5SE	-1.1	-3.5	-1.5	0.6
1.09	-4.35	-2.4	179.19	15SE	-1.1	-3.5	-1.6	0.7
358.12	-3.71	-2.3	189.08	25SE	-0.8	-3.4	-1.6	0.7
355.93	-2.98	-2.0	199.04	5OC	-0.4	-3.4	-1.7	0.8
354.97	-2.25	-1.6	209.04	15OC	-0.3	-3.4	-1.7	0.8
355.33	-1.58	-1.3	219.10	25OC	-0.2	-3.4	-1.8	0.8
356.91	-1.02	-1.0	229.21	4NO	-0.1	-3.4	-1.9	0.8
359.52	-0.54	-0.7	239.35	14NO	0.8	-3.4	-1.9	0.8
2.94	-0.16	-0.4	249.52	24NO	2.8	-3.4	-2.0	0.8
7.01	0.16	-0.1	259.71	4DE	0.4	-3.4	-2.0	0.8
11.57	0.41	0.1	269.91	14DE	-0.1	-3.4	-2.1	0.8
16.50	0.62	0.3	280.10	24DE	-0.2	-3.4	-2.1	0.8
				1137				
21.73	0.79	0.5	290.29	3JA	-0.3	-3.4	-2.1	0.8
27.19	0.92	0.7	300.45	13JA	-0.5	-3.4	-2.1	0.8
32.81	1.03	0.9	310.58	23JA	-0.9	-3.4	-2.1	0.8
38.57	1.11	1.1	320.67	2FE	-1.3	-3.5	-2.1	0.9
44.43	1.18	1.2	330.71	12FE	-1.1	-3.5	-2.0	0.9
50.36	1.24	1.3	340.70	22FE	-0.1	-3.5	-2.0	0.9
56.36	1.28	1.4	350.63	4MR	1.7	-3.4	-1.9	0.9
62.39	1.30	1.5	0.50	14MR	3.3	-3.4	-1.9	0.9
68.46	1.32	1.6	10.32	24MR	1.8	-3.4	-1.8	0.9
74.55	1.34	1.7	20.09	3AP	1.0	-3.4	-1.7	0.9
80.67	1.34	1.8	29.79	13AP	0.6	-3.3	-1.7	0.8
86.80	1.34	1.8	39.46	23AP	0.1	-3.3	-1.6	0.8
92.94	1.33	1.9	49.08	3MY	-0.7	-3.3	-1.5	0.8
99.10	1.32	1.9	58.66	13MY	-1.6	-3.3	-1.5	0.7
105.28	1.30	2.0	68.22	23MY	-1.4	-3.3	-1.4	0.6
111.47	1.28	2.0	77.76	2JN	-0.4	-3.4	-1.4	0.6
117.68	1.25	2.0	87.29	12JN	0.3	-3.4	-1.4	0.5
123.91	1.22	2.0	96.82	22JN	0.8	-3.4	-1.3	0.4
130.17	1.19	2.0	106.36	2JL	1.3	-3.4	-1.3	0.4
136.45	1.15	2.0	115.91	12JL	2.3	-3.5	-1.3	0.3
142.77	1.11	2.0	125.49	22JL	3.0	-3.5	-1.3	0.4
149.12	1.07	2.0	135.10	1AU	1.0	-3.6	-1.3	0.4
155.50	1.02	2.0	144.76	11AU	-0.5	-3.7	-1.3	0.5
161.93	0.96	2.0	154.46	21AU	-1.2	-3.8	-1.3	0.5
168.41	0.91	1.9	164.21	31AU	-1.3	-3.8	-1.3	0.6
174.93	0.85	1.9	174.02	10SE	-0.7	-3.9	-1.3	0.6
181.51	0.78	1.9	183.89	20SE	-0.3	-4.1	-1.4	0.7
188.14	0.71	1.9	193.81	30SE	-0.2	-4.2	-1.4	0.7
194.82	0.64	1.9	203.79	10OC	-0.1	-4.3	-1.4	0.8
201.57	0.56	1.9	213.82	20OC	0.1	-4.4	-1.5	0.8
208.38	0.48	1.9	223.90	30OC	1.0	-4.4	-1.5	0.8
215.25	0.39	1.8	234.02	9NO	2.6	-4.2	-1.6	0.9
222.18	0.30	1.8	244.18	19NO	0.2	-3.7	-1.7	0.9
229.17	0.20	1.8	254.36	29NO	-0.3	-2.9	-1.7	0.9
236.23	0.09	1.7	264.56	9DE	-0.3	-3.8	-1.8	0.9
243.35	-0.02	1.7	274.76	19DE	-0.4	-4.2	-1.9	0.9
250.54	-0.13	1.7	284.95	29DE	-0.6	-4.4	-1.9	0.9

♂ LONG	♂ LAT	♂ MAG	⊙ LONG	16.00UT	☿	♀	♃	♄
				1138	MAGNITUDES			
257.78	-0.26	1.6	295.12	8JA	-0.9	-4.4	-2.0	0.8
265.07	-0.38	1.5	305.27	18JA	-1.1	-4.3	-2.0	0.8
272.42	-0.51	1.5	315.38	28JA	-1.0	-4.2	-2.0	0.9
279.82	-0.65	1.4	325.45	7FE	-0.1	-4.1	-2.0	0.9
287.27	-0.78	1.4	335.47	17FE	2.0	-3.9	-2.0	1.0
294.75	-0.92	1.3	345.43	27FE	2.6	-3.8	-2.0	1.0
302.25	-1.06	1.2	355.34	9MR	1.3	-3.7	-2.0	1.0
309.78	-1.19	1.2	5.19	19MR	0.7	-3.7	-1.9	1.0
317.32	-1.33	1.1	14.98	29MR	0.4	-3.6	-1.9	1.0
324.85	-1.45	1.0	24.72	8AP	-0.0	-3.5	-1.8	1.0
332.36	-1.57	1.0	34.41	18AP	-0.7	-3.5	-1.8	1.0
339.83	-1.68	0.9	44.04	28AP	-1.6	-3.4	-1.7	0.9
347.25	-1.78	0.8	53.65	8MY	-1.4	-3.4	-1.6	0.9
354.60	-1.86	0.7	63.22	18MY	-0.4	-3.3	-1.6	0.8
1.84	-1.93	0.7	72.77	28MY	0.4	-3.3	-1.5	0.8
8.97	-1.98	0.6	82.30	7JN	1.0	-3.3	-1.5	0.7
15.94	-2.01	0.5	91.83	17JN	1.7	-3.3	-1.4	0.7
22.74	-2.02	0.4	101.36	27JN	2.9	-3.3	-1.4	0.6
29.31	-2.01	0.3	110.90	7JL	2.5	-3.3	-1.3	0.5
35.62	-1.98	0.2	120.47	17JL	0.9	-3.3	-1.3	0.5
41.62	-1.92	0.1	130.06	27JL	-0.5	-3.4	-1.3	0.4
47.23	-1.83	0.0	139.69	6AU	-1.3	-3.4	-1.2	0.5
52.37	-1.72	-0.1	149.37	16AU	-1.3	-3.4	-1.2	0.5
56.95	-1.57	-0.3	159.09	26AU	-0.7	-3.5	-1.2	0.6
60.82	-1.38	-0.5	168.87	5SE	-0.3	-3.5	-1.2	0.6
63.82	-1.15	-0.6	178.71	15SE	-0.0	-3.5	-1.2	0.7
65.75	-0.85	-0.8	188.60	25SE	0.1	-3.5	-1.2	0.7
66.40	-0.50	-1.1	198.55	5OC	0.3	-3.4	-1.3	0.8
65.61	-0.08	-1.3	208.56	15OC	1.3	-3.4	-1.3	0.8
63.40	0.40	-1.5	218.61	25OC	2.3	-3.4	-1.3	0.9
60.05	0.89	-1.6	228.71	4NO	0.0	-3.4	-1.4	0.9
56.24	1.35	-1.7	238.85	14NO	-0.4	-3.3	-1.4	0.9
52.78	1.72	-1.4	249.02	24NO	-0.5	-3.3	-1.5	0.9
50.31	1.98	-1.1	259.21	4DE	-0.5	-3.3	-1.5	1.0
49.17	2.14	-0.8	269.41	14DE	-0.7	-3.3	-1.6	1.0
49.36	2.22	-0.5	279.61	24DE	-0.9	-3.4	-1.7	1.0
				1139				
50.72	2.25	-0.2	289.79	3JA	-0.9	-3.4	-1.7	1.0
53.06	2.24	0.1	299.96	13JA	-0.9	-3.4	-1.8	0.9
56.16	2.21	0.3	310.08	23JA	0.0	-3.4	-1.8	0.9
59.87	2.16	0.6	320.18	2FE	2.3	-3.4	-1.9	1.0
64.06	2.10	0.8	330.22	12FE	2.0	-3.5	-1.9	1.0
68.61	2.04	0.9	340.21	22FE	0.9	-3.5	-2.0	1.1
73.45	1.98	1.1	350.15	4MR	0.5	-3.6	-2.0	1.1
78.53	1.91	1.2	0.03	14MR	0.3	-3.6	-2.0	1.1
83.80	1.84	1.4	9.85	24MR	-0.1	-3.7	-2.0	1.1
89.21	1.77	1.5	19.62	3AP	-0.7	-3.8	-2.0	1.1
94.76	1.70	1.6	29.33	13AP	-1.6	-3.9	-1.9	1.1
100.40	1.63	1.7	38.99	23AP	-1.4	-3.9	-1.9	1.1
106.14	1.56	1.7	48.62	3MY	-0.4	-4.0	-1.8	1.1
111.97	1.49	1.8	58.20	13MY	0.5	-4.1	-1.8	1.0
117.86	1.42	1.9	67.76	23MY	1.3	-4.2	-1.7	1.0
123.82	1.35	1.9	77.30	2JN	2.3	-4.2	-1.7	1.0
129.85	1.27	1.9	86.83	12JN	3.4	-4.0	-1.6	0.9
135.94	1.20	2.0	96.36	22JN	2.0	-3.6	-1.5	0.8
142.10	1.12	2.0	105.90	2JL	0.7	-3.0	-1.5	0.8
148.31	1.04	2.0	115.45	12JL	-0.6	-3.5	-1.4	0.7
154.60	0.96	2.0	125.02	22JL	-1.4	-4.0	-1.4	0.6
160.95	0.88	2.0	134.64	1AU	-1.3	-4.2	-1.3	0.6
167.36	0.80	2.0	144.29	11AU	-0.6	-4.2	-1.3	0.6
173.85	0.71	1.9	153.98	21AU	-0.2	-4.2	-1.3	0.6
180.42	0.62	1.9	163.74	31AU	0.1	-4.1	-1.2	0.6
187.06	0.53	1.9	173.54	10SE	0.2	-4.0	-1.2	0.7
193.77	0.44	1.8	183.40	20SE	0.5	-3.9	-1.2	0.7
200.57	0.35	1.8	193.33	30SE	1.7	-3.9	-1.2	0.8
207.45	0.25	1.7	203.30	10OC	2.1	-3.8	-1.2	0.8
214.41	0.15	1.7	213.33	20OC	-0.1	-3.7	-1.2	0.9
221.45	0.05	1.7	223.42	30OC	-0.6	-3.6	-1.2	0.9
228.57	-0.05	1.7	233.53	9NO	-0.6	-3.6	-1.3	1.0
235.78	-0.15	1.6	243.69	19NO	-0.6	-3.5	-1.3	1.0
243.07	-0.26	1.6	253.87	29NO	-0.8	-3.5	-1.3	1.0
250.43	-0.36	1.6	264.07	9DE	-0.8	-3.5	-1.4	1.1
257.86	-0.47	1.6	274.26	19DE	-0.8	-3.4	-1.4	1.1
265.37	-0.57	1.6	284.46	29DE	-0.7	-3.4	-1.5	1.1

♂ LONG	♂ LAT	♂ MAG	⊙ LONG	16.00UT	☿	♀	♃	♄
				1140	MAGNITUDES			
272.93	-0.67	1.5	294.63	8JA	0.1	-3.4	-1.6	1.1
280.56	-0.77	1.5	304.78	18JA	2.6	-3.3	-1.6	1.1
288.23	-0.87	1.5	314.90	28JA	1.5	-3.3	-1.7	1.1
295.94	-0.96	1.5	324.97	7FE	0.6	-3.3	-1.8	1.1
303.68	-1.04	1.4	334.99	17FE	0.3	-3.3	-1.8	1.1
311.44	-1.12	1.4	344.95	27FE	0.1	-3.3	-1.9	1.2
319.21	-1.18	1.4	354.86	8MR	-0.2	-3.3	-1.9	1.2
326.98	-1.24	1.4	4.71	18MR	-0.8	-3.4	-2.0	1.2
334.73	-1.28	1.3	14.51	28MR	-1.6	-3.4	-2.0	1.3
342.45	-1.31	1.3	24.24	7AP	-1.4	-3.4	-2.0	1.3
350.13	-1.33	1.3	33.93	17AP	-0.3	-3.5	-2.0	1.3
357.75	-1.33	1.2	43.58	27AP	0.7	-3.5	-2.0	1.3
5.30	-1.31	1.2	53.18	7MY	1.7	-3.5	-2.0	1.3
12.78	-1.29	1.2	62.75	17MY	3.1	-3.4	-2.0	1.2
20.16	-1.24	1.2	72.30	27MY	2.9	-3.4	-1.9	1.2
27.42	-1.19	1.1	81.83	6JN	1.6	-3.4	-1.9	1.2
34.58	-1.11	1.1	91.36	16JN	0.5	-3.3	-1.8	1.1
41.59	-1.02	1.1	100.89	26JN	-0.6	-3.3	-1.7	1.0
48.46	-0.92	1.0	110.43	6JL	-1.5	-3.3	-1.7	1.0
55.17	-0.80	1.0	120.00	16JL	-1.3	-3.3	-1.6	0.9
61.69	-0.67	1.0	129.59	26JL	-0.6	-3.3	-1.6	0.8
68.00	-0.52	0.9	139.22	5AU	-0.1	-3.3	-1.5	0.8
74.08	-0.36	0.8	148.89	15AU	0.2	-3.4	-1.4	0.7
79.89	-0.17	0.7	158.62	25AU	0.4	-3.4	-1.4	0.7
85.39	0.03	0.7	168.39	4SE	0.8	-3.4	-1.4	0.8
90.52	0.25	0.5	178.23	14SE	2.1	-3.5	-1.3	0.8
95.20	0.51	0.4	188.12	24SE	1.9	-3.5	-1.3	0.8
99.35	0.79	0.3	198.06	4OC	-0.2	-3.6	-1.3	0.9
102.83	1.11	0.1	208.07	14OC	-0.7	-3.6	-1.3	0.9
105.48	1.47	-0.1	218.12	24OC	-0.8	-3.7	-1.3	1.0
107.15	1.88	-0.3	228.22	3NO	-0.8	-3.8	-1.3	1.0
107.61	2.32	-0.5	238.36	13NO	-0.7	-3.9	-1.3	1.1
106.71	2.79	-0.7	248.53	23NO	-0.6	-4.0	-1.3	1.1
104.45	3.25	-0.9	258.72	3DE	-0.6	-4.1	-1.3	1.2
101.04	3.63	-1.1	268.92	13DE	-0.6	-4.2	-1.3	1.2
97.09	3.87	-1.2	279.12	23DE	0.2	-4.3	-1.4	1.2
				1141				
93.35	3.96	-1.0	289.30	2JA	2.9	-4.4	-1.4	1.2
90.48	3.89	-0.7	299.47	12JA	1.1	-4.4	-1.5	1.3
88.86	3.72	-0.5	309.60	22JA	0.4	-4.2	-1.5	1.3
88.57	3.50	-0.2	319.69	1FE	0.2	-3.7	-1.6	1.3
89.46	3.26	0.1	329.74	11FE	0.0	-3.3	-1.7	1.3
91.37	3.02	0.3	339.73	21FE	-0.3	-3.8	-1.7	1.3
94.09	2.79	0.5	349.67	3MR	-0.8	-4.2	-1.8	1.3
97.46	2.57	0.7	359.55	13MR	-1.6	-4.3	-1.9	1.4
101.36	2.38	0.9	9.38	23MR	-1.3	-4.3	-1.9	1.4
105.66	2.19	1.1	19.14	2AP	-0.3	-4.2	-2.0	1.4
110.30	2.02	1.2	28.86	12AP	0.9	-4.1	-2.0	1.4
115.22	1.86	1.3	38.53	22AP	2.3	-4.0	-2.1	1.4
120.36	1.71	1.4	48.15	2MY	3.8	-3.9	-2.1	1.4
125.70	1.56	1.5	57.74	12MY	2.2	-3.8	-2.1	1.3
131.20	1.43	1.6	67.30	22MY	1.2	-3.7	-2.1	1.3
136.85	1.30	1.6	76.84	1JN	0.4	-3.6	-2.1	1.3
142.64	1.17	1.7	86.37	11JN	-0.6	-3.5	-2.1	1.2
148.55	1.05	1.7	95.90	21JN	-1.5	-3.5	-2.0	1.2
154.58	0.93	1.8	105.43	1JL	-1.4	-3.4	-2.0	1.1
160.71	0.81	1.8	114.99	11JL	-0.5	-3.4	-1.9	1.1
166.96	0.69	1.8	124.56	21JL	-0.0	-3.4	-1.9	1.0
173.31	0.58	1.8	134.17	31JL	0.3	-3.4	-1.8	1.0
179.76	0.46	1.8	143.82	10AU	0.6	-3.4	-1.7	0.9
186.32	0.35	1.8	153.52	20AU	1.1	-3.3	-1.7	0.9
192.97	0.24	1.8	163.26	30AU	2.5	-3.3	-1.6	0.9
199.74	0.13	1.8	173.07	9SE	1.7	-3.4	-1.6	0.9
206.60	0.02	1.7	182.93	19SE	-0.3	-3.4	-1.5	0.9
213.57	-0.08	1.7	192.84	29SE	-0.9	-3.4	-1.5	1.0
220.64	-0.19	1.7	202.82	9OC	-0.9	-3.4	-1.4	1.0
227.80	-0.29	1.6	212.84	19OC	-0.9	-3.4	-1.4	1.1
235.07	-0.39	1.6	222.92	29OC	-0.6	-3.4	-1.4	1.1
242.43	-0.49	1.6	233.04	8NO	-0.5	-3.5	-1.4	1.2
249.87	-0.58	1.5	243.19	18NO	-0.5	-3.5	-1.4	1.2
257.41	-0.67	1.5	253.37	28NO	-0.4	-3.5	-1.3	1.3
265.01	-0.75	1.4	263.57	8DE	0.4	-3.5	-1.4	1.3
272.69	-0.83	1.4	273.77	18DE	3.0	-3.4	-1.4	1.4
280.43	-0.90	1.4	283.96	28DE	0.8	-3.4	-1.4	1.4

♂ LONG	♂ LAT	♂ MAG	⊙ LONG	16.00UT	☿	♀	♃	♄
				1142	MAGNITUDES			
288.22	-0.96	1.4	294.14	7JA	0.2	-3.4	-1.4	1.4
296.04	-1.01	1.4	304.29	17JA	0.0	-3.4	-1.4	1.4
303.90	-1.05	1.4	314.40	27JA	-0.1	-3.3	-1.5	1.3
311.77	-1.08	1.4	324.48	6FE	-0.4	-3.3	-1.5	1.3
319.64	-1.10	1.4	334.50	16FE	-0.8	-3.3	-1.6	1.2
327.50	-1.11	1.4	344.47	26FE	-1.5	-3.3	-1.6	1.2
335.34	-1.11	1.4	354.38	8MR	-1.3	-3.3	-1.7	1.1
343.14	-1.09	1.4	4.23	18MR	-0.3	-3.4	-1.8	1.2
350.90	-1.06	1.4	14.03	28MR	1.2	-3.4	-1.9	1.2
358.60	-1.02	1.4	23.78	7AP	2.9	-3.4	-1.9	1.2
6.24	-0.97	1.4	33.46	17AP	2.9	-3.4	-2.0	1.1
13.80	-0.91	1.4	43.11	27AP	1.6	-3.5	-2.1	1.1
21.28	-0.83	1.5	52.72	7MY	0.9	-3.5	-2.1	1.1
28.66	-0.75	1.5	62.29	17MY	0.2	-3.6	-2.2	1.1
35.95	-0.66	1.5	71.84	27MY	-0.6	-3.6	-2.2	1.0
43.14	-0.55	1.5	81.38	6JN	-1.6	-3.7	-2.2	1.0
50.22	-0.45	1.5	90.90	16JN	-1.4	-3.8	-2.3	1.0
57.19	-0.33	1.5	100.44	26JN	-0.5	-3.9	-2.3	0.9
64.04	-0.21	1.5	109.98	6JL	0.1	-4.0	-2.2	0.9
70.77	-0.08	1.5	119.54	16JL	0.5	-4.1	-2.2	0.8
77.37	0.06	1.4	129.13	26JL	0.8	-4.2	-2.2	0.8
83.83	0.20	1.4	138.76	5AU	1.5	-4.2	-2.1	0.7
90.15	0.35	1.4	148.43	15AU	3.0	-4.2	-2.0	0.7
96.30	0.51	1.3	158.15	25AU	1.4	-4.1	-2.0	0.6
102.28	0.68	1.3	167.92	4SE	-0.4	-3.6	-1.9	0.6
108.05	0.86	1.2	177.75	14SE	-1.0	-3.3	-1.9	0.6
113.61	1.05	1.2	187.64	24SE	-1.1	-3.8	-1.8	0.6
118.89	1.25	1.1	197.58	4OC	-0.8	-4.2	-1.7	0.7
123.86	1.48	1.0	207.58	14OC	-0.5	-4.3	-1.7	0.8
128.45	1.72	0.8	217.63	24OC	-0.4	-4.3	-1.6	0.8
132.59	1.98	0.7	227.73	3NO	-0.3	-4.3	-1.6	0.9
136.17	2.28	0.5	237.86	13NO	-0.2	-4.2	-1.5	1.0
139.05	2.60	0.3	248.03	23NO	0.6	-4.1	-1.5	1.0
141.07	2.96	0.1	258.22	3DE	3.0	-4.0	-1.5	1.0
142.06	3.34	-0.1	268.42	13DE	0.6	-3.9	-1.5	1.1
141.82	3.73	-0.4	278.62	23DE	0.0	-3.8	-1.5	1.1
				1143				
140.26	4.10	-0.6	288.81	2JA	-0.1	-3.7	-1.5	1.1
137.45	4.40	-0.8	298.97	12JA	-0.2	-3.6	-1.5	1.1
133.74	4.55	-1.0	309.11	22JA	-0.4	-3.6	-1.5	1.1
129.80	4.53	-0.9	319.21	1FE	-0.9	-3.5	-1.5	1.1
126.33	4.34	-0.7	329.25	11FE	-1.3	-3.5	-1.5	1.0
123.88	4.03	-0.5	339.26	21FE	-1.2	-3.4	-1.6	1.0
122.73	3.66	-0.3	349.20	3MR	-0.2	-3.4	-1.6	0.9
122.83	3.28	-0.0	359.08	13MR	1.4	-3.3	-1.6	0.9
124.07	2.92	0.2	8.91	23MR	3.5	-3.3	-1.7	0.9
126.25	2.58	0.4	18.68	2AP	2.1	-3.3	-1.8	0.9
129.20	2.27	0.6	28.40	12AP	1.2	-3.3	-1.8	0.9
132.78	1.99	0.8	38.07	22AP	0.7	-3.3	-1.9	0.9
136.87	1.73	0.9	47.69	2MY	0.1	-3.3	-2.0	0.9
141.36	1.50	1.0	57.28	12MY	-0.7	-3.3	-2.0	0.9
146.21	1.28	1.1	66.84	22MY	-1.6	-3.4	-2.1	0.9
151.35	1.09	1.2	76.38	1JN	-1.4	-3.4	-2.2	0.8
156.74	0.90	1.3	85.91	11JN	-0.5	-3.4	-2.2	0.8
162.35	0.73	1.4	95.44	21JN	0.2	-3.5	-2.3	0.8
168.16	0.56	1.4	104.97	1JL	0.6	-3.5	-2.3	0.7
174.15	0.41	1.5	114.52	11JL	1.1	-3.5	-2.4	0.7
180.30	0.26	1.5	124.10	21JL	1.9	-3.4	-2.4	0.6
186.62	0.12	1.5	133.71	31JL	3.2	-3.4	-2.4	0.6
193.08	-0.01	1.5	143.35	10AU	1.2	-3.4	-2.4	0.5
199.69	-0.14	1.6	153.05	20AU	-0.4	-3.4	-2.3	0.5
206.43	-0.26	1.6	162.79	30AU	-1.2	-3.3	-2.3	0.4
213.30	-0.37	1.6	172.59	9SE	-1.2	-3.3	-2.2	0.4
220.31	-0.48	1.6	182.45	19SE	-0.8	-3.3	-2.2	0.3
227.43	-0.58	1.5	192.36	29SE	-0.4	-3.3	-2.1	0.3
234.67	-0.67	1.5	202.33	9OC	-0.2	-3.3	-2.0	0.4
242.03	-0.76	1.5	212.36	19OC	-0.2	-3.4	-2.0	0.4
249.49	-0.84	1.5	222.43	29OC	-0.0	-3.4	-1.9	0.5
257.04	-0.91	1.5	232.54	8NO	0.8	-3.4	-1.8	0.6
264.68	-0.96	1.5	242.70	18NO	2.8	-3.4	-1.8	0.6
272.39	-1.01	1.4	252.87	28NO	0.3	-3.5	-1.7	0.7
280.17	-1.05	1.4	263.07	8DE	-0.2	-3.5	-1.7	0.7
288.01	-1.08	1.4	273.27	18DE	-0.2	-3.6	-1.6	0.8
295.88	-1.09	1.4	283.46	28DE	-0.3	-3.7	-1.6	0.8

♂ LONG	LAT	MAG	☉ LONG	16.00UT	☿	♀	♃	♄
				1144				
303.77	-1.10	1.4	293.64	7JA	-0.5	-3.7	-1.6	0.8
311.68	-1.09	1.3	303.80	17JA	-0.9	-3.8	-1.6	0.8
319.59	-1.07	1.3	313.91	27JA	-1.2	-3.9	-1.5	0.8
327.48	-1.04	1.3	323.99	6FE	-1.1	-4.0	-1.5	0.8
335.34	-0.99	1.3	334.02	16FE	-0.2	-4.1	-1.5	0.8
343.16	-0.94	1.3	343.99	26FE	1.7	-4.2	-1.6	0.8
350.93	-0.88	1.4	353.90	7MR	3.1	-4.2	-1.6	0.7
358.64	-0.81	1.4	3.76	17MR	1.6	-4.3	-1.6	0.7
6.28	-0.73	1.4	13.56	27MR	0.9	-4.2	-1.6	0.6
13.84	-0.64	1.5	23.31	6AP	0.5	-3.9	-1.7	0.6
21.32	-0.55	1.5	33.00	16AP	0.0	-3.3	-1.7	0.7
28.71	-0.45	1.5	42.65	26AP	-0.7	-3.1	-1.8	0.7
36.02	-0.35	1.6	52.26	6MY	-1.6	-3.8	-1.8	0.7
43.23	-0.25	1.6	61.83	16MY	-1.4	-4.1	-1.9	0.7
50.35	-0.14	1.6	71.38	26MY	-0.5	-4.2	-2.0	0.7
57.38	-0.03	1.7	80.92	5JN	0.3	-4.2	-2.0	0.6
64.31	0.08	1.7	90.44	15JN	0.8	-4.1	-2.1	0.6
71.15	0.20	1.7	99.97	25JN	1.5	-4.0	-2.2	0.6
77.90	0.31	1.7	109.52	5JL	2.5	-3.9	-2.2	0.6
84.56	0.43	1.7	119.08	15JL	2.9	-3.8	-2.3	0.5
91.12	0.54	1.7	128.66	25JL	1.0	-3.7	-2.4	0.5
97.59	0.66	1.7	138.29	4AU	-0.5	-3.7	-2.4	0.4
103.96	0.78	1.7	147.96	14AU	-1.3	-3.6	-2.4	0.4
110.22	0.90	1.7	157.68	24AU	-1.3	-3.6	-2.5	0.3
116.39	1.02	1.7	167.45	3SE	-0.7	-3.5	-2.5	0.3
122.43	1.15	1.7	177.27	13SE	-0.3	-3.5	-2.4	0.2
128.35	1.28	1.6	187.16	23SE	-0.1	-3.4	-2.4	0.1
134.13	1.41	1.5	197.10	3OC	-0.0	-3.4	-2.4	0.1
139.74	1.55	1.5	207.10	13OC	0.2	-3.4	-2.3	0.1
145.17	1.70	1.4	217.14	23OC	1.1	-3.4	-2.2	0.2
150.38	1.86	1.3	227.24	2NO	2.6	-3.4	-2.2	0.2
155.32	2.03	1.2	237.37	12NO	0.2	-3.4	-2.1	0.3
159.96	2.21	1.0	247.54	22NO	-0.3	-3.4	-2.0	0.4
164.20	2.40	0.9	257.73	2DE	-0.4	-3.4	-2.0	0.4
167.97	2.61	0.7	267.92	12DE	-0.4	-3.4	-1.9	0.5
171.15	2.84	0.5	278.12	22DE	-0.6	-3.4	-1.8	0.5
				1145				
173.59	3.08	0.3	288.31	1JA	-0.9	-3.4	-1.8	0.6
175.13	3.33	0.0	298.48	11JA	-1.0	-3.4	-1.7	0.6
175.59	3.58	-0.2	308.61	21JA	-0.9	-3.4	-1.7	0.6
174.81	3.80	-0.5	318.71	31JA	-0.1	-3.5	-1.6	0.6
172.74	3.95	-0.8	328.76	10FE	2.0	-3.5	-1.6	0.6
169.57	3.99	-1.0	338.76	20FE	2.4	-3.5	-1.6	0.6
165.75	3.87	-1.1	348.71	2MR	1.1	-3.4	-1.6	0.6
162.00	3.60	-1.0	358.59	12MR	0.7	-3.4	-1.6	0.6
158.97	3.21	-0.8	8.43	22MR	0.3	-3.4	-1.6	0.5
157.11	2.77	-0.6	18.20	1AP	-0.1	-3.4	-1.6	0.5
156.56	2.32	-0.4	27.92	11AP	-0.7	-3.3	-1.6	0.4
157.25	1.90	-0.1	37.59	21AP	-1.6	-3.3	-1.6	0.5
159.03	1.51	0.1	47.22	1MY	-1.4	-3.3	-1.7	0.5
161.71	1.17	0.2	56.81	11MY	-0.4	-3.3	-1.7	0.5
165.13	0.86	0.4	66.37	21MY	0.4	-3.3	-1.7	0.5
169.16	0.59	0.5	75.92	31MY	1.1	-3.4	-1.8	0.5
173.70	0.34	0.7	85.45	10JN	1.9	-3.4	-1.8	0.5
178.66	0.12	0.8	94.98	20JN	3.2	-3.4	-1.9	0.5
183.99	-0.07	0.9	104.51	30JN	2.4	-3.4	-2.0	0.5
189.63	-0.25	0.9	114.06	10JL	0.8	-3.5	-2.0	0.5
195.54	-0.42	1.0	123.63	20JL	-0.5	-3.5	-2.1	0.4
201.69	-0.56	1.1	133.24	30JL	-1.4	-3.6	-2.2	0.4
208.07	-0.69	1.1	142.88	9AU	-1.3	-3.7	-2.2	0.3
214.64	-0.81	1.1	152.58	19AU	-0.7	-3.8	-2.3	0.3
221.41	-0.92	1.2	162.32	29AU	-0.2	-3.8	-2.3	0.2
228.33	-1.01	1.2	172.11	8SE	0.0	-3.9	-2.4	0.2
235.42	-1.09	1.2	181.97	18SE	0.1	-4.1	-2.4	0.1
242.65	-1.15	1.2	191.88	28SE	0.4	-4.2	-2.4	0.0
250.01	-1.20	1.3	201.85	8OC	1.4	-4.3	-2.4	-0.0
257.48	-1.24	1.3	211.87	18OC	2.4	-4.3	-2.4	-0.1
265.06	-1.26	1.3	221.94	28OC	0.0	-4.3	-2.4	-0.1
272.73	-1.27	1.3	232.05	7NO	-0.5	-4.2	-2.4	-0.0
280.47	-1.27	1.3	242.21	17NO	-0.5	-3.6	-2.3	0.1
288.27	-1.25	1.3	252.38	27NO	-0.6	-2.9	-2.3	0.1
296.10	-1.22	1.3	262.57	7DE	-0.7	-3.8	-2.2	0.2
303.97	-1.18	1.4	272.78	17DE	-0.8	-4.2	-2.1	0.3
311.84	-1.12	1.4	282.97	27DE	-0.9	-4.4	-2.0	0.3

♂ LONG	LAT	MAG	☉ LONG	16.00UT	☿	♀	♃	♄
				1146				
319.71	-1.06	1.4	293.15	6JA	-0.8	-4.4	-2.0	0.4
327.56	-0.99	1.4	303.30	16JA	0.0	-4.3	-1.9	0.4
335.38	-0.91	1.4	313.42	26JA	2.4	-4.2	-1.8	0.5
343.15	-0.82	1.5	323.50	5FE	1.8	-4.1	-1.8	0.5
350.87	-0.72	1.5	333.53	15FE	0.8	-3.9	-1.7	0.5
358.52	-0.63	1.5	343.50	25FE	0.5	-3.8	-1.7	0.5
6.10	-0.52	1.5	353.42	7MR	0.2	-3.7	-1.6	0.5
13.61	-0.42	1.5	3.28	17MR	-0.1	-3.7	-1.6	0.5
21.03	-0.31	1.5	13.09	27MR	-0.7	-3.6	-1.6	0.4
28.37	-0.21	1.6	22.83	6AP	-1.6	-3.5	-1.6	0.4
35.63	-0.10	1.6	32.53	16AP	-1.4	-3.5	-1.5	0.3
42.80	0.00	1.6	42.18	26AP	-0.4	-3.4	-1.5	0.3
49.88	0.11	1.6	51.79	6MY	0.6	-3.4	-1.5	0.3
56.89	0.21	1.7	61.37	16MY	1.5	-3.3	-1.6	0.4
63.81	0.31	1.7	70.92	26MY	2.6	-3.3	-1.6	0.4
70.65	0.41	1.8	80.46	5JN	3.3	-3.3	-1.6	0.4
77.42	0.50	1.8	89.99	15JN	1.9	-3.3	-1.6	0.4
84.11	0.60	1.8	99.52	25JN	0.6	-3.3	-1.7	0.4
90.74	0.69	1.9	109.06	5JL	-0.5	-3.3	-1.7	0.4
97.30	0.78	1.9	118.62	15JL	-1.4	-3.3	-1.8	0.4
103.79	0.86	1.9	128.20	25JL	-1.4	-3.4	-1.8	0.4
110.22	0.94	1.9	137.83	4AU	-0.6	-3.4	-1.9	0.3
116.59	1.03	1.9	147.49	14AU	-0.1	-3.4	-1.9	0.3
122.91	1.11	1.9	157.21	24AU	0.1	-3.5	-2.0	0.2
129.16	1.18	1.9	166.97	3SE	0.3	-3.5	-2.1	0.2
135.35	1.26	1.9	176.80	13SE	0.6	-3.5	-2.1	0.1
141.47	1.34	1.9	186.67	23SE	1.8	-3.5	-2.2	0.1
147.52	1.41	1.8	196.61	3OC	2.1	-3.4	-2.3	-0.0
153.49	1.49	1.8	206.61	13OC	-0.1	-3.4	-2.3	-0.1
159.38	1.56	1.7	216.65	23OC	-0.7	-3.4	-2.3	-0.1
165.16	1.63	1.7	226.74	2NO	-0.7	-3.4	-2.4	-0.2
170.83	1.71	1.6	236.88	12NO	-0.7	-3.3	-2.4	-0.2
176.36	1.79	1.5	247.04	22NO	-0.8	-3.3	-2.4	-0.1
181.74	1.86	1.4	257.23	2DE	-0.7	-3.3	-2.3	-0.0
186.91	1.94	1.2	267.43	12DE	-0.7	-3.3	-2.3	0.0
191.84	2.02	1.1	277.62	22DE	-0.7	-3.4	-2.2	0.1
				1147				
196.49	2.09	0.9	287.81	1JA	0.1	-3.4	-2.2	0.2
200.78	2.17	0.7	297.98	11JA	2.7	-3.4	-2.1	0.2
204.62	2.24	0.5	308.12	21JA	1.4	-3.4	-2.0	0.3
207.91	2.30	0.3	318.22	31JA	0.5	-3.4	-1.9	0.3
210.50	2.35	0.0	328.28	10FE	0.3	-3.5	-1.9	0.3
212.24	2.37	-0.2	338.28	20FE	0.1	-3.5	-1.8	0.4
212.93	2.35	-0.5	348.23	2MR	-0.2	-3.6	-1.7	0.4
212.42	2.28	-0.8	358.12	12MR	-0.8	-3.6	-1.7	0.4
210.64	2.12	-1.1	7.95	22MR	-1.6	-3.7	-1.6	0.4
207.74	1.86	-1.4	17.73	1AP	-1.4	-3.8	-1.6	0.4
204.15	1.49	-1.5	27.45	11AP	-0.4	-3.9	-1.5	0.3
200.58	1.05	-1.4	37.12	21AP	0.8	-3.9	-1.5	0.3
197.69	0.58	-1.3	46.76	1MY	1.9	-4.0	-1.5	0.3
195.98	0.13	-1.1	56.35	11MY	3.4	-4.1	-1.5	0.2
195.62	-0.27	-0.9	65.91	21MY	2.6	-4.2	-1.5	0.2
196.56	-0.61	-0.6	75.46	31MY	1.4	-4.2	-1.4	0.3
198.65	-0.90	-0.3	84.99	10JN	0.5	-4.0	-1.5	0.3
201.71	-1.14	-0.3	94.52	20JN	-0.6	-3.6	-1.5	0.3
205.56	-1.33	-0.1	104.05	30JN	-1.5	-2.9	-1.5	0.3
210.08	-1.49	0.0	113.60	10JL	-1.4	-3.5	-1.5	0.3
215.15	-1.61	0.2	123.17	20JL	-0.6	-4.0	-1.5	0.3
220.67	-1.71	0.3	132.78	30JL	-0.1	-4.2	-1.6	0.3
226.58	-1.78	0.4	142.42	9AU	0.2	-4.2	-1.6	0.3
232.82	-1.83	0.5	152.10	19AU	0.5	-4.2	-1.7	0.3
239.34	-1.86	0.5	161.84	29AU	0.9	-4.1	-1.7	0.2
246.10	-1.86	0.6	171.64	8SE	2.2	-4.0	-1.8	0.2
253.07	-1.85	0.7	181.49	18SE	1.9	-3.9	-1.8	0.1
260.22	-1.82	0.8	191.40	28SE	-0.2	-3.9	-1.9	0.1
267.53	-1.77	0.8	201.36	8OC	-0.8	-3.8	-2.0	0.0
274.95	-1.71	0.9	211.38	18OC	-0.8	-3.7	-2.0	-0.1
282.48	-1.63	0.9	221.45	28OC	-0.8	-3.6	-2.1	-0.1
290.08	-1.54	1.0	231.56	7NO	-0.7	-3.6	-2.1	-0.2
297.74	-1.44	1.1	241.71	17NO	-0.6	-3.5	-2.2	-0.3
305.43	-1.33	1.1	251.89	27NO	-0.6	-3.5	-2.2	-0.2
313.15	-1.21	1.2	262.08	7DE	-0.5	-3.5	-2.2	-0.2
320.86	-1.09	1.2	272.28	17DE	0.3	-3.4	-2.2	-0.1
328.55	-0.96	1.3	282.48	27DE	2.9	-3.4	-2.2	-0.0

Left Table

♂ LONG	LAT	MAG	☉ LONG	16.00UT	☿	♀	♃	♄
				1148				
336.22	-0.83	1.3	292.66	6JA	1.0	-3.4	-2.2	0.0
343.84	-0.70	1.4	302.81	16JA	0.3	-3.3	-2.1	0.1
351.40	-0.57	1.4	312.94	26JA	0.1	-3.3	-2.1	0.2
358.91	-0.44	1.5	323.01	5FE	-0.0	-3.3	-2.0	0.2
6.35	-0.32	1.5	333.04	15FE	-0.3	-3.3	-2.0	0.3
13.72	-0.19	1.6	343.02	25FE	-0.8	-3.3	-1.9	0.3
21.01	-0.07	1.6	352.94	6MR	-1.5	-3.3	-1.8	0.3
28.22	0.04	1.7	2.80	16MR	-1.3	-3.4	-1.7	0.3
35.35	0.15	1.7	12.61	26MR	-0.3	-3.4	-1.7	0.3
42.41	0.25	1.7	22.36	5AP	1.0	-3.4	-1.6	0.3
49.39	0.35	1.8	32.06	15AP	2.5	-3.5	-1.6	0.3
56.30	0.44	1.8	41.71	25AP	3.5	-3.5	-1.5	0.3
63.14	0.53	1.8	51.32	5MY	2.0	-3.5	-1.5	0.3
69.91	0.61	1.8	60.90	15MY	1.1	-3.4	-1.4	0.2
76.62	0.69	1.8	70.46	25MY	0.3	-3.4	-1.4	0.2
83.27	0.76	1.8	79.99	4JN	-0.6	-3.4	-1.4	0.2
89.87	0.83	1.8	89.52	14JN	-1.5	-3.3	-1.4	0.3
96.42	0.90	1.9	99.05	24JN	-1.4	-3.3	-1.3	0.3
102.92	0.95	1.9	108.59	4JL	-0.6	-3.3	-1.3	0.3
109.38	1.01	1.9	118.15	14JL	0.0	-3.3	-1.3	0.3
115.81	1.06	2.0	127.74	24JL	0.4	-3.3	-1.4	0.3
122.21	1.10	2.0	137.35	3AU	0.7	-3.3	-1.4	0.3
128.57	1.15	2.0	147.02	13AU	1.2	-3.4	-1.4	0.3
134.91	1.18	2.0	156.73	23AU	2.7	-3.4	-1.4	0.3
141.23	1.22	2.0	166.49	2SE	1.7	-3.4	-1.5	0.3
147.52	1.25	2.0	176.32	12SE	-0.3	-3.5	-1.5	0.2
153.79	1.28	2.0	186.20	22SE	-0.9	-3.5	-1.6	0.2
160.03	1.30	2.0	196.13	2OC	-1.0	-3.6	-1.6	0.1
166.25	1.32	1.9	206.12	12OC	-0.9	-3.6	-1.7	0.1
172.45	1.33	1.9	216.17	22OC	-0.6	-3.7	-1.7	-0.0
178.61	1.35	1.9	226.25	1NO	-0.4	-3.8	-1.8	-0.1
184.75	1.35	1.8	236.39	11NO	-0.4	-3.9	-1.9	-0.1
190.84	1.35	1.7	246.55	21NO	-0.4	-4.0	-1.9	-0.2
196.88	1.35	1.6	256.73	1DE	0.4	-4.1	-2.0	-0.3
202.86	1.34	1.6	266.93	11DE	3.1	-4.2	-2.0	-0.2
208.77	1.32	1.4	277.13	21DE	0.8	-4.3	-2.1	-0.2
214.60	1.29	1.3	287.32	31DE	0.1	-4.4	-2.1	-0.1
				1149				
220.33	1.24	1.2	297.49	10JA	-0.0	-4.3	-2.1	-0.0
225.92	1.19	1.1	307.63	20JA	-0.1	-4.2	-2.1	0.0
231.37	1.11	0.9	317.73	30JA	-0.4	-3.7	-2.1	0.1
236.62	1.02	0.7	327.79	9FE	-0.8	-3.3	-2.1	0.2
241.64	0.90	0.5	337.80	19FE	-1.4	-3.8	-2.0	0.2
246.36	0.74	0.3	347.74	1MR	-1.2	-4.2	-2.0	0.3
250.72	0.55	0.1	357.64	11MR	-0.3	-4.3	-1.9	0.3
254.60	0.31	-0.2	7.48	21MR	1.2	-4.3	-1.8	0.3
257.90	0.00	-0.5	17.25	31MR	3.1	-4.2	-1.8	0.3
260.45	-0.37	-0.7	26.98	10AP	2.6	-4.1	-1.7	0.3
262.08	-0.83	-1.1	36.66	20AP	1.5	-4.0	-1.6	0.3
262.61	-1.38	-1.4	46.29	30AP	0.8	-3.9	-1.6	0.3
261.91	-2.01	-1.7	55.89	10MY	0.2	-3.8	-1.5	0.3
260.01	-2.68	-2.0	65.45	20MY	-0.6	-3.7	-1.5	0.3
257.22	-3.33	-2.2	74.99	30MY	-1.6	-3.6	-1.4	0.2
254.12	-3.87	-2.2	84.53	9JN	-1.4	-3.5	-1.4	0.2
251.48	-4.24	-2.1	94.06	19JN	-0.5	-3.5	-1.3	0.2
249.88	-4.42	-1.9	103.59	29JN	0.1	-3.4	-1.3	0.3
249.62	-4.44	-1.7	113.14	9JL	0.5	-3.4	-1.3	0.3
250.75	-4.36	-1.4	122.71	19JL	0.9	-3.4	-1.3	0.3
253.10	-4.19	-1.2	132.31	29JL	1.6	-3.4	-1.3	0.4
256.51	-3.98	-1.0	141.95	8AU	3.1	-3.3	-1.3	0.4
260.78	-3.75	-0.8	151.64	18AU	1.4	-3.3	-1.3	0.4
265.73	-3.49	-0.6	161.37	28AU	-0.4	-3.3	-1.3	0.4
271.23	-3.23	-0.4	171.17	7SE	-1.1	-3.4	-1.3	0.4
277.17	-2.96	-0.3	181.01	17SE	-1.1	-3.4	-1.3	0.3
283.45	-2.68	-0.1	190.92	27SE	-0.8	-3.4	-1.3	0.3
290.00	-2.41	0.1	200.88	7OC	-0.5	-3.4	-1.4	0.2
296.76	-2.15	0.2	210.90	17OC	-0.3	-3.4	-1.4	0.2
303.68	-1.89	0.3	220.96	27OC	-0.1	-3.5	-1.5	0.1
310.71	-1.64	0.5	231.07	6NO	-0.1	-3.5	-1.5	0.0
317.83	-1.40	0.6	241.22	16NO	0.6	-3.5	-1.6	-0.0
324.99	-1.17	0.7	251.39	26NO	3.0	-3.5	-1.7	-0.1
332.18	-0.95	0.8	261.59	6DE	0.5	-3.5	-1.7	-0.2
339.38	-0.74	0.9	271.78	16DE	-0.1	-3.4	-1.8	-0.2
346.57	-0.55	1.0	281.98	26DE	-0.2	-3.4	-1.9	-0.2

Right Table

♂ LONG	LAT	MAG	☉ LONG	16.00UT	☿	♀	♃	♄
				1150				
353.74	-0.37	1.2	292.16	5JA	-0.2	-3.4	-1.9	-0.1
0.88	-0.20	1.2	302.32	15JA	-0.5	-3.4	-2.0	-0.1
7.97	-0.05	1.3	312.44	25JA	-0.9	-3.3	-2.0	0.0
15.02	0.09	1.4	322.52	4FE	-1.3	-3.3	-2.0	0.1
22.01	0.22	1.5	332.55	14FE	-1.1	-3.3	-2.0	0.1
28.95	0.34	1.6	342.53	24FE	-0.2	-3.3	-2.0	0.2
35.84	0.45	1.6	352.46	6MR	1.5	-3.4	-2.0	0.2
42.67	0.55	1.7	2.33	16MR	3.4	-3.4	-2.0	0.3
49.44	0.63	1.8	12.13	26MR	1.9	-3.4	-1.9	0.3
56.16	0.71	1.8	21.89	5AP	1.1	-3.4	-1.9	0.3
62.82	0.79	1.9	31.59	15AP	0.6	-3.4	-1.8	0.4
69.44	0.85	1.9	41.25	25AP	0.1	-3.5	-1.7	0.4
76.01	0.91	1.9	50.86	5MY	-0.6	-3.5	-1.7	0.4
82.54	0.96	1.9	60.44	15MY	-1.6	-3.6	-1.6	0.4
89.04	1.00	2.0	69.99	25MY	-1.4	-3.6	-1.5	0.4
95.50	1.04	2.0	79.53	4JN	-0.5	-3.7	-1.5	0.3
101.94	1.07	2.0	89.06	14JN	0.2	-3.8	-1.4	0.3
108.35	1.10	2.0	98.59	24JN	0.7	-3.9	-1.4	0.3
114.74	1.12	2.0	108.13	4JL	1.2	-4.0	-1.3	0.3
121.12	1.14	2.0	117.69	14JL	2.1	-4.1	-1.3	0.4
127.49	1.15	1.9	127.27	24JL	3.1	-4.2	-1.3	0.4
133.86	1.15	2.0	136.89	3AU	1.2	-4.2	-1.2	0.4
140.23	1.16	2.0	146.55	13AU	-0.4	-4.2	-1.2	0.5
146.59	1.16	2.0	156.26	23AU	-1.2	-4.0	-1.2	0.5
152.97	1.15	2.0	166.03	2SE	-1.3	-3.6	-1.2	0.5
159.35	1.14	2.0	175.84	12SE	-0.8	-3.3	-1.2	0.5
165.75	1.12	2.0	185.72	22SE	-0.4	-3.8	-1.2	0.4
172.16	1.10	2.0	195.65	2OC	-0.2	-4.2	-1.2	0.4
178.59	1.08	2.0	205.64	12OC	-0.1	-4.3	-1.3	0.4
185.03	1.05	2.0	215.68	22OC	0.0	-4.3	-1.3	0.3
191.49	1.01	1.9	225.77	1NO	0.9	-4.3	-1.3	0.3
197.97	0.96	1.9	235.90	11NO	2.9	-4.2	-1.4	0.2
204.46	0.91	1.8	246.06	21NO	0.3	-4.1	-1.4	0.1
210.97	0.85	1.8	256.24	1DE	-0.2	-4.0	-1.5	0.1
217.50	0.78	1.7	266.44	11DE	-0.3	-3.9	-1.5	-0.0
224.03	0.71	1.6	276.64	21DE	-0.4	-3.8	-1.6	-0.1
230.58	0.62	1.6	286.83	31DE	-0.5	-3.7	-1.7	-0.1
				1151				
237.13	0.51	1.5	297.00	10JA	-0.9	-3.6	-1.7	-0.1
243.68	0.40	1.4	307.14	20JA	-1.1	-3.6	-1.8	-0.0
250.23	0.27	1.3	317.25	30JA	-1.0	-3.5	-1.9	0.0
256.77	0.12	1.1	327.31	9FE	-0.2	-3.5	-1.9	0.1
263.30	-0.05	1.0	337.32	19FE	1.8	-3.4	-2.0	0.2
269.81	-0.24	0.9	347.27	1MR	2.8	-3.4	-2.0	0.2
276.27	-0.45	0.7	357.17	11MR	1.4	-3.3	-2.0	0.3
282.69	-0.69	0.6	7.01	21MR	0.8	-3.3	-2.0	0.3
289.04	-0.96	0.4	16.79	31MR	0.4	-3.3	-2.0	0.4
295.29	-1.26	0.3	26.52	10AP	0.0	-3.3	-2.0	0.4
301.41	-1.59	0.1	36.20	20AP	-0.7	-3.3	-1.9	0.4
307.35	-1.95	-0.1	45.83	30AP	-1.6	-3.3	-1.9	0.5
313.07	-2.35	-0.3	55.42	10MY	-1.4	-3.3	-1.8	0.5
318.48	-2.80	-0.5	64.99	20MY	-0.5	-3.4	-1.8	0.5
323.49	-3.28	-0.8	74.54	30MY	0.3	-3.4	-1.7	0.5
327.98	-3.80	-1.0	84.07	9JN	0.9	-3.4	-1.6	0.5
331.79	-4.36	-1.3	93.60	19JN	1.6	-3.5	-1.6	0.5
334.72	-4.94	-1.5	103.13	29JN	2.8	-3.5	-1.5	0.4
336.60	-5.51	-1.8	112.67	9JL	2.7	-3.5	-1.5	0.4
337.19	-6.03	-2.1	122.24	19JL	1.0	-3.4	-1.4	0.5
336.45	-6.42	-2.3	131.84	29JL	-0.5	-3.4	-1.4	0.5
334.53	-6.59	-2.5	141.48	8AU	-1.3	-3.4	-1.3	0.5
331.89	-6.45	-2.6	151.17	18AU	-1.3	-3.4	-1.3	0.6
329.29	-6.01	-2.5	160.90	28AU	-0.7	-3.3	-1.3	0.6
327.43	-5.34	-2.2	170.68	7SE	-0.3	-3.3	-1.2	0.6
326.75	-4.56	-1.9	180.53	17SE	-0.1	-3.3	-1.2	0.6
327.39	-3.78	-1.6	190.43	27SE	0.1	-3.3	-1.2	0.6
329.24	-3.05	-1.3	200.39	7OC	0.2	-3.3	-1.2	0.6
332.13	-2.40	-1.0	210.41	17OC	1.2	-3.4	-1.2	0.6
335.84	-1.83	-0.7	220.47	27OC	2.6	-3.4	-1.2	0.5
340.19	-1.34	-0.5	230.58	6NO	0.1	-3.4	-1.3	0.5
345.05	-0.92	-0.2	240.73	16NO	-0.4	-3.4	-1.3	0.4
350.29	-0.57	0.0	250.90	26NO	-0.4	-3.5	-1.3	0.3
355.82	-0.27	0.2	261.09	6DE	-0.5	-3.5	-1.4	0.3
1.58	-0.01	0.4	271.29	16DE	-0.6	-3.6	-1.4	0.2
7.50	0.21	0.6	281.49	26DE	-0.9	-3.7	-1.5	0.1

♂ LONG	LAT	MAG	☉ LONG	16.00UT 1152	☿	♀	♃	♄
13.54	0.39	0.8	291.67	5JA	-1.0	-3.7	-1.5	0.1
19.68	0.55	0.9	301.83	15JA	-0.9	-3.8	-1.6	0.0
25.88	0.68	1.1	311.95	25JA	-0.1	-3.9	-1.7	0.1
32.13	0.79	1.2	322.04	4FE	2.1	-4.0	-1.7	0.1
38.40	0.89	1.3	332.08	14FE	2.2	-4.1	-1.8	0.2
44.70	0.97	1.4	342.06	24FE	1.0	-4.2	-1.9	0.2
51.00	1.03	1.5	351.98	5MR	0.6	-4.2	-1.9	0.3
57.30	1.09	1.6	1.86	15MR	0.3	-4.3	-2.0	0.4
63.60	1.13	1.7	11.66	25MR	-0.1	-4.2	-2.0	0.4
69.90	1.17	1.8	21.42	4AP	-0.7	-3.9	-2.0	0.5
76.19	1.19	1.8	31.13	14AP	-1.6	-3.3	-2.0	0.5
82.47	1.21	1.9	40.78	24AP	-1.4	-3.1	-2.0	0.6
88.75	1.22	1.9	50.40	4MY	-0.4	-3.8	-2.0	0.6
95.03	1.23	2.0	59.98	14MY	0.5	-4.1	-2.0	0.6
101.30	1.23	2.0	69.53	24MY	1.2	-4.2	-2.0	0.6
107.57	1.23	2.0	79.07	3JN	2.2	-4.2	-1.9	0.6
113.84	1.22	2.0	88.60	13JN	3.4	-4.1	-1.9	0.6
120.13	1.20	2.0	98.13	23JN	2.2	-4.0	-1.8	0.6
126.42	1.19	2.0	107.67	3JL	0.8	-3.9	-1.7	0.6
132.73	1.16	2.0	117.22	13JL	-0.5	-3.8	-1.7	0.6
139.06	1.14	2.0	126.80	23JL	-1.4	-3.7	-1.6	0.6
145.42	1.11	2.0	136.42	2AU	-1.4	-3.7	-1.5	0.7
151.80	1.07	2.0	146.08	12AU	-0.7	-3.6	-1.5	0.7
158.21	1.03	1.9	155.79	22AU	-0.2	-3.5	-1.4	0.7
164.66	0.99	1.9	165.55	1SE	0.1	-3.5	-1.4	0.8
171.14	0.94	1.9	175.37	11SE	0.2	-3.5	-1.4	0.8
177.67	0.89	1.9	185.23	21SE	0.5	-3.4	-1.3	0.8
184.24	0.83	1.9	195.17	1OC	1.5	-3.4	-1.3	0.8
190.86	0.77	1.9	205.15	11OC	2.4	-3.4	-1.3	0.8
197.53	0.70	1.9	215.19	21OC	-0.0	-3.4	-1.3	0.8
204.25	0.63	1.9	225.28	31OC	-0.6	-3.4	-1.3	0.8
211.01	0.55	1.9	235.41	10NO	-0.6	-3.4	-1.3	0.7
217.84	0.46	1.8	245.56	20NO	-0.6	-3.4	-1.3	0.7
224.71	0.37	1.8	255.75	30NO	-0.7	-3.4	-1.3	0.6
231.64	0.27	1.8	265.95	10DE	-0.8	-3.4	-1.3	0.6
238.62	0.16	1.7	276.14	20DE	-0.8	-3.4	-1.4	0.5
245.66	0.05	1.6	286.34	30DE	-0.8	-3.4	-1.4	0.4
				1153				
252.74	-0.07	1.6	296.51	9JA	0.0	-3.4	-1.5	0.4
259.88	-0.20	1.5	306.65	19JA	2.4	-3.4	-1.5	0.3
267.06	-0.34	1.5	316.76	29JA	1.7	-3.5	-1.6	0.2
274.29	-0.49	1.4	326.82	8FE	0.7	-3.5	-1.6	0.3
281.55	-0.64	1.3	336.83	18FE	0.4	-3.5	-1.7	0.3
288.85	-0.80	1.2	346.79	28FE	0.2	-3.4	-1.8	0.4
296.18	-0.96	1.1	356.69	10MR	-0.2	-3.4	-1.8	0.4
303.52	-1.13	1.0	6.53	20MR	-0.7	-3.4	-1.9	0.5
310.87	-1.30	1.0	16.31	30MR	-1.6	-3.4	-2.0	0.6
318.22	-1.47	0.9	26.04	9AP	-1.4	-3.3	-2.0	0.6
325.55	-1.64	0.8	35.72	19AP	-0.4	-3.3	-2.1	0.7
332.84	-1.80	0.7	45.36	29AP	0.6	-3.3	-2.1	0.7
340.07	-1.96	0.6	54.96	9MY	1.6	-3.3	-2.1	0.8
347.21	-2.11	0.5	64.53	19MY	2.9	-3.3	-2.1	0.8
354.24	-2.25	0.4	74.08	29MY	3.1	-3.4	-2.1	0.8
1.12	-2.38	0.3	83.61	8JN	1.7	-3.4	-2.1	0.8
7.82	-2.49	0.2	93.13	18JN	0.6	-3.4	-2.1	0.9
14.28	-2.58	0.0	102.67	28JN	-0.5	-3.4	-2.0	0.9
20.43	-2.65	-0.1	112.21	8JL	-1.4	-3.5	-2.0	0.9
26.23	-2.69	-0.2	121.78	18JL	-1.4	-3.5	-1.9	0.9
31.57	-2.71	-0.4	131.38	28JL	-0.6	-3.6	-1.9	0.9
36.34	-2.70	-0.5	141.01	7AU	-0.1	-3.7	-1.8	0.9
40.42	-2.65	-0.7	150.69	17AU	0.2	-3.8	-1.7	0.9
43.60	-2.56	-0.9	160.43	27AU	0.4	-3.8	-1.7	1.0
45.71	-2.41	-1.1	170.21	6SE	0.7	-4.0	-1.6	1.0
46.52	-2.18	-1.4	180.05	16SE	1.9	-4.1	-1.6	1.0
45.88	-1.86	-1.6	189.95	26SE	2.1	-4.2	-1.5	1.1
43.80	-1.44	-1.8	199.91	6OC	-0.1	-4.3	-1.5	1.1
40.61	-0.94	-1.9	209.92	16OC	-0.7	-4.3	-1.4	1.1
36.98	-0.39	-1.9	219.49	26OC	-0.7	-4.3	-1.4	1.1
33.75	0.13	-1.7	230.09	5NO	-0.7	-4.1	-1.4	1.1
31.54	0.58	-1.4	240.23	15NO	-0.7	-3.6	-1.4	1.0
30.67	0.94	-1.0	250.41	25NO	-0.6	-3.0	-1.4	1.0
31.12	1.20	-0.7	260.60	5DE	-0.7	-3.9	-1.4	1.0
32.74	1.39	-0.4	270.79	15DE	-0.6	-4.1	-1.4	0.9
35.31	1.53	-0.1	280.99	25DE	0.1	-4.4	-1.4	0.8

♂ LONG	LAT	MAG	☉ LONG	16.00UT 1154	☿	♀	♃	♄
38.63	1.61	0.2	291.18	4JA	2.7	-4.4	-1.4	0.8
42.54	1.67	0.4	301.34	14JA	1.3	-4.3	-1.4	0.7
46.91	1.70	0.6	311.47	24JA	0.5	-4.2	-1.5	0.6
51.63	1.72	0.8	321.55	3FE	0.2	-4.1	-1.5	0.6
56.63	1.72	1.0	331.59	13FE	0.1	-3.9	-1.6	0.5
61.85	1.71	1.1	341.58	23FE	-0.2	-3.8	-1.6	0.6
67.24	1.69	1.3	351.51	5MR	-0.8	-3.7	-1.7	0.6
72.76	1.67	1.4	1.38	15MR	-1.5	-3.7	-1.7	0.6
78.40	1.64	1.5	11.20	25MR	-1.4	-3.6	-1.8	0.7
84.12	1.60	1.6	20.95	4AP	-0.4	-3.5	-1.9	0.8
89.92	1.56	1.7	30.66	14AP	0.8	-3.5	-2.0	0.8
95.79	1.52	1.8	40.32	24AP	2.1	-3.4	-2.0	0.9
101.70	1.48	1.8	49.94	4MY	3.7	-3.4	-2.1	0.9
107.67	1.43	1.9	59.52	14MY	2.4	-3.3	-2.2	1.0
113.69	1.38	1.9	69.08	24MY	1.3	-3.3	-2.2	1.0
119.75	1.33	2.0	78.61	3JN	0.4	-3.3	-2.2	1.0
125.86	1.27	2.0	88.15	13JN	-0.5	-3.3	-2.3	1.1
132.02	1.21	2.0	97.68	23JN	-1.5	-3.3	-2.3	1.1
138.22	1.15	2.0	107.21	3JL	-1.4	-3.3	-2.3	1.1
144.46	1.09	2.0	116.77	13JL	-0.6	-3.3	-2.3	1.1
150.76	1.03	2.0	126.35	23JL	-0.0	-3.4	-2.2	1.1
157.12	0.96	2.0	135.96	2AU	0.3	-3.4	-2.2	1.1
163.53	0.89	2.0	145.62	12AU	0.6	-3.4	-2.1	1.1
170.00	0.81	2.0	155.32	22AU	1.0	-3.5	-2.1	1.1
176.53	0.74	1.9	165.07	1SE	2.3	-3.5	-2.0	1.2
183.13	0.66	1.9	174.89	11SE	1.9	-3.5	-1.9	1.2
189.79	0.57	1.8	184.76	21SE	-0.2	-3.5	-1.9	1.3
196.53	0.49	1.8	194.68	1OC	-0.9	-3.4	-1.8	1.3
203.34	0.40	1.8	204.66	11OC	-0.9	-3.4	-1.7	1.4
210.22	0.31	1.8	214.70	21OC	-0.9	-3.4	-1.7	1.4
217.18	0.21	1.7	224.78	31OC	-0.6	-3.4	-1.6	1.3
224.21	0.11	1.7	234.91	10NO	-0.5	-3.3	-1.6	1.3
231.32	0.01	1.7	245.06	20NO	-0.5	-3.3	-1.6	1.3
238.51	-0.10	1.7	255.24	30NO	-0.4	-3.3	-1.5	1.2
245.76	-0.20	1.7	265.44	10DE	0.3	-3.3	-1.5	1.2
253.09	-0.31	1.6	275.64	20DE	2.9	-3.4	-1.5	1.1
260.48	-0.42	1.6	285.83	30DE	1.0	-3.4	-1.5	1.1
				1155				
267.94	-0.54	1.6	296.01	9JA	0.2	-3.4	-1.5	1.0
275.46	-0.65	1.5	306.16	19JA	0.1	-3.4	-1.5	1.0
283.03	-0.76	1.5	316.26	29JA	-0.1	-3.5	-1.5	1.0
290.65	-0.87	1.5	326.33	8FE	-0.3	-3.5	-1.5	0.9
298.30	-0.98	1.4	336.35	18FE	-0.8	-3.5	-1.5	0.8
305.98	-1.08	1.4	346.31	28FE	-1.5	-3.6	-1.6	0.8
313.68	-1.17	1.3	356.21	10MR	-1.3	-3.6	-1.6	0.9
321.38	-1.25	1.3	6.06	20MR	-0.4	-3.7	-1.7	0.9
329.08	-1.33	1.3	15.84	30MR	1.0	-3.8	-1.7	0.9
336.76	-1.39	1.2	25.58	9AP	2.7	-3.9	-1.8	1.0
344.40	-1.44	1.2	35.26	19AP	3.1	-4.0	-1.9	1.1
351.99	-1.48	1.1	44.90	29AP	1.8	-4.0	-1.9	1.1
359.53	-1.50	1.1	54.50	9MY	1.0	-4.1	-2.0	1.2
6.98	-1.50	1.0	64.07	19MY	0.3	-4.2	-2.1	1.2
14.34	-1.49	1.0	73.62	29MY	-0.6	-4.2	-2.1	1.3
21.59	-1.46	1.0	83.15	8JN	-1.5	-4.0	-2.2	1.3
28.70	-1.42	0.9	92.68	18JN	-1.4	-3.6	-2.3	1.3
35.67	-1.35	0.9	102.21	28JN	-0.6	-2.9	-2.3	1.4
42.47	-1.27	0.8	111.75	8JL	0.0	-3.5	-2.4	1.4
49.07	-1.17	0.7	121.32	18JL	0.4	-4.0	-2.4	1.4
55.46	-1.05	0.7	130.91	28JL	0.8	-4.2	-2.4	1.4
61.59	-0.91	0.6	140.55	7AU	1.3	-4.2	-2.4	1.4
67.42	-0.75	0.5	150.23	17AU	2.8	-4.2	-2.4	1.3
72.90	-0.57	0.4	159.96	27AU	1.6	-4.1	-2.3	1.3
77.96	-0.36	0.3	169.74	6SE	-0.3	-4.0	-2.3	1.2
82.51	-0.12	0.2	179.57	16SE	-1.0	-3.9	-2.2	1.2
86.44	0.16	-0.0	189.47	26SE	-1.0	-3.8	-2.2	1.2
89.60	0.48	-0.2	199.43	6OC	-0.9	-3.8	-2.1	1.2
91.83	0.85	-0.4	209.43	16OC	-0.5	-3.7	-2.0	1.2
92.91	1.26	-0.6	219.49	26OC	-0.4	-3.6	-2.0	1.2
92.66	1.72	-0.8	229.60	5NO	-0.3	-3.6	-1.9	1.2
90.99	2.20	-1.0	239.74	15NO	-0.3	-3.5	-1.8	1.2
88.03	2.66	-1.2	249.91	25NO	0.5	-3.5	-1.8	1.1
84.24	3.04	-1.3	260.10	5DE	3.1	-3.5	-1.7	1.1
80.35	3.28	-1.2	270.30	15DE	0.7	-3.4	-1.7	1.0
77.12	3.38	-1.0	280.49	25DE	0.0	-3.4	-1.6	1.0

♂ LONG	♂ LAT	♂ MAG	⊙ LONG	16.00UT	☿	♀	♃	♄
				1156	MAGNITUDES			
75.07	3.36	-0.7	290.68	4JA	-0.1	-3.4	-1.6	1.0
74.35	3.25	-0.4	300.84	14JA	-0.2	-3.3	-1.6	0.9
74.89	3.10	-0.1	310.97	24JA	-0.4	-3.3	-1.6	0.9
76.52	2.94	0.2	321.06	3FE	-0.8	-3.3	-1.6	0.8
79.03	2.77	0.4	331.10	13FE	-1.4	-3.3	-1.6	0.8
82.24	2.60	0.6	341.09	23FE	-1.2	-3.3	-1.6	0.7
86.01	2.44	0.8	351.03	4MR	-0.3	-3.3	-1.6	0.7
90.22	2.29	1.0	0.90	14MR	1.3	-3.4	-1.6	0.7
94.77	2.15	1.1	10.72	24MR	3.3	-3.4	-1.6	0.8
99.62	2.02	1.3	20.48	3AP	2.3	-3.4	-1.7	0.9
104.69	1.89	1.4	30.19	13AP	1.3	-3.5	-1.7	0.9
109.95	1.77	1.5	39.85	23AP	0.7	-3.5	-1.7	1.0
115.38	1.65	1.6	49.47	3MY	0.2	-3.5	-1.8	1.1
120.94	1.54	1.7	59.05	13MY	-0.6	-3.4	-1.9	1.1
126.63	1.43	1.7	68.61	23MY	-1.5	-3.4	-1.9	1.2
132.43	1.32	1.8	78.15	2JN	-1.4	-3.4	-2.0	1.2
138.33	1.21	1.8	87.68	12JN	-0.5	-3.3	-2.1	1.2
144.33	1.11	1.8	97.21	22JN	0.1	-3.3	-2.1	1.3
150.42	1.01	1.9	106.75	2JL	0.6	-3.3	-2.2	1.3
156.60	0.90	1.9	116.30	12JL	1.0	-3.3	-2.3	1.3
162.87	0.80	1.9	125.88	22JL	1.8	-3.3	-2.3	1.2
169.23	0.70	1.9	135.49	1AU	3.2	-3.3	-2.4	1.2
175.68	0.60	1.9	145.14	11AU	1.4	-3.4	-2.4	1.2
182.22	0.49	1.9	154.85	21AU	-0.4	-3.4	-2.4	1.2
188.85	0.39	1.8	164.60	31AU	-1.1	-3.4	-2.5	1.1
195.57	0.29	1.8	174.41	10SE	-1.2	-3.5	-2.5	1.1
202.38	0.19	1.8	184.28	20SE	-0.8	-3.5	-2.4	1.0
209.29	0.08	1.8	194.20	30SE	-0.4	-3.6	-2.4	1.0
216.29	-0.02	1.7	204.18	10OC	-0.3	-3.6	-2.4	1.1
223.39	-0.12	1.7	214.21	20OC	-0.2	-3.7	-2.3	1.1
230.58	-0.23	1.6	224.29	30OC	-0.1	-3.8	-2.2	1.0
237.85	-0.33	1.6	234.41	9NO	0.7	-3.9	-2.2	1.0
245.22	-0.43	1.5	244.57	19NO	3.1	-4.0	-2.1	1.0
252.66	-0.52	1.5	254.75	29NO	0.5	-4.1	-2.0	1.0
260.19	-0.62	1.5	264.95	9DE	-0.1	-4.2	-2.0	1.0
267.79	-0.71	1.5	275.15	19DE	-0.2	-4.3	-1.9	0.9
275.45	-0.79	1.5	285.34	29DE	-0.3	-4.4	-1.8	0.9
				1157				
283.17	-0.87	1.5	295.51	8JA	-0.5	-4.3	-1.8	0.9
290.93	-0.94	1.4	305.66	18JA	-0.9	-4.1	-1.7	0.8
298.73	-1.01	1.4	315.77	28JA	-1.2	-3.7	-1.7	0.8
306.56	-1.06	1.4	325.84	7FE	-1.1	-3.3	-1.6	0.7
314.40	-1.11	1.4	335.86	17FE	-0.3	-3.9	-1.6	0.7
322.24	-1.14	1.4	345.82	27FE	1.6	-4.2	-1.6	0.6
330.07	-1.16	1.4	355.72	9MR	3.3	-4.3	-1.6	0.6
337.87	-1.17	1.4	5.58	19MR	1.7	-4.3	-1.6	0.5
345.64	-1.16	1.4	15.36	29MR	1.0	-4.2	-1.6	0.6
353.37	-1.15	1.4	25.10	8AP	0.5	-4.1	-1.6	0.6
1.03	-1.11	1.4	34.79	18AP	0.1	-4.0	-1.6	0.7
8.62	-1.07	1.4	44.43	28AP	-0.6	-3.9	-1.6	0.8
16.14	-1.02	1.4	54.03	8MY	-1.5	-3.8	-1.7	0.9
23.57	-0.95	1.4	63.61	18MY	-1.5	-3.7	-1.7	0.9
30.90	-0.87	1.4	73.15	28MY	-0.5	-3.6	-1.7	1.0
38.12	-0.78	1.4	82.69	7JN	0.2	-3.5	-1.8	1.0
45.24	-0.67	1.4	92.22	17JN	0.8	-3.5	-1.8	1.0
52.23	-0.56	1.3	101.75	27JN	1.4	-3.4	-1.9	1.1
59.09	-0.44	1.3	111.29	7JL	2.3	-3.4	-2.0	1.1
65.82	-0.31	1.3	120.86	17JL	3.0	-3.4	-2.0	1.1
72.40	-0.17	1.3	130.45	27JL	1.2	-3.4	-2.1	1.1
78.82	-0.02	1.2	140.08	6AU	-0.4	-3.3	-2.2	1.1
85.06	0.14	1.2	149.76	16AU	-1.2	-3.3	-2.2	1.1
91.10	0.31	1.1	159.48	26AU	-1.3	-3.3	-2.3	1.0
96.93	0.50	1.1	169.26	5SE	-0.8	-3.3	-2.4	1.0
102.50	0.70	1.0	179.10	15SE	-0.3	-3.4	-2.4	1.0
107.77	0.91	0.9	188.99	25SE	-0.1	-3.4	-2.4	0.9
112.70	1.15	0.8	198.94	5OC	-0.0	-3.4	-2.4	0.9
117.20	1.41	0.7	208.95	15OC	0.1	-3.4	-2.4	0.9
121.20	1.70	0.5	219.00	25OC	0.9	-3.4	-2.4	0.9
124.56	2.02	0.3	229.11	4NO	2.9	-3.5	-2.4	0.9
127.14	2.38	0.2	239.25	14NO	0.3	-3.5	-2.3	0.9
128.77	2.77	-0.1	249.41	24NO	-0.3	-3.5	-2.3	0.9
129.26	3.19	-0.3	259.60	4DE	-0.4	-3.5	-2.2	0.9
128.44	3.62	-0.5	269.80	14DE	-0.4	-3.4	-2.1	0.9
126.28	4.02	-0.8	280.00	24DE	-0.6	-3.4	-2.1	0.9

♂ LONG	♂ LAT	♂ MAG	⊙ LONG	16.00UT	☿	♀	♃	♄
				1158	MAGNITUDES			
122.98	4.32	-0.9	290.18	3JA	-0.9	-3.4	-2.0	0.8
119.05	4.46	-1.0	300.35	13JA	-1.1	-3.4	-1.9	0.8
115.23	4.43	-0.9	310.47	23JA	-1.0	-3.3	-1.9	0.7
112.19	4.25	-0.7	320.57	2FE	-0.2	-3.3	-1.8	0.7
110.35	3.96	-0.4	330.61	12FE	1.8	-3.3	-1.7	0.6
109.82	3.64	-0.1	340.60	22FE	2.6	-3.3	-1.7	0.6
110.49	3.30	0.1	350.54	4MR	1.2	-3.4	-1.6	0.5
112.21	2.98	0.3	0.42	14MR	0.7	-3.4	-1.6	0.5
114.77	2.68	0.5	10.24	24MR	0.4	-3.4	-1.6	0.4
118.02	2.41	0.7	20.00	3AP	-0.0	-3.4	-1.6	0.4
121.83	2.16	0.9	29.72	13AP	-0.7	-3.4	-1.5	0.5
126.07	1.93	1.0	39.38	23AP	-1.5	-3.5	-1.5	0.5
130.69	1.72	1.1	49.00	3MY	-1.5	-3.5	-1.5	0.6
135.61	1.53	1.3	58.59	13MY	-0.5	-3.6	-1.5	0.7
140.78	1.34	1.3	68.15	23MY	0.4	-3.6	-1.5	0.7
146.18	1.17	1.4	77.69	2JN	1.0	-3.7	-1.6	0.8
151.77	1.01	1.5	87.22	12JN	1.8	-3.8	-1.6	0.9
157.54	0.86	1.5	96.75	22JN	3.0	-3.9	-1.6	0.9
163.47	0.71	1.6	106.29	2JL	2.5	-4.0	-1.7	0.9
169.54	0.57	1.6	115.84	12JL	0.9	-4.1	-1.7	1.0
175.75	0.44	1.6	125.42	22JL	-0.4	-4.2	-1.8	1.0
182.10	0.31	1.7	135.03	1AU	-1.3	-4.2	-1.8	1.0
188.58	0.18	1.7	144.68	11AU	-1.4	-4.2	-1.9	1.0
195.18	0.06	1.7	154.38	21AU	-0.7	-4.0	-1.9	1.0
201.91	-0.06	1.7	164.13	31AU	-0.3	-3.6	-2.0	1.0
208.75	-0.17	1.7	173.94	10SE	-0.0	-3.3	-2.1	0.9
215.71	-0.28	1.6	183.80	20SE	0.1	-3.8	-2.1	0.9
222.79	-0.39	1.6	193.72	30SE	0.3	-4.2	-2.2	0.8
229.98	-0.49	1.6	203.69	10OC	1.2	-4.3	-2.3	0.8
237.28	-0.58	1.6	213.72	20OC	2.7	-4.3	-2.3	0.8
244.67	-0.67	1.5	223.81	30OC	0.1	-4.3	-2.3	0.8
252.17	-0.76	1.5	233.92	9NO	-0.5	-4.2	-2.3	0.8
259.75	-0.83	1.5	244.08	19NO	-0.5	-4.1	-2.3	0.9
267.41	-0.90	1.5	254.26	29NO	-0.5	-4.0	-2.3	0.9
275.13	-0.95	1.4	264.46	9DE	-0.7	-3.9	-2.3	0.8
282.92	-1.00	1.4	274.65	19DE	-0.8	-3.8	-2.2	0.8
290.76	-1.04	1.4	284.85	29DE	-0.9	-3.7	-2.2	0.8
				1159				
298.63	-1.06	1.3	295.02	8JA	-0.9	-3.6	-2.1	0.8
306.52	-1.08	1.3	305.17	18JA	-0.1	-3.6	-2.1	0.8
314.42	-1.08	1.3	315.29	28JA	2.1	-3.5	-2.0	0.7
322.32	-1.07	1.3	325.35	7FE	2.0	-3.5	-1.9	0.7
330.20	-1.05	1.3	335.37	17FE	0.9	-3.4	-1.8	0.6
338.05	-1.02	1.4	345.34	27FE	0.5	-3.4	-1.8	0.6
345.85	-0.98	1.4	355.25	9MR	0.2	-3.3	-1.7	0.5
353.61	-0.93	1.4	5.10	19MR	-0.1	-3.3	-1.7	0.5
1.30	-0.86	1.4	14.90	29MR	-0.7	-3.3	-1.6	0.4
8.92	-0.79	1.5	24.63	8AP	-1.5	-3.3	-1.6	0.4
16.46	-0.71	1.5	34.32	18AP	-1.4	-3.3	-1.5	0.4
23.92	-0.62	1.5	43.97	28AP	-0.5	-3.3	-1.5	0.4
31.29	-0.53	1.5	53.57	8MY	0.5	-3.3	-1.5	0.5
38.57	-0.43	1.6	63.14	18MY	1.3	-3.4	-1.4	0.5
45.76	-0.32	1.6	72.69	28MY	2.4	-3.4	-1.4	0.6
52.84	-0.21	1.6	82.22	7JN	3.4	-3.4	-1.4	0.7
59.83	-0.10	1.6	91.75	17JN	2.0	-3.5	-1.4	0.7
66.72	0.02	1.6	101.29	27JN	0.7	-3.5	-1.4	0.8
73.50	0.14	1.6	110.83	7JL	-0.5	-3.5	-1.5	0.8
80.18	0.26	1.6	120.39	17JL	-1.4	-3.4	-1.5	0.9
86.76	0.39	1.6	129.99	27JL	-1.4	-3.4	-1.5	0.9
93.22	0.52	1.6	139.61	6AU	-0.7	-3.4	-1.6	0.9
99.57	0.65	1.6	149.29	16AU	-0.2	-3.4	-1.6	0.9
105.79	0.79	1.6	159.01	26AU	0.1	-3.3	-1.7	0.9
111.89	0.93	1.6	168.78	5SE	0.3	-3.3	-1.7	0.9
117.84	1.08	1.5	178.62	15SE	0.5	-3.3	-1.8	0.9
123.62	1.23	1.4	188.51	25SE	1.6	-3.3	-1.8	0.8
129.23	1.39	1.4	198.45	5OC	2.4	-3.3	-1.9	0.8
134.62	1.56	1.3	208.46	15OC	-0.0	-3.4	-2.0	0.8
139.76	1.75	1.2	218.51	25OC	-0.6	-3.4	-2.0	0.7
144.61	1.94	1.1	228.61	4NO	-0.7	-3.4	-2.1	0.8
149.10	2.16	0.9	238.75	14NO	-0.7	-3.4	-2.1	0.8
153.15	2.39	0.8	248.92	24NO	-0.8	-3.5	-2.2	0.8
156.66	2.64	0.6	259.11	4DE	-0.7	-3.5	-2.2	0.8
159.50	2.92	0.4	269.31	14DE	-0.7	-3.6	-2.2	0.8
161.50	3.22	0.2	279.51	24DE	-0.7	-3.7	-2.2	0.8

♂ LONG	♂ LAT	♂ MAG	⊙ LONG	16.00UT 1160	☿	♀	♃	♄
162.51	3.54	-0.1	289.69	3JA	0.0	-3.7	-2.2	0.8
162.32	3.84	-0.3	299.86	13JA	2.4	-3.8	-2.2	0.8
160.84	4.12	-0.6	309.99	23JA	1.6	-3.9	-2.1	0.7
158.12	4.30	-0.8	320.08	2FE	0.6	-4.0	-2.1	0.7
154.49	4.34	-1.0	330.13	12FE	0.3	-4.1	-2.0	0.7
150.58	4.22	-1.0	340.12	22FE	0.1	-4.2	-1.9	0.6
147.09	3.94	-0.8	350.06	3MR	-0.2	-4.2	-1.9	0.6
144.60	3.56	-0.6	359.94	13MR	-0.7	-4.3	-1.8	0.5
143.38	3.13	-0.4	9.77	23MR	-1.5	-4.2	-1.7	0.4
143.44	2.70	-0.1	19.53	2AP	-1.4	-3.9	-1.7	0.4
144.65	2.30	0.1	29.25	12AP	-0.4	-3.3	-1.6	0.3
146.83	1.94	0.3	38.91	22AP	0.7	-3.2	-1.5	0.3
149.82	1.61	0.5	48.53	2MY	1.8	-3.8	-1.5	0.3
153.46	1.31	0.6	58.13	12MY	3.2	-4.1	-1.4	0.4
157.64	1.05	0.7	67.68	22MY	2.8	-4.2	-1.4	0.4
162.26	0.81	0.9	77.22	1JN	1.6	-4.2	-1.4	0.5
167.26	0.59	1.0	86.76	11JN	0.6	-4.1	-1.4	0.6
172.57	0.39	1.0	96.28	21JN	-0.5	-4.0	-1.3	0.6
178.16	0.20	1.1	105.82	1JL	-1.4	-3.9	-1.3	0.7
184.00	0.03	1.2	115.37	11JL	-1.4	-3.8	-1.3	0.7
190.06	-0.13	1.2	124.95	21JL	-0.6	-3.7	-1.3	0.8
196.32	-0.27	1.3	134.56	31JL	-0.1	-3.7	-1.3	0.8
202.77	-0.41	1.3	144.21	10AU	0.2	-3.6	-1.4	0.8
209.39	-0.53	1.3	153.91	20AU	0.4	-3.5	-1.4	0.8
216.17	-0.65	1.3	163.66	30AU	0.8	-3.5	-1.4	0.8
223.11	-0.75	1.4	173.46	9SE	2.0	-3.5	-1.5	0.8
230.19	-0.84	1.4	183.32	19SE	2.1	-3.4	-1.5	0.8
237.40	-0.93	1.4	193.24	29SE	-0.1	-3.4	-1.6	0.8
244.74	-1.00	1.4	203.21	9OC	-0.8	-3.4	-1.6	0.8
252.19	-1.06	1.4	213.24	19OC	-0.8	-3.4	-1.7	0.7
259.74	-1.10	1.4	223.32	29OC	-0.8	-3.4	-1.7	0.7
267.39	-1.14	1.4	233.44	8NO	-0.7	-3.4	-1.8	0.7
275.11	-1.16	1.4	243.59	18NO	-0.6	-3.4	-1.9	0.7
282.90	-1.17	1.4	253.77	28NO	-0.6	-3.4	-1.9	0.8
290.74	-1.17	1.4	263.96	8DE	-0.5	-3.4	-2.0	0.8
298.61	-1.16	1.4	274.16	18DE	0.1	-3.4	-2.0	0.8
306.51	-1.13	1.4	284.35	28DE	2.7	-3.4	-2.1	0.8
				1161				
314.41	-1.09	1.4	294.53	7JA	1.2	-3.4	-2.1	0.8
322.30	-1.04	1.4	304.68	17JA	0.4	-3.4	-2.1	0.8
330.16	-0.98	1.4	314.80	27JA	0.2	-3.5	-2.1	0.7
338.00	-0.91	1.4	324.87	6FE	0.0	-3.5	-2.1	0.7
345.78	-0.84	1.4	334.89	16FE	-0.3	-3.5	-2.0	0.7
353.51	-0.75	1.4	344.86	26FE	-0.7	-3.4	-2.0	0.6
1.17	-0.66	1.4	354.77	8MR	-1.5	-3.4	-1.9	0.6
8.76	-0.57	1.4	4.62	18MR	-1.4	-3.4	-1.9	0.5
16.27	-0.47	1.4	14.42	28MR	-0.4	-3.4	-1.8	0.5
23.70	-0.37	1.5	24.16	7AP	0.9	-3.3	-1.7	0.4
31.04	-0.27	1.5	33.85	17AP	2.3	-3.3	-1.7	0.3
38.29	-0.16	1.6	43.50	27AP	3.7	-3.3	-1.6	0.3
45.46	-0.05	1.6	53.10	7MY	2.1	-3.3	-1.5	0.2
52.54	0.05	1.7	62.67	17MY	1.2	-3.3	-1.5	0.3
59.53	0.16	1.7	72.22	27MY	0.4	-3.4	-1.4	0.4
66.44	0.26	1.7	81.76	6JN	-0.5	-3.4	-1.4	0.4
73.27	0.36	1.8	91.29	16JN	-1.5	-3.4	-1.3	0.5
80.01	0.47	1.8	100.82	26JN	-1.4	-3.4	-1.3	0.5
86.68	0.57	1.8	110.36	6JL	-0.6	-3.5	-1.3	0.6
93.27	0.67	1.8	119.92	16JL	-0.0	-3.5	-1.3	0.7
99.79	0.76	1.9	129.52	26JL	0.3	-3.6	-1.2	0.7
106.23	0.86	1.9	139.15	5AU	0.6	-3.7	-1.2	0.7
112.59	0.96	1.9	148.82	15AU	1.1	-3.8	-1.2	0.8
118.89	1.05	1.9	158.54	25AU	2.5	-3.9	-1.2	0.8
125.10	1.15	1.8	168.31	4SE	1.9	-4.0	-1.3	0.8
131.23	1.24	1.8	178.14	14SE	-0.2	-4.1	-1.3	0.8
137.28	1.34	1.8	188.03	24SE	-0.9	-4.2	-1.3	0.8
143.23	1.44	1.7	197.98	4OC	-1.0	-4.3	-1.3	0.8
149.07	1.54	1.7	207.97	14OC	-0.9	-4.3	-1.4	0.8
154.79	1.64	1.6	218.03	24OC	-0.6	-4.3	-1.4	0.7
160.37	1.74	1.5	228.13	3NO	-0.5	-4.1	-1.5	0.7
165.78	1.85	1.4	238.26	13NO	-0.4	-3.6	-1.6	0.7
171.00	1.96	1.3	248.43	23NO	-0.4	-3.0	-1.6	0.7
175.98	2.08	1.2	258.62	3DE	0.3	-3.9	-1.7	0.7
180.68	2.20	1.0	268.81	13DE	3.0	-4.3	-1.7	0.7
185.02	2.33	0.9	279.01	23DE	0.9	-4.4	-1.8	0.8

♂ LONG	♂ LAT	♂ MAG	⊙ LONG	16.00UT 1162	☿	♀	♃	♄
188.93	2.46	0.7	289.20	2JA	0.2	-4.4	-1.9	0.8
192.30	2.60	0.5	299.36	12JA	0.0	-4.3	-1.9	0.8
195.00	2.74	0.2	309.50	22JA	-0.1	-4.2	-2.0	0.8
196.85	2.87	-0.0	319.59	1FE	-0.3	-4.0	-2.0	0.8
197.70	2.99	-0.3	329.64	11FE	-0.8	-3.9	-2.0	0.7
197.36	3.06	-0.6	339.64	21FE	-1.4	-3.8	-2.0	0.7
195.75	3.07	-0.9	349.58	3MR	-1.3	-3.7	-2.0	0.7
192.95	2.97	-1.1	359.47	13MR	-0.4	-3.7	-2.0	0.6
189.35	2.74	-1.3	9.30	23MR	1.1	-3.6	-1.9	0.6
185.61	2.39	-1.2	19.06	2AP	2.9	-3.5	-1.9	0.5
182.41	1.95	-1.1	28.78	12AP	2.8	-3.5	-1.8	0.4
180.30	1.49	-0.9	38.45	22AP	1.6	-3.4	-1.8	0.4
179.51	1.05	-0.7	48.08	2MY	0.9	-3.4	-1.7	0.3
180.02	0.64	-0.5	57.67	12MY	0.3	-3.3	-1.7	0.3
181.68	0.28	-0.3	67.23	22MY	-0.6	-3.3	-1.6	0.2
184.34	-0.04	-0.1	76.77	1JN	-1.5	-3.3	-1.5	0.3
187.81	-0.31	0.1	86.30	11JN	-1.5	-3.3	-1.5	0.4
191.95	-0.54	0.2	95.83	21JN	-0.6	-3.3	-1.4	0.4
196.65	-0.74	0.3	105.36	1JL	0.1	-3.3	-1.4	0.5
201.82	-0.91	0.5	114.91	11JL	0.5	-3.3	-1.3	0.5
207.39	-1.06	0.5	124.49	21JL	0.9	-3.4	-1.3	0.6
213.29	-1.19	0.6	134.09	31JL	1.5	-3.4	-1.3	0.7
219.49	-1.29	0.7	143.74	10AU	2.9	-3.4	-1.2	0.7
225.95	-1.38	0.8	153.44	20AU	1.6	-3.5	-1.2	0.7
232.65	-1.45	0.8	163.18	30AU	-0.3	-3.5	-1.2	0.8
239.54	-1.50	0.9	172.98	9SE	-1.0	-3.5	-1.2	0.8
246.62	-1.53	0.9	182.84	19SE	-1.1	-3.5	-1.2	0.8
253.85	-1.54	1.0	192.75	29SE	-0.9	-3.4	-1.2	0.8
261.22	-1.54	1.0	202.72	9OC	-0.5	-3.4	-1.2	0.8
268.71	-1.52	1.1	212.75	19OC	-0.3	-3.4	-1.3	0.8
276.30	-1.49	1.1	222.82	29OC	-0.3	-3.4	-1.3	0.7
283.96	-1.44	1.1	232.94	8NO	-0.2	-3.3	-1.3	0.7
291.69	-1.38	1.2	243.09	18NO	0.5	-3.3	-1.4	0.7
299.46	-1.31	1.2	253.27	28NO	3.2	-3.3	-1.4	0.7
307.26	-1.23	1.3	263.47	8DE	0.6	-3.3	-1.5	0.7
315.06	-1.14	1.3	273.67	18DE	-0.0	-3.4	-1.6	0.7
322.85	-1.04	1.3	283.86	28DE	-0.1	-3.4	-1.6	0.8
				1163				
330.62	-0.93	1.4	294.04	7JA	-0.2	-3.4	-1.7	0.8
338.36	-0.82	1.4	304.19	17JA	-0.4	-3.4	-1.8	0.8
346.04	-0.71	1.5	314.30	27JA	-0.8	-3.5	-1.8	0.8
353.67	-0.59	1.5	324.38	6FE	-1.3	-3.5	-1.9	0.8
1.23	-0.48	1.5	334.41	16FE	-1.2	-3.5	-1.9	0.8
8.72	-0.36	1.6	344.38	26FE	-0.3	-3.6	-2.0	0.7
16.13	-0.25	1.6	354.29	8MR	1.3	-3.6	-2.0	0.7
23.46	-0.13	1.6	4.15	18MR	3.4	-3.7	-2.0	0.7
30.71	-0.02	1.7	13.95	28MR	2.1	-3.8	-2.0	0.6
37.88	0.08	1.7	23.69	7AP	1.2	-3.9	-2.0	0.5
44.97	0.19	1.7	33.39	17AP	0.7	-4.0	-2.0	0.5
51.98	0.29	1.7	43.03	27AP	0.2	-4.0	-1.9	0.4
58.91	0.38	1.7	52.64	7MY	-0.6	-4.1	-1.9	0.4
65.77	0.47	1.7	62.22	17MY	-1.5	-4.2	-1.8	0.3
72.55	0.56	1.7	71.77	27MY	-1.5	-4.2	-1.8	0.2
79.27	0.65	1.8	81.30	6JN	-0.6	-4.0	-1.7	0.3
85.93	0.72	1.8	90.83	16JN	0.2	-3.6	-1.6	0.3
92.53	0.80	1.9	100.36	26JN	0.7	-2.9	-1.6	0.4
99.08	0.87	1.9	109.90	6JL	1.1	-3.5	-1.5	0.4
105.57	0.94	1.9	119.46	16JL	2.0	-4.0	-1.5	0.5
112.02	1.00	2.0	129.05	26JL	3.2	-4.2	-1.4	0.6
118.43	1.06	2.0	138.68	5AU	1.4	-4.2	-1.4	0.6
124.79	1.12	2.0	148.35	15AU	-0.3	-4.2	-1.3	0.7
131.12	1.17	2.0	158.06	25AU	-1.2	-4.1	-1.3	0.7
137.40	1.22	2.0	167.84	4SE	-1.2	-4.0	-1.3	0.7
143.65	1.27	2.0	177.66	14SE	-0.8	-3.9	-1.2	0.8
149.86	1.32	2.0	187.54	24SE	-0.4	-3.8	-1.2	0.8
156.04	1.36	1.9	197.49	4OC	-0.2	-3.8	-1.2	0.8
162.16	1.40	1.9	207.49	14OC	-0.1	-3.7	-1.2	0.8
168.25	1.43	1.8	217.54	24OC	-0.0	-3.6	-1.2	0.8
174.27	1.47	1.8	227.64	3NO	0.7	-3.6	-1.2	0.8
180.23	1.50	1.7	237.77	13NO	3.1	-3.5	-1.3	0.7
186.13	1.52	1.6	247.93	23NO	0.4	-3.5	-1.3	0.7
191.93	1.55	1.5	258.12	3DE	-0.2	-3.5	-1.3	0.7
197.63	1.56	1.4	268.32	13DE	-0.3	-3.4	-1.4	0.7
203.21	1.58	1.3	278.52	23DE	-0.3	-3.4	-1.4	0.7

♂ LONG	♂ LAT	♂ MAG	☉ LONG	16.00UT	☿	♀	♃	♄
LONG	LAT	MAG	LONG	1164	MAGNITUDES			
208.64	1.58	1.2	288.71	2JA	-0.5	-3.4	-1.5	0.8
213.88	1.57	1.0	298.87	12JA	-0.9	-3.3	-1.6	0.8
218.90	1.55	0.9	309.01	22JA	-1.1	-3.3	-1.6	0.8
223.65	1.52	0.7	319.11	1FE	-1.1	-3.3	-1.7	0.8
228.05	1.47	0.5	329.16	11FE	-0.3	-3.3	-1.8	0.8
232.03	1.39	0.2	339.16	21FE	1.6	-3.3	-1.8	0.8
235.47	1.27	-0.0	349.11	2MR	3.0	-3.3	-1.9	0.8
238.23	1.12	-0.3	358.99	12MR	1.5	-3.4	-1.9	0.7
240.16	0.90	-0.6	8.82	22MR	0.9	-3.4	-2.0	0.7
241.06	0.62	-0.9	18.59	1AP	0.5	-3.4	-2.0	0.7
240.78	0.25	-1.2	28.31	11AP	0.1	-3.5	-2.1	0.6
239.24	-0.20	-1.5	37.98	21AP	-0.6	-3.5	-2.1	0.6
236.59	-0.73	-1.8	47.61	1MY	-1.5	-3.5	-2.1	0.5
233.29	-1.27	-2.0	57.20	11MY	-1.5	-3.4	-2.0	0.4
230.00	-1.79	-1.8	66.76	21MY	-0.5	-3.4	-2.0	0.4
227.45	-2.21	-1.7	76.31	31MY	0.3	-3.4	-2.0	0.3
226.13	-2.54	-1.5	85.83	10JN	0.9	-3.3	-1.9	0.2
226.17	-2.75	-1.3	95.36	20JN	1.5	-3.3	-1.9	0.3
227.55	-2.88	-1.1	104.90	30JN	2.6	-3.3	-1.8	0.3
230.09	-2.95	-0.9	114.44	10JL	2.9	-3.3	-1.7	0.4
233.61	-2.96	-0.7	124.02	20JL	1.1	-3.3	-1.7	0.5
237.93	-2.94	-0.5	133.63	30JL	-0.4	-3.3	-1.6	0.5
242.91	-2.89	-0.3	143.27	9AU	-1.3	-3.4	-1.5	0.6
248.42	-2.81	-0.2	152.96	19AU	-1.4	-3.4	-1.5	0.6
254.38	-2.70	-0.1	162.70	29AU	-0.7	-3.4	-1.4	0.7
260.68	-2.58	0.1	172.50	8SE	-0.3	-3.5	-1.4	0.7
267.28	-2.45	0.2	182.35	18SE	-0.1	-3.5	-1.4	0.7
274.11	-2.30	0.3	192.27	28SE	0.0	-3.6	-1.3	0.8
281.14	-2.13	0.4	202.23	8OC	0.2	-3.6	-1.3	0.8
288.31	-1.96	0.5	212.26	18OC	1.0	-3.7	-1.3	0.8
295.59	-1.79	0.6	222.33	28OC	2.9	-3.8	-1.3	0.8
302.96	-1.61	0.7	232.45	7NO	0.2	-3.9	-1.3	0.8
310.38	-1.43	0.8	242.60	17NO	-0.4	-4.0	-1.3	0.8
317.84	-1.25	0.9	252.78	27NO	-0.4	-4.1	-1.3	0.7
325.31	-1.07	1.0	262.97	7DE	-0.5	-4.2	-1.3	0.7
332.77	-0.90	1.1	273.17	17DE	-0.6	-4.3	-1.4	0.7
340.21	-0.73	1.2	283.37	27DE	-0.9	-4.4	-1.4	0.7
				1165				
347.62	-0.56	1.3	293.54	6JA	-1.0	-4.3	-1.4	0.8
354.98	-0.41	1.3	303.70	16JA	-0.9	-4.1	-1.5	0.8
2.29	-0.26	1.4	313.82	26JA	-0.2	-3.6	-1.5	0.8
9.54	-0.12	1.5	323.89	5FE	1.9	-3.3	-1.6	0.8
16.73	0.01	1.5	333.92	15FE	2.4	-3.9	-1.7	0.8
23.86	0.13	1.6	343.90	25FE	1.1	-4.2	-1.7	0.8
30.91	0.25	1.7	353.81	7MR	0.6	-4.3	-1.8	0.8
37.90	0.36	1.7	3.67	17MR	0.3	-4.3	-1.9	0.8
44.82	0.45	1.8	13.48	27MR	-0.0	-4.2	-1.9	0.8
51.67	0.54	1.8	23.22	6AP	-0.6	-4.1	-2.0	0.7
58.46	0.63	1.8	32.92	16AP	-1.5	-4.0	-2.1	0.7
65.19	0.70	1.9	42.57	26AP	-1.5	-3.9	-2.1	0.6
71.87	0.77	1.9	52.18	6MY	-0.5	-3.8	-2.1	0.6
78.49	0.84	1.9	61.76	16MY	0.4	-3.7	-2.2	0.5
85.07	0.90	1.9	71.31	26MY	1.1	-3.6	-2.2	0.4
91.60	0.95	1.9	80.84	5JN	2.0	-3.5	-2.2	0.4
98.10	0.99	1.9	90.37	15JN	3.3	-3.5	-2.1	0.3
104.56	1.03	1.9	99.90	25JN	2.4	-3.4	-2.1	0.3
110.99	1.07	1.9	109.44	5JL	0.9	-3.4	-2.1	0.3
117.41	1.10	1.9	119.01	15JL	-0.4	-3.4	-2.0	0.4
123.80	1.13	2.0	128.59	25JL	-1.3	-3.4	-1.9	0.4
130.17	1.15	2.0	138.21	4AU	-1.4	-3.3	-1.9	0.5
136.54	1.17	2.0	147.88	14AU	-0.7	-3.3	-1.8	0.5
142.89	1.18	2.0	157.60	24AU	-0.2	-3.3	-1.8	0.6
149.24	1.19	2.0	167.36	3SE	0.0	-3.3	-1.7	0.7
155.59	1.19	2.0	177.19	13SE	0.2	-3.4	-1.6	0.7
161.94	1.19	2.0	187.07	23SE	0.4	-3.4	-1.6	0.7
168.29	1.19	2.0	197.00	3OC	1.3	-3.4	-1.5	0.8
174.64	1.18	2.0	207.00	13OC	2.7	-3.4	-1.5	0.8
180.98	1.16	2.0	217.05	23OC	0.1	-3.4	-1.5	0.8
187.34	1.14	1.9	227.14	2NO	-0.5	-3.5	-1.4	0.8
193.69	1.11	1.9	237.27	12NO	-0.6	-3.5	-1.4	0.8
200.03	1.08	1.8	247.44	22NO	-0.6	-3.5	-1.4	0.8
206.38	1.04	1.7	257.62	2DE	-0.7	-3.5	-1.4	0.8
212.71	0.99	1.7	267.82	12DE	-0.8	-3.4	-1.4	0.8
219.03	0.93	1.6	278.02	22DE	-0.8	-3.4	-1.4	0.7

♂ LONG	♂ LAT	♂ MAG	☉ LONG	16.00UT	☿	♀	♃	♄
LONG	LAT	MAG	LONG	1166	MAGNITUDES			
225.32	0.85	1.5	288.21	1JA	-0.8	-3.4	-1.4	0.7
231.60	0.77	1.4	298.38	11JA	-0.1	-3.4	-1.4	0.8
237.83	0.67	1.3	308.51	21JA	2.2	-3.3	-1.5	0.8
244.03	0.55	1.1	318.62	31JA	1.9	-3.3	-1.5	0.9
250.16	0.41	1.0	328.67	10FE	0.8	-3.3	-1.5	0.9
256.22	0.25	0.8	338.67	20FE	0.4	-3.3	-1.6	0.9
262.19	0.06	0.7	348.62	2MR	0.2	-3.4	-1.7	0.9
268.05	-0.16	0.5	358.51	12MR	-0.1	-3.4	-1.7	0.9
273.75	-0.42	0.3	8.34	22MR	-0.7	-3.4	-1.8	0.9
279.27	-0.72	0.1	18.12	1AP	-1.5	-3.4	-1.9	0.8
284.54	-1.07	-0.1	27.85	11AP	-1.5	-3.4	-1.9	0.8
289.50	-1.47	-0.3	37.52	21AP	-0.5	-3.5	-2.0	0.8
294.05	-1.94	-0.6	47.15	1MY	0.5	-3.5	-2.1	0.7
298.07	-2.47	-0.9	56.74	11MY	1.5	-3.6	-2.1	0.7
301.43	-3.08	-1.1	66.30	21MY	2.7	-3.6	-2.2	0.6
303.93	-3.76	-1.4	75.85	31MY	3.3	-3.7	-2.2	0.5
305.37	-4.50	-1.7	85.38	10JN	1.8	-3.8	-2.3	0.5
305.62	-5.26	-2.0	94.91	20JN	0.7	-3.9	-2.3	0.4
304.60	-5.95	-2.3	104.44	30JN	-0.4	-4.0	-2.3	0.3
302.53	-6.48	-2.6	113.99	10JL	-1.4	-4.1	-2.3	0.3
299.93	-6.73	-2.6	123.56	20JL	-1.4	-4.2	-2.3	0.4
297.50	-6.67	-2.5	133.17	30JL	-0.7	-4.2	-2.2	0.4
295.93	-6.32	-2.3	142.81	9AU	-0.2	-4.2	-2.2	0.5
295.61	-5.79	-2.0	152.50	19AU	0.1	-4.0	-2.1	0.5
296.60	-5.16	-1.7	162.24	29AU	0.3	-3.5	-2.1	0.6
298.84	-4.52	-1.4	172.03	8SE	0.6	-3.3	-2.0	0.6
302.09	-3.90	-1.2	181.88	18SE	1.7	-3.8	-1.9	0.7
306.17	-3.32	-0.9	191.79	28SE	2.4	-4.2	-1.9	0.7
310.92	-2.79	-0.7	201.75	8OC	-0.0	-4.3	-1.8	0.8
316.16	-2.31	-0.4	211.77	18OC	-0.7	-4.3	-1.7	0.8
321.78	-1.88	-0.2	221.84	28OC	-0.7	-4.3	-1.7	0.8
327.71	-1.50	-0.0	231.95	7NO	-0.7	-4.2	-1.6	0.8
333.84	-1.15	0.2	242.10	17NO	-0.8	-4.1	-1.6	0.9
340.15	-0.84	0.3	252.28	27NO	-0.7	-4.0	-1.6	0.9
346.58	-0.57	0.5	262.47	7DE	-0.7	-3.9	-1.5	0.8
353.09	-0.33	0.7	272.67	17DE	-0.6	-3.8	-1.5	0.8
359.65	-0.12	0.8	282.87	27DE	0.0	-3.7	-1.5	0.8
				1167				
6.26	0.07	1.0	293.05	6JA	2.5	-3.6	-1.5	0.8
12.87	0.24	1.1	303.20	16JA	1.4	-3.6	-1.5	0.8
19.49	0.38	1.2	313.33	26JA	0.5	-3.5	-1.5	0.9
26.11	0.51	1.3	323.41	5FE	0.3	-3.5	-1.5	0.9
32.71	0.62	1.4	333.44	15FE	0.1	-3.4	-1.5	0.9
39.29	0.72	1.5	343.42	25FE	-0.2	-3.4	-1.6	1.0
45.85	0.81	1.6	353.34	7MR	-0.7	-3.3	-1.6	1.0
52.37	0.88	1.7	3.20	17MR	-1.5	-3.3	-1.7	1.0
58.87	0.94	1.7	13.01	27MR	-1.4	-3.3	-1.7	1.0
65.35	1.00	1.8	22.76	6AP	-0.5	-3.3	-1.8	0.9
71.79	1.04	1.9	32.46	16AP	0.7	-3.3	-1.8	0.9
78.21	1.08	1.9	42.11	26AP	1.9	-3.3	-1.9	0.9
84.61	1.11	1.9	51.72	6MY	3.5	-3.3	-2.0	0.8
90.98	1.14	2.0	61.30	16MY	2.6	-3.4	-2.0	0.8
97.34	1.15	2.0	70.85	26MY	1.4	-3.4	-2.1	0.7
103.69	1.17	2.0	80.39	5JN	0.5	-3.4	-2.2	0.7
110.03	1.17	2.0	89.91	15JN	-0.5	-3.5	-2.2	0.6
116.36	1.17	2.0	99.45	25JN	-1.4	-3.5	-2.3	0.5
122.69	1.17	2.0	108.98	5JL	-1.5	-3.5	-2.3	0.5
129.03	1.16	2.0	118.54	15JL	-0.6	-3.4	-2.4	0.4
135.38	1.15	2.0	128.13	25JL	-0.1	-3.4	-2.4	0.4
141.73	1.13	2.0	137.75	4AU	0.3	-3.4	-2.4	0.5
148.11	1.11	2.0	147.41	14AU	0.5	-3.4	-2.4	0.5
154.51	1.09	2.0	157.13	24AU	0.9	-3.3	-2.4	0.6
160.93	1.06	2.0	166.89	3SE	2.1	-3.3	-2.3	0.6
167.37	1.02	2.0	176.71	13SE	2.1	-3.3	-2.3	0.7
173.85	0.98	2.0	186.59	23SE	-0.1	-3.3	-2.2	0.7
180.36	0.94	2.0	196.52	3OC	-0.8	-3.3	-2.2	0.8
186.91	0.89	2.0	206.51	13OC	-0.9	-3.4	-2.1	0.8
193.50	0.83	1.9	216.56	23OC	-0.8	-3.4	-2.0	0.8
200.12	0.77	1.9	226.64	2NO	-0.7	-3.4	-2.0	0.9
206.78	0.70	1.9	236.78	12NO	-0.5	-3.4	-1.9	0.9
213.48	0.63	1.8	246.94	22NO	-0.5	-3.5	-1.8	0.9
220.23	0.55	1.8	257.12	2DE	-0.5	-3.5	-1.8	0.9
227.01	0.45	1.8	267.32	12DE	0.2	-3.6	-1.7	0.9
233.84	0.35	1.7	277.52	22DE	2.7	-3.7	-1.7	0.9

♂ LONG	LAT	MAG	⊙ LONG	16.00UT	☿	♀	♃	♄
				1168	MAGNITUDES			
240.70	0.24	1.6	287.71	1JA	1.1	-3.7	-1.6	0.9
247.60	0.13	1.6	297.88	11JA	0.3	-3.8	-1.6	0.9
254.54	-0.00	1.5	308.02	21JA	0.1	-3.9	-1.6	0.9
261.52	-0.15	1.4	318.12	31JA	-0.0	-4.0	-1.6	0.9
268.53	-0.30	1.3	328.18	10FE	-0.3	-4.1	-1.6	1.0
275.56	-0.46	1.2	338.19	20FE	-0.7	-4.2	-1.6	1.0
282.62	-0.64	1.1	348.14	1MR	-1.5	-4.3	-1.6	1.0
289.69	-0.83	1.0	358.03	11MR	-1.3	-4.3	-1.6	1.1
296.77	-1.03	0.9	7.87	21MR	-0.4	-4.2	-1.6	1.1
303.85	-1.24	0.8	17.65	31MR	0.9	-3.9	-1.6	1.1
310.92	-1.46	0.7	27.38	10AP	2.5	-3.2	-1.7	1.1
317.94	-1.68	0.5	37.05	20AP	3.4	-3.2	-1.7	1.0
324.91	-1.92	0.4	46.68	30AP	1.9	-3.8	-1.8	1.0
331.80	-2.16	0.3	56.28	10MY	1.1	-4.1	-1.8	1.0
338.56	-2.40	0.1	65.85	20MY	0.4	-4.2	-1.9	0.9
345.17	-2.64	-0.0	75.39	30MY	-0.5	-4.2	-1.9	0.9
351.56	-2.87	-0.2	84.92	9JN	-1.5	-4.1	-2.0	0.8
357.67	-3.10	-0.3	94.45	19JN	-1.5	-4.0	-2.1	0.8
3.44	-3.33	-0.5	103.98	29JN	-0.6	-3.9	-2.1	0.7
8.75	-3.53	-0.7	113.53	9JL	0.0	-3.8	-2.2	0.6
13.47	-3.72	-0.8	123.10	19JL	0.4	-3.7	-2.3	0.6
17.47	-3.88	-1.0	132.70	29JL	0.7	-3.7	-2.3	0.5
20.53	-4.00	-1.3	142.34	8AU	1.2	-3.6	-2.4	0.5
22.45	-4.06	-1.5	152.03	18AU	2.6	-3.5	-2.4	0.6
23.04	-4.04	-1.7	161.76	28AU	1.8	-3.5	-2.5	0.6
22.16	-3.88	-2.0	171.56	7SE	-0.2	-3.5	-2.5	0.7
19.93	-3.56	-2.1	181.41	17SE	-1.0	-3.4	-2.5	0.7
16.81	-3.06	-2.3	191.31	27SE	-1.0	-3.4	-2.4	0.8
13.53	-2.44	-2.2	201.27	7OC	-0.9	-3.4	-2.4	0.8
10.91	-1.77	-1.9	211.29	17OC	-0.6	-3.4	-2.3	0.9
9.44	-1.14	-1.5	221.35	27OC	-0.4	-3.4	-2.3	0.9
9.32	-0.58	-1.2	231.47	6NO	-0.4	-3.4	-2.2	0.9
10.47	-0.12	-0.9	241.61	16NO	-0.3	-3.4	-2.2	1.0
12.69	0.25	-0.6	251.79	26NO	0.3	-3.4	-2.1	1.0
15.78	0.54	-0.3	261.98	6DE	3.0	-3.4	-2.0	1.0
19.57	0.77	-0.0	272.18	16DE	0.8	-3.4	-1.9	1.0
23.87	0.95	0.2	282.37	26DE	0.1	-3.4	-1.9	1.0
				1169				
28.58	1.09	0.4	292.55	5JA	-0.0	-3.4	-1.8	1.0
33.62	1.19	0.6	302.71	15JA	-0.1	-3.4	-1.8	1.0
38.89	1.27	0.8	312.83	25JA	-0.4	-3.5	-1.7	1.0
44.35	1.33	1.0	322.91	4FE	-0.8	-3.5	-1.7	1.0
49.96	1.38	1.1	332.95	14FE	-1.4	-3.5	-1.6	1.1
55.67	1.41	1.3	342.92	24FE	-1.3	-3.4	-1.6	1.1
61.48	1.43	1.4	352.85	6MR	-0.4	-3.4	-1.6	1.2
67.36	1.44	1.5	2.72	16MR	1.1	-3.4	-1.6	1.2
73.29	1.44	1.6	12.52	26MR	3.1	-3.4	-1.6	1.2
79.27	1.43	1.7	22.28	5AP	2.5	-3.3	-1.6	1.2
85.28	1.42	1.8	31.98	15AP	1.4	-3.3	-1.6	1.2
91.33	1.41	1.8	41.63	25AP	0.8	-3.3	-1.6	1.2
97.40	1.39	1.9	51.25	5MY	0.2	-3.3	-1.6	1.2
103.50	1.36	1.9	60.83	15MY	-0.5	-3.3	-1.7	1.2
109.62	1.33	2.0	70.38	25MY	-1.5	-3.4	-1.7	1.1
115.78	1.30	2.0	79.92	4JN	-1.5	-3.4	-1.7	1.1
121.95	1.26	2.0	89.45	14JN	-0.6	-3.4	-1.8	1.0
128.16	1.22	2.0	98.98	24JN	0.1	-3.4	-1.8	0.9
134.40	1.17	2.0	108.52	4JL	0.5	-3.5	-1.9	0.9
140.68	1.13	2.0	118.08	14JL	0.9	-3.5	-2.0	0.8
146.99	1.08	2.0	127.66	24JL	1.6	-3.6	-2.0	0.7
153.35	1.02	2.0	137.28	3AU	3.1	-3.7	-2.1	0.7
159.75	0.96	2.0	146.94	13AU	1.6	-3.8	-2.2	0.6
166.20	0.90	2.0	156.65	23AU	-0.3	-3.9	-2.2	0.7
172.71	0.84	1.9	166.42	2SE	-1.1	-4.0	-2.3	0.7
179.26	0.77	1.9	176.23	12SE	-1.2	-4.1	-2.3	0.7
185.88	0.70	1.8	186.11	22SE	-0.8	-4.2	-2.4	0.8
192.56	0.62	1.8	196.04	2OC	-0.5	-4.3	-2.4	0.8
199.30	0.54	1.8	206.03	12OC	-0.3	-4.3	-2.4	0.9
206.10	0.46	1.8	216.07	22OC	-0.2	-4.3	-2.4	0.9
212.98	0.37	1.8	226.16	1NO	-0.1	-4.1	-2.4	1.0
219.91	0.27	1.8	236.29	11NO	0.6	-3.5	-2.4	1.0
226.92	0.17	1.8	246.45	21NO	3.2	-3.1	-2.3	1.1
234.00	0.07	1.7	256.63	1DE	0.6	-3.9	-2.3	1.1
241.14	-0.04	1.7	266.83	11DE	-0.1	-4.3	-2.2	1.1
248.35	-0.15	1.7	277.03	21DE	-0.2	-4.4	-2.1	1.2
255.62	-0.26	1.6	287.22	31DE	-0.3	-4.4	-2.0	1.2

♂ LONG	LAT	MAG	⊙ LONG	16.00UT	☿	♀	♃	♄
				1170	MAGNITUDES			
262.95	-0.38	1.6	297.39	10JA	-0.5	-4.3	-2.0	1.2
270.34	-0.51	1.5	307.53	20JA	-0.8	-4.2	-1.9	1.2
277.78	-0.63	1.5	317.63	30JA	-1.2	-4.0	-1.8	1.2
285.27	-0.76	1.4	327.69	9FE	-1.2	-3.9	-1.8	1.2
292.81	-0.88	1.4	337.70	19FE	-0.4	-3.8	-1.7	1.2
300.37	-1.01	1.3	347.66	1MR	1.4	-3.7	-1.7	1.3
307.96	-1.13	1.3	357.55	11MR	3.4	-3.7	-1.6	1.3
315.57	-1.25	1.2	7.39	21MR	1.9	-3.6	-1.6	1.3
323.17	-1.36	1.2	17.18	31MR	1.1	-3.5	-1.6	1.4
330.76	-1.46	1.1	26.90	10AP	0.6	-3.5	-1.5	1.4
338.33	-1.55	1.0	36.58	20AP	0.1	-3.4	-1.5	1.4
345.85	-1.63	1.0	46.22	30AP	-0.6	-3.4	-1.5	1.4
353.32	-1.69	0.9	55.81	10MY	-1.5	-3.3	-1.5	1.4
0.71	-1.74	0.9	65.38	20MY	-1.5	-3.3	-1.5	1.4
8.00	-1.77	0.8	74.93	30MY	-0.6	-3.3	-1.5	1.3
15.17	-1.78	0.7	84.46	9JN	0.2	-3.3	-1.6	1.3
22.19	-1.78	0.7	93.99	19JN	0.7	-3.3	-1.6	1.2
29.05	-1.75	0.6	103.52	29JN	1.3	-3.3	-1.6	1.2
35.70	-1.71	0.5	113.07	9JL	2.2	-3.3	-1.7	1.1
42.12	-1.64	0.4	122.64	19JL	3.2	-3.4	-1.7	1.0
48.25	-1.55	0.3	132.24	29JL	1.3	-3.4	-1.8	1.0
54.05	-1.43	0.2	141.88	8AU	-0.3	-3.4	-1.8	0.9
59.45	-1.29	0.1	151.56	18AU	-1.2	-3.5	-1.9	0.8
64.36	-1.11	-0.0	161.29	28AU	-1.3	-3.5	-1.9	0.8
68.68	-0.90	-0.2	171.08	7SE	-0.8	-3.5	-2.0	0.8
72.27	-0.65	-0.4	180.93	17SE	-0.4	-3.5	-2.1	0.9
74.95	-0.35	-0.5	190.83	27SE	-0.2	-3.4	-2.1	0.9
76.54	-0.00	-0.8	200.78	7OC	-0.1	-3.4	-2.2	1.0
76.81	0.41	-1.0	210.80	17OC	0.1	-3.4	-2.2	1.0
75.64	0.87	-1.2	220.86	27OC	0.8	-3.4	-2.3	1.1
73.10	1.36	-1.4	230.97	6NO	3.2	-3.3	-2.3	1.1
69.53	1.83	-1.5	241.12	16NO	0.4	-3.3	-2.3	1.2
65.66	2.21	-1.5	251.29	26NO	-0.3	-3.3	-2.3	1.2
62.28	2.47	-1.2	261.48	6DE	-0.3	-3.3	-2.3	1.3
59.98	2.62	-0.9	271.68	16DE	-0.4	-3.4	-2.3	1.3
59.02	2.66	-0.6	281.88	26DE	-0.5	-3.4	-2.2	1.3
				1171				
59.37	2.65	-0.3	292.06	5JA	-0.9	-3.4	-2.2	1.3
60.85	2.59	-0.0	302.22	15JA	-1.1	-3.4	-2.1	1.4
63.27	2.51	0.2	312.34	25JA	-1.0	-3.5	-2.0	1.4
66.42	2.42	0.5	322.42	4FE	-0.3	-3.5	-1.9	1.4
70.16	2.32	0.7	332.46	14FE	1.6	-3.5	-1.9	1.4
74.36	2.22	0.9	342.44	24FE	2.8	-3.6	-1.8	1.3
78.91	2.13	1.0	352.37	6MR	1.4	-3.7	-1.7	1.3
83.75	2.03	1.2	2.24	16MR	0.8	-3.7	-1.7	1.3
88.82	1.94	1.3	12.05	26MR	0.4	-3.8	-1.6	1.3
94.08	1.85	1.4	21.81	5AP	0.0	-3.9	-1.6	1.3
99.49	1.76	1.5	31.51	15AP	-0.6	-4.0	-1.5	1.3
105.03	1.67	1.6	41.17	25AP	-1.5	-4.0	-1.5	1.2
110.68	1.58	1.7	50.79	5MY	-1.5	-4.1	-1.5	1.2
116.43	1.50	1.8	60.37	15MY	-0.5	-4.2	-1.4	1.2
122.27	1.41	1.8	69.92	25MY	0.3	-4.2	-1.4	1.2
128.19	1.33	1.9	79.46	4JN	0.9	-4.0	-1.4	1.1
134.18	1.24	1.9	88.99	14JN	1.7	-3.5	-1.4	1.1
140.25	1.16	1.9	98.52	24JN	2.8	-2.9	-1.4	1.0
146.39	1.07	1.9	108.06	4JL	2.7	-3.5	-1.4	1.0
152.61	0.98	1.9	117.62	14JL	1.1	-4.0	-1.5	0.9
158.90	0.89	1.9	127.20	24JL	-0.4	-4.2	-1.5	0.9
165.26	0.80	1.9	136.82	3AU	-1.3	-4.2	-1.5	0.8
171.70	0.71	1.9	146.48	13AU	-1.4	-4.2	-1.5	0.8
178.22	0.62	1.9	156.18	23AU	-0.7	-4.1	-1.6	0.7
184.82	0.53	1.9	165.94	2SE	-0.3	-4.0	-1.6	0.7
191.50	0.43	1.9	175.76	12SE	-0.0	-3.9	-1.7	0.7
198.26	0.34	1.8	185.63	22SE	0.1	-3.8	-1.8	0.8
205.11	0.24	1.8	195.56	2OC	0.3	-3.8	-1.8	0.9
212.04	0.14	1.7	205.55	12OC	1.1	-3.7	-1.9	0.9
219.06	0.04	1.7	215.58	22OC	3.0	-3.6	-2.0	1.0
226.17	-0.07	1.6	225.67	1NO	0.2	-3.6	-2.0	1.1
233.37	-0.17	1.6	235.80	11NO	-0.4	-3.5	-2.1	1.1
240.64	-0.27	1.6	245.95	21NO	-0.5	-3.5	-2.1	1.2
248.00	-0.37	1.6	256.14	1DE	-0.5	-3.5	-2.2	1.2
255.44	-0.48	1.6	266.34	11DE	-0.6	-3.4	-2.2	1.2
262.94	-0.58	1.6	276.53	21DE	-0.9	-3.4	-2.2	1.2
270.52	-0.67	1.5	286.73	31DE	-0.9	-3.4	-2.2	1.2

Left Table

♂ LONG	LAT	MAG	☉ LONG	16.00UT	☿	♀	♃	♄
				1172		MAGNITUDES		
278.15	-0.77	1.5	296.90	10JA	-0.9	-3.3	-2.2	1.2
285.84	-0.86	1.5	307.04	20JA	-0.2	-3.3	-2.1	1.2
293.57	-0.94	1.5	317.15	30JA	1.9	-3.3	-2.1	1.2
301.34	-1.02	1.4	327.21	9FE	2.2	-3.3	-2.0	1.1
309.12	-1.08	1.4	337.22	19FE	1.0	-3.3	-2.0	1.1
316.93	-1.14	1.4	347.18	29FE	0.5	-3.3	-1.9	1.1
324.72	-1.19	1.4	357.08	10MR	0.3	-3.4	-1.8	1.0
332.51	-1.22	1.4	6.91	20MR	-0.1	-3.4	-1.8	1.0
340.27	-1.25	1.3	16.70	30MR	-0.6	-3.4	-1.7	1.0
348.00	-1.26	1.3	26.43	9AP	-1.5	-3.5	-1.6	1.0
355.67	-1.25	1.3	36.11	19AP	-1.5	-3.5	-1.6	1.0
3.28	-1.23	1.3	45.75	29AP	-0.5	-3.5	-1.5	1.0
10.81	-1.20	1.3	55.35	9MY	0.4	-3.4	-1.5	1.0
18.26	-1.15	1.3	64.91	19MY	1.2	-3.4	-1.4	1.0
25.62	-1.09	1.2	74.46	29MY	2.2	-3.4	-1.4	0.9
32.86	-1.02	1.2	83.99	8JN	3.4	-3.3	-1.4	0.9
39.99	-0.93	1.2	93.52	18JN	2.2	-3.3	-1.3	0.9
46.98	-0.83	1.2	103.05	28JN	0.8	-3.3	-1.3	0.8
53.83	-0.71	1.2	112.60	8JL	-0.4	-3.3	-1.3	0.8
60.53	-0.59	1.1	122.17	18JL	-1.3	-3.3	-1.3	0.7
67.05	-0.45	1.1	131.77	28JL	-1.5	-3.3	-1.3	0.7
73.39	-0.29	1.0	141.40	7AU	-0.7	-3.4	-1.3	0.6
79.51	-0.13	1.0	151.08	17AU	-0.2	-3.4	-1.4	0.6
85.38	0.06	0.9	160.82	27AU	0.1	-3.4	-1.4	0.5
90.98	0.26	0.8	170.60	6SE	0.2	-3.5	-1.4	0.5
96.25	0.48	0.7	180.44	16SE	0.5	-3.5	-1.5	0.4
101.12	0.72	0.6	190.35	26SE	1.4	-3.6	-1.5	0.4
105.52	1.00	0.5	200.30	6OC	2.7	-3.6	-1.6	0.5
109.35	1.30	0.3	210.31	16OC	0.1	-3.7	-1.6	0.6
112.46	1.64	0.1	220.38	26OC	-0.6	-3.8	-1.7	0.7
114.70	2.02	-0.1	230.48	5NO	-0.6	-3.9	-1.7	0.7
115.87	2.45	-0.3	240.62	15NO	-0.6	-4.0	-1.8	0.8
115.79	2.90	-0.5	250.80	25NO	-0.7	-4.1	-1.9	0.8
114.34	3.35	-0.7	260.99	5DE	-0.7	-4.2	-1.9	0.9
111.59	3.76	-0.9	271.19	15DE	-0.8	-4.3	-2.0	0.9
107.89	4.06	-1.1	281.38	25DE	-0.7	-4.4	-2.0	0.9
				1173				
103.92	4.19	-1.1	291.56	4JA	-0.1	-4.3	-2.1	1.0
100.42	4.16	-0.8	301.72	14JA	2.2	-4.1	-2.1	1.0
97.97	4.00	-0.6	311.85	24JA	1.7	-3.6	-2.1	1.0
96.83	3.76	-0.3	321.94	3FE	0.7	-3.4	-2.1	0.9
96.95	3.48	-0.0	331.98	13FE	0.4	-3.9	-2.1	0.9
98.22	3.20	0.2	341.96	23FE	0.2	-4.2	-2.0	0.9
100.41	2.93	0.4	351.89	5MR	-0.1	-4.3	-2.0	0.8
103.36	2.68	0.6	1.76	15MR	-0.7	-4.3	-1.9	0.8
106.92	2.45	0.8	11.58	25MR	-1.5	-4.2	-1.8	0.7
110.96	2.24	1.0	21.33	4AP	-1.5	-4.1	-1.8	0.7
115.39	2.04	1.1	31.04	14AP	-0.5	-4.0	-1.7	0.8
120.14	1.86	1.2	40.70	24AP	0.6	-3.9	-1.6	0.8
125.15	1.69	1.4	50.32	4MY	1.6	-3.8	-1.6	0.8
130.38	1.53	1.4	59.90	14MY	2.9	-3.7	-1.5	0.8
135.81	1.38	1.5	69.46	24MY	3.0	-3.6	-1.5	0.7
141.40	1.24	1.6	79.00	3JN	1.7	-3.5	-1.4	0.7
147.15	1.10	1.6	88.53	13JN	0.7	-3.5	-1.4	0.7
153.03	0.96	1.7	98.06	23JN	-0.4	-3.4	-1.3	0.7
159.04	0.83	1.7	107.59	3JL	-1.4	-3.4	-1.3	0.6
165.17	0.71	1.7	117.15	13JL	-1.5	-3.4	-1.3	0.6
171.43	0.59	1.8	126.74	23JL	-0.7	-3.4	-1.3	0.5
177.79	0.46	1.8	136.35	2AU	-0.1	-3.3	-1.2	0.5
184.27	0.35	1.8	146.01	12AU	0.2	-3.3	-1.2	0.4
190.86	0.23	1.8	155.71	22AU	0.4	-3.3	-1.2	0.4
197.55	0.12	1.7	165.47	1SE	0.7	-3.3	-1.2	0.3
204.36	0.01	1.7	175.28	11SE	1.8	-3.4	-1.3	0.3
211.28	-0.10	1.7	185.15	21SE	2.4	-3.4	-1.3	0.2
218.30	-0.21	1.7	195.08	1OC	-0.0	-3.4	-1.3	0.2
225.42	-0.31	1.7	205.06	11OC	-0.7	-3.4	-1.4	0.2
232.65	-0.41	1.6	215.10	21OC	-0.8	-3.4	-1.4	0.3
239.98	-0.51	1.6	225.18	31OC	-0.8	-3.5	-1.4	0.3
247.40	-0.60	1.5	235.31	10NO	-0.7	-3.5	-1.5	0.4
254.91	-0.69	1.5	245.46	20NO	-0.6	-3.5	-1.6	0.5
262.50	-0.77	1.5	255.64	30NO	-0.6	-3.5	-1.6	0.5
270.16	-0.84	1.4	265.84	10DE	-0.6	-3.4	-1.7	0.6
277.89	-0.91	1.4	276.04	20DE	0.0	-3.4	-1.8	0.6
285.67	-0.96	1.3	286.23	30DE	2.5	-3.4	-1.8	0.7

Right Table

♂ LONG	LAT	MAG	☉ LONG	16.00UT	☿	♀	♃	♄
				1174		MAGNITUDES		
293.50	-1.01	1.3	296.40	9JA	1.3	-3.4	-1.9	0.7
301.36	-1.05	1.3	306.55	19JA	0.4	-3.3	-1.9	0.7
309.24	-1.08	1.4	316.65	29JA	0.2	-3.3	-2.0	0.7
317.13	-1.09	1.4	326.72	8FE	0.0	-3.3	-2.0	0.7
325.01	-1.09	1.4	336.73	18FE	-0.2	-3.3	-2.0	0.7
332.87	-1.09	1.4	346.69	28FE	-0.7	-3.4	-2.0	0.7
340.70	-1.06	1.4	356.60	10MR	-1.4	-3.4	-2.0	0.6
348.49	-1.03	1.4	6.44	20MR	-1.4	-3.4	-2.0	0.6
356.22	-0.99	1.4	16.23	30MR	-0.5	-3.4	-1.9	0.6
3.89	-0.93	1.4	25.96	9AP	0.7	-3.4	-1.9	0.5
11.49	-0.87	1.5	35.64	19AP	2.1	-3.5	-1.8	0.5
19.01	-0.79	1.5	45.28	29AP	3.8	-3.5	-1.8	0.6
26.44	-0.71	1.5	54.89	9MY	2.3	-3.6	-1.7	0.6
33.78	-0.61	1.5	64.45	19MY	1.3	-3.6	-1.6	0.6
41.02	-0.51	1.5	74.00	29MY	0.5	-3.7	-1.6	0.6
48.16	-0.41	1.5	83.54	8JN	-0.5	-3.8	-1.5	0.6
55.20	-0.29	1.5	93.06	18JN	-1.4	-3.9	-1.5	0.5
62.12	-0.17	1.5	102.59	28JN	-1.5	-4.0	-1.4	0.5
68.94	-0.05	1.5	112.14	8JL	-0.6	-4.1	-1.4	0.5
75.63	0.08	1.5	121.71	18JL	-0.1	-4.2	-1.3	0.5
82.21	0.21	1.5	131.30	28JL	0.3	-4.2	-1.3	0.4
88.66	0.35	1.5	140.94	7AU	0.6	-4.2	-1.2	0.4
94.97	0.50	1.5	150.62	17AU	1.0	-4.0	-1.2	0.3
101.13	0.65	1.4	160.35	27AU	2.3	-3.5	-1.2	0.3
107.13	0.81	1.4	170.13	6SE	2.1	-3.3	-1.2	0.2
112.95	0.98	1.3	179.97	16SE	-0.1	-3.9	-1.2	0.1
118.57	1.16	1.3	189.87	26SE	-0.9	-4.2	-1.2	0.1
123.95	1.35	1.2	199.82	6OC	-0.9	-4.3	-1.2	0.0
129.04	1.55	1.1	209.83	16OC	-0.9	-4.3	-1.2	-0.0
133.81	1.78	1.0	219.89	26OC	-0.6	-4.2	-1.3	0.0
138.17	2.02	0.8	229.99	5NO	-0.5	-4.2	-1.3	0.1
142.04	2.29	0.7	240.13	15NO	-0.5	-4.1	-1.4	0.2
145.31	2.58	0.5	250.31	25NO	-0.4	-4.0	-1.4	0.2
147.82	2.91	0.3	260.50	5DE	0.2	-3.9	-1.5	0.3
149.40	3.25	0.0	270.69	15DE	2.8	-3.8	-1.5	0.4
149.88	3.62	-0.2	280.89	25DE	1.0	-3.7	-1.6	0.4
				1175				
149.09	3.98	-0.4	291.08	4JA	0.2	-3.6	-1.6	0.5
146.99	4.29	-0.7	301.24	14JA	0.0	-3.6	-1.7	0.5
143.75	4.49	-0.9	311.37	24JA	-0.1	-3.5	-1.8	0.5
139.86	4.53	-1.0	321.45	3FE	-0.3	-3.5	-1.8	0.5
136.04	4.39	-0.9	331.49	13FE	-0.7	-3.4	-1.9	0.5
132.95	4.11	-0.7	341.48	23FE	-1.4	-3.4	-1.9	0.5
131.03	3.74	-0.4	351.42	5MR	-1.3	-3.3	-2.0	0.5
130.41	3.34	-0.2	1.29	15MR	-0.5	-3.3	-2.0	0.5
131.01	2.95	0.0	11.11	25MR	0.9	-3.3	-2.0	0.5
132.67	2.58	0.3	20.87	4AP	2.7	-3.3	-2.0	0.4
135.20	2.24	0.5	30.58	14AP	3.0	-3.3	-2.0	0.4
138.45	1.93	0.6	40.24	24AP	1.7	-3.3	-2.0	0.4
142.28	1.65	0.8	49.86	4MY	1.0	-3.3	-1.9	0.4
146.58	1.40	0.9	59.44	14MY	0.3	-3.4	-1.9	0.4
151.28	1.17	1.0	69.00	24MY	-0.5	-3.4	-1.8	0.4
156.30	0.96	1.1	78.54	3JN	-1.5	-3.4	-1.7	0.4
161.61	0.77	1.2	88.07	13JN	-1.5	-3.5	-1.7	0.4
167.17	0.59	1.3	97.60	23JN	-0.6	-3.5	-1.6	0.4
172.94	0.42	1.3	107.13	3JL	0.0	-3.5	-1.6	0.4
178.91	0.26	1.4	116.69	13JL	0.4	-3.4	-1.5	0.4
185.07	0.11	1.4	126.27	23JL	0.8	-3.4	-1.4	0.4
191.40	-0.03	1.4	135.88	2AU	1.4	-3.4	-1.4	0.3
197.88	-0.17	1.5	145.53	12AU	2.8	-3.4	-1.4	0.3
204.52	-0.29	1.5	155.24	22AU	1.8	-3.3	-1.3	0.2
211.30	-0.41	1.5	164.99	1SE	-0.2	-3.3	-1.3	0.2
218.22	-0.52	1.5	174.80	11SE	-1.0	-3.3	-1.3	0.1
225.27	-0.62	1.5	184.67	21SE	-1.1	-3.3	-1.2	0.1
232.44	-0.72	1.5	194.59	1OC	-0.9	-3.3	-1.2	-0.0
239.74	-0.80	1.5	204.57	11OC	-0.5	-3.4	-1.2	-0.1
247.15	-0.88	1.5	214.61	21OC	-0.4	-3.4	-1.2	-0.1
254.60	-0.95	1.5	224.69	31OC	-0.3	-3.4	-1.2	-0.2
262.26	-1.00	1.4	234.81	10NO	-0.2	-3.4	-1.2	-0.2
269.94	-1.05	1.4	244.97	20NO	0.4	-3.5	-1.3	-0.0
277.70	-1.08	1.4	255.15	30NO	3.0	-3.5	-1.3	0.0
285.51	-1.10	1.4	265.35	10DE	0.7	-3.6	-1.4	0.1
293.37	-1.11	1.4	275.55	20DE	0.0	-3.7	-1.4	0.2
301.25	-1.11	1.4	285.74	30DE	-0.1	-3.7	-1.5	0.2

♂ LONG	♂ LAT	♂ MAG	☉ LONG	16.00UT	☿	♀	♃	♄
				1176	MAGNITUDES			
309.16	-1.10	1.4	295.91	9JA	-0.2	-3.8	-1.5	0.3
317.06	-1.07	1.4	306.06	19JA	-0.4	-3.9	-1.6	0.3
324.96	-1.04	1.4	316.17	29JA	-0.8	-4.0	-1.6	0.4
332.83	-0.99	1.4	326.24	8FE	-1.3	-4.1	-1.7	0.4
340.66	-0.93	1.3	336.25	18FE	-1.2	-4.2	-1.8	0.4
348.45	-0.87	1.3	346.21	28FE	-0.4	-4.3	-1.8	0.4
356.17	-0.79	1.3	356.12	9MR	1.2	-4.3	-1.9	0.4
3.84	-0.71	1.4	5.97	19MR	3.3	-4.2	-2.0	0.4
11.42	-0.62	1.4	15.76	29MR	2.3	-3.9	-2.0	0.4
18.93	-0.53	1.5	25.49	8AP	1.3	-3.2	-2.0	0.3
26.35	-0.43	1.5	35.18	18AP	0.7	-3.2	-2.1	0.3
33.68	-0.33	1.6	44.82	28AP	0.2	-3.9	-2.1	0.3
40.93	-0.22	1.6	54.42	8MY	-0.5	-4.1	-2.1	0.2
48.08	-0.12	1.6	63.99	18MY	-1.5	-4.2	-2.0	0.3
55.14	-0.01	1.7	73.54	28MY	-1.5	-4.2	-2.0	0.3
62.12	0.10	1.7	83.07	7JN	-0.6	-4.1	-2.0	0.3
69.00	0.21	1.7	92.60	17JN	0.1	-4.0	-1.9	0.4
75.80	0.32	1.8	102.13	27JN	0.6	-3.9	-1.9	0.4
82.51	0.43	1.8	111.68	7JL	1.1	-3.8	-1.8	0.4
89.13	0.54	1.8	121.24	17JL	1.8	-3.7	-1.7	0.4
95.66	0.65	1.8	130.84	27JL	3.2	-3.7	-1.7	0.3
102.11	0.76	1.8	140.47	6AU	1.5	-3.6	-1.6	0.3
108.47	0.87	1.8	150.15	16AU	-0.2	-3.5	-1.5	0.3
114.73	0.98	1.8	159.87	26AU	-1.1	-3.5	-1.5	0.2
120.90	1.09	1.7	169.65	5SE	-1.2	-3.5	-1.4	0.2
126.96	1.21	1.7	179.49	15SE	-0.8	-3.4	-1.4	0.1
132.91	1.33	1.7	189.38	25SE	-0.4	-3.4	-1.4	0.1
138.74	1.45	1.6	199.33	5OC	-0.2	-3.4	-1.3	-0.0
144.42	1.57	1.5	209.34	15OC	-0.2	-3.4	-1.3	-0.1
149.94	1.71	1.5	219.40	25OC	-0.0	-3.4	-1.3	-0.2
155.26	1.84	1.4	229.50	4NO	0.6	-3.4	-1.3	-0.2
160.36	1.99	1.3	239.64	14NO	3.3	-3.4	-1.3	-0.3
165.18	2.14	1.1	249.81	24NO	0.5	-3.4	-1.3	-0.2
169.66	2.31	1.0	260.00	4DE	-0.2	-3.4	-1.3	-0.1
173.73	2.49	0.8	270.20	14DE	-0.3	-3.4	-1.3	-0.1
177.30	2.68	0.6	280.40	24DE	-0.3	-3.4	-1.4	0.0
				1177				
180.23	2.88	0.4	290.58	3JA	-0.5	-3.4	-1.4	0.1
182.37	3.09	0.2	300.75	13JA	-0.8	-3.4	-1.5	0.1
183.56	3.30	-0.1	310.87	23JA	-1.2	-3.5	-1.5	0.2
183.60	3.50	-0.4	320.97	2FE	-1.1	-3.5	-1.6	0.3
182.37	3.64	-0.6	331.01	12FE	-0.4	-3.5	-1.6	0.3
179.91	3.70	-0.9	341.00	22FE	1.4	-3.4	-1.7	0.3
176.46	3.62	-1.1	350.93	4MR	3.2	-3.4	-1.8	0.3
172.61	3.40	-1.1	0.81	14MR	1.7	-3.4	-1.8	0.3
169.08	3.05	-1.0	10.63	24MR	0.9	-3.4	-1.9	0.3
166.46	2.62	-0.8	20.40	3AP	0.5	-3.3	-2.0	0.3
165.11	2.16	-0.6	30.11	13AP	0.1	-3.3	-2.0	0.3
165.05	1.72	-0.4	39.77	23AP	-0.6	-3.3	-2.1	0.3
166.20	1.32	-0.2	49.39	3MY	-1.5	-3.3	-2.1	0.2
168.38	0.96	0.0	58.98	13MY	-1.5	-3.3	-2.2	0.2
171.41	0.64	0.2	68.53	23MY	-0.6	-3.4	-2.2	0.2
175.15	0.35	0.4	78.08	2JN	0.2	-3.4	-2.2	0.2
179.47	0.10	0.5	87.61	12JN	0.8	-3.4	-2.2	0.3
184.26	-0.12	0.6	97.13	22JN	1.4	-3.4	-2.1	0.3
189.47	-0.32	0.7	106.67	2JL	2.4	-3.5	-2.1	0.3
195.02	-0.49	0.8	116.22	12JL	3.1	-3.5	-2.1	0.3
200.88	-0.65	0.9	125.80	22JL	1.3	-3.6	-2.0	0.3
207.01	-0.79	0.9	135.42	1AU	-0.3	-3.7	-1.9	0.3
213.37	-0.91	1.0	145.07	11AU	-1.2	-3.8	-1.9	0.3
219.95	-1.02	1.0	154.77	21AU	-1.4	-3.9	-1.8	0.3
226.73	-1.11	1.1	164.52	31AU	-0.8	-4.0	-1.8	0.3
233.69	-1.19	1.1	174.32	10SE	-0.4	-4.1	-1.7	0.2
240.80	-1.25	1.1	184.19	20SE	-0.1	-4.2	-1.6	0.2
248.06	-1.30	1.2	194.11	30SE	-0.0	-4.3	-1.6	0.1
255.45	-1.33	1.2	204.09	10OC	0.1	-4.3	-1.5	0.0
262.96	-1.35	1.2	214.12	20OC	0.9	-4.3	-1.5	-0.0
270.57	-1.35	1.2	224.20	30OC	3.2	-4.1	-1.5	-0.1
278.25	-1.34	1.3	234.32	9NO	0.3	-3.5	-1.4	-0.2
286.01	-1.31	1.3	244.47	19NO	-0.3	-3.1	-1.4	-0.3
293.81	-1.27	1.3	254.66	29NO	-0.4	-3.9	-1.4	-0.3
301.65	-1.22	1.3	264.85	9DE	-0.4	-4.3	-1.4	-0.2
309.50	-1.16	1.3	275.05	19DE	-0.6	-4.4	-1.4	-0.2
317.35	-1.09	1.4	285.24	29DE	-0.9	-4.3	-1.4	-0.1

♂ LONG	♂ LAT	♂ MAG	☉ LONG	16.00UT	☿	♀	♃	♄
				1178	MAGNITUDES			
325.19	-1.01	1.4	295.42	8JA	-1.0	-4.3	-1.4	-0.0
333.00	-0.92	1.4	305.57	18JA	-1.0	-4.2	-1.5	0.1
340.77	-0.83	1.4	315.68	28JA	-0.3	-4.0	-1.5	0.1
348.49	-0.73	1.5	325.75	7FE	1.7	-3.9	-1.5	0.2
356.15	-0.62	1.5	335.77	17FE	2.6	-3.8	-1.6	0.2
3.74	-0.52	1.5	345.74	27FE	1.2	-3.7	-1.6	0.3
11.25	-0.41	1.5	355.64	9MR	0.7	-3.7	-1.7	0.3
18.69	-0.30	1.6	5.49	19MR	0.4	-3.6	-1.7	0.3
26.04	-0.19	1.6	15.29	29MR	0.0	-3.5	-1.8	0.3
33.31	-0.08	1.6	25.03	8AP	-0.6	-3.5	-1.9	0.3
40.49	0.02	1.6	34.72	18AP	-1.5	-3.4	-2.0	0.3
47.59	0.13	1.6	44.36	28AP	-1.5	-3.4	-2.0	0.3
54.61	0.23	1.6	53.96	8MY	-0.6	-3.3	-2.1	0.3
61.55	0.33	1.7	63.53	18MY	0.3	-3.3	-2.2	0.2
68.41	0.42	1.7	73.09	28MY	1.0	-3.3	-2.2	0.2
75.20	0.52	1.8	82.62	7JN	1.8	-3.3	-2.3	0.2
81.92	0.60	1.8	92.15	17JN	3.1	-3.3	-2.3	0.2
88.57	0.69	1.9	101.68	27JN	2.5	-3.3	-2.3	0.3
95.15	0.77	1.9	111.22	7JL	1.0	-3.3	-2.3	0.3
101.68	0.85	1.9	120.78	17JL	-0.3	-3.4	-2.3	0.3
108.15	0.93	1.9	130.38	27JL	-1.3	-3.4	-2.3	0.3
114.57	1.01	2.0	140.00	6AU	-1.4	-3.4	-2.2	0.3
120.93	1.08	2.0	149.68	16AU	-0.7	-3.5	-2.2	0.3
127.24	1.15	2.0	159.40	26AU	-0.3	-3.5	-2.1	0.3
133.50	1.21	1.9	169.18	5SE	-0.0	-3.5	-2.1	0.3
139.70	1.28	1.9	179.01	15SE	0.1	-3.5	-2.0	0.3
145.85	1.34	1.9	188.90	25SE	0.3	-3.4	-1.9	0.2
151.93	1.41	1.9	198.84	5OC	1.2	-3.4	-1.9	0.2
157.96	1.47	1.8	208.85	15OC	3.0	-3.4	-1.8	0.1
163.90	1.53	1.8	218.90	25OC	0.2	-3.4	-1.7	0.0
169.77	1.59	1.7	229.00	4NO	-0.5	-3.3	-1.7	-0.0
175.53	1.64	1.6	239.14	14NO	-0.5	-3.3	-1.6	-0.1
181.19	1.70	1.5	249.31	24NO	-0.6	-3.3	-1.6	-0.2
186.70	1.75	1.4	259.50	4DE	-0.7	-3.3	-1.6	-0.2
192.06	1.81	1.3	269.70	14DE	-0.8	-3.4	-1.5	-0.3
197.21	1.86	1.2	279.90	24DE	-0.9	-3.4	-1.5	-0.2
				1179				
202.13	1.90	1.0	290.08	3JA	-0.8	-3.4	-1.5	-0.1
206.75	1.94	0.8	300.25	13JA	-0.2	-3.4	-1.5	-0.1
211.01	1.98	0.6	310.38	23JA	1.9	-3.5	-1.5	0.0
214.81	2.00	0.4	320.47	2FE	2.1	-3.5	-1.5	0.1
218.04	2.00	0.2	330.52	12FE	0.9	-3.5	-1.5	0.1
220.56	1.98	-0.1	340.52	22FE	0.5	-3.6	-1.6	0.2
222.21	1.92	-0.4	350.45	4MR	0.2	-3.7	-1.6	0.2
222.79	1.81	-0.7	0.34	14MR	-0.1	-3.7	-1.6	0.3
222.15	1.63	-1.0	10.16	24MR	-0.6	-3.8	-1.7	0.3
220.26	1.36	-1.3	19.93	3AP	-1.4	-3.9	-1.7	0.3
217.29	0.99	-1.6	29.64	13AP	-1.5	-4.0	-1.8	0.3
213.74	0.55	-1.7	39.31	23AP	-0.6	-4.1	-1.8	0.3
210.31	0.07	-1.5	48.93	3MY	0.5	-4.1	-1.9	0.3
207.67	-0.40	-1.4	58.52	13MY	1.4	-4.2	-2.0	0.3
206.27	-0.81	-1.2	68.08	23MY	2.5	-4.2	-2.0	0.3
206.24	-1.15	-1.0	77.62	2JN	3.4	-4.0	-2.1	0.3
207.49	-1.43	-0.8	87.15	12JN	2.0	-3.5	-2.2	0.3
209.89	-1.65	-0.6	96.68	22JN	0.8	-2.8	-2.3	0.2
213.22	-1.82	-0.4	106.21	2JL	-0.4	-3.6	-2.3	0.3
217.34	-1.94	-0.2	115.77	12JL	-1.3	-4.0	-2.4	0.3
222.10	-2.03	-0.1	125.34	22JL	-1.5	-4.2	-2.4	0.4
227.40	-2.09	0.0	134.95	1AU	-0.7	-4.2	-2.4	0.4
233.14	-2.13	0.2	144.60	11AU	-0.2	-4.2	-2.4	0.4
239.25	-2.14	0.3	154.30	21AU	0.1	-4.1	-2.4	0.4
245.68	-2.13	0.4	164.04	31AU	0.3	-4.0	-2.4	0.4
252.37	-2.09	0.5	173.85	10SE	0.6	-3.9	-2.3	0.4
259.29	-2.04	0.5	183.71	20SE	1.5	-3.8	-2.3	0.4
266.39	-1.97	0.6	193.62	30SE	2.7	-3.8	-2.2	0.3
273.65	-1.89	0.7	203.60	10OC	0.1	-3.7	-2.2	0.3
281.03	-1.79	0.8	213.63	20OC	-0.6	-3.6	-2.1	0.3
288.51	-1.68	0.8	223.70	30OC	-0.7	-3.6	-2.0	0.2
296.07	-1.55	0.9	233.83	9NO	-0.7	-3.5	-2.0	0.1
303.68	-1.42	1.0	243.98	19NO	-0.8	-3.5	-1.9	0.0
311.31	-1.29	1.1	254.16	29NO	-0.7	-3.4	-1.8	-0.0
318.96	-1.15	1.1	264.35	9DE	-0.7	-3.4	-1.8	-0.1
326.60	-1.01	1.2	274.55	19DE	-0.7	-3.4	-1.7	-0.2
334.22	-0.86	1.3	284.74	29DE	-0.1	-3.4	-1.7	-0.2

♂ LONG	♂ LAT	♂ MAG	⊙ LONG	16.00UT	☿	♀	♃	♄
				1180	MAGNITUDES			
341.80	-0.72	1.3	294.92	8JA	2.2	-3.4	-1.6	-0.1
349.34	-0.58	1.4	305.07	18JA	1.6	-3.3	-1.6	-0.1
356.81	-0.44	1.4	315.19	28JA	0.6	-3.3	-1.6	0.0
4.23	-0.31	1.5	325.26	7FE	0.3	-3.3	-1.6	0.1
11.58	-0.18	1.6	335.28	17FE	0.1	-3.3	-1.6	0.1
18.85	-0.06	1.6	345.25	27FE	-0.2	-3.4	-1.6	0.2
26.06	0.06	1.7	355.16	8MR	-0.7	-3.4	-1.6	0.2
33.18	0.17	1.7	5.01	18MR	-1.4	-3.4	-1.6	0.3
40.23	0.28	1.7	14.81	28MR	-1.5	-3.4	-1.6	0.3
47.21	0.38	1.8	24.56	7AP	-0.5	-3.5	-1.6	0.4
54.11	0.47	1.8	34.24	17AP	0.6	-3.5	-1.7	0.4
60.95	0.56	1.8	43.89	27AP	1.8	-3.5	-1.7	0.4
67.72	0.64	1.8	53.50	7MY	3.3	-3.4	-1.8	0.4
74.43	0.71	1.9	63.07	17MY	2.8	-3.4	-1.8	0.4
81.08	0.78	1.9	72.62	27MY	1.5	-3.4	-1.9	0.4
87.69	0.85	1.9	82.15	6JN	0.6	-3.3	-1.9	0.4
94.24	0.91	1.9	91.68	16JN	-0.4	-3.3	-2.0	0.4
100.75	0.96	1.9	101.21	26JN	-1.4	-3.3	-2.1	0.4
107.22	1.01	1.9	110.75	6JL	-1.5	-3.3	-2.2	0.4
113.66	1.05	1.9	120.31	16JL	-0.7	-3.3	-2.2	0.4
120.07	1.09	2.0	129.91	26JL	-0.1	-3.3	-2.3	0.4
126.45	1.13	2.0	139.54	5AU	0.2	-3.4	-2.3	0.5
132.81	1.16	2.0	149.20	15AU	0.5	-3.4	-2.4	0.5
139.16	1.19	2.0	158.93	25AU	0.8	-3.4	-2.4	0.5
145.48	1.22	2.0	168.70	4SE	1.9	-3.5	-2.5	0.5
151.79	1.24	2.0	178.53	14SE	2.4	-3.5	-2.5	0.5
158.09	1.25	2.0	188.42	24SE	-0.0	-3.6	-2.5	0.5
164.37	1.26	2.0	198.36	4OC	-0.8	-3.7	-2.4	0.5
170.63	1.27	2.0	208.36	14OC	-0.9	-3.7	-2.4	0.5
176.88	1.27	1.9	218.42	24OC	-0.8	-3.8	-2.3	0.4
183.11	1.27	1.9	228.51	3NO	-0.7	-3.9	-2.3	0.4
189.32	1.27	1.8	238.65	13NO	-0.6	-4.0	-2.2	0.3
195.51	1.25	1.8	248.82	23NO	-0.5	-4.1	-2.1	0.3
201.66	1.23	1.7	259.00	3DE	-0.5	-4.2	-2.1	0.2
207.77	1.20	1.6	269.20	13DE	0.1	-4.3	-2.0	0.1
213.83	1.16	1.5	279.40	23DE	2.5	-4.4	-1.9	0.0
				1181				
219.83	1.11	1.4	289.59	2JA	1.2	-4.3	-1.9	-0.0
225.76	1.05	1.3	299.75	12JA	0.3	-4.1	-1.8	-0.0
231.59	0.97	1.1	309.89	22JA	0.1	-3.6	-1.7	0.0
237.32	0.87	1.0	319.98	1FE	-0.0	-3.4	-1.7	0.1
242.91	0.76	0.8	330.03	11FE	-0.3	-3.9	-1.7	0.1
248.33	0.61	0.6	340.03	21FE	-0.7	-4.2	-1.6	0.2
253.53	0.43	0.4	349.97	3MR	-1.4	-4.3	-1.6	0.3
258.48	0.22	0.2	359.85	13MR	-1.4	-4.3	-1.6	0.3
263.09	-0.05	-0.0	9.68	23MR	-0.5	-4.2	-1.6	0.4
267.27	-0.37	-0.3	19.45	2AP	0.8	-4.1	-1.6	0.4
270.92	-0.75	-0.6	29.17	12AP	2.3	-4.0	-1.6	0.5
273.87	-1.22	-0.8	38.84	22AP	3.6	-3.8	-1.6	0.5
275.98	-1.77	-1.2	48.46	2MY	2.1	-3.8	-1.6	0.5
277.04	-2.41	-1.5	58.05	12MY	1.2	-3.7	-1.6	0.5
276.89	-3.12	-1.8	67.62	22MY	0.4	-3.6	-1.7	0.6
275.53	-3.86	-2.1	77.16	1JN	-0.4	-3.5	-1.7	0.6
273.15	-4.54	-2.3	86.69	11JN	-1.4	-3.5	-1.7	0.6
270.25	-5.07	-2.4	96.22	21JN	-1.5	-3.4	-1.8	0.6
267.60	-5.37	-2.3	105.75	1JL	-0.7	-3.4	-1.8	0.5
265.83	-5.44	-2.1	115.30	11JL	-0.0	-3.4	-1.9	0.5
265.34	-5.31	-1.9	124.88	21JL	0.3	-3.4	-2.0	0.5
266.23	-5.05	-1.6	134.49	31JL	0.7	-3.3	-2.0	0.6
268.38	-4.72	-1.4	144.14	10AU	1.1	-3.3	-2.1	0.6
271.63	-4.35	-1.1	153.83	20AU	2.4	-3.3	-2.2	0.7
275.76	-3.97	-0.9	163.57	30AU	2.0	-3.3	-2.2	0.7
280.59	-3.58	-0.7	173.37	9SE	-0.1	-3.4	-2.3	0.7
285.99	-3.21	-0.5	183.23	19SE	-0.9	-3.4	-2.3	0.7
291.81	-2.85	-0.3	193.14	29SE	-1.0	-3.4	-2.4	0.7
297.97	-2.50	-0.2	203.12	9OC	-0.9	-3.4	-2.4	0.7
304.40	-2.17	0.0	213.14	19OC	-0.6	-3.4	-2.4	0.7
311.02	-1.85	0.2	223.21	29OC	-0.4	-3.5	-2.4	0.7
317.78	-1.56	0.3	233.33	8NO	-0.4	-3.5	-2.4	0.6
324.65	-1.29	0.5	243.49	18NO	-0.3	-3.5	-2.3	0.6
331.59	-1.03	0.6	253.66	28NO	0.2	-3.5	-2.3	0.5
338.57	-0.79	0.7	263.86	8DE	2.8	-3.4	-2.2	0.4
345.57	-0.58	0.9	274.06	18DE	0.9	-3.4	-2.2	0.4
352.57	-0.38	1.0	284.25	28DE	0.1	-3.4	-2.1	0.3

♂ LONG	♂ LAT	♂ MAG	⊙ LONG	16.00UT	☿	♀	♃	♄
				1182	MAGNITUDES			
359.57	-0.19	1.1	294.43	7JA	-0.0	-3.4	-2.0	0.2
6.54	-0.03	1.2	304.58	17JA	-0.1	-3.3	-1.9	0.2
13.48	0.12	1.3	314.69	27JA	-0.3	-3.3	-1.9	0.2
20.39	0.26	1.4	324.77	6FE	-0.7	-3.3	-1.8	0.2
27.25	0.38	1.5	334.79	16FE	-1.4	-3.3	-1.8	0.3
34.06	0.49	1.6	344.76	26FE	-1.3	-3.4	-1.7	0.3
40.83	0.59	1.6	354.68	8MR	-0.5	-3.4	-1.6	0.4
47.55	0.68	1.7	4.54	18MR	1.0	-3.4	-1.6	0.4
54.23	0.76	1.8	14.33	28MR	2.9	-3.4	-1.6	0.5
60.85	0.83	1.8	24.08	7AP	2.7	-3.4	-1.5	0.5
67.44	0.89	1.9	33.77	17AP	1.5	-3.5	-1.5	0.6
73.98	0.95	1.9	43.42	27AP	0.9	-3.5	-1.5	0.6
80.49	0.99	1.9	53.03	7MY	0.3	-3.6	-1.5	0.7
86.96	1.03	2.0	62.61	17MY	-0.5	-3.6	-1.5	0.7
93.40	1.07	2.0	72.16	27MY	-1.4	-3.7	-1.5	0.7
99.82	1.09	2.0	81.70	6JN	-1.6	-3.8	-1.5	0.7
106.22	1.12	2.0	91.22	16JN	-0.6	-3.9	-1.5	0.8
112.60	1.13	2.0	100.75	26JN	0.0	-4.0	-1.6	0.8
118.97	1.14	2.0	110.30	6JL	0.5	-4.1	-1.6	0.8
125.33	1.15	2.0	119.86	16JL	0.9	-4.2	-1.7	0.8
131.69	1.15	2.0	129.45	26JL	1.5	-4.2	-1.7	0.7
138.06	1.15	2.0	139.08	5AU	2.9	-4.2	-1.8	0.8
144.43	1.14	2.0	148.74	15AU	1.8	-4.0	-1.8	0.8
150.80	1.13	2.0	158.46	25AU	-0.2	-3.5	-1.9	0.9
157.20	1.12	2.0	168.23	4SE	-1.0	-3.3	-1.9	0.9
163.61	1.10	2.0	178.06	14SE	-1.2	-3.9	-2.0	0.9
170.03	1.07	2.0	187.94	24SE	-0.9	-4.2	-2.1	0.9
176.48	1.04	2.0	197.88	4OC	-0.5	-4.3	-2.1	1.0
182.96	1.00	2.0	207.88	14OC	-0.3	-4.3	-2.2	1.0
189.45	0.96	2.0	217.93	24OC	-0.2	-4.2	-2.2	0.9
195.98	0.91	1.9	228.03	3NO	-0.2	-4.1	-2.3	0.9
202.52	0.86	1.9	238.16	13NO	0.4	-4.1	-2.3	0.9
209.10	0.80	1.8	248.32	23NO	3.1	-4.0	-2.3	0.9
215.70	0.73	1.8	258.51	3DE	0.7	-3.9	-2.3	0.8
222.32	0.65	1.7	268.71	13DE	-0.1	-3.8	-2.3	0.7
228.97	0.56	1.7	278.91	23DE	-0.2	-3.7	-2.2	0.7
				1183				
235.64	0.46	1.6	289.10	2JA	-0.2	-3.6	-2.2	0.6
242.33	0.35	1.5	299.26	12JA	-0.4	-3.6	-2.1	0.5
249.04	0.22	1.4	309.40	22JA	-0.8	-3.5	-2.1	0.5
255.76	0.08	1.3	319.50	1FE	-1.2	-3.5	-2.0	0.4
262.49	-0.07	1.2	329.55	11FE	-1.2	-3.4	-1.9	0.4
269.22	-0.25	1.1	339.55	21FE	-0.4	-3.4	-1.8	0.4
275.95	-0.44	1.0	349.49	3MR	1.2	-3.4	-1.8	0.5
282.67	-0.65	0.8	359.38	13MR	3.4	-3.3	-1.7	0.5
289.37	-0.88	0.7	9.21	23MR	2.0	-3.3	-1.7	0.6
296.03	-1.13	0.5	18.98	2AP	1.1	-3.3	-1.6	0.7
302.63	-1.40	0.4	28.70	12AP	0.7	-3.3	-1.5	0.7
309.15	-1.70	0.2	38.37	22AP	0.2	-3.3	-1.5	0.8
315.55	-2.02	0.1	48.00	2MY	-0.5	-3.3	-1.5	0.8
321.79	-2.36	-0.1	57.59	12MY	-1.4	-3.4	-1.4	0.9
327.82	-2.72	-0.3	67.15	22MY	-1.6	-3.4	-1.4	0.9
333.56	-3.11	-0.5	76.70	1JN	-0.6	-3.4	-1.4	0.9
338.93	-3.51	-0.7	86.23	11JN	0.1	-3.5	-1.4	1.0
343.81	-3.93	-0.9	95.76	21JN	0.7	-3.5	-1.4	1.0
348.05	-4.37	-1.1	105.29	1JL	1.2	-3.5	-1.4	1.0
351.48	-4.80	-1.4	114.84	11JL	2.0	-3.4	-1.4	1.0
353.88	-5.20	-1.6	124.41	21JL	3.2	-3.4	-1.4	1.0
355.05	-5.55	-1.9	134.02	31JL	1.5	-3.4	-1.5	1.0
354.85	-5.78	-2.1	143.66	10AU	-0.2	-3.4	-1.5	1.0
353.29	-5.82	-2.4	153.36	20AU	-1.1	-3.3	-1.5	1.0
350.72	-5.60	-2.5	163.10	30AU	-1.3	-3.3	-1.6	1.1
347.82	-5.11	-2.5	172.90	9SE	-0.8	-3.3	-1.6	1.1
345.33	-4.43	-2.2	182.75	19SE	-0.4	-3.3	-1.7	1.2
343.90	-3.65	-1.9	192.66	29SE	-0.2	-3.3	-1.8	1.2
343.76	-2.88	-1.6	202.63	9OC	-0.1	-3.4	-1.8	1.2
344.90	-2.18	-1.3	212.65	19OC	0.0	-3.4	-1.9	1.2
347.16	-1.57	-1.0	222.73	29OC	0.7	-3.4	-2.0	1.2
350.34	-1.05	-0.7	232.84	8NO	3.4	-3.5	-2.0	1.2
354.24	-0.62	-0.4	242.99	18NO	0.5	-3.5	-2.1	1.2
358.71	-0.26	-0.1	253.17	28NO	-0.2	-3.5	-2.1	1.2
3.60	0.04	0.1	263.36	8DE	-0.3	-3.6	-2.2	1.1
8.82	0.28	0.3	273.56	18DE	-0.4	-3.7	-2.2	1.1
14.30	0.48	0.5	283.76	28DE	-0.5	-3.7	-2.2	1.0

♂			☉	16.00UT	☿	♀	♃	♄
LONG	LAT	MAG	LONG	1184	MAGNITUDES			
19.97	0.65	0.7	293.93	7JA	-0.8	-3.8	-2.2	1.0
25.79	0.79	0.9	304.09	17JA	-1.1	-3.9	-2.1	0.9
31.72	0.90	1.0	314.21	27JA	-1.1	-4.0	-2.1	0.8
37.73	1.00	1.2	324.28	6FE	-0.4	-4.1	-2.1	0.7
43.79	1.07	1.3	334.31	16FE	1.5	-4.2	-2.0	0.7
49.91	1.14	1.4	344.28	26FE	3.0	-4.3	-1.9	0.7
56.05	1.18	1.5	354.20	7MR	1.5	-4.3	-1.9	0.7
62.22	1.22	1.6	4.06	17MR	0.8	-4.2	-1.8	0.8
68.40	1.25	1.7	13.86	27MR	0.5	-3.9	-1.7	0.8
74.59	1.27	1.8	23.61	6AP	0.1	-3.2	-1.7	0.9
80.78	1.28	1.8	33.31	16AP	-0.5	-3.3	-1.6	0.9
86.99	1.29	1.9	42.96	26AP	-1.4	-3.9	-1.5	1.0
93.19	1.29	1.9	52.56	6MY	-1.6	-4.1	-1.5	1.0
99.41	1.28	2.0	62.14	16MY	-0.6	-4.2	-1.4	1.1
105.63	1.27	2.0	71.69	26MY	0.2	-4.2	-1.4	1.1
111.86	1.25	2.0	81.23	5JN	0.9	-4.1	-1.4	1.2
118.10	1.23	2.0	90.76	15JN	1.5	-4.0	-1.3	1.2
124.36	1.21	2.0	100.29	25JN	2.6	-3.9	-1.3	1.2
130.64	1.18	2.0	109.83	5JL	2.9	-3.8	-1.3	1.3
136.94	1.15	2.0	119.39	15JL	1.2	-3.7	-1.3	1.3
143.27	1.11	2.0	128.98	25JL	-0.3	-3.6	-1.3	1.3
149.63	1.07	2.0	138.60	4AU	-1.2	-3.6	-1.3	1.3
156.02	1.03	2.0	148.28	14AU	-1.4	-3.5	-1.3	1.3
162.45	0.98	1.9	157.99	24AU	-0.8	-3.5	-1.3	1.3
168.92	0.93	1.9	167.76	3SE	-0.3	-3.5	-1.4	1.3
175.44	0.87	1.9	177.58	13SE	-0.1	-3.4	-1.4	1.3
182.00	0.81	1.9	187.46	23SE	0.1	-3.4	-1.5	1.3
188.62	0.75	1.9	197.40	3OC	0.2	-3.4	-1.5	1.3
195.29	0.67	1.9	207.40	13OC	0.9	-3.4	-1.6	1.3
202.01	0.60	1.9	217.44	23OC	3.3	-3.4	-1.6	1.3
208.79	0.52	1.9	227.54	2NO	0.3	-3.4	-1.7	1.3
215.63	0.43	1.8	237.67	12NO	-0.4	-3.4	-1.7	1.2
222.52	0.34	1.8	247.83	22NO	-0.5	-3.4	-1.8	1.2
229.48	0.24	1.8	258.02	2DE	-0.5	-3.4	-1.9	1.2
236.49	0.14	1.7	268.22	12DE	-0.6	-3.4	-1.9	1.1
243.57	0.03	1.7	278.41	22DE	-0.9	-3.4	-2.0	1.1
				1185				
250.70	-0.09	1.6	288.60	1JA	-0.9	-3.4	-2.0	1.0
257.88	-0.21	1.6	298.77	11JA	-0.9	-3.4	-2.1	1.0
265.12	-0.34	1.5	308.91	21JA	-0.3	-3.5	-2.1	0.9
272.41	-0.48	1.5	319.00	31JA	1.7	-3.5	-2.1	0.9
279.75	-0.62	1.4	329.06	10FE	2.4	-3.5	-2.1	0.8
287.13	-0.76	1.3	339.06	20FE	1.1	-3.4	-2.0	0.8
294.54	-0.91	1.3	349.01	2MR	0.6	-3.4	-2.0	0.8
301.98	-1.06	1.2	358.90	12MR	0.3	-3.4	-1.9	0.8
309.44	-1.21	1.1	8.73	22MR	-0.0	-3.4	-1.9	0.9
316.90	-1.36	1.0	18.51	1AP	-0.6	-3.3	-1.8	1.0
324.36	-1.50	0.9	28.23	11AP	-1.4	-3.3	-1.7	1.0
331.79	-1.64	0.9	37.90	21AP	-1.5	-3.3	-1.7	1.1
339.19	-1.77	0.8	47.53	1MY	-0.6	-3.3	-1.6	1.2
346.52	-1.89	0.7	57.13	11MY	0.3	-3.3	-1.6	1.2
353.77	-1.99	0.6	66.69	21MY	1.1	-3.4	-1.5	1.3
0.92	-2.08	0.5	76.23	31MY	2.1	-3.4	-1.4	1.3
7.92	-2.16	0.4	85.76	10JN	3.3	-3.4	-1.4	1.3
14.76	-2.21	0.3	95.29	20JN	2.4	-3.4	-1.3	1.3
21.38	-2.25	0.2	104.83	30JN	1.0	-3.5	-1.3	1.3
27.74	-2.26	0.1	114.38	10JL	-0.3	-3.5	-1.3	1.3
33.79	-2.25	0.0	123.95	20JL	-1.3	-3.6	-1.3	1.3
39.46	-2.21	-0.1	133.56	30JL	-1.5	-3.7	-1.2	1.3
44.65	-2.15	-0.3	143.20	9AU	-0.8	-3.8	-1.2	1.3
49.26	-2.04	-0.4	152.89	19AU	-0.2	-3.9	-1.2	1.2
53.15	-1.90	-0.6	162.63	29AU	0.0	-4.0	-1.2	1.2
56.14	-1.71	-0.8	172.43	8SE	0.2	-4.1	-1.2	1.1
58.05	-1.46	-1.0	182.27	18SE	0.4	-4.2	-1.3	1.1
58.64	-1.15	-1.2	192.19	28SE	1.3	-4.3	-1.3	1.1
57.78	-0.75	-1.4	202.15	8OC	3.0	-4.3	-1.3	1.1
55.51	-0.28	-1.6	212.17	18OC	0.2	-4.3	-1.4	1.1
52.15	0.22	-1.8	222.24	28OC	-0.6	-4.1	-1.4	1.1
48.41	0.72	-1.8	232.35	7NO	-0.6	-3.4	-1.5	1.1
45.10	1.14	-1.5	242.50	17NO	-0.6	-3.2	-1.5	1.1
42.84	1.47	-1.2	252.68	27NO	-0.7	-4.0	-1.6	1.1
41.93	1.70	-0.9	262.87	7DE	-0.8	-4.3	-1.6	1.0
42.34	1.85	-0.6	273.07	17DE	-0.8	-4.4	-1.7	1.0
43.90	1.94	-0.3	283.26	27DE	-0.8	-4.3	-1.8	0.9

♂			☉	16.00UT	☿	♀	♃	♄
LONG	LAT	MAG	LONG	1186	MAGNITUDES			
46.41	1.98	0.0	293.44	6JA	-0.2	-4.3	-1.8	0.9
49.66	1.99	0.3	303.59	16JA	2.0	-4.2	-1.9	0.9
53.50	1.98	0.5	313.72	26JA	1.9	-4.0	-1.9	0.8
57.80	1.96	0.7	323.80	5FE	0.8	-3.9	-2.0	0.8
62.45	1.93	0.9	333.82	15FE	0.4	-3.8	-2.0	0.7
67.38	1.89	1.1	343.80	25FE	0.2	-3.7	-2.0	0.7
72.53	1.84	1.2	353.72	7MR	-0.1	-3.7	-2.0	0.6
77.85	1.79	1.3	3.59	17MR	-0.6	-3.6	-2.0	0.6
83.32	1.74	1.5	13.39	27MR	-1.4	-3.5	-2.0	0.7
88.91	1.69	1.6	23.14	6AP	-1.5	-3.5	-1.9	0.7
94.59	1.63	1.7	32.84	16AP	-0.6	-3.4	-1.9	0.8
100.36	1.57	1.7	42.50	26AP	0.5	-3.4	-1.8	0.9
106.20	1.51	1.8	52.11	6MY	1.5	-3.3	-1.7	0.9
112.10	1.45	1.9	61.68	16MY	2.7	-3.3	-1.7	1.0
118.07	1.39	1.9	71.24	26MY	3.2	-3.3	-1.6	1.1
124.09	1.32	1.9	80.77	5JN	1.8	-3.3	-1.6	1.1
130.16	1.25	2.0	90.30	15JN	0.7	-3.3	-1.5	1.1
136.30	1.19	2.0	99.83	25JN	-0.3	-3.3	-1.4	1.2
142.48	1.12	2.0	109.37	5JL	-1.3	-3.3	-1.4	1.2
148.73	1.04	2.0	118.93	15JL	-1.5	-3.4	-1.3	1.2
155.03	0.97	2.0	128.52	25JL	-0.7	-3.4	-1.3	1.2
161.40	0.89	2.0	138.14	4AU	-0.2	-3.4	-1.3	1.2
167.83	0.81	2.0	147.80	14AU	0.1	-3.5	-1.2	1.1
174.33	0.73	1.9	157.52	24AU	0.4	-3.5	-1.2	1.1
180.89	0.65	1.9	167.28	3SE	0.7	-3.5	-1.2	1.1
187.53	0.56	1.9	177.10	13SE	1.6	-3.5	-1.2	1.0
194.24	0.48	1.8	186.98	23SE	2.6	-3.4	-1.2	1.0
201.02	0.38	1.8	196.92	3OC	0.1	-3.4	-1.2	1.0
207.89	0.29	1.7	206.91	13OC	-0.7	-3.4	-1.2	1.0
214.83	0.19	1.7	216.95	23OC	-0.8	-3.4	-1.3	1.0
221.85	0.09	1.7	227.04	2NO	-0.7	-3.4	-1.3	1.0
228.96	-0.01	1.7	237.17	12NO	-0.8	-3.3	-1.3	1.0
236.14	-0.11	1.7	247.34	22NO	-0.6	-3.3	-1.4	1.0
243.39	-0.22	1.7	257.52	2DE	-0.6	-3.3	-1.4	1.0
250.73	-0.32	1.6	267.72	12DE	-0.6	-3.4	-1.5	0.9
258.13	-0.43	1.6	277.92	22DE	-0.1	-3.4	-1.5	0.9
				1187				
265.60	-0.54	1.6	288.11	1JA	2.2	-3.4	-1.6	0.9
273.13	-0.65	1.6	298.28	11JA	1.5	-3.4	-1.7	0.8
280.72	-0.75	1.5	308.42	21JA	0.5	-3.5	-1.7	0.8
288.36	-0.85	1.5	318.52	31JA	0.2	-3.5	-1.8	0.7
296.04	-0.95	1.5	328.57	10FE	0.1	-3.5	-1.9	0.7
303.75	-1.04	1.4	338.58	20FE	-0.2	-3.6	-1.9	0.6
311.48	-1.12	1.4	348.53	2MR	-0.7	-3.7	-2.0	0.6
319.23	-1.20	1.4	358.42	12MR	-1.4	-3.7	-2.0	0.5
326.97	-1.26	1.3	8.26	22MR	-1.5	-3.8	-2.0	0.5
334.69	-1.32	1.3	18.04	1AP	-0.6	-3.9	-2.0	0.5
342.39	-1.36	1.3	27.76	11AP	0.6	-4.0	-2.0	0.6
350.05	-1.38	1.2	37.44	21AP	1.9	-4.1	-2.0	0.6
357.65	-1.40	1.2	47.07	1MY	3.6	-4.1	-2.0	0.7
5.18	-1.39	1.2	56.67	11MY	2.5	-4.2	-1.9	0.8
12.63	-1.37	1.1	66.23	21MY	1.4	-4.2	-1.9	0.8
19.98	-1.34	1.1	75.77	31MY	0.6	-4.0	-1.8	0.9
27.22	-1.29	1.1	85.31	10JN	-0.4	-3.5	-1.7	0.9
34.33	-1.22	1.0	94.83	20JN	-1.4	-2.8	-1.7	1.0
41.30	-1.14	1.0	104.36	30JN	-1.6	-3.6	-1.6	1.0
48.11	-1.04	0.9	113.91	10JL	-0.7	-4.0	-1.5	1.0
54.73	-0.92	0.9	123.49	20JL	-0.1	-4.2	-1.5	1.0
61.16	-0.79	0.8	133.09	30JL	0.3	-4.2	-1.4	1.0
67.35	-0.64	0.8	142.73	9AU	0.5	-4.2	-1.4	1.0
73.27	-0.47	0.7	152.42	19AU	0.9	-4.1	-1.3	1.0
78.88	-0.28	0.6	162.15	29AU	2.1	-4.0	-1.3	1.0
84.12	-0.06	0.5	171.95	8SE	2.3	-3.9	-1.3	1.0
88.91	0.18	0.4	181.79	18SE	-0.0	-3.8	-1.3	0.9
93.18	0.45	0.2	191.70	28SE	-0.8	-3.8	-1.2	0.9
96.78	0.76	0.0	201.66	8OC	-0.9	-3.7	-1.2	0.8
99.57	1.12	-0.1	211.68	18OC	-0.9	-3.6	-1.2	0.9
101.36	1.51	-0.3	221.75	28OC	-0.7	-3.6	-1.2	0.9
101.96	1.95	-0.6	231.86	7NO	-0.5	-3.5	-1.3	0.9
101.20	2.43	-0.8	242.01	17NO	-0.5	-3.5	-1.3	0.9
99.04	2.90	-1.0	252.18	27NO	-0.4	-3.4	-1.3	0.9
95.71	3.31	-1.2	262.38	7DE	0.1	-3.4	-1.3	0.9
91.78	3.60	-1.2	272.58	17DE	2.5	-3.4	-1.4	0.9
88.01	3.73	-1.1	282.77	27DE	1.2	-3.4	-1.4	0.8

♂			⊙	16.00UT	☿	♀	♃	♄
LONG	LAT	MAG	LONG	1188	MAGNITUDES			
85.09	3.72	-0.8	292.95	6JA	0.3	-3.4	-1.5	0.8
83.42	3.60	-0.5	303.11	16JA	0.1	-3.3	-1.5	0.8
83.07	3.41	-0.2	313.23	26JA	-0.1	-3.3	-1.6	0.7
83.93	3.20	0.0	323.31	5FE	-0.3	-3.3	-1.7	0.7
85.81	2.99	0.3	333.34	15FE	-0.7	-3.3	-1.7	0.6
88.52	2.78	0.5	343.32	25FE	-1.4	-3.4	-1.8	0.6
91.88	2.59	0.7	353.25	6MR	-1.4	-3.4	-1.9	0.5
95.76	2.41	0.9	3.11	16MR	-0.5	-3.4	-1.9	0.5
100.06	2.24	1.1	12.92	26MR	0.8	-3.4	-2.0	0.4
104.69	2.08	1.2	22.67	5AP	2.5	-3.5	-2.0	0.4
109.60	1.93	1.3	32.37	15AP	3.3	-3.5	-2.1	0.4
114.72	1.78	1.4	42.02	25AP	1.9	-3.5	-2.1	0.5
120.04	1.65	1.5	51.64	5MY	1.1	-3.4	-2.1	0.5
125.52	1.52	1.6	61.22	15MY	0.4	-3.4	-2.1	0.6
131.14	1.40	1.7	70.77	25MY	-0.4	-3.4	-2.0	0.7
136.89	1.28	1.7	80.31	4JN	-1.4	-3.3	-2.0	0.7
142.76	1.16	1.8	89.83	14JN	-1.6	-3.3	-2.0	0.8
148.73	1.04	1.8	99.36	24JN	-0.7	-3.3	-1.9	0.8
154.81	0.93	1.8	108.91	4JL	-0.0	-3.3	-1.8	0.9
160.99	0.82	1.8	118.46	14JL	0.4	-3.3	-1.8	0.9
167.27	0.71	1.8	128.05	24JL	0.7	-3.3	-1.7	0.9
173.65	0.60	1.8	137.67	3AU	1.3	-3.4	-1.7	0.9
180.12	0.49	1.8	147.33	13AU	2.6	-3.4	-1.6	0.9
186.69	0.39	1.8	157.04	23AU	2.0	-3.4	-1.5	0.9
193.36	0.28	1.8	166.80	2SE	-0.1	-3.5	-1.5	0.9
200.12	0.17	1.8	176.62	12SE	-1.0	-3.5	-1.4	0.9
206.99	0.07	1.8	186.50	22SE	-1.1	-3.6	-1.4	0.9
213.95	-0.04	1.7	196.43	2OC	-0.9	-3.7	-1.4	0.8
221.01	-0.14	1.7	206.42	12OC	-0.6	-3.7	-1.3	0.8
228.17	-0.25	1.6	216.46	22OC	-0.4	-3.8	-1.3	0.8
235.42	-0.35	1.6	226.55	1NO	-0.3	-3.9	-1.3	0.8
242.76	-0.45	1.5	236.68	11NO	-0.3	-4.0	-1.3	0.8
250.20	-0.54	1.5	246.84	21NO	0.3	-4.1	-1.3	0.8
257.71	-0.63	1.4	257.03	1DE	2.8	-4.2	-1.3	0.8
265.30	-0.72	1.4	267.22	11DE	0.9	-4.3	-1.3	0.8
272.96	-0.80	1.4	277.43	21DE	0.1	-4.4	-1.4	0.8
280.68	-0.87	1.4	287.62	31DE	-0.1	-4.3	-1.4	0.8
				1189				
288.44	-0.94	1.4	297.78	10JA	-0.2	-4.1	-1.4	0.8
296.25	-1.00	1.4	307.93	20JA	-0.4	-3.5	-1.5	0.7
304.09	-1.05	1.4	318.03	30JA	-0.8	-3.4	-1.5	0.7
311.94	-1.09	1.4	328.09	9FE	-1.3	-3.9	-1.6	0.7
319.81	-1.11	1.4	338.10	19FE	-1.3	-4.2	-1.6	0.6
327.66	-1.13	1.4	348.05	1MR	-0.5	-4.3	-1.7	0.6
335.49	-1.13	1.4	357.94	11MR	1.0	-4.3	-1.8	0.5
343.29	-1.12	1.4	7.78	21MR	3.1	-4.2	-1.9	0.4
351.04	-1.10	1.4	17.57	31MR	2.5	-4.1	-1.9	0.4
358.74	-1.07	1.4	27.29	10AP	1.4	-4.0	-2.0	0.3
6.38	-1.02	1.4	36.97	20AP	0.8	-3.8	-2.1	0.3
13.94	-0.96	1.4	46.61	30AP	0.3	-3.8	-2.1	0.4
21.41	-0.89	1.4	56.20	10MY	-0.5	-3.7	-2.1	0.4
28.80	-0.81	1.4	65.77	20MY	-1.4	-3.6	-2.2	0.5
36.09	-0.72	1.4	75.31	30MY	-1.6	-3.5	-2.2	0.5
43.27	-0.62	1.4	84.84	9JN	-0.7	-3.5	-2.2	0.6
50.34	-0.51	1.4	94.38	19JN	0.1	-3.4	-2.2	0.7
57.29	-0.39	1.4	103.91	29JN	0.5	-3.4	-2.2	0.7
64.12	-0.27	1.4	113.45	9JL	1.0	-3.4	-2.1	0.8
70.82	-0.14	1.4	123.03	19JL	1.7	-3.4	-2.1	0.8
77.38	0.01	1.4	132.63	29JL	3.1	-3.3	-2.0	0.8
83.80	0.15	1.3	142.26	8AU	1.7	-3.3	-1.9	0.9
90.05	0.31	1.3	151.95	18AU	-0.2	-3.3	-1.9	0.9
96.11	0.48	1.2	161.68	28AU	-1.1	-3.3	-1.8	0.9
101.98	0.66	1.2	171.47	7SE	-1.2	-3.4	-1.8	0.9
107.62	0.85	1.1	181.32	17SE	-0.9	-3.4	-1.7	0.8
113.00	1.05	1.0	191.22	27SE	-0.5	-3.4	-1.6	0.8
118.07	1.28	0.9	201.18	7OC	-0.3	-3.4	-1.6	0.8
122.76	1.52	0.8	211.19	17OC	-0.2	-3.4	-1.5	0.7
127.00	1.79	0.7	221.26	27OC	-0.1	-3.5	-1.5	0.7
130.69	2.09	0.5	231.36	6NO	0.5	-3.5	-1.5	0.7
133.70	2.41	0.3	241.51	16NO	3.1	-3.5	-1.4	0.7
135.87	2.78	0.1	251.68	26NO	0.6	-3.5	-1.4	0.8
137.02	3.17	-0.1	261.87	6DE	-0.1	-3.4	-1.4	0.8
136.95	3.58	-0.4	272.08	16DE	-0.2	-3.4	-1.4	0.8
135.56	3.97	-0.6	282.27	26DE	-0.3	-3.4	-1.4	0.8

♂			⊙	16.00UT	☿	♀	♃	♄
LONG	LAT	MAG	LONG	1190	MAGNITUDES			
132.88	4.30	-0.8	292.45	5JA	-0.5	-3.4	-1.4	0.8
129.25	4.51	-1.0	302.61	15JA	-0.8	-3.3	-1.4	0.8
125.28	4.54	-1.0	312.74	25JA	-1.2	-3.3	-1.5	0.7
121.71	4.40	-0.8	322.82	4FE	-1.1	-3.3	-1.5	0.7
119.13	4.12	-0.5	332.86	14FE	-0.5	-3.3	-1.5	0.7
117.82	3.78	-0.3	342.84	24FE	1.2	-3.4	-1.6	0.6
117.79	3.42	-0.0	352.76	6MR	3.3	-3.4	-1.6	0.6
118.90	3.06	0.2	2.64	16MR	1.8	-3.4	-1.7	0.5
120.98	2.73	0.4	12.44	26MR	1.0	-3.4	-1.8	0.5
123.84	2.43	0.6	22.20	5AP	0.6	-3.4	-1.8	0.4
127.34	2.15	0.8	31.91	15AP	0.1	-3.5	-1.9	0.3
131.35	1.90	0.9	41.56	25AP	-0.5	-3.5	-2.0	0.3
135.78	1.67	1.0	51.18	5MY	-1.4	-3.6	-2.0	0.3
140.56	1.46	1.2	60.76	15MY	-1.6	-3.6	-2.1	0.3
145.62	1.26	1.3	70.31	25MY	-0.6	-3.7	-2.2	0.4
150.94	1.08	1.3	79.85	4JN	0.2	-3.8	-2.2	0.4
156.47	0.91	1.4	89.38	14JN	0.7	-3.9	-2.3	0.5
162.20	0.74	1.5	98.91	24JN	1.3	-4.0	-2.3	0.6
168.10	0.59	1.5	108.45	4JL	2.2	-4.1	-2.3	0.6
174.17	0.44	1.5	118.01	14JL	3.2	-4.2	-2.3	0.7
180.39	0.30	1.6	127.59	24JL	1.4	-4.2	-2.3	0.7
186.75	0.17	1.6	137.21	3AU	-0.2	-4.2	-2.3	0.8
193.25	0.04	1.6	146.87	13AU	-1.2	-3.9	-2.2	0.8
199.88	-0.08	1.6	156.57	23AU	-1.4	-3.5	-2.2	0.8
206.65	-0.20	1.6	166.33	2SE	-0.8	-3.3	-2.1	0.8
213.54	-0.32	1.6	176.15	12SE	-0.4	-3.9	-2.1	0.8
220.55	-0.42	1.6	186.02	22SE	-0.1	-4.2	-2.0	0.8
227.68	-0.52	1.6	195.95	2OC	-0.0	-4.3	-1.9	0.8
234.93	-0.62	1.6	205.94	12OC	0.1	-4.3	-1.9	0.8
242.28	-0.71	1.5	215.97	22OC	0.7	-4.2	-1.8	0.7
249.73	-0.79	1.5	226.06	1NO	3.5	-4.1	-1.7	0.7
257.28	-0.86	1.5	236.19	11NO	0.5	-4.0	-1.7	0.7
264.91	-0.92	1.5	246.35	21NO	-0.3	-4.0	-1.6	0.7
272.62	-0.98	1.4	256.53	1DE	-0.4	-3.9	-1.6	0.7
280.39	-1.02	1.4	266.73	11DE	-0.4	-3.8	-1.6	0.8
288.22	-1.05	1.4	276.92	21DE	-0.5	-3.7	-1.5	0.8
296.09	-1.08	1.4	287.12	31DE	-0.9	-3.6	-1.5	0.8
				1191				
303.98	-1.09	1.3	297.29	10JA	-1.0	-3.6	-1.5	0.8
311.88	-1.08	1.3	307.43	20JA	-1.0	-3.5	-1.5	0.8
319.79	-1.07	1.3	317.54	30JA	-0.4	-3.5	-1.5	0.7
327.68	-1.05	1.3	327.60	9FE	1.5	-3.4	-1.5	0.7
335.55	-1.01	1.3	337.61	19FE	2.8	-3.4	-1.5	0.7
343.37	-0.96	1.3	347.57	1MR	1.3	-3.4	-1.6	0.6
351.15	-0.91	1.4	357.47	11MR	0.7	-3.3	-1.6	0.6
358.87	-0.84	1.4	7.31	21MR	0.4	-3.3	-1.6	0.5
6.52	-0.77	1.4	17.10	31MR	0.0	-3.3	-1.7	0.5
14.09	-0.68	1.5	26.83	10AP	-0.5	-3.3	-1.7	0.4
21.58	-0.59	1.5	36.51	20AP	-1.4	-3.3	-1.8	0.4
28.98	-0.50	1.5	46.15	30AP	-1.6	-3.3	-1.9	0.3
36.30	-0.40	1.6	55.74	10MY	-0.6	-3.4	-1.9	0.2
43.52	-0.29	1.6	65.31	20MY	0.3	-3.4	-2.0	0.3
50.65	-0.19	1.6	74.86	30MY	1.0	-3.4	-2.1	0.3
57.68	-0.07	1.7	84.39	9JN	1.7	-3.5	-2.1	0.4
64.62	0.04	1.7	93.91	19JN	2.9	-3.5	-2.2	0.4
71.46	0.16	1.7	103.45	29JN	2.7	-3.5	-2.3	0.5
78.20	0.27	1.7	112.99	9JL	1.1	-3.4	-2.3	0.6
84.85	0.39	1.7	122.56	19JL	-0.2	-3.4	-2.4	0.6
91.39	0.51	1.7	132.16	29JL	-1.2	-3.4	-2.4	0.7
97.84	0.64	1.7	141.79	8AU	-1.5	-3.4	-2.4	0.7
104.17	0.76	1.7	151.48	18AU	-0.8	-3.3	-2.4	0.7
110.40	0.89	1.7	161.21	28AU	-0.3	-3.3	-2.4	0.8
116.50	1.02	1.6	170.99	7SE	-0.0	-3.3	-2.4	0.8
122.48	1.16	1.6	180.83	17SE	0.1	-3.3	-2.3	0.8
128.31	1.30	1.5	190.74	27SE	0.3	-3.3	-2.3	0.8
133.98	1.45	1.5	200.69	7OC	1.0	-3.4	-2.2	0.8
139.45	1.60	1.4	210.70	17OC	3.3	-3.4	-2.2	0.8
144.71	1.77	1.3	220.77	27OC	0.3	-3.4	-2.1	0.7
149.71	1.94	1.2	230.87	6NO	-0.5	-3.5	-2.0	0.7
154.39	2.13	1.0	241.01	16NO	-0.5	-3.5	-1.9	0.7
158.69	2.34	0.9	251.19	26NO	-0.5	-3.6	-1.9	0.7
162.51	2.56	0.7	261.38	6DE	-0.7	-3.6	-1.8	0.7
165.76	2.80	0.5	271.58	16DE	-0.8	-3.7	-1.8	0.7
168.28	3.07	0.3	281.78	26DE	-0.9	-3.8	-1.7	0.8

♂ LONG	LAT	MAG	☉ LONG	16.00UT	☿	♀	♃	♄
				1192		MAGNITUDES		
169.91	3.34	0.1	291.96	5JA	-0.9	-3.8	-1.7	0.8
170.46	3.62	-0.2	302.12	15JA	-0.3	-3.9	-1.6	0.8
169.78	3.88	-0.4	312.25	25JA	1.7	-4.0	-1.6	0.8
167.81	4.07	-0.7	322.33	4FE	2.3	-4.1	-1.6	0.8
164.70	4.16	-0.9	332.37	14FE	1.0	-4.2	-1.6	0.7
160.89	4.09	-1.1	342.36	24FE	0.5	-4.3	-1.6	0.7
157.09	3.86	-1.0	352.28	5MR	0.3	-4.3	-1.6	0.7
153.95	3.50	-0.8	2.16	15MR	-0.0	-4.2	-1.6	0.6
151.96	3.07	-0.6	11.97	25MR	-0.6	-3.9	-1.6	0.6
151.28	2.63	-0.3	21.73	4AP	-1.4	-3.2	-1.6	0.5
151.83	2.21	-0.1	31.44	14AP	-1.6	-3.3	-1.6	0.5
153.48	1.82	0.1	41.10	24AP	-0.6	-3.9	-1.7	0.4
156.05	1.47	0.3	50.71	4MY	0.4	-4.1	-1.7	0.3
159.35	1.16	0.4	60.29	14MY	1.2	-4.2	-1.8	0.3
163.28	0.88	0.6	69.85	24MY	2.3	-4.2	-1.8	0.2
167.71	0.62	0.7	79.39	3JN	3.5	-4.1	-1.9	0.3
172.57	0.40	0.8	88.92	13JN	2.2	-4.0	-2.0	0.3
177.79	0.19	0.9	98.45	23JN	0.9	-3.9	-2.0	0.4
183.31	0.00	1.0	107.98	3JL	-0.3	-3.8	-2.1	0.4
189.10	-0.17	1.1	117.54	13JL	-1.3	-3.7	-2.2	0.5
195.14	-0.33	1.1	127.12	23JL	-1.5	-3.6	-2.2	0.6
201.40	-0.47	1.2	136.74	2AU	-0.8	-3.5	-2.3	0.6
207.85	-0.60	1.2	146.40	12AU	-0.2	-3.5	-2.4	0.7
214.50	-0.72	1.2	156.10	22AU	0.1	-3.5	-2.4	0.7
221.31	-0.82	1.3	165.86	1SE	0.3	-3.5	-2.4	0.7
228.28	-0.92	1.3	175.67	11SE	0.5	-3.4	-2.5	0.8
235.40	-1.00	1.3	185.54	21SE	1.4	-3.4	-2.5	0.8
242.65	-1.07	1.3	195.47	1OC	2.9	-3.4	-2.4	0.8
250.03	-1.13	1.3	205.45	11OC	0.2	-3.4	-2.4	0.8
257.52	-1.17	1.3	215.49	21OC	-0.6	-3.4	-2.4	0.8
265.11	-1.20	1.3	225.57	31OC	-0.7	-3.4	-2.3	0.7
272.78	-1.22	1.3	235.70	10NO	-0.7	-3.4	-2.3	0.7
280.53	-1.22	1.4	245.86	20NO	-0.8	-3.4	-2.2	0.7
288.34	-1.21	1.4	256.04	30NO	-0.7	-3.4	-2.1	0.6
296.19	-1.19	1.4	266.23	10DE	-0.7	-3.4	-2.0	0.7
304.06	-1.16	1.4	276.43	20DE	-0.7	-3.4	-2.0	0.7
311.95	-1.12	1.4	286.62	30DE	-0.2	-3.4	-1.9	0.7
				1193				
319.83	-1.06	1.4	296.80	9JA	2.0	-3.5	-1.8	0.8
327.70	-1.00	1.4	306.94	19JA	1.8	-3.5	-1.8	0.8
335.53	-0.92	1.4	317.04	29JA	0.7	-3.5	-1.7	0.8
343.32	-0.84	1.4	327.11	8FE	0.3	-3.5	-1.7	0.8
351.06	-0.75	1.4	337.12	18FE	0.1	-3.4	-1.6	0.8
358.73	-0.66	1.5	347.08	28FE	-0.1	-3.4	-1.6	0.8
6.34	-0.56	1.5	356.98	10MR	-0.6	-3.4	-1.6	0.7
13.87	-0.46	1.5	6.83	20MR	-1.4	-3.4	-1.6	0.7
21.31	-0.35	1.5	16.61	30MR	-1.5	-3.3	-1.6	0.6
28.67	-0.25	1.5	26.35	9AP	-0.6	-3.3	-1.6	0.6
35.95	-0.14	1.5	36.03	19AP	0.5	-3.3	-1.6	0.5
43.14	-0.04	1.6	45.67	29AP	1.6	-3.3	-1.6	0.5
50.24	0.07	1.6	55.27	9MY	3.0	-3.3	-1.6	0.4
57.26	0.17	1.7	64.84	19MY	3.0	-3.4	-1.6	0.3
64.19	0.27	1.7	74.39	29MY	1.7	-3.4	-1.7	0.3
71.05	0.37	1.8	83.92	8JN	0.7	-3.4	-1.7	0.2
77.83	0.47	1.8	93.45	18JN	-0.3	-3.4	-1.7	0.3
84.53	0.57	1.8	102.98	28JN	-1.3	-3.5	-1.8	0.3
91.16	0.66	1.9	112.53	8JL	-1.6	-3.5	-1.9	0.4
97.71	0.76	1.9	122.10	18JL	-0.7	-3.6	-1.9	0.5
104.20	0.85	1.9	131.69	28JL	-0.2	-3.7	-2.0	0.5
110.62	0.94	1.9	141.33	7AU	0.2	-3.8	-2.0	0.6
116.98	1.02	1.9	151.01	17AU	0.4	-3.9	-2.1	0.6
123.27	1.11	1.9	160.74	27AU	0.8	-4.0	-2.2	0.7
129.49	1.19	1.9	170.52	6SE	1.7	-4.1	-2.2	0.7
135.64	1.28	1.9	180.36	16SE	2.6	-4.2	-2.3	0.7
141.71	1.36	1.8	190.26	26SE	0.1	-4.3	-2.3	0.8
147.70	1.45	1.8	200.21	6OC	-0.8	-4.3	-2.4	0.8
153.59	1.53	1.7	210.22	16OC	-0.8	-4.3	-2.4	0.8
159.38	1.62	1.7	220.28	26OC	-0.8	-4.0	-2.4	0.8
165.05	1.70	1.6	230.39	5NO	-0.7	-3.4	-2.4	0.8
170.58	1.79	1.5	240.52	15NO	-0.6	-3.2	-2.4	0.7
175.94	1.88	1.4	250.70	25NO	-0.6	-4.0	-2.3	0.7
181.10	1.98	1.3	260.89	5DE	-0.5	-4.4	-2.3	0.7
186.01	2.07	1.1	271.08	15DE	-0.0	-4.4	-2.2	0.7
190.63	2.17	1.0	281.28	25DE	2.2	-4.3	-2.1	0.7

♂ LONG	LAT	MAG	☉ LONG	16.00UT	☿	♀	♃	♄
				1194		MAGNITUDES		
194.88	2.27	0.8	291.47	4JA	1.4	-4.3	-2.1	0.8
198.68	2.37	0.6	301.62	14JA	0.4	-4.1	-2.0	0.8
201.91	2.46	0.3	311.75	24JA	0.2	-4.0	-1.9	0.8
204.44	2.55	0.1	321.84	3FE	0.0	-3.9	-1.9	0.8
206.09	2.62	-0.2	331.88	13FE	-0.2	-3.8	-1.8	0.8
206.69	2.66	-0.5	341.87	23FE	-0.7	-3.7	-1.7	0.8
206.07	2.64	-0.8	351.80	5MR	-1.4	-3.7	-1.7	0.8
204.18	2.55	-1.0	1.68	15MR	-1.4	-3.6	-1.6	0.8
201.18	2.34	-1.3	11.50	25MR	-0.6	-3.5	-1.6	0.7
197.53	2.01	-1.4	21.26	4AP	0.7	-3.5	-1.6	0.7
193.92	1.60	-1.3	30.96	14AP	2.1	-3.4	-1.5	0.6
191.02	1.14	-1.2	40.63	24AP	3.9	-3.4	-1.5	0.6
189.31	0.68	-0.9	50.25	4MY	2.3	-3.3	-1.5	0.5
188.93	0.27	-0.7	59.83	14MY	1.3	-3.3	-1.5	0.5
189.84	-0.10	-0.5	69.39	24MY	0.5	-3.3	-1.5	0.4
191.89	-0.42	-0.3	78.93	3JN	-0.4	-3.3	-1.5	0.3
194.88	-0.68	-0.2	88.46	13JN	-1.4	-3.3	-1.5	0.3
198.66	-0.91	-0.0	97.99	23JN	-1.6	-3.3	-1.5	0.3
203.09	-1.10	0.1	107.53	3JL	-0.7	-3.3	-1.6	0.3
208.06	-1.26	0.3	117.08	13JL	-0.1	-3.4	-1.6	0.4
213.48	-1.39	0.4	126.66	23JL	0.3	-3.4	-1.7	0.4
219.28	-1.49	0.5	136.28	2AU	0.6	-3.4	-1.7	0.5
225.41	-1.58	0.5	145.93	12AU	1.0	-3.5	-1.8	0.5
231.83	-1.64	0.6	155.63	22AU	2.2	-3.5	-1.8	0.6
238.50	-1.68	0.7	165.38	1SE	2.3	-3.5	-1.9	0.6
245.37	-1.70	0.8	175.19	11SE	-0.0	-3.5	-1.9	0.7
252.44	-1.70	0.8	185.06	21SE	-0.9	-3.4	-2.0	0.7
259.68	-1.69	0.9	194.98	1OC	-1.0	-3.4	-2.1	0.8
267.05	-1.66	0.9	204.96	11OC	-0.9	-3.4	-2.1	0.8
274.53	-1.61	1.0	215.00	21OC	-0.6	-3.4	-2.2	0.8
282.11	-1.55	1.0	225.08	31OC	-0.5	-3.3	-2.2	0.8
289.77	-1.48	1.1	235.20	10NO	-0.4	-3.3	-2.3	0.8
297.47	-1.39	1.1	245.36	20NO	-0.4	-3.3	-2.3	0.8
305.21	-1.30	1.2	255.54	30NO	0.1	-3.3	-2.3	0.8
312.97	-1.19	1.2	265.73	10DE	2.5	-3.4	-2.3	0.7
320.72	-1.08	1.3	275.94	20DE	1.1	-3.4	-2.3	0.7
328.46	-0.97	1.3	286.13	30DE	0.2	-3.4	-2.2	0.7
				1195				
336.17	-0.85	1.4	296.30	9JA	0.0	-3.4	-2.2	0.8
343.83	-0.72	1.4	306.45	19JA	-0.1	-3.5	-2.1	0.8
351.44	-0.60	1.5	316.56	29JA	-0.3	-3.5	-2.0	0.8
358.99	-0.48	1.5	326.62	8FE	-0.7	-3.5	-2.0	0.8
6.47	-0.35	1.5	336.64	18FE	-1.3	-3.6	-1.9	0.9
13.88	-0.23	1.6	346.60	28FE	-1.3	-3.7	-1.8	0.9
21.20	-0.12	1.6	356.51	10MR	-0.6	-3.7	-1.7	0.9
28.46	-0.01	1.7	6.36	20MR	0.9	-3.8	-1.7	0.8
35.62	0.10	1.7	16.14	30MR	2.7	-3.9	-1.6	0.8
42.71	0.21	1.7	25.88	9AP	3.0	-4.0	-1.6	0.8
49.73	0.31	1.7	35.57	19AP	1.7	-4.1	-1.5	0.7
56.66	0.40	1.8	45.21	29AP	1.0	-4.1	-1.5	0.7
63.53	0.49	1.8	54.81	9MY	0.4	-4.2	-1.5	0.6
70.32	0.58	1.8	64.38	19MY	-0.4	-4.2	-1.4	0.5
77.05	0.66	1.8	73.93	29MY	-1.4	-4.0	-1.4	0.5
83.72	0.74	1.8	83.46	8JN	-1.6	-3.5	-1.4	0.4
90.33	0.81	1.8	92.99	18JN	-0.7	-2.8	-1.4	0.4
96.89	0.87	1.9	102.52	28JN	-0.0	-3.6	-1.4	0.3
103.41	0.94	1.9	112.07	8JL	0.4	-4.0	-1.4	0.3
109.87	1.00	1.9	121.63	18JL	0.8	-4.2	-1.4	0.4
116.30	1.05	2.0	131.23	28JL	1.4	-4.2	-1.4	0.4
122.69	1.10	2.0	140.86	7AU	2.7	-4.2	-1.5	0.5
129.05	1.15	2.0	150.54	17AU	1.9	-4.1	-1.5	0.5
135.38	1.19	2.0	160.26	27AU	-0.1	-4.0	-1.5	0.6
141.67	1.23	2.0	170.04	6SE	-1.0	-3.9	-1.6	0.6
147.94	1.27	2.0	179.88	16SE	-1.1	-3.8	-1.6	0.7
154.18	1.30	2.0	189.77	26SE	-0.9	-3.7	-1.7	0.7
160.38	1.33	2.0	199.73	6OC	-0.5	-3.7	-1.8	0.8
166.56	1.36	1.9	209.73	16OC	-0.3	-3.6	-1.8	0.8
172.69	1.38	1.9	219.79	26OC	-0.3	-3.6	-1.9	0.8
178.79	1.40	1.8	229.89	5NO	-0.2	-3.5	-2.0	0.8
184.84	1.42	1.8	240.03	15NO	0.3	-3.5	-2.0	0.8
190.83	1.43	1.7	250.20	25NO	2.8	-3.4	-2.1	0.8
196.75	1.43	1.6	260.39	5DE	0.8	-3.4	-2.1	0.8
202.59	1.43	1.5	270.59	15DE	-0.0	-3.4	-2.2	0.8
208.34	1.42	1.4	280.79	25DE	-0.1	-3.4	-2.2	0.8

Left table

♂ LONG	LAT	MAG	☉ LONG	16.00UT	☿	♀	♃	♄
				1196				
213.97	1.40	1.2	290.97	4JA	-0.2	-3.4	-2.2	0.8
219.46	1.37	1.1	301.14	14JA	-0.4	-3.3	-2.2	0.8
224.78	1.32	0.9	311.26	24JA	-0.8	-3.3	-2.1	0.8
229.88	1.26	0.8	321.36	3FE	-1.3	-3.3	-2.1	0.9
234.72	1.17	0.6	331.40	13FE	-1.2	-3.3	-2.0	0.9
239.24	1.06	0.3	341.39	23FE	-0.5	-3.4	-2.0	0.9
243.34	0.91	0.1	351.32	4MR	1.1	-3.4	-1.9	0.9
246.92	0.72	-0.1	1.20	14MR	3.2	-3.4	-1.8	0.9
249.86	0.47	-0.4	11.02	24MR	2.2	-3.4	-1.8	0.9
251.97	0.15	-0.7	20.78	3AP	1.2	-3.5	-1.7	0.9
253.09	-0.25	-1.0	30.49	13AP	0.7	-3.5	-1.6	0.9
253.05	-0.73	-1.4	40.16	23AP	0.2	-3.5	-1.6	0.9
251.75	-1.30	-1.7	49.78	3MY	-0.5	-3.4	-1.5	0.8
249.33	-1.91	-2.0	59.36	13MY	-1.4	-3.4	-1.5	0.7
246.21	-2.51	-2.1	68.92	23MY	-1.6	-3.4	-1.4	0.7
243.04	-3.02	-2.0	78.46	2JN	-0.7	-3.3	-1.4	0.6
240.58	-3.39	-1.9	87.99	12JN	0.1	-3.3	-1.3	0.5
239.31	-3.62	-1.7	97.52	22JN	0.6	-3.3	-1.3	0.5
239.41	-3.72	-1.4	107.06	2JL	1.1	-3.3	-1.3	0.4
240.87	-3.72	-1.2	116.61	12JL	1.9	-3.3	-1.3	0.3
243.50	-3.66	-1.0	126.19	22JL	3.2	-3.3	-1.3	0.4
247.12	-3.55	-0.8	135.80	1AU	1.6	-3.4	-1.3	0.4
251.55	-3.41	-0.6	145.46	11AU	-0.1	-3.4	-1.3	0.5
256.63	-3.25	-0.5	155.16	21AU	-1.1	-3.4	-1.3	0.5
262.23	-3.06	-0.3	164.91	31AU	-1.3	-3.5	-1.3	0.6
268.26	-2.87	-0.2	174.71	10SE	-0.9	-3.5	-1.4	0.6
274.63	-2.66	-0.0	184.58	20SE	-0.4	-3.6	-1.4	0.7
281.27	-2.45	0.1	194.50	30SE	-0.2	-3.7	-1.5	0.7
288.12	-2.23	0.3	204.48	10OC	-0.1	-3.7	-1.5	0.8
295.13	-2.01	0.4	214.51	20OC	-0.0	-3.8	-1.6	0.8
302.27	-1.79	0.5	224.59	30OC	0.6	-3.9	-1.6	0.8
309.49	-1.57	0.6	234.71	9NO	3.1	-4.0	-1.7	0.9
316.78	-1.36	0.7	244.86	19NO	0.6	-4.1	-1.7	0.9
324.10	-1.15	0.8	255.05	29NO	-0.2	-4.2	-1.8	0.9
331.43	-0.95	0.9	265.24	9DE	-0.3	-4.3	-1.9	0.9
338.76	-0.76	1.0	275.44	19DE	-0.3	-4.4	-1.9	0.9
346.07	-0.58	1.1	285.63	29DE	-0.5	-4.3	-2.0	0.9
				1197				
353.35	-0.41	1.2	295.81	8JA	-0.8	-4.1	-2.0	0.8
0.58	-0.25	1.3	305.95	18JA	-1.1	-3.5	-2.1	0.8
7.77	-0.10	1.4	316.07	28JA	-1.1	-3.4	-2.1	0.9
14.90	0.04	1.5	326.13	7FE	-0.5	-4.0	-2.1	0.9
21.97	0.16	1.5	336.15	17FE	1.3	-4.2	-2.0	1.0
28.98	0.28	1.6	346.12	27FE	3.2	-4.3	-2.0	1.0
35.93	0.39	1.7	356.03	9MR	1.6	-4.3	-2.0	1.0
42.82	0.49	1.7	5.88	19MR	0.9	-4.2	-1.9	1.0
49.64	0.58	1.8	15.67	29MR	0.5	-4.1	-1.9	1.0
56.41	0.66	1.8	25.41	8AP	0.1	-3.9	-1.8	1.0
63.12	0.74	1.9	35.10	18AP	-0.5	-3.8	-1.7	1.0
69.78	0.81	1.9	44.74	28AP	-1.4	-3.7	-1.7	0.9
76.38	0.87	1.9	54.34	8MY	-1.6	-3.7	-1.6	0.9
82.95	0.92	1.9	63.92	18MY	-0.7	-3.6	-1.5	0.9
89.47	0.97	2.0	73.47	28MY	0.2	-3.5	-1.5	0.8
95.96	1.01	2.0	83.00	7JN	0.8	-3.5	-1.4	0.7
102.41	1.05	2.0	92.53	17JN	1.4	-3.4	-1.4	0.7
108.84	1.08	2.0	102.06	27JN	2.4	-3.4	-1.3	0.6
115.25	1.11	1.9	111.60	7JL	3.1	-3.4	-1.3	0.5
121.64	1.13	1.9	121.17	17JL	1.3	-3.4	-1.3	0.5
128.02	1.14	2.0	130.77	27JL	-0.2	-3.3	-1.2	0.4
134.38	1.16	2.0	140.39	6AU	-1.2	-3.3	-1.2	0.5
140.75	1.16	2.0	150.07	16AU	-1.4	-3.3	-1.2	0.5
147.11	1.17	2.0	159.80	26AU	-0.8	-3.3	-1.2	0.6
153.47	1.17	2.0	169.57	5SE	-0.4	-3.4	-1.2	0.6
159.85	1.16	2.0	179.41	15SE	-0.1	-3.4	-1.2	0.7
166.22	1.15	2.0	189.30	25SE	0.0	-3.4	-1.3	0.7
172.61	1.13	2.0	199.24	5OC	0.2	-3.4	-1.3	0.8
179.01	1.11	2.0	209.25	15OC	0.8	-3.4	-1.3	0.8
185.41	1.09	2.0	219.30	25OC	3.5	-3.5	-1.4	0.9
191.82	1.05	1.9	229.40	4NO	0.4	-3.5	-1.4	0.9
198.25	1.02	1.9	239.54	14NO	-0.4	-3.5	-1.5	0.9
204.68	0.97	1.8	249.71	24NO	-0.4	-3.5	-1.5	0.9
211.11	0.91	1.8	259.89	4DE	-0.5	-3.4	-1.6	1.0
217.55	0.85	1.7	270.09	14DE	-0.6	-3.4	-1.7	1.0
223.99	0.77	1.6	280.29	24DE	-0.9	-3.4	-1.7	1.0

Right table

♂ LONG	LAT	MAG	☉ LONG	16.00UT	☿	♀	♃	♄
				1198				
230.43	0.69	1.5	290.47	3JA	-1.0	-3.4	-1.8	1.0
236.86	0.59	1.4	300.64	13JA	-1.0	-3.3	-1.8	1.0
243.28	0.47	1.3	310.77	23JA	-0.4	-3.3	-1.9	0.9
249.67	0.34	1.2	320.86	2FE	1.5	-3.3	-1.9	1.0
256.04	0.19	1.1	330.91	12FE	2.7	-3.3	-2.0	1.0
262.36	0.02	0.9	340.90	22FE	1.2	-3.4	-2.0	1.1
268.64	-0.18	0.8	350.84	4MR	0.6	-3.4	-2.0	1.1
274.84	-0.40	0.6	0.72	14MR	0.4	-3.4	-2.0	1.1
280.95	-0.66	0.4	10.55	24MR	0.0	-3.4	-2.0	1.1
286.95	-0.95	0.2	20.31	3AP	-0.5	-3.4	-2.0	1.1
292.78	-1.28	0.1	30.03	13AP	-1.4	-3.5	-1.9	1.1
298.40	-1.65	-0.2	39.69	23AP	-1.6	-3.5	-1.9	1.1
303.76	-2.08	-0.4	49.32	3MY	-0.7	-3.6	-1.8	1.1
308.76	-2.55	-0.6	58.91	13MY	0.3	-3.7	-1.7	1.1
313.29	-3.08	-0.9	68.47	23MY	1.0	-3.7	-1.7	1.0
317.23	-3.66	-1.1	78.01	2JN	1.9	-3.8	-1.6	1.0
320.39	-4.30	-1.4	87.54	12JN	3.2	-3.9	-1.5	0.9
322.59	-4.97	-1.7	97.06	22JN	2.5	-4.0	-1.5	0.9
323.62	-5.65	-2.0	106.60	2JL	1.1	-4.1	-1.4	0.8
323.35	-6.25	-2.2	116.16	12JL	-0.2	-4.2	-1.4	0.7
321.86	-6.69	-2.5	125.73	22JL	-1.2	-4.2	-1.3	0.6
319.48	-6.87	-2.6	135.34	1AU	-1.5	-4.2	-1.3	0.6
316.86	-6.71	-2.6	144.99	11AU	-0.8	-3.9	-1.3	0.6
314.77	-6.26	-2.4	154.69	21AU	-0.3	-3.4	-1.2	0.6
313.73	-5.61	-2.1	164.44	31AU	0.0	-3.4	-1.2	0.6
314.00	-4.86	-1.8	174.24	10SE	0.2	-3.9	-1.2	0.7
315.54	-4.12	-1.5	184.10	20SE	0.4	-4.2	-1.2	0.7
318.19	-3.42	-1.2	194.02	30SE	1.1	-4.3	-1.2	0.8
321.76	-2.79	-1.0	204.00	10OC	3.2	-4.3	-1.2	0.8
326.04	-2.23	-0.7	214.02	20OC	0.3	-4.2	-1.2	0.9
330.88	-1.74	-0.4	224.10	30OC	-0.5	-4.1	-1.3	0.9
336.14	-1.31	-0.2	234.22	9NO	-0.6	-4.0	-1.3	1.0
341.74	-0.94	0.0	244.37	19NO	-0.6	-3.9	-1.3	1.0
347.57	-0.62	0.2	254.55	29NO	-0.7	-3.9	-1.4	1.1
353.60	-0.34	0.4	264.75	9DE	-0.8	-3.8	-1.4	1.1
359.76	-0.10	0.6	274.95	19DE	-0.8	-3.7	-1.5	1.1
6.03	0.11	0.7	285.14	29DE	-0.8	-3.6	-1.6	1.1
				1199				
12.36	0.29	0.9	295.32	8JA	-0.3	-3.6	-1.6	1.1
18.74	0.45	1.0	305.47	18JA	1.7	-3.5	-1.7	1.1
25.15	0.59	1.2	315.58	28JA	2.1	-3.5	-1.8	1.1
31.58	0.70	1.3	325.66	7FE	0.8	-3.4	-1.8	1.1
38.01	0.80	1.4	335.67	17FE	0.4	-3.4	-1.9	1.1
44.44	0.88	1.5	345.64	27FE	0.2	-3.4	-1.9	1.2
50.86	0.96	1.6	355.56	9MR	-0.1	-3.3	-2.0	1.2
57.26	1.02	1.7	5.41	19MR	-0.6	-3.3	-2.0	1.3
63.65	1.07	1.7	15.20	29MR	-1.4	-3.3	-2.0	1.3
70.03	1.11	1.8	24.95	8AP	-1.6	-3.3	-2.0	1.3
76.39	1.14	1.9	34.63	18AP	-0.7	-3.3	-2.0	1.3
82.73	1.16	1.9	44.28	28AP	0.4	-3.3	-2.0	1.3
89.06	1.18	2.0	53.89	8MY	1.4	-3.4	-1.9	1.3
95.38	1.19	2.0	63.46	18MY	2.5	-3.4	-1.9	1.3
101.69	1.20	2.0	73.01	28MY	3.4	-3.4	-1.8	1.2
107.99	1.20	2.0	82.54	7JN	2.0	-3.5	-1.8	1.2
114.29	1.20	2.0	92.07	17JN	0.8	-3.5	-1.7	1.1
120.60	1.19	2.0	101.60	27JN	-0.3	-3.5	-1.7	1.1
126.92	1.18	2.0	111.14	7JL	-1.3	-3.4	-1.6	1.0
133.24	1.16	2.0	120.70	17JL	-1.6	-3.4	-1.5	0.9
139.58	1.14	2.0	130.30	27JL	-0.8	-3.4	-1.5	0.9
145.94	1.11	2.0	139.93	6AU	-0.2	-3.4	-1.4	0.8
152.32	1.08	2.0	149.60	16AU	0.1	-3.3	-1.4	0.7
158.74	1.05	1.9	159.32	26AU	0.3	-3.3	-1.3	0.7
165.18	1.01	1.9	169.09	5SE	0.6	-3.3	-1.3	0.8
171.65	0.96	2.0	178.92	15SE	1.5	-3.3	-1.3	0.8
178.17	0.92	2.0	188.81	25SE	2.9	-3.3	-1.3	0.9
184.72	0.86	2.0	198.76	5OC	0.2	-3.4	-1.2	0.9
191.32	0.80	1.9	208.76	15OC	-0.7	-3.4	-1.2	1.0
197.96	0.74	1.9	218.81	25OC	-0.7	-3.4	-1.2	1.0
204.64	0.67	1.9	228.91	4NO	-0.7	-3.5	-1.2	1.1
211.38	0.59	1.9	239.04	14NO	-0.8	-3.5	-1.3	1.1
218.16	0.51	1.8	249.22	24NO	-0.7	-3.6	-1.3	1.1
224.99	0.42	1.8	259.40	4DE	-0.6	-3.6	-1.3	1.2
231.86	0.32	1.7	269.60	14DE	-0.6	-3.7	-1.4	1.2
238.79	0.21	1.7	279.80	24DE	-0.2	-3.8	-1.4	1.2

♂ LONG	♂ LAT	♂ MAG	⊙ LONG	16.00UT	☿	♀	♃	♄
LONG	LAT	MAG	LONG	1200	MAGNITUDES			
245.76	0.10	1.6	289.99	3JA	2.0	-3.8	-1.4	1.3
252.78	-0.03	1.6	300.15	13JA	1.7	-3.9	-1.5	1.3
259.84	-0.16	1.5	310.29	23JA	0.6	-4.0	-1.6	1.3
266.94	-0.30	1.4	320.38	2FE	0.3	-4.1	-1.6	1.3
274.09	-0.45	1.3	330.43	12FE	0.1	-4.2	-1.7	1.3
281.26	-0.61	1.2	340.43	22FE	-0.2	-4.3	-1.8	1.3
288.47	-0.78	1.2	350.36	3MR	-0.6	-4.3	-1.8	1.3
295.70	-0.96	1.1	0.25	13MR	-1.3	-4.2	-1.9	1.4
302.94	-1.14	1.0	10.08	23MR	-1.5	-3.9	-2.0	1.4
310.18	-1.33	0.9	19.84	2AP	-0.6	-3.2	-2.0	1.4
317.42	-1.52	0.8	29.56	12AP	0.5	-3.3	-2.1	1.4
324.62	-1.71	0.6	39.23	22AP	1.8	-3.9	-2.1	1.4
331.77	-1.91	0.5	48.85	2MY	3.3	-4.2	-2.1	1.3
338.86	-2.10	0.4	58.44	12MY	2.7	-4.2	-2.1	1.3
345.83	-2.28	0.3	68.01	22MY	1.5	-4.2	-2.1	1.3
352.67	-2.46	0.2	77.54	1JN	0.6	-4.1	-2.1	1.2
359.34	-2.62	0.1	87.08	11JN	-0.3	-4.0	-2.0	1.2
5.76	-2.77	-0.1	96.61	21JN	-1.3	-3.9	-2.0	1.1
11.90	-2.91	-0.2	106.14	1JL	-1.6	-3.8	-1.9	1.1
17.67	-3.02	-0.4	115.69	11JL	-0.7	-3.7	-1.8	1.0
22.97	-3.11	-0.5	125.27	21JL	-0.1	-3.6	-1.8	1.0
27.68	-3.18	-0.7	134.87	31JL	0.2	-3.6	-1.7	0.9
31.67	-3.20	-0.9	144.52	10AU	0.5	-3.5	-1.7	0.9
34.72	-3.19	-1.1	154.22	20AU	0.9	-3.5	-1.6	0.9
36.66	-3.11	-1.3	163.96	30AU	1.9	-3.5	-1.5	0.8
37.26	-2.94	-1.5	173.76	9SE	2.6	-3.4	-1.5	0.9
36.38	-2.67	-1.8	183.62	19SE	0.1	-3.4	-1.4	0.9
34.12	-2.27	-2.0	193.54	29SE	-0.8	-3.4	-1.4	1.0
30.88	-1.76	-2.1	203.51	9OC	-0.9	-3.4	-1.4	1.0
27.39	-1.17	-2.0	213.54	19OC	-0.9	-3.4	-1.4	1.1
24.48	-0.59	-1.7	223.61	29OC	-0.7	-3.4	-1.3	1.2
22.69	-0.06	-1.4	233.73	8NO	-0.5	-3.4	-1.3	1.2
22.25	0.37	-1.1	243.88	18NO	-0.5	-3.4	-1.3	1.3
23.11	0.71	-0.8	254.06	28NO	-0.5	-3.4	-1.3	1.3
25.07	0.97	-0.4	264.26	8DE	-0.0	-3.4	-1.3	1.3
27.93	1.16	-0.2	274.46	18DE	2.2	-3.4	-1.4	1.4
31.51	1.30	0.1	284.65	28DE	1.3	-3.4	-1.4	1.4
				1201				
35.63	1.40	0.3	294.83	7JA	0.3	-3.5	-1.4	1.4
40.18	1.48	0.6	304.98	17JA	0.1	-3.5	-1.5	1.3
45.05	1.53	0.8	315.09	27JA	-0.0	-3.5	-1.5	1.3
50.19	1.56	0.9	325.17	6FE	-0.2	-3.5	-1.6	1.3
55.52	1.57	1.1	335.19	16FE	-0.7	-3.4	-1.6	1.2
61.01	1.58	1.2	345.16	26FE	-1.3	-3.4	-1.7	1.2
66.63	1.58	1.4	355.07	8MR	-1.4	-3.4	-1.7	1.1
72.34	1.57	1.5	4.93	18MR	-0.6	-3.4	-1.8	1.1
78.13	1.55	1.6	14.73	28MR	0.7	-3.3	-1.9	1.1
83.99	1.52	1.7	24.47	7AP	2.3	-3.3	-2.0	1.1
89.90	1.50	1.7	34.17	17AP	3.6	-3.3	-2.0	1.1
95.86	1.46	1.8	43.81	27AP	2.0	-3.3	-2.1	1.1
101.86	1.43	1.9	53.42	7MY	1.2	-3.3	-2.1	1.1
107.90	1.39	1.9	63.00	17MY	0.5	-3.4	-2.2	1.1
113.97	1.35	2.0	72.54	27MY	-0.4	-3.4	-2.2	1.0
120.08	1.30	2.0	82.08	6JN	-1.3	-3.4	-2.2	1.0
126.23	1.25	2.0	91.61	16JN	-1.6	-3.4	-2.2	0.9
132.42	1.20	2.0	101.14	26JN	-0.7	-3.5	-2.2	0.9
138.65	1.15	2.0	110.68	6JL	-0.1	-3.5	-2.2	0.9
144.92	1.09	2.0	120.24	16JL	0.4	-3.6	-2.1	0.8
151.24	1.03	2.0	129.83	26JL	0.7	-3.7	-2.1	0.7
157.60	0.97	2.0	139.46	5AU	1.2	-3.8	-2.0	0.7
164.02	0.90	2.0	149.13	15AU	2.4	-3.9	-1.9	0.6
170.49	0.83	1.9	158.85	25AU	2.2	-4.0	-1.9	0.6
177.03	0.76	1.9	168.62	4SE	0.0	-4.1	-1.8	0.6
183.62	0.68	1.9	178.45	14SE	-0.9	-4.2	-1.8	0.5
190.28	0.61	1.8	188.33	24SE	-1.0	-4.3	-1.7	0.6
197.01	0.52	1.8	198.28	4OC	-1.0	-4.3	-1.6	0.7
203.80	0.44	1.8	208.27	14OC	-0.6	-4.3	-1.6	0.7
210.66	0.35	1.8	218.32	24OC	-0.4	-4.0	-1.5	0.8
217.60	0.25	1.8	228.42	3NO	-0.3	-3.4	-1.5	0.9
224.61	0.15	1.8	238.55	13NO	-0.3	-3.3	-1.5	0.9
231.69	0.05	1.7	248.72	23NO	0.2	-4.0	-1.4	1.0
238.84	-0.05	1.7	258.91	3DE	2.5	-4.3	-1.4	1.0
246.06	-0.16	1.7	269.10	13DE	1.0	-4.4	-1.4	1.0
253.35	-0.27	1.7	279.30	23DE	0.1	-4.3	-1.4	1.1

♂ LONG	♂ LAT	♂ MAG	⊙ LONG	16.00UT	☿	♀	♃	♄
LONG	LAT	MAG	LONG	1202	MAGNITUDES			
260.71	-0.39	1.6	289.49	2JA	-0.1	-4.3	-1.4	1.1
268.12	-0.51	1.6	299.66	12JA	-0.1	-4.1	-1.4	1.1
275.60	-0.62	1.5	309.79	22JA	-0.3	-4.0	-1.5	1.1
283.13	-0.74	1.5	319.89	1FE	-0.7	-3.9	-1.5	1.0
290.70	-0.86	1.4	329.94	11FE	-1.3	-3.8	-1.5	1.0
298.31	-0.97	1.4	339.94	21FE	-1.3	-3.7	-1.6	1.0
305.95	-1.08	1.3	349.89	3MR	-0.6	-3.7	-1.6	0.9
313.61	-1.18	1.3	359.77	13MR	0.9	-3.6	-1.7	0.9
321.27	-1.28	1.2	9.60	23MR	2.8	-3.5	-1.7	0.8
328.93	-1.37	1.2	19.38	2AP	2.7	-3.5	-1.8	0.9
336.57	-1.44	1.2	29.09	12AP	1.5	-3.4	-1.9	0.9
344.18	-1.50	1.1	38.77	22AP	0.9	-3.4	-1.9	0.9
351.74	-1.55	1.1	48.40	2MY	0.3	-3.3	-2.0	0.9
359.23	-1.59	1.0	57.98	12MY	-0.4	-3.3	-2.1	0.8
6.64	-1.61	1.0	67.55	22MY	-1.3	-3.3	-2.1	0.8
13.95	-1.61	0.9	77.09	1JN	-1.6	-3.3	-2.2	0.8
21.14	-1.59	0.9	86.62	11JN	-0.7	-3.3	-2.3	0.8
28.20	-1.55	0.8	96.15	21JN	0.0	-3.3	-2.3	0.7
35.09	-1.50	0.7	105.68	1JL	0.5	-3.3	-2.3	0.7
41.79	-1.43	0.7	115.23	11JL	0.9	-3.4	-2.3	0.7
48.27	-1.33	0.6	124.81	21JL	1.5	-3.4	-2.3	0.6
54.50	-1.22	0.5	134.41	31JL	2.9	-3.4	-2.3	0.6
60.44	-1.08	0.4	144.06	10AU	1.9	-3.5	-2.3	0.5
66.02	-0.91	0.3	153.75	20AU	-0.1	-3.5	-2.2	0.4
71.18	-0.72	0.2	163.49	30AU	-1.0	-3.5	-2.2	0.4
75.84	-0.50	0.1	173.29	9SE	-1.2	-3.5	-2.1	0.3
79.87	-0.25	-0.1	183.14	19SE	-0.9	-3.4	-2.1	0.3
83.14	0.05	-0.3	193.06	29SE	-0.5	-3.4	-2.0	0.3
85.46	0.40	-0.5	203.02	9OC	-0.3	-3.4	-1.9	0.3
86.63	0.80	-0.7	213.05	19OC	-0.2	-3.4	-1.9	0.4
86.47	1.25	-0.9	223.12	29OC	-0.1	-3.3	-1.8	0.5
84.88	1.74	-1.1	233.23	8NO	0.4	-3.3	-1.7	0.5
81.97	2.21	-1.3	243.38	18NO	2.8	-3.3	-1.7	0.6
78.21	2.62	-1.4	253.56	28NO	0.8	-3.3	-1.6	0.7
74.35	2.91	-1.3	263.75	8DE	-0.1	-3.4	-1.6	0.7
71.13	3.06	-1.1	273.95	18DE	-0.2	-3.4	-1.6	0.8
69.09	3.09	-0.8	284.15	28DE	-0.3	-3.4	-1.5	0.8
				1203				
68.40	3.04	-0.5	294.32	7JA	-0.4	-3.4	-1.5	0.8
68.97	2.94	-0.2	304.48	17JA	-0.8	-3.5	-1.5	0.8
70.63	2.81	0.1	314.60	27JA	-1.2	-3.5	-1.5	0.8
73.18	2.68	0.4	324.67	6FE	-1.2	-3.6	-1.5	0.8
76.42	2.54	0.6	334.70	16FE	-0.5	-3.6	-1.5	0.8
80.23	2.40	0.8	344.68	26FE	1.1	-3.7	-1.6	0.8
84.46	2.27	1.0	354.59	8MR	3.3	-3.7	-1.6	0.7
89.04	2.15	1.1	4.46	18MR	2.0	-3.8	-1.6	0.7
93.91	2.03	1.2	14.26	28MR	1.1	-3.9	-1.7	0.6
98.99	1.92	1.4	24.00	7AP	0.6	-4.0	-1.7	0.6
104.26	1.81	1.5	33.70	17AP	0.2	-4.1	-1.8	0.6
109.69	1.70	1.6	43.35	27AP	-0.4	-4.1	-1.8	0.6
115.25	1.60	1.7	52.96	7MY	-1.3	-4.2	-1.9	0.6
120.93	1.50	1.7	62.54	17MY	-1.6	-4.2	-2.0	0.6
126.71	1.40	1.8	72.09	27MY	-0.7	-4.0	-2.0	0.6
132.58	1.30	1.8	81.62	6JN	0.1	-3.4	-2.1	0.6
138.55	1.20	1.9	91.15	16JN	0.7	-2.8	-2.2	0.6
144.60	1.10	1.9	100.68	26JN	1.2	-3.6	-2.2	0.6
150.73	1.01	1.9	110.22	6JL	2.1	-4.0	-2.3	0.6
156.95	0.91	1.9	119.78	16JL	3.2	-4.2	-2.3	0.5
163.25	0.82	1.9	129.37	26JL	1.6	-4.2	-2.4	0.5
169.63	0.72	1.9	138.99	5AU	-0.1	-4.1	-2.4	0.4
176.09	0.62	1.9	148.66	15AU	-1.1	-4.1	-2.4	0.4
182.64	0.52	1.9	158.38	25AU	-1.3	-4.0	-2.4	0.3
189.27	0.42	1.9	168.14	4SE	-0.9	-3.9	-2.4	0.2
196.00	0.32	1.8	177.97	14SE	-0.4	-3.8	-2.4	0.2
202.81	0.22	1.8	187.85	24SE	-0.2	-3.7	-2.3	0.1
209.71	0.12	1.8	197.79	4OC	-0.1	-3.7	-2.3	0.1
216.70	0.02	1.7	207.79	14OC	0.1	-3.6	-2.2	0.1
223.79	-0.08	1.7	217.83	24OC	0.6	-3.6	-2.2	0.1
230.96	-0.19	1.6	227.93	3NO	3.2	-3.5	-2.1	0.2
238.22	-0.29	1.5	238.06	13NO	0.6	-3.5	-2.0	0.3
245.56	-0.39	1.5	248.22	23NO	-0.3	-3.4	-1.9	0.3
252.99	-0.49	1.5	258.41	3DE	-0.4	-3.4	-1.9	0.4
260.49	-0.58	1.5	268.61	13DE	-0.4	-3.4	-1.8	0.5
268.06	-0.68	1.5	278.80	23DE	-0.5	-3.4	-1.8	0.5

♂ LONG	LAT	MAG	☉ LONG	16.00UT	☿	♀	♃	♄
					MAGNITUDES			
				1204				
275.70	-0.77	1.5	288.99	2JA	-0.8	-3.4	-1.7	0.6
283.39	-0.85	1.5	299.16	12JA	-1.0	-3.3	-1.7	0.6
291.13	-0.93	1.5	309.30	22JA	-1.0	-3.3	-1.6	0.6
298.91	-1.00	1.5	319.40	1FE	-0.5	-3.3	-1.6	0.6
306.72	-1.06	1.4	329.45	11FE	1.3	-3.3	-1.6	0.6
314.54	-1.11	1.4	339.45	21FE	3.0	-3.4	-1.6	0.6
322.37	-1.15	1.4	349.40	2MR	1.5	-3.4	-1.6	0.6
330.18	-1.18	1.4	359.29	12MR	0.8	-3.4	-1.6	0.6
337.98	-1.20	1.4	9.12	22MR	0.5	-3.4	-1.6	0.5
345.74	-1.20	1.4	18.90	1AP	0.1	-3.5	-1.6	0.5
353.45	-1.19	1.4	28.62	11AP	-0.5	-3.5	-1.6	0.4
1.11	-1.17	1.4	38.29	21AP	-1.3	-3.5	-1.7	0.4
8.70	-1.13	1.3	47.92	1MY	-1.6	-3.4	-1.7	0.5
16.21	-1.08	1.3	57.52	11MY	-0.7	-3.4	-1.7	0.5
23.62	-1.02	1.3	67.08	21MY	0.2	-3.4	-1.8	0.5
30.95	-0.94	1.3	76.62	31MY	0.9	-3.3	-1.8	0.5
38.16	-0.85	1.3	86.16	10JN	1.6	-3.3	-1.9	0.5
45.25	-0.75	1.3	95.68	20JN	2.7	-3.3	-2.0	0.5
52.22	-0.64	1.3	105.22	30JN	2.9	-3.3	-2.0	0.5
59.05	-0.52	1.2	114.77	10JL	1.3	-3.3	-2.1	0.4
65.73	-0.39	1.2	124.34	20JL	-0.2	-3.3	-2.2	0.4
72.25	-0.24	1.2	133.94	30JL	-1.2	-3.4	-2.2	0.4
78.60	-0.09	1.1	143.59	9AU	-1.5	-3.4	-2.3	0.3
84.74	0.08	1.1	153.27	19AU	-0.8	-3.4	-2.4	0.3
90.67	0.26	1.0	163.02	29AU	-0.3	-3.5	-2.4	0.2
96.34	0.46	0.9	172.81	8SE	-0.1	-3.5	-2.4	0.2
101.73	0.67	0.9	182.66	18SE	0.1	-3.6	-2.5	0.1
106.76	0.91	0.7	192.57	28SE	0.2	-3.7	-2.5	0.0
111.37	1.17	0.6	202.54	8OC	0.9	-3.7	-2.4	-0.0
115.49	1.45	0.5	212.56	18OC	3.5	-3.8	-2.4	-0.1
118.98	1.77	0.3	222.63	28OC	0.4	-3.9	-2.4	-0.1
121.70	2.13	0.1	232.74	7NO	-0.4	-4.0	-2.3	-0.0
123.49	2.52	-0.1	242.89	17NO	-0.5	-4.1	-2.2	0.0
124.15	2.95	-0.3	253.07	27NO	-0.5	-4.3	-2.2	0.1
123.50	3.40	-0.5	263.26	7DE	-0.6	-4.3	-2.1	0.2
121.50	3.82	-0.8	273.46	17DE	-0.8	-4.4	-2.0	0.2
118.30	4.15	-1.0	283.65	27DE	-0.9	-4.3	-2.0	0.3
				1205				
114.41	4.35	-1.1	293.83	6JA	-0.9	-4.1	-1.9	0.4
110.54	4.37	-0.9	303.98	16JA	-0.4	-3.5	-1.8	0.4
107.39	4.23	-0.7	314.11	26JA	1.5	-3.4	-1.8	0.4
105.42	3.99	-0.4	324.18	5FE	2.5	-4.0	-1.7	0.5
104.75	3.68	-0.2	334.21	15FE	1.1	-4.3	-1.7	0.5
105.32	3.37	0.1	344.19	25FE	0.6	-4.3	-1.6	0.5
106.95	3.06	0.3	354.11	7MR	0.3	-4.3	-1.6	0.5
109.43	2.78	0.5	3.97	17MR	-0.0	-4.2	-1.6	0.4
112.62	2.51	0.7	13.78	27MR	-0.5	-4.1	-1.6	0.4
116.37	2.27	0.9	23.53	6AP	-1.3	-3.9	-1.6	0.4
120.57	2.05	1.0	33.23	16AP	-1.6	-3.8	-1.6	0.3
125.14	1.85	1.2	42.88	26AP	-0.7	-3.7	-1.6	0.3
130.01	1.66	1.3	52.49	6MY	0.3	-3.7	-1.6	0.3
135.13	1.48	1.4	62.07	16MY	1.1	-3.6	-1.6	0.3
140.47	1.32	1.4	71.63	26MY	2.1	-3.5	-1.6	0.4
146.01	1.16	1.5	81.16	5JN	3.4	-3.5	-1.7	0.4
151.71	1.01	1.6	90.69	15JN	2.3	-3.4	-1.7	0.4
157.56	0.87	1.6	100.22	25JN	1.0	-3.4	-1.7	0.4
163.56	0.73	1.7	109.76	5JL	-0.2	-3.4	-1.8	0.4
169.69	0.60	1.7	119.32	15JL	-1.2	-3.3	-1.9	0.4
175.96	0.47	1.7	128.91	25JL	-1.6	-3.3	-1.9	0.4
182.34	0.34	1.7	138.53	4AU	-0.8	-3.3	-2.0	0.3
188.85	0.22	1.7	148.20	14AU	-0.3	-3.3	-2.0	0.3
195.47	0.10	1.7	157.91	24AU	0.0	-3.3	-2.1	0.2
202.21	-0.01	1.7	167.67	3SE	0.2	-3.4	-2.2	0.2
209.07	-0.13	1.7	177.49	13SE	0.5	-3.4	-2.2	0.1
216.04	-0.23	1.7	187.38	23SE	1.2	-3.4	-2.3	0.1
223.11	-0.34	1.6	197.31	3OC	3.2	-3.4	-2.3	-0.0
230.30	-0.44	1.6	207.30	13OC	0.3	-3.4	-2.4	-0.1
237.59	-0.54	1.6	217.35	23OC	-0.6	-3.5	-2.4	-0.1
244.97	-0.63	1.6	227.44	2NO	-0.7	-3.5	-2.4	-0.2
252.46	-0.71	1.5	237.57	12NO	-0.6	-3.5	-2.4	-0.2
260.03	-0.79	1.5	247.73	22NO	-0.7	-3.5	-2.4	-0.1
267.67	-0.86	1.4	257.91	2DE	-0.7	-3.4	-2.3	-0.1
275.39	-0.92	1.4	268.11	12DE	-0.7	-3.4	-2.2	0.0
283.16	-0.98	1.4	278.31	22DE	-0.7	-3.4	-2.2	0.1

♂ LONG	LAT	MAG	☉ LONG	16.00UT	☿	♀	♃	♄
					MAGNITUDES			
				1206				
290.99	-1.02	1.3	288.49	1JA	-0.3	-3.4	-2.1	0.1
298.85	-1.05	1.3	298.66	11JA	1.7	-3.3	-2.0	0.2
306.73	-1.07	1.3	308.80	21JA	2.0	-3.3	-2.0	0.3
314.62	-1.08	1.3	318.90	31JA	0.7	-3.3	-1.9	0.3
322.52	-1.08	1.3	328.96	10FE	0.4	-3.3	-1.8	0.3
330.39	-1.07	1.4	338.97	20FE	0.2	-3.4	-1.8	0.4
338.24	-1.04	1.4	348.91	2MR	-0.1	-3.4	-1.7	0.4
346.05	-1.01	1.4	358.81	12MR	-0.6	-3.4	-1.7	0.4
353.81	-0.96	1.4	8.64	22MR	-1.3	-3.4	-1.6	0.4
1.51	-0.90	1.4	18.42	1AP	-1.5	-3.5	-1.6	0.4
9.13	-0.83	1.5	28.15	11AP	-0.7	-3.5	-1.5	0.3
16.68	-0.76	1.5	37.82	21AP	0.4	-3.5	-1.5	0.3
24.15	-0.67	1.5	47.45	1MY	1.5	-3.6	-1.5	0.3
31.53	-0.58	1.5	57.05	11MY	2.8	-3.7	-1.5	0.2
38.82	-0.48	1.5	66.62	21MY	3.2	-3.7	-1.5	0.2
46.00	-0.37	1.6	76.16	31MY	1.8	-3.8	-1.5	0.3
53.09	-0.26	1.6	85.70	10JN	0.8	-3.9	-1.5	0.3
60.08	-0.15	1.6	95.22	20JN	-0.3	-4.0	-1.5	0.3
66.96	-0.03	1.6	104.76	30JN	-1.3	-4.1	-1.5	0.3
73.74	0.10	1.6	114.31	10JL	-1.6	-4.2	-1.6	0.3
80.40	0.22	1.6	123.88	20JL	-0.8	-4.2	-1.6	0.3
86.95	0.35	1.6	133.48	30JL	-0.2	-4.2	-1.6	0.3
93.38	0.49	1.6	143.12	9AU	0.2	-3.9	-1.7	0.3
99.69	0.63	1.5	152.81	19AU	0.4	-3.4	-1.8	0.3
105.86	0.77	1.5	162.55	29AU	0.7	-3.4	-1.8	0.2
111.88	0.93	1.5	172.34	8SE	1.6	-3.9	-1.9	0.2
117.73	1.08	1.4	182.19	18SE	2.9	-4.2	-1.9	0.1
123.41	1.25	1.4	192.09	28SE	0.2	-4.3	-2.0	0.1
128.87	1.43	1.3	202.06	8OC	-0.7	-4.3	-2.1	0.0
134.08	1.61	1.2	212.07	18OC	-0.8	-4.2	-2.1	-0.1
139.00	1.82	1.1	222.14	28OC	-0.8	-4.1	-2.2	-0.1
143.56	2.04	0.9	232.25	7NO	-0.7	-4.0	-2.2	-0.2
147.69	2.28	0.8	242.40	17NO	-0.6	-3.9	-2.3	-0.3
151.29	2.54	0.6	252.57	27NO	-0.6	-3.9	-2.3	-0.3
154.23	2.83	0.4	262.77	7DE	-0.6	-3.8	-2.3	-0.2
156.35	3.15	0.2	272.96	17DE	-0.1	-3.7	-2.3	-0.1
157.48	3.48	-0.0	283.16	27DE	2.0	-3.6	-2.2	-0.0
				1207				
157.44	3.81	-0.3	293.34	6JA	1.6	-3.6	-2.2	0.0
156.10	4.12	-0.5	303.50	16JA	0.5	-3.5	-2.1	0.1
153.49	4.35	-0.8	313.62	26JA	0.2	-3.5	-2.1	0.2
149.92	4.44	-1.0	323.70	5FE	0.1	-3.4	-2.0	0.2
145.99	4.37	-1.0	333.73	15FE	-0.2	-3.4	-1.9	0.3
142.42	4.13	-0.8	343.71	25FE	-0.6	-3.4	-1.9	0.3
139.78	3.77	-0.6	353.64	7MR	-1.3	-3.3	-1.8	0.3
138.41	3.36	-0.4	3.50	17MR	-1.5	-3.3	-1.7	0.3
138.32	2.95	-0.1	13.31	27MR	-0.7	-3.3	-1.7	0.3
139.39	2.55	0.1	23.06	6AP	0.6	-3.3	-1.6	0.3
141.45	2.18	0.3	32.76	16AP	1.9	-3.3	-1.5	0.3
144.32	1.85	0.5	42.41	26AP	3.7	-3.3	-1.5	0.3
147.86	1.55	0.6	52.03	6MY	2.4	-3.4	-1.5	0.3
151.94	1.29	0.8	61.60	16MY	1.4	-3.4	-1.4	0.2
156.47	1.04	0.9	71.16	26MY	0.6	-3.4	-1.4	0.2
161.37	0.82	1.0	80.70	5JN	-0.3	-3.5	-1.4	0.2
166.58	0.62	1.1	90.23	15JN	-1.3	-3.5	-1.4	0.3
172.08	0.43	1.2	99.76	25JN	-1.6	-3.5	-1.4	0.3
177.81	0.25	1.2	109.30	5JL	-0.8	-3.4	-1.4	0.3
183.76	0.09	1.3	118.85	15JL	-0.1	-3.4	-1.4	0.3
189.91	-0.06	1.3	128.44	25JL	0.3	-3.4	-1.4	0.3
196.25	-0.20	1.4	138.06	4AU	0.6	-3.4	-1.4	0.3
202.75	-0.34	1.4	147.72	14AU	1.0	-3.3	-1.5	0.3
209.42	-0.46	1.4	157.43	24AU	2.0	-3.3	-1.5	0.3
216.24	-0.57	1.4	167.20	3SE	2.5	-3.3	-1.5	0.3
223.20	-0.68	1.4	177.01	13SE	0.1	-3.3	-1.6	0.3
230.30	-0.77	1.4	186.89	23SE	-0.8	-3.3	-1.6	0.2
237.53	-0.86	1.4	196.83	3OC	-1.0	-3.4	-1.7	0.1
244.88	-0.93	1.4	206.81	13OC	-0.9	-3.4	-1.8	0.1
252.33	-1.00	1.4	216.86	23OC	-0.7	-3.4	-1.8	0.0
259.89	-1.05	1.4	226.95	2NO	-0.5	-3.5	-1.9	-0.1
267.54	-1.09	1.4	237.07	12NO	-0.4	-3.5	-2.0	-0.1
275.27	-1.12	1.4	247.24	22NO	-0.4	-3.6	-2.0	-0.2
283.06	-1.14	1.4	257.42	2DE	0.0	-3.6	-2.1	-0.3
290.90	-1.14	1.4	267.61	12DE	2.2	-3.7	-2.1	-0.3
298.77	-1.14	1.4	277.82	22DE	1.2	-3.8	-2.1	-0.2

Left table (1208 / 1209):

♂ LONG	LAT	MAG	⊙ LONG	16.00UT	☿	♀	♃	♄
				1208	MAGNITUDES			
306.67	-1.12	1.4	288.01	1JA	0.2	-3.8	-2.2	-0.1
314.58	-1.09	1.4	298.17	11JA	0.0	-3.9	-2.1	-0.0
322.47	-1.05	1.4	308.32	21JA	-0.1	-4.0	-2.1	0.0
330.35	-0.99	1.4	318.42	31JA	-0.3	-4.1	-2.1	0.1
338.20	-0.93	1.4	328.48	10FE	-0.7	-4.2	-2.1	0.1
345.99	-0.86	1.4	338.49	20FE	-1.3	-4.3	-2.0	0.2
353.74	-0.78	1.4	348.44	1MR	-1.4	-4.3	-1.9	0.2
1.41	-0.70	1.4	358.33	11MR	-0.6	-4.2	-1.9	0.3
9.02	-0.61	1.4	8.17	21MR	0.7	-3.8	-1.8	0.3
16.55	-0.51	1.4	17.95	31MR	2.4	-3.2	-1.7	0.3
23.99	-0.41	1.5	27.68	10AP	3.2	-3.4	-1.7	0.3
31.35	-0.31	1.5	37.36	20AP	1.8	-3.9	-1.6	0.3
38.62	-0.20	1.6	46.99	30AP	1.0	-4.2	-1.5	0.3
45.80	-0.10	1.6	56.58	10MY	0.4	-4.2	-1.5	0.3
52.89	0.01	1.7	66.15	20MY	-0.3	-4.2	-1.4	0.3
59.90	0.12	1.7	75.70	30MY	-1.3	-4.1	-1.4	0.3
66.82	0.22	1.7	85.23	9JN	-1.7	-4.0	-1.4	0.2
73.65	0.33	1.8	94.76	19JN	-0.8	-3.9	-1.3	0.2
80.40	0.43	1.8	104.29	29JN	-0.1	-3.8	-1.3	0.3
87.06	0.54	1.8	113.84	9JL	0.4	-3.7	-1.3	0.3
93.65	0.64	1.8	123.41	19JL	0.8	-3.6	-1.3	0.3
100.16	0.75	1.8	133.01	29JL	1.3	-3.6	-1.3	0.4
106.58	0.85	1.8	142.65	8AU	2.5	-3.5	-1.3	0.4
112.93	0.95	1.8	152.34	18AU	2.2	-3.5	-1.3	0.4
119.19	1.05	1.8	162.07	28AU	0.0	-3.4	-1.3	0.4
125.36	1.16	1.8	171.86	7SE	-1.0	-3.4	-1.3	0.4
131.45	1.26	1.8	181.71	17SE	-1.1	-3.4	-1.4	0.3
137.43	1.36	1.7	191.61	27SE	-1.0	-3.4	-1.4	0.3
143.30	1.47	1.7	201.57	7OC	-0.6	-3.4	-1.5	0.3
149.04	1.58	1.6	211.59	17OC	-0.4	-3.4	-1.5	0.2
154.64	1.70	1.5	221.65	27OC	-0.3	-3.4	-1.6	0.1
160.07	1.82	1.4	231.76	6NO	-0.2	-3.4	-1.6	0.1
165.31	1.94	1.3	241.91	16NO	0.2	-3.4	-1.7	-0.0
170.30	2.07	1.2	252.08	26NO	2.5	-3.4	-1.8	-0.1
175.01	2.21	1.1	262.27	6DE	1.0	-3.4	-1.8	-0.2
179.36	2.36	0.9	272.47	16DE	0.0	-3.4	-1.9	-0.2
183.27	2.51	0.7	282.66	26DE	-0.1	-3.4	-1.9	-0.2
				1209				
186.64	2.68	0.5	292.84	5JA	-0.2	-3.5	-2.0	-0.1
189.34	2.85	0.3	303.00	15JA	-0.4	-3.5	-2.0	-0.1
191.20	3.01	0.0	313.13	25JA	-0.7	-3.5	-2.0	-0.0
192.05	3.17	-0.2	323.21	4FE	-1.3	-3.5	-2.1	0.1
191.71	3.29	-0.5	333.25	14FE	-1.3	-3.4	-2.0	0.1
190.09	3.35	-0.8	343.22	24FE	-0.6	-3.4	-2.0	0.2
187.29	3.30	-1.1	353.15	6MR	0.9	-3.4	-2.0	0.2
183.67	3.12	-1.2	3.02	16MR	3.0	-3.4	-1.9	0.3
179.88	2.81	-1.2	12.83	26MR	2.4	-3.3	-1.9	0.3
176.63	2.40	-1.0	22.58	5AP	1.3	-3.3	-1.8	0.4
174.45	1.94	-0.8	32.29	15AP	0.8	-3.3	-1.8	0.4
173.58	1.49	-0.6	41.94	25AP	0.3	-3.3	-1.7	0.4
173.99	1.07	-0.4	51.56	5MY	-0.4	-3.3	-1.6	0.4
175.57	0.69	-0.2	61.14	15MY	-1.3	-3.4	-1.6	0.4
178.13	0.36	-0.0	70.69	25MY	-1.7	-3.4	-1.5	0.4
181.50	0.07	0.2	80.23	4JN	-0.7	-3.4	-1.4	0.4
185.54	-0.18	0.3	89.76	14JN	0.0	-3.4	-1.4	0.3
190.13	-0.40	0.4	99.29	24JN	0.6	-3.5	-1.4	0.3
195.18	-0.59	0.5	108.83	4JL	1.0	-3.5	-1.3	0.3
200.63	-0.76	0.6	118.39	14JL	1.7	-3.6	-1.3	0.4
206.42	-0.91	0.7	127.97	24JL	3.1	-3.7	-1.3	0.4
212.50	-1.04	0.8	137.59	3AU	1.8	-3.8	-1.2	0.4
218.84	-1.15	0.9	147.26	13AU	-0.0	-3.9	-1.2	0.5
225.41	-1.24	0.9	156.96	23AU	-1.1	-4.0	-1.2	0.5
232.20	-1.31	1.0	166.72	2SE	-1.3	-4.1	-1.2	0.5
239.17	-1.37	1.0	176.54	12SE	-0.9	-4.2	-1.2	0.5
246.30	-1.41	1.0	186.41	22SE	-0.5	-4.3	-1.2	0.5
253.58	-1.44	1.1	196.35	2OC	-0.2	-4.3	-1.3	0.4
261.00	-1.45	1.1	206.33	12OC	-0.1	-4.3	-1.3	0.4
268.52	-1.44	1.1	216.37	22OC	-0.1	-4.0	-1.3	0.4
276.14	-1.42	1.2	226.46	1NO	0.4	-3.3	-1.4	0.3
283.84	-1.39	1.2	236.58	11NO	2.8	-3.3	-1.4	0.2
291.60	-1.34	1.2	246.74	21NO	0.8	-4.0	-1.5	0.2
299.40	-1.28	1.3	256.93	1DE	-0.1	-4.3	-1.5	0.1
307.22	-1.21	1.3	267.12	11DE	-0.3	-4.4	-1.6	0.0
315.05	-1.13	1.3	277.32	21DE	-0.3	-4.3	-1.7	-0.1
322.87	-1.04	1.4	287.51	31DE	-0.5	-4.2	-1.7	-0.1

Right table (1210 / 1211):

♂ LONG	LAT	MAG	⊙ LONG	16.00UT	☿	♀	♃	♄
				1210	MAGNITUDES			
330.67	-0.94	1.4	297.68	10JA	-0.8	-4.1	-1.8	-0.1
338.43	-0.84	1.4	307.82	20JA	-1.1	-4.0	-1.9	-0.0
346.15	-0.73	1.5	317.93	30JA	-1.1	-3.9	-1.9	0.0
353.81	-0.62	1.5	327.99	9FE	-0.5	-3.8	-2.0	0.1
1.40	-0.51	1.5	338.00	19FE	1.1	-3.7	-2.0	0.2
8.92	-0.40	1.6	347.96	1MR	3.3	-3.7	-2.0	0.2
16.36	-0.29	1.6	357.86	11MR	1.8	-3.6	-2.0	0.3
23.72	-0.18	1.6	7.70	21MR	1.0	-3.5	-2.0	0.3
31.00	-0.07	1.6	17.48	31MR	0.6	-3.5	-2.0	0.4
38.20	0.04	1.7	27.21	10AP	0.2	-3.4	-1.9	0.4
45.31	0.15	1.7	36.89	20AP	-0.4	-3.4	-1.9	0.4
52.35	0.25	1.7	46.53	30AP	-1.3	-3.3	-1.8	0.5
59.30	0.34	1.7	56.13	10MY	-1.7	-3.3	-1.8	0.5
66.17	0.44	1.7	65.69	20MY	-0.7	-3.3	-1.7	0.5
72.98	0.53	1.7	75.24	30MY	0.1	-3.3	-1.6	0.5
79.72	0.61	1.8	84.77	9JN	0.7	-3.3	-1.6	0.5
86.39	0.70	1.8	94.30	19JN	1.3	-3.3	-1.5	0.5
93.00	0.78	1.9	103.84	29JN	2.3	-3.3	-1.5	0.5
99.55	0.85	1.9	113.38	9JL	3.2	-3.4	-1.4	0.4
106.04	0.92	1.9	122.95	19JL	1.5	-3.4	-1.4	0.5
112.49	0.99	2.0	132.55	29JL	-0.1	-3.4	-1.3	0.5
118.89	1.06	2.0	142.18	8AU	-1.1	-3.5	-1.3	0.6
125.24	1.12	2.0	151.87	18AU	-1.4	-3.5	-1.3	0.6
131.55	1.18	2.0	161.60	28AU	-0.9	-3.5	-1.2	0.6
137.81	1.24	2.0	171.38	7SE	-0.4	-3.5	-1.2	0.6
144.03	1.29	2.0	181.23	17SE	-0.1	-3.4	-1.2	0.6
150.21	1.34	1.9	191.13	27SE	-0.0	-3.4	-1.2	0.6
156.33	1.39	1.9	201.08	7OC	0.1	-3.4	-1.2	0.6
162.40	1.44	1.9	211.10	17OC	0.7	-3.3	-1.2	0.6
168.41	1.49	1.8	221.16	27OC	3.2	-3.3	-1.2	0.6
174.35	1.53	1.7	231.27	6NO	0.6	-3.3	-1.3	0.5
180.21	1.57	1.7	241.41	16NO	-0.3	-3.3	-1.3	0.5
185.98	1.61	1.6	251.59	26NO	-0.4	-3.3	-1.3	0.4
191.63	1.64	1.5	261.77	6DE	-0.4	-3.4	-1.4	0.3
197.15	1.67	1.4	271.97	16DE	-0.6	-3.4	-1.5	0.3
202.51	1.70	1.2	282.17	26DE	-0.8	-3.4	-1.5	0.2
				1211				
207.68	1.71	1.1	292.35	5JA	-1.0	-3.4	-1.6	0.1
212.61	1.72	0.9	302.51	15JA	-1.0	-3.5	-1.6	0.0
217.24	1.72	0.7	312.64	25JA	-0.5	-3.5	-1.7	0.1
221.51	1.70	0.5	322.72	4FE	1.3	-3.6	-1.8	0.1
225.33	1.67	0.3	332.76	14FE	2.9	-3.6	-1.8	0.2
228.57	1.60	0.0	342.75	24FE	1.3	-3.7	-1.9	0.2
231.10	1.50	-0.2	352.67	6MR	0.7	-3.7	-1.9	0.3
232.74	1.35	-0.5	2.55	16MR	0.4	-3.8	-2.0	0.4
233.31	1.13	-0.8	12.36	26MR	0.1	-3.9	-2.0	0.4
232.67	0.83	-1.2	22.12	5AP	-0.5	-4.0	-2.0	0.5
230.77	0.45	-1.5	31.82	15AP	-1.3	-4.1	-2.0	0.5
227.83	-0.02	-1.7	41.48	25AP	-1.6	-4.2	-2.0	0.6
224.37	-0.52	-1.8	51.10	5MY	-0.7	-4.2	-2.0	0.6
221.08	-1.02	-1.7	60.68	15MY	0.2	-4.2	-1.9	0.6
218.65	-1.45	-1.5	70.24	25MY	1.0	-4.0	-1.9	0.6
217.49	-1.81	-1.3	79.77	4JN	1.8	-3.4	-1.8	0.7
217.70	-2.07	-1.1	89.31	14JN	3.0	-2.8	-1.8	0.7
219.21	-2.27	-0.9	98.83	24JN	2.7	-3.6	-1.7	0.7
221.84	-2.40	-0.7	108.37	4JL	1.2	-4.0	-1.6	0.7
225.40	-2.48	-0.5	117.93	14JL	-0.1	-4.2	-1.6	0.7
229.75	-2.52	-0.3	127.51	24JL	-1.2	-4.2	-1.5	0.6
234.72	-2.54	-0.2	137.12	3AU	-1.5	-4.1	-1.5	0.7
240.21	-2.52	-0.1	146.78	13AU	-0.8	-4.1	-1.4	0.7
246.14	-2.48	0.1	156.49	23AU	-0.3	-4.0	-1.4	0.8
252.42	-2.41	0.2	166.24	2SE	-0.0	-3.9	-1.3	0.8
258.99	-2.33	0.3	176.06	12SE	0.1	-3.8	-1.3	0.8
265.81	-2.23	0.4	185.93	22SE	0.3	-3.7	-1.3	0.8
272.83	-2.12	0.5	195.86	2OC	1.0	-3.7	-1.3	0.9
280.02	-1.99	0.6	205.84	12OC	3.4	-3.6	-1.3	0.8
287.33	-1.85	0.7	215.88	22OC	0.4	-3.6	-1.2	0.8
294.74	-1.70	0.7	225.96	1NO	-0.5	-3.5	-1.3	0.8
302.22	-1.55	0.8	236.09	11NO	-0.6	-3.5	-1.3	0.8
309.75	-1.39	0.9	246.25	21NO	-0.6	-3.4	-1.3	0.7
317.30	-1.23	1.0	256.43	1DE	-0.7	-3.4	-1.3	0.7
324.85	-1.06	1.1	266.63	11DE	-0.8	-3.4	-1.3	0.6
332.40	-0.90	1.2	276.83	21DE	-0.8	-3.4	-1.4	0.5
339.92	-0.75	1.2	287.02	31DE	-0.8	-3.4	-1.4	0.5

♂			⊙	16.00UT	☿	♀	♃	♄
LONG	LAT	MAG	LONG	1212	MAGNITUDES			
347.40	-0.59	1.3	297.19	10JA	-0.4	-3.3	-1.5	0.4
354.84	-0.45	1.4	307.34	20JA	1.5	-3.3	-1.5	0.3
2.22	-0.30	1.4	317.44	30JA	2.3	-3.3	-1.6	0.3
9.53	-0.17	1.5	327.51	9FE	0.9	-3.3	-1.7	0.3
16.78	-0.04	1.6	337.52	19FE	0.5	-3.4	-1.7	0.4
23.96	0.08	1.6	347.48	29FE	0.3	-3.4	-1.8	0.4
31.07	0.20	1.7	357.38	10MR	-0.0	-3.4	-1.9	0.5
38.10	0.30	1.7	7.22	20MR	-0.5	-3.4	-1.9	0.5
45.06	0.40	1.8	17.01	30MR	-1.3	-3.5	-2.0	0.6
51.96	0.50	1.8	26.74	9AP	-1.6	-3.5	-2.0	0.6
58.79	0.58	1.8	36.42	19AP	-0.7	-3.4	-2.1	0.7
65.55	0.66	1.9	46.06	29AP	0.3	-3.4	-2.1	0.7
72.26	0.74	1.9	55.66	9MY	1.3	-3.4	-2.1	0.8
78.91	0.80	1.9	65.23	19MY	2.3	-3.4	-2.1	0.8
85.51	0.86	1.9	74.78	29MY	3.5	-3.3	-2.1	0.8
92.07	0.92	1.9	84.31	8JN	2.1	-3.3	-2.1	0.9
98.58	0.97	1.9	93.84	18JN	0.9	-3.3	-2.0	0.9
105.06	1.02	1.9	103.37	28JN	-0.2	-3.3	-2.0	0.9
111.50	1.06	1.9	112.92	8JL	-1.2	-3.3	-1.9	0.9
117.92	1.09	1.9	122.48	18JL	-1.6	-3.3	-1.8	0.9
124.31	1.12	2.0	132.08	28JL	-0.8	-3.4	-1.8	0.9
130.69	1.15	2.0	141.72	7AU	-0.2	-3.4	-1.7	0.9
137.05	1.17	2.0	151.39	17AU	0.1	-3.4	-1.7	0.9
143.39	1.19	2.0	161.12	27AU	0.3	-3.5	-1.6	1.0
149.73	1.20	2.0	170.91	6SE	0.5	-3.5	-1.5	1.0
156.06	1.21	2.0	180.74	16SE	1.3	-3.6	-1.5	1.1
162.38	1.22	2.0	190.64	26SE	3.2	-3.7	-1.4	1.1
168.70	1.22	2.0	200.60	6OC	0.3	-3.7	-1.4	1.1
175.01	1.22	2.0	210.61	16OC	-0.6	-3.8	-1.4	1.1
181.32	1.21	1.9	220.67	26OC	-0.7	-3.9	-1.4	1.1
187.61	1.19	1.9	230.78	5NO	-0.7	-4.0	-1.3	1.1
193.90	1.17	1.8	240.92	15NO	-0.8	-4.1	-1.3	1.1
200.17	1.14	1.8	251.09	25NO	-0.7	-4.3	-1.3	1.0
206.43	1.11	1.7	261.28	5DE	-0.7	-4.3	-1.3	1.0
212.66	1.06	1.6	271.48	15DE	-0.7	-4.4	-1.4	0.9
218.86	1.01	1.5	281.68	25DE	-0.2	-4.3	-1.4	0.9
				1213				
225.02	0.94	1.4	291.86	4JA	1.7	-4.0	-1.4	0.8
231.13	0.86	1.3	302.02	14JA	1.9	-3.4	-1.4	0.7
237.19	0.76	1.2	312.15	24JA	0.6	-3.5	-1.5	0.7
243.17	0.65	1.0	322.24	3FE	0.3	-4.0	-1.5	0.6
249.05	0.51	0.9	332.27	13FE	0.1	-4.3	-1.6	0.5
254.83	0.34	0.7	342.26	23FE	-0.1	-4.3	-1.6	0.6
260.45	0.15	0.5	352.20	5MR	-0.6	-4.3	-1.7	0.6
265.90	-0.08	0.3	2.07	15MR	-1.3	-4.2	-1.8	0.7
271.12	-0.36	0.1	11.89	25MR	-1.5	-4.1	-1.9	0.7
276.04	-0.68	-0.1	21.65	4AP	-0.7	-3.9	-1.9	0.8
280.59	-1.06	-0.4	31.35	14AP	0.4	-3.8	-2.0	0.8
284.66	-1.51	-0.6	41.02	24AP	1.6	-3.7	-2.1	0.9
288.11	-2.04	-0.9	50.64	4MY	3.1	-3.7	-2.1	0.9
290.79	-2.65	-1.2	60.22	14MY	2.9	-3.6	-2.2	1.0
292.50	-3.35	-1.5	69.78	24MY	1.6	-3.5	-2.2	1.0
293.06	-4.10	-1.8	79.32	3JN	0.7	-3.5	-2.2	1.1
292.39	-4.87	-2.1	88.85	13JN	-0.2	-3.4	-2.2	1.1
290.56	-5.56	-2.4	98.38	23JN	-1.3	-3.4	-2.2	1.1
287.98	-6.06	-2.6	107.92	3JL	-1.6	-3.4	-2.2	1.2
285.32	-6.30	-2.5	117.47	13JL	-0.8	-3.3	-2.2	1.2
283.27	-6.24	-2.3	127.05	23JL	-0.2	-3.3	-2.1	1.2
282.39	-5.95	-2.1	136.66	2AU	0.2	-3.3	-2.1	1.2
282.85	-5.52	-1.8	146.32	12AU	0.4	-3.3	-2.0	1.2
284.61	-5.01	-1.5	156.02	22AU	0.8	-3.3	-1.9	1.2
287.51	-4.48	-1.3	165.78	1SE	1.7	-3.4	-1.9	1.2
291.34	-3.95	-1.0	175.58	11SE	2.8	-3.4	-1.8	1.3
295.90	-3.45	-0.8	185.45	21SE	0.2	-3.4	-1.8	1.3
301.06	-2.99	-0.6	195.37	1OC	-0.8	-3.4	-1.7	1.4
306.66	-2.55	-0.4	205.35	11OC	-0.9	-3.5	-1.6	1.4
312.60	-2.15	-0.2	215.39	21OC	-0.8	-3.5	-1.6	1.4
318.81	-1.78	0.0	225.47	31OC	-0.7	-3.5	-1.6	1.3
325.21	-1.44	0.2	235.59	10NO	-0.5	-3.5	-1.5	1.3
331.75	-1.14	0.4	245.75	20NO	-0.5	-3.5	-1.5	1.3
338.41	-0.86	0.5	255.93	30NO	-0.5	-3.4	-1.5	1.2
345.12	-0.61	0.7	266.13	10DE	-0.1	-3.4	-1.4	1.2
351.89	-0.39	0.8	276.33	20DE	2.0	-3.4	-1.4	1.1
358.68	-0.18	0.9	286.52	30DE	1.5	-3.4	-1.4	1.1

♂			⊙	16.00UT	☿	♀	♃	♄
LONG	LAT	MAG	LONG	1214	MAGNITUDES			
5.47	-0.00	1.1	296.69	9JA	0.4	-3.3	-1.5	1.0
12.26	0.16	1.2	306.84	19JA	0.1	-3.3	-1.5	1.0
19.03	0.31	1.3	316.95	29JA	0.0	-3.3	-1.5	0.9
25.77	0.44	1.4	327.02	8FE	-0.2	-3.3	-1.5	0.9
32.49	0.55	1.5	337.03	18FE	-0.6	-3.4	-1.6	0.9
39.17	0.65	1.6	346.99	28FE	-1.3	-3.4	-1.6	0.9
45.82	0.74	1.6	356.90	10MR	-1.4	-3.4	-1.7	0.9
52.43	0.82	1.7	6.75	20MR	-0.7	-3.4	-1.7	0.9
59.00	0.88	1.8	16.54	30MR	0.6	-3.5	-1.8	1.0
65.54	0.94	1.8	26.27	9AP	2.1	-3.5	-1.8	1.0
72.04	0.99	1.9	35.96	19AP	3.8	-3.5	-1.9	1.1
78.51	1.04	1.9	45.60	29AP	2.2	-3.6	-2.0	1.1
84.95	1.07	2.0	55.20	9MY	1.2	-3.7	-2.0	1.2
91.36	1.10	2.0	64.77	19MY	0.5	-3.7	-2.1	1.2
97.76	1.12	2.0	74.32	29MY	-0.3	-3.8	-2.2	1.3
104.13	1.14	2.0	83.85	8JN	-1.3	-3.9	-2.2	1.3
110.50	1.15	2.0	93.38	18JN	-1.7	-4.0	-2.3	1.4
116.85	1.16	2.0	102.91	28JN	-0.8	-4.1	-2.3	1.4
123.20	1.16	2.0	112.46	8JL	-0.1	-4.2	-2.4	1.4
129.55	1.16	2.0	122.03	18JL	0.3	-4.2	-2.4	1.4
135.90	1.15	2.0	131.62	28JL	0.6	-4.1	-2.4	1.4
142.26	1.14	2.0	141.25	7AU	1.1	-3.9	-2.3	1.3
148.64	1.12	2.0	150.93	17AU	2.2	-3.4	-2.3	1.3
155.03	1.10	2.0	160.65	27AU	2.4	-3.4	-2.3	1.2
161.45	1.08	2.0	170.44	6SE	0.1	-3.9	-2.2	1.2
167.88	1.05	2.0	180.27	16SE	-0.9	-4.2	-2.1	1.2
174.34	1.01	2.0	190.16	26SE	-1.0	-4.3	-2.1	1.2
180.83	0.97	2.0	200.12	6OC	-1.0	-4.3	-2.0	1.2
187.35	0.92	2.0	210.12	16OC	-0.6	-4.2	-1.9	1.2
193.91	0.87	1.9	220.18	26OC	-0.4	-4.1	-1.9	1.2
200.49	0.82	1.9	230.28	5NO	-0.4	-4.0	-1.8	1.2
207.11	0.75	1.9	240.42	15NO	-0.3	-3.9	-1.7	1.1
213.76	0.68	1.8	250.59	25NO	0.1	-3.8	-1.7	1.1
220.45	0.60	1.8	260.78	5DE	2.2	-3.8	-1.7	1.1
227.18	0.51	1.7	270.98	15DE	1.2	-3.7	-1.6	1.0
233.93	0.41	1.7	281.18	25DE	0.2	-3.6	-1.6	1.0
				1215				
240.72	0.30	1.6	291.36	4JA	-0.0	-3.6	-1.6	0.9
247.54	0.18	1.5	301.53	14JA	-0.1	-3.5	-1.5	0.9
254.38	0.05	1.4	311.66	24JA	-0.3	-3.5	-1.5	0.8
261.26	-0.09	1.3	321.75	3FE	-0.7	-3.4	-1.5	0.8
268.15	-0.25	1.2	331.79	13FE	-1.3	-3.4	-1.5	0.8
275.06	-0.42	1.1	341.78	23FE	-1.3	-3.4	-1.6	0.7
281.99	-0.61	1.0	351.72	5MR	-0.7	-3.3	-1.6	0.7
288.92	-0.81	0.9	1.60	15MR	0.8	-3.3	-1.6	0.7
295.85	-1.03	0.8	11.41	25MR	2.6	-3.3	-1.6	0.8
302.76	-1.26	0.6	21.18	4AP	2.9	-3.3	-1.7	0.8
309.63	-1.51	0.5	30.89	14AP	1.6	-3.3	-1.7	0.9
316.45	-1.77	0.4	40.55	24AP	0.9	-3.3	-1.8	1.0
323.18	-2.04	0.2	50.18	4MY	0.4	-3.4	-1.8	1.0
329.80	-2.32	0.1	59.76	14MY	-0.3	-3.4	-1.9	1.1
336.25	-2.61	-0.1	69.32	24MY	-1.3	-3.4	-2.0	1.1
342.49	-2.91	-0.2	78.86	3JN	-1.7	-3.5	-2.1	1.2
348.45	-3.21	-0.4	88.39	13JN	-0.8	-3.5	-2.1	1.2
354.04	-3.52	-0.6	97.92	23JN	-0.0	-3.5	-2.2	1.2
359.15	-3.82	-0.8	107.45	3JL	0.4	-3.4	-2.3	1.2
3.64	-4.11	-1.0	117.01	13JL	0.8	-3.4	-2.3	1.2
7.35	-4.38	-1.2	126.58	23JL	1.4	-3.4	-2.4	1.2
10.07	-4.61	-1.5	136.20	2AU	2.7	-3.3	-2.4	1.2
11.59	-4.78	-1.7	145.85	12AU	2.1	-3.3	-2.4	1.2
11.73	-4.84	-1.9	155.55	22AU	0.1	-3.3	-2.5	1.1
10.42	-4.74	-2.2	165.30	1SE	-1.0	-3.3	-2.4	1.1
7.93	-4.44	-2.3	175.11	11SE	-1.2	-3.3	-2.4	1.1
4.83	-3.92	-2.4	184.97	21SE	-1.0	-3.3	-2.4	1.0
1.89	-3.25	-2.2	194.89	1OC	-0.5	-3.4	-2.3	1.0
359.85	-2.53	-1.9	204.87	11OC	-0.3	-3.4	-2.3	1.0
359.07	-1.83	-1.6	214.90	21OC	-0.2	-3.4	-2.2	1.0
359.61	-1.20	-1.2	224.98	31OC	-0.2	-3.5	-2.1	1.0
1.36	-0.68	-0.9	235.10	10NO	0.3	-3.5	-2.1	1.0
4.08	-0.25	-0.6	245.26	20NO	2.5	-3.6	-2.0	1.0
7.59	0.10	-0.3	255.44	30NO	0.9	-3.6	-1.9	1.0
11.72	0.38	-0.1	265.63	10DE	-0.0	-3.7	-1.9	1.0
16.31	0.60	0.2	275.83	20DE	-0.2	-3.8	-1.8	0.9
21.28	0.78	0.4	286.02	30DE	-0.2	-3.8	-1.8	0.9

♂ LONG	♂ LAT	♂ MAG	☉ LONG	16.00UT	☿	♀	♃	♄
				1216	MAGNITUDES			
26.53	0.93	0.6	296.20	9JA	-0.4	-3.9	-1.7	0.8
31.98	1.04	0.8	306.35	19JA	-0.7	-4.0	-1.7	0.8
37.61	1.13	1.0	316.46	29JA	-1.2	-4.1	-1.6	0.8
43.36	1.21	1.1	326.53	8FE	-1.2	-4.2	-1.6	0.7
49.21	1.26	1.2	336.55	18FE	-0.6	-4.3	-1.6	0.7
55.13	1.30	1.4	346.51	28FE	0.9	-4.3	-1.6	0.6
61.11	1.33	1.5	356.42	9MR	3.1	-4.2	-1.6	0.6
67.13	1.35	1.6	6.27	19MR	2.2	-3.8	-1.6	0.5
73.19	1.36	1.7	16.06	29MR	1.2	-3.2	-1.6	0.6
79.27	1.37	1.7	25.80	8AP	0.7	-3.4	-1.6	0.6
85.37	1.36	1.8	35.49	18AP	0.2	-3.9	-1.6	0.7
91.49	1.35	1.9	45.13	28AP	-0.4	-4.2	-1.7	0.8
97.63	1.34	1.9	54.74	8MY	-1.3	-4.2	-1.7	0.8
103.79	1.32	2.0	64.31	18MY	-1.7	-4.2	-1.8	0.9
109.96	1.30	2.0	73.86	28MY	-0.8	-4.1	-1.8	0.9
116.15	1.27	2.0	83.39	7JN	0.0	-4.0	-1.9	1.0
122.37	1.24	2.0	92.92	17JN	0.6	-3.9	-1.9	1.0
128.60	1.21	2.0	102.45	27JN	1.1	-3.8	-2.0	1.1
134.87	1.17	2.0	111.99	7JL	1.9	-3.7	-2.1	1.1
141.16	1.13	2.0	121.56	17JL	3.2	-3.6	-2.1	1.1
147.49	1.08	2.0	131.16	27JL	1.7	-3.6	-2.2	1.1
153.86	1.03	2.0	140.78	6AU	-0.0	-3.5	-2.3	1.1
160.26	0.98	2.0	150.46	16AU	-1.1	-3.5	-2.3	1.1
166.72	0.92	1.9	160.19	26AU	-1.3	-3.4	-2.4	1.0
173.22	0.86	1.9	169.96	5SE	-0.9	-3.4	-2.4	1.0
179.77	0.80	1.9	179.80	15SE	-0.5	-3.4	-2.4	1.0
186.37	0.73	1.9	189.69	25SE	-0.2	-3.4	-2.5	0.9
193.04	0.65	1.9	199.64	5OC	-0.1	-3.4	-2.5	0.9
199.76	0.58	1.9	209.64	15OC	0.0	-3.4	-2.4	0.9
206.54	0.49	1.9	219.70	25OC	0.5	-3.4	-2.4	0.9
213.39	0.41	1.8	229.79	4NO	2.9	-3.4	-2.4	0.9
220.30	0.31	1.8	239.94	14NO	0.7	-3.4	-2.3	0.9
227.27	0.22	1.8	250.10	24NO	-0.2	-3.4	-2.2	0.9
234.31	0.11	1.8	260.29	4DE	-0.3	-3.4	-2.2	0.9
241.41	0.01	1.7	270.49	14DE	-0.4	-3.4	-2.1	0.9
248.57	-0.11	1.7	280.68	24DE	-0.5	-3.4	-2.0	0.9
				1217				
255.79	-0.23	1.6	290.87	3JA	-0.8	-3.5	-1.9	0.8
263.08	-0.35	1.6	301.03	13JA	-1.1	-3.5	-1.9	0.8
270.42	-0.48	1.5	311.16	23JA	-1.1	-3.5	-1.8	0.7
277.80	-0.61	1.5	321.25	2FE	-0.5	-3.5	-1.8	0.7
285.24	-0.74	1.4	331.30	12FE	1.1	-3.4	-1.7	0.6
292.72	-0.87	1.3	341.29	22FE	3.2	-3.4	-1.7	0.6
300.22	-1.01	1.3	351.23	4MR	1.6	-3.4	-1.6	0.5
307.76	-1.14	1.2	1.11	14MR	0.9	-3.4	-1.6	0.5
315.30	-1.27	1.2	10.93	24MR	0.5	-3.3	-1.6	0.4
322.85	-1.39	1.1	20.70	3AP	0.1	-3.3	-1.6	0.4
330.39	-1.51	1.0	30.42	13AP	-0.4	-3.3	-1.6	0.5
337.89	-1.62	0.9	40.08	23AP	-1.3	-3.3	-1.6	0.5
345.36	-1.71	0.9	49.70	3MY	-1.7	-3.3	-1.6	0.6
352.76	-1.80	0.8	59.29	13MY	-0.8	-3.4	-1.6	0.7
0.07	-1.86	0.7	68.85	23MY	0.1	-3.4	-1.6	0.7
7.29	-1.91	0.7	78.39	2JN	0.8	-3.4	-1.6	0.8
14.36	-1.94	0.6	87.93	12JN	1.5	-3.4	-1.7	0.8
21.28	-1.96	0.5	97.45	22JN	2.5	-3.5	-1.7	0.9
28.00	-1.95	0.4	106.99	2JL	3.1	-3.5	-1.7	0.9
34.49	-1.92	0.3	116.55	12JL	1.4	-3.6	-1.8	0.9
40.71	-1.86	0.2	126.12	22JL	-0.1	-3.7	-1.9	1.0
46.59	-1.78	0.1	135.73	1AU	-1.1	-3.8	-1.9	1.0
52.07	-1.68	-0.0	145.39	11AU	-1.4	-3.9	-2.0	1.0
57.05	-1.54	-0.1	155.08	21AU	-0.9	-4.0	-2.0	1.0
61.44	-1.36	-0.3	164.83	31AU	-0.4	-4.1	-2.1	0.9
65.08	-1.15	-0.5	174.64	10SE	-0.1	-4.2	-2.2	0.9
67.81	-0.88	-0.7	184.49	20SE	0.1	-4.3	-2.2	0.9
69.41	-0.56	-0.9	194.41	30SE	0.2	-4.3	-2.3	0.8
69.69	-0.17	-1.1	204.39	10OC	0.8	-4.3	-2.3	0.8
68.51	0.28	-1.3	214.42	20OC	3.2	-4.0	-2.4	0.8
65.96	0.77	-1.5	224.49	30OC	0.6	-3.3	-2.4	0.8
62.40	1.25	-1.6	234.61	9NO	-0.4	-3.4	-2.4	0.8
58.60	1.67	-1.6	244.77	19NO	-0.5	-4.0	-2.4	0.8
55.31	2.00	-1.3	254.95	29NO	-0.5	-4.3	-2.3	0.8
53.15	2.21	-1.0	265.14	9DE	-0.6	-4.4	-2.3	0.8
52.34	2.32	-0.7	275.34	19DE	-0.8	-4.3	-2.2	0.8
52.82	2.36	-0.4	285.53	29DE	-0.9	-4.2	-2.2	0.8

♂ LONG	♂ LAT	♂ MAG	☉ LONG	16.00UT	☿	♀	♃	♄
				1218	MAGNITUDES			
54.44	2.35	-0.1	295.71	8JA	-0.9	-4.1	-2.1	0.8
56.97	2.32	0.2	305.85	18JA	-0.5	-4.0	-2.0	0.8
60.22	2.26	0.4	315.97	28JA	1.3	-3.9	-1.9	0.7
64.05	2.20	0.6	326.04	7FE	2.7	-3.8	-1.9	0.7
68.32	2.13	0.8	336.06	17FE	1.2	-3.7	-1.8	0.6
72.94	2.06	1.0	346.03	27FE	0.6	-3.6	-1.7	0.6
77.84	1.98	1.1	355.94	9MR	0.3	-3.6	-1.7	0.5
82.96	1.91	1.3	5.79	19MR	0.0	-3.5	-1.6	0.5
88.26	1.83	1.4	15.59	29MR	-0.5	-3.5	-1.6	0.4
93.70	1.76	1.5	25.33	8AP	-1.3	-3.4	-1.5	0.4
99.26	1.68	1.6	35.02	18AP	-1.6	-3.4	-1.5	0.3
104.93	1.61	1.7	44.67	28AP	-0.8	-3.3	-1.5	0.4
110.69	1.53	1.8	54.28	8MY	0.2	-3.3	-1.5	0.5
116.53	1.46	1.8	63.85	18MY	1.1	-3.3	-1.5	0.5
122.44	1.38	1.9	73.40	28MY	2.0	-3.3	-1.5	0.6
128.43	1.31	1.9	82.93	7JN	3.2	-3.3	-1.5	0.7
134.47	1.23	1.9	92.46	17JN	2.5	-3.3	-1.5	0.7
140.59	1.15	2.0	101.99	27JN	1.1	-3.3	-1.5	0.8
146.77	1.07	2.0	111.54	7JL	-0.1	-3.4	-1.5	0.8
153.01	0.99	2.0	121.10	17JL	-1.2	-3.4	-1.6	0.8
159.32	0.90	2.0	130.69	27JL	-1.6	-3.4	-1.6	0.9
165.70	0.82	2.0	140.32	6AU	-0.9	-3.5	-1.6	0.9
172.15	0.73	1.9	149.99	16AU	-0.3	-3.5	-1.7	0.9
178.68	0.65	1.9	159.71	26AU	0.0	-3.5	-1.7	0.9
185.28	0.56	1.9	169.49	5SE	0.2	-3.5	-1.8	0.9
191.95	0.46	1.9	179.32	15SE	0.4	-3.4	-1.9	0.9
198.71	0.37	1.8	189.21	25SE	1.1	-3.4	-1.9	0.8
205.55	0.27	1.8	199.15	5OC	3.4	-3.4	-2.0	0.8
212.48	0.18	1.7	209.15	15OC	0.4	-3.3	-2.1	0.8
219.48	0.08	1.7	219.20	25OC	-0.5	-3.3	-2.1	0.7
226.57	-0.03	1.6	229.30	4NO	-0.6	-3.3	-2.2	0.7
233.74	-0.13	1.6	239.43	14NO	-0.6	-3.3	-2.2	0.8
241.00	-0.23	1.6	249.61	24NO	-0.7	-3.3	-2.2	0.8
248.33	-0.34	1.6	259.79	4DE	-0.7	-3.4	-2.3	0.8
255.74	-0.44	1.6	269.99	14DE	-0.8	-3.4	-2.3	0.8
263.22	-0.54	1.6	280.19	24DE	-0.8	-3.4	-2.2	0.8
				1219				
270.76	-0.64	1.6	290.37	3JA	-0.3	-3.4	-2.2	0.8
278.37	-0.74	1.5	300.54	13JA	1.5	-3.5	-2.2	0.8
286.02	-0.84	1.5	310.67	23JA	2.2	-3.5	-2.1	0.7
293.73	-0.93	1.5	320.77	2FE	0.8	-3.6	-2.0	0.7
301.46	-1.01	1.4	330.81	12FE	0.4	-3.6	-2.0	0.7
309.23	-1.09	1.4	340.81	22FE	0.2	-3.7	-1.9	0.6
317.00	-1.15	1.4	350.75	4MR	-0.1	-3.7	-1.8	0.6
324.78	-1.21	1.4	0.63	14MR	-0.5	-3.8	-1.8	0.5
332.54	-1.25	1.3	10.46	24MR	-1.3	-3.9	-1.7	0.4
340.29	-1.29	1.3	20.23	3AP	-1.6	-4.0	-1.6	0.4
347.99	-1.31	1.3	29.95	13AP	-0.8	-4.1	-1.6	0.3
355.65	-1.31	1.3	39.62	23AP	0.3	-4.2	-1.5	0.3
3.24	-1.30	1.2	49.24	3MY	1.4	-4.2	-1.5	0.3
10.76	-1.27	1.2	58.83	13MY	2.6	-4.2	-1.4	0.4
18.20	-1.23	1.2	68.39	23MY	3.4	-3.9	-1.4	0.4
25.53	-1.18	1.2	77.93	2JN	2.0	-3.4	-1.4	0.5
32.75	-1.11	1.2	87.46	12JN	0.9	-2.8	-1.4	0.6
39.84	-1.02	1.1	96.99	22JN	-0.2	-3.6	-1.4	0.6
46.79	-0.93	1.1	106.53	2JL	-1.2	-4.1	-1.4	0.7
53.60	-0.81	1.1	116.08	12JL	-1.6	-4.2	-1.4	0.7
60.23	-0.69	1.0	125.66	22JL	-0.8	-4.2	-1.4	0.8
66.67	-0.54	1.0	135.26	1AU	-0.2	-4.1	-1.4	0.8
72.90	-0.39	0.9	144.91	11AU	0.1	-4.1	-1.4	0.8
78.88	-0.21	0.8	154.61	21AU	0.3	-4.0	-1.4	0.8
84.58	-0.02	0.7	164.35	31AU	0.6	-3.9	-1.5	0.8
89.96	0.19	0.6	174.16	10SE	1.4	-3.8	-1.5	0.8
94.95	0.43	0.5	184.02	20SE	3.1	-3.7	-1.6	0.8
99.47	0.70	0.4	193.93	30SE	0.3	-3.7	-1.6	0.8
103.42	0.99	0.3	203.90	10OC	-0.7	-3.6	-1.7	0.8
106.66	1.33	0.1	213.93	20OC	-0.8	-3.6	-1.8	0.7
109.04	1.71	-0.1	224.00	30OC	-0.8	-3.5	-1.8	0.7
110.36	2.13	-0.3	234.12	9NO	-0.8	-3.5	-1.9	0.7
110.43	2.58	-0.6	244.27	19NO	-0.6	-3.4	-2.0	0.7
109.13	3.05	-0.8	254.45	29NO	-0.6	-3.4	-2.0	0.7
106.49	3.48	-1.0	264.65	9DE	-0.6	-3.4	-2.1	0.8
102.86	3.82	-1.1	274.85	19DE	-0.2	-3.4	-2.1	0.8
98.89	4.00	-1.1	285.04	29DE	1.7	-3.4	-2.1	0.8

♂ LONG	♂ LAT	♂ MAG	☉ LONG	16.00UT 1220	☿	♀	♃	♄
					MAGNITUDES			
95.33	4.02	-0.9	295.22	8JA	1.8	-3.3	-2.1	0.8
92.80	3.91	-0.6	305.37	18JA	0.5	-3.3	-2.1	0.8
91.56	3.70	-0.4	315.48	28JA	0.2	-3.3	-2.1	0.7
91.61	3.46	-0.1	325.55	7FE	0.1	-3.3	-2.1	0.7
92.81	3.20	0.2	335.58	17FE	-0.2	-3.4	-2.0	0.7
94.96	2.96	0.4	345.55	27FE	-0.6	-3.4	-2.0	0.6
97.88	2.72	0.6	355.46	8MR	-1.3	-3.4	-1.9	0.6
101.41	2.51	0.8	5.32	18MR	-1.5	-3.4	-1.8	0.5
105.42	2.31	1.0	15.11	28MR	-0.7	-3.5	-1.8	0.5
109.83	2.12	1.1	24.86	7AP	0.5	-3.5	-1.7	0.4
114.55	1.95	1.2	34.55	17AP	1.8	-3.4	-1.6	0.4
119.53	1.79	1.4	44.19	27AP	3.4	-3.4	-1.6	0.3
124.74	1.63	1.5	53.80	7MY	2.6	-3.4	-1.5	0.2
130.13	1.49	1.5	63.38	17MY	1.5	-3.4	-1.5	0.3
135.68	1.35	1.6	72.93	27MY	0.7	-3.3	-1.4	0.3
141.38	1.22	1.7	82.47	6JN	-0.2	-3.3	-1.4	0.4
147.21	1.09	1.7	91.99	16JN	-1.2	-3.3	-1.3	0.5
153.16	0.97	1.7	101.52	26JN	-1.7	-3.3	-1.3	0.5
159.23	0.84	1.8	111.07	6JL	-0.8	-3.3	-1.3	0.6
165.42	0.73	1.8	120.63	16JL	-0.2	-3.3	-1.3	0.6
171.70	0.61	1.8	130.22	26JL	0.2	-3.4	-1.3	0.7
178.10	0.49	1.8	139.85	5AU	0.5	-3.4	-1.3	0.7
184.60	0.38	1.8	149.52	15AU	0.9	-3.4	-1.3	0.8
191.20	0.27	1.8	159.24	25AU	1.9	-3.5	-1.3	0.8
197.91	0.16	1.8	169.01	4SE	2.7	-3.5	-1.3	0.8
204.72	0.05	1.8	178.84	14SE	0.2	-3.6	-1.3	0.8
211.64	-0.06	1.7	188.72	24SE	-0.8	-3.7	-1.4	0.8
218.66	-0.16	1.7	198.67	4OC	-0.9	-3.7	-1.4	0.8
225.78	-0.27	1.7	208.67	14OC	-0.9	-3.8	-1.5	0.8
232.99	-0.37	1.6	218.71	24OC	-0.7	-3.9	-1.5	0.7
240.31	-0.47	1.6	228.81	3NO	-0.5	-4.0	-1.6	0.7
247.72	-0.56	1.5	238.95	13NO	-0.5	-4.2	-1.6	0.7
255.21	-0.65	1.5	249.11	23NO	-0.4	-4.3	-1.7	0.7
262.79	-0.73	1.4	259.30	3DE	-0.0	-4.4	-1.8	0.7
270.43	-0.81	1.4	269.49	13DE	2.0	-4.4	-1.8	0.7
278.15	-0.88	1.4	279.69	23DE	1.4	-4.3	-1.9	0.8
				1221				
285.92	-0.94	1.4	289.88	2JA	0.3	-4.0	-1.9	0.8
293.73	-1.00	1.4	300.04	12JA	0.1	-3.4	-2.0	0.8
301.58	-1.04	1.4	310.18	22JA	-0.0	-3.5	-2.0	0.8
309.44	-1.07	1.4	320.28	1FE	-0.2	-4.0	-2.0	0.8
317.32	-1.10	1.4	330.33	11FE	-0.6	-4.3	-2.0	0.7
325.19	-1.11	1.4	340.32	21FE	-1.3	-4.3	-2.0	0.7
333.05	-1.10	1.4	350.27	3MR	-1.4	-4.3	-2.0	0.7
340.87	-1.09	1.4	0.15	13MR	-0.7	-4.2	-2.0	0.6
348.66	-1.06	1.4	9.98	23MR	0.6	-4.1	-1.9	0.6
356.40	-1.03	1.4	19.76	2AP	2.2	-3.9	-1.9	0.5
4.07	-0.98	1.4	29.47	12AP	3.5	-3.8	-1.8	0.4
11.67	-0.92	1.4	39.15	22AP	2.0	-3.7	-1.7	0.4
19.19	-0.84	1.5	48.78	2MY	1.1	-3.7	-1.7	0.3
26.62	-0.76	1.5	58.36	12MY	0.5	-3.6	-1.6	0.3
33.96	-0.67	1.5	67.93	22MY	-0.3	-3.5	-1.5	0.2
41.20	-0.57	1.5	77.47	1JN	-1.3	-3.5	-1.5	0.3
48.34	-0.47	1.5	87.00	11JN	-1.7	-3.4	-1.4	0.3
55.37	-0.35	1.5	96.53	21JN	-0.8	-3.4	-1.4	0.4
62.28	-0.23	1.5	106.07	1JL	-0.1	-3.4	-1.3	0.5
69.08	-0.10	1.5	115.62	11JL	0.4	-3.3	-1.3	0.5
75.75	0.03	1.5	125.19	21JL	0.7	-3.3	-1.3	0.6
82.29	0.17	1.5	134.80	31JL	1.2	-3.3	-1.2	0.6
88.69	0.31	1.4	144.45	10AU	2.4	-3.3	-1.2	0.7
94.95	0.47	1.4	154.14	20AU	2.4	-3.3	-1.2	0.7
101.03	0.63	1.3	163.89	30AU	0.1	-3.4	-1.2	0.8
106.94	0.80	1.3	173.68	9SE	-0.9	-3.4	-1.2	0.8
112.64	0.98	1.2	183.54	19SE	-1.1	-3.4	-1.2	0.8
118.10	1.17	1.1	193.45	29SE	-1.0	-3.4	-1.2	0.8
123.28	1.38	1.0	203.42	9OC	-0.6	-3.5	-1.3	0.8
128.14	1.61	0.9	213.44	19OC	-0.4	-3.5	-1.3	0.8
132.59	1.85	0.8	223.52	29OC	-0.3	-3.5	-1.3	0.7
136.57	2.13	0.6	233.63	8NO	-0.3	-3.5	-1.4	0.7
139.94	2.43	0.5	243.78	18NO	0.1	-3.5	-1.4	0.7
142.58	2.76	0.3	253.96	28NO	2.3	-3.4	-1.5	0.7
144.31	3.12	0.1	264.15	8DE	1.1	-3.4	-1.6	0.7
144.94	3.50	-0.2	274.35	18DE	0.1	-3.4	-1.6	0.7
144.32	3.88	-0.4	284.54	28DE	-0.1	-3.4	-1.7	0.8

♂ LONG	♂ LAT	♂ MAG	☉ LONG	16.00UT 1222	☿	♀	♃	♄
					MAGNITUDES			
142.37	4.22	-0.7	294.72	7JA	-0.2	-3.3	-1.7	0.8
139.24	4.47	-0.9	304.87	17JA	-0.3	-3.3	-1.8	0.8
135.39	4.56	-1.0	314.99	27JA	-0.7	-3.3	-1.9	0.8
131.51	4.47	-0.9	325.06	6FE	-1.2	-3.3	-1.9	0.8
128.30	4.23	-0.7	335.09	16FE	-1.3	-3.4	-2.0	0.8
126.23	3.89	-0.4	345.06	26FE	-0.7	-3.4	-2.0	0.7
125.45	3.51	-0.2	354.98	8MR	0.8	-3.4	-2.0	0.7
125.91	3.12	0.0	4.84	18MR	2.8	-3.4	-2.0	0.7
127.45	2.76	0.3	14.64	28MR	2.6	-3.5	-2.0	0.6
129.87	2.43	0.5	24.39	7AP	1.4	-3.5	-2.0	0.6
133.02	2.12	0.6	34.09	17AP	0.8	-3.5	-1.9	0.5
136.77	1.85	0.8	43.73	27AP	0.3	-3.6	-1.9	0.4
140.99	1.60	0.9	53.34	7MY	-0.3	-3.7	-1.8	0.4
145.60	1.37	1.1	62.92	17MY	-1.3	-3.7	-1.8	0.3
150.55	1.16	1.2	72.47	27MY	-1.7	-3.8	-1.7	0.3
155.77	0.96	1.2	82.01	6JN	-0.8	-3.9	-1.6	0.2
161.24	0.78	1.3	91.54	16JN	-0.0	-4.0	-1.6	0.3
166.93	0.61	1.4	101.07	26JN	0.5	-4.1	-1.5	0.4
172.81	0.45	1.4	110.61	6JL	0.9	-4.2	-1.4	0.4
178.87	0.30	1.5	120.17	16JL	1.6	-4.2	-1.4	0.5
185.09	0.16	1.5	129.76	26JL	2.9	-4.1	-1.3	0.5
191.47	0.02	1.5	139.38	5AU	2.0	-3.9	-1.3	0.6
198.00	-0.11	1.5	149.05	15AU	0.1	-3.3	-1.3	0.6
204.66	-0.23	1.5	158.77	25AU	-1.0	-3.4	-1.2	0.7
211.47	-0.35	1.5	168.54	4SE	-1.2	-3.9	-1.2	0.7
218.41	-0.46	1.5	178.36	14SE	-1.0	-4.2	-1.2	0.8
225.47	-0.56	1.5	188.24	24SE	-0.5	-4.3	-1.2	0.8
232.65	-0.66	1.5	198.18	4OC	-0.3	-4.3	-1.2	0.8
239.95	-0.75	1.5	208.18	14OC	-0.2	-4.2	-1.2	0.8
247.36	-0.83	1.5	218.23	24OC	-0.1	-4.1	-1.2	0.8
254.86	-0.90	1.5	228.32	3NO	0.3	-4.0	-1.3	0.8
262.46	-0.96	1.5	238.46	13NO	2.6	-3.9	-1.3	0.7
270.14	-1.01	1.4	248.62	23NO	0.9	-3.8	-1.3	0.7
277.89	-1.05	1.4	258.81	3DE	-0.1	-3.8	-1.4	0.7
285.70	-1.08	1.4	269.01	13DE	-0.2	-3.7	-1.4	0.7
293.55	-1.09	1.4	279.20	23DE	-0.3	-3.6	-1.5	0.7
				1223				
301.44	-1.10	1.4	289.39	2JA	-0.4	-3.6	-1.5	0.8
309.35	-1.09	1.4	299.56	12JA	-0.7	-3.5	-1.6	0.8
317.25	-1.08	1.3	309.69	22JA	-1.1	-3.5	-1.7	0.8
325.15	-1.05	1.3	319.79	1FE	-1.2	-3.4	-1.7	0.8
333.03	-1.01	1.3	329.85	11FE	-0.6	-3.4	-1.8	0.8
340.87	-0.95	1.3	339.85	21FE	0.9	-3.4	-1.9	0.8
348.66	-0.89	1.3	349.79	3MR	3.2	-3.3	-1.9	0.8
356.40	-0.82	1.4	359.68	13MR	1.9	-3.3	-2.0	0.8
4.07	-0.75	1.4	9.51	23MR	1.1	-3.3	-2.0	0.7
11.67	-0.66	1.4	19.29	2AP	0.6	-3.3	-2.0	0.7
19.19	-0.57	1.5	29.01	12AP	0.2	-3.3	-2.0	0.6
26.63	-0.47	1.5	38.68	22AP	-0.4	-3.3	-2.0	0.6
33.97	-0.37	1.6	48.31	2MY	-1.3	-3.4	-2.0	0.5
41.23	-0.27	1.6	57.91	12MY	-1.7	-3.4	-2.0	0.4
48.40	-0.16	1.6	67.47	22MY	-0.8	-3.4	-1.9	0.4
55.47	-0.05	1.7	77.01	1JN	0.1	-3.5	-1.9	0.3
62.45	0.06	1.7	86.54	11JN	0.7	-3.5	-1.8	0.3
69.34	0.17	1.7	96.07	21JN	1.2	-3.5	-1.8	0.3
76.14	0.28	1.7	105.60	1JL	2.1	-3.4	-1.7	0.3
82.84	0.40	1.7	115.15	11JL	3.3	-3.4	-1.6	0.4
89.45	0.51	1.8	124.72	21JL	1.7	-3.4	-1.6	0.5
95.97	0.63	1.8	134.33	31JL	0.0	-3.3	-1.5	0.5
102.40	0.74	1.7	143.98	10AU	-1.1	-3.3	-1.5	0.6
108.72	0.86	1.7	153.66	20AU	-1.4	-3.3	-1.4	0.6
114.95	0.98	1.7	163.41	30AU	-0.9	-3.3	-1.4	0.7
121.07	1.10	1.7	173.20	9SE	-0.4	-3.3	-1.3	0.7
127.07	1.22	1.6	183.05	19SE	-0.2	-3.3	-1.3	0.7
132.94	1.35	1.6	192.96	29SE	-0.0	-3.4	-1.3	0.8
138.66	1.49	1.5	202.93	9OC	0.1	-3.4	-1.3	0.8
144.22	1.63	1.5	212.95	19OC	0.6	-3.4	-1.3	0.8
149.58	1.77	1.4	223.02	29OC	2.9	-3.5	-1.3	0.8
154.71	1.93	1.3	233.14	8NO	0.7	-3.5	-1.3	0.8
159.57	2.09	1.1	243.28	18NO	-0.3	-3.6	-1.3	0.8
164.09	2.27	1.0	253.46	28NO	-0.4	-3.6	-1.3	0.7
168.20	2.47	0.8	263.66	8DE	-0.4	-3.7	-1.3	0.7
171.81	2.67	0.6	273.85	18DE	-0.5	-3.8	-1.4	0.7
174.79	2.90	0.4	284.05	28DE	-0.8	-3.9	-1.4	0.7

♂ LONG	LAT	MAG	⊙ LONG	16.00UT	☿	♀	♃	♄
				1224	MAGNITUDES			
176.99	3.13	0.2	294.23	7JA	-1.0	-3.9	-1.4	0.8
178.24	3.37	-0.0	304.38	17JA	-1.0	-4.0	-1.5	0.8
178.35	3.60	-0.3	314.51	27JA	-0.5	-4.1	-1.6	0.8
177.20	3.79	-0.6	324.58	6FE	1.1	-4.2	-1.6	0.8
174.79	3.90	-0.8	334.61	16FE	3.0	-4.3	-1.7	0.8
171.37	3.88	-1.1	344.59	26FE	1.4	-4.3	-1.8	0.8
167.50	3.70	-1.1	354.50	7MR	0.8	-4.2	-1.8	0.8
163.90	3.38	-0.9	4.36	17MR	0.4	-3.8	-1.9	0.8
161.19	2.97	-0.7	14.17	27MR	0.1	-3.2	-2.0	0.8
159.71	2.52	-0.5	23.92	6AP	-0.4	-3.4	-2.0	0.7
159.53	2.08	-0.3	33.62	16AP	-1.2	-4.0	-2.1	0.7
160.56	1.66	-0.1	43.27	26AP	-1.7	-4.2	-2.1	0.6
162.63	1.29	0.1	52.88	6MY	-0.8	-4.2	-2.1	0.6
165.54	0.96	0.3	62.46	16MY	0.1	-4.2	-2.1	0.5
169.17	0.67	0.4	72.01	26MY	0.9	-4.1	-2.1	0.5
173.38	0.41	0.6	81.55	5JN	1.6	-4.0	-2.1	0.4
178.06	0.18	0.7	91.07	15JN	2.8	-3.9	-2.1	0.3
183.15	-0.03	0.8	100.61	25JN	2.9	-3.8	-2.0	0.3
188.59	-0.22	0.9	110.14	5JL	1.3	-3.7	-2.0	0.3
194.33	-0.39	0.9	119.70	15JL	-0.1	-3.6	-1.9	0.4
200.34	-0.54	1.0	129.30	25JL	-1.1	-3.6	-1.8	0.4
206.59	-0.68	1.1	138.92	4AU	-1.5	-3.5	-1.8	0.5
213.04	-0.80	1.1	148.58	14AU	-0.9	-3.5	-1.7	0.5
219.70	-0.91	1.1	158.30	24AU	-0.4	-3.4	-1.7	0.6
226.54	-1.01	1.2	168.06	3SE	-0.1	-3.4	-1.6	0.6
233.54	-1.09	1.2	177.88	13SE	0.1	-3.4	-1.5	0.7
240.70	-1.16	1.2	187.76	23SE	0.3	-3.4	-1.5	0.7
247.99	-1.21	1.2	197.70	3OC	0.9	-3.4	-1.5	0.8
255.40	-1.25	1.3	207.69	13OC	3.2	-3.4	-1.4	0.8
262.93	-1.27	1.3	217.74	23OC	0.6	-3.4	-1.4	0.8
270.55	-1.29	1.3	227.83	2NO	-0.4	-3.4	-1.4	0.8
278.25	-1.28	1.3	237.96	12NO	-0.5	-3.4	-1.4	0.8
286.02	-1.27	1.3	248.13	22NO	-0.5	-3.4	-1.3	0.8
293.84	-1.24	1.3	258.31	2DE	-0.6	-3.4	-1.4	0.8
301.69	-1.20	1.4	268.51	12DE	-0.8	-3.4	-1.4	0.8
309.56	-1.15	1.4	278.71	22DE	-0.8	-3.4	-1.4	0.8
				1225				
317.43	-1.09	1.4	288.89	1JA	-0.9	-3.5	-1.4	0.7
325.29	-1.01	1.4	299.06	11JA	-0.4	-3.5	-1.4	0.8
333.12	-0.93	1.4	309.20	21JA	1.3	-3.5	-1.5	0.8
340.91	-0.85	1.4	319.30	31JA	2.5	-3.5	-1.5	0.9
348.65	-0.75	1.5	329.36	10FE	1.0	-3.4	-1.6	0.9
356.34	-0.65	1.5	339.36	20FE	0.5	-3.4	-1.6	0.9
3.95	-0.55	1.5	349.31	2MR	0.3	-3.4	-1.7	0.9
11.49	-0.45	1.5	359.20	12MR	0.0	-3.4	-1.7	0.9
18.95	-0.34	1.5	9.04	22MR	-0.5	-3.3	-1.8	0.9
26.32	-0.23	1.6	18.81	1AP	-1.2	-3.3	-1.9	0.9
33.61	-0.13	1.6	28.54	11AP	-1.6	-3.3	-2.0	0.8
40.82	-0.02	1.6	38.21	21AP	-0.8	-3.3	-2.0	0.8
47.94	0.09	1.6	47.84	1MY	0.2	-3.3	-2.1	0.7
54.98	0.19	1.6	57.44	11MY	1.1	-3.4	-2.1	0.7
61.94	0.29	1.7	67.01	21MY	2.2	-3.4	-2.2	0.6
68.81	0.39	1.7	76.55	31MY	3.5	-3.4	-2.2	0.6
75.61	0.48	1.8	86.08	10JN	2.3	-3.5	-2.2	0.5
82.34	0.58	1.8	95.61	20JN	1.1	-3.5	-2.3	0.4
89.00	0.67	1.9	105.14	30JN	-0.1	-3.6	-2.2	0.4
95.59	0.75	1.9	114.69	10JL	-1.2	-3.6	-2.2	0.3
102.12	0.84	1.9	124.27	20JL	-1.6	-3.7	-2.2	0.4
108.58	0.92	1.9	133.87	30JL	-0.9	-3.8	-2.1	0.4
114.99	1.00	1.9	143.51	9AU	-0.3	-3.9	-2.1	0.5
121.34	1.08	1.9	153.20	19AU	0.1	-4.0	-2.0	0.5
127.62	1.15	1.9	162.93	29AU	0.3	-4.1	-2.0	0.6
133.85	1.23	1.9	172.73	8SE	0.5	-4.2	-1.9	0.6
140.02	1.30	1.9	182.58	18SE	1.2	-4.3	-1.8	0.7
146.12	1.37	1.9	192.48	28SE	3.4	-4.3	-1.8	0.7
152.15	1.44	1.8	202.45	8OC	0.4	-4.3	-1.7	0.8
158.10	1.51	1.8	212.46	18OC	-0.6	-3.9	-1.6	0.8
163.96	1.58	1.7	222.53	28OC	-0.7	-3.3	-1.6	0.8
169.71	1.65	1.6	232.64	7NO	-0.7	-3.4	-1.6	0.9
175.35	1.72	1.5	242.79	17NO	-0.7	-4.1	-1.5	0.9
180.84	1.79	1.4	252.96	27NO	-0.7	-4.3	-1.5	0.9
186.17	1.86	1.3	263.16	7DE	-0.7	-4.4	-1.5	0.9
191.29	1.93	1.2	273.36	17DE	-0.7	-4.3	-1.5	0.8
196.16	2.00	1.0	283.55	27DE	-0.3	-4.2	-1.5	0.8

♂ LONG	LAT	MAG	⊙ LONG	16.00UT	☿	♀	♃	♄
				1226	MAGNITUDES			
200.73	2.07	0.9	293.74	6JA	1.5	-4.1	-1.5	0.8
204.91	2.13	0.7	303.89	16JA	2.1	-4.0	-1.5	0.8
208.64	2.19	0.5	314.01	26JA	0.7	-3.9	-1.5	0.9
211.77	2.23	0.2	324.10	5FE	0.3	-3.8	-1.5	0.9
214.16	2.25	-0.0	334.13	15FE	0.2	-3.7	-1.5	0.9
215.66	2.25	-0.3	344.10	25FE	-0.1	-3.6	-1.6	1.0
216.07	2.20	-0.6	354.03	7MR	-0.5	-3.6	-1.6	1.0
215.25	2.08	-0.9	3.89	17MR	-1.2	-3.5	-1.7	1.0
213.18	1.87	-1.2	13.70	27MR	-1.6	-3.5	-1.7	1.0
210.06	1.56	-1.5	23.46	6AP	-0.8	-3.4	-1.8	1.0
206.42	1.16	-1.5	33.15	16AP	0.4	-3.4	-1.9	0.9
202.97	0.70	-1.4	42.81	26AP	1.5	-3.3	-1.9	0.9
200.36	0.24	-1.2	52.42	6MY	2.9	-3.3	-2.0	0.9
199.01	-0.19	-1.0	62.00	16MY	3.1	-3.3	-2.1	0.8
199.01	-0.57	-0.8	71.55	26MY	1.8	-3.3	-2.1	0.7
200.28	-0.88	-0.6	81.09	5JN	0.8	-3.3	-2.2	0.7
202.66	-1.14	-0.4	90.62	15JN	-0.2	-3.3	-2.3	0.6
205.96	-1.35	-0.3	100.15	25JN	-1.2	-3.3	-2.3	0.6
210.02	-1.52	-0.1	109.69	5JL	-1.7	-3.4	-2.4	0.5
214.71	-1.65	0.0	119.24	15JL	-0.8	-3.4	-2.4	0.4
219.93	-1.75	0.2	128.83	25JL	-0.2	-3.4	-2.4	0.4
225.58	-1.83	0.3	138.45	4AU	0.2	-3.5	-2.4	0.5
231.61	-1.88	0.4	148.11	14AU	0.4	-3.5	-2.3	0.5
237.95	-1.91	0.5	157.82	24AU	0.7	-3.5	-2.3	0.6
244.56	-1.92	0.5	167.59	3SE	1.6	-3.5	-2.3	0.6
251.40	-1.91	0.6	177.40	13SE	3.0	-3.4	-2.2	0.7
258.44	-1.88	0.7	187.28	23SE	0.3	-3.4	-2.1	0.7
265.65	-1.83	0.8	197.22	3OC	-0.7	-3.4	-2.1	0.8
273.00	-1.77	0.8	207.20	13OC	-0.9	-3.3	-2.0	0.8
280.47	-1.69	0.9	217.24	23OC	-0.8	-3.3	-1.9	0.8
288.02	-1.60	0.9	227.33	2NO	-0.7	-3.3	-1.9	0.9
295.64	-1.49	1.0	237.46	12NO	-0.6	-3.3	-1.8	0.9
303.31	-1.38	1.1	247.62	22NO	-0.5	-3.3	-1.7	0.9
311.01	-1.26	1.1	257.81	2DE	-0.5	-3.4	-1.7	0.9
318.71	-1.14	1.2	268.00	12DE	-0.2	-3.4	-1.7	0.9
326.41	-1.01	1.2	278.20	22DE	1.7	-3.4	-1.6	0.9
				1227				
334.08	-0.87	1.3	288.40	1JA	1.7	-3.4	-1.6	0.9
341.71	-0.74	1.4	298.56	11JA	0.5	-3.5	-1.6	0.9
349.30	-0.61	1.4	308.71	21JA	0.2	-3.5	-1.6	0.9
356.83	-0.48	1.5	318.81	31JA	0.0	-3.6	-1.5	0.9
4.29	-0.35	1.5	328.87	10FE	-0.2	-3.6	-1.5	1.0
11.69	-0.22	1.6	338.88	20FE	-0.6	-3.7	-1.6	1.0
19.01	-0.10	1.6	348.83	2MR	-1.2	-3.7	-1.6	1.1
26.25	0.01	1.7	358.73	12MR	-1.5	-3.8	-1.6	1.1
33.41	0.13	1.7	8.57	22MR	-0.8	-3.9	-1.6	1.1
40.50	0.23	1.7	18.35	1AP	0.5	-4.0	-1.7	1.1
47.51	0.33	1.8	28.07	11AP	1.9	-4.1	-1.7	1.1
54.45	0.43	1.8	37.75	21AP	3.8	-4.2	-1.8	1.1
61.32	0.52	1.8	47.39	1MY	2.4	-4.2	-1.8	1.0
68.11	0.60	1.8	56.98	11MY	1.4	-4.2	-1.9	1.0
74.85	0.68	1.8	66.55	21MY	0.6	-3.9	-1.9	1.0
81.52	0.75	1.8	76.09	31MY	-0.2	-3.4	-2.0	0.9
88.14	0.82	1.8	85.62	10JN	-1.2	-2.8	-2.1	0.8
94.71	0.88	1.8	95.15	20JN	-1.7	-3.6	-2.1	0.8
101.23	0.94	1.9	104.68	30JN	-0.8	-4.1	-2.2	0.7
107.72	0.99	1.9	114.23	10JL	-0.2	-4.2	-2.3	0.6
114.16	1.04	2.0	123.80	20JL	0.3	-4.2	-2.3	0.6
120.57	1.09	2.0	133.40	30JL	0.6	-4.1	-2.4	0.5
126.95	1.13	2.0	143.04	9AU	1.0	-4.0	-2.4	0.5
133.31	1.17	2.0	152.73	19AU	2.0	-4.0	-2.4	0.6
139.63	1.20	2.0	162.46	29AU	2.7	-3.9	-2.5	0.6
145.94	1.23	2.0	172.25	8SE	0.2	-3.8	-2.4	0.7
152.23	1.26	2.0	182.10	18SE	-0.8	-3.7	-2.4	0.7
158.49	1.28	2.0	192.00	28SE	-1.0	-3.7	-2.4	0.8
164.74	1.30	2.0	201.96	8OC	-1.0	-3.6	-2.3	0.8
170.96	1.31	1.9	211.98	18OC	-0.7	-3.6	-2.3	0.9
177.15	1.32	1.9	222.04	28OC	-0.5	-3.5	-2.2	0.9
183.32	1.33	1.8	232.15	7NO	-0.4	-3.5	-2.1	1.0
189.45	1.33	1.8	242.30	17NO	-0.4	-3.4	-2.1	1.0
195.54	1.32	1.7	252.47	27NO	0.0	-3.4	-2.0	1.0
201.58	1.31	1.6	262.66	7DE	2.0	-3.4	-1.9	1.0
207.57	1.29	1.5	272.86	17DE	1.4	-3.4	-1.9	1.0
213.48	1.26	1.4	283.05	27DE	0.2	-3.4	-1.8	1.1

♂			☉	16.00UT	☿	♀	♃	♄
LONG	LAT	MAG	LONG	1228	MAGNITUDES			
219.30	1.22	1.3	293.24	6JA	0.0	-3.3	-1.8	1.1
225.02	1.16	1.2	303.40	16JA	-0.1	-3.3	-1.7	1.0
230.61	1.09	1.0	313.52	26JA	-0.3	-3.3	-1.7	1.0
236.04	1.00	0.8	323.60	5FE	-0.6	-3.3	-1.6	1.0
241.28	0.89	0.7	333.64	15FE	-1.2	-3.4	-1.6	1.1
246.27	0.74	0.5	343.62	25FE	-1.4	-3.4	-1.6	1.1
250.96	0.57	0.2	353.54	6MR	-0.7	-3.4	-1.6	1.2
255.26	0.35	0.0	3.41	16MR	0.6	-3.4	-1.6	1.2
259.09	0.07	-0.3	13.22	26MR	2.4	-3.5	-1.6	1.2
262.30	-0.27	-0.5	22.98	5AP	3.1	-3.5	-1.6	1.2
264.74	-0.68	-0.8	32.68	15AP	1.8	-3.4	-1.6	1.2
266.23	-1.19	-1.1	42.34	25AP	1.0	-3.4	-1.6	1.2
266.59	-1.78	-1.5	51.95	5MY	0.4	-3.4	-1.7	1.2
265.71	-2.44	-1.8	61.53	15MY	-0.3	-3.4	-1.7	1.2
263.67	-3.13	-2.1	71.09	25MY	-1.2	-3.3	-1.8	1.1
260.81	-3.77	-2.3	80.63	4JN	-1.7	-3.3	-1.8	1.1
257.78	-4.26	-2.2	90.16	14JN	-0.8	-3.3	-1.9	1.0
255.32	-4.57	-2.1	99.68	24JN	-0.1	-3.3	-1.9	1.0
253.96	-4.68	-1.9	109.22	4JL	0.4	-3.3	-2.0	0.9
253.98	-4.64	-1.6	118.78	14JL	0.8	-3.3	-2.1	0.8
255.35	-4.49	-1.4	128.36	24JL	1.3	-3.4	-2.1	0.8
257.93	-4.28	-1.2	137.98	3AU	2.5	-3.4	-2.2	0.7
261.52	-4.02	-1.0	147.64	13AU	2.3	-3.4	-2.3	0.6
265.93	-3.75	-0.8	157.35	23AU	0.2	-3.5	-2.3	0.7
271.00	-3.46	-0.6	167.11	2SE	-0.9	-3.5	-2.4	0.7
276.60	-3.17	-0.4	176.93	12SE	-1.2	-3.6	-2.4	0.8
282.60	-2.88	-0.2	186.80	22SE	-1.0	-3.7	-2.4	0.8
288.93	-2.59	-0.1	196.73	2OC	-0.6	-3.7	-2.5	0.9
295.51	-2.30	0.1	206.72	12OC	-0.3	-3.8	-2.4	0.9
302.29	-2.03	0.2	216.76	22OC	-0.3	-3.9	-2.4	1.0
309.21	-1.76	0.4	226.85	1NO	-0.2	-4.0	-2.4	1.0
316.23	-1.50	0.5	236.97	11NO	0.2	-4.2	-2.3	1.1
323.33	-1.26	0.6	247.13	21NO	2.3	-4.3	-2.3	1.1
330.47	-1.03	0.7	257.32	1DE	1.1	-4.4	-2.2	1.1
337.63	-0.81	0.9	267.51	11DE	0.0	-4.4	-2.1	1.2
344.79	-0.61	1.0	277.71	21DE	-0.2	-4.3	-2.1	1.2
351.95	-0.42	1.1	287.90	31DE	-0.2	-4.0	-2.0	1.2
				1229				
359.07	-0.25	1.2	298.07	10JA	-0.4	-3.4	-1.9	1.2
6.16	-0.09	1.3	308.21	20JA	-0.7	-3.5	-1.9	1.2
13.21	0.06	1.4	318.32	30JA	-1.2	-4.1	-1.8	1.2
20.21	0.20	1.5	328.38	9FE	-1.2	-4.3	-1.7	1.2
27.16	0.32	1.5	338.39	19FE	-0.7	-4.3	-1.7	1.2
34.06	0.43	1.6	348.34	1MR	0.8	-4.3	-1.6	1.3
40.90	0.53	1.7	358.24	11MR	2.9	-4.2	-1.6	1.3
47.68	0.62	1.7	8.08	21MR	2.4	-4.1	-1.6	1.4
54.42	0.71	1.8	17.87	31MR	1.3	-3.9	-1.6	1.4
61.10	0.78	1.8	27.60	10AP	0.8	-3.8	-1.5	1.4
67.73	0.85	1.9	37.28	20AP	0.3	-3.7	-1.5	1.4
74.32	0.90	1.9	46.92	30AP	-0.3	-3.7	-1.5	1.4
80.86	0.95	1.9	56.52	10MY	-1.2	-3.6	-1.5	1.4
87.36	1.00	2.0	66.08	20MY	-1.7	-3.5	-1.6	1.4
93.84	1.04	2.0	75.63	30MY	-0.8	-3.5	-1.6	1.4
100.28	1.07	2.0	85.16	9JN	-0.0	-3.4	-1.6	1.3
106.70	1.10	2.0	94.69	19JN	0.6	-3.4	-1.6	1.3
113.10	1.12	2.0	104.23	29JN	1.0	-3.4	-1.7	1.2
119.48	1.13	2.0	113.77	9JL	1.8	-3.3	-1.7	1.1
125.85	1.15	2.0	123.34	19JL	3.1	-3.3	-1.8	1.1
132.22	1.15	2.0	132.94	29JL	1.9	-3.3	-1.9	1.0
138.58	1.15	2.0	142.58	8AU	0.1	-3.3	-1.9	0.9
144.95	1.15	2.0	152.26	18AU	-1.0	-3.3	-2.0	0.9
151.33	1.15	2.0	162.00	28AU	-1.3	-3.4	-2.0	0.8
157.71	1.14	2.0	171.78	7SE	-1.0	-3.4	-2.1	0.8
164.10	1.12	2.0	181.62	17SE	-0.5	-3.4	-2.2	0.9
170.51	1.10	2.0	191.52	27SE	-0.2	-3.4	-2.2	0.9
176.94	1.07	2.0	201.48	7OC	-0.1	-3.5	-2.3	1.0
183.38	1.04	2.0	211.49	17OC	-0.0	-3.5	-2.3	1.0
189.84	1.00	2.0	221.55	27OC	0.4	-3.5	-2.4	1.1
196.32	0.96	1.9	231.66	6NO	2.6	-3.5	-2.4	1.1
202.82	0.91	1.9	241.80	16NO	0.9	-3.5	-2.4	1.2
209.33	0.85	1.8	251.98	26NO	-0.2	-3.4	-2.3	1.2
215.87	0.79	1.8	262.16	6DE	-0.3	-3.4	-2.3	1.3
222.42	0.71	1.7	272.36	16DE	-0.3	-3.4	-2.3	1.3
228.98	0.62	1.6	282.56	26DE	-0.5	-3.4	-2.2	1.3

♂			☉	16.00UT	☿	♀	♃	♄
LONG	LAT	MAG	LONG	1230	MAGNITUDES			
235.55	0.52	1.5	292.74	5JA	-0.8	-3.3	-2.1	1.4
242.13	0.41	1.4	302.90	15JA	-1.1	-3.3	-2.1	1.4
248.72	0.29	1.3	313.03	25JA	-1.1	-3.3	-2.0	1.4
255.30	0.15	1.2	323.11	4FE	-0.6	-3.3	-1.9	1.4
261.88	-0.01	1.1	333.15	14FE	1.0	-3.4	-1.8	1.3
268.45	-0.19	1.0	343.13	24FE	3.2	-3.4	-1.8	1.3
274.99	-0.39	0.8	353.06	6MR	1.7	-3.4	-1.7	1.3
281.49	-0.61	0.7	2.93	16MR	0.9	-3.4	-1.7	1.3
287.95	-0.86	0.5	12.75	26MR	0.6	-3.5	-1.6	1.3
294.33	-1.14	0.4	22.50	5AP	0.2	-3.5	-1.6	1.3
300.61	-1.44	0.2	32.21	15AP	-0.4	-3.5	-1.5	1.2
306.75	-1.78	0.0	41.87	25AP	-1.2	-3.6	-1.5	1.2
312.72	-2.15	-0.2	51.49	5MY	-1.7	-3.7	-1.5	1.2
318.45	-2.56	-0.4	61.07	15MY	-0.8	-3.7	-1.5	1.2
323.86	-3.00	-0.6	70.63	25MY	0.1	-3.8	-1.4	1.1
328.85	-3.47	-0.8	80.17	4JN	0.7	-3.9	-1.4	1.1
333.30	-3.98	-1.0	89.70	14JN	1.4	-4.0	-1.5	1.1
337.03	-4.52	-1.3	99.23	24JN	2.3	-4.1	-1.5	1.0
339.88	-5.06	-1.6	108.76	4JL	3.2	-4.2	-1.5	1.0
341.60	-5.59	-1.8	118.32	14JL	1.6	-4.2	-1.5	0.9
342.04	-6.05	-2.1	127.91	24JL	0.0	-4.1	-1.5	0.9
341.14	-6.36	-2.3	137.52	3AU	-1.1	-3.8	-1.6	0.8
339.07	-6.44	-2.5	147.18	13AU	-1.4	-3.3	-1.6	0.8
336.37	-6.21	-2.6	156.89	23AU	-0.9	-3.4	-1.7	0.7
333.79	-5.69	-2.4	166.64	2SE	-0.4	-4.0	-1.7	0.7
332.01	-4.98	-2.2	176.46	12SE	-0.1	-4.2	-1.8	0.7
331.46	-4.19	-1.9	186.32	22SE	0.0	-4.3	-1.9	0.8
332.22	-3.41	-1.6	196.25	2OC	0.2	-4.3	-1.9	0.8
334.17	-2.70	-1.2	206.24	12OC	0.7	-4.2	-2.0	0.9
337.13	-2.07	-0.9	216.27	22OC	2.9	-4.1	-2.1	1.0
340.89	-1.53	-0.7	226.35	1NO	0.7	-4.0	-2.1	1.0
345.27	-1.07	-0.4	236.48	11NO	-0.3	-3.9	-2.2	1.1
350.14	-0.68	-0.2	246.64	21NO	-0.5	-3.8	-2.2	1.1
355.38	-0.35	0.1	256.82	1DE	-0.5	-3.8	-2.2	1.2
0.91	-0.07	0.3	267.02	11DE	-0.6	-3.7	-2.2	1.2
6.65	0.17	0.5	277.22	21DE	-0.8	-3.6	-2.2	1.2
12.54	0.37	0.7	287.41	31DE	-0.9	-3.6	-2.2	1.2
				1231				
18.56	0.53	0.8	297.58	10JA	-1.0	-3.5	-2.2	1.2
24.68	0.68	1.0	307.72	20JA	-0.5	-3.5	-2.1	1.2
30.85	0.79	1.1	317.83	30JA	1.1	-3.4	-2.1	1.2
37.06	0.89	1.2	327.90	9FE	2.9	-3.4	-2.0	1.1
43.31	0.98	1.4	337.91	19FE	1.3	-3.4	-1.9	1.1
49.58	1.04	1.5	347.87	1MR	0.7	-3.3	-1.9	1.0
55.85	1.10	1.6	357.77	11MR	0.4	-3.3	-1.8	1.0
62.13	1.14	1.7	7.61	21MR	0.1	-3.3	-1.7	1.0
68.41	1.18	1.7	17.40	31MR	-0.4	-3.3	-1.7	1.0
74.68	1.21	1.8	27.13	10AP	-1.2	-3.3	-1.6	1.0
80.95	1.23	1.9	36.81	20AP	-1.7	-3.3	-1.5	1.0
87.22	1.24	1.9	46.45	30AP	-0.8	-3.4	-1.5	1.0
93.48	1.24	1.9	56.05	10MY	0.2	-3.4	-1.4	1.0
99.74	1.25	2.0	65.62	20MY	1.0	-3.4	-1.4	0.9
106.00	1.24	2.0	75.17	30MY	1.8	-3.5	-1.4	0.9
112.27	1.23	2.0	84.70	9JN	3.0	-3.5	-1.4	0.9
118.54	1.22	2.0	94.23	19JN	2.7	-3.5	-1.3	0.8
124.83	1.20	2.0	103.76	29JN	1.3	-3.4	-1.3	0.8
131.13	1.17	2.0	113.31	9JL	-0.0	-3.4	-1.3	0.7
137.44	1.15	2.0	122.87	19JL	-1.1	-3.4	-1.3	0.7
143.78	1.12	2.0	132.47	29JL	-1.5	-3.3	-1.4	0.6
150.15	1.08	2.0	142.11	8AU	-0.9	-3.3	-1.4	0.6
156.54	1.04	2.0	151.79	18AU	-0.3	-3.3	-1.4	0.5
162.97	1.00	1.9	161.52	28AU	-0.0	-3.3	-1.4	0.5
169.44	0.95	1.9	171.30	7SE	0.2	-3.3	-1.5	0.4
175.94	0.90	1.9	181.14	17SE	0.4	-3.3	-1.5	0.4
182.50	0.84	1.9	191.04	27SE	1.0	-3.4	-1.6	0.4
189.09	0.78	1.9	201.00	7OC	3.3	-3.4	-1.6	0.5
195.74	0.71	1.9	211.00	17OC	0.6	-3.4	-1.7	0.5
202.44	0.64	1.9	221.06	27OC	-0.5	-3.5	-1.8	0.6
209.19	0.56	1.9	231.17	6NO	-0.6	-3.5	-1.8	0.7
215.99	0.48	1.9	241.31	16NO	-0.6	-3.6	-1.9	0.7
222.85	0.39	1.8	251.48	26NO	-0.7	-3.6	-2.0	0.8
229.76	0.29	1.8	261.67	6DE	-0.8	-3.7	-2.0	0.9
236.73	0.18	1.7	271.87	16DE	-0.8	-3.8	-2.1	0.9
243.75	0.07	1.7	282.07	26DE	-0.8	-3.9	-2.1	0.9

♂ LONG	♂ LAT	♂ MAG	⊙ LONG	16.00UT 1232	☿	♀	♃	♄
250.82	-0.05	1.6	292.25	5JA	-0.4	-3.9	-2.1	0.9
257.95	-0.17	1.6	302.41	15JA	1.3	-4.0	-2.1	0.9
265.13	-0.30	1.5	312.54	25JA	2.4	-4.1	-2.1	0.9
272.35	-0.45	1.4	322.63	4FE	0.9	-4.2	-2.1	0.9
279.61	-0.59	1.4	332.66	14FE	0.5	-4.3	-2.0	0.9
286.92	-0.75	1.3	342.65	24FE	0.2	-4.3	-2.0	0.9
294.25	-0.90	1.2	352.58	5MR	-0.0	-4.2	-1.9	0.8
301.61	-1.07	1.1	2.45	15MR	-0.5	-3.8	-1.9	0.8
308.99	-1.23	1.0	12.27	25MR	-1.2	-3.2	-1.8	0.7
316.36	-1.39	0.9	22.03	4AP	-1.6	-3.5	-1.7	0.7
323.73	-1.56	0.8	31.74	14AP	-0.8	-4.0	-1.7	0.7
331.07	-1.72	0.8	41.40	24AP	0.3	-4.2	-1.6	0.7
338.36	-1.87	0.7	51.02	4MY	1.3	-4.2	-1.5	0.7
345.59	-2.01	0.6	60.60	14MY	2.4	-4.2	-1.5	0.7
352.72	-2.15	0.5	70.16	24MY	3.5	-4.1	-1.4	0.7
359.72	-2.27	0.4	79.70	3JN	2.1	-4.0	-1.4	0.7
6.56	-2.37	0.3	89.23	13JN	1.0	-3.9	-1.3	0.7
13.20	-2.46	0.1	98.76	23JN	-0.1	-3.8	-1.3	0.7
19.58	-2.53	0.0	108.30	3JL	-1.2	-3.7	-1.3	0.6
25.66	-2.57	-0.1	117.86	13JL	-1.6	-3.6	-1.3	0.6
31.35	-2.59	-0.2	127.44	23JL	-0.9	-3.6	-1.3	0.5
36.55	-2.58	-0.4	137.05	2AU	-0.3	-3.5	-1.3	0.5
41.17	-2.53	-0.6	146.71	12AU	0.1	-3.5	-1.3	0.4
45.04	-2.45	-0.7	156.42	22AU	0.3	-3.4	-1.3	0.4
47.98	-2.32	-0.9	166.17	1SE	0.6	-3.4	-1.3	0.3
49.80	-2.12	-1.2	175.98	11SE	1.3	-3.4	-1.3	0.2
50.27	-1.85	-1.4	185.85	21SE	3.3	-3.4	-1.3	0.2
49.28	-1.49	-1.6	195.77	1OC	0.4	-3.4	-1.4	0.2
46.89	-1.03	-1.8	205.75	11OC	-0.6	-3.4	-1.4	0.2
43.51	-0.51	-1.9	215.79	21OC	-0.8	-3.4	-1.5	0.2
39.87	0.02	-1.9	225.87	31OC	-0.7	-3.4	-1.5	0.3
36.79	0.51	-1.6	235.99	10NO	-0.8	-3.4	-1.6	0.4
34.84	0.91	-1.3	246.15	20NO	-0.6	-3.4	-1.6	0.5
34.24	1.21	-0.9	256.33	30NO	-0.6	-3.4	-1.7	0.5
34.94	1.43	-0.6	266.52	10DE	-0.6	-3.4	-1.8	0.6
36.76	1.58	-0.3	276.73	20DE	-0.3	-3.4	-1.8	0.6
39.49	1.68	-0.0	286.91	30DE	1.5	-3.5	-1.9	0.7
				1233				
42.93	1.74	0.2	297.09	9JA	2.0	-3.5	-1.9	0.7
46.93	1.77	0.5	307.23	19JA	0.6	-3.5	-2.0	0.7
51.36	1.78	0.7	317.34	29JA	0.3	-3.5	-2.0	0.7
56.13	1.78	0.9	327.41	8FE	0.1	-3.4	-2.0	0.7
61.16	1.77	1.0	337.42	18FE	-0.1	-3.4	-2.0	0.7
66.40	1.75	1.2	347.38	28FE	-0.5	-3.4	-2.0	0.7
71.81	1.72	1.3	357.28	10MR	-1.2	-3.4	-2.0	0.6
77.34	1.69	1.4	7.13	20MR	-1.5	-3.3	-1.9	0.6
82.99	1.65	1.5	16.92	30MR	-0.8	-3.3	-1.9	0.5
88.72	1.61	1.6	26.66	9AP	0.4	-3.3	-1.8	0.5
94.52	1.56	1.7	36.34	19AP	1.6	-3.3	-1.8	0.5
100.39	1.51	1.8	45.98	29AP	3.2	-3.3	-1.7	0.5
106.32	1.46	1.8	55.58	9MY	2.8	-3.4	-1.6	0.5
112.30	1.41	1.9	65.16	19MY	1.6	-3.4	-1.6	0.6
118.32	1.36	1.9	74.70	29MY	0.7	-3.4	-1.5	0.6
124.39	1.30	2.0	84.24	8JN	-0.2	-3.5	-1.5	0.5
130.52	1.24	2.0	93.77	18JN	-1.2	-3.5	-1.4	0.5
136.68	1.18	2.0	103.30	28JN	-1.7	-3.6	-1.4	0.5
142.90	1.11	2.0	112.84	8JL	-0.9	-3.6	-1.3	0.5
149.17	1.05	2.0	122.41	18JL	-0.2	-3.7	-1.3	0.5
155.49	0.98	2.0	132.01	28JL	0.2	-3.8	-1.3	0.4
161.87	0.91	2.0	141.64	7AU	0.5	-3.9	-1.2	0.4
168.31	0.83	2.0	151.32	17AU	0.8	-4.0	-1.2	0.3
174.81	0.76	1.9	161.05	27AU	1.7	-4.1	-1.2	0.2
181.37	0.68	1.9	170.83	6SE	3.0	-4.2	-1.2	0.2
188.01	0.59	1.9	180.67	16SE	0.3	-4.3	-1.2	0.1
194.71	0.51	1.8	190.56	26SE	-0.8	-4.3	-1.2	0.1
201.49	0.42	1.8	200.51	6OC	-0.9	-4.2	-1.2	0.0
208.34	0.33	1.7	210.52	16OC	-0.9	-3.9	-1.3	-0.0
215.26	0.23	1.7	220.58	26OC	-0.7	-3.2	-1.3	-0.0
222.26	0.13	1.7	230.68	5NO	-0.5	-3.5	-1.3	0.1
229.34	0.03	1.7	240.82	15NO	-0.5	-4.1	-1.4	0.1
236.49	-0.07	1.7	250.99	25NO	-0.5	-4.3	-1.4	0.2
243.72	-0.18	1.7	261.18	5DE	-0.1	-4.4	-1.5	0.3
251.02	-0.29	1.7	271.38	15DE	1.7	-4.3	-1.6	0.3
258.39	-0.40	1.6	281.57	25DE	1.6	-4.2	-1.6	0.4

♂ LONG	♂ LAT	♂ MAG	⊙ LONG	16.00UT 1234	☿	♀	♃	♄
265.82	-0.51	1.6	291.76	4JA	0.4	-4.1	-1.7	0.4
273.32	-0.62	1.6	301.92	14JA	0.1	-4.0	-1.8	0.5
280.87	-0.73	1.5	312.05	24JA	-0.0	-3.9	-1.8	0.5
288.47	-0.83	1.5	322.14	3FE	-0.2	-3.8	-1.9	0.5
296.12	-0.94	1.4	332.18	13FE	-0.6	-3.7	-1.9	0.5
303.79	-1.04	1.4	342.17	23FE	-1.2	-3.6	-2.0	0.5
311.49	-1.13	1.4	352.11	5MR	-1.4	-3.6	-2.0	0.5
319.20	-1.22	1.3	1.98	15MR	-0.8	-3.5	-2.0	0.5
326.91	-1.29	1.3	11.80	25MR	0.5	-3.5	-2.0	0.5
334.61	-1.36	1.2	21.57	4AP	2.1	-3.4	-2.0	0.4
342.28	-1.41	1.2	31.28	14AP	3.7	-3.4	-2.0	0.4
349.91	-1.45	1.2	40.94	24AP	2.1	-3.3	-1.9	0.3
357.48	-1.47	1.1	50.56	4MY	1.2	-3.3	-1.9	0.4
4.98	-1.48	1.1	60.15	14MY	0.6	-3.3	-1.8	0.4
12.40	-1.47	1.0	69.71	24MY	-0.2	-3.3	-1.7	0.4
19.72	-1.44	1.0	79.25	3JN	-1.2	-3.3	-1.7	0.4
26.91	-1.40	1.0	88.78	13JN	-1.8	-3.3	-1.6	0.4
33.98	-1.34	0.9	98.30	23JN	-0.9	-3.3	-1.6	0.4
40.88	-1.26	0.9	107.84	3JL	-0.1	-3.4	-1.5	0.4
47.61	-1.17	0.8	117.39	13JL	0.3	-3.4	-1.4	0.4
54.14	-1.05	0.7	126.97	23JL	0.7	-3.4	-1.4	0.4
60.44	-0.92	0.7	136.59	2AU	1.1	-3.5	-1.3	0.3
66.46	-0.77	0.6	146.24	12AU	2.2	-3.5	-1.3	0.3
72.19	-0.59	0.5	155.94	22AU	2.6	-3.5	-1.3	0.2
77.54	-0.39	0.4	165.69	1SE	0.3	-3.5	-1.2	0.2
82.44	-0.17	0.3	175.50	11SE	-0.9	-3.4	-1.2	0.1
86.82	0.09	0.1	185.36	21SE	-1.1	-3.4	-1.2	0.1
90.54	0.39	-0.0	195.29	1OC	-1.0	-3.4	-1.2	-0.0
93.44	0.73	-0.2	205.26	11OC	-0.6	-3.3	-1.2	-0.1
95.36	1.11	-0.4	215.30	21OC	-0.4	-3.3	-1.2	-0.1
96.07	1.55	-0.6	225.38	31OC	-0.3	-3.3	-1.2	-0.2
95.43	2.02	-0.9	235.50	10NO	-0.3	-3.3	-1.3	-0.1
93.37	2.50	-1.1	245.65	20NO	0.1	-3.3	-1.3	-0.1
90.11	2.94	-1.2	255.84	30NO	2.0	-3.4	-1.3	-0.0
86.21	3.27	-1.3	266.03	10DE	1.3	-3.4	-1.4	0.1
82.43	3.45	-1.1	276.23	20DE	0.1	-3.4	-1.4	0.1
79.48	3.49	-0.9	286.42	30DE	-0.1	-3.4	-1.5	0.2
				1235				
77.78	3.42	-0.6	296.59	9JA	-0.1	-3.5	-1.6	0.3
77.41	3.28	-0.3	306.74	19JA	-0.3	-3.5	-1.6	0.3
78.26	3.11	-0.0	316.86	29JA	-0.6	-3.6	-1.7	0.4
80.14	2.93	0.2	326.92	8FE	-1.2	-3.6	-1.8	0.4
82.85	2.75	0.5	336.94	18FE	-1.3	-3.7	-1.8	0.4
86.22	2.57	0.7	346.91	28FE	-0.7	-3.8	-1.9	0.4
90.11	2.41	0.9	356.81	10MR	0.6	-3.8	-1.9	0.4
94.41	2.26	1.0	6.66	20MR	2.6	-3.9	-2.0	0.4
99.05	2.11	1.2	16.46	30MR	2.8	-4.0	-2.0	0.4
103.95	1.97	1.3	26.19	9AP	1.6	-4.1	-2.0	0.3
109.07	1.84	1.4	35.88	19AP	0.9	-4.2	-2.0	0.3
114.38	1.72	1.5	45.52	29AP	0.4	-4.2	-2.0	0.3
119.85	1.60	1.6	55.12	9MY	-0.3	-4.2	-2.0	0.2
125.45	1.48	1.7	64.70	19MY	-1.2	-3.9	-2.0	0.3
131.17	1.37	1.7	74.25	29MY	-1.8	-3.3	-1.9	0.3
137.00	1.26	1.8	83.78	8JN	-0.9	-2.8	-1.9	0.3
142.94	1.15	1.8	93.31	18JN	-0.1	-3.7	-1.8	0.3
148.97	1.04	1.8	102.84	28JN	0.4	-4.1	-1.8	0.4
155.10	0.94	1.9	112.38	8JL	0.9	-4.2	-1.7	0.4
161.31	0.83	1.9	121.95	18JL	1.5	-4.2	-1.6	0.3
167.62	0.73	1.9	131.54	28JL	2.7	-4.1	-1.6	0.3
174.02	0.62	1.9	141.17	7AU	2.2	-4.0	-1.5	0.3
180.51	0.52	1.9	150.85	17AU	0.2	-4.0	-1.5	0.3
187.09	0.42	1.9	160.57	27AU	-1.0	-3.9	-1.4	0.2
193.76	0.31	1.8	170.35	6SE	-1.2	-3.8	-1.4	0.2
200.53	0.21	1.8	180.19	16SE	-1.0	-3.7	-1.3	0.1
207.40	0.11	1.8	190.08	26SE	-0.5	-3.7	-1.3	0.1
214.35	0.00	1.7	200.02	6OC	-0.3	-3.6	-1.3	-0.0
221.41	-0.10	1.7	210.03	16OC	-0.2	-3.6	-1.3	-0.1
228.55	-0.20	1.6	220.09	26OC	-0.1	-3.5	-1.3	-0.2
235.79	-0.31	1.6	230.19	5NO	0.3	-3.5	-1.3	-0.2
243.12	-0.41	1.5	240.33	15NO	2.3	-3.4	-1.3	-0.3
250.53	-0.50	1.5	250.50	25NO	1.1	-3.4	-1.3	-0.2
258.02	-0.60	1.5	260.68	5DE	-0.0	-3.4	-1.3	-0.1
265.59	-0.69	1.5	270.89	15DE	-0.2	-3.4	-1.3	-0.1
273.23	-0.77	1.5	281.08	25DE	-0.3	-3.4	-1.4	-0.0

♂			☉	16.00UT	☿	♀	♃	♄
LONG	LAT	MAG	LONG	1236		MAGNITUDES		
280.92	-0.85	1.5	291.26	4JA	-0.4	-3.3	-1.4	0.1
288.67	-0.92	1.4	301.43	14JA	-0.7	-3.3	-1.5	0.1
296.46	-0.99	1.4	311.56	24JA	-1.1	-3.3	-1.5	0.2
304.28	-1.04	1.4	321.65	3FE	-1.2	-3.3	-1.6	0.2
312.12	-1.09	1.4	331.70	13FE	-0.7	-3.4	-1.7	0.3
319.96	-1.12	1.4	341.69	23FE	0.8	-3.4	-1.7	0.3
327.80	-1.15	1.4	351.62	4MR	3.1	-3.4	-1.8	0.3
335.62	-1.16	1.4	1.51	14MR	2.1	-3.4	-1.9	0.3
343.42	-1.16	1.4	11.33	24MR	1.1	-3.5	-1.9	0.3
351.16	-1.14	1.4	21.09	3AP	0.7	-3.5	-2.0	0.3
358.86	-1.11	1.4	30.81	13AP	0.3	-3.4	-2.0	0.3
6.49	-1.07	1.4	40.47	23AP	-0.3	-3.4	-2.1	0.3
14.05	-1.02	1.4	50.09	3MY	-1.2	-3.4	-2.1	0.2
21.52	-0.95	1.4	59.68	13MY	-1.8	-3.3	-2.1	0.2
28.90	-0.88	1.4	69.24	23MY	-0.9	-3.3	-2.1	0.2
36.18	-0.79	1.4	78.78	2JN	-0.0	-3.3	-2.1	0.2
43.35	-0.69	1.4	88.31	12JN	0.6	-3.3	-2.1	0.3
50.41	-0.58	1.4	97.84	22JN	1.1	-3.3	-2.1	0.3
57.35	-0.46	1.4	107.37	2JL	1.9	-3.3	-2.0	0.3
64.15	-0.34	1.3	116.93	12JL	3.2	-3.3	-2.0	0.3
70.81	-0.20	1.3	126.51	22JL	1.8	-3.4	-1.9	0.3
77.32	-0.05	1.3	136.12	1AU	0.1	-3.4	-1.8	0.3
83.67	0.10	1.2	145.77	11AU	-1.0	-3.4	-1.8	0.3
89.84	0.27	1.2	155.47	21AU	-1.4	-3.5	-1.7	0.3
95.81	0.44	1.1	165.21	31AU	-1.0	-3.5	-1.7	0.3
101.54	0.63	1.1	175.02	10SE	-0.5	-3.6	-1.6	0.2
107.02	0.84	1.0	184.88	20SE	-0.2	-3.7	-1.5	0.2
112.18	1.06	0.9	194.80	30SE	-0.1	-3.7	-1.5	0.1
116.98	1.30	0.8	204.78	10OC	0.0	-3.8	-1.5	0.0
121.33	1.57	0.6	214.80	20OC	0.5	-3.9	-1.4	-0.0
125.14	1.87	0.5	224.88	30OC	2.6	-4.0	-1.4	-0.1
128.28	2.20	0.3	235.01	9NO	0.9	-4.2	-1.4	-0.2
130.60	2.57	0.1	245.16	19NO	-0.2	-4.3	-1.4	-0.3
131.90	2.97	-0.1	255.34	29NO	-0.4	-4.4	-1.4	-0.3
132.01	3.39	-0.4	265.53	9DE	-0.4	-4.4	-1.4	-0.2
130.79	3.81	-0.6	275.73	19DE	-0.5	-4.3	-1.4	-0.2
128.26	4.17	-0.8	285.93	29DE	-0.8	-4.0	-1.4	-0.1
				1237				
124.70	4.42	-1.0	296.10	8JA	-1.0	-3.3	-1.4	-0.0
120.73	4.50	-1.0	306.25	18JA	-1.0	-3.5	-1.5	0.0
117.08	4.41	-0.8	316.36	28JA	-0.6	-4.1	-1.5	0.1
114.37	4.18	-0.6	326.44	7FE	1.0	-4.3	-1.5	0.2
112.91	3.86	-0.3	336.45	17FE	3.2	-4.3	-1.6	0.2
112.74	3.52	-0.1	346.42	27FE	1.6	-4.3	-1.7	0.3
113.74	3.18	0.2	356.33	9MR	0.8	-4.2	-1.7	0.3
115.72	2.86	0.4	6.18	19MR	0.5	-4.1	-1.8	0.3
118.50	2.56	0.6	15.98	29MR	0.1	-3.9	-1.8	0.3
121.93	2.29	0.8	25.72	8AP	-0.4	-3.8	-1.9	0.3
125.87	2.04	0.9	35.41	18AP	-1.2	-3.7	-2.0	0.3
130.24	1.82	1.1	45.06	28AP	-1.7	-3.7	-2.1	0.3
134.95	1.61	1.2	54.66	8MY	-0.9	-3.6	-2.1	0.3
139.96	1.42	1.3	64.23	18MY	0.1	-3.5	-2.2	0.2
145.21	1.24	1.4	73.79	28MY	0.8	-3.5	-2.2	0.2
150.67	1.07	1.4	83.32	7JN	1.5	-3.4	-2.3	0.2
156.33	0.91	1.5	92.85	17JN	2.6	-3.4	-2.3	0.2
162.15	0.76	1.5	102.38	27JN	3.1	-3.4	-2.3	0.3
168.14	0.61	1.6	111.93	7JL	1.5	-3.3	-2.3	0.3
174.27	0.47	1.6	121.49	17JL	-0.0	-3.3	-2.2	0.3
180.54	0.34	1.6	131.08	27JL	-1.1	-3.3	-2.2	0.3
186.94	0.21	1.7	140.71	6AU	-1.5	-3.3	-2.1	0.4
193.47	0.09	1.7	150.38	16AU	-0.9	-3.3	-2.1	0.4
200.13	-0.03	1.7	160.11	26AU	-0.4	-3.4	-2.0	0.3
206.92	-0.15	1.7	169.88	5SE	-0.1	-3.4	-2.0	0.3
213.82	-0.26	1.6	179.71	15SE	0.1	-3.4	-1.9	0.3
220.84	-0.37	1.6	189.60	25SE	0.2	-3.4	-1.8	0.2
227.97	-0.47	1.6	199.54	5OC	0.8	-3.5	-1.8	0.2
235.21	-0.57	1.6	209.54	15OC	3.0	-3.5	-1.7	0.1
242.56	-0.66	1.6	219.60	25OC	0.7	-3.5	-1.7	0.1
250.01	-0.74	1.5	229.69	4NO	-0.4	-3.5	-1.6	-0.0
257.54	-0.82	1.5	239.83	14NO	-0.5	-3.4	-1.6	-0.1
265.17	-0.88	1.5	250.00	24NO	-0.5	-3.4	-1.5	-0.2
272.87	-0.94	1.4	260.18	4DE	-0.6	-3.4	-1.5	-0.2
280.63	-0.99	1.4	270.38	14DE	-0.8	-3.4	-1.5	-0.3
288.45	-1.03	1.4	280.58	24DE	-0.9	-3.4	-1.5	-0.2

♂			☉	16.00UT	☿	♀	♃	♄
LONG	LAT	MAG	LONG	1238		MAGNITUDES		
296.31	-1.06	1.3	290.76	3JA	-0.9	-3.3	-1.5	-0.1
304.19	-1.08	1.3	300.93	13JA	-0.5	-3.3	-1.5	-0.1
312.09	-1.08	1.3	311.07	23JA	1.1	-3.3	-1.5	-0.0
320.00	-1.08	1.3	321.16	2FE	2.8	-3.3	-1.5	0.1
327.89	-1.06	1.3	331.21	12FE	1.1	-3.4	-1.5	0.1
335.75	-1.03	1.3	341.20	22FE	0.6	-3.4	-1.6	0.2
343.58	-0.99	1.4	351.14	4MR	0.3	-3.4	-1.6	0.2
351.36	-0.94	1.4	1.03	14MR	0.0	-3.4	-1.6	0.3
359.09	-0.88	1.4	10.85	24MR	-0.4	-3.5	-1.7	0.3
6.74	-0.81	1.4	20.62	3AP	-1.2	-3.5	-1.8	0.3
14.33	-0.73	1.5	30.34	13AP	-1.7	-3.5	-1.8	0.3
21.83	-0.64	1.5	40.01	23AP	-0.9	-3.6	-1.9	0.3
29.24	-0.55	1.5	49.63	3MY	0.2	-3.7	-2.0	0.3
36.57	-0.45	1.6	59.22	13MY	1.0	-3.7	-2.0	0.3
43.80	-0.34	1.6	68.78	23MY	2.0	-3.8	-2.1	0.3
50.93	-0.23	1.6	78.32	2JN	3.3	-3.9	-2.2	0.3
57.97	-0.12	1.6	87.85	12JN	2.5	-4.0	-2.2	0.3
64.90	-0.01	1.6	97.39	22JN	1.2	-4.1	-2.3	0.2
71.74	0.11	1.7	106.92	2JL	-0.0	-4.2	-2.3	0.3
78.48	0.23	1.7	116.47	12JL	-1.1	-4.2	-2.4	0.3
85.11	0.36	1.7	126.05	22JL	-1.6	-4.1	-2.4	0.4
91.63	0.48	1.7	135.66	1AU	-0.9	-3.8	-2.4	0.4
98.05	0.61	1.6	145.31	11AU	-0.3	-3.3	-2.4	0.4
104.34	0.75	1.6	155.00	21AU	0.0	-3.4	-2.4	0.4
110.52	0.88	1.6	164.75	31AU	0.2	-4.0	-2.3	0.4
116.56	1.02	1.5	174.55	10SE	0.4	-4.2	-2.3	0.4
122.45	1.17	1.5	184.41	20SE	1.1	-4.3	-2.2	0.4
128.18	1.32	1.4	194.32	30SE	3.3	-4.3	-2.1	0.4
133.71	1.48	1.4	204.29	10OC	0.6	-4.2	-2.1	0.3
139.03	1.66	1.3	214.32	20OC	-0.5	-4.1	-2.0	0.3
144.08	1.84	1.2	224.39	30OC	-0.7	-4.0	-1.9	0.2
148.83	2.04	1.0	234.51	9NO	-0.7	-3.9	-1.9	0.1
153.19	2.25	0.9	244.66	19NO	-0.7	-3.8	-1.8	0.1
157.09	2.49	0.7	254.84	29NO	-0.7	-3.8	-1.7	-0.0
160.41	2.74	0.5	265.04	9DE	-0.7	-3.7	-1.7	-0.1
163.02	3.02	0.3	275.24	19DE	-0.7	-3.6	-1.7	-0.1
164.75	3.32	0.1	285.43	29DE	-0.4	-3.6	-1.6	-0.2
				1239				
165.42	3.62	-0.1	295.61	8JA	1.3	-3.5	-1.6	-0.1
164.86	3.91	-0.4	305.76	18JA	2.3	-3.5	-1.6	-0.1
163.01	4.15	-0.7	315.87	28JA	0.8	-3.4	-1.6	-0.0
159.98	4.28	-0.9	325.95	7FE	0.4	-3.4	-1.6	0.1
156.20	4.26	-1.0	335.97	17FE	0.2	-3.4	-1.6	0.1
152.34	4.08	-0.9	345.94	27FE	-0.1	-3.3	-1.6	0.2
149.10	3.75	-0.8	355.86	9MR	-0.5	-3.3	-1.6	0.2
146.97	3.35	-0.5	5.71	19MR	-1.2	-3.3	-1.6	0.3
146.14	2.91	-0.3	15.51	29MR	-1.6	-3.3	-1.6	0.3
146.55	2.49	-0.1	25.25	8AP	-0.9	-3.3	-1.7	0.4
148.07	2.10	0.1	34.95	18AP	0.3	-3.3	-1.7	0.4
150.51	1.75	0.3	44.59	28AP	1.4	-3.4	-1.8	0.4
153.71	1.43	0.5	54.20	8MY	2.7	-3.4	-1.8	0.4
157.53	1.14	0.7	63.77	18MY	3.3	-3.4	-1.9	0.4
161.86	0.88	0.8	73.32	28MY	1.9	-3.5	-1.9	0.4
166.61	0.65	0.9	82.86	7JN	0.9	-3.5	-2.0	0.4
171.72	0.44	1.0	92.39	17JN	-0.1	-3.5	-2.1	0.4
177.14	0.25	1.1	101.92	27JN	-1.1	-3.4	-2.2	0.4
182.82	0.07	1.1	111.46	7JL	-1.7	-3.4	-2.2	0.4
188.75	-0.10	1.2	121.02	17JL	-0.9	-3.4	-2.3	0.4
194.89	-0.25	1.2	130.61	27JL	-0.3	-3.3	-2.3	0.5
201.23	-0.39	1.3	140.24	6AU	0.1	-3.3	-2.4	0.5
207.75	-0.52	1.3	149.91	16AU	0.4	-3.3	-2.4	0.5
214.44	-0.63	1.3	159.63	26AU	0.7	-3.3	-2.5	0.5
221.29	-0.74	1.3	169.40	5SE	1.4	-3.3	-2.5	0.6
228.30	-0.84	1.4	179.23	15SE	3.2	-3.3	-2.4	0.6
235.44	-0.92	1.4	189.11	25SE	0.5	-3.4	-2.4	0.5
242.71	-1.00	1.4	199.06	5OC	-0.7	-3.4	-2.4	0.5
250.10	-1.06	1.4	209.06	15OC	-0.8	-3.4	-2.3	0.5
257.60	-1.11	1.4	219.10	25OC	-0.8	-3.5	-2.3	0.5
265.20	-1.14	1.4	229.20	4NO	-0.8	-3.5	-2.2	0.4
272.89	-1.17	1.4	239.34	14NO	-0.6	-3.6	-2.1	0.3
280.64	-1.18	1.4	249.50	24NO	-0.6	-3.6	-2.1	0.3
288.45	-1.18	1.4	259.69	4DE	-0.6	-3.7	-2.0	0.2
296.31	-1.17	1.4	269.89	14DE	-0.2	-3.8	-1.9	0.1
304.19	-1.15	1.4	280.08	24DE	1.5	-3.9	-1.9	0.1

461

♂ LONG	♂ LAT	♂ MAG	☉ LONG	16.00UT 1240	☿	♀	♃	♄
					MAGNITUDES			
312.09	-1.11	1.4	290.27	3JA	1.9	-4.0	-1.8	-0.0
319.98	-1.06	1.4	300.44	13JA	0.5	-4.1	-1.7	-0.0
327.86	-1.00	1.4	310.57	23JA	0.2	-4.1	-1.7	0.0
335.71	-0.94	1.4	320.67	2FE	0.1	-4.2	-1.7	0.1
343.52	-0.86	1.4	330.72	12FE	-0.2	-4.3	-1.6	0.1
351.27	-0.78	1.4	340.72	22FE	-0.5	-4.3	-1.6	0.2
358.96	-0.69	1.4	350.66	3MR	-1.2	-4.2	-1.6	0.3
6.59	-0.59	1.4	0.54	13MR	-1.5	-3.8	-1.6	0.3
14.13	-0.50	1.5	10.37	23MR	-0.8	-3.2	-1.6	0.4
21.60	-0.39	1.5	20.15	2AP	0.4	-3.5	-1.6	0.4
28.98	-0.29	1.5	29.86	12AP	1.8	-4.0	-1.6	0.5
36.27	-0.18	1.5	39.54	22AP	3.5	-4.2	-1.6	0.5
43.48	-0.08	1.6	49.17	2MY	2.6	-4.2	-1.6	0.5
50.60	0.03	1.6	58.75	12MY	1.5	-4.2	-1.7	0.5
57.63	0.13	1.7	68.32	22MY	0.7	-4.1	-1.7	0.6
64.58	0.24	1.7	77.86	1JN	-0.2	-4.0	-1.8	0.6
71.45	0.34	1.8	87.39	11JN	-1.2	-3.9	-1.8	0.6
78.23	0.44	1.8	96.92	21JN	-1.8	-3.8	-1.9	0.6
84.94	0.54	1.8	106.46	1JL	-0.9	-3.7	-1.9	0.6
91.56	0.64	1.8	116.01	11JL	-0.2	-3.6	-2.0	0.5
98.12	0.74	1.9	125.58	21JL	0.2	-3.6	-2.1	0.5
104.60	0.83	1.9	135.19	31JL	0.5	-3.5	-2.1	0.6
111.01	0.93	1.9	144.84	10AU	0.9	-3.5	-2.2	0.6
117.34	1.02	1.9	154.53	20AU	1.8	-3.4	-2.3	0.7
123.60	1.11	1.9	164.28	30AU	2.9	-3.4	-2.3	0.7
129.79	1.20	1.8	174.07	9SE	0.4	-3.4	-2.4	0.7
135.89	1.30	1.8	183.93	19SE	-0.8	-3.4	-2.4	0.7
141.90	1.39	1.8	193.84	29SE	-1.0	-3.4	-2.4	0.7
147.82	1.48	1.7	203.81	9OC	-0.9	-3.4	-2.4	0.7
153.62	1.58	1.7	213.84	19OC	-0.7	-3.4	-2.4	0.7
159.30	1.68	1.6	223.91	29OC	-0.5	-3.4	-2.4	0.7
164.84	1.78	1.5	234.02	8NO	-0.4	-3.4	-2.4	0.6
170.20	1.88	1.4	244.17	18NO	-0.4	-3.4	-2.3	0.6
175.36	1.99	1.3	254.35	28NO	-0.1	-3.4	-2.3	0.5
180.27	2.10	1.1	264.54	8DE	1.7	-3.4	-2.2	0.5
184.88	2.22	1.0	274.74	18DE	1.5	-3.4	-2.1	0.4
189.11	2.34	0.8	284.93	28DE	0.3	-3.5	-2.0	0.3
				1241				
192.89	2.47	0.6	295.11	7JA	0.0	-3.5	-2.0	0.3
196.10	2.59	0.4	305.26	17JA	-0.1	-3.5	-1.9	0.2
198.58	2.72	0.2	315.38	27JA	-0.2	-3.4	-1.8	0.2
200.20	2.83	-0.1	325.45	6FE	-0.6	-3.4	-1.8	0.2
200.75	2.92	-0.4	335.48	16FE	-1.2	-3.4	-1.7	0.3
200.08	2.96	-0.7	345.45	26FE	-1.4	-3.4	-1.7	0.3
198.14	2.91	-1.0	355.37	8MR	-0.8	-3.4	-1.6	0.4
195.08	2.76	-1.2	5.23	18MR	0.5	-3.3	-1.6	0.4
191.39	2.48	-1.3	15.03	28MR	2.2	-3.3	-1.6	0.5
187.73	2.10	-1.2	24.77	7AP	3.4	-3.3	-1.5	0.5
184.80	1.65	-1.1	34.47	17AP	1.9	-3.3	-1.5	0.6
183.05	1.19	-0.9	44.12	27AP	1.1	-3.3	-1.5	0.6
182.62	0.76	-0.6	53.73	7MY	0.5	-3.4	-1.5	0.7
183.47	0.37	-0.4	63.31	17MY	-0.2	-3.4	-1.5	0.7
185.44	0.03	-0.2	72.86	27MY	-1.2	-3.4	-1.6	0.7
188.35	-0.26	-0.1	82.39	6JN	-1.8	-3.5	-1.6	0.8
192.04	-0.51	0.1	91.93	16JN	-0.9	-3.5	-1.6	0.8
196.37	-0.73	0.2	101.46	26JN	-0.1	-3.6	-1.6	0.8
201.23	-0.91	0.4	111.00	6JL	0.4	-3.6	-1.7	0.8
206.54	-1.07	0.5	120.56	16JL	0.7	-3.7	-1.7	0.8
212.23	-1.20	0.6	130.15	26JL	1.2	-3.8	-1.8	0.8
218.24	-1.31	0.6	139.77	5AU	2.3	-3.9	-1.8	0.8
224.55	-1.40	0.7	149.44	15AU	2.5	-4.0	-1.9	0.8
231.10	-1.47	0.8	159.16	25AU	0.3	-4.1	-2.0	0.9
237.88	-1.52	0.8	168.93	4SE	-0.9	-4.2	-2.0	0.9
244.85	-1.56	0.9	178.76	14SE	-1.1	-4.3	-2.1	0.9
251.99	-1.58	0.9	188.64	24SE	-1.0	-4.3	-2.2	1.0
259.28	-1.58	1.0	198.58	4OC	-0.6	-4.2	-2.2	1.0
266.71	-1.56	1.0	208.57	14OC	-0.4	-3.9	-2.3	1.0
274.24	-1.53	1.1	218.62	24OC	-0.3	-3.2	-2.3	1.0
281.86	-1.48	1.1	228.71	3NO	-0.2	-3.5	-2.3	1.0
289.56	-1.42	1.1	238.85	13NO	0.1	-4.1	-2.4	0.9
297.30	-1.35	1.2	249.01	23NO	2.0	-4.3	-2.3	0.9
305.08	-1.27	1.2	259.20	3DE	1.3	-4.4	-2.3	0.8
312.87	-1.18	1.3	269.39	13DE	0.1	-4.3	-2.3	0.8
320.66	-1.08	1.3	279.59	23DE	-0.1	-4.2	-2.2	0.7

♂ LONG	♂ LAT	♂ MAG	☉ LONG	16.00UT 1242	☿	♀	♃	♄
					MAGNITUDES			
328.43	-0.97	1.4	289.78	2JA	-0.2	-4.1	-2.2	0.6
336.18	-0.86	1.4	299.95	12JA	-0.3	-4.0	-2.1	0.6
343.87	-0.74	1.4	310.08	22JA	-0.7	-3.9	-2.0	0.5
351.52	-0.63	1.5	320.18	1FE	-1.2	-3.8	-1.9	0.4
359.11	-0.51	1.5	330.23	11FE	-1.3	-3.7	-1.9	0.4
6.62	-0.39	1.5	340.23	21FE	-0.8	-3.6	-1.8	0.5
14.06	-0.28	1.6	350.18	3MR	0.7	-3.6	-1.7	0.5
21.43	-0.16	1.6	0.07	13MR	2.7	-3.5	-1.7	0.6
28.71	-0.05	1.6	9.90	23MR	2.6	-3.5	-1.6	0.6
35.91	0.06	1.7	19.68	2AP	1.4	-3.4	-1.6	0.7
43.03	0.17	1.7	29.40	12AP	0.8	-3.4	-1.5	0.7
50.07	0.27	1.7	39.07	22AP	0.4	-3.3	-1.5	0.8
57.04	0.36	1.7	48.70	2MY	-0.3	-3.3	-1.5	0.8
63.92	0.46	1.7	58.29	12MY	-1.2	-3.3	-1.5	0.9
70.74	0.54	1.7	67.86	22MY	-1.8	-3.3	-1.4	0.9
77.49	0.63	1.8	77.40	1JN	-0.9	-3.3	-1.4	1.0
84.17	0.71	1.8	86.93	11JN	-0.1	-3.3	-1.4	1.0
90.80	0.78	1.8	96.46	21JN	0.5	-3.3	-1.4	1.0
97.37	0.85	1.9	106.00	1JL	1.0	-3.4	-1.5	1.0
103.89	0.92	1.9	115.55	11JL	1.6	-3.4	-1.5	1.0
110.36	0.98	2.0	125.12	21JL	2.9	-3.4	-1.5	1.0
116.79	1.04	2.0	134.73	31JL	2.1	-3.5	-1.5	1.0
123.17	1.10	2.0	144.37	10AU	0.2	-3.5	-1.6	1.0
129.52	1.15	2.0	154.06	20AU	-1.0	-3.5	-1.6	1.0
135.83	1.20	2.0	163.80	30AU	-1.3	-3.4	-1.7	1.1
142.10	1.25	2.0	173.59	9SE	-1.0	-3.4	-1.7	1.2
148.34	1.29	2.0	183.44	19SE	-0.5	-3.4	-1.8	1.2
154.54	1.33	2.0	193.36	29SE	-0.3	-3.4	-1.9	1.2
160.70	1.37	1.9	203.32	9OC	-0.1	-3.3	-1.9	1.3
166.82	1.40	1.9	213.34	19OC	-0.1	-3.3	-2.0	1.3
172.89	1.43	1.8	223.41	29OC	0.3	-3.3	-2.1	1.3
178.90	1.46	1.8	233.53	8NO	2.3	-3.3	-2.1	1.3
184.86	1.49	1.7	243.67	18NO	1.0	-3.3	-2.2	1.2
190.73	1.50	1.6	253.85	28NO	-0.1	-3.4	-2.2	1.2
196.52	1.52	1.5	264.04	8DE	-0.3	-3.4	-2.2	1.2
202.20	1.53	1.4	274.24	18DE	-0.3	-3.4	-2.2	1.1
207.76	1.53	1.3	284.44	28DE	-0.4	-3.4	-2.2	1.1
				1243				
213.15	1.52	1.1	294.62	7JA	-0.7	-3.5	-2.2	1.0
218.36	1.50	1.0	304.77	17JA	-1.1	-3.5	-2.1	0.9
223.34	1.47	0.8	314.89	27JA	-1.1	-3.6	-2.1	0.9
228.03	1.42	0.6	324.97	6FE	-0.7	-3.6	-2.0	0.8
232.37	1.35	0.4	335.00	16FE	0.8	-3.7	-2.0	0.7
236.25	1.24	0.2	344.97	26FE	3.1	-3.8	-1.9	0.7
239.58	1.10	-0.1	354.89	8MR	1.9	-3.8	-1.8	0.8
242.20	0.91	-0.4	4.75	18MR	1.0	-3.9	-1.8	0.8
243.94	0.65	-0.7	14.56	28MR	0.6	-4.0	-1.7	0.9
244.62	0.32	-1.0	24.31	7AP	0.2	-4.1	-1.6	0.9
244.10	-0.09	-1.3	34.00	17AP	-0.3	-4.2	-1.6	1.0
242.32	-0.59	-1.6	43.66	27AP	-1.2	-4.2	-1.5	1.0
239.52	-1.14	-1.9	53.27	7MY	-1.8	-4.2	-1.5	1.1
236.18	-1.69	-2.0	62.84	17MY	-0.9	-3.9	-1.4	1.1
233.02	-2.19	-1.9	72.40	27MY	0.0	-3.3	-1.4	1.2
230.74	-2.57	-1.7	81.93	6JN	0.7	-2.9	-1.4	1.2
229.72	-2.85	-1.5	91.46	16JN	1.3	-3.7	-1.3	1.2
230.09	-3.01	-1.3	100.99	26JN	2.2	-4.1	-1.3	1.3
231.77	-3.10	-1.0	110.53	6JL	3.3	-4.2	-1.3	1.3
234.57	-3.12	-0.8	120.09	16JL	1.7	-4.2	-1.3	1.3
238.31	-3.10	-0.7	129.68	26JL	0.1	-4.1	-1.3	1.3
242.82	-3.04	-0.5	139.30	5AU	-1.0	-4.0	-1.4	1.3
247.95	-2.96	-0.3	148.97	15AU	-1.4	-3.9	-1.4	1.3
253.60	-2.85	-0.2	158.69	25AU	-1.0	-3.9	-1.4	1.3
259.65	-2.72	-0.0	168.45	4SE	-0.4	-3.8	-1.4	1.3
266.05	-2.58	0.1	178.27	14SE	-0.1	-3.7	-1.5	1.3
272.72	-2.42	0.2	188.16	24SE	0.0	-3.7	-1.5	1.3
279.61	-2.25	0.3	198.09	4OC	0.1	-3.6	-1.6	1.3
286.67	-2.07	0.4	208.08	14OC	0.6	-3.5	-1.6	1.3
293.87	-1.89	0.5	218.13	24OC	2.6	-3.5	-1.7	1.3
301.18	-1.70	0.6	228.22	3NO	0.9	-3.5	-1.8	1.2
308.55	-1.51	0.7	238.35	13NO	-0.3	-3.4	-1.8	1.2
315.97	-1.32	0.8	248.52	23NO	-0.4	-3.4	-1.9	1.2
323.41	-1.14	0.9	258.70	3DE	-0.4	-3.4	-2.0	1.1
330.85	-0.96	1.0	268.90	13DE	-0.5	-3.4	-2.0	1.1
338.29	-0.78	1.1	279.10	23DE	-0.8	-3.4	-2.1	1.1

♂ LONG	LAT	MAG	☉ LONG	16.00UT	☿	♀	♃	♄
				1244				
345.69	-0.61	1.2	289.28	2JA	-0.9	-3.3	-2.1	1.0
353.06	-0.45	1.3	299.45	12JA	-1.0	-3.3	-2.1	1.0
0.38	-0.30	1.4	309.59	22JA	-0.6	-3.3	-2.1	0.9
7.64	-0.15	1.4	319.69	1FE	1.0	-3.4	-2.1	0.9
14.85	-0.02	1.5	329.75	11FE	3.1	-3.4	-2.1	0.8
21.99	0.11	1.6	339.75	21FE	1.4	-3.4	-2.0	0.8
29.06	0.23	1.6	349.70	2MR	0.7	-3.4	-2.0	0.8
36.07	0.34	1.7	359.59	12MR	0.4	-3.4	-1.9	0.8
43.01	0.44	1.7	9.43	22MR	0.1	-3.5	-1.8	0.9
49.88	0.53	1.8	19.20	1AP	-0.4	-3.5	-1.8	0.9
56.69	0.62	1.8	28.93	11AP	-1.2	-3.4	-1.7	1.0
63.46	0.69	1.9	38.60	21AP	-1.7	-3.4	-1.6	1.1
70.13	0.77	1.9	48.23	1MY	-0.9	-3.4	-1.6	1.1
76.77	0.83	1.9	57.83	11MY	0.1	-3.4	-1.5	1.2
83.37	0.89	1.9	67.39	21MY	0.9	-3.3	-1.5	1.2
89.91	0.94	1.9	76.93	31MY	1.7	-3.3	-1.4	1.3
96.42	0.99	1.9	86.47	10JN	2.8	-3.3	-1.3	1.3
102.89	1.03	1.9	95.99	20JN	2.9	-3.3	-1.3	1.3
109.34	1.06	1.9	105.53	30JN	1.4	-3.3	-1.3	1.3
115.76	1.09	1.9	115.08	10JL	0.1	-3.3	-1.3	1.3
122.15	1.12	1.9	124.65	20JL	-1.1	-3.4	-1.3	1.3
128.53	1.14	2.0	134.25	30JL	-1.5	-3.4	-1.2	1.3
134.90	1.16	2.0	143.90	9AU	-0.9	-3.4	-1.2	1.3
141.26	1.17	2.0	153.59	19AU	-0.4	-3.5	-1.2	1.2
147.62	1.18	2.0	163.32	29AU	-0.0	-3.5	-1.3	1.2
153.97	1.18	2.0	173.12	8SE	0.1	-3.6	-1.3	1.1
160.32	1.18	2.0	182.97	18SE	0.3	-3.7	-1.3	1.1
166.67	1.18	2.0	192.87	28SE	0.8	-3.7	-1.3	1.1
173.03	1.17	2.0	202.84	8OC	3.0	-3.8	-1.4	1.1
179.39	1.15	2.0	212.86	18OC	0.7	-3.9	-1.4	1.1
185.75	1.13	1.9	222.92	28OC	-0.4	-4.1	-1.5	1.1
192.11	1.10	1.9	233.04	7NO	-0.6	-4.2	-1.5	1.1
198.48	1.07	1.8	243.18	17NO	-0.6	-4.3	-1.6	1.1
204.84	1.03	1.8	253.36	27NO	-0.7	-4.4	-1.6	1.0
211.20	0.98	1.7	263.55	7DE	-0.8	-4.4	-1.7	1.0
217.54	0.92	1.6	273.75	17DE	-0.8	-4.3	-1.8	1.0
223.88	0.85	1.5	283.94	27DE	-0.8	-3.9	-1.8	0.9
				1245				
230.19	0.77	1.4	294.12	6JA	-0.5	-3.3	-1.9	0.9
236.48	0.67	1.3	304.28	16JA	1.1	-3.6	-2.0	0.8
242.74	0.56	1.2	314.40	26JA	2.6	-4.1	-2.0	0.8
248.95	0.43	1.1	324.48	5FE	1.0	-4.3	-2.0	0.7
255.10	0.27	0.9	334.51	15FE	0.5	-4.3	-2.0	0.7
261.18	0.10	0.8	344.49	25FE	0.3	-4.3	-2.0	0.7
267.17	-0.11	0.6	354.41	7MR	0.0	-4.2	-2.0	0.6
273.04	-0.35	0.4	4.28	17MR	-0.4	-4.0	-2.0	0.6
278.77	-0.62	0.3	14.08	27MR	-1.2	-3.9	-1.9	0.6
284.29	-0.94	0.0	23.84	6AP	-1.7	-3.8	-1.9	0.7
289.56	-1.31	-0.2	33.54	16AP	-0.9	-3.7	-1.8	0.8
294.52	-1.73	-0.4	43.19	26AP	0.2	-3.7	-1.8	0.9
299.05	-2.21	-0.7	52.81	6MY	1.1	-3.6	-1.7	0.9
303.04	-2.77	-0.9	62.38	16MY	2.2	-3.5	-1.6	1.0
306.34	-3.39	-1.2	71.94	26MY	3.6	-3.5	-1.6	1.0
308.75	-4.09	-1.5	81.48	5JN	2.3	-3.4	-1.5	1.1
310.09	-4.82	-1.8	91.00	15JN	1.1	-3.4	-1.4	1.1
310.21	-5.56	-2.1	100.53	25JN	-0.0	-3.4	-1.4	1.1
309.06	-6.21	-2.4	110.08	5JL	-1.1	-3.3	-1.3	1.2
306.91	-6.66	-2.6	119.63	15JL	-1.6	-3.3	-1.3	1.2
304.29	-6.82	-2.6	129.22	25JL	-0.9	-3.3	-1.3	1.2
301.95	-6.66	-2.5	138.84	4AU	-0.3	-3.3	-1.2	1.1
300.53	-6.23	-2.2	148.50	14AU	0.1	-3.3	-1.2	1.1
300.37	-5.63	-2.0	158.22	24AU	0.3	-3.4	-1.2	1.1
301.53	-4.97	-1.7	167.98	3SE	0.5	-3.4	-1.2	1.0
303.90	-4.30	-1.4	177.80	13SE	1.2	-3.4	-1.2	1.0
307.25	-3.67	-1.1	187.68	23SE	3.3	-3.4	-1.2	1.0
311.41	-3.08	-0.9	197.61	3OC	0.6	-3.5	-1.2	1.0
316.20	-2.56	-0.6	207.60	13OC	-0.6	-3.5	-1.3	1.0
321.47	-2.08	-0.4	217.64	23OC	-0.7	-3.5	-1.3	1.0
327.11	-1.66	-0.2	227.73	2NO	-0.7	-3.5	-1.3	1.0
333.03	-1.28	0.0	237.86	12NO	-0.8	-3.4	-1.4	1.0
339.16	-0.95	0.2	248.02	22NO	-0.7	-3.4	-1.4	1.0
345.45	-0.66	0.4	258.21	2DE	-0.6	-3.4	-1.5	0.9
351.85	-0.40	0.6	268.40	12DE	-0.7	-3.4	-1.5	0.9
358.33	-0.17	0.7	278.60	22DE	-0.4	-3.4	-1.6	0.9

♂ LONG	LAT	MAG	☉ LONG	16.00UT	☿	♀	♃	♄
				1246				
4.86	0.03	0.9	288.79	1JA	1.3	-3.3	-1.7	0.9
11.43	0.21	1.0	298.96	11JA	2.2	-3.3	-1.7	0.8
18.01	0.36	1.1	309.10	21JA	0.7	-3.3	-1.8	0.8
24.59	0.50	1.2	319.20	31JA	0.3	-3.3	-1.8	0.7
31.17	0.62	1.4	329.26	10FE	0.1	-3.4	-1.9	0.7
37.73	0.72	1.5	339.27	20FE	-0.1	-3.4	-1.9	0.6
44.28	0.81	1.5	349.22	2MR	-0.5	-3.4	-2.0	0.6
50.80	0.88	1.6	359.11	12MR	-1.2	-3.4	-2.0	0.5
57.29	0.95	1.7	8.95	22MR	-1.6	-3.5	-2.0	0.5
63.76	1.00	1.8	18.73	1AP	-0.9	-3.5	-2.0	0.5
70.20	1.05	1.8	28.46	11AP	0.3	-3.6	-2.0	0.5
76.62	1.09	1.9	38.14	21AP	1.5	-3.6	-1.9	0.6
83.02	1.12	1.9	47.77	1MY	2.9	-3.7	-1.9	0.7
89.39	1.14	2.0	57.37	11MY	3.1	-3.7	-1.8	0.7
95.75	1.16	2.0	66.94	21MY	1.7	-3.8	-1.8	0.8
102.10	1.17	2.0	76.48	31MY	0.8	-3.9	-1.7	0.9
108.43	1.18	2.0	86.01	10JN	-0.1	-4.0	-1.7	0.9
114.76	1.18	2.0	95.54	20JN	-1.1	-4.1	-1.6	0.9
121.09	1.18	2.0	105.07	30JN	-1.7	-4.2	-1.5	1.0
127.42	1.17	2.0	114.62	10JL	-0.9	-4.2	-1.5	1.0
133.76	1.16	2.0	124.19	20JL	-0.3	-4.1	-1.4	1.0
140.11	1.14	2.0	133.79	30JL	0.2	-3.8	-1.4	1.0
146.47	1.12	2.0	143.43	9AU	0.4	-3.3	-1.3	1.0
152.86	1.09	1.9	153.12	19AU	0.7	-3.5	-1.3	1.0
159.26	1.06	2.0	162.85	29AU	1.5	-4.0	-1.3	1.0
165.70	1.03	2.0	172.64	8SE	3.2	-4.2	-1.2	1.0
172.16	0.99	2.0	182.49	18SE	0.5	-4.3	-1.2	0.9
178.66	0.94	2.0	192.39	28SE	-0.7	-4.3	-1.2	0.9
185.19	0.89	2.0	202.35	8OC	-0.9	-4.2	-1.2	0.8
191.77	0.84	2.0	212.37	18OC	-0.9	-4.1	-1.2	0.8
198.37	0.78	1.9	222.44	28OC	-0.7	-4.0	-1.2	0.9
205.03	0.71	1.9	232.55	7NO	-0.5	-3.9	-1.3	0.9
211.72	0.64	1.9	242.70	17NO	-0.5	-3.8	-1.3	0.9
218.45	0.56	1.8	252.87	27NO	-0.5	-3.8	-1.3	0.9
225.23	0.47	1.7	263.06	7DE	-0.2	-3.7	-1.4	0.9
232.05	0.37	1.7	273.26	17DE	1.5	-3.6	-1.4	0.9
238.91	0.27	1.7	283.45	27DE	1.8	-3.6	-1.5	0.8
				1247				
245.81	0.15	1.6	293.63	6JA	0.4	-3.5	-1.5	0.8
252.75	0.02	1.5	303.79	16JA	0.1	-3.5	-1.6	0.8
259.73	-0.11	1.4	313.91	26JA	0.0	-3.4	-1.6	0.7
266.75	-0.26	1.4	324.00	5FE	-0.2	-3.4	-1.7	0.7
273.79	-0.42	1.3	334.03	15FE	-0.6	-3.4	-1.8	0.6
280.86	-0.58	1.2	344.01	25FE	-1.2	-3.3	-1.8	0.6
287.96	-0.76	1.1	353.94	7MR	-1.5	-3.3	-1.9	0.5
295.07	-0.95	1.0	3.81	17MR	-0.9	-3.3	-2.0	0.5
302.19	-1.15	0.9	13.61	27MR	0.4	-3.3	-2.0	0.4
309.30	-1.36	0.7	23.37	6AP	1.9	-3.3	-2.0	0.4
316.38	-1.58	0.6	33.07	16AP	3.8	-3.3	-2.1	0.4
323.42	-1.80	0.5	42.73	26AP	2.3	-3.4	-2.1	0.5
330.41	-2.03	0.4	52.34	6MY	1.3	-3.4	-2.0	0.5
337.29	-2.25	0.3	61.92	16MY	0.6	-3.4	-2.0	0.6
344.04	-2.48	0.1	71.48	26MY	-0.2	-3.5	-2.0	0.6
350.62	-2.70	-0.0	81.02	5JN	-1.1	-3.5	-1.9	0.7
356.79	-2.92	-0.2	90.55	15JN	-1.8	-3.5	-1.9	0.8
3.03	-3.12	-0.3	100.07	25JN	-0.9	-3.4	-1.8	0.8
8.71	-3.31	-0.5	109.61	5JL	-0.2	-3.4	-1.8	0.9
13.91	-3.49	-0.7	119.17	15JL	0.3	-3.4	-1.7	0.9
18.50	-3.64	-0.9	128.75	25JL	0.6	-3.3	-1.6	0.9
22.32	-3.76	-1.1	138.37	4AU	1.0	-3.3	-1.6	0.9
25.16	-3.83	-1.3	148.03	14AU	2.0	-3.3	-1.5	0.9
26.83	-3.84	-1.5	157.74	24AU	2.8	-3.3	-1.5	0.9
27.11	-3.75	-1.7	167.50	3SE	0.4	-3.3	-1.4	0.9
25.92	-3.53	-2.0	177.32	13SE	-0.8	-3.3	-1.4	0.9
23.45	-3.15	-2.1	187.19	23SE	-1.0	-3.4	-1.3	0.9
20.19	-2.61	-2.2	197.12	3OC	-1.0	-3.4	-1.3	0.8
16.95	-1.98	-2.1	207.11	13OC	-0.7	-3.4	-1.3	0.8
14.47	-1.33	-1.8	217.15	23OC	-0.4	-3.5	-1.3	0.8
13.22	-0.74	-1.5	227.24	2NO	-0.4	-3.5	-1.3	0.8
13.32	-0.23	-1.1	237.37	12NO	-0.3	-3.6	-1.3	0.8
14.65	0.18	-0.8	247.53	22NO	-0.0	-3.6	-1.3	0.8
17.03	0.51	-0.5	257.71	2DE	1.8	-3.7	-1.3	0.8
20.24	0.76	-0.2	267.91	12DE	1.5	-3.8	-1.3	0.8
24.10	0.96	0.1	278.11	22DE	0.2	-3.9	-1.4	0.8

♂ LONG	♂ LAT	♂ MAG	☉ LONG	16.00UT	☿	♀	♃	♄
				1248	MAGNITUDES			
28.46	1.11	0.3	288.30	1JA	-0.0	-4.0	-1.4	0.8
33.22	1.22	0.5	298.47	11JA	-0.1	-4.1	-1.4	0.8
38.28	1.31	0.7	308.61	21JA	-0.3	-4.2	-1.5	0.7
43.57	1.37	0.9	318.72	31JA	-0.6	-4.2	-1.6	0.7
49.04	1.41	1.0	328.78	10FE	-1.2	-4.3	-1.6	0.7
54.65	1.45	1.2	338.78	20FE	-1.4	-4.3	-1.7	0.6
60.37	1.46	1.3	348.74	1MR	-0.8	-4.2	-1.8	0.6
66.17	1.47	1.4	358.64	11MR	0.5	-3.8	-1.8	0.5
72.05	1.47	1.5	8.48	21MR	2.4	-3.2	-1.9	0.5
77.97	1.47	1.6	18.26	31MR	3.1	-3.5	-2.0	0.4
83.95	1.45	1.7	27.99	10AP	1.7	-4.0	-2.0	0.3
89.95	1.44	1.8	37.67	20AP	1.0	-4.2	-2.1	0.3
96.00	1.41	1.9	47.31	30AP	0.5	-4.2	-2.1	0.3
102.07	1.38	1.9	56.90	10MY	-0.2	-4.2	-2.1	0.4
108.16	1.35	1.9	66.47	20MY	-1.1	-4.1	-2.2	0.5
114.29	1.32	2.0	76.02	30MY	-1.8	-4.0	-2.2	0.5
120.45	1.28	2.0	85.55	9JN	-0.9	-3.9	-2.1	0.6
126.63	1.24	2.0	95.08	19JN	-0.1	-3.8	-2.1	0.7
132.85	1.19	2.0	104.61	29JN	0.4	-3.7	-2.1	0.7
139.10	1.14	2.0	114.16	9JL	0.8	-3.6	-2.0	0.8
145.39	1.09	2.0	123.73	19JL	1.4	-3.6	-2.0	0.8
151.72	1.04	2.0	133.33	29JL	2.5	-3.5	-1.9	0.8
158.10	0.98	2.0	142.97	8AU	2.4	-3.5	-1.8	0.8
164.52	0.92	2.0	152.65	18AU	0.3	-3.4	-1.8	0.9
171.00	0.85	1.9	162.38	28AU	-0.9	-3.4	-1.7	0.9
177.53	0.78	1.9	172.17	7SE	-1.2	-3.4	-1.7	0.9
184.12	0.71	1.9	182.01	17SE	-1.0	-3.4	-1.6	0.8
190.77	0.64	1.8	191.92	27SE	-0.6	-3.4	-1.5	0.8
197.48	0.56	1.8	201.87	7OC	-0.3	-3.4	-1.5	0.8
204.25	0.47	1.8	211.88	17OC	-0.2	-3.4	-1.5	0.7
211.10	0.38	1.8	221.95	27OC	-0.2	-3.4	-1.4	0.7
218.01	0.29	1.8	232.05	6NO	0.2	-3.4	-1.4	0.7
224.99	0.19	1.8	242.20	16NO	2.0	-3.4	-1.4	0.7
232.04	0.09	1.8	252.37	26NO	1.2	-3.4	-1.4	0.8
239.15	-0.01	1.7	262.56	6DE	0.0	-3.4	-1.4	0.8
246.34	-0.12	1.7	272.76	16DE	-0.2	-3.4	-1.4	0.8
253.59	-0.24	1.7	282.96	26DE	-0.2	-3.5	-1.4	0.8
				1249				
260.90	-0.35	1.6	293.14	5JA	-0.4	-3.5	-1.4	0.8
268.27	-0.47	1.6	303.30	15JA	-0.7	-3.5	-1.4	0.8
275.70	-0.60	1.5	313.42	25JA	-1.1	-3.4	-1.5	0.7
283.18	-0.72	1.5	323.51	4FE	-1.2	-3.4	-1.5	0.7
290.71	-0.84	1.4	333.54	14FE	-0.8	-3.4	-1.6	0.7
298.27	-0.96	1.4	343.53	24FE	0.7	-3.4	-1.6	0.6
305.86	-1.08	1.3	353.45	6MR	2.8	-3.4	-1.7	0.6
313.47	-1.20	1.3	3.33	16MR	2.3	-3.3	-1.7	0.5
321.09	-1.31	1.2	13.14	26MR	1.2	-3.3	-1.8	0.5
328.70	-1.41	1.1	22.89	5AP	0.7	-3.3	-1.9	0.4
336.30	-1.50	1.1	32.60	15AP	0.3	-3.3	-1.9	0.3
343.86	-1.57	1.0	42.26	25AP	-0.3	-3.4	-2.0	0.3
351.37	-1.64	1.0	51.87	5MY	-1.1	-3.4	-2.1	0.3
358.82	-1.69	0.9	61.46	15MY	-1.8	-3.4	-2.1	0.3
6.17	-1.72	0.9	71.02	25MY	-0.9	-3.4	-2.2	0.4
13.42	-1.74	0.8	80.55	4JN	-0.1	-3.5	-2.2	0.4
20.54	-1.73	0.7	90.08	14JN	0.5	-3.5	-2.3	0.5
27.51	-1.71	0.7	99.61	24JN	1.1	-3.6	-2.3	0.6
34.30	-1.67	0.6	109.15	4JL	1.8	-3.6	-2.3	0.6
40.87	-1.60	0.5	118.71	14JL	3.1	-3.7	-2.3	0.7
47.19	-1.52	0.4	128.29	24JL	2.0	-3.8	-2.2	0.7
53.23	-1.41	0.3	137.91	3AU	0.2	-3.9	-2.2	0.8
58.91	-1.27	0.2	147.57	13AU	-1.0	-4.0	-2.1	0.8
64.16	-1.11	0.1	157.28	23AU	-1.3	-4.1	-2.1	0.8
68.91	-0.91	-0.0	167.03	2SE	-1.0	-4.2	-2.0	0.8
73.03	-0.68	-0.2	176.85	12SE	-0.5	-4.3	-2.0	0.8
76.38	-0.41	-0.4	186.72	22SE	-0.2	-4.3	-1.9	0.8
78.77	-0.08	-0.6	196.64	2OC	-0.1	-4.2	-1.8	0.8
80.01	0.30	-0.8	206.63	12OC	0.0	-3.9	-1.8	0.8
79.90	0.73	-1.0	216.66	22OC	0.4	-3.2	-1.7	0.7
78.35	1.21	-1.2	226.75	1NO	2.3	-3.5	-1.7	0.7
75.46	1.70	-1.4	236.88	11NO	1.0	-4.1	-1.6	0.7
71.74	2.14	-1.5	247.03	21NO	-0.2	-4.3	-1.6	0.7
67.91	2.48	-1.4	257.21	1DE	-0.3	-4.4	-1.5	0.7
64.75	2.69	-1.1	267.41	11DE	-0.4	-4.3	-1.5	0.8
62.78	2.78	-0.8	277.61	21DE	-0.5	-4.2	-1.5	0.8
62.16	2.78	-0.5	287.80	31DE	-0.7	-4.1	-1.5	0.8

♂ LONG	♂ LAT	♂ MAG	☉ LONG	16.00UT	☿	♀	♃	♄
				1250	MAGNITUDES			
62.80	2.73	-0.2	297.98	10JA	-1.0	-4.0	-1.5	0.8
64.54	2.65	0.0	308.12	20JA	-1.1	-3.9	-1.5	0.8
67.15	2.55	0.3	318.22	30JA	-0.7	-3.8	-1.5	0.8
70.46	2.44	0.5	328.29	9FE	0.8	-3.7	-1.5	0.7
74.31	2.33	0.7	338.30	19FE	3.2	-3.6	-1.5	0.7
78.60	2.23	0.9	348.26	1MR	1.7	-3.6	-1.6	0.7
83.22	2.12	1.1	358.16	11MR	0.9	-3.5	-1.6	0.6
88.12	2.02	1.2	8.01	21MR	0.5	-3.5	-1.7	0.5
93.23	1.92	1.4	17.79	31MR	0.2	-3.4	-1.7	0.5
98.53	1.82	1.5	27.53	10AP	-0.3	-3.4	-1.8	0.4
103.97	1.73	1.6	37.21	20AP	-1.1	-3.3	-1.8	0.4
109.54	1.64	1.6	46.84	30AP	-1.8	-3.3	-1.9	0.3
115.21	1.55	1.7	56.45	10MY	-0.9	-3.3	-2.0	0.3
120.99	1.46	1.8	66.01	20MY	0.0	-3.3	-2.1	0.2
126.85	1.37	1.8	75.56	30MY	0.7	-3.3	-2.1	0.3
132.79	1.28	1.9	85.10	9JN	1.4	-3.3	-2.2	0.4
138.81	1.19	1.9	94.62	19JN	2.4	-3.3	-2.3	0.4
144.91	1.10	1.9	104.15	29JN	3.2	-3.4	-2.3	0.5
151.08	1.01	1.9	113.70	9JL	1.6	-3.4	-2.4	0.6
157.33	0.92	1.9	123.27	19JL	0.1	-3.4	-2.4	0.6
163.65	0.83	1.9	132.86	29JL	-1.0	-3.5	-2.4	0.7
170.05	0.74	1.9	142.50	8AU	-1.5	-3.5	-2.4	0.7
176.52	0.64	1.9	152.18	18AU	-1.0	-3.5	-2.4	0.7
183.08	0.55	1.9	161.91	28AU	-0.4	-3.4	-2.4	0.8
189.72	0.45	1.9	171.69	7SE	-0.1	-3.4	-2.3	0.8
196.44	0.36	1.8	181.53	17SE	0.1	-3.4	-2.3	0.8
203.24	0.26	1.8	191.43	27SE	0.2	-3.4	-2.2	0.8
210.14	0.16	1.7	201.39	7OC	0.7	-3.3	-2.1	0.8
217.12	0.06	1.7	211.39	17OC	2.7	-3.3	-2.1	0.8
224.19	-0.04	1.6	221.45	27OC	0.9	-3.3	-2.0	0.7
231.34	-0.15	1.6	231.56	6NO	-0.3	-3.3	-1.9	0.7
238.58	-0.25	1.6	241.70	16NO	-0.5	-3.3	-1.9	0.7
245.90	-0.35	1.6	251.87	26NO	-0.5	-3.4	-1.8	0.7
253.30	-0.45	1.6	262.06	6DE	-0.6	-3.4	-1.7	0.7
260.78	-0.55	1.6	272.26	16DE	-0.8	-3.4	-1.7	0.7
268.33	-0.65	1.5	282.46	26DE	-0.9	-3.4	-1.7	0.8
				1251				
275.94	-0.74	1.5	292.64	5JA	-0.9	-3.5	-1.6	0.8
283.61	-0.83	1.5	302.80	15JA	-0.6	-3.5	-1.6	0.8
291.32	-0.91	1.5	312.93	25JA	1.0	-3.6	-1.6	0.8
299.08	-0.99	1.5	323.02	4FE	2.9	-3.6	-1.6	0.8
306.86	-1.06	1.4	333.05	14FE	1.3	-3.7	-1.6	0.8
314.66	-1.12	1.4	343.04	24FE	0.6	-3.8	-1.6	0.7
322.47	-1.17	1.4	352.98	6MR	0.4	-3.8	-1.6	0.7
330.27	-1.21	1.4	2.85	16MR	0.1	-3.9	-1.6	0.6
338.05	-1.23	1.4	12.67	26MR	-0.4	-4.0	-1.6	0.6
345.79	-1.24	1.3	22.43	5AP	-1.1	-4.1	-1.6	0.5
353.50	-1.24	1.3	32.13	15AP	-1.7	-4.2	-1.7	0.5
1.15	-1.22	1.3	41.80	25AP	-0.9	-4.2	-1.7	0.4
8.72	-1.19	1.3	51.42	5MY	0.1	-4.2	-1.8	0.3
16.22	-1.15	1.3	61.00	15MY	1.0	-3.9	-1.8	0.3
23.63	-1.09	1.3	70.56	25MY	1.9	-3.3	-1.9	0.2
30.93	-1.02	1.3	80.10	4JN	3.1	-2.9	-2.0	0.3
38.12	-0.94	1.2	89.62	14JN	2.7	-3.7	-2.0	0.3
45.19	-0.84	1.2	99.15	24JN	1.3	-4.1	-2.1	0.4
52.13	-0.73	1.2	108.69	4JL	0.1	-4.2	-2.2	0.4
58.92	-0.60	1.2	118.24	14JL	-1.1	-4.2	-2.2	0.5
65.54	-0.47	1.1	127.83	24JL	-1.6	-4.1	-2.3	0.6
71.99	-0.32	1.1	137.44	3AU	-1.0	-4.0	-2.4	0.6
78.25	-0.16	1.0	147.10	13AU	-0.4	-3.9	-2.4	0.7
84.28	0.01	1.0	156.80	23AU	-0.0	-3.9	-2.4	0.7
90.06	0.21	0.9	166.56	2SE	0.2	-3.8	-2.5	0.7
95.55	0.41	0.8	176.37	12SE	0.4	-3.7	-2.5	0.8
100.69	0.64	0.7	186.24	22SE	0.9	-3.6	-2.4	0.8
105.42	0.90	0.6	196.16	2OC	3.0	-3.6	-2.4	0.8
109.65	1.18	0.4	206.14	12OC	0.7	-3.5	-2.4	0.8
113.27	1.50	0.3	216.18	22OC	-0.5	-3.5	-2.3	0.8
116.13	1.85	0.1	226.26	1NO	-0.7	-3.5	-2.3	0.7
118.07	2.25	-0.1	236.38	11NO	-0.6	-3.4	-2.2	0.7
118.89	2.68	-0.3	246.54	21NO	-0.7	-3.4	-2.1	0.7
118.41	3.13	-0.6	256.72	1DE	-0.7	-3.4	-2.0	0.7
116.56	3.57	-0.8	266.91	11DE	-0.7	-3.4	-2.0	0.7
113.47	3.95	-1.0	277.12	21DE	-0.8	-3.4	-1.9	0.7
109.61	4.19	-1.1	287.30	31DE	-0.5	-3.3	-1.8	0.8

Left table (1252 / 1253)

♂ LONG	♂ LAT	♂ MAG	☉ LONG	16.00UT	☿	♀	♃	♄
105.70	4.26	-1.0	297.48	10JA	1.1	-3.3	-1.8	0.8
102.46	4.17	-0.8	307.62	20JA	2.5	-3.3	-1.7	0.8
100.38	3.97	-0.5	317.73	30JA	0.9	-3.4	-1.7	0.8
99.61	3.70	-0.2	327.80	9FE	0.4	-3.4	-1.6	0.8
100.08	3.41	0.0	337.81	19FE	0.2	-3.4	-1.6	0.8
101.63	3.12	0.3	347.77	29FE	-0.0	-3.4	-1.6	0.8
104.05	2.85	0.5	357.68	10MR	-0.4	-3.4	-1.6	0.7
107.19	2.60	0.7	7.52	20MR	-1.1	-3.5	-1.6	0.7
110.90	2.37	0.9	17.31	30MR	-1.7	-3.5	-1.6	0.6
115.05	2.15	1.0	27.05	9AP	-0.9	-3.4	-1.6	0.6
119.58	1.96	1.2	36.73	19AP	0.2	-3.4	-1.6	0.5
124.42	1.78	1.3	46.37	29AP	1.2	-3.4	-1.6	0.5
129.50	1.60	1.4	55.97	9MY	2.5	-3.4	-1.6	0.4
134.80	1.44	1.5	65.55	19MY	3.5	-3.3	-1.7	0.3
140.28	1.29	1.5	75.09	29MY	2.1	-3.3	-1.7	0.3
145.92	1.15	1.6	84.63	8JN	1.0	-3.3	-1.8	0.2
151.72	1.01	1.6	94.16	18JN	-0.0	-3.3	-1.8	0.3
157.65	0.88	1.7	103.69	28JN	-1.1	-3.3	-1.9	0.3
163.72	0.75	1.7	113.23	8JL	-1.7	-3.3	-1.9	0.4
169.90	0.62	1.7	122.80	18JL	-1.0	-3.4	-2.0	0.5
176.20	0.50	1.8	132.39	28JL	-0.3	-3.4	-2.1	0.5
182.62	0.38	1.8	142.03	7AU	0.1	-3.4	-2.1	0.6
189.15	0.26	1.8	151.71	17AU	0.3	-3.5	-2.2	0.6
195.79	0.14	1.7	161.43	27AU	0.6	-3.5	-2.3	0.7
202.54	0.03	1.7	171.22	6SE	1.3	-3.6	-2.3	0.7
209.40	-0.08	1.7	181.05	16SE	3.3	-3.7	-2.4	0.7
216.37	-0.19	1.7	190.95	26SE	0.6	-3.8	-2.4	0.8
223.45	-0.29	1.7	200.90	6OC	-0.6	-3.8	-2.4	0.8
230.63	-0.39	1.6	210.91	16OC	-0.8	-3.9	-2.4	0.8
237.91	-0.49	1.6	220.97	26OC	-0.8	-4.1	-2.4	0.8
245.28	-0.58	1.6	231.07	5NO	-0.8	-4.2	-2.4	0.8
252.75	-0.67	1.5	241.21	15NO	-0.6	-4.3	-2.3	0.7
260.31	-0.75	1.5	251.38	25NO	-0.6	-4.4	-2.3	0.7
267.94	-0.83	1.4	261.57	5DE	-0.6	-4.4	-2.2	0.7
275.64	-0.89	1.4	271.77	15DE	-0.3	-4.3	-2.2	0.7
283.40	-0.95	1.4	281.96	25DE	1.3	-3.9	-2.1	0.7

1253

♂ LONG	♂ LAT	♂ MAG	☉ LONG	16.00UT	☿	♀	♃	♄
291.22	-1.00	1.3	292.15	4JA	2.1	-3.3	-2.0	0.8
299.06	-1.04	1.3	302.31	14JA	0.6	-3.6	-1.9	0.8
306.94	-1.07	1.3	312.43	24JA	0.2	-4.1	-1.9	0.8
314.82	-1.09	1.4	322.52	3FE	0.1	-4.3	-1.8	0.8
322.71	-1.09	1.4	332.57	13FE	-0.1	-4.3	-1.7	0.8
330.58	-1.08	1.4	342.55	23FE	-0.5	-4.3	-1.7	0.8
338.43	-1.07	1.4	352.49	5MR	-1.1	-4.2	-1.6	0.8
346.23	-1.04	1.4	2.37	15MR	-1.6	-4.0	-1.6	0.8
353.99	-0.99	1.4	12.18	25MR	-0.9	-3.9	-1.6	0.7
1.70	-0.94	1.4	21.95	4AP	0.3	-3.8	-1.6	0.7
9.33	-0.88	1.5	31.66	14AP	1.6	-3.7	-1.5	0.7
16.88	-0.80	1.5	41.32	24AP	3.2	-3.7	-1.5	0.6
24.36	-0.72	1.5	50.95	4MY	2.8	-3.6	-1.5	0.5
31.74	-0.63	1.5	60.53	14MY	1.6	-3.5	-1.5	0.5
39.03	-0.53	1.5	70.09	24MY	0.8	-3.5	-1.5	0.4
46.22	-0.43	1.5	79.64	3JN	-0.1	-3.4	-1.5	0.3
53.31	-0.32	1.5	89.16	13JN	-1.1	-3.4	-1.6	0.3
60.29	-0.20	1.6	98.69	23JN	-1.8	-3.4	-1.6	0.3
67.17	-0.08	1.6	108.23	3JL	-0.9	-3.3	-1.6	0.3
73.93	0.05	1.6	117.79	13JL	-0.2	-3.3	-1.7	0.4
80.57	0.18	1.5	127.37	23JL	0.2	-3.3	-1.7	0.4
87.09	0.32	1.5	136.98	2AU	0.5	-3.3	-1.8	0.5
93.48	0.46	1.5	146.63	12AU	0.8	-3.3	-1.9	0.5
99.73	0.61	1.5	156.33	22AU	1.7	-3.4	-1.9	0.6
105.83	0.76	1.4	166.09	1SE	3.1	-3.4	-2.0	0.6
111.77	0.92	1.4	175.89	11SE	0.5	-3.4	-2.0	0.7
117.52	1.09	1.3	185.76	21SE	-0.7	-3.4	-2.1	0.7
123.05	1.27	1.2	195.68	1OC	-1.0	-3.5	-2.2	0.8
128.34	1.46	1.2	205.66	11OC	-0.9	-3.5	-2.2	0.8
133.34	1.67	1.0	215.69	21OC	-0.7	-3.5	-2.3	0.8
137.98	1.90	0.9	225.77	31OC	-0.5	-3.5	-2.3	0.8
142.20	2.14	0.8	235.89	10NO	-0.4	-3.4	-2.3	0.8
145.89	2.42	0.6	246.04	20NO	-0.4	-3.4	-2.3	0.8
148.94	2.71	0.4	256.23	30NO	-0.1	-3.4	-2.3	0.8
151.19	3.04	0.2	266.42	10DE	1.5	-3.4	-2.3	0.7
152.46	3.39	-0.0	276.62	20DE	1.7	-3.4	-2.3	0.7
152.57	3.75	-0.3	286.81	30DE	0.4	-3.3	-2.2	0.7

Right table (1254 / 1255)

♂ LONG	♂ LAT	♂ MAG	☉ LONG	16.00UT	☿	♀	♃	♄
151.38	4.09	-0.5	296.98	9JA	0.1	-3.3	-2.1	0.8
148.91	4.36	-0.8	307.13	19JA	-0.0	-3.3	-2.1	0.8
145.42	4.50	-0.9	317.24	29JA	-0.2	-3.3	-2.0	0.8
141.48	4.48	-1.0	327.31	8FE	-0.6	-3.4	-1.9	0.9
137.81	4.28	-0.8	337.32	18FE	-1.2	-3.4	-1.8	0.9
135.04	3.96	-0.6	347.29	28FE	-1.4	-3.4	-1.8	0.9
133.51	3.57	-0.4	357.19	10MR	-0.9	-3.4	-1.7	0.9
133.26	3.16	-0.1	7.04	20MR	0.4	-3.5	-1.7	0.8
134.20	2.77	0.1	16.84	30MR	2.0	-3.5	-1.6	0.8
136.14	2.40	0.3	26.57	9AP	3.6	-3.6	-1.6	0.8
138.90	2.07	0.5	36.26	19AP	2.1	-3.6	-1.5	0.7
142.34	1.78	0.7	45.91	29AP	1.2	-3.7	-1.5	0.7
146.33	1.51	0.8	55.51	9MY	0.6	-3.7	-1.5	0.6
150.76	1.26	0.9	65.08	19MY	-0.2	-3.8	-1.4	0.6
155.58	1.04	1.0	74.63	29MY	-1.1	-3.9	-1.4	0.5
160.70	0.83	1.1	84.17	8JN	-1.8	-4.0	-1.4	0.4
166.10	0.64	1.2	93.70	18JN	-0.9	-4.1	-1.4	0.4
171.74	0.46	1.3	103.23	28JN	-0.2	-4.2	-1.4	0.3
177.59	0.30	1.3	112.77	8JL	0.3	-4.2	-1.4	0.3
183.64	0.14	1.4	122.34	18JL	0.7	-4.1	-1.5	0.4
189.87	-0.00	1.4	131.94	28JL	1.1	-3.8	-1.5	0.4
196.26	-0.14	1.4	141.56	7AU	2.2	-3.2	-1.5	0.5
202.81	-0.27	1.5	151.24	17AU	2.7	-3.3	-1.6	0.5
209.52	-0.39	1.5	160.97	27AU	0.4	-4.0	-1.6	0.6
216.36	-0.50	1.5	170.74	6SE	-0.8	-4.2	-1.7	0.6
223.35	-0.61	1.5	180.58	16SE	-1.1	-4.3	-1.7	0.7
230.46	-0.71	1.5	190.47	26SE	-1.0	-4.3	-1.8	0.7
237.70	-0.79	1.5	200.42	6OC	-0.6	-4.2	-1.9	0.8
245.05	-0.87	1.5	210.42	16OC	-0.4	-4.1	-1.9	0.8
252.51	-0.94	1.5	220.48	26OC	-0.3	-4.0	-2.0	0.8
260.07	-1.00	1.4	230.58	5NO	-0.3	-3.9	-2.1	0.8
267.72	-1.05	1.4	240.72	15NO	0.0	-3.8	-2.1	0.8
275.44	-1.08	1.4	250.89	25NO	1.8	-3.7	-2.2	0.8
283.23	-1.11	1.4	261.07	5DE	1.5	-3.7	-2.2	0.8
291.07	-1.12	1.4	271.27	15DE	0.1	-3.6	-2.2	0.8
298.95	-1.12	1.4	281.47	25DE	-0.1	-3.6	-2.2	0.8

1255

♂ LONG	♂ LAT	♂ MAG	☉ LONG	16.00UT	☿	♀	♃	♄
306.85	-1.11	1.4	291.65	4JA	-0.2	-3.5	-2.2	0.8
314.76	-1.09	1.4	301.82	14JA	-0.3	-3.5	-2.2	0.8
322.66	-1.05	1.4	311.95	24JA	-0.6	-3.4	-2.1	0.8
330.54	-1.01	1.4	322.04	3FE	-1.2	-3.4	-2.1	0.9
338.40	-0.95	1.4	332.09	13FE	-1.3	-3.4	-2.0	0.9
346.20	-0.89	1.4	342.08	23FE	-0.8	-3.3	-1.9	0.9
353.96	-0.81	1.4	352.01	5MR	0.5	-3.3	-1.9	0.9
1.65	-0.73	1.4	1.89	15MR	2.5	-3.3	-1.8	0.9
9.27	-0.64	1.4	11.71	25MR	2.8	-3.3	-1.7	0.9
16.81	-0.55	1.4	21.48	4AP	1.5	-3.3	-1.7	0.9
24.27	-0.45	1.5	31.19	14AP	0.9	-3.3	-1.6	0.9
31.64	-0.35	1.5	40.86	24AP	0.4	-3.4	-1.5	0.8
38.93	-0.25	1.6	50.48	4MY	-0.2	-3.4	-1.5	0.8
46.13	-0.14	1.6	60.07	14MY	-1.1	-3.4	-1.4	0.7
53.23	-0.03	1.7	69.63	24MY	-1.8	-3.5	-1.4	0.7
60.24	0.07	1.7	79.17	3JN	-0.9	-3.5	-1.4	0.6
67.17	0.18	1.7	88.70	13JN	-0.1	-3.5	-1.4	0.5
74.01	0.29	1.7	98.23	23JN	0.4	-3.4	-1.3	0.5
80.76	0.40	1.8	107.76	3JL	0.9	-3.4	-1.3	0.4
87.42	0.51	1.8	117.32	13JL	1.5	-3.4	-1.3	0.4
94.00	0.62	1.8	126.90	23JL	2.7	-3.3	-1.3	0.4
100.49	0.73	1.8	136.51	2AU	2.3	-3.3	-1.3	0.4
106.90	0.84	1.8	146.16	12AU	0.3	-3.3	-1.3	0.5
113.22	0.94	1.8	155.86	22AU	-0.9	-3.3	-1.4	0.5
119.44	1.05	1.8	165.61	1SE	-1.3	-3.3	-1.4	0.6
125.57	1.16	1.7	175.42	11SE	-1.0	-3.3	-1.4	0.6
131.59	1.28	1.7	185.27	21SE	-0.6	-3.4	-1.5	0.7
137.50	1.39	1.7	195.19	1OC	-0.3	-3.4	-1.5	0.7
143.28	1.51	1.6	205.17	11OC	-0.2	-3.4	-1.6	0.8
148.91	1.64	1.5	215.20	21OC	-0.1	-3.5	-1.6	0.8
154.37	1.76	1.4	225.28	31OC	0.2	-3.5	-1.7	0.8
159.63	1.90	1.3	235.40	10NO	2.0	-3.6	-1.8	0.9
164.64	2.04	1.2	245.55	20NO	1.2	-3.6	-1.8	0.9
169.37	2.20	1.1	255.73	30NO	-0.1	-3.7	-1.9	0.9
173.74	2.36	0.9	265.93	10DE	-0.3	-3.8	-2.0	0.9
177.68	2.53	0.7	276.12	20DE	-0.3	-3.9	-2.0	0.9
181.08	2.72	0.5	286.32	30DE	-0.4	-4.0	-2.1	0.9

♂ LONG	♂ LAT	♂ MAG	☉ LONG	16.00UT	☿	♀	♃	♄
				1256	MAGNITUDES			
183.80	2.91	0.3	296.49	9JA	-0.7	-4.1	-2.1	0.9
185.69	3.11	0.1	306.64	19JA	-1.1	-4.2	-2.1	0.9
186.57	3.31	-0.2	316.75	29JA	-1.2	-4.2	-2.1	0.9
186.26	3.47	-0.5	326.82	8FE	-0.8	-4.3	-2.1	0.9
184.68	3.58	-0.7	336.84	18FE	0.7	-4.3	-2.0	1.0
181.89	3.58	-1.0	346.81	28FE	3.0	-4.2	-2.0	1.0
178.27	3.45	-1.2	356.72	9MR	2.1	-3.8	-1.9	1.0
174.45	3.18	-1.1	6.57	19MR	1.1	-3.2	-1.9	1.0
171.13	2.79	-1.0	16.37	29MR	0.7	-3.6	-1.8	1.0
168.86	2.35	-0.7	26.11	8AP	0.3	-4.0	-1.7	1.0
167.89	1.89	-0.5	35.79	18AP	-0.3	-4.2	-1.7	1.0
168.20	1.46	-0.3	45.44	28AP	-1.1	-4.2	-1.6	1.0
169.67	1.08	-0.1	55.05	8MY	-1.8	-4.1	-1.6	0.9
172.12	0.73	0.1	64.62	18MY	-1.0	-4.1	-1.5	0.9
175.38	0.43	0.2	74.17	28MY	-0.1	-4.0	-1.4	0.8
179.32	0.16	0.4	83.70	7JN	0.6	-3.9	-1.4	0.8
183.79	-0.08	0.5	93.23	17JN	1.2	-3.8	-1.4	0.7
188.73	-0.28	0.6	102.77	27JN	2.0	-3.7	-1.3	0.6
194.07	-0.47	0.7	112.31	7JL	3.2	-3.6	-1.3	0.6
199.73	-0.63	0.8	121.87	17JL	1.9	-3.6	-1.3	0.5
205.69	-0.78	0.9	131.47	27JL	0.2	-3.5	-1.2	0.5
211.92	-0.91	0.9	141.10	6AU	-1.0	-3.5	-1.2	0.5
218.37	-1.02	1.0	150.77	16AU	-1.4	-3.4	-1.2	0.5
225.04	-1.11	1.0	160.50	26AU	-1.0	-3.4	-1.2	0.6
231.89	-1.20	1.1	170.27	5SE	-0.5	-3.4	-1.3	0.6
238.91	-1.26	1.1	180.10	15SE	-0.2	-3.4	-1.3	0.7
246.10	-1.31	1.1	189.99	25SE	-0.0	-3.4	-1.3	0.7
253.42	-1.34	1.2	199.94	5OC	0.1	-3.4	-1.3	0.8
260.86	-1.36	1.2	209.94	15OC	0.5	-3.4	-1.4	0.8
268.42	-1.37	1.2	220.00	25OC	2.4	-3.4	-1.4	0.9
276.06	-1.36	1.2	230.09	4NO	1.0	-3.4	-1.5	0.9
283.78	-1.34	1.3	240.23	14NO	-0.2	-3.4	-1.5	0.9
291.56	-1.30	1.3	250.40	24NO	-0.4	-3.4	-1.6	1.0
299.38	-1.25	1.3	260.58	4DE	-0.4	-3.4	-1.7	1.0
307.22	-1.19	1.3	270.78	14DE	-0.5	-3.4	-1.7	1.0
315.08	-1.12	1.4	280.98	24DE	-0.8	-3.5	-1.8	1.0
				1257				
322.92	-1.04	1.4	291.16	3JA	-1.0	-3.5	-1.9	1.0
330.75	-0.95	1.4	301.32	13JA	-1.0	-3.5	-1.9	1.0
338.54	-0.86	1.4	311.46	23JA	-0.7	-3.4	-2.0	1.0
346.28	-0.76	1.5	321.55	2FE	0.8	-3.4	-2.0	1.0
353.96	-0.65	1.5	331.59	12FE	3.1	-3.4	-2.0	1.0
1.58	-0.54	1.5	341.59	22FE	1.5	-3.4	-2.0	1.1
9.13	-0.44	1.5	351.53	4MR	0.8	-3.4	-2.0	1.1
16.60	-0.33	1.6	1.41	14MR	0.5	-3.3	-2.0	1.1
23.99	-0.22	1.6	11.24	24MR	0.1	-3.3	-2.0	1.2
31.29	-0.11	1.6	21.00	3AP	-0.3	-3.3	-1.9	1.2
38.51	-0.00	1.6	30.72	13AP	-1.1	-3.3	-1.9	1.2
45.65	0.10	1.6	40.39	23AP	-1.8	-3.4	-1.8	1.1
52.71	0.21	1.6	50.01	3MY	-1.0	-3.4	-1.7	1.1
59.68	0.31	1.7	59.60	13MY	0.0	-3.4	-1.7	1.1
66.58	0.40	1.7	69.17	23MY	0.8	-3.4	-1.6	1.0
73.40	0.49	1.8	78.71	2JN	1.6	-3.5	-1.5	1.0
80.15	0.58	1.8	88.24	12JN	2.6	-3.5	-1.5	0.9
86.83	0.67	1.8	97.77	22JN	3.1	-3.6	-1.4	0.9
93.44	0.75	1.9	107.30	2JL	1.5	-3.6	-1.4	0.8
100.00	0.83	1.9	116.86	12JL	0.1	-3.7	-1.3	0.7
106.50	0.91	1.9	126.43	22JL	-1.0	-3.8	-1.3	0.7
112.94	0.98	1.9	136.04	1AU	-1.5	-3.9	-1.3	0.6
119.33	1.05	2.0	145.69	11AU	-1.0	-4.0	-1.2	0.6
125.67	1.12	2.0	155.39	21AU	-0.4	-4.1	-1.2	0.6
131.96	1.19	2.0	165.14	31AU	-0.1	-4.2	-1.2	0.7
138.20	1.25	1.9	174.94	10SE	0.1	-4.3	-1.2	0.7
144.38	1.31	1.9	184.80	20SE	0.3	-4.3	-1.2	0.8
150.51	1.37	1.9	194.71	30SE	0.7	-4.2	-1.2	0.8
156.58	1.43	1.9	204.69	10OC	2.7	-3.8	-1.2	0.9
162.58	1.49	1.8	214.72	20OC	0.9	-3.2	-1.3	0.9
168.50	1.54	1.7	224.79	30OC	-0.4	-3.6	-1.3	1.0
174.35	1.59	1.7	234.91	9NO	-0.6	-4.1	-1.3	1.0
180.09	1.65	1.6	245.06	19NO	-0.6	-4.3	-1.4	1.0
185.71	1.70	1.5	255.24	29NO	-0.6	-4.4	-1.4	1.1
191.19	1.74	1.4	265.43	9DE	-0.8	-4.3	-1.5	1.1
196.50	1.79	1.3	275.63	19DE	-0.8	-4.2	-1.5	1.1
201.61	1.83	1.1	285.82	29DE	-0.8	-4.1	-1.6	1.1

♂ LONG	♂ LAT	♂ MAG	☉ LONG	16.00UT	☿	♀	♃	♄
				1258	MAGNITUDES			
206.47	1.86	0.9	296.00	8JA	-0.6	-4.0	-1.7	1.1
211.01	1.89	0.8	306.15	18JA	0.9	-3.9	-1.7	1.1
215.18	1.91	0.6	316.26	28JA	2.8	-3.8	-1.8	1.1
218.86	1.91	0.3	326.34	7FE	1.1	-3.7	-1.9	1.1
221.95	1.89	0.1	336.36	17FE	0.6	-3.6	-1.9	1.1
224.29	1.84	-0.2	346.33	27FE	0.3	-3.6	-2.0	1.2
225.70	1.75	-0.5	356.24	9MR	0.0	-3.5	-2.0	1.2
226.02	1.60	-0.8	6.10	19MR	-0.4	-3.5	-2.0	1.3
225.10	1.37	-1.1	15.90	29MR	-1.1	-3.4	-2.0	1.3
222.93	1.05	-1.4	25.64	8AP	-1.7	-3.4	-2.0	1.3
219.78	0.64	-1.6	35.33	18AP	-1.0	-3.3	-2.0	1.3
216.21	0.17	-1.7	44.98	28AP	0.1	-3.3	-1.9	1.3
212.93	-0.31	-1.5	54.59	8MY	1.0	-3.3	-1.9	1.3
210.59	-0.75	-1.4	64.16	18MY	2.1	-3.3	-1.8	1.3
209.54	-1.13	-1.1	73.71	28MY	3.4	-3.3	-1.8	1.3
209.84	-1.44	-0.9	83.25	7JN	2.5	-3.3	-1.7	1.2
211.42	-1.68	-0.7	92.78	17JN	1.2	-3.3	-1.7	1.2
214.08	-1.87	-0.5	102.30	27JN	0.1	-3.4	-1.6	1.1
217.65	-2.01	-0.4	111.85	7JL	-1.0	-3.4	-1.5	1.0
221.97	-2.11	-0.2	121.41	17JL	-1.6	-3.4	-1.5	1.0
226.90	-2.17	-0.1	131.00	27JL	-1.0	-3.5	-1.4	0.9
232.34	-2.21	0.1	140.63	6AU	-0.4	-3.5	-1.4	0.8
238.20	-2.22	0.2	150.30	16AU	0.0	-3.5	-1.3	0.8
244.42	-2.21	0.3	160.02	26AU	0.2	-3.4	-1.3	0.7
250.94	-2.17	0.4	169.79	5SE	0.5	-3.4	-1.3	0.8
257.71	-2.12	0.5	179.62	15SE	1.0	-3.4	-1.2	0.8
264.69	-2.05	0.5	189.50	25SE	3.1	-3.4	-1.2	0.9
271.85	-1.96	0.6	199.45	5OC	0.7	-3.3	-1.2	0.9
279.16	-1.86	0.7	209.45	15OC	-0.5	-3.3	-1.2	1.0
286.57	-1.75	0.8	219.50	25OC	-0.7	-3.3	-1.2	1.0
294.08	-1.62	0.9	229.60	4NO	-0.7	-3.3	-1.2	1.1
301.64	-1.49	0.9	239.73	14NO	-0.7	-3.3	-1.3	1.1
309.25	-1.35	1.0	249.90	24NO	-0.7	-3.4	-1.3	1.2
316.88	-1.21	1.1	260.09	4DE	-0.7	-3.4	-1.3	1.2
324.51	-1.06	1.1	270.28	14DE	-0.7	-3.4	-1.4	1.2
332.13	-0.91	1.2	280.48	24DE	-0.4	-3.4	-1.4	1.3
				1259				
339.71	-0.77	1.3	290.67	3JA	1.1	-3.5	-1.5	1.3
347.26	-0.62	1.3	300.83	13JA	2.4	-3.5	-1.5	1.3
354.76	-0.48	1.4	310.97	23JA	0.8	-3.6	-1.6	1.3
2.19	-0.34	1.5	321.07	2FE	0.3	-3.6	-1.7	1.3
9.56	-0.21	1.5	331.11	12FE	0.2	-3.7	-1.7	1.3
16.87	-0.09	1.6	341.11	22FE	-0.1	-3.8	-1.8	1.3
24.09	0.03	1.6	351.06	4MR	-0.5	-3.8	-1.9	1.3
31.25	0.15	1.7	0.94	14MR	-1.1	-3.9	-1.9	1.4
38.33	0.26	1.7	10.77	24MR	-1.6	-4.0	-2.0	1.4
45.33	0.36	1.8	20.54	3AP	-0.9	-4.1	-2.0	1.4
52.26	0.45	1.8	30.26	13AP	0.2	-4.2	-2.1	1.4
59.13	0.54	1.8	39.93	23AP	1.3	-4.2	-2.1	1.3
65.93	0.62	1.8	49.56	3MY	2.7	-4.2	-2.1	1.3
72.66	0.70	1.9	59.14	13MY	3.3	-3.9	-2.1	1.3
79.34	0.77	1.9	68.71	23MY	1.9	-3.3	-2.0	1.2
85.96	0.83	1.9	78.25	2JN	0.9	-2.9	-2.0	1.2
92.53	0.89	1.9	87.78	12JN	-0.0	-3.7	-1.9	1.2
99.06	0.95	1.9	97.31	22JN	-1.1	-4.1	-1.9	1.1
105.55	1.00	1.9	106.84	2JL	-1.7	-4.2	-1.8	1.1
112.00	1.04	1.9	116.39	12JL	-1.0	-4.2	-1.8	1.0
118.43	1.08	2.0	125.97	22JL	-0.3	-4.1	-1.7	1.0
124.82	1.12	2.0	135.58	1AU	0.1	-4.0	-1.6	0.9
131.19	1.15	2.0	145.22	11AU	0.4	-3.9	-1.6	0.9
137.55	1.18	2.0	154.92	21AU	0.7	-3.9	-1.5	0.8
143.88	1.20	2.0	164.66	31AU	1.4	-3.8	-1.5	0.8
150.20	1.22	2.0	174.46	10SE	3.3	-3.7	-1.4	0.8
156.51	1.24	2.0	184.32	20SE	0.6	-3.6	-1.4	0.9
162.80	1.25	2.0	194.23	30SE	-0.7	-3.6	-1.3	1.0
169.09	1.25	2.0	204.20	10OC	-0.9	-3.5	-1.3	1.0
175.35	1.26	1.9	214.23	20OC	-0.8	-3.5	-1.3	1.1
181.61	1.25	1.9	224.30	30OC	-0.8	-3.5	-1.3	1.2
187.84	1.25	1.9	234.41	9NO	-0.6	-3.4	-1.3	1.2
194.06	1.23	1.8	244.57	19NO	-0.5	-3.4	-1.3	1.3
200.24	1.21	1.7	254.75	29NO	-0.5	-3.4	-1.3	1.3
206.40	1.18	1.6	264.94	9DE	-0.3	-3.4	-1.3	1.3
212.52	1.14	1.6	275.14	19DE	1.3	-3.4	-1.4	1.3
218.58	1.09	1.4	285.33	29DE	2.0	-3.3	-1.4	1.3

1260

♂ LONG	LAT	MAG	☉ LONG	16.00UT	☿	♀	♃	♄
224.58	1.03	1.3	295.51	8JA	0.5	-3.3	-1.4	1.3
230.52	0.96	1.2	305.66	18JA	0.2	-3.3	-1.5	1.3
236.35	0.87	1.1	315.78	28JA	0.0	-3.4	-1.5	1.3
242.08	0.76	0.9	325.85	7FE	-0.2	-3.4	-1.6	1.2
247.67	0.62	0.7	335.88	17FE	-0.5	-3.4	-1.6	1.2
253.08	0.46	0.6	345.85	27FE	-1.1	-3.4	-1.7	1.1
258.28	0.26	0.4	355.76	8MR	-1.5	-3.4	-1.8	1.1
263.20	0.02	0.1	5.62	18MR	-0.9	-3.5	-1.9	1.1
267.78	-0.27	-0.1	15.42	28MR	0.3	-3.5	-1.9	1.1
271.93	-0.62	-0.4	25.17	7AP	1.7	-3.4	-2.0	1.1
275.52	-1.04	-0.6	34.86	17AP	3.5	-3.4	-2.0	1.1
278.39	-1.54	-0.9	44.51	27AP	2.5	-3.4	-2.1	1.1
280.39	-2.13	-1.2	54.12	7MY	1.4	-3.4	-2.1	1.1
281.32	-2.81	-1.6	63.70	17MY	0.7	-3.3	-2.2	1.0
281.03	-3.54	-1.9	73.25	27MY	-0.1	-3.3	-2.2	1.0
279.53	-4.29	-2.2	82.78	6JN	-1.1	-3.3	-2.2	1.0
277.06	-4.94	-2.4	92.31	16JN	-1.8	-3.3	-2.2	0.9
274.21	-5.41	-2.4	101.84	26JN	-1.0	-3.3	-2.1	0.9
271.68	-5.64	-2.3	111.38	6JL	-0.2	-3.3	-2.1	0.8
270.11	-5.62	-2.1	120.95	16JL	0.2	-3.4	-2.0	0.8
269.87	-5.42	-1.8	130.53	26JL	0.6	-3.4	-2.0	0.7
270.98	-5.10	-1.6	140.16	5AU	0.9	-3.4	-1.9	0.7
273.34	-4.71	-1.4	149.83	15AU	1.8	-3.5	-1.8	0.6
276.75	-4.30	-1.1	159.54	25AU	3.0	-3.5	-1.8	0.6
281.00	-3.89	-0.9	169.31	4SE	0.5	-3.6	-1.7	0.5
285.93	-3.48	-0.7	179.14	14SE	-0.8	-3.7	-1.7	0.5
291.40	-3.08	-0.5	189.02	24SE	-1.0	-3.8	-1.6	0.6
297.27	-2.70	-0.3	198.96	4OC	-1.0	-3.8	-1.6	0.6
303.46	-2.35	-0.1	208.96	14OC	-0.7	-3.9	-1.5	0.7
309.90	-2.01	0.0	219.01	24OC	-0.5	-4.1	-1.5	0.8
316.51	-1.69	0.2	229.10	3NO	-0.4	-4.2	-1.4	0.8
323.27	-1.40	0.4	239.24	13NO	-0.4	-4.3	-1.4	0.9
330.12	-1.13	0.5	249.40	23NO	-0.1	-4.4	-1.4	0.9
337.03	-0.88	0.7	259.59	3DE	1.5	-4.4	-1.4	1.0
343.98	-0.64	0.8	269.79	13DE	1.7	-4.3	-1.4	1.0
350.94	-0.43	0.9	279.98	23DE	0.3	-3.9	-1.4	1.0

1261

♂ LONG	LAT	MAG	☉ LONG	16.00UT	☿	♀	♃	♄
357.90	-0.24	1.0	290.17	2JA	-0.0	-3.2	-1.4	1.0
4.86	-0.07	1.1	300.34	12JA	-0.1	-3.6	-1.4	1.0
11.79	0.09	1.2	310.47	22JA	-0.2	-4.1	-1.5	1.0
18.68	0.24	1.3	320.57	1FE	-0.6	-4.3	-1.5	1.0
25.54	0.36	1.4	330.63	11FE	-1.1	-4.3	-1.5	1.0
32.36	0.48	1.5	340.63	21FE	-1.4	-4.3	-1.6	0.9
39.13	0.58	1.6	350.57	3MR	-0.9	-4.2	-1.6	0.9
45.86	0.67	1.7	0.46	13MR	0.4	-4.0	-1.7	0.9
52.54	0.76	1.7	10.29	23MR	2.2	-3.9	-1.8	0.8
59.17	0.83	1.8	20.07	2AP	3.3	-3.8	-1.8	0.8
65.77	0.89	1.8	29.79	12AP	1.8	-3.7	-1.9	0.8
72.32	0.94	1.9	39.46	22AP	1.1	-3.7	-2.0	0.8
78.83	0.99	1.9	49.09	2MY	0.5	-3.6	-2.0	0.8
85.31	1.03	2.0	58.69	12MY	-0.2	-3.5	-2.1	0.8
91.76	1.07	2.0	68.25	22MY	-1.1	-3.5	-2.2	0.8
98.19	1.09	2.0	77.79	1JN	-1.8	-3.4	-2.2	0.8
104.59	1.12	2.0	87.32	11JN	-1.0	-3.4	-2.3	0.8
110.97	1.13	2.0	96.85	21JN	-0.2	-3.4	-2.3	0.7
117.34	1.15	2.0	106.39	1JL	0.4	-3.3	-2.3	0.7
123.71	1.15	2.0	115.94	11JL	0.7	-3.3	-2.3	0.6
130.07	1.15	2.0	125.51	21JL	1.3	-3.3	-2.3	0.6
136.43	1.15	2.0	135.12	31JL	2.3	-3.3	-2.2	0.5
142.79	1.14	2.0	144.76	10AU	2.6	-3.3	-2.2	0.5
149.17	1.13	2.0	154.45	20AU	0.4	-3.4	-2.1	0.4
155.55	1.12	2.0	164.19	30AU	-0.9	-3.4	-2.1	0.4
161.96	1.10	2.0	173.99	9SE	-1.2	-3.4	-2.0	0.3
168.38	1.07	2.0	183.84	19SE	-1.1	-3.4	-2.0	0.3
174.82	1.04	2.0	193.75	29SE	-0.6	-3.5	-1.9	0.2
181.29	1.00	2.0	203.72	9OC	-0.3	-3.5	-1.8	0.3
187.78	0.96	2.0	213.73	19OC	-0.2	-3.5	-1.8	0.4
194.30	0.91	1.9	223.81	29OC	-0.2	-3.5	-1.7	0.4
200.84	0.86	1.9	233.92	8NO	0.1	-3.4	-1.7	0.5
207.41	0.80	1.9	244.07	18NO	1.8	-3.4	-1.6	0.6
214.01	0.73	1.8	254.24	28NO	1.4	-3.4	-1.6	0.6
220.64	0.65	1.8	264.43	8DE	0.1	-3.4	-1.5	0.7
227.29	0.57	1.7	274.63	18DE	-0.2	-3.4	-1.5	0.7
233.97	0.47	1.6	284.83	28DE	-0.2	-3.3	-1.5	0.8

1262

♂ LONG	LAT	MAG	☉ LONG	16.00UT	☿	♀	♃	♄
240.67	0.36	1.5	295.01	7JA	-0.3	-3.3	-1.5	0.8
247.39	0.24	1.5	305.16	17JA	-0.6	-3.3	-1.5	0.8
254.13	0.11	1.4	315.28	27JA	-1.1	-3.3	-1.5	0.8
260.88	-0.04	1.3	325.36	6FE	-1.2	-3.4	-1.5	0.8
267.65	-0.20	1.2	335.39	16FE	-0.8	-3.4	-1.5	0.8
274.42	-0.38	1.0	345.37	26FE	0.5	-3.4	-1.6	0.7
281.19	-0.58	0.9	355.28	8MR	2.6	-3.4	-1.6	0.7
287.95	-0.80	0.8	5.15	18MR	2.5	-3.5	-1.6	0.7
294.68	-1.03	0.6	14.95	28MR	1.4	-3.5	-1.7	0.6
301.37	-1.29	0.5	24.70	7AP	0.8	-3.6	-1.7	0.6
308.00	-1.56	0.3	34.40	17AP	0.4	-3.6	-1.8	0.6
314.54	-1.86	0.2	44.05	27AP	-0.2	-3.7	-1.9	0.6
320.95	-2.18	0.0	53.66	7MY	-1.1	-3.7	-1.9	0.6
327.20	-2.51	-0.2	63.24	17MY	-1.9	-3.8	-2.0	0.6
333.22	-2.87	-0.3	72.79	27MY	-1.0	-3.9	-2.1	0.6
338.94	-3.24	-0.5	82.33	6JN	-0.1	-4.0	-2.2	0.6
344.26	-3.63	-0.7	91.86	16JN	0.5	-4.1	-2.3	0.6
349.07	-4.03	-0.9	101.39	26JN	1.0	-4.2	-2.3	0.6
353.21	-4.43	-1.2	110.93	6JL	1.7	-4.2	-2.3	0.5
356.50	-4.82	-1.4	120.49	16JL	2.9	-4.1	-2.4	0.5
358.73	-5.17	-1.7	130.08	26JL	2.2	-3.8	-2.4	0.5
359.69	-5.45	-1.9	139.70	5AU	0.3	-3.2	-2.4	0.4
359.25	-5.61	-2.2	149.37	15AU	-0.9	-3.5	-2.4	0.4
357.48	-5.56	-2.4	159.08	25AU	-1.3	-4.0	-2.4	0.3
354.78	-5.26	-2.5	168.84	4SE	-1.1	-4.2	-2.4	0.2
351.84	-4.71	-2.4	178.67	14SE	-0.5	-4.3	-2.3	0.2
349.45	-3.99	-2.2	188.55	24SE	-0.2	-4.2	-2.3	0.1
348.16	-3.21	-1.9	198.48	4OC	-0.1	-4.2	-2.2	0.1
348.18	-2.47	-1.5	208.48	14OC	-0.0	-4.1	-2.1	0.0
349.47	-1.80	-1.2	218.52	24OC	0.3	-4.0	-2.1	0.1
351.85	-1.23	-0.9	228.61	3NO	2.1	-3.9	-2.0	0.2
355.11	-0.75	-0.6	238.74	13NO	1.2	-3.8	-1.9	0.2
359.07	-0.35	-0.3	248.91	23NO	-0.1	-3.7	-1.9	0.3
3.58	-0.03	-0.1	259.09	3DE	-0.3	-3.7	-1.8	0.4
8.49	0.24	0.2	269.29	13DE	-0.3	-3.6	-1.7	0.4
13.73	0.46	0.4	279.49	23DE	-0.5	-3.6	-1.7	0.5

1263

♂ LONG	LAT	MAG	☉ LONG	16.00UT	☿	♀	♃	♄
19.21	0.64	0.6	289.67	2JA	-0.7	-3.5	-1.7	0.5
24.87	0.79	0.7	299.84	12JA	-1.0	-3.5	-1.6	0.6
30.68	0.91	0.9	309.98	22JA	-1.1	-3.4	-1.6	0.6
36.59	1.01	1.1	320.08	1FE	-0.7	-3.4	-1.6	0.6
42.58	1.09	1.2	330.14	11FE	0.6	-3.4	-1.6	0.6
48.63	1.15	1.3	340.14	21FE	-0.3	-3.3	-1.6	0.6
54.73	1.20	1.4	350.09	3MR	1.9	-3.3	-1.6	0.5
60.85	1.24	1.5	359.99	13MR	1.0	-3.3	-1.6	0.5
67.00	1.27	1.6	9.82	23MR	0.6	-3.3	-1.6	0.5
73.16	1.29	1.7	19.60	2AP	0.2	-3.3	-1.6	0.5
79.33	1.30	1.8	29.32	12AP	-0.3	-3.3	-1.7	0.4
85.52	1.31	1.8	39.00	22AP	-1.1	-3.4	-1.7	0.4
91.71	1.31	1.9	48.63	2MY	-1.8	-3.4	-1.7	0.4
97.90	1.30	1.9	58.22	12MY	-1.0	-3.5	-1.8	0.5
104.11	1.29	2.0	67.79	22MY	-0.0	-3.5	-1.9	0.5
110.32	1.27	2.0	77.33	1JN	0.7	-3.5	-1.9	0.5
116.55	1.25	2.0	86.86	11JN	1.3	-3.5	-2.0	0.5
122.80	1.22	2.0	96.39	21JN	2.2	-3.4	-2.0	0.5
129.06	1.20	2.0	105.92	1JL	3.3	-3.4	-2.1	0.5
135.35	1.16	2.0	115.47	11JL	1.8	-3.4	-2.2	0.4
141.66	1.13	2.0	125.04	21JL	0.2	-3.3	-2.3	0.4
148.00	1.08	2.0	134.65	31JL	-1.0	-3.3	-2.3	0.4
154.37	1.04	2.0	144.29	10AU	-1.4	-3.3	-2.4	0.3
160.78	0.99	2.0	153.98	20AU	-1.0	-3.3	-2.4	0.3
167.23	0.94	1.9	163.72	30AU	-0.5	-3.3	-2.5	0.2
173.73	0.88	1.9	173.51	9SE	-0.1	-3.3	-2.5	0.2
180.27	0.82	1.9	183.36	19SE	0.0	-3.4	-2.5	0.1
186.86	0.76	1.9	193.27	29SE	0.2	-3.4	-2.4	0.0
193.51	0.69	1.9	203.23	9OC	0.6	-3.4	-2.4	-0.0
200.21	0.61	1.9	213.25	19OC	2.4	-3.5	-2.4	-0.1
206.97	0.53	1.9	223.32	29OC	1.0	-3.5	-2.3	-0.1
213.79	0.45	1.9	233.43	8NO	-0.3	-3.6	-2.2	-0.1
220.66	0.36	1.8	243.58	18NO	-0.5	-3.6	-2.2	0.0
227.60	0.26	1.8	253.75	28NO	-0.5	-3.7	-2.1	0.1
234.60	0.16	1.8	263.94	8DE	-0.6	-3.8	-2.0	0.2
241.65	0.05	1.7	274.14	18DE	-0.8	-3.9	-2.0	0.2
248.77	-0.06	1.7	284.33	28DE	-0.9	-4.0	-1.9	0.3

♂			☉	16.00UT	☿	♀	♃	♄
LONG	LAT	MAG	LONG	1264	MAGNITUDES			
255.94	-0.18	1.6	294.52	7JA	-0.9	-4.1	-1.8	0.3
263.17	-0.31	1.6	304.67	17JA	-0.6	-4.2	-1.8	0.4
270.45	-0.44	1.5	314.79	27JA	0.8	-4.3	-1.7	0.4
277.78	-0.58	1.4	324.87	6FE	3.1	-4.3	-1.7	0.4
285.15	-0.72	1.4	334.90	16FE	1.4	-4.3	-1.6	0.5
292.56	-0.86	1.3	344.88	26FE	0.7	-4.2	-1.6	0.5
300.00	-1.01	1.2	354.80	7MR	0.4	-3.8	-1.6	0.4
307.47	-1.15	1.2	4.66	17MR	0.1	-3.2	-1.6	0.4
314.95	-1.29	1.1	14.47	27MR	-0.3	-3.6	-1.6	0.4
322.43	-1.43	1.0	24.23	6AP	-1.1	-4.0	-1.6	0.4
329.89	-1.57	0.9	33.92	16AP	-1.8	-4.2	-1.6	0.3
337.33	-1.70	0.8	43.58	26AP	-1.0	-4.2	-1.6	0.3
344.71	-1.81	0.8	53.20	6MY	0.0	-4.1	-1.6	0.3
352.03	-1.92	0.7	62.77	16MY	0.9	-4.1	-1.6	0.3
359.26	-2.00	0.6	72.33	26MY	1.7	-4.0	-1.7	0.3
6.36	-2.08	0.5	81.87	5JN	2.9	-3.9	-1.7	0.4
13.31	-2.13	0.4	91.39	15JN	2.9	-3.8	-1.8	0.4
20.09	-2.17	0.3	100.93	25JN	1.4	-3.7	-1.8	0.4
26.63	-2.18	0.2	110.47	5JL	0.1	-3.6	-1.9	0.4
32.90	-2.17	0.1	120.02	15JL	-1.0	-3.6	-1.9	0.4
38.85	-2.14	0.0	129.61	25JL	-1.6	-3.5	-2.0	0.4
44.38	-2.07	-0.1	139.24	4AU	-1.0	-3.5	-2.1	0.3
49.42	-1.98	-0.3	148.90	14AU	-0.4	-3.4	-2.1	0.3
53.85	-1.85	-0.4	158.61	24AU	-0.0	-3.4	-2.2	0.2
57.51	-1.68	-0.6	168.37	3SE	0.2	-3.4	-2.3	0.2
60.24	-1.45	-0.8	178.19	13SE	0.3	-3.4	-2.3	0.1
61.82	-1.17	-1.0	188.07	23SE	0.8	-3.4	-2.4	0.1
62.05	-0.81	-1.2	198.01	3OC	2.8	-3.4	-2.4	-0.0
60.82	-0.38	-1.5	207.99	13OC	0.9	-3.4	-2.4	-0.1
58.21	0.10	-1.6	218.04	23OC	-0.4	-3.4	-2.4	-0.1
54.67	0.61	-1.8	228.13	2NO	-0.6	-3.4	-2.4	-0.2
50.95	1.08	-1.7	238.25	12NO	-0.6	-3.4	-2.4	-0.2
47.82	1.46	-1.4	248.42	22NO	-0.7	-3.4	-2.3	-0.2
45.87	1.73	-1.1	258.60	2DE	-0.8	-3.4	-2.3	-0.1
45.27	1.91	-0.8	268.79	12DE	-0.8	-3.4	-2.2	-0.0
45.95	2.02	-0.5	278.99	22DE	-0.8	-3.5	-2.1	0.1
				1265				
47.74	2.07	-0.2	289.18	1JA	-0.5	-3.5	-2.1	0.1
50.43	2.08	0.1	299.35	11JA	0.9	-3.5	-2.0	0.2
53.81	2.07	0.4	309.49	21JA	2.7	-3.4	-1.9	0.2
57.76	2.05	0.6	319.59	31JA	1.0	-3.4	-1.8	0.3
62.13	2.01	0.8	329.64	10FE	0.5	-3.4	-1.8	0.3
66.84	1.96	1.0	339.65	20FE	0.3	-3.4	-1.7	0.3
71.82	1.91	1.1	349.60	2MR	0.0	-3.4	-1.7	0.4
77.00	1.86	1.3	359.50	12MR	-0.4	-3.3	-1.6	0.4
82.35	1.80	1.4	9.34	22MR	-1.1	-3.3	-1.6	0.4
87.84	1.74	1.5	19.12	1AP	-1.7	-3.3	-1.6	0.3
93.44	1.68	1.6	28.84	11AP	-1.0	-3.3	-1.5	0.3
99.14	1.62	1.7	38.52	21AP	0.1	-3.4	-1.5	0.3
104.92	1.55	1.8	48.16	1MY	1.1	-3.4	-1.5	0.3
110.77	1.49	1.8	57.75	11MY	2.3	-3.4	-1.5	0.2
116.69	1.42	1.9	67.32	21MY	3.6	-3.4	-1.5	0.2
122.67	1.35	1.9	76.87	31MY	2.2	-3.5	-1.5	0.3
128.71	1.28	1.9	86.40	10JN	1.1	-3.5	-1.5	0.3
134.80	1.21	2.0	95.93	20JN	0.1	-3.6	-1.6	0.3
140.96	1.14	2.0	105.46	30JN	-1.0	-3.6	-1.6	0.3
147.17	1.07	2.0	115.01	10JL	-1.7	-3.7	-1.6	0.3
153.44	0.99	2.0	124.59	20JL	-1.0	-3.8	-1.7	0.3
159.77	0.92	2.0	134.19	30JL	-0.3	-3.9	-1.7	0.3
166.16	0.84	2.0	143.83	9AU	0.1	-4.0	-1.8	0.3
172.62	0.75	2.0	153.51	19AU	0.3	-4.1	-1.9	0.3
179.15	0.67	1.9	163.25	29AU	0.5	-4.2	-1.9	0.3
185.75	0.58	1.9	173.04	8SE	1.2	-4.3	-2.0	0.2
192.43	0.50	1.8	182.89	18SE	3.1	-4.3	-2.0	0.1
199.18	0.40	1.8	192.79	28SE	0.8	-4.2	-2.1	0.1
206.01	0.31	1.8	202.75	8OC	-0.6	-3.8	-2.2	0.0
212.92	0.21	1.7	212.77	18OC	-0.8	-3.2	-2.2	-0.1
219.90	0.12	1.7	222.83	28OC	-0.8	-3.6	-2.3	-0.1
226.97	0.01	1.7	232.94	7NO	-0.8	-4.1	-2.3	-0.2
234.12	-0.09	1.7	243.09	17NO	-0.6	-4.3	-2.3	-0.3
241.35	-0.19	1.7	253.26	27NO	-0.6	-4.4	-2.3	-0.3
248.65	-0.30	1.6	263.45	7DE	-0.6	-4.3	-2.3	-0.2
256.03	-0.40	1.6	273.65	17DE	-0.4	-4.2	-2.3	-0.1
263.48	-0.51	1.6	283.84	27DE	1.1	-4.1	-2.2	-0.1

♂			☉	16.00UT	☿	♀	♃	♄
LONG	LAT	MAG	LONG	1266	MAGNITUDES			
270.99	-0.62	1.6	294.02	6JA	2.3	-4.0	-2.2	0.0
278.57	-0.72	1.5	304.18	16JA	0.7	-3.9	-2.1	0.1
286.19	-0.82	1.5	314.30	26JA	0.3	-3.8	-2.0	0.1
293.86	-0.92	1.5	324.38	5FE	0.1	-3.7	-2.0	0.2
301.57	-1.01	1.4	334.42	15FE	-0.1	-3.6	-1.9	0.2
309.30	-1.09	1.4	344.40	25FE	-0.5	-3.6	-1.8	0.3
317.04	-1.17	1.4	354.32	7MR	-1.1	-3.5	-1.7	0.3
324.80	-1.23	1.3	4.19	17MR	-1.6	-3.5	-1.7	0.3
332.54	-1.29	1.3	14.00	27MR	-1.0	-3.4	-1.6	0.3
340.26	-1.33	1.3	23.76	6AP	0.2	-3.4	-1.6	0.3
347.94	-1.36	1.3	33.46	16AP	1.5	-3.3	-1.5	0.3
355.58	-1.37	1.2	43.11	26AP	3.0	-3.3	-1.5	0.3
3.15	-1.37	1.2	52.73	6MY	3.0	-3.3	-1.5	0.3
10.65	-1.35	1.2	62.31	16MY	1.7	-3.3	-1.4	0.2
18.06	-1.32	1.1	71.87	26MY	0.8	-3.3	-1.4	0.2
25.36	-1.27	1.1	81.41	5JN	-0.0	-3.3	-1.4	0.2
32.55	-1.21	1.1	90.94	15JN	-1.0	-3.3	-1.4	0.2
39.60	-1.13	1.0	100.46	25JN	-1.7	-3.4	-1.4	0.3
46.50	-1.04	1.0	110.01	5JL	-1.0	-3.4	-1.4	0.3
53.24	-0.92	0.9	119.56	15JL	-0.3	-3.4	-1.4	0.3
59.78	-0.80	0.9	129.15	25JL	0.2	-3.5	-1.5	0.3
66.12	-0.65	0.8	138.77	4AU	0.5	-3.5	-1.5	0.3
72.21	-0.49	0.8	148.43	14AU	0.8	-3.5	-1.5	0.3
78.03	-0.31	0.7	158.13	24AU	1.5	-3.4	-1.6	0.3
83.52	-0.11	0.6	167.90	3SE	3.2	-3.4	-1.6	0.3
88.62	0.12	0.5	177.71	13SE	0.6	-3.4	-1.7	0.3
93.26	0.37	0.3	187.59	23SE	-0.7	-3.4	-1.7	0.2
97.33	0.66	0.2	197.52	3OC	-0.9	-3.3	-1.8	0.2
100.69	0.99	0.0	207.51	13OC	-0.9	-3.3	-1.9	0.1
103.21	1.36	-0.2	217.54	23OC	-0.8	-3.3	-1.9	0.0
104.67	1.77	-0.4	227.63	2NO	-0.5	-3.3	-2.0	-0.0
104.88	2.22	-0.6	237.76	12NO	-0.5	-3.3	-2.1	-0.1
103.71	2.70	-0.8	247.92	22NO	-0.5	-3.4	-2.1	-0.2
101.18	3.15	-1.0	258.10	2DE	-0.2	-3.4	-2.1	-0.3
97.61	3.53	-1.2	268.30	12DE	1.3	-3.4	-2.2	-0.3
93.65	3.76	-1.2	278.49	22DE	2.0	-3.4	-2.2	-0.2
				1267				
90.05	3.83	-1.0	288.69	1JA	0.4	-3.5	-2.2	-0.1
87.45	3.77	-0.7	298.86	11JA	0.1	-3.5	-2.2	-0.1
86.15	3.61	-0.4	309.00	21JA	-0.0	-3.6	-2.1	0.0
86.15	3.40	-0.1	319.11	31JA	-0.2	-3.6	-2.1	0.1
87.31	3.17	0.1	329.16	10FE	-0.5	-3.7	-2.0	0.1
89.43	2.95	0.4	339.17	20FE	-1.1	-3.8	-2.0	0.2
92.33	2.74	0.6	349.13	2MR	-1.5	-3.8	-1.9	0.2
95.84	2.54	0.8	359.02	12MR	-0.9	-3.9	-1.8	0.3
99.84	2.35	0.9	8.86	22MR	0.3	-4.0	-1.8	0.3
104.24	2.18	1.1	18.65	1AP	1.8	-4.1	-1.7	0.3
108.95	2.01	1.2	28.38	11AP	3.8	-4.2	-1.6	0.3
113.92	1.86	1.3	38.05	21AP	2.2	-4.2	-1.6	0.3
119.10	1.72	1.5	47.69	1MY	1.3	-4.2	-1.5	0.3
124.47	1.58	1.5	57.29	11MY	0.6	-3.9	-1.5	0.3
129.99	1.45	1.6	66.86	21MY	-0.1	-3.2	-1.4	0.3
135.66	1.33	1.7	76.40	31MY	-1.0	-2.9	-1.4	0.3
141.45	1.20	1.7	85.93	10JN	-1.8	-3.7	-1.4	0.2
147.35	1.08	1.8	95.46	20JN	-1.0	-4.1	-1.3	0.2
153.37	0.97	1.8	105.00	30JN	-0.2	-4.2	-1.3	0.3
159.48	0.85	1.8	114.54	10JL	0.3	-4.2	-1.3	0.3
165.70	0.74	1.8	124.11	20JL	0.6	-4.1	-1.3	0.4
172.03	0.63	1.8	133.72	30JL	1.0	-4.0	-1.3	0.4
178.44	0.52	1.8	143.35	9AU	2.0	-3.9	-1.3	0.4
184.96	0.41	1.8	153.04	19AU	2.9	-3.8	-1.3	0.4
191.58	0.30	1.8	162.77	29AU	0.5	-3.8	-1.4	0.4
198.29	0.20	1.8	172.56	8SE	-0.8	-3.7	-1.4	0.4
205.11	0.09	1.8	182.40	18SE	-1.1	-3.6	-1.4	0.3
212.02	-0.02	1.7	192.31	28SE	-1.0	-3.6	-1.5	0.3
219.04	-0.12	1.7	202.26	8OC	-0.7	-3.5	-1.5	0.3
226.15	-0.22	1.7	212.28	18OC	-0.4	-3.5	-1.6	0.2
233.36	-0.33	1.6	222.34	28OC	-0.3	-3.5	-1.6	0.1
240.66	-0.42	1.6	232.45	7NO	-0.3	-3.4	-1.7	0.1
248.05	-0.52	1.5	242.59	17NO	-0.0	-3.4	-1.8	0.0
255.53	-0.61	1.5	252.77	27NO	1.5	-3.4	-1.8	-0.1
263.08	-0.70	1.4	262.95	7DE	1.7	-3.4	-1.9	-0.1
270.72	-0.78	1.4	273.15	17DE	0.2	-3.4	-2.0	-0.2
278.41	-0.86	1.4	283.35	27DE	-0.1	-3.3	-2.0	-0.2

Left table — ♂ / ☉ / 16.00UT — year 1268

♂ LONG	LAT	MAG	☉ LONG	16.00UT 1268	☿	♀	♃	♄
286.16	-0.92	1.4	293.53	6JA	-0.1	-3.3	-2.0	-0.1
293.96	-0.98	1.4	303.69	16JA	-0.3	-3.3	-2.1	-0.1
301.79	-1.03	1.4	313.81	26JA	-0.6	-3.4	-2.1	-0.0
309.64	-1.07	1.4	323.89	5FE	-1.1	-3.4	-2.1	0.1
317.51	-1.10	1.4	333.93	15FE	-1.3	-3.4	-2.0	0.1
325.37	-1.12	1.4	343.92	25FE	-0.9	-3.4	-2.0	0.2
333.22	-1.13	1.4	353.84	6MR	0.4	-3.4	-2.0	0.2
341.04	-1.12	1.4	3.71	16MR	2.3	-3.5	-1.9	0.3
348.82	-1.10	1.4	13.53	26MR	3.0	-3.5	-1.8	0.3
356.55	-1.07	1.4	23.28	5AP	1.6	-3.4	-1.8	0.4
4.22	-1.02	1.4	32.99	15AP	1.0	-3.4	-1.7	0.4
11.82	-0.97	1.4	42.64	25AP	0.5	-3.4	-1.7	0.4
19.34	-0.90	1.4	52.26	5MY	-0.2	-3.4	-1.6	0.4
26.78	-0.82	1.4	61.84	15MY	-1.0	-3.3	-1.5	0.4
34.11	-0.73	1.4	71.40	25MY	-1.9	-3.3	-1.5	0.4
41.35	-0.64	1.4	80.93	4JN	-1.0	-3.3	-1.4	0.4
48.48	-0.53	1.4	90.47	14JN	-0.2	-3.3	-1.4	0.3
55.50	-0.41	1.4	100.00	24JN	0.4	-3.3	-1.3	0.3
62.40	-0.29	1.4	109.53	4JL	0.8	-3.3	-1.3	0.3
69.17	-0.16	1.4	119.09	14JL	1.4	-3.4	-1.3	0.4
75.81	-0.03	1.4	128.68	24JL	2.5	-3.4	-1.2	0.4
82.31	0.12	1.4	138.29	3AU	2.5	-3.4	-1.2	0.4
88.65	0.27	1.3	147.96	13AU	0.4	-3.5	-1.2	0.5
94.83	0.43	1.3	157.66	23AU	-0.9	-3.5	-1.2	0.5
100.83	0.60	1.2	167.42	2SE	-1.2	-3.6	-1.2	0.5
106.62	0.78	1.2	177.24	12SE	-1.1	-3.7	-1.3	0.5
112.17	0.98	1.1	187.11	22SE	-0.6	-3.8	-1.3	0.5
117.44	1.19	1.0	197.04	2OC	-0.3	-3.8	-1.3	0.5
122.39	1.42	0.9	207.02	12OC	-0.2	-4.0	-1.3	0.4
126.95	1.67	0.8	217.06	22OC	-0.1	-4.1	-1.4	0.4
131.03	1.94	0.6	227.14	1NO	0.2	-4.2	-1.4	0.3
134.52	2.24	0.5	237.27	11NO	1.8	-4.3	-1.5	0.3
137.29	2.58	0.3	247.43	21NO	1.4	-4.4	-1.5	0.2
139.17	2.95	0.1	257.61	1DE	0.0	-4.4	-1.6	0.1
139.97	3.34	-0.2	267.81	11DE	-0.2	-4.3	-1.7	0.0
139.51	3.74	-0.4	278.00	21DE	-0.3	-3.9	-1.7	-0.0
137.73	4.12	-0.7	288.19	31DE	-0.4	-3.2	-1.8	-0.1

1269

♂ LONG	LAT	MAG	☉ LONG	16.00UT	☿	♀	♃	♄
134.72	4.41	-0.9	298.37	10JA	-0.7	-3.7	-1.9	-0.1
130.91	4.55	-1.0	308.51	20JA	-1.1	-4.1	-1.9	-0.0
126.99	4.51	-0.9	318.61	30JA	-1.2	-4.3	-2.0	0.0
123.67	4.31	-0.7	328.68	9FE	-0.8	-4.3	-2.0	0.1
121.45	4.00	-0.5	338.69	19FE	0.5	-4.2	-2.0	0.2
120.51	3.64	-0.2	348.64	1MR	2.7	-4.2	-2.0	0.2
120.83	3.27	0.0	358.55	11MR	2.3	-4.0	-2.0	0.3
122.25	2.92	0.3	8.39	21MR	1.2	-3.9	-2.0	0.3
124.57	2.59	0.5	18.17	31MR	0.7	-3.8	-1.9	0.4
127.64	2.29	0.6	27.91	10AP	0.3	-3.7	-1.9	0.4
131.30	2.02	0.8	37.59	20AP	-0.2	-3.7	-1.8	0.5
135.45	1.77	1.0	47.23	30AP	-1.0	-3.6	-1.8	0.5
139.99	1.55	1.1	56.83	10MY	-1.9	-3.5	-1.7	0.5
144.86	1.34	1.2	66.40	20MY	-1.0	-3.5	-1.6	0.5
150.01	1.14	1.3	75.94	30MY	-0.1	-3.4	-1.6	0.5
155.41	0.96	1.4	85.48	9JN	0.5	-3.4	-1.5	0.5
161.01	0.79	1.4	95.00	19JN	1.1	-3.4	-1.5	0.5
166.81	0.63	1.5	104.54	29JN	1.9	-3.3	-1.4	0.5
172.78	0.48	1.5	114.09	9JL	3.1	-3.3	-1.4	0.5
178.90	0.34	1.5	123.65	19JL	2.1	-3.3	-1.3	0.5
185.18	0.20	1.6	133.25	29JL	0.3	-3.3	-1.3	0.5
191.61	0.07	1.6	142.89	8AU	-0.9	-3.3	-1.3	0.6
198.17	-0.06	1.6	152.57	18AU	-1.4	-3.4	-1.2	0.6
204.86	-0.18	1.6	162.30	28AU	-1.1	-3.4	-1.2	0.6
211.69	-0.29	1.6	172.09	7SE	-0.5	-3.4	-1.2	0.6
218.64	-0.40	1.6	181.93	17SE	-0.2	-3.4	-1.2	0.7
225.71	-0.51	1.6	191.83	27SE	-0.1	-3.5	-1.2	0.7
232.90	-0.60	1.6	201.78	7OC	0.0	-3.5	-1.2	0.6
240.19	-0.69	1.5	211.79	17OC	0.4	-3.5	-1.2	0.6
247.60	-0.77	1.5	221.85	27OC	2.1	-3.5	-1.3	0.6
255.10	-0.85	1.5	231.96	6NO	1.2	-3.4	-1.3	0.5
262.69	-0.91	1.5	242.10	16NO	-0.2	-3.4	-1.3	0.5
270.37	-0.97	1.5	252.27	26NO	-0.4	-3.4	-1.4	0.4
278.11	-1.02	1.4	262.46	6DE	-0.4	-3.4	-1.4	0.4
285.91	-1.05	1.4	272.65	16DE	-0.5	-3.4	-1.5	0.3
293.76	-1.07	1.4	282.85	26DE	-0.7	-3.3	-1.6	0.2

Right table — ♂ / ☉ / 16.00UT — year 1270

♂ LONG	LAT	MAG	☉ LONG	16.00UT 1270	☿	♀	♃	♄
301.64	-1.09	1.4	293.04	5JA	-1.0	-3.3	-1.6	0.1
309.54	-1.09	1.3	303.19	15JA	-1.0	-3.3	-1.7	0.1
317.45	-1.08	1.3	313.32	25JA	-0.7	-3.3	-1.8	0.1
325.35	-1.05	1.3	323.41	4FE	0.6	-3.4	-1.8	0.1
333.23	-1.02	1.3	333.44	14FE	3.1	-3.4	-1.9	0.2
341.08	-0.98	1.3	343.43	24FE	1.7	-3.4	-1.9	0.3
348.88	-0.92	1.4	353.36	6MR	0.9	-3.4	-2.0	0.3
356.63	-0.86	1.4	3.23	16MR	0.5	-3.5	-2.0	0.4
4.31	-0.78	1.4	13.05	26MR	0.2	-3.5	-2.0	0.4
11.92	-0.70	1.5	22.81	5AP	-0.3	-3.6	-2.0	0.5
19.45	-0.61	1.5	32.52	15AP	-1.0	-3.6	-2.0	0.5
26.90	-0.52	1.5	42.18	25AP	-1.8	-3.7	-2.0	0.6
34.26	-0.42	1.6	51.80	5MY	-1.0	-3.8	-1.9	0.6
41.53	-0.32	1.6	61.38	15MY	-0.0	-3.8	-1.9	0.6
48.70	-0.21	1.6	70.94	25MY	0.7	-3.9	-1.8	0.7
55.78	-0.10	1.6	80.48	4JN	1.4	-4.0	-1.8	0.7
62.77	0.01	1.7	90.01	14JN	2.5	-4.1	-1.7	0.7
69.66	0.13	1.7	99.54	24JN	3.2	-4.2	-1.6	0.7
76.45	0.24	1.7	109.08	4JL	1.7	-4.2	-1.6	0.7
83.15	0.36	1.7	118.63	14JL	0.2	-4.1	-1.5	0.7
89.75	0.48	1.7	128.22	24JL	-1.0	-3.7	-1.5	0.7
96.25	0.60	1.7	137.83	3AU	-1.5	-3.2	-1.4	0.7
102.65	0.72	1.7	147.49	13AU	-1.0	-3.5	-1.4	0.7
108.94	0.85	1.7	157.19	23AU	-0.5	-4.0	-1.3	0.8
115.12	0.98	1.7	166.95	2SE	-0.1	-4.2	-1.3	0.8
121.18	1.11	1.6	176.76	12SE	0.1	-4.3	-1.3	0.8
127.10	1.24	1.6	186.63	22SE	0.2	-4.2	-1.2	0.9
132.88	1.38	1.5	196.55	2OC	0.6	-4.2	-1.2	0.9
138.48	1.53	1.4	206.53	12OC	2.5	-4.1	-1.2	0.9
143.89	1.68	1.4	216.57	22OC	1.1	-4.0	-1.2	0.9
149.07	1.85	1.3	226.65	1NO	-0.3	-3.9	-1.2	0.8
153.98	2.02	1.1	236.78	11NO	-0.5	-3.8	-1.3	0.8
158.55	2.21	1.0	246.94	21NO	-0.5	-3.7	-1.3	0.8
162.71	2.42	0.8	257.12	1DE	-0.6	-3.7	-1.3	0.7
166.38	2.64	0.7	267.31	11DE	-0.8	-3.6	-1.4	0.6
169.43	2.88	0.5	277.51	21DE	-0.8	-3.6	-1.4	0.6
171.70	3.14	0.2	287.70	31DE	-0.9	-3.5	-1.4	0.5

1271

♂ LONG	LAT	MAG	☉ LONG	16.00UT	☿	♀	♃	♄
173.04	3.40	-0.0	297.87	10JA	-0.6	-3.5	-1.5	0.4
173.25	3.67	-0.3	308.02	20JA	0.8	-3.4	-1.6	0.4
172.19	3.90	-0.5	318.13	30JA	3.0	-3.4	-1.6	0.3
169.87	4.05	-0.8	328.19	9FE	1.2	-3.4	-1.7	0.3
166.49	4.08	-1.0	338.21	19FE	0.6	-3.3	-1.8	0.4
162.62	3.95	-1.1	348.17	1MR	0.3	-3.3	-1.8	0.4
158.95	3.67	-0.9	358.07	11MR	0.1	-3.3	-1.9	0.5
156.12	3.28	-0.7	7.92	21MR	-0.4	-3.3	-2.0	0.5
154.51	2.84	-0.5	17.70	31MR	-1.1	-3.3	-2.0	0.6
154.19	2.40	-0.3	27.44	10AP	-1.7	-3.3	-2.0	0.6
155.09	1.98	-0.1	37.13	20AP	-1.0	-3.4	-2.1	0.7
157.03	1.61	0.1	46.76	30AP	0.0	-3.4	-2.1	0.7
159.83	1.27	0.3	56.36	10MY	0.9	-3.4	-2.1	0.8
163.35	0.97	0.5	65.94	20MY	1.9	-3.5	-2.1	0.8
167.44	0.70	0.6	75.48	30MY	3.2	-3.5	-2.0	0.9
172.02	0.46	0.7	85.02	9JN	2.6	-3.5	-2.0	0.9
177.00	0.24	0.8	94.54	19JN	1.3	-3.4	-1.9	0.9
182.33	0.04	0.9	104.07	29JN	0.1	-3.4	-1.9	0.9
187.95	-0.14	1.0	113.62	9JL	-1.0	-3.4	-1.8	0.9
193.85	-0.30	1.1	123.19	19JL	-1.6	-3.3	-1.7	0.9
199.97	-0.45	1.1	132.78	29JL	-1.0	-3.3	-1.7	0.9
206.31	-0.58	1.2	142.42	8AU	-0.4	-3.3	-1.6	0.9
212.85	-0.71	1.2	152.10	18AU	-0.0	-3.3	-1.6	0.9
219.56	-0.82	1.2	161.82	28AU	0.2	-3.3	-1.5	1.0
226.45	-0.91	1.3	171.60	7SE	0.4	-3.3	-1.5	1.1
233.49	-1.00	1.3	181.44	17SE	0.9	-3.4	-1.4	1.1
240.67	-1.07	1.3	191.34	27SE	2.8	-3.4	-1.4	1.1
247.98	-1.13	1.3	201.29	7OC	0.9	-3.4	-1.4	1.1
255.42	-1.18	1.3	211.30	17OC	-0.5	-3.5	-1.3	1.1
262.96	-1.21	1.3	221.36	27OC	-0.7	-3.5	-1.3	1.1
270.59	-1.23	1.3	231.46	6NO	-0.7	-3.6	-1.3	1.1
278.31	-1.24	1.3	241.60	16NO	-0.7	-3.6	-1.3	1.1
286.05	-1.23	1.4	251.77	26NO	-0.7	-3.7	-1.3	1.1
293.92	-1.21	1.4	261.97	6DE	-0.7	-3.8	-1.3	1.0
301.78	-1.18	1.4	272.16	16DE	-0.7	-3.9	-1.3	1.0
309.66	-1.14	1.4	282.36	26DE	-0.5	-4.0	-1.4	0.9

♂ LONG	♂ LAT	♂ MAG	☉ LONG	16.00UT 1272	☿	♀	♃	♄
					MAGNITUDES			
317.54	-1.08	1.4	292.55	5JA	0.9	-4.1	-1.4	0.8
325.42	-1.02	1.4	302.71	15JA	2.6	-4.2	-1.5	0.8
333.26	-0.95	1.4	312.83	25JA	0.9	-4.3	-1.5	0.7
341.08	-0.86	1.4	322.92	4FE	0.4	-4.3	-1.6	0.6
348.84	-0.78	1.4	332.96	14FE	0.2	-4.3	-1.6	0.6
356.54	-0.68	1.5	342.95	24FE	-0.0	-4.2	-1.7	0.6
4.17	-0.58	1.5	352.89	5MR	-0.4	-3.7	-1.7	0.6
11.73	-0.48	1.5	2.76	15MR	-1.1	-3.2	-1.8	0.7
19.21	-0.38	1.5	12.58	25MR	-1.7	-3.6	-1.9	0.7
26.61	-0.27	1.5	22.34	4AP	-1.0	-4.1	-2.0	0.8
33.93	-0.17	1.5	32.05	14AP	0.1	-4.2	-2.0	0.9
41.15	-0.06	1.5	41.71	24AP	1.2	-4.2	-2.1	0.9
48.29	0.04	1.6	51.34	4MY	2.5	-4.1	-2.1	1.0
55.35	0.15	1.6	60.92	14MY	3.5	-4.1	-2.2	1.0
62.32	0.25	1.7	70.48	24MY	2.0	-3.9	-2.2	1.1
69.22	0.35	1.7	80.02	3JN	1.0	-3.9	-2.2	1.1
76.03	0.45	1.8	89.55	13JN	0.1	-3.8	-2.2	1.1
82.76	0.55	1.8	99.08	23JN	-1.0	-3.7	-2.2	1.2
89.43	0.64	1.8	108.62	3JL	-1.7	-3.6	-2.1	1.2
96.02	0.73	1.9	118.17	13JL	-1.0	-3.6	-2.1	1.2
102.55	0.82	1.9	127.75	23JL	-0.3	-3.5	-2.0	1.2
109.00	0.91	1.9	137.37	2AU	0.1	-3.5	-2.0	1.2
115.40	0.99	1.9	147.02	12AU	0.4	-3.4	-1.9	1.2
121.72	1.08	1.9	156.72	22AU	0.6	-3.4	-1.8	1.2
127.99	1.16	1.9	166.48	1SE	1.3	-3.4	-1.8	1.2
134.18	1.24	1.9	176.28	11SE	3.2	-3.4	-1.7	1.3
140.30	1.32	1.9	186.15	21SE	0.8	-3.4	-1.7	1.3
146.35	1.40	1.8	196.07	1OC	-0.6	-3.4	-1.6	1.4
152.31	1.48	1.8	206.05	11OC	-0.9	-3.4	-1.6	1.4
158.18	1.56	1.7	216.08	21OC	-0.8	-3.4	-1.5	1.3
163.93	1.64	1.6	226.16	31OC	-0.8	-3.4	-1.5	1.3
169.56	1.73	1.6	236.28	10NO	-0.6	-3.4	-1.5	1.3
175.05	1.81	1.5	246.44	20NO	-0.5	-3.4	-1.4	1.2
180.36	1.90	1.3	256.62	30NO	-0.5	-3.4	-1.4	1.2
185.46	1.98	1.2	266.81	10DE	-0.3	-3.4	-1.4	1.2
190.31	2.07	1.1	277.01	20DE	1.1	-3.5	-1.4	1.1
194.84	2.16	0.9	287.21	30DE	2.2	-3.5	-1.4	1.1
				1273				
198.98	2.25	0.7	297.38	9JA	0.6	-3.5	-1.4	1.0
202.65	2.34	0.5	307.53	19JA	0.2	-3.4	-1.5	1.0
205.71	2.42	0.3	317.64	29JA	0.1	-3.4	-1.5	0.9
208.04	2.49	0.0	327.70	8FE	-0.1	-3.4	-1.5	0.9
209.44	2.53	-0.3	337.72	18FE	-0.5	-3.4	-1.6	0.9
209.74	2.54	-0.5	347.69	28FE	-1.1	-3.4	-1.6	0.8
208.80	2.48	-0.8	357.59	10MR	-1.5	-3.3	-1.7	0.9
206.61	2.33	-1.1	7.44	20MR	-1.0	-3.3	-1.7	1.0
203.39	2.07	-1.4	17.23	30MR	0.2	-3.3	-1.8	1.0
199.69	1.71	-1.4	26.97	9AP	1.6	-3.3	-1.9	1.1
196.21	1.27	-1.3	36.65	19AP	3.3	-3.4	-1.9	1.1
193.60	0.81	-1.1	46.30	29AP	2.7	-3.4	-2.0	1.2
192.25	0.37	-0.9	55.90	9MY	1.5	-3.4	-2.1	1.2
192.24	-0.03	-0.7	65.47	19MY	0.8	-3.4	-2.1	1.3
193.48	-0.37	-0.5	75.02	29MY	-0.0	-3.5	-2.2	1.3
195.81	-0.66	-0.3	84.55	8JN	-1.0	-3.5	-2.3	1.4
199.04	-0.90	-0.1	94.08	18JN	-1.8	-3.6	-2.3	1.4
203.03	-1.10	0.0	103.62	28JN	-1.0	-3.6	-2.3	1.4
207.64	-1.27	0.1	113.16	8JL	-0.3	-3.7	-2.3	1.4
212.75	-1.41	0.3	122.73	18JL	0.2	-3.8	-2.3	1.4
218.31	-1.52	0.4	132.32	28JL	0.5	-3.9	-2.3	1.4
224.23	-1.61	0.5	141.95	7AU	0.9	-4.0	-2.3	1.3
230.47	-1.67	0.5	151.63	17AU	1.7	-4.1	-2.2	1.3
236.98	-1.72	0.6	161.36	27AU	3.2	-4.2	-2.2	1.2
243.73	-1.74	0.7	171.13	6SE	0.7	-4.3	-2.1	1.2
250.69	-1.75	0.8	180.97	16SE	-0.7	-4.3	-2.0	1.2
257.82	-1.74	0.8	190.86	26SE	-1.0	-4.2	-2.0	1.2
265.11	-1.71	0.9	200.81	6OC	-1.0	-3.8	-1.9	1.2
272.53	-1.66	0.9	210.82	16OC	-0.7	-3.1	-1.8	1.2
280.06	-1.60	1.0	220.87	26OC	-0.5	-3.7	-1.8	1.2
287.66	-1.53	1.0	230.97	5NO	-0.4	-4.2	-1.7	1.1
295.34	-1.44	1.1	241.11	15NO	-0.4	-4.4	-1.7	1.1
303.06	-1.35	1.1	251.28	25NO	-0.2	-4.4	-1.6	1.1
310.80	-1.24	1.2	261.46	5DE	1.3	-4.3	-1.6	1.1
318.55	-1.13	1.2	271.67	15DE	1.9	-4.2	-1.6	1.0
326.29	-1.01	1.3	281.86	25DE	0.3	-4.1	-1.5	1.0

♂ LONG	♂ LAT	♂ MAG	☉ LONG	16.00UT 1274	☿	♀	♃	♄
					MAGNITUDES			
334.00	-0.89	1.3	292.04	4JA	0.0	-4.0	-1.5	0.9
341.68	-0.76	1.4	302.21	14JA	-0.1	-3.9	-1.5	0.9
349.31	-0.64	1.4	312.34	24JA	-0.2	-3.8	-1.5	0.8
356.88	-0.51	1.5	322.43	3FE	-0.5	-3.7	-1.5	0.8
4.39	-0.39	1.5	332.48	13FE	-1.1	-3.6	-1.5	0.7
11.83	-0.26	1.6	342.47	23FE	-1.4	-3.6	-1.6	0.7
19.19	-0.15	1.6	352.41	5MR	-0.9	-3.5	-1.6	0.7
26.47	-0.03	1.6	2.29	15MR	0.3	-3.5	-1.6	0.7
33.67	0.08	1.7	12.11	25MR	2.0	-3.4	-1.7	0.7
40.79	0.19	1.7	21.87	4AP	3.6	-3.4	-1.7	0.8
47.84	0.29	1.7	31.59	14AP	2.0	-3.3	-1.8	0.9
54.80	0.38	1.8	41.25	24AP	1.2	-3.3	-1.8	0.9
61.69	0.48	1.8	50.88	4MY	0.6	-3.3	-1.9	1.0
68.52	0.56	1.8	60.47	14MY	-0.1	-3.3	-2.0	1.1
75.27	0.64	1.8	70.02	24MY	-1.0	-3.3	-2.0	1.1
81.97	0.72	1.8	79.56	3JN	-1.8	-3.3	-2.1	1.2
88.61	0.79	1.8	89.09	13JN	-1.0	-3.3	-2.2	1.2
95.19	0.86	1.9	98.62	23JN	-0.2	-3.4	-2.2	1.2
101.72	0.92	1.9	108.16	3JL	0.3	-3.4	-2.3	1.2
108.21	0.98	1.9	117.71	13JL	0.7	-3.4	-2.4	1.2
114.66	1.03	2.0	127.29	23JL	1.2	-3.5	-2.4	1.2
121.07	1.08	2.0	136.90	2AU	2.2	-3.5	-2.4	1.2
127.44	1.13	2.0	146.55	12AU	2.8	-3.5	-2.4	1.2
133.79	1.17	2.0	156.25	22AU	0.6	-3.4	-2.4	1.1
140.10	1.21	2.0	166.00	1SE	-0.8	-3.4	-2.4	1.1
146.39	1.25	2.0	175.81	11SE	-1.1	-3.4	-2.4	1.0
152.65	1.28	2.0	185.66	21SE	-1.1	-3.4	-2.3	1.0
158.88	1.31	2.0	195.59	1OC	-0.6	-3.3	-2.3	1.0
165.08	1.33	1.9	205.56	11OC	-0.4	-3.3	-2.2	1.0
171.25	1.36	1.9	215.59	21OC	-0.3	-3.3	-2.1	1.0
177.38	1.37	1.9	225.67	31OC	-0.2	-3.3	-2.1	1.0
183.47	1.39	1.8	235.79	10NO	0.0	-3.3	-2.0	1.0
189.51	1.39	1.7	245.94	20NO	1.6	-3.4	-1.9	1.0
195.50	1.40	1.6	256.12	30NO	1.6	-3.4	-1.9	1.0
201.42	1.39	1.6	266.32	10DE	0.1	-3.4	-1.8	1.0
207.25	1.38	1.4	276.51	20DE	-0.1	-3.4	-1.7	0.9
212.99	1.36	1.3	286.71	30DE	-0.2	-3.5	-1.7	0.9
				1275				
218.61	1.33	1.2	296.88	9JA	-0.3	-3.5	-1.7	0.8
224.08	1.29	1.0	307.03	19JA	-0.6	-3.6	-1.6	0.8
229.37	1.22	0.9	317.14	29JA	-1.1	-3.6	-1.6	0.7
234.44	1.14	0.7	327.22	8FE	-1.3	-3.7	-1.6	0.7
239.25	1.04	0.5	337.23	18FE	-0.9	-3.8	-1.6	0.6
243.71	0.90	0.3	347.20	28FE	0.4	-3.9	-1.6	0.6
247.74	0.72	0.0	357.11	10MR	2.4	-3.9	-1.6	0.6
251.24	0.50	-0.2	6.96	20MR	2.7	-4.0	-1.6	0.5
254.06	0.21	-0.5	16.76	30MR	1.5	-4.1	-1.6	0.5
256.02	-0.14	-0.8	26.50	9AP	0.9	-4.2	-1.6	0.6
256.97	-0.59	-1.1	36.19	19AP	0.4	-4.2	-1.7	0.7
256.72	-1.12	-1.5	45.84	29AP	-0.2	-4.2	-1.7	0.7
255.22	-1.72	-1.8	55.44	9MY	-1.0	-3.9	-1.8	0.8
252.66	-2.35	-2.0	65.01	19MY	-1.9	-3.2	-1.8	0.9
249.50	-2.94	-2.2	74.56	29MY	-1.0	-3.0	-1.9	0.9
246.45	-3.41	-2.0	84.10	8JN	-0.2	-3.7	-1.9	1.0
244.22	-3.73	-1.9	93.62	18JN	0.4	-4.1	-2.0	1.0
243.23	-3.90	-1.7	103.16	28JN	0.9	-4.2	-2.1	1.0
243.64	-3.94	-1.4	112.70	8JL	1.5	-4.2	-2.1	1.1
245.35	-3.89	-1.2	122.26	18JL	2.8	-4.1	-2.2	1.1
248.22	-3.79	-1.0	131.86	28JL	2.4	-4.0	-2.3	1.1
252.03	-3.64	-0.8	141.49	7AU	0.4	-3.9	-2.3	1.1
256.62	-3.46	-0.6	151.16	17AU	-0.9	-3.8	-2.4	1.0
261.83	-3.26	-0.4	160.88	27AU	-1.3	-3.8	-2.4	1.0
267.54	-3.05	-0.3	170.66	6SE	-1.1	-3.7	-2.5	1.0
273.65	-2.83	-0.1	180.49	16SE	-0.6	-3.6	-2.5	0.9
280.08	-2.60	0.0	190.38	26SE	-0.3	-3.6	-2.5	0.9
286.76	-2.37	0.1	200.33	6OC	-0.1	-3.5	-2.4	0.9
293.64	-2.14	0.3	210.33	16OC	-0.1	-3.5	-2.4	0.9
300.67	-1.90	0.4	220.38	26OC	0.2	-3.5	-2.4	0.9
307.81	-1.68	0.5	230.48	5NO	1.8	-3.4	-2.3	0.9
315.03	-1.45	0.6	240.62	15NO	1.4	-3.4	-2.2	0.9
322.30	-1.23	0.7	250.79	25NO	-0.1	-3.4	-2.2	0.9
329.60	-1.03	0.9	260.97	5DE	-0.3	-3.4	-2.1	0.9
336.90	-0.83	1.0	271.17	15DE	-0.3	-3.4	-2.0	0.9
344.20	-0.64	1.1	281.37	25DE	-0.4	-3.3	-1.9	0.9

♂			⊙	16.00UT	☿	♀	♃	♄
LONG	LAT	MAG	LONG	1276	MAGNITUDES			
351.47	-0.46	1.2	291.55	4JA	-0.7	-3.3	-1.9	0.8
358.70	-0.29	1.2	301.71	14JA	-1.0	-3.3	-1.8	0.8
5.90	-0.14	1.3	311.85	24JA	-1.1	-3.4	-1.8	0.7
13.04	0.00	1.4	321.94	3FE	-0.8	-3.4	-1.7	0.7
20.12	0.14	1.5	331.99	13FE	0.5	-3.4	-1.7	0.6
27.15	0.26	1.6	341.98	23FE	2.8	-3.4	-1.6	0.6
34.11	0.37	1.6	351.92	4MR	2.1	-3.5	-1.6	0.5
41.02	0.48	1.7	1.80	14MR	1.1	-3.5	-1.6	0.5
47.86	0.57	1.7	11.63	24MR	0.6	-3.5	-1.6	0.4
54.64	0.65	1.8	21.40	3AP	0.3	-3.4	-1.6	0.4
61.37	0.73	1.8	31.11	13AP	-0.2	-3.4	-1.6	0.4
68.05	0.80	1.9	40.78	23AP	-1.0	-3.4	-1.6	0.5
74.67	0.86	1.9	50.40	3MY	-1.9	-3.4	-1.6	0.6
81.25	0.92	1.9	59.99	13MY	-1.0	-3.3	-1.6	0.6
87.78	0.97	1.9	69.56	23MY	-0.1	-3.3	-1.6	0.7
94.28	1.01	2.0	79.10	2JN	0.6	-3.3	-1.7	0.8
100.75	1.04	2.0	88.63	12JN	1.2	-3.3	-1.7	0.8
107.18	1.08	2.0	98.16	22JN	2.1	-3.3	-1.8	0.9
113.60	1.10	2.0	107.69	2JL	3.3	-3.3	-1.8	0.9
119.99	1.12	1.9	117.25	12JL	1.9	-3.4	-1.9	0.9
126.37	1.14	1.9	126.82	22JL	0.3	-3.4	-1.9	1.0
132.74	1.15	2.0	136.43	1AU	-0.9	-3.4	-2.0	1.0
139.11	1.16	2.0	146.08	11AU	-1.4	-3.5	-2.1	1.0
145.48	1.16	2.0	155.78	21AU	-1.1	-3.5	-2.2	1.0
151.84	1.16	2.0	165.52	31AU	-0.5	-3.6	-2.2	0.9
158.21	1.15	2.0	175.33	10SE	-0.2	-3.7	-2.3	0.9
164.59	1.14	2.0	185.19	20SE	0.0	-3.8	-2.3	0.9
170.97	1.13	2.0	195.10	30SE	0.1	-3.9	-2.4	0.8
177.37	1.11	2.0	205.08	10OC	0.5	-4.0	-2.4	0.8
183.78	1.08	2.0	215.11	20OC	2.2	-4.1	-2.4	0.8
190.20	1.05	1.9	225.18	30OC	1.2	-4.2	-2.4	0.8
196.63	1.01	1.9	235.30	9NO	-0.2	-4.3	-2.4	0.8
203.07	0.96	1.9	245.45	19NO	-0.5	-4.4	-2.4	0.8
209.53	0.91	1.8	255.63	29NO	-0.5	-4.4	-2.3	0.8
215.99	0.85	1.7	265.82	9DE	-0.5	-4.3	-2.2	0.8
222.45	0.78	1.7	276.02	19DE	-0.8	-3.8	-2.2	0.8
228.91	0.69	1.6	286.21	29DE	-0.9	-3.2	-2.1	0.8
				1277				
235.38	0.60	1.5	296.39	8JA	-1.0	-3.7	-2.0	0.8
241.83	0.49	1.4	306.54	18JA	-0.7	-4.2	-2.0	0.8
248.28	0.36	1.3	316.65	28JA	0.6	-4.3	-1.9	0.7
254.71	0.22	1.1	326.72	7FE	3.1	-4.3	-1.8	0.7
261.10	0.06	1.0	336.75	17FE	1.5	-4.3	-1.8	0.6
267.46	-0.13	0.9	346.71	27FE	0.8	-4.2	-1.7	0.6
273.77	-0.34	0.7	356.63	9MR	0.4	-4.0	-1.7	0.5
280.00	-0.58	0.6	6.48	19MR	0.2	-3.9	-1.6	0.5
286.15	-0.85	0.4	16.28	29MR	-0.3	-3.8	-1.6	0.4
292.17	-1.15	0.2	26.03	8AP	-1.0	-3.7	-1.5	0.4
298.02	-1.49	0.0	35.72	18AP	-1.8	-3.6	-1.5	0.3
303.67	-1.88	-0.2	45.36	28AP	-1.0	-3.6	-1.5	0.4
309.03	-2.31	-0.4	54.98	8MY	-0.0	-3.5	-1.5	0.4
314.03	-2.79	-0.7	64.55	18MY	0.8	-3.5	-1.5	0.5
318.55	-3.32	-0.9	74.10	28MY	1.6	-3.4	-1.5	0.6
322.45	-3.90	-1.2	83.64	7JN	2.7	-3.4	-1.5	0.6
325.55	-4.53	-1.4	93.17	17JN	3.1	-3.4	-1.5	0.7
327.66	-5.18	-1.7	102.70	27JN	1.6	-3.3	-1.6	0.7
328.56	-5.82	-2.0	112.24	7JL	0.2	-3.3	-1.6	0.8
328.16	-6.36	-2.3	121.80	17JL	-0.9	-3.3	-1.6	0.8
326.55	-6.73	-2.5	131.40	27JL	-1.5	-3.3	-1.7	0.9
324.08	-6.80	-2.7	141.03	6AU	-1.1	-3.3	-1.7	0.9
321.48	-6.56	-2.6	150.69	16AU	-0.4	-3.4	-1.8	0.9
319.46	-6.03	-2.4	160.41	26AU	-0.1	-3.4	-1.8	0.9
318.54	-5.32	-2.1	170.19	5SE	0.1	-3.4	-1.9	0.9
318.93	-4.56	-1.8	180.01	15SE	0.3	-3.4	-2.0	0.9
320.58	-3.81	-1.5	189.90	25SE	0.7	-3.5	-2.0	0.8
323.31	-3.11	-1.2	199.85	5OC	2.5	-3.5	-2.1	0.8
326.94	-2.49	-0.9	209.84	15OC	1.1	-3.5	-2.2	0.8
331.25	-1.95	-0.6	219.89	25OC	-0.4	-3.5	-2.2	0.7
336.10	-1.48	-0.4	229.99	4NO	-0.6	-3.4	-2.3	0.7
341.37	-1.07	-0.1	240.12	14NO	-0.6	-3.4	-2.3	0.8
346.95	-0.72	0.1	250.29	24NO	-0.6	-3.4	-2.3	0.8
352.78	-0.42	0.3	260.48	4DE	-0.8	-3.4	-2.3	0.8
358.78	-0.15	0.5	270.67	14DE	-0.8	-3.3	-2.3	0.8
4.91	0.07	0.6	280.87	24DE	-0.8	-3.3	-2.2	0.8

♂			⊙	16.00UT	☿	♀	♃	♄
LONG	LAT	MAG	LONG	1278	MAGNITUDES			
11.15	0.26	0.8	291.06	3JA	-0.6	-3.3	-2.2	0.8
17.45	0.43	0.9	301.22	13JA	0.8	-3.3	-2.1	0.8
23.79	0.57	1.1	311.35	23JA	2.9	-3.3	-2.1	0.7
30.17	0.69	1.2	321.45	2FE	1.1	-3.4	-2.0	0.7
36.56	0.80	1.3	331.50	12FE	0.5	-3.4	-1.9	0.7
42.96	0.89	1.4	341.50	22FE	0.3	-3.4	-1.9	0.6
49.36	0.96	1.5	351.44	4MR	0.0	-3.4	-1.8	0.6
55.74	1.02	1.6	1.32	14MR	-0.4	-3.5	-1.7	0.5
62.12	1.07	1.7	11.15	24MR	-1.0	-3.5	-1.7	0.5
68.49	1.12	1.8	20.93	3AP	-1.7	-3.6	-1.6	0.4
74.83	1.15	1.8	30.64	13AP	-1.0	-3.6	-1.6	0.3
81.17	1.17	1.9	40.31	23AP	0.0	-3.7	-1.5	0.3
87.49	1.19	1.9	49.94	3MY	1.0	-3.8	-1.5	0.3
93.80	1.21	2.0	59.53	13MY	2.1	-3.8	-1.4	0.3
100.11	1.21	2.0	69.10	23MY	3.5	-3.9	-1.4	0.4
106.41	1.21	2.0	78.64	2JN	2.4	-4.0	-1.4	0.5
112.70	1.21	2.0	88.17	12JN	1.2	-4.1	-1.4	0.5
119.00	1.20	2.0	97.70	22JN	0.1	-4.2	-1.4	0.6
125.31	1.19	2.0	107.24	2JL	-1.0	-4.2	-1.4	0.7
131.62	1.17	2.0	116.79	12JL	-1.6	-4.1	-1.4	0.7
137.95	1.15	2.0	126.36	22JL	-1.1	-3.7	-1.4	0.8
144.30	1.12	2.0	135.97	1AU	-0.4	-3.2	-1.5	0.8
150.67	1.09	2.0	145.62	11AU	0.0	-3.5	-1.5	0.8
157.07	1.05	1.9	155.31	21AU	0.3	-4.0	-1.5	0.8
163.49	1.01	1.9	165.06	31AU	0.5	-4.2	-1.6	0.8
169.95	0.97	1.9	174.85	10SE	1.0	-4.3	-1.6	0.8
176.45	0.92	2.0	184.71	20SE	2.9	-4.2	-1.7	0.8
182.99	0.87	2.0	194.62	30SE	0.9	-4.2	-1.7	0.8
189.56	0.81	1.9	204.59	10OC	-0.5	-4.1	-1.8	0.8
196.19	0.75	1.9	214.62	20OC	-0.8	-4.0	-1.9	0.7
202.86	0.68	1.9	224.69	30OC	-0.7	-3.9	-1.9	0.7
209.57	0.60	1.9	234.80	9NO	-0.8	-3.8	-2.0	0.7
216.34	0.52	1.9	244.96	19NO	-0.7	-3.7	-2.0	0.7
223.16	0.43	1.8	255.13	29NO	-0.6	-3.7	-2.1	0.7
230.02	0.34	1.8	265.33	9DE	-0.6	-3.6	-2.1	0.8
236.93	0.23	1.7	275.53	19DE	-0.4	-3.5	-2.2	0.8
243.90	0.12	1.7	285.72	29DE	0.9	-3.5	-2.2	0.8
				1279				
250.91	0.00	1.6	295.90	8JA	2.5	-3.5	-2.2	0.8
257.97	-0.13	1.5	306.05	18JA	0.8	-3.4	-2.1	0.8
265.07	-0.26	1.5	316.17	28JA	0.3	-3.4	-2.1	0.8
272.21	-0.41	1.4	326.24	7FE	0.1	-3.4	-2.1	0.7
279.39	-0.56	1.3	336.27	17FE	-0.1	-3.3	-2.0	0.7
286.61	-0.73	1.2	346.24	27FE	-0.4	-3.3	-1.9	0.6
293.85	-0.89	1.1	356.15	9MR	-1.0	-3.3	-1.9	0.6
301.12	-1.07	1.0	6.01	19MR	-1.6	-3.3	-1.8	0.5
308.39	-1.25	0.9	15.81	29MR	-1.0	-3.3	-1.7	0.5
315.66	-1.43	0.8	25.56	8AP	0.1	-3.3	-1.7	0.4
322.92	-1.62	0.7	35.25	18AP	1.3	-3.4	-1.6	0.4
330.13	-1.80	0.6	44.90	28AP	2.8	-3.4	-1.5	0.3
337.30	-1.99	0.5	54.51	8MY	3.2	-3.4	-1.5	0.2
344.38	-2.16	0.4	64.09	18MY	1.8	-3.5	-1.4	0.3
351.34	-2.33	0.3	73.64	28MY	0.9	-3.5	-1.4	0.3
358.16	-2.48	0.2	83.17	7JN	0.1	-3.5	-1.4	0.4
4.78	-2.63	0.1	92.70	17JN	-1.0	-3.4	-1.3	0.5
11.15	-2.75	-0.1	102.23	27JN	-1.7	-3.4	-1.3	0.5
17.21	-2.86	-0.2	111.77	7JL	-1.1	-3.4	-1.3	0.6
22.88	-2.95	-0.4	121.34	17JL	-0.3	-3.3	-1.3	0.6
28.05	-3.01	-0.5	130.93	27JL	0.1	-3.3	-1.3	0.7
32.62	-3.03	-0.7	140.55	6AU	0.4	-3.3	-1.3	0.7
36.40	-3.02	-0.9	150.22	16AU	0.7	-3.3	-1.3	0.8
39.23	-2.96	-1.1	159.94	26AU	1.4	-3.3	-1.3	0.8
40.89	-2.83	-1.3	169.71	5SE	3.2	-3.3	-1.4	0.8
41.16	-2.62	-1.6	179.54	15SE	0.8	-3.4	-1.4	0.8
39.97	-2.29	-1.8	189.42	25SE	-0.6	-3.4	-1.4	0.8
37.43	-1.85	-2.0	199.36	5OC	-0.9	-3.4	-1.5	0.8
34.04	-1.31	-2.1	209.36	15OC	-0.9	-3.5	-1.5	0.8
30.58	-0.73	-1.9	219.40	25OC	-0.8	-3.5	-1.6	0.7
27.82	-0.18	-1.6	229.50	4NO	-0.6	-3.6	-1.6	0.7
26.26	0.29	-1.3	239.63	14NO	-0.5	-3.6	-1.7	0.7
26.07	0.67	-1.0	249.79	24NO	-0.5	-3.7	-1.8	0.7
27.13	0.97	-0.7	259.98	4DE	-0.3	-3.8	-1.8	0.7
29.27	1.18	-0.4	270.18	14DE	1.1	-3.9	-1.9	0.7
32.27	1.34	-0.1	280.37	24DE	2.2	-4.0	-2.0	0.8

♂ LONG	♂ LAT	♂ MAG	☉ LONG	16.00UT 1280	☿	♀	♃	♄
35.93	1.45	0.2	290.56	3JA	0.5	-4.1	-2.0	0.8
40.13	1.53	0.4	300.73	13JA	0.1	-4.2	-2.0	0.8
44.72	1.58	0.6	310.86	23JA	0.0	-4.3	-2.1	0.8
49.63	1.61	0.8	320.96	2FE	-0.2	-4.3	-2.1	0.8
54.79	1.63	1.0	331.02	12FE	-0.5	-4.3	-2.1	0.7
60.14	1.63	1.1	341.01	22FE	-1.1	-4.2	-2.0	0.7
65.64	1.62	1.3	350.96	3MR	-1.5	-3.7	-2.0	0.7
71.26	1.61	1.4	0.85	13MR	-1.0	-3.2	-1.9	0.6
76.97	1.59	1.5	10.68	23MR	0.2	-3.7	-1.9	0.6
82.77	1.56	1.6	20.45	2AP	1.7	-4.1	-1.8	0.5
88.62	1.53	1.7	30.17	12AP	3.6	-4.2	-1.8	0.5
94.53	1.50	1.8	39.84	22AP	2.4	-4.2	-1.7	0.4
100.49	1.46	1.8	49.47	2MY	1.4	-4.1	-1.6	0.3
106.50	1.42	1.9	59.07	12MY	0.7	-4.1	-1.6	0.3
112.54	1.38	1.9	68.63	22MY	-0.0	-3.9	-1.5	0.2
118.62	1.33	2.0	78.17	1JN	-1.0	-3.8	-1.4	0.3
124.74	1.28	2.0	87.71	11JN	-1.8	-3.8	-1.4	0.3
130.90	1.23	2.0	97.23	21JN	-1.1	-3.7	-1.3	0.4
137.10	1.17	2.0	106.77	1JL	-0.3	-3.6	-1.3	0.5
143.34	1.11	2.0	116.32	11JL	0.2	-3.5	-1.3	0.5
149.63	1.05	2.0	125.90	21JL	0.6	-3.5	-1.3	0.6
155.97	0.99	2.0	135.51	31JL	1.0	-3.5	-1.2	0.6
162.36	0.92	2.0	145.15	10AU	1.8	-3.4	-1.2	0.7
168.80	0.85	2.0	154.84	20AU	3.1	-3.4	-1.2	0.7
175.30	0.78	1.9	164.59	30AU	0.7	-3.4	-1.2	0.8
181.87	0.70	1.9	174.38	9SE	-0.7	-3.4	-1.2	0.8
188.49	0.62	1.9	184.23	19SE	-1.1	-3.4	-1.2	0.8
195.19	0.54	1.8	194.15	29SE	-1.0	-3.4	-1.3	0.8
201.95	0.45	1.8	204.11	9OC	-0.7	-3.4	-1.3	0.8
208.78	0.36	1.8	214.13	19OC	-0.4	-3.4	-1.3	0.8
215.69	0.27	1.8	224.21	29OC	-0.3	-3.4	-1.4	0.7
222.66	0.17	1.8	234.32	8NO	-0.3	-3.4	-1.4	0.7
229.71	0.07	1.7	244.47	18NO	-0.1	-3.4	-1.5	0.7
236.84	-0.03	1.7	254.64	28NO	1.3	-3.4	-1.6	0.7
244.03	-0.14	1.7	264.83	8DE	1.9	-3.5	-1.6	0.7
251.29	-0.25	1.7	275.03	18DE	0.3	-3.5	-1.7	0.7
258.63	-0.36	1.6	285.23	28DE	-0.0	-3.5	-1.7	0.8
				1281				
266.02	-0.47	1.6	295.40	7JA	-0.1	-3.5	-1.8	0.8
273.48	-0.59	1.6	305.55	17JA	-0.3	-3.4	-1.9	0.8
280.99	-0.70	1.5	315.68	27JA	-0.6	-3.4	-1.9	0.8
288.55	-0.82	1.5	325.75	6FE	-1.1	-3.4	-2.0	0.8
296.15	-0.93	1.4	335.78	16FE	-1.4	-3.4	-2.0	0.8
303.79	-1.04	1.4	345.75	26FE	-1.0	-3.4	-2.0	0.8
311.45	-1.14	1.3	355.67	8MR	0.3	-3.3	-2.0	0.7
319.12	-1.24	1.3	5.53	18MR	2.1	-3.3	-2.0	0.7
326.79	-1.32	1.2	15.34	28MR	3.2	-3.3	-2.0	0.6
334.45	-1.40	1.2	25.08	7AP	1.8	-3.3	-1.9	0.6
342.09	-1.46	1.1	34.78	17AP	1.0	-3.4	-1.9	0.5
349.69	-1.51	1.1	44.43	27AP	0.5	-3.4	-1.8	0.4
357.22	-1.55	1.0	54.04	7MY	-0.1	-3.4	-1.8	0.4
4.69	-1.57	1.0	63.62	17MY	-1.0	-3.4	-1.7	0.3
12.07	-1.57	1.0	73.18	27MY	-1.8	-3.5	-1.6	0.3
19.34	-1.56	0.9	82.71	6JN	-1.1	-3.5	-1.6	0.2
26.48	-1.53	0.8	92.24	16JN	-0.2	-3.6	-1.5	0.3
33.47	-1.48	0.8	101.77	26JN	0.3	-3.6	-1.4	0.4
40.30	-1.41	0.7	111.31	6JL	0.8	-3.7	-1.4	0.4
46.93	-1.32	0.7	120.87	16JL	1.3	-3.8	-1.3	0.5
53.32	-1.21	0.6	130.47	26JL	2.4	-3.9	-1.3	0.5
59.46	-1.08	0.5	140.09	5AU	2.7	-4.0	-1.3	0.6
65.29	-0.92	0.4	149.75	15AU	0.6	-4.1	-1.2	0.6
70.74	-0.74	0.3	159.47	25AU	-0.8	-4.2	-1.2	0.7
75.76	-0.53	0.2	169.24	4SE	-1.2	-4.3	-1.2	0.7
80.24	-0.29	0.0	179.06	14SE	-1.1	-4.3	-1.2	0.8
84.06	-0.02	-0.1	188.94	24SE	-0.6	-4.2	-1.2	0.8
87.07	0.30	-0.3	198.88	4OC	-0.3	-3.8	-1.2	0.8
89.09	0.68	-0.5	208.87	14OC	-0.2	-3.1	-1.2	0.8
89.90	1.10	-0.7	218.92	24OC	-0.2	-3.7	-1.2	0.8
89.34	1.57	-0.9	229.01	3NO	0.1	-4.2	-1.3	0.8
87.36	2.05	-1.1	239.14	13NO	1.6	-4.4	-1.3	0.7
84.15	2.51	-1.3	249.31	23NO	1.6	-4.4	-1.4	0.7
80.28	2.88	-1.4	259.49	3DE	0.1	-4.3	-1.4	0.7
76.51	3.12	-1.2	269.69	13DE	-0.2	-4.2	-1.5	0.7
73.57	3.21	-1.0	279.89	23DE	-0.2	-4.1	-1.5	0.7

♂ LONG	♂ LAT	♂ MAG	☉ LONG	16.00UT 1282	☿	♀	♃	♄
71.88	3.20	-0.7	290.07	2JA	-0.4	-4.0	-1.6	0.8
71.52	3.11	-0.4	300.24	12JA	-0.6	-3.9	-1.6	0.8
72.40	2.98	-0.1	310.38	22JA	-1.1	-3.8	-1.7	0.8
74.30	2.83	0.2	320.48	1FE	-1.2	-3.7	-1.8	0.8
77.04	2.68	0.4	330.53	11FE	-0.9	-3.6	-1.8	0.8
80.44	2.53	0.6	340.54	21FE	0.4	-3.6	-1.9	0.8
84.35	2.39	0.8	350.48	3MR	2.5	-3.5	-1.9	0.8
88.68	2.25	1.0	0.38	13MR	2.5	-3.5	-2.0	0.8
93.34	2.12	1.2	10.21	23MR	1.3	-3.4	-2.0	0.7
98.26	2.00	1.3	19.98	2AP	0.8	-3.4	-2.0	0.7
103.39	1.88	1.4	29.71	12AP	0.4	-3.3	-2.0	0.6
108.70	1.76	1.5	39.38	22AP	-0.2	-3.3	-2.0	0.6
114.16	1.65	1.6	49.01	2MY	-1.0	-3.3	-2.0	0.5
119.76	1.55	1.7	58.61	12MY	-1.9	-3.3	-1.9	0.4
125.46	1.44	1.7	68.17	22MY	-1.1	-3.3	-1.9	0.4
131.27	1.34	1.8	77.72	1JN	-0.2	-3.3	-1.8	0.3
137.18	1.24	1.8	87.25	11JN	0.5	-3.3	-1.8	0.3
143.17	1.14	1.9	96.78	21JN	1.0	-3.4	-1.7	0.3
149.25	1.04	1.9	106.31	1JL	1.7	-3.4	-1.6	0.3
155.42	0.94	1.9	115.86	11JL	3.0	-3.4	-1.6	0.4
161.67	0.84	1.9	125.43	21JL	2.2	-3.5	-1.5	0.4
168.00	0.75	1.9	135.04	31JL	0.5	-3.5	-1.5	0.5
174.42	0.65	1.9	144.68	10AU	-0.9	-3.5	-1.4	0.6
180.92	0.55	1.9	154.37	20AU	-1.3	-3.4	-1.4	0.6
187.51	0.45	1.9	164.11	30AU	-1.1	-3.4	-1.3	0.7
194.19	0.35	1.8	173.90	9SE	-0.6	-3.4	-1.3	0.7
200.96	0.25	1.8	183.75	19SE	-0.2	-3.4	-1.3	0.7
207.81	0.14	1.8	193.66	29SE	-0.1	-3.3	-1.2	0.8
214.77	0.04	1.7	203.63	9OC	0.0	-3.3	-1.2	0.8
221.81	-0.06	1.7	213.64	19OC	0.3	-3.3	-1.2	0.8
228.94	-0.16	1.6	223.71	29OC	1.9	-3.3	-1.2	0.8
236.16	-0.26	1.6	233.82	8NO	1.4	-3.3	-1.3	0.8
243.46	-0.37	1.5	243.97	18NO	-0.1	-3.4	-1.3	0.8
250.86	-0.46	1.5	254.15	28NO	-0.4	-3.4	-1.3	0.8
258.33	-0.56	1.5	264.34	8DE	-0.4	-3.4	-1.3	0.7
265.87	-0.65	1.5	274.54	18DE	-0.5	-3.4	-1.4	0.7
273.48	-0.74	1.5	284.73	28DE	-0.7	-3.5	-1.4	0.7
				1283				
281.16	-0.83	1.5	294.91	7JA	-1.0	-3.5	-1.5	0.8
288.88	-0.91	1.5	305.07	17JA	-1.1	-3.6	-1.5	0.8
296.65	-0.98	1.5	315.19	27JA	-0.8	-3.6	-1.6	0.8
304.45	-1.04	1.4	325.27	6FE	0.5	-3.7	-1.7	0.8
312.27	-1.09	1.4	335.30	16FE	2.9	-3.8	-1.7	0.9
320.10	-1.13	1.4	345.27	26FE	1.9	-3.9	-1.8	0.9
327.92	-1.17	1.4	355.20	8MR	0.9	-3.9	-1.9	0.8
335.73	-1.18	1.4	5.06	18MR	0.6	-4.0	-1.9	0.8
343.51	-1.19	1.4	14.86	28MR	0.2	-4.1	-2.0	0.8
351.25	-1.18	1.4	24.62	7AP	-0.2	-4.2	-2.0	0.8
358.94	-1.16	1.4	34.31	17AP	-1.0	-4.2	-2.1	0.7
6.57	-1.13	1.4	43.97	27AP	-1.9	-4.1	-2.1	0.7
14.12	-1.08	1.4	53.58	7MY	-1.1	-3.9	-2.1	0.6
21.59	-1.02	1.3	63.16	17MY	-0.1	-3.2	-2.1	0.5
28.96	-0.95	1.3	72.71	27MY	0.7	-3.0	-2.1	0.5
36.23	-0.86	1.3	82.25	6JN	1.3	-3.8	-2.0	0.4
43.39	-0.76	1.3	91.78	16JN	2.3	-4.1	-2.0	0.3
50.43	-0.66	1.3	101.31	26JN	3.3	-4.2	-1.9	0.3
57.33	-0.54	1.3	110.85	6JL	1.8	-4.2	-1.9	0.3
64.10	-0.41	1.3	120.41	16JL	0.3	-4.1	-1.8	0.4
70.72	-0.27	1.2	130.00	26JL	-0.9	-4.0	-1.7	0.4
77.17	-0.12	1.2	139.62	5AU	-1.5	-3.9	-1.7	0.5
83.44	0.04	1.1	149.28	15AU	-1.1	-3.8	-1.6	0.5
89.51	0.21	1.1	158.99	25AU	-0.5	-3.8	-1.6	0.6
95.35	0.40	1.0	168.76	4SE	-0.1	-3.7	-1.5	0.6
100.92	0.60	0.9	178.58	14SE	0.1	-3.6	-1.5	0.7
106.19	0.82	0.8	188.46	24SE	0.2	-3.6	-1.4	0.7
111.09	1.07	0.7	198.39	4OC	0.6	-3.5	-1.4	0.8
115.56	1.33	0.6	208.38	14OC	2.2	-3.5	-1.4	0.8
119.49	1.63	0.4	218.43	24OC	1.2	-3.5	-1.3	0.8
122.76	1.96	0.3	228.52	3NO	-0.3	-3.4	-1.3	0.8
125.22	2.33	0.1	238.65	13NO	-0.5	-3.4	-1.3	0.8
126.68	2.73	-0.1	248.81	23NO	-0.5	-3.4	-1.3	0.8
126.96	3.16	-0.4	259.00	3DE	-0.6	-3.4	-1.3	0.8
125.91	3.60	-0.6	269.19	13DE	-0.8	-3.4	-1.3	0.8
123.52	4.00	-0.8	279.39	23DE	-0.9	-3.3	-1.4	0.8

♂ LONG	LAT	MAG	☉ LONG	16.00UT 1284	☿	♀	♃	♄
120.06	4.29	-1.0	289.58	2JA	-0.9	-3.3	-1.4	0.8
116.10	4.42	-1.0	299.75	12JA	-0.7	-3.4	-1.4	0.8
112.37	4.38	-0.9	309.89	22JA	0.6	-3.4	-1.5	0.8
109.54	4.19	-0.6	319.99	1FE	3.1	-3.4	-1.5	0.9
107.94	3.91	-0.4	330.04	11FE	1.4	-3.4	-1.6	0.9
107.65	3.59	-0.1	340.05	21FE	0.7	-3.4	-1.6	0.9
108.55	3.27	0.1	350.00	2MR	0.4	-3.5	-1.7	0.9
110.44	2.96	0.4	359.89	12MR	0.1	-3.5	-1.8	0.9
113.14	2.67	0.6	9.73	22MR	-0.3	-3.5	-1.9	0.9
116.51	2.41	0.8	19.51	1AP	-1.0	-3.4	-1.9	0.9
120.40	2.17	0.9	29.23	11AP	-1.8	-3.4	-2.0	0.8
124.71	1.95	1.1	38.91	21AP	-1.1	-3.4	-2.1	0.8
129.38	1.75	1.2	48.55	1MY	-0.0	-3.4	-2.1	0.8
134.33	1.56	1.3	58.14	11MY	0.9	-3.3	-2.2	0.7
139.52	1.39	1.4	67.71	21MY	1.8	-3.3	-2.2	0.6
144.93	1.22	1.5	77.25	31MY	3.0	-3.3	-2.2	0.6
150.52	1.06	1.5	86.78	10JN	2.8	-3.3	-2.2	0.5
156.28	0.92	1.6	96.31	20JN	1.4	-3.3	-2.2	0.4
162.19	0.77	1.6	105.85	30JN	0.2	-3.3	-2.2	0.4
168.24	0.64	1.7	115.39	10JL	-0.9	-3.4	-2.1	0.3
174.43	0.50	1.7	124.97	20JL	-1.6	-3.4	-2.1	0.4
180.74	0.37	1.7	134.57	30JL	-1.1	-3.4	-2.0	0.4
187.18	0.25	1.7	144.21	9AU	-0.4	-3.5	-2.0	0.5
193.74	0.13	1.7	153.90	19AU	-0.0	-3.5	-1.9	0.5
200.42	0.01	1.7	163.63	29AU	0.2	-3.6	-1.8	0.5
207.21	-0.10	1.7	173.42	8SE	0.4	-3.7	-1.8	0.6
214.12	-0.21	1.7	183.27	18SE	0.8	-3.8	-1.7	0.7
221.14	-0.32	1.7	193.17	28SE	2.6	-3.9	-1.7	0.7
228.27	-0.42	1.6	203.14	8OC	1.1	-4.0	-1.6	0.8
235.52	-0.52	1.6	213.16	18OC	-0.4	-4.1	-1.6	0.8
242.86	-0.61	1.6	223.22	28OC	-0.7	-4.2	-1.5	0.8
250.29	-0.70	1.5	233.33	7NO	-0.7	-4.3	-1.5	0.9
257.83	-0.77	1.5	243.48	17NO	-0.7	-4.4	-1.5	0.9
265.44	-0.85	1.5	253.65	27NO	-0.7	-4.4	-1.4	0.9
273.12	-0.91	1.4	263.84	7DE	-0.7	-4.2	-1.4	0.9
280.88	-0.97	1.4	274.04	17DE	-0.7	-3.8	-1.4	0.9
288.68	-1.01	1.4	284.23	27DE	-0.5	-3.1	-1.4	0.8
				1285				
296.53	-1.05	1.3	294.42	6JA	0.8	-3.7	-1.4	0.8
304.41	-1.07	1.3	304.57	16JA	2.8	-4.2	-1.5	0.8
312.30	-1.08	1.3	314.69	26JA	1.0	-4.3	-1.5	0.9
320.20	-1.08	1.3	324.78	5FE	0.4	-4.3	-1.5	0.9
328.09	-1.07	1.3	334.81	15FE	0.2	-4.3	-1.5	1.0
335.95	-1.05	1.4	344.79	25FE	0.0	-4.2	-1.6	1.0
343.78	-1.01	1.4	354.72	7MR	-0.4	-4.0	-1.6	1.0
351.57	-0.97	1.4	4.58	17MR	-1.0	-3.9	-1.7	1.0
359.29	-0.91	1.4	14.39	27MR	-1.7	-3.8	-1.8	1.0
6.96	-0.85	1.5	24.15	6AP	-1.1	-3.7	-1.8	1.0
14.55	-0.77	1.5	33.85	16AP	0.0	-3.6	-1.9	1.0
22.06	-0.69	1.5	43.50	26AP	1.1	-3.6	-2.0	0.9
29.48	-0.60	1.5	53.12	6MY	2.3	-3.5	-2.0	0.9
36.81	-0.50	1.5	62.70	16MY	3.7	-3.5	-2.1	0.8
44.04	-0.39	1.6	72.25	26MY	2.2	-3.4	-2.2	0.8
51.18	-0.29	1.6	81.79	5JN	1.1	-3.4	-2.2	0.7
58.22	-0.17	1.6	91.32	15JN	0.1	-3.4	-2.3	0.6
65.15	-0.05	1.6	100.85	25JN	-0.9	-3.3	-2.3	0.6
71.99	0.07	1.6	110.39	5JL	-1.7	-3.3	-2.3	0.5
78.71	0.19	1.6	119.95	15JL	-1.1	-3.3	-2.3	0.4
85.32	0.32	1.6	129.54	25JL	-0.4	-3.3	-2.3	0.4
91.82	0.45	1.6	139.16	4AU	0.1	-3.4	-2.3	0.5
98.20	0.59	1.6	148.82	14AU	0.3	-3.4	-2.3	0.5
104.45	0.73	1.6	158.53	24AU	0.6	-3.4	-2.2	0.6
110.56	0.87	1.5	168.29	3SE	1.1	-3.4	-2.2	0.6
116.52	1.02	1.5	178.11	13SE	3.0	-3.4	-2.1	0.7
122.32	1.18	1.4	187.98	23SE	1.0	-3.5	-2.0	0.7
127.92	1.35	1.3	197.91	3OC	-0.5	-3.5	-2.0	0.8
133.30	1.53	1.3	207.90	13OC	-0.8	-3.5	-1.9	0.8
138.42	1.72	1.2	217.93	23OC	-0.8	-3.4	-1.8	0.9
143.24	1.92	1.0	228.02	2NO	-0.8	-3.4	-1.8	0.9
147.67	2.14	0.9	238.15	12NO	-0.6	-3.4	-1.7	0.9
151.65	2.39	0.7	248.31	22NO	-0.6	-3.4	-1.7	0.9
155.06	2.65	0.6	258.49	2DE	-0.6	-3.4	-1.6	0.9
157.77	2.94	0.4	268.69	12DE	-0.4	-3.3	-1.6	1.0
159.63	3.26	0.1	278.89	22DE	0.9	-3.3	-1.6	1.0

♂ LONG	LAT	MAG	☉ LONG	16.00UT 1286	☿	♀	♃	♄
160.43	3.58	-0.1	289.08	1JA	2.5	-3.3	-1.5	0.9
160.01	3.90	-0.4	299.25	11JA	0.7	-3.3	-1.5	0.9
158.29	4.18	-0.6	309.39	21JA	0.2	-3.4	-1.5	0.9
155.35	4.36	-0.8	319.50	31JA	0.1	-3.4	-1.5	1.0
151.61	4.40	-1.0	329.55	10FE	-0.1	-3.4	-1.5	1.0
147.71	4.26	-0.9	339.56	20FE	-0.4	-3.4	-1.6	1.0
144.36	3.97	-0.7	349.52	2MR	-1.0	-3.4	-1.6	1.1
142.09	3.58	-0.5	359.42	12MR	-1.6	-3.5	-1.6	1.1
141.10	3.16	-0.3	9.26	22MR	-1.1	-3.5	-1.6	1.1
141.37	2.75	-0.1	19.04	1AP	0.1	-3.6	-1.7	1.1
142.76	2.36	0.2	28.77	11AP	1.4	-3.6	-1.7	1.1
145.08	2.00	0.3	38.45	21AP	3.0	-3.7	-1.8	1.1
148.17	1.68	0.5	48.09	1MY	2.9	-3.8	-1.9	1.1
151.88	1.39	0.7	57.68	11MY	1.7	-3.8	-1.9	1.0
156.11	1.13	0.8	67.25	21MY	0.9	-3.9	-2.0	1.0
160.77	0.89	0.9	76.80	31MY	0.0	-4.0	-2.1	0.9
165.78	0.68	1.0	86.33	10JN	-0.9	-4.1	-2.1	0.9
171.09	0.48	1.1	95.86	20JN	-1.7	-4.2	-2.2	0.8
176.68	0.30	1.2	105.39	30JN	-1.1	-4.2	-2.3	0.7
182.49	0.13	1.2	114.94	10JL	-0.3	-4.1	-2.3	0.7
188.52	-0.03	1.3	124.51	20JL	0.2	-3.7	-2.4	0.6
194.75	-0.18	1.3	134.11	30JL	0.5	-3.1	-2.4	0.5
201.16	-0.31	1.4	143.75	9AU	0.8	-3.5	-2.4	0.5
207.73	-0.44	1.4	153.43	19AU	1.5	-4.0	-2.4	0.6
214.47	-0.56	1.4	163.16	29AU	3.2	-4.2	-2.4	0.6
221.35	-0.66	1.4	172.95	8SE	0.8	-4.3	-2.4	0.7
228.38	-0.76	1.4	182.79	18SE	-0.6	-4.2	-2.4	0.7
235.54	-0.85	1.4	192.70	28SE	-1.0	-4.2	-2.3	0.8
242.82	-0.93	1.4	202.65	8OC	-1.0	-4.1	-2.3	0.8
250.22	-0.99	1.4	212.67	18OC	-0.8	-4.0	-2.2	0.9
257.73	-1.05	1.4	222.73	28OC	-0.5	-3.9	-2.1	0.9
265.31	-1.09	1.4	232.84	7NO	-0.4	-3.8	-2.0	1.0
273.02	-1.12	1.4	242.98	17NO	-0.4	-3.7	-2.0	1.0
280.78	-1.14	1.4	253.16	27NO	-0.2	-3.7	-1.9	1.0
288.60	-1.15	1.4	263.34	7DE	1.1	-3.6	-1.8	1.0
296.46	-1.15	1.4	273.54	17DE	2.1	-3.5	-1.8	1.1
304.34	-1.13	1.4	283.74	27DE	0.4	-3.5	-1.7	1.1
				1287				
312.25	-1.10	1.4	293.92	6JA	0.1	-3.5	-1.7	1.1
320.15	-1.06	1.4	304.08	16JA	-0.0	-3.4	-1.7	1.1
328.04	-1.01	1.4	314.20	26JA	-0.2	-3.4	-1.6	1.1
335.90	-0.95	1.4	324.29	5FE	-0.5	-3.4	-1.6	1.1
343.72	-0.88	1.4	334.32	15FE	-1.0	-3.3	-1.6	1.1
351.49	-0.80	1.4	344.31	25FE	-1.4	-3.3	-1.6	1.1
359.19	-0.72	1.4	354.23	7MR	-1.0	-3.3	-1.6	1.2
6.83	-0.63	1.4	4.11	17MR	0.2	-3.3	-1.6	1.2
14.40	-0.53	1.4	13.92	27MR	1.8	-3.3	-1.6	1.2
21.88	-0.43	1.5	23.67	6AP	3.7	-3.3	-1.6	1.3
29.28	-0.33	1.5	33.38	16AP	2.2	-3.4	-1.6	1.3
36.59	-0.23	1.5	43.04	26AP	1.2	-3.4	-1.7	1.2
43.81	-0.12	1.6	52.65	6MY	0.6	-3.4	-1.7	1.2
50.95	-0.02	1.6	62.24	16MY	-0.0	-3.5	-1.8	1.2
57.99	0.09	1.7	71.79	26MY	-1.0	-3.5	-1.8	1.2
64.95	0.20	1.7	81.33	5JN	-1.8	-3.5	-1.9	1.1
71.83	0.30	1.8	90.86	15JN	-1.1	-3.4	-1.9	1.1
78.62	0.41	1.8	100.39	25JN	-0.3	-3.4	-2.0	1.0
85.32	0.51	1.8	109.93	5JL	0.3	-3.4	-2.1	0.9
91.95	0.61	1.8	119.49	15JL	0.6	-3.3	-2.1	0.9
98.50	0.72	1.8	129.07	25JL	1.1	-3.3	-2.2	0.8
104.97	0.82	1.8	138.68	4AU	2.0	-3.3	-2.3	0.7
111.36	0.92	1.8	148.35	14AU	3.0	-3.3	-2.3	0.7
117.67	1.02	1.8	158.05	24AU	0.7	-3.3	-2.4	0.7
123.89	1.12	1.8	167.81	3SE	-0.7	-3.3	-2.4	0.7
130.03	1.22	1.8	177.63	13SE	-1.1	-3.4	-2.5	0.8
136.08	1.32	1.8	187.50	23SE	-1.1	-3.4	-2.5	0.8
142.03	1.42	1.7	197.42	3OC	-0.7	-3.4	-2.5	0.9
147.86	1.53	1.7	207.41	13OC	-0.4	-3.5	-2.4	0.9
153.56	1.63	1.6	217.45	23OC	-0.3	-3.5	-2.4	1.0
159.11	1.74	1.5	227.53	2NO	-0.2	-3.6	-2.3	1.0
164.49	1.86	1.4	237.66	12NO	-0.0	-3.6	-2.3	1.1
169.65	1.98	1.3	247.82	22NO	1.3	-3.7	-2.2	1.1
174.57	2.11	1.2	258.00	2DE	1.9	-3.8	-2.1	1.1
179.19	2.24	1.0	268.20	12DE	0.2	-3.9	-2.1	1.2
183.42	2.38	0.8	278.39	22DE	-0.1	-4.0	-2.0	1.2

♂ LONG	♂ LAT	♂ MAG	⊙ LONG	16.00UT	☿	♀	♃	♄
				1288	MAGNITUDES			
187.20	2.53	0.7	288.58	1JA	-0.2	-4.1	-1.9	1.2
190.41	2.69	0.4	298.76	11JA	-0.3	-4.2	-1.9	1.2
192.89	2.84	0.2	308.90	21JA	-0.6	-4.3	-1.8	1.2
194.50	3.00	-0.1	319.00	31JA	-1.1	-4.3	-1.7	1.2
195.04	3.13	-0.3	329.07	10FE	-1.3	-4.3	-1.7	1.2
194.36	3.22	-0.6	339.08	20FE	-0.9	-4.1	-1.7	1.2
192.41	3.23	-0.9	349.03	1MR	0.3	-3.7	-1.6	1.3
189.33	3.13	-1.1	358.93	11MR	2.2	-3.2	-1.6	1.3
185.60	2.90	-1.3	8.77	21MR	3.0	-3.7	-1.6	1.4
181.90	2.54	-1.2	18.56	31MR	1.6	-4.1	-1.6	1.4
178.91	2.11	-1.0	28.30	10AP	0.9	-4.2	-1.6	1.4
177.10	1.65	-0.8	37.98	20AP	0.5	-4.2	-1.6	1.4
176.60	1.21	-0.6	47.61	30AP	-0.1	-4.1	-1.6	1.4
177.36	0.80	-0.3	57.22	10MY	-1.0	-4.0	-1.6	1.4
179.25	0.45	-0.2	66.78	20MY	-1.8	-3.9	-1.6	1.4
182.06	0.14	0.0	76.33	30MY	-1.1	-3.8	-1.6	1.3
185.65	-0.13	0.2	85.87	9JN	-0.2	-3.8	-1.7	1.3
189.87	-0.37	0.3	95.39	19JN	0.4	-3.7	-1.7	1.2
194.62	-0.57	0.4	104.93	29JN	0.9	-3.6	-1.8	1.2
199.82	-0.75	0.5	114.48	9JL	1.4	-3.5	-1.8	1.2
205.39	-0.91	0.6	124.04	19JL	2.6	-3.5	-1.9	1.1
211.28	-1.04	0.7	133.64	29JL	2.6	-3.5	-1.9	1.0
217.47	-1.16	0.8	143.28	8AU	0.6	-3.4	-2.0	1.0
223.91	-1.25	0.8	152.96	18AU	-0.8	-3.4	-2.1	0.9
230.57	-1.33	0.9	162.69	28AU	-1.3	-3.4	-2.1	0.8
237.43	-1.39	1.0	172.48	7SE	-1.1	-3.4	-2.2	0.9
244.47	-1.44	1.0	182.32	17SE	-0.6	-3.4	-2.3	0.9
251.67	-1.46	1.0	192.22	27SE	-0.3	-3.4	-2.3	0.9
259.01	-1.48	1.1	202.17	7OC	-0.2	-3.4	-2.4	1.0
266.48	-1.47	1.1	212.18	17OC	-0.1	-3.4	-2.4	1.0
274.04	-1.45	1.1	222.24	27OC	0.2	-3.4	-2.4	1.1
281.70	-1.42	1.2	232.35	6NO	1.6	-3.4	-2.4	1.1
289.42	-1.37	1.2	242.49	16NO	1.6	-3.4	-2.4	1.2
297.19	-1.32	1.2	252.66	26NO	0.0	-3.4	-2.3	1.2
305.00	-1.24	1.3	262.85	6DE	-0.3	-3.5	-2.3	1.3
312.82	-1.16	1.3	273.05	16DE	-0.3	-3.5	-2.2	1.3
320.64	-1.07	1.3	283.24	26DE	-0.4	-3.5	-2.2	1.4
				1289				
328.44	-0.98	1.4	293.43	5JA	-0.6	-3.5	-2.1	1.4
336.22	-0.87	1.4	303.58	15JA	-1.0	-3.4	-2.0	1.4
343.95	-0.77	1.4	313.71	25JA	-1.1	-3.4	-1.9	1.4
351.63	-0.66	1.5	323.80	4FE	-0.9	-3.4	-1.9	1.4
359.24	-0.54	1.5	333.83	14FE	0.4	-3.4	-1.8	1.3
6.79	-0.43	1.5	343.82	24FE	2.6	-3.4	-1.7	1.3
14.27	-0.32	1.6	353.75	6MR	2.2	-3.3	-1.7	1.2
21.66	-0.20	1.6	3.62	16MR	1.2	-3.3	-1.6	1.2
28.97	-0.09	1.6	13.44	26MR	0.7	-3.3	-1.6	1.2
36.20	0.02	1.6	23.20	5AP	0.3	-3.3	-1.6	1.2
43.35	0.12	1.7	32.90	15AP	-0.2	-3.4	-1.5	1.2
50.42	0.22	1.7	42.57	25AP	-1.0	-3.4	-1.5	1.2
57.41	0.32	1.7	52.19	5MY	-1.9	-3.4	-1.5	1.2
64.31	0.42	1.7	61.77	15MY	-1.1	-3.4	-1.5	1.2
71.15	0.51	1.7	71.33	25MY	-0.2	-3.5	-1.5	1.1
77.92	0.60	1.8	80.87	4JN	0.5	-3.5	-1.5	1.1
84.62	0.68	1.8	90.40	14JN	1.1	-3.6	-1.5	1.0
91.26	0.76	1.9	99.93	24JN	1.9	-3.6	-1.5	1.0
97.83	0.83	1.9	109.47	4JL	3.2	-3.7	-1.5	0.9
104.36	0.90	1.9	119.02	14JL	2.1	-3.8	-1.6	0.9
110.83	0.97	1.9	128.61	24JL	0.5	-3.9	-1.6	0.8
117.25	1.04	2.0	138.22	3AU	-0.8	-4.0	-1.7	0.8
123.63	1.10	2.0	147.88	13AU	-1.4	-4.1	-1.7	0.7
129.97	1.16	2.0	157.59	23AU	-1.1	-4.2	-1.8	0.7
136.26	1.21	2.0	167.34	2SE	-0.5	-4.3	-1.8	0.7
142.50	1.26	2.0	177.15	12SE	-0.2	-4.3	-1.9	0.7
148.71	1.31	2.0	187.02	22SE	-0.0	-4.1	-2.0	0.7
154.87	1.36	1.9	196.94	2OC	0.1	-3.7	-2.0	0.8
160.98	1.41	1.9	206.93	12OC	0.4	-3.2	-2.1	0.9
167.03	1.45	1.8	216.96	22OC	1.9	-3.7	-2.2	0.9
173.02	1.49	1.8	227.04	1NO	1.4	-4.2	-2.2	1.0
178.95	1.53	1.7	237.17	11NO	-0.2	-4.4	-2.2	1.0
184.79	1.56	1.6	247.33	21NO	-0.4	-4.4	-2.3	1.1
190.53	1.59	1.5	257.51	1DE	-0.4	-4.3	-2.3	1.1
196.17	1.62	1.4	267.70	11DE	-0.5	-4.2	-2.3	1.2
201.66	1.64	1.3	277.90	21DE	-0.7	-4.1	-2.3	1.2
206.98	1.65	1.2	288.09	31DE	-0.9	-4.0	-2.2	1.2

♂ LONG	♂ LAT	♂ MAG	⊙ LONG	16.00UT	☿	♀	♃	♄
				1290	MAGNITUDES			
212.11	1.66	1.0	298.26	10JA	-1.0	-3.9	-2.2	1.2
216.98	1.66	0.8	308.41	20JA	-0.8	-3.8	-2.1	1.2
221.55	1.64	0.7	318.52	30JA	0.5	-3.7	-2.0	1.1
225.74	1.60	0.4	328.58	9FE	3.0	-3.6	-2.0	1.1
229.44	1.54	0.2	338.60	19FE	1.7	-3.6	-1.9	1.1
232.55	1.45	-0.0	348.55	1MR	0.8	-3.5	-1.8	1.0
234.91	1.32	-0.3	358.46	11MR	0.5	-3.5	-1.8	1.0
236.34	1.13	-0.6	8.31	21MR	0.2	-3.4	-1.7	0.9
236.67	0.87	-0.9	18.09	31MR	-0.3	-3.4	-1.6	1.0
235.75	0.52	-1.3	27.83	10AP	-1.0	-3.3	-1.6	1.0
233.61	0.09	-1.6	37.51	20AP	-1.8	-3.3	-1.5	1.0
230.52	-0.40	-1.8	47.15	30AP	-1.1	-3.3	-1.5	0.9
227.05	-0.92	-1.8	56.75	10MY	-0.1	-3.3	-1.4	0.9
223.93	-1.40	-1.7	66.33	20MY	0.7	-3.3	-1.4	0.9
221.79	-1.81	-1.5	75.87	30MY	1.5	-3.3	-1.4	0.9
220.96	-2.12	-1.3	85.41	9JN	2.5	-3.3	-1.4	0.9
221.51	-2.35	-1.1	94.94	19JN	3.2	-3.4	-1.4	0.8
223.33	-2.50	-0.9	104.47	29JN	1.7	-3.4	-1.4	0.8
226.21	-2.59	-0.7	114.01	9JL	0.3	-3.4	-1.4	0.7
230.01	-2.64	-0.5	123.58	19JL	-0.9	-3.5	-1.4	0.7
234.54	-2.66	-0.3	133.18	29JL	-1.5	-3.5	-1.4	0.6
239.67	-2.64	-0.2	142.81	8AU	-1.1	-3.5	-1.4	0.6
245.30	-2.59	-0.1	152.49	18AU	-0.5	-3.4	-1.5	0.5
251.34	-2.53	0.1	162.22	28AU	-0.1	-3.4	-1.5	0.5
257.72	-2.44	0.2	172.00	7SE	0.1	-3.4	-1.6	0.4
264.38	-2.34	0.3	181.84	17SE	0.3	-3.4	-1.6	0.4
271.27	-2.22	0.4	191.73	27SE	0.6	-3.3	-1.7	0.4
278.34	-2.08	0.5	201.69	7OC	2.3	-3.3	-1.7	0.4
285.56	-1.94	0.6	211.69	17OC	1.3	-3.3	-1.8	0.5
292.90	-1.79	0.7	221.75	27OC	-0.3	-3.3	-1.9	0.6
300.33	-1.63	0.8	231.85	6NO	-0.6	-3.3	-1.9	0.7
307.81	-1.46	0.9	241.99	16NO	-0.6	-3.4	-2.0	0.7
315.34	-1.29	0.9	252.16	26NO	-0.6	-3.4	-2.0	0.8
322.87	-1.13	1.0	262.35	6DE	-0.8	-3.4	-2.1	0.8
330.41	-0.96	1.1	272.55	16DE	-0.8	-3.4	-2.1	0.9
337.93	-0.80	1.2	282.75	26DE	-0.8	-3.5	-2.1	0.9
				1291				
345.41	-0.64	1.3	292.93	5JA	-0.6	-3.5	-2.2	0.9
352.86	-0.49	1.3	303.09	15JA	0.6	-3.6	-2.1	0.9
0.25	-0.34	1.4	313.22	25JA	3.0	-3.6	-2.1	0.9
7.58	-0.20	1.5	323.31	4FE	1.2	-3.7	-2.1	0.9
14.85	-0.07	1.5	333.35	14FE	0.6	-3.8	-2.0	0.9
22.04	0.06	1.6	343.34	24FE	0.3	-3.9	-2.0	0.8
29.17	0.17	1.6	353.27	6MR	0.1	-3.9	-1.9	0.8
36.23	0.29	1.7	3.15	16MR	-0.3	-4.0	-1.8	0.7
43.22	0.39	1.7	12.97	26MR	-1.0	-4.1	-1.8	0.7
50.14	0.48	1.8	22.73	5AP	-1.8	-4.2	-1.7	0.7
56.99	0.57	1.8	32.44	15AP	-1.1	-4.2	-1.6	0.7
63.78	0.65	1.8	42.10	25AP	-0.0	-4.1	-1.6	0.7
70.50	0.73	1.9	51.72	5MY	0.9	-3.9	-1.5	0.7
77.17	0.79	1.9	61.31	15MY	1.9	-3.2	-1.5	0.7
83.79	0.86	1.9	70.86	25MY	3.3	-3.0	-1.4	0.7
90.36	0.91	1.9	80.41	4JN	2.6	-3.8	-1.4	0.7
96.89	0.96	1.9	89.94	14JN	1.3	-4.1	-1.3	0.7
103.38	1.01	1.9	99.46	24JN	0.2	-4.2	-1.3	0.6
109.84	1.05	1.9	109.01	4JL	-0.9	-4.2	-1.3	0.6
116.26	1.08	1.9	118.56	14JL	-1.6	-4.1	-1.3	0.6
122.67	1.11	2.0	128.14	24JL	-1.1	-4.0	-1.3	0.5
129.05	1.14	2.0	137.76	3AU	-0.4	-3.9	-1.3	0.5
135.42	1.16	2.0	147.41	13AU	-0.0	-3.8	-1.3	0.4
141.77	1.18	2.0	157.11	23AU	0.2	-3.8	-1.3	0.4
148.12	1.19	2.0	166.87	2SE	0.4	-3.7	-1.3	0.3
154.45	1.20	2.0	176.67	12SE	0.9	-3.6	-1.4	0.2
160.78	1.21	2.0	186.54	22SE	2.7	-3.6	-1.4	0.2
167.11	1.21	2.0	196.47	2OC	1.1	-3.5	-1.4	0.1
173.43	1.20	2.0	206.44	12OC	-0.5	-3.5	-1.5	0.2
179.75	1.19	2.0	216.48	22OC	-0.7	-3.5	-1.5	0.2
186.06	1.18	1.9	226.56	1NO	-0.7	-3.4	-1.6	0.3
192.37	1.16	1.9	236.68	11NO	-0.7	-3.4	-1.7	0.4
198.66	1.13	1.8	246.83	21NO	-0.7	-3.4	-1.7	0.4
204.94	1.09	1.7	257.02	1DE	-0.6	-3.4	-1.8	0.5
211.21	1.05	1.7	267.21	11DE	-0.7	-3.4	-1.8	0.5
217.45	1.00	1.6	277.41	21DE	-0.5	-3.4	-1.9	0.6
223.66	0.93	1.5	287.60	31DE	0.8	-3.3	-2.0	0.6

♂			☉	16.00UT	☿	♀	♃	♄
LONG	LAT	MAG	LONG	1292	MAGNITUDES			
229.84	0.85	1.4	297.77	10JA	2.8	-3.4	-2.0	0.7
235.96	0.76	1.3	307.92	20JA	0.9	-3.4	-2.0	0.7
242.03	0.65	1.1	318.03	30JA	0.4	-3.4	-2.1	0.7
248.02	0.52	1.0	328.09	9FE	0.2	-3.4	-2.1	0.7
253.92	0.37	0.8	338.11	19FE	-0.0	-3.4	-2.0	0.7
259.70	0.19	0.7	348.07	29FE	-0.4	-3.5	-2.0	0.6
265.33	-0.03	0.5	357.98	10MR	-1.0	-3.5	-2.0	0.6
270.79	-0.28	0.3	7.83	20MR	-1.6	-3.5	-1.9	0.6
276.00	-0.57	0.1	17.62	30MR	-1.1	-3.4	-1.9	0.5
280.91	-0.92	-0.2	27.35	9AP	0.0	-3.4	-1.8	0.5
285.44	-1.33	-0.4	37.04	19AP	1.2	-3.4	-1.7	0.5
289.47	-1.81	-0.7	46.68	29AP	2.6	-3.4	-1.7	0.5
292.88	-2.36	-1.0	56.28	9MY	3.5	-3.3	-1.6	0.5
295.48	-3.00	-1.3	65.86	19MY	2.0	-3.3	-1.5	0.5
297.08	-3.71	-1.6	75.41	29MY	1.0	-3.3	-1.5	0.5
297.52	-4.48	-1.9	84.94	8JN	0.1	-3.3	-1.4	0.5
296.71	-5.23	-2.2	94.47	18JN	-0.9	-3.3	-1.4	0.5
294.78	-5.88	-2.4	104.00	28JN	-1.7	-3.3	-1.3	0.5
292.18	-6.32	-2.6	113.54	8JL	-1.1	-3.4	-1.3	0.5
289.58	-6.46	-2.5	123.11	18JL	-0.4	-3.4	-1.3	0.4
287.70	-6.31	-2.3	132.71	28JL	0.1	-3.4	-1.2	0.4
287.01	-5.95	-2.0	142.34	7AU	0.4	-3.5	-1.2	0.4
287.65	-5.45	-1.8	152.02	17AU	0.6	-3.5	-1.2	0.3
289.60	-4.89	-1.5	161.74	27AU	1.3	-3.6	-1.2	0.2
292.63	-4.33	-1.3	171.52	6SE	3.1	-3.7	-1.2	0.2
296.57	-3.78	-1.0	181.36	16SE	1.0	-3.8	-1.2	0.1
301.22	-3.27	-0.8	191.25	26SE	-0.6	-3.9	-1.3	0.1
306.42	-2.79	-0.5	201.20	6OC	-0.9	-4.0	-1.3	-0.0
312.05	-2.35	-0.3	211.21	16OC	-0.9	-4.1	-1.3	-0.1
318.02	-1.95	-0.1	221.26	26OC	-0.8	-4.2	-1.4	-0.0
324.22	-1.59	0.0	231.36	5NO	-0.6	-4.3	-1.4	0.0
330.62	-1.26	0.2	241.51	15NO	-0.5	-4.4	-1.4	0.1
337.15	-0.96	0.4	251.67	25NO	-0.5	-4.4	-1.5	0.2
343.77	-0.69	0.6	261.86	5DE	-0.3	-4.2	-1.6	0.2
350.46	-0.45	0.7	272.06	15DE	0.9	-3.8	-1.6	0.3
357.19	-0.23	0.8	282.25	25DE	2.4	-3.1	-1.7	0.4
				1293				
3.94	-0.04	1.0	292.44	4JA	0.6	-3.7	-1.8	0.4
10.69	0.13	1.1	302.60	14JA	0.2	-4.2	-1.8	0.5
17.44	0.28	1.2	312.73	24JA	0.0	-4.3	-1.9	0.5
24.17	0.42	1.3	322.82	3FE	-0.1	-4.3	-1.9	0.5
30.87	0.54	1.4	332.87	13FE	-0.4	-4.3	-2.0	0.5
37.55	0.64	1.5	342.85	23FE	-1.0	-4.2	-2.0	0.5
44.19	0.73	1.6	352.79	5MR	-1.5	-4.0	-2.0	0.5
50.80	0.81	1.7	2.67	15MR	-1.1	-3.9	-2.0	0.5
57.38	0.88	1.7	12.49	25MR	0.1	-3.8	-2.0	0.4
63.91	0.94	1.8	22.26	4AP	1.5	-3.7	-2.0	0.4
70.42	1.00	1.9	31.97	14AP	3.3	-3.6	-1.9	0.4
76.89	1.04	1.9	41.64	24AP	2.6	-3.6	-1.9	0.3
83.33	1.08	1.9	51.26	4MY	1.5	-3.5	-1.8	0.4
89.75	1.10	2.0	60.85	14MY	0.8	-3.5	-1.7	0.4
96.15	1.13	2.0	70.41	24MY	0.0	-3.4	-1.7	0.4
102.52	1.15	2.0	79.95	3JN	-0.9	-3.4	-1.6	0.4
108.88	1.16	2.0	89.48	13JN	-1.7	-3.4	-1.5	0.4
115.23	1.16	2.0	99.01	23JN	-1.1	-3.3	-1.5	0.4
121.58	1.17	2.0	108.55	3JL	-0.3	-3.3	-1.4	0.4
127.93	1.16	2.0	118.10	13JL	0.2	-3.3	-1.4	0.4
134.27	1.15	2.0	127.68	23JL	0.5	-3.3	-1.3	0.4
140.63	1.14	2.0	137.29	2AU	0.9	-3.3	-1.3	0.3
147.00	1.13	2.0	146.95	12AU	1.7	-3.4	-1.3	0.3
153.38	1.10	2.0	156.64	22AU	3.2	-3.4	-1.2	0.2
159.78	1.08	2.0	166.39	1SE	0.8	-3.4	-1.2	0.2
166.21	1.05	2.0	176.20	11SE	-0.7	-3.4	-1.2	0.1
172.66	1.01	2.0	186.06	21SE	-1.0	-3.5	-1.2	0.1
179.14	0.97	2.0	195.98	1OC	-1.0	-3.5	-1.2	-0.0
185.65	0.93	2.0	205.96	11OC	-0.7	-3.5	-1.2	-0.1
192.20	0.88	2.0	215.99	21OC	-0.5	-3.5	-1.2	-0.1
198.77	0.82	1.9	226.07	31OC	-0.4	-3.4	-1.3	-0.2
205.39	0.76	1.9	236.19	10NO	-0.3	-3.4	-1.3	-0.2
212.03	0.69	1.9	246.34	20NO	-0.2	-3.4	-1.3	-0.1
218.72	0.61	1.8	256.52	30NO	1.1	-3.4	-1.4	-0.0
225.44	0.52	1.8	266.71	10DE	2.1	-3.3	-1.4	0.1
232.19	0.43	1.7	276.91	20DE	0.4	-3.3	-1.5	0.1
238.98	0.32	1.6	287.10	30DE	-0.0	-3.3	-1.5	0.2

♂			☉	16.00UT	☿	♀	♃	♄
LONG	LAT	MAG	LONG	1294	MAGNITUDES			
245.80	0.21	1.6	297.28	9JA	-0.1	-3.3	-1.6	0.2
252.66	0.08	1.5	307.42	19JA	-0.2	-3.4	-1.7	0.3
259.54	-0.06	1.4	317.54	29JA	-0.5	-3.4	-1.7	0.3
266.45	-0.21	1.3	327.61	8FE	-1.0	-3.4	-1.8	0.4
273.38	-0.37	1.2	337.62	18FE	-1.4	-3.4	-1.9	0.4
280.34	-0.55	1.1	347.59	28FE	-1.0	-3.4	-1.9	0.4
287.30	-0.74	1.0	357.50	10MR	0.2	-3.5	-2.0	0.4
294.27	-0.95	0.9	7.35	20MR	1.9	-3.5	-2.0	0.4
301.23	-1.17	0.7	17.15	30MR	3.5	-3.6	-2.0	0.4
308.16	-1.40	0.6	26.89	9AP	1.9	-3.6	-2.0	0.3
315.06	-1.64	0.5	36.58	19AP	1.1	-3.7	-2.0	0.3
321.90	-1.90	0.3	46.22	29AP	0.6	-3.8	-2.0	0.3
328.64	-2.16	0.2	55.83	9MY	-0.0	-3.8	-2.0	0.2
335.26	-2.44	0.0	65.40	19MY	-0.9	-3.9	-1.9	0.3
341.71	-2.72	-0.1	74.95	29MY	-1.8	-4.0	-1.9	0.3
347.92	-3.00	-0.3	84.49	8JN	-1.1	-4.1	-1.8	0.3
353.83	-3.28	-0.4	94.01	18JN	-0.3	-4.2	-1.7	0.3
359.35	-3.55	-0.6	103.55	28JN	0.3	-4.2	-1.7	0.3
4.36	-3.82	-0.8	113.09	8JL	0.7	-4.0	-1.6	0.4
8.73	-4.07	-1.0	122.65	18JL	1.2	-3.7	-1.6	0.3
12.27	-4.30	-1.2	132.25	28JL	2.2	-3.1	-1.5	0.3
14.78	-4.48	-1.5	141.88	7AU	2.9	-3.6	-1.4	0.3
16.06	-4.59	-1.7	151.55	17AU	0.7	-4.0	-1.4	0.3
15.91	-4.58	-2.0	161.27	27AU	-0.7	-4.2	-1.4	0.2
14.34	-4.41	-2.2	171.05	6SE	-1.2	-4.3	-1.3	0.2
11.66	-4.04	-2.3	180.88	16SE	-1.1	-4.2	-1.3	0.1
8.47	-3.47	-2.4	190.77	26SE	-0.7	-4.1	-1.3	0.1
5.62	-2.79	-2.1	200.72	6OC	-0.4	-4.1	-1.3	-0.0
3.73	-2.08	-1.8	210.72	16OC	-0.2	-4.0	-1.2	-0.1
3.14	-1.41	-1.5	220.78	26OC	-0.2	-3.9	-1.2	-0.1
3.88	-0.84	-1.1	230.87	5NO	0.0	-3.8	-1.3	-0.2
5.76	-0.36	-0.8	241.01	15NO	1.4	-3.7	-1.3	-0.3
8.61	0.02	-0.5	251.18	25NO	1.9	-3.7	-1.3	-0.2
12.21	0.34	-0.2	261.37	5DE	0.1	-3.6	-1.3	-0.2
16.40	0.58	0.0	271.57	15DE	-0.2	-3.5	-1.4	-0.1
21.04	0.78	0.3	281.77	25DE	-0.2	-3.5	-1.4	-0.0
				1295				
26.03	0.94	0.5	291.95	4JA	-0.3	-3.5	-1.4	0.1
31.29	1.06	0.7	302.11	14JA	-0.6	-3.4	-1.5	0.1
36.76	1.16	0.8	312.25	24JA	-1.0	-3.4	-1.6	0.2
42.38	1.23	1.0	322.34	3FE	-1.2	-3.4	-1.6	0.2
48.13	1.29	1.2	332.38	13FE	-0.9	-3.3	-1.7	0.3
53.97	1.33	1.3	342.38	23FE	0.3	-3.3	-1.8	0.3
59.88	1.36	1.4	352.32	5MR	2.3	-3.3	-1.8	0.3
65.85	1.38	1.5	2.20	15MR	2.7	-3.3	-1.9	0.3
71.86	1.39	1.6	12.02	25MR	1.4	-3.3	-2.0	0.3
77.90	1.39	1.7	21.79	4AP	0.8	-3.3	-2.0	0.3
83.97	1.39	1.8	31.50	14AP	0.4	-3.4	-2.1	0.3
90.07	1.38	1.8	41.17	24AP	-0.1	-3.4	-2.1	0.3
96.18	1.36	1.9	50.79	4MY	-0.9	-3.4	-2.1	0.2
102.32	1.34	1.9	60.38	14MY	-1.8	-3.5	-2.1	0.2
108.47	1.32	2.0	69.95	24MY	-1.1	-3.5	-2.1	0.2
114.64	1.29	2.0	79.49	3JN	-0.2	-3.5	-2.1	0.2
120.83	1.26	2.0	89.02	13JN	0.4	-3.4	-2.0	0.3
127.05	1.22	2.0	98.55	23JN	0.9	-3.4	-2.0	0.3
133.30	1.18	2.0	108.08	3JL	1.6	-3.4	-1.9	0.3
139.57	1.14	2.0	117.63	13JL	2.8	-3.3	-1.9	0.3
145.88	1.09	2.0	127.21	23JL	2.4	-3.3	-1.8	0.3
152.22	1.04	2.0	136.82	2AU	0.6	-3.3	-1.7	0.3
158.60	0.99	2.0	146.47	12AU	-0.8	-3.3	-1.7	0.3
165.03	0.93	2.0	156.17	22AU	-1.3	-3.3	-1.6	0.3
171.51	0.87	1.9	165.91	1SE	-1.1	-3.3	-1.6	0.3
178.03	0.81	1.9	175.72	11SE	-0.6	-3.4	-1.5	0.2
184.61	0.74	1.9	185.58	21SE	-0.3	-3.4	-1.5	0.2
191.25	0.67	1.9	195.49	1OC	-0.1	-3.4	-1.4	0.1
197.95	0.59	1.9	205.47	11OC	-0.0	-3.5	-1.4	0.0
204.70	0.51	1.9	215.50	21OC	0.2	-3.5	-1.4	-0.0
211.52	0.42	1.8	225.57	31OC	1.6	-3.6	-1.3	-0.1
218.41	0.33	1.8	235.69	10NO	1.6	-3.6	-1.3	-0.2
225.35	0.24	1.8	245.85	20NO	-0.1	-3.7	-1.3	-0.2
232.37	0.14	1.8	256.02	30NO	-0.3	-3.8	-1.3	-0.3
239.45	0.03	1.7	266.22	10DE	-0.4	-3.9	-1.3	-0.3
246.59	-0.08	1.7	276.42	20DE	-0.4	-4.0	-1.4	-0.2
253.80	-0.20	1.7	286.61	30DE	-0.7	-4.1	-1.4	-0.1

1296

♂ LONG	LAT	MAG	☉ LONG	16.00UT	☿	♀	♃	♄
261.06	-0.32	1.6	296.79	9JA	-1.0	-4.2	-1.4	-0.0
268.39	-0.44	1.6	306.94	19JA	-1.1	-4.3	-1.5	0.0
275.76	-0.57	1.5	317.05	29JA	-0.8	-4.3	-1.5	0.1
283.19	-0.70	1.4	327.12	8FE	0.4	-4.3	-1.6	0.2
290.66	-0.83	1.4	337.14	18FE	2.7	-4.1	-1.6	0.2
298.17	-0.96	1.3	347.11	28FE	2.0	-3.7	-1.7	0.2
305.71	-1.09	1.3	357.02	9MR	1.0	-3.2	-1.7	0.3
313.26	-1.22	1.2	6.88	19MR	0.6	-3.7	-1.8	0.3
320.83	-1.34	1.1	16.67	29MR	0.3	-4.1	-1.9	0.3
328.39	-1.45	1.1	26.42	8AP	-0.2	-4.2	-2.0	0.3
335.93	-1.56	1.0	36.11	18AP	-0.9	-4.2	-2.0	0.3
343.43	-1.65	0.9	45.75	28AP	-1.8	-4.1	-2.1	0.3
350.88	-1.73	0.9	55.36	8MY	-1.1	-4.0	-2.1	0.3
358.26	-1.80	0.8	64.94	18MY	-0.2	-3.9	-2.2	0.2
5.55	-1.85	0.7	74.49	28MY	0.6	-3.8	-2.2	0.2
12.72	-1.88	0.7	84.03	7JN	1.2	-3.8	-2.2	0.2
19.74	-1.90	0.6	93.55	17JN	2.1	-3.7	-2.2	0.2
26.60	-1.89	0.5	103.08	27JN	3.3	-3.6	-2.2	0.3
33.25	-1.86	0.4	112.63	7JL	2.0	-3.5	-2.2	0.3
39.65	-1.81	0.3	122.19	17JL	0.4	-3.5	-2.1	0.3
45.77	-1.74	0.2	131.78	27JL	-0.8	-3.5	-2.1	0.3
51.53	-1.64	0.1	141.41	6AU	-1.4	-3.4	-2.0	0.4
56.87	-1.51	-0.0	151.08	16AU	-1.1	-3.4	-2.0	0.4
61.70	-1.34	-0.2	160.80	26AU	-0.5	-3.4	-1.9	0.3
65.88	-1.15	-0.3	170.58	5SE	-0.2	-3.4	-1.9	0.3
69.29	-0.90	-0.5	180.41	15SE	0.0	-3.4	-1.8	0.3
71.73	-0.61	-0.7	190.29	25SE	0.2	-3.4	-1.7	0.3
72.99	-0.25	-0.9	200.24	5OC	0.5	-3.4	-1.7	0.2
72.89	0.16	-1.1	210.24	15OC	2.0	-3.4	-1.6	0.1
71.34	0.63	-1.3	220.29	25OC	1.5	-3.4	-1.6	0.1
68.46	1.13	-1.5	230.38	4NO	-0.2	-3.4	-1.5	0.0
64.76	1.59	-1.6	240.52	14NO	-0.5	-3.4	-1.5	-0.1
60.99	1.98	-1.6	250.68	24NO	-0.5	-3.4	-1.5	-0.1
57.93	2.25	-1.2	260.87	4DE	-0.6	-3.5	-1.5	-0.2
56.09	2.40	-0.9	271.07	14DE	-0.7	-3.5	-1.4	-0.3
55.59	2.47	-0.6	281.26	24DE	-0.9	-3.5	-1.4	-0.2

1297

♂ LONG	LAT	MAG	☉ LONG	16.00UT	☿	♀	♃	♄
56.37	2.47	-0.3	291.45	3JA	-0.9	-3.5	-1.4	-0.2
58.21	2.44	-0.0	301.61	13JA	-0.7	-3.4	-1.5	-0.1
60.92	2.38	0.2	311.75	23JA	0.5	-3.4	-1.5	-0.0
64.32	2.31	0.5	321.85	2FE	3.0	-3.4	-1.5	0.1
68.26	2.23	0.7	331.89	12FE	1.5	-3.4	-1.5	0.1
72.61	2.15	0.9	341.89	22FE	0.7	-3.4	-1.6	0.2
77.30	2.07	1.0	351.83	4MR	0.4	-3.3	-1.6	0.2
82.24	1.99	1.2	1.72	14MR	0.2	-3.3	-1.7	0.3
87.40	1.90	1.3	11.54	24MR	-0.3	-3.3	-1.7	0.3
92.73	1.82	1.4	21.32	3AP	-0.9	-3.3	-1.8	0.3
98.20	1.74	1.5	31.03	13AP	-1.8	-3.4	-1.9	0.3
103.79	1.66	1.6	40.70	23AP	-1.1	-3.4	-1.9	0.3
109.48	1.58	1.7	50.33	3MY	-0.1	-3.4	-2.0	0.3
115.25	1.50	1.8	59.92	13MY	0.8	-3.4	-2.1	0.3
121.11	1.42	1.8	69.48	23MY	1.6	-3.5	-2.1	0.3
127.05	1.34	1.9	79.03	2JN	2.8	-3.5	-2.2	0.3
133.05	1.26	1.9	88.56	12JN	3.0	-3.6	-2.3	0.3
139.12	1.18	1.9	98.09	22JN	1.6	-3.6	-2.3	0.2
145.26	1.10	2.0	107.63	2JL	0.3	-3.7	-2.3	0.3
151.46	1.01	2.0	117.17	12JL	-0.9	-3.8	-2.4	0.3
157.73	0.93	2.0	126.75	22JL	-1.5	-3.9	-2.4	0.4
164.08	0.84	2.0	136.36	1AU	-1.1	-4.0	-2.3	0.4
170.49	0.76	1.9	146.01	11AU	-0.5	-4.1	-2.3	0.4
176.97	0.67	1.9	155.70	21AU	-0.1	-4.2	-2.3	0.4
183.54	0.58	1.9	165.45	31AU	0.2	-4.3	-2.2	0.4
190.17	0.49	1.9	175.25	10SE	0.3	-4.3	-2.2	0.4
196.89	0.39	1.8	185.10	20SE	0.7	-4.1	-2.1	0.4
203.69	0.30	1.8	195.02	30SE	2.4	-3.7	-2.0	0.4
210.57	0.20	1.7	204.98	10OC	1.3	-3.2	-2.0	0.3
217.54	0.10	1.7	215.01	20OC	-0.4	-3.7	-1.9	0.3
224.59	-0.00	1.6	225.08	30OC	-0.6	-4.2	-1.8	0.2
231.73	-0.10	1.6	235.20	9NO	-0.6	-4.4	-1.8	0.2
238.94	-0.21	1.6	245.35	19NO	-0.7	-4.4	-1.7	0.1
246.24	-0.31	1.6	255.53	29NO	-0.8	-4.3	-1.7	0.0
253.62	-0.41	1.6	265.72	9DE	-0.7	-4.2	-1.6	-0.1
261.07	-0.52	1.6	275.92	19DE	-0.8	-4.1	-1.6	-0.1
268.59	-0.62	1.6	286.11	29DE	-0.6	-4.0	-1.6	-0.2

1298

♂ LONG	LAT	MAG	☉ LONG	16.00UT	☿	♀	♃	♄
276.17	-0.72	1.5	296.29	8JA	0.6	-3.9	-1.5	-0.1
283.81	-0.81	1.5	306.44	18JA	3.0	-3.8	-1.5	-0.1
291.50	-0.90	1.5	316.56	28JA	1.1	-3.7	-1.5	-0.0
299.22	-0.98	1.5	326.63	7FE	0.5	-3.6	-1.5	0.1
306.98	-1.06	1.4	336.66	17FE	0.3	-3.6	-1.5	0.1
314.76	-1.13	1.4	346.63	27FE	0.0	-3.5	-1.6	0.2
322.54	-1.18	1.4	356.54	9MR	-0.3	-3.5	-1.6	0.2
330.32	-1.23	1.4	6.41	19MR	-1.0	-3.4	-1.6	0.3
338.08	-1.26	1.3	16.21	29MR	-1.7	-3.4	-1.7	0.3
345.81	-1.29	1.3	25.95	8AP	-1.1	-3.3	-1.7	0.4
353.50	-1.29	1.3	35.65	18AP	-0.1	-3.3	-1.8	0.4
1.14	-1.28	1.3	45.29	28AP	1.0	-3.3	-1.8	0.4
8.70	-1.26	1.3	54.90	8MY	2.1	-3.3	-1.9	0.4
16.18	-1.23	1.2	64.48	18MY	3.6	-3.3	-1.9	0.4
23.57	-1.17	1.2	74.03	28MY	2.4	-3.3	-2.0	0.4
30.86	-1.11	1.2	83.57	7JN	1.2	-3.3	-2.1	0.4
38.02	-1.03	1.2	93.10	17JN	0.2	-3.4	-2.2	0.4
45.06	-0.93	1.1	102.63	27JN	-0.9	-3.4	-2.2	0.4
51.95	-0.82	1.1	112.17	7JL	-1.6	-3.4	-2.3	0.4
58.68	-0.70	1.1	121.73	17JL	-1.1	-3.5	-2.3	0.4
65.24	-0.56	1.0	131.32	27JL	-0.4	-3.5	-2.4	0.5
71.60	-0.41	1.0	140.94	6AU	0.0	-3.5	-2.4	0.5
77.74	-0.25	0.9	150.62	16AU	0.3	-3.4	-2.4	0.5
83.63	-0.06	0.8	160.33	26AU	0.5	-3.4	-2.4	0.6
89.23	0.14	0.7	170.10	5SE	1.0	-3.4	-2.4	0.6
94.48	0.36	0.6	179.93	15SE	2.8	-3.4	-2.4	0.6
99.33	0.61	0.5	189.81	25SE	1.1	-3.3	-2.4	0.6
103.68	0.89	0.4	199.75	5OC	-0.5	-3.3	-2.3	0.5
107.42	1.20	0.2	209.75	15OC	-0.8	-3.3	-2.2	0.5
110.42	1.55	0.0	219.79	25OC	-0.8	-3.3	-2.2	0.5
112.51	1.94	-0.2	229.89	4NO	-0.8	-3.3	-2.1	0.4
113.47	2.37	-0.4	240.02	14NO	-0.7	-3.4	-2.0	0.4
113.15	2.83	-0.6	250.18	24NO	-0.6	-3.4	-2.0	0.3
111.45	3.29	-0.8	260.37	4DE	-0.6	-3.4	-1.9	0.2
108.47	3.70	-1.0	270.57	14DE	-0.4	-3.4	-1.8	0.2
104.66	3.98	-1.2	280.76	24DE	0.8	-3.5	-1.8	0.1

1299

♂ LONG	LAT	MAG	☉ LONG	16.00UT	☿	♀	♃	♄
100.73	4.11	-1.1	290.95	3JA	2.7	-3.5	-1.7	0.0
97.41	4.07	-0.8	301.12	13JA	0.8	-3.6	-1.7	-0.0
95.23	3.90	-0.6	311.25	23JA	0.3	-3.6	-1.7	0.0
94.36	3.67	-0.3	321.35	2FE	0.1	-3.7	-1.6	0.1
94.76	3.41	-0.0	331.41	12FE	-0.1	-3.8	-1.6	0.1
96.24	3.14	0.2	341.40	22FE	-0.4	-3.9	-1.6	0.2
98.62	2.89	0.5	351.35	4MR	-1.0	-4.0	-1.6	0.3
101.72	2.65	0.7	1.24	14MR	-1.6	-4.0	-1.6	0.3
105.40	2.43	0.8	11.07	24MR	-1.1	-4.1	-1.6	0.4
109.53	2.23	1.0	20.85	3AP	0.0	-4.2	-1.6	0.4
114.03	2.05	1.2	30.57	13AP	1.3	-4.2	-1.6	0.5
118.83	1.87	1.3	40.24	23AP	2.8	-4.1	-1.6	0.5
123.88	1.71	1.4	49.87	3MY	3.1	-3.8	-1.7	0.5
129.15	1.56	1.5	59.46	13MY	1.8	-3.1	-1.7	0.6
134.59	1.41	1.6	69.02	23MY	0.9	-3.1	-1.8	0.6
140.19	1.27	1.6	78.57	2JN	0.1	-3.8	-1.8	0.6
145.94	1.14	1.7	88.10	12JN	-0.9	-4.1	-1.9	0.6
151.82	1.01	1.7	97.62	22JN	-1.7	-4.2	-1.9	0.6
157.82	0.88	1.7	107.16	2JL	-1.1	-4.2	-2.0	0.6
163.93	0.76	1.8	116.71	12JL	-0.4	-4.1	-2.1	0.6
170.16	0.64	1.8	126.28	22JL	0.1	-4.0	-2.2	0.5
176.49	0.52	1.8	135.89	1AU	0.4	-3.9	-2.2	0.6
182.93	0.41	1.8	145.54	11AU	0.7	-3.8	-2.3	0.6
189.48	0.29	1.8	155.23	21AU	1.4	-3.8	-2.3	0.7
196.14	0.18	1.8	164.97	31AU	3.1	-3.7	-2.4	0.7
202.90	0.07	1.8	174.77	10SE	1.0	-3.6	-2.4	0.7
209.76	-0.04	1.7	184.62	20SE	-0.6	-3.6	-2.5	0.8
216.73	-0.14	1.7	194.54	30SE	-1.0	-3.5	-2.5	0.8
223.80	-0.25	1.7	204.50	10OC	-0.9	-3.5	-2.4	0.8
230.97	-0.35	1.6	214.52	20OC	-0.8	-3.5	-2.4	0.7
238.25	-0.45	1.6	224.60	30OC	-0.5	-3.4	-2.4	0.7
245.61	-0.54	1.6	234.71	9NO	-0.4	-3.4	-2.3	0.7
253.06	-0.63	1.5	244.86	19NO	-0.4	-3.4	-2.3	0.6
260.61	-0.71	1.5	255.00	29NO	-0.3	-3.4	-2.2	0.6
268.22	-0.79	1.4	265.23	9DE	0.9	-3.4	-2.1	0.5
275.91	-0.86	1.4	275.42	19DE	2.4	-3.4	-2.0	0.4
283.65	-0.93	1.4	285.62	29DE	0.5	-3.3	-2.0	0.4

Left table (1300 / 1301):

♂ LONG	LAT	MAG	☉ LONG	16.00UT	☿	♀	♃	♄
				1300		MAGNITUDES		
291.45	-0.98	1.4	295.79	8JA	0.1	-3.4	-1.9	0.3
299.28	-1.03	1.4	305.95	18JA	-0.0	-3.4	-1.8	0.2
307.14	-1.06	1.4	316.07	28JA	-0.2	-3.4	-1.8	0.2
315.02	-1.09	1.4	326.14	7FE	-0.5	-3.4	-1.7	0.2
322.89	-1.10	1.4	336.17	17FE	-1.0	-3.4	-1.7	0.3
330.76	-1.10	1.4	346.14	27FE	-1.5	-3.5	-1.6	0.3
338.60	-1.09	1.4	356.06	8MR	-1.1	-3.5	-1.6	0.4
346.41	-1.07	1.4	5.92	18MR	0.1	-3.5	-1.6	0.4
354.17	-1.03	1.4	15.72	28MR	1.6	-3.4	-1.6	0.5
1.87	-0.98	1.4	25.47	7AP	3.6	-3.4	-1.6	0.6
9.51	-0.92	1.4	35.17	17AP	2.3	-3.4	-1.6	0.6
17.07	-0.85	1.5	44.82	27AP	1.3	-3.4	-1.6	0.7
24.54	-0.78	1.5	54.43	7MY	0.7	-3.3	-1.6	0.7
31.93	-0.69	1.5	64.01	17MY	0.0	-3.3	-1.6	0.7
39.22	-0.59	1.5	73.56	27MY	-0.9	-3.3	-1.6	0.8
46.41	-0.49	1.5	83.10	6JN	-1.8	-3.3	-1.6	0.8
53.50	-0.37	1.5	92.63	16JN	-1.1	-3.3	-1.7	0.8
60.47	-0.26	1.5	102.16	26JN	-0.3	-3.4	-1.7	0.8
67.33	-0.13	1.5	111.70	6JL	0.2	-3.4	-1.8	0.8
74.07	-0.00	1.5	121.26	16JL	0.6	-3.4	-1.8	0.8
80.69	0.13	1.5	130.85	26JL	1.3	-3.4	-1.9	0.8
87.17	0.28	1.5	140.47	5AU	1.8	-3.5	-1.9	0.8
93.51	0.42	1.4	150.14	15AU	3.2	-3.5	-2.0	0.8
99.70	0.58	1.4	159.86	25AU	0.8	-3.6	-2.1	0.9
105.72	0.74	1.3	169.62	4SE	-0.7	-3.7	-2.1	0.9
111.55	0.91	1.3	179.45	14SE	-1.1	-3.8	-2.2	1.0
117.17	1.10	1.2	189.33	24SE	-1.1	-3.9	-2.3	1.0
122.54	1.29	1.1	199.27	4OC	-0.7	-4.0	-2.3	1.0
127.62	1.51	1.0	209.26	14OC	-0.4	-4.1	-2.4	1.0
132.35	1.74	0.9	219.31	24OC	-0.3	-4.2	-2.4	1.0
136.66	1.99	0.8	229.40	3NO	-0.3	-4.3	-2.4	1.0
140.46	2.26	0.6	239.53	13NO	-0.1	-4.4	-2.4	1.0
143.62	2.57	0.4	249.70	23NO	1.1	-4.4	-2.3	0.9
146.01	2.91	0.2	259.88	3DE	2.1	-4.2	-2.3	0.9
147.42	3.27	-0.0	270.08	13DE	0.3	-3.7	-2.3	0.8
147.69	3.64	-0.2	280.27	23DE	-0.1	-3.1	-2.2	0.7
				1301				
146.67	4.01	-0.5	290.46	2JA	-0.1	-3.8	-2.1	0.7
144.34	4.32	-0.7	300.63	12JA	-0.3	-4.2	-2.0	0.6
140.94	4.51	-0.9	310.76	22JA	-0.5	-4.3	-2.0	0.5
137.01	4.54	-1.0	320.86	1FE	-1.0	-4.3	-1.9	0.5
133.26	4.39	-0.8	330.92	11FE	-1.3	-4.3	-1.8	0.4
130.36	4.10	-0.6	340.92	21FE	-1.0	-4.1	-1.8	0.5
128.67	3.73	-0.4	350.86	3MR	0.2	-4.0	-1.7	0.5
128.27	3.34	-0.1	0.76	13MR	2.0	-3.9	-1.7	0.6
129.07	2.96	0.1	10.59	23MR	3.2	-3.8	-1.6	0.6
130.88	2.60	0.3	20.37	2AP	1.7	-3.7	-1.6	0.7
133.54	2.27	0.5	30.09	12AP	1.0	-3.6	-1.5	0.7
136.89	1.97	0.7	39.77	22AP	0.5	-3.6	-1.5	0.8
140.79	1.71	0.8	49.40	2MY	-0.1	-3.5	-1.5	0.8
145.14	1.46	1.0	59.00	12MY	-0.9	-3.5	-1.5	0.9
149.87	1.24	1.1	68.56	22MY	-1.8	-3.4	-1.5	0.9
154.91	1.03	1.2	78.10	1JN	-1.2	-3.4	-1.5	1.0
160.23	0.84	1.3	87.64	11JN	-0.3	-3.3	-1.5	1.0
165.78	0.66	1.3	97.17	21JN	0.3	-3.3	-1.5	1.0
171.54	0.50	1.4	106.70	1JL	0.8	-3.3	-1.5	1.0
177.49	0.34	1.4	116.25	11JL	1.3	-3.3	-1.5	1.1
183.61	0.19	1.5	125.83	21JL	2.4	-3.3	-1.6	1.1
189.90	0.05	1.5	135.43	31JL	2.8	-3.3	-1.6	1.1
196.35	-0.08	1.5	145.08	10AU	0.7	-3.4	-1.7	1.1
202.94	-0.21	1.5	154.76	20AU	-0.7	-3.4	-1.7	1.1
209.67	-0.33	1.5	164.50	30AU	-1.2	-3.4	-1.8	1.1
216.54	-0.44	1.5	174.30	9SE	-0.6	-3.4	-1.8	1.2
223.54	-0.55	1.5	184.14	19SE	-0.6	-3.5	-1.9	1.2
230.66	-0.64	1.5	194.05	29SE	-0.3	-3.5	-2.0	1.3
237.91	-0.73	1.5	204.02	9OC	-0.2	-3.5	-2.0	1.3
245.26	-0.82	1.5	214.03	19OC	-0.1	-3.5	-2.1	1.3
252.72	-0.89	1.5	224.10	29OC	0.1	-3.4	-2.2	1.3
260.28	-0.95	1.5	234.22	8NO	1.4	-3.4	-2.2	1.3
267.93	-1.00	1.5	244.36	18NO	1.9	-3.4	-2.2	1.3
275.64	-1.05	1.4	254.53	28NO	0.1	-3.4	-2.3	1.2
283.43	-1.08	1.4	264.73	8DE	-0.2	-3.3	-2.3	1.2
291.26	-1.10	1.4	274.92	18DE	-0.3	-3.3	-2.3	1.2
299.14	-1.10	1.4	285.12	28DE	-0.4	-3.3	-2.2	1.1

Right table (1302 / 1303):

♂ LONG	LAT	MAG	☉ LONG	16.00UT	☿	♀	♃	♄
				1302		MAGNITUDES		
307.04	-1.10	1.4	295.30	7JA	-0.6	-3.3	-2.2	1.0
314.95	-1.08	1.4	305.45	17JA	-1.0	-3.4	-2.1	1.0
322.85	-1.06	1.3	315.57	27JA	-1.2	-3.4	-2.1	0.9
330.74	-1.02	1.3	325.65	6FE	-0.9	-3.4	-2.0	0.8
338.60	-0.97	1.3	335.68	16FE	0.3	-3.4	-1.9	0.8
346.42	-0.91	1.3	345.66	26FE	2.4	-3.4	-1.9	0.7
354.18	-0.84	1.3	355.58	8MR	2.4	-3.5	-1.8	0.8
1.88	-0.76	1.4	5.44	18MR	1.3	-3.5	-1.7	0.8
9.52	-0.68	1.4	15.25	28MR	0.7	-3.6	-1.7	0.9
17.07	-0.59	1.5	25.00	7AP	0.4	-3.6	-1.6	0.9
24.54	-0.50	1.5	34.70	17AP	-0.1	-3.7	-1.5	1.0
31.93	-0.40	1.5	44.35	27AP	-0.9	-3.8	-1.5	1.0
39.23	-0.29	1.6	53.97	7MY	-1.8	-3.8	-1.5	1.1
46.44	-0.19	1.6	63.55	17MY	-1.2	-3.9	-1.4	1.1
53.55	-0.08	1.6	73.10	27MY	-0.2	-4.0	-1.4	1.2
60.58	0.03	1.7	82.64	6JN	0.5	-4.1	-1.4	1.2
67.51	0.14	1.7	92.17	16JN	1.0	-4.2	-1.4	1.3
74.35	0.25	1.7	101.70	26JN	1.8	-4.2	-1.4	1.3
81.10	0.37	1.7	111.24	6JL	3.0	-4.0	-1.4	1.3
87.76	0.48	1.8	120.80	16JL	2.3	-3.6	-1.4	1.3
94.33	0.59	1.8	130.39	26JL	0.6	-3.1	-1.4	1.4
100.80	0.71	1.8	140.01	5AU	-0.8	-3.6	-1.4	1.4
107.19	0.82	1.8	149.67	15AU	-1.4	-4.0	-1.4	1.4
113.47	0.94	1.7	159.39	25AU	-1.2	-4.2	-1.5	1.3
119.65	1.06	1.7	169.15	4SE	-0.6	-4.3	-1.5	1.3
125.73	1.18	1.7	178.97	14SE	-0.2	-4.2	-1.6	1.3
131.68	1.30	1.6	188.85	24SE	-0.1	-4.1	-1.6	1.3
137.50	1.42	1.6	198.79	4OC	0.0	-4.1	-1.7	1.3
143.17	1.56	1.5	208.78	14OC	0.3	-4.0	-1.7	1.3
148.67	1.69	1.4	218.82	24OC	1.7	-3.9	-1.8	1.2
153.96	1.84	1.3	228.91	3NO	1.7	-3.8	-1.9	1.2
159.01	1.99	1.2	239.04	13NO	-0.1	-3.7	-1.9	1.2
163.78	2.16	1.1	249.20	23NO	-0.4	-3.7	-2.0	1.2
168.18	2.33	0.9	259.39	3DE	-0.4	-3.6	-2.0	1.1
172.16	2.52	0.8	269.58	13DE	-0.5	-3.5	-2.1	1.1
175.60	2.73	0.6	279.78	23DE	-0.7	-3.5	-2.1	1.0
				1303				
178.37	2.95	0.4	289.97	2JA	-0.9	-3.5	-2.1	1.0
180.32	3.17	0.1	300.13	12JA	-1.0	-3.4	-2.1	0.9
181.26	3.40	-0.1	310.28	22JA	-0.8	-3.4	-2.1	0.9
181.02	3.61	-0.4	320.38	1FE	0.3	-3.4	-2.1	0.8
179.50	3.76	-0.7	330.43	11FE	2.8	-3.3	-2.1	0.8
176.75	3.82	-0.9	340.44	21FE	1.9	-3.3	-2.0	0.8
173.15	3.74	-1.1	350.39	3MR	0.9	-3.3	-1.9	0.7
169.29	3.50	-1.1	0.28	13MR	0.5	-3.3	-1.9	0.8
165.89	3.14	-0.9	10.12	23MR	0.2	-3.3	-1.8	0.8
163.52	2.71	-0.7	19.90	2AP	-0.2	-3.3	-1.7	0.9
162.42	2.26	-0.5	29.62	12AP	-0.9	-3.4	-1.7	1.0
162.61	1.83	-0.3	39.30	22AP	-1.8	-3.4	-1.6	1.0
163.97	1.43	-0.1	48.93	2MY	-1.2	-3.4	-1.5	1.1
166.31	1.08	0.1	58.53	12MY	-0.2	-3.5	-1.5	1.2
169.45	0.76	0.3	68.10	22MY	0.6	-3.5	-1.4	1.2
173.28	0.48	0.4	77.64	1JN	1.4	-3.5	-1.4	1.3
177.64	0.23	0.6	87.17	11JN	2.4	-3.4	-1.4	1.3
182.47	0.01	0.7	96.70	21JN	3.3	-3.4	-1.3	1.3
187.69	-0.19	0.8	106.23	1JL	1.9	-3.4	-1.3	1.3
193.24	-0.36	0.9	115.78	11JL	0.4	-3.3	-1.3	1.3
199.08	-0.52	0.9	125.36	21JL	-0.8	-3.3	-1.3	1.3
205.18	-0.67	1.0	134.96	31JL	-1.5	-3.3	-1.3	1.3
211.51	-0.79	1.0	144.60	10AU	-1.2	-3.3	-1.3	1.2
218.06	-0.91	1.1	154.29	20AU	-0.5	-3.3	-1.3	1.2
224.79	-1.01	1.1	164.02	30AU	-0.1	-3.3	-1.3	1.2
231.70	-1.09	1.2	173.81	9SE	0.1	-3.4	-1.3	1.1
238.77	-1.16	1.2	183.67	19SE	0.2	-3.4	-1.4	1.1
245.99	-1.22	1.2	193.57	29SE	0.6	-3.4	-1.4	1.1
253.34	-1.26	1.2	203.53	9OC	2.0	-3.5	-1.4	1.1
260.81	-1.29	1.3	213.55	19OC	1.5	-3.5	-1.5	1.1
268.39	-1.30	1.3	223.61	29OC	-0.3	-3.6	-1.5	1.1
276.05	-1.30	1.3	233.72	8NO	-0.6	-3.6	-1.6	1.1
283.79	-1.29	1.3	243.87	18NO	-0.6	-3.7	-1.7	1.1
291.58	-1.26	1.3	254.04	28NO	-0.6	-3.8	-1.7	1.0
299.42	-1.22	1.3	264.23	8DE	-0.8	-3.9	-1.8	1.0
307.28	-1.17	1.4	274.43	18DE	-0.8	-4.0	-1.9	1.0
315.15	-1.11	1.4	284.63	28DE	-0.9	-4.1	-1.9	0.9

Left Table

♂ LONG	♂ LAT	♂ MAG	☉ LONG	16.00UT 1304	☿	♀	♃	♄
323.01	-1.04	1.4	294.81	7JA	-0.7	-4.2	-2.0	0.9
330.86	-0.96	1.4	304.96	17JA	0.5	-4.3	-2.0	0.8
338.67	-0.87	1.4	315.08	27JA	3.0	-4.3	-2.0	0.8
346.43	-0.78	1.5	325.17	6FE	1.4	-4.3	-2.0	0.7
354.14	-0.68	1.5	335.20	16FE	0.6	-4.1	-2.0	0.7
1.79	-0.58	1.5	345.18	26FE	0.4	-3.7	-2.0	0.6
9.36	-0.47	1.5	355.10	7MR	0.1	-3.2	-2.0	0.6
16.85	-0.37	1.5	4.97	17MR	-0.3	-3.7	-1.9	0.6
24.27	-0.26	1.6	14.77	27MR	-0.9	-4.1	-1.9	0.6
31.60	-0.15	1.6	24.53	6AP	-1.8	-4.2	-1.8	0.7
38.84	-0.04	1.6	34.23	16AP	-1.2	-4.2	-1.8	0.8
46.00	0.06	1.6	43.89	26AP	-0.1	-4.1	-1.7	0.8
53.07	0.17	1.6	53.50	6MY	0.8	-4.0	-1.6	0.9
60.07	0.27	1.7	63.08	16MY	1.8	-3.9	-1.6	1.0
66.98	0.37	1.7	72.64	26MY	3.1	-3.8	-1.5	1.0
73.81	0.46	1.8	82.18	5JN	2.8	-3.8	-1.5	1.1
80.58	0.55	1.8	91.71	15JN	1.5	-3.7	-1.4	1.1
87.27	0.64	1.8	101.23	25JN	0.3	-3.6	-1.4	1.1
93.89	0.73	1.9	110.78	5JL	-0.8	-3.5	-1.3	1.1
100.45	0.81	1.9	120.34	15JL	-1.6	-3.5	-1.3	1.1
106.94	0.90	1.9	129.92	25JL	-1.2	-3.4	-1.3	1.1
113.38	0.97	1.9	139.54	4AU	-0.5	-3.4	-1.2	1.1
119.76	1.05	1.9	149.21	14AU	-0.0	-3.4	-1.2	1.1
126.08	1.12	1.9	158.92	24AU	0.2	-3.4	-1.2	1.1
132.34	1.20	1.9	168.68	3SE	0.4	-3.4	-1.2	1.0
138.55	1.27	1.9	178.50	13SE	0.8	-3.4	-1.2	1.0
144.69	1.34	1.9	188.37	23SE	2.4	-3.4	-1.2	0.9
150.77	1.40	1.9	198.31	3OC	1.3	-3.4	-1.3	0.9
156.77	1.47	1.8	208.30	13OC	-0.4	-3.4	-1.3	1.0
162.70	1.54	1.8	218.33	23OC	-0.7	-3.4	-1.3	1.0
168.53	1.60	1.7	228.42	2NO	-0.7	-3.4	-1.4	1.0
174.26	1.67	1.6	238.55	12NO	-0.7	-3.4	-1.4	1.0
179.86	1.73	1.5	248.71	22NO	-0.7	-3.4	-1.5	1.0
185.32	1.79	1.4	258.89	2DE	-0.7	-3.5	-1.5	0.9
190.59	1.86	1.3	269.09	12DE	-0.7	-3.5	-1.6	0.9
195.66	1.92	1.1	279.28	22DE	-0.5	-3.5	-1.7	0.9
				1305				
200.46	1.98	1.0	289.47	1JA	0.6	-3.5	-1.7	0.9
204.94	2.03	0.8	299.64	11JA	2.9	-3.4	-1.8	0.8
209.03	2.08	0.6	309.78	21JA	1.0	-3.4	-1.8	0.8
212.62	2.12	0.4	319.89	31JA	0.4	-3.4	-1.9	0.7
215.59	2.15	0.1	329.94	10FE	0.2	-3.4	-1.9	0.7
217.80	2.15	-0.1	339.95	20FE	0.0	-3.4	-2.0	0.6
219.05	2.11	-0.4	349.91	2MR	-0.3	-3.3	-2.0	0.6
219.18	2.02	-0.7	359.80	12MR	-0.9	-3.3	-2.0	0.5
218.06	1.86	-1.0	9.64	22MR	-1.7	-3.3	-2.0	0.5
215.70	1.60	-1.3	19.43	1AP	-1.2	-3.3	-2.0	0.5
212.41	1.24	-1.5	29.15	11AP	-0.1	-3.4	-1.9	0.5
208.76	0.81	-1.5	38.83	21AP	1.1	-3.4	-1.9	0.6
205.48	0.34	-1.4	48.47	1MY	2.4	-3.4	-1.8	0.6
203.19	-0.11	-1.2	58.06	11MY	3.7	-3.4	-1.8	0.7
202.18	-0.51	-1.0	67.63	21MY	2.2	-3.5	-1.7	0.8
202.53	-0.85	-0.8	77.18	31MY	1.1	-3.5	-1.7	0.8
204.12	-1.14	-0.6	86.71	10JN	0.2	-3.6	-1.6	0.9
206.77	-1.37	-0.4	96.24	20JN	-0.8	-3.6	-1.5	0.9
210.30	-1.55	-0.2	105.78	30JN	-1.6	-3.7	-1.5	1.0
214.56	-1.69	-0.1	115.32	10JL	-1.2	-3.8	-1.4	1.0
219.42	-1.80	0.0	124.89	20JL	-0.4	-3.9	-1.4	1.0
224.78	-1.88	0.2	134.50	30JL	0.1	-4.0	-1.3	1.0
230.56	-1.94	0.3	144.13	9AU	0.3	-4.1	-1.3	1.0
236.70	-1.97	0.4	153.82	19AU	0.6	-4.2	-1.3	1.0
243.14	-1.98	0.5	163.56	29AU	1.2	-4.3	-1.2	1.0
249.83	-1.97	0.5	173.34	8SE	2.9	-4.3	-1.2	0.9
256.75	-1.94	0.6	183.19	18SE	1.1	-4.1	-1.2	0.9
263.86	-1.89	0.7	193.09	28SE	-0.5	-3.7	-1.2	0.9
271.12	-1.83	0.8	203.05	8OC	-0.9	-3.2	-1.2	0.8
278.51	-1.75	0.8	213.06	18OC	-0.8	-3.8	-1.2	0.8
286.01	-1.66	0.9	223.13	28OC	-0.8	-4.2	-1.2	0.9
293.59	-1.55	1.0	233.23	7NO	-0.6	-4.4	-1.3	0.9
301.22	-1.44	1.0	243.38	17NO	-0.5	-4.3	-1.3	0.9
308.89	-1.32	1.1	253.55	27NO	-0.5	-4.3	-1.3	0.9
316.58	-1.19	1.1	263.74	7DE	-0.4	-4.2	-1.4	0.9
324.27	-1.06	1.2	273.94	17DE	0.8	-4.1	-1.4	0.9
331.94	-0.92	1.3	284.14	27DE	2.7	-4.0	-1.5	0.8

Right Table

♂ LONG	♂ LAT	♂ MAG	☉ LONG	16.00UT 1306	☿	♀	♃	♄
339.59	-0.78	1.3	294.32	6JA	0.7	-3.9	-1.6	0.8
347.19	-0.65	1.4	304.47	16JA	0.2	-3.8	-1.6	0.8
354.74	-0.51	1.4	314.60	26JA	0.1	-3.7	-1.7	0.7
2.22	-0.38	1.5	324.68	5FE	-0.1	-3.6	-1.8	0.7
9.64	-0.25	1.5	334.72	15FE	-0.4	-3.6	-1.8	0.6
16.99	-0.13	1.6	344.70	25FE	-1.0	-3.5	-1.9	0.6
24.26	-0.01	1.6	354.63	7MR	-1.5	-3.5	-1.9	0.5
31.46	0.10	1.7	4.50	17MR	-1.1	-3.4	-2.0	0.5
38.58	0.21	1.7	14.31	27MR	0.0	-3.4	-2.0	0.4
45.62	0.31	1.7	24.06	6AP	1.4	-3.3	-2.0	0.4
52.59	0.41	1.8	33.77	16AP	3.1	-3.3	-2.0	0.4
59.49	0.50	1.8	43.43	26AP	2.8	-3.3	-2.0	0.4
66.31	0.58	1.8	53.04	6MY	1.6	-3.3	-2.0	0.5
73.07	0.66	1.8	62.63	16MY	0.9	-3.3	-2.0	0.6
79.77	0.74	1.8	72.18	26MY	0.1	-3.3	-1.9	0.6
86.41	0.81	1.8	81.72	5JN	-0.9	-3.3	-1.9	0.7
93.00	0.87	1.8	91.25	15JN	-1.7	-3.4	-1.8	0.7
99.55	0.93	1.9	100.78	25JN	-1.2	-3.4	-1.7	0.8
106.05	0.98	1.9	110.32	5JL	-0.4	-3.4	-1.7	0.8
112.51	1.03	1.9	119.88	15JL	0.1	-3.5	-1.6	0.9
118.93	1.08	2.0	129.46	25JL	0.5	-3.5	-1.6	0.9
125.33	1.12	2.0	139.07	4AU	0.8	-3.5	-1.5	0.9
131.70	1.15	2.0	148.74	14AU	1.5	-3.4	-1.4	0.9
138.04	1.19	2.0	158.44	24AU	3.2	-3.4	-1.4	0.9
144.36	1.22	2.0	168.20	3SE	1.0	-3.4	-1.3	0.9
150.66	1.24	2.0	178.02	13SE	-0.6	-3.4	-1.3	0.9
156.94	1.26	2.0	187.89	23SE	-1.0	-3.3	-1.3	0.9
163.20	1.28	2.0	197.82	3OC	-1.0	-3.3	-1.3	0.8
169.45	1.29	2.0	207.80	13OC	-0.8	-3.3	-1.3	0.8
175.66	1.30	1.9	217.84	23OC	-0.5	-3.3	-1.3	0.7
181.86	1.30	1.9	227.93	2NO	-0.4	-3.3	-1.3	0.8
188.03	1.30	1.8	238.06	12NO	-0.4	-3.4	-1.3	0.8
194.16	1.30	1.8	248.21	22NO	-0.2	-3.4	-1.3	0.8
200.25	1.28	1.7	258.39	2DE	1.0	-3.4	-1.3	0.8
206.29	1.26	1.6	268.59	12DE	2.4	-3.4	-1.3	0.8
212.27	1.23	1.5	278.79	22DE	0.4	-3.5	-1.4	0.8
				1307				
218.18	1.19	1.4	288.98	1JA	0.0	-3.5	-1.4	0.8
224.00	1.14	1.2	299.15	11JA	-0.1	-3.6	-1.5	0.8
229.71	1.07	1.1	309.30	21JA	-0.2	-3.7	-1.5	0.8
235.29	0.99	1.0	319.40	31JA	-0.5	-3.7	-1.6	0.7
240.71	0.88	0.8	329.46	10FE	-1.0	-3.8	-1.7	0.7
245.93	0.75	0.6	339.47	20FE	-1.4	-3.9	-1.7	0.6
250.90	0.58	0.4	349.40	2MR	-1.1	-4.0	-1.8	0.6
255.55	0.38	0.2	359.33	12MR	0.1	-4.0	-1.9	0.5
259.81	0.13	-0.1	9.17	22MR	1.7	-4.1	-1.9	0.5
263.57	-0.18	-0.3	18.96	1AP	3.6	-4.2	-2.0	0.4
266.69	-0.55	-0.6	28.69	11AP	2.1	-4.2	-2.0	0.3
269.02	-1.01	-0.9	38.37	21AP	1.2	-4.1	-2.1	0.3
270.37	-1.55	-1.2	48.01	1MY	0.6	-3.8	-2.1	0.3
270.56	-2.18	-1.6	57.61	11MY	0.0	-3.1	-2.1	0.4
269.52	-2.88	-1.9	67.17	21MY	-0.9	-3.1	-2.1	0.5
267.34	-3.57	-2.2	76.72	31MY	-1.8	-3.8	-2.1	0.5
264.44	-4.18	-2.3	86.25	10JN	-1.2	-4.1	-2.1	0.6
261.50	-4.63	-2.2	95.78	20JN	-0.3	-4.2	-2.0	0.6
259.22	-4.86	-2.1	105.31	30JN	0.3	-4.2	-2.0	0.7
258.12	-4.90	-1.9	114.86	10JL	0.7	-4.1	-1.9	0.7
258.41	-4.80	-1.6	124.43	20JL	1.1	-4.0	-1.9	0.8
260.02	-4.59	-1.4	134.03	30JL	2.0	-3.9	-1.8	0.8
262.82	-4.33	-1.2	143.67	9AU	3.1	-3.8	-1.7	0.8
266.58	-4.03	-1.0	153.34	19AU	0.8	-3.7	-1.7	0.9
271.13	-3.72	-0.8	163.08	29AU	-0.7	-3.7	-1.6	0.9
276.31	-3.40	-0.6	172.86	8SE	-1.1	-3.6	-1.6	0.9
281.99	-3.09	-0.4	182.70	18SE	-1.1	-3.6	-1.5	0.8
288.06	-2.78	-0.2	192.60	28SE	-0.7	-3.5	-1.5	0.8
294.43	-2.47	-0.1	202.56	8OC	-0.4	-3.5	-1.4	0.8
301.04	-2.18	0.1	212.57	18OC	-0.3	-3.5	-1.4	0.7
307.82	-1.89	0.2	222.63	28OC	-0.2	-3.4	-1.4	0.7
314.75	-1.62	0.4	232.74	7NO	-0.0	-3.4	-1.4	0.7
321.76	-1.36	0.5	242.88	17NO	1.2	-3.4	-1.3	0.7
328.83	-1.12	0.7	253.06	27NO	2.1	-3.4	-1.3	0.8
335.95	-0.89	0.8	263.25	7DE	0.2	-3.4	-1.3	0.8
343.08	-0.67	0.9	273.44	17DE	-0.1	-3.4	-1.4	0.8
350.20	-0.48	1.0	283.64	27DE	-0.2	-3.4	-1.4	0.8

♂			☉	16.00UT	☿ ♀ ♃ ♄		♂			☉	16.00UT	☿ ♀ ♃ ♄
LONG	LAT	MAG	LONG	1308	MAGNITUDES		LONG	LAT	MAG	LONG	1310	MAGNITUDES
357.31	-0.29	1.1	293.82	6JA	-0.3 -3.4 -1.4 0.8		17.55	0.52	0.7	293.32	5JA	-0.9 -3.3 -1.6 0.8
4.39	-0.12	1.2	303.98	16JA	-0.6 -3.4 -1.4 0.8		23.56	0.67	0.9	303.48	15JA	-0.8 -3.4 -1.6 0.8
11.44	0.03	1.3	314.11	26JA	-1.0 -3.4 -1.5 0.7		29.64	0.79	1.0	313.61	25JA	0.3 -3.4 -1.5 0.8
18.44	0.17	1.4	324.20	5FE	-1.3 -3.4 -1.5 0.7		35.79	0.90	1.2	323.70	4FE	2.8 -3.4 -1.5 0.8
25.39	0.30	1.5	334.23	15FE	-0.3 -3.4 -1.6 0.7		41.98	0.98	1.3	333.74	14FE	1.7 -3.4 -1.5 0.8
32.30	0.41	1.6	344.22	25FE	0.1 -3.5 -1.6 0.6		48.20	1.06	1.4	343.73	24FE	0.8 -3.4 -1.6 0.7
39.15	0.52	1.6	354.15	6MR	2.1 -3.5 -1.7 0.6		54.44	1.11	1.5	353.67	6MR	0.5 -3.5 -1.6 0.7
45.95	0.61	1.7	4.02	16MR	2.9 -3.5 -1.8 0.5		60.69	1.16	1.6	3.54	16MR	0.2 -3.5 -1.6 0.7
52.69	0.70	1.8	13.83	26MR	1.5 -3.4 -1.8 0.5		66.95	1.20	1.7	13.36	26MR	-0.2 -3.6 -1.6 0.6
59.38	0.77	1.8	23.59	5AP	0.9 -3.4 -1.9 0.4		73.20	1.22	1.8	23.12	5AP	-0.9 -3.6 -1.7 0.5
66.03	0.84	1.9	33.30	15AP	0.5 -3.4 -2.0 0.3		79.46	1.24	1.8	32.83	15AP	-1.8 -3.7 -1.7 0.5
72.63	0.90	1.9	42.96	25AP	-0.1 -3.4 -2.1 0.3		85.71	1.25	1.9	42.50	25AP	-1.2 -3.8 -1.8 0.4
79.18	0.95	1.9	52.58	5MY	-0.9 -3.3 -2.1 0.2		91.96	1.26	1.9	52.12	5MY	-0.2 -3.9 -1.8 0.4
85.70	1.00	2.0	62.16	15MY	-1.8 -3.3 -2.2 0.3		98.21	1.26	2.0	61.70	15MY	0.7 -3.9 -1.9 0.3
92.18	1.03	2.0	71.72	25MY	-1.2 -3.3 -2.2 0.4		104.46	1.25	2.0	71.26	25MY	1.5 -4.0 -2.0 0.2
98.63	1.07	2.0	81.26	4JN	-0.3 -3.3 -2.2 0.4		110.72	1.24	2.0	80.80	4JN	2.6 -4.1 -2.0 0.3
105.05	1.09	2.0	90.78	14JN	0.4 -3.3 -2.2 0.5		116.98	1.23	2.0	90.33	14JN	3.2 -4.2 -2.1 0.3
111.46	1.12	2.0	100.32	24JN	0.9 -3.4 -2.2 0.6		123.25	1.21	2.0	99.86	24JN	1.7 -4.2 -2.2 0.4
117.85	1.13	2.0	109.85	4JL	1.5 -3.4 -2.2 0.6		129.54	1.19	2.0	109.40	4JL	0.4 -4.0 -2.2 0.4
124.22	1.14	2.0	119.41	14JL	2.6 -3.4 -2.2 0.7		135.84	1.16	2.0	118.95	14JL	-0.8 -3.6 -2.3 0.5
130.59	1.15	2.0	128.99	24JL	2.6 -3.4 -2.2 0.7		142.16	1.13	2.0	128.53	24JL	-1.5 -3.1 -2.4 0.6
136.95	1.15	2.0	138.61	3AU	0.7 -3.5 -2.1 0.8		148.51	1.09	2.0	138.15	3AU	-1.2 -3.6 -2.4 0.6
143.32	1.15	2.0	148.27	13AU	-0.7 -3.5 -2.0 0.8		154.89	1.05	2.0	147.80	13AU	-0.5 -4.0 -2.4 0.7
149.69	1.14	2.0	157.97	23AU	-1.3 -3.6 -2.0 0.8		161.30	1.01	2.0	157.50	23AU	-0.1 -4.2 -2.4 0.7
156.07	1.13	2.0	167.73	2SE	-1.2 -3.7 -1.9 0.8		167.75	0.96	1.9	167.26	2SE	0.1 -4.3 -2.4 0.7
162.46	1.12	2.0	177.54	12SE	-0.6 -3.8 -1.8 0.8		174.24	0.91	1.9	177.06	12SE	0.3 -4.2 -2.4 0.8
168.87	1.10	2.0	187.41	22SE	-0.3 -3.9 -1.8 0.8		180.78	0.85	1.9	186.93	22SE	0.7 -4.1 -2.4 0.8
175.29	1.07	2.0	197.33	2OC	-0.1 -4.0 -1.7 0.8		187.36	0.79	1.9	196.86	2OC	2.1 -4.0 -2.4 0.8
181.73	1.04	2.0	207.32	12OC	-0.0 -4.1 -1.7 0.8		193.98	0.72	1.9	206.83	12OC	1.5 -4.0 -2.3 0.8
188.19	1.00	2.0	217.35	22OC	0.2 -4.2 -1.6 0.7		200.66	0.65	1.9	216.87	22OC	-0.3 -3.9 -2.2 0.8
194.67	0.96	1.9	227.43	1NO	1.4 -4.3 -1.6 0.7		207.39	0.57	1.9	226.95	1NO	-0.6 -3.8 -2.2 0.8
201.17	0.91	1.9	237.56	11NO	1.9 -4.4 -1.5 0.7		214.18	0.49	1.9	237.07	11NO	-0.6 -3.7 -2.1 0.7
207.69	0.85	1.9	247.72	21NO	0.0 -4.4 -1.5 0.7		221.02	0.40	1.8	247.22	21NO	-0.6 -3.7 -2.0 0.7
214.23	0.79	1.8	257.90	1DE	-0.3 -4.2 -1.5 0.7		227.91	0.31	1.8	257.41	1DE	-0.8 -3.6 -2.0 0.7
220.78	0.71	1.7	268.09	11DE	-0.3 -3.7 -1.5 0.8		234.86	0.20	1.8	267.60	11DE	-0.7 -3.5 -1.9 0.8
227.36	0.63	1.7	278.29	21DE	-0.4 -3.1 -1.5 0.8		241.87	0.10	1.7	277.80	21DE	-0.8 -3.5 -1.8 0.7
233.95	0.54	1.6	288.48	31DE	-0.6 -3.8 -1.4 0.8		248.93	-0.02	1.7	287.99	31DE	-0.6 -3.4 -1.8 0.8
				1309							1311	
240.55	0.43	1.5	298.66	10JA	-1.0 -4.2 -1.5 0.8		256.04	-0.14	1.6	298.16	10JA	0.5 -3.4 -1.7 0.8
247.16	0.31	1.4	308.80	20JA	-1.1 -4.4 -1.5 0.8		263.21	-0.27	1.5	308.31	20JA	3.0 -3.4 -1.7 0.8
253.77	0.17	1.3	318.91	30JA	-0.9 -4.3 -1.5 0.8		270.42	-0.41	1.5	318.42	30JA	1.2 -3.4 -1.6 0.8
260.39	0.02	1.2	328.97	9FE	0.2 -4.3 -1.5 0.7		277.68	-0.55	1.4	328.48	9FE	0.5 -3.3 -1.6 0.8
267.00	-0.14	1.1	338.99	19FE	2.5 -4.1 -1.6 0.7		284.99	-0.70	1.3	338.50	19FE	0.3 -3.3 -1.6 0.8
273.60	-0.33	0.9	348.95	1MR	2.2 -4.0 -1.6 0.7		292.32	-0.85	1.2	348.47	1MR	0.1 -3.3 -1.6 0.8
280.18	-0.54	0.8	358.85	11MR	1.1 -3.9 -1.6 0.6		299.69	-1.01	1.2	358.37	11MR	-0.3 -3.3 -1.6 0.7
286.71	-0.78	0.7	8.70	21MR	0.7 -3.8 -1.7 0.6		307.08	-1.16	1.1	8.22	21MR	-0.9 -3.3 -1.6 0.7
293.20	-1.03	0.5	18.48	31MR	0.3 -3.7 -1.8 0.5		314.47	-1.32	1.0	18.01	31MR	-1.7 -3.3 -1.6 0.7
299.61	-1.32	0.3	28.22	10AP	-0.2 -3.6 -1.8 0.4		321.87	-1.48	0.9	27.75	10AP	-1.2 -3.4 -1.6 0.6
305.92	-1.63	0.2	37.91	20AP	-0.9 -3.6 -1.9 0.4		329.25	-1.64	0.8	37.44	20AP	-0.1 -3.4 -1.6 0.5
312.09	-1.97	-0.0	47.54	30AP	-1.8 -3.5 -2.0 0.3		336.59	-1.78	0.7	47.08	30AP	0.9 -3.4 -1.7 0.5
318.07	-2.34	-0.2	57.15	10MY	-1.2 -3.5 -2.0 0.3		343.88	-1.92	0.6	56.68	10MY	2.0 -3.5 -1.7 0.4
323.79	-2.75	-0.4	66.72	20MY	-0.2 -3.4 -2.1 0.2		351.09	-2.05	0.5	66.25	20MY	3.4 -3.5 -1.7 0.3
329.19	-3.18	-0.6	76.26	30MY	0.5 -3.4 -2.2 0.3		358.19	-2.17	0.5	75.80	30MY	2.6 -3.5 -1.8 0.3
334.15	-3.65	-0.8	85.80	9JN	1.1 -3.3 -2.2 0.4		5.15	-2.27	0.4	85.33	9JN	1.3 -3.4 -1.8 0.2
338.53	-4.14	-1.1	95.33	19JN	2.0 -3.3 -2.3 0.4		11.95	-2.35	0.2	94.86	19JN	0.3 -3.4 -1.9 0.3
342.19	-4.65	-1.3	104.86	29JN	3.2 -3.3 -2.3 0.5		18.52	-2.42	0.1	104.40	29JN	-0.8 -3.4 -2.0 0.3
344.91	-5.16	-1.6	114.41	9JL	2.2 -3.3 -2.4 0.5		24.83	-2.46	0.0	113.94	9JL	-1.6 -3.3 -2.0 0.4
346.48	-5.64	-1.9	123.97	19JL	0.6 -3.3 -2.4 0.6		30.80	-2.48	-0.1	123.51	19JL	-1.2 -3.3 -2.1 0.5
346.74	-6.03	-2.1	133.57	29JL	-0.8 -3.3 -2.4 0.7		36.37	-2.47	-0.2	133.10	29JL	-0.5 -3.3 -2.2 0.5
345.65	-6.26	-2.4	143.21	8AU	-1.4 -3.4 -2.4 0.7		41.43	-2.43	-0.4	142.73	8AU	-0.0 -3.3 -2.2 0.6
343.43	-6.24	-2.5	152.88	18AU	-1.2 -3.4 -2.3 0.7		45.86	-2.35	-0.6	152.41	18AU	0.3 -3.3 -2.3 0.6
340.68	-5.92	-2.4	162.61	28AU	-0.6 -3.4 -2.3 0.8		49.51	-2.23	-0.8	162.14	28AU	0.5 -3.3 -2.3 0.7
338.12	-5.34	-2.4	172.39	7SE	-0.2 -3.4 -2.2 0.8		52.19	-2.06	-1.0	171.91	7SE	0.9 -3.4 -2.4 0.7
336.46	-4.59	-2.1	182.23	17SE	-0.0 -3.5 -2.2 0.8		53.69	-1.83	-1.2	181.75	17SE	2.5 -3.4 -2.4 0.7
336.05	-3.79	-1.8	192.12	27SE	0.1 -3.5 -2.1 0.8		53.82	-1.51	-1.4	191.64	27SE	1.3 -3.4 -2.4 0.8
336.94	-3.03	-1.5	202.08	7OC	0.4 -3.5 -2.0 0.8		52.47	-1.11	-1.6	201.59	7OC	-0.4 -3.5 -2.4 0.8
339.01	-2.34	-1.2	212.09	17OC	1.8 -3.5 -2.0 0.8		49.78	-0.63	-1.8	211.60	17OC	-0.8 -3.5 -2.4 0.8
342.05	-1.74	-0.9	222.14	27OC	1.7 -3.4 -1.9 0.7		46.25	-0.10	-1.9	221.66	27OC	-0.8 -3.6 -2.4 0.8
345.86	-1.23	-0.6	232.25	6NO	-0.1 -3.4 -1.8 0.7		42.65	0.41	-1.8	231.76	6NO	-0.8 -3.6 -2.4 0.8
350.29	-0.80	-0.3	242.39	16NO	-0.5 -3.4 -1.8 0.7		39.76	0.86	-1.5	241.90	16NO	-0.7 -3.7 -2.3 0.8
355.18	-0.44	-0.1	252.55	26NO	-0.5 -3.4 -1.7 0.7		38.09	1.21	-1.2	252.06	26NO	-0.6 -3.8 -2.2 0.7
0.42	-0.13	0.1	262.74	6DE	-0.5 -3.3 -1.7 0.7		37.77	1.46	-0.8	262.25	6DE	-0.6 -3.9 -2.2 0.7
5.95	0.12	0.3	272.94	16DE	-0.7 -3.3 -1.6 0.7		38.72	1.63	-0.5	272.45	16DE	-0.5 -4.0 -2.1 0.7
11.67	0.34	0.5	283.14	26DE	-0.9 -3.3 -1.6 0.8		40.75	1.74	-0.2	282.65	26DE	0.6 -4.1 -2.0 0.7

♂			⊙	16.00UT	☿	♀	♃	♄
LONG	LAT	MAG	LONG	1312	MAGNITUDES			
43.63	1.81	0.1	292.83	5JA	2.9	-4.2	-2.0	0.8
47.19	1.85	0.3	302.99	15JA	0.9	-4.3	-1.9	0.8
51.28	1.86	0.5	313.12	25JA	0.3	-4.3	-1.8	0.8
55.77	1.85	0.7	323.21	4FE	0.2	-4.3	-1.8	0.8
60.59	1.84	0.9	333.25	14FE	-0.0	-4.1	-1.7	0.8
65.66	1.81	1.1	343.24	24FE	-0.4	-3.7	-1.7	0.8
70.93	1.78	1.2	353.18	5MR	-0.9	-3.3	-1.6	0.8
76.35	1.74	1.4	3.06	15MR	-1.6	-3.8	-1.6	0.8
81.90	1.70	1.5	12.88	25MR	-1.2	-4.1	-1.6	0.8
87.56	1.65	1.6	22.64	4AP	-0.1	-4.2	-1.6	0.7
93.30	1.60	1.7	32.36	14AP	1.2	-4.2	-1.5	0.7
99.11	1.55	1.7	42.02	24AP	2.6	-4.1	-1.5	0.6
104.99	1.50	1.8	51.65	4MY	3.4	-4.0	-1.5	0.6
110.92	1.44	1.9	61.24	14MY	1.9	-3.9	-1.6	0.5
116.91	1.39	1.9	70.79	24MY	1.0	-3.8	-1.6	0.4
122.95	1.33	1.9	80.34	3JN	0.2	-3.7	-1.6	0.4
129.04	1.27	2.0	89.87	13JN	-0.8	-3.7	-1.6	0.3
135.17	1.20	2.0	99.40	23JN	-1.7	-3.6	-1.7	0.3
141.36	1.14	2.0	108.94	3JL	-1.2	-3.5	-1.7	0.3
147.59	1.07	2.0	118.49	13JL	-0.4	-3.5	-1.8	0.4
153.88	1.00	2.0	128.07	23JL	0.1	-3.4	-1.8	0.4
160.23	0.93	2.0	137.68	2AU	0.4	-3.4	-1.9	0.5
166.64	0.85	2.0	147.34	12AU	0.7	-3.4	-2.0	0.5
173.10	0.77	2.0	157.03	22AU	1.3	-3.4	-2.0	0.6
179.64	0.69	1.9	166.79	1SE	3.0	-3.4	-2.1	0.6
186.24	0.61	1.9	176.59	11SE	1.2	-3.4	-2.2	0.7
192.91	0.53	1.8	186.45	21SE	-0.5	-3.4	-2.2	0.7
199.65	0.44	1.8	196.38	1OC	-0.9	-3.4	-2.3	0.8
206.47	0.35	1.7	206.35	11OC	-0.9	-3.4	-2.3	0.8
213.36	0.25	1.7	216.38	21OC	-0.8	-3.4	-2.3	0.8
220.33	0.16	1.7	226.46	31OC	-0.6	-3.4	-2.4	0.8
227.37	0.05	1.7	236.58	10NO	-0.5	-3.4	-2.4	0.8
234.50	-0.05	1.7	246.73	20NO	-0.5	-3.4	-2.4	0.8
241.70	-0.15	1.7	256.91	30NO	-0.3	-3.5	-2.3	0.8
248.97	-0.26	1.7	267.11	10DE	0.8	-3.5	-2.3	0.8
256.31	-0.37	1.6	277.30	20DE	2.6	-3.5	-2.2	0.7
263.73	-0.48	1.6	287.49	30DE	0.6	-3.5	-2.2	0.7
				1313				
271.20	-0.59	1.6	297.67	9JA	0.1	-3.4	-2.1	0.8
278.74	-0.69	1.5	307.81	19JA	0.0	-3.4	-2.0	0.8
286.33	-0.80	1.5	317.93	29JA	-0.1	-3.4	-1.9	0.8
293.96	-0.90	1.5	327.99	8FE	-0.4	-3.4	-1.9	0.9
301.63	-1.00	1.4	338.01	18FE	-1.0	-3.4	-1.8	0.9
309.33	-1.10	1.4	347.98	28FE	-1.5	-3.3	-1.7	0.9
317.04	-1.18	1.3	357.89	10MR	-1.1	-3.3	-1.7	0.9
324.76	-1.26	1.3	7.73	20MR	-0.0	-3.3	-1.6	0.9
332.48	-1.32	1.3	17.53	30MR	1.5	-3.3	-1.6	0.8
340.17	-1.37	1.2	27.27	9AP	3.3	-3.4	-1.5	0.8
347.82	-1.41	1.2	36.96	19AP	2.5	-3.4	-1.5	0.8
355.44	-1.44	1.2	46.61	29AP	1.5	-3.4	-1.5	0.7
2.98	-1.45	1.1	56.21	9MY	0.8	-3.4	-1.5	0.6
10.45	-1.44	1.1	65.78	19MY	0.1	-3.5	-1.5	0.6
17.83	-1.42	1.0	75.34	29MY	-0.8	-3.5	-1.5	0.5
25.09	-1.38	1.0	84.87	8JN	-1.7	-3.6	-1.5	0.5
32.24	-1.32	1.0	94.40	18JN	-1.2	-3.6	-1.5	0.4
39.24	-1.25	0.9	103.94	28JN	-0.4	-3.7	-1.5	0.3
46.07	-1.16	0.9	113.48	8JL	0.2	-3.8	-1.5	0.3
52.72	-1.05	0.8	123.04	18JL	0.5	-3.9	-1.5	0.4
59.17	-0.93	0.7	132.64	28JL	0.9	-4.0	-1.6	0.4
65.37	-0.78	0.7	142.27	7AU	1.7	-4.1	-1.6	0.5
71.30	-0.61	0.6	151.94	17AU	3.2	-4.2	-1.7	0.5
76.90	-0.43	0.5	161.67	27AU	1.0	-4.3	-1.7	0.6
82.11	-0.21	0.4	171.44	6SE	-0.6	-4.3	-1.8	0.6
86.86	0.03	0.3	181.28	16SE	-1.1	-4.1	-1.8	0.7
91.04	0.30	0.1	191.17	26SE	-1.1	-3.6	-1.9	0.7
94.53	0.62	-0.1	201.11	6OC	-0.7	-3.2	-2.0	0.8
97.16	0.97	-0.2	211.11	16OC	-0.5	-3.8	-2.0	0.8
98.74	1.38	-0.5	221.17	26OC	-0.3	-4.2	-2.1	0.8
99.08	1.83	-0.7	231.27	5NO	-0.3	-4.4	-2.2	0.8
98.03	2.31	-0.9	241.40	15NO	-0.1	-4.4	-2.2	0.8
95.59	2.78	-1.1	251.57	25NO	1.0	-4.3	-2.2	0.8
92.08	3.19	-1.3	261.76	5DE	2.4	-4.2	-2.2	0.8
88.13	3.47	-1.3	271.95	15DE	0.4	-4.1	-2.2	0.8
84.51	3.59	-1.1	282.15	25DE	-0.0	-4.0	-2.2	0.8

♂			⊙	16.00UT	☿	♀	♃	♄
LONG	LAT	MAG	LONG	1314	MAGNITUDES			
81.88	3.58	-0.8	292.34	4JA	-0.1	-3.9	-2.2	0.8
80.54	3.46	-0.5	302.50	14JA	-0.2	-3.8	-2.2	0.8
80.52	3.30	-0.2	312.63	24JA	-0.5	-3.7	-2.1	0.8
81.67	3.10	0.1	322.72	3FE	-1.0	-3.6	-2.0	0.9
83.78	2.91	0.3	332.77	13FE	-1.3	-3.6	-2.0	0.9
86.68	2.72	0.5	342.77	23FE	-1.1	-3.5	-1.9	0.9
90.20	2.54	0.7	352.70	5MR	0.1	-3.4	-1.8	0.9
94.20	2.37	0.9	2.58	15MR	1.8	-3.4	-1.8	0.9
98.60	2.21	1.1	12.41	25MR	3.4	-3.4	-1.7	0.9
103.31	2.06	1.2	22.18	4AP	1.9	-3.3	-1.6	0.9
108.28	1.92	1.3	31.89	14AP	1.1	-3.3	-1.6	0.9
113.45	1.79	1.4	41.56	24AP	0.6	-3.3	-1.5	0.9
118.81	1.66	1.5	51.18	4MY	-0.0	-3.3	-1.5	0.8
124.31	1.54	1.6	60.77	14MY	-0.8	-3.3	-1.4	0.8
129.95	1.42	1.7	70.34	24MY	-1.7	-3.3	-1.4	0.7
135.71	1.30	1.7	79.87	3JN	-1.2	-3.3	-1.4	0.6
141.58	1.19	1.8	89.41	13JN	-0.3	-3.4	-1.4	0.6
147.55	1.08	1.8	98.94	23JN	0.3	-3.4	-1.4	0.5
153.62	0.97	1.8	108.47	3JL	0.7	-3.4	-1.3	0.4
159.78	0.86	1.9	118.03	13JL	1.2	-3.5	-1.4	0.4
166.03	0.76	1.9	127.61	23JL	2.2	-3.5	-1.4	0.4
172.38	0.65	1.9	137.21	2AU	3.0	-3.5	-1.4	0.4
178.82	0.55	1.9	146.87	12AU	0.8	-3.4	-1.4	0.5
185.35	0.44	1.9	156.56	22AU	-0.7	-3.4	-1.4	0.5
191.98	0.34	1.8	166.31	1SE	-1.2	-3.4	-1.5	0.6
198.70	0.23	1.8	176.11	11SE	-1.2	-3.4	-1.5	0.6
205.51	0.13	1.8	185.97	21SE	-0.7	-3.3	-1.6	0.7
212.42	0.02	1.7	195.89	1OC	-0.4	-3.3	-1.6	0.7
219.43	-0.08	1.7	205.86	11OC	-0.2	-3.3	-1.7	0.8
226.53	-0.18	1.7	215.89	21OC	-0.1	-3.3	-1.7	0.8
233.73	-0.28	1.6	225.96	31OC	0.0	-3.3	-1.8	0.9
241.01	-0.38	1.6	236.08	10NO	1.2	-3.4	-1.9	0.9
248.39	-0.48	1.5	246.24	20NO	2.1	-3.4	-1.9	0.9
255.85	-0.57	1.5	256.41	30NO	0.2	-3.4	-2.0	0.9
263.39	-0.66	1.5	266.61	10DE	-0.2	-3.4	-2.0	0.9
270.99	-0.75	1.5	276.81	20DE	-0.3	-3.5	-2.1	0.9
278.67	-0.83	1.5	287.00	30DE	-0.3	-3.5	-2.1	0.9
				1315				
286.40	-0.90	1.4	297.18	9JA	-0.6	-3.6	-2.1	0.9
294.18	-0.97	1.4	307.32	19JA	-1.0	-3.7	-2.1	0.9
301.99	-1.03	1.4	317.44	29JA	-1.2	-3.7	-2.1	0.9
309.83	-1.07	1.4	327.51	8FE	-1.0	-3.8	-2.1	1.0
317.68	-1.11	1.4	337.53	18FE	0.1	-3.9	-2.0	1.0
325.53	-1.13	1.4	347.50	28FE	2.2	-4.0	-2.0	1.0
333.36	-1.15	1.4	357.41	10MR	2.7	-4.1	-1.9	1.0
341.18	-1.15	1.4	7.27	20MR	1.4	-4.1	-1.8	1.0
348.95	-1.14	1.4	17.06	30MR	0.8	-4.2	-1.8	1.0
356.68	-1.11	1.4	26.81	9AP	0.4	-4.2	-1.7	1.0
4.35	-1.07	1.4	36.50	19AP	-0.1	-4.1	-1.6	1.0
11.95	-1.02	1.4	46.14	29AP	-0.8	-3.8	-1.6	1.0
19.46	-0.96	1.4	55.75	9MY	-1.8	-3.1	-1.5	0.9
26.89	-0.88	1.4	65.32	19MY	-1.2	-3.1	-1.5	0.9
34.23	-0.80	1.4	74.87	29MY	-0.3	-3.8	-1.4	0.8
41.46	-0.70	1.4	84.41	8JN	0.4	-4.1	-1.4	0.8
48.58	-0.60	1.4	93.94	18JN	1.0	-4.2	-1.3	0.7
55.58	-0.48	1.4	103.47	28JN	1.6	-4.2	-1.3	0.7
62.45	-0.36	1.4	113.01	8JL	2.8	-4.1	-1.3	0.6
69.19	-0.23	1.3	122.58	18JL	2.5	-4.0	-1.3	0.5
75.79	-0.09	1.3	132.17	28JL	0.7	-3.9	-1.3	0.5
82.24	0.06	1.3	141.80	7AU	-0.7	-3.8	-1.3	0.5
88.52	0.22	1.2	151.47	17AU	-1.3	-3.7	-1.3	0.5
94.61	0.39	1.2	161.19	27AU	-1.2	-3.7	-1.3	0.6
100.49	0.57	1.1	170.97	6SE	-0.6	-3.6	-1.3	0.6
106.14	0.77	1.1	180.80	16SE	-0.3	-3.6	-1.3	0.7
111.51	0.98	1.0	190.69	26SE	-0.1	-3.5	-1.4	0.7
116.56	1.21	0.9	200.63	6OC	0.0	-3.5	-1.4	0.8
121.22	1.46	0.7	210.63	16OC	0.3	-3.4	-1.4	0.8
125.41	1.73	0.6	220.68	26OC	1.5	-3.4	-1.5	0.9
129.02	2.04	0.4	230.78	5NO	1.9	-3.4	-1.5	0.9
131.92	2.38	0.3	240.91	15NO	-0.0	-3.4	-1.6	0.9
133.95	2.75	0.0	251.08	25NO	-0.4	-3.4	-1.7	1.0
134.91	3.16	-0.2	261.27	5DE	-0.4	-3.4	-1.7	1.0
134.63	3.57	-0.4	271.46	15DE	-0.5	-3.4	-1.8	1.0
133.01	3.97	-0.6	281.66	25DE	-0.7	-3.4	-1.9	1.0

♂ LONG	♂ LAT	♂ MAG	⊙ LONG	16.00UT	☿	♀	♃	♄
LONG	LAT	MAG	LONG	1316	MAGNITUDES			
130.13	4.30	-0.9	291.85	4JA	-1.0	-3.4	-1.9	1.0
126.38	4.49	-1.0	302.01	14JA	-1.0	-3.4	-2.0	1.0
122.43	4.51	-0.9	312.14	24JA	-0.9	-3.4	-2.0	1.0
119.00	4.36	-0.7	322.24	3FE	0.2	-3.4	-2.0	1.0
116.64	4.08	-0.5	332.28	13FE	2.6	-3.4	-2.0	1.0
115.56	3.75	-0.2	342.28	23FE	2.0	-3.5	-2.0	1.1
115.75	3.39	0.0	352.22	4MR	1.0	-3.5	-2.0	1.1
117.06	3.05	0.2	2.10	14MR	0.6	-3.5	-2.0	1.1
119.29	2.73	0.5	11.93	24MR	0.3	-3.4	-1.9	1.2
122.27	2.44	0.6	21.70	3AP	-0.2	-3.4	-1.9	1.2
125.86	2.17	0.8	31.42	13AP	-0.9	-3.4	-1.8	1.2
129.94	1.93	1.0	41.09	23AP	-1.8	-3.4	-1.8	1.2
134.42	1.71	1.1	50.72	3MY	-1.2	-3.3	-1.7	1.1
139.23	1.50	1.2	60.30	13MY	-0.2	-3.3	-1.6	1.1
144.32	1.31	1.3	69.87	23MY	0.6	-3.3	-1.6	1.1
149.64	1.13	1.4	79.41	2JN	1.3	-3.3	-1.5	1.0
155.18	0.96	1.4	88.94	12JN	2.2	-3.3	-1.4	1.0
160.89	0.81	1.5	98.47	22JN	3.3	-3.4	-1.4	0.9
166.78	0.66	1.6	108.01	2JL	2.0	-3.4	-1.3	0.8
172.82	0.51	1.6	117.56	12JL	0.5	-3.4	-1.3	0.8
179.01	0.37	1.6	127.14	22JL	-0.7	-3.4	-1.3	0.7
185.34	0.24	1.6	136.75	1AU	-1.4	-3.5	-1.2	0.6
191.80	0.11	1.6	146.39	11AU	-1.2	-3.5	-1.2	0.6
198.39	-0.01	1.6	156.09	21AU	-0.6	-3.6	-1.2	0.6
205.11	-0.13	1.6	165.84	31AU	-0.2	-3.7	-1.2	0.7
211.95	-0.24	1.6	175.63	10SE	0.0	-3.8	-1.2	0.7
218.91	-0.35	1.6	185.50	20SE	0.2	-3.9	-1.2	0.8
225.98	-0.45	1.6	195.41	30SE	0.5	-4.0	-1.2	0.8
233.17	-0.55	1.6	205.38	10OC	1.8	-4.1	-1.3	0.9
240.47	-0.64	1.6	215.41	20OC	1.7	-4.2	-1.3	0.9
247.87	-0.73	1.5	225.48	30OC	-0.2	-4.3	-1.3	1.0
255.36	-0.80	1.5	235.59	9NO	-0.5	-4.4	-1.4	1.0
262.95	-0.87	1.5	245.75	19NO	-0.5	-4.4	-1.4	1.0
270.61	-0.93	1.5	255.92	29NO	-0.6	-4.2	-1.5	1.1
278.34	-0.98	1.4	266.11	9DE	-0.7	-3.7	-1.5	1.1
286.14	-1.03	1.4	276.31	19DE	-0.8	-3.1	-1.6	1.1
293.98	-1.06	1.4	286.51	29DE	-0.9	-3.8	-1.7	1.1
				1317				
301.85	-1.08	1.3	296.68	8JA	-0.7	-4.2	-1.7	1.1
309.75	-1.08	1.3	306.83	18JA	0.3	-4.4	-1.8	1.1
317.66	-1.08	1.3	316.95	28JA	2.9	-4.3	-1.9	1.1
325.55	-1.06	1.3	327.02	7FE	1.5	-4.3	-1.9	1.1
333.44	-1.04	1.3	337.05	17FE	0.7	-4.1	-2.0	1.2
341.29	-1.00	1.3	347.02	27FE	0.4	-4.0	-2.0	1.2
349.09	-0.95	1.4	356.93	9MR	0.1	-3.9	-2.0	1.3
356.85	-0.89	1.4	6.79	19MR	-0.2	-3.8	-2.0	1.3
4.54	-0.82	1.4	16.59	29MR	-0.9	-3.7	-2.0	1.3
12.16	-0.74	1.5	26.33	8AP	-1.8	-3.6	-2.0	1.3
19.70	-0.66	1.5	36.03	18AP	-1.2	-3.6	-1.9	1.3
27.16	-0.57	1.5	45.68	28AP	-0.2	-3.5	-1.9	1.3
34.53	-0.47	1.5	55.29	8MY	0.8	-3.5	-1.8	1.3
41.80	-0.36	1.6	64.86	18MY	1.7	-3.4	-1.8	1.3
48.99	-0.26	1.6	74.41	28MY	2.9	-3.4	-1.7	1.3
56.07	-0.15	1.6	83.95	7JN	3.0	-3.3	-1.6	1.2
63.06	-0.03	1.6	93.48	17JN	1.6	-3.3	-1.6	1.2
69.95	0.08	1.7	103.01	27JN	0.4	-3.3	-1.5	1.1
76.74	0.20	1.7	112.55	7JL	-0.8	-3.3	-1.5	1.1
83.43	0.33	1.7	122.12	17JL	-1.5	-3.3	-1.4	1.0
90.01	0.45	1.7	131.71	27JL	-1.2	-3.3	-1.4	0.9
96.49	0.58	1.7	141.33	6AU	-0.5	-3.4	-1.3	0.8
102.85	0.71	1.6	151.01	16AU	-0.1	-3.4	-1.3	0.8
109.10	0.84	1.6	160.72	26AU	0.2	-3.4	-1.3	0.8
115.22	0.97	1.6	170.49	5SE	0.4	-3.4	-1.2	0.8
121.21	1.11	1.5	180.32	15SE	0.7	-3.5	-1.2	0.8
127.04	1.26	1.5	190.20	25SE	2.2	-3.5	-1.2	0.9
132.71	1.41	1.4	200.15	5OC	1.5	-3.5	-1.2	0.9
138.17	1.57	1.3	210.14	15OC	-0.3	-3.5	-1.2	1.0
143.41	1.75	1.2	220.19	25OC	-0.7	-3.4	-1.2	1.0
148.37	1.93	1.1	230.29	4NO	-0.7	-3.4	-1.3	1.1
153.00	2.13	1.0	240.42	14NO	-0.7	-3.4	-1.3	1.1
157.23	2.34	0.9	250.58	24NO	-0.7	-3.4	-1.3	1.2
160.97	2.58	0.7	260.77	4DE	-0.7	-3.3	-1.4	1.2
164.09	2.83	0.5	270.97	14DE	-0.7	-3.3	-1.4	1.2
166.46	3.11	0.3	281.16	24DE	-0.6	-3.3	-1.5	1.3

♂ LONG	♂ LAT	♂ MAG	⊙ LONG	16.00UT	☿	♀	♃	♄
LONG	LAT	MAG	LONG	1318	MAGNITUDES			
167.90	3.40	0.0	291.35	3JA	0.5	-3.3	-1.5	1.3
168.22	3.69	-0.2	301.52	13JA	3.0	-3.4	-1.6	1.3
167.29	3.96	-0.5	311.65	23JA	1.1	-3.4	-1.7	1.3
165.07	4.16	-0.7	321.75	2FE	0.5	-3.4	-1.7	1.3
161.76	4.24	-0.9	331.80	12FE	0.2	-3.4	-1.8	1.3
157.89	4.16	-1.0	341.80	22FE	0.0	-3.4	-1.9	1.3
154.15	3.91	-0.9	351.74	4MR	-0.3	-3.5	-1.9	1.4
151.20	3.55	-0.7	1.63	14MR	-0.9	-3.5	-2.0	1.4
149.45	3.13	-0.5	11.46	24MR	-1.7	-3.6	-2.0	1.4
148.98	2.69	-0.2	21.23	3AP	-1.2	-3.6	-2.0	1.4
149.75	2.28	-0.0	30.95	13AP	-0.1	-3.7	-2.0	1.3
151.56	1.90	0.2	40.62	23AP	1.0	-3.8	-2.0	1.3
154.24	1.55	0.4	50.26	3MY	2.2	-3.9	-2.0	1.3
157.65	1.25	0.5	59.85	13MY	3.7	-3.9	-2.0	1.3
161.64	0.97	0.7	69.41	23MY	2.3	-4.0	-2.0	1.2
166.11	0.72	0.8	78.96	2JN	1.2	-4.1	-1.9	1.2
170.99	0.50	0.9	88.49	12JN	0.3	-4.2	-1.9	1.1
176.21	0.30	1.0	98.01	22JN	-0.8	-4.2	-1.8	1.1
181.72	0.11	1.1	107.55	2JL	-1.6	-4.0	-1.7	1.1
187.50	-0.06	1.1	117.10	12JL	-1.2	-3.6	-1.7	1.0
193.51	-0.22	1.2	126.67	22JL	-0.5	-3.1	-1.6	1.0
199.74	-0.36	1.2	136.28	1AU	0.0	-3.6	-1.5	0.9
206.15	-0.50	1.3	145.93	11AU	0.3	-4.1	-1.5	0.9
212.75	-0.62	1.3	155.62	21AU	0.5	-4.2	-1.4	0.8
219.52	-0.73	1.3	165.36	31AU	1.0	-4.3	-1.4	0.8
226.44	-0.83	1.3	175.16	10SE	2.6	-4.2	-1.4	0.8
233.50	-0.92	1.3	185.01	20SE	1.3	-4.1	-1.3	0.9
240.71	-0.99	1.4	194.93	30SE	-0.4	-4.0	-1.3	0.9
248.04	-1.06	1.4	204.89	10OC	-0.8	-4.0	-1.3	1.0
255.49	-1.11	1.4	214.91	20OC	-0.8	-3.9	-1.3	1.1
263.04	-1.15	1.4	224.99	30OC	-0.8	-3.8	-1.3	1.1
270.68	-1.18	1.4	235.10	9NO	-0.6	-3.7	-1.3	1.2
278.40	-1.19	1.4	245.25	19NO	-0.5	-3.7	-1.3	1.2
286.19	-1.19	1.4	255.43	29NO	-0.6	-3.6	-1.3	1.3
294.02	-1.18	1.4	265.62	9DE	-0.4	-3.5	-1.3	1.3
301.89	-1.16	1.4	275.82	19DE	0.6	-3.5	-1.4	1.3
309.78	-1.13	1.4	286.02	29DE	2.9	-3.4	-1.4	1.3
				1319				
317.68	-1.08	1.4	296.19	8JA	0.8	-3.4	-1.4	1.3
325.56	-1.02	1.4	306.34	18JA	0.3	-3.4	-1.5	1.3
333.43	-0.96	1.4	316.47	28JA	0.1	-3.4	-1.6	1.3
341.25	-0.88	1.4	326.54	7FE	-0.1	-3.3	-1.6	1.2
349.03	-0.80	1.4	336.56	17FE	-0.4	-3.3	-1.7	1.2
356.75	-0.71	1.4	346.54	27FE	-0.9	-3.3	-1.8	1.1
4.40	-0.62	1.4	356.46	9MR	-1.6	-3.3	-1.8	1.1
11.98	-0.52	1.5	6.31	19MR	-1.2	-3.3	-1.9	1.1
19.48	-0.42	1.5	16.12	29MR	-0.1	-3.3	-2.0	1.1
26.90	-0.31	1.5	25.86	8AP	1.2	-3.4	-2.0	1.1
34.23	-0.21	1.5	35.56	18AP	2.9	-3.4	-2.1	1.1
41.48	-0.10	1.6	45.21	28AP	3.1	-3.4	-2.1	1.1
48.64	0.00	1.6	54.82	8MY	1.8	-3.5	-2.1	1.0
55.71	0.11	1.7	64.40	18MY	0.9	-3.5	-2.1	1.0
62.70	0.21	1.7	73.95	28MY	0.2	-3.5	-2.1	1.0
69.60	0.31	1.7	83.49	7JN	-0.8	-3.4	-2.1	1.0
76.43	0.42	1.8	93.02	17JN	-1.7	-3.4	-2.1	0.9
83.17	0.52	1.8	102.55	27JN	-1.2	-3.4	-2.0	0.9
89.84	0.61	1.8	112.09	7JL	-0.4	-3.3	-2.0	0.8
96.43	0.71	1.9	121.65	17JL	0.1	-3.3	-1.9	0.8
102.95	0.80	1.9	131.24	27JL	0.4	-3.3	-1.9	0.7
109.40	0.90	1.9	140.86	6AU	0.8	-3.3	-1.8	0.7
115.77	0.99	1.9	150.53	16AU	1.4	-3.3	-1.7	0.6
122.08	1.08	1.9	160.25	26AU	3.1	-3.3	-1.7	0.6
128.31	1.17	1.9	170.01	5SE	1.2	-3.4	-1.6	0.5
134.46	1.26	1.8	179.84	15SE	-0.5	-3.4	-1.6	0.5
140.53	1.35	1.8	189.72	25SE	-1.0	-3.4	-1.5	0.5
146.51	1.44	1.8	199.66	5OC	-1.0	-3.5	-1.5	0.6
152.40	1.53	1.7	209.65	15OC	-0.8	-3.5	-1.4	0.7
158.17	1.62	1.6	219.70	25OC	-0.5	-3.6	-1.4	0.7
163.81	1.71	1.6	229.79	4NO	-0.4	-3.7	-1.4	0.8
169.30	1.81	1.5	239.92	14NO	-0.4	-3.7	-1.4	0.9
174.60	1.91	1.4	250.09	24NO	-0.3	-3.8	-1.4	0.9
179.70	2.01	1.2	260.27	4DE	0.8	-3.9	-1.4	1.0
184.53	2.12	1.1	270.47	14DE	2.6	-4.0	-1.4	1.0
189.05	2.23	0.9	280.67	24DE	0.5	-4.1	-1.4	1.0

Left table (1320–1321)

♂ LONG	♂ LAT	♂ MAG	☉ LONG	16.00UT 1320	☿	♀	♃	♄
193.17	2.34	0.8	290.85	3JA	0.1	-4.2	-1.4	1.0
196.81	2.46	0.5	301.02	13JA	-0.0	-4.3	-1.4	1.0
199.84	2.57	0.3	311.16	23JA	-0.2	-4.4	-1.5	1.0
202.12	2.68	0.1	321.26	2FE	-0.4	-4.3	-1.5	1.0
203.46	2.77	-0.2	331.31	12FE	-0.9	-4.1	-1.6	1.0
203.70	2.83	-0.5	341.32	22FE	-1.4	-3.6	-1.6	0.9
202.70	2.83	-0.8	351.26	3MR	-1.1	-3.3	-1.7	0.9
200.43	2.74	-1.1	1.15	13MR	-0.0	-3.8	-1.7	0.8
197.16	2.53	-1.3	10.99	23MR	1.6	-4.1	-1.8	0.8
193.41	2.21	-1.3	20.76	2AP	3.6	-4.2	-1.9	0.8
189.89	1.79	-1.2	30.48	12AP	2.3	-4.2	-1.9	0.8
187.26	1.33	-1.0	40.16	22AP	1.3	-4.1	-2.0	0.8
185.87	0.88	-0.8	49.79	2MY	0.7	-4.0	-2.1	0.8
185.82	0.47	-0.6	59.38	12MY	0.1	-3.9	-2.1	0.8
187.01	0.11	-0.4	68.95	22MY	-0.8	-3.8	-2.2	0.8
189.27	-0.21	-0.2	78.49	1JN	-1.7	-3.7	-2.2	0.8
192.42	-0.48	-0.0	88.03	11JN	-1.2	-3.7	-2.3	0.7
196.32	-0.71	0.1	97.55	21JN	-0.4	-3.6	-2.3	0.7
200.82	-0.90	0.2	107.09	1JL	0.2	-3.5	-2.3	0.7
205.83	-1.07	0.4	116.64	11JL	0.6	-3.5	-2.2	0.6
211.28	-1.21	0.5	126.21	21JL	1.0	-3.4	-2.2	0.6
217.09	-1.33	0.6	135.82	31JL	1.9	-3.4	-2.2	0.5
223.21	-1.42	0.6	145.46	10AU	3.2	-3.4	-2.1	0.5
229.62	-1.50	0.7	155.15	20AU	1.0	-3.4	-2.0	0.4
236.26	-1.55	0.8	164.89	30AU	-0.6	-3.4	-2.0	0.4
243.12	-1.59	0.8	174.69	9SE	-1.1	-3.4	-1.9	0.3
250.16	-1.61	0.9	184.54	19SE	-1.1	-3.4	-1.9	0.3
257.37	-1.61	0.9	194.44	29SE	-0.7	-3.4	-1.8	0.2
264.72	-1.60	1.0	204.41	90C	-0.4	-3.4	-1.7	0.3
272.19	-1.57	1.0	214.43	190C	-0.3	-3.4	-1.7	0.3
279.76	-1.52	1.1	224.50	290C	-0.2	-3.4	-1.6	0.4
287.42	-1.47	1.1	234.61	8NO	-0.1	-3.4	-1.6	0.5
295.13	-1.39	1.2	244.76	18NO	1.0	-3.4	-1.5	0.5
302.88	-1.31	1.2	254.93	28NO	2.4	-3.5	-1.5	0.6
310.66	-1.22	1.2	265.12	8DE	0.3	-3.5	-1.5	0.7
318.45	-1.12	1.3	275.32	18DE	-0.1	-3.5	-1.5	0.7
326.23	-1.01	1.3	285.51	28DE	-0.2	-3.5	-1.5	0.7
				1321				
333.98	-0.90	1.4	295.70	7JA	-0.3	-3.4	-1.5	0.8
341.70	-0.78	1.4	305.85	17JA	-0.5	-3.4	-1.5	0.8
349.36	-0.66	1.5	315.97	27JA	-1.0	-3.4	-1.5	0.8
356.97	-0.54	1.5	326.05	6FE	-1.3	-3.4	-1.5	0.8
4.52	-0.42	1.5	336.08	16FE	-1.0	-3.3	-1.5	0.8
11.99	-0.30	1.6	346.05	26FE	0.0	-3.3	-1.6	0.7
19.38	-0.19	1.6	355.98	8MR	1.9	-3.3	-1.6	0.7
26.70	-0.07	1.6	5.84	18MR	3.1	-3.3	-1.7	0.6
33.93	0.04	1.7	15.64	28MR	1.7	-3.3	-1.7	0.6
41.09	0.14	1.7	25.40	7AP	1.0	-3.4	-1.8	0.6
48.16	0.24	1.7	35.09	17AP	0.5	-3.4	-1.8	0.6
55.15	0.34	1.7	44.75	27AP	-0.0	-3.4	-1.9	0.6
62.07	0.44	1.7	54.36	7MY	-0.8	-3.4	-2.0	0.6
68.92	0.53	1.8	63.94	17MY	-1.7	-3.5	-2.1	0.6
75.70	0.61	1.8	73.49	27MY	-1.2	-3.5	-2.1	0.6
82.41	0.69	1.8	83.03	6JN	-0.3	-3.6	-2.2	0.6
89.06	0.77	1.8	92.56	16JN	0.3	-3.6	-2.3	0.6
95.65	0.84	1.9	102.09	26JN	0.8	-3.7	-2.3	0.6
102.20	0.90	1.9	111.63	6JL	1.4	-3.8	-2.3	0.5
108.69	0.97	1.9	121.19	16JL	2.4	-3.9	-2.4	0.5
115.14	1.03	2.0	130.78	26JL	2.8	-4.0	-2.4	0.4
121.55	1.08	2.0	140.40	5AU	0.8	-4.1	-2.4	0.4
127.91	1.13	2.0	150.07	15AU	-0.6	-4.2	-2.4	0.3
134.25	1.18	2.0	159.78	25AU	-1.2	-4.3	-2.3	0.3
140.54	1.23	2.0	169.55	4SE	-1.2	-4.2	-2.3	0.2
146.81	1.27	2.0	179.36	14SE	-0.7	-4.1	-2.2	0.2
153.03	1.31	2.0	189.24	24SE	-0.3	-3.6	-2.2	0.1
159.23	1.34	1.9	199.18	40C	-0.2	-3.2	-2.1	0.1
165.38	1.37	1.9	209.17	140C	-0.1	-3.8	-2.0	0.0
171.49	1.40	1.9	219.21	240C	0.1	-4.2	-2.0	0.1
177.55	1.43	1.8	229.30	3NO	1.3	-4.4	-1.9	0.1
183.55	1.45	1.7	239.43	13NO	2.1	-4.3	-1.8	0.2
189.50	1.47	1.7	249.59	23NO	0.1	-4.3	-1.8	0.3
195.36	1.48	1.6	259.78	3DE	-0.3	-4.2	-1.7	0.3
201.13	1.49	1.5	269.97	13DE	-0.3	-4.1	-1.7	0.4
206.80	1.48	1.4	280.17	23DE	-0.4	-4.0	-1.6	0.5

Right table (1322–1323)

♂ LONG	♂ LAT	♂ MAG	☉ LONG	16.00UT 1322	☿	♀	♃	♄
212.33	1.48	1.2	290.36	2JA	-0.6	-3.9	-1.6	0.5
217.70	1.46	1.1	300.53	12JA	-1.0	-3.8	-1.6	0.5
222.88	1.42	0.9	310.67	22JA	-1.1	-3.7	-1.6	0.6
227.81	1.37	0.7	320.77	1FE	-0.9	-3.6	-1.6	0.6
232.45	1.31	0.5	330.82	11FE	0.1	-3.6	-1.5	0.6
236.72	1.21	0.3	340.83	21FE	2.3	-3.5	-1.6	0.6
240.51	1.08	0.1	350.78	3MR	2.4	-3.5	-1.6	0.6
243.73	0.91	-0.2	0.68	13MR	1.2	-3.4	-1.6	0.5
246.20	0.68	-0.5	10.52	23MR	0.7	-3.4	-1.6	0.5
247.76	0.38	-0.8	20.29	2AP	0.4	-3.3	-1.6	0.5
248.23	0.01	-1.1	30.02	12AP	-0.1	-3.3	-1.7	0.4
247.46	-0.45	-1.4	39.70	22AP	-0.8	-3.3	-1.7	0.4
245.48	-0.99	-1.7	49.33	2MY	-1.7	-3.3	-1.8	0.4
242.54	-1.56	-2.0	58.92	12MY	-1.3	-3.3	-1.8	0.4
239.19	-2.11	-2.0	68.49	22MY	-0.3	-3.3	-1.9	0.5
236.20	-2.58	-1.8	78.03	1JN	0.5	-3.3	-2.0	0.5
234.18	-2.92	-1.7	87.57	11JN	1.1	-3.4	-2.0	0.5
233.48	-3.14	-1.5	97.10	21JN	1.8	-3.4	-2.1	0.5
234.17	-3.26	-1.2	106.63	1JL	3.1	-3.4	-2.2	0.5
236.12	-3.30	-1.0	116.18	11JL	2.3	-3.5	-2.3	0.4
239.16	-3.28	-0.8	125.75	21JL	0.7	-3.5	-2.3	0.4
243.11	-3.22	-0.6	135.35	31JL	-0.7	-3.5	-2.4	0.4
247.78	-3.13	-0.5	144.99	10AU	-1.4	-3.4	-2.4	0.3
253.06	-3.01	-0.3	154.68	20AU	-1.2	-3.4	-2.4	0.3
258.82	-2.87	-0.2	164.41	30AU	-0.6	-3.4	-2.5	0.2
264.97	-2.72	-0.0	174.21	9SE	-0.2	-3.4	-2.5	0.1
271.44	-2.55	0.1	184.06	19SE	-0.0	-3.3	-2.4	0.1
278.17	-2.37	0.2	193.96	29SE	0.1	-3.3	-2.4	0.0
285.10	-2.19	0.3	203.92	90C	0.3	-3.3	-2.4	-0.0
292.20	-2.00	0.4	213.94	190C	1.6	-3.3	-2.3	-0.1
299.42	-1.80	0.6	224.00	290C	1.9	-3.3	-2.2	-0.1
306.73	-1.60	0.7	234.11	8NO	-0.1	-3.4	-2.2	-0.1
314.10	-1.41	0.8	244.26	18NO	-0.4	-3.4	-2.1	-0.0
321.51	-1.21	0.9	254.40	28NO	-0.5	-3.4	-2.0	0.1
328.93	-1.02	1.0	264.62	8DE	-0.5	-3.4	-1.9	0.1
336.35	-0.84	1.0	274.82	18DE	-0.7	-3.5	-1.9	0.2
343.74	-0.67	1.1	285.02	28DE	-0.9	-3.5	-1.8	0.3
				1323				
351.11	-0.50	1.2	295.20	7JA	-1.0	-3.6	-1.8	0.3
358.44	-0.34	1.3	305.35	17JA	-0.8	-3.7	-1.7	0.4
5.71	-0.19	1.4	315.47	27JA	0.2	-3.7	-1.7	0.4
12.93	-0.05	1.5	325.56	6FE	2.6	-3.8	-1.6	0.4
20.09	0.08	1.5	335.59	16FE	1.8	-3.9	-1.6	0.4
27.18	0.20	1.6	345.57	26FE	0.9	-4.0	-1.6	0.4
34.21	0.32	1.7	355.49	8MR	0.5	-4.1	-1.6	0.4
41.17	0.42	1.7	5.36	18MR	0.2	-4.1	-1.6	0.4
48.07	0.52	1.8	15.17	28MR	-0.2	-4.2	-1.6	0.4
54.90	0.60	1.8	24.92	7AP	-0.8	-4.2	-1.6	0.4
61.67	0.68	1.8	34.63	17AP	-1.7	-4.1	-1.6	0.3
68.38	0.76	1.9	44.28	27AP	-1.3	-3.8	-1.6	0.3
75.04	0.82	1.9	53.90	7MY	-0.3	-3.1	-1.7	0.3
81.65	0.88	1.9	63.48	17MY	0.6	-3.8	-1.7	0.3
88.21	0.93	1.9	73.03	27MY	1.4	-3.8	-1.7	0.3
94.73	0.98	1.9	82.57	6JN	2.4	-4.1	-1.8	0.4
101.22	1.02	1.9	92.10	16JN	3.3	-4.2	-1.8	0.4
107.67	1.06	1.9	101.63	26JN	1.9	-4.2	-1.9	0.4
114.10	1.09	1.9	111.17	6JL	0.5	-4.1	-2.0	0.4
120.51	1.11	1.9	120.73	16JL	-0.7	-4.0	-2.0	0.4
126.89	1.14	2.0	130.31	26JL	-1.5	-3.9	-2.1	0.3
133.27	1.15	2.0	139.93	5AU	-1.2	-3.8	-2.2	0.3
139.63	1.16	2.0	149.60	15AU	-0.6	-3.7	-2.2	0.3
145.99	1.17	2.0	159.30	25AU	-0.1	-3.7	-2.3	0.2
152.35	1.18	2.0	169.07	4SE	0.1	-3.6	-2.4	0.2
158.70	1.17	2.0	178.89	14SE	0.2	-3.6	-2.4	0.1
165.06	1.17	2.0	188.76	24SE	0.6	-3.5	-2.4	0.1
171.42	1.16	2.0	198.70	40C	1.9	-3.5	-2.4	-0.0
177.79	1.14	2.0	208.68	140C	1.7	-3.4	-2.4	-0.1
184.16	1.12	2.0	218.72	240C	-0.2	-3.4	-2.4	-0.1
190.53	1.09	1.9	228.81	3NO	-0.6	-3.4	-2.4	-0.2
196.91	1.06	1.9	238.94	13NO	-0.6	-3.4	-2.3	-0.2
203.29	1.02	1.8	249.10	23NO	-0.6	-3.4	-2.3	-0.2
209.67	0.97	1.8	259.29	3DE	-0.8	-3.4	-2.2	-0.1
216.04	0.91	1.7	269.48	13DE	-0.8	-3.4	-2.1	-0.0
222.41	0.85	1.6	279.68	23DE	-0.8	-3.4	-2.1	0.0

♂ LONG	♂ LAT	♂ MAG	☉ LONG	16.00UT	☿	♀	♃	♄
				1324	MAGNITUDES			
228.76	0.77	1.5	289.87	2JA	-0.7	-3.4	-2.0	0.1
235.10	0.68	1.4	300.03	12JA	0.3	-3.4	-1.9	0.2
241.41	0.57	1.3	310.17	22JA	2.9	-3.4	-1.9	0.2
247.69	0.44	1.2	320.28	1FE	1.4	-3.4	-1.8	0.3
253.93	0.30	1.0	330.34	11FE	0.6	-3.4	-1.7	0.3
260.11	0.13	0.9	340.34	21FE	0.3	-3.5	-1.7	0.3
266.22	-0.06	0.7	350.30	2MR	0.1	-3.5	-1.6	0.4
272.23	-0.28	0.6	0.19	12MR	-0.3	-3.5	-1.6	0.4
278.13	-0.53	0.4	10.03	22MR	-0.9	-3.4	-1.6	0.4
283.87	-0.83	0.2	19.81	1AP	-1.7	-3.4	-1.6	0.3
289.41	-1.16	-0.0	29.54	11AP	-1.3	-3.4	-1.5	0.3
294.70	-1.55	-0.2	39.22	21AP	-0.2	-3.4	-1.5	0.3
299.66	-1.99	-0.5	48.86	1MY	0.8	-3.3	-1.5	0.3
304.18	-2.49	-0.7	58.45	11MY	1.8	-3.3	-1.5	0.2
308.15	-3.06	-1.0	68.02	21MY	3.2	-3.3	-1.6	0.2
311.39	-3.69	-1.3	77.57	31MY	2.8	-3.3	-1.6	0.2
313.74	-4.38	-1.6	87.10	10JN	1.5	-3.3	-1.6	0.3
314.99	-5.11	-1.9	96.63	20JN	0.4	-3.4	-1.6	0.3
314.97	-5.82	-2.1	106.17	30JN	-0.7	-3.4	-1.7	0.3
313.73	-6.41	-2.4	115.71	10JL	-1.5	-3.4	-1.7	0.3
311.50	-6.78	-2.6	125.28	20JL	-1.2	-3.4	-1.8	0.3
308.88	-6.85	-2.6	134.89	30JL	-0.5	-3.5	-1.8	0.3
306.62	-6.59	-2.4	144.52	9AU	-0.0	-3.5	-1.9	0.3
305.32	-6.09	-2.2	154.21	19AU	0.2	-3.6	-2.0	0.3
305.30	-5.43	-1.9	163.95	29AU	0.4	-3.7	-2.0	0.3
306.60	-4.74	-1.6	173.73	8SE	0.8	-3.8	-2.1	0.2
309.06	-4.05	-1.4	183.58	18SE	2.3	-3.9	-2.2	0.2
312.50	-3.41	-1.1	193.48	28SE	1.5	-4.0	-2.2	0.1
316.71	-2.83	-0.8	203.44	8OC	-0.3	-4.1	-2.3	0.0
321.53	-2.31	-0.6	213.45	18OC	-0.7	-4.2	-2.3	-0.0
326.81	-1.84	-0.3	223.52	28OC	-0.7	-4.3	-2.3	-0.1
332.45	-1.43	-0.1	233.62	7NO	-0.8	-4.4	-2.4	-0.2
338.35	-1.07	0.1	243.77	17NO	-0.7	-4.4	-2.4	-0.3
344.46	-0.75	0.3	253.94	27NO	-0.6	-4.2	-2.3	-0.3
350.72	-0.47	0.4	264.13	7DE	-0.6	-3.6	-2.3	-0.2
357.09	-0.23	0.6	274.33	17DE	-0.5	-3.1	-2.3	-0.1
3.54	-0.01	0.8	284.52	27DE	0.5	-3.9	-2.2	-0.1
				1325				
10.03	0.17	0.9	294.70	6JA	3.0	-4.2	-2.1	-0.0
16.56	0.34	1.0	304.86	16JA	1.0	-4.4	-2.1	0.1
23.11	0.48	1.2	314.99	26JA	0.4	-4.3	-2.0	0.1
29.65	0.60	1.3	325.07	5FE	0.2	-4.3	-1.9	0.2
36.19	0.71	1.4	335.10	15FE	0.0	-4.1	-1.8	0.2
42.72	0.80	1.5	345.09	25FE	-0.3	-4.0	-1.8	0.3
49.22	0.88	1.6	355.01	7MR	-0.9	-3.9	-1.7	0.3
55.71	0.95	1.7	4.88	17MR	-1.6	-3.8	-1.7	0.3
62.17	1.01	1.7	14.69	27MR	-1.2	-3.7	-1.6	0.3
68.61	1.06	1.8	24.45	6AP	-0.2	-3.6	-1.6	0.3
75.03	1.09	1.9	34.16	16AP	1.0	-3.6	-1.5	0.3
81.42	1.13	1.9	43.81	26AP	2.4	-3.5	-1.5	0.3
87.80	1.15	1.9	53.43	6MY	3.6	-3.4	-1.5	0.3
94.15	1.17	2.0	63.01	16MY	2.1	-3.4	-1.5	0.2
100.50	1.18	2.0	72.57	26MY	1.1	-3.4	-1.4	0.2
106.83	1.19	2.0	82.11	5JN	0.3	-3.3	-1.4	0.2
113.15	1.19	2.0	91.64	15JN	-0.7	-3.3	-1.4	0.2
119.48	1.18	2.0	101.17	25JN	-1.6	-3.3	-1.5	0.3
125.80	1.18	2.0	110.71	5JL	-1.2	-3.3	-1.5	0.3
132.13	1.16	2.0	120.27	15JL	-0.5	-3.3	-1.5	0.3
138.47	1.15	2.0	129.85	25JL	0.0	-3.3	-1.5	0.3
144.82	1.12	2.0	139.47	4AU	0.4	-3.4	-1.6	0.3
151.20	1.10	2.0	149.13	14AU	0.6	-3.4	-1.6	0.3
157.59	1.07	1.9	158.84	24AU	1.2	-3.4	-1.7	0.3
164.01	1.03	2.0	168.60	3SE	2.8	-3.4	-1.7	0.3
170.47	0.99	2.0	178.41	13SE	1.3	-3.5	-1.8	0.3
176.95	0.95	2.0	188.28	23SE	-0.4	-3.5	-1.8	0.2
183.47	0.90	2.0	198.21	3OC	-0.9	-3.5	-1.9	0.2
190.03	0.84	2.0	208.20	13OC	-0.9	-3.5	-2.0	0.1
196.62	0.78	1.9	218.23	23OC	-0.9	-3.4	-2.0	0.0
203.26	0.72	1.9	228.32	2NO	-0.6	-3.4	-2.1	-0.0
209.94	0.65	1.9	238.45	12NO	-0.5	-3.4	-2.1	-0.1
216.66	0.57	1.9	248.60	22NO	-0.5	-3.4	-2.2	-0.2
223.43	0.48	1.8	258.78	2DE	-0.4	-3.3	-2.2	-0.3
230.24	0.39	1.8	268.98	12DE	0.6	-3.3	-2.2	-0.3
237.10	0.28	1.7	279.18	22DE	2.8	-3.3	-2.2	-0.2

♂ LONG	♂ LAT	♂ MAG	☉ LONG	16.00UT	☿	♀	♃	♄
				1326	MAGNITUDES			
244.00	0.17	1.6	289.37	1JA	0.7	-3.3	-2.2	-0.1
250.94	0.05	1.6	299.54	11JA	0.2	-3.4	-2.2	-0.1
257.92	-0.08	1.5	309.68	21JA	0.0	-3.4	-2.1	0.0
264.94	-0.22	1.4	319.79	31JA	-0.1	-3.4	-2.1	0.1
271.99	-0.37	1.3	329.85	10FE	-0.4	-3.4	-2.0	0.1
279.08	-0.53	1.2	339.86	20FE	-0.9	-3.4	-1.9	0.2
286.19	-0.70	1.1	349.81	2MR	-1.5	-3.5	-1.9	0.2
293.33	-0.89	1.0	359.72	12MR	-1.2	-3.5	-1.8	0.3
300.47	-1.08	0.9	9.56	22MR	-0.1	-3.6	-1.7	0.3
307.63	-1.28	0.8	19.34	1AP	1.3	-3.6	-1.7	0.3
314.76	-1.48	0.7	29.08	11AP	3.1	-3.7	-1.6	0.3
321.87	-1.69	0.6	38.75	21AP	2.8	-3.8	-1.5	0.3
328.94	-1.91	0.5	48.39	1MY	1.6	-3.9	-1.5	0.3
335.93	-2.12	0.4	58.00	11MY	0.9	-3.9	-1.5	0.3
342.81	-2.33	0.2	67.56	21MY	0.2	-4.0	-1.4	0.3
349.55	-2.54	0.1	77.11	31MY	-0.8	-4.1	-1.4	0.3
356.10	-2.74	-0.0	86.65	10JN	-1.7	-4.2	-1.4	0.2
2.41	-2.94	-0.2	96.17	20JN	-1.3	-4.2	-1.3	0.2
8.41	-3.12	-0.3	105.71	30JN	-0.4	-4.0	-1.3	0.3
14.00	-3.28	-0.5	115.26	10JL	0.1	-3.6	-1.3	0.3
19.09	-3.42	-0.7	124.82	20JL	0.5	-3.1	-1.3	0.4
23.53	-3.53	-0.9	134.42	30JL	0.9	-3.6	-1.3	0.4
27.16	-3.61	-1.1	144.06	9AU	1.6	-4.1	-1.4	0.4
29.78	-3.63	-1.3	153.74	19AU	3.2	-4.2	-1.4	0.4
31.18	-3.58	-1.5	163.47	29AU	1.2	-4.2	-1.4	0.4
31.15	-3.43	-1.8	173.26	8SE	-0.5	-4.2	-1.5	0.4
29.68	-3.15	-2.0	183.10	18SE	-1.0	-4.1	-1.5	0.4
26.97	-2.72	-2.1	193.00	28SE	-1.0	-4.0	-1.6	0.3
23.61	-2.15	-2.2	202.96	8OC	-0.8	-3.9	-1.6	0.3
20.43	-1.52	-2.0	212.97	18OC	-0.5	-3.9	-1.7	0.2
18.11	-0.90	-1.7	223.03	28OC	-0.4	-3.8	-1.7	0.2
17.08	-0.35	-1.4	233.14	7NO	-0.3	-3.7	-1.8	0.1
17.38	0.10	-1.0	243.28	17NO	-0.2	-3.6	-1.9	0.0
18.89	0.47	-0.7	253.45	27NO	0.8	-3.6	-1.9	-0.1
21.41	0.75	-0.4	263.64	7DE	2.6	-3.5	-2.0	-0.1
24.72	0.96	-0.1	273.83	17DE	0.5	-3.5	-2.0	-0.2
28.66	1.13	0.1	284.03	27DE	-0.0	-3.4	-2.1	-0.2
				1327				
33.08	1.25	0.4	294.21	6JA	-0.1	-3.4	-2.1	-0.1
37.87	1.34	0.6	304.37	16JA	-0.2	-3.4	-2.1	-0.1
42.95	1.41	0.8	314.50	26JA	-0.5	-3.4	-2.1	-0.0
48.25	1.45	0.9	324.58	5FE	-0.9	-3.3	-2.1	0.1
53.73	1.49	1.1	334.62	15FE	-1.4	-3.3	-2.0	0.1
59.34	1.50	1.2	344.61	25FE	-1.1	-3.3	-2.0	0.2
65.05	1.51	1.4	354.54	7MR	-0.0	-3.3	-1.9	0.2
70.85	1.51	1.5	4.40	17MR	1.6	-3.3	-1.9	0.3
76.72	1.50	1.6	14.22	27MR	3.6	-3.3	-1.8	0.3
82.64	1.49	1.7	23.98	6AP	2.0	-3.4	-1.7	0.4
88.61	1.47	1.7	33.68	16AP	1.2	-3.4	-1.7	0.4
94.61	1.44	1.8	43.35	26AP	0.6	-3.4	-1.6	0.4
100.65	1.41	1.9	52.96	6MY	0.1	-3.5	-1.5	0.4
106.72	1.38	1.9	62.55	16MY	-0.8	-3.5	-1.5	0.4
112.82	1.34	2.0	72.11	26MY	-1.7	-3.5	-1.4	0.4
118.95	1.30	2.0	81.64	5JN	-1.3	-3.4	-1.4	0.4
125.11	1.26	2.0	91.17	15JN	-0.4	-3.4	-1.4	0.4
131.30	1.21	2.0	100.71	25JN	0.2	-3.4	-1.3	0.3
137.53	1.16	2.0	110.24	5JL	0.7	-3.3	-1.3	0.3
143.80	1.11	2.0	119.80	15JL	1.1	-3.3	-1.3	0.4
150.10	1.05	2.0	129.39	25JL	2.0	-3.3	-1.3	0.4
156.45	1.00	2.0	139.00	4AU	3.1	-3.3	-1.2	0.5
162.85	0.93	2.0	148.66	14AU	1.0	-3.3	-1.2	0.5
169.30	0.87	2.0	158.37	24AU	-0.6	-3.3	-1.3	0.5
175.80	0.80	1.9	168.12	3SE	-1.2	-3.4	-1.3	0.5
182.36	0.73	1.9	177.93	13SE	-1.2	-3.4	-1.3	0.5
188.98	0.65	1.8	187.80	23SE	-0.7	-3.4	-1.3	0.5
195.66	0.57	1.8	197.73	3OC	-0.4	-3.5	-1.4	0.5
202.41	0.49	1.8	207.71	13OC	-0.2	-3.5	-1.4	0.4
209.22	0.40	1.8	217.75	23OC	-0.2	-3.6	-1.4	0.4
216.10	0.31	1.8	227.83	2NO	-0.0	-3.7	-1.5	0.3
223.06	0.21	1.8	237.96	12NO	1.0	-3.7	-1.6	0.3
230.08	0.12	1.8	248.11	22NO	2.4	-3.8	-1.6	0.2
237.17	0.01	1.7	258.29	2DE	0.2	-3.9	-1.7	0.1
244.32	-0.10	1.7	268.49	12DE	-0.2	-4.0	-1.7	0.1
251.55	-0.21	1.7	278.69	22DE	-0.2	-4.1	-1.8	-0.0

1328

♂ LONG	LAT	MAG	☉ LONG	16.00UT	☿	♀	♃	♄
258.84	-0.32	1.6	288.87	1JA	-0.3	-4.2	-1.9	-0.1
266.20	-0.44	1.6	299.05	11JA	-0.5	-4.3	-1.9	-0.1
273.61	-0.56	1.5	309.19	21JA	-1.0	-4.4	-2.0	-0.0
281.07	-0.68	1.5	319.30	31JA	-1.2	-4.3	-2.0	0.0
288.59	-0.80	1.4	329.36	10FE	-1.0	-4.1	-2.0	0.1
296.15	-0.92	1.4	339.38	20FE	0.0	-3.6	-2.0	0.2
303.74	-1.04	1.3	349.33	1MR	2.0	-3.3	-2.0	0.2
311.35	-1.15	1.3	359.24	11MR	2.9	-3.8	-2.0	0.3
318.98	-1.26	1.2	9.08	21MR	1.5	-4.2	-2.0	0.3
326.61	-1.36	1.2	18.87	31MR	0.9	-4.3	-1.9	0.4
334.23	-1.45	1.1	28.60	10AP	0.5	-4.2	-1.8	0.4
341.82	-1.53	1.1	38.29	20AP	-0.0	-4.1	-1.8	0.5
349.37	-1.59	1.0	47.92	30AP	-0.8	-4.0	-1.7	0.5
356.87	-1.64	1.0	57.53	10MY	-1.7	-3.9	-1.7	0.5
4.29	-1.68	0.9	67.10	20MY	-1.3	-3.8	-1.6	0.5
11.61	-1.69	0.8	76.64	30MY	-0.3	-3.7	-1.5	0.5
18.82	-1.69	0.8	86.18	9JN	0.4	-3.7	-1.5	0.5
25.89	-1.67	0.7	95.71	19JN	0.9	-3.6	-1.4	0.5
32.79	-1.63	0.7	105.24	29JN	1.5	-3.5	-1.4	0.5
39.51	-1.57	0.6	114.79	9JL	2.6	-3.5	-1.3	0.5
46.00	-1.49	0.5	124.36	19JL	2.7	-3.4	-1.3	0.5
52.24	-1.39	0.4	133.96	29JL	0.8	-3.4	-1.3	0.5
58.16	-1.26	0.3	143.59	8AU	-0.6	-3.4	-1.2	0.6
63.71	-1.10	0.2	153.27	18AU	-1.3	-3.4	-1.2	0.6
68.83	-0.92	0.1	163.00	28AU	-1.2	-3.4	-1.2	0.6
73.40	-0.71	-0.1	172.79	7SE	-0.7	-3.4	-1.2	0.7
77.31	-0.45	-0.2	182.63	17SE	-0.3	-3.4	-1.2	0.7
80.41	-0.15	-0.4	192.52	27SE	-0.1	-3.4	-1.2	0.7
82.51	0.20	-0.6	202.48	7OC	-0.0	-3.4	-1.2	0.7
83.39	0.60	-0.8	212.49	17OC	0.2	-3.4	-1.3	0.6
82.89	1.06	-1.0	222.54	27OC	1.3	-3.4	-1.3	0.6
80.95	1.55	-1.2	232.65	6NO	2.1	-3.4	-1.3	0.6
77.78	2.03	-1.4	242.79	16NO	0.0	-3.4	-1.4	0.5
73.94	2.43	-1.5	252.96	26NO	-0.3	-3.5	-1.4	0.4
70.20	2.72	-1.3	263.15	6DE	-0.4	-3.5	-1.5	0.3
67.31	2.87	-1.0	273.34	16DE	-0.4	-3.5	-1.6	0.3
65.68	2.92	-0.7	283.53	26DE	-0.6	-3.5	-1.6	0.2

1329

♂ LONG	LAT	MAG	☉ LONG	16.00UT	☿	♀	♃	♄
65.38	2.88	-0.4	293.72	5JA	-1.0	-3.4	-1.7	0.2
66.32	2.80	-0.1	303.86	15JA	-1.1	-3.4	-1.7	0.1
68.28	2.69	0.1	314.00	25JA	-0.9	-3.4	-1.8	0.1
71.07	2.57	0.4	324.09	4FE	0.1	-3.4	-1.9	0.2
74.53	2.45	0.6	334.13	14FE	2.4	-3.3	-1.9	0.2
78.49	2.34	0.8	344.12	24FE	2.2	-3.3	-2.0	0.3
82.86	2.22	1.0	354.05	6MR	1.1	-3.3	-2.0	0.3
87.55	2.11	1.1	3.93	16MR	0.6	-3.3	-2.0	0.4
92.50	2.00	1.3	13.74	26MR	0.3	-3.3	-2.0	0.4
97.65	1.90	1.4	23.51	5AP	-0.1	-3.4	-2.0	0.5
102.98	1.79	1.5	33.22	15AP	-0.8	-3.4	-2.0	0.5
108.45	1.69	1.6	42.88	25AP	-1.7	-3.4	-1.9	0.6
114.05	1.60	1.7	52.50	5MY	-1.3	-3.4	-1.9	0.6
119.75	1.50	1.7	62.08	15MY	-0.3	-3.5	-1.8	0.6
125.55	1.41	1.8	71.64	25MY	0.5	-3.5	-1.8	0.7
131.44	1.32	1.8	81.19	4JN	1.2	-3.6	-1.7	0.7
137.40	1.23	1.9	90.71	14JN	2.0	-3.6	-1.7	0.7
143.45	1.13	1.9	100.24	24JN	3.3	-3.7	-1.6	0.7
149.58	1.04	1.9	109.79	4JL	2.2	-3.8	-1.5	0.7
155.78	0.95	1.9	119.34	14JL	0.7	-3.9	-1.5	0.7
162.05	0.86	1.9	128.92	24JL	-0.7	-4.0	-1.4	0.7
168.41	0.76	1.9	138.54	3AU	-1.4	-4.1	-1.4	0.7
174.84	0.67	1.9	148.19	13AU	-1.3	-4.2	-1.3	0.7
181.35	0.57	1.9	157.89	23AU	-0.6	-4.3	-1.3	0.8
187.95	0.48	1.9	167.65	2SE	-0.2	-4.2	-1.3	0.8
194.63	0.38	1.9	177.46	12SE	0.0	-4.1	-1.2	0.9
201.40	0.28	1.8	187.32	22SE	0.1	-3.6	-1.2	0.9
208.25	0.18	1.8	197.25	2OC	0.4	-3.2	-1.2	0.9
215.19	0.08	1.7	207.23	12OC	1.6	-3.8	-1.2	0.9
222.22	-0.02	1.7	217.26	22OC	1.9	-4.2	-1.2	0.9
229.33	-0.12	1.6	227.35	1NO	-0.1	-4.4	-1.3	0.8
236.53	-0.22	1.6	237.46	11NO	-0.5	-4.3	-1.3	0.8
243.82	-0.33	1.6	247.62	21NO	-0.5	-4.3	-1.3	0.8
251.19	-0.43	1.6	257.80	1DE	-0.6	-4.2	-1.3	0.7
258.63	-0.53	1.6	267.99	11DE	-0.7	-4.1	-1.4	0.7
266.15	-0.62	1.5	278.19	21DE	-0.8	-4.0	-1.4	0.6
273.74	-0.72	1.5	288.39	31DE	-0.9	-3.9	-1.5	0.5

1330

♂ LONG	LAT	MAG	☉ LONG	16.00UT	☿	♀	♃	♄
281.39	-0.81	1.5	298.56	10JA	-0.8	-3.8	-1.5	0.5
289.09	-0.89	1.5	308.70	20JA	0.2	-3.7	-1.6	0.4
296.83	-0.97	1.5	318.82	30JA	2.7	-3.6	-1.7	0.3
304.60	-1.04	1.4	328.88	9FE	1.7	-3.6	-1.7	0.3
312.40	-1.10	1.4	338.90	19FE	0.8	-3.5	-1.8	0.4
320.21	-1.15	1.4	348.86	1MR	0.4	-3.4	-1.9	0.4
328.02	-1.19	1.4	358.76	11MR	0.2	-3.4	-1.9	0.5
335.81	-1.21	1.4	8.61	21MR	-0.2	-3.4	-2.0	0.5
343.58	-1.23	1.4	18.40	31MR	-0.8	-3.3	-2.0	0.6
351.31	-1.23	1.3	28.14	10AP	-1.7	-3.3	-2.0	0.7
358.99	-1.21	1.3	37.82	20AP	-1.3	-3.3	-2.1	0.7
6.61	-1.19	1.3	47.47	30AP	-0.3	-3.3	-2.0	0.8
14.15	-1.15	1.3	57.07	10MY	0.7	-3.3	-2.0	0.8
21.61	-1.09	1.3	66.64	20MY	1.5	-3.3	-2.0	0.8
28.97	-1.02	1.3	76.19	30MY	2.7	-3.3	-2.0	0.9
36.22	-0.94	1.3	85.72	9JN	3.2	-3.4	-1.9	0.9
43.36	-0.85	1.2	95.25	19JN	1.7	-3.4	-1.8	0.9
50.37	-0.74	1.2	104.78	29JN	0.5	-3.4	-1.8	0.9
57.24	-0.62	1.2	114.33	9JL	-0.7	-3.5	-1.7	0.9
63.97	-0.49	1.2	123.90	19JL	-1.5	-3.5	-1.7	0.9
70.52	-0.35	1.1	133.49	29JL	-1.3	-3.5	-1.6	0.9
76.90	-0.19	1.1	143.12	8AU	-0.6	-3.4	-1.5	0.9
83.08	-0.03	1.0	152.80	18AU	-0.1	-3.4	-1.5	1.0
89.02	0.16	1.0	162.53	28AU	0.1	-3.4	-1.4	1.0
94.70	0.35	0.9	172.30	7SE	0.3	-3.4	-1.4	1.1
100.08	0.57	0.8	182.14	17SE	0.7	-3.3	-1.4	1.1
105.09	0.81	0.7	192.04	27SE	2.0	-3.3	-1.3	1.1
109.67	1.07	0.5	201.98	7OC	1.7	-3.3	-1.3	1.2
113.72	1.37	0.4	211.99	17OC	-0.2	-3.3	-1.3	1.2
117.12	1.70	0.2	222.05	27OC	-0.7	-3.3	-1.3	1.2
119.73	2.06	0.0	232.15	6NO	-0.7	-3.4	-1.3	1.2
121.35	2.47	-0.2	242.29	16NO	-0.7	-3.4	-1.3	1.1
121.79	2.91	-0.4	252.46	26NO	-0.8	-3.4	-1.3	1.1
120.91	3.36	-0.6	262.65	6DE	-0.7	-3.5	-1.3	1.1
118.67	3.78	-0.8	272.85	16DE	-0.7	-3.5	-1.3	1.0
115.30	4.11	-1.0	283.04	26DE	-0.6	-3.5	-1.4	0.9

1331

♂ LONG	LAT	MAG	☉ LONG	16.00UT	☿	♀	♃	♄
111.35	4.29	-1.1	293.23	5JA	0.3	-3.6	-1.4	0.9
107.56	4.30	-0.9	303.39	15JA	2.9	-3.7	-1.5	0.8
104.63	4.16	-0.7	313.52	25JA	1.3	-3.7	-1.5	0.7
102.91	3.91	-0.4	323.61	4FE	0.5	-3.8	-1.6	0.7
102.51	3.62	-0.1	333.65	14FE	0.3	-3.9	-1.6	0.6
103.31	3.32	0.1	343.64	24FE	0.1	-4.0	-1.7	0.6
105.12	3.03	0.3	353.58	6MR	-0.3	-4.1	-1.8	0.7
107.77	2.76	0.6	3.46	16MR	-0.8	-4.2	-1.9	0.7
111.08	2.50	0.7	13.27	26MR	-1.7	-4.2	-1.9	0.8
114.92	2.27	0.9	23.04	5AP	-1.3	-4.2	-2.0	0.8
119.20	2.06	1.1	32.75	15AP	-0.2	-4.1	-2.0	0.9
123.82	1.87	1.2	42.41	25AP	0.9	-3.8	-2.1	0.9
128.73	1.69	1.3	52.04	5MY	2.0	-3.1	-2.1	1.0
133.88	1.52	1.4	61.63	15MY	3.5	-3.3	-2.1	1.0
139.24	1.36	1.5	71.18	25MY	2.5	-3.8	-2.2	1.1
144.77	1.20	1.5	80.72	4JN	1.3	-4.1	-2.1	1.1
150.47	1.06	1.6	90.26	14JN	0.4	-4.2	-2.1	1.1
156.32	0.92	1.7	99.78	24JN	-0.7	-4.2	-2.1	1.2
162.30	0.79	1.7	109.32	4JL	-1.6	-4.1	-2.0	1.2
168.41	0.66	1.7	118.88	14JL	-1.3	-4.0	-2.0	1.2
174.64	0.53	1.7	128.45	24JL	-0.5	-3.9	-1.9	1.2
180.99	0.41	1.7	138.07	3AU	-0.0	-3.8	-1.9	1.2
187.46	0.29	1.7	147.72	13AU	0.3	-3.7	-1.8	1.2
194.04	0.17	1.7	157.42	23AU	0.5	-3.7	-1.7	1.2
200.74	0.06	1.7	167.17	2SE	1.0	-3.6	-1.7	1.3
207.54	-0.06	1.7	176.98	12SE	2.4	-3.6	-1.6	1.3
214.45	-0.16	1.7	186.84	22SE	1.5	-3.5	-1.6	1.3
221.48	-0.27	1.7	196.76	2OC	-0.4	-3.5	-1.5	1.3
228.61	-0.37	1.6	206.74	12OC	-0.8	-3.4	-1.5	1.3
235.84	-0.47	1.6	216.77	22OC	-0.8	-3.4	-1.4	1.3
243.17	-0.56	1.6	226.85	1NO	-0.8	-3.4	-1.4	1.3
250.60	-0.65	1.5	236.97	11NO	-0.7	-3.4	-1.4	1.3
258.12	-0.73	1.5	247.13	21NO	-0.6	-3.4	-1.4	1.2
265.72	-0.81	1.5	257.31	1DE	-0.6	-3.4	-1.4	1.2
273.39	-0.88	1.4	267.50	11DE	-0.5	-3.4	-1.4	1.1
281.13	-0.94	1.4	277.70	21DE	0.5	-3.4	-1.4	1.1
288.92	-0.99	1.3	287.89	31DE	3.0	-3.4	-1.4	1.1

$\male$ LONG	LAT	MAG	$\odot$ LONG	16.00UT 1332	$\mercury$	$\venus$	$\jupiter$	$\saturn$
					MAGNITUDES			
296.76	-1.03	1.3	298.07	10JA	0.9	-3.4	-1.4	1.0
304.62	-1.06	1.3	308.21	20JA	0.3	-3.4	-1.5	1.0
312.51	-1.08	1.3	318.33	30JA	0.1	-3.4	-1.5	0.9
320.40	-1.09	1.4	328.39	9FE	-0.0	-3.4	-1.5	0.9
328.28	-1.08	1.4	338.41	19FE	-0.3	-3.5	-1.6	0.8
336.14	-1.07	1.4	348.38	29FE	-0.9	-3.5	-1.6	0.8
343.97	-1.04	1.4	358.28	10MR	-1.6	-3.5	-1.7	0.9
351.76	-1.00	1.4	8.13	20MR	-1.2	-3.4	-1.8	0.9
359.49	-0.95	1.4	17.93	30MR	-0.2	-3.4	-1.8	1.0
7.16	-0.89	1.4	27.67	9AP	1.1	-3.4	-1.9	1.1
14.75	-0.82	1.5	37.35	19AP	2.6	-3.4	-2.0	1.1
22.27	-0.74	1.5	47.00	29AP	3.3	-3.3	-2.1	1.2
29.69	-0.65	1.5	56.60	9MY	1.9	-3.3	-2.1	1.2
37.03	-0.55	1.5	66.17	19MY	1.0	-3.3	-2.2	1.3
44.27	-0.45	1.5	75.72	29MY	0.2	-3.3	-2.2	1.3
51.41	-0.34	1.5	85.26	8JN	-0.7	-3.3	-2.3	1.4
58.44	-0.22	1.6	94.79	18JN	-1.6	-3.4	-2.3	1.4
65.37	-0.10	1.6	104.32	28JN	-1.3	-3.4	-2.3	1.4
72.19	0.02	1.6	113.86	8JL	-0.5	-3.4	-2.3	1.4
78.90	0.15	1.6	123.43	18JL	0.1	-3.4	-2.3	1.4
85.49	0.28	1.6	133.03	28JL	0.4	-3.5	-2.2	1.3
91.95	0.42	1.5	142.66	7AU	0.7	-3.5	-2.2	1.3
98.29	0.56	1.5	152.33	17AU	1.3	-3.6	-2.1	1.3
104.48	0.71	1.5	162.06	27AU	2.9	-3.7	-2.1	1.2
110.52	0.86	1.4	171.83	6SE	1.3	-3.8	-2.0	1.2
116.39	1.03	1.4	181.66	16SE	-0.4	-3.9	-1.9	1.2
122.07	1.20	1.3	191.56	26SE	-1.0	-4.0	-1.9	1.2
127.52	1.38	1.2	201.50	6OC	-1.0	-4.1	-1.8	1.2
132.72	1.58	1.1	211.50	16OC	-0.8	-4.2	-1.7	1.2
137.61	1.79	1.0	221.56	26OC	-0.6	-4.3	-1.7	1.1
142.13	2.01	0.9	231.66	5NO	-0.4	-4.4	-1.6	1.1
146.20	2.26	0.7	241.79	15NO	-0.4	-4.4	-1.6	1.1
149.71	2.54	0.6	251.96	25NO	-0.3	-4.1	-1.6	1.1
152.53	2.84	0.4	262.15	5DE	0.6	-3.6	-1.5	1.0
154.51	3.17	0.1	272.35	15DE	2.8	-3.1	-1.5	1.0
155.45	3.51	-0.1	282.55	25DE	0.6	-3.9	-1.5	1.0
				1333				
155.18	3.86	-0.3	292.73	4JA	0.1	-4.3	-1.5	0.9
153.61	4.17	-0.6	302.89	14JA	-0.0	-4.4	-1.5	0.9
150.79	4.40	-0.8	313.03	24JA	-0.1	-4.3	-1.5	0.8
147.11	4.48	-1.0	323.12	3FE	-0.4	-4.2	-1.5	0.8
143.17	4.39	-0.9	333.16	13FE	-0.9	-4.1	-1.5	0.7
139.72	4.14	-0.8	343.16	23FE	-1.4	-4.0	-1.6	0.7
137.30	3.79	-0.5	353.09	5MR	-1.2	-3.9	-1.6	0.7
136.14	3.38	-0.3	2.98	15MR	-0.1	-3.8	-1.6	0.6
136.27	2.97	-0.1	12.80	25MR	1.4	-3.7	-1.7	0.7
137.52	2.58	0.2	22.57	4AP	3.3	-3.6	-1.8	0.8
139.72	2.23	0.4	32.28	14AP	2.5	-3.6	-1.8	0.8
142.71	1.91	0.5	41.95	24AP	1.4	-3.5	-1.9	0.9
146.33	1.62	0.7	51.57	4MY	0.8	-3.4	-1.9	1.0
150.46	1.36	0.8	61.16	14MY	0.1	-3.4	-2.0	1.0
155.02	1.12	1.0	70.73	24MY	-0.7	-3.4	-2.1	1.1
159.94	0.90	1.1	80.27	3JN	-1.7	-3.3	-2.2	1.1
165.16	0.70	1.1	89.80	13JN	-1.3	-3.3	-2.2	1.2
170.65	0.51	1.2	99.33	23JN	-0.4	-3.3	-2.3	1.2
176.37	0.34	1.3	108.86	3JL	0.2	-3.3	-2.3	1.2
182.30	0.18	1.3	118.42	13JL	0.6	-3.3	-2.4	1.2
188.42	0.03	1.4	128.00	23JL	1.0	-3.3	-2.4	1.2
194.71	-0.11	1.4	137.61	2AU	1.7	-3.4	-2.4	1.2
201.18	-0.25	1.4	147.26	12AU	3.2	-3.4	-2.4	1.2
207.80	-0.37	1.5	156.96	22AU	1.2	-3.4	-2.4	1.1
214.56	-0.49	1.5	166.70	1SE	-0.5	-3.4	-2.3	1.1
221.47	-0.59	1.5	176.51	11SE	-1.1	-3.5	-2.3	1.0
228.52	-0.69	1.5	186.37	21SE	-1.1	-3.5	-2.2	1.0
235.69	-0.78	1.5	196.28	1OC	-0.8	-3.5	-2.2	1.0
242.99	-0.86	1.5	206.26	11OC	-0.4	-3.5	-2.1	1.0
250.39	-0.93	1.5	216.28	21OC	-0.3	-3.4	-2.0	1.0
257.90	-0.99	1.4	226.36	31OC	-0.3	-3.4	-2.0	1.0
265.51	-1.04	1.4	236.48	10NO	-0.1	-3.4	-1.9	1.0
273.19	-1.08	1.4	246.63	20NO	0.8	-3.4	-1.8	1.0
280.95	-1.11	1.4	256.80	30NO	2.6	-3.3	-1.8	1.0
288.77	-1.12	1.4	267.00	10DE	0.4	-3.3	-1.7	0.9
296.63	-1.13	1.4	277.20	20DE	-0.1	-3.3	-1.7	0.9
304.52	-1.12	1.4	287.39	30DE	-0.2	-3.3	-1.6	0.9

$\male$ LONG	LAT	MAG	$\odot$ LONG	16.00UT 1334	$\mercury$	$\venus$	$\jupiter$	$\saturn$
					MAGNITUDES			
312.42	-1.10	1.4	297.56	9JA	-0.3	-3.4	-1.6	0.8
320.33	-1.06	1.4	307.71	19JA	-0.5	-3.4	-1.6	0.8
328.22	-1.02	1.4	317.83	29JA	-0.9	-3.4	-1.6	0.7
336.09	-0.97	1.4	327.90	8FE	-1.3	-3.4	-1.6	0.7
343.92	-0.90	1.4	337.92	18FE	-1.1	-3.4	-1.6	0.6
351.70	-0.83	1.4	347.89	28FE	-0.1	-3.5	-1.6	0.6
359.43	-0.75	1.4	357.80	10MR	1.7	-3.5	-1.6	0.5
7.08	-0.66	1.4	7.66	20MR	3.4	-3.6	-1.6	0.5
14.66	-0.57	1.4	17.45	30MR	1.8	-3.6	-1.6	0.5
22.16	-0.48	1.5	27.20	9AP	1.0	-3.7	-1.7	0.6
29.57	-0.37	1.5	36.89	19AP	0.6	-3.8	-1.7	0.6
36.90	-0.27	1.6	46.53	29AP	0.0	-3.9	-1.8	0.7
44.14	-0.17	1.6	56.14	9MY	-0.7	-4.0	-1.8	0.8
51.29	-0.06	1.6	65.72	19MY	-1.7	-4.0	-1.9	0.8
58.34	0.05	1.7	75.27	29MY	-1.3	-4.1	-1.9	0.9
65.31	0.16	1.7	84.80	8JN	-0.4	-4.2	-2.0	0.9
72.19	0.27	1.7	94.33	18JN	0.3	-4.2	-2.1	1.0
78.99	0.37	1.8	103.86	28JN	0.7	-4.0	-2.1	1.0
85.70	0.48	1.8	113.41	8JL	1.3	-3.5	-2.2	1.0
92.32	0.59	1.8	122.97	18JL	2.2	-3.0	-2.3	1.1
98.86	0.70	1.8	132.56	28JL	3.0	-3.6	-2.3	1.1
105.31	0.80	1.8	142.19	7AU	1.0	-4.1	-2.4	1.1
111.68	0.91	1.8	151.86	17AU	-0.6	-4.2	-2.4	1.0
117.96	1.02	1.8	161.58	27AU	-1.2	-4.2	-2.5	1.0
124.15	1.12	1.8	171.36	6SE	-1.2	-4.2	-2.5	1.0
130.23	1.23	1.7	181.19	16SE	-0.7	-4.1	-2.5	0.9
136.22	1.34	1.7	191.07	26SE	-0.3	-4.0	-2.4	0.9
142.08	1.45	1.6	201.02	6OC	-0.2	-3.9	-2.4	0.9
147.81	1.57	1.6	211.02	16OC	-0.1	-3.9	-2.4	0.9
153.39	1.69	1.5	221.07	26OC	0.1	-3.8	-2.3	0.9
158.79	1.82	1.4	231.17	5NO	1.1	-3.7	-2.2	0.9
163.98	1.95	1.3	241.30	15NO	2.4	-3.6	-2.2	0.9
168.93	2.09	1.2	251.47	25NO	0.2	-3.6	-2.1	0.9
173.56	2.24	1.0	261.66	5DE	-0.2	-3.5	-2.0	0.9
177.82	2.40	0.9	271.85	15DE	-0.3	-3.5	-1.9	0.9
181.62	2.57	0.7	282.05	25DE	-0.4	-3.4	-1.9	0.8
				1335				
184.84	2.75	0.5	292.24	4JA	-0.6	-3.4	-1.8	0.8
187.35	2.93	0.3	302.40	14JA	-1.0	-3.4	-1.8	0.8
188.98	3.12	0.0	312.53	24JA	-1.1	-3.4	-1.7	0.7
189.54	3.29	-0.3	322.63	3FE	-1.0	-3.3	-1.7	0.7
188.89	3.43	-0.5	332.67	13FE	0.0	-3.3	-1.6	0.6
186.96	3.49	-0.8	342.67	23FE	2.1	-3.3	-1.6	0.6
183.89	3.45	-1.1	352.61	5MR	2.6	-3.3	-1.6	0.5
180.15	3.26	-1.2	2.50	15MR	1.3	-3.3	-1.6	0.5
176.41	2.94	-1.1	12.32	25MR	0.8	-3.3	-1.6	0.4
173.35	2.52	-0.9	22.10	4AP	0.4	-3.4	-1.6	0.4
171.44	2.07	-0.7	31.81	14AP	-0.1	-3.4	-1.6	0.4
170.84	1.62	-0.5	41.48	24AP	-0.8	-3.4	-1.6	0.5
171.51	1.21	-0.3	51.11	4MY	-1.7	-3.5	-1.6	0.6
173.28	0.84	-0.1	60.70	14MY	-1.3	-3.5	-1.7	0.6
175.99	0.51	0.1	70.26	24MY	-0.4	-3.5	-1.7	0.7
179.47	0.22	0.3	79.80	3JN	0.4	-3.4	-1.7	0.7
183.58	-0.03	0.4	89.33	13JN	1.0	-3.4	-1.8	0.8
188.22	-0.25	0.5	98.86	23JN	1.7	-3.4	-1.9	0.8
193.30	-0.45	0.6	108.40	3JL	2.9	-3.3	-1.9	0.9
198.75	-0.62	0.7	117.95	13JL	2.5	-3.3	-2.0	0.9
204.53	-0.77	0.8	127.53	23JL	0.8	-3.3	-2.0	0.9
210.60	-0.90	0.9	137.14	2AU	-0.6	-3.3	-2.1	0.9
216.91	-1.02	0.9	146.78	12AU	-1.3	-3.3	-2.2	1.0
223.45	-1.12	1.0	156.48	22AU	-1.3	-3.3	-2.2	0.9
230.20	-1.20	1.0	166.23	1SE	-0.6	-3.4	-2.3	0.9
237.12	-1.27	1.1	176.02	11SE	-0.3	-3.4	-2.4	0.9
244.22	-1.33	1.1	185.88	21SE	-0.1	-3.4	-2.4	0.9
251.47	-1.36	1.1	195.80	1OC	0.1	-3.5	-2.4	0.8
258.84	-1.39	1.2	205.77	11OC	0.3	-3.5	-2.4	0.8
266.34	-1.39	1.2	215.80	21OC	1.4	-3.6	-2.4	0.8
273.94	-1.39	1.2	225.87	31OC	2.2	-3.7	-2.4	0.8
281.61	-1.36	1.2	235.98	10NO	0.0	-3.7	-2.4	0.8
289.36	-1.33	1.3	246.14	20NO	-0.4	-3.8	-2.3	0.8
297.15	-1.28	1.3	256.31	30NO	-0.4	-3.9	-2.3	0.8
304.98	-1.22	1.3	266.50	10DE	-0.5	-4.0	-2.2	0.8
312.82	-1.15	1.3	276.71	20DE	-0.7	-4.1	-2.1	0.8
320.67	-1.07	1.4	286.90	30DE	-0.9	-4.2	-2.1	0.8

♂ LONG	♂ LAT	♂ MAG	⊙ LONG	16.00UT 1336	☿ MAGNITUDES	♀	♃	♄
328.50	-0.98	1.4	297.07	9JA	-1.0	-4.3	-2.0	0.8
336.30	-0.89	1.4	307.22	19JA	-0.9	-4.4	-1.9	0.8
344.06	-0.79	1.4	317.34	29JA	0.1	-4.3	-1.8	0.7
351.76	-0.68	1.5	327.41	8FE	2.4	-4.1	-1.8	0.7
359.41	-0.57	1.5	337.44	18FE	2.0	-3.6	-1.7	0.6
6.99	-0.46	1.5	347.40	28FE	1.0	-3.3	-1.7	0.6
14.49	-0.35	1.6	357.32	9MR	0.6	-3.8	-1.6	0.5
21.91	-0.24	1.6	7.18	19MR	0.3	-4.2	-1.6	0.5
29.25	-0.13	1.6	16.97	29MR	-0.1	-4.3	-1.6	0.4
36.51	-0.03	1.6	26.72	8AP	-0.8	-4.2	-1.5	0.4
43.68	0.08	1.6	36.42	18AP	-1.7	-4.1	-1.5	0.3
50.77	0.18	1.6	46.06	28AP	-1.3	-4.0	-1.5	0.4
57.78	0.28	1.7	55.67	8MY	-0.3	-3.9	-1.5	0.4
64.71	0.38	1.7	65.25	18MY	0.5	-3.8	-1.5	0.5
71.57	0.47	1.7	74.80	28MY	1.3	-3.7	-1.5	0.6
78.35	0.56	1.8	84.34	7JN	2.2	-3.7	-1.6	0.6
85.06	0.65	1.8	93.87	17JN	3.4	-3.6	-1.6	0.7
91.71	0.73	1.9	103.40	27JN	2.0	-3.5	-1.6	0.7
98.29	0.81	1.9	112.94	7JL	0.6	-3.5	-1.7	0.8
104.82	0.89	1.9	122.51	17JL	-0.6	-3.4	-1.7	0.8
111.29	0.96	1.9	132.10	27JL	-1.4	-3.4	-1.8	0.8
117.71	1.03	2.0	141.73	6AU	-1.3	-3.4	-1.8	0.9
124.08	1.10	2.0	151.40	16AU	-0.6	-3.4	-1.9	0.9
130.39	1.16	2.0	161.11	26AU	-0.2	-3.4	-2.0	0.9
136.66	1.22	2.0	170.89	5SE	0.1	-3.4	-2.0	0.9
142.88	1.28	1.9	180.71	15SE	0.2	-3.4	-2.1	0.9
149.05	1.34	1.9	190.60	25SE	0.5	-3.4	-2.1	0.8
155.16	1.39	1.9	200.54	5OC	1.7	-3.4	-2.2	0.8
161.21	1.45	1.8	210.54	15OC	2.0	-3.4	-2.3	0.8
167.19	1.50	1.8	220.58	25OC	-0.1	-3.4	-2.3	0.7
173.10	1.55	1.7	230.68	4NO	-0.6	-3.4	-2.3	0.7
178.92	1.60	1.6	240.81	14NO	-0.6	-3.4	-2.3	0.8
184.63	1.64	1.6	250.97	24NO	-0.6	-3.5	-2.3	0.8
190.22	1.69	1.5	261.16	4DE	-0.7	-3.5	-2.3	0.8
195.67	1.73	1.3	271.36	14DE	-0.8	-3.5	-2.3	0.8
200.94	1.76	1.2	281.55	24DE	-0.8	-3.5	-2.2	0.8
				1337				
206.00	1.79	1.1	291.74	3JA	-0.7	-3.4	-2.2	0.8
210.79	1.82	0.9	301.91	13JA	0.2	-3.4	-2.1	0.8
215.26	1.83	0.7	312.04	23JA	2.7	-3.4	-2.0	0.7
219.33	1.83	0.5	322.14	2FE	1.5	-3.4	-2.0	0.7
222.90	1.81	0.3	332.19	12FE	0.7	-3.3	-1.9	0.7
225.83	1.77	0.0	342.18	22FE	0.4	-3.3	-1.8	0.6
227.99	1.69	-0.3	352.13	4MR	0.1	-3.3	-1.8	0.6
229.17	1.56	-0.6	2.01	14MR	-0.2	-3.3	-1.7	0.5
229.22	1.37	-0.9	11.84	24MR	-0.8	-3.3	-1.6	0.5
228.02	1.09	-1.2	21.62	3AP	-1.7	-3.4	-1.6	0.4
225.60	0.72	-1.5	31.34	13AP	-1.3	-3.4	-1.5	0.3
222.32	0.28	-1.7	41.01	23AP	-0.3	-3.4	-1.5	0.3
218.76	-0.20	-1.7	50.64	3MY	0.7	-3.4	-1.5	0.3
215.68	-0.68	-1.5	60.23	13MY	1.7	-3.5	-1.5	0.3
213.64	-1.10	-1.3	69.79	23MY	3.0	-3.5	-1.4	0.4
212.93	-1.45	-1.1	79.34	2JN	3.0	-3.6	-1.4	0.5
213.58	-1.72	-0.9	88.87	12JN	1.6	-3.6	-1.4	0.5
215.46	-1.93	-0.7	98.40	22JN	0.5	-3.7	-1.4	0.6
218.38	-2.08	-0.5	107.94	2JL	-0.7	-3.8	-1.4	0.6
222.17	-2.19	-0.4	117.49	12JL	-1.5	-3.9	-1.5	0.7
226.68	-2.26	-0.2	127.07	22JL	-1.3	-4.0	-1.5	0.7
231.77	-2.30	-0.1	136.68	1AU	-0.5	-4.1	-1.5	0.8
237.35	-2.31	0.1	146.32	11AU	-0.1	-4.2	-1.6	0.8
243.34	-2.29	0.2	156.01	21AU	0.2	-4.3	-1.6	0.8
249.66	-2.26	0.3	165.76	31AU	0.4	-4.2	-1.7	0.8
256.27	-2.21	0.4	175.55	10SE	0.8	-4.0	-1.7	0.8
263.12	-2.13	0.5	185.41	20SE	2.1	-3.6	-1.8	0.8
270.17	-2.04	0.6	195.32	30SE	1.7	-3.3	-1.8	0.8
277.38	-1.94	0.6	205.29	10OC	-0.3	-3.9	-1.9	0.8
284.72	-1.82	0.7	215.31	20OC	-0.7	-4.2	-2.0	0.7
292.16	-1.69	0.8	225.38	30OC	-0.7	-4.4	-2.0	0.7
299.68	-1.56	0.9	235.49	9NO	-0.7	-4.3	-2.1	0.7
307.25	-1.42	0.9	245.64	19NO	-0.7	-4.3	-2.1	0.7
314.86	-1.27	1.0	255.82	29NO	-0.6	-4.2	-2.2	0.7
322.48	-1.12	1.1	266.01	9DE	-0.7	-4.1	-2.2	0.8
330.08	-0.97	1.2	276.21	19DE	-0.6	-4.0	-2.2	0.8
337.67	-0.81	1.2	286.40	29DE	0.3	-3.9	-2.2	0.8

♂ LONG	♂ LAT	♂ MAG	⊙ LONG	16.00UT 1338	☿ MAGNITUDES	♀	♃	♄
345.23	-0.67	1.3	296.58	8JA	3.0	-3.8	-2.2	0.8
352.73	-0.52	1.4	306.73	18JA	1.1	-3.7	-2.1	0.8
0.19	-0.38	1.4	316.85	28JA	0.4	-3.6	-2.1	0.8
7.58	-0.24	1.5	326.93	7FE	0.2	-3.6	-2.0	0.7
14.90	-0.12	1.5	336.95	17FE	0.0	-3.5	-2.0	0.7
22.15	0.01	1.6	346.93	27FE	-0.3	-3.4	-1.9	0.7
29.33	0.13	1.7	356.84	9MR	-0.8	-3.4	-1.8	0.6
36.43	0.24	1.7	6.70	19MR	-1.6	-3.4	-1.8	0.5
43.46	0.34	1.7	16.51	29MR	-1.3	-3.3	-1.7	0.5
50.42	0.44	1.8	26.25	8AP	-0.2	-3.3	-1.6	0.4
57.31	0.53	1.8	35.95	18AP	0.9	-3.3	-1.6	0.4
64.13	0.61	1.8	45.60	28AP	2.2	-3.3	-1.5	0.3
70.89	0.69	1.9	55.21	8MY	3.8	-3.3	-1.5	0.3
77.58	0.76	1.9	64.79	18MY	2.3	-3.3	-1.4	0.3
84.23	0.82	1.9	74.34	28MY	1.2	-3.3	-1.4	0.3
90.82	0.88	1.9	83.88	7JN	0.3	-3.4	-1.4	0.4
97.36	0.94	1.9	93.41	17JN	-0.7	-3.4	-1.3	0.5
103.87	0.99	1.9	102.94	27JN	-1.6	-3.4	-1.3	0.5
110.34	1.03	1.9	112.48	7JL	-1.3	-3.5	-1.3	0.6
116.77	1.07	1.9	122.04	17JL	-0.5	-3.5	-1.3	0.6
123.18	1.11	2.0	131.63	27JL	0.0	-3.5	-1.3	0.7
129.56	1.14	2.0	141.26	6AU	0.3	-3.5	-1.3	0.7
135.93	1.17	2.0	150.92	16AU	0.6	-3.4	-1.4	0.8
142.27	1.19	2.0	160.64	26AU	1.1	-3.4	-1.4	0.8
148.60	1.21	2.0	170.41	5SE	2.5	-3.4	-1.4	0.8
154.92	1.22	2.0	180.23	15SE	1.5	-3.3	-1.5	0.8
161.23	1.23	2.0	190.12	25SE	-0.4	-3.3	-1.5	0.8
167.52	1.24	2.0	200.05	5OC	-0.9	-3.3	-1.6	0.8
173.81	1.24	2.0	210.05	15OC	-0.9	-3.3	-1.6	0.8
180.08	1.24	1.9	220.09	25OC	-0.9	-3.3	-1.7	0.7
186.33	1.23	1.9	230.18	4NO	-0.6	-3.3	-1.7	0.7
192.58	1.21	1.8	240.32	14NO	-0.5	-3.4	-1.8	0.7
198.79	1.19	1.8	250.48	24NO	-0.5	-3.4	-1.9	0.7
204.99	1.16	1.7	260.66	4DE	-0.4	-3.5	-1.9	0.7
211.15	1.13	1.6	270.86	14DE	0.5	-3.5	-2.0	0.7
217.27	1.08	1.5	281.06	24DE	3.0	-3.5	-2.0	0.8
				1339				
223.34	1.02	1.4	291.24	3JA	0.8	-3.6	-2.1	0.8
229.35	0.95	1.3	301.41	13JA	0.2	-3.7	-2.1	0.8
235.28	0.86	1.2	311.55	23JA	0.1	-3.7	-2.1	0.8
241.12	0.76	1.0	321.65	2FE	-0.1	-3.8	-2.1	0.8
246.85	0.63	0.9	331.70	12FE	-0.4	-3.9	-2.1	0.8
252.44	0.48	0.7	341.71	22FE	-0.9	-4.0	-2.0	0.7
257.85	0.29	0.5	351.65	4MR	-1.5	-4.1	-2.0	0.7
263.03	0.07	0.3	1.54	14MR	-1.2	-4.2	-1.9	0.6
267.94	-0.20	0.1	11.37	24MR	-0.2	-4.2	-1.8	0.6
272.50	-0.51	-0.2	21.15	3AP	1.2	-4.2	-1.8	0.5
276.60	-0.89	-0.4	30.87	13AP	2.9	-4.1	-1.7	0.5
280.12	-1.35	-0.7	40.55	23AP	3.0	-3.8	-1.6	0.4
282.92	-1.88	-1.0	50.17	3MY	1.7	-3.0	-1.6	0.3
284.81	-2.51	-1.3	59.77	13MY	0.9	-3.2	-1.5	0.3
285.59	-3.21	-1.6	69.33	23MY	0.2	-3.9	-1.5	0.2
285.17	-3.96	-1.9	78.88	2JN	-0.7	-4.1	-1.4	0.3
283.53	-4.70	-2.2	88.41	12JN	-1.6	-4.2	-1.4	0.3
281.00	-5.32	-2.5	97.94	22JN	-1.3	-4.2	-1.3	0.4
278.18	-5.72	-2.4	107.47	2JL	-0.5	-4.1	-1.3	0.5
275.79	-5.86	-2.3	117.03	12JL	0.1	-4.0	-1.3	0.5
274.45	-5.76	-2.1	126.60	22JL	0.5	-3.9	-1.3	0.6
274.43	-5.49	-1.8	136.20	1AU	0.8	-3.8	-1.2	0.6
275.76	-5.10	-1.6	145.85	11AU	1.4	-3.7	-1.2	0.7
278.32	-4.67	-1.3	155.54	21AU	3.0	-3.7	-1.2	0.7
281.87	-4.22	-1.1	165.28	31AU	1.3	-3.6	-1.2	0.7
286.25	-3.78	-0.9	175.08	10SE	-0.4	-3.6	-1.3	0.8
291.28	-3.35	-0.7	184.93	20SE	-1.0	-3.5	-1.3	0.8
296.80	-2.93	-0.5	194.84	30SE	-1.0	-3.5	-1.3	0.8
302.72	-2.55	-0.3	204.81	10OC	-0.8	-3.4	-1.4	0.8
308.94	-2.18	-0.1	214.82	20OC	-0.5	-3.4	-1.4	0.8
315.38	-1.84	0.1	224.89	30OC	-0.4	-3.4	-1.4	0.8
322.00	-1.52	0.2	235.01	9NO	-0.4	-3.4	-1.5	0.7
328.74	-1.23	0.4	245.15	19NO	-0.2	-3.4	-1.6	0.7
335.57	-0.96	0.5	255.33	29NO	0.7	-3.4	-1.6	0.7
342.45	-0.72	0.7	265.52	9DE	2.9	-3.4	-1.7	0.7
349.37	-0.49	0.8	275.72	19DE	0.6	-3.4	-1.8	0.7
356.29	-0.29	0.9	285.91	29DE	0.0	-3.4	-1.8	0.8

♂ LONG	♂ LAT	♂ MAG	☉ LONG	16.00UT 1340	☿	♀	♃	♄
					MAGNITUDES			
3.22	-0.11	1.1	296.09	8JA	-0.1	-3.4	-1.9	0.8
10.13	0.06	1.2	306.24	18JA	-0.2	-3.4	-1.9	0.8
17.01	0.21	1.3	316.36	28JA	-0.4	-3.4	-2.0	0.8
23.87	0.34	1.4	326.44	7FE	-0.9	-3.4	-2.0	0.8
30.68	0.46	1.5	336.47	17FE	-1.4	-3.5	-2.0	0.8
37.45	0.57	1.6	346.44	27FE	-1.2	-3.5	-2.0	0.8
44.18	0.67	1.6	356.36	8MR	-0.1	-3.5	-2.0	0.7
50.87	0.75	1.7	6.22	18MR	1.5	-3.4	-2.0	0.7
57.51	0.82	1.8	16.03	28MR	3.5	-3.4	-1.9	0.6
64.11	0.89	1.8	25.78	7AP	2.2	-3.4	-1.9	0.6
70.67	0.95	1.9	35.48	17AP	1.3	-3.4	-1.8	0.5
77.19	0.99	1.9	45.13	27AP	0.7	-3.3	-1.8	0.5
83.68	1.03	1.9	54.74	7MY	0.1	-3.3	-1.7	0.4
90.13	1.07	2.0	64.32	17MY	-0.7	-3.3	-1.6	0.3
96.56	1.10	2.0	73.88	27MY	-1.6	-3.3	-1.6	0.3
102.96	1.12	2.0	83.41	6JN	-1.3	-3.3	-1.5	0.2
109.35	1.14	2.0	92.94	16JN	-0.4	-3.4	-1.5	0.3
115.72	1.15	2.0	102.47	26JN	0.2	-3.4	-1.4	0.4
122.08	1.15	2.0	112.02	6JL	0.6	-3.4	-1.4	0.4
128.44	1.16	2.0	121.57	16JL	1.1	-3.4	-1.3	0.5
134.80	1.15	2.0	131.16	26JL	1.9	-3.5	-1.3	0.5
141.16	1.15	2.0	140.79	5AU	3.2	-3.5	-1.2	0.6
147.53	1.13	2.0	150.45	15AU	1.1	-3.6	-1.2	0.6
153.91	1.12	2.0	160.17	25AU	-0.5	-3.7	-1.2	0.7
160.31	1.10	2.0	169.93	4SE	-1.1	-3.8	-1.2	0.7
166.72	1.07	2.0	179.75	14SE	-1.2	-3.9	-1.2	0.8
173.16	1.04	2.0	189.64	24SE	-0.8	-4.0	-1.2	0.8
179.62	1.00	2.0	199.57	4OC	-0.4	-4.1	-1.2	0.8
186.10	0.96	2.0	209.56	14OC	-0.3	-4.2	-1.2	0.8
192.61	0.92	2.0	219.61	24OC	-0.2	-4.3	-1.3	0.8
199.15	0.86	1.9	229.70	3NO	-0.1	-4.4	-1.3	0.8
205.72	0.80	1.9	239.83	13NO	0.9	-4.3	-1.3	0.7
212.32	0.74	1.9	249.99	23NO	2.6	-4.1	-1.4	0.7
218.95	0.66	1.8	260.17	3DE	0.3	-3.6	-1.4	0.7
225.60	0.58	1.7	270.37	13DE	-0.1	-3.1	-1.5	0.7
232.29	0.48	1.7	280.57	23DE	-0.2	-3.9	-1.6	0.7
				1341				
238.99	0.38	1.6	290.75	2JA	-0.3	-4.3	-1.6	0.8
245.73	0.27	1.5	300.92	12JA	-0.5	-4.4	-1.7	0.8
252.48	0.14	1.4	311.06	22JA	-0.9	-4.3	-1.8	0.8
259.25	-0.00	1.3	321.16	1FE	-1.2	-4.2	-1.8	0.8
266.05	-0.16	1.2	331.21	11FE	-1.1	-4.1	-1.9	0.8
272.85	-0.33	1.1	341.22	21FE	-0.1	-4.0	-1.9	0.8
279.65	-0.52	1.0	351.17	3MR	1.8	-3.9	-2.0	0.8
286.46	-0.72	0.9	1.06	13MR	3.1	-3.8	-2.0	0.8
293.25	-0.94	0.7	10.90	23MR	1.6	-3.7	-2.0	0.7
300.01	-1.18	0.6	20.68	2AP	0.9	-3.6	-2.0	0.7
306.74	-1.44	0.5	30.40	12AP	0.5	-3.6	-2.0	0.6
313.39	-1.72	0.3	40.08	22AP	0.0	-3.5	-2.0	0.6
319.95	-2.02	0.2	49.72	2MY	-0.7	-3.4	-1.9	0.5
326.38	-2.33	-0.0	59.31	12MY	-1.7	-3.4	-1.9	0.5
332.62	-2.66	-0.2	68.88	22MY	-1.3	-3.4	-1.8	0.4
338.62	-3.00	-0.4	78.42	1JN	-0.4	-3.3	-1.7	0.3
344.32	-3.36	-0.6	87.95	11JN	0.3	-3.3	-1.7	0.3
349.58	-3.72	-0.8	97.49	21JN	0.8	-3.3	-1.6	0.3
354.31	-4.09	-1.0	107.02	1JL	1.4	-3.3	-1.6	0.3
358.34	-4.45	-1.2	116.57	11JL	2.5	-3.3	-1.5	0.4
1.48	-4.80	-1.4	126.14	21JL	2.9	-3.3	-1.4	0.4
3.52	-5.10	-1.7	135.74	31JL	1.0	-3.3	-1.4	0.5
4.25	-5.32	-1.9	145.38	10AU	-0.6	-3.4	-1.3	0.6
3.57	-5.39	-2.2	155.07	20AU	-1.3	-3.4	-1.3	0.6
1.59	-5.26	-2.4	164.81	30AU	-1.3	-3.4	-1.3	0.7
358.76	-4.88	-2.5	174.60	9SE	-0.7	-3.5	-1.3	0.7
355.81	-4.28	-2.4	184.45	19SE	-0.3	-3.5	-1.2	0.7
353.53	-3.54	-2.1	194.35	29SE	-0.1	-3.5	-1.2	0.8
352.38	-2.78	-1.8	204.32	9OC	-0.0	-3.5	-1.2	0.8
352.57	-2.06	-1.5	214.34	19OC	0.1	-3.4	-1.2	0.8
354.01	-1.43	-1.1	224.40	29OC	1.1	-3.4	-1.2	0.8
356.50	-0.90	-0.8	234.51	8NO	2.4	-3.4	-1.3	0.8
359.85	-0.46	-0.5	244.66	18NO	0.1	-3.4	-1.3	0.8
3.88	-0.10	-0.3	254.83	28NO	-0.3	-3.3	-1.3	0.8
8.42	0.19	-0.0	265.02	8DE	-0.4	-3.3	-1.3	0.7
13.36	0.43	0.2	275.22	18DE	-0.4	-3.3	-1.4	0.7
18.61	0.63	0.4	285.41	28DE	-0.6	-3.3	-1.4	0.7

♂ LONG	♂ LAT	♂ MAG	☉ LONG	16.00UT 1342	☿	♀	♃	♄
					MAGNITUDES			
24.09	0.79	0.6	295.59	7JA	-0.9	-3.4	-1.5	0.8
29.75	0.92	0.8	305.75	17JA	-1.1	-3.4	-1.6	0.8
35.54	1.02	1.0	315.87	27JA	-1.0	-3.4	-1.6	0.8
41.44	1.11	1.1	325.95	6FE	-0.0	-3.4	-1.7	0.9
47.42	1.17	1.2	335.98	16FE	2.1	-3.5	-1.8	0.9
53.45	1.22	1.4	345.96	26FE	2.4	-3.5	-1.8	0.9
59.53	1.26	1.5	355.89	8MR	1.2	-3.5	-1.9	0.9
65.64	1.29	1.6	5.75	18MR	0.7	-3.6	-2.0	0.8
71.77	1.31	1.7	15.56	28MR	0.4	-3.6	-2.0	0.8
77.91	1.32	1.7	25.31	7AP	-0.1	-3.7	-2.0	0.8
84.07	1.33	1.8	35.02	17AP	-0.7	-3.8	-2.1	0.7
90.24	1.33	1.9	44.67	27AP	-1.7	-3.9	-2.1	0.7
96.42	1.32	1.9	54.29	7MY	-1.3	-4.0	-2.1	0.6
102.61	1.31	2.0	63.87	17MY	-0.4	-4.0	-2.0	0.5
108.81	1.29	2.0	73.42	27MY	0.4	-4.1	-2.0	0.5
115.02	1.27	2.0	82.96	6JN	1.1	-4.2	-2.0	0.4
121.25	1.24	2.0	92.49	16JN	1.9	-4.2	-1.9	0.4
127.50	1.21	2.0	102.02	26JN	3.1	-4.0	-1.8	0.3
133.77	1.18	2.0	111.56	6JL	2.4	-3.5	-1.8	0.3
140.06	1.14	2.0	121.12	16JL	0.8	-3.0	-1.7	0.4
146.38	1.10	2.0	130.70	26JL	-0.6	-3.6	-1.7	0.4
152.73	1.05	2.0	140.33	5AU	-1.4	-4.1	-1.6	0.5
159.12	1.00	2.0	149.99	15AU	-1.3	-4.2	-1.5	0.5
165.55	0.95	2.0	159.69	25AU	-0.6	-4.2	-1.5	0.6
172.03	0.89	1.9	169.46	4SE	-0.2	-4.2	-1.4	0.6
178.55	0.83	1.9	179.28	14SE	-0.0	-4.1	-1.4	0.7
185.11	0.77	1.9	189.15	24SE	0.1	-4.0	-1.4	0.7
191.74	0.70	1.9	199.09	4OC	0.4	-3.9	-1.3	0.8
198.42	0.63	1.9	209.08	14OC	1.4	-3.9	-1.3	0.8
205.15	0.55	1.9	219.12	24OC	2.2	-3.8	-1.3	0.8
211.94	0.46	1.9	229.21	3NO	-0.0	-3.7	-1.3	0.8
218.80	0.37	1.8	239.33	13NO	-0.5	-3.6	-1.3	0.8
225.71	0.28	1.8	249.49	23NO	-0.5	-3.6	-1.3	0.8
232.69	0.18	1.8	259.68	3DE	-0.5	-3.5	-1.3	0.8
239.72	0.07	1.7	269.88	13DE	-0.7	-3.5	-1.3	0.8
246.82	-0.04	1.7	280.07	23DE	-0.9	-3.4	-1.4	0.8
				1343				
253.98	-0.15	1.6	290.26	2JA	-0.9	-3.4	-1.4	0.8
261.19	-0.28	1.6	300.43	12JA	-0.8	-3.4	-1.4	0.8
268.46	-0.40	1.5	310.57	22JA	0.1	-3.4	-1.5	0.8
275.78	-0.54	1.5	320.68	1FE	2.5	-3.3	-1.6	0.9
283.15	-0.67	1.4	330.73	11FE	1.9	-3.3	-1.6	0.9
290.56	-0.81	1.3	340.74	21FE	0.8	-3.3	-1.7	0.9
298.00	-0.95	1.3	350.70	3MR	0.5	-3.3	-1.7	0.9
305.48	-1.10	1.2	0.59	13MR	0.2	-3.3	-1.8	0.9
312.97	-1.23	1.1	10.43	23MR	-0.2	-3.3	-1.9	0.9
320.47	-1.37	1.1	20.21	2AP	-0.8	-3.4	-2.0	0.9
327.96	-1.50	1.0	29.94	12AP	-1.6	-3.4	-2.0	0.9
335.43	-1.63	0.9	39.61	22AP	-1.3	-3.4	-2.1	0.8
342.86	-1.74	0.8	49.25	2MY	-0.3	-3.5	-2.1	0.8
350.24	-1.84	0.8	58.85	12MY	0.6	-3.5	-2.1	0.7
357.54	-1.93	0.7	68.41	22MY	1.4	-3.5	-2.2	0.7
4.73	-2.00	0.6	77.96	1JN	2.5	-3.4	-2.2	0.6
11.79	-2.06	0.5	87.49	11JN	3.3	-3.4	-2.2	0.5
18.70	-2.09	0.4	97.02	21JN	1.9	-3.4	-2.1	0.5
25.40	-2.10	0.3	106.56	1JL	0.6	-3.3	-2.1	0.4
31.87	-2.10	0.2	116.10	11JL	-0.6	-3.3	-2.1	0.3
38.05	-2.06	0.1	125.67	21JL	-1.4	-3.3	-2.0	0.4
43.88	-2.00	0.0	135.28	31JL	-1.3	-3.3	-1.9	0.4
49.28	-1.92	-0.1	144.91	10AU	-0.6	-3.3	-1.9	0.5
54.16	-1.80	-0.3	154.60	20AU	-0.1	-3.3	-1.8	0.5
58.39	-1.64	-0.4	164.33	30AU	0.1	-3.4	-1.7	0.6
61.83	-1.43	-0.6	174.12	9SE	0.3	-3.4	-1.7	0.7
64.28	-1.18	-0.8	183.97	19SE	0.6	-3.4	-1.6	0.7
65.52	-0.86	-1.0	193.87	29SE	1.8	-3.5	-1.6	0.7
65.39	-0.47	-1.3	203.83	9OC	2.0	-3.5	-1.5	0.8
63.78	-0.01	-1.5	213.84	19OC	-0.2	-3.6	-1.5	0.8
60.87	0.49	-1.6	223.91	29OC	-0.6	-3.7	-1.5	0.8
57.20	0.98	-1.8	234.02	8NO	-0.6	-3.7	-1.4	0.9
53.53	1.41	-1.6	244.16	18NO	-0.7	-3.8	-1.4	0.9
50.63	1.75	-1.3	254.34	28NO	-0.8	-3.9	-1.4	0.9
48.98	1.97	-1.0	264.52	8DE	-0.7	-4.0	-1.4	0.9
48.67	2.10	-0.7	274.73	18DE	-0.8	-4.1	-1.4	0.9
49.63	2.17	-0.4	284.92	28DE	-0.7	-4.2	-1.4	0.9

Left table (1344–1345):

♂ LONG	LAT	MAG	☉ LONG	16.00UT	☿	♀	♃	♄
				1344	MAGNITUDES			
51.63	2.19	-0.1	295.10	7JA	0.2	-4.3	-1.4	0.8
54.48	2.17	0.2	305.26	17JA	2.8	-4.4	-1.4	0.8
58.00	2.14	0.4	315.38	27JA	1.4	-4.3	-1.5	0.9
62.04	2.10	0.6	325.46	6FE	0.6	-4.1	-1.5	0.9
66.48	2.05	0.8	335.50	16FE	0.3	-3.6	-1.6	1.0
71.25	1.99	1.0	345.48	26FE	0.1	-3.3	-1.6	1.0
76.27	1.93	1.2	355.40	7MR	-0.2	-3.9	-1.7	1.0
81.48	1.86	1.3	5.28	17MR	-0.8	-4.2	-1.7	1.0
86.86	1.80	1.4	15.09	27MR	-1.6	-4.3	-1.8	1.0
92.37	1.73	1.5	24.84	6AP	-1.3	-4.2	-1.9	1.0
97.99	1.67	1.6	34.55	16AP	-0.3	-4.1	-1.9	1.0
103.71	1.60	1.7	44.20	26AP	0.8	-4.0	-2.0	0.9
109.50	1.53	1.8	53.82	6MY	1.9	-3.9	-2.1	0.9
115.37	1.46	1.8	63.40	16MY	3.3	-3.8	-2.1	0.8
121.30	1.39	1.9	72.96	26MY	2.7	-3.7	-2.2	0.8
127.30	1.32	1.9	82.50	5JN	1.5	-3.7	-2.2	0.7
133.35	1.24	1.9	92.03	15JN	0.4	-3.6	-2.3	0.7
139.47	1.17	2.0	101.56	25JN	-0.6	-3.5	-2.3	0.6
145.64	1.09	2.0	111.10	5JL	-1.5	-3.5	-2.3	0.5
151.87	1.02	2.0	120.66	15JL	-1.3	-3.4	-2.3	0.5
158.17	0.94	2.0	130.24	25JL	-0.6	-3.4	-2.3	0.4
164.52	0.86	2.0	139.86	4AU	-0.1	-3.4	-2.2	0.5
170.95	0.78	2.0	149.52	14AU	0.2	-3.4	-2.2	0.5
177.44	0.69	1.9	159.23	24AU	0.5	-3.4	-2.1	0.6
184.01	0.60	1.9	168.99	3SE	0.9	-3.4	-2.1	0.6
190.64	0.52	1.9	178.81	13SE	2.2	-3.4	-2.0	0.7
197.36	0.42	1.8	188.68	23SE	1.8	-3.4	-1.9	0.7
204.15	0.33	1.8	198.61	3OC	-0.3	-3.4	-1.9	0.8
211.02	0.24	1.7	208.59	13OC	-0.8	-3.4	-1.8	0.8
217.97	0.14	1.7	218.63	23OC	-0.8	-3.4	-1.7	0.9
225.00	0.04	1.7	228.71	2NO	-0.8	-3.4	-1.7	0.9
232.12	-0.06	1.7	238.84	12NO	-0.7	-3.4	-1.6	0.9
239.31	-0.17	1.7	249.00	22NO	-0.6	-3.5	-1.6	0.9
246.58	-0.27	1.6	259.18	2DE	-0.6	-3.5	-1.6	1.0
253.93	-0.38	1.6	269.38	12DE	-0.5	-3.5	-1.5	1.0
261.35	-0.48	1.6	279.57	22DE	0.3	-3.5	-1.5	1.0
				1345				
268.83	-0.59	1.6	289.76	1JA	3.0	-3.4	-1.5	1.0
276.39	-0.69	1.5	299.93	11JA	1.0	-3.4	-1.5	1.0
283.99	-0.79	1.5	310.07	21JA	0.3	-3.4	-1.5	0.9
291.65	-0.89	1.5	320.18	31JA	0.2	-3.4	-1.5	1.0
299.35	-0.98	1.5	330.24	10FE	-0.0	-3.3	-1.5	1.0
307.07	-1.06	1.4	340.25	20FE	-0.3	-3.3	-1.6	1.0
314.82	-1.14	1.4	350.21	2MR	-0.8	-3.3	-1.6	1.1
322.58	-1.20	1.4	0.11	12MR	-1.6	-3.3	-1.6	1.1
330.33	-1.26	1.3	9.95	22MR	-1.3	-3.4	-1.7	1.1
338.08	-1.30	1.3	19.73	1AP	-0.3	-3.4	-1.7	1.1
345.79	-1.33	1.3	29.47	11AP	1.0	-3.4	-1.8	1.1
353.45	-1.35	1.2	39.15	21AP	2.4	-3.4	-1.8	1.1
1.07	-1.35	1.2	48.78	1MY	3.6	-3.4	-1.9	1.1
8.61	-1.34	1.2	58.39	11MY	2.0	-3.5	-2.0	1.0
16.07	-1.31	1.2	67.95	21MY	1.1	-3.5	-2.0	1.0
23.44	-1.26	1.1	77.50	31MY	0.3	-3.6	-2.1	0.9
30.69	-1.20	1.1	87.04	10JN	-0.7	-3.6	-2.2	0.9
37.82	-1.13	1.1	96.56	20JN	-1.6	-3.7	-2.3	0.8
44.82	-1.04	1.0	106.10	30JN	-1.3	-3.8	-2.3	0.8
51.65	-0.93	1.0	115.65	10JL	-0.5	-3.9	-2.4	0.7
58.32	-0.81	0.9	125.21	20JL	0.0	-4.0	-2.4	0.6
64.79	-0.67	0.9	134.82	30JL	0.4	-4.1	-2.4	0.6
71.04	-0.51	0.8	144.45	9AU	0.7	-4.2	-2.4	0.6
77.04	-0.34	0.7	154.13	19AU	1.2	-4.3	-2.4	0.6
82.75	-0.15	0.7	163.87	29AU	2.7	-4.2	-2.4	0.6
88.11	0.06	0.6	173.65	8SE	1.5	-4.0	-2.3	0.7
93.07	0.30	0.4	183.49	18SE	-0.4	-3.5	-2.3	0.7
97.54	0.57	0.3	193.39	28SE	-0.9	-3.3	-2.2	0.8
101.41	0.87	0.2	203.35	8OC	-0.9	-3.9	-2.2	0.8
104.54	1.21	-0.0	213.36	18OC	-0.9	-4.2	-2.1	0.9
106.75	1.60	-0.2	223.42	28OC	-0.6	-4.4	-2.0	0.9
107.86	2.03	-0.4	233.53	7NO	-0.5	-4.3	-2.0	1.0
107.69	2.49	-0.7	243.67	17NO	-0.4	-4.3	-1.9	1.0
106.11	2.97	-0.9	253.84	27NO	-0.4	-4.2	-1.8	1.0
103.24	3.40	-1.1	264.03	7DE	0.5	-4.1	-1.8	1.1
99.48	3.72	-1.2	274.22	17DE	3.0	-4.0	-1.7	1.1
95.53	3.90	-1.1	284.42	27DE	0.7	-3.9	-1.7	1.1

Right table (1346–1347):

♂ LONG	LAT	MAG	☉ LONG	16.00UT	☿	♀	♃	♄
				1346	MAGNITUDES			
92.17	3.91	-0.9	294.60	6JA	0.1	-3.8	-1.6	1.1
89.92	3.79	-0.6	304.76	16JA	0.0	-3.7	-1.6	1.1
88.98	3.60	-0.3	314.89	26JA	-0.1	-3.6	-1.6	1.1
89.32	3.37	-0.1	324.97	5FE	-0.4	-3.6	-1.6	1.1
90.76	3.13	0.2	335.01	15FE	-0.9	-3.5	-1.6	1.1
93.11	2.90	0.4	345.00	25FE	-1.5	-3.4	-1.6	1.2
96.19	2.68	0.6	354.93	7MR	-1.2	-3.4	-1.6	1.2
99.84	2.48	0.8	4.80	17MR	-0.2	-3.4	-1.6	1.2
103.96	2.29	1.0	14.62	27MR	1.3	-3.3	-1.6	1.3
108.45	2.11	1.1	24.37	6AP	3.1	-3.3	-1.6	1.3
113.23	1.95	1.3	34.08	16AP	2.7	-3.3	-1.7	1.3
118.27	1.80	1.4	43.74	26AP	1.5	-3.3	-1.7	1.3
123.51	1.65	1.5	53.36	6MY	0.8	-3.3	-1.8	1.3
128.92	1.51	1.6	62.94	16MY	0.2	-3.3	-1.8	1.2
134.44	1.38	1.6	72.50	26MY	-0.7	-3.3	-1.9	1.2
140.20	1.25	1.7	82.04	5JN	-1.6	-3.4	-1.9	1.1
146.03	1.13	1.7	91.57	15JN	-1.3	-3.4	-2.0	1.1
151.98	1.01	1.8	101.10	25JN	-0.5	-3.4	-2.1	1.0
158.03	0.89	1.8	110.64	5JL	0.1	-3.5	-2.2	1.0
164.19	0.78	1.8	120.19	15JL	0.5	-3.5	-2.2	0.9
170.46	0.66	1.8	129.78	25JL	0.9	-3.5	-2.3	0.8
176.82	0.55	1.8	139.39	4AU	1.6	-3.4	-2.3	0.8
183.28	0.44	1.8	149.05	14AU	3.1	-3.4	-2.4	0.7
189.85	0.33	1.8	158.76	24AU	1.3	-3.4	-2.4	0.7
196.51	0.22	1.8	168.51	3SE	-0.4	-3.4	-2.5	0.7
203.28	0.11	1.8	178.32	13SE	-1.1	-3.3	-2.5	0.8
210.14	0.01	1.8	188.20	23SE	-1.1	-3.3	-2.5	0.8
217.11	-0.10	1.7	198.12	3OC	-0.8	-3.3	-2.4	0.9
224.17	-0.20	1.7	208.10	13OC	-0.5	-3.3	-2.4	0.9
231.34	-0.30	1.6	218.14	23OC	-0.3	-3.4	-2.3	1.0
238.59	-0.40	1.6	228.22	2NO	-0.3	-3.4	-2.3	1.0
245.94	-0.50	1.6	238.35	12NO	-0.2	-3.4	-2.2	1.1
253.38	-0.59	1.5	248.51	22NO	0.7	-3.4	-2.1	1.1
260.90	-0.68	1.5	258.68	2DE	2.9	-3.5	-2.1	1.2
268.50	-0.76	1.4	268.88	12DE	0.5	-3.5	-2.0	1.2
276.17	-0.84	1.4	279.08	22DE	-0.0	-3.5	-1.9	1.2
				1347				
283.90	-0.91	1.4	289.27	1JA	-0.1	-3.6	-1.9	1.2
291.68	-0.97	1.4	299.44	11JA	-0.2	-3.7	-1.8	1.2
299.49	-1.02	1.4	309.58	21JA	-0.5	-3.7	-1.7	1.2
307.34	-1.06	1.4	319.69	31JA	-0.9	-3.8	-1.7	1.2
315.20	-1.09	1.4	329.75	10FE	-1.3	-3.9	-1.7	1.2
323.07	-1.11	1.4	339.77	20FE	-1.1	-4.0	-1.6	1.3
330.93	-1.12	1.4	349.72	2MR	-0.2	-4.1	-1.6	1.3
338.76	-1.11	1.4	359.63	12MR	1.6	-4.2	-1.6	1.3
346.57	-1.10	1.4	9.47	22MR	3.5	-4.2	-1.6	1.4
354.32	-1.07	1.4	19.26	1AP	2.0	-4.2	-1.6	1.4
2.03	-1.03	1.4	28.99	11AP	1.1	-4.1	-1.6	1.4
9.67	-0.97	1.4	38.68	21AP	0.6	-3.8	-1.6	1.5
17.23	-0.91	1.4	48.32	1MY	0.1	-3.0	-1.6	1.4
24.70	-0.83	1.4	57.92	11MY	-0.7	-3.3	-1.6	1.4
32.09	-0.75	1.4	67.49	21MY	-1.6	-3.9	-1.7	1.4
39.38	-0.65	1.4	77.04	31MY	-1.4	-4.1	-1.7	1.3
46.57	-0.55	1.5	86.57	10JN	-0.4	-4.2	-1.7	1.3
53.64	-0.44	1.5	96.10	20JN	0.2	-4.1	-1.8	1.2
60.60	-0.32	1.4	105.63	30JN	0.7	-4.1	-1.9	1.2
67.45	-0.19	1.4	115.18	10JL	1.2	-4.0	-1.9	1.1
74.16	-0.06	1.4	124.75	20JL	2.1	-3.9	-2.0	1.1
80.74	0.08	1.4	134.34	30JL	3.1	-3.8	-2.0	1.0
87.18	0.23	1.4	143.98	9AU	1.1	-3.7	-2.1	1.0
93.46	0.39	1.3	153.66	19AU	-0.5	-3.7	-2.2	0.9
99.57	0.55	1.3	163.39	29AU	-1.2	-3.6	-2.2	0.9
105.49	0.72	1.2	173.18	8SE	-1.2	-3.6	-2.3	0.9
111.20	0.91	1.2	183.02	18SE	-0.7	-3.5	-2.4	0.9
116.66	1.11	1.1	192.91	28SE	-0.4	-3.5	-2.4	1.0
121.83	1.32	1.0	202.87	8OC	-0.2	-3.4	-2.4	1.0
126.65	1.55	0.9	212.88	18OC	-0.1	-3.4	-2.4	1.1
131.07	1.81	0.8	222.93	28OC	0.0	-3.4	-2.4	1.1
134.97	2.09	0.6	233.04	7NO	0.9	-3.4	-2.4	1.2
138.25	2.40	0.4	243.18	17NO	2.7	-3.4	-2.4	1.2
140.77	2.74	0.2	253.35	27NO	-0.3	-3.4	-2.3	1.3
142.34	3.12	-0.0	263.54	7DE	-0.2	-3.4	-2.2	1.3
142.77	3.51	-0.2	273.73	17DE	-0.3	-3.4	-2.2	1.3
141.93	3.90	-0.5	283.93	27DE	-0.3	-3.4	-2.1	1.4

Left table (1348–1349):

♂ LONG	♂ LAT	♂ MAG	☉ LONG	16.00UT 1348	☿	♀	♃	♄
139.75	4.24	-0.7	294.11	6JA	-0.5	-3.4	-2.0	1.4
136.45	4.48	-0.9	304.27	16JA	-0.9	-3.4	-2.0	1.4
132.53	4.56	-1.0	314.40	26JA	-1.2	-3.4	-1.9	1.4
128.72	4.46	-0.9	324.48	5FE	-1.0	-3.4	-1.8	1.4
125.69	4.21	-0.6	334.52	15FE	-0.1	-3.5	-1.8	1.3
123.84	3.87	-0.4	344.51	25FE	1.9	-3.5	-1.7	1.2
123.29	3.50	-0.1	354.44	6MR	2.9	-3.5	-1.7	1.2
123.95	3.12	0.1	4.31	16MR	1.5	-3.4	-1.6	1.2
125.66	2.77	0.3	14.13	26MR	0.8	-3.4	-1.6	1.2
128.22	2.45	0.5	23.90	5AP	0.5	-3.4	-1.6	1.2
131.48	2.15	0.7	33.60	15AP	-0.0	-3.4	-1.5	1.2
135.30	1.89	0.8	43.26	25AP	-0.7	-3.3	-1.5	1.2
139.57	1.64	1.0	52.89	5MY	-1.6	-3.3	-1.5	1.2
144.23	1.42	1.1	62.47	15MY	-1.4	-3.3	-1.5	1.1
149.19	1.22	1.2	72.03	25MY	-0.4	-3.3	-1.5	1.1
154.43	1.03	1.3	81.57	4JN	0.3	-3.3	-1.5	1.1
159.90	0.85	1.4	91.10	14JN	0.9	-3.4	-1.6	1.0
165.58	0.68	1.4	100.63	24JN	1.6	-3.4	-1.6	1.0
171.44	0.53	1.5	110.17	4JL	2.7	-3.4	-1.6	0.9
177.48	0.38	1.5	119.73	14JL	2.7	-3.4	-1.7	0.9
183.67	0.23	1.5	129.31	24JL	0.9	-3.5	-1.7	0.8
190.01	0.10	1.6	138.93	3AU	-0.5	-3.5	-1.8	0.8
196.50	-0.03	1.6	148.58	13AU	-1.3	-3.6	-1.8	0.7
203.12	-0.15	1.6	158.28	23AU	-1.3	-3.7	-1.9	0.7
209.88	-0.27	1.6	168.04	2SE	-0.7	-3.8	-1.9	0.7
216.76	-0.38	1.6	177.85	12SE	-0.3	-3.9	-2.0	0.6
223.78	-0.49	1.6	187.72	22SE	-0.1	-4.0	-2.1	0.7
230.91	-0.59	1.6	197.64	2OC	0.0	-4.1	-2.1	0.8
238.15	-0.68	1.5	207.62	12OC	0.2	-4.2	-2.2	0.8
245.51	-0.76	1.5	217.65	22OC	1.2	-4.3	-2.2	0.9
252.97	-0.84	1.5	227.74	1NO	2.4	-4.4	-2.3	1.0
260.52	-0.90	1.5	237.85	11NO	0.1	-4.3	-2.3	1.0
268.15	-0.96	1.5	248.01	21NO	-0.4	-4.1	-2.3	1.1
275.87	-1.01	1.4	258.19	1DE	-0.4	-3.5	-2.3	1.1
283.64	-1.05	1.4	268.38	11DE	-0.5	-3.2	-2.3	1.1
291.47	-1.07	1.4	278.58	21DE	-0.6	-3.9	-2.3	1.2
299.34	-1.09	1.4	288.77	31DE	-0.9	-4.3	-2.2	1.2
				1349				
307.24	-1.09	1.4	298.94	10JA	-1.0	-4.4	-2.1	1.2
315.15	-1.08	1.3	309.09	20JA	-0.9	-4.3	-2.1	1.1
323.05	-1.06	1.3	319.20	30JA	-0.0	-4.2	-2.0	1.1
330.94	-1.03	1.3	329.26	9FE	2.2	-4.1	-1.9	1.1
338.80	-0.99	1.3	339.28	19FE	2.2	-4.0	-1.9	1.1
346.63	-0.93	1.3	349.24	1MR	1.1	-3.9	-1.8	1.0
354.40	-0.87	1.4	359.15	11MR	0.6	-3.8	-1.7	0.9
2.11	-0.80	1.4	9.00	21MR	0.3	-3.7	-1.7	0.9
9.76	-0.72	1.4	18.79	31MR	-0.1	-3.6	-1.6	0.9
17.32	-0.63	1.5	28.52	10AP	-0.7	-3.6	-1.6	0.9
24.81	-0.54	1.5	38.21	20AP	-1.6	-3.5	-1.5	0.9
32.21	-0.44	1.5	47.85	30AP	-1.4	-3.4	-1.5	0.9
39.52	-0.34	1.6	57.45	10MY	-0.4	-3.4	-1.5	0.9
46.74	-0.23	1.6	67.03	20MY	0.5	-3.4	-1.4	0.9
53.86	-0.12	1.6	76.58	30MY	1.2	-3.3	-1.4	0.9
60.89	-0.01	1.7	86.11	9JN	2.1	-3.3	-1.4	0.8
67.83	0.10	1.7	95.64	19JN	3.3	-3.3	-1.4	0.8
74.67	0.22	1.7	105.18	29JN	2.2	-3.3	-1.4	0.8
81.42	0.33	1.7	114.72	9JL	0.7	-3.3	-1.4	0.7
88.07	0.45	1.7	124.29	19JL	-0.6	-3.3	-1.4	0.7
94.62	0.57	1.7	133.89	29JL	-1.4	-3.3	-1.5	0.6
101.07	0.69	1.7	143.52	8AU	-1.3	-3.4	-1.5	0.6
107.43	0.81	1.7	153.20	18AU	-0.6	-3.4	-1.5	0.5
113.67	0.93	1.7	162.92	28AU	-0.2	-3.4	-1.6	0.5
119.81	1.06	1.7	172.70	7SE	0.0	-3.5	-1.6	0.4
125.81	1.19	1.6	182.54	17SE	0.2	-3.5	-1.7	0.4
131.68	1.32	1.6	192.43	27SE	0.4	-3.5	-1.8	0.4
137.40	1.46	1.5	202.38	7OC	1.5	-3.5	-1.8	0.4
142.95	1.61	1.4	212.39	17OC	2.2	-3.4	-1.9	0.5
148.28	1.76	1.3	222.44	27OC	-0.1	-3.4	-2.0	0.6
153.38	1.92	1.2	232.54	6NO	-0.5	-3.4	-2.0	0.6
158.19	2.10	1.1	242.68	16NO	-0.6	-3.4	-2.1	0.7
162.65	2.29	1.0	252.85	26NO	-0.6	-3.3	-2.1	0.8
166.68	2.49	0.8	263.04	6DE	-0.7	-3.3	-2.2	0.8
170.17	2.71	0.6	273.24	16DE	-0.8	-3.3	-2.2	0.8
173.00	2.95	0.4	283.43	26DE	-0.8	-3.3	-2.2	0.9

Right table (1350–1351):

♂ LONG	♂ LAT	♂ MAG	☉ LONG	16.00UT 1350	☿	♀	♃	♄
175.03	3.20	0.2	293.61	5JA	-0.8	-3.4	-2.2	0.9
176.05	3.46	-0.1	303.78	15JA	0.1	-3.4	-2.2	0.9
175.90	3.70	-0.3	313.90	25JA	2.5	-3.4	-2.1	0.9
174.47	3.90	-0.6	323.99	4FE	1.7	-3.4	-2.1	0.9
171.80	4.00	-0.9	334.04	14FE	0.7	-3.5	-2.0	0.9
168.22	3.97	-1.1	344.03	24FE	0.4	-3.5	-1.9	0.8
164.34	3.78	-1.0	353.96	6MR	-0.2	-3.5	-1.9	0.8
160.86	3.45	-0.9	3.84	16MR	-0.2	-3.6	-1.8	0.7
158.37	3.04	-0.7	13.66	26MR	-0.8	-3.6	-1.7	0.7
157.14	2.60	-0.4	23.42	5AP	-1.6	-3.7	-1.7	0.7
157.20	2.16	-0.2	33.14	15AP	-1.4	-3.8	-1.6	0.7
158.43	1.76	-0.0	42.80	25AP	-0.4	-3.9	-1.5	0.7
160.64	1.40	0.2	52.42	5MY	0.6	-4.0	-1.5	0.7
163.68	1.07	0.4	62.01	15MY	1.6	-4.1	-1.4	0.7
167.39	0.78	0.5	71.57	25MY	2.8	-4.1	-1.4	0.7
171.64	0.53	0.6	81.11	4JN	3.1	-4.2	-1.4	0.7
176.36	0.30	0.7	90.65	14JN	1.7	-4.2	-1.3	0.7
181.47	0.09	0.8	100.17	24JN	0.6	-4.0	-1.3	0.6
186.90	-0.10	0.9	109.71	4JL	-0.6	-3.5	-1.3	0.6
192.63	-0.27	1.0	119.27	14JL	-1.5	-3.0	-1.3	0.6
198.62	-0.43	1.1	128.85	24JL	-1.3	-3.7	-1.3	0.5
204.83	-0.57	1.1	138.46	3AU	-0.6	-4.1	-1.3	0.5
211.25	-0.69	1.2	148.12	13AU	-0.1	-4.2	-1.3	0.4
217.87	-0.81	1.2	157.81	23AU	0.2	-4.2	-1.3	0.3
224.66	-0.91	1.2	167.57	2SE	0.3	-4.2	-1.4	0.3
231.61	-1.00	1.2	177.37	12SE	0.7	-4.1	-1.4	0.2
238.72	-1.07	1.3	187.23	22SE	1.9	-4.0	-1.5	0.2
245.97	-1.13	1.3	197.16	2OC	2.0	-3.9	-1.5	0.1
253.34	-1.18	1.3	207.14	12OC	-0.2	-3.8	-1.6	0.1
260.83	-1.22	1.3	217.16	22OC	-0.7	-3.8	-1.6	0.2
268.42	-1.24	1.3	227.24	1NO	-0.7	-3.7	-1.7	0.3
276.09	-1.25	1.3	237.37	11NO	-0.7	-3.6	-1.7	0.3
283.84	-1.25	1.3	247.52	21NO	-0.7	-3.6	-1.8	0.4
291.65	-1.23	1.3	257.70	1DE	-0.7	-3.5	-1.9	0.5
299.50	-1.20	1.4	267.89	11DE	-0.7	-3.5	-1.9	0.5
307.37	-1.16	1.4	278.09	21DE	-0.6	-3.4	-2.0	0.6
315.25	-1.11	1.4	288.28	31DE	0.2	-3.4	-2.0	0.6
				1351				
323.13	-1.04	1.4	298.46	10JA	2.8	-3.4	-2.1	0.6
330.99	-0.97	1.4	308.60	20JA	1.3	-3.4	-2.1	0.7
338.82	-0.89	1.4	318.71	30JA	0.5	-3.3	-2.1	0.7
346.61	-0.80	1.4	328.78	9FE	0.2	-3.3	-2.1	0.7
354.33	-0.71	1.5	338.80	19FE	0.1	-3.3	-2.0	0.7
2.00	-0.61	1.5	348.76	1MR	-0.2	-3.3	-2.0	0.6
9.60	-0.51	1.5	358.67	11MR	-0.8	-3.3	-1.9	0.6
17.11	-0.40	1.5	8.52	21MR	-1.6	-3.3	-1.9	0.6
24.55	-0.30	1.5	18.32	31MR	-1.3	-3.4	-1.8	0.5
31.90	-0.19	1.5	28.05	10AP	-0.3	-3.4	-1.8	0.5
39.16	-0.08	1.5	37.74	20AP	0.8	-3.4	-1.7	0.5
46.34	0.02	1.6	47.39	30AP	2.0	-3.5	-1.6	0.5
53.44	0.13	1.6	56.99	10MY	3.6	-3.5	-1.6	0.5
60.45	0.23	1.7	66.56	20MY	2.5	-3.5	-1.5	0.5
67.37	0.33	1.7	76.11	30MY	1.3	-3.4	-1.4	0.5
74.22	0.43	1.8	85.65	9JN	0.4	-3.4	-1.4	0.5
80.99	0.52	1.8	95.17	19JN	-0.6	-3.4	-1.4	0.5
87.69	0.62	1.8	104.71	29JN	-1.5	-3.3	-1.3	0.5
94.32	0.71	1.9	114.25	9JL	-1.4	-3.3	-1.3	0.5
100.87	0.80	1.9	123.82	19JL	-0.6	-3.3	-1.3	0.4
107.37	0.88	1.9	133.41	29JL	-0.0	-3.3	-1.2	0.4
113.79	0.97	1.9	143.04	8AU	0.3	-3.3	-1.2	0.4
120.15	1.05	1.9	152.72	18AU	0.5	-3.3	-1.2	0.3
126.45	1.13	1.9	162.45	28AU	1.0	-3.4	-1.2	0.2
132.69	1.21	1.9	172.22	7SE	2.3	-3.4	-1.2	0.2
138.86	1.29	1.9	182.06	17SE	1.8	-3.4	-1.3	0.1
144.95	1.36	1.8	191.95	27SE	-0.3	-3.5	-1.3	0.1
150.97	1.44	1.8	201.90	7OC	-0.8	-3.5	-1.3	-0.0
156.90	1.52	1.8	211.90	17OC	-0.9	-3.6	-1.4	-0.1
162.74	1.59	1.7	221.96	27OC	-0.8	-3.7	-1.4	-0.1
168.46	1.67	1.6	232.05	6NO	-0.6	-3.7	-1.5	0.0
174.06	1.75	1.5	242.19	16NO	-0.5	-3.8	-1.5	0.1
179.50	1.82	1.3	252.36	26NO	-0.5	-3.9	-1.6	0.2
184.76	1.90	1.3	262.55	6DE	-0.5	-4.0	-1.6	0.2
189.80	1.98	1.2	272.74	16DE	0.3	-4.1	-1.7	0.3
194.57	2.06	1.0	282.94	26DE	3.0	-4.2	-1.8	0.4

♂ LONG	LAT	MAG	☉ LONG	16.00UT 1352	☿	♀	♃	♄
199.01	2.14	0.8	293.12	5JA	0.9	-4.3	-1.8	0.4
203.04	2.22	0.7	303.28	15JA	0.3	-4.4	-1.9	0.4
206.57	2.30	0.4	313.42	25JA	0.1	-4.3	-1.9	0.5
209.46	2.36	0.2	323.51	4FE	-0.0	-4.1	-2.0	0.5
211.57	2.41	-0.1	333.55	14FE	-0.3	-3.6	-2.0	0.5
212.71	2.42	-0.3	343.55	24FE	-0.8	-3.4	-2.0	0.5
212.71	2.39	-0.6	353.48	5MR	-1.5	-3.9	-2.0	0.5
211.45	2.29	-0.9	3.36	15MR	-1.3	-4.2	-2.0	0.5
208.96	2.09	-1.2	13.19	25MR	-0.3	-4.3	-2.0	0.4
205.57	1.79	-1.5	22.95	4AP	1.1	-4.2	-1.9	0.4
201.87	1.39	-1.4	32.67	14AP	2.7	-4.1	-1.9	0.4
198.57	0.94	-1.3	42.34	24AP	3.2	-4.0	-1.8	0.3
196.29	0.48	-1.1	51.96	4MY	1.8	-3.9	-1.7	0.3
195.30	0.06	-0.9	61.55	14MY	1.0	-3.8	-1.7	0.4
195.65	-0.31	-0.7	71.11	24MY	0.3	-3.7	-1.6	0.4
197.21	-0.63	-0.5	80.65	3JN	-0.6	-3.7	-1.6	0.4
199.81	-0.89	-0.3	90.18	13JN	-1.6	-3.6	-1.5	0.4
203.28	-1.11	-0.1	99.71	23JN	-1.4	-3.5	-1.4	0.4
207.47	-1.29	0.0	109.25	3JL	-0.5	-3.5	-1.4	0.4
212.24	-1.43	0.2	118.80	13JL	0.1	-3.4	-1.3	0.4
217.51	-1.55	0.3	128.38	23JL	0.4	-3.4	-1.3	0.4
223.19	-1.65	0.4	137.99	2AU	0.7	-3.4	-1.3	0.3
229.23	-1.71	0.5	147.65	12AU	1.3	-3.4	-1.2	0.3
235.57	-1.76	0.5	157.35	22AU	2.8	-3.4	-1.2	0.2
242.17	-1.79	0.6	167.09	1SE	1.5	-3.3	-1.2	0.2
249.00	-1.80	0.7	176.90	11SE	-0.4	-3.3	-1.2	0.1
256.03	-1.79	0.8	186.76	21SE	-1.0	-3.4	-1.2	0.1
263.23	-1.76	0.8	196.68	1OC	-1.0	-3.4	-1.2	-0.0
270.57	-1.71	0.9	206.65	11OC	-0.9	-3.4	-1.2	-0.1
278.03	-1.65	0.9	216.68	21OC	-0.5	-3.4	-1.3	-0.1
285.59	-1.58	1.0	226.76	31OC	-0.4	-3.4	-1.3	-0.2
293.22	-1.49	1.0	236.88	10NO	-0.4	-3.4	-1.3	-0.2
300.91	-1.39	1.1	247.03	20NO	-0.3	-3.5	-1.4	-0.1
308.63	-1.29	1.2	257.20	30NO	0.5	-3.5	-1.4	-0.0
316.37	-1.17	1.2	267.40	10DE	3.0	-3.5	-1.5	0.0
324.11	-1.05	1.3	277.60	20DE	0.7	-3.5	-1.5	0.1
331.83	-0.93	1.3	287.78	30DE	0.1	-3.4	-1.6	0.2

1353

♂ LONG	LAT	MAG	☉ LONG	16.00UT 1353	☿	♀	♃	♄
339.52	-0.80	1.4	297.96	9JA	-0.0	-3.4	-1.7	0.2
347.17	-0.67	1.4	308.11	19JA	-0.2	-3.4	-1.7	0.3
354.76	-0.55	1.5	318.22	29JA	-0.4	-3.4	-1.8	0.3
2.29	-0.42	1.5	328.29	8FE	-0.9	-3.3	-1.9	0.4
9.76	-0.30	1.5	338.31	18FE	-1.4	-3.3	-1.9	0.4
17.14	-0.17	1.6	348.28	28FE	-1.2	-3.3	-2.0	0.4
24.46	-0.06	1.6	358.19	10MR	-0.2	-3.3	-2.0	0.4
31.69	0.06	1.7	8.04	20MR	1.3	-3.4	-2.0	0.4
38.85	0.16	1.7	17.84	30MR	3.3	-3.4	-2.0	0.4
45.92	0.27	1.7	27.58	9AP	2.4	-3.4	-2.0	0.3
52.92	0.37	1.8	37.27	19AP	1.4	-3.4	-2.0	0.3
59.85	0.46	1.8	46.92	29AP	0.8	-3.4	-2.0	0.3
66.70	0.55	1.8	56.53	9MY	0.2	-3.5	-1.9	0.2
73.48	0.63	1.8	66.10	19MY	-0.6	-3.5	-1.9	0.2
80.20	0.71	1.8	75.65	29MY	-1.6	-3.6	-1.8	0.3
86.87	0.78	1.8	85.19	8JN	-1.4	-3.6	-1.7	0.3
93.47	0.84	1.8	94.72	18JN	-0.5	-3.7	-1.7	0.3
100.03	0.91	1.9	104.25	28JN	0.1	-3.8	-1.6	0.3
106.54	0.97	1.9	113.80	8JL	0.6	-3.9	-1.5	0.3
113.00	1.02	1.9	123.36	18JL	1.0	-4.0	-1.5	0.3
119.43	1.07	2.0	132.95	28JL	1.7	-4.1	-1.4	0.3
125.82	1.12	2.0	142.58	7AU	3.2	-4.2	-1.4	0.3
132.18	1.16	2.0	152.25	17AU	1.3	-4.2	-1.3	0.3
138.51	1.20	2.0	161.97	27AU	-0.4	-4.2	-1.3	0.2
144.82	1.23	2.0	171.75	6SE	-1.1	-4.0	-1.3	0.2
151.10	1.26	2.0	181.58	16SE	-1.2	-3.5	-1.3	0.1
157.35	1.29	2.0	191.47	26SE	-0.8	-3.3	-1.2	0.1
163.57	1.31	2.0	201.42	6OC	-0.4	-3.9	-1.2	-0.0
169.77	1.33	1.9	211.41	16OC	-0.3	-4.2	-1.2	-0.1
175.93	1.35	1.9	221.47	26OC	-0.2	-4.3	-1.2	-0.1
182.06	1.36	1.8	231.57	5NO	-0.1	-4.3	-1.2	-0.2
188.15	1.36	1.8	241.70	15NO	0.7	-4.3	-1.3	-0.3
194.19	1.37	1.7	251.87	25NO	2.9	-4.2	-1.3	-0.2
200.17	1.36	1.6	262.06	5DE	0.4	-4.1	-1.3	-0.2
206.07	1.35	1.5	272.25	15DE	-0.1	-4.0	-1.4	-0.1
211.90	1.33	1.4	282.45	25DE	-0.2	-3.9	-1.4	-0.0

♂ LONG	LAT	MAG	☉ LONG	16.00UT 1354	☿	♀	♃	♄
217.63	1.30	1.3	292.63	4JA	-0.3	-3.8	-1.5	0.0
223.23	1.25	1.1	302.80	14JA	-0.5	-3.7	-1.5	0.1
228.69	1.20	1.0	312.93	24JA	-0.9	-3.6	-1.6	0.2
233.95	1.12	0.8	323.03	3FE	-1.3	-3.6	-1.7	0.2
239.00	1.02	0.6	333.07	13FE	-1.1	-3.5	-1.7	0.3
243.76	0.89	0.4	343.07	23FE	-0.2	-3.4	-1.8	0.3
248.17	0.73	0.2	353.01	5MR	1.6	-3.4	-1.9	0.3
252.13	0.53	-0.0	2.89	15MR	3.3	-3.4	-1.9	0.3
255.54	0.27	-0.3	12.72	25MR	1.8	-3.3	-2.0	0.3
258.23	-0.05	-0.6	22.49	4AP	1.0	-3.3	-2.0	0.3
260.06	-0.46	-0.9	32.20	14AP	0.6	-3.3	-2.1	0.3
260.82	-0.94	-1.2	41.87	24AP	0.1	-3.3	-2.1	0.3
260.36	-1.52	-1.5	51.50	4MY	-0.7	-3.3	-2.1	0.3
258.69	-2.15	-1.9	61.09	14MY	-1.6	-3.3	-2.1	0.2
255.99	-2.79	-2.1	70.65	24MY	-1.4	-3.3	-2.0	0.2
252.83	-3.36	-2.2	80.19	3JN	-0.5	-3.4	-2.0	0.2
249.93	-3.79	-2.0	89.72	13JN	0.2	-3.4	-2.0	0.2
247.93	-4.05	-1.9	99.25	23JN	0.8	-3.4	-1.9	0.3
247.24	-4.15	-1.6	108.79	3JL	1.3	-3.5	-1.8	0.3
247.93	-4.14	-1.4	118.34	13JL	2.3	-3.5	-1.8	0.3
249.91	-4.04	-1.2	127.92	23JL	3.0	-3.5	-1.7	0.3
253.00	-3.88	-1.0	137.53	2AU	1.1	-3.4	-1.7	0.3
257.00	-3.69	-0.8	147.17	12AU	-0.5	-3.4	-1.6	0.3
261.74	-3.48	-0.6	156.87	22AU	-1.2	-3.4	-1.5	0.3
267.08	-3.25	-0.3	166.62	1SE	-1.3	-3.4	-1.5	0.3
272.88	-3.02	-0.3	176.41	11SE	-0.7	-3.3	-1.4	0.2
279.07	-2.77	-0.1	186.27	21SE	-0.3	-3.3	-1.4	0.2
285.56	-2.52	0.0	196.19	1OC	-0.2	-3.3	-1.4	0.1
292.28	-2.28	0.2	206.16	11OC	-0.1	-3.3	-1.3	0.1
299.18	-2.03	0.3	216.19	21OC	0.1	-3.3	-1.3	-0.0
306.23	-1.79	0.4	226.26	31OC	1.0	-3.4	-1.3	-0.1
313.37	-1.55	0.5	236.38	10NO	2.7	-3.4	-1.3	-0.2
320.58	-1.32	0.7	246.53	20NO	0.2	-3.4	-1.3	-0.2
327.83	-1.11	0.8	256.71	30NO	-0.3	-3.5	-1.3	-0.3
335.10	-0.90	0.9	266.90	10DE	-0.3	-3.5	-1.3	-0.3
342.37	-0.70	1.0	277.10	20DE	-0.4	-3.6	-1.4	-0.2
349.63	-0.51	1.1	287.29	30DE	-0.6	-3.6	-1.4	-0.1

1355

♂ LONG	LAT	MAG	☉ LONG	16.00UT 1355	☿	♀	♃	♄
356.85	-0.34	1.2	297.47	9JA	-0.9	-3.7	-1.4	-0.1
4.05	-0.18	1.3	307.62	19JA	-1.1	-3.7	-1.5	0.0
11.19	-0.03	1.4	317.74	29JA	-1.0	-3.8	-1.5	0.1
18.28	0.11	1.4	327.81	8FE	-0.1	-3.9	-1.6	0.1
25.32	0.24	1.5	337.84	18FE	1.9	-4.0	-1.6	0.2
32.30	0.35	1.6	347.80	28FE	2.6	-4.1	-1.7	0.2
39.22	0.46	1.7	357.72	10MR	1.3	-4.2	-1.8	0.3
46.08	0.56	1.7	7.57	20MR	0.7	-4.2	-1.8	0.3
52.88	0.64	1.8	17.37	30MR	0.4	-4.2	-1.9	0.3
59.63	0.72	1.8	27.11	9AP	-0.0	-4.1	-2.0	0.3
66.32	0.79	1.9	36.81	19AP	-0.7	-3.8	-2.0	0.3
72.96	0.86	1.9	46.46	29AP	-1.6	-3.0	-2.1	0.3
79.55	0.91	1.9	56.06	9MY	-1.4	-3.3	-2.1	0.3
86.10	0.96	1.9	65.64	19MY	-0.4	-3.9	-2.2	0.3
92.61	1.00	2.0	75.19	29MY	0.4	-4.1	-2.2	0.2
99.08	1.04	2.0	84.73	8JN	1.0	-4.2	-2.2	0.2
105.53	1.07	2.0	94.26	18JN	1.7	-4.1	-2.2	0.2
111.95	1.10	2.0	103.79	28JN	2.9	-4.1	-2.1	0.3
118.35	1.12	2.0	113.33	8JL	2.6	-4.0	-2.1	0.3
124.74	1.14	1.9	122.90	18JL	0.9	-3.9	-2.1	0.3
131.11	1.15	2.0	132.48	28JL	-0.5	-3.8	-2.0	0.3
137.48	1.16	2.0	142.11	7AU	-1.3	-3.7	-1.9	0.4
143.85	1.16	2.0	151.78	17AU	-1.3	-3.7	-1.9	0.4
150.21	1.16	2.0	161.50	27AU	-0.7	-3.6	-1.8	0.3
156.58	1.15	2.0	171.27	6SE	-0.3	-3.5	-1.7	0.3
162.96	1.14	2.0	181.10	16SE	-0.0	-3.5	-1.7	0.3
169.34	1.12	2.0	190.98	26SE	0.1	-3.5	-1.6	0.3
175.74	1.10	2.0	200.93	6OC	0.3	-3.4	-1.6	0.2
182.15	1.07	2.0	210.93	16OC	1.3	-3.4	-1.5	0.2
188.57	1.04	2.0	220.97	26OC	2.5	-3.4	-1.5	0.1
195.01	1.00	1.9	231.07	5NO	0.1	-3.4	-1.5	0.0
201.46	0.96	1.8	241.21	15NO	-0.4	-3.4	-1.4	-0.1
207.92	0.91	1.8	251.37	25NO	-0.5	-3.4	-1.4	-0.1
214.39	0.85	1.8	261.56	5DE	-0.5	-3.4	-1.4	-0.2
220.87	0.78	1.7	271.76	15DE	-0.7	-3.4	-1.4	-0.3
227.36	0.70	1.6	281.95	25DE	-0.9	-3.4	-1.4	-0.2

♂			☉	16.00UT	☿	♀	♃	♄
LONG	LAT	MAG	LONG	1356	MAGNITUDES			
233.85	0.60	1.5	292.14	4JA	-0.9	-3.4	-1.4	-0.2
240.33	0.50	1.4	302.30	14JA	-0.9	-3.4	-1.4	-0.1
246.82	0.38	1.3	312.44	24JA	-0.0	-3.4	-1.5	-0.0
253.29	0.25	1.2	322.53	3FE	2.2	-3.4	-1.5	0.0
259.74	0.09	1.1	332.59	13FE	2.0	-3.5	-1.5	0.1
266.17	-0.08	1.0	342.58	23FE	0.9	-3.5	-1.6	0.2
272.56	-0.28	0.8	352.53	4MR	0.5	-3.5	-1.6	0.2
278.89	-0.50	0.7	2.41	14MR	0.3	-3.4	-1.7	0.3
285.16	-0.75	0.5	12.24	24MR	-0.1	-3.4	-1.8	0.3
291.33	-1.03	0.3	22.01	3AP	-0.7	-3.4	-1.8	0.3
297.37	-1.35	0.1	31.73	13AP	-1.6	-3.4	-1.9	0.3
303.25	-1.70	-0.0	41.40	23AP	-1.4	-3.3	-2.0	0.4
308.91	-2.10	-0.3	51.03	3MY	-0.4	-3.3	-2.0	0.4
314.28	-2.53	-0.5	60.62	13MY	0.5	-3.3	-2.1	0.3
319.27	-3.02	-0.7	70.18	23MY	1.3	-3.3	-2.2	0.3
323.76	-3.55	-1.0	79.73	2JN	2.3	-3.3	-2.2	0.3
327.61	-4.13	-1.2	89.26	12JN	3.4	-3.4	-2.3	0.3
330.63	-4.74	-1.5	98.79	22JN	2.0	-3.4	-2.3	0.3
332.62	-5.36	-1.8	108.33	2JL	0.7	-3.4	-2.3	0.3
333.39	-5.95	-2.0	117.88	12JL	-0.5	-3.4	-2.3	0.3
332.85	-6.43	-2.3	127.45	22JL	-1.4	-3.5	-2.3	0.4
331.09	-6.71	-2.5	137.06	1AU	-1.4	-3.6	-2.3	0.4
328.56	-6.69	-2.7	146.71	11AU	-0.6	-3.6	-2.2	0.4
325.97	-6.35	-2.5	156.40	21AU	-0.2	-3.7	-2.2	0.4
324.03	-5.75	-2.3	166.15	31AU	0.1	-3.8	-2.1	0.4
323.25	-5.00	-2.0	175.94	10SE	0.2	-3.9	-2.1	0.4
323.77	-4.22	-1.7	185.80	20SE	0.5	-4.0	-2.0	0.4
325.53	-3.47	-1.4	195.71	30SE	1.6	-4.1	-1.9	0.4
328.36	-2.79	-1.1	205.68	10OC	2.2	-4.2	-1.9	0.3
332.04	-2.19	-0.8	215.70	20OC	-0.1	-4.3	-1.8	0.3
336.40	-1.67	-0.6	225.77	30OC	-0.6	-4.4	-1.7	0.2
341.28	-1.22	-0.3	235.88	9NO	-0.6	-4.3	-1.7	0.2
346.55	-0.83	-0.1	246.03	19NO	-0.6	-4.1	-1.6	0.1
352.13	-0.50	0.1	256.21	29NO	-0.7	-3.5	-1.6	0.0
357.94	-0.22	0.3	266.40	9DE	-0.7	-3.2	-1.6	-0.0
3.92	0.02	0.5	276.60	19DE	-0.8	-4.0	-1.5	-0.1
10.03	0.23	0.7	286.80	29DE	-0.7	-4.3	-1.5	-0.2
				1357				
16.24	0.41	0.8	296.97	8JA	0.1	-4.4	-1.5	-0.1
22.51	0.56	1.0	307.12	18JA	2.6	-4.3	-1.5	-0.1
28.82	0.69	1.1	317.24	28JA	1.6	-4.2	-1.5	-0.0
35.17	0.80	1.3	327.32	7FE	0.6	-4.1	-1.5	0.1
41.53	0.89	1.4	337.34	17FE	0.3	-4.0	-1.5	0.1
47.89	0.97	1.5	347.32	27FE	0.1	-3.9	-1.6	0.2
54.26	1.03	1.6	357.24	9MR	-0.2	-3.8	-1.6	0.2
60.62	1.08	1.6	7.10	19MR	-0.7	-3.7	-1.6	0.3
66.97	1.13	1.7	16.90	29MR	-1.6	-3.6	-1.7	0.3
73.31	1.16	1.8	26.65	8AP	-1.4	-3.6	-1.7	0.4
79.63	1.19	1.9	36.34	18AP	-0.4	-3.5	-1.8	0.4
85.95	1.20	1.9	46.00	28AP	0.7	-3.4	-1.9	0.4
92.25	1.22	1.9	55.60	8MY	1.7	-3.4	-1.9	0.4
98.55	1.22	2.0	65.18	18MY	3.1	-3.4	-2.0	0.4
104.84	1.22	2.0	74.74	28MY	2.9	-3.3	-2.1	0.4
111.13	1.22	2.0	84.27	7JN	1.6	-3.3	-2.1	0.4
117.42	1.21	2.0	93.80	17JN	0.5	-3.3	-2.2	0.4
123.72	1.20	2.0	103.34	27JN	-0.6	-3.3	-2.3	0.4
130.02	1.18	2.0	112.87	7JL	-1.5	-3.3	-2.3	0.4
136.34	1.15	2.0	122.44	17JL	-1.4	-3.3	-2.4	0.4
142.68	1.13	2.0	132.03	27JL	-0.6	-3.3	-2.4	0.5
149.03	1.10	2.0	141.65	6AU	-0.1	-3.4	-2.4	0.5
155.42	1.06	2.0	151.32	16AU	0.2	-3.4	-2.4	0.5
161.83	1.02	1.9	161.04	26AU	0.4	-3.4	-2.4	0.6
168.27	0.98	1.9	170.80	5SE	0.8	-3.5	-2.4	0.6
174.75	0.93	1.9	180.63	15SE	2.0	-3.5	-2.3	0.6
181.27	0.88	1.9	190.51	25SE	2.0	-3.5	-2.3	0.6
187.84	0.82	1.9	200.44	5OC	-0.2	-3.5	-2.2	0.6
194.44	0.76	1.9	210.44	15OC	-0.8	-3.4	-2.2	0.5
201.10	0.69	1.9	220.49	25OC	-0.8	-3.4	-2.1	0.5
207.79	0.61	1.9	230.58	4NO	-0.8	-3.4	-2.0	0.4
214.55	0.53	1.9	240.71	14NO	-0.7	-3.4	-1.9	0.4
221.34	0.45	1.8	250.87	24NO	-0.6	-3.3	-1.9	0.3
228.19	0.35	1.8	261.05	4DE	-0.6	-3.3	-1.8	0.3
235.10	0.25	1.7	271.25	14DE	-0.6	-3.3	-1.8	0.2
242.05	0.14	1.7	281.45	24DE	0.2	-3.3	-1.7	0.1

♂			☉	16.00UT	☿	♀	♃	♄
LONG	LAT	MAG	LONG	1358	MAGNITUDES			
249.05	0.03	1.6	291.63	3JA	2.8	-3.4	-1.7	0.0
256.10	-0.10	1.6	301.80	13JA	1.2	-3.4	-1.6	-0.0
263.19	-0.23	1.5	311.94	23JA	0.4	-3.4	-1.6	0.0
270.33	-0.37	1.4	322.04	2FE	0.2	-3.4	-1.6	0.1
277.52	-0.52	1.4	332.09	12FE	0.0	-3.5	-1.6	0.1
284.74	-0.67	1.3	342.10	22FE	-0.3	-3.5	-1.6	0.2
291.99	-0.84	1.2	352.04	4MR	-0.8	-3.5	-1.6	0.3
299.27	-1.00	1.1	1.93	14MR	-1.6	-3.6	-1.6	0.3
306.56	-1.18	1.0	11.77	24MR	-1.3	-3.7	-1.6	0.4
313.86	-1.36	0.9	21.54	3AP	-0.3	-3.7	-1.6	0.4
321.15	-1.53	0.8	31.27	13AP	0.9	-3.8	-1.7	0.5
328.42	-1.71	0.7	40.94	23AP	2.2	-3.9	-1.7	0.5
335.64	-1.88	0.6	50.57	3MY	3.8	-4.0	-1.7	0.5
342.80	-2.05	0.5	60.17	13MY	2.2	-4.1	-1.8	0.6
349.86	-2.21	0.4	69.73	23MY	1.2	-4.1	-1.8	0.6
356.81	-2.36	0.3	79.27	2JN	0.4	-4.2	-1.9	0.6
3.58	-2.50	0.2	88.81	12JN	-0.6	-4.2	-2.0	0.6
10.15	-2.62	0.0	98.34	22JN	-1.5	-4.0	-2.0	0.6
16.45	-2.72	-0.1	107.87	2JL	-1.4	-3.5	-2.1	0.6
22.42	-2.80	-0.2	117.42	12JL	-0.6	-3.0	-2.2	0.6
27.98	-2.85	-0.4	126.99	22JL	-0.0	-3.7	-2.2	0.6
33.02	-2.88	-0.5	136.60	1AU	0.3	-4.1	-2.3	0.6
37.40	-2.87	-0.7	146.25	11AU	0.6	-4.2	-2.4	0.7
40.98	-2.82	-0.9	155.93	21AU	1.1	-4.2	-2.4	0.7
43.55	-2.72	-1.1	165.67	31AU	2.5	-4.2	-2.4	0.7
44.89	-2.54	-1.3	175.47	10SE	1.7	-4.1	-2.5	0.8
44.83	-2.28	-1.6	185.32	20SE	-0.3	-4.0	-2.5	0.8
43.30	-1.90	-1.8	195.23	30SE	-0.9	-3.9	-2.5	0.8
40.49	-1.43	-2.0	205.20	10OC	-0.9	-3.8	-2.4	0.8
37.00	-0.87	-2.1	215.21	20OC	-0.9	-3.8	-2.4	0.8
33.59	-0.31	-1.8	225.28	30OC	-0.6	-3.7	-2.3	0.7
31.03	0.20	-1.5	235.40	9NO	-0.5	-3.6	-2.3	0.7
29.75	0.63	-1.2	245.54	19NO	-0.5	-3.6	-2.2	0.6
29.81	0.95	-0.9	255.72	29NO	-0.4	-3.5	-2.1	0.6
31.09	1.20	-0.6	265.91	9DE	0.4	-3.5	-2.0	0.5
33.41	1.37	-0.3	276.11	19DE	3.0	-3.4	-2.0	0.5
36.54	1.50	0.0	286.30	29DE	0.9	-3.4	-1.9	0.4
				1359				
40.31	1.58	0.3	296.48	8JA	0.2	-3.4	-1.8	0.3
44.58	1.64	0.5	306.63	18JA	0.0	-3.4	-1.8	0.2
49.23	1.67	0.7	316.75	28JA	-0.1	-3.3	-1.7	0.2
54.18	1.68	0.9	326.83	7FE	-0.3	-3.3	-1.7	0.2
59.36	1.68	1.0	336.86	17FE	-0.8	-3.3	-1.7	0.3
64.73	1.68	1.2	346.83	27FE	-1.5	-3.3	-1.6	0.3
70.24	1.66	1.3	356.76	9MR	-1.3	-3.3	-1.6	0.4
75.87	1.64	1.4	6.61	19MR	-0.3	-3.3	-1.6	0.5
81.59	1.61	1.5	16.42	29MR	1.1	-3.4	-1.6	0.5
87.38	1.57	1.6	26.17	8AP	2.9	-3.4	-1.6	0.6
93.24	1.54	1.7	35.87	18AP	2.9	-3.4	-1.6	0.6
99.16	1.50	1.8	45.52	28AP	1.7	-3.5	-1.6	0.7
105.12	1.45	1.8	55.14	8MY	0.9	-3.5	-1.6	0.7
111.12	1.41	1.9	64.71	18MY	0.3	-3.5	-1.6	0.7
117.17	1.36	1.9	74.27	28MY	-0.6	-3.4	-1.7	0.8
123.26	1.30	2.0	83.81	7JN	-1.6	-3.4	-1.7	0.8
129.39	1.25	2.0	93.34	17JN	-1.4	-3.4	-1.7	0.8
135.56	1.19	2.0	102.87	27JN	-0.5	-3.3	-1.8	0.8
141.78	1.13	2.0	112.41	7JL	0.1	-3.3	-1.9	0.8
148.04	1.07	2.0	121.97	17JL	0.5	-3.3	-1.9	0.8
154.35	1.01	2.0	131.56	27JL	0.8	-3.3	-2.0	0.8
160.71	0.94	2.0	141.18	6AU	1.4	-3.3	-2.0	0.8
167.12	0.87	2.0	150.84	16AU	3.0	-3.3	-2.1	0.8
173.60	0.80	2.0	160.56	26AU	1.5	-3.4	-2.2	0.9
180.13	0.72	1.9	170.33	5SE	-0.4	-3.4	-2.2	0.9
186.72	0.64	1.9	180.14	15SE	-1.0	-3.4	-2.3	1.0
193.39	0.56	1.8	190.03	25SE	-1.1	-3.5	-2.4	1.0
200.12	0.47	1.8	199.96	5OC	-0.8	-3.5	-2.4	1.0
206.92	0.38	1.8	209.95	15OC	-0.5	-3.6	-2.4	1.0
213.80	0.29	1.8	220.00	25OC	-0.4	-3.7	-2.4	1.0
220.74	0.19	1.8	230.09	4NO	-0.3	-3.7	-2.4	1.0
227.76	0.10	1.7	240.22	14NO	-0.2	-3.8	-2.4	1.0
234.86	-0.01	1.7	250.38	24NO	0.6	-3.9	-2.3	0.9
242.03	-0.11	1.7	260.56	4DE	3.1	-4.0	-2.3	0.9
249.27	-0.22	1.7	270.76	14DE	0.6	-4.1	-2.2	0.8
256.58	-0.33	1.6	280.96	24DE	0.0	-4.2	-2.1	0.8

♂ LONG	♂ LAT	♂ MAG	⊙ LONG	16.00UT	☿	♀	♃	♄
				1360	MAGNITUDES			
263.95	-0.44	1.6	291.14	3JA	-0.1	-4.3	-2.1	0.7
271.39	-0.56	1.6	301.31	13JA	-0.2	-4.4	-2.0	0.6
278.89	-0.67	1.5	311.45	23JA	-0.4	-4.3	-1.9	0.6
286.43	-0.78	1.5	321.55	2FE	-0.9	-4.1	-1.9	0.5
294.03	-0.89	1.5	331.60	12FE	-1.3	-3.5	-1.8	0.5
301.66	-1.00	1.4	341.61	22FE	-1.2	-3.4	-1.7	0.5
309.31	-1.10	1.4	351.56	3MR	-0.3	-3.9	-1.7	0.5
316.99	-1.20	1.3	1.45	13MR	1.4	-4.2	-1.6	0.6
324.68	-1.28	1.3	11.29	23MR	3.5	-4.3	-1.6	0.6
332.35	-1.36	1.2	21.06	2AP	2.2	-4.2	-1.6	0.7
340.01	-1.42	1.2	30.79	12AP	1.2	-4.1	-1.5	0.8
347.64	-1.48	1.1	40.47	22AP	0.7	-4.0	-1.5	0.8
355.21	-1.51	1.1	50.10	2MY	0.1	-3.9	-1.5	0.9
2.73	-1.54	1.0	59.69	12MY	-0.6	-3.8	-1.5	0.9
10.16	-1.54	1.0	69.27	22MY	-1.6	-3.7	-1.5	0.9
17.49	-1.53	1.0	78.81	1JN	-1.4	-3.7	-1.5	1.0
24.71	-1.50	0.9	88.34	11JN	-0.5	-3.6	-1.5	1.0
31.80	-1.45	0.9	97.88	21JN	0.2	-3.5	-1.5	1.0
38.72	-1.39	0.8	107.41	1JL	0.6	-3.5	-1.6	1.1
45.48	-1.30	0.7	116.96	11JL	1.1	-3.4	-1.6	1.1
52.02	-1.20	0.7	126.53	21JL	1.9	-3.4	-1.7	1.1
58.33	-1.07	0.6	136.14	31JL	3.2	-3.4	-1.7	1.1
64.36	-0.93	0.5	145.78	10AU	1.3	-3.4	-1.8	1.1
70.07	-0.76	0.4	155.47	20AU	-0.4	-3.4	-1.8	1.1
75.39	-0.56	0.3	165.20	30AU	-1.1	-3.3	-1.9	1.1
80.25	-0.34	0.2	175.00	9SE	-1.2	-3.3	-1.9	1.2
84.53	-0.08	0.0	184.85	19SE	-0.8	-3.4	-2.0	1.2
88.13	0.22	-0.1	194.75	29SE	-0.4	-3.4	-2.1	1.3
90.87	0.56	-0.3	204.71	9OC	-0.2	-3.4	-2.1	1.3
92.56	0.95	-0.5	214.73	19OC	-0.2	-3.4	-2.2	1.3
93.00	1.39	-0.8	224.79	29OC	-0.0	-3.4	-2.2	1.3
92.04	1.87	-1.0	234.91	8NO	0.8	-3.4	-2.3	1.3
89.67	2.36	-1.2	245.05	18NO	2.9	-3.5	-2.3	1.3
86.22	2.79	-1.3	255.22	28NO	0.4	-3.5	-2.3	1.3
82.28	3.11	-1.4	265.41	8DE	-0.2	-3.5	-2.3	1.2
78.67	3.29	-1.1	275.61	18DE	-0.2	-3.5	-2.3	1.2
76.04	3.33	-0.9	285.80	28DE	-0.3	-3.4	-2.2	1.1
				1361				
74.70	3.27	-0.6	295.98	7JA	-0.5	-3.4	-2.2	1.1
74.69	3.15	-0.3	306.14	17JA	-0.9	-3.4	-2.1	1.0
75.85	2.99	0.0	316.26	27JA	-1.2	-3.4	-2.0	0.9
77.99	2.83	0.3	326.34	6FE	-1.1	-3.3	-2.0	0.9
80.91	2.67	0.5	336.37	16FE	-0.2	-3.3	-1.9	0.8
84.45	2.51	0.7	346.35	26FE	1.7	-3.3	-1.8	0.8
88.48	2.36	0.9	356.27	8MR	3.1	-3.3	-1.8	0.8
92.90	2.22	1.1	6.14	18MR	1.6	-3.4	-1.7	0.8
97.62	2.08	1.2	15.94	28MR	0.9	-3.4	-1.6	0.9
102.60	1.96	1.3	25.70	7AP	0.5	-3.4	-1.6	0.9
107.78	1.83	1.4	35.40	17AP	0.0	-3.4	-1.5	1.0
113.13	1.72	1.5	45.05	27AP	-0.7	-3.4	-1.5	1.0
118.63	1.60	1.6	54.67	7MY	-1.6	-3.5	-1.5	1.1
124.26	1.49	1.7	64.25	17MY	-1.4	-3.5	-1.4	1.2
129.99	1.39	1.8	73.80	27MY	-0.5	-3.6	-1.4	1.2
135.83	1.28	1.8	83.35	6JN	0.3	-3.7	-1.4	1.2
141.77	1.18	1.8	92.88	16JN	0.8	-3.7	-1.4	1.3
147.80	1.08	1.9	102.41	26JN	1.4	-3.8	-1.4	1.3
153.91	0.98	1.9	111.95	6JL	2.5	-3.9	-1.4	1.3
160.11	0.87	1.9	121.51	16JL	2.9	-4.0	-1.4	1.4
166.39	0.77	1.9	131.10	26JL	1.1	-4.1	-1.4	1.4
172.77	0.67	1.9	140.72	5AU	-0.5	-4.2	-1.5	1.4
179.22	0.57	1.9	150.38	15AU	-1.3	-4.2	-1.5	1.3
185.76	0.47	1.9	160.09	25AU	-1.3	-4.2	-1.5	1.3
192.40	0.37	1.9	169.86	4SE	-0.7	-4.0	-1.6	1.2
199.12	0.27	1.8	179.67	14SE	-0.3	-3.5	-1.6	1.2
205.93	0.17	1.8	189.55	24SE	-0.1	-3.3	-1.7	1.2
212.84	0.06	1.8	199.48	4OC	-0.0	-3.9	-1.8	1.2
219.84	-0.04	1.7	209.47	14OC	0.2	-4.3	-1.8	1.2
226.93	-0.14	1.7	219.51	24OC	1.0	-4.3	-1.9	1.2
234.11	-0.24	1.6	229.60	3NO	2.7	-4.3	-2.0	1.2
241.38	-0.34	1.6	239.73	13NO	0.2	-4.3	-2.0	1.2
248.73	-0.44	1.5	249.89	23NO	-0.3	-4.2	-2.1	1.1
256.17	-0.54	1.5	260.07	3DE	-0.4	-4.1	-2.1	1.1
263.69	-0.63	1.5	270.27	13DE	-0.4	-4.0	-2.2	1.1
271.27	-0.72	1.5	280.46	23DE	-0.6	-3.9	-2.2	1.0

♂ LONG	♂ LAT	♂ MAG	⊙ LONG	16.00UT	☿	♀	♃	♄
				1362	MAGNITUDES			
278.93	-0.81	1.5	290.65	2JA	-0.9	-3.8	-2.2	1.0
286.63	-0.88	1.5	300.82	12JA	-1.0	-3.7	-2.2	0.9
294.39	-0.96	1.5	310.96	22JA	-1.0	-3.6	-2.1	0.9
302.18	-1.02	1.4	321.07	1FE	-0.1	-3.6	-2.1	0.8
310.00	-1.07	1.4	331.12	11FE	2.0	-3.5	-2.0	0.8
317.83	-1.12	1.4	341.13	21FE	2.4	-3.4	-2.0	0.8
325.67	-1.15	1.4	351.08	3MR	1.1	-3.4	-1.9	0.7
333.49	-1.17	1.4	0.98	13MR	0.7	-3.4	-1.8	0.7
341.29	-1.18	1.4	10.81	23MR	0.3	-3.3	-1.8	0.8
349.06	-1.17	1.4	20.60	2AP	-0.0	-3.3	-1.7	0.9
356.78	-1.16	1.4	30.32	12AP	-0.7	-3.3	-1.6	1.0
4.44	-1.12	1.4	40.00	22AP	-1.6	-3.3	-1.6	1.0
12.03	-1.08	1.4	49.64	2MY	-1.4	-3.3	-1.5	1.1
19.55	-1.02	1.4	59.23	12MY	-0.5	-3.3	-1.5	1.1
26.97	-0.95	1.4	68.80	22MY	0.4	-3.3	-1.4	1.2
34.29	-0.87	1.3	78.35	1JN	1.1	-3.4	-1.4	1.2
41.51	-0.78	1.3	87.88	11JN	1.9	-3.4	-1.4	1.3
48.62	-0.67	1.3	97.41	21JN	3.2	-3.4	-1.3	1.3
55.60	-0.56	1.3	106.95	1JL	2.4	-3.5	-1.3	1.3
62.44	-0.43	1.3	116.49	11JL	0.9	-3.5	-1.3	1.3
69.15	-0.30	1.3	126.06	21JL	-0.5	-3.5	-1.3	1.3
75.69	-0.15	1.2	135.67	31JL	-1.3	-3.4	-1.3	1.3
82.07	0.00	1.2	145.31	10AU	-1.4	-3.4	-1.3	1.2
88.27	0.17	1.1	154.99	20AU	-0.7	-3.4	-1.3	1.2
94.25	0.35	1.1	164.73	30AU	-0.2	-3.4	-1.3	1.1
100.00	0.54	1.0	174.52	9SE	0.0	-3.3	-1.4	1.1
105.47	0.75	0.9	184.36	19SE	0.1	-3.3	-1.4	1.1
110.62	0.98	0.8	194.27	29SE	0.4	-3.3	-1.5	1.1
115.39	1.23	0.7	204.22	9OC	1.3	-3.3	-1.5	1.1
119.69	1.50	0.6	214.24	19OC	2.5	-3.3	-1.6	1.1
123.43	1.81	0.4	224.31	29OC	0.0	-3.4	-1.6	1.1
126.47	2.15	0.2	234.41	8NO	-0.5	-3.4	-1.7	1.1
128.64	2.53	0.0	244.56	18NO	-0.5	-3.4	-1.7	1.0
129.77	2.94	-0.2	254.73	28NO	-0.6	-3.5	-1.8	1.0
129.66	3.37	-0.4	264.92	8DE	-0.7	-3.5	-1.9	1.0
128.21	3.79	-0.7	275.12	18DE	-0.8	-3.6	-1.9	1.0
125.46	4.15	-0.9	285.31	28DE	-0.9	-3.6	-2.0	0.9
				1363				
121.78	4.39	-1.0	295.49	7JA	-0.8	-3.7	-2.0	0.9
117.80	4.46	-1.0	305.65	17JA	-0.0	-3.7	-2.1	0.8
114.28	4.36	-0.8	315.77	27JA	2.3	-3.8	-2.1	0.8
111.79	4.12	-0.5	325.85	6FE	1.9	-3.9	-2.1	0.7
110.58	3.81	-0.3	335.89	16FE	0.8	-4.0	-2.0	0.7
110.65	3.48	-0.0	345.87	26FE	0.5	-4.1	-2.0	0.6
111.85	3.15	0.2	355.79	8MR	0.2	-4.2	-2.0	0.6
113.99	2.84	0.4	5.66	18MR	-0.1	-4.2	-1.9	0.6
116.90	2.56	0.6	15.47	28MR	-0.7	-4.2	-1.9	0.6
120.43	2.30	0.8	25.23	7AP	-1.6	-4.1	-1.8	0.7
124.45	2.06	1.0	34.93	17AP	-1.4	-3.8	-1.7	0.7
128.88	1.85	1.1	44.59	27AP	-0.4	-3.0	-1.7	0.8
133.63	1.65	1.2	54.20	7MY	0.6	-3.3	-1.6	0.9
138.67	1.46	1.3	63.79	17MY	1.4	-3.9	-1.5	0.9
143.94	1.29	1.4	73.34	27MY	2.6	-4.1	-1.5	1.0
149.41	1.12	1.5	82.88	6JN	3.3	-4.2	-1.4	1.0
155.06	0.97	1.5	92.41	16JN	1.9	-4.1	-1.4	1.1
160.87	0.82	1.6	101.94	26JN	0.7	-4.1	-1.3	1.1
166.84	0.68	1.6	111.48	6JL	-0.5	-4.0	-1.3	1.1
172.94	0.54	1.7	121.04	16JL	-1.4	-3.9	-1.3	1.1
179.18	0.41	1.7	130.63	26JL	-1.4	-3.8	-1.2	1.1
185.55	0.28	1.7	140.25	5AU	-0.6	-3.7	-1.2	1.1
192.04	0.16	1.7	149.91	15AU	-0.2	-3.7	-1.2	1.1
198.66	0.04	1.7	159.62	25AU	0.1	-3.6	-1.2	1.1
205.39	-0.08	1.7	169.38	4SE	0.3	-3.5	-1.2	1.0
212.24	-0.19	1.7	179.20	14SE	0.6	-3.5	-1.2	1.0
219.21	-0.30	1.7	189.07	24SE	1.7	-3.5	-1.3	0.9
226.28	-0.40	1.6	199.00	4OC	2.2	-3.4	-1.3	0.9
233.47	-0.50	1.6	208.99	14OC	-0.1	-3.4	-1.3	0.9
240.76	-0.59	1.6	219.03	24OC	-0.7	-3.4	-1.4	0.9
248.15	-0.68	1.6	229.11	3NO	-0.7	-3.4	-1.4	1.0
255.64	-0.76	1.5	239.24	13NO	-0.7	-3.4	-1.5	0.9
263.22	-0.83	1.5	249.40	23NO	-0.8	-3.4	-1.5	0.9
270.87	-0.90	1.4	259.58	3DE	-0.7	-3.4	-1.6	0.9
278.59	-0.95	1.4	269.78	13DE	-0.7	-3.4	-1.7	0.9
286.38	-1.00	1.4	279.97	23DE	-0.7	-3.4	-1.7	0.9

♂ LONG	♂ LAT	♂ MAG	☉ LONG	16.00UT	☿ MAGNITUDES	♀	♃	♄
				1364				
294.21	-1.04	1.3	290.16	2JA	0.1	-3.4	-1.8	0.8
302.07	-1.06	1.3	300.33	12JA	2.6	-3.4	-1.8	0.8
309.96	-1.08	1.3	310.47	22JA	1.4	-3.4	-1.9	0.8
317.86	-1.08	1.3	320.57	1FE	0.5	-3.4	-1.9	0.7
325.76	-1.07	1.3	330.63	11FE	0.3	-3.5	-2.0	0.7
333.64	-1.05	1.3	340.64	21FE	0.1	-3.5	-2.0	0.6
341.49	-1.02	1.4	350.60	2MR	-0.2	-3.5	-2.0	0.6
349.30	-0.98	1.4	0.50	12MR	-0.7	-3.4	-2.0	0.5
357.06	-0.92	1.4	10.33	22MR	-1.5	-3.4	-2.0	0.5
4.76	-0.86	1.4	20.12	1AP	-1.4	-3.4	-2.0	0.4
12.38	-0.79	1.5	29.85	11AP	-0.4	-3.4	-1.9	0.5
19.93	-0.70	1.5	39.53	21AP	0.7	-3.3	-1.9	0.6
27.40	-0.61	1.5	49.17	1MY	1.9	-3.3	-1.8	0.6
34.78	-0.52	1.5	58.77	11MY	3.4	-3.3	-1.7	0.7
42.06	-0.42	1.6	68.33	21MY	2.6	-3.3	-1.7	0.8
49.25	-0.31	1.6	77.88	31MY	1.5	-3.3	-1.6	0.8
56.34	-0.20	1.6	87.42	10JN	0.5	-3.4	-1.5	0.9
63.32	-0.08	1.6	96.94	20JN	-0.5	-3.4	-1.5	0.9
70.21	0.04	1.6	106.48	30JN	-1.5	-3.4	-1.4	0.9
76.99	0.16	1.6	116.03	10JL	-1.4	-3.4	-1.4	1.0
83.66	0.29	1.6	125.60	20JL	-0.6	-3.5	-1.3	1.0
90.23	0.42	1.6	135.20	30JL	-0.1	-3.6	-1.3	1.0
96.67	0.55	1.6	144.84	9AU	0.2	-3.6	-1.3	1.0
103.00	0.69	1.6	154.52	19AU	0.5	-3.7	-1.2	1.0
109.20	0.83	1.6	164.25	29AU	0.9	-3.8	-1.2	1.0
115.25	0.97	1.5	174.04	8SE	2.1	-3.9	-1.2	0.9
121.15	1.12	1.5	183.88	18SE	2.0	-4.0	-1.2	0.9
126.88	1.28	1.4	193.79	28SE	-0.2	-4.1	-1.2	0.9
132.41	1.45	1.3	203.74	8OC	-0.8	-4.2	-1.2	0.8
137.71	1.63	1.2	213.75	18OC	-0.8	-4.3	-1.2	0.8
142.74	1.82	1.1	223.82	28OC	-0.8	-4.4	-1.3	0.8
147.44	2.02	1.0	233.93	7NO	-0.7	-4.3	-1.3	0.9
151.74	2.25	0.9	244.07	17NO	-0.6	-4.1	-1.3	0.9
155.56	2.49	0.7	254.24	27NO	-0.6	-3.4	-1.4	0.9
158.78	2.76	0.5	264.43	7DE	-0.5	-3.2	-1.4	0.9
161.25	3.05	0.3	274.62	17DE	0.2	-4.0	-1.5	0.8
162.81	3.36	0.1	284.82	27DE	2.9	-4.3	-1.5	0.8
				1365				
163.26	3.68	-0.2	295.00	6JA	1.1	-4.4	-1.6	0.8
162.47	3.98	-0.4	305.16	16JA	0.3	-4.3	-1.7	0.8
160.36	4.22	-0.7	315.28	26JA	0.1	-4.2	-1.7	0.7
157.14	4.35	-0.9	325.37	5FE	-0.0	-4.1	-1.8	0.7
153.29	4.32	-1.0	335.40	15FE	-0.3	-4.0	-1.9	0.6
149.48	4.12	-0.9	345.39	25FE	-0.8	-3.9	-1.9	0.6
146.41	3.79	-0.7	355.32	7MR	-1.5	-3.8	-2.0	0.5
144.50	3.38	-0.5	5.19	17MR	-1.3	-3.7	-2.0	0.5
143.89	2.95	-0.2	15.00	27MR	-0.4	-3.6	-2.0	0.4
144.51	2.54	-0.0	24.76	6AP	0.9	-3.6	-2.0	0.4
146.19	2.16	0.2	34.47	16AP	2.4	-3.5	-2.0	0.4
148.76	1.81	0.4	44.13	26AP	3.5	-3.4	-2.0	0.4
152.06	1.50	0.6	53.74	6MY	2.0	-3.4	-1.9	0.5
155.95	1.22	0.7	63.33	16MY	1.1	-3.4	-1.9	0.5
160.32	0.97	0.8	72.89	26MY	0.4	-3.3	-1.8	0.6
165.10	0.74	0.9	82.43	5JN	-0.6	-3.3	-1.8	0.7
170.21	0.53	1.0	91.95	15JN	-1.5	-3.3	-1.7	0.7
175.63	0.34	1.1	101.49	25JN	-1.4	-3.3	-1.7	0.8
181.30	0.17	1.2	111.02	5JL	-0.6	-3.3	-1.6	0.8
187.20	0.00	1.2	120.58	15JL	0.0	-3.3	-1.5	0.9
193.31	-0.15	1.3	130.17	25JL	0.4	-3.3	-1.5	0.9
199.62	-0.29	1.3	139.78	4AU	0.7	-3.4	-1.4	0.9
206.10	-0.42	1.3	149.44	14AU	1.2	-3.4	-1.4	0.9
212.74	-0.54	1.4	159.15	24AU	2.6	-3.4	-1.3	0.9
219.55	-0.65	1.4	168.90	3SE	1.7	-3.5	-1.3	0.9
226.50	-0.75	1.4	178.72	13SE	-0.3	-3.5	-1.3	0.9
233.59	-0.84	1.4	188.59	23SE	-0.9	-3.5	-1.3	0.9
240.81	-0.92	1.4	198.51	3OC	-1.0	-3.5	-1.2	0.8
248.15	-0.99	1.4	208.50	13OC	-0.9	-3.4	-1.2	0.8
255.61	-1.05	1.4	218.54	23OC	-0.6	-3.4	-1.2	0.7
263.16	-1.09	1.4	228.62	2NO	-0.4	-3.4	-1.2	0.8
270.81	-1.13	1.4	238.74	12NO	-0.4	-3.4	-1.3	0.8
278.53	-1.15	1.4	248.90	22NO	-0.3	-3.3	-1.3	0.8
286.32	-1.16	1.4	259.08	2DE	0.4	-3.3	-1.3	0.8
294.16	-1.16	1.4	269.28	12DE	3.1	-3.3	-1.3	0.8
302.04	-1.14	1.4	279.48	22DE	0.8	-3.3	-1.4	0.8

♂ LONG	♂ LAT	♂ MAG	☉ LONG	16.00UT	☿ MAGNITUDES	♀	♃	♄
				1366				
309.93	-1.12	1.4	289.66	1JA	0.1	-3.4	-1.4	0.8
317.83	-1.08	1.4	299.84	11JA	-0.0	-3.4	-1.5	0.8
325.73	-1.03	1.4	309.98	21JA	-0.1	-3.4	-1.6	0.8
333.60	-0.97	1.4	320.08	31JA	-0.4	-3.4	-1.6	0.7
341.44	-0.90	1.4	330.15	10FE	-0.8	-3.5	-1.7	0.7
349.23	-0.83	1.4	340.16	20FE	-1.4	-3.5	-1.8	0.6
356.97	-0.74	1.4	350.12	2MR	-1.3	-3.5	-1.8	0.6
4.64	-0.65	1.4	0.02	12MR	-0.3	-3.6	-1.9	0.5
12.24	-0.56	1.4	9.87	22MR	1.2	-3.7	-2.0	0.5
19.75	-0.46	1.4	19.65	1AP	3.1	-3.7	-2.0	0.4
27.19	-0.36	1.5	29.39	11AP	2.6	-3.8	-2.0	0.3
34.54	-0.25	1.5	39.07	21AP	1.5	-3.9	-2.1	0.3
41.81	-0.15	1.6	48.71	1MY	0.8	-4.0	-2.1	0.3
48.98	-0.04	1.6	58.31	11MY	0.2	-4.1	-2.1	0.4
56.07	0.07	1.7	67.88	21MY	-0.6	-4.1	-2.1	0.4
63.07	0.17	1.7	77.42	31MY	-1.5	-4.2	-2.0	0.5
69.99	0.28	1.7	86.96	10JN	-1.4	-4.1	-2.0	0.6
76.82	0.38	1.8	96.49	20JN	-0.5	-3.9	-2.0	0.6
83.57	0.49	1.8	106.02	30JN	0.1	-3.4	-1.9	0.7
90.23	0.59	1.8	115.57	10JL	0.5	-3.0	-1.8	0.7
96.82	0.69	1.8	125.14	20JL	0.9	-3.7	-1.8	0.8
103.33	0.79	1.8	134.73	30JL	1.6	-4.1	-1.7	0.8
109.77	0.89	1.9	144.37	9AU	3.1	-4.2	-1.7	0.8
116.12	0.98	1.8	154.05	19AU	1.5	-4.2	-1.6	0.8
122.40	1.08	1.8	163.78	29AU	-0.3	-4.2	-1.5	0.9
128.60	1.18	1.8	173.56	8SE	-1.1	-4.1	-1.5	0.8
134.70	1.27	1.8	183.40	18SE	-1.1	-4.0	-1.4	0.8
140.71	1.37	1.7	193.30	28SE	-0.8	-3.9	-1.4	0.8
146.62	1.47	1.7	203.26	8OC	-0.5	-3.8	-1.4	0.8
152.41	1.57	1.6	213.27	18OC	-0.3	-3.8	-1.3	0.8
158.07	1.68	1.6	223.32	28OC	-0.2	-3.7	-1.3	0.7
163.57	1.79	1.5	233.43	7NO	-0.1	-3.6	-1.3	0.7
168.89	1.90	1.4	243.57	17NO	0.6	-3.6	-1.3	0.7
173.99	2.01	1.3	253.74	27NO	3.1	-3.5	-1.3	0.8
178.83	2.14	1.1	263.93	7DE	0.5	-3.5	-1.3	0.8
183.35	2.27	1.0	274.13	17DE	-0.1	-3.4	-1.3	0.8
187.47	2.40	0.8	284.32	27DE	-0.2	-3.4	-1.4	0.8
				1367				
191.10	2.54	0.6	294.51	6JA	-0.2	-3.4	-1.4	0.8
194.12	2.69	0.4	304.67	16JA	-0.5	-3.4	-1.4	0.8
196.39	2.83	0.1	314.79	26JA	-0.9	-3.3	-1.5	0.8
197.72	2.96	-0.1	324.88	5FE	-1.3	-3.3	-1.6	0.7
197.94	3.07	-0.4	334.92	15FE	-1.2	-3.3	-1.6	0.7
196.92	3.12	-0.7	344.91	25FE	-0.3	-3.3	-1.7	0.6
194.63	3.08	-1.0	354.84	7MR	1.4	-3.3	-1.7	0.6
191.33	2.93	-1.2	4.71	17MR	3.4	-3.3	-1.8	0.5
187.54	2.65	-1.3	14.53	27MR	1.9	-3.4	-1.9	0.5
183.98	2.26	-1.1	24.29	6AP	1.1	-3.4	-1.9	0.4
181.30	1.81	-0.9	34.00	16AP	0.6	-3.4	-2.0	0.4
179.84	1.36	-0.7	43.66	26AP	0.1	-3.5	-2.1	0.3
179.72	0.93	-0.5	53.28	6MY	-0.6	-3.5	-2.1	0.2
180.83	0.55	-0.3	62.86	16MY	-1.5	-3.4	-2.2	0.3
182.99	0.21	-0.1	72.42	26MY	-1.5	-3.4	-2.2	0.4
186.05	-0.08	0.0	81.96	5JN	-0.5	-3.4	-2.2	0.4
189.85	-0.33	0.2	91.49	15JN	0.2	-3.4	-2.2	0.5
194.25	-0.55	0.3	101.02	25JN	0.7	-3.3	-2.2	0.5
199.15	-0.74	0.4	110.56	5JL	1.2	-3.3	-2.2	0.6
204.48	-0.90	0.6	120.11	15JL	2.1	-3.3	-2.1	0.7
210.17	-1.04	0.6	129.70	25JL	3.2	-3.3	-2.1	0.7
216.18	-1.16	0.7	139.31	4AU	1.2	-3.3	-2.0	0.7
222.46	-1.26	0.8	148.97	14AU	-0.4	-3.3	-1.9	0.8
228.99	-1.35	0.8	158.67	24AU	-1.2	-3.4	-1.9	0.8
235.74	-1.41	0.9	168.43	3SE	-1.3	-3.4	-1.8	0.8
242.68	-1.46	0.9	178.23	13SE	-0.8	-3.4	-1.7	0.8
249.79	-1.49	1.0	188.10	23SE	-0.4	-3.5	-1.7	0.8
257.05	-1.50	1.0	198.03	3OC	-0.2	-3.5	-1.6	0.8
264.45	-1.50	1.1	208.01	13OC	-0.1	-3.6	-1.6	0.8
271.96	-1.49	1.1	218.04	23OC	0.0	-3.7	-1.5	0.7
279.57	-1.45	1.1	228.13	2NO	0.8	-3.7	-1.5	0.7
287.26	-1.41	1.2	238.25	12NO	3.0	-3.8	-1.5	0.7
295.00	-1.35	1.2	248.40	22NO	0.3	-3.9	-1.4	0.7
302.79	-1.28	1.2	258.59	2DE	-0.2	-4.0	-1.4	0.7
310.60	-1.20	1.3	268.78	12DE	-0.3	-4.1	-1.4	0.7
318.41	-1.11	1.3	278.98	22DE	-0.4	-4.2	-1.4	0.8

1368

♂ LONG LAT MAG	⊙ LONG	16.00UT	☿	♀	♃	♄
				MAGNITUDES		
326.22 -1.01 1.4	289.17	1JA	-0.5	-4.3	-1.4	0.8
334.01 -0.91 1.4	299.34	11JA	-0.9	-4.4	-1.4	0.8
341.75 -0.80 1.4	309.49	21JA	-1.1	-4.3	-1.5	0.8
349.46 -0.69 1.5	319.60	31JA	-1.0	-4.1	-1.5	0.8
357.10 -0.57 1.5	329.66	10FE	-0.2	-3.5	-1.5	0.7
4.67 -0.46 1.5	339.68	20FE	1.7	-3.4	-1.6	0.7
12.18 -0.34 1.6	349.64	1MR	2.9	-3.9	-1.6	0.7
19.60 -0.23 1.6	359.54	11MR	1.4	-4.2	-1.7	0.6
26.95 -0.12 1.6	9.39	21MR	0.8	-4.3	-1.7	0.6
34.21 -0.01 1.6	19.18	31MR	0.4	-4.2	-1.8	0.5
41.40 0.10 1.7	28.91	10AP	0.0	-4.1	-1.9	0.4
48.50 0.20 1.7	38.60	20AP	-0.6	-4.0	-1.9	0.4
55.52 0.30 1.7	48.24	30AP	-1.5	-3.9	-2.0	0.3
62.46 0.40 1.7	57.84	10MY	-1.5	-3.8	-2.1	0.3
69.33 0.49 1.7	67.42	20MY	-0.5	-3.7	-2.1	0.3
76.12 0.58 1.7	76.97	30MY	0.3	-3.7	-2.2	0.3
82.85 0.66 1.8	86.50	9JN	0.9	-3.6	-2.3	0.4
89.52 0.74 1.8	96.03	19JN	1.6	-3.5	-2.3	0.4
96.12 0.81 1.9	105.56	29JN	2.7	-3.5	-2.3	0.5
102.67 0.89 1.9	115.11	9JL	2.7	-3.4	-2.3	0.5
109.17 0.95 1.9	124.68	19JL	1.0	-3.4	-2.3	0.6
115.62 1.02 2.0	134.27	29JL	-0.4	-3.4	-2.3	0.7
122.02 1.08 2.0	143.91	8AU	-1.3	-3.4	-2.3	0.7
128.38 1.13 2.0	153.59	18AU	-1.4	-3.4	-2.2	0.7
134.69 1.19 2.0	163.31	28AU	-0.7	-3.3	-2.2	0.8
140.97 1.24 2.0	173.09	7SE	-0.3	-3.3	-2.1	0.8
147.20 1.29 2.0	182.93	17SE	-0.1	-3.4	-2.1	0.8
153.39 1.33 1.9	192.82	27SE	0.1	-3.4	-2.0	0.8
159.54 1.38 1.9	202.77	7OC	0.2	-3.4	-1.9	0.8
165.64 1.42 1.9	212.78	17OC	1.1	-3.4	-1.9	0.8
171.68 1.45 1.8	222.83	27OC	2.7	-3.4	-1.8	0.7
177.66 1.49 1.8	232.94	6NO	0.2	-3.4	-1.7	0.7
183.57 1.52 1.7	243.08	16NO	-0.4	-3.5	-1.7	0.7
189.39 1.55 1.6	253.24	26NO	-0.5	-3.5	-1.6	0.7
195.12 1.57 1.5	263.43	6DE	-0.5	-3.5	-1.6	0.7
200.73 1.59 1.4	273.63	16DE	-0.6	-3.5	-1.6	0.7
206.19 1.60 1.3	283.82	26DE	-0.9	-3.4	-1.5	0.8

1369

♂ LONG LAT MAG	⊙ LONG	16.00UT	☿	♀	♃	♄
211.49 1.60 1.1	294.01	5JA	-1.0	-3.4	-1.5	0.8
216.57 1.60 1.0	304.17	15JA	-0.9	-3.4	-1.5	0.8
221.39 1.58 0.8	314.30	25JA	-0.1	-3.4	-1.5	0.8
225.89 1.54 0.6	324.39	4FE	2.0	-3.3	-1.5	0.8
229.99 1.49 0.4	334.43	14FE	2.2	-3.3	-1.5	0.8
233.59 1.40 0.1	344.42	24FE	1.0	-3.3	-1.6	0.7
236.56 1.28 -0.1	354.35	6MR	0.6	-3.3	-1.6	0.7
238.75 1.11 -0.4	4.23	16MR	0.3	-3.4	-1.6	0.7
239.97 0.88 -0.7	14.05	26MR	-0.1	-3.4	-1.7	0.6
240.05 0.58 -1.0	23.82	5AP	-0.7	-3.4	-1.7	0.6
238.88 0.18 -1.4	33.53	15AP	-1.5	-3.4	-1.8	0.5
236.52 -0.29 -1.6	43.19	25AP	-1.5	-3.4	-1.8	0.4
233.31 -0.81 -1.9	52.82	5MY	-0.5	-3.5	-1.9	0.4
229.87 -1.32 -1.8	62.40	15MY	0.4	-3.5	-1.9	0.3
226.95 -1.78 -1.7	71.96	25MY	1.2	-3.6	-2.0	0.3
225.10 -2.15 -1.5	81.50	4JN	2.1	-3.7	-2.1	0.3
224.61 -2.42 -1.3	91.04	14JN	3.4	-3.7	-2.2	0.3
225.48 -2.60 -1.1	100.56	24JN	2.2	-3.8	-2.2	0.4
227.57 -2.71 -0.9	110.10	4JL	0.8	-3.9	-2.3	0.4
230.71 -2.77 -0.7	119.66	14JL	-0.5	-4.0	-2.3	0.5
234.71 -2.79 -0.5	129.24	24JL	-1.4	-4.1	-2.4	0.5
239.41 -2.77 -0.3	138.85	3AU	-1.4	-4.2	-2.4	0.6
244.69 -2.72 -0.2	148.51	13AU	-0.7	-4.2	-2.4	0.7
250.45 -2.65 -0.1	158.20	23AU	-0.2	-4.2	-2.4	0.7
256.59 -2.56 0.1	167.96	2SE	0.1	-4.0	-2.4	0.7
263.06 -2.45 0.2	177.77	12SE	0.2	-3.5	-2.4	0.8
269.79 -2.32 0.3	187.63	22SE	0.5	-3.4	-2.3	0.8
276.73 -2.18 0.4	197.55	2OC	1.4	-3.9	-2.3	0.8
283.85 -2.04 0.5	207.53	12OC	2.5	-4.3	-2.2	0.8
291.10 -1.88 0.6	217.55	22OC	0.0	-4.3	-2.1	0.8
298.46 -1.71 0.7	227.64	1NO	-0.6	-4.3	-2.1	0.8
305.90 -1.54 0.8	237.76	11NO	-0.6	-4.2	-2.0	0.7
313.38 -1.37 0.9	247.91	21NO	-0.6	-4.2	-1.9	0.7
320.89 -1.19 1.0	258.09	1DE	-0.7	-4.1	-1.9	0.7
328.41 -1.02 1.0	268.28	11DE	-0.8	-4.0	-1.8	0.7
335.92 -0.85 1.1	278.48	21DE	-0.8	-3.9	-1.8	0.7
343.41 -0.69 1.2	288.67	31DE	-0.8	-3.8	-1.7	0.8

1370

♂ LONG LAT MAG	⊙ LONG	16.00UT	☿	♀	♃	♄
				MAGNITUDES		
350.86 -0.53 1.3	298.85	10JA	-0.0	-3.7	-1.7	0.8
358.26 -0.38 1.4	308.99	20JA	2.3	-3.6	-1.6	0.8
5.61 -0.24 1.4	319.11	30JA	1.7	-3.6	-1.6	0.8
12.90 -0.10 1.5	329.17	9FE	0.7	-3.5	-1.6	0.8
20.12 0.03 1.6	339.19	19FE	0.4	-3.4	-1.6	0.8
27.27 0.15 1.6	349.16	1MR	0.2	-3.4	-1.6	0.8
34.35 0.26 1.7	359.07	11MR	-0.2	-3.4	-1.6	0.7
41.36 0.37 1.7	8.91	21MR	-0.7	-3.3	-1.6	0.7
48.31 0.47 1.8	18.71	31MR	-1.5	-3.3	-1.6	0.7
55.18 0.56 1.8	28.45	10AP	-1.4	-3.3	-1.6	0.6
61.99 0.64 1.8	38.14	20AP	-0.4	-3.3	-1.7	0.6
68.74 0.71 1.9	47.78	30AP	0.6	-3.3	-1.7	0.5
75.43 0.78 1.9	57.39	10MY	1.6	-3.3	-1.7	0.4
82.07 0.85 1.9	66.96	20MY	2.9	-3.3	-1.8	0.4
88.66 0.90 1.9	76.51	30MY	3.1	-3.4	-1.8	0.3
95.20 0.95 1.9	86.04	9JN	1.7	-3.4	-1.9	0.2
101.70 1.00 1.9	95.57	19JN	0.6	-3.4	-2.0	0.3
108.17 1.04 1.9	105.10	29JN	-0.5	-3.5	-2.0	0.3
114.61 1.08 1.9	114.65	9JL	-1.4	-3.5	-2.1	0.4
121.02 1.11 1.9	124.21	19JL	-1.4	-3.5	-2.2	0.4
127.42 1.13 2.0	133.81	29JL	-0.6	-3.4	-2.2	0.5
133.79 1.15 2.0	143.44	8AU	-0.1	-3.4	-2.3	0.6
140.15 1.17 2.0	153.11	18AU	0.2	-3.4	-2.4	0.6
146.50 1.18 2.0	162.84	28AU	0.4	-3.3	-2.4	0.7
152.85 1.19 2.0	172.61	7SE	0.7	-3.3	-2.4	0.7
159.18 1.20 2.0	182.45	17SE	1.8	-3.3	-2.5	0.7
165.52 1.20 2.0	192.34	27SE	2.2	-3.3	-2.5	0.8
171.85 1.19 2.0	202.29	7OC	-0.1	-3.4	-2.4	0.8
178.18 1.18 2.0	212.29	17OC	-0.7	-3.3	-2.4	0.8
184.51 1.16 1.9	222.35	27OC	-0.7	-3.4	-2.4	0.8
190.83 1.14 1.9	232.44	6NO	-0.7	-3.4	-2.3	0.8
197.15 1.12 1.8	242.58	16NO	-0.7	-3.4	-2.2	0.8
203.45 1.08 1.8	252.75	26NO	-0.6	-3.5	-2.2	0.7
209.75 1.04 1.7	262.94	6DE	-0.6	-3.5	-2.1	0.7
216.03 0.99 1.6	273.13	16DE	-0.6	-3.6	-2.0	0.7
222.29 0.92 1.5	283.33	26DE	0.1	-3.6	-2.0	0.7

1371

♂ LONG LAT MAG	⊙ LONG	16.00UT	☿	♀	♃	♄
228.51 0.85 1.4	293.51	5JA	2.6	-3.7	-1.9	0.8
234.70 0.76 1.3	303.68	15JA	1.3	-3.8	-1.8	0.8
240.85 0.66 1.2	313.81	25JA	0.5	-3.8	-1.8	0.8
246.93 0.53 1.1	323.90	4FE	0.2	-3.9	-1.7	0.8
252.94 0.39 0.9	333.94	14FE	0.1	-4.0	-1.7	0.8
258.85 0.22 0.8	343.94	24FE	-0.2	-4.1	-1.6	0.8
264.65 0.02 0.6	353.87	6MR	-0.7	-4.2	-1.6	0.8
270.30 -0.21 0.4	3.75	16MR	-1.5	-4.2	-1.6	0.8
275.76 -0.48 0.2	13.58	26MR	-1.4	-4.3	-1.6	0.8
280.98 -0.80 -0.0	23.34	5AP	-0.4	-4.1	-1.6	0.7
285.89 -1.17 -0.2	33.06	15AP	0.8	-3.7	-1.6	0.7
290.41 -1.60 -0.5	42.73	25AP	2.1	-3.0	-1.6	0.6
294.42 -2.11 -0.8	52.35	5MY	3.7	-3.3	-1.6	0.6
297.77 -2.69 -1.0	61.94	15MY	2.4	-3.9	-1.6	0.5
300.30 -3.35 -1.3	71.50	25MY	1.3	-4.2	-1.6	0.4
301.81 -4.07 -1.7	81.04	4JN	0.5	-4.2	-1.7	0.4
302.14 -4.84 -2.0	90.57	14JN	-0.5	-4.1	-1.7	0.3
301.21 -5.57 -2.2	100.10	24JN	-1.5	-4.1	-1.7	0.3
299.21 -6.17 -2.5	109.64	4JL	-1.4	-4.0	-1.8	0.3
296.58 -6.52 -2.6	119.20	14JL	-0.6	-3.9	-1.9	0.4
294.06 -6.57 -2.5	128.77	24JL	-0.1	-3.8	-1.9	0.4
292.32 -6.34 -2.3	138.38	3AU	0.3	-3.7	-2.0	0.5
291.79 -5.89 -2.0	148.04	13AU	0.6	-3.6	-2.0	0.5
292.62 -5.34 -1.8	157.74	23AU	1.0	-3.6	-2.1	0.6
294.71 -4.74 -1.5	167.48	2SE	2.3	-3.5	-2.2	0.6
297.86 -4.15 -1.2	177.29	12SE	2.0	-3.5	-2.2	0.7
301.88 -3.59 -1.0	187.15	22SE	-0.2	-3.5	-2.3	0.7
306.59 -3.06 -0.7	197.07	2OC	-0.9	-3.4	-2.3	0.8
311.83 -2.58 -0.5	207.04	12OC	-0.9	-3.4	-2.4	0.8
317.48 -2.14 -0.3	217.07	22OC	-0.9	-3.4	-2.4	0.8
323.44 -1.75 -0.1	227.15	1NO	-0.6	-3.4	-2.4	0.8
329.64 -1.39 0.1	237.27	11NO	-0.5	-3.4	-2.4	0.8
336.01 -1.07 0.3	247.42	21NO	-0.4	-3.4	-2.4	0.8
342.51 -0.78 0.4	257.59	1DE	-0.4	-3.4	-2.3	0.8
349.11 -0.52 0.6	267.79	11DE	0.2	-3.4	-2.3	0.8
355.76 -0.29 0.7	277.99	21DE	2.9	-3.4	-2.2	0.7
2.45 -0.09 0.9	288.18	31DE	1.0	-3.4	-2.1	0.7

♂			⊙	16.00UT	☿	♀	♃	♄
LONG	LAT	MAG	LONG	1372	MAGNITUDES			
9.16	0.10	1.0	298.35	10JA	0.2	-3.4	-2.0	0.8
15.88	0.26	1.1	308.50	20JA	0.1	-3.4	-2.0	0.8
22.58	0.40	1.3	318.61	30JA	-0.1	-3.4	-1.9	0.8
29.27	0.52	1.4	328.68	9FE	-0.3	-3.5	-1.8	0.9
35.93	0.63	1.5	338.70	19FE	-0.8	-3.5	-1.8	0.9
42.57	0.73	1.5	348.67	29FE	-1.5	-3.5	-1.7	0.9
49.17	0.81	1.6	358.58	10MR	-1.3	-3.4	-1.7	0.9
55.75	0.88	1.7	8.43	20MR	-0.4	-3.4	-1.6	0.9
62.28	0.95	1.8	18.23	30MR	1.0	-3.4	-1.6	0.8
68.79	1.00	1.8	27.97	9AP	2.7	-3.4	-1.5	0.8
75.27	1.04	1.9	37.66	19AP	3.1	-3.3	-1.5	0.8
81.71	1.08	1.9	47.30	29AP	1.8	-3.3	-1.5	0.7
88.13	1.11	2.0	56.91	9MY	1.0	-3.3	-1.5	0.7
94.53	1.13	2.0	66.49	19MY	0.3	-3.3	-1.5	0.6
100.90	1.15	2.0	76.04	29MY	-0.5	-3.3	-1.5	0.5
107.27	1.16	2.0	85.58	8JN	-1.5	-3.4	-1.5	0.5
113.62	1.17	2.0	95.11	18JN	-1.5	-3.4	-1.5	0.4
119.96	1.17	2.0	104.64	28JN	-0.6	-3.4	-1.5	0.3
126.30	1.17	2.0	114.19	8JL	0.0	-3.4	-1.6	0.3
132.65	1.16	2.0	123.75	18JL	0.4	-3.5	-1.6	0.4
138.99	1.15	2.0	133.34	28JL	0.8	-3.6	-1.7	0.4
145.36	1.13	2.0	142.97	7AU	1.3	-3.6	-1.7	0.5
151.73	1.11	2.0	152.64	17AU	2.8	-3.7	-1.8	0.5
158.12	1.08	2.0	162.37	27AU	1.7	-3.8	-1.8	0.6
164.54	1.05	2.0	172.14	6SE	-0.3	-3.9	-1.9	0.6
170.98	1.02	2.0	181.97	16SE	-1.0	-4.0	-1.9	0.7
177.45	0.98	2.0	191.86	26SE	-1.0	-4.1	-2.0	0.7
183.95	0.93	2.0	201.81	6OC	-0.9	-4.2	-2.1	0.8
190.48	0.88	2.0	211.81	16OC	-0.5	-4.3	-2.1	0.8
197.05	0.82	1.9	221.86	26OC	-0.4	-4.4	-2.2	0.8
203.65	0.76	1.9	231.96	5NO	-0.3	-4.3	-2.2	0.8
210.29	0.69	1.9	242.09	15NO	-0.3	-4.0	-2.3	0.8
216.97	0.62	1.8	252.26	25NO	0.4	-3.4	-2.3	0.8
223.68	0.53	1.8	262.44	5DE	3.1	-3.3	-2.3	0.8
230.44	0.44	1.7	272.64	15DE	0.7	-4.0	-2.3	0.8
237.23	0.34	1.7	282.84	25DE	0.0	-4.3	-2.2	0.8
				1373				
244.05	0.23	1.6	293.02	4JA	-0.1	-4.4	-2.2	0.8
250.91	0.11	1.5	303.18	14JA	-0.2	-4.3	-2.1	0.8
257.80	-0.03	1.4	313.32	24JA	-0.4	-4.2	-2.1	0.9
264.73	-0.17	1.4	323.41	3FE	-0.8	-4.1	-2.0	0.9
271.68	-0.33	1.3	333.46	13FE	-1.3	-4.0	-1.9	0.9
278.65	-0.50	1.2	343.45	23FE	-1.2	-3.9	-1.9	0.9
285.65	-0.68	1.1	353.40	5MR	-0.4	-3.8	-1.8	1.0
292.65	-0.88	0.9	3.28	15MR	1.2	-3.7	-1.7	1.0
299.66	-1.08	0.8	13.10	25MR	3.3	-3.6	-1.7	0.9
306.65	-1.30	0.7	22.87	4AP	2.3	-3.6	-1.6	0.9
313.62	-1.53	0.6	32.59	14AP	1.3	-3.5	-1.6	0.9
320.55	-1.77	0.4	42.26	24AP	0.8	-3.5	-1.5	0.9
327.40	-2.02	0.3	51.89	4MY	0.2	-3.4	-1.5	0.8
334.16	-2.28	0.2	61.48	14MY	-0.6	-3.4	-1.4	0.8
340.78	-2.54	0.0	71.04	24MY	-1.5	-3.3	-1.4	0.7
347.21	-2.80	-0.1	80.58	3JN	-1.5	-3.3	-1.4	0.7
353.39	-3.06	-0.3	90.11	13JN	-0.6	-3.3	-1.4	0.6
359.26	-3.31	-0.5	99.65	23JN	0.1	-3.3	-1.4	0.5
4.70	-3.56	-0.6	109.18	3JL	0.6	-3.3	-1.4	0.5
9.62	-3.80	-0.8	118.73	13JL	1.0	-3.3	-1.4	0.4
13.85	-4.01	-1.0	128.31	23JL	1.8	-3.3	-1.4	0.4
17.22	-4.19	-1.3	137.92	2AU	3.2	-3.4	-1.4	0.4
19.53	-4.31	-1.5	147.57	12AU	1.4	-3.4	-1.5	0.5
20.54	-4.36	-1.7	157.27	22AU	-0.3	-3.4	-1.5	0.5
20.13	-4.28	-2.0	167.01	1SE	-1.1	-3.5	-1.5	0.6
18.32	-4.04	-2.2	176.81	11SE	-1.2	-3.5	-1.6	0.7
15.45	-3.61	-2.3	186.67	21SE	-0.8	-3.5	-1.6	0.7
12.22	-3.01	-2.3	196.59	1OC	-0.4	-3.5	-1.7	0.7
9.45	-2.32	-2.0	206.56	11OC	-0.3	-3.4	-1.8	0.8
7.72	-1.64	-1.7	216.58	21OC	-0.2	-3.4	-1.8	0.8
7.32	-1.02	-1.4	226.66	31OC	-0.1	-3.4	-1.9	0.9
8.22	-0.49	-1.0	236.77	10NO	0.6	-3.4	-2.0	0.9
10.24	-0.06	-0.7	246.92	20NO	3.2	-3.3	-2.0	0.9
13.19	0.29	-0.4	257.10	30NO	0.5	-3.3	-2.1	0.9
16.87	0.56	-0.2	267.29	10DE	-0.1	-3.3	-2.1	0.9
21.10	0.78	0.1	277.49	20DE	-0.2	-3.3	-2.1	0.9
25.78	0.94	0.3	287.68	30DE	-0.3	-3.4	-2.2	0.9

♂			⊙	16.00UT	☿	♀	♃	♄
LONG	LAT	MAG	LONG	1374	MAGNITUDES			
30.79	1.08	0.5	297.86	9JA	-0.5	-3.4	-2.2	0.9
36.06	1.18	0.7	308.01	19JA	-0.9	-3.4	-2.1	0.9
41.53	1.26	0.9	318.12	29JA	-1.2	-3.4	-2.1	0.9
47.15	1.32	1.1	328.19	8FE	-1.1	-3.5	-2.1	1.0
52.89	1.36	1.2	338.22	18FE	-0.3	-3.5	-2.0	1.0
58.72	1.39	1.3	348.19	28FE	1.5	-3.5	-1.9	1.0
64.62	1.41	1.4	358.10	10MR	3.3	-3.6	-1.9	1.0
70.58	1.42	1.5	7.96	20MR	1.7	-3.7	-1.8	1.1
76.58	1.42	1.6	17.76	30MR	1.0	-3.7	-1.7	1.1
82.61	1.42	1.7	27.50	9AP	0.6	-3.8	-1.7	1.0
88.67	1.41	1.8	37.20	19AP	0.1	-3.9	-1.6	1.0
94.76	1.39	1.9	46.84	29AP	-0.6	-4.0	-1.5	1.0
100.86	1.37	1.9	56.45	9MY	-1.5	-4.1	-1.5	1.0
106.99	1.34	1.9	66.03	19MY	-1.5	-4.1	-1.4	0.9
113.14	1.31	2.0	75.58	29MY	-0.5	-4.2	-1.4	0.9
119.32	1.28	2.0	85.12	8JN	0.2	-4.1	-1.4	0.8
125.51	1.24	2.0	94.65	18JN	0.8	-3.9	-1.3	0.7
131.74	1.20	2.0	104.18	28JN	1.3	-3.4	-1.3	0.7
137.99	1.16	2.0	113.72	8JL	2.3	-3.0	-1.3	0.6
144.28	1.11	2.0	123.29	18JL	3.1	-3.7	-1.3	0.5
150.60	1.06	2.0	132.88	28JL	1.2	-4.1	-1.3	0.5
156.96	1.01	2.0	142.51	7AU	-0.4	-4.2	-1.3	0.5
163.36	0.95	2.0	152.18	17AU	-1.2	-4.2	-1.3	0.5
169.81	0.89	2.0	161.89	27AU	-1.3	-4.2	-1.3	0.6
176.31	0.82	1.9	171.67	6SE	-0.8	-4.1	-1.3	0.6
182.87	0.75	1.9	181.50	16SE	-0.4	-4.0	-1.4	0.7
189.47	0.68	1.9	191.38	26SE	-0.1	-3.9	-1.4	0.7
196.14	0.61	1.9	201.32	6OC	-0.0	-3.8	-1.5	0.8
202.87	0.52	1.9	211.32	16OC	0.1	-3.8	-1.5	0.8
209.67	0.44	1.8	221.37	26OC	0.9	-3.7	-1.6	0.9
216.52	0.35	1.8	231.47	5NO	3.0	-3.6	-1.6	0.9
223.45	0.26	1.8	241.60	15NO	0.3	-3.6	-1.7	1.0
230.43	0.16	1.8	251.76	25NO	-0.3	-3.5	-1.8	1.0
237.49	0.05	1.8	261.95	5DE	-0.4	-3.5	-1.8	1.0
244.61	-0.05	1.7	272.15	15DE	-0.4	-3.4	-1.9	1.0
251.79	-0.17	1.7	282.34	25DE	-0.6	-3.4	-1.9	1.0
				1375				
259.04	-0.28	1.6	292.53	4JA	-0.9	-3.4	-2.0	1.0
266.34	-0.41	1.6	302.69	14JA	-1.0	-3.4	-2.0	1.0
273.71	-0.53	1.5	312.83	24JA	-1.0	-3.3	-2.1	1.0
281.12	-0.66	1.5	322.92	3FE	-0.2	-3.3	-2.1	1.0
288.59	-0.78	1.4	332.97	13FE	1.8	-3.3	-2.1	1.1
296.09	-0.91	1.4	342.97	23FE	2.6	-3.3	-2.0	1.1
303.63	-1.04	1.3	352.92	5MR	1.3	-3.3	-2.0	1.1
311.19	-1.16	1.2	2.80	15MR	0.7	-3.4	-2.0	1.2
318.77	-1.28	1.2	12.63	25MR	0.4	-3.4	-1.9	1.2
326.35	-1.40	1.1	22.40	4AP	-0.0	-3.4	-1.8	1.2
333.91	-1.50	1.1	32.12	14AP	-0.6	-3.4	-1.8	1.2
341.46	-1.60	1.0	41.79	24AP	-1.5	-3.5	-1.7	1.2
348.95	-1.68	0.9	51.42	4MY	-1.5	-3.5	-1.6	1.2
356.39	-1.74	0.9	61.01	14MY	-0.5	-3.4	-1.6	1.1
3.75	-1.79	0.8	70.57	24MY	0.3	-3.4	-1.5	1.1
11.00	-1.83	0.7	80.12	3JN	1.0	-3.4	-1.5	1.0
18.13	-1.84	0.7	89.65	13JN	1.8	-3.4	-1.4	1.0
25.10	-1.84	0.6	99.18	23JN	3.0	-3.3	-1.4	0.9
31.89	-1.81	0.5	108.72	3JL	2.6	-3.3	-1.3	0.9
38.47	-1.77	0.4	118.27	13JL	1.0	-3.3	-1.3	0.8
44.79	-1.70	0.3	127.84	23JL	-0.4	-3.3	-1.3	0.7
50.80	-1.60	0.2	137.45	2AU	-1.3	-3.3	-1.2	0.7
56.44	-1.48	0.1	147.10	12AU	-1.4	-3.3	-1.2	0.6
61.64	-1.33	-0.0	156.79	22AU	-0.7	-3.4	-1.2	0.6
66.29	-1.14	-0.2	166.54	1SE	-0.3	-3.4	-1.2	0.7
70.28	-0.92	-0.3	176.33	11SE	-0.0	-3.4	-1.2	0.7
73.44	-0.64	-0.5	186.19	21SE	0.1	-3.5	-1.2	0.8
75.58	-0.32	-0.7	196.11	1OC	0.3	-3.5	-1.3	0.8
76.50	0.06	-0.9	206.07	11OC	1.2	-3.6	-1.3	0.9
76.02	0.50	-1.2	216.10	21OC	2.8	-3.7	-1.3	0.9
74.09	0.99	-1.4	226.17	31OC	0.1	-3.7	-1.4	1.0
70.93	1.48	-1.5	236.28	10NO	-0.5	-3.8	-1.4	1.0
67.13	1.92	-1.6	246.43	20NO	-0.5	-3.9	-1.5	1.1
63.45	2.26	-1.4	256.61	30NO	-0.5	-4.0	-1.5	1.1
60.66	2.47	-1.1	266.80	10DE	-0.7	-4.1	-1.6	1.1
59.13	2.58	-0.8	277.00	20DE	-0.8	-4.2	-1.7	1.1
58.95	2.60	-0.5	287.19	30DE	-0.9	-4.3	-1.7	1.1

♂ LONG	♂ LAT	♂ MAG	⊙ LONG	16.00UT 1376	☿ MAGNITUDES	♀	♃	♄
59.99	2.57	-0.2	297.37	9JA	-0.9	-4.4	-1.8	1.2
62.05	2.51	0.1	307.52	19JA	-0.1	-4.3	-1.9	1.2
64.93	2.43	0.3	317.64	29JA	2.1	-4.0	-1.9	1.2
68.46	2.34	0.5	327.71	8FE	2.1	-3.5	-2.0	1.1
72.49	2.25	0.8	337.73	18FE	0.9	-3.4	-2.0	1.2
76.93	2.16	0.9	347.71	28FE	0.5	-4.0	-2.0	1.2
81.67	2.07	1.1	357.62	9MR	0.2	-4.2	-2.0	1.3
86.66	1.98	1.2	7.48	19MR	-0.1	-4.3	-2.0	1.3
91.86	1.89	1.4	17.28	29MR	-0.7	-4.2	-2.0	1.3
97.21	1.80	1.5	27.03	8AP	-1.5	-4.1	-1.9	1.4
102.71	1.72	1.6	36.73	18AP	-1.5	-4.0	-1.9	1.4
108.32	1.63	1.7	46.38	28AP	-0.5	-3.9	-1.8	1.4
114.03	1.55	1.7	55.99	8MY	0.5	-3.8	-1.8	1.4
119.82	1.46	1.8	65.56	18MY	1.3	-3.7	-1.7	1.3
125.70	1.38	1.8	75.12	28MY	2.4	-3.6	-1.6	1.3
131.66	1.30	1.9	84.65	7JN	3.4	-3.6	-1.6	1.3
137.68	1.21	1.9	94.19	17JN	2.0	-3.5	-1.5	1.2
143.77	1.13	1.9	103.72	27JN	0.8	-3.5	-1.5	1.1
149.93	1.04	2.0	113.26	7JL	-0.4	-3.4	-1.4	1.1
156.17	0.96	2.0	122.82	17JL	-1.4	-3.4	-1.4	1.0
162.47	0.87	2.0	132.42	27JL	-1.4	-3.4	-1.3	0.9
168.84	0.78	2.0	142.04	6AU	-0.7	-3.4	-1.3	0.9
175.29	0.69	1.9	151.71	16AU	-0.2	-3.3	-1.3	0.8
181.81	0.60	1.9	161.43	26AU	0.1	-3.3	-1.2	0.8
188.41	0.51	1.9	171.19	5SE	0.3	-3.3	-1.2	0.8
195.09	0.41	1.9	181.02	15SE	0.5	-3.4	-1.2	0.8
201.85	0.32	1.8	190.90	25SE	1.5	-3.4	-1.2	0.9
208.69	0.22	1.8	200.84	5OC	2.5	-3.4	-1.2	0.9
215.62	0.12	1.7	210.84	15OC	0.0	-3.4	-1.2	1.0
222.63	0.02	1.7	220.89	25OC	-0.6	-3.4	-1.2	1.0
229.73	-0.08	1.6	230.98	4NO	-0.7	-3.4	-1.3	1.1
236.91	-0.18	1.6	241.11	14NO	-0.7	-3.5	-1.3	1.1
244.17	-0.29	1.6	251.27	24NO	-0.8	-3.5	-1.3	1.2
251.52	-0.39	1.6	261.45	4DE	-0.7	-3.5	-1.4	1.2
258.94	-0.49	1.6	271.65	14DE	-0.7	-3.5	-1.4	1.3
266.43	-0.59	1.6	281.85	24DE	-0.7	-3.4	-1.5	1.3
				1377				
273.98	-0.69	1.5	292.03	3JA	-0.0	-3.4	-1.6	1.3
281.60	-0.78	1.5	302.20	13JA	2.3	-3.4	-1.6	1.3
289.27	-0.87	1.5	312.34	23JA	1.6	-3.4	-1.7	1.3
296.99	-0.96	1.5	322.43	2FE	0.6	-3.3	-1.8	1.3
304.74	-1.03	1.4	332.49	12FE	0.3	-3.3	-1.8	1.3
312.51	-1.10	1.4	342.49	22FE	0.1	-3.3	-1.9	1.3
320.30	-1.16	1.4	352.43	4MR	-0.2	-3.3	-1.9	1.3
328.09	-1.21	1.4	2.32	14MR	-0.7	-3.4	-2.0	1.4
335.86	-1.24	1.4	12.15	24MR	-1.5	-3.4	-2.0	1.3
343.61	-1.27	1.3	21.93	3AP	-1.4	-3.4	-2.0	1.3
351.33	-1.28	1.3	31.65	13AP	-0.5	-3.4	-2.0	1.3
358.99	-1.27	1.3	41.32	23AP	0.6	-3.4	-2.0	1.3
6.60	-1.25	1.3	50.95	3MY	1.7	-3.5	-2.0	1.3
14.13	-1.22	1.3	60.55	13MY	3.2	-3.5	-1.9	1.3
21.57	-1.17	1.2	70.11	23MY	2.8	-3.6	-1.9	1.2
28.91	-1.11	1.2	79.66	2JN	1.6	-3.7	-1.8	1.2
36.15	-1.03	1.2	89.19	12JN	0.6	-3.7	-1.8	1.1
43.26	-0.94	1.2	98.72	22JN	-0.5	-3.8	-1.7	1.1
50.24	-0.83	1.1	108.26	2JL	-1.4	-3.9	-1.7	1.0
57.06	-0.72	1.1	117.81	12JL	-1.5	-4.0	-1.6	1.0
63.73	-0.58	1.1	127.38	22JL	-0.6	-4.1	-1.5	0.9
70.21	-0.44	1.0	136.99	1AU	-0.1	-4.2	-1.5	0.9
76.50	-0.28	1.0	146.64	11AU	0.2	-4.2	-1.4	0.8
82.55	-0.10	0.9	156.32	21AU	0.4	-4.2	-1.4	0.8
88.34	0.09	0.8	166.06	31AU	0.8	-3.9	-1.3	0.8
93.83	0.30	0.7	175.86	10SE	1.9	-3.4	-1.3	0.8
98.95	0.53	0.6	185.71	20SE	2.2	-3.4	-1.3	0.8
103.65	0.79	0.5	195.62	30SE	-0.1	-4.0	-1.3	0.9
107.82	1.08	0.3	205.59	10OC	-0.8	-4.3	-1.2	1.0
111.35	1.40	0.2	215.61	20OC	-0.8	-4.3	-1.2	1.1
114.09	1.77	0.0	225.68	30OC	-0.8	-4.3	-1.2	1.1
115.86	2.17	-0.2	235.79	9NO	-0.7	-4.2	-1.3	1.2
116.47	2.61	-0.4	245.94	19NO	-0.6	-4.1	-1.3	1.2
115.74	3.08	-0.7	256.11	29NO	-0.6	-4.0	-1.3	1.3
113.64	3.52	-0.9	266.31	9DE	-0.5	-3.9	-1.3	1.3
110.36	3.89	-1.1	276.50	19DE	0.1	-3.9	-1.4	1.3
106.44	4.12	-1.1	286.70	29DE	2.6	-3.8	-1.4	1.3

♂ LONG	♂ LAT	♂ MAG	⊙ LONG	16.00UT 1378	☿ MAGNITUDES	♀	♃	♄
102.61	4.18	-1.0	296.88	8JA	1.2	-3.7	-1.5	1.3
99.59	4.08	-0.7	307.03	18JA	0.4	-3.6	-1.5	1.3
97.77	3.88	-0.5	317.15	28JA	0.2	-3.6	-1.6	1.2
97.27	3.62	-0.2	327.23	7FE	0.0	-3.5	-1.7	1.2
97.99	3.34	0.1	337.25	17FE	-0.3	-3.4	-1.7	1.2
99.74	3.07	0.3	347.23	27FE	-0.7	-3.4	-1.8	1.1
102.34	2.81	0.5	357.15	9MR	-1.5	-3.4	-1.9	1.0
105.61	2.57	0.7	7.01	19MR	-1.4	-3.3	-1.9	1.1
109.42	2.36	0.9	16.82	29MR	-0.4	-3.3	-2.0	1.1
113.67	2.15	1.1	26.57	8AP	0.8	-3.3	-2.0	1.1
118.26	1.97	1.2	36.26	18AP	2.3	-3.3	-2.1	1.1
123.14	1.79	1.3	45.92	28AP	3.7	-3.3	-2.1	1.0
128.26	1.63	1.4	55.53	8MY	2.2	-3.3	-2.1	1.0
133.57	1.47	1.5	65.10	18MY	1.2	-3.3	-2.1	1.0
139.07	1.33	1.6	74.66	28MY	0.4	-3.4	-2.1	1.0
144.72	1.19	1.6	84.20	7JN	-0.5	-3.4	-2.1	0.9
150.51	1.05	1.7	93.73	17JN	-1.5	-3.4	-2.0	0.9
156.43	0.92	1.7	103.26	27JN	-1.5	-3.5	-2.0	0.9
162.48	0.80	1.7	112.80	7JL	-0.6	-3.5	-1.9	0.8
168.64	0.68	1.8	122.36	17JL	-0.0	-3.5	-1.8	0.8
174.91	0.56	1.8	131.95	27JL	0.3	-3.4	-1.8	0.7
181.29	0.44	1.8	141.57	6AU	0.6	-3.4	-1.7	0.7
187.78	0.32	1.8	151.23	16AU	1.1	-3.4	-1.7	0.6
194.38	0.21	1.8	160.95	26AU	2.4	-3.3	-1.6	0.6
201.09	0.10	1.8	170.72	5SE	1.9	-3.3	-1.5	0.5
207.90	-0.01	1.7	180.54	15SE	-0.2	-3.3	-1.5	0.5
214.81	-0.12	1.7	190.42	25SE	-0.9	-3.3	-1.4	0.5
221.84	-0.22	1.7	200.35	5OC	-1.0	-3.3	-1.4	0.6
228.96	-0.33	1.7	210.34	15OC	-0.9	-3.3	-1.4	0.6
236.18	-0.43	1.6	220.39	25OC	-0.6	-3.4	-1.4	0.7
243.51	-0.52	1.6	230.48	4NO	-0.5	-3.4	-1.3	0.8
250.92	-0.61	1.5	240.61	14NO	-0.4	-3.4	-1.3	0.8
258.42	-0.70	1.5	250.78	24NO	-0.4	-3.5	-1.3	0.9
266.01	-0.78	1.4	260.96	4DE	0.3	-3.5	-1.3	0.9
273.66	-0.85	1.4	271.15	14DE	2.9	-3.6	-1.3	1.0
281.39	-0.91	1.4	281.36	24DE	0.9	-3.6	-1.4	1.0
				1379				
289.16	-0.97	1.4	291.54	3JA	0.2	-3.7	-1.4	1.0
296.98	-1.02	1.4	301.71	13JA	0.0	-3.8	-1.4	1.0
304.84	-1.05	1.4	311.85	23JA	-0.1	-3.8	-1.5	1.0
312.71	-1.08	1.4	321.95	2FE	-0.3	-3.9	-1.5	1.0
320.59	-1.10	1.4	332.00	12FE	-0.8	-4.0	-1.6	1.0
328.47	-1.10	1.4	342.01	22FE	-1.4	-4.1	-1.6	0.9
336.32	-1.09	1.4	351.96	4MR	-1.3	-4.2	-1.7	0.9
344.15	-1.07	1.4	1.85	14MR	-0.4	-4.3	-1.8	0.8
351.94	-1.03	1.4	11.68	24MR	1.0	-4.3	-1.8	0.8
359.67	-0.99	1.4	21.46	3AP	2.9	-4.1	-1.9	0.8
7.34	-0.93	1.4	31.18	13AP	2.8	-3.7	-2.0	0.8
14.94	-0.87	1.5	40.86	23AP	1.6	-3.0	-2.0	0.8
22.46	-0.79	1.5	50.49	3MY	0.9	-3.4	-2.1	0.8
29.89	-0.70	1.5	60.09	13MY	0.3	-3.9	-2.2	0.8
37.23	-0.61	1.5	69.65	23MY	-0.5	-4.2	-2.2	0.8
44.47	-0.51	1.5	79.20	2JN	-1.5	-4.2	-2.2	0.8
51.60	-0.40	1.5	88.73	12JN	-1.5	-4.1	-2.2	0.7
58.64	-0.28	1.5	98.26	22JN	-0.6	-4.1	-2.2	0.7
65.55	-0.16	1.5	107.79	2JL	0.1	-4.0	-2.2	0.7
72.36	-0.03	1.5	117.34	12JL	0.5	-3.9	-2.2	0.6
79.05	0.10	1.5	126.92	22JL	0.8	-3.8	-2.1	0.6
85.60	0.24	1.5	136.52	1AU	1.5	-3.7	-2.1	0.5
92.03	0.38	1.5	146.16	11AU	2.9	-3.6	-2.0	0.5
98.31	0.53	1.4	155.85	21AU	1.7	-3.6	-1.9	0.4
104.43	0.69	1.4	165.59	31AU	-0.2	-3.5	-1.9	0.4
110.38	0.86	1.3	175.38	10SE	-1.0	-3.5	-1.8	0.3
116.14	1.03	1.3	185.23	20SE	-1.1	-3.5	-1.7	0.3
121.68	1.22	1.2	195.14	30SE	-0.9	-3.4	-1.7	0.2
126.96	1.41	1.1	205.10	10OC	-0.5	-3.4	-1.6	0.2
131.93	1.63	1.0	215.12	20OC	-0.3	-3.4	-1.6	0.3
136.53	1.86	0.9	225.19	30OC	-0.3	-3.4	-1.5	0.4
140.70	2.12	0.7	235.30	9NO	-0.2	-3.4	-1.5	0.5
144.31	2.40	0.6	245.45	19NO	0.5	-3.4	-1.5	0.5
147.25	2.71	0.4	255.62	29NO	3.2	-3.4	-1.5	0.6
149.36	3.05	0.2	265.81	9DE	0.7	-3.4	-1.4	0.6
150.45	3.41	-0.1	276.01	19DE	-0.0	-3.4	-1.4	0.7
150.35	3.78	-0.3	286.20	29DE	-0.1	-3.4	-1.4	0.7

♂ LONG	♂ LAT	♂ MAG	⊙ LONG	16.00UT 1380	☿	♀	♃	♄
					MAGNITUDES			
148.93	4.13	-0.6	296.38	8JA	-0.2	-3.4	-1.4	0.7
146.25	4.39	-0.8	306.54	18JA	-0.4	-3.4	-1.5	0.8
142.63	4.53	-1.0	316.66	28JA	-0.8	-3.4	-1.5	0.8
138.67	4.49	-0.9	326.74	7FE	-1.3	-3.5	-1.5	0.8
135.12	4.28	-0.8	336.77	17FE	-1.2	-3.5	-1.5	0.7
132.55	3.95	-0.5	346.74	27FE	-0.4	-3.5	-1.6	0.7
131.24	3.57	-0.3	356.67	8MR	1.3	-3.4	-1.6	0.7
131.22	3.17	-0.1	6.53	18MR	3.4	-3.4	-1.7	0.6
132.34	2.79	0.2	16.34	28MR	2.1	-3.4	-1.8	0.6
134.42	2.44	0.4	26.09	7AP	1.2	-3.4	-1.8	0.6
137.31	2.11	0.6	35.79	17AP	0.7	-3.3	-1.9	0.6
140.83	1.83	0.7	45.44	27AP	0.2	-3.3	-2.0	0.6
144.88	1.56	0.9	55.06	7MY	-0.6	-3.3	-2.0	0.6
149.36	1.32	1.0	64.64	17MY	-1.5	-3.3	-2.1	0.6
154.19	1.11	1.1	74.19	27MY	-1.5	-3.3	-2.2	0.6
159.33	0.90	1.2	83.73	6JN	-0.6	-3.4	-2.2	0.6
164.73	0.72	1.3	93.26	16JN	0.1	-3.4	-2.3	0.6
170.35	0.54	1.3	102.79	26JN	0.6	-3.4	-2.3	0.6
176.19	0.38	1.4	112.34	6JL	1.1	-3.5	-2.3	0.5
182.21	0.23	1.4	121.90	16JL	2.0	-3.5	-2.4	0.5
188.40	0.08	1.5	131.48	26JL	3.2	-3.6	-2.3	0.4
194.76	-0.06	1.5	141.11	5AU	1.4	-3.6	-2.3	0.4
201.27	-0.19	1.5	150.77	15AU	-0.3	-3.7	-2.3	0.3
207.92	-0.31	1.5	160.48	25AU	-1.1	-3.8	-2.2	0.3
214.72	-0.42	1.5	170.25	4SE	-1.3	-3.9	-2.2	0.2
221.65	-0.53	1.5	180.06	14SE	-0.8	-4.0	-2.1	0.2
228.71	-0.63	1.5	189.94	24SE	-0.4	-4.1	-2.1	0.1
235.89	-0.72	1.5	199.87	4OC	-0.2	-4.2	-2.0	0.0
243.19	-0.80	1.5	209.86	14OC	-0.1	-4.3	-1.9	-0.0
250.60	-0.88	1.5	219.90	24OC	-0.0	-4.4	-1.9	0.0
258.11	-0.94	1.5	229.99	3NO	0.7	-4.3	-1.8	0.1
265.71	-1.00	1.5	240.12	13NO	3.2	-4.0	-1.7	0.2
273.39	-1.04	1.4	250.28	23NO	0.4	-3.3	-1.7	0.3
281.15	-1.08	1.4	260.46	3DE	-0.2	-3.3	-1.6	0.3
288.96	-1.10	1.4	270.66	13DE	-0.3	-4.0	-1.6	0.4
296.82	-1.11	1.4	280.85	23DE	-0.3	-4.3	-1.6	0.4
				1381				
304.71	-1.11	1.4	291.04	2JA	-0.5	-4.4	-1.6	0.5
312.61	-1.09	1.4	301.21	12JA	-0.9	-4.3	-1.5	0.5
320.52	-1.07	1.4	311.35	22JA	-1.1	-4.2	-1.5	0.6
328.42	-1.03	1.3	321.46	1FE	-1.1	-4.1	-1.5	0.6
336.30	-0.98	1.3	331.51	11FE	-0.3	-4.0	-1.5	0.6
344.14	-0.93	1.3	341.52	21FE	1.5	-3.9	-1.5	0.6
351.93	-0.86	1.3	351.47	3MR	3.1	-3.8	-1.6	0.6
359.66	-0.78	1.4	1.37	13MR	1.5	-3.7	-1.6	0.5
7.33	-0.70	1.4	11.21	23MR	0.9	-3.6	-1.6	0.5
14.92	-0.61	1.4	20.99	2AP	0.5	-3.6	-1.7	0.5
22.43	-0.52	1.5	30.72	12AP	0.1	-3.5	-1.7	0.4
29.86	-0.42	1.5	40.40	22AP	-0.6	-3.4	-1.8	0.4
37.20	-0.32	1.6	50.03	2MY	-1.5	-3.4	-1.8	0.4
44.45	-0.21	1.6	59.63	12MY	-1.5	-3.4	-1.9	0.4
51.61	-0.10	1.6	69.19	22MY	-0.6	-3.3	-2.0	0.4
58.68	0.01	1.7	78.74	1JN	0.2	-3.3	-2.0	0.5
65.66	0.12	1.7	88.27	11JN	0.9	-3.3	-2.1	0.5
72.54	0.23	1.7	97.80	21JN	1.5	-3.3	-2.2	0.5
79.34	0.34	1.7	107.34	1JL	2.6	-3.3	-2.2	0.4
86.04	0.45	1.8	116.89	11JL	2.9	-3.3	-2.3	0.4
92.66	0.56	1.8	126.46	21JL	1.2	-3.3	-2.4	0.4
99.18	0.67	1.8	136.06	31JL	-0.3	-3.4	-2.4	0.4
105.62	0.79	1.8	145.70	10AU	-1.2	-3.4	-2.4	0.3
111.96	0.90	1.8	155.39	20AU	-1.4	-3.4	-2.4	0.3
118.20	1.01	1.7	165.12	30AU	-0.8	-3.5	-2.4	0.2
124.34	1.13	1.7	174.91	9SE	-0.3	-3.5	-2.4	0.1
130.37	1.25	1.7	184.76	19SE	-0.1	-3.5	-2.4	0.1
136.28	1.37	1.6	194.66	29SE	0.0	-3.5	-2.3	0.0
142.05	1.49	1.6	204.62	9OC	0.2	-3.4	-2.3	-0.0
147.67	1.62	1.5	214.63	19OC	1.0	-3.4	-2.2	-0.1
153.10	1.76	1.4	224.70	29OC	3.1	-3.4	-2.1	-0.2
158.33	1.90	1.3	234.80	8NO	0.3	-3.4	-2.1	-0.1
163.30	2.05	1.2	244.95	18NO	-0.4	-3.3	-2.0	-0.0
167.96	2.22	1.1	255.12	28NO	-0.4	-3.3	-1.9	0.0
172.26	2.39	0.9	265.31	8DE	-0.5	-3.3	-1.9	0.1
176.09	2.58	0.7	275.51	18DE	-0.6	-3.3	-1.8	0.2
179.35	2.78	0.5	285.70	28DE	-0.9	-3.4	-1.8	0.2

♂ LONG	♂ LAT	♂ MAG	⊙ LONG	16.00UT 1382	☿	♀	♃	♄
					MAGNITUDES			
181.91	2.99	0.3	295.88	7JA	-1.0	-3.4	-1.7	0.3
183.59	3.20	0.0	306.04	17JA	-0.9	-3.4	-1.7	0.4
184.21	3.41	-0.2	316.16	27JA	-0.2	-3.4	-1.6	0.4
183.62	3.59	-0.5	326.24	6FE	1.8	-3.5	-1.6	0.4
181.74	3.71	-0.8	336.28	16FE	2.4	-3.5	-1.6	0.4
178.71	3.71	-1.0	346.26	26FE	1.1	-3.5	-1.6	0.4
174.96	3.57	-1.1	356.18	8MR	0.6	-3.6	-1.6	0.4
171.17	3.29	-1.0	6.06	18MR	0.3	-3.7	-1.6	0.4
168.03	2.89	-0.9	15.87	28MR	-0.0	-3.7	-1.6	0.4
166.01	2.45	-0.7	25.62	7AP	-0.6	-3.8	-1.6	0.4
165.29	2.00	-0.4	35.33	17AP	-1.5	-3.9	-1.6	0.3
165.84	1.58	-0.2	44.98	27AP	-1.5	-4.0	-1.7	0.3
167.50	1.20	-0.0	54.60	7MY	-0.5	-4.1	-1.7	0.3
170.10	0.86	0.2	64.18	17MY	0.4	-4.1	-1.7	0.3
173.47	0.56	0.3	73.74	27MY	1.1	-4.2	-1.8	0.3
177.47	0.30	0.5	83.28	6JN	2.0	-4.1	-1.9	0.3
182.00	0.06	0.6	92.81	16JN	3.2	-3.9	-1.9	0.4
186.96	-0.15	0.7	102.34	26JN	2.4	-3.4	-2.0	0.4
192.30	-0.34	0.8	111.88	6JL	0.9	-3.0	-2.1	0.4
197.96	-0.50	0.9	121.44	16JL	-0.4	-3.7	-2.1	0.4
203.90	-0.65	0.9	131.02	26JL	-1.3	-4.1	-2.2	0.3
210.09	-0.79	1.0	140.64	5AU	-1.4	-4.2	-2.3	0.3
216.52	-0.90	1.0	150.30	15AU	-0.7	-4.2	-2.3	0.3
223.14	-1.01	1.1	160.01	25AU	-0.2	-4.2	-2.4	0.2
229.95	-1.09	1.1	169.77	4SE	0.0	-4.1	-2.4	0.2
236.93	-1.17	1.2	179.59	14SE	0.2	-4.0	-2.4	0.1
244.07	-1.23	1.2	189.46	24SE	0.4	-3.9	-2.5	0.1
251.35	-1.27	1.2	199.39	4OC	1.3	-3.8	-2.5	-0.0
258.76	-1.30	1.2	209.38	14OC	2.8	-3.8	-2.4	-0.1
266.28	-1.32	1.2	219.41	24OC	0.1	-3.7	-2.4	-0.1
273.89	-1.32	1.3	229.50	3NO	-0.5	-3.6	-2.4	-0.2
281.59	-1.31	1.3	239.63	13NO	-0.6	-3.6	-2.3	-0.3
289.35	-1.29	1.3	249.79	23NO	-0.6	-3.5	-2.2	-0.2
297.16	-1.25	1.3	259.97	3DE	-0.7	-3.5	-2.2	-0.1
305.01	-1.20	1.3	270.17	13DE	-0.8	-3.4	-2.1	-0.0
312.87	-1.14	1.4	280.36	23DE	-0.8	-3.4	-2.0	0.0
				1383				
320.74	-1.07	1.4	290.55	2JA	-0.8	-3.4	-1.9	0.1
328.59	-0.99	1.4	300.72	12JA	-0.1	-3.4	-1.9	0.2
336.41	-0.90	1.4	310.86	22JA	2.1	-3.3	-1.8	0.2
344.19	-0.81	1.4	320.96	1FE	1.9	-3.3	-1.8	0.3
351.93	-0.71	1.5	331.03	11FE	0.8	-3.3	-1.7	0.3
359.60	-0.61	1.5	341.03	21FE	0.4	-3.3	-1.7	0.3
7.20	-0.50	1.5	350.99	3MR	0.2	-3.3	-1.6	0.3
14.73	-0.39	1.5	0.89	13MR	-0.1	-3.4	-1.6	0.4
22.17	-0.28	1.6	10.73	23MR	-0.7	-3.4	-1.6	0.4
29.54	-0.18	1.6	20.51	2AP	-1.5	-3.4	-1.6	0.3
36.82	-0.07	1.6	30.24	12AP	-1.5	-3.4	-1.6	0.3
44.02	0.04	1.6	39.92	22AP	-0.5	-3.5	-1.6	0.3
51.13	0.14	1.6	49.56	2MY	0.5	-3.5	-1.6	0.3
58.16	0.24	1.6	59.16	12MY	1.5	-3.4	-1.6	0.2
65.11	0.34	1.7	68.73	22MY	2.6	-3.4	-1.6	0.2
71.98	0.44	1.7	78.27	1JN	3.3	-3.4	-1.6	0.2
78.77	0.53	1.8	87.81	11JN	1.9	-3.4	-1.7	0.3
85.50	0.62	1.8	97.34	21JN	0.7	-3.3	-1.7	0.3
92.15	0.71	1.9	106.87	1JL	-0.4	-3.3	-1.7	0.3
98.74	0.79	1.9	116.42	11JL	-1.4	-3.3	-1.8	0.3
105.27	0.87	1.9	125.99	21JL	-1.5	-3.3	-1.9	0.3
111.73	0.95	1.9	135.59	31JL	-0.7	-3.3	-1.9	0.3
118.14	1.03	1.9	145.23	10AU	-0.2	-3.3	-2.0	0.3
124.50	1.10	1.9	154.91	20AU	0.1	-3.4	-2.1	0.3
130.79	1.17	1.9	164.64	30AU	0.3	-3.4	-2.1	0.3
137.03	1.24	1.9	174.43	9SE	0.6	-3.4	-2.2	0.2
143.22	1.30	1.9	184.27	19SE	1.6	-3.5	-2.2	0.2
149.34	1.37	1.9	194.18	29SE	2.5	-3.5	-2.3	0.1
155.39	1.43	1.8	204.13	9OC	-0.0	-3.6	-2.3	0.0
161.38	1.49	1.8	214.14	19OC	-0.7	-3.7	-2.4	-0.0
167.28	1.55	1.7	224.21	29OC	-0.7	-3.8	-2.4	-0.1
173.08	1.61	1.7	234.31	8NO	-0.7	-3.8	-2.4	-0.2
178.78	1.67	1.6	244.45	18NO	-0.8	-3.9	-2.4	-0.3
184.35	1.73	1.5	254.63	28NO	-0.7	-4.0	-2.3	-0.3
189.76	1.79	1.4	264.82	8DE	-0.7	-4.1	-2.3	-0.2
195.00	1.84	1.2	275.01	18DE	-0.6	-4.3	-2.2	-0.2
200.01	1.90	1.1	285.21	28DE	-0.0	-4.3	-2.2	-0.1

♂ LONG	♂ LAT	♂ MAG	☉ LONG	16.00UT	☿	♀	♃	♄
				1384	MAGNITUDES			
204.74	1.95	0.9	295.39	7JA	2.4	-4.4	-2.1	-0.0
209.14	1.99	0.7	305.54	17JA	1.5	-4.3	-2.0	0.1
213.12	2.02	0.5	315.67	27JA	0.5	-4.0	-1.9	0.1
216.58	2.05	0.3	325.75	6FE	0.3	-3.5	-1.9	0.2
219.38	2.05	0.1	335.79	16FE	0.1	-3.5	-1.8	0.2
221.38	2.02	-0.2	345.77	26FE	-0.2	-4.0	-1.7	0.3
222.38	1.95	-0.5	355.70	7MR	-0.7	-4.2	-1.7	0.3
222.22	1.82	-0.8	5.57	17MR	-1.5	-4.3	-1.6	0.3
220.79	1.61	-1.1	15.39	27MR	-1.4	-4.2	-1.6	0.3
218.19	1.31	-1.4	25.14	6AP	-0.5	-4.1	-1.6	0.3
214.76	0.91	-1.6	34.85	16AP	0.7	-4.0	-1.5	0.3
211.14	0.46	-1.5	44.51	26AP	1.9	-3.9	-1.5	0.3
208.08	-0.01	-1.4	54.13	6MY	3.5	-3.8	-1.5	0.3
206.10	-0.45	-1.2	63.71	16MY	2.6	-3.7	-1.5	0.2
205.46	-0.82	-1.0	73.27	26MY	1.4	-3.6	-1.5	0.2
206.16	-1.13	-0.8	82.81	5JN	0.5	-3.6	-1.5	0.2
208.05	-1.38	-0.6	92.34	15JN	-0.4	-3.5	-1.5	0.2
210.97	-1.58	-0.4	101.88	25JN	-1.4	-3.5	-1.5	0.3
214.72	-1.74	-0.2	111.41	5JL	-1.5	-3.4	-1.5	0.3
219.17	-1.86	-0.1	120.97	15JL	-0.7	-3.4	-1.6	0.3
224.20	-1.94	0.0	130.56	25JL	-0.1	-3.4	-1.6	0.3
229.71	-2.00	0.2	140.17	4AU	0.3	-3.4	-1.6	0.3
235.61	-2.04	0.3	149.83	14AU	0.5	-3.3	-1.7	0.3
241.86	-2.05	0.4	159.54	24AU	0.9	-3.3	-1.8	0.3
248.39	-2.04	0.5	169.30	3SE	2.1	-3.3	-1.8	0.3
255.17	-2.01	0.5	179.11	13SE	2.2	-3.4	-1.9	0.3
262.16	-1.96	0.6	188.98	23SE	-0.1	-3.4	-1.9	0.2
269.33	-1.90	0.7	198.91	3OC	-0.8	-3.4	-2.0	0.2
276.64	-1.82	0.8	208.89	13OC	-0.9	-3.4	-2.1	0.1
284.07	-1.72	0.8	218.93	23OC	-0.9	-3.4	-2.1	0.0
291.60	-1.62	0.9	229.01	2NO	-0.7	-3.4	-2.2	-0.0
299.19	-1.50	1.0	239.13	12NO	-0.5	-3.5	-2.2	-0.1
306.83	-1.38	1.0	249.29	22NO	-0.5	-3.5	-2.3	-0.2
314.50	-1.24	1.1	259.47	2DE	-0.5	-3.5	-2.3	-0.2
322.18	-1.11	1.2	269.67	12DE	0.1	-3.5	-2.3	-0.3
329.85	-0.97	1.2	279.87	22DE	2.6	-3.4	-2.2	-0.2
				1385				
337.49	-0.83	1.3	290.05	1JA	1.1	-3.4	-2.2	-0.1
345.10	-0.69	1.3	300.22	11JA	0.3	-3.4	-2.2	-0.1
352.67	-0.55	1.4	310.37	21JA	0.1	-3.4	-2.1	-0.0
0.17	-0.42	1.5	320.47	31JA	-0.0	-3.3	-2.0	0.1
7.61	-0.29	1.5	330.53	10FE	-0.3	-3.3	-2.0	0.1
14.99	-0.16	1.6	340.55	20FE	-0.7	-3.3	-1.9	0.2
22.28	-0.04	1.6	350.50	2MR	-1.4	-3.3	-1.8	0.2
29.51	0.08	1.7	0.41	12MR	-1.4	-3.4	-1.8	0.3
36.65	0.19	1.7	10.25	22MR	-0.5	-3.4	-1.7	0.3
43.72	0.29	1.7	20.03	1AP	0.9	-3.4	-1.6	0.3
50.72	0.39	1.8	29.77	11AP	2.4	-3.4	-1.6	0.3
57.64	0.48	1.8	39.45	21AP	3.4	-3.4	-1.5	0.3
64.49	0.57	1.8	49.09	1MY	1.9	-3.5	-1.5	0.3
71.28	0.65	1.8	58.69	11MY	1.1	-3.5	-1.4	0.3
78.00	0.72	1.8	68.27	21MY	0.4	-3.6	-1.4	0.3
84.66	0.79	1.9	77.81	31MY	-0.5	-3.7	-1.4	0.3
91.28	0.86	1.9	87.35	10JN	-1.4	-3.7	-1.4	0.2
97.84	0.92	1.9	96.88	20JN	-1.5	-3.8	-1.4	0.2
104.35	0.97	1.9	106.41	30JN	-0.6	-3.9	-1.4	0.3
110.83	1.02	1.9	115.96	10JL	-0.0	-4.0	-1.4	0.3
117.27	1.06	2.0	125.53	20JL	0.4	-4.1	-1.4	0.4
123.68	1.10	2.0	135.13	30JL	0.7	-4.2	-1.4	0.4
130.07	1.14	2.0	144.77	9AU	1.2	-4.2	-1.4	0.4
136.42	1.17	2.0	154.45	19AU	2.6	-4.2	-1.5	0.4
142.76	1.20	2.0	164.17	29AU	1.9	-3.9	-1.5	0.4
149.07	1.22	2.0	173.96	8SE	-0.2	-3.4	-1.5	0.4
155.37	1.24	2.0	183.80	18SE	-1.0	-3.4	-1.6	0.4
161.65	1.26	2.0	193.70	28SE	-1.0	-4.0	-1.6	0.3
167.91	1.27	2.0	203.65	8OC	-0.9	-4.3	-1.7	0.3
174.15	1.28	1.9	213.66	18OC	-0.6	-4.3	-1.8	0.2
180.37	1.28	1.9	223.72	28OC	-0.4	-4.3	-1.8	0.2
186.56	1.28	1.9	233.82	7NO	-0.4	-4.2	-1.9	0.1
192.73	1.27	1.8	243.96	17NO	-0.3	-4.1	-2.0	0.0
198.86	1.26	1.7	254.13	27NO	0.3	-4.0	-2.0	-0.0
204.95	1.24	1.6	264.32	7DE	2.9	-3.9	-2.1	-0.1
210.99	1.21	1.6	274.52	17DE	0.8	-3.9	-2.1	-0.2
216.97	1.17	1.4	284.71	27DE	0.1	-3.8	-2.1	-0.2

♂ LONG	♂ LAT	♂ MAG	☉ LONG	16.00UT	☿	♀	♃	♄
				1386	MAGNITUDES			
222.88	1.12	1.3	294.90	6JA	-0.1	-3.7	-2.1	-0.1
228.69	1.05	1.2	305.06	16JA	-0.1	-3.6	-2.1	-0.1
234.40	0.97	1.1	315.18	26JA	-0.4	-3.6	-2.1	-0.0
239.97	0.87	0.9	325.27	5FE	-0.8	-3.5	-2.1	0.1
245.38	0.75	0.7	335.31	15FE	-1.4	-3.4	-2.0	0.1
250.58	0.60	0.5	345.30	25FE	-1.3	-3.4	-2.0	0.2
255.52	0.41	0.3	355.23	7MR	-0.4	-3.4	-1.9	0.2
260.14	0.18	0.1	5.10	17MR	1.1	-3.3	-1.8	0.3
264.35	-0.10	-0.1	14.92	27MR	3.1	-3.3	-1.8	0.3
268.04	-0.44	-0.4	24.68	6AP	2.5	-3.3	-1.7	0.4
271.07	-0.85	-0.7	34.39	16AP	1.4	-3.3	-1.6	0.4
273.28	-1.35	-1.0	44.05	26AP	0.8	-3.3	-1.6	0.4
274.48	-1.93	-1.3	53.67	6MY	0.3	-3.3	-1.5	0.4
274.50	-2.60	-1.7	63.25	16MY	-0.5	-3.3	-1.5	0.4
273.29	-3.31	-2.0	72.81	26MY	-1.5	-3.4	-1.4	0.4
270.99	-4.01	-2.2	82.35	5JN	-1.5	-3.4	-1.4	0.4
268.07	-4.58	-2.4	91.88	15JN	-0.6	-3.4	-1.3	0.4
265.22	-4.97	-2.3	101.41	25JN	0.1	-3.5	-1.3	0.3
263.15	-5.13	-2.1	110.95	5JL	0.5	-3.5	-1.3	0.3
262.32	-5.09	-1.9	120.51	15JL	0.9	-3.5	-1.3	0.4
262.87	-4.92	-1.6	130.09	25JL	1.6	-3.4	-1.3	0.4
264.73	-4.66	-1.4	139.71	4AU	3.1	-3.4	-1.3	0.5
267.73	-4.35	-1.2	149.36	14AU	1.6	-3.4	-1.3	0.5
271.66	-4.01	-0.9	159.07	24AU	-0.2	-3.3	-1.3	0.5
276.35	-3.67	-0.7	168.82	3SE	-1.1	-3.3	-1.3	0.5
281.64	-3.33	-0.5	178.63	13SE	-1.2	-3.3	-1.3	0.5
287.39	-2.99	-0.4	188.50	23SE	-0.9	-3.3	-1.4	0.5
293.52	-2.66	-0.2	198.43	3OC	-0.5	-3.3	-1.4	0.5
299.93	-2.34	-0.0	208.40	13OC	-0.3	-3.3	-1.5	0.4
306.56	-2.04	0.1	218.44	23OC	-0.2	-3.4	-1.5	0.4
313.36	-1.74	0.3	228.52	2NO	-0.1	-3.4	-1.6	0.4
320.27	-1.47	0.4	238.64	12NO	0.5	-3.4	-1.6	0.3
327.27	-1.21	0.6	248.80	22NO	3.2	-3.5	-1.7	0.2
334.33	-0.97	0.7	258.98	2DE	0.6	-3.5	-1.8	0.2
341.41	-0.74	0.8	269.17	12DE	-0.1	-3.6	-1.8	0.1
348.51	-0.53	0.9	279.37	22DE	-0.2	-3.6	-1.9	0.0
				1387				
355.59	-0.34	1.0	289.56	1JA	-0.3	-3.7	-1.9	-0.1
2.66	-0.16	1.1	299.73	11JA	-0.5	-3.8	-2.0	-0.1
9.70	-0.00	1.3	309.88	21JA	-0.8	-3.8	-2.0	-0.0
16.70	0.14	1.3	319.99	31JA	-1.2	-3.9	-2.0	0.0
23.65	0.28	1.4	330.05	10FE	-1.2	-4.0	-2.0	0.1
30.56	0.40	1.5	340.07	20FE	-0.4	-4.1	-2.0	0.2
37.42	0.51	1.6	350.03	2MR	1.3	-4.2	-2.0	0.2
44.23	0.60	1.7	359.93	12MR	3.4	-4.3	-2.0	0.3
50.98	0.69	1.7	9.78	22MR	1.9	-4.3	-1.9	0.3
57.69	0.77	1.8	19.57	1AP	1.1	-4.1	-1.9	0.4
64.34	0.84	1.8	29.30	11AP	0.6	-3.7	-1.8	0.4
70.95	0.90	1.9	38.99	21AP	0.1	-3.0	-1.7	0.5
77.51	0.95	1.9	48.63	1MY	-0.5	-3.4	-1.7	0.5
84.04	1.00	1.9	58.23	11MY	-1.5	-3.9	-1.6	0.5
90.53	1.03	2.0	67.80	21MY	-1.5	-4.2	-1.5	0.5
96.98	1.07	2.0	77.35	31MY	-0.6	-4.2	-1.5	0.5
103.41	1.09	2.0	86.88	10JN	0.2	-4.1	-1.4	0.5
109.82	1.12	2.0	96.42	20JN	0.7	-4.1	-1.4	0.5
116.21	1.13	2.0	105.95	30JN	1.2	-4.0	-1.3	0.5
122.59	1.14	2.0	115.49	10JL	2.2	-3.9	-1.3	0.5
128.96	1.15	2.0	125.06	20JL	3.2	-3.8	-1.3	0.5
135.32	1.15	2.0	134.66	30JL	1.4	-3.7	-1.2	0.5
141.69	1.15	2.0	144.29	9AU	-0.3	-3.6	-1.2	0.6
148.06	1.14	2.0	153.98	19AU	-1.2	-3.6	-1.2	0.6
154.43	1.13	2.0	163.70	29AU	-1.3	-3.5	-1.2	0.6
160.82	1.11	2.0	173.48	8SE	-0.8	-3.5	-1.2	0.7
167.22	1.09	2.0	183.32	18SE	-0.4	-3.5	-1.2	0.7
173.64	1.07	2.0	193.22	28SE	-0.2	-3.4	-1.2	0.7
180.07	1.04	2.0	203.17	8OC	-0.1	-3.4	-1.3	0.7
186.53	1.00	2.0	213.18	18OC	0.1	-3.4	-1.3	0.7
193.00	0.96	2.0	223.23	28OC	0.8	-3.4	-1.3	0.6
199.50	0.91	1.9	233.33	7NO	3.3	-3.4	-1.4	0.6
206.03	0.85	1.9	243.48	17NO	0.4	-3.4	-1.4	0.5
212.57	0.79	1.8	253.64	27NO	-0.3	-3.4	-1.5	0.5
219.13	0.72	1.8	263.83	7DE	-0.3	-3.4	-1.5	0.4
225.72	0.64	1.7	274.03	17DE	-0.4	-3.4	-1.6	0.3
232.32	0.55	1.6	284.22	27DE	-0.5	-3.4	-1.7	0.3

1388

♂ LONG	LAT	MAG	⊙ LONG	16.00UT	☿	♀	♃	♄
238.93	0.44	1.6	294.40	6JA	-0.9	-3.4	-1.7	0.2
245.57	0.33	1.5	304.57	16JA	-1.1	-3.4	-1.8	0.1
252.20	0.20	1.4	314.69	26JA	-1.0	-3.4	-1.9	0.1
258.85	0.06	1.3	324.78	5FE	-0.3	-3.5	-1.9	0.2
265.50	-0.10	1.1	334.82	15FE	1.6	-3.5	-2.0	0.2
272.15	-0.28	1.0	344.81	25FE	2.9	-3.5	-2.0	0.3
278.78	-0.48	0.9	354.74	6MR	1.4	-3.4	-2.0	0.3
285.39	-0.70	0.8	4.62	16MR	0.8	-3.4	-2.0	0.4
291.96	-0.94	0.6	14.44	26MR	0.4	-3.4	-2.0	0.4
298.48	-1.20	0.5	24.20	5AP	0.0	-3.4	-2.0	0.5
304.92	-1.49	0.3	33.92	15AP	-0.6	-3.3	-1.9	0.5
311.25	-1.81	0.1	43.58	25AP	-1.5	-3.3	-1.9	0.6
317.44	-2.15	-0.1	53.20	5MY	-1.5	-3.3	-1.8	0.6
323.42	-2.52	-0.2	62.79	15MY	-0.6	-3.3	-1.8	0.7
329.14	-2.92	-0.4	72.35	25MY	0.3	-3.3	-1.7	0.7
334.52	-3.35	-0.7	81.89	4JN	0.9	-3.4	-1.6	0.7
339.43	-3.80	-0.9	91.42	14JN	1.7	-3.4	-1.6	0.7
343.75	-4.27	-1.1	100.95	24JN	2.8	-3.4	-1.5	0.7
347.31	-4.75	-1.4	110.49	4JL	2.7	-3.5	-1.4	0.7
349.88	-5.22	-1.6	120.05	14JL	1.1	-3.5	-1.4	0.7
351.29	-5.64	-1.9	129.62	24JL	-0.3	-3.6	-1.3	0.7
351.35	-5.96	-2.1	139.24	3AU	-1.3	-3.6	-1.3	0.7
350.05	-6.11	-2.4	148.90	13AU	-1.4	-3.7	-1.3	0.8
347.70	-6.00	-2.5	158.59	23AU	-0.8	-3.8	-1.2	0.8
344.88	-5.60	-2.6	168.35	2SE	-0.3	-3.9	-1.2	0.8
342.38	-4.96	-2.3	178.16	12SE	-0.0	-4.0	-1.2	0.9
340.84	-4.18	-2.1	188.02	22SE	0.1	-4.1	-1.2	0.9
340.57	-3.39	-1.7	197.94	2OC	0.3	-4.2	-1.2	0.9
341.60	-2.65	-1.4	207.92	12OC	1.0	-4.3	-1.2	0.9
343.78	-1.99	-1.1	217.95	22OC	3.1	-4.4	-1.2	0.9
346.91	-1.42	-0.8	228.03	1NO	0.2	-4.3	-1.2	0.9
350.79	-0.94	-0.5	238.16	11NO	-0.4	-4.0	-1.3	0.8
355.55	-0.54	-0.3	248.31	21NO	-0.5	-3.3	-1.3	0.8
0.16	-0.21	-0.0	258.49	1DE	-0.5	-3.4	-1.4	0.8
5.42	0.07	0.2	268.68	11DE	-0.6	-4.0	-1.4	0.7
10.94	0.31	0.4	278.88	21DE	-0.9	-4.3	-1.5	0.6
16.66	0.50	0.6	289.07	31DE	-0.9	-4.4	-1.5	0.6

1389

♂ LONG	LAT	MAG	⊙ LONG	16.00UT	☿	♀	♃	♄
22.53	0.66	0.8	299.24	10JA	-0.9	-4.3	-1.6	0.5
28.51	0.79	0.9	309.39	20JA	-0.2	-4.2	-1.7	0.4
34.58	0.90	1.1	319.50	30JA	1.8	-4.1	-1.7	0.4
40.71	0.99	1.2	329.57	9FE	2.3	-4.0	-1.8	0.4
46.87	1.07	1.3	339.58	19FE	1.0	-3.9	-1.9	0.4
53.07	1.13	1.4	349.55	1MR	0.5	-3.8	-1.9	0.4
59.29	1.18	1.5	359.45	11MR	0.3	-3.7	-2.0	0.5
65.52	1.21	1.6	9.30	21MR	-0.1	-3.6	-2.0	0.6
71.75	1.24	1.7	19.10	31MR	-0.6	-3.6	-2.0	0.6
77.99	1.26	1.8	28.84	10AP	-1.4	-3.5	-2.0	0.7
84.22	1.27	1.8	38.52	20AP	-1.5	-3.4	-2.0	0.7
90.46	1.28	1.9	48.17	30AP	-0.6	-3.4	-2.0	0.8
96.70	1.28	1.9	57.77	10MY	0.4	-3.4	-2.0	0.8
102.94	1.27	2.0	67.34	20MY	1.2	-3.3	-1.9	0.8
109.18	1.26	2.0	76.90	30MY	2.2	-3.3	-1.9	0.9
115.43	1.24	2.0	86.43	9JN	3.4	-3.3	-1.8	0.9
121.69	1.22	2.0	95.96	19JN	2.2	-3.3	-1.8	0.9
127.96	1.20	2.0	105.49	29JN	0.9	-3.3	-1.7	1.0
134.25	1.17	2.0	115.04	9JL	-0.4	-3.3	-1.6	1.0
140.55	1.14	2.0	124.60	19JL	-1.3	-3.3	-1.6	1.0
146.89	1.10	2.0	134.20	29JL	-1.5	-3.4	-1.5	1.0
153.25	1.06	2.0	143.83	8AU	-0.7	-3.4	-1.5	1.0
159.64	1.02	2.0	153.50	18AU	-0.2	-3.4	-1.4	1.0
166.07	0.97	1.9	163.23	28AU	0.1	-3.5	-1.4	1.0
172.54	0.92	1.9	173.01	7SE	0.2	-3.5	-1.3	1.1
179.05	0.86	1.9	182.84	17SE	0.5	-3.5	-1.3	1.1
185.61	0.80	1.9	192.74	27SE	1.4	-3.5	-1.3	1.1
192.22	0.73	1.9	202.68	7OC	2.8	-3.4	-1.3	1.2
198.88	0.66	1.9	212.69	17OC	0.1	-3.4	-1.3	1.2
205.59	0.58	1.9	222.74	27OC	-0.6	-3.4	-1.3	1.2
212.35	0.50	1.9	232.84	6NO	-0.6	-3.4	-1.3	1.2
219.17	0.42	1.8	242.98	16NO	-0.6	-3.3	-1.3	1.2
226.05	0.32	1.8	253.15	26NO	-0.7	-3.3	-1.3	1.1
232.98	0.23	1.8	263.33	6DE	-0.7	-3.4	-1.3	1.1
239.97	0.12	1.7	273.53	16DE	-0.8	-3.3	-1.3	1.0
247.02	0.01	1.7	283.73	26DE	-0.7	-3.4	-1.4	1.0

1390

♂ LONG	LAT	MAG	⊙ LONG	16.00UT	☿	♀	♃	♄
254.12	-0.11	1.6	293.91	5JA	-0.1	-3.4	-1.4	0.9
261.27	-0.24	1.6	304.07	15JA	2.1	-3.4	-1.5	0.8
268.48	-0.37	1.5	314.20	25JA	1.8	-3.4	-1.5	0.8
275.73	-0.51	1.4	324.29	4FE	0.7	-3.5	-1.6	0.7
283.04	-0.65	1.4	334.34	14FE	0.4	-3.5	-1.7	0.6
290.37	-0.80	1.3	344.33	24FE	0.2	-3.6	-1.7	0.6
297.75	-0.95	1.2	354.27	6MR	-0.1	-3.6	-1.8	0.7
305.15	-1.10	1.1	4.15	16MR	-0.6	-3.7	-1.9	0.7
312.56	-1.26	1.1	13.97	26MR	-1.4	-3.7	-1.9	0.8
319.98	-1.41	1.0	23.74	5AP	-1.5	-3.8	-2.0	0.8
327.39	-1.56	0.9	33.45	15AP	-0.5	-3.9	-2.1	0.9
334.78	-1.70	0.8	43.12	25AP	0.5	-4.0	-2.1	0.9
342.12	-1.84	0.7	52.74	5MY	1.6	-4.1	-2.1	1.0
349.40	-1.96	0.6	62.33	15MY	2.9	-4.1	-2.1	1.0
356.59	-2.08	0.5	71.89	25MY	3.1	-4.2	-2.1	1.1
3.66	-2.18	0.4	81.43	4JN	1.7	-4.1	-2.1	1.1
10.58	-2.26	0.3	90.96	14JN	0.7	-3.9	-2.1	1.2
17.32	-2.32	0.2	100.50	24JN	-0.4	-3.3	-2.0	1.2
23.82	-2.36	0.1	110.03	4JL	-1.4	-3.0	-2.0	1.2
30.04	-2.38	0.0	119.59	14JL	-1.5	-3.7	-1.9	1.2
35.91	-2.37	-0.1	129.16	24JL	-0.7	-4.1	-1.8	1.2
41.35	-2.33	-0.3	138.77	3AU	-0.1	-4.2	-1.8	1.3
46.25	-2.26	-0.4	148.43	13AU	0.2	-4.2	-1.7	1.3
50.49	-2.15	-0.6	158.13	23AU	0.4	-4.2	-1.6	1.3
53.91	-2.00	-0.8	167.87	2SE	0.7	-4.1	-1.6	1.3
56.31	-1.79	-1.0	177.68	12SE	1.8	-4.0	-1.5	1.3
57.49	-1.52	-1.2	187.54	22SE	2.5	-3.9	-1.5	1.3
57.25	-1.16	-1.4	197.46	2OC	0.0	-3.8	-1.4	1.3
55.54	-0.72	-1.6	207.44	12OC	-0.7	-3.7	-1.4	1.3
52.58	-0.22	-1.8	217.46	22OC	-0.8	-3.7	-1.4	1.3
48.94	0.30	-1.9	227.54	1NO	-0.8	-3.6	-1.4	1.3
45.41	0.78	-1.7	237.66	11NO	-0.7	-3.6	-1.3	1.2
42.76	1.18	-1.4	247.81	21NO	-0.6	-3.5	-1.3	1.2
41.37	1.48	-1.1	257.99	1DE	-0.6	-3.5	-1.3	1.2
41.33	1.68	-0.7	268.19	11DE	-0.6	-3.4	-1.3	1.1
42.53	1.81	-0.4	278.38	21DE	0.0	-3.4	-1.4	1.1
44.75	1.89	-0.1	288.57	31DE	2.4	-3.4	-1.4	1.0

1391

♂ LONG	LAT	MAG	⊙ LONG	16.00UT	☿	♀	♃	♄
47.78	1.93	0.1	298.75	10JA	1.4	-3.4	-1.4	1.0
51.46	1.94	0.4	308.90	20JA	0.4	-3.3	-1.5	0.9
55.63	1.93	0.6	319.01	30JA	0.2	-3.3	-1.5	0.9
60.19	1.91	0.8	329.08	9FE	0.0	-3.3	-1.6	0.9
65.06	1.88	1.0	339.10	19FE	-0.2	-3.3	-1.6	0.8
70.16	1.84	1.1	349.07	1MR	-0.7	-3.3	-1.7	0.8
75.46	1.80	1.3	358.98	11MR	-1.4	-3.4	-1.7	0.8
80.90	1.75	1.4	8.83	21MR	-1.4	-3.4	-1.8	0.9
86.47	1.70	1.5	18.62	31MR	-0.5	-3.4	-1.9	1.0
92.13	1.65	1.6	28.37	10AP	0.7	-3.4	-1.9	1.1
97.89	1.60	1.7	38.05	20AP	2.1	-3.5	-2.0	1.1
103.71	1.54	1.8	47.70	30AP	3.8	-3.5	-2.1	1.2
109.59	1.48	1.8	57.31	10MY	2.3	-3.5	-2.1	1.2
115.54	1.42	1.9	66.88	20MY	1.3	-3.4	-2.2	1.3
121.54	1.36	1.9	76.43	30MY	0.5	-3.4	-2.2	1.3
127.59	1.30	2.0	85.97	9JN	-0.4	-3.4	-2.2	1.4
133.69	1.23	2.0	95.49	19JN	-1.4	-3.3	-2.2	1.4
139.84	1.16	2.0	105.03	29JN	-1.5	-3.3	-2.2	1.4
146.04	1.09	2.0	114.57	9JL	-0.7	-3.3	-2.2	1.4
152.30	1.02	2.0	124.14	19JL	-0.1	-3.3	-2.1	1.3
158.61	0.95	2.0	133.73	29JL	0.3	-3.3	-2.1	1.3
164.99	0.87	2.0	143.36	8AU	0.6	-3.3	-2.1	1.3
171.42	0.79	2.0	153.03	18AU	1.0	-3.4	-2.0	1.3
177.92	0.71	1.9	162.75	28AU	2.2	-3.4	-1.9	1.2
184.48	0.63	1.9	172.53	7SE	2.2	-3.4	-1.9	1.2
191.12	0.55	1.9	182.36	17SE	-0.1	-3.5	-1.8	1.1
197.83	0.46	1.8	192.25	27SE	-0.9	-3.5	-1.8	1.1
204.61	0.37	1.8	202.20	7OC	-0.9	-3.6	-1.7	1.1
211.46	0.27	1.7	212.20	17OC	-0.9	-3.7	-1.6	1.1
218.40	0.18	1.7	222.25	27OC	-0.6	-3.8	-1.6	1.1
225.41	0.08	1.7	232.35	6NO	-0.5	-3.8	-1.6	1.1
232.50	-0.02	1.7	242.48	16NO	-0.4	-3.9	-1.5	1.1
239.67	-0.13	1.7	252.65	26NO	-0.4	-4.0	-1.5	1.1
246.91	-0.23	1.7	262.84	6DE	0.2	-4.2	-1.5	1.0
254.23	-0.34	1.6	273.03	16DE	2.7	-4.3	-1.5	1.0
261.61	-0.45	1.6	283.23	26DE	1.0	-4.3	-1.4	1.0

♂ LONG	♂ LAT	♂ MAG	⊙ LONG	16.00UT	☿	♀	♃	♄
				1392	MAGNITUDES			
269.07	-0.56	1.6	293.42	5JA	0.2	-4.4	-1.5	0.9
276.58	-0.66	1.6	303.58	15JA	0.0	-4.3	-1.5	0.9
284.16	-0.77	1.5	313.71	25JA	-0.1	-4.0	-1.5	0.8
291.78	-0.87	1.5	323.81	4FE	-0.3	-3.4	-1.5	0.8
299.44	-0.97	1.4	333.85	14FE	-0.7	-3.5	-1.5	0.7
307.13	-1.06	1.4	343.85	24FE	-1.4	-4.0	-1.6	0.7
314.85	-1.15	1.4	353.79	5MR	-1.3	-4.2	-1.6	0.6
322.58	-1.22	1.3	3.67	15MR	-0.5	-4.3	-1.7	0.6
330.30	-1.29	1.3	13.50	25MR	0.9	-4.2	-1.7	0.7
338.02	-1.34	1.3	23.27	4AP	2.6	-4.1	-1.8	0.7
345.70	-1.38	1.2	32.98	14AP	3.1	-4.0	-1.9	0.8
353.35	-1.41	1.2	42.65	24AP	1.7	-3.9	-1.9	0.9
0.93	-1.42	1.2	52.28	4MY	1.0	-3.8	-2.0	1.0
8.45	-1.42	1.1	61.86	14MY	0.4	-3.7	-2.1	1.0
15.89	-1.40	1.1	71.43	24MY	-0.5	-3.6	-2.1	1.1
23.22	-1.36	1.0	80.97	3JN	-1.4	-3.6	-2.2	1.1
30.43	-1.31	1.0	90.50	13JN	-1.6	-3.5	-2.3	1.1
37.52	-1.24	1.0	100.04	23JN	-0.7	-3.5	-2.3	1.2
44.46	-1.16	0.9	109.57	3JL	0.0	-3.4	-2.3	1.2
51.22	-1.05	0.9	119.12	13JL	0.4	-3.4	-2.4	1.2
57.80	-0.93	0.8	128.70	23JL	0.8	-3.4	-2.4	1.2
64.16	-0.79	0.7	138.31	2AU	1.4	-3.4	-2.4	1.2
70.27	-0.63	0.7	147.96	12AU	2.7	-3.3	-2.3	1.1
76.09	-0.46	0.6	157.66	22AU	1.9	-3.3	-2.3	1.1
81.56	-0.25	0.5	167.41	1SE	-0.2	-3.3	-2.3	1.1
86.63	-0.03	0.4	177.21	11SE	-1.0	-3.4	-2.2	1.0
91.22	0.23	0.2	187.07	21SE	-1.1	-3.4	-2.1	1.0
95.20	0.52	0.1	196.98	1OC	-0.9	-3.4	-2.1	1.0
98.45	0.85	-0.1	206.95	11OC	-0.5	-3.4	-2.0	1.0
100.79	1.23	-0.3	216.98	21OC	-0.3	-3.4	-1.9	1.0
102.03	1.65	-0.5	227.05	31OC	-0.3	-3.5	-1.9	1.0
101.98	2.11	-0.7	237.16	10NO	-0.2	-3.5	-1.8	1.0
100.52	2.59	-0.9	247.32	20NO	0.4	-3.5	-1.7	1.0
97.74	3.05	-1.1	257.49	30NO	3.0	-3.5	-1.7	1.0
94.03	3.41	-1.3	267.68	10DE	0.8	-3.5	-1.7	0.9
90.08	3.64	-1.2	277.88	20DE	0.0	-3.4	-1.6	0.9
86.69	3.70	-1.0	288.07	30DE	-0.1	-3.4	-1.6	0.9
				1393				
84.39	3.64	-0.7	298.25	9JA	-0.2	-3.4	-1.6	0.8
83.41	3.49	-0.4	308.40	19JA	-0.4	-3.4	-1.5	0.8
83.72	3.29	-0.1	318.51	29JA	-0.8	-3.3	-1.5	0.7
85.14	3.09	0.1	328.59	8FE	-1.3	-3.3	-1.5	0.7
87.49	2.88	0.4	338.61	18FE	-1.2	-3.3	-1.6	0.6
90.56	2.68	0.6	348.58	28FE	-0.5	-3.3	-1.6	0.6
94.22	2.50	0.8	358.49	10MR	1.1	-3.4	-1.6	0.5
98.34	2.32	1.0	8.35	20MR	3.2	-3.4	-1.6	0.5
102.83	2.16	1.1	18.15	30MR	2.3	-3.4	-1.7	0.5
107.61	2.01	1.3	27.89	9AP	1.3	-3.4	-1.7	0.6
112.63	1.86	1.4	37.59	19AP	0.7	-3.5	-1.7	0.6
117.86	1.73	1.5	47.23	29AP	0.2	-3.5	-1.8	0.7
123.26	1.60	1.6	56.84	9MY	-0.5	-3.5	-1.9	0.8
128.81	1.47	1.6	66.42	19MY	-1.4	-3.6	-1.9	0.8
134.48	1.35	1.7	75.97	29MY	-1.6	-3.7	-2.0	0.9
140.28	1.24	1.8	85.51	8JN	-0.6	-3.7	-2.1	0.9
146.18	1.12	1.8	95.04	18JN	0.1	-3.8	-2.1	1.0
152.19	1.01	1.8	104.57	28JN	0.6	-3.9	-2.2	1.0
158.30	0.90	1.8	114.11	8JL	1.0	-4.0	-2.3	1.0
164.50	0.79	1.9	123.68	18JL	1.8	-4.1	-2.3	1.0
170.79	0.68	1.9	133.27	28JL	3.2	-4.2	-2.4	1.0
177.18	0.58	1.9	142.90	7AU	1.6	-4.2	-2.4	1.0
183.66	0.47	1.9	152.57	17AU	-0.2	-4.2	-2.4	1.0
190.24	0.36	1.8	162.29	27AU	-1.1	-3.9	-2.5	1.0
196.91	0.26	1.8	172.06	6SE	-1.2	-3.4	-2.5	1.0
203.68	0.15	1.8	181.89	16SE	-0.9	-3.4	-2.4	0.9
210.54	0.05	1.8	191.77	26SE	-0.4	-4.0	-2.4	0.9
217.50	-0.06	1.7	201.72	6OC	-0.2	-4.3	-2.3	0.8
224.56	-0.16	1.7	211.72	16OC	-0.1	-4.3	-2.3	0.9
231.71	-0.26	1.6	221.76	26OC	-0.0	-4.3	-2.2	0.9
238.95	-0.36	1.6	231.86	5NO	0.6	-4.2	-2.1	0.9
246.29	-0.46	1.5	241.99	15NO	3.3	-4.1	-2.1	0.9
253.71	-0.55	1.5	252.15	25NO	0.6	-4.0	-2.0	0.9
261.21	-0.64	1.5	262.34	5DE	-0.2	-3.9	-1.9	0.9
268.79	-0.73	1.5	272.54	15DE	-0.3	-3.8	-1.9	0.9
276.44	-0.81	1.4	282.73	25DE	-0.3	-3.8	-1.8	0.8

♂ LONG	♂ LAT	♂ MAG	⊙ LONG	16.00UT	☿	♀	♃	♄
				1394	MAGNITUDES			
284.15	-0.88	1.4	292.92	4JA	-0.5	-3.7	-1.8	0.8
291.90	-0.95	1.4	303.08	14JA	-0.8	-3.6	-1.7	0.8
299.70	-1.01	1.4	313.22	24JA	-1.2	-3.6	-1.7	0.7
307.53	-1.06	1.4	323.31	3FE	-1.1	-3.5	-1.6	0.7
315.38	-1.10	1.4	333.36	13FE	-0.4	-3.4	-1.6	0.6
323.23	-1.12	1.4	343.36	23FE	1.4	-3.4	-1.6	0.6
331.08	-1.14	1.4	353.31	5MR	3.2	-3.4	-1.6	0.5
338.91	-1.14	1.4	3.19	15MR	1.7	-3.3	-1.6	0.5
346.71	-1.13	1.4	13.02	25MR	0.9	-3.3	-1.6	0.4
354.46	-1.11	1.4	22.80	4AP	0.5	-3.3	-1.6	0.4
2.16	-1.07	1.4	32.51	14AP	0.1	-3.3	-1.6	0.4
9.80	-1.02	1.4	42.18	24AP	-0.5	-3.3	-1.6	0.5
17.36	-0.96	1.4	51.81	4MY	-1.4	-3.3	-1.7	0.5
24.83	-0.89	1.4	61.41	14MY	-1.6	-3.3	-1.7	0.6
32.22	-0.81	1.4	70.97	24MY	-0.6	-3.4	-1.8	0.7
39.50	-0.72	1.4	80.51	3JN	0.2	-3.4	-1.8	0.7
46.68	-0.62	1.4	90.04	13JN	0.8	-3.4	-1.9	0.8
53.74	-0.50	1.4	99.57	23JN	1.4	-3.5	-1.9	0.8
60.69	-0.38	1.4	109.11	3JL	2.4	-3.5	-2.0	0.9
67.51	-0.25	1.4	118.66	13JL	3.1	-3.5	-2.1	0.9
74.19	-0.12	1.4	128.23	23JL	1.3	-3.4	-2.1	0.9
80.72	0.03	1.3	137.85	2AU	-0.3	-3.4	-2.2	0.9
87.10	0.18	1.3	147.49	12AU	-1.2	-3.4	-2.3	0.9
93.31	0.35	1.2	157.18	22AU	-1.4	-3.3	-2.3	0.9
99.33	0.52	1.2	166.93	1SE	-0.8	-3.3	-2.4	0.9
105.12	0.70	1.1	176.73	11SE	-0.4	-3.3	-2.4	0.9
110.68	0.90	1.0	186.58	21SE	-0.1	-3.3	-2.4	0.9
115.95	1.12	1.0	196.50	1OC	-0.0	-3.3	-2.5	0.8
120.87	1.35	0.8	206.46	11OC	0.1	-3.3	-2.5	0.8
125.39	1.61	0.7	216.49	21OC	0.8	-3.4	-2.4	0.8
129.41	1.89	0.6	226.56	31OC	3.4	-3.4	-2.4	0.8
132.82	2.21	0.4	236.67	10NO	0.4	-3.4	-2.3	0.8
135.47	2.55	0.2	246.82	20NO	-0.3	-3.5	-2.3	0.8
137.19	2.93	-0.0	257.00	30NO	-0.4	-3.5	-2.2	0.8
137.80	3.34	-0.2	267.19	10DE	-0.4	-3.6	-2.1	0.8
137.12	3.75	-0.5	277.39	20DE	-0.6	-3.6	-2.1	0.8
135.10	4.13	-0.7	287.58	30DE	-0.9	-3.7	-2.0	0.8
				1395				
131.91	4.41	-0.9	297.76	9JA	-1.0	-3.8	-1.9	0.8
128.03	4.54	-1.0	307.91	19JA	-1.0	-3.8	-1.9	0.8
124.15	4.49	-0.9	318.03	29JA	-0.3	-3.9	-1.8	0.7
121.00	4.28	-0.7	328.10	8FE	1.6	-4.0	-1.7	0.7
119.01	3.97	-0.4	338.12	18FE	2.7	-4.1	-1.7	0.6
118.31	3.62	-0.2	348.10	28FE	1.2	-4.2	-1.7	0.6
118.85	3.26	0.1	358.01	10MR	0.7	-4.3	-1.6	0.5
120.44	2.92	0.3	7.87	20MR	0.4	-4.3	-1.6	0.5
122.91	2.60	0.5	17.67	30MR	0.0	-4.1	-1.6	0.4
126.09	2.31	0.7	27.42	9AP	-0.6	-3.7	-1.6	0.4
129.84	2.05	0.9	37.12	19AP	-1.4	-3.0	-1.5	0.3
134.05	1.81	1.0	46.77	29AP	-1.6	-3.4	-1.6	0.4
138.64	1.59	1.1	56.38	9MY	-0.6	-4.0	-1.6	0.4
143.54	1.39	1.2	65.95	19MY	0.3	-4.2	-1.6	0.5
148.71	1.20	1.3	75.51	29MY	1.0	-4.2	-1.6	0.5
154.11	1.02	1.4	85.04	8JN	1.8	-4.1	-1.6	0.6
159.71	0.86	1.5	94.58	18JN	3.1	-4.1	-1.7	0.7
165.49	0.70	1.5	104.11	28JN	2.6	-4.0	-1.7	0.7
171.44	0.55	1.6	113.65	8JL	1.0	-3.9	-1.7	0.8
177.54	0.41	1.6	123.21	18JL	-0.3	-3.8	-1.8	0.8
183.80	0.27	1.6	132.80	28JL	-1.3	-3.7	-1.9	0.8
190.18	0.14	1.6	142.43	7AU	-1.5	-3.6	-1.9	0.9
196.70	0.02	1.6	152.10	17AU	-0.8	-3.6	-2.0	0.9
203.35	-0.10	1.6	161.82	27AU	-0.3	-3.5	-2.1	0.9
210.13	-0.22	1.6	171.58	6SE	-0.0	-3.5	-2.1	0.9
217.03	-0.33	1.6	181.41	16SE	0.1	-3.5	-2.2	0.8
224.05	-0.43	1.6	191.30	26SE	0.3	-3.4	-2.2	0.8
231.18	-0.53	1.6	201.23	6OC	1.1	-3.4	-2.3	0.8
238.43	-0.63	1.6	211.23	16OC	3.1	-3.4	-2.3	0.8
245.78	-0.71	1.5	221.28	26OC	0.2	-3.4	-2.4	0.7
253.23	-0.79	1.5	231.37	5NO	-0.5	-3.4	-2.4	0.7
260.77	-0.86	1.5	241.50	15NO	-0.6	-3.4	-2.4	0.7
268.40	-0.92	1.5	251.67	25NO	-0.6	-3.4	-2.4	0.8
276.11	-0.98	1.4	261.85	5DE	-0.7	-3.4	-2.3	0.8
283.87	-1.02	1.4	272.05	15DE	-0.8	-3.4	-2.3	0.8
291.70	-1.05	1.4	282.24	25DE	-0.8	-3.4	-2.2	0.8

1396

♂ LONG	LAT	MAG	☉ LONG	16.00UT 1396	☿	♀	♃	♄
299.56	-1.07	1.4	292.43	4JA	-0.8	-3.4	-2.1	0.8
307.45	-1.08	1.3	302.59	14JA	-0.2	-3.4	-2.1	0.8
315.35	-1.08	1.3	312.73	24JA	1.9	-3.5	-2.0	0.7
323.25	-1.07	1.3	322.82	3FE	2.1	-3.5	-1.9	0.7
331.14	-1.04	1.3	332.88	13FE	0.9	-3.5	-1.9	0.7
339.01	-1.01	1.3	342.87	23FE	0.5	-3.5	-1.8	0.6
346.83	-0.96	1.4	352.82	4MR	0.2	-3.4	-1.7	0.6
354.61	-0.90	1.4	2.71	14MR	-0.1	-3.4	-1.7	0.5
2.33	-0.84	1.4	12.54	24MR	-0.6	-3.4	-1.6	0.5
9.98	-0.76	1.4	22.31	3AP	-1.4	-3.4	-1.6	0.4
17.56	-0.68	1.5	32.04	13AP	-1.5	-3.3	-1.5	0.3
25.06	-0.59	1.5	41.71	23AP	-0.6	-3.3	-1.5	0.3
32.47	-0.49	1.5	51.34	3MY	0.4	-3.3	-1.5	0.3
39.79	-0.39	1.6	60.94	13MY	1.3	-3.3	-1.5	0.3
47.02	-0.28	1.6	70.50	23MY	2.5	-3.3	-1.5	0.4
54.15	-0.17	1.6	80.04	2JN	3.4	-3.4	-1.5	0.5
61.18	-0.06	1.6	89.58	12JN	2.0	-3.4	-1.5	0.5
68.12	0.06	1.7	99.11	22JN	0.8	-3.4	-1.5	0.6
74.96	0.17	1.7	108.64	2JL	-0.3	-3.5	-1.5	0.6
81.70	0.29	1.7	118.20	12JL	-1.3	-3.5	-1.5	0.7
88.34	0.42	1.7	127.77	22JL	-1.5	-3.6	-1.6	0.7
94.88	0.54	1.7	137.38	1AU	-0.7	-3.6	-1.6	0.8
101.30	0.67	1.7	147.03	11AU	-0.2	-3.7	-1.6	0.8
107.62	0.79	1.6	156.71	21AU	0.1	-3.8	-1.7	0.8
113.82	0.93	1.6	166.46	31AU	0.3	-3.9	-1.7	0.8
119.89	1.06	1.6	176.25	10SE	0.6	-4.0	-1.8	0.8
125.82	1.20	1.5	186.10	20SE	1.5	-4.1	-1.9	0.8
131.60	1.35	1.5	196.01	30SE	2.8	-4.2	-1.9	0.8
137.20	1.50	1.4	205.98	10OC	0.1	-4.3	-2.0	0.8
142.59	1.66	1.3	216.00	20OC	-0.6	-4.4	-2.1	0.7
147.74	1.83	1.2	226.07	30OC	-0.7	-4.3	-2.1	0.7
152.61	2.02	1.1	236.19	9NO	-0.7	-4.0	-2.2	0.7
157.12	2.21	1.0	246.33	19NO	-0.8	-3.2	-2.2	0.7
161.21	2.43	0.8	256.50	29NO	-0.7	-3.4	-2.2	0.7
164.78	2.67	0.6	266.70	9DE	-0.7	-4.1	-2.3	0.8
167.69	2.92	0.4	276.89	19DE	-0.7	-4.3	-2.2	0.8
169.80	3.19	0.2	287.09	29DE	-0.1	-4.4	-2.2	0.8

1397

♂ LONG	LAT	MAG	☉ LONG	16.00UT 1397	☿	♀	♃	♄
170.93	3.47	-0.0	297.27	8JA	2.1	-4.3	-2.2	0.8
170.90	3.75	-0.3	307.42	18JA	1.7	-4.2	-2.1	0.8
169.58	3.99	-0.6	317.54	28JA	0.6	-4.1	-2.1	0.8
167.00	4.14	-0.8	327.61	7FE	0.3	-4.0	-2.0	0.7
163.47	4.16	-1.0	337.64	17FE	0.1	-3.9	-1.9	0.7
159.56	4.02	-1.0	347.62	27FE	-0.2	-3.8	-1.9	0.7
156.00	3.73	-0.8	357.54	9MR	-0.6	-3.7	-1.8	0.6
153.38	3.33	-0.6	7.39	19MR	-1.4	-3.6	-1.7	0.6
152.00	2.90	-0.4	17.20	29MR	-1.5	-3.6	-1.7	0.5
151.93	2.47	-0.2	26.95	8AP	-0.6	-3.5	-1.6	0.4
153.02	2.06	0.0	36.65	18AP	0.6	-3.4	-1.6	0.4
155.11	1.69	0.2	46.30	28AP	1.7	-3.4	-1.5	0.3
158.04	1.36	0.4	55.91	8MY	3.2	-3.4	-1.5	0.3
161.63	1.07	0.6	65.49	18MY	2.8	-3.3	-1.4	0.3
165.78	0.80	0.7	75.05	28MY	1.6	-3.3	-1.4	0.3
170.40	0.56	0.8	84.59	7JN	0.6	-3.3	-1.4	0.4
175.40	0.35	0.9	94.11	17JN	-0.4	-3.4	-1.4	0.4
180.73	0.15	1.0	103.65	27JN	-1.4	-3.3	-1.4	0.5
186.34	-0.03	1.1	113.19	7JL	-1.5	-3.3	-1.3	0.6
192.21	-0.19	1.1	122.75	17JL	-0.7	-3.3	-1.4	0.6
198.31	-0.34	1.2	132.34	27JL	-0.1	-3.4	-1.4	0.7
204.61	-0.48	1.2	141.97	6AU	0.2	-3.4	-1.4	0.7
211.11	-0.60	1.3	151.63	16AU	0.5	-3.4	-1.4	0.7
217.78	-0.72	1.3	161.35	26AU	0.8	-3.4	-1.4	0.8
224.62	-0.82	1.3	171.11	5SE	1.9	-3.5	-1.5	0.8
231.61	-0.91	1.3	180.93	15SE	2.4	-3.5	-1.5	0.8
238.74	-0.99	1.3	190.81	25SE	0.0	-3.5	-1.6	0.8
246.01	-1.06	1.3	200.75	5OC	-0.8	-3.4	-1.6	0.8
253.40	-1.11	1.4	210.74	15OC	-0.9	-3.4	-1.7	0.8
260.90	-1.16	1.4	220.79	25OC	-0.8	-3.4	-1.8	0.7
268.50	-1.19	1.4	230.88	4NO	-0.7	-3.3	-1.8	0.7
276.18	-1.20	1.4	241.00	14NO	-0.5	-3.3	-1.9	0.7
283.94	-1.21	1.4	251.17	24NO	-0.5	-3.3	-2.0	0.7
291.75	-1.20	1.4	261.35	4DE	-0.5	-3.3	-2.0	0.7
299.61	-1.18	1.4	271.54	14DE	0.0	-3.3	-2.1	0.7
307.49	-1.14	1.4	281.74	24DE	2.4	-3.4	-2.1	0.8

1398

♂ LONG	LAT	MAG	☉ LONG	16.00UT 1398	☿	♀	♃	♄
315.38	-1.10	1.4	291.93	3JA	1.3	-3.4	-2.1	0.8
323.27	-1.05	1.4	302.10	13JA	0.3	-3.4	-2.1	0.8
331.15	-0.98	1.4	312.24	23JA	0.1	-3.4	-2.1	0.8
338.99	-0.91	1.4	322.33	2FE	-0.0	-3.5	-2.1	0.8
346.79	-0.82	1.4	332.39	12FE	-0.2	-3.5	-2.1	0.8
354.54	-0.74	1.4	342.39	22FE	-0.7	-3.6	-2.0	0.7
2.22	-0.64	1.4	352.34	4MR	-1.4	-3.6	-1.9	0.7
9.84	-0.54	1.5	2.23	14MR	-1.4	-3.7	-1.9	0.6
17.38	-0.44	1.5	12.07	24MR	-0.6	-3.7	-1.8	0.6
24.83	-0.34	1.5	21.85	3AP	0.8	-3.8	-1.7	0.5
32.20	-0.23	1.5	31.57	13AP	2.3	-3.9	-1.7	0.5
39.49	-0.13	1.5	41.25	23AP	3.7	-4.0	-1.6	0.4
46.68	-0.02	1.6	50.88	3MY	2.1	-4.1	-1.5	0.3
53.80	0.08	1.6	60.47	13MY	1.2	-4.2	-1.5	0.3
60.82	0.19	1.7	70.04	23MY	0.5	-4.2	-1.4	0.2
67.76	0.29	1.7	79.58	2JN	-0.4	-4.1	-1.4	0.3
74.62	0.39	1.8	89.12	12JN	-1.4	-3.9	-1.4	0.3
81.40	0.49	1.8	98.65	22JN	-1.6	-3.3	-1.3	0.4
88.10	0.59	1.8	108.18	2JL	-0.7	-3.0	-1.3	0.4
94.73	0.68	1.9	117.73	12JL	-0.1	-3.7	-1.3	0.5
101.29	0.78	1.9	127.31	22JL	0.3	-4.1	-1.3	0.6
107.77	0.87	1.9	136.91	1AU	0.7	-4.2	-1.3	0.6
114.19	0.96	1.9	146.55	11AU	1.1	-4.2	-1.3	0.7
120.53	1.05	1.9	156.25	21AU	2.4	-4.1	-1.3	0.7
126.80	1.13	1.9	165.98	31AU	2.1	-4.1	-1.3	0.7
133.00	1.22	1.9	175.78	10SE	-0.1	-4.0	-1.3	0.8
139.13	1.31	1.8	185.63	20SE	-0.9	-3.9	-1.3	0.8
145.17	1.39	1.8	195.53	30SE	-1.0	-3.8	-1.4	0.8
151.12	1.48	1.8	205.50	10OC	-0.9	-3.7	-1.4	0.8
156.97	1.56	1.7	215.52	20OC	-0.6	-3.7	-1.5	0.8
162.70	1.65	1.6	225.58	30OC	-0.4	-3.6	-1.5	0.8
168.30	1.74	1.5	235.69	9NO	-0.4	-3.6	-1.6	0.7
173.74	1.83	1.4	245.84	19NO	-0.3	-3.5	-1.6	0.7
178.99	1.93	1.3	256.01	29NO	0.2	-3.5	-1.7	0.7
184.02	2.02	1.2	266.20	9DE	2.7	-3.4	-1.8	0.7
188.77	2.12	1.0	276.40	19DE	1.0	-3.4	-1.8	0.7
193.19	2.23	0.9	286.59	29DE	0.1	-3.4	-1.9	0.8

1399

♂ LONG	LAT	MAG	☉ LONG	16.00UT 1399	☿	♀	♃	♄
197.20	2.33	0.7	296.77	8JA	-0.0	-3.4	-1.9	0.8
200.69	2.44	0.5	306.93	18JA	-0.1	-3.3	-2.0	0.8
203.54	2.54	0.2	317.05	28JA	-0.3	-3.3	-2.0	0.8
205.59	2.62	-0.0	327.13	7FE	-0.7	-3.3	-2.0	0.8
206.66	2.69	-0.3	337.16	17FE	-1.3	-3.3	-2.0	0.8
206.59	2.71	-0.6	347.13	27FE	-1.3	-3.3	-2.0	0.8
205.24	2.67	-0.9	357.06	9MR	-0.5	-3.4	-2.0	0.7
202.68	2.53	-1.1	6.92	19MR	0.9	-3.4	-2.0	0.7
199.23	2.28	-1.4	16.73	29MR	2.8	-3.4	-1.9	0.6
195.48	1.91	-1.3	26.48	8AP	2.8	-3.4	-1.8	0.6
192.16	1.47	-1.2	36.18	18AP	1.5	-3.5	-1.8	0.5
189.85	1.02	-1.0	45.83	28AP	0.9	-3.5	-1.7	0.5
188.83	0.58	-0.8	55.45	8MY	0.3	-3.4	-1.6	0.4
189.14	0.19	-0.6	65.03	18MY	-0.4	-3.4	-1.6	0.3
190.65	-0.15	-0.4	74.58	28MY	-1.4	-3.4	-1.5	0.3
193.18	-0.44	-0.2	84.12	7JN	-1.6	-3.4	-1.5	0.2
196.57	-0.69	-0.0	93.65	17JN	-0.7	-3.3	-1.4	0.3
200.66	-0.90	0.1	103.18	27JN	0.0	-3.3	-1.4	0.3
205.34	-1.07	0.3	112.72	7JL	0.5	-3.3	-1.3	0.4
210.50	-1.22	0.4	122.28	17JL	0.9	-3.3	-1.3	0.5
216.07	-1.35	0.5	131.87	27JL	1.5	-3.3	-1.3	0.5
222.00	-1.45	0.6	141.49	6AU	2.9	-3.3	-1.2	0.6
228.23	-1.53	0.6	151.16	16AU	1.8	-3.4	-1.2	0.6
234.51	-1.59	0.7	160.87	26AU	-0.1	-3.4	-1.2	0.7
241.46	-1.63	0.8	170.63	5SE	-1.0	-3.4	-1.2	0.7
248.39	-1.65	0.8	180.45	15SE	-1.2	-3.5	-1.2	0.7
255.50	-1.65	0.9	190.33	25SE	-0.9	-3.5	-1.2	0.8
262.77	-1.64	0.9	200.27	5OC	-0.5	-3.6	-1.2	0.8
270.18	-1.61	1.0	210.26	15OC	-0.3	-3.7	-1.3	0.8
277.69	-1.57	1.0	220.30	25OC	-0.2	-3.8	-1.3	0.8
285.30	-1.51	1.1	230.39	4NO	-0.2	-3.8	-1.3	0.8
292.97	-1.44	1.1	240.52	14NO	0.4	-3.9	-1.4	0.8
300.70	-1.35	1.2	250.68	24NO	3.0	-4.1	-1.4	0.7
308.47	-1.26	1.2	260.86	4DE	0.7	-4.2	-1.5	0.7
316.25	-1.16	1.3	271.05	14DE	-0.1	-4.3	-1.6	0.7
324.02	-1.05	1.3	281.25	24DE	-0.2	-4.3	-1.6	0.7

♂ LONG	♂ LAT	♂ MAG	⊙ LONG	16.00UT	☿	♀	♃	♄
				1400	MAGNITUDES			
331.78	-0.94	1.3	291.44	3JA	-0.2	-4.4	-1.7	0.8
339.51	-0.82	1.4	301.61	13JA	-0.4	-4.3	-1.8	0.8
347.20	-0.70	1.4	311.74	23JA	-0.8	-4.0	-1.8	0.8
354.83	-0.58	1.5	321.85	2FE	-1.2	-3.4	-1.9	0.8
2.40	-0.46	1.5	331.90	12FE	-1.2	-3.5	-1.9	0.8
9.90	-0.33	1.5	341.91	22FE	-0.5	-4.0	-2.0	0.8
17.33	-0.22	1.6	351.86	3MR	1.2	-4.3	-2.0	0.8
24.68	-0.10	1.6	1.75	13MR	3.3	-4.3	-2.0	0.8
31.95	0.01	1.7	11.59	23MR	2.0	-4.2	-2.0	0.7
39.13	0.12	1.7	21.37	2AP	1.1	-4.1	-2.0	0.7
46.24	0.22	1.7	31.10	12AP	0.7	-4.0	-2.0	0.7
53.27	0.32	1.7	40.78	22AP	0.2	-3.9	-1.9	0.6
60.22	0.42	1.7	50.41	2MY	-0.5	-3.8	-1.9	0.5
67.10	0.51	1.8	60.01	12MY	-1.4	-3.7	-1.8	0.5
73.90	0.59	1.8	69.58	22MY	-1.6	-3.6	-1.7	0.4
80.64	0.67	1.8	79.13	1JN	-0.7	-3.6	-1.7	0.3
87.32	0.75	1.8	88.66	11JN	0.1	-3.5	-1.6	0.3
93.94	0.82	1.8	98.19	21JN	0.7	-3.5	-1.5	0.3
100.51	0.89	1.9	107.72	1JL	1.2	-3.4	-1.5	0.3
107.02	0.95	1.9	117.27	11JL	2.0	-3.4	-1.4	0.4
113.49	1.01	2.0	126.84	21JL	3.2	-3.4	-1.4	0.4
119.92	1.06	2.0	136.45	31JL	1.5	-3.4	-1.3	0.5
126.31	1.11	2.0	146.09	10AU	-0.2	-3.3	-1.3	0.6
132.66	1.16	2.0	155.78	20AU	-1.1	-3.3	-1.3	0.6
138.98	1.21	2.0	165.51	30AU	-1.3	-3.3	-1.2	0.7
145.26	1.25	2.0	175.30	9SE	-0.8	-3.4	-1.2	0.7
151.51	1.28	2.0	185.15	19SE	-0.4	-3.4	-1.2	0.7
157.73	1.32	2.0	195.05	29SE	-0.2	-3.4	-1.2	0.8
163.91	1.35	1.9	205.01	9OC	-0.1	-3.4	-1.2	0.8
170.06	1.37	1.9	215.03	19OC	0.0	-3.4	-1.2	0.8
176.16	1.40	1.9	225.10	29OC	0.6	-3.5	-1.2	0.8
182.21	1.42	1.8	235.20	8NO	3.3	-3.5	-1.3	0.8
188.21	1.43	1.7	245.35	18NO	0.5	-3.5	-1.3	0.8
194.14	1.44	1.6	255.52	28NO	-0.2	-3.5	-1.3	0.8
199.99	1.45	1.5	265.71	8DE	-0.3	-3.5	-1.4	0.7
205.75	1.44	1.4	275.91	18DE	-0.4	-3.4	-1.4	0.7
211.40	1.43	1.3	286.10	28DE	-0.5	-3.4	-1.5	0.7
				1401				
216.91	1.41	1.2	296.28	7JA	-0.8	-3.4	-1.5	0.8
222.25	1.38	1.0	306.44	17JA	-1.1	-3.4	-1.6	0.8
227.39	1.33	0.9	316.56	27JA	-1.1	-3.3	-1.7	0.8
232.29	1.27	0.7	326.64	6FE	-0.4	-3.3	-1.7	0.9
236.87	1.18	0.5	336.67	16FE	1.4	-3.3	-1.8	0.9
241.06	1.06	0.2	346.65	26FE	3.1	-3.3	-1.9	0.9
244.77	0.90	0.0	356.57	8MR	1.5	-3.4	-1.9	0.9
247.87	0.70	-0.3	6.44	18MR	0.8	-3.4	-2.0	0.8
250.19	0.43	-0.6	16.25	28MR	0.5	-3.4	-2.0	0.8
251.58	0.09	-0.9	26.01	7AP	0.1	-3.4	-2.0	0.8
251.83	-0.33	-1.2	35.71	17AP	-0.5	-3.5	-2.0	0.7
250.84	-0.84	-1.5	45.37	27AP	-1.4	-3.5	-2.0	0.7
248.66	-1.41	-1.8	54.98	7MY	-1.6	-3.5	-2.0	0.6
245.61	-1.99	-2.0	64.57	17MY	-0.6	-3.6	-2.0	0.6
242.31	-2.53	-2.0	74.12	27MY	0.2	-3.7	-1.9	0.5
239.50	-2.95	-1.8	83.66	6JN	0.9	-3.7	-1.9	0.4
237.76	-3.25	-1.7	93.20	16JN	1.5	-3.8	-1.8	0.4
237.38	-3.41	-1.4	102.72	26JN	2.6	-3.9	-1.8	0.3
238.36	-3.48	-1.2	112.26	6JL	2.9	-4.0	-1.7	0.3
240.58	-3.47	-1.0	121.83	16JL	1.2	-4.1	-1.6	0.4
243.86	-3.41	-0.8	131.41	26JL	-0.2	-4.2	-1.6	0.4
247.99	-3.31	-0.6	141.03	5AU	-1.2	-4.2	-1.5	0.5
252.83	-3.19	-0.5	150.69	15AU	-1.4	-4.2	-1.5	0.5
258.24	-3.04	-0.3	160.40	25AU	-0.8	-3.9	-1.4	0.6
264.10	-2.88	-0.2	170.16	4SE	-0.3	-3.3	-1.4	0.6
270.34	-2.70	-0.0	179.98	14SE	-0.1	-3.5	-1.3	0.7
276.88	-2.51	0.1	189.85	24SE	0.1	-4.0	-1.3	0.7
283.66	-2.32	0.2	199.78	4OC	0.2	-4.3	-1.3	0.8
290.64	-2.11	0.4	209.77	14OC	0.9	-4.3	-1.3	0.8
297.76	-1.91	0.5	219.81	24OC	3.4	-4.3	-1.3	0.8
304.99	-1.70	0.6	229.89	3NO	0.4	-4.2	-1.3	0.8
312.30	-1.50	0.7	240.02	13NO	-0.4	-4.1	-1.3	0.8
319.66	-1.29	0.8	250.18	23NO	-0.5	-4.0	-1.3	0.8
327.05	-1.10	0.9	260.36	3DE	-0.5	-3.9	-1.3	0.8
334.44	-0.91	1.0	270.56	13DE	-0.6	-3.8	-1.3	0.8
341.83	-0.72	1.1	280.76	23DE	-0.9	-3.8	-1.4	0.8

♂ LONG	♂ LAT	♂ MAG	⊙ LONG	16.00UT	☿	♀	♃	♄
				1402	MAGNITUDES			
349.19	-0.55	1.2	290.95	2JA	-0.9	-3.7	-1.4	0.8
356.52	-0.38	1.3	301.12	12JA	-0.9	-3.6	-1.5	0.8
3.80	-0.23	1.3	311.26	22JA	-0.3	-3.6	-1.5	0.8
11.03	-0.08	1.4	321.36	1FE	1.6	-3.5	-1.6	0.9
18.21	0.05	1.5	331.42	11FE	2.5	-3.4	-1.6	0.9
25.31	0.18	1.6	341.43	21FE	1.1	-3.4	-1.7	0.9
32.36	0.30	1.6	351.39	3MR	0.6	-3.4	-1.8	0.9
39.34	0.40	1.7	1.29	13MR	0.3	-3.3	-1.9	0.9
46.26	0.50	1.7	11.12	23MR	-0.0	-3.3	-1.9	0.9
53.11	0.59	1.8	20.91	2AP	-0.6	-3.3	-2.0	0.9
59.90	0.67	1.8	30.64	12AP	-1.4	-3.3	-2.0	0.9
66.63	0.75	1.9	40.32	22AP	-1.6	-3.3	-2.1	0.8
73.31	0.81	1.9	49.95	2MY	-0.6	-3.3	-2.1	0.8
79.94	0.87	1.9	59.55	12MY	0.3	-3.3	-2.1	0.7
86.52	0.93	1.9	69.12	22MY	1.1	-3.4	-2.1	0.7
93.05	0.98	1.9	78.67	1JN	2.0	-3.4	-2.1	0.6
99.55	1.02	2.0	88.20	11JN	3.3	-3.4	-2.1	0.5
106.02	1.05	2.0	97.73	21JN	2.4	-3.5	-2.1	0.5
112.45	1.08	1.9	107.26	1JL	1.0	-3.5	-2.0	0.4
118.87	1.11	1.9	116.81	11JL	-0.3	-3.5	-2.0	0.4
125.26	1.13	1.9	126.38	21JL	-1.3	-3.4	-1.9	0.4
131.64	1.15	2.0	135.98	31JL	-1.5	-3.4	-1.8	0.4
138.01	1.16	2.0	145.62	10AU	-0.8	-3.4	-1.8	0.5
144.37	1.17	2.0	155.30	20AU	-0.3	-3.3	-1.7	0.5
150.73	1.17	2.0	165.03	30AU	0.0	-3.3	-1.6	0.6
157.09	1.17	2.0	174.82	9SE	0.2	-3.3	-1.6	0.7
163.45	1.16	2.0	184.66	19SE	0.4	-3.3	-1.5	0.7
169.81	1.15	2.0	194.57	29SE	1.2	-3.3	-1.5	0.7
176.18	1.13	2.0	204.52	9OC	3.1	-3.3	-1.5	0.8
182.56	1.11	2.0	214.53	19OC	0.2	-3.4	-1.4	0.8
188.94	1.09	1.9	224.60	29OC	-0.6	-3.4	-1.4	0.9
195.32	1.05	1.9	234.71	8NO	-0.6	-3.4	-1.4	0.9
201.72	1.01	1.9	244.85	18NO	-0.6	-3.5	-1.4	0.9
208.11	0.97	1.8	255.02	28NO	-0.7	-3.5	-1.4	0.9
214.51	0.91	1.7	265.21	8DE	-0.8	-3.6	-1.4	0.9
220.90	0.84	1.7	275.41	18DE	-0.8	-3.6	-1.4	0.9
227.29	0.77	1.6	285.61	28DE	-0.8	-3.7	-1.4	0.9
				1403				
233.66	0.68	1.5	295.79	7JA	-0.2	-3.8	-1.4	0.9
240.02	0.58	1.4	305.94	17JA	1.9	-3.9	-1.4	0.9
246.35	0.46	1.3	316.07	27JA	2.0	-3.9	-1.5	0.9
252.65	0.32	1.1	326.15	6FE	0.8	-4.0	-1.5	0.9
258.91	0.17	1.0	336.19	16FE	0.4	-4.1	-1.6	1.0
265.12	-0.01	0.8	346.18	26FE	0.2	-4.2	-1.6	1.0
271.25	-0.22	0.7	356.10	8MR	-0.1	-4.3	-1.7	1.0
277.29	-0.45	0.5	5.97	18MR	-0.6	-4.3	-1.8	1.0
283.20	-0.72	0.3	15.79	28MR	-1.4	-4.1	-1.8	1.0
288.96	-1.03	0.1	25.54	7AP	-1.5	-3.7	-1.9	1.0
294.53	-1.39	-0.1	35.25	17AP	-0.6	-3.0	-2.0	1.0
299.82	-1.79	-0.3	44.91	27AP	0.5	-3.5	-2.0	1.0
304.78	-2.24	-0.5	54.52	7MY	1.5	-4.0	-2.1	0.9
309.29	-2.76	-0.8	64.10	17MY	2.7	-4.2	-2.2	0.9
313.22	-3.34	-1.0	73.66	27MY	3.3	-4.2	-2.2	0.8
316.41	-3.97	-1.3	83.20	6JN	1.8	-4.1	-2.2	0.7
318.68	-4.66	-1.6	92.73	16JN	0.8	-4.0	-2.3	0.7
319.80	-5.37	-1.9	102.27	26JN	-0.3	-4.0	-2.3	0.6
319.67	-6.04	-2.2	111.80	6JL	-1.3	-3.9	-2.3	0.5
318.31	-6.57	-2.4	121.36	16JL	-1.6	-3.8	-2.2	0.5
316.00	-6.86	-2.6	130.95	26JL	-0.7	-3.7	-2.2	0.4
313.40	-6.82	-2.6	140.56	5AU	-0.2	-3.6	-2.1	0.5
311.21	-6.47	-2.4	150.22	15AU	0.1	-3.6	-2.1	0.5
310.05	-5.90	-2.2	159.93	25AU	0.4	-3.5	-2.0	0.6
310.19	-5.20	-1.9	169.69	4SE	0.7	-3.5	-2.0	0.6
311.62	-4.48	-1.6	179.50	14SE	1.6	-3.5	-1.9	0.7
314.19	-3.78	-1.3	189.37	24SE	2.7	-3.4	-1.8	0.7
317.71	-3.14	-1.0	199.30	4OC	0.1	-3.4	-1.8	0.8
321.97	-2.56	-0.8	209.28	14OC	-0.7	-3.4	-1.7	0.8
326.82	-2.05	-0.5	219.32	24OC	-0.8	-3.4	-1.6	0.9
332.12	-1.60	-0.3	229.40	3NO	-0.8	-3.4	-1.6	0.9
337.76	-1.20	-0.1	239.53	13NO	-0.8	-3.4	-1.6	0.9
343.65	-0.86	0.1	249.69	23NO	-0.6	-3.4	-1.5	1.0
349.75	-0.55	0.3	259.87	3DE	-0.6	-3.4	-1.5	1.0
355.98	-0.29	0.5	270.06	13DE	-0.6	-3.4	-1.5	1.0
2.32	-0.06	0.7	280.26	23DE	-0.1	-3.4	-1.5	1.0

1404

♂ LONG	LAT	MAG	⊙ LONG	16.00UT	☿	♀	♃	♄
8.73	0.14	0.8	290.45	2JA	2.1	-3.4	-1.5	1.0
15.19	0.31	1.0	300.62	12JA	1.5	-3.4	-1.5	1.0
21.68	0.46	1.1	310.77	22JA	0.5	-3.5	-1.5	0.9
28.19	0.59	1.2	320.87	1FE	0.2	-3.5	-1.5	1.0
34.70	0.71	1.3	330.93	11FE	0.1	-3.5	-1.5	1.0
41.20	0.80	1.4	340.95	21FE	-0.2	-3.5	-1.6	1.1
47.69	0.89	1.5	350.90	2MR	-0.6	-3.4	-1.6	1.1
54.16	0.95	1.6	0.80	12MR	-1.4	-3.4	-1.6	1.1
60.62	1.01	1.7	10.65	22MR	-1.5	-3.4	-1.7	1.1
67.05	1.06	1.8	20.43	1AP	-0.6	-3.4	-1.8	1.1
73.46	1.10	1.8	30.16	11AP	0.6	-3.3	-1.8	1.1
79.85	1.13	1.9	39.85	21AP	1.9	-3.3	-1.9	1.1
86.22	1.16	1.9	49.48	1MY	3.6	-3.3	-2.0	1.1
92.58	1.18	2.0	59.09	11MY	2.5	-3.3	-2.0	1.1
98.92	1.19	2.0	68.66	21MY	1.4	-3.3	-2.1	1.0
105.24	1.20	2.0	78.20	31MY	0.6	-3.4	-2.2	1.0
111.56	1.20	2.0	87.74	10JN	-0.4	-3.4	-2.2	0.9
117.88	1.19	2.0	97.27	20JN	-1.4	-3.4	-2.3	0.8
124.20	1.18	2.0	106.80	30JN	-1.6	-3.5	-2.3	0.8
130.52	1.17	2.0	116.35	10JL	-0.7	-3.5	-2.4	0.7
136.86	1.15	2.0	125.92	20JL	-0.1	-3.6	-2.4	0.6
143.20	1.13	2.0	135.52	30JL	0.3	-3.6	-2.4	0.6
149.57	1.10	2.0	145.16	9AU	0.5	-3.7	-2.4	0.6
155.95	1.07	1.9	154.84	19AU	0.9	-3.8	-2.3	0.6
162.36	1.04	1.9	164.56	29AU	2.0	-3.9	-2.3	0.6
168.80	1.00	2.0	174.35	8SE	2.4	-4.0	-2.3	0.7
175.27	0.95	2.0	184.19	18SE	0.0	-4.1	-2.2	0.7
181.77	0.90	2.0	194.09	28SE	-0.8	-4.2	-2.1	0.8
188.32	0.85	2.0	204.04	8OC	-0.9	-4.3	-2.1	0.8
194.90	0.79	1.9	214.05	18OC	-0.9	-4.4	-2.0	0.9
201.52	0.73	1.9	224.11	28OC	-0.7	-4.3	-1.9	0.9
208.19	0.66	1.9	234.22	7NO	-0.5	-3.9	-1.9	1.0
214.90	0.58	1.9	244.36	17NO	-0.5	-3.2	-1.8	1.0
221.66	0.49	1.8	254.52	27NO	-0.4	-3.4	-1.7	1.0
228.46	0.40	1.8	264.71	7DE	0.1	-4.1	-1.7	1.1
235.30	0.30	1.7	274.91	17DE	2.4	-4.3	-1.7	1.1
242.19	0.20	1.7	285.10	27DE	1.2	-4.4	-1.6	1.1

1405

♂ LONG	LAT	MAG	⊙ LONG	16.00UT	☿	♀	♃	♄
249.13	0.08	1.6	295.29	6JA	0.3	-4.3	-1.6	1.1
256.10	-0.05	1.5	305.45	16JA	-0.1	-4.2	-1.6	1.1
263.12	-0.18	1.5	315.57	26JA	-0.1	-4.1	-1.6	1.1
270.18	-0.33	1.4	325.66	5FE	-0.3	-4.0	-1.6	1.1
277.27	-0.48	1.3	335.70	15FE	-0.7	-3.9	-1.6	1.1
284.40	-0.65	1.2	345.69	25FE	-1.4	-3.8	-1.6	1.2
291.55	-0.82	1.1	355.62	7MR	-1.4	-3.7	-1.6	1.2
298.72	-1.00	1.0	5.49	17MR	-0.6	-3.6	-1.6	1.2
305.90	-1.20	0.9	15.31	27MR	0.8	-3.6	-1.6	1.3
313.08	-1.39	0.8	25.08	6AP	2.4	-3.5	-1.7	1.3
320.23	-1.59	0.7	34.78	16AP	3.3	-3.4	-1.7	1.3
327.36	-1.80	0.6	44.44	26AP	1.9	-3.4	-1.8	1.3
334.43	-2.00	0.5	54.07	6MY	1.1	-3.4	-1.8	1.3
341.42	-2.20	0.3	63.65	16MY	0.4	-3.3	-1.9	1.2
348.29	-2.40	0.2	73.21	26MY	-0.4	-3.3	-1.9	1.2
355.01	-2.59	0.1	82.75	5JN	-1.4	-3.3	-2.0	1.2
1.52	-2.77	-0.1	92.28	15JN	-1.6	-3.3	-2.1	1.1
7.78	-2.94	-0.2	101.81	25JN	-0.7	-3.3	-2.2	1.1
13.69	-3.09	-0.4	111.35	5JL	-0.0	-3.3	-2.2	1.0
19.18	-3.22	-0.5	120.90	15JL	0.4	-3.3	-2.3	0.9
24.15	-3.33	-0.7	130.49	25JL	0.7	-3.4	-2.4	0.9
28.42	-3.40	-0.9	140.10	4AU	1.3	-3.4	-2.4	0.8
31.85	-3.43	-1.1	149.76	14AU	2.5	-3.4	-2.4	0.7
34.23	-3.41	-1.3	159.46	24AU	2.1	-3.5	-2.5	0.7
35.32	-3.31	-1.5	169.22	3SE	-0.1	-3.5	-2.5	0.8
34.99	-3.10	-1.8	179.02	13SE	-1.0	-3.5	-2.5	0.8
33.21	-2.76	-2.0	188.89	23SE	-1.1	-3.4	-2.4	0.8
30.28	-2.28	-2.1	198.82	3OC	-0.9	-3.4	-2.4	0.9
26.87	-1.70	-2.2	208.80	13OC	-0.6	-3.4	-2.3	0.9
23.76	-1.08	-1.9	218.83	23OC	-0.4	-3.4	-2.3	1.0
21.65	-0.49	-1.6	228.91	2NO	-0.3	-3.3	-2.2	1.0
20.85	0.01	-1.3	239.03	12NO	-0.3	-3.3	-2.1	1.1
21.37	0.41	-0.9	249.19	22NO	0.3	-3.3	-2.1	1.1
23.07	0.73	-0.6	259.37	2DE	2.7	-3.3	-2.0	1.2
25.74	0.97	-0.3	269.56	12DE	0.9	-3.3	-1.9	1.2
29.15	1.15	-0.0	279.76	22DE	0.1	-3.4	-1.9	1.2

1406

♂ LONG	LAT	MAG	⊙ LONG	16.00UT	☿	♀	♃	♄
33.17	1.28	0.2	289.95	1JA	-0.1	-3.4	-1.8	1.2
37.65	1.38	0.4	300.12	11JA	-0.2	-3.4	-1.7	1.3
42.48	1.45	0.6	310.27	21JA	-0.4	-3.4	-1.7	1.3
47.59	1.50	0.8	320.38	31JA	-0.7	-3.5	-1.7	1.3
52.90	1.53	1.0	330.44	10FE	-1.3	-3.5	-1.6	1.3
58.39	1.55	1.1	340.46	20FE	-1.3	-3.6	-1.6	1.3
64.01	1.55	1.3	350.42	2MR	-0.5	-3.6	-1.6	1.3
69.72	1.55	1.4	0.32	12MR	1.0	-3.7	-1.6	1.4
75.52	1.54	1.5	10.17	22MR	3.0	-3.7	-1.6	1.4
81.39	1.52	1.6	19.96	1AP	2.5	-3.8	-1.6	1.4
87.31	1.50	1.7	29.69	11AP	1.4	-3.9	-1.6	1.5
93.27	1.47	1.8	39.38	21AP	0.8	-4.0	-1.6	1.5
99.27	1.44	1.8	49.02	1MY	0.3	-4.1	-1.6	1.4
105.31	1.41	1.9	58.62	11MY	-0.4	-4.2	-1.7	1.4
111.38	1.37	1.9	68.20	21MY	-1.4	-4.2	-1.7	1.4
117.48	1.33	2.0	77.75	31MY	-1.6	-4.1	-1.8	1.3
123.61	1.28	2.0	87.28	10JN	-0.7	-3.9	-1.8	1.3
129.78	1.23	2.0	96.81	20JN	0.0	-3.3	-1.9	1.2
135.99	1.18	2.0	106.34	30JN	0.5	-3.1	-1.9	1.2
142.23	1.13	2.0	115.89	10JL	1.0	-3.0	-2.0	1.1
148.51	1.07	2.0	125.46	20JL	1.7	-4.1	-2.1	1.1
154.83	1.01	2.0	135.05	30JL	3.0	-4.2	-2.1	1.0
161.21	0.95	2.0	144.69	9AU	1.8	-4.2	-2.2	1.0
167.63	0.88	2.0	154.37	19AU	-0.1	-4.1	-2.3	0.9
174.10	0.82	1.9	164.09	29AU	-1.1	-4.1	-2.3	0.9
180.63	0.74	1.9	173.87	8SE	-1.2	-4.0	-2.4	0.9
187.23	0.67	1.9	183.71	18SE	-0.9	-3.9	-2.4	0.9
193.88	0.59	1.8	193.61	28SE	-0.5	-3.8	-2.4	1.0
200.60	0.51	1.8	203.56	8OC	-0.3	-3.7	-2.5	1.0
207.38	0.42	1.8	213.57	18OC	-0.2	-3.7	-2.4	1.1
214.24	0.33	1.8	223.62	28OC	-0.1	-3.6	-2.4	1.1
221.16	0.23	1.8	233.72	7NO	0.5	-3.6	-2.4	1.2
228.16	0.14	1.8	243.87	17NO	3.0	-3.5	-2.3	1.2
235.22	0.03	1.7	254.03	27NO	0.7	-3.5	-2.3	1.3
242.35	-0.07	1.7	264.22	7DE	-0.1	-3.4	-2.2	1.3
249.56	-0.18	1.7	274.42	17DE	-0.2	-3.4	-2.1	1.4
256.82	-0.29	1.7	284.61	27DE	-0.3	-3.4	-2.0	1.4

1407

♂ LONG	LAT	MAG	⊙ LONG	16.00UT	☿	♀	♃	♄
264.16	-0.41	1.6	294.79	6JA	-0.5	-3.4	-2.0	1.4
271.55	-0.53	1.6	304.96	16JA	-0.8	-3.3	-1.9	1.4
279.00	-0.65	1.5	315.08	26JA	-1.2	-3.3	-1.8	1.4
286.51	-0.76	1.5	325.17	5FE	-1.1	-3.3	-1.8	1.3
294.06	-0.88	1.4	335.21	15FE	-0.5	-3.3	-1.7	1.3
301.64	-1.00	1.4	345.20	25FE	1.2	-3.3	-1.7	1.2
309.26	-1.11	1.3	355.13	7MR	3.3	-3.4	-1.6	1.2
316.89	-1.21	1.3	5.01	17MR	1.8	-3.4	-1.6	1.2
324.53	-1.31	1.2	14.83	27MR	1.0	-3.4	-1.6	1.2
332.17	-1.40	1.2	24.59	6AP	0.6	-3.4	-1.6	1.2
339.78	-1.48	1.1	34.31	16AP	0.2	-3.5	-1.5	1.2
347.37	-1.55	1.1	43.97	26AP	-0.5	-3.5	-1.5	1.2
354.90	-1.60	1.0	53.59	6MY	-1.4	-3.4	-1.5	1.2
2.37	-1.63	1.0	63.18	16MY	-1.6	-3.4	-1.5	1.1
9.75	-1.65	0.9	72.74	26MY	-0.7	-3.4	-1.6	1.1
17.04	-1.65	0.8	82.28	5JN	0.1	-3.4	-1.6	1.1
24.19	-1.64	0.8	91.81	15JN	0.7	-3.3	-1.6	1.0
31.20	-1.60	0.7	101.34	25JN	1.3	-3.3	-1.7	1.0
38.04	-1.54	0.7	110.88	5JL	2.2	-3.3	-1.7	0.9
44.68	-1.47	0.6	120.44	15JL	3.2	-3.3	-1.7	0.9
51.09	-1.37	0.5	130.02	25JL	1.5	-3.3	-1.8	0.8
57.22	-1.25	0.4	139.63	4AU	-0.2	-3.3	-1.9	0.8
63.02	-1.10	0.3	149.29	14AU	-1.2	-3.4	-1.9	0.7
68.45	-0.93	0.2	158.99	24AU	-1.4	-3.4	-2.0	0.7
73.40	-0.73	0.1	168.74	3SE	-0.8	-3.4	-2.1	0.6
77.79	-0.49	-0.1	178.55	13SE	-0.4	-3.5	-2.1	0.6
81.48	-0.22	-0.2	188.41	23SE	-0.1	-3.5	-2.2	0.7
84.31	0.11	-0.4	198.34	3OC	-0.0	-3.6	-2.2	0.7
86.08	0.48	-0.6	208.31	13OC	0.1	-3.7	-2.3	0.8
86.60	0.91	-0.8	218.34	23OC	0.7	-3.8	-2.3	0.9
85.70	1.39	-1.1	228.42	2NO	3.4	-3.9	-2.4	0.9
83.38	1.88	-1.3	238.55	12NO	0.5	-3.9	-2.4	1.0
79.96	2.33	-1.4	248.70	22NO	-0.3	-4.1	-2.4	1.0
76.06	2.70	-1.5	258.88	2DE	-0.4	-4.2	-2.3	1.1
72.48	2.93	-1.2	269.07	12DE	-0.4	-4.3	-2.3	1.1
69.88	3.03	-0.9	279.27	22DE	-0.5	-4.4	-2.2	1.1

♂ / ☉ / Magnitudes — 1408–1409

♂ LONG	LAT	MAG	☉ LONG	16.00UT 1408	☿	♀	♃	♄
68.59	3.02	-0.6	289.46	1JA	-0.8	-4.4	-2.2	1.1
68.63	2.95	-0.3	299.63	11JA	-1.0	-4.3	-2.1	1.1
69.84	2.84	-0.1	309.77	21JA	-1.0	-4.0	-2.0	1.1
72.03	2.72	0.2	319.89	31JA	-0.4	-3.4	-2.0	1.1
75.00	2.59	0.4	329.95	10FE	1.4	-3.6	-1.9	1.1
78.59	2.45	0.7	339.97	20FE	2.9	-4.0	-1.8	1.0
82.66	2.33	0.9	349.93	1MR	1.3	-4.3	-1.8	1.0
87.11	2.20	1.0	359.84	11MR	0.7	-4.3	-1.7	0.9
91.87	2.09	1.2	9.69	21MR	0.4	-4.2	-1.6	0.9
96.87	1.97	1.3	19.48	31MR	0.1	-4.1	-1.6	0.9
102.07	1.86	1.4	29.22	10AP	-0.5	-4.0	-1.5	0.9
107.43	1.76	1.5	38.91	20AP	-1.4	-3.9	-1.5	0.9
112.93	1.65	1.6	48.55	30AP	-1.6	-3.8	-1.5	0.9
118.56	1.55	1.7	58.16	10MY	-0.7	-3.7	-1.5	0.9
124.29	1.46	1.7	67.73	20MY	0.2	-3.6	-1.4	0.9
130.11	1.36	1.8	77.28	30MY	0.9	-3.6	-1.4	0.9
136.02	1.26	1.8	86.82	9JN	1.7	-3.5	-1.4	0.8
142.02	1.17	1.9	96.35	19JN	2.9	-3.5	-1.4	0.8
148.09	1.07	1.9	105.88	29JN	2.8	-3.4	-1.5	0.8
154.25	0.98	1.9	115.43	9JL	1.2	-3.4	-1.5	0.7
160.48	0.88	1.9	124.99	19JL	-0.2	-3.4	-1.5	0.7
166.79	0.79	1.9	134.59	29JL	-1.2	-3.4	-1.5	0.6
173.18	0.69	1.9	144.22	8AU	-1.5	-3.3	-1.6	0.6
179.65	0.60	1.9	153.90	18AU	-0.8	-3.3	-1.6	0.5
186.20	0.50	1.9	163.63	28AU	-0.3	-3.3	-1.7	0.5
192.84	0.40	1.9	173.40	7SE	-0.0	-3.4	-1.7	0.4
199.56	0.30	1.8	183.24	17SE	0.1	-3.4	-1.8	0.4
206.37	0.20	1.8	193.13	27SE	0.3	-3.4	-1.9	0.3
213.27	0.10	1.8	203.08	7OC	1.0	-3.4	-1.9	0.4
220.26	0.00	1.7	213.08	17OC	3.4	-3.4	-2.0	0.5
227.34	-0.10	1.7	223.14	27OC	0.3	-3.5	-2.1	0.5
234.50	-0.20	1.6	233.23	6NO	-0.5	-3.5	-2.1	0.6
241.75	-0.30	1.6	243.37	16NO	-0.5	-3.5	-2.2	0.7
249.08	-0.40	1.6	253.54	26NO	-0.5	-3.5	-2.2	0.7
256.50	-0.50	1.5	263.72	6DE	-0.6	-3.5	-2.2	0.8
263.99	-0.60	1.5	273.92	16DE	-0.8	-3.4	-2.2	0.8
271.55	-0.69	1.5	284.12	26DE	-0.9	-3.4	-2.2	0.9
				1409				
279.18	-0.78	1.5	294.30	5JA	-0.9	-3.4	-2.2	0.9
286.86	-0.87	1.5	304.46	15JA	-0.3	-3.4	-2.2	0.9
294.59	-0.94	1.5	314.59	25JA	1.6	-3.3	-2.1	0.9
302.36	-1.01	1.4	324.68	4FE	2.3	-3.3	-2.0	0.9
310.15	-1.07	1.4	334.72	14FE	1.0	-3.3	-1.9	0.8
317.97	-1.13	1.4	344.72	24FE	0.5	-3.3	-1.9	0.8
325.78	-1.17	1.4	354.65	6MR	0.3	-3.4	-1.8	0.8
333.59	-1.20	1.4	4.53	16MR	-0.0	-3.4	-1.8	0.7
341.38	-1.21	1.4	14.36	26MR	-0.6	-3.4	-1.7	0.7
349.13	-1.21	1.4	24.12	5AP	-1.4	-3.4	-1.6	0.7
356.85	-1.20	1.3	33.83	15AP	-1.6	-3.5	-1.6	0.7
4.50	-1.18	1.3	43.50	25AP	-0.7	-3.5	-1.5	0.7
12.08	-1.14	1.3	53.12	5MY	0.4	-3.5	-1.5	0.7
19.58	-1.09	1.3	62.71	15MY	1.2	-3.6	-1.4	0.7
27.00	-1.02	1.3	72.28	25MY	2.3	-3.7	-1.4	0.7
34.31	-0.95	1.3	81.82	4JN	3.5	-3.7	-1.4	0.7
41.51	-0.86	1.3	91.35	14JN	2.2	-3.8	-1.4	0.7
48.59	-0.75	1.3	100.88	24JN	0.9	-3.9	-1.3	0.6
55.54	-0.64	1.2	110.42	4JL	-0.3	-4.0	-1.3	0.6
62.35	-0.51	1.2	119.98	14JL	-1.3	-4.1	-1.3	0.6
69.01	-0.37	1.2	129.56	24JL	-1.5	-4.2	-1.3	0.5
75.49	-0.23	1.1	139.17	3AU	-0.8	-4.2	-1.4	0.5
81.79	-0.06	1.1	148.82	13AU	-0.2	-4.1	-1.4	0.4
87.88	0.11	1.0	158.52	23AU	0.1	-3.9	-1.4	0.3
93.73	0.30	1.0	168.27	2SE	0.3	-3.3	-1.4	0.3
99.31	0.51	0.9	178.08	12SE	0.5	-3.5	-1.5	0.2
104.56	0.73	0.8	187.94	22SE	1.3	-4.0	-1.5	0.2
109.44	0.98	0.7	197.85	2OC	3.0	-4.3	-1.6	0.1
113.86	1.25	0.5	207.83	12OC	0.2	-4.3	-1.6	0.1
117.72	1.55	0.2	217.86	22OC	-0.6	-4.3	-1.7	0.2
120.90	1.89	0.2	227.93	1NO	-0.7	-4.2	-1.8	0.2
123.22	2.27	-0.0	238.06	11NO	-0.7	-4.1	-1.8	0.3
124.52	2.69	-0.2	248.21	21NO	-0.8	-4.0	-1.9	0.4
124.57	3.13	-0.4	258.38	1DE	-0.7	-3.9	-2.0	0.4
123.27	3.57	-0.7	268.58	11DE	-0.7	-3.8	-2.0	0.5
120.66	3.97	-0.9	278.77	21DE	-0.7	-3.8	-2.1	0.6
117.05	4.25	-1.1	288.96	31DE	-0.2	-3.7	-2.1	0.6

♂ / ☉ / Magnitudes — 1410–1411

♂ LONG	LAT	MAG	☉ LONG	16.00UT 1410	☿	♀	♃	♄
113.07	4.37	-1.0	299.14	10JA	1.9	-3.6	-2.1	0.6
109.48	4.31	-0.8	309.29	20JA	1.8	-3.5	-2.1	0.7
106.87	4.12	-0.6	319.40	30JA	0.7	-3.5	-2.1	0.7
105.53	3.84	-0.3	329.47	9FE	0.3	-3.4	-2.1	0.7
105.49	3.53	-0.1	339.49	19FE	0.1	-3.4	-2.0	0.7
106.60	3.22	0.2	349.45	1MR	-0.1	-3.4	-2.0	0.6
108.67	2.93	0.4	359.37	11MR	-0.6	-3.3	-1.9	0.6
111.52	2.66	0.6	9.22	21MR	-1.4	-3.3	-1.8	0.6
114.99	2.41	0.8	19.01	31MR	-1.5	-3.3	-1.8	0.5
118.97	2.18	1.0	28.75	10AP	-0.6	-3.3	-1.7	0.5
123.35	1.97	1.1	38.44	20AP	0.5	-3.3	-1.6	0.5
128.06	1.77	1.2	48.09	30AP	1.6	-3.3	-1.6	0.5
133.05	1.59	1.3	57.70	10MY	3.0	-3.3	-1.5	0.5
138.27	1.42	1.4	67.27	20MY	3.0	-3.4	-1.5	0.5
143.69	1.26	1.5	76.82	30MY	1.7	-3.4	-1.4	0.5
149.28	1.11	1.6	86.36	9JN	0.7	-3.4	-1.4	0.5
155.04	0.97	1.6	95.88	19JN	-0.3	-3.5	-1.3	0.5
160.93	0.83	1.7	105.42	29JN	-1.3	-3.5	-1.3	0.5
166.97	0.69	1.7	114.96	9JL	-1.6	-3.5	-1.3	0.5
173.13	0.56	1.7	124.53	19JL	-0.8	-3.4	-1.3	0.4
179.41	0.44	1.7	134.12	29JL	-0.2	-3.4	-1.2	0.4
185.81	0.32	1.7	143.75	8AU	0.2	-3.4	-1.2	0.4
192.33	0.20	1.7	153.43	18AU	0.4	-3.3	-1.2	0.3
198.96	0.08	1.7	163.15	28AU	0.8	-3.3	-1.3	0.2
205.71	-0.03	1.7	172.93	7SE	1.7	-3.3	-1.3	0.2
212.57	-0.14	1.7	182.76	17SE	2.7	-3.3	-1.3	0.1
219.53	-0.25	1.7	192.65	27SE	0.1	-3.3	-1.3	0.1
226.61	-0.35	1.7	202.59	7OC	-0.8	-3.3	-1.4	-0.0
233.79	-0.45	1.6	212.59	17OC	-0.8	-3.4	-1.4	-0.1
241.08	-0.55	1.6	222.65	27OC	-0.8	-3.4	-1.5	-0.1
248.46	-0.63	1.6	232.75	6NO	-0.7	-3.4	-1.5	0.0
255.94	-0.72	1.5	242.88	16NO	-0.6	-3.5	-1.6	0.1
263.50	-0.80	1.5	253.05	26NO	-0.6	-3.5	-1.6	0.1
271.14	-0.86	1.4	263.23	6DE	-0.5	-3.6	-1.7	0.2
278.85	-0.93	1.4	273.43	16DE	-0.1	-3.6	-1.8	0.3
286.62	-0.98	1.4	283.62	26DE	2.1	-3.7	-1.8	0.3
				1411				
294.44	-1.02	1.3	293.81	5JA	1.4	-3.8	-1.9	0.4
302.30	-1.05	1.3	303.97	15JA	0.4	-3.9	-1.9	0.4
310.18	-1.08	1.3	314.10	25JA	0.2	-3.9	-2.0	0.5
318.07	-1.09	1.3	324.20	4FE	0.0	-4.0	-2.0	0.5
325.96	-1.08	1.4	334.24	14FE	-0.2	-4.1	-2.0	0.5
333.84	-1.07	1.4	344.24	24FE	-0.6	-4.2	-2.0	0.5
341.69	-1.04	1.4	354.18	6MR	-1.3	-4.3	-2.0	0.5
349.50	-1.01	1.4	4.06	16MR	-1.4	-4.3	-2.0	0.5
357.26	-0.96	1.4	13.88	26MR	-0.6	-4.1	-1.9	0.4
4.97	-0.90	1.4	23.65	5AP	0.6	-3.7	-1.9	0.4
12.60	-0.83	1.5	33.37	15AP	2.1	-3.0	-1.8	0.4
20.15	-0.75	1.5	43.04	25AP	3.9	-3.5	-1.8	0.3
27.63	-0.66	1.5	52.66	5MY	2.3	-4.0	-1.7	0.3
35.01	-0.57	1.5	62.25	15MY	1.3	-4.2	-1.6	0.4
42.30	-0.47	1.5	71.81	25MY	0.5	-4.2	-1.6	0.4
49.49	-0.36	1.5	81.36	4JN	-0.3	-4.1	-1.5	0.4
56.58	-0.25	1.6	90.89	14JN	-1.3	-4.0	-1.4	0.4
63.56	-0.13	1.6	100.42	24JN	-1.6	-3.9	-1.4	0.4
70.44	-0.01	1.6	109.96	4JL	-0.7	-3.9	-1.3	0.4
77.21	0.12	1.6	119.51	14JL	-0.1	-3.8	-1.3	0.4
83.86	0.25	1.6	129.09	24JL	0.3	-3.7	-1.3	0.4
90.40	0.38	1.6	138.70	3AU	0.6	-3.6	-1.2	0.3
96.81	0.52	1.5	148.35	13AU	1.0	-3.6	-1.2	0.3
103.08	0.66	1.5	158.05	23AU	2.2	-3.5	-1.2	0.3
109.22	0.81	1.5	167.79	2SE	2.4	-3.5	-1.2	0.2
115.20	0.97	1.4	177.60	12SE	0.0	-3.4	-1.2	0.1
121.00	1.13	1.4	187.46	22SE	-0.9	-3.4	-1.2	0.1
126.60	1.30	1.3	197.37	2OC	-1.0	-3.4	-1.2	0.0
131.97	1.49	1.2	207.35	12OC	-0.9	-3.4	-1.2	-0.1
137.07	1.68	1.1	217.38	22OC	-0.6	-3.4	-1.3	-0.1
141.84	1.90	1.0	227.45	1NO	-0.4	-3.4	-1.3	-0.2
146.23	2.13	0.9	237.56	11NO	-0.4	-3.4	-1.4	-0.2
150.14	2.38	0.7	247.72	21NO	-0.4	-3.4	-1.4	-0.1
153.45	2.66	0.5	257.89	1DE	0.1	-3.4	-1.5	-0.0
156.04	2.96	0.3	268.08	11DE	2.4	-3.4	-1.5	0.0
157.72	3.29	0.1	278.28	21DE	1.1	-3.4	-1.6	0.1
158.32	3.63	-0.2	288.47	31DE	0.2	-3.4	-1.6	0.2

♂			☉	16.00UT	☿	♀	♃	♄
LONG	LAT	MAG	LONG	1412	MAGNITUDES			
157.67	3.96	-0.4	298.65	10JA	0.0	-3.4	-1.7	0.2
155.71	4.24	-0.7	308.80	20JA	-0.1	-3.5	-1.8	0.3
152.59	4.42	-0.9	318.91	30JA	-0.3	-3.5	-1.8	0.3
148.76	4.44	-1.0	328.98	9FE	-0.7	-3.5	-1.9	0.4
144.90	4.29	-0.9	339.00	19FE	-1.3	-3.4	-1.9	0.4
141.71	3.99	-0.7	348.97	29FE	-1.3	-3.4	-2.0	0.4
139.65	3.60	-0.5	358.88	10MR	-0.6	-3.4	-2.0	0.4
138.88	3.19	-0.2	8.74	20MR	0.8	-3.4	-2.0	0.4
139.36	2.78	0.0	18.53	30MR	2.6	-3.4	-2.0	0.4
140.91	2.40	0.2	28.28	9AP	3.0	-3.3	-2.0	0.3
143.36	2.05	0.4	37.97	19AP	1.7	-3.3	-1.9	0.3
146.56	1.74	0.6	47.62	29AP	1.0	-3.3	-1.9	0.3
150.34	1.46	0.7	57.23	9MY	0.4	-3.3	-1.8	0.2
154.62	1.20	0.9	66.80	19MY	-0.4	-3.3	-1.8	0.2
159.30	0.97	1.0	76.35	29MY	-1.3	-3.4	-1.7	0.3
164.33	0.76	1.1	85.89	8JN	-1.6	-3.4	-1.7	0.3
169.65	0.57	1.2	95.42	18JN	-0.7	-3.4	-1.6	0.3
175.22	0.39	1.2	104.95	28JN	-0.0	-3.5	-1.5	0.3
181.02	0.22	1.3	114.50	8JL	0.4	-3.5	-1.5	0.3
187.03	0.06	1.3	124.06	18JL	0.8	-3.6	-1.4	0.3
193.22	-0.08	1.4	133.65	28JL	1.4	-3.6	-1.4	0.3
199.59	-0.22	1.4	143.28	7AU	2.7	-3.7	-1.3	0.3
206.12	-0.35	1.4	152.96	17AU	2.0	-3.8	-1.3	0.3
212.81	-0.47	1.4	162.67	27AU	-0.0	-3.9	-1.3	0.2
219.64	-0.58	1.5	172.45	6SE	-1.0	-4.0	-1.2	0.2
226.62	-0.68	1.5	182.28	16SE	-1.1	-4.1	-1.2	0.1
233.72	-0.77	1.5	192.16	26SE	-0.9	-4.2	-1.2	0.1
240.96	-0.86	1.5	202.11	6OC	-0.5	-4.3	-1.2	0.0
248.31	-0.93	1.5	212.11	16OC	-0.3	-4.3	-1.2	-0.1
255.77	-0.99	1.4	222.16	26OC	-0.3	-4.3	-1.2	-0.1
263.33	-1.04	1.4	232.26	5NO	-0.2	-3.9	-1.2	-0.2
270.97	-1.08	1.4	242.39	15NO	0.3	-3.1	-1.3	-0.3
278.70	-1.11	1.4	252.55	25NO	2.7	-3.5	-1.3	-0.3
286.49	-1.13	1.4	262.74	5DE	0.9	-4.1	-1.4	-0.2
294.33	-1.13	1.4	272.94	15DE	-0.0	-4.3	-1.4	-0.1
302.20	-1.13	1.4	283.13	25DE	-0.1	-4.4	-1.5	-0.0
				1413				
310.10	-1.11	1.4	293.32	4JA	-0.2	-4.3	-1.5	0.0
318.01	-1.08	1.4	303.48	14JA	-0.4	-4.2	-1.6	0.1
325.91	-1.04	1.4	313.61	24JA	-0.7	-4.1	-1.6	0.2
333.79	-0.98	1.4	323.71	3FE	-1.2	-4.0	-1.7	0.2
341.64	-0.92	1.4	333.76	13FE	-1.2	-3.9	-1.8	0.3
349.44	-0.85	1.4	343.75	23FE	-0.6	-3.8	-1.8	0.3
357.19	-0.77	1.4	353.70	5MR	1.0	-3.7	-1.9	0.3
4.88	-0.69	1.4	3.58	15MR	3.2	-3.6	-2.0	0.3
12.49	-0.59	1.4	13.41	25MR	2.2	-3.6	-2.0	0.3
20.03	-0.50	1.4	23.18	4AP	1.2	-3.5	-2.0	0.3
27.48	-0.40	1.5	32.90	14AP	0.7	-3.4	-2.0	0.3
34.85	-0.29	1.5	42.57	24AP	0.2	-3.4	-2.1	0.3
42.13	-0.19	1.6	52.20	4MY	-0.4	-3.4	-2.0	0.3
49.32	-0.08	1.6	61.79	14MY	-1.3	-3.3	-2.0	0.2
56.42	0.02	1.7	71.36	24MY	-1.6	-3.3	-2.0	0.2
63.43	0.13	1.7	80.90	3JN	-0.7	-3.3	-1.9	0.2
70.36	0.24	1.7	90.43	13JN	0.1	-3.3	-1.9	0.2
77.19	0.35	1.8	99.96	23JN	0.6	-3.3	-1.8	0.3
83.94	0.45	1.8	109.50	3JL	1.1	-3.3	-1.8	0.3
90.61	0.56	1.8	119.05	13JL	1.8	-3.3	-1.7	0.3
97.19	0.67	1.8	128.62	23JL	3.2	-3.4	-1.6	0.3
103.69	0.77	1.8	138.24	2AU	1.7	-3.4	-1.6	0.3
110.11	0.87	1.8	147.88	12AU	-0.1	-3.4	-1.5	0.3
116.44	0.98	1.8	157.57	22AU	-1.1	-3.5	-1.5	0.3
122.69	1.08	1.8	167.32	1SE	-1.3	-3.5	-1.4	0.3
128.83	1.19	1.8	177.12	11SE	-0.9	-3.5	-1.4	0.2
134.89	1.29	1.7	186.97	21SE	-0.4	-3.4	-1.3	0.2
140.83	1.40	1.7	196.89	1OC	-0.2	-3.4	-1.3	0.1
146.65	1.51	1.6	206.86	11OC	-0.1	-3.4	-1.3	0.1
152.34	1.63	1.6	216.88	21OC	-0.0	-3.4	-1.3	-0.0
157.86	1.74	1.5	226.95	31OC	0.5	-3.3	-1.3	-0.1
163.20	1.87	1.4	237.07	10NO	3.0	-3.3	-1.3	-0.2
168.32	2.00	1.3	247.22	20NO	0.7	-3.3	-1.3	-0.2
173.18	2.13	1.1	257.40	30NO	-0.2	-3.3	-1.3	-0.3
177.71	2.28	1.0	267.59	10DE	-0.3	-3.3	-1.3	-0.3
181.85	2.43	0.8	277.78	20DE	-0.3	-3.4	-1.4	-0.2
185.50	2.60	0.6	287.98	30DE	-0.5	-3.4	-1.4	-0.1

♂			☉	16.00UT	☿	♀	♃	♄
LONG	LAT	MAG	LONG	1414	MAGNITUDES			
188.53	2.77	0.4	298.15	9JA	-0.8	-3.4	-1.4	-0.1
190.82	2.94	0.2	308.30	19JA	-1.1	-3.4	-1.5	0.0
192.17	3.11	-0.1	318.42	29JA	-1.1	-3.5	-1.5	0.1
192.41	3.26	-0.4	328.49	8FE	-0.5	-3.5	-1.6	0.1
191.40	3.36	-0.6	338.52	18FE	1.2	-3.6	-1.7	0.2
189.12	3.38	-0.9	348.49	28FE	3.2	-3.6	-1.7	0.2
185.82	3.28	-1.1	358.41	10MR	1.6	-3.7	-1.8	0.3
182.01	3.04	-1.2	8.27	20MR	0.9	-3.7	-1.9	0.3
178.39	2.68	-1.1	18.07	30MR	0.5	-3.8	-1.9	0.3
175.64	2.24	-0.9	27.81	9AP	0.1	-3.9	-2.0	0.3
174.09	1.79	-0.7	37.51	19AP	-0.5	-4.0	-2.1	0.3
173.87	1.35	-0.4	47.16	29AP	-1.3	-4.1	-2.1	0.3
174.88	0.96	-0.2	56.77	9MY	-1.6	-4.2	-2.1	0.3
176.95	0.61	-0.0	66.35	19MY	-0.7	-4.2	-2.2	0.3
179.91	0.30	0.1	75.90	29MY	0.2	-4.1	-2.2	0.2
183.59	0.03	0.3	85.43	8JN	0.8	-3.9	-2.1	0.2
187.88	-0.21	0.4	94.96	18JN	1.4	-3.2	-2.1	0.2
192.67	-0.42	0.5	104.50	28JN	2.4	-3.1	-2.1	0.3
197.88	-0.60	0.6	114.04	8JL	3.1	-3.8	-2.0	0.3
203.46	-0.76	0.7	123.60	18JL	1.4	-4.1	-2.0	0.3
209.35	-0.90	0.8	133.19	28JL	-0.1	-4.2	-1.9	0.3
215.51	-1.02	0.9	142.82	7AU	-1.2	-4.2	-1.8	0.4
221.92	-1.13	0.9	152.49	17AU	-1.4	-4.1	-1.8	0.4
228.55	-1.21	1.0	162.20	27AU	-0.8	-4.1	-1.7	0.4
235.37	-1.29	1.0	171.97	6SE	-0.4	-4.0	-1.7	0.3
242.37	-1.34	1.1	181.80	16SE	-0.1	-3.9	-1.6	0.3
249.54	-1.38	1.1	191.68	26SE	0.0	-3.8	-1.5	0.3
256.84	-1.41	1.1	201.62	6OC	0.2	-3.7	-1.5	0.2
264.28	-1.42	1.2	211.62	16OC	0.8	-3.7	-1.5	0.2
271.82	-1.41	1.2	221.67	26OC	3.4	-3.6	-1.4	0.1
279.46	-1.39	1.2	231.76	5NO	0.5	-3.6	-1.4	0.0
287.17	-1.36	1.2	241.89	15NO	-0.4	-3.5	-1.4	-0.0
294.94	-1.31	1.3	252.06	25NO	-0.4	-3.5	-1.4	-0.1
302.75	-1.25	1.3	262.24	5DE	-0.5	-3.4	-1.4	-0.2
310.58	-1.18	1.3	272.44	15DE	-0.6	-3.4	-1.4	-0.3
318.42	-1.10	1.3	282.64	25DE	-0.8	-3.4	-1.4	-0.2
				1415				
326.26	-1.02	1.4	292.82	4JA	-0.9	-3.4	-1.4	-0.2
334.07	-0.92	1.4	302.99	14JA	-1.0	-3.3	-1.4	-0.1
341.84	-0.82	1.4	313.13	24JA	-0.4	-3.3	-1.5	-0.0
349.57	-0.71	1.5	323.22	3FE	1.4	-3.3	-1.5	0.0
357.24	-0.60	1.5	333.28	13FE	2.7	-3.3	-1.6	0.1
4.85	-0.49	1.5	343.27	23FE	1.2	-3.3	-1.6	0.2
12.38	-0.38	1.5	353.22	5MR	0.6	-3.4	-1.7	0.2
19.84	-0.27	1.6	3.11	15MR	0.4	-3.4	-1.7	0.3
27.21	-0.16	1.6	12.94	25MR	0.0	-3.4	-1.8	0.3
34.50	-0.05	1.6	22.71	4AP	-0.5	-3.4	-1.9	0.3
41.71	0.06	1.6	32.43	14AP	-1.3	-3.5	-1.9	0.3
48.84	0.16	1.6	42.10	24AP	-1.6	-3.5	-2.0	0.4
55.88	0.26	1.7	51.73	4MY	-0.7	-3.4	-2.1	0.4
62.85	0.36	1.7	61.33	14MY	0.3	-3.4	-2.1	0.4
69.73	0.45	1.7	70.89	24MY	1.0	-3.4	-2.2	0.3
76.54	0.54	1.7	80.43	3JN	1.9	-3.4	-2.2	0.3
83.29	0.63	1.8	89.97	13JN	3.1	-3.3	-2.3	0.3
89.96	0.71	1.8	99.50	23JN	2.6	-3.3	-2.3	0.3
96.58	0.79	1.9	109.03	3JL	1.1	-3.3	-2.3	0.3
103.13	0.87	1.9	118.59	13JL	-0.2	-3.3	-2.3	0.3
109.63	0.94	1.9	128.16	23JL	-1.2	-3.3	-2.2	0.4
116.08	1.01	2.0	137.76	2AU	-1.5	-3.3	-2.2	0.4
122.47	1.07	2.0	147.41	12AU	-0.8	-3.4	-2.1	0.4
128.82	1.14	2.0	157.10	22AU	-0.3	-3.4	-2.1	0.4
135.11	1.20	2.0	166.84	1SE	0.0	-3.4	-2.0	0.4
141.36	1.25	2.0	176.64	11SE	0.2	-3.5	-2.0	0.4
147.56	1.31	1.9	186.49	21SE	0.4	-3.5	-1.9	0.4
153.71	1.36	1.9	196.40	1OC	1.1	-3.6	-1.8	0.4
159.81	1.41	1.9	206.37	11OC	3.3	-3.7	-1.8	0.4
165.84	1.46	1.8	216.39	21OC	0.3	-3.8	-1.7	0.3
171.81	1.51	1.8	226.46	31OC	-0.5	-3.9	-1.7	0.3
177.70	1.55	1.7	236.58	10NO	-0.6	-4.0	-1.6	0.2
183.49	1.59	1.6	246.72	20NO	-0.6	-4.1	-1.6	0.1
189.18	1.63	1.5	256.90	30NO	-0.7	-4.2	-1.5	0.0
194.75	1.67	1.4	267.09	10DE	-0.8	-4.3	-1.5	-0.0
200.16	1.70	1.3	277.29	20DE	-0.8	-4.4	-1.5	-0.1
205.39	1.73	1.2	287.48	30DE	-0.8	-4.4	-1.5	-0.2

♂ LONG	♂ LAT	♂ MAG	☉ LONG	16.00UT	☿	♀	♃	♄
				1416	MAGNITUDES			
210.39	1.74	1.0	297.66	9JA	-0.3	-4.3	-1.5	-0.1
215.12	1.76	0.8	307.81	19JA	1.7	-4.0	-1.5	-0.1
219.52	1.75	0.6	317.93	29JA	2.2	-3.4	-1.5	-0.0
223.48	1.74	0.4	328.01	8FE	0.8	-3.6	-1.5	0.1
226.93	1.70	0.2	338.03	18FE	0.4	-4.1	-1.5	0.1
229.71	1.63	-0.1	348.01	28FE	0.2	-4.3	-1.6	0.2
231.66	1.52	-0.4	357.93	9MR	-0.1	-4.3	-1.6	0.2
232.61	1.35	-0.7	7.79	19MR	-0.6	-4.2	-1.7	0.3
232.38	1.11	-1.0	17.59	29MR	-1.3	-4.1	-1.7	0.3
230.89	0.79	-1.3	27.34	8AP	-1.6	-4.0	-1.8	0.4
228.26	0.38	-1.6	37.04	18AP	-0.7	-3.9	-1.8	0.4
224.86	-0.10	-1.8	46.69	28AP	0.4	-3.8	-1.9	0.4
221.37	-0.59	-1.7	56.31	8MY	1.3	-3.7	-2.0	0.4
218.51	-1.05	-1.5	65.88	18MY	2.5	-3.6	-2.1	0.5
216.79	-1.44	-1.3	75.44	28MY	3.4	-3.6	-2.1	0.5
216.43	-1.75	-1.1	84.98	7JN	2.0	-3.5	-2.2	0.4
217.42	-1.98	-0.9	94.50	17JN	0.9	-3.5	-2.3	0.4
219.58	-2.15	-0.7	104.04	27JN	-0.2	-3.4	-2.3	0.4
222.75	-2.27	-0.5	113.58	7JL	-1.3	-3.4	-2.4	0.4
226.76	-2.35	-0.3	123.14	17JL	-1.6	-3.4	-2.4	0.4
231.44	-2.39	-0.2	132.73	27JL	-0.8	-3.3	-2.4	0.5
236.69	-2.40	-0.1	142.36	6AU	-0.2	-3.3	-2.4	0.5
242.40	-2.39	0.1	152.02	16AU	0.1	-3.3	-2.4	0.5
248.49	-2.35	0.2	161.74	26AU	0.3	-3.3	-2.4	0.6
254.92	-2.30	0.3	171.50	5SE	0.6	-3.4	-2.3	0.6
261.61	-2.22	0.4	181.32	15SE	1.4	-3.4	-2.3	0.6
268.53	-2.13	0.5	191.21	25SE	3.0	-3.4	-2.2	0.6
275.63	-2.02	0.6	201.14	5OC	0.2	-3.4	-2.1	0.6
282.88	-1.90	0.6	211.13	15OC	-0.7	-3.4	-2.1	0.5
290.26	-1.77	0.7	221.18	25OC	-0.8	-3.5	-2.0	0.5
297.72	-1.63	0.8	231.27	4NO	-0.7	-3.5	-1.9	0.5
305.25	-1.49	0.9	241.40	14NO	-0.8	-3.5	-1.9	0.4
312.83	-1.33	1.0	251.56	24NO	-0.6	-3.5	-1.8	0.3
320.43	-1.18	1.0	261.74	4DE	-0.6	-3.4	-1.7	0.3
328.02	-1.02	1.1	271.94	14DE	-0.6	-3.4	-1.7	0.2
335.61	-0.87	1.2	282.14	24DE	-0.2	-3.4	-1.7	0.1
				1417				
343.17	-0.71	1.3	292.32	3JA	1.9	-3.4	-1.6	0.1
350.69	-0.56	1.3	302.49	13JA	1.7	-3.4	-1.6	-0.0
358.16	-0.42	1.4	312.63	23JA	0.6	-3.3	-1.6	0.0
5.56	-0.28	1.5	322.72	2FE	0.3	-3.3	-1.6	0.1
12.91	-0.15	1.5	332.78	12FE	0.1	-3.3	-1.6	0.1
20.19	-0.02	1.6	342.79	22FE	-0.2	-3.3	-1.6	0.2
27.39	0.10	1.6	352.73	4MR	-0.6	-3.4	-1.6	0.3
34.52	0.21	1.7	2.62	14MR	-1.3	-3.4	-1.6	0.3
41.58	0.32	1.7	12.46	24MR	-1.5	-3.4	-1.6	0.4
48.56	0.42	1.8	22.24	3AP	-0.7	-3.4	-1.6	0.4
55.48	0.51	1.8	31.96	13AP	0.5	-3.5	-1.7	0.5
62.32	0.60	1.8	41.64	23AP	1.7	-3.5	-1.7	0.5
69.11	0.67	1.8	51.27	3MY	3.3	-3.5	-1.8	0.5
75.83	0.75	1.9	60.87	13MY	2.7	-3.6	-1.8	0.6
82.49	0.81	1.9	70.44	23MY	1.5	-3.7	-1.9	0.6
89.10	0.87	1.9	79.98	2JN	0.7	-3.7	-2.0	0.6
95.66	0.93	1.9	89.51	12JN	-0.3	-3.8	-2.0	0.6
102.19	0.98	1.9	99.04	22JN	-1.3	-3.9	-2.1	0.6
108.67	1.02	1.9	108.58	2JL	-1.6	-4.0	-2.2	0.6
115.12	1.06	1.9	118.13	12JL	-0.8	-4.1	-2.2	0.6
121.54	1.10	2.0	127.70	22JL	-0.2	-4.2	-2.3	0.6
127.93	1.13	2.0	137.30	1AU	0.2	-4.2	-2.4	0.6
134.30	1.16	2.0	146.95	11AU	0.5	-4.1	-2.4	0.7
140.66	1.18	2.0	156.64	21AU	0.8	-3.8	-2.4	0.7
147.00	1.20	2.0	166.37	31AU	1.8	-3.3	-2.5	0.7
153.33	1.21	2.0	176.17	10SE	2.6	-3.5	-2.5	0.8
159.65	1.22	2.0	186.02	20SE	0.1	-4.0	-2.4	0.8
165.95	1.22	2.0	195.92	30SE	-0.8	-4.3	-2.4	0.8
172.25	1.22	2.0	205.89	10OC	-0.9	-4.3	-2.4	0.8
178.54	1.22	2.0	215.91	20OC	-0.9	-4.3	-2.3	0.8
184.82	1.21	1.9	225.97	30OC	-0.7	-4.2	-2.3	0.7
191.08	1.20	1.9	236.08	9NO	-0.5	-4.1	-2.2	0.7
197.33	1.17	1.8	246.23	19NO	-0.5	-4.0	-2.1	0.7
203.56	1.15	1.7	256.40	29NO	-0.5	-3.9	-2.0	0.6
209.76	1.11	1.7	266.59	9DE	-0.0	-3.8	-2.0	0.5
215.93	1.06	1.6	276.79	19DE	2.1	-3.8	-1.9	0.5
222.06	1.01	1.5	286.98	29DE	1.4	-3.7	-1.8	0.4

♂ LONG	♂ LAT	♂ MAG	☉ LONG	16.00UT	☿	♀	♃	♄
				1418	MAGNITUDES			
228.14	0.94	1.4	297.16	8JA	0.3	-3.6	-1.8	0.3
234.16	0.85	1.2	307.32	18JA	0.1	-3.5	-1.7	0.3
240.10	0.75	1.1	317.44	28JA	-0.0	-3.5	-1.7	0.2
245.95	0.63	1.0	327.52	7FE	-0.2	-3.4	-1.7	0.2
251.68	0.49	0.8	337.55	17FE	-0.6	-3.4	-1.6	0.3
257.27	0.32	0.6	347.52	27FE	-1.3	-3.4	-1.6	0.3
262.68	0.12	0.4	357.45	9MR	-1.4	-3.3	-1.6	0.4
267.86	-0.13	0.2	7.31	19MR	-0.7	-3.3	-1.6	0.5
272.76	-0.42	0.0	17.12	29MR	0.7	-3.3	-1.6	0.5
277.29	-0.76	-0.2	26.87	8AP	2.2	-3.3	-1.6	0.6
281.35	-1.17	-0.5	36.57	18AP	3.6	-3.3	-1.6	0.6
284.83	-1.66	-0.8	46.23	28AP	2.0	-3.3	-1.6	0.7
287.55	-2.23	-1.1	55.84	8MY	1.2	-3.3	-1.6	0.7
289.33	-2.88	-1.4	65.42	18MY	0.5	-3.4	-1.7	0.7
290.00	-3.61	-1.7	74.98	28MY	-0.3	-3.4	-1.7	0.8
289.43	-4.37	-2.0	84.52	7JN	-1.3	-3.4	-1.8	0.8
287.68	-5.08	-2.3	94.05	17JN	-1.7	-3.5	-1.8	0.8
285.11	-5.66	-2.5	103.57	27JN	-0.8	-3.5	-1.9	0.8
282.33	-5.99	-2.4	113.12	7JL	-0.1	-3.5	-1.9	0.8
280.09	-6.04	-2.3	122.68	17JL	0.4	-3.4	-2.0	0.8
278.95	-5.86	-2.1	132.26	27JL	0.7	-3.4	-2.1	0.8
279.14	-5.51	-1.8	141.89	6AU	1.2	-3.4	-2.1	0.8
280.68	-5.07	-1.6	151.55	16AU	2.3	-3.3	-2.2	0.9
283.39	-4.60	-1.3	161.26	26AU	2.3	-3.3	-2.3	0.9
287.08	-4.11	-1.1	171.03	5SE	0.1	-3.3	-2.3	1.0
291.56	-3.64	-0.8	180.84	15SE	-0.9	-3.3	-2.4	1.0
296.66	-3.19	-0.6	190.72	25SE	-1.1	-3.3	-2.4	1.0
302.23	-2.77	-0.4	200.66	5OC	-1.0	-3.3	-2.4	1.0
308.18	-2.37	-0.2	210.65	15OC	-0.6	-3.4	-2.4	1.0
314.41	-2.00	-0.1	220.69	25OC	-0.4	-3.4	-2.4	1.0
320.86	-1.66	0.1	230.78	4NO	-0.3	-3.4	-2.4	1.0
327.46	-1.35	0.3	240.91	14NO	-0.3	-3.5	-2.4	1.0
334.18	-1.06	0.4	251.07	24NO	0.2	-3.5	-2.3	1.0
340.98	-0.80	0.6	261.25	4DE	2.4	-3.6	-2.2	0.9
347.83	-0.56	0.7	271.44	14DE	1.1	-3.6	-2.2	0.9
354.71	-0.35	0.9	281.64	24DE	0.1	-3.7	-2.1	0.8
				1419				
1.60	-0.15	1.0	291.83	3JA	-0.1	-3.8	-2.0	0.7
8.48	0.02	1.1	302.00	13JA	-0.1	-3.9	-1.9	0.7
15.34	0.18	1.2	312.13	23JA	-0.3	-3.9	-1.9	0.6
22.19	0.32	1.3	322.24	2FE	-0.7	-4.0	-1.8	0.5
28.99	0.45	1.4	332.29	12FE	-1.3	-4.1	-1.8	0.5
35.76	0.56	1.5	342.30	22FE	-1.3	-4.2	-1.7	0.5
42.50	0.66	1.6	352.25	4MR	-0.6	-4.3	-1.7	0.5
49.19	0.74	1.7	2.14	14MR	0.8	-4.3	-1.6	0.6
55.83	0.82	1.7	11.98	24MR	2.8	-4.1	-1.6	0.7
62.44	0.89	1.8	21.76	3AP	2.7	-3.7	-1.6	0.7
69.01	0.95	1.8	31.49	13AP	1.5	-3.0	-1.5	0.8
75.53	0.99	1.9	41.17	23AP	0.9	-3.5	-1.5	0.8
82.03	1.04	1.9	50.80	3MY	0.3	-4.0	-1.5	0.9
88.49	1.07	2.0	60.40	13MY	-0.4	-4.2	-1.5	0.9
94.92	1.10	2.0	69.97	23MY	-1.3	-4.2	-1.5	1.0
101.33	1.12	2.0	79.52	2JN	-1.7	-4.1	-1.6	1.0
107.72	1.14	2.0	89.05	12JN	-0.7	-4.0	-1.6	1.0
114.09	1.15	2.0	98.58	22JN	-0.0	-3.9	-1.6	1.1
120.45	1.16	2.0	108.12	2JL	0.5	-3.9	-1.6	1.1
126.81	1.16	2.0	117.66	12JL	0.9	-3.8	-1.7	1.1
133.16	1.15	2.0	127.24	22JL	1.5	-3.7	-1.7	1.1
139.52	1.15	2.0	136.84	1AU	2.9	-3.6	-1.8	1.1
145.88	1.13	2.0	146.48	11AU	1.9	-3.6	-1.9	1.1
152.26	1.12	2.0	156.17	21AU	-0.0	-3.5	-1.9	1.1
158.65	1.10	2.0	165.90	31AU	-1.0	-3.5	-2.0	1.1
165.05	1.07	2.0	175.69	10SE	-1.2	-3.4	-2.0	1.2
171.48	1.04	2.0	185.54	20SE	-0.9	-3.4	-2.1	1.2
177.94	1.00	2.0	195.45	30SE	-0.5	-3.4	-2.2	1.3
184.41	0.96	2.0	205.40	10OC	-0.3	-3.4	-2.2	1.3
190.92	0.92	2.0	215.42	20OC	-0.2	-3.4	-2.3	1.3
197.45	0.87	1.9	225.49	30OC	-0.1	-3.4	-2.3	1.3
204.02	0.81	1.9	235.59	9NO	0.4	-3.4	-2.3	1.3
210.61	0.74	1.9	245.74	19NO	2.7	-3.4	-2.3	1.3
217.24	0.67	1.8	255.91	29NO	0.8	-3.4	-2.3	1.3
223.89	0.59	1.8	266.10	9DE	-0.1	-3.4	-2.3	1.2
230.58	0.50	1.7	276.30	19DE	-0.2	-3.4	-2.3	1.2
237.30	0.40	1.6	286.49	29DE	-0.3	-3.4	-2.2	1.1

♂ LONG	♂ LAT	♂ MAG	☉ LONG	16.00UT	☿	♀	♃	♄
				1420	MAGNITUDES			
244.04	0.29	1.6	296.67	8JA	-0.4	-3.4	-2.1	1.1
250.81	0.16	1.5	306.83	18JA	-0.8	-3.5	-2.1	1.0
257.60	0.03	1.4	316.95	28JA	-1.2	-3.5	-2.0	0.9
264.41	-0.12	1.3	327.03	7FE	-1.2	-3.5	-1.9	0.9
271.24	-0.28	1.2	337.06	17FE	-0.6	-3.4	-1.9	0.8
278.08	-0.46	1.1	347.04	27FE	1.0	-3.4	-1.8	0.8
284.93	-0.65	1.0	356.96	8MR	3.3	-3.4	-1.7	0.8
291.78	-0.86	0.8	6.83	18MR	2.0	-3.4	-1.7	0.9
298.61	-1.09	0.7	16.64	28MR	1.1	-3.4	-1.6	0.9
305.41	-1.33	0.6	26.39	7AP	0.6	-3.3	-1.6	1.0
312.17	-1.59	0.4	36.10	17AP	0.2	-3.3	-1.5	1.0
318.84	-1.87	0.3	45.76	27AP	-0.4	-3.3	-1.5	1.1
325.43	-2.16	0.1	55.37	7MY	-1.3	-3.3	-1.5	1.1
331.86	-2.47	-0.0	64.96	17MY	-1.7	-3.3	-1.4	1.2
338.10	-2.78	-0.2	74.51	27MY	-0.7	-3.4	-1.4	1.2
344.10	-3.11	-0.4	84.05	6JN	0.1	-3.4	-1.4	1.3
349.75	-3.44	-0.6	93.58	16JN	0.7	-3.4	-1.4	1.3
354.96	-3.78	-0.8	103.11	26JN	1.2	-3.5	-1.4	1.3
359.60	-4.12	-1.0	112.65	6JL	2.0	-3.5	-1.5	1.4
3.51	-4.44	-1.2	122.22	16JL	3.2	-3.6	-1.5	1.4
6.49	-4.74	-1.5	131.80	26JL	1.6	-3.6	-1.5	1.4
8.34	-4.98	-1.7	141.42	5AU	-0.1	-3.7	-1.5	1.4
8.84	-5.13	-1.9	151.09	15AU	-1.1	-3.8	-1.6	1.3
7.92	-5.13	-2.2	160.79	25AU	-1.3	-3.9	-1.6	1.3
5.75	-4.92	-2.4	170.55	4SE	-0.9	-4.0	-1.7	1.2
2.80	-4.47	-2.5	180.37	14SE	-0.4	-4.1	-1.7	1.2
359.86	-3.83	-2.3	190.25	24SE	-0.2	-4.2	-1.8	1.2
357.67	-3.09	-2.0	200.18	4OC	-0.1	-4.3	-1.9	1.2
356.68	-2.34	-1.7	210.17	14OC	0.1	-4.3	-1.9	1.2
357.03	-1.66	-1.4	220.20	24OC	0.6	-4.3	-2.0	1.2
358.59	-1.08	-1.1	230.29	3NO	3.1	-3.9	-2.1	1.2
1.19	-0.59	-0.7	240.42	13NO	0.6	-3.1	-2.1	1.2
4.62	-0.19	-0.5	250.57	23NO	-0.3	-3.5	-2.2	1.1
8.69	0.14	-0.2	260.75	3DE	-0.4	-4.1	-2.2	1.1
13.26	0.40	0.1	270.95	13DE	-0.4	-4.3	-2.2	1.1
18.22	0.61	0.3	281.15	23DE	-0.5	-4.4	-2.2	1.0
				1421				
23.47	0.79	0.5	291.33	2JA	-0.8	-4.3	-2.2	1.0
28.95	0.92	0.7	301.51	12JA	-1.0	-4.2	-2.2	0.9
34.61	1.04	0.9	311.64	22JA	-1.0	-4.1	-2.1	0.9
40.39	1.12	1.0	321.75	1FE	-0.5	-4.0	-2.1	0.8
46.28	1.19	1.2	331.81	11FE	1.2	-3.9	-2.0	0.8
52.23	1.25	1.3	341.82	21FE	3.1	-3.8	-1.9	0.7
58.25	1.29	1.4	351.77	3MR	1.5	-3.7	-1.9	0.7
64.31	1.32	1.5	1.67	13MR	0.8	-3.6	-1.8	0.7
70.40	1.34	1.6	11.51	23MR	0.5	-3.6	-1.7	0.8
76.51	1.35	1.7	21.29	2AP	0.1	-3.5	-1.7	0.9
82.64	1.35	1.8	31.02	12AP	-0.5	-3.4	-1.6	0.9
88.79	1.35	1.8	40.70	22AP	-1.3	-3.4	-1.5	1.0
94.95	1.34	1.9	50.34	2MY	-1.7	-3.4	-1.5	1.1
101.12	1.33	1.9	59.94	12MY	-0.7	-3.3	-1.5	1.1
107.30	1.31	2.0	69.51	22MY	0.2	-3.3	-1.4	1.2
113.50	1.28	2.0	79.06	1JN	0.9	-3.3	-1.4	1.2
119.71	1.26	2.0	88.59	11JN	1.6	-3.3	-1.4	1.2
125.94	1.23	2.0	98.12	21JN	2.7	-3.3	-1.3	1.3
132.19	1.19	2.0	107.66	1JL	2.9	-3.3	-1.3	1.3
138.46	1.15	2.0	117.20	11JL	1.3	-3.3	-1.3	1.3
144.76	1.11	2.0	126.77	21JL	-0.1	-3.4	-1.3	1.3
151.10	1.07	2.0	136.38	31JL	-1.2	-3.4	-1.3	1.2
157.47	1.02	2.0	146.02	10AU	-1.5	-3.4	-1.3	1.2
163.87	0.96	2.0	155.70	20AU	-0.9	-3.5	-1.4	1.2
170.33	0.91	1.9	165.43	30AU	-0.3	-3.5	-1.4	1.1
176.82	0.85	1.9	175.22	9SE	-0.1	-3.5	-1.4	1.1
183.37	0.78	1.9	185.06	19SE	0.1	-3.4	-1.5	1.0
189.97	0.71	1.9	194.96	29SE	0.2	-3.4	-1.5	1.1
196.62	0.64	1.9	204.92	9OC	0.9	-3.4	-1.6	1.1
203.33	0.56	1.9	214.93	19OC	3.4	-3.4	-1.6	1.1
210.10	0.48	1.9	225.00	29OC	0.5	-3.3	-1.7	1.1
216.93	0.39	1.8	235.10	8NO	-0.4	-3.3	-1.8	1.0
223.82	0.30	1.8	245.24	18NO	-0.5	-3.3	-1.8	1.0
230.77	0.20	1.8	255.42	28NO	-0.5	-3.3	-1.9	1.0
237.79	0.10	1.8	265.60	8DE	-0.6	-3.3	-2.0	1.0
244.86	-0.01	1.7	275.80	18DE	-0.8	-3.4	-2.0	0.9
252.00	-0.13	1.7	286.00	28DE	-0.9	-3.4	-2.1	0.9

♂ LONG	♂ LAT	♂ MAG	☉ LONG	16.00UT	☿	♀	♃	♄
				1422	MAGNITUDES			
259.20	-0.25	1.6	296.17	7JA	-0.9	-3.4	-2.1	0.9
266.45	-0.37	1.6	306.33	17JA	-0.4	-3.4	-2.1	0.8
273.76	-0.50	1.5	316.46	27JA	1.4	-3.5	-2.1	0.8
281.12	-0.63	1.5	326.54	6FE	2.5	-3.5	-2.1	0.7
288.52	-0.77	1.4	336.57	16FE	1.1	-3.6	-2.0	0.7
295.97	-0.90	1.3	346.56	26FE	0.6	-3.6	-2.0	0.6
303.45	-1.04	1.3	356.49	8MR	0.3	-3.7	-1.9	0.6
310.95	-1.18	1.2	6.36	18MR	-0.0	-3.8	-1.9	0.5
318.46	-1.31	1.1	16.17	28MR	-0.5	-3.8	-1.8	0.6
325.98	-1.44	1.0	25.93	7AP	-1.3	-3.9	-1.8	0.6
333.48	-1.56	1.0	35.63	17AP	-1.6	-4.0	-1.7	0.7
340.96	-1.67	0.9	45.29	27AP	-0.7	-4.1	-1.6	0.8
348.39	-1.77	0.8	54.91	7MY	0.3	-4.2	-1.6	0.8
355.75	-1.86	0.7	64.49	17MY	1.1	-4.2	-1.5	0.9
3.03	-1.93	0.7	74.05	27MY	2.1	-4.1	-1.4	1.0
10.19	-1.98	0.6	83.59	6JN	3.4	-3.8	-1.4	1.0
17.21	-2.02	0.5	93.12	16JN	2.4	-3.2	-1.4	1.1
24.06	-2.03	0.4	102.65	26JN	1.0	-3.1	-1.3	1.1
30.69	-2.03	0.3	112.19	6JL	-0.2	-3.8	-1.3	1.1
37.08	-2.00	0.2	121.75	16JL	-1.2	-4.1	-1.3	1.1
43.16	-1.94	0.1	131.34	26JL	-1.6	-4.2	-1.2	1.1
48.87	-1.86	-0.0	140.95	5AU	-0.8	-4.2	-1.2	1.1
54.14	-1.75	-0.1	150.61	15AU	-0.3	-4.1	-1.2	1.1
58.85	-1.60	-0.3	160.32	25AU	0.0	-4.0	-1.2	1.1
62.88	-1.41	-0.5	170.08	4SE	0.2	-4.0	-1.3	1.0
66.07	-1.18	-0.6	179.89	14SE	0.5	-3.9	-1.3	1.0
68.22	-0.89	-0.8	189.77	24SE	1.2	-3.8	-1.3	0.9
69.13	-0.53	-1.1	199.69	4OC	3.3	-3.7	-1.3	0.9
68.62	-0.11	-1.3	209.68	14OC	0.3	-3.7	-1.4	0.9
66.65	0.36	-1.5	219.72	24OC	-0.6	-3.6	-1.4	0.9
63.49	0.87	-1.7	229.80	3NO	-0.7	-3.6	-1.5	0.9
59.72	1.34	-1.7	239.93	13NO	-0.7	-3.5	-1.5	0.9
56.15	1.73	-1.5	250.08	23NO	-0.7	-3.5	-1.6	0.9
53.51	2.01	-1.2	260.26	3DE	-0.7	-3.4	-1.7	0.9
52.16	2.18	-0.9	270.46	13DE	-0.7	-3.4	-1.7	0.9
52.15	2.27	-0.6	280.66	23DE	-0.7	-3.4	-1.8	0.9
				1423				
53.36	2.30	-0.3	290.84	2JA	-0.3	-3.4	-1.8	0.8
55.56	2.28	0.0	301.01	12JA	1.7	-3.3	-1.9	0.8
58.57	2.25	0.3	311.16	22JA	2.0	-3.3	-2.0	0.8
62.21	2.20	0.5	321.26	1FE	0.7	-3.3	-2.0	0.7
66.34	2.14	0.7	331.32	11FE	0.4	-3.3	-2.0	0.7
70.85	2.07	0.9	341.34	21FE	0.2	-3.3	-2.0	0.6
75.67	2.01	1.1	351.29	3MR	-0.1	-3.4	-2.0	0.6
80.72	1.94	1.2	1.19	13MR	-0.6	-3.4	-2.0	0.5
85.97	1.86	1.3	11.03	23MR	-1.3	-3.4	-2.0	0.5
91.37	1.79	1.5	20.82	2AP	-1.6	-3.4	-1.9	0.4
96.90	1.72	1.6	30.55	12AP	-0.7	-3.5	-1.9	0.5
102.54	1.65	1.6	40.23	22AP	0.4	-3.5	-1.8	0.5
108.27	1.57	1.7	49.87	2MY	1.5	-3.4	-1.7	0.6
114.08	1.50	1.8	59.47	12MY	2.8	-3.4	-1.7	0.7
119.96	1.43	1.8	69.04	22MY	3.2	-3.4	-1.6	0.7
125.91	1.35	1.9	78.59	1JN	1.8	-3.4	-1.5	0.8
131.92	1.28	1.9	88.12	11JN	0.8	-3.3	-1.5	0.8
137.99	1.20	2.0	97.65	21JN	-0.2	-3.3	-1.4	0.9
144.12	1.12	2.0	107.19	1JL	-1.3	-3.3	-1.4	0.9
150.32	1.04	2.0	116.74	11JL	-1.6	-3.3	-1.3	1.0
156.58	0.96	2.0	126.30	21JL	-0.8	-3.3	-1.3	1.0
162.90	0.88	2.0	135.90	31JL	-0.2	-3.3	-1.3	1.0
169.29	0.80	2.0	145.54	10AU	0.2	-3.4	-1.2	1.0
175.74	0.71	1.9	155.22	20AU	0.4	-3.4	-1.2	1.0
182.27	0.63	1.9	164.95	30AU	0.7	-3.4	-1.2	1.0
188.87	0.54	1.9	174.74	9SE	1.6	-3.5	-1.2	0.9
195.55	0.44	1.9	184.58	19SE	3.0	-3.5	-1.2	0.9
202.30	0.35	1.8	194.48	29SE	0.2	-3.6	-1.2	0.9
209.14	0.26	1.8	204.44	9OC	-0.7	-3.7	-1.2	0.8
216.05	0.16	1.7	214.45	19OC	-0.8	-3.8	-1.3	0.8
223.05	0.06	1.7	224.51	29OC	-0.8	-3.9	-1.3	0.8
230.13	-0.04	1.7	234.61	8NO	-0.8	-4.0	-1.3	0.8
237.28	-0.14	1.7	244.75	18NO	-0.6	-4.1	-1.4	0.9
244.52	-0.25	1.6	254.92	28NO	-0.6	-4.2	-1.4	0.9
251.84	-0.35	1.6	265.11	8DE	-0.6	-4.3	-1.5	0.9
259.23	-0.46	1.6	275.31	18DE	-0.1	-4.4	-1.5	0.8
266.69	-0.56	1.6	285.50	28DE	1.9	-4.4	-1.6	0.8

♂ LONG	♂ LAT	♂ MAG	⊙ LONG	16.00UT 1424	☿	♀	♃	♄
					MAGNITUDES			
274.22	-0.66	1.6	295.69	7JA	1.6	-4.3	-1.7	0.8
281.80	-0.76	1.5	305.84	17JA	0.5	-3.9	-1.7	0.8
289.44	-0.86	1.5	315.97	27JA	0.2	-3.3	-1.8	0.7
297.13	-0.95	1.5	326.05	6FE	0.0	-3.6	-1.9	0.7
304.85	-1.03	1.4	336.09	16FE	-0.2	-4.1	-1.9	0.6
312.59	-1.11	1.4	346.07	26FE	-0.6	-4.3	-2.0	0.6
320.35	-1.17	1.4	356.01	7MR	-1.3	-4.3	-2.0	0.5
328.12	-1.23	1.4	5.88	17MR	-1.5	-4.2	-2.0	0.5
335.87	-1.28	1.3	15.69	27MR	-0.7	-4.1	-2.0	0.4
343.60	-1.31	1.3	25.46	6AP	0.5	-4.0	-2.0	0.4
351.30	-1.33	1.3	35.16	16AP	1.9	-3.9	-2.0	0.3
358.95	-1.33	1.2	44.82	26AP	3.7	-3.8	-1.9	0.4
6.53	-1.32	1.2	54.45	6MY	2.4	-3.7	-1.9	0.5
14.04	-1.30	1.2	64.03	16MY	1.4	-3.6	-1.8	0.5
21.47	-1.25	1.2	73.59	26MY	0.6	-3.6	-1.8	0.6
28.79	-1.20	1.1	83.13	5JN	-0.3	-3.5	-1.7	0.7
35.99	-1.13	1.1	92.66	15JN	-1.3	-3.5	-1.7	0.7
43.07	-1.04	1.1	102.19	25JN	-1.7	-3.4	-1.6	0.8
50.00	-0.94	1.0	111.73	5JL	-0.8	-3.4	-1.5	0.8
56.77	-0.82	1.0	121.29	15JL	-0.1	-3.4	-1.5	0.9
63.36	-0.69	0.9	130.87	25JL	0.3	-3.3	-1.4	0.9
69.75	-0.54	0.9	140.49	4AU	0.6	-3.3	-1.4	0.9
75.91	-0.37	0.8	150.15	14AU	1.0	-3.3	-1.3	0.9
81.82	-0.19	0.7	159.85	24AU	2.0	-3.3	-1.3	0.9
87.41	0.01	0.7	169.61	3SE	2.6	-3.3	-1.3	0.9
92.65	0.24	0.5	179.42	13SE	0.2	-3.4	-1.2	0.9
97.46	0.49	0.4	189.29	23SE	-0.8	-3.4	-1.2	0.8
101.75	0.77	0.3	199.22	3OC	-1.0	-3.4	-1.2	0.8
105.41	1.09	0.1	209.19	13OC	-0.9	-3.4	-1.2	0.8
108.29	1.44	-0.1	219.23	23OC	-0.7	-3.5	-1.2	0.7
110.20	1.84	-0.3	229.31	2NO	-0.5	-3.5	-1.2	0.8
110.95	2.29	-0.5	239.43	12NO	-0.4	-3.5	-1.3	0.8
110.37	2.76	-0.7	249.59	22NO	-0.4	-3.5	-1.3	0.8
108.40	3.22	-0.9	259.77	2DE	0.0	-3.4	-1.3	0.8
105.22	3.62	-1.1	269.96	12DE	2.1	-3.4	-1.4	0.8
101.32	3.89	-1.2	280.16	22DE	1.3	-3.4	-1.4	0.8
				1425				
97.46	4.01	-1.1	290.35	1JA	0.2	-3.4	-1.5	0.8
94.37	3.96	-0.8	300.52	11JA	0.0	-3.4	-1.5	0.8
92.47	3.80	-0.5	310.66	21JA	-0.1	-3.3	-1.6	0.8
91.90	3.58	-0.2	320.77	31JA	-0.3	-3.3	-1.7	0.7
92.57	3.33	0.0	330.83	10FE	-0.7	-3.3	-1.7	0.7
94.27	3.08	0.3	340.85	20FE	-1.3	-3.3	-1.8	0.6
96.84	2.84	0.5	350.81	2MR	-1.4	-3.4	-1.9	0.6
100.09	2.62	0.7	0.71	12MR	-0.7	-3.4	-1.9	0.5
103.88	2.41	0.9	10.56	22MR	0.7	-3.4	-2.0	0.5
108.11	2.22	1.0	20.35	1AP	2.4	-3.4	-2.0	0.4
112.68	2.04	1.2	30.08	11AP	3.2	-3.5	-2.0	0.3
117.54	1.88	1.3	39.77	21AP	1.8	-3.5	-2.1	0.3
122.63	1.72	1.4	49.41	1MY	1.0	-3.5	-2.1	0.3
127.93	1.58	1.5	59.01	11MY	0.4	-3.6	-2.0	0.4
133.39	1.44	1.6	68.58	21MY	-0.3	-3.7	-2.0	0.4
139.00	1.30	1.6	78.13	31MY	-1.3	-3.7	-2.0	0.5
144.75	1.18	1.7	87.67	10JN	-1.7	-3.8	-1.9	0.6
150.62	1.05	1.7	97.20	20JN	-0.8	-3.9	-1.9	0.6
156.61	0.93	1.8	106.73	30JN	-0.1	-4.0	-1.8	0.7
162.71	0.81	1.8	116.28	10JL	0.4	-4.1	-1.8	0.7
168.91	0.69	1.8	125.85	20JL	0.8	-4.2	-1.7	0.8
175.22	0.58	1.8	135.44	30JL	1.3	-4.2	-1.6	0.8
181.62	0.47	1.8	145.08	9AU	2.5	-4.1	-1.6	0.8
188.14	0.36	1.8	154.76	19AU	2.2	-3.8	-1.5	0.8
194.75	0.25	1.8	164.48	29AU	0.1	-3.3	-1.5	0.8
201.46	0.14	1.8	174.27	8SE	-0.9	-3.5	-1.4	0.8
208.28	0.03	1.8	184.11	18SE	-1.1	-4.0	-1.4	0.8
215.19	-0.08	1.7	194.00	28SE	-1.0	-4.3	-1.3	0.8
222.21	-0.18	1.7	203.95	8OC	-0.6	-4.3	-1.3	0.8
229.33	-0.28	1.7	213.96	18OC	-0.4	-4.3	-1.3	0.8
236.54	-0.38	1.6	224.02	28OC	-0.3	-4.2	-1.3	0.7
243.85	-0.48	1.6	234.12	7NO	-0.2	-4.1	-1.3	0.7
251.25	-0.57	1.5	244.26	17NO	0.2	-4.0	-1.3	0.7
258.74	-0.66	1.5	254.43	27NO	2.4	-3.9	-1.3	0.8
266.30	-0.74	1.4	264.62	7DE	1.0	-3.8	-1.3	0.8
273.94	-0.82	1.4	274.82	17DE	0.0	-3.8	-1.3	0.8
281.65	-0.89	1.4	285.01	27DE	-0.1	-3.7	-1.4	0.8

♂ LONG	♂ LAT	♂ MAG	⊙ LONG	16.00UT 1426	☿	♀	♃	♄
					MAGNITUDES			
289.40	-0.95	1.4	295.19	6JA	-0.2	-3.6	-1.4	0.8
297.21	-1.00	1.4	305.36	16JA	-0.4	-3.5	-1.5	0.8
305.05	-1.05	1.4	315.48	26JA	-0.7	-3.5	-1.5	0.8
312.91	-1.08	1.4	325.57	5FE	-1.2	-3.4	-1.6	0.7
320.78	-1.10	1.4	335.61	15FE	-1.3	-3.4	-1.6	0.7
328.64	-1.11	1.4	345.60	25FE	-0.6	-3.4	-1.7	0.6
336.49	-1.11	1.4	355.53	7MR	0.9	-3.3	-1.8	0.6
344.32	-1.10	1.4	5.41	17MR	3.0	-3.3	-1.8	0.5
352.10	-1.07	1.4	15.22	27MR	2.4	-3.3	-1.9	0.5
359.83	-1.03	1.4	24.99	6AP	1.3	-3.3	-2.0	0.4
7.50	-0.98	1.4	34.70	16AP	0.8	-3.3	-2.0	0.4
15.10	-0.92	1.4	44.36	26AP	0.3	-3.3	-2.1	0.3
22.62	-0.84	1.4	53.98	6MY	-0.4	-3.3	-2.1	0.3
30.05	-0.76	1.5	63.57	16MY	-1.3	-3.4	-2.2	0.3
37.39	-0.67	1.5	73.13	26MY	-1.7	-3.4	-2.2	0.3
44.63	-0.57	1.5	82.67	5JN	-0.8	-3.4	-2.2	0.4
51.76	-0.46	1.5	92.20	15JN	0.0	-3.5	-2.1	0.5
58.78	-0.34	1.5	101.73	25JN	0.5	-3.5	-2.1	0.5
65.69	-0.22	1.5	111.27	5JL	1.0	-3.5	-2.1	0.6
72.48	-0.09	1.4	120.83	15JL	1.7	-3.4	-2.0	0.7
79.13	0.05	1.4	130.40	25JL	3.0	-3.4	-2.0	0.7
85.65	0.19	1.4	140.02	4AU	1.9	-3.4	-1.9	0.7
92.02	0.34	1.4	149.68	14AU	0.0	-3.3	-1.8	0.8
98.24	0.50	1.3	159.37	24AU	-1.0	-3.3	-1.8	0.8
104.28	0.67	1.3	169.13	3SE	-1.3	-3.3	-1.7	0.8
110.13	0.85	1.2	178.94	13SE	-0.9	-3.3	-1.7	0.8
115.75	1.03	1.2	188.80	23SE	-0.5	-3.3	-1.6	0.8
121.12	1.24	1.1	198.73	3OC	-0.2	-3.3	-1.5	0.8
126.18	1.45	1.0	208.70	13OC	-0.1	-3.4	-1.5	0.8
130.88	1.69	0.8	218.73	23OC	-0.1	-3.4	-1.5	0.7
135.14	1.95	0.7	228.82	2NO	0.4	-3.4	-1.4	0.7
138.87	2.24	0.5	238.94	12NO	2.7	-3.5	-1.4	0.7
141.93	2.55	0.4	249.09	22NO	0.8	-3.5	-1.4	0.7
144.18	2.90	0.2	259.27	2DE	-0.1	-3.6	-1.4	0.7
145.42	3.28	-0.1	269.47	12DE	-0.3	-3.6	-1.4	0.7
145.49	3.66	-0.3	279.66	22DE	-0.3	-3.7	-1.4	0.8
				1427				
144.24	4.04	-0.5	289.86	1JA	-0.5	-3.8	-1.4	0.8
141.69	4.35	-0.8	300.03	11JA	-0.8	-3.9	-1.4	0.8
138.15	4.53	-1.0	310.17	21JA	-1.1	-4.0	-1.5	0.8
134.18	4.54	-1.0	320.29	31JA	-1.1	-4.0	-1.5	0.8
130.54	4.38	-0.8	330.35	10FE	-0.6	-4.1	-1.5	0.7
127.84	4.09	-0.6	340.37	20FE	1.0	-4.2	-1.6	0.7
126.37	3.72	-0.3	350.33	2MR	3.3	-4.3	-1.6	0.7
126.20	3.34	-0.1	0.24	12MR	1.8	-4.3	-1.7	0.6
127.19	2.97	0.2	10.08	22MR	1.0	-4.1	-1.8	0.6
129.16	2.62	0.4	19.88	1AP	0.6	-3.7	-1.8	0.5
131.95	2.30	0.6	29.61	11AP	0.2	-3.0	-1.9	0.5
135.39	2.01	0.7	39.30	21AP	-0.4	-3.6	-2.0	0.4
139.35	1.75	0.9	48.94	1MY	-1.3	-4.0	-2.0	0.3
143.76	1.51	1.0	58.55	11MY	-1.7	-4.2	-2.1	0.3
148.51	1.30	1.1	68.12	21MY	-0.8	-4.2	-2.2	0.2
153.57	1.10	1.2	77.67	31MY	0.1	-4.1	-2.2	0.3
158.89	0.91	1.3	87.20	10JN	0.7	-4.0	-2.3	0.3
164.44	0.73	1.4	96.73	20JN	1.3	-3.9	-2.3	0.4
170.19	0.57	1.4	106.27	30JN	2.3	-3.8	-2.3	0.5
176.11	0.42	1.5	115.81	10JL	3.2	-3.8	-2.3	0.5
182.21	0.27	1.5	125.38	20JL	1.5	-3.7	-2.3	0.6
188.47	0.13	1.5	134.98	30JL	-0.0	-3.6	-2.2	0.6
194.87	-0.00	1.6	144.61	9AU	-1.1	-3.6	-2.2	0.7
201.42	-0.13	1.6	154.29	19AU	-1.4	-3.5	-2.1	0.7
208.10	-0.25	1.6	164.01	29AU	-0.9	-3.5	-2.1	0.8
214.92	-0.36	1.6	173.79	8SE	-0.4	-3.4	-2.0	0.8
221.87	-0.47	1.6	183.63	18SE	-0.1	-3.4	-2.0	0.8
228.94	-0.57	1.6	193.52	28SE	0.0	-3.4	-1.9	0.8
236.12	-0.66	1.5	203.47	8OC	0.1	-3.4	-1.8	0.8
243.42	-0.75	1.5	213.47	18OC	0.7	-3.4	-1.8	0.8
250.83	-0.83	1.5	223.53	28OC	3.1	-3.4	-1.7	0.7
258.34	-0.90	1.5	233.63	7NO	0.6	-3.4	-1.7	0.7
265.94	-0.96	1.5	243.77	17NO	-0.3	-3.4	-1.6	0.7
273.62	-1.00	1.4	253.94	27NO	-0.4	-3.4	-1.6	0.7
281.36	-1.04	1.4	264.12	7DE	-0.4	-3.4	-1.5	0.7
289.17	-1.07	1.4	274.32	17DE	-0.6	-3.4	-1.5	0.7
297.03	-1.09	1.4	284.51	27DE	-0.8	-3.4	-1.5	0.8

♂ LONG	LAT	MAG	⊙ LONG	16.00UT	☿	♀	♃	♄
				1428	MAGNITUDES			
304.91	-1.09	1.4	294.69	6JA	-1.0	-3.4	-1.5	0.8
312.82	-1.09	1.3	304.86	16JA	-1.0	-3.5	-1.5	0.8
320.73	-1.07	1.3	314.99	26JA	-0.5	-3.5	-1.5	0.8
328.62	-1.04	1.3	325.08	5FE	1.2	-3.5	-1.5	0.8
336.50	-1.00	1.3	335.12	15FE	2.9	-3.4	-1.5	0.8
344.35	-0.95	1.3	345.11	25FE	1.3	-3.4	-1.6	0.7
352.15	-0.89	1.3	355.05	6MR	0.7	-3.4	-1.6	0.7
359.89	-0.82	1.4	4.93	16MR	0.4	-3.4	-1.6	0.7
7.57	-0.74	1.4	14.75	26MR	0.1	-3.4	-1.7	0.6
15.17	-0.65	1.5	24.51	5AP	-0.5	-3.3	-1.7	0.6
22.70	-0.56	1.5	34.23	15AP	-1.3	-3.3	-1.8	0.5
30.14	-0.46	1.5	43.89	25AP	-1.7	-3.3	-1.9	0.4
37.49	-0.36	1.6	53.51	5MY	-0.8	-3.3	-1.9	0.4
44.75	-0.26	1.6	63.11	15MY	0.2	-3.3	-2.0	0.3
51.92	-0.15	1.6	72.67	25MY	0.9	-3.4	-2.1	0.3
58.99	-0.04	1.7	82.21	4JN	1.8	-3.4	-2.1	0.2
65.98	0.07	1.7	91.74	14JN	3.0	-3.4	-2.2	0.3
72.87	0.19	1.7	101.27	24JN	2.7	-3.5	-2.3	0.4
79.66	0.30	1.7	110.81	4JL	1.2	-3.5	-2.3	0.4
86.36	0.42	1.7	120.36	14JL	-0.1	-3.6	-2.4	0.5
92.97	0.53	1.7	129.94	24JL	-1.2	-3.6	-2.4	0.5
99.47	0.65	1.7	139.55	3AU	-1.5	-3.7	-2.4	0.6
105.89	0.77	1.7	149.21	13AU	-0.9	-3.8	-2.4	0.6
112.19	0.89	1.7	158.91	23AU	-0.3	-3.9	-2.4	0.7
118.39	1.01	1.7	168.66	2SE	-0.0	-4.0	-2.4	0.7
124.48	1.14	1.7	178.47	12SE	0.1	-4.1	-2.3	0.8
130.44	1.27	1.6	188.33	22SE	0.3	-4.2	-2.3	0.8
136.26	1.40	1.6	198.24	2OC	1.0	-4.3	-2.2	0.8
141.92	1.53	1.5	208.22	12OC	3.4	-4.3	-2.1	0.8
147.40	1.68	1.4	218.25	22OC	0.5	-4.2	-2.1	0.8
152.67	1.83	1.3	228.32	1NO	-0.5	-3.9	-2.0	0.8
157.68	1.99	1.2	238.45	11NO	-0.6	-3.1	-1.9	0.7
162.39	2.17	1.1	248.60	21NO	-0.6	-3.6	-1.9	0.7
166.73	2.35	0.9	258.77	1DE	-0.7	-4.1	-1.8	0.7
170.61	2.56	0.7	268.97	11DE	-0.8	-4.4	-1.7	0.7
173.93	2.77	0.5	279.17	21DE	-0.8	-4.4	-1.7	0.7
176.55	3.01	0.3	289.35	31DE	-0.8	-4.3	-1.7	0.8
				1429				
178.30	3.25	0.1	299.53	10JA	-0.4	-4.2	-1.6	0.8
179.01	3.49	-0.2	309.68	20JA	1.4	-4.1	-1.6	0.8
178.50	3.71	-0.4	319.79	30JA	2.4	-4.0	-1.6	0.8
176.70	3.87	-0.7	329.86	9FE	0.9	-3.9	-1.6	0.8
173.73	3.93	-0.9	339.88	19FE	0.5	-3.8	-1.6	0.8
170.00	3.84	-1.1	349.85	1MR	0.3	-3.7	-1.6	0.8
166.16	3.59	-1.0	359.76	11MR	-0.0	-3.6	-1.6	0.8
162.92	3.23	-0.8	9.61	21MR	-0.5	-3.6	-1.6	0.7
160.78	2.79	-0.6	19.40	31MR	-1.3	-3.5	-1.6	0.7
159.93	2.35	-0.4	29.15	10AP	-1.6	-3.4	-1.7	0.6
160.36	1.92	-0.2	38.84	20AP	-0.8	-3.4	-1.7	0.6
161.89	1.54	0.0	48.48	30AP	0.3	-3.4	-1.7	0.5
164.37	1.19	0.2	58.09	10MY	1.2	-3.3	-1.8	0.4
167.63	0.88	0.4	67.66	20MY	2.3	-3.3	-1.8	0.4
171.52	0.60	0.5	77.21	30MY	3.5	-3.3	-1.9	0.3
175.94	0.36	0.7	86.75	9JN	2.2	-3.3	-2.0	0.3
180.79	0.14	0.8	96.28	19JN	1.0	-3.3	-2.0	0.3
186.01	-0.06	0.9	105.81	29JN	-0.2	-3.3	-2.1	0.3
191.56	-0.24	0.9	115.36	9JL	-1.2	-3.3	-2.2	0.4
197.39	-0.40	1.0	124.92	19JL	-1.6	-3.4	-2.3	0.4
203.46	-0.55	1.1	134.51	29JL	-0.8	-3.4	-2.3	0.5
209.76	-0.68	1.1	144.15	8AU	-0.3	-3.4	-2.4	0.6
216.26	-0.80	1.2	153.82	18AU	0.1	-3.5	-2.4	0.6
222.95	-0.90	1.2	163.54	28AU	0.3	-3.5	-2.4	0.7
229.82	-1.00	1.2	173.32	7SE	0.5	-3.5	-2.5	0.7
236.84	-1.07	1.2	183.15	17SE	1.3	-3.4	-2.5	0.7
244.01	-1.14	1.3	193.04	27SE	3.2	-3.4	-2.4	0.8
251.32	-1.19	1.3	202.99	7OC	0.4	-3.4	-2.4	0.8
258.74	-1.23	1.3	212.98	17OC	-0.6	-3.4	-2.4	0.8
266.28	-1.25	1.3	223.04	27OC	-0.7	-3.3	-2.3	0.8
273.91	-1.27	1.3	233.14	6NO	-0.7	-3.3	-2.2	0.8
281.62	-1.26	1.3	243.27	16NO	-0.8	-3.3	-2.2	0.8
289.40	-1.25	1.3	253.44	26NO	-0.7	-3.3	-2.1	0.7
297.22	-1.22	1.3	263.62	6DE	-0.7	-3.3	-2.0	0.7
305.08	-1.18	1.4	273.82	16DE	-0.7	-3.4	-2.0	0.7
312.96	-1.13	1.4	284.01	26DE	-0.3	-3.4	-1.9	0.7

♂ LONG	LAT	MAG	⊙ LONG	16.00UT	☿	♀	♃	♄
				1430	MAGNITUDES			
320.84	-1.07	1.4	294.20	5JA	1.7	-3.4	-1.8	0.8
328.70	-1.00	1.4	304.36	15JA	1.9	-3.4	-1.8	0.8
336.55	-0.92	1.4	314.49	25JA	0.6	-3.5	-1.7	0.8
344.35	-0.83	1.4	324.59	4FE	0.3	-3.5	-1.7	0.8
352.10	-0.74	1.5	334.63	14FE	0.1	-3.6	-1.6	0.8
359.80	-0.64	1.5	344.63	24FE	-0.1	-3.6	-1.6	0.8
7.42	-0.53	1.5	354.57	6MR	-0.6	-3.7	-1.6	0.8
14.97	-0.43	1.5	4.45	16MR	-1.3	-3.8	-1.6	0.8
22.45	-0.32	1.5	14.27	26MR	-1.5	-3.8	-1.6	0.8
29.84	-0.22	1.5	24.04	5AP	-0.7	-3.9	-1.6	0.7
37.14	-0.11	1.5	33.76	15AP	0.4	-4.0	-1.6	0.7
44.36	-0.00	1.6	43.43	25AP	1.6	-4.1	-1.6	0.6
51.49	0.10	1.6	53.06	5MY	3.1	-4.2	-1.6	0.6
58.54	0.20	1.6	62.64	15MY	2.9	-4.2	-1.6	0.5
65.50	0.31	1.7	72.21	25MY	1.7	-4.1	-1.7	0.4
72.39	0.40	1.7	81.75	4JN	0.7	-3.8	-1.7	0.4
79.19	0.50	1.8	91.28	14JN	-0.2	-3.2	-1.8	0.3
85.93	0.59	1.8	100.81	24JN	-1.2	-3.1	-1.8	0.3
92.59	0.68	1.9	110.35	4JL	-1.7	-3.8	-1.9	0.3
99.18	0.77	1.9	119.90	14JL	-0.8	-4.1	-1.9	0.4
105.70	0.86	1.9	129.48	24JL	-0.2	-4.2	-2.0	0.4
112.16	0.94	1.9	139.09	3AU	0.2	-4.2	-2.1	0.5
118.56	1.02	1.9	148.74	13AU	0.4	-4.1	-2.1	0.5
124.90	1.10	1.9	158.44	23AU	0.8	-4.0	-2.2	0.6
131.17	1.18	1.9	168.18	2SE	1.7	-4.0	-2.3	0.6
137.38	1.25	1.9	177.98	12SE	2.9	-3.9	-2.3	0.7
143.52	1.33	1.9	187.85	22SE	0.3	-3.8	-2.4	0.7
149.59	1.40	1.8	197.76	2OC	-0.8	-3.7	-2.4	0.8
155.58	1.47	1.8	207.73	12OC	-0.9	-3.7	-2.4	0.8
161.49	1.54	1.7	217.76	22OC	-0.9	-3.6	-2.4	0.8
167.29	1.61	1.7	227.84	1NO	-0.7	-3.6	-2.4	0.8
172.99	1.69	1.6	237.95	11NO	-0.5	-3.5	-2.4	0.8
178.54	1.76	1.5	248.11	21NO	-0.5	-3.5	-2.3	0.8
183.94	1.83	1.4	258.28	1DE	-0.5	-3.4	-2.3	0.8
189.15	1.90	1.3	268.47	11DE	-0.1	-3.4	-2.2	0.8
194.12	1.98	1.1	278.67	21DE	1.9	-3.4	-2.1	0.7
198.82	2.05	1.0	288.86	31DE	1.5	-3.4	-2.1	0.7
				1431				
203.17	2.12	0.8	299.04	10JA	0.4	-3.3	-2.0	0.8
207.09	2.18	0.6	309.19	20JA	0.1	-3.3	-1.9	0.8
210.47	2.24	0.4	319.30	30JA	0.0	-3.3	-1.9	0.8
213.19	2.29	0.1	329.37	9FE	-0.2	-3.3	-1.8	0.9
215.07	2.31	-0.2	339.40	19FE	-0.6	-3.3	-1.7	0.9
215.95	2.30	-0.4	349.36	1MR	-1.3	-3.4	-1.7	0.9
215.64	2.23	-0.7	359.27	11MR	-1.5	-3.4	-1.6	0.9
214.07	2.08	-1.0	9.13	21MR	-0.7	-3.4	-1.6	0.9
211.33	1.83	-1.3	18.92	31MR	0.6	-3.4	-1.6	0.9
207.80	1.48	-1.5	28.67	10AP	2.1	-3.5	-1.5	0.8
204.14	1.06	-1.4	38.36	20AP	3.8	-3.5	-1.5	0.8
201.08	0.60	-1.3	48.01	30AP	2.2	-3.4	-1.5	0.7
199.12	0.15	-1.1	57.62	10MY	1.3	-3.4	-1.5	0.7
198.50	-0.25	-0.9	67.20	20MY	0.6	-3.4	-1.5	0.6
199.20	-0.59	-0.6	76.74	30MY	-0.3	-3.4	-1.5	0.5
201.06	-0.88	-0.5	86.28	9JN	-1.3	-3.3	-1.5	0.5
203.93	-1.11	-0.3	95.81	19JN	-1.7	-3.3	-1.6	0.4
207.63	-1.31	-0.1	105.34	29JN	-0.8	-3.3	-1.6	0.4
212.00	-1.46	0.0	114.89	9JL	-0.1	-3.3	-1.6	0.3
216.94	-1.59	0.2	124.45	19JL	0.3	-3.3	-1.7	0.4
222.35	-1.69	0.3	134.04	29JL	0.6	-3.3	-1.7	0.4
228.16	-1.76	0.4	143.67	8AU	1.1	-3.4	-1.8	0.5
234.31	-1.81	0.5	153.35	18AU	2.2	-3.4	-1.9	0.5
240.75	-1.84	0.5	163.06	28AU	2.5	-3.4	-1.9	0.6
247.44	-1.85	0.6	172.84	7SE	0.2	-3.5	-2.0	0.6
254.35	-1.84	0.7	182.67	17SE	-0.9	-3.5	-2.0	0.7
261.45	-1.81	0.8	192.55	27SE	-1.0	-3.6	-2.1	0.7
268.70	-1.77	0.8	202.50	7OC	-1.0	-3.7	-2.2	0.8
276.10	-1.71	0.9	212.50	17OC	-0.6	-3.8	-2.2	0.8
283.59	-1.63	0.9	222.55	27OC	-0.4	-3.9	-2.3	0.8
291.18	-1.55	1.0	232.64	6NO	-0.4	-4.0	-2.3	0.8
298.83	-1.45	1.1	242.78	16NO	-0.3	-4.1	-2.3	0.9
306.53	-1.34	1.1	252.94	26NO	0.1	-4.2	-2.3	0.9
314.25	-1.22	1.2	263.13	6DE	2.2	-4.3	-2.3	0.8
321.98	-1.10	1.2	273.33	16DE	1.2	-4.4	-2.3	0.8
329.69	-0.97	1.3	283.52	26DE	0.2	-4.4	-2.2	0.8

Left table (1432–1433):

♂ LONG	♂ LAT	♂ MAG	☉ LONG	16.00UT 1432	☿	♀	♃	♄
337.39	-0.84	1.3	293.71	5JA	-0.0	-4.3	-2.2	0.8
345.05	-0.71	1.4	303.87	15JA	-0.1	-3.9	-2.1	0.8
352.66	-0.58	1.4	314.00	25JA	-0.3	-3.3	-2.0	0.9
0.21	-0.45	1.5	324.10	4FE	-0.7	-3.6	-2.0	0.9
7.69	-0.33	1.5	334.14	14FE	-1.3	-4.1	-1.9	0.9
15.11	-0.20	1.6	344.14	24FE	-1.3	-4.3	-1.8	0.9
22.45	-0.08	1.6	354.08	5MR	-0.7	-4.3	-1.8	1.0
29.71	0.03	1.6	3.97	15MR	0.7	-4.2	-1.7	1.0
36.90	0.14	1.7	13.79	25MR	2.6	-4.1	-1.6	1.0
44.00	0.25	1.7	23.57	4AP	2.9	-4.0	-1.6	0.9
51.03	0.35	1.7	33.29	14AP	1.6	-3.9	-1.5	0.9
57.98	0.44	1.8	42.96	24AP	0.9	-3.8	-1.5	0.9
64.87	0.53	1.8	52.59	4MY	0.4	-3.7	-1.5	0.8
71.68	0.61	1.8	62.18	14MY	-0.3	-3.6	-1.4	0.8
78.42	0.69	1.8	71.74	24MY	-1.3	-3.6	-1.4	0.7
85.11	0.76	1.8	81.29	3JN	-1.7	-3.5	-1.4	0.7
91.74	0.83	1.8	90.82	13JN	-0.8	-3.5	-1.4	0.6
98.31	0.89	1.8	100.35	23JN	-0.1	-3.4	-1.4	0.5
104.84	0.95	1.9	109.89	3JL	0.4	-3.4	-1.4	0.5
111.33	1.01	1.9	119.44	13JL	0.8	-3.4	-1.4	0.4
117.77	1.05	2.0	129.02	23JL	1.4	-3.3	-1.5	0.4
124.19	1.10	2.0	138.63	2AU	2.7	-3.3	-1.5	0.4
130.56	1.14	2.0	148.28	12AU	2.1	-3.3	-1.5	0.5
136.91	1.18	2.0	157.97	22AU	0.1	-3.3	-1.6	0.5
143.23	1.21	2.0	167.72	1SE	-1.0	-3.3	-1.6	0.6
149.53	1.24	2.0	177.51	11SE	-1.2	-3.4	-1.7	0.7
155.80	1.27	2.0	187.37	21SE	-1.0	-3.4	-1.7	0.7
162.04	1.29	2.0	197.29	1OC	-0.5	-3.4	-1.8	0.8
168.26	1.31	2.0	207.25	11OC	-0.3	-3.4	-1.9	0.8
174.46	1.32	1.9	217.27	21OC	-0.2	-3.5	-1.9	0.8
180.62	1.33	1.9	227.35	31OC	-0.2	-3.5	-2.0	0.9
186.74	1.34	1.8	237.46	10NO	0.3	-3.5	-2.1	0.9
192.82	1.34	1.8	247.61	20NO	2.4	-3.5	-2.1	0.9
198.86	1.33	1.7	257.79	30NO	1.0	-3.4	-2.2	0.9
204.83	1.32	1.6	267.98	10DE	-0.0	-3.4	-2.2	0.9
210.74	1.30	1.5	278.17	20DE	-0.2	-3.4	-2.2	0.9
216.56	1.27	1.4	288.37	30DE	-0.2	-3.4	-2.2	0.9
				1433				
222.27	1.22	1.2	298.54	9JA	-0.4	-3.4	-2.2	0.9
227.86	1.17	1.1	308.69	19JA	-0.7	-3.3	-2.1	0.9
233.29	1.10	0.9	318.81	29JA	-1.2	-3.3	-2.1	0.9
238.54	1.00	0.8	328.88	8FE	-1.2	-3.3	-2.0	1.0
243.55	0.88	0.6	338.91	18FE	-0.6	-3.3	-2.0	1.0
248.27	0.74	0.4	348.88	28FE	0.9	-3.4	-1.9	1.0
252.63	0.55	0.1	358.79	10MR	3.1	-3.4	-1.8	1.1
256.53	0.31	-0.1	8.65	20MR	2.2	-3.4	-1.8	1.1
259.83	0.02	-0.4	18.45	30MR	1.2	-3.4	-1.7	1.1
262.41	-0.34	-0.7	28.20	9AP	0.7	-3.5	-1.6	1.1
264.08	-0.78	-1.0	37.89	19AP	0.3	-3.5	-1.6	1.0
264.66	-1.32	-1.3	47.54	29AP	-0.4	-3.6	-1.5	1.0
264.01	-1.93	-1.6	57.15	9MY	-1.3	-3.6	-1.5	1.0
262.15	-2.59	-1.9	66.73	19MY	-1.7	-3.7	-1.4	0.9
259.36	-3.23	-2.2	76.29	29MY	-0.8	-3.8	-1.4	0.9
256.23	-3.77	-2.2	85.82	8JN	0.0	-3.8	-1.4	0.8
253.48	-4.14	-2.0	95.35	18JN	0.6	-3.9	-1.3	0.8
251.74	-4.34	-1.9	104.89	28JN	1.1	-4.0	-1.3	0.7
251.34	-4.38	-1.6	114.43	8JL	1.9	-4.1	-1.3	0.6
252.31	-4.30	-1.4	123.99	18JL	3.2	-4.2	-1.3	0.6
254.55	-4.15	-1.2	133.59	28JL	1.8	-4.2	-1.3	0.5
257.85	-3.95	-1.0	143.21	7AU	0.0	-4.1	-1.3	0.5
262.03	-3.73	-0.8	152.88	17AU	-1.1	-3.8	-1.3	0.6
266.91	-3.48	-0.6	162.60	27AU	-1.3	-3.3	-1.4	0.6
272.36	-3.22	-0.4	172.37	6SE	-0.9	-3.6	-1.4	0.6
278.26	-2.96	-0.3	182.20	16SE	-0.5	-4.1	-1.4	0.7
284.52	-2.70	-0.1	192.08	26SE	-0.2	-4.3	-1.5	0.8
291.06	-2.43	0.0	202.02	6OC	-0.1	-4.3	-1.5	0.8
297.82	-2.17	0.2	212.02	16OC	0.0	-4.3	-1.6	0.8
304.74	-1.91	0.3	222.06	26OC	0.5	-4.2	-1.6	0.9
311.79	-1.66	0.4	232.15	5NO	2.8	-4.1	-1.7	0.9
318.51	-1.42	0.6	242.29	15NO	0.8	-4.0	-1.8	1.0
326.13	-1.19	0.7	252.45	25NO	-0.2	-3.9	-1.8	1.0
333.36	-0.97	0.8	262.63	5DE	-0.3	-3.8	-1.9	1.0
340.60	-0.76	0.9	272.83	15DE	-0.4	-3.7	-2.0	1.0
347.84	-0.57	1.0	283.03	25DE	-0.5	-3.7	-2.0	1.0

Right table (1434–1435):

♂ LONG	♂ LAT	♂ MAG	☉ LONG	16.00UT 1434	☿	♀	♃	♄
355.05	-0.39	1.1	293.21	4JA	-0.8	-3.6	-2.0	1.0
2.24	-0.22	1.2	303.38	14JA	-1.1	-3.5	-2.1	1.0
9.39	-0.06	1.3	313.51	24JA	-1.1	-3.5	-2.1	1.0
16.48	0.08	1.4	323.61	3FE	-0.6	-3.4	-2.1	1.0
23.53	0.21	1.5	333.66	13FE	1.1	-3.4	-2.1	1.1
30.52	0.33	1.6	343.66	23FE	3.2	-3.4	-2.0	1.1
37.45	0.44	1.6	353.61	5MR	1.6	-3.3	-2.0	1.1
44.32	0.54	1.7	3.50	15MR	0.9	-3.3	-1.9	1.2
51.14	0.63	1.7	13.33	25MR	0.5	-3.3	-1.9	1.2
57.90	0.71	1.8	23.10	4AP	0.1	-3.3	-1.8	1.2
64.60	0.79	1.8	32.82	14AP	-0.4	-3.3	-1.7	1.2
71.26	0.85	1.9	42.49	24AP	-1.2	-3.3	-1.7	1.2
77.86	0.91	1.9	52.12	4MY	-1.7	-3.3	-1.6	1.2
84.42	0.96	1.9	61.72	14MY	-0.8	-3.4	-1.5	1.2
90.94	1.00	2.0	71.28	24MY	0.1	-3.4	-1.5	1.1
97.43	1.04	2.0	80.82	3JN	0.8	-3.4	-1.4	1.1
103.88	1.07	2.0	90.36	13JN	1.5	-3.5	-1.4	1.0
110.31	1.10	2.0	99.89	23JN	2.5	-3.5	-1.3	0.9
116.71	1.12	2.0	109.42	3JL	3.1	-3.5	-1.3	0.9
123.11	1.13	2.0	118.98	13JL	1.5	-3.4	-1.3	0.8
129.48	1.15	1.9	128.55	23JL	-0.0	-3.4	-1.3	0.7
135.85	1.15	2.0	138.16	2AU	-1.1	-3.4	-1.2	0.7
142.22	1.15	2.0	147.81	12AU	-1.5	-3.3	-1.2	0.6
148.58	1.15	2.0	157.49	22AU	-0.9	-3.3	-1.2	0.6
154.95	1.14	2.0	167.24	1SE	-0.4	-3.3	-1.2	0.7
161.33	1.13	2.0	177.04	11SE	-0.1	-3.3	-1.3	0.7
167.71	1.12	2.0	186.89	21SE	0.1	-3.3	-1.3	0.8
174.11	1.10	2.0	196.80	1OC	0.2	-3.3	-1.3	0.8
180.52	1.07	2.0	206.77	11OC	0.8	-3.4	-1.3	0.9
186.94	1.04	2.0	216.79	21OC	3.1	-3.4	-1.4	0.9
193.38	1.00	1.9	226.86	31OC	0.6	-3.4	-1.4	1.0
199.83	0.96	1.9	236.97	10NO	-0.4	-3.5	-1.5	1.0
206.30	0.90	1.9	247.12	20NO	-0.5	-3.5	-1.5	1.1
212.78	0.85	1.8	257.29	30NO	-0.5	-3.6	-1.6	1.1
219.27	0.78	1.7	267.49	10DE	-0.6	-3.6	-1.7	1.1
225.77	0.70	1.7	277.68	20DE	-0.8	-3.7	-1.7	1.1
232.29	0.61	1.6	287.87	30DE	-0.9	-3.8	-1.8	1.2
				1435				
238.80	0.51	1.5	298.05	9JA	-0.9	-3.9	-1.9	1.2
245.31	0.40	1.4	308.20	19JA	-0.5	-4.0	-1.9	1.2
251.82	0.27	1.3	318.32	29JA	1.2	-4.1	-2.0	1.2
258.32	0.12	1.2	328.40	8FE	2.7	-4.1	-2.0	1.2
264.81	-0.04	1.1	338.42	18FE	1.2	-4.2	-2.0	1.2
271.27	-0.23	0.9	348.40	28FE	0.6	-4.3	-2.0	1.2
277.68	-0.43	0.8	358.32	10MR	0.3	-4.3	-2.0	1.3
284.05	-0.67	0.6	8.18	20MR	0.0	-4.1	-2.0	1.3
290.35	-0.93	0.5	17.98	30MR	-0.5	-3.6	-1.9	1.3
296.54	-1.22	0.3	27.73	9AP	-1.2	-3.0	-1.9	1.4
302.61	-1.55	0.1	37.43	19AP	-1.7	-3.6	-1.8	1.4
308.51	-1.91	-0.1	47.08	29AP	-0.8	-4.0	-1.8	1.4
314.17	-2.31	-0.3	56.69	9MY	0.2	-4.2	-1.7	1.4
319.54	-2.75	-0.5	66.27	19MY	1.0	-4.2	-1.7	1.4
324.52	-3.24	-0.8	75.82	29MY	1.9	-4.1	-1.6	1.3
328.97	-3.76	-1.0	85.36	8JN	3.2	-4.0	-1.5	1.3
332.76	-4.32	-1.3	94.89	18JN	2.5	-3.9	-1.5	1.2
335.68	-4.91	-1.5	104.42	28JN	1.2	-3.8	-1.4	1.2
337.54	-5.50	-1.8	113.97	8JL	-0.1	-3.8	-1.4	1.1
338.17	-6.04	-2.1	123.53	18JL	-1.2	-3.7	-1.3	1.0
337.45	-6.45	-2.3	133.12	28JL	-1.6	-3.6	-1.3	1.0
335.55	-6.63	-2.5	142.74	7AU	-0.9	-3.6	-1.3	0.9
332.96	-6.52	-2.6	152.41	17AU	-0.3	-3.5	-1.2	0.8
330.38	-6.09	-2.5	162.13	27AU	0.0	-3.5	-1.2	0.8
328.55	-5.43	-2.3	171.90	6SE	0.2	-3.4	-1.2	0.8
327.90	-4.66	-2.0	181.72	16SE	0.4	-3.4	-1.2	0.9
328.56	-3.87	-1.7	191.60	26SE	1.1	-3.4	-1.2	0.9
330.45	-3.13	-1.4	201.54	6OC	3.4	-3.4	-1.2	0.9
333.36	-2.47	-1.1	211.53	16OC	0.5	-3.4	-1.2	1.0
337.11	-1.89	-0.8	221.58	26OC	-0.5	-3.4	-1.3	1.0
341.51	-1.39	-0.5	231.67	5NO	-0.6	-3.4	-1.3	1.1
346.41	-0.96	-0.3	241.80	15NO	-0.6	-3.4	-1.3	1.1
351.69	-0.60	-0.0	251.96	25NO	-0.7	-3.4	-1.4	1.2
357.27	-0.29	0.2	262.14	5DE	-0.7	-3.4	-1.4	1.2
3.06	-0.03	0.4	272.34	15DE	-0.7	-3.4	-1.5	1.3
9.03	0.20	0.6	282.54	25DE	-0.8	-3.4	-1.6	1.3

Left table (1436 / 1437):

♂ LONG	LAT	MAG	☉ LONG	16.00UT	☿	♀	♃	♄
15.12	0.39	0.7	292.72	4JA	-0.4	-3.4	-1.6	1.3
21.30	0.55	0.9	302.89	14JA	1.4	-3.5	-1.7	1.3
27.54	0.68	1.0	313.02	24JA	2.2	-3.5	-1.8	1.4
33.83	0.80	1.2	323.12	3FE	0.8	-3.4	-1.8	1.4
40.14	0.89	1.3	333.17	13FE	0.4	-3.4	-1.9	1.4
46.47	0.97	1.4	343.18	23FE	0.2	-3.4	-1.9	1.3
52.81	1.04	1.5	353.12	4MR	-0.1	-3.4	-2.0	1.3
59.15	1.10	1.6	3.01	14MR	-0.5	-3.4	-2.0	1.3
65.48	1.14	1.7	12.85	24MR	-1.2	-3.4	-2.0	1.3
71.80	1.17	1.8	22.62	3AP	-1.6	-3.3	-2.0	1.3
78.12	1.20	1.8	32.35	13AP	-0.8	-3.3	-2.0	1.3
84.42	1.22	1.9	42.03	23AP	0.3	-3.3	-2.0	1.3
90.72	1.23	1.9	51.65	3MY	1.3	-3.3	-1.9	1.3
97.01	1.24	2.0	61.25	13MY	2.6	-3.3	-1.9	1.2
103.29	1.24	2.0	70.82	23MY	3.4	-3.4	-1.8	1.2
109.57	1.23	2.0	80.36	2JN	2.0	-3.4	-1.8	1.2
115.85	1.22	2.0	89.90	12JN	0.9	-3.4	-1.7	1.1
122.14	1.21	2.0	99.43	22JN	-0.1	-3.5	-1.6	1.1
128.44	1.19	2.0	108.96	2JL	-1.2	-3.5	-1.6	1.0
134.74	1.16	2.0	118.51	12JL	-1.7	-3.6	-1.5	1.0
141.07	1.14	2.0	128.09	22JL	-0.9	-3.6	-1.5	0.9
147.41	1.11	2.0	137.69	1AU	-0.2	-3.7	-1.4	0.9
153.78	1.07	2.0	147.34	11AU	0.1	-3.8	-1.4	0.8
160.17	1.03	2.0	157.03	21AU	0.4	-3.9	-1.3	0.8
166.60	0.99	1.9	166.77	31AU	0.6	-4.0	-1.3	0.8
173.06	0.94	1.9	176.56	10SE	1.4	-4.1	-1.3	0.8
179.57	0.88	1.9	186.41	20SE	3.2	-4.2	-1.2	0.8
186.11	0.83	1.9	196.32	30SE	0.4	-4.3	-1.2	0.9
192.70	0.76	1.9	206.28	10OC	-0.7	-4.3	-1.2	1.0
199.33	0.70	1.9	216.30	20OC	-0.8	-4.2	-1.2	1.0
206.02	0.62	1.9	226.37	30OC	-0.8	-3.8	-1.2	1.1
212.75	0.55	1.9	236.48	9NO	-0.8	-3.0	-1.3	1.1
219.53	0.46	1.9	246.63	19NO	-0.6	-3.6	-1.3	1.2
226.37	0.37	1.8	256.80	29NO	-0.6	-4.2	-1.3	1.2
233.26	0.27	1.8	266.99	9DE	-0.6	-4.4	-1.3	1.3
240.20	0.17	1.7	277.19	19DE	-0.2	-4.4	-1.4	1.3
247.18	0.05	1.7	287.38	29DE	1.7	-4.3	-1.4	1.3

1437

♂ LONG	LAT	MAG	☉ LONG	16.00UT	☿	♀	♃	♄
254.23	-0.07	1.6	297.56	8JA	1.8	-4.2	-1.5	1.3
261.32	-0.19	1.5	307.72	18JA	0.5	-4.1	-1.6	1.3
268.45	-0.33	1.5	317.83	28JA	0.2	-4.0	-1.6	1.2
275.64	-0.47	1.4	327.91	7FE	0.1	-3.9	-1.7	1.2
282.86	-0.62	1.3	337.94	17FE	-0.1	-3.8	-1.8	1.1
290.11	-0.78	1.2	347.92	27FE	-0.6	-3.7	-1.8	1.1
297.40	-0.94	1.2	357.84	9MR	-1.2	-3.6	-1.9	1.0
304.71	-1.11	1.1	7.71	19MR	-1.5	-3.6	-2.0	1.0
312.03	-1.28	1.0	17.51	29MR	-0.8	-3.5	-2.0	1.0
319.36	-1.45	0.9	27.26	8AP	0.4	-3.4	-2.0	1.0
326.66	-1.62	0.8	36.97	18AP	1.7	-3.4	-2.1	1.0
333.94	-1.79	0.7	46.62	28AP	3.4	-3.4	-2.1	1.0
341.17	-1.95	0.6	56.23	8MY	2.6	-3.3	-2.1	1.0
348.32	-2.11	0.5	65.81	18MY	1.5	-3.3	-2.1	1.0
355.36	-2.25	0.4	75.37	28MY	0.7	-3.3	-2.0	1.0
2.27	-2.38	0.3	84.91	7JN	-0.2	-3.3	-2.0	0.9
9.00	-2.49	0.1	94.44	17JN	-1.2	-3.3	-1.9	0.9
15.51	-2.59	0.0	103.97	27JN	-1.7	-3.3	-1.9	0.8
21.74	-2.67	-0.1	113.51	7JL	-0.8	-3.3	-1.8	0.8
27.61	-2.72	-0.2	123.07	17JL	-0.2	-3.4	-1.7	0.8
33.05	-2.74	-0.4	132.65	27JL	0.2	-3.4	-1.7	0.7
37.93	-2.74	-0.6	142.28	6AU	0.5	-3.4	-1.6	0.6
42.14	-2.69	-0.7	151.94	16AU	0.9	-3.5	-1.6	0.6
45.49	-2.60	-0.9	161.65	26AU	1.8	-3.5	-1.5	0.5
47.79	-2.45	-1.1	171.42	5SE	2.8	-3.5	-1.5	0.5
48.82	-2.23	-1.4	181.24	15SE	0.3	-3.4	-1.4	0.5
48.41	-1.92	-1.6	191.11	25SE	-0.8	-3.4	-1.4	0.5
46.54	-1.50	-1.8	201.05	5OC	-0.9	-3.4	-1.4	0.5
43.51	-1.00	-1.9	211.04	15OC	-0.9	-3.4	-1.3	0.6
39.93	-0.44	-2.0	221.08	25OC	-0.7	-3.3	-1.3	0.7
36.62	0.09	-1.8	231.17	4NO	-0.5	-3.3	-1.3	0.7
34.29	0.56	-1.4	241.30	14NO	-0.4	-3.3	-1.3	0.8
33.27	0.93	-1.1	251.46	24NO	-0.4	-3.3	-1.3	0.9
33.58	1.21	-0.8	261.65	4DE	-0.1	-3.3	-1.3	0.9
35.09	1.41	-0.5	271.84	14DE	1.9	-3.4	-1.3	0.9
37.57	1.55	-0.2	282.04	24DE	1.5	-3.4	-1.4	1.0

Right table (1438 / 1439):

♂ LONG	LAT	MAG	☉ LONG	16.00UT	☿	♀	♃	♄
40.83	1.64	0.1	292.23	3JA	0.3	-3.4	-1.4	1.0
44.69	1.70	0.3	302.39	13JA	0.1	-3.4	-1.4	1.0
49.03	1.73	0.5	312.53	23JA	-0.0	-3.5	-1.5	1.0
53.73	1.74	0.7	322.64	2FE	-0.2	-3.5	-1.5	1.0
58.72	1.74	0.9	332.69	12FE	-0.6	-3.6	-1.6	0.9
63.93	1.73	1.1	342.70	22FE	-1.2	-3.6	-1.7	0.9
69.32	1.71	1.2	352.65	4MR	-1.4	-3.7	-1.7	0.9
74.84	1.69	1.4	2.54	14MR	-0.7	-3.8	-1.8	0.8
80.48	1.65	1.5	12.38	24MR	0.6	-3.8	-1.9	0.8
86.20	1.62	1.6	22.16	3AP	2.2	-3.9	-1.9	0.8
92.00	1.58	1.7	31.88	13AP	3.5	-4.0	-2.0	0.8
97.87	1.53	1.7	41.56	23AP	2.0	-4.1	-2.1	0.8
103.79	1.49	1.8	51.20	3MY	1.1	-4.2	-2.1	0.8
109.75	1.44	1.9	60.79	13MY	0.5	-4.2	-2.2	0.8
115.77	1.39	1.9	70.36	23MY	-0.3	-4.1	-2.2	0.8
121.82	1.33	2.0	79.91	2JN	-1.2	-3.8	-2.2	0.8
127.92	1.28	2.0	89.44	12JN	-1.7	-3.2	-2.2	0.7
134.06	1.22	2.0	98.97	22JN	-0.8	-3.1	-2.2	0.7
140.25	1.16	2.0	108.51	2JL	-0.1	-3.8	-2.1	0.7
146.48	1.09	2.0	118.05	12JL	0.4	-4.1	-2.1	0.6
152.76	1.03	2.0	127.62	22JL	0.7	-4.2	-2.0	0.6
159.09	0.96	2.0	137.23	1AU	1.2	-4.2	-2.0	0.5
165.47	0.89	2.0	146.87	11AU	2.3	-4.1	-1.9	0.5
171.91	0.81	2.0	156.56	21AU	2.4	-4.0	-1.8	0.4
178.41	0.74	1.9	166.29	31AU	0.2	-3.9	-1.8	0.4
184.98	0.66	1.9	176.08	10SE	-0.9	-3.9	-1.7	0.3
191.61	0.58	1.9	185.93	20SE	-1.1	-3.8	-1.7	0.2
198.31	0.49	1.8	195.84	30SE	-1.0	-3.7	-1.6	0.2
205.08	0.40	1.8	205.79	10OC	-0.6	-3.7	-1.6	0.2
211.92	0.31	1.8	215.81	20OC	-0.4	-3.6	-1.5	0.3
218.83	0.21	1.8	225.88	30OC	-0.3	-3.5	-1.5	0.4
225.82	0.12	1.7	235.99	9NO	-0.3	-3.5	-1.4	0.4
232.89	0.02	1.7	246.13	19NO	0.1	-3.5	-1.4	0.5
240.02	-0.09	1.7	256.31	29NO	2.2	-3.4	-1.4	0.6
247.23	-0.19	1.7	266.49	9DE	1.2	-3.4	-1.4	0.6
254.52	-0.30	1.7	276.69	19DE	0.1	-3.4	-1.4	0.7
261.87	-0.41	1.6	286.89	29DE	-0.1	-3.4	-1.4	0.7

1439

♂ LONG	LAT	MAG	☉ LONG	16.00UT	☿	♀	♃	♄
269.29	-0.53	1.6	297.07	8JA	-0.2	-3.3	-1.4	0.7
276.77	-0.64	1.6	307.22	18JA	-0.3	-3.3	-1.4	0.7
284.30	-0.75	1.5	317.35	28JA	-0.7	-3.3	-1.5	0.8
291.88	-0.86	1.5	327.42	7FE	-1.2	-3.3	-1.5	0.7
299.50	-0.96	1.4	337.46	17FE	-1.3	-3.3	-1.6	0.7
307.16	-1.06	1.4	347.44	27FE	-0.7	-3.4	-1.6	0.7
314.84	-1.16	1.3	357.36	9MR	0.7	-3.4	-1.7	0.7
322.53	-1.24	1.3	7.23	19MR	2.7	-3.4	-1.7	0.6
330.22	-1.32	1.3	17.04	29MR	2.6	-3.5	-1.8	0.6
337.90	-1.39	1.2	26.79	8AP	1.4	-3.5	-1.9	0.5
345.56	-1.44	1.2	36.50	18AP	0.8	-3.5	-1.9	0.6
353.17	-1.48	1.1	46.15	28AP	0.3	-3.4	-2.0	0.6
0.72	-1.50	1.1	55.76	8MY	-0.3	-3.4	-2.1	0.6
8.21	-1.51	1.0	65.35	18MY	-1.2	-3.4	-2.1	0.6
15.60	-1.50	1.0	74.90	28MY	-1.7	-3.4	-2.2	0.6
22.90	-1.48	1.0	84.44	7JN	-0.8	-3.3	-2.2	0.6
30.07	-1.43	0.9	93.97	17JN	-0.0	-3.3	-2.3	0.6
37.09	-1.37	0.9	103.50	27JN	0.5	-3.3	-2.3	0.6
43.95	-1.29	0.8	113.04	7JL	0.9	-3.3	-2.3	0.5
50.63	-1.19	0.7	122.60	17JL	1.6	-3.3	-2.3	0.5
57.09	-1.07	0.7	132.19	27JL	2.9	-3.3	-2.3	0.4
63.31	-0.93	0.6	141.81	6AU	2.1	-3.4	-2.3	0.4
69.23	-0.77	0.5	151.47	16AU	0.1	-3.4	-2.2	0.3
74.82	-0.59	0.4	161.18	26AU	-1.0	-3.4	-2.1	0.3
80.00	-0.38	0.3	170.94	5SE	-1.2	-3.5	-2.1	0.2
84.69	-0.14	0.2	180.76	15SE	-1.0	-3.5	-2.0	0.2
88.78	0.14	0.0	190.63	25SE	-0.5	-3.6	-2.0	0.1
92.15	0.46	-0.2	200.57	5OC	-0.3	-3.7	-1.9	0.0
94.60	0.82	-0.4	210.56	15OC	-0.2	-3.8	-1.8	-0.0
95.95	1.23	-0.6	220.59	25OC	-0.1	-3.9	-1.8	0.0
96.01	1.69	-0.8	230.68	4NO	0.3	-4.0	-1.7	0.1
94.64	2.18	-1.0	240.81	14NO	2.5	-4.1	-1.7	0.2
91.93	2.65	-1.2	250.96	24NO	1.0	-4.2	-1.6	0.2
88.26	3.05	-1.3	261.15	4DE	-0.1	-4.3	-1.6	0.3
84.33	3.32	-1.3	271.34	14DE	-0.2	-4.4	-1.5	0.4
80.93	3.44	-1.0	281.54	24DE	-0.3	-4.4	-1.5	0.4

♂			☉	16.00UT	☿	♀	♃	♄
LONG	LAT	MAG	LONG	1440	MAGNITUDES			
78.62	3.43	-0.8	291.73	3JA	-0.4	-4.3	-1.5	0.5
77.64	3.32	-0.5	301.90	13JA	-0.7	-3.9	-1.5	0.5
77.95	3.17	-0.2	312.04	23JA	-1.1	-3.3	-1.5	0.5
79.38	3.00	0.1	322.14	2FE	-1.2	-3.7	-1.5	0.6
81.74	2.82	0.3	332.20	12FE	-0.6	-4.1	-1.5	0.6
84.84	2.65	0.6	342.21	22FE	0.9	-4.3	-1.5	0.6
88.51	2.49	0.8	352.16	3MR	3.2	-4.3	-1.6	0.5
92.65	2.33	0.9	2.06	13MR	2.0	-4.2	-1.6	0.5
97.15	2.18	1.1	11.90	23MR	1.1	-4.1	-1.7	0.5
101.94	2.04	1.2	21.68	2AP	0.6	-4.0	-1.7	0.5
106.97	1.91	1.4	31.42	12AP	0.2	-3.9	-1.8	0.4
112.20	1.79	1.5	41.09	22AP	-0.4	-3.8	-1.8	0.4
117.59	1.67	1.6	50.73	2MY	-1.2	-3.7	-1.9	0.4
123.13	1.55	1.6	60.33	12MY	-1.7	-3.6	-2.0	0.4
128.79	1.44	1.7	69.90	22MY	-0.8	-3.6	-2.0	0.4
134.56	1.33	1.8	79.45	1JN	0.0	-3.5	-2.1	0.5
140.43	1.22	1.8	88.98	11JN	0.7	-3.5	-2.2	0.5
146.40	1.11	1.8	98.51	21JN	1.2	-3.4	-2.2	0.5
152.46	1.01	1.9	108.05	1JL	2.1	-3.4	-2.3	0.4
158.61	0.91	1.9	117.60	11JL	3.3	-3.4	-2.3	0.4
164.84	0.80	1.9	127.16	21JL	1.7	-3.3	-2.4	0.4
171.16	0.70	1.9	136.77	31JL	0.1	-3.3	-2.4	0.4
177.57	0.60	1.9	146.41	10AU	-1.1	-3.3	-2.4	0.3
184.07	0.50	1.9	156.09	20AU	-1.4	-3.3	-2.4	0.3
190.65	0.39	1.9	165.83	30AU	-0.9	-3.3	-2.4	0.2
197.33	0.29	1.8	175.61	9SE	-0.4	-3.4	-2.4	0.1
204.10	0.19	1.8	185.45	19SE	-0.2	-3.4	-2.3	0.1
210.96	0.09	1.8	195.36	29SE	-0.0	-3.4	-2.3	0.0
217.91	-0.02	1.7	205.32	9OC	0.1	-3.4	-2.2	-0.0
224.96	-0.12	1.7	215.33	19OC	0.6	-3.5	-2.1	-0.1
232.10	-0.22	1.6	225.39	29OC	2.8	-3.5	-2.1	-0.2
239.33	-0.32	1.6	235.49	8NO	0.8	-3.5	-2.0	-0.1
246.64	-0.42	1.5	245.63	18NO	-0.3	-3.5	-1.9	-0.0
254.04	-0.52	1.5	255.81	28NO	-0.4	-3.4	-1.9	0.0
261.52	-0.61	1.5	266.00	8DE	-0.4	-3.4	-1.8	0.1
269.08	-0.70	1.5	276.19	18DE	-0.5	-3.4	-1.7	0.2
276.71	-0.78	1.5	286.39	28DE	-0.8	-3.4	-1.7	0.2
				1441				
284.39	-0.86	1.5	296.57	7JA	-1.0	-3.4	-1.7	0.3
292.13	-0.93	1.5	306.72	17JA	-1.0	-3.3	-1.6	0.3
299.91	-1.00	1.4	316.85	27JA	-0.6	-3.3	-1.6	0.4
307.72	-1.06	1.4	326.93	6FE	1.1	-3.3	-1.6	0.4
315.55	-1.10	1.4	336.96	16FE	3.1	-3.3	-1.6	0.4
323.39	-1.14	1.4	346.95	26FE	1.4	-3.4	-1.6	0.4
331.22	-1.16	1.4	356.88	8MR	0.8	-3.4	-1.6	0.4
339.04	-1.17	1.4	6.75	18MR	0.4	-3.4	-1.6	0.4
346.83	-1.17	1.4	16.56	28MR	0.1	-3.4	-1.6	0.4
354.57	-1.15	1.4	26.32	7AP	-0.4	-3.5	-1.6	0.4
2.27	-1.12	1.4	36.02	17AP	-1.2	-3.5	-1.7	0.3
9.90	-1.08	1.4	45.69	27AP	-1.7	-3.6	-1.7	0.3
17.45	-1.02	1.4	55.30	7MY	-0.8	-3.6	-1.8	0.3
24.93	-0.96	1.4	64.89	17MY	0.1	-3.7	-1.8	0.3
32.30	-0.88	1.4	74.45	27MY	0.9	-3.8	-1.9	0.3
39.58	-0.79	1.4	83.98	6JN	1.6	-3.8	-1.9	0.3
46.75	-0.69	1.3	93.51	16JN	2.8	-3.9	-2.0	0.4
53.79	-0.58	1.3	103.05	26JN	2.9	-4.0	-2.1	0.4
60.71	-0.45	1.3	112.59	6JL	1.4	-4.1	-2.1	0.4
67.50	-0.32	1.3	122.14	16JL	-0.0	-4.2	-2.2	0.4
74.14	-0.18	1.3	131.73	26JL	-1.1	-4.2	-2.3	0.3
80.62	-0.03	1.2	141.35	5AU	-1.5	-4.1	-2.3	0.3
86.93	0.13	1.2	151.01	15AU	-0.9	-3.8	-2.4	0.3
93.05	0.30	1.1	160.72	25AU	-0.4	-3.3	-2.4	0.2
98.94	0.48	1.1	170.47	4SE	-0.1	-3.6	-2.5	0.2
104.60	0.68	1.0	180.29	14SE	0.1	-4.1	-2.5	0.1
109.97	0.90	0.9	190.16	24SE	0.3	-4.3	-2.5	0.1
115.00	1.13	0.8	200.09	4OC	0.8	-4.3	-2.4	0.0
119.62	1.39	0.7	210.07	14OC	3.1	-4.3	-2.4	-0.1
123.76	1.67	0.5	220.11	24OC	0.6	-4.2	-2.4	-0.1
127.29	1.99	0.4	230.19	3NO	-0.4	-4.1	-2.3	-0.2
130.08	2.34	0.2	240.32	13NO	-0.6	-4.0	-2.2	-0.3
131.96	2.72	-0.0	250.48	23NO	-0.6	-3.9	-2.2	-0.2
132.73	3.14	-0.2	260.65	3DE	-0.6	-3.8	-2.1	-0.1
132.23	3.57	-0.5	270.85	13DE	-0.8	-3.7	-2.0	-0.1
130.37	3.97	-0.7	281.05	23DE	-0.8	-3.7	-1.9	0.0

♂			☉	16.00UT	☿	♀	♃	♄
LONG	LAT	MAG	LONG	1442	MAGNITUDES			
127.30	4.29	-0.9	291.23	2JA	-0.9	-3.6	-1.9	0.1
123.45	4.47	-1.0	301.40	12JA	-0.5	-3.5	-1.8	0.1
119.54	4.47	-0.9	311.55	22JA	1.2	-3.5	-1.8	0.2
116.28	4.31	-0.7	321.65	1FE	2.6	-3.4	-1.7	0.3
114.14	4.03	-0.5	331.71	11FE	1.0	-3.4	-1.7	0.3
113.31	3.70	-0.2	341.73	21FE	0.5	-3.4	-1.6	0.3
113.73	3.36	0.0	351.68	3MR	0.3	-3.3	-1.6	0.3
115.22	3.03	0.3	1.58	13MR	0.0	-3.3	-1.6	0.4
117.61	2.73	0.5	11.43	23MR	-0.5	-3.3	-1.6	0.3
120.71	2.44	0.7	21.21	2AP	-1.2	-3.3	-1.6	0.3
124.40	2.19	0.9	30.94	12AP	-1.7	-3.3	-1.6	0.3
128.55	1.95	1.0	40.63	22AP	-0.8	-3.3	-1.6	0.3
133.08	1.74	1.1	50.26	2MY	0.2	-3.3	-1.6	0.3
137.93	1.54	1.2	59.87	12MY	1.1	-3.4	-1.6	0.2
143.04	1.36	1.3	69.44	22MY	2.2	-3.4	-1.6	0.2
148.38	1.18	1.4	78.98	1JN	3.5	-3.4	-1.7	0.2
153.91	1.02	1.5	88.52	11JN	2.3	-3.5	-1.7	0.3
159.63	0.87	1.5	98.05	21JN	1.1	-3.5	-1.8	0.3
165.50	0.72	1.6	107.58	1JL	-0.1	-3.5	-1.8	0.3
171.52	0.58	1.6	117.13	11JL	-1.2	-3.4	-1.9	0.3
177.68	0.44	1.7	126.70	21JL	-1.6	-3.4	-1.9	0.3
183.98	0.31	1.7	136.30	31JL	-0.9	-3.4	-2.0	0.3
190.40	0.19	1.7	145.94	10AU	-0.3	-3.3	-2.1	0.3
196.95	0.06	1.7	155.62	20AU	0.1	-3.3	-2.2	0.3
203.62	-0.05	1.7	165.35	30AU	0.3	-3.3	-2.2	0.3
210.41	-0.17	1.7	175.13	9SE	0.5	-3.3	-2.3	0.2
217.32	-0.28	1.7	184.97	19SE	1.2	-3.3	-2.3	0.2
224.34	-0.38	1.6	194.87	29SE	3.4	-3.3	-2.4	0.1
231.48	-0.48	1.6	204.83	9OC	0.5	-3.4	-2.4	0.0
238.72	-0.58	1.6	214.84	19OC	-0.6	-3.4	-2.4	-0.0
246.07	-0.66	1.6	224.90	29OC	-0.7	-3.4	-2.4	-0.1
253.51	-0.75	1.5	235.00	8NO	-0.7	-3.5	-2.4	-0.2
261.05	-0.82	1.5	245.14	18NO	-0.7	-3.5	-2.4	-0.2
268.66	-0.89	1.5	255.31	28NO	-0.7	-3.6	-2.3	-0.3
276.36	-0.95	1.4	265.50	8DE	-0.7	-3.6	-2.3	-0.2
284.11	-0.99	1.4	275.70	18DE	-0.7	-3.7	-2.2	-0.2
291.93	-1.03	1.4	285.89	28DE	-0.3	-3.8	-2.1	-0.1
				1443				
299.78	-1.06	1.3	296.08	7JA	1.4	-3.9	-2.0	-0.0
307.66	-1.08	1.3	306.23	17JA	2.1	-4.0	-2.0	0.0
315.56	-1.08	1.3	316.36	27JA	0.7	-4.1	-1.9	0.1
323.46	-1.08	1.3	326.44	6FE	0.3	-4.2	-1.8	0.2
331.34	-1.06	1.3	336.48	16FE	0.2	-4.2	-1.8	0.2
339.21	-1.03	1.4	346.46	26FE	-0.1	-4.3	-1.7	0.3
347.04	-0.99	1.4	356.40	8MR	-0.5	-4.3	-1.7	0.3
354.82	-0.93	1.4	6.27	18MR	-1.2	-4.1	-1.6	0.3
2.55	-0.87	1.4	16.08	28MR	-1.6	-3.6	-1.6	0.3
10.21	-0.80	1.4	25.84	7AP	-0.8	-3.0	-1.6	0.3
17.79	-0.72	1.5	35.55	17AP	0.3	-3.6	-1.5	0.3
25.30	-0.63	1.5	45.21	27AP	1.5	-4.0	-1.5	0.3
32.72	-0.54	1.5	54.83	7MY	2.9	-4.2	-1.5	0.3
40.05	-0.44	1.6	64.42	17MY	3.1	-4.2	-1.5	0.2
47.28	-0.33	1.6	73.98	27MY	1.8	-4.1	-1.5	0.2
54.42	-0.22	1.6	83.52	6JN	0.8	-4.0	-1.5	0.2
61.46	-0.11	1.6	93.05	16JN	-0.1	-3.9	-1.5	0.2
68.39	0.01	1.6	102.58	26JN	-1.2	-3.8	-1.6	0.3
75.23	0.13	1.6	112.12	6JL	-1.7	-3.8	-1.6	0.3
81.96	0.26	1.6	121.68	16JL	-0.9	-3.7	-1.6	0.3
88.58	0.38	1.6	131.26	26JL	-0.2	-3.6	-1.7	0.3
95.09	0.51	1.6	140.88	5AU	0.2	-3.6	-1.7	0.3
101.49	0.64	1.6	150.54	15AU	0.4	-3.5	-1.8	0.3
107.76	0.78	1.6	160.24	25AU	0.7	-3.5	-1.9	0.3
113.90	0.92	1.6	170.00	4SE	1.5	-3.4	-1.9	0.3
119.90	1.07	1.5	179.81	14SE	3.1	-3.4	-2.0	0.3
125.74	1.22	1.5	189.68	24SE	0.4	-3.4	-2.0	0.2
131.40	1.38	1.4	199.61	4OC	-0.7	-3.4	-2.1	0.2
136.86	1.54	1.3	209.59	14OC	-0.9	-3.4	-2.2	0.1
142.08	1.72	1.2	219.62	24OC	-0.8	-3.4	-2.2	0.1
147.00	1.91	1.1	229.71	3NO	-0.8	-3.4	-2.3	-0.0
151.59	2.12	1.0	239.82	13NO	-0.6	-3.4	-2.3	-0.1
155.75	2.35	0.8	249.98	23NO	-0.5	-3.4	-2.3	-0.2
159.39	2.59	0.6	260.16	3DE	-0.5	-3.4	-2.3	-0.2
162.40	2.86	0.4	270.35	13DE	-0.2	-3.4	-2.3	-0.3
164.62	3.15	0.2	280.55	23DE	1.7	-3.4	-2.3	-0.2

♂ / ☉ / Magnitudes (1444–1447)

1444–1445

♂ LONG	♂ LAT	♂ MAG	☉ LONG	16.00UT	☿	♀	♃	♄
165.86	3.45	-0.0	290.74	2JA	1.7	-3.4	-2.2	-0.2
165.96	3.76	-0.3	300.91	12JA	0.5	-3.5	-2.1	-0.1
164.77	4.04	-0.5	311.06	22JA	0.2	-3.5	-2.1	-0.0
162.31	4.24	-0.8	321.16	1FE	0.0	-3.5	-2.0	0.1
158.84	4.31	-1.0	331.22	11FE	-0.2	-3.4	-1.9	0.1
154.92	4.21	-1.0	341.24	21FE	-0.6	-3.4	-1.9	0.2
151.28	3.96	-0.8	351.20	2MR	-1.2	-3.4	-1.8	0.2
148.53	3.59	-0.6	1.10	12MR	-1.5	-3.4	-1.7	0.3
147.00	3.17	-0.4	10.94	22MR	-0.8	-3.4	-1.7	0.3
146.77	2.74	-0.2	20.73	1AP	0.5	-3.3	-1.6	0.3
147.73	2.34	0.0	30.46	11AP	1.9	-3.3	-1.6	0.3
149.69	1.97	0.2	40.15	21AP	3.7	-3.3	-1.5	0.3
152.50	1.63	0.4	49.79	1MY	2.4	-3.3	-1.5	0.3
155.99	1.33	0.6	59.39	11MY	1.4	-3.4	-1.5	0.3
160.04	1.06	0.7	68.97	21MY	0.6	-3.4	-1.4	0.3
164.56	0.82	0.9	78.52	31MY	-0.2	-3.4	-1.4	0.3
169.45	0.60	1.0	88.05	10JN	-1.2	-3.4	-1.4	0.2
174.68	0.39	1.0	97.58	20JN	-1.7	-3.5	-1.4	0.2
180.19	0.21	1.1	107.12	30JN	-0.9	-3.5	-1.4	0.3
185.95	0.04	1.2	116.66	10JL	-0.2	-3.6	-1.4	0.3
191.94	-0.12	1.2	126.24	20JL	0.3	-3.6	-1.4	0.4
198.13	-0.26	1.3	135.83	30JL	0.6	-3.7	-1.5	0.4
204.51	-0.40	1.3	145.47	9AU	1.0	-3.8	-1.5	0.4
211.06	-0.52	1.3	155.15	19AU	2.0	-3.9	-1.5	0.4
217.78	-0.64	1.4	164.88	29AU	2.7	-4.0	-1.6	0.4
224.65	-0.74	1.4	174.66	8SE	0.3	-4.1	-1.6	0.4
231.67	-0.83	1.4	184.50	18SE	-0.8	-4.2	-1.7	0.4
238.83	-0.92	1.4	194.39	28SE	-1.0	-4.3	-1.7	0.3
246.11	-0.99	1.4	204.35	8OC	-1.0	-4.3	-1.8	0.3
253.51	-1.05	1.4	214.36	18OC	-0.7	-4.2	-1.9	0.2
261.02	-1.10	1.4	224.41	28OC	-0.4	-3.8	-1.9	0.2
268.62	-1.13	1.4	234.51	7NO	-0.4	-3.0	-2.0	0.1
276.31	-1.16	1.4	244.65	17NO	-0.4	-3.6	-2.1	0.0
284.07	-1.17	1.4	254.82	27NO	-0.0	-4.2	-2.1	-0.0
291.89	-1.17	1.4	265.01	7DE	1.9	-4.4	-2.1	-0.1
299.75	-1.16	1.4	275.21	17DE	1.4	-4.4	-2.2	-0.2
307.64	-1.13	1.4	285.40	27DE	0.2	-4.3	-2.2	-0.2
				1445				
315.54	-1.10	1.4	295.58	6JA	-0.0	-4.2	-2.2	-0.1
323.44	-1.05	1.4	305.74	16JA	-0.1	-4.1	-2.2	-0.1
331.32	-0.99	1.4	315.87	26JA	-0.4	-4.0	-2.1	-0.0
339.18	-0.92	1.4	325.96	5FE	-0.6	-3.9	-2.1	0.0
346.99	-0.85	1.4	336.00	15FE	-1.2	-3.8	-2.0	0.1
354.75	-0.76	1.4	345.98	25FE	-1.4	-3.7	-1.9	0.2
2.45	-0.67	1.4	355.92	7MR	-0.8	-3.6	-1.9	0.2
10.09	-0.58	1.4	5.80	17MR	0.6	-3.6	-1.8	0.3
17.64	-0.48	1.4	15.61	27MR	2.4	-3.5	-1.7	0.3
25.11	-0.38	1.4	25.37	6AP	3.2	-3.4	-1.7	0.4
32.50	-0.28	1.5	35.09	16AP	1.8	-3.4	-1.6	0.4
39.81	-0.17	1.5	44.75	26AP	1.0	-3.4	-1.5	0.4
47.02	-0.06	1.6	54.37	6MY	0.5	-3.3	-1.5	0.4
54.15	0.04	1.6	63.96	16MY	-0.3	-3.3	-1.4	0.4
61.19	0.15	1.7	73.52	26MY	-1.2	-3.3	-1.4	0.4
68.14	0.25	1.7	83.06	5JN	-1.8	-3.3	-1.4	0.4
75.01	0.36	1.8	92.59	15JN	-0.9	-3.3	-1.3	0.4
81.79	0.46	1.8	102.12	25JN	-0.1	-3.3	-1.3	0.3
88.50	0.56	1.8	111.66	5JL	0.4	-3.3	-1.3	0.3
95.13	0.66	1.8	121.22	15JL	0.8	-3.4	-1.3	0.4
101.68	0.76	1.8	130.81	25JL	1.3	-3.4	-1.3	0.4
108.15	0.86	1.9	140.41	4AU	2.5	-3.4	-1.3	0.5
114.55	0.95	1.9	150.07	14AU	2.3	-3.5	-1.3	0.5
120.87	1.05	1.9	159.77	24AU	0.2	-3.5	-1.3	0.5
127.12	1.14	1.8	169.53	3SE	-0.9	-3.5	-1.4	0.5
133.28	1.23	1.8	179.33	13SE	-1.2	-3.5	-1.4	0.5
139.35	1.33	1.8	189.20	23SE	-1.0	-3.4	-1.4	0.5
145.33	1.42	1.7	199.12	3OC	-0.6	-3.4	-1.5	0.5
151.20	1.52	1.7	209.10	13OC	-0.3	-3.4	-1.5	0.5
156.95	1.62	1.6	219.13	23OC	-0.2	-3.3	-1.6	0.4
162.56	1.72	1.5	229.21	2NO	-0.2	-3.3	-1.7	0.4
168.01	1.82	1.3	239.33	12NO	-0.2	-3.3	-1.7	0.3
173.27	1.93	1.3	249.48	22NO	2.2	-3.3	-1.8	0.2
178.30	2.04	1.2	259.67	2DE	1.2	-3.3	-1.8	0.2
183.05	2.16	1.1	269.86	12DE	0.0	-3.4	-1.9	0.1
187.47	2.28	0.9	280.05	22DE	-0.2	-3.4	-2.0	0.0

1446–1447

♂ LONG	♂ LAT	♂ MAG	☉ LONG	16.00UT	☿	♀	♃	♄
191.47	2.41	0.7	290.25	1JA	-0.2	-3.4	-2.0	-0.0
194.95	2.54	0.5	300.42	11JA	-0.4	-3.4	-2.0	-0.1
197.78	2.67	0.3	310.56	21JA	-0.7	-3.5	-2.1	-0.0
199.81	2.80	0.0	320.67	31JA	-1.2	-3.5	-2.1	0.0
200.86	2.91	-0.2	330.74	10FE	-1.2	-3.6	-2.1	0.1
200.75	2.99	-0.5	340.75	20FE	-0.7	-3.6	-2.0	0.2
199.38	3.00	-0.8	350.72	2MR	0.7	-3.7	-2.0	0.2
196.78	2.91	-1.1	0.62	12MR	2.9	-3.8	-2.0	0.3
193.29	2.71	-1.3	10.47	22MR	2.4	-3.8	-1.9	0.3
189.50	2.38	-1.2	20.27	1AP	1.3	-3.9	-1.8	0.4
186.14	1.96	-1.1	30.00	11AP	0.8	-4.0	-1.8	0.4
183.77	1.51	-0.9	39.69	21AP	0.3	-4.1	-1.7	0.5
182.69	1.06	-0.7	49.33	1MY	-0.3	-4.2	-1.6	0.5
182.94	0.66	-0.5	58.93	11MY	-1.2	-4.2	-1.6	0.5
184.37	0.29	-0.3	68.51	21MY	-1.8	-4.1	-1.5	0.5
186.82	-0.02	-0.1	78.06	31MY	-0.9	-3.8	-1.5	0.5
190.12	-0.29	0.1	87.59	10JN	-0.0	-3.1	-1.4	0.5
194.10	-0.52	0.2	97.12	20JN	0.5	-3.1	-1.4	0.5
198.67	-0.72	0.3	106.66	30JN	1.0	-3.8	-1.3	0.5
203.73	-0.90	0.5	116.20	10JL	1.7	-4.1	-1.3	0.5
209.18	-1.05	0.6	125.77	20JL	3.1	-4.2	-1.3	0.5
215.00	-1.17	0.6	135.37	30JL	2.0	-4.2	-1.2	0.5
221.11	-1.28	0.7	145.00	9AU	0.1	-4.1	-1.2	0.6
227.49	-1.36	0.8	154.68	19AU	-1.0	-4.0	-1.2	0.6
234.11	-1.43	0.8	164.40	29AU	-1.3	-3.9	-1.2	0.7
240.93	-1.48	0.9	174.18	8SE	-1.0	-3.9	-1.2	0.7
247.95	-1.52	0.9	184.02	18SE	-0.5	-3.8	-1.3	0.7
255.13	-1.53	1.0	193.91	28SE	-0.2	-3.7	-1.3	0.7
262.45	-1.53	1.0	203.86	8OC	-0.1	-3.7	-1.3	0.7
269.90	-1.52	1.1	213.87	18OC	-0.0	-3.6	-1.3	0.7
277.46	-1.49	1.1	223.92	28OC	0.4	-3.5	-1.4	0.6
285.10	-1.45	1.1	234.02	7NO	2.5	-3.5	-1.4	0.6
292.81	-1.39	1.2	244.16	17NO	0.9	-3.5	-1.5	0.5
300.58	-1.32	1.2	254.33	27NO	-0.2	-3.4	-1.6	0.5
308.37	-1.24	1.3	264.51	7DE	-0.3	-3.4	-1.6	0.4
316.18	-1.15	1.3	274.71	17DE	-0.3	-3.4	-1.7	0.3
323.99	-1.05	1.3	284.91	27DE	-0.5	-3.4	-1.7	0.3
				1447				
331.78	-0.95	1.4	295.09	6JA	-0.7	-3.3	-1.8	0.2
339.55	-0.84	1.4	305.25	16JA	-1.1	-3.3	-1.9	0.1
347.27	-0.72	1.4	315.38	26JA	-1.1	-3.3	-1.9	0.1
354.93	-0.61	1.5	325.47	5FE	-0.6	-3.3	-2.0	0.2
2.54	-0.49	1.5	335.51	15FE	0.9	-3.3	-2.0	0.2
10.08	-0.37	1.5	345.50	25FE	3.2	-3.4	-2.0	0.3
17.53	-0.26	1.6	355.44	7MR	1.8	-3.4	-2.0	0.3
24.91	-0.14	1.6	5.32	17MR	0.9	-3.4	-2.0	0.4
32.22	-0.03	1.6	15.14	27MR	0.6	-3.5	-2.0	0.4
39.43	0.08	1.7	24.90	6AP	0.2	-3.5	-1.9	0.5
46.57	0.18	1.7	34.62	16AP	-0.4	-3.5	-1.9	0.6
53.62	0.28	1.7	44.28	26AP	-1.2	-3.4	-1.8	0.6
60.60	0.38	1.7	53.90	6MY	-1.8	-3.4	-1.8	0.6
67.50	0.47	1.7	63.49	16MY	-0.9	-3.4	-1.7	0.7
74.32	0.56	1.7	73.05	26MY	0.0	-3.4	-1.6	0.7
81.08	0.64	1.8	82.59	5JN	0.7	-3.3	-1.6	0.7
87.77	0.72	1.8	92.12	15JN	1.4	-3.3	-1.5	0.7
94.40	0.80	1.9	101.66	25JN	2.3	-3.3	-1.5	0.7
100.98	0.87	1.9	111.19	5JL	3.2	-3.3	-1.4	0.7
107.50	0.94	1.9	120.75	15JL	1.6	-3.3	-1.4	0.7
113.97	1.00	2.0	130.33	25JL	0.1	-3.3	-1.3	0.7
120.39	1.06	2.0	139.94	4AU	-1.1	-3.4	-1.3	0.7
126.78	1.11	2.0	149.60	14AU	-1.4	-3.4	-1.2	0.8
133.11	1.17	2.0	159.30	24AU	-0.9	-3.4	-1.2	0.8
139.41	1.22	2.0	169.04	3SE	-0.4	-3.5	-1.2	0.9
145.68	1.26	2.0	178.85	13SE	-0.1	-3.6	-1.2	0.9
151.89	1.31	2.0	188.72	23SE	0.0	-3.6	-1.2	0.9
158.07	1.35	1.9	198.64	3OC	0.2	-3.7	-1.2	0.9
164.21	1.38	1.9	208.62	13OC	0.6	-3.8	-1.2	0.9
170.29	1.42	1.9	218.64	23OC	2.8	-3.9	-1.2	0.9
176.32	1.45	1.8	228.72	2NO	0.8	-4.0	-1.3	0.9
182.29	1.48	1.7	238.84	12NO	-0.3	-4.1	-1.3	0.9
188.18	1.50	1.7	248.99	22NO	-0.5	-4.2	-1.3	0.8
193.99	1.52	1.6	259.17	2DE	-0.5	-4.3	-1.4	0.8
199.70	1.54	1.5	269.37	12DE	-0.6	-4.4	-1.4	0.7
205.28	1.55	1.4	279.56	22DE	-0.8	-4.4	-1.5	0.7

♂ LONG	♂ LAT	♂ MAG	☉ LONG	16.00UT	☿	♀	♃	♄
				1448	MAGNITUDES			
210.71	1.55	1.2	289.75	1JA	-0.9	-4.3	-1.6	0.6
215.97	1.54	1.1	299.93	11JA	-0.9	-3.9	-1.6	0.5
221.01	1.52	0.9	310.07	21JA	-0.6	-3.3	-1.7	0.4
225.77	1.49	0.7	320.18	31JA	1.1	-3.7	-1.8	0.4
230.21	1.44	0.5	330.25	10FE	2.9	-4.1	-1.8	0.4
234.22	1.36	0.3	340.27	20FE	1.3	-4.3	-1.9	0.4
237.71	1.25	0.0	350.23	1MR	0.7	-4.3	-1.9	0.4
240.54	1.10	-0.2	0.15	11MR	0.4	-4.2	-2.0	0.5
242.55	0.89	-0.5	9.99	21MR	0.1	-4.1	-2.0	0.6
243.56	0.62	-0.8	19.79	31MR	-0.4	-4.0	-2.0	0.6
243.39	0.27	-1.1	29.53	10AP	-1.2	-3.9	-2.0	0.7
241.97	-0.17	-1.5	39.22	20AP	-1.7	-3.8	-2.0	0.7
239.41	-0.68	-1.7	48.86	30AP	-0.9	-3.7	-2.0	0.8
236.10	-1.22	-1.9	58.47	10MY	0.1	-3.6	-1.9	0.8
232.75	-1.73	-1.8	68.05	20MY	0.9	-3.6	-1.9	0.9
230.05	-2.15	-1.7	77.60	30MY	1.8	-3.5	-1.8	0.9
228.51	-2.48	-1.5	87.14	9JN	3.0	-3.5	-1.8	0.9
228.36	-2.70	-1.3	96.66	19JN	2.7	-3.4	-1.7	0.9
229.54	-2.84	-1.0	106.20	29JN	1.3	-3.4	-1.6	1.0
231.91	-2.91	-0.8	115.74	9JL	-0.0	-3.4	-1.6	1.0
235.28	-2.93	-0.7	125.31	19JL	-1.1	-3.3	-1.5	1.0
239.47	-2.91	-0.5	134.90	29JL	-1.6	-3.3	-1.5	1.0
244.35	-2.86	-0.3	144.54	8AU	-0.9	-3.3	-1.4	1.0
249.77	-2.79	-0.2	154.21	18AU	-0.3	-3.3	-1.4	1.0
255.64	-2.69	-0.0	163.93	28AU	-0.0	-3.3	-1.3	1.0
261.88	-2.57	0.1	173.71	7SE	0.2	-3.4	-1.3	1.1
268.43	-2.44	0.2	183.54	17SE	0.4	-3.4	-1.3	1.1
275.22	-2.30	0.3	193.43	27SE	0.9	-3.4	-1.2	1.2
282.22	-2.14	0.4	203.38	7OC	3.2	-3.4	-1.2	1.2
289.38	-1.97	0.5	213.38	17OC	0.6	-3.5	-1.2	1.2
296.65	-1.80	0.6	223.43	27OC	-0.5	-3.5	-1.2	1.2
304.03	-1.63	0.7	233.53	6NO	-0.6	-3.5	-1.2	1.2
311.47	-1.45	0.8	243.67	16NO	-0.6	-3.5	-1.3	1.2
318.94	-1.27	0.9	253.83	26NO	-0.7	-3.4	-1.3	1.1
326.44	-1.09	1.0	264.02	6DE	-0.8	-3.4	-1.3	1.1
333.94	-0.91	1.1	274.21	16DE	-0.8	-3.4	-1.4	1.1
341.42	-0.74	1.1	284.41	26DE	-0.8	-3.4	-1.4	1.0
				1449				
348.87	-0.58	1.2	294.60	5JA	-0.4	-3.4	-1.5	0.9
356.28	-0.42	1.3	304.75	15JA	1.2	-3.3	-1.5	0.9
3.64	-0.27	1.4	314.89	25JA	2.5	-3.3	-1.6	0.8
10.94	-0.13	1.4	324.98	4FE	0.9	-3.3	-1.6	0.7
18.18	0.00	1.5	335.02	14FE	0.5	-3.3	-1.7	0.6
25.36	0.13	1.6	345.02	24FE	0.2	-3.4	-1.8	0.6
32.46	0.24	1.6	354.96	6MR	-0.0	-3.4	-1.9	0.7
39.50	0.35	1.7	4.84	16MR	-0.5	-3.4	-1.9	0.7
46.46	0.45	1.7	14.66	26MR	-1.2	-3.4	-2.0	0.8
53.36	0.54	1.8	24.43	5AP	-1.6	-3.5	-2.0	0.8
60.20	0.63	1.8	34.15	15AP	-0.9	-3.5	-2.1	0.9
66.97	0.70	1.8	43.82	25AP	0.2	-3.6	-2.1	0.9
73.68	0.77	1.9	53.44	5MY	1.2	-3.6	-2.1	1.0
80.34	0.84	1.9	63.03	15MY	2.4	-3.7	-2.1	1.0
86.94	0.90	1.9	72.60	25MY	3.6	-3.8	-2.1	1.1
93.50	0.95	1.9	82.14	4JN	2.1	-3.8	-2.0	1.1
100.02	0.99	1.9	91.67	14JN	1.0	-3.9	-2.0	1.2
106.51	1.03	1.9	101.20	24JN	-0.1	-4.0	-1.9	1.2
112.96	1.07	1.9	110.74	4JL	-1.1	-4.1	-1.9	1.2
119.38	1.10	1.9	120.29	14JL	-1.7	-4.2	-1.8	1.3
125.78	1.13	2.0	129.87	24JL	-0.9	-4.2	-1.7	1.3
132.16	1.15	2.0	139.48	3AU	-0.3	-4.1	-1.7	1.3
138.53	1.16	2.0	149.13	13AU	0.1	-3.7	-1.6	1.3
144.89	1.18	2.0	158.83	23AU	0.3	-3.2	-1.6	1.3
151.24	1.18	2.0	168.57	2SE	0.6	-3.6	-1.5	1.3
157.58	1.19	2.0	178.38	12SE	1.3	-4.1	-1.5	1.3
163.92	1.19	2.0	188.24	22SE	3.3	-4.3	-1.4	1.3
170.26	1.18	2.0	198.15	2OC	0.5	-4.3	-1.4	1.3
176.59	1.17	2.0	208.13	12OC	-0.6	-4.3	-1.4	1.3
182.93	1.15	2.0	218.16	22OC	-0.8	-4.2	-1.3	1.3
189.27	1.13	1.9	228.23	1NO	-0.7	-4.1	-1.3	1.3
195.60	1.10	1.9	238.35	11NO	-0.8	-4.0	-1.3	1.2
201.93	1.07	1.8	248.50	21NO	-0.6	-3.9	-1.3	1.2
208.25	1.03	1.8	258.68	1DE	-0.6	-3.8	-1.3	1.2
214.56	0.98	1.7	268.87	11DE	-0.6	-3.7	-1.3	1.1
220.85	0.92	1.6	279.07	21DE	-0.3	-3.7	-1.4	1.1
227.12	0.84	1.5	289.26	31DE	1.4	-3.6	-1.4	1.0

♂ LONG	♂ LAT	♂ MAG	☉ LONG	16.00UT	☿	♀	♃	♄
				1450	MAGNITUDES			
233.36	0.76	1.4	299.44	10JA	2.0	-3.5	-1.4	1.0
239.57	0.66	1.3	309.59	20JA	0.6	-3.5	-1.5	0.9
245.73	0.55	1.2	319.70	30JA	0.3	-3.4	-1.5	0.9
251.83	0.41	1.0	329.77	9FE	0.1	-3.4	-1.6	0.8
257.85	0.25	0.9	339.79	19FE	-0.1	-3.4	-1.6	0.8
263.78	0.07	0.7	349.76	1MR	-0.5	-3.3	-1.7	0.8
269.59	-0.15	0.5	359.67	11MR	-1.2	-3.3	-1.8	0.8
275.25	-0.40	0.4	9.53	21MR	-1.6	-3.3	-1.8	0.9
280.72	-0.69	0.1	19.32	31MR	-0.8	-3.3	-1.9	1.0
285.95	-1.03	-0.1	29.06	10AP	0.3	-3.3	-2.0	1.0
290.85	-1.42	-0.3	38.76	20AP	1.6	-3.3	-2.0	1.1
295.35	-1.88	-0.6	48.40	30AP	3.2	-3.3	-2.1	1.2
299.34	-2.41	-0.8	58.01	10MY	2.9	-3.4	-2.1	1.2
302.64	-3.01	-1.1	67.59	20MY	1.6	-3.4	-2.2	1.3
305.09	-3.68	-1.4	77.14	30MY	0.8	-3.4	-2.2	1.3
306.51	-4.42	-1.7	86.67	9JN	-0.1	-3.5	-2.2	1.3
306.71	-5.17	-2.0	96.21	19JN	-1.2	-3.5	-2.2	1.3
305.67	-5.87	-2.3	105.73	29JN	-1.7	-3.5	-2.2	1.4
303.58	-6.41	-2.5	115.28	9JL	-0.9	-3.4	-2.1	1.3
300.95	-6.68	-2.6	124.85	19JL	-0.2	-3.4	-2.1	1.3
298.51	-6.63	-2.5	134.43	29JL	0.2	-3.4	-2.0	1.3
296.92	-6.31	-2.2	144.07	8AU	0.5	-3.3	-2.0	1.3
296.58	-5.80	-2.0	153.74	18AU	0.8	-3.3	-1.9	1.2
297.58	-5.19	-1.7	163.45	28AU	1.7	-3.3	-1.8	1.2
299.80	-4.56	-1.4	173.23	7SE	3.0	-3.3	-1.8	1.1
303.07	-3.95	-1.2	183.06	17SE	0.4	-3.3	-1.7	1.1
307.18	-3.37	-0.9	192.94	27SE	-0.7	-3.3	-1.7	1.1
311.94	-2.84	-0.7	202.89	7OC	-0.9	-3.4	-1.6	1.1
317.22	-2.36	-0.5	212.89	17OC	-0.9	-3.4	-1.6	1.1
322.88	-1.93	-0.3	222.94	27OC	-0.7	-3.4	-1.5	1.1
328.84	-1.54	-0.1	233.04	6NO	-0.5	-3.5	-1.5	1.1
335.03	-1.19	0.1	243.17	16NO	-0.5	-3.5	-1.5	1.1
341.39	-0.88	0.3	253.33	26NO	-0.5	-3.6	-1.4	1.1
347.86	-0.60	0.5	263.52	6DE	-0.1	-3.6	-1.4	1.0
354.43	-0.35	0.6	273.72	16DE	1.7	-3.7	-1.4	1.0
1.04	-0.13	0.8	283.91	26DE	1.7	-3.8	-1.4	1.0
				1451				
7.70	0.06	0.9	294.10	5JA	0.4	-3.9	-1.4	0.9
14.37	0.23	1.1	304.27	15JA	0.1	-4.0	-1.4	0.9
21.04	0.38	1.2	314.40	25JA	-0.0	-4.1	-1.5	0.8
27.70	0.51	1.3	324.49	4FE	-0.2	-4.2	-1.5	0.8
34.35	0.62	1.4	334.54	14FE	-0.6	-4.2	-1.5	0.7
40.97	0.72	1.5	344.54	24FE	-1.2	-4.3	-1.6	0.7
47.57	0.81	1.6	354.48	6MR	-1.4	-4.3	-1.6	0.6
54.14	0.88	1.7	4.37	16MR	-0.8	-4.1	-1.7	0.6
60.67	0.95	1.7	14.19	26MR	0.5	-3.6	-1.8	0.7
67.18	1.00	1.8	23.96	5AP	2.0	-3.1	-1.8	0.7
73.66	1.05	1.8	33.68	15AP	3.7	-3.6	-1.9	0.8
80.10	1.09	1.9	43.35	25AP	2.1	-4.1	-2.0	0.9
86.53	1.12	1.9	52.98	5MY	1.2	-4.2	-2.0	0.9
92.92	1.14	2.0	62.57	15MY	0.6	-4.2	-2.1	1.0
99.30	1.16	2.0	72.13	25MY	-0.2	-4.1	-2.2	1.0
105.66	1.17	2.0	81.68	4JN	-1.2	-4.0	-2.2	1.1
112.01	1.17	2.0	91.21	14JN	-1.8	-3.9	-2.3	1.1
118.35	1.18	2.0	100.74	24JN	-0.9	-3.8	-2.3	1.2
124.69	1.17	2.0	110.28	4JL	-0.2	-3.8	-2.3	1.2
131.03	1.16	2.0	119.83	14JL	0.3	-3.7	-2.3	1.2
137.37	1.15	2.0	129.41	24JL	0.6	-3.6	-2.3	1.2
143.73	1.13	2.0	139.02	3AU	1.1	-3.6	-2.3	1.2
150.09	1.11	2.0	148.66	13AU	2.1	-3.5	-2.3	1.1
156.48	1.08	2.0	158.36	23AU	2.7	-3.5	-2.2	1.1
162.89	1.05	2.0	168.11	2SE	0.3	-3.4	-2.2	1.1
169.31	1.02	2.0	177.90	12SE	-0.9	-3.4	-2.1	1.0
175.77	0.98	2.0	187.76	22SE	-1.1	-3.4	-2.0	1.0
182.26	0.93	2.0	197.68	2OC	-1.0	-3.4	-2.0	1.0
188.78	0.88	2.0	207.64	12OC	-0.6	-3.4	-1.9	1.0
195.34	0.83	2.0	217.67	22OC	-0.4	-3.4	-1.8	1.0
201.93	0.77	1.9	227.74	1NO	-0.3	-3.4	-1.8	1.0
208.56	0.70	1.9	237.85	11NO	-0.3	-3.4	-1.7	1.0
215.23	0.63	1.9	248.00	21NO	0.0	-3.4	-1.7	1.0
221.94	0.54	1.8	258.18	1DE	1.9	-3.4	-1.6	1.0
228.68	0.45	1.8	268.37	11DE	1.4	-3.4	-1.6	0.9
235.47	0.36	1.7	278.57	21DE	0.2	-3.4	-1.6	0.9
242.29	0.25	1.6	288.76	31DE	-0.1	-3.4	-1.5	0.9

♂ LONG	♂ LAT	♂ MAG	⊙ LONG	16.00UT	☿	♀	♃	♄
				1452	MAGNITUDES			
249.15	0.13	1.6	298.94	10JA	-0.1	-3.5	-1.5	0.8
256.05	0.00	1.5	309.09	20JA	-0.3	-3.5	-1.5	0.8
262.97	-0.14	1.4	319.21	30JA	-0.6	-3.5	-1.5	0.7
269.93	-0.29	1.3	329.28	9FE	-1.2	-3.4	-1.5	0.7
276.93	-0.45	1.2	339.30	19FE	-1.3	-3.4	-1.5	0.6
283.94	-0.62	1.1	349.27	29FE	-0.8	-3.4	-1.6	0.6
290.97	-0.81	1.0	359.19	10MR	0.6	-3.4	-1.6	0.5
298.01	-1.01	0.9	9.05	20MR	2.5	-3.4	-1.6	0.5
305.04	-1.21	0.8	18.85	30MR	2.9	-3.3	-1.7	0.5
312.07	-1.43	0.7	28.59	9AP	1.6	-3.3	-1.7	0.5
319.07	-1.66	0.5	38.29	19AP	0.9	-3.3	-1.8	0.6
326.01	-1.90	0.4	47.94	29AP	0.4	-3.3	-1.9	0.7
332.87	-2.14	0.3	57.54	9MY	-0.3	-3.4	-1.9	0.7
339.63	-2.38	0.1	67.12	19MY	-1.2	-3.4	-2.0	0.8
346.23	-2.62	-0.0	76.68	29MY	-1.8	-3.4	-2.1	0.9
352.64	-2.87	-0.2	86.21	8JN	-0.9	-3.4	-2.1	0.9
358.78	-3.10	-0.3	95.74	18JN	-0.1	-3.5	-2.2	1.0
4.58	-3.33	-0.5	105.28	28JN	0.4	-3.5	-2.3	1.0
9.94	-3.55	-0.7	114.82	8JL	0.9	-3.6	-2.3	1.0
14.74	-3.75	-0.9	124.38	18JL	1.5	-3.6	-2.4	1.0
18.82	-3.92	-1.1	133.98	28JL	2.7	-3.7	-2.4	1.0
22.01	-4.05	-1.3	143.60	7AU	2.3	-3.8	-2.4	1.0
24.07	-4.12	-1.5	153.27	17AU	0.2	-3.9	-2.4	1.0
24.81	-4.11	-1.7	162.99	27AU	-0.9	-4.0	-2.4	1.0
24.11	-3.96	-2.0	172.76	6SE	-1.2	-4.1	-2.4	1.0
22.03	-3.65	-2.2	182.59	16SE	-1.0	-4.2	-2.4	0.9
19.00	-3.17	-2.3	192.47	26SE	-0.5	-4.3	-2.3	0.9
15.76	-2.54	-2.2	202.41	6OC	-0.3	-4.3	-2.3	0.8
13.09	-1.86	-1.9	212.41	16OC	-0.2	-4.2	-2.2	0.9
11.56	-1.21	-1.6	222.46	26OC	-0.1	-3.8	-2.1	0.9
11.37	-0.63	-1.3	232.55	5NO	0.2	-3.0	-2.0	0.9
12.45	-0.16	-1.0	242.68	15NO	2.2	-3.7	-2.0	0.9
14.63	0.23	-0.6	252.84	25NO	1.1	-4.2	-1.9	0.9
17.69	0.53	-0.4	263.02	5DE	-0.0	-4.4	-1.9	0.9
21.45	0.77	-0.1	273.22	15DE	-0.2	-4.4	-1.8	0.9
25.75	0.95	0.2	283.42	25DE	-0.3	-4.3	-1.7	0.8
				1453				
30.47	1.10	0.4	293.60	4JA	-0.4	-4.2	-1.7	0.8
35.50	1.21	0.6	303.77	14JA	-0.7	-4.1	-1.7	0.8
40.79	1.29	0.8	313.90	24JA	-1.1	-4.0	-1.6	0.7
46.26	1.35	1.0	324.00	3FE	-1.2	-3.9	-1.6	0.7
51.89	1.39	1.1	334.05	13FE	-0.7	-3.8	-1.6	0.6
57.62	1.42	1.2	344.05	23FE	0.7	-3.7	-1.6	0.6
63.45	1.44	1.4	354.00	5MR	3.0	-3.6	-1.6	0.5
69.34	1.45	1.5	3.89	15MR	2.1	-3.6	-1.6	0.5
75.29	1.45	1.6	13.72	25MR	1.2	-3.5	-1.6	0.4
81.28	1.45	1.7	23.49	4AP	0.7	-3.4	-1.6	0.4
87.30	1.43	1.7	33.22	14AP	0.3	-3.4	-1.6	0.4
93.36	1.42	1.8	42.89	24AP	-0.3	-3.4	-1.7	0.5
99.44	1.39	1.9	52.52	4MY	-1.2	-3.3	-1.7	0.5
105.54	1.37	1.9	62.11	14MY	-1.8	-3.3	-1.8	0.6
111.67	1.34	2.0	71.68	24MY	-0.9	-3.3	-1.8	0.6
117.82	1.30	2.0	81.22	3JN	-0.0	-3.3	-1.9	0.7
123.99	1.26	2.0	90.76	13JN	0.6	-3.3	-1.9	0.8
130.20	1.22	2.0	100.28	23JN	1.1	-3.3	-2.0	0.8
136.43	1.18	2.0	109.82	3JL	1.9	-3.3	-2.1	0.9
142.69	1.13	2.0	119.37	13JL	3.2	-3.4	-2.2	0.9
148.99	1.08	2.0	128.94	23JL	1.9	-3.4	-2.2	0.9
155.33	1.02	2.0	138.55	2AU	0.2	-3.4	-2.3	0.9
161.71	0.96	2.0	148.20	12AU	-1.0	-3.5	-2.3	0.9
168.14	0.90	2.0	157.89	22AU	-1.4	-3.5	-2.4	0.9
174.61	0.84	1.9	167.63	1SE	-1.0	-3.5	-2.4	0.9
181.14	0.77	1.9	177.43	11SE	-0.5	-3.4	-2.5	0.9
187.72	0.70	1.9	187.28	21SE	-0.2	-3.4	-2.5	0.9
194.37	0.62	1.8	197.19	1OC	-0.0	-3.4	-2.5	0.8
201.07	0.54	1.8	207.16	11OC	0.1	-3.4	-2.4	0.8
207.84	0.46	1.8	217.18	21OC	0.5	-3.3	-2.4	0.8
214.67	0.37	1.8	227.25	31OC	2.5	-3.3	-2.3	0.8
221.57	0.27	1.8	237.36	10NO	0.9	-3.3	-2.3	0.8
228.53	0.18	1.8	247.51	20NO	-0.2	-3.3	-2.2	0.8
235.56	0.08	1.8	257.68	30NO	-0.4	-3.3	-2.1	0.8
242.66	-0.03	1.7	267.88	10DE	-0.4	-3.4	-2.1	0.8
249.82	-0.14	1.7	278.07	20DE	-0.5	-3.4	-2.0	0.8
257.04	-0.26	1.7	288.26	30DE	-0.8	-3.4	-1.9	0.8

♂ LONG	♂ LAT	♂ MAG	⊙ LONG	16.00UT	☿	♀	♃	♄
				1454	MAGNITUDES			
264.33	-0.37	1.6	298.44	9JA	-1.0	-3.4	-1.9	0.8
271.68	-0.50	1.6	308.59	19JA	-1.0	-3.5	-1.8	0.7
279.08	-0.62	1.5	318.71	29JA	-0.6	-3.5	-1.8	0.7
286.53	-0.74	1.5	328.79	8FE	0.9	-3.6	-1.7	0.7
294.03	-0.87	1.4	338.81	18FE	3.2	-3.6	-1.7	0.6
301.57	-0.99	1.3	348.79	28FE	1.6	-3.7	-1.6	0.6
309.13	-1.12	1.3	358.71	10MR	0.8	-3.8	-1.6	0.5
316.71	-1.23	1.2	8.57	20MR	0.5	-3.8	-1.6	0.5
324.30	-1.34	1.2	18.37	30MR	0.2	-3.9	-1.6	0.4
331.89	-1.45	1.1	28.12	9AP	-0.4	-4.0	-1.6	0.3
339.46	-1.54	1.0	37.82	19AP	-1.2	-4.1	-1.6	0.3
346.99	-1.62	1.0	47.47	29AP	-1.8	-4.2	-1.6	0.3
354.47	-1.69	0.9	57.08	9MY	-0.9	-4.2	-1.6	0.4
1.88	-1.74	0.9	66.66	19MY	0.1	-4.1	-1.6	0.5
9.21	-1.78	0.8	76.22	29MY	0.8	-3.8	-1.6	0.5
16.42	-1.79	0.7	85.76	8JN	1.5	-3.1	-1.7	0.6
23.49	-1.79	0.7	95.28	18JN	2.6	-3.1	-1.7	0.7
30.41	-1.77	0.6	104.82	28JN	3.1	-3.8	-1.8	0.7
37.12	-1.72	0.5	114.36	8JL	1.5	-4.1	-1.8	0.8
43.62	-1.66	0.4	123.92	18JL	0.1	-4.2	-1.9	0.8
49.84	-1.57	0.3	133.51	28JL	-1.1	-4.2	-2.0	0.8
55.73	-1.45	0.2	143.14	7AU	-1.5	-4.1	-2.0	0.8
61.25	-1.31	0.1	152.80	17AU	-1.0	-4.0	-2.1	0.9
66.29	-1.14	-0.0	162.52	27AU	-0.4	-3.9	-2.2	0.9
70.76	-0.93	-0.2	172.29	6SE	-0.1	-3.9	-2.2	0.9
74.53	-0.68	-0.4	182.11	16SE	0.1	-3.8	-2.3	0.8
77.42	-0.38	-0.5	191.99	26SE	0.2	-3.7	-2.3	0.8
79.24	-0.03	-0.7	201.93	6OC	0.7	-3.6	-2.4	0.8
79.80	0.38	-1.0	211.92	16OC	2.9	-3.6	-2.4	0.8
78.91	0.84	-1.2	221.97	26OC	0.8	-3.5	-2.4	0.7
76.62	1.34	-1.4	232.06	5NO	-0.4	-3.5	-2.4	0.7
73.22	1.82	-1.5	242.19	15NO	-0.5	-3.5	-2.4	0.7
69.35	2.22	-1.6	252.35	25NO	-0.5	-3.4	-2.3	0.8
65.84	2.51	-1.3	262.53	5DE	-0.6	-3.4	-2.3	0.8
63.33	2.67	-1.0	272.73	15DE	-0.8	-3.4	-2.2	0.8
62.14	2.72	-0.7	282.93	25DE	-0.9	-3.4	-2.2	0.8
				1455				
62.28	2.71	-0.4	293.11	4JA	-0.9	-3.3	-2.1	0.8
63.59	2.65	-0.1	303.28	14JA	-0.5	-3.3	-2.0	0.8
65.87	2.56	0.2	313.42	24JA	1.1	-3.3	-1.9	0.7
68.92	2.47	0.4	323.51	3FE	2.8	-3.3	-1.9	0.7
72.57	2.37	0.6	333.56	13FE	1.2	-3.3	-1.8	0.7
76.71	2.26	0.8	343.57	23FE	0.6	-3.4	-1.7	0.6
81.21	2.16	1.0	353.51	5MR	0.3	-3.4	-1.7	0.6
86.02	2.06	1.1	3.40	15MR	0.0	-3.4	-1.6	0.5
91.06	1.97	1.3	13.24	25MR	-0.4	-3.5	-1.6	0.5
96.29	1.87	1.4	23.01	4AP	-1.2	-3.5	-1.6	0.4
101.68	1.78	1.5	32.74	14AP	-1.7	-3.5	-1.5	0.3
107.20	1.69	1.6	42.41	24AP	-0.9	-3.4	-1.5	0.3
112.83	1.60	1.7	52.04	4MY	0.1	-3.4	-1.5	0.3
118.56	1.51	1.8	61.64	14MY	1.0	-3.4	-1.5	0.3
124.38	1.42	1.8	71.21	24MY	2.0	-3.4	-1.5	0.4
130.28	1.33	1.9	80.75	3JN	3.3	-3.3	-1.5	0.4
136.26	1.25	1.9	90.28	13JN	2.5	-3.3	-1.5	0.5
142.31	1.16	1.9	99.82	23JN	1.2	-3.3	-1.5	0.6
148.42	1.07	1.9	109.35	3JL	0.0	-3.3	-1.6	0.6
154.61	0.98	2.0	118.90	13JL	-1.1	-3.3	-1.6	0.7
160.87	0.90	2.0	128.48	23JL	-1.6	-3.4	-1.6	0.7
167.21	0.81	2.0	138.08	2AU	-0.9	-3.4	-1.7	0.8
173.61	0.71	1.9	147.73	12AU	-0.3	-3.4	-1.7	0.8
180.09	0.62	1.9	157.42	22AU	0.0	-3.4	-1.8	0.8
186.65	0.53	1.9	167.16	1SE	0.2	-3.5	-1.9	0.8
193.29	0.43	1.9	176.95	11SE	0.4	-3.5	-1.9	0.8
200.01	0.34	1.8	186.80	21SE	1.0	-3.6	-2.0	0.8
206.82	0.24	1.8	196.71	1OC	3.2	-3.7	-2.0	0.8
213.71	0.14	1.7	206.67	11OC	0.6	-3.8	-2.1	0.8
220.68	0.04	1.7	216.69	21OC	-0.5	-3.9	-2.2	0.7
227.74	-0.06	1.6	226.76	31OC	-0.7	-4.0	-2.2	0.7
234.89	-0.16	1.6	236.87	10NO	-0.7	-4.1	-2.3	0.7
242.12	-0.26	1.6	247.02	20NO	-0.7	-4.2	-2.3	0.7
249.43	-0.36	1.6	257.19	30NO	-0.7	-4.3	-2.3	0.7
256.82	-0.47	1.6	267.38	10DE	-0.7	-4.4	-2.3	0.8
264.28	-0.57	1.6	277.58	20DE	-0.7	-4.4	-2.3	0.8
271.82	-0.66	1.5	287.77	30DE	-0.4	-4.2	-2.2	0.8

1456

♂ LONG	LAT	MAG	☉ LONG	16.00UT	☿	♀	♃	♄
279.42	-0.76	1.5	297.95	9JA	1.2	-3.8	-2.2	0.8
287.07	-0.85	1.5	308.10	19JA	2.3	-3.3	-2.1	0.8
294.77	-0.93	1.5	318.22	29JA	0.8	-3.7	-2.1	0.8
302.52	-1.01	1.5	328.30	8FE	0.4	-4.2	-2.0	0.7
310.29	-1.08	1.4	338.33	18FE	0.2	-4.3	-1.9	0.7
318.07	-1.14	1.4	348.30	28FE	-0.1	-4.3	-1.8	0.7
325.87	-1.18	1.4	358.23	9MR	-0.5	-4.2	-1.8	0.6
333.66	-1.22	1.4	8.09	19MR	-1.2	-4.1	-1.7	0.6
341.43	-1.25	1.4	17.89	29MR	-1.6	-4.0	-1.6	0.5
349.17	-1.26	1.3	27.65	8AP	-0.9	-3.9	-1.6	0.4
356.87	-1.26	1.3	37.35	18AP	0.2	-3.8	-1.5	0.4
4.51	-1.24	1.3	47.00	28AP	1.3	-3.7	-1.5	0.3
12.08	-1.21	1.3	56.62	8MY	2.6	-3.6	-1.5	0.3
19.57	-1.16	1.3	66.20	18MY	3.4	-3.6	-1.4	0.2
26.96	-1.10	1.2	75.75	28MY	1.9	-3.5	-1.4	0.3
34.26	-1.03	1.2	85.29	7JN	0.9	-3.5	-1.4	0.4
41.44	-0.94	1.2	94.82	17JN	-0.1	-3.4	-1.4	0.4
48.49	-0.84	1.2	104.35	27JN	-1.1	-3.4	-1.4	0.5
55.40	-0.73	1.1	113.90	7JL	-1.7	-3.4	-1.4	0.6
62.16	-0.60	1.1	123.46	17JL	-0.9	-3.3	-1.4	0.6
68.75	-0.46	1.1	133.05	27JL	-0.3	-3.3	-1.4	0.7
75.16	-0.31	1.0	142.67	6AU	0.1	-3.3	-1.4	0.7
81.35	-0.14	1.0	152.34	16AU	0.4	-3.3	-1.5	0.7
87.32	0.04	0.9	162.05	26AU	0.6	-3.3	-1.5	0.8
93.00	0.24	0.8	171.82	5SE	1.4	-3.4	-1.6	0.8
98.37	0.46	0.7	181.63	15SE	3.3	-3.4	-1.6	0.8
103.36	0.71	0.6	191.51	25SE	0.5	-3.4	-1.7	0.8
107.90	0.97	0.5	201.45	5OC	-0.7	-3.4	-1.7	0.8
111.88	1.27	0.3	211.44	15OC	-0.8	-3.5	-1.8	0.8
115.19	1.61	0.2	221.48	25OC	-0.8	-3.5	-1.9	0.7
117.65	1.99	-0.0	231.57	4NO	-0.8	-3.5	-1.9	0.7
119.09	2.40	-0.3	241.69	14NO	-0.6	-3.5	-2.0	0.7
119.32	2.85	-0.5	251.85	24NO	-0.6	-3.4	-2.0	0.7
118.18	3.31	-0.7	262.04	4DE	-0.6	-3.4	-2.1	0.7
115.70	3.73	-0.9	272.23	14DE	-0.3	-3.4	-2.1	0.7
112.17	4.06	-1.1	282.43	24DE	1.4	-3.4	-2.2	0.8

1457

♂ LONG	LAT	MAG	☉ LONG	16.00UT	☿	♀	♃	♄
108.19	4.22	-1.1	292.62	3JA	1.9	-3.4	-2.2	0.8
104.53	4.22	-0.9	302.78	13JA	0.5	-3.3	-2.2	0.8
101.82	4.07	-0.6	312.92	23JA	0.2	-3.3	-2.1	0.8
100.38	3.83	-0.4	323.02	2FE	0.1	-3.3	-2.1	0.8
100.25	3.55	-0.1	333.07	12FE	-0.1	-3.3	-2.0	0.8
101.28	3.27	0.1	343.08	22FE	-0.5	-3.4	-2.0	0.7
103.29	2.99	0.4	353.03	4MR	-1.2	-3.4	-1.9	0.7
106.09	2.73	0.6	2.92	14MR	-1.5	-3.4	-1.8	0.6
109.52	2.49	0.8	12.76	24MR	-0.9	-3.4	-1.8	0.6
113.47	2.27	0.9	22.54	3AP	0.4	-3.5	-1.7	0.5
117.82	2.07	1.1	32.27	13AP	1.7	-3.5	-1.6	0.5
122.49	1.88	1.2	41.95	23AP	3.5	-3.6	-1.6	0.4
127.45	1.71	1.3	51.58	3MY	2.6	-3.6	-1.5	0.4
132.63	1.54	1.4	61.18	13MY	1.5	-3.7	-1.5	0.3
138.01	1.39	1.5	70.75	23MY	0.7	-3.8	-1.4	0.2
143.56	1.24	1.6	80.29	2JN	-0.1	-3.8	-1.4	0.3
149.26	1.10	1.6	89.82	12JN	-1.1	-3.9	-1.3	0.3
155.10	0.97	1.7	99.36	22JN	-1.8	-4.0	-1.3	0.4
161.06	0.84	1.7	108.89	2JL	-0.9	-4.1	-1.3	0.4
167.15	0.71	1.7	118.44	12JL	-0.2	-4.2	-1.3	0.5
173.36	0.59	1.8	128.02	22JL	0.2	-4.2	-1.3	0.6
179.68	0.47	1.8	137.62	1AU	0.5	-4.1	-1.3	0.6
186.11	0.35	1.8	147.26	11AU	0.9	-3.7	-1.3	0.7
192.65	0.24	1.8	156.95	21AU	1.8	-3.2	-1.3	0.7
199.30	0.12	1.8	166.69	31AU	3.0	-3.6	-1.3	0.7
206.05	0.01	1.7	176.48	10SE	0.4	-4.1	-1.4	0.8
212.92	-0.10	1.7	186.33	20SE	-0.8	-4.3	-1.4	0.8
219.89	-0.20	1.7	196.23	30SE	-1.0	-4.3	-1.4	0.8
226.96	-0.30	1.7	206.19	10OC	-1.0	-4.3	-1.5	0.8
234.14	-0.40	1.6	216.21	20OC	-0.7	-4.2	-1.5	0.8
241.41	-0.50	1.6	226.27	30OC	-0.5	-4.1	-1.6	0.8
248.78	-0.59	1.6	236.38	9NO	-0.4	-4.0	-1.7	0.7
256.25	-0.68	1.5	246.53	19NO	-0.4	-3.9	-1.7	0.7
263.80	-0.76	1.5	256.70	29NO	-0.1	-3.8	-1.8	0.7
271.42	-0.83	1.4	266.89	9DE	1.7	-3.7	-1.8	0.7
279.12	-0.90	1.4	277.09	19DE	1.6	-3.7	-1.9	0.7
286.87	-0.96	1.4	287.28	29DE	0.3	-3.6	-2.0	0.8

1458

♂ LONG	LAT	MAG	☉ LONG	16.00UT	☿	♀	♃	♄
294.68	-1.01	1.4	297.46	8JA	0.0	-3.5	-2.0	0.8
302.52	-1.04	1.4	307.61	18JA	-0.1	-3.5	-2.0	0.8
310.39	-1.07	1.4	317.73	28JA	-0.2	-3.4	-2.1	0.8
318.27	-1.09	1.4	327.81	7FE	-0.6	-3.4	-2.1	0.8
326.16	-1.09	1.4	337.85	17FE	-1.2	-3.4	-2.0	0.8
334.03	-1.09	1.4	347.83	27FE	-1.4	-3.3	-2.0	0.8
341.88	-1.07	1.4	357.75	9MR	-0.8	-3.3	-2.0	0.7
349.69	-1.04	1.4	7.62	19MR	0.5	-3.3	-1.9	0.7
357.45	-0.99	1.4	17.43	29MR	2.2	-3.3	-1.9	0.7
5.15	-0.94	1.4	27.18	8AP	3.4	-3.3	-1.8	0.6
12.79	-0.88	1.4	36.88	18AP	1.9	-3.3	-1.7	0.5
20.35	-0.80	1.5	46.54	28AP	1.1	-3.3	-1.7	0.5
27.83	-0.72	1.5	56.15	8MY	0.5	-3.4	-1.6	0.4
35.21	-0.62	1.5	65.73	18MY	-0.2	-3.4	-1.5	0.3
42.50	-0.52	1.5	75.29	28MY	-1.1	-3.4	-1.5	0.3
49.70	-0.42	1.5	84.83	7JN	-1.8	-3.5	-1.4	0.2
56.78	-0.30	1.5	94.36	17JN	-0.9	-3.5	-1.4	0.3
63.76	-0.18	1.5	103.89	27JN	-0.2	-3.5	-1.3	0.3
70.63	-0.06	1.5	113.43	7JL	0.4	-3.4	-1.3	0.4
77.38	0.07	1.5	122.99	17JL	0.7	-3.4	-1.3	0.5
84.00	0.20	1.5	132.58	27JL	1.2	-3.4	-1.2	0.5
90.51	0.34	1.5	142.20	6AU	2.3	-3.3	-1.2	0.6
96.87	0.49	1.5	151.86	16AU	2.6	-3.3	-1.2	0.6
103.09	0.64	1.4	161.57	26AU	0.3	-3.3	-1.2	0.7
109.15	0.80	1.4	171.33	5SE	-0.9	-3.3	-1.2	0.7
115.03	0.97	1.3	181.16	15SE	-1.1	-3.3	-1.2	0.7
120.71	1.14	1.3	191.03	25SE	-1.0	-3.3	-1.2	0.8
126.16	1.33	1.2	200.96	5OC	-0.6	-3.4	-1.3	0.8
131.34	1.53	1.1	210.95	15OC	-0.4	-3.4	-1.3	0.8
136.20	1.75	1.0	220.99	25OC	-0.3	-3.4	-1.3	0.8
140.68	1.99	0.8	231.08	4NO	-0.2	-3.5	-1.4	0.8
144.68	2.25	0.7	241.21	14NO	0.1	-3.5	-1.4	0.8
148.10	2.53	0.5	251.36	24NO	1.9	-3.6	-1.5	0.7
150.80	2.85	0.3	261.54	4DE	1.3	-3.6	-1.6	0.7
152.62	3.19	0.1	271.74	14DE	0.1	-3.7	-1.6	0.7
153.37	3.55	-0.1	281.93	24DE	-0.1	-3.8	-1.7	0.7

1459

♂ LONG	LAT	MAG	☉ LONG	16.00UT	☿	♀	♃	♄
152.89	3.90	-0.4	292.12	3JA	-0.2	-3.9	-1.8	0.8
151.08	4.22	-0.6	302.29	13JA	-0.3	-4.0	-1.8	0.8
148.07	4.44	-0.8	312.43	23JA	-0.6	-4.1	-1.9	0.8
144.28	4.51	-1.0	322.53	2FE	-1.2	-4.2	-1.9	0.8
140.37	4.41	-0.9	332.59	12FE	-1.3	-4.3	-2.0	0.8
137.07	4.15	-0.7	342.60	22FE	-0.8	-4.3	-2.0	0.8
134.85	3.79	-0.5	352.55	4MR	0.6	-4.3	-2.0	0.8
133.93	3.39	-0.2	2.45	14MR	2.7	-4.1	-2.0	0.8
134.26	2.99	0.0	12.29	24MR	2.6	-3.6	-2.0	0.8
135.68	2.61	0.2	22.07	3AP	1.4	-3.1	-2.0	0.7
138.03	2.27	0.4	31.80	13AP	0.8	-3.7	-1.9	0.7
141.12	1.96	0.6	41.48	23AP	0.4	-4.1	-1.9	0.6
144.81	1.67	0.8	51.11	3MY	-0.3	-4.2	-1.8	0.5
149.00	1.42	0.9	60.71	13MY	-1.1	-4.2	-1.7	0.5
153.60	1.19	1.0	70.28	23MY	-1.8	-4.1	-1.7	0.4
158.54	0.97	1.1	79.83	2JN	-0.9	-4.0	-1.6	0.4
163.77	0.78	1.2	89.36	12JN	-0.1	-3.9	-1.5	0.3
169.25	0.59	1.3	98.89	22JN	0.5	-3.8	-1.5	0.3
174.96	0.42	1.3	108.43	2JL	1.0	-3.7	-1.4	0.3
180.87	0.26	1.4	117.98	12JL	1.6	-3.7	-1.4	0.4
186.95	0.11	1.4	127.55	22JL	2.9	-3.6	-1.3	0.4
193.22	-0.03	1.5	137.15	1AU	2.1	-3.6	-1.3	0.5
199.64	-0.16	1.5	146.79	11AU	0.2	-3.5	-1.3	0.6
206.21	-0.29	1.5	156.48	21AU	-1.0	-3.5	-1.2	0.6
212.93	-0.40	1.5	166.21	31AU	-1.3	-3.4	-1.2	0.7
219.79	-0.51	1.5	176.00	10SE	-1.0	-3.4	-1.2	0.7
226.78	-0.61	1.5	185.85	20SE	-0.5	-3.4	-1.2	0.7
233.90	-0.71	1.5	195.75	30SE	-0.3	-3.4	-1.2	0.8
241.05	-0.79	1.5	205.71	10OC	-0.1	-3.4	-1.2	0.8
248.50	-0.87	1.5	215.72	20OC	-0.1	-3.4	-1.2	0.8
255.96	-0.94	1.5	225.79	30OC	0.3	-3.4	-1.2	0.8
263.52	-0.99	1.5	235.89	9NO	2.2	-3.4	-1.3	0.8
271.16	-1.04	1.4	246.04	19NO	1.1	-3.4	-1.3	0.8
278.89	-1.08	1.4	256.21	29NO	-0.1	-3.4	-1.4	0.8
286.67	-1.10	1.4	266.40	9DE	-0.3	-3.4	-1.4	0.8
294.51	-1.11	1.4	276.59	19DE	-0.3	-3.4	-1.5	0.7
302.38	-1.11	1.4	286.79	29DE	-0.4	-3.4	-1.5	0.7

Left table (1460 / 1461)

♂ LONG	LAT	MAG	☉ LONG	16.00UT	☿	♀	♃	♄
310.28	-1.10	1.4	296.97	8JA	-0.7	-3.5	-1.6	0.8
318.19	-1.08	1.4	307.12	18JA	-1.1	-3.5	-1.7	0.8
326.10	-1.04	1.4	317.25	28JA	-1.1	-3.5	-1.7	0.8
333.98	-1.00	1.3	327.32	7FE	-0.7	-3.4	-1.8	0.9
341.84	-0.94	1.3	337.36	17FE	0.7	-3.4	-1.9	0.9
349.65	-0.88	1.3	347.34	27FE	3.1	-3.4	-1.9	0.9
357.41	-0.80	1.3	357.27	8MR	1.9	-3.4	-2.0	0.9
5.11	-0.72	1.4	7.13	18MR	1.0	-3.4	-2.0	0.8
12.74	-0.63	1.4	16.95	28MR	0.6	-3.3	-2.0	0.8
20.29	-0.54	1.5	26.70	7AP	0.2	-3.3	-2.0	0.8
27.76	-0.44	1.5	36.41	17AP	-0.3	-3.3	-2.0	0.7
35.14	-0.34	1.5	46.07	27AP	-1.1	-3.3	-2.0	0.7
42.44	-0.23	1.6	55.68	7MY	-1.8	-3.4	-2.0	0.6
49.64	-0.13	1.6	65.75	17MY	-0.9	-3.4	-1.9	0.6
56.75	-0.02	1.7	74.83	27MY	-0.0	-3.4	-1.9	0.5
63.77	0.09	1.7	84.37	6JN	0.7	-3.4	-1.8	0.4
70.71	0.20	1.7	93.90	16JN	1.3	-3.5	-1.7	0.4
77.55	0.31	1.7	103.43	26JN	2.1	-3.5	-1.7	0.3
84.30	0.42	1.8	112.97	6JL	3.3	-3.6	-1.6	0.3
90.96	0.53	1.8	122.53	16JL	1.8	-3.6	-1.6	0.4
97.54	0.64	1.8	132.12	26JL	0.2	-3.7	-1.5	0.4
104.02	0.75	1.8	141.73	5AU	-1.0	-3.8	-1.4	0.5
110.42	0.86	1.8	151.39	15AU	-1.4	-3.9	-1.4	0.5
116.72	0.97	1.8	161.10	25AU	-1.0	-4.0	-1.4	0.6
122.92	1.09	1.7	170.86	4SE	-0.5	-4.1	-1.3	0.6
129.02	1.20	1.7	180.68	14SE	-0.1	-4.2	-1.3	0.7
135.01	1.32	1.7	190.55	24SE	0.0	-4.3	-1.3	0.7
140.87	1.43	1.6	200.48	4OC	0.1	-4.3	-1.2	0.8
146.59	1.56	1.6	210.47	14OC	0.6	-4.2	-1.2	0.8
152.15	1.68	1.5	220.50	24OC	2.5	-3.7	-1.2	0.8
157.52	1.82	1.4	230.59	3NO	0.9	-3.0	-1.2	0.8
162.68	1.96	1.3	240.71	13NO	-0.3	-3.7	-1.3	0.8
167.56	2.11	1.2	250.87	23NO	-0.4	-4.2	-1.3	0.8
172.12	2.27	1.0	261.05	3DE	-0.5	-4.4	-1.3	0.8
176.29	2.44	0.8	271.25	13DE	-0.5	-4.4	-1.3	0.8
179.97	2.62	0.7	281.44	23DE	-0.8	-4.3	-1.4	0.8

1461

♂ LONG	LAT	MAG	☉ LONG	16.00UT	☿	♀	♃	♄
183.05	2.81	0.5	291.63	2JA	-0.9	-4.2	-1.4	0.8
185.37	3.02	0.2	301.80	12JA	-1.0	-4.1	-1.5	0.8
186.76	3.22	-0.0	311.94	22JA	-0.6	-4.0	-1.5	0.8
187.06	3.41	-0.3	322.05	1FE	0.9	-3.9	-1.6	0.9
186.10	3.56	-0.6	332.11	11FE	3.1	-3.8	-1.7	0.9
183.87	3.63	-0.8	342.12	21FE	1.4	-3.7	-1.7	0.9
180.59	3.58	-1.1	352.07	3MR	0.7	-3.6	-1.8	0.9
176.76	3.38	-1.1	1.98	13MR	0.4	-3.6	-1.9	0.9
173.09	3.05	-1.0	11.82	23MR	0.1	-3.5	-1.9	0.9
170.24	2.63	-0.8	21.60	2AP	-0.4	-3.4	-2.0	0.9
168.59	2.18	-0.6	31.34	12AP	-1.1	-3.4	-2.0	0.9
168.26	1.74	-0.4	41.02	22AP	-1.8	-3.4	-2.1	0.8
169.15	1.34	-0.2	50.65	2MY	-0.9	-3.3	-2.1	0.8
171.11	0.97	0.0	60.26	12MY	0.1	-3.3	-2.1	0.7
173.95	0.65	0.2	69.83	22MY	0.9	-3.3	-2.1	0.7
177.53	0.37	0.3	79.37	1JN	1.7	-3.3	-2.1	0.6
181.71	0.11	0.5	88.91	11JN	2.8	-3.3	-2.0	0.6
186.39	-0.11	0.6	98.44	21JN	2.9	-3.3	-2.0	0.5
191.48	-0.31	0.7	107.97	1JL	1.4	-3.3	-1.9	0.4
196.94	-0.48	0.8	117.52	11JL	0.1	-3.4	-1.9	0.4
202.71	-0.64	0.9	127.09	21JL	-1.1	-3.4	-1.8	0.4
208.75	-0.78	0.9	136.69	31JL	-1.5	-3.4	-1.7	0.4
215.03	-0.90	1.0	146.33	10AU	-1.0	-3.5	-1.7	0.5
221.54	-1.01	1.0	156.01	20AU	-0.4	-3.5	-1.6	0.5
228.24	-1.10	1.1	165.74	30AU	-0.0	-3.5	-1.6	0.6
235.13	-1.17	1.1	175.52	9SE	0.1	-3.4	-1.5	0.7
242.19	-1.24	1.1	185.36	19SE	0.3	-3.4	-1.5	0.7
249.39	-1.29	1.2	195.26	29SE	0.8	-3.4	-1.4	0.7
256.73	-1.32	1.2	205.20	9OC	2.9	-3.4	-1.4	0.8
264.19	-1.34	1.2	215.23	19OC	0.8	-3.3	-1.4	0.8
271.76	-1.34	1.2	225.29	29OC	-0.4	-3.3	-1.3	0.9
279.41	-1.33	1.3	235.40	8NO	-0.6	-3.3	-1.3	0.9
287.14	-1.31	1.3	245.54	18NO	-0.6	-3.3	-1.3	0.9
294.93	-1.28	1.3	255.71	28NO	-0.7	-3.3	-1.3	0.9
302.76	-1.23	1.3	265.90	8DE	-0.8	-3.4	-1.3	0.9
310.61	-1.17	1.3	276.09	18DE	-0.8	-3.4	-1.4	0.9
318.47	-1.10	1.4	286.29	28DE	-0.8	-3.4	-1.4	0.9

Right table (1462 / 1463)

♂ LONG	LAT	MAG	☉ LONG	16.00UT	☿	♀	♃	♄
326.33	-1.02	1.4	296.47	7JA	-0.5	-3.4	-1.4	0.9
334.16	-0.93	1.4	306.63	17JA	1.1	-3.5	-1.4	0.9
341.96	-0.84	1.4	316.75	27JA	2.7	-3.5	-1.5	0.9
349.71	-0.74	1.5	326.84	6FE	1.0	-3.6	-1.5	0.9
357.41	-0.63	1.5	336.88	16FE	0.5	-3.6	-1.6	1.0
5.04	-0.53	1.5	346.86	26FE	0.3	-3.7	-1.7	1.0
12.60	-0.42	1.5	356.79	8MR	0.0	-3.8	-1.7	1.0
20.08	-0.31	1.5	6.66	18MR	-0.4	-3.9	-1.8	1.0
27.48	-0.20	1.6	16.48	28MR	-1.1	-3.9	-1.9	1.0
34.80	-0.09	1.6	26.24	7AP	-1.7	-4.0	-1.9	1.0
42.03	0.01	1.6	35.95	17AP	-0.9	-4.1	-2.0	1.0
49.18	0.12	1.6	45.61	27AP	0.2	-4.2	-2.1	1.0
56.25	0.22	1.6	55.23	7MY	1.1	-4.2	-2.1	0.9
63.23	0.32	1.7	64.81	17MY	2.2	-4.1	-2.2	0.9
70.14	0.42	1.7	74.37	27MY	3.6	-3.8	-2.2	0.8
76.97	0.51	1.8	83.91	6JN	2.3	-3.1	-2.2	0.8
83.72	0.60	1.8	93.44	16JN	1.1	-3.2	-2.2	0.7
90.41	0.69	1.8	102.97	26JN	0.0	-3.8	-2.2	0.6
97.03	0.77	1.9	112.51	6JL	-1.1	-4.1	-2.2	0.6
103.58	0.85	1.9	122.06	16JL	-1.6	-4.2	-2.1	0.5
110.08	0.93	1.9	131.65	26JL	-1.0	-4.2	-2.1	0.5
116.52	1.00	1.9	141.27	5AU	-0.3	-4.1	-2.0	0.5
122.91	1.07	1.9	150.92	15AU	0.1	-4.0	-2.0	0.5
129.24	1.14	2.0	160.63	25AU	0.3	-3.9	-1.9	0.6
135.51	1.21	1.9	170.39	4SE	0.5	-3.8	-1.8	0.6
141.73	1.27	1.9	180.20	14SE	1.1	-3.8	-1.8	0.7
147.90	1.33	1.9	190.07	24SE	3.2	-3.7	-1.7	0.7
154.00	1.39	1.9	199.99	4OC	0.7	-3.6	-1.7	0.8
160.03	1.45	1.8	209.97	14OC	-0.6	-3.6	-1.6	0.8
165.99	1.51	1.8	220.01	24OC	-0.8	-3.5	-1.6	0.9
171.87	1.57	1.7	230.13	3NO	-0.7	-3.5	-1.5	0.9
177.65	1.62	1.6	240.22	13NO	-0.8	-3.5	-1.5	0.9
183.32	1.68	1.5	250.37	23NO	-0.7	-3.4	-1.5	1.0
188.85	1.73	1.4	260.56	3DE	-0.6	-3.4	-1.4	1.0
194.23	1.78	1.3	270.75	13DE	-0.7	-3.4	-1.4	1.0
199.42	1.82	1.2	280.95	23DE	-0.4	-3.4	-1.4	1.0

1463

♂ LONG	LAT	MAG	☉ LONG	16.00UT	☿	♀	♃	♄
204.36	1.87	1.0	291.14	2JA	1.2	-3.3	-1.4	1.0
209.03	1.90	0.9	301.31	12JA	2.2	-3.3	-1.4	1.0
213.34	1.93	0.7	311.45	22JA	0.7	-3.3	-1.5	1.0
217.20	1.95	0.5	321.56	1FE	0.3	-3.3	-1.5	1.0
220.52	1.95	0.2	331.62	11FE	0.1	-3.4	-1.5	1.0
223.15	1.93	-0.0	341.64	21FE	-0.1	-3.4	-1.6	1.1
224.93	1.88	-0.3	351.60	3MR	-0.5	-3.4	-1.6	1.1
225.68	1.77	-0.6	1.50	13MR	-1.1	-3.4	-1.7	1.1
225.62	1.60	-0.9	11.34	23MR	-1.6	-3.5	-1.7	1.1
223.50	1.34	-1.2	21.13	2AP	-0.9	-3.5	-1.8	1.2
220.66	0.99	-1.5	30.86	12AP	0.2	-3.5	-1.9	1.1
217.13	0.56	-1.6	40.55	22AP	1.5	-3.4	-1.9	1.1
213.61	0.09	-1.5	50.19	2MY	2.9	-3.4	-2.0	1.1
210.79	-0.37	-1.4	59.79	12MY	3.1	-3.4	-2.1	1.1
209.14	-0.78	-1.2	69.36	22MY	1.8	-3.4	-2.1	1.0
208.87	-1.12	-1.0	78.91	1JN	0.9	-3.3	-2.2	1.0
209.90	-1.40	-0.7	88.44	11JN	-0.1	-3.3	-2.3	0.9
212.08	-1.62	-0.6	97.97	21JN	-1.1	-3.3	-2.3	0.9
215.25	-1.79	-0.4	107.51	1JL	-1.7	-3.3	-2.3	0.8
219.21	-1.92	-0.2	117.05	11JL	-0.9	-3.3	-2.4	0.7
223.84	-2.01	-0.1	126.62	21JL	-0.3	-3.4	-2.4	0.7
229.03	-2.07	0.1	136.22	31JL	0.2	-3.4	-2.3	0.6
234.66	-2.11	0.2	145.85	10AU	0.4	-3.4	-2.3	0.6
240.69	-2.12	0.3	155.54	20AU	0.7	-3.4	-2.3	0.6
247.04	-2.11	0.4	165.26	30AU	1.5	-3.5	-2.2	0.7
253.66	-2.08	0.5	175.00	9SE	3.2	-3.5	-2.2	0.7
260.52	-2.03	0.5	184.89	19SE	0.5	-3.6	-2.1	0.8
267.58	-1.97	0.6	194.78	29SE	-0.7	-3.7	-2.0	0.8
274.80	-1.89	0.7	204.73	9OC	-0.9	-3.8	-2.0	0.9
282.15	-1.79	0.8	214.74	19OC	-0.9	-3.9	-1.9	0.9
289.62	-1.68	0.8	224.80	29OC	-0.8	-4.0	-1.8	0.9
297.17	-1.56	0.9	234.90	8NO	-0.5	-4.1	-1.8	1.0
304.77	-1.44	1.0	245.04	18NO	-0.5	-4.2	-1.7	1.0
312.42	-1.30	1.0	255.21	28NO	-0.5	-4.3	-1.7	1.1
320.08	-1.16	1.1	265.40	8DE	-0.2	-4.4	-1.6	1.1
327.75	-1.02	1.2	275.60	18DE	1.4	-4.4	-1.6	1.1
335.39	-0.88	1.2	285.79	28DE	1.9	-4.2	-1.6	1.1

Left Table — 1464

♂ LONG	LAT	MAG	☉ LONG	16.00UT	☿	♀	♃	♄
343.01	-0.74	1.3	295.97	7JA	0.4	-3.8	-1.5	1.1
350.59	-0.59	1.4	306.13	17JA	0.1	-3.3	-1.5	1.1
358.11	-0.46	1.4	316.26	27JA	0.0	-3.7	-1.5	1.1
5.57	-0.32	1.5	326.35	6FE	-0.2	-4.2	-1.5	1.1
12.97	-0.19	1.5	336.39	16FE	-0.5	-4.3	-1.5	1.1
20.29	-0.07	1.6	346.38	26FE	-1.1	-4.3	-1.6	1.2
27.55	0.05	1.6	356.31	7MR	-1.5	-4.2	-1.6	1.2
34.72	0.16	1.7	6.19	17MR	-0.9	-4.1	-1.6	1.3
41.82	0.27	1.7	16.00	27MR	0.4	-4.0	-1.7	1.3
48.84	0.37	1.7	25.77	6AP	1.9	-3.9	-1.7	1.3
55.79	0.47	1.8	35.48	16AP	3.8	-3.8	-1.8	1.3
62.67	0.55	1.8	45.14	26AP	2.3	-3.7	-1.8	1.3
69.48	0.64	1.8	54.76	6MY	1.3	-3.6	-1.9	1.3
76.23	0.71	1.8	64.35	16MY	0.6	-3.6	-1.9	1.3
82.92	0.78	1.9	73.91	26MY	-0.1	-3.5	-2.0	1.2
89.55	0.85	1.9	83.45	5JN	-1.1	-3.4	-2.1	1.2
96.13	0.90	1.9	92.98	15JN	-1.8	-3.4	-2.2	1.1
102.67	0.96	1.9	102.51	25JN	-0.9	-3.4	-2.2	1.1
109.16	1.01	1.9	112.05	5JL	-0.2	-3.4	-2.3	1.0
115.62	1.05	1.9	121.61	15JL	0.3	-3.3	-2.3	0.9
122.04	1.09	2.0	131.19	25JL	0.6	-3.3	-2.4	0.9
128.44	1.13	2.0	140.81	4AU	1.0	-3.3	-2.4	0.8
134.81	1.16	2.0	150.46	14AU	2.0	-3.3	-2.4	0.7
141.16	1.19	2.0	160.16	24AU	2.9	-3.4	-2.4	0.7
147.49	1.21	2.0	169.92	3SE	0.4	-3.4	-2.4	0.8
153.80	1.23	2.0	179.73	13SE	-0.8	-3.4	-2.4	0.8
160.09	1.24	2.0	189.59	23SE	-1.1	-3.4	-2.4	0.9
166.37	1.25	2.0	199.52	3OC	-1.0	-3.4	-2.3	0.9
172.63	1.26	2.0	209.49	13OC	-0.7	-3.5	-2.3	1.0
178.87	1.26	1.9	219.52	23OC	-0.4	-3.5	-2.2	1.0
185.09	1.26	1.9	229.60	2NO	-0.3	-3.5	-2.1	1.1
191.29	1.25	1.8	239.72	12NO	-0.3	-3.5	-2.0	1.1
197.46	1.24	1.8	249.88	22NO	-0.0	-3.4	-2.0	1.1
203.60	1.22	1.7	260.06	2DE	1.7	-3.4	-1.9	1.2
209.69	1.19	1.6	270.25	12DE	1.6	-3.4	-1.8	1.2
215.74	1.15	1.5	280.44	22DE	0.2	-3.4	-1.8	1.2

1465

♂ LONG	LAT	MAG	☉ LONG	16.00UT	☿	♀	♃	♄
221.72	1.10	1.4	290.64	1JA	-0.0	-3.4	-1.7	1.3
227.63	1.04	1.3	300.81	11JA	-0.1	-3.3	-1.7	1.3
233.45	0.96	1.2	310.95	21JA	-0.3	-3.3	-1.7	1.3
239.15	0.86	1.0	321.06	31JA	-0.6	-3.3	-1.6	1.3
244.72	0.75	0.8	331.13	10FE	-1.2	-3.4	-1.6	1.3
250.12	0.61	0.7	341.14	20FE	-1.4	-3.4	-1.6	1.3
255.30	0.43	0.5	351.11	2MR	-0.8	-3.4	-1.6	1.3
260.22	0.22	0.3	1.01	12MR	0.5	-3.4	-1.6	1.4
264.81	-0.03	0.0	10.86	22MR	2.3	-3.4	-1.6	1.4
268.98	-0.34	-0.2	20.66	1AP	3.1	-3.5	-1.6	1.5
272.60	-0.71	-0.5	30.39	11AP	1.7	-3.5	-1.6	1.5
275.56	-1.16	-0.8	40.08	21AP	1.0	-3.6	-1.7	1.4
277.65	-1.70	-1.1	49.73	1MY	0.5	-3.6	-1.7	1.4
278.71	-2.32	-1.4	59.33	11MY	-0.2	-3.7	-1.7	1.4
278.58	-3.02	-1.7	68.90	21MY	-1.1	-3.8	-1.8	1.3
277.21	-3.74	-2.0	78.45	31MY	-1.8	-3.8	-1.8	1.3
274.82	-4.42	-2.3	87.99	10JN	-0.9	-3.9	-1.9	1.3
271.90	-4.96	-2.4	97.52	20JN	-0.1	-4.0	-2.0	1.2
269.16	-5.27	-2.3	107.05	30JN	0.4	-4.1	-2.0	1.2
267.30	-5.36	-2.1	116.60	10JL	0.8	-4.2	-2.1	1.1
266.71	-5.25	-1.8	126.16	20JL	1.3	-4.2	-2.2	1.1
267.50	-5.01	-1.6	135.76	30JL	2.5	-4.1	-2.2	1.0
269.58	-4.69	-1.4	145.39	9AU	2.5	-3.7	-2.3	1.0
272.76	-4.34	-1.1	155.07	19AU	0.4	-3.2	-2.4	0.9
276.84	-3.97	-0.9	164.80	29AU	-0.9	-3.6	-2.4	0.9
281.64	-3.59	-0.7	174.57	8SE	-1.2	-4.1	-2.4	0.9
287.01	-3.23	-0.5	184.41	18SE	-1.1	-4.3	-2.5	0.9
292.83	-2.87	-0.3	194.31	28SE	-0.6	-4.3	-2.5	1.0
299.00	-2.52	-0.2	204.25	8OC	-0.3	-4.3	-2.5	1.0
305.43	-2.20	0.0	214.26	18OC	-0.2	-4.2	-2.4	1.1
312.07	-1.88	0.2	224.31	28OC	-0.2	-4.1	-2.4	1.1
318.87	-1.59	0.3	234.41	7NO	0.2	-4.0	-2.3	1.2
325.77	-1.32	0.5	244.55	17NO	1.9	-3.9	-2.3	1.2
332.75	-1.06	0.6	254.72	27NO	1.3	-3.8	-2.2	1.3
339.78	-0.82	0.7	264.90	7DE	0.0	-3.7	-2.1	1.3
346.83	-0.60	0.8	275.10	17DE	-0.2	-3.7	-2.1	1.4
353.89	-0.40	1.0	285.30	27DE	-0.2	-3.6	-2.0	1.4

Right Table — 1466

♂ LONG	LAT	MAG	☉ LONG	16.00UT	☿	♀	♃	♄
0.93	-0.21	1.1	295.48	6JA	-0.4	-3.5	-1.9	1.4
7.95	-0.04	1.2	305.64	16JA	-0.7	-3.5	-1.9	1.4
14.95	0.11	1.3	315.77	26JA	-1.1	-3.4	-1.8	1.4
21.90	0.25	1.4	325.86	5FE	-1.2	-3.4	-1.7	1.3
28.81	0.38	1.5	335.90	15FE	-0.8	-3.4	-1.7	1.3
35.68	0.49	1.5	345.89	25FE	0.6	-3.3	-1.7	1.2
42.49	0.59	1.6	355.83	7MR	2.8	-3.3	-1.6	1.2
49.26	0.68	1.7	5.71	17MR	2.3	-3.3	-1.6	1.2
55.97	0.76	1.7	15.53	27MR	1.2	-3.3	-1.6	1.2
62.64	0.83	1.8	25.29	6AP	0.7	-3.3	-1.6	1.2
69.26	0.89	1.9	35.01	16AP	0.3	-3.3	-1.6	1.2
75.84	0.95	1.9	44.68	26AP	-0.3	-3.3	-1.6	1.2
82.37	0.99	1.9	54.30	6MY	-1.1	-3.4	-1.6	1.1
88.87	1.03	2.0	63.89	16MY	-1.8	-3.4	-1.6	1.1
95.34	1.07	2.0	73.45	26MY	-0.9	-3.4	-1.6	1.1
101.77	1.10	2.0	82.99	5JN	-0.1	-3.5	-1.6	1.1
108.19	1.12	2.0	92.52	15JN	0.5	-3.5	-1.7	1.0
114.58	1.13	2.0	102.05	25JN	1.1	-3.5	-1.7	1.0
120.96	1.14	2.0	111.59	5JL	1.8	-3.4	-1.8	0.9
127.33	1.15	2.0	121.15	15JL	3.1	-3.4	-1.8	0.9
133.69	1.15	2.0	130.72	25JL	2.0	-3.4	-1.9	0.8
140.05	1.15	2.0	140.33	4AU	0.3	-3.3	-2.0	0.8
146.42	1.14	2.0	149.99	14AU	-1.0	-3.3	-2.0	0.7
152.79	1.13	2.0	159.69	24AU	-1.3	-3.3	-2.1	0.7
159.17	1.11	2.0	169.44	3SE	-1.0	-3.3	-2.2	0.6
165.57	1.09	2.0	179.25	13SE	-0.5	-3.3	-2.2	0.6
171.98	1.07	2.0	189.11	23SE	-0.2	-3.3	-2.3	0.6
178.41	1.03	2.0	199.03	3OC	-0.1	-3.4	-2.3	0.7
184.87	1.00	2.0	209.01	13OC	0.0	-3.4	-2.4	0.8
191.34	0.96	2.0	219.03	23OC	0.4	-3.4	-2.4	0.9
197.84	0.91	1.9	229.11	2NO	2.2	-3.5	-2.4	0.9
204.36	0.85	1.9	239.23	12NO	1.1	-3.5	-2.4	1.0
210.90	0.79	1.9	249.39	22NO	-0.2	-3.6	-2.4	1.0
217.47	0.72	1.8	259.56	2DE	-0.4	-3.6	-2.3	1.1
224.07	0.64	1.7	269.76	12DE	-0.4	-3.7	-2.3	1.1
230.68	0.56	1.7	279.95	22DE	-0.5	-3.8	-2.2	1.1

1467

♂ LONG	LAT	MAG	☉ LONG	16.00UT	☿	♀	♃	♄
237.31	0.46	1.6	290.14	1JA	-0.7	-3.9	-2.1	1.1
243.96	0.35	1.5	300.32	11JA	-1.0	-4.0	-2.1	1.1
250.62	0.22	1.4	310.46	21JA	-1.1	-4.1	-2.0	1.1
257.30	0.09	1.3	320.57	31JA	-0.7	-4.2	-1.9	1.1
263.99	-0.06	1.2	330.64	10FE	0.7	-4.3	-1.8	1.1
270.67	-0.23	1.1	340.66	20FE	3.1	-4.3	-1.8	1.0
277.36	-0.42	1.0	350.62	2MR	1.7	-4.3	-1.7	1.0
284.01	-0.63	0.9	0.54	12MR	0.9	-4.1	-1.7	0.9
290.68	-0.85	0.7	10.38	22MR	0.5	-3.6	-1.6	0.9
297.29	-1.10	0.6	20.18	1AP	0.2	-3.1	-1.6	0.9
303.84	-1.37	0.4	29.92	11AP	-0.3	-3.7	-1.5	0.9
310.31	-1.66	0.3	39.61	21AP	-1.1	-4.1	-1.5	0.9
316.67	-1.98	0.1	49.25	1MY	-1.8	-4.2	-1.5	0.9
322.87	-2.32	-0.1	58.86	11MY	-1.0	-4.2	-1.5	0.9
328.87	-2.69	-0.3	68.43	21MY	-0.0	-4.1	-1.5	0.9
334.59	-3.08	-0.5	77.99	31MY	0.7	-4.0	-1.5	0.9
339.93	-3.49	-0.7	87.53	10JN	1.4	-3.9	-1.5	0.8
344.81	-3.92	-0.9	97.05	20JN	2.4	-3.8	-1.5	0.8
349.06	-4.36	-1.1	106.59	30JN	3.2	-3.7	-1.5	0.8
352.50	-4.80	-1.4	116.14	10JL	1.7	-3.7	-1.5	0.7
354.95	-5.22	-1.6	125.70	20JL	0.2	-3.6	-1.6	0.7
356.18	-5.59	-1.9	135.30	30JL	-1.0	-3.5	-1.6	0.6
356.03	-5.84	-2.2	144.93	9AU	-1.5	-3.5	-1.7	0.6
354.56	-5.90	-2.4	154.60	19AU	-1.0	-3.5	-1.7	0.5
352.06	-5.70	-2.5	164.33	29AU	-0.4	-3.4	-1.8	0.5
349.20	-5.22	-2.5	174.11	8SE	-0.1	-3.4	-1.8	0.4
346.76	-4.54	-2.3	183.93	18SE	0.1	-3.4	-1.9	0.4
345.32	-3.76	-2.0	193.83	28SE	0.2	-3.4	-2.0	0.3
345.19	-2.98	-1.7	203.78	8OC	0.6	-3.4	-2.0	0.4
346.34	-2.26	-1.3	213.77	18OC	2.6	-3.4	-2.1	0.5
348.62	-1.64	-1.0	223.83	28OC	0.9	-3.4	-2.2	0.5
351.82	-1.10	-0.7	233.93	7NO	-0.3	-3.4	-2.2	0.6
355.74	-0.66	-0.5	244.06	17NO	-0.5	-3.4	-2.2	0.7
0.23	-0.29	-0.2	254.23	27NO	-0.5	-3.4	-2.3	0.7
5.16	0.01	0.0	264.41	7DE	-0.6	-3.4	-2.3	0.8
10.42	0.27	0.3	274.61	17DE	-0.8	-3.4	-2.3	0.8
15.93	0.48	0.5	284.80	27DE	-0.9	-3.4	-2.2	0.8

♂			☉	16.00UT	☿	♀	♃	♄	♂			☉	16.00UT	☿	♀	♃	♄
LONG	LAT	MAG	LONG	1468		MAGN	ITUDES		LONG	LAT	MAG	LONG	1470		MAGN	ITUDES	
21.64	0.65	0.6	294.99	6JA	-0.9	-3.5	-2.2	0.9	48.82	2.02	-0.0	294.49	5JA	2.2	-3.4	-2.0	0.4
27.49	0.79	0.8	305.15	16JA	-0.6	-3.5	-2.1	0.9	51.99	2.03	0.2	304.66	15JA	0.6	-3.5	-2.0	0.4
33.45	0.91	1.0	315.28	26JA	0.9	-3.5	-2.1	0.9	55.77	2.02	0.5	314.79	25JA	0.2	-3.5	-2.0	0.5
39.49	1.00	1.1	325.37	5FE	3.0	-3.4	-2.0	0.9	60.02	1.99	0.7	324.88	4FE	0.1	-3.6	-2.0	0.5
45.60	1.08	1.3	335.41	15FE	1.3	-3.4	-1.9	0.8	64.64	1.96	0.9	334.93	14FE	-0.1	-3.6	-2.0	0.5
51.74	1.14	1.4	345.41	25FE	0.6	-3.4	-1.9	0.8	69.54	1.91	1.0	344.92	24FE	-0.5	-3.7	-2.0	0.5
57.91	1.19	1.5	355.34	6MR	0.4	-3.4	-1.8	0.8	74.68	1.87	1.2	354.87	6MR	-1.1	-3.8	-2.0	0.5
64.11	1.23	1.6	5.22	16MR	0.1	-3.4	-1.7	0.7	80.00	1.81	1.3	4.75	16MR	-1.6	-3.9	-2.0	0.5
70.31	1.26	1.7	15.05	26MR	-0.4	-3.3	-1.7	0.7	85.45	1.76	1.4	14.58	26MR	-0.9	-3.9	-1.9	0.4
76.53	1.28	1.7	24.82	5AP	-1.1	-3.3	-1.6	0.7	91.03	1.70	1.5	24.35	5AP	0.3	-4.0	-1.8	0.4
82.74	1.29	1.8	34.53	15AP	-1.7	-3.3	-1.6	0.7	96.71	1.64	1.6	34.07	15AP	1.6	-4.1	-1.8	0.4
88.96	1.29	1.9	44.20	25AP	-1.0	-3.3	-1.5	0.7	102.47	1.58	1.7	43.74	25AP	3.2	-4.2	-1.7	0.3
95.19	1.29	1.9	53.83	5MY	0.1	-3.4	-1.5	0.7	108.30	1.52	1.8	53.37	5MY	2.8	-4.2	-1.6	0.3
101.41	1.29	2.0	63.42	15MY	0.9	-3.4	-1.4	0.7	114.20	1.46	1.8	62.96	15MY	1.6	-4.1	-1.6	0.4
107.64	1.27	2.0	72.98	25MY	1.8	-3.4	-1.4	0.7	120.15	1.39	1.9	72.52	25MY	0.8	-3.8	-1.5	0.4
113.88	1.26	2.0	82.52	4JN	3.1	-3.4	-1.4	0.7	126.16	1.33	1.9	82.07	4JN	-0.1	-3.0	-1.5	0.4
120.12	1.24	2.0	92.06	14JN	2.7	-3.5	-1.4	0.7	132.23	1.26	2.0	91.60	14JN	-1.1	-3.2	-1.4	0.4
126.38	1.21	2.0	101.59	24JN	1.3	-3.5	-1.4	0.6	138.34	1.19	2.0	101.12	24JN	-1.8	-3.8	-1.4	0.4
132.66	1.18	2.0	111.13	4JL	0.1	-3.6	-1.4	0.6	144.51	1.12	2.0	110.67	4JL	-1.0	-4.1	-1.3	0.4
138.95	1.15	2.0	120.68	14JL	-1.0	-3.6	-1.4	0.6	150.74	1.05	2.0	120.22	14JL	-0.3	-4.2	-1.3	0.4
145.27	1.11	2.0	130.26	24JL	-1.6	-3.7	-1.4	0.5	157.02	0.97	2.0	129.79	24JL	0.2	-4.2	-1.3	0.4
151.61	1.07	2.0	139.87	3AU	-1.0	-3.8	-1.4	0.5	163.36	0.89	2.0	139.41	3AU	0.5	-4.1	-1.2	0.3
157.99	1.03	2.0	149.52	13AU	-0.4	-3.9	-1.4	0.4	169.76	0.81	2.0	149.05	13AU	0.8	-4.0	-1.2	0.3
164.40	0.98	2.0	159.22	23AU	-0.0	-4.0	-1.5	0.3	176.22	0.73	2.0	158.75	23AU	1.7	-3.9	-1.2	0.3
170.85	0.93	1.9	168.97	2SE	0.2	-4.1	-1.5	0.3	182.75	0.65	1.9	168.50	2SE	3.2	-3.8	-1.2	0.2
177.34	0.87	1.9	178.77	12SE	0.4	-4.2	-1.6	0.2	189.35	0.56	1.9	178.30	12SE	0.6	-3.8	-1.2	0.1
183.88	0.81	1.9	188.64	22SE	0.9	-4.3	-1.6	0.2	196.02	0.48	1.8	188.15	22SE	-0.7	-3.7	-1.2	0.1
190.46	0.74	1.9	198.55	2OC	3.0	-4.3	-1.7	0.1	202.77	0.39	1.8	198.07	2OC	-1.0	-3.6	-1.3	0.0
197.10	0.67	1.9	208.52	12OC	0.8	-4.2	-1.7	0.1	209.59	0.29	1.7	208.04	12OC	-0.9	-3.6	-1.3	-0.1
203.79	0.60	1.9	218.55	22OC	-0.5	-3.7	-1.8	0.2	216.49	0.20	1.7	218.06	22OC	-0.7	-3.5	-1.3	-0.1
210.53	0.52	1.9	228.63	1NO	-0.7	-3.0	-1.9	0.2	223.47	0.10	1.7	228.14	1NO	-0.5	-3.5	-1.4	-0.2
217.33	0.43	1.9	238.74	11NO	-0.6	-3.7	-1.9	0.3	230.52	-0.00	1.7	238.25	11NO	-0.4	-3.5	-1.4	-0.2
224.18	0.34	1.8	248.89	21NO	-0.7	-4.2	-2.0	0.4	237.66	-0.10	1.7	248.40	21NO	-0.4	-3.4	-1.5	-0.1
231.10	0.24	1.8	259.07	1DE	-0.7	-4.4	-2.0	0.4	244.87	-0.21	1.7	258.58	1DE	-0.2	-3.4	-1.5	-0.1
238.07	0.14	1.8	269.26	11DE	-0.7	-4.4	-2.1	0.5	252.16	-0.31	1.7	268.77	11DE	1.4	-3.4	-1.6	0.0
245.10	0.03	1.7	279.46	21DE	-0.7	-4.3	-2.1	0.5	259.52	-0.42	1.6	278.97	21DE	1.8	-3.4	-1.6	0.1
252.19	-0.08	1.7	289.65	31DE	-0.5	-4.2	-2.1	0.6	266.94	-0.53	1.6	289.16	31DE	0.4	-3.3	-1.7	0.2
				1469									1471				
259.33	-0.20	1.6	299.82	10JA	1.1	-4.1	-2.1	0.6	274.44	-0.63	1.6	299.33	10JA	0.1	-3.3	-1.8	0.2
266.53	-0.33	1.5	309.97	20JA	2.6	-4.0	-2.1	0.6	281.99	-0.74	1.5	309.48	20JA	-0.0	-3.3	-1.8	0.3
273.77	-0.47	1.5	320.08	30JA	0.9	-3.9	-2.1	0.7	289.60	-0.84	1.5	319.60	30JA	-0.2	-3.3	-1.9	0.3
281.07	-0.61	1.4	330.16	9FE	0.4	-3.8	-2.1	0.7	297.25	-0.94	1.5	329.67	9FE	-0.5	-3.4	-1.9	0.3
288.41	-0.75	1.3	340.18	19FE	0.2	-3.7	-2.0	0.6	304.93	-1.03	1.4	339.69	19FE	-1.1	-3.4	-2.0	0.4
295.78	-0.90	1.3	350.14	1MR	-0.0	-3.6	-1.9	0.6	312.65	-1.11	1.4	349.66	1MR	-1.4	-3.4	-2.0	0.4
303.19	-1.04	1.2	0.06	11MR	-0.4	-3.5	-1.9	0.6	320.38	-1.19	1.4	359.58	11MR	-0.9	-3.4	-2.0	0.4
310.62	-1.19	1.1	9.91	21MR	-1.1	-3.5	-1.8	0.6	328.11	-1.26	1.3	9.43	21MR	0.4	-3.5	-2.0	0.4
318.06	-1.34	1.0	19.71	31MR	-1.7	-3.4	-1.7	0.5	335.84	-1.31	1.3	19.23	31MR	2.0	-3.5	-2.0	0.4
325.50	-1.49	1.0	29.45	10AP	-0.9	-3.4	-1.7	0.5	343.55	-1.36	1.3	28.98	10AP	3.7	-3.5	-1.9	0.3
332.93	-1.63	0.9	39.15	20AP	0.2	-3.4	-1.6	0.5	351.22	-1.38	1.2	38.67	20AP	2.1	-3.4	-1.9	0.3
340.32	-1.76	0.8	48.79	30AP	1.2	-3.3	-1.6	0.5	358.85	-1.40	1.2	48.32	30AP	1.2	-3.4	-1.8	0.3
347.66	-1.88	0.7	58.40	10MY	2.4	-3.3	-1.5	0.5	6.41	-1.40	1.2	57.93	10MY	0.6	-3.4	-1.8	0.2
354.93	-1.99	0.6	67.98	20MY	3.5	-3.3	-1.4	0.5	13.90	-1.38	1.1	67.51	20MY	-0.1	-3.4	-1.7	0.2
2.09	-2.09	0.5	77.53	30MY	2.1	-3.3	-1.4	0.5	21.29	-1.35	1.1	77.06	30MY	-1.1	-3.3	-1.7	0.3
9.13	-2.17	0.4	87.06	9JN	1.0	-3.3	-1.4	0.5	28.58	-1.30	1.0	86.60	9JN	-1.8	-3.3	-1.6	0.3
16.01	-2.23	0.3	96.60	19JN	0.0	-3.3	-1.3	0.5	35.75	-1.23	1.0	96.13	19JN	-1.0	-3.3	-1.5	0.3
22.69	-2.26	0.2	106.13	29JN	-1.1	-3.3	-1.3	0.5	42.77	-1.15	1.0	105.66	29JN	-0.2	-3.3	-1.5	0.3
29.12	-2.28	0.1	115.67	9JL	-1.7	-3.4	-1.3	0.5	49.64	-1.05	0.9	115.20	9JL	0.3	-3.3	-1.4	0.3
35.25	-2.27	0.0	125.24	19JL	-1.0	-3.4	-1.3	0.4	56.34	-0.94	0.9	124.77	19JL	0.7	-3.4	-1.4	0.3
41.01	-2.24	-0.1	134.83	29JL	-0.3	-3.4	-1.3	0.4	62.84	-0.80	0.8	134.36	29JL	1.1	-3.4	-1.3	0.3
46.31	-2.17	-0.3	144.46	8AU	0.1	-3.5	-1.3	0.4	69.11	-0.65	0.7	143.99	8AU	2.1	-3.4	-1.3	0.3
51.05	-2.08	-0.4	154.13	18AU	0.3	-3.5	-1.3	0.3	75.12	-0.48	0.7	153.66	18AU	2.8	-3.4	-1.3	0.3
55.10	-1.94	-0.6	163.85	28AU	0.6	-3.5	-1.3	0.2	80.83	-0.29	0.6	163.38	28AU	0.5	-3.5	-1.2	0.2
58.28	-1.75	-0.8	173.63	7SE	1.3	-3.4	-1.3	0.2	86.18	-0.08	0.5	173.15	7SE	-0.8	-3.6	-1.2	0.2
60.40	-1.50	-1.0	183.46	17SE	3.3	-3.4	-1.4	0.1	91.11	0.16	0.4	182.98	17SE	-1.1	-3.6	-1.2	0.1
61.24	-1.19	-1.2	193.34	27SE	0.7	-3.4	-1.4	0.1	95.52	0.43	0.2	192.86	27SE	-1.1	-3.7	-1.2	0.1
60.64	-0.79	-1.4	203.29	7OC	-0.6	-3.4	-1.4	-0.0	99.30	0.74	0.1	202.80	7OC	-0.6	-3.8	-1.2	0.0
58.60	-0.33	-1.6	213.29	17OC	-0.8	-3.3	-1.5	-0.1	102.30	1.09	-0.1	212.80	17OC	-0.4	-3.9	-1.2	-0.1
55.40	0.19	-1.8	223.34	27OC	-0.8	-3.3	-1.5	-0.1	104.33	1.48	-0.3	222.85	27OC	-0.3	-4.0	-1.2	-0.1
51.68	0.69	-1.8	233.43	6NO	-0.8	-3.3	-1.6	-0.0	105.22	1.92	-0.5	232.94	6NO	-0.3	-4.1	-1.3	-0.2
48.28	1.14	-1.6	243.57	16NO	-0.6	-3.3	-1.7	0.1	104.77	2.39	-0.8	243.08	16NO	0.0	-4.2	-1.3	-0.3
45.87	1.48	-1.3	253.73	26NO	-0.6	-3.4	-1.7	0.1	102.92	2.87	-1.0	253.24	26NO	1.7	-4.3	-1.3	-0.3
44.76	1.73	-1.0	263.92	6DE	-0.6	-3.4	-1.8	0.2	99.82	3.30	-1.2	263.42	6DE	1.5	-4.4	-1.4	-0.2
45.00	1.88	-0.6	274.11	16DE	-0.3	-3.4	-1.9	0.3	95.96	3.62	-1.3	273.62	16DE	0.1	-4.4	-1.4	-0.1
46.42	1.97	-0.3	284.31	26DE	1.2	-3.4	-1.9	0.3	92.09	3.78	-1.1	283.82	26DE	-0.1	-4.2	-1.5	-0.0

LONG	LAT	MAG	LONG	16.00UT	☿	♀	♃	♄
88.95	3.79	-0.9	294.00	5JA	-0.2	-3.8	-1.6	0.0
87.00	3.67	-0.6	304.17	15JA	-0.3	-3.2	-1.6	0.1
86.39	3.49	-0.3	314.30	25JA	-0.6	-3.8	-1.7	0.2
87.01	3.28	-0.0	324.39	4FE	-1.1	-4.2	-1.8	0.2
88.70	3.05	0.2	334.44	14FE	-1.3	-4.3	-1.8	0.2
91.26	2.84	0.5	344.44	24FE	-0.9	-4.3	-1.9	0.3
94.49	2.64	0.7	354.38	5MR	0.5	-4.2	-1.9	0.3
98.28	2.45	0.9	4.27	15MR	2.5	-4.1	-2.0	0.3
102.51	2.27	1.0	14.10	25MR	2.8	-4.0	-2.0	0.3
107.07	2.10	1.2	23.88	4AP	1.5	-3.9	-2.0	0.3
111.92	1.95	1.3	33.60	14AP	0.9	-3.8	-2.0	0.3
117.01	1.80	1.4	43.27	24AP	0.4	-3.7	-2.0	0.3
122.28	1.66	1.5	52.90	4MY	-0.2	-3.6	-2.0	0.3
127.72	1.53	1.6	62.50	14MY	-1.1	-3.6	-2.0	0.2
133.31	1.41	1.7	72.06	24MY	-1.9	-3.5	-1.9	0.2
139.03	1.28	1.7	81.60	3JN	-1.0	-3.4	-1.9	0.2
144.86	1.16	1.8	91.14	13JN	-0.1	-3.4	-1.8	0.2
150.80	1.05	1.8	100.67	23JN	0.4	-3.4	-1.7	0.3
156.84	0.93	1.8	110.20	3JL	0.9	-3.4	-1.7	0.3
162.99	0.82	1.8	119.76	13JL	1.5	-3.3	-1.6	0.3
169.23	0.71	1.9	129.33	23JL	2.7	-3.3	-1.6	0.3
175.56	0.60	1.9	138.94	2AU	2.3	-3.4	-1.5	0.3
181.99	0.50	1.9	148.59	12AU	0.4	-3.3	-1.4	0.3
188.52	0.39	1.8	158.28	22AU	-0.9	-3.3	-1.4	0.3
195.14	0.28	1.8	168.00	1SE	-1.3	-3.4	-1.4	0.3
201.86	0.18	1.8	177.82	11SE	-1.1	-3.4	-1.3	0.2
208.67	0.07	1.8	187.67	21SE	-0.6	-3.4	-1.3	0.2
215.59	-0.03	1.7	197.59	1OC	-0.3	-3.4	-1.3	0.1
222.60	-0.14	1.7	207.56	11OC	-0.2	-3.5	-1.3	0.1
229.71	-0.24	1.7	217.57	21OC	-0.1	-3.5	-1.3	0.0
236.91	-0.34	1.6	227.65	31OC	0.2	-3.5	-1.3	-0.1
244.20	-0.44	1.6	237.76	10NO	2.0	-3.5	-1.3	-0.1
251.59	-0.53	1.5	247.91	20NO	1.3	-3.4	-1.3	-0.2
259.05	-0.62	1.5	258.08	30NO	-0.1	-3.4	-1.3	-0.3
266.60	-0.71	1.4	268.27	10DE	-0.3	-3.4	-1.3	-0.3
274.22	-0.79	1.4	278.47	20DE	-0.3	-3.4	-1.4	-0.2
281.90	-0.86	1.4	288.66	30DE	-0.4	-3.3	-1.4	-0.1
				1473				
289.64	-0.93	1.4	298.84	9JA	-0.7	-3.3	-1.5	-0.1
297.43	-0.99	1.4	308.99	19JA	-1.1	-3.3	-1.5	0.0
305.25	-1.04	1.4	319.11	29JA	-1.2	-3.3	-1.6	0.1
313.09	-1.08	1.4	329.18	8FE	-0.8	-3.4	-1.6	0.1
320.95	-1.11	1.4	339.21	18FE	0.6	-3.4	-1.7	0.2
328.80	-1.13	1.4	349.18	28FE	2.9	-3.4	-1.8	0.2
336.64	-1.13	1.4	359.10	10MR	2.1	-3.4	-1.8	0.3
344.46	-1.13	1.4	8.96	20MR	1.1	-3.4	-1.9	0.3
352.24	-1.11	1.4	18.76	30MR	0.7	-3.5	-2.0	0.3
359.97	-1.07	1.4	28.51	9AP	0.3	-3.5	-2.0	0.3
7.64	-1.03	1.4	38.21	19AP	-0.3	-3.6	-2.1	0.3
15.24	-0.97	1.4	47.86	29AP	-1.1	-3.6	-2.1	0.3
22.75	-0.90	1.4	57.47	9MY	-1.8	-3.7	-2.1	0.3
30.19	-0.82	1.4	67.05	19MY	-1.0	-3.8	-2.1	0.3
37.52	-0.73	1.4	76.60	29MY	-0.1	-3.9	-2.1	0.2
44.76	-0.63	1.4	86.14	8JN	0.6	-3.9	-2.1	0.2
51.88	-0.52	1.4	95.67	18JN	1.2	-4.0	-2.0	0.2
58.89	-0.41	1.4	105.21	28JN	2.0	-4.1	-2.0	0.3
65.78	-0.28	1.4	114.75	8JL	3.2	-4.2	-1.9	0.3
72.54	-0.15	1.4	124.31	18JL	1.9	-4.2	-1.9	0.3
79.16	-0.01	1.4	133.90	28JL	0.3	-4.0	-1.8	0.4
85.63	0.14	1.3	143.53	7AU	-1.0	-3.7	-1.7	0.4
91.94	0.30	1.3	153.19	17AU	-1.4	-3.2	-1.7	0.4
98.08	0.47	1.2	162.91	27AU	-1.0	-3.7	-1.6	0.4
104.01	0.65	1.2	172.68	6SE	-0.5	-4.1	-1.6	0.3
109.73	0.83	1.1	182.50	16SE	-0.2	-4.3	-1.5	0.3
115.19	1.04	1.0	192.38	26SE	-0.0	-4.3	-1.5	0.3
120.35	1.26	0.9	202.32	6OC	0.1	-4.2	-1.4	0.3
125.15	1.50	0.8	212.31	16OC	0.5	-4.2	-1.4	0.2
129.52	1.76	0.7	222.36	26OC	2.3	-4.1	-1.4	0.1
133.36	2.05	0.5	232.45	5NO	1.1	-4.0	-1.4	0.0
136.55	2.37	0.4	242.58	15NO	-0.2	-3.9	-1.3	-0.0
138.94	2.72	0.2	252.75	25NO	-0.4	-3.8	-1.3	-0.1
140.34	3.11	-0.1	262.93	5DE	-0.4	-3.7	-1.3	-0.2
140.58	3.51	-0.3	273.12	15DE	-0.5	-3.7	-1.4	-0.2
139.50	3.91	-0.5	283.32	25DE	-0.8	-3.6	-1.4	-0.2

LONG	LAT	MAG	LONG	16.00UT	☿	♀	♃	♄
137.11	4.26	-0.8	293.51	4JA	-1.0	-3.5	-1.4	-0.2
133.65	4.49	-1.0	303.67	14JA	-1.0	-3.5	-1.4	-0.1
129.69	4.55	-1.0	313.81	24JA	-0.7	-3.4	-1.5	-0.0
125.97	4.44	-0.8	323.91	3FE	0.7	-3.4	-1.5	0.0
123.12	4.18	-0.3	333.96	13FE	3.1	-3.4	-1.6	0.1
121.51	3.84	-0.3	343.97	23FE	1.6	-3.3	-1.6	0.2
121.19	3.48	-0.1	353.91	5MR	0.8	-3.3	-1.7	0.2
122.05	3.12	0.1	3.80	15MR	0.5	-3.3	-1.8	0.3
123.92	2.77	0.4	13.64	25MR	0.2	-3.3	-1.8	0.3
126.62	2.46	0.6	23.41	4AP	-0.3	-3.3	-1.9	0.3
129.98	2.18	0.7	33.13	14AP	-1.1	-3.3	-2.0	0.3
133.87	1.92	0.9	42.81	24AP	-1.8	-3.3	-2.0	0.4
138.21	1.69	1.0	52.44	4MY	-1.0	-3.4	-2.1	0.4
142.89	1.47	1.1	62.03	14MY	-0.0	-3.4	-2.2	0.4
147.89	1.27	1.2	71.60	24MY	0.8	-3.4	-2.2	0.3
153.14	1.09	1.3	81.14	3JN	1.5	-3.5	-2.2	0.3
158.61	0.91	1.4	90.68	13JN	2.6	-3.5	-2.2	0.3
164.28	0.75	1.5	100.21	23JN	3.1	-3.5	-2.2	0.3
170.12	0.60	1.5	109.74	3JL	1.5	-3.4	-2.2	0.3
176.13	0.45	1.6	119.29	13JL	0.2	-3.4	-2.2	0.3
182.29	0.31	1.6	128.87	23JL	-1.0	-3.4	-2.2	0.4
188.60	0.17	1.6	138.47	2AU	-1.5	-3.3	-2.1	0.4
195.04	0.05	1.6	148.12	12AU	-1.0	-3.3	-2.0	0.4
201.62	-0.08	1.6	157.81	22AU	-0.4	-3.3	-2.0	0.4
208.33	-0.19	1.6	167.54	1SE	-0.1	-3.3	-1.9	0.5
215.16	-0.31	1.6	177.34	11SE	0.1	-3.3	-1.8	0.4
222.12	-0.41	1.6	187.19	21SE	0.3	-3.3	-1.8	0.4
229.20	-0.51	1.6	197.10	1OC	0.7	-3.4	-1.7	0.4
236.39	-0.61	1.6	207.06	11OC	2.6	-3.4	-1.7	0.4
243.69	-0.70	1.6	217.08	21OC	0.9	-3.4	-1.6	0.3
251.09	-0.78	1.5	227.15	31OC	-0.4	-3.5	-1.6	0.3
258.59	-0.85	1.5	237.26	10NO	-0.6	-3.5	-1.5	0.2
266.19	-0.91	1.5	247.41	20NO	-0.6	-3.6	-1.5	0.1
273.86	-0.97	1.5	257.58	30NO	-0.6	-3.7	-1.5	0.1
281.60	-1.01	1.4	267.78	10DE	-0.8	-3.7	-1.5	-0.0
289.40	-1.05	1.4	277.97	20DE	-0.8	-3.8	-1.5	-0.1
297.24	-1.07	1.4	288.16	30DE	-0.8	-3.9	-1.4	-0.1
				1475				
305.12	-1.08	1.3	298.35	9JA	-0.6	-4.0	-1.5	-0.1
313.03	-1.08	1.3	308.50	19JA	0.9	-4.1	-1.5	-0.1
320.93	-1.07	1.3	318.62	29JA	2.9	-4.2	-1.5	-0.0
328.83	-1.05	1.3	328.70	8FE	1.1	-4.3	-1.5	0.1
336.71	-1.02	1.3	338.73	18FE	0.6	-4.3	-1.5	0.1
344.56	-0.97	1.3	348.70	28FE	0.3	-4.3	-1.6	0.2
352.37	-0.92	1.4	358.62	10MR	-0.0	-4.1	-1.6	0.2
0.11	-0.85	1.4	8.48	20MR	-0.4	-3.6	-1.7	0.3
7.80	-0.78	1.4	18.29	30MR	-1.1	-3.1	-1.8	0.3
15.42	-0.69	1.5	28.04	9AP	-1.7	-3.7	-1.8	0.4
22.95	-0.60	1.5	37.74	19AP	-1.0	-4.1	-1.9	0.4
30.40	-0.51	1.5	47.39	29AP	0.1	-4.2	-2.0	0.4
37.76	-0.41	1.6	57.01	9MY	1.0	-4.2	-2.0	0.5
45.04	-0.30	1.6	66.59	19MY	2.0	-4.1	-2.1	0.5
52.21	-0.20	1.6	76.14	29MY	3.4	-4.0	-2.2	0.5
59.30	-0.08	1.6	85.68	8JN	2.5	-3.9	-2.2	0.5
66.28	0.03	1.7	95.21	18JN	1.2	-3.8	-2.3	0.4
73.17	0.15	1.7	104.74	28JN	0.1	-3.7	-2.3	0.4
79.96	0.27	1.7	114.29	8JL	-1.0	-3.7	-2.4	0.4
86.65	0.38	1.7	123.85	18JL	-1.6	-3.6	-2.4	0.4
93.24	0.51	1.7	133.43	28JL	-1.0	-3.5	-2.4	0.5
99.73	0.63	1.7	143.06	7AU	-0.4	-3.5	-2.4	0.5
106.11	0.75	1.7	152.72	17AU	0.0	-3.5	-2.3	0.5
112.38	0.88	1.7	162.44	27AU	0.2	-3.4	-2.3	0.6
118.53	1.01	1.6	172.21	6SE	0.5	-3.4	-2.2	0.6
124.55	1.15	1.5	182.02	16SE	1.0	-3.4	-2.2	0.6
130.43	1.29	1.5	191.90	26SE	3.0	-3.4	-2.1	0.6
136.15	1.43	1.5	201.84	6OC	0.8	-3.4	-2.0	0.6
141.68	1.58	1.4	211.83	16OC	-0.5	-3.4	-2.0	0.6
147.00	1.74	1.3	221.87	26OC	-0.7	-3.4	-1.9	0.5
152.06	1.91	1.2	231.96	5NO	-0.7	-3.4	-1.8	0.5
156.82	2.10	1.1	242.09	15NO	-0.7	-3.4	-1.8	0.4
161.22	2.30	0.9	252.25	25NO	-0.7	-3.4	-1.7	0.4
165.17	2.51	0.8	262.43	5DE	-0.7	-3.4	-1.7	0.3
168.55	2.74	0.6	272.62	15DE	-0.7	-3.4	-1.6	0.2
171.25	3.00	0.4	282.82	25DE	-0.4	-3.4	-1.6	0.1

1476 / 1477

♂ LONG	♂ LAT	♂ MAG	☉ LONG	16.00UT 1476	☿	♀	♃	♄
173.09	3.26	0.1	293.01	4JA	1.0	-3.5	-1.6	0.1
173.90	3.53	-0.1	303.17	14JA	2.5	-3.5	-1.6	0.0
173.51	3.79	-0.4	313.31	24JA	0.8	-3.5	-1.5	0.0
171.81	3.99	-0.6	323.42	3FE	0.3	-3.4	-1.5	0.1
168.92	4.10	-0.9	333.47	13FE	0.2	-3.4	-1.5	0.1
165.21	4.06	-1.1	343.47	23FE	-0.1	-3.4	-1.6	0.2
161.33	3.86	-1.0	353.43	4MR	-0.4	-3.4	-1.6	0.3
158.00	3.52	-0.8	3.32	14MR	-1.1	-3.3	-1.6	0.3
155.73	3.10	-0.6	13.15	24MR	-1.6	-3.3	-1.6	0.4
154.74	2.66	-0.4	22.93	3AP	-1.0	-3.3	-1.7	0.4
155.02	2.24	-0.1	32.66	13AP	0.2	-3.3	-1.7	0.5
156.43	1.84	0.1	42.34	23AP	1.3	-3.3	-1.8	0.5
158.79	1.49	0.3	51.97	3MY	2.7	-3.4	-1.8	0.6
161.93	1.17	0.4	61.57	13MY	3.3	-3.4	-1.9	0.6
165.71	0.89	0.6	71.14	23MY	1.9	-3.4	-2.0	0.6
170.02	0.63	0.7	80.68	2JN	0.9	-3.4	-2.0	0.6
174.77	0.41	0.8	90.21	12JN	0.0	-3.5	-2.1	0.6
179.88	0.20	0.9	99.75	22JN	-1.0	-3.5	-2.2	0.6
185.32	0.01	1.0	109.28	2JL	-1.7	-3.6	-2.2	0.6
191.03	-0.16	1.1	118.83	12JL	-1.0	-3.6	-2.3	0.6
196.98	-0.32	1.1	128.41	22JL	-0.3	-3.7	-2.4	0.6
203.17	-0.46	1.2	138.01	1AU	0.1	-3.8	-2.4	0.6
209.55	-0.59	1.2	147.65	11AU	0.4	-3.9	-2.4	0.7
216.12	-0.71	1.2	157.34	21AU	0.7	-4.0	-2.5	0.7
222.87	-0.81	1.3	167.08	31AU	1.4	-4.1	-2.5	0.7
229.77	-0.91	1.3	176.87	10SE	3.3	-4.2	-2.4	0.8
236.83	-0.99	1.3	186.72	20SE	0.7	-4.3	-2.4	0.8
244.02	-1.06	1.3	196.62	30SE	-0.6	-4.3	-2.4	0.8
251.35	-1.12	1.3	206.58	10OC	-0.9	-4.1	-2.3	0.8
258.79	-1.16	1.3	216.60	20OC	-0.9	-3.7	-2.2	0.8
266.34	-1.19	1.3	226.67	30OC	-0.8	-3.0	-2.2	0.8
273.98	-1.21	1.4	236.77	9NO	-0.6	-3.8	-2.1	0.7
281.70	-1.22	1.4	246.92	19NO	-0.5	-4.2	-2.0	0.7
289.49	-1.21	1.4	257.09	29NO	-0.5	-4.4	-2.0	0.6
297.32	-1.19	1.4	267.28	9DE	-0.3	-4.4	-1.9	0.6
305.19	-1.16	1.4	277.48	19DE	1.2	-4.3	-1.8	0.5
313.08	-1.12	1.4	287.67	29DE	2.1	-4.2	-1.8	0.4
				1477				
320.97	-1.07	1.4	297.85	8JA	0.5	-4.1	-1.7	0.3
328.85	-1.00	1.4	308.00	18JA	0.2	-4.0	-1.7	0.3
336.71	-0.93	1.4	318.12	28JA	0.0	-3.9	-1.7	0.2
344.53	-0.85	1.4	328.20	7FE	-0.1	-3.8	-1.6	0.2
352.30	-0.76	1.4	338.24	17FE	-0.5	-3.7	-1.6	0.3
0.01	-0.67	1.4	348.22	27FE	-1.1	-3.6	-1.6	0.3
7.66	-0.57	1.5	358.14	9MR	-1.5	-3.5	-1.6	0.4
15.23	-0.47	1.5	8.01	19MR	-1.0	-3.5	-1.6	0.5
22.72	-0.36	1.5	17.82	29MR	0.3	-3.4	-1.6	0.5
30.13	-0.26	1.5	27.57	8AP	1.7	-3.4	-1.6	0.6
37.46	-0.15	1.5	37.28	18AP	3.5	-3.4	-1.6	0.6
44.69	-0.05	1.6	46.93	28AP	2.5	-3.3	-1.7	0.7
51.84	0.06	1.6	56.54	8MY	1.4	-3.3	-1.7	0.7
58.91	0.16	1.7	66.13	18MY	0.7	-3.3	-1.7	0.8
65.89	0.27	1.7	75.68	28MY	-0.1	-3.3	-1.8	0.8
72.79	0.37	1.7	85.22	7JN	-1.0	-3.3	-1.8	0.8
79.60	0.47	1.8	94.76	17JN	-1.8	-3.3	-1.9	0.8
86.34	0.56	1.8	104.28	27JN	-1.0	-3.3	-2.0	0.8
93.01	0.66	1.8	113.82	7JL	-0.3	-3.4	-2.0	0.8
99.60	0.75	1.9	123.39	17JL	0.2	-3.4	-2.1	0.8
106.12	0.84	1.9	132.97	27JL	0.6	-3.4	-2.2	0.8
112.57	0.93	1.9	142.59	6AU	0.9	-3.5	-2.2	0.9
118.96	1.02	1.9	152.26	16AU	1.8	-3.5	-2.3	0.9
125.27	1.10	1.9	161.96	26AU	3.1	-3.5	-2.4	0.9
131.51	1.19	1.9	171.73	5SE	0.6	-3.4	-2.4	1.0
137.69	1.27	1.9	181.55	15SE	-0.7	-3.4	-2.5	1.0
143.78	1.35	1.8	191.42	25SE	-1.0	-3.4	-2.5	1.0
149.79	1.43	1.8	201.35	5OC	-1.0	-3.4	-2.5	1.0
155.71	1.51	1.7	211.34	15OC	-0.7	-3.3	-2.4	1.1
161.52	1.60	1.7	221.38	25OC	-0.5	-3.3	-2.4	1.1
167.22	1.68	1.6	231.47	4NO	-0.4	-3.3	-2.4	1.0
172.78	1.77	1.5	241.60	14NO	-0.4	-3.3	-2.3	1.0
178.17	1.85	1.4	251.75	24NO	-0.1	-3.3	-2.3	1.0
183.37	1.94	1.3	261.93	4DE	1.4	-3.4	-2.2	0.9
188.33	2.03	1.2	272.13	14DE	1.8	-3.4	-2.1	0.9
193.00	2.12	1.0	282.32	24DE	0.3	-3.4	-2.0	0.8

1478 / 1479

♂ LONG	♂ LAT	♂ MAG	☉ LONG	16.00UT 1478	☿	♀	♃	♄
197.32	2.22	0.8	292.51	3JA	-0.0	-3.4	-2.0	0.8
201.21	2.31	0.6	302.68	13JA	-0.1	-3.5	-1.9	0.7
204.55	2.40	0.4	312.82	23JA	-0.2	-3.5	-1.8	0.6
207.21	2.49	0.2	322.92	2FE	-0.6	-3.6	-1.8	0.5
209.02	2.55	-0.1	332.98	12FE	-1.1	-3.6	-1.7	0.5
209.82	2.59	-0.4	342.99	22FE	-1.4	-3.7	-1.7	0.5
209.43	2.58	-0.7	352.94	4MR	-0.9	-3.8	-1.6	0.6
207.76	2.49	-1.0	2.84	14MR	0.4	-3.9	-1.6	0.6
204.93	2.30	-1.2	12.68	24MR	2.1	-3.9	-1.6	0.7
201.34	2.00	-1.4	22.46	3AP	3.3	-4.0	-1.6	0.7
197.64	1.60	-1.3	32.19	13AP	1.8	-4.1	-1.6	0.8
194.56	1.15	-1.2	41.87	23AP	1.1	-4.2	-1.5	0.8
192.58	0.70	-1.0	51.51	3MY	0.5	-4.2	-1.6	0.9
191.94	0.28	-0.7	61.11	13MY	-0.1	-4.1	-1.6	0.9
192.60	-0.08	-0.5	70.67	23MY	-1.0	-3.8	-1.6	1.0
194.41	-0.40	-0.3	80.22	2JN	-1.8	-3.0	-1.6	1.0
197.21	-0.67	-0.2	89.76	12JN	-1.0	-3.2	-1.6	1.0
200.82	-0.89	-0.0	99.28	22JN	-0.2	-3.9	-1.7	1.1
205.11	-1.08	0.1	108.82	2JL	0.3	-4.1	-1.7	1.1
209.95	-1.24	0.3	118.37	12JL	0.7	-4.2	-1.8	1.1
215.25	-1.37	0.4	127.94	22JL	1.2	-4.2	-1.8	1.1
220.95	-1.47	0.5	137.54	1AU	2.3	-4.1	-1.9	1.1
227.00	-1.56	0.6	147.18	11AU	2.7	-4.0	-2.0	1.1
233.33	-1.62	0.6	156.87	21AU	0.5	-3.9	-2.0	1.1
239.92	-1.66	0.7	166.60	31AU	-0.8	-3.8	-2.1	1.2
246.73	-1.69	0.7	176.39	10SE	-1.2	-3.8	-2.2	1.2
253.74	-1.69	0.8	186.23	20SE	-1.1	-3.7	-2.2	1.3
260.92	-1.68	0.9	196.14	30SE	-0.6	-3.6	-2.3	1.3
268.24	-1.66	0.9	206.10	10OC	-0.3	-3.6	-2.3	1.3
275.69	-1.61	1.0	216.11	20OC	-0.2	-3.5	-2.4	1.3
283.24	-1.55	1.0	226.18	30OC	-0.2	-3.5	-2.4	1.3
290.88	-1.48	1.1	236.28	9NO	0.1	-3.5	-2.4	1.3
298.57	-1.40	1.1	246.42	19NO	1.7	-3.4	-2.4	1.3
306.31	-1.31	1.2	256.60	29NO	1.5	-3.4	-2.3	1.2
314.08	-1.20	1.2	266.79	9DE	0.1	-3.4	-2.3	1.2
321.85	-1.09	1.3	276.98	19DE	-0.2	-3.4	-2.2	1.1
329.61	-0.98	1.3	287.18	29DE	-0.2	-3.3	-2.2	1.1
				1479				
337.34	-0.86	1.4	297.36	8JA	-0.3	-3.3	-2.1	1.1
345.04	-0.74	1.4	307.51	18JA	-0.6	-3.3	-2.0	1.0
352.69	-0.61	1.4	317.64	28JA	-1.1	-3.3	-2.0	1.0
0.28	-0.49	1.5	327.72	7FE	-1.2	-3.4	-1.9	0.9
7.81	-0.37	1.5	337.75	17FE	-0.9	-3.4	-1.8	0.8
15.26	-0.25	1.6	347.73	27FE	0.5	-3.4	-1.8	0.8
22.64	-0.13	1.6	357.66	9MR	2.6	-3.4	-1.7	0.8
29.94	-0.01	1.6	7.52	19MR	2.5	-3.5	-1.6	0.9
37.16	0.10	1.7	17.34	29MR	1.4	-3.5	-1.6	0.9
44.30	0.20	1.7	27.09	8AP	0.8	-3.5	-1.6	1.0
51.36	0.30	1.7	36.80	18AP	0.4	-3.4	-1.5	1.0
58.34	0.40	1.7	46.46	28AP	-0.2	-3.4	-1.5	1.1
65.25	0.49	1.8	56.07	8MY	-1.0	-3.4	-1.5	1.1
72.08	0.58	1.8	65.66	18MY	-1.9	-3.4	-1.5	1.2
78.85	0.66	1.8	75.22	28MY	-1.0	-3.3	-1.5	1.2
85.55	0.73	1.8	84.75	7JN	-0.1	-3.3	-1.5	1.3
92.20	0.80	1.8	94.28	17JN	0.5	-3.3	-1.5	1.3
98.79	0.87	1.9	103.82	27JN	1.0	-3.3	-1.5	1.3
105.32	0.93	1.9	113.36	7JL	1.7	-3.3	-1.5	1.4
111.82	0.99	1.9	122.92	17JL	2.9	-3.4	-1.5	1.4
118.26	1.05	2.0	132.50	27JL	2.2	-3.4	-1.6	1.4
124.67	1.10	2.0	142.12	6AU	0.4	-3.4	-1.6	1.4
131.04	1.14	2.0	151.78	16AU	-0.9	-3.4	-1.7	1.3
137.38	1.19	2.0	161.49	26AU	-1.3	-3.5	-1.7	1.3
143.68	1.23	2.0	171.25	5SE	-1.1	-3.6	-1.8	1.2
149.96	1.26	2.0	181.07	15SE	-0.5	-3.6	-1.8	1.2
156.20	1.29	2.0	190.94	25SE	-0.2	-3.7	-1.9	1.2
162.41	1.32	2.0	200.87	5OC	-0.1	-3.8	-2.0	1.2
168.58	1.35	1.9	210.85	15OC	-0.0	-3.9	-2.0	1.2
174.72	1.37	1.9	220.89	25OC	0.3	-4.0	-2.1	1.2
180.81	1.39	1.8	230.98	4NO	2.0	-4.1	-2.2	1.2
186.86	1.40	1.8	241.10	14NO	1.3	-4.2	-2.2	1.2
192.85	1.41	1.7	251.26	24NO	-0.1	-4.3	-2.2	1.1
198.77	1.41	1.6	261.44	4DE	-0.3	-4.4	-2.2	1.1
204.61	1.41	1.5	271.64	14DE	-0.4	-4.4	-2.3	1.1
210.35	1.40	1.4	281.83	24DE	-0.4	-4.2	-2.2	1.0

♂ LONG	♂ LAT	♂ MAG	☉ LONG	16.00UT 1480	☿	♀	♃	♄
					MAGNITUDES			
215.98	1.38	1.3	292.02	3JA	-0.7	-3.7	-2.2	1.0
221.47	1.34	1.1	302.19	13JA	-1.0	-3.2	-2.2	0.9
226.79	1.30	1.0	312.33	23JA	-1.1	-3.8	-2.1	0.9
231.90	1.24	0.8	322.43	2FE	-0.8	-4.2	-2.0	0.8
236.75	1.15	0.6	332.49	12FE	0.6	-4.3	-2.0	0.8
241.27	1.04	0.4	342.50	22FE	3.0	-4.3	-1.9	0.7
245.40	0.90	0.2	352.46	3MR	1.9	-4.2	-1.8	0.7
249.01	0.71	-0.1	2.36	13MR	1.0	-4.1	-1.8	0.7
251.98	0.47	-0.4	12.20	23MR	0.6	-4.0	-1.7	0.8
254.16	0.17	-0.6	21.98	2AP	0.2	-3.9	-1.6	0.8
255.35	-0.22	-1.0	31.72	12AP	-0.3	-3.8	-1.6	0.9
255.39	-0.68	-1.3	41.40	22AP	-1.0	-3.7	-1.5	1.0
254.18	-1.23	-1.6	51.04	2MY	-1.8	-3.6	-1.5	1.0
251.81	-1.83	-1.9	60.64	12MY	-1.0	-3.6	-1.4	1.1
248.69	-2.42	-2.1	70.21	22MY	-0.1	-3.5	-1.4	1.1
245.46	-2.93	-2.0	79.76	1JN	0.7	-3.4	-1.4	1.2
242.85	-3.31	-1.8	89.29	11JN	1.3	-3.4	-1.4	1.2
241.41	-3.55	-1.6	98.82	21JN	2.2	-3.4	-1.4	1.2
241.34	-3.66	-1.4	108.36	1JL	3.3	-3.4	-1.4	1.3
242.62	-3.67	-1.2	117.91	11JL	1.8	-3.3	-1.4	1.3
245.11	-3.62	-1.0	127.48	21JL	0.3	-3.3	-1.4	1.2
248.61	-3.52	-0.8	137.08	31JL	-0.9	-3.3	-1.4	1.2
252.93	-3.39	-0.6	146.72	10AU	-1.4	-3.3	-1.4	1.2
257.92	-3.23	-0.5	156.40	20AU	-1.0	-3.3	-1.4	1.2
263.45	-3.06	-0.3	166.13	30AU	-0.5	-3.4	-1.5	1.1
269.42	-2.86	-0.1	175.92	9SE	-0.1	-3.4	-1.5	1.1
275.75	-2.66	-0.0	185.76	19SE	0.0	-3.4	-1.6	1.0
282.36	-2.46	0.1	195.66	29SE	0.2	-3.4	-1.6	1.0
289.19	-2.24	0.2	205.62	9OC	0.5	-3.5	-1.7	1.0
296.20	-2.03	0.4	215.63	19OC	2.3	-3.5	-1.7	1.0
303.33	-1.81	0.5	225.68	29OC	1.1	-3.5	-1.8	1.0
310.57	-1.59	0.6	235.79	8NO	-0.3	-3.5	-1.9	1.0
317.88	-1.38	0.7	245.93	18NO	-0.5	-3.4	-1.9	1.0
325.23	-1.17	0.8	256.10	28NO	-0.5	-3.4	-2.0	1.0
332.60	-0.97	0.9	266.29	8DE	-0.6	-3.4	-2.0	1.0
339.97	-0.78	1.0	276.48	18DE	-0.8	-3.4	-2.1	0.9
347.32	-0.60	1.1	286.68	28DE	-0.9	-3.3	-2.1	0.9
				1481				
354.64	-0.43	1.2	296.86	7JA	-0.9	-3.3	-2.1	0.9
1.92	-0.27	1.3	307.02	17JA	-0.7	-3.3	-2.1	0.8
9.16	-0.12	1.4	317.14	27JA	0.7	-3.3	-2.1	0.8
16.34	0.02	1.4	327.23	6FE	3.1	-3.4	-2.1	0.7
23.47	0.16	1.5	337.26	16FE	1.4	-3.4	-2.0	0.7
30.53	0.28	1.6	347.25	26FE	0.7	-3.4	-2.0	0.6
37.52	0.39	1.6	357.18	8MR	0.4	-3.4	-1.9	0.6
44.46	0.49	1.7	7.05	18MR	0.1	-3.4	-1.9	0.5
51.33	0.58	1.8	16.86	28MR	-0.3	-3.5	-1.8	0.6
58.14	0.66	1.8	26.63	7AP	-1.1	-3.5	-1.7	0.6
64.89	0.74	1.8	36.33	17AP	-1.8	-3.6	-1.6	0.7
71.58	0.81	1.9	45.99	27AP	-1.0	-3.6	-1.6	0.8
78.22	0.87	1.9	55.61	7MY	-0.0	-3.7	-1.5	0.8
84.82	0.92	1.9	65.20	17MY	0.9	-3.8	-1.5	0.9
91.37	0.97	1.9	74.76	27MY	1.7	-3.9	-1.4	0.9
97.88	1.01	2.0	84.30	6JN	2.9	-3.9	-1.4	1.0
104.35	1.05	2.0	93.83	16JN	2.9	-4.0	-1.3	1.0
110.80	1.08	2.0	103.36	26JN	1.4	-4.1	-1.3	1.1
117.22	1.11	1.9	112.90	6JL	0.2	-4.2	-1.3	1.1
123.62	1.13	1.9	122.46	16JL	-1.0	-4.2	-1.3	1.1
130.00	1.14	2.0	132.04	26JL	-1.6	-4.0	-1.3	1.1
136.38	1.15	2.0	141.66	5AU	-1.0	-3.6	-1.3	1.1
142.74	1.16	2.0	151.32	15AU	-0.4	-3.2	-1.3	1.1
149.10	1.16	2.0	161.02	25AU	-0.0	-3.7	-1.3	1.0
155.46	1.16	2.0	170.78	4SE	0.2	-4.1	-1.3	1.0
161.82	1.15	2.0	180.59	14SE	0.3	-4.3	-1.3	1.0
168.19	1.14	2.0	190.46	24SE	0.8	-4.3	-1.4	0.9
174.56	1.13	2.0	200.39	4OC	2.7	-4.2	-1.4	0.9
180.94	1.11	2.0	210.37	14OC	1.0	-4.2	-1.4	0.9
187.33	1.08	2.0	220.41	24OC	-0.4	-4.1	-1.5	0.9
193.72	1.05	1.9	230.49	3NO	-0.6	-4.0	-1.5	0.9
200.12	1.01	1.9	240.61	13NO	-0.6	-3.9	-1.6	0.9
206.53	0.96	1.8	250.77	23NO	-0.7	-3.8	-1.7	0.9
212.94	0.90	1.8	260.95	3DE	-0.8	-3.7	-1.7	0.9
219.35	0.84	1.7	271.14	13DE	-0.7	-3.7	-1.8	0.9
225.77	0.77	1.6	281.34	23DE	-0.8	-3.6	-1.9	0.9

♂ LONG	♂ LAT	♂ MAG	☉ LONG	16.00UT 1482	☿	♀	♃	♄
					MAGNITUDES			
232.17	0.68	1.5	291.53	2JA	-0.5	-3.5	-1.9	0.8
238.57	0.59	1.4	301.70	12JA	0.9	-3.5	-2.0	0.8
244.95	0.47	1.3	311.84	22JA	2.8	-3.4	-2.0	0.8
251.30	0.34	1.2	321.95	1FE	1.0	-3.4	-2.0	0.7
257.63	0.20	1.1	332.01	11FE	0.5	-3.4	-2.0	0.7
263.91	0.03	0.9	342.03	21FE	0.3	-3.3	-2.0	0.6
270.14	-0.16	0.8	351.98	3MR	0.0	-3.3	-2.0	0.5
276.29	-0.38	0.6	1.88	13MR	-0.4	-3.3	-2.0	0.5
282.36	-0.63	0.5	11.73	23MR	-1.1	-3.3	-1.9	0.5
288.29	-0.92	0.3	21.52	2AP	-1.7	-3.3	-1.9	0.4
294.08	-1.24	0.1	31.25	12AP	-1.0	-3.3	-1.8	0.5
299.65	-1.61	-0.1	40.94	22AP	0.1	-3.3	-1.8	0.5
304.95	-2.03	-0.3	50.58	2MY	1.1	-3.4	-1.7	0.6
309.90	-2.49	-0.6	60.18	12MY	2.3	-3.4	-1.6	0.7
314.39	-3.02	-0.8	69.75	22MY	3.6	-3.4	-1.6	0.7
318.28	-3.60	-1.1	79.30	1JN	2.3	-3.5	-1.5	0.8
321.41	-4.24	-1.4	88.83	11JN	1.1	-3.5	-1.4	0.8
323.57	-4.92	-1.7	98.36	21JN	0.1	-3.5	-1.4	0.9
324.58	-5.60	-2.0	107.89	1JL	-1.0	-3.4	-1.3	0.9
324.32	-6.22	-2.2	117.44	11JL	-1.7	-3.4	-1.3	0.9
322.81	-6.68	-2.5	127.01	21JL	-1.0	-3.4	-1.3	1.0
320.44	-6.87	-2.6	136.61	31JL	-0.4	-3.3	-1.2	1.0
317.84	-6.74	-2.6	146.25	10AU	0.1	-3.3	-1.2	1.0
315.75	-6.31	-2.4	155.93	20AU	0.3	-3.3	-1.2	1.0
314.74	-5.67	-2.1	165.66	30AU	0.5	-3.3	-1.2	1.0
315.02	-4.93	-1.8	175.44	9SE	1.1	-3.3	-1.2	0.9
316.59	-4.19	-1.5	185.28	19SE	3.1	-3.3	-1.2	0.9
319.28	-3.49	-1.3	195.18	29SE	0.8	-3.4	-1.2	0.9
322.87	-2.86	-1.0	205.13	9OC	-0.6	-3.4	-1.3	0.8
327.19	-2.29	-0.7	215.14	19OC	-0.8	-3.4	-1.3	0.8
332.07	-1.79	-0.5	225.20	29OC	-0.8	-3.5	-1.3	0.8
337.38	-1.36	-0.2	235.30	8NO	-0.8	-3.5	-1.4	0.8
343.02	-0.98	-0.0	245.44	18NO	-0.6	-3.6	-1.4	0.8
348.91	-0.65	0.2	255.61	28NO	-0.6	-3.7	-1.5	0.9
354.98	-0.36	0.4	265.80	8DE	-0.6	-3.7	-1.5	0.8
1.20	-0.11	0.5	275.99	18DE	-0.4	-3.8	-1.6	0.8
7.51	0.10	0.7	286.19	28DE	1.0	-3.9	-1.7	0.8
				1483				
13.89	0.29	0.9	296.37	7JA	2.4	-4.0	-1.7	0.8
20.32	0.45	1.0	306.53	17JA	0.7	-4.1	-1.8	0.8
26.78	0.58	1.1	316.65	27JA	0.3	-4.2	-1.9	0.7
33.24	0.70	1.3	326.74	6FE	0.1	-4.3	-1.9	0.7
39.72	0.80	1.4	336.78	16FE	-0.1	-4.3	-1.9	0.6
46.19	0.89	1.5	346.77	26FE	-0.4	-4.3	-2.0	0.6
52.64	0.96	1.6	356.70	8MR	-1.1	-4.1	-2.0	0.5
59.08	1.02	1.6	6.58	18MR	-1.6	-3.6	-2.0	0.5
65.51	1.07	1.7	16.39	28MR	-1.0	-3.2	-2.0	0.4
71.91	1.11	1.8	26.15	7AP	0.2	-3.7	-2.0	0.4
78.30	1.14	1.8	35.86	17AP	1.4	-4.1	-1.9	0.3
84.66	1.17	1.9	45.52	27AP	3.0	-4.2	-1.9	0.4
91.01	1.19	1.9	55.15	7MY	3.0	-4.2	-1.8	0.5
97.35	1.20	2.0	64.73	17MY	1.7	-4.1	-1.8	0.5
103.67	1.21	2.0	74.29	27MY	0.9	-4.0	-1.7	0.6
109.98	1.21	2.0	83.83	6JN	0.0	-3.9	-1.6	0.6
116.30	1.20	2.0	93.37	16JN	-1.0	-3.8	-1.6	0.7
122.61	1.19	2.0	102.90	26JN	-1.7	-3.7	-1.5	0.8
128.92	1.18	2.0	112.44	6JL	-1.0	-3.7	-1.5	0.8
135.25	1.16	2.0	121.99	16JL	-0.3	-3.6	-1.4	0.8
141.58	1.14	2.0	131.57	26JL	0.2	-3.5	-1.4	0.9
147.93	1.11	2.0	141.19	5AU	0.5	-3.5	-1.3	0.9
154.30	1.08	2.0	150.85	15AU	0.8	-3.5	-1.3	0.9
160.70	1.04	1.9	160.55	25AU	1.5	-3.4	-1.3	0.9
167.13	1.00	1.9	170.31	4SE	3.2	-3.4	-1.2	0.9
173.58	0.96	2.0	180.12	14SE	0.7	-3.4	-1.2	0.9
180.07	0.91	2.0	189.98	24SE	-0.7	-3.4	-1.2	0.8
186.60	0.86	2.0	199.91	4OC	-0.9	-3.4	-1.2	0.8
193.17	0.80	2.0	209.89	14OC	-0.9	-3.4	-1.2	0.8
199.78	0.74	1.9	219.92	24OC	-0.8	-3.4	-1.2	0.7
206.43	0.67	1.9	230.00	3NO	-0.5	-3.4	-1.2	0.7
213.13	0.59	1.9	240.12	13NO	-0.5	-3.4	-1.3	0.8
219.87	0.51	1.9	250.28	23NO	-0.4	-3.4	-1.3	0.8
226.66	0.42	1.8	260.46	3DE	-0.2	-3.4	-1.4	0.8
233.49	0.32	1.8	270.65	13DE	1.2	-3.4	-1.4	0.8
240.38	0.22	1.7	280.84	23DE	2.0	-3.5	-1.5	0.8

♂			⊙	16.00UT	☿	♀	♃	♄
LONG	LAT	MAG	LONG	1484	MAGNITUDES			
247.31	0.10	1.7	291.04	2JA	0.4	-3.5	-1.5	0.8
254.28	-0.02	1.6	301.21	12JA	0.1	-3.5	-1.6	0.8
261.29	-0.15	1.5	311.35	22JA	-0.0	-3.5	-1.6	0.7
268.35	-0.29	1.4	321.46	1FE	-0.2	-3.4	-1.7	0.7
275.45	-0.44	1.4	331.52	11FE	-0.5	-3.4	-1.8	0.7
282.58	-0.60	1.3	341.54	21FE	-1.1	-3.4	-1.8	0.6
289.74	-0.76	1.2	351.50	2MR	-1.5	-3.4	-1.9	0.6
296.93	-0.94	1.1	1.40	12MR	-1.0	-3.3	-2.0	0.5
304.13	-1.12	1.0	11.25	22MR	0.3	-3.3	-2.0	0.5
311.34	-1.31	0.9	21.04	1AP	1.8	-3.3	-2.0	0.4
318.54	-1.50	0.8	30.78	11AP	3.7	-3.3	-2.0	0.3
325.73	-1.70	0.7	40.46	21AP	2.2	-3.3	-2.0	0.3
332.86	-1.89	0.5	50.11	1MY	1.3	-3.4	-2.0	0.3
339.94	-2.08	0.4	59.71	11MY	0.6	-3.4	-2.0	0.4
346.92	-2.27	0.3	69.28	21MY	-0.1	-3.4	-2.0	0.4
353.77	-2.45	0.2	78.84	31MY	-1.0	-3.4	-1.9	0.5
0.45	-2.62	0.1	88.37	10JN	-1.8	-3.5	-1.9	0.5
6.92	-2.78	-0.1	97.90	20JN	-1.0	-3.5	-1.8	0.6
13.10	-2.92	-0.2	107.44	30JN	-0.2	-3.6	-1.7	0.7
18.93	-3.04	-0.4	116.98	10JL	0.3	-3.6	-1.7	0.7
24.32	-3.14	-0.5	126.55	20JL	0.6	-3.7	-1.6	0.8
29.13	-3.21	-0.7	136.15	30JL	1.0	-3.8	-1.5	0.8
33.24	-3.25	-0.9	145.78	9AU	2.0	-3.9	-1.5	0.8
36.45	-3.23	-1.1	155.46	19AU	3.0	-4.0	-1.4	0.8
38.55	-3.16	-1.3	165.19	29AU	0.6	-4.1	-1.4	0.8
39.35	-3.00	-1.6	174.97	8SE	-0.8	-4.2	-1.4	0.8
38.69	-2.74	-1.8	184.80	18SE	-1.1	-4.3	-1.3	0.8
36.60	-2.35	-2.0	194.70	28SE	-1.1	-4.3	-1.3	0.8
33.49	-1.83	-2.1	204.65	8OC	-0.7	-4.1	-1.3	0.8
30.03	-1.24	-2.1	214.65	18OC	-0.4	-3.7	-1.3	0.7
27.04	-0.65	-1.8	224.71	28OC	-0.3	-3.1	-1.3	0.7
25.16	-0.10	-1.5	234.81	7NO	-0.3	-3.8	-1.3	0.7
24.60	0.34	-1.2	244.95	17NO	-0.0	-4.2	-1.3	0.7
25.35	0.70	-0.8	255.12	27NO	1.5	-4.4	-1.3	0.7
27.24	0.97	-0.5	265.30	7DE	1.7	-4.4	-1.3	0.8
30.04	1.17	-0.2	275.50	17DE	0.2	-4.3	-1.3	0.8
33.57	1.32	0.0	285.70	27DE	-0.1	-4.2	-1.4	0.8
				1485				
37.67	1.42	0.3	295.88	6JA	-0.1	-4.1	-1.4	0.8
42.20	1.50	0.5	306.04	16JA	-0.3	-4.0	-1.5	0.8
47.07	1.55	0.7	316.17	26JA	-0.6	-3.9	-1.5	0.7
52.20	1.58	0.9	326.25	5FE	-1.1	-3.8	-1.6	0.7
57.54	1.60	1.0	336.30	15FE	-1.3	-3.7	-1.7	0.7
63.04	1.60	1.2	346.29	25FE	-0.9	-3.6	-1.7	0.6
68.66	1.59	1.3	356.22	7MR	0.4	-3.5	-1.8	0.6
74.38	1.58	1.4	6.10	17MR	2.2	-3.5	-1.9	0.5
80.18	1.56	1.5	15.92	27MR	3.0	-3.4	-1.9	0.5
86.04	1.54	1.6	25.69	6AP	1.6	-3.4	-2.0	0.4
91.96	1.51	1.7	35.40	16AP	1.0	-3.4	-2.1	0.4
97.92	1.48	1.8	45.07	26AP	0.5	-3.3	-2.1	0.3
103.92	1.44	1.9	54.69	6MY	-0.1	-3.3	-2.1	0.3
109.96	1.40	1.9	64.28	16MY	-1.0	-3.3	-2.1	0.3
116.04	1.35	1.9	73.84	26MY	-1.9	-3.3	-2.1	0.3
122.14	1.31	2.0	83.38	5JN	-1.0	-3.3	-2.1	0.4
128.28	1.26	2.0	92.91	15JN	-0.2	-3.3	-2.1	0.5
134.46	1.21	2.0	102.44	25JN	0.4	-3.3	-2.0	0.5
140.67	1.15	2.0	111.98	5JL	0.8	-3.4	-2.0	0.6
146.93	1.09	2.0	121.54	15JL	1.4	-3.4	-1.9	0.6
153.23	1.03	2.0	131.12	25JL	2.5	-3.4	-1.9	0.7
159.57	0.97	2.0	140.73	4AU	2.5	-3.5	-1.8	0.7
165.97	0.90	2.0	150.38	14AU	0.5	-3.5	-1.7	0.8
172.41	0.83	2.0	160.08	24AU	-0.8	-3.5	-1.7	0.8
178.91	0.76	1.9	169.83	3SE	-1.2	-3.4	-1.6	0.8
185.48	0.68	1.9	179.64	13SE	-1.1	-3.4	-1.6	0.8
192.10	0.61	1.8	189.50	23SE	-0.6	-3.4	-1.5	0.8
198.79	0.52	1.8	199.42	3OC	-0.3	-3.4	-1.5	0.8
205.54	0.44	1.8	209.40	13OC	-0.2	-3.3	-1.4	0.8
212.36	0.35	1.8	219.43	23OC	-0.1	-3.3	-1.4	0.7
219.26	0.25	1.8	229.50	2NO	0.2	-3.3	-1.4	0.7
226.22	0.16	1.8	239.63	12NO	1.7	-3.3	-1.4	0.7
233.26	0.06	1.8	249.78	22NO	1.5	-3.3	-1.4	0.7
240.36	-0.05	1.7	259.96	2DE	0.0	-3.4	-1.4	0.7
247.54	-0.16	1.7	270.15	12DE	-0.2	-3.4	-1.4	0.7
254.79	-0.27	1.7	280.35	22DE	-0.3	-3.4	-1.4	0.8

♂			⊙	16.00UT	☿	♀	♃	♄
LONG	LAT	MAG	LONG	1486	MAGNITUDES			
262.10	-0.38	1.6	290.54	1JA	-0.4	-3.4	-1.4	0.8
269.48	-0.49	1.6	300.71	11JA	-0.6	-3.5	-1.4	0.8
276.91	-0.61	1.6	310.86	21JA	-1.1	-3.5	-1.5	0.8
284.40	-0.73	1.5	320.97	31JA	-1.2	-3.6	-1.5	0.8
291.94	-0.84	1.5	331.04	10FE	-0.8	-3.6	-1.5	0.7
299.52	-0.96	1.4	341.06	20FE	0.5	-3.7	-1.6	0.7
307.13	-1.07	1.4	351.02	2MR	2.7	-3.8	-1.7	0.7
314.77	-1.17	1.3	0.93	12MR	2.3	-3.9	-1.7	0.6
322.42	-1.27	1.3	10.78	22MR	1.2	-4.0	-1.8	0.6
330.07	-1.36	1.2	20.57	1AP	0.7	-4.0	-1.9	0.5
337.71	-1.44	1.2	30.32	11AP	0.3	-4.1	-1.9	0.5
345.32	-1.50	1.1	40.00	21AP	-0.2	-4.2	-2.0	0.4
352.90	-1.55	1.1	49.65	1MY	-1.0	-4.2	-2.1	0.3
0.41	-1.59	1.0	59.25	11MY	-1.9	-4.1	-2.1	0.3
7.85	-1.61	1.0	68.83	21MY	-1.0	-3.7	-2.2	0.2
15.21	-1.62	0.9	78.38	31MY	-0.1	-3.0	-2.2	0.3
22.44	-1.60	0.8	87.91	10JN	0.5	-3.2	-2.2	0.3
29.54	-1.57	0.8	97.44	20JN	1.1	-3.9	-2.3	0.4
36.50	-1.52	0.7	106.98	30JN	1.8	-4.1	-2.3	0.5
43.26	-1.44	0.7	116.52	10JL	3.1	-4.2	-2.2	0.5
49.82	-1.35	0.6	126.09	20JL	2.1	-4.2	-2.2	0.6
56.14	-1.24	0.5	135.68	30JL	0.4	-4.1	-2.2	0.6
62.16	-1.10	0.4	145.32	9AU	-0.9	-4.0	-2.1	0.7
67.85	-0.94	0.3	154.99	19AU	-1.4	-3.9	-2.0	0.7
73.13	-0.75	0.2	164.71	29AU	-1.1	-3.8	-2.0	0.8
77.92	-0.53	0.1	174.49	8SE	-0.5	-3.8	-1.9	0.8
82.12	-0.27	-0.1	184.32	18SE	-0.2	-3.7	-1.8	0.8
85.58	0.03	-0.3	194.21	28SE	-0.0	-3.6	-1.8	0.8
88.12	0.37	-0.5	204.16	8OC	0.1	-3.6	-1.7	0.8
89.56	0.77	-0.7	214.16	18OC	0.4	-3.5	-1.7	0.8
89.69	1.22	-0.9	224.22	28OC	2.0	-3.5	-1.6	0.8
88.39	1.71	-1.1	234.32	7NO	1.3	-3.5	-1.6	0.7
85.73	2.19	-1.3	244.45	17NO	-0.2	-3.4	-1.5	0.7
82.09	2.62	-1.4	254.62	27NO	-0.4	-3.4	-1.5	0.6
78.19	2.94	-1.4	264.81	7DE	-0.4	-3.4	-1.5	0.7
74.82	3.11	-1.1	275.00	17DE	-0.5	-3.4	-1.5	0.7
72.53	3.16	-0.8	285.20	27DE	-0.7	-3.4	-1.5	0.8
				1487				
71.59	3.11	-0.5	295.38	6JA	-1.0	-3.3	-1.5	0.8
71.95	3.01	-0.2	305.54	16JA	-1.0	-3.3	-1.5	0.8
73.42	2.87	0.0	315.68	26JA	-0.7	-3.3	-1.5	0.8
75.83	2.73	0.3	325.77	5FE	0.6	-3.4	-1.5	0.8
78.97	2.59	0.5	335.81	15FE	3.0	-3.4	-1.5	0.8
82.68	2.45	0.7	345.81	25FE	1.7	-3.4	-1.6	0.7
86.85	2.31	0.9	355.74	7MR	0.9	-3.4	-1.6	0.7
91.38	2.18	1.1	5.62	17MR	0.5	-3.5	-1.7	0.7
96.20	2.06	1.2	15.45	27MR	0.2	-3.5	-1.7	0.6
101.25	1.94	1.3	25.21	6AP	-0.3	-3.5	-1.8	0.6
106.49	1.83	1.5	34.93	16AP	-1.0	-3.4	-1.8	0.5
111.89	1.72	1.6	44.60	26AP	-1.8	-3.4	-1.9	0.4
117.43	1.61	1.6	54.22	6MY	-1.0	-3.4	-2.0	0.4
123.08	1.51	1.7	63.81	16MY	-0.1	-3.4	-2.1	0.3
128.84	1.40	1.8	73.37	26MY	0.7	-3.3	-2.1	0.3
134.69	1.30	1.8	82.91	5JN	1.4	-3.3	-2.2	0.2
140.63	1.21	1.9	92.44	15JN	2.4	-3.3	-2.3	0.3
146.66	1.11	1.9	101.98	25JN	3.2	-3.3	-2.3	0.4
152.76	1.01	1.9	111.51	5JL	1.7	-3.3	-2.3	0.4
158.94	0.91	1.9	121.07	15JL	0.3	-3.4	-2.4	0.5
165.21	0.82	1.9	130.65	25JL	-0.9	-3.4	-2.4	0.5
171.55	0.72	1.9	140.26	4AU	-1.5	-3.4	-2.4	0.6
177.98	0.62	1.9	149.91	14AU	-1.1	-3.4	-2.4	0.6
184.49	0.52	1.9	159.61	24AU	-0.5	-3.5	-2.3	0.7
191.08	0.43	1.9	169.36	3SE	-0.1	-3.6	-2.3	0.7
197.76	0.33	1.8	179.16	13SE	0.1	-3.6	-2.2	0.8
204.53	0.23	1.8	189.02	23SE	0.2	-3.7	-2.2	0.8
211.38	0.13	1.8	198.94	3OC	0.6	-3.8	-2.1	0.8
218.33	0.02	1.7	208.91	13OC	2.4	-3.9	-2.0	0.8
225.36	-0.08	1.7	218.94	23OC	1.1	-4.0	-2.0	0.8
232.49	-0.18	1.6	229.01	2NO	-0.3	-4.1	-1.9	0.8
239.70	-0.28	1.6	239.13	12NO	-0.5	-4.2	-1.8	0.7
246.99	-0.38	1.6	249.29	22NO	-0.5	-4.3	-1.8	0.7
254.37	-0.48	1.5	259.46	2DE	-0.6	-4.4	-1.7	0.7
261.83	-0.57	1.5	269.65	12DE	-0.8	-4.4	-1.7	0.7
269.37	-0.67	1.5	279.85	22DE	-0.8	-4.2	-1.6	0.7

1488

♂ LONG	LAT	MAG	☉ LONG	16.00UT	☿	♀	♃	♄
276.96	-0.76	1.5	290.04	1JA	-0.9	-3.7	-1.6	0.8
284.63	-0.84	1.5	300.21	11JA	-0.6	-3.2	-1.6	0.8
292.34	-0.92	1.5	310.36	21JA	0.7	-3.8	-1.6	0.8
300.09	-0.99	1.5	320.47	31JA	3.0	-4.2	-1.5	0.8
307.88	-1.05	1.4	330.55	10FE	1.3	-4.3	-1.5	0.8
315.69	-1.11	1.4	340.57	20FE	0.6	-4.3	-1.6	0.8
323.51	-1.15	1.4	350.53	1MR	0.3	-4.2	-1.6	0.8
331.33	-1.18	1.4	0.45	11MR	0.1	-4.1	-1.6	0.8
339.14	-1.20	1.4	10.30	21MR	-0.3	-4.0	-1.6	0.7
346.91	-1.20	1.4	20.10	31MR	-1.0	-3.9	-1.6	0.7
354.65	-1.20	1.4	29.84	10AP	-1.8	-3.8	-1.7	0.6
2.34	-1.17	1.3	39.54	20AP	-1.0	-3.7	-1.7	0.6
9.96	-1.14	1.3	49.18	30AP	0.0	-3.6	-1.8	0.5
17.51	-1.09	1.3	58.79	10MY	0.9	-3.6	-1.8	0.4
24.97	-1.03	1.3	68.37	20MY	1.9	-3.5	-1.9	0.4
32.34	-0.95	1.3	77.92	30MY	3.2	-3.4	-2.0	0.3
39.60	-0.86	1.3	87.45	9JN	2.7	-3.4	-2.0	0.3
46.75	-0.77	1.3	96.99	19JN	1.3	-3.4	-2.1	0.3
53.77	-0.65	1.3	106.52	29JN	0.2	-3.4	-2.2	0.3
60.66	-0.53	1.2	116.06	9JL	-1.0	-3.3	-2.3	0.4
67.41	-0.40	1.2	125.63	19JL	-1.6	-3.3	-2.3	0.4
73.99	-0.26	1.2	135.22	29JL	-1.1	-3.3	-2.4	0.5
80.41	-0.10	1.1	144.85	8AU	-0.4	-3.3	-2.4	0.6
86.63	0.07	1.1	154.53	18AU	-0.0	-3.3	-2.4	0.6
92.63	0.25	1.0	164.24	28AU	0.2	-3.4	-2.5	0.7
98.39	0.44	0.9	174.02	7SE	0.4	-3.4	-2.5	0.7
103.86	0.66	0.9	183.85	17SE	0.9	-3.4	-2.4	0.7
109.00	0.89	0.8	193.74	27SE	2.8	-3.4	-2.4	0.8
113.74	1.14	0.6	203.68	7OC	1.0	-3.5	-2.4	0.8
117.99	1.43	0.5	213.68	17OC	-0.5	-3.5	-2.3	0.8
121.65	1.74	0.3	223.73	27OC	-0.7	-3.5	-2.2	0.8
124.58	2.09	0.2	233.83	6NO	-0.7	-3.5	-2.2	0.8
126.61	2.48	-0.0	243.96	16NO	-0.7	-3.4	-2.1	0.8
127.55	2.90	-0.3	254.12	26NO	-0.7	-3.4	-2.0	0.8
127.21	3.34	-0.5	264.31	6DE	-0.7	-3.4	-2.0	0.7
125.52	3.77	-0.7	274.50	16DE	-0.7	-3.4	-1.9	0.7
122.57	4.13	-0.9	284.69	26DE	-0.5	-3.3	-1.8	0.7

1489

♂ LONG	LAT	MAG	☉ LONG	16.00UT	☿	♀	♃	♄
118.77	4.36	-1.1	294.88	5JA	0.9	-3.3	-1.8	0.8
114.83	4.41	-1.0	305.04	15JA	2.7	-3.3	-1.7	0.8
111.47	4.29	-0.8	315.17	25JA	0.9	-3.3	-1.7	0.8
109.21	4.06	-0.5	325.27	4FE	0.4	-3.4	-1.6	0.8
108.26	3.76	-0.2	335.32	14FE	0.2	-3.4	-1.6	0.8
108.56	3.44	0.0	345.31	24FE	-0.0	-3.4	-1.6	0.8
109.96	3.12	0.3	355.26	6MR	-0.4	-3.4	-1.6	0.8
112.28	2.83	0.5	5.14	16MR	-1.0	-3.4	-1.6	0.8
115.32	2.55	0.7	14.97	26MR	-1.7	-3.5	-1.6	0.8
118.95	2.31	0.8	24.74	5AP	-1.0	-3.5	-1.6	0.7
123.06	2.08	1.0	34.46	15AP	0.1	-3.6	-1.6	0.7
127.54	1.87	1.1	44.13	25AP	1.2	-3.6	-1.6	0.6
132.34	1.68	1.2	53.76	5MY	2.5	-3.7	-1.7	0.6
137.41	1.50	1.4	63.35	15MY	3.5	-3.8	-1.7	0.5
142.69	1.33	1.4	72.91	25MY	2.1	-3.9	-1.7	0.5
148.17	1.17	1.5	82.46	4JN	1.0	-3.9	-1.8	0.4
153.82	1.02	1.6	91.99	14JN	0.1	-4.0	-1.8	0.3
159.63	0.87	1.6	101.52	24JN	-1.0	-4.1	-1.9	0.3
165.58	0.73	1.7	111.06	4JL	-1.7	-4.2	-2.0	0.3
171.67	0.60	1.7	120.61	14JL	-1.1	-4.2	-2.0	0.4
177.88	0.47	1.7	130.19	24JL	-0.3	-4.0	-2.1	0.4
184.21	0.35	1.7	139.80	3AU	0.1	-3.6	-2.2	0.5
190.67	0.23	1.7	149.45	13AU	0.4	-3.2	-2.2	0.5
197.24	0.11	1.7	159.14	23AU	0.6	-3.7	-2.3	0.6
203.93	-0.01	1.7	168.89	2SE	1.3	-4.1	-2.4	0.6
210.73	-0.12	1.7	178.69	12SE	3.1	-4.3	-2.4	0.7
217.64	-0.23	1.7	188.54	22SE	0.9	-4.3	-2.4	0.7
224.67	-0.33	1.7	198.46	2OC	-0.6	-4.2	-2.4	0.8
231.80	-0.43	1.6	208.43	12OC	-0.9	-4.1	-2.4	0.8
239.03	-0.53	1.6	218.45	22OC	-0.8	-4.1	-2.4	0.8
246.37	-0.62	1.6	228.53	1NO	-0.8	-4.0	-2.4	0.8
253.81	-0.70	1.5	238.64	11NO	-0.6	-3.9	-2.4	0.8
261.33	-0.78	1.5	248.79	21NO	-0.5	-3.8	-2.3	0.8
268.94	-0.85	1.5	258.97	1DE	-0.5	-3.7	-2.2	0.8
276.62	-0.91	1.4	269.16	11DE	-0.3	-3.7	-2.2	0.8
284.36	-0.97	1.4	279.36	21DE	1.0	-3.6	-2.1	0.8
292.16	-1.01	1.3	289.55	31DE	2.3	-3.5	-2.0	0.7

1490

♂ LONG	LAT	MAG	☉ LONG	16.00UT	☿	♀	♃	♄
300.00	-1.05	1.3	299.72	10JA	0.6	-3.5	-1.9	0.8
307.87	-1.07	1.3	309.87	20JA	0.2	-3.4	-1.9	0.8
315.76	-1.08	1.3	319.99	30JA	0.1	-3.4	-1.8	0.9
323.65	-1.08	1.3	330.06	9FE	-0.1	-3.4	-1.8	0.9
331.54	-1.07	1.4	340.08	19FE	-0.5	-3.3	-1.7	0.9
339.40	-1.05	1.4	350.05	1MR	-1.1	-3.3	-1.7	0.9
347.23	-1.01	1.4	359.97	11MR	-1.5	-3.3	-1.6	0.9
355.02	-0.97	1.4	9.82	21MR	-1.0	-3.3	-1.6	0.9
2.75	-0.91	1.4	19.63	31MR	0.2	-3.3	-1.6	0.9
10.42	-0.84	1.4	29.37	10AP	1.5	-3.3	-1.5	0.8
18.01	-0.77	1.5	39.06	20AP	3.3	-3.3	-1.5	0.8
25.52	-0.68	1.5	48.71	30AP	2.7	-3.4	-1.5	0.7
32.95	-0.59	1.5	58.32	10MY	1.5	-3.4	-1.5	0.7
40.28	-0.49	1.5	67.90	20MY	0.8	-3.4	-1.6	0.6
47.52	-0.38	1.5	77.45	30MY	0.0	-3.5	-1.6	0.6
54.66	-0.27	1.6	86.99	9JN	-1.0	-3.5	-1.6	0.5
61.70	-0.16	1.6	96.52	19JN	-1.8	-3.5	-1.6	0.4
68.63	-0.04	1.6	106.05	29JN	-1.1	-3.4	-1.7	0.4
75.46	0.09	1.6	115.60	9JL	-0.3	-3.4	-1.7	0.3
82.17	0.21	1.6	125.16	19JL	0.2	-3.4	-1.8	0.4
88.77	0.35	1.6	134.75	29JL	0.5	-3.3	-1.8	0.4
95.25	0.48	1.6	144.38	8AU	0.9	-3.3	-1.9	0.5
101.61	0.62	1.5	154.05	18AU	1.7	-3.3	-2.0	0.5
107.83	0.76	1.5	163.77	28AU	3.2	-3.3	-2.0	0.6
113.91	0.91	1.5	173.54	7SE	0.7	-3.3	-2.1	0.6
119.82	1.07	1.4	183.37	17SE	-0.7	-3.3	-2.2	0.7
125.55	1.23	1.4	193.25	27SE	-1.0	-3.4	-2.2	0.7
131.08	1.41	1.3	203.19	7OC	-1.0	-3.4	-2.3	0.8
136.36	1.59	1.2	213.19	17OC	-0.7	-3.4	-2.3	0.8
141.36	1.79	1.1	223.24	27OC	-0.5	-3.5	-2.3	0.8
146.02	2.00	1.0	233.33	6NO	-0.4	-3.5	-2.4	0.8
150.27	2.24	0.8	243.47	16NO	-0.4	-3.6	-2.4	0.9
154.00	2.49	0.6	253.63	26NO	-0.2	-3.7	-2.3	0.9
157.11	2.77	0.5	263.81	6DE	1.2	-3.7	-2.3	0.8
159.44	3.08	0.2	274.01	16DE	2.0	-3.8	-2.3	0.8
160.81	3.40	0.0	284.21	26DE	0.4	-3.9	-2.2	0.8

1491

♂ LONG	LAT	MAG	☉ LONG	16.00UT	☿	♀	♃	♄
161.06	3.73	-0.2	294.39	5JA	0.0	-4.0	-2.1	0.8
160.02	4.04	-0.5	304.56	15JA	-0.1	-4.1	-2.1	0.8
157.68	4.28	-0.7	314.69	25JA	-0.2	-4.2	-2.0	0.9
154.30	4.41	-0.9	324.78	4FE	-0.5	-4.3	-1.9	0.9
150.38	4.36	-1.0	334.83	14FE	-1.1	-4.3	-1.9	0.9
146.66	4.15	-0.8	344.83	24FE	-1.4	-4.3	-1.8	1.0
143.77	3.81	-0.6	354.77	6MR	-1.0	-4.1	-1.7	1.0
142.09	3.41	-0.4	4.66	16MR	0.2	-3.5	-1.7	1.0
141.71	2.99	-0.2	14.49	26MR	1.9	-3.2	-1.6	1.0
142.52	2.59	0.1	24.26	5AP	3.6	-3.8	-1.6	1.0
144.36	2.21	0.3	33.99	15AP	2.0	-4.1	-1.5	0.9
147.06	1.88	0.5	43.66	25AP	1.2	-4.2	-1.5	0.9
150.44	1.57	0.6	53.28	5MY	0.6	-4.2	-1.5	0.9
154.40	1.30	0.8	62.88	15MY	-0.1	-4.1	-1.5	0.8
158.81	1.05	0.9	72.44	25MY	-1.0	-4.0	-1.4	0.7
163.61	0.83	1.0	81.99	4JN	-1.8	-3.9	-1.4	0.7
168.75	0.62	1.1	91.53	14JN	-1.1	-3.8	-1.4	0.6
174.16	0.43	1.2	101.05	24JN	-0.2	-3.7	-1.5	0.6
179.82	0.26	1.2	110.59	4JL	0.3	-3.7	-1.5	0.5
185.70	0.10	1.3	120.15	14JL	0.7	-3.6	-1.5	0.4
191.78	-0.05	1.3	129.72	24JL	1.2	-3.5	-1.5	0.4
198.05	-0.20	1.4	139.33	3AU	2.1	-3.5	-1.6	0.5
204.49	-0.33	1.4	148.98	13AU	2.9	-3.5	-1.6	0.5
211.09	-0.45	1.4	158.67	23AU	0.6	-3.4	-1.7	0.6
217.85	-0.56	1.4	168.42	2SE	-0.8	-3.4	-1.7	0.6
224.75	-0.67	1.4	178.22	12SE	-1.2	-3.4	-1.8	0.7
231.79	-0.76	1.4	188.07	22SE	-1.1	-3.4	-1.8	0.7
238.96	-0.85	1.4	197.98	2OC	-0.7	-3.4	-1.9	0.8
246.26	-0.92	1.4	207.95	12OC	-0.4	-3.4	-2.0	0.8
253.66	-0.99	1.4	217.97	22OC	-0.3	-3.4	-2.0	0.8
261.18	-1.04	1.4	228.04	1NO	-0.2	-3.4	-2.1	0.9
268.78	-1.08	1.4	238.15	11NO	0.0	-3.4	-2.1	0.9
276.47	-1.12	1.4	248.30	21NO	1.5	-3.4	-2.2	0.9
284.23	-1.13	1.4	258.48	1DE	1.7	-3.4	-2.2	0.9
292.05	-1.14	1.4	268.67	11DE	0.1	-3.4	-2.2	0.9
299.91	-1.14	1.4	278.86	21DE	-0.1	-3.5	-2.2	0.9
307.80	-1.12	1.4	289.05	31DE	-0.2	-3.5	-2.2	0.9

♂ LONG	LAT	MAG	☉ LONG	16.00UT 1492	☿	♀	♃	♄
315.71	-1.09	1.4	299.23	10JA	-0.3	-3.5	-2.2	0.9
323.61	-1.05	1.4	309.38	20JA	-0.6	-3.5	-2.1	0.9
331.50	-1.00	1.4	319.50	30JA	-1.1	-3.4	-2.1	0.9
339.37	-0.94	1.4	329.57	9FE	-1.3	-3.4	-2.0	1.0
347.19	-0.87	1.4	339.59	19FE	-0.9	-3.4	-1.9	1.0
354.97	-0.79	1.4	349.57	29FE	0.3	-3.4	-1.9	1.0
2.68	-0.71	1.4	359.49	10MR	2.4	-3.3	-1.8	1.1
10.33	-0.62	1.4	9.34	20MR	2.8	-3.3	-1.7	1.1
17.90	-0.52	1.4	19.15	30MR	1.5	-3.3	-1.7	1.1
25.39	-0.42	1.5	28.89	9AP	0.9	-3.3	-1.6	1.1
32.80	-0.32	1.5	38.59	19AP	0.4	-3.3	-1.6	1.1
40.12	-0.21	1.6	48.24	29AP	-0.2	-3.4	-1.5	1.0
47.35	-0.11	1.6	57.85	9MY	-1.0	-3.4	-1.5	1.0
54.49	0.00	1.6	67.43	19MY	-1.9	-3.4	-1.4	1.0
61.54	0.11	1.7	76.99	29MY	-1.1	-3.4	-1.4	0.9
68.50	0.22	1.7	86.53	8JN	-0.2	-3.5	-1.4	0.8
75.38	0.32	1.7	96.06	18JN	0.4	-3.5	-1.4	0.8
82.17	0.43	1.8	105.59	28JN	0.9	-3.6	-1.3	0.7
88.88	0.53	1.8	115.14	8JL	1.5	-3.6	-1.3	0.6
95.50	0.64	1.8	124.70	18JL	2.7	-3.7	-1.3	0.6
102.05	0.74	1.8	134.29	28JL	2.4	-3.8	-1.4	0.5
108.51	0.84	1.8	143.92	7AU	0.5	-3.9	-1.4	0.5
114.89	0.94	1.8	153.58	17AU	-0.8	-4.0	-1.4	0.6
121.18	1.05	1.8	163.30	27AU	-1.3	-4.1	-1.4	0.6
127.39	1.15	1.8	173.07	6SE	-1.1	-4.2	-1.5	0.7
133.50	1.25	1.8	182.89	16SE	-0.6	-4.3	-1.5	0.7
139.51	1.35	1.7	192.78	26SE	-0.3	-4.3	-1.6	0.8
145.41	1.46	1.7	202.72	6OC	-0.1	-4.1	-1.6	0.8
151.19	1.56	1.6	212.71	16OC	-0.0	-3.6	-1.7	0.9
156.83	1.68	1.5	222.76	26OC	0.2	-3.1	-1.7	0.9
162.30	1.79	1.5	232.84	5NO	1.8	-3.8	-1.8	0.9
167.58	1.91	1.4	242.97	15NO	1.5	-4.2	-1.9	1.0
172.63	2.04	1.2	253.14	25NO	-0.0	-4.4	-1.9	1.0
177.40	2.17	1.1	263.32	5DE	-0.3	-4.4	-2.0	1.0
181.83	2.31	0.9	273.51	15DE	-0.3	-4.3	-2.0	1.0
185.83	2.46	0.8	283.71	25DE	-0.4	-4.2	-2.1	1.0
				1493				
189.32	2.61	0.6	293.90	4JA	-0.7	-4.1	-2.1	1.0
192.17	2.78	0.3	304.06	14JA	-1.0	-4.0	-2.1	1.0
194.20	2.94	0.1	314.20	24JA	-1.1	-3.9	-2.1	1.0
195.26	3.09	-0.2	324.30	3FE	-0.8	-3.8	-2.1	1.0
195.17	3.21	-0.4	334.35	13FE	0.5	-3.7	-2.1	1.1
193.80	3.27	-0.7	344.35	23FE	2.8	-3.6	-2.0	1.1
191.21	3.24	-1.0	354.30	5MR	2.1	-3.5	-2.0	1.2
187.71	3.09	-1.2	4.19	15MR	1.1	-3.5	-1.9	1.2
183.88	2.80	-1.2	14.02	25MR	0.6	-3.4	-1.8	1.2
180.46	2.40	-1.0	23.80	4AP	0.3	-3.4	-1.8	1.2
178.02	1.96	-0.8	33.52	14AP	-0.2	-3.4	-1.7	1.2
176.86	1.51	-0.6	43.20	24AP	-1.0	-3.3	-1.6	1.2
177.01	1.09	-0.4	52.82	4MY	-1.9	-3.3	-1.6	1.2
178.35	0.71	-0.2	62.42	14MY	-1.1	-3.3	-1.5	1.2
180.70	0.38	-0.0	71.99	24MY	-0.1	-3.3	-1.4	1.1
183.89	0.09	0.2	81.53	3JN	0.6	-3.3	-1.4	1.1
187.77	-0.17	0.3	91.06	13JN	1.2	-3.3	-1.4	1.0
192.24	-0.39	0.4	100.60	23JN	2.0	-3.3	-1.3	1.0
197.17	-0.58	0.5	110.13	3JL	3.3	-3.4	-1.3	0.9
202.51	-0.75	0.6	119.68	13JL	2.0	-3.4	-1.3	0.8
208.21	-0.90	0.7	129.26	23JL	0.4	-3.4	-1.3	0.8
214.20	-1.02	0.8	138.86	2AU	-0.9	-3.5	-1.3	0.7
220.46	-1.13	0.9	148.51	12AU	-1.4	-3.5	-1.3	0.6
226.96	-1.22	0.9	158.20	22AU	-1.1	-3.5	-1.3	0.7
233.67	-1.30	1.0	167.94	1SE	-0.5	-3.4	-1.3	0.7
240.57	-1.36	1.0	177.74	11SE	-0.2	-3.4	-1.3	0.7
247.65	-1.40	1.0	187.59	21SE	0.0	-3.4	-1.3	0.8
254.87	-1.43	1.1	197.49	1OC	0.1	-3.4	-1.4	0.8
262.24	-1.44	1.1	207.46	11OC	0.5	-3.3	-1.4	0.9
269.73	-1.44	1.1	217.48	21OC	2.1	-3.3	-1.4	0.9
277.31	-1.42	1.2	227.55	31OC	1.3	-3.3	-1.5	1.0
284.98	-1.39	1.2	237.66	10NO	-0.2	-3.3	-1.6	1.0
292.73	-1.34	1.2	247.81	20NO	-0.5	-3.3	-1.6	1.1
300.51	-1.29	1.3	257.98	30NO	-0.5	-3.4	-1.7	1.1
308.33	-1.22	1.3	268.17	10DE	-0.5	-3.4	-1.7	1.1
316.17	-1.14	1.3	278.37	20DE	-0.7	-3.4	-1.8	1.2
324.01	-1.05	1.4	288.56	30DE	-0.9	-3.5	-1.9	1.2

♂ LONG	LAT	MAG	☉ LONG	16.00UT 1494	☿	♀	♃	♄
331.83	-0.95	1.4	298.74	9JA	-1.0	-3.5	-1.9	1.2
339.62	-0.85	1.4	308.89	19JA	-0.7	-3.5	-2.0	1.2
347.37	-0.75	1.4	319.01	29JA	0.6	-3.6	-2.0	1.2
355.07	-0.64	1.5	329.09	8FE	3.1	-3.7	-2.0	1.2
2.70	-0.52	1.5	339.11	18FE	1.5	-3.7	-2.0	1.2
10.26	-0.41	1.5	349.09	28FE	0.8	-3.8	-2.0	1.2
17.75	-0.30	1.6	359.01	10MR	0.4	-3.9	-2.0	1.3
25.17	-0.19	1.6	8.87	20MR	0.2	-4.0	-2.0	1.3
32.49	-0.07	1.6	18.68	30MR	-0.3	-4.0	-1.9	1.4
39.74	0.03	1.6	28.43	9AP	-1.0	-4.1	-1.9	1.4
46.90	0.14	1.7	38.13	19AP	-1.8	-4.1	-1.8	1.4
53.98	0.24	1.7	47.78	29AP	-1.1	-4.2	-1.7	1.4
60.98	0.34	1.7	57.39	9MY	-0.1	-4.1	-1.7	1.4
67.89	0.43	1.7	66.97	19MY	0.8	-3.7	-1.6	1.4
74.74	0.52	1.7	76.53	29MY	1.6	-2.9	-1.5	1.3
81.51	0.61	1.8	86.07	8JN	2.7	-3.3	-1.5	1.3
88.22	0.69	1.8	95.60	18JN	3.1	-3.9	-1.4	1.3
94.86	0.77	1.9	105.13	28JN	1.6	-4.1	-1.4	1.2
101.44	0.85	1.9	114.67	8JL	0.3	-4.2	-1.3	1.1
107.97	0.92	1.9	124.23	18JL	-0.9	-4.2	-1.3	1.1
114.44	0.99	1.9	133.82	28JL	-1.5	-4.1	-1.3	1.0
120.86	1.05	2.0	143.45	7AU	-1.1	-4.0	-1.2	0.9
127.23	1.11	2.0	153.11	17AU	-0.5	-3.9	-1.2	0.9
133.55	1.17	2.0	162.82	27AU	-0.1	-3.8	-1.2	0.8
139.83	1.23	2.0	172.59	6SE	0.1	-3.8	-1.2	0.8
146.06	1.28	2.0	182.41	16SE	0.3	-3.7	-1.2	0.9
152.25	1.33	1.9	192.29	26SE	0.7	-3.6	-1.2	0.9
158.38	1.38	1.9	202.23	6OC	2.4	-3.6	-1.2	1.0
164.46	1.43	1.9	212.22	16OC	1.1	-3.5	-1.3	1.0
170.48	1.47	1.8	222.27	26OC	-0.4	-3.5	-1.3	1.1
176.43	1.51	1.8	232.36	5NO	-0.6	-3.5	-1.3	1.1
182.29	1.55	1.7	242.48	15NO	-0.6	-3.4	-1.4	1.2
188.07	1.58	1.6	252.65	25NO	-0.7	-3.4	-1.4	1.2
193.73	1.62	1.5	262.83	5DE	-0.8	-3.4	-1.5	1.2
199.27	1.64	1.4	273.02	15DE	-0.8	-3.4	-1.5	1.3
204.65	1.67	1.3	283.22	25DE	-0.8	-3.4	-1.6	1.3
				1495				
209.83	1.68	1.1	293.41	4JA	-0.6	-3.3	-1.7	1.3
214.78	1.69	0.9	303.57	14JA	0.7	-3.3	-1.7	1.4
219.45	1.69	0.8	313.71	24JA	2.9	-3.3	-1.8	1.4
223.76	1.67	0.6	323.81	3FE	1.1	-3.4	-1.9	1.4
227.63	1.63	0.3	333.86	13FE	0.5	-3.4	-1.9	1.4
230.94	1.57	0.1	343.87	23FE	0.3	-3.4	-2.0	1.3
233.56	1.47	-0.2	353.82	5MR	0.0	-3.4	-2.0	1.3
235.31	1.32	-0.5	3.71	15MR	-0.3	-3.5	-2.0	1.3
236.02	1.11	-0.8	13.54	25MR	-1.0	-3.5	-2.0	1.3
235.52	0.83	-1.1	23.32	4AP	-1.7	-3.5	-2.0	1.3
233.77	0.46	-1.4	33.05	14AP	-1.1	-3.4	-2.0	1.3
230.93	0.01	-1.7	42.72	24AP	-0.0	-3.4	-1.9	1.3
227.46	-0.48	-1.8	52.36	4MY	1.0	-3.4	-1.9	1.3
224.09	-0.97	-1.7	61.95	14MY	2.1	-3.4	-1.8	1.2
221.47	-1.41	-1.5	71.52	24MY	3.5	-3.3	-1.8	1.2
220.08	-1.76	-1.3	81.07	3JN	2.4	-3.3	-1.7	1.2
220.08	-2.03	-1.1	90.60	13JN	1.2	-3.3	-1.6	1.1
221.37	-2.23	-0.9	100.13	23JN	0.2	-3.3	-1.6	1.1
223.82	-2.36	-0.7	109.67	3JL	-0.9	-3.3	-1.5	1.0
227.23	-2.45	-0.5	119.21	13JL	-1.6	-3.4	-1.5	1.0
231.43	-2.50	-0.3	128.79	23JL	-1.1	-3.4	-1.4	0.9
236.28	-2.51	-0.2	138.39	2AU	-0.4	-3.4	-1.4	0.9
241.67	-2.50	-0.1	148.03	12AU	0.0	-3.4	-1.3	0.8
247.51	-2.46	0.1	157.73	22AU	0.3	-3.5	-1.3	0.8
253.71	-2.40	0.2	167.46	1SE	0.5	-3.6	-1.3	0.7
260.22	-2.32	0.3	177.25	11SE	1.0	-3.6	-1.2	0.7
266.99	-2.23	0.4	187.11	21SE	2.8	-3.7	-1.2	0.8
273.97	-2.12	0.5	197.01	1OC	1.0	-3.8	-1.2	0.9
281.12	-1.99	0.6	206.97	11OC	-0.5	-3.9	-1.2	0.9
288.41	-1.86	0.7	216.99	21OC	-0.8	-4.0	-1.2	1.0
295.82	-1.71	0.7	227.06	31OC	-0.7	-4.1	-1.2	1.1
303.29	-1.56	0.8	237.17	10NO	-0.8	-4.2	-1.3	1.1
310.83	-1.40	0.9	247.32	20NO	-0.7	-4.3	-1.3	1.2
318.40	-1.24	1.0	257.49	30NO	-0.6	-4.4	-1.3	1.2
325.98	-1.08	1.1	267.67	10DE	-0.6	-4.4	-1.4	1.2
333.56	-0.92	1.1	277.87	20DE	-0.4	-4.2	-1.4	1.3
341.12	-0.76	1.2	288.07	30DE	0.9	-3.7	-1.5	1.3

♂ LONG	♂ LAT	♂ MAG	☉ LONG	16.00UT 1496	☿	♀	♃	♄
					MAGNITUDES			
348.64	-0.61	1.3	298.24	9JA	2.6	-3.2	-1.5	1.3
356.12	-0.46	1.4	308.40	19JA	0.8	-3.9	-1.6	1.2
3.55	-0.32	1.4	318.52	29JA	0.3	-4.2	-1.7	1.2
10.91	-0.18	1.5	328.60	8FE	0.1	-4.3	-1.7	1.2
18.22	-0.05	1.5	338.63	18FE	-0.1	-4.3	-1.8	1.1
25.44	0.07	1.6	348.61	28FE	-0.4	-4.2	-1.9	1.1
32.60	0.19	1.6	358.53	9MR	-1.0	-4.1	-1.9	1.0
39.69	0.30	1.7	8.40	19MR	-1.6	-4.0	-2.0	1.0
46.70	0.40	1.7	18.20	29MR	-1.1	-3.9	-2.0	1.0
53.64	0.49	1.8	27.96	8AP	0.1	-3.8	-2.0	1.0
60.51	0.58	1.8	37.66	18AP	1.3	-3.7	-2.1	1.0
67.32	0.66	1.8	47.32	28AP	2.8	-3.6	-2.0	1.0
74.06	0.74	1.9	56.93	8MY	3.2	-3.6	-2.0	1.0
80.75	0.80	1.9	66.51	18MY	1.9	-3.5	-2.0	1.0
87.38	0.86	1.9	76.07	28MY	0.9	-3.4	-2.0	1.0
93.96	0.92	1.9	85.61	7JN	0.1	-3.4	-1.9	0.9
100.50	0.97	1.9	95.14	17JN	-0.9	-3.4	-1.8	0.9
107.00	1.02	1.9	104.67	27JN	-1.7	-3.3	-1.8	0.8
113.46	1.06	1.9	114.21	7JL	-1.1	-3.3	-1.7	0.8
119.89	1.09	1.9	123.78	17JL	-0.3	-3.3	-1.7	0.7
126.30	1.12	2.0	133.36	27JL	0.1	-3.3	-1.6	0.7
132.68	1.15	2.0	142.98	6AU	0.4	-3.3	-1.5	0.6
139.05	1.17	2.0	152.65	16AU	0.7	-3.3	-1.5	0.6
145.39	1.19	2.0	162.36	26AU	1.4	-3.4	-1.4	0.5
151.73	1.20	2.0	172.12	5SE	3.2	-3.4	-1.4	0.5
158.06	1.21	2.0	181.94	15SE	0.9	-3.4	-1.4	0.5
164.38	1.21	2.0	191.81	25SE	-0.6	-3.4	-1.3	0.5
170.69	1.21	2.0	201.75	5OC	-0.9	-3.5	-1.3	0.5
176.99	1.21	2.0	211.74	15OC	-0.9	-3.5	-1.3	0.6
183.28	1.20	1.9	221.77	25OC	-0.8	-3.5	-1.3	0.7
189.56	1.18	1.9	231.86	4NO	-0.6	-3.5	-1.3	0.7
195.83	1.16	1.9	241.99	14NO	-0.5	-3.4	-1.3	0.8
202.09	1.13	1.8	252.15	24NO	-0.5	-3.4	-1.3	0.8
208.33	1.09	1.7	262.33	4DE	-0.3	-3.4	-1.3	0.9
214.54	1.05	1.6	272.53	14DE	1.0	-3.4	-1.3	0.9
220.71	0.99	1.5	282.72	24DE	2.3	-3.3	-1.4	1.0
				1497				
226.86	0.93	1.4	292.91	3JA	0.5	-3.3	-1.4	1.0
232.94	0.85	1.3	303.08	13JA	0.1	-3.3	-1.5	1.0
238.97	0.75	1.2	313.22	23JA	0.0	-3.3	-1.5	1.0
244.92	0.64	1.1	323.32	2FE	-0.2	-3.4	-1.6	1.0
250.78	0.51	0.9	333.38	12FE	-0.5	-3.4	-1.6	0.9
256.52	0.35	0.8	343.38	22FE	-1.0	-3.4	-1.7	0.9
262.12	0.16	0.6	353.34	4MR	-1.5	-3.4	-1.8	0.9
267.52	-0.07	0.4	3.23	14MR	-1.0	-3.4	-1.8	0.8
272.70	-0.33	0.2	13.07	24MR	0.2	-3.5	-1.9	0.8
277.59	-0.65	-0.1	22.85	3AP	1.7	-3.5	-2.0	0.8
282.10	-1.02	-0.3	32.58	13AP	3.5	-3.6	-2.0	0.8
286.13	-1.46	-0.6	42.26	23AP	2.4	-3.6	-2.1	0.8
289.56	-1.97	-0.9	51.90	3MY	1.4	-3.7	-2.1	0.8
292.20	-2.57	-1.2	61.50	13MY	0.7	-3.8	-2.1	0.8
293.88	-3.25	-1.5	71.06	23MY	-0.0	-3.9	-2.1	0.8
294.42	-4.00	-1.8	80.61	2JN	-0.9	-4.0	-2.1	0.7
293.72	-4.76	-2.1	90.15	12JN	-1.8	-4.0	-2.1	0.7
291.88	-5.45	-2.4	99.67	22JN	-1.1	-4.1	-2.1	0.7
289.26	-5.96	-2.5	109.21	2JL	-0.3	-4.2	-2.0	0.7
286.55	-6.21	-2.5	118.76	12JL	0.2	-4.2	-2.0	0.6
284.47	-6.18	-2.3	128.33	22JL	0.6	-4.0	-1.9	0.6
283.52	-5.91	-2.0	137.93	1AU	1.0	-3.6	-1.9	0.5
283.93	-5.50	-1.8	147.57	11AU	1.8	-3.2	-1.8	0.5
285.65	-5.01	-1.5	157.26	21AU	3.1	-3.7	-1.7	0.4
288.52	-4.49	-1.3	166.99	31AU	0.7	-4.1	-1.7	0.4
292.34	-3.98	-1.0	176.78	10SE	-0.7	-4.3	-1.6	0.3
296.90	-3.48	-0.8	186.62	20SE	-1.1	-4.3	-1.6	0.2
302.06	-3.02	-0.6	196.53	30SE	-1.0	-4.2	-1.5	0.2
307.68	-2.59	-0.4	206.49	10OC	-0.7	-4.1	-1.5	0.2
313.65	-2.19	-0.2	216.50	20OC	-0.4	-4.1	-1.4	0.3
319.89	-1.82	-0.0	226.57	30OC	-0.3	-4.0	-1.4	0.4
326.33	-1.48	0.2	236.67	9NO	-0.3	-3.9	-1.4	0.4
332.92	-1.17	0.3	246.82	19NO	-0.1	-3.8	-1.4	0.5
339.61	-0.89	0.5	256.99	29NO	1.3	-3.7	-1.4	0.6
346.39	-0.64	0.6	267.18	9DE	2.0	-3.7	-1.4	0.6
353.20	-0.41	0.8	277.38	19DE	0.3	-3.6	-1.4	0.7
0.04	-0.20	0.9	287.57	29DE	-0.0	-3.5	-1.4	0.7

♂ LONG	♂ LAT	♂ MAG	☉ LONG	16.00UT 1498	☿	♀	♃	♄
					MAGNITUDES			
6.89	-0.01	1.0	297.75	8JA	-0.1	-3.5	-1.4	0.7
13.72	0.15	1.1	307.91	18JA	-0.3	-3.4	-1.4	0.7
20.55	0.30	1.3	318.03	28JA	-0.5	-3.4	-1.5	0.7
27.34	0.43	1.4	328.11	7FE	-1.1	-3.4	-1.5	0.7
34.11	0.55	1.4	338.15	17FE	-1.4	-3.3	-1.6	0.7
40.84	0.65	1.5	348.13	27FE	-1.0	-3.3	-1.6	0.7
47.53	0.74	1.6	358.05	9MR	0.2	-3.3	-1.7	0.7
54.18	0.82	1.7	7.92	19MR	2.1	-3.3	-1.8	0.6
60.79	0.89	1.8	17.74	29MR	3.3	-3.3	-1.8	0.6
67.36	0.95	1.8	27.49	8AP	1.8	-3.3	-1.9	0.5
73.89	1.00	1.9	37.19	18AP	1.0	-3.3	-2.0	0.6
80.39	1.04	1.9	46.85	28AP	0.5	-3.4	-2.0	0.6
86.86	1.07	1.9	56.47	8MY	-0.1	-3.4	-2.1	0.6
93.30	1.10	2.0	66.05	18MY	-1.0	-3.4	-2.2	0.6
99.71	1.13	2.0	75.61	28MY	-1.8	-3.5	-2.2	0.6
106.10	1.14	2.0	85.15	7JN	-1.1	-3.5	-2.2	0.6
112.48	1.15	2.0	94.68	17JN	-0.2	-3.4	-2.3	0.6
118.84	1.16	2.0	104.21	27JN	0.3	-3.4	-2.3	0.5
125.19	1.16	2.0	113.75	7JL	0.8	-3.4	-2.3	0.5
131.55	1.16	2.0	123.31	17JL	1.3	-3.4	-2.2	0.5
137.90	1.15	2.0	132.90	27JL	2.3	-3.3	-2.2	0.4
144.26	1.14	2.0	142.51	6AU	2.7	-3.3	-2.2	0.4
150.63	1.12	2.0	152.17	16AU	0.6	-3.3	-2.1	0.3
157.01	1.10	2.0	161.88	26AU	-0.8	-3.3	-2.0	0.3
163.41	1.07	2.0	171.64	5SE	-1.2	-3.3	-2.0	0.2
169.83	1.04	2.0	181.46	15SE	-1.1	-3.3	-1.9	0.2
176.27	1.01	2.0	191.33	25SE	-0.6	-3.4	-1.9	0.1
182.74	0.97	2.0	201.26	5OC	-0.3	-3.4	-1.8	0.0
189.24	0.92	2.0	211.25	15OC	-0.2	-3.4	-1.7	-0.0
195.76	0.87	2.0	221.28	25OC	-0.1	-3.5	-1.7	0.0
202.32	0.81	1.9	231.37	4NO	0.1	-3.5	-1.6	0.1
208.91	0.75	1.9	241.49	14NO	1.5	-3.6	-1.6	0.2
215.53	0.67	1.9	251.65	24NO	1.7	-3.7	-1.5	0.2
222.18	0.60	1.8	261.83	4DE	0.1	-3.7	-1.5	0.3
228.87	0.51	1.8	272.03	14DE	-0.2	-3.8	-1.5	0.4
235.59	0.41	1.7	282.23	24DE	-0.3	-3.9	-1.5	0.4
				1499				
242.33	0.30	1.6	292.41	3JA	-0.4	-4.0	-1.5	0.5
249.11	0.19	1.5	302.59	13JA	-0.6	-4.1	-1.5	0.5
255.91	0.06	1.5	312.73	23JA	-1.1	-4.2	-1.5	0.5
262.74	-0.08	1.4	322.83	2FE	-1.2	-4.3	-1.5	0.6
269.59	-0.24	1.3	332.89	12FE	-0.9	-4.3	-1.5	0.6
276.46	-0.41	1.2	342.90	22FE	0.3	-4.3	-1.6	0.6
283.34	-0.59	1.0	352.85	4MR	2.5	-4.1	-1.6	0.5
290.23	-0.79	0.9	2.76	14MR	2.5	-3.5	-1.6	0.5
297.10	-1.01	0.8	12.60	24MR	1.3	-3.2	-1.7	0.5
303.97	-1.24	0.7	22.38	3AP	0.8	-3.8	-1.7	0.5
310.80	-1.48	0.5	32.11	13AP	0.4	-4.1	-1.8	0.4
317.58	-1.74	0.4	41.80	23AP	-0.2	-4.2	-1.9	0.4
324.29	-2.01	0.2	51.43	3MY	-1.0	-4.2	-1.9	0.4
330.88	-2.30	0.1	61.03	13MY	-1.9	-4.1	-2.0	0.4
337.31	-2.59	-0.1	70.60	23MY	-1.1	-4.0	-2.1	0.4
343.55	-2.89	-0.2	80.15	2JN	-0.2	-3.9	-2.1	0.4
349.51	-3.20	-0.4	89.69	12JN	0.5	-3.8	-2.2	0.5
355.11	-3.51	-0.6	99.22	22JN	1.0	-3.7	-2.3	0.4
0.26	-3.82	-0.8	108.75	2JL	1.7	-3.7	-2.3	0.4
4.80	-4.12	-1.0	118.30	12JL	2.9	-3.6	-2.4	0.4
8.57	-4.41	-1.2	127.87	22JL	2.3	-3.5	-2.4	0.4
11.39	-4.65	-1.5	137.47	1AU	0.5	-3.5	-2.4	0.4
13.01	-4.84	-1.7	147.11	11AU	-0.8	-3.5	-2.4	0.3
13.26	-4.91	-2.0	156.79	21AU	-1.3	-3.4	-2.4	0.3
12.10	-4.83	-2.2	166.52	31AU	-1.1	-3.4	-2.3	0.2
9.71	-4.54	-2.4	176.31	10SE	-0.6	-3.4	-2.3	0.2
6.68	-4.04	-2.4	186.15	20SE	-0.2	-3.4	-2.2	0.1
3.77	-3.37	-2.2	196.05	30SE	-0.1	-3.4	-2.2	0.0
1.70	-2.63	-2.0	206.01	10OC	0.0	-3.4	-2.1	-0.0
0.89	-1.91	-1.6	216.02	20OC	0.3	-3.4	-2.0	-0.1
1.40	-1.27	-1.3	226.08	30OC	1.8	-3.4	-2.0	-0.2
3.10	-0.73	-1.0	236.19	9NO	1.5	-3.4	-1.9	-0.1
5.82	-0.29	-0.7	246.33	19NO	-0.1	-3.4	-1.8	-0.1
9.32	0.07	-0.4	256.49	29NO	-0.4	-3.4	-1.8	0.0
13.45	0.36	-0.1	266.68	9DE	-0.4	-3.4	-1.7	0.1
18.07	0.60	0.1	276.88	19DE	-0.5	-3.5	-1.7	0.2
23.04	0.78	0.4	287.07	29DE	-0.7	-3.5	-1.6	0.2

1500

♂ LONG	LAT	MAG	☉ LONG	16.00UT	☿	♀	♃	♄
28.31	0.93	0.6	297.25	8JA	-1.0	-3.5	-1.6	0.3
33.79	1.05	0.7	307.41	18JA	-1.0	-3.5	-1.6	0.3
39.44	1.14	0.9	317.53	28JA	-0.8	-3.4	-1.6	0.4
45.22	1.22	1.1	327.62	7FE	0.4	-3.4	-1.6	0.4
51.09	1.27	1.2	337.65	17FE	2.9	-3.4	-1.6	0.4
57.03	1.31	1.3	347.64	27FE	1.9	-3.4	-1.6	0.4
63.03	1.34	1.4	357.57	8MR	1.0	-3.3	-1.6	0.4
69.08	1.36	1.5	7.44	18MR	0.6	-3.3	-1.6	0.4
75.15	1.37	1.6	17.26	28MR	0.2	-3.3	-1.6	0.4
81.25	1.38	1.7	27.02	7AP	-0.2	-3.3	-1.7	0.4
87.37	1.37	1.8	36.72	17AP	-1.0	-3.3	-1.7	0.3
93.50	1.36	1.9	46.38	27AP	-1.9	-3.4	-1.7	0.3
99.65	1.35	1.9	56.01	7MY	-1.1	-3.4	-1.8	0.2
105.82	1.33	1.9	65.59	17MY	-0.1	-3.4	-1.9	0.3
111.99	1.30	2.0	75.15	27MY	0.6	-3.4	-1.9	0.3
118.19	1.28	2.0	84.69	6JN	1.3	-3.5	-2.0	0.3
124.40	1.24	2.0	94.22	16JN	2.3	-3.5	-2.1	0.4
130.63	1.21	2.0	103.75	26JN	3.3	-3.6	-2.1	0.4
136.89	1.17	2.0	113.29	6JL	1.9	-3.6	-2.2	0.4
143.17	1.13	2.0	122.85	16JL	0.4	-3.7	-2.3	0.4
149.49	1.08	2.0	132.43	26JL	-0.9	-3.8	-2.3	0.3
155.84	1.03	2.0	142.05	5AU	-1.5	-3.9	-2.4	0.3
162.23	0.98	2.0	151.71	15AU	-1.1	-4.0	-2.4	0.3
168.66	0.92	2.0	161.42	25AU	-0.5	-4.1	-2.5	0.2
175.13	0.86	1.9	171.18	4SE	-0.1	-4.2	-2.5	0.2
181.65	0.79	1.9	180.99	14SE	0.1	-4.3	-2.5	0.1
188.23	0.72	1.9	190.86	24SE	0.2	-4.3	-2.4	0.1
194.86	0.65	1.9	200.78	4OC	0.5	-4.1	-2.4	0.0
201.54	0.57	1.9	210.76	14OC	2.1	-3.6	-2.3	-0.1
208.29	0.49	1.9	220.80	24OC	1.3	-3.1	-2.3	-0.1
215.10	0.41	1.9	230.88	3NO	-0.3	-3.8	-2.2	-0.2
221.96	0.32	1.8	241.00	13NO	-0.5	-4.2	-2.2	-0.3
228.89	0.22	1.8	251.16	23NO	-0.5	-4.4	-2.1	-0.1
235.89	0.12	1.8	261.34	3DE	-0.6	-4.4	-2.0	-0.1
242.94	0.01	1.7	271.53	13DE	-0.8	-4.3	-1.9	-0.1
250.06	-0.10	1.7	281.73	23DE	-0.8	-4.2	-1.9	0.0

1501

♂ LONG	LAT	MAG	☉ LONG	16.00UT	☿	♀	♃	♄
257.24	-0.22	1.7	291.92	2JA	-0.9	-4.1	-1.8	0.1
264.48	-0.34	1.6	302.09	12JA	-0.7	-4.0	-1.8	0.1
271.77	-0.46	1.5	312.23	22JA	0.6	-3.9	-1.7	0.2
279.12	-0.59	1.5	322.34	1FE	3.1	-3.8	-1.7	0.2
286.52	-0.72	1.4	332.40	11FE	1.4	-3.7	-1.6	0.3
293.96	-0.86	1.4	342.41	21FE	0.7	-3.6	-1.6	0.3
301.43	-0.99	1.3	352.37	3MR	0.4	-3.5	-1.6	0.3
308.94	-1.13	1.2	2.27	13MR	0.1	-3.5	-1.6	0.3
316.46	-1.26	1.2	12.12	23MR	-0.3	-3.4	-1.6	0.3
323.99	-1.38	1.1	21.91	2AP	-1.0	-3.4	-1.6	0.3
331.52	-1.50	1.0	31.64	12AP	-1.8	-3.4	-1.6	0.3
339.03	-1.61	1.0	41.33	22AP	-1.1	-3.3	-1.6	0.3
346.49	-1.71	0.9	50.97	2MY	-0.1	-3.3	-1.6	0.3
353.91	-1.79	0.8	60.57	12MY	0.8	-3.3	-1.7	0.2
1.25	-1.86	0.7	70.15	22MY	1.7	-3.3	-1.7	0.2
8.49	-1.92	0.7	79.69	1JN	3.0	-3.3	-1.7	0.2
15.61	-1.95	0.6	89.23	11JN	2.9	-3.3	-1.8	0.3
22.58	-1.97	0.5	98.76	21JN	1.5	-3.3	-1.8	0.3
29.36	-1.97	0.4	108.29	1JL	0.3	-3.4	-1.9	0.3
35.92	-1.94	0.3	117.84	11JL	-0.9	-3.4	-2.0	0.3
42.22	-1.89	0.2	127.41	21JL	-1.6	-3.4	-2.0	0.3
48.19	-1.81	0.1	137.01	31JL	-1.1	-3.5	-2.1	0.3
53.78	-1.70	-0.0	146.64	10AU	-0.4	-3.5	-2.2	0.3
58.89	-1.57	-0.2	156.32	20AU	-0.0	-3.5	-2.2	0.3
63.42	-1.39	-0.3	166.05	30AU	0.2	-3.4	-2.3	0.3
67.24	-1.18	-0.5	175.83	9SE	0.4	-3.4	-2.4	0.2
70.17	-0.91	-0.7	185.68	19SE	0.8	-3.4	-2.4	0.2
72.01	-0.59	-0.9	195.57	29SE	2.5	-3.4	-2.4	0.1
72.56	-0.21	-1.1	205.52	9OC	1.2	-3.3	-2.4	0.0
71.66	0.24	-1.3	215.53	19OC	-0.4	-3.3	-2.4	-0.0
69.35	0.74	-1.5	225.59	29OC	-0.7	-3.3	-2.4	-0.1
65.95	1.23	-1.7	235.69	8NO	-0.7	-3.3	-2.4	-0.2
62.13	1.68	-1.7	245.83	18NO	-0.7	-3.3	-2.3	-0.2
58.73	2.02	-1.4	256.00	28NO	-0.7	-3.4	-2.3	-0.3
56.37	2.24	-1.1	266.19	8DE	-0.7	-3.4	-2.2	-0.2
55.34	2.36	-0.8	276.39	18DE	-0.7	-3.4	-2.1	-0.2
55.63	2.41	-0.5	286.58	28DE	-0.5	-3.5	-2.1	-0.1

1502

♂ LONG	LAT	MAG	☉ LONG	16.00UT	☿	♀	♃	♄
57.08	2.40	-0.2	296.76	7JA	0.7	-3.5	-2.0	-0.0
59.49	2.37	0.1	306.92	17JA	2.9	-3.5	-1.9	0.0
62.65	2.31	0.3	317.04	27JA	1.0	-3.6	-1.9	0.1
66.40	2.24	0.6	327.13	6FE	0.4	-3.7	-1.8	0.2
70.62	2.17	0.8	337.17	16FE	0.2	-3.7	-1.7	0.2
75.20	2.09	1.0	347.15	26FE	0.0	-3.8	-1.7	0.3
80.07	2.01	1.1	357.09	8MR	-0.4	-3.9	-1.6	0.3
85.17	1.94	1.3	6.96	18MR	-1.0	-4.0	-1.6	0.3
90.44	1.86	1.4	16.78	28MR	-1.7	-4.1	-1.6	0.3
95.87	1.78	1.5	26.54	7AP	-1.1	-4.1	-1.6	0.3
101.42	1.70	1.6	36.26	17AP	-0.0	-4.2	-1.5	0.3
107.08	1.62	1.7	45.91	27AP	1.1	-4.2	-1.5	0.3
112.82	1.55	1.7	55.54	7MY	2.3	-4.1	-1.5	0.3
118.65	1.47	1.8	65.13	17MY	3.7	-3.7	-1.5	0.2
124.55	1.39	1.9	74.68	27MY	2.2	-2.9	-1.5	0.2
130.51	1.31	1.9	84.23	6JN	1.1	-3.3	-1.6	0.2
136.55	1.23	1.9	93.76	16JN	0.2	-3.9	-1.6	0.2
142.64	1.15	2.0	103.29	26JN	-0.9	-4.1	-1.6	0.3
148.79	1.07	2.0	112.83	6JL	-1.7	-4.2	-1.7	0.3
155.02	0.99	2.0	122.39	16JL	-1.1	-4.2	-1.7	0.3
161.30	0.91	2.0	131.97	26JL	-0.4	-4.1	-1.8	0.3
167.65	0.82	2.0	141.59	5AU	0.1	-4.0	-1.8	0.3
174.07	0.73	2.0	151.24	15AU	0.3	-3.9	-1.9	0.3
180.56	0.65	1.9	160.94	25AU	0.6	-3.8	-2.0	0.3
187.12	0.56	1.9	170.70	4SE	1.1	-3.7	-2.0	0.3
193.76	0.47	1.9	180.51	14SE	2.9	-3.7	-2.1	0.3
200.48	0.37	1.8	190.37	24SE	1.0	-3.6	-2.2	0.2
207.28	0.28	1.8	200.30	4OC	-0.5	-3.6	-2.2	0.1
214.16	0.18	1.7	210.28	14OC	-0.8	-3.5	-2.3	0.1
221.11	0.08	1.7	220.31	24OC	-0.8	-3.5	-2.3	0.1
228.16	-0.02	1.7	230.39	3NO	-0.8	-3.5	-2.3	-0.0
235.28	-0.12	1.7	240.51	13NO	-0.6	-3.4	-2.3	-0.1
242.49	-0.22	1.6	250.67	23NO	-0.6	-3.4	-2.3	-0.2
249.78	-0.33	1.6	260.85	3DE	-0.6	-3.4	-2.3	-0.2
257.14	-0.43	1.6	271.04	13DE	-0.4	-3.4	-2.3	-0.3
264.57	-0.53	1.6	281.24	23DE	0.9	-3.4	-2.2	-0.2

1503

♂ LONG	LAT	MAG	☉ LONG	16.00UT	☿	♀	♃	♄
272.08	-0.63	1.6	291.43	2JA	2.5	-3.3	-2.2	-0.2
279.65	-0.73	1.5	301.60	12JA	0.7	-3.3	-2.1	-0.1
287.27	-0.83	1.5	311.74	22JA	0.2	-3.3	-2.0	-0.0
294.94	-0.92	1.5	321.85	1FE	0.1	-3.4	-2.0	0.1
302.65	-1.00	1.5	331.91	11FE	-0.1	-3.4	-1.9	0.1
310.40	-1.08	1.4	341.93	21FE	-0.4	-3.4	-1.8	0.2
318.16	-1.15	1.4	351.89	3MR	-1.0	-3.4	-1.8	0.2
325.93	-1.20	1.4	1.79	13MR	-1.6	-3.5	-1.7	0.3
333.70	-1.25	1.3	11.64	23MR	-1.1	-3.5	-1.6	0.3
341.45	-1.29	1.3	21.43	2AP	0.1	-3.5	-1.6	0.3
349.17	-1.31	1.3	31.17	12AP	1.4	-3.4	-1.6	0.3
356.84	-1.31	1.3	40.85	22AP	3.0	-3.4	-1.5	0.3
4.47	-1.31	1.2	50.50	2MY	2.9	-3.4	-1.5	0.3
12.02	-1.28	1.2	60.10	12MY	1.7	-3.4	-1.5	0.3
19.49	-1.24	1.2	69.67	22MY	0.9	-3.3	-1.4	0.3
26.87	-1.19	1.2	79.22	1JN	0.1	-3.3	-1.4	0.3
34.14	-1.12	1.1	88.76	11JN	-0.9	-3.3	-1.4	0.3
41.29	-1.04	1.1	98.29	21JN	-1.7	-3.3	-1.4	0.2
48.30	-0.94	1.1	107.83	1JL	-1.1	-3.3	-1.4	0.3
55.16	-0.83	1.0	117.37	11JL	-0.3	-3.4	-1.5	0.3
61.86	-0.70	1.0	126.94	21JL	0.1	-3.4	-1.5	0.4
68.38	-0.56	0.9	136.54	31JL	0.5	-3.4	-1.5	0.4
74.68	-0.40	0.9	146.17	10AU	0.8	-3.4	-1.6	0.4
80.75	-0.23	0.8	155.85	20AU	1.5	-3.5	-1.6	0.4
86.55	-0.04	0.7	165.58	30AU	3.2	-3.6	-1.7	0.4
92.03	0.18	0.6	175.36	9SE	0.9	-3.6	-1.7	0.4
97.13	0.41	0.5	185.20	19SE	-0.6	-3.7	-1.8	0.4
101.78	0.68	0.4	195.09	29SE	-1.0	-3.8	-1.8	0.3
105.89	0.97	0.3	205.04	9OC	-1.0	-3.9	-1.9	0.3
109.32	1.30	0.1	215.05	19OC	-0.8	-4.0	-2.0	0.3
111.92	1.67	-0.1	225.10	29OC	-0.5	-4.1	-2.0	0.2
113.51	2.09	-0.3	235.20	8NO	-0.4	-4.2	-2.1	0.1
113.83	2.54	-0.5	245.34	18NO	-0.4	-4.3	-2.1	0.1
112.90	3.01	-0.7	255.51	28NO	-0.2	-4.4	-2.2	-0.0
110.55	3.46	-1.0	265.69	8DE	1.1	-4.4	-2.2	-0.1
107.09	3.82	-1.1	275.89	18DE	2.2	-4.2	-2.2	-0.2
103.12	4.03	-1.2	286.08	28DE	0.4	-3.6	-2.2	-0.2

♂			⊙	16.00UT	☿	♀	♃	♄
LONG	LAT	MAG	LONG	1504	MAGNITUDES			
99.42	4.08	-1.0	296.26	7JA	0.1	-3.2	-2.2	-0.1
96.63	3.98	-0.7	306.42	17JA	-0.0	-3.9	-2.2	-0.1
95.11	3.78	-0.4	316.55	27JA	-0.2	-4.2	-2.1	-0.0
94.90	3.53	-0.2	326.64	6FE	-0.5	-4.3	-2.1	0.0
95.87	3.27	0.1	336.68	16FE	-1.0	-4.3	-2.0	0.1
97.84	3.02	0.3	346.67	26FE	-1.4	-4.2	-1.9	0.2
100.61	2.77	0.6	356.60	7MR	-1.0	-4.1	-1.8	0.2
104.01	2.55	0.8	6.48	17MR	0.1	-4.0	-1.8	0.3
107.94	2.34	0.9	16.31	27MR	1.8	-3.9	-1.7	0.3
112.26	2.15	1.1	26.07	6AP	3.7	-3.8	-1.6	0.4
116.92	1.97	1.2	35.78	16AP	2.2	-3.7	-1.6	0.4
121.85	1.80	1.3	45.45	26AP	1.3	-3.6	-1.5	0.4
127.00	1.65	1.4	55.07	6MY	0.6	-3.6	-1.5	0.4
132.35	1.50	1.5	64.66	16MY	-0.0	-3.5	-1.4	0.4
137.86	1.36	1.6	74.22	26MY	-0.9	-3.4	-1.4	0.4
143.52	1.23	1.7	83.76	5JN	-1.8	-3.4	-1.4	0.4
149.31	1.10	1.7	93.30	15JN	-1.1	-3.4	-1.3	0.4
155.23	0.97	1.7	102.83	25JN	-0.3	-3.3	-1.3	0.4
161.26	0.85	1.8	112.37	5JL	0.3	-3.3	-1.3	0.3
167.40	0.73	1.8	121.93	15JL	0.6	-3.3	-1.3	0.4
173.64	0.61	1.8	131.51	25JL	1.1	-3.3	-1.3	0.4
179.99	0.50	1.8	141.12	4AU	2.0	-3.3	-1.3	0.5
186.45	0.38	1.8	150.78	14AU	3.0	-3.3	-1.4	0.5
193.00	0.27	1.8	160.48	24AU	0.8	-3.4	-1.4	0.5
199.66	0.16	1.8	170.23	3SE	-0.7	-3.4	-1.4	0.5
206.43	0.05	1.8	180.04	13SE	-1.1	-3.4	-1.5	0.5
213.29	-0.05	1.7	189.90	23SE	-1.1	-3.4	-1.5	0.5
220.26	-0.16	1.7	199.82	3OC	-0.7	-3.5	-1.6	0.5
227.33	-0.26	1.7	209.80	13OC	-0.4	-3.5	-1.6	0.5
234.50	-0.36	1.6	219.82	23OC	-0.3	-3.5	-1.7	0.4
241.76	-0.46	1.6	229.90	2NO	-0.2	-3.5	-1.7	0.4
249.12	-0.55	1.6	240.02	12NO	-0.0	-3.4	-1.8	0.3
256.57	-0.64	1.5	250.17	22NO	1.3	-3.4	-1.9	0.3
264.10	-0.72	1.5	260.35	2DE	1.9	-3.4	-1.9	0.2
271.71	-0.80	1.4	270.54	12DE	0.2	-3.4	-2.0	0.1
279.39	-0.87	1.4	280.74	22DE	-0.1	-3.3	-2.0	0.0
				1505				
287.13	-0.94	1.4	290.93	1JA	-0.2	-3.3	-2.1	-0.0
294.92	-0.99	1.4	301.10	11JA	-0.3	-3.3	-2.1	-0.1
302.74	-1.04	1.4	311.25	21JA	-0.6	-3.3	-2.1	-0.0
310.60	-1.07	1.4	321.36	31JA	-1.0	-3.4	-2.1	0.0
318.47	-1.09	1.4	331.43	10FE	-1.3	-3.4	-2.1	0.1
326.35	-1.11	1.4	341.44	20FE	-1.0	-3.4	-2.0	0.2
334.21	-1.11	1.4	351.40	2MR	0.2	-3.4	-2.0	0.2
342.05	-1.09	1.4	1.32	12MR	2.2	-3.4	-1.9	0.3
349.86	-1.07	1.4	11.16	22MR	3.0	-3.5	-1.9	0.3
357.62	-1.03	1.4	20.96	1AP	1.6	-3.5	-1.8	0.4
5.33	-0.98	1.4	30.70	11AP	0.9	-3.6	-1.7	0.4
12.96	-0.92	1.4	40.39	21AP	0.5	-3.6	-1.7	0.5
20.53	-0.85	1.4	50.03	1MY	-0.1	-3.7	-1.6	0.5
28.01	-0.77	1.5	59.64	11MY	-0.9	-3.8	-1.5	0.5
35.39	-0.68	1.5	69.21	21MY	-1.8	-3.9	-1.5	0.5
42.68	-0.58	1.5	78.76	31MY	-1.1	-4.0	-1.4	0.5
49.87	-0.48	1.5	88.30	10JN	-0.2	-4.0	-1.4	0.5
56.95	-0.36	1.5	97.83	20JN	0.4	-4.1	-1.3	0.5
63.92	-0.24	1.5	107.36	30JN	0.8	-4.2	-1.3	0.5
70.77	-0.11	1.5	116.91	10JL	1.4	-4.2	-1.3	0.5
77.50	0.02	1.5	126.48	20JL	2.5	-4.0	-1.3	0.5
84.10	0.16	1.4	136.07	30JL	2.6	-3.5	-1.2	0.5
90.55	0.30	1.4	145.71	9AU	0.6	-3.2	-1.2	0.6
96.86	0.46	1.4	155.38	19AU	-0.8	-3.7	-1.2	0.6
103.01	0.62	1.3	165.11	29AU	-1.3	-4.1	-1.2	0.7
108.98	0.79	1.3	174.89	8SE	-1.1	-4.3	-1.3	0.7
114.75	0.97	1.2	184.72	18SE	-0.6	-4.3	-1.3	0.7
120.28	1.16	1.1	194.61	28SE	-0.3	-4.2	-1.3	0.7
125.55	1.36	1.1	204.56	8OC	-0.2	-4.1	-1.4	0.7
130.50	1.58	0.9	214.56	18OC	-0.1	-4.0	-1.4	0.7
135.07	1.82	0.8	224.61	28OC	0.2	-4.0	-1.4	0.6
139.17	2.09	0.7	234.71	7NO	1.5	-3.9	-1.5	0.6
142.70	2.38	0.5	244.85	17NO	1.7	-3.8	-1.6	0.6
145.54	2.70	0.3	255.02	27NO	0.0	-3.7	-1.6	0.5
147.50	3.05	0.1	265.20	7DE	-0.3	-3.6	-1.7	0.4
148.41	3.43	-0.1	275.39	17DE	-0.3	-3.6	-1.8	0.4
148.09	3.81	-0.4	285.59	27DE	-0.4	-3.5	-1.8	0.3

♂			⊙	16.00UT	☿	♀	♃	♄
LONG	LAT	MAG	LONG	1506	MAGNITUDES			
146.44	4.16	-0.6	295.78	6JA	-0.6	-3.5	-1.9	0.2
143.56	4.42	-0.8	305.93	16JA	-1.0	-3.4	-1.9	0.2
139.82	4.55	-1.0	316.07	26JA	-1.1	-3.4	-2.0	0.1
135.88	4.49	-0.9	326.16	5FE	-0.9	-3.4	-2.0	0.2
132.46	4.27	-0.7	336.20	15FE	0.3	-3.3	-2.0	0.2
130.10	3.95	-0.5	346.19	25FE	2.6	-3.3	-2.0	0.3
129.02	3.56	-0.2	356.13	7MR	2.3	-3.3	-2.0	0.3
129.20	3.17	-0.0	6.01	17MR	1.2	-3.3	-2.0	0.4
130.50	2.80	0.2	15.83	27MR	0.7	-3.3	-1.9	0.5
132.74	2.46	0.4	25.60	6AP	0.3	-3.3	-1.9	0.5
135.73	2.15	0.6	35.31	16AP	-0.2	-3.3	-1.8	0.6
139.34	1.87	0.8	44.98	26AP	-0.9	-3.4	-1.8	0.6
143.45	1.61	0.9	54.61	6MY	-1.8	-3.4	-1.7	0.6
147.97	1.38	1.0	64.19	16MY	-1.1	-3.4	-1.6	0.7
152.83	1.17	1.1	73.76	26MY	-0.2	-3.5	-1.6	0.7
157.98	0.97	1.2	83.30	5JN	0.5	-3.5	-1.5	0.7
163.38	0.79	1.3	92.83	15JN	1.1	-3.4	-1.5	0.7
169.00	0.62	1.4	102.36	25JN	1.9	-3.4	-1.4	0.7
174.81	0.46	1.4	111.90	5JL	3.2	-3.4	-1.4	0.7
180.81	0.31	1.5	121.45	15JL	2.2	-3.4	-1.3	0.7
186.97	0.16	1.5	131.04	25JL	0.5	-3.3	-1.3	0.7
193.29	0.03	1.5	140.65	4AU	-0.8	-3.3	-1.3	0.7
199.76	-0.10	1.5	150.30	14AU	-1.4	-3.3	-1.2	0.8
206.37	-0.23	1.6	160.00	24AU	-1.1	-3.3	-1.2	0.8
213.11	-0.34	1.6	169.75	3SE	-0.6	-3.3	-1.2	0.9
219.99	-0.45	1.6	179.55	13SE	-0.2	-3.3	-1.2	0.9
227.00	-0.55	1.6	189.42	23SE	-0.0	-3.4	-1.2	0.9
234.12	-0.65	1.5	199.33	3OC	0.1	-3.4	-1.2	0.9
241.37	-0.74	1.5	209.31	13OC	0.4	-3.4	-1.2	0.9
248.72	-0.82	1.5	219.34	23OC	1.8	-3.5	-1.3	0.9
256.18	-0.89	1.5	229.41	2NO	1.5	-3.5	-1.3	0.9
263.74	-0.95	1.5	239.53	12NO	-0.2	-3.6	-1.3	0.9
271.38	-1.00	1.5	249.68	22NO	-0.4	-3.7	-1.4	0.8
279.10	-1.04	1.4	259.86	2DE	-0.4	-3.7	-1.4	0.8
286.88	-1.07	1.4	270.05	12DE	-0.5	-3.8	-1.5	0.7
294.71	-1.09	1.4	280.25	22DE	-0.7	-3.9	-1.6	0.7
				1507				
302.58	-1.10	1.4	290.44	1JA	-0.9	-4.0	-1.6	0.6
310.48	-1.09	1.4	300.61	11JA	-1.0	-4.1	-1.7	0.5
318.39	-1.08	1.3	310.76	21JA	-0.8	-4.2	-1.8	0.5
326.30	-1.05	1.3	320.87	31JA	0.4	-4.3	-1.8	0.4
334.19	-1.01	1.3	330.94	10FE	2.9	-4.3	-1.9	0.4
342.05	-0.96	1.3	340.96	20FE	1.7	-4.3	-1.9	0.4
349.87	-0.90	1.3	350.92	2MR	0.8	-4.0	-2.0	0.5
357.64	-0.83	1.4	0.84	12MR	0.5	-3.5	-2.0	0.5
5.35	-0.76	1.4	10.69	22MR	0.2	-3.3	-2.0	0.6
12.99	-0.67	1.4	20.48	1AP	-0.2	-3.8	-2.0	0.6
20.56	-0.58	1.5	30.23	11AP	-0.9	-4.1	-2.0	0.7
28.04	-0.48	1.5	39.92	21AP	-1.8	-4.2	-2.0	0.7
35.43	-0.38	1.5	49.56	1MY	-1.1	-4.2	-1.9	0.8
42.74	-0.28	1.6	59.17	11MY	-0.1	-4.1	-1.9	0.8
49.96	-0.17	1.6	68.75	21MY	0.7	-4.0	-1.8	0.9
57.08	-0.06	1.6	78.30	31MY	1.5	-3.9	-1.7	0.9
64.11	0.05	1.7	87.84	10JN	2.5	-3.8	-1.7	0.9
71.04	0.16	1.7	97.37	20JN	3.2	-3.7	-1.6	1.0
77.88	0.27	1.7	106.90	30JN	1.7	-3.7	-1.6	1.0
84.63	0.39	1.7	116.45	10JL	0.4	-3.6	-1.5	1.0
91.29	0.50	1.7	126.01	20JL	-0.8	-3.5	-1.4	1.0
97.85	0.62	1.7	135.61	30JL	-1.5	-3.5	-1.4	1.0
104.32	0.73	1.7	145.24	9AU	-1.1	-3.4	-1.3	1.0
110.68	0.85	1.7	154.91	19AU	-0.5	-3.4	-1.3	1.0
116.95	0.97	1.7	164.63	29AU	-0.1	-3.4	-1.3	1.0
123.11	1.09	1.7	174.41	8SE	0.1	-3.4	-1.2	1.1
129.15	1.21	1.6	184.24	18SE	0.3	-3.4	-1.2	1.1
135.06	1.34	1.6	194.13	28SE	0.6	-3.4	-1.2	1.2
140.83	1.47	1.5	204.08	8OC	2.2	-3.4	-1.2	1.2
146.43	1.61	1.5	214.08	18OC	1.3	-3.4	-1.2	1.2
151.84	1.75	1.4	224.13	28OC	-0.3	-3.4	-1.2	1.2
157.03	1.90	1.3	234.23	7NO	-0.6	-3.4	-1.2	1.2
161.96	2.06	1.2	244.36	17NO	-0.6	-3.4	-1.3	1.2
166.56	2.23	1.0	254.52	27NO	-0.6	-3.4	-1.3	1.2
170.78	2.42	0.9	264.71	7DE	-0.8	-3.4	-1.3	1.1
174.50	2.62	0.7	274.90	17DE	-0.8	-3.5	-1.4	1.1
177.63	2.83	0.5	285.10	27DE	-0.8	-3.5	-1.4	1.0

♂			☉	16.00UT	☿	♀	♃	♄
LONG	LAT	MAG	LONG	1508			MAGNITUDES	
180.02	3.06	0.3	295.28	6JA	-0.6	-3.5	-1.5	0.9
181.49	3.29	0.0	305.44	16JA	0.6	-3.5	-1.6	0.9
181.86	3.51	-0.2	315.57	26JA	3.0	-3.4	-1.6	0.8
180.99	3.71	-0.5	325.67	5FE	1.3	-3.4	-1.7	0.7
178.83	3.82	-0.8	335.71	15FE	0.6	-3.4	-1.8	0.7
175.59	3.83	-1.0	345.71	25FE	0.3	-3.4	-1.8	0.7
171.75	3.68	-1.1	355.65	6MR	0.1	-3.3	-1.9	0.7
168.03	3.38	-1.0	5.53	16MR	-0.3	-3.3	-1.9	0.7
165.08	2.98	-0.8	15.35	26MR	-1.0	-3.3	-2.0	0.8
163.30	2.54	-0.6	25.13	5AP	-1.8	-3.3	-2.0	0.8
162.84	2.10	-0.3	34.84	15AP	-1.1	-3.3	-2.0	0.9
163.60	1.69	-0.1	44.51	25AP	-0.1	-3.4	-2.1	1.0
165.44	1.31	0.1	54.14	5MY	0.9	-3.4	-2.0	1.0
168.17	0.98	0.3	63.73	15MY	1.9	-3.4	-2.0	1.1
171.63	0.68	0.4	73.29	25MY	3.3	-3.4	-2.0	1.1
175.70	0.42	0.5	82.84	4JN	2.6	-3.5	-1.9	1.1
180.26	0.19	0.7	92.37	14JN	1.4	-3.5	-1.9	1.2
185.25	-0.02	0.8	101.90	24JN	0.3	-3.6	-1.8	1.2
190.59	-0.21	0.9	111.44	4JL	-0.9	-3.7	-1.8	1.2
196.24	-0.38	0.9	120.99	14JL	-1.6	-3.7	-1.7	1.3
202.16	-0.53	1.0	130.57	24JL	-1.1	-3.8	-1.7	1.3
208.33	-0.67	1.1	140.19	3AU	-0.4	-3.9	-1.6	1.3
214.71	-0.79	1.1	149.83	13AU	-0.0	-4.0	-1.5	1.3
221.29	-0.90	1.1	159.53	23AU	0.2	-4.1	-1.5	1.3
228.06	-0.99	1.2	169.28	2SE	0.4	-4.2	-1.4	1.3
235.00	-1.08	1.2	179.08	12SE	0.9	-4.3	-1.4	1.3
242.09	-1.14	1.2	188.94	22SE	2.6	-4.3	-1.4	1.3
249.33	-1.20	1.3	198.85	2OC	1.2	-4.1	-1.3	1.3
256.69	-1.24	1.3	208.82	12OC	-0.4	-3.6	-1.3	1.3
264.17	-1.27	1.3	218.85	22OC	-0.7	-3.2	-1.3	1.3
271.76	-1.28	1.3	228.92	1NO	-0.7	-3.9	-1.3	1.2
279.43	-1.28	1.3	239.04	11NO	-0.7	-4.3	-1.3	1.2
287.17	-1.27	1.3	249.19	21NO	-0.7	-4.4	-1.3	1.2
294.97	-1.24	1.3	259.37	1DE	-0.6	-4.4	-1.3	1.2
302.81	-1.20	1.3	269.56	11DE	-0.7	-4.3	-1.3	1.1
310.68	-1.15	1.4	279.75	21DE	-0.5	-4.2	-1.4	1.1
318.56	-1.09	1.4	289.95	31DE	0.7	-4.1	-1.4	1.0
				1509				
326.43	-1.02	1.4	300.12	10JA	2.8	-4.0	-1.4	1.0
334.28	-0.94	1.4	310.27	20JA	0.9	-3.9	-1.5	0.9
342.10	-0.86	1.4	320.39	30JA	0.4	-3.8	-1.5	0.9
349.87	-0.76	1.4	330.46	9FE	0.2	-3.7	-1.6	0.8
357.59	-0.66	1.5	340.48	19FE	-0.0	-3.6	-1.7	0.8
5.25	-0.56	1.5	350.45	1MR	-0.4	-3.5	-1.7	0.8
12.83	-0.46	1.5	0.36	11MR	-1.0	-3.5	-1.8	0.8
20.34	-0.35	1.5	10.22	21MR	-1.6	-3.4	-1.9	0.9
27.76	-0.24	1.5	20.02	31MR	-1.1	-3.4	-1.9	0.9
35.10	-0.13	1.5	29.76	10AP	-0.0	-3.4	-2.0	1.0
42.36	-0.03	1.6	39.46	20AP	1.2	-3.3	-2.1	1.1
49.53	0.08	1.6	49.11	30AP	2.5	-3.3	-2.1	1.1
56.61	0.18	1.6	58.71	10MY	3.5	-3.3	-2.1	1.2
63.62	0.28	1.7	68.29	20MY	2.0	-3.3	-2.2	1.2
70.54	0.38	1.7	77.85	30MY	1.0	-3.3	-2.2	1.3
77.38	0.48	1.8	87.38	9JN	0.2	-3.3	-2.2	1.3
84.15	0.57	1.8	96.91	19JN	-0.9	-3.3	-2.1	1.3
90.84	0.66	1.8	106.44	29JN	-1.7	-3.4	-2.1	1.3
97.47	0.75	1.9	115.99	9JL	-1.1	-3.4	-2.0	1.3
104.03	0.83	1.9	125.55	19JL	-0.4	-3.4	-2.0	1.3
110.52	0.92	1.9	135.14	29JL	0.1	-3.5	-1.9	1.3
116.95	0.99	1.9	144.77	8AU	0.4	-3.5	-1.9	1.3
123.32	1.07	1.9	154.44	18AU	0.6	-3.5	-1.8	1.2
129.63	1.15	1.9	164.16	28AU	1.3	-3.4	-1.7	1.2
135.88	1.22	1.9	173.93	7SE	3.0	-3.4	-1.7	1.1
142.07	1.29	1.9	183.76	17SE	1.0	-3.4	-1.6	1.1
148.19	1.36	1.9	193.64	27SE	-0.5	-3.4	-1.6	1.1
154.23	1.43	1.8	203.58	7OC	-0.9	-3.3	-1.5	1.1
160.20	1.50	1.8	213.58	17OC	-0.9	-3.3	-1.5	1.1
166.08	1.57	1.7	223.63	27OC	-0.8	-3.3	-1.4	1.1
171.85	1.63	1.6	233.72	6NO	-0.6	-3.3	-1.4	1.1
177.51	1.70	1.6	243.86	16NO	-0.5	-3.3	-1.4	1.1
183.03	1.77	1.5	254.02	26NO	-0.5	-3.4	-1.4	1.0
188.38	1.83	1.4	264.21	6DE	-0.3	-3.4	-1.4	1.0
193.53	1.90	1.2	274.41	16DE	0.9	-3.4	-1.4	1.0
198.44	1.96	1.1	284.60	26DE	2.5	-3.5	-1.4	0.9

♂			☉	16.00UT	☿	♀	♃	♄
LONG	LAT	MAG	LONG	1510			MAGNITUDES	
203.06	2.02	0.9	294.78	5JA	0.6	-3.5	-1.4	0.9
207.31	2.08	0.7	304.95	15JA	0.2	-3.5	-1.4	0.9
211.11	2.13	0.5	315.09	25JA	0.0	-3.6	-1.5	0.8
214.34	2.17	0.3	325.18	4FE	-0.1	-3.7	-1.5	0.8
216.87	2.20	0.0	335.23	14FE	-0.4	-3.7	-1.6	0.7
218.52	2.19	-0.2	345.23	24FE	-1.0	-3.8	-1.6	0.7
219.12	2.15	-0.5	355.17	6MR	-1.5	-3.9	-1.7	0.6
218.50	2.03	-0.8	5.06	16MR	-1.1	-4.0	-1.7	0.6
216.62	1.84	-1.1	14.89	26MR	0.1	-4.1	-1.8	0.6
213.64	1.55	-1.4	24.66	5AP	1.5	-4.1	-1.9	0.7
210.01	1.16	-1.5	34.38	15AP	3.3	-4.2	-1.9	0.8
206.66	0.71	-1.4	44.05	25AP	2.6	-4.2	-2.0	0.8
203.66	0.26	-1.2	53.68	5MY	1.5	-4.1	-2.1	0.9
202.04	-0.17	-1.0	63.27	15MY	0.8	-3.7	-2.1	1.0
201.79	-0.54	-0.8	72.84	25MY	0.1	-2.9	-2.2	1.0
202.82	-0.86	-0.6	82.38	4JN	-0.9	-3.3	-2.2	1.1
204.99	-1.11	-0.4	91.92	14JN	-1.7	-3.9	-2.3	1.1
208.11	-1.32	-0.3	101.44	24JN	-1.1	-4.1	-2.3	1.1
212.01	-1.49	-0.1	110.98	4JL	-0.3	-4.2	-2.3	1.2
216.57	-1.63	0.0	120.53	14JL	0.2	-4.2	-2.3	1.2
221.67	-1.73	0.2	130.11	24JL	0.5	-4.1	-2.3	1.2
227.21	-1.81	0.3	139.72	3AU	0.9	-4.0	-2.2	1.1
233.15	-1.86	0.4	149.37	13AU	1.7	-3.9	-2.2	1.1
239.40	-1.89	0.5	159.06	23AU	3.2	-3.8	-2.1	1.1
245.94	-1.91	0.5	168.80	2SE	0.9	-3.7	-2.1	1.1
252.72	-1.90	0.6	178.60	12SE	-0.6	-3.7	-2.0	1.0
259.70	-1.87	0.7	188.45	22SE	-1.0	-3.6	-1.9	1.0
266.86	-1.83	0.8	198.37	2OC	-1.0	-3.6	-1.9	0.9
274.17	-1.77	0.8	208.34	12OC	-0.7	-3.5	-1.8	1.0
281.61	-1.69	0.9	218.36	22OC	-0.5	-3.5	-1.7	1.0
289.14	-1.60	0.9	228.43	1NO	-0.4	-3.4	-1.7	1.0
296.75	-1.50	1.0	238.54	11NO	-0.3	-3.4	-1.6	1.0
304.42	-1.39	1.1	248.69	21NO	-0.2	-3.4	-1.6	1.0
312.12	-1.27	1.1	258.87	1DE	1.1	-3.4	-1.6	0.9
319.84	-1.15	1.2	269.06	11DE	2.2	-3.4	-1.5	0.9
327.55	-1.02	1.2	279.25	21DE	0.4	-3.4	-1.5	0.9
335.25	-0.89	1.3	289.45	31DE	-0.0	-3.3	-1.5	0.9
				1511				
342.92	-0.76	1.3	299.63	10JA	-0.1	-3.3	-1.5	0.8
350.54	-0.62	1.4	309.77	20JA	-0.2	-3.4	-1.5	0.8
358.11	-0.49	1.4	319.89	30JA	-0.5	-3.4	-1.5	0.7
5.63	-0.36	1.5	329.97	9FE	-1.0	-3.4	-1.5	0.7
13.07	-0.23	1.5	339.99	19FE	-1.4	-3.4	-1.5	0.6
20.44	-0.11	1.6	349.97	1MR	-0.9	-3.4	-1.6	0.6
27.73	0.01	1.6	359.88	11MR	0.1	-3.5	-1.6	0.5
34.94	0.12	1.7	9.74	21MR	1.9	-3.5	-1.7	0.5
42.08	0.22	1.7	19.54	31MR	3.5	-3.5	-1.7	0.5
49.14	0.33	1.7	29.29	10AP	1.9	-3.4	-1.8	0.5
56.12	0.42	1.8	38.98	20AP	1.1	-3.4	-1.8	0.6
63.03	0.51	1.8	48.64	30AP	0.6	-3.4	-1.9	0.7
69.87	0.60	1.8	58.25	10MY	-0.0	-3.4	-2.0	0.7
76.64	0.68	1.8	67.82	20MY	-0.9	-3.3	-2.0	0.8
83.35	0.75	1.8	77.38	30MY	-1.8	-3.3	-2.1	0.8
90.00	0.82	1.8	86.92	9JN	-1.2	-3.3	-2.2	0.9
96.60	0.88	1.8	96.44	19JN	-0.3	-3.3	-2.2	0.9
103.15	0.94	1.9	105.98	29JN	0.3	-3.3	-2.3	1.0
109.65	0.99	1.9	115.52	9JL	0.7	-3.4	-2.4	1.0
116.12	1.04	1.9	125.08	19JL	1.2	-3.4	-2.4	1.0
122.55	1.09	2.0	134.68	29JL	2.2	-3.4	-2.4	1.0
128.94	1.13	2.0	144.30	8AU	2.9	-3.5	-2.4	1.0
135.30	1.16	2.0	153.97	18AU	0.8	-3.5	-2.4	1.0
141.64	1.20	2.0	163.69	28AU	-0.7	-3.6	-2.4	1.0
147.95	1.22	2.0	173.46	7SE	-1.2	-3.6	-2.3	1.0
154.24	1.25	2.0	183.28	17SE	-1.1	-3.7	-2.3	0.9
160.51	1.27	2.0	193.17	27SE	-0.7	-3.8	-2.2	0.9
166.75	1.29	2.0	203.10	7OC	-0.4	-3.9	-2.2	0.8
172.97	1.30	1.9	213.10	17OC	-0.2	-4.0	-2.1	0.8
179.16	1.31	1.9	223.14	27OC	-0.2	-4.1	-2.0	0.9
185.32	1.31	1.9	233.23	6NO	0.0	-4.2	-2.0	0.9
191.44	1.31	1.8	243.36	16NO	1.3	-4.3	-1.9	0.9
197.52	1.30	1.7	253.53	26NO	1.9	-4.4	-1.8	0.9
203.56	1.29	1.6	263.71	6DE	0.1	-4.4	-1.8	0.9
209.53	1.27	1.5	273.90	16DE	-0.2	-4.1	-1.7	0.9
215.43	1.24	1.4	284.10	26DE	-0.2	-3.6	-1.7	0.8

♂ LONG	♂ LAT	♂ MAG	☉ LONG	16.00UT	☿	♀	♃	♄
				1512	MAGNITUDES			
221.24	1.20	1.3	294.29	5JA	-0.3	-3.2	-1.6	0.8
226.95	1.14	1.2	304.45	15JA	-0.6	-3.9	-1.6	0.8
232.53	1.07	1.0	314.59	25JA	-1.0	-4.2	-1.6	0.7
237.95	0.99	0.9	324.69	4FE	-1.2	-4.3	-1.6	0.7
243.17	0.88	0.7	334.74	14FE	-1.0	-4.3	-1.6	0.6
248.16	0.74	0.5	344.74	24FE	0.2	-4.2	-1.6	0.6
252.84	0.57	0.3	354.69	5MR	2.3	-4.1	-1.6	0.5
257.14	0.35	0.1	4.58	15MR	2.7	-4.0	-1.6	0.5
260.97	0.09	-0.2	14.41	25MR	1.4	-3.9	-1.6	0.4
264.19	-0.24	-0.5	24.19	4AP	0.8	-3.8	-1.6	0.4
266.65	-0.64	-0.8	33.91	14AP	0.4	-3.7	-1.7	0.4
268.16	-1.13	-1.1	43.59	24AP	-0.1	-3.6	-1.7	0.4
268.56	-1.70	-1.4	53.22	4MY	-0.9	-3.6	-1.8	0.5
267.73	-2.35	-1.7	62.81	14MY	-1.8	-3.5	-1.8	0.6
265.70	-3.03	-2.0	72.38	24MY	-1.2	-3.4	-1.9	0.6
262.84	-3.66	-2.2	81.92	3JN	-0.2	-3.4	-1.9	0.7
259.75	-4.16	-2.2	91.46	13JN	0.4	-3.4	-2.0	0.8
257.18	-4.47	-2.0	100.99	23JN	0.9	-3.3	-2.1	0.8
255.70	-4.60	-1.8	110.52	3JL	1.6	-3.3	-2.2	0.8
255.57	-4.57	-1.6	120.08	13JL	2.8	-3.3	-2.2	0.9
256.82	-4.44	-1.4	129.65	23JL	2.5	-3.3	-2.3	0.9
259.29	-4.24	-1.2	139.26	2AU	0.6	-3.3	-2.4	0.9
262.78	-4.00	-1.0	148.90	12AU	-0.8	-3.3	-2.4	0.9
267.12	-3.73	-0.8	158.59	22AU	-1.3	-3.4	-2.4	0.9
272.13	-3.46	-0.6	168.33	1SE	-1.2	-3.4	-2.5	0.9
277.68	-3.17	-0.4	178.13	11SE	-0.6	-3.4	-2.5	0.9
283.66	-2.89	-0.2	187.98	21SE	-0.3	-3.4	-2.5	0.9
289.97	-2.60	-0.1	197.89	1OC	-0.1	-3.5	-2.4	0.8
296.55	-2.32	0.1	207.85	11OC	-0.0	-3.5	-2.4	0.8
303.34	-2.05	0.2	217.87	21OC	0.2	-3.5	-2.3	0.7
310.27	-1.78	0.3	227.94	31OC	1.6	-3.5	-2.3	0.8
317.32	-1.53	0.5	238.05	10NO	1.7	-3.4	-2.2	0.8
324.45	-1.28	0.6	248.20	20NO	-0.0	-3.4	-2.1	0.8
331.63	-1.05	0.7	258.37	30NO	-0.3	-3.4	-2.1	0.8
338.83	-0.83	0.8	268.56	10DE	-0.4	-3.4	-2.0	0.8
346.04	-0.63	0.9	278.76	20DE	-0.4	-3.3	-1.9	0.8
353.24	-0.44	1.1	288.94	30DE	-0.7	-3.3	-1.9	0.8
				1513				
0.42	-0.26	1.2	299.12	9JA	-1.0	-3.3	-1.8	0.8
7.56	-0.10	1.3	309.28	19JA	-1.1	-3.3	-1.8	0.7
14.66	0.05	1.3	319.39	29JA	-0.9	-3.4	-1.7	0.7
21.71	0.19	1.4	329.47	8FE	0.3	-3.4	-1.7	0.7
28.71	0.31	1.5	339.50	18FE	2.7	-3.4	-1.6	0.6
35.66	0.43	1.6	349.47	28FE	2.1	-3.4	-1.6	0.6
42.54	0.53	1.6	359.40	10MR	1.0	-3.4	-1.6	0.5
49.38	0.62	1.7	9.26	20MR	0.6	-3.5	-1.6	0.5
56.15	0.71	1.8	19.06	30MR	0.3	-3.5	-1.6	0.4
62.87	0.78	1.8	28.82	9AP	-0.2	-3.6	-1.6	0.3
69.54	0.85	1.9	38.52	19AP	-0.9	-3.6	-1.6	0.3
76.16	0.90	1.9	48.17	29AP	-1.8	-3.7	-1.6	0.3
82.73	0.96	1.9	57.79	9MY	-1.2	-3.8	-1.6	0.4
89.27	1.00	1.9	67.37	19MY	-0.2	-3.9	-1.7	0.4
95.76	1.04	2.0	76.92	29MY	0.6	-4.0	-1.7	0.5
102.23	1.07	2.0	86.46	8JN	1.2	-4.1	-1.8	0.6
108.66	1.10	2.0	95.99	18JN	2.1	-4.1	-1.8	0.6
115.07	1.12	2.0	105.52	28JN	3.3	-4.2	-1.9	0.7
121.47	1.13	2.0	115.07	8JL	2.0	-4.2	-1.9	0.7
127.85	1.14	2.0	124.63	18JL	0.5	-4.0	-2.0	0.8
134.22	1.15	1.9	134.22	28JL	-0.8	-3.5	-2.1	0.8
140.58	1.15	2.0	143.84	7AU	-1.4	-3.2	-2.1	0.8
146.95	1.15	2.0	153.51	17AU	-1.2	-3.7	-2.2	0.9
153.32	1.14	2.0	163.22	27AU	-0.5	-4.1	-2.3	0.9
159.69	1.13	2.0	172.99	6SE	-0.2	-4.3	-2.3	0.9
166.07	1.11	2.0	182.81	16SE	0.0	-4.3	-2.4	0.8
172.47	1.09	2.0	192.68	26SE	0.2	-4.2	-2.4	0.8
178.88	1.07	2.0	202.62	6OC	0.5	-4.1	-2.4	0.8
185.30	1.03	2.0	212.61	16OC	1.9	-4.0	-2.4	0.8
191.74	1.00	2.0	222.66	26OC	1.5	-3.9	-2.4	0.7
198.20	0.95	1.9	232.75	5NO	-0.2	-3.9	-2.4	0.7
204.67	0.90	1.9	242.87	15NO	-0.5	-3.8	-2.4	0.7
211.16	0.84	1.8	253.04	25NO	-0.5	-3.7	-2.3	0.8
217.66	0.78	1.8	263.22	5DE	-0.6	-3.6	-2.3	0.8
224.18	0.70	1.7	273.41	15DE	-0.7	-3.6	-2.2	0.8
230.71	0.62	1.6	283.61	25DE	-0.9	-3.5	-2.1	0.8

♂ LONG	♂ LAT	♂ MAG	☉ LONG	16.00UT	☿	♀	♃	♄
				1514	MAGNITUDES			
237.25	0.52	1.6	293.80	4JA	-0.9	-3.5	-2.0	0.8
243.79	0.41	1.5	303.96	14JA	-0.7	-3.4	-2.0	0.8
250.34	0.29	1.4	314.10	24JA	0.4	-3.4	-1.9	0.7
256.88	0.15	1.3	324.20	3FE	3.0	-3.4	-1.8	0.7
263.41	-0.00	1.1	334.25	13FE	1.5	-3.3	-1.8	0.7
269.93	-0.18	1.0	344.26	23FE	0.7	-3.3	-1.7	0.6
276.42	-0.37	0.9	354.21	5MR	0.4	-3.3	-1.7	0.6
282.88	-0.59	0.7	4.10	15MR	0.2	-3.3	-1.6	0.5
289.28	-0.84	0.6	13.93	25MR	-0.3	-3.3	-1.6	0.5
295.61	-1.11	0.4	23.71	4AP	-0.9	-3.3	-1.6	0.4
301.84	-1.41	0.2	33.44	14AP	-1.8	-3.3	-1.5	0.3
307.94	-1.74	0.1	43.12	24AP	-1.2	-3.4	-1.5	0.3
313.85	-2.11	-0.1	52.75	4MY	-0.1	-3.4	-1.5	0.2
319.53	-2.51	-0.3	62.34	14MY	0.8	-3.4	-1.5	0.3
324.90	-2.95	-0.6	71.91	24MY	1.6	-3.5	-1.5	0.4
329.86	-3.43	-0.8	81.46	3JN	2.8	-3.5	-1.5	0.4
334.28	-3.95	-1.0	90.99	13JN	3.0	-3.4	-1.6	0.5
338.00	-4.49	-1.3	100.53	23JN	1.6	-3.4	-1.6	0.6
340.83	-5.05	-1.6	110.06	3JL	0.4	-3.4	-1.6	0.6
342.57	-5.59	-1.8	119.61	13JL	-0.8	-3.4	-1.7	0.7
343.03	-6.07	-2.1	129.18	23JL	-1.5	-3.3	-1.7	0.7
342.15	-6.40	-2.4	138.79	2AU	-1.2	-3.3	-1.8	0.8
340.14	-6.50	-2.6	148.43	12AU	-0.5	-3.3	-1.8	0.8
337.47	-6.29	-2.6	158.12	22AU	-0.1	-3.3	-1.9	0.8
334.92	-5.79	-2.5	167.86	1SE	0.2	-3.3	-2.0	0.8
333.18	-5.08	-2.2	177.65	11SE	0.3	-3.3	-2.0	0.8
332.64	-4.29	-1.9	187.50	21SE	0.7	-3.4	-2.1	0.8
333.43	-3.50	-1.6	197.41	1OC	2.3	-3.4	-2.2	0.8
335.41	-2.78	-1.3	207.37	11OC	1.4	-3.4	-2.2	0.8
338.40	-2.14	-1.0	217.38	21OC	-0.3	-3.5	-2.3	0.7
342.19	-1.59	-0.7	227.45	31OC	-0.7	-3.5	-2.3	0.7
346.62	-1.11	-0.4	237.56	10NO	-0.6	-3.6	-2.3	0.7
351.53	-0.71	-0.2	247.71	20NO	-0.7	-3.7	-2.3	0.7
356.81	-0.38	0.0	257.88	30NO	-0.8	-3.7	-2.3	0.7
2.38	-0.09	0.2	268.07	10DE	-0.7	-3.8	-2.3	0.7
8.16	0.15	0.4	278.26	20DE	-0.8	-3.9	-2.3	0.8
14.11	0.36	0.6	288.46	30DE	-0.6	-4.0	-2.2	0.8
				1515				
20.17	0.53	0.8	298.63	9JA	0.5	-4.1	-2.2	0.8
26.32	0.67	0.9	308.79	19JA	3.0	-4.2	-2.1	0.8
32.54	0.80	1.1	318.91	29JA	1.1	-4.3	-2.0	0.8
38.79	0.90	1.2	328.99	8FE	0.5	-4.3	-1.9	0.7
45.08	0.98	1.3	339.02	18FE	0.3	-4.3	-1.9	0.7
51.38	1.05	1.4	349.00	28FE	0.0	-4.0	-1.8	0.7
57.68	1.11	1.5	358.92	10MR	-0.3	-3.5	-1.7	0.6
63.99	1.15	1.6	8.78	20MR	-0.9	-3.3	-1.7	0.6
70.30	1.19	1.7	18.59	30MR	-1.7	-3.8	-1.6	0.5
76.60	1.21	1.8	28.34	9AP	-1.2	-4.1	-1.6	0.4
82.89	1.23	1.8	38.05	19AP	-0.1	-4.2	-1.5	0.4
89.18	1.25	1.9	47.70	29AP	1.0	-4.2	-1.5	0.3
95.46	1.25	1.9	57.32	9MY	2.1	-4.1	-1.5	0.3
101.74	1.25	2.0	66.90	19MY	3.6	-4.0	-1.4	0.2
108.01	1.24	2.0	76.46	29MY	2.4	-3.9	-1.4	0.3
114.28	1.23	2.0	85.99	8JN	1.2	-3.8	-1.4	0.4
120.56	1.22	2.0	95.53	18JN	0.2	-3.7	-1.4	0.4
126.84	1.20	2.0	105.06	28JN	-0.8	-3.7	-1.4	0.5
133.14	1.17	2.0	114.60	8JL	-1.6	-3.6	-1.4	0.5
139.45	1.15	2.0	124.17	18JL	-1.2	-3.5	-1.5	0.6
145.78	1.11	2.0	133.75	28JL	-0.4	-3.5	-1.5	0.7
152.13	1.08	2.0	143.37	7AU	0.0	-3.4	-1.5	0.7
158.51	1.04	2.0	153.04	17AU	0.3	-3.4	-1.6	0.7
164.92	0.99	1.9	162.75	27AU	0.5	-3.4	-1.6	0.8
171.37	0.95	1.9	172.51	6SE	1.0	-3.4	-1.7	0.8
177.85	0.89	1.9	182.34	16SE	2.7	-3.4	-1.7	0.8
184.38	0.84	1.9	192.21	26SE	1.2	-3.4	-1.8	0.8
190.95	0.77	1.9	202.14	6OC	-0.5	-3.4	-1.8	0.8
197.56	0.71	1.9	212.13	16OC	-0.8	-3.4	-1.9	0.8
204.23	0.64	1.9	222.17	26OC	-0.8	-3.4	-2.0	0.7
210.94	0.56	1.9	232.26	5NO	-0.8	-3.4	-2.0	0.7
217.71	0.48	1.9	242.39	15NO	-0.6	-3.4	-2.1	0.7
224.53	0.39	1.8	252.54	25NO	-0.6	-3.4	-2.1	0.7
231.40	0.29	1.8	262.72	5DE	-0.6	-3.4	-2.2	0.7
238.32	0.19	1.8	272.92	15DE	-0.4	-3.5	-2.2	0.7
245.30	0.08	1.7	283.11	25DE	0.7	-3.5	-2.2	0.8

♂ LONG	♂ LAT	♂ MAG	⊙ LONG	16.00UT	☿	♀	♃	♄
				1516	MAGNITUDES			
252.33	-0.04	1.6	293.30	4JA	2.8	-3.5	-2.2	0.8
259.41	-0.16	1.6	303.47	14JA	0.8	-3.5	-2.2	0.8
266.54	-0.29	1.5	313.60	24JA	0.3	-3.4	-2.1	0.8
273.72	-0.43	1.4	323.71	3FE	0.1	-3.4	-2.1	0.8
280.94	-0.58	1.4	333.77	13FE	-0.1	-3.4	-2.0	0.8
288.21	-0.73	1.3	343.77	23FE	-0.4	-3.4	-2.0	0.7
295.50	-0.89	1.2	353.72	4MR	-1.0	-3.3	-1.9	0.7
302.83	-1.05	1.1	3.62	14MR	-1.6	-3.3	-1.8	0.7
310.17	-1.21	1.0	13.45	24MR	-1.1	-3.3	-1.7	0.6
317.52	-1.38	0.9	23.24	3AP	-0.0	-3.3	-1.7	0.5
324.87	-1.54	0.9	32.97	13AP	1.3	-3.3	-1.6	0.5
332.19	-1.70	0.8	42.64	23AP	2.8	-3.4	-1.6	0.4
339.48	-1.86	0.7	52.28	3MY	3.2	-3.4	-1.5	0.4
346.70	-2.01	0.6	61.88	13MY	1.8	-3.4	-1.5	0.3
353.85	-2.14	0.5	71.45	23MY	1.0	-3.4	-1.4	0.2
0.87	-2.27	0.4	81.00	2JN	0.1	-3.5	-1.4	0.3
7.74	-2.38	0.3	90.53	12JN	-0.8	-3.5	-1.4	0.3
14.43	-2.47	0.1	100.06	22JN	-1.7	-3.6	-1.3	0.4
20.87	-2.54	0.0	109.60	2JL	-1.2	-3.7	-1.3	0.4
27.01	-2.59	-0.1	119.15	12JL	-0.4	-3.7	-1.3	0.5
32.79	-2.61	-0.2	128.72	22JL	0.1	-3.8	-1.3	0.6
38.10	-2.61	-0.4	138.33	1AU	0.4	-3.9	-1.3	0.6
42.83	-2.57	-0.6	147.97	11AU	0.7	-4.0	-1.3	0.7
46.84	-2.49	-0.7	157.65	21AU	1.4	-4.1	-1.4	0.7
49.96	-2.36	-0.9	167.39	31AU	3.1	-4.2	-1.4	0.7
51.98	-2.17	-1.2	177.18	10SE	1.1	-4.3	-1.4	0.8
52.69	-1.90	-1.4	187.02	20SE	-0.6	-4.3	-1.5	0.8
51.93	-1.54	-1.6	196.93	30SE	-1.0	-4.1	-1.5	0.8
49.75	-1.09	-1.8	206.89	10OC	-0.9	-3.5	-1.6	0.8
46.51	-0.57	-1.9	216.90	20OC	-0.8	-3.2	-1.6	0.8
42.88	-0.02	-1.9	226.97	30OC	-0.5	-3.9	-1.7	0.8
39.72	0.48	-1.7	237.07	9NO	-0.4	-4.3	-1.7	0.7
37.62	0.90	-1.3	247.21	19NO	-0.4	-4.4	-1.8	0.7
36.85	1.22	-1.0	257.39	29NO	-0.3	-4.4	-1.9	0.7
37.42	1.45	-0.7	267.57	9DE	0.9	-4.3	-1.9	0.7
39.12	1.60	-0.4	277.77	19DE	2.5	-4.2	-2.0	0.7
41.76	1.71	-0.1	287.96	29DE	0.5	-4.1	-2.0	0.7
				1517				
45.14	1.77	0.2	298.14	8JA	0.1	-4.0	-2.1	0.8
49.09	1.80	0.4	308.30	18JA	-0.0	-3.9	-2.1	0.8
53.49	1.81	0.6	318.42	28JA	-0.2	-3.8	-2.1	0.8
58.25	1.81	0.8	328.50	7FE	-0.5	-3.7	-2.1	0.8
63.26	1.80	1.0	338.53	17FE	-1.0	-3.6	-2.0	0.8
68.50	1.77	1.1	348.52	27FE	-1.5	-3.5	-2.0	0.8
73.90	1.74	1.3	358.44	9MR	-1.1	-3.5	-2.0	0.7
79.43	1.71	1.4	8.31	19MR	0.0	-3.4	-1.9	0.7
85.08	1.67	1.5	18.12	29MR	1.6	-3.4	-1.8	0.7
90.81	1.62	1.6	27.88	8AP	3.5	-3.4	-1.8	0.6
96.61	1.57	1.7	37.58	18AP	2.4	-3.3	-1.7	0.5
102.48	1.52	1.8	47.24	28AP	1.4	-3.3	-1.6	0.5
108.40	1.47	1.8	56.86	8MY	0.7	-3.3	-1.6	0.4
114.38	1.42	1.9	66.44	18MY	0.0	-3.3	-1.5	0.4
120.40	1.36	1.9	76.00	28MY	-0.9	-3.3	-1.5	0.3
126.46	1.30	2.0	85.54	7JN	-1.7	-3.3	-1.4	0.2
132.57	1.24	2.0	95.07	17JN	-1.2	-3.3	-1.4	0.3
138.72	1.18	2.0	104.60	27JN	-0.3	-3.4	-1.3	0.3
144.92	1.11	2.0	114.14	7JL	0.2	-3.4	-1.3	0.4
151.17	1.05	2.0	123.70	17JL	0.6	-3.4	-1.3	0.5
157.47	0.98	2.0	133.29	27JL	1.0	-3.5	-1.2	0.5
163.83	0.91	2.0	142.91	6AU	1.8	-3.5	-1.2	0.6
170.24	0.83	2.0	152.57	16AU	3.2	-3.5	-1.2	0.6
176.71	0.76	2.0	162.28	26AU	0.9	-3.4	-1.2	0.7
183.24	0.68	1.9	172.04	5SE	-0.6	-3.4	-1.2	0.7
189.84	0.59	1.9	181.85	15SE	-1.1	-3.4	-1.3	0.7
196.50	0.51	1.8	191.73	25SE	-1.1	-3.4	-1.3	0.8
203.24	0.42	1.8	201.66	5OC	-0.7	-3.3	-1.3	0.8
210.05	0.33	1.8	211.64	15OC	-0.4	-3.3	-1.4	0.8
216.93	0.23	1.8	221.68	25OC	-0.3	-3.3	-1.4	0.8
223.88	0.14	1.7	231.76	4NO	-0.3	-3.3	-1.4	0.8
230.92	0.04	1.7	241.89	14NO	-0.1	-3.3	-1.5	0.8
238.02	-0.06	1.7	252.05	24NO	1.1	-3.4	-1.6	0.7
245.20	-0.17	1.7	262.23	4DE	2.2	-3.4	-1.6	0.7
252.46	-0.28	1.7	272.42	14DE	0.3	-3.4	-1.7	0.7
259.78	-0.39	1.6	282.62	24DE	-0.1	-3.5	-1.8	0.7

♂ LONG	♂ LAT	♂ MAG	⊙ LONG	16.00UT	☿	♀	♃	♄
				1518	MAGNITUDES			
267.17	-0.50	1.6	292.81	3JA	-0.1	-3.5	-1.8	0.8
274.63	-0.61	1.6	302.98	13JA	-0.3	-3.5	-1.9	0.8
282.15	-0.72	1.5	313.12	23JA	-0.5	-3.6	-1.9	0.8
289.71	-0.82	1.5	323.22	2FE	-1.0	-3.7	-2.0	0.8
297.33	-0.93	1.5	333.28	12FE	-1.3	-3.7	-2.0	0.8
304.98	-1.03	1.4	343.29	22FE	-1.0	-3.8	-2.0	0.8
312.66	-1.12	1.4	353.24	4MR	0.1	-3.9	-2.0	0.8
320.36	-1.21	1.3	3.14	14MR	2.0	-4.0	-2.0	0.8
328.06	-1.28	1.3	12.99	24MR	3.2	-4.1	-2.0	0.8
335.75	-1.35	1.2	22.77	3AP	1.7	-4.1	-1.9	0.7
343.43	-1.41	1.2	32.50	13AP	1.0	-4.2	-1.9	0.7
351.08	-1.45	1.2	42.18	23AP	0.5	-4.2	-1.8	0.6
358.67	-1.47	1.1	51.82	3MY	-0.0	-4.1	-1.7	0.6
6.21	-1.48	1.1	61.42	13MY	-0.9	-3.7	-1.7	0.5
13.66	-1.48	1.0	70.99	23MY	-1.8	-2.8	-1.6	0.4
21.02	-1.45	1.0	80.53	2JN	-1.2	-3.3	-1.6	0.4
28.26	-1.41	1.0	90.07	12JN	-0.3	-3.9	-1.5	0.3
35.38	-1.35	0.9	99.60	22JN	0.3	-4.2	-1.4	0.3
42.35	-1.28	0.9	109.13	2JL	0.8	-4.2	-1.4	0.3
49.14	-1.18	0.8	118.69	12JL	1.3	-4.1	-1.3	0.4
55.74	-1.07	0.7	128.25	22JL	2.4	-4.1	-1.3	0.4
62.12	-0.94	0.7	137.86	1AU	2.8	-4.0	-1.3	0.5
68.24	-0.79	0.6	147.50	11AU	0.8	-3.9	-1.2	0.5
74.05	-0.61	0.5	157.18	21AU	-0.7	-3.8	-1.2	0.6
79.52	-0.41	0.4	166.91	31AU	-1.2	-3.7	-1.2	0.7
84.55	-0.19	0.3	176.70	10SE	-1.2	-3.7	-1.2	0.7
89.07	0.07	0.1	186.54	20SE	-0.7	-3.6	-1.2	0.7
92.96	0.36	-0.0	196.44	30SE	-0.3	-3.6	-1.2	0.8
96.07	0.70	-0.2	206.41	10OC	-0.2	-3.5	-1.2	0.8
98.23	1.08	-0.4	216.41	20OC	-0.1	-3.5	-1.2	0.8
99.23	1.52	-0.6	226.48	30OC	0.1	-3.4	-1.3	0.8
98.89	1.99	-0.8	236.58	9NO	1.3	-3.4	-1.3	0.8
97.13	2.47	-1.0	246.72	19NO	2.0	-3.4	-1.4	0.8
94.10	2.93	-1.2	256.89	29NO	0.1	-3.4	-1.4	0.8
90.27	3.28	-1.3	267.08	9DE	-0.2	-3.4	-1.5	0.8
86.41	3.50	-1.2	277.28	19DE	-0.3	-3.4	-1.5	0.7
83.25	3.56	-1.0	287.47	29DE	-0.4	-3.3	-1.6	0.7
				1519				
81.29	3.50	-0.7	297.65	8JA	-0.6	-3.3	-1.6	0.8
80.66	3.36	-0.4	307.81	18JA	-1.0	-3.4	-1.7	0.8
81.28	3.18	-0.1	317.93	28JA	-1.2	-3.4	-1.8	0.8
82.98	2.99	0.2	328.02	7FE	-0.9	-3.4	-1.8	0.9
85.54	2.80	0.4	338.05	17FE	0.2	-3.4	-1.9	0.9
88.79	2.62	0.6	348.03	27FE	2.4	-3.4	-1.9	0.9
92.60	2.45	0.8	357.96	9MR	2.5	-3.5	-2.0	0.9
96.83	2.29	1.0	7.83	19MR	1.3	-3.5	-2.0	0.9
101.41	2.14	1.1	17.64	29MR	0.7	-3.5	-2.0	0.8
106.26	2.00	1.3	27.40	8AP	0.4	-3.4	-2.0	0.8
111.35	1.86	1.4	37.11	18AP	-0.1	-3.4	-2.0	0.8
116.62	1.73	1.5	46.77	28AP	-0.9	-3.4	-1.9	0.7
122.05	1.61	1.6	56.39	8MY	-1.8	-3.4	-1.9	0.6
127.62	1.49	1.7	65.97	18MY	-1.2	-3.3	-1.9	0.6
133.31	1.38	1.7	75.53	28MY	-0.3	-3.3	-1.8	0.5
139.12	1.26	1.8	85.07	7JN	0.5	-3.3	-1.7	0.5
145.02	1.15	1.8	94.60	17JN	1.0	-3.3	-1.7	0.4
151.02	1.05	1.8	104.13	27JN	1.8	-3.3	-1.6	0.3
157.12	0.94	1.9	113.68	7JL	3.0	-3.4	-1.5	0.3
163.30	0.83	1.9	123.23	17JL	2.3	-3.4	-1.5	0.4
169.57	0.73	1.9	132.82	27JL	0.6	-3.4	-1.4	0.4
175.93	0.63	1.9	142.44	6AU	-0.7	-3.5	-1.4	0.5
182.38	0.52	1.9	152.09	16AU	-1.4	-3.5	-1.3	0.5
188.92	0.42	1.9	161.80	26AU	-1.2	-3.6	-1.3	0.6
195.55	0.32	1.8	171.56	5SE	-0.6	-3.6	-1.3	0.6
202.27	0.21	1.8	181.37	15SE	-0.2	-3.7	-1.2	0.7
209.09	0.11	1.8	191.25	25SE	-0.0	-3.8	-1.2	0.7
216.00	0.01	1.8	201.17	5OC	0.1	-3.9	-1.2	0.8
223.00	-0.10	1.7	211.16	15OC	0.3	-4.0	-1.2	0.8
230.10	-0.20	1.7	221.19	25OC	1.6	-4.1	-1.2	0.8
237.28	-0.30	1.6	231.28	4NO	1.7	-4.2	-1.2	0.8
244.56	-0.40	1.6	241.40	14NO	-0.1	-4.3	-1.3	0.8
251.93	-0.49	1.5	251.56	24NO	-0.4	-4.4	-1.3	0.8
259.37	-0.59	1.5	261.74	4DE	-0.4	-4.4	-1.3	0.8
266.90	-0.68	1.5	271.93	14DE	-0.5	-4.1	-1.4	0.8
274.50	-0.76	1.5	282.13	24DE	-0.7	-3.6	-1.4	0.8

♂ LONG	LAT	MAG	☉ LONG	16.00UT	☿	♀	♃	♄
				1520	MAGNITUDES			
282.16	-0.84	1.5	292.32	3JA	-0.9	-3.3	-1.5	0.8
289.87	-0.91	1.5	302.48	13JA	-1.0	-3.9	-1.5	0.8
297.64	-0.98	1.4	312.63	23JA	-0.8	-4.3	-1.6	0.8
305.44	-1.04	1.4	322.73	2FE	0.3	-4.3	-1.6	0.9
313.26	-1.08	1.4	332.79	12FE	2.7	-4.3	-1.7	0.9
321.10	-1.12	1.4	342.81	22FE	1.9	-4.2	-1.8	0.9
328.94	-1.15	1.4	352.76	3MR	0.9	-4.1	-1.9	0.9
336.77	-1.16	1.4	2.66	13MR	0.5	-4.0	-1.9	0.9
344.58	-1.16	1.4	12.51	23MR	0.2	-3.9	-2.0	0.9
352.35	-1.14	1.4	22.30	2AP	-0.2	-3.8	-2.0	0.9
0.08	-1.12	1.4	32.03	12AP	-0.9	-3.7	-2.0	0.9
7.74	-1.08	1.4	41.72	22AP	-1.8	-3.6	-2.1	0.9
15.34	-1.03	1.4	51.35	2MY	-1.2	-3.5	-2.1	0.8
22.85	-0.96	1.4	60.96	12MY	-0.2	-3.5	-2.1	0.8
30.28	-0.89	1.4	70.53	22MY	0.6	-3.4	-2.0	0.7
37.61	-0.80	1.4	80.08	1JN	1.4	-3.4	-2.0	0.6
44.83	-0.70	1.4	89.61	11JN	2.3	-3.4	-1.9	0.6
51.95	-0.59	1.4	99.15	21JN	3.3	-3.3	-1.9	0.5
58.94	-0.48	1.3	108.68	1JL	1.9	-3.3	-1.8	0.4
65.80	-0.35	1.3	118.22	11JL	0.5	-3.3	-1.8	0.4
72.52	-0.21	1.3	127.80	21JL	-0.8	-3.3	-1.7	0.4
79.10	-0.07	1.3	137.39	31JL	-1.5	-3.3	-1.6	0.4
85.51	0.09	1.2	147.03	10AU	-1.2	-3.3	-1.6	0.5
91.75	0.25	1.2	156.71	20AU	-0.5	-3.4	-1.5	0.5
97.78	0.43	1.1	166.44	30AU	-0.1	-3.4	-1.5	0.6
103.60	0.62	1.1	176.22	9SE	0.1	-3.4	-1.4	0.7
109.16	0.82	1.0	186.07	19SE	0.2	-3.4	-1.4	0.7
114.42	1.04	0.9	195.96	29SE	0.6	-3.5	-1.4	0.8
119.33	1.28	0.8	205.92	9OC	2.0	-3.5	-1.3	0.8
123.80	1.55	0.7	215.93	19OC	1.6	-3.5	-1.3	0.8
127.76	1.84	0.5	225.98	29OC	-0.2	-3.5	-1.3	0.9
131.08	2.16	0.3	236.09	8NO	-0.6	-3.4	-1.3	0.9
133.61	2.52	0.1	246.23	18NO	-0.6	-3.4	-1.3	0.9
135.18	2.91	-0.1	256.39	28NO	-0.6	-3.4	-1.3	0.9
135.59	3.33	-0.3	266.58	8DE	-0.8	-3.4	-1.3	0.9
134.68	3.75	-0.5	276.78	18DE	-0.8	-3.3	-1.3	0.9
132.44	4.13	-0.8	286.97	28DE	-0.8	-3.3	-1.4	0.9
				1521				
129.08	4.40	-1.0	297.15	7JA	-0.7	-3.3	-1.4	0.9
125.13	4.52	-1.0	307.31	17JA	0.4	-3.3	-1.5	0.9
121.35	4.45	-0.9	317.44	27JA	3.0	-3.4	-1.5	0.9
118.37	4.24	-0.6	327.52	6FE	1.4	-3.4	-1.6	0.9
116.62	3.93	-0.4	337.57	16FE	0.6	-3.4	-1.6	1.0
116.17	3.58	-0.1	347.55	26FE	0.4	-3.4	-1.7	1.0
116.91	3.24	0.1	357.48	8MR	0.1	-3.4	-1.8	1.0
118.68	2.91	0.3	7.36	18MR	-0.3	-3.5	-1.8	1.0
121.29	2.60	0.6	17.17	28MR	-0.9	-3.5	-1.9	1.0
124.58	2.32	0.7	26.93	7AP	-1.8	-3.6	-2.0	1.0
128.41	2.07	0.9	36.65	17AP	-1.2	-3.6	-2.0	1.0
132.68	1.84	1.0	46.31	27AP	-0.2	-3.7	-2.1	1.0
137.32	1.63	1.2	55.93	7MY	0.8	-3.8	-2.1	0.9
142.25	1.43	1.3	65.52	17MY	1.8	-3.9	-2.2	0.9
147.44	1.25	1.4	75.07	27MY	3.1	-4.0	-2.2	0.8
152.84	1.08	1.4	84.62	6JN	2.8	-4.1	-2.2	0.8
158.44	0.92	1.5	94.15	16JN	1.5	-4.1	-2.2	0.7
164.21	0.76	1.5	103.68	26JN	0.3	-4.2	-2.1	0.6
170.14	0.62	1.6	113.22	6JL	-0.8	-4.2	-2.1	0.6
176.22	0.48	1.6	122.78	16JL	-1.6	-4.0	-2.1	0.5
182.44	0.35	1.6	132.36	26JL	-1.2	-3.5	-2.0	0.5
188.79	0.22	1.7	141.97	5AU	-0.5	-3.2	-1.9	0.5
195.27	0.09	1.7	151.63	15AU	-0.0	-3.7	-1.9	0.5
201.87	-0.03	1.7	161.33	25AU	0.2	-4.1	-1.8	0.6
208.60	-0.14	1.7	171.09	4SE	0.4	-4.3	-1.7	0.6
215.45	-0.25	1.7	180.90	14SE	0.8	-4.3	-1.7	0.7
222.41	-0.36	1.6	190.76	24SE	2.4	-4.2	-1.6	0.7
229.49	-0.46	1.6	200.69	4OC	1.4	-4.1	-1.6	0.8
236.68	-0.56	1.6	210.67	14OC	-0.4	-4.0	-1.5	0.8
243.98	-0.65	1.6	220.70	24OC	-0.7	-3.9	-1.5	0.9
251.38	-0.73	1.5	230.78	3NO	-0.7	-3.9	-1.5	0.9
258.87	-0.81	1.5	240.91	13NO	-0.7	-3.8	-1.4	0.9
266.45	-0.88	1.5	251.06	23NO	-0.7	-3.7	-1.4	1.0
274.11	-0.93	1.5	261.24	3DE	-0.7	-3.6	-1.4	1.0
281.84	-0.99	1.4	271.43	13DE	-0.7	-3.6	-1.4	1.0
289.63	-1.03	1.4	281.63	23DE	-0.5	-3.5	-1.4	1.0

♂ LONG	LAT	MAG	☉ LONG	16.00UT	☿	♀	♃	♄
				1522	MAGNITUDES			
297.47	-1.06	1.4	291.82	2JA	0.5	-3.5	-1.4	1.0
305.34	-1.07	1.3	301.99	12JA	3.0	-3.4	-1.4	1.0
313.23	-1.08	1.3	312.14	22JA	1.0	-3.4	-1.5	1.0
321.14	-1.08	1.3	322.25	1FE	0.4	-3.4	-1.5	1.0
329.03	-1.06	1.3	332.31	11FE	0.2	-3.3	-1.5	1.0
336.91	-1.03	1.3	342.32	21FE	0.0	-3.3	-1.6	1.1
344.77	-0.99	1.4	352.29	3MR	-0.3	-3.3	-1.6	1.1
352.57	-0.94	1.4	2.19	13MR	-0.9	-3.3	-1.7	1.1
0.33	-0.88	1.4	12.04	23MR	-1.7	-3.3	-1.8	1.2
8.02	-0.81	1.4	21.83	2AP	-1.2	-3.3	-1.8	1.2
15.64	-0.74	1.5	31.56	12AP	-0.1	-3.3	-1.9	1.2
23.19	-0.65	1.5	41.25	22AP	1.1	-3.4	-2.0	1.2
30.65	-0.56	1.5	50.89	2MY	2.4	-3.4	-2.0	1.1
38.02	-0.46	1.5	60.49	12MY	3.7	-3.4	-2.1	1.1
45.30	-0.35	1.6	70.06	22MY	2.2	-3.5	-2.2	1.1
52.48	-0.24	1.6	79.62	1JN	1.1	-3.5	-2.2	1.0
59.57	-0.13	1.6	89.15	11JN	0.2	-3.4	-2.3	1.0
66.56	-0.02	1.6	98.68	21JN	-0.8	-3.4	-2.3	0.9
73.44	0.10	1.6	108.22	1JL	-1.6	-3.4	-2.3	0.8
80.23	0.23	1.6	117.76	11JL	-1.2	-3.4	-2.3	0.8
86.91	0.35	1.6	127.33	21JL	-0.4	-3.3	-2.3	0.7
93.48	0.48	1.6	136.93	31JL	0.0	-3.3	-2.3	0.6
99.94	0.60	1.6	146.56	10AU	0.3	-3.3	-2.2	0.6
106.29	0.74	1.6	156.24	20AU	0.6	-3.3	-2.2	0.6
112.51	0.87	1.6	165.97	30AU	1.1	-3.3	-2.1	0.7
118.59	1.01	1.5	175.75	9SE	2.8	-3.3	-2.1	0.7
124.54	1.16	1.5	185.58	19SE	1.2	-3.4	-2.0	0.8
130.32	1.31	1.4	195.48	29SE	-0.5	-3.4	-1.9	0.8
135.91	1.47	1.4	205.43	9OC	-0.9	-3.4	-1.9	0.9
141.29	1.63	1.3	215.43	19OC	-0.9	-3.5	-1.8	0.9
146.42	1.81	1.2	225.49	29OC	-0.9	-3.5	-1.7	1.0
151.24	2.01	1.1	235.59	8NO	-0.6	-3.6	-1.7	1.0
155.71	2.21	0.9	245.73	18NO	-0.5	-3.7	-1.6	1.0
159.72	2.44	0.8	255.90	28NO	-0.5	-3.7	-1.6	1.1
163.19	2.69	0.6	266.08	8DE	-0.4	-3.8	-1.6	1.1
165.98	2.95	0.4	276.28	18DE	0.7	-3.9	-1.5	1.1
167.93	3.24	0.2	286.48	28DE	2.7	-4.0	-1.5	1.1
				1523				
168.85	3.54	-0.1	296.66	7JA	0.7	-4.1	-1.5	1.1
168.59	3.82	-0.3	306.82	17JA	0.2	-4.2	-1.5	1.1
167.01	4.07	-0.6	316.95	27JA	0.1	-4.3	-1.5	1.1
164.22	4.22	-0.8	327.03	6FE	-0.1	-4.3	-1.5	1.1
160.56	4.23	-1.0	337.08	16FE	-0.4	-4.3	-1.5	1.1
156.65	4.08	-1.0	347.07	26FE	-0.9	-4.0	-1.6	1.2
153.22	3.78	-0.8	357.00	8MR	-1.5	-3.5	-1.6	1.2
150.80	3.38	-0.6	6.88	18MR	-1.1	-3.3	-1.6	1.3
149.66	2.95	-0.3	16.70	28MR	-0.0	-3.9	-1.7	1.3
149.80	2.52	-0.1	26.46	7AP	1.4	-4.2	-1.7	1.3
151.07	2.13	0.1	36.18	17AP	3.1	-4.2	-1.8	1.3
153.31	1.77	0.3	45.84	27AP	2.8	-4.2	-1.9	1.3
156.34	1.45	0.5	55.46	7MY	1.6	-4.1	-1.9	1.3
160.02	1.16	0.6	65.05	17MY	0.9	-4.0	-2.0	1.3
164.22	0.90	0.8	74.61	27MY	0.1	-3.9	-2.1	1.3
168.87	0.66	0.9	84.15	6JN	-0.8	-3.8	-2.1	1.2
173.88	0.45	1.0	93.69	16JN	-1.7	-3.7	-2.2	1.2
179.21	0.25	1.1	103.22	26JN	-1.2	-3.6	-2.3	1.1
184.81	0.08	1.1	112.75	6JL	-0.4	-3.6	-2.3	1.0
190.66	-0.09	1.2	122.31	16JL	0.1	-3.5	-2.4	1.0
196.73	-0.24	1.2	131.89	26JL	0.5	-3.5	-2.4	0.9
202.99	-0.38	1.3	141.51	5AU	0.8	-3.4	-2.4	0.8
209.45	-0.51	1.3	151.17	15AU	1.5	-3.4	-2.4	0.8
216.08	-0.62	1.3	160.87	25AU	3.2	-3.4	-2.4	0.7
222.86	-0.73	1.4	170.62	4SE	1.1	-3.4	-2.4	0.8
229.80	-0.83	1.4	180.43	14SE	-0.6	-3.4	-2.3	0.8
236.88	-0.91	1.4	190.29	24SE	-1.0	-3.4	-2.3	0.9
244.10	-0.98	1.4	200.21	4OC	-1.0	-3.4	-2.2	0.9
251.44	-1.05	1.4	210.19	14OC	-0.8	-3.4	-2.2	1.0
258.89	-1.10	1.4	220.22	24OC	-0.5	-3.4	-2.1	1.0
266.45	-1.14	1.4	230.29	3NO	-0.4	-3.4	-2.0	1.1
274.09	-1.16	1.4	240.42	13NO	-0.4	-3.4	-2.0	1.1
281.82	-1.18	1.4	250.57	23NO	-0.2	-3.4	-1.9	1.2
289.61	-1.18	1.4	260.74	3DE	0.9	-3.4	-1.8	1.2
297.45	-1.17	1.4	270.94	13DE	2.4	-3.5	-1.8	1.2
305.32	-1.15	1.4	281.13	23DE	0.4	-3.5	-1.7	1.3

♂ LONG	♂ LAT	♂ MAG	☉ LONG	16.00UT	☿	♀	♃	♄
				1524	MAGNITUDES			
313.22	-1.11	1.4	291.32	2JA	0.0	-3.5	-1.7	1.3
321.12	-1.07	1.4	301.50	12JA	-0.1	-3.5	-1.6	1.3
329.01	-1.01	1.4	311.64	22JA	-0.2	-3.4	-1.6	1.3
336.88	-0.94	1.4	321.75	1FE	-0.5	-3.4	-1.6	1.3
344.71	-0.87	1.4	331.82	11FE	-1.0	-3.4	-1.6	1.3
352.50	-0.79	1.4	341.83	21FE	-1.4	-3.4	-1.6	1.3
0.23	-0.70	1.4	351.80	2MR	-1.1	-3.3	-1.6	1.3
7.89	-0.60	1.4	1.71	12MR	0.0	-3.3	-1.6	1.4
15.48	-0.51	1.4	11.55	22MR	1.7	-3.3	-1.6	1.4
22.99	-0.40	1.4	21.35	1AP	3.6	-3.3	-1.6	1.5
30.42	-0.30	1.5	31.09	11AP	2.1	-3.3	-1.7	1.4
37.77	-0.19	1.5	40.78	21AP	1.2	-3.4	-1.7	1.4
45.02	-0.09	1.6	50.42	1MY	0.6	-3.4	-1.7	1.4
52.19	0.02	1.6	60.03	11MY	0.0	-3.4	-1.8	1.4
59.27	0.12	1.7	69.60	21MY	-0.8	-3.4	-1.8	1.3
66.26	0.23	1.7	79.15	31MY	-1.7	-3.5	-1.9	1.3
73.17	0.33	1.7	88.69	10JN	-1.2	-3.5	-2.0	1.2
80.00	0.44	1.8	98.22	20JN	-0.3	-3.6	-2.0	1.2
86.74	0.54	1.8	107.75	30JN	0.2	-3.7	-2.1	1.2
93.41	0.63	1.8	117.30	10JL	0.7	-3.7	-2.2	1.1
100.00	0.73	1.8	126.87	20JL	1.1	-3.8	-2.2	1.1
106.51	0.83	1.9	136.46	30JL	2.0	-3.9	-2.3	1.0
112.95	0.92	1.9	146.10	9AU	3.1	-4.0	-2.4	1.0
119.32	1.01	1.9	155.77	19AU	0.9	-4.1	-2.4	0.9
125.61	1.10	1.9	165.50	29AU	-0.6	-4.2	-2.4	0.9
131.82	1.20	1.8	175.28	8SE	-1.2	-4.3	-2.5	0.9
137.95	1.29	1.8	185.11	18SE	-1.1	-4.3	-2.5	1.0
143.99	1.38	1.8	195.00	28SE	-0.7	-4.0	-2.5	1.0
149.93	1.47	1.7	204.95	8OC	-0.4	-3.5	-2.4	1.1
155.76	1.56	1.7	214.95	18OC	-0.3	-3.3	-2.4	1.1
161.47	1.66	1.6	225.01	28OC	-0.2	-3.9	-2.3	1.2
167.04	1.75	1.5	235.11	7NO	-0.0	-4.3	-2.3	1.2
172.44	1.85	1.4	245.24	17NO	1.1	-4.4	-2.2	1.3
177.64	1.96	1.3	255.41	27NO	2.2	-4.3	-2.1	1.3
182.60	2.06	1.2	265.59	7DE	0.2	-4.3	-2.1	1.3
187.27	2.17	1.0	275.78	17DE	-0.1	-4.2	-2.0	1.4
191.58	2.29	0.9	285.98	27DE	-0.2	-4.1	-1.9	1.4
				1525				
195.45	2.41	0.7	296.16	6JA	-0.3	-4.0	-1.9	1.4
198.78	2.53	0.5	306.32	16JA	-0.5	-3.9	-1.8	1.4
201.41	2.65	0.2	316.46	26JA	-1.0	-3.8	-1.7	1.4
203.20	2.76	-0.0	326.55	5FE	-1.3	-3.7	-1.7	1.3
203.96	2.84	-0.3	336.59	15FE	-1.0	-3.6	-1.7	1.3
203.53	2.88	-0.6	346.58	25FE	0.1	-3.5	-1.6	1.2
201.81	2.85	-0.9	356.52	7MR	2.1	-3.5	-1.6	1.2
198.95	2.72	-1.1	6.40	17MR	2.9	-3.4	-1.6	1.2
195.31	2.46	-1.3	16.23	27MR	1.5	-3.4	-1.6	1.2
191.58	2.09	-1.2	25.99	6AP	0.9	-3.4	-1.6	1.2
188.45	1.66	-1.1	35.71	16AP	0.5	-3.3	-1.6	1.2
186.42	1.20	-0.9	45.38	26AP	-0.1	-3.3	-1.6	1.2
185.73	0.77	-0.7	55.00	6MY	-0.8	-3.3	-1.6	1.1
186.32	0.39	-0.4	64.59	16MY	-1.8	-3.3	-1.6	1.1
188.06	0.05	-0.2	74.15	26MY	-1.2	-3.3	-1.7	1.1
190.78	-0.25	-0.1	83.70	5JN	-0.3	-3.3	-1.7	1.0
194.29	-0.50	0.1	93.23	15JN	0.4	-3.3	-1.8	1.0
198.47	-0.71	0.2	102.76	25JN	0.9	-3.4	-1.8	1.0
203.21	-0.89	0.4	112.30	5JL	1.5	-3.4	-1.9	0.9
208.40	-1.05	0.5	121.85	15JL	2.6	-3.4	-1.9	0.9
213.99	-1.18	0.6	131.43	25JL	2.7	-3.5	-2.0	0.8
219.91	-1.29	0.6	141.04	4AU	0.7	-3.5	-2.1	0.8
226.13	-1.39	0.7	150.69	14AU	-0.7	-3.5	-2.1	0.7
232.60	-1.46	0.8	160.39	24AU	-1.3	-3.4	-2.2	0.7
239.31	-1.51	0.8	170.14	3SE	-1.2	-3.4	-2.3	0.6
246.21	-1.55	0.9	179.94	13SE	-0.6	-3.4	-2.3	0.6
253.29	-1.57	0.9	189.81	23SE	-0.3	-3.4	-2.4	0.6
260.54	-1.57	1.0	199.72	3OC	-0.1	-3.3	-2.4	0.7
267.91	-1.56	1.0	209.70	13OC	-0.0	-3.3	-2.4	0.8
275.41	-1.53	1.1	219.73	23OC	0.2	-3.3	-2.4	0.8
283.00	-1.48	1.1	229.80	2NO	1.4	-3.3	-2.4	0.9
290.67	-1.43	1.1	239.92	12NO	2.0	-3.3	-2.4	1.0
298.41	-1.36	1.2	250.07	22NO	0.0	-3.4	-2.3	1.0
306.18	-1.28	1.2	260.25	2DE	-0.3	-3.4	-2.3	1.1
313.97	-1.19	1.3	270.44	12DE	-0.3	-3.4	-2.2	1.1
321.78	-1.09	1.3	280.64	22DE	-0.4	-3.5	-2.2	1.1

♂ LONG	♂ LAT	♂ MAG	☉ LONG	16.00UT	☿	♀	♃	♄
				1526	MAGNITUDES			
329.57	-0.98	1.3	290.83	1JA	-0.6	-3.5	-2.1	1.1
337.34	-0.87	1.4	301.00	11JA	-1.0	-3.5	-2.0	1.1
345.08	-0.76	1.4	311.15	21JA	-1.1	-3.6	-1.9	1.1
352.76	-0.64	1.5	321.26	31JA	-0.9	-3.7	-1.9	1.1
0.39	-0.52	1.5	331.33	10FE	0.2	-3.7	-1.8	1.1
7.95	-0.40	1.5	341.35	20FE	2.4	-3.8	-1.7	1.0
15.44	-0.29	1.6	351.31	2MR	2.3	-3.9	-1.7	1.0
22.85	-0.17	1.6	1.23	12MR	1.1	-4.0	-1.6	0.9
30.19	-0.06	1.6	11.08	22MR	0.7	-4.1	-1.6	0.9
37.44	0.05	1.7	20.87	1AP	0.3	-4.2	-1.6	0.9
44.61	0.16	1.7	30.62	11AP	-0.1	-4.2	-1.5	0.9
51.69	0.26	1.7	40.31	21AP	-0.9	-4.2	-1.5	0.9
58.70	0.36	1.7	49.95	1MY	-1.8	-4.1	-1.5	0.9
65.63	0.45	1.7	59.56	11MY	-1.2	-3.7	-1.5	0.9
72.49	0.54	1.7	69.14	21MY	-0.3	-2.8	-1.5	0.9
79.28	0.62	1.7	78.69	31MY	0.5	-3.4	-1.5	0.9
86.00	0.70	1.8	88.23	10JN	1.1	-3.9	-1.5	0.8
92.66	0.78	1.8	97.76	20JN	2.0	-4.2	-1.6	0.8
99.26	0.85	1.9	107.29	30JN	3.2	-4.2	-1.6	0.8
105.80	0.92	1.9	116.84	10JL	2.2	-4.1	-1.6	0.7
112.30	0.98	1.9	126.40	20JL	0.6	-4.1	-1.7	0.7
118.75	1.04	2.0	136.00	30JL	-0.7	-4.0	-1.7	0.6
125.15	1.09	2.0	145.63	9AU	-1.4	-3.9	-1.8	0.6
131.51	1.15	2.0	155.30	19AU	-1.2	-3.8	-1.8	0.5
137.84	1.19	2.0	165.02	29AU	-0.6	-3.7	-1.9	0.4
144.12	1.24	2.0	174.80	8SE	-0.2	-3.7	-1.9	0.4
150.37	1.28	2.0	184.63	18SE	-0.0	-3.6	-2.0	0.4
156.57	1.32	2.0	194.52	28SE	0.1	-3.6	-2.1	0.3
162.74	1.36	1.9	204.47	8OC	0.4	-3.5	-2.1	0.4
168.86	1.39	1.9	214.47	18OC	1.7	-3.5	-2.2	0.4
174.94	1.42	1.8	224.51	28OC	1.8	-3.4	-2.2	0.5
180.96	1.44	1.8	234.61	7NO	-0.1	-3.4	-2.3	0.6
186.91	1.47	1.7	244.75	17NO	-0.5	-3.4	-2.3	0.6
192.79	1.48	1.6	254.91	27NO	-0.5	-3.4	-2.3	0.7
198.58	1.50	1.5	265.10	7DE	-0.5	-3.4	-2.3	0.8
204.26	1.50	1.4	275.29	17DE	-0.7	-3.4	-2.3	0.8
209.82	1.50	1.3	285.49	27DE	-0.9	-3.3	-2.2	0.8
				1527				
215.23	1.50	1.2	295.67	6JA	-0.9	-3.3	-2.2	0.8
220.45	1.48	1.0	305.83	16JA	-0.8	-3.4	-2.1	0.9
225.45	1.44	0.8	315.96	26JA	0.3	-3.4	-2.1	0.9
230.16	1.39	0.7	326.06	5FE	2.8	-3.4	-2.0	0.9
234.52	1.32	0.5	336.10	15FE	1.7	-3.4	-1.9	0.8
238.45	1.22	0.2	346.10	25FE	0.8	-3.4	-1.8	0.8
241.82	1.08	-0.0	356.04	7MR	0.5	-3.5	-1.8	0.8
244.51	0.90	-0.3	5.92	17MR	0.2	-3.5	-1.7	0.7
246.35	0.65	-0.6	15.74	27MR	-0.2	-3.5	-1.6	0.7
247.13	0.33	-0.9	25.52	6AP	-0.9	-3.4	-1.6	0.6
246.73	-0.07	-1.2	35.23	16AP	-1.8	-3.4	-1.5	0.7
245.08	-0.55	-1.5	44.90	26AP	-1.2	-3.4	-1.5	0.7
242.33	-1.08	-1.8	54.53	6MY	-0.2	-3.4	-1.5	0.7
238.99	-1.63	-1.9	64.12	16MY	0.7	-3.3	-1.4	0.7
235.74	-2.12	-1.8	73.68	26MY	1.5	-3.3	-1.4	0.7
233.28	-2.51	-1.7	83.23	5JN	2.6	-3.3	-1.4	0.7
232.07	-2.79	-1.5	92.76	15JN	3.2	-3.3	-1.4	0.6
232.23	-2.97	-1.2	102.29	25JN	1.7	-3.3	-1.4	0.6
233.72	-3.06	-1.0	111.83	5JL	0.5	-3.4	-1.4	0.6
236.36	-3.09	-0.8	121.38	15JL	-0.8	-3.4	-1.4	0.6
239.95	-3.07	-0.6	130.96	25JL	-1.5	-3.4	-1.4	0.5
244.34	-3.02	-0.5	140.58	4AU	-1.2	-3.5	-1.5	0.5
249.37	-2.94	-0.3	150.22	14AU	-0.5	-3.5	-1.5	0.4
254.92	-2.83	-0.2	159.92	24AU	-0.1	-3.6	-1.5	0.3
260.91	-2.71	-0.0	169.67	3SE	0.1	-3.6	-1.6	0.3
267.25	-2.57	0.1	179.47	13SE	0.3	-3.7	-1.6	0.2
273.87	-2.42	0.2	189.33	23SE	0.6	-3.8	-1.7	0.2
280.72	-2.26	0.3	199.25	3OC	2.1	-3.9	-1.8	0.1
287.77	-2.08	0.4	209.21	13OC	1.6	-4.0	-1.8	0.1
294.95	-1.90	0.5	219.24	23OC	-0.3	-4.1	-1.9	0.1
302.25	-1.72	0.6	229.31	2NO	-0.6	-4.2	-2.0	0.2
309.64	-1.53	0.7	239.43	12NO	-0.6	-4.3	-2.0	0.3
317.07	-1.34	0.8	249.58	22NO	-0.7	-4.4	-2.1	0.4
324.54	-1.16	0.9	259.76	2DE	-0.8	-4.3	-2.1	0.4
332.01	-0.98	1.0	269.94	12DE	-0.7	-4.1	-2.2	0.5
339.48	-0.80	1.1	280.14	22DE	-0.8	-3.5	-2.2	0.5

♂ LONG	♂ LAT	♂ MAG	☉ LONG	16.00UT 1528	☿	♀	♃	♄
					MAGNITUDES			
346.93	-0.63	1.2	290.33	1JA	-0.6	-3.3	-2.2	0.6
354.34	-0.47	1.3	300.51	11JA	0.4	-4.0	-2.2	0.6
1.70	-0.31	1.3	310.66	21JA	3.0	-4.3	-2.1	0.6
9.02	-0.17	1.4	320.77	31JA	1.3	-4.4	-2.1	0.6
16.27	-0.03	1.5	330.84	10FE	0.5	-4.3	-2.0	0.6
23.46	0.10	1.5	340.86	20FE	0.3	-4.2	-2.0	0.6
30.59	0.22	1.6	350.83	1MR	0.1	-4.1	-1.9	0.6
37.64	0.33	1.7	0.74	11MR	-0.3	-4.0	-1.8	0.6
44.63	0.43	1.7	10.60	21MR	-0.9	-3.9	-1.8	0.6
51.55	0.53	1.8	20.40	31MR	-1.7	-3.8	-1.7	0.5
58.40	0.61	1.8	30.14	10AP	-1.2	-3.7	-1.6	0.5
65.20	0.69	1.8	39.84	20AP	-0.2	-3.6	-1.6	0.5
71.93	0.77	1.9	49.49	30AP	0.9	-3.5	-1.5	0.5
78.60	0.83	1.9	59.10	10MY	2.0	-3.5	-1.5	0.5
85.23	0.89	1.9	68.68	20MY	3.4	-3.4	-1.4	0.5
91.80	0.94	1.9	78.23	30MY	2.6	-3.4	-1.4	0.5
98.33	0.99	1.9	87.77	9JN	1.4	-3.4	-1.4	0.5
104.83	1.03	1.9	97.30	19JN	0.3	-3.3	-1.3	0.5
111.29	1.06	1.9	106.83	29JN	-0.8	-3.3	-1.3	0.5
117.73	1.09	1.9	116.38	9JL	-1.6	-3.3	-1.3	0.5
124.14	1.12	1.9	125.94	19JL	-1.2	-3.3	-1.3	0.4
130.53	1.14	2.0	135.54	29JL	-0.5	-3.3	-1.3	0.4
136.90	1.16	2.0	145.16	8AU	-0.0	-3.3	-1.3	0.4
143.26	1.17	2.0	154.84	18AU	0.3	-3.4	-1.3	0.3
149.62	1.17	2.0	164.56	28AU	0.5	-3.4	-1.3	0.2
155.96	1.18	2.0	174.33	7SE	0.9	-3.4	-1.4	0.2
162.31	1.18	2.0	184.16	17SE	2.5	-3.4	-1.4	0.1
168.65	1.17	2.0	194.04	27SE	1.4	-3.5	-1.5	0.1
175.00	1.16	2.0	203.98	7OC	-0.4	-3.5	-1.5	-0.0
181.34	1.14	2.0	213.98	17OC	-0.8	-3.5	-1.6	-0.1
187.69	1.12	2.0	224.03	27OC	-0.8	-3.5	-1.6	-0.1
194.03	1.09	1.9	234.12	6NO	-0.8	-3.4	-1.7	-0.0
200.38	1.06	1.9	244.26	16NO	-0.7	-3.4	-1.7	0.0
206.72	1.02	1.8	254.42	26NO	-0.6	-3.4	-1.8	0.1
213.05	0.97	1.7	264.60	6DE	-0.6	-3.4	-1.9	0.2
219.37	0.91	1.7	274.80	16DE	-0.5	-3.3	-1.9	0.3
225.68	0.84	1.6	284.99	26DE	0.5	-3.3	-2.0	0.3
				1529				
231.97	0.76	1.5	295.17	5JA	2.9	-3.3	-2.0	0.4
238.23	0.67	1.4	305.34	15JA	0.9	-3.3	-2.1	0.4
244.45	0.56	1.3	315.47	25JA	0.3	-3.4	-2.1	0.4
250.62	0.43	1.1	325.57	4FE	0.2	-3.4	-2.1	0.5
256.74	0.28	1.0	335.62	14FE	-0.0	-3.4	-2.1	0.5
262.78	0.11	0.8	345.61	24FE	-0.3	-3.4	-2.0	0.5
268.73	-0.09	0.7	355.56	6MR	-0.9	-3.5	-2.0	0.5
274.56	-0.33	0.5	5.45	16MR	-1.6	-3.5	-1.9	0.5
280.23	-0.59	0.3	15.27	26MR	-1.2	-3.5	-1.9	0.4
285.71	-0.91	0.1	25.04	5AP	-0.1	-3.6	-1.8	0.4
290.94	-1.26	-0.1	34.77	15AP	1.1	-3.6	-1.7	0.4
295.84	-1.68	-0.4	44.44	25AP	2.6	-3.7	-1.7	0.3
300.33	-2.16	-0.6	54.07	5MY	3.4	-3.8	-1.6	0.3
304.27	-2.70	-0.9	63.66	15MY	2.0	-3.9	-1.5	0.3
307.52	-3.32	-1.2	73.22	25MY	1.0	-4.0	-1.5	0.4
309.90	-4.00	-1.5	82.77	4JN	0.2	-4.1	-1.4	0.4
311.20	-4.74	-1.8	92.31	14JN	-0.8	-4.1	-1.4	0.4
311.29	-5.48	-2.1	101.83	24JN	-1.6	-4.2	-1.3	0.4
310.13	-6.14	-2.4	111.37	4JL	-1.2	-4.2	-1.3	0.4
307.95	-6.60	-2.6	120.93	14JL	-0.4	-3.9	-1.3	0.4
305.33	-6.78	-2.6	130.50	24JL	0.1	-3.5	-1.2	0.4
302.97	-6.64	-2.5	140.11	3AU	0.4	-3.2	-1.2	0.3
301.53	-6.23	-2.2	149.76	13AU	0.7	-3.8	-1.2	0.3
301.37	-5.66	-2.0	159.45	23AU	1.3	-4.1	-1.2	0.3
302.53	-5.01	-1.7	169.20	2SE	2.9	-4.3	-1.2	0.2
304.90	-4.35	-1.4	179.00	12SE	1.2	-4.3	-1.2	0.1
308.28	-3.72	-1.1	188.85	22SE	-0.5	-4.2	-1.3	0.1
312.46	-3.14	-0.9	198.76	2OC	-0.9	-4.1	-1.3	0.0
317.28	-2.61	-0.7	208.73	12OC	-0.9	-4.0	-1.3	-0.1
322.58	-2.13	-0.4	218.75	22OC	-0.8	-3.9	-1.4	-0.1
328.26	-1.70	-0.2	228.83	1NO	-0.6	-3.9	-1.4	-0.2
334.23	-1.32	-0.0	238.94	11NO	-0.5	-3.8	-1.5	-0.2
340.40	-0.99	0.2	249.09	21NO	-0.5	-3.7	-1.5	-0.1
346.74	-0.69	0.4	259.26	1DE	-0.3	-3.6	-1.6	-0.1
353.19	-0.42	0.5	269.45	11DE	0.7	-3.6	-1.6	0.0
359.72	-0.19	0.7	279.65	21DE	2.7	-3.5	-1.7	0.1
6.30	0.02	0.8	289.84	31DE	0.6	-3.5	-1.8	0.1

♂ LONG	♂ LAT	♂ MAG	☉ LONG	16.00UT 1530	☿	♀	♃	♄
					MAGNITUDES			
12.92	0.20	1.0	300.02	10JA	0.1	-3.4	-1.8	0.2
19.55	0.36	1.1	310.17	20JA	0.0	-3.4	-1.9	0.3
26.18	0.49	1.2	320.28	30JA	-0.1	-3.4	-1.9	0.3
32.81	0.61	1.3	330.36	9FE	-0.4	-3.3	-2.0	0.3
39.41	0.72	1.4	340.38	19FE	-0.9	-3.3	-2.0	0.4
46.00	0.81	1.5	350.36	1MR	-1.5	-3.3	-2.0	0.4
52.56	0.89	1.6	0.27	11MR	-1.1	-3.3	-2.0	0.4
59.09	0.95	1.7	10.13	21MR	-0.1	-3.3	-2.0	0.4
65.59	1.01	1.8	19.93	31MR	1.4	-3.3	-2.0	0.4
72.07	1.05	1.8	29.68	10AP	3.3	-3.3	-1.9	0.3
78.52	1.09	1.9	39.37	20AP	2.6	-3.4	-1.9	0.3
84.94	1.12	1.9	49.03	30AP	1.5	-3.4	-1.8	0.3
91.33	1.15	1.9	58.64	10MY	0.8	-3.4	-1.7	0.2
97.71	1.16	2.0	68.21	20MY	0.1	-3.5	-1.7	0.2
104.07	1.18	2.0	77.77	30MY	-0.8	-3.5	-1.6	0.3
110.42	1.18	2.0	87.31	9JN	-1.7	-3.4	-1.5	0.3
116.76	1.18	2.0	96.83	19JN	-1.2	-3.4	-1.5	0.3
123.09	1.18	2.0	106.37	29JN	-0.4	-3.4	-1.4	0.3
129.42	1.17	2.0	115.91	9JL	0.2	-3.4	-1.4	0.3
135.76	1.16	2.0	125.47	19JL	0.5	-3.3	-1.3	0.3
142.10	1.14	2.0	135.07	29JL	0.9	-3.3	-1.3	0.3
148.46	1.12	2.0	144.69	8AU	1.7	-3.3	-1.3	0.3
154.83	1.09	2.0	154.36	18AU	3.2	-3.3	-1.2	0.3
161.23	1.06	2.0	164.08	28AU	1.1	-3.3	-1.2	0.3
167.65	1.02	2.0	173.85	7SE	-0.6	-3.3	-1.2	0.2
174.09	0.98	2.0	183.67	17SE	-1.1	-3.4	-1.2	0.2
180.57	0.94	2.0	193.56	27SE	-1.1	-3.4	-1.2	0.1
187.08	0.89	2.0	203.50	7OC	-0.8	-3.4	-1.2	0.0
193.62	0.83	2.0	213.49	17OC	-0.5	-3.5	-1.2	-0.0
200.21	0.77	1.9	223.54	27OC	-0.3	-3.5	-1.3	-0.1
206.82	0.71	1.9	233.63	6NO	-0.3	-3.6	-1.3	-0.2
213.48	0.63	1.9	243.76	16NO	-0.1	-3.7	-1.3	-0.3
220.18	0.55	1.8	253.93	26NO	0.9	-3.7	-1.4	-0.3
226.92	0.47	1.8	264.11	6DE	2.4	-3.8	-1.4	-0.2
233.70	0.37	1.7	274.30	16DE	0.4	-3.9	-1.5	-0.1
240.52	0.27	1.7	284.50	26DE	-0.0	-4.0	-1.5	-0.1
				1531				
247.38	0.15	1.6	294.68	5JA	-0.1	-4.1	-1.6	0.0
254.28	0.03	1.5	304.85	15JA	-0.2	-4.2	-1.7	0.1
261.21	-0.10	1.5	314.99	25JA	-0.5	-4.3	-1.7	0.1
268.18	-0.25	1.4	325.08	4FE	-1.0	-4.3	-1.8	0.2
275.18	-0.40	1.3	335.13	14FE	-1.3	-4.3	-1.9	0.2
282.21	-0.57	1.2	345.13	24FE	-1.1	-4.0	-1.9	0.3
289.26	-0.74	1.1	355.08	6MR	0.0	-3.5	-2.0	0.3
296.32	-0.93	1.0	4.97	16MR	1.8	-3.3	-2.0	0.3
303.40	-1.13	0.9	14.80	26MR	3.4	-3.9	-2.0	0.3
310.47	-1.34	0.8	24.57	5AP	1.9	-4.2	-2.0	0.3
317.52	-1.56	0.6	34.30	15AP	1.1	-4.2	-2.0	0.3
324.54	-1.78	0.5	43.97	25AP	0.6	-4.2	-2.0	0.3
331.49	-2.01	0.4	53.60	5MY	0.0	-4.1	-1.9	0.3
338.37	-2.24	0.3	63.20	15MY	-0.8	-4.0	-1.9	0.2
345.11	-2.47	0.1	72.76	25MY	-1.7	-3.9	-1.8	0.2
351.70	-2.69	-0.0	82.31	4JN	-1.2	-3.8	-1.8	0.2
358.07	-2.91	-0.2	91.84	14JN	-0.4	-3.7	-1.7	0.2
4.16	-3.13	-0.3	101.37	24JN	0.3	-3.6	-1.7	0.3
9.89	-3.33	-0.5	110.91	4JL	0.7	-3.6	-1.6	0.3
15.15	-3.51	-0.7	120.46	14JL	1.2	-3.5	-1.5	0.3
19.82	-3.67	-0.9	130.04	24JL	2.2	-3.5	-1.5	0.3
23.74	-3.80	-1.1	139.64	3AU	3.0	-3.4	-1.4	0.3
26.72	-3.88	-1.3	149.29	13AU	0.9	-3.4	-1.4	0.3
28.53	-3.90	-1.5	158.98	23AU	-0.6	-3.4	-1.3	0.3
28.99	-3.82	-1.8	168.72	2SE	-1.2	-3.4	-1.3	0.3
27.98	-3.61	-2.0	178.52	12SE	-1.2	-3.4	-1.3	0.2
25.65	-3.24	-2.2	188.38	22SE	-0.7	-3.4	-1.3	0.2
22.49	-2.71	-2.3	198.28	2OC	-0.4	-3.4	-1.2	0.1
19.25	-2.07	-2.1	208.25	12OC	-0.2	-3.4	-1.2	0.1
16.72	-1.41	-1.9	218.27	22OC	-0.1	-3.4	-1.2	0.0
15.40	-0.80	-1.5	228.34	1NO	0.0	-3.4	-1.2	-0.1
15.41	-0.27	-1.2	238.45	11NO	1.2	-3.4	-1.3	-0.1
16.68	0.16	-0.9	248.60	21NO	2.2	-3.4	-1.3	-0.2
19.01	0.49	-0.6	258.77	1DE	0.2	-3.4	-1.3	-0.3
22.18	0.76	-0.3	268.96	11DE	-0.2	-3.5	-1.3	-0.3
26.03	0.96	-0.0	279.16	21DE	-0.3	-3.5	-1.4	-0.2
30.39	1.12	0.2	289.35	31DE	-0.3	-3.5	-1.4	-0.1

♂ LONG	LAT	MAG	⊙ LONG	16.00UT 1532	☿	♀	♃	♄
35.14	1.23	0.5	299.53	10JA	-0.6	-3.5	-1.5	-0.1
40.20	1.32	0.7	309.68	20JA	-1.0	-3.4	-1.5	0.0
45.50	1.39	0.8	319.79	30JA	-1.2	-3.4	-1.6	0.1
50.98	1.43	1.0	329.87	9FE	-1.0	-3.4	-1.7	0.1
56.61	1.46	1.2	339.90	19FE	0.1	-3.4	-1.7	0.2
62.34	1.48	1.3	349.87	29FE	2.1	-3.3	-1.8	0.2
68.16	1.49	1.4	359.79	10MR	2.7	-3.3	-1.9	0.3
74.05	1.49	1.5	9.65	20MR	1.4	-3.3	-1.9	0.3
79.99	1.48	1.6	19.45	30MR	0.8	-3.3	-2.0	0.3
85.97	1.47	1.7	29.21	9AP	0.4	-3.3	-2.0	0.3
91.99	1.45	1.8	38.90	19AP	-0.1	-3.4	-2.1	0.3
98.04	1.42	1.8	48.56	29AP	-0.8	-3.4	-2.1	0.3
104.12	1.39	1.9	58.17	9MY	-1.7	-3.4	-2.1	0.3
110.22	1.36	1.9	67.75	19MY	-1.3	-3.4	-2.1	0.3
116.34	1.32	2.0	77.31	29MY	-0.3	-3.5	-2.0	0.2
122.50	1.28	2.0	86.85	8JN	0.4	-3.5	-2.0	0.2
128.68	1.24	2.0	96.38	18JN	1.0	-3.6	-2.0	0.2
134.88	1.19	2.0	105.91	28JN	1.6	-3.7	-1.9	0.3
141.13	1.15	2.0	115.46	8JL	2.8	-3.7	-1.8	0.3
147.40	1.09	2.0	125.02	18JL	2.5	-3.8	-1.8	0.3
153.72	1.04	2.0	134.60	28JL	0.7	-3.9	-1.7	0.4
160.07	0.98	2.0	144.23	7AU	-0.7	-4.0	-1.6	0.4
166.47	0.92	2.0	153.90	17AU	-1.3	-4.1	-1.6	0.4
172.92	0.85	2.0	163.61	27AU	-1.2	-4.2	-1.5	0.4
179.43	0.78	1.9	173.38	6SE	-0.6	-4.3	-1.5	0.4
185.98	0.71	1.9	183.20	16SE	-0.3	-4.2	-1.4	0.3
192.60	0.64	1.8	193.08	26SE	-0.1	-4.0	-1.4	0.3
199.27	0.56	1.8	203.02	6OC	0.0	-3.5	-1.4	0.2
206.01	0.47	1.8	213.01	16OC	0.3	-3.3	-1.3	0.2
212.81	0.39	1.8	223.05	26OC	1.4	-3.9	-1.3	0.1
219.68	0.29	1.8	233.14	5NO	2.0	-4.3	-1.3	0.1
226.62	0.20	1.8	243.27	15NO	-0.0	-4.4	-1.3	-0.0
233.62	0.10	1.8	253.43	25NO	-0.4	-4.3	-1.3	-0.1
240.70	-0.01	1.7	263.62	5DE	-0.4	-4.3	-1.3	-0.2
247.84	-0.12	1.7	273.81	15DE	-0.5	-4.2	-1.3	-0.2
255.04	-0.23	1.7	284.01	25DE	-0.7	-4.1	-1.4	-0.2

1533

♂ LONG	LAT	MAG	⊙ LONG	16.00UT	☿	♀	♃	♄
262.31	-0.34	1.6	294.19	4JA	-0.9	-4.0	-1.4	-0.2
269.64	-0.46	1.6	304.36	14JA	-1.0	-3.9	-1.4	-0.1
277.03	-0.58	1.5	314.50	24JA	-0.9	-3.8	-1.5	-0.0
284.47	-0.71	1.5	324.60	3FE	0.2	-3.7	-1.5	0.0
291.96	-0.83	1.4	334.65	13FE	2.5	-3.6	-1.6	0.1
299.49	-0.95	1.4	344.65	23FE	2.1	-3.5	-1.7	0.2
307.05	-1.07	1.3	354.60	5MR	1.0	-3.5	-1.7	0.2
314.64	-1.19	1.3	4.49	15MR	0.6	-3.4	-1.8	0.3
322.24	-1.30	1.2	14.33	25MR	0.3	-3.4	-1.9	0.3
329.85	-1.40	1.1	24.11	4AP	-0.2	-3.4	-1.9	0.3
337.44	-1.49	1.1	33.83	14AP	-0.8	-3.3	-2.0	0.3
345.01	-1.57	1.0	43.51	24AP	-1.7	-3.3	-2.1	0.4
352.53	-1.64	1.0	53.14	4MY	-1.3	-3.3	-2.1	0.4
359.99	-1.69	0.9	62.74	14MY	-0.3	-3.3	-2.2	0.4
7.38	-1.73	0.8	72.31	24MY	0.6	-3.3	-2.2	0.4
14.67	-1.75	0.8	81.85	3JN	1.3	-3.3	-2.2	0.3
21.83	-1.75	0.7	91.38	13JN	2.2	-3.3	-2.2	0.3
28.86	-1.73	0.7	100.92	23JN	3.3	-3.4	-2.2	0.3
35.70	-1.68	0.6	110.45	3JL	2.0	-3.4	-2.2	0.3
42.35	-1.62	0.5	120.00	13JL	0.6	-3.4	-2.1	0.3
48.75	-1.54	0.4	129.58	23JL	-0.7	-3.5	-2.1	0.4
54.87	-1.43	0.3	139.18	2AU	-1.4	-3.5	-2.0	0.4
60.65	-1.30	0.2	148.82	12AU	-1.2	-3.5	-1.9	0.4
66.03	-1.13	0.1	158.51	22AU	-0.6	-3.4	-1.9	0.5
70.91	-0.94	-0.0	168.25	1SE	-0.2	-3.4	-1.8	0.5
75.19	-0.71	-0.2	178.04	11SE	0.0	-3.4	-1.7	0.5
78.72	-0.43	-0.4	187.89	21SE	0.2	-3.3	-1.7	0.4
81.34	-0.11	-0.6	197.80	1OC	0.5	-3.3	-1.6	0.4
82.84	0.27	-0.8	207.76	11OC	1.8	-3.3	-1.6	0.4
83.01	0.70	-1.0	217.78	21OC	1.8	-3.3	-1.5	0.3
81.74	1.18	-1.2	227.84	31OC	-0.2	-3.3	-1.5	0.3
79.11	1.68	-1.4	237.95	10NO	-0.5	-3.3	-1.5	0.2
75.50	2.14	-1.5	248.10	20NO	-0.5	-3.4	-1.4	0.1
71.64	2.50	-1.5	258.27	30NO	-0.6	-3.4	-1.4	0.1
68.32	2.73	-1.2	268.46	10DE	-0.7	-3.4	-1.4	0.0
66.12	2.83	-0.9	278.66	20DE	-0.8	-3.5	-1.4	-0.1
65.27	2.84	-0.6	288.85	30DE	-0.9	-3.5	-1.4	-0.1

♂ LONG	LAT	MAG	⊙ LONG	16.00UT 1534	☿	♀	♃	♄
65.70	2.79	-0.3	299.03	9JA	-0.7	-3.6	-1.4	-0.1
67.27	2.71	-0.0	309.19	19JA	0.3	-3.6	-1.5	-0.1
69.75	2.60	0.2	319.30	29JA	2.8	-3.7	-1.5	-0.0
72.95	2.49	0.5	329.38	8FE	1.5	-3.7	-1.5	0.1
76.73	2.38	0.7	339.42	18FE	0.7	-3.8	-1.6	0.1
80.96	2.26	0.9	349.39	28FE	0.4	-3.9	-1.6	0.2
85.53	2.15	1.0	359.31	10MR	0.2	-4.0	-1.7	0.2
90.39	2.05	1.2	9.18	20MR	-0.2	-4.1	-1.7	0.3
95.48	1.94	1.3	18.99	30MR	-0.9	-4.2	-1.8	0.3
100.74	1.84	1.4	28.74	9AP	-1.7	-4.2	-1.9	0.4
106.16	1.75	1.5	38.44	19AP	-1.3	-4.2	-1.9	0.4
111.71	1.65	1.6	48.10	29AP	-0.2	-4.1	-2.0	0.4
117.37	1.56	1.7	57.71	9MY	0.7	-3.6	-2.1	0.5
123.13	1.47	1.8	67.29	19MY	1.6	-2.8	-2.1	0.5
128.97	1.37	1.8	76.85	29MY	2.9	-3.4	-2.2	0.5
134.89	1.28	1.9	86.39	8JN	3.0	-3.9	-2.2	0.5
140.89	1.19	1.9	95.92	18JN	1.6	-4.2	-2.3	0.5
146.96	1.10	1.9	105.45	28JN	0.4	-4.2	-2.3	0.4
153.11	1.01	1.9	114.99	8JL	-0.7	-4.1	-2.3	0.4
159.32	0.92	2.0	124.55	18JL	-1.5	-4.1	-2.3	0.4
165.61	0.83	2.0	134.14	28JL	-1.2	-4.0	-2.3	0.5
171.98	0.74	1.9	143.76	7AU	-0.5	-3.9	-2.3	0.5
178.42	0.65	1.9	153.43	17AU	-0.1	-3.8	-2.2	0.6
184.94	0.55	1.9	163.14	27AU	0.2	-3.7	-2.2	0.6
191.53	0.46	1.9	172.90	6SE	0.4	-3.7	-2.1	0.6
198.21	0.36	1.9	182.72	16SE	0.7	-3.6	-2.1	0.6
204.98	0.26	1.8	192.60	26SE	2.2	-3.6	-2.0	0.6
211.83	0.16	1.8	202.53	6OC	1.6	-3.5	-1.9	0.6
218.76	0.06	1.7	212.52	16OC	-0.3	-3.5	-1.9	0.6
225.78	-0.04	1.7	222.56	26OC	-0.7	-3.4	-1.8	0.5
232.89	-0.14	1.6	232.65	5NO	-0.7	-3.4	-1.7	0.5
240.08	-0.24	1.6	242.78	15NO	-0.7	-3.4	-1.7	0.4
247.35	-0.34	1.6	252.93	25NO	-0.7	-3.4	-1.6	0.4
254.71	-0.44	1.6	263.12	5DE	-0.7	-3.4	-1.6	0.3
262.14	-0.54	1.6	273.31	15DE	-0.7	-3.4	-1.6	0.2
269.65	-0.64	1.6	283.51	25DE	-0.6	-3.3	-1.5	0.2

1535

♂ LONG	LAT	MAG	⊙ LONG	16.00UT	☿	♀	♃	♄
277.22	-0.73	1.5	293.70	4JA	0.4	-3.3	-1.5	0.1
284.86	-0.82	1.5	303.86	14JA	3.0	-3.4	-1.5	0.0
292.54	-0.90	1.5	314.00	24JA	1.1	-3.4	-1.5	0.0
300.27	-0.98	1.5	324.10	3FE	0.5	-3.4	-1.5	0.1
308.04	-1.05	1.4	334.16	13FE	0.2	-3.4	-1.5	0.1
315.82	-1.11	1.4	344.16	23FE	0.0	-3.4	-1.6	0.2
323.62	-1.16	1.4	354.12	5MR	-0.3	-3.5	-1.6	0.3
331.42	-1.20	1.4	4.01	15MR	-0.9	-3.5	-1.6	0.3
339.21	-1.23	1.4	13.85	25MR	-1.7	-3.5	-1.7	0.4
346.97	-1.24	1.3	23.63	4AP	-1.2	-3.4	-1.7	0.4
354.70	-1.24	1.3	33.36	14AP	-0.2	-3.4	-1.8	0.5
2.37	-1.23	1.3	43.04	24AP	1.0	-3.4	-1.8	0.5
9.98	-1.20	1.3	52.67	4MY	2.2	-3.4	-1.9	0.6
17.52	-1.16	1.3	62.27	14MY	3.7	-3.3	-1.9	0.6
24.97	-1.10	1.3	71.84	24MY	2.3	-3.3	-2.0	0.6
32.32	-1.03	1.2	81.39	3JN	1.2	-3.3	-2.1	0.6
39.56	-0.95	1.2	90.92	13JN	0.3	-3.3	-2.1	0.6
46.69	-0.85	1.2	100.45	23JN	-0.7	-3.3	-2.2	0.6
53.68	-0.74	1.2	109.99	3JL	-1.6	-3.4	-2.3	0.6
60.53	-0.62	1.1	119.54	13JL	-1.2	-3.4	-2.3	0.6
67.22	-0.48	1.1	129.11	23JL	-0.5	-3.4	-2.4	0.6
73.74	-0.34	1.1	138.71	2AU	0.0	-3.5	-2.4	0.6
80.07	-0.18	1.0	148.35	12AU	0.3	-3.5	-2.4	0.7
86.18	-0.00	1.0	158.04	22AU	0.5	-3.6	-2.4	0.7
92.05	0.19	0.9	167.78	1SE	1.0	-3.6	-2.4	0.8
97.63	0.40	0.8	177.57	11SE	2.6	-3.7	-2.4	0.8
102.88	0.63	0.7	187.41	21SE	1.4	-3.8	-2.3	0.8
107.73	0.88	0.6	197.32	1OC	-0.4	-3.9	-2.3	0.8
112.10	1.16	0.4	207.28	11OC	-0.8	-4.0	-2.2	0.8
115.89	1.47	0.3	217.29	21OC	-0.8	-4.1	-2.2	0.8
118.95	1.82	0.1	227.36	31OC	-0.8	-4.2	-2.1	0.8
121.13	2.21	-0.1	237.46	10NO	-0.6	-4.3	-2.0	0.7
122.23	2.63	-0.3	247.60	20NO	-0.5	-4.4	-1.9	0.7
122.06	3.08	-0.5	257.78	30NO	-0.5	-4.3	-1.9	0.6
120.52	3.53	-0.8	267.96	10DE	-0.4	-4.1	-1.8	0.6
117.69	3.93	-1.0	278.16	20DE	0.6	-3.5	-1.8	0.5
113.95	4.20	-1.1	288.35	30DE	2.9	-3.3	-1.7	0.4

♂ LONG	♂ LAT	♂ MAG	☉ LONG	16.00UT	☿	♀	♃	♄
				1536	MAGNITUDES			
109.99	4.30	-1.0	298.53	9JA	0.8	-4.0	-1.7	0.4
106.55	4.24	-0.8	308.68	19JA	0.3	-4.3	-1.6	0.3
104.18	4.04	-0.6	318.81	29JA	0.1	-4.4	-1.6	0.2
103.12	3.77	-0.3	328.89	8FE	-0.1	-4.3	-1.6	0.3
103.33	3.48	-0.0	338.92	18FE	-0.4	-4.2	-1.6	0.3
104.66	3.18	0.2	348.91	28FE	-0.9	-4.1	-1.6	0.4
106.91	2.90	0.4	358.83	9MR	-1.6	-4.0	-1.6	0.4
109.90	2.64	0.6	8.70	19MR	-1.2	-3.9	-1.6	0.5
113.49	2.40	0.8	18.51	29MR	-0.1	-3.8	-1.6	0.5
117.56	2.18	1.0	28.27	8AP	1.2	-3.7	-1.6	0.6
122.01	1.98	1.1	37.97	18AP	2.8	-3.6	-1.7	0.6
126.77	1.79	1.3	47.63	28AP	3.1	-3.5	-1.7	0.7
131.80	1.62	1.4	57.25	8MY	1.8	-3.5	-1.7	0.7
137.04	1.46	1.5	66.83	18MY	0.9	-3.4	-1.8	0.8
142.48	1.30	1.5	76.39	28MY	0.2	-3.4	-1.8	0.8
148.08	1.16	1.6	85.93	7JN	-0.8	-3.4	-1.9	0.8
153.83	1.02	1.6	95.46	17JN	-1.6	-3.3	-2.0	0.8
159.72	0.88	1.7	104.99	27JN	-1.3	-3.3	-2.0	0.8
165.73	0.75	1.7	114.53	7JL	-0.4	-3.3	-2.1	0.9
171.87	0.62	1.7	124.09	17JL	0.1	-3.3	-2.2	0.9
178.13	0.50	1.8	133.68	27JL	0.4	-3.3	-2.2	0.9
184.49	0.38	1.8	143.30	6AU	0.8	-3.3	-2.3	0.8
190.97	0.26	1.8	152.96	16AU	1.4	-3.4	-2.4	0.9
197.57	0.15	1.8	162.67	26AU	3.0	-3.4	-2.4	0.9
204.26	0.04	1.8	172.43	5SE	1.2	-3.4	-2.4	1.0
211.07	-0.07	1.7	182.25	15SE	-0.5	-3.4	-2.5	1.0
217.99	-0.18	1.7	192.12	25SE	-1.0	-3.5	-2.5	1.0
225.01	-0.28	1.7	202.05	5OC	-1.0	-3.5	-2.5	1.1
232.14	-0.38	1.7	212.04	15OC	-0.8	-3.5	-2.4	1.1
239.37	-0.48	1.6	222.07	25OC	-0.5	-3.5	-2.4	1.1
246.70	-0.57	1.6	232.16	4NO	-0.4	-3.4	-2.3	1.1
254.12	-0.66	1.5	242.28	14NO	-0.4	-3.4	-2.3	1.0
261.63	-0.74	1.5	252.44	24NO	-0.3	-3.4	-2.2	1.0
269.22	-0.82	1.5	262.62	4DE	0.7	-3.4	-2.1	1.0
276.88	-0.88	1.4	272.81	14DE	2.7	-3.3	-2.0	0.9
284.61	-0.94	1.4	283.01	24DE	0.5	-3.3	-2.0	0.8
				1537				
292.40	-0.99	1.3	293.19	3JA	0.1	-3.3	-1.9	0.8
300.22	-1.03	1.3	303.36	13JA	-0.0	-3.3	-1.8	0.7
308.08	-1.06	1.4	313.50	23JA	-0.2	-3.4	-1.8	0.6
315.96	-1.08	1.4	323.61	2FE	-0.4	-3.4	-1.7	0.6
323.85	-1.09	1.4	333.67	12FE	-0.9	-3.4	-1.7	0.5
331.73	-1.09	1.4	343.68	22FE	-1.4	-3.4	-1.6	0.5
339.59	-1.07	1.4	353.63	4MR	-1.1	-3.5	-1.6	0.6
347.42	-1.04	1.4	3.53	14MR	-0.1	-3.5	-1.6	0.6
355.21	-1.00	1.4	13.37	24MR	1.5	-3.5	-1.6	0.7
2.94	-0.95	1.4	23.15	3AP	3.5	-3.6	-1.6	0.7
10.61	-0.89	1.4	32.89	13AP	2.3	-3.7	-1.6	0.8
18.21	-0.81	1.5	42.57	23AP	1.3	-3.7	-1.6	0.8
25.73	-0.73	1.5	52.21	3MY	0.7	-3.8	-1.6	0.9
33.16	-0.64	1.5	61.81	13MY	0.1	-3.9	-1.6	0.9
40.50	-0.54	1.5	71.38	23MY	-0.8	-4.0	-1.6	1.0
47.74	-0.44	1.5	80.93	2JN	-1.7	-4.1	-1.7	1.0
54.88	-0.33	1.5	90.47	12JN	-1.3	-4.1	-1.7	1.1
61.91	-0.21	1.5	99.99	22JN	-0.4	-4.2	-1.8	1.1
68.84	-0.09	1.5	109.53	2JL	0.2	-4.1	-1.8	1.1
75.65	0.04	1.5	119.08	12JL	0.6	-3.9	-1.9	1.1
82.35	0.17	1.5	128.65	22JL	1.0	-3.4	-1.9	1.1
88.92	0.31	1.5	138.25	1AU	1.8	-3.2	-2.0	1.1
95.37	0.45	1.5	147.89	11AU	3.2	-3.8	-2.1	1.1
101.67	0.60	1.5	157.57	21AU	1.1	-4.1	-2.1	1.1
107.83	0.75	1.4	167.30	31AU	-0.6	-4.3	-2.2	1.2
113.82	0.91	1.4	177.09	10SE	-1.1	-4.3	-2.3	1.2
119.63	1.08	1.3	186.93	20SE	-1.1	-4.2	-2.3	1.3
125.24	1.25	1.3	196.83	30SE	-0.7	-4.1	-2.4	1.3
130.60	1.44	1.2	206.80	10OC	-0.4	-4.0	-2.4	1.3
135.68	1.65	1.1	216.80	20OC	-0.3	-3.9	-2.4	1.4
140.42	1.87	0.9	226.87	30OC	-0.2	-3.8	-2.4	1.3
144.75	2.11	0.8	236.97	9NO	-0.1	-3.8	-2.4	1.3
148.59	2.37	0.6	247.11	19NO	0.9	-3.7	-2.4	1.3
151.81	2.66	0.5	257.28	29NO	2.5	-3.6	-2.3	1.2
154.26	2.98	0.3	267.47	9DE	0.3	-3.6	-2.3	1.2
155.78	3.32	0.0	277.67	19DE	-0.1	-3.5	-2.2	1.1
156.17	3.67	-0.2	287.86	29DE	-0.2	-3.5	-2.1	1.1

♂ LONG	♂ LAT	♂ MAG	☉ LONG	16.00UT	☿	♀	♃	♄
				1538	MAGNITUDES			
155.29	4.01	-0.5	298.04	8JA	-0.3	-3.4	-2.1	1.0
153.10	4.29	-0.7	308.20	18JA	-0.5	-3.4	-2.0	1.0
149.81	4.46	-0.9	318.32	28JA	-1.0	-3.4	-1.9	1.0
145.90	4.47	-1.0	328.40	7FE	-1.3	-3.3	-1.8	0.9
142.11	4.30	-0.8	338.44	17FE	-1.1	-3.3	-1.8	0.9
139.09	4.00	-0.6	348.42	27FE	-0.0	-3.3	-1.7	0.8
137.25	3.62	-0.4	358.35	9MR	1.9	-3.3	-1.7	0.8
136.72	3.21	-0.2	8.22	19MR	3.2	-3.3	-1.6	0.9
137.38	2.81	0.1	18.03	29MR	1.7	-3.3	-1.6	0.9
139.10	2.44	0.3	27.79	8AP	1.0	-3.3	-1.6	1.0
141.69	2.10	0.5	37.50	18AP	0.5	-3.4	-1.5	1.0
144.97	1.80	0.6	47.16	28AP	-0.0	-3.4	-1.5	1.1
148.83	1.52	0.8	56.78	8MY	-0.8	-3.4	-1.5	1.1
153.16	1.27	0.9	66.36	18MY	-1.7	-3.5	-1.5	1.2
157.87	1.05	1.0	75.92	28MY	-1.3	-3.5	-1.5	1.2
162.92	0.84	1.1	85.46	7JN	-0.4	-3.4	-1.5	1.3
168.24	0.65	1.2	94.99	17JN	0.3	-3.4	-1.5	1.3
173.80	0.47	1.3	104.53	27JN	0.8	-3.4	-1.5	1.4
179.59	0.30	1.3	114.07	7JL	1.4	-3.4	-1.6	1.4
185.57	0.15	1.4	123.62	17JL	2.4	-3.3	-1.6	1.4
191.73	0.00	1.4	133.21	27JL	2.9	-3.3	-1.7	1.4
198.07	-0.13	1.5	142.83	6AU	0.9	-3.3	-1.7	1.3
204.55	-0.26	1.5	152.49	16AU	-0.6	-3.3	-1.8	1.3
211.20	-0.38	1.5	162.20	26AU	-1.2	-3.3	-1.8	1.3
217.98	-0.50	1.5	171.95	5SE	-1.2	-3.3	-1.9	1.2
224.91	-0.60	1.5	181.77	15SE	-0.7	-3.4	-1.9	1.2
231.96	-0.70	1.5	191.64	25SE	-0.3	-3.4	-2.0	1.2
239.14	-0.78	1.5	201.57	5OC	-0.2	-3.4	-2.1	1.2
246.44	-0.86	1.5	211.55	15OC	-0.1	-3.5	-2.1	1.2
253.85	-0.93	1.5	221.58	25OC	0.1	-3.5	-2.2	1.2
261.37	-0.99	1.5	231.67	4NO	1.2	-3.6	-2.2	1.2
268.97	-1.04	1.4	241.79	14NO	2.2	-3.7	-2.3	1.1
276.66	-1.08	1.4	251.95	24NO	0.1	-3.8	-2.3	1.1
284.42	-1.10	1.4	262.13	4DE	-0.3	-3.8	-2.3	1.1
292.23	-1.12	1.4	272.32	14DE	-0.3	-3.9	-2.3	1.0
300.09	-1.12	1.4	282.52	24DE	-0.4	-4.0	-2.2	1.0
				1539				
307.98	-1.11	1.4	292.71	3JA	-0.6	-4.1	-2.2	1.0
315.89	-1.09	1.4	302.87	13JA	-1.0	-4.2	-2.2	0.9
323.80	-1.06	1.4	313.02	23JA	-1.1	-4.3	-2.1	0.9
331.70	-1.01	1.4	323.12	2FE	-1.0	-4.3	-2.0	0.8
339.57	-0.96	1.3	333.18	12FE	0.1	-4.3	-1.9	0.8
347.40	-0.89	1.3	343.19	22FE	2.2	-4.0	-1.9	0.7
355.19	-0.82	1.3	353.15	4MR	2.5	-3.5	-1.8	0.7
2.91	-0.74	1.3	3.05	14MR	1.2	-3.4	-1.7	0.7
10.57	-0.65	1.4	12.90	24MR	0.7	-3.9	-1.7	0.7
18.16	-0.56	1.4	22.68	3AP	0.4	-4.2	-1.6	0.8
25.66	-0.46	1.5	32.41	13AP	-0.1	-4.2	-1.6	0.9
33.09	-0.36	1.5	42.10	23AP	-0.8	-4.2	-1.5	1.0
40.42	-0.26	1.6	51.74	3MY	-1.7	-4.1	-1.5	1.0
47.66	-0.15	1.6	61.34	13MY	-1.3	-4.0	-1.4	1.1
54.82	-0.04	1.6	70.91	23MY	-0.3	-3.9	-1.4	1.1
61.88	0.07	1.7	80.46	2JN	0.4	-3.8	-1.4	1.2
68.85	0.18	1.7	90.00	12JN	1.1	-3.7	-1.4	1.2
75.74	0.29	1.7	99.53	22JN	1.8	-3.6	-1.4	1.2
82.53	0.39	1.8	109.06	2JL	3.0	-3.6	-1.4	1.2
89.24	0.50	1.8	118.61	12JL	2.4	-3.5	-1.4	1.2
95.86	0.61	1.8	128.19	22JL	0.7	-3.5	-1.4	1.2
102.39	0.72	1.8	137.78	1AU	-0.7	-3.4	-1.4	1.2
108.83	0.83	1.8	147.42	11AU	-1.4	-3.4	-1.5	1.2
115.19	0.94	1.8	157.11	21AU	-1.3	-3.4	-1.5	1.2
121.45	1.05	1.8	166.84	31AU	-0.6	-3.4	-1.5	1.1
127.61	1.16	1.7	176.62	10SE	-0.2	-3.4	-1.6	1.1
133.67	1.27	1.7	186.46	20SE	-0.0	-3.4	-1.6	1.0
139.61	1.38	1.7	196.36	30SE	0.1	-3.4	-1.7	1.0
145.43	1.50	1.6	206.31	10OC	0.3	-3.4	-1.8	1.0
151.10	1.62	1.5	216.32	20OC	1.5	-3.4	-1.8	1.0
156.60	1.74	1.5	226.38	30OC	2.0	-3.4	-1.9	1.0
161.91	1.87	1.4	236.48	9NO	-0.1	-3.4	-2.0	1.0
166.99	2.01	1.2	246.62	19NO	-0.4	-3.4	-2.0	1.0
171.78	2.16	1.1	256.79	29NO	-0.5	-3.4	-2.1	1.0
176.24	2.31	1.0	266.98	9DE	-0.5	-3.5	-2.1	1.0
180.27	2.48	0.8	277.17	19DE	-0.7	-3.5	-2.1	0.9
183.79	2.66	0.6	287.36	29DE	-0.9	-3.5	-2.2	0.9

1540 / 1541

♂ LONG	♂ LAT	♂ MAG	☉ LONG	16.00UT	☿	♀	♃	♄
186.67	2.84	0.4	297.55	8JA	-1.0	-3.5	-2.2	0.9
188.74	3.03	0.1	307.70	18JA	-0.8	-3.4	-2.1	0.8
189.85	3.22	-0.1	317.83	28JA	-0.2	-3.4	-2.1	0.8
189.80	3.39	-0.4	327.91	7FE	2.6	-3.4	-2.1	0.7
188.48	3.50	-0.7	337.95	17FE	1.9	-3.4	-2.0	0.7
185.92	3.52	-0.9	347.93	27FE	0.9	-3.3	-2.0	0.6
182.42	3.42	-1.1	357.87	8MR	0.5	-3.3	-1.9	0.6
178.57	3.17	-1.1	7.74	18MR	0.2	-3.3	-1.8	0.5
175.09	2.80	-1.0	17.55	28MR	-0.2	-3.3	-1.7	0.5
172.55	2.36	-0.8	27.32	7AP	-0.8	-3.3	-1.7	0.6
171.29	1.91	-0.6	37.03	17AP	-1.7	-3.4	-1.6	0.7
171.33	1.48	-0.3	46.69	27AP	-1.3	-3.4	-1.6	0.7
172.55	1.09	-0.1	56.31	7MY	-0.3	-3.4	-1.5	0.8
174.79	0.75	0.1	65.90	17MY	0.6	-3.4	-1.4	0.9
177.87	0.44	0.2	75.46	27MY	1.4	-3.5	-1.4	0.9
181.64	0.17	0.4	85.00	6JN	2.4	-3.5	-1.4	1.0
185.99	-0.06	0.5	94.53	16JN	3.3	-3.6	-1.3	1.0
190.81	-0.27	0.6	104.06	26JN	1.9	-3.7	-1.3	1.0
196.04	-0.46	0.7	113.61	6JL	0.6	-3.7	-1.3	1.1
201.61	-0.62	0.8	123.16	16JL	-0.7	-3.8	-1.3	1.1
207.49	-0.77	0.9	132.75	26JL	-1.5	-3.9	-1.3	1.1
213.62	-0.89	0.9	142.37	5AU	-1.3	-4.0	-1.3	1.1
220.00	-1.01	1.0	152.02	15AU	-0.6	-4.1	-1.3	1.1
226.59	-1.10	1.0	161.73	25AU	-0.1	-4.2	-1.3	1.0
233.38	-1.18	1.1	171.49	4SE	0.1	-4.3	-1.3	1.0
240.34	-1.25	1.1	181.30	14SE	0.3	-4.2	-1.4	1.0
247.46	-1.30	1.1	191.16	24SE	0.6	-4.0	-1.4	0.9
254.73	-1.34	1.2	201.09	4OC	1.9	-3.4	-1.5	0.9
262.13	-1.36	1.2	211.07	14OC	1.8	-3.3	-1.5	0.9
269.64	-1.36	1.2	221.10	24OC	-0.2	-4.0	-1.6	0.9
277.25	-1.36	1.2	231.18	3NO	-0.6	-4.3	-1.6	0.9
284.94	-1.34	1.3	241.30	13NO	-0.6	-4.4	-1.7	0.9
292.70	-1.30	1.3	251.45	23NO	-0.6	-4.3	-1.7	0.9
300.51	-1.26	1.3	261.64	3DE	-0.8	-4.3	-1.8	0.9
308.35	-1.20	1.3	271.83	13DE	-0.8	-4.2	-1.9	0.9
316.21	-1.13	1.3	282.02	23DE	-0.8	-4.1	-1.9	0.9
				1541				
324.07	-1.05	1.4	292.21	2JA	-0.7	-4.0	-2.0	0.8
331.91	-0.96	1.4	302.38	12JA	0.3	-3.9	-2.0	0.8
339.72	-0.87	1.4	312.53	22JA	2.9	-3.8	-2.0	0.7
347.50	-0.77	1.4	322.64	1FE	1.4	-3.7	-2.1	0.7
355.22	-0.66	1.5	332.70	11FE	0.6	-3.6	-2.1	0.6
2.88	-0.56	1.5	342.71	21FE	0.3	-3.5	-2.0	0.6
10.47	-0.45	1.5	352.68	3MR	0.1	-3.5	-2.0	0.5
17.99	-0.34	1.5	2.58	13MR	-0.2	-3.4	-2.0	0.5
25.43	-0.23	1.6	12.42	23MR	-0.8	-3.4	-1.9	0.4
32.78	-0.12	1.6	22.22	2AP	-1.7	-3.4	-1.8	0.4
40.05	-0.01	1.6	31.95	12AP	-1.3	-3.3	-1.8	0.4
47.23	0.10	1.6	41.64	22AP	-0.2	-3.3	-1.7	0.5
54.34	0.20	1.6	51.28	2MY	0.8	-3.3	-1.6	0.6
61.35	0.30	1.6	60.88	12MY	1.8	-3.3	-1.6	0.6
68.29	0.40	1.7	70.45	22MY	3.2	-3.3	-1.5	0.7
75.16	0.49	1.7	80.00	1JN	2.8	-3.3	-1.5	0.8
81.94	0.58	1.8	89.54	11JN	1.5	-3.3	-1.4	0.8
88.66	0.67	1.8	99.07	21JN	0.4	-3.4	-1.4	0.9
95.31	0.75	1.9	108.60	1JL	-0.7	-3.4	-1.3	0.9
101.90	0.83	1.9	118.15	11JL	-1.5	-3.4	-1.3	0.9
108.42	0.91	1.9	127.72	21JL	-1.3	-3.5	-1.3	1.0
114.89	0.98	1.9	137.32	31JL	-0.5	-3.5	-1.2	1.0
121.30	1.05	2.0	146.95	10AU	-0.1	-3.5	-1.2	1.0
127.66	1.12	2.0	156.63	20AU	0.2	-3.4	-1.2	1.0
133.97	1.18	2.0	166.36	30AU	0.4	-3.4	-1.2	0.9
140.22	1.24	1.9	176.14	9SE	0.8	-3.3	-1.2	0.9
146.42	1.30	1.9	185.98	19SE	2.3	-3.3	-1.2	0.9
152.56	1.36	1.9	195.88	29SE	1.6	-3.3	-1.3	0.8
158.64	1.42	1.9	205.82	9OC	-0.3	-3.3	-1.3	0.8
164.66	1.47	1.8	215.83	19OC	-0.8	-3.3	-1.3	0.8
170.60	1.52	1.8	225.89	29OC	-0.8	-3.3	-1.4	0.8
176.45	1.57	1.7	235.99	8NO	-0.8	-3.3	-1.4	0.8
182.21	1.62	1.6	246.13	18NO	-0.7	-3.4	-1.5	0.8
187.85	1.67	1.5	256.30	28NO	-0.6	-3.4	-1.5	0.8
193.34	1.71	1.4	266.48	8DE	-0.6	-3.4	-1.6	0.8
198.68	1.76	1.3	276.68	18DE	-0.5	-3.5	-1.7	0.8
203.81	1.79	1.1	286.87	28DE	0.4	-3.5	-1.7	0.8

1542 / 1543

♂ LONG	♂ LAT	♂ MAG	☉ LONG	16.00UT	☿	♀	♃	♄
208.70	1.83	1.0	297.05	7JA	3.0	-3.6	-1.8	0.8
213.29	1.85	0.8	307.21	17JA	1.0	-3.6	-1.8	0.8
217.51	1.87	0.6	317.34	27JA	0.4	-3.7	-1.9	0.7
221.26	1.87	0.4	327.43	6FE	0.2	-3.7	-1.9	0.7
224.44	1.85	0.2	337.47	16FE	0.0	-3.8	-2.0	0.6
226.88	1.80	-0.1	347.46	26FE	-0.3	-3.9	-2.0	0.6
228.44	1.71	-0.4	357.39	8MR	-0.9	-4.0	-2.0	0.5
228.92	1.57	-0.7	7.27	18MR	-1.6	-4.1	-2.0	0.5
228.17	1.35	-1.0	17.09	28MR	-1.2	-4.2	-2.0	0.4
226.17	1.04	-1.3	26.85	7AP	-0.2	-4.2	-1.9	0.4
223.12	0.65	-1.6	36.56	17AP	1.0	-4.2	-1.9	0.3
219.55	0.20	-1.6	46.23	27AP	2.4	-4.1	-1.8	0.4
216.16	-0.28	-1.5	55.85	7MY	3.6	-3.6	-1.8	0.4
213.60	-0.72	-1.3	65.44	17MY	2.1	-2.8	-1.7	0.5
212.31	-1.10	-1.1	75.00	27MY	1.1	-3.4	-1.6	0.6
212.38	-1.41	-0.9	84.54	6JN	0.3	-4.0	-1.6	0.6
213.73	-1.65	-0.7	94.07	16JN	-0.7	-4.2	-1.5	0.7
216.20	-1.84	-0.5	103.60	26JN	-1.6	-4.2	-1.5	0.7
219.60	-1.98	-0.4	113.14	6JL	-1.3	-4.1	-1.4	0.8
223.77	-2.08	-0.2	122.70	16JL	-0.5	-4.1	-1.4	0.8
228.58	-2.15	-0.1	132.28	26JL	0.0	-4.0	-1.3	0.9
233.91	-2.19	0.1	141.89	5AU	0.4	-3.9	-1.3	0.9
239.68	-2.20	0.2	151.55	15AU	0.6	-3.8	-1.3	0.9
245.82	-2.19	0.3	161.25	25AU	1.2	-3.7	-1.2	0.9
252.26	-2.16	0.4	171.00	4SE	2.7	-3.7	-1.2	0.9
258.97	-2.11	0.5	180.82	14SE	1.4	-3.6	-1.2	0.9
265.90	-2.04	0.6	190.68	24SE	-0.4	-3.6	-1.2	0.8
273.02	-1.96	0.6	200.60	4OC	-0.9	-3.5	-1.2	0.8
280.29	-1.86	0.7	210.58	14OC	-0.9	-3.5	-1.2	0.8
287.68	-1.75	0.8	220.61	24OC	-0.9	-3.4	-1.2	0.7
295.15	-1.63	0.9	230.69	3NO	-0.6	-3.4	-1.3	0.7
302.74	-1.50	0.9	240.81	13NO	-0.5	-3.4	-1.3	0.8
310.35	-1.36	1.0	250.96	23NO	-0.5	-3.4	-1.3	0.8
317.99	-1.22	1.1	261.14	3DE	-0.4	-3.4	-1.4	0.8
325.65	-1.08	1.1	271.34	13DE	0.6	-3.4	-1.4	0.8
333.29	-0.93	1.2	281.53	23DE	2.9	-3.4	-1.5	0.8
				1543				
340.91	-0.78	1.3	291.72	2JA	0.7	-3.4	-1.6	0.8
348.50	-0.64	1.3	301.89	12JA	0.2	-3.4	-1.6	0.8
356.04	-0.50	1.4	312.04	22JA	0.0	-3.4	-1.7	0.7
3.52	-0.36	1.4	322.15	1FE	-0.1	-3.4	-1.8	0.7
10.94	-0.22	1.5	332.21	11FE	-0.4	-3.4	-1.9	0.7
18.29	-0.10	1.6	342.23	21FE	-0.9	-3.4	-1.9	0.6
25.57	0.03	1.6	352.19	3MR	-1.5	-3.5	-1.9	0.6
32.78	0.14	1.7	2.10	13MR	-1.2	-3.5	-2.0	0.5
39.90	0.25	1.7	11.94	23MR	-0.1	-3.5	-2.0	0.5
46.96	0.35	1.7	21.74	2AP	1.3	-3.4	-2.0	0.4
53.94	0.45	1.8	31.48	12AP	3.1	-3.4	-2.0	0.3
60.84	0.54	1.8	41.16	22AP	2.8	-3.4	-2.0	0.3
67.68	0.62	1.8	50.81	2MY	1.6	-3.4	-2.0	0.3
74.45	0.70	1.8	60.42	12MY	0.9	-3.3	-1.9	0.3
81.16	0.77	1.9	69.99	22MY	0.2	-3.3	-1.9	0.4
87.82	0.83	1.9	79.54	1JN	-0.7	-3.3	-1.8	0.5
94.42	0.89	1.9	89.08	11JN	-1.6	-3.3	-1.8	0.5
100.98	0.95	1.9	98.60	21JN	-1.3	-3.3	-1.7	0.6
107.49	1.00	1.9	108.14	1JL	-0.4	-3.4	-1.7	0.7
113.96	1.04	1.9	117.69	11JL	0.1	-3.4	-1.6	0.7
120.40	1.08	1.9	127.25	21JL	0.5	-3.4	-1.5	0.7
126.81	1.12	2.0	136.85	31JL	0.9	-3.5	-1.5	0.8
133.19	1.15	2.0	146.48	10AU	1.5	-3.5	-1.4	0.8
139.55	1.17	2.0	156.16	20AU	3.1	-3.6	-1.4	0.8
145.89	1.20	2.0	165.88	30AU	1.2	-3.6	-1.3	0.8
152.21	1.22	2.0	175.66	9SE	-0.5	-3.7	-1.3	0.8
158.52	1.23	2.0	185.50	19SE	-1.0	-3.8	-1.3	0.8
164.81	1.24	2.0	195.39	29SE	-1.0	-3.9	-1.3	0.8
171.08	1.24	2.0	205.34	9OC	-0.8	-4.0	-1.2	0.8
177.35	1.25	2.0	215.34	19OC	-0.5	-4.1	-1.2	0.7
183.59	1.24	1.9	225.40	29OC	-0.4	-4.2	-1.2	0.7
189.81	1.23	1.9	235.50	8NO	-0.3	-4.3	-1.3	0.7
196.02	1.22	1.8	245.63	18NO	-0.2	-4.4	-1.3	0.7
202.19	1.20	1.7	255.80	28NO	0.8	-4.3	-1.3	0.7
208.33	1.17	1.7	265.99	8DE	2.7	-4.1	-1.3	0.8
214.43	1.13	1.6	276.18	18DE	0.5	-3.4	-1.4	0.8
220.48	1.08	1.5	286.38	28DE	-0.0	-3.3	-1.4	0.8

♂ LONG	♂ LAT	♂ MAG	⊙ LONG	16.00UT	☿	♀	♃	♄
				1544	MAGNITUDES			
226.46	1.02	1.4	296.56	7JA	-0.1	-4.0	-1.5	0.8
232.37	0.95	1.2	306.72	17JA	-0.2	-4.3	-1.5	0.8
238.19	0.86	1.1	316.85	27JA	-0.5	-4.4	-1.6	0.8
243.89	0.75	1.0	326.94	6FE	-0.9	-4.3	-1.6	0.7
249.46	0.62	0.8	336.98	16FE	-1.4	-4.2	-1.7	0.7
254.85	0.46	0.6	346.98	26FE	-1.1	-4.1	-1.8	0.6
260.02	0.26	0.4	356.91	7MR	-0.1	-4.0	-1.8	0.6
264.92	0.03	0.2	6.79	17MR	1.6	-3.9	-1.9	0.5
269.48	-0.25	-0.0	16.62	27MR	3.6	-3.8	-2.0	0.5
273.60	-0.59	-0.3	26.38	6AP	2.0	-3.7	-2.0	0.4
277.18	-1.00	-0.6	36.09	16AP	1.2	-3.6	-2.1	0.4
280.04	-1.48	-0.9	45.76	26AP	0.6	-3.5	-2.1	0.3
282.02	-2.06	-1.2	55.39	6MY	0.1	-3.5	-2.1	0.3
282.95	-2.71	-1.5	64.98	16MY	-0.7	-3.4	-2.1	0.3
282.66	-3.44	-1.8	74.54	26MY	-1.7	-3.4	-2.1	0.3
281.16	-4.17	-2.1	84.08	5JN	-1.3	-3.4	-2.0	0.4
278.68	-4.82	-2.4	93.61	15JN	-0.4	-3.3	-2.0	0.5
275.77	-5.31	-2.4	103.15	25JN	0.2	-3.3	-2.0	0.5
273.18	-5.54	-2.3	112.69	5JL	0.7	-3.3	-1.9	0.6
271.52	-5.55	-2.1	122.24	15JL	1.1	-3.3	-1.8	0.6
271.17	-5.36	-1.8	131.82	25JL	2.0	-3.3	-1.8	0.7
272.21	-5.06	-1.6	141.43	4AU	3.1	-3.3	-1.7	0.7
274.49	-4.69	-1.3	151.08	14AU	1.0	-3.4	-1.6	0.8
277.84	-4.30	-1.1	160.79	24AU	-0.6	-3.4	-1.6	0.8
282.06	-3.89	-0.9	170.53	3SE	-1.2	-3.4	-1.5	0.8
286.96	-3.49	-0.7	180.34	13SE	-1.2	-3.4	-1.5	0.8
292.41	-3.10	-0.5	190.20	23SE	-0.7	-3.5	-1.4	0.8
298.29	-2.73	-0.3	200.12	3OC	-0.4	-3.5	-1.4	0.8
304.49	-2.38	-0.1	210.09	13OC	-0.2	-3.5	-1.4	0.8
310.94	-2.04	0.0	220.12	23OC	-0.2	-3.5	-1.4	0.7
317.59	-1.72	0.2	230.19	2NO	-0.0	-3.4	-1.3	0.7
324.38	-1.43	0.3	240.31	12NO	1.0	-3.4	-1.3	0.7
331.26	-1.16	0.5	250.47	22NO	2.5	-3.4	-1.3	0.7
338.22	-0.90	0.6	260.64	2DE	0.3	-3.4	-1.3	0.7
345.21	-0.67	0.8	270.83	12DE	-0.2	-3.3	-1.3	0.7
352.23	-0.45	0.9	281.03	22DE	-0.2	-3.3	-1.4	0.8
				1545				
359.25	-0.26	1.0	291.22	1JA	-0.3	-3.3	-1.4	0.8
6.25	-0.08	1.1	301.40	11JA	-0.5	-3.3	-1.4	0.8
13.23	0.08	1.2	311.55	21JA	-1.0	-3.4	-1.5	0.8
20.18	0.23	1.3	321.65	31JA	-1.2	-3.4	-1.5	0.8
27.09	0.36	1.4	331.72	10FE	-1.0	-3.4	-1.6	0.7
33.95	0.48	1.5	341.75	20FE	-0.0	-3.4	-1.6	0.7
40.77	0.58	1.6	351.71	2MR	1.9	-3.5	-1.7	0.7
47.54	0.67	1.6	1.62	12MR	2.9	-3.5	-1.8	0.6
54.27	0.76	1.7	11.48	22MR	1.5	-3.5	-1.8	0.6
60.95	0.83	1.8	21.27	1AP	0.9	-3.6	-1.9	0.5
67.58	0.89	1.8	31.01	11AP	0.5	-3.7	-2.0	0.5
74.16	0.95	1.9	40.70	21AP	-0.0	-3.7	-2.0	0.4
80.71	1.00	1.9	50.35	1MY	-0.8	-3.8	-2.1	0.3
87.21	1.04	1.9	59.96	11MY	-1.7	-3.9	-2.1	0.3
93.69	1.07	2.0	69.53	21MY	-1.3	-4.0	-2.2	0.2
100.13	1.10	2.0	79.08	31MY	-0.4	-4.1	-2.2	0.3
106.55	1.12	2.0	88.62	10JN	0.3	-4.1	-2.2	0.3
112.95	1.13	2.0	98.15	20JN	0.9	-4.2	-2.2	0.4
119.33	1.15	2.0	107.68	30JN	1.5	-4.1	-2.2	0.5
125.70	1.15	2.0	117.23	10JL	2.6	-3.9	-2.2	0.5
132.07	1.15	2.0	126.80	20JL	2.7	-3.4	-2.1	0.6
138.43	1.15	2.0	136.39	30JL	0.9	-3.2	-2.1	0.6
144.79	1.14	2.0	146.02	9AU	-0.6	-3.8	-2.0	0.7
151.16	1.13	2.0	155.69	19AU	-1.3	-4.1	-1.9	0.7
157.54	1.11	2.0	165.41	29AU	-1.3	-4.3	-1.9	0.7
163.93	1.09	2.0	175.19	8SE	-0.7	-4.2	-1.8	0.8
170.33	1.06	2.0	185.02	18SE	-0.3	-4.2	-1.7	0.8
176.76	1.03	2.0	194.91	28SE	-0.1	-4.1	-1.7	0.8
183.21	1.00	2.0	204.86	8OC	-0.0	-4.0	-1.6	0.8
189.68	0.96	2.0	214.86	18OC	0.2	-3.9	-1.6	0.8
196.17	0.91	2.0	224.91	28OC	1.3	-3.8	-1.5	0.8
202.69	0.85	1.9	235.01	7NO	2.2	-3.8	-1.5	0.7
209.23	0.79	1.9	245.14	17NO	0.1	-3.7	-1.5	0.7
215.80	0.73	1.8	255.30	27NO	-0.3	-3.6	-1.5	0.6
222.39	0.65	1.8	265.49	7DE	-0.4	-3.6	-1.4	0.7
229.01	0.56	1.7	275.69	17DE	-0.4	-3.5	-1.4	0.7
235.65	0.47	1.7	285.88	27DE	-0.6	-3.5	-1.4	0.8

♂ LONG	♂ LAT	♂ MAG	⊙ LONG	16.00UT	☿	♀	♃	♄
				1546	MAGNITUDES			
242.31	0.36	1.6	296.07	6JA	-0.9	-3.4	-1.4	0.8
248.99	0.25	1.5	306.23	16JA	-1.0	-3.4	-1.5	0.8
255.69	0.12	1.4	316.36	26JA	-0.9	-3.4	-1.5	0.8
262.40	-0.03	1.3	326.46	5FE	0.1	-3.3	-1.5	0.8
269.12	-0.19	1.2	336.50	15FE	2.3	-3.3	-1.5	0.8
275.84	-0.37	1.1	346.50	25FE	2.2	-3.3	-1.6	0.8
282.57	-0.56	0.9	356.44	7MR	1.1	-3.3	-1.6	0.7
289.27	-0.77	0.8	6.32	17MR	0.6	-3.3	-1.7	0.7
295.96	-1.01	0.7	16.14	27MR	0.3	-3.3	-1.7	0.6
302.60	-1.26	0.5	25.92	6AP	-0.1	-3.3	-1.8	0.6
309.18	-1.53	0.4	35.63	16AP	-0.8	-3.4	-1.9	0.5
315.68	-1.83	0.2	45.30	26AP	-1.7	-3.4	-2.0	0.5
322.06	-2.15	0.0	54.93	6MY	-1.3	-3.4	-2.0	0.4
328.27	-2.48	-0.1	64.52	16MY	-0.3	-3.5	-2.1	0.3
334.27	-2.84	-0.3	74.08	26MY	0.5	-3.5	-2.2	0.3
339.97	-3.22	-0.5	83.62	5JN	1.2	-3.4	-2.2	0.2
345.29	-3.61	-0.7	93.15	15JN	2.0	-3.4	-2.3	0.3
350.11	-4.02	-0.9	102.68	25JN	3.3	-3.4	-2.3	0.3
354.27	-4.43	-1.2	112.22	5JL	2.2	-3.4	-2.3	0.4
357.60	-4.83	-1.4	121.78	15JL	0.7	-3.3	-2.3	0.5
359.89	-5.20	-1.7	131.35	25JL	-0.6	-3.3	-2.3	0.5
0.92	-5.50	-1.9	140.97	4AU	-1.4	-3.3	-2.3	0.6
0.57	-5.67	-2.2	150.61	14AU	-1.3	-3.3	-2.3	0.6
358.89	-5.65	-2.4	160.31	24AU	-0.6	-3.3	-2.2	0.7
356.26	-5.36	-2.5	170.06	3SE	-0.2	-3.3	-2.2	0.7
353.37	-4.82	-2.5	179.86	13SE	0.0	-3.4	-2.1	0.7
351.00	-4.11	-2.2	189.72	23SE	0.2	-3.4	-2.1	0.8
349.70	-3.32	-1.9	199.64	3OC	0.4	-3.4	-2.0	0.8
349.72	-2.56	-1.6	209.60	13OC	1.6	-3.5	-1.9	0.8
351.00	-1.88	-1.3	219.63	23OC	2.0	-3.5	-1.9	0.8
353.38	-1.29	-1.0	229.71	2NO	-0.1	-3.6	-1.8	0.8
356.66	-0.80	-0.7	239.82	12NO	-0.5	-3.7	-1.7	0.7
0.64	-0.39	-0.4	249.97	22NO	-0.5	-3.8	-1.7	0.7
5.17	-0.05	-0.1	260.15	2DE	-0.6	-3.8	-1.6	0.7
10.11	0.22	0.1	270.34	12DE	-0.7	-3.9	-1.6	0.7
15.37	0.45	0.3	280.53	22DE	-0.8	-4.0	-1.6	0.7
				1547				
20.89	0.64	0.5	290.73	1JA	-0.9	-4.1	-1.6	0.8
26.58	0.79	0.7	300.90	11JA	-0.8	-4.2	-1.5	0.8
32.42	0.92	0.9	311.05	21JA	0.2	-4.3	-1.5	0.8
38.36	1.02	1.0	321.17	31JA	2.6	-4.3	-1.5	0.8
44.39	1.10	1.2	331.23	10FE	1.7	-4.3	-1.5	0.8
50.47	1.16	1.3	341.26	20FE	0.8	-4.0	-1.5	0.8
56.59	1.21	1.4	351.23	2MR	0.4	-3.4	-1.6	0.8
62.74	1.25	1.5	1.14	12MR	0.2	-3.4	-1.6	0.8
68.91	1.28	1.6	11.00	22MR	-0.2	-3.9	-1.6	0.7
75.10	1.30	1.7	20.80	1AP	-0.8	-4.2	-1.7	0.7
81.29	1.31	1.8	30.54	11AP	-1.7	-4.2	-1.7	0.6
87.49	1.31	1.8	40.23	21AP	-1.3	-4.2	-1.8	0.6
93.70	1.31	1.9	49.88	1MY	-0.3	-4.1	-1.8	0.5
99.91	1.31	1.9	59.49	11MY	0.6	-4.0	-1.9	0.5
106.12	1.29	2.0	69.07	21MY	1.5	-3.9	-2.0	0.4
112.35	1.28	2.0	78.62	31MY	2.7	-3.8	-2.0	0.3
118.58	1.25	2.0	88.16	10JN	3.2	-3.7	-2.1	0.3
124.83	1.23	2.0	97.69	20JN	1.8	-3.6	-2.2	0.3
131.09	1.20	2.0	107.22	30JN	0.5	-3.6	-2.2	0.3
137.36	1.16	2.0	116.77	10JL	-0.7	-3.5	-2.3	0.4
143.67	1.13	2.0	126.34	20JL	-1.5	-3.5	-2.4	0.4
149.99	1.08	2.0	135.93	30JL	-1.3	-3.4	-2.4	0.5
156.35	1.04	2.0	145.55	9AU	-0.6	-3.4	-2.4	0.6
162.75	0.99	2.0	155.23	19AU	-0.1	-3.4	-2.4	0.6
169.18	0.94	1.9	164.95	29AU	0.1	-3.4	-2.4	0.7
175.65	0.88	1.9	174.72	8SE	0.3	-3.4	-2.4	0.7
182.16	0.82	1.9	184.55	18SE	0.7	-3.4	-2.4	0.7
188.73	0.75	1.9	194.44	28SE	1.9	-3.4	-2.3	0.8
195.34	0.68	1.9	204.38	8OC	1.8	-3.4	-2.3	0.8
202.01	0.61	1.9	214.38	18OC	-0.2	-3.4	-2.2	0.8
208.73	0.53	1.9	224.42	28OC	-0.7	-3.4	-2.1	0.8
215.51	0.45	1.9	234.51	7NO	-0.7	-3.4	-2.1	0.8
222.34	0.36	1.8	244.65	17NO	-0.7	-3.4	-2.0	0.8
229.24	0.26	1.8	254.81	27NO	-0.8	-3.4	-1.9	0.8
236.19	0.16	1.8	264.99	7DE	-0.7	-3.5	-1.9	0.7
243.20	0.06	1.7	275.19	17DE	-0.7	-3.5	-1.8	0.7
250.27	-0.06	1.7	285.38	27DE	-0.6	-3.5	-1.8	0.7

♂ LONG	♂ LAT	♂ MAG	⊙ LONG	16.00UT	☿	♀	♃	♄
				1548	MAGNITUDES			
257.40	-0.18	1.6	295.57	6JA	0.3	-3.5	-1.7	0.8
264.58	-0.30	1.6	305.73	16JA	2.9	-3.4	-1.7	0.8
271.82	-0.43	1.5	315.86	26JA	1.3	-3.4	-1.6	0.8
279.10	-0.56	1.5	325.96	5FE	0.5	-3.4	-1.6	0.8
286.44	-0.70	1.4	336.01	15FE	0.3	-3.4	-1.6	0.8
293.81	-0.85	1.3	346.00	25FE	0.1	-3.3	-1.6	0.8
301.22	-0.99	1.2	355.95	6MR	-0.3	-3.3	-1.6	0.8
308.65	-1.14	1.2	5.84	16MR	-0.8	-3.3	-1.6	0.8
316.10	-1.28	1.1	15.66	26MR	-1.7	-3.3	-1.6	0.8
323.57	-1.42	1.0	25.44	5AP	-1.3	-3.4	-1.6	0.8
331.02	-1.56	0.9	35.16	15AP	-0.3	-3.4	-1.6	0.7
338.45	-1.69	0.9	44.83	25AP	0.9	-3.4	-1.7	0.7
345.84	-1.80	0.8	54.46	5MY	2.0	-3.4	-1.7	0.6
353.17	-1.91	0.7	64.05	15MY	3.5	-3.4	-1.7	0.5
0.41	-2.01	0.6	73.62	25MY	2.5	-3.5	-1.8	0.5
7.55	-2.08	0.5	83.16	4JN	1.4	-3.5	-1.9	0.4
14.55	-2.14	0.4	92.70	14JN	0.4	-3.6	-1.9	0.3
21.37	-2.18	0.3	102.23	24JN	-0.7	-3.7	-2.0	0.3
27.98	-2.20	0.2	111.76	4JL	-1.5	-3.7	-2.0	0.3
34.32	-2.19	0.1	121.32	14JL	-1.3	-3.8	-2.1	0.4
40.35	-2.16	-0.0	130.89	24JL	-0.5	-3.9	-2.2	0.4
45.99	-2.10	-0.1	140.50	3AU	-0.0	-4.0	-2.3	0.5
51.15	-2.01	-0.3	150.15	13AU	0.3	-4.1	-2.3	0.5
55.72	-1.88	-0.4	159.85	23AU	0.5	-4.2	-2.4	0.6
59.55	-1.71	-0.6	169.59	2SE	0.9	-4.3	-2.4	0.6
62.47	-1.49	-0.8	179.39	12SE	2.4	-4.2	-2.4	0.7
64.28	-1.21	-1.0	189.24	22SE	1.6	-4.0	-2.5	0.7
64.77	-0.85	-1.2	199.16	2OC	-0.3	-3.4	-2.5	0.8
63.80	-0.42	-1.5	209.13	12OC	-0.8	-3.4	-2.4	0.8
61.42	0.06	-1.6	219.15	22OC	-0.8	-4.0	-2.4	0.8
58.01	0.58	-1.8	229.22	1NO	-0.8	-4.3	-2.4	0.8
54.27	1.06	-1.8	239.33	11NO	-0.7	-4.4	-2.3	0.8
51.04	1.46	-1.5	249.48	21NO	-0.6	-4.3	-2.2	0.8
48.89	1.76	-1.2	259.66	1DE	-0.6	-4.3	-2.2	0.8
48.10	1.95	-0.8	269.85	11DE	-0.5	-4.2	-2.1	0.8
48.62	2.06	-0.5	280.04	21DE	0.4	-4.0	-2.0	0.8
50.26	2.11	-0.2	290.23	31DE	3.0	-3.9	-1.9	0.7
				1549				
52.84	2.13	0.0	300.41	10JA	0.9	-3.9	-1.9	0.8
56.16	2.11	0.3	310.56	20JA	0.3	-3.8	-1.8	0.8
60.04	2.08	0.5	320.67	30JA	0.1	-3.7	-1.8	0.9
64.37	2.04	0.7	330.75	9FE	-0.0	-3.6	-1.7	0.9
69.05	1.99	0.9	340.77	19FE	-0.3	-3.5	-1.7	0.9
74.00	1.94	1.1	350.75	1MR	-0.9	-3.5	-1.6	0.9
79.17	1.88	1.2	0.66	11MR	-1.6	-3.4	-1.6	0.9
84.51	1.82	1.4	10.52	21MR	-1.3	-3.4	-1.6	0.9
89.99	1.76	1.5	20.32	31MR	-0.2	-3.4	-1.6	0.9
95.58	1.70	1.6	30.07	10AP	1.1	-3.3	-1.6	0.8
101.27	1.63	1.7	39.77	20AP	2.6	-3.3	-1.6	0.8
107.04	1.56	1.7	49.42	30AP	3.3	-3.3	-1.6	0.8
112.89	1.50	1.8	59.03	10MY	1.9	-3.3	-1.6	0.7
118.80	1.43	1.9	68.61	20MY	1.0	-3.3	-1.6	0.6
124.76	1.36	1.9	78.16	30MY	0.3	-3.3	-1.6	0.6
130.79	1.29	1.9	87.70	9JN	-0.7	-3.3	-1.7	0.5
136.87	1.22	2.0	97.23	19JN	-1.6	-3.4	-1.7	0.4
143.00	1.14	2.0	106.77	29JN	-1.3	-3.4	-1.7	0.4
149.19	1.07	2.0	116.31	9JL	-0.5	-3.4	-1.8	0.3
155.44	0.99	2.0	125.87	19JL	0.1	-3.5	-1.9	0.4
161.74	0.92	2.0	135.46	29JL	0.4	-3.5	-1.9	0.4
168.11	0.84	2.0	145.09	8AU	0.7	-3.5	-2.0	0.5
174.54	0.75	2.0	154.76	18AU	1.3	-3.4	-2.1	0.5
181.04	0.67	1.9	164.48	28AU	2.8	-3.4	-2.1	0.6
187.60	0.58	1.9	174.24	7SE	1.4	-3.4	-2.2	0.6
194.24	0.50	1.9	184.07	17SE	-0.4	-3.3	-2.2	0.7
200.95	0.41	1.8	193.95	27SE	-1.0	-3.3	-2.3	0.7
207.74	0.31	1.8	203.89	7OC	-1.0	-3.3	-2.3	0.8
214.60	0.22	1.7	213.88	17OC	-0.8	-3.3	-2.4	0.8
221.55	0.12	1.7	223.93	27OC	-0.6	-3.3	-2.4	0.8
228.57	0.02	1.7	234.02	6NO	-0.4	-3.3	-2.4	0.9
235.67	-0.08	1.7	244.15	16NO	-0.4	-3.4	-2.4	0.9
242.85	-0.18	1.7	254.32	26NO	-0.3	-3.4	-2.3	0.9
250.11	-0.29	1.7	264.50	6DE	0.6	-3.4	-2.3	0.9
257.44	-0.39	1.6	274.69	16DE	2.9	-3.5	-2.2	0.9
264.85	-0.50	1.6	284.89	26DE	0.6	-3.5	-2.2	0.8

♂ LONG	♂ LAT	♂ MAG	⊙ LONG	16.00UT	☿	♀	♃	♄
				1550	MAGNITUDES			
272.32	-0.61	1.6	295.07	5JA	0.1	-3.6	-2.1	0.8
279.85	-0.71	1.6	305.24	15JA	-0.0	-3.6	-2.0	0.8
287.44	-0.81	1.5	315.38	25JA	-0.1	-3.7	-2.0	0.9
295.08	-0.91	1.5	325.47	4FE	-0.4	-3.7	-1.9	0.9
302.76	-1.00	1.5	335.52	14FE	-0.9	-3.8	-1.8	0.9
310.47	-1.08	1.4	345.52	24FE	-1.4	-3.9	-1.8	1.0
318.20	-1.16	1.4	355.47	6MR	-1.2	-4.0	-1.7	1.0
325.94	-1.23	1.4	5.36	16MR	-0.2	-4.1	-1.6	1.0
333.68	-1.28	1.3	15.19	26MR	1.4	-4.2	-1.6	1.0
341.41	-1.33	1.3	24.96	5AP	3.3	-4.2	-1.6	1.0
349.11	-1.36	1.3	34.69	15AP	2.5	-4.2	-1.5	0.9
356.76	-1.37	1.2	44.36	25AP	1.4	-4.1	-1.5	0.9
4.36	-1.38	1.2	53.99	5MY	0.8	-3.6	-1.5	0.9
11.89	-1.36	1.2	63.59	15MY	0.1	-2.7	-1.5	0.8
19.34	-1.33	1.1	73.15	25MY	-0.7	-3.4	-1.5	0.8
26.69	-1.29	1.1	82.70	4JN	-1.6	-4.0	-1.5	0.7
33.93	-1.22	1.1	92.23	14JN	-1.3	-4.2	-1.5	0.6
41.03	-1.15	1.0	101.76	24JN	-0.5	-4.2	-1.5	0.6
48.00	-1.05	1.0	111.30	4JL	0.2	-4.1	-1.5	0.5
54.80	-0.94	0.9	120.85	14JL	0.6	-4.1	-1.6	0.4
61.42	-0.81	0.9	130.43	24JL	1.0	-4.0	-1.6	0.4
67.83	-0.67	0.8	140.03	3AU	1.7	-3.9	-1.6	0.5
74.01	-0.51	0.7	149.68	13AU	3.2	-3.8	-1.7	0.5
79.92	-0.33	0.7	159.38	23AU	1.2	-3.7	-1.8	0.6
85.51	-0.13	0.6	169.11	2SE	-0.5	-3.7	-1.8	0.6
90.73	0.10	0.5	178.91	12SE	-1.1	-3.6	-1.9	0.7
95.49	0.35	0.3	188.77	22SE	-1.1	-3.5	-1.9	0.7
99.72	0.64	0.2	198.67	2OC	-0.8	-3.5	-2.0	0.8
103.27	0.96	0.0	208.64	12OC	-0.4	-3.5	-2.1	0.8
106.00	1.33	-0.2	218.66	22OC	-0.3	-3.4	-2.1	0.8
107.72	1.74	-0.4	228.73	1NO	-0.3	-3.4	-2.2	0.9
108.24	2.19	-0.6	238.84	11NO	-0.1	-3.4	-2.2	0.9
107.38	2.67	-0.8	248.99	21NO	0.8	-3.4	-2.3	0.9
105.14	3.13	-1.0	259.16	1DE	2.7	-3.4	-2.3	0.9
101.75	3.53	-1.2	269.35	11DE	0.4	-3.4	-2.3	0.9
97.80	3.79	-1.2	279.55	21DE	-0.1	-3.4	-2.3	0.9
94.08	3.89	-1.0	289.74	31DE	-0.2	-3.4	-2.2	0.9
				1551				
91.23	3.84	-0.8	299.92	10JA	-0.3	-3.4	-2.2	0.9
89.65	3.68	-0.5	310.07	20JA	-0.5	-3.4	-2.1	0.9
89.39	3.47	-0.2	320.18	30JA	-0.9	-3.4	-2.1	0.9
90.33	3.24	0.1	330.26	9FE	-1.3	-3.4	-2.0	1.0
92.27	3.01	0.3	340.29	19FE	-1.1	-3.4	-1.9	1.0
95.03	2.79	0.5	350.26	1MR	-0.1	-3.5	-1.8	1.1
98.43	2.58	0.7	0.18	11MR	1.7	-3.5	-1.8	1.1
102.34	2.39	0.9	10.04	21MR	3.4	-3.5	-1.7	1.1
106.66	2.21	1.1	19.84	31MR	1.8	-3.4	-1.6	1.1
111.31	2.04	1.2	29.59	10AP	1.1	-3.4	-1.6	1.1
116.23	1.88	1.3	39.29	20AP	0.6	-3.4	-1.5	1.1
121.37	1.74	1.4	48.94	30AP	0.0	-3.4	-1.5	1.0
126.70	1.60	1.5	58.56	10MY	-0.7	-3.3	-1.5	1.0
132.18	1.46	1.6	68.14	20MY	-1.7	-3.3	-1.4	1.0
137.81	1.33	1.7	77.69	30MY	-1.3	-3.3	-1.4	0.9
143.57	1.21	1.7	87.23	9JN	-0.4	-3.3	-1.4	0.9
149.44	1.09	1.8	96.76	19JN	0.3	-3.3	-1.4	0.8
155.41	0.97	1.8	106.30	29JN	0.7	-3.4	-1.4	0.7
161.50	0.86	1.8	115.84	9JL	1.3	-3.4	-1.4	0.7
167.68	0.75	1.8	125.40	19JL	2.2	-3.4	-1.4	0.6
173.96	0.63	1.9	134.99	29JL	3.0	-3.5	-1.4	0.5
180.34	0.52	1.9	144.62	8AU	1.0	-3.5	-1.4	0.5
186.81	0.41	1.8	154.29	18AU	-0.5	-3.6	-1.5	0.6
193.38	0.31	1.8	164.00	28AU	-1.2	-3.6	-1.5	0.6
200.05	0.20	1.8	173.77	7SE	-1.2	-3.7	-1.5	0.7
206.82	0.09	1.8	183.59	17SE	-0.7	-3.8	-1.6	0.7
213.68	-0.01	1.8	193.47	27SE	-0.4	-3.9	-1.6	0.8
220.65	-0.11	1.7	203.41	7OC	-0.2	-4.0	-1.7	0.8
227.71	-0.22	1.7	213.40	17OC	-0.1	-4.1	-1.8	0.9
234.87	-0.32	1.6	223.44	27OC	0.1	-4.2	-1.8	0.9
242.12	-0.42	1.6	233.54	6NO	1.0	-4.3	-1.9	0.9
249.47	-0.51	1.5	243.66	16NO	2.5	-4.4	-2.0	1.0
256.90	-0.60	1.5	253.82	26NO	0.2	-4.3	-2.0	1.0
264.41	-0.69	1.4	264.01	6DE	-0.2	-4.0	-2.1	1.0
272.00	-0.77	1.4	274.20	16DE	-0.3	-3.4	-2.1	1.0
279.66	-0.85	1.4	284.39	26DE	-0.4	-3.4	-2.1	1.1

♂ LONG	LAT	MAG	☉ LONG	16.00UT	☿	♀	♃	♄
				1552	MAGNITUDES			
287.38	-0.91	1.4	294.58	5JA	-0.6	-4.0	-2.1	1.1
295.15	-0.98	1.4	304.74	15JA	-0.9	-4.3	-2.1	1.1
302.96	-1.03	1.4	314.88	25JA	-1.1	-4.4	-2.1	1.0
310.80	-1.07	1.4	324.98	4FE	-1.0	-4.3	-2.1	1.0
318.66	-1.10	1.4	335.03	14FE	-0.0	-4.2	-2.0	1.1
326.52	-1.12	1.4	345.04	24FE	2.0	-4.1	-2.0	1.1
334.38	-1.13	1.4	354.99	5MR	2.7	-4.0	-1.9	1.2
342.21	-1.12	1.4	4.88	15MR	1.3	-3.9	-1.9	1.2
350.01	-1.10	1.4	14.71	25MR	0.8	-3.8	-1.8	1.2
357.77	-1.07	1.4	24.49	4AP	0.4	-3.7	-1.7	1.2
5.47	-1.03	1.4	34.21	14AP	-0.0	-3.6	-1.7	1.2
13.11	-0.98	1.4	43.89	24AP	-0.7	-3.5	-1.6	1.2
20.67	-0.91	1.4	53.53	4MY	-1.7	-3.5	-1.5	1.2
28.15	-0.83	1.4	63.12	14MY	-1.3	-3.4	-1.5	1.2
35.53	-0.74	1.4	72.69	24MY	-0.4	-3.4	-1.4	1.1
42.82	-0.65	1.4	82.24	3JN	0.4	-3.4	-1.4	1.1
50.00	-0.54	1.4	91.77	13JN	1.0	-3.3	-1.3	1.0
57.07	-0.43	1.4	101.30	23JN	1.7	-3.3	-1.3	1.0
64.03	-0.30	1.4	110.84	3JL	2.9	-3.3	-1.3	0.9
70.86	-0.17	1.4	120.39	13JL	2.6	-3.3	-1.3	0.9
77.55	-0.04	1.4	129.97	23JL	0.8	-3.3	-1.3	0.8
84.11	0.11	1.4	139.57	2AU	-0.6	-3.3	-1.3	0.7
90.52	0.26	1.3	149.22	12AU	-1.3	-3.4	-1.3	0.7
96.76	0.42	1.3	158.91	22AU	-1.3	-3.4	-1.3	0.7
102.82	0.59	1.2	168.65	1SE	-0.7	-3.4	-1.3	0.7
108.68	0.77	1.2	178.44	11SE	-0.3	-3.4	-1.3	0.7
114.31	0.96	1.1	188.29	21SE	-0.1	-3.5	-1.4	0.8
119.66	1.17	1.0	198.19	1OC	0.1	-3.5	-1.4	0.8
124.71	1.40	0.9	208.16	11OC	0.3	-3.5	-1.5	0.9
129.38	1.64	0.8	218.18	21OC	1.3	-3.5	-1.5	0.9
133.59	1.91	0.7	228.24	31OC	2.3	-3.4	-1.6	1.0
137.24	2.20	0.5	238.35	10NO	0.0	-3.4	-1.6	1.0
140.20	2.53	0.3	248.49	20NO	-0.4	-3.4	-1.7	1.1
142.31	2.89	0.1	258.66	30NO	-0.4	-3.4	-1.8	1.1
143.38	3.28	-0.1	268.85	10DE	-0.5	-3.3	-1.8	1.1
143.23	3.68	-0.4	279.05	20DE	-0.7	-3.3	-1.9	1.2
141.75	4.06	-0.6	289.24	30DE	-0.9	-3.3	-1.9	1.2
				1553				
139.01	4.36	-0.8	299.42	9JA	-1.0	-3.3	-2.0	1.2
135.33	4.54	-1.0	309.57	19JA	-0.9	-3.4	-2.0	1.2
131.37	4.53	-1.0	319.69	29JA	0.0	-3.4	-2.0	1.2
127.85	4.36	-0.8	329.77	8FE	2.4	-3.4	-2.1	1.2
125.34	4.06	-0.5	339.80	18FE	2.1	-3.4	-2.0	1.2
124.11	3.70	-0.3	349.78	28FE	1.0	-3.5	-2.0	1.3
124.16	3.33	-0.0	359.70	10MR	0.6	-3.5	-2.0	1.3
125.33	2.97	0.2	9.57	20MR	0.3	-3.5	-1.9	1.3
127.47	2.63	0.4	19.37	30MR	-0.1	-3.6	-1.9	1.4
130.37	2.32	0.6	29.12	9AP	-0.8	-3.7	-1.8	1.4
133.90	2.05	0.8	38.83	19AP	-1.7	-3.7	-1.8	1.4
137.94	1.79	0.9	48.48	29AP	-1.3	-3.8	-1.7	1.4
142.38	1.56	1.1	58.10	9MY	-0.3	-3.9	-1.6	1.4
147.18	1.35	1.2	67.68	19MY	0.5	-4.0	-1.6	1.4
152.26	1.15	1.3	77.23	29MY	1.3	-4.1	-1.5	1.4
157.58	0.97	1.3	86.77	8JN	2.2	-4.1	-1.4	1.3
163.12	0.80	1.4	96.31	18JN	3.4	-4.2	-1.4	1.3
168.86	0.64	1.5	105.84	28JN	2.1	-4.1	-1.3	1.2
174.77	0.49	1.5	115.38	8JL	0.7	-3.9	-1.3	1.2
180.84	0.34	1.6	124.94	18JL	-0.6	-3.4	-1.3	1.1
187.06	0.21	1.6	134.53	28JL	-1.4	-3.2	-1.2	1.0
193.43	0.07	1.6	144.15	7AU	-1.3	-3.8	-1.2	0.9
199.93	-0.05	1.6	153.82	17AU	-0.6	-4.1	-1.2	0.9
206.57	-0.17	1.6	163.53	27AU	-0.2	-4.3	-1.2	0.8
213.34	-0.29	1.6	173.29	6SE	0.1	-4.2	-1.2	0.8
220.23	-0.39	1.6	183.12	16SE	0.2	-4.2	-1.2	0.9
227.25	-0.50	1.6	192.99	26SE	0.5	-4.1	-1.2	0.9
234.38	-0.59	1.6	202.93	6OC	1.7	-4.0	-1.3	1.0
241.63	-0.68	1.6	212.92	16OC	2.1	-3.9	-1.3	1.0
248.98	-0.76	1.5	222.96	26OC	-0.1	-3.8	-1.3	1.1
256.43	-0.84	1.5	233.05	5NO	-0.6	-3.8	-1.4	1.1
263.98	-0.90	1.5	243.17	15NO	-0.6	-3.7	-1.4	1.2
271.61	-0.96	1.5	253.33	25NO	-0.6	-3.6	-1.5	1.2
279.32	-1.01	1.4	263.51	5DE	-0.7	-3.6	-1.5	1.3
287.10	-1.04	1.4	273.71	15DE	-0.8	-3.5	-1.6	1.3
294.92	-1.07	1.4	283.90	25DE	-0.8	-3.5	-1.7	1.3

♂ LONG	LAT	MAG	☉ LONG	16.00UT	☿	♀	♃	♄
				1554	MAGNITUDES			
302.79	-1.08	1.4	294.09	4JA	-0.7	-3.4	-1.7	1.4
310.69	-1.09	1.3	304.26	14JA	0.1	-3.4	-1.8	1.4
318.59	-1.08	1.3	314.39	24JA	2.7	-3.4	-1.9	1.4
326.50	-1.06	1.3	324.50	3FE	1.6	-3.3	-1.9	1.4
334.39	-1.03	1.3	334.55	13FE	0.7	-3.3	-2.0	1.4
342.26	-0.98	1.3	344.56	23FE	0.4	-3.3	-2.0	1.3
350.09	-0.93	1.3	354.51	5MR	0.1	-3.3	-2.0	1.3
357.87	-0.86	1.4	4.40	15MR	-0.2	-3.3	-2.0	1.3
5.58	-0.79	1.4	14.24	25MR	-0.8	-3.3	-2.0	1.3
13.23	-0.71	1.4	24.02	4AP	-1.6	-3.3	-2.0	1.3
20.81	-0.62	1.5	33.75	14AP	-1.3	-3.4	-1.9	1.3
28.30	-0.53	1.5	43.42	24AP	-0.3	-3.4	-1.9	1.3
35.71	-0.43	1.5	53.06	4MY	0.7	-3.4	-1.8	1.2
43.03	-0.33	1.6	62.66	14MY	1.7	-3.5	-1.8	1.2
50.25	-0.22	1.6	72.23	24MY	3.0	-3.5	-1.7	1.2
57.38	-0.11	1.6	81.77	3JN	3.0	-3.4	-1.6	1.1
64.41	0.00	1.6	91.31	13JN	1.6	-3.4	-1.6	1.1
71.35	0.12	1.7	100.84	23JN	0.5	-3.4	-1.5	1.1
78.19	0.24	1.7	110.37	3JL	-0.6	-3.4	-1.5	1.0
84.94	0.35	1.7	119.92	13JL	-1.5	-3.3	-1.4	1.0
91.58	0.47	1.7	129.49	23JL	-1.3	-3.3	-1.4	0.9
98.13	0.59	1.7	139.10	2AU	-0.6	-3.3	-1.3	0.9
104.57	0.72	1.7	148.74	12AU	-0.1	-3.3	-1.3	0.8
110.90	0.84	1.7	158.43	22AU	0.2	-3.3	-1.3	0.8
117.13	0.97	1.7	168.17	1SE	0.4	-3.3	-1.2	0.7
123.22	1.10	1.6	177.96	11SE	0.8	-3.4	-1.2	0.7
129.19	1.23	1.6	187.80	21SE	2.0	-3.4	-1.2	0.8
135.02	1.37	1.5	197.71	1OC	1.8	-3.4	-1.2	0.8
140.67	1.51	1.5	207.67	11OC	-0.2	-3.5	-1.2	0.9
146.14	1.66	1.4	217.68	21OC	-0.7	-3.5	-1.2	1.0
151.38	1.82	1.3	227.75	31OC	-0.7	-3.6	-1.2	1.0
156.35	1.99	1.2	237.86	10NO	-0.7	-3.7	-1.3	1.1
161.01	2.17	1.0	248.00	20NO	-0.7	-3.8	-1.3	1.2
165.28	2.37	0.9	258.18	30NO	-0.6	-3.8	-1.4	1.2
169.07	2.59	0.7	268.36	10DE	-0.7	-3.9	-1.4	1.2
172.27	2.82	0.5	278.56	20DE	-0.6	-4.0	-1.5	1.2
174.73	3.06	0.3	288.75	30DE	0.3	-4.1	-1.5	1.2
				1555				
176.28	3.32	0.1	298.93	9JA	2.9	-4.2	-1.6	1.2
176.76	3.58	-0.2	309.08	19JA	1.2	-4.3	-1.6	1.2
175.99	3.81	-0.5	319.21	29JA	0.4	-4.4	-1.7	1.2
173.93	3.98	-0.7	329.28	8FE	0.2	-4.3	-1.8	1.2
170.75	4.03	-0.9	339.32	18FE	0.0	-4.0	-1.8	1.1
166.92	3.93	-1.1	349.30	28FE	-0.3	-3.4	-1.9	1.1
163.14	3.67	-0.9	359.22	10MR	-0.8	-3.4	-2.0	1.0
160.08	3.30	-0.8	9.09	20MR	-1.6	-4.0	-2.0	1.0
158.17	2.87	-0.5	18.90	30MR	-1.3	-4.2	-2.0	1.0
157.57	2.43	-0.3	28.65	9AP	-0.3	-4.2	-2.0	1.0
158.20	2.01	-0.1	38.36	19AP	0.9	-4.2	-2.0	1.0
159.91	1.63	0.1	48.02	29AP	2.2	-4.1	-2.0	1.0
162.52	1.29	0.3	57.63	9MY	3.8	-4.0	-2.0	1.0
165.87	0.98	0.5	67.21	19MY	2.3	-3.9	-1.9	1.0
169.83	0.71	0.6	76.77	29MY	1.2	-3.8	-1.9	0.9
174.29	0.47	0.7	86.31	8JN	0.4	-3.7	-1.8	0.9
179.16	0.25	0.8	95.84	18JN	-0.6	-3.6	-1.8	0.9
184.39	0.05	0.9	105.38	28JN	-1.6	-3.6	-1.7	0.8
189.93	-0.13	1.0	114.92	8JL	-1.3	-3.5	-1.6	0.8
195.74	-0.29	1.1	124.48	18JL	-0.5	-3.5	-1.6	0.7
201.79	-0.44	1.1	134.07	28JL	-0.0	-3.4	-1.5	0.7
208.05	-0.57	1.2	143.69	7AU	0.3	-3.4	-1.5	0.6
214.51	-0.69	1.2	153.35	17AU	0.6	-3.4	-1.4	0.6
221.16	-0.80	1.2	163.06	27AU	1.1	-3.4	-1.4	0.5
227.98	-0.90	1.3	172.82	6SE	2.5	-3.4	-1.3	0.5
234.95	-0.99	1.3	182.64	16SE	1.6	-3.4	-1.3	0.4
242.08	-1.06	1.3	192.51	26SE	-0.3	-3.4	-1.3	0.4
249.34	-1.12	1.3	202.44	6OC	-0.9	-3.4	-1.3	0.5
256.72	-1.17	1.3	212.43	16OC	-0.9	-3.4	-1.3	0.6
264.21	-1.20	1.3	222.47	26OC	-0.9	-3.4	-1.2	0.6
271.81	-1.22	1.3	232.55	5NO	-0.6	-3.4	-1.3	0.7
279.49	-1.23	1.3	242.68	15NO	-0.5	-3.4	-1.3	0.8
287.24	-1.23	1.4	252.84	25NO	-0.5	-3.4	-1.3	0.8
295.05	-1.21	1.4	263.02	5DE	-0.4	-3.5	-1.3	0.9
302.90	-1.18	1.4	273.21	15DE	0.4	-3.5	-1.3	0.9
310.78	-1.14	1.4	283.41	25DE	3.0	-3.5	-1.4	0.9

♂ LONG	♂ LAT	♂ MAG	☉ LONG	16.00UT 1556	☿	♀	♃	♄
318.67	-1.09	1.4	293.59	4JA	0.8	-3.5	-1.4	1.0
326.55	-1.03	1.4	303.77	14JA	0.2	-3.4	-1.5	1.0
334.42	-0.95	1.4	313.90	24JA	0.1	-3.4	-1.5	1.0
342.26	-0.87	1.4	324.00	3FE	-0.1	-3.4	-1.6	1.0
350.05	-0.79	1.4	334.06	13FE	-0.4	-3.4	-1.7	0.9
357.79	-0.69	1.4	344.07	23FE	-0.8	-3.3	-1.7	0.9
5.46	-0.60	1.5	354.02	4MR	-1.5	-3.3	-1.8	0.9
13.07	-0.49	1.5	3.92	14MR	-1.2	-3.3	-1.9	0.8
20.59	-0.39	1.5	13.76	24MR	-0.2	-3.3	-1.9	0.8
28.04	-0.28	1.5	23.54	3AP	1.2	-3.4	-2.0	0.8
35.40	-0.18	1.5	33.28	13AP	2.8	-3.4	-2.0	0.8
42.68	-0.07	1.5	42.96	23AP	3.0	-3.4	-2.1	0.8
49.87	0.04	1.6	52.59	3MY	1.7	-3.4	-2.1	0.8
56.97	0.14	1.6	62.20	13MY	0.9	-3.4	-2.1	0.8
63.99	0.24	1.7	71.77	23MY	0.2	-3.5	-2.1	0.8
70.93	0.35	1.7	81.31	2JN	-0.7	-3.5	-2.1	0.7
77.79	0.44	1.8	90.85	12JN	-1.6	-3.6	-2.1	0.7
84.56	0.54	1.8	100.38	22JN	-1.3	-3.7	-2.0	0.7
91.26	0.64	1.8	109.91	2JL	-0.5	-3.7	-2.0	0.7
97.89	0.73	1.9	119.47	12JL	0.1	-3.8	-1.9	0.6
104.45	0.82	1.9	129.04	22JL	0.5	-3.9	-1.8	0.6
110.94	0.90	1.9	138.64	1AU	0.8	-4.0	-1.8	0.5
117.36	0.99	1.9	148.28	11AU	1.4	-4.1	-1.7	0.5
123.71	1.07	1.9	157.96	21AU	3.0	-4.2	-1.6	0.4
130.00	1.15	1.9	167.70	31AU	1.4	-4.3	-1.6	0.3
136.22	1.23	1.9	177.49	10SE	-0.4	-4.2	-1.5	0.3
142.36	1.31	1.9	187.33	20SE	-1.0	-3.9	-1.5	0.2
148.43	1.39	1.8	197.23	30SE	-1.0	-3.4	-1.4	0.2
154.41	1.47	1.8	207.19	10OC	-0.8	-3.4	-1.4	0.2
160.30	1.55	1.7	217.20	20OC	-0.5	-4.0	-1.4	0.3
166.08	1.62	1.7	227.26	30OC	-0.4	-4.3	-1.4	0.3
171.74	1.70	1.6	237.37	9NO	-0.3	-4.4	-1.3	0.4
177.25	1.78	1.5	247.51	19NO	-0.2	-4.3	-1.3	0.5
182.60	1.87	1.4	257.68	29NO	0.6	-4.2	-1.3	0.5
187.73	1.95	1.2	267.87	9DE	2.9	-4.1	-1.3	0.6
192.62	2.03	1.1	278.06	19DE	0.6	-4.0	-1.4	0.6
197.21	2.12	0.9	288.26	29DE	0.0	-3.9	-1.4	0.7
				1557				
201.43	2.20	0.8	298.44	8JA	-0.1	-3.8	-1.4	0.7
205.18	2.28	0.6	308.59	18JA	-0.2	-3.8	-1.4	0.7
208.36	2.36	0.3	318.72	28JA	-0.4	-3.7	-1.5	0.7
210.82	2.42	0.1	328.80	7FE	-0.9	-3.6	-1.5	0.7
212.39	2.47	-0.2	338.84	17FE	-1.4	-3.5	-1.6	0.7
212.89	2.47	-0.5	348.82	27FE	-1.2	-3.5	-1.7	0.7
212.17	2.42	-0.8	358.75	9MR	-0.2	-3.4	-1.7	0.7
210.19	2.29	-1.1	8.62	19MR	1.4	-3.4	-1.8	0.6
207.11	2.05	-1.3	18.43	29MR	3.5	-3.4	-1.9	0.6
203.43	1.70	-1.4	28.19	8AP	2.2	-3.3	-1.9	0.5
199.85	1.28	-1.3	37.89	18AP	1.3	-3.3	-2.0	0.5
197.02	0.83	-1.1	47.55	28AP	0.7	-3.3	-2.1	0.6
195.41	0.39	-0.9	57.17	8MY	0.1	-3.3	-2.1	0.6
195.13	-0.01	-0.7	66.76	18MY	-0.7	-3.3	-2.2	0.6
196.13	-0.35	-0.5	76.32	28MY	-1.6	-3.3	-2.2	0.6
198.25	-0.64	-0.3	85.86	7JN	-1.4	-3.3	-2.2	0.6
201.30	-0.88	-0.1	95.39	17JN	-0.5	-3.4	-2.2	0.6
205.12	-1.09	0.0	104.92	27JN	0.2	-3.4	-2.2	0.5
209.59	-1.25	0.1	114.46	7JL	0.6	-3.4	-2.2	0.5
214.59	-1.39	0.3	124.02	17JL	1.1	-3.5	-2.2	0.5
220.03	-1.50	0.4	133.60	27JL	1.9	-3.5	-2.1	0.4
225.86	-1.59	0.5	143.22	6AU	3.2	-3.5	-2.1	0.4
232.01	-1.66	0.6	152.88	16AU	1.2	-3.4	-2.0	0.3
238.44	-1.70	0.6	162.59	26AU	-0.5	-3.4	-1.9	0.3
245.12	-1.73	0.7	172.34	5SE	-1.1	-3.4	-1.9	0.2
252.01	-1.74	0.8	182.16	15SE	-1.2	-3.3	-1.8	0.2
259.09	-1.73	0.8	192.03	25SE	-0.8	-3.3	-1.8	0.1
266.33	-1.70	0.9	201.96	5OC	-0.4	-3.3	-1.7	0.0
273.71	-1.66	0.9	211.94	15OC	-0.2	-3.3	-1.6	-0.0
281.20	-1.60	1.0	221.98	25OC	-0.2	-3.3	-1.6	0.0
288.79	-1.53	1.0	232.06	4NO	-0.1	-3.3	-1.5	0.1
296.45	-1.45	1.1	242.18	14NO	0.8	-3.4	-1.5	0.1
304.16	-1.35	1.1	252.34	24NO	2.7	-3.4	-1.5	0.2
311.91	-1.25	1.2	262.52	4DE	0.3	-3.4	-1.5	0.3
319.67	-1.14	1.2	272.71	14DE	-0.1	-3.5	-1.4	0.3
327.43	-1.02	1.3	282.91	24DE	-0.2	-3.5	-1.4	0.4

♂ LONG	♂ LAT	♂ MAG	☉ LONG	16.00UT 1558	☿	♀	♃	♄
335.17	-0.90	1.3	293.10	3JA	-0.3	-3.6	-1.4	0.5
342.88	-0.78	1.4	303.27	13JA	-0.5	-3.6	-1.5	0.5
350.55	-0.65	1.4	313.41	23JA	-0.9	-3.7	-1.5	0.5
358.16	-0.52	1.5	323.52	2FE	-1.2	-3.8	-1.5	0.5
5.71	-0.40	1.5	333.58	12FE	-1.1	-3.8	-1.5	0.6
13.20	-0.28	1.5	343.59	22FE	-0.1	-3.9	-1.6	0.6
20.60	-0.16	1.6	353.55	4MR	1.8	-4.0	-1.6	0.5
27.93	-0.04	1.6	3.45	14MR	3.1	-4.1	-1.7	0.5
35.18	0.07	1.7	13.29	24MR	1.6	-4.2	-1.7	0.5
42.36	0.18	1.7	23.08	3AP	0.9	-4.2	-1.8	0.4
49.45	0.28	1.7	32.81	13AP	0.5	-4.2	-1.8	0.4
56.46	0.38	1.7	42.50	23AP	0.0	-4.1	-1.9	0.4
63.40	0.47	1.8	52.13	3MY	-0.7	-3.6	-2.0	0.4
70.26	0.56	1.8	61.73	13MY	-1.6	-2.7	-2.1	0.4
77.06	0.64	1.8	71.31	23MY	-1.4	-3.5	-2.1	0.4
83.79	0.72	1.8	80.85	2JN	-0.4	-4.0	-2.2	0.4
90.46	0.79	1.8	90.39	12JN	0.3	-4.2	-2.2	0.4
97.07	0.86	1.8	99.92	22JN	0.8	-4.2	-2.3	0.4
103.63	0.92	1.9	109.45	2JL	1.4	-4.1	-2.3	0.4
110.14	0.98	1.9	119.00	12JL	2.4	-4.0	-2.4	0.4
116.61	1.03	2.0	128.57	22JL	2.9	-4.0	-2.4	0.4
123.04	1.08	2.0	138.17	1AU	1.0	-3.9	-2.4	0.4
129.43	1.13	2.0	147.81	11AU	-0.5	-3.8	-2.3	0.3
135.79	1.17	2.0	157.49	21AU	-1.3	-3.7	-2.3	0.3
142.11	1.21	2.0	167.22	31AU	-1.3	-3.7	-2.2	0.2
148.40	1.24	2.0	177.01	10SE	-0.7	-3.6	-2.2	0.2
154.67	1.27	2.0	186.85	20SE	-0.3	-3.5	-2.1	0.1
160.90	1.30	2.0	196.75	30SE	-0.1	-3.5	-2.1	0.0
167.10	1.32	2.0	206.70	10OC	-0.0	-3.5	-2.0	-0.0
173.27	1.34	1.9	216.71	20OC	0.1	-3.4	-1.9	-0.1
179.40	1.36	1.9	226.77	30OC	1.1	-3.4	-1.9	-0.2
185.49	1.37	1.8	236.87	9NO	2.5	-3.4	-1.8	-0.1
191.53	1.38	1.7	247.02	19NO	0.2	-3.4	-1.7	-0.1
197.51	1.38	1.7	257.18	29NO	-0.3	-3.4	-1.7	0.0
203.43	1.37	1.6	267.37	9DE	-0.4	-3.4	-1.6	0.1
209.26	1.36	1.5	277.57	19DE	-0.4	-3.4	-1.6	0.2
214.99	1.34	1.4	287.76	29DE	-0.6	-3.4	-1.6	0.2
				1559				
220.60	1.31	1.2	297.94	8JA	-0.9	-3.4	-1.6	0.3
226.07	1.26	1.1	308.10	18JA	-1.1	-3.4	-1.5	0.3
231.36	1.20	0.9	318.22	28JA	-1.0	-3.4	-1.5	0.4
236.44	1.12	0.7	328.31	7FE	-0.1	-3.4	-1.5	0.4
241.25	1.02	0.6	338.35	17FE	2.1	-3.4	-1.5	0.4
245.72	0.89	0.3	348.33	27FE	2.5	-3.5	-1.6	0.4
249.77	0.72	0.1	358.26	9MR	1.2	-3.5	-1.6	0.4
253.29	0.50	-0.2	8.14	19MR	0.7	-3.5	-1.6	0.4
256.15	0.22	-0.4	17.95	29MR	0.4	-3.4	-1.6	0.4
258.17	-0.12	-0.7	27.71	8AP	-0.1	-3.4	-1.7	0.4
259.18	-0.55	-1.1	37.42	18AP	-0.7	-3.4	-1.7	0.3
259.01	-1.06	-1.4	47.08	28AP	-1.6	-3.4	-1.8	0.3
257.58	-1.64	-1.7	56.70	8MY	-1.4	-3.3	-1.9	0.2
255.06	-2.26	-2.0	66.29	18MY	-0.4	-3.3	-1.9	0.3
251.89	-2.85	-2.1	75.85	28MY	0.4	-3.3	-2.0	0.3
248.76	-3.33	-2.0	85.39	7JN	1.1	-3.3	-2.1	0.3
246.38	-3.65	-1.8	94.92	17JN	1.9	-3.3	-2.1	0.4
245.23	-3.83	-1.6	104.45	27JN	3.1	-3.4	-2.2	0.4
245.46	-3.88	-1.4	113.99	7JL	2.4	-3.4	-2.3	0.4
247.02	-3.85	-1.2	123.55	17JL	0.8	-3.4	-2.3	0.4
249.75	-3.75	-1.0	133.13	27JL	-0.6	-3.5	-2.4	0.3
253.45	-3.61	-0.8	142.75	6AU	-1.4	-3.5	-2.4	0.3
257.94	-3.44	-0.6	152.41	16AU	-1.3	-3.6	-2.4	0.3
263.07	-3.25	-0.4	162.11	26AU	-0.7	-3.6	-2.5	0.3
268.71	-3.05	-0.3	171.87	5SE	-0.2	-3.7	-2.4	0.2
274.78	-2.83	-0.1	181.68	15SE	-0.0	-3.8	-2.4	0.1
281.17	-2.61	0.0	191.55	25SE	0.1	-3.9	-2.4	0.1
287.83	-2.38	0.1	201.48	5OC	0.4	-4.0	-2.3	0.0
294.71	-2.15	0.3	211.46	15OC	1.4	-4.1	-2.3	-0.1
301.73	-1.92	0.4	221.49	25OC	2.3	-4.2	-2.2	-0.1
308.89	-1.70	0.5	231.57	4NO	-0.0	-4.3	-2.1	-0.2
316.13	-1.47	0.6	241.69	14NO	-0.5	-4.4	-2.1	-0.3
323.42	-1.26	0.7	251.84	24NO	-0.5	-4.3	-2.0	-0.2
330.76	-1.05	0.8	262.03	4DE	-0.5	-4.0	-1.9	-0.1
338.10	-0.85	0.9	272.22	14DE	-0.7	-3.3	-1.9	-0.1
345.43	-0.66	1.0	282.41	24DE	-0.9	-3.4	-1.8	-0.0

♂ LONG	♂ LAT	♂ MAG	⊙ LONG	16.00UT	☿	♀	♃	♄
				1560	MAGNITUDES			
352.75	-0.48	1.1	292.60	3JA	-0.9	-4.0	-1.8	0.1
0.03	-0.31	1.2	302.77	13JA	-0.8	-4.3	-1.7	0.1
7.27	-0.15	1.3	312.91	23JA	0.0	-4.4	-1.7	0.2
14.47	-0.01	1.4	323.02	2FE	2.4	-4.3	-1.6	0.2
21.60	0.13	1.5	333.09	12FE	1.9	-4.2	-1.6	0.3
28.68	0.25	1.5	343.10	22FE	0.8	-4.1	-1.6	0.3
35.69	0.37	1.6	353.06	3MR	0.5	-4.0	-1.6	0.3
42.65	0.47	1.7	2.97	13MR	0.2	-3.9	-1.6	0.3
49.53	0.57	1.7	12.81	23MR	-0.1	-3.8	-1.6	0.3
56.36	0.65	1.8	22.61	2AP	-0.7	-3.7	-1.6	0.3
63.13	0.73	1.8	32.34	12AP	-1.6	-3.6	-1.6	0.3
69.84	0.80	1.9	42.03	22AP	-1.4	-3.5	-1.6	0.3
76.50	0.86	1.9	51.67	2MY	-0.4	-3.5	-1.7	0.3
83.11	0.92	1.9	61.27	12MY	0.6	-3.4	-1.7	0.2
89.68	0.97	1.9	70.85	22MY	1.4	-3.4	-1.8	0.2
96.20	1.01	2.0	80.40	1JN	2.5	-3.4	-1.8	0.2
102.69	1.05	2.0	89.93	11JN	3.3	-3.3	-1.9	0.3
109.14	1.08	2.0	99.46	21JN	1.9	-3.3	-1.9	0.3
115.57	1.10	2.0	109.00	1JL	0.6	-3.3	-2.0	0.3
121.98	1.12	1.9	118.54	11JL	-0.6	-3.3	-2.1	0.3
128.37	1.14	1.9	128.11	21JL	-1.4	-3.3	-2.1	0.3
134.74	1.15	2.0	137.71	31JL	-1.3	-3.3	-2.2	0.3
141.11	1.16	2.0	147.35	10AU	-0.6	-3.4	-2.3	0.3
147.47	1.16	2.0	157.03	20AU	-0.1	-3.4	-2.3	0.3
153.84	1.16	2.0	166.76	30AU	0.1	-3.4	-2.4	0.3
160.20	1.15	2.0	176.53	9SE	0.3	-3.4	-2.4	0.2
166.57	1.14	2.0	186.37	19SE	0.6	-3.5	-2.4	0.2
172.94	1.12	2.0	196.27	29SE	1.7	-3.5	-2.5	0.1
179.32	1.10	2.0	206.22	9OC	2.1	-3.5	-2.5	0.1
185.72	1.07	2.0	216.22	19OC	-0.1	-3.5	-2.4	-0.0
192.12	1.04	2.0	226.28	29OC	-0.6	-3.4	-2.4	-0.1
198.53	1.00	1.9	236.38	8NO	-0.6	-3.4	-2.3	-0.2
204.95	0.95	1.9	246.52	18NO	-0.7	-3.4	-2.3	-0.2
211.37	0.90	1.8	256.69	28NO	-0.8	-3.4	-2.2	-0.3
217.80	0.84	1.8	266.87	8DE	-0.7	-3.3	-2.1	-0.2
224.24	0.77	1.7	277.07	18DE	-0.8	-3.3	-2.1	-0.2
230.67	0.69	1.6	287.26	28DE	-0.7	-3.3	-2.0	-0.1
				1561				
237.10	0.59	1.5	297.44	7JA	0.1	-3.3	-1.9	-0.0
243.52	0.49	1.4	307.60	17JA	2.7	-3.4	-1.9	0.0
249.93	0.36	1.3	317.73	27JA	1.4	-3.4	-1.8	0.1
256.31	0.22	1.2	327.81	6FE	0.6	-3.4	-1.7	0.2
262.67	0.06	1.0	337.86	16FE	0.3	-3.4	-1.7	0.2
268.98	-0.12	0.9	347.85	26FE	0.1	-3.5	-1.7	0.3
275.24	-0.32	0.8	357.78	8MR	-0.2	-3.5	-1.6	0.3
281.43	-0.55	0.6	7.66	18MR	-0.8	-3.5	-1.6	0.3
287.52	-0.82	0.4	17.48	28MR	-1.6	-3.6	-1.6	0.3
293.49	-1.12	0.2	27.24	7AP	-1.3	-3.7	-1.6	0.3
299.29	-1.45	0.0	36.95	17AP	-0.3	-3.7	-1.6	0.3
304.88	-1.83	-0.2	46.62	27AP	0.8	-3.8	-1.6	0.3
310.19	-2.26	-0.4	56.24	7MY	1.8	-3.9	-1.6	0.3
315.14	-2.74	-0.6	65.83	17MY	3.3	-4.0	-1.6	0.3
319.61	-3.27	-0.9	75.39	27MY	2.7	-4.1	-1.6	0.2
323.47	-3.85	-1.1	84.93	6JN	1.5	-4.1	-1.6	0.2
326.53	-4.48	-1.4	94.47	16JN	0.5	-4.2	-1.7	0.2
328.60	-5.14	-1.7	104.00	26JN	-0.6	-4.1	-1.7	0.3
329.50	-5.78	-2.0	113.54	6JL	-1.5	-3.9	-1.7	0.3
329.09	-6.34	-2.3	123.10	16JL	-1.3	-3.3	-1.8	0.3
327.48	-6.73	-2.5	132.68	26JL	-0.6	-3.2	-1.9	0.3
325.04	-6.83	-2.7	142.29	5AU	-0.1	-3.8	-1.9	0.4
322.44	-6.60	-2.6	151.95	15AU	0.2	-4.1	-2.0	0.4
320.45	-6.09	-2.4	161.65	25AU	0.5	-4.2	-2.1	0.3
319.55	-5.39	-2.1	171.40	4SE	0.9	-4.2	-2.1	0.3
319.96	-4.63	-1.8	181.21	14SE	2.2	-4.2	-2.2	0.3
321.65	-3.88	-1.5	191.07	24SE	1.8	-4.1	-2.2	0.2
324.41	-3.18	-1.2	200.99	4OC	-0.2	-4.0	-2.3	0.2
328.07	-2.56	-0.9	210.97	14OC	-0.8	-3.9	-2.3	0.1
332.43	-2.01	-0.7	221.00	24OC	-0.8	-3.8	-2.4	0.1
337.32	-1.53	-0.4	231.08	3NO	-0.8	-3.8	-2.4	-0.0
342.64	-1.11	-0.2	241.20	13NO	-0.7	-3.7	-2.4	-0.1
348.27	-0.75	0.0	251.35	23NO	-0.6	-3.6	-2.4	-0.1
354.14	-0.44	0.2	261.53	3DE	-0.6	-3.6	-2.3	-0.2
0.19	-0.17	0.4	271.73	13DE	-0.5	-3.5	-2.3	-0.3
6.38	0.06	0.6	281.92	23DE	0.3	-3.5	-2.2	-0.2

♂ LONG	♂ LAT	♂ MAG	⊙ LONG	16.00UT	☿	♀	♃	♄
				1562	MAGNITUDES			
12.66	0.25	0.8	292.11	2JA	2.9	-3.4	-2.1	-0.2
19.01	0.42	0.9	302.28	12JA	1.1	-3.4	-2.1	-0.1
25.40	0.57	1.0	312.43	22JA	0.3	-3.4	-2.0	-0.0
31.82	0.69	1.2	322.54	1FE	0.2	-3.3	-1.9	0.0
38.26	0.80	1.3	332.60	11FE	-0.0	-3.3	-1.9	0.1
44.70	0.89	1.4	342.62	21FE	-0.3	-3.3	-1.8	0.2
51.13	0.97	1.5	352.58	3MR	-0.8	-3.3	-1.7	0.2
57.55	1.03	1.6	2.49	13MR	-1.6	-3.3	-1.7	0.3
63.96	1.08	1.7	12.33	23MR	-1.3	-3.3	-1.6	0.3
70.36	1.12	1.7	22.13	2AP	-0.3	-3.4	-1.6	0.3
76.74	1.15	1.8	31.87	12AP	1.0	-3.4	-1.5	0.3
83.10	1.18	1.9	41.55	22AP	2.4	-3.4	-1.5	0.3
89.44	1.20	1.9	51.20	2MY	3.6	-3.4	-1.5	0.3
95.77	1.21	2.0	60.81	12MY	2.1	-3.5	-1.5	0.3
102.09	1.22	2.0	70.38	22MY	1.1	-3.5	-1.5	0.3
108.40	1.22	2.0	79.93	1JN	0.3	-3.4	-1.5	0.3
114.71	1.21	2.0	89.47	11JN	-0.6	-3.4	-1.5	0.3
121.01	1.20	2.0	99.00	21JN	-1.6	-3.4	-1.5	0.2
127.32	1.19	2.0	108.53	1JL	-1.4	-3.4	-1.5	0.3
133.63	1.17	2.0	118.08	11JL	-0.5	-3.3	-1.5	0.3
139.96	1.15	2.0	127.65	21JL	0.0	-3.3	-1.6	0.4
146.30	1.12	2.0	137.24	31JL	0.4	-3.3	-1.6	0.4
152.66	1.09	2.0	146.88	10AU	0.7	-3.3	-1.6	0.4
159.04	1.05	2.0	156.55	20AU	1.2	-3.3	-1.7	0.4
165.45	1.01	1.9	166.28	30AU	2.6	-3.3	-1.8	0.4
171.89	0.97	1.9	176.06	9SE	1.6	-3.4	-1.8	0.4
178.37	0.92	2.0	185.89	19SE	-0.3	-3.4	-1.9	0.4
184.88	0.86	2.0	195.79	29SE	-0.9	-3.4	-1.9	0.4
191.43	0.81	2.0	205.73	9OC	-0.9	-3.5	-2.0	0.3
198.02	0.74	1.9	215.73	19OC	-0.9	-3.5	-2.1	0.3
204.66	0.67	1.9	225.79	29OC	-0.6	-3.6	-2.1	0.2
211.34	0.60	1.9	235.89	8NO	-0.5	-3.7	-2.2	0.1
218.07	0.52	1.9	246.02	18NO	-0.4	-3.8	-2.2	0.1
224.85	0.43	1.8	256.19	28NO	-0.4	-3.8	-2.2	-0.0
231.68	0.34	1.8	266.38	8DE	0.4	-3.9	-2.3	-0.1
238.55	0.24	1.7	276.57	18DE	3.0	-4.0	-2.3	-0.1
245.47	0.13	1.7	286.77	28DE	0.8	-4.1	-2.2	-0.2
				1563				
252.44	0.01	1.6	296.95	7JA	0.1	-4.2	-2.2	-0.1
259.45	-0.12	1.6	307.11	17JA	0.0	-4.3	-2.1	-0.1
266.51	-0.25	1.5	317.24	27JA	-0.1	-4.4	-2.1	-0.0
273.61	-0.40	1.4	327.33	6FE	-0.4	-4.3	-2.0	0.0
280.75	-0.55	1.3	337.37	16FE	-0.8	-4.0	-2.0	0.1
287.92	-0.71	1.2	347.36	26FE	-1.5	-3.4	-1.9	0.2
295.12	-0.88	1.1	357.30	8MR	-1.2	-3.5	-1.8	0.2
302.35	-1.05	1.0	7.18	18MR	-0.3	-4.0	-1.7	0.3
309.59	-1.23	1.0	17.00	28MR	1.2	-4.2	-1.7	0.3
316.83	-1.42	0.8	26.77	7AP	3.1	-4.3	-1.6	0.4
324.06	-1.60	0.7	36.48	17AP	2.7	-4.2	-1.6	0.4
331.26	-1.79	0.6	46.15	27AP	1.5	-4.1	-1.5	0.4
338.41	-1.97	0.5	55.77	7MY	0.9	-4.0	-1.5	0.4
345.48	-2.15	0.4	65.36	17MY	0.2	-3.9	-1.4	0.4
352.46	-2.32	0.3	74.93	27MY	-0.6	-3.8	-1.4	0.4
359.29	-2.48	0.2	84.47	6JN	-1.6	-3.7	-1.4	0.4
5.94	-2.63	0.0	94.00	16JN	-1.4	-3.6	-1.4	0.4
12.35	-2.76	-0.1	103.54	26JN	-0.5	-3.6	-1.4	0.4
18.46	-2.88	-0.2	113.07	6JL	0.1	-3.5	-1.4	0.3
24.21	-2.97	-0.4	122.63	16JL	0.5	-3.5	-1.4	0.4
29.47	-3.04	-0.5	132.21	26JL	0.9	-3.4	-1.4	0.4
34.14	-3.07	-0.7	141.83	5AU	1.6	-3.4	-1.4	0.5
38.07	-3.06	-0.9	151.48	15AU	3.1	-3.4	-1.4	0.5
41.05	-3.01	-1.1	161.18	25AU	1.4	-3.4	-1.5	0.5
42.89	-2.88	-1.3	170.93	4SE	-0.4	-3.4	-1.5	0.5
43.38	-2.68	-1.6	180.74	14SE	-1.1	-3.4	-1.5	0.5
42.38	-2.36	-1.8	190.60	24SE	-1.1	-3.4	-1.6	0.5
40.03	-1.92	-2.0	200.52	4OC	-0.8	-3.4	-1.6	0.5
36.75	-1.38	-2.1	210.49	14OC	-0.5	-3.4	-1.7	0.5
33.28	-0.79	-2.0	220.52	24OC	-0.3	-3.4	-1.8	0.4
30.45	-0.23	-1.7	230.59	3NO	-0.3	-3.4	-1.8	0.4
28.77	0.27	-1.4	240.71	13NO	-0.2	-3.4	-1.9	0.3
28.45	0.66	-1.1	250.86	23NO	0.6	-3.4	-2.0	0.3
29.41	0.97	-0.7	261.04	3DE	3.0	-3.5	-2.0	0.2
31.46	1.19	-0.4	271.23	13DE	0.5	-3.5	-2.1	0.1
34.39	1.35	-0.1	281.43	23DE	-0.0	-3.5	-2.1	0.1

♂ LONG	♂ LAT	♂ MAG	☉ LONG	16.00UT 1564	☿	♀	♃	♄
					MAGNITUDES			
38.02	1.47	0.1	291.61	2JA	-0.1	-3.5	-2.1	-0.0
42.18	1.55	0.4	301.79	12JA	-0.2	-3.4	-2.1	-0.1
46.77	1.60	0.6	311.93	22JA	-0.5	-3.4	-2.1	-0.0
51.67	1.63	0.8	322.04	1FE	-0.9	-3.4	-2.1	0.0
56.82	1.65	0.9	332.11	11FE	-1.3	-3.4	-2.1	0.1
62.18	1.65	1.1	342.13	21FE	-1.2	-3.3	-2.0	0.2
67.68	1.64	1.2	352.09	2MR	-0.2	-3.3	-2.0	0.2
73.30	1.63	1.4	2.01	12MR	1.5	-3.3	-1.9	0.3
79.02	1.61	1.5	11.86	22MR	3.5	-3.3	-1.8	0.3
84.82	1.58	1.6	21.65	1AP	2.0	-3.4	-1.8	0.4
90.68	1.55	1.7	31.39	11AP	1.1	-3.4	-1.7	0.4
96.60	1.51	1.7	41.09	21AP	0.6	-3.4	-1.6	0.5
102.56	1.47	1.8	50.73	1MY	0.1	-3.4	-1.6	0.5
108.56	1.43	1.9	60.34	11MY	-0.7	-3.4	-1.5	0.5
114.61	1.38	1.9	69.92	21MY	-1.6	-3.5	-1.5	0.5
120.68	1.33	2.0	79.47	31MY	-1.4	-3.5	-1.4	0.6
126.80	1.28	2.0	89.01	10JN	-0.5	-3.6	-1.4	0.6
132.95	1.23	2.0	98.54	20JN	0.2	-3.7	-1.3	0.5
139.13	1.17	2.0	108.07	30JN	0.7	-3.7	-1.3	0.5
145.36	1.11	2.0	117.62	10JL	1.2	-3.8	-1.3	0.5
151.63	1.05	2.0	127.19	20JL	2.1	-3.9	-1.3	0.5
157.95	0.99	2.0	136.78	30JL	3.2	-4.0	-1.3	0.6
164.32	0.92	2.0	146.41	9AU	1.2	-4.1	-1.3	0.6
170.74	0.85	2.0	156.09	19AU	-0.5	-4.2	-1.3	0.6
177.21	0.78	1.9	165.81	29AU	-1.2	-4.3	-1.3	0.7
183.74	0.70	1.9	175.59	8SE	-1.2	-4.2	-1.3	0.7
190.33	0.62	1.9	185.42	18SE	-0.8	-3.9	-1.3	0.7
196.99	0.54	1.8	195.31	28SE	-0.4	-3.4	-1.4	0.7
203.71	0.45	1.8	205.26	8OC	-0.2	-3.4	-1.4	0.7
210.51	0.37	1.8	215.26	18OC	-0.1	-4.0	-1.5	0.7
217.37	0.27	1.8	225.30	28OC	0.0	-4.3	-1.5	0.7
224.30	0.18	1.8	235.40	7NO	0.9	-4.4	-1.6	0.6
231.31	0.08	1.8	245.54	17NO	2.8	-4.3	-1.6	0.6
238.38	-0.02	1.7	255.70	27NO	0.3	-4.2	-1.7	0.5
245.53	-0.13	1.7	265.89	7DE	-0.2	-4.1	-1.8	0.5
252.75	-0.24	1.7	276.08	17DE	-0.3	-4.0	-1.8	0.4
260.04	-0.35	1.7	286.27	27DE	-0.3	-3.9	-1.9	0.3
				1565				
267.39	-0.46	1.6	296.46	6JA	-0.5	-3.8	-1.9	0.2
274.81	-0.58	1.6	306.62	16JA	-0.9	-3.8	-2.0	0.2
282.28	-0.69	1.5	316.75	26JA	-1.2	-3.7	-2.0	0.1
289.81	-0.81	1.5	326.84	5FE	-1.0	-3.6	-2.0	0.2
297.38	-0.92	1.4	336.89	15FE	-0.1	-3.5	-2.0	0.2
304.99	-1.03	1.4	346.88	25FE	1.8	-3.5	-2.0	0.3
312.63	-1.13	1.3	356.82	7MR	2.9	-3.4	-2.0	0.3
320.28	-1.23	1.3	6.71	17MR	1.5	-3.4	-2.0	0.4
327.95	-1.32	1.2	16.53	27MR	0.8	-3.4	-1.9	0.5
335.61	-1.39	1.2	26.30	6AP	0.5	-3.3	-1.9	0.5
343.25	-1.46	1.1	36.02	16AP	0.0	-3.3	-1.8	0.6
350.86	-1.51	1.1	45.68	26AP	-0.7	-3.3	-1.7	0.6
358.42	-1.55	1.0	55.31	6MY	-1.6	-3.3	-1.7	0.6
5.91	-1.58	1.0	64.90	16MY	-1.4	-3.3	-1.6	0.7
13.32	-1.58	0.9	74.47	26MY	-0.4	-3.3	-1.5	0.7
20.63	-1.57	0.9	84.01	5JN	0.3	-3.3	-1.5	0.7
27.82	-1.54	0.8	93.54	15JN	0.9	-3.4	-1.4	0.7
34.87	-1.49	0.8	103.07	25JN	1.6	-3.4	-1.4	0.7
41.76	-1.42	0.7	112.61	5JL	2.7	-3.4	-1.3	0.8
48.46	-1.33	0.7	122.17	15JL	2.7	-3.5	-1.3	0.7
54.94	-1.23	0.6	131.74	25JL	1.0	-3.5	-1.3	0.7
61.16	-1.09	0.5	141.36	4AU	-0.5	-3.5	-1.2	0.7
67.08	-0.94	0.4	151.01	14AU	-1.3	-3.4	-1.2	0.8
72.65	-0.76	0.3	160.70	24AU	-1.3	-3.4	-1.2	0.8
77.78	-0.56	0.2	170.46	3SE	-0.7	-3.4	-1.2	0.9
82.41	-0.32	0.0	180.25	13SE	-0.3	-3.3	-1.2	0.9
86.40	-0.04	-0.1	190.11	23SE	-0.1	-3.3	-1.2	0.9
89.61	0.28	-0.3	200.03	3OC	0.0	-3.3	-1.2	0.9
91.87	0.65	-0.5	210.00	13OC	0.2	-3.3	-1.3	0.9
92.96	1.07	-0.7	220.03	23OC	1.2	-3.3	-1.3	0.9
92.70	1.54	-0.9	230.10	2NO	2.5	-3.3	-1.3	0.9
91.02	2.03	-1.1	240.22	12NO	0.1	-3.4	-1.4	0.9
88.04	2.50	-1.3	250.37	22NO	-0.4	-3.4	-1.4	0.9
84.24	2.89	-1.4	260.55	2DE	-0.4	-3.4	-1.5	0.8
80.40	3.15	-1.3	270.73	12DE	-0.5	-3.5	-1.6	0.8
77.26	3.27	-1.0	280.93	22DE	-0.6	-3.5	-1.6	0.7

♂ LONG	♂ LAT	♂ MAG	☉ LONG	16.00UT 1566	☿	♀	♃	♄
					MAGNITUDES			
75.31	3.26	-0.7	291.12	1JA	-0.9	-3.6	-1.7	0.6
74.71	3.18	-0.4	301.29	11JA	-1.0	-3.6	-1.8	0.5
75.36	3.05	-0.2	311.44	21JA	-0.9	-3.7	-1.8	0.5
77.09	2.89	0.1	321.56	31JA	-0.1	-3.8	-1.9	0.4
79.69	2.73	0.4	331.63	10FE	2.1	-3.8	-1.9	0.4
82.98	2.58	0.6	341.65	20FE	2.3	-3.9	-2.0	0.4
86.82	2.43	0.8	351.62	2MR	1.1	-4.0	-2.0	0.5
91.08	2.29	1.0	1.53	12MR	0.6	-4.1	-2.0	0.5
95.68	2.15	1.1	11.39	22MR	0.3	-4.2	-2.0	0.6
100.56	2.02	1.3	21.18	1AP	-0.1	-4.2	-2.0	0.6
105.65	1.90	1.4	30.93	11AP	-0.7	-4.2	-2.0	0.7
110.93	1.78	1.5	40.62	21AP	-1.6	-4.1	-1.9	0.7
116.37	1.67	1.6	50.27	1MY	-1.4	-3.6	-1.9	0.8
121.93	1.56	1.7	59.88	11MY	-0.4	-2.7	-1.8	0.8
127.62	1.45	1.7	69.45	21MY	0.5	-3.5	-1.7	0.9
133.40	1.35	1.8	79.01	31MY	1.2	-4.0	-1.7	0.9
139.28	1.25	1.8	88.54	10JN	2.1	-4.2	-1.6	0.9
145.25	1.14	1.9	98.08	20JN	3.3	-4.2	-1.6	1.0
151.30	1.04	1.9	107.61	30JN	2.2	-4.1	-1.5	1.0
157.44	0.95	1.9	117.15	10JL	0.8	-4.0	-1.4	1.0
163.65	0.85	1.9	126.72	20JL	-0.5	-4.0	-1.4	1.0
169.95	0.75	1.9	136.31	30JL	-1.4	-3.9	-1.3	1.0
176.33	0.65	1.9	145.94	9AU	-1.3	-3.8	-1.3	1.0
182.80	0.55	1.9	155.62	19AU	-0.7	-3.7	-1.3	1.0
189.35	0.45	1.9	165.34	29AU	-0.2	-3.6	-1.2	1.1
195.98	0.35	1.9	175.11	8SE	0.0	-3.6	-1.2	1.1
202.71	0.25	1.8	184.94	18SE	0.2	-3.5	-1.2	1.1
209.52	0.15	1.8	194.83	28SE	0.4	-3.5	-1.2	1.2
216.42	0.05	1.8	204.77	8OC	1.5	-3.5	-1.2	1.2
223.41	-0.05	1.7	214.77	18OC	2.3	-3.4	-1.2	1.2
230.50	-0.16	1.7	224.82	28OC	-0.0	-3.4	-1.2	1.2
237.67	-0.26	1.6	234.91	7NO	-0.5	-3.4	-1.3	1.2
244.93	-0.36	1.5	245.05	17NO	-0.6	-3.4	-1.3	1.2
252.27	-0.46	1.5	255.21	27NO	-0.6	-3.4	-1.3	1.2
259.70	-0.55	1.5	265.39	7DE	-0.7	-3.4	-1.4	1.1
267.20	-0.64	1.5	275.59	17DE	-0.8	-3.4	-1.4	1.1
274.77	-0.73	1.5	285.78	27DE	-0.8	-3.4	-1.5	1.0
				1567				
282.41	-0.82	1.5	295.97	6JA	-0.8	-3.4	-1.5	1.0
290.10	-0.90	1.5	306.13	16JA	0.0	-3.4	-1.6	0.9
297.84	-0.97	1.5	316.26	26JA	2.4	-3.4	-1.7	0.8
305.62	-1.03	1.5	326.35	5FE	1.7	-3.4	-1.7	0.8
313.43	-1.09	1.4	336.40	15FE	0.7	-3.4	-1.8	0.7
321.25	-1.13	1.4	346.40	25FE	0.4	-3.5	-1.9	0.7
329.07	-1.16	1.4	356.34	7MR	0.2	-3.5	-1.9	0.7
336.89	-1.18	1.4	6.23	17MR	-0.2	-3.5	-2.0	0.8
344.68	-1.19	1.4	16.05	27MR	-0.7	-3.4	-2.0	0.8
352.44	-1.19	1.4	25.82	6AP	-1.6	-3.4	-2.0	0.9
0.16	-1.17	1.4	35.55	16AP	-1.4	-3.4	-2.0	0.9
7.81	-1.13	1.4	45.21	26AP	-0.4	-3.4	-2.0	1.0
15.40	-1.09	1.3	54.84	6MY	0.6	-3.3	-2.0	1.0
22.91	-1.03	1.3	64.44	16MY	1.5	-3.3	-2.0	1.1
30.33	-0.96	1.3	74.00	26MY	2.8	-3.3	-1.9	1.1
37.65	-0.87	1.3	83.54	5JN	3.2	-3.3	-1.9	1.2
44.86	-0.78	1.3	93.08	15JN	1.7	-3.3	-1.8	1.2
51.96	-0.67	1.3	102.61	25JN	0.6	-3.4	-1.8	1.2
58.92	-0.55	1.3	112.15	5JL	-0.6	-3.4	-1.7	1.3
65.75	-0.42	1.2	121.70	15JL	-1.5	-3.4	-1.6	1.3
72.43	-0.28	1.2	131.28	25JL	-1.4	-3.5	-1.6	1.3
78.95	-0.13	1.2	140.89	4AU	-0.6	-3.5	-1.5	1.3
85.29	0.03	1.1	150.54	14AU	-0.1	-3.6	-1.5	1.3
91.43	0.20	1.1	160.23	24AU	0.2	-3.6	-1.4	1.3
97.35	0.39	1.0	169.98	3SE	0.3	-3.7	-1.4	1.3
103.01	0.59	0.9	179.78	13SE	0.7	-3.8	-1.3	1.3
108.38	0.81	0.8	189.63	23SE	1.9	-3.9	-1.3	1.3
113.39	1.05	0.7	199.55	3OC	2.1	-4.0	-1.3	1.3
117.98	1.31	0.6	209.52	13OC	-0.2	-4.1	-1.3	1.3
122.06	1.60	0.5	219.54	23OC	-0.7	-4.2	-1.3	1.3
125.51	1.93	0.3	229.61	2NO	-0.7	-4.3	-1.3	1.2
128.18	2.29	0.1	239.73	12NO	-0.7	-4.4	-1.3	1.2
129.91	2.68	-0.1	249.88	22NO	-0.7	-4.3	-1.3	1.2
130.49	3.11	-0.3	260.05	2DE	-0.7	-4.0	-1.3	1.1
129.75	3.55	-0.6	270.24	12DE	-0.7	-3.3	-1.3	1.1
127.67	3.95	-0.8	280.44	22DE	-0.6	-3.4	-1.4	1.1

♂ LONG	♂ LAT	♂ MAG	☉ LONG	16.00UT	☿ MAGNITUDES	♀	♃	♄
				1568				
124.41	4.27	-1.0	290.63	1JA	0.2	-4.1	-1.4	1.0
120.49	4.44	-1.1	300.81	11JA	2.7	-4.3	-1.4	1.0
116.65	4.42	-0.9	310.95	21JA	1.3	-4.4	-1.5	0.9
113.57	4.25	-0.7	321.07	31JA	0.5	-4.3	-1.6	0.9
111.68	3.98	-0.4	331.14	10FE	0.2	-4.2	-1.6	0.8
111.10	3.66	-0.2	341.16	20FE	0.1	-4.1	-1.7	0.8
111.74	3.33	0.1	351.14	1MR	-0.2	-4.0	-1.8	0.8
113.42	3.01	0.3	1.05	11MR	-0.8	-3.9	-1.8	0.8
115.96	2.72	0.5	10.91	21MR	-1.6	-3.8	-1.9	0.9
119.18	2.45	0.7	20.71	31MR	-1.4	-3.7	-2.0	0.9
122.96	2.20	0.9	30.46	10AP	-0.4	-3.6	-2.0	1.0
127.18	1.97	1.0	40.15	20AP	0.8	-3.5	-2.1	1.1
131.77	1.77	1.2	49.81	30AP	2.0	-3.5	-2.1	1.1
136.65	1.58	1.3	59.42	10MY	3.6	-3.4	-2.1	1.2
141.79	1.40	1.4	68.99	20MY	2.5	-3.4	-2.1	1.2
147.14	1.23	1.5	78.55	30MY	1.3	-3.4	-2.1	1.3
152.68	1.07	1.5	88.09	9JN	0.4	-3.3	-2.1	1.3
158.38	0.92	1.6	97.62	19JN	-0.6	-3.3	-2.1	1.3
164.24	0.78	1.6	107.15	29JN	-1.5	-3.3	-2.0	1.3
170.25	0.64	1.7	116.70	9JL	-1.4	-3.3	-2.0	1.3
176.38	0.51	1.7	126.26	19JL	-0.6	-3.3	-1.9	1.3
182.65	0.38	1.7	135.85	29JL	-0.0	-3.3	-1.8	1.3
189.03	0.25	1.7	145.48	8AU	0.3	-3.4	-1.8	1.3
195.54	0.13	1.7	155.15	18AU	0.5	-3.4	-1.7	1.2
202.17	0.02	1.7	164.87	28AU	1.0	-3.4	-1.6	1.2
208.91	-0.10	1.7	174.63	7SE	2.3	-3.4	-1.6	1.1
215.76	-0.21	1.7	184.46	17SE	1.8	-3.5	-1.5	1.1
222.73	-0.31	1.7	194.35	27SE	-0.3	-3.5	-1.5	1.1
229.81	-0.41	1.6	204.28	7OC	-0.8	-3.5	-1.4	1.1
237.00	-0.51	1.6	214.28	17OC	-0.9	-3.5	-1.4	1.1
244.29	-0.60	1.6	224.33	27OC	-0.8	-3.4	-1.4	1.1
251.68	-0.69	1.6	234.42	6NO	-0.6	-3.4	-1.4	1.1
259.16	-0.77	1.5	244.55	16NO	-0.5	-3.4	-1.4	1.1
266.73	-0.84	1.5	254.71	26NO	-0.5	-3.4	-1.3	1.0
274.38	-0.90	1.4	264.89	6DE	-0.5	-3.3	-1.3	1.0
282.10	-0.96	1.4	275.09	16DE	0.3	-3.3	-1.4	1.0
289.87	-1.00	1.4	285.29	26DE	3.0	-3.3	-1.4	0.9
				1569				
297.70	-1.04	1.3	295.47	5JA	1.0	-3.3	-1.4	0.9
305.56	-1.07	1.3	305.63	15JA	0.3	-3.4	-1.4	0.8
313.45	-1.08	1.3	315.77	25JA	0.1	-3.4	-1.5	0.8
321.34	-1.08	1.3	325.86	4FE	-0.0	-3.4	-1.5	0.7
329.24	-1.07	1.3	335.92	14FE	-0.3	-3.4	-1.6	0.7
337.11	-1.05	1.4	345.92	24FE	-0.8	-3.5	-1.6	0.7
344.96	-1.02	1.4	355.86	6MR	-1.5	-3.5	-1.7	0.6
352.77	-0.98	1.4	5.75	16MR	-1.3	-3.5	-1.8	0.6
0.53	-0.92	1.4	15.58	26MR	-0.3	-3.6	-1.8	0.6
8.23	-0.86	1.4	25.36	5AP	1.0	-3.7	-1.9	0.7
15.86	-0.78	1.5	35.08	15AP	2.6	-3.7	-2.0	0.8
23.41	-0.70	1.5	44.75	25AP	3.2	-3.8	-2.0	0.8
30.88	-0.61	1.5	54.38	5MY	1.8	-3.9	-2.1	0.9
38.26	-0.51	1.5	63.98	15MY	1.0	-4.0	-2.2	1.0
45.54	-0.41	1.5	73.55	25MY	0.3	-4.1	-2.2	1.0
52.73	-0.30	1.6	83.09	4JN	-0.6	-4.2	-2.2	1.1
59.82	-0.18	1.6	92.62	14JN	-1.5	-4.2	-2.2	1.1
66.81	-0.06	1.6	102.16	24JN	-1.4	-4.1	-2.3	1.1
73.69	0.06	1.6	111.69	4JL	-0.5	-3.9	-2.2	1.1
80.46	0.18	1.6	121.24	14JL	0.0	-3.3	-2.2	1.1
87.13	0.31	1.6	130.82	24JL	0.4	-3.2	-2.2	1.1
93.67	0.44	1.6	140.42	3AU	0.7	-3.8	-2.1	1.1
100.10	0.58	1.6	150.07	13AU	1.3	-4.1	-2.1	1.1
106.40	0.72	1.5	159.76	23AU	2.8	-4.2	-2.0	1.1
112.57	0.86	1.5	169.50	2SE	1.6	-4.2	-1.9	1.0
118.58	1.01	1.5	179.30	12SE	-0.3	-4.2	-1.9	1.0
124.43	1.17	1.4	189.15	22SE	-1.0	-4.1	-1.8	1.0
130.09	1.33	1.3	199.06	2OC	-1.0	-4.0	-1.8	0.9
135.54	1.51	1.3	209.03	12OC	-0.9	-3.9	-1.7	1.0
140.74	1.69	1.2	219.05	22OC	-0.5	-3.8	-1.6	1.0
145.64	1.89	1.1	229.12	1NO	-0.4	-3.8	-1.6	1.0
150.18	2.11	0.9	239.23	11NO	-0.4	-3.7	-1.6	1.0
154.27	2.34	0.8	249.38	21NO	-0.3	-3.6	-1.5	1.0
157.83	2.60	0.6	259.55	1DE	0.5	-3.6	-1.5	0.9
160.73	2.88	0.4	269.75	11DE	3.1	-3.5	-1.5	0.9
162.79	3.19	0.2	279.94	21DE	0.7	-3.5	-1.5	0.9
163.85	3.50	-0.1	290.13	31DE	0.1	-3.4	-1.5	0.9

♂ LONG	♂ LAT	♂ MAG	☉ LONG	16.00UT	☿ MAGNITUDES	♀	♃	♄
				1570				
163.72	3.82	-0.3	300.31	10JA	-0.0	-3.4	-1.5	0.8
162.29	4.10	-0.6	310.46	20JA	-0.2	-3.4	-1.5	0.8
159.62	4.30	-0.8	320.58	30JA	-0.4	-3.3	-1.5	0.7
156.01	4.37	-1.0	330.66	9FE	-0.8	-3.3	-1.5	0.7
152.08	4.26	-1.0	340.68	19FE	-1.4	-3.3	-1.5	0.6
148.55	3.99	-0.8	350.66	1MR	-1.2	-3.3	-1.6	0.6
145.99	3.62	-0.6	0.58	11MR	-0.3	-3.3	-1.6	0.5
144.70	3.20	-0.3	10.44	21MR	1.3	-3.3	-1.7	0.5
144.68	2.78	-0.1	20.24	31MR	3.3	-3.4	-1.7	0.4
145.82	2.39	0.1	29.99	10AP	2.4	-3.4	-1.8	0.5
147.94	2.03	0.3	39.69	20AP	1.4	-3.4	-1.9	0.6
150.85	1.70	0.5	49.34	30AP	0.8	-3.5	-1.9	0.6
154.42	1.41	0.7	58.95	10MY	0.2	-3.5	-2.0	0.7
158.53	1.14	0.8	68.53	20MY	-0.6	-3.5	-2.1	0.8
163.08	0.90	0.9	78.08	30MY	-1.6	-3.4	-2.2	0.8
167.99	0.69	1.0	87.63	9JN	-1.4	-3.4	-2.2	0.9
173.22	0.49	1.1	97.15	19JN	-0.5	-3.4	-2.3	0.9
178.73	0.30	1.2	106.69	29JN	0.1	-3.4	-2.3	1.0
184.47	0.13	1.2	116.23	9JL	0.6	-3.3	-2.4	1.0
190.43	-0.02	1.3	125.79	19JL	1.0	-3.3	-2.4	1.0
196.59	-0.17	1.3	135.38	29JL	1.7	-3.3	-2.4	1.0
202.93	-0.30	1.4	145.01	8AU	3.2	-3.3	-2.4	1.0
209.44	-0.43	1.4	154.67	18AU	1.4	-3.3	-2.3	1.0
216.11	-0.55	1.4	164.39	28AU	-0.4	-3.3	-2.3	1.0
222.93	-0.65	1.4	174.16	7SE	-1.1	-3.4	-2.3	0.9
229.90	-0.75	1.4	183.98	17SE	-1.2	-3.4	-2.2	0.9
237.00	-0.84	1.4	193.86	27SE	-0.8	-3.4	-2.1	0.9
244.23	-0.92	1.4	203.80	7OC	-0.4	-3.5	-2.1	0.8
251.58	-0.98	1.4	213.79	17OC	-0.3	-3.5	-2.0	0.8
259.04	-1.04	1.4	223.83	27OC	-0.2	-3.6	-1.9	0.8
266.60	-1.09	1.4	233.93	6NO	-0.1	-3.7	-1.9	0.9
274.25	-1.12	1.4	244.05	16NO	0.7	-3.8	-1.8	0.9
281.97	-1.14	1.4	254.21	26NO	3.0	-3.9	-1.7	0.9
289.77	-1.15	1.4	264.40	6DE	0.5	-3.9	-1.7	0.9
297.61	-1.15	1.4	274.59	16DE	-0.1	-4.1	-1.7	0.8
305.48	-1.13	1.4	284.79	26DE	-0.2	-4.2	-1.6	0.8
				1571				
313.38	-1.11	1.4	294.98	5JA	-0.3	-4.3	-1.6	0.8
321.29	-1.07	1.4	305.14	15JA	-0.5	-4.3	-1.6	0.8
329.19	-1.02	1.4	315.28	25JA	-0.9	-4.4	-1.6	0.7
337.07	-0.96	1.4	325.38	4FE	-1.2	-4.3	-1.5	0.7
344.91	-0.89	1.4	335.43	14FE	-1.1	-4.0	-1.6	0.6
352.71	-0.81	1.4	345.43	24FE	-0.2	-3.4	-1.6	0.6
0.46	-0.73	1.4	355.38	6MR	1.6	-3.5	-1.6	0.5
8.14	-0.64	1.4	5.27	16MR	3.3	-4.0	-1.6	0.5
15.74	-0.54	1.4	15.11	26MR	1.8	-4.2	-1.6	0.4
23.27	-0.44	1.4	24.88	5AP	1.0	-4.3	-1.7	0.4
30.72	-0.34	1.5	34.61	15AP	0.6	-4.2	-1.7	0.4
38.08	-0.24	1.5	44.28	25AP	0.1	-4.1	-1.8	0.4
45.35	-0.13	1.6	53.92	5MY	-0.6	-4.0	-1.8	0.5
52.53	-0.02	1.6	63.51	15MY	-1.6	-3.9	-1.9	0.6
59.62	0.08	1.7	73.08	25MY	-1.4	-3.8	-1.9	0.6
66.63	0.19	1.7	82.63	4JN	-0.5	-3.7	-2.0	0.7
73.55	0.30	1.7	92.16	14JN	0.2	-3.6	-2.1	0.7
80.38	0.40	1.8	101.69	24JN	0.8	-3.6	-2.2	0.8
87.13	0.51	1.8	111.23	4JL	1.3	-3.5	-2.2	0.8
93.80	0.61	1.8	120.78	14JL	2.3	-3.5	-2.3	0.9
100.38	0.71	1.8	130.36	24JL	3.1	-3.4	-2.3	0.9
106.89	0.81	1.8	139.96	3AU	1.1	-3.4	-2.4	0.9
113.31	0.91	1.8	149.61	13AU	-0.4	-3.4	-2.4	0.9
119.65	1.01	1.8	159.30	23AU	-1.2	-3.4	-2.5	0.9
125.91	1.11	1.8	169.04	2SE	-1.3	-3.4	-2.5	0.9
132.08	1.21	1.8	178.83	12SE	-0.7	-3.4	-2.5	0.9
138.16	1.31	1.8	188.68	22SE	-0.4	-3.4	-2.4	0.9
144.13	1.41	1.7	198.59	2OC	-0.2	-3.4	-2.4	0.8
150.00	1.51	1.7	208.55	12OC	-0.1	-3.4	-2.3	0.8
155.73	1.61	1.6	218.57	22OC	0.1	-3.4	-2.3	0.7
161.32	1.72	1.5	228.63	1NO	0.9	-3.4	-2.2	0.8
166.74	1.83	1.4	238.74	11NO	2.8	-3.4	-2.1	0.8
171.96	1.95	1.3	248.89	21NO	0.3	-3.4	-2.1	0.8
176.93	2.07	1.2	259.06	1DE	-0.3	-3.5	-2.0	0.8
181.61	2.20	1.1	269.25	11DE	-0.3	-3.5	-1.9	0.8
185.93	2.33	0.9	279.44	21DE	-0.4	-3.5	-1.9	0.8
189.81	2.48	0.7	289.63	31DE	-0.6	-3.5	-1.8	0.8

♂ LONG	♂ LAT	♂ MAG	☉ LONG	16.00UT 1572	☿	♀	♃	♄
193.14	2.62	0.5	299.81	10JA	-0.9	-3.4	-1.8	0.8
195.78	2.77	0.3	309.96	20JA	-1.1	-3.4	-1.7	0.7
197.57	2.92	0.0	320.08	30JA	-1.0	-3.4	-1.7	0.7
198.34	3.05	-0.3	330.16	9FE	-0.2	-3.4	-1.6	0.7
197.90	3.14	-0.5	340.19	19FE	1.9	-3.3	-1.6	0.6
196.19	3.16	-0.8	350.17	29FE	2.7	-3.3	-1.6	0.6
193.32	3.08	-1.1	0.09	10MR	1.3	-3.3	-1.6	0.5
189.66	2.87	-1.2	9.95	20MR	0.7	-3.3	-1.6	0.5
185.89	2.53	-1.2	19.76	30MR	0.4	-3.4	-1.6	0.4
182.70	2.12	-1.0	29.51	9AP	-0.0	-3.4	-1.6	0.3
180.60	1.66	-0.8	39.22	19AP	-0.7	-3.4	-1.6	0.3
179.82	1.22	-0.6	48.87	29AP	-1.6	-3.4	-1.6	0.3
180.32	0.82	-0.4	58.49	9MY	-1.4	-3.5	-1.7	0.4
181.97	0.46	-0.2	68.07	19MY	-0.5	-3.5	-1.7	0.4
184.58	0.15	0.0	77.62	29MY	0.4	-3.5	-1.8	0.5
187.99	-0.12	0.2	87.16	8JN	1.0	-3.6	-1.8	0.6
192.07	-0.35	0.3	96.70	18JN	1.7	-3.7	-1.9	0.6
196.69	-0.56	0.4	106.23	28JN	2.9	-3.7	-1.9	0.7
201.77	-0.74	0.5	115.77	8JL	2.6	-3.8	-2.0	0.7
207.24	-0.89	0.6	125.34	18JL	0.9	-3.9	-2.1	0.8
213.04	-1.03	0.7	134.92	28JL	-0.5	-4.0	-2.1	0.8
219.14	-1.14	0.8	144.55	7AU	-1.3	-4.1	-2.2	0.8
225.49	-1.24	0.9	154.21	17AU	-1.4	-4.2	-2.3	0.8
232.08	-1.32	0.9	163.92	27AU	-0.7	-4.3	-2.3	0.8
238.87	-1.38	1.0	173.69	6SE	-0.3	-4.2	-2.4	0.8
245.85	-1.42	1.0	183.51	16SE	-0.0	-3.9	-2.4	0.8
252.99	-1.45	1.0	193.38	26SE	0.1	-3.3	-2.5	0.8
260.28	-1.47	1.1	203.32	6OC	0.3	-3.5	-2.5	0.8
267.70	-1.47	1.1	213.31	16OC	1.2	-4.0	-2.5	0.7
275.23	-1.45	1.1	223.35	26OC	2.6	-4.3	-2.4	0.7
282.85	-1.42	1.2	233.44	5NO	0.1	-4.4	-2.4	0.7
290.55	-1.38	1.2	243.57	15NO	-0.4	-4.3	-2.3	0.7
298.31	-1.32	1.2	253.72	25NO	-0.5	-4.2	-2.3	0.7
306.11	-1.25	1.3	263.91	5DE	-0.5	-4.1	-2.2	0.8
313.94	-1.17	1.3	274.10	15DE	-0.7	-4.0	-2.1	0.8
321.77	-1.08	1.3	284.29	25DE	-0.9	-3.9	-2.1	0.8
				1573				
329.59	-0.99	1.4	294.48	4JA	-0.9	-3.8	-2.0	0.8
337.39	-0.89	1.4	304.65	14JA	-0.9	-3.8	-1.9	0.8
345.15	-0.78	1.4	314.78	24JA	-0.1	-3.7	-1.9	0.7
352.87	-0.67	1.5	324.89	3FE	2.2	-3.6	-1.8	0.7
0.53	-0.55	1.5	334.94	13FE	2.1	-3.5	-1.7	0.7
8.12	-0.44	1.5	344.95	23FE	0.9	-3.5	-1.7	0.6
15.64	-0.33	1.6	354.90	5MR	0.5	-3.4	-1.6	0.6
23.09	-0.21	1.6	4.79	15MR	0.3	-3.4	-1.6	0.5
30.45	-0.10	1.6	14.63	25MR	-0.1	-3.4	-1.6	0.5
37.73	0.01	1.6	24.41	4AP	-0.7	-3.3	-1.6	0.4
44.92	0.12	1.6	34.14	14AP	-1.6	-3.3	-1.6	0.3
52.04	0.22	1.7	43.82	24AP	-1.4	-3.3	-1.5	0.3
59.07	0.32	1.7	53.45	4MY	-0.4	-3.3	-1.5	0.2
66.03	0.41	1.7	63.05	14MY	0.5	-3.3	-1.6	0.3
72.90	0.50	1.7	72.62	24MY	1.3	-3.3	-1.6	0.3
79.71	0.59	1.7	82.17	3JN	2.3	-3.3	-1.6	0.4
86.44	0.68	1.8	91.70	13JN	3.4	-3.4	-1.6	0.5
93.11	0.75	1.8	101.23	23JN	2.1	-3.4	-1.7	0.5
99.72	0.83	1.9	110.77	3JL	0.7	-3.4	-1.7	0.6
106.28	0.90	1.9	120.32	13JL	-0.5	-3.5	-1.8	0.7
112.77	0.97	1.9	129.89	23JL	-1.4	-3.5	-1.8	0.7
119.22	1.03	2.0	139.50	2AU	-1.4	-3.5	-1.9	0.7
125.62	1.09	2.0	149.14	12AU	-0.7	-3.4	-1.9	0.8
131.97	1.15	2.0	158.82	22AU	-0.2	-3.4	-2.0	0.8
138.27	1.20	2.0	168.56	1SE	0.1	-3.4	-2.1	0.8
144.53	1.26	2.0	178.35	11SE	0.2	-3.3	-2.1	0.8
150.75	1.30	2.0	188.20	21SE	0.5	-3.3	-2.2	0.8
156.92	1.35	1.9	198.10	1OC	1.6	-3.3	-2.3	0.8
163.03	1.39	1.9	208.06	11OC	2.3	-3.3	-2.3	0.8
169.10	1.43	1.9	218.07	21OC	-0.0	-3.3	-2.3	0.7
175.10	1.47	1.8	228.14	31OC	-0.6	-3.3	-2.4	0.7
181.03	1.51	1.7	238.25	10NO	-0.6	-3.4	-2.4	0.7
186.88	1.54	1.7	248.39	20NO	-0.6	-3.4	-2.4	0.7
192.63	1.57	1.6	258.57	30NO	-0.7	-3.4	-2.3	0.7
198.27	1.59	1.5	268.75	10DE	-0.7	-3.5	-2.3	0.7
203.78	1.61	1.3	278.95	20DE	-0.8	-3.5	-2.3	0.8
209.12	1.62	1.2	289.14	30DE	-0.7	-3.6	-2.2	0.8

♂ LONG	♂ LAT	♂ MAG	☉ LONG	16.00UT 1574	☿	♀	♃	♄
214.26	1.63	1.1	299.32	9JA	0.0	-3.6	-2.1	0.8
219.16	1.62	0.9	309.47	19JA	2.5	-3.7	-2.1	0.8
223.76	1.61	0.7	319.60	29JA	1.6	-3.8	-2.0	0.8
227.99	1.57	0.5	329.67	8FE	0.6	-3.8	-1.9	0.7
231.76	1.51	0.3	339.71	18FE	0.3	-3.9	-1.8	0.7
234.94	1.42	0.0	349.69	28FE	0.1	-4.0	-1.8	0.7
237.39	1.29	-0.3	359.61	10MR	-0.2	-4.1	-1.7	0.6
238.94	1.11	-0.6	9.48	20MR	-0.7	-4.2	-1.7	0.6
239.40	0.86	-0.9	19.29	30MR	-1.6	-4.2	-1.6	0.5
238.64	0.53	-1.2	29.04	9AP	-1.4	-4.2	-1.6	0.4
236.64	0.11	-1.5	38.75	19AP	-0.4	-4.0	-1.5	0.4
233.62	-0.37	-1.7	48.40	29AP	0.7	-3.6	-1.5	0.3
230.14	-0.88	-1.8	58.02	9MY	1.7	-2.7	-1.5	0.3
226.89	-1.36	-1.7	67.60	19MY	3.1	-3.5	-1.5	0.2
224.56	-1.76	-1.5	77.16	29MY	2.9	-4.0	-1.5	0.3
223.51	-2.08	-1.3	86.70	8JN	1.6	-4.2	-1.5	0.3
223.83	-2.31	-1.1	96.23	18JN	0.6	-4.2	-1.5	0.4
225.44	-2.46	-0.9	105.77	28JN	-0.5	-4.1	-1.5	0.5
228.16	-2.56	-0.7	115.31	8JL	-1.5	-4.0	-1.5	0.5
231.79	-2.61	-0.5	124.87	18JL	-1.4	-3.9	-1.5	0.6
236.19	-2.63	-0.3	134.46	28JL	-0.6	-3.9	-1.6	0.6
241.21	-2.61	-0.2	144.08	7AU	-0.1	-3.8	-1.6	0.7
246.74	-2.57	-0.0	153.74	17AU	0.2	-3.7	-1.6	0.7
252.70	-2.51	0.1	163.45	27AU	0.4	-3.6	-1.7	0.8
259.00	-2.43	0.2	173.21	6SE	0.8	-3.6	-1.8	0.8
265.60	-2.33	0.3	183.03	16SE	2.0	-3.5	-1.8	0.8
272.44	-2.22	0.4	192.91	26SE	2.1	-3.5	-1.9	0.8
279.48	-2.09	0.5	202.83	6OC	-0.2	-3.5	-1.9	0.8
286.67	-1.95	0.6	212.82	16OC	-0.8	-3.4	-2.0	0.8
294.00	-1.80	0.7	222.86	26OC	-0.8	-3.4	-2.1	0.7
301.42	-1.64	0.8	232.95	5NO	-0.8	-3.4	-2.1	0.7
308.91	-1.48	0.8	243.07	15NO	-0.7	-3.4	-2.2	0.7
316.45	-1.31	0.9	253.23	25NO	-0.6	-3.4	-2.2	0.6
324.00	-1.15	1.0	263.41	5DE	-0.6	-3.4	-2.2	0.7
331.57	-0.98	1.1	273.61	15DE	-0.6	-3.4	-2.2	0.7
339.12	-0.82	1.2	283.80	25DE	0.2	-3.4	-2.2	0.7
				1575				
346.64	-0.66	1.2	293.99	4JA	2.8	-3.4	-2.2	0.8
354.13	-0.50	1.3	304.16	14JA	1.2	-3.4	-2.2	0.8
1.57	-0.35	1.4	314.30	24JA	0.4	-3.4	-2.1	0.8
8.94	-0.21	1.4	324.40	3FE	0.2	-3.4	-2.1	0.8
16.26	-0.08	1.5	334.46	13FE	0.0	-3.4	-2.0	0.8
23.51	0.05	1.6	344.46	23FE	-0.3	-3.5	-1.9	0.7
30.69	0.17	1.6	354.41	5MR	-0.8	-3.5	-1.9	0.7
37.80	0.28	1.7	4.31	15MR	-1.5	-3.4	-1.8	0.7
44.83	0.38	1.7	14.15	25MR	-1.4	-3.4	-1.7	0.6
51.80	0.48	1.8	23.93	4AP	-0.4	-3.4	-1.6	0.5
58.69	0.57	1.8	33.66	14AP	0.9	-3.4	-1.6	0.5
65.52	0.65	1.8	43.34	24AP	2.2	-3.4	-1.5	0.4
72.29	0.72	1.9	52.98	4MY	3.8	-3.3	-1.5	0.4
78.99	0.79	1.9	62.58	14MY	2.2	-3.3	-1.4	0.3
85.64	0.86	1.9	72.15	24MY	1.2	-3.3	-1.4	0.2
92.24	0.91	1.9	81.70	3JN	0.4	-3.3	-1.4	0.2
98.80	0.96	1.9	91.24	13JN	-0.6	-3.3	-1.4	0.3
105.31	1.01	1.9	100.77	23JN	-1.5	-3.4	-1.3	0.4
111.79	1.05	1.9	110.30	3JL	-1.4	-3.4	-1.3	0.4
118.23	1.08	1.9	119.85	13JL	-0.6	-3.4	-1.3	0.5
124.65	1.11	2.0	129.42	23JL	-0.0	-3.5	-1.3	0.5
131.04	1.14	2.0	139.03	2AU	0.3	-3.5	-1.4	0.6
137.42	1.16	2.0	148.67	12AU	0.6	-3.6	-1.4	0.6
143.77	1.18	2.0	158.35	22AU	1.1	-3.6	-1.4	0.7
150.12	1.19	2.0	168.09	1SE	2.4	-3.7	-1.4	0.7
156.45	1.20	2.0	177.88	11SE	1.8	-3.8	-1.5	0.8
162.78	1.20	2.0	187.72	21SE	-0.2	-3.9	-1.5	0.8
169.10	1.20	2.0	197.62	1OC	-0.9	-4.0	-1.6	0.8
175.41	1.19	2.0	207.58	11OC	-0.9	-4.1	-1.6	0.8
181.72	1.18	2.0	217.59	21OC	-0.9	-4.2	-1.7	0.8
188.02	1.17	1.9	227.65	31OC	-0.6	-4.3	-1.8	0.8
194.31	1.14	1.9	237.76	10NO	-0.5	-4.4	-1.8	0.7
200.58	1.12	1.8	247.90	20NO	-0.5	-4.3	-1.9	0.7
206.85	1.08	1.8	258.07	30NO	-0.4	-4.0	-2.0	0.7
213.09	1.04	1.7	268.26	10DE	0.3	-3.2	-2.0	0.7
219.31	0.98	1.6	278.45	20DE	3.0	-3.5	-2.1	0.7
225.50	0.92	1.5	288.65	30DE	0.9	-4.1	-2.1	0.7

♂ LONG	♂ LAT	♂ MAG	☉ LONG	16.00UT 1576	☿	♀	♃	♄
					MAGNITUDES			
231.65	0.84	1.4	298.83	9JA	0.2	-4.3	-2.1	0.8
237.75	0.75	1.3	308.98	19JA	0.0	-4.4	-2.1	0.8
243.79	0.65	1.2	319.10	29JA	-0.1	-4.3	-2.1	0.8
249.75	0.52	1.0	329.19	8FE	-0.3	-4.2	-2.1	0.8
255.61	0.37	0.9	339.22	18FE	-0.8	-4.1	-2.0	0.8
261.36	0.19	0.7	349.20	28FE	-1.5	-4.0	-2.0	0.8
266.96	-0.01	0.5	359.13	9MR	-1.3	-3.9	-1.9	0.7
272.37	-0.26	0.3	9.00	19MR	-0.3	-3.8	-1.9	0.7
277.55	-0.55	0.1	18.82	29MR	1.1	-3.7	-1.8	0.7
282.42	-0.88	-0.1	28.58	8AP	2.8	-3.6	-1.7	0.6
286.91	-1.28	-0.4	38.28	18AP	2.9	-3.5	-1.7	0.5
290.91	-1.75	-0.6	47.94	28AP	1.7	-3.5	-1.6	0.5
294.27	-2.29	-0.9	57.56	8MY	0.9	-3.4	-1.5	0.4
296.84	-2.92	-1.2	67.14	18MY	0.3	-3.4	-1.5	0.4
298.42	-3.62	-1.5	76.70	28MY	-0.6	-3.4	-1.4	0.3
298.83	-4.38	-1.9	86.24	7JN	-1.5	-3.3	-1.4	0.2
298.00	-5.13	-2.2	95.77	17JN	-1.4	-3.3	-1.3	0.3
296.05	-5.78	-2.4	105.31	27JN	-0.6	-3.3	-1.3	0.3
293.41	-6.23	-2.6	114.85	7JL	0.1	-3.3	-1.3	0.4
290.78	-6.39	-2.5	124.41	17JL	0.5	-3.3	-1.3	0.4
288.85	-6.27	-2.3	133.99	27JL	0.8	-3.3	-1.3	0.5
288.11	-5.92	-2.0	143.61	6AU	1.4	-3.4	-1.2	0.6
288.73	-5.44	-1.8	153.27	16AU	2.9	-3.4	-1.3	0.6
290.63	-4.90	-1.5	162.98	26AU	1.6	-3.4	-1.3	0.7
293.66	-4.35	-1.3	172.74	5SE	-0.3	-3.4	-1.3	0.7
297.59	-3.81	-1.0	182.55	15SE	-1.0	-3.5	-1.3	0.7
302.24	-3.31	-0.8	192.43	25SE	-1.1	-3.5	-1.3	0.8
307.46	-2.83	-0.6	202.35	5OC	-0.9	-3.5	-1.4	0.8
313.12	-2.39	-0.4	212.33	15OC	-0.5	-3.4	-1.4	0.8
319.10	-1.99	-0.2	222.37	25OC	-0.3	-3.4	-1.5	0.8
325.35	-1.62	0.0	232.46	4NO	-0.3	-3.4	-1.5	0.8
331.78	-1.29	0.2	242.58	14NO	-0.2	-3.4	-1.6	0.8
338.36	-0.99	0.4	252.74	24NO	0.5	-3.4	-1.6	0.7
345.03	-0.72	0.5	262.91	4DE	3.1	-3.3	-1.7	0.7
351.77	-0.47	0.7	273.10	14DE	0.6	-3.3	-1.8	0.7
358.55	-0.25	0.8	283.30	24DE	0.0	-3.3	-1.8	0.7
				1577				
5.35	-0.06	0.9	293.49	3JA	-0.1	-3.3	-1.9	0.8
12.16	0.12	1.1	303.66	13JA	-0.2	-3.4	-1.9	0.8
18.95	0.27	1.2	313.80	23JA	-0.4	-3.4	-2.0	0.8
25.73	0.41	1.3	323.91	2FE	-0.8	-3.4	-2.0	0.8
32.48	0.53	1.4	333.96	12FE	-1.3	-3.4	-2.0	0.8
39.21	0.64	1.5	343.98	22FE	-1.2	-3.5	-2.0	0.8
45.90	0.73	1.6	353.93	4MR	-0.3	-3.5	-2.0	0.8
52.54	0.82	1.6	3.83	14MR	1.3	-3.6	-2.0	0.8
59.16	0.89	1.7	13.68	24MR	3.4	-3.6	-1.9	0.8
65.73	0.95	1.8	23.47	3AP	2.2	-3.7	-1.9	0.7
72.27	1.00	1.8	33.20	13AP	1.2	-3.7	-1.8	0.7
78.77	1.04	1.9	42.88	23AP	0.7	-3.8	-1.8	0.6
85.24	1.08	1.9	52.52	3MY	0.2	-3.9	-1.7	0.6
91.68	1.11	2.0	62.12	13MY	-0.6	-4.0	-1.6	0.5
98.10	1.13	2.0	71.70	23MY	-1.5	-4.1	-1.6	0.4
104.49	1.15	2.0	81.24	2JN	-1.5	-4.2	-1.5	0.4
110.86	1.16	2.0	90.78	12JN	-0.5	-4.2	-1.5	0.3
117.23	1.16	2.0	100.31	22JN	0.2	-4.1	-1.4	0.3
123.58	1.17	2.0	109.84	2JL	0.6	-3.8	-1.4	0.3
129.93	1.16	2.0	119.39	12JL	1.1	-3.3	-1.3	0.4
136.28	1.15	2.0	128.96	22JL	1.9	-3.2	-1.3	0.4
142.63	1.14	2.0	138.56	1AU	3.2	-3.8	-1.2	0.5
148.99	1.12	2.0	148.20	11AU	1.3	-4.2	-1.2	0.5
155.37	1.10	2.0	157.89	21AU	-0.4	-4.2	-1.2	0.6
161.76	1.07	2.0	167.61	31AU	-1.1	-4.2	-1.2	0.6
168.17	1.04	2.0	177.40	10SE	-1.2	-4.2	-1.2	0.7
174.60	1.01	2.0	187.24	20SE	-0.8	-4.1	-1.2	0.7
181.06	0.97	2.0	197.14	30SE	-0.4	-4.0	-1.2	0.8
187.55	0.92	2.0	207.10	10OC	-0.2	-3.9	-1.2	0.8
194.07	0.87	2.0	217.11	20OC	-0.2	-3.8	-1.3	0.8
200.62	0.81	1.9	227.16	30OC	-0.0	-3.7	-1.3	0.8
207.20	0.75	1.9	237.27	9NO	0.7	-3.7	-1.3	0.8
213.81	0.68	1.9	247.41	19NO	3.0	-3.6	-1.4	0.8
220.46	0.60	1.8	257.58	29NO	0.4	-3.6	-1.4	0.8
227.15	0.52	1.8	267.77	9DE	-0.2	-3.5	-1.5	0.8
233.86	0.43	1.7	277.96	19DE	-0.3	-3.5	-1.6	0.7
240.61	0.32	1.7	288.15	29DE	-0.3	-3.4	-1.6	0.7

♂ LONG	♂ LAT	♂ MAG	☉ LONG	16.00UT 1578	☿	♀	♃	♄
					MAGNITUDES			
247.39	0.21	1.6	298.34	8JA	-0.5	-3.4	-1.7	0.8
254.20	0.08	1.5	308.49	18JA	-0.9	-3.4	-1.8	0.8
261.04	-0.05	1.4	318.62	28JA	-1.2	-3.3	-1.8	0.8
267.91	-0.20	1.3	328.70	7FE	-1.1	-3.3	-1.9	0.9
274.79	-0.36	1.2	338.74	17FE	-0.2	-3.3	-1.9	0.9
281.70	-0.54	1.1	348.72	27FE	1.6	-3.3	-2.0	0.9
288.62	-0.72	1.0	358.66	9MR	3.1	-3.3	-2.0	0.9
295.54	-0.93	0.9	8.53	19MR	1.6	-3.3	-2.0	0.9
302.45	-1.14	0.8	18.34	29MR	0.9	-3.4	-2.0	0.8
309.35	-1.38	0.6	28.10	8AP	0.5	-3.4	-2.0	0.8
316.21	-1.62	0.5	37.81	18AP	0.1	-3.4	-1.9	0.8
323.01	-1.87	0.4	47.47	28AP	-0.6	-3.5	-1.9	0.7
329.73	-2.14	0.2	57.10	8MY	-1.5	-3.5	-1.8	0.7
336.33	-2.42	0.1	66.68	18MY	-1.5	-3.5	-1.8	0.6
342.76	-2.70	-0.1	76.24	28MY	-0.5	-3.4	-1.7	0.5
348.98	-2.98	-0.3	85.78	7JN	0.3	-3.4	-1.7	0.5
354.90	-3.27	-0.4	95.31	17JN	0.8	-3.4	-1.6	0.4
0.45	-3.56	-0.6	104.84	27JN	1.4	-3.4	-1.5	0.3
5.51	-3.83	-0.8	114.38	7JL	2.5	-3.3	-1.5	0.3
9.93	-4.10	-1.0	123.94	17JL	2.9	-3.3	-1.4	0.4
13.56	-4.34	-1.3	133.52	27JL	1.1	-3.3	-1.4	0.4
16.17	-4.53	-1.5	143.14	6AU	-0.4	-3.3	-1.3	0.5
17.56	-4.65	-1.7	152.80	16AU	-1.2	-3.3	-1.3	0.5
17.56	-4.66	-2.0	162.50	26AU	-1.3	-3.3	-1.3	0.6
16.13	-4.50	-2.2	172.26	5SE	-0.7	-3.4	-1.2	0.6
13.56	-4.14	-2.4	182.07	15SE	-0.3	-3.4	-1.2	0.7
10.44	-3.59	-2.4	191.94	25SE	-0.1	-3.4	-1.2	0.7
7.58	-2.90	-2.2	201.87	5OC	-0.0	-3.5	-1.2	0.8
5.67	-2.17	-1.9	211.85	15OC	0.2	-3.5	-1.2	0.8
5.04	-1.49	-1.5	221.88	25OC	1.0	-3.6	-1.2	0.8
5.72	-0.90	-1.2	231.97	4NO	2.8	-3.7	-1.2	0.8
7.58	-0.40	-0.9	242.09	14NO	0.2	-3.8	-1.3	0.8
10.40	-0.00	-0.6	252.24	24NO	-0.3	-3.9	-1.3	0.8
13.99	0.32	-0.3	262.43	4DE	-0.4	-4.0	-1.3	0.8
18.19	0.58	-0.0	272.62	14DE	-0.4	-4.1	-1.4	0.8
22.83	0.78	0.2	282.81	24DE	-0.6	-4.2	-1.4	0.8
				1579				
27.84	0.94	0.4	293.00	3JA	-0.9	-4.3	-1.5	0.8
33.12	1.07	0.6	303.17	13JA	-1.0	-4.3	-1.6	0.8
38.60	1.17	0.8	313.31	23JA	-1.0	-4.4	-1.6	0.8
44.25	1.24	1.0	323.42	2FE	-0.2	-4.3	-1.7	0.9
50.02	1.30	1.1	333.48	12FE	1.9	-3.9	-1.8	0.9
55.88	1.34	1.2	343.49	22FE	2.5	-3.4	-1.8	0.9
61.81	1.37	1.4	353.46	4MR	1.2	-3.5	-1.9	0.9
67.80	1.39	1.5	3.36	14MR	0.7	-4.0	-1.9	0.9
73.83	1.40	1.6	13.20	24MR	0.4	-4.2	-2.0	0.9
79.89	1.40	1.7	22.99	3AP	-0.0	-4.3	-2.0	0.9
85.98	1.40	1.7	32.73	13AP	-0.7	-4.2	-2.0	0.9
92.08	1.39	1.8	42.41	23AP	-1.5	-4.1	-2.0	0.9
98.21	1.37	1.9	52.06	3MY	-1.5	-4.0	-2.0	0.8
104.35	1.35	1.9	61.66	13MY	-0.5	-3.9	-2.0	0.8
110.51	1.33	2.0	71.23	23MY	0.4	-3.8	-2.0	0.7
116.68	1.30	2.0	80.78	2JN	1.1	-3.7	-1.9	0.6
122.88	1.26	2.0	90.32	12JN	1.9	-3.6	-1.9	0.6
129.09	1.23	2.0	99.85	22JN	3.2	-3.6	-1.8	0.5
135.33	1.19	2.0	109.39	2JL	2.4	-3.5	-1.7	0.5
141.59	1.14	2.0	118.93	12JL	0.9	-3.5	-1.7	0.4
147.89	1.09	2.0	128.50	22JL	-0.4	-3.4	-1.6	0.4
154.22	1.04	2.0	138.10	1AU	-1.3	-3.4	-1.6	0.4
160.58	0.99	2.0	147.74	11AU	-1.4	-3.4	-1.5	0.5
166.99	0.93	2.0	157.42	21AU	-0.7	-3.4	-1.5	0.5
173.44	0.87	1.9	167.15	31AU	-0.2	-3.4	-1.4	0.6
179.94	0.81	1.9	176.93	10SE	0.0	-3.4	-1.4	0.7
186.49	0.74	1.9	186.77	20SE	0.2	-3.4	-1.3	0.7
193.09	0.67	1.9	196.66	30SE	0.4	-3.4	-1.3	0.8
199.75	0.59	1.9	206.61	10OC	1.3	-3.4	-1.3	0.8
206.47	0.51	1.9	216.62	20OC	2.6	-3.4	-1.3	0.8
213.25	0.42	1.9	226.68	30OC	0.1	-3.4	-1.3	0.9
220.10	0.33	1.8	236.78	9NO	-0.5	-3.4	-1.3	0.9
227.00	0.24	1.8	246.92	19NO	-0.5	-3.4	-1.3	0.9
233.97	0.14	1.8	257.09	29NO	-0.6	-3.5	-1.3	0.9
241.01	0.04	1.8	267.27	9DE	-0.7	-3.5	-1.3	0.9
248.11	-0.07	1.7	277.47	19DE	-0.8	-3.5	-1.3	0.9
255.27	-0.19	1.7	287.66	29DE	-0.9	-3.5	-1.4	0.9

♂ LONG	♂ LAT	♂ MAG	⊙ LONG	16.00UT	☿	♀	♃	♄
				1580	MAGNITUDES			
262.49	-0.31	1.6	297.84	8JA	-0.8	-3.4	-1.4	0.9
269.77	-0.43	1.6	308.00	18JA	-0.1	-3.4	-1.5	0.9
277.10	-0.55	1.5	318.13	28JA	2.2	-3.4	-1.5	0.9
284.49	-0.68	1.5	328.21	7FE	1.9	-3.4	-1.6	1.0
291.92	-0.81	1.4	338.25	17FE	0.8	-3.3	-1.7	1.0
299.40	-0.94	1.3	348.24	27FE	0.5	-3.3	-1.7	1.0
306.90	-1.07	1.3	358.17	8MR	0.2	-3.3	-1.8	1.0
314.43	-1.20	1.2	8.05	18MR	-0.1	-3.3	-1.9	1.0
321.98	-1.32	1.1	17.87	28MR	-0.7	-3.4	-1.9	1.0
329.52	-1.44	1.1	27.63	7AP	-1.5	-3.4	-2.0	1.0
337.06	-1.55	1.0	37.34	17AP	-1.4	-3.4	-2.0	1.0
344.56	-1.65	0.9	47.01	27AP	-0.5	-3.4	-2.1	1.0
352.03	-1.73	0.9	56.63	7MY	0.5	-3.5	-2.1	1.0
359.43	-1.80	0.8	66.22	17MY	1.4	-3.5	-2.1	0.9
6.74	-1.86	0.7	75.78	27MY	2.6	-3.5	-2.1	0.9
13.95	-1.89	0.7	85.32	6JN	3.3	-3.6	-2.1	0.8
21.02	-1.91	0.6	94.86	16JN	1.9	-3.7	-2.1	0.7
27.93	-1.91	0.5	104.39	26JN	0.7	-3.8	-2.1	0.7
34.64	-1.88	0.4	113.93	6JL	-0.5	-3.8	-2.0	0.6
41.12	-1.83	0.3	123.49	16JL	-1.4	-3.9	-2.0	0.5
47.32	-1.76	0.2	133.07	26JL	-1.4	-4.0	-1.9	0.5
53.18	-1.66	0.1	142.68	5AU	-0.7	-4.1	-1.8	0.5
58.63	-1.53	-0.0	152.34	15AU	-0.2	-4.2	-1.8	0.5
63.59	-1.37	-0.2	162.04	25AU	0.1	-4.3	-1.7	0.6
67.93	-1.17	-0.3	171.79	4SE	0.3	-4.2	-1.6	0.6
71.52	-0.93	-0.5	181.60	14SE	0.6	-3.9	-1.6	0.7
74.17	-0.64	-0.7	191.47	24SE	1.7	-3.3	-1.5	0.7
75.69	-0.29	-0.9	201.39	4OC	2.3	-3.5	-1.5	0.8
75.86	0.13	-1.1	211.37	14OC	-0.1	-4.0	-1.5	0.8
74.58	0.60	-1.3	221.40	24OC	-0.7	-4.3	-1.4	0.9
71.94	1.10	-1.5	231.47	3NO	-0.7	-4.4	-1.4	0.9
68.34	1.58	-1.7	241.60	13NO	-0.7	-4.3	-1.4	0.9
64.55	1.99	-1.6	251.75	23NO	-0.8	-4.2	-1.4	1.0
61.34	2.28	-1.3	261.93	3DE	-0.7	-4.1	-1.4	1.0
59.28	2.45	-1.0	272.12	13DE	-0.7	-4.0	-1.4	1.0
58.57	2.52	-0.7	282.32	23DE	-0.7	-3.9	-1.4	1.0
				1581				
59.15	2.53	-0.4	292.51	2JA	0.0	-3.8	-1.4	1.0
60.84	2.49	-0.1	302.68	12JA	2.5	-3.8	-1.4	1.0
63.44	2.43	0.2	312.82	22JA	1.5	-3.7	-1.5	1.0
66.75	2.36	0.4	322.93	1FE	0.5	-3.6	-1.5	1.0
70.61	2.27	0.6	333.00	11FE	0.3	-3.5	-1.5	1.0
74.92	2.19	0.8	343.02	21FE	0.1	-3.5	-1.6	1.1
79.56	2.10	1.0	352.98	3MR	-0.2	-3.4	-1.7	1.1
84.48	2.01	1.2	2.89	13MR	-0.7	-3.4	-1.7	1.1
89.62	1.93	1.3	12.73	23MR	-1.5	-3.4	-1.8	1.2
94.92	1.84	1.4	22.53	2AP	-1.4	-3.3	-1.9	1.2
100.38	1.76	1.5	32.27	12AP	-0.4	-3.3	-1.9	1.2
105.95	1.68	1.6	41.95	22AP	0.7	-3.3	-2.0	1.2
111.63	1.59	1.7	51.60	2MY	1.9	-3.3	-2.1	1.1
117.39	1.51	1.8	61.20	12MY	3.4	-3.3	-2.1	1.1
123.23	1.43	1.8	70.77	22MY	2.7	-3.3	-2.2	1.1
129.15	1.35	1.9	80.32	1JN	1.5	-3.3	-2.2	1.0
135.14	1.27	1.9	89.86	11JN	0.5	-3.4	-2.3	1.0
141.19	1.18	1.9	99.39	21JN	-0.5	-3.4	-2.3	0.9
147.30	1.10	2.0	108.93	1JL	-1.5	-3.4	-2.3	0.8
153.48	1.02	2.0	118.47	11JL	-1.4	-3.5	-2.3	0.8
159.73	0.93	2.0	128.04	21JL	-0.6	-3.5	-2.2	0.7
166.04	0.84	2.0	137.64	31JL	-0.1	-3.5	-2.2	0.6
172.42	0.76	2.0	147.27	10AU	0.2	-3.4	-2.1	0.6
178.87	0.67	1.9	156.94	20AU	0.5	-3.4	-2.1	0.6
185.40	0.58	1.9	166.67	30AU	0.9	-3.4	-2.0	0.7
192.00	0.49	1.9	176.45	9SE	2.1	-3.3	-1.9	0.7
198.67	0.39	1.9	186.28	19SE	2.1	-3.3	-1.9	0.8
205.43	0.30	1.8	196.18	29SE	-0.2	-3.3	-1.8	0.8
212.27	0.20	1.8	206.13	9OC	-0.8	-3.3	-1.8	0.9
219.19	0.10	1.7	216.13	19OC	-0.8	-3.3	-1.7	0.9
226.20	0.00	1.7	226.18	29OC	-0.8	-3.3	-1.6	1.0
233.28	-0.10	1.6	236.28	8NO	-0.7	-3.4	-1.6	1.0
240.45	-0.20	1.6	246.42	18NO	-0.6	-3.4	-1.6	1.0
247.71	-0.30	1.6	256.59	28NO	-0.5	-3.4	-1.5	1.1
255.04	-0.41	1.6	266.77	8DE	-0.5	-3.5	-1.5	1.1
262.44	-0.51	1.6	276.97	18DE	0.2	-3.5	-1.5	1.1
269.92	-0.61	1.6	287.16	28DE	2.8	-3.6	-1.5	1.1

♂ LONG	♂ LAT	♂ MAG	⊙ LONG	16.00UT	☿	♀	♃	♄
				1582	MAGNITUDES			
277.46	-0.71	1.6	297.35	7JA	1.1	-3.6	-1.5	1.1
285.07	-0.80	1.5	307.50	17JA	0.3	-3.7	-1.5	1.1
292.73	-0.89	1.5	317.64	27JA	0.1	-3.8	-1.5	1.1
300.43	-0.97	1.5	327.73	6FE	-0.0	-3.8	-1.5	1.1
308.16	-1.05	1.4	337.77	16FE	-0.3	-3.9	-1.5	1.2
315.93	-1.12	1.4	347.76	26FE	-0.8	-4.0	-1.6	1.2
323.70	-1.18	1.4	357.70	8MR	-1.5	-4.1	-1.6	1.2
331.48	-1.23	1.4	7.57	18MR	-1.3	-4.2	-1.7	1.3
339.25	-1.26	1.3	17.40	28MR	-0.4	-4.2	-1.7	1.3
346.99	-1.29	1.3	27.16	7AP	0.9	-4.2	-1.8	1.3
354.69	-1.30	1.3	36.88	17AP	2.4	-4.0	-1.8	1.3
2.35	-1.29	1.3	46.55	27AP	3.5	-3.6	-1.9	1.3
9.95	-1.27	1.2	56.17	7MY	2.0	-2.8	-2.0	1.3
17.47	-1.23	1.2	65.75	17MY	1.1	-3.5	-2.0	1.3
24.90	-1.18	1.2	75.32	27MY	0.4	-4.0	-2.1	1.3
32.23	-1.12	1.2	84.86	6JN	-0.5	-4.2	-2.2	1.2
39.45	-1.04	1.1	94.39	16JN	-1.5	-4.2	-2.2	1.2
46.54	-0.95	1.1	103.93	26JN	-1.5	-4.1	-2.3	1.1
53.49	-0.84	1.1	113.46	6JL	-0.6	-4.0	-2.3	1.1
60.29	-0.72	1.0	123.02	16JL	0.0	-3.9	-2.4	1.0
66.92	-0.58	1.0	132.60	26JL	0.4	-3.9	-2.4	0.9
73.35	-0.43	0.9	142.21	5AU	0.7	-3.8	-2.4	0.9
79.57	-0.26	0.9	151.87	15AU	1.2	-3.7	-2.4	0.8
85.55	-0.08	0.8	161.57	25AU	2.6	-3.6	-2.4	0.8
91.24	0.12	0.7	171.32	4SE	1.8	-3.6	-2.3	0.8
96.60	0.35	0.6	181.12	14SE	-0.2	-3.5	-2.3	0.8
101.56	0.59	0.5	190.99	24SE	-0.9	-3.5	-2.2	0.9
106.05	0.87	0.4	200.91	4OC	-1.0	-3.5	-2.1	0.9
109.96	1.17	0.2	210.88	14OC	-0.9	-3.4	-2.1	1.0
113.16	1.52	0.1	220.91	24OC	-0.6	-3.4	-2.0	1.0
115.47	1.90	-0.1	230.98	3NO	-0.4	-3.4	-1.9	1.1
116.72	2.33	-0.3	241.10	13NO	-0.4	-3.4	-1.9	1.1
116.71	2.79	-0.6	251.26	23NO	-0.3	-3.4	-1.8	1.2
115.31	3.25	-0.8	261.43	3DE	0.4	-3.4	-1.7	1.2
112.60	3.68	-1.0	271.62	13DE	3.0	-3.4	-1.7	1.2
108.92	3.99	-1.1	281.82	23DE	0.8	-3.4	-1.7	1.3
				1583				
104.95	4.14	-1.1	292.01	2JA	0.1	-3.4	-1.6	1.3
101.45	4.13	-0.9	302.18	12JA	-0.0	-3.4	-1.6	1.3
99.00	3.98	-0.6	312.33	22JA	-0.1	-3.4	-1.6	1.3
97.85	3.75	-0.3	322.44	1FE	-0.4	-3.4	-1.6	1.3
97.98	3.48	-0.1	332.51	11FE	-0.8	-3.4	-1.6	1.3
99.25	3.21	0.2	342.53	21FE	-1.4	-3.5	-1.6	1.3
101.46	2.94	0.4	352.49	3MR	-1.3	-3.5	-1.6	1.3
104.42	2.70	0.6	2.40	13MR	-0.4	-3.4	-1.6	1.4
107.98	2.47	0.8	12.25	23MR	1.1	-3.4	-1.6	1.4
112.02	2.26	1.0	22.05	2AP	3.1	-3.4	-1.6	1.4
116.45	2.07	1.1	31.79	12AP	2.6	-3.4	-1.7	1.4
121.19	1.89	1.3	41.48	22AP	1.5	-3.4	-1.7	1.4
126.19	1.73	1.4	51.12	2MY	0.8	-3.3	-1.8	1.4
131.40	1.57	1.5	60.73	12MY	0.2	-3.3	-1.8	1.4
136.80	1.42	1.5	70.31	22MY	-0.6	-3.3	-1.9	1.3
142.36	1.28	1.6	79.86	1JN	-1.5	-3.3	-2.0	1.3
148.07	1.14	1.7	89.40	11JN	-1.5	-3.3	-2.0	1.2
153.91	1.01	1.7	98.93	21JN	-0.6	-3.4	-2.1	1.2
159.86	0.89	1.7	108.46	1JL	0.1	-3.4	-2.2	1.1
165.94	0.76	1.8	118.01	11JL	0.5	-3.4	-2.2	1.1
172.12	0.64	1.8	127.57	21JL	0.9	-3.5	-2.3	1.0
178.41	0.53	1.8	137.17	31JL	1.6	-3.5	-2.4	1.0
184.81	0.41	1.8	146.80	10AU	3.1	-3.6	-2.4	1.0
191.31	0.30	1.8	156.48	20AU	1.5	-3.6	-2.4	0.9
197.92	0.19	1.8	166.20	30AU	-0.3	-3.7	-2.5	0.9
204.63	0.08	1.8	175.98	9SE	-1.1	-3.8	-2.5	0.9
211.44	-0.03	1.8	185.81	19SE	-1.1	-3.9	-2.5	1.0
218.36	-0.14	1.7	195.70	29SE	-0.8	-4.0	-2.4	1.0
225.38	-0.24	1.7	205.65	9OC	-0.5	-4.1	-2.4	1.1
232.50	-0.34	1.7	215.65	19OC	-0.3	-4.2	-2.3	1.1
239.72	-0.44	1.6	225.70	29OC	-0.2	-4.3	-2.3	1.2
247.03	-0.53	1.6	235.80	8NO	-0.1	-4.4	-2.2	1.2
254.44	-0.62	1.5	245.93	18NO	0.6	-4.3	-2.1	1.3
261.93	-0.71	1.5	256.09	28NO	3.2	-3.9	-2.1	1.3
269.51	-0.78	1.4	266.28	8DE	0.6	-3.2	-2.0	1.4
277.16	-0.86	1.4	276.47	18DE	-0.1	-3.5	-1.9	1.4
284.87	-0.92	1.4	286.66	28DE	-0.2	-4.1	-1.9	1.4

Left table (1584–1585):

♂ LONG	LAT	MAG	☉ LONG	16.00UT 1584	☿	♀	24	♄
292.64	-0.98	1.4	296.85	7JA	-0.2	-4.3	-1.8	1.4
300.45	-1.02	1.4	307.01	17JA	-0.4	-4.4	-1.7	1.4
308.29	-1.06	1.4	317.14	27JA	-0.8	-4.3	-1.7	1.3
316.16	-1.09	1.4	327.23	6FE	-1.3	-4.2	-1.7	1.3
324.04	-1.10	1.4	337.28	16FE	-1.2	-4.1	-1.6	1.3
331.91	-1.10	1.4	347.27	26FE	-0.3	-4.0	-1.6	1.2
339.77	-1.09	1.4	357.21	7MR	1.4	-3.9	-1.6	1.2
347.59	-1.07	1.4	7.09	17MR	3.4	-3.8	-1.6	1.2
355.38	-1.04	1.4	16.92	27MR	1.9	-3.7	-1.6	1.2
3.11	-0.99	1.4	26.69	6AP	1.1	-3.6	-1.6	1.2
10.78	-0.93	1.4	36.41	16AP	0.6	-3.5	-1.6	1.2
18.39	-0.86	1.4	46.08	26AP	0.1	-3.5	-1.6	1.1
25.91	-0.79	1.5	55.71	6MY	-0.6	-3.4	-1.7	1.1
33.34	-0.70	1.5	65.29	16MY	-1.5	-3.4	-1.7	1.1
40.68	-0.60	1.5	74.86	26MY	-1.5	-3.4	-1.7	1.1
47.92	-0.50	1.5	84.41	5JN	-0.5	-3.3	-1.8	1.0
55.06	-0.39	1.5	93.94	15JN	0.2	-3.3	-1.8	1.0
62.09	-0.27	1.5	103.47	25JN	0.7	-3.3	-1.9	1.0
69.00	-0.14	1.5	113.01	5JL	1.2	-3.3	-1.9	0.9
75.79	-0.01	1.5	122.56	15JL	2.1	-3.3	-2.0	0.9
82.47	0.12	1.5	132.14	25JL	3.2	-3.3	-2.1	0.8
89.00	0.27	1.4	141.75	4AU	1.3	-3.4	-2.1	0.8
95.40	0.41	1.4	151.40	14AU	-0.4	-3.4	-2.2	0.7
101.65	0.57	1.4	161.10	24AU	-1.2	-3.4	-2.3	0.7
107.73	0.73	1.3	170.85	3SE	-1.3	-3.4	-2.3	0.6
113.62	0.90	1.3	180.65	13SE	-0.8	-3.5	-2.4	0.6
119.31	1.08	1.2	190.51	23SE	-0.4	-3.5	-2.4	0.6
124.76	1.28	1.1	200.43	3OC	-0.2	-3.5	-2.4	0.7
129.92	1.48	1.0	210.39	13OC	-0.1	-3.4	-2.5	0.7
134.76	1.71	0.9	220.42	23OC	0.0	-3.4	-2.4	0.8
139.19	1.96	0.8	230.50	2NO	0.8	-3.4	-2.4	0.9
143.12	2.23	0.6	240.61	12NO	3.1	-3.4	-2.4	0.9
146.46	2.52	0.5	250.76	22NO	0.4	-3.4	-2.3	1.0
149.04	2.85	0.3	260.94	2DE	-0.2	-3.3	-2.3	1.0
150.70	3.20	0.0	271.12	12DE	-0.3	-3.3	-2.2	1.1
151.26	3.57	-0.2	281.32	22DE	-0.4	-3.3	-2.1	1.1

1585

♂ LONG	LAT	MAG	☉ LONG	16.00UT 1585	☿	♀	24	♄
150.54	3.94	-0.4	291.51	1JA	-0.5	-3.3	-2.0	1.1
148.51	4.26	-0.7	301.68	11JA	-0.9	-3.4	-2.0	1.1
145.32	4.48	-0.9	311.83	21JA	-1.1	-3.4	-1.9	1.1
141.44	4.54	-1.0	321.95	31JA	-1.0	-3.4	-1.8	1.1
137.59	4.42	-0.9	332.01	10FE	-0.3	-3.4	-1.8	1.1
134.44	4.15	-0.7	342.04	20FE	1.7	-3.5	-1.7	1.0
132.45	3.79	-0.4	352.01	2MR	2.9	-3.5	-1.7	1.0
131.76	3.40	-0.2	1.92	12MR	1.4	-3.6	-1.6	0.9
132.29	3.01	0.1	11.77	22MR	0.8	-3.6	-1.6	0.9
133.89	2.64	0.3	21.57	1AP	0.4	-3.7	-1.6	0.9
136.36	2.30	0.5	31.31	11AP	0.0	-3.7	-1.6	0.9
139.55	2.00	0.7	41.01	21AP	-0.6	-3.8	-1.5	0.9
143.33	1.73	0.8	50.66	1MY	-1.5	-3.9	-1.5	0.9
147.57	1.48	0.9	60.27	11MY	-1.5	-4.0	-1.5	0.9
152.21	1.25	1.1	69.85	21MY	-0.5	-4.1	-1.5	0.9
157.17	1.04	1.2	79.40	31MY	0.3	-4.2	-1.6	0.8
162.41	0.85	1.3	88.94	10JN	0.9	-4.2	-1.6	0.8
167.89	0.67	1.3	98.47	20JN	1.6	-4.1	-1.6	0.8
173.59	0.50	1.4	108.00	30JN	2.7	-3.8	-1.7	0.7
179.47	0.34	1.4	117.55	10JL	2.8	-3.2	-1.7	0.7
185.54	0.20	1.5	127.11	20JL	1.1	-3.2	-1.7	0.7
191.77	0.06	1.5	136.71	30JL	-0.4	-3.8	-1.8	0.6
198.15	-0.08	1.5	146.33	9AU	-1.3	-4.2	-1.9	0.6
204.68	-0.20	1.5	156.01	19AU	-1.4	-4.4	-1.9	0.5
211.36	-0.32	1.5	165.73	29AU	-0.7	-4.2	-2.0	0.4
218.17	-0.43	1.6	175.50	8SE	-0.3	-4.2	-2.1	0.4
225.11	-0.54	1.5	185.33	18SE	-0.1	-4.1	-2.1	0.3
232.18	-0.63	1.5	195.22	28SE	0.1	-4.0	-2.2	0.3
239.36	-0.72	1.5	205.16	8OC	0.2	-3.9	-2.2	0.4
246.67	-0.81	1.5	215.16	18OC	1.1	-3.8	-2.3	0.4
254.08	-0.88	1.5	225.21	28OC	2.9	-3.7	-2.3	0.5
261.59	-0.94	1.5	235.30	7NO	0.2	-3.6	-2.3	0.6
269.19	-1.00	1.5	245.44	17NO	-0.4	-3.6	-2.4	0.6
276.87	-1.04	1.4	255.60	27NO	-0.5	-3.6	-2.3	0.7
284.62	-1.07	1.4	265.78	7DE	-0.5	-3.5	-2.3	0.7
292.44	-1.09	1.4	275.98	17DE	-0.6	-3.5	-2.3	0.8
300.29	-1.10	1.4	286.17	27DE	-0.9	-3.4	-2.2	0.8

Right table (1586–1587):

♂ LONG	LAT	MAG	☉ LONG	16.00UT 1586	☿	♀	24	♄
308.18	-1.10	1.4	296.36	6JA	-1.0	-3.4	-2.2	0.8
316.08	-1.09	1.4	306.52	16JA	-0.9	-3.4	-2.1	0.8
323.99	-1.06	1.3	316.65	26JA	-0.2	-3.4	-2.0	0.9
331.89	-1.02	1.3	326.74	5FE	2.0	-3.3	-1.9	0.8
339.77	-0.98	1.3	336.79	15FE	2.3	-3.3	-1.9	0.8
347.61	-0.92	1.3	346.79	25FE	1.0	-3.3	-1.8	0.8
355.40	-0.85	1.3	356.73	7MR	0.6	-3.3	-1.7	0.8
3.14	-0.77	1.4	6.62	17MR	0.3	-3.3	-1.7	0.7
10.81	-0.69	1.4	16.44	27MR	-0.1	-3.4	-1.6	0.7
18.41	-0.60	1.4	26.21	6AP	-0.6	-3.4	-1.6	0.6
25.93	-0.51	1.5	35.93	16AP	-1.5	-3.4	-1.5	0.7
33.36	-0.41	1.5	45.60	26AP	-1.5	-3.5	-1.5	0.7
40.71	-0.30	1.6	55.23	6MY	-0.5	-3.5	-1.5	0.7
47.97	-0.20	1.6	64.83	16MY	0.4	-3.5	-1.5	0.7
55.13	-0.09	1.6	74.39	26MY	1.2	-3.4	-1.4	0.7
62.20	0.02	1.7	83.93	5JN	2.1	-3.4	-1.4	0.7
69.18	0.13	1.7	93.47	15JN	3.4	-3.4	-1.4	0.6
76.07	0.25	1.7	103.00	25JN	2.2	-3.4	-1.4	0.6
82.87	0.36	1.7	112.54	5JL	0.8	-3.3	-1.5	0.6
89.57	0.47	1.7	122.09	15JL	-0.4	-3.3	-1.5	0.6
96.18	0.59	1.8	131.67	25JL	-1.4	-3.3	-1.5	0.5
102.70	0.70	1.8	141.28	4AU	-1.4	-3.3	-1.5	0.5
109.12	0.81	1.7	150.93	14AU	-0.7	-3.3	-1.6	0.4
115.44	0.93	1.7	160.62	24AU	-0.2	-3.4	-1.6	0.3
121.67	1.05	1.7	170.37	3SE	0.1	-3.4	-1.7	0.3
127.78	1.16	1.7	180.17	13SE	0.2	-3.4	-1.7	0.2
133.77	1.29	1.6	190.02	23SE	0.5	-3.4	-1.8	0.2
139.63	1.41	1.6	199.94	3OC	1.4	-3.5	-1.9	0.1
145.34	1.54	1.5	209.91	13OC	2.6	-3.6	-1.9	0.1
150.89	1.67	1.4	219.93	23OC	0.0	-3.6	-2.0	0.1
156.24	1.81	1.4	230.01	2NO	-0.6	-3.7	-2.1	0.2
161.35	1.96	1.2	240.12	12NO	-0.6	-3.8	-2.1	0.3
166.18	2.12	1.1	250.27	22NO	-0.6	-3.9	-2.2	0.3
170.68	2.29	1.0	260.44	2DE	-0.7	-4.0	-2.2	0.4
174.76	2.47	0.8	270.63	12DE	-0.8	-4.1	-2.2	0.5
178.32	2.67	0.6	280.83	22DE	-0.8	-4.2	-2.2	0.5

1587

♂ LONG	LAT	MAG	☉ LONG	16.00UT 1587	☿	♀	24	♄
181.25	2.88	0.4	291.02	1JA	-0.8	-4.3	-2.2	0.6
183.38	3.10	0.2	301.20	11JA	-0.1	-4.3	-2.2	0.6
184.56	3.32	-0.1	311.34	21JA	2.2	-4.4	-2.1	0.6
184.58	3.52	-0.3	321.46	31JA	1.8	-4.3	-2.1	0.6
183.33	3.68	-0.6	331.53	10FE	0.7	-3.9	-2.0	0.6
180.84	3.75	-0.9	341.55	20FE	0.4	-3.9	-2.0	0.6
177.36	3.70	-1.1	351.52	2MR	0.2	-3.6	-1.9	0.6
173.49	3.49	-1.1	1.44	12MR	-0.1	-4.0	-1.8	0.6
169.93	3.15	-0.9	11.29	22MR	-0.7	-4.2	-1.7	0.6
167.29	2.73	-0.7	21.10	1AP	-1.5	-4.3	-1.7	0.5
165.91	2.28	-0.5	30.84	11AP	-1.4	-4.2	-1.6	0.5
165.82	1.85	-0.3	40.54	21AP	-0.5	-4.1	-1.6	0.4
166.92	1.45	-0.1	50.19	1MY	0.6	-4.0	-1.5	0.5
169.04	1.09	0.1	59.80	11MY	1.6	-3.9	-1.5	0.5
172.01	0.78	0.3	69.38	21MY	2.8	-3.8	-1.4	0.5
175.68	0.49	0.4	78.94	31MY	3.1	-3.7	-1.4	0.5
179.91	0.25	0.6	88.47	10JN	1.7	-3.6	-1.4	0.5
184.62	0.02	0.7	98.00	20JN	0.7	-3.6	-1.3	0.5
189.74	-0.17	0.8	107.54	30JN	-0.5	-3.5	-1.3	0.5
195.19	-0.35	0.9	117.08	10JL	-1.4	-3.5	-1.3	0.5
200.95	-0.51	0.9	126.65	20JL	-1.4	-3.4	-1.3	0.4
206.97	-0.65	1.0	136.24	30JL	-0.7	-3.4	-1.3	0.4
213.22	-0.78	1.1	145.87	9AU	-0.1	-3.4	-1.4	0.4
219.69	-0.89	1.1	155.54	19AU	0.2	-3.4	-1.4	0.3
226.36	-0.99	1.1	165.26	29AU	0.4	-3.4	-1.4	0.3
233.20	-1.08	1.2	175.03	8SE	0.7	-3.3	-1.4	0.2
240.21	-1.15	1.2	184.86	18SE	1.8	-3.4	-1.5	0.1
247.37	-1.21	1.2	194.74	28SE	2.3	-3.4	-1.5	0.1
254.67	-1.25	1.2	204.68	8OC	-0.1	-3.4	-1.6	0.0
262.09	-1.28	1.3	214.68	18OC	-0.7	-3.4	-1.6	-0.1
269.62	-1.30	1.3	224.72	28OC	-0.8	-3.4	-1.7	-0.1
277.25	-1.30	1.3	234.81	7NO	-0.7	-3.4	-1.8	-0.0
284.96	-1.29	1.3	244.94	17NO	-0.7	-3.5	-1.8	0.0
292.73	-1.26	1.3	255.11	27NO	-0.6	-3.5	-1.9	0.1
300.55	-1.23	1.3	265.29	7DE	-0.6	-3.5	-2.0	0.2
308.41	-1.18	1.4	275.48	17DE	-0.6	-3.5	-2.0	0.2
316.28	-1.12	1.4	285.68	27DE	0.1	-3.4	-2.1	0.3

♂			☉	16.00UT	☿	♀	♃	♄
LONG	LAT	MAG	LONG	1588	MAGNITUDES			
324.16	-1.05	1.4	295.86	6JA	2.5	-3.4	-2.1	0.4
332.02	-0.97	1.4	306.02	16JA	1.4	-3.4	-2.1	0.4
339.85	-0.88	1.4	316.16	26JA	0.5	-3.4	-2.1	0.4
347.65	-0.79	1.4	326.25	5FE	0.2	-3.4	-2.1	0.5
355.39	-0.69	1.5	336.30	15FE	0.1	-3.3	-2.1	0.5
3.07	-0.59	1.5	346.31	25FE	-0.2	-3.3	-2.0	0.5
10.69	-0.48	1.5	356.25	6MR	-0.7	-3.3	-2.0	0.5
18.23	-0.38	1.5	6.14	16MR	-1.5	-3.3	-1.9	0.5
25.69	-0.27	1.5	15.97	26MR	-1.4	-3.4	-1.8	0.4
33.07	-0.16	1.6	25.74	5AP	-0.5	-3.4	-1.8	0.4
40.36	-0.05	1.6	35.46	15AP	0.8	-3.4	-1.7	0.4
47.57	0.05	1.6	45.14	25AP	2.0	-3.4	-1.6	0.3
54.69	0.16	1.6	54.77	5MY	3.7	-3.5	-1.6	0.3
61.73	0.26	1.6	64.36	15MY	2.4	-3.5	-1.5	0.3
68.68	0.36	1.7	73.93	25MY	1.3	-3.5	-1.5	0.4
75.56	0.46	1.7	83.47	4JN	0.5	-3.6	-1.4	0.4
82.36	0.55	1.8	93.01	14JN	-0.5	-3.7	-1.4	0.4
89.09	0.64	1.8	102.54	24JN	-1.5	-3.8	-1.3	0.4
95.75	0.73	1.9	112.08	4JL	-1.5	-3.8	-1.3	0.4
102.34	0.81	1.9	121.63	14JL	-0.6	-3.9	-1.3	0.4
108.86	0.89	1.9	131.21	24JL	-0.1	-4.0	-1.3	0.4
115.32	0.97	1.9	140.82	3AU	0.3	-4.1	-1.2	0.3
121.73	1.05	1.9	150.47	13AU	0.6	-4.2	-1.2	0.3
128.07	1.12	1.9	160.16	23AU	1.0	-4.3	-1.2	0.3
134.36	1.19	1.9	169.90	2SE	2.2	-4.2	-1.3	0.2
140.58	1.26	1.9	179.70	12SE	2.0	-3.9	-1.3	0.2
146.74	1.33	1.9	189.55	22SE	-0.2	-3.3	-1.3	0.1
152.84	1.39	1.9	199.46	2OC	-0.9	-3.5	-1.3	0.0
158.86	1.46	1.8	209.43	12OC	-0.9	-4.1	-1.4	-0.1
164.80	1.52	1.8	219.45	22OC	-0.9	-4.3	-1.4	-0.1
170.65	1.58	1.7	229.52	1NO	-0.6	-4.4	-1.5	-0.2
176.39	1.64	1.6	239.63	11NO	-0.5	-4.3	-1.5	-0.2
182.02	1.71	1.5	249.78	21NO	-0.5	-4.2	-1.6	-0.1
187.50	1.77	1.4	259.95	1DE	-0.4	-4.1	-1.6	-0.1
192.80	1.83	1.3	270.14	11DE	0.2	-4.0	-1.7	-0.0
197.90	1.88	1.2	280.34	21DE	2.8	-3.9	-1.8	0.1
202.74	1.94	1.0	290.53	31DE	1.0	-3.8	-1.8	0.1
				1589				
207.28	1.99	0.9	300.70	10JA	0.2	-3.7	-1.9	0.2
211.43	2.04	0.7	310.85	20JA	0.1	-3.7	-1.9	0.3
215.10	2.07	0.4	320.97	30JA	-0.1	-3.6	-2.0	0.3
218.17	2.09	0.2	331.05	9FE	-0.3	-3.5	-2.0	0.3
220.50	2.09	-0.1	341.07	19FE	-0.8	-3.5	-2.0	0.4
221.91	2.06	-0.3	351.05	1MR	-1.4	-3.4	-2.0	0.4
222.22	1.97	-0.6	0.97	11MR	-1.3	-3.4	-2.0	0.4
221.29	1.82	-0.9	10.83	21MR	-0.4	-3.4	-2.0	0.4
219.12	1.58	-1.2	20.63	31MR	1.0	-3.3	-1.9	0.4
215.93	1.24	-1.5	30.38	10AP	2.6	-3.3	-1.9	0.3
212.27	0.82	-1.5	40.08	20AP	3.2	-3.3	-1.8	0.3
208.86	0.36	-1.4	49.73	30AP	1.8	-3.3	-1.7	0.3
206.34	-0.08	-1.2	59.34	10MY	1.0	-3.3	-1.7	0.2
205.09	-0.49	-1.0	68.92	20MY	0.3	-3.3	-1.6	0.2
205.19	-0.83	-0.8	78.47	30MY	-0.5	-3.3	-1.5	0.3
206.55	-1.11	-0.6	88.02	9JN	-1.5	-3.4	-1.5	0.3
209.01	-1.34	-0.4	97.54	19JN	-1.5	-3.4	-1.4	0.3
212.36	-1.53	-0.2	107.08	29JN	-0.6	-3.4	-1.4	0.3
216.47	-1.67	-0.1	116.62	9JL	0.0	-3.5	-1.3	0.3
221.21	-1.78	0.1	126.18	19JL	0.4	-3.5	-1.3	0.3
226.45	-1.86	0.2	135.77	29JL	0.8	-3.5	-1.3	0.3
232.14	-1.92	0.3	145.40	8AU	1.3	-3.4	-1.2	0.3
238.18	-1.95	0.4	155.07	18AU	2.7	-3.4	-1.2	0.3
244.54	-1.97	0.5	164.78	28AU	1.8	-3.4	-1.2	0.3
251.17	-1.96	0.6	174.55	7SE	-0.2	-3.3	-1.2	0.2
258.03	-1.93	0.6	184.37	17SE	-1.0	-3.3	-1.2	0.2
265.08	-1.89	0.7	194.26	27SE	-1.1	-3.3	-1.2	0.1
272.30	-1.83	0.8	204.20	7OC	-0.9	-3.3	-1.2	0.0
279.66	-1.75	0.8	214.19	17OC	-0.5	-3.3	-1.2	-0.0
287.13	-1.66	0.9	224.23	27OC	-0.4	-3.3	-1.3	-0.1
294.69	-1.56	1.0	234.32	6NO	-0.3	-3.4	-1.3	-0.2
302.32	-1.45	1.0	244.45	16NO	-0.3	-3.4	-1.4	-0.3
310.00	-1.33	1.1	254.61	26NO	0.4	-3.4	-1.4	-0.3
317.70	-1.20	1.1	264.80	6DE	3.1	-3.5	-1.5	-0.2
325.41	-1.07	1.2	274.99	16DE	0.7	-3.5	-1.5	-0.1
333.10	-0.94	1.2	285.19	26DE	0.0	-3.6	-1.6	-0.1

♂			☉	16.00UT	☿	♀	♃	♄
LONG	LAT	MAG	LONG	1590	MAGNITUDES			
340.78	-0.80	1.3	295.37	5JA	-0.1	-3.6	-1.7	0.0
348.42	-0.66	1.4	305.53	15JA	-0.2	-3.7	-1.7	0.1
356.01	-0.53	1.4	315.67	25JA	-0.4	-3.8	-1.8	0.1
3.54	-0.40	1.5	325.77	4FE	-0.8	-3.9	-1.8	0.2
11.01	-0.27	1.5	335.82	14FE	-1.3	-3.9	-1.9	0.2
18.41	-0.14	1.6	345.82	24FE	-1.2	-4.0	-1.9	0.3
25.73	-0.02	1.6	355.77	6MR	-0.4	-4.1	-2.0	0.3
32.98	0.09	1.7	5.66	16MR	1.2	-4.2	-2.0	0.3
40.14	0.20	1.7	15.50	26MR	3.2	-4.2	-2.0	0.3
47.23	0.31	1.7	25.27	5AP	2.4	-4.2	-2.0	0.3
54.25	0.40	1.8	35.00	15AP	1.3	-4.0	-2.0	0.3
61.19	0.50	1.8	44.67	25AP	0.8	-3.5	-1.9	0.3
68.05	0.58	1.8	54.31	5MY	0.2	-2.8	-1.9	0.3
74.85	0.66	1.8	63.90	15MY	-0.5	-3.5	-1.8	0.2
81.59	0.74	1.8	73.47	25MY	-1.5	-4.0	-1.8	0.2
88.26	0.80	1.8	83.01	4JN	-1.5	-4.2	-1.7	0.2
94.88	0.87	1.8	92.55	14JN	-0.6	-4.2	-1.6	0.2
101.45	0.93	1.8	102.08	24JN	0.1	-4.1	-1.6	0.3
107.98	0.98	1.9	111.62	4JL	0.6	-4.0	-1.5	0.3
114.46	1.03	1.9	121.16	14JL	1.0	-3.9	-1.5	0.3
120.90	1.07	2.0	130.74	24JL	1.8	-3.9	-1.4	0.3
127.31	1.11	2.0	140.35	3AU	3.2	-3.8	-1.4	0.3
133.69	1.15	2.0	149.99	13AU	1.5	-3.7	-1.3	0.3
140.04	1.18	2.0	159.68	23AU	-0.3	-3.6	-1.3	0.3
146.37	1.21	2.0	169.42	2SE	-1.1	-3.6	-1.3	0.3
152.67	1.23	2.0	179.22	12SE	-1.2	-3.5	-1.2	0.3
158.96	1.25	2.0	189.07	22SE	-0.8	-3.5	-1.2	0.2
165.22	1.27	2.0	198.98	2OC	-0.4	-3.5	-1.2	0.2
171.46	1.28	2.0	208.94	12OC	-0.3	-3.4	-1.2	0.1
177.67	1.29	1.9	218.96	22OC	-0.2	-3.4	-1.2	0.0
183.86	1.29	1.9	229.03	1NO	-0.1	-3.4	-1.2	-0.0
190.02	1.29	1.8	239.14	11NO	0.6	-3.4	-1.3	-0.1
196.14	1.28	1.8	249.29	21NO	3.2	-3.4	-1.3	-0.2
202.22	1.27	1.7	259.46	1DE	0.5	-3.4	-1.3	-0.3
208.25	1.24	1.6	269.65	11DE	-0.1	-3.4	-1.4	-0.3
214.22	1.21	1.5	279.85	21DE	-0.2	-3.4	-1.4	-0.2
220.12	1.17	1.4	290.04	31DE	-0.3	-3.4	-1.5	-0.1
				1591				
225.93	1.12	1.3	300.21	10JA	-0.5	-3.4	-1.5	-0.1
231.63	1.05	1.1	310.37	20JA	-0.8	-3.4	-1.6	-0.0
237.19	0.97	1.0	320.48	30JA	-1.2	-3.4	-1.7	0.1
242.60	0.87	0.8	330.56	9FE	-1.1	-3.5	-1.7	0.1
247.81	0.74	0.6	340.59	19FE	-0.3	-3.5	-1.8	0.2
252.76	0.58	0.4	350.56	1MR	1.4	-3.5	-1.9	0.2
257.41	0.38	0.2	0.48	11MR	3.3	-3.4	-1.9	0.3
261.66	0.14	-0.0	10.35	21MR	1.7	-3.4	-2.0	0.3
265.42	-0.16	-0.3	20.15	31MR	1.0	-3.4	-2.0	0.3
268.56	-0.52	-0.6	29.90	10AP	0.6	-3.4	-2.0	0.3
270.89	-0.96	-0.9	39.61	20AP	0.1	-3.3	-2.1	0.3
272.27	-1.49	-1.2	49.26	30AP	-0.6	-3.3	-2.1	0.3
272.50	-2.10	-1.5	58.87	10MY	-1.5	-3.3	-2.0	0.3
271.48	-2.78	-1.8	68.46	20MY	-1.5	-3.3	-2.0	0.3
269.32	-3.46	-2.1	78.01	30MY	-0.6	-3.3	-2.0	0.2
266.41	-4.07	-2.3	87.55	9JN	0.2	-3.3	-1.9	0.2
263.39	-4.52	-2.2	97.08	19JN	0.8	-3.4	-1.9	0.2
261.01	-4.77	-2.0	106.61	29JN	1.3	-3.4	-1.8	0.3
259.78	-4.83	-1.8	116.16	9JL	2.3	-3.4	-1.7	0.3
259.93	-4.74	-1.6	125.72	19JL	3.1	-3.5	-1.7	0.3
261.44	-4.55	-1.4	135.31	29JL	1.2	-3.5	-1.6	0.4
264.12	-4.30	-1.2	144.93	8AU	-0.3	-3.6	-1.6	0.4
267.80	-4.01	-0.9	154.60	18AU	-1.2	-3.6	-1.5	0.4
272.29	-3.71	-0.7	164.31	28AU	-1.3	-3.7	-1.5	0.4
277.41	-3.41	-0.6	174.08	7SE	-0.8	-3.8	-1.4	0.4
283.06	-3.10	-0.4	183.90	17SE	-0.4	-3.9	-1.4	0.3
289.11	-2.79	-0.2	193.77	27SE	-0.1	-4.0	-1.3	0.3
295.47	-2.49	-0.1	203.71	7OC	-0.0	-4.1	-1.3	0.3
302.08	-2.20	0.1	213.70	17OC	0.1	-4.2	-1.3	0.2
308.88	-1.92	0.2	223.74	27OC	0.9	-4.3	-1.3	0.1
315.82	-1.64	0.4	233.83	6NO	3.1	-4.4	-1.3	0.1
322.86	-1.39	0.5	243.96	16NO	0.3	-4.3	-1.3	-0.0
329.97	-1.14	0.6	254.12	26NO	-0.3	-3.9	-1.3	-0.1
337.12	-0.91	0.8	264.30	6DE	-0.4	-3.1	-1.3	-0.2
344.30	-0.70	0.9	274.50	16DE	-0.4	-3.5	-1.3	-0.2
351.47	-0.49	1.0	284.69	26DE	-0.6	-4.1	-1.4	-0.2

♂ LONG	♂ LAT	♂ MAG	⊙ LONG	16.00UT	☿	♀	♃	♄
				1592	MAGNITUDES			
358.63	-0.31	1.1	294.88	5JA	-0.9	-4.3	-1.4	-0.2
5.77	-0.14	1.2	305.04	15JA	-1.0	-4.4	-1.5	-0.1
12.86	0.02	1.3	315.18	25JA	-1.0	-4.3	-1.5	-0.0
19.92	0.16	1.4	325.28	4FE	-0.3	-4.2	-1.6	0.0
26.92	0.29	1.5	335.34	14FE	1.7	-4.1	-1.6	0.1
33.88	0.41	1.5	345.34	24FE	2.7	-4.0	-1.7	0.2
40.78	0.52	1.6	355.29	5MR	1.3	-3.9	-1.8	0.2
47.62	0.61	1.7	5.19	15MR	0.7	-3.8	-1.8	0.3
54.41	0.70	1.7	15.02	25MR	0.4	-3.7	-1.9	0.3
61.14	0.77	1.8	24.80	4AP	0.0	-3.6	-2.0	0.3
67.83	0.84	1.8	34.53	14AP	-0.6	-3.5	-2.0	0.4
74.46	0.90	1.9	44.21	24AP	-1.5	-3.5	-2.1	0.4
81.05	0.95	1.9	53.85	4MY	-1.5	-3.4	-2.1	0.4
87.59	1.00	1.9	63.44	14MY	-0.5	-3.4	-2.1	0.4
94.10	1.04	2.0	73.01	24MY	0.3	-3.4	-2.2	0.4
100.57	1.07	2.0	82.56	3JN	1.0	-3.3	-2.2	0.3
107.02	1.10	2.0	92.09	13JN	1.8	-3.3	-2.1	0.3
113.43	1.12	2.0	101.62	23JN	3.0	-3.3	-2.1	0.3
119.83	1.13	2.0	111.16	3JL	2.6	-3.3	-2.1	0.3
126.22	1.14	2.0	120.71	13JL	1.0	-3.3	-2.0	0.3
132.59	1.15	2.0	130.28	23JL	-0.4	-3.3	-2.0	0.4
138.96	1.15	2.0	139.89	2AU	-1.3	-3.4	-1.9	0.4
145.32	1.15	2.0	149.53	12AU	-1.4	-3.4	-1.8	0.4
151.69	1.14	2.0	159.21	22AU	-0.7	-3.4	-1.8	0.5
158.06	1.13	2.0	168.95	1SE	-0.3	-3.4	-1.7	0.5
164.44	1.11	2.0	178.74	11SE	-0.0	-3.5	-1.7	0.5
170.83	1.09	2.0	188.59	21SE	0.1	-3.5	-1.6	0.4
177.24	1.06	2.0	198.50	1OC	0.3	-3.5	-1.5	0.4
183.66	1.03	2.0	208.45	11OC	1.2	-3.4	-1.5	0.4
190.10	0.99	2.0	218.47	21OC	2.9	-3.4	-1.5	0.3
196.56	0.95	2.0	228.54	31OC	0.2	-3.4	-1.4	0.3
203.03	0.90	1.9	238.64	10NO	-0.5	-3.4	-1.4	0.2
209.52	0.84	1.9	248.78	20NO	-0.5	-3.3	-1.4	0.2
216.03	0.78	1.8	258.96	30NO	-0.5	-3.3	-1.4	0.1
222.56	0.71	1.8	269.15	10DE	-0.7	-3.3	-1.4	0.0
229.10	0.63	1.7	279.34	20DE	-0.8	-3.3	-1.4	-0.1
235.65	0.53	1.6	289.54	30DE	-0.9	-3.3	-1.4	-0.1
				1593				
242.22	0.43	1.5	299.71	9JA	-0.9	-3.4	-1.4	-0.1
248.79	0.31	1.4	309.87	19JA	-0.2	-3.4	-1.4	-0.1
255.36	0.18	1.3	319.99	29JA	2.0	-3.4	-1.5	-0.0
261.94	0.03	1.2	330.07	8FE	2.1	-3.4	-1.5	0.1
268.50	-0.13	1.1	340.10	18FE	0.9	-3.5	-1.6	0.1
275.06	-0.32	1.0	350.09	28FE	0.5	-3.5	-1.6	0.2
281.58	-0.52	0.8	0.01	10MR	0.3	-3.6	-1.7	0.2
288.07	-0.75	0.7	9.87	20MR	-0.1	-3.6	-1.8	0.3
294.51	-1.01	0.5	19.69	30MR	-0.6	-3.7	-1.8	0.3
300.87	-1.29	0.4	29.44	9AP	-1.5	-3.7	-1.9	0.4
307.13	-1.59	0.2	39.14	19AP	-1.5	-3.8	-2.0	0.4
313.25	-1.93	0.0	48.80	29AP	-0.5	-3.9	-2.0	0.4
319.18	-2.30	-0.2	58.41	9MY	0.5	-4.0	-2.1	0.5
324.87	-2.71	-0.4	68.00	19MY	1.3	-4.1	-2.2	0.5
330.23	-3.14	-0.6	77.56	29MY	2.4	-4.2	-2.2	0.5
335.16	-3.61	-0.8	87.09	8JN	3.4	-4.2	-2.3	0.5
339.54	-4.11	-1.1	96.63	18JN	2.1	-4.1	-2.3	0.5
343.18	-4.63	-1.3	106.16	28JN	0.8	-3.8	-2.3	0.4
345.91	-5.15	-1.6	115.70	8JL	-0.4	-3.2	-2.3	0.4
347.51	-5.65	-1.9	125.26	18JL	-1.4	-3.3	-2.3	0.4
347.80	-6.06	-2.1	134.85	28JL	-1.5	-3.9	-2.2	0.5
346.76	-6.31	-2.4	144.47	7AU	-0.7	-4.2	-2.2	0.5
344.59	-6.31	-2.6	154.13	17AU	-0.2	-4.2	-2.1	0.6
341.87	-6.01	-2.6	163.84	27AU	0.1	-4.2	-2.1	0.6
339.37	-5.44	-2.4	173.60	6SE	0.3	-4.1	-2.0	0.6
337.72	-4.70	-2.2	183.42	16SE	0.5	-4.1	-2.0	0.6
337.32	-3.89	-1.9	193.30	26SE	1.5	-4.0	-1.9	0.6
338.24	-3.12	-1.5	203.22	6OC	2.6	-3.9	-1.8	0.6
340.32	-2.42	-1.2	213.21	16OC	0.0	-3.8	-1.8	0.6
343.39	-1.81	-0.9	223.25	26OC	-0.6	-3.7	-1.7	0.5
347.24	-1.28	-0.6	233.34	5NO	-0.7	-3.7	-1.7	0.5
351.69	-0.84	-0.4	243.46	15NO	-0.7	-3.6	-1.6	0.4
356.62	-0.47	-0.1	253.62	25NO	-0.8	-3.6	-1.6	0.4
1.91	-0.16	0.1	263.80	5DE	-0.7	-3.5	-1.5	0.3
7.47	0.11	0.3	274.00	15DE	-0.7	-3.5	-1.5	0.2
13.24	0.33	0.5	284.19	25DE	-0.7	-3.4	-1.5	0.2

♂ LONG	♂ LAT	♂ MAG	⊙ LONG	16.00UT	☿	♀	♃	♄
				1594	MAGNITUDES			
19.16	0.51	0.7	294.38	4JA	-0.1	-3.4	-1.5	0.1
25.20	0.67	0.8	304.55	14JA	2.3	-3.4	-1.5	0.0
31.33	0.80	1.0	314.69	24JA	1.6	-3.4	-1.5	0.0
37.51	0.90	1.1	324.79	3FE	0.6	-3.3	-1.5	0.1
43.74	0.99	1.3	334.85	13FE	0.3	-3.3	-1.5	0.1
50.00	1.06	1.4	344.86	23FE	0.1	-3.3	-1.6	0.2
56.27	1.12	1.5	354.81	5MR	-0.2	-3.3	-1.6	0.3
62.55	1.17	1.6	4.71	15MR	-0.7	-3.3	-1.6	0.3
68.84	1.20	1.7	14.55	25MR	-1.5	-3.4	-1.7	0.4
75.12	1.23	1.7	24.33	4AP	-1.5	-3.4	-1.7	0.4
81.40	1.25	1.8	34.06	14AP	-0.5	-3.4	-1.8	0.5
87.67	1.26	1.9	43.74	24AP	0.6	-3.5	-1.9	0.5
93.94	1.27	1.9	53.38	4MY	1.7	-3.5	-1.9	0.6
100.20	1.27	2.0	62.98	14MY	3.1	-3.5	-2.0	0.6
106.47	1.26	2.0	72.55	24MY	2.9	-3.4	-2.1	0.6
112.73	1.25	2.0	82.09	3JN	1.6	-3.4	-2.1	0.6
119.00	1.23	2.0	91.63	13JN	0.6	-3.4	-2.2	0.6
125.27	1.21	2.0	101.16	23JN	-0.4	-3.4	-2.3	0.6
131.55	1.19	2.0	110.69	3JL	-1.4	-3.3	-2.3	0.6
137.85	1.16	2.0	120.25	13JL	-1.5	-3.3	-2.4	0.6
144.17	1.13	2.0	129.82	23JL	-0.7	-3.3	-2.4	0.6
150.51	1.09	2.0	139.42	2AU	-0.1	-3.3	-2.4	0.6
156.88	1.05	2.0	149.06	12AU	0.2	-3.3	-2.4	0.7
163.27	1.00	2.0	158.74	22AU	0.4	-3.4	-2.4	0.7
169.70	0.96	1.9	168.48	1SE	0.8	-3.4	-2.4	0.8
176.17	0.90	1.9	178.27	11SE	1.9	-3.4	-2.3	0.8
182.67	0.85	1.9	188.11	21SE	2.3	-3.4	-2.3	0.8
189.22	0.78	1.9	198.01	1OC	-0.1	-3.5	-2.2	0.8
195.82	0.72	1.9	207.97	11OC	-0.8	-3.6	-2.1	0.8
202.47	0.65	1.9	217.98	21OC	-0.8	-3.6	-2.1	0.8
209.16	0.57	1.9	228.04	31OC	-0.8	-3.7	-2.0	0.8
215.91	0.49	1.9	238.15	10NO	-0.7	-3.8	-1.9	0.8
222.71	0.40	1.9	248.29	20NO	-0.6	-3.9	-1.9	0.7
229.56	0.31	1.8	258.46	30NO	-0.6	-4.0	-1.8	0.7
236.47	0.21	1.8	268.65	10DE	-0.5	-4.1	-1.7	0.6
243.43	0.10	1.7	278.84	20DE	0.1	-4.2	-1.7	0.5
250.45	-0.01	1.7	289.04	30DE	2.5	-4.3	-1.7	0.5
				1595				
257.52	-0.13	1.6	299.22	9JA	1.3	-4.3	-1.6	0.4
264.64	-0.26	1.6	309.37	19JA	0.4	-4.4	-1.6	0.3
271.81	-0.39	1.5	319.50	29JA	0.2	-4.3	-1.6	0.2
279.03	-0.53	1.4	329.58	8FE	0.0	-3.9	-1.6	0.3
286.29	-0.68	1.3	339.61	18FE	-0.2	-3.3	-1.6	0.3
293.59	-0.83	1.3	349.60	28FE	-0.7	-3.6	-1.6	0.4
300.92	-0.99	1.2	359.53	10MR	-1.4	-4.0	-1.6	0.4
308.27	-1.15	1.1	9.39	20MR	-1.4	-4.2	-1.6	0.5
315.64	-1.31	1.0	19.21	30MR	-0.5	-4.3	-1.6	0.5
323.01	-1.46	0.9	28.97	9AP	0.8	-4.2	-1.7	0.6
330.37	-1.62	0.8	38.67	19AP	2.2	-4.1	-1.7	0.6
337.71	-1.77	0.7	48.33	29AP	3.8	-4.0	-1.7	0.7
345.00	-1.92	0.6	57.95	9MY	2.2	-3.9	-1.8	0.7
352.21	-2.05	0.6	67.53	19MY	1.2	-3.8	-1.9	0.8
359.34	-2.17	0.5	77.09	29MY	0.4	-3.7	-1.9	0.8
6.33	-2.27	0.3	86.64	8JN	-0.5	-3.6	-2.0	0.8
13.16	-2.36	0.2	96.17	18JN	-1.4	-3.6	-2.0	0.8
19.79	-2.43	0.1	105.70	28JN	-1.5	-3.5	-2.1	0.9
26.15	-2.48	0.0	115.24	8JL	-0.6	-3.5	-2.2	0.9
32.21	-2.50	-0.1	124.80	18JL	-0.0	-3.4	-2.3	0.9
37.87	-2.50	-0.3	134.39	28JL	0.3	-3.4	-2.3	0.9
43.04	-2.46	-0.4	144.01	7AU	0.6	-3.4	-2.4	0.9
47.60	-2.39	-0.6	153.67	17AU	1.1	-3.4	-2.4	0.9
51.41	-2.27	-0.8	163.38	27AU	2.4	-3.4	-2.5	0.9
54.27	-2.11	-1.0	173.14	6SE	2.0	-3.3	-2.5	1.0
55.99	-1.87	-1.2	182.95	16SE	-0.1	-3.4	-2.5	1.0
56.36	-1.56	-1.4	192.82	26SE	-0.9	-3.4	-2.5	1.0
55.25	-1.16	-1.6	202.75	6OC	-1.0	-3.4	-2.4	1.1
52.77	-0.68	-1.8	212.73	16OC	-0.9	-3.4	-2.4	1.1
49.33	-0.15	-1.9	222.77	26OC	-0.6	-3.4	-2.3	1.1
45.72	0.38	-1.8	232.85	5NO	-0.4	-3.4	-2.3	1.1
42.73	0.84	-1.6	242.97	15NO	-0.4	-3.5	-2.2	1.0
40.89	1.21	-1.2	253.13	25NO	-0.4	-3.5	-2.1	1.0
40.41	1.48	-0.9	263.31	5DE	0.2	-3.5	-2.0	1.0
41.22	1.66	-0.6	273.50	15DE	2.8	-3.5	-2.0	0.9
43.13	1.77	-0.3	283.70	25DE	0.9	-3.4	-1.9	0.9

♂ LONG	♂ LAT	♂ MAG	⊙ LONG	16.00UT	☿	♀	♃	♄
				1596	MAGNITUDES			
45.93	1.84	-0.0	293.88	4JA	0.2	-3.4	-1.8	0.8
49.42	1.88	0.2	304.05	14JA	0.0	-3.4	-1.8	0.7
53.47	1.89	0.5	314.19	24JA	-0.1	-3.4	-1.7	0.6
57.94	1.88	0.7	324.30	3FE	-0.3	-3.4	-1.7	0.6
62.73	1.87	0.9	334.35	13FE	-0.8	-3.3	-1.7	0.5
67.79	1.84	1.0	344.37	23FE	-1.4	-3.3	-1.6	0.5
73.05	1.80	1.2	354.32	4MR	-1.3	-3.3	-1.6	0.6
78.46	1.76	1.3	4.22	14MR	-0.4	-3.3	-1.6	0.6
84.01	1.72	1.4	14.07	24MR	1.0	-3.4	-1.6	0.7
89.67	1.67	1.5	23.85	3AP	2.8	-3.4	-1.6	0.7
95.40	1.62	1.6	33.59	13AP	2.8	-3.4	-1.6	0.8
101.22	1.57	1.7	43.27	23AP	1.6	-3.4	-1.6	0.8
107.09	1.51	1.8	52.91	3MY	0.9	-3.5	-1.6	0.9
113.02	1.45	1.8	62.51	13MY	0.3	-3.5	-1.7	0.9
119.00	1.39	1.9	72.09	23MY	-0.5	-3.5	-1.7	1.0
125.03	1.33	1.9	81.63	2JN	-1.5	-3.6	-1.7	1.0
131.11	1.27	2.0	91.17	12JN	-1.5	-3.7	-1.8	1.1
137.23	1.21	2.0	100.70	22JN	-0.6	-3.8	-1.8	1.1
143.40	1.14	2.0	110.24	2JL	0.0	-3.8	-1.9	1.1
149.62	1.07	2.0	119.79	12JL	0.5	-3.9	-1.9	1.1
155.89	1.00	2.0	129.36	22JL	0.8	-4.0	-2.0	1.1
162.21	0.93	2.0	138.96	1AU	1.5	-4.1	-2.1	1.1
168.59	0.85	2.0	148.60	11AU	2.9	-4.2	-2.2	1.2
175.03	0.77	2.0	158.28	21AU	1.7	-4.2	-2.2	1.1
181.53	0.69	1.9	168.01	31AU	-0.2	-4.2	-2.3	1.2
188.09	0.61	1.9	177.80	10SE	-1.0	-3.8	-2.3	1.2
194.73	0.53	1.9	187.64	20SE	-1.1	-3.3	-2.4	1.3
201.43	0.44	1.8	197.53	30SE	-0.9	-3.6	-2.4	1.3
208.21	0.35	1.8	207.49	10OC	-0.5	-4.1	-2.4	1.3
215.06	0.25	1.7	217.50	20OC	-0.3	-4.3	-2.4	1.3
221.98	0.16	1.7	227.56	30OC	-0.3	-4.3	-2.4	1.3
228.98	0.06	1.7	237.66	9NO	-0.2	-4.3	-2.4	1.3
236.06	-0.04	1.7	247.80	19NO	0.4	-4.2	-2.3	1.3
243.21	-0.15	1.7	257.97	29NO	3.1	-4.1	-2.3	1.2
250.44	-0.25	1.7	268.16	9DE	0.7	-4.0	-2.2	1.2
257.74	-0.36	1.7	278.36	19DE	-0.0	-3.9	-2.2	1.1
265.11	-0.47	1.6	288.54	29DE	-0.1	-3.8	-2.1	1.1
				1597				
272.54	-0.58	1.6	298.73	8JA	-0.2	-3.7	-2.0	1.0
280.04	-0.68	1.6	308.88	18JA	-0.4	-3.7	-1.9	1.0
287.59	-0.79	1.5	319.01	28JA	-0.8	-3.6	-1.9	0.9
295.19	-0.89	1.5	329.09	7FE	-1.3	-3.5	-1.8	0.9
302.83	-0.99	1.4	339.13	17FE	-1.2	-3.5	-1.7	0.9
310.51	-1.09	1.4	349.11	27FE	-0.4	-3.4	-1.7	0.8
318.21	-1.17	1.4	359.04	9MR	1.2	-3.4	-1.6	0.8
325.91	-1.25	1.3	8.92	19MR	3.4	-3.4	-1.6	0.9
333.62	-1.32	1.3	18.73	29MR	2.1	-3.3	-1.6	0.9
341.32	-1.37	1.2	28.49	8AP	1.2	-3.3	-1.6	1.0
348.99	-1.41	1.2	38.20	18AP	0.7	-3.3	-1.5	1.0
356.61	-1.44	1.2	47.86	28AP	0.2	-3.3	-1.5	1.1
4.19	-1.45	1.1	57.49	8MY	-0.5	-3.3	-1.5	1.1
11.69	-1.45	1.1	67.07	18MY	-1.5	-3.3	-1.5	1.2
19.10	-1.43	1.0	76.63	28MY	-1.5	-3.3	-1.5	1.2
26.42	-1.39	1.0	86.17	7JN	-0.6	-3.4	-1.6	1.3
33.61	-1.34	1.0	95.71	17JN	0.1	-3.4	-1.6	1.3
40.67	-1.27	0.9	105.24	27JN	0.6	-3.4	-1.6	1.4
47.57	-1.18	0.9	114.78	7JL	1.1	-3.5	-1.6	1.4
54.29	-1.07	0.8	124.34	17JL	1.9	-3.5	-1.7	1.4
60.81	-0.94	0.7	133.92	27JL	3.2	-3.5	-1.7	1.4
67.10	-0.80	0.7	143.54	6AU	1.5	-3.4	-1.8	1.3
73.11	-0.63	0.6	153.20	16AU	-0.3	-3.4	-1.9	1.3
78.81	-0.44	0.5	162.90	26AU	-1.1	-3.4	-1.9	1.2
84.14	-0.23	0.4	172.66	5SE	-1.3	-3.3	-2.0	1.2
89.02	0.01	0.3	182.47	15SE	-0.8	-3.3	-2.1	1.2
93.36	0.28	0.1	192.34	25SE	-0.4	-3.3	-2.1	1.2
97.03	0.59	-0.0	202.26	5OC	-0.2	-3.3	-2.2	1.2
99.88	0.95	-0.2	212.24	15OC	-0.1	-3.3	-2.2	1.2
101.72	1.35	-0.4	222.28	25OC	-0.0	-3.3	-2.3	1.2
102.35	1.80	-0.7	232.36	4NO	0.7	-3.4	-2.3	1.2
101.61	2.28	-0.9	242.48	14NO	3.3	-3.4	-2.3	1.1
99.46	2.76	-1.1	252.63	24NO	0.5	-3.4	-2.3	1.1
96.13	3.18	-1.3	262.82	4DE	-0.2	-3.5	-2.3	1.1
92.22	3.49	-1.3	273.01	14DE	-0.3	-3.5	-2.3	1.0
88.48	3.64	-1.1	283.20	24DE	-0.3	-3.6	-2.2	1.0

♂ LONG	♂ LAT	♂ MAG	⊙ LONG	16.00UT	☿	♀	♃	♄
				1598	MAGNITUDES			
85.61	3.65	-0.9	293.39	3JA	-0.5	-3.6	-2.2	0.9
84.00	3.54	-0.6	303.56	13JA	-0.9	-3.7	-2.1	0.9
83.72	3.37	-0.3	313.70	23JA	-1.1	-3.8	-2.1	0.8
84.65	3.17	-0.0	323.81	2FE	-1.1	-3.9	-2.0	0.8
86.60	2.97	0.3	333.87	12FE	-0.3	-3.9	-1.9	0.8
89.35	2.77	0.5	343.88	22FE	1.5	-4.0	-1.8	0.7
92.76	2.59	0.7	353.85	4MR	3.1	-4.1	-1.8	0.7
96.69	2.41	0.9	3.75	14MR	1.5	-4.2	-1.7	0.7
101.01	2.24	1.0	13.59	24MR	0.9	-4.3	-1.6	0.7
105.67	2.09	1.2	23.38	3AP	0.5	-4.2	-1.6	0.8
110.59	1.94	1.3	33.11	13AP	0.1	-4.0	-1.6	0.9
115.72	1.80	1.4	42.80	23AP	-0.6	-3.5	-1.5	0.9
121.04	1.67	1.5	52.44	3MY	-1.5	-2.8	-1.5	1.0
126.52	1.55	1.6	62.04	13MY	-1.5	-3.6	-1.5	1.1
132.12	1.43	1.7	71.62	23MY	-0.6	-4.0	-1.4	1.1
137.85	1.31	1.7	81.17	2JN	0.2	-4.2	-1.4	1.2
143.69	1.20	1.8	90.70	12JN	0.8	-4.2	-1.4	1.2
149.63	1.08	1.8	100.24	22JN	1.5	-4.1	-1.4	1.2
155.66	0.98	1.8	109.77	2JL	2.5	-4.0	-1.4	1.2
161.79	0.87	1.9	119.32	12JL	2.9	-3.9	-1.4	1.2
168.01	0.76	1.9	128.89	22JL	1.2	-3.8	-1.5	1.2
174.32	0.65	1.9	138.49	1AU	-0.3	-3.8	-1.5	1.2
180.72	0.55	1.9	148.13	11AU	-1.2	-3.7	-1.5	1.2
187.21	0.44	1.9	157.81	21AU	-1.4	-3.6	-1.6	1.1
193.79	0.34	1.9	167.54	31AU	-0.8	-3.6	-1.6	1.1
200.47	0.24	1.8	177.32	10SE	-0.3	-3.5	-1.7	1.1
207.24	0.13	1.8	187.16	20SE	-0.1	-3.5	-1.7	1.0
214.10	0.03	1.8	197.06	30SE	0.0	-3.5	-1.8	1.0
221.06	-0.07	1.7	207.01	10OC	0.2	-3.4	-1.9	1.0
228.11	-0.17	1.7	217.01	20OC	0.9	-3.4	-1.9	1.0
235.26	-0.28	1.6	227.07	30OC	3.2	-3.4	-2.0	1.0
242.49	-0.37	1.6	237.17	9NO	0.3	-3.4	-2.1	1.0
249.82	-0.47	1.5	247.31	19NO	-0.4	-3.4	-2.1	1.0
257.24	-0.56	1.5	257.48	29NO	-0.4	-3.4	-2.2	1.0
264.73	-0.65	1.5	267.66	9DE	-0.5	-3.4	-2.2	1.0
272.30	-0.74	1.5	277.86	19DE	-0.6	-3.4	-2.2	0.9
279.94	-0.82	1.5	288.05	29DE	-0.9	-3.4	-2.2	0.9
				1599				
287.63	-0.89	1.5	298.23	8JA	-1.0	-3.4	-2.2	0.8
295.38	-0.96	1.4	308.39	18JA	-0.9	-3.4	-2.2	0.8
303.18	-1.02	1.4	318.52	28JA	-0.3	-3.4	-2.1	0.7
311.00	-1.07	1.4	328.60	7FE	1.7	-3.5	-2.1	0.7
318.84	-1.11	1.4	338.64	17FE	2.5	-3.5	-2.0	0.6
326.68	-1.13	1.4	348.63	27FE	1.1	-3.5	-1.9	0.6
334.53	-1.15	1.4	358.56	9MR	0.6	-3.4	-1.9	0.6
342.35	-1.15	1.4	8.44	19MR	0.3	-3.4	-1.8	0.5
350.14	-1.14	1.4	18.25	29MR	-0.0	-3.4	-1.7	0.5
357.89	-1.12	1.4	28.01	8AP	-0.6	-3.4	-1.6	0.6
5.59	-1.08	1.4	37.73	18AP	-1.5	-3.3	-1.6	0.7
13.22	-1.03	1.4	47.39	28AP	-1.5	-3.3	-1.5	0.7
20.78	-0.97	1.4	57.01	8MY	-0.6	-3.3	-1.5	0.8
28.26	-0.90	1.4	66.60	18MY	0.4	-3.3	-1.4	0.9
35.64	-0.81	1.4	76.16	28MY	1.1	-3.3	-1.4	0.9
42.92	-0.72	1.4	85.70	7JN	2.0	-3.3	-1.4	1.0
50.09	-0.61	1.4	95.24	17JN	3.2	-3.4	-1.3	1.0
57.14	-0.50	1.4	104.77	27JN	2.4	-3.4	-1.3	1.0
64.08	-0.37	1.3	114.31	7JL	1.0	-3.4	-1.3	1.1
70.88	-0.24	1.3	123.87	17JL	-0.3	-3.5	-1.3	1.1
77.54	-0.10	1.3	133.45	27JL	-1.3	-3.5	-1.3	1.1
84.05	0.05	1.3	143.07	6AU	-1.5	-3.6	-1.3	1.1
90.39	0.21	1.2	152.73	16AU	-0.7	-3.6	-1.3	1.1
96.55	0.38	1.2	162.43	26AU	-0.3	-3.7	-1.4	1.0
102.51	0.56	1.1	172.18	5SE	0.0	-3.8	-1.4	1.0
108.23	0.75	1.1	182.00	15SE	0.2	-3.9	-1.4	1.0
113.69	0.96	1.0	191.86	25SE	0.4	-4.0	-1.5	0.9
118.83	1.19	0.9	201.78	5OC	1.2	-4.1	-1.5	0.9
123.61	1.43	0.8	211.76	15OC	2.9	-4.2	-1.6	0.9
127.93	1.70	0.6	221.79	25OC	0.2	-4.3	-1.6	0.9
131.70	2.00	0.5	231.87	4NO	-0.5	-4.4	-1.7	0.9
134.79	2.33	0.3	241.99	14NO	-0.6	-4.3	-1.8	0.9
137.05	2.70	0.1	252.14	24NO	-0.6	-3.9	-1.8	0.9
138.28	3.10	-0.1	262.32	4DE	-0.7	-3.1	-1.9	0.9
138.31	3.51	-0.4	272.51	14DE	-0.8	-3.6	-2.0	0.9
137.00	3.92	-0.6	282.71	24DE	-0.8	-4.1	-2.0	0.9

♂			⊙	16.00UT	☿	♀	♃	♄
LONG	LAT	MAG	LONG	1600	MAGNITUDES			
134.39	4.26	-0.8	292.90	3JA	-0.8	-4.3	-2.0	0.8
130.79	4.48	-1.0	303.07	13JA	-0.2	-4.4	-2.1	0.8
126.82	4.53	-1.0	313.21	23JA	2.0	-4.3	-2.1	0.7
123.21	4.41	-0.8	323.32	2FE	2.0	-4.2	-2.1	0.7
120.57	4.15	-0.6	333.39	12FE	0.8	-4.1	-2.1	0.6
119.20	3.81	-0.3	343.40	22FE	0.4	-4.0	-2.0	0.6
119.11	3.45	-0.1	353.36	3MR	0.2	-3.9	-2.0	0.5
120.17	3.10	0.2	3.27	13MR	-0.1	-3.8	-1.9	0.5
122.21	2.78	0.4	13.12	23MR	-0.6	-3.7	-1.9	0.4
125.03	2.47	0.6	22.91	2AP	-1.4	-3.6	-1.8	0.4
128.49	2.20	0.8	32.65	12AP	-1.5	-3.5	-1.7	0.4
132.46	1.95	0.9	42.33	22AP	-0.6	-3.5	-1.7	0.5
136.84	1.72	1.1	51.98	2MY	0.5	-3.5	-1.6	0.6
141.58	1.51	1.2	61.59	12MY	1.4	-3.4	-1.5	0.6
146.59	1.32	1.3	71.16	22MY	2.6	-3.4	-1.5	0.7
151.85	1.14	1.4	80.71	1JN	3.3	-3.3	-1.4	0.7
157.33	0.97	1.4	90.25	11JN	1.9	-3.3	-1.4	0.8
162.99	0.81	1.5	99.78	21JN	0.7	-3.3	-1.3	0.9
168.82	0.66	1.6	109.31	1JL	-0.4	-3.3	-1.3	0.9
174.81	0.52	1.6	118.86	11JL	-1.4	-3.3	-1.3	0.9
180.95	0.38	1.6	128.43	21JL	-1.5	-3.3	-1.3	0.9
187.22	0.25	1.6	138.03	31JL	-0.7	-3.4	-1.2	1.0
193.63	0.12	1.7	147.66	10AU	-0.2	-3.4	-1.2	1.0
200.16	-0.00	1.7	157.34	20AU	0.1	-3.4	-1.2	1.0
206.82	-0.12	1.7	167.07	30AU	0.3	-3.4	-1.2	0.9
213.61	-0.23	1.7	176.85	9SE	0.6	-3.5	-1.3	0.9
220.51	-0.34	1.6	186.68	19SE	1.6	-3.5	-1.3	0.9
227.53	-0.44	1.6	196.57	29SE	2.6	-3.5	-1.3	0.8
234.67	-0.54	1.6	206.52	9OC	0.0	-3.4	-1.3	0.8
241.91	-0.63	1.6	216.53	19OC	-0.7	-3.4	-1.4	0.8
249.26	-0.72	1.6	226.58	29OC	-0.7	-3.4	-1.4	0.8
256.71	-0.79	1.5	236.68	8NO	-0.7	-3.4	-1.5	0.8
264.25	-0.86	1.5	246.81	18NO	-0.8	-3.3	-1.5	0.8
271.87	-0.92	1.5	256.98	28NO	-0.7	-3.3	-1.6	0.8
279.57	-0.98	1.4	267.17	8DE	-0.7	-3.3	-1.7	0.8
287.34	-1.02	1.4	277.36	18DE	-0.6	-3.3	-1.7	0.8
295.15	-1.05	1.4	287.56	28DE	-0.0	-3.3	-1.8	0.8
				1601				
303.01	-1.07	1.3	297.74	7JA	2.3	-3.4	-1.9	0.8
310.90	-1.08	1.3	307.90	17JA	1.5	-3.4	-1.9	0.8
318.80	-1.08	1.3	318.03	27JA	0.5	-3.4	-2.0	0.7
326.71	-1.07	1.3	328.11	6FE	0.3	-3.4	-2.0	0.7
334.60	-1.04	1.3	338.15	16FE	0.1	-3.5	-2.0	0.6
342.47	-1.00	1.3	348.15	26FE	-0.2	-3.5	-2.0	0.6
350.30	-0.96	1.4	358.08	8MR	-0.7	-3.6	-2.0	0.5
358.08	-0.90	1.4	7.96	18MR	-1.4	-3.6	-2.0	0.5
5.81	-0.83	1.4	17.78	28MR	-1.4	-3.7	-1.9	0.4
13.47	-0.75	1.4	27.55	7AP	-0.5	-3.7	-1.9	0.4
21.05	-0.67	1.5	37.26	17AP	0.6	-3.8	-1.8	0.3
28.56	-0.58	1.5	46.93	27AP	1.9	-3.9	-1.8	0.4
35.97	-0.48	1.5	56.55	7MY	3.5	-4.0	-1.7	0.4
43.30	-0.38	1.6	66.14	17MY	2.6	-4.1	-1.7	0.5
50.53	-0.27	1.6	75.71	27MY	1.4	-4.2	-1.6	0.6
57.67	-0.16	1.6	85.25	6JN	0.6	-4.2	-1.5	0.6
64.70	-0.04	1.6	94.78	16JN	-0.4	-4.1	-1.5	0.7
71.64	0.08	1.6	104.31	26JN	-1.4	-3.8	-1.4	0.7
78.48	0.20	1.6	113.85	6JL	-1.5	-3.2	-1.4	0.8
85.21	0.32	1.7	123.41	16JL	-0.7	-3.3	-1.3	0.8
91.85	0.44	1.7	132.99	26JL	-0.1	-3.9	-1.3	0.8
98.37	0.57	1.6	142.60	5AU	0.3	-4.2	-1.3	0.9
104.78	0.70	1.6	152.26	15AU	0.5	-4.2	-1.2	0.9
111.07	0.83	1.6	161.96	25AU	0.9	-4.2	-1.2	0.9
117.24	0.96	1.6	171.71	4SE	2.0	-4.1	-1.2	0.9
123.28	1.10	1.5	181.52	14SE	2.3	-4.1	-1.2	0.9
129.16	1.25	1.5	191.38	24SE	-0.1	-4.0	-1.2	0.8
134.88	1.40	1.4	201.30	4OC	-0.8	-3.9	-1.2	0.8
140.40	1.55	1.4	211.28	14OC	-0.9	-3.8	-1.2	0.8
145.70	1.72	1.3	221.31	24OC	-0.9	-3.7	-1.3	0.7
150.74	1.90	1.2	231.38	3NO	-0.7	-3.7	-1.3	0.7
155.46	2.09	1.0	241.50	13NO	-0.5	-3.6	-1.3	0.8
159.79	2.30	0.9	251.65	23NO	-0.5	-3.6	-1.4	0.8
163.65	2.53	0.7	261.83	3DE	-0.5	-3.5	-1.4	0.8
166.93	2.77	0.5	272.02	13DE	0.1	-3.5	-1.5	0.8
169.48	3.04	0.3	282.22	23DE	2.6	-3.4	-1.5	0.8

♂			⊙	16.00UT	☿	♀	♃	♄
LONG	LAT	MAG	LONG	1602	MAGNITUDES			
171.15	3.32	0.1	292.41	2JA	1.2	-3.4	-1.6	0.8
171.74	3.61	-0.2	302.58	12JA	0.3	-3.4	-1.7	0.8
171.10	3.87	-0.4	312.73	22JA	0.1	-3.4	-1.7	0.7
169.16	4.08	-0.7	322.83	1FE	-0.0	-3.3	-1.8	0.7
166.06	4.18	-0.9	332.90	11FE	-0.3	-3.3	-1.9	0.7
162.25	4.13	-1.0	342.92	21FE	-0.7	-3.3	-1.9	0.6
158.42	3.92	-0.9	352.88	3MR	-1.4	-3.3	-2.0	0.6
155.24	3.58	-0.7	2.79	13MR	-1.4	-3.3	-2.0	0.5
153.19	3.16	-0.5	12.64	23MR	-0.5	-3.4	-2.0	0.5
152.43	2.72	-0.3	22.44	2AP	0.8	-3.4	-2.0	0.4
152.92	2.31	-0.1	32.18	12AP	2.4	-3.4	-2.0	0.3
154.50	1.92	0.1	41.87	22AP	3.4	-3.5	-2.0	0.3
157.00	1.57	0.3	51.51	2MY	1.9	-3.5	-1.9	0.3
160.23	1.26	0.5	61.12	12MY	1.1	-3.5	-1.9	0.3
164.09	0.98	0.6	70.70	22MY	0.4	-3.4	-1.8	0.4
168.44	0.73	0.8	80.24	1JN	-0.4	-3.4	-1.8	0.5
173.21	0.51	0.9	89.78	11JN	-1.4	-3.4	-1.7	0.5
178.34	0.30	1.0	99.31	21JN	-1.5	-3.4	-1.6	0.6
183.77	0.12	1.1	108.85	1JL	-0.7	-3.3	-1.6	0.6
189.46	-0.05	1.1	118.39	11JL	-0.0	-3.3	-1.5	0.7
195.40	-0.21	1.2	127.96	21JL	0.4	-3.3	-1.5	0.7
201.55	-0.36	1.2	137.55	31JL	0.7	-3.3	-1.4	0.8
207.89	-0.49	1.3	147.19	10AU	1.2	-3.3	-1.4	0.8
214.42	-0.61	1.3	156.86	20AU	2.5	-3.4	-1.3	0.8
221.12	-0.72	1.3	166.59	30AU	2.0	-3.4	-1.3	0.8
227.98	-0.82	1.3	176.37	9SE	-0.1	-3.4	-1.3	0.8
234.98	-0.91	1.4	186.20	19SE	-1.0	-3.4	-1.2	0.8
242.13	-0.98	1.4	196.09	29SE	-1.0	-3.5	-1.2	0.8
249.41	-1.05	1.4	206.04	9OC	-0.9	-3.6	-1.2	0.8
256.80	-1.10	1.4	216.04	19OC	-0.6	-3.6	-1.2	0.7
264.31	-1.14	1.4	226.09	29OC	-0.4	-3.7	-1.2	0.7
271.91	-1.17	1.4	236.19	8NO	-0.4	-3.8	-1.2	0.7
279.60	-1.19	1.4	246.33	18NO	-0.3	-3.9	-1.3	0.7
287.35	-1.19	1.4	256.49	28NO	0.3	-4.0	-1.3	0.7
295.17	-1.18	1.4	266.68	8DE	2.8	-4.1	-1.3	0.8
303.03	-1.16	1.4	276.87	18DE	0.9	-4.2	-1.4	0.8
310.91	-1.13	1.4	287.06	28DE	0.1	-4.3	-1.4	0.8
				1603				
318.81	-1.09	1.4	297.25	7JA	-0.1	-4.4	-1.5	0.8
326.71	-1.03	1.4	307.41	17JA	-0.1	-4.4	-1.5	0.8
334.58	-0.97	1.4	317.54	27JA	-0.4	-4.2	-1.6	0.8
342.43	-0.89	1.4	327.63	6FE	-0.8	-3.9	-1.7	0.7
350.24	-0.81	1.4	337.67	16FE	-1.3	-3.3	-1.7	0.7
358.00	-0.72	1.4	347.66	26FE	-1.3	-3.6	-1.8	0.6
5.69	-0.63	1.4	357.60	8MR	-0.5	-4.1	-1.9	0.6
13.31	-0.53	1.4	7.48	18MR	1.0	-4.2	-1.9	0.5
20.86	-0.43	1.4	17.31	28MR	3.0	-4.3	-2.0	0.5
28.33	-0.32	1.5	27.08	7AP	2.6	-4.2	-2.0	0.4
35.71	-0.22	1.5	36.79	17AP	1.4	-4.1	-2.1	0.4
43.00	-0.11	1.5	46.46	27AP	0.8	-4.0	-2.1	0.3
50.21	-0.01	1.6	56.09	7MY	0.3	-3.9	-2.1	0.3
57.33	0.10	1.6	65.68	17MY	-0.5	-3.8	-2.0	0.3
64.37	0.21	1.7	75.24	27MY	-1.4	-3.7	-2.0	0.3
71.32	0.31	1.7	84.79	6JN	-1.6	-3.6	-2.0	0.4
78.18	0.41	1.8	94.32	16JN	-0.6	-3.6	-1.9	0.4
84.97	0.51	1.8	103.85	26JN	0.1	-3.5	-1.9	0.5
91.67	0.61	1.8	113.39	6JL	0.5	-3.5	-1.8	0.6
98.30	0.70	1.8	122.95	16JL	0.9	-3.4	-1.7	0.6
104.86	0.80	1.9	132.53	26JL	1.6	-3.4	-1.7	0.7
111.34	0.89	1.9	142.14	5AU	3.0	-3.4	-1.6	0.7
117.74	0.98	1.9	151.79	15AU	1.7	-3.4	-1.6	0.8
124.08	1.07	1.9	161.49	25AU	-0.2	-3.3	-1.5	0.8
130.34	1.16	1.9	171.24	4SE	-1.1	-3.3	-1.5	0.8
136.51	1.25	1.8	181.04	14SE	-1.2	-3.3	-1.4	0.8
142.61	1.33	1.8	190.90	24SE	-0.9	-3.4	-1.4	0.8
148.62	1.42	1.8	200.82	4OC	-0.5	-3.4	-1.3	0.8
154.53	1.51	1.7	210.79	14OC	-0.3	-3.4	-1.3	0.8
160.33	1.60	1.7	220.82	24OC	-0.2	-3.4	-1.3	0.7
165.99	1.69	1.6	230.89	3NO	-0.1	-3.4	-1.3	0.7
171.51	1.78	1.5	241.01	13NO	0.5	-3.5	-1.3	0.7
176.86	1.88	1.4	251.16	23NO	3.1	-3.5	-1.3	0.7
182.00	1.98	1.3	261.34	3DE	0.6	-3.5	-1.3	0.7
186.89	2.08	1.1	271.52	13DE	-0.1	-3.5	-1.3	0.7
191.47	2.18	1.0	281.72	23DE	-0.2	-3.4	-1.4	0.8

♂			⊙	16.00UT	☿	♀	♃	♄
LONG	LAT	MAG	LONG	1604	MAGNITUDES			
195.67	2.29	0.8	291.91	2JA	-0.3	-3.4	-1.4	0.8
199.40	2.40	0.6	302.08	12JA	-0.4	-3.4	-1.4	0.8
202.56	2.51	0.4	312.23	22JA	-0.8	-3.4	-1.5	0.8
204.98	2.61	0.1	322.34	1FE	-1.2	-3.4	-1.5	0.8
206.52	2.70	-0.1	332.41	11FE	-1.2	-3.3	-1.6	0.7
206.97	2.75	-0.4	342.43	21FE	-0.4	-3.3	-1.7	0.7
206.20	2.76	-0.7	352.40	2MR	1.3	-3.3	-1.7	0.7
204.17	2.68	-1.0	2.31	12MR	3.4	-3.3	-1.8	0.6
201.04	2.49	-1.2	12.17	22MR	1.9	-3.4	-1.9	0.6
197.32	2.19	-1.3	21.96	1AP	1.1	-3.4	-1.9	0.5
193.69	1.79	-1.2	31.71	11AP	0.6	-3.4	-2.0	0.5
190.83	1.35	-1.0	41.40	21AP	0.2	-3.4	-2.1	0.4
189.17	0.90	-0.8	51.05	1MY	-0.5	-3.5	-2.1	0.3
188.84	0.49	-0.6	60.66	11MY	-1.4	-3.5	-2.1	0.3
189.77	0.12	-0.4	70.24	21MY	-1.6	-3.6	-2.2	0.2
191.82	-0.19	-0.2	79.79	31MY	-0.6	-3.6	-2.2	0.3
194.78	-0.46	-0.0	89.32	10JN	0.2	-3.7	-2.2	0.3
198.51	-0.69	0.1	98.86	20JN	0.7	-3.8	-2.2	0.4
202.88	-0.89	0.2	108.39	30JN	1.2	-3.8	-2.1	0.4
207.77	-1.06	0.4	117.93	10JL	2.1	-3.9	-2.1	0.5
213.10	-1.20	0.5	127.50	20JL	3.2	-4.0	-2.0	0.6
218.81	-1.31	0.6	137.10	30JL	1.4	-4.1	-2.0	0.6
224.85	-1.41	0.6	146.72	9AU	-0.2	-4.2	-1.9	0.7
231.17	-1.48	0.7	156.40	19AU	-1.2	-4.2	-1.8	0.7
237.73	-1.54	0.8	166.12	29AU	-1.3	-4.1	-1.8	0.7
244.52	-1.58	0.8	175.89	8SE	-0.8	-3.8	-1.7	0.8
251.50	-1.60	0.9	185.72	18SE	-0.4	-3.3	-1.7	0.8
258.65	-1.60	0.9	195.61	28SE	-0.2	-3.6	-1.6	0.8
265.95	-1.59	1.0	205.55	8OC	-0.1	-4.1	-1.6	0.8
273.39	-1.57	1.0	215.55	18OC	0.1	-4.3	-1.5	0.8
280.93	-1.52	1.1	225.60	28OC	0.7	-4.3	-1.5	0.8
288.55	-1.47	1.1	235.69	7NO	3.4	-4.3	-1.4	0.7
296.25	-1.40	1.2	245.83	17NO	0.4	-4.2	-1.4	0.7
304.00	-1.32	1.2	255.99	27NO	-0.3	-4.1	-1.4	0.7
311.78	-1.23	1.2	266.18	7DE	-0.4	-4.0	-1.4	0.7
319.58	-1.13	1.3	276.38	17DE	-0.4	-3.9	-1.4	0.7
327.37	-1.02	1.3	286.57	27DE	-0.5	-3.8	-1.4	0.7
				1605				
335.15	-0.91	1.4	296.75	6JA	-0.9	-3.7	-1.4	0.8
342.90	-0.79	1.4	306.92	16JA	-1.1	-3.7	-1.4	0.8
350.60	-0.68	1.4	317.05	26JA	-1.0	-3.6	-1.5	0.8
358.25	-0.56	1.5	327.14	5FE	-0.4	-3.5	-1.5	0.8
5.84	-0.44	1.5	337.19	15FE	1.5	-3.5	-1.5	0.8
13.35	-0.32	1.5	347.19	25FE	2.9	-3.4	-1.6	0.8
20.80	-0.20	1.6	357.13	7MR	1.4	-3.4	-1.7	0.7
28.16	-0.08	1.6	7.02	17MR	0.8	-3.4	-1.7	0.7
35.44	0.03	1.6	16.84	27MR	0.4	-3.3	-1.8	0.6
42.65	0.14	1.7	26.61	6AP	0.0	-3.3	-1.9	0.6
49.77	0.24	1.7	36.33	16AP	-0.6	-3.3	-1.9	0.5
56.81	0.34	1.7	46.00	26AP	-1.4	-3.3	-2.0	0.5
63.77	0.43	1.7	55.63	6MY	-1.6	-3.3	-2.1	0.4
70.66	0.52	1.7	65.22	16MY	-0.6	-3.3	-2.1	0.3
77.48	0.61	1.7	74.78	26MY	0.3	-3.3	-2.2	0.3
84.23	0.69	1.8	84.33	5JN	0.9	-3.4	-2.2	0.2
90.91	0.76	1.8	93.86	15JN	1.6	-3.4	-2.3	0.3
97.54	0.83	1.8	103.39	25JN	2.8	-3.4	-2.3	0.3
104.11	0.90	1.9	112.93	5JL	2.8	-3.5	-2.3	0.4
110.63	0.96	1.9	122.49	15JL	1.1	-3.5	-2.3	0.5
117.10	1.02	2.0	132.06	25JL	-0.3	-3.5	-2.3	0.5
123.52	1.08	2.0	141.67	4AU	-1.2	-3.4	-2.2	0.6
129.91	1.13	2.0	151.32	14AU	-1.4	-3.4	-2.2	0.6
136.25	1.17	2.0	161.01	24AU	-0.8	-3.4	-2.1	0.7
142.56	1.22	2.0	170.76	3SE	-0.3	-3.3	-2.1	0.7
148.83	1.26	2.0	180.56	13SE	-0.0	-3.3	-2.0	0.7
155.07	1.30	2.0	190.42	23SE	0.1	-3.3	-2.0	0.8
161.27	1.33	2.0	200.33	3OC	0.3	-3.3	-1.9	0.8
167.42	1.36	1.9	210.30	13OC	1.0	-3.3	-1.8	0.8
173.53	1.39	1.9	220.32	23OC	3.2	-3.3	-1.8	0.8
179.60	1.41	1.8	230.40	2NO	0.3	-3.4	-1.7	0.8
185.61	1.43	1.8	240.51	12NO	-0.4	-3.4	-1.7	0.7
191.55	1.45	1.7	250.66	22NO	-0.5	-3.4	-1.6	0.7
197.41	1.46	1.6	260.83	2DE	-0.5	-3.5	-1.6	0.7
203.19	1.46	1.5	271.03	12DE	-0.6	-3.5	-1.5	0.7
208.86	1.46	1.4	281.22	22DE	-0.8	-3.6	-1.5	0.7

♂			⊙	16.00UT	☿	♀	♃	♄
LONG	LAT	MAG	LONG	1606	MAGNITUDES			
214.39	1.45	1.3	291.41	1JA	-0.9	-3.6	-1.5	0.7
219.77	1.43	1.1	301.59	11JA	-0.9	-3.7	-1.5	0.8
224.96	1.40	1.0	311.74	21JA	-0.3	-3.8	-1.5	0.8
229.91	1.35	0.8	321.86	31JA	1.8	-3.9	-1.5	0.8
234.56	1.28	0.6	331.93	10FE	2.3	-3.9	-1.5	0.8
238.85	1.19	0.4	341.95	20FE	1.0	-4.0	-1.5	0.8
242.69	1.06	0.1	351.92	2MR	0.5	-4.1	-1.6	0.8
245.94	0.89	-0.1	1.84	12MR	0.3	-4.2	-1.6	0.8
248.49	0.67	-0.4	11.69	22MR	-0.0	-4.3	-1.7	0.7
250.13	0.39	-0.7	21.50	1AP	-0.6	-4.2	-1.7	0.7
250.70	0.03	-1.0	31.24	11AP	-1.4	-4.0	-1.8	0.6
250.06	-0.42	-1.3	40.93	21AP	-1.5	-3.5	-1.8	0.6
248.16	-0.94	-1.6	50.59	1MY	-0.6	-2.8	-1.9	0.5
245.27	-1.50	-1.9	60.20	11MY	0.4	-3.6	-2.0	0.5
241.91	-2.04	-2.0	69.77	21MY	1.2	-4.0	-2.0	0.4
238.80	-2.51	-1.8	79.33	31MY	2.2	-4.2	-2.1	0.3
236.61	-2.86	-1.6	88.86	10JN	3.4	-4.2	-2.2	0.3
235.71	-3.08	-1.4	98.39	20JN	2.2	-4.1	-2.2	0.3
236.20	-3.21	-1.2	107.93	30JN	0.9	-4.0	-2.3	0.3
237.98	-3.25	-1.0	117.47	10JL	-0.3	-3.9	-2.3	0.4
240.86	-3.24	-0.8	127.04	20JL	-1.3	-3.8	-2.4	0.4
244.67	-3.19	-0.6	136.63	30JL	-1.5	-3.8	-2.4	0.5
249.24	-3.10	-0.5	146.26	9AU	-0.7	-3.7	-2.4	0.5
254.42	-2.99	-0.3	155.93	19AU	-0.2	-3.6	-2.4	0.6
260.10	-2.86	-0.2	165.65	29AU	0.1	-3.6	-2.4	0.7
266.19	-2.71	-0.0	175.42	8SE	0.2	-3.5	-2.4	0.7
272.61	-2.55	0.1	185.24	18SE	0.5	-3.5	-2.3	0.7
279.30	-2.38	0.2	195.13	28SE	1.3	-3.5	-2.3	0.8
286.21	-2.20	0.3	205.07	8OC	2.9	-3.4	-2.2	0.8
293.29	-2.01	0.4	215.07	18OC	0.1	-3.4	-2.1	0.8
300.50	-1.82	0.5	225.12	28OC	-0.6	-3.4	-2.1	0.8
307.82	-1.62	0.6	235.20	7NO	-0.6	-3.4	-2.0	0.8
315.20	-1.43	0.7	245.34	17NO	-0.6	-3.4	-1.9	0.8
322.63	-1.23	0.8	255.50	27NO	-0.7	-3.4	-1.9	0.8
330.08	-1.04	0.9	265.68	7DE	-0.7	-3.4	-1.8	0.7
337.53	-0.86	1.0	275.88	17DE	-0.7	-3.4	-1.7	0.7
344.97	-0.68	1.1	286.07	27DE	-0.7	-3.4	-1.7	0.7
				1607				
352.39	-0.52	1.2	296.25	6JA	-0.2	-3.4	-1.7	0.8
359.76	-0.35	1.3	306.42	16JA	2.0	-3.4	-1.6	0.8
7.08	-0.20	1.4	316.55	26JA	1.8	-3.4	-1.6	0.8
14.35	-0.06	1.4	326.65	5FE	0.7	-3.5	-1.6	0.8
21.56	0.07	1.5	336.70	15FE	0.4	-3.5	-1.6	0.9
28.70	0.20	1.6	346.70	25FE	0.2	-3.5	-1.6	0.9
35.78	0.31	1.6	356.64	7MR	-0.1	-3.4	-1.6	0.8
42.79	0.42	1.7	6.53	17MR	-0.6	-3.4	-1.6	0.8
49.73	0.51	1.7	16.36	27MR	-1.4	-3.4	-1.6	0.8
56.60	0.60	1.8	26.13	6AP	-1.5	-3.4	-1.6	0.8
63.42	0.68	1.8	35.86	16AP	-0.6	-3.3	-1.7	0.7
70.17	0.76	1.9	45.53	26AP	0.5	-3.3	-1.7	0.7
76.86	0.82	1.9	55.16	6MY	1.6	-3.3	-1.8	0.6
83.51	0.88	1.9	64.76	16MY	2.9	-3.3	-1.8	0.5
90.10	0.93	1.9	74.32	26MY	3.1	-3.3	-1.9	0.5
96.65	0.98	1.9	83.86	5JN	1.7	-3.3	-1.9	0.4
103.15	1.02	1.9	93.40	15JN	0.7	-3.4	-2.0	0.4
109.63	1.06	1.9	102.93	25JN	-0.4	-3.4	-2.1	0.3
116.07	1.09	1.9	112.47	5JL	-1.4	-3.4	-2.1	0.3
122.49	1.11	1.9	122.02	15JL	-1.5	-3.5	-2.2	0.4
128.89	1.13	1.9	131.60	25JL	-0.7	-3.5	-2.3	0.4
135.27	1.15	2.0	141.20	4AU	-0.2	-3.6	-2.3	0.5
141.64	1.16	2.0	150.85	14AU	0.2	-3.6	-2.4	0.5
148.00	1.17	2.0	160.55	24AU	0.4	-3.7	-2.4	0.6
154.35	1.17	2.0	170.29	3SE	0.7	-3.8	-2.5	0.6
160.70	1.17	2.0	180.09	13SE	1.7	-3.9	-2.5	0.7
167.05	1.16	2.0	189.94	23SE	2.6	-4.0	-2.5	0.7
173.40	1.15	2.0	199.85	3OC	0.0	-4.1	-2.4	0.8
179.75	1.13	2.0	209.82	13OC	-0.7	-4.2	-2.4	0.8
186.11	1.11	2.0	219.84	23OC	-0.8	-4.3	-2.4	0.8
192.46	1.08	1.9	229.91	2NO	-0.8	-4.4	-2.3	0.8
198.82	1.05	1.9	240.02	12NO	-0.7	-4.3	-2.2	0.8
205.18	1.01	1.8	250.17	22NO	-0.6	-3.9	-2.2	0.8
211.54	0.96	1.8	260.34	2DE	-0.6	-3.0	-2.1	0.8
217.89	0.90	1.7	270.53	12DE	-0.6	-3.6	-2.0	0.8
224.23	0.84	1.6	280.73	22DE	-0.0	-4.2	-2.0	0.8

♂ LONG	♂ LAT	♂ MAG	☉ LONG	16.00UT 1608	☿	♀	♃	♄
					MAGNITUDES			
230.55	0.76	1.5	290.91	1JA	2.3	-4.4	-1.9	0.7
236.86	0.67	1.4	301.09	11JA	1.4	-4.4	-1.8	0.8
243.14	0.56	1.3	311.24	21JA	0.4	-4.3	-1.8	0.8
249.38	0.44	1.2	321.36	31JA	0.2	-4.2	-1.7	0.9
255.58	0.30	1.1	331.44	10FE	0.0	-4.1	-1.7	0.9
261.72	0.14	0.9	341.46	20FE	-0.2	-4.0	-1.6	0.9
267.78	-0.05	0.8	351.43	1MR	-0.7	-3.9	-1.6	0.9
273.75	-0.26	0.6	1.36	11MR	-1.4	-3.8	-1.6	0.9
279.60	-0.51	0.4	11.21	21MR	-1.4	-3.7	-1.6	0.9
285.29	-0.80	0.2	21.02	31MR	-0.6	-3.6	-1.6	0.9
290.79	-1.13	0.0	30.77	10AP	0.7	-3.5	-1.6	0.8
296.02	-1.50	-0.2	40.47	20AP	2.0	-3.5	-1.6	0.8
300.93	-1.94	-0.4	50.12	30AP	3.8	-3.4	-1.6	0.8
305.39	-2.43	-0.7	59.74	10MY	2.3	-3.4	-1.6	0.7
309.31	-2.99	-1.0	69.31	20MY	1.3	-3.4	-1.6	0.6
312.51	-3.62	-1.2	78.87	30MY	0.5	-3.3	-1.7	0.6
314.81	-4.31	-1.5	88.41	9JN	-0.4	-3.3	-1.7	0.5
316.01	-5.04	-1.8	97.94	19JN	-1.4	-3.3	-1.8	0.5
315.98	-5.75	-2.1	107.47	29JN	-1.6	-3.3	-1.8	0.4
314.70	-6.35	-2.4	117.02	9JL	-0.7	-3.3	-1.9	0.3
312.46	-6.74	-2.6	126.58	19JL	-0.1	-3.3	-2.0	0.4
309.84	-6.83	-2.6	136.17	29JL	0.3	-3.4	-2.0	0.4
307.56	-6.59	-2.4	145.80	8AU	0.6	-3.4	-2.1	0.5
306.27	-6.11	-2.2	155.46	18AU	1.0	-3.4	-2.2	0.5
306.26	-5.47	-1.9	165.18	28AU	2.2	-3.4	-2.2	0.6
307.56	-4.79	-1.7	174.95	7SE	2.2	-3.5	-2.3	0.6
310.05	-4.11	-1.4	184.77	17SE	-0.0	-3.5	-2.3	0.7
313.51	-3.47	-1.1	194.65	27SE	-0.9	-3.5	-2.4	0.7
317.76	-2.89	-0.8	204.59	7OC	-1.0	-3.4	-2.4	0.8
322.61	-2.36	-0.6	214.58	17OC	-0.9	-3.4	-2.4	0.8
327.93	-1.89	-0.4	224.62	27OC	-0.6	-3.4	-2.4	0.8
333.62	-1.48	-0.2	234.71	6NO	-0.5	-3.4	-2.4	0.9
339.57	-1.11	0.0	244.84	16NO	-0.4	-3.3	-2.4	0.9
345.73	-0.78	0.2	255.00	26NO	-0.4	-3.3	-2.3	0.9
352.05	-0.50	0.4	265.19	6DE	0.1	-3.3	-2.3	0.9
358.47	-0.25	0.6	275.38	16DE	2.6	-3.3	-2.2	0.9
4.96	-0.03	0.7	285.57	26DE	1.1	-3.3	-2.1	0.8
				1609				
11.51	0.16	0.9	295.76	5JA	0.2	-3.4	-2.1	0.8
18.09	0.33	1.0	305.92	15JA	0.0	-3.4	-2.0	0.8
24.68	0.48	1.1	316.06	25JA	-0.1	-3.4	-1.9	0.9
31.27	0.60	1.3	326.16	4FE	-0.3	-3.4	-1.8	0.9
37.86	0.71	1.4	336.21	14FE	-0.7	-3.5	-1.8	0.9
44.43	0.81	1.5	346.21	24FE	-1.4	-3.5	-1.7	1.0
50.98	0.89	1.6	356.16	6MR	-1.3	-3.6	-1.7	1.0
57.50	0.95	1.6	6.05	16MR	-0.5	-3.6	-1.6	1.0
64.00	1.01	1.7	15.88	26MR	0.9	-3.7	-1.6	1.0
70.47	1.06	1.8	25.66	5AP	2.6	-3.8	-1.6	1.0
76.92	1.10	1.8	35.39	15AP	3.1	-3.8	-1.5	1.0
83.34	1.13	1.9	45.06	25AP	1.7	-3.9	-1.5	0.9
89.74	1.15	1.9	54.70	5MY	1.0	-4.0	-1.5	0.9
96.11	1.17	2.0	64.29	15MY	0.4	-4.1	-1.5	0.8
102.47	1.18	2.0	73.86	25MY	-0.4	-4.2	-1.5	0.8
108.82	1.19	2.0	83.41	4JN	-1.4	-4.2	-1.5	0.7
115.15	1.19	2.0	92.94	14JN	-1.6	-4.1	-1.5	0.6
121.48	1.19	2.0	102.48	24JN	-0.7	-3.8	-1.6	0.6
127.81	1.18	2.0	112.01	4JL	-0.0	-3.2	-1.6	0.5
134.14	1.16	2.0	121.56	14JL	0.4	-3.3	-1.6	0.5
140.48	1.14	2.0	131.14	24JL	0.8	-3.9	-1.7	0.4
146.83	1.12	2.0	140.74	3AU	1.4	-4.2	-1.7	0.5
153.19	1.10	2.0	150.39	13AU	2.7	-4.2	-1.8	0.5
159.58	1.06	1.9	160.08	23AU	1.9	-4.2	-1.9	0.6
165.98	1.03	2.0	169.82	2SE	-0.1	-4.1	-1.9	0.6
172.41	0.99	2.0	179.61	12SE	-1.0	-4.1	-2.0	0.7
178.88	0.94	2.0	189.46	22SE	-1.1	-4.0	-2.0	0.7
185.37	0.89	2.0	199.37	2OC	-0.9	-3.9	-2.1	0.8
191.91	0.84	2.0	209.33	12OC	-0.5	-3.8	-2.2	0.8
198.47	0.78	2.0	219.35	22OC	-0.3	-3.7	-2.2	0.8
205.08	0.72	1.9	229.42	1NO	-0.3	-3.7	-2.3	0.9
211.73	0.64	1.9	239.53	11NO	-0.2	-3.6	-2.3	0.9
218.42	0.57	1.9	249.68	21NO	0.3	-3.6	-2.3	0.9
225.15	0.48	1.8	259.85	1DE	2.9	-3.5	-2.3	0.9
231.92	0.39	1.8	270.04	11DE	0.8	-3.5	-2.3	0.9
238.74	0.29	1.7	280.24	21DE	0.0	-3.4	-2.3	1.0
245.59	0.18	1.7	290.42	31DE	-0.1	-3.4	-2.2	0.9

♂ LONG	♂ LAT	♂ MAG	☉ LONG	16.00UT 1610	☿	♀	♃	♄
					MAGNITUDES			
252.49	0.06	1.6	300.60	10JA	-0.2	-3.4	-2.2	0.9
259.43	-0.07	1.5	310.76	20JA	-0.4	-3.4	-2.1	0.9
266.40	-0.21	1.4	320.87	30JA	-0.8	-3.3	-2.0	0.9
273.41	-0.36	1.4	330.95	9FE	-1.3	-3.3	-2.0	1.0
280.45	-0.52	1.3	340.98	19FE	-1.2	-3.3	-1.9	1.0
287.52	-0.69	1.2	350.95	1MR	-0.5	-3.3	-1.8	1.1
294.62	-0.87	1.1	0.87	11MR	1.1	-3.3	-1.7	1.1
301.72	-1.06	1.0	10.74	21MR	3.2	-3.4	-1.7	1.1
308.83	-1.25	0.8	20.54	31MR	2.3	-3.4	-1.6	1.1
315.94	-1.46	0.7	30.29	10AP	1.3	-3.4	-1.6	1.1
323.02	-1.67	0.6	40.00	20AP	0.7	-3.5	-1.5	1.1
330.05	-1.89	0.5	49.65	30AP	0.2	-3.5	-1.5	1.1
337.03	-2.10	0.4	59.26	10MY	-0.5	-3.5	-1.5	1.0
343.90	-2.32	0.2	68.85	20MY	-1.4	-3.4	-1.4	1.0
350.64	-2.53	0.1	78.40	30MY	-1.6	-3.4	-1.4	0.9
357.21	-2.74	-0.0	87.94	9JN	-0.7	-3.4	-1.4	0.9
3.54	-2.94	-0.2	97.48	19JN	0.1	-3.4	-1.4	0.8
9.58	-3.12	-0.3	107.01	29JN	0.6	-3.3	-1.4	0.7
15.23	-3.29	-0.5	116.55	9JL	1.0	-3.3	-1.4	0.7
20.38	-3.44	-0.7	126.11	19JL	1.8	-3.3	-1.4	0.6
24.92	-3.56	-0.9	135.70	29JL	3.2	-3.3	-1.5	0.5
28.67	-3.65	-1.1	145.33	8AU	1.6	-3.3	-1.5	0.5
31.42	-3.68	-1.3	154.99	18AU	-0.2	-3.4	-1.5	0.6
32.98	-3.64	-1.5	164.70	28AU	-1.1	-3.4	-1.6	0.6
33.15	-3.50	-1.8	174.47	7SE	-1.2	-3.4	-1.6	0.7
31.84	-3.23	-2.0	184.29	17SE	-0.9	-3.4	-1.7	0.7
29.29	-2.81	-2.2	194.17	27SE	-0.4	-3.5	-1.7	0.8
26.01	-2.25	-2.3	204.10	7OC	-0.2	-3.6	-1.8	0.8
22.80	-1.61	-2.1	214.10	17OC	-0.1	-3.6	-1.9	0.9
20.43	-0.97	-1.8	224.13	27OC	-0.0	-3.7	-1.9	0.9
19.30	-0.40	-1.4	234.22	6NO	0.5	-3.8	-2.0	1.0
19.52	0.07	-1.1	244.35	16NO	3.2	-3.9	-2.1	1.0
20.96	0.45	-0.8	254.51	26NO	0.6	-4.0	-2.1	1.0
23.42	0.74	-0.5	264.69	6DE	-0.2	-4.1	-2.1	1.0
26.70	0.97	-0.2	274.89	16DE	-0.3	-4.2	-2.2	1.1
30.61	1.14	0.1	285.08	26DE	-0.3	-4.3	-2.2	1.1
				1611				
35.02	1.26	0.3	295.27	5JA	-0.5	-4.4	-2.2	1.1
39.81	1.36	0.5	305.43	15JA	-0.8	-4.4	-2.2	1.1
44.89	1.43	0.7	315.57	25JA	-1.1	-4.2	-2.1	1.1
50.20	1.47	0.9	325.67	4FE	-1.1	-3.9	-2.1	1.0
55.69	1.50	1.1	335.73	14FE	-0.4	-3.3	-2.0	1.1
61.31	1.52	1.2	345.73	24FE	1.3	-3.6	-2.0	1.1
67.04	1.53	1.3	355.68	6MR	3.3	-4.1	-1.9	1.2
72.86	1.53	1.4	5.57	16MR	1.7	-4.3	-1.8	1.2
78.74	1.52	1.6	15.41	26MR	0.9	-4.3	-1.8	1.2
84.67	1.50	1.6	25.19	5AP	0.5	-4.2	-1.7	1.2
90.65	1.48	1.7	34.92	15AP	0.1	-4.1	-1.6	1.2
96.66	1.45	1.8	44.59	25AP	-0.5	-4.0	-1.6	1.2
102.70	1.42	1.9	54.23	5MY	-1.4	-3.9	-1.5	1.2
108.78	1.39	1.9	63.83	15MY	-1.6	-3.8	-1.5	1.2
114.88	1.35	1.9	73.39	25MY	-0.6	-3.7	-1.4	1.2
121.01	1.31	2.0	82.94	4JN	0.2	-3.6	-1.4	1.1
127.16	1.26	2.0	92.48	14JN	0.8	-3.6	-1.3	1.1
133.35	1.22	2.0	102.01	24JN	1.4	-3.5	-1.3	1.0
139.57	1.16	2.0	111.55	4JL	2.4	-3.5	-1.3	0.9
145.82	1.11	2.0	121.10	14JL	3.1	-3.4	-1.3	0.9
152.11	1.05	2.0	130.67	24JL	1.3	-3.4	-1.3	0.8
158.44	1.00	2.0	140.28	3AU	-0.2	-3.4	-1.3	0.7
164.82	0.93	2.0	149.92	13AU	-1.2	-3.4	-1.3	0.7
171.24	0.87	2.0	159.61	23AU	-1.4	-3.3	-1.3	0.7
177.72	0.80	1.9	169.35	2SE	-0.8	-3.3	-1.4	0.7
184.24	0.73	1.9	179.14	12SE	-0.4	-3.3	-1.4	0.8
190.83	0.65	1.9	188.99	22SE	-0.1	-3.4	-1.4	0.8
197.48	0.57	1.8	198.90	2OC	0.0	-3.4	-1.5	0.9
204.19	0.49	1.8	208.86	12OC	0.1	-3.4	-1.5	0.9
210.96	0.40	1.8	218.87	22OC	0.8	-3.4	-1.6	1.0
217.80	0.31	1.8	228.94	1NO	3.5	-3.4	-1.6	1.0
224.71	0.22	1.8	239.04	11NO	0.4	-3.5	-1.7	1.0
231.69	0.12	1.8	249.18	21NO	-0.3	-3.5	-1.8	1.1
238.73	0.02	1.8	259.36	1DE	-0.4	-3.5	-1.8	1.1
245.85	-0.09	1.7	269.54	11DE	-0.4	-3.5	-1.9	1.2
253.03	-0.20	1.7	279.74	21DE	-0.6	-3.4	-2.0	1.2
260.27	-0.31	1.7	289.93	31DE	-0.9	-3.4	-2.0	1.2

♂ LONG	♂ LAT	♂ MAG	☉ LONG	16.00UT 1612	☿	♀	♃	♄
					MAGNITUDES			
267.58	-0.43	1.6	300.10	10JA	-1.0	-3.4	-2.0	1.2
274.96	-0.55	1.6	310.26	20JA	-1.0	-3.4	-2.1	1.2
282.38	-0.67	1.5	320.38	30JA	-0.4	-3.4	-2.1	1.2
289.86	-0.79	1.5	330.46	9FE	1.5	-3.3	-2.1	1.2
297.38	-0.91	1.4	340.49	19FE	2.7	-3.3	-2.0	1.2
304.95	-1.03	1.4	350.47	29FE	1.2	-3.3	-2.0	1.3
312.53	-1.14	1.3	0.39	10MR	0.7	-3.3	-2.0	1.3
320.15	-1.25	1.2	10.26	20MR	0.4	-3.4	-1.9	1.4
327.76	-1.35	1.2	20.07	30MR	0.0	-3.4	-1.8	1.4
335.38	-1.44	1.1	29.82	9AP	-0.5	-3.4	-1.8	1.4
342.98	-1.52	1.1	39.53	19AP	-1.4	-3.4	-1.7	1.4
350.54	-1.59	1.0	49.18	29AP	-1.6	-3.5	-1.6	1.4
358.05	-1.64	1.0	58.80	9MY	-0.6	-3.5	-1.6	1.4
5.49	-1.68	0.9	68.38	19MY	0.3	-3.6	-1.5	1.4
12.85	-1.70	0.8	77.94	29MY	1.0	-3.6	-1.5	1.4
20.10	-1.70	0.8	87.48	8JN	1.8	-3.7	-1.4	1.3
27.22	-1.69	0.7	97.01	18JN	3.1	-3.8	-1.4	1.3
34.18	-1.65	0.6	106.55	28JN	2.6	-3.8	-1.3	1.2
40.97	-1.59	0.6	116.09	8JL	1.1	-3.9	-1.3	1.2
47.54	-1.51	0.5	125.65	18JL	-0.3	-4.0	-1.3	1.1
53.85	-1.41	0.4	135.24	28JL	-1.3	-4.1	-1.2	1.0
59.87	-1.28	0.3	144.86	7AU	-1.5	-4.2	-1.2	1.0
65.53	-1.13	0.2	154.53	17AU	-0.8	-4.2	-1.2	0.9
70.76	-0.94	0.1	164.24	27AU	-0.3	-4.1	-1.2	0.8
75.48	-0.73	-0.1	174.00	6SE	-0.0	-3.8	-1.2	0.9
79.56	-0.48	-0.2	183.82	16SE	0.2	-3.3	-1.3	0.9
82.85	-0.18	-0.4	193.70	26SE	0.3	-3.6	-1.3	0.9
85.18	0.17	-0.6	203.62	6OC	1.1	-4.1	-1.3	1.0
86.33	0.57	-0.8	213.61	16OC	3.2	-4.3	-1.3	1.0
86.12	1.03	-1.0	223.65	26OC	0.3	-4.3	-1.4	1.1
84.47	1.52	-1.2	233.74	5NO	-0.5	-4.3	-1.4	1.1
81.52	2.01	-1.4	243.86	15NO	-0.6	-4.2	-1.5	1.2
77.76	2.44	-1.5	254.02	25NO	-0.6	-4.1	-1.5	1.2
73.96	2.75	-1.4	264.20	5DE	-0.7	-4.0	-1.6	1.3
70.87	2.92	-1.1	274.40	15DE	-0.8	-3.9	-1.7	1.3
69.00	2.98	-0.8	284.59	25DE	-0.8	-3.8	-1.7	1.3
				1613				
68.47	2.94	-0.5	294.77	4JA	-0.8	-3.7	-1.8	1.4
69.20	2.86	-0.2	304.94	14JA	-0.3	-3.7	-1.9	1.4
71.00	2.75	0.1	315.08	24JA	1.8	-3.6	-1.9	1.4
73.67	2.63	0.3	325.18	3FE	2.1	-3.5	-2.0	1.4
77.02	2.50	0.5	335.24	13FE	0.9	-3.5	-2.0	1.4
80.91	2.38	0.7	345.25	23FE	0.5	-3.4	-2.0	1.3
85.22	2.26	0.9	355.20	5MR	0.2	-3.4	-2.0	1.3
89.86	2.14	1.1	5.10	15MR	-0.1	-3.4	-2.0	1.3
94.77	2.03	1.2	14.94	25MR	-0.6	-3.3	-2.0	1.3
99.90	1.92	1.4	24.72	4AP	-1.4	-3.3	-1.9	1.3
105.20	1.81	1.5	34.45	14AP	-1.5	-3.3	-1.9	1.3
110.65	1.71	1.6	44.13	24AP	-0.6	-3.3	-1.8	1.3
116.22	1.61	1.7	53.77	4MY	0.4	-3.3	-1.8	1.3
121.91	1.51	1.7	63.37	14MY	1.3	-3.3	-1.7	1.2
127.69	1.42	1.8	72.94	24MY	2.4	-3.3	-1.6	1.2
133.55	1.32	1.8	82.48	3JN	3.5	-3.4	-1.6	1.1
139.50	1.23	1.9	92.02	13JN	2.0	-3.4	-1.5	1.1
145.52	1.14	1.9	101.55	23JN	0.8	-3.4	-1.5	1.0
151.62	1.04	1.9	111.08	3JL	-0.3	-3.5	-1.4	1.0
157.79	0.95	1.9	120.64	13JL	-1.3	-3.5	-1.4	1.0
164.04	0.86	1.9	130.21	23JL	-1.5	-3.5	-1.3	0.9
170.36	0.76	1.9	139.81	2AU	-0.8	-3.4	-1.3	0.9
176.76	0.67	1.9	149.45	12AU	-0.2	-3.4	-1.2	0.8
183.23	0.58	1.9	159.14	22AU	0.1	-3.4	-1.2	0.8
189.79	0.48	1.9	168.87	1SE	0.3	-3.3	-1.2	0.7
196.43	0.38	1.9	178.66	11SE	0.6	-3.3	-1.2	0.7
203.15	0.28	1.8	188.51	21SE	1.4	-3.3	-1.2	0.8
209.96	0.19	1.8	198.41	1OC	2.9	-3.3	-1.2	0.8
216.85	0.09	1.7	208.37	11OC	0.1	-3.3	-1.2	0.9
223.83	-0.01	1.7	218.38	21OC	-0.6	-3.3	-1.2	1.0
230.90	-0.12	1.6	228.44	31OC	-0.7	-3.4	-1.3	1.0
238.05	-0.22	1.6	238.55	10NO	-0.7	-3.4	-1.3	1.1
245.29	-0.32	1.6	248.69	20NO	-0.8	-3.4	-1.3	1.1
252.61	-0.42	1.6	258.86	30NO	-0.7	-3.5	-1.4	1.2
260.01	-0.52	1.6	269.05	10DE	-0.7	-3.5	-1.4	1.2
267.49	-0.61	1.6	279.24	20DE	-0.7	-3.6	-1.5	1.2
275.04	-0.71	1.5	289.44	30DE	-0.1	-3.6	-1.6	1.2

♂ LONG	♂ LAT	♂ MAG	☉ LONG	16.00UT 1614	☿	♀	♃	♄
					MAGNITUDES			
282.65	-0.80	1.5	299.62	9JA	2.0	-3.7	-1.6	1.2
290.31	-0.88	1.5	309.77	19JA	1.7	-3.8	-1.7	1.2
298.03	-0.96	1.5	319.89	29JA	0.6	-3.9	-1.8	1.2
305.78	-1.03	1.5	329.98	8FE	0.3	-4.0	-1.8	1.2
313.56	-1.09	1.4	340.01	18FE	0.1	-4.0	-1.9	1.1
321.37	-1.14	1.4	349.99	28FE	-0.2	-4.1	-1.9	1.1
329.17	-1.18	1.4	359.92	10MR	-0.6	-4.2	-2.0	1.0
336.97	-1.21	1.4	9.78	20MR	-1.4	-4.3	-2.0	1.0
344.75	-1.23	1.4	19.60	30MR	-1.5	-4.2	-2.0	1.0
352.50	-1.23	1.3	29.36	9AP	-0.6	-4.0	-2.0	1.0
0.20	-1.22	1.3	39.06	19AP	0.6	-3.5	-2.0	1.0
7.85	-1.19	1.3	48.72	29AP	1.7	-2.9	-2.0	1.0
15.43	-1.16	1.3	58.34	9MY	3.2	-3.6	-1.9	1.0
22.92	-1.10	1.3	67.92	19MY	2.8	-4.0	-1.9	1.0
30.33	-1.03	1.3	77.48	29MY	1.6	-4.2	-1.8	0.9
37.64	-0.95	1.2	87.02	8JN	0.6	-4.2	-1.8	0.9
44.83	-0.86	1.2	96.55	18JN	-0.3	-4.1	-1.7	0.9
51.90	-0.75	1.2	106.08	28JN	-1.3	-4.0	-1.6	0.8
58.83	-0.64	1.2	115.63	8JL	-1.6	-3.9	-1.6	0.8
65.62	-0.51	1.2	125.18	18JL	-0.7	-3.8	-1.5	0.7
72.24	-0.36	1.1	134.77	28JL	-0.1	-3.8	-1.5	0.7
78.69	-0.21	1.1	144.39	7AU	0.2	-3.7	-1.4	0.6
84.94	-0.04	1.0	154.05	17AU	0.5	-3.6	-1.4	0.6
90.96	0.14	1.0	163.76	27AU	0.8	-3.6	-1.3	0.5
96.73	0.34	0.9	173.52	6SE	1.9	-3.5	-1.3	0.5
102.20	0.56	0.8	183.34	16SE	2.5	-3.5	-1.3	0.4
107.33	0.79	0.7	193.21	26SE	0.1	-3.4	-1.2	0.4
112.03	1.05	0.6	203.14	6OC	-0.8	-3.4	-1.2	0.5
116.23	1.34	0.4	213.13	16OC	-0.9	-3.4	-1.2	0.6
119.81	1.66	0.3	223.17	26OC	-0.8	-3.4	-1.2	0.6
122.62	2.03	0.1	233.25	5NO	-0.7	-3.4	-1.2	0.7
124.50	2.42	-0.1	243.37	15NO	-0.5	-3.4	-1.3	0.8
125.23	2.86	-0.4	253.53	25NO	-0.5	-3.4	-1.3	0.8
124.67	3.31	-0.6	263.71	5DE	-0.5	-3.4	-1.3	0.9
122.73	3.74	-0.8	273.90	15DE	0.0	-3.4	-1.4	0.9
119.58	4.10	-1.0	284.10	25DE	2.3	-3.4	-1.4	0.9
				1615				
115.70	4.31	-1.1	294.28	4JA	1.3	-3.4	-1.5	1.0
111.81	4.35	-1.0	304.45	14JA	0.4	-3.4	-1.5	1.0
108.64	4.23	-0.7	314.59	24JA	0.1	-3.4	-1.6	1.0
106.64	3.99	-0.5	324.70	3FE	-0.0	-3.5	-1.6	0.9
105.95	3.70	-0.2	334.75	13FE	-0.2	-3.5	-1.7	0.9
106.49	3.39	0.1	344.77	23FE	-0.7	-3.5	-1.8	0.9
108.10	3.09	0.3	354.72	5MR	-1.4	-3.4	-1.8	0.8
110.58	2.81	0.5	4.62	15MR	-1.4	-3.4	-1.9	0.8
113.75	2.55	0.7	14.46	25MR	-0.6	-3.4	-2.0	0.7
117.49	2.31	0.9	24.24	4AP	0.7	-3.4	-2.0	0.7
121.67	2.09	1.0	33.98	14AP	2.2	-3.3	-2.0	0.8
126.22	1.89	1.2	43.66	24AP	3.7	-3.3	-2.1	0.8
131.06	1.70	1.3	53.30	4MY	2.1	-3.3	-2.1	0.8
136.16	1.53	1.4	62.90	14MY	1.2	-3.3	-2.1	0.8
141.46	1.37	1.5	72.47	24MY	0.5	-3.3	-2.1	0.8
146.95	1.21	1.5	82.02	3JN	-0.4	-3.3	-2.0	0.7
152.61	1.07	1.6	91.55	13JN	-1.4	-3.4	-2.0	0.7
158.41	0.93	1.7	101.09	23JN	-1.6	-3.4	-1.9	0.7
164.34	0.79	1.7	110.62	3JL	-0.7	-3.4	-1.9	0.7
170.41	0.66	1.7	120.17	13JL	-0.1	-3.5	-1.8	0.6
176.59	0.53	1.7	129.74	23JL	0.3	-3.5	-1.7	0.6
182.90	0.41	1.8	139.34	2AU	0.7	-3.6	-1.7	0.5
189.32	0.29	1.8	148.98	12AU	1.1	-3.6	-1.6	0.5
195.85	0.17	1.8	158.67	22AU	2.3	-3.7	-1.6	0.4
202.49	0.06	1.8	168.40	1SE	2.2	-3.8	-1.5	0.3
209.24	-0.05	1.7	178.18	11SE	-0.0	-3.9	-1.5	0.3
216.11	-0.16	1.7	188.03	21SE	-0.9	-4.0	-1.4	0.2
223.08	-0.26	1.7	197.93	1OC	-1.0	-4.1	-1.4	0.2
230.15	-0.36	1.7	207.88	11OC	-1.0	-4.2	-1.4	0.2
237.34	-0.46	1.6	217.89	21OC	-0.6	-4.3	-1.3	0.3
244.62	-0.55	1.6	227.95	31OC	-0.4	-4.4	-1.3	0.3
252.00	-0.64	1.6	238.06	10NO	-0.4	-4.2	-1.3	0.4
259.47	-0.73	1.5	248.20	20NO	-0.3	-3.8	-1.3	0.5
267.02	-0.80	1.5	258.37	30NO	0.2	-3.0	-1.3	0.5
274.66	-0.87	1.4	268.55	10DE	2.6	-3.6	-1.3	0.6
282.36	-0.93	1.4	278.75	20DE	1.0	-4.2	-1.3	0.6
290.12	-0.98	1.4	288.94	30DE	0.1	-4.4	-1.4	0.7

♂ LONG	♂ LAT	♂ MAG	☉ LONG	16.00UT 1616	☿	♀	♃	♄
					MAGNITUDES			
297.93	-1.03	1.3	299.12	9JA	-0.0	-4.4	-1.4	0.7
305.78	-1.06	1.3	309.28	19JA	-0.1	-4.3	-1.5	0.7
313.66	-1.08	1.3	319.40	29JA	-0.3	-4.2	-1.5	0.7
321.54	-1.09	1.4	329.49	8FE	-0.7	-4.1	-1.6	0.7
329.43	-1.09	1.4	339.53	18FE	-1.3	-4.0	-1.6	0.7
337.30	-1.07	1.4	349.51	28FE	-1.3	-3.9	-1.7	0.7
345.15	-1.04	1.4	359.44	9MR	-0.6	-3.8	-1.8	0.7
352.96	-1.01	1.4	9.31	19MR	0.9	-3.7	-1.8	0.6
0.72	-0.96	1.4	19.12	29MR	2.8	-3.6	-1.9	0.6
8.42	-0.90	1.4	28.89	8AP	2.8	-3.5	-2.0	0.5
16.06	-0.83	1.5	38.60	18AP	1.6	-3.5	-2.0	0.5
23.61	-0.75	1.5	48.25	28AP	0.9	-3.4	-2.1	0.6
31.09	-0.66	1.5	57.88	8MY	0.3	-3.4	-2.1	0.6
38.47	-0.56	1.5	67.46	18MY	-0.4	-3.4	-2.2	0.6
45.76	-0.46	1.5	77.02	28MY	-1.4	-3.3	-2.2	0.6
52.95	-0.35	1.5	86.57	7JN	-1.6	-3.3	-2.2	0.6
60.04	-0.24	1.5	96.10	17JN	-0.7	-3.3	-2.2	0.6
67.02	-0.12	1.5	105.63	27JN	0.0	-3.3	-2.2	0.5
73.90	0.01	1.5	115.17	7JL	0.5	-3.3	-2.1	0.5
80.65	0.14	1.5	124.73	17JL	0.9	-3.3	-2.1	0.5
87.29	0.27	1.5	134.31	27JL	1.5	-3.4	-2.0	0.4
93.81	0.41	1.5	143.93	6AU	2.9	-3.4	-2.0	0.4
100.20	0.55	1.5	153.59	16AU	1.9	-3.4	-1.9	0.3
106.45	0.70	1.5	163.29	26AU	-0.1	-3.4	-1.8	0.3
112.54	0.85	1.4	173.05	5SE	-1.0	-3.5	-1.8	0.2
118.47	1.01	1.4	182.86	15SE	-1.2	-3.5	-1.7	0.2
124.21	1.18	1.3	192.73	25SE	-0.9	-3.5	-1.7	0.1
129.73	1.36	1.2	202.66	5OC	-0.5	-3.4	-1.6	0.0
135.01	1.55	1.1	212.64	15OC	-0.3	-3.4	-1.6	-0.0
139.98	1.76	1.0	222.67	25OC	-0.2	-3.4	-1.5	-0.0
144.60	1.98	0.9	232.75	4NO	-0.2	-3.4	-1.5	0.1
148.79	2.23	0.8	242.87	14NO	0.4	-3.3	-1.4	0.1
152.45	2.49	0.6	253.03	24NO	2.9	-3.3	-1.4	0.2
155.45	2.78	0.4	263.21	4DE	0.8	-3.3	-1.4	0.3
157.64	3.10	0.2	273.40	14DE	-0.1	-3.3	-1.4	0.3
158.84	3.44	-0.0	283.60	24DE	-0.2	-3.3	-1.4	0.4
				1617				
158.87	3.78	-0.3	293.79	3JA	-0.2	-3.4	-1.4	0.4
157.60	4.10	-0.5	303.96	13JA	-0.4	-3.4	-1.4	0.5
155.05	4.34	-0.8	314.10	23JA	-0.8	-3.4	-1.5	0.5
151.51	4.45	-1.0	324.21	2FE	-1.2	-3.4	-1.5	0.5
147.57	4.40	-1.0	334.27	12FE	-1.2	-3.5	-1.5	0.5
143.94	4.17	-0.8	344.28	22FE	-0.5	-3.5	-1.6	0.5
141.24	3.83	-0.6	354.24	4MR	1.1	-3.6	-1.6	0.5
139.79	3.43	-0.3	4.14	14MR	3.3	-3.6	-1.7	0.5
139.62	3.02	-0.1	13.99	24MR	2.1	-3.7	-1.7	0.5
140.63	2.62	0.1	23.78	3AP	1.1	-3.8	-1.8	0.4
142.62	2.26	0.3	33.51	13AP	0.7	-3.8	-1.9	0.4
145.43	1.93	0.5	43.20	23AP	0.2	-3.9	-2.0	0.4
148.91	1.64	0.7	52.84	3MY	-0.5	-4.0	-2.0	0.4
152.92	1.37	0.8	62.44	13MY	-1.4	-4.1	-2.1	0.4
157.38	1.13	0.9	72.01	23MY	-1.6	-4.2	-2.2	0.4
162.20	0.91	1.1	81.56	2JN	-0.7	-4.2	-2.2	0.4
167.34	0.71	1.1	91.10	12JN	0.1	-4.1	-2.3	0.4
172.75	0.52	1.2	100.63	22JN	0.6	-3.8	-2.3	0.4
178.40	0.35	1.3	110.17	2JL	1.2	-3.1	-2.3	0.4
184.26	0.19	1.3	119.71	12JL	2.0	-3.3	-2.3	0.4
190.31	0.04	1.4	129.28	22JL	3.2	-3.9	-2.3	0.4
196.55	-0.11	1.4	138.88	1AU	1.6	-4.2	-2.3	0.4
202.95	-0.24	1.4	148.52	11AU	-0.1	-4.2	-2.3	0.3
209.50	-0.36	1.5	158.20	21AU	-1.1	-4.2	-2.2	0.3
216.21	-0.48	1.5	167.93	31AU	-1.3	-4.1	-2.2	0.2
223.06	-0.58	1.5	177.71	10SE	-0.9	-4.0	-2.1	0.2
230.04	-0.68	1.5	187.55	20SE	-0.4	-4.0	-2.0	0.1
237.16	-0.77	1.5	197.45	30SE	-0.2	-3.9	-2.0	0.0
244.40	-0.85	1.5	207.40	10OC	-0.1	-3.8	-1.9	-0.0
251.76	-0.92	1.5	217.40	20OC	0.0	-3.7	-1.8	-0.1
259.22	-0.99	1.5	227.46	30OC	0.6	-3.7	-1.8	-0.2
266.78	-1.04	1.4	237.56	9NO	3.2	-3.6	-1.7	-0.1
274.43	-1.08	1.4	247.70	19NO	0.6	-3.6	-1.7	-0.1
282.16	-1.10	1.4	257.87	29NO	-0.2	-3.5	-1.6	-0.0
289.94	-1.12	1.4	268.05	9DE	-0.3	-3.5	-1.6	0.1
297.79	-1.13	1.4	278.25	19DE	-0.4	-3.4	-1.6	0.1
305.66	-1.12	1.4	288.45	29DE	-0.5	-3.4	-1.5	0.2

♂ LONG	♂ LAT	♂ MAG	☉ LONG	16.00UT 1618	☿	♀	♃	♄
					MAGNITUDES			
313.56	-1.10	1.4	298.63	8JA	-0.8	-3.4	-1.5	0.3
321.48	-1.07	1.4	308.79	18JA	-1.1	-3.4	-1.5	0.3
329.38	-1.03	1.4	318.91	28JA	-1.1	-3.3	-1.5	0.4
337.26	-0.97	1.4	329.00	7FE	-0.4	-3.3	-1.5	0.4
345.12	-0.91	1.4	339.04	17FE	1.3	-3.3	-1.5	0.4
352.93	-0.84	1.4	349.03	27FE	3.1	-3.3	-1.6	0.4
0.68	-0.76	1.4	358.96	9MR	1.5	-3.3	-1.6	0.4
8.38	-0.68	1.4	8.84	19MR	0.8	-3.4	-1.6	0.4
16.00	-0.58	1.4	18.65	29MR	0.5	-3.4	-1.7	0.4
23.54	-0.49	1.4	28.41	8AP	0.1	-3.4	-1.7	0.4
31.00	-0.39	1.5	38.13	18AP	-0.5	-3.5	-1.8	0.3
38.38	-0.28	1.5	47.79	28AP	-1.4	-3.5	-1.8	0.3
45.66	-0.17	1.6	57.41	8MY	-1.6	-3.5	-1.9	0.2
52.86	-0.07	1.6	67.00	18MY	-0.7	-3.4	-2.0	0.3
59.96	0.04	1.7	76.56	28MY	0.2	-3.4	-2.0	0.3
66.98	0.15	1.7	86.10	7JN	0.9	-3.4	-2.1	0.3
73.91	0.26	1.7	95.63	17JN	1.5	-3.4	-2.2	0.4
80.74	0.37	1.7	105.16	27JN	2.6	-3.3	-2.3	0.4
87.49	0.48	1.8	114.70	7JL	3.0	-3.3	-2.3	0.4
94.16	0.58	1.8	124.26	17JL	1.3	-3.3	-2.4	0.4
100.73	0.69	1.8	133.84	27JL	-0.2	-3.3	-2.4	0.4
107.23	0.80	1.8	143.46	6AU	-1.2	-3.3	-2.4	0.3
113.63	0.90	1.8	153.12	16AU	-1.4	-3.4	-2.4	0.3
119.94	1.01	1.8	162.82	26AU	-0.8	-3.4	-2.4	0.3
126.17	1.11	1.8	172.57	5SE	-0.3	-3.4	-2.4	0.2
132.29	1.22	1.7	182.39	15SE	-0.1	-3.4	-2.4	0.1
138.31	1.33	1.7	192.25	25SE	0.1	-3.5	-2.3	0.1
144.21	1.44	1.7	202.17	5OC	0.2	-3.6	-2.3	0.0
149.98	1.55	1.6	212.15	15OC	0.9	-3.6	-2.2	-0.1
155.60	1.67	1.5	222.18	25OC	3.5	-3.7	-2.1	-0.1
161.04	1.79	1.4	232.26	4NO	0.4	-3.8	-2.0	-0.2
166.29	1.92	1.3	242.38	14NO	-0.4	-3.9	-2.0	-0.3
171.28	2.06	1.2	252.53	24NO	-0.5	-4.0	-1.9	-0.2
175.99	2.20	1.1	262.71	4DE	-0.5	-4.1	-1.8	-0.2
180.33	2.35	0.9	272.91	14DE	-0.6	-4.2	-1.8	-0.1
184.23	2.51	0.7	283.10	24DE	-0.8	-4.3	-1.7	-0.0
				1619				
187.59	2.68	0.5	293.29	3JA	-0.9	-4.4	-1.7	0.1
190.25	2.86	0.3	303.46	13JA	-0.9	-4.4	-1.7	0.1
192.08	3.04	0.1	313.60	23JA	-0.4	-4.2	-1.6	0.2
192.88	3.21	-0.2	323.71	2FE	1.6	-3.8	-1.6	0.2
192.48	3.35	-0.5	333.78	12FE	2.5	-3.3	-1.6	0.3
190.81	3.42	-0.7	343.79	22FE	1.1	-3.7	-1.6	0.3
187.95	3.39	-1.0	353.75	4MR	0.6	-4.1	-1.6	0.3
184.29	3.23	-1.2	3.66	14MR	0.3	-4.3	-1.6	0.3
180.48	2.93	-1.1	13.51	24MR	-0.0	-4.3	-1.6	0.3
177.21	2.53	-0.9	23.30	3AP	-0.5	-4.2	-1.6	0.3
175.03	2.08	-0.7	33.04	13AP	-1.4	-4.1	-1.6	0.3
174.14	1.64	-0.5	42.73	23AP	-1.6	-4.0	-1.7	0.3
174.54	1.23	-0.3	52.37	3MY	-0.7	-3.9	-1.7	0.3
176.08	0.85	-0.1	61.98	13MY	0.3	-3.8	-1.8	0.2
178.59	0.53	0.1	71.55	23MY	1.1	-3.7	-1.8	0.2
181.89	0.24	0.2	81.10	2JN	2.0	-3.6	-1.9	0.2
185.86	-0.02	0.4	90.64	12JN	3.3	-3.6	-1.9	0.3
190.37	-0.24	0.5	100.17	22JN	2.4	-3.5	-2.0	0.3
195.33	-0.43	0.6	109.71	2JL	1.0	-3.5	-2.1	0.3
200.68	-0.61	0.7	119.25	12JL	-0.2	-3.4	-2.1	0.3
206.37	-0.76	0.8	128.82	22JL	-1.3	-3.4	-2.2	0.3
212.34	-0.89	0.9	138.42	1AU	-1.5	-3.4	-2.3	0.3
218.58	-1.01	0.9	148.05	11AU	-0.8	-3.4	-2.3	0.3
225.04	-1.11	1.0	157.73	21AU	-0.3	-3.3	-2.4	0.3
231.71	-1.19	1.0	167.46	31AU	0.0	-3.3	-2.4	0.3
238.57	-1.26	1.1	177.24	10SE	0.2	-3.3	-2.5	0.2
245.61	-1.31	1.1	187.07	20SE	0.4	-3.4	-2.5	0.2
252.79	-1.35	1.1	196.97	30SE	1.2	-3.4	-2.5	0.1
260.12	-1.38	1.2	206.92	10OC	3.2	-3.4	-2.4	0.1
267.57	-1.39	1.2	216.92	20OC	0.3	-3.4	-2.4	-0.0
275.13	-1.38	1.2	226.98	30OC	-0.6	-3.4	-2.4	-0.1
282.78	-1.36	1.2	237.08	9NO	-0.6	-3.5	-2.3	-0.2
290.50	-1.33	1.3	247.21	19NO	-0.6	-3.5	-2.2	-0.2
298.28	-1.28	1.3	257.38	29NO	-0.7	-3.5	-2.2	-0.3
306.10	-1.23	1.3	267.56	9DE	-0.7	-3.5	-2.1	-0.3
313.94	-1.16	1.3	277.75	19DE	-0.8	-3.4	-2.0	-0.2
321.80	-1.08	1.4	287.95	29DE	-0.8	-3.4	-1.9	-0.1

♂ LONG	♂ LAT	♂ MAG	⊙ LONG	16.00UT 1620	☿	♀	♃	♄
					MAGNITUDES			
329.64	-0.99	1.4	298.13	8JA	-0.3	-3.4	-1.9	-0.0
337.47	-0.90	1.4	308.29	18JA	1.8	-3.4	-1.8	0.0
345.26	-0.80	1.4	318.42	28JA	2.0	-3.4	-1.8	0.1
353.00	-0.69	1.5	328.50	7FE	0.8	-3.3	-1.7	0.2
0.69	-0.59	1.5	338.54	17FE	0.4	-3.3	-1.7	0.2
8.31	-0.48	1.5	348.54	27FE	0.2	-3.3	-1.6	0.2
15.86	-0.37	1.5	358.47	8MR	-0.1	-3.3	-1.6	0.3
23.33	-0.25	1.6	8.35	18MR	-0.6	-3.4	-1.6	0.3
30.72	-0.14	1.6	18.17	28MR	-1.4	-3.4	-1.6	0.3
38.02	-0.03	1.6	27.94	7AP	-1.5	-3.4	-1.6	0.3
45.24	0.07	1.6	37.65	17AP	-0.7	-3.4	-1.6	0.3
52.38	0.18	1.6	47.32	27AP	0.4	-3.5	-1.6	0.3
59.44	0.28	1.6	56.94	7MY	1.5	-3.5	-1.6	0.3
66.41	0.38	1.7	66.53	17MY	2.7	-3.6	-1.6	0.3
73.31	0.47	1.7	76.10	27MY	3.3	-3.6	-1.6	0.2
80.13	0.56	1.8	85.64	6JN	1.9	-3.7	-1.7	0.2
86.88	0.65	1.8	95.17	16JN	0.8	-3.8	-1.7	0.2
93.56	0.73	1.8	104.71	26JN	-0.3	-3.9	-1.8	0.3
100.18	0.81	1.9	114.25	6JL	-1.3	-3.9	-1.8	0.3
106.73	0.88	1.9	123.80	16JL	-1.6	-4.0	-1.9	0.3
113.23	0.96	1.9	133.39	26JL	-0.8	-4.1	-2.0	0.3
119.67	1.03	1.9	143.00	5AU	-0.2	-4.2	-2.0	0.4
126.06	1.09	2.0	152.65	15AU	0.1	-4.2	-2.1	0.4
132.40	1.16	2.0	162.35	25AU	0.4	-4.1	-2.2	0.3
138.68	1.22	2.0	172.10	4SE	0.7	-3.8	-2.2	0.3
144.92	1.27	1.9	181.91	14SE	1.6	-3.2	-2.3	0.3
151.10	1.33	1.9	191.77	24SE	2.8	-3.6	-2.3	0.3
157.22	1.38	1.9	201.69	4OC	0.2	-4.1	-2.4	0.2
163.28	1.43	1.9	211.67	14OC	-0.7	-4.3	-2.4	0.1
169.28	1.48	1.8	221.70	24OC	-0.8	-4.3	-2.4	0.1
175.19	1.53	1.7	231.77	3NO	-0.8	-4.3	-2.4	0.0
181.03	1.58	1.7	241.89	13NO	-0.8	-4.2	-2.4	-0.1
186.75	1.62	1.6	252.05	23NO	-0.6	-4.1	-2.4	-0.1
192.36	1.66	1.5	262.22	3DE	-0.6	-4.0	-2.3	-0.2
197.83	1.70	1.4	272.41	13DE	-0.6	-3.9	-2.2	-0.3
203.12	1.73	1.2	282.61	23DE	-0.1	-3.8	-2.2	-0.2
				1621				
208.20	1.76	1.1	292.80	2JA	2.0	-3.7	-2.1	-0.2
213.03	1.78	0.9	302.97	12JA	1.6	-3.7	-2.0	-0.1
217.55	1.79	0.7	313.12	22JA	0.5	-3.6	-2.0	-0.0
221.67	1.79	0.5	323.22	1FE	0.2	-3.5	-1.9	0.0
225.31	1.77	0.3	333.29	11FE	0.1	-3.5	-1.8	0.1
228.33	1.73	0.1	343.31	21FE	-0.2	-3.4	-1.8	0.2
230.60	1.65	-0.2	353.27	3MR	-0.6	-3.4	-1.7	0.2
231.93	1.53	-0.5	3.18	13MR	-1.3	-3.4	-1.7	0.3
232.13	1.34	-0.8	13.03	23MR	-1.5	-3.3	-1.6	0.3
231.10	1.08	-1.1	22.83	2AP	-0.6	-3.3	-1.6	0.3
228.84	0.73	-1.4	32.57	12AP	0.6	-3.3	-1.5	0.3
225.62	0.30	-1.7	42.26	22AP	1.9	-3.3	-1.5	0.3
222.05	-0.18	-1.7	51.90	2MY	3.6	-3.3	-1.5	0.3
218.81	-0.65	-1.5	61.51	12MY	2.5	-3.3	-1.5	0.3
216.56	-1.07	-1.3	71.09	22MY	1.4	-3.3	-1.5	0.3
215.62	-1.41	-1.1	80.64	1JN	0.6	-3.4	-1.5	0.3
216.03	-1.69	-0.9	90.18	11JN	-0.3	-3.4	-1.5	0.3
217.70	-1.90	-0.7	99.71	21JN	-1.3	-3.4	-1.5	0.2
220.44	-2.05	-0.5	109.24	1JL	-1.6	-3.5	-1.6	0.3
224.07	-2.16	-0.4	118.79	11JL	-0.7	-3.5	-1.6	0.3
228.44	-2.23	-0.2	128.36	21JL	-0.1	-3.5	-1.6	0.4
233.41	-2.27	-0.1	137.95	31JL	0.3	-3.4	-1.7	0.4
238.88	-2.29	0.1	147.58	10AU	0.5	-3.4	-1.7	0.4
244.78	-2.28	0.2	157.26	20AU	0.9	-3.4	-1.8	0.4
251.02	-2.25	0.3	166.98	30AU	2.0	-3.3	-1.9	0.4
257.56	-2.20	0.4	176.76	9SE	2.5	-3.3	-1.9	0.4
264.36	-2.13	0.5	186.59	19SE	0.1	-3.3	-2.0	0.4
271.35	-2.04	0.6	196.48	29SE	-0.8	-3.3	-2.0	0.4
278.52	-1.94	0.6	206.43	9OC	-0.9	-3.3	-2.1	0.3
285.84	-1.83	0.7	216.43	19OC	-0.9	-3.3	-2.2	0.3
293.26	-1.70	0.8	226.48	29OC	-0.7	-3.4	-2.2	0.2
300.78	-1.57	0.9	236.58	8NO	-0.5	-3.4	-2.3	0.2
308.35	-1.43	0.9	246.72	18NO	-0.5	-3.4	-2.3	0.1
315.96	-1.28	1.0	256.88	28NO	-0.4	-3.5	-2.3	0.0
323.60	-1.13	1.1	267.07	8DE	0.0	-3.5	-2.3	-0.1
331.23	-0.98	1.2	277.26	18DE	2.3	-3.6	-2.3	-0.1
338.85	-0.83	1.2	287.45	28DE	1.2	-3.6	-2.2	-0.2

♂ LONG	♂ LAT	♂ MAG	⊙ LONG	16.00UT 1622	☿	♀	♃	♄
					MAGNITUDES			
346.44	-0.68	1.3	297.64	7JA	0.3	-3.7	-2.2	-0.1
353.99	-0.54	1.3	307.80	17JA	0.1	-3.8	-2.1	-0.1
1.49	-0.39	1.4	317.93	27JA	-0.1	-3.9	-2.1	-0.0
8.93	-0.26	1.5	328.02	6FE	-0.3	-4.0	-2.0	0.0
16.30	-0.13	1.5	338.06	16FE	-0.7	-4.1	-1.9	0.1
23.60	-0.00	1.6	348.05	26FE	-1.3	-4.1	-1.8	0.2
30.83	0.12	1.6	358.00	8MR	-1.4	-4.2	-1.8	0.2
37.98	0.23	1.7	7.87	18MR	-0.6	-4.3	-1.7	0.3
45.06	0.33	1.7	17.70	28MR	0.7	-4.2	-1.7	0.3
52.07	0.43	1.8	27.47	7AP	2.4	-4.0	-1.6	0.4
59.00	0.52	1.8	37.18	17AP	3.3	-3.5	-1.5	0.4
65.86	0.61	1.8	46.85	27AP	1.9	-2.9	-1.5	0.4
72.66	0.69	1.8	56.48	7MY	1.1	-3.6	-1.5	0.4
79.39	0.76	1.9	66.07	17MY	0.4	-4.1	-1.4	0.4
86.07	0.82	1.9	75.63	27MY	-0.4	-4.2	-1.4	0.4
92.69	0.88	1.9	85.18	6JN	-1.3	-4.2	-1.4	0.4
99.27	0.94	1.9	94.71	16JN	-1.6	-4.1	-1.4	0.4
105.79	0.99	1.9	104.24	26JN	-0.7	-4.0	-1.4	0.4
112.28	1.03	1.9	113.78	6JL	-0.0	-3.9	-1.4	0.3
118.74	1.07	1.9	123.33	16JL	0.4	-3.8	-1.4	0.4
125.16	1.11	2.0	132.92	26JL	0.7	-3.8	-1.4	0.4
131.55	1.14	2.0	142.53	5AU	1.2	-3.7	-1.5	0.5
137.92	1.16	2.0	152.18	15AU	2.5	-3.6	-1.5	0.5
144.28	1.18	2.0	161.88	25AU	2.1	-3.6	-1.5	0.5
150.61	1.20	2.0	171.63	4SE	-0.0	-3.5	-1.6	0.5
156.93	1.22	2.0	181.43	14SE	-1.0	-3.5	-1.6	0.5
163.23	1.22	2.0	191.29	24SE	-1.1	-3.4	-1.7	0.5
169.52	1.23	2.0	201.21	4OC	-1.0	-3.4	-1.7	0.5
175.80	1.23	2.0	211.18	14OC	-0.6	-3.4	-1.8	0.5
182.06	1.22	1.9	221.21	24OC	-0.4	-3.4	-1.9	0.5
188.31	1.22	1.9	231.29	3NO	-0.3	-3.4	-1.9	0.4
194.53	1.20	1.9	241.40	13NO	-0.3	-3.4	-2.0	0.3
200.74	1.18	1.8	251.55	23NO	0.2	-3.4	-2.1	0.3
206.92	1.15	1.7	261.73	3DE	2.6	-3.4	-2.1	0.2
213.06	1.11	1.6	271.92	13DE	1.0	-3.4	-2.1	0.1
219.17	1.06	1.5	282.11	23DE	0.1	-3.4	-2.2	0.1
				1623				
225.22	1.01	1.4	292.30	2JA	-0.1	-3.4	-2.2	-0.0
231.21	0.94	1.3	302.47	12JA	-0.2	-3.4	-2.2	-0.1
237.12	0.85	1.2	312.62	22JA	-0.4	-3.4	-2.1	-0.0
242.94	0.75	1.1	322.74	1FE	-0.7	-3.5	-2.1	0.0
248.64	0.62	0.9	332.80	11FE	-1.3	-3.5	-2.1	0.1
254.21	0.47	0.7	342.82	21FE	-1.3	-3.5	-2.0	0.2
259.59	0.29	0.5	352.79	3MR	-0.6	-3.4	-1.9	0.2
264.75	0.08	0.3	2.70	13MR	0.9	-3.4	-1.9	0.3
269.63	-0.18	0.1	12.55	23MR	3.0	-3.4	-1.8	0.3
274.16	-0.49	-0.1	22.35	2AP	2.5	-3.4	-1.7	0.4
278.24	-0.85	-0.4	32.09	12AP	1.4	-3.3	-1.7	0.4
281.75	-1.29	-0.7	41.79	22AP	0.8	-3.3	-1.6	0.5
284.53	-1.82	-0.9	51.43	2MY	0.3	-3.3	-1.5	0.5
286.41	-2.42	-1.3	61.04	12MY	-0.4	-3.3	-1.5	0.5
287.19	-3.11	-1.6	70.62	22MY	-1.3	-3.3	-1.4	0.5
286.76	-3.85	-1.9	80.17	1JN	-1.6	-3.3	-1.4	0.6
285.13	-4.58	-2.2	89.71	11JN	-0.7	-3.4	-1.4	0.6
282.57	-5.20	-2.4	99.24	21JN	0.0	-3.4	-1.3	0.6
279.70	-5.62	-2.4	108.78	1JL	0.5	-3.4	-1.3	0.5
277.25	-5.78	-2.3	118.32	11JL	1.0	-3.5	-1.3	0.5
275.81	-5.70	-2.1	127.89	21JL	1.7	-3.5	-1.3	0.5
275.71	-5.44	-1.8	137.48	31JL	3.0	-3.6	-1.3	0.6
276.97	-5.08	-1.6	147.11	10AU	1.8	-3.6	-1.3	0.6
279.45	-4.66	-1.3	156.79	20AU	-0.1	-3.7	-1.3	0.6
282.97	-4.23	-1.1	166.51	30AU	-1.1	-3.8	-1.3	0.7
287.31	-3.79	-0.9	176.29	9SE	-1.2	-3.9	-1.4	0.7
292.31	-3.37	-0.7	186.12	19SE	-0.9	-4.0	-1.4	0.7
297.84	-2.96	-0.5	196.00	29SE	-0.5	-4.1	-1.4	0.7
303.76	-2.58	-0.3	205.95	9OC	-0.3	-4.2	-1.5	0.7
309.99	-2.21	-0.1	215.95	19OC	-0.2	-4.3	-1.5	0.7
316.47	-1.87	0.1	226.00	29OC	-0.1	-4.4	-1.6	0.7
323.11	-1.56	0.2	236.09	8NO	0.4	-4.2	-1.7	0.6
329.89	-1.26	0.4	246.23	18NO	2.9	-3.8	-1.7	0.6
336.76	-0.99	0.5	256.39	28NO	0.7	-2.9	-1.8	0.5
343.69	-0.74	0.7	266.57	8DE	-0.1	-3.7	-1.8	0.5
350.65	-0.52	0.8	276.77	18DE	-0.2	-4.2	-1.9	0.4
357.63	-0.31	0.9	286.96	28DE	-0.3	-4.4	-2.0	0.3

Left (1624–1625)

♂ LONG	LAT	MAG	☉ LONG	16.00UT	☿	♀	♃	♄
				1624		MAGNITUDES		
4.60	-0.12	1.0	297.14	7JA	-0.4	-4.4	-2.0	0.3
11.56	0.05	1.1	307.31	17JA	-0.8	-4.3	-2.0	0.2
18.50	0.20	1.3	317.44	27JA	-1.2	-4.2	-2.1	0.1
25.40	0.34	1.4	327.53	6FE	-1.1	-4.1	-2.1	0.2
32.26	0.46	1.4	337.58	16FE	-0.5	-4.0	-2.0	0.2
39.08	0.57	1.5	347.57	26FE	1.1	-3.9	-2.0	0.3
45.86	0.67	1.6	357.51	7MR	3.3	-3.8	-2.0	0.3
52.59	0.75	1.7	7.40	17MR	1.8	-3.7	-1.9	0.4
59.27	0.83	1.7	17.23	27MR	1.0	-3.6	-1.9	0.5
65.91	0.89	1.8	27.00	6AP	0.6	-3.5	-1.8	0.5
72.50	0.95	1.8	36.72	16AP	0.2	-3.5	-1.7	0.6
79.06	1.00	1.9	46.39	26AP	-0.5	-3.4	-1.7	0.6
85.57	1.04	1.9	56.02	6MY	-1.3	-3.4	-1.6	0.7
92.05	1.07	2.0	65.61	16MY	-1.6	-3.4	-1.6	0.7
98.50	1.10	2.0	75.17	26MY	-0.7	-3.3	-1.5	0.7
104.92	1.12	2.0	84.72	5JN	0.1	-3.3	-1.4	0.7
111.33	1.14	2.0	94.25	15JN	0.7	-3.3	-1.4	0.8
117.71	1.15	2.0	103.78	25JN	1.3	-3.3	-1.3	0.8
124.08	1.15	2.0	113.32	5JL	2.2	-3.3	-1.3	0.8
130.44	1.16	2.0	122.88	15JL	3.2	-3.3	-1.3	0.8
136.80	1.15	2.0	132.45	25JL	1.5	-3.4	-1.3	0.8
143.16	1.14	2.0	142.07	4AU	-0.1	-3.4	-1.2	0.7
149.53	1.13	2.0	151.72	14AU	-1.1	-3.4	-1.2	0.8
155.90	1.11	2.0	161.41	24AU	-1.4	-3.4	-1.2	0.8
162.28	1.09	2.0	171.16	3SE	-0.9	-3.5	-1.2	0.9
168.68	1.06	2.0	180.96	13SE	-0.4	-3.5	-1.2	0.9
175.10	1.03	2.0	190.81	23SE	-0.1	-3.5	-1.2	0.9
181.54	1.00	2.0	200.73	3OC	-0.0	-3.4	-1.3	1.0
188.01	0.96	2.0	210.70	13OC	0.1	-3.4	-1.3	1.0
194.49	0.91	2.0	220.72	23OC	0.7	-3.4	-1.3	1.0
201.01	0.86	1.9	230.79	2NO	3.3	-3.4	-1.4	0.9
207.55	0.80	1.9	240.91	12NO	0.5	-3.3	-1.4	0.9
214.12	0.73	1.9	251.05	22NO	-0.3	-3.3	-1.5	0.9
220.71	0.66	1.8	261.23	2DE	-0.4	-3.3	-1.6	0.8
227.33	0.57	1.8	271.42	12DE	-0.4	-3.3	-1.6	0.8
233.98	0.48	1.7	281.61	22DE	-0.5	-3.3	-1.7	0.7
				1625				
240.65	0.38	1.6	291.81	1JA	-0.8	-3.4	-1.8	0.6
247.34	0.27	1.5	301.98	11JA	-1.0	-3.4	-1.8	0.6
254.05	0.14	1.5	312.13	21JA	-1.0	-3.4	-1.9	0.5
260.78	0.00	1.4	322.25	31JA	-0.4	-3.4	-1.9	0.4
267.53	-0.15	1.3	332.32	10FE	1.3	-3.5	-2.0	0.4
274.28	-0.32	1.1	342.34	20FE	2.9	-3.5	-2.0	0.4
281.04	-0.50	1.0	352.31	2MR	1.4	-3.6	-2.0	0.5
287.80	-0.70	0.9	2.23	12MR	0.7	-3.6	-2.0	0.5
294.54	-0.92	0.8	12.08	22MR	0.4	-3.7	-2.0	0.6
301.26	-1.16	0.6	21.88	1AP	0.1	-3.8	-2.0	0.6
307.93	-1.42	0.5	31.63	11AP	-0.5	-3.8	-1.9	0.7
314.55	-1.69	0.3	41.32	21AP	-1.3	-3.9	-1.9	0.7
321.07	-1.99	0.2	50.98	1MY	-1.6	-4.0	-1.8	0.8
327.46	-2.30	0.0	60.59	11MY	-0.7	-4.1	-1.7	0.8
333.68	-2.63	-0.2	70.16	21MY	0.2	-4.2	-1.7	0.9
339.67	-2.98	-0.4	79.72	31MY	0.9	-4.2	-1.6	0.9
345.34	-3.34	-0.5	89.25	10JN	1.7	-4.1	-1.6	1.0
350.62	-3.71	-0.8	98.78	20JN	2.9	-3.7	-1.5	1.0
355.36	-4.09	-1.0	108.32	30JN	2.8	-3.1	-1.4	1.0
359.42	-4.46	-1.2	117.86	10JL	1.2	-3.3	-1.4	1.0
2.62	-4.82	-1.4	127.43	20JL	-0.2	-3.9	-1.3	1.0
4.72	-5.14	-1.7	137.02	30JL	-1.2	-4.2	-1.3	1.0
5.55	-5.38	-2.0	146.65	9AU	-1.5	-4.2	-1.3	1.0
4.97	-5.47	-2.2	156.32	19AU	-0.8	-4.2	-1.2	1.0
3.08	-5.36	-2.4	166.04	29AU	-0.3	-4.1	-1.2	1.1
0.33	-4.99	-2.5	175.81	8SE	-0.0	-4.0	-1.2	1.1
357.43	-4.40	-2.4	185.64	18SE	0.1	-4.0	-1.2	1.2
355.14	-3.66	-2.2	195.53	28SE	0.3	-3.9	-1.2	1.2
354.00	-2.88	-1.8	205.46	8OC	1.0	-3.8	-1.2	1.2
354.17	-2.15	-1.5	215.46	18OC	3.5	-3.7	-1.2	1.2
355.60	-1.50	-1.2	225.51	28OC	0.4	-3.7	-1.2	1.2
358.10	-0.96	-0.9	235.60	7NO	-0.5	-3.6	-1.3	1.2
1.46	-0.50	-0.6	245.73	17NO	-0.5	-3.5	-1.3	1.2
5.49	-0.13	-0.3	255.90	27NO	-0.5	-3.5	-1.4	1.2
10.06	0.17	-0.1	266.08	7DE	-0.6	-3.5	-1.4	1.2
15.02	0.42	0.2	276.28	17DE	-0.8	-3.4	-1.5	1.1
20.30	0.63	0.4	286.47	27DE	-0.9	-3.4	-1.5	1.1

Right (1626–1627)

♂ LONG	LAT	MAG	☉ LONG	16.00UT	☿	♀	♃	♄
				1626		MAGNITUDES		
25.81	0.79	0.6	296.65	6JA	-0.9	-3.4	-1.6	1.0
31.50	0.92	0.8	306.82	16JA	-0.4	-3.4	-1.6	0.9
37.33	1.03	0.9	316.95	26JA	1.6	-3.3	-1.7	0.8
43.25	1.12	1.1	327.04	5FE	2.4	-3.3	-1.8	0.8
49.26	1.18	1.2	337.09	15FE	1.0	-3.3	-1.8	0.7
55.32	1.24	1.3	347.09	25FE	0.5	-3.3	-1.9	0.7
61.42	1.27	1.4	357.03	7MR	0.3	-3.3	-1.9	0.7
67.56	1.30	1.5	6.92	17MR	-0.0	-3.4	-2.0	0.8
73.71	1.32	1.6	16.75	27MR	-0.5	-3.4	-2.0	0.8
79.88	1.33	1.7	26.52	6AP	-1.3	-3.4	-2.0	0.9
86.05	1.34	1.8	36.25	16AP	-1.6	-3.5	-2.0	0.9
92.24	1.33	1.9	45.92	26AP	-0.7	-3.5	-2.0	1.0
98.43	1.33	1.9	55.55	6MY	0.3	-3.5	-2.0	1.0
104.63	1.31	1.9	65.14	16MY	1.2	-3.4	-1.9	1.1
110.84	1.29	2.0	74.71	26MY	2.3	-3.4	-1.9	1.1
117.05	1.27	2.0	84.25	5JN	3.5	-3.4	-1.8	1.2
123.28	1.24	2.0	93.79	15JN	2.2	-3.4	-1.7	1.2
129.53	1.21	2.0	103.32	25JN	1.0	-3.3	-1.7	1.2
135.79	1.18	2.0	112.85	5JL	-0.2	-3.3	-1.6	1.3
142.07	1.14	2.0	122.41	15JL	-1.3	-3.3	-1.6	1.3
148.39	1.10	2.0	131.99	25JL	-1.6	-3.3	-1.5	1.3
154.72	1.05	2.0	141.59	4AU	-0.8	-3.3	-1.4	1.3
161.10	1.00	2.0	151.24	14AU	-0.2	-3.4	-1.4	1.3
167.51	0.95	2.0	160.94	24AU	0.1	-3.4	-1.4	1.3
173.96	0.89	1.9	170.68	3SE	0.3	-3.4	-1.3	1.3
180.45	0.83	1.9	180.48	13SE	0.5	-3.5	-1.3	1.3
186.99	0.77	1.9	190.33	23SE	1.3	-3.5	-1.3	1.3
193.58	0.70	1.9	200.24	3OC	3.1	-3.6	-1.2	1.3
200.23	0.62	1.9	210.21	13OC	0.3	-3.6	-1.2	1.3
206.93	0.55	1.9	220.23	23OC	-0.6	-3.7	-1.2	1.2
213.68	0.46	1.9	230.30	2NO	-0.7	-3.8	-1.2	1.2
220.50	0.37	1.9	240.42	12NO	-0.7	-3.9	-1.2	1.2
227.37	0.28	1.8	250.57	22NO	-0.8	-4.0	-1.3	1.2
234.31	0.18	1.8	260.74	2DE	-0.7	-4.1	-1.3	1.1
241.30	0.08	1.8	270.93	12DE	-0.7	-4.2	-1.3	1.1
248.35	-0.03	1.7	281.13	22DE	-0.7	-4.3	-1.4	1.1
				1627				
255.46	-0.15	1.7	291.31	1JA	-0.2	-4.4	-1.4	1.0
262.63	-0.27	1.6	301.49	11JA	1.8	-4.4	-1.5	1.0
269.85	-0.39	1.6	311.64	21JA	1.9	-4.2	-1.5	0.9
277.13	-0.52	1.5	321.76	31JA	0.7	-3.8	-1.6	0.9
284.46	-0.66	1.4	331.83	10FE	0.3	-3.3	-1.7	0.8
291.83	-0.80	1.4	341.86	20FE	0.1	-3.7	-1.7	0.8
299.24	-0.94	1.3	351.83	2MR	-0.1	-4.1	-1.8	0.7
306.68	-1.08	1.2	1.75	12MR	-0.6	-4.3	-1.9	0.8
314.14	-1.22	1.2	11.60	22MR	-1.3	-4.3	-1.9	0.8
321.62	-1.36	1.1	21.41	1AP	-1.5	-4.2	-2.0	0.9
329.09	-1.49	1.0	31.16	11AP	-0.7	-4.1	-2.0	1.0
336.55	-1.62	0.9	40.85	21AP	0.5	-4.0	-2.1	1.0
343.94	-1.73	0.8	50.51	1MY	1.6	-3.9	-2.1	1.1
351.37	-1.84	0.8	60.12	11MY	3.0	-3.8	-2.1	1.2
358.68	-1.93	0.7	69.70	21MY	3.0	-3.7	-2.1	1.2
5.90	-2.01	0.6	79.25	31MY	1.7	-3.6	-2.1	1.3
13.00	-2.06	0.5	88.80	10JN	0.7	-3.6	-2.0	1.3
19.95	-2.10	0.4	98.33	20JN	-0.3	-3.5	-2.0	1.3
26.71	-2.12	0.3	107.86	30JN	-1.3	-3.4	-1.9	1.3
33.25	-2.12	0.2	117.41	10JL	-1.6	-3.4	-1.9	1.3
39.50	-2.09	0.1	126.97	20JL	-0.8	-3.4	-1.8	1.3
45.43	-2.03	-0.0	136.56	30JL	-0.2	-3.4	-1.7	1.3
50.93	-1.94	-0.1	146.19	9AU	0.2	-3.4	-1.7	1.2
55.94	-1.83	-0.3	155.85	19AU	0.4	-3.3	-1.6	1.2
60.32	-1.67	-0.4	165.57	29AU	0.7	-3.3	-1.6	1.2
63.93	-1.47	-0.6	175.34	8SE	1.7	-3.3	-1.5	1.1
66.58	-1.21	-0.8	185.16	18SE	2.8	-3.4	-1.5	1.1
68.07	-0.89	-1.0	195.04	28SE	0.2	-3.4	-1.4	1.1
68.19	-0.51	-1.3	204.99	8OC	-0.7	-3.4	-1.4	1.1
66.85	-0.05	-1.5	214.98	18OC	-0.8	-3.4	-1.4	1.1
64.16	0.45	-1.7	225.02	28OC	-0.8	-3.4	-1.3	1.1
60.58	0.96	-1.8	235.11	7NO	-0.7	-3.5	-1.3	1.1
56.88	1.41	-1.7	245.24	17NO	-0.6	-3.5	-1.3	1.1
53.84	1.76	-1.4	255.40	27NO	-0.5	-3.5	-1.3	1.0
51.98	2.00	-1.1	265.59	7DE	-0.5	-3.5	-1.3	1.0
51.50	2.14	-0.7	275.78	17DE	-0.1	-3.4	-1.3	1.0
52.28	2.21	-0.4	285.97	27DE	2.0	-3.4	-1.4	0.9

Left table (1628–1629)

♂ LONG	♂ LAT	♂ MAG	☉ LONG	16.00UT 1628	☿	♀	♃	♄
54.15	2.23	-0.1	296.16	6JA	1.5	-3.4	-1.4	0.9
56.90	2.22	0.1	306.32	16JA	0.4	-3.4	-1.4	0.8
60.34	2.18	0.4	316.46	26JA	0.2	-3.4	-1.5	0.8
64.33	2.14	0.6	326.56	5FE	0.0	-3.3	-1.5	0.7
68.73	2.08	0.8	336.60	15FE	-0.2	-3.3	-1.6	0.7
73.46	2.02	1.0	346.61	25FE	-0.6	-3.3	-1.6	0.6
78.46	1.96	1.1	356.55	6MR	-1.3	-3.3	-1.7	0.6
83.66	1.89	1.3	6.44	16MR	-1.5	-3.4	-1.8	0.6
89.03	1.82	1.4	16.28	26MR	-0.7	-3.4	-1.9	0.6
94.53	1.75	1.5	26.05	5AP	0.6	-3.4	-1.9	0.7
100.14	1.68	1.6	35.78	15AP	2.0	-3.4	-2.0	0.7
105.84	1.61	1.7	45.45	25AP	3.9	-3.5	-2.1	0.8
111.63	1.54	1.8	55.09	5MY	2.3	-3.5	-2.1	0.9
117.49	1.47	1.8	64.68	15MY	1.3	-3.6	-2.2	0.9
123.41	1.40	1.9	74.25	25MY	0.5	-3.6	-2.2	1.0
129.39	1.32	1.9	83.80	4JN	-0.3	-3.7	-2.2	1.0
135.45	1.25	1.9	93.33	14JN	-1.3	-3.8	-2.2	1.1
141.53	1.17	2.0	102.86	24JN	-1.6	-3.9	-2.2	1.1
147.68	1.10	2.0	112.40	4JL	-0.8	-3.9	-2.2	1.1
153.89	1.02	2.0	121.95	14JL	-0.1	-4.0	-2.1	1.1
160.16	0.94	2.0	131.53	24JL	0.3	-4.1	-2.1	1.1
166.49	0.86	2.0	141.14	3AU	0.6	-4.2	-2.0	1.1
172.88	0.78	2.0	150.78	13AU	1.0	-4.2	-2.0	1.1
179.35	0.69	2.0	160.47	23AU	2.1	-4.1	-1.9	1.1
185.88	0.60	1.9	170.21	2SE	2.4	-3.7	-1.8	1.0
192.47	0.52	1.9	180.00	12SE	0.1	-3.2	-1.8	1.0
199.15	0.43	1.8	189.86	22SE	-0.9	-3.7	-1.7	0.9
205.90	0.33	1.8	199.76	2OC	-1.0	-4.1	-1.7	0.9
212.73	0.24	1.8	209.73	12OC	-1.0	-4.3	-1.6	0.9
219.64	0.14	1.7	219.75	22OC	-0.6	-4.3	-1.6	0.9
226.62	0.04	1.7	229.81	1NO	-0.4	-4.3	-1.5	1.0
233.69	-0.06	1.7	239.92	11NO	-0.4	-4.2	-1.5	1.0
240.84	-0.16	1.7	250.07	21NO	-0.4	-4.1	-1.5	0.9
248.06	-0.26	1.7	260.24	1DE	0.1	-4.0	-1.4	0.9
255.36	-0.37	1.6	270.43	11DE	2.3	-3.9	-1.4	0.9
262.74	-0.47	1.6	280.63	21DE	1.2	-3.8	-1.4	0.9
270.18	-0.58	1.6	290.82	31DE	0.2	-3.7	-1.4	0.8
				1629				
277.70	-0.68	1.6	301.00	10JA	-0.0	-3.7	-1.4	0.8
285.27	-0.78	1.5	311.15	20JA	-0.1	-3.6	-1.5	0.8
292.89	-0.87	1.5	321.27	30JA	-0.3	-3.5	-1.5	0.7
300.56	-0.97	1.5	331.35	9FE	-0.7	-3.5	-1.5	0.7
308.27	-1.05	1.4	341.38	19FE	-1.3	-3.4	-1.6	0.6
316.00	-1.13	1.4	351.35	1MR	-1.4	-3.4	-1.6	0.6
323.74	-1.20	1.4	1.27	11MR	-0.6	-3.4	-1.7	0.5
331.50	-1.25	1.3	11.14	21MR	0.8	-3.3	-1.7	0.5
339.24	-1.30	1.3	20.94	31MR	2.6	-3.3	-1.8	0.4
346.96	-1.33	1.3	30.69	10AP	3.0	-3.3	-1.8	0.5
354.64	-1.35	1.2	40.39	20AP	1.7	-3.3	-1.9	0.6
2.28	-1.36	1.2	50.04	30AP	1.0	-3.3	-2.0	0.6
9.85	-1.34	1.2	59.66	10MY	0.4	-3.3	-2.1	0.7
17.35	-1.32	1.2	69.24	20MY	-0.4	-3.3	-2.1	0.8
24.76	-1.28	1.1	78.80	30MY	-1.3	-3.4	-2.2	0.8
32.06	-1.22	1.1	88.33	9JN	-1.7	-3.4	-2.2	0.9
39.25	-1.14	1.1	97.87	19JN	-0.7	-3.4	-2.3	0.9
46.29	-1.05	1.0	107.40	29JN	-0.0	-3.5	-2.3	0.9
53.20	-0.95	1.0	116.94	9JL	0.4	-3.5	-2.3	1.0
59.93	-0.82	0.9	126.51	19JL	0.8	-3.4	-2.3	1.0
66.47	-0.69	0.9	136.09	29JL	1.4	-3.4	-2.3	1.0
72.80	-0.53	0.8	145.72	8AU	2.7	-3.4	-2.3	1.0
78.88	-0.36	0.7	155.39	18AU	2.1	-3.4	-2.3	1.0
84.68	-0.17	0.7	165.10	28AU	0.0	-3.3	-2.2	1.0
90.16	0.05	0.6	174.86	7SE	-1.0	-3.3	-2.2	0.9
95.24	0.28	0.4	184.69	17SE	-1.1	-3.3	-2.1	0.9
99.84	0.55	0.3	194.56	27SE	-1.0	-3.3	-2.0	0.9
103.88	0.85	0.2	204.50	7OC	-0.5	-3.3	-2.0	0.8
107.20	1.19	-0.0	214.49	17OC	-0.3	-3.3	-1.9	0.8
109.65	1.57	-0.2	224.53	27OC	-0.3	-3.4	-1.8	0.8
111.03	1.99	-0.4	234.62	6NO	-0.2	-3.4	-1.8	0.9
111.16	2.45	-0.6	244.75	16NO	0.3	-3.4	-1.7	0.9
109.90	2.93	-0.8	254.90	26NO	2.6	-3.5	-1.7	0.9
107.29	3.38	-1.1	265.09	6DE	0.9	-3.5	-1.6	0.9
103.67	3.73	-1.2	275.28	16DE	-0.0	-3.6	-1.6	0.8
99.71	3.94	-1.2	285.47	26DE	-0.2	-3.6	-1.6	0.8

Right table (1630–1631)

♂ LONG	♂ LAT	♂ MAG	☉ LONG	16.00UT 1630	☿	♀	♃	♄
96.16	3.97	-1.0	295.66	5JA	-0.2	-3.7	-1.5	0.8
93.65	3.87	-0.7	305.83	15JA	-0.4	-3.8	-1.5	0.7
92.43	3.68	-0.4	315.96	25JA	-0.7	-3.9	-1.5	0.7
92.51	3.44	-0.1	326.07	4FE	-1.2	-4.0	-1.5	0.7
93.75	3.20	0.1	336.12	14FE	-1.2	-4.1	-1.5	0.6
95.93	2.96	0.4	346.12	24FE	-0.6	-4.1	-1.6	0.6
98.87	2.73	0.6	356.08	6MR	1.0	-4.2	-1.6	0.5
102.42	2.52	0.8	5.97	16MR	3.1	-4.3	-1.6	0.5
106.45	2.32	1.0	15.80	26MR	2.2	-4.2	-1.7	0.4
110.87	2.14	1.1	25.59	5AP	1.2	-4.0	-1.7	0.4
115.59	1.97	1.2	35.31	15AP	0.7	-3.5	-1.7	0.3
120.58	1.81	1.4	44.99	25AP	0.3	-2.9	-1.8	0.4
125.77	1.67	1.5	54.62	5MY	-0.4	-3.7	-1.9	0.5
131.15	1.52	1.5	64.22	15MY	-1.3	-4.1	-1.9	0.5
136.68	1.39	1.6	73.79	25MY	-1.7	-4.2	-2.0	0.6
142.35	1.26	1.7	83.34	4JN	-0.7	-4.2	-2.1	0.7
148.14	1.13	1.7	92.87	14JN	0.0	-4.1	-2.1	0.7
154.05	1.01	1.8	102.40	24JN	0.6	-4.0	-2.2	0.8
160.07	0.89	1.8	111.94	4JL	1.1	-3.9	-2.3	0.8
166.20	0.78	1.8	121.49	14JL	1.8	-3.8	-2.3	0.9
172.42	0.66	1.8	131.06	24JL	3.2	-3.8	-2.4	0.9
178.74	0.55	1.8	140.67	3AU	1.7	-3.7	-2.4	0.9
185.16	0.44	1.8	150.31	13AU	-0.0	-3.6	-2.4	0.9
191.68	0.33	1.8	160.00	23AU	-1.1	-3.6	-2.4	0.9
198.30	0.22	1.8	169.74	2SE	-1.3	-3.5	-2.4	0.9
205.02	0.12	1.8	179.53	12SE	-0.9	-3.5	-2.4	0.9
211.83	0.01	1.8	189.38	22SE	-0.5	-3.4	-2.4	0.8
218.75	-0.09	1.7	199.29	2OC	-0.2	-3.4	-2.3	0.8
225.76	-0.20	1.7	209.25	12OC	-0.1	-3.4	-2.3	0.8
232.87	-0.30	1.7	219.26	22OC	-0.0	-3.4	-2.2	0.7
240.08	-0.39	1.6	229.33	1NO	0.5	-3.4	-2.1	0.8
247.38	-0.49	1.6	239.43	11NO	2.9	-3.4	-2.0	0.8
254.77	-0.58	1.5	249.58	21NO	0.7	-3.4	-2.0	0.8
262.25	-0.67	1.5	259.75	1DE	-0.2	-3.4	-1.9	0.8
269.81	-0.75	1.4	269.94	11DE	-0.3	-3.4	-1.8	0.8
277.44	-0.83	1.4	280.13	21DE	-0.3	-3.4	-1.8	0.8
285.13	-0.90	1.4	290.32	31DE	-0.5	-3.4	-1.7	0.8
				1631				
292.88	-0.96	1.4	300.50	10JA	-0.8	-3.4	-1.7	0.8
300.68	-1.01	1.4	310.65	20JA	-1.1	-3.4	-1.7	0.7
308.51	-1.06	1.4	320.77	30JA	-1.1	-3.5	-1.6	0.7
316.36	-1.09	1.4	330.85	9FE	-0.5	-3.5	-1.6	0.7
324.22	-1.11	1.4	340.88	19FE	1.2	-3.5	-1.6	0.6
332.09	-1.12	1.4	350.86	1MR	3.2	-3.5	-1.6	0.6
339.93	-1.12	1.4	0.78	11MR	1.7	-3.4	-1.6	0.5
347.75	-1.10	1.4	10.65	21MR	0.9	-3.4	-1.6	0.4
355.54	-1.07	1.4	20.46	31MR	0.5	-3.4	-1.6	0.4
3.27	-1.03	1.4	30.21	10AP	0.1	-3.3	-1.6	0.3
10.94	-0.98	1.4	39.92	20AP	-0.4	-3.3	-1.7	0.3
18.54	-0.92	1.4	49.58	30AP	-1.3	-3.3	-1.7	0.3
26.06	-0.84	1.4	59.19	10MY	-1.7	-3.3	-1.7	0.4
33.50	-0.76	1.4	68.77	20MY	-0.7	-3.3	-1.8	0.4
40.83	-0.66	1.4	78.33	30MY	0.1	-3.3	-1.8	0.5
48.07	-0.56	1.4	87.87	9JN	0.8	-3.4	-1.9	0.6
55.20	-0.45	1.4	97.40	19JN	1.4	-3.4	-2.0	0.6
62.22	-0.33	1.4	106.94	29JN	2.4	-3.4	-2.0	0.7
69.11	-0.20	1.4	116.48	9JL	3.1	-3.5	-2.1	0.7
75.89	-0.07	1.4	126.04	19JL	1.4	-3.5	-2.2	0.8
82.52	0.07	1.4	135.63	29JL	-0.1	-3.6	-2.2	0.8
89.02	0.22	1.4	145.25	8AU	-1.2	-3.7	-2.3	0.8
95.36	0.38	1.3	154.92	18AU	-1.4	-3.7	-2.4	0.8
101.53	0.54	1.2	164.63	28AU	-0.9	-3.8	-2.4	0.8
107.52	0.71	1.2	174.39	7SE	-0.4	-3.9	-2.4	0.8
113.30	0.90	1.2	184.21	17SE	-0.1	-4.0	-2.5	0.8
118.84	1.09	1.1	194.09	27SE	0.0	-4.1	-2.5	0.8
124.10	1.30	1.0	204.02	7OC	0.2	-4.3	-2.5	0.8
129.02	1.53	0.9	214.01	17OC	0.8	-4.3	-2.4	0.7
133.55	1.78	0.8	224.04	27OC	3.3	-4.3	-2.4	0.7
137.60	2.06	0.6	234.13	6NO	0.5	-4.2	-2.3	0.7
141.05	2.36	0.4	244.26	16NO	-0.4	-3.8	-2.3	0.7
143.77	2.69	0.3	254.41	26NO	-0.5	-2.8	-2.2	0.7
145.58	3.05	0.0	264.59	6DE	-0.5	-3.7	-2.1	0.8
146.30	3.44	-0.2	274.79	16DE	-0.6	-4.2	-2.1	0.8
145.76	3.83	-0.4	284.98	26DE	-0.8	-4.4	-2.0	0.8

♂			⊙	16.00UT	☿	♀	♃	♄
LONG	LAT	MAG	LONG	1632	MAGNITUDES			
143.89	4.19	-0.7	295.16	5JA	-0.9	-4.4	-1.9	0.8
140.81	4.45	-0.9	305.33	15JA	-0.9	-4.3	-1.9	0.8
136.98	4.56	-1.0	315.47	25JA	-0.4	-4.2	-1.8	0.7
133.08	4.49	-0.9	325.57	4FE	1.4	-4.1	-1.7	0.7
129.81	4.26	-0.7	335.63	14FE	2.7	-4.0	-1.7	0.7
127.67	3.93	-0.4	345.64	24FE	1.2	-3.9	-1.7	0.6
126.82	3.55	-0.2	355.59	5MR	0.6	-3.8	-1.6	0.6
127.22	3.18	0.0	5.49	15MR	0.4	-3.7	-1.6	0.5
128.70	2.82	0.3	15.33	25MR	0.0	-3.6	-1.6	0.5
131.07	2.48	0.5	25.11	4AP	-0.5	-3.5	-1.6	0.4
134.18	2.18	0.7	34.84	14AP	-1.3	-3.5	-1.6	0.3
137.87	1.91	0.8	44.52	24AP	-1.6	-3.4	-1.6	0.3
142.04	1.66	1.0	54.16	4MY	-0.7	-3.4	-1.6	0.2
146.61	1.44	1.1	63.76	14MY	0.2	-3.4	-1.6	0.3
151.50	1.23	1.2	73.33	24MY	1.0	-3.3	-1.6	0.3
156.66	1.04	1.3	82.88	3JN	1.9	-3.3	-1.7	0.4
162.06	0.86	1.4	92.41	13JN	3.1	-3.3	-1.7	0.5
167.68	0.69	1.4	101.94	23JN	2.6	-3.3	-1.7	0.5
173.48	0.53	1.5	111.48	3JL	1.1	-3.3	-1.8	0.6
179.46	0.38	1.5	121.03	13JL	-0.2	-3.3	-1.8	0.6
185.59	0.24	1.5	130.60	23JL	-1.2	-3.4	-1.9	0.7
191.87	0.10	1.6	140.20	2AU	-1.5	-3.4	-2.0	0.7
198.30	-0.02	1.6	149.85	12AU	-0.8	-3.4	-2.0	0.8
204.87	-0.15	1.6	159.53	22AU	-0.3	-3.4	-2.1	0.8
211.57	-0.26	1.6	169.27	1SE	0.0	-3.5	-2.2	0.8
218.40	-0.37	1.6	179.06	11SE	0.2	-3.5	-2.2	0.8
225.35	-0.48	1.6	188.90	21SE	0.4	-3.5	-2.3	0.8
232.43	-0.58	1.6	198.80	1OC	1.1	-3.4	-2.3	0.8
239.62	-0.67	1.6	208.76	11OC	3.4	-3.4	-2.4	0.8
246.92	-0.75	1.5	218.77	21OC	0.4	-3.4	-2.4	0.7
254.33	-0.83	1.5	228.83	31OC	-0.5	-3.4	-2.4	0.7
261.83	-0.90	1.5	238.94	10NO	-0.6	-3.3	-2.4	0.7
269.43	-0.95	1.5	249.08	20NO	-0.6	-3.3	-2.4	0.7
277.10	-1.00	1.5	259.25	30NO	-0.7	-3.3	-2.3	0.7
284.85	-1.04	1.4	269.44	10DE	-0.8	-3.3	-2.3	0.7
292.65	-1.07	1.4	279.63	20DE	-0.8	-3.3	-2.2	0.8
300.50	-1.09	1.4	289.83	30DE	-0.8	-3.4	-2.2	0.8
				1633				
308.38	-1.09	1.4	300.01	9JA	-0.3	-3.4	-2.1	0.8
316.29	-1.08	1.3	310.16	19JA	1.6	-3.4	-2.0	0.8
324.19	-1.06	1.3	320.28	29JA	2.2	-3.4	-1.9	0.8
332.09	-1.03	1.3	330.36	8FE	0.9	-3.5	-1.9	0.7
339.97	-0.99	1.3	340.39	18FE	0.4	-3.5	-1.8	0.7
347.82	-0.94	1.3	350.38	28FE	0.2	-3.6	-1.7	0.7
355.62	-0.88	1.4	0.31	10MR	-0.1	-3.6	-1.7	0.6
3.37	-0.81	1.4	10.17	20MR	-0.5	-3.7	-1.6	0.6
11.05	-0.73	1.4	19.99	30MR	-1.3	-3.8	-1.6	0.5
18.66	-0.64	1.5	29.74	9AP	-1.6	-3.8	-1.6	0.4
26.19	-0.55	1.5	39.45	19AP	-0.7	-3.9	-1.5	0.4
33.64	-0.45	1.5	49.11	29AP	0.4	-4.0	-1.5	0.3
40.99	-0.35	1.6	58.73	9MY	1.3	-4.1	-1.5	0.3
48.26	-0.24	1.6	68.31	19MY	2.5	-4.2	-1.5	0.2
55.44	-0.13	1.6	77.87	29MY	3.4	-4.2	-1.5	0.3
62.52	-0.02	1.6	87.41	8JN	2.0	-4.1	-1.5	0.3
69.50	0.09	1.7	96.94	18JN	0.9	-3.7	-1.5	0.4
76.39	0.21	1.7	106.48	28JN	-0.2	-3.1	-1.5	0.5
83.18	0.32	1.7	116.02	8JL	-1.2	-3.3	-1.6	0.5
89.88	0.44	1.7	125.58	18JL	-1.6	-3.9	-1.6	0.6
96.48	0.56	1.7	135.17	28JL	-0.8	-4.2	-1.6	0.6
102.98	0.68	1.7	144.79	7AU	-0.2	-4.2	-1.7	0.7
109.37	0.80	1.7	154.44	17AU	0.1	-4.2	-1.7	0.7
115.66	0.92	1.7	164.16	27AU	0.3	-4.1	-1.8	0.7
121.83	1.05	1.7	173.92	6SE	0.6	-4.0	-1.9	0.8
127.88	1.18	1.6	183.73	16SE	1.4	-3.9	-1.9	0.8
133.80	1.31	1.6	193.60	26SE	3.1	-3.9	-2.0	0.8
139.57	1.44	1.5	203.53	6OC	0.3	-3.8	-2.0	0.8
145.16	1.59	1.4	213.52	16OC	-0.7	-3.7	-2.1	0.8
150.56	1.74	1.3	223.56	26OC	-0.8	-3.7	-2.2	0.7
155.72	1.89	1.2	233.64	5NO	-0.7	-3.6	-2.2	0.7
160.60	2.06	1.1	243.76	15NO	-0.8	-3.5	-2.3	0.7
165.15	2.25	1.0	253.92	25NO	-0.6	-3.5	-2.3	0.6
169.28	2.44	0.8	264.10	5DE	-0.6	-3.5	-2.3	0.7
172.90	2.65	0.7	274.29	15DE	-0.6	-3.4	-2.3	0.7
175.90	2.88	0.4	284.49	25DE	-0.2	-3.4	-2.3	0.7

♂			⊙	16.00UT	☿	♀	♃	♄
LONG	LAT	MAG	LONG	1634	MAGNITUDES			
178.11	3.12	0.2	294.67	4JA	1.8	-3.4	-2.2	0.8
179.37	3.37	-0.0	304.84	14JA	1.8	-3.4	-2.2	0.8
179.49	3.61	-0.3	314.99	24JA	0.6	-3.3	-2.1	0.8
178.35	3.81	-0.5	325.09	3FE	0.3	-3.3	-2.0	0.8
175.93	3.93	-0.8	335.14	13FE	0.1	-3.3	-2.0	0.8
172.50	3.93	-1.0	345.16	23FE	-0.1	-3.3	-1.9	0.7
168.62	3.77	-1.0	355.11	5MR	-0.6	-3.3	-1.8	0.7
164.99	3.46	-0.9	5.01	15MR	-1.3	-3.4	-1.8	0.7
162.23	3.06	-0.7	14.85	25MR	-1.5	-3.4	-1.7	0.6
160.71	2.62	-0.5	24.63	4AP	-0.7	-3.4	-1.6	0.5
160.48	2.19	-0.3	34.36	14AP	0.5	-3.5	-1.6	0.5
161.45	1.78	-0.0	44.05	24AP	1.7	-3.5	-1.5	0.4
163.46	1.42	0.2	53.69	4MY	3.3	-3.5	-1.5	0.4
166.31	1.09	0.3	63.29	14MY	2.7	-3.4	-1.4	0.3
169.86	0.80	0.5	72.86	24MY	1.5	-3.4	-1.4	0.2
173.99	0.54	0.6	82.41	3JN	0.7	-3.4	-1.4	0.2
178.59	0.31	0.7	91.94	13JN	-0.3	-3.4	-1.4	0.3
183.59	0.10	0.8	101.48	23JN	-1.3	-3.3	-1.4	0.4
188.94	-0.09	0.9	111.01	3JL	-1.7	-3.3	-1.4	0.4
194.57	-0.26	1.0	120.56	13JL	-0.8	-3.3	-1.4	0.5
200.48	-0.42	1.1	130.13	23JL	-0.2	-3.3	-1.4	0.5
206.61	-0.56	1.1	139.73	2AU	0.2	-3.3	-1.4	0.6
212.96	-0.68	1.2	149.37	12AU	0.5	-3.4	-1.4	0.6
219.51	-0.80	1.2	159.06	22AU	0.8	-3.4	-1.5	0.7
226.23	-0.90	1.2	168.79	1SE	1.8	-3.4	-1.5	0.7
233.12	-0.98	1.3	178.58	11SE	2.7	-3.5	-1.6	0.7
240.17	-1.06	1.3	188.42	21SE	0.2	-3.5	-1.6	0.8
247.36	-1.12	1.3	198.32	1OC	-0.8	-3.6	-1.7	0.8
254.68	-1.17	1.3	208.28	11OC	-0.9	-3.6	-1.7	0.8
262.12	-1.21	1.3	218.29	21OC	-0.9	-3.7	-1.8	0.8
269.66	-1.24	1.3	228.34	31OC	-0.7	-3.8	-1.9	0.8
277.30	-1.25	1.3	238.45	10NO	-0.5	-3.9	-1.9	0.7
285.02	-1.24	1.3	248.59	20NO	-0.5	-4.0	-2.0	0.7
292.81	-1.23	1.4	258.76	30NO	-0.5	-4.1	-2.0	0.7
300.64	-1.20	1.4	268.95	10DE	-0.0	-4.2	-2.1	0.7
308.51	-1.16	1.4	279.14	20DE	2.1	-4.3	-2.1	0.7
316.39	-1.11	1.4	289.33	30DE	1.4	-4.4	-2.1	0.7
				1635				
324.28	-1.05	1.4	299.51	9JA	0.3	-4.4	-2.2	0.8
332.16	-0.98	1.4	309.67	19JA	0.1	-4.2	-2.1	0.8
340.01	-0.90	1.4	319.79	29JA	-0.0	-3.8	-2.1	0.8
347.82	-0.81	1.4	329.88	8FE	-0.2	-3.3	-2.1	0.8
355.58	-0.72	1.4	339.91	18FE	-0.6	-3.7	-2.0	0.8
3.28	-0.62	1.5	349.89	28FE	-1.3	-4.1	-2.0	0.8
10.92	-0.52	1.5	359.82	10MR	-1.4	-4.3	-1.9	0.7
18.48	-0.41	1.5	9.70	20MR	-0.7	-4.3	-1.8	0.7
25.96	-0.31	1.5	19.51	30MR	0.6	-4.2	-1.8	0.7
33.36	-0.20	1.5	29.27	9AP	2.2	-4.1	-1.7	0.6
40.68	-0.09	1.5	38.98	19AP	3.6	-4.0	-1.6	0.6
47.90	0.01	1.5	48.64	29AP	2.0	-3.9	-1.6	0.5
55.05	0.12	1.6	58.26	9MY	1.2	-3.8	-1.5	0.4
62.10	0.22	1.7	67.85	19MY	0.5	-3.7	-1.5	0.4
69.07	0.32	1.7	77.40	29MY	-0.3	-3.6	-1.4	0.3
75.96	0.42	1.7	86.95	8JN	-1.3	-3.6	-1.4	0.2
82.78	0.52	1.8	96.48	18JN	-1.7	-3.5	-1.3	0.3
89.51	0.61	1.8	106.01	28JN	-0.8	-3.4	-1.3	0.3
96.17	0.70	1.8	115.56	8JL	-0.1	-3.4	-1.3	0.4
102.77	0.79	1.9	125.12	18JL	0.3	-3.4	-1.3	0.4
109.29	0.88	1.9	134.70	28JL	0.7	-3.4	-1.3	0.5
115.75	0.96	1.9	144.32	7AU	1.1	-3.3	-1.3	0.6
122.13	1.04	1.9	153.98	17AU	2.3	-3.3	-1.3	0.6
128.46	1.12	1.9	163.69	27AU	2.4	-3.3	-1.3	0.7
134.72	1.20	1.9	173.45	6SE	0.1	-3.3	-1.3	0.7
140.91	1.28	1.9	183.26	16SE	-0.9	-3.4	-1.4	0.7
147.02	1.35	1.9	193.13	26SE	-1.1	-3.4	-1.4	0.8
153.06	1.43	1.8	203.06	6OC	-1.0	-3.4	-1.4	0.8
159.01	1.50	1.8	213.04	16OC	-0.6	-3.4	-1.5	0.8
164.87	1.57	1.7	223.07	26OC	-0.4	-3.4	-1.5	0.8
170.62	1.65	1.6	233.15	5NO	-0.3	-3.5	-1.6	0.8
176.24	1.72	1.5	243.27	15NO	-0.3	-3.5	-1.7	0.8
181.71	1.80	1.5	253.42	25NO	0.1	-3.5	-1.7	0.7
187.00	1.87	1.3	263.61	5DE	2.3	-3.5	-1.8	0.7
192.08	1.95	1.2	273.80	15DE	1.1	-3.4	-1.9	0.7
196.90	2.02	1.1	283.99	25DE	0.1	-3.4	-1.9	0.7

1636 / 1637

♂ LONG	♂ LAT	♂ MAG	☉ LONG	16.00UT 1636	☿	♀	♃	♄
201.40	2.10	0.9	294.18	4JA	-0.1	-3.4	-2.0	0.7
205.50	2.17	0.7	304.35	14JA	-0.1	-3.4	-2.0	0.8
209.12	2.24	0.5	314.49	24JA	-0.3	-3.4	-2.0	0.8
212.13	2.30	0.3	324.60	3FE	-0.7	-3.3	-2.0	0.8
214.38	2.34	0.0	334.65	13FE	-1.3	-3.3	-2.0	0.8
215.70	2.36	-0.3	344.67	23FE	-1.3	-3.3	-2.0	0.8
215.90	2.33	-0.6	354.63	4MR	-0.7	-3.3	-2.0	0.8
214.86	2.24	-0.9	4.53	14MR	0.8	-3.4	-2.0	0.8
212.57	2.06	-1.1	14.37	24MR	2.8	-3.4	-1.9	0.8
209.28	1.77	-1.4	24.16	3AP	2.7	-3.4	-1.8	0.7
205.57	1.39	-1.4	33.89	13AP	1.5	-3.4	-1.8	0.7
202.13	0.95	-1.3	43.58	23AP	0.9	-3.5	-1.7	0.6
199.61	0.50	-1.1	53.22	3MY	0.3	-3.5	-1.7	0.6
198.36	0.08	-0.9	62.82	13MY	-0.4	-3.6	-1.6	0.5
198.45	-0.29	-0.7	72.40	23MY	-1.3	-3.6	-1.5	0.4
199.78	-0.61	-0.5	81.95	2JN	-1.7	-3.7	-1.5	0.4
202.18	-0.87	-0.3	91.48	12JN	-0.8	-3.8	-1.4	0.3
205.47	-1.09	-0.1	101.02	22JN	-0.0	-3.9	-1.4	0.3
209.50	-1.27	0.0	110.55	2JL	0.5	-4.0	-1.3	0.3
214.14	-1.42	0.2	120.10	12JL	0.9	-4.1	-1.3	0.4
219.29	-1.53	0.3	129.67	22JL	1.5	-4.1	-1.3	0.4
224.87	-1.63	0.4	139.27	1AU	2.9	-4.2	-1.2	0.5
230.81	-1.70	0.5	148.91	11AU	2.0	-4.2	-1.2	0.5
237.06	-1.75	0.6	158.59	21AU	0.0	-4.1	-1.2	0.6
243.59	-1.78	0.6	168.32	31AU	-1.0	-3.7	-1.2	0.6
250.36	-1.78	0.7	178.10	10SE	-1.2	-3.2	-1.2	0.7
257.32	-1.78	0.8	187.94	20SE	-0.9	-3.7	-1.2	0.7
264.47	-1.75	0.8	197.84	30SE	-0.5	-4.1	-1.2	0.8
271.77	-1.71	0.9	207.79	10OC	-0.3	-4.3	-1.3	0.8
279.20	-1.65	0.9	217.80	20OC	-0.2	-4.3	-1.3	0.8
286.73	-1.58	1.0	227.86	30OC	-0.1	-4.3	-1.3	0.8
294.35	-1.50	1.0	237.96	9NO	0.3	-4.2	-1.4	0.8
302.03	-1.40	1.1	248.10	19NO	2.6	-4.1	-1.4	0.8
309.75	-1.30	1.1	258.27	29NO	0.9	-4.0	-1.5	0.8
317.50	-1.19	1.2	268.45	9DE	-0.1	-3.9	-1.6	0.8
325.25	-1.07	1.2	278.65	19DE	-0.2	-3.8	-1.6	0.7
333.00	-0.94	1.3	288.84	29DE	-0.3	-3.7	-1.7	0.7
				1637				
340.72	-0.82	1.3	299.02	8JA	-0.4	-3.7	-1.8	0.8
348.40	-0.69	1.4	309.18	18JA	-0.7	-3.6	-1.8	0.8
356.03	-0.56	1.4	319.31	28JA	-1.2	-3.5	-1.9	0.8
3.61	-0.43	1.5	329.39	7FE	-1.2	-3.5	-1.9	0.9
11.12	-0.31	1.5	339.43	17FE	-0.6	-3.4	-2.0	0.9
18.56	-0.18	1.6	349.42	27FE	1.0	-3.4	-2.0	0.9
25.92	-0.07	1.6	359.35	9MR	3.2	-3.4	-2.0	0.9
33.20	0.05	1.6	9.23	19MR	2.0	-3.3	-2.0	0.9
40.41	0.16	1.7	19.04	29MR	1.1	-3.3	-2.0	0.8
47.53	0.26	1.7	28.80	8AP	0.6	-3.3	-1.9	0.8
54.58	0.36	1.7	38.52	18AP	0.2	-3.3	-1.9	0.8
61.54	0.45	1.8	48.18	28AP	-0.4	-3.3	-1.8	0.7
68.44	0.54	1.8	57.80	8MY	-1.3	-3.3	-1.8	0.7
75.26	0.63	1.8	67.39	18MY	-1.7	-3.3	-1.7	0.6
82.02	0.70	1.8	76.95	28MY	-0.8	-3.4	-1.7	0.5
88.71	0.78	1.8	86.49	7JN	0.1	-3.4	-1.6	0.5
95.35	0.84	1.8	96.02	17JN	0.6	-3.5	-1.5	0.4
101.93	0.91	1.9	105.55	27JN	1.2	-3.5	-1.5	0.3
108.47	0.96	1.9	115.09	7JL	2.0	-3.5	-1.4	0.3
114.96	1.02	1.9	124.65	17JL	3.2	-3.4	-1.4	0.4
121.40	1.07	2.0	134.23	27JL	1.7	-3.4	-1.3	0.4
127.81	1.11	2.0	143.85	6AU	-0.0	-3.4	-1.3	0.5
134.18	1.15	2.0	153.51	16AU	-1.1	-3.4	-1.3	0.5
140.53	1.19	2.0	163.21	26AU	-1.3	-3.3	-1.2	0.6
146.84	1.22	2.0	172.96	5SE	-0.9	-3.3	-1.2	0.6
153.12	1.25	2.0	182.78	15SE	-0.4	-3.3	-1.2	0.7
159.37	1.28	2.0	192.64	25SE	-0.2	-3.3	-1.2	0.7
165.60	1.30	2.0	202.57	5OC	-0.1	-3.3	-1.2	0.8
171.80	1.32	1.9	212.55	15OC	0.1	-3.4	-1.2	0.8
177.96	1.33	1.9	222.58	25OC	0.6	-3.4	-1.2	0.8
184.09	1.34	1.9	232.66	4NO	3.0	-3.4	-1.3	0.8
190.17	1.35	1.8	242.78	14NO	0.7	-3.4	-1.3	0.8
196.20	1.35	1.7	252.93	24NO	-0.3	-3.5	-1.3	0.8
202.18	1.34	1.6	263.11	4DE	-0.4	-3.5	-1.4	0.8
208.08	1.33	1.5	273.31	14DE	-0.4	-3.6	-1.4	0.8
213.90	1.31	1.4	283.50	24DE	-0.5	-3.6	-1.5	0.8

1638 / 1639

♂ LONG	♂ LAT	♂ MAG	☉ LONG	16.00UT 1638	☿	♀	♃	♄
219.62	1.28	1.3	293.69	3JA	-0.8	-3.7	-1.5	0.8
225.22	1.23	1.2	303.86	13JA	-1.0	-3.8	-1.6	0.8
230.67	1.17	1.0	314.00	23JA	-1.0	-3.9	-1.7	0.8
235.94	1.10	0.9	324.11	2FE	-0.5	-4.0	-1.7	0.9
240.98	1.00	0.7	334.18	12FE	1.2	-4.1	-1.8	0.9
245.74	0.88	0.5	344.19	22FE	3.1	-4.2	-1.9	0.9
250.16	0.72	0.3	354.15	4MR	1.5	-4.2	-1.9	0.9
254.14	0.52	0.0	4.06	14MR	0.8	-4.3	-2.0	1.0
257.57	0.28	-0.2	13.90	24MR	0.5	-4.2	-2.0	0.9
260.31	-0.04	-0.5	23.69	3AP	0.1	-4.0	-2.0	0.9
262.17	-0.42	-0.8	33.43	13AP	-0.4	-3.5	-2.0	0.9
263.00	-0.89	-1.2	43.11	23AP	-1.3	-3.0	-2.0	0.9
262.62	-1.45	-1.5	52.76	3MY	-1.7	-3.7	-2.0	0.8
260.99	-2.07	-1.8	62.36	13MY	-0.8	-4.1	-2.0	0.8
258.34	-2.70	-2.1	71.93	23MY	0.2	-4.2	-1.9	0.7
255.15	-3.27	-2.1	81.49	2JN	0.9	-4.2	-1.9	0.7
252.14	-3.70	-2.0	91.02	12JN	1.6	-4.1	-1.8	0.6
250.01	-3.97	-1.8	100.55	22JN	2.7	-4.0	-1.7	0.5
249.14	-4.09	-1.6	110.09	2JL	3.0	-3.9	-1.7	0.5
249.68	-4.08	-1.4	119.64	12JL	1.3	-3.8	-1.6	0.4
251.51	-3.99	-1.2	129.20	22JL	-0.1	-3.7	-1.5	0.4
254.47	-3.85	-1.0	138.80	1AU	-1.2	-3.7	-1.5	0.4
258.36	-3.67	-0.8	148.44	11AU	-1.5	-3.6	-1.4	0.5
263.01	-3.47	-0.6	158.12	21AU	-0.9	-3.6	-1.4	0.5
268.27	-3.25	-0.4	167.85	31AU	-0.4	-3.5	-1.3	0.6
274.03	-3.02	-0.3	177.63	10SE	-0.1	-3.5	-1.3	0.6
280.17	-2.78	-0.1	187.46	20SE	0.1	-3.4	-1.3	0.7
286.63	-2.54	0.0	197.36	30SE	0.2	-3.4	-1.3	0.7
293.34	-2.29	0.2	207.31	10OC	0.8	-3.4	-1.2	0.8
300.24	-2.05	0.3	217.31	20OC	3.3	-3.4	-1.2	0.8
307.29	-1.81	0.4	227.37	30OC	0.5	-3.4	-1.2	0.9
314.45	-1.58	0.5	237.47	9NO	-0.4	-3.4	-1.2	0.9
321.68	-1.35	0.6	247.61	19NO	-0.5	-3.4	-1.3	0.9
328.97	-1.13	0.8	257.78	29NO	-0.5	-3.4	-1.3	0.9
336.27	-0.92	0.9	267.96	9DE	-0.6	-3.4	-1.3	0.9
343.59	-0.72	1.0	278.15	19DE	-0.8	-3.4	-1.4	0.9
350.89	-0.53	1.1	288.35	29DE	-0.9	-3.4	-1.4	0.9
				1639				
358.17	-0.36	1.2	298.53	8JA	-0.9	-3.4	-1.4	0.9
5.40	-0.19	1.3	308.69	18JA	-0.4	-3.4	-1.5	0.9
12.60	-0.04	1.3	318.82	28JA	1.4	-3.5	-1.6	0.9
19.75	0.10	1.4	328.90	7FE	2.6	-3.5	-1.6	1.0
26.83	0.23	1.5	338.94	17FE	1.1	-3.5	-1.7	1.0
33.86	0.35	1.6	348.94	27FE	0.6	-3.4	-1.8	1.0
40.83	0.46	1.6	358.87	9MR	0.3	-3.4	-1.8	1.0
47.74	0.55	1.7	8.74	19MR	0.0	-3.4	-1.9	1.1
54.59	0.64	1.7	18.57	29MR	-0.5	-3.4	-2.0	1.1
61.37	0.72	1.8	28.33	8AP	-1.3	-3.3	-2.0	1.0
68.10	0.79	1.8	38.04	18AP	-1.6	-3.3	-2.1	1.0
74.78	0.86	1.9	47.71	28AP	-0.8	-3.3	-2.1	1.0
81.41	0.91	1.9	57.33	8MY	0.3	-3.3	-2.1	1.0
87.98	0.96	1.9	66.92	18MY	1.1	-3.3	-2.1	0.9
94.52	1.01	1.9	76.48	28MY	2.1	-3.4	-2.1	0.9
101.04	1.04	2.0	86.03	7JN	3.4	-3.4	-2.0	0.8
107.49	1.07	2.0	95.56	17JN	2.4	-3.4	-2.0	0.7
113.93	1.10	2.0	105.09	27JN	1.1	-3.4	-2.0	0.7
120.34	1.12	2.0	114.63	7JL	-0.1	-3.5	-1.9	0.6
126.73	1.14	1.9	124.19	17JL	-1.2	-3.5	-1.9	0.5
133.12	1.15	1.9	133.77	27JL	-1.6	-3.6	-1.8	0.5
139.49	1.15	2.0	143.38	6AU	-0.8	-3.7	-1.7	0.5
145.85	1.15	2.0	153.04	16AU	-0.3	-3.7	-1.7	0.5
152.22	1.15	2.0	162.74	26AU	0.0	-3.8	-1.6	0.6
158.58	1.14	2.0	172.49	5SE	0.2	-3.9	-1.6	0.6
164.95	1.13	2.0	182.30	15SE	0.5	-4.0	-1.5	0.7
171.32	1.12	2.0	192.17	25SE	1.2	-4.1	-1.5	0.7
177.70	1.09	2.0	202.08	5OC	3.4	-4.3	-1.4	0.8
184.10	1.07	2.0	212.06	15OC	-0.4	-4.3	-1.4	0.8
190.50	1.03	2.0	222.09	25OC	-0.6	-4.3	-1.4	0.9
196.92	1.00	1.9	232.17	4NO	-0.7	-4.2	-1.3	0.9
203.34	0.95	1.9	242.29	14NO	-0.7	-3.7	-1.3	1.0
209.78	0.90	1.8	252.44	24NO	-0.7	-2.8	-1.3	1.0
216.22	0.84	1.8	262.61	4DE	-0.7	-3.7	-1.3	1.0
222.67	0.77	1.7	272.81	14DE	-0.7	-4.2	-1.3	1.0
229.13	0.69	1.6	283.01	24DE	-0.7	-4.4	-1.4	1.0

1640

♂ LONG LAT MAG	☉ LONG	16.00UT	☿ ♀ ♃ ♄ MAGNITUDES
235.58 0.60 1.6	293.19	3JA	-0.3 -4.4 -1.4 1.0
242.04 0.50 1.5	303.37	13JA	1.6 -4.3 -1.4 1.0
248.48 0.38 1.4	313.51	23JA	2.1 -4.2 -1.5 1.0
254.92 0.25 1.3	323.62	2FE	0.7 -4.1 -1.5 1.0
261.33 0.10 1.1	333.69	12FE	0.4 -4.0 -1.6 1.0
267.71-0.07 1.0	343.71	22FE	0.2 -3.9 -1.6 1.1
274.05-0.26 0.9	353.67	3MR	-0.1 -3.8 -1.7 1.1
280.34-0.48 0.7	3.58	13MR	-0.5 -3.7 -1.7 1.2
286.56-0.73 0.5	13.43	23MR	-1.3 -3.6 -1.8 1.2
292.68-1.00 0.4	23.22	2AP	-1.6 -3.5 -1.9 1.2
298.67-1.31 0.2	32.96	12AP	-0.7 -3.5 -2.0 1.2
304.49-1.66 -0.0	42.66	22AP	0.4 -3.4 -2.0 1.2
310.10-2.05 -0.2	52.30	2MY	1.5 -3.4 -2.1 1.2
315.42-2.49 -0.4	61.91	12MY	2.8 -3.4 -2.1 1.1
320.36-2.97 -0.7	71.48	22MY	3.2 -3.3 -2.2 1.1
324.81-3.50 -0.9	81.03	1JN	1.8 -3.3 -2.2 1.0
328.62-4.08 -1.2	90.57	11JN	0.8 -3.3 -2.2 1.0
331.61-4.69 -1.5	100.10	21JN	-0.2 -3.3 -2.2 0.9
333.59-5.32 -1.8	109.63	1JL	-1.2 -3.3 -2.2 0.9
334.35-5.93 -2.0	119.18	11JL	-1.7 -3.3 -2.2 0.8
333.80-6.43 -2.3	128.75	21JL	-0.8 -3.4 -2.1 0.7
332.07-6.72 -2.5	138.34	31JL	-0.2 -3.4 -2.1 0.7
329.56-6.73 -2.7	147.98	10AU	0.1 -3.4 -2.0 0.6
326.99-6.41 -2.6	157.65	20AU	0.4 -3.4 -2.0 0.6
325.08-5.82 -2.3	167.37	30AU	0.7 -3.5 -1.9 0.7
324.31-5.09 -2.0	177.15	9SE	1.5 -3.5 -1.8 0.7
324.87-4.30 -1.7	186.99	19SE	3.0 -3.5 -1.8 0.8
326.66-3.55 -1.4	196.87	29SE	0.3 -3.4 -1.7 0.8
329.51-2.87 -1.2	206.82	9OC	-0.7 -3.4 -1.7 0.9
333.24-2.26 -0.9	216.82	19OC	-0.8 -3.4 -1.6 0.9
337.63-1.73 -0.6	226.87	29OC	-0.8 -3.4 -1.6 1.0
342.55-1.27 -0.4	236.97	8NO	-0.8 -3.3 -1.5 1.0
347.87-0.87 -0.1	247.11	18NO	-0.6 -3.3 -1.5 1.0
353.49-0.53 0.1	257.27	28NO	-0.6 -3.3 -1.5 1.1
359.35-0.24 0.3	267.46	8DE	-0.6 -3.3 -1.5 1.1
5.38 0.01 0.5	277.65	18DE	-0.2 -3.3 -1.4 1.1
11.54 0.22 0.6	287.85	28DE	1.8 -3.4 -1.4 1.1

1641

♂ LONG LAT MAG	☉ LONG	16.00UT	☿ ♀ ♃ ♄ MAGNITUDES
17.79 0.40 0.8	298.03	7JA	1.7 -3.4 -1.4 1.2
24.11 0.56 1.0	308.19	17JA	0.5 -3.4 -1.5 1.2
30.47 0.69 1.1	318.32	27JA	0.2 -3.4 -1.5 1.2
36.86 0.80 1.2	328.42	6FE	0.0 -3.5 -1.5 1.1
43.26 0.89 1.3	338.46	16FE	-0.2 -3.5 -1.5 1.2
49.66 0.97 1.4	348.45	26FE	-0.6 -3.6 -1.6 1.2
56.06 1.04 1.5	358.39	8MR	-1.3 -3.6 -1.6 1.3
62.46 1.09 1.6	8.27	18MR	-1.5 -3.7 -1.7 1.3
68.84 1.13 1.7	18.10	28MR	-0.7 -3.8 -1.7 1.3
75.20 1.17 1.8	27.87	7AP	0.5 -3.8 -1.8 1.3
81.56 1.19 1.8	37.58	17AP	1.9 -3.9 -1.9 1.4
87.89 1.21 1.9	47.25	27AP	3.6 -4.0 -1.9 1.4
94.22 1.22 1.9	56.88	7MY	2.5 -4.1 -2.0 1.3
100.53 1.23 2.0	66.46	17MY	1.4 -4.2 -2.1 1.3
106.84 1.23 2.0	76.03	27MY	0.6 -4.2 -2.2 1.3
113.14 1.22 2.0	85.57	6JN	-0.2 -4.1 -2.2 1.2
119.44 1.21 2.0	95.10	16JN	-1.2 -3.7 -2.3 1.2
125.74 1.20 2.0	104.64	26JN	-1.7 -3.0 -2.3 1.1
132.04 1.18 2.0	114.18	6JL	-0.8 -3.3 -2.3 1.1
138.36 1.15 2.0	123.73	16JL	-0.2 -3.9 -2.4 1.0
144.69 1.13 2.0	133.31	26JL	0.3 -4.2 -2.4 0.9
151.04 1.09 2.0	142.92	5AU	0.6 -4.2 -2.3 0.9
157.40 1.06 2.0	152.57	15AU	0.9 -4.2 -2.3 0.8
163.80 1.02 1.9	162.27	25AU	2.0 -4.1 -2.3 0.8
170.23 0.97 1.9	172.02	4SE	2.7 -4.0 -2.2 0.8
176.69 0.93 1.9	181.82	14SE	0.2 -3.9 -2.2 0.8
183.18 0.87 2.0	191.68	24SE	-0.8 -3.9 -2.1 0.9
189.72 0.81 2.0	201.60	4OC	-1.0 -3.8 -2.0 0.9
196.29 0.75 1.9	211.57	14OC	-0.9 -3.7 -2.0 1.0
202.92 0.68 1.9	221.60	24OC	-0.7 -3.6 -1.9 1.0
209.58 0.61 1.9	231.68	3NO	-0.5 -3.6 -1.8 1.1
216.30 0.53 1.9	241.79	13NO	-0.4 -3.5 -1.8 1.1
223.06 0.45 1.9	251.94	23NO	-0.4 -3.5 -1.7 1.2
229.87 0.35 1.8	262.12	3DE	-0.0 -3.5 -1.7 1.2
236.73 0.25 1.8	272.31	13DE	2.1 -3.4 -1.6 1.2
243.64 0.15 1.7	282.51	23DE	1.3 -3.4 -1.6 1.3

1642

♂ LONG LAT MAG	☉ LONG	16.00UT	☿ ♀ ♃ ♄ MAGNITUDES
250.60 0.03 1.7	292.70	2JA	0.2 -3.4 -1.6 1.3
257.60-0.09 1.6	302.87	12JA	0.0 -3.4 -1.6 1.3
264.65-0.22 1.5	313.02	22JA	-0.1 -3.3 -1.5 1.3
271.75-0.36 1.5	323.13	1FE	-0.3 -3.3 -1.5 1.3
278.89-0.50 1.4	333.20	11FE	-0.6 -3.3 -1.5 1.3
286.07-0.66 1.3	343.22	21FE	-1.3 -3.3 -1.6 1.3
293.28-0.82 1.2	353.19	3MR	-1.4 -3.3 -1.6 1.4
300.51-0.99 1.1	3.10	13MR	-0.7 -3.4 -1.6 1.4
307.77-1.16 1.0	12.95	23MR	0.7 -3.4 -1.6 1.4
315.04-1.34 0.9	22.75	2AP	2.4 -3.4 -1.7 1.4
322.30-1.52 0.8	32.49	12AP	3.2 -3.5 -1.7 1.4
329.55-1.69 0.7	42.19	22AP	1.8 -3.5 -1.8 1.4
336.76-1.87 0.6	51.83	2MY	1.1 -3.5 -1.8 1.4
343.92-2.04 0.5	61.44	12MY	0.4 -3.4 -1.9 1.3
350.99-2.21 0.4	71.02	22MY	-0.3 -3.4 -2.0 1.3
357.94-2.36 0.3	80.57	1JN	-1.3 -3.4 -2.0 1.3
4.74-2.50 0.2	90.10	11JN	-1.7 -3.4 -2.1 1.2
11.35-2.63 0.0	99.64	21JN	-0.8 -3.3 -2.2 1.2
17.70-2.73 -0.1	109.17	1JL	-0.1 -3.3 -2.2 1.1
23.73-2.82 -0.2	118.72	11JL	0.4 -3.3 -2.3 1.1
29.37-2.88 -0.4	128.28	21JL	0.8 -3.3 -2.4 1.0
34.50-2.91 -0.6	137.88	31JL	1.3 -3.3 -2.4 1.0
39.02-2.91 -0.7	147.51	10AU	2.5 -3.4 -2.4 0.9
42.74-2.86 -0.9	157.18	20AU	2.3 -3.4 -2.5 0.9
45.48-2.76 -1.1	166.90	30AU	0.1 -3.4 -2.5 0.9
47.03-2.60 -1.4	176.68	9SE	-0.9 -3.5 -2.4 0.9
47.18-2.33 -1.6	186.51	19SE	-1.1 -3.5 -2.4 0.9
45.87-1.97 -1.8	196.40	29SE	-1.0 -3.6 -2.4 1.0
43.24-1.49 -2.0	206.34	9OC	-0.6 -3.6 -2.3 1.1
39.81-0.94 -2.1	216.34	19OC	-0.4 -3.7 -2.3 1.1
36.40-0.36 -1.9	226.39	29OC	-0.3 -3.8 -2.2 1.2
33.74 0.17 -1.6	236.48	8NO	-0.2 -3.9 -2.1 1.2
32.31 0.61 -1.3	246.62	18NO	0.2 -4.0 -2.0 1.3
32.24 0.95 -1.0	256.78	28NO	2.3 -4.1 -2.0 1.3
33.41 1.21 -0.6	266.96	8DE	1.1 -4.2 -1.9 1.4
35.64 1.39 -0.3	277.16	18DE	0.0 -4.3 -1.8 1.4
38.71 1.52 -0.1	287.35	28DE	-0.1 -4.4 -1.8 1.4

1643

♂ LONG LAT MAG	☉ LONG	16.00UT	☿ ♀ ♃ ♄ MAGNITUDES
42.43 1.61 0.2	297.53	7JA	-0.2 -4.4 -1.7 1.4
46.67 1.66 0.4	307.70	17JA	-0.4 -4.2 -1.7 1.4
51.30 1.69 0.6	317.83	27JA	-0.7 -3.8 -1.7 1.3
56.24 1.71 0.8	327.92	6FE	-1.2 -3.3 -1.6 1.3
61.42 1.71 1.0	337.97	16FE	-1.3 -3.8 -1.6 1.2
66.79 1.70 1.1	347.96	26FE	-0.7 -4.2 -1.6 1.2
72.30 1.68 1.3	357.90	8MR	0.8 -4.3 -1.6 1.1
77.93 1.65 1.4	7.79	18MR	2.9 -4.3 -1.6 1.1
83.66 1.62 1.5	17.61	28MR	2.4 -4.2 -1.6 1.1
89.46 1.59 1.6	27.39	7AP	1.3 -4.1 -1.6 1.1
95.32 1.55 1.7	37.11	17AP	0.8 -4.0 -1.6 1.1
101.24 1.51 1.8	46.78	27AP	0.3 -3.9 -1.7 1.1
107.20 1.46 1.8	56.41	7MY	-0.3 -3.8 -1.7 1.1
113.21 1.41 1.9	66.00	17MY	-1.3 -3.7 -1.7 1.1
119.25 1.36 1.9	75.56	27MY	-1.7 -3.6 -1.8 1.1
125.33 1.31 2.0	85.11	6JN	-0.8 -3.5 -1.8 1.0
131.46 1.25 2.0	94.65	16JN	-0.0 -3.5 -1.9 1.0
137.62 1.20 2.0	104.18	26JN	0.5 -3.4 -2.0 0.9
143.82 1.13 2.0	113.72	6JL	1.0 -3.4 -2.0 0.9
150.06 1.07 2.0	123.27	16JL	1.7 -3.4 -2.1 0.8
156.35 1.01 2.0	132.85	26JL	3.0 -3.4 -2.2 0.8
162.69 0.94 2.0	142.46	5AU	1.9 -3.3 -2.2 0.7
169.08 0.87 2.0	152.11	15AU	0.1 -3.3 -2.3 0.7
175.53 0.79 2.0	161.80	25AU	-1.0 -3.3 -2.4 0.6
182.03 0.72 1.9	171.55	4SE	-1.3 -3.3 -2.4 0.6
188.59 0.64 1.9	181.35	14SE	-0.9 -3.4 -2.4 0.6
195.22 0.56 1.9	191.21	24SE	-0.5 -3.4 -2.5 0.6
201.91 0.47 1.8	201.13	4OC	-0.2 -3.4 -2.5 0.7
208.68 0.38 1.8	211.10	14OC	-0.1 -3.4 -2.5 0.7
215.51 0.29 1.8	221.12	24OC	-0.0 -3.4 -2.4 0.8
222.41 0.20 1.8	231.19	3NO	0.4 -3.5 -2.4 0.9
229.39 0.10 1.8	241.30	13NO	2.6 -3.5 -2.3 0.9
236.44-0.00 1.7	251.45	23NO	0.8 -3.5 -2.3 1.0
243.56-0.11 1.7	261.63	3DE	-0.1 -3.5 -2.2 1.0
250.75-0.21 1.7	271.81	13DE	-0.3 -3.4 -2.1 1.1
258.01-0.32 1.7	282.01	23DE	-0.3 -3.4 -2.0 1.1

♂			☉	16.00UT	☿	♀	♃	♄
LONG	LAT	MAG	LONG	1644		MAGN	ITUDES	
265.35	-0.43	1.6	292.20	2JA	-0.5	-3.4	-2.0	1.1
272.74	-0.55	1.6	302.37	12JA	-0.8	-3.4	-1.9	1.1
280.19	-0.66	1.6	312.52	22JA	-1.1	-3.4	-1.8	1.1
287.71	-0.77	1.5	322.63	1FE	-1.1	-3.3	-1.8	1.1
295.26	-0.88	1.5	332.70	11FE	-0.6	-3.3	-1.7	1.0
302.86	-0.99	1.4	342.72	21FE	1.0	-3.3	-1.7	1.0
310.50	-1.09	1.4	352.70	2MR	3.3	-3.3	-1.6	1.0
318.16	-1.19	1.3	2.61	12MR	1.8	-3.3	-1.6	0.9
325.83	-1.27	1.3	12.47	22MR	1.0	-3.4	-1.6	0.9
333.50	-1.35	1.2	22.27	1AP	0.6	-3.4	-1.6	0.9
341.16	-1.42	1.2	32.02	11AP	0.2	-3.4	-1.6	0.9
348.79	-1.48	1.1	41.71	21AP	-0.4	-3.5	-1.6	0.9
356.39	-1.52	1.1	51.36	1MY	-1.3	-3.5	-1.6	0.9
3.92	-1.54	1.0	60.97	11MY	-1.7	-3.6	-1.6	0.9
11.39	-1.55	1.0	70.55	21MY	-0.8	-3.6	-1.6	0.9
18.76	-1.54	0.9	80.11	31MY	0.1	-3.7	-1.6	0.8
26.02	-1.51	0.9	89.65	10JN	0.7	-3.8	-1.6	0.8
33.16	-1.47	0.8	99.18	20JN	1.3	-3.9	-1.7	0.8
40.15	-1.40	0.8	108.72	30JN	2.2	-4.0	-1.7	0.7
46.97	-1.32	0.7	118.26	10JL	3.2	-4.1	-1.8	0.7
53.59	-1.22	0.7	127.82	20JL	1.6	-4.1	-1.8	0.7
59.98	-1.09	0.6	137.42	30JL	-0.0	-4.2	-1.9	0.6
66.10	-0.95	0.5	147.05	9AU	-1.1	-4.2	-2.0	0.5
71.91	-0.78	0.4	156.72	19AU	-1.4	-4.1	-2.0	0.5
77.34	-0.58	0.3	166.44	29AU	-0.9	-3.7	-2.1	0.4
82.33	-0.36	0.2	176.21	8SE	-0.4	-3.2	-2.2	0.4
86.77	-0.10	0.0	186.03	18SE	-0.1	-3.7	-2.2	0.3
90.54	0.19	-0.1	195.92	28SE	0.0	-4.1	-2.3	0.3
93.50	0.53	-0.3	205.86	8OC	0.1	-4.3	-2.3	0.3
95.44	0.92	-0.5	215.85	18OC	0.7	-4.3	-2.4	0.4
96.16	1.36	-0.7	225.90	28OC	3.0	-4.3	-2.4	0.5
95.51	1.84	-1.0	235.99	7NO	0.7	-4.2	-2.4	0.5
93.43	2.34	-1.2	246.13	17NO	-0.3	-4.1	-2.4	0.6
90.15	2.79	-1.3	256.29	27NO	-0.4	-4.0	-2.4	0.7
86.27	3.13	-1.4	266.47	7DE	-0.4	-3.9	-2.3	0.7
82.54	3.34	-1.2	276.67	17DE	-0.6	-3.8	-2.3	0.8
79.68	3.40	-0.9	286.86	27DE	-0.8	-3.7	-2.2	0.8
				1645				
78.09	3.34	-0.6	297.04	6JA	-1.0	-3.7	-2.1	0.8
77.83	3.22	-0.3	307.21	16JA	-1.0	-3.6	-2.0	0.8
78.78	3.06	-0.1	317.34	26JA	-0.5	-3.5	-2.0	0.8
80.75	2.89	0.2	327.43	5FE	1.2	-3.5	-1.9	0.8
83.54	2.72	0.4	337.48	15FE	2.9	-3.4	-1.8	0.8
86.98	2.56	0.7	347.48	25FE	1.3	-3.4	-1.8	0.8
90.93	2.40	0.8	357.42	7MR	0.7	-3.4	-1.7	0.8
95.29	2.25	1.0	7.31	17MR	0.4	-3.3	-1.7	0.7
99.96	2.11	1.2	17.14	27MR	0.1	-3.3	-1.6	0.7
104.90	1.98	1.3	26.91	6AP	-0.4	-3.3	-1.6	0.6
110.04	1.85	1.4	36.64	16AP	-1.2	-3.3	-1.5	0.6
115.37	1.73	1.5	46.31	26AP	-1.7	-3.3	-1.5	0.7
120.83	1.62	1.6	55.94	6MY	-0.8	-3.3	-1.5	0.7
126.43	1.50	1.7	65.54	16MY	0.2	-3.3	-1.5	0.7
132.15	1.40	1.7	75.10	26MY	0.9	-3.4	-1.5	0.7
137.96	1.29	1.8	84.65	5JN	1.7	-3.4	-1.5	0.7
143.87	1.18	1.8	94.18	15JN	2.9	-3.5	-1.5	0.6
149.87	1.08	1.9	103.71	25JN	2.8	-3.5	-1.5	0.6
155.95	0.98	1.9	113.25	5JL	1.3	-3.5	-1.5	0.6
162.12	0.88	1.9	122.81	15JL	-0.1	-3.4	-1.5	0.5
168.37	0.78	1.9	132.38	25JL	-1.2	-3.4	-1.6	0.5
174.70	0.68	1.9	141.99	4AU	-1.5	-3.4	-1.6	0.5
181.12	0.57	1.9	151.64	14AU	-0.9	-3.4	-1.7	0.4
187.63	0.47	1.9	161.33	24AU	-0.3	-3.3	-1.7	0.3
194.22	0.37	1.9	171.07	3SE	-0.0	-3.3	-1.8	0.3
200.90	0.27	1.8	180.87	13SE	0.1	-3.3	-1.8	0.2
207.67	0.17	1.8	190.73	23SE	0.3	-3.3	-1.9	0.2
214.52	0.07	1.8	200.64	3OC	0.9	-3.3	-2.0	0.1
221.48	-0.03	1.7	210.61	13OC	3.3	-3.4	-2.0	0.1
228.52	-0.13	1.7	220.63	23OC	0.5	-3.4	-2.1	0.1
235.65	-0.23	1.6	230.70	2NO	-0.5	-3.4	-2.2	0.2
242.87	-0.33	1.6	240.81	12NO	-0.6	-3.4	-2.2	0.3
250.18	-0.43	1.5	250.96	22NO	-0.6	-3.5	-2.2	0.3
257.57	-0.53	1.5	261.13	2DE	-0.7	-3.5	-2.3	0.4
265.04	-0.62	1.5	271.32	12DE	-0.8	-3.6	-2.3	0.5
272.59	-0.71	1.5	281.52	22DE	-0.8	-3.7	-2.3	0.5

♂			☉	16.00UT	☿	♀	♃	♄
LONG	LAT	MAG	LONG	1646		MAGN	ITUDES	
280.20	-0.80	1.5	291.71	1JA	-0.8	-3.7	-2.2	0.6
287.88	-0.87	1.5	301.88	11JA	-0.4	-3.8	-2.2	0.6
295.60	-0.95	1.5	312.03	21JA	1.4	-3.9	-2.1	0.6
303.37	-1.01	1.5	322.15	31JA	2.4	-4.0	-2.1	0.6
311.18	-1.07	1.4	332.22	10FE	0.9	-4.1	-2.0	0.6
319.00	-1.11	1.4	342.25	20FE	0.5	-4.2	-1.9	0.6
326.83	-1.15	1.4	352.22	2MR	-0.2	-4.2	-1.9	0.6
334.65	-1.17	1.4	2.14	12MR	-0.0	-4.3	-1.8	0.6
342.46	-1.18	1.4	11.99	22MR	-0.5	-4.2	-1.7	0.6
350.25	-1.18	1.4	21.79	1AP	-1.2	-4.0	-1.7	0.5
357.99	-1.16	1.4	31.55	11AP	-1.6	-3.4	-1.6	0.5
5.68	-1.13	1.4	41.24	21AP	-0.8	-3.0	-1.5	0.4
13.31	-1.09	1.4	50.89	1MY	0.3	-3.7	-1.5	0.5
20.86	-1.03	1.3	60.51	11MY	1.2	-4.1	-1.5	0.5
28.32	-0.96	1.3	70.08	21MY	2.3	-4.2	-1.4	0.5
35.70	-0.88	1.3	79.64	31MY	3.5	-4.2	-1.4	0.5
42.96	-0.79	1.3	89.18	10JN	2.2	-4.1	-1.4	0.5
50.12	-0.69	1.3	98.71	20JN	1.0	-4.0	-1.4	0.5
57.16	-0.57	1.3	108.25	30JN	-0.1	-3.9	-1.4	0.5
64.06	-0.45	1.3	117.79	10JL	-1.2	-3.8	-1.4	0.5
70.82	-0.31	1.2	127.36	20JL	-1.6	-3.7	-1.4	0.4
77.44	-0.17	1.2	136.95	30JL	-0.9	-3.7	-1.4	0.4
83.88	-0.01	1.2	146.58	9AU	-0.3	-3.6	-1.4	0.4
90.14	0.16	1.1	156.25	19AU	0.1	-3.6	-1.4	0.3
96.21	0.34	1.1	165.96	29AU	0.3	-3.5	-1.5	0.3
102.03	0.53	1.0	175.74	8SE	0.5	-3.5	-1.5	0.2
107.59	0.73	0.9	185.56	18SE	1.3	-3.4	-1.6	0.1
112.84	0.96	0.8	195.44	28SE	3.3	-3.4	-1.6	0.1
117.72	1.21	0.7	205.38	8OC	0.4	-3.4	-1.7	0.0
122.16	1.48	0.6	215.37	18OC	-0.6	-3.4	-1.7	-0.1
126.05	1.78	0.5	225.42	28OC	-0.7	-3.4	-1.8	-0.1
129.27	2.11	0.3	235.51	7NO	-0.7	-3.4	-1.9	-0.0
131.67	2.48	0.1	245.64	17NO	-0.8	-3.4	-1.9	0.0
133.07	2.88	-0.2	255.80	27NO	-0.7	-3.4	-2.0	0.1
133.26	3.31	-0.4	265.98	7DE	-0.7	-3.4	-2.0	0.2
132.13	3.74	-0.6	276.17	17DE	-0.7	-3.4	-2.1	0.2
129.66	4.12	-0.8	286.37	27DE	-0.3	-3.4	-2.1	0.3
				1647				
126.14	4.38	-1.0	296.55	6JA	1.6	-3.4	-2.1	0.4
122.17	4.48	-1.0	306.71	16JA	2.0	-3.4	-2.1	0.4
118.48	4.41	-0.8	316.85	26JA	0.6	-3.5	-2.1	0.4
115.73	4.19	-0.6	326.95	5FE	0.3	-3.5	-2.1	0.5
114.22	3.88	-0.3	336.99	15FE	0.1	-3.5	-2.0	0.5
114.01	3.55	-0.1	347.00	25FE	-0.1	-3.4	-2.0	0.5
114.97	3.21	0.2	356.94	7MR	-0.5	-3.4	-1.9	0.5
116.92	2.89	0.4	6.83	17MR	-1.2	-3.4	-1.9	0.5
119.67	2.60	0.6	16.66	27MR	-1.6	-3.4	-1.8	0.4
123.07	2.33	0.8	26.44	6AP	-0.8	-3.3	-1.7	0.4
126.98	2.09	0.9	36.16	16AP	0.4	-3.3	-1.7	0.4
131.32	1.87	1.1	45.84	26AP	1.6	-3.3	-1.6	0.3
136.00	1.66	1.2	55.47	6MY	3.1	-3.3	-1.5	0.3
140.97	1.47	1.3	65.06	16MY	2.9	-3.4	-1.5	0.3
146.17	1.30	1.4	74.63	26MY	1.7	-3.4	-1.4	0.4
151.59	1.13	1.5	84.18	5JN	0.8	-3.4	-1.4	0.4
157.19	0.97	1.5	93.71	15JN	-0.2	-3.4	-1.3	0.4
162.95	0.82	1.6	103.25	25JN	-1.2	-3.4	-1.3	0.4
168.87	0.68	1.6	112.79	5JL	-1.7	-3.5	-1.3	0.4
174.92	0.54	1.7	122.34	15JL	-0.8	-3.5	-1.3	0.4
181.11	0.41	1.7	131.92	25JL	-0.2	-3.6	-1.3	0.4
187.43	0.29	1.7	141.52	4AU	0.2	-3.7	-1.3	0.3
193.87	0.16	1.7	151.17	14AU	0.4	-3.7	-1.3	0.3
200.43	0.04	1.7	160.86	24AU	0.8	-3.8	-1.3	0.3
207.11	-0.07	1.7	170.60	3SE	1.7	-3.9	-1.3	0.2
213.91	-0.18	1.7	180.40	13SE	3.0	-4.0	-1.3	0.2
220.82	-0.29	1.7	190.25	23SE	0.3	-4.1	-1.3	0.1
227.84	-0.39	1.7	200.16	3OC	-0.7	-4.3	-1.4	0.0
234.98	-0.49	1.6	210.12	13OC	-0.9	-4.3	-1.4	-0.0
242.22	-0.58	1.6	220.15	23OC	-0.9	-4.3	-1.5	-0.1
249.56	-0.67	1.6	230.21	2NO	-0.7	-4.2	-1.5	-0.2
257.00	-0.75	1.5	240.32	12NO	-0.5	-3.7	-1.6	-0.2
264.53	-0.82	1.5	250.47	22NO	-0.5	-2.8	-1.7	-0.2
272.14	-0.89	1.5	260.64	2DE	-0.5	-3.8	-1.7	-0.1
279.83	-0.95	1.4	270.83	12DE	-0.1	-4.2	-1.8	-0.0
287.59	-1.00	1.4	281.02	22DE	1.8	-4.4	-1.9	0.1

♂ LONG	♂ LAT	♂ MAG	⊙ LONG	16.00UT	☿	♀	♃	♄
				1648	MAGNITUDES			
295.39	-1.03	1.4	291.21	1JA	1.6	-4.4	-1.9	0.1
303.24	-1.06	1.3	301.39	11JA	0.4	-4.3	-2.0	0.2
311.12	-1.08	1.3	311.54	21JA	0.1	-4.2	-2.0	0.2
319.02	-1.08	1.3	321.66	31JA	0.0	-4.1	-2.0	0.3
326.92	-1.07	1.3	331.73	10FE	-0.2	-4.0	-2.0	0.3
334.81	-1.06	1.3	341.76	20FE	-0.6	-3.9	-2.0	0.4
342.67	-1.03	1.4	351.74	1MR	-1.2	-3.8	-2.0	0.4
350.51	-0.98	1.4	1.66	11MR	-1.5	-3.7	-2.0	0.4
358.29	-0.93	1.4	11.52	21MR	-0.8	-3.6	-1.9	0.4
6.02	-0.87	1.4	21.32	31MR	0.5	-3.5	-1.9	0.4
13.69	-0.80	1.4	31.08	10AP	2.0	-3.5	-1.8	0.3
21.28	-0.71	1.5	40.78	20AP	3.9	-3.4	-1.8	0.3
28.79	-0.62	1.5	50.43	30AP	2.2	-3.4	-1.7	0.3
36.22	-0.53	1.5	60.05	10MY	1.3	-3.4	-1.6	0.2
43.55	-0.43	1.5	69.63	20MY	0.6	-3.3	-1.6	0.2
50.79	-0.32	1.6	79.18	30MY	-0.2	-3.3	-1.5	0.3
57.93	-0.21	1.6	88.72	9JN	-1.2	-3.3	-1.4	0.3
64.96	-0.09	1.6	98.26	19JN	-1.7	-3.3	-1.4	0.3
71.90	0.03	1.6	107.79	29JN	-0.8	-3.3	-1.3	0.3
78.73	0.15	1.6	117.33	9JL	-0.1	-3.3	-1.3	0.3
85.45	0.28	1.6	126.90	19JL	0.3	-3.4	-1.3	0.4
92.06	0.41	1.6	136.48	29JL	0.6	-3.4	-1.2	0.3
98.56	0.54	1.6	146.11	8AU	1.1	-3.4	-1.2	0.3
104.93	0.68	1.6	155.78	18AU	2.1	-3.4	-1.2	0.3
111.18	0.82	1.5	165.49	28AU	2.6	-3.5	-1.2	0.3
117.29	0.96	1.5	175.26	7SE	0.2	-3.5	-1.2	0.2
123.24	1.11	1.5	185.08	17SE	-0.9	-3.5	-1.2	0.2
129.02	1.27	1.4	194.96	27SE	-1.0	-3.4	-1.2	0.1
134.62	1.43	1.3	204.89	7OC	-1.0	-3.4	-1.3	0.0
139.98	1.60	1.2	214.89	17OC	-0.6	-3.4	-1.3	-0.0
145.08	1.79	1.1	224.92	27OC	-0.4	-3.4	-1.3	-0.1
149.87	1.99	1.0	235.02	6NO	-0.4	-3.3	-1.4	-0.2
154.28	2.21	0.9	245.14	16NO	-0.3	-3.3	-1.4	-0.2
158.22	2.45	0.7	255.30	26NO	0.0	-3.3	-1.5	-0.3
161.60	2.70	0.5	265.48	6DE	2.1	-3.3	-1.5	-0.2
164.25	2.99	0.3	275.68	16DE	1.3	-3.3	-1.6	-0.1
166.04	3.29	0.1	285.87	26DE	0.2	-3.4	-1.6	-0.1
				1649				
166.76	3.60	-0.1	296.06	5JA	-0.0	-3.4	-1.7	0.0
166.25	3.90	-0.4	306.22	15JA	-0.1	-3.4	-1.8	0.1
164.45	4.15	-0.6	316.36	25JA	-0.3	-3.4	-1.8	0.1
161.45	4.30	-0.9	326.46	4FE	-0.7	-3.5	-1.9	0.2
157.67	4.30	-1.0	336.51	14FE	-1.2	-3.5	-1.9	0.2
153.79	4.13	-0.9	346.51	24FE	-1.3	-3.6	-2.0	0.3
150.50	3.82	-0.7	356.46	6MR	-0.7	-3.6	-2.0	0.3
148.30	3.42	-0.5	6.36	16MR	0.7	-3.7	-2.0	0.3
147.39	2.99	-0.3	16.19	26MR	2.5	-3.8	-2.0	0.3
147.74	2.58	-0.0	25.97	5AP	2.9	-3.9	-2.0	0.3
149.19	2.19	0.2	35.70	15AP	1.6	-3.9	-1.9	0.3
151.56	1.84	0.4	45.37	25AP	0.9	-4.0	-1.9	0.3
154.69	1.52	0.5	55.01	5MY	0.4	-4.1	-1.8	0.3
158.44	1.24	0.7	64.61	15MY	-0.3	-4.2	-1.8	0.2
162.69	0.98	0.8	74.17	25MY	-1.2	-4.2	-1.7	0.2
167.37	0.75	0.9	83.72	4JN	-1.7	-4.1	-1.6	0.2
172.40	0.54	1.0	93.26	14JN	-0.8	-3.7	-1.6	0.2
177.72	0.35	1.1	102.79	24JN	-0.1	-3.0	-1.5	0.3
183.32	0.17	1.2	112.33	4JL	0.4	-3.4	-1.5	0.3
189.15	0.01	1.2	121.88	14JL	0.8	-3.9	-1.4	0.3
195.19	-0.14	1.3	131.45	24JL	1.4	-4.2	-1.4	0.3
201.42	-0.28	1.3	141.06	3AU	2.7	-4.2	-1.3	0.3
207.83	-0.41	1.4	150.70	13AU	2.2	-4.2	-1.3	0.3
214.41	-0.53	1.4	160.39	23AU	0.1	-4.1	-1.3	0.3
221.15	-0.64	1.4	170.13	2SE	-1.0	-4.0	-1.2	0.3
228.04	-0.74	1.4	179.92	12SE	-1.2	-3.9	-1.2	0.3
235.07	-0.83	1.4	189.77	22SE	-1.0	-3.9	-1.2	0.2
242.24	-0.91	1.4	199.68	2OC	-0.5	-3.8	-1.2	0.2
249.53	-0.98	1.4	209.64	12OC	-0.3	-3.7	-1.2	0.1
256.93	-1.04	1.4	219.65	22OC	-0.2	-3.6	-1.2	0.0
264.44	-1.09	1.4	229.72	1NO	-0.2	-3.6	-1.2	-0.0
272.05	-1.12	1.4	239.83	11NO	0.2	-3.5	-1.3	-0.1
279.74	-1.15	1.4	249.97	21NO	2.3	-3.5	-1.3	-0.2
287.50	-1.16	1.4	260.15	1DE	1.0	-3.5	-1.3	-0.3
295.32	-1.16	1.4	270.33	11DE	-0.0	-3.4	-1.4	-0.3
303.18	-1.14	1.4	280.53	21DE	-0.2	-3.4	-1.4	-0.2
311.07	-1.12	1.4	290.72	31DE	-0.2	-3.4	-1.5	-0.1

♂ LONG	♂ LAT	♂ MAG	⊙ LONG	16.00UT	☿	♀	♃	♄
				1650	MAGNITUDES			
318.97	-1.08	1.4	300.90	10JA	-0.4	-3.4	-1.6	-0.1
326.87	-1.04	1.4	311.05	20JA	-0.7	-3.3	-1.6	-0.0
334.76	-0.98	1.4	321.17	30JA	-1.2	-3.3	-1.7	0.1
342.62	-0.91	1.4	331.25	9FE	-1.2	-3.3	-1.8	0.1
350.44	-0.84	1.4	341.28	19FE	-0.7	-3.3	-1.8	0.2
358.21	-0.75	1.4	351.26	1MR	0.8	-3.3	-1.9	0.2
5.92	-0.66	1.4	1.18	11MR	3.1	-3.4	-1.9	0.3
13.56	-0.57	1.4	11.04	21MR	2.2	-3.4	-2.0	0.3
21.13	-0.47	1.4	20.85	31MR	1.2	-3.4	-2.0	0.3
28.61	-0.37	1.5	30.60	10AP	0.7	-3.5	-2.0	0.3
36.01	-0.26	1.5	40.31	20AP	0.3	-3.5	-2.0	0.3
43.32	-0.16	1.5	49.96	30AP	-0.3	-3.5	-2.0	0.3
50.55	-0.05	1.6	59.58	10MY	-1.2	-3.4	-2.0	0.3
57.68	0.06	1.6	69.16	20MY	-1.7	-3.4	-2.0	0.3
64.73	0.17	1.7	78.72	30MY	-0.8	-3.4	-1.9	0.3
71.69	0.27	1.7	88.26	9JN	0.0	-3.4	-1.8	0.2
78.57	0.38	1.8	97.79	19JN	0.6	-3.3	-1.8	0.2
85.36	0.48	1.8	107.32	29JN	1.1	-3.3	-1.7	0.3
92.06	0.58	1.8	116.86	9JL	1.9	-3.3	-1.7	0.3
98.69	0.68	1.8	126.43	19JL	3.2	-3.3	-1.6	0.3
105.24	0.78	1.8	136.02	29JL	1.8	-3.3	-1.5	0.4
111.71	0.88	1.8	145.64	8AU	0.1	-3.4	-1.5	0.4
118.10	0.98	1.8	155.30	18AU	-1.0	-3.4	-1.4	0.4
124.40	1.07	1.8	165.01	28AU	-1.3	-3.4	-1.4	0.4
130.63	1.17	1.8	174.78	7SE	-1.0	-3.5	-1.4	0.4
136.76	1.26	1.8	184.60	17SE	-0.5	-3.5	-1.3	0.3
142.80	1.36	1.8	194.47	27SE	-0.2	-3.6	-1.3	0.3
148.74	1.46	1.7	204.41	7OC	-0.1	-3.6	-1.3	0.3
154.56	1.56	1.7	214.40	17OC	0.0	-3.7	-1.3	0.2
160.25	1.66	1.6	224.44	27OC	0.5	-3.8	-1.3	0.1
165.79	1.76	1.5	234.52	6NO	2.7	-3.9	-1.3	0.1
171.16	1.87	1.4	244.65	16NO	0.8	-4.0	-1.3	0.0
176.30	1.98	1.3	254.81	26NO	-0.2	-4.1	-1.3	-0.1
181.20	2.10	1.2	264.99	6DE	-0.3	-4.2	-1.3	-0.1
185.79	2.22	1.0	275.19	16DE	-0.4	-4.3	-1.3	-0.2
189.99	2.35	0.8	285.38	26DE	-0.5	-4.4	-1.4	-0.2
				1651				
193.73	2.48	0.6	295.56	5JA	-0.8	-4.4	-1.4	-0.2
196.88	2.62	0.4	305.73	15JA	-1.0	-4.2	-1.5	-0.1
199.30	2.76	0.2	315.87	25JA	-1.1	-3.8	-1.5	-0.0
200.83	2.88	-0.1	325.97	4FE	-0.6	-3.3	-1.6	0.0
201.28	2.99	-0.3	336.03	14FE	1.0	-3.8	-1.7	0.1
200.50	3.04	-0.6	346.03	24FE	3.2	-4.2	-1.7	0.2
198.46	3.02	-0.9	355.98	6MR	1.6	-4.3	-1.8	0.2
195.31	2.89	-1.1	5.88	16MR	0.9	-4.3	-1.9	0.3
191.56	2.63	-1.3	15.72	26MR	0.5	-4.2	-1.9	0.3
187.89	2.26	-1.1	25.50	5AP	0.1	-4.1	-2.0	0.3
184.97	1.82	-1.0	35.23	15AP	-0.4	-4.0	-2.0	0.4
183.24	1.37	-0.7	44.91	25AP	-1.2	-3.9	-2.1	0.4
182.83	0.95	-0.5	54.54	5MY	-1.7	-3.8	-2.1	0.4
183.68	0.56	-0.3	64.15	15MY	-0.8	-3.7	-2.1	0.4
185.64	0.23	-0.1	73.71	25MY	0.1	-3.6	-2.1	0.4
188.50	-0.07	0.0	83.26	4JN	0.8	-3.5	-2.1	0.4
192.13	-0.32	0.2	92.80	14JN	1.5	-3.5	-2.1	0.3
196.39	-0.54	0.3	102.33	24JN	2.5	-3.4	-2.0	0.3
201.17	-0.73	0.5	111.87	4JL	3.1	-3.4	-2.0	0.3
206.39	-0.89	0.6	121.42	14JL	1.5	-3.4	-1.9	0.3
211.98	-1.03	0.6	130.99	24JL	0.0	-3.4	-1.9	0.4
217.90	-1.15	0.7	140.59	3AU	-1.1	-3.3	-1.8	0.4
224.10	-1.25	0.8	150.24	13AU	-1.5	-3.3	-1.7	0.4
230.54	-1.33	0.9	159.92	23AU	-0.9	-3.3	-1.7	0.5
237.21	-1.40	0.9	169.66	2SE	-0.4	-3.3	-1.6	0.5
244.09	-1.45	1.0	179.45	12SE	-0.1	-3.4	-1.6	0.5
251.14	-1.48	1.0	189.29	22SE	0.1	-3.4	-1.5	0.5
258.34	-1.50	1.0	199.19	2OC	0.2	-3.4	-1.5	0.4
265.70	-1.50	1.1	209.16	12OC	0.7	-3.4	-1.4	0.4
273.17	-1.48	1.1	219.17	22OC	3.0	-3.4	-1.4	0.4
280.74	-1.45	1.1	229.23	1NO	0.7	-3.5	-1.4	0.3
288.40	-1.41	1.2	239.34	11NO	-0.4	-3.5	-1.4	0.2
296.13	-1.35	1.2	249.48	21NO	-0.5	-3.5	-1.3	0.2
303.90	-1.29	1.2	259.65	1DE	-0.5	-3.5	-1.3	0.1
311.71	-1.21	1.3	269.84	11DE	-0.6	-3.4	-1.4	0.0
319.54	-1.12	1.3	280.03	21DE	-0.8	-3.4	-1.4	-0.0
327.36	-1.02	1.3	290.22	31DE	-0.9	-3.4	-1.4	-0.1

Finding Home

MARY,
I HOPE YOU FIND SOME INSPIRATION
IN THESE PAGES FOR YOUR OWN
JOURNEY HOME!

K

FINDING HOME

THE HOUSES OF PURSLEY DIXON

BY KEN PURSLEY
WRITTEN WITH JACQUELINE TERREBONNE
FOREWORD BY SUZANNE KASLER

RIZZOLI
NEW YORK

New York Paris London Milan

TO SARAH, CRAIG, BRONWYN, AND DAY,
FOR SHINING THE LIGHT THAT GUIDES ME.
MY LIFE'S JOURNEY IS RICHER
BECAUSE OF YOU.

Table of Contents

FOREWORD

For me, great interiors start with great architecture—it really is the most important part of a home. One of the highlights of my career has been working with talented architects like Pursley Dixon. It has been so special to collaborate with Ken Pursley and his partner, Craig Dixon, on houses on the coast, in the low country, and in the city. They have a holistic approach to their homes and a sensitivity to interior spaces, and they know how people want to live today.

In 2014, we were paired together to create the *Southern Living* Idea House in Palmetto Bluff, South Carolina. Pursley Dixon had decided to set the porch down several feet from the first floor. As Ken explained to me how this shift in heights allowed for interior views that were uninterrupted by outdoor furniture, I instantly bonded with him. He perfectly understood the importance of supporting the overall experience with these subtle gestures. As a designer, that unique sensitivity plays an immeasurable role in achieving my goal of conceiving houses that live comfortably but beautifully.

Working together from the ground up on a seaside home, I gained an even deeper understanding of Ken's quiet yet powerful way with interior spaces. Little things add up, such as how the living room's sculptural plaster ceiling evokes a quiet calm in the space, and how he manipulated views from the rooms to highlight the water beyond. But as the breadth of styles in this book proves, the firm can apply those same principles and that rigorous thinking just as graciously to rustic mountain getaways and elegant city homes.

Great architecture often pushes the designer, giving them the opportunity to do things they might not do otherwise. It is a rare occurrence when I come across a home that surprises me with its innovative design. When that falls into place, the decorating does not have to be as hard, allowing layers of nuance to emerge.

The collaboration involved in imbuing a house with a positive kind of energy that you can really feel as you move through it is unique and special. Pursley Dixon captures something even more elusive by creating homes that spark an unforgettable emotional connection.

Pursley Dixon's design philosophy comes to light in this beautiful book, and I'm elated for the firm to share its work that will resonate with and inspire so many.

—*Suzanne Kasler*

INTRODUCTION

Most mornings, I start the day with a cup of tea in my favorite mug. As I go through this daily ritual, I like to reflect on the image staring back at me—or, more precisely, staring back at himself. The artwork depicted on the mug is Norman Rockwell's *Triple Self-Portrait,* which features Rockwell painting an idealized version of himself while looking in the mirror. The illustration makes me smile and think about how we see ourselves, how we'd like others to see us, and how we truly appear. Those differing perspectives are the same self-truths

I encounter daily when joining my clients on their journeys to find home.

Conceiving a house isn't as simple as studying room arrangements or figuring out the best views. The process is more nuanced with every design decision building upon the ultimate mission of reflecting the clients back to themselves. Like the Rockwell painting, this translation of their personalities into architecture is not an exact portrait, because building that perfect home captures not who we are, but who we dream to be.

About a year ago, I took a month off from my day-to-day duties at Pursley Dixon Architecture to reflect on the meaning of finding home and how we interpret that into architectural tenets at our firm in Charlotte, North Carolina. Ensconced in a guesthouse on the eastern shore of Maryland that I had designed earlier in my career, I could map out the values that distinguish us from others while getting a tangible sense of how far we've come in our nearly two decades. I left feeling invigorated and filled with the stories and ideas that comprise this monograph of our work.

Like most stories, each project starts with the setting. When I begin to sketch ideas, I think first about how the space will be used and how the house will connect with the land around it. I like to contemplate what makes a person feel welcome and comfortable and develop my design from that foundation rather than chasing the latest trend. I trace my philosophy back to a simple extrasensory exercise from my architecture-school days

at Auburn University. My professor asked the class to go outside at night, close our eyes, and report back on what we heard. As with many assignments in college, I greeted this one with a dose of skepticism. Putting my reservations aside, I fetched a lawn chair, found a patch of grass, and settled in. As I closed my eyes, two distinct sounds came forward. The first, and more dominant, was a car. It started quietly, reached a crescendo, then faded into the night. The noise captured my attention, but it was fleeting. The second sound was less apparent: the hum of crickets. Their musical contribution to the night air was more consistent and melodic than the car's, but also more elusive and challenging to hear. I came to the small but powerful realization that had I only fixated on the dominant sound, I would have missed that evening symphony.

In our architecture studio, I often relate our design principles back to that distinction between the crickets and the car. The car represents the industry trends. They come in quietly, crescendo, and fade away when the next fashionable thing comes along. I encourage my team to be aware of changes in our industry, but not have that be their primary focus. Instead, I ask them to listen for the crickets, those design gestures that are saturated with integrity and authenticity. The teachings of proportions, spatial arrangements, and building orientation are not as shiny or sexy as chasing the latest fashion, but if carefully understood and thoughtfully executed, they will lead to more enduring results.

For me, the foundation of those principles was established throughout my time at Auburn. My tenure in the architecture program began the summer after my freshman year with "Summer Option." This end-of-year intensive served as a boot camp of sorts, intended to weed out the uncommitted. Just as a pressure cooker melds flavors, I forged many close bonds over those months. Among them was the bond I formed with my future business partner, Craig Dixon. As we stayed up late listening to John Denver with piles of chipboard, plaster, and balsa wood surrounding us, neither of us could foresee how our paths would later intersect.

Architecture, like most trades, is a mentor-based profession. You learn your craft from those who have come before you. When young practitioners ask me about the best path for success, I tell them it is simple: work for the best firm you can find that does the kind of work you want to do. In my case, that meant working for Bobby McAlpine in Montgomery, Alabama. It was a graduate school of sorts, building on foundations formed when he was my professor at Auburn. I still remember the first day of class when Bobby explained his views on life and architecture. He spoke with a majestic air about buildings being "honest, functional, and real." I could not write fast enough to capture

all the wisdom. Later, as my boss, Bobby continued to guide me, but he also gave me the space needed to grow and flourish.

After eight wonderful years at the McAlpine firm, I decided it was time to put down roots elsewhere. Several folks enthusiastically encouraged me to move to Charlotte—which in 2002 was pretty monotone—in hopes of adding a fresh voice architecturally. As my commissions began to mount, the need for additional help became apparent. As luck would have it, my former Summer Option studio mate, Craig Dixon, was living in Charlotte and he seemed the perfect fit. After a lively dinner, with more than a few glasses of wine, I convinced him he should join me. We had moved far beyond our chipboard modeling days, but that same spirit and enthusiasm for our creations was as alive as ever. We worked under the umbrella of Pursley Architecture for seven years, and then, to honor Craig's contributions, I made him a partner and changed our name to Pursley Dixon. Our collaboration of nearly a decade has exceeded all my expectations, and I could not ask for a better teammate or friend. To this day, I am in awe that our studio consists of fourteen people, and I relish the time I spend with my work family, whether visiting construction sites or grabbing an after-work cocktail.

When asked to define the style of our houses, I often pause. Not because I don't know what characterizes them, but because there's no simple, one-word answer. Modern, Craftsman, Colonial, Georgian—the residences we create don't fit neatly into the standard playbook. Instead, I prefer to think more in terms of linked architectural moments that speak to one another and work together. If you work within a singular style, the design solutions are governed by an expected and prescribed palette. I find it a shame to stifle creative possibilities and choose to remain open to stylistic freedom with project specific interpretations. I opt to expand the architectural tools at my disposal, and script solutions that blend styles as needed, while not losing sight of their historical roots.

On the surface, all of the houses in this book may appear to be different, but they're linked on an innate and authentic level. Whether located in the mountains, by the water, or in town, as the chapters are themed, they are connected by their sense of proportion, attention to detail, and understated elegance. While we like to weave together traditional and modern styles, there's a certain element of Southern culture that threads through all our designs as well. Not Southern in its historical references, but in its cultural values—a nod to the qualities and ideals that tie people together in this region. There's a humility and warmth to how our houses greet neighbors on the street. It's essential to find a way for each house to be distinctive but still compatible with the neighborhood, because there's a fine line between fitting in and being overlooked. As you spend more time in the homes our studio designs and really get to know them, they open up past the polite conversation and let you in on their eccentricities and secrets. Little touches of humanity and hospitality are waiting to be discovered in each of them. It could be a sleeping nook that perches you

right in the highest branches of a tree or a secret stairwell that leads to a dramatic overlook. We also find joy in carving out inviting places to entertain—kitchens that open up to the rest of the house, gathering spaces that are forever the life of the party.

A carpenter once told me that he could tell our houses are drawn by hand because of the shapes, curves, and movement. A house conceived on a computer can't offer the same fluidity and finesse. These nuances make the finished product look and feel right, hand-crafted rather than mass produced. The artisans we collaborate with bring an equal passion and skill to their realization. Without their commitment, the homes would just be ideas on paper.

To translate these sentiments and values into homes, we honor timeless constants instead of looking to trends. Our firm was founded on a simple principle: "Build beautiful things." Each house has its own thoughtful narrative that pulls from styles and references far beyond the obvious. We develop unique languages that each say, "Welcome home!" in a distinctive voice. While we always want to create architectural works that function well and can weather storms, we know that inherent beauty is what sparks an emotional connection that lasts a lifetime. It's in that connection that one truly finds home.

ABOVE: A cottage I designed fifteen years ago on the eastern shore of Maryland provided a place of solitude for reflecting on the firm's work while creating the blueprint for this monograph.

IN THE
MOUNTAINS

FINDING HOME
IN THE MOUNTAINS

Sitting at the base of a tree is very different from sitting high in its branches. From the top of a tree, your view of the world is amplified. You're simultaneously humbled by the vastness of the landscape and invigorated by the dramatic change in elevation. In this heightened state, you feel alive. Living in the mountains provides a similar experience. As architects, we seek to harness this feeling of exhilaration as we respond to the diverse beauty found atop these aerial perches. In some cases, a floorplan cranes its neck, turning and angling to get the best view. Other times, the structure balances on the shoulders of rocks reaching out over a cliff, seemingly defying gravity. Nature extends the invitation for these dynamic relationships, and we answer the call by making the most out of the extraordinary opportunities presented.

To fully appreciate the surroundings, all the senses must be engaged. Sweeping views capture the eye, but hearing a waterfall cascade or feeling the warmth of a fire adds an undeniable richness to the setting. To deepen the immersion in these exceptional places, the materials we construct with reflect a departure from the urban and set a more organic tone. As a general rule, we like to use items one might find on a walk through the property. Building from indigenous trees and boulders lets the home exist in harmony with the site. Selected for its ability to mellow over time, this palette is carefully curated to work in concert with the natural conditions. That's when the magic of being fully absorbed in a place happens, and architecture reveals its power to make all other places and worries disappear.

Every mountain setting necessitates a different architectural embrace. An extraordinary view may call for a dining porch open on three sides to the majesty of a valley. Sometimes the earth slopes away dramatically, and a house must be devised that follows this drop with ease and elegance. In another case, a house takes on a broken L shape to offer continual glimpses of the waterfall that is the cornerstone of the property. The gestures can also be as simple as planning for which way the breeze is going to flow through a living space. Architecture amplifies nature, both making the terrain livable and accentuating the awe-inspiring majesty of the world around us.

EXPLORATION

Gazing out from a mountain overlook elicits many responses. Feelings of excitement and exhilaration might be laced with an undercurrent of apprehension. All of the aforementioned emotions struck me as I reflected on the recently cleared lot for this home in Blowing Rock, North Carolina. I remember standing on the precipice overlooking the steep embankment and thinking, Where is this house going to go? And therein lay the challenge and the opportunity. The house we ultimately created maximizes the drama of the landscape by incorporating a sense of discovery and surprise within its unique design.

Having worked with these clients before, we knew their refined taste would mesh in a unique way with the traditional mountain aesthetic, and they gave us incredible leeway to push things in a more rustic direction without losing sight of their cultivated sensibilities. That translated into gilded frames against fieldstone walls and articulated beams mixed with rough boulders—exactly the kind of tension that makes a place singular and unforgettable.

Just as a film never begins at the climax, neither should a home. The entry and front facade welcome visitors with a warm, yet modest, greeting—the dormers peel open from under the cedar shakes, the roofline lilts toward the earth, and the windowsill just grazes the garden bed. Walking down into the entry feels like stepping into a cave—you're surrounded by stone and almost completely encased to make what lies ahead all the more dramatic.

Once inside the living room, the astonishing view comes into full focus with a sheer expanse of mountains that often leaves first-time visitors awestruck. The soaring tiers of porches are thoughtfully stepped down from the main level or placed adjacent to primary living spaces in order not to interrupt the magnitude of the panorama. A dining porch cantilevered over the vista, open on three sides, gives the homeowners, who are warm and engaging hosts, a one-of-a-kind space to serve dinner. Those staying the night also marvel at the bridge to the guest bedroom, which both provides an adventure on the walk to bed and carves out some personal space.

But the literal crowning achievement comes in the form of a daring perch soaring at the pinnacle of the roof. Only reached by ascending a secret stairwell, it's the perfect setting to raise a sunset cocktail to the majestic Blue Ridge Mountains. The architectural journey of discovery and surprise woven throughout this home has been leading to this moment, where the house celebrates the unbridled power of the boundless landscape.

PREVIOUS PAGES: Nestling the main floor windows against the ground reduces the overall appearance of the house's scale, while a cupola, concealing a viewing porch access, climbs toward the sky. ABOVE, FROM LEFT: The eaves of a wood-shingled roof gracefully undulate upward, allowing space for passage. A bracket emerges from a column of boulders with a lissome arch. Cut limestone tames a wall of natural fieldstone. Carved, untreated oak marries strength and elegance. FOLLOWING PAGES: Shaped timbers and steel-framed doors frame the view. Lowering the porch three steps opens up the sight line and exposes more of the mountain range.

The soaring tiers of porches are thoughtfully stepped down from the main level or placed adjacent to primary living spaces in order not to interrupt the magnitude of the panorama.

PREVIOUS PAGES: A dining porch cantilevered over the vista and open on three sides gives the homeowners a one-of-a-kind space to enjoy dinner. The ceiling timbers, rough boulders, and planked flooring connect the architectural vocabulary of the indoors and outdoors. OPPOSITE: A cantilevered porch off the main living room above provides shade for the lower terrace, which connects to the guest quarters by a suspended wooden bridge. FOLLOWING PAGES: A floor-to-ceiling wall of glass in a guest bedroom gives the effect of sleeping among the trees, while casement windows open to bring in mountain breezes. PAGES 32–33: The house takes advantage of its spectacular panorama of Pisgah National Forest through a series of porches on multiple levels as well as a dramatic rooftop terrace.

Connection

Western architecture is often associated with big boulders and overscaled timbers, but here we decided to forgo those clichés in pursuit of something more restrained. We referenced the lessons of Modernism, such as a connection to the outdoors and an emphasis on space and light, then filtered them through our own lens. At the same time, we made sure the home still addressed the rustic Western environment by instilling its design with a layer of pragmatism—substituting extraneous flourishes for thoughtful details that hold up well against the harsh climate and active use. Since our team designed both the architecture and the interiors, there's a real cohesion to that vision in every detail.

We began the design process by siting the home to maximize the views at every opportunity. The structure is almost completely surrounded by the glorious sweep of mountains, which are magnified by their reflection in the tranquil pond. As a result, there's no real facade. A porte cochere acts as the de facto entry, linking the house's three pavilions. Once the door pivots open, you're immediately greeted by the landscape, in the form of a perfectly framed, two-story window capturing the mountains. As you move through the house, the vistas

continue to unfold. In the main living area, a wall of glass slides into a hidden pocket, bringing all the invigorating elements of the mountain setting indoors.

But not everything revolves around swaths of dramatic glass; some moments are smaller and more subtle. In the dining room, a diminutive window provides a sense of intimacy with the range's tallest peak. A system of apertures throughout the interiors allows for glimpses into other spaces and to the outside. For example, a slim cutout in a wall of stone in the foyer peers through the kitchen and to the living space beyond. There's even an outdoor shower, which frames the perfect slice of a special vista. These moments of majesty and discovery come from careful planning for the site's spectacular surroundings along with a desire to infuse the interiors with an equal sense of awe.

The choice of materials throughout emphasizes this play between interior and exterior while still turning those Western banalities on their head. Durable materials and simple rooflines reflect a practical side, while sliding walls and tactile fabrics add intrigue and style. Inside and out, corral board, highly weathered wood that's upcycled after years of managing livestock and snowdrifts, adds grit and texture. Blackened steel also gives a

contemporary edge as cladding on the wall above the kitchen space as well as razor-thin bookshelves nestled into a fireplace mass. But there's also a soft side: at the core of the house, an internal light well connects the upper-story bedrooms and casts rays down to the main floor, so things never feel dark. Textural materials in a tonal palette, such as Bel-gian linen slipcovers, alpaca throws, and touches of napa leather, mellow edges.

By eschewing Western stereotypes and get-ting into the essence of what makes a house feel like a home, we ended up with a place that is totally comfortable with itself and exists harmo-niously in its extraordinary surroundings.

PREVIOUS PAGES: Rustic Modernism makes the most of this spectacular setting through clean lines mixed with earthy materials. OPPOSITE: A pivot door, flanked by frameless glass, rotates open to reveal a sight line straight through the house to the extraordinary expanse beyond. ABOVE: A slot opening cut through a stone wall provides a peek from the entry hall to the kitchen and living room beyond. FOLLOWING PAGES: When the retractable doors are tucked out of sight, the living room completely opens up to nature—becoming a true indoor-outdoor living area.

PREVIOUS PAGES: A cantilevered concrete slab island anchors the kitchen, while a custom weathered wood credenza conceals most of the storage. OPPOSITE: Varying ceiling heights between the kitchen and dining areas provide intimacy and spaciousness while delineating each area in the vast room. The blackened-steel wall above conceals a guest quarters, complete with its own private window looking down into the space. ABOVE: Blackened-steel details carry the theme of Modernism throughout the house.

ABOVE, FROM LEFT: A crisp metal-framed aperture cuts a sight line through rough stone. A handrail, wrapped in leather, brings an urbane sensibility to rough-hewn oak stairs. A custom-designed light fixture does the same in the entry. A cantilevered I-beam supports the expansive concrete island.

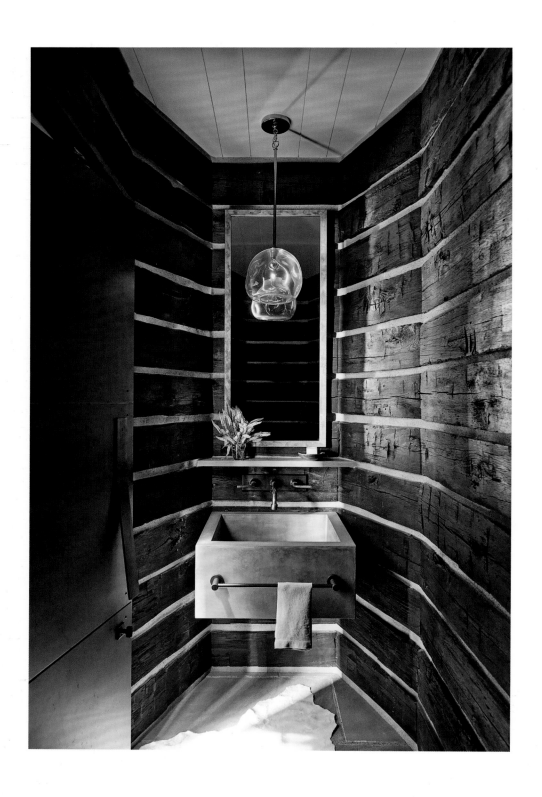

ABOVE: Dark rough-hewn beams separated by pioneer-inspired mortar joints animate the faceted powder room for a moment of surprise. OPPOSITE: Off the entry, a sunken study with lowered ceilings provides a cozy, intimate gathering space away from the open floor plan's larger living area.

ABOVE: On the second floor, a sitting area tucked in at the top of the stairs offers a view of the pond and a quiet place to retreat. OPPOSITE: The space is centered on a sizable leather-trimmed light well, through which sunshine flows to the entryway and the ground floor below.

LEFT: In the upstairs sitting room, a bookcase, fashioned from upcycled corral boards with steel shelving, swings open to reveal a hidden bunk room. The additional accommodations can be sealed off when not in use to diminish the perceived scale of the house.

53

ABOVE: Bedrooms take on a slightly more refined air. Plastered walls curve upward to meet a wood-paneled ceiling. Built-in daybeds have become another firm signature beloved by clients, who are always looking for a sunny space to lounge as well as options for fitting in extra guests. OPPOSITE: In a bath, whimsical mirrors on pullies can be adjusted to the correct height for adults and children.

PREVIOUS PAGES AND ABOVE: An outdoor fireplace mirrors the living room's indoor one while sharing the same rock-clad chimney. OPPOSITE: The simple pleasures of a plein air shower are enhanced by a cutout in the corral boards, which perfectly frames the mountain peak in the distance. FOLLOWING PAGES: A mix of stone techniques creates the ideal gathering spot near the water. Local fieldstone forms the firepit while organically shaped flagstone slabs anchor the sitting area.

GRACE

Some houses want to be more private, not showing you everything at first glance. This mountain escape in Linville, North Carolina, is just that kind of place, from its meandering floorplan to the intriguing mix of materials. By partitioning the footprint into several modest structures and linking them together with varied textures of earth-toned cladding, we were able to devise an approachable demeanor, masking the breadth of discovery that lies within.

Since the house is hidden from the street, there was an opportunity to create a striking bay window to fill the living space with light and offer grand views from inside to the wooded forecourt. In the adjacent entry alcove, Cedar wraps around the frame, proving that even small architectural moments can leave a big impression. We created that detail as a nod to the Shingle Style cottages that architect Bruce Price designed as charming turn-of-the-century retreats in Tuxedo Park, New York. Those retreats, like this one, are a magical merger of elegance and whimsy, providing familiarity without being predictable.

Once through the Gothic door, the house unfolds in a genteel fashion, with a modest entrance hall that greets you before you enter the house's multifaceted main room. We divided this dynamic volume into two sections: the dining room has a vaulted two-story ceiling that opens up to sweeping views and gives the house air, while the living area has lower ceilings that provide cozier opportunities for enjoying fires and relaxing.

The idea of this amalgamation of volumes plays into the narrative for this new home. We imagined that it was actually built over time—an exterior porch space closed in here, a room added on later there. Some things look new and others look old, thereby playing into those sorts of incongruent yet delightful moments you often find in generational houses that have evolved organically over time. That baseline offers the ability to change materials from room to room and, as a result, you get these interesting shifts in the architectural story throughout the house. Overall, the palette of the interior spaces takes on a rustic language with some primitive touches: rough-hewn beams, rugged flagstone flooring, and plank walls of varying widths. But those unpretentious details are tempered by more polished ones—most prominently, an abstracted board-and-batten texture with a sculptural half round found in the main living space.

These types of joyful discoveries are as essential to us as the floorplan itself. We weave them into all parts of the home. A wooden library has bark panels on the ceiling, a curiosity to make one wonder if this was originally a porch. In the kitchen, the counter legs are fashioned out of logs. Since they're tucked below the concrete slab, their uniqueness is subtle, inviting a moment of unexpected revelation.

On the second floor, you walk through the stone chimney to arrive in a sleeping nook that looks out across the dining room and offers a different perspective of the view beyond.

The more time you spend within these walls, the more the character of the place is revealed. Like any strong relationship, the experience of the house grows richer with each visit, as it quietly shares its nurturing and playful personality.

PREVIOUS PAGES: Changing the dimension, direction, and materials of classic wood elements adds beauty and rhythm in subtle, unexpected ways. ABOVE: In the small structure that houses the primary bath, cedar shingles and pine skirting are stained varied earth tones to blend quietly with their surroundings. OPPOSITE: Inside, wide planks support a stair rail with gaps of light to accentuate the cadence. Horizontal slats in varying widths add a subtle pattern to the walls, while overhead a uniform board application brings order.

PREVIOUS PAGES: The verticality of the double-height dining room is accentuated by the board-and-batten paneling. RIGHT: Reclaimed-oak ceiling beams create intimacy in the living room and separate the space from the soaring dining area. The smooth decorative walls contrast with the rough nature of the beams and the bark-paneled sliding door.

ABOVE: Throughout the house, elements of bark connect the design to its mountain
location. OPPOSITE: The sliding door that leads into the study harmonizes
with the ceiling clad in square panels of bark running in alternating directions.
FOLLOWING PAGES: In the kitchen, timber logs anchor the concrete counter.

ABOVE: The primary bath is reached through a series of transitions, peacefully isolating it from the rest of the house. OPPOSITE: On the second level, a cozy nook, entered through a passageway of stones, overlooks the dining area with windows that can be left open to capture cross breezes or closed to control sound. FOLLOWING PAGES: A stone banquette maximizes seating on a covered porch while strategically placed columns frame the mountain view.

COMMUNITY

Cheerful and *mountain house* have never exactly been synonymous. Although we love working with mighty boulders and rough timbers, sometimes the client and the location prescribe a more upbeat and refined approach. Certainly, this Linville, North Carolina, home has some touches of those classic materials, but the overall feeling of the architecture comes across much lighter and airier, reflecting the couple's bright personalities.

Stone, bark, and cedar all make appearances here, but in a sophisticated manner that elevates them beyond primitive craft. For example, poplar-bark siding, which carries forward the mountain-resort area's tradition of chestnut bark–clad structures from the early 1900s, lines the house, while the contrast of white-painted trim provides a crisp delineation from the earth-toned material. In line with the clients' wishes for refulgence, the interior wood tones were lifted and the windows made sizable for maximum light. We also pulled back the eaves, so the sun shines through instead of being blocked by deep overhangs. The whole second floor is tucked into the roofline to keep the scale more friendly, with thoughtfully arranged dormers

peeking wide-eyed from the shingled roof.

Situated on a former rail route that is now one of the primary arteries through the resort town, the house has a layered entry sequence that removes it from the street. Channeling architect Frank Lloyd Wright, who insisted on navigating at least three turns before reaching the heart of a home, we planned a series of architectural moments that usher guests strategically through to the living space. You enter to the left of the facade's central bay, featuring full-width, chalet-inspired Swiss diamond windows, before veering into a crisp foyer of light-stained oak and Pennsylvania bluestone. From here you ascend three steps and walk around the rear of the fireplace, all before being welcomed into the family room. This not only makes for a memorable experience, it also creates an architectural layer that buffers the space, and those using it, from the distractions of the roadway beyond.

However, the home's spirit truly comes alive on the gracious porch, which animates the rear elevation overlooking the golf course. Complete with a large outdoor fireplace, the portico functions as the de facto living room. An ancient-looking chimney, assembled by a skilled local

mason with artful proclivities, adds a whimsical feature, especially given that you can walk straight through its massing. Another anachronistic touch, harkening back to the late nineteenth century when this community was founded, is a wooden gate that swings open to the fairway, prompting neighbors to say hello while playing a round of golf and perhaps make plans to drop in later for a visit. Reflecting the Southern values of hospitality and neighborliness, this home embodies what makes this Blue Ridge Mountain enclave so special.

PREVIOUS PAGES: Charming touches, such as the single shuttered window and tapered wooden brackets, lend a sense of warmth and welcome. OPPOSITE: Dormers peel up from the wood-shingled roof to bring light into the upper floor. ABOVE: A rambling stone chimney, assembled by a local mason with artful proclivities, warms the covered porch. FOLLOWING PAGES: The terrace overlooks a fairway of the golf course—creating the ideal place to say hello to friends, admire their game, and invite them for a post-round visit.

Transformation

When it comes to mountain retreats, creating a successful connection with nature is essential. Few homes we've designed execute that objective better than this one on Lake Glenville near Cashiers, North Carolina. With a lake house nestled in the mountains, the opportunities for exploring the outdoors were boundless. The owners are an active family, so this getaway's kinetic spirit, with its walls folding open to the woods and water below, suits them perfectly.

Just entering the home is an adventure unto itself. You cross a ravine on a bridge that leads to the front door, whose sliver of a window gives a peek at the lake beyond. Once inside, there's a progression of steps—three down here, another three down there. This gradual descent into the main living area allows visitors always to keep the view in sight as they slowly move closer to the water's edge.

In the generous main room, a wall of folding glass doors can be fully retracted, forging an unencumbered connection with the outdoors. On temperate summer evenings, a twenty-foot-wide insect screen can be rolled down from a carefully concealed pocket to convert the room into a dramatic living pavilion with all the sights, smells, and sounds of the environment adorning the space. All of the interior materials, from a driftwood-lined entry hall to a fieldstone-clad fireplace, were carefully chosen to mirror the surrounding woods. That, to me, is the allure of this transformative architecture: the ability to seamlessly peel back the skin of the building, allowing the full enjoyment of the environment without sacrificing the comfort of your favorite chair.

Since the kitchen is central to the main living space, we concealed most of the appliances and storage. Wanting to minimize the visual clutter, we chose to create a "non-kitchen," quietly integrating the necessities of the space behind paneling that matches the adjacent wall material. Using wood for the island's countertop and a large timber to mimic a hearth around the kitchen sink adds to the unified look. Woodwork became a theme throughout other areas as well—whether it was the exposed, hand-hewn oak in the entry foyer or sculptural bark on the walls of the powder room, the textures add a layer of intrigue to the architectural palette.

While all of these details reflect the home's relationship with nature, nothing can compete with nature itself. The two porches that flank the main structure were conceived to be as connected

PREVIOUS PAGES: At dusk, the
house's poplar bark siding helps it
to disappear into its surroundings.
RIGHT: Even the great outdoors
can benefit from modern
inventions. In the living room,
a retractable wall of doors truly
brings the outdoors in. Artist
Stuart Coleman Budd was
commissioned to paint an
evocative landscape that serves as a
four-panel bifold television cover.

as possible with the outdoors, but in different ways. The screened porch amply accommodates sitting and dining, while also opening up to the kitchen to allow a fluid link between the spaces. The uncovered porch is more dramatic: cantilevered high among the trees, its platform reaches out from the house to place you in the heart of the forest. Here, you can intimately connect to the magic of this place, which seamlessly blends the rustic and the poetic.

ABOVE: To integrate the kitchen and screened porch, large casement windows are folded out of view. OPPOSITE: Custom cabinetry is cleverly concealed within horizontal planks, virtually disappearing for an orderly look.

ABOVE: A windowed stair floods light into multiple floors, while a cascading custom
iron handrail and a coordinating sconce add dramatic flair. OPPOSITE: In the game
room, natural textures are offset by plush fabrics in an intimate alcove.

REJUVENATION

Of all the houses we've designed, this North Carolina getaway is one that touched my heart deeply. The rolling hills and massive boulders leading up the drive give clues to the true natural wonder of the nine-acre property. However, it's only once you pass through the screened entry porch that its full majesty comes into play—with a towering waterfall cascading into the creek bed below.

When the owners shared their desire for a whimsical tree house compound with our studio, I knew immediately we were kindred spirits. I envisioned a house that would sit on the property, balanced on a steep outcropping and towering on stilts high in the trees. The owners shared fantastical images of romantic sleeping follies and living porches dangling from ropes to further fuel my imagination. I even spent time researching Indonesian and Balinese architecture to introduce a mildly exotic flair. I wanted to create an environment that was like no place the owners had ever been, yet, paradoxically, they would feel they had always known.

While there is excitement in architectural exploration, I still wanted to craft a comfortable place where people could feel restored and connect with the natural beauty. To ground those daydreams, many of the elements took on earthy characteristics. The dark bronze roof and mossy-green exterior cladding give the house a chameleonlike quality, helping it meld quietly into its densely wooded backdrop. By placing most of the structure on pine poles, we created the illusion of the home stepping gently into the surrounding forest. Pilings normally seen on coastal homes bolster up the rear while mimicking the vertical rhythm of the surrounding trees. When it came to the stonework, the homeowners got involved, personally joining our team on visits to the best local sources to select the most interesting fieldstone for the fireplaces, the steps to the front door, and paths down through the forest. There's a real artistry in how each one fits together, weaving the various sizes and shapes into a powerful tapestry.

It's not a large house, so to expand the perceived interior scale we carved out spaces within spaces. For example, in the living room, you have the option to lounge next to a generous picture window, which strategically frames a collection of boulders and plants. Or you can choose to sit in the window seat and gaze out to the woods and waterfall in the distance. To add a layer of comfort to the internal living experience, the exterior eave hangs just low enough that you feel protected, like the brim of a sun hat, without blocking the mesmerizing view.

As the hub of the space, the kitchen connects seamlessly with the adjacent living room. Sliding doors and folding windows open up to the uncovered and screened porch, erasing any division between indoor and out. As the owners are wonderful cooks, this layout allows them to be the center of activity, wherever family and friends choose to unwind.

The screened-in living room gets the most use, though. Nestled between the main house and the owners' bedroom suite, this outdoor living space functions as a bar, dining area, napping destination, and gathering spot for fireside chats. It's also the ideal backdrop for enjoying the symphonic sounds of nature while sipping a favorite bourbon. This is when architecture is at its best: all the dreams and aspirations are realized, your family and friends are gathered, and the resulting joy floats effortlessly in the mountain air.

PAGES 100–101: A gracious screened porch connects the main house with the primary bedroom suite.
PREVIOUS PAGES, ABOVE, AND OPPOSITE: The living room offers multiple seating options, creating spaces within spaces. A banquette defines a cozy place for lounging in the far corner. A combination of single-paned and divided windows blurs the lines between indoors and out, giving the room the feeling of an open porch.

PAGES 106-107: With its central location, the kitchen easily connects to the screened and open porches. PREVIOUS PAGES: The porch off the kitchen, like the living room, offers multiple options for dining and sitting. Large windows fold open, allowing a convenient link for serving food or drink. ABOVE AND OPPOSITE: There's a moment of discovery when guests stumble upon this stone path. The walkway climbs up toward the screened porch, its massive hearth anchored by three boulders. The stone walls and steps were sourced locally and artfully arranged by talented craftspeople from the region.

It's not a large house, so to expand the perceived interior scale we carved out spaces within spaces. For example, in the living room, you have the option to lounge next to a generous picture window, which strategically frames a collection of boulders and plants. Or you can choose to sit in the window seat and gaze out to the woods and waterfall in the distance.

OPPOSITE: Built by the owners' son, the hanging daybed provides a comfortable setting for an afternoon nap in front of the floor-to-ceiling screened opening.

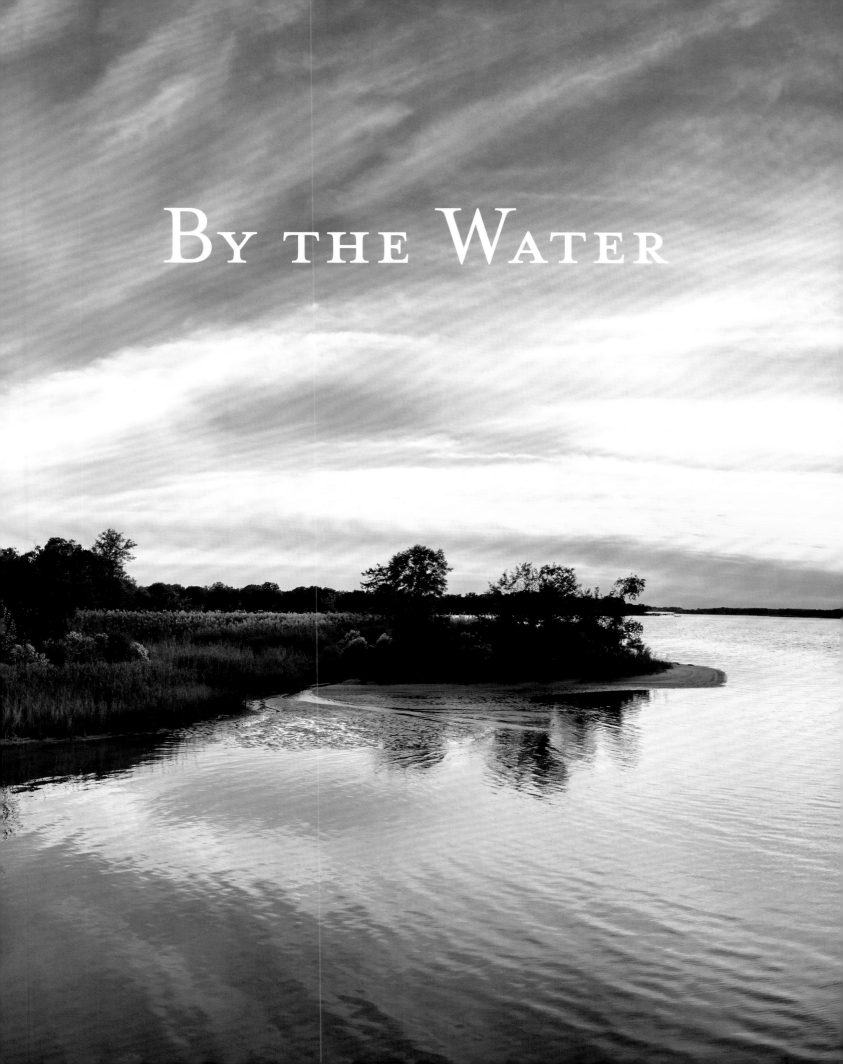

BY THE WATER

Finding Home
by the Water

Water stirs a strong emotional connection within us all. As we observe the turbulence of an ocean wave or the stillness of a tranquil pond, we are reminded that nature is dynamic. I often consider these environmental complexities when designing houses on the water's edge. Much like setting up for a day at the beach, I want enough "architecture" to be comfortable, but not so much as to inhibit my connection with the place I've come to enjoy. Like a properly positioned umbrella, sliding glass walls and thoughtfully outfitted porches can be adjusted to celebrate a beautiful day, or shuttered when Mother Nature is feeling grumpy. Creating a charming vessel that provides shelter is important, but it is the deeper emotional and physical link to the water that is at the root of every design decision we make.

While water is the common thread for the houses featured in this chapter, each location calls for a unique architectural interpretation. For a house situated off the Gulf of Mexico, the rear opens up to make the most of the marine vistas with sweeping verandas and covered porches, perfect for admiring the boats as they pass by. In South Florida, a certain harshness in the climate demands building with stucco-on-block construction rather than wood, which can rot in the damp heat. The house's rooms all face inward to the pool, while the shady courtyard keeps things cool and provides privacy. In the Carolina countryside, a farmhouse-inspired escape takes into account the stillness of the property's pond with an expansive bay window large enough to settle into and connect with the wetland's deep calm.

Since not all water is created equal, the architecture's responsiveness to each body's characteristics must be taken into account. To devise spaces that feed into our natural instinct to connect with water's calming qualities, I plan for as many rooms as possible to interact with it. Large swaths of glass foster that relationship, as do palettes and materials that complement the setting. For example, an elegant beach house may call for bleached oak and creamy marbles, while a more laid-back country house requires rough-sawn beams and stair rails fashioned from rope. These layers of decisions give personality and warmth, and they ultimately make each home a unique reflection of the family who lives there.

TRANQUILITY

When I first visited the town of Boca Grande on Florida's Gulf Coast, I was pleasantly surprised to discover a place that reflects a bygone era, capturing a nostalgic simplicity missing from our fast-paced digital lifestyles. The cadence here is relaxed, and I wanted to reflect that in the home we created for a family who has been visiting this understated enclave for years. Given the stylistic diversity of the area, I looked toward the British West Indies to inspire my design. Balanced forms composed of masonry, stucco, and cypress are the building blocks of this waterside escape. Open-air living amenities, such as outdoor rooms, a kitted-out cabana, and porches with deep overhangs, not only give the house its character but also provide ample opportunities to enjoy the temperate climate.

By exercising restraint, the design features used have an even greater impact. On the front facade, a mirrored pair of subtle S curves are carved into the stucco and clad with cypress boards, guiding visitors to the oak-paneled entry door. On the more sprawling exterior overlooking the water, a unique bracket shape on the column has a distinctive positive and negative space—giving interest while also framing the vistas.

Inside the home, the architectural story continues with a reverence for both unadorned space and axial views. Because of the monolithic quality of the masonry structure, I explored the idea of the house as a carved plaster box. The envelope that forms the central living space is crisp and white, as if chiseled from a solid block to reveal its shape. Additional elements such as the wooden door, stairwell, and paneling are treated as an additive layer, offering texture in contrast to the austerity of the plasterwork. In the discreetly grand entry, storage is tucked into the woodwork below the symmetrical stairs and slivered sidelights flank the front door. Creating quiet yet thoughtful touches like these makes the house memorable and gives me such delight as an architect.

Throughout the interiors, the connection to the water was central to my thinking.

PAGES 120–121: Cypress siding under the eaves and in the recessed entry provides a textural relief from the stucco facade. PREVIOUS PAGES: Symmetrical stairwells surround the paneled oak entry door, which is flanked by slot sidelights. ABOVE, FROM LEFT: The architectural details and curved gestures that are signatures of the firm get an island-influenced makeover for this Florida home. From outdoor balustrades to stylized facades, carved doors, and indoor handrails, thoughtful moments of surprise and whimsy abound.

We worked in tandem with the incredibly talented Atlanta-based interior designer Suzanne Kasler to tie the interiors back to the architecture with clean, sculptural furniture and a muted color palette that reflects the sand and ocean outside the generous windows. Reached by ascending the double staircase, the second-floor landing is centered on the primary bedroom and flanked with whimsical bookcases. With its vaulted ceiling and private balcony, this space makes the ideal setting for the owners to escape and peacefully enjoy the panoramic views.

The home's straightforward elegance channels the understated spirit of Boca Grande. Looking back, this is precisely the quality I admire so much in these clients, and what originally attracted them to this region. It's not always the loudest among us who has the most to say, and I think this is as true of architecture as it is of people.

RIGHT: A serene living space begins with architectural finesse. A shaped plaster ceiling conceals the ventilation system, resulting in a seamless backdrop unmarred by visual clutter. A pair of symmetrical doorways bring color and harmony to the space.

OPPOSITE: Exposed wood ceiling beams add warmth to the kitchen's palette of creams and beiges. Storage becomes nearly invisible when placed behind horizontally laid wood slats. ABOVE: Just beyond the island, a curved banquette softens a square nook accented by vertical paneling and a lowered ceiling.

ABOVE AND OPPOSITE: Bookcases in a signature architectural shape repeated throughout the house flank the entry to the primary bedroom, which is illuminated by a wall of floor-to-ceiling windows along with a door to a private balcony overlooking the pool and waterway beyond. A vaulted ceiling and virtually all-white palette enhance the bright nature of the space.

ABOVE: By plastering the pool in a color that matches the bay the two bodies of water meld compositionally. OPPOSITE: The house takes on a grander countenance in the rear compared to the more private entry facade. FOLLOWING PAGES: A series of balconies, porches, and stairs connect the house and its terrace to the aquatic view, while a detached pavilion provides a comfortable spot to rest waterside.

DISCOVERY

Lately, more clients have requested their vacation home take on the character of a personal resort. It's an easy request to understand, because if you are creating a sanctuary for yourself, why not make it a reflection of places that have brought you pleasure and relaxation? This reimagining of an existing home for a family, who migrates to Wellington, Florida, to ride horses every winter, provided the perfect opportunity to create such an oasis.

As with any great hospitality experience, setting the tone is key from the moment you enter. On the exterior approach, the pacing of the pavers is echoed in the rhythm of the pivoting gate's slats, which are designed to let breezes through while blocking enough to keep what lies beyond a bit mysterious. Once you enter, a grand courtyard reveals itself. The outdoor space connects to almost every room through a multitude of sliding doors, and tucked in a corner, a hedged alcove offers a fireside nook with a cozier atmosphere. To build on the overall relaxing ambience of the clients' dream retreat, we recast the roof in a warm gray and outlined the windows in black, complementing the shimmering pool's new charcoal tinted base and the surrounding fieldstone-colored tiles.

Inside, the style aligns with the clean, uncluttered exterior. Working with Bronwyn Ford of our interior design team, we created a warm, modern palette that was carefully curated to not be too clinical or fussy. Calm, cool, collected furnishings are rendered in a soothing spectrum of grays, reinforcing the overall serene atmosphere.

The architecture, with its variation in ceiling heights, forms a rhythm of highs and lows. To remedy the existing voluminous fourteen-foot-tall spaces, we lowered ceilings in the bar, television lounge, and master bath hall to create juxtapositions of scale. In the television lounge, we pulled the ceiling down to nine feet and painted it a darker tone, creating a cozy cocoon—perfect for a movie marathon or an afternoon nap. Conversely, we left the height in the open kitchen, where the extra volume is welcome when everyone inevitably gathers there.

Reimagining structures like this one has made me realize that what is taken away is often more impactful than what is added. Whether it is extraneous moldings or an overly lofted ceiling, by carefully deciding what to keep and what to change, a home can be dramatically transformed—in this case, into a personal getaway tucked away from the world, existing as its own private paradise.

RIGHT: Once inside the
courtyard, the canopied front
door provides access to the house,
but all the rooms lining the
enclosure have a fluid connection
as well. The curved stucco wall
covered in star jasmine keeps the
outdoor grill out of sight.

RIGHT: A sliding upholstered
door conceals the lounge,
making it the perfect escape from
the sun. Lowered ceilings, a
darker palette, and textures such
as leather and shearling complete
this cozy environment.

ABOVE: A sliver of a window divides wooden slats to bring a ray of light into the entryway.
OPPOSITE: A sense of calm permeates the dining room, which is grounded by a dining table
with a bleached-oak base and a custom-designed concrete top.

OPPOSITE AND ABOVE: Small gestures make a big impact even in a seemingly streamlined kitchen. Pecky cypress supports a quartz island top with stainless steel panels beneath for added interest. A footrail offers a place to kick up and settle in, while a back ledge above the counter plays a balancing act between storage and display. FOLLOWING PAGES: An arched doorway frames a vestibule outside the primary bedroom, buffering it from the rest of the home. In the bathroom, a mirrored wall concealing storage expands on the notion of the traditional medicine cabinet.

RIGHT: Tucked off the main courtyard, this open-air room provides the best of outdoor living: generous seating with a statement fireplace and a view of the nearby pool.

Authenticity

If our firm were a rock band, this farmhouse would be our first big hit. Its publication and subsequent social-media exposure has garnered us many fans through the years; however, when I initially visited the site in rural South Carolina, I didn't think so much of the setting's potential. It was a sticky day in July, and while walking the grounds with the clients, I remember surveying the property—an algae-slicked pond, a dilapidated ranch house, a colony of menacing things with wings—and wondering to myself, Of all the beautiful countryside in the Carolinas, why here? Hiding my concerns, I set out to design a structure that would give meaning to its surroundings. The home I imagined would be down to earth, reflecting the owners yet aspirational enough to elevate the setting.

To strike the right chord, I knew I needed a structure with personality but not blind to its agrarian roots. When considering a design direction, an unlikely source of inspiration came to mind. I had recently traveled to Cumberland Island off the coast of Georgia, once a winter retreat for the Carnegie family, and was particularly fascinated by an old icehouse that was still intact on the property. It had a distinctive, continuous vent along the ridgeline and vertical stacked windows on the gable ends running from ground to peak, giving this humble structure an almost chapel-like quality. Those two elements became the seeds from which my new design would grow.

The result was a three-story structure, larger and more modern than its inspiration, but in keeping with the spirit. Each elevation is scaled down by having a one-story component—on the pond side there's a generous bay window; on the approach, a wide porch with a sweeping cantilevered awning. From here, family and guests can enjoy a shady respite while taking in the dreamy grounds, which were adeptly transformed by talented landscape architects Ben Page and Ed Tessier. Sliding walls and large swaths of glass are knitted into its farmhouse vernacular, providing a dining room that seamlessly opens up to this alfresco space. Above, the clerestory accommodates a nest of sleeping bunks, while the primary bedroom suite is located in a separate structure isolated from the buzz of the main household.

The architecture may have given this plot of land a sense of place, but the family gave it life. Five children grew up running through that screened porch and, like a proud parent, I take great delight in watching this homestead evolve in response to the next generation. The property has now become an innovative organic farming operation, and my heart is warmed by the knowledge that what I helped conceive so many years ago will continue to grow in its mission to provide joy and inspire others.

PREVIOUS PAGES: A gracious screened porch serves as the entry to the home and connects the main living structure with the primary bedroom suite. ABOVE AND OPPOSITE: Nods to farmhouse living are woven throughout elements of the home. In the main living area, simple rope serves as handrails and also suspends the chandelier above the dining table. Oak floors are left untreated, while beams and stair supports have a light stain. A large sliding door opens to connect with the outdoors.

PREVIOUS PAGES: A generous
bay window with floor to ceiling
glass provides an architectural
home for an expansive sofa.
RIGHT: A bunk room nestled
into the clerestory offers
everything guests need: reading
lamps, bed-curtains for privacy
and light control, closets, and
even extra pillows. It's no wonder
this space has become one of the
most coveted spots in the house.

IN TOWN

Finding Home in Town

Southern culture is revered for many things—music, cuisine, cocktails—but perhaps its hospitality above all else. The houses in the region often follow suit, presenting a genteel and welcoming temperament to the street. I like to carry on this tradition of greeting others with a friendly face in the homes we design while simultaneously layering in unexpected drama behind the cordial facades. These architectural decisions shape the personality of the house and bring an element of discovery and adventure for those invited inside.

These dynamics especially come into play for the houses in town, where I prefer to keep a quietness to the street with just enough mystery built in to intrigue people. As a general rule, the entry rarely presides as the central focal point; instead we often place a living space in the middle of the elevation and locate the entry door in a lesser position to the side or around a corner. With this arrangement, the more communal area can now inhabit the best location in the house, rather than a hallway or foyer used for coming and going. Impressive foyers may dazzle visitors, but it's the rooms where we gather together that give a home a deeper purpose.

Different clients want different levels of formality, and we relish hitting the right note for each of them. To give an air that's slightly more dressed up for the city, we turn to refined materials, sculpted moldings, and gracious living spaces, yet we still like to underplay the scale to keep the house from becoming ostentatious. When renovating a house, we acknowledge the context of what came before while reengineering the way it lives for today. New additions may be conceived to meet the needs of a growing family, but these structures generally share a common language with the original. Reflecting our Southern roots, we want the dialogue between old and new to be courteous yet animated, expressing the best of both eras. Whether a renovation or new construction, we strive to design homes that will be good neighbors, welcomed and beloved by those surrounding them.

REBIRTH

Merging classical and modern sensibilities has become a calling card for our studio. The challenge and the opportunity in combining these seemingly opposed architectural directions is to arrive at a cohesive result that celebrates the best of both. For the renovation of this 1920s home, we carefully wove together the old and the new to create a respectful dialogue between the original grande dame and its youthful additions. The complementary solution breathes fresh life into this historic estate, representing the best of both eras and, ultimately, creating a better place to dwell.

Respecting the existing Georgian-style structure, with its handsome limed brick and clay-tile roof, was important, but as we weighed the design possibilities, it became clear that what looked good from the outside wasn't living well on the inside. The pre-existing pond facade was beautifully proportioned, but the lack of windows impeded the views of the expansive nine-acre property, divorcing the house from the landscape. Taking some liberties, we inserted a double-height bay of windows there instead, creating a cathedral-like living room. Since the space was conceived from the inside out, we looked to modern architecture for a window concept with minimal corner mullions, allowing for a panoramic view of the idyllic pond.

Indoors, new elements generate a less formal tone for the house, making the atmosphere more comfortable and approachable to fit the clients' personalities. The sweeping stair nods to the former shape while contemporary open treads allow light to pour through. A new internal fireplace separates the cozy dining space from the soaring living room but has an open hearth to maintain a connection. A hidden television lifts from a custom-designed cabinet that's partially recessed into the floor in front of the window. The flat-screen rises up in front of the view, so the seating group always faces in the direction of the pond.

A dialogue between cultures also comes into play with influences inspired by the couple's extensive travels to Africa. The clients came to our design team with images and experiences that we knitted into the Georgian elements in unexpected ways. The stair hall chandelier adds a textural surprise with its strips of leather, rather than traditional sparkling crystal, and is based on a snapshot from Maasai Mara National Reserve in southwest Kenya that one homeowner shared with us.

South of the main house, a new outdoor room with a thatched roof provides a welcoming vantage for enjoying the pond. Large enough for dining and seating areas as well as a wood-burning fireplace, the space has become a popular destination for the family to gather and relax. By placing the porch adjacent to the house,

RIGHT: A two-story porch and brick facade were removed to allow for the addition of a double-height room with walls of windows.

ABOVE: Following the footprint of the original staircase, the reimagined version allows for light to flow through open treads. Additionally, the plaster balustrade creates a sculptural moment accented by a custom steel handrail that highlights its sinuous path. OPPOSITE: To preserve the view from the living room, the television lifts out of the custom cabinetry.

it captures the views while not being the primary focal point of the property.

From its Georgian roots, the home has been transformed into a merger of worlds—old and new, classical and modern. The new language we constructed proves the power of a holistic living environment where the architecture, landscape, and interiors truly align. That sense of coming together can be felt in every square inch, humming with a peaceful energy that recharges and invigorates anyone who spends time here.

ABOVE: Separated by a double-sided fireplace, the dining room takes on a different personality from the airy living room through the use of whitewashed oak beams on its lowered ceiling. OPPOSITE: In contrast to the light and bright character of the rest of the house, the lounge, complete with dark upholstered walls and a blackened-steel fire surround, becomes a quiet retreat completely tucked away.

OPPOSITE: The breakfast room opens onto the thatched porch—perfect for inviting morning strolls and evening breezes. ABOVE: This seemingly classic kitchen is modernized by elements of surprise. A floating-slat pantry door combines glass and white oak. Vinyl upholstered cabinet fronts bring together practicality and style.

ABOVE AND OPPOSITE: The primitive thatched roof on this pavilion serves as an unexpected counterpoint to the structure's classically inspired architecture. The pattern on its apex denotes a unique signature from the dedicated artisan who crafted it. FOLLOWING PAGES: A pondside folly anchors the landscape composition, providing storage for canoes and a dock for fishing.

CONFIDENCE

Far more than creating livable floor plans and striking facades, balancing a couple's disparate requests can be one of the greatest challenges of an architect. I revel in melding diverse dreams into a unified, successful vision, and this home is one of those perfect combinations. From the beginning of our process, the husband wanted a grand stone estate with an austere edge. On the other hand, the wife's tastes leaned more toward a welcoming, light-filled cottage. The challenge was to blend the two directions by listening and then interpreting, adding our own imaginative influences to complete the composition.

As you enter the property, a lilting roofline offers a humble greeting before revealing the full magnitude of the entry facade. By carefully curating the stone veneer and providing gracious bands of glass, we were able to create a dwelling that is mature in its nature yet stays light and approachable in its temperament. The unique stone cladding actually came by way of a bit of serendipity. While reviewing some salvaged wood and reclaimed brick from a local supplier, I noticed a few blackened stones lying about. They were from the shuttered Old Crow Distillery in Versailles, Kentucky, and the surface was darkened by ash from decades of charring bourbon barrels. However, when we flipped them over, there was a silky buff stone side that I knew was exactly what this house needed. As we laid them, we left mortar on the face, giving the structure a softer, weathered look.

Once inside, you see straight through the house, across the living room, and out to a lush landscape. Avoiding the tropes of a central stairway, which can often block light and views, we placed dual staircases in windowed towers flanking the entrance hall. Their placement allows light to flood both floor levels while providing for fluid movement. Collaborating with the talented interior designer Phoebe Howard, we created a serene palette of colors and textures that conjures a soothing yet refined atmosphere. For example, the interior walls are mostly wood, but paneled to make them a bit more formal. The ceilings are beamed but simply detailed and painted smooth instead of being overtly rustic or highly ornamented. The family room, in contrast, has an interior stone

PREVIOUS PAGES: Lilting rooflines flank the towers that house the structure's symmetrically placed stairwells. OPPOSITE: Architecture draws you through this entire house—not just leading you from room to room, but setting the tone and mood as well. The painted-wood paneling in the main entry hall, with its linear designs and stately columns, takes on a more formal tone. ABOVE: Conversely, an entryway to the study is understated with a lower arched doorway beneath a sweeping plaster stair.

RIGHT: A rustic stone wall matches the refinement of the interiors when balanced with an arched paned window and cut limestone. Part of the overall vision at the project's conception, the opposites work in harmony, balancing each other.

wall, an unexpected element that draws the interior and exterior personalities together.

The pool house and garden offer a surprise destination unseen from the main house. The central room of this folly can be opened up as needed to enjoy nice weather or closed and air-conditioned to provide year-round use. With the clients' children now out of the house, it has become a de facto man cave, a place to enjoy a cigar and savor a glass of Burgundy while taking in the beautifully layered gardens designed by the gifted landscape architect Ben Page.

Architecture doesn't have to be about compromise to achieve divergent goals. Instead, we look to different viewpoints to inform and improve one another, creating a dynamic opportunity that elevates the final outcome. This house strikes the seamless balance of approachability and permanence, heavy and light, masculine and feminine—perfectly pleasing both the lady and the gentleman of the house.

ABOVE AND OPPOSITE: The strength of the stone facade is balanced by moments of charm and grace. A pierced-wood gate makes for a friendly welcome, while an arched window outlined in trained Carolina jasmine softens the stone mass.

ABOVE: Multiple spaces surrounding the pool area are used for more than sunny-day swimming. A deep porch serves as a covered outdoor living area, ideal for enjoying fresh air while avoiding the sun. OPPOSITE: The bifold pool house doors open to create an open-air pavilion. FOLLOWING PAGES: The roof lines of the pool house mimic those of the main house with modest edges surrounding a grander central core.

TRADITION

As a perpetual student of architecture, there are some architects' works that have shaped my approach and provided a consistent source of inspiration. One of my favorites is Harrie T. Lindeberg. He was known for designing evocative country estates for some of the most influential families of early twentieth-century America, and I've long admired his gift for creating intimacy on a grand scale. So I knew it was kismet when interior designer and homeowner Andrea Burridge shared photographs of a Lake Forest, Illinois house she loved. Working together, my associate Mark Kline and I identified the home as one by Lindeberg and eagerly dove into the client's project—crafting a house in the nearby community of Hinsdale with a reverence for the past and its beautiful traditions, yet distinct in its interpretation.

Planned around a central courtyard, the stately brick home maintains an air of mystery behind its gated brick walls. A cobblestone motor court is flanked by a pair of whimsical parking structures, and a bluestone pedestrian path leads to an arched opening that frames the front door beyond. By placing the formal entry off-center, you move through the fairly symmetrical home in an asymmetrical manner, another hallmark of Lindeberg's designs as well as those of his English

contemporary, Edwin Lutyens. The horizontal-paneled stair foyer welcomes guests and serves as the central connection to all the spaces of the home, which we conceived to provide natural light on at least two sides, allowing for maximum ventilation and views. The natural progression of the main level unfolds through the gallery, which runs parallel to both the living room and a reflecting pool outside. The limestone slab flooring amplifies the sound of the water, and the indoor-outdoor connection is heightened when the windows are open. Pocket screens emerge out of the walls to provide protection but tuck in when not needed to keep views crisp.

Collaborating with Burridge on the interiors, we married the modern and traditional in the living room's scheme. Flush plank white oak, bleached to a warm hue, surrounds the chimney. Direct set glass forgoes mullions and a typical window frame to keep the visual lines uncomplicated. Additional touches of rusticity, such as rough-hewn ceiling beams and a blackened-steel hearth set within a carved-limestone fireplace surround, soften the tone and bring down the formality.

But it is the kitchen that truly functions as the heart of this home. A beam-and-plaster ceiling gives rhythm and texture to the handmade

PREVIOUS PAGES, ABOVE, AND OPPOSITE: The gracious walled courtyard creates intentional drama and intrigue, surprising visitors with a series of discoveries hidden from the street before they reach the front door, including boxwood-lined hedges that frame a blanket of lavender. The centrally placed reflecting pool provides the soothing sound of splashing water, and a corbel balcony breaks away from the facade and frames the limestone entry surround.

Christopher Peacock cabinetry of brushed and mirrored stainless steel. A backsplash of alternating polished and honed marble has a similar effect, as does the pantry door with white oak slats played against panels of plate glass. The room is adjacent to a three-season space with French doors and an elegant arched transom. Inspired by classic sleeping porches, the room can be closed off from the rest of the house as weather dictates. Such flexible spaces, designed to capture breezes and collect the warming sun, are as relevant today as they were more than a hundred years ago.

As a whole, this house reflects the timeless continuum of architecture. It acknowledges the legacy of the architects who have come before us without being dogmatic or replicative. It's not afraid to add a fresh voice but does so with maturity, never losing sight of its fundamentals. Imbued with classical principles and a sensitivity for the modern lifestyle, this home translates tradition into a living language for today.

ABOVE: A round window animates the rear of the home while bringing light to the basement stair. OPPOSITE: The theme of brightness continues inside with a white palette of horizontal wood-paneled walls and subtly patterned limestone floors.

ABOVE: The house oscillates between levels of formality. A reserved limestone-clad hall with a series of French doors runs the length of the front of the house parallel to the reflecting pool. RIGHT: In the living room, horizontal painted-wood paneling mixed with bleached-oak slats offers a twist on the traditional.

LEFT: There's a careful balance between matte and glossy finishes in the kitchen, from the mirrored stainless-steel hood to the flat, chalky plaster ceiling, and the rough oak beams to the smooth wood counters. The backsplash pairs both honed and polished marble to create an understated pattern. FOLLOWING PAGES: The classically influenced French doors and transom windows contrast with the modern slatted glass pantry door and cantilevered concrete hearth.

MODESTY

Seeing a new home erected in an established neighborhood often makes people uneasy. The concern is that what goes up will be overscaled, stylistically misplaced, or otherwise ruinous to the streetscape. As I designed this new home in Charlotte, I strove to overturn those stereotypes. I wanted the home to be a welcome addition to the neighborhood, reflecting the gracious personalities of the owners—stylish but friendly, approachable yet distinctive. Essentially, the neighbor everyone dreams of.

Those visiting for the first time are generally surprised that this unpretentious facade actually conceals a 10,000-square-foot home. The modest approach to the street comes from tucking the second floor into the roofline and allowing the plan to open up toward the expansive backyard. The style adds to its humble nature as well. Reflecting a modern Belgian influence, ornamentation takes a back seat to crisp, clean lines. Painted brick mixes with sculpted wood beams, cast-stone chimney caps, and a cedar-shake roof along with such contemporary touches as a cantilevered concrete bench. Lines, curves, and materiality, all working in a perfectly imperfect balance.

Tucked under the sloping eave on the right, the front door offers a warm welcome before turning to enter the formal living room. The same graceful swoop of the roofline accents the double-height space, which can also be appreciated from the balcony above. The more casual rooms give way to a palette of less refined materials—steel beams, rough timbers, and exposed brick—that feel utterly collected. We even cut the kitchen's bleached-oak floor into four-foot-long strips to form a dynamic patchwork in tune with the mix of stained and painted wood on the ceiling. Every element was chosen to imbue the space with a sense of weathered comfort that compels you to kick up your feet and relax a bit.

As regular entertainers, the clients wanted lots of options for hosting casual gatherings of friends, and the outdoor living area has proven to be the ideal setting for dinner parties, glasses

of wine, football games, and so much more. An outdoor fireplace and heaters in the ceilings mean those invitations come year-round. A sweeping, cantilevered roof keeps views to the pool unobstructed by columns. Similarly, the rooms that overlook the backyard have broad swaths of glass with paned windows for ventilation and clear openings for panoramas.

One of the least romantic but impactful aspects of designing a house is deciding where the cars go. Here, we burrowed the garage into the center of the house. This means you don't see the garage as a separate structure on the front, nor does the backyard have to be given over to parking and the driveway. By hiding the cars out of plain sight, the house seems more demure. With good design as with good neighbors, humility often makes the best company.

PREVIOUS PAGES: Influenced by a Belgian aesthetic, the house relies little on ornamentation. Instead, the architecture brings every element down to its essence for a subtle yet powerful effect. RIGHT: In the formal living room, the curve in the wall comes from the line of the exterior roof. The curvaceous balcony projecting from the second floor takes advantage of the triple-sash windows in addition to bringing unexpected visual interest.

OPPOSITE: In the family room, architectural devices deliver both form and function. A blackened-steel I-beam supports antique oak ceiling joists. A large sliding-glass door can be pocketed for direct access to the covered porch. ABOVE: A deep brick archway houses a raised fireplace and serves as a passageway into the kitchen.

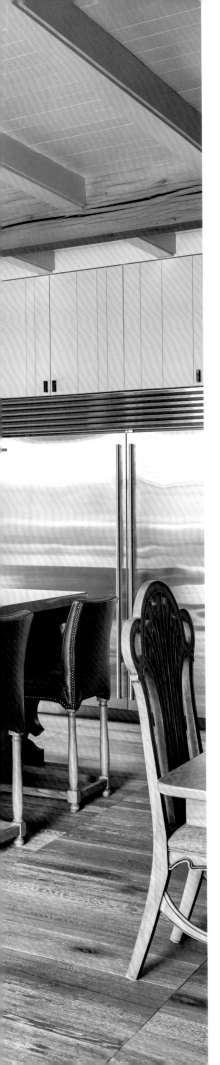

LEFT: The kitchen has a collected approach stemming from an unexpected mix of imperfect materials working together. The bleached-oak floor was laid in four-foot strips to create a patchwork. The ceiling combines painted and stained beams. A pewter range hood with an uneven patina runs over the cooktop, the vinyl-upholstered door leading into the pantry, and the cabinetry. FOLLOWING PAGES: In contrast to the home's modest street facade, the impressive rear view offers an expansive area for entertaining tucked under a cantilevered roof.

RESTRAINT

Some houses are quieter than others. They mind their manners by dutifully lining the block and saying very little. Their appropriateness lets them blend in, never calling too much attention to themselves. When clients asked us to renovate one such elegant and understated home in the Eastover neighborhood of Charlotte, we were intrigued by the opportunity. A symmetrical Georgian-style structure with agreeable proportions and fitting details gave us complete freedom to create a rich, interesting world behind the exterior's polite camouflage.

Choosing to keep the existing facade intact, I knew we could be more imaginative behind the scenes. In response, we devised a complementary brick-box addition to the rear of the house and added a more modern wooden link that unites these two design languages. In doing so, we doubled the square footage of the home without undermining the traditional character of the street.

At first glance, the entry door seems to reflect the Georgian vernacular, but upon more careful inspection, it reveals itself as a pivoting door, a subtle indication of the unique interiors within. By adding planks of wood around the existing stair rail, we freshened up the look with minimal structural intervention. The interior palette has a similar dynamic tension. We left the charming plaster-paneled living room intact, while Heather Smith of Circa Interiors animated the space with rustic and refined treasures that reflect the clients' far-flung travels as well as their Southern roots.

For the new portions of the home, we maintained the same level of integrity and detailing as the original while imbuing each with an element of distinction. The outdoor porch's arch echoes some of the original brickwork, but we made the space modern with expanses of glass framed by blackened steel, which pivot open, creating a fluid connection with the family room. That introspective and cozy space, the response to an unexpected client request for a dark room, is centered on a dramatic double fireplace with a concealed television nestled between the flues. Above, a bridge leads to the primary bedroom suite on the upper floor with a gap just wide enough for the drapery to cascade down three floors. By separating the owners' bedroom from the rest of the home, there's a real sense of escape and privacy, and its location above the porch lends to the narrative of inhabiting a glamorous tree house.

I love the process of discovery that comes with designing each project and this one proved no different. While a true reflection of the homeowners who are utterly stylish and never afraid to forge their own path, the home remains understated and unfolds gradually as a portrait of the people who live here.

PREVIOUS PAGES: When renovating, saving elements that are inherently beautiful provides a link to the past and gives the structure an air of maturity. We left the front facade virtually untouched aside from adding a pivoting door and painting the brick and shutters in a monotone scheme. RIGHT: Likewise, this living room, with its classic woodwork, just needed new decor. Only the fireplace surround was updated to bring a fresh viewpoint.

RIGHT: The existing house and the expansive addition are visually separated by an antique timber that delineates the kitchen from the new living room. On the opposite side, a shaped-plaster surround echoes the form to contain the range and adjacent cabinets.
FOLLOWING PAGES: Twin fireplaces flank a recessed television hidden from view.

ABOVE AND OPPOSITE: In the primary suite's bathroom, a floating mirror and custom-designed pedestal vanity allow windows to wrap the space. Enclosing the double shower in floor-to-ceiling glass keeps the full visual volume of the space.

ABOVE: The material and detailing of the two-story extension at the rear of the home reflect the original Georgian-style house. OPPOSITE AND FOLLOWING PAGES: By situating the new primary suite on the second floor of the volume, the space below becomes a generous outdoor terrace with room for dining and sitting fireside. The configuration also makes way for a choreographed connection to the garden.

MY JOURNEY

When my wife, Sarah, and I decided we had outgrown our home, I had only two requirements for our next one. The first was quite pragmatic: more bathrooms. But the second was more idealistic: I wanted a house I could learn from. Our previous homes had been charming, congenial Colonial styles, where we happily raised our family, but they didn't have much to teach me as an architect. I felt a yearning to engage in a spirited dialogue with the walls that surrounded us—challenging me to think differently and inspiring me to expand the limits of my work.

I found just the architectural mentor I was looking for in a flat-roofed, midcentury gem. It was in need of a bit of polish, but underneath the layers of mauve carpet and questionable wallpaper, I could make out the exceptional bones. In 1952, New York–based architect Saul Edelbaum had designed the house for arts patrons Anita and Herman Blumenthal, avid collectors and forward thinkers in the community. At the time, Charlotte wasn't known for its groundbreaking design aesthetic, and this International-style dwelling must have seemed otherworldly, especially among its more conventional neighbors.

From the street, the house presents itself as a blank slate. Compared to the neighborhood's other homes that seem to cheerfully wave at passersby, it's much more evasive, hiding the allures of its personality for only those who take the time to get to know it. There's a modesty to that which I appreciate. Inside, a soft palette of wood panels, painted brick, and stone pavers brings warmth to this strain of Modernism. Naturally, glass still plays a major role with expansive floor-to-ceiling windows forging a seamless connection to the outdoors, perfect for appreciating the change of seasons in the lush backyard. Carefully placed overhangs shade the interior from the harsh summer sun, yet allow light to slip under the brow and heat the room in the winter. Just watching how the sunlight interacts with the space has inspired and informed my own designs.

Knowing I wanted to maximize time spent together as a family, I pulled many of the cooking functions out of the existing kitchen and into the loftlike main living area. A marble screen acts as a divider, subtly separating culinary activities from the adjacent sitting areas. There's a flush cooktop, a sink, and a sizable workspace for preparing meals, while more intensive tasks

and cleanup are performed out of sight in the original kitchen. Appropriately renamed the working pantry, this behind-the-scenes space houses the refrigerator, dishwasher, and ovens, keeping the public space orderly. The layout works wonders for entertaining, too. As the chef, you can still entertain guests and socialize while cooking dinner, yet dirty plates can disappear backstage throughout the evening. In the dining area, on the opposite end, the wood-paneled wall already existed, but I added steel shelving and a quirky library ladder to display a collection of books, prints, and curiosities. I fashioned the table as a modern barrel crafted from extra tongue-and-groove oak floorboards and surrounded it with a mix of skirted antique and tubular steel chairs.

As a preservationist as well as an architect, I find making the decisions of what to keep and what to update a delicate balance. As much as I want to express my own ideas, there's a history here that I feel compelled to protect. To that end, I decided to leave the original study intact. The wrap-around fluorescent lighting meticulously integrates with the built-in cabinetry, complete with custom-shaped wooden pulls, providing a tactile link to this bygone era.

After living in this home for eight years, I've learned many lessons. It has instilled in me a more acute sense of how the desire to maintain the history of a home and the instinct to reshape it can actually complement and elevate each other. Personally, I've come to appreciate Modernism more deeply and to understand its oft-overlooked warm and sympathetic side. As with any healthy relationship, I am forever shaped by our time together and grateful for the positive influence this experience continues to have on my work.

PREVIOUS PAGES: The living room demonstrates the open beauty of loftlike living; the kitchen is discreetly tucked behind the Calacatta marble divider. OPPOSITE: The fireplace hearth and display shelf, with floating slabs of Crab Orchard stone, reflect the home's vintage midcentury design.

ABOVE AND OPPOSITE: Knowing a full wall of bookshelves would be challenging to fill, I instead assigned the middle portion to showcase a collection of family photos and etchings. A custom-designed library ladder was fashioned from a salvaged eighteenth-century roof truss, while the dining table was designed using excess flooring material from the renovation.

PREVIOUS PAGES: The kitchen in the open living area serves as the entertaining hub.
A custom designed butcher block rests on the blackened steel island. ABOVE: In a house
of glass walls, hanging art takes a bit of ingenuity. Here, a painting floats in front of
felt curtains. OPPOSITE: Backstage tasks can be performed in the house's original
kitchen, which acts as a working pantry. A beamed ceiling reveals a lightwell above.

ABOVE: Modern houses require organic touches to add warmth. A landscape portrait by photographer Mark Dauber, petrified-wood lamp, and reeded-bamboo cabinet enliven a corner. OPPOSITE: Once a screened porch, the sitting room proves that modern spaces have an agnostic sensibility that provides the perfect backdrop for mixing styles. Here, the club chair is French Art Deco, the table is Shaker, and the other pieces are discoveries from markets such as Brimfield and Round Top.

ABOVE AND OPPOSITE: Keeping select period details allows the house to remain connected to its past while bringing it into the present. The study boasts the wealth of these preserved elements, including integrated fluorescent lighting, a custom stereo system, and amber-colored cabinetry with its original shellac finish. FOLLOWING PAGES: In the primary bedroom, the period's lower ceiling heights are visually raised with the choice of white paint, the use of flush mount lighting, and the addition of full-length draperies.

PROJECT CREDITS

EXPLORATION *Page 18*
ARCHITECTURE: Ken Pursley and Craig Dixon
INTERIOR DESIGN: Bronwyn Ford, Pursley Dixon
Ford Interior Design
LANDSCAPE DESIGN: Edward Snyder, Greenleaf
Services

CONNECTION *Page 34*
ARCHITECTURE: Ken Pursley and Mark Kline
INTERIOR DESIGN: Bronwyn Ford, Pursley Dixon
Ford Interior Design
LANDSCAPE ARCHITECTURE: Heath G. Kuszak
and R. Jason Snider, Agrostis

GRACE *Page 62*
ARCHITECTURE: Ken Pursley and Drew Button
INTERIOR DESIGN: Patrick Lewis, Patrick Lewis
Interiors
LANDSCAPE ARCHITECTURE: Cathy Davis, High
Country Consulting Landscape Architecture

COMMUNITY *Page 80*
ARCHITECTURE: Ken Pursley and Drew Button
LANDSCAPE ARCHITECTURE: Cathy Davis, High
Country Consulting Landscape Architecture

TRANSFORMATION *Page 88*
ARCHITECTURE: Ken Pursley and Mark Kline
INTERIOR DESIGN: Kathy Smith, Kathy Smith
Interiors
LANDSCAPE DESIGN: Robert Mathis, Nature View
Landscapes and Excavations

REJUVENATION *Page 98*
ARCHITECTURE: Ken Pursley and Craig Dixon

TRANQUILITY *Page 118*
ARCHITECTURE: Ken Pursley and Mark Kline
INTERIOR DESIGN: Suzanne Kasler, Suzanne
Kasler Interiors
LANDSCAPE ARCHITECTURE: Laurie Durden,
Laurie Durden Garden Design

DISCOVERY *Page 136*
RENOVATION ARCHITECTURE: Ken Pursley
and Jon Hindman
INTERIOR DESIGN: Bronwyn Ford, Pursley Dixon
Ford Interior Design

LANDSCAPE DESIGN: Brian DeRubeis,
Hadden Landscape

AUTHENTICITY *Page 150*
ARCHITECTURE: Ken Pursley, Craig Dixon,
and Mark Kline
LANDSCAPE ARCHITECTURE: Ben Page and Ed
Tessier, Page Duke Landscape Architects

REBIRTH *Page 164*
RENOVATION ARCHITECTURE: Ken Pursley and
Drew Button
INTERIOR DESIGN: Bronwyn Ford, Pursley Dixon
Ford Interior Design
LANDSCAPE ARCHITECTURE: Ben Page and Ed
Tessier, Page Duke Landscape Architects

CONFIDENCE *Page 178*
ARCHITECTURE: Ken Pursley and Craig Dixon
INTERIOR DESIGN: Phoebe Howard, Mrs. Howard
LANDSCAPE ARCHITECTURE: Ben Page and Ed
Tessier, Page Duke Landscape Architects

TRADITION *Page 192*
ARCHITECTURE: Ken Pursley and Mark Kline
INTERIOR DESIGN: Andrea Burridge, AXB
Interiors
LANDSCAPE ARCHITECTURE: Craig Bergmann,
Craig Bergmann Landscape Design

MODESTY *Page 206*
ARCHITECTURE: Ken Pursley and Mark Kline
INTERIOR DESIGN: Amanda Swaringen,
Carolina Design Associates
LANDSCAPE ARCHITECTURE: Carole Joyner,
Joyner Benfield

RESTRAINT *Page 218*
RENOVATION ARCHITECTURE: Ken Pursley and
Drew Button
INTERIOR DESIGN: Heather Smith, Circa
Interiors

MY JOURNEY *Page 234*
RENOVATION ARCHITECTURE: Ken Pursley
INTERIOR DESIGN: Ken Pursley and Bronwyn
Ford, Pursley Dixon Ford Interior Design

ACKNOWLEDGMENTS

Creating a successful architecture practice involves vision, effort, and a healthy dose of serendipity. Much like the way Barney Fife from *The Andy Griffith Show* would catch a crook, I have backed into and tripped over many of my accomplishments, being fortunate enough to land on good people. I would like to express my gratitude to a few of them:

To my Pursley Dixon family, especially my business partner and dear friend Craig Dixon, for your kind nature and steady hand in steering the ship. To my interior design partner Bronwyn Ford, for making our work, and me, whole. To my business manager Day Palmer, you are therapist, cheerleader, bus driver, wrangler, hammer, and glue, all while keeping a smile on your face and reminding me life should be fun. To the talented Mark Kline and Aaron Cote, whose sketches adorn each chapter, I am grateful for the trail of beauty you have forged. To the rest of our crew, Scott Hambrick, Maribeth Prosser, Polly Finn, Bailey Newsome, Alex Ancona, Emily Bell, Hope Wietman, Julie Reo, and Abbey Allen, you make each day a joy. I look forward to all that is on our horizon.

To the book team, I have learned that it's not what you do, but who you do it with that makes a venture memorable. I couldn't have made this trek with a better group. To Jill Cohen and her staff, you are masters of your craft and I am appreciative beyond words for your sage council. To Jackie Terrebonne, our cocktail-hour Zoom calls were a haven of peace during a turbulent time, and I will miss them. Thank you for being a wonderful and patient writing partner. To Kathleen Jayes, for polishing the words in this book to a lustrous glow. To Charles Miers and Rizzoli, for taking a chance on a group of unknowns from North Carolina; may your hopes be exceeded. To Doug Turshen and David Huang, your talent for shaping random images into a visual narrative is second to none. I have long admired your handiwork and am honored to be a part of your literary legacy.

To the photographers who provide the windows to our work, I have enjoyed seeing our designs blossom through your lens. To my friend Chris Edwards, whose beautiful images appear throughout the book, thank you. To Bill Abranowicz, Roger Davies (and Huey Tran), Emily Followill, Tria Giovan, Michael Robinson, Reed Brown, and a host of others, I am indebted.

To Eleanor Roper, Anita Sarsidi, and Helen Crowther for bringing the chic factor to every shoot.

The environments we create would be hollow vessels without the brilliant interior designers and landscape architects we have been fortunate enough to work with. To Suzanne Kasler, for her supportive words and fruitful collaborations, I am particularly appreciative. To Phoebe Howard, Kathy Smith, Patrick Lewis, Andrea Burridge, Amanda Swaringen, and Heather Smith, thank you for elevating the spaces in this monograph with your

vision. To Cindy Smith and Jane Schwab, your support and friendship have meant everything. Ben Page and Ed Tessier have designed landscapes for our homes from the firm's earliest days. I have been privileged to watch these masters execute their skills, shaping nature in perfect response to the architecture. I am also grateful for partnerships with the superbly talented Laurie Durden, Craig Bergmann, Ed Snyder, Cathy Davis, Carole Joyner, Heath Kuszak, and Jason Snider.

To the artisans, craftspeople, and builders who bring our work to life, thank you for sharing your gifts. I offer a special nod to Jeff Franz, Damon Rumsch, Cliff Newbury, Charlie Thomas, Scott and Steven Whitlock, Tyler Mahan, Matthew Collins, Joe Balderson, Jesse Stover, Eric Morley, Eric Cockrell, David Purser, Jesse Morgan, Gina Willis, Jeremy Hubbard, and Tony Valdez. And to the late Randy Carder, you are missed.

I owe deep gratitude to all the publishers, editors, and writers who have shined a light on our work through the years—in particular to *Garden & Gun* and *Southern Living* magazines for being ardent supporters of our studio from the beginning and continuing to share the well of talent in the South.

I'm sure my former English teachers are utterly perplexed by the notion of me authoring a book. To them and to the other educators and mentors who have shaped me along the way, I am indebted—in particular to Nancy Kopfle, Bill Gwin, Paul Zorr, Gaines Blackwell, and Ryland Koets for giving me the road map to pursue what I love. Also, to Bobby McAlpine and Greg Tankersley for unselfishly sharing your gifts and, in turn, helping me discover mine.

To Turner and Judy Smith, thank you for providing the cottage by the shore we dreamed up so many years ago. It was the perfect setting to untangle the wires of my life and plot things to come. Our conversations and time together will forever hold a special place in my heart.

To Laura Lawrence and the Sedgewood Social Club, your joy and laughter proved the perfect antidote to get us through COVID with our sanity intact. To Brian Speas and Jay Everette, for encouraging me to stretch the limits of my work and being impassioned champions of our efforts.

To my family, particularly my mom, Emmaline, and my late father, Rock, for giving me the freedom to explore the boundaries of my imagination. Your unconditional love is the foundation on which all else is built. And to my wife, Sarah, I was reminded in a bold way that every day with you is a gift. Ryland, Grace, and I are immensely blessed to be the recipients of your loving radiance.

Last but not least, to our courageous clients: this book would not exist without the trust you have placed in our studio. Because of you, each creative journey has been memorable. I will forever relish our time together, locking arms as we traveled through the unknown on our joyful adventures to find home.

PHOTOGRAPHY CREDITS
Tria Giovan Photography: Front endpaper, 137–149
Emily Followill Photography: 2–3, 164, 166–167, 174–177, 253
Chris Edwards: 5, 90–97, 150–161, 168–173, 207–215, 216–227, 229–233
Roger Davies: 6, 35–61, 114–115
Erica George Dines: 8
Ken Pursley: 13
William Abranowicz: 14–16, 19–31, 63–113, 117–135, 235–251, back endpaper
Eric Morley: 32–33
John Bessler: 89
Annie Schlechter: 162
Bella Loren: 179, 183
Reed Brown: 180–181, 186–187, 190–191
Josh Savage Gibson: 182, 184–185
Michael Carroll: 188–189
Michael Robinson: 193–199, 201–205
Karen Knecht: 200
Jeff Herr: 228

Triple Self-Portrait printed by permission of the Norman Rockwell Family Agency
Copyright ©1960 the Norman Rockwell Family Entities: 10

First published in the United States of America in 2021 by
Rizzoli International Publications, Inc.
300 Park Avenue South
New York, NY 10010
www.rizzoliusa.com

Copyright © 2021 Pursley Dixon Architecture, Inc.
Foreword: Suzanne Kasler
Text: Jacqueline Terrebonne

Publisher: Charles Miers
Senior Editor: Kathleen Jayes
Design: Doug Turshen with David Huang
Production Manager: Barbara Sadick
Managing Editor: Lynn Scrabis

Developed in collaboration with Jill Cohen Associates, LLC.

Printed in China

2021 2022 2023 2024 / 10 9 8 7 6 5 4 3 2 1

ISBN: 978-0-8478-7082-0
Library of Congress Control Number: 2021938774

Visit us online:
Facebook.com/RizzoliNewYork
Twitter: @Rizzoli_Books
Instagram.com/RizzoliBooks
Pinterest.com/RizzoliBooks
Youtube.com/user/RizzoliNY
Issuu.com/Rizzoli